Have you considered publishing your journal with Oxford University Press?

Throughout the world...

Oxford University Press is the largest and most international of university presses, employing some 2500 people in over thirty branches and offices internationally. With such a broad reach, we can offer you unparalleled marketing and distribution opportunities, to help ensure your journal reaches the widest possible audience.

In every discipline...

Publishing journals has been part of our business since the beginning of the century. Today we publish over 150 journals worldwide, the vast majority on behalf of professional societies and learned organisations, encompassing the entire range of academic and professional fields, from biomedicine to literary studies; dance and drama to business strategy; molecular biology to archaeology.

The best in publishing...

We pride ourselves on producing your journal to the highest possible standards. Using the latest technological developments, we can ensure rapid publication, superb presentation, and speedy distribution to your readers.

Why not consider using us as the publisher of your society's journal?
For more information, write to:

Martin Richardson
Director
Journals Department
Oxford University Press
Walton Street
Oxford OX2 6DP
United Kingdom
Tel: +44 (0)865 56767
Fax: +44 (0)865 257773
e-mail: richarm@oup.co.uk

ASSOCIAZIONE PEDAGOGICA ITALIANA (As. Pe. I.)
SIRA SERENELLA MACCHIETTI
Presidente Nazionale
Facoltà di Magistero Università di Siena
52100 AREZZO

L'Associazione Pedagogica Italiana (As.Pe.I.) è una delle più antiche e illustri associazioni culturali esistenti in Italia. E' stata fondata a Firenze negli anni '50 da Giovanni Calò, grande esponente della pedagogia del tempo; ne furono soci fondatori, insieme a Calò, Aldo Agazzi, Cecilia Motzo Dentice d'Accadia, Santino Caramella, Giuseppe Flores D'Arcais, Lamberto Borghi, Giovanni Maria Bertin: A.Calò succedettero nella Presidenza Nazionale la Prof.ssa Cecilia Dentice d'Accadia, i Prof.ri Aldo Agazzi, Mauro Laeng, Mario Mencarelli, Lamberto Borghi, Claudio Volpi; è attualmente Presidente la Prof.ssa Sira Serenella Macchietti dell'Università di Siena.

All'As.Pe.I. aderiscono professori universitari di discipline pedagogiche, insegnanti di istituti scolastici di ogni ordine e grado, educatori e pedagogisti che operano nel territorio. Suo scopo è la promozione del miglioramento e del rinnovamento della scuola in genere e di ogni altra istituzione a carattere educativo, nonchè la valorizzazione, il potenziamento e lo sviluppo degli studi e delle ricerche pedagogiche (art. 1 dello Statuto). "L'Associazione non ha carattere sindacale, nè collocazione ideologica, politica e religiosa, ma si avvale del dialogo fra uomini di diverso orientamento per realizzare i pròpri scopi e non persegue scopi di lucro" (art. 3 dello Statuto).

Nell'ultimo triennio l'Associazione si è particolarmente impegnata nella promozione di indagini e ricerche a carattere nazionale su queste tematiche: la nuova scheda di valutazione nella scuola media; la formazione universitaria degli insegnanti della scuola materna ed elementare; la formazione universitaria dell'educatore professionale extrascolastico; le riviste pedagogiche-didattiche.

Le Sezioni, diffuse in tutto il territorio nazionale, effettuano ricerche relative a vari settori del "sapere pedagogico" ed esprimono un particolare impegno nella formazione in servizio degli insegnanti della scuola di ogni ordine e grado.

Ogni due o tre anni l'Associazione tiene un Congresso Nazionale in cui vengono affrontate questioni pedagogiche ed educative di grande rilevanza ed attualità. L'ultimo Congresso Verso un'educazione interculturale: temi, problemi, prospettive si è tenuto a Bergamo dal 21 al 24 aprile 1993.

Le notizie sulla vita dell'Associazione sono presentate nel Bollettino As.Pe.I. che viene pubblicato con cadenza trimestrale.

53043 Chiusi Città (SI) - Via Cassia Aurelia II, n. 32 - Telefono 0578/20451

saur

Handbook of International Documentation and Information

Vol. 13

World Guide to Scientific Associations and Learned Societies

6th Edition

Q10
W67
1994

K·G·Saur München · New Providence · London · Paris 1994

Editor / Redaktion:

Michael Zils

Die Deutsche Bibliothek – CIP-Einheitsaufnahme

World guide to scientific associations and learned societies /
[ed.: Michael Zils]. – 6. ed. – München ; New Providence ;
London ; Paris : Saur, 1994
(Handbook of international documentation and information ; Vol. 13)
ISBN 3-598-20580-5
NE: Zils, Michael [Hrsg.]; GT

Copyright © 1994 by K.G. Saur Verlag GmbH & Co KG, München
Printed and bound in the Federal Republic of Germany.
All rights reserved. No part of this publication may be reproduced,
stored in a retrieval system, or transmitted in any form or by any means,
electronic, mechanical, photocopying, recording, or otherwise,
without permission in writing from the publisher.

Digiset data preparation and automatic data processing
by Microcomposition, Wittmar
Printed and Bound: Strauss Offsetdruck, Mörlenbach
ISBN 3-598-20580-5
ISSN 0939-1959

Foreword

The new sixth edition of the »World Guide to Scientific Associations and Learned Societies« provides description of some 17,200 associations and societies from the fields of science, culture and technology.

The work is arranged alphabetically by name within an A–Z of countries (following English spelling). The sixth edition also contains an Alphabetical Index of Association Names, a Subject Index, and a Publications Index. The manuscript was revised with the help of an internationally sent questionnaire, as well as a wide range of national and international source works. Besides corrections, deletions and new additions, the new edition also takes into account geopolitical changes in the former Soviet Union, Czechoslovakia, Yugoslavia and Germany, so that associations and societies will be found under the new state names. Details on organizations in Bosnia-Herzegovia could not be published in this edition due to indeterminate data.

Each entry contains the following information (as far as data was available): association name, address, telephone, telex, telegramme address, telefax, year of foundation, number of members, president, general secretary, area of activity and details on periodical publications. The editorial deadline was the 31st December 1993. New entries and amendments which reached us after this date could unfortunately not be included in the sixth edition.

The use of electronic media in the production of the work means that we are able to offer address collections for all subject groups. Please contact us if you are interested in this offer.

We would like to thank all associations and societies for their friendly cooperation. Please let us know of any necessary amendments or further societies which should be included in the World Guide to Scientific Associations and Learned Societies. Your information helps to improve still further upon the accuracy and reliability of this reference work.

Spring 1994 The Editorial

Vorwort

Die vorliegende 6. Ausgabe des »Internationalen Verzeichnisses wissenschaftlicher Verbände und Gesellschaften« enthält ca. 17,200 Verbände und Gesellschaften aus den Bereichen Wissenschaft, Kultur und Technik.

Das Werk ist nach dem englischen Länderalphabet und innerhalb der Länder namensalphabetisch geordnet. Außerdem enthält die 6. Ausgabe ein alphabetisches Verzeichnis aller Verbandsnamen, ein Register nach Fachgebieten und ein Register der periodischen Publikationen. Die Manuskript-Bearbeitung erfolgte anhand eines weltweiten Fragebogenversands sowie zahlreicher internationaler und nationaler Quellenwerke. Außer den daraus resultierenden Korrekturen, Löschungen und Neuaufnahmen wurden die bisherigen Bundesstaaten Sowjetunion, Tschechoslowakei und Jugoslawien in die entsprechenden neuen Nachfolgestaaten aufgeteilt. Aufgrund der Kriegsereignisse in Bosnien-Herzegowina wurde auf den Abdruck dieses Staates verzichtet, da zur Zeit aus diesem Land keine gesicherten Inormationen erhältlich sind.

Die Einzeleinträge enthalten – sofern bekannt – folgende Detailinformationen: Verbandsname, Adresse, Telefon, Telex, Telegrammanschrift, Telefax, Gründungsjahr, Mitgliederzahl, Präsident, Generalsekretär, Tätigkeitsgebiete und periodische Publikationen mit Hinweisen auf die Erscheinungsweise. Redaktionsschluß war am 31. Dezember 1993. Neueintragungen und Änderungen, die uns nach diesem Datum erreichten, konnten für die vorliegende 6. Ausgabe leider nicht mehr berücksichtigt werden.

Die elektronische Speicherung der Daten ermöglicht es uns, Adressenkollektionen aller Fachgruppen anzubieten. Bei Interesse bitten wir um Ihre Kontaktaufnahme.

Wir danken allen Verbänden und Gesellschaften für ihre freundliche Unterstützung unserer redaktionellen Arbeit. Bitte machen Sie uns auf Fehler oder fehlende Verbände aufmerksam. Ihre entsprechenden Hinweise tragen dazu bei, die Genauigkeit und Zuverlässigkeit dieses Nachschlagewerks noch zu erhöhen.

Frühjar 1994 — Die Redaktion

Contents

Areas of Specialization Abbreviations XI

Afghanistan 3
Albania 3
Algeria 3
Andorra 3
Angola 3
Antigua and Barbuda 3
Argentina 3
Armenia 6
Australia 6
Austria 12
Azerbaijan 21

Bahamas 21
Bahrain 21
Bangladesh 22
Barbados 22
Belarus 22
Belgium 22
Belize 29
Benin 29
Bermuda 29
Birma s. Myanmar
Bolivia 30
Botswana 30
Brazil 30
Bulgaria 33
Burkina Faso 34

Cameroon 34
Canada 34
Chile 53
China, People's Republic 55
China, Republic 57
Colombia 58
Congo 60
Costa Rica 60
Croatia 60
Cuba 60
Cyprus 61
Czech Republic 61

Denmark 62
Dominican Republic 66

Ecuador 67
Egypt 67
El Salvador 68

Inhaltsverzeichnis

Abkürzungsverzeichnis der Fachgebiete XIII

Afghanistan 3
Ägypten 67
Albanien 3
Algerien 3
Andorra 3
Angola 3
Antigua und Barbuda 3
Argentinen 3
Armenien 6
Aserbaidschan 21
Äthiopien 68
Australien 6

Bahamas 21
Bahrain 21
Bangladesh 22
Barbados 22
Belarus 22
Belize 29
Benin 29
Bermuda 29
Bolivien 30
Botswana 30
Brasilien 30
Bulgarien 33
Burkina Faso 34
Burma s. Myanmar

Chile 53
China, Republik 57
China, Volksrepublik 55
Costa Rica 60

Dänemark 62
Deutschland 87
Dominikanische Republik 66

Ecuador 67
Elfenbeinküste 139
El Salvador 68
Estland 68

Fidschi 68
Finnland 68
Frankreich 71
Französisch-Polynesien 87

Estonia 68
Ethiopia 68

Fiji 68
Finland 68
France 71
French Polynesia 87

Gabon 86
Gambia 86
Georgia 86
Germany 87
Ghana 112
Gibraltar 112
Greece 112
Grenada 113
Guam 113
Guatemala 113
Guyana 114

Haiti 114
Honduras 114
Hong Kong 114
Hungary 114

Iceland 116
India 116
Indonesia 119
Iran 120
Iraq 120
Ireland 120
Israel 122
Italy 123
Ivory Coast 139

Jamaica 139
Japan 140
Jordan 147

Kazakhstan 147
Kenya 147
Korea, Democratic People's
Republic 147
Korea, Republic 148
Kuwait 148
Kyrgyzstan 148

Laos 148
Latvia 148
Lebanon 149
Liberia 149
Libya 149
Liechtenstein 149

Gabun 86
Gambia 86
Georgien 86
Ghana 112
Gibraltar 112
Grenada 113
Griechenland 112
Guam 113
Guatemala 113
Guyana 114

Haiti 114
Honduras 114
Hong Kong 114

Indien 116
Indonesien 119
Irak 120
Iran 120
Irland 120
Island 116
Israel 122
Italien 123

Jamaika 139
Japan 140
Jordanien 147
Jugoslawien 313
Jungferninseln (U.S.A.) 313

Kamerun 34
Kanada 34
Kasachstan 147
Kenia 147
Kirgisien 148
Kolumbien 58
Kongo 60
Korea, Demokratische
Volksrepublik 147
Korea, Republik 148
Kroatien 60
Kuba 60
Kuwait 148

Laos 148
Lettland 148
Libanon 149
Liberia 149
Libyen 149
Liechtenstein 149
Litauen 149
Luxemburg 149

Lithuania 149
Luxembourg 149

Macao 149
Macedonia 149
Madagascar 150
Malawi 150
Malaysia 150
Mali 150
Malta 151
Mauritania 151
Mauritius 151
Mexico 151
Moldova, Republic of 152
Monaco 153
Mongolia 153
Morocco 153
Myanmar 153

Namibia 153
Nepal 153
Netherlands 153
Netherlands Antilles 159
New Caledonia 159
New Zealand 159
Nicaragua 160
Niger 160
Nigeria 160
Norway 161

Oman 163

Pakistan 163
Panama 164
Papua New Guinea 164
Paraguay 164
Peru 164
Philippines 165
Poland 166
Portugal 168
Puerto Rico 169

Réunion 169
Romania 170
Russia 171
Rwanda 174

St. Lucia 174
San Marino 174
Saudi Arabia 174
Senegal 174
Sierra Leone 175
Singapore 175

Macau 149
Madagaskar 150
Malawi 150
Malaysia 150
Mali 150
Malta 151
Marokko 153
Mauretanien 151
Mauritius 151
Mazedonien 149
Mexiko 151
Moldovien 152
Monaco 153
Mongolei 153
Myanmar 153

Namibia 153
Nepal 153
Neukaledonien 159
Neuseeland 159
Nicaragua 160
Niederländische Antillen 159
Niederlande 153
Niger 160
Nigeria 160
Norwegen 161

Österreich 12
Oman 163

Pakistan 163
Panama 164
Papua-Neuguinea 164
Paraguay 164
Peru 164
Philippinen 165
Polen 166
Portugal 168
Puerto Rico 169

Reunion 169
Ruanda 174
Rumänien 170
Russland 171

Sambia 314
St. Lucia 174
San Marino 174
Saudi-Arabien 174
Schweden 182
Schweitz 185
Senegal 174
Sierra Leone 175

Slovakia 175
Slovenia 176
South Africa 176
Spain 178
Sri Lanka 181
Sudan 182
Suriname 182
Swaziland 182
Sweden 182
Switzerland 185
Syria 193

Tajikistan 193
Tanzania 193
Thailand 193
Togo 194
Trinidad and Tobago 194
Tunisia 194
Turkey 195
Turkmenistan 195

Uganda 195
Ukraine 196
United Kingdom 196
Uruguay 232
U.S.A. 233
Uzbekistan 311

Vatican City 311
Venezuela 311
Vietnam 312
Virgin Islands (U.S.) 313

Yugoslavia 313

Zaire 314
Zambia 314
Zimbabwe 314

INDEX

Alphabetical Index to Association Names 317

German-English Concordance to Areas of Specialization . . . 399

Subject Index 401

Publications Index 501

Simbabwe 314
Singapur 175
Slovenien 176
Slowakei 175
Spanien 178
Sri Lanka 181
Sudan 182
Südafrika 176
Surinam 182
Swasiland 182
Syrien 193

Tadschikistan 193
Tansania 193
Thailand 193
Togo 194
Trinidad und Tobago 194
Tschechien 61
Türkei 195
Tunesien 194
Turkmenistan 195

Uganda 195
Ukraine 196
Ungarn 114
Uruguay 232
U.S.A. 233
Usbekistan 311

Vatikanstadt 311
Venezuela 311
Vereinigtes Königreich 196
Vietnam 312

Weißrußland 22

Zaire 314
Zypern 61

REGISTER

Alphabetisches Verzeichnis der Gesellschaftsnamen 317

Deutsch-Englische Konkordanz der Fachgebiete 399

Register nach Fachgebieten . . . 401

Register der periodischen Publikationen 501

Areas of Specialization Abbreviations

Acoustics	Acoustics
Adult Educ	Adult Education
Advert	Advertising
Aero	Aeronautics, Aviation, Space Technology
Agri	Agriculture
Air Cond	Air Conditioning
Anat	Anatomy
Anesthetics	Anesthetics
Animal Husb	Animal Husbandry
Anthro	Anthropology
Antique	Antiquities
Archeol	Archeology
Archit	Architecture
Archives	Archives
Arts	Arts
Astronomy	Astronomy
Astrophys	Astrophysics
Auto Eng	Automotive Engineering
Behav Sci	Behavioral Sciences
Biochem	Biochemistry
Bio	Biology
Biophys	Biophysics
Botany	Botany
Botany, Specific	Botany, Specific
Business Admin	Business Administration, Management
Cardiol	Cardiology
Cart	Cartography
Cell Biol & Cancer Res	Cell Biology, Cancer Research
Chem	Chemistry
Cinema	Cinematography
Civil Eng	Civil Engineering
Coffee, Tea, Cocoa	Coffee, Tea, Cocoa
Comm Sci	Communication Science
Commerce	Commerce
Comm	Communications
Computer & Info Sci	Computer and Information Science, Data Processing
Criminology	Criminology
Crop Husb	Crop Husbandry
Cultur Hist	Cultural History, History of Civilization
Cybernetics	Cybernetics
Dairy Sci	Dairy Science
Dent	Dentistry
Depth Psych	Depth Psychology
Derm	Dermatology
Develop Areas	Developing Areas
Diabetes	Diabetes
Doc	Documentation
Ecology	Ecology
Econ	Economics
Educ	Education
Educ Handic	Education of the Handicapped
Electric Eng	Electrical Engineering
Electrochem	Electrochemistry
Electronic Eng	Electronic Engineering
Endocrinology	Endocrinology
Energy	Energy
Eng	Engineering
Entomology	Entomology
Ethnology	Ethnology
Family Plan	Family Planning
Finance	Finance
Fine Arts	Fine Arts
Fisheries	Fisheries
Folklore	Folklore
Food	Food
Forensic Med	Forensic Medicine
Forestry	Forestry
Futurology	Futurology
Gastroenter	Gastroenterology
Genealogy	Genealogy, Heraldry
Genetics	Genetics
Geography	Geography
Geology	Geology
Geomorph	Geomorphology
Geophys	Geophysics
Geriatrics	Geriatrics
Graphic & Dec Arts, Design	Graphic and Decorative Arts, Design
Graphology	Graphology
Gynecology	Gynecology
Hematology	Hematology
Hist	History
Home Econ	Home Economics
Homeopathy	Homeopathy
Hort	Horticulture
Humanities	Humanities, general
Hydrology	Hydrology
Hygiene	Hygiene
Immunology	Immunology
Insurance	Insurance
Int'l Relat	International Relations
Intern Med	Internal Medicine

Journalism	Journalism
Law	Law
Libraries & Bk Sci	Librarianship and Book Science
Ling	Linguistics
Lit	Literature
Logic	Logic
Logopedics	Logopedics
Mach Eng	Machine Engineering
Marketing	Marketing
Mass Media	Mass Media
Materials Sci	Materials Science
Math	Mathematics
Med	Medicine
Metallurgy	Metallurgy
Microbio	Microbiology
Military Sci	Military Science
Mineralogy	Mineralogy
Mining	Mining
Music	Musicology
Nat Res	Natural Resources
Nat Sci	Natural Sciences, general
Navig	Navigation
Neurology	Neurology
Nucl Med	Nuclear Medicine
Nucl Res	Nuclear Research
Numismatics	Numismatics
Nutrition	Nutrition
Oceanography	Oceanography, Marine Sciences
Ophthal	Ophthalmology
Optics	Optics
Ornithology	Ornithology
Orthopedics	Orthopedics
Otorhinolaryn-gology	Otorhinolaryn-gology
Paleontology	Paleontology
Parapsych	Parapsychology
Pathology	Pathology
Pediatrics	Pediatrics
Perf Arts	Performing Arts, Theater
Petrochem	Petrochemistry
Pharmacol	Pharmacology
Philos	Philosophy
Photo	Photography
Physical Therapy	Physical Therapy
Physics	Physics
Physiology	Physiology
Poli Sci	Political Science
Preserv Hist Monuments	Preservation of Historical Monuments, Restoration
Prom Peace	Promotion of Peace
Psych	Psychology
Psychiatry	Psychiatry
Psychoan	Psychoanalysis
Public Admin	Public Administration
Public Health	Public Health
Pulmon Dis	Pulmonary Disease
Radiology	Radiology
Rehabil	Rehabilitation
Rel & Theol	Religions and Theology
Rheuma	Rheumatology
Safety	Safetey and Protection, Safety Engineering
Sci	Science, general
Soc Sci	Social Sciences
Socialism	Socialism
Sociology	Sociology
Speleology	Speleology
Sports	Sports
Standards	Standardization
Stats	Statistics
Stomatology	Stomatology
Surgery	Surgery
Surveying	Surveying, Photo-grammetry
Textiles	Textiles
Therapeutics	Therapeutics
Tobacco	Tobacco
Toxicology	Toxicology
Transport	Transport and Traffic
Traumatology	Traumatology
Travel	Travel and Tourism
Trop Med	Tropical Medicine
Urban Plan	Urban and Regional Planning
Urology	Urology
Venereology	Venereology
Vet Med	Veterinary Medicine
Water Res	Water Resources
Wines	Wines and Wine Making
X-Ray Tech	X-Ray Technology
Zoology	Zoology

Abkürzungsverzeichnis der Fachgebiete

Acoustics	Akustik
Adult Educ	Erwachsenenbildung
Advert	Werbung
Aero	Luftfahrt, Raumfahrttechnik
Agri	Landwirtschaft
Air Cond	Klimatechnik
Anat	Anatomie
Anesthetics	Anästhesiologie
Animal Husb	Tierzüchtung
Anthro	Anthropologie
Antique	Antiquitäten
Archeol	Archäologie
Archit	Architektur
Archives	Archivwesen
Arts	Kunst
Astronomy	Astronomie
Astrophys	Astrophysik
Auto Eng	Kraftfahrzeugbau
Behav Sci	Verhaltensforschung
Biochem	Biochemie
Bio	Biologie
Biophys	Biophysik
Botany	Botanik
Botany, Specific	Botanik, Systematische
Business Admin	Unternehmensführung, Betriebswirtschaft
Cardiol	Kardiologie
Cart	Kartographie
Cell Biol & Cancer Res	Zellbiologie, Krebsforschung
Chem	Chemie
Cinema	Filmkunst
Civil Eng	Bautechnik
Coffee, Tea Cocoa	Kaffee, Tee Kakao
Comm Sci	Kommunikationswissenschaft
Commerce	Handel
Comm	Nachrichtentechnik
Computer & Info Sci	Informatik, Datenverarbeitung
Criminology	Kriminologie
Crop Husb	Nutzpflanzenzüchtung
Cultur Hist	Kulturgeschichte
Cybernetics	Kybernetik
Dairy Sci	Milchwirtschaft
Dent	Zahnheilkunde
Depth Psych	Tiefenpsychologie
Derm	Dermatologie
Develop Areas	Entwicklungshilfe
Diabetes	Diabetes
Doc	Dokumentation
Ecology	Ökologie
Econ	Wirtschaft
Educ	Erziehung und Ausbildung
Educ Handic	Behindertenpädagogik
Electric Eng	Elektrotechnik
Electrochem	Elektrochemie
Electronic Eng	Elektronik
Endocrinology	Endokrinologie
Energy	Energiewesen
Eng	Ingenieurwesen
Entomology	Insektenkunde
Ethnology	Völkerkunde
Family Plan	Familienplanung
Finance	Finanzen
Fine Arts	Malerei, Bildhauerei
Fisheries	Fischerei
Folklore	Volkskunde
Food	Nahrungsmittel
Forensic Med	Gerichtsmedizin
Forestry	Forstwissenschaft
Futurology	Futurologie
Gastroenter	Gastroenterologie
Genealogy	Genealogie, Heraldik
Genetics	Genetik
Geography	Geographie
Geology	Geologie
Geomorph	Geomorphologie
Geophys	Geophysik
Geriatrics	Geriatrie
Graphic & Dec Arts, Design	Graphische und Dekorative Künste, Design
Graphology	Graphologie
Gynecology	Gynäkologie
Hematology	Hämatologie
Hist	Geschichte
Home Econ	Hauswirtschaft
Homeopathy	Homöopathie
Hort	Gartenbau
Humanities	Geisteswissenschaften, allgemeine
Hydrology	Hydrologie
Hygiene	Hygiene
Immunology	Immunologie
Insurance	Versicherung
Int'l Relat	Internationale Beziehungen
Intern Med	Innere Medizin
Journalism	Publizistik

Law	Recht
Libraries & Bk Sci	Bibliotheks- und Buchwesen
Ling	Linguistik
Lit	Literatur
Logic	Logik
Logopedics	Logopädie
Mach Eng	Maschinenbau
Marketing	Marktforschung
Mass Media	Massenmedien
Materials Sci	Werkstoffkunde
Math	Mathematik
Med	Medizin
Metallurgy	Metallurgie
Microbio	Mikrobiologie
Military Sci	Militärwissenschaft
Mineralogy	Mineralogie
Mining	Bergbau
Music	Musikwissenschaft
Nat Res	Rohstoffe
Nat Sci	Naturwissenschaften, allgemeine
Navig	Navigation
Neurology	Neurologie
Nucl Med	Nuklearmedizin
Nucl Res	Kernforschung
Numismatics	Numismatik
Nutrition	Ernährung
Oceanography	Meereskunde
Ophthal	Augenheilkunde
Optics	Optik
Ornithology	Ornithologie
Orthopedics	Orthopädie
Otorhinolaryngology	Hals-Nasen-Ohrenheilkunde
Paleontology	Paläontologie
Parapsych	Parapsychologie
Pathology	Pathologie
Pediatrics	Pädiatrie
Perf Arts	Darstellende Künste, Theater
Petrochem	Petrochemie
Pharmacol	Pharmakologie
Philos	Philosophie
Photo	Photographie
Physical Therapy	Physiotherapie
Physics	Physik
Physiology	Physiologie
Poli Sci	Politologie
Preserv Hist Monuments	Denkmalschutz, Restaurierung
Prom Peace	Friedensforschung
Psych	Psychologie
Psychiatry	Psychiatrie
Psychoan	Psychoanalyse
Public Admin	Verwaltung
Public Health	Gesundheitswesen
Pulmon Dis	Pulmologie
Radiology	Radiologie
Rehabil	Rehabilitation
Rel & Theol	Religionsphilosophie, Theologie
Rheuma	Rheumatologie
Safety	Unfallverhütung, Sicherheitstechnik
Sci	Wissenschaft, allgemeine
Soc Sci	Sozialwissenschaften
Socialism	Sozialismus
Sociology	Soziologie
Speleology	Höhlenkunde
Sports	Sport
Standards	Normung
Stats	Statistik
Stomatology	Stomatologie
Surgery	Chirurgie
Surveying	Landvermessung, Photogrammetrie
Textiles	Textilien
Therapeutics	Therapeutik
Tobacco	Tabak
Toxicology	Toxikologie
Transport	Verkehrswesen
Traumatology	Traumatologie
Travel	Reisen und Tourismus
Trop Med	Tropenmedizin
Urban Plan	Stadt- und Regionalplanung
Urology	Urologie
Venereology	Venerologie
Vet Med	Veterinärmedizin
Water Res	Wasserversorgung
Wines	Wein und Weinbau
X-Ray Tech	Röntgenologie
Zoology	Zoologie

World Guide to Scientific Associations and Learned Societies

Afghanistan

Afghanistan Academy of Sciences, Sher Alikhan St, Kabul
T: (93) 20350
Founded: 1979
Focus: Sci
Periodical
Pashtu Quarterly (weekly) 00001

Asia Foundation, POB 257, Kabul
Founded: 1955
Focus: Educ; Econ 00002

Albania

Albanian Academy of Sciences, Rruga Myslym Shyri 7, Tirana
T: (042) 6049; Tx: 2214
Founded: 1972; Members: 30
Pres: Prof. Aleks Buda; Gen Secr: Prof. Luan Omari
Focus: Sci
Periodicals
Ceshtje te folklorit shqiptar: Problems of the Albanian Folklore (weekly)
Ceshtje te gramatikes se shqipes se sotme: Problems of the Grammar of the Modern Albanian (weekly)
Dialektologjia Shqiptare: Albanian Dialectology (weekly)
Ethnographie Albanaise (weekly)
Etnografia Shqiptare (weekly)
Gjuha jone (weekly)
Iliria: Illyria (weekly)
Informatika dhe Matematika Llogaritese: Informatics and Computer Nathematics (weekly)
Kerkime dhe Studime Hidraulike: Hydraulic Studies (weekly)
Kultura Popullore: Folk Culture (weekly)
Punime te Institutit te Fizikes Berthamore: Studies of the Institute of Nuclear Physics (weekly)
Studia Albanica (weekly)
Studime Filologjike: Philological Studies (weekly)
Studime Gjeografike: Geographic Studies (weekly)
Studime Historike: Historical Studies (weekly)
Studime Meteorologjike e Hidrologjike: Meteorological and Hydrological Studies (weekly)
Studime per Letersine Shqiptare: Studies on Albanian Literature (weekly)
Vjetar i Punimeve Shkencore te Qendres se Kerkimeve Biologjike: Yearbook of the Biological Research Centre (weekly) 00003

Council of Libraries, Rruga Abdi Toptani 3, Tirana
Founded: 1977
Focus: Libraries & Bk Sci 00004

Lidhjy e Shkrimtareve dhe e Artisteve te Shqiperise, Z. Baboci 37, Tirana
Founded: 1957; Members: 1750
Pres: D. Agolli
Focus: Lit
Periodicals
Drita: Journal (weekly)
Les Lettres Albanaises (weekly)
Nentori: Review (weekly) 00005

PEN Centre of Albania, c/o Lidhjy e Shkrimtareve dhe e Artisteve te Shqiperise, Z. Baboci 37, Tirana
Pres: Besnik Mustafaj; Gen Secr: Piro Misho; Sabri Hamiti
Focus: Lit 00006

Shoqata e Gjeologeve te Shqiperise / Geologists' Association of Albania, Bloku Vasil Shanto, Tirana
T: (042) 22511; Tx: 4204
Founded: 1989; Members: 870
Pres: Aleksander Cina; Gen Secr: Afat Serjani
Focus: Geology 00007

Algeria

Arab Scientific and Technical Information Network (ASTINET), c/o Centre d'Etude et de Recherche en Informatique Scientifique et Technique, 6 Pl El-Qods, BP 47, Algiers
T: (02) 600054; Tx: 66036; Fax: (02) 593871
Founded: 1989
Gen Secr: M. Benhamadi
Focus: Nat Sci; Eng 00008

Centre de Coordination des Etudes et des Recherches sur les Infrastructures et les Equipements, 1 Rue Bachir Attar, Alger
Focus: Econ 00009

Centre de Recherches Anthropologiques, Préhistoriques et Ethnographiques (CRAPE), 3 Blvd Franklin Roosevelt, Alger
Founded: 1957
Focus: Anthro; Ethnology 00010

Centre de Recherches en Economie Appliquée pour le Développement (C.R.E.A.D.), Rue Djanel Eddine El-Afghani, El-Hannadia-Bouzaréah, BP 29, Alger
Fax: (02) 44277987
Focus: Econ
Periodicals
Bulletin Bibliographique
Les Cahiers du CREAD 00011

Centre de Recherches Océanographiques et des Pêches (CROP), Jetée Nord, Amirauté, Alger
Focus: Oceanography 00012

Centre de Recherches sur les Resources Biologiques Terrestres (CRBT), 2 Rue Didouche Mourad, Alger
Fax: (02) 642705
Focus: Bio
Periodical
Biocenoses 00013

Centre des Sciences et de la Technologie Nucléaires (CSTN), 3 Blvd Frantz Fanon, BP 1147, Alger
Focus: Nucl Res 00014

Centre National d'Astronomie, d'Astrophysique et de Géophysique (CNAAG), Observatoire de Bouzaréah, Alger
Focus: Astronomy; Geophys 00015

Centre National de Documentation et de Recherche en Pédagogie (CNDRP), c/o Université, 2 Rue Didouche Mourad, Alger
Focus: Doc; Educ 00016

Centre National de Recherches et d'Application des Géosciences (CRAG), 2 Rue Didouche Mourad, Alger
Focus: Geology 00017

Centre National de Recherches et Expérimentations Forestières (CNREF), BP 63, El Mouradia, Alger
Founded: 1911
Focus: Forestry
Periodicals
Les Annales du CNREF
Notes Techniques Forestières 00018

Centre National de Recherche sur les Zones Arides (CNRZA), c/o Université, 2 Rue Didouche Mourad, Alger
Focus: Geophys 00019

Centre National de Traduction et de Terminologie Arabe (CNTTA), 3 Blvd Franklin Roosevelt, Alger
Focus: Ling 00020

Centre National d'Etudes et de Recherches pour l'Aménagement du Territoire (CNERAT), 3 Rue Professor Vincent, Telemly, Alger
Focus: Ecology 00021

El-Djazairia El-Mossilia, 1 Rue Hamitouche, Alger
Founded: 1930; Members: 452
Pres: Ali Benmerabet; Gen Secr: Abdelhadi Meraoubi
Focus: Music 00022

Société Archéologique du Département de Constantine, c/o Musée Gustave Mercier, Constantine
Founded: 1852; Members: 250
Focus: Archeol
Periodical
Recueil des Notices et Mémoires . . . 00023

Société Historique Algérienne, c/o Faculté des Lettres, Université, Alger
Founded: 1963; Members: 600
Focus: Hist 00024

Union des Ecrivains Algériens, 12 Rue Ali Boumendjel, Alger
Founded: 1963; Members: 60
Focus: Lit 00025

Union Médicale Algérienne, 3 Blvd Zirout Yousef, Alger
Focus: Med
Periodical
Algérie Médicale 00026

Andorra

Associació Cultural i Artística Els Esquirols, Sala Parroquial, Plaça de l'Església, La Massana
Focus: Cultur Hist; Arts 00027

Associació per la Defensa de la Natura, Apartat de Correus 96, Andorra la Vella
T: 43248
Founded: 1986
Pres: Alain Moreno; Gen Secr: A. Blanc
Focus: Ecology 00028

Cercle de les Arts i de les Lletres, Av Carlemany 24, Escaldes-Engordany
T: 21233
Founded: 1968
Pres: Lídia Burgués Martisella
Focus: Arts; Lit 00029

Comité Andorrà de Ciències Històriques, Apartat de Correus 113, Andorra la Vella
T: 24077
Founded: 1979; Members: 32
Pres: Lídia Armengol Vila; Gen Secr: Xavier Llovera Massana
Focus: Hist
Periodical
Butlletí Andorrà de Ciències Històriques . 00030

International Council of Museums, Andorran National Committee, c/o Patrimoni Artístic Nacional, Edif Prada Casadet, Prat de la Creu, Andorra la Vella
T: 24077; Fax: 60768
Founded: 1988
Pres: Xavier Llovera; Gen Secr: Bernadeta Garralla Coma
Focus: Arts 00031

Societat Andorrana de Ciències, Carrer de la Borda, Edif Rosella, Apt 3, Andorra la Vella
T: 29729
Founded: 1983
Pres: J. Vilanova
Focus: Sci 00032

Angola

Direcção Provincial dos Servicos de Geologia e Minas de Angola, CP 1260-C, Luanda
Founded: 1914; Members: 506
Focus: Geology; Mining
Periodicals
Boletim
Memória 00033

União dos Escritores Angolanos, CP 2767-C, Luanda
T: (02) 322155; Tx: 3056
Founded: 1975; Members: 75
Gen Secr: Luandino Vieira
Focus: Lit
Periodical
Lavra & Oficina (weekly) 00034

Antigua and Barbuda

Caribbean Regional Council for Adult Education (CARCAE), POB 1810, Saint John's
Founded: 1978
Gen Secr: Edris Bird
Focus: Adult Educ
Periodical
CARCAE Newsletter (weekly) 00035

Library Association of Antigua and Barbuda, c/o Economic Affairs Secretariat, Organization of Eastern Caribbean States, POB 822, Saint John's
T: 4623500; Tx: 2157; Fax: 4621537
Founded: 1983; Members: 40
Pres: Sue Evan-Wong; Gen Secr: D. Josiah
Focus: Libraries & Bk Sci 00036

Argentina

Academia Argentina de Cirugía, Santa Fé 1171, 1059 Buenos Aires
T: (01) 411633
Founded: 1911
Pres: Dr. Santiago G. Perera; Gen Secr: Dr. Enrique M. Beveraggi
Focus: Surgery 00037

Academia Argentina de Letras, Sánchez de Bustamante 2663, 1425 Buenos Aires
T: (01) 8023814; Fax: (01) 8028390
Founded: 1931
Pres: Raúl H. Castagnino; Gen Secr: Jorgelina Loubet
Focus: Lit
Periodical
Boletín (weekly) 00038

Academia de Ciencias Médicas, CC 130, 5000 Córdoba
T: (051) 690051
Founded: 1975; Members: 350
Pres: Dr. Hector Buteler Riu; Gen Secr: Dr. Benito Monis
Focus: Med 00039

Academia Nacional de Agronomía y Veterinaría, Av Alvear 1711, 1014 Buenos Aires
T: (01) 444168
Founded: 1909; Members: 35
Pres: Dr. Norberto Ras; Gen Secr: Dr. Alberto Cano
Focus: Agri; Vet Med
Periodical
Anales (weekly) 00040

Academia Nacional de Bellas Artes, Sánchez de Bustamante 2663, 1425 Buenos Aires
T: (01) 8022469
Founded: 1936; Members: 60
Pres: Alfredo C. Casares; Gen Secr: Ary Brizzi
Focus: Arts 00041

Academia Nacional de Ciencias de Buenos Aires, Av Alvear 1711, 1014 Buenos Aires
T: (01) 413066
Founded: 1935; Members: 35
Pres: Dr. Osvaldo Fustinoni; Gen Secr: Dr. Juan Carlos Agulla
Focus: Sci 00042

Academia Nacional de Ciencias de Córdoba, Av Velez Sarsfield 229, Casilla de Correo 36, 5000 Córdoba
T: (051) 229687; Fax: (051) 229687
Founded: 1869; Members: 35
Pres: Dr. Albert P. Maiztegui; Gen Secr: Dr. Alfredo Cocucci
Focus: Nat Sci
Periodicals
Actas (weekly)
Boletín (weekly)
Miscelanea (weekly) 00043

Academia Nacional de Ciencias Económicas, Av Alvear 1790, 1014 Buenos Aires
T: (01) 422344
Founded: 1914; Members: 35
Pres: Dr. Vicente Vázquez Presedo; Gen Secr: Dr. Enrique J. Loncán
Focus: Econ
Periodical
Anales 00044

Academia Nacional de Ciencias Exactas, Físicas y Naturales (ANCEFN) / The National Academy of Exact, Physical and Natural Sciences of Argentina, Av Alvear 1711, 1014 Buenos Aires
T: (01) 8112998; Fax: (01) 8116951
Founded: 1874; Members: 34
Pres: Dr. Andrés O.M. Stoppani; Gen Secr: Horacio C. Reggini
Focus: Math; Physics; Nat Sci
Periodical
Anales (weekly) 00045

Academia Nacional de Ciencias Morales y Políticas, Av Alvear 1711, 1014 Buenos Aires
T: (01) 412049
Founded: 1938; Members: 35
Pres: Dr. Segundo V. Linares Quintana; Gen Secr: Dr. Carlos M. Bidegain
Focus: Philos; Poli Sci 00046

Academia Nacional de Derecho y Ciencias Sociales, Av Alvear 1711, 1014 Buenos Aires
T: (01) 8213522
Founded: 1874; Members: 25
Pres: Dr. Segundo V. Linares Quintana; Gen Secr: Dr. A.G. Padilla; L. Moreno
Focus: Law; Sociology
Periodical
Anales 00047

Academia Nacional de Derecho y Ciencias Sociales (Córdoba), Artigas 74, 5000 Córdoba
T: (051) 214929
Founded: 1941; Members: 83
Pres: Dr. Pedro J. Frías; Gen Secr: Dr. Olsen A. Ghirardi
Focus: Law; Sociology
Periodicals
Anales
Ediciones Academia 00048

Academia Nacional de Geografía, Av Alvear 1711, 1014 Buenos Aires
T: (01) 413066
Founded: 1956; Members: 30
Pres: Lorenzo Dagnino Pastore; Gen Secr: Aristides Incarnato
Focus: Geography
Periodical
Anales 00049

Academia Nacional de la Historia, Balcarce 139, 1064 Buenos Aires
T: (01) 3315147; Fax: (01) 3314633
Members: 40
Pres: Dr. Ricardo Zorraquin Becú; Gen Secr: Miguel Angel de Marco
Focus: Hist
Periodicals
Boletín (weekly)
Investigaciones y Ensayos (weekly) . . . 00050

Academia Nacional de Medicina, Las Heras 3092, 1425 Buenos Aires
T: (01) 8056890; Fax: (01) 8053415
Founded: 1822; Members: 35
Pres: Dr. Leoncio Arrighi; Gen Secr: Roberto A. Votta
Focus: Med
Periodical
Boletín (weekly) 00051

American Association of Jurists (AAJ), Paraná 257, 1017 Buenos Aires
T: (01) 402724; Fax: (01) 3256454
Founded: 1975
Gen Secr: Nora Morales
Focus: Law 00052

Asociación Archivística Argentina, Leandro N. Alem 250, Buenos Aires
T: (01) 335531
Founded: 1968; Members: 160
Pres: Luis F. Piazzali; Gen Secr: Dr. Augusto Fernández Pinto
Focus: Archives
Periodical
Boletín 00053

Asociacíon Argentina Amigos de la Astronomía (A.A.A.A.), Casilla de Correo 369, Correo Central, 1000 Buenos Aires
T: (01) 883366; Fax: (01) 883366
Founded: 1929; Members: 1000
Pres: Gloria Roitman
Focus: Astronomy

Periodical
Revista Astronómica 00054

Asociación Argentina de Alergía e Inmunología, Moreno, 1091 Buenos Aires
T: (01) 387683
Founded: 1949
Focus: Immunology 00055

Asociación Argentina de Astronomía, c/o Observatorio Astronómico, 1900 La Plata
T: (021) 27308
Founded: 1958; Members: 120
Pres: Raúl Colomb; Gen Secr: Hugo Levato
Focus: Astronomy
Periodical
Boletín (weekly) 00056

Asociación Argentina de Bibliotecas y Centros de Información Cientificis y Tecnical, Santa Fé, 1145 Buenos Aires
T: (01) 411405
Focus: Libraries & Bk Sci 00057

Asociación Argentina de Bibliotecas y Centros de Información Científicos y Técnicos, Santa Fé 1145, 1059 Buenos Aires
T: (01) 3938406
Founded: 1937; Members: 84
Focus: Libraries & Bk Sci; Computer & Info Sci 00058

Asociación Argentina de Biología y Medicina Nuclear, Santa Fé 1145, 1059 Buenos Aires
T: (01) 3935682
Founded: 1963; Members: 190
Pres: Dr. Juan J. O'Farrell; Gen Secr: Dr. Carlos Cañellas
Focus: Bio; Nucl Med 00059

Asociación Argentina de Ciencias Naturales, Av Angel Gallardo 470, 1405 Buenos Aires
T: (01) 9828370; Fax: (01) 9824494
Founded: 1912; Members: 500
Pres: Juan Carlos Giacchi; Gen Secr: Graciela Esnal
Focus: Nat Sci
Periodical
Revista Physis (weekly) 00060

Asociación Argentina de Cirugía, Marcelo T. de Alvear 2415, 1122 Buenos Aires
T: (01) 832905; Fax: (01) 836458
Founded: 1930; Members: 4000
Pres: Dr. Santiago G. Perera; Gen Secr: Dr. Guillermo A. Flaherty
Focus: Surgery
Periodicals
Anuario
Boletín
Revista Argentina de Cirugía (8 times annually) 00061

Asociación Argentina de Ecología, CC 1025, Correo Central, 5000 Córdoba
T: (051) 22284 ext 49
Founded: 1972; Members: 600
Pres: Dr. Raúl A. Montenegro
Focus: Ecology
Periodical
Bulletin (3 times annually) 00062

Asociación Argentina de Estudios Americanos, Maipú 672, 1424 Buenos Aires
T: (01) 3924971
Focus: Ethnology 00063

Asociación Argentina de Farmacia y Bioquímica Industrial, Uruguay 469, 1015 Buenos Aires
T: (01) 498900; Fax: (01) 498794
Founded: 1952; Members: 1650
Pres: Dr. Enrique F. Piccinni; Gen Secr: Dr. Raúl Alberto Revilla
Focus: Pharmacol; Biochem 00064

Asociación Argentina de Geofísicos y Geodestas / Argentine Association of Geophysicists and Geodesists, CC 106, 1428 Buenos Aires
Founded: 1959
Pres: Dr. María L. Altinger; Gen Secr: Dr. María C. Pomposiello
Focus: Surveying; Geology
Periodicals
Boletín (3 times annually)
Geoacta (biennial) 00065

Asociación Argentina de la Ciencia del Suelo, J.R. de Velasco 847, 1414 Buenos Aires
T: (01) 7818968
Founded: 1958; Members: 610
Pres: Carlos O. Miaczynski; Gen Secr: H.E. del Campo
Focus: Agri
Periodicals
Boletín (3 times annually)
Ciencia del Suelo (weekly) 00066

Asociación Argentina de la Cultura Inglesa, Suipacha 1333, 1011 Buenos Aires
T: (01) 3934864
Founded: 1927
Focus: Lit 00067

Asociación Argentina del Frío, Av Mayo 1123, 1085 Buenos Aires
T: (01) 387544
Founded: 1932; Members: 178
Pres: Enzo Locatelli; Gen Secr: Norberto M. Miranda
Focus: Eng; Air Cond
Periodical
Clima (weekly) 00068

Asociación Argentina de Micología, c/o Dept de Microbiología, Facultad de Medicina, Universidad de Buenos Aires, Paraguay 2155, 1121 Buenos Aires
T: (01) 9627274; Fax: (01) 9625404
Founded: 1960; Members: 150
Pres: Dr. Jorge Luis Finquelievich; Gen Secr: Dr. Iris Nora Tiraboschi
Focus: Botany, Specific
Periodical
Revista (3 times annually) 00069

Asociación Argentina de Ortopedía y Traumatología, Vicente López 1878, 1128 Buenos Aires
T: (01) 8012320; Fax: (01) 8017703
Founded: 1936; Members: 1550
Pres: Dr. Carlos A.N. Firpo; Gen Secr: Enrique J.C. Lafrenz
Focus: Orthopedics; Traumatology
Periodical
Revista de la Asociación de Ortopedía y Traumatología (weekly) 00070

Asociación Argentina para el Estudio Científico de la Deficiencia Mental, Pacheco de Melo 2483, 1425 Buenos Aires
Focus: Psychiatry 00071

Asociación Argentina para el Progreso de las Ciencias, Av Alvear 1711, 1014 Buenos Aires
T: (01) 416951; Fax: (01) 7868578
Founded: 1933; Members: 135
Pres: Dr. Héctor N. Torres; Gen Secr: Dr. Máximo Barón
Focus: Nat Sci
Periodical
Ciencia e Investigación 00072

Asociación Bernardino Rivadavia, Av Colón 31, 8000 Bahía Blanca
T: (091) 27492
Founded: 1882
Pres: Prof. Raúl Oscar Gouamalusse; Gen Secr: María Luisa Gastaminza
Focus: Libraries & Bk Sci
Periodical
Boletín Informativo (weekly) 00073

Asociación Bioquímica Argentina, Venezuela 1823, 1096 Buenos Aires
T: (01) 382907
Founded: 1934
Focus: Biochem 00074

Asociación Científica Argentino-Alemana, Santa Fé 1145, 1059 Buenos Aires
Founded: 1954
Focus: Sci 00075

Asociación Dante Alighieri, Tucumán 1646, 1050 Buenos Aires
T: (01) 402480
Founded: 1896; Members: 3077
Pres: Dr. Dionisio Petriella; Gen Secr: Dr. Juan A. Stefani
Focus: Lit; Ling; Arts 00076

Asociación de Bibliotecarios Graduados de la República Argentina (ABGRA), Montevideo 581, 1019 Buenos Aires
T: (01) 409728; Tx: 18201
Founded: 1953; Members: 1200
Pres: Stella Maris Fernández; Gen Secr: Eduardo del Caño
Focus: Libraries & Bk Sci
Periodicals
Bibliotecología y Documentación
Boletín Informativo
Trabajos de Congresos (weekly) 00077

Asociación de Ciencias Naturales del Litoral, José Macía 1933, 3016 Santo Tomé
Founded: 1962; Members: 500
Pres: Dr. Hetty Bertoldi de Pomar
Focus: Nat Sci
Periodicals
Climax
Revista de la Asociación de Ciencias Naturales del Litoral (weekly) 00078

Asociación Electrotécnica Argentina, Posadas 1659, 1112 Buenos Aires
T: (01) 8043454
Founded: 1913; Members: 2000
Pres: Ernesto H. Rodil
Focus: Electric Eng 00079

Asociación Electrotécnica Argentina / Argentine Electrotechnical Association, Posadas 1659, 1112 Buenos Aires
Founded: 1913; Members: 930
Focus: Electric Eng
Periodicals
Revista Electrotécnica
Revista Electrotécnica 00080

Asociación Física Argentina, CC 5, 1894 Villa Elisa
Founded: 1944
Pres: Ernesto E. Galloni; Gen Secr: Dr. Carlos Bollini
Focus: Physics
Periodical
Revista 00081

Asociación Geológica Argentina / Argentinian Geological Association, Maipú 645, 1006 Buenos Aires
T: (01) 3222820
Founded: 1945; Members: 1600
Pres: Dr. A.C. Riccardi; Gen Secr: Dr. M.K. de Brodtkorb
Focus: Geology
Periodical
Revista de la Asociación Geológica Argentina: RAGA (weekly) 00082

Asociación Iberoamericana de Estudio de los Problemas del Alcohol y la Droga (AIEPAD) / Ibero-American Association for the Study of Alcohol and Drug Problems, Casilla 1264, 5000 Córdoba
Focus: Soc Sci 00083

Asociación Interamericana de Escritores, Casilla de Correo 4852, 1000 Buenos Aires
Focus: Lit 00084

Asociación Internacional de Hidatidología, Florida 460, 1005 Buenos Aires
T: (01) 3222070; Tx: 2314
Founded: 1941; Members: 600
Focus: Med; Physiology; Pathology
Periodical
Boletín de Hidatidosis (3 times annually) 00085

Asociación Latinoamericana de Centros de Educación (ALCECOOP) / Latin American Association of Cooperative Centres of Education, Laprida 1062, Casilla 135, 2000 Rosario
T: (041) 48106; Tx: 41939; Fax: (041) 245832
Founded: 1989
Focus: Educ 00086

Asociación Latinoamericana de Sociedades de Biología y Medicina Nuclear, Santa Fé 1145, 1059 Buenos Aires
Founded: 1964
Focus: Bio; Nucl Med 00087

Asociación Latinoamericana de Sociología (ALAS), Trejo 241, 5000 Córdoba
Founded: 1950
Focus: Sociology 00088

Asociación Latinoamericana para la Calidad (ALC) / Latin American Association for Quality, c/o IACC, Carlos Calvo 1551, 1102 Buenos Aires
Focus: Materials Sci 00089

Asociación Médica Argentina, Santa Fé 1171, 1059 Buenos Aires
T: (01) 411633
Founded: 1891
Pres: Dr. Carlos Reussi; Gen Secr: Laura Astarloa
Focus: Med
Periodical
Revista 00090

Asociación Odontológica Argentina, Junín 959, 1113 Buenos Aires
T: (01) 9616062
Founded: 1896; Members: 8200
Pres: Jorge H. Canzani; Gen Secr: A. Pereira
Focus: Dent
Periodical
Revista (weekly) 00091

Asociación Ornitológica del Plata, 25 de Mayo 749, 1002 Buenos Aires
T: (01) 3128958; Fax: (01) 3128958
Founded: 1916; Members: 1000
Pres: Manuel Nores; Gen Secr: Diego Gallegos-Luque
Focus: Ornithology
Periodicals
El Hornero (weekly)
Nuestras Aves (weekly) 00092

Asociación pala la Lucha la Parálisis Infantil, Salguero 1639, 1425 Buenos Aires
T: (01) 841034
Founded: 1943; Members: 30
Pres: Verónica S.M. de Busto
Focus: Pediatrics 00093

Asociación Paleontológica Argentina (APA), Maipu 645, 1006 Buenos Aires
T: (01) 3922820
Founded: 1955; Members: 650
Pres: Dr. Edgardo Romero; Gen Secr: Dr. Eduardo Ottone
Focus: Paleontology
Periodical
Ameghiniana (weekly) 00094

Asociación Psicoanalitica Argentina, Rodriguez Peña 1674, 1021 Buenos Aires
T: (01) 8123518, 8140757
Founded: 1943; Members: 932
Focus: Psychoan 00095

Asociación Química Argentina, Sánchez de Bustamente 1749, 1425 Buenos Aires
T: (01) 834886
Founded: 1912; Members: 5000
Pres: Dr. Mario A. Crivelli; Gen Secr: Dr. César Roitman
Focus: Chem
Periodicals
Anales (weekly)
Boletín (10 times annually)
Industría y Química (5 times annually) . 00096

Center for Integrated Social Development, c/o CIDES, Blvd 27 de Febrero 1905, 2000 Rosario
Founded: 1961
Gen Secr: Clemente R. Laurito
Focus: Soc Sci 00097

Centro Argentino de Espeleología, Av de Nayo 651, 1428 Buenos Aires
T: (01) 3316798
Founded: 1970; Members: 58
Pres: Julio Goyén Aguado; Gen Secr: Roberto Oscar Bermejo
Focus: Geology 00098

Centro Argentino de Información Cientifica y Tecnológica (CAICYT), Moreno 431-433, 1091 Buenos Aires
T: (01) 305749
Focus: Eng; Sci; Doc 00099

Centro Argentino de Ingenieros (CAI) / Argentine Centre of Engineering, Cerrito 1250, 1010 Buenos Aires
T: (01) 414133
Founded: 1895; Members: 10122
Pres: Simón Aisiks
Focus: Eng
Periodical
CAI Informa (3 times annually) 00100

Centro de Documentación Bibliotecológica, c/o Universidad Nacional del Sur, Av Alem 1253, 8000 Bahía Blanca
Founded: 1962
Focus: Libraries & Bk Sci; Doc
Periodicals
Documentación Bibliotecológica (weekly)
Memoria Anual (weekly)
Ultimas Adquisiciones (weekly) 00101

Centro de Estudios Económicos Sociales, Libertad 1050, 1012 Buenos Aires
Focus: Econ; Sociology 00102

Centro de Estudios Urbanos y Regionales (CEUR), Bartolomé Mitre 2212, 1039 Buenos Aires
Founded: 1977
Focus: Urban Plan 00103

Centro de Investigación de Biologia Marina, 1650 San Martín
Founded: 1960; Members: 30
Focus: Bio; Oceanography 00104

Centro de Investigación Documentaria, Casilla 1359, 1000 Buenos Aires
Tx: 021859
Focus: Doc 00105

Centro de Investigaciones Bella Vista, José Manuel Estrada 66, 3432 Bella Vista
Founded: 1954
Focus: Bio; Botany 00106

Centro de Investigaciones de Recursos Naturales, 1712 Castelar
Founded: 1944
Pres: Dr. C. Scoppa
Focus: Ecology
Periodicals
Boletines Climaticos (weekly)
Floras (weekly)
Serie Carta de Suelos (weekly)
Serie Fitogeográfica (weekly) 00107

Centro de Investigaciones Económicas, c/o Instituto Torcuato di Tella, 11 de Septiembre 2139, 1428 Buenos Aires
T: (01) 7815013
Founded: 1960
Focus: Econ 00108

Centro de Medicina Nuclear, c/o Instituto Modelo, Av Amancio Alcorta 1402, 1283 Buenos Aires
Founded: 1958
Focus: Nucl Med 00109

Centro de Navegación Transatlántica, Maipú 521, 1006 Buenos Aires
T: (01) 3940520; Tx: 25572
Focus: Navig
Periodical
River Plate Shipping Guide (every 3-4 years) 00110

Centro Lincoln, Florida 935, 1005 Buenos Aires
T: (01) 323428
Founded: 1950
Focus: Lit 00111

Centro Nacional de Documentación e Información Educativa, c/o Ministerio de Cultura y Educación, Madero 235, 1408 Buenos Aires
Focus: Educ; Doc 00112

Círculo Médico de Córdoba, Colón 637, 5000 Córdoba
Founded: 1910
Focus: Med 00113

Círculo Médico de Rosario, Santa Fé 1798, 2000 Rosario
T: (041) 210120; Fax: (041) 259089
Founded: 1910; Members: 1600
Focus: Med
Periodical
Revista Medica de Rosario (weekly) . . 00114

Colegio de Abogados de la Ciudad de Buenos Aires, Montevideo 640, 1019 Buenos Aires
T: (01) 422690; Fax: (01) 422690
Founded: 1913; Members: 1600
Pres: Dr. Ernesto O'Farrell; Gen Secr: Dr. Julio Cesar Saguier
Focus: Law
Periodicals
Revista
Revista 00115

Colegio de Graduados en Ciencias Económicas, Viamonte 1592, 1055 Buenos Aires
T: (01) 450242
Founded: 1891
Pres: Rubén Veiga
Focus: Econ
Periodicals
Boletín Informativo
Revista de Ciencias Económicas 00116

Comisión de Investigaciones Científicas de la Provincia de Buenos Aires, Calle 526 entre 10-11, 1900 La Plata
T: (021) 217374; Tx: 31149
Founded: 1956
Focus: Sci
Periodicals
Informes
Monografías 00117

Comisión Nacional de Energía Atómica, Av del Libertador 8250, 1429 Buenos Aires
T: (01) 707711; Tx: 21388
Founded: 1950
Focus: Nucl Res
Periodical
Informes 00118

Comisión Nacional de Investigaciones Espaciales (CNIE) / National Commission on Space Research, Dorrego 4018, 1425 Buenos Aires
T: (01) 7743234; Tx: 17511; Fax: (01) 548041021
Founded: 1960
Focus: Aero; Nat Res 00119

Comisión Nacional de Museos y de Monumentos y Lugares Históricos, Av de Mayo 556, 1084 Buenos Aires
T: (01) 3316151
Founded: 1938
Pres: Jorge Enrique Hardoy; Gen Secr: Matilde I. Orueta
Focus: Preserv Hist Monuments; Arts
Periodical
Boletín 00120

Comisión Nacional Protectora de Bibliotecas Populares / National Commission for the Protection of Public Libraries, Ayacucho 1578, 1112 Buenos Aires
T: (01) 8036545
Founded: 1870
Pres: Alberto F.J. Vanasco; Gen Secr: Prof. Daniel R. Ríos
Focus: Econ
Periodical
Boletín 00121

Comisión Panamericana de Normas Técnicas (COPANT) / Pan American Standards Commission, Lima 629, 1073 Buenos Aires
T: (01) 375123; CA: COPANTEC
Founded: 1961; Members: 21
Focus: Standards
Periodical
Noticias COPANT: Catalogue of Standards (weekly) 00122

Comité Nacional de Cristalografía, Av Libertador San Martín 8250, 1429 Buenos Aires
Founded: 1958
Focus: Chem; Geology; Physics 00123

Confederación Sudamericana de Medicina del Deporte (CONSUMED), Casilla 994, 1000 Buenos Aires
Founded: 1986
Pres: Prof. Dr. Carlos d' Angelo
Focus: Med 00124

Conference of Latin American Data-Processing Authorities, Puerredón 524, Buenos Aires
Focus: Computer & Info Sci 00125

Congreso Internacional de Americanistas, c/o Museo La Plata, Paseo del Bosque, 1900 La Plata
Founded: 1875
Focus: Lit 00126

Consejo Federal de Inversiones, Alsina 1407, 1088 Buenos Aires
Focus: Geophys 00127

Consejo Internacional de Administración Científica, Tucumán 1668, 1050 Buenos Aires
Focus: Sci 00128

Consejo Inter-Universitario Nacional, c/o Universidad, Viamonte 444, 1053 Buenos Aires
Focus: Adult Educ 00129

Consejo Nacional de Investigaciones Científicas y Técnicas (CONICET) / National Council of Scientific and Technical Research, Rivadavia 1917, 1033 Buenos Aires
T: (01) 9537230; Fax: (01) 9534345
Founded: 1958; Members: 8100
Pres: Prof. Dr. R.F. Matera; Gen Secr: Alejandro Andreini
Focus: Eng; Sci
Periodical
Boletín SECYT-CONICET (weekly) . . . 00130

Departamento de Estudios Etnográficos y Coloniales / Dept of Ethnographical and Colonial Studies, Calle 25 de Mayo 1470, 3000 Santa Fé
Founded: 1940
Focus: Ethnology
Periodical
America (weekly) 00131

Departamento de Estudios Históricos Navales / Department of Naval History Studies, Av Almirante Brown 401, 1155 Buenos Aires
T: (01) 3621248; Tx: 21338
Founded: 1957
Focus: Hist; Navig 00132

Departamento de Microbiología (INTA), Villa Udaondo, 1712 Castelar
Founded: 1956
Focus: Microbio 00133

Federación Argentina de Asociaciones de Anestesiología, J.F. Aranguren 1323, 1405 Buenos Aires
T: (01) 4312463; Fax: (01) 4312463
Founded: 1970; Members: 1113
Pres: Dr. Gustavo A. Elena; Gen Secr: Dr. Alfredo Parietti
Focus: Anesthetics
Periodicals
Boletín Informativo
Revista Argentina de Anestesiología . . 00134

Federación del Patronato del Enfermo de Lepra de la República Argentina, Beruti 2373-2377, 1117 Buenos Aires
T: (01) 8215531
Founded: 1930
Focus: Derm
Periodical
Temars de Leprologia (weekly) 00135

Federación Interamericana del Instituto de Enseñanza Publicitaria, Av de Mayo 621, 1084 Buenos Aires
Focus: Advert 00136

Federación Lanera Argentina (FLA), Av Paseo Colón 823, 1063 Buenos Aires
T: (01) 3614604; Tx: 22021; Fax: (01) 3616517
Founded: 1929; Members: 105
Focus: Marketing
Periodicals
Boletín Lanero (daily)
Informe Mensual Estadístico (weekly) . . 00137

Federación Universitaria Argentina (FUA), c/o Universidad, Viamonte 444, 1053 Buenos Aires
Focus: Adult Educ 00138

Fondo Nacional de las Artes / National Arts Foundation, Alsina 673, 1087 Buenos Aires
T: (01) 301591
Founded: 1958
Pres: Dr. J. Harvey
Focus: Arts
Periodicals
Anuario del Teatro Argentino
Bibliografía Argentina de Artes y Letras (weekly)
Informativo (weekly) 00139

Fundación Miguel Lillo, Miguel Lillo 251, 4000 San Miguel de Tucumán
Founded: 1931
Focus: Botany; Geology; Zoology
Periodicals
Acta Geologica Lilloana (weekly)
Acta Zoologica Lilloana (weekly)
Genera et Species Animalium Argentinorum (weekly)
Genera et Species Plantarum Argentinorum (weekly)
Lilloa (weekly) 00140

Grupo Latinoamericano de R.I.L.E.M. (GLAR), Av L. N. Alem 1067, 1001 Buenos Aires
T: (01) 3133013; Tx: 021859
Founded: 1963; Members: 61
Focus: Civil Eng; Materials Sci
Periodical
GLARILEM Informations (3-4 times annually) 00141

Instituto Bonaerense de Numismática y Antigüedades, San Martín 336, 1004 Buenos Aires
T: (01) 492659
Founded: 1872; Members: 50
Pres: Humberto F. Burzio
Focus: Numismatics; Antique
Periodical
Boletín 00142

Instituto de Literatura, Calle 47 No 625, 1900 La Plata
Founded: 1968
Focus: Lit
Periodicals
Boletín
Investigaciones 00143

Instituto Nacional de Investigación y Desarrollo Pesquero (INIDEP), Casilla de Correo 175, 7600 Mar del Plata
T: (023) 517818; Tx: 39975; CA: INIDEP
Founded: 1960; Members: 200
Focus: Fisheries; Oceanography; Food
Periodicals
Boletín Informativo (weekly)
Contribuciones INIDEP (weekly)
Memoria Anual (weekly)
Revista de Investigación y Desarrollo Pesquero (weekly) 00144

Instituto Nacional de Tecnologia Industrial (INTI) / National Institute of Industrial Technology, Av Leandro N. Alem 1067, 1001 Buenos Aires
Tx: 021859; Fax: (01) 3133966
Founded: 1957; Members: 1800
Focus: Eng; Fisheries; Food; Textiles . . 00145

Instituto Nacional de Vitivinicultura / National Institute of Viti-Viniculture, Av San Martin 430, 5500 Mendoza
T: (061) 248288; Tx: 55343; Fax: (061) 340283
Founded: 1959; Members: 530
Pres: Eduardo Martinez
Focus: Agri
Periodicals
Estadistica Vitivinicola (weekly)
Exportaciones Argentinas de Productos Vitivinicolas (weekly)
Sintesis Básica (weekly) 00146

Instituto para la Integración de América Latina, Cerrito 264, 1010 Buenos Aires
Founded: 1965; Members: 30
Focus: Finance; Poli Sci 00147

Junta de Historia Eclesiástica Argentina, Reconquista 269, 1003 Buenos Aires
T: (01) 3316239
Founded: 1942; Members: 100
Pres: Prof. Carlos M. Gelly y Obes; Gen Secr: Alberto de Paula
Focus: Rel & Theol
Periodicals
Boletin
Revista Archivum (weekly) 00148

Liga Argentina contra la Tuberculosis (LALAT), Santa Fé 4292, 1425 Buenos Aires
T: (01) 7742811
Founded: 1901
Pres: Dr. Rodolfo Cucchiani Acevedo; Gen Secr: Dr. G. Quintela Novoa
Focus: Pulmon Dis
Periodicals
Doble Cruz (weekly)
Revista Argentina del Toiax (weekly) . . 00149

Organización de Universidades Católicas de América Latina (ODUCAL), Juncal 1912, 1116 Buenos Aires
Founded: 1953
Focus: Sci 00150

PEN Club Argentino, Rivadavia 4060, 1205 Buenos Aires
Founded: 1930; Members: 100
Pres: Miguel A. Olivera
Focus: Lit
Periodical
Boletín 00151

Servicio de Endocrinología y Metabolismo, Martínez de Hoz y Marconi, Villa Sarmiento, 1408 Buenos Aires
Founded: 1969
Focus: Endocrinology 00152

Sociedad Argentina de Alergia e Inmunopatología (SAAI), Santa Fé 1171, 1059 Buenos Aires
T: (01) 411633
Founded: 1940; Members: 84
Focus: Immunology
Periodicals
An Up-to-Date in Pollenosis and Drug Allergy (weekly)
Clinical Reports of Allergy and Immunology (weekly)
Monographs of Allergy and Clinical Immunology (weekly) 00153

Sociedad Argentina de Anatomía Normal y Patológica, Santa Fé 1171, 1059 Buenos Aires
Founded: 1933; Members: 200
Focus: Anat; Pathology 00154

Sociedad Argentina de Antropología, Moreno 350, 1091 Buenos Aires
Founded: 1936; Members: 250
Pres: C.A. Aschero; Gen Secr: V.I. Williams
Focus: Anthro
Periodical
Relaciones (weekly) 00155

Sociedad Argentina de Autores y Compositores de Música (SADAIC), Lavalle 1547, 1048 Buenos Aires
T: (01) 404868; Fax: (01) 111985
Founded: 1936
Pres: Ariel Ramirez
Focus: Music; Lit
Periodical
Catálogos Bibliográficos 00156

Sociedad Argentina de Biofísica / Argentine Biophysical Society, Santa Fé 1145, 1059 Buenos Aires
Founded: 1972
Focus: Biophys 00157

Sociedad Argentina de Biología, Obligado 2490, 1428 Buenos Aires
T: (01) 7832869
Founded: 1920; Members: 180
Pres: Dr. Enrique T. Segura; Gen Secr: Dr. José Lino Barañao
Focus: Bio
Periodical
Revista (weekly) 00158

Sociedad Argentina de Botánica, c/o Instituto de Botánica Darwinion, Lavardén 200, 1642 San Isidro
Founded: 1945
Focus: Botany
Periodical
Boletín (weekly) 00159

Sociedad Argentina de Cardiología, Paraná 489, 1017 Buenos Aires
T: (01) 464221
Focus: Cardiol 00160

Sociedad Argentina de' Ciencias Fisiológicas (SACF) / Argentine Society of Physiological Sciences, Solís 453, 1078 Buenos Aires
T: (01) 3831110; Fax: (01) 3810323
Founded: 1950; Members: 180
Pres: Prof. Dr. Pedro Aramendía; Gen Secr: Dr. Ricardo A. Quinteiro
Focus: Physiology 00161

Sociedad Argentina de Ciencias Neurológicas, Psiquiátricas y Neuroquirúrgicas, Santa Fé 1171, 1059 Buenos Aires
T: (01) 411633
Founded: 1920; Members: 400
Pres: Prof. Dr. Diego Brage; Gen Secr: Prof. Dr. Carlos Márquez
Focus: Surgery; Neurology; Psychiatry
Periodical
Revista (weekly) 00162

Sociedad Argentina de Criminología, Libertad 555, 1012 Buenos Aires
Founded: 1933
Focus: Criminology 00163

Sociedad Argentina de Dermatología, B. Mitre 2517, 1039 Buenos Aires
T: (01) 489967
Founded: 1934; Members: 600
Pres: Dr. Sergio G. Stringa; Gen Secr: Dr. Carlos Bianchi
Focus: Derm 00164

Sociedad Argentina de Endocrinología y Metabolismo / Argentine Society of Endocrinology and Metabolism, Larrea 705, 1030 Buenos Aires
T: (01) 9627311; Fax: (01) 9627311
Founded: 1941; Members: 500
Pres: Dr. Isaac Sinay; Gen Secr: Dr. Alberto Chervin
Focus: Endocrinology
Periodical
Revista Argentina de Endocrinologia y Metabolismo (weekly) 00165

Sociedad Argentina de Escritores (SADE), Uruguay 1371, 1016 Buenos Aires
T: (01) 413520
Focus: Lit 00166

Sociedad Argentina de Estudios Geográficos / Argentine Society of Geographical Studies, Rodriguez Peña 158, 1020 Buenos Aires
T: (01) 402076
Founded: 1922; Members: 4000
Pres: Dr. Raúl C. Rey Balmaceda; Gen Secr: Prof. Raquel Barrera de Mesiano
Focus: Geography
Periodical
Boletín (weekly) 00167

Sociedad Argentina de Farmacología y Terapéutica, Santa Fé 1171, 1059 Buenos Aires
T: (01) 411633
Founded: 1929; Members: 100
Pres: Prof. Dr. Manuel Litter; Gen Secr: Dr. José A.L. Chiesa
Focus: Pharmacol; Therapeutics 00168

Sociedad Argentina de Fisiología Vegetal, CC 25, 1712 Castelar
T: 6211534
Founded: 1958; Members: 140
Pres: Enrique M. Sívori
Focus: Botany 00169

Sociedad Argentina de Gastroenterología, Santa Fé 1171, 1059 Buenos Aires
T: (01) 411633
Founded: 1927; Members: 900
Pres: Dr. Erman E. Crosetti; Gen Secr: Dr. Leonardo Pinchuk
Focus: Gastroenter
Periodical
Acta Gastroenterológica Latinoamericana . 00170

Sociedad Argentina de Gerontología y Geriatría, Santa Fé 1171, 1059 Buenos Aires
T: (01) 411633
Founded: 1950; Members: 160
Pres: Prof. Dr. Osvaldo Fustinoni; Gen Secr: Dr. Juan Carlos Borkowski
Focus: Geriatrics
Periodical
Sinopsis de Geriatría (weekly) 00171

Sociedad Argentina de Hematología (SAH), Av Angel Gallardo 899, 1405 Buenos Aires
Founded: 1948; Members: 350
Pres: Dr. Guillermo Carlos Vilaseca; Gen Secr: Dr. Eduardo Dibar
Focus: Hematology *00172*

Sociedad Argentina de Investigación Clínica, c/o Instituto de Investigaciones Médicas, Hospital Tornu, Donato Alvarez 3150, 1427 Buenos Aires
T: (01) 520061
Founded: 1960
Pres: Dr. Felisa C. Molinas; Gen Secr: Dr. Enrique S. Borda
Focus: Med; Surgery
Periodical
Medicina (weekly) *00173*

Sociedad Argentina de Investigación Operativa (SADIO), c/o Centro Argentino de Ingenieros, Cerrito 1250, 1010 Buenos Aires
Focus: Business Admin *00174*

Sociedad Argentina de Leprología, CC 2899, 1000 Buenos Aires
T: (01) 831815
Founded: 1954
Gen Secr: J.C. Gatti
Focus: Derm
Periodical
Leprología (2-3 times annually) *00175*

Sociedad Argentina de Medicina Nuclear (SADMN), Salta 990, 1074 Buenos Aires
Founded: 1963
Focus: Nucl Med *00176*

Sociedad Argentina de Medicina Social, Santa Fé 1171, 1059 Buenos Aires
Founded: 1939
Focus: Hygiene *00177*

Sociedad Argentina de Neurología, Psiquiatría y Neurocirugía, Santa Fé 1171, 1059 Buenos Aires
Founded: 1921
Focus: Neurology; Psychiatry; Surgery . . *00178*

Sociedad Argentina de Oftalmología, Santa Fé 1171, 1059 Buenos Aires
T: (01) 2410392
Founded: 1920; Members: 2000
Pres: Dr. Roberto Sampoalesi; Gen Secr: Dr. José A. Badia
Focus: Ophthal
Periodical
Archivos de Oftalmología de Buenos Aires (weekly) *00179*

Sociedad Argentina de Patología, Santa Fé 1171, 1059 Buenos Aires
T: (01) 411633
Founded: 1933; Members: 220
Pres: Dr. Alberto Sundblad; Gen Secr: Dr. Mabel Pomar de Gil
Focus: Pathology
Periodical
Archivos *00180*

Sociedad Argentina de Pediatría, Coronel Díaz 1971, 1425 Buenos Aires
T: (01) 8242063
Founded: 1911; Members: 7500
Pres: Dr. Carlos A. Gianantonio; Gen Secr: Dr. María Luisa Ageitos
Focus: Pediatrics
Periodical
Archivos Argentinos de Pediatría (weekly) *00181*

Sociedad Argentina de Psicología, Callao 435, 1022 Buenos Aires
T: (01) 4323760
Founded: 1930
Pres: Juan Cuatrecasas
Focus: Psych *00182*

Sociedad Argentina de Radiología, Santa Fé 1171, 1059 Buenos Aires
Founded: 1917
Focus: Radiology *00183*

Sociedad Argentina de Sociología, Trejo 241, 5000 Córdoba
T: (051) 45901
Founded: 1950
Pres: Prof. Alfredo Poviña; Gen Secr: Prof. Odorico Pires Pinto
Focus: Sociology *00184*

Sociedad Argentina de Socorros, 3080 Esperanza
Focus: Safety *00185*

Sociedad Argentina para el Estudio de la Esterilidad, Santa Fé 1171, 1059 Buenos Aires
Founded: 1945
Focus: Gynecology *00186*

Sociedad Argentina para el Estudio del Cáncer, Av San Martín 5481, 1417 Buenos Aires
Focus: Cell Biol & Cancer Res *00187*

Sociedad Central de Arquitectos, Montevideo 938, 1019 Buenos Aires
T: (01) 422375
Founded: 1886; Members: 8500
Pres: Francisco J. García Vázquez; Gen Secr: Julio Keselman
Focus: Archit *00188*

Sociedad Científica Argentina, Santa Fé 1145, 1059 Buenos Aires
T: (01) 3938406
Founded: 1872; Members: 954
Pres: Lucio R. Ballester
Focus: Sci *00189*

Sociedad de Biología de Córdoba, Colón 637, 5000 Córdoba
Founded: 1934
Focus: Bio *00190*

Sociedad de Cirugía de Buenos Aires, Santa Fé 1171, 1059 Buenos Aires
T: (01) 440664
Pres: Ivan Goñi Moreno; Gen Secr: Guillermo I. Belleville
Focus: Surgery *00191*

Sociedad de Dermatología y Sifilografía, Santa Fé 1171, 1059 Buenos Aires
Founded: 1934
Focus: Derm *00192*

Sociedad de Medicina Interna de Buenos Aires, Santa Fé 1171, 1059 Buenos Aires
Founded: 1925
Focus: Intern Med *00193*

Sociedad de Medicina Interna de Córdoba, Colón 637, 5000 Córdoba
Focus: Intern Med *00194*

Sociedad de Psicología Médica, Psicoanálisis y Medicina Psicosomática, Santa Fé 1171, 1059 Buenos Aires
T: (01) 440664
Founded: 1940; Members: 80
Pres: Dr. F.A. Gioannini; Gen Secr: Dr. M.A. Rubinstein
Focus: Psychiatry; Psych *00195*

Sociedad de Tisiología y Neumonología del Hospital Tornu y Dispensarios, c/o Hospital Tornu, Donato Alvarez 3000, 1427 Buenos Aires
Founded: 1932
Focus: Pulmon Dis *00196*

Sociedad Entomológica Argentina (SEA) / Argentine Entomological Society, c/o Museo de la Plata, 1900 La Plata
T: (021) 257527
Founded: 1926; Members: 400
Pres: Dr. Norma B. Diaz; Gen Secr: Dr. Amalia A. Lanteri
Focus: Entomology
Periodical
Revista de la Sociedad Entomológica Argentina (weekly) *00197*

Sociedad General de Autores de la Argentina, J.A. Pacheco de Melo 1820, 1126 Buenos Aires
T: (01) 421227
Founded: 1910; Members: 2000
Pres: Roberto A. Talice; Gen Secr: Emilio Villalba Wesh
Focus: Lit
Periodical
Boletín (weekly) *00198*

Sociedad Hebraica Argentina, Sarmiento 2233, 1044 Buenos Aires
T: (01) 480178
Focus: Lit *00199*

Sociedad Latinoamericana de Ciencias de los Suelos, c/o Facultad de Ciencias Agrarias, Chacras de Corita, 5500 Mendoza
Focus: Agri *00200*

Sociedad Rural Argentina, Florida 460, 1005 Buenos Aires
T: (01) 3220468
Founded: 1866; Members: 12000
Pres: Dr. Eduardo A.C. de Zavalia
Focus: Agri; Vet Med *00201*

Unión Matemática Argentina, FAMDF, Valparaiso y R. Martinez, Ciudad Universitaria, 5000 Cordoba
T: (051) 690068; Fax: (051) 210669
Founded: 1936; Members: 650
Pres: Dr. Juan Tirao; Gen Secr: Jorge Vargas
Focus: Math
Periodical
Revista de la Unión Matemática Argentina *00202*

Armenia

Armenian Academy of Sciences, Pr Marshala Bagramyana 24, 375019 Yerewan
T: 524580; Tx: 243344
Founded: 1943; Members: 129
Pres: V.A. Ambartsumyan; Gen Secr: D.M. Sedrakian
Focus: Sci
Periodicals
Atrofizika: Astrophysics
Biologicheskii Zhurnal Armenii: Biological Journal of Armenia
Doklady: Report
Istoriko-Filologicheskii Zhurnal: Historical and Philological Journal
Izvestiya: Bulletin
Khimicheskii Zhurnal Armenii: Chemical Journal of Armenia
Neirokhimiya: Neurochemistry
Soobshcheniya Byurakanskoi Observatorii: Report of the Byurakan Astrophysical Observatory
Vestnik Obshchestvennykh Nauk: Herald of Social Sciences
Zhurnal Experimentalnoi i Klinicheskoi Medisiny: Journal of Experimental and Clinical Medicine
. . . . *00203*

Commission on Computer Technology, Pr Marshala Bagramyana 24, 375019 Yerevan
Pres: F.T. Sarkisyan
Focus: Computer & Info Sci *00204*

Committee on UNESCO Programme „Man and Biosphere", Ul Abouyana 68, 375025 Yerevan
T: 563570
Pres: A.B. Bagdasaryan
Focus: Bio *00205*

Council on Informatics, Ul Paruyra 1, 375044 Yerevan
Pres: R.R. Varshamov
Focus: Computer & Info Sci *00206*

Council on Use of Atomic Energy and Technology, Ul Markaryana 2, 375036 Yerevan
T: 355063
Pres: A.M. Petrosyants
Focus: Nucl Res *00207*

Interdepartmental Council for Study of Ethnic Problems, Pr Marshala Bagramyana 24, 375019 Yerevan
T: 581823
Pres: K.A. Barsegyan
Focus: Ethnology *00208*

Interdepartmental Council on Physicochemical Biology and Biotechnology, Pr Marshala Bagramyana 24, 375019 Yerevan
T: 281814
Pres: A.A. Galoyan
Focus: Bio; Biophys *00209*

Yerevan Academy of National Economy, Ul Abovyana 52, 375025 Yerevan
T: 560413; Tx: 411626
Members: 1975
Focus: Econ
Periodical
Economics *00210*

Australia

Academy of the Social Sciences in Australia (ASSA), GPOB 1956, Canberra, ACT 2601
Founded: 1971; Members: 174
Pres: Prof. P.W. Sheehan; Gen Secr: Prof. O. McDonagh
Focus: Sociology
Periodicals
Newsletter (weekly)
Report (weekly) *00211*

Accounting Association of Australia and New Zealand (A.A.A.N.Z.), c/o Dept of Accounting and Finance, Monash University, Clayton, Vic 3168
Focus: Business Admin
Periodical
Accounting and Finance (weekly) . . . *00212*

Agricultural Engineering Society, Australia (AESA), 191 Royal Parade, Parkville, Vic 3052
Fax: (03) 7420202
Founded: 1950; Members: 550
Focus: Agri; Eng
Periodical
Agricultural Engineering Australia (weekly) *00213*

Anthropological Society of New South Wales, c/o Anthropological Dept, University of Sydney, Sydney, NSW 2006
Founded: 1928; Members: 170
Pres: P. Hinton; Gen Secr: P.J.F. Newton
Focus: Anthro
Periodical
Mankind (3 times annually) *00214*

Anthropological Society of Queensland, c/o Geology Dept, University of Queensland, Saint Lucia, Qld 4067
Founded: 1948
Focus: Anthro *00215*

Anthropological Society of South Australia, c/o Conservation Centre, 120 Wakefield St, Adelaide, SA 5000
T: (08) 2235155
Founded: 1926; Members: 125
Pres: S. Hemming; Gen Secr: R. Allison
Focus: Anthro; Archeol; Hist
Periodicals
Journal of the Anthropological Society of South Australia (weekly)
Newsletter of the Anthropological Society of South Australia (9 times annually) *00216*

Anthropological Society of Western Australia (A.S.W.A.), c/o Dept of Anthropology, University of Western Australia, Nedlands, WA 6009
T: 3802853
Founded: 1958; Members: 150
Focus: Anthro
Periodical
Anthropology News: Proceedings (10 times annually) *00217*

Archaeological and Anthropological Society of Victoria, POB 328C, Melbourne, Vic 3001
Founded: 1934
Focus: Anthro; Archeol
Periodical
Artefact (weekly) *00218*

Arts Council of Australia, 80 George St, The Rocks, Sydney, NSW 2000
Founded: 1946; Members: 20000
Pres: David Hamer; Gen Secr: Jill Nash
Focus: Arts *00219*

Asian and Pacific Association for Social Work Education (APASWE), c/o Dept of Social Work, La Trobe University, Bundoora 3083
T: (03) 4792570 ext 2854; Tx: 33143; Fax: (03) 4785814
Founded: 1974
Pres: Prof. David R. Cox; Gen Secr: Janet George
Focus: Soc Sci; Educ
Periodical
APASWE Newsletter *00220*

Asian-Australasian Society of Neurological Surgeons, 201 Wickham Terrace, Brisbane 4001
T: (07) 8393393
Founded: 1964; Members: 18
Gen Secr: Dr. Leigh Atkinson
Focus: Neurology; Surgery *00221*

Asian Coordinating Group for Chemistry (ACGC), c/o Dept of Chemistry, University of Western Australia, Nedlands 6009
T: (09) 3803152; Tx: 92992; Fax: (09) 3801005
Founded: 1984
Gen Secr: Prof. J.R. Cannon
Focus: Chem *00222*

Asian Network for Analytical and Inorganic Chemistry (ANAIC), c/o School of Mathematical and Physical Sciences, Murdoch University, Perth 6150
T: (09) 3322211, 3322152; Tx: 92711; Fax: (09) 3322507
Founded: 1984; Members: 18
Gen Secr: Dr. J. Webb
Focus: Chem
Periodical
Network News *00223*

Asian Pacific Confederation of Chemical Engineering (APCChE), c/o Institution of Engineers in Australia, 11 National Circuit, Barton 2600
T: (06) 706555; Fax: (06) 2731488
Founded: 1975
Gen Secr: Robert T. Flower
Focus: Eng
Periodical
APCChE Newsletter *00224*

Asian-Pacific Corrosion Control Organization (APCC), Science Centre, 35-43 Clarence St, Sydney 2000
T: (02) 2659177
Focus: Metallurgy *00225*

Asian Pacific Federation of Clinical Biochemistry (APFCB), c/o Clinical Chemistry Div, Institute of Medical and Veterinary Science, POB 14, Rundle Mall Post Office, Adelaide 5000
T: (08) 2287563; Fax: (08) 2287538
Founded: 1982
Pres: Prof. K. Miyai; Gen Secr: T.D. Geary
Focus: Chem
Periodical
APFCB News (weekly) *00226*

Asian Pan-Pacific Society for Paediatric Gastroenterology and Nutrition (APPSGAN), c/o Health Dept of Western Australia, 189 Royal St, East Perth 6004
T: (09) 2222491; Fax: (09) 2222481
Founded: 1976
Pres: Dr. Michael Gracey
Focus: Nutrition; Pediatrics; Gastroenter . *00227*

Asia Pacific Food Analysis Network (APFAN), c/o Div of Botany and Zoology, Australian National University, POB 4, Canberra, ACT 2601
T: (06) 2490775; Tx: 62760; Fax: (06) 2495573
Founded: 1989
Gen Secr: Dr. J. Howard Bradbury
Focus: Food *00228*

Association of Australasian and Pacific Area Police Medical Officers, c/o Office of Forensic Medicine, POB 2763, Melbourne, 3000
T: (03) 6671657; Fax: (03) 6671709
Pres: Dr. W. Ryan
Focus: Forensic Med *00229*

Association of Medical Directors of the Australian Pharmaceutical Industry, 12 Wharf Rd, West Ryde, NSW 2114
T: 850122
Founded: 1962; Members: 40
Focus: Med; Pharmacol *00230*

Astronomical Society of Australia (ASA), c/o Physics Dept, Macquarie University, Macquarie, NSW 2109
T: (02) 8058933; Fax: (02) 8058983
Founded: 1966; Members: 250
Pres: Dr. D.A. Allen; Gen Secr: Dr. L.R. Allen; Dr. M.L. Duldig
Focus: Astronomy; Astrophys
Periodical
Proceedings (biennial) *00231*

Astronomical Society of Queensland, 14 Suelin St, Boondall, Qld 4034
Founded: 1927
Focus: Astronomy 00232

Astronomical Society of South Australia, GPOB 199, Adelaide, SA 5001
T: (08) 3381231
Founded: 1892; Members: 350
Pres: Prof. G. Goodwin; Gen Secr: A.C. Beresford
Focus: Astronomy
Periodicals
Astronomical Data (weekly)
The Bulletin (weekly) 00233

Astronomical Society of Tasmania, c/o Queen Victoria Museum, Wellington St, Launceston, Tas 7250
Founded: 1930; Members: 93
Pres: M. George
Focus: Astronomy
Periodical
Bulletin (weekly) 00234

Astronomical Society of Victoria (ASV), GPOB 1059J, Melbourne, Vic 3001
Founded: 1922; Members: 680
Pres: K.M. Harrison; Gen Secr: Dr. S.J. Roberts
Focus: Astronomy
Periodicals
ASV Newsletter (weekly)
Yearbook (weekly) 00235

Astronomical Society of Western Australia, GPOB 1460, Perth, WA 6001
Founded: 1950; Members: 125
Pres: F. Ward; Gen Secr: A. Murray
Focus: Astronomy
Periodical
Journal (weekly) 00236

Australasian Association for the History, Philosophy and Social Studies of Science (AAHPSSS), c/o History Dept, University, Sydney, NSW 2006
T: (02) 6923610
Founded: 1967; Members: 150
Focus: Hist; Philos; Sociology 00237

Australasian Association of Philosophy (AAP), c/o Dept of Philosophy, La Trobe University, Melbourne, Vic 3083
Founded: 1923; Members: 400
Pres: Robert Pargetter; Gen Secr: G. Marshall
Focus: Philos
Periodical
Australasian Journal of Philosophy (weekly) 00238

The Australasian Ceramic Society, c/o Dept of Materials Science, University of Technology, Broadway, POB 123, Sydney, NSW 2007
T: (02) 3301784; Fax: (02) 3301755
Founded: 1961; Members: 550
Pres: R.G. Bowman; Gen Secr: Dr. K.D. Reeve
Focus: Materials Sci
Periodical
Journal (weekly) 00239

Australasian College of Dermatologists, 136 Pittwater Rd, Gladesville, NSW 2111
T: 8796177; Fax: 8161174
Founded: 1966; Members: 229
Pres: Dr. J.B. Rohr; Gen Secr: Dr. G.B. Crosland
Focus: Derm
Periodical
The Australasian Journal of Dermatology (3 times annually) 00240

Australasian College of Physical Scientists and Engineers in Medicine, c/o Dept of Physical Sciences, Peter MacCallum Cancer Institute, 481 Little Lonsdale St, Melbourne, Vic 3000
T: (03) 6415569; Tx: 34567; Fax: (03) 6703357
Founded: 1977; Members: 300
Pres: E.R. Scull; Gen Secr: Dr. J.R. Coles
Focus: Med 00241

Australasian College of Venereologists, c/o Sydney Sexual Health Centre, Sydney Hospital, Macquarie St, Sydney, NSW 2000
T: (02) 2328319; Fax: (02) 2233183
Founded: 1988
Pres: Dr. C.R. Philpot; Gen Secr: Dr. J. Moran
Focus: Venereology
Periodical
Venereology 00242

Australasian Corrosion Association (A.C.A.), c/o Australasian Corrosion Centre, POB 250, Clayton, Vic 3168
T: 35440066; Fax: 35435905
Founded: 1955; Members: 1200
Pres: L.H. Boulton; Gen Secr: Sally Nugent
Focus: Metallurgy
Periodicals
ACA Conference Proceedings (weekly)
Corrosion Australasia (weekly) 00243

The Australasian Institute of Mining and Metallurgy, 191 Royal Parade, POB 122, Parkville, Vic 3052
Founded: 1893; Members: 6400
Focus: Mining; Metallurgy 00244

Australasian Pharmaceutical Science Association (APSA), c/o Dept of Pharmacy, University of Queensland, Saint Lucia, Qld 4067
T: (07) 3773191; Fax: (07) 3715896
Founded: 1960; Members: 150
Pres: Dr. I.G. Tucker; Gen Secr: Prof. E.J. Triggs
Focus: Pharmacol
Periodical
Newsletter (2-4 times annually) 00245

Australasian Plant Pathology Society (APPS), c/o Adelaide Botanic Garden, North Terrace, Adelaide, SA 6000
T: (08) 2282327
Founded: 1969; Members: 350
Focus: Botany 00246

Australasian Political Studies Association (APSA), c/o Federalism Research Centre, Australian National University, Canberra, ACT 2601
T: (06) 2493668
Founded: 1952
Pres: Leslie Holmes; Gen Secr: Christine Fletcher
Focus: Poli Sci
Periodicals
APSA Newsletter (5 times annually)
Politics (weekly) 00247

Australasian Universities Language and Literature Association (AULLA), c/o Dept of French, University of Sydney, Sydney 2006
T: (02) 6922397
Founded: 1950
Gen Secr: Dr. Robert White
Focus: Ling; Lit
Periodical
AUMLA (weekly) 00248

Australia Council, 181 Lawson St, Redfern, NWS 2016
T: (02) 9509000; Tx: 26023; Fax: (02) 9509111
Founded: 1968
Pres: Rodney Hall
Focus: Arts 00249

Australian Academy of Science (AAS), GPOB 783, Canberra, ACT 2601
T: (06) 2475777; Tx: 82406; Fax: (06) 2574620
Founded: 1954; Members: 208
Pres: Prof. D.P. Craig; Gen Secr: P. Vallee
Focus: Sci
Periodicals
Biology in Action (weekly)
Historical Records of Australian Science (weekly)
Yearbook (weekly) 00250

Australian Academy of Technological Sciences and Engineering, Ian McLennan House, 197 Royal Parade, Parkville, Vic 3052
T: (03) 3470622; Tx: 33552; Fax: (03) 3478237
Founded: 1976; Members: 185
Pres: Sir Rupert Myers; Gen Secr: H.C. Coe
Focus: Eng
Periodicals
Annual Report (weekly)
Focus Newsletter
Handbook
Symposia Series 00251

Australian Academy of the Humanities (A.A.H.), GPOB 93, Canberra, ACT 2601
T: (06) 2487744; Fax: (06) 2486287
Founded: 1969; Members: 208
Pres: Prof. G.E.O. Schulz; Gen Secr: Prof. D.J. Mulvaney
Focus: Humanities
Periodical
Proceedings 00252

Australian Acoustical Society (A.A.S.), Private Bag 1, Darlinghurst, NSW 2010
T: (02) 3316920; Fax: (02) 3317296
Founded: 1971; Members: 390
Pres: S.E. Samuels; Gen Secr: R.A. Piesse
Focus: Acoustics
Periodical
Acoustics Australia (3 times annually) . . 00253

The Australian Agricultural Council, c/o Commonwealth Dept of Primary Industries and Energy, Canberra, ACT 2600
Founded: 1934; Members: 9
Focus: Agri 00254

Australian Agricultural Economics Society (A.A.E.S.), c/o Dept of Agricultural Economics and Business Management, University of New England, Armidale, NSW 2351
T: (067) 732915; Fax: (067) 711
Founded: 1957; Members: 1300
Pres: Dr. B.S. Fisher; Gen Secr: Dr. V.E. Wright
Focus: Agri
Periodical
Australian Journal of Agricultural Economics (3 times annually) 00255

Australian and New Zealand Association for Canadian Studies, c/o School of Earth Sciences, Macquarie University, North Ryde, NSW 2113
T: 8888000; Tx: 22377
Focus: Hist; Geography 00256

Australian and New Zealand Association for Medieval and Renaissance Studies, c/o Dept of English, Sydney University, Sydney, NSW 2006
Founded: 1968; Members: 300
Gen Secr: D.P. Speed
Focus: Hist; Lit
Periodical
Parergon (weekly) 00257

Australian and New Zealand Association for the Advancement of Science (ANZAAS), c/o University of Sydney, 118 Darlington Rd, Sydney, NSW 2006
Founded: 1886; Members: 1000
Gen Secr: Dr. D.J. O'Connor
Focus: Sci 00258

Australian and New Zealand Society of Nuclear Medicine, c/o Science Centre, Private Bag 1, Darlinghurst, NSW 2010
T: (02) 3316920
Founded: 1969; Members: 500
Pres: A. Van der Schaaf; Gen Secr: J.G. Chan
Focus: Nucl Med
Periodical
Newsletter of the Australian and New Zealand Society of Nuclear Medicine (weekly) . . 00259

Australian and New Zealand Solar Energy Society, POB 124, Caulfield East, Vic 3145
T: (03) 5717557; Fax: (03) 5635173
Founded: 1962; Members: 800
Focus: Energy
Periodical
Solar Progress incorporating South Wind (weekly) 00260

Australian Association for the Study of Religions (AASR), c/o Dept of Studies in Religion, University of Queensland, Saint Lucia, Qld 4067
T: (07) 3772154
Founded: 1975; Members: 320
Pres: Dr. Edgar Conrad; Gen Secr: Scott Cowdell
Focus: Rel & Theol; Hist
Periodical
Australian Religion Studies Review (weekly) 00261

Australian Association of Adult and Community Education, POB 1346, Canberra, ACT 2601
T: (06) 2478383; Fax: (06) 2574498
Founded: 1961; Members: 500
Pres: Ann Whyte; Gen Secr: M. Hannan
Focus: Adult Educ
Periodicals
Australian Journal of Adult Education (3 times annually)
Newsletter (weekly) 00262

Australian Association of Clinical Biochemists (AACB), POB 118, Maylands, WA 6051
T: (09) 3285720; Fax: (09) 2275040
Founded: 1961; Members: 1500
Pres: L. Wyndham; Gen Secr: L. Watkinson
Focus: Chem
Periodicals
The Clinical Biochemist Newsletter (weekly)
The Clinical Biochemist Review (weekly) 00263

Australian Association of Gerontology (A.A.G.), 191 Royal Parade, Parkville, Vic 3052
T: (03) 3472570; Fax: (03) 3477819
Founded: 1964; Members: 1000
Focus: Geriatrics
Periodical
Proceedings (weekly) 00264

Australian Association of Neurologists (A.A.N.), c/o Dept of Neurology, Royal Melbourne Hospital, Melbourne, Vic 3050
T: (03) 3427272
Founded: 1950; Members: 360
Pres: Dr. J. King; Gen Secr: Dr. C. Kilpatrick
Focus: Neurology
Periodical
Clinical and Experimental Neurology (weekly) 00265

Australian Association of Occupational Therapists (A.A.O.T.), 97 Grange Rd, Alphington, Vic 3078
Founded: 1945; Members: 7
Pres: Margaret Hiep; Gen Secr: Helen McGee
Focus: Therapeutics
Periodical
Australian Occupational Therapy Journal (weekly) 00266

Australian Association of Social Workers, POB 1059, North Richmond, Vic 3121
T: 416960, 414797
Founded: 1946
Focus: Sociology 00267

Australian Atomic Energy Commission Research Establishment, Private Mail Bag, Sutherland, NSW 2232
Founded: 1955
Focus: Nucl Res
Periodical
Annual Report (weekly) 00268

Australian Bar Association, Owen Dixon Chambers, 205 William St, Melbourne, Vic 3000
T: (03) 6087111; Tx: 36782; Fax: (03) 6702959
Founded: 1962; Members: 2350
Pres: G.L. Davies; Gen Secr: D.M. Brennan
Focus: Law 00269

Australian Biochemical Society, c/o Dr. R.J. Porra, C.S.I.R.O.-Div of Plant Industry, GPOB 1600, Canberra, ACT 2601
T: (06) 465465; Tx: 62351; Fax: (06) 473785
Founded: 1955; Members: 1000
Focus: Biochem
Periodical
Proceedings (weekly) 00270

Australian Bird Study Association, GPOB A313, Sydney South, NSW 2001
Focus: Ornithology
Periodical
Corelle (5 times annually) 00271

Australian Cancer Society, 500 George St, Sydney, NSW 2000
Founded: 1961
Pres: Dr. K.J. Donald; Gen Secr: L.A. Wright
Focus: Cell Biol & Cancer Res
Periodical
Cancer Forum (3 times annually) . . . 00272

Australian Clay Minerals Society (ACMS), c/o C.S.I.R.O.-Div of Soils, GPOB 639, Canberra, ACT 2601
Founded: 1962; Members: 47
Focus: Mineralogy
Periodical
Proceedings National Conference 00273

Australian College of Education, 42 Geils Court, Deakin, ACT 2600
T: (06) 2811677
Founded: 1959; Members: 6300
Gen Secr: Beverley J. Pope
Focus: Educ
Periodicals
ACE News (weekly)
Unicorn (weekly) 00274

The Australian College of Obstetricians and Gynaecologists (A.C.O.G.), 8 La Trobe St, Melbourne, Vic 3000
T: (03) 3470231
Founded: 1978; Members: 870
Focus: Gynecology 00275

Australian College of Paediatrics (A.C.P.), POB 30, Parkville, Vic 3052
T: 3298181; Fax: 3297030
Founded: 1956; Members: 700
Pres: Dr. W.G. Grigor; Gen Secr: Dr. L.E.G. Sloan
Focus: Pediatrics
Periodical
Australian Paediatric Journal (weekly) . . 00276

Australian Committee of Directors of Principals (ACDP), Churchill House, 218 Northbourne Av, Braddon, ACT 2601
T: (062) 478880; CA: ACOMDAP; Fax: (062) 573079
Founded: 1983; Members: 40
Pres: Dr. B. Twistlethwayte; Gen Secr: J.R. Scutt
Focus: Educ
Periodical
Newsletter (4-5 times annually) 00277

Australian Conservation Foundation, 340 Gore St, Fitzroy, Vic 3065
T: (03) 4161455; Fax: (03) 4160767
Founded: 1965; Members: 40000
Pres: Peter Garrett; Gen Secr: Dr. Phillip Toyne
Focus: Nat Res; Ecology
Periodicals
Habitat Australia
Newsletter 00278

The Australian Council for Educational Research (ACER), 9 Frederick St, Hawthorn, Vic 3122
Founded: 1930; Members: 25
Pres: Prof. P.H. Karmel; Gen Secr: Dr. Barry McGaw
Focus: Educ
Periodicals
Australian Education Index (weekly)
Australian Education Review (weekly)
Australian Journal of Education (3 times annually)
Bibliography of Educational Theses in Australia (weekly)
Newsletter (3 times annually) 00279

The Australian Council for Health, Physical Education and Recreation (ACHPER), POB 304, Hindmarsh, SA 5007
Founded: 1952; Members: 1700
Pres: E. Murphy; Gen Secr: Lyall Cowell
Focus: Public Health; Sports
Periodicals
ACHPER National Journal (weekly)
Australian Journal of Science and Medicine in Sport 00280

Australian Council of National Trusts, POB 1002, Civic Square, ACT 2608
T: (06) 2476766; Fax: (06) 2491395
Founded: 1965; Members: 80000
Gen Secr: Duncan Marshall
Focus: Ecology; Preserv Hist Monuments
Periodical
Heritage Australia 00281

Australian Dental Association (A.D.A.), 75 Lithgow St, POB 520, Saint Leonards, NSW 2065
T: (02) 9064412; Fax: (02) 9064676
Founded: 1928; Members: 6939
Pres: Dr. John Havelock Southwick; Gen Secr: Colin H. Wall
Focus: Dent
Periodical
Australian Dental Journal (weekly) . . . 00282

Australian Dental Association, Northern Territory Branch, POB 4496, Darwin, NT 5794
Focus: Dent 00283

Australian Dental Association, Queensland Branch, 61 Brookes St, Bowen Hills, Qld 4006
T: (07) 2529866; Fax: (07) 2524488
Focus: Dent 00284

Australian Dental Association, South Australian Branch, POB 858, Unley, SA 5061
T: (08) 2728111
Focus: Dent 00285

Australian Dental Association, Tasmania Branch, 130 Main Rd, New Town, Tas 7008
T: (002) 295917
Focus: Dent 00286

Australian Dental Association, Victorian Branch, POB 434, Toorak, Vic 3142
T: (03) 2408318; Fax: (010) 6129231261
Focus: Dent 00287

Australian Dental Association, Western Australia Branch, 14 Altona St, West Perth, WA 6005
T: (09) 3217880
Focus: Dent 00288

The Australian Entomological Society, c/o Plant Research Institute, Swan St, Burnley, Vic 3121
Fax: (03) 8195653
Founded: 1965; Members: 769
Focus: Entomology
Periodicals
Journal of the Australian Entomological Society (weekly)
Myrmecia (weekly) 00289

Australian Fabian Society, GPOB 2707X, Melbourne, Vic 3001
Focus: Socialism; Poli Sci
Periodical
Fabian Pamphlets 00290

Australian Federation for Medical and Biological Engineering, c/o Dept of Electrical Engineering, University of Melbourne, Parkville, Vic 3052
T: 3416686
Founded: 1959
Focus: Eng; Med; Bio 00291

Australian Geography Teachers' Association, c/o Adelaide College of T.A.F.E., 208 Currie St, Adelaide, SA 5000
T: (08) 2123622
Founded: 1967
Focus: Geography; Educ
Periodical
Geographical Education (weekly) 00292

Australian Geomechanics Society, c/o The Institution of Engineers, Australia, 11 National Circuit, Barton, ACT 2600
T: (062) 706555; Tx: 726548; Fax: (062) 731488
Founded: 1970; Members: 638
Pres: Dr. N.S. Mattes; Gen Secr: R. Bushnell
Focus: Mining; Eng; Geology; Mineralogy
Periodicals
Australasian Geomechanics Computing Newsletter (weekly)
Australian Geomechanics (weekly) . . . 00293

Australian Institute for Non-Destructive Testing (AINDT), 191 Royal Parade, Parkville, Vic 3052
T: 3471166
Founded: 1963; Members: 700
Focus: Materials Sci
Periodical
Non-Destructive Testing (10 times annually)
. . . . 00294

Australian Institute of Aboriginal Studies, GPOB 553, Canberra City, ACT 2601
Founded: 1961
Focus: Ethnology
Periodicals
Australian Aboriginal Studies
Bibliographies
Manuals
Newsletter (weekly)
Regional and Research Studies 00295

Australian Institute of Agricultural Science (AJA5), 191 Royal Parade, Ste 303, Parkville, Vic 3052
T: (03) 3471277; Fax: (03) 3471792
Founded: 1935; Members: 3100
Gen Secr: Simon L. Field
Focus: Agri
Periodical
Agricultural Science (weekly) 00296

Australian Institute of Archaeology, 174 Collins St, Melbourne, Vic 3000
T: (03) 6503477
Founded: 1946
Pres: J.A. Thompson
Focus: Archeol
Periodical
Buried History (weekly) 00297

Australian Institute of Cartographers, GPOB 1817, Brisbane, Qld 4001
Founded: 1953; Members: 1502
Gen Secr: D.V. Diggles
Focus: Cart
Periodical
Cartography (weekly) 00298

Australian Institute of Credit Management, POB 558, Artarmon, NSW 2064
Founded: 1967; Members: 3000
Pres: W.L. Duncam; Gen Secr: L.J. Wilson
Focus: Finance
Periodical
Credit Review (weekly) 00299

Australian Institute of Energy, POB 230, Wahroonga, NSW 2076
T: (02) 4491800
Founded: 1978; Members: 1500
Pres: Ian Smith; Gen Secr: Ian Maitland
Focus: Energy
Periodical
News Journal (weekly) 00300

Australian Institute of Food Science and Technology, POB 277, Pymble, NSW 2073
Founded: 1967; Members: 1754
Pres: Alan Mortimer; Gen Secr: Judy Fairbrother
Focus: Food; Eng; Nutrition
Periodical
Food Technology in Australia (weekly) . 00301

Australian Institute of Homoeopathy, POB 122, Roseville, NSW 2069
Founded: 1946
Pres: Roger Denne; Gen Secr: Joan Brandt
Focus: Med
Periodical
Newsletter 00302

Australian Institute of Industrial Psychology, 33-35 Bligh St, Sydney, NSW 2000
T: (02) 2325063
Founded: 1927
Focus: Psych 00303

Australian Institute of International Affairs (A.I.I.A.), Queen Victoria Terrace, POB E181, Canberra, ACT 2600
T: (06) 2822133; Fax: (06) 2852334
Founded: 1933; Members: 2000
Pres: R.J. White; Gen Secr: S. Allica
Focus: Poli Sci
Periodical
Australian Outlook (3 times annually) . . 00304

Australian Institute of Management, 10 Saint Leonards Av, Saint Kilda, Vic 3182
T: (03) 5348181; Fax: (03) 5345050
Founded: 1941; Members: 25500
Focus: Business Admin 00305

Australian Institute of Marine Science, PMB 3, Townsville, Qld 4810
Fax: (077) 725852
Founded: 1972
Gen Secr: J. Bradridge
Focus: Oceanography
Periodical
Annual Report: Projected Research Activities
. . . . 00306

Australian Institute of Mining and Metallurgy (AIMM), Clunies Ross House, 191 Royal Parade, Parkville, Vic 3052
T: (03) 3473166; Fax: (03) 3478525
Founded: 1893; Members: 8400
Pres: Sir Bruce Watson; Gen Secr: J.M. Webber
Focus: Mining; Metallurgy
Periodicals
Bulletin
Proceedings 00307

Australian Institute of Navigation (A.I.N.), GPOB 2250, Sydney, NSW 2001
T: (02) 944399
Founded: 1949; Members: 600
Pres: D. Pyett; Gen Secr: A.P. Davey
Focus: Navig 00308

Australian Institute of Nuclear Science and Engineering (AINSE), PMB 1, Menai, NSW 2234
T: (02) 7173376; Fax: (02) 7179268
Founded: 1958
Gen Secr: Dr. R. Gammon
Focus: Nucl Res
Periodical
AINSE Annual Report (weekly) 00309

Australian Institute of Physics (AIP), 191 Royal Parade, Parkville, Vic 3052
T: (03) 3476872; Fax: (03) 3491409
Founded: 1963; Members: 2400
Pres: Prof. A. Thomas; Gen Secr: Dr. J. Riley
Focus: Physics
Periodical
The Australian Physicist (weekly) . . . 00310

Australian Institute of Political Science (A.I.P.S.), 149 Castlereagh St, Ste 303B, Sydney, NSW 2000
T: (02) 2648923
Founded: 1932; Members: 3000
Pres: J. Best
Focus: Poli Sci
Periodical
The Australian Quarterly (weekly) . . . 00311

Australian Institute of Quantity Surveyors, 27-29 Napier Close, Deakin, ACT 2600
T: (06) 2822222; Fax: (06) 2852427
Founded: 1971; Members: 1500
Pres: A.J. Ott; Gen Secr: A.G. Hordern
Focus: Surveying
Periodical
The Building Economist (weekly) 00312

Australian Institute of Radiography (A.I.R.), 212 Clarendon St, East Melbourne, Vic 3002
T: (03) 4193336
Founded: 1950; Members: 2300
Pres: Gillian Tickall; Gen Secr: E.M. Hughes
Focus: Radiology
Periodical
The Radiographer (weekly) 00313

The Australian Institute of Refrigeration, Air Conditioning and Heating (AIRAH), 191 Royal Parade, Parkville, Vic 3052
T: (03) 3474941
Founded: 1920; Members: 2050
Focus: Electric Eng
Periodical
Australian Refrigeration, Air Conditioning and Heating: AIRAH Journal (weekly) 00314

Australian Institute of Urban Studies (AIUS), POB 809, Canberra, ACT 2601
Founded: 1966; Members: 750
Pres: Justice Rae Else-Mitchell; Gen Secr: Anthony D. Winter
Focus: Urban Plan
Periodicals
Australian Institute of Urban Studies (weekly)
Australian Urban Studies (weekly) . . . 00315

Australian Institute of Valuers and Land Economists, 6 Campion St, Deakin, ACT 2600
T: (06) 2822411; Fax: (06) 2852194
Founded: 1926; Members: 7000
Pres: R. Westwood; Gen Secr: Dr. G.R. Webb
Focus: Finance
Periodical
The Valuer and Land Economist (weekly) 00316

Australian Law Librarians' Group, 367 Collins St, POB 92, Melbourne, Vic 3000
Focus: Libraries & Bk Sci; Law
Periodical
Australian Law Librarians' Group Newsletter (weekly) 00317

Australian Library and Information Association, Queen Victoria Terrace, POB E441, Canberra, ACT 2600
Founded: 1936; Members: 9500
Pres: Jenny Cram; Gen Secr: S. Kosse
Focus: Libraries & Bk Sci
Periodicals
Australian Academic Research Libraries (weekly)
Australian Library Journal (weekly)
Australian Special Libraries News (weekly)
Cataloguing Australia (weekly)
Incite (weekly)
Orana (weekly) 00318

Australian Mammal Society, c/o C.S.I.R.O.-Div of Wildlife and Ecology, POB 84, Lyneham, ACT 2602
Founded: 1959; Members: 560
Pres: Dr. A. Cockburn; Gen Secr: P. Catling
Focus: Zoology
Periodical
Australian Mammalogy (weekly) 00319

The Australian Mathematical Society (A.M.S.), c/o Dept of Mathematics, University of Tasmania, Hobart, Tas 7001
T: (002) 202442; Fax: (002) 202867
Founded: 1956; Members: 900
Pres: Prof. J.H. Rubinstein; Gen Secr: Prof. D. Elliott
Focus: Math
Periodicals
Bulletin (weekly)
Gazette (weekly)
Journal: Series A (weekly)
Journal: Series B (weekly) 00320

Australian Mathematics Competition Committee (AMC), POB 1, Belconnen, ACT 2616
T: 522227
Founded: 1978
Pres: P.J. O'Halloran; Gen Secr: S. Bakker
Focus: Math
Periodical
Australian Mathematics Competition for the Westpac Awards: Solutions and Statistics (weekly)
. . . . 00321

Australian Medical Association (AMA), 42 Macquarie St, Barton, ACT 2600
T: (06) 2705400; Tx: 24814; Fax: (06) 2705499
Founded: 1962; Members: 18000
Gen Secr: Alan Passmore
Focus: Med
Periodicals
Australian Medicine
Medical Journal of Australia 00322

Australian National Committee on Large Dams (ANCOLD), c/o Queensland Water Resources Commission, GPOB 2454, Brisbane, Qld 4001
T: (07) 2247281
Founded: 1960; Members: 112
Focus: Civil Eng
Periodical
ANCOLD Bulletin (3 times annually) . . 00323

Australian Numismatic Society (ANS), Royal Exchange, POB R4, Sydney, NSW 2000
T: (02) 4195790
Founded: 1913; Members: 300
Pres: T.E. Hanley; Gen Secr: F.S. Dobbin
Focus: Numismatics
Periodical
Report (weekly) 00324

Australian Optometrical Association, POB 185, Carlton South, Vic 3053
T: (03) 6636833
Founded: 1918; Members: 2000
Pres: Stephen Leslie; Gen Secr: Dr. Joseph Chakman
Focus: Optics
Periodical
Clinical & Experimental Optometry (weekly)
. . . . 00325

Australian Orthopaedic Association, 229 Macquarie St, Sydney, NSW 2000
T: (02) 2333018; Fax: (02) 2218301
Founded: 1936
Focus: Orthopedics 00326

Australian Petroleum Exploration Association (APEA), 60 Margaret St, Sydney, NSW 2000
T: (02) 276718
Founded: 1959; Members: 219
Focus: Mining
Periodical
APEA Journal (weekly) 00327

Australian Physiological and Pharmacological Society, c/o Dept of Physiology, University of Melbourne, Melbourne, Vic 3052
T: (03) 3445846; Fax: (03) 3445818
Founded: 1960; Members: 460
Gen Secr: Prof. T.O. Morgan
Focus: Physiology; Pharmacol
Periodical
Proceedings (weekly) 00328

Australian Physiotherapy Association, POB 119, Concord, NSW 2137
Founded: 1905; Members: 2300
Pres: Virginia Binns; Gen Secr: Andrew McKinnon
Focus: Therapeutics
Periodical
NSW Physiotherapy Bulletin (weekly) . . 00329

Australian Postgraduate Federation in Medicine, 22 Lascelles Av, Camperdown, VIC 3147
T: (03) 2408671
Founded: 1962; Members: 40
Focus: Med; Adult Educ 00330

Australian Psychological Society, 191 Royal Parade, Parkville, Vic 3052
T: (03) 3472622; Fax: (03) 3474841
Founded: 1966; Members: 6500
Pres: F.D. Naylor; Gen Secr: A.F. Bennett
Focus: Psych
Periodicals
Australian Journal of Psychology (3 times annually)
Australian Psychologist (3 times annually)
Bulletin of the Australian Psychological Society (weekly) 00331

Australian Remedial Education Association (AREA), GPOB 334, Camberwell, Vic 3124
T: 821659
Founded: 1964; Members: 560
Focus: Educ 00332

Australian Research Council, c/o Dept of Employment Education and Training, 1 Farrell Pl, Canberra, ACT 2601
Tx: 61906
Founded: 1965
Pres: Prof. M.A. Brennan
Focus: Educ 00333

Australian Research Grants Committee, c/o Dept of Science and Environment, POB 449, Woden, ACT 2606
Founded: 1965; Members: 19
Focus: Nat Sci 00334

Australian Road Research Board, 500 Burwood Hwy, Vermont South, Melbourne, Vic 3133
T: (03) 2351555; Tx: 33113; Fax: (03) 2338878
Founded: 1960
Gen Secr: M.G. Lay
Focus: Civil Eng; Transport; Textiles
Periodicals
ARR Research Reports
Australian Road Research in Progress
Australian Road Research Journal (weekly)
Proceedings of Biennial Conference . . 00335

Australian School Library Association (ASLA), POB 287, Alderley, Qld 4051
T: 8782712
Founded: 1969; Members: 8
Focus: Libraries & Bk Sci
Periodical
School Libraries in Australia (weekly) . . 00336

Australian Science Teachers Association (A.S.T.A.), 15 Sussex St, Warradale, SA 5046
T: 2966472
Founded: 1950; Members: 4000
Focus: Sci; Educ 00337

Australian Society for Biochemistry and Molecular Biology (ASBMB), POB 62, Edwardstown, SA 5039
T: (08) 2770055; Fax: (08) 2771489
Founded: 1955; Members: 1150
Pres: Prof. W.H. Sawyer
Focus: Biochem; Bio

Periodical
Proceedings (weekly) 00338

Australian Society for Fish Biology, c/o Marine Science Laboratories, POB 114, Queenscliff, Vic 3225
Founded: 1971; Members: 530
Pres: Dr. J. Glaister; Gen Secr: Dr. D. Gwyther
Focus: Zoology
Periodical
Newsletter (weekly) 00339

Australian Society for Limnology, c/o Museum of Victoria, 71 Victoria Crescent, Abbotsford, Vic 3067
T: (03) 4195200; Fax: (03) 4160475
Founded: 1961; Members: 370
Gen Secr: Dr. Richard Marchant
Focus: Hydrology
Periodicals
Bulletin (weekly)
Newsletter (weekly)
Special Publications (weekly) 00340

Australian Society for Medical Research, 145 Maquarie St, Sydney, NSW 2000
Founded: 1961
Focus: Med
Periodical
Proceedings (weekly) 00341

Australian Society for Microbiology (A.S.M.), 191 Royal Parade, Parkville, Vic 3052
T: (03) 3481441; Fax: (03) 3481221
Founded: 1959; Members: 3000
Pres: Prof. J. Mackenzie
Focus: Microbio
Periodical
Australian Microbiologist (5 times annually) 00342

Australian Society for Operations Research (ASOR), GPOB 2021S, Melbourne, Vic 3001
Founded: 1972; Members: 500
Pres: P.A. Evans; Gen Secr: C. Malonas
Focus: Computer & Info Sci
Periodical
ASOR Bulletin (weekly) 00343

Australian Society for Parasitology, c/o ACTS, GPOB 2200, Canberra, ACT 2601
T: (06) 2573299; Fax: (06) 2573256
Founded: 1964; Members: 450
Focus: Immunology 00344

Australian Society for Parasitology, c/o CTS, GPOB 2200, Canberra, ACT 2601
Fax: (06) 573256
Founded: 1964; Members: 450
Focus: Microbio
Periodicals
Abstracts of Annual Conference
International Journal for Parasitology
International Journal for Parasitology (8 times annually) 00345

Australian Society for Reproductive Biology, c/o Dept of Biology Science, University of Newcastle, Newcastle, NSW 2308
T: (049) 685511
Founded: 1969
Focus: Bio 00346

Australian Society of Anaesthesists (ASA), POB 345, Paddington, NSW 2021
T: 3313211
Founded: 1934; Members: 1500
Focus: Anesthetics
Periodical
Anaesthesia and Intensive Care (weekly) 00347

Australian Society of Animal Production (ASAP), c/o Dept of Agriculture, POB 361, Mount Barker, SA 5251
T: (08) 3910155
Founded: 1956; Members: 2000
Pres: Dr. J.C. Radcliffe; Gen Secr: L.D. McLaren
Focus: Animal Husb
Periodical
Proceedings (biennial) 00348

Australian Society of Authors (A.S.A.), POB 1566, Strawberry Hills, NSW 2012
T: (02) 3180877; Fax: (02) 3180530
Founded: 1963; Members: 2180
Gen Secr: Gail Cork
Focus: Lit
Periodicals
The Acquisition of Film Rights in Literary Properties
The Australian Author (weekly)
Australian Book Contracts 00349

Australian Society of Clinical Hypnotherapy, POB 44, Eastwood, NSW 2122
T: (02) 8741276
Founded: 1974; Members: 400
Pres: Norman McMaster; Gen Secr: Patricia Weir
Focus: Therapeutics
Periodical
The Australian Journal of Clinical Hypnotherapy and Hypnosis (weekly) 00350

Australian Society of Cosmetic Chemists, POB 301, Kingsgrove, NSW 2208
Founded: 1964
Focus: Chem
Periodical
Cosmetic Science in Australia (weekly) . 00351

Australian Society of Endodontology (A.S.E.), 11 Kings Rd, Subiaco, WA 6008
T: 3813820
Founded: 1967; Members: 400
Focus: Dent 00352

Australian Society of Herpetologists, c/o John Wombey, C.S.I.R.O. Div of Wildlife and Ecology, POB 84, Lyneham, ACT 2602
Founded: 1964
Focus: Zoology
Periodical
Newsletter (weekly) 00353

Australian Society of Indexers (AusSI), GPOB 1251L, Melbourne 3001
Focus: Libraries & Bk Sci
Periodical
Newsletter (weekly) 00354

Australian Society of Orthodontists, 167 Saint George Terrace, Perth, WA 6000
Founded: 1927
Focus: Dent 00355

Australian Society of Periodontology, 145 Collins St, Melbourne, Vic 3000
T: (03) 633616
Founded: 1961
Focus: Dent 00356

Australian Society of Plant Physiologists (A.S.P.P.), c/o Dr. I.F. Wardlaw, C.S.I.R.O. Div of Plant Industry, GPOB 1600, Canberra, ACT 2601
T: (06) 465838
Founded: 1958; Members: 641
Focus: Botany 00357

Australian Society of Prosthodontists (A.S.P.), c/o School of Dental Science, 711 Elizabeth St, Melbourne, WA 3000
T: (03) 3474222
Founded: 1961; Members: 250
Focus: Dent 00358

Australian Society of Soil Science (ASSSI), c/o C.S.I.R.O. Cunningham Laboratory, 306 Carmidy Rd, Saint Lucia, QND 4061
T: (01) 37700209; Tx: AA 42159
Founded: 1956; Members: 104
Focus: Agri 00359

Australian Sociological Association, c/o Dept of Sociology, La Trobe University, Bundoora, Vic 3083
Founded: 1963; Members: 550
Pres: Kathy Richmond; Gen Secr: Ray Jureidini
Focus: Sociology 00360

Australian Veterinary Association (AVA), 134-136 Hampden Rd, Artarmon, NSW 2064
T: 4112733
Founded: 1921; Members: 3500
Pres: D.R. Kerr; Gen Secr: P.E. Greenwood
Focus: Vet Med
Periodicals
Australian Advances in Veterinary Science (weekly)
Australian Veterinary Journal (weekly)
AVA Yearbook (weekly) 00361

Australian Vice-Chancellors' Committee, GPOB 1142, Canberra, ACT 2601
Tx: 62180
Founded: 1920; Members: 26
Pres: Prof. K.R. McKinnon; Gen Secr: F.S. Hambly
Focus: Educ
Periodicals
Information Summaries (weekly)
Occasional Papers
Univation (3 times annually) 00362

Australian Wool Corporation, Wool House, 369 Royal Parade, Parkville, Vic 3052
Fax: (03) 3419273
Founded: 1936
Pres: H.S. Beggs
Focus: Textiles
Periodical
Annual Report (weekly) 00363

Australian Zinc Development Association, 95 Collins St, Melbourne, Vic 3000
Focus: Metallurgy 00364

Australia-Pacific Society for Management Studies, c/o School of Business, Bond University, Gold Coast, Qld 4229
T: (075) 952281; Fax: (075) 951160
Founded: 1989
Focus: Business Admin
Periodical
Bond Management Review (weekly) . . 00365

Bibliographical Society of Australia and New Zealand, c/o Margaret Dent, National Library of Australia, Canberra, ACT 2600
Founded: 1969
Pres: W.D. Thorn
Focus: Libraries & Bk Sci
Periodicals
Broadsheet (3-4 times annually)
Bulletin (weekly) 00366

Bio-Rhythm Research and Information Centre, 22 Mungala St, Rochdale, Brisbane, Qld 4123
T: (07) 3414928
Founded: 1954
Focus: Bio
Periodical
The Basic Biorhythm Story 00367

Bird Observers Club of Australia (B.O.C.), 183 Springvale Rd, Nunawading, Vic 3131
T: 8775342
Founded: 1905; Members: 3300
Focus: Ornithology
Periodicals
The Australian Bird Watcher (weekly)
The Bird Observer (weekly) 00368

Botany 2000-Asia, c/o Dept of Conservation and Land Management, Western Australian Herbarium, George St, Kensington
T: (09) 3670515; Tx: 94616; Fax: (09) 3670515
Founded: 1989
Focus: Botany 00369

The Cardiac Society of Australia and New Zealand, 145 Macquarie St, Sydney, NSW 2000
T: (02) 274461; Fax: (02) 231320
Founded: 1951; Members: 1100
Pres: Dr. John Waddy; Gen Secr: Dr. David Richards
Focus: Cardiol 00370

Chartered Institute of Transport in Australia, Queen Victoria Bldgs, POB Q398, Queen Victoria P.O., Sydney, NSW 2000
T: (02) 2646413; Fax: (02) 2671682
Founded: 1935; Members: 2500
Gen Secr: K.L. Duncan
Focus: Transport
Periodical
Australian Transport (weekly) 00371

Clean Air Society of Australia and New Zealand, POB 191, Eastwood, NSW 2122
T: 20557
Founded: 1966
Focus: Ecology 00372

Commission for the Conservation of Antarctic Marine Living Resources (CCAMLR), 25 Old Wharf, Hobart 7000
T: (02) 310366; Tx: 57236; Fax: (02) 232714
Founded: 1982; Members: 20
Gen Secr: Dr. D. Powell
Focus: Ecology
Periodicals
CCAMLR Newsletter (weekly)
CCAMLR Statistical Bulletin 00373

Committee for Economic Development of Australia (CEDA), CEDA House, 123 Lonsdale St, Melbourne, Vic 3000
T: (03) 6623544; Fax: (03) 6637271
Founded: 1960; Members: 740
Pres: R.A. Johnston; Gen Secr: P. Grey
Focus: Econ
Periodicals
PEN: Public Economic Newsletter (weekly)
PIPS: Public Information Paper (weekly) . 00374

Commonwealth Council for Educational Administration (CCEA), c/o University of New England, Armidale, NSW 2351
T: (067) 732543; Tx: 66050; CA: CCEA UNE Armidale; Fax: (067) 733122
Founded: 1970; Members: 5000
Gen Secr: John Weeks
Focus: Educ; Public Admin
Periodicals
CCEA Newsletter (weekly)
Studies in Educational Administration (weekly) 00375

Commonwealth Institute of Valuers, 119 York St, Sydney
Founded: 1926; Members: 5 000
Focus: Finance
Periodical
The Valuer (weekly) 00376

Commonwealth Scientific and Industrial Research Organisation (C.S.I.R.O.), POB 225, Dickson, ACT 2602
T: (062) 484211; Tx: 62528; Fax: (062) 473832
Founded: 1926; Members: 7000
Pres: N.K. Wran; Gen Secr: N.K. Boardman
Focus: Sci; Eng
Periodicals
Australian Forest Research (weekly)
Australian Journal of Agricultural Research (weekly)
Australian Journal of Biological Sciences (weekly)
Australian Journal of Botany (weekly)
Australian Journal of Chemistry (weekly)
Australian Journal of Experimental Agriculture and Animal Husbandry (weekly)
Australian Journal of Marine and Freshwater Research (weekly)
Australian Journal of Physics (weekly)
Australian Journal of Plant Physiology (weekly)
Australian Journal of Soil Research (weekly)
Australian Journal of Zoology (weekly)
Australian Wildlife Research
Brunonia (weekly)
Food Research Quarterly (weekly) . . . 00377

Commonwealth Veterinary Association (CVA), RMB N141, Pryors Rd, Scotsburn 3352
Founded: 1968; Members: 47
Gen Secr: Dr. W.J. Pryor
Focus: Vet Med
Periodical
CVA News (weekly) 00378

The Contact Lens Society of Australia (C.L.S.A.), 818 Australia Sq Tower, 264 George St, Sydney, NSW 2000
T: (02) 273997
Founded: 1962; Members: 560
Pres: M. Knipe; Gen Secr: Kenneth W. Bell
Focus: Ophthal 00379

Contemporary Art Society of Australia, GPOB 3271, Sydney, NSW 2001
Founded: 1939
Focus: Arts 00380

Council for International Congresses of Entomology, c/o Div of Entomology, C.S.I.R.O., GPOB 1700, Canberra, ACT 2601
Members: 23
Pres: Dr. R. Galun; Gen Secr: Dr. M. Whitter
Focus: Entomology 00381

Council of Adult Education, 256 Flinders St, Melbourne, Vic 3000
T: (03) 6520611; Tx: 30625; Fax: (03) 6546759
Founded: 1947
Pres: Lloyd Smith; Gen Secr: Dr. Shirley Randell
Focus: Adult Educ
Periodical
Groupe Affairs (3 times annually) . . . 00382

Council of Australian Food Technology Associations, POB 1310, North Sydney, NSW 2059
T: (02) 9637670
Founded: 1949
Focus: Eng; Food
Periodical
Food Australia (weekly) 00383

Dairy Industry Association of Australia, POB 20, Highett, Vic 3190
T: (03) 2526000; Fax: (03) 2626555
Founded: 1946; Members: 1200
Pres: M. Seuret; Gen Secr: Helen Dornom
Focus: Dairy Sci
Periodical
Australian Journal of Dairy Technology . 00384

Dietitians Association of Australia (D.A.A.), Queen Victoria Terrace, POB E187, Canberra, ACT 2600
T: (06) 814750
Founded: 1983; Members: 800
Focus: Nutrition
Periodical
Annual Conference Proceedings 00385

Eastern Dredging Association (EADA), GPOB 1818, Brisbane 4001
Founded: 1982
Gen Secr: John E. Dobson 00386

Ecological Society of Australia (E.S.A.), POB 1564, Canberra, ACT 2601
Founded: 1960; Members: 800
Pres: Dr. I. Noble; Gen Secr: Dr. T. Norton
Focus: Ecology
Periodical
Australian Journal of Ecology (weekly) . 00387

Economic Society of Australia, c/o Prof. A.D. Woodland, Dept of Econometrics, Sydney University, Sydney, NSW 2006
T: (02) 6923069; Fax: (02) 5523105
Founded: 1925; Members: 3200
Pres: F. Argy; Gen Secr: W. Shields
Focus: Econ
Periodicals
Economic Papers (weekly)
The Economic Record (weekly) 00388

Endocrine Society of Australia (ESA), c/o Dept of Endocrinology, Royal Prince Alfred Hospital, Camperdown, NSW 2050
T: (02) 5168071
Founded: 1958; Members: 500
Focus: Endocrinology
Periodical
Proceedings (weekly) 00389

English Association – Sydney Branch, POB 187, Rozelle, NSW 2039
Founded: 1923; Members: 120
Pres: D.L. Robinson; Gen Secr: C.A. Lee
Focus: Ling
Periodical
Southerly (weekly) 00390

Entomological Society of New South Wales, c/o Dept of Entomology, The Australian Museum, 6-8 College St, Sydney, NSW 2001
Founded: 1952; Members: 194
Pres: G.R. Brown; Gen Secr: M.J. Fletcher
Focus: Entomology
Periodical
General and Applied Entomology (weekly) 00391

Entomological Society of Queensland, c/o Entomology Dept, University of Queensland, Saint Lucia, Brisbane, Qld 4072
Founded: 1923; Members: 300
Pres: Dr. G.G. White; Gen Secr: Dr. J. King
Focus: Entomology
Periodicals
Australian Entomological Magazine (weekly)
News Bulletin (10 times annually) . . . 00392

Ergonomics Society of Australia, POB 1337, Fortitude Valley, Qld 4006
T: (07) 2521046; Tx: 140472; Fax: (07) 2523850
Pres: J. Frith; Gen Secr: Bill Green
Periodical
Conference Proceedings (weekly) 00393

Federation of Australian Scientific and Technological Societies, GPOB 2181, Canberra, ACT 2601
T: (06) 2473554; Fax: (06) 2496419
Founded: 1985; Members: 60
Pres: Prof. Tony Wicken; Gen Secr: Dr. David Widdup
Focus: Sci; Eng *00394*

Fellowship of Australian Writers, GPOB 3448, Sydney, NSW 2001
Founded: 1928; Members: 3500
Pres: Hilarie Lindsay; Gen Secr: Gaye Manley
Focus: Lit
Periodical
Bulletin (weekly) *00395*

Field Naturalists' Club of Victoria (F.N.C.V.), c/o National Herbarium, Birdwood Av, South Yarra, Vic 3141
Founded: 1880; Members: 700
Pres: A. Farnworth
Focus: Hist; Nat Sci; Ecology
Periodical
The Victorian Naturalist (weekly) *00396*

The Field Naturalists' Society of South Australia (FNSSA), GPOB 1594, Adelaide, SA 5001
Founded: 1883; Members: 380
Focus: Hist; Nat Sci; Ecology
Periodical
The South Australian Naturalist (weekly) . *00397*

Food Technology Association of New South Wales, 60 York St, Sydney, NSW 2000
T: (02) 2900700
Founded: 1946
Focus: Eng; Nutrition *00398*

Food Technology Association of Queensland, 375 Wickham Terrace, Brisbane, Qld 4000
Founded: 1947
Focus: Nutrition *00399*

Food Technology Association of Tasmania, 191 Liverpool St, Hobart, Tas 7000
T: (002) 345933
Founded: 1952
Focus: Nutrition *00400*

Food Technology Association of Western Australia, 212-220 Adelaide Terrace, POB 6209, Perth, WA 6000
Founded: 1954
Focus: Eng; Nutrition *00401*

Geographical Society of New South Wales, POB 602, Gladesville, NSW 2111
T: (02) 8173647
Founded: 1927; Members: 500
Gen Secr: J. Dodson
Focus: Geography
Periodical
Australian Geographer (weekly) *00402*

Geography Teachers' Association of New South Wales, POB 187, Rozelle, NSW 2039
T: 275589
Founded: 1936
Focus: Geography; Educ *00403*

Geological Society of Australia, 301 George St, Sydney, NSW 2000
Founded: 1953; Members: 3300
Focus: Geology
Periodicals
Journal (weekly)
Meeting Abstracts (weekly)
Special Publications *00404*

Geological Survey of New South Wales, c/o Dept of Minerals and Energy, Sydney
Founded: 1874
Focus: Geology
Periodicals
Bulletin
Geological Maps and Notes
Geological Memoirs
Metallogenic Maps and Notes
Mineral Industry
Mineral Resources
Palaeontological Memoirs
Quarterly Notes (weekly)
Records *00405*

Geological Survey of Western Australia, Mineral House, 100 Plain St, East Perth, WA 6004
Fax: (09) 2223633
Founded: 1888; Members: 110
Focus: Geology
Periodicals
Bulletin
Information Pamphlets
Memoir
Record
Report *00406*

Gosford District Historical Research and Heritage Association, 230 Scenic Rd, Killcare Heights, Hardy's Bay, NSW 2256
Focus: Hist *00407*

Grassland Society of Victoria (G.S.V.), 191 Royal Parade, Parkville, Vic 3052
T: (03) 3471277
Founded: 1959; Members: 700
Focus: Agri
Periodicals
Conference Proceedings (weekly)
Newsletter (weekly) *00408*

Horological Guild of Australasia, 228 Pitt St, Sydney, NSW 2000
T: (02) 9828561
Founded: 1951
Focus: Eng *00409*

Indo-Pacific Prehistory Association (IPPA), c/o Dept of Prehistory and Anthropology, Faculty of Arts, Australian National University, POB 4, Canberra, ACT 2601
T: (06) 494395
Founded: 1929; Members: 250
Focus: Anthro; Hist; Archeol
Periodical
Bulletin (weekly) *00410*

The Institute of Actuaries of Australia, 5 Moorina Rd, Saint Ives, NSW 2075
T: (02) 443311
Founded: 1897; Members: 910
Focus: Insurance
Periodicals
General Insurance Seminar (weekly)
Transactions (weekly) *00411*

Institute of Australian Geographers (I.A.G.), c/o AIUS, GPOB 809, Canberra, ACT 2601
Founded: 1958; Members: 400
Pres: Dr. R.L. Heathcote; Gen Secr: Dr. C. Adrian
Focus: Geography
Periodicals
Australian Geographical Studies (weekly)
Newsletter (weekly) *00412*

Institute of Materials Handling, 79 Buckhurst St, South Melbourne, Vic 3205
T: (03) 6991333
Founded: 1953
Focus: Materials Sci *00413*

Institute of Mental Health Research and Postgraduate Training, 35 Poplar Rd, Parkville, Vic 3052
Founded: 1955
Focus: Psychiatry
Periodical
Statistical Bulletin (weekly) *00414*

Institute of Metals and Materials Australasia, 191 Royal Parade, Parkville, Vic 3052
T: (03) 3472544; Fax: (03) 3481208
Founded: 1946; Members: 1500
Gen Secr: Dr. I.J. McKee
Focus: Metallurgy
Periodicals
Materials Australasia (weekly)
Materials Forum (weekly) *00415*

Institute of Municipal Management, 27-33 Raglan St, Ste 4, South Melbourne, Vic 3205
T: (03) 6965799; Fax: (03) 6904217
Founded: 1936; Members: 2700
Gen Secr: Barrie Beattie
Focus: Public Admin
Periodical
Local Government Management *00416*

Institute of Photographic Technology, GPOB 5385BB, Melbourne, Vic 3001
Founded: 1945
Focus: Eng; Photo *00417*

Institute of Public Affairs (IPA), 128-135 Jolimont Rd, Jolimont, Vic 3002
Founded: 1943; Members: 4250
Pres: C.B. Goode; Gen Secr: John Hyde
Focus: Econ; Business Admin
Periodicals
IPA Facts (weekly)
IPA Review (weekly)
Policy Issues *00418*

Institute of Transport, 22 Karne St, Beverly Hills, NSW 2209
Founded: 1919
Focus: Transport *00419*

Institutiom of Surveyors – Australia, 27-29 Napier Close, Deakin, Canberra, ACT 2600
T: (06) 2822282; Fax: (06) 2822576
Founded: 1952; Members: 3500
Gen Secr: C.A. Fuller
Focus: Eng *00420*

Institution of Engineers – Australia, 11 National Circuit, Barton, ACT 2600
T: (06) 2706555; Fax: (06) 2731488
Founded: 1919; Members: 60000
Focus: Eng
Periodicals
Australian Geomechanics News (weekly)
Chemical Engineering in Australia (weekly)
Civil Engineering Transactions (weekly)
Conference Volumes (weekly)
Engineers Australia (weekly)
General Engineering Transactions (weekly)
Journal of Electrical & Electronics Engineering in Australia (weekly)
Mechanical Engineering Transactions (weekly)
. . . . *00421*

Institution of Radio and Electronics Engineers – Australia, c/o Science Centre, 35-43 Clarence St, Sydney, NSW 2000
Founded: 1932; Members: 3000
Pres: Prof. G.A. Rigby; Gen Secr: Heather Harriman
Focus: Electronic Eng
Periodicals
Journal of Electrical and Electronics Engineering
Monitor (weekly)
Proceedings of the I.R.E.E. (weekly) . . *00422*

International Association of Environmental Mutagen Societies, c/o Dept of Haematology, Flinders University, Adelaide, SA 5042
T: (08) 2044431; Fax: (08) 2045450
Founded: 1973; Members: 2500
Focus: Bio *00423*

International Association of Trichologists (IAT), 185 Elizabeth St, Ste 610, Sydney, NSW 2000
T: (2) 2671384
Founded: 1973; Members: 210
Focus: Med; Immunology *00424*

International Development Program of Australian Universities and Colleges, GPOB 2006, Canberra, ACT 2601
T: (06) 2858222; Tx: 61940; Fax: (06) 2853036
Founded: 1984
Gen Secr: Dr. D.G. Blight
Focus: Educ
Periodicals
Newsletter (weekly)
Report (weekly) *00425*

Inter-Union Commission on Frequency Allocations for Radio Astronomy and Space Science, c/o Dr. B.J. Robinson, C.S.I.R.O., POB 76, Epping, NSW 2121
T: (02) 8680222; Tx: 26230; Fax: (02) 8680457
Founded: 1950
Pres: Dr. B.J. Robinson
Focus: Astronomy; Comm *00426*

Law Council of Australia, 19 Torrens St, Braddon, ACT 2601
T: (06) 2473788; Fax: (06) 2480639
Founded: 1933; Members: 27000
Pres: David Miles; Gen Secr: Peter G. Levy
Focus: Law
Periodical
Australian Law News (weekly) *00427*

Law Society of New South Wales, 170 Phillip St, Sydney, NSW 2000
T: (02) 2200333; Fax: (02) 2315809
Founded: 1884; Members: 11600
Gen Secr: J.R. Hunt
Focus: Law
Periodicals
Caveat (11 times annually)
Law Society Journal (11 times annually) *00428*

Linnean Society of New South Wales, 6-24 Cliff St, POB 457, Milsons Point, NSW 2061
T: 843616
Founded: 1894; Members: 300
Pres: Dr. P.J. Myerslough; Gen Secr: Barbara Stoddard
Focus: Hist; Nat Sci *00429*

Malacological Society of Australia, c/o Dept of Invertebrate Zoology, Museum of Victoria, 285-321 Russell St, Melbourne, Vic 3000
T: (03) 6699880
Founded: 1955; Members: 400
Pres: Dr. N. Coleman; Gen Secr: S. Boyd
Focus: Zoology
Periodicals
Australian Shell News (weekly)
Journal (weekly) *00430*

Medical Foundation, c/o Coppleson Institute, University of Sydney, Sydney, NSW 2006
Fax: (02) 6924160
Founded: 1958
Pres: A.J. Williams; Gen Secr: V.S.M. Harper
Focus: Med *00431*

Medical Society of Victoria, 293 Royal Parade, Parkville, Vic 3052
T: 3478722
Founded: 1852
Focus: Med *00432*

Medico Legal Society of New South Wales, POB 121, Saint Leonards, NSW 2065
T: 4398822
Founded: 1947
Focus: Law
Periodical
Proceedings *00433*

Metropolitan Research Trust, POB 597, Canberra City, ACT 2601
Founded: 1968
Focus: Urban Plan *00434*

Museums Association of Australia, c/o Performing Arts Museum, Victoria Arts Centre, 100 Saint Kilda Rd, Melbourne, Vic 3004
T: (03) 6848325
Founded: 1937; Members: 1600
Pres: M. Anderson
Focus: Arts
Periodical
Museums Australia (weekly) *00435*

Musicological Society of Australia, GPOB 2404, Canberra, ACT 2601
Founded: 1963; Members: 300
Pres: Andrew McCredie; Gen Secr: K. Coaldrake
Focus: Music
Periodicals
Musicology Australia (weekly)
Newsletter (3 times annually) *00436*

National Association of Testing Authorities, Australia (NATA), 688 Pacific Hwy, Chatswood, NSW 2067
T: 4114000
Founded: 1947; Members: 1325
Focus: Materials Sci
Periodicals
Annual Report (weekly)
Directory of Laboratories (weekly)
NATA News (weekly) *00437*

National Health and Medical Research Council, POB 9848, Canberra, ACT 2601
Tx: 62149; Fax: (062) 896957
Founded: 1936; Members: 28
Focus: Public Health; Med
Periodicals
Council Session Reports
Medical Research Projects (weekly) . . *00438*

Native Plants Preservation Society of Victoria, 3 Denham Pl, Toorak, Vic 3142
Founded: 1952
Focus: Ecology *00439*

Natural Resources Conservation League of Victoria (N.R.C.L.), 593 Springvale Rd, Springvale South, Vic 3172
T: (03) 5469744; Fax: (03) 5478791
Founded: 1944; Members: 4400
Pres: Helen Harrison; Gen Secr: L. Schultz
Focus: Ecology
Periodical
Trees and Natural Resources (weekly) . *00440*

North Queensland Naturalists Club (N.Q.N.C.), POB 991, Cairns, Qld 4870
Founded: 1932; Members: 100
Focus: Hist; Nat Sci; Ecology *00441*

Oaks Historical Society, Strathmore Rd, The Oaks, NSW 2570
T: (046) 571366
Focus: Hist *00442*

Opticians and Optometrists Association of New South Wales, 235 Elizabeth St, Sydney, NSW 2000
T: (02) 265284
Founded: 1926
Focus: Optics; Ophthal *00443*

Paediatric Society of Victoria, c/o Royal Children's Hospital, Flemington Rd, Parkville, Vic 3052
T: 3475522
Founded: 1906
Focus: Pediatrics *00444*

PEN International (Sydney Centre), POB 153, Woollahra, NSW 2025
T: 325668
Founded: 1926; Members: 160
Pres: Stella Wilkes; Gen Secr: Susan Yorke
Focus: Lit
Periodical
Newsletter (weekly) *00445*

Plant Protection Society of Western Australia, POB 190, Victoria Park, WA 6100
T: (09) 3683333; Fax: (09) 3681205
Founded: 1976; Members: 97
Focus: Ecology
Periodical
Proceedings (weekly) *00446*

Primary English Teaching Association (PETA), POB 167, Rozelle, NSW 2039
T: (02) 8182591, 8182346
Focus: Educ *00447*

Queensland Institute for Educational Research, c/o Faculty of Education, University of Queensland, Saint Lucia, Qld 4067
T: 3706969
Founded: 1930
Focus: Educ *00448*

The Queensland Naturalists' Club (Q.N.C.), GPOB 1220, Brisbane, Qld 4001
T: (07) 3722028
Founded: 1906; Members: 440
Focus: Hist; Nat Sci; Ecology
Periodical
The Queensland Naturalist (biennial) . . *00449*

Royal Aeronautical Society (Australian Division), POB 573, Mascot, NSW 2020
Founded: 1927
Pres: R. Noble; Gen Secr: R.D. Barkla
Focus: Aero *00450*

Royal Agricultural and Horticultural Society of South Australia, Royal Showground, Wayville, SA 5034
T: 874421, 514951
Focus: Hort; Agri *00451*

Royal Agricultural Society of Tasmania, Rothmans Bldg, Royal Showgrounds, Glenorchy, Tas 7010
T: 726812
Founded: 1862
Focus: Agri *00452*

The Royal Agricultural Society of Western Australia, POB 135, Claremont, WA 6010
T: 311933
Founded: 1829; Members: 3600
Focus: Agri *00453*

Royal Art Society of New South Wales, 25-27 Walker St, North Sydney, NSW 2060
Founded: 1880; Members: 640
Pres: Frederic Bates; Gen Secr: Marjorie McLachlan
Focus: Arts
Periodical
Newsletter (weekly) 00454

Royal Australasian College of Dental Surgeons, 64 Castlereagh St, Sydney, NSW 2000
T: (02) 2323800; Fax: (02) 2218108
Founded: 1965; Members: 920
Gen Secr: K.H. Wendon
Focus: Surgery
Periodical
Annals 00455

Royal Australasian College of Physicians (RACP), 145 Macquarie St, Sydney, NSW 2000
T: (02) 2565444; Fax: (02) 2523310
Founded: 1938; Members: 5915
Pres: Prof. A.K. Cohen; Gen Secr: D.W. Swinbourne
Focus: Med
Periodical
Australian and New Zealand Journal of Medicine (weekly) 00456

Royal Australasian College of Radiologists, 37 Lower Fort St, Millers Point, NSW 2000
T: (02) 2477797; Fax: (02) 2514629
Founded: 1935; Members: 1770
Gen Secr: Dr. Peter J.A. Carr
Focus: Radiology
Periodical
Australasian Radiology (weekly) 00457

Royal Australasian College of Surgeons (R.A.C.S.), Spring St, Melbourne, Vic 3000
T: (03) 6621033; Fax: (03) 6634075
Founded: 1927; Members: 2600
Gen Secr: P.H. Carter
Focus: Surgery
Periodical
Australian and New Zealand Journal of Surgery (weekly) 00458

Royal Australasian Ornithologists Union (R.A.O.U.), 21 Gladstone St, Moonee Ponds, Vic 3039
T: (03) 3701422; Fax: (03) 3709194
Founded: 1901; Members: 3000
Pres: Prof. B.G. Collins; Gen Secr: B. Wilson
Focus: Ornithology
Periodicals
Newsletter (weekly)
WA Bird Notes (weekly) 00459

The Royal Australian and New-Zealand College of Psychiatrists (R.A.N.Z.C.P.), Maudsley House, 101 Rathdowne St, Carlton, Vic 3053
T: (03) 6635466; Fax: (03) 6631998
Founded: 1963; Members: 1850
Pres: Dr. Norman James; Gen Secr: Dr. Michael Epstein
Focus: Psychiatry
Periodical
The Australian and New Zealand Journal of Psychiatry (weekly) 00460

Royal Australian Chemical Institute (R.A.C.I.), 191 Royal Parade, Parkville, Vic 3052
T: (03) 3471577; Fax: (03) 3491409
Founded: 1917; Members: 8000
Gen Secr: Dr. F. Bryant
Focus: Chem
Periodical
Chemistry in Australia (weekly) 00461

Royal Australian College of General Practitioners (RACGP), 43 Lower Fort St, Sydney, NSW 2000
T: (02) 273244, 278845
Founded: 1958; Members: 3324
Focus: Med 00462

Royal Australian College of Ophthalmologists, 27 Commonwealth St, Sydney, NSW 2010
T: (02) 2677006; Fax: (02) 2676534
Founded: 1969; Members: 983
Gen Secr: Dr. Stephen Cains
Focus: Ophthal
Periodical
Australian and New Zealand Journal of Ophthalmology (weekly) 00463

Royal Australian Historical Society, History House, 133 Macquarie St, Sydney, NSW 2000
T: (02) 278001
Founded: 1901; Members: 3000
Gen Secr: Patricia Parker
Focus: Hist
Periodical
Journal (weekly) 00464

Royal Australian Planning Institute (RAPI), 615 Burwood Rd, Hawthorn, Vic 3122
T: (03) 8190728; Fax: (03) 8190676
Founded: 1951; Members: 2300
Pres: Peter D. Webb; Gen Secr: Gaye McKenzie
Focus: Urban Plan
Periodical
Australian Planner (weekly) 00465

Royal College of Nursing – Australia, 2 Slater St, Melbourne, Vic 3004
T: (03) 8202055; Fax: (03) 8671315
Founded: 1949; Members: 4000
Focus: Med 00466

Royal College of Pathologists of Australasia, Durham Hall, 207 Albion St, Surrey Hills, NSW 2010
Founded: 1956; Members: 1500
Pres: Dr. D.W. Fortune; Gen Secr: Dr. J.V. Wells
Focus: Pathology
Periodical
Pathology 00467

Royal Geographical Society of Queensland (R.G.S.A.), 112 Brookes St, Fortitude Valley, Qld 4006
T: (07) 2523856; Fax: (07) 2524986
Founded: 1885; Members: 580
Pres: P. Feeney; Gen Secr: Keith Smith
Focus: Geography
Periodical
Queensland Geographical Journal (weekly) 00468

The Royal Historical Society of Queensland (R.H.S.Q.), Commissariat Stores, 115 William St, POB 57, Brisbane North Quay, Qld 4002
T: (07) 2214198
Founded: 1913; Members: 600
Pres: Dr. R.F.J. Wood; Gen Secr: Ruth Seffrin
Focus: Hist
Periodicals
Journal of The Royal Historical Society of Queensland (weekly)
The Royal Historical Society of Queensland Bulletin (weekly) 00469

Royal Historical Society of Victoria (RHSV), Royal Mint, 280 William St, Melbourne, Vic 3000
T: (03) 6701219; Fax: (03) 6701241
Founded: 1909; Members: 1600
Pres: Prof. W. Bate; Gen Secr: Dr. Leonie Foster
Focus: Hist
Periodicals
History News (weekly)
Journal (weekly) 00470

Royal Horticultural Society of New South Wales, GPOB 4728, Sydney, NSW 2001
Founded: 1862; Members: 510
Focus: Hort 00471

Royal Melbourne Institute of Technology, GPOB 2476V, Melbourne, Vic 3001
Focus: Eng 00472

Royal Meteorological Society, Australian Branch, c/o Clunies Ross Memorial Foundation, 191 Royal Parade, Parkville, Vic 3058
T: 3476077
Founded: 1973; Members: 300
Focus: Geophys 00473

Royal Queensland Art Society, GPOB 1602, Brisbane, Qld 4001
T: (07) 8313455
Founded: 1887; Members: 850
Gen Secr: Karen Kane
Focus: Arts 00474

Royal Society of Canberra, POB 130, Canberra, ACT 2601
Founded: 1930
Focus: Sci 00475

Royal Society of New South Wales, 134 Herring Rd, POB 1525, North Ryde, NSW 2113
T: (02) 8874448
Founded: 1821; Members: 305
Pres: Dr. F.L. Sutherland; Gen Secr: J.R. Hardie
Focus: Sci
Periodicals
Journal
Newsletter (weekly)
Proceedings (weekly) 00476

Royal Society of Queensland, POB 21, Saint Lucia, Qld 4067
T: 8701697
Founded: 1884; Members: 250
Pres: Dr. R. Hynes; Gen Secr: P. Young
Focus: Sci
Periodicals
Newsletter (weekly)
Proceedings (weekly)
Symposia 00477

Royal Society of South Australia, c/o South Australian Museum, North Terrace, Adelaide, SA 5000
Founded: 1853; Members: 300
Focus: Sci
Periodicals
Regional Natural Histories (weekly)
Transactions (weekly) 00478

Royal Society of Tasmania, GPOB 1166M, Hobart, Tas 7001
T: (002) 231422
Founded: 1843; Members: 480
Gen Secr: D.R. Gregg
Focus: Sci
Periodical
Papers and Proceedings of the Royal Society of Tasmania (weekly) 00479

Royal Society of Victoria, 9 Victoria St, Melbourne, Vic 3000
T: (03) 6635259; Fax: (03) 6632301
Founded: 1854; Members: 590
Pres: Dr. G.F. Watson; Gen Secr: Dr. T.A. Darragh
Focus: Sci
Periodical
Proceedings (weekly) 00480

Royal Society of Western Australia, c/o Western Australian Museum, Francis St, Perth, WA 6000
Fax: (09) 3104929
Founded: 1913; Members: 350
Pres: Dr. M. Candy; Gen Secr: L. Thomas; B. Gozzard
Focus: Sci
Periodical
Journal of the Royal Society of Western Australia (weekly) 00481

Royal South Australian Society of Arts, c/o Kintore Gallery, Institute Bldg, North Terrace, Adelaide, SA 5000
Founded: 1856; Members: 702
Pres: Stephanie Schrapel
Focus: Arts
Periodical
Kalori (weekly) 00482

Royal Western Australian Historical Society, Stirling House, 49 Broadway, Nedlands, WA 6009
T: 3863841
Founded: 1926; Members: 1000
Gen Secr: Judy Campbell
Focus: Hist
Periodicals
Early Days (weekly)
Newsletter (weekly) 00483

Royal Zoological Society of New South Wales, POB 20, Mosman, NSW 2088
Founded: 1879; Members: 1500
Gen Secr: Arthur White
Focus: Zoology
Periodical
Australian Zoologist (weekly) 00484

Royal Zoological Society of South Australia, c/o Adelaide Zoo, Frome Rd, Adelaide, SA 5000
T: (08) 2673255; Fax: (08) 2390637
Founded: 1878; Members: 3810
Pres: Dr. P.A.S. Davidson; Gen Secr: E.J. McAllister
Focus: Zoology
Periodical
Annual Report (weekly) 00485

Safe Water Association of New South Wales, Clarence St, POB C9, Sydney, NSW 2000
Focus: Water Res
Periodical
The Australian Fluoridation News (weekly) 00486

School Library Association of the Northern Territory (SLANT), POB 3162, Darwin, NT 0801
Members: 90
Pres: Susan Ross; Gen Secr: Celia Otley
Focus: Libraries & Bk Sci 00487

Science Teachers Association of New South Wales, POB 187, Rozelle, 2039
T: (02) 8182591
Founded: 1949; Members: 1500
Focus: Sci; Educ
Periodical
SEN-Science Education News (weekly) . 00488

Science Teachers Association of Queensland (S.T.A.Q.), POB 84, Spring Hill, Qld 4000
Founded: 1945; Members: 530
Focus: Sci; Educ 00489

Science Teachers' Association of Victoria, 191 Royal Parade, Parkville, Vic 3052
T: 3472537
Founded: 1943
Focus: Sci; Educ 00490

Society for Social Responsibility in Science (A.C.T.) (SSRS), POB 48, Canberra, ACT 2601
T: (06) 465682
Founded: 1970; Members: 150
Focus: Sociology 00491

Society of Australian Genealogists, Richmond Villa, 120 Kent St, Sydney, NSW 2000
T: (02) 273953
Founded: 1932; Members: 8000
Pres: E.C. Best; Gen Secr: H.E. Garnsey
Focus: Genealogy
Periodical
Descent (weekly) 00492

Society of Automotive Engineers – Australasia (SAE), 191 Royal Parade, Parkville, Vic 3052
T: (03) 3472220; Fax: (03) 3470464
Founded: 1927; Members: 3200
Pres: John Marshall; Gen Secr: Phil Bourke
Focus: Auto Eng
Periodicals
Recommended Practices
SAE-Australasia (weekly)
SAE-News (weekly) 00493

Society of Leather Technologists and Chemists, c/o Austral Finishes Pty Ltd, 19 Wilson St, Botany, NSW 2019
Founded: 1897; Members: 71
Focus: Eng; Chem 00494

Sociological Association of Australia and New Zealand, c/o Dept of Social Sciences, Mitchell College of Advanced Education, Bathurst, NSW 2795
Founded: 1963; Members: 600
Pres: Prof. Robert Lingard; Gen Secr: Eva Barboza
Focus: Sociology
Periodical
The Australian and New Zealand Journal of Sociology (3 times annually) 00495

South Australian Ornithological Association (S.A.O.A.), c/o South Australian Museum, North Terrace, Adelaide, SA 5000
Founded: 1899; Members: 650
Pres: Andrew Black; Gen Secr: Jean Wilson
Focus: Ornithology
Periodicals
Newsletter (weekly)
The South Australian Ornithologist (weekly)
. . . . 00496

South Australian Science Teachers Association, 163a Greenhill Rd, Parkside, SA 5063
Founded: 1941; Members: 850
Focus: Sci; Educ
Periodical
Sasta Journal (3 times annually) . . . 00497

Standards Association of Australia, 80 Arthur St, POB 458, North Sydney, NSW 2060
T: (02) 9634111; Tx: 26514; Fax: (02) 9593896
Founded: 1922
Gen Secr: W.S. Horwood
Focus: Standards
Periodical
The Australian Standard (weekly) . . . 00498

Statistical Society of Australia (S.S.A.), GPOB 573, Canberra, ACT 2601
Founded: 1959; Members: 900
Pres: T.C. Brown; Gen Secr: H. MacGillivray
Focus: Stats
Periodicals
Australian Journal of Statistics (3 times annually)
Newsletter (weekly) 00499

Sydney University Chemical Engineering Association (S.U.C.E.A.), c/o Dept of Chemical Engineering, University of Sydney, Sydney, NSW 2006
T: (02) 6608455
Members: 500
Focus: Chem; Eng 00500

Sydney University Chemical Society, c/o School of Chemistry, University, Sydney, NSW 2006
T: (02) 6922732; Fax: (02) 6923329
Founded: 1929
Focus: Chem 00501

Sydney University Medical Society, Blackburn Bldg, University, Sydney, NSW 2006
T: (02) 6600522
Founded: 1886
Focus: Med 00502

Sydney University Psychological Society, c/o Psychology Dept, University, Sydney, NSW 2006
T: (02) 6600522
Founded: 1929
Focus: Psych 00503

Sydney University Veterinary Society, c/o Faculty of Veterinary Science, University, Sydney, NSW 2006
Focus: Vet Med 00504

Tasmanian Geological Survey, c/o Dept of Mines, POB 56, Rosny, Tas 7018
Founded: 1860
Focus: Geology
Periodicals
Explanatory Report
Geological Survey Bulletin
Geological Survey Paper
Technical Report 00505

Tasmanian Historical Research Association, POB 441, Sandy Bay, Tas 7005
Founded: 1951; Members: 450
Pres: T.D. Sprod; Gen Secr: R. Kelly
Focus: Hist
Periodical
Papers and Proceedings (weekly) . . . 00506

Tasmanian University Agricultural Science Society, c/o Agricultural Faculty, University of Tasmania, GPOB 252, Hobart, Tas 7001
T: (002) 230561
Founded: 1962; Members: 45
Focus: Agri 00507

Technical Association of the Australian and New Zealand Pulp and Paper Industry (APPITA), 191 Royal Parade, Melbourne, Vic 3052
T: (03) 3472377; Tx: 10104
Founded: 1946; Members: 1200
Focus: Materials Sci
Periodical
APPITA Journal (weekly) 00508

Telecommunication Society of Australia (T.S.A.), GPOB 4050, Melbourne, Vic 3001
T: (03) 6307650
Founded: 1874; Members: 6000
Focus: Electric Eng 00509

Textile Society of Australia, 594 Saint Kilda Rd, Melbourne, Vic 3004
T: (03) 510221
Founded: 1945; Members: 279
Focus: Textiles 00510

Town and Country Planning Association of Victoria, 324 William St, Melbourne, Vic 3000
T: (03) 3295516
Founded: 1918
Focus: Urban Plan 00511

The Tropical Grassland Society of Australia, c/o C.S.I.R.O., Carmody Rd, Saint Lucia, Qld 4067
T: 3770209
Founded: 1963; Members: 850
Focus: Agri
Periodical
Tropical Grasslands (weekly) 00512

University Geographical Society, c/o Dept of Geography, University of Western Australia, Nedlands, WA 6009
T: (09) 3802698; Tx: 92992
Founded: 1964; Members: 140
Focus: Geography
Periodicals
Australind (weekly)
Real World 00513

University of New South Wales Chemical Engineering Association, c/o University of New South Wales, POB 1, Kensington, NSW 2033
T: 6630351
Founded: 1962
Focus: Eng 00514

Victorian Artists' Society, 430 Albert St, East Melbourne, Vic 3002
T: (03) 6621484; Fax: (03) 6622343
Founded: 1870; Members: 1000
Pres: A.W. Harding; Gen Secr: Beverley Snelling
Focus: Arts
Periodicals
Annual Report (weekly)
News Letter 00515

Victorian Post-Secondary Education Commission, Invergowrie, 21 Coppin Grove, Hawthorn, Vic 3122
Founded: 1978
Pres: Dr. R.B. Cullen
Focus: Educ 00516

Victorian Public Interest Research Group, 47 Charles St, Fitzroy, Vic 3065
Focus: Econ 00517

Victorian Society of Pathology and Experimental Medicine, 191 Royal Parade, Parkville, Vic 3052
T: 3476077
Founded: 1936
Focus: Med; Pathology 00518

Water Authorities Association of Victoria, 468 Saint Kilda Rd, Melbourne, Vic 3004
T: (03) 2675266
Founded: 1988
Focus: Civil Eng 00519

Water Research Foundation of Australia, POB 47, Kingsford, NSW 2032
Founded: 1956; Members: 600
Focus: Water Res
Periodical
Newsletter 00520

Weed Science Society of New South Wales, POB K287, Haymarket, NSW 2000
T: 6310655
Founded: 1966; Members: 125
Focus: Botany 00521

Weed Science Society of South Australia, c/o Ian Clunies Ross Centre, Australian Mineral Foundation, Conyngham St, Glenside, SA 5065
T: (085) 797821
Founded: 1970; Members: 50
Focus: Botany
Periodical
Newsletter (9 times annually) 00522

Western Australian Mental Health Association, 311-313 Hay St, Subiaco, WA 6008
T: 811986
Founded: 1960; Members: 300
Focus: Psychiatry 00523

Western Australian Naturalists' Club, 63-65 Merriwa St, POB 156, Nedlands, WA 6009
Founded: 1924; Members: 545
Focus: Hist; Nat Sci; Ecology
Periodical
Australian Naturalist (3-4 times annually) 00524

Western Australian Shell Club (W.A.S.C.), c/o W.A. Museum, Francis St, Perth, WA 6000
T: (09) 2714639
Founded: 1965; Members: 94
Focus: Ecology
Periodical
W.A. Shell Collector (weekly) 00525

Wildlife Conservation Society, Ettamogah Sanctuary, Hume Hwy, Lavington, NSW 2641
T: 251473
Founded: 1967
Focus: Ecology 00526

Wildlife Preservation Society of Australia (W.L.P.S. of A.), POB 3428, Sydney, NSW 2001
Founded: 1909; Members: 500
Pres: Vincent Serventy; Gen Secr: Sandy Johnson
Focus: Ecology
Periodical
Australian Wildlife Newsletter (weekly) . . 00527

Wildlife Preservation Society of Queensland (W.P.S.Q.), 8 Clifton St, Brisbane, Qld 4000
Founded: 1962; Members: 1500
Focus: Ecology 00528

World's Poultry Science Association, Australian Branch, c/o Poultry Science Research Station, Seven Hills, NSW 2147
T: 6226322
Founded: 1956; Members: 500
Focus: Zoology 00529

Zoological Board of Victoria, POB 74, Parkville, Vic 3052
T: (03) 3471522; Fax: (03) 3470309
Founded: 1937; Members: 12
Pres: P.E. Feilman; Gen Secr: I.D. Reid
Focus: Zoology 00530

Austria

Adalbert Stifter-Gesellschaft (A.St.G.), c/o Historisches Museum, Karlspl, 1040 Wien
T: (0222) 374354
Founded: 1918; Members: 150
Focus: Lit; Fine Arts 00531

Ärztegemeinschaft im Katholischen Akademieverband der Erzdiözese Wien, Währinger Str 2, 1090 Wien
T: (0222) 345493
Focus: Med 00532

Ärztegesellschaft Innsbruck, c/o Medizinische Fakultät, Universität, 6020 Innsbruck
Founded: 1894; Members: 900
Pres: Prof. Dr. G. Bartsch; Gen Secr: Dr. C. Radmayr
Focus: Med
Periodicals
Münchener Medizinische Wochenschrift (weekly)
Wiener Klinische Wochenschrift (weekly) . 00533

Ärztekammer für Kärnten, Bahnhofstr 22, 9020 Klagenfurt
T: (0463) 514222
Focus: Med
Periodical
Mitteilungen der Ärztekammer für Kärnten (weekly) 00534

Ärztekammer für Niederösterreich, Wipplingerstr 2, 1010 Wien
T: (0222) 633611
Focus: Med 00535

Ärztekammer für Wien, Weihburggasse 10-12, 1010 Wien
T: (0222) 51501223; Tx: 133921; Fax: (0222) 51501289
Focus: Med
Periodical
Wiener Arzt (11 times annually) 00536

Ärztliche Gesellschaft für Physiotherapie, Österreichischer Kneippärztebund, c/o Dr. Hans Kramer, Auzeile 22a, 2620 Neunkirchen
T: 2798
Focus: Therapeutics 00537

AGEMUS Arbeitsgemeinschaft Evolution, Menschheitszukunft und Sinnfragen, c/o Dr. Pretzmann, Burgring 7, 1014 Wien
T: (0222) 52177330; Fax: (0222) 935254
Founded: 1982; Members: 158
Gen Secr: Dr. Ina Wukowitz-Kumbaratschi
Focus: Soc Sci
Periodical
AGEMUS Nachrichten Wien: Internes Informationsorgan der Arbeitsgemeinschaft (weekly) 00538

Akademie der bildenden Künste in Wien, Schillerpl 3, 1010 Wien
T: (0222) 58816
Focus: Arts 00539

Akademie für Allgemeinmedizin, Universitätspl 4, 8010 Graz
T: (031581) 323
Founded: 1978
Focus: Med 00540

Akademie für Sozialarbeit der Stadt Wien, Freytaggasse 32, 1210 Wien
T: (0222) 387251; Fax: (0222) 305622
Focus: Soc Sci 00541

Akademische Arbeitsgemeinschaft für Volkskunde, Hanuschgasse 3, 1010 Wien
T: (0222) 5123837
Founded: 1967; Members: 80
Pres: Prof. Dr. Hermann Steininger
Focus: Ethnology; Folklore 00542

Akademisch-soziale Arbeitsgemeinschaft Österreichs, Schreyvogelgasse 3, 1010 Wien
T: (0222) 633589, 636844, 638419
Focus: Sociology 00543

Anthropologische Gesellschaft in Wien (AGW), Burgring 7, 1014 Wien
Founded: 1870; Members: 298
Pres: Prof. Dr. Karl Wernhart; Gen Secr: Dr. Herbert Kritscher
Focus: Anthro; Ethnology; Hist
Periodicals
Anthropologische Forschungen (weekly)
Mitteilungen der Anthropologischen Gesellschaft in Wien (weekly)
Prähistorische Forschungen (weekly)
Völkerkundliche Veröffentlichungen (weekly)
Volkskundliche Veröffentlichungen (weekly) 00544

Anthroposophische Gesellschaft in Österreich, Siebensterngasse 27, 1070 Wien
T: (0222) 932198
Founded: 1913
Focus: Anthro; Philos 00545

Anthroposophische Gesellschaft in Wien, Tilgnerstr 3, 1040 Wien
T: (0222) 653207
Focus: Anthro 00546

Anton-Bruckner-Institut Linz, Untere Donaulände 7, 4010 Linz
T: (0732) 7612268; Fax: (0732) 5133851
Founded: 1978; Members: 8
Pres: Prof. Dr. Othmar Wessely; Gen Secr: Dr. Elisabeth Maier
Focus: Music
Periodical
Bruckner-Jahrbuch (weekly) 00547

Arbeitsgemeinschaft audiovisueller Archive Österreichs (AGAVA), Liebiggasse 5, 1010 Wien
T: (0222) 401032734; Fax: (0222) 4030465
Founded: 1976; Members: 38
Pres: Prof. Dr. Gerhard Jagschitz; Gen Secr: Dr. Dietrich Schüller
Focus: Archives
Periodical
Das audiovisuelle Archiv (weekly) . . . 00548

Arbeitsgemeinschaft der Musikerzieher Österreichs (AGMÖ), c/o Bundesgymnasium Wien 3, Kundmanngasse 22, 1030 Wien
T: (0222) 7123364
Founded: 1947; Members: 1680
Pres: Dr. Wolf Peschl; Gen Secr: Dr. Karl Schnürl
Focus: Music; Educ
Periodical
Musikerziehung: Zeitschrift der Musikerzieher Österreichs 00549

Arbeitsgemeinschaft für Deutschdidaktik, c/o Institut für Germanistik, Universität für Bildungswissenschaften, Universitätsstr 65-67, 9022 Klagenfurt
T: (0463) 2700458; Fax: (0463) 27006110
Founded: 1976
Pres: Werner Wintersteiner; Gen Secr: Prof. Dr. Robert Saxer
Focus: Educ; Ling; Lit
Periodical
Informationen zur Deutschdidaktik (weekly) 00550

Arbeitsgemeinschaft für Historische Sozialkunde, c/o Institut für Wirtschafts- und Sozialgeschichte, Universität, Dr.-Karl-Lueger-Ring 1, 1010 Wien
T: (0222) 427611
Focus: Hist; Sociology
Periodical
Beiträge zur historischen Sozialkunde (weekly) 00551

Arbeitsgemeinschaft für interdisziplinäre angewandte Sozialforschung (AIAS), Lerchenfelder Str 36, 1080 Wien
Founded: 1968; Members: 28
Focus: Sociology
Periodical
Angewandte Sozialforschung (weekly) . . 00552

Arbeitsgemeinschaft für klinische Ernährung, c/o 1. Universitätsklinik für Gastroenterologie und Hepatologie, Lazarettgasse 14, 1090 Wien
Focus: Nutrition 00553

Arbeitsgemeinschaft für Neurobiologie, c/o Neurologisches Institut, Schwarzspanierstr 17, 1090 Wien
T: (0222) 431526
Founded: 1948; Members: 150
Focus: Bio; Neurology 00554

Arbeitsgemeinschaft für Präventivpsychologie, Stadiongasse 6-8, 1010 Wien
T: (0222) 425204
Members: 350
Pres: Dr. Christian Konrad; Gen Secr: Dr. Anneliese Fuchs
Focus: Psych
Periodical
APP-Info: Vereinszeitschrift der Arge für Präventivpsychologie (weekly) 00555

Arbeitsgemeinschaft für Psychotechnik in Österreich (APÖ), Vegagasse 4, 1190 Wien
T: (0222) 341130
Founded: 1926; Members: 30
Focus: Psych
Periodical
Mensch und Arbeit (weekly) 00556

Arbeitsgemeinschaft für theoretische und klinische Leistungsmedizin der Hochschullehrer Österreichs (ATKL), c/o Physiologisches Institut, Universität, Harrachgasse 21, 8010 Graz
T: (0316) 3804263; Fax: (0316) 383976
Founded: 1980; Members: 31
Pres: Dr. Günther Schwaberger
Focus: Med 00557

Arbeitsgemeinschaft für Verhaltensmodifikation (AVM), c/o Institut für Psychologie, Universität, Hellbrunnerstr 34, 5020 Salzburg
T: (0662) 80445122; Fax: (0662) 80446636
Founded: 1975; Members: 489
Pres: Dr. Anton Laireiter; Gen Secr: Dr. Mercedes Zsifkovics
Focus: Psych 00558

Arbeitsgemeinschaft für Wissenschaft und Politik, Innrain 52, 6020 Innsbruck
Fax: (0512) 5073467
Founded: 1971; Members: 50
Pres: Prof. Dr. Hans Köchler
Focus: Sci; Poli Sci 00559

Arbeitsgemeinschaft Österreichischer Entomologen (AÖG), Ludo-Hartmann-Pl 7, 1160 Wien
T: (0222) 4030338
Founded: 1949; Members: 250
Pres: Dr. Alexander Dostal; Gen Secr: Gottfried Novak
Focus: Entomology
Periodical
Zeitschrift der Arbeitsgemeinschaft Österreichischer Entomologen (weekly) 00560

Arbeitsgemeinschaft Personalwesen im Österreichischen Produktivitäts- und Wirtschaftlichkeits-Zentrum, Rockhgasse 6, Postf 131, 1014 Wien
Focus: Business Admin 00561

Arbeitsgemeinschaft Personenzentrierte Psychotherapie und Gesprächsführung, Postf 524, 1171 Wien
T: (0222) 4095573; Fax: (0222) 4095573
Focus: Psych; Therapeutics 00562

Arbeitsgemeinschaft zur Erforschung der Ärztlichen Allgemeinpraxis, c/o Niederösterreichisches Institut für Allgemeinmedizin, 3595 Brunn
T: (02989) 2249; Fax: (02989) 224918
Focus: Med 00563

Arbeitskreis der Wiener Altgermanisten, c/o Institut für Germanistik, Universität, Universitätsstr 7, 1010 Wien
T: (0222) 43002389
Founded: 1959; Members: 188
Focus: Ling; Lit 00564

Arbeitskreis der Wiener Skandinavisten, c/o Institut für Germanistik, Universität, Universitätsstr 7, 1010 Wien
Founded: 1976; Members: 20
Focus: Ling; Lit 00565

Arbeitskreis für neue Methoden in der Regionalforschung (AMR), Gustav-Tschermak-Gasse 36, 1190 Wien
T: (0222) 6387470
Founded: 1971; Members: 80
Focus: Urban Plan 00566

Arbeitskreis für Tibetische und Buddhistische Studien, Maria-Theresien-Str 3, 1090 Wien
T: (0222) 347493
Founded: 1977; Members: 20
Focus: Ethnology; Rel & Theol
Periodical
Wiener Studien zur Tibetologie und Buddhismuskunde 00567

Archaeoloische Gesellschaft Steiermark (AGST), Plüddemanngasse 95a, 8010 Graz
Founded: 1979; Members: 104
Pres: Prof. Dr. Manfred Hainzmann; Dr. Diether Kramer; Dr. Erwin Pochmarski
Focus: Archeol
Periodicals
Mitteilungen der Archäologischen Gesellschaft Steiermark
Nachrichtenblatt AGST 00568

Arthur Schnitzler-Institut, Herrengasse 5, 1010 Wien
T: (0222) 5338159; Fax: (0222) 5334067
Founded: 1974
Pres: Dr. Wolfgang Kraus; Gen Secr: Helmuth A. Niederle
Focus: Lit 00569

Asian Regional Cooperative Project on Food Irradiation (RPFI), c/o FAO/IAEA Div of Nuclear Techniques in Food and Agriculture, Food Preservation Section, Postf 100, 1400 Wien
T: (0222) 23600; Tx: 112645; Fax: (0222) 234564
Founded: 1980; Members: 13
Focus: Food 00570

Austria Esperanto Federacio / Austrian Esperanto Society, Fünfhausgasse 16, 1150 Wien
T: (0222) 834475
Founded: 1901; Members: 1000
Pres: Helga Farukvoye; Gen Secr: Emil Vokal
Focus: Ling

Periodicals
Austria-Esperanto-Revuo (weekly)
La Socialisto (weekly) *00571*

Austrotransplant – Österreichische Gesellschaft für Transplantation, Transfusion und Genetik, c/o Institut für Blutgruppenserologie, Spitalgasse 4, 1090 Wien
Founded: 1970; Members: 102
Focus: Genetics; Surgery *00572*

Autorenkreis Linz, c/o Margret Czerni, Schubertstr 7, 4020 Linz
Pres: Margret Czerni
Focus: Lit *00573*

Berufsverband freiberuflich tätiger Tierärzte Österreichs (BFÖ), c/o Dr. Gerald Lamprecht, Falkenburg 114, 8952 Irdning
T: (03682) 22937
Founded: 1958; Members: 264
Focus: Vet Med
Periodical
Der Österreichische Freiberufstierarzt (weekly)
. . . . *00574*

Berufsverband Österreichischer Psychologen (B.Ö.P.), Salzgries 10, 1010 Wien
T: (0222) 5335265
Founded: 1953; Members: 823
Focus: Psych
Periodical
Psychologie in Österreich (weekly) . . . *00575*

Bundeskammer der Tierärzte Österreichs, Biberstr 22, 1010 Wien
T: (0222) 5121766; Fax: (0222) 5121470
Focus: Vet Med
Periodical
Österreichische Tierärztezeitung (11 times annually)
. . . . *00576*

Chemisch-Physikalische Gesellschaft in Wien, Strudlhofgasse 4, 1090 Wien
T: (0222) 342630; Tx: 116222
Founded: 1869; Members: 260
Pres: Prof. P. Weinzierl; Gen Secr: F. Vesely
Focus: Chem; Physics
Periodical
Bulletin (weekly) *00577*

Commission Internationale de l'Eclairage (CIE) / International Commission on Illumination, Kegelgasse 27, 1030 Wien
T: (0222) 753187; Tx: 11151; Fax: (0222) 7130838
Founded: 1900
Gen Secr: Dr. János Schanda
Focus: Eng
Periodical
CIE (weekly) *00578*

Commission Internationale d'Histoire du Sel (CIHS), c/o Universität, Innrain 52, 6020 Innsbruck
T: 20053305
Founded: 1987
Gen Secr: Prof. Dr. Rudolf Palme
Focus: Cultur Hist *00579*

Deutscher Rechtshistorikertag, c/o Institut für Österreichische Rechtsgeschichte, Universitätspl 3, 8010 Graz
Focus: Law; Hist *00580*

Diplomatische Akademie Wien / Diplomatic Academy of Vienna, Favoritenstr 15, 1040 Wien
T: (0222) 5042265
Founded: 1964
Gen Secr: Dr. Alfred Missong
Focus: Poli Sci
Periodical
Jahrbuch (weekly) *00581*

Dokumentationsarchiv des österreichischen Widerstandes, Wipplingerstr 8, 1010 Wien
T: (0222) 53436739
Founded: 1963; Members: 203
Pres: D.H. Pfoch; Gen Secr: Dr. Wolfgang Neugebauer
Focus: Hist
Periodicals
Jahrbuch (weekly)
Mitteilungen des Dokumentationsarchives des österreichischen Widerstandes *00582*

Dokumentationsstelle für neuere österreichische Literatur, Seidengasse 13, 1070 Wien
T: (0222) 5262044
Founded: 1966
Focus: Lit
Periodical
Zirkular (3 times annually) *00583*

Eranos Vindobonensis, c/o Institut für Klassische Philologie, Universität, Dr.-Karl-Lueger-Ring 1, 1010 Wien
Founded: 1885; Members: 91
Gen Secr: Dr. Paul Raimund Lorenz
Focus: Ling; Lit; Archeol; Hist *00584*

Europäisches Zentrum für Wohlfahrtspolitik und Sozialforschung, Berggasse 17, 1090 Wien
T: (0222) 31945050; CA: Eurosocial Wien; Fax: (0222) 319450519
Founded: 1974
Gen Secr: Prof. Dr. B. Marin
Focus: Poli Sci; Sociology
Periodicals
Eurosocial Reports
Journal für Sozialforschung (weekly)
Newsletter/Bulletin d'Information/Nachrichten (weekly)
. . . . *00585*

European Association for Catholic Adult Education, Kapuzinerstr 84, 4020 Linz
T: (0732) 274441 ext 238
Founded: 1963
Focus: Adult Educ
Periodical
FEECA Bulletin (weekly) *00586*

European Bone Marrow Transplant Group (EBMT), Interconvention, Austria Centre, 1450 Wien
Gen Secr: Prof. Dr. J.W.H. Leer
Focus: Surgery *00587*

European Centre for Social Welfare Policy and Research (ECSWPR), Berggasse 17, 1090 Wien
Fax: (0222) 314505
Founded: 1974
Gen Secr: Prof. Dr. Bernd Marin
Focus: Soc Sci
Periodicals
Eurosocial Bulletin (weekly)
Journal für Sozialforschung (weekly)
Sovetskaya Tyurkologiya: Soviet Turkology *00588*

European Chapter of Combinatorial Optimization (ECCO), c/o Institut für Mathematik, Technische Universität Graz, Steirergasse 30, 8010 Graz
T: (0316) 8737127; Fax: (0316) 827685
Gen Secr: Günter Rote
Focus: Math *00589*

European Cooperation in Social Science Information and Documentation (ECSSID), c/o Vienna Centre, Grünangergasse 2, Postf 974, 1010 Wien
T: (0222) 5124333; Fax: (0222) 5125366
Gen Secr: Gerhard Auinger
Focus: Soc Sci
Periodical
ECSSID Bulletin *00590*

European Coordination Centre for Research and Documentation in Social Sciences, c/o Vienna Centre, Grünangergasse 2, 1010 Wien
T: (0222) 5124333; Fax: (0222) 512536616
Founded: 1963
Gen Secr: Dr. L. Kiuzadjan
Focus: Soc Sci
Periodical
Vienna Centre Newsletter (3 times annually)
. . . . *00591*

Evangelische Akademie in Wien, Schwarzspanierstr 13, Postf 15, 1096 Wien
T: (0222) 426106
Founded: 1955
Pres: Prof. Dr. Ulrich Trinks
Focus: Rel & Theol; Adult Educ . . . *00592*

Föderation der Internationalen Donausymposia über Diabetes mellitus, c/o Klinik für Innere Medizin III, Währinger Gürtel 18-20, 1090 Wien
Focus: Diabetes *00593*

Forschungsgemeinschaft für Erkrankungen des Bewegungsapparates, Hochstr 31, 1238 Wien
T: (0222) 881371
Founded: 1966
Focus: Orthopedics *00594*

Forschungsgesellschaft für das Verkehrs- und Straßenwesen im ÖIAV (FVS), Eschenbachgasse 9, 1010 Wien
T: (0222) 587353631; Fax: (0222) 58735365
Founded: 1950; Members: 964
Pres: Herbert Kainbacher; Gen Secr: Herbert Stundner
Focus: Transport; Civil Eng
Periodical
Schriftenreihe der FVS (weekly) *00595*

Forschungsgesellschaft für Psycho-Elektronik und Kybernetik, Sechskrügelgasse 2, 1030 Wien
T: (0222) 71247383
Founded: 1971; Members: 127
Focus: Electronic Eng; Cybernetics
Periodical
Psychotronics (weekly) *00596*

Forschungsgesellschaft für Wohnen, Bauen und Planen (FGW), Löwengasse 47, 1030 Wien
T: (0222) 726251, 726741
Founded: 1956; Members: 700
Focus: Civil Eng; Urban Plan
Periodicals
Österreichische Dokumentation für Wohnen, Bauen und Planen (ODW) (weekly)
Wohnbauforschung in Österreich (WBFÖ) (weekly)
. . . . *00597*

Franz Schmidt-Gesellschaft, c/o Musikverein, Bösendorferstr 12, 1010 Wien
T: (0222) 505172171
Founded: 1951; Members: 160
Pres: Prof. Albert Moser; Gen Secr: Dr. Carmen Ottner
Focus: Music
Periodical
Studien zu Franz Schmidt (weekly) . . . *00598*

Friedrich Hebbel-Gesellschaft, Frauengasse 14, 1170 Wien
T: (0222) 4374224
Founded: 1957; Members: 102
Pres: Ida Koller-Andorf; Gen Secr: Kurt Kerschbaum
Focus: Lit *00599*

Gebrüder Schrammel-Gesellschaft, Elterleinpl 14, 1170 Wien
Founded: 1967
Focus: Music *00600*

Gemeinnütziger Verein zur Durchführung von Lehr- und Forschungsaufgaben an der Wirtschaftsuniversität Wien, Augasse 2-6, 1090 Wien
T: (0222) 340525693
Founded: 1987
Focus: Econ *00601*

Geschichtsverein für Kärnten, Museumgasse 2, 9020 Klagenfurt
T: (0463) 53630573; Fax: (0463) 53630570
Founded: 1844; Members: 2050
Focus: Hist
Periodical
Carinthia I: Zeitschrift für geschichtliche Landeskunde von Kärnten (weekly) . . . *00602*

Gesellschaft der Ärzte in Vorarlberg, Schulgasse 17, 6850 Dornbirn
Founded: 1951; Members: 452
Focus: Med *00603*

Gesellschaft der Ärzte in Wien, Frankgasse 8, Postf 147, 1096 Wien
T: (0222) 424777; Fax: (0222) 4023090
Founded: 1837; Members: 1752
Focus: Med
Periodical
Wiener Klinische Wochenschrift (weekly) . *00604*

Gesellschaft der Chirurgen in Wien, c/o Chirurgische Universitätsklinik, Währinger Gürtel 18-20, 1090 Wien
Fax: (0222) 404004004
Founded: 1837; Members: 137
Gen Secr: Prof. Dr. R. Jakesz
Focus: Surgery *00605*

Gesellschaft der Freunde der Biologischen Station Wilhelminenberg, Savoyenstr 1, 1160 Wien
Founded: 1957; Members: 790
Focus: Bio *00606*

Gesellschaft der Freunde der Neuen Galerie, Sackgasse 16, 8010 Graz
T: (0316) 829155, 829186
Founded: 1946; Members: 100
Gen Secr: Dr. Werner Fenz
Focus: Arts *00607*

Gesellschaft der Freunde der Österreichischen Nationalbibliothek, Josefspl 1, 1015 Wien
T: (0222) 53410
Founded: 1921; Members: 130
Focus: Libraries & Bk Sci
Periodical
Biblos: Österreichische Zeitschrift für Buch- und Bibliothekswesen, Dokumentation, Bibliographie und Bibliophilie (weekly) *00608*

Gesellschaft der Freunde des Kunsthistorischen Institutes der Karl-Franzens-Universität in Graz, Universitätspl 3, 8010 Graz
T: (0316) 3802395
Founded: 1964; Members: 114
Focus: Arts *00609*

Gesellschaft der Geologie- und Bergbaustudenten in Österreich, c/o Institut für Geologie, Universität, Universitätsstr 7, 1010 Wien
T: (0222) 43002518
Founded: 1948; Members: 329
Focus: Geology; Mining *00610*

Gesellschaft der Kunstfreunde, Neudeggergasse 8, 1080 Wien
T: (0222) 423125
Focus: Arts *00611*

Gesellschaft der Musikfreunde in Wien (GdM), Bösendorferstr 12, 1010 Wien
T: (0222) 5058681; Fax: (0222) 5059409
Founded: 1812; Members: 10000
Focus: Music
Periodical
Musikfreunde (8 times annually) *00612*

Gesellschaft für Arbeitsrecht und Sozialrecht, c/o Institut für Arbeitsrecht und Sozialrecht, Auhof, 4040 Linz
Founded: 1965
Focus: Law *00613*

Gesellschaft für biologische und psychosomatische Medizin, Peterspl 4, 1010 Wien
Focus: Psychiatry; Neurology *00614*

Gesellschaft für Chemiewirtschaft, Salesianergasse 208, 1030 Wien
T: (0222) 725611360
Founded: 1949; Members: 251
Focus: Chem; Econ *00615*

Gesellschaft für das Recht der Ostkirchen, c/o Institut für Kirchenrecht, Freyung 6, 1010 Wien
T: (0222) 5333861; Fax: (0222) 5351019
Founded: 1971; Members: 200
Focus: Law; Rel & Theol
Periodical
Kanon: Jahrbuch der Gesellschaft für das Recht der Ostkirchen *00616*

Gesellschaft für die Geschichte des Protestantismus in Österreich (GPrÖ), c/o Evangelisch-theologische Fakultät, Universität, Rooseveltpl 10, 1090 Wien
T: (0222) 43598113; Fax: (0222) 4020595
Founded: 1880; Members: 328
Pres: Prof. Dr. Peter F. Barton; Gen Secr: Dr. Karl Schwarz
Focus: Hist; Rel & Theol
Periodical
Jahrbuch für die Geschichte des Protestantismus Oesterreich *00617*

Gesellschaft für Forschungen zur Aufführungspraxis (GfA), Leonhardstr 15, 8010 Graz
T: (0316) 74025
Founded: 1972; Members: 10
Focus: Perf Arts *00618*

Gesellschaft für Ganzheitsforschung (G.f.G.), c/o Wirtschaftsuniversität Wien, Augasse 2-6, 1090 Wien
T: (0222) 313364524; Fax: (0222) 31336727
Founded: 1956; Members: 120
Pres: Prof. Dr. J. Hanns Pichler; Gen Secr: Dr. Hubert Verhonig
Focus: Philos
Periodical
Zeitschrift für Ganzheitsforschung (weekly) *00619*

Gesellschaft für Geschichte der Neuzeit, c/o Prof. Dr. Fritz Fellner, Ignaz-Rieder-Kai 19, 5020 Salzburg
T: (0662) 80444757
Founded: 1981; Members: 67
Pres: Prof. Dr. Fritz Fellner; Gen Secr: Dr. Brigitte Mazohl-Wallnig
Focus: Hist *00620*

Gesellschaft für Innere Medizin in Wien, Garnisongasse 13, 1090 Wien
Founded: 1902; Members: 375
Focus: Intern Med *00621*

Gesellschaft für Input-Output-Analyse, Postf 108, 1033 Wien
Founded: 1988; Members: 288
Pres: Dr. Jiri Skolka; Gen Secr: Dr. Norbert Rainer
Focus: Computer & Info Sci
Periodical
Economic Systems Research (weekly) . . *00622*

Gesellschaft für Klassische Philologie in Innsbruck, c/o Institut für Klassische Philologie, Neue Universität, Innrain 52, 6020 Innsbruck
T: (05222) 7243663
Founded: 1958; Members: 200
Pres: Eva Dragosits; Gen Secr: Veronika Bohle
Focus: Ling; Lit
Periodical
Acta Philologica Aenipotana *00623*

Gesellschaft für Kulturpsychologie, Schwesternweg 9, 5020 Salzburg
T: (0662) 80445111
Founded: 1987; Members: 26
Focus: Psych *00624*

Gesellschaft für Logotherapie und Existenzanalyse, Eduard-Sueß-Gasse 10, 1150 Wien
T: (0222) 9859566; Fax: (0222) 9824845
Founded: 1984; Members: 600
Pres: Dr. A. Längle; Gen Secr: Dr. Liselotte Tutsch
Focus: Therapeutics; Psychoan
Periodical
Bulletin *00625*

Gesellschaft für Manuelle Lymphdrainage nach Dr. Vodder, Alleestr 30, 6344 Walchsee
T: (05374) 52450; Fax: (05374) 52454
Founded: 1973; Members: 800
Focus: Cell Biol & Cancer Res *00626*

Gesellschaft für österreichische Kulturgeschichte, Haydngasse 1, 7000 Eisenstadt
Founded: 1968
Focus: Hist
Periodical
Jahrbuch für österreichische Kulturgeschichte (weekly) *00627*

Gesellschaft für Ost- und Südostkunde / Association for Eastern and Southeastern European Studies, Wienerstr 131, 4024 Linz
T: (0732) 273380
Founded: 1955; Members: 90
Pres: Karl Ochsner; Gen Secr: Franz Templ
Focus: Ethnology
Periodicals
Das Kleine Ostpanorama (weekly)
Ostpanorama-Sonderausgabe (weekly) . . *00628*

Gesellschaft für Phänomenologie und Kritische Anthropologie, c/o Prof. Dr. Benedikt, Chimanistr 27, 1190 Wien
Focus: Anthro *00629*

Gesellschaft für Photographie und Geschichte, c/o Institut für Zeitgeschichte, Universität, Rotenhausgasse 6, 1090 Wien
T: (0222) 426280
Founded: 1988
Focus: Photo; Hist
Periodical
Photographie und Gesellschaft (weekly) . 00630

Gesellschaft für politische Aufklärung, c/o Institut für Höhere Studien und wissenschaftliche Forschung, Stumpergasse 56, 1060 Wien
T: (0222) 59991172; Fax: (0222) 5970635
Founded: 1982; Members: 24
Pres: Prof. Dr. Anton Pelinka
Focus: Poli Sci
Periodical
Informationen der Gesellschaft für politische Aufklärung (weekly) 00631

Gesellschaft für Salzburger Landeskunde (GSLK), Michael-Pacher-Str 40, 5020 Salzburg
T: (0662) 80424511
Founded: 1860; Members: 1480
Focus: Hist
Periodical
Mitteilungen (weekly) 00632

Gesellschaft für Soziologie an der Universität Graz, Heinrichstr 106, 8010 Graz
Founded: 1951
Focus: Sociology 00633

Gesellschaft für Strategische Unternehmensführung, c/o Institut für Unternehmensführung, Universität, Rennweg 25, 6020 Innsbruck
T: (0512) 57700111; Fax: (0512) 57700117
Founded: 1982; Members: 80
Pres: Prof. Dr. Hans H. Hinterhuber; Gen Secr: M. Stumpf
Focus: Business Admin
Periodical
Informationsdienst 00634

Gesellschaft für vergleichende Felsbildforschung (GE.FE.BI.), Geidorfgürtel 40, 8010 Graz
T: (0316) 329845
Founded: 1977; Members: 42
Pres: Dr. Lothar Wanke; Gen Secr: Michaela Loidl
Focus: Arts
Periodical
Jahrbuch (weekly) 00635

Gesellschaft für vergleichende Kunstforschung, c/o Institut für Kunstgeschichte, Universitätsstr 7, 1010 Wien
T: (0222) 43002509; Fax: (0222) 4028510
Founded: 1934; Members: 324
Gen Secr: Prof. Dr. Walter Krause
Focus: Arts
Periodical
Mitteilungen der Gesellschaft für vergleichende Kunstforschung 00636

Gesellschaft österreichischer Chemiker (VÖCH), Nibelungengasse 11, 1010 Wien
Founded: 1897; Members: 1720
Pres: Prof. Dr. K. Schlögl
Focus: Chem
Periodical
Monatshefte für Chemie (weekly) . . . 00637

Gesellschaft Österreichischer Nervenärzte und Psychiater, c/o Neurologisches Krankenhaus Rosenhügel, Riedelgasse 5, 1130 Wien
T: (0222) 882515270
Founded: 1957; Members: 300
Pres: Prof. Dr. Kurt Jellinger; Gen Secr: Prof. Dr. H. Binder; Dr. H.G. Zapotoczky
Focus: Psychiatry; Neurology
Periodical
Neuropsychiatrie (weekly) 00638

Gesellschaft zum Studium und zur Erneuerung der Struktur der Rechtsordnung, Schottenbastei 10-16, 1010 Wien
T: (0222) 43003114
Founded: 1970; Members: 89
Focus: Law 00639

Gesellschaft zur Erforschung slawischer Sprachen und Kulturen, Heinrichstr 26, 8010 Graz
Founded: 1973
Focus: Ethnology; Ling 00640

Gesellschaft zur Errichtung der Akademie für Allgemeinmedizin, Universitätspl 4, 8010 Graz
T: (0316) 31581323
Founded: 1978
Focus: Med 00641

Gesellschaft zur Förderung der industriellen Pflanzenbaus, Getreidemarkt 9, 1060 Wien
Founded: 1964
Focus: Agri 00642

Gesellschaft zur Förderung Slawistischer Studien, Teschnergasse 4, 1180 Wien
T: (0222) 401032941
Founded: 1983
Focus: Ling; Lit
Periodical
Wiener Slawistischer Almanach (WSA) (weekly) 00643

Gesellschaft zur Förderung von Nordamerika-Studien an der Universität Wien, Lammgasse 8, 1080 Wien
T: (0222) 438679
Founded: 1985; Members: 20
Focus: Ethnology 00644

Gesellschaft zur Herausgabe von Denkmälern der Tonkunst in Österreich, c/o Institut für Musikwissenschaft, Universitätsstr 7, 1010 Wien
T: (0222) 401032629; Fax: (0222) 4020533
Founded: 1893; Members: 230
Pres: Dr. Heinrich Haerdtl
Focus: Music
Periodicals
Studien zur Musikwissenschaft
Wiener Beiträge zur Musikwissenschaft . 00645

Gregor Mendel-Gesellschaft Wien, c/o Institut für Pflanzenbau und Pflanzenzüchtung, Gregor-Mendel-Str 33, 1180 Wien
T: (0222) 342500
Founded: 1972
Focus: Genetics 00646

Grillparzer-Gesellschaft, Gumpendorfer Str 15, 1060 Wien
T: (0222) 5861090
Founded: 1890; Members: 200
Pres: Dr. Leopold O. Knobloch; Gen Secr: Dr. Robert Pichl
Focus: Lit
Periodical
Grillparzer-Jahrbuch (weekly) 00647

Haus des Meeres Vivarium Wien, Fritz-Grünbaum-Pl 1, 1060 Wien
T: (0222) 5871417; Fax: (0222) 5860617
Founded: 1957; Members: 30
Focus: Bio; Oceanography 00648

Heraldisch-Genealogische Gesellschaft Adler, Haarhof 4a, 1010 Wien
Founded: 1870; Members: 850
Pres: Dr. Georg Johannes Kugler; Gen Secr: Dr. Andreas Cornaro
Focus: Genealogy
Periodical
Adler: Zeitschrift für Genealogie und Heraldik (weekly) 00649

Historische Landeskommission für Steiermark, Karmeliterpl 3, 8010 Graz
Founded: 1892; Members: 30
Pres: Dr. Josef Krainer; Gen Secr: Prof. Dr. Othmar Pickl
Focus: Hist
Periodical
Forschungen zur geschichtlichen Landeskunde der Steiermark 00650

Historischer Verein für Steiermark, Hamerlinggasse 3, 8010 Graz
T: (0316) 8772366
Founded: 1850; Members: 1400
Focus: Hist
Periodicals
Blätter für Heimatkunde (weekly)
Zeitschrift des Historischen Vereins für Steiermark (weekly) 00651

Innsbrucker Arbeitskreis für Psychoanalyse, Psychoanalytisches Forschungs- und Ausbildungsinstitut, Collingasse 7, 6020 Innsbruck
T: (05222) 582827, 577905
Founded: 1976; Members: 26
Focus: Psychoan 00652

Innsbrucker Germanistische Arbeitsgemeinschaft (IGA), c/o Institut für Germanistik, Innrain 52, 6020 Innsbruck
T: (0512) 7243453
Members: 20
Focus: Ling; Lit 00653

Innsbrucker Gesellschaft zur Pflege der Geisteswissenschaften (IGG), Innrain 52, 6020 Innsbruck
T: (0512) 5073535
Founded: 1949; Members: 80
Pres: Prof. Dr. Peter W. Haider; Gen Secr: Dr. Günther Lorenz
Focus: Humanities
Periodical
Innsbrucker Beiträge zur Kulturwissenschaft (weekly) 00654

Innsbrucker Sprachwissenschaftliche Gesellschaft (ISG), Innrain 52, 6020 Innsbruck
T: (0512) 5073570
Founded: 1970; Members: 96
Pres: Dr. Peter Anreiter; Gen Secr: Dr. Barbara Stefan
Focus: Ling 00655

Institut für Grenzgebiete der Wissenschaft (IGW), Maximilianstr 8, Postf 8, 6010 Innsbruck
T: (0512) 574772; Fax: (0512) 586463
Founded: 1980; Members: 370
Focus: Sci
Periodicals
Ethica: Wissenschaft und Verantwortung
Grenzgebiete der Wissenschaft (weekly) . 00656

Institut für Österreichische Musikdokumentation (IÖM), Augustinerstr 1, 1010 Wien
T: (0222) 5335087; Fax: (0222) 53410310
Founded: 1972; Members: 8
Pres: Dr. Heinz-Georg Kamler; Gen Secr: Dr. Günter Brosche
Focus: Music 00657

Institut für Österreichkunde, Hanuschgasse 3, 1010 Wien
T: (0222) 5127932
Founded: 1957; Members: 700
Pres: Prof. Dr. Ernst Bruckmüller; Gen Secr: Prof. Hermann Möcker
Focus: Hist; Poli Sci; Lit; Ling; Geography; Econ
Periodical
Österreich in Geschichte und Literatur (mit Geographie) (weekly) 00658

Institut für Sozialdienste, Schedlerstr 10, 6900 Bregenz
T: (05574) 451870; Fax: (05574) 4518721
Founded: 1962; Members: 43
Pres: Prof. Hans Sperandio; Gen Secr: Manfred Dörler
Focus: Soc Sci
Periodical
IfS-Information (weekly) 00659

Institut für Soziales Design – Entwicklung und Forschung (ISD), Grenzackerstr 7-11, 1100 Wien
T: (0222) 6030772
Founded: 1975; Members: 20
Pres: Peter Pruner; Dieter Berdel
Focus: Graphic & Dec Arts, Design . . 00660

Institut für Wissenschaft und Kunst (IWK), Berggasse 17, 1090 Wien
T: (0222) 344342
Founded: 1946; Members: 800
Pres: Dr. Johann Altenhuber; Gen Secr: Dr. Helga Kaschl
Focus: Sci; Arts
Periodical
Mitteilungen (weekly) 00661

International Association of Schools of Social Work (IASSW), Palais Palfy, Josefpl 6, 1010 Wien
T: (0222) 5134297; Fax: (0222) 5138468
Founded: 1929; Members: 500
Gen Secr: Dr. Vera Mehta
Focus: Adult Educ; Sociology
Periodicals
International Social Work (weekly)
Newsletter (3 times annually) 00662

International Atomic Energy Agency (IAEA), c/o Vienna International Centre, Wagramerstr 5, Postf 100, 1400 Wien
T: (0222) 23600; Tx: 112645; Fax: (0222) 234564
Founded: 1957; Members: 113
Gen Secr: Dr. Hans Blix
Focus: Energy; Nucl Res
Periodicals
IAEA Bulletin (weekly)
IAEA Newsbrief (weekly)
Meetings on Atomic Energy (weekly) . . 00663

International Council on Social Welfare (ICSW), Koestlergasse 1, 1060 Wien
Fax: (0222) 5879951
Founded: 1928
Gen Secr: Sirpa Utriainen
Focus: Soc Sci
Periodical
ICSW-Information 00664

Internationale Albrechtsberger-Gesellschaft, Rathaus, 3400 Klosteneuburg
T: (02243) 6795222
Founded: 1962; Members: 70
Focus: Music 00665

Internationale Bruckner-Gesellschaft, Rathauspl 3, 1010 Wien
T: (0222) 427140
Focus: Music
Periodical
Mitteilungsblatt der IBG (weekly) 00666

Internationale Chopin-Gesellschaft in Wien, Biberstr 4, 1010 Wien
T: (0222) 5122374
Founded: 1952; Members: 240
Focus: Music
Periodical
Wiener Chopin-Blätter (2-3 times annually) 00667

Internationale Coronelli-Gesellschaft für Globen- und Instrumentenkunde, Dominikanerbastei 21, 1010 Wien
T: (0222) 5333285; Fax: (0222) 5320824
Founded: 1952; Members: 280
Pres: Rudolf Schmidt; Gen Secr: Dr. Johannes Dörflinger
Focus: Cart
Periodical
Der Globusfreund (weekly) 00668

Internationale Gesellschaft für Getreidewissenschaft und -technologie / International Association for Cereal Science and Technology, Wienerstr 22, Postf 77, 2320 Schwechat
T: 777202; Fax: 777204
Founded: 1955; Members: 33
Pres: Dr. P. Steyn; Gen Secr: Dr. H. Glattes
Focus: Biochem
Periodical
Newsletter (weekly) 00669

Internationale Gesellschaft für Ingenieurpädagogik, c/o Universität für Bildungswissenschaften, Universitätsstr 65-67, 9022 Klagenfurt
T: (0463) 2700371
Founded: 1972; Members: 300
Pres: Prof. Dr. Adolf Melezinek; Gen Secr: Hartmut Weidner
Focus: Eng; Educ
Periodical
IGIP-Report (3 times annually) 00670

Internationale Gesellschaft für Jazzforschung (IGJ), c/o Institut für Jazzforschung, Hochschule für Musik und darstellende Kunst, Leonhardstr 15, 8010 Graz
T: (0316) 3891221
Founded: 1969; Members: 286
Pres: Prof. Dr. Franz Kerschbaumer; Gen Secr: Dr. Elisabeth Kolleritsch
Focus: Music
Periodicals
Beiträge zur Jazzforschung: Studies in Jazz Research (weekly)
Jazzforschung: Jazz Research (weekly) . 00671

Internationale Gesellschaft zur Erforschung und Förderung der Blasmusik (IGEB) / International Society for the Promotion and Investigation of Band Music, Leonhardstr 15, 8010 Graz
T: (0316) 3891123
Founded: 1974
Pres: Prof. Dr. Wolfgang Suppan; Gen Secr: Dr. Bernhard Habla
Focus: Music
Periodicals
Alta Musica: Jahrbücher (weekly)
Mitteilungsblatt (3 times annually) . . . 00672

Internationale Gustav Mahler Gesellschaft (IGMG), Wiedner Gürtel 6, 1040 Wien
T: (0222) 5057330
Founded: 1955; Members: 450
Pres: Dr. Rainer Bischof; Gen Secr: Emmy Hauswirth
Focus: Music
Periodical
Nachrichten zur Mahler-Forschung: News about Mahler Research (weekly) 00673

Internationale Hugo Wolf-Gesellschaft, Latschkagasse 4, 1090 Wien
T: (0222) 3101388
Founded: 1956; Members: 35
Pres: Prof. Leopold Spitzer; Gen Secr: Prof. Dr. Hans Jancik
Focus: Music 00674

Internationale Nestroy-Gesellschaft, Gentzgasse 10, 1180 Wien
T: (0222) 4707067
Founded: 1973; Members: 245
Pres: Prof. Dr. Heinrich Kraus; Gen Secr: Karl Zimmel
Focus: Perf Arts; Lit
Periodical
Nestroyana: Blätter der Internationalen Nestroy-Gesellschaft (weekly) 00675

Internationale Paracelsus-Gesellschaft (IPG), Ignaz-Harrer-Str 79, 5020 Salzburg
Founded: 1950; Members: 250
Pres: Prof. Dr. Kurt Goldammer
Focus: Med; Philos; Rel & Theol
Periodicals
Paracelsusbrief: Mitteilungsblatt (weekly)
Salzburger Beiträge zur Paracelsusforschung 00676

Internationaler Verband Forstlicher Forschungsanstalten / Union Internationale des Instituts de Recherche Forestières / International Union of Forestry Research Organisations, Schönbrunn, Tirolergarten, 1131 Wien
T: (0222) 8770151; Fax: (0222) 829355
Founded: 1891; Members: 15000
Pres: Dr. Mohd N. Sallah; Gen Secr: Heinrich Schmutzenhofer
Focus: Forestry
Periodical
IUFRO-NEWS (weekly) 00677

Internationale Schönberg-Gesellschaft, Bernhardgasse 6, 2340 Mödling
T: (02236) 42223
Founded: 1972
Pres: Prof. Dr. Rudolf Stephan; Gen Secr: Dr. Sigrid Wiesmann
Focus: Music
Periodical
Mitteilungen (weekly) 00678

Internationales Forschungszentrum für Grundfragen der Wissenschaften, Mönchsberg 2a, 5020 Salzburg
T: (0662) 842521; Fax: (0662) 84252118
Founded: 1961; Members: 24
Pres: Prof. Dr. Franz-Martin Schmölz; Gen Secr: Prof. Dr. Paul Weingartner
Focus: Sci
Periodicals
Erkenntnis
Journal of Symbolic Logic
Kairos
Synthese
Zeitgeschichte 00679

Internationales Institut für den Frieden / International Institute for Peace, c/o Dr. Peter Stania, Möllwaldpl 5, 1040 Wien
T: (0222) 5046437; CA: paxinstitu vienna; Fax: (0222) 5053236
Founded: 1957
Pres: Erwin Lanc
Focus: Poli Sci; Soc Sci; Nat Sci
Periodicals
IIP Monitor (weekly)
Peace in the Sciences: Occasional Papers (weekly) 00680

Internationales Institut für Jugendliteratur und Leseforschung, Mayerhofgasse 6, 1040 Wien
T: (0222) 5050359
Founded: 1965; Members: 500
Focus: Lit
Periodicals
1000 + 1 Buch (weekly)
Bookbird (weekly)
PA-Kontakte (weekly) 00681

Internationales Musikzentrum (IMZ), Lothringerstr 20, 1030 Wien
T: (0222) 7130777; Tx: 75311745
Founded: 1961; Members: 250
Pres: Avril MacRory; Gen Secr: Eric Marinitsch
Focus: Music
Periodical
Music in the Media: IMZ-Bulletin (10 times annually) 00682

Internationale Stiftung Mozarteum, Schwarzstr 26, 5020 Salzburg
T: (0662) 8394010
Founded: 1841
Focus: Music
Periodical
Mitteilungen der Internationalen Stiftung Mozarteum (weekly) 00683

Internationale Tagung der Historiker der Arbeiterbewegung (ITH) / International Conference of Historians of the Labour Movement, Altes Rathaus, Wipplingerstr 8, 1010 Wien
T: (0222) 53436210
Founded: 1964
Gen Secr: Dr. Barry McLoughlin
Focus: Hist; Socialism
Periodicals
Auswahl der Konferenzbeiträge (weekly)
Mitteilungsblatt (weekly) 00684

Internationale Vereinigung für Selbstmordprophylaxe (IASP) / International Association for Suicide Prevention, c/o Institut für medizinische Psychology, Severingasse 9, 1090 Wien
T: (0222) 408356824; Fax: (0222) 408356812
Founded: 1960; Members: 300
Pres: Prof. David Lester; Gen Secr: Prof. Gernot Sonneck
Focus: Psych; Psychiatry
Periodical
Crisis: Journal of Crisis Intervention and Suicide Prevention (weekly) 00685

International Federation of Automatic Control (IFAC), Schlosspl 12, 2361 Laxenburg
T: (02236) 71447; Tx: 79248; Fax: (02236) 72859
Founded: 1957; Members: 44
Gen Secr: Dr. G. Hencsey
Focus: Eng
Periodicals
Automatica (weekly)
Newsletter (weekly) 00686

International Institute for Applied Systems Analysis (IIASA), Schlosspl 1, 2361 Laxenburg
T: (02236) 715210; Tx: 079137; Fax: (02236) 71313
Founded: 1972
Focus: Ecology; Econ; Eng
Periodical
Options (weekly) 00687

International Law Association, Österreichischer Zweigverein, Rotenturmstr 13, 1010 Wien
Founded: 1947
Focus: Law 00688

International Society of Soil Science, c/o Universität für Bodenkultur, Gregor-Mendel-Str 33, 1180 Wien
T: (0222) 3106026; Tx: 111010; Fax: (0222) 3106027
Founded: 1924; Members: 7500
Gen Secr: Prof. W.E.H. Blum
Focus: Agri
Periodical
Bulletin (weekly) 00689

Joanneum-Verein, Raubergasse 10, 8010 Graz
T: (0316) 8772461
Founded: 1819; Members: 415
Focus: Arts 00690

Johann-Joseph-Fux-Gesellschaft, c/o Institut für Musikethnologie, Hochschule für Musik und darstellende Kunst, Leonhardstr 15, 8010 Graz
T: (0316) 3891122
Founded: 1955; Members: 75
Pres: Prof. Dr. Berthold Sutter; Gen Secr: Prof. Dr. Wolfgang Suppan
Focus: Lit
Periodical
Jahresgabe der Johann-Joseph-Fux-Gesellschaft (weekly) 00691

Johann Strauß-Gesellschaft, Kleeblattgasse 9, 1010 Wien
T: (0222) 5339194
Founded: 1936; Members: 200
Pres: Prof. Franz Mailer
Focus: Music
Periodicals
Mitteilungen (3 times annually)
Wiener Bonbons (3 times annually) . . 00692

Kärntner Juristische Gesellschaft (KJG), Dobernigstr 2, 9010 Klagenfurt
T: (0463) 5840288
Founded: 1950; Members: 480
Focus: Law 00693

Klagenfurter Sprachwissenschaftliche Gesellschaft, Universitätsstr 67, 9020 Klagenfurt
T: (0463) 2700347; Fax: (0463) 2700100
Founded: 1977; Members: 40
Pres: Prof. Dr. Heinz-Dieter Pohl
Focus: Ling
Periodical
Klagenfurter Beiträge zur Sprachwissenschaft (weekly) 00694

Kommission für Neuere Geschichte Österreichs, c/o Institut für Geschichte, Universität Salzburg, Rudolfskai 42, 5020 Salzburg
T: (0662) 80444750; Fax: (0662) 8044413
Founded: 1902; Members: 25
Pres: Prof. Dr. Fritz Fellner; Gen Secr: Dr. Franz Adlgasser
Focus: Hist 00695

Kontext – Institut für Kommunikations- und Textanalysen, Margaretenpl 4, 1050 Wien
T: (0222) 5457764
Founded: 1988; Members: 20
Pres: Johanna Lalouschek; Gen Secr: Dr. Florian Menz
Focus: Comm Sci; Ling 00696

Krahuletz-Gesellschaft, c/o Krahuletz-Museum, 3730 Eggenburg
Founded: 1899
Focus: Arts 00697

Kulturgeschichtliche Gesellschaft am Landesmuseum Joanneum, Neutorgasse 45, 8010 Graz
T: (0316) 80174780
Founded: 1948
Focus: Hist 00698

Kunsthistorische Gesellschaft, c/o Institut für Kunstgeschichte, Universitätsstr 7, 1010 Wien
T: (0222) 401032617
Founded: 1956; Members: 152
Pres: Prof. Dr. Gerhard Schmidt
Focus: Hist; Arts 00699

Kunsthistorische Gesellschaft an der Universität Graz, c/o Kunsthistorisches Institut, Universität, 8010 Graz
Founded: 1959; Members: 43
Focus: Hist; Arts 00700

Kuratorium für Verkehrssicherheit, Ölzeltgasse 3, 1030 Wien
T: (0222) 717700; Fax: (0222) 717709
Pres: Dr. Ernst Baumgartner; Gen Secr: Franz M. Bogner
Focus: Transport 00701

Landesverein für Höhlenkunde in Wien und Niederösterreich, Obere Donaustr 97, 1020 Wien
Founded: 1937
Focus: Speleology
Periodical
Die Höhle 00702

Ludwig Boltzmann-Gesellschaft, Österreichische Vereinigung zur Förderung der wissenschaftlichen Forschung (LBG), Hofburg, Postf 33, 1014 Wien
T: (0222) 5338024, 5336082
Founded: 1960; Members: 80
Focus: Sci 00703

Mathematisch-Physikalische Gesellschaft in Innsbruck, Technikerstr 25, 6020 Innsbruck
Fax: (0512) 2185173
Founded: 1936; Members: 130
Focus: Math; Physics 00704

Mediacult, Internationales Forschungsinstitut für Medien, Kommunikation und kulturelle Entwicklung, Schönburgstr 27, 1040 Wien
T: (0222) 5041316; Fax: (0222) 50413164
Founded: 1969; Members: 64
Pres: John P.L. Roberts; Gen Secr: Alfred Smudits
Focus: Music; Perf Arts; Sociology; Mass Media; Comm Sci
Periodical
Mediacult Newsletter (weekly) 00705

Medizinische Gesellschaft für Oberösterreich, Dinghoferstr 4, 4020 Linz
T: (0732) 77837164; Fax: (0732) 77837158
Founded: 1946; Members: 2686
Focus: Med 00706

Mikrographische Gesellschaft, Postf 255, 1084 Wien
T: (0222) 8138446
Founded: 1910; Members: 80
Focus: Eng
Periodical
Informationsblatt (3 times annually) . . . 00707

Montanhistorischer Verein für Österreich, Peter-Tunner-Str 15, Postf 1, 8704 Leoben-Donawitz
T: (03842) 40702377
Founded: 1976; Members: 750
Pres: Prof. Dr. Karl Stadlober; Gen Secr: Anton Manfreda
Focus: Mining; Hist
Periodical
Leobener Grüne Hefte, Neue Folge (weekly) 00708

Musealverein in Hallstatt, c/o Museum, 4830 Hallstatt
Founded: 1884
Focus: Arts 00709

Museums-Verein Stillfried, 2262 Stillfried
Founded: 1914; Members: 120
Focus: Arts 00710

Nationalökonomische Gesellschaft (NÖG) / Austrian Economic Association, c/o Institut für Wirtschaftswissenschaften, Universität, Hohenstaufengasse 9, 1010 Wien
T: (0222) 401032676
Founded: 1917; Members: 270
Pres: Prof. Dr. Erich Streissler; Gen Secr: Prof. Dr. Fritz Breuss
Focus: Econ
Periodical
Empirica – Austrian Economic Papers (weekly) 00711

Naturwissenschaftlicher Verein für Kärnten, Museumgasse 2, 9020 Klagenfurt
T: (04222) 53630574
Founded: 1848; Members: 1300
Pres: Dr. Hans Sampl
Focus: Nat Sci
Periodicals
Carinthia II: Naturwissenschaftliche Beiträge zur Heimatkunde Kärntens (weekly)
Karinthin (weekly) 00712

Naturwissenschaftlicher Verein für Steiermark, c/o Universitätsbibliothek, Universitätspl 3, 8010 Graz
Founded: 1862; Members: 500
Pres: Prof. Dr. Georg Heinrich
Focus: Nat Sci
Periodical
Mitteilungen des Naturwissenschaftlichen Vereins für Steiermark (weekly) 00713

Naturwissenschaftlich-medizinischer Verein in Innsbruck, Technikerstr 25, 6020 Innsbruck
T: (0512) 2185354, 2185341
Founded: 1870; Members: 300
Focus: Med; Nat Sci
Periodical
Berichte Naturwissenschaftlich-medizinischer Verein in Innsbruck (weekly) 00714

Naturwissenschaftlich-Medizinische Vereinigung in Salzburg, Schopperstr 13, 5020 Salzburg
T: (0662) 510204
Founded: 1974; Members: 78
Focus: Med
Periodical
Berichte der naturwissenschaftlich-medizinischen Vereinigung Salzburg 00715

Niederösterreichisches Bildungs- und Heimatwerk, Arbeitsgemeinschaft für Volkskunde, Strauchgasse 3, 1010 Wien
Founded: 1960
Focus: Ethnology 00716

Oberösterreichischer Musealverein / Gesellschaft für Landeskunde, Lamdstr 31, 4020 Linz
T: (0732) 770218
Founded: 1833; Members: 1200
Pres: Dr. Georg Heilingsetzer
Focus: Arts
Periodicals
Beiträge zur Landeskunde von Oberösterreich: I. Historische Reihe
Beiträge zur Landeskunde von Oberösterreich: II. Naturwissenschaftliche Reihe
Jahrbuch: Teil I Abhandlungen, Teil II Berichte
Mitteilungen (weekly)
Schriftenreihe 00717

Ökosoziales Forum, Franz-Josefs-Kai 13, 1010 Wien
T: (0222) 5330797; Fax: (0222) 533079790
Founded: 1969
Focus: Ecology
Periodical
Agrarische Rundschau (weekly) 00718

Österreichische Ärztegesellschaft für Psychotherapie, Mariannengasse 10, 1090 Wien
Founded: 1950; Members: 30
Pres: Prof. V.E. Frankl
Focus: Therapeutics; Psych 00719

Österreichische Ärztegesellschaft zur Bekämpfung der cystischen Fibrose, c/o Universitäts-Kinderklinik, Währinger Gürtel 18-20, 1090 Wien
T: (0222) 48993232
Founded: 1968; Members: 30
Focus: Cell Biol & Cancer Res 00720

Österreichische Ärztekammer (ÖÄK), Weihburggasse 10-12, Postf 213, 1010 Wien
T: (0222) 514060; Fax: (0222) 5140642
Founded: 1950; Members: 29000
Focus: Med
Periodicals
medizin populär: Patienteninformation der Österreichischen Ärztekammer (10 times annually)
Österreichische Ärztezeitung: Organ der Österreichischen Ärztekammer (weekly) . 00721

Österreichische ärztliche Gesellschaft für medizinisches und technisches Ozon, c/o Prof. Dr. Ottokar Rokitansky, Walfischgasse 14, 1010 Wien
T: (0222) 5131948; Fax: (0222) 5131948
Focus: Therapeutics 00722

Österreichische Akademie der Wissenschaften (ÖAW), Dr.-Ignaz-Seipel-Pl 2, 1010 Wien
T: (0222) 51581; Tx: 12628; Fax: (0222) 5139541
Founded: 1847; Members: 529
Pres: Prof. Dr. Werner Welzig; Gen Secr: Prof. Dr. Karl Schlögl
Focus: Sci
Periodicals
Anzeiger der mathematisch-naturwissenschaftlichen Klasse (weekly)
Anzeiger der philosophisch-historischen Klasse (weekly)
Jahrbuch der österreichischen Byzantinistik (weekly)
Römische Historische Mitteilungen (weekly)
Sitzungsbericht I und II (weekly)
Sprachkunst: Beiträge zur Literaturwissenschaft (weekly)
Wiener Slavistisches Jahrbuch (weekly)
Wiener Studien: Zeitschrift für Klassische Philologie und Patristik (weekly)
Wiener Zeitschrift für die Kunde Südasiens und Archiv für indische Philosophie (weekly)
Zeitschrift des Instituts für Demographie: Demographische Informationen 00723

Österreichische Arbeitsgemeinschaft für morphologische und funktionelle Atherosklerosefoschung, Schwarzspanierstr 17, 1090 Wien
T: (0222) 431526395
Founded: 1975; Members: 125
Focus: Physiology; Pathology 00724

Österreichische Arbeitsgemeinschaft für Neuropsychiatrie und Psychologie des Kindes- und Jugendalters und verwandter Berufe (ÖANP), Währinger Gürtel 18-20, 1090 Wien
T: (0222) 48003012
Founded: 1973; Members: 400
Focus: Psychiatry; Psych 00725

Österreichische Arbeitsgemeinschaft für Rehabilitation / Dachorganisation der österreichischen Behindertenverbände, Brigittenauer Lände 42, 1200 Wien
T: (0222) 3326101; Fax: (0222) 3309314
Founded: 1950
Focus: Rehabil
Periodical
Monatsbericht: Sozialpolitische Rundschau (weekly) 00726

Österreichische Arbeitsgemeinschaft für Volksgesundheit (ÖAV), Stubenring 6, 1010 Wien
T: (0222) 529661
Founded: 1947; Members: 110
Focus: Public Health
Periodical
ÖAV-Informationen (weekly) 00727

Österreichische Arbeitskreise für Tiefenpsychologie (ÖAKT), c/o Psychologisches Institut, Universität, Akademiestr 20, 5020 Salzburg
T: (0662) 33501305
Founded: 1947; Members: 66
Focus: Psych 00728

Österreichische Bankwissenschaftliche Gesellschaft an der Wirtschaftsuniversität Wien (BWG), Strauchgasse 3, 1010 Wien
T: (0222) 5336545; Fax: (0222) 53127247
Founded: 1952
Focus: Finance
Periodicals
Bank-Archiv (weekly)
Bankwissenschaftliche Schriftenreihe (weekly) 00729

Österreichische Bibelgesellschaft, Breite Gasse 8, 1070 Wien
T: (0222) 5238240
Focus: Rel & Theol 00730

Österreichische Biochemische Gesellschaft, c/o Institut für Biochemie, Universität Wien, 1090 Wien
T: (0222) 436141; Fax: (0222) 434687
Founded: 1954; Members: 650
Pres: Prof. Dr. H. Tuppy; Gen Secr: Dr. Dieter Blaas
Focus: Biochem 00731

Österreichische Biophysikalische Gesellschaft, c/o Institut für allgemeine und experimentelle Pathologie, Währinger Str 13, 1090 Wien
T: (0222) 431526333
Founded: 1961; Members: 150
Focus: Biophys 00732

Österreichische Bodenkundliche Gesellschaft (ÖBG), c/o Universität für Bodenkultur, Gregor-Mendel-Str 33, 1180 Wien
T: (0222) 476540
Founded: 1954; Members: 269
Pres: Dr. Walter Kilian; Gen Secr: Dr. Franz Mutsch
Focus: Geology
Periodical
Mitteilungen der Österreichischen Bodenkundlichen Gesellschaft (weekly) 00733

Österreichische Byzantinische Gesellschaft, Postgasse 7-9, 1010 Wien
T: (0222) 5120217
Founded: 1946; Members: 150
Pres: Prof. Dr. Herbert Hunger
Focus: Hist
Periodical
Mitteilungen aus der österreichischen Byzantinistik und Neogräzistik (weekly) 00734

Österreichische Computer-Gesellschaft (OCG), Wollzeile 1, 1010 Wien
T: (0222) 5120235; Fax: (0222) 5120235
Founded: 1975; Members: 1000
Focus: Computer & Info Sci
Periodicals
ocg-Kommunikativ (weekly)
ocg-Schriftenreihe (weekly) 00735

Österreichische Dentistenkammer (ÖDK), Kohlmarkt 11, 1010 Wien
T: (0222) 5337064; Fax: (0222) 5350758
Founded: 1949; Members: 388
Focus: Dent
Periodical
Österreichische Zahnärzte-Zeitung (weekly) 00736

Österreichische Diabetikervereinigung, Obere Augartenstr 26-28, 1020 Wien
T: (0222) 3323217; Fax: (0222) 3326828
Founded: 1978; Members: 4400
Focus: Diabetes
Periodical
Mein Leben (weekly) 00737

Österreichische Entomologische Gesellschaft, c/o Institut für Naturschutz und Landschaftspflege, Heinrichstr 5, 8010 Graz
T: (0316) 36068; Fax: (0316) 360685
Founded: 1975; Members: 121
Pres: Prof. Dr. Horst Aspöck; Gen Secr: Dr. Johann Gepp
Focus: Entomology 00738

Österreichische Ethnologische Gesellschaft (ÖEG), c/o Museum für Völkerkunde, Neue Hofburg, Heldenpl, 1014 Wien
T: (0222) 52177; Fax: (0222) 5355320
Founded: 1950; Members: 180
Focus: Ethnology
Periodical
Wiener Völkerkundliche Mitteilungen (WVM) (weekly) 00739

Österreichische Ethnomedizinische Gesellschaft, c/o Institut für Geschichte der Medizin, Währingerstr 25, 1090 Wien
Founded: 1978; Members: 84
Focus: Ethnology; Med 00740

Österreichische Exlibris-Gesellschaft, c/o Gertrud Slattner, Ferdinandstr 14, 1020 Wien
T: (0222) 2615213
Founded: 1903; Members: 239
Focus: Lit
Periodical
Mitteilungen der Österreichischen Exlibris-Gesellschaft (3 times annually) 00741

Österreichische Forschungsgemeinschaft, Berggasse 25, 1090 Wien
T: (0222) 3195770; Fax: (0222) 319577020
Focus: Sci 00742

Österreichische Forschungsstiftung für Entwicklungshilfe, Berggasse 7, 1090 Wien
T: (0222) 3174010; Fax: (0222) 3174015
Founded: 1967
Focus: Develop Areas
Periodicals
Ausgewählte neue Literatur zur Entwicklungspolitik (weekly)
Handbook of Austrian Development Aid . 00743

Österreichische Gartenbau-Gesellschaft, Parkring 12, 1010 Wien
T: (0222) 5128416, 5128808; Fax: (0222) 512841617
Focus: Hort
Periodical
Garten (11 times annually) 00744

Österreichische Geographische Gesellschaft (ÖGG), Karl-Schweighofer-Gasse 3, 1071 Wien
T: (0222) 5237974
Founded: 1856; Members: 1100
Pres: Dr. Walter Petrowitz; Gen Secr: Prof. Dr. Ingrid Kretschmer
Focus: Geography
Periodical
Mitteilungen der Österreichischen Geographischen Gesellschaft (weekly) 00745

Österreichische Geographische Gesellschaft, Zweigverein Innsbruck, Innrain 52, 6020 Innsbruck
T: (0512) 5073112, 5073101; Fax: (0512) 5073132
Founded: 1971; Members: 420
Pres: Prof. Dr. Axel Borsdorf; Gen Secr: Dr. Josef Aistleitner
Focus: Geography
Periodical
Jahresbericht 00746

Österreichische Geologische Gesellschaft (ÖGG), c/o Geologische Bundesanstalt, Rasumofskygasse 23, 1031 Wien
T: (0222) 712567456
Founded: 1907; Members: 750
Pres: Prof. Dr. Walter J. Schmidt; Gen Secr: Dr. Werner R. Janoschen
Focus: Geology
Periodical
Mitteilungen der Österreichischen Geologischen Gesellschaft (weekly) 00747

Österreichische Gesellschaft der Tierärzte, Linke Bahngasse 11, 1030 Wien
T: (0222) 711550; Fax: (0222) 7136875
Founded: 1919; Members: 1030
Pres: Dr. C. Stanek
Focus: Vet Med
Periodical
Wiener Tierärztliche Monatsschrift (weekly) 00748

Österreichische Gesellschaft für ärztliche Hypnose und autogenes Training, Pyrkergasse 23, 1190 Wien
T: (0222) 362366
Founded: 1969; Members: 170
Focus: Psych 00749

Österreichische Gesellschaft für Agrar- und Umweltrecht, c/o Institut für Wirtschaft, Politik und Recht, Universität für Bodenkultur, Gregor-Mendel-Str 33, 1180 Wien
T: (0222) 47654; Fax: (0222) 3105175
Founded: 1967; Members: 127
Pres: Dr. Ernst Massauer; Gen Secr: Dr. Hellmuth Gatterbauer
Focus: Law
Periodical
Agrarische Rundschau (weekly) 00750

Österreichische Gesellschaft für Akupunktur und Aurikulotherapie, c/o Kaiserin-Elisabeth-Spital, Huglgasse 1-3, 1150 Wien
T: (0222) 981042610; Fax: (0222) 98104460
Founded: 1954; Members: 2000
Focus: Anesthetics; Med; Physical Therapy
Periodical
Deutsche Zeitschrift für Akupunktur (weekly) 00751

Österreichische Gesellschaft für Allgemeinmedizin (ÖGAM), Bahnhofstr 22, 9020 Klagenfurt
T: (0463) 55449
Focus: Med 00752

Österreichische Gesellschaft für Amerikastudien / Austrian Association for American Studies, Akademiestr 24c, 5020 Salzburg
Founded: 1970; Members: 85
Pres: Heinz Tschachler
Focus: Lit; Hist; Cultur Hist 00753

Österreichische Gesellschaft für Anästhesiologie, Reanimation und Intensivtherapie, Spitalgasse 23, 1090 Wien
T: (0222) 404002502
Founded: 1951; Members: 800
Pres: Prof. W. Haider; Gen Secr: Dr. W. Zwölfer
Focus: Anesthetics; Therapeutics
Periodical
Der Anaesthesist 00754

Österreichische Gesellschaft für angewandte Fremdenverkehrswissenschaft (ÖGFW), Türkenschanzstr 18, 1018 Wien
T: (0222) 346656
Focus: Econ 00755

Österreichische Gesellschaft für angewandte Zytologie, Auenbruggerpl 26, 8036 Graz
T: (0222) 48003694
Founded: 1964; Members: 262
Pres: Dr. Leopoldine Pokieser; Gen Secr: Prof. Dr. Gerhard Breitenecker
Focus: Cell Biol & Cancer Res; Physiology
Periodical
Acta Cytologica (weekly) 00756

Österreichische Gesellschaft für Angiologie, c/o 1. Medizinische Universitätsklinik, Lazarettgasse 14, 1090 Wien
Founded: 1966; Members: 350
Focus: Med 00757

Österreichische Gesellschaft für Arbeitsmedizin, c/o Arbeitsmedizinisches Zentrum, 6060 Hall
T: (05223) 7304
Founded: 1954; Members: 350
Pres: Dr. Egmont Baumgartner
Focus: Med; Hygiene 00758

Österreichische Gesellschaft für Arbeitsrecht und Sozialrecht, c/o Institut für Arbeitsrecht und Sozialrecht, Universität, 4040 Linz-Auhof
T: (0732) 2468255
Founded: 1965; Members: 86
Focus: Law 00759

Österreichische Gesellschaft für Archäologie (ÖGA), c/o Institut für Alte Geschichte, Altertumskunde und Epigraphik, Dr.-Karl-Lueger-Ring 1, 1010 Wien
T: (0222) 401032200; Fax: (0222) 401032118
Founded: 1972; Members: 210
Pres: Dr. Günther Dembski; Gen Secr: Dr. Kurt Gschwantler
Focus: Archeol
Periodical
Römisches Österreich (weekly) 00760

Österreichische Gesellschaft für Architektur, Seidengasse 26, 1070 Wien
T: (0222) 5268395; Fax: (0222) 5268395
Focus: Archit
Periodical
Umbau (weekly) 00761

Österreichische Gesellschaft für Artificial Intelligence, c/o Institut für Medizinische Kybernetik und Artificial Intelligence, Universität, Freyung 6, 1010 Wien
T: (0222) 53532810
Founded: 1981; Members: 240
Focus: Computer & Info Sci
Periodical
ÖGAI Journal (weekly) 00762

Österreichische Gesellschaft für Aussenpolitik und Internationale Beziehungen, Hofburg, Schweizer Hof, Brunnenstiege, 1010 Wien
Founded: 1958; Members: 500
Pres: Dr. Hans Haumer; Gen Secr: Dr. Hedwig Wolfram
Focus: Poli Sci
Periodical
Österreichisches Jahrbuch für Internationale Politik (weekly) 00763

Österreichische Gesellschaft für Autogenes Training und allgemeine Psychotherapie, Schelleingasse 8, 1040 Wien
T: (0222) 5054454
Founded: 1969; Members: 680
Focus: Psychoan; Therapeutics 00764

Österreichische Gesellschaft für Balneologie und medizinische Klimatologie, c/o Institut für medizinische Physiologie, Universität Wien, Schwarzspanierstr 17, 1090 Wien
Founded: 1956; Members: 136
Pres: Dr. W. Marktl
Focus: Physiology
Periodical
Zeitschrift für Physikalische Medizin, Balneologie und Medizinische Klimatologie 00765

Österreichische Gesellschaft für Biomedizinische Technik, c/o Technische Universität, Gusshausstr 27-29, 1040 Wien
T: (0222) 588013842; Fax: (0222) 5052666
Founded: 1975; Members: 202
Focus: Med 00766

Österreichische Gesellschaft für Bionome Psychotherapie, Pyrkergasse 23, 1190 Wien
Founded: 1978
Focus: Psych; Therapeutics 00767

Österreichische Gesellschaft für China-Forschung (ÖGCF), Wickenburggasse 4, 1080 Wien
T: (0222) 439793; Fax: (0222) 439794
Focus: Sociology; Nat Sci
Periodical
China-Report (weekly) 00768

Österreichische Gesellschaft für Chirurgie, Frankgasse 8, Postf 80, 1096 Wien
T: (0222) 4087920; Fax: (0222) 4081328
Founded: 1958; Members: 967
Pres: Prof. Dr. Oskar Boeckl; Gen Secr: Prof. Dr. Erich Wayand
Focus: Surgery
Periodicals
Acta Chirurgica Austriaca (weekly)
Mitteilungen der Österreichischen Gesellschaft für Chirurgie mit den assoziierten Fachgesellschaften (weekly) 00769

Österreichische Gesellschaft für Chirurgische Forschung, c/o Zentrum für Biomedizinische Forschung, Allgemeines Krankenhaus, Währinger Gürtel 18-20, 1090 Wien
T: (0222) 404005221; Fax: (0222) 404005229
Founded: 1976; Members: 240
Pres: Dr. S. Uranös
Focus: Surgery
Periodical
Wissenschaftlicher Bericht des jährlich abgehaltenen Seminars (weekly) 00770

Österreichische Gesellschaft für Christliche Kunst, Kreuzherrengasse 1, 1040 Wien
Founded: 1909; Members: 150
Pres: Dr. Alfred Sammer
Focus: Arts 00771

Österreichische Gesellschaft für Denkmal- und Ortsbildpflege, Künstlerhaus, Karlspl 5, 1010 Wien
T: (0222) 5879663
Founded: 1947; Members: 670
Focus: Preserv Hist Monuments
Periodicals
Steine Sprechen (weekly)
Steinschlag (weekly) 00772

Österreichische Gesellschaft für Dermatologie und Venerologie, c/o Fachbereich Dermatologie, Universität, Auenbruggerpl 8, 8036 Graz
Pres: Prof. Dr. H. Kerl; Gen Secr: Prof. Dr. S. Hödl
Focus: Derm 00773

Österreichische Gesellschaft für Dokumentation und Information (ÖGDI), Heinestr 38, 1021 Wien
T: (0222) 267535; Tx: 115960
Founded: 1951; Members: 180
Pres: H. Jobst
Focus: Doc; Computer & Info Sci
Periodical
Fakten, Daten, Zitate: Das Informationsangebot für Wissenschaft & Wirtschaft (weekly) . . . 00774

Österreichische Gesellschaft für Elektroencephalographie und klinische Neurophysiologie, c/o Universitätsklinik für Neurologie, Anichstr 35, 6020 Innsbruck
Founded: 1953; Members: 350
Pres: Prof. Dr. H. Lechner; Gen Secr: Prof. Dr. G. Ladurner; Prof. Dr. G. Bauer
Focus: Physiology
Periodical
EEG/EMG 00775

Österreichische Gesellschaft für Elektronenmikroskopie, c/o Zentrum für Elektronenmikroskopie Graz, Steyrergasse 17, 8010 Graz
T: (0316) 825363; Fax: (0316) 811596
Founded: 1965; Members: 238
Pres: Dr. Wolfgang Geymayer; Gen Secr: Prof. Dr. Peter Golob
Focus: Optics 00776

Österreichische Gesellschaft für Erdölwissenschaften (ÖGEW), Erdbergstr 72, 1031 Wien
T: (0222) 7132348
Founded: 1960; Members: 600
Pres: Engelbert Pott; Gen Secr: Dr. Herbert Lang
Focus: Geology
Periodical
Erdöl Erdgas Kohle (weekly) 00777

Österreichische Gesellschaft für Ernährungsforschung (ÖGE), Zaunergasse 1-3, Postf 74, 1037 Wien
Fax: (0222) 7131802
Founded: 1950; Members: 276
Pres: Prof. Dr. Josef Leibetseder
Focus: Food
Periodical
ernährung-nutrition 00778

Österreichische Gesellschaft für Filmwissenschaft, Kommunikations- und Medienforschung (ÖGFKM), Rauhensteingasse 5, 1010 Wien
T: (0222) 5129936; Fax: (0222) 5135330
Founded: 1952; Members: 250
Pres: Prof. Dr. Giselher Guttmann; Gen Secr: Dr. Josef Schuchnig
Focus: Cinema; Comm Sci
Periodicals
Filmkunst (weekly)
Film/Video Manual (3 times annually)
Mitteilungen der ÖGFKM (8-10 times annually) 00779

Österreichische Gesellschaft für Friedensforschung, Landstrasser Hauptstr 24, 1030 Wien
Focus: Poli Sci; Prom Peace 00780

Österreichische Gesellschaft für Gastroenterologie und Hepatologie (ÖGGH), c/o Universitätsklinik für Innere Medizin IV, Währinger Gürtel 18-20, 1090 Wien
T: (0222) 404004744; Fax: (0222) 404004735
Founded: 1968; Members: 633
Pres: Prof. Dr. Arnulf Fritsch; Gen Secr: Dr. Rainer Schöfl
Focus: Gastroenter
Periodical
Zeitschrift für Gastroenterologie 00781

Österreichische Gesellschaft für Gefässchirurgie, c/o Landeskrankenanstalten Salzburg, Müllner Hauptstr 48, 5020 Salzburg
Focus: Surgery 00782

Österreichische Gesellschaft für Geriatrie und Gerontologie, c/o Krankenhaus der Barmherzigen Brüder, Medizinische Abteilung, Grosse Mohrengasse 9, 1020 Wien
T: (0222) 21121395
Founded: 1955; Members: 121
Pres: Prof. Dr. R. Willvonseder; Gen Secr: Prof. Dr. Monika Skalicky
Focus: Geriatrics
Periodicals
Aktuelle Gerontologie (weekly)
Scriptum Geriatricum (weekly) 00783

Österreichische Gesellschaft für gerichtliche Medizin, Sensengasse 2, 1090 Wien
T: (0222) 4024051; Fax: (0222) 422726
Founded: 1965; Members: 64
Focus: Forensic Med
Periodical
Beiträge zur Gerichtlichen Medizin (weekly) 00784

Österreichische Gesellschaft für Geschichte der Pharmazie, Spitalgasse 31, 1090 Wien
T: (0222) 5580555
Founded: 1950; Members: 146
Focus: Hist; Pharmacol 00785

Österreichische Gesellschaft für Gesetzgebungslehre (ÖGGL), c/o Institut für Verfassungs- und Verwaltungsrecht, Kapitelgasse 5-7, 5020 Salzburg
T: (0662) 80443630; Fax: (0662) 8044303
Founded: 1982; Members: 150
Pres: Prof. Dr. Heinz Schäffer; Gen Secr: Dr. Gerhard Holzinger
Focus: Law 00786

Österreichische Gesellschaft für Gruppendynamik und Organisationsberatung, c/o Institut für Philosophie, Universität, Universitätsstr 67, 9022 Klagenfurt
T: (0463) 37509
Founded: 1973; Members: 80
Pres: Dr. Ewald Krainz; Gen Secr: Dr. Veronika Dalheimer
Focus: Philos 00787

Österreichische Gesellschaft für Gynäkologie und Geburtshilfe, Spitalgasse 23, 1090 Wien
T: (0222) 404002915; Fax: (0222) 404002911
Founded: 1980; Members: 398
Focus: Gynecology 00788

Österreichische Gesellschaft für Hals-, Nasen- und Ohrenheilkunde, Kopf- und Halschirurgie, c/o Universitäts-HNO-Klinik, Währinger Gürtel 18-20, 1090 Wien
Founded: 1892; Members: 270
Focus: Otorhinolaryngology; Surgery
Periodical
Laryngologie, Rhinologie, Otologie (weekly) 00789

Österreichische Gesellschaft für Hochschuldidaktik, Strozzigasse 2, 1080 Wien
Fax: (0222) 425618248
Founded: 1977; Members: 150
Focus: Educ
Periodical
Zeitschrift für Hochschuldidaktik: Beiträge zu Studium, Wissenschaft und Beruf . . . 00790

Österreichische Gesellschaft für Holzforschung (ÖGH), Arsenal, Franz-Grill-Str 7, 1031 Wien
T: (0222) 79826230
Founded: 1947; Members: 300
Pres: Christian Reinberg; Gen Secr: Helmuth Resch
Focus: Forestry
Periodical
Holzforschung und Holzverwertung: Mitteilungen (weekly) 00791

Österreichische Gesellschaft für Humanökologie, Karlspl 13, 1040 Wien
T: (0222) 653785799
Founded: 1976
Focus: Ecology 00792

Österreichische Gesellschaft für Hygiene, Mikrobiologie und Präventivmedizin (ÖGHMP), Kinderspitalgasse 15, 1095 Wien
T: (0222) 40490; Fax: (0222) 40490
Founded: 1925; Members: 430
Pres: Prof. Dr. M. Rotter; Gen Secr: Prof. Dr. G. Wewaka
Focus: Hygiene; Med; Microbio 00793

Österreichische Gesellschaft für Industrielle Strahltechnik (ÖGIST), Costenoblegasse 2, 1130 Wien
T: (0222) 588013869
Founded: 1979; Members: 6
Focus: Radiology
Periodical
Mitteilungen der ÖGIST 00794

Österreichische Gesellschaft für Informatik (ÖGI), Altenbergerstr 69, 4040 Linz
T: (0732) 24689238
Founded: 1975; Members: 200
Pres: Prof. Dr. Arno Schulz
Focus: Computer & Info Sci 00795

Österreichische Gesellschaft für Innere Medizin, c/o Allgemeines Krankenhaus, Währinger Gürtel 18-20, 1090 Wien
Founded: 1886; Members: 400
Pres: Prof. Dr. A. Gangl
Focus: Intern Med
Periodical
Wiener Zeitschrift für Innere Medizin . . 00796

Österreichische Gesellschaft für internistische Intensivmedizin, c/o Intensivstation, Klinik für Innere Medizin IV, Währinger Gürtel 18-20, 1090 Wien
Founded: 1970; Members: 290
Pres: Prof. Dr. W. Waldhäusl; Gen Secr: Prof. Dr. K. Lenz
Focus: Intern Med 00797

Österreichische Gesellschaft für Kinderchirurgie, c/o Universitätsklinik für Kinderchirurgie, Auenbruggerpl 34, 8036 Graz
Founded: 1966; Members: 70
Focus: Surgery 00798

Österreichische Gesellschaft für Kinderphilosophie (KiPhi), Schmiedgasse 12, 8010 Graz
Fax: (0316) 811513
Founded: 1985
Pres: Dr. Daniela G. Camhy
Focus: Philos
Periodical
Info Kinderphilosophie (weekly) 00799

Österreichische Gesellschaft für Kinder- und Jugendheilkunde, c/o G. von Preyer'sches Kinderspital, Schrankenberggasse 31, 1100 Wien
T: (0222) 60113201; Fax: (0222) 60113311
Founded: 1962; Members: 800
Pres: Prof. Dr. W. Stögmann; Gen Secr: Dr. F. Paky
Focus: Pediatrics
Periodical
Pädiatrie und Pädologie 00800

Österreichische Gesellschaft für Kirchenrecht, c/o Institut für Kirchenrecht, Freyung 6, 1010 Wien
T: (0222) 5339861; Fax: (0222) 5351019
Founded: 1949; Members: 225
Pres: Prof. Dr. Marianne Meinhart; Gen Secr: Prof. Dr. Bruno Primetshofer
Focus: Law; Rel & Theol
Periodical
Österreichisches Archiv für Kirchenrecht (weekly) 00801

Österreichische Gesellschaft für Klinische Chemie (ÖGKC), Währingerstr 10, 1090 Wien
T: (0222) 5072295
Founded: 1968; Members: 505
Pres: Prof. Dr. Mathias Müller; Gen Secr: Prof. Dr. Wolfgang Doloczfcuk
Focus: Chem
Periodical
Berichte der ÖGKC 00802

Österreichische Gesellschaft für Kommunikationsfragen (ÖKG) / Austrian Society for Communication, Bankgasse 8, 1010 Wien
T: (0222) 3197977
Founded: 1976; Members: 110
Pres: Dr. Kurt Luger; Gen Secr: Dr. Rudolf Renger
Focus: Comm Sci; Journalism
Periodical
Medien-Journal (weekly) 00803

Österreichische Gesellschaft für Laboratoriumsmedizin, Franz-Josefs-Kai 65, 1010 Wien
T: (0222) 349130
Founded: 1973; Members: 72
Focus: Med 00804

Österreichische Gesellschaft für Literatur, Herrengasse 5, 1010 Wien
T: (0222) 5338159; Fax: (0222) 5334067
Founded: 1961
Pres: Dr. Wolfgang Kraus
Focus: Lit 00805

Österreichische Gesellschaft für Logopädie, Phoniatrie und Pädoaudiologie, c/o 1. HNO-Universitätsklinik, Lazarettgasse 14, 1090 Wien
T: (0222) 48003316
Founded: 1951; Members: 100
Focus: Logopedy; Otorhinolaryngology . . 00806

Österreichische Gesellschaft für Lungenerkrankungen und Tuberkulose, c/o Wiener Medizinische Akademie, Alserstr 4, 1090 Wien
T: (0222) 42138321
Founded: 1950; Members: 480
Focus: Pulmon Dis
Periodical
Jahresberichte der wissenschaftlichen Veranstaltungen 00807

Österreichische Gesellschaft für Medizinsoziologie (ÖGMS), Stumpergasse 56, 1060 Wien
T: (0222) 59991243; Fax: (0222) 5971871
Founded: 1978; Members: 15
Pres: Prof. Dr. Jürgen M. Pelikan; Gen Secr: Wolfgang Dür
Focus: Med; Sociology 00808

Österreichische Gesellschaft für Meteorologie, Hohe Warte 38, 1190 Wien
T: (0222) 364453; CA: Meteor Wien; Fax: (0222) 36911233
Founded: 1865; Members: 230
Pres: Prof. Dr. Helmut Pichler; Gen Secr: Dr. E. Dreiseitl
Focus: Geophys
Periodicals
Meteorologische Zeitschrift (weekly)
Wetter und Leben: Zeitschrift für Angewandte Meteorologie (weekly) 00809

Österreichische Gesellschaft für Mikrochemie und analytische Chemie (ÖGMAC), c/o Institut für analytische Chemie und Mikrochemie, Technischen Universität, Getreidemarkt 9, 1060 Wien
T: (0222) 4322258801, 48474940
Founded: 1946; Members: 220
Focus: Chem 00810

Österreichische Gesellschaft für Musik (ÖGfM), Hanuschgasse 3, 1010 Wien
T: (0222) 5123143; Fax: (0222) 5124299
Founded: 1963; Members: 1000
Pres: Prof. Gerhard Wimberger; Gen Secr: Dr. Harald Goertz
Focus: Music
Periodical
Beiträge (weekly) 00811

Österreichische Gesellschaft für Nephrologie, c/o Krankenhaus der Elisabethinen, 4020 Linz
T: (0732) 278021
Focus: Med; Geriatrics
Periodical
Nephrologische Nachrichten (weekly) . . 00812

Österreichische Gesellschaft für Neugriechische Studien, c/o Institut für Byzantinistik und Neogräzistik, Postgasse 7, 1010 Wien
T: (0222) 5120217; Fax: (0222) 5127023
Founded: 1988; Members: 36
Pres: Prof. Dr. Gunnar Hering; Gen Secr: Prof. Dr. Max D. Peyfuss
Focus: Ling; Hist 00813

Österreichische Gesellschaft für Neurochirurgie, Währinger Gürtel 18-20, 1090 Wien
Fax: (0222) 404004566
Founded: 1961; Members: 85
Focus: Surgery 00814

Österreichische Gesellschaft für Neuropathologie, c/o Neurologisches Institut, Universität, Schwarzspanierstr 17, 1090 Wien
T: (0222) 84163431
Founded: 1964; Members: 48
Pres: Prof. Dr. H. Lassmann
Focus: Pathology
Periodical
Current Topics in Neuropathology (weekly) 00815

Österreichische Gesellschaft für Nuclearmedizin, Postf 35, 1025 Wien
T: (0222) 43956518; Fax: (0222) 43956514
Founded: 1968; Members: 209
Focus: Nucl Med 00816

Österreichische Gesellschaft für Operations Research (ÖGOR), Argentinierstr 8, 1040 Wien
Founded: 1978
Focus: Business Admin; Math
Periodical
ÖGOR-Nachrichten 00817

Österreichische Gesellschaft für Orthopädie und Orthopädische Chirurgie, c/o Orthopädische Universitätsklinik, Garnisongasse 13, 1090 Wien
Founded: 1948; Members: 276
Focus: Orthopedics; Surgery 00818

Österreichische Gesellschaft für Parapsychologie (ÖGP), c/o Technische Universität, Gusshausstr 27, 1040 Wien
T: (0222) 588013940
Founded: 1927; Members: 260
Pres: Prof. Dr. Hellmut Hofmann; Gen Secr: Prof. Dr. I. Schenk
Focus: Parapsych 00819

Österreichische Gesellschaft für Pathologie, c/o Pathologisch-anatomisches Institut, Universität, Währinger Gürtel 18-20, 1090 Wien
T: (0222) 404003651; Fax: (0222) 423402
Founded: 1921; Members: 250
Pres: Prof. Dr. W. Dutz; Gen Secr: Dr. F. Wrba
Focus: Pathology 00820

Österreichische Gesellschaft für Perinatale Medizin, c/o Universitäts-Frauenklinik, Spitalgasse 23, 1090 Wien
T: (0222) 404002915; Fax: (0222) 404002911
Founded: 1972; Members: 350
Pres: Prof. Dr. K. Baumgarten; Gen Secr: Prof. Dr. W. Gruber
Focus: Gynecology 00821

Österreichische Gesellschaft für Philosophie, c/o Dr. Johannes Brandl, Institut für Philosophie, Universität Salzburg, Franziskanergasse 1, 5020 Salzburg
Focus: Philos
Periodical
Philosophie-Österreich (weekly) 00822

Österreichische Gesellschaft für Physikalische Medizin und Rehabilitation, c/o Institut für Physikalische Medizin, Hanuschkrankenhaus, Heinrich-Collin-Str 30, 1140 Wien
T: (0222) 949701; Fax: (0222) 949264
Founded: 1951; Members: 180
Pres: Dr. O. Rathkolb; Gen Secr: Dr. K. Ammer
Focus: Rehabil; Physiology; Physical Therapy
Periodical
Österreichische Zeitschrift für Physikalische Medizin 00823

Österreichische Gesellschaft für Politikwissenschaft, Stumpergasse 56, 1060 Wien
T: (0222) 59991
Founded: 1972; Members: 350
Focus: Poli Sci
Periodicals
ÖGPW-Rundbrief (weekly)
Österreichische Zeitschrift für Politikwissenschaft; ÖZP (weekly) 00824

Österreichische Gesellschaft für Psychische Hygiene, c/o Psychiatrische Universitätsklinik, Lazarettgasse 14, 1090 Wien
T: (0222) 48003567
Founded: 1948; Members: 85
Focus: Hygiene
Periodical
Wiener Zeitschrift für Nervenheilkunde . . 00825

Österreichische Gesellschaft für Raumforschung und Raumplanung (ÖGRR), Karlspl 13, 1040 Wien
T: (0222) 5877749
Founded: 1954; Members: 1000
Focus: Urban Plan
Periodical
Schriftenreihe 00826

Österreichische Gesellschaft für Rechtsvergleichung (ÖGfRV), c/o Institut für Rechtsvergleichung, Universität, Schottenbastei 10-16, 1010 Wien
T: (0222) 401033240
Founded: 1959; Members: 183
Pres: Prof. Dr. Fritz Schwind
Focus: Law
Periodicals
Wiener Rechtswissenschaftliche Studien
Zeitschrift für Rechtsvergleichung, Internationales Privatrecht und Europarecht 00827

Österreichische Gesellschaft für Religionswissenschaft, Hans-Pfitzner-Str 5, 5020 Salzburg
Founded: 1962
Focus: Rel & Theol 00828

Österreichische Gesellschaft für Schweisstechnik, Arsenal, 1030 Wien
T: (0222) 782168
Founded: 1947
Focus: Eng
Periodical
Schweisstechnik (weekly) 00829

Österreichische Gesellschaft für Semiotik (ÖGS), c/o Institut für Romanistik, Universität, Schlickgasse 4, 1090 Wien
T: (0222) 316146
Founded: 1976; Members: 174
Focus: Ling
Periodicals
Angewandte Semiotik
Semiotische Berichte: Zeitschrift für Semiotik 00830

Österreichische Gesellschaft für Sexualforschung, Währinger Str 79, 1018 Wien
T: (0222) 421320
Founded: 1954
Focus: Sociology 00831

Österreichische Gesellschaft für Soziologie (ÖGS), Neutorgasse 12, 1013 Wien
T: (0222) 5332878; Fax: (0222) 5336592
Founded: 1950; Members: 373
Pres: Prof. Dr. Rudolf Richter
Focus: Sociology
Periodical
Österreichische Zeitschrift für Soziologie (weekly) 00832

Österreichische Gesellschaft für Sprachheilpädagogik, Erlachgasse 91, 1100 Wien
T: (0222) 6033796
Founded: 1969; Members: 1500
Pres: Inge Frühwirth; Gen Secr: Konrad Kregcjk
Focus: Educ; Rehabil; Educ Handic
Periodical
Der Sprachheilpädagoge (weekly) . . . 00833

Österreichische Gesellschaft für Statistik (ÖGS), Hintere Zollamtstr 2b, Postf 90, 1033 Wien
T: (0222) 71128
Founded: 1951; Members: 450
Focus: Computer & Info Sci; Stats
Periodical
Österreichische Zeitschrift für Statistik und Informatik (weekly) 00834

Österreichische Gesellschaft für Strassenwesen (ÖGS), Marxergasse 10, 1030 Wien
T: (0222) 736296
Founded: 1950; Members: 120
Focus: Transport
Periodical
Die Straße im Scheinwetter: Monatsberichte (weekly) 00835

Österreichische Gesellschaft für Tropenmedizin und Parasitologie, Kinderspitalgasse 15, 1095 Wien
T: (0222) 40490
Founded: 1967; Members: 200
Pres: Prof. Dr. E. Kutzer; Gen Secr: Dr. H. Auer
Focus: Trop Med
Periodical
Mitteilungen (weekly) 00836

Österreichische Gesellschaft für Unfallchirurgie (ÖGfU), Donaueschingenstr 13, 1200 Wien
T: (0222) 33110
Founded: 1965; Members: 900
Pres: Dr. V. Vecsei; Gen Secr: Dr. W. Buchinger
Focus: Surgery
Periodical
Kongreßberichte: Hefte zur Unfallkunde (weekly) 00837

Österreichische Gesellschaft für Unternehmensgeschichte, Augasse 2-6, 1090 Wien
T: (0222) 2536004711
Members: 120

Pres: Prof. Dr. Alois Mosser; Gen Secr: Oskar Ermann
Focus: Hist; Econ 00838

Österreichische Gesellschaft für Urologie, c/o Urologische Universitätsklinik, Alserstr 4, 1090 Wien
T: (0222) 48002616
Founded: 1947; Members: 286
Pres: Prof. Dr. M. Marberger
Focus: Urology 00839

Österreichische Gesellschaft für Ur- und Frühgeschichte, c/o Institut für Ur- und Frühgeschichte, Franz-Klein-Gasse 1, 1190 Wien
T: (0222) 31352377; Fax: (0222) 31352350
Founded: 1950; Members: 997
Focus: Hist
Periodical
Archäologie Österreichs (weekly) 00840

Österreichische Gesellschaft für Vakuumtechnik, c/o Institut für Allgemeine Physik, Karlspl 13, 1040 Wien
Founded: 1969; Members: 170
Focus: Eng 00841

Österreichische Gesellschaft für Versicherungsfachwissen, Schwarzenbergpl 7, 1030 Wien
T: (0222) 7132135
Focus: Insurance
Periodical
Die Versicherungsrundschau (weekly) . . 00842

Österreichische Gesellschaft für Vogelkunde (ÖGV), c/o Naturhistorisches Museum, Burgring 7, Postf 417, 1014 Wien
T: (0222) 934651; Fax: (0222) 935254
Founded: 1953; Members: 1400
Pres: Dr. Kurt Bauer; Gen Secr: Andreas Ranner
Focus: Ornithology
Periodical
Egretta: Vogelkundliche Nachrichten aus Österreich (weekly) 00843

Österreichische Gesellschaft für Warenkunde und Technologie (ÖGWT), c/o Institut für Technologie und Warenwirtschaftslehre, Wirtschaftsuniversität, Augasse 2-6, 1090 Wien
T: (0222) 313364806; Fax: (0222) 31336706
Founded: 1957; Members: 83
Pres: Prof. Dr. Josef Hölzl; Gen Secr: Dr. Ingrid Wagner
Focus: Business Admin; Eng
Periodicals
Bioware: Zeitschrift für Biologie und Warenlehre (3 times annually)
Forum Ware: Die Ware und ihre Bedeutung für Mensch, Wirtschaft und Natur (weekly) . 00844

Österreichische Gesellschaft für Weltraumforschung, Postf 67, 6020 Innsbruck
Founded: 1952
Focus: Astronomy 00845

Österreichische Gesellschaft für Wirtschaftspolitik, Franz-Klein-Gasse 1, 1190 Wien
Founded: 1947
Focus: Poli Sci 00846

Österreichische Gesellschaft für Wirtschaftsraumforschung (ÖGW), c/o Institut für Wirtschafts- und Sozialgeographie, Augasse 2-6, 1090 Wien
T: (0222) 347541813
Founded: 1962; Members: 75
Focus: Econ
Periodical
Wiener Geographische Schriften (weekly) 00847

Österreichische Gesellschaft für Wirtschaftssoziologie, c/o Wirtschaftsuniversität, Augasse 2-6, 1090 Wien
T: (0222) 340525742
Founded: 1973; Members: 25
Focus: Sociology
Periodical
Studien zur Soziologie (weekly) 00848

Österreichische Gesellschaft für Wissenschaftsgeschichte, Postgasse 7-9, 1010 Wien
Founded: 1980; Members: 200
Focus: Hist; Nat Sci
Periodical
Mitteilungen (weekly) 00849

Österreichische Gesellschaft für Zahn-, Mund- und Kieferheilkunde, c/o Bundesfachgruppe für Zahn-, Mund- und Kieferheilkunde der Österreichischen Ärztekammer, Weihburggasse 10-12, 1010 Wien
T: (0222) 51501233; Fax: (0222) 512512667
Founded: 1861; Members: 1950
Pres: Prof. Dr. G. Watzek; Gen Secr: Dr. E. Stoiber
Focus: Dent; Stomatology
Periodical
Zeitschrift für Stomatologie (8 times annually) 00850

Österreichische Gesellschaft für Zeitgeschichte, Rotenhausgasse 6, 1090 Wien
Founded: 1960; Members: 10
Focus: Hist
Periodicals
Jahrbuch für Zeitgeschichte (weekly)
Materialien zur Zeitgeschichte (weekly)
Veröffentlichungen zur Zeitgeschichte (weekly) 00851

Österreichische Gesellschaft und Institut für Umweltschutz, Umwelttechnologie und Umweltwissenschaften, Marc-Aurel-Str 5, 1010 Wien
T: (0222) 634124
Focus: Ecology; Eng; Nat Sci 00852

Österreichische Gesellschaft zum Studium der Sterilität und Fertilität, c/o 1. Universitäts-Frauenklinik, Spitalgasse 23, 1090 Wien
T: (0222) 404002851
Founded: 1956; Members: 114
Pres: Prof. Dr. G. Tscherne; Gen Secr: Prof. Dr. J.C. Huber
Focus: Gynecology 00853

Österreichische Gesellschaft zur Bekämpfung der Cystischen Fibrose, c/o Universitäts-Kinderklinik, Währinger Gürtel 74-76, 1090 Wien
Founded: 1968
Focus: Physiology; Pathology 00854

Österreichische Gesellschaft zur Erforschung des 18. Jahrhunderts, c/o Institut für Geschichte, Universität, Dr.-Karl-Lueger-Ring 1, 1010 Wien
T: (0222) 43002280
Members: 140
Focus: Hist 00855

Österreichische Gesellschaft zur Föderung von Umweltschutz und Energieforschung (ÖGEFUE), Obere Weissgerberstr 16, 1030 Wien
T: (0222) 7358234
Founded: 1973; Members: 3500
Focus: Ecology; Energy
Periodical
Umwelt-Report (weekly) 00856

Österreichische Gesellschaft zur Förderung medizin-meteorologischer Forschung in Österreich, Hohe Warte 38, 1190 Wien
T: (0222) 364453
Members: 180
Focus: Med 00857

Österreichische Hämophilie-Gesellschaft (ÖHG) / Austrian Hemophilia Society, Obere Augartenstr. 26-29, 1020 Wien
T: (0222) 3303257
Founded: 1965; Members: 700
Pres: Helmut Heisig
Focus: Hematology
Periodical
Mitgliederzeitschrift der Österreichischen Hämophilie Gesellschaft (weekly) 00858

Österreichische Himalaya-Gesellschaft (ÖHG), Prinz-Eugen-Str 12, 1040 Wien
T: (0222) 6540553
Focus: Geology 00859

Österreichische Humanistische Gesellschaft für die Steiermark, c/o Institut für Ökumenische Theologie und Patrologie, Universitätspl 3, 8010 Graz
T: (0316) 3803180
Founded: 1946; Members: 350
Pres: Prof. Dr. Johannes B. Bauer; Gen Secr: Dr. Erwin Pochmarski
Focus: Humanities 00860

Österreichische Kardiologische Gesellschaft, c/o Kardiologische Universitätsklinik, Garnisongasse 13, 1090 Wien
T: (0222) 48002201
Founded: 1968; Members: 611
Focus: Cardiol
Periodical
Acta medica austriaca (weekly) 00861

Österreichische Kommission für Internationale Erdmessung, c/o Bundesamt für Eich- und Vermessungswesen, Schiffamtsgasse 1-3, 1025 Wien
T: (0222) 211763201; Fax: (0222) 2161062
Founded: 1863; Members: 20
Pres: Prof. Dr. H. Moritz; Gen Secr: Dr. E. Erker
Focus: Surveying
Periodical
Geodätische Arbeiten Österreichs für die Internationale Erdmessung (weekly) . . . 00862

Österreichische Krebshilfe – Österreichische Krebsgesellschaft, Spitalgasse 19, 1090 Wien
T: (0222) 426363, 425139; Fax: (0222) 425139
Founded: 1910; Members: 1500
Focus: Cell Biol & Cancer Res
Periodical
Onkologie (weekly) 00863

Österreichische Krebshilfe Steiermark, Plöddemanngasse 51, 8010 Graz
T: (0316) 474433
Founded: 1974; Members: 200
Pres: Prof. Dr. Peter Steindorfer
Focus: Cell Biol & Cancer Res 00864

Österreichische Kulturgemeinschaft, Gunoldstr 14, 1019 Wien
T: (0222) 365168
Focus: Ethnology 00865

Österreichische Ludwig-Wittgenstein-Gesellschaft, 2880 Kirchberg
T: (02641) 2280
Focus: Philos 00866

Österreichische Mathematische Gesellschaft (ÖMG), Wiedner Hauptstr 8-10, 1040 Wien
T: (0222) 588015451
Founded: 1903; Members: 1200
Pres: Prof. Dr. L. Reich
Focus: Math
Periodical
International Mathematical News (3 times annually) 00867

Österreichische medizinische Gesellschaft für Neuraltherapie nach Huneke-Regulationsforschung (ÖNR), Wastiangasse 7, 8010 Graz
T: (05354) 2120; Fax: (05354) 472575
Founded: 1971; Members: 715
Pres: Dr. Winfried Muhri
Focus: Therapeutics
Periodical
Ganzheitsmedizin: Zeitschrift für Regulationsmedizin (weekly) 00868

Österreichische Mineralogische Gesellschaft (ÖMG), c/o Naturhistorisches Museum, Burgring 7, Postf 417, 1014 Wien
T: (0222) 43152177
Founded: 1901; Members: 333
Pres: Prof. Dr. Georg Hainkes; Gen Secr: Dr. Franz Brandstätter
Focus: Mineralogy
Periodical
Mitteilungen der ÖMG (weekly) 00869

Österreichische Multiple Sklerosis Gesellschaft, Währinger Gürtel 18-20, 1090 Wien
Fax: (0222) 404003141
Focus: Neurology
Periodical
Neue Horizonte (weekly) 00870

Österreichische Mykologische Gesellschaft (ÖMG), c/o Institut für Botanik, Universität, Rennweg 14, 1030 Wien
T: (0222) 787101; Fax: (0222) 787101131
Founded: 1919; Members: 200
Pres: Prof. Dr. Meinhard Moser; Gen Secr: Irmgard Krisai-Greilhuber
Focus: Botany
Periodical
Österreichische Zeitschrift für Pilzkunde (weekly) 00871

Österreichische Numismatische Gesellschaft (ÖNG), Burgring 5, 1010 Wien
T: (0222) 52177380; Fax: (0222) 932770
Founded: 1870; Members: 385
Pres: Prof. Dr. Bernhard Koch; Gen Secr: Dr. Helmut Jungwirth
Focus: Numismatics
Periodicals
Mitteilungen der Österreichischen Numismatischen Gesellschaft (weekly)
Numismatische Zeitschrift (weekly) . . . 00872

Österreichische Ophthalmologische Gesellschaft (ÖOG), Schlösselgasse 9, 1080 Wien
T: (0222) 4028540
Founded: 1955; Members: 540
Focus: Ophthal
Periodical
Spectrum der Augenheilkunde (weekly) . 00873

Österreichische Orchideen-Gesellschaft, Postf 300, 1222 Wien
T: (0222) 5128416
Focus: Botany
Periodical
Der Orchideenkurier (weekly) 00874

Österreichische Orient-Gesellschaft Hammer-Purgstall, Dominikanerbastei 6, 1010 Wien
T: (0222) 5128936; Fax: (0222) 512893617
Members: 163
Pres: Prof. Dr. Hans Weis; Gen Secr: Dr. Siegfried Haas
Focus: Ethnology; Ling 00875

Österreichische Pädagogische Gesellschaft, c/o Institut für Erziehungswissenschaften, Garnisongasse 3, Postf 26, 1090 Wien
T: (0222) 436171; Fax: (0222) 43617131
Founded: 1966; Members: 300
Pres: Dr. Josef Kurzreiter; Gen Secr: Dr. Dr. Nikolaus Severinski
Focus: Educ
Periodical
Jahrbuch der Österreichischen Pädagogischen Gesellschaft (weekly) 00876

Österreichische Paläontologische Gesellschaft (ÖPG), Universitätsstr 7, 1010 Wien
T: (0222) 43002494
Founded: 1966; Members: 165
Focus: Paleontology 00877

Österreichische Pharmazeutische Gesellschaft (ÖPhG), c/o Institut für Pharmakognosie, Währingerstr 25, 1090 Wien
Founded: 1979; Members: 330
Focus: Pharmacol
Periodical
Scientia Pharmaceutica (weekly) 00878

Österreichische Physikalische Gesellschaft (ÖPG), c/o Dr. Ewald Schachinger, Institut für Theoretische Physik, Technische Universität Graz, Petersgasse 16, 8010 Graz
T: (0316) 8738176
Founded: 1950; Members: 1000
Pres: Prof. Dr. Heinrich Mitter; Gen Secr: Dr. Ewald Schachinger
Focus: Physics 00879

Österreichische Physiologische Gesellschaft, Schwarzspanierstr 17, 1090 Wien
T: (0222) 40480239; Fax: (0222) 4028822
Founded: 1974; Members: 83
Focus: Physiology 00880

Österreichischer Alpenverein, Wissenschaftlicher Unterausschuss (ÖAV), Wilhelm-Greil-Str 15, 6010 Innsbruck
T: (0512) 23171
Founded: 1862; Members: 4
Focus: Geology 00881

Österreichischer Arbeitskreis für Gruppentherapie und Gruppendynamik (ÖAGG), Heiligenstädter Str 7, 1190 Wien
T: (0222) 3117062
Founded: 1959; Members: 938
Focus: Therapeutics
Periodicals
Feedback (weekly)
Gruppenpsychotherapie und Gruppendynamik (weekly) 00882

Österreichischer Arbeitskreis für Soziologie des Sports und der Leibeserziehung / Austrian Committee for Sociology of Sport, c/o Prof. Dr. Roland Bässler, Auf der Schmelz 6, 1150 Wien
T: (0222) 9822661112; Fax: (0222) 9822661131
Founded: 1969; Members: 20
Pres: Prof. Dr. Reinhard Bachleitner; Gen Secr: Prof. Dr. Roland Bässler
Focus: Sociology; Sports
Periodical
Sportsoziologie: Informationsschrift des Österreichischen Arbeitskreises für Soziologie des Sports und der Leibeserziehung (weekly) 00883

Österreichischer Arbeitskreis für Stadtgeschichtsforschung, Römerstr 14, Postf 320, 4010 Linz
T: (0732) 781064
Founded: 1969; Members: 104
Focus: Hist
Periodicals
Arbeitsbehelfe
Österreichische Städtebibliographie
Pro Civitate Austriae 00884

Österreichischer Arbeitsring für Lärmbekämpfung (ÖAL), Jägerstr 71, 1200 Wien
T: (0222) 3329236
Founded: 1958
Focus: Ecology
Periodical
ÖAL-Richtlinien (weekly) 00885

Österreichischer Astronomischer Verein (AV), Hasenwartgasse 32, 1238 Wien
T: (0222) 8893541; Fax: (0222) 8893541
Founded: 1924; Members: 1500
Pres: Johann Albrecht; Gen Secr: Prof. Hermann Mucke
Focus: Astronomy
Periodical
Österreichischer Himmelskalender (weekly) 00886

Österreichische Raumordnungskonferenz (ÖROK) / Austrian Conference on Regional Planning, Annagasse 5, 1010 Wien
T: (0222) 5134888; Fax: (0222) 5134890
Founded: 1971
Gen Secr: Dr. Eduard Kunze; Rudolf Schicker
Focus: Urban Plan
Periodical
ÖROK-Schriftenreihe (weekly) 00887

Österreichischer Burgenlandbund, Arbeitsgemeinschaft für Burgenländische Geschichte und Persönlichkeiten-Deutschtum in Ungarn, c/o Dr. Emmerich Karl Horvath, Gartengasse 12, 7000 Eisenstadt
T: (02682) 2893
Founded: 1958; Members: 8
Gen Secr: Dr. Emmerich Karl Horvath
Focus: Hist 00888

Österreichischer Burgenverein / Verein zur Erhaltung Historischer Bauten, Schloss Parz, 4710 Grieskirchen
T: (07248) 8771
Founded: 1955; Members: 700
Focus: Hist
Periodical
ARX: Burgen und Schlösser in Bayern, Österreich und Südtirol (weekly) 00889

Österreichische Rektorenkonferenz (ÖRK), Liechtensteinstr 22a, 1090 Wien
T: (0222) 31056500; Fax: (0222) 310565622
Founded: 1955; Members: 36
Pres: Prof. Dr. Alfred Ebenbauer; Gen Secr: Dr. Alexandra Suess
Focus: Educ 00890

Österreichischer Fachverband für Volkskunde (ÖFVK), Hans-Sachs-Gasse 3, 8010 Graz
Founded: 1958; Members: 100
Focus: Ethnology 00891

Österreichischer Fernschulverband, Margaretenstr 65, 1050 Wien
T: (0222) 5879650
Focus: Educ 00892

Österreichischer Ingenieur- und Architektenverein (ÖIAV), Eschenbachgasse 9, 1010 Wien
T: (0222) 5873536
Founded: 1848; Members: 7000
Pres: Karl Rabus; Gen Secr: Friedrich Smola
Focus: Archit
Periodical
Österreichische Ingenieur- und Architekten-Zeitschrift (weekly) 00893

Österreichischer Juristentag, Postf 3, 1016 Wien
T: (0222) 5261100
Founded: 1960; Members: 1500
Focus: Law
Periodical
Verhandlungen des Österreichischen Juristentages 00894

Österreichischer Kunsthistorikerverband, c/o Bundesdenkmalamt, Hofburg, 1010 Wien
T: (0222) 5353964
Members: 450
Focus: Arts; Hist 00895

Österreichischer Lehrerverband (ÖLV), Grillparzerstr 7, 1010 Wien
T: (0222) 439352
Founded: 1867; Members: 7
Focus: Educ 00896

Österreichischer Museumsbund, Burgring 5, 1010 Wien
T: (0222) 934541
Focus: Arts 00897

Österreichischer Naturschutzbund (ÖNB), Arenbergstr 10, 5020 Salzburg
T: (0662) 75492
Founded: 1913; Members: 60000
Pres: Prof. Dr. Eberhard Stüber; Gen Secr: Birgit Mair-Markart
Focus: Ecology
Periodicals
Natur und Land (weekly)
ÖNB-Kurier (weekly)
Steirischer Naturschutzbrief (weekly)
Wiener Naturschutznachrichten (weekly) . 00898

Österreichische Röntgengesellschaft – Gesellschaft für medizinische Radiologie und Nuklearmedizin, Rotenhausgasse 6, 1090 Wien
T: (0222) 435222
Founded: 1935; Members: 700
Pres: Prof. Dr. H. Schreyer; Gen Secr: Prof. Dr. J. Lammer
Focus: Radiology; X-Ray Tech; Nucl Med
Periodical
ÖRG-Mitteilungen 00899

Österreichischer PEN-Club, Bankgasse 8, 1010 Wien
T: (0222) 5334459; Fax: (0222) 5328749
Founded: 1922
Pres: Prof. Dr. Alexander Giese
Focus: Lit
Periodical
PEN-Nachrichten (weekly) 00900

Österreichischer Schriftstellerverband, Kettenbrückengasse 11, 1050 Wien
T: (0222) 564151
Focus: Lit
Periodical
Literarisches Österreich (weekly) 00901

Österreichischer Sportlehrerverband (ÖSLV), Prinz Eugen-Str 12, Haus des Sports, 1040 Wien
Focus: Sports
Periodical
Informationsblatt des ÖSLV (weekly) . . 00902

Österreichischer Stahlbauverband, Larochegasse 28, 1130 Wien
T: (0222) 8776170; Fax: (0222) 8774836
Focus: Civil Eng
Periodicals
Stahlbau-Rundschau (weekly)
Stahlbau-Rundschau-Mitteilungen (weekly) . 00903

Österreichischer Verband für Elektrotechnik (ÖVE) / Austrian Electrotechnical Association, Eschenbachgasse 9, 1010 Wien
T: (0222) 58763730; Fax: (0222) 567408
Focus: Electric Eng
Periodical
e & i Elektrotechnik und Informationstechnik (weekly) 00904

Österreichischer Verband für Strahlenschutz (ÖVS), c/o Forschungszentrum Seibersdorf, 2444 Seibersdorf
T: (02254) 7802500; Fax: (02254) 74060
Founded: 1966; Members: 307
Pres: Dr. N. Vana; Gen Secr: Dr. A. Hefner
Focus: Ecology; Radiology
Periodicals
Mitteilungsblatt des ÖVS (weekly)
Tagungsberichte 00905

Österreichischer Verein für Individualpsychologie, Währinger Gürtel 18-20, 1090 Wien
Founded: 1912; Members: 250
Pres: Dr. Günther Ratzka; Gen Secr: Dr. Eva Presslich
Focus: Psych
Periodical
Zeitschrift für Individualpsychologie (weekly) 00906

Österreichischer Verein für Vermessungswesen und Photogrammetrie, Schiffamtsgasse 1-3, 1025 Wien
T: (0222) 3576112701
Founded: 1903; Members: 630
Pres: August Hochwartner; Gen Secr: Gerhard Muggenhuber
Focus: Surveying
Periodical
Österreichische Zeitschrift für Vermessungswesen und Photogrammetrie (weekly) 00907

Österreichisches College: Collegegemeinschaft Wien, c/o Institut für Philosophie, Universitätsstr 7, 1010 Wien
T: (0222) 43002402
Founded: 1945; Members: 52
Focus: Educ 00908

Österreichisches Forschungsinstitut für Sparkassenwesen, Grimmelshausengasse 1, 1030 Wien
T: (0222) 71169290
Founded: 1951; Members: 140
Gen Secr: Dr. Gustav Raab
Focus: Finance 00909

Österreichisches Giesserei-Institut, Verein für Praktische Giessereiforschung (ÖGI), Parkstr 21, 8700 Leoben
T: (03842) 431010, 430720; Fax: (03842) 4310116
Founded: 1952; Members: 124
Pres: Dr. Robert Sponer; Gen Secr: Erich Nechtelberger
Focus: Metallurgy
Periodical
Gießerei-Rundschau (weekly) 00910

Österreichisches Institut für Bibliographie, Rathauspl 4, 1010 Wien
Founded: 1949
Focus: Libraries & Bk Sci 00911

Österreichisches Institut für Wirtschaftsforschung, Arsenal, Postf 91, 1103 Wien
T: (0222) 7826010; Fax: (0222) 789386
Focus: Econ
Periodicals
Empirica: Austrian Economic Papers (2-4 times annually)
Monatsberichte des Österreichischen Institutes für Wirtschaftsforschung (weekly) 00912

Österreichisches Lateinamerika-Institut, Schmerlingpl 8, 1010 Wien
T: (0222) 933315
Founded: 1965; Members: 1500
Focus: Ethnology; Develop Areas; Poli Sci; Sociology
Periodical
Zeitschrift für Lateinamerika (weekly) . . 00913

Österreichisches Normungsinstitut (ON), Heinestr 38, Postf 130, 1021 Wien
T: (0222) 267535; Tx: 115960; CA: austria norm; Fax: (0222) 267552
Founded: 1920; Members: 460
Pres: Prof. Dr. Karl Korinek; Gen Secr: Dr. Gerhard Hartmann
Focus: Standards
Periodical
ÖNORM: Fachzeitschrift des ON Österreichisches Normungsinstitut (weekly) 00914

Österreichisches Ost- und Südosteuropa-Institut / Austrian Institute of East and South-East European Studies, Josefspl 6, 1010 Wien
T: (0222) 5121895
Founded: 1958
Focus: Hist; Geography; Econ; Cultur Hist; Educ; Law; Poli Sci
Periodicals
Dokumentation der Gesetze und Verordnungen Osteuropas (DGVO) (weekly)
Österreichische Osthefte (weekly)
Ost-Dokumentation Bildungswesen (weekly)
Ost-Dokumentation Wirtschaft
Presseschau Ostwirtschaft (weekly) . . . 00915

Österreichische Sportwissenschaftliche Gesellschaft, c/o Institut für Sportwissenschaften, Akademiestr 26, 5020 Salzburg
T: (0662) 80444857
Members: 140
Gen Secr: Dr. Rudolf Stadler
Focus: Sports 00916

Österreichisches Produktivitäts- und Wirtschaftlichkeits-Zentrum (ÖPWZ), Rockhgasse 6, 1014 Wien
T: (0222) 5338636; Fax: (0222) 533863636
Founded: 1950; Members: 7
Focus: Business Admin 00917

Österreichisches Studienzentrum für Frieden und Konfliktforschung, 7461 Burg Schlaining
T: (03355) 2498, 2522; Fax: (03355) 2662
Founded: 1983
Pres: Dr. Gerald Mader
Focus: Prom Peace 00918

Österreichische Statistische Gesellschaft, Zollamtsstr 2b, 1030 Wien
T: (0222) 711287712
Founded: 1951; Members: 595
Gen Secr: Prof. Dr. Alfred Franz
Focus: Stats
Periodical
Österreichische Zeitschrift für Statistika und Informatik (weekly) 00919

Österreichische Studiengesellschaft für Kinderpsychoanalyse, Konstanze-Weber-Gasse 45, 5020 Salzburg
T: (0662) 845138
Founded: 1976; Members: 15
Focus: Psychoan
Periodical
Studien zur Kinderpsychoanalyse (weekly) 00920

Österreichische Studiengesellschaft für Kybernetik (ÖSGK), Schottengasse 3, 1010 Wien
T: (0222) 53532810
Founded: 1970; Members: 210
Focus: Cybernetics
Periodicals
Berichte: Reports (weekly)
Cybernetics and Systems: An International Journal (weekly) 00921

Österreichisches Volksliedwerk / Verband der Volksliedwerke der Bundesländer, Fuhrmannsgasse 18, 1080 Wien
T: (022) 420140
Founded: 1974; Members: 9
Focus: Music
Periodical
Jahrbuch (weekly) 00922

Österreichische Tribologische Gesellschaft (ÖTG) / Wissenschaftlicher Verein – Arbeitsgemeinschaft für Reibungs- und Verschleissfragen / Austrian Society of Tribology, Gußhausstr 27-29, 1040 Wien
T: (0222) 5053400; Fax: (0222) 5053400
Founded: 1976; Members: 36
Pres: Prof. Dr. Friedrich Franek
Focus: Eng; Physics; Chem; Econ . . . 00923

Österreichische Vereinigung der Zellstoff- und Papierchemiker und -techniker (ÖZEPA), Gumpendorfer Str 6, 1060 Wien
T: (0222) 58886207; Fax: (0222) 58886222
Members: 282
Focus: Chem; Eng 00924

Österreichische Vereinigung für politische Wissenschaften, Judenpl 11, 1014 Wien
Founded: 1951
Focus: Poli Sci 00925

Österreichische Verkehrswissenschaftliche Gesellschaft (ÖVG), Elisabethstr 9, 1010 Wien
T: (0222) 5879727
Founded: 1926; Members: 940
Focus: Transport
Periodicals
ÖVG-Spezial
Verkehrsannalen 00926

Österreichische Verwaltungswissenschaftliche Vereinigung, Judenpl 11, 1014 Wien
Founded: 1948
Focus: Business Admin 00927

Österreichische Werbewissenschaftliche Gesellschaft, Augasse 2-6, 1090 Wien
T: (0222) 313364617
Founded: 1955; Members: 1500
Focus: Advert
Periodical
Werbeforschung & Praxis (5 times annually) 00928

Österreichische wissenschaftliche Gesellschaft für Prophylaktische Medizin und Sozialhygiene, Berggasse 4, 1090 Wien
T: (0222) 343685
Founded: 1954; Members: 180
Pres: Prof. Dr. Oswald Jahn; Gen Secr: Dr. Otto Voelkel
Focus: Med; Hygiene
Periodical
Sozialmedizinische Schriftenreihe 00929

Orientalische Gesellschaft (OG), Universitätsstr 7, 1010 Wien
T: (0222) 43002593
Founded: 1952; Members: 74
Pres: Prof. Dr. Hermann Hunger; Gen Secr: Prof. Dr. Arne A. Ambros
Focus: Ling; Ethnology 00930

Philosophische Gesellschaft an der Universität Graz, c/o Philosophisches Institut, Heinrichstr 26, 8010 Graz
Focus: Philos 00931

Philosophische Gesellschaft Innsbruck, c/o Institut für Philosophie, Innrain 20, 6020 Innsbruck
T: (05222) 5073462
Founded: 1974; Members: 30
Pres: Prof. Dr. Reinhard Kleinknecht; Gen Secr: Prof. Dr. Rainer Thurnher
Focus: Philos 00932

Philosophische Gesellschaft in Salzburg, c/o Institut für Philosophie, Franziskanergasse 1, 5020 Salzburg
T: (0662) 80444070
Founded: 1966; Members: 91
Pres: Prof. Dr. Gerhard Zecha; Gen Secr: Dr. Alfons Süßbauer
Focus: Philos 00933

Philosophische Gesellschaft Klagenfurt, c/o Universität für Bildungswissenschaften, Universitätsstr 67, 9020 Klagenfurt
T: (04222) 23730504
Founded: 1973; Members: 65
Focus: Philos 00934

Philosophische Gesellschaft Wien, c/o Institut für Philosophie, Universitätsstr 7, 1010 Wien
T: (0222) 43002402, 401032402
Founded: 1954; Members: 100
Pres: Prof. Dr. Hans-Dieter Klein
Focus: Philos 00935

Pro Austria / Gesellschaft zur Erforschung und Förderung der österreichischen Bundesstaatsidee und des österreichischen Nationalbewusstseins, Josefspl 1, 1010 Wien
T: (0222) 521684245
Founded: 1969
Focus: Ethnology 00936

Rudolf Kassner-Gesellschaft, c/o Prof. Dr. Viktor Suchy, Schanzstr 33, 1140 Wien
T: (0222) 9820394
Founded: 1961; Members: 19
Pres: Prof. Dr. Viktor Suchy
Focus: Lit 00937

Salzburger Ärztegesellschaft, c/o Landeskrankenanstalten, 5020 Salzburg
T: (0662) 44824427
Founded: 1964; Members: 810
Pres: Prof. Dr. Friedrich Sandhofer; Gen Secr: Dr. Josef Koller
Focus: Med 00938

Salzburger Arbeitskreis für Psychoanalyse, Getreidegasse 16, 5020 Salzburg
T: (0662) 842205
Members: 32
Focus: Psychoan 00939

Salzburger Institut für juristische Information und Fortbildung (SIJIF), Künstlerhausgasse 4, 5020 Salzburg
T: (0662) 843663; Fax: (0662) 847986
Founded: 1976
Pres: Dr. Rudolf Zitta
Focus: Law 00940

Salzburger Juristische Gesellschaft, Churfürststr 1, 5020 Salzburg
T: (0662) 80443511; Fax: (0662) 8044302
Founded: 1970; Members: 163
Pres: Prof. Dr. Rolf Ostheim; Gen Secr: Prof. Dr. Wolfgang Schuhmacher
Focus: Law 00941

Salzburger Kulturvereinigung (SKV), Waagpl 1a, 5010 Salzburg
T: (0662) 845346; Fax: (0662) 891110
Founded: 1947; Members: 10250
Focus: Ethnology 00942

SIGMA – Salzburger Gesellschaft für Semiologie, c/o Dr. Sigrid Schmid, Institut für Germanistik, Akademiestr 20, 5020 Salzburg
T: (0662) 80444358; Fax: (0662) 8044612
Founded: 1986; Members: 21
Pres: Dr. Georg Schmid; Gen Secr: Dr. Hans Petschar
Focus: Ethnology; Lit; Hist 00943

Sigmund Freud-Gesellschaft, Berggasse 19, 1090 Wien
T: (0222) 3181596; Fax: (0222) 340278
Founded: 1968; Members: 1100
Focus: Psych; Psychoan
Periodical
Sigmund Freud House Bulletin (weekly) . 00944

Societas Linguistica Europaea, Universitätsstr 7, 1010 Wien
T: (0222) 401032655
Founded: 1966; Members: 1200
Pres: Prof. Dr. Peter Trudgill; Gen Secr: Prof. Dr. Dieter Kastovsky
Focus: Ling
Periodicals
Folia Linguistica
Folia Linguistica Historica 00945

Sonnblick-Verein, Hohe Warte 38, 1190 Wien
T: (0222) 364453
Founded: 1892; Members: 350
Pres: Dr. H. Kienzl; Gen Secr: Dr. Otto Motschka
Focus: Geophys
Periodical
Jahresbericht (weekly) 00946

Sozialwissenschaftliche Arbeitsgemeinschaft (SWA), Johannesgasse 4, 1010 Wien
T: (0222) 524850
Founded: 1953
Focus: Sociology 00947

Sozialwissenschaftliche Studiengesellschaft (SWS), Maria-Theresien-Str 9, 1090 Wien
T: (0222) 343127; Fax: (0222) 3102238
Founded: 1960; Members: 40
Focus: Sociology
Periodical
SWS-Rundschau (weekly) 00948

Sportwissenschaftliche Gesellschaft der Universität Graz, Conrad-von-Hötzendorf-Str 11, 8010 Graz
T: (0316) 73312
Founded: 1959
Focus: Sports 00949

Steirische Gesellschaft für Psychologie, c/o Prof. Dr. Helmuth P. Huber, Universitätspl 2, 8010 Graz
T: (0316) 3805113
Founded: 1980; Members: 183
Pres: Prof. Dr. Helmuth P. Huber; Gen Secr: Gerhild Bachmann
Focus: Psych 00950

SYNEMA – Gesellschaft für Film und Medien, Neubaugasse 36, 1070 Wien
T: (0222) 933797; Fax: (0222) 933797
Founded: 1981; Members: 60
Pres: Dr. Georg Haberl; Gen Secr: Gottfried Schlemmer
Focus: Cinema; Mass Media
Periodical
Kinoschriften 00951

Technisch-Wissenschaftlicher Verein Bergmännischer Verband Österreichs (BVÖ), c/o Montanuniversität, Franz-Josef-Str 18, 8700 Leoben
T: (03842) 45279; Fax: (03842) 402530
Founded: 1950; Members: 853
Focus: Mining
Periodical
Berg- und Hüttenmännische Monatshefte (weekly) 00952

Technisch-wissenschaftlicher Verein Eisenhütte Österreich, c/o Institut für Eisenhüttenkunde, Montanuniversität, 8700 Leoben
T: (03842) 45189; Fax: (03842) 46852
Founded: 1925; Members: 1097
Pres: Heribert Kreulitsch
Focus: Metallurgy
Periodical
Berg- und Hüttenmännische Monatshefte (weekly) 00953

Tiroler Gesellschaft zur Förderung der Alterswisschaft und des Seniorenstudiums an der Universität Innsbruck, Josef-Him-Str 7, 6020 Innsbruck
T: (05222) 20750
Founded: 1981; Members: 228
Focus: Geriatrics 00954

Tiroler Juristische Gesellschaft, c/o Universität, Innrain 80, 6020 Innsbruck
T: (05222) 5072561
Founded: 1968; Members: 218
Focus: Law 00955

Verband der Akademikerinnen Österreichs (VAÖ), Reitschulgasse 2, 1010 Wien
T: (0222) 5339080
Founded: 1922; Members: 650
Pres: Ingrid Fleischmann; Gen Secr: Dr. Gabriele Huber
Focus: Sci
Periodical
VAÖ Mitteilungen (weekly) 00956

Verband der diplomierten Physiotherapeuten Österreichs, Gießergasse 6, 1090 Wien
T: (0222) 4087577
Founded: 1962; Members: 2000
Focus: Physical Therapy
Periodical
Physiotherapie (weekly) 00957

Verband der Geistig Schaffenden Österreichs, Kärntnerstr 51, 1010 Wien
T: (0222) 525865, 5125865
Focus: Music; Arts
Periodical
Der Geistig Schaffende (weekly) 00958

Verband der Marktforscher Österreichs (VMÖ), Postf 182, 1013 Wien
T: (0222) 63391102
Founded: 1963; Members: 150
Focus: Marketing 00959

Verband der Österreicher zur Wahrung der Geschichte Österreichs, Huttengasse 8, 1014 Wien
T: (0222) 9247102
Founded: 1964; Members: 433
Focus: Hist 00960

Verband der österreichischen Neuphilologen (VÖN), Universitätsstr 6, 1090 Wien
Fax: (0222) 4038384
Founded: 1955
Focus: Ling; Lit
Periodical
Moderne Sprachen (weekly) 00961

Verband der Russischlehrer Österreichs (VRÖ), Josefspl 6, 1010 Wien
Focus: Educ; Ling
Periodical
Mitteilungen für Lehrer slawischer Fremdsprachen (weekly) 00962

Verband der wissenschaftlichen Gesellschaften Österreichs (VWGÖ), Lindengasse 37, 1070 Wien
T: (0222) 934756
Founded: 1949; Members: 321
Gen Secr: Dr. Rainer Zitta
Focus: Sci 00963

Verband für medizinischen Strahlenschutz in Österreich (VMSÖ), Embelgasse 52, 1050 Wien
T: (0222) 555332
Founded: 1974; Members: 500
Focus: Radiology
Periodical
Informationsblatt (1-2 times annually) . . 00964

Verband niederösterreichischer Volkshochschulen, Zelinkagasse 2, 1010 Wien
T: (0222) 5335379, 5355605
Focus: Adult Educ 00965

Verband österreichischer Archivare (VÖA), Postf 164, 1014 Wien
T: (0222) 932740
Founded: 1967; Members: 300
Focus: Archives
Periodical
Scrinium-Zeitschrift (weekly) 00966

Verband Österreichischer Bildungswerke, Heinrichgasse 4, 1010 Wien
T: (0222) 636547
Focus: Educ 00967

Verband Österreichischer Geschichtsvereine, c/o Allgemeines Verwaltungsarchiv, Nottendorfer Gasse 2, 1030 Wien
T: (0222) 786641350
Founded: 1949; Members: 133
Pres: Prof. Dr. Karl Heinz Burmeister; Gen Secr: Dr. Erwin A. Schmidl
Focus: Hist 00968

Verband Österreichischer Höhlenforscher, Obere Donaustr 97, 1020 Wien
Founded: 1949; Members: 24
Pres: Heinz Ilming; Gen Secr: Günter Stummer
Focus: Speleology
Periodicals
Die Höhle: Zeitschrift für Karst- und Höhlenkunde (weekly)
Wissenschaftliche Beihefte zur Zeitschrift Die Höhle (weekly) 00969

Verband Österreichischer Kurärzte, Josefspl 6, 1010 Wien
T: (0222) 5121904
Focus: Med 00970

Verband Österreichischer Privat-Museen, 4654 Bad Winsbach-Neydharting
Founded: 1950; Members: 156
Focus: Arts
Periodical
Die Österreichische Kunstforschung (weekly) 00971

Verband Österreichischer Sportärzte, Auf der Schmelz 6, 1150 Wien
T: (0222) 9822661174
Founded: 1950; Members: 714
Focus: Med 00972

Verband Österreichischer Volksbüchereien und Volksbibliothekare, Lange Gasse 37, 1080 Wien
T: (0222) 439722
Founded: 1948; Members: 780
Focus: Libraries & Bk Sci 00973

Verband Österreichischer Volkshochschulen, Rudolfspl 8, 1010 Wien
T: (0222) 630245
Focus: Adult Educ
Periodical
Die Österreichische Volkshochschule (weekly) 00974

Verband Österreichischer Wirtschaftsakademiker (VÖWA), Teinfaltstr 1, 1010 Wien
T: (0222) 5336876; Fax: (0222) 5327669
Founded: 1927; Members: 1900
Focus: Econ
Periodical
Wirtschaftskurier (weekly) 00975

Verein der Freunde der im Mittelalter von Österreich aus besiedelten Sprachinseln, Semperstr 29, 1180 Wien
T: (0222) 4796083
Founded: 1972; Members: 210
Pres: Prof. Dr. Maria Hornung; Gen Secr: Dr. Manfred Skopec
Focus: Ling
Periodical
Beiträge zur Sprachinselforschung (weekly) 00976

Verein der Freunde des Radwerkes IV in Vordernberg, Peter-Tunner-Str 2, 8794 Vordernberg
T: (03849) 283
Founded: 1956; Members: 120
Pres: Dr. Ernst Steyrer
Focus: Hist 00977

Verein der Mundartfreunde Österreichs, Postgasse 7-9, 1010 Wien
Members: 150
Focus: Lit
Periodical
Mitteilungen der Mundartfreunde Österreichs (weekly) 00978

Verein der Museumsfreunde in Wien, Praterstr 30, 1020 Wien
T: (0222) 269371; Fax: (0222) 214890004
Founded: 1911; Members: 2900
Pres: Herbert Schimetschek
Focus: Arts
Periodical
Mitteilungen des Vereins der Museumsfreunde in Wien (10 times annually) 00979

Verein Forschung für das graphische Gewerbe (VFG), Leyserstr 6, 1140 Wien
T: (0222) 92265421
Founded: 1966; Members: 61
Focus: Eng 00980

Verein Freunde der Archäologie, Plüddemanngasse 95a, 8010 Graz
Founded: 1979
Focus: Archeol
Periodical
Mitteilungen des Vereins Freunde der Archäologie (weekly) 00981

Verein Freunde der Völkerkunde, c/o Museum für Völkerkunde, Neue Hofburg, 1014 Wien
T: (0222) 52177; Fax: (0222) 5355320
Founded: 1933; Members: 185
Pres: Dr. Friedrich Berg; Gen Secr: Dr. Walter Warthol
Focus: Ethnology
Periodical
Archiv für Völkerkunde (weekly) 00982

Verein für Geschichte der Arbeiterbewegung, Rechte Wienzeile 97, 1050 Wien
T: (0222) 5467870; Fax: (0222) 5463097
Founded: 1959; Members: 382
Focus: Hist; Sociology; Socialism
Periodical
Archiv: Mitteilungsblatt (weekly) 00983

Verein für Geschichte der Stadt Wien (VGStW), c/o Wiener Stadt- und Landesarchiv, Rathaus, 1082 Wien
T: (0222) 428002701
Founded: 1853; Members: 879
Focus: Hist
Periodicals
Forschungen und Beiträge zur Wiener Stadtgeschichte (weekly)
Jahrbuch des Vereins für Geschichte der Stadt Wien (weekly)
Wiener Geschichtsblätter (weekly) . . . 00984

Verein für Landeskunde von Niederösterreich, Herrengasse 11, 1014 Wien
T: (0222) 6357112042
Founded: 1864; Members: 1300
Pres: Prof. Dr. Helmuth Feigl; Gen Secr: Dr. Silvia Petrin
Focus: Hist
Periodicals
Jahrbuch für Landeskunde von Niederösterreich (weekly)
Unsere Heimat (weekly) 00985

Verein für Psychiatrie und Neurologie, c/o Psychiatrische Universitätsklinik, Währinger Gürtel 18-20, 1090 Wien
T: (0222) 404003520
Founded: 1868; Members: 400
Pres: Prof. Dr. B. Saletu; Gen Secr: Dr. V. Passweg
Focus: Psychiatry; Neurology 00986

Verein für Sozial- und Wirtschaftspolitik, Ebendorferstr 6, 1010 Wien
T: (0222) 422674; Fax: (0222) 4023777
Founded: 1953; Members: 283
Focus: Poli Sci
Periodical
Gesellschaft und Politik (weekly) 00987

Verein für Volkskunde in Wien, Laudongasse 15-19, 1080 Wien
T: (0222) 43890528; Fax: (0222) 4085342
Founded: 1894; Members: 900
Pres: Dr. Klaus Beitl; Gen Secr: Dr. Franz Grieshofer; Dr. Margot Schindler
Focus: Ethnology
Periodicals
Oesterreichische Zeitschrift für Volkskunde (weekly)
Volkskunde in Österreich: Nachrichtenblatt des Vereins für Volkskunde (10 times annually) 00988

Vereinigung bildender Künstler (Wiener Secession), Friedrichstr 12, 1010 Wien
Fax: (0222) 587530734
Founded: 1897; Members: 209
Pres: Prof. Adolf Krischanitz
Focus: Arts 00989

Vereinigung Burgenländischer Geographen, 7551 Stegersbach
T: (03326) 2711; Fax: (03326) 257622
Founded: 1975; Members: 555
Focus: Geography
Periodical
Geographisches Jahrbuch Burgenland (weekly) 00990

Vereinigung der Humanistischen Gesellschaften Österreichs, c/o Institut für Klassische Philologie, Universität, Universitätspl 3, 8010 Graz
Pres: Prof. Dr. W. Pötscher
Focus: Philos 00991

Vereinigung der kooperativen Forschungsinstitute der Österreichischen Wirtschaft, Franz-Grill-Str 1, 1030 Wien
T: (0222) 782772; Fax: (0222) 7986457
Founded: 1954; Members: 34
Focus: Econ
Periodical
ACR-Info (3-4 times annually) 00992

Vereinigung für Angewandte Lagerstättenforschung, c/o Montanuniversität, Franz-Josef-Str 19, 8700 Leoben
Founded: 1975
Focus: Geology 00993

Vereinigung für Hydrogeologische Forschungen in Graz, Rechbauerstr 12, 8010 Graz
T: (0316) 8020373
Founded: 1962; Members: 300
Pres: Prof. Dr. G. Riedmüller; Gen Secr: Prof. Dr. H. Zojer
Focus: Geology
Periodical
Steirische Beiträge zur Hydrogeologie . . 00994

Vereinigung für wissenschaftliche Grundlagenforschung, Heinrichstr 26, 8010 Graz
Founded: 1964; Members: 39
Pres: Prof. Dr. Johann C. Marek; Gen Secr: Prof. Dr. Wolfgang L. Gombocz
Focus: Sci; Philos 00995

Vereinigung Österreichischer Ärzte (VÖÄ), Weihburggasse 10-12, 1010 Wien
T: (0222) 527958
Founded: 1945; Members: 1800
Focus: Med
Periodical
Der Österreichische Arzt (weekly) . . . 00996

Vereinigung Österreichischer Bibliothekarinnen und Bibliothekare (VÖBB), Innrain 50, 6010 Innsbruck
T: (0512) 5072087; Fax: (0512) 5072307
Founded: 1896; Members: 1020
Pres: Dr. Walter Neuhauser; Gen Secr: Eva Ramminger
Focus: Libraries & Bk Sci
Periodical
BIBLOS: Mitteilungen der Vereinigung Österreichischer Bibliothekarinnen und Bibliothekare 00997

Verein Montandenkmal Altböckstein, Postfach 78, 8700 Leoben
Founded: 1979; Members: 120
Pres: Eva Sika
Focus: Preserv Hist Monuments
Periodicals
Böcksteiner Montana
Jahresberichte (weekly) 00998

Verein Muttersprache, Postf 27, 2103 Lang-Enzersdorf
T: (0222) 4058837; Fax: (0222) 2788501
Founded: 1949; Members: 1890
Focus: Ling
Periodicals
Schriftenreihe Wissenschaft (weekly)
Wiener Sprachblätter (weekly) 00999

Verein Österreichischer Lebensmittel- und Biotechnologen (V.Ö.L.B.), c/o Institut für Lebensmitteltechnologie, Peter-Jordan-Str 82, 1190 Wien
Founded: 1968; Members: 400
Pres: Prof. Dr. Josef Weiss
Focus: Food; Eng 01000

Verein österreichischer Ledertechniker (VÖLT), Rosensteingasse 79, 1170 Wien
T: (0222) 461480; Fax: (0222) 454068
Founded: 1950; Members: 250
Pres: Prof. Karl Heinz Munz; Gen Secr: Prof. Christian Anderson
Focus: Eng 01001

Verein Österreichischer Textilchemiker und Coloristen (VÖTC), Postf 2, 6850 Dornbirn
Focus: Chem 01002

Verein Tiroler Landesmuseum Ferdinandeum, Museumstr 15, 6020 Innsbruck
T: (0512) 59489
Founded: 1823; Members: 2800
Pres: Prof. Dr. Josef Riedmann; Gen Secr: Dr. Gert Ammann
Focus: Arts; Archeol; Hist; Nat Sci
Periodical
Veröffentlichungen des Tiroler Landesmuseums Ferdinandeum (weekly) 01003

Verein Wiener Frauenverlag, Lange Gasse 51, 1080 Wien
T: (0222) 4390412
Founded: 1980; Members: 42
Focus: Lit 01004

Verein zur Förderung des physikalischen und chemischen Unterrichts, c/o Institut für theoretische Physik, Strudlhofgasse 4, 1090 Wien
T: (0222) 31367
Founded: 1893; Members: 918
Focus: Chem; Educ; Physics
Periodical
Plus Lucis (weekly) 01005

Verein zur Verbreitung naturwissenschaftlicher Kenntnisse, Althanstr 14, 1091 Wien
Founded: 1860
Focus: Nat Sci
Periodical
Schriften (weekly) 01006

Volkswirtschaftliche Gesellschaft Österreich / Verband für Bildungswesen, Bauernmarkt 21, 1010 Wien
T: (0222) 5351185
Focus: Econ
Periodical
Wirtschaft in der Praxis (5-8 times annually) 01007

Vorarlberger Landesmuseumsverein (VLMV), Marktstr 33, 6050 Dornbirn
T: (05572) 23235
Founded: 1857; Members: 1200
Pres: Dr. Edwin Oberhauser; Gen Secr: Dr. Walter Krieg
Focus: Hist; Sci
Periodical
Jahrbuch des Vorarlberger Landesmuseumsvereins (weekly) 01008

Waldviertler Heimatbund, Wissenschaftliche Sektion, Postf 100, 3580 Horn
T: (02982) 3991
Founded: 1951; Members: 1000
Pres: Prof. Dr. Erich Rabl
Focus: Hist
Periodical
Das Waldviertel: Zeitschrift für Heimat- und Regionalkunde des Waldviertels und der Wachau (weekly) 01009

Walter Buchebner Gesellschaft, Wiener Str 35, 8680 Mürzzuschlag
T: (03852) 56200; Fax: (03852) 56209
Founded: 1977
Pres: Robert Lotter
Focus: Lit; Arts; Music
Periodical
Was: Zeitschrift für Kultur und Politik (weekly) 01010

Wiener Arbeitskreis für Psychoanalyse, Postf 76, 1000 Wien
T: (0222) 7153957
Founded: 1947; Members: 23
Pres: Dr. Josef Shaked; Gen Secr: Dr. Brigitte Garger-Grossmann
Focus: Psychoan 01011

Wiener Beethoven-Gesellschaft (WBG), Probusgasse 6, 1190 Wien
T: (0222) 3714085
Founded: 1954; Members: 495
Pres: Dr. Albert Mitringer; Gen Secr: Walther Brauneis
Focus: Music
Periodical
Mitteilungsblatt der Wiener Beethoven-Gesellschaft (weekly) 01012

Wiener Bibliophilen-Gesellschaft (WBG), c/o Walter R. Schaden, Sonnenfelsgasse 4, 1010 Wien
Founded: 1912
Focus: Libraries & Bk Sci 01013

Wiener Gesellschaft der Hals-Nasen-Ohren-Ärzte, Lazarettgasse 14, 1090 Wien
T: (0222) 48003306
Focus: Otorhinolaryngology 01014

Wiener Gesellschaft für Innere Medizin, Währinger Gürtel 18-20, 1090 Wien
T: (0222) 404004409; Fax: (0222) 4026930
Members: 223
Pres: Prof. Dr. Klaus Lechner
Focus: Intern Med 01015

Wiener Gesellschaft für Theaterforschung, c/o Institut für Theaterwissenschaft, Hofburg, 1010 Wien
T: (0222) 5350599
Founded: 1944; Members: 100
Focus: Perf Arts
Periodicals
Quellen zur Theatergeschichte (weekly)
Theater in Österreich: Verzeichnis der Inszenierungen (theadok) (weekly) 01016

Wiener Goethe-Verein, Stallburggasse 2, 1010 Wien
T: (0222) 401032542
Founded: 1878; Members: 320
Pres: Prof. Dr. Herbert Zeman
Focus: Lit
Periodical
Jahrbuch des Wiener Goethe-Vereins (weekly) 01017

Wiener Humanistische Gesellschaft, c/o Institut für Klassische Philologie, Universität, Dr. Karl-Lueger-Ring 1, 1010 Wien
T: (0222) 401032326
Founded: 1946; Members: 600
Pres: Prof. Dr. Heinrich Stremitzer; Gen Secr: Angelika Huber
Focus: Ling; Lit; Hist
Periodical
Wiener Humanistische Blätter (weekly) . 01018

Wiener Institut für Entwicklungsfragen und Zusammenarbeit / Vienna Institute for Development and Cooperation, Weyrgasse 5, 1030 Wien
T: (0222) 7133594; Fax: (0222) 713359473
Founded: 1962
Pres: Dr. Franz Vranitzky; Gen Secr: Dr. Erich Andrlik
Focus: Econ; Develop Areas; Poli Sci; Int'l Relat
Periodical
Report Series (weekly) 01019

Wiener Institut für Internationale Wirtschaftsvergleiche, Postfach 87, 1103 Wien
T: (0222) 7826010; Fax: (0222) 787120
Founded: 1972
Focus: Econ
Periodicals
Comecon Data
East-West European Economic Interaction/Workshop Papers
Forschungsberichte: Reprint-Serie
Studien über Wirtschafts- und Systemvergleiche 01020

Wiener Juristische Gesellschaft (WJG), Tuchlauben 13, Postf 41, 1014 Wien
T: (0222) 53437; Fax: (0222) 5332521
Founded: 1867; Members: 550
Pres: Prof. Dr. Walter Barfuß
Focus: Law 01021

Wiener Konzerthausgesellschaft, Lothringerstr 20, 1030 Wien
T: (0222) 71246860; Tx: 3222302; Fax: (0222) 7131709
Founded: 1913; Members: 4000
Pres: Dr. Heinrich Haerdtl; Gen Secr: Karsten Witt
Focus: Music
Periodical
Konzerthaus-Nachrichten (8 times annually) 01022

Wiener Kulturkreis, Prinz-Eugen-Str 3, 1030 Wien
T: (0222) 784350
Founded: 1947; Members: 750
Focus: Arts; Sci 01023

Wiener Medizinische Akademie für Ärztliche Fortbildung und Forschung, c/o Allgemeines Krankenhaus, Alserstr 4, 1090 Wien
T: (0222) 421383
Founded: 1896; Members: 320
Pres: Prof. Dr. O. Mayrhofer; Gen Secr: Prof. Dr. H. Ludwig
Focus: Med; Adult Educ 01024

Wiener Psychoanalytische Vereinigung, Gonzagagasse 11, 1010 Wien
T: (0222) 5330767; Fax: (0222) 5330767
Founded: 1910; Members: 68
Focus: Psychoan
Periodical
Bulletin der Wiener Psychoanalytischen Vereinigung (weekly) 01025

Wiener Rechtsgeschichtliche Gesellschaft, Schottenbastei 10-16, 1010 Wien
T: (0222) 43003260
Founded: 1975; Members: 486
Focus: Law 01026

Wiener Schubertbund, Lothringerstr 20, 1030 Wien
T: (0222) 732429
Focus: Music 01027

Wiener Sprachgesellschaft (WSG), c/o Institut für Sprachwissenschaft, Dr.-Karl-Lueger-Ring 1, 1010 Wien
T: (0222) 4039080
Founded: 1947; Members: 200
Focus: Ling
Periodical
Die Sprache: Zeitschrift für Sprachwissenschaft (weekly) 01028

Wiener Volkswirtschaftliche Gesellschaft, Fischhof 3, 1010 Wien
T: (0222) 5330871; Fax: (0222) 5330688
Focus: Econ
Periodical
Die Aussprache (weekly) 01029

Wissenschaftliche Ärztegesellschaft Innsbruck, c/o Landeskrankenhaus, Anichstr 35, 6020 Innsbruck
T: (0512) 28711900
Founded: 1894; Members: 900
Focus: Med 01030

Wissenschaftliche Arbeitsgemeinschaft für Leibeserziehung und Sportmedizin in Innsbruck, Fürstenweg 185, 6020 Innsbruck
T: (0522) 5076530
Founded: 1969; Members: 24
Focus: Med; Sports 01031

Wissenschaftliche Gesellschaft der Ärzte in der Steiermark, Auenbruggerpl 25, 8036 Graz
T: (0316) 382653; Fax: (0316) 3853062
Founded: 1863; Members: 1500
Pres: Prof. Dr. H.C. Zapotoczky; Prof. Dr. K. Schauenstein
Focus: Med
Periodical
Abstracts über die wissenschaftlichen Sitzungen 01032

Wissenschaftliche Gesellschaft für Sport und Leibeserziehung am Institut für Sportwissenschaften der Universität Salzburg, Akademiestr 26, 5020 Salzburg
T: (0662) 44511252
Founded: 1970; Members: 50
Pres: Prof. Dr. Stefan Größing
Focus: Sports
Periodical
Salzburger Beiträge zum Sport unserer Zeit 01033

Zentralvereinigung der Architekten Österreichs, Salvatorgasse 10, 1010 Wien
T: (0222) 5334429
Founded: 1907; Members: 700
Pres: Prof. Eugen Wörle
Focus: Archit 01034

Zoologisch-Botanische Gesellschaft in Österreich, c/o Biozentrum, Althanstr 14, Postf 287, 1091 Wien
T: (0222) 314510
Founded: 1851; Members: 355
Pres: Prof. Dr. Walter Fiedler; Gen Secr: Dr. Wolfgang Punz
Focus: Botany; Zoology
Periodicals
Abhandlungen der Zoologisch-Botanischen Gesellschaft (weekly)
Koleopterologische Rundschau (weekly)
Verhandlungen der Zoologisch-Botanischen Gesellschaft (weekly) 01035

Azerbaijan

Azerbaijan Academy of Sciences, Kommunisticheskaya ul 10, 370601 Baku
Members: 102
Pres: E.Y. Salaev; Gen Secr: A.A.K. Nadirov
Focus: Sci
Periodicals
Azerbaidzhanskii Khimicheskii Zhurnal: Azerbaijan Chemical Journal
Doklady: Report
Izvestiya: Bulletin
Sibirskii Vestnik Selskokhozyaistvennoi Nauki: Siberian Agricultural Science Journal
Tsirkulyar Shemakhinskoi Astrofizicheskoi Observatorii: Newsletter of the Shemakha Astrophysics Observatory 01036

Azerbaijan Mathematics Society, Ul Kommunisticheskaya 10, 370601 Baku
Pres: M.A. Dzhavadov
Focus: Math 01037

Azerbaijan Petroleum Academy, Pr Azadlyk 20, 370601 Baku
T: 934557
Focus: Petrochem 01038

Azerbaijan Physical Society, Pr Narimanova 33, 370143 Baku
Pres: G.M. Abdullaev
Focus: Physics 01039

Commision on Mining, Ul Nizami 67, Baku
Pres: A.D. Sultanov
Focus: Mining 01040

Commission for the Study of Productive Forces and Natural Resources, Pr Narimanova 31, 370143 Baku
T: 387573
Focus: Nat Sci; Eng 01041

Commission on International Scientific Contacts, Kommunisticheskaya ul 10, 370601 Baku
Pres: M.F. Nagev
Focus: Sci 01042

Commission on Mountain Mud Flows, Ul Narimanova 31, 370143 Baku
Pres: S.G. Rustamov
Focus: Hydrology 01043

Commission on Nature Conservation, Kommunisticheskaya ul 10, 370601 Baku
Pres: G.A. Aliev
Focus: Ecology 01044

Commission on the Caspian Sea, Ul Narimanova 31, 370143 Baku
Pres: K.K. Gyul
Focus: Hydrology 01045

Council on Exploitation of Sciemtific Equipment, Ul Kommunisticheskaya 10, 370601 Baku
Pres: V.Y. Akhundov
Focus: Eng 01046

Science Production Association Biotech, Pr Metbuat 2, 370073 Baku
T: 381287
Gen Secr: N.K.O. Mekhtiev
Focus: Eng 01047

Scientific Council on Complex Problems (Cybernetics), Ul Kommunisticheskaya 10, 370601 Baku
Pres: B.A. Azimov
Focus: Cybernetics 01048

Terminological Committee, Pr Narimanova 31, 370143 Baku
T: 395713
Pres: M.S. Kasumov
Focus: Nat Sci 01049

Bahamas

Association of Caribbean Historians, c/o Dept of Archives, POB 6341, Nassau
T: 3232175
Founded: 1974
Gen Secr: Dr. D. Gail Saunders
Focus: Hist
Periodical
Association of Caribbean Historians Bulletin (weekly) 01050

Bahamas Historical Society, POB N143, Nassau
Founded: 1959; Members: 314
Pres: June Maura; Gen Secr: Anne Lawlor
Focus: Hist
Periodical
Journal (weekly) 01051

Bahamas National Trust, POB N4105, Nassau
T: (809) 3231317
Founded: 1958; Members: 2100
Focus: Ecology
Periodicals
The Bahamas Naturalist (weekly)
Currents (weekly) 01052

Bahrain

Arab Regional Office of the United Schools International, POB 726, Gudaibiya
T: 258839; Tx: 9094; Fax: 272252
Founded: 1982
Gen Secr: Jiya Lal Jain
Focus: Educ
Periodical
Workshop of Peace (weekly) 01053

Bahrain Arts Society, POB 26264, Manama
T: 230367
Founded: 1983; Members: 150
Pres: Rashid Khalifa Al-Khalifa
Focus: Arts 01054

Bahrain Bar Society, POB 5025, Manama
T: 720566
Founded: 1977
Pres: Ali Abdulla Alayoobi
Focus: Law 01055

Bahrain Computer Society, POB 26089, Manama
T: 245366
Founded: 1981; Members: 114
Pres: Mohammed Ahmed Al-Amer
Focus: Computer & Info Sci 01056

Bahrain Contemporary Art Association, POB 26232, Manama
T: 728046
Founded: 1970; Members: 60
Pres: Abdul Karim Al-Orrayed
Focus: Arts 01057

Bahrain Medical Society, POB 26136, Manama
T: 742666; Fax: 715559
Founded: 1972; Members: 310
Pres: Dr. Ali M. Matar; Gen Secr: Dr. Ahmed J. Jamal
Focus: Med
Periodicals
Journal of the Bahrain Medical Society (3 times annually)
Journal of the Bahrain Medical Society . 01058

Bahrain Society of Engineers, POB 835, Manama
T: 727100; Tx: 7592; Fax: 729819
Founded: 1972; Members: 950
Pres: Emad A.R. Almoayed; Gen Secr: Ali Redha Hussain
Focus: Eng
Periodical
Al-Mohandes (weekly) 01059

Bahrain Society of Sociologists, POB 26488, Manama
T: 727483
Founded: 1979; Members: 65
Pres: Dr. Baquer Salman Al-Najjar
Focus: Sociology 01060

Bahrain Writers and Literators Association, POB 1010, Manama
Founded: 1969; Members: 35
Pres: Q. Haddad; Gen Secr: Abdul Qadir Aqeel
Focus: Lit 01061

Bahrein Historical and Archaeological Society, POB 5087, Manama
T: 727895
Founded: 1953; Members: 342
Pres: Issa Al-Khalifa; Gen Secr: Robert Jarman
Focus: Archeol; Hist
Periodical
Dilmun (weekly) 01062

The Islamic Association, POB 22484, Manama
T: 345405; Fax: 345407
Founded: 1979; Members: 110
Pres: Abdulrahman E. Abdul Salam
Focus: Rel & Theol 01063

Bangladesh

Association of Universities of Bangladesh, 28 Shyamolee, Dhaka 7
Gen Secr: M.K. Hussain Sirkar
Focus: Educ 01064

Bangla Academy, Burdwan House, Dhaka 2
Founded: 1972
Gen Secr: Dr. A.H. Siddiqui
Focus: Sci
Periodicals
Dhan Shaliker Desh (weekly)
Journal (weekly)
Research Journal
Science Journal (weekly) 01065

Bangladesh Academy of Sciences, 3/8 Asad Av, Mohammadpur, Dhaka 7
T: (02) 310425
Founded: 1973; Members: 37
Pres: Dr. M.O. Ghani; Gen Secr: Dr. S.D. Chaudhury
Focus: Sci 01066

Bangladesh Council of Scientific and Industrial Research, Mirpur Rd, Dhanmondi, Dhaka 5
T: (02) 505686
Founded: 1955
Pres: Prof. Dr. S.S.M.A. Khorasani; Gen Secr: K.C. Sikdar
Focus: Sci; Eng 01067

Bangladesh Economic Association, c/o Economics Dept, Dhaka University, Dhaka
Founded: 1958
Pres: Dr. M. Huq; Gen Secr: Dr. S.R. Bose
Focus: Econ 01068

Library Association of Bangladesh, c/o Central Public Library, Shanbagh, Dhaka 2
T: (02) 504269
Founded: 1956
Pres: A. Zakiuddin; Gen Secr: A, Sultanuddin
Focus: Libraries & Bk Sci
Periodical
The Eastern Librarian (weekly) 01069

Society of Arts, Literature and Welfare, Society Park, K.C. Dey Rd, Chittagong
Founded: 1942; Members: 500
Gen Secr: Husain Musharraf
Focus: Arts; Lit 01070

Zoological Society of Bangladesh, c/o Dept of Zoology, Dhaka University, Dhaka 1000
Focus: Zoology
Periodical
Bangladesh Journal of Zoology (weekly) . 01071

Barbados

Barbados Association of Medical Practitioners, c/o Queen Elizabeth Hospital, Martindales Rd, Saint Michael
Founded: 1973; Members: 75
Pres: Prof. E.R. Walrond; Gen Secr: Dr. G. Dixit
Focus: Med
Periodical
Newsletter (weekly) 01072

Barbados Astronomical Society (BAS), Brittons Hill, Saint Michael, POB 41B, Bridgetown
T: 4277424
Founded: 1956; Members: 75
Pres: Janice Stahl; Gen Secr: Lynda Hamilton
Focus: Astronomy
Periodical
Journal (weekly) 01073

Barbados Museum and Historical Society (BMHS), Saint Ann's Garrison, Saint Michael, Bridgetown
T: 4270201, 4361956
Founded: 1933; Members: 700
Pres: Jack Q.C. Dear; Gen Secr: Alissandra Cummins
Focus: Hist; Arts; Geology; Zoology; Archeol
Periodicals
Journal of BMHS (weekly)
Newsletter (weekly) 01074

Barbados Pharmaceutical Society, Saint Michael, POB 820E, Bridgetown
Founded: 1948; Members: 120
Pres: Peter Bourne; Gen Secr: Cheryl-Ann Yearwood
Focus: Pharmacol 01075

Caribbean Appropriate Technology Centre (CATC), c/o CCC, POB 616, Bridgetown
T: (042) 72681; Tx: 2335
Founded: 1981
Focus: Eng 01076

Caribbean Association of Law Librarians (CAROLL), c/o Faculty of Law Library, University of the West Indies, Cave Hill
Gen Secr: John Dyrud
Focus: Libraries & Bk Sci 01077

Caribbean Atlantic Regional Dental Association (CARDA), c/o Rosedale Dental Centre, Upper Collymore Rock, Saint Michael
Founded: 1977
Pres: Dr. Victor H. Eastmond
Focus: Dent 01078

Caribbean Centre for Development Administration (CARICAD), 27 Block C, The Garrison, Saint Michael
Founded: 1980
Gen Secr: Selwyn Patrick Smith
Focus: Business Admin
Periodical
CARICAD Newsletter (weekly) 01079

Caribbean Conservation Association (CCA), Savannah Lodge, Saint Michael, Bridgetown
T: 4265373; CA: Concarb Barbados
Founded: 1967; Members: 250
Pres: Yves Renard; Gen Secr: Michael King
Focus: Ecology; Cultur Hist
Periodical
Caribbean Conservation News (weekly) . 01080

Caribbean Examinations Council (CXC), The Garrison, Saint Michael
Founded: 1972; Members: 15
Gen Secr: W.W. Beckles
Focus: Educ 01081

Caribbean Institute for Meteorology and Hydrology (CIMH), POB 130, Bridgetown
Founded: 1967; Members: 16
Pres: Dr. Colin A. Depradine
Focus: Geophys; Hydrology
Periodical
Climatic Data (weekly) 01082

Caribbean Law Institute, c/o Faculty of Law, University of the West Indies, POB 64, Bridgetown
Focus: Law 01083

Caribbean Management Development Association (CMDA), 27 Block C, Garrison, Saint Michael
Focus: Business Admin 01084

Caribbean Network of Educational Innovation for Development (CARNEID), Garrison Garden, POB 423, Saint Michael
Founded: 1979
Gen Secr: Hubert J. Charles
Focus: Educ
Periodical
CARNEID Newsletter (weekly) 01085

Caribbean Policy Development Centre (CPDC), c/o CCC, POB 616, Bridgetown
Founded: 1991
Gen Secr: Joan French
Focus: Poli Sci 01086

Library Association of Barbados, POB 827E, Bridgetown
Founded: 1968; Members: 60
Pres: Elizabeth Campbell; Gen Secr: Alan Moss
Focus: Libraries & Bk Sci
Periodicals
Bulletin (weekly)
Update: A Newsletter (weekly) 01087

Belarus

Belarussian Academy of Sciences, Skaryny pr 66, 220072 Minsk
T: 395836; Tx: 252277
Founded: 1929; Members: 152
Pres: L.M. Sushchenya
Focus: Sci
Periodicals
Differentsialnye Uravneniya: Differential Equations (weekly)
Doklady: Report
Inzhenerno-Fizicheskii Zhurnal: Engineerring Physics Journal
Trenie i Iznos: Friction and Wear (weekly)
Vestsi: Bulletin
Zhurnal Prikladnoi Spektroskopii: Journal of Applied Spectroscopy 01088

Belorussian Agricultural Academy, Mogilev oblast, 213410 Gorkii
T: 21545
Founded: 1840
Focus: Agri 01089

Belgium

Académie Européenne d'Allergologie et d'Immunologie Clinique, 52 Blvd de la Cambre, 1050 Bruxelles
T: (02) 6472407
Focus: Immunology 01090

Académie Européenne des Ecrivains Publics (AEP) / European Academy of Public Writers, 1-13 Rue Defacqz, 1050 Bruxelles
T: (02) 5380473
Focus: Lit 01091

Académie Royale d'Archéologie de Belgique, c/o Musée Bellevue, 7 Pl des Palais, 1000 Bruxelles
Founded: 1842; Members: 100
Pres: Paul Beckhout; Gen Secr: Claire Dumortier
Focus: Archeol
Periodical
Revue Belge d'Archéologie et d'Histoire de l'Art (weekly) 01092

Académie Royale de Langue et de Littérature Françaises, 1 Rue Ducale, 1000 Bruxelles
T: (02) 5115687
Founded: 1920; Members: 40
Pres: P. Jones; Gen Secr: Georges Gion
Focus: Ling; Lit
Periodicals
Annuaire (weekly)
Bulletin 01093

Académie Royale de Médecine de Belgique (ARMB), 1 Rue Ducale, 1000 Bruxelles
T: (02) 5112471; Fax: (02) 5020712
Founded: 1841; Members: 299
Gen Secr: Prof. A. de Scoville
Focus: Med
Periodical
Monthly Bulletin (weekly) 01094

Académie Royale des Sciences, des Lettres et des Beaux-Arts de Belgique, 1 Rue Ducale, 1000 Bruxelles
T: (02) 5144064; Fax: (02) 5020424
Founded: 1772; Members: 300
Pres: Pierre Harmel; Gen Secr: P. Roberts-Jones
Focus: Arts; Lit; Sci
Periodicals
Annuaire de l'Académie (weekly)
Bulletin de la Classe des Lettres (10)
Bulletin de la Classe des Sciences (10 times annually)
Mémoires de la Classe des Beaux-Arts (weekly)
Mémoires de la Classe des Lettres (weekly)
. . . . 01095

Académie Royale des Sciences d'Outre-Mer, 1 Rue Defacqz, 1050 Bruxelles
T: (02) 5380211
Founded: 1928; Members: 250
Gen Secr: Prof. J.-J. Symoens
Focus: Poli Sci; Med; Nat Sci
Periodical
Bulletin des Sciences (weekly) 01096

Advanced Informatics in Medicine in Europe (AIM), c/o Commission of the European Communities, DG XIII, AIM Programme, 200 Rue de la Loi, 1049 Bruxelles
T: (02) 2351111; Tx: 21877
Founded: 1989
Gen Secr: Michel Carpentier
Focus: Computer & Info Sci 01097

Aire Méditerranéenne et Latinoaméricaine (AMELA), 50 Av Franklin Roosevelt, 1050 Bruxelles
Founded: 1979
Gen Secr: Isabelle Stengens
Focus: Ecology 01098

Alzheimer Europe (AE), c/o ECAS, 98 Rue du Trône, 1050 Bruxelles
T: (02) 5124444; Fax: (02) 5126673
Founded: 1990
Pres: Michael Coote
Focus: Med
Periodical
Alzheimer Europe Annual Newsletter (weekly)
. . . . 01099

American Management Association / International (AMAI), 15 Rue Caroly, 1040 Bruxelles
Founded: 1923; Members: 80000
Gen Secr: Domenico A. Fanelli
Focus: Econ
Periodical
Management Review 01100

Architects Council of Europe (ACE), 207 Av Louise, 1050 Bruxelles
T: (02) 6450905, 6450982; Fax: (02) 6464266
Founded: 1990; Members: 17
Focus: Archit
Periodical
ACE Newsletter (weekly) 01101

Archives et Bibliothèques de Belgique / Archief- en Biblioteekwezen in Belgie, 4 Blvd de l'Empereur, 1000 Bruxelles
T: (02) 5195351
Founded: 1907; Members: 410
Gen Secr: T. Verschaffel
Focus: Libraries & Bk Sci; Archives
Periodical
Archives et Bibliothèques de Belgique: Archief- en Bibliotheekwezen in Belgie (weekly) . . 01102

ASAB-VEBI, 39 Rue Joseph Cuylits, 1180 Bruxelles
T: (02) 3470700; Fax: (02) 3432194
Founded: 1964; Members: 2600
Focus: Computer & Info Sci
Periodical
Computer Organ Inform (10 times annually)
. . . . 01103

ASM International-European Council, 19 Rue de l'Orme, 1040 Bruxelles
T: (02) 7331264, 7341240; Tx: 61473; Fax: (02) 7346702
Founded: 1985
Gen Secr: Alain Galaski
Focus: Eng 01104

Association Actuarielle Internationale (AAI), c/o CGER, 48 Rue du Fossé-aux-Loups, 1000 Bruxelles
T: (02) 2136976; Tx: 61189; Fax: (02) 2136799
Founded: 1895
Gen Secr: W.P. Lenaerts
Focus: Insurance
Periodicals
ASTIN Bulletin (weekly)
Bulletin AAI (weekly)
Index AAI (weekly) 01105

Association Belge de Documentation (ABD), 110 Blvd Louis Schmidt, 1040 Bruxelles
T: (02) 7368251
Founded: 1947; Members: 310
Pres: Jean-Louis Janssens; Gen Secr: Philippe Laurent
Focus: Doc
Periodicals
ABD Flash (weekly)
Cahiers de la Documentation (weekly) . 01106

Association Belge de Droit Rural (A.B.D.R.-B.V.A.R.) / Belgische Vereniging voor Agrarisch Recht, c/o Vlaamse Landmaatschappij, 72 Av de la Toison d'Or, 1060 Bruxelles
T: (02) 5388160; Fax: (02) 5388426
Founded: 1961; Members: 455
Pres: R. de Meyer; Gen Secr: M. Heyerick
Focus: Law
Periodical
Revue de Droit Rural: Tijdschrift voor Agrarisch Recht (weekly) 01107

Association Belge de l'Eclairage (ABE), 120 Rue Saint-Denis, 1190 Bruxelles
Founded: 1957
Focus: Electric Eng 01108

Association Belge de Photographie et de Cinématographie (ABPC), 57 Rue Claessens, 1020 Bruxelles
Founded: 1874; Members: 130
Pres: J. Peeters
Focus: Cinema; Photo
Periodical
Informations (weekly) 01109

Association Belge de Radioprotection / Belgische Vereniging voor Stralingsbescherming, 14 Rue Juliette Wytsman, 1050 Bruxelles
Fax: (02) 6606322
Founded: 1963; Members: 260
Focus: Radiology
Periodical
Annales: Annalen (weekly) 01110

Association Belge des Analystes Financiers, c/o Générale de Banque, 3 Montagne du Parc, 1000 Bruxelles
T: (02) 5163564; Fax: (02) 5163763
Founded: 1959; Members: 424
Focus: Finance 01111

Association Belge d'Hygiène et de Médecine Sociale, c/o National Institute of Hygiene and Epidemiology, 14 Rue Juliette Weytsman, 1050 Bruxelles
Founded: 1938; Members: 160
Gen Secr: Dr. G. Thiers
Focus: Hygiene 01112

Association Centrale des Assistants Sociaux (ACAS), 24 Rue de l'Abbaye, 1050 Bruxelles
T: (02) 6483684
Founded: 1925; Members: 300
Focus: Soc Sci
Periodical
Bulletin d'Information Professionelle, BIP (weekly)
. . . . 01113

Association de l'Europe Occidentale pour la Psychologie Aéronautique, 35 Rue Cardinal Mercier, 1000 Bruxelles
T: (02) 5119060
Founded: 1956; Members: 100
Focus: Psych 01114

Association des Bibliothécaires et du Personnel des Bibliothèques des Ministères de Belgique, 22 Rue des Petits Carmes, 1000 Bruxelles
Focus: Libraries & Bk Sci 01115

Association des Diplomés en Histoire de l'Art et Archéologie de l'Université Catholique de Louvain, c/o Collège Erasme, 31 Pl Blaise Pascal, 1348 Louvain-la-Neuve
Founded: 1965; Members: 250
Focus: Archeol; Hist; Arts; Cultur Hist
Periodical
Revue des Archéologues et Historiens d'Art de Louvain (weekly) 01116

Association des Ecrivains Belges de Langue Française (A.E.B.), 150 Chaussée de Wavre, 1050 Bruxelles
T: (02) 5122968
Founded: 1902; Members: 500
Pres: Roger Foulon; Gen Secr: Jean Lacroix
Focus: Lit
Periodical
Nos Lettres (weekly) 01117

Association des Sociétés Scientifiques Médicales Belges, 138a Av Circulaire, 1180 Bruxelles
T: (02) 3745158
Focus: Med 01118

Association d'Instituts Européens de Conjoncture Economique (AIECE) / Association of European Conjuncture Institutes, 3 Pl Montesquieu, BP 4, 1348 Louvain-la-Neuve
T: (010) 474152

Founded: 1957; Members: 36
Gen Secr: Paul Olbrechts
Focus: Econ
Periodical
Report (weekly) 01119

Association Euro-Arabe de Juristes, 60 Rue Mignot-Delstanche, 1060 Bruxelles
Focus: Law 01120

Association Européenne Camac / European Camac Association, c/o UCL, FYAM, 2 Chemin du Cyclotron, 1348 Louvain-la-Neuve
T: (010) 4363244; Tx: 59065
Founded: 1975; Members: 354
Gen Secr: Dr. Jacques Steyaert
Focus: Eng
Periodical
ECA Newsletter (3 times annually) . . . 01121

Association Européenne contre les Maladies à Virus, c/o Institut Pasteur de Brabant, 1040 Bruxelles
Founded: 1951
Focus: Immunology 01122

Association Européenne de l'Ethnie Française (AEEF), 22 Rue Henri Vieuxtemps, 1070 Bruxelles
Founded: 1959
Focus: Ethnology
Periodical
L'Ethnie Française d'Europe (weekly) . . 01123

Association Européenne des Affaires Internationales / European International Business Association, c/o EIASM, 13 Rue d'Egmont, 1050 Bruxelles
T: (02) 5119116; Fax: (02) 5121929
Founded: 1974
Focus: Adult Educ
Periodical
Newsletter (weekly) 01124

Association Européenne des Barreaux des Courts Suprêmes, Bergstr 16, 1851 Grimbergen
Founded: 1990
Gen Secr: Georges Van Hecke
Focus: Law 01125

Association Européenne des Centres de Lutte contre les Poisons (A.E.C.A.P.) / European Association of Poisons Control Centres, 1 Rue Joseph Stallaert, 1060 Bruxelles
T: (02) 3451818, 3457473; Fax: (02) 3475860
Founded: 1966; Members: 177
Pres: Hans Persson
Focus: Toxicology
Periodicals
Bulletin de Médecine Légale et de Toxicologie Médicale
European Journal of Toxicology
Newsletter (weekly)
Reports of Meetings 01126

Association Européenne des Centres Nationaux de Productivité / European Association of National Productivitn Centres, 60 Rue de la Concorde, 1050 Bruxelles
T: (02) 5117100
Founded: 1966; Members: 17
Gen Secr: A.C. Hubert
Focus: Econ
Periodicals
EPI (weekly)
Europroductivity (weekly) 01127

Association Européenne des Institutions d'Aménagement Rural, c/o Vlaamse Landmaatschappij, 72 Av de la Toison d'Or, 1060 Bruxelles
T: (02) 5388160; Fax: (02) 5388426
Founded: 1965
Pres: F.P. Luttmer; Gen Secr: M. Dubois
Focus: Urban Plan 01128

Association Européenne F. Matthias Alexander (AEFMA), 14 Av du Prince Héritier, 1200 Bruxelles
T: (02) 7710471
Founded: 1983
Gen Secr: Colette Dethioux
Focus: Educ 01129

Association Européenne pour l'Analyse Transculturelle de Groupe, 43 Rue du Taciturne, 1040 Bruxelles
Focus: Ethnology 01130

Association Européenne pour le Droit Bancaire et Financier (AEDBF), 103 Blvd Auguste Reyers, 1040 Bruxelles
T: (02) 7364057
Founded: 1990
Pres: M. Buyle
Focus: Law 01131

Association for European Training for Employees in Technology / Association pour la Formation Européenne des Travailleurs aux Technologies, 38 Rue Fossé-aux-Loups, 1000 Bruxelles
T: (02) 2182865; Fax: (02) 2182836
Founded: 1986
Gen Secr: Antonio Miniutti
Focus: Educ; Eng 01132

Association for Teacher Education in Europe (ATEE), 60 Rue de la Concorde, 1050 Bruxelles
T: (02) 5143340; Tx: 21504; Fax: (02) 5341172
Founded: 1976
Gen Secr: Yves Beervaert
Focus: Educ
Periodicals
ATEE News
European Journal of Teacher Education (3 times annually) 01133

Association for the Epidemiological Study and Assessment of Disasters in Developing Countries, 30 Clos Chapelle-aux-Champs, 1200 Bruxelles
T: (02) 7643327, 7643823; Tx: 23722
Founded: 1977
Gen Secr: Fabienne Keymeulen
Focus: Med 01134

Association Internationale de Cybernétique (AIC) / International Association for Cybernetics, Palais des Expositions, Pl André Rijckmans, 5000 Namur
T: (081) 735209; Tx: 59101; Fax: (081) 742945
Founded: 1957; Members: 200
Pres: Jean Ramaekers; Gen Secr: Carine Aigret
Focus: Cybernetics
Periodicals
Cybernetica (weekly)
Proceedings of International Congresses on Cybernetics (every 3 years) 01135

Association Internationale de Droit Economique (AIDE), c/o CRIDE, UCL, 3 Pl Montesquieu, 1348 Louvain-la-Neuve
Founded: 1982
Focus: Law 01136

Association Internationale des Ecoles Supérieures d'Education Physique (AIESEP), c/o Institut Supérieur d'Education Physique, Université, 21 Sart-Tilman, 4000 Liège
Fax: (041) 562901
Founded: 1962
Pres: Prof. Dr. John Cheffers; Gen Secr: Prof. Dr. Maurice Pieron
Focus: Sports
Periodical
Proceedings of AIESEP (weekly) 01137

Association Internationale des Etudes Coptes, c/o Institut Orientaliste, Collège Erasme, 1348 Louvain-la-Neuve
T: (010) 473199
Gen Secr: Prof. Julien Ries
Focus: Rel & Theol 01138

Association Internationale des Juristes Démocrates (AIJD) / International Association of Democratic Lawyers, 263 Av Albert, 1180 Bruxelles
T: (02) 3451471
Founded: 1946
Focus: Law
Periodical
International Review of Contemporary Law (weekly) 01139

Association Internationale des Laboratoires Textiles Lainiers, 19 Rue du Luxembourg, 1040 Bruxelles
T: (02) 5130620; Tx: 26885; CA: Interwoollabs Brussels; Fax: (02) 5140665
Founded: 1969; Members: 106
Pres: Umberto Fracassi; Gen Secr: Dimitri Orekhoff
Focus: Textiles 01140

Association Internationale des Mathématiques et Calculateurs en Simulation (IMACS) / International Association for Mathematics and Computers in Simulation, c/o Institut Montefiore, Bât. B28 Sart-Tilman, 4000 Liège
T: (041) 563710
Founded: 1955; Members: 1250
Pres: Prof. R. Vichnevetsky; Gen Secr: Prof. J. Robert
Focus: Math; Computer & Info Sci
Periodical
Mathematics and Computer in Simulation: Applied Numerical Mathematics (weekly) 01141

Association Internationale des Métiers et Enseignements d'Art (AIMEA), 32 Av J.G. van Goolen, 1200 Bruxelles
T: (02) 7703706
Founded: 1938
Focus: Arts; Educ 01142

Association Internationale pour la Coopération et le Développement en Afrique Australe (ACODA), Schoonzichtlaan 46, 3009 Winksele
Founded: 1990
Gen Secr: Prof. Pol Marck
Focus: Develop Areas 01143

Association Internationale pour le Progrès Social (AIPS), 7 Sq Wiser, 1040 Bruxelles
T: (02) 2310662
Founded: 1925
Gen Secr: Jean-Paul Promper
Focus: Sociology
Periodical
Bulletin d'Information 01144

Association Internationale pour l'Utilisation des Langues Régionales à l'Ecole (SCOLARE), 9 Rue du Beau-Mur, 4030 Liège
T: (041) 415072
Focus: Ling 01145

Association Laïque pour l'Education et la Formation Professionnelle des Adolescents en Europe (ALEFPA-Europe), 117 Av Sainte-Anne, 1640 Bruxelles
T: (02) 3581165
Founded: 1987
Gen Secr: Liliane Zahlen
Focus: Adult Educ 01146

Association Médicale Européenne, 34 Blvd Général Jacques, 1050 Bruxelles
Founded: 1991
Focus: Med 01147

Association Mondiale des Sciences de l'Education (A.M.S.E.) / World Association for Educational Research, W.A.E.R., Henri Dunantlaan 1, 9000 Gent
T: (091) 254100
Founded: 1953; Members: 310
Pres: Prof. Dr. M. Cipro; Gen Secr: Prof. Dr. M.-L. van Herreweghe
Focus: Educ
Periodical
Communicationes (weekly) 01148

Association Nationale pour la Protection contre l'Incendie (ANPI), Parc Scientifique, Ottignies
T: (010) 418712
Founded: 1957
Focus: Safety
Periodicals
Belgisch Brandtijdschrift (weekly)
Revue Belge du Feu (weekly) 01149

Association pour la Promotion des Publications Scientifiques (A.P.P.S.), 26 Av de l'Amarante, 1020 Bruxelles
T: (02) 2682933; Fax: (02) 2682514
Founded: 1981; Members: 90
Pres: Dr. Jean C. Baudet; Gen Secr: Marianne Allard
Focus: Libraries & Bk Sci
Periodical
Ingénieur et Industrie (10 times annually) 01150

Association pour les Etudes et Recherches de Zoologie Appliquée et de Phytopathologie (A.E.R.Z.A.P.) / Vereniging voor Studie en Onderzoek over Fitopathologie en Toegepaste Zoologie, c/o Laboratoire de Phytopathologie, Université Catholique, 2 Pl Croix du Sud, 1348 Louvain-la-Neuve
Founded: 1944; Members: 155
Focus: Pathology; Zoology
Periodical
Parasitica (weekly) 01151

Association pour l'Etude et l'Evaluation Epidémiologiques des Désastres dans les Pays en Voie de Développement, 30 Clos Chapelle-aux-Champs, EPID 3034, 1200 Bruxelles
T: (02) 7643327; Tx: 23722; CA: Ecolsante
Founded: 1977; Members: 24
Gen Secr: C. Boucquey-Hachez
Focus: Develop Areas; Trop Med; Public Health 01152

Association pour l'Etude Taxonomique de la Flore d'Afrique Tropicale (AETFAT), 136 Rue de la Hulpe, 1331 Rosières
T: (02) 6532457, 6490030
Founded: 1951; Members: 704
Focus: Botany 01153

Association pour l' Introduction de la Nomenclature Biologique Nouvelle / Vereniging voor het Invoeren van Nieuwe Biologische Nomenklatuur / Association for the Introduction of New Biological Nomenclature, Hertendreef 12, 2920 Kalmthout
Founded: 1971
Pres: Herman de Vos; Gen Secr: W.M.A. de Smet
Focus: Standards; Bio 01154

Association Professionnelle Belge des Pédiatres, 20 Av de la Couronne, 1050 Bruxelles
T: (02) 6492147
Founded: 1954; Members: 365
Focus: Pediatrics 01155

Association Professionnelle des Bibliothécaires et Documentalistes (APBD), BP 31, 1070 Bruxelles
T: (02) 4289273, 5216208
Founded: 1975; Members: 350
Pres: Jean Lheureux; Gen Secr: Georges Lecocq
Focus: Libraries & Bk Sci; Doc
Periodical
Bloc-Notes (10 times annually) 01156

Association Royale des Actuaires Belges (ARAB) / Koninklijke Vereniging der Belgische Actuarissen, 48 Rue du Fossé-aux-Loups, 1000 Bruxelles
T: (02) 2184490
Founded: 1895; Members: 461
Focus: Insurance
Periodical
Bulletin (weekly) 01157

Association Royale des Demeures Historiques de Belgique, 24 Rue Vergote, 1200 Bruxelles
T: (02) 7350965
Founded: 1934; Members: 2000
Focus: Preserv Hist Monuments
Periodical
La Maison d'Hier et d'Aujourd'hui (weekly) 01158

Aviation Research and Development Institute (ARDI), 50 Rue P. Van Obberghen, 1140 Bruxelles
Founded: 1989
Gen Secr: Adrianus Groenewege
Focus: Aero 01159

Belgische Neus-, Keel- en Oorheelkundige Vereniging, c/o Dr. W. Ampe, Sint Jaanstr 13, 8000 Brugge
Members: 320
Focus: Otorhinolaryngology 01160

Belgische Vereniging voor Aardrijkskundige Studies / Société Belge d'Etudes Géographiques, W. de Croylaan 42, 3001 Heverlee
T: (016) 286611; Fax: (016) 200720
Founded: 1931; Members: 400
Focus: Geography
Periodical
Tijdschrift: Bulletin (weekly) 01161

Belgische Vereniging voor Toegepaste Linguistiek (BVTL) / Association Belge de Linguistique Appliquée; ABLA, c/o Instituut voor Taalonderwijs, Vrije Universiteit Brussel, Pleinlaan 2, 1000 Bruxelles
Founded: 1974; Members: 185
Focus: Ling
Periodical
ABLA-Papers (weekly) 01162

Belgische Vereniging voor Tropische Geneeskunde / Société Belge de Médecine Tropicale, Nationalestr 155, 2000 Antwerpen
Founded: 1920; Members: 540
Gen Secr: Prof. L. Eyckmans
Focus: Trop Med
Periodical
Annales (weekly) 01163

Belgische Wetenschappelijke Vereniging voor Neurochirurgie, Bisschopsdreef 53, 8000 Brugge
Founded: 1977; Members: 50
Focus: Surgery; Neurology 01164

Belgisch Instituut voor Arbeidsverhoudingen, c/o Collegium Falconis, Tiensestr 41, 3000 Leuven
Focus: Law 01165

Benelux Association of Energy Economists (BAEE), 39 Rue de la Régence, 1000 Bruxelles
T: (02) 5193823; Tx: 61540; Fax: (02) 5134206
Founded: 1984
Gen Secr: Jan Daneels
Focus: Energy 01166

Benelux Phlebology Society, 180 Av du Roi Albert, 1080 Bruxelles
Founded: 1961
Gen Secr: Dr. I. Staelens
Focus: Med 01167

Biotechnology Research for Innovation, Development and Growth in Europe (BRIDGE), 200 Rue de la Loi, 1049 Bruxelles
T: (02) 2354044; Tx: 21877; Fax: (02) 2350145
Founded: 1989
Gen Secr: D. de Nettancourt
Focus: Eng 01168

Bisnuth Institute, Borgtstr 301, 1850 Grimbergen
T: (02) 2524747; Tx: 27000; Fax: (02) 2522775
Founded: 1972
Focus: Mineralogy 01169

Board of Governors of the European Schools, 200 Rue de la Loi, 1049 Bruxelles
T: (02) 2351111; Tx: 21877; Fax: (02) 2301930
Founded: 1957; Members: 12
Gen Secr: J. Olsen
Focus: Educ
Periodical
Pedagogical Bulletin (weekly) 01170

Branche Belge de la Société de Chimie Industrielle, 3 Rue Ravenstein, 1000 Bruxelles
Fax: (02) 5145735
Pres: Lucien Clerbois
Focus: Chem 01171

Brite-EuRam II, c/o Commission of the European Communities, DG XII-C, 75 Rue Montoyer, 1040 Bruxelles
T: (02) 2361232; Tx: 21877; Fax: (02) 2358046
Founded: 1990
Gen Secr: H. Pero
Focus: Eng 01172

Bureau International d'Audiophonologie / International Office for Audiophonology, 29 Rue J. B. Vandercammen, 1160 Bruxelles
T: (02) 6731036
Founded: 1964; Members: 269
Gen Secr: Albert Roussel
Focus: Med; Ling
Periodical
Rapport Annuel: Recommandations (weekly) 01173

Bureau International de la Récupération (BIR), 24 Rue du Lombard, 1000 Bruxelles
T: (02) 5142180; Tx: 61965; Fax: (02) 5141226
Founded: 1948
Gen Secr: Francis Veys
Focus: Ecology 01174

Bureau International Technique de l'ABS (BIT), 4 Av E. Van Nieuwenhuyse, 1160 Bruxelles
T: (02) 6767211; Tx: 62444; Fax: (02) 6767301
Founded: 1974
Focus: Eng 01175

Bureau International Technique des Polyesters Insaturés, 4 Av E. Van Nieuwenhuyse, 1160 Bruxelles
T: (02) 6767211; Tx: 62444; Fax: (02) 6767301
Founded: 1969
Focus: Eng 01176

Bureau International Technique du Spathfluor, 4 Av E. Van Nieuwenhuyse, 1160 Bruxelles
T: (02) 6767211; Tx: 62444; Fax: (02) 6767300
Founded: 1974
Gen Secr: B. Jeensen
Focus: Eng 01177

Cadmium Pigments Association, c/o CEFIC, 4 Av E. Van Nieuwenhuyse, 1160 Bruxelles
T: (02) 6767211; Tx: 62444; Fax: (02) 6767301
Gen Secr: Dr. M. Papez
Focus: Metallurgy 01178

Catholic International Education Office / Internationales Katholisches Büro für Unterricht und Erziehung, 60 Rue des Eburons, 1040 Bruxelles
T: (02) 2307252
Founded: 1952
Gen Secr: Paulus Adams
Focus: Educ 01179

Catholic Office for Information on European Problems, 13 Av Père Damien, 1150 Bruxelles
T: (02) 7713678; Fax: (02) 7707654
Focus: Soc Sci 01180

CELEX, c/o Eurobases, EC Commission, 200 Rue de la Loi, 1049 Bruxelles
T: (02) 2350001; Tx: 21877; Fax: (02) 2360624
Founded: 1970; Members: 12
Focus: Law 01181

CEN/CENELEC, 38 Rue de Stassart, 1050 Bruxelles
T: (02) 5196852; Fax: (02) 5196819
Gen Secr: P.S. Sanson
Focus: Standards
Periodical
CEN/CENELEC Review 01182

Central Bureau for Nuclear Measurements (CBNM), Steenweg naar Retie, 2440 Geel
T: (014) 571211; Tx: 33589; Fax: (014) 584273
Founded: 1960
Gen Secr: Dr. Herbert H. Hansen
Focus: Nucl Res 01183

Centre Belge d'Etude de la Corrosion (CEBELCOR) / Belgian Center for Corrosion Study, 2 Av Paul Héger-Grille, 1050 Bruxelles
T: (02) 6493516; Fax: (02) 6493107
Founded: 1951
Focus: Metallurgy
Periodical
Rapports Techniques CEBELCOR (weekly) 01184

Centre Belge pour la Gestion de la Qualité, 21 Rue des Drapiers, 1050 Bruxelles
Focus: Materials Sci 01185

Centre Européen des Silicones (CES), 4 Av E. Van Nieuwenhuyse, 1160 Bruxelles
T: (02) 6767211; Tx: 62444; Fax: (02) 6767301
Founded: 1968
Gen Secr: René M. Van Sloten
Focus: Chem 01186

Centre Européen pour l'Etude de l'Argumentation (CEAA), c/o Institut de Philosophie, 143 Av A. Buyl, 1050 Bruxelles
T: (02) 6422628
Gen Secr: Patricia Boeyen
Focus: Philos 01187

Centre for Research on European Women (CREW), 21 Rue de la Tourelle, 1040 Bruxelles
T: (02) 2305158; Fax: (02) 2306230
Founded: 1980
Focus: Soc Sci
Periodical
CREW Report (weekly) 01188

Centre for Research on the Epidemiology of Disasters (CRED), c/o Ecole de Santé Publique, 30 Clos Chapelle-aux-Champs, 1200 Bruxelles
T: (02) 7643327; Tx: 23722; Fax: (02) 7643328
Founded: 1973
Gen Secr: Dr. Michel F. Lechat
Focus: Med 01189

Centre International de Documentation Marguerite Yourcenar (CIDMY), c/o Archives de Bruxelles, 65 Rue des Tanneurs, 1000 Bruxelles
T: (02) 5027475; Fax: (02) 5027475
Founded: 1989
Gen Secr: Michèle Goslar
Focus: Sci 01190

Centre International de Formation et de Recherche en Population et Développement en Association avec les Nations Unies (CIDEP), 1 Pl Montesquieu, 1348, Louvain-la-Neuve
T: (010) 474106; Tx: 62966
Founded: 1984
Focus: Soc Sci; Develop Areas 01191

Centre International de Liaison des Ecoles de Cinéma et de Télévision (CILECT), 8 Rue Thérésienne, 1000 Bruxelles
T: (02) 5119839; Fax: (02) 5110279
Founded: 1955
Gen Secr: Henry Verhasselt
Focus: Educ; Cinema
Periodical
CILECT Newsletter 01192

Centre International de Recherches et d'Information sur l'Economie Publique, Sociale et Coopérative (CIRIEC), c/o Université de Liège, Bar. B 31, 4000 Liège
T: (041) 521032
Founded: 1948
Pres: Dr. Anton Rauter; Gen Secr: Prof. Dr. Guy Quaden
Focus: Econ
Periodical
Annals of Public and Cooperative Economics (weekly) 01193

Centre International de Recherches Glyptographiques (CIRG), 13 Rue Mathias de la Bruyère, 1440 Braine-le-Château
T: (02) 3660529
Founded: 1981
Gen Secr: J.L. Van Belle
Focus: Hist 01194

Centre International des Langues, Littératures et Traditions d'Afrique au Service du Développement (CILTADE), 30 Av des Clos, 1348 Louvain-la-Neuve
T: (010) 450665
Founded: 1986
Gen Secr: Dr. C. F.-N. Madiya
Focus: Ling; Lit; Hist 01195

Centre International d'Etude de la Peinture Médiévale des Bassins de l'Exant et de la Meuse / International Centre for the Study of Medieval Painting in the Schelde and the Meuse Valleys, 1 Parc du Cinquantenaire, 1040 Bruxelles
Founded: 1950; Members: 15
Pres: H. Pauwels
Focus: Fine Arts
Periodicals
Contributions à l'Etude des Primitifs Flamands (weekly)
Corpus de la Peinture des Anciens Pays-Bas Méridionaux au 15e Siècle
Répertoire des Peintures Flamandes du 15e Siècle 01196

Centre International d'Etude de Tantale et de Niobium, 40 Rue Washington, 1050 Bruxelles
T: (02) 6495158; Tx: 65080; CA: Tictan Brussels; Fax: (02) 6493269
Founded: 1974
Focus: Eng
Periodical
Bulletin (weekly) 01197

Centre International d'Etudes de la Formation Religieuse / International Centre for Studies in Religious Education, 184-186 Rue Washington, 1050 Bruxelles
T: (02) 3441882; Fax: (02) 3465745
Founded: 1935; Members: 7
Focus: Educ; Rel & Theol
Periodicals
Lumen Vitae: International Review of Religious Education (weekly)
Lumen Vitae: Revue Internationale de la Formation Religieuse (weekly) 01198

Centre International d'Etudes, de Recherche et d'Action pour le Développemnt (CINTERAD), 186 Blvd Louis Schmidt, 1050 Bruxelles
Founded: Gbossa, Lambert; Members: 13
Focus: Develop Areas 01199

Centre International d'Etudes du Lindane (CIEL), 4 Av E. Van Nieuwenhuyse, 1160 Bruxelles
T: (02) 6767241; Tx: 62444; Fax: (02) 6767301
Founded: 1969
Gen Secr: Rodolphe Schmitt
Focus: Chem 01200

Centre National de Documentation Scientifique et Technique (CNDST) / Nationaal Centrum voor Wetenschappelijke en Technische Documentatie; NCWTD / National Science Information Center, 4 Blvd de l'Empereur, 1000 Bruxelles
T: (02) 5195640; Tx: 21157; CA: BIBREG.Bru; Fax: (02) 5195679
Founded: 1964; Members: 10
Pres: G. Renson; Gen Secr: I. Clemens
Focus: Doc 01201

Centre National de Documentation Scientifique et Technique (CNDST), c/o Bibliothèque Royale Albert Ier, 4 Blvd de l'Empereur, 1000 Bruxelles
T: (02) 5195640; Fax: (02) 5195679
Founded: 1981
Focus: Doc
Periodicals
Information Resources (weekly)
Inventaire des Bibliothèques et Centres de Documentation Belges (weekly) 01202

Centre National de Recherches de Logique, 32 Rue de la Pêcherie, 1180 Bruxelles
Founded: 1951
Focus: Logic
Periodical
Logique et Analyse (weekly) 01203

Centre of Promotion of Catholic Education in Europe / Centre de Promotion de l'Enseignement Catholique en Europe, 42 Rue de l'Industrie, 1040 Bruxelles
T: (02) 5117501; Fax: (02) 5145989
Founded: 1984
Gen Secr: R. de Zutter
Focus: Educ 01204

Centre pour l'Etude des Problèmes du Monde Musulman Contemporain / Study Centre for Problems of the Contemporary Muslim World, 44 Av Jeanne, 1050 Bruxelles
T: (02) 6423359
Founded: 1958
Gen Secr: Destrée. A.
Focus: Rel & Theol; Anthro
Periodical
Le Monde Muselman Contemporain-Initiations 01205

Cercle Archéologique de Mons (CAM), c/o Maison Losseau, 37 Rue de Nimy, 7000 Mons
T: (065) 331765
Founded: 1856; Members: 350
Focus: Archeol 01206

Cercle Benelux d'Histoire de la Pharmacie, Romeinselaan 6, 8500 Kortrijk
T: (056) 225499
Founded: 1950
Pres: Bernard Mattelaer
Focus: Pharmacol; Hist 01207

Cercle d'Etudes Numismatiques (CEN), 4 Blvd de l'Empereur, 1000 Bruxelles
T: (02) 5136180
Founded: 1964; Members: 465
Focus: Numismatics
Periodical
Traveaux 01208

Chambre Belge des Pédicures Médicaux (CBPM), 5 Av des Cormorans, 1150 Bruxelles
T: (02) 7713864
Founded: 1939; Members: 200
Focus: Orthopedics 01209

Chambre Belge des Traducteurs, Interprètes et Philologues (CBTIF), c/o M. Lepeer, 110 Av de Heyn, 1090 Bruxelles
Founded: 1955; Members: 250
Focus: Ling; Lit 01210

Chambre des Architectes de Belgique (CAB), 36 Rue des Echevins, 1050 Bruxelles
T: (02) 6472671
Founded: 1963
Focus: Archit 01211

Charles Darwin Foundation for the Galapagos Isles, 1 Rue Ducale, 1000 Bruxelles
Gen Secr: Alfredo Carrasco
Focus: Bio
Periodical
Noticias de Galápagos 01212

China-Europe Management Institute (CEMI), c/o EFMD, 40 Rue Washington, 1050 Bruxelles
T: (02) 6480385; Tx: 222362; Fax: (02) 8418412
Founded: 1989
Gen Secr: Jean-Claude Cuzzi
Focus: Business Admin 01213

Coimbra Group, c/o UCL, 15 Av de l'Assomption, 1200 Bruxelles
T: (02) 7642209; Fax: (02) 7642299
Founded: 1985; Members: 30
Gen Secr: Simon-Pierre Nothomb
Focus: Educ 01214

Collège Européen des Technologies, Rue Netzert, 6700 Arlorn
Focus: Eng 01215

Collège Européen d'Hygiène et de Médecine Naturelles (CEHMN), 42 Rue des Lilas, 7850 Enghien
Gen Secr: Frans Meersseman
Focus: Med; Hygiene 01216

College of Europe, Dyver 10-11, 8000 Brugge
T: (050) 335334; Tx: 81457; Fax: (050) 343158
Founded: 1949
Pres: Dr. Werner Ungerer
Focus: Educ
Periodical
Information (weekly) 01217

Comité Belge d'Histoire des Sciences, c/o Bibliothèque Royale Albert Ier, 4 Blvd de l'Empereur, 1000 Bruxelles
T: (02) 5195740; Fax: (02) 5195533
Founded: 1933; Members: 40
Gen Secr: Hossam Elkhadem
Focus: Hist; Sci 01218

Comité d'Associations Européennes de Médecins Catholiques (CAEMC), 38 Rue des Deux-Eglises, 1040 Bruxelles
Gen Secr: Dr. R. Verly
Focus: Med 01219

Comité de Liaison des Architectes de l'Europe Unie, 158 Rue de Livourne, 1050 Bruxelles
Founded: 1959
Gen Secr: Dan Craet
Focus: Archit 01220

Comité des Cancérologues de la Communauté Européenne, 75 Av Hippocrate, BP 7550, 1200 Bruxelles
T: (02) 7623292; Tx: 23722; Fax: (02) 7645322
Gen Secr: Prof. Christian de Duve
Focus: Cell Biol & Cancer Res 01221

Comité d'Etude de la Corrosion et de la Protection des Canalisations (CEOCOR), c/o Institut de Chimie et Métallurgie, 2 Rue Armand Stevart, 4000 Liège
T: (041) 520180; Tx: 41397
Founded: 1963; Members: 120
Focus: Materials Sci; Metallurgy 01222

Comité d'Etude des Producteurs de Charbon d'Europe Occidentale (CEPCEO), 168 Av de Tervueren, 1150 Bruxelles
T: (02) 7719974
Founded: 1953
Gen Secr: Rogier Jean van der Stichelen
Focus: Mining
Periodical
CEPCEO Report 01223

Comité d'Etudes Economiques de l'Industrie du Gaz (COMETEC-GAZ), 4 Av Palmerston, 1040 Bruxelles
T: (02) 7348188
Founded: 1954
Focus: Econ 01224

Comité Economique et Social des Communautés Européennes (CES) / Economic and Social Committee of the European Communities, 2 Rue Ravenstein, 1000 Bruxelles
T: (02) 5199011; Tx: 25983; Fax: (02) 5134893
Members: 189
Focus: Econ; Sociology
Periodical
CES Bulletin (weekly) 01225

Comité Européen de Normalisation Electrotechnique (CENELEC), 2 Rue de Bréderode, 1000 Bruxelles
T: (02) 5196811; Tx: 26257; Fax: (02) 5196819
Founded: 1973; Members: 14
Gen Secr: H.K. Tronnier
Focus: Standards; Electric Eng 01226

Comité Européen de Recherche et de Développement (CERD), 200 Rue de la Loi, 1040 Bruxelles
T: (02) 2351111 ext 3774
Founded: 1973
Gen Secr: U. Colombo
Focus: Econ 01227

Comité International de Médecine Militaire / International Committee of Military Medicine, c/o Hôpital Militaire, 79 Rue Saint-Laurent, 4000 Liège
T: (041) 222183; Fax: (041) 222150
Founded: 1921; Members: 33
Gen Secr: Dr. M. Colls
Focus: Med; Pharmacol; Military Sci
Periodical
Revue Internationale des Services de Santé des Forces Armées: International Review of the Armed Forces Medical Services (weekly) 01228

Comité International de Recherche et d'Etude de Facteurs de l'Ambiance (CIFA), 27 Drève des Wégélias, 1170 Bruxelles
T: (02) 6490030
Focus: Ecology 01229

Comité International de Standardisation en Biologie Humaine (CISBH), c/o Ecole de Santé Publique, 100 Rue Belliard, 1040 Bruxelles
T: (02) 7360063
Founded: 1968; Members: 200
Focus: Standards; Bio 01230

Comité International d'Etude des Géants Processionnels, 27 Drève des Wégélias, 1170 Bruxelles
Founded: 1954; Members: 8
Focus: Hist 01231

Comité National Belge de l'Organisation Scientifique (CNBOS), 93 Rue de Stassart, 1050 Bruxelles
T: (02) 5112844
Founded: 1926
Focus: Metallurgy 01232

Comité National pour l'Etude et la Prévention de l'Alcoolisme et des Autres Toxicomanies (CNA), 78 Rue du Rempart-des-Moines, 1000 Bruxelles
T: (02) 5111188
Founded: 1949; Members: 21
Focus: Sociology 01233

Commission Belge de Bibliographie, 4 Blvd de l'Empereur, 1000 Bruxelles
T: (02) 510465
Founded: 1951; Members: 68
Pres: C.F. Verbeke
Focus: Libraries & Bk Sci
Periodicals
Bibliographica Belgica
Bulletin (weekly) 01234

Commission Mixte sur les Aspects Internationaux de l'Arriération Mentale, 12 Rue Forestière, 1050 Bruxelles
Founded: 1970
Focus: Psychiatry 01235

Commission of the European Communities Liaison Committee of Historians / Groupe de Liaison des Historiens près la Commission des Communautés Européennes, c/o Unité d'Histoire Contemporaine, Collège Erasme, UCL, 1 Pl Blaise Pascal, 1348 Louvain-la-Neuve
Founded: 1982
Gen Secr: Michel Dumoulin
Focus: Hist
Periodical
Lettre d'Information des Historiens de l'Europe Contemporaine 01236

Commission Permanente des Groupements Professionnels d'Assistants Sociaux, c/o Comité Belge de Service Social, 3 Rue de la Vièrge Noire, 1000 Bruxelles
T: (02) 5118481
Focus: Sociology 01237

Committee on Higher Education in the European Community (CHEEC), c/o HEURAS, 60 Rue de la Concorde, 1050 Bruxelles
T: (02) 5143340; Tx: 21004; Fax: (02) 5141172
Focus: Educ 01238

Committee on the Challenges of Modern Society (CCMS), c/o North Atlantic Treaty Organization, 1110 Bruxelles
Founded: 1969; Members: 15
Focus: Ecology; Sci 01239

Common Office for European Training / Office Commun de Formation Européenne, 43-45 Rue Notre-Dame, 7000 Mons
T: (065) 364259; Fax: (065) 317443
Founded: 1986
Gen Secr: Ivanka Zonta
Focus: Educ 01240

Community Network for European Education and Training (COMNET), 6 Quai Banning, 4000 Liège
T: (041) 528085; Tx: 41488; Fax: (041) 534097
Founded: 1988; Members: 90
Gen Secr: Prof. Victor de Kosinsky
Focus: Educ
Periodical
COMNET Newsletter 01241

Comparative Education Society in Europe (CESE), 60 Rue de la Concorde, 1050 Bruxelles
T: (02) 5143340; Fax: (02) 5141172
Founded: 1961
Pres: Prof. Henk Van Daele
Focus: Educ
Periodical
CESE Newsletter 01242

Confédération d'Associations d'Ecoles Indépendantes de la Communauté Européenne (CADEICA), 15 Av Montana, 1180 Bruxelles
Founded: 1989
Gen Secr: C. Moraiti Kartali
Focus: Educ 01243

Confédération Européenne des Syndicats Nationaux et Associations Professionnelles de Pédiatres, 20 Av de la Couronne, 1050 Bruxelles
T: (02) 6492147
Founded: 1959
Focus: Pediatrics 01244

Confederation of European Specialists in Paediatrics (CESP), 20 Av de la Couronne, 1050 Bruxelles
T: (02) 6492147; CA: 616052; Fax: (02) 6492690
Founded: 1959
Pres: Prof. Georges Van den Berghe . 01245

Conférence des Recteurs des Universités Belges, 5 Rue d'Egmont, 1050 Bruxelles
T: (02) 5125815; Tx: 25498
Founded: 1973
Focus: Sci 01246

Conférence Diplomatique de Droit Maritime International, c/o Ministère des Affaires Etrangères et du Commerce Extérieur, 2 Rue Quatre Bras, 1000 Bruxelles
T: (02) 5136240
Founded: 1905
Focus: Law 01247

Conférence Internationale Permanente de Directeurs d'Instituts Universitaires pour la Formation de Traducteurs et d'Interprètes (CIUTI), Schildersstr 41, 2000 Antwerpen
T: (03) 2389832
Founded: 1961; Members: 17
Gen Secr: Prof. D. Godfrind
Focus: Educ; Ling 01248

Conferentie voor Regionale Ontwikkeling in Noord-West-Europa / Conférence des Regions de l'Europe du Nord-Ouest / Conference of Regions of North-West-Europe, Postbus 107, 8000 Brugge 1
Founded: 1955; Members: 5
Pres: Sir Jack Stewart Clark; Gen Secr: Prof. Dr. I.B.F. Kormoss
Focus: Urban Plan
Periodical
Proceedings of Seminars Organized (weekly) 01249

Conseil Consultatif Economique et Social de l'Union Economique Benelux (CCES), 21 Av de la Joyeuse Entrée, 1040 Bruxelles
T: (02) 2338904
Founded: 1960
Gen Secr: L. Denayer
Focus: Econ; Sociology
Periodical
Report of Council Meetings 01250

Conseil des Recteurs des Institutions Universitaires Francophones de Belgique, 5 Rue d'Egmont, 1050 Bruxelles
T: (02) 5049300; Tx: 25498; Fax: (02) 5140006
Focus: Educ 01251

Conseil Interuniversitaire de la Communauté Française, 5 Rue d'Egmont, 1050 Bruxelles
T: (02) 5049211; Tx: 25498; Fax: (02) 5049292
Founded: 1980
Focus: Adult Educ 01252

Conseil National de l'Ordre des Médecins, 32 Pl de Jamblinne de Meux, 1040 Bruxelles
T: (02) 7368291
Founded: 1947
Focus: Med
Periodical
Bulletin du Conseil National: Tijdschrift Nationale Raad (weekly) 01253

Conseil Supérieur de Statistique (C.S.S.), 44 Rue de Louvain, 1000 Bruxelles
T: (02) 5486211
Founded: 1841; Members: 36
Pres: M. Despontin; Gen Secr: H. Schepers
Focus: Stats 01254

Consortium of Institutions of Higher Education in Health and Rehabilitation in Europe (COHEHRE), c/o HIBP Sint-Vincentius, Sint-Lievenspoortstr 143, 9000 Gent
T: (091) 240209; Fax: (091) 239902
Founded: 1990
Gen Secr: Rudy Van Renterghem
Focus: Educ; Public Health; Rehabil
Periodical
COHEHRE Newsletter (weekly) 01255

Cooperation for Open Systems Interconnection Networking in Europe (COSINE), c/o CEC, 200 Rue de la Loi, 1049 Bruxelles
T: (02) 2362075; Fax: (02) 2353821
Founded: 1985; Members: 19
Focus: Computer & Info Sci 01256

Coordination et Promotion de l'Enseignement de la Réligion en Europe (COOPERE), c/o COMECE, 13 Av Père Damien, 1150 Bruxelles
T: (02) 7713678; Fax: (02) 7707654
Founded: 1989
Focus: Rel & Theol; Educ 01257

C.R.I.F., Technologiepark 9, 9052 Zwijnaarde
T: (09) 2645689; Fax: (09) 2645848
Founded: 1929; Members: 200
Pres: Dr. F. Defrancq
Focus: Metallurgy 01258

Dedicated Road Infrastructure for Vehicle Safety in Europe (DRIVE), 200 Rue de la Loi, 1049 Bruxelles
T: (02) 2363449; Tx: 22045; Fax: (02) 2362391
Founded: 1988; Members: 17
Gen Secr: F. Karamitsos
Focus: Transport 01259

Education Information Network in the European Community (EURYDICE), c/o Unité Européenne, 17 Rue Archimède, 1040 Bruxelles
T: (02) 2300382; Tx: 65398; Fax: (02) 2306562
Founded: 1976
Gen Secr: José Antonio Fernandez
Focus: Educ
Periodical
EURYDICE Info (2-3 times annually) . . 01260

Euratom Scientific and Technical Committee, c/o CEC, 200 Rue de la Loi, 1049 Bruxelles
T: (02) 2354494; Tx: 21877; Fax: (02) 2350150
Gen Secr: Gianpiero Alessi
Focus: Nucl Res 01261

Eureka Organization, 19h Av des Arts, 1040 Bruxelles
T: (02) 2170030; Fax: (02) 2187906
Founded: 1985; Members: 19
Gen Secr: Prof. Olaf Meyer
Focus: Eng 01262

Euro-China Research Association in Management (ECRAM), c/o EFMD, 40 Rue Washington, 1050 Bruxelles
T: (02) 6480385; Tx: 65080; Fax: (02) 6460768
Focus: Business Admin 01263

Euro-China Research Centre for Business Cooperation (ECRCBC), 3 Av d'Uccle, 1190 Bruxelles
T: (02) 3445907; Fax: (02) 3473096
Founded: 1990
Gen Secr: Hong Zhi
Focus: Business Admin 01264

Euro Chlor, c/o CEFIC, 4 Av E. Van Nieuwenhuyse, 1160 Bruxelles
T: (02) 6767211; Tx: 62444; Fax: (02) 6767300
Gen Secr: Dr. A. Seys
Focus: Chem 01265

Europa Esperanto-Centro (EEC), Pr. Leopoldstr 51, 8310 Brugge
T: (030) 354935
Founded: 1977
Gen Secr: G. Maertens
Focus: Ling 01266

European Academy for Film and Television / Académie Européenne du Cinéma et de la Télévision, 69 Rue Verte, 1210 Bruxelles
T: (02) 2186607; Fax: (02) 2175572
Founded: 1990; Members: 18
Gen Secr: Dimitri Balachoff
Focus: Mass Media; Cinema 01267

European Accounting Association (EAA) / Association Européenne de Comptabilité, c/o EIASM, 13 Rue d'Egmont, 1050 Bruxelles
T: (02) 5119111; Fax: (02) 5121929
Founded: 1977
Focus: Business Admin
Periodical
EAA Newsletter 01268

European Alliance of Safe Meat, 436 Blvd de Lambermont, 1030 Bruxelles
Founded: 1991
Focus: Food 01269

European Anthropological Association / Association Européenne d'Anthropologie, c/o Laboratoire Anthropogenetika, Pleinlaan 2, 1080 Bruxelles
Founded: 1975
Gen Secr: Prof. Charles Susanne
Focus: Anthro
Periodicals
International Journal of Anthropology
Newsletter 01270

European Association against Fibre Pollution (EAFP) / Association Européenne pour la Réduction de la Pollution due aux Fibres, 31 Rue Montoyer, 1040 Bruxelles
T: (02) 5136085; Fax: (02) 5143386
Founded: 1988
Pres: Mark Dubrulle
Focus: Ecology 01271

European Association against Pigmentary Dystrophy and its Syndromes / Association Européenne de Lutte contre la Rétinite Pigmentaire et ses Syndromes, 3 Blvd de la Cambre, 1050 Bruxelles
T: (02) 6402504; Fax: (02) 6402504
Founded: 1985
Focus: Derm 01272

European Association for Audiovisual Media Education / Association Européenne pour l'Education aux Médias Audiovisuels, 162 Rue du Midi, 1000 Bruxelles
T: (02) 5020643
Founded: 1989
Pres: Didier Schretter
Focus: Educ
Periodical
AEEMA Newsletter (weekly) 01273

European Association for Burgundy Studies (EABS) / Association Européenne d'Etudes Bourguignonnes, c/o Facultés Universitaires Saint-Louis, 43 Blvd du Jardin Botanique, 1000 Bruxelles
T: (02) 2117811; Fax: (02) 2117997
Gen Secr: Prof. Jean-Marie Cauchies
Focus: Cultur Hist 01274

European Association for Country Planning Institutions / Association Européenne des Institutions d'Aménagement Rural, 72 Av de la Toison d'Or, 1060 Bruxelles
T: (02) 5388160
Founded: 1965; Members: 16
Gen Secr: M. Dubois
Focus: Urban Plan 01275

European Association for Health Information and Libraries (EAHIL) / Association Européenne pour l'Information et les Bibliothèques de Santé, 60 Rue de la Concorde, 1050 Bruxelles
T: (02) 5118063; Fax: (02) 5141172
Founded: 1987
Pres: Monique Cleland
Focus: Libraries & Bk Sci; Computer & Info Sci
Periodical
Newsletter to European Health Librarians 01276

European Association for Information on Local Development / Association Européenne pour l'Information sur le Développement Local, 34 Rue Breydel, 1040 Bruxelles
T: (02) 2305234; Fax: (02) 2303482
Gen Secr: Viviane Schreiber
Focus: Develop Areas 01277

European Association for Institutions of Higher Education (EURASHE) / Association Européenne des Institutions d'Enseignement Supérieur, c/o HEURAS, 51 Rue de la Concorde, 1050 Bruxelles
T: (02) 5121734; Tx: 21004; Fax: (02) 5123265
Founded: 1990
Focus: Educ 01278

European Association for Research in Industrial Economics (EARIE) / Association Européenne de la Recherche en Economie Industrielle, c/o EIASM, 13 Rue d'Egmont, 1050 Bruxelles
T: (02) 5119116; Fax: (02) 5121929
Founded: 1974
Focus: Business Admin 01279

European Association for Research into Adapted Physical Activity / Association Européenne de Recherche en Avtivité Adaptée, c/o Institut Supérieur d'Education Physique, Université Libre de Bruxelles, 28 Av Paul Héger, 1050 Bruxelles
Founded: 1987
Gen Secr: Dr. Jean-Claude de Potter
Focus: Sports 01280

European Association for Textile Polyolefins (EATP) / Association Européenne des Polyoléfines Textiles, 4 Av E. Van Nieuwenhuyse, 1160 Bruxelles
T: (02) 6767472; Tx: 62444; Fax: (02) 6767474
Founded: 1971
Gen Secr: J.P. Peckstadt
Focus: Textiles 01281

European Association for the Promotion of Poetry, Blijde Inkomststr 9, 3000 Leuven
T: (016) 235351; Fax: (016) 234949
Founded: 1979
Focus: Lit
Periodical
PI (European Poetry Quarterly) (weekly) . 01282

European Association for the Promotion of the Hand Hygiene / Association Européenne pour la Promotion de l'Hygiène des Mains, 46 Rue Lieutenant Liedel, 1070 Bruxelles
T: (02) 5212099
Founded: 1983; Members: 24
Gen Secr: Annie Lalé
Focus: Hygiene 01283

European Association for the Study of Safety Problems in the Production and Use of Propellant Powders (EASSP) / Association Européenne pour l'Etude des Problèmes de Sécurité dans la Fabrication et l'Emploi des Poudres Propulsives, 4 Av E. Van Nieuwenhuyse, 1160 Bruxelles
T: (02) 6767211
Founded: 1977
Focus: Eng 01284

European Association for the Teaching of Legal Theory (EATLT) / Association Européenne pour l'Enseignment de la Théorie du Droit, 43 Blvd du Jardin Botanique, 1000 Bruxelles
Focus: Law; Educ 01285

European Association of Agricultural Economists (EAAE) / Association Européenne d'Economistes Agricoles, 82 Rue de Trèves, 1040 Bruxelles
Founded: 1975; Members: 1000
Gen Secr: Prof. Laurent Martens
Focus: Agri
Periodical
EAAE Newsletter 01286

European Association of Architectural Education / Association Européenne pour l'Enseignement de l'Architecture, c/o Unité Architecture, Université Catholique de Louvain, 1 Pl du Levant, 1348 Louvain-la-Neuve
T: (010) 478078
Founded: 1975; Members: 91
Gen Secr: Nicole Mouzon
Focus: Educ; Archit
Periodical
EAAE News Sheet (2-3 times annually) . 01287

European Association of Centers of Medical Ethics / Association Européenne des Centres d'Ethique Médicale, 5 Promenade de l'Alma, BP 4534, 1200 Bruxelles
T: (02) 7620297
Founded: 1985
Gen Secr: Jean-F. Malherbe
Focus: Philos 01288

European Association of Classification Societies (EurACS), c/o Germanischer Lloyd, Victor Govaerslaan 16, 2060 Antwerpen
T: (031) 6460416; Tx: 34447; Fax: (031) 6460064
Founded: 1979
Gen Secr: G. Heinacher
Focus: Standards
Periodical
EurACs 01289

European Association of Education and Research in Public Relations (CERP-EDUCATION), c/o Hibo, Sint Pietersnieuwstr 160, 9000 Gent
T: (091) 250092; Fax: (091) 330949
Members: 89
Pres: J. Willems
Focus: Educ 01290

European Association of Experimental Social Psychology / Association Européenne de Psychologie Sociale Expérimentale, c/o Faculté de Psychologie, Université de Louvain, 20 Voie du Roman Pays, 1348 Louvain-la-Neuve
T: (010) 474036; Fax: (010) 472999
Founded: 1964; Members: 348
Gen Secr: Dr. Bernard Rimé
Focus: Psych
Periodical
European Journal of Social Psychology . 01291

European Association of Hearing Aid Audiologists / Association Européenne des Audioprothésistes, 16 Av Général Jourdan, 6220 Fleurus
Founded: 1968
Gen Secr: Serge Wittemans
Focus: Otorhinolaryngology 01292

European Association of Historical Associations, c/o Faculté de Philosophie et Lettres, Université Catholique de Louvain, 1348 Louvain-la-Neuve
Gen Secr: Claude Bruneel
Focus: Hist 01293

European Association of Metals / Association Européenne des Métaux, 47 Rue Montoyer, 1040 Bruxelles
T: (02) 5138634; Tx: 22077; Fax: (02) 5117553
Founded: 1957; Members: 21
Gen Secr: Jacques Spaas
Focus: Metallurgy 01294

European Association of Podologists / Association Européenne des Podologues, 5 Av des Cormorans, 1150 Bruxelles
T: (02) 7713864
Gen Secr: G. Demoulin-Delhaye
Focus: Med 01295

European Association of Radiology (EAR), c/o Dept of Radiology, University Hospital, Herestr 49, 3000 Leuven
T: (016) 213771; Fax: (016) 213769
Founded: 1962; Members: 100
Gen Secr: Prof. Dr. Albert Baert
Focus: Radiology
Periodical
Newsletter 01296

European Association of Users of Satellites in Training and Education Programmes (EUROSTEP), c/o OXFAM Belgium, 39 Rue du Conseil, 1050 Bruxelles
T: (02) 5129990
Founded: 1990
Pres: Pierre Galand
Focus: Educ 01297

European Association of Veterinary Anatomists, c/o Fakulteit Diergeneeskunde, Casinopl 24, 9000 Gent
T: (09) 2233765; Fax: (09) 2332234
Founded: 1963
Focus: Vet Med
Periodical
Anatomia, Histologia, Embryologia: Journal of Veterinary Medicine (4-6 times annually) . 01298

European Association on Catalysis / Association Européenne de Catalyse, c/o Dept de Chimie, Laboratoire de Catalyse, Faculté Universitaire de Namur, 61 Rue de Bruxelles, 5000 Namur
T: (081) 724554; Tx: 59222; Fax: (081) 230391
Founded: 1980
Gen Secr: Eric Derouane
Focus: Chem 01299

European Bitumen Association (EUROBITUME), c/o Centre d'Information du Bitume, 351 Blvd Emile Bockstael, 1020 Bruxelles
T: (02) 4782589
Founded: 1969
Gen Secr: W. Depicker
Focus: Materials Sci 01300

European Brain Injury Society (EBIS) / Association Européenne d'Etude des Traumatisés Crâniens et de leur Réinsertion, 17 Rue de Londres, 1050 Bruxelles
Founded: 1990
Focus: Neurology 01301

European Bureau for Conservation and Development (EBCD), 9 Rue de la Science, 1040 Bruxelles
T: (02) 2303070; Fax: (02) 2308272
Founded: 1989
Gen Secr: Despina Symons
Focus: Ecology 01302

European Bureau of Lesser-Used Languages / Bureau Européen des Langues moins Répandues, 1 Rue Montoyer, BP 63, 1040 Bruxelles
T: (02) 4285614
Founded: 1982
Gen Secr: Dr. Yvo Peeters
Focus: Ling
Periodical
Contact Bulletin (3 times annually) . . . 01303

European Burns Association (EBA), c/o Akademisch Ziekenhius Sint-Pieter, Katholiek Universiteit, 3000 Leuven
Gen Secr: Dr. Willy Boeckx
Focus: Derm 01304

European Business Ethics Network (EBEN), c/o EFMD, 40 Rue Washington, 1050 Bruxelles
T: (02) 6480385; Tx: 65080; Fax: (02) 6460768
Founded: 1987; Members: 227
Focus: Philos 01305

European Cancer Prevention Organizazion / ECP, 1307 Chaussée de Waterloo, 1180 Bruxelles
Founded: 1981; Members: 120
Focus: Cell Biol & Cancer Res 01306

European Capitals Universities Network (UNICA), c/o Rectorat de l'ULB, 28 Av Franklin Roosevelt, 1050 Bruxelles
T: (02) 6503152; Fax: (02) 6504243
Founded: 1990
Gen Secr: Chantal Zoller
Focus: Educ 01307

European Carbon Black Centre, 73 Av du Karreveld, 1180 Bruxelles
Founded: 1966
Focus: Energy 01308

European Centre for Ethnolinguistic Cartography, 13 Rue Lenoir, 1090 Bruxelles
Founded: 1986
Focus: Cart 01309

European Centre for Geodynamics and Seismology (ECGS), c/o Observatoire Royal, 3 Av Circulaire, 1180 Bruxelles
T: (02) 3730211; Tx: 21565
Founded: 1988; Members: 9
Gen Secr: Dr. B. Ducarme
Focus: Geology 01310

European Centre for Plastics in the Environment, c/o APME, 4 Av Van Nieuwenhuyse, 1160 Bruxelles
T: (02) 6767211; Tx: 62444; Fax: (02) 6767300
Focus: Ecology 01311

European Centre for Strategic Management of Universities (ESMU), 14 Rue Montoyer, 1040 Bruxelles
T: (02) 5138622; Fax: (02) 5125743
Founded: 1986
Gen Secr: E. Prosser
Focus: Public Admin 01312

European Centre of Ophthalmology, 32 Rue du Méridien, 1030 Bruxelles
Founded: 1990
Focus: Ophthal 01313

European Centre of Studies on Linear Alkylbenzene, c/o ECOSOL, 4 Av E. Van Nieuwenhuyse, 1160 Bruxelles
T: (02) 6767247; Tx: 62444; Fax: (02) 6767301
Founded: 1986
Gen Secr: René Van Sloten
Focus: Chem 01314

European Chemical Industry Ecology and Toxicology Centre (ECETOC), 4 Av E. Van Nieuwenhuyse, 1160 Bruxelles
T: (02) 6753600; Fax: (02) 6753625
Founded: 1978
Gen Secr: Dr. D.A. Stringer
Focus: Ecology; Toxicology 01315

European Chlorinated Solvents Association, 4 Av E. Van Nieuwenhuyse, 1160 Bruxelles
T: (02) 6767211; Tx: 62444; Fax: (02) 6767301
Gen Secr: N. Ginn
Focus: Chem 01316

European Coil Coating Association (ECCA), 47 Rue Montoyer, 1040 Bruxelles
T: (02) 5136052; Tx: 20689; Fax: (02) 5114361
Founded: 1967
Gen Secr: P. Franck
Focus: Metallurgy 01317

European Committee for Catholic Education, 42 Rue de l'Industrie, 1040 Bruxelles
T: (02) 5118232; Fax: (02) 5145989
Founded: 1974
Gen Secr: R. de Zutter
Focus: Educ 01318

European Committee for EC Agricultural Engineers, c/o CEPFAR, 23-25 Rue de la Science, 1040 Bruxelles
T: (02) 2303263; Tx: 25816; Fax: (02) 2311845
Founded: 1987
Gen Secr: Anton Hardt
Focus: Eng; Agri 01319

European Committee for Electrotechnical Standardization, 35 Rue de Stassart, 1050 Bruxelles
T: (02) 5196871; Tx: 172210097; Fax: (02) 5196919
Founded: 1973
Gen Secr: Stephen Marriott
Focus: Standards 01320

European Committee for Interoperable Systems (ECIS), c/o Bates & Wacker, 9 Rue du Moniteur, 1000 Bruxelles
T: (02) 2190305; Fax: (02) 2193215
Founded: 1989
Gen Secr: Philippe Wacker
Focus: Computer & Info Sci 01321

European Committee for Iron and Steel Standardization (ECISS), c/o CEN, 36 Rue de Stassart, 1050 Bruxelles
T: (02) 5196811; Tx: 172210097; Fax: (02) 5196819
Focus: Standards 01322

European Committee for Treatment and Research in Multiple Sclerosis, c/o Belgisch Centrum voor Multiple Sclerosis, Vanheylenstr 16, 1910 Melsbroek
T: (02) 7518030
Pres: Prof. M. Gonzette
Focus: Med 01323

European Committee of Organic Surfactants and their Intermediates, 4 Av E. Van Nieuwenhuyse, 1160 Bruxelles
T: (02) 6767211; Tx: 62444; Fax: (02) 6767300
Founded: 1974
Gen Secr: J.G. Aalbers
Focus: Chem 01324

European Committee on Computational Methods in Applied Sciences (ECCOMAS), c/o VUB, Triomflaan 43, 1160 Bruxelles
T: (02) 6412878; Fax: (02) 6412880
Gen Secr: Prof. C. Hirsch
Focus: Computer & Info Sci 01325

European Community Biologists Association (ECBA), 175 Rue des Brasseurs, 5000 Namur
T: (081) 241133; Fax: (081) 241164
Founded: 1975; Members: 17
Pres: Dr. R.H. Priestley
Focus: Bio 01326

European Community Network of the National Academic Recognition Information Centres (NARIC), c/o Erasmus Bureau, 70 Rue Montoyer, 1040 Bruxelles
T: (02) 2330111; Tx: 63528; Fax: (02) 2330150
Founded: 1984; Members: 20
Focus: Sci 01327

European Community University Professors in Ophthalmology (EUPO), c/o UZ Gent, De Pintelaan 185, 9000 Gent
T: (091) 402319; Fax: (091) 404963
Founded: 1988; Members: 196
Gen Secr: Prof. Dr. J.J de Laey
Focus: Ophthal 01328

European Consulting Engineering Network, 170 Blvd Léopold II, 1080 Bruxelles
T: (02) 4271716; Fax: (02) 4201072
Founded: 1990
Gen Secr: P. Fusager
Focus: Eng 01329

European Consumer Law Group (ECLG), c/o Faculté de Droit, UCL, 2 Pl Montesquieu, 1348 Louvain-la-Neuve
Founded: 1977; Members: 25
Gen Secr: T. Bourgoignie
Focus: Law 01330

European Control Data Users Group (ECODU), c/o Rekenzentrum, Limburgs Universitaire Centrum, Universitaire Campus, 3590 Diepenbeek
T: (011) 286111
Founded: 1965
Gen Secr: Marc L. Thoelen
Focus: Computer & Info Sci 01331

European Convention for Constructional Steelwork (ECCS), 32-36 Av des Ombrages, 1200 Bruxelles
T: (02) 7620429; Fax: (02) 7620935
Founded: 1955
Gen Secr: J. Van Neste
Focus: Civil Eng 01332

European Cooperation in the Field of Scientific and Technical Research, c/o Council of the European Communities, 170 Rue de la Loi, 1049 Bruxelles
T: (02) 2346111; Tx: 21711; Fax: (02) 2347381
Founded: 1970; Members: 24
Gen Secr: N. Ersbøll
Focus: Sci; Eng 01333

European Council for Rural Law, Drongenstationstr 101, 9031 Gent
T: (091) 271857
Founded: 1957
Gen Secr: Marc Heyerick
Focus: Law 01334

European Council of Integrated Medicine (ECIM), 71 Rue des Echevins, 1050 Bruxelles
T: (02) 6483480
Gen Secr: G.J. Van Lamoen
Focus: Med 01335

European Crop Protection Association (ECPA), 79a Av Albert Lancaster, 1180 Bruxelles
T: (02) 3756860; Tx: 62120; Fax: (02) 3752793
Members: 30
Gen Secr: K.P. Vlahodimos
Focus: Crop Husb 01336

European Down's Syndrome Association (EDSA), 41 Rue Victor Close, 4800 Verviers
T: (087) 220534; Fax: (087) 220716
Founded: 1987; Members: 11
Gen Secr: Richard Bonjean
Focus: Psych 01337

European Economic Association (EEA), 2b Van Evenstr, 3000 Leuven
T: (016) 283070; Fax: (016) 293361
Founded: 1985; Members: 1693
Pres: Martin F. Hellwig; Gen Secr: Prof. Anton Barten
Focus: Econ 01338

European Economic Research and Advisory Consortium (ERECO), 48 Rue du Cardinal, 1040 Bruxelles
T: (02) 2302212; Fax: (02) 2306499
Founded: 1990
Gen Secr: Jean-Marie Poutrel
Focus: Econ 01339

European Educational Association for News Distribution, 20 Allée des Bouleaux, 6280 Gerpinnes
T: (071) 216153; Fax: (071) 217713
Founded: 1971
Gen Secr: Michel Paunet
Focus: Educ 01340

European Environment Agency (EEA), c/o CEC, Directorate General XI, 200 Rue de la Loi, 1049 Bruxelles
T: (02) 2351111; Tx: 21877; Fax: (02) 2350138
Founded: 1990
Focus: Ecology 01341

European Environmental Bureau (EEB), 22-26 Rue de la Victoire, 1060 Bruxelles
T: (02) 5141250; Tx: 62720; Fax: (02) 5140937
Founded: 1974; Members: 128
Gen Secr: Raymond Van Ermen
Focus: Ecology 01342

European Environment Information and Observation Network, c/o CEC, Diretorate-General XI, 200 Rue de la Loi, 1049 Bruxelles
T: (02) 2968814; Tx: 21877; Fax: (02) 2969562
Founded: 1990
Focus: Ecology 01343

European Extruded Polystyrene Insulation Board Association (EXIBA), 4 Av E. Van Nieuwenhuyse, 1160 Bruxelles
T: (02) 6767240; Fax: (02) 6767301
Founded: 1987; Members: 9
Gen Secr: B. Jensen
Focus: Materials Sci 01344

European Family Therapy Association (EFTA), 1 Rue Defacqz, 1050 Bruxelles
T: (02) 6401372; Tx: 65080
Gen Secr: Edith Goldbeter-Merinfeld
Focus: Therapeutics 01345

European Federation for the Education of Children of Occupational Travellers (EFECOT), 42 Rue de l'Industrie, 1040 Bruxelles
T: (02) 5118232; Fax: (02) 5145989
Founded: 1988
Gen Secr: L. Knaepkens
Focus: Educ 01346

European Federation of Animal Health, 1 Rue Defacqz, 1050 Bruxelles
T: (02) 5372125; Fax: (02) 5370049
Founded: 1987; Members: 33
Gen Secr: Dr. J. Vanhemelrijk
Focus: Vet Med
Periodical
Animal Health Focus (3 times annually) . 01347

European Federation of Child Neurology Societies (EFCNS), c/o Service de Neurologie Pédiatrique, UCL-Cliniques Universitaires Saint-Luc, 10 Av Hippocrate, 1200 Bruxelles
T: (02) 7641303; Fax: (02) 7645231
Founded: 1970
Gen Secr: Prof. Philippe Evrard
Focus: Neurology 01348

European Federation of Energy Management Associations (EFEM), c/o Energik, Brouwersvliet 15, 2000 Antwerpen
T: (03) 2311660; Tx: 72893
Founded: 1986
Gen Secr: Luc Beernaert
Focus: Business Admin 01349

European Finance Association, 13 Rue d'Egmont, 1050 Bruxelles
T: (02) 5119116; Fax: (02) 5121929
Focus: Educ; Finance
Periodical
Newsletter (weekly) 01350

European Power Electronics and Drives Association, 2 Blvd de la Plaine, 1050 Bruxelles
T: (02) 6412819
Founded: 1989; Members: 300
Pres: J.C. Sabonnadière
Focus: Electronic Eng
Periodical
EPE Journal (weekly) 01351

European Technical Association for Protective Coatings (ETAPC), Korenstr 6-8, 2170 Antwerpen
T: (03) 6463373; Fax: (03) 6454905
Founded: 1972
Pres: E. van Hoydonck
Focus: Materials Sci 01352

European Trade Union Institute (ETUI), 155 Blvd Emile Jacqmain, 1210 Bruxelles
T: (02) 2240470; Fax: (02) 2240502
Focus: Socialism; Sociology; Econ; Poli Sci
Periodicals
ETUI Documentation Centre Bulletin (weekly)
ETUI Newsletter (weekly) 01353

Fédération Belge d'Education Physique (FBEP), c/o Free-University, 1050 Bruxelles
T: (02) 3581446
Founded: 1923; Members: 3000
Focus: Sports 01354

Fédération Belge des Chambres Syndicales de Médecins (FBCSM), 15 Rue du Château, 1420 Braine l'Allend
T: 3843930
Founded: 1963; Members: 4000
Focus: Med 01355

Fédération Belge des Sociétés Scientifiques, 31 Rue Vautier, 1040 Bruxelles
T: (02) 6480475
Founded: 1949
Focus: Sci 01356

Fédération Européenne de Médecine Physique et Réadaption / European Federation for Physical Medicine and Rehabilitation, c/o Kliniek voor Fysiotherapie en Orthopedie, De Pintelaan 185, 9000 Gent
T: (09) 2402234; Fax: (09) 2404975
Founded: 1963
Gen Secr: Dr. G. Vanderstraeten
Focus: Med; Physical Therapy; Rehabil
Periodicals
Europa Medicophysica
European Journal of Physical Medicine and Rehabilitation (weekly) 01357

Fédération Internationale des Instituts de Recherches Socio-Religieuses (FERES), 5 Av Sainte-Gertrude, 1348 Louvain-la-Neuve
T: (010) 450822
Founded: 1958; Members: 38
Pres: F. Houtart
Focus: Sociology; Rel & Theol
Periodical
Social Compass: International Review of Sociology of Religion (weekly) 01358

Fédération Nationale de l'Enseignement Moyen Catholique, 5 Rue Guimard, 1040 Bruxelles
T: (02) 5136860
Focus: Educ 01359

Fédération Royale des Sociétés d'Architectes de Belgique (FAB), 537 Blvd de Smet de Naeyer, 1020 Bruxelles
Members: 16
Gen Secr: E. Draps
Focus: Archit 01360

Fondation Archéologique / Archaeological Foundation, c/o Université Libre de Bruxelles, 50 Av Franklin Roosevelt, 1050 Bruxelles
T: (02) 6502421; Fax: (02) 6502450
Founded: 1930
Focus: Archeol
Periodical
Culture et Cité 01361

Fondation Born-Bunge pour la Recherche / Born-Bunge Research Foundation, Universiteitsplein 1, 2610 Wilrijk
Founded: 1954
Focus: Neurology; Cardiol 01362

Fondation Egyptologique Reine Elisabeth, 10 Parc du Cinquantenaire, 1040 Bruxelles
T: (02) 7417364
Founded: 1923; Members: 650
Pres: Comte d' Arschot; Gen Secr: A. Mekhitarian
Focus: Anthro; Ethnology; Archeol
Periodicals
Bibliographie Papyrologique (weekly)
Chronique d'Egypte (weekly) 01363

Fondation Fernand Lazard, 11 Rue d'Egmont, 1050 Bruxelles
Founded: 1949
Pres: Julien Fautrez
Focus: Educ 01364

Fondation Francqui, 11 Rue d'Egmont, 1050 Bruxelles
Founded: 1932
Pres: Jacques Groothaert; Gen Secr: Luc Eyckmans
Focus: Educ
Periodical
Rapport (5 times annually) 01365

Fondation Médicale Reine Elisabeth, 3 Av J.J. Crocq, 1020 Bruxelles
T: (02) 4783556
Founded: 1926
Pres: Prof. Dr. T. de Barsy; Gen Secr: Prof. Dr. R. Six
Focus: Med 01366

Fondation Universitaire, 11 Rue d'Egmont, 1050 Bruxelles
Founded: 1920
Focus: Adult Educ; Sci
Periodicals
Bureau de Statistiques Universitaires: Dienst voor Universitaire Statistiek (weekly)
Cahiers de la Fondation Universitaire: Cahiers van de Universitaire Stichting (weekly) . . . 01367

Groupe Belge d'Etude de l'Arriération Mentale, 66 Rue de la Limite, 1030 Bruxelles
Focus: Educ Handic
Periodical
Proceedings (weekly) 01368

Groupement Belge des Omnipraticiens (GBO), 76 Rue du Tabellion, 1050 Bruxelles
T: (02) 5387365; Fax: (02) 5385105
Founded: 1968
Pres: Dr. Philippe Vandermeeren; Gen Secr: Dr. Paul Vandaele
Focus: Med
Periodicals
GBO (10 times annually)
VBO (10 times annually) 01369

Groupement des Unions Professionnelles Belges de Médecins Spécialistes (GBS), 20 Av de la Couronne, 1050 Bruxelles
T: (02) 6492147; Fax: (02) 6492690
Members: 21
Focus: Med
Periodical
Le Médecin Spécialiste (10 times annually) 01370

Groupement International pour la Recherche Scientifique en Odontologie et en Stomatologie (GIRSO), 322 Rue Haute, 1000 Bruxelles
T: (02) 5380000
Founded: 1956; Members: 200
Focus: Dent; Stomatology 01371

Groupement International pour la Recherche Scientifique en Stomatologie et Odontologie (GIRSO), 19 Av Jeanne, 1050 Bruxelles
T: (02) 6482481
Founded: 1956; Members: 200
Gen Secr: Dr. Roland Rodembourg
Focus: Stomatology; Dent
Periodical
Bulletin du GIRSO (weekly) 01372

Icon / Association on Marginal Literature and Art, Lobergenbos 27, 3200 Leuven
Gen Secr: Jozef Peeters
Focus: Lit; Arts
Periodical
Cahier (weekly) 01373

INBEL, Institut Belge d'Information et de Documentation, 155 Rue de la Loi, 1040 Bruxelles
T: (02) 2874111; Fax: (02) 2874100
Founded: 1962
Pres: G. Demez
Focus: Doc
Periodical
FAITS/FEITEN (weekly) 01374

Institut Archéologique du Luxembourg (IAL), 13 Rue des Martyrs, 6700 Arlon
T: (063) 221236
Founded: 1847; Members: 735
Focus: Archeol; Hist; Folklore; Genealogy; Arts; Numismatics
Periodicals
Annales de l'Institut Archéologique du Luxembourg (weekly)
Bulletin (weekly) 01375

Institut Archéologique Liégeois (IAL), c/o Musée Curtius, 13 Quai de Maastricht, 4000 Liège
Founded: 1850; Members: 490
Pres: Pierre Colman; Gen Secr: B. Dumont
Focus: Archeol; Hist
Periodical
Bulletin de l'Institut Archéologique Liègeois (weekly) 01376

Institut Belge de Droit Comparé, 14 Rue Bosquet, 1060 Bruxelles
T: (02) 5380618
Founded: 1907; Members: 400
Pres: M. Vauthier; Gen Secr: Paul Landrien
Focus: Law
Periodical
La Revue de Droit International et de Droit Comparé (weekly) 01377

Institut Belge de la Soudure (IBS), 21 Rue des Drapiers, 1050 Bruxelles
T: (02) 5122892; Fax: (02) 5127457
Founded: 1942; Members: 315
Focus: Eng
Periodical
Revue de la Soudure-Lastijdschrift (weekly) 01378

Institut Belge de Normalisation (IBN), 29 Av de la Brabançonne, 1040 Bruxelles
T: (02) 7349205; Fax: (02) 7334264
Founded: 1946; Members: 625
Focus: Standards
Periodical
Revue IBN (10 times annually) 01379

Institut Belge de Régulation et d'Automatisme (IBRA), 3 Rue Ravenstein, 1000 Bruxelles
T: (02) 5117004; Fax: (02) 5117004
Founded: 1955; Members: 500
Focus: Eng
Periodicals
Automatique (3 times annually)
Journal A.: Benelux Quarterly Journal on Automatic Control 01380

Institut Belge des Hautes Etudes Chinoises, c/o Musées Royaux d'Art et d'Histoire, 10 Parc du Cinquantenaire, 1040 Bruxelles
Fax: (02) 7337735
Founded: 1929; Members: 250
Pres: M. de Noorhout
Focus: Ling; Philos; Hist; Rel & Theol
Periodical
Mélanges Chinois et Bouddhiques (weekly) 01381

Institut Belge des Sciences Administratives (I.B.S.A.), 15 Rue du Gouvernement Provisoire, 1000 Bruxelles
T: (02) 2191750
Founded: 1936; Members: 150
Pres: A. François; Gen Secr: H. van Hassel
Focus: Business Admin
Periodical
Administration Publique (weekly) 01382

Institut Européen d'Ecologie et de Cancérologie, 24bis Rue des Fripiers, 1000 Bruxelles
T: (02) 2190830
Founded: 1965
Focus: Ecology; Cell Biol & Cancer Res
Periodical
Medicine-Biology-Environment (weekly) . . 01383

Institut Européen des Armes de Chasse et de Sport, 3 Rue Charles Morren, 4000 Liège
T: (041) 533986, 713778; Fax: (041) 533989
Founded: 1977; Members: 8
Pres: J. Chatain; Gen Secr: H. Heidebroek
Focus: Eng 01384

Institut Européen Interuniversitaire de l'Action Sociale (IEIAS), 179 Rue du Debarcadère, 6001 Marcinelle
T: (071) 366273, 432920
Founded: 1970
Focus: Sociology; Soc Sci
Periodical
COMM: Community Work and Communication (3 times annually) 01385

Institut Historique Belge de Rome (I.H.B.R.), 3 Av du Derby, 1050 Bruxelles
T: (02) 6492381
Founded: 1902
Focus: Hist 01386

Institut International de Recherches Betteravières (I.I.R.B.) / International Institute for Sugar Beeb Research, 47 Rue Montoyer, 1040 Bruxelles
T: (02) 5091530/33; Tx: 21287; Fax: (02) 5126506
Founded: 1932; Members: 507
Pres: J.A. Esteban Baselga; Gen Secr: L. Weickmans
Focus: Agri
Periodical
Proceedings (weekly) 01387

Institut International des Sciences Administratives (IISA) / International Institute of Administrative Sciences, 1 Rue Defacqz, 1050 Bruxelles
T: (02) 5389165; CA: INTERADMIN; Fax: (02) 5379702
Founded: 1930; Members: 48
Pres: Guy Braibant; Gen Secr: Turkia Ould Daddah
Focus: Public Admin
Periodicals
Infoadmin (weekly)
International Review of Administrative Sciences (weekly) 01388

Institut International du Fer et de l'Acier / International Iron and Steel Institute, 120 Rue Colonel Bourg, 1140 Bruxelles
T: (02) 7359075; Tx: 22639; Fax: (02) 7358012
Founded: 1967; Members: 150
Gen Secr: L.J. Holschuh
Focus: Metallurgy
Periodicals
Bulletin
Conference Proceedings
Statistical and Economic Reports
Technical Surveys
World Steel in Figures 01389

Institut National de Cinématographie Scientifique de Belgique, 29 Rue Vautier, 1040 Bruxelles
T: (02) 6480475
Pres: Prof. G. Thines; Gen Secr: A. Quintart
Focus: Cinema 01390

Institut Royal Belge du Pétrole (IBP), 4 Rue de la Science, 1040 Bruxelles
T: (02) 5728767; Fax: (02) 4480825
Founded: 1937; Members: 612
Focus: Materials Sci
Periodical
Annales/Annalen (weekly) 01391

Institut Royal des Relations Internationales (IRRI), 65 Rue Belliard, 1040 Bruxelles
T: (02) 2302230; Fax: (02) 2305230
Founded: 1947; Members: 1200
Focus: Poli Sci; Law
Periodical
Studia Diplomatica (weekly) 01392

International Committee of Dialectologists, c/o Centre International de Dialectologie Générale, Université Catholique, Ravenstr 46, 3000 Leuven
Focus: Ling 01393

International Council of Onomastic Sciences (ICOS), c/o International Centre of Onomastics, Blijde Inkomststr 21, 3000 Leuven
T: (016) 284819; Fax: (016) 285025
Founded: 1949
Gen Secr: Prof. Dr. W.F.H. Nicolaissen
Focus: Ling
Periodical
Onoma 01394

International Diabetes Federation (IDF), 40 Rue Washington, 1050 Bruxelles
T: (02) 6474414; Tx: 65080; Fax: (02) 6408565
Founded: 1949; Members: 1675
Pres: Wendell Mayes
Focus: Diabetes
Periodicals
IDF Bulletin (3 times annually)
IDF Newsletter (3 times annually) . . . 01395

International Educational and Cultural Association (ITECA), Merellaan 10, 1150 Bruxelles
T: (02) 6738437
Focus: Educ 01396

International Institute for Organizational and Social Development (I.O.D.), Predikherenberg 55, 3010 Leuven
T: (016) 251671; Fax: (016) 251680
Founded: 1970
Pres: Dr. L.S. Vansina
Focus: Business Admin; Sociology . . . 01397

International Institute of Cellular and Molecular Pathology, 75 Av Hippocrate, 1200 Bruxelles
T: (02) 7623292; Tx: 23722; Fax: (02) 7647573
Founded: 1974
Gen Secr: R. Bouckaert
Focus: Pathology
Periodical
Scientific Report (weekly) 01398

International League of Societies for Persons with Mental Handicap, 248 Av Louise, 1050 Bruxelles
T: (02) 6476180; Fax: (02) 6472969
Founded: 1960; Members: 149
Pres: Victor Wahlstrom
Focus: Rehabil
Periodical
News of the International League (weekly) 01399

International PEN Club, Flemish Centre, Albert Heyrbautlaan 48, 1710 Dilbeek
Gen Secr: Willem M. Roggeman
Focus: Lit
Periodical
PEN-Tijdingen (weekly) 01400

International PEN Club, French Speaking Branch, 10 Av des Cerfs, 1950 Kraainem
Fax: (02) 7314847
Founded: 1922; Members: 230
Pres: Georges Sion; Gen Secr: Huguette de Broqueville
Focus: Lit 01401

International Rhinologic Society (IRS), c/o ENT Dept A.Z-V.U.B., Laarbeeklaan 101, 1090 Bruxelles
T: (02) 4776002; Fax: (02) 4775800
Founded: 1965; Members: 1000
Gen Secr: P.A.R. Clement
Focus: Otorhinolaryngology
Periodicals
Journal (weekly)
Journal of Rhinology (weekly) 01402

International Secretariat for University Study of Education, Baertsoenkaai 3, 9000 Gent
Focus: Educ 01403

International Union for the Scientific Study of Population, 34 Rue des Augustins, 4000 Liège
T: (041) 224080; Fax: (041) 223847
Founded: 1928; Members: 1900
Gen Secr: Bruno Remiche
Focus: Sociology
Periodical
UIESP Newsletter 01404

Jeunesse Intellectuelle, 11 Rue d'Egmont, 1050 Bruxelles
Founded: 1945
Pres: G. de Landsheere
Focus: Educ 01405

Koninklijke Academie en Nationaal Hoger Instituut voor Schone Kunsten (KASKA-NHISKA), Mutsaertstr 31, 2000 Antwerpen
T: (03) 2314186, 2324161, 2335619
Founded: 1663; Members: 87
Focus: Arts 01406

Koninklijke Academie voor Geneeskunde van België, Hertogsstr 1, 1000 Bruxelles
Founded: 1938; Members: 156
Pres: Prof. Dr. E. Carmeliet; Gen Secr: Prof. Dr. A. Lacquet
Focus: Med
Periodical
Verhandelingen (weekly) 01407

Koninklijke Academie voor Nederlandse Taal- en Letterkunde (KANTL), Koningstr 18, 9000 Gent
T: (09) 2252774; Fax: (09) 2232718
Founded: 1886; Members: 30
Gen Secr: Prof. Dr. Ada Deprez
Focus: Ling; Lit
Periodical
Verslagen en Mededelingen (2-3 times annually) 01408

Koninklijke Academie voor Wetenschappen, Letteren en Schone Kunsten van België, Hertogsstr 1, 1000 Bruxelles
T: (02) 5112623; Fax: (02) 5110143
Members: 120
Focus: Fine Arts; Law; Philos; Humanities; Hist; Rel & Theol; Music; Nat Sci; Eng; Ling; Lit

Periodicals
Academiae Analecta: Mededelingen van de K.A.W.L.S.K. Klasse der Letteren (weekly)
Academiae Analecta: Mededelingen van de K.A.W.L.S.K. Klasse der Schone Kunsten (weekly)
Academiae Analecta: Mededelingen van de K.A.W.L.S.K. Klasse der Wetenschappen (weekly)
Jaarboek (weekly) 01409

Koninklijke Maatschappij voor Dierkunde van Antwerpen / Royal Zoological Society of Antwerp, Koningin Astridplein 26, 2018 Antwerpen
T: (03) 2311640
Founded: 1843; Members: 25000
Focus: Zoology
Periodical
Zoo (weekly) 01410

Koninklijke Vereniging voor Natuur- en Stedeschoon, Koningin Astridplein 24, 2018 Antwerpen
T: (03) 2323531
Founded: 1910; Members: 1250
Focus: Ecology; Urban Plan
Periodical
Natuur- en Stedeschoon (weekly) . . . 01411

Koninklijke Vereniging voor Vogel- en Natuurstudie de Wielewaal, Graatakker 11, 2300 Turnhout
T: (014) 412252; Fax: (014) 412252
Founded: 1933; Members: 7000
Focus: Ornithology
Periodicals
Oriolus (weekly)
Wielewaal (weekly) 01412

Ligue Nationale Belge contre l'Epilepsie, 135 Av Albert, 1060 Bruxelles
T: (02) 3443293
Founded: 1955
Focus: Pathology 01413

Les Naturalistes Belges, 29 Rue Vautier, 1040 Bruxelles
Founded: 1918; Members: 1000
Pres: Alain Quintart
Focus: Nat Sci
Periodical
Bulletin (weekly) 01414

Oeuvre Belge du Cancer (OBC), 21 Rue des Deux Eglises, 1040 Bruxelles
T: (02) 2185191
Founded: 1950
Focus: Cell Biol & Cancer Res 01415

Office Généalogique et Héraldique de Belgique, c/o Musées Royaux d'Art et d'Histoire, 10 Parc du Cinquantenaire, 1040 Bruxelles
Founded: 1942
Focus: Genealogy
Periodicals
Le Héraut (weekly)
Le Parchemin (weekly)
Le Recueil Généalogique et Héraldique de Belgique (weekly) 01416

Office International de l'Enseignement Catholique (OIEC), 60 Rue des Eburons, 1040 Bruxelles
T: (02) 2307252
Founded: 1952; Members: 84
Pres: John Meyers; Gen Secr: Paulus Adams
Focus: Educ; Rel & Theol
Periodical
Bulletin de l'OIEC (weekly) 01417

Ordre des Architectes / Orde van Architecten, Livornostr 160, 1050 Bruxelles
T: (02) 6470669; Fax: (02) 6463818
Founded: 1963; Members: 8000
Pres: J. Ketelaer
Focus: Archit 01418

Ruusbroecgenootschap, Prinsstr 13, 2000 Antwerpen
T: (03) 2204367; Fax: (03) 2204420
Founded: 1925; Members: 7
Pres: Dr. G. de Baere
Focus: Rel & Theol
Periodical
Ons Geestelijk Erf (weekly) 01419

Societas Logopedica Latina, 69 Rue de Bruxelles, 3000 Leuven
Founded: 1967; Members: 300
Focus: Logopedy 01420

Société Archéologique de Namur, Hôtel de Gaiffier d'Hestroy, 24 Rue de Fer, 5000 Namur
T: (081) 224362
Founded: 1845; Members: 400
Pres: Eugène Nemery de Bellevaux; Gen Secr: Jacques Toussaint
Focus: Archeol; Hist
Periodical
Annales de la Société Archéologique de Namur (ASAN) (weekly) 01421

Société Astronomique de Liège (SAL), c/o Institut d'Astrophysique, 5 Av de Cointe, 4200 Cointe-Ougrée
T: (041) 529980; Tx: 41264
Founded: 1938; Members: 616
Pres: A. Lausberg; Gen Secr: D. Gering
Focus: Astronomy
Periodical
Le Ciel (weekly) 01422

Société Belge d'Allergologie et d'immunologie Clinique, c/o Academic Hospital, De Pintelaan 185, 9000 Gent
Founded: 1946; Members: 250
Focus: Immunology 01423

Société Belge d'Anesthésie et de Réanimation / Belgische Vereniging voor Anesthesie en Reanimatie / Belgian Society of Anaesthesia and Resuscitation, 7 Rue de Rosières, 1301 Bierges
T: (010) 224742; Fax: (010) 241832
Founded: 1963; Members: 705
Pres: Prof. Dr. G. Rolly; Gen Secr: Prof. S.L. Scholtes
Focus: Anesthetics
Periodical
Acta Anaesthesiologica Belgica (weekly) . 01424

Société Belge de Biochimie et de Biologie Moléculaire, 75 Av Hippocrate, 1200 Bruxelles
T: (02) 7647451; Fax: (02) 7647573
Founded: 1951; Members: 1050
Gen Secr: Philippe Hoet
Focus: Biochem; Biophys 01425

Société Belge de Biologie Clinique, c/o Laboratoire de l'Institut Gailly, 1 Blvd Zoë Drion 1, 6000 Charleroi
Founded: 1949
Focus: Bio 01426

Société Belge de Cardiologie, 43 Rue des Champs-Elysées, 1050 Bruxelles
Fax: (02) 4776381
Focus: Cardiol
Periodicals
Acta Cardiologica Belgica (weekly)
News Bulletin (weekly) 01427

Société Belge de Chirurgie Orthopédique et de Traumatologie (SOBCOT), c/o Dr. Roger Lemaire, 44 Rue du Centre, 4130 Hony-Esneux
Founded: 1921; Members: 378
Pres: Lemaire. Dr. Roger; Gen Secr: Patrick van Elegem
Focus: Surgery
Periodical
ACTA Orthopaedica Belgica (3 times annually) 01428

Société Belge de Gastro-Entérologie (SBGE), c/o Dr. A. Delcourt, 10 Rue P.-E. Janson, 1050 Bruxelles
Founded: 1928; Members: 500
Focus: Gastroenter 01429

Société Belge de Géologie (SBG), 13 Rue Jenner, 1040 Bruxelles
T: (02) 6476400; Fax: (02) 6477359
Founded: 1887; Members: 430
Pres: Eric Groessens; Gen Secr: M. Dusar
Focus: Geology
Periodical
Géologie (weekly) 01430

Société Belge d'Electroencéphalographie et de Neurophysiologie Clinique, c/o Centre William Lennox, 1340 Ottignies
T: 417810
Founded: 1974; Members: 150
Focus: Physiology; Neurology 01431

Société Belge de Logique et de Philosophie des Sciences, 317 Av Charles Woeste, 1090 Bruxelles
Founded: 1948; Members: 120
Gen Secr: G. Hirsch
Focus: Philos 01432

Société Belge de Médecine Interne (SBMI), c/o Laboratorium voor Experimentele Geneeskunde, 115 Blvd de Waterloo, 1000 Bruxelles
Founded: 1946; Members: 400
Focus: Intern Med 01433

Société Belge d'EMG, c/o Prof. J. E. Desmedt, 115 Blvd de Waterloo, 1000 Bruxelles
Members: 70
Focus: Med 01434

Société Belge de Microscopie Electronique (SBME), c/o CRM, 69 Rue du Val-Benot, 4000 Liège
T: (041) 527050
Founded: 1966; Members: 160
Focus: Electronic Eng 01435

Société Belge de Musicologie, 30 Rue de la Régence, 1000 Bruxelles
Founded: 1946; Members: 300
Pres: R. Wangermee; Gen Secr: H. Vanhulst
Focus: Music
Periodical
Revue Belge de Musicology (weekly) . . 01436

Société Belge de Neurologie, 97 Rue aux Laines, 1000 Bruxelles
Founded: 1896; Members: 200
Focus: Neurology 01437

Société Belge de Pédiatrie, 20 Av de la Couronne, 1050 Bruxelles
T: (02) 6492147
Founded: 1926; Members: 331
Focus: Pediatrics 01438

Société Belge de Philosophie, c/o Institut de Philosophie, 143 Av Adolphe Buyl, 1050 Bruxelles
Fax: (02) 6503647
Founded: 1920; Members: 100
Pres: Anne-Marie Roviello; Gen Secr: Jean-N. Missa
Focus: Philos 01439

Société Belge de Photogrammétrie, de Télédétection et de Cartographie, C.A.E.-Tour Finances, 50 Blvd du Jardin Botanique, 1010 Bruxelles
Founded: 1931; Members: 169
Pres: Urbain van Twembeke; Gen Secr: José van Hemelrijck
Focus: Surveying; Cart
Periodical
Bulletin (weekly) 01440

Société Belge de Physiologie et de Pharmacologie, c/o Prof. T. Godfraind, 54 Av Hippocrate, 1200 Bruxelles
Focus: Physiology; Pharmacol 01441

Société Belge de Physique (S.B.P.) / Belgische Natuurkundige Vereniging / Belgian Physical Society, c/o Belgisch Instituut voor Ruimte-Aéronomie, Ringlaan 3, 1180 Bruxelles
Founded: 1929; Members: 470
Focus: Physics
Periodicals
Physicalia (weekly)
Physicalia Magazine (weekly) 01442

Société Belge de Pneumologie, 56 Rue de la Concorde, 1050 Bruxelles
T: (02) 5125451; Fax: (02) 5114614
Founded: 1924; Members: 360
Focus: Pulmon Dis 01443

Société Belge de Psychoanalyse / Belgische Vereniging voor Psychoanalyse, 53 Allée du Vieux Frêne, 6270 Loverval
T: (071) 430639
Founded: 1946; Members: 98
Focus: Psychoan
Periodical
Revue Belge de Psychoanalyse (weekly) 01444

Société Belge de Psychologie (SBP), Dieweg 47, 1180 Bruxelles
T: (02) 3742896
Founded: 1946; Members: 450
Focus: Psych 01445

Société Belge d'Ergologie (SBE), 46 Av Michel Ange, 1040 Bruxelles
T: (02) 7330952
Founded: 1935; Members: 2000
Focus: Eng 01446

Société Belge des Auteurs, Compositeurs et Editeurs (SABAM), 75-77 Rue d'Arlon, 1040 Bruxelles
T: (02) 2302660; Tx: 65854; Fax: (02) 2311800
Founded: 1922; Members: 11000
Pres: Vic Legley; Gen Secr: Jan Corbet
Focus: Lit; Music
Periodical
Bulletin (3 times annually) 01447

Société Belge d'Etudes Byzantines, 4 Blvd de l'Empereur, 1000 Bruxelles
Founded: 1956; Members: 74
Pres: Alice Leroy-Molinghen; Gen Secr: Jacques Noret
Focus: Hist
Periodical
Byzantion: Revue Internationale des Etudes Byzantines (weekly) 01448

Société Belge de Vacuologie et de Vacuotechnique, 30 Av de la Renaissance, 1040 Bruxelles
Founded: 1963; Members: 62
Focus: Physics; Eng 01449

Société Belge d'Histoire des Hôpitaux, c/o Archives de l'Assistance Publique, 298a Rue Haute, 1000 Bruxelles
Founded: 1963
Focus: Hist 01450

Société Belge d'Ophthalmologie, c/o Hôpital Erasme, 808 Rte de Lennick, 1070 Bruxelles
Fax: (02) 5554504
Founded: 1896; Members: 550
Gen Secr: Dr. A. Zanen
Focus: Ophthal
Periodical
Bulletin (weekly) 01451

Société Beneluxienne de Métallurgie, 43 Rue des Champs-Elysees, 1050 Bruxelles
Focus: Metallurgy 01452

Société Centrale d'Architecture de Belgique (S.C.A.B.), 3 Rue Ravenstein, 1000 Bruxelles
T: (02) 5113492
Founded: 1872; Members: 180
Pres: Alberto Vanderauwers; Gen Secr: Ghislain Ladrière
Focus: Archit
Periodical
Bulletin Mensuel (weekly) 01453

Société de Langue et de Littérature Wallonnes, c/o Université de Liège, 7 Pl du XX Août, 4000 Liège
T: (041) 525809
Founded: 1856; Members: 400
Pres: M.-T. Counet; Gen Secr: Victor George
Focus: Ling; Lit
Periodicals
Bulletin
Dialectes de Wallonie (weekly) 01454

Société des Auteurs et Compositeurs Dramatiques (S.A.C.D.), 29 Av Jeanne, 1050 Bruxelles
T: (02) 6275211; Fax: (02) 6275252
Focus: Lit; Music 01455

Société des Bollandistes, 24 Blvd Saint-Michel, 1040 Bruxelles
Founded: 1615; Members: 5
Pres: Paul Devos; Gen Secr: J. van der Straeten
Focus: Hist; Rel & Theol
Periodical
Analecta Bollandiana (weekly) 01456

Société des Physiciens des Hôpitaux d'Expression Française, c/o Service de Radiothérapie, Hôpital de Bavière, 66 Blvd de la Constitution, 4020 Liège
Focus: Physiology 01457

Société d'Etudes Latines de Bruxelles, 18 Av Van Cutsem, 7500 Tournai
T: (069) 214713
Founded: 1937; Members: 750
Pres: C. Deroux; Gen Secr: J. Dumortier-Bibauw
Focus: Ling; Hist; Archeol
Periodical
Latomus: Revue d'Etudes Latines . . . 01458

Société Européenne de Radiobiologie, c/o Institut de Chimie, Sart-Tilman, 4000 Liège
T: (041) 563360; Tx: 41397
Founded: 1959
Gen Secr: Prof. R. Goutier
Focus: Bio; Radiology
Periodical
International Journal of Radiation Biology 01459

Société Européenne pour la Formation des Ingénieurs / European Society for Engineering Education, 51 Rue de la Concorde, 1050 Bruxelles
T: (02) 5121734; CA: 21504; Fax: (02) 5123265
Founded: 1973
Pres: Yves Beernaert; Gen Secr: Dr. D. Jenkins
Focus: Adult Educ; Eng
Periodicals
European Journal of Engineering Education (weekly)
SEFI-News (4 times annually) 01460

Société Géologique de Belgique (S.G.B.), c/o Université, 7 Pl du XX Août, 4000 Liège
T: (041) 665338; Fax: (041) 665700
Founded: 1874; Members: 458
Pres: J. Bouckaert; Gen Secr: E. Poty
Focus: Geology
Periodical
Annales (weekly) 01461

Société Internationale de Chirurgie Orthopédique et de Traumatologie (SICOT), 40 Rue Washington, 1050 Bruxelles
T: (02) 6486823; Fax: (02) 6498601
Founded: 1929; Members: 2496
Gen Secr: Prof. M. Hinsenkamp
Focus: Surgery; Traumatology
Periodical
International Orthopaedics (weekly) . . . 01462

Société Internationale de Droit Pénal Militaire et de Droit de la Guerre, c/o Auditorat Général près de la Cour Militaire, Palais de Justice, 1000 Bruxelles
Fax: (02) 5086085
Founded: 1956
Gen Secr: F. Gorle
Focus: Law
Periodical
The Military Law and Law of War Review (weekly) 01463

Société Internationale pour la Recherche sur les Maladies de Civilisation et sur l'Environnement (SIRMCE) / International Society for Research on Civilization Diseases and on Environment, 20 Sq Larousse, 1060 Bruxelles
T: (02) 3430461
Founded: 1973; Members: 700
Gen Secr: Raimond Sanchez-Perez
Focus: Med; Ecology; Public Health
Periodical
Congress Proceedings 01464

Société Internationale pour l'Etude de la Philosophie Médiévale (S.I.E.P.M.), 1 Chemin d'Aristote, 1348 Louvain-la-Neuve
T: (010) 474807; Fax: (010) 474807
Founded: 1959; Members: 538
Pres: Albert Zimmermann; Gen Secr: Jacqueline Hamesse
Focus: Philos
Periodical
Bulletin de Philosophie Médievale (weekly) 01465

Société Mathématique de Belgique, 317 Av Charles Woeste, 1090 Bruxelles
Founded: 1921; Members: 600
Gen Secr: Guy Hirsch
Focus: Math
Periodical
Bulletin (weekly) 01466

Société Nationale de Laiterie, 95-99 Rue Froissart, 1040 Bruxelles
Founded: 1898
Pres: R. Desmedt; Gen Secr: E. Dobbelaere
Focus: Dairy Sci 01467

Société Philosophique de Louvain, c/o Institut Supérieur de Philosophie, Collège Thomas More, 1 Chemin d'Aristote, 1348 Louvain-la-Neuve
T: (010) 474613
Founded: 1888; Members: 135
Pres: Ghislaine Florival
Focus: Philos 01468

Société Royale Belge d'Anthropologie et de Préhistoire, 29 Rue Vautier, 1040 Bruxelles
T: (02) 6274381; Fax: (02) 6464433
Founded: 1882; Members: 135
Pres: Stéphane Louryan; Gen Secr: Rosine Orban
Focus: Anthro; Hist; Archeol
Periodicals
Anthropologie et Préhistoire (weekly)
Hominid Remains: An Up-date (weekly) . 01469

Société Royale Belge d'Astronomie, de Météorologie et de Physique du Globe, 3 Av Circulaire, 1180 Bruxelles
T: (02) 3730253; Fax: (02) 3749822
Founded: 1895; Members: 750
Pres: R. Dejaiffe; Gen Secr: T. Camelbeeck
Focus: Astronomy; Geophys
Periodical
Ciel et Terre (weekly) 01470

Société Royale Belge de Chirurgie / Royal Belgian Society of Surgery, 138a Av Circulaire, 1180 Bruxelles
Founded: 1893; Members: 1000
Focus: Surgery
Periodical
Acta Chirurgica Belgica (weekly) 01471

Société Royale Belge de Dermatologie et de Vénérologie, c/o Dr. J. Delescluse, 14 Av Emile Demot, 1050 Bruxelles
T: (02) 6475399; Fax: (02) 6419227
Founded: 1901; Members: 207
Pres: Prof. J.M. Lachapelle; Gen Secr: Dr. J. Delescluse
Focus: Derm; Venereology 01472

Société Royale Belge de Géographie (SRBG), c/o Institut de Géographie, Campus de la Plaine ULB, Blvd du Triomphe, BP 246, 1050 Bruxelles
T: (02) 6400015 ext 5073
Founded: 1876; Members: 400
Pres: Y. Verhasselt; Gen Secr: J.-P. Grimmeau
Focus: Geography
Periodical
Revue Belge de Géographie (weekly) . . 01473

Société Royale Belge de Gynécologie et d'Obstétrique, 309 Av Molière, 1060 Bruxelles
Founded: 1889
Focus: Gynecology 01474

Société Royale Belge de Médecine Physique et de Réhabilitation, 37 Blvd Louis Smidt, 1040 Bruxelles
Founded: 1937; Members: 262
Focus: Rehabil; Physiology 01475

Société Royale Belge d'Entomologie, 29 Rue Vautier, 1040 Bruxelles
Founded: 1855; Members: 180
Gen Secr: P. Grootaert
Focus: Entomology
Periodicals
Bulletin et Annales
Catalogue des Coléoptères de Belgique
Mémoires 01476

Société Royale Belge de Rheumatologie, 33 Rue du Saphir, 1040 Bruxelles
T: (02) 7369439
Founded: 1926; Members: 250
Focus: Rheuma 01477

Société Royale Belge des Electriciens (SRBE), 2 Blvd de la Plaine, 1050 Bruxelles
T: (02) 6412819; Fax: (02) 6412282
Founded: 1884; Members: 1600
Pres: G. Maggetto
Focus: Electric Eng
Periodical
Revue E Tijdschrift (weekly) 01478

Société Royale Belge de Stomatologie et Chirurgie Maxillo-Faciale, c/o Dr. E. de Wolf, Koningstr 16, 9000 Gent
Focus: Stomatology; Surgery 01479

Société Royale d'Archéologie de Bruxelles, c/o Musée de la Porte de Hal, 1000 Bruxelles
Founded: 1887; Members: 1700
Pres: A. de Greef; Gen Secr: Comte J. de Borchgrave d'Altena
Focus: Archeol
Periodicals
Annales
Bulletin 01480

Société Royale de Chimie, Blvd du Triomphe, ULB, Campus Plaine, CP 206/4, 1050 Bruxelles
Founded: 1887; Members: 1000
Focus: Chem
Periodicals
Bulletin (weekly)
Chimie Nouvelle (weekly) 01481

Société Royale d'Economie Politique de Belgique, c/o CIFOP, 1b Av Général Michel, 6000 Charleroi
T: (071) 327394; Fax: (071) 328676
Founded: 1855; Members: 600
Pres: R. Lamy; Gen Secr: A. Schleiper
Focus: Poli Sci
Periodical
Comptes Rendus des Travaux (5-6 times annually) 01482

Société Royale de Médecine Mentale de Belgique, 6 Rue Jean Paquot, 1050 Bruxelles
T: (02) 6482110
Founded: 1869; Members: 179
Focus: Psychiatry
Periodical
Acta Psychiatrica Belgica (weekly) . . . 01483

Société Royale des Amis du Musée Royal de l'Armée et d'Histoire Militaire, c/o Musée Royal de l'Armée, 3 Parc du Cinquantenaire, 1040 Bruxelles
T: (02) 7334493
Founded: 1925; Members: 550
Focus: Military Sci; Hist 01484

Société Royale des Beaux-Arts, 25 Av J. Lambeaux, 1000 Bruxelles
Founded: 1893
Pres: Baron Albert Houtart; Gen Secr: P.P. Hamesse
Focus: Fine Arts 01485

Société Royale des Bibliophiles et Iconophiles de Belgique, 4 Blvd de l'Empereur, 1000 Bruxelles
Founded: 1910; Members: 150
Focus: Libraries & Bk Sci
Periodical
Le Livre et l'Estampe (weekly) 01486

Société Royale des Sciences de Liège, 15 Av des Tilleuls, 4000 Liège
Founded: 1835; Members: 220
Focus: Sci
Periodical
Bulletin (weekly) 01487

Société Royale des Sciences Médicales et Naturelles de Bruxelles, 115 Blvd de Waterloo, 1000 Bruxelles
Founded: 1822; Members: 150
Focus: Med; Nat Sci 01488

Société Royale Zoologique de Belgique / Koninklijke Belgische Vereniging voor Dierkunde, 50 Av F.D. Roosevelt, 1050 Bruxelles
T: (02) 6492055
Founded: 1863; Members: 400
Pres: J. Hulselmans; Gen Secr: F. de Vree
Focus: Zoology
Periodical
Belgian Journal of Zoology (weekly) . . 01489

Société Scientifique de Bruxelles, 61 Rue de Bruxelles, 5000 Namur
T: (081) 230391; Fax: (081) 724502
Founded: 1875; Members: 220
Pres: J. Delhalle; Gen Secr: C. Courtoy
Focus: Sci
Periodical
Revue des Questions Scientifiques . . . 01490

Société Technique et Chimique de Sucrerie de Belgique, 182 Av de Tervueren, 1150 Bruxelles
T: (02) 7710130
Founded: 1931; Members: 170
Focus: Eng; Chem 01491

Union Académique Internationale (UAI) / International Union of Academies, 1 Rue Ducale, 1000 Bruxelles
T: (02) 5126079
Founded: 1919; Members: 34
Pres: R. Ris; Gen Secr: P. Jones
Focus: Sci
Periodicals
Compte Rendu des Sessions (weekly)
Mémoires de la Classe des Sciences (weekly) 01492

Union Belge des Géomètres-Experts Immobiliers (UBG) / Belgische Unie van Landmeters en Meetkundigen-Schatters van Onroerende Goederen, 76 Rue du Nord, 1000 Bruxelles
T: (02) 2180713; Fax: (02) 2193147
Founded: 1946; Members: 700
Pres: J. van den Boogaerde; Gen Secr: J.P. Cloquet
Focus: Surveying; Civil Eng
Periodical
Landmeter Expert Vastgoed (10 times annually) 01493

Union Belge des Journalistes et Ecrivains du Tourisme (U.B.J.E.T.) / Belgische Vereniging van Toeristische Schrijvers en Journalisten, Vlierkenstr 147, 1800 Vilvoorde
T: (02) 2516125
Founded: 1951; Members: 151
Gen Secr: F. Weemaels
Focus: Lit
Periodicals
Nieuws (weekly)
Nouvelles (weekly) 01494

Union des Associations d'Assistants Sociaux Francophones (U.F.A.S.), 3c Rue de la Vierge Noire, 1000 Bruxelles
Members: 3000
Focus: Sociology 01495

Union Européenne des Médecins Spécialistes (UEMS), 20 Av de la Couronne, 1050 Bruxelles
T: (02) 6495164; Fax: (02) 6403730
Founded: 1958
Focus: Med
Periodical
UEMS Bulletin (weekly) 01496

Union Internationale des Sciences Préhistoriques et Protohistoriques (UISPP), c/o Faculté des Lettres, Université, Blandijnberg 2, 9000 Gent
T: (091) 257571
Founded: 1931
Gen Secr: Prof. Dr. J.A.E. Nenquin
Focus: Hist 01497

Union Internationale des Services Médicaux des Chemins de Fer (UIMC), 85 Rue de France, 1070 Bruxelles
T: (02) 5252550; Fax: (02) 5252501
Founded: 1949; Members: 36
Pres: Dr. Endre Tari
Focus: Med 01498

Union Professionnelle Belge des Médécins Ophthalmologistes (UPBMO), 20 Av de la Couronne, 1050 Bruxelles
Founded: 1928; Members: 435
Focus: Ophthal 01499

Union Radio-Scientifique Internationale (URSI), c/o Observatoire Royal de Belgique, 3 Av Circulaire, 1180 Bruxelles
T: (02) 3741308
Founded: 1913; Members: 36
Gen Secr: Y. Stevanovitch
Focus: Electric Eng
Periodicals
Proceedings of General Assemblies
Review of Radio Science
URSI Information Bulletin (weekly) . . . 01500

Union Royale Belge pour les Pays d'Outre-Mer et l'Europe Unie, 22 Rue de Stassart, 1050 Bruxelles
Founded: 1912; Members: 29
Pres: Raoul Suain
Focus: Geography 01501

Union Scientifique Continentale du Verre (U.S.C.V.), 10 Blvd Defontaine, 6000 Charleroi
T: (071) 310041; Fax: (071) 334480
Founded: 1950; Members: 111
Pres: J.P. Mazeau; Gen Secr: P. Eloy
Focus: Materials Sci 01502

Union Syndicale Vétérinaire Belge (USVB), 41 Av Fonsny, 1060 Bruxelles
T: (02) 5381754
Founded: 1934; Members: 1220
Focus: Vet Med 01503

Vereniging der Antwerpsche Bibliophielen, c/o Museum Plantin-Moretus, Vrijdagmarkt 22, 2000 Antwerpen
T: (03) 2330294; Fax: (03) 2262516
Founded: 1877
Focus: Libraries & Bk Sci
Periodical
De Gulden Passer 01504

Vereniging Leraars Aardrijkskunde, De Croylaan 42, 3001 Heverlee
Founded: 1977; Members: 1169
Focus: Educ; Geography
Periodical
De Aardrijkskunde (weekly) 01505

Vereniging van Religieus-Wetenschappelijke Bibliothecarissen (V.R.B.), Minderbroederstr 5, 3800 Sint Truiden
Founded: 1965; Members: 65
Pres: E. d' Hondt; Gen Secr: Kris van de Casteele
Focus: Libraries & Bk Sci
Periodical
V.R.B.-Informatie (weekly) 01506

Vlaamse Chemische Vereniging (VCV), Krijgslaan 281, 9000 Gent
T: (091) 225715 ext 2827
Founded: 1939; Members: 1800
Pres: Dr. J. Engelmann; Gen Secr: Dr. D. de Keukeleire
Focus: Chem
Periodicals
Bulletin des Sociétés Chimiques Belges (weekly)
Chemie Magazine (weekly) 01507

Vlaamse Interuniversitaire Raad / Flemish Interuniversity Council, Egmontstr 5, 1050 Bruxelles
T: (02) 5129110; Fax: (02) 5122996
Founded: 1976
Pres: Léon de Meyer; Gen Secr: Jef van der Perre
Focus: Adult Educ 01508

Vlaamse Museumvereniging / Flemish Museums Association, c/o Y. Morel, Koninklijk Museum voor Schone Kunsten, Plaatsnijdersstr 2, 2000 Antwerpen
Founded: 1968; Members: 240
Pres: A. de Vos
Focus: Archives
Periodicals
Museumbrief (3 times annually)
Museumleven (weekly) 01509

Vlaamse Vereniging voor Familiekunde, Van Heybeeckstr 3, 2170 Merksen
T: (03) 6469988; Fax: (03) 6444620
Founded: 1964; Members: 3000
Focus: Genealogy
Periodical
Vlaamse Stam: Tijdschrift voor Familiegeschiedenis (weekly) 01510

Vlaams Kinesitherapeuten Verbond (VKV), Boeschepestr 70, 8970 Poperinge
T: 333186
Founded: 1963; Members: 3200
Focus: Therapeutics 01511

Von Karman Institute for Fluid Dynamics / Postgraduate Teaching and Research Organization, 72 Chaussée de Waterloo, 1640 Rhode-Saint-Genese
T: (02) 3581901; Fax: (02) 3582885
Founded: 1956; Members: 80
Gen Secr: John F. Wendt
Focus: Aero
Periodicals
Lecture Series Notes
Technical Notes 01512

Belize

National Library Service, c/o Bliss Institute, Southern Foreshore, POB 287, Belize City
Focus: Libraries & Bk Sci 01513

Benin

African Geographers' Association, BP 7060, Cotonou
T: 331917; Tx: 5010; Fax: 331981
Founded: 1981; Members: 45
Gen Secr: Dr. K.S. Adam
Focus: Geography 01514

African Union for the Management of Development Banks (UAMDB), BP 2045, Cotonou
Tx: 5238
Focus: Finance 01515

Association des Pédiatres d'Afrique Noire Francophone (APANF), BP 523, Cotonou
Founded: 1981
Gen Secr: Prof. Félix Hazoumé
Focus: Pediatrics 01516

Bermuda

Amalgamated Bermuda Union of Teachers, POB 726, Hamilton
Founded: 1963
Focus: Educ 01517

Astronomical Society of Bermuda, POB 1054, Hamilton 5
Pres: M.E.B. Nash; Gen Secr: C. McGonagle
Focus: Astronomy
Periodicals
Bermuda Sky Watch
Cosmos 01518

Bermuda Audubon Society, c/o King Edward VII Memorial Hospital, POB 1328, Hamilton 5
Members: 250
Pres: Helge Trapnell
Focus: Ecology; Educ
Periodical
Newsletter (4-6 times annually) 01519

Bermuda Historical Society, Par-la-Ville, Hamilton
Members: 120
Pres: Sir Dudley Spurling
Focus: Hist
Periodical
Bermuda Books Out of Print 01520

Bermuda Medical Society, c/o King Edward VII Memorial Hospital, Paget
T: 2962345
Founded: 1970; Members: 60
Pres: Dr. James King; Gen Secr: Dr. Ian Fulton
Focus: Med 01521

The Bermuda National Trust, POB 61, Hamilton
Founded: 1970; Members: 2000
Pres: Dennis Sherwin; Gen Secr: W.S. Zuill
Focus: Humanities
Periodical
Newsletter (weekly) 01522

Bermuda Society of Arts, c/o Art Gallery, City Hall, POB 1202, Hamilton 5
Founded: 1956; Members: 400
Pres: Chris Wilcox; Gen Secr: Kathryn Peterson
Focus: Arts 01523

Bermuda Technical Society, POB 151, Hamilton 5
Members: 85
Pres: W.R. Holland; Gen Secr: Allan Blundell
Focus: Eng 01524

Bermuda Tuberculosis, Cancer and Health Association, POB 1562, Hamilton 5
T: 57738
Focus: Pulmon Dis; Public Health . . . 01525

Royal Commonwealth Society, Bermuda Branch, POB 673, Devonshire 4
Pres: B. Darling-Meade; Gen Secr: Jent M.
Focus: Humanities 01526

Saint George's Historical Society, Corner Duke of Kent St and Featherbed Alley, Saint George
Members: 109
Pres: W.S. Olivey; Gen Secr: M.V. Ridgeway
Focus: Hist 01527

Bolivia

Academia Boliviana, Casilla de Correo 4145, La Paz
Founded: 1927; Members: 26
Pres: Juan Quirós; Gen Secr: Carlos Castañon Barrientos
Focus: Sci
Periodical
Revista 01528

Academia Nacional de Ciencias de Bolivia, Av 16 de Julio 1732, Casilla de Correo 5829, La Paz
Founded: 1960; Members: 32
Pres: Dr. L.F. Hartmann; Gen Secr: M. Zubieta
Focus: Sci
Periodicals
Boletín Informativo (weekly)
Publicaciones (weekly)
Revista (weekly) 01529

Academia Nacional de la Historia, Av Abel Iturralde 205, La Paz
Founded: 1929; Members: 18
Pres: Dr. David Alvestegui; Gen Secr: Dr. Humberto Vázquez-Machicado
Focus: Hist 01530

Amigos de la Ciudad, Plaza del Teatro Núñez del Prado 576, Casilla de Correo 911, La Paz
Founded: 1916; Members: 518
Focus: Urban Plan 01531

Asociación de Arquitectos de Bolivia, Casilla de Correo 1498, La Paz
Founded: 1942; Members: 35
Pres: Hugo López Videla; Gen Secr: Fernando Calderón
Focus: Archit 01532

Asociación de Ingenieros y Geólogos de Yacimientos Petrolíferos Fiscales Bolivianos (AIGYPFB), Casilla de Correo 401, La Paz
Founded: 1959; Members: 210
Focus: Geology; Energy 01533

Ateneo de Medicina de Sucre, Sucre
Focus: Med 01534

Centro Intelectual Galindo, Oficina Mapiri, La Paz
Founded: 1924
Focus: Sci 01535

Centro Nacional de Documentación Científica y Tecnológica, Plaza del Obelisco, Edificio Facultad de Tecnologia, Casilla de Correo 3283, La Paz
T: 359586, 359587
Founded: 1967
Focus: Doc
Periodicals
Actualidades (weekly)
Serie Bibliográfica (4-5 times annually) . 01536

Centro Nacional de Documentación e Información Educativa, c/o Ministerio de Educación y Cultura, La Paz
Founded: 1967
Focus: Educ 01537

Centro Pedagógico y Cultural de Portales, Av Potosí, Casilla de Correo 544, Cochabamba
T: 1137
Founded: 1969
Focus: Educ 01538

Círculo de Bellas Artes, Plaza Teatro, La Paz
Founded: 1912
Pres: Ernesto Peñaranda
Focus: Arts 01539

Comisión Bolivana de Energía Nuclear, Av 6 de Agosto 2905, Casilla de Correo 4821, La Paz
Founded: 1960
Focus: Nucl Res 01540

Comisión de Planeamiento y Coordinación de las Universidades Bolvianas, c/o Universidad Mayor de San Simón, Cochabamba
Focus: Sci 01541

Confederación Boliviana de Odontólogos, Casilla de Correo 2203, La Paz
Founded: 1940
Focus: Dent 01542

Confederación Universitaria Boliviana (CUB), c/o Universidad Mayor de San Andrés, Villazón 465, La Paz
Focus: Adult Educ 01543

Consejo Nacional de Educación, Yanacocha 475, La Paz
Focus: Educ 01544

Fundación Universitaria Simon I. Patiño, Casilla de Correo 3687, Cochabamba
Founded: 1931; Members: 5
Pres: Rafael Anaya Arze
Focus: Adult Educ 01545

Instituto Boliviano del Petróleo (IBP), Casilla de Correo 4722, La Paz
Founded: 1959; Members: 50
Pres: José Patiño; Gen Secr: Reynaldo Salgueiro Pabón
Focus: Geology; Energy
Periodicals
Boletín del Instituto Boliviano del Petróleo
Manual de Signos Convencionales . . . 01546

PEN Club de Bolivia-Centro Internacional de Escritores, Goitia 17, Casilla de Correo 149, La Paz
Founded: 1931; Members: 47
Gen Secr: Yolanda Bedregal de Cónitzer
Focus: Lit 01547

Sociedad Arqueológica de Bolivia, Av Chacaltaya 500, Casilla de Correo 1487, La Paz
Focus: Archeol 01548

Sociedad Boliviana de Cirugía, Casilla de Correo 1252, La Paz
Founded: 1939; Members: 185
Focus: Surgery 01549

Sociedad Boliviana de Salud Pública, Plaza Franz Tamayo, Casilla de Correo 151, La Paz
Founded: 1963
Focus: Public Health 01550

Sociedad de Estudios Geográficos e Históricos, Plaza 24 de Setiembre, Santa Cruz de la Sierra
Founded: 1903
Focus: Geography; Hist 01551

Sociedad de Pediatría de Cochabamba, Casilla de Correo 1429, Cochabamba
Founded: 1945; Members: 14
Focus: Pediatrics 01552

Sociedad Geográfica de La Paz, Tiahuanaco 12, Casilla de Correo 632, La Paz
Founded: 1889; Members: 580
Focus: Geography 01553

Sociedad Geográfica Sucre, Plaza 25 de Mayo, Casilla de Correo 101, Sucre
Founded: 1887; Members: 21
Focus: Geography 01554

Sociedad Geográfica y de Historia Potosío, Casa Nacional de Moneda, Casilla de Correo 39, Potosí
Founded: 1905; Members: 20
Focus: Geography; Hist 01555

Sociedad Geológica Boliviana, Edificio Miniminas, Casilla de Correo 2729, La Paz
Founded: 1961
Focus: Geology 01556

Sociedad Rural Boliviana, Comercio 979, Casilla de Correo 786, La Paz
Founded: 1934; Members: 31
Pres: Carlos Montes y Montes
Focus: Agri
Periodicals
IFAP News
Universitas 01557

Botswana

The Botswana Society, c/o National Museum and Art Gallery, Independence Av, POB 71, Gaborone
T: 351500; Fax: 359321
Pres: Gobe Matenge; Gen Secr: Angela Scales
Focus: Sci
Periodical
Botswana Notes and Records (weekly) . 01559

Brazil

Academia Alagoana de Letras, 57000 Maceió, AL
Focus: Lit 01560

Academia Amazonense de Letras, Rua Ramos Ferreira 1009, 69000 Manaus, AM
Founded: 1918; Members: 40
Focus: Lit
Periodical
Revista 01561

Academia Brasileira de Ciência da Administração / Brazilian Academy of Administration, Av 13 de Maio 23, Salas 1117-1119, Rio de Janeiro, RJ
Founded: 1973; Members: 40
Focus: Public Admin 01562

Academia Brasileira de Ciências / Brazilian Academy of Sciences, CP 229, 20001-970 Rio de Janeiro, RJ
T: (021) 2204794; Tx: (021) 2123087; Fax: (011) 2404695
Founded: 1916; Members: 286
Pres: Prof. E.M. Krieger; Gen Secr: A. Guidão Gomes
Focus: Sci
Periodicals
Anais da Academia Brasileira de Ciências (weekly)
Revista Brasileira de Biologia (weekly) . 01563

Academia Brasileira de Letras, Av Presidente Wilson 203, 20030 Rio de Janeiro, RJ
Founded: 1897
Focus: Lit
Periodical
Revista 01564

Academia Cachoeirense de Letras, Prâca J. Monteiro 105, 29300 Cachoeira de Itapemerim, ES
Founded: 1962; Members: 40
Focus: Lit 01565

Academia Campinense de Letras, Rua Marechal Deodoro 525, 13100 Campinas, SP
Focus: Lit
Periodical
Publicações (weekly) 01566

Academia Catarinense de Letras, Rua Vidal Ramos, Edifício José Daux, 88000 Florianópolis, SC
Founded: 1920; Members: 60
Focus: Lit
Periodical
Revista 01567

Academia Cearense de Letras, Palácio Senador Alencar, Rua São Paulo 51, 60030 Fortaleza, CE
Founded: 1894; Members: 61
Pres: Claudio Martins; Gen Secr: I. de Santiago Espindola
Focus: Lit
Periodical
Revista 01568

Academia de Letras, 58000 Jõao Pessoa, PB
Focus: Lit 01569

Academia de Letras da Bahia, Praça 15 de Novembro 15a, 40000 Salvador, BA
Founded: 1917; Members: 65
Focus: Lit
Periodical
Revista 01570

Academia de Letras de Piauí, 64000 Teresina, PI
Focus: Lit
Periodical
Revista 01571

Academia de Letras e Artes do Planalto / Planalto Academy of Arts and Letters, Rua Joseph de Mello Alvares 57, 77220 Luziânia, GO
Founded: 1976; Members: 30
Focus: Arts; Lit 01572

Academia de Medicina de São Paulo, Rua Teodoro Sampaio 115, 05405 São Paulo, SP
Founded: 1895
Pres: Dr. Arthur B. Garrido Junior; Gen Secr: Dr. José R. Louza; Dr. S. Atlas
Focus: Med 01573

Academia Matogrossense de Letras, Rua 13 de Junho 173, 78000 Cuiabá, MT
Founded: 1921
Focus: Lit
Periodical
Revista 01574

Academia Miniera de Letras, Rua Carijos 150, 30000 Belo Horizonte, MG
Focus: Lit 01575

Academia Nacional de Farmacia, Rua dos Andradas 96, 20000 Rio de Janeiro, RJ
Founded: 1937
Focus: Pharmacol
Periodical
Boletim 01576

Academia Nacional de Medicina, Av General Justo 365, CP 459, 20021 Rio de Janeiro, RJ
Founded: 1829; Members: 100
Pres: José C. de Paula Lopes Pontes; Gen Secr: José Cardoso de Castro
Focus: Med
Periodical
Boletim 01577

Academia Paraibana de Letras, Rua Duque de Caxias 25, CP 334, 58000 João Pessoa, PB
Members: 40
Focus: Lit
Periodicals
Boletim Informativo
Discursos e Ensaios
Revista 01578

Academia Paulista de Letras (APL), Largo do Arouche 312, 01219 São Paulo, SP
T: (011) 2207222
Founded: 1909; Members: 40
Pres: Antônio Soares Amôra
Focus: Lit
Periodical
Revista No. 107 01579

Academia Pernambucana de Letras, Av Rui Barbosa 1596, Graças, 50000 Recife, PE
Founded: 1901; Members: 40
Focus: Lit
Periodical
Revista 01580

Academia Riograndense de Letras, Rua Cândido Silveira 43, 90000 Pôrto Alegre, RS
Focus: Lit
Periodical
Revista 01581

Asociación Interamericana de Ingenieria Sanitaria y Ambiental / Inter-American Association of Sanitary and Environmental Engineering, Rua Nicolau Gagliard 354, 05429 Saõ Paulo, SP
T: (011) 2124080; Tx: 81453; Fax: (011) 8142441
Founded: 1946; Members: 10000
Focus: Eng
Periodicals
Desafio (weekly)
Ingenieria Sanitaria (weekly) 01582

Asociación Latinoamericana de Escuelas de Trabajo Social (ALAETS) / Latin American Association of Schools of Social Work, c/o Dep de Servicio Social, Universidade Federal do Maranho, 65000 Saõ Luís, MA
Founded: 1965
Pres: Rodrigo Facio; Gen Secr: Josefa Batista Lopes
Focus: Soc Sci; Educ 01583

Asociación Latinoamericana de Paleobotánica y Palinología (ALPP) / Latin American Association of Paleobotany and Palynology, c/o Instituto de Geociencias, Universidade Federal do RGS, CP 15001, Porto Alegre, RS
Founded: 1972
Gen Secr: Miriam Cazzulo Klepzig
Focus: Paleontology; Botany
Periodical
Boletín de la Asociación Latinoamericana de Paleobotánicy y Palinología (weekly) . . 01584

Asociación Latinoamericana para la Producción Animal (ALPA) / Latin American Association for Animal Production, Av Benito Goncalves 7712, CP 776, 90001 Porto Alegre, RS
T: (0512) 365011; Tx: 520253; Fax: (0512) 272295
Founded: 1966; Members: 23
Gen Secr: Prof. Sergio Nicolaiewsky
Focus: Animal Husb 01585

Associação Bahiana de Medicina, Av 7 de Setembro 48, 40000 Salvador, BA
Founded: 1894
Focus: Med 01586

Associação Brasileira de Educadores Lassalistas, Av José Pereira Lopes 252, CP 352, 13560 São Carlos, SP
T: (0162) 710121; Fax: (0162) 717100
Focus: Educ 01587

Associação Brasileira de Engenharia Sanitária e Ambiental (ABES), Av Beira Mar 296, 20021-060 Rio de Janeiro, RJ
T: (021) 2103221; Tx: 31902; CA: ABESRIO; Fax: (021) 2626838
Founded: 1966; Members: 8000
Focus: Ecology
Periodicals
Catálogo Brasileiro de Engenharia Sanitária e Ambiental – CABES (weekly)
Jornal de ABES (weekly)
Revista Bio (weekly) 01588

Associação Brasileira de Escolas Superiores Católicas (ABESC), SGAN, Quadra 916, Módulo B, 70790-160 Brasília, DF
Pres: Prof. Ivo Mauri
Focus: Educ 01589

Associação Brasileira de Farmacêuticos, Rua Andradas 96, 20051 Rio de Janeiro
T: (021) 2630791
Founded: 1916; Members: 2000
Focus: Pharmacol
Periodical
Revista Brasileira de Farmacia 01590

Associação Brasileira de Imprensa, Rua Araujo Porto Alegre 71, 20000 Rio de Janairo, RJ
Focus: Libraries & Bk Sci 01591

Associação Brasileira de Mecânica dos Solos (ABMS), CP 7141, 01000 São Paulo, SP
Focus: Eng
Periodical
Solos e Rochas (weekly) 01592

Associação Brasileira de Química, Av Rio Branco 156, CP 550, 20000 Rio de Janeiro, RJ
Founded: 1951; Members: 3500
Focus: Chem 01593

Associação Católica Interamericana de Filosofía, Rua Marquês de São Vicente 293, 22451 Rio de Janeiro, FJ
T: (021) 2744596
Founded: 1972
Focus: Philos 01594

Associação de Educação Católica do Brasil, SCS Q3, Bloco A, 79, Edif João Paulo II, 70300 Brasília, DF
Founded: 1945; Members: 1750
Pres: P. Agustín Castejón
Focus: Educ
Periodical
Revista de Educação AEC 01595

Associação de Ensino, Rua Hygino Muzzi Filho 1001, CP 554, 17500 Marília, SP
Focus: Educ 01596

Associação de Ensino, Av Costabile Romano 2201, 14100 Ribeirão Preto, SP
T: 257306
Focus: Educ 01597

Associação de Ensino e Cultura Urubupungá, Av Cel. Jonas Alves de Mello 1660, CP 126, 15370 Pereira Barreto, SP
T: 2044
Focus: Educ 01598

Associação dos Advogados, Largo de São Francisco 34, 01005 São Paulo, SP
T: (011) 2588355
Focus: Law 01599

Associação Educacional Presidente Kennedy, Rua Barão de Maurá 600, 07000 Guarulhos, SP
T: 2095681
Focus: Educ 01600

Associação Internacional de Críticos de Arte, Rua Visconde de Piraja 228, App 802, 20000 Rio de Janeiro, RJ
Focus: Arts 01601

Associação Itaquerense de Ensino, Rua Carolina Fonseca 548, 08200 Itaquera, SP
T: 2976065
Focus: Educ 01602

Associação Médica Brasileira, Rua São Carlos do Prinhal 324, CP 8904, 01333 São Paulo
T: (011) 2893511; Tx: 1135300
Founded: 1951; Members: 35000
Focus: Med
Periodicals
Journal da Associação Médica Brasileira (weekly)
Revista da Associação Médica Brasileira (weekly) 01603

Associação Pananericana de Medicina Social / Pan American Association of Social Medicine, Moura 81, Tijuca, Rio de Janeiro, RJ
Focus: Med 01604

Associação Paulista de Bibliotecários (APB), Rua Maestro Cardim 94, Liberdade, 01323-000 São Paulo
T: (011) 2853831; Fax: (011) 2853831
Founded: 1938; Members: 1500
Pres: Marta Ligia P. Valentim; Gen Secr: Marcia D. Segato
Focus: Libraries & Bk Sci
Periodical
Palavra Chave: Key Word (weekly) . . 01605

Associação Paulista de Medicina, Av Brigadeiro Luíz António 278, CP 2103, 01000 São Paulo, SP
Founded: 1930; Members: 25000
Pres: Dr. Celso C. Campos Guerra; Gen Secr: Dr. Roberto Simão Mathias
Focus: Med
Periodicals
Jornal da Associação Paulista de Medicina (weekly)
Revista Paulista de Medicina (weekly) . 01606

Associação Pernambucana de Bibliotecarios, Rua das Fronteiras 255, 50000 Recife, PE
Focus: Libraries & Bk Sci 01607

Associação Prudentina de Educação e Cultura, Rua José Bongiovani 700, 19100 Presidente Prudente
Focus: Educ; Physical Therapy; Hygiene; Vet Med; Business Admin 01608

Associação Rio-Grandense de Bibliotecarios (ARB), CP 2344, 90020-122 Pôrto Alegre, RS
T: (051) 2258194
Founded: 1951; Members: 500
Pres: Luciana Franke Nebel
Focus: Libraries & Bk Sci 01609

Associação Tibirica de Educação, Largo São Bento, 01029 São Paulo, SP
T: (011) 2273377
Focus: Educ 01610

Associação Universitaria Santa Ursula, Rua Fernando Ferrari 75, 20000 Rio de Janeiro, RJ
Focus: Educ 01611

Associaçcão de Geografia Teorética, CP 178, 13500 Rio Claro, SP
Focus: Geography
Periodicals
Boletim de Geografia Teorética (weekly)
Geografia (weekly) 01612

Association of Amazonian Universities (UNAMAZ) / Association des Universités de l'Amazonie, CP 558, 66010 Belém, Paraná
T: (091) 2299310; Tx: 1013; Fax: (091) 2294339, 2299677
Founded: 1987; Members: 57
Focus: Educ 01613

Association of Latin American Lawyers for the Defense of Human Rights, Av São Luiz 131, 01046 São Paulo, SP
Founded: 1979
Focus: Law 01614

Centro Académico Hugo Simas, Rua Marechal Floriano Peixoto 524, 80000 Curitiba, PI
Focus: Sci 01615

Centro Brasileiro de Estudos, Rua Sacramento 108, 13100 Campinas, SP
Focus: Sci 01616

Centro Brasileiro de Pesquisas Fisicas, Av Wenceslau Bráz 71, 20000 Rio de Janeiro, RJ
Focus: Physics 01617

Centro Cultural de Botucatu, Rua Cesário Alvim 296, 18600 Botucatu, SP
Focus: Sci 01618

Centro de Análise Conjuntura Econômica / Centre for Analysis of Economic Affairs, Av Gomes Freire 647, 20231 Rio de Janeiro, RJ
Focus: Econ 01619

Centro de Aperfeiçoamento e Especialização Médica, Rua Sacadura Cabral 178, 20000 Rio de Janeiro, RJ
Focus: Metallurgy 01620

Centro de Biomédica de Campina Grande, 58100 Campina Grande, PB
Focus: Bio; Med 01621

Centro de Ciências, Letras e Artes, Rua Bernardino de Campos 989, 13100 Campinas, SP
Focus: Lit 01622

Centro de Estudios de Demografía Histórica de América Latina, c/o Faculdade de Filosofía, Letras e Ciencias Humanas, Universidade de São Paulo, CP 8105, 05508 São Paulo
T: (011) 2102122 ext 616
Founded: 1986
Gen Secr: Dr. Maria Luiza Marcílio
Focus: Hist
Periodical
Estudios CEDHAL (weekly) 01623

Centro de Estudos e Pesquisas em Administração, Av João Pessoa 52, 90000 Porto Alegre
T: (0512) 243837
Founded: 1959
Focus: Public Admin 01624

Centro de Pesquisa Agropecuária dos Cerrados (CPAC) / Center for Agricultural Research for the Brazilian Savannahs, Estrada Brasilia-Fortaleza Km 18, CP 08223, 73301-970 Planaltina, DF
T: (061) 3891171; Fax: (061) 3892953
Focus: Agri
Periodicals
Anual do Centro de Pesquisa Agropecuária dos Cerrados (weekly)
Boletim de Pesquisa (weekly)
Circular Técnica (weekly)
Comunicado Técnico (weekly)
Documentos (weekly)
Pesquisa em Andamento (weekly) . . . 01625

Centro de Pesquisa Agropecuária do Trópico-Árido (EMBRAPA), BR-428, Km 152, s/n, Zona Rural, Apdo 23, 56300 Petrolina, PE
T: (081) 9614411; Tx: 810016; Fax: (081) 9615681
Founded: 1975
Focus: Agri
Periodicals
Annual Research Report (weekly)
Research Bulletin (weekly)
Technical Bulletin (weekly)
Technical Information (weekly) 01626

Centro de Pesquisa Agropecuária do Trópico Umido (CPATU), Travessa Dr. Eneas Pinheiro, Bairro do Marco, CP 48, 66000 Belém
Focus: Agri
Periodicals
Boletim de Pesquisa
Circular Técnica
Documentos
Relatório Décimo Anual CPATU 01627

Centro de Pesquisas de Geografia do Brasil, Largo de São Francisco, 20000 Rio de Janeiro, RJ
Focus: Geography 01628

Centro de Pesquisas Folclóricas, c/o Escola de Música, Rua do Passeio 98, 20021 Rio de Janeiro, RJ
Pres: Rosa Maria Zamith
Focus: Ethnology
Periodical
Revista Brasileira de Música 01629

Centro Latino Americano de Física (CLAF) / Latin American Centre for Physics, Av Venceslau Braz 71, 22290-970 Rio de Janeiro, RJ
T: (021) 2955096, 2955145; Tx: 2122563; CA: CELAFI; Fax: (021) 5412047
Founded: 1962
Gen Secr: Prof. Juan José Giambiagi
Focus: Physics
Periodical
Noticia (weekly) 01630

Centro Nacional de Informação Científica em Microbiologia, Av Pasteur 250, 20000 Rio de Janeiro, RJ
Focus: Microbio 01631

Centro Nacional de Pesquisa de Arroz e Feijão (EMBRAPA), Rodovia GYN 10, km 12, CP 179, 74001-970 Goiânia, GO
Fax: (062) 2613880
Focus: Agri; Food
Periodicals
Série Documentos (weekly)
Série Técnica (weekly) 01632

Centro Nacional de Pesquisa de Mandioca e Fruticultura (EMBRAPA), Rua Lauro Passos, CP 007, 44380 Cruz das Almas, BA
T: 7212120
Focus: Agri; Food
Periodicals
Boletim Agrometeorológico (weekly)
Boletim de Pesquisa (weekly)
Boletim Informativo (weekly)
Circular Técnica (weekly)
Comunicado Técnico (weekly)
Documentos (weekly)
Pesquisa em Andamento (weekly)
Relatório Técnico Anual (weekly)
Revista Brasileira de Mandioca (weekly) . 01633

Centro Nacional de Pesquisa de Milho e Sorgo, CP 151, 35701-970 Sete Lagoas, MG
T: (031) 9215644; Tx: (031) 2099; Fax: (031) 9219252
Founded: 1974
Focus: Agri
Periodicals
Circular Técnica
Documentos
Relatório 01634

Centro Nacional de Pesquisa de Seringueira (EMBRAPA), Rodovia AM 10, km 30, CP 319, 69000 Manaus, AM
T: (092) 2341417
Focus: Sci
Periodicals
Boletim de Pesquisa
Circular Técnica
Resumos Informativos
Série Documento 01635

Centro Nacional de Pesquisa de Soja (EMBRAPA), Rodovia Celso Garcia Cid, Km 375, CP 1061, 86100 Londrina
Focus: Food; Agri 01636

Centro Nacional de Pesquisa do Algodão (EMBRAPA), CP 174, 58100 Campina Grande, PB
Focus: Agri 01637

Centro Regional de Pesquisas Educacionais do Sul, Av Oswaldo Aranha 271, CP 2872, 90000 Porto Alegre, RS
Focus: Educ 01638

Comissão Brasileira de Documentâcão Agricola (CBDA), c/o Ministério da Agricultura, Anexo I, Bloco H, Ala Oeste, Esplanada dos Ministérios, CP 102432, 70043 Brasília, DF
T: (061) 2251101; Tx: 1871
Focus: Doc
Periodical
Agricolas (weekly) 01639

Comissão Nacional de Folclore, Av Marechal Floriano 196, 20080-002 Rio de Janeiro, RJ
Focus: Ethnology; Folklore
Periodical
Boletim da Comissão Nacional de Folclore (weekly) 01640

Confederación Panamericana de Medicina Deportiva (COPAMADE), Rua Felipe Becker 95, 91330 Porto Alegre, R.S.
T: (0512) 348083; Tx: 1055
Founded: 1975; Members: 27
Pres: Dr. Eduardo Henrique de Rosa
Focus: Med 01641

Congresso da América do Sul da Zoologia, CP 7172, 01000 São Paulo, SP
Focus: Zoology 01642

Conselho Federal de Biblioteconomia (CFB), CRN Quadra 702/703, Bloco G, Edifício Coencisa Sobreloja, 70000 Brasília, DF
Founded: 1966
Focus: Libraries & Bk Sci 01643

Conselho Federal de Educação, Rua da Imprensa 16, 20000 Rio de Janeiro, RJ
Focus: Educ 01644

Conselho Nacional de Desenvolvimento Científico e Tecnológico / National Council of Scientific and Technological Development, Av W3 Norte, Q. 507/B, CP 111142, 70740 Brasília, DF
T: (061) 2741155; Tx: 1089
Founded: 1951
Focus: Eng
Periodical
Revista Brasileira de Tecnologia (weekly) 01645

Conselho Nacional Serviço Social, Imprensa 16, 20000 Rio de Janeiro, RJ
Focus: Sociology 01646

Coordenação de Folclore e Cultura Popular, Rua do Catete 179, 22220-000 Rio de Janeiro, RJ
T: (021) 2850441
Gen Secr: Claudia Marcia Ferreira
Focus: Ethnology
Periodicals
Bibliografia Folclórica
Cuadernos de Folclore
Documentário Sonoro
Série Encontros e Estudos
Série Referência 01647

Empresa Brasileira de Pesquisa Agropecuária (EMBRAPA) / Centro Nacional de Pesquisa de Gado de Corte / Brazilian Agricultural Research Corporation, Rodovia BR 262 km 04, CP 154, 79106-000 Campo Grande
T: (067) 7631030; Fax: (067) 7632245
Founded: 1976; Members: 47
Pres: Ivo Martins Cezar; Gen Secr: Geraldo Ramos de Figueiredo
Focus: Agri; Vet Med
Periodicals
Boletim Agrometeorológico (weekly)
Boletim de Pesquisa (weekly)
Circular Técnica (weekly)
Comunicado Técnico (weekly)
Documentos (weekly)
Pesquisa em Andamento (weekly)
Relatório Técnico Anual (weekly) 01648

Federação Brasileira de Associaeões de Bibliotecarios (FEBAB), Rua Avanhandava 40, 01306 São Paulo
Founded: 1959; Members: 27
Pres: Salvadore Nascimento Mirian
Focus: Libraries & Bk Sci
Periodicals
Jornal da Febab (weekly)
Revista Brasileira de Biblioteconomia e Documentação (weekly) 01649

Fundação Antonio Prudente (FAP), c/o Instituto Central-Hospital A. C. Camargo, Rua Prof. Antonio Prudente 211, CP 5271, 01509 São Paulo, SP
T: (011) 2788811
Focus: Cell Biol & Cancer Res
Periodical
Acta Oncologica Brasileira (weekly) . . . 01650

Fundação Centro de Pesquisas e Estudos (CPE) / Foundation Centre of Research and Study, Centro Administrativo, Av Luíz Viana Filho, Paralela, 40000 Salvador, BA
Founded: 1960
Focus: Adult Educ; Public Admin
Periodical
Planejamento (weekly) 01651

Fundação de Estudos Sociais do Paraná, Rua Cametro 216, 80000 Curitiba, PI
Focus: Sociology 01652

Fundação Educacional de Fortaleza, Av Capistrano de Abreu 5609, 60000 Fortaleza, CE
Founded: 1974
Focus: Educ 01653

Fundação Getulio Vargas (FGV), Praia de Botafogo 190, 22253-900 Rio de Janeiro, RJ
Fax: (021) 5536372
Founded: 1944; Members: 1440
Pres: Prof. Jorge Oscar Mello Flores
Focus: Sci
Periodicals
Correio da Unesco (weekly)
Finanças e Desenvolvimento (weekly)
Revista Brasileira de Economia (weekly)
Revista de Administraçao de Empresas (weekly)
Revista de Administração Pública (weekly) 01654

Fundação Instituto Brasileiro de Geografia e Estatística, Av Franklin Roosevelt 166, 20021 Rio de Janeiro, RJ
T: (021) 3917788, 2304747; CA: IBGE
Focus: Geography; Stats
Periodicals
Revista Brasileira de Estatística (weekly)
Revista Brasileira de Geografia (weekly) . 01655

Fundação Instituto Tecnológico do Estado de Pernambuco (ITEP), Av Conde da Boa Vista 428, Recife, PE
Founded: 1942; Members: 181
Focus: Eng
Periodical
Boletim Técnico (weekly) 01656

Fundação João Pineiro, Alameda das Acácias 70, CP 2210, 31270 Belo Horizonte
T: (031) 4411133; Tx: 1302
Founded: 1970
Focus: Public Admin; Econ; Urban Plan; Poli Sci; Hist; Soc Sci
Periodical
Análise & Conjuntura (weekly) 01657

Fundação Joaquim Nabuco (FUNDAJ) / Joaquim Nabuco Foundation, Av 17 de Agosto 2187, Casa Forte, 52061-070 Recife, PE
T: (081) 4415500; Tx: (081) 1180; CA: JONABUCO; Fax: (081) 4415600
Founded: 1949; Members: 477
Pres: Fernando de Mello Freyre
Focus: Anthro; Econ; Sociology
Periodicals
Cadernos de Estudos Sociais (weekly)
Revista Ciência e Trópico (weekly) . . . 01658

Fundação Moinho Santista / Moinho Santista Foundation, Av M.C. Aguiar 215, Bloco A, São Paulo, SP
Founded: 1955
Focus: Arts; Lit; Sci 01659

Fundação Nacional do Indio (FUNAI) / Indian National Foundation, SEUPS Quadra 702, Edificio LEX, 70330 Brasília
T: (061) 2268211; Tx: 1344
Founded: 1967
Focus: Anthro; Ethnology 01660

Grémio Literario Carlos Ferreira, Praça Barão do Rio 141, 13900 Amparo, SP
Focus: Lit 01661

Grémio Literario e Comercial Portugués, Rua Senador Manuel Barata 237, 66000 Belém, PA
Focus: Lit 01662

Instituto Archeológico, Histórico e Geográfico Pernambucano, Rua do Hospicio 130, 50000 Recife, PE
Founded: 1862; Members: 185
Focus: Hist; Archeol; Lit; Geography; Ethnology; Sociology 01663

Instituto Brasileiro de Economia / Brazilian Institute of Economics, c/o Getúlio Vergas Foundation, CP 9052, 22250 Rio de Janeiro, RJ
Founded: 1951
Focus: Econ
Periodicals
Agroanalysis (weekly)
National Accounts (weekly)
Revista Brasileira de Economia (weekly)
Revista Conjuntura Economia (weekly) . 01664

Instituto Brasileiro de Educâcão, Ciéncia e Cultura (IBECC), Av Marechal Floriano 196, 20080-002 Rio de Janeiro, RJ
T: (021) 5162458; Fax: (021) 5516897
Members: 46
Focus: Educ
Periodical
Correio do IBECC (weekly) 01665

Instituto de Botânica / Botanical Institute, Av Miguel Estefano 3687, CP 4005, 01000 São Paulo, SP
T: (011) 2753322; Tx: 381123694; Fax: (011) 5773678
Founded: 1938; Members: 81
Focus: Botany
Periodicals
Boletin do Instituto de Botánica (weekly)
Hoehnea (weekly)
Monographs (weekly) 01666

Instituto de Engenharia de São Paulo, Palacio Maná, 01000 São Paulo, SP
Founded: 1917; Members: 7500
Focus: Eng 01667

Instituto de Planejamento de Pernambuco, Rua Gervásio Pires 399, Boa Vista, 50000 Recife, PE
T: (081) 2315005; Tx: 3167; Fax: (081) 2313379
Founded: 1952; Members: 226
Focus: Agri; Ecology; Econ; Sociology
Periodicals
Desempenho da Economia de Pernambuco (weekly)
Revista Pernambucana de Desenvolvimento (weekly)
Sondagem Conjuntural (weekly) 01668

Instituto do Ceará, Rua Barã do Rio Branco 1594, 60025-061 Fortaleza, CE
Founded: 1887; Members: 40
Pres: Prof. Geraldo da Silva Nobre; Gen Secr: Prof. Vinicius Barros Leal
Focus: Hist; Geography; Anthro
Periodical
Revista do Instituto do Ceará (weekly) . 01669

Instituto dos Advogados Brasileiros, Av Marechal Câmera 210, 20000 Rio de Janeiro, RJ
Founded: 1843; Members: 600
Focus: Law 01670

Instituto Genealógico Brasileiro, Rua Dr. Zuquim 1525, 01000 São Paulo, SP
Focus: Genealogy 01671

Instituto Geográfico e Histórico da Bahia (I.G.H.B.), Av Sete de Setembro 94A, 40000 Salvador, BA
Founded: 1894
Focus: Geography; Hist
Periodical
Revista do I.G.H.B. (weekly) 01672

Instituto Geográfico e Histórico do Amazonas, Rua B. Ramos 131, 69000 Manaus, AM
Founded: 1917; Members: 194
Focus: Geography; Hist 01673

Instituto Histórico de Alagoas, Rua J. Pessoa 382, 57000 Maceió, AL
Founded: 1869; Members: 40
Focus: Hist 01674

Instituto Histórico e Geográfico Brasileiro, Av A. Severo 8, 20021 Rio de Janeiro, RJ
Founded: 1838
Pres: Prof. Américo Jacobina Lacombe; Gen Secr: Prof. Vicente Costa Santos
Focus: Geography; Hist
Periodical
Revista (weekly) 01675

Instituto Histórico e Geográfico de Goiás, 74000 Goiânia, GO
Focus: Geography; Hist 01676

Instituto Histórico e Geográfico de Santa Catarina (IHGSC), Praca XV de Novembro, Palácio Cruz e Sousa, CP 1582, 88000 Florianópolis
Founded: 1896; Members: 100
Pres: Prof. Dr. Victor A. Peluso; Gen Secr: Prof. Jali Meirinho
Focus: Geography; Hist
Periodical
Revista do IHGSC (weekly) 01677

Instituto Histórico e Geográfico de Santos, Av Concelheiro Nébias 689, Boqueirão, 11100 Santos, SP
Focus: Hist; Geography 01678

Instituto Histórico e Geográfico de São Paulo, Rua B. Constant 158, 01005-000 São Paulo, SP
T: (011) 2323582
Founded: 1894
Focus: Geography; Hist
Periodical
Revista (weekly) 01679

Instituto Histórico e Geográfico de Sergipe, Rua de Itabaianinha 41, 49 000 Aracajú, SE
Founded: 1912
Focus: Geography; Hist 01680

Instituto Histórico e Geográfico do Maranhão, Rua Santa Rita 230, 65000 São Luís, MA
Founded: 1925
Focus: Geography; Hist
Periodical
Revista 01681

Instituto Histórico e Geográfico do Pará, Rua d'Aveiro, Cidade Irmã 62, CP 547, 66000 Belém, PA
Founded: 1917
Focus: Geography; Hist
Periodical
Leusla (weekly) 01682

Instituto Histórico e Geográfico do Rio Grande do Norte (I.H.G./R.N.), Rua da Conceição 622, 59000 Natal
Founded: 1902
Pres: Dr. Enélio Lima Petrovich
Focus: Geography; Hist
Periodical
Revista do I.H.G./R.N. (weekly) 01683

Instituto Histórico e Geográfico do Rio Grande do Sul, Rua Riachuelo 1317, 90010 Porto Alegre
Founded: 1920
Pres: Dr. Luis Alberto Cibils
Focus: Geography; Hist
Periodical
Revista (weekly) 01684

Instituto Histórico e Geográfico Paraíbano, Rua Barão do Abiaí 64, CP 37, 58000 João Pessoa, PB
Founded: 1905; Members: 70
Focus: Geography; Hist 01685

Instituto Histórico, Geográfico e Etnográfico Paranaense, Rua J. Loureiro 43, 80000 Curitiba, PI
Founded: 1900
Focus: Ethnology; Geography; Hist . . . 01686

Instituto Nacional de Estudos e Pesquisas Educacionais / National Institute for Educational Research, CP 04366, 70300 Brasília, DF
Focus: Educ 01687

Instituto Nacional de Pesquisas da Amazonia, Rua André Araújo 1756, CP 478, 69083-970 Manaus, AM
T: (092) 6423377
Focus: Ecology; Botany; Bio; Forestry; Chem; Nutrition; Eng
Periodical
Acta Amazonica (weekly) 01688

Instituto Nacional de Pesquisas Hidroviarias, Rua General Gurjão 166, 20000 Rio de Janeiro, RJ
Focus: Water Res 01689

Instituto Nacional de Tecnologia (INT) / National Institute of Technology, Av Venezuela 82, 20081 Rio de Janeiro, RJ
T: (021) 2231320; Tx: 30056
Founded: 1921
Focus: Chem; Eng
Periodicals
Corrosão & Protecão Boletim Informativo (weekly)
Informativo INT (weekly) 01690

Instituto Nacional do Cancer, Praça Cruz Vermelha 23, 20000 Rio de Janeiro, RJ
Focus: Cell Biol & Cancer Res 01691

Instituto Nacional do Livro, Av W/3 Sul, Entre Quadras 506, 70000 Brasília, DF
Focus: Libraries & Bk Sci 01692

Instituto Nacional do Livro, Rua Pedro Lessa 36, 20000 Rio de Janeiro, RJ
Focus: Libraries & Bk Sci 01693

International Communication Agency (ICA), Av Paulista 2439, 01000 São Paulo, SP
Focus: Lit 01694

International Mathematical Union (IMU), c/o IMPA, Estrada D. Castouna 110, 22460 Rio de Janeiro, RJ
T: (021) 2949032; Fax: (021) 5124112
Founded: 1952; Members: 55
Pres: Prof. J.L. Leons; Gen Secr: Prof. Jacob Palis
Focus: Math
Periodical
IMU Bulletin 01695

International Seaweed Association (ISA), c/o Enrico C. Oliveira, Instituto de Biociências, USP, CP 11461, 05422-970 São Paulo, SP
T: (011) 2114773; Fax: (011) 8154272
Founded: 1977; Members: 750
Pres: Mark A. Ragan; Gen Secr: Enrico C. Oliveira
Focus: Botany
Periodical
Proceedings 01696

Istituto Histórico e Geográfico do Espirito Santo, Av República, CP, 29000 Vitória, ES
Founded: 1916
Focus: Geography; Hist
Periodical
Do Ihges (weekly) 01697

Latin American Centre for Physics, c/o Centro Brasileiro de Pesquisas Físicas, Av Wenceslau Bráz 71, 22290 Rio de Janeiro, RJ
T: (021) 2955096
Founded: 1962; Members: 13
Focus: Physics
Periodical
Noticia (weekly) 01698

Organização Guarão de Ensino, Av Pedro de Toledo, 12500 Guaratinguetá, SP
T: 224416
Focus: Educ 01699

PEN Clube do Brasil-Associação Universal de Escritores, Praia do Flamengo 172, 20000 Rio de Janeiro, RJ
Founded: 1936; Members: 225
Focus: Lit
Periodical
Boletim 01700

Secção de Farmacia Galénica, c/o Universidade do Paraná, 80000 Curitiba, PI
Founded: 1960
Focus: Pharmacol 01701

Secretaria do Patrimônio Histórico e Artístico Nacional (SPHAN), Rua da Imprensa 16, 20000 Rio de Janeiro
Tx: 21170
Founded: 1937
Focus: Hist; Arts
Periodical
Revista do Patrimônio Histórico e Artístico Nacional (weekly) 01702

Sociedade Botânica do Brasil, Edificio Venâncio II, 70302 Brasília, DF
Founded: 1950; Members: 1234
Focus: Botany 01703

Sociedade Brasileira de Autores Teatrais, Av Almirante Barroso 97, 20031 Rio de Janeiro
Founded: 1917; Members: 5000
Pres: Daniel da Silva Rocha; Gen Secr: Danilo Martins Rocha
Focus: Perf Arts
Periodical
Revista de Teatro (weekly) 01704

Sociedade Brasileira de Belas Artes, Rua Araújo Pôrto Alegre 70, 20000 Rio de Janeiro, RJ
Focus: Arts 01705

Sociedade Brasileira de Cartografia, Rua México 41, 20000 Rio de Janeiro, RJ
Founded: 1958; Members: 2500
Focus: Cart 01706

Sociedade Brasileira de Cultura, Alameda Eduardo Prado 705, 01218 São Paulo, SP
Focus: Cultur Hist
Periodicals
Convivium: Revista de Investigação e Cultura (weekly)
Política e Estratégia: Revista de Política Internacional e Assuntos Militares . . . 01707

Sociedade Brasileira de Dermatologia (SBD) / Brazilian Society of Dermatology, CP 389, 20001-970 Rio de Janeiro, RJ
T: (021) 2222648; Fax: (021) 2210774
Founded: 1912; Members: 3000
Focus: Derm
Periodical
Anais Brasileiros de Dermatologia (weekly) 01708

Sociedade Brasileira de Entomologia (SBE), CP 9063, 01065-970 São Paulo, SP
T: (011) 2743455 ext 60
Founded: 1937; Members: 500
Pres: Sergio Antonio Vanin; Gen Secr: Pedro Gnaspini Netto
Focus: Entomology
Periodical
Revista Brasileira de Entomologia (weekly) 01709

Sociedade Brasileira de Filosofia, Praça da República 54, 20000 Rio de Janeiro, RJ
Founded: 1927; Members: 105
Focus: Philos 01710

Sociedade Brasileira de Genética / Brazilian Genetics Society, c/o Departamento de Genética, Faculdade de Medicina, USP, 14049 Ribeirão Preto
T: (016) 6331610; Tx: 166354
Founded: 1955; Members: 2000
Focus: Genetics
Periodical
Revista Brasileira de Genética: Brazilian Journal of Genetics (weekly) 01711

Sociedade Brasileira de Geografia, Prâca da República 54, 20000 Rio de Janeiro, RJ
Founded: 1883; Members: 284
Focus: Geography 01712

Sociedade Brasileira de Geologia, c/o Instituto de Geociências, USP, CP 20897, 01498 São Paulo, SP
T: (011) 2126166
Founded: 1945; Members: 4000
Focus: Geology
Periodicals
Revista Brasileira de Geociências (weekly)
Revista Ciências da Terra (weekly) . . 01713

Sociedade Brasileira de Instrução, Rua da Matriz 82, 22260-100 Botafogo, RJ
T: (021) 2867146
Focus: Sociology; Poli Sci
Periodicals
Caderno de Conjuntura (weekly)
DADOS (3 times annually)
Indice de Ciências Sociais (weekly)
Série Estudos (weekly) 01714

Sociedade Brasileira de Microbiologia, Av Pasteur 250, 20000 Rio de Janeiro, RJ
Founded: 1956
Focus: Microbio 01715

Sociedade Brasileira de Romanistas, Av Rio Branco 123, 20000 Rio de Janeiro, RJ
Focus: Ling 01716

Sociedade Civil de Educação São Marcos, Av Nazaré 900, 04262 São Paulo, SP
T: (011) 2745711
Focus: Educ 01717

Sociedade de Biologia do Brasil, CP 1587, 20000 Rio de Janeiro, RJ
Founded: 1947
Focus: Bio 01718

Sociedade de Biologia do Rio Grande do Sul, Av Candido de Godoy 112, 90000 Pôrto Alegre, RS
Focus: Bio 01719

Sociedade de Cultura e Educação do Litoral Sul, Rua São Francisco Xavier 165, 11900 Registro, SP
T: 22360
Focus: Educ 01720

Sociedade de Engenharia do Rio Grande do Sul (SERGS), Travessa Eng. Acilino Carvalho 33, 90000 Pôrto Alegre, RS
T: (0512) 246133
Founded: 1930; Members: 3660
Focus: Eng 01721

Sociedade de Ensino Piratininga, Av Angélica 381, 01277 São Paulo, SP
T: (011) 666628
Focus: Educ 01722

Sociedade de Farmácia e Química de São Paulo, Av Brigadeiro Luiz Antonio 393, 01317 São Paulo, SP
Founded: 1924; Members: 300
Pres: Dr. Luiz Marques de Sá; Gen Secr: Dr. Elfriede Marianne Bacchi
Focus: Bio; Chem; Pharmacol
Periodical
Anais de Farmácia e Química (weekly) . 01723

Sociedade de Medicina de Alagoas, Rua do Comércio 150, 57000 Maceió, AL
Founded: 1917
Focus: Med 01724

Sociedade de Pediatria da Bahia, Av Joanna Angélica 75, 40000 Salvador, BA
Members: 43
Focus: Pediatrics 01725

Sociedade Geográfica Brasileira (SGB), Rua 24 de Maio 104, 01041 São Paulo, SP
T: (011) 327802
Founded: 1948
Focus: Geography 01726

Sociedade Nacional de Agricultura, Av General Justo 171, CP 1245, 20021-130 Rio de Janeiro
T: (021) 2404573; Fax: (021) 2404189
Founded: 1897
Pres: Octavio Mello Alvarenga; Gen Secr: Elvo Santoro
Focus: Agri
Periodical
A Lavoura (3 times annually) 01727

Sociedade Paranaense de Matemática, CP 1261, 80001-970 Curitiba, PR
Founded: 1953
Focus: Math

Periodical
Boletim da Sociedade Paranaense de Matemática (weekly) 01728

Sociedade Propagadora Esdeva, Rua Halfeld 1179, 36100 Juiz de Fora, MG
Focus: Sci 01729

Sociedade Visconde de São Leopoldo, Rua Carvalho de Mendonça 140, 11100 Santos, SP
T: 345187
Focus: Sci 01730

Unidade de Pesquisa de Ambito Estadual em Barreiras (EMBRAPA), Rodovia Barreiras, São Desiderio, Km 15, 47800 Barreiras, BA
Focus: Sci 01731

Unidade de Pesquisa de Ambito Estadual em Campos (EMBRAPA), Av Francisco Lamego 134, 28100 Campos, RJ
Focus: Sci 01732

Unidade de Pesquisa de Ambito Estadual em Corumba (EMBRAPA), Rua Antonio Maria 786, 79300 Corumba, MT
Focus: Sci 01733

Unidade de Pesquisa de Ambito Estadual em Dourados (EMBRAPA), Rua Joachim Teixeira Alves 2190, 79800 Dourados, MT
Focus: Sci 01734

Unidade de Pesquisa de Ambito Estadual em Itaguai (EMBRAPA), Antiga Rodoviario São Paulo, Km 47, 26000 Nova Iguaçu, RJ
Focus: Sci 01735

Unidade de Pesquisa de Ambito Estadual em Itapirema (EMBRAPA), 55900 Goiânia, GO
Focus: Sci 01736

Unidade de Pesquisa de Ambito Estadual em Manaus (EMBRAPA), Rodovia Torquato Tapajos, Km 30, CP 455, 69000 Manaus, AM
Focus: Sci 01737

Unidade de Pesquisa de Ambito Estadual em Ponta Grossa (EMBRAPA), Av Presidente Kennedy, Rodovia do Café, Km 104, 84100 Ponta Grossa
Focus: Sci
Periodicals
Boletim Técnico
Circular
Informe da Pesquisa
Relatório de Atividades
Relatório Técnico 01738

Unidade de Pesquisa de Ambito Estadual em Porto Velho (EMBRAPA), c/o Laboratorio de Solos, 78900 Porto Velho, RO
Focus: Sci 01739

Unidade de Pesquisa de Ambito Estadual em Teresina (EMBRAPA), Av Perimetral 5650, 64000 Teresina, PI
T: 2620
Focus: Sci 01740

Bulgaria

Academy of Agricultural Sciences, Bul Dragan Cankov 6, Sofia
Founded: 1961; Members: 30
Focus: Agri 01741

Bulgarian Academy of Sciences, 7 Noemvri 1, 1040 Sofia
T: (02) 84141; Tx: 22424
Founded: 1869; Members: 100
Pres: B. Sendov; Gen Secr: Prof. Vasil Hristov
Focus: Sci 01742

Bulgarian Association of Penal Law, Benkovski 3, 1000 Sofia
T: (02) 84121
Founded: 1961; Members: 70
Gen Secr: Prof. V. Karakashev
Focus: Law 01743

Bulgarian Astronautical Society, V. Kolarov 36, 1000 Sofia
T: (02) 875619
Founded: 1957; Members: 100
Pres: K. Serafimov; Gen Secr: M. Velkov
Focus: Aero 01744

Bulgarian Biochemical and Biophysical Society, c/o Institute of Molecular Biology, Bulgarian Academy of Sciences, Acad. G. Bonchev Str Bl 21, Sofia 1113
Members: 250
Focus: Biochem; Biophys 01745

Bulgarian Botanical Society, Acad. B. Bonchev St Bl 1, 1113 Sofia
T: (02) 720951
Founded: 1923; Members: 150
Pres: Prof. Velcho Velchev; Gen Secr: Dr. Tenjo Mesinev
Focus: Botany 01746

Bulgarian Dermatological Society (BDD), Bul G. Sofiisky 1, 1431 Sofia
Founded: 1920; Members: 40
Pres: Prof. N. Botev-Zlatkov; Gen Secr: Prof. V. Goranov
Focus: Derm
Periodical
Dermatologia i Venerologia (weekly) . . 01747

Bulgarian Geographical Society, Bul Ruski 15, 1000 Sofia
T: (02) 872845
Founded: 1918
Pres: Prof. T. Hristov; Gen Secr: Dr. M. Glovnja
Focus: Geography 01748

Bulgarian Geological Society (BGS), Bul Ruski 15, POB 228, 1080 Sofia
T: (02) 833977
Founded: 1925; Members: 1208
Pres: Dr. M. Staykov; Gen Secr: M. Evstatiev
Focus: Geology
Periodical
Review of the Bulgarian Geological Society (3 times annually) 01749

The Bulgarian Gynecological and Obstetrical Society, Zdrave 2, 1431 Sofia
T: (02) 521026
Founded: 1961; Members: 600
Pres: Prof. Dr. Kiril Mirkov; Gen Secr: Prof. Alexander Hadjiev
Focus: Gynecology
Periodicals
Akuserstvo i Ginekologija (weekly)
Problems of Obstetrics and Gynecology (weekly) 01750

Bulgarian Historical Society, A. Zhdanov 5, 1000 Sofia
Founded: 1901
Pres: D. Kosev; Gen Secr: Prof. S. Trifonov
Focus: Hist 01751

Bulgarian Nutrition Society, c/o Institute of Gastroentereology and Nutrition, D. Nesterov Str 5, 1431 Sofia
Gen Secr: Dr. Donna Baynova
Focus: Food 01752

Bulgarian Philosophical Society, Benkovski 6, 1000 Sofia
T: (02) 884035
Founded: 1968
Pres: S. Ganovski; Gen Secr: V. Prodanov
Focus: Philos 01753

Bulgarian Scientific Pharmaceutical Association, Dunav 2, Sofia
Focus: Pharmacol
Periodical
Farmacija (weekly) 01754

Bulgarian Society for Microbiology, Tolbukhin 18, Sofia
Focus: Microbio 01755

Bulgarian Society of Anaesthesiology and Resuscitation, Bul G. Sofiisky 1, Sofia
Members: 200
Focus: Anesthetics 01756

Bulgarian Society of Cardiology, Mico Papo 65, 1309 Sofia
Members: 70
Pres: Prof. Ilia Tomov Iliev; Gen Secr: Prof. Tihomir Radomirov Daskalov
Focus: Cardiol; Rheuma 01757

Bulgarian Society of Electroencephalography, Electromyography and Clinical Neurophysiology, c/o First Neurological Clinic, Medical Academy, G. Sofiliski Bul 1, 1431 Sofia
Members: 20
Focus: Physiology; Neurology 01758

Bulgarian Society of Natural History, D. Cankov 8, 1421 Sofia
T: (02) 666594
Founded: 1896
Pres: B. Botev; Gen Secr: I. Jakov
Focus: Nat Sci 01759

Bulgarian Society of Neurosurgery (BSN), Bul G. Sofiisky 1, 1431 Sofia
T: (02) 5321316
Founded: 1975; Members: 83
Focus: Surgery
Periodical
Journal of Neurology, Psychiatry and Neurosurgery (weekly) 01760

Bulgarian Society of Physiological Sciences, c/o Dept of Physiology, Medical Academy, 1431 Sofia
Focus: Physiology 01761

Bulgarian Sociological Society, Moskovska 27b, 1000 Sofia
T: (02) 884181
Founded: 1959
Pres: N. Jahiel; Gen Secr: B. Cakalov
Focus: Sociology 01762

Bulgarian Soil Society, Shosé Bankya 7, POB 1369, 1080 Sofia
T: (02) 242257; Tx: 22701
Founded: 1959; Members: 200
Pres: Prof. M. Jolevski; Gen Secr: Dr. R. Dilkova
Focus: Agri 01763

Bulgarian Union of Public Libraries, Ul Alabin 31, Sofia
Focus: Libraries & Bk Sci 01764

Carpathian Balkan Geological Association (CBGA), c/o University of Sofia, 1504 Sofia
Gen Secr: Dr. Kristina Kolcheva
Focus: Geology 01765

Central Council of Scientific and Technical Unions, Ravovski 108, 1000 Sofia
T: (02) 898379; Tx: 22185
Founded: 1949
Pres: I. Popov; Gen Secr: M. Trifonova
Focus: Sci; Eng 01766

Federation of European Biochemical Societies (FEBS), c/o Medical Academy, Blvd VI Zaimov, Sofia
Founded: 1964; Members: 25
Focus: Chem; Biochem 01767

Medical Academy, Bul D. Nestorov 15, 1431 Sofia
T: (02) 58121; Tx: 22712; Fax: (02) 598116
Founded: 1918
Focus: Med 01768

National Scientific Society of Biomedical Physics and Engineering, c/o Medical Academy, Dragan Tzankov 8, 1000 Sofia
T: (02) 668718
Founded: 1971; Members: 86
Pres: Prof. Dr. Ivan Daskalov; Gen Secr: Prof. Dr. Marko Markov
Focus: Physics; Eng 01769

Research Association for the Fine Arts, c/o Bulgarian Academy of Sciences, 7 Noemvri 1, 1040 Sofia
T: (02) 84141
Founded: 1972
Focus: Arts
Periodicals
Arhitekturata i zizenata sreda na coveka
Balgarski folklor
Muzikoznanie
Problemi na balgarskija folklor
Problemi na izkustvoto 01770

Research Association on Fundamental Problems of Technical Sciences, c/o Bulgarian Academy of Sciences, 7 Noemvri 1, 1040 Sofia
T: (02) 84141
Founded: 1977
Focus: Eng
Periodicals
Fizko-himiceska Mehanika
Materialoznanie i Technologija
Problemi na Tehniceskata Kibernetika
Tehniceska Misal
Vodni Problemi 01771

Scientific and Technical Union of Agricultural Specialists, Rakovski 108, 1000 Sofia
Founded: 1965
Focus: Agri 01772

Scientific and Technical Union of Transport, Rakovski 108, 1000 Sofia
T: (02) 872371
Founded: 1965
Pres: E. Zahariev; Gen Secr: S. Sapundjiev
Focus: Transport 01773

Scientific Association of the Bulgarian Neurologists, c/o II. Neurological Clinic, Medical Academy, Belo-More 8, 1040 Sofia
Founded: 1960; Members: 120
Focus: Neurology 01774

Scientific Information Centre at the Bulgarian Academy of Sciences, Ul 7 Noemvri 1, 1040 Sofia
T: (02) 84141 ext 336
Founded: 1959; Members: 76
Pres: S. Gabrovska
Focus: Math; Nat Sci; Sociology
Periodicals
Biology (weekly)
Bulletin Catalogue of the Inventions of BAS (weekly)
Chemistry (weekly)
Culture (weekly)
Economics and Law (weekly)
Forthcoming Scientific Congresses, Conferences and Symposia abroad and in Bulgaria, Series A: Natural and Mathematical Sciences (weekly)
Forthcoming Scientific Congresses, Conferences and Symposia abroad and in Bulgaria, Series B: Social Sciences (weekly)
Geosciences (weekly)
History, Archaeology and Ethnography (weekly)
Linguistics and Literature (weekly)
Mathematical and Physical Sciences (weekly)
Psychology and Pedagogics (weekly)
Reports from International Scientific Congresses and Conferences, Series B: Social Sciences (weekly)
Scientific Communism, Philosophy, Sociology, Science of Science and Scientific Information (weekly)
Scientific Information serving in Natural, Mathematical and Social Sciences (weekly)
Topical Problems of Science: A Survey with Subtitles corresponding to the Material included (weekly)
Topical Problems of Science and Science Policy (weekly) 01775

Société Bulgare de Gastroentérologie, c/o Dr. B. Damianov, Belo-More 8, Sofia
Members: 50
Focus: Gastroenter 01776

Society of Bulgarian Anatomists, Histologists and Embryologists, Zdrave Str 2, 1431 Sofia
Founded: 1962; Members: 135
Pres: Prof. Dr. Vanko Vankov; Gen Secr: Prof. Dr. Wladimir Ovtscharoff
Focus: Anat; Med; Anthro; Endocrinology
Periodical
Experimental Medicine and Morphology (weekly) 01777

Society of Bulgarian Chemists, A. Ivanov 1, Sofia
Founded: 1971
Focus: Chem 01778

Society of Bulgarian Physicists, A. Ivanov 5, 1126 Sofia
T: (02) 627660
Founded: 1971
Pres: M. Borisov; Gen Secr: T. Troev
Focus: Physics 01779

Society of Bulgarian Psychologists, Tolbouhin St 18, POB 1333, Sofia
T: (02) 872171
Founded: 1968
Pres: S. Ganovski; Gen Secr: L. Mavrov
Focus: Psych
Periodical
Psihologija (weekly) 01780

Society of Foreign Language Teachers, Bul Ruski 15, Sofia
Founded: 1970
Focus: Ling; Educ 01781

Society of Sport Medicine, T. Kirkova 1, 1000 Sofia
T: (02) 881511
Founded: 1953; Members: 200
Pres: Prof. V. Slancev; Gen Secr: T. Todorov
Focus: Med; Physical Therapy 01782

Union of Architects in Bulgaria, E. Georgiev 1, 1504 Sofia
T: (02) 442673; Tx: 23569
Founded: 1965
Pres: G. Stoilov
Focus: Archit
Periodical
Architecture and Society 01783

Union of Bulgarian Composers, Ivan Vazov 2, 1000 Sofia
T: (02) 881560
Founded: 1947; Members: 223
Pres: Prof. Alexander Rajcev
Focus: Music 01784

Union of Bulgarian Film Makers, Bul Ruski 8a, 1000 Sofia
T: (02) 878956
Founded: 1934
Pres: G. Stojanov
Focus: Cinema
Periodicals
Filmovi Novini
Kinoizkustvo 01785

Union of Bulgarian Mathematicians, Acad. G. St Bl 8, Bonchev, POB 373, 1113 Sofia
T: (02) 721189, 756029; Tx: 22628
Founded: 1971; Members: 8000
Pres: L. Iliev; Gen Secr: G. Gerov
Focus: Math
Periodical
Mathematics and Mathematical Education (weekly) 01786

Union of Bulgarian Writers, Angel Kancev 5, 1040 Sofia
T: (02) 874711
Founded: 1913
Pres: L. Levchey
Focus: Lit 01787

Union of Chemistry and the Chemical Industry, Rakovski 108, 1000 Sofia
T: (02) 875812
Founded: 1965
Pres: G. Bliznakov; Gen Secr: S. Dzalev
Focus: Chem 01788

Union of Civil Engineering, Rakovski 108, 1000 Sofia
Founded: 1965
Pres: Prof. S. Simeonov; Gen Secr: M. Ruseva
Focus: Civil Eng 01789

Union of Economics, Rakovski 108, 1000 Sofia
Founded: 1970
Pres: I. Iliev; Gen Secr: D. Pravov
Focus: Econ 01790

Union of Electronics, Electrical Engineering and Communications, Rakovski 108, 1000 Sofia
T: (02) 884158
Founded: 1965
Pres: S. Popadijn; Gen Secr: I. Ivancev
Focus: Energy; Electric Eng 01791

Union of Energetics, Rakovski 108, 1000 Sofia
T: (02) 884158
Founded: 1965
Pres: S. Misev; Gen Secr: G. Grigorov
Focus: Eng; Electric Eng; Comm
Periodical
Energetika 01792

Union of Forest Engineering, Rakovski 108, 1000 Sofia
T: (02) 883683
Founded: 1965
Pres: I. Gruev; Gen Secr: S. Ivanov
Focus: Forestry
Periodical
Darvoobrabotvasta i mebelna Promislenost *01793*

Union of Geodesy and Cartography, Rakovski 108, 1000 Sofia
Founded: 1949
Pres: Prof. V. Peevski; Gen Secr: S. Bogdanov
Focus: Surveying *01794*

Union of Mechanical Engineering, Rakovski 108, 1000 Sofia
T: (02) 877290
Founded: 1965
Pres: Prof. V. Diviziev; Gen Secr: D. Damjanov
Focus: Eng *01795*

Union of Mining Engineering, Geology and Metallurgy, Rakovski 108, 1000 Sofia
T: (02) 875727
Founded: 1965
Gen Secr: B. Petkov
Focus: Mining; Geology; Metallurgy
Periodicals
Metalurgija
Rudodobiv
Vaglista *01796*

Union of Rural Economics, Rakovski 108, 1000 Sofia
T: (02) 876513
Founded: 1965
Pres: K. Malinov; Gen Secr: P. Lekov
Focus: Agri
Periodical
Bjuletin-Vnedreni Novosti *01797*

Union of Scientific Medical Societies in Bulgaria, Serdica 2, 1000 Sofia
T: (02) 883111
Founded: 1968; Members: 46
Focus: Med
Periodical
Information Bulletin (weekly) *01798*

Union of Scientific Workers in Bulgaria, Oborishte Str 35, 1504 Sofia
T: (02) 441157, 441590, 446297
Founded: 1944; Members: 11200
Pres: Alexander Yankov; Gen Secr: Alexander Assenov
Focus: Sci
Periodicals
Medical Archives (weekly)
Scientific Life (weekly) *01799*

Union of Textiles, Clothing and Leather, Rakovski 108, 1000 Sofia
T: (02) 881641
Founded: 1965
Pres: Prof. G. Damjanov; Gen Secr: I. Mechev
Focus: Textiles *01800*

Union of the Food Industry, Rakovski 108, 1000 Sofia
T: (02) 874744
Founded: 1965
Pres: R. Bojadzhiev; Gen Secr: V. Fumadjiev
Focus: Eng; Food *01801*

Union of Translators' in Bulgaria, Neofit Rilsky 5, 1000 Sofia
T: (02) 522438
Founded: 1974; Members: 850
Pres: Naiden Vulchev; Gen Secr: Barbara Müller
Focus: Ling
Periodicals
The Art of Translation: Collection of Theoretical Articles (weekly)
Fakel: Magazine (weekly)
News: Bulletin (weekly)
Panorama: Magazine (weekly) *01802*

Union of Water Works, Rakovski 108, 1000 Sofia
Founded: 1965
Pres: A. Pasev; Gen Secr: M. Sjarova
Focus: Civil Eng *01803*

Burkina Faso

African and Malagasy Council on Higher Education, POB 134, Ouagadougou
Founded: 1968
Focus: Educ *01804*

Association des Techniciens et Ingénieurs Sanitaires Africains (ATISA), c/o CEPA, BP 7112, Ouagadougou
Focus: Eng *01805*

Centre d'Epidémiologie, Statistique et Information Sanitaire (CESIS), c/o OCCGE, 01 BP 153, Bobo-Dioulasso
T: 981372; Tx: 8269; Fax: 981372
Focus: Public Health *01806*

Centre Muraz, BP 153, Bobo-Dioulasso
T: 982875; Tx: 8260
Founded: 1939
Focus: Med *01807*

Centre of Economic and Social Studies and Experiments in Western Africa, 01 BP 305, Bobo-Dioulasso
T: (09) 980935; Tx: 8257
Founded: 1978
Pres: F. Tevi Sedalo
Focus: Soc Sci; Econ
Periodical
Construire Ensemble (weekly) *01808*

Coordination and Cooperation Organization for the Control of the Major Endemioc Diseases, c/o OCCGE, 01 BP 153, Bobo-Dioulasso
T: 981372; Tx: 8260; Fax: 981372
Founded: 1960; Members: 8
Gen Secr: Aminata Dao
Focus: Med
Periodicals
Bulletin Mensuel d'Informations (weekly)
Communiqué Bibliographique (weekly)
SERITEC (weekly) *01809*

Organization for Co-ordination and Co-operation in the Control of Major Endemic Diseases (OCCGE), BP 153, Bobo-Dioulasso
Focus: Zoology
Periodicals
Bulletin OCCGE Information (weekly)
Seritec (weekly) *01810*

Cameroon

Association des Institutions d'Enseignement Théologique en Afrique Centrale (ASTHEOL-CENTRAL), c/o Faculté de Théologie Protestante, BP 4011, Yaoundé
Focus: Rel & Theol; Educ *01811*

Association Oecuménique des Théologiens Africains (AOTA), BP 1539, Yaoundé
Gen Secr: Engelbert Mveng
Focus: Rel & Theol *01812*

Compagnie Française pour le Développement des Fibres Textiles (CFDT), BP 1699, Douala
Focus: Textiles *01813*

Canada

Académie des Lettres du Québec, 5724 Chemin de la Côte Saint-Antoine, Montréal, Qué. H4A 1R9
T: (514) 488-5883
Founded: 1944; Members: 36
Pres: Jean-Guy Pilon; Gen Secr: Jean-Pierre Duquette
Focus: Ling; Lit
Periodical
Les Cahiers de l'Académie (weekly) . . *01814*

Academy of Canadian Cinema / Académie du Cinéma Canadien, 653 Yonge St, Toronto, Ont. M4Y 1Z9
T: (416) 967-0315
Founded: 1979; Members: 600
Focus: Cinema *01815*

Academy of Medicine, 288 Bloor St W, Toronto, Ont. M5S 1V8
T: (416) 922-1134
Founded: 1907; Members: 2800
Gen Secr: Dr. D.J. McKnight
Focus: Med
Periodical
Bulletin *01816*

Accident Prevention Association of Manitoba, 266 Kilbride Av, Winnipeg, Man. R2V 1A2
T: (204) 339-7796
Focus: Safety *01817*

Acupuncture Foundation of Canada, 7321 Victoria Park Av, Unit 18, Markham, Ont. L3R 2Z8
T: (416) 474-0383
Founded: 1974; Members: 600
Pres: Dr. Gilbert Gagne
Focus: Med
Periodical
Newsletter (weekly) *01818*

Addiction Research Foundation, 33 Russell St, Toronto, Ont. M5S 2S1
T: (416) 595-6000
Founded: 1949
Pres: Mark Taylor
Focus: Immunology
Periodicals
Ontario Report (weekly)
The Journal (weekly) *01819*

Administrative Sciences Association of Canada / Association des Sciences Administratives du Canada, c/o Département des Sciences Administratives, CP 8888, Succ A, Montréal, Qué. H3C 3P8
T: (514) 282-3893; Fax: (514) 282-3343
Founded: 1971; Members: 700
Gen Secr: Jean Pasquero
Focus: Public Admin
Periodicals
Canadian Journal of Administrative Sciences (weekly)
Newsletter (weekly)
Proceedings (weekly) *01820*

African Literature Association (ALA), c/o Dept of Comparative Literature, Edmonton, Alta. T6G 2E6
T: (403) 432-5535
Founded: 1974; Members: 1000
Gen Secr: Stephen Arnold
Focus: Lit
Periodicals
ALA Bulletin (weekly)
Annual Selected Conference Paper . . . *01821*

Agency for Tele-Education in Canada, c/o OECA, Broadcast Relations, POB 200, Station Q, Toronto, Ont. M4T 2T1
T: (416) 484-2639
Focus: Educ *01822*

Agricultural Institute of Canada / Institut Agricole du Canada, 151 Slater St, Ste 907, Ottawa, Ont. K1P 5H4
T: (613) 232-9459; Fax: (613) 594-5190
Founded: 1920; Members: 6500
Focus: Agri
Periodicals
Canadian Journal of Animal Science
Canadian Journal of Plant Science
Canadian Journal of Soil Science . . . *01823*

Alberta Association of College Librarians, c/o Southern Alberta Institute of Technology, 1301 16 Av NW, Calgary, Alta. T2M 0L4
Pres: Ronald Peters
Focus: Libraries & Bk Sci *01824*

Alberta Association of Library Technicians, POB 700, Edmonton, Alta. T5J 2L4
Founded: 1975; Members: 243
Focus: Libraries & Bk Sci *01825*

Alberta Association of Rehabilitation Centres, 2725 12 St NE, POB 105, Calgary, Alta. T2E 7J2
T: (403) 276-8182
Founded: 1973; Members: 25
Gen Secr: Carol Anderson
Focus: Rehabil *01826*

Alberta Association of Social Workers, 11831 123 St, Ste 100, Edmonton, Alta. T5L 0G7
T: (403) 454-1426
Founded: 1962; Members: 610
Gen Secr: Mark Nicoll
Focus: Sociology
Periodical
The Advocate (weekly) *01827*

Alberta Association on Gerontology, c/o 7-122 Clinical Sciences Bldg, University of Alberta, Edmonton, Alta. T6G 2J8
T: (403) 492-5975
Founded: 1980
Pres: Ruth Wolfe
Focus: Geriatrics *01828*

Alberta Chiropractic Association, 12536 125 St, Edmonton, Alta. T5L 0T3
T: (403) 453-2438
Focus: Orthopedics *01829*

Alberta Construction Association, POB 3830, Station D, Edmonton, Alta. T5L 4J8
T: (403) 455-1122; Fax: (403) 455-1120
Founded: 1958; Members: 1600
Pres: Ronald W. Jones
Focus: Civil Eng
Periodicals
Alberta Construction (weekly)
Membership Roster (weekly) *01830*

Alberta Dental Association, 8230 105 St, Ste 101, Edmonton, Alta. T6E 5H9
Gen Secr: Dr. B.P. Martinello
Focus: Dent *01831*

Alberta Educational Communications Corporation, 16930 114 Av, Edmonton, Alta. T5M 3S2
T: (403) 451-3160; Tx: 0373948; Fax: (403) 452-7233
Founded: 1973
Pres: Peter L. Senchuk
Focus: Educ; Comm *01832*

Alberta Forest Development Research Trust Fund, Postal Bag 6343, Spruce Grove, Alta. T7X 2Y4
T: (403) 962-8700; Fax: (403) 962-5414
Founded: 1974
Pres: K.D. Higginbotham
Focus: Forestry *01833*

Alberta Genealogical Society, POB 12015, Edmonton, Alta. T5J 3L2
Founded: 1973; Members: 800
Pres: Dolores Christie
Focus: Genealogy
Periodical
Relatively Speaking (weekly) *01834*

Alberta Health Records Association, POB 1752, Edmonton, Alta. T5J 2P1
Founded: 1960; Members: 455
Pres: Sandra Moffatt
Focus: Public Health *01835*

Alberta Heritage Foundation for Medical Research, 10235 101 St, Ste 1200, Edmonton, Alta. T5J 3G1
T: (403) 423-5727
Founded: 1979
Pres: E.A. Geddes
Focus: Med *01836*

Alberta Historical Resources Foundation, 102 Eighth Av SE, Calgary, Alta. T2G 0K6
T: (403) 297-7320
Founded: 1973; Members: 3000
Gen Secr: Trudy Cowan
Focus: Hist *01837*

Alberta Law Foundation, 407 Eighth Av SW, Ste 205, Calgary, Alta. T2P 1E3
T: (403) 264-4701
Founded: 1973
Gen Secr: Owen G. Snider
Focus: Law *01838*

Alberta Lung Association, 10618 124 St, Edmonton, Alta. T5N 3X4
T: (403) 482-6527; Fax: (403) 482-6748
Founded: 1939
Gen Secr: Gary Lathan
Focus: Pulmon Dis *01839*

Alberta Medical Council, c/o Dr. J.A. Noakes, 106 Rideau Medical Dental Centre, Calgary, Alta. T2S 1V8
Focus: Med *01840*

Alberta Museums Association, POB 4036, Station C, Calgary, Alta. T2T 5M9
T: (403) 264-8300
Founded: 1971; Members: 170
Pres: Morris Flewwelling
Focus: Archives *01841*

Alberta Psychiatric Association, c/o University of Alberta Hospital, 155 Clinical Sciences Bldg, Edmonton, Alta. T6G 2J3
T: (403) 432-6565
Focus: Psychiatry *01842*

Alberta Public Health Association, POB 459, Olds, Alta. T0M 1P0
T: (403) 556-8441
Members: 350
Gen Secr: Mildred Harris
Focus: Public Health *01843*

Alberta Registered Dietitians Association, POB 2208, Station M, Calgary, Alta. T2P 2M4
Founded: 1959; Members: 330
Pres: Sandra Follett
Focus: Nutrition *01844*

Alberta Research Council, 250 Karl Clark Rd, POB 8330, Station F, Edmonton, Alb. T6H 5X2
T: (403) 450-5111; Tx: 0372147; Fax: (403) 461-2651
Founded: 1921
Pres: C.E.O. Bowman
Focus: Geology *01845*

Alberta Safety Council, 10526 Jasper Av, Ste 201, Edmonton, Alta. T5J 1Z4
T: (403) 428-7555
Focus: Safety *01846*

Alberta School Trustees' Association, 12310 105 Av, Edmonton, Alta. T5N 0Y4
T: (403) 482-7311
Gen Secr: Dr. Larry Ferguson
Focus: Educ *01847*

Alberta Society of Professional Biologists, POB 566, Edmonton, Alta. T5J 2K8
T: (403) 429-9110
Founded: 1975
Pres: Donald C. Thompson
Focus: Bio *01848*

Alberta Sulphur Research, 2500 University Dr NW, Calgary, Alta. T2N 1N4
T: (403) 220-5372; Fax: (403) 289-9488
Founded: 1964; Members: 25
Focus: Chem; Nat Res
Periodical
Bulletin (weekly) *01849*

Alberta Teachers' Association, 11010 142 St, Edmonton, Alta. T5N 2R1
T: (403) 453-2411; Fax: (403) 455-6481
Founded: 1918
Gen Secr: Dr. J.S. Buski
Focus: Educ
Periodicals
The ATA Magazine (weekly)
The ATA News (weekly) *01850*

Alberta Thoracic Society, 2E4-32 Walter C. McKenzie Center, Edmonton, Alta. T6G 2B7
T: (403) 432-4862
Founded: 1960; Members: 65
Pres: Dr. D. Stollerys
Focus: Pulmon Dis *01851*

Alberta Veterinary Medical Association, 8615 149 St, Edmonton, Alta. T5P 1B3
T: (403) 489-5007
Founded: 1905; Members: 575
Gen Secr: Hans E. Flatla
Focus: Vet Med
Periodical
Newsletter *01852*

Algoma Lung Association, 212 Queen St E, Ste 103, Sault Sainte Marie, Ont. P6A 5X8
T: (705) 256-2335
Gen Secr: Jean Shaw
Focus: Pulmon Dis *01853*

Allergy Association of Calgary, POB 1116, Station 6, Calgary, Alta. T3A 3G120
T: (403) 277-4999
Members: 120
Pres: Phyllis Kane
Focus: Immunology *01854*

Allergy Foundation of Canada, POB 1904, Saskatoon, Sask. S7K 3S5
T: (306) 664-4618
Founded: 1972; Members: 650
Pres: Sandy Woynarski
Focus: Immunology
Periodical
Allergy Alert (weekly) 01855

Allergy Information Association, 65 Tromley Dr, Ste 10, Etobicoke, Ont. M9B 5Y7
Founded: 1964; Members: 4000
Pres: M. Susan Daglish
Focus: Immunology
Periodical
Allergy Quarterly (weekly) 01856

Alzheimer Society of Manitoba, 205 Edmonton St, Winnipeg, Man. R3C 1R4
Members: 600
Focus: Geriatrics
Periodical
Newsletter (10 times annually) 01857

American and Canadian Underwater Certifications, Upper Canada Pl, 460 Brant St, Ste 201, Burlington, Ont. L7R 4B6
T: (416) 639-2357
Founded: 1986
Pres: R.W. Crankwright
Focus: Sports 01858

American Association of Physics Teachers (Ontario), c/o Frontenac County Board of Education, Bay Ridge Secondary School, Kingston, Ont.
T: (613) 389-8932
Founded: 1979; Members: 300
Pres: Brenda Molloy
Focus: Educ; Physics 01859

American Dialect Society (ADS), c/o Dept of English, University of Western Ontario, London, Ont. N5A 3K7
T: (591) 679-3707
Founded: 1889; Members: 465
Gen Secr: H. Rex Wilson
Focus: Ling 01860

American Institute of Iranian Studies (AIIS), c/o West Asian Dept, Royal Ontario Museum, 100 Queen's Park, Toronto, Ont. M5S 2C6
Founded: 1969; Members: 120
Pres: Cuyler T. Young; Gen Secr: Louis D. Levine
Focus: Ethnology 01861

American Osler Society (AOS), 3H56-HSC, McMaster University, Hamilton, Ont. L8N 3Z5
T: (416) 525-9140
Founded: 1970; Members: 70
Gen Secr: Charles G. Roland
Focus: Med; Cultur Hist 01862

American Society for Aesthetics (ASA), 4-108 Humanities Centre, University of Alberta, Edmonton, Alta. T6G 2E5
T: (403) 492-4102
Founded: 1942; Members: 1000
Pres: Arthur Danto; Gen Secr: Roger Shiner
Focus: Arts
Periodicals
ASA Newsletter
Journal of Aesthetics and Art Criticism (weekly) 01863

American Society for Information Science (Western Canada Chapter), 11546 77 Av, Edmonton, Alta. T6G 0M1
T: (604) 384-2444
Focus: Computer & Info Sci 01864

American Society for Metals, c/o Al J. Delaine, Debro Industries, 180 Cree Crescent, Winnipeg, Man. R3J 3W1
Focus: Metallurgy 01865

Amyotrophic Lateral Sclerosis Society of British Columbia, 411 Dunsmuir St, Vancouver, B.C. V6B 1X4
T: (604) 685-0737
Founded: 1981; Members: 200
Pres: Roy Slater
Focus: Pathology 01866

Amyotrophic Lateral Sclerosis Society of Canada / Société Canadienne de la Sclérose Laterale Amyotrophique, 90 Adelaide St E, Toronto, Ont. M5C 2R4
T: (416) 362-0269
Founded: 1977; Members: 4000
Focus: Neurology
Periodicals
ALS News (weekly)
Info ALS (weekly) 01867

Anthroposophical Society in Canada, 81 Lawton Blvd, Toronto, Ont. M4V 1Z6
T: (416) 481-2886
Founded: 1953; Members: 470
Focus: Philos 01868

Arbitrators' Institute of Canada / Institut des Arbitres du Canada, 234 Eglinton Av E, Ste 411, Toronto, Ont. M4P 1K5
T: (416) 487-8433
Founded: 1974; Members: 270
Gen Secr: J.A. Tuck
Focus: Law 01869

Architectural Intitute of British Columbia, 970 Richards St, Vancouver, B.C. V6B 3C1
T: (604) 683-8588
Founded: 1914; Members: 900
Gen Secr: Cheryl Williams
Focus: Archit 01870

Arctic Institute of North America (AINA), 2500 University Dr NW, Calgary, Alta. T2N 1N4
T: (403) 220-7515; Fax: (403) 282-7298
Founded: 1945; Members: 2000
Gen Secr: Michael P. Robinson
Focus: Geography
Periodicals
Information North (weekly)
Journal Arctic (weekly) 01871

Assocation of Canadian Industrial Designers (ACID) / Association des Désigners Industriels du Canada, c/o Resource Centre, 168 Bedford Rd, Toronto, Ont. M5R 2K9
Founded: 1948; Members: 250
Pres: Alex Manu
Focus: Eng; Graphic & Dec Arts, Design 01872

Association Canadienne d'Education de Langue Française (ACELF), 980 Chemin Saint-Louis, Sillery, Qué. G1S 1C7
T: (418) 681-4661
Founded: 1947; Members: 500
Focus: Educ; Ling
Periodical
Revue de l'ACELF 01873

Association Canadienne-Française pour l'Avancement des Sciences, 425 Rue De La Gauchetière Est, Montréal, Qué. H2L 2M7
T: (514) 849-0045; Fax: (514) 849-5558
Founded: 1923; Members: 45
Gen Secr: Françoise Braun
Focus: Nat Sci 01874

Association Canadienne-Française pour l'Avancement des Sciences, 2730 Côte Sainte Catherine, C.P. 6060, Montréal, Qué. H3T 1B7
T: (514) 342-1411
Founded: 1923; Members: 4000
Gen Secr: Guy Arbour
Focus: Sci
Periodicals
Annales (weekly)
Interface (weekly) 01875

Association de l'Enseignement du Nouveau-Québec / Northern Quebec Teaching Association, 2336 Chemin Sainte-Foy, Sainte-Foy, Qué. G1V 1S5
T: (418) 658-5711
Founded: 1970; Members: 500
Focus: Educ 01876

Association de Paralysie Cérébrale du Québec / Cerebral Palsy Association of Quebec, 525 Blvd Hamel Est, Bureau A-50, Québec, Qué. G1M 2S8
T: (418) 529-5371
Founded: 1949; Members: 4000
Gen Secr: Claude Desjardins
Focus: Neurology 01877

Association des Chirurgiens Généraux de la Province de Québec / Association of General Surgeons of the Province of Quebec, 2 Complexe Desjardins, CP 216, Succ Desjardins, Montréal, Qué. H5B 1G8
T: (514) 849-8887
Founded: 1965; Members: 610
Gen Secr: Paul Roy
Focus: Surgery 01878

Association des Collèges du Québec, 1940 Blvd Henri Bourassa Est, Montréal, Qué. H2B 1S2
T: (514) 381-8891
Founded: 1968; Members: 25
Gen Secr: Jean-Marie Saint-Germain
Focus: Educ 01879

Association des Diplômés de Polytechnique, CP 6079, Montréal, Qué. H3C 3A7
T: (514) 344-4764
Founded: 1910; Members: 8376
Gen Secr: Lucille Charbonneau
Focus: Educ
Periodical
Journal d'Ingénieur (weekly) 01880

Association de Securité des Exploitations Forestières du Québec / Quebec Logging Safety Association, 425 Saint Amable, Québec, Qué. G1R 5E4
T: (418) 522-1638
Founded: 1969; Members: 110
Focus: Safety; Forestry 01881

Association de Securité des Industriels Forestiers du Québec / Quebec Forest Industrial Safety Association, 3350 Blvd Wilfrid-Hamel, Ste 300, Québec, Qué. G1P 2J9
T: (418) 872-6126
Founded: 1934; Members: 2200
Gen Secr: J. Turgeon
Focus: Safety; Forestry 01882

Association de Securité des Pâtes et Papiers du Québec / Quebec Pulp and Paper Safety Association, 425 Saint Amable, Ste 1325, Québec, Qué. G1R 5E4
T: (418) 522-1638
Founded: 1932; Members: 82
Gen Secr: J. Aurele Saint Pierre
Focus: Safety; Materials Sci 01883

Association des Enseignants Franco-Ontariens, 681 Chemin Belfast, Ottawa, Ont. K1G 0Z4
T: (613) 230-9583
Founded: 1939; Members: 5200
Gen Secr: Jacques Schryburt
Focus: Educ 01884

Association des Enseignants Francophones du Nouveau-Brunswick, CP 712, Fredericton, N.B. E3B 5B4
T: (506) 454-2654
Founded: 1970; Members: 2700
Gen Secr: Ronald LeBreton
Focus: Educ 01885

Association des Institutions de Niveaux Préscolaire et Elementaire du Québec, 1141 Blvd Saint-Joseph Est, Montréal, Qué. H2J 1L3
T: (514) 381-8891
Founded: 1970; Members: 57
Gen Secr: André Paré
Focus: Educ 01886

Association des Institutions d'Enseignement Secondaire, 1940 Blvd Henri Bourassa Est, Montréal, Qué. H2B 1S2
T: (514) 381-8891; Fax: (514) 381-4086
Founded: 1968; Members: 102
Gen Secr: Paul-Emile Tingras
Focus: Educ 01887

Association des Médecins de Langue Française du Canada, 1440 Rue Sainte-Catherine Ouest, Ste 510, Montréal, Qué. H3G 2P9
T: (514) 866-2053
Founded: 1902; Members: 4000
Gen Secr: Dr. Raymond Robillard
Focus: Med 01888

Association des Médecins du Travail du Québec / Quebec Occupational Medical Association, 2 Complexe Desjardins, CP 216, Succ Desjardins, Montréal, Qué. H4A 3B5
T: (514) 286-1220
Members: 250
Pres: Dr. Claude Lapierre
Focus: Hygiene
Periodical
Bulletin (weekly) 01889

Association des Professeurs de Français de la Saskatchewan / Saskatchewan Association of Teachers of French, 2317 Arlington Av, Saskatoon, Sask. S7J 2H8
T: (306) 373-1660
Founded: 1967; Members: 239
Focus: Educ; Ling 01890

Association des Professeurs de Français des Universités et Collèges Canadiens / Association of Canadian University and College Teachers of French, c/o Dept of French, University of Victoria, POB 1700, Victoria, B.C. V8W 2Y2
T: (604) 721-7373
Founded: 1958; Members: 350
Pres: Danielle Thaler
Focus: Educ; Ling 01891

Association des Professeurs en Alimentation du Québec, 600 Rue Fullum, Montréal, Qué. H2K 4L1
T: (514) 873-3196
Pres: Armand Durandon
Focus: Food 01892

Association des Psychiatres du Québec, CP 216, Succ Desjardins, Montréal, Qué. H5B 1G8
T: (514) 845-3259
Focus: Psychiatry 01893

Association des Services de Réhabilitation Sociale du Québec / Quebec Association of Social Rehabilitation Agencies, 4217 Rue d'Iberville, Montréal, Qué. H2H 2L5
T: (514) 521-3733
Founded: 1962; Members: 26
Focus: Rehabil 01894

Association des Universités Partiellement ou Entièrement de Langue Française (AUPELF), c/o Université de Montréal, BP 6128, Montréal, Qué. H3C 3J7
T: (514) 343-6630
Founded: 1961; Members: 122
Focus: Sci
Periodicals
Etudes créoles (weekly)
Idées (weekly)
La Revue de l'AUPELF (weekly)
Universités (weekly) 01895

Association du Planning Familial des Sept-Iles, 421 Arnaud, Sept-Iles, Qué. G4R 3B3
T: (418) 968-8948
Members: 10
Pres: Carmen Lajoie
Focus: Family Plan 01896

Association du Syndrôme de Turner du Québec, 535 De La Vérendrye, Boucherville, Qué. J4B 2Y2
T: (514) 449-2212
Pres: Francine Beland
Focus: Med; Pathology 01897

Association for 18th Century Studies, c/o TSH-626 McMaster University, Hamilton, Ont. L8S 4M2
T: (416) 525-9140
Founded: 1968; Members: 3000
Pres: Dr. James King
Focus: Hist 01898

Association for Canadian Studies (ACS) / Association d'Etudes Canadiennes, CP 8888, Succ A, Montréal, Qué. H3C 3P8
T: (514) 987-7784; Fax: (514) 987-8210
Founded: 1973; Members: 900
Focus: Hist; Sociology
Periodicals
ACS Newsletter/Bulletin de l'AEC (weekly)
Canadian Issues (weekly) 01899

Association for Computing Machinery, c/o INRS Telecommunications, 3 Pl du Commerce, Verdun, Qué. H3E 1H6
T: (514) 761-5831
Focus: Computer & Info Sci 01900

Association for Media and Technology in Education / Association des Media et de la Technologie en Education au Canada, POB 53, Station R, Toronto, Ont. M4G 3Z3
Founded: 1975; Members: 450
Focus: Mass Media; Educ
Periodical
Media Message (weekly) 01901

Association for Social Psychology, 812 O'Connor Dr, POB 152, Station O, Toronto, Ont. M4A 2N3
Founded: 1961
Pres: K. Carsen; Gen Secr: Ann Skirko
Focus: Psych 01902

Association for the Advancement of Christian Scholarship, 229 College St, Toronto, Ont. M5T 1R4
T: (416) 979-2331
Founded: 1956; Members: 1800
Pres: Dr. B. Zulstra
Focus: Rel & Theol
Periodicals
Anakainosis (weekly)
Perspective (weekly) 01903

Association for the Advancement of Scandinavian Studies in Canada, c/o Dept of Drama, University of Guelph, Guelph, Ont. N1G 2W1
T: (519) 824-4120; Fax: (519) 837-1315
Founded: 1982; Members: 140
Pres: Dr. Harry Lane; Gen Secr: Patricia Appavoo
Focus: Ling; Lit
Periodical
Newsbulletin 01904

Association for the Advancement of Science in Canada / Association pour l'Avancement des Sciences au Canada, 2450 Lancaster Rd, Ste 36, Ottawa, Ont. K1B 5N3
T: (613) 521-2557; Fax: (613) 731-9782
Founded: 1982; Members: 750
Focus: Sci 01905

Association for the Study of Canadian Radio and Television / Association pour les Etudes sur la Radiotélévision Canadienne, 1455 de Maisonneuve W, Rm N-312, Montréal, Qué. H3G 1M8
T: (514) 879-5901
Founded: 1977; Members: 200
Focus: Mass Media 01906

Association Française des Conseils Scolaires de l'Ontario, 50 Rue Vaughan, Ottawa, Ont. K1M 1X1
T: (613) 745-3195
Founded: 1944; Members: 800
Gen Secr: Ginette M. Gratton
Focus: Public Admin; Educ
Periodicals
Cahier des actualités (weekly)
Infoscolaire (weekly) 01907

Association Francophone Internationale des Directeurs d'Etablissements Scolaires (AFIDES) / International Association of French-Speaking Directors of Educational Institutions, 500 Blvd Crémazie Est, Montréal, Qué. H2P 1E7
T: (0514) 3837335; Fax: (0514) 3842139
Founded: 1983
Gen Secr: R. de Guire
Focus: Educ
Periodical
La Revue des Echanges (weekly) . . . 01908

Association Internationale de Pédagogie Universitaire, c/o Service Pédagogique, Université de Montréal, CP 6128, Succ A, Montréal, Qué. H3C 3J7
T: (514) 343-7087; Fax: (514) 343-2107
Founded: 1979; Members: 800
Gen Secr: Jean-Marie Van der Maren
Focus: Educ
Periodical
Pédagogiques (weekly) 01909

Association Internationale du Théâtre pour l'Enfance et la Jeunesse, POB 190, Station E, Montréal, Qué. H2T 3A7
T: (514) 672-7512
Founded: 1975
Focus: Perf Arts 01910

Association Mathématique du Québec, 490 Chemin du Golf, L'Assomption, Qué. J0K 1G0
T: (514) 581-5153
Founded: 1958; Members: 900

Pres: Louise Trudel
Focus: Math
Periodical
Bulletin de l'AMQ (weekly) *01911*

Association Médicale du Québec / Quebec Medical Association, 1010 Rue Sherbrooke Ouest, Ste 1010, Montréal, Qué. H3A 2R7
T: (514) 282-1443
Founded: 1922; Members: 7500
Gen Secr: Dr. Gerald Caron
Focus: Med *01912*

Association Mondiale des Médecins Francophones (AMMF), 545 Blvd Saint-Laurent, Ottawa, Ont K1K 4H9
T: (06132) 7494001
Founded: 1973; Members: 303
Gen Secr: Dr. Pierre Sarda
Focus: Med
Periodical
AMMF Bulletin *01913*

Association Museums of New Brunswick / Association Musées du Nouveau-Brunswick, 277 Douglas Av, Saint John, N.B. E2K 1E5
T: (506) 642-5850
Founded: 1976; Members: 70
Focus: Archives *01914*

Association of British Columbia Archivists (A.B.C.A.), 2075 Wesbrooke Mall, POB 3859, Main Post Office, Vancouver, B.C. V6B 3Z3
Focus: Archives
Periodical
A.B.C.A. Newsletter (weekly) *01915*

Association of British Columbia School Superintendents, RR 1, 6205 89 St, Osoyoos, B.C. V0H 1V0
T: (604) 495-7672
Members: 185
Gen Secr: B.G. Webber
Focus: Educ *01916*

Association of Canadian Archivists (ACA), POB 2596, Station D, Ottawa, Ont. K1P 5W6
T: (613) 830-9663
Founded: 1975; Members: 587
Pres: Jay Atherton; Gen Secr: Gina Meacoe
Focus: Archives
Periodicals
ACA Bulletin (weekly)
Archivaria (biennial) *01917*

Association of Canadian Bible Colleges, 100 Fergus Av, Kitchener, Ont. N2A 2H2
T: (519) 742-3572
Focus: Rel & Theol *01918*

Association of Canadian Career Colleges / Association Canadienne des Collèges Carrière, POB 340, Brantford, Ont. N3T 5N3
T: (519) 753-8689
Founded: 1964
Focus: Educ *01919*

Association of Canadian College and University Information Bureaus / Association des Bureaux d'Information des Collèges et Universités du Canada, c/o Public Relations and Development, Mount Saint Vincent University, Halifax, N.S. B3M 2J6
T: (902) 443-4450
Founded: 1968; Members: 5
Focus: Sci *01920*

Association of Canadian Community Colleges / Association des Collèges Communautaires du Canada, 110 Eglinton Av W, Toronto, Ont. M4R 1A3
T: (416) 489-5925; Tx: 06217566; Fax: (416) 489-5080
Founded: 1972; Members: 155
Gen Secr: Tom Norton
Focus: Educ
Periodicals
Bulletin (weekly)
College Canada
Communiqué (weekly)
Newsletter *01921*

Association of Canadian Faculties of Dentistry / Association des Facultés Dentaires du Canada, c/o Faculty of Dentistry, University of Alberta, 5059 Dentistry Pharmacy Bldg, Edmonton, Alta. T6G 2N8
T: (403) 432-5762
Founded: 1967; Members: 450
Focus: Dent *01922*

Association of Canadian Industrial Designers / Association des Designers Industriels du Canada, c/o Humber College, 205 Humber College Blvd, Rexdale, Ont. M9W 5L7
Founded: 1948; Members: 250
Pres: Ken Cummings; Gen Secr: L.E. Windsor
Focus: Graphic & Dec Arts, Design . . *01923*

Association of Canadian Map Libraries (ACML) / Association des Cartothèques Canadiennes, c/o National Map Collection, Public Archives of Canada, 395 Wellington St, Ottawa, Ont. K1A ON3
T: (613) 995-1077
Founded: 1967; Members: 260
Focus: Libraries & Bk Sci; Cart *01924*

Association of Canadian Medical Colleges / Association des Facultés de Médecine du Canada, 151 Slater St, Ste 1006, Ottawa, Ont. K1P 5N1
T: (613) 237-0070
Founded: 1943
Focus: Med
Periodical
Forum (weekly) *01925*

Association of Canadian Universities for Northern Studies / Association Universitaire Canadienne d'Etudes Nordiques, 130 Albert St, Ste 1915, Ottawa, Ont. K1P 5G4
T: (613) 238-3525
Founded: 1977
Focus: Hist; Sociology; Lit
Periodical
Northline/Point Nord (weekly) *01926*

Association of Canadian University Planning Programs, c/o Technical University of Nova Scotia, POB 1000, Halifax, N.S. B3J 2X4
T: (902) 420-7699; Fax: (902) 423-6672
Founded: 1978; Members: 16
Pres: Prof. F. Palermo; Gen Secr: Prof. W. Smith
Focus: Urban Plan *01927*

Association of Concern for Ultimate Reality and Meaning, 15 Saint Mary St, Ste 209, Toronto, Ont. M4Y 2R5
T: (416) 922-2476
Founded: 1976; Members: 10
Gen Secr: Tibor Horvath
Focus: Philos
Periodical
Ultimate Reality and Meaning *01928*

Association of Deans of Pharmacy of Canada / Association des Doyens de Pharmacie du Canada, c/o Faculty of Pharmacy, University of Toronto, Toronto, Ont. M5S 2S2
T: (416) 978-2880
Founded: 1965
Focus: Pharmacol *01929*

Association of Directors of Journalism Programs in Canadian Universities / Association des Directeurs et Coordonnateurs des Programmes de Journalisme des Universités Canadienne, c/o The School of Journalism, Middlesex College, University of Western Ontario, London, Ont. N6A 5B7
T: (519) 679-3377
Focus: Journalism *01930*

Association of Educators of Gifted, Talented and Creative Children in B.C., 2030 Orland Dr, Coquitlam, B.C. V3J 3J4
T: (604) 936-1184
Founded: 1977; Members: 800
Pres: Linda Lewis
Focus: Educ *01931*

Association of Exploration Geochemists, POB 523, Rexdale, Ont. M9W 5L4
Founded: 1970
Focus: Geology *01932*

Association of Faculties of Pharmacy of Canada / Association des Facultés de Pharmacie du Canada, c/o Faculty of Pharmaceutical Science, University of British Columbia, Vancouver, B.C. V6T 1W5
T: (604) 228-3815
Founded: 1944; Members: 143
Focus: Pharmacol *01933*

Association of Large School Boards in Ontario / Association des Grands Conseils Scolaires de l'Ontario, 252 Bloor St W, Ste 12-205, Toronto, Ont. M5S 1V5
T: (416) 922-4858
Founded: 1976
Focus: Public Admin; Educ *01934*

Association of Manitoba Museums (AMM), 422-167 Lombard Av, Winnipeg, Man. R3B 0T6
T: (204) 947-1782
Founded: 1971; Members: 330
Pres: George E. Lammers
Focus: Arts
Periodicals
AMM Newsletter (weekly)
Dawson & Hind (3 times annually) . . . *01935*

Association of Nova Scotia Education Administrators / Association de Directeurs d'Education en Nouvelle-Ecosse, 266 Whitney Av, Sydney, N.S. B1P 5A6
T: (902) 539-0940
Members: 50
Pres: Ernest J. Chiasson
Focus: Educ *01936*

Association of Occupational Therapists of Manitoba, 1061 Ingersoll St, Winnipeg, Man. R3E 2M1
T: (204) 237-1762
Founded: 1971; Members: 178
Gen Secr: Ann Ogrodnik
Focus: Therapeutics *01937*

Association of Parliamentary Librarians in Canada (APLIC) / Association des Bibliothécaires Parlementaires du Canada, c/o Library of Parliament, Parliament Bldgs, Ottawa, Ont. K1A 0A9
T: (613) 992-2427; Fax: (613) 992-1269
Founded: 1975
Focus: Libraries & Bk Sci
Periodical
APLIC Bulletin: Bulletin ABPAC (weekly) *01938*

Association of Psychologists of Nova Scotia, R.R. 1, Eureka, Millstream, N.S. B0K 1B0
T: (902) 424-2081
Founded: 1965; Members: 150
Pres: John Service
Focus: Psych *01939*

Association of Quebec Regional English Media / Association des Medias Régionaux Anglophones du Québec, c/o Extension Dept, Macdonald College, POB 284, Anne de Bellevue, Qué. H9X 1C0
T: (514) 457-2000; Tx: 05821788
Founded: 1979; Members: 35
Focus: Mass Media *01940*

Association of Registrars of Universities and Colleges of Canada / Association des Régistraires des Universités et Collèges du Canada, 151 Slater St, Ottawa, Ont. K1P 5N1
T: (613) 563-1236
Founded: 1959; Members: 200
Focus: Educ *01941*

Association of School Business Officials of Saskatchewan, 33 Darlington St W, Yorkton, Sask. S3N 0E3
T: (306) 783-8576
Gen Secr: W.G. Fleming
Focus: Public Admin *01942*

Association of Schools of Optometry of Canada / Association des Ecoles d'Optométrie du Canada, c/o School of Optometry, Faculty of Science, University of Waterloo, Waterloo, Ont. N2L 3G1
T: (519) 885-1211
Focus: Ophthal *01943*

Association of Small Public Libraries of Ontario, c/o Public Library, 2 Library Ln, Tillsonburg, Ont. N4G 3P9
T: (519) 842-5571
Founded: 1981
Focus: Libraries & Bk Sci *01944*

Association of Universities and Colleges of Canada, 151 Slater St, Ste 1200, Ottawa, Ont. K1P 5N1
T: (613) 563-1236
Founded: 1911; Members: 89
Pres: Dr. Claude Lajeunesse; Gen Secr: Dr. Kenneth Ozmon
Focus: Sci
Periodical
University Affairs (10 times annually) . . *01945*

Association of University Forestry Schools of Canada / Association de E1coles Forestières Universitaires du Canada, c/o Lakehead University, Thunder Bay, Ont. P7B 5E1
T: (807) 345-2121; Tx: 0734594
Founded: 1973
Focus: Forestry *01946*

Association of University of New Brunswick Teachers, POB 4400, Fredericton, N.B. E3B 5A3
T: (506) 453-4661
Pres: Dr. Gerald M. Clarke
Focus: Educ *01947*

Association of Wholly or Partially French Language Universities (AUPELF), c/o University of Montréal, BP 6128, Montréal, Qué. H3C 3J7
T: (514) 343-6630
Founded: 1961; Members: 122
Gen Secr: Maurice Beutler
Focus: Ling; Doc; Sci
Periodicals
Etudes créoles (weekly)
Perspectives universitaires (weekly)
Universités (weekly) *01948*

Association of Workers' Compensation Boards of Canada / Association des Commissions des Accidents du Travail du Canada, 255 Bamburgh Circle, Ste 1501, Scarborough, Ont. M1W 3T6
T: (416) 495-7524
Founded: 1919
Gen Secr: Kenneth B. Harding
Focus: Safety *01949*

Association Paritaire de Prévention pour la Santé et la Securité du Travail, 50 Pl Crémaize, Ste 812, Montréal, Qué. H2P 2T5
T: (514) 389-8295
Focus: Safety; Public Health *01950*

Association pour l'Avancement des Sciences et des Techniques de la Documentation (ASTED), 1030 Rue Cherrier, Montréal, Qué. H2L 1H9
T: (514) 522-7833; Fax: (514) 521-9561
Founded: 1973; Members: 900
Pres: Richard Paré; Gen Secr: Louis Cabral
Focus: Doc
Periodical
Documentation et Bibliothèques (weekly) . *01951*

Association Professionnelle des Géologues et des Géophysiciens du Québec / Association of Professional Geologists and Geophysicists of Quebec, CP 470, Succ Tour de la Bourse, Montréal, Qué. H4Z 1J2
Founded: 1968; Members: 400
Focus: Geology
Periodical
Bulletin de l'Association Professionnelle des Géologues et des Géophysiciens du Québec (weekly) *01952*

Association Pulmonaire du Québec / Québec Lung Association, 264 Rue Chénier, Québec, Qué. G1K 1R2
T: (418) 524-4254, 524-2735
Founded: 1938; Members: 300
Gen Secr: Guy Paré
Focus: Pulmon Dis *01953*

Association Québécoise des Archivistes Médicales, 4357 Pl Viger, CP 1020, Rock-Forest, Qué. J0B 2J0
T: (819) 567-6935
Founded: 1960; Members: 800
Focus: Archives *01954*

Association Québécoise des Techniques de l'Eau, 407 Saint-Laurent, Bureau 500, Montréal, Qué. H2Y 2Y5
T: (514) 874-3700
Founded: 1962; Members: 1600
Gen Secr: N. Pageau-Goyette
Focus: Water Res
Periodicals
Effluent (10 times annually)
Sciences et Techniques de l'Eau (weekly) *01955*

Asthma Association of Canada, Toronto-Dominion Centre, POB 192, Toronto, Ont. M5K 1H6
T: (416) 481-9627
Focus: Rheuma *01956*

Atlantic Cerebral Palsy Association, 24 Bristol St, Fredericton, N.B. E3B 4W3
T: (506) 454-2804
Founded: 1973; Members: 300
Focus: Neurology *01957*

Atlantic Conference on Learning Disabilities, POB 2036, DEPS, Dartmouth, N.S. B2W 3X8
T: (902) 469-7282, 469-8903
Founded: 1974; Members: 20
Pres: Judy Pelletier
Focus: Educ Handic *01958*

Atlantic Petroleum Association, POB 1145, Halifax, N.S. B3J 2X1
T: (902) 463-8975
Founded: 1972; Members: 6
Gen Secr: Stan Miller
Focus: Petrochem; Nat Res *01959*

Atlantic Planners Institute, POB 2012, Halifax, N.S. B3J 2Z1
Focus: Urban Plan
Periodical
The Atlantic Planners Pen (weekly) . . *01960*

Atlantic Provinces Council on the Sciences / Conseil des Provinces Atlantiques pour les Sciences, c/o Science Dept, Memorial University of Newfoundland, St. John's, Nfld. A1B 3X7
Tx: (709) 737-8153
Founded: 1962
Focus: Sci *01961*

Atlantic Provinces Library Association (APLA), c/o School of Library Science, Dalhousie University, Halifax, N.S. B3H 4M8
Focus: Libraries & Bk Sci
Periodical
APLA Bulletin (weekly) *01962*

Banting Research Foundation, c/o Faculty of Medicine, University of Toronto, McMurrich Bldg, Toronto, Ont. M5S 1A8
T: (416) 978-4251
Pres: R.H. Pearce
Focus: Med *01963*

Bibliographical Society of Canada / Société Bibliographique du Canada, POB 575, Postal Station P, Toronto, Ont. M5S 2T1
Founded: 1946; Members: 500
Pres: Sandra Alston; Gen Secr: Anne Dondertman
Focus: Libraries & Bk Sci
Periodicals
Bulletin (weekly)
Papers/Cahiers (weekly) *01964*

Biological Council of Canada / Conseil Canadian de Biologie, c/o Dept of Biology, University of Ottawa, Ottawa, Ont. K1N 6N5
T: (613) 564-2336
Founded: 1965
Gen Secr: Dr. T.W. Moon
Focus: Bio *01965*

Biomass Energy Institute, 1329 Niakwa Rd, Winnipeg, Man. R2J 3T4
T: (204) 257-3891
Founded: 1971; Members: 500
Gen Secr: Beth Candlish
Focus: Energy *01966*

Birks Family Foundation, 1240 Sq Phillips, Montréal, Qué. H3B 3H4
Gen Secr: O.R. Macklem
Focus: Sci *01967*

Bobechko Foundation, 176 Saint George St, Toronto, Ont. M5R 2M7
Gen Secr: W.P. Bobechko
Focus: Sci *01968*

Brewing and Malting Barley Research Institute, 167 Lombard Av, Ste 206, Winnipeg, Man. R3B 0T6
Tx: (204) 942-1407
Founded: 1948
Gen Secr: Dr. N.T. Kendall
Focus: Nutrition 01969

British Columbia and Yukon Heart Foundation, 1212 Broadway W, Vancouver, B.C. V6H 3V2
T: (604) 736-4404
Founded: 1955; Members: 26
Focus: Cardiol 01970

British Columbia Association of Podiatry (B.C.A.P.), 2309 41 Av W, Ste 115, Vancouver, B.C. V6M 2A3
T: (604) 261-2265
Founded: 1968
Focus: Orthopedics
Periodical
B.C.A.P. Journal (weekly) 01971

British Columbia Association of Speech / Language Pathologists and Audiologists, 1616 Seventh Av W, Vancouver, B.C. V6J 1S5
T: (604) 734-5577
Founded: 1957; Members: 243
Focus: Logopedy 01972

British Columbia Cancer Foundation, 601 10 Av W, Vancouver, B.C. V5Z 1L3
T: (604) 877-6010
Founded: 1935; Members: 200
Focus: Cell Biol & Cancer Res
Periodical
Cancer Research News (weekly) 01973

British Columbia Corrections Association, 1090 Georgia St W, Ste 1300, Vancouver, B.C. V6E 3X9
T: (604) 687-2242
Members: 105
Focus: Criminology
Periodical
The Courier 01974

British Columbia Council for Leadership in Educational Administration, 5840 Cedarbridge Way, Richmond, B.C. V6X 2A7
T: (604) 273-5461
Gen Secr: Robert M. Taylor
Focus: Educ 01975

British Columbia Drama Association, 835 Humbolt St, Ste 400, Victoria, B.C. V8V 2M4
T: (604) 381-2443
Founded: 1932; Members: 102
Gen Secr: W.J. Langlois
Focus: Perf Arts 01976

British Columbia Epilepsy Society, 1195 Eighth Av W, Vancouver, B.C. V6H 1C5
T: (604) 734-2221
Founded: 1959; Members: 800
Focus: Pathology 01977

British Columbia Genealogical Society, POB 94371, Richmond, B.C. V6Y 2A8
T: (604) 274-3659
Focus: Genealogy
Periodicals
British Columbia Generalogist (weekly)
Newsletter (weekly) 01978

British Columbia Historical Association, 450 Stewart Av, Ste 211, Nanaimo, B.C. V9S 5E9
Pres: Barbara Stannard
Focus: Hist 01979

British Columbia Library Trustees' Association, c/o Dunsmuir Secondary School, 3391 Painter Rd, Victoria, B.C. V9C 2J5
T: (604) 478-5548
Focus: Libraries & Bk Sci 01980

British Columbia Lung Association, 906 Broadway W, Vancouver, B.C. V5Z 1K7
T: (604) 731-4961
Founded: 1922
Focus: Pulmon Dis 01981

British Columbia Medical Association, 115-1665 Broadway W, Vancouver, B.C. V6J 5A4
T: (604) 736-5551
Founded: 1936; Members: 4766
Focus: Med 01982

British Columbia Museums Association, 514 Government St, Victoria, B.C. V8V 4X4
T: (604) 387-3315; Fax: (604) 356-8197
Founded: 1959; Members: 500
Gen Secr: Richard A. Duckles
Focus: Archives
Periodical
Museum Round-up (weekly) 01983

British Columbia Parents in Crisis Society, 13-250 Willingdon Av S, Burnaby, B.C. V5C 5C9
T: (604) 299-0521
Founded: 1974; Members: 100
Focus: Educ
Periodical
Newsletter (weekly) 01984

British Columbia Parkinsons's Disease Association, 1195 Eighth Av W, Vancouver, B.C. V6H 1C5
T: (604) 734-2221
Members: 85
Focus: Med; Neurology 01985

British Columbia Pharmacists' Society, 1200 73 Av W, Ste 604, Vancouver, B.C. V6P 6G5
T: (604) 263-2766; Fax: (604) 263-2786
Founded: 1968; Members: 1673
Gen Secr: Frank Merlin Archer
Focus: Pharmacol
Periodical
In Pharmation: B.C. Pharmacist (weekly) 01986

British Columbia Pollution Control Association, 100 Kapilano, Park Royal, West Vancouver, B.C. V7T 1A2
Focus: Ecology 01987

British Columbia Research Council, 3650 Wesbrook Mall, Vancouver, B.C. V6S 2L2
T: (604) 224-4331; Tx: 04507748;
CA: RESEARCHBC
Founded: 1944; Members: 72
Gen Secr: Alan Mode
Focus: Eng; Sci
Periodicals
Annual Report (weekly)
List of Publications (weekly) 01988

British Columbia Society of Occupational Therapists (B.C.S.O.T.), 222-4585 Canada Way, Burnaby, B.C. V5G 4L6
T: (604) 294-2717; Fax: (604) 294-9191
Founded: 1958; Members: 620
Pres: K. Scalzo; Gen Secr: Kim Hudson
Focus: Therapeutics
Periodical
OT Line (weekly) 01989

British Columbia Speleological Federation, POB 733, Gold River, B.C. V0P 1G0
Pres: Paul Griffiths
Focus: Speleology 01990

British Columbia Teacher-Librarians' Association, 2235 Burrard St, Vancouver, B.C. V6J 3H9
T: (604) 731-8121
Members: 925
Pres: Liz Austrom
Focus: Libraries & Bk Sci 01991

British Columbia Teachers' Federation, 2235 Burrard St, Vancouver, B.C. V6J 3H9
T: (604) 731-8121
Founded: 1917; Members: 38064
Gen Secr: Robert M. Buzza
Focus: Educ 01992

British Columbia Thoracic Society, 906 Broadway W, Vancouver, B.C. V5Z 1K7
T: (604) 731-4961
Gen Secr: Scott McDonald
Focus: Pulmon Dis 01993

Broadcast Education Association of Canada / Association Canadienne des Educateurs en Radiodiffusion, c/o Broadcast Communications, BCIT, 3700 Willingdon Av, Burnaby, B.C. V5G 3H2
Founded: 1977; Members: 82
Focus: Educ 01994

Broadcast Research Council of Canada, 2 Bloor St W, Ste 100, Toronto, Ont. M4W 3E2
T: (416) 445-9800
Founded: 1965; Members: 120
Focus: Mass Media 01995

Brodie Club, c/o Dept of Anthropology, University of Toronto, Toronto, Ont. M5S 1A1
Tx: (416) 978-5260
Founded: 1922; Members: 30
Focus: Anthro 01996

Bruce, Dufferin, Grey Lung Association, 836 Second Av E, Ste 1, Owen Sound, Ont. N4K 2H3
T: (519) 371-2321
Gen Secr: Helen Cowan
Focus: Pulmon Dis 01997

Calgary Society for the Treatment of Autism, 404 94 Av SE, Calgary, Alta. T2J 0E8
T: (403) 253-2291
Focus: Psych 01998

Calgary Zoological Society, POB 3036, Station B, Calgary, Alta. T2M 4R8
Tx: (403) 265-9310
Focus: Zoology 01999

Canada Council / Conseil des Arts du Canada, 99 Metcalfe St, POB 1047, Ottawa, Ont. K1P 5V8
T: (613) 598-4365; Fax: (613) 598-4390
Founded: 1957; Members: 21
Pres: Allan Gotlieb; Gen Secr: Joyce Zemans
Focus: Arts
Periodical
Annual Report (weekly) 02000

Canada Safety Council / Conseil Canadienne de la Sécurité, 1765 Saint Laurent Blvd, Ottawa, Ont. K1G 3V4
T: (613) 521-6881
Founded: 1968; Members: 25
Focus: Safety
Periodicals
Council Update (weekly)
Living Safety (weekly)
News Brake (weekly)
Public Focus (weekly)
Safety Canada (weekly)
Safetylines (weekly) 02001

Canadian Academic Accounting Association, c/o Faculty of Management, University of Toronto, Toronto, Ont. M5S 1V4
Founded: 1976; Members: 600
Focus: Business Admin
Periodicals
Canadian Accounting Education and Research News (weekly)
Contemporary Accounting Research (biennial)
. . . . 02002

Canadian Academy of Engineering, 130 Albert St, Ste 1414, Ottawa, K1P 5G4
T: (613) 235-9056
Founded: 1988; Members: 150
Pres: Dr. Camille Dagenais; Gen Secr: Dr. L.M. Nadeau
Focus: Eng
Periodical
Newsbrief (weekly) 02003

Canadian Academy of Medical Illustrators, 15 Brendan Rd, Toronto, Ont. M4G 2W9
T: (416) 486-3611
Founded: 1965; Members: 50
Focus: Graphic & Dec Arts, Design; Med 02004

Canadian Academy of Periondontology, 170 Saint George St, Ste 818, Toronto, Ont. M5R 2M8
T: (416) 925-0293
Founded: 1954; Members: 100
Pres: Dr. Frederick C. Lackie
Focus: Dent 02005

Canadian Academy of Recording Arts and Sciences, 89 Bloor St E, Toronto, Ont. M4W 1A9
T: (416) 9225029
Founded: 1974; Members: 1000
Gen Secr: Daisy C. Falle
Focus: Doc
Periodical
Newsletter (weekly) 02006

Canadian Academy of Restorative Dentistry, 558 Village Green Av, London, Ont. N6K 1G4
T: (519) 472-1930
Focus: Dent 02007

Canadian Academy of Sport Medicine / Académie Canadienne de Medécine Sportive, 333 River Rd, Vanier City, Ont. K1L 8H9
Members: 360
Focus: Physical Therapy 02008

Canadian Academy of the History of Pharmacy / Académie Canadienne d'Histoire de la Pharmacie, c/o Faculty of Pharmacy, University of Toronto, Toronto, Ont. M5S 1A1
T: (416) 978-5626; Fax: (416) 978-8511
Founded: 1955; Members: 200
Focus: Pharmacol 02009

Canadian Academy of Urological Surgeons, 888 Eighth Av W, Ste 312, Vancouver, B.C. V5Z 3Y1
Focus: Surgery; Urology 02010

Canadian Acid Precipitation Foundation, 112 Saint Clair Av W, Toronto, Ont. M4V 2Y3
Fax: (416) 923-4911
Focus: Ecology 02011

Canadian Acoustical Association / Association Canadienne d'Acoustique, POB 1351, Station F, Ottawa, Ont. K4T 2V9
Founded: 1977; Members: 400
Focus: Acoustics
Periodical
Canadian Acoustics/Acoustique Canadienne (weekly) 02012

Canadian Active Health Foundation, 246 Broadway E, Ste 1, Vancouver, B.C. V5T 1W3
T: (604) 872-2210
Focus: Public Health 02013

Canadian Advanced Technology Association, 275 Slater St, Ste 803, Ottawa, Ont. K1P 5H9
T: (613) 236-6550
Founded: 1978; Members: 26
Gen Secr: Robert Long
Focus: Eng 02014

Canadian Advertising Research Foundation (CARF) / Fondation Canadienne de Recherche en Publicité, 180 Bloor St W, Ste 803, Toronto, Ont. M5S 2V6
T: (416) 964-3832
Members: 118
Focus: Advert
Periodical
CARF Newsletter 02015

Canadian Aeronautics and Space Institute / Institut Aéronautique et Spatial du Canada, 130 Slater St, Ste 818, Ottawa, Ont. K1P 6E2
Founded: 1954; Members: 2300
Pres: Dr. G. Marsters; Gen Secr: A.J.S. Timmins
Focus: Aero
Periodical
Canadian Aeronautics and Space Journal (weekly)
. . . . 02016

Canadian Agricultural Economics and Farm Management Society, 151 Slater St, Suite 907, Ottawa, Ont. K1P 5H4
T: (613) 2329459
Founded: 1952; Members: 500
Pres: George Lee
Focus: Agri
Periodical
Canadian Journal of Agricultural Economics (weekly) 02017

Canadian Alarm and Security Association (CANASA) / Association Canadienne de l'Alarme et de la Securité, 2175 Sheppard Av E, Ste 110, Willowdale, Ont. M2J 1W8
Fax: (416) 491-1670
Founded: 1975; Members: 260
Gen Secr: Jennifer Rogers
Focus: Safety
Periodical
CANASA Newsletter (2-3 times annually) 02018

Canadian Aldeburgh Foundation, 34 Glenallan Rd, Toronto, Ont. M4N 1G8
T: (416) 481-1964
Focus: Music 02019

Canadian Amphibian and Reptile Conservation Society, 9 Mississauga Rd N, Ste 1, Mississauga, Ont. L5H 2H5
T: (416) 274-0914
Founded: 1960; Members: 130
Pres: Wayne F. Weller
Focus: Zoology 02020

Canadian Anaesthetists' Society / Société Canadienne des Anesthésistes, 94 Cumberland St, Ste 901, Toronto, Ont. M5R 1A3
Founded: 1943; Members: 1930
Focus: Anesthetics 02021

Canadian Apparel Manufacturers Institute, 141 Laurier W, Ste 804, Ottawa Ont. K1P 5J3
T: (613) 238-7744
Focus: Textiles 02022

Canadian Arctic Resources Committee, 111 Sparks St, Ottawa, Ont. K1P 5B5
T: (613) 236-7379
Founded: 1971
Gen Secr: Stephen Hazell
Focus: Nat Res
Periodicals
Northern Decisions (weekly)
Northern Perspectives (weekly) 02023

Canadian Art Museums Directors' Organization / Organisation des Directeurs des Musées d'Art Canadiens, c/o Art Gallery of Greater Victoria, 1040 Moss St, Victoria, B.C. V8V 4P1
T: (604) 384-4101
Founded: 1964; Members: 35
Focus: Fine Arts 02024

Canadian Association for Adult Education, 29 Prince Arthur Av, Toronto, Ont. MSR 1B2
T: (416) 964-0559
Founded: 1935; Members: 1000
Pres: Patrick Flanagan; Gen Secr: Ian Morrison
Focus: Adult Educ
Periodicals
Learning (weekly)
Resources Kit (weekly) 02025

Canadian Association for American Studies (CAAS), c/o Dept. of History, Simon Fraser University, Burnaby, B.C.
Founded: 1964; Members: 200
Pres: Michael Fellman
Focus: Hist
Periodicals
Canadian Review of American Studies (weekly)
Newsletter (weekly) 02026

Canadian Association for Clinical Micro and Infectious Diseases / Association Canadienne de Microbiologie Clinique et des Maladies, c/o Connaught Research Institute, 1755 Steeles Av W, Willowdale, Ont. M2R 3T4
T: (416) 667-2800
Founded: 1980; Members: 400
Focus: Immunology 02027

Canadian Association for Corporate Growth, POB 87, Commerce Court Post Office, Toronto, Ont. M5L 1C6
T: (416) 864-9520
Founded: 1972
Gen Secr: Barbara Corder
Focus: Econ 02028

Canadian Association for Dental Research / Association Canadienne de Recherches Dentaires, c/o Faculty of Dentistry, University of Western Ontario, London, Ont. N6A 5C1
T: (519) 679-2111 ext 6137
Founded: 1976; Members: 195
Focus: Dent 02029

Canadian Association for Distance Education, c/o Wilfrid Laurier University, 75 University Av W, Waterloo, Ont. N2L 3C5
T: (519) 884-1970
Founded: 1983
Pres: Marion Croft
Focus: Educ 02030

Canadian Association for Educational Psychology / Association Canadienne en Psychopédagogie, Faculty of Education, University of Western Ontario, London, Ont. N6G 1G7
T: (519) 679-3259
Founded: 1974; Members: 150
Focus: Educ; Psych 02031

Canadian Association for Future Studies / Association Canadienne des Etudes Prospectives, 3764 Côte des Neiges, Montréal, Qué. H3H 1V6
T: (514) 937-2212
Founded: 1976; Members: 300
Focus: Futurology
Periodical
Futures Canada (weekly) 02032

Canadian Association for Health Physical Education and Recreation / Association Canadienne pour la Santé, l'Education Physique et la Récréation, 333 River Rd, Tower B, Vanier, Ont. K1L 8B9
T: (613) 746-5909; Tx: 0533660
Founded: 1933; Members: 2500
Focus: Public Health 02033

Canadian Association for Information Science / Association Canadienne des Sciences de l'Information, 47 Gore St E, Perth, Ont. K7H 1H6
T: (613) 267-3499
Founded: 1970; Members: 600
Focus: Computer & Info Sci
Periodicals
CAIS Newsletter: Nouvelles ACSI (weekly)
Canadian Journal of Information Science: Revue canadienne des sciences de l'information (weekly)
Proceedings of the Annual Conference: Procesverbaux de la conférence (weekly) 02034

Canadian Association for Laboratory Animal Science / Association Canadienne pour la Technologie des Animaux de Laboratoire, c/o M524 Biological Sciences Bldg, University of Alberta, Edmonton, Alta. T6G 2E9
T: (403) 492-5793
Founded: 1962; Members: 500
Focus: Vet Med
Periodical
Newsletter (weekly) 02035

Canadian Association for Latin American and Caribbean Studies (C.A.L.A.C.S.) / Association Canadienne des Etudes Latino-Américaines et Caraïbes, c/o University of Ottawa, Ottawa, Ont. K1N 6N5
T: (613) 564-5457
Founded: 1969
Focus: Sci
Periodicals
The Canadian Journal of Latin American and Caribbean Studies (biennial)
Newsletter 02036

Canadian Association for Narcolepsy / Association Canadienne de la Narcolepsie, POB 223, Station S, Toronto, Ont. M5M 4L7
T: (416) 398-1627
Founded: 1978; Members: 1500
Pres: Dr. Fred Demanuele
Focus: Anesthetics
Periodical
Keeping Awake (weekly) 02037

Canadian Association for Pastoral Education / Association Canadienne pour l'Education Pastorale, 40 Saint Clair Av E, Toronto, Ont. M4T 1M9
T: (416) 921-4152
Focus: Doc 02038

Canadian Association for Production and Inventory Control, 123 Burnhamthorpe Rd, Islington, Ont. M9H 1A5
T: (416) 234-5774
Focus: Materials Sci 02039

Canadian Association for Scottish Studies / Société des Etudes Ecossaises, c/o Dept of History, University of Guelph, Guelph, Ont. N1G 2W1
T: (519) 824-4120 ext 3201
Founded: 1969; Members: 217
Focus: Ethnology 02040

Canadian Association for the Advancement of Netherlandic Studies / Association Canadienne pour l'Avancement des Etudes Néerlandaises, c/o Dept of French, University of Windsor, Windsor, Ont. N9B 3P4
T: (519) 253-4232 ext 2066
Founded: 1971; Members: 300
Focus: Ling; Lit; Cultur Hist; Sociology; Poli Sci
Periodical
Canadian Journal of Netherlandic Studies (weekly) 02041

Canadian Association for the Social Studies, POB 752, Fredericton, N.B. E3R 5R6
Focus: Sociology 02042

Canadian Association for the Study of Educational Administration / Association Canadienne pour l'Etude de l'Administration Scolaire, c/o Education Dept, Dalhousie University, Halifax, N.S. B3H 3J5
T: (902) 425-5430
Founded: 1974; Members: 225
Focus: Educ 02043

Canadian Association for University Continuing Education / Association pour l'Education Permanente dans des Universités du Canada, 151 Slater St, Ottawa, Ont. K1P 5N1
T: (613) 563-1236; Tx: 0533329
Founded: 1954; Members: 70
Focus: Sci; Educ 02044

Canadian Association of African Studies / Association Canadienne des Etudes Africaines, c/o Innis College, 2 Sussex Av, Toronto, Ont. M5S 1J5
T: (416) 978-3424; Fax: (416) 978-5503
Founded: 1970; Members: 310
Pres: Serge Genest
Focus: Ethnology
Periodicals
Canadian Journal of African Studies (3 times annually)
Newsletter (weekly) 02045

Canadian Association of Anatomists (CAA) / Association Canadienne des Anatomistes, c/o Dept of Anatomy, University of Western Ontario, London, Ont. N6A 5C1
Founded: 1956; Members: 265
Pres: Dr. P.K. Lala
Focus: Anat
Periodical
CAA Bulletin (weekly) 02046

Canadian Association of Business Education Teachers, 439 Eighth St SW, Medicine Hat, Alta. T1A 4M1
T: (403) 526-5198
Focus: Educ 02047

Canadian Association of Deans of Arts and Sciences / Association Canadienne des Doyens des Facultés des Lettres et des Sciences, c/o University of Prince Edward Island, Charlottetown, P.E.I. C1A 4P3
T: (902) 892-4121
Focus: Sci; Arts; Lit 02048

Canadian Association of Deans of Education / Association Canadienne des Doyens d'Education, c/o Faculty of Education, University of British Columbia, 2125 Main Mall, Vancouver, B.C. V6T 1Z5
T: (604) 228-5211
Focus: Educ 02049

Canadian Association of Foundations of Education / Association Canadienne pour l'Etude des Fondements de l'Education, c/o Queen's University, Kingston, Ont. K7L 5C4
Founded: 1967; Members: 204
Focus: Educ 02050

Canadian Association of Gastroenterology / Association Canadienne de Gastroentérologie, c/o Sir Mortimer B. Davis Jewish General Hospital, 3755 Cote St, Catherine Rd, Montréal, Qué. H3T 1E2
T: (613) 544-0499
Founded: 1961; Members: 254
Focus: Gastroenter 02051

Canadian Association of Geographers (CAG/ACG) / Association Canadienne des Géographes, c/o McGill University, Burnside Hall, 805 Sherbrooke St W, Montréal, Qué. H3A 2K6
T: (514) 398-4946; Fax: (514) 398-7437
Founded: 1951; Members: 1400
Pres: Dr. Olav Slaymaker; Gen Secr: Dr. Mark Rosenberg
Focus: Geography
Periodicals
The Canadian Geographer (weekly)
The Operational Geographer (weekly)
The Directory (weekly) 02052

Canadian Association of Immersion Teachers / Association Canadienne des Professeurs d'Immersion, POB 8843, Alta Vista PO, Ottawa, Ont. K1G 3H8
T: (613) 731-8661
Members: 1000
Focus: Educ 02053

Canadian Association of Independent Schools, 10 Elm Av, Toronto, Ont. M4W 1N4
T: (416) 920-9741
Founded: 1979
Pres: Allison Roach
Focus: Educ 02054

Canadian Association of Latin American Studies / Association Canadienne des Etudes Latino-Américaines, c/o Dept of Hispanic Studies, University of British Columbia, Vancouver, B.C. V6T 1W5
Founded: 1969; Members: 200
Pres: J.C.M. Ogelsby
Focus: Ling; Lit
Periodicals
Canadian Journal of Latin American Studies (weekly)
Newsletter (weekly) 02055

Canadian Association of Law Libraries (CALL), c/o York University Law Library, 4700 Keele St, North York, Ont. M3J 1P3
Founded: 1961
Pres: Patricia M. Young; Gen Secr: Neil Campbell
Focus: Libraries & Bk Sci
Periodical
CALL Newsletter (5 times annually) . . 02056

Canadian Association of Library Schools, c/o School of Library Service, Dalhousie University, Halifax, N.S. B3H 4H8
T: (902) 424-3656
Founded: 1967; Members: 60
Pres: L.J. Amey
Focus: Libraries & Bk Sci 02057

Canadian Association of Marketing Research Organizations, 69 Sherbourne St, Ste 315, Toronto, Ont. M5A 3X7
T: (416) 364-1223
Members: 23
Pres: Michael Adams
Focus: Marketing 02058

Canadian Association of Music Libraries (CAML) / Association Canadienne des Bibliothèques Musicales, c/o National Library, Music Division, 395 Wellington St, Ottawa, Ont. K1A 0N4
T: (613) 996-3530
Founded: 1971; Members: 80
Focus: Libraries & Bk Sci; Music
Periodical
CAML Newsletter (3 times annually) . . 02059

Canadian Association of Optometrists / Association Canadienne des Optométristes, 301-1785 Alta Vista Dr, Ottawa, Ont. K1G 3Y6
T: (613) 738-4412; Fax: (613) 738-7161
Founded: 1948; Members: 1400
Pres: Dr. Jean-M. Rodrigue; Gen Secr: Gérard Lambert
Focus: Ophthal
Periodical
The Canadian Journal of Optometry (weekly) 02060

Canadian Association of Paediatric Surgeons / Association de la Chirurgie Infantile Canadienne, c/o Victoria Hospital, South St, London, Ont. N6A 4G5
T: (519) 432-5977
Founded: 1968; Members: 60
Focus: Pediatrics; Surgery 02061

Canadian Association of Parapsychologists, 69 Gilmour St, Ottawa, Ont. K2P 0N1
T: (613) 563-4264
Focus: Parapsych 02062

Canadian Association of Pathologists / Association Canadienne des Pathologistes, c/o Dept of Pathology, Hotel Dieu Hospital, Kingston, Ont. K7L 3H6
T: (613) 549-2080
Founded: 1949; Members: 650
Focus: Pathology 02063

Canadian Association of Physical Medicine and Rehabilitation / Association Canadienne de Médecine Physique et de Réadaptation, c/o University Hospital, 339 Windmere Rd, London, Ont. N6A 5A5
T: (519) 663-3513
Founded: 1962; Members: 230
Focus: Physical Therapy
Periodical
Newsletter (weekly) 02064

Canadian Association of Physicists / Association Canadienne des Physiciens, 151 Slater St, Ste 903, Ottawa, Ont. K1P 5H3
T: (613) 237-3392; Fax: (613) 238-1677
Founded: 1958; Members: 2000
Pres: R.M. Lees; Gen Secr: Mona L. Jento
Focus: Physics
Periodical
Physics in Canada (weekly) 02065

Canadian Association of Radiologists / Association Canadienne des Radiologistes, 1440 Sainte Catherine St W, Ste 806, Montréal, Qué. H3G 1R8
Founded: 1937; Members: 1700
Focus: Nucl Med 02066

Canadian Association of Research Libraries / Association des Bibliothèques de Recherche du Canada, 7 Noel Av, Toronto, Ont. M4G 1B2
T: (416) 425-8145
Focus: Libraries & Bk Sci 02067

Canadian Association of Social Workers / Association Canadienne des Travailleurs Sociaux, 55 Parkdale Av, Ottawa, Ont. K1Y 1E5
T: (613) 728-1865
Founded: 1926; Members: 9000
Focus: Sociology 02068

Canadian Association of Sport Sciences / Association Canadienne des Sciences du Sport, 1600 James Naismith Dr, Gloucester, Ont. K1B 5N4
T: (613) 748-5671; Fax: (613) 748-5729
Founded: 1967; Members: 400-500
Focus: Sports
Periodical
Canadian Journal of Applied Sport Sciences (weekly) 02069

Canadian Association of Toy Libraries and Parent Resources Centres / Association Canadienne des Ludothèques et des Centres de Resources pour la Famille, POB 4478, Station E, Ottawa, Ont. K1S 5B4
T: (613) 563-0438
Founded: 1975
Focus: Libraries & Bk Sci
Periodical
Play and Parenting Connections (weekly) 02070

Canadian Association of University Development Officers, 151 Slater St, Ottawa, Ont. K1P 5N1
T: (613) 563-1236
Pres: Kenneth Clements
Focus: Educ 02071

Canadian Association of University Schools of Nursing (CAUSN) / Association Canadienne des Ecoles Universitaires de Nursing, 151 Slater St, Ste 1200, Ottawa, Ont. K1P 5N1
T: (613) 830-2532
Founded: 1942
Focus: Educ
Periodicals
The Canadian Journal of Nursing/Revue canadienne de recherche en sciences infirmières (weekly)
CAUSN Newsletter/Bulletin d'information (weekly) 02072

Canadian Association of University Schools of Rehabilitation / Association Canadienne des Ecoles Universitaires de Réadaptation, c/o School of Physical and Occupational Therapy, McGill University, 3654 Drummond St, Montréal, Qué. H3G 1Y5
T: (514) 392-5487
Founded: 1965
Focus: Educ; Rehabil 02073

Canadian Association of University Teachers / Association Canadienne des Professeurs d'Université, 75 Albert St, Ste 1001, Ottawa, Ont. K1P 5E7
T: (613) 237-6885
Founded: 1951; Members: 27000
Focus: Educ 02074

Canadian Association of University Teachers of German / Association des Professeurs d'Allemand des Universités Canadiennes, c/o Dept of Germanic Languages, University of Alberta, Edmonton, Alta. T6G 2E6
T: (403) 492-5930
Founded: 1961; Members: 300
Focus: Educ; Ling
Periodical
Seminar (weekly) 02075

Canadian Association on Water Pollution Research and Control / Association Canadienne pour la Recherche sur la Pollution de l'Eau et sa Maîtrise, c/o Prof. R. Gehr, Dept of Civil Engineering and Applied Mechanics, McGill University, 817 Sherbrooke St W, Montréal, Qué. H3A 2K6
T: (514) 398-6861; Tx: 05268510; Fax: (514) 398-7361
Members: 320
Focus: Water Res; Hygiene; Ecology
Periodical
Water Pollution Research Journal of Canada (weekly) 02076

Canadian Astronomical Society / Société Canadienne d'Astronomie, c/o Herzberg Institute of Astrophysics, 100 Sussex Dr, Ottawa, Ont. K1A 0R6
T: (613) 226-5297, 993-6060
Founded: 1971; Members: 360
Focus: Astronomy 02077

Canadian Athletic Therapists Association / Association Canadienne des Thérapeutes du Sport, 1600 James Naismith Dr, Alorchester, Ont. K1B 5N4
Members: 500
Focus: Physical Therapy
Periodicals
The Journal of the Canadian Athletic (weekly)
Therapists Association (weekly) 02078

Canadian Authors Association, 275 Slate St, Ste 500, Ottawa, Ont. K1P 5H9
Fax: (613) 235-8237
Founded: 1921; Members: 900
Pres: Mary E. Dawe
Focus: Lit
Periodical
Canadian Author & Bookman (weekly) . 02079

Canadian Aviation Historical Society (C.A.H.S.), POB 224, Willowdale, Ont. M2N 5S8
Founded: 1963; Members: 1000
Pres: William J. Wheeler; Gen Secr: N. Ennott
Focus: Aero; Hist
Periodical
C.A.H.S. Journal (weekly) 02080

Canadian Bar Association (CBA/ABC) / Association du Barreau Canadien, 50 O'Connor St, Ste 902, Ottawa, Ont. K1P 6L2
T: (613) 237-2925; Fax: (613) 237-0185
Founded: 1914; Members: 38000
Pres: J.J. Camp
Focus: Law
Periodicals
Canadian Bar Review (weekly)
National (weekly) 02081

Canadian Biochemical Society (CBS) / Société Canadienne de Biochimie, c/o Dept of Biochemistry, University of Western Ontario, London, Ont. N6A 5C1
T: (519) 661-3055
Founded: 1958; Members: 900
Pres: Dr. P.C. Choy; Gen Secr: Dr. S.D. Dunn
Focus: Biochem
Periodical
Bulletin (weekly) 02082

Canadian Botanical Association (CBA) / Association Botanique du Canada, c/o Dept of Biology, Carleton University, Ottawa, Ont. K1S 5B6
T: (613) 231-3813

Founded: 1964; Members: 400
Focus: Botany
Periodical
CBA/ABC Bulletin (weekly) 02083

Canadian Bureau for International Education / Bureau Canadien de l'Education Internationale, 85 Albert St, Ottawa, Ont. K1P 6A4
T: (613) 237-4820; Tx: 0533255; Fax: (613) 237-1073
Founded: 1966; Members: 90
Pres: James W. Fox
Focus: Educ
Periodicals
Annual Report (weekly)
International Education Magazine (weekly) 02084

Canadian Bureau for the Advancement of Music / Bureau Canadien pour l'Avancement de Musique, Exhibition Pl, Toronto, Ont. M6K 3C3
T: (416) 598-2798
Founded: 1922
Focus: Music 02085

Canadian Canon Law Society / Société Canadienne de Droit Canonique, 223 Main St, Ottawa, Ont. K1S 1C4
T: (613) 236-1393 ext 215; Fax: (613) 236-5278
Founded: 1966; Members: 450
Focus: Rel & Theol; Law
Periodical
Newsletter/Bulletin de Nouvelles (weekly) 02086

Canadian Cardiovascular Society / Société Canadienne de Cardiologie, 360 Victoria Av, Rm 401, Westmount, Qué. H3Z 2N4
T: (514) 482-3407
Founded: 1947; Members: 855
Focus: Cardiol 02087

Canadian Cartographic Association / Association Canadienne de Cartographie, c/o Roger Wheate, University of Calgary, Calgary, Alta. T2N 1N4
T: (403) 220-4892; Fax: (403) 282-7298
Founded: 1975; Members: 400
Focus: Cart
Periodical
Cartographica (weekly) 02088

Canadian Catholic Historical Association (English Section), c/o Saint Peter's Seminary, 1040 Waterloo St, London, Ont. N6A 3Y1
Founded: 1933
Focus: Hist; Rel & Theol 02089

Canadian Celtic Arts Association, c/o Saint Michael's College, 81 Saint Mary St, Toronto, Ont. M4T 1B4
T: (416) 926-1300
Founded: 1977; Members: 2000
Gen Secr: Robert O'Driscoll
Focus: Arts 02090

Canadian Centre for Toxicology / Centre Canadien de Toxicologie, 645 Gordon St, Guelph, Ont. N1G 2W1
T: (519) 837-3320
Founded: 1983
Focus: Toxicology 02091

Canadian Ceramic Society, 2175 Sheppard Av E, Ste 110, Willowdale, Ont. M2J 1W8
T: (416) 491-2886
Founded: 1932; Members: 600
Gen Secr: J. Milne
Focus: Materials Sci
Periodical
Canadian Ceramics (weekly) 02092

Canadian Certified General Accountants Research Foundation / Fondation de Recherche de l'Association des Comptables Généraux Licenciés du Canada, 1176 W Georgia St, Ste 740, Vancouver, B.C. V6E 4A2
T: (604) 669-3555; Fax: (604) 689-5845
Founded: 1979; Members: 850
Focus: Business Admin 02093

Canadian Children's Book Centre, 229 College St, Toronto, Ont. M5T 1R4
T: (416) 597-1331
Founded: 1976; Members: 1000
Gen Secr: Debbie Rogosin
Focus: Libraries & Bk Sci
Periodical
Children's Book News (weekly) 02094

Canadian Chiropractic Association / Association Chiropratique Canadienne, 1396 Ellington Av W, Toronto, Ont. M6C 2E4
T: (416) 718-5656; Fax: (416) 781-7344
Founded: 1953; Members: 1950
Focus: Surgery
Periodical
Journal of the CCA (weekly) 02095

Canadian Classification Research Group / Groupe Canadien pour la Recherche en Classification, c/o School of Library and Information Science, Universitv of Western Ontario, London, Ont. N6G 1H1
T: (519) 679-3542
Founded: 1973; Members: 33
Focus: Standards; Doc 02096

Canadian Coalition for Nuclear Responsibility / Regroupement pour la Surveillance du Nucléaire, CP 236, Succ Snowdon, Montréal, Qué. H3X 3T4
T: (514) 842-1471
Founded: 1975
Focus: Nucl Res 02097

Canadian Coalition on Acid Rain, 112 Saint Clair Av W, Ste 504, Toronto, Ont. M4B 2Y3
T: (416) 968-2135
Founded: 1980
Focus: Ecology 02098

Canadian College of Medical Geneticists / Collège Canadien de Généticians Médicaux, c/o Alberta Children's Hospital, 1820 Richmond Rd SW, Calgary, Alta. T2T 5C7
T: (403) 229-7373
Founded: 1975; Members: 90
Focus: Genetics; Med 02099

Canadian College of Teachers / Collège Canadien des Enseignants, 7546 Tenth Av, Edmonton, Alta. T6K 2T6
T: (403) 462-5103
Founded: 1958; Members: 1200
Focus: Educ 02100

Canadian Committee on Early Childhood / Organisation Canadienne pour l'Education Préscolaire, 6228 127 St, Edmonton, Alta. T5H 3W9
Focus: Educ 02101

Canadian Conference of the Arts / Conférence Canadienne des Arts, 126 York St, Ste 400, Ottawa, Ont. K1N 5T5
T: (613) 238-3561
Founded: 1945; Members: 500
Focus: Arts 02102

Canadian Conference on Continuing Education in Pharmacy, 35 Hazen Av, Renforth, Saint John, N.B. E2N 1N9
T: (506) 847-7671
Founded: 1973
Pres: Kay McLean
Focus: Pharmacol 02103

Canadian Construction Association / Association Canadienne de la Construction, 85 Albert St, Ottawa, Ont. K1P 6A4
T: (613) 236-9455; Tx: 0534436
Founded: 1914; Members: 25000
Focus: Civil Eng
Periodical
Construction (weekly) 02104

Canadian Council for European Affairs / Conseil Canadien des Affaires Européennes, c/o Dept of Political Studies, University of Saskatchewan, Saskatoon, Sask. S7N 0W0
T: (306) 966-5231; Fax: (306) 966-8839
Founded: 1980
Pres: Eldon P. Black; Gen Secr: Prof. H.J. Michelmann
Focus: Int'l Relat
Periodical
Journal of European Integration 02105

Canadian Council of Library Schools / Conseil Canadien des Ecoles de Bibliothécaires, c/o School of Librarianship, University of British Columbia, Vancouver, B.C. V6T 1W5
T: (604) 228-4991
Founded: 1972
Focus: Libraries & Bk Sci 02106

Canadian Council of Professional Engineers, 116 Albert St, Ste 401, Ottawa, Ont. K1P 5G3
T: (613) 232-2474; Fax: (613) 230-5759
Founded: 1936; Members: 151000
Gen Secr: D.G. Laplante
Focus: Eng 02107

Canadian Council of Teachers of English (CCTE), 1243 Wood Pl, Oakville, Ont. L6L 2R4
T: (416) 827-4850
Founded: 1967; Members: 1200
Focus: Educ; Ling
Periodicals
CCTE Newsletter (weekly)
English Quarterly (weekly) 02108

Canadian Council of University Biology Chairmen / Conseil Universitaire des Directeurs de Biologie du Canada, Dept of Biology, University of Saskatchewan, Saskatoon, Sask. S7N 0W0
T: (306) 343-5980
Focus: Bio 02109

Canadian Council of University Physical Education Administrators / Conseil Canadien des Administrateurs Universitaires en Education Physique, c/o Faculty of Physical Education, University of Alberta, Edmonton, Alta. T6G 2E9
T: (403) 432-3364
Focus: Sports 02110

Canadian Council on Social Development / Conseil Canadien de Développement Social, POB 3505, Station C, Ottawa, Ont. K1Y 4G1
T: (613) 728-1865
Founded: 1920; Members: 900
Focus: Sociology 02111

Canadian Credit Institute Educational Foundation / Fondation Scolaire de l'Institut Canadien du Crédit, 5090 Explorer Dr, Ste 501, Mississauga, Ont. L4W 3T9
T: (416) 629-9805; Fax: (416) 629-9809
Founded: 1967; Members: 150
Focus: Finance 02112

Canadian Criminal Justice Association / Association Canadienne de Justice Pénale, 55 Parkdale Av, Ottawa, Ont. K1Y 1E5
T: (613) 725-3715
Founded: 1919; Members: 1000
Focus: Criminology
Periodicals
Canadian Journal of Criminology: Revue canadienne de criminologie (weekly)
Directory Services for Victims of Crime: Répertoire Services aux victimes d'actes criminels (weekly)
Justice Directory of Services: Répertoire des services (weekly)
Justice Report: Actualités-Justice (weekly) 02113

Canadian Dental Association / Association Dentaire Canadienne, 1815 Alta Vista Dr, Ottawa, Ont. K1G 3Y6
T: (613) 523-1770; Fax: (613) 523-7736
Founded: 1902; Members: 11436
Gen Secr: A.J. Neilson
Focus: Dent
Periodical
Canadian Dental Association Journal (weekly) 02114

Canadian Dental Research Foundation, 1815 Alta Vista Dr, Ottawa, Ont. K1G 3Y6
T: (613) 523-1770; Fax: (613) 523-7736
Founded: 1902
Focus: Dent 02115

Canadian Dermatological Association / Association Canadienne de Dermatologie, 1160 Burrard St, Ste 302, Vancouver, B.C. V6Z 2E8
T: (604) 681-8494
Members: 400
Focus: Derm 02116

Canadian Diabetes Association, 78 Bond St, Toronto, Ont. M5B 2J8
T: (416) 362-4440
Founded: 1953; Members: 40000
Gen Secr: Elwood Springman
Focus: Cardiol 02117

Canadian Dietetic Association / Association Canadienne des Diététistes, 385 Yonge St, Ste 304, Toronto, Ont. M5B 1S1
T: (416) 596-0857
Founded: 1935; Members: 3509
Focus: Nutrition 02118

Canadian Drilling Research Association, 237 Fourth St SW, Rm 3232, Calgary, Alta. T2P 0H6
T: (403) 237-2938
Focus: Eng 02119

Canadian Economics Association, c/o Dept of Economics, McGill University, Montréal, Qué.
Gen Secr: Prof. C. Green
Focus: Econ
Periodical
Canadian Journal of Economics 02120

Canadian Educational Researchers' Association / Association Canadienne des Chercheurs en Education, 14 Henderson Av, Ottawa, Ont. K1N 7P1
Founded: 1976; Members: 311
Focus: Educ 02121

Canadian Education Association (CEA/ACE) / Association Canadienne d'Education, 252 Bloor St W, Ste 8-200, Toronto, Ont. M5S 1V5
T: (416) 924-7721; Fax: (416) 924-3188
Founded: 1891; Members: 1200
Gen Secr: Robert E. Blair
Focus: Educ
Periodicals
CEA Handbook
CEA Newsletter: Nouvelles
Education Canada (weekly) 02122

Canadian Electrical Association / Association Canadienne de l'Electricité, 1 Westmount Sq, Ste 1600, Montréal, Qué. H3Z 2P9
T: (514) 937-6181; Fax: (514) 937-6498
Founded: 1891; Members: 2609
Pres: W.S. Read; Gen Secr: I.M. Phillips
Focus: Electric Eng
Periodicals
Bulletin
Research Reports 02123

Canadian Energy Research Institute, 3512 33 St NW, Calgary, Alta. T2L 2A6
T: (403) 282-1231
Founded: 1975
Gen Secr: Dr. G.E. Angevine
Focus: Energy 02124

Canadian Esperanto Association / Association Canadienne d'Espéranto, CP 126, Succ Beaubien, Montréal, Qué. H2G 3C8
T: (514) 495-8442
Founded: 1954; Members: 300
Focus: Ling 02125

Canadian Esperanto Youth / Jeunesse Espérantiste Canadienne, CP 126, Succ Beaubien, Montréal, Qué. H2G 3C8
T: (514) 495-8442
Founded: 1976
Pres: Paulo Hopkins
Focus: Ling 02126

Canadian Ethnic Studies Association (CESA) / Société Canadienne d'Etudes Ethniques, 43 Queen's Park Crescent E, Toronto, Ont. M5S 2C3
T: (416) 979-2973
Founded: 1968; Members: 1300
Focus: Ethnology
Periodicals
Canadian Ethnic Studies
CESA Bulletin (weekly)
Proceedings (weekly) 02127

Canadian Federation for the Humanities (CFH) / Fédération Canadienne des Etudes Humaines, 151 Slater St, Ste 407, Ottawa, Ont. K1P 5H3
T: (613) 236-4686; Tx: 0534862; Fax: (613) 238-6114
Founded: 1943; Members: 8000
Focus: Humanities
Periodical
Bulletin (weekly) 02128

Canadian Federation of Biological Societies (CFBS) / Fédération Canadienne des Sociétés de Biologie, 575 King Edward, Ottawa, Ont. K1N 7N5
T: (613) 234-9555
Founded: 1957; Members: 5000
Focus: Bio
Periodicals
CFBS Newsletter (weekly)
Programme & Proceedings (weekly) . . 02129

Canadian Federation of Deans of Management and Administrative Studies / Fédération Canadienne de Doyens de Gestion et d'Administration, c/o Faculty of Administration, University of Ottawa, Ottawa, Ont. K1N 9B5
T: (613) 231-3301
Founded: 1957
Focus: Public Admin 02130

Canadian Federation of University Women / Fédération Canadienne des Femmes Diplômées des Universités, c/o Malaspina College, 900 Fifth St, Nanaimo, B.C. V9R 5S5
T: (604) 753-6233
Founded: 1919; Members: 12500
Focus: Educ 02131

Canadian Fertility and Andrology Society / Société Canadienne de Fertilité et d'Andrologie, 2065 Alexandre de Seve, Ste 409, Montréal, Qué. H2L 2W5
T: (514) 524-9009
Founded: 1977; Members: 250
Focus: Venereology 02132

Canadian Fertilizer Institute / L'Institut Canadien des Engrais, 280 Albert St, Ste 301, Ottawa, Ont. K1P 5G8
T: (613) 230-9020; Tx: 0533803
Founded: 1958
Focus: Agri 02133

Canadian Film Institute / Institut Canadien du Film, 2 Daly, Ottawa, Ont. K1N 6E2
T: (613) 232-6727; Tx: 0636700474; Fax: (613) 232-6315
Founded: 1935
Pres: Serge Losique
Focus: Cinema
Periodical
The Guide to Film, Television and Communications Courses in Canada (biennial) 02134

Canadian Fire Safety Association / Association Canadienne de Sécurité Incendie, 2175 Sheppard Av E, Ste 110, Willowdale, Ont. M2J 1W8
T: (416) 492-9417
Founded: 1971; Members: 500
Focus: Safety 02135

Canadian Flexible Packaging Institute / Institut Canadien de l'Emballage Souple, 1 Yonge St, Ste 1400, Toronto, Ont. M5E 1J9
T: (416) 363-7261
Focus: Eng 02136

Canadian Folk Arts Council / Conseil Canadien des Arts Populaires, 263 Adelaide St W, Toronto, Ont. M5H 1Y2
T: (416) 977-8311
Founded: 1964; Members: 130
Focus: Folklore; Arts 02137

Canadian Folk Music Society / Société Canadienne de Musique Folklorique, 1314 Shelbourne St SW, Calgary, Alta. T3C 2K8
T: (402) 244-2804
Founded: 1956; Members: 800
Focus: Music; Folklore 02138

Canadian Forestry Association, 185 Somerset St W, Ste 203, Ottawa, Ont. K2P 0J2
T: (613) 232-1815; Fax: (613) 232-4210
Founded: 1900; Members: 35000
Pres: Walter Giles
Focus: Forestry
Periodical
What They Say About Forestry on the Hill (weekly) 02139

Canadian Foundation for Economic Education / Fondation Canadienne d'Education Economique, 252 Bloor St W, Ste 7205, Toronto, Ont. M5S 1V5
T: (416) 968-2236

Founded: 1974; Members: 2000
Focus: Econ; Educ 02140

Canadian Foundation for the Advancement of Pharmacy, 123 Edward St, Ste 303, Toronto, Ont. M5G 1E2
T: (416) 979-2024
Founded: 1945
Gen Secr: Baxter G. Rowsell
Focus: Pharmacol 02141

Canadian Foundation for the Study of Infant Deaths, POB 190, Station R, Toronto, Ont. M4G 3Z9
T: (416) 488-3260
Founded: 1973; Members: 200
Gen Secr: Beverley DeBruyn
Focus: Pediatrics 02142

Canadian Gas Association / Association Canadienne du Gaz, 55 Scarsdale Rd, Don Mills, Ont. M3B 2R3
T: (416) 447-6465; Tx: 06966824
Founded: 1907; Members: 650
Focus: Energy
Periodical
CGA News (weekly) 02143

Canadian Gemmological Association, POB 1106, Station Q, Toronto, Ont. M4T 2P2
T: (416) 652-3137; Fax: (416) 730-8105
Founded: 1957; Members: 400
Pres: Warren Boyd
Focus: Mineralogy
Periodical
Canadian Gemmologist (weekly) 02144

Canadian General Standards Board, 111 Pl du Portage, Hull, Qué. K1A 1G6
T: (613) 990-7897
Focus: Standards 02145

Canadian Geoscience Council / Conseil Géoscientifique Canadien, c/o Dept of Earth Sciences, University of Waterloo, Waterloo, Ont. N2L 3G1
T: (519) 855-1211 ext 3231
Founded: 1972
Gen Secr: A.V. Morgan
Focus: Geology 02146

Canadian Geotechnical Society / Société Canadienne Géotechnique, 170 Attwell Dr, Ste 602, Rexdale, Ont. M9W 5Z5
T: (416) 674-0366; Fax: (416) 674-9507
Founded: 1972; Members: 1254
Focus: Eng 02147

Canadian Geriatrics Research Society / Société Canadienne de Recherche en Geriatrie, 351 Christie St, Toronto, Ont. M6G 3C3
T: (416) 537-6000
Founded: 1955
Focus: Geriatrics
Periodical
Geriascope (3 times annually) 02148

Canadian Graphic Arts Institute, 19 Duncan St, Toronto, Ont. M5H 3H1
T: (416) 593-4018
Focus: Graphic & Dec Arts, Design . . 02149

Canadian Group Psychotherapy Association / Association Canadienne de Psychothérapie de Groupe, 3666 McTavish St, Montréal, Qué. H3A 1Y2
T: (416) 842-1231 ext 1771
Founded: 1980; Members: 225
Focus: Therapeutics; Psych 02150

Canadian Health Care Material Management Association, POB 30, High River, Alta. T0L 1B0
T: (403) 652-2321
Focus: Public Health 02151

Canadian Health Education Society (CHES) / Société Canadienne en Education Sanitaire, 253 College St, POB 306, Toronto, Ont. M5T 1R5
Founded: 1963; Members: 120
Focus: Educ; Public Health
Periodical
CHES Newsletter (weekly) 02152

Canadian Health Libraries Association / Association des Bibliothèques de la Santé du Canada, POB 983, Station B, Ottawa, Ont. K1A 5R1
T: (613) 787-2743
Founded: 1976; Members: 325
Focus: Libraries & Bk Sci; Public Health 02153

Canadian Hearing Society Foundation, 60 Bedford Rd, Toronto, Ont. M5R 2K2
T: (416) 964-9595
Founded: 1979
Gen Secr: Joan Beattie
Focus: Logopedy 02154

Canadian Heart Foundation / Fondation Canadienne des Maladies du Coeur, 1 Nicholas St, Ste 1200, Ottawa, Ont. K1N 7B7
T: (613) 237-4361
Founded: 1956
Focus: Cardiol 02155

Canadian Hematology Society / Société Canadienne d'Hématologie, c/o Cross Cancer Institute, 11560 University Av, Edmonton, Alta. T6G 1Z2
T: (403) 432-8717
Members: 200
Focus: Hematology 02156

Canadian Historical Association, 395 Wellington St, Ottawa, Ont. K1A 0N3
T: (613) 233-7885
Founded: 1922; Members: 1800
Gen Secr: John Lutz; Denise Rioux
Focus: Hist
Periodicals
Canada's Ethnic Groups Series
Historical Booklets (weekly)
Historical Papers (weekly) 02157

Canadian Hypnotherapy Association / Association des Hypnologues du Canada, 1538 Sherbrooke St W, Ste 710, Montréal, Qué. H3G 1L5
T: (514) 935-7755
Gen Secr: Nathan Schiff
Focus: Psych 02158

Canadian Image Processing and Pattern Recognition Society / Société Canadienne de Traitement de l'Image et de Reconnaissance des Structures, 243 College St, Toronto, Ont. M5T 2Y1
T: (416) 593-4040
Founded: 1980; Members: 200
Focus: Computer & Info Sci 02159

Canadian Industrial Arts Association, 3750 Willingdon Av, Burnaby, B.C. V5G 3G9
T: (604) 434-8777
Founded: 1965; Members: 250
Pres: Robert J. Leduc
Focus: Eng 02160

Canadian Industrial Computer Society, c/o R.E. Gagne, 12 Kindle Court, Ottawa, Ont. K1J 6E2
T: (613) 993-2030
Gen Secr: R.E. Gagne
Focus: Computer & Info Sci 02161

Canadian Institute for Environmental Law and Policy / Institut Canadien du Droit et de la Politique de l'Environnement, 517 College St, Ste 400, Toronto, Ont. M6G 1A8
T: (416) 923-3529; Fax: (416) 960-9392
Founded: 1970; Members: 800
Gen Secr: Barbara Heidenreich
Focus: Law; Poli Sci 02162

Canadian Institute for Historical Microreproductions / Institut Canadien de Microréproductions Historiques, POB 2428, Station D, Ottawa, Ont. K1P 5W5
T: (613) 235-2628
Founded: 1978
Focus: Doc 02163

Canadian Institute for Middle East Research, POB 970, Station G, Calgary, Alta. T3A 3G2
T: (403) 247-1155
Focus: Sociology; Poli Sci 02164

Canadian Institute for Organization Management, 200 Elgin St, Ste 301, Ottawa, Ont. K2P 1J7
T: (613) 238-4000
Founded: 1946
Gen Secr: Roger Stanion
Focus: Business Admin 02165

Canadian Institute for Studies in Telecommunications / Institut Canadien pour les Etudes des Télécommunications, 35 Pl Bergeron, Pierrefonds, Qué. H8Y 1P4
T: (514) 684-2751
Founded: 1977
Focus: Comm 02166

Canadian Institute for the Administration of Justice, c/o Faculté de Droit, Université de Montréal, CP 6128, Succ A, Montréal, Qué. H3C 3J7
Gen Secr: Joyce Whitman
Focus: Law 02167

Canadian Institute of Chartered Accountants, 150 Bloor St W, Toronto, Ont. M5S 2Y2
T: (416) 962-1242; Fax: (416) 962-3375
Founded: 1902; Members: 50565
Pres: K.C. Fincham
Focus: Business Admin
Periodicals
CA Magazine (weekly)
Members Directory (biennial) 02168

Canadian Institute of Child Health / Institut Canadien de la Santé Infantile, 17 York St, Ste 105, Ottawa, Ont. K1N 5S7
T: (613) 238-8425
Founded: 1977
Focus: Pediatrics
Periodical
Child Health (weekly) 02169

Canadian Institute of Cost Reduction, 2468 Wyndale Crescent, Ottawa, Ont. K1H 7A6
T: (613) 731-0900
Founded: 1973
Gen Secr: James V. Thoppil
Focus: Finance 02170

Canadian Institute of Credit and Financial Management, 5090 Explorer Dr, Ste 501, Mississauga, Ont. L4W 3T9
T: (416) 629-9805; Fax: (416) 629-9809
Founded: 1928; Members: 2500
Focus: Finance
Periodical
The Credit and Financial Journal (weekly) 02171

Canadian Institute of Facial Plastic Surgery / Institut Canadien de Chirurgie Plastique Faciale, 170 Saint George St, Ste 728, Toronto, Ont. M5R 2M8
Focus: Surgery 02172

Canadian Institute of Financial Planning, 70 Bond St, Ste 400, Toronto, Ont. M5B 1X2
T: (416) 363-2159
Focus: Finance 02173

Canadian Institute of Food Science and Technology (CIFST) / Institut Canadien de Science et Technologie Alimentaire, 1335 Carling Av, Ste 309, Ottawa, Ont. K1Z 8N8
T: (613) 724-7752; Fax: (613) 724-7754
Founded: 1951; Members: 2500
Focus: Food
Periodical
CIFST Journal (5 times annually) . . . 02174

Canadian Institute of Forestry / Institut Forestier du Canada, 151 Slater St, Ste 1005, Ottawa, Ont. K1P 5H3
T: (613) 234-2242; Fax: (613) 234-6181
Founded: 1908; Members: 2400
Focus: Forestry
Periodical
The Forestry Chronicle (weekly) 02175

Canadian Institute of Hypnotism / Institut Canadienne d'Hypnotisme, 3465 Côte des Neiges Rd, Ste 91, POB 60, Sainte Anne de Bellevue, Qué. H9X 3L4
T: (514) 457-3377
Founded: 1953; Members: 420
Focus: Psychoan 02176

Canadian Institute of International Affairs, 15 King's College Circle, Toronto, Ont. M5S 2V9
T: (416) 979-1851; Fax: (416) 979-8575
Founded: 1928; Members: 2100
Pres: Dr. John English; Gen Secr: David A.T. Stafford
Focus: Poli Sci
Periodicals
Behind the Headlines (weekly)
Etudes Internationales (weekly)
International Journal (weekly) 02177

Canadian Institute of Management / Institut Canadien de Gestion, 2175 Sheppard Av E, Ste 110, Willowdale, Ont. M2J 1W8
T: (416) 493-0155
Founded: 1942; Members: 5000
Focus: Business Admin
Periodical
Canadian Manager Magazine 02178

Canadian Institute of Metalworking, 1280 Main St W, Hamilton, Ont. L8S 4K1
T: (416) 528-2777
Founded: 1970
Gen Secr: James E. Crozier
Focus: Materials Sci 02179

Canadian Institute of Mining and Metallurgy (CIM), 3400 de Maisonneuve Blvd W, Ste 1210, Montréal, Qué. H3Z 3B8
T: (514) 939-2710; Fax: (514) 935-2714
Founded: 1898; Members: 10500
Gen Secr: Pierre Michaud
Focus: Mining; Metallurgy
Periodicals
CIM Bulletin (weekly)
Directory (weekly)
Journal of Canadian Petroleum Technology (weekly) 02180

Canadian Institute of Planners / Institut Canadien des Urbanistes, 30-46 Elgin St, Ottawa, Ont. K1P 5K6
Founded: 1923; Members: 3000
Pres: David Witty; Gen Secr: David H. Sherwood
Focus: Urban Plan
Periodical
Plan Canada (10 times annually) . . . 02181

Canadian Institute of Religion and Gerontology, 40 Saint Clair Av E, Ste 203, Toronto, Ont. M4T 1M9
T: (416) 924-5865
Founded: 1974; Members: 1000
Gen Secr: Michael Guinan
Focus: Rel & Theol; Geriatrics 02182

Canadian Institute of Stress, 24 Humbervale Blvd Dr, Toronto, Ont. M8Y 3P2
T: (416) 236-1433
Focus: Public Health 02183

Canadian Institute of Surveying (CISM), POB 5378, Station F, Ottawa, Ont. K2C 3J1
T: (613) 224-9851
Founded: 1882; Members: 2650
Focus: Surveying
Periodical
CISM Journal (weekly) 02184

Canadian Institute of Treated Wood / Institut Canadien des Bois Traités, 75 Albert St, Ste 506, Ottawa, Ont. K1P 5E7
T: (613) 234-9456; Fax: (613) 234-1228
Founded: 1954
Focus: Forestry; Ecology 02185

Canadian Institute of Ukrainian Studies, c/o 352 Athabasca Hall, University of Alberta, Edmonton, Alta. T6G 2E8
T: (403) 492-2972; Fax: (403) 492-4967
Founded: 1976
Gen Secr: Bohdan A. Krawchenko
Focus: Ethnology
Periodical
Journal of Ukrainian Studies 02186

Canadian Law and Society Association, c/o Osgoode Hall Law School, 4700 Keele St, Downsview, Ont. M3J 2R5
Founded: 1983
Focus: Law 02187

Canadian Law Information Council / Conseil Canadien de la Documentation Juridique, 161 Laurier Av W, Ste 5, Ottawa, Ont. K1P 5J2
T: (613) 236-9766
Founded: 1973; Members: 60
Focus: Law 02188

Canadian Library Association (CLA), 200 Elgin St, Ottawa, Ont. K2P 1L5
T: (613) 232-9625; Fax: (613) 563-9895
Founded: 1946; Members: 5000
Pres: Margaret Andrews; Gen Secr: Karen Adams
Focus: Libraries & Bk Sci
Periodicals
Canadian Library Journal (weekly)
Canadian Materials
Feliciter (weekly) 02189

Canadian Lime Institute, c/o Beachville Lime Co Ltd, Ingerson, Ont. N5E 3K5
T: (519) 423-6283
Gen Secr: Bruce Yearhay
Focus: Materials Sci 02190

Canadian Linguistic Association (CLA) / Association Canadienne de Linguistique, c/o Experimental Phonetics Laboratory, New College, University of Toronto, Toronto, Ont. M5S 1A1
T: (416) 599-0973
Founded: 1954; Members: 731
Pres: Marie-T. Vinet; Gen Secr: P. Bhatt
Focus: Ling
Periodical
The Canadian Journal of Linguistics (weekly) 02191

Canadian Lung Association (CLA), 75 Albert St, Ste 908, Ottawa, Ont. K1P 5E7
T: (613) 237-1208; Fax: (613) 563-3362
Founded: 1900
Pres: R.W. King; Gen Secr: Bob Mech
Focus: Pulmon Dis
Periodical
Canadian Lung Association Bulletin (3 times annually) 02192

Canadian Man-Computer Communications Society / Société Canadienne du Dialogue Hommemachine, 243 College St, Toronto, Ont. M5T 2Y1
T: (416) 593-4040
Founded: 1976; Members: 455
Focus: Computer & Info Sci 02193

Canadian Maritime Law Association / Association Canadienne de Droit Maritime, 363 Saint François Xavier St, Ste 300, Montréal, Qué. H2Y 3P9
T: (514) 845-9151; Tx: 0525266; Fax: (514) 845-9161
Founded: 1951; Members: 311
Focus: Law 02194

Canadian Mathematical Society / Société Mathématique du Canada, 577 King Edward, Ottawa, Ont. K1N 6N5
T: (613) 564-2223; Fax: (613) 789-1539
Founded: 1945; Members: 1250
Pres: Dr. S. Riemenschneider; Gen Secr: Dr. Graham P. Wright
Focus: Math
Periodicals
Applied Mathematics Notes (weekly)
Canadian Mathematical Bulletin (weekly) . 02195

Canadian Medical and Biological Engineering Society / Société Canadienne de Génie Biomedical, Bldg M-50, National Research Council, Ste 307, Ottawa, Ont. K1A 0R8
T: (613) 993-1686; Tx: 0534134
Founded: 1966; Members: 400
Focus: Med; Bio; Eng 02196

Canadian Medical Association / Association Médicale Canadienne, POB 8650, Ottawa, Ont. K1G 0G8
T: (613) 731-9331; Fax: (613) 731-9013
Founded: 1867; Members: 46000
Gen Secr: Léo-Paul Landry
Focus: Med
Periodicals
Canadian Journal of Surgery: Journal canadien de chirurgie (weekly)
CMA Journal: Journal de l'Association médicale canadienne (weekly)
Net Worth: Valeurs nettes (weekly) . . . 02197

Canadian Medical Protective Association / Association Canadienne de Protection Médicale, POB 8225, Ottawa, Ont. K1G 3H7
T: (613) 236-2100
Founded: 1910; Members: 37000
Gen Secr: Stuart P. Lee
Focus: Public Health 02198

Canadian Meteorological and Oceanographic Society / La Société Canadienne de Météorologie et d'Océanographie, POB 334, Newmarket, Ont. L3Y 4X7
T: (416) 898-1040; Fax: (416) 898-7937

Founded: 1977; Members: 1000
Pres: Dr. Louis Hobson; Gen Secr: Uri Schwarz
Focus: Oceanography; Geophys
Periodicals
Atmosphere-Ocean (weekly)
Chinook (weekly)
Climatological Bulletin (tri-annually) . . . 02199

Canadian Metric Association / Association Métrique Canadienne, POB 35, Fonthill, Ont. L0S 1E0
T: (416) 358-0171
Founded: 1969; Members: 300
Focus: Ling; Lit 02200

Canadian Museums Association (CMA) / Association des Musées Canadiens, 280 Metcalfe St, Ste 400, Ottawa, Ont. K2P 1R7
T: (613) 233-5653; Fax: (613) 233-5438
Founded: 1947; Members: 2000
Gen Secr: John G. McAvity
Focus: Arts
Periodicals
Muse (weekly)
Museogramme (weekly) 02201

Canadian Music Centre / Centre de Musique Canadienne, 20 Saint Joseph St, Toronto, Ont. M4Y 1J9
T: (416) 961-6601; Fax: (416) 961-7198
Founded: 1959; Members: 310
Gen Secr: Simone Auger
Focus: Music
Periodical
Acquisitions (weekly) 02202

Canadian Music Council (CMC) / Conseil Canadien de la Musique, 189 Laurier Av East, Ottawa, Ont. K1N 6P1
T: (613) 238-5893; Fax: (613) 235-7425
Founded: 1949; Members: 300
Gen Secr: Myra Grimley Dahl
Focus: Music
Periodical
Musicanada (weekly) 02203

Canadian National Committee for The International Geographical Union, c/o Royal Canadian Geographical Society, 39 McArthur Av, Vanier, Ont. K1L 8L7
T: (613) 745-4629
Founded: 1950; Members: 12
Focus: Geography 02204

Canadian National Committee of World Energy Conference, 130 Albert St, Ste 305, Ottawa, Ont. K1P 5G4
T: (613) 993-4624; Fax: (613) 993-7679
Founded: 1924; Members: 125
Gen Secr: C.H. Smith
Focus: Energy
Periodical
Proceedings of Canadian National Energy Forums (weekly) 02205

Canadian National Institute for the Blind / Institut National Canadien pour les Aveugles, 1931 Bayview Av, Toronto, Ont. M4G 4C8
T: (416) 486-2636
Founded: 1918; Members: 596
Focus: Educ Handic 02206

Canadian Natural Hygiene Society, 14 Lynn Haven Rd, Toronto, Ont. M6A 2K8
T: (416) 781-0359
Focus: Hygiene 02207

Canadian Nautical Research Society, POB 7008, Station J, Ottawa, Ont. K2A 3Z6
Focus: Navig
Periodicals
Argonauta (weekly)
Canadian Maritime Bibliography (weekly) . 02208

The Canadian Neurological Society / Société Canadienne de Neurologie, 1650 Cedar Av, Montréal, Qué. H3G 1A4
T: (514) 937-4833
Founded: 1948; Members: 400
Focus: Neurology 02209

Canadian Neurosurgical Society, c/o Hospital for Sick Children, 555 University Av, Toronto, Ont. M5G 1X8
T: (416) 598-1210
Pres: Dr. B.K.W. Wein; Gen Secr: Dr. H.J. Hoffman
Focus: Neurology; Surgery
Periodical
Canadian Journal of Neurological Science (9 times annually) 02210

Canadian Nuclear Association / Association Nucléaire Canadienne, 111 Elizabeth St, Toronto, Ont. M5G 1P7
T: (416) 977-6152; Tx: 0623741
Founded: 1960
Focus: Nucl Res
Periodicals
Bulletin of the Canadian Nuclear Society (weekly)
CNA Annual International Conference Proceedings (weekly)
Nuclear Canada (9 times annually)
Nuclear Canada Yearbook (weekly) . . . 02211

Canadian Numismatic Association, POB 226, Barrie, Ont. L4M 4T2
T: (705) 737-0845
Founded: 1950; Members: 2500
Gen Secr: K.B. Prophet
Focus: Numismatics
Periodical
The Canadian Numismatic Journal (weekly)
. . . . 02212

Canadian Operational Research Society (CORS), POB 2225, Stn D, Ottawa, Ont. K1P 5W4
T: (613) 992-4079
Founded: 1958; Members: 760
Pres: Dr. J.A. Buzacott; Gen Secr: Dr. M.A. Weinberger
Focus: Business Admin
Periodicals
Bulletin (9 times annually)
Infor (weekly) 02213

Canadian Oral History Association / Société Canadienne d'Histoire Orale, POB 301, Station A, Ottawa, Ont. K1N 8V3
T: (613) 995-1311
Founded: 1974; Members: 300
Focus: Hist 02214

Canadian Orthoptic Society, 610 Champagner, Montréal, Qué. H2V 3P6
Focus: Ophthal 02215

Canadian Osteogenesis Imperfecta Society / Société Canadienne Osteogenesis Imperfecta, POB 607, Station U, Toronto, Ont. M8Z 5Y9
Pres: Lorna Smith
Focus: Med; Pathology 02216

Canadian Osteopathic Aid Society / Sociétè Canadienne d'Assistance Ostéopathique, 575 Waterloo St, London, Ont. N6B 2R2
T: (519) 439-5521
Founded: 1960; Members: 600
Focus: Med; Pathology 02217

Canadian Osteopathic Association / Société Canadienne Ostéopathique, 575 Waterloo St, London, Ont. N6B 2R2
Tx: (519) 439-5521
Founded: 1926; Members: 31
Focus: Med; Pathology 02218

Canadian Paediatric Society / Société Canadienne de Pédiatrie, c/o Children's Hospital of Eastern Ontario, 401 Smyth Rd, Ottawa, Ont. K1H 8L1
T: (613) 737-2728; Fax: (613) 737-2794
Founded: 1923; Members: 1834
Pres: Dr. Claude Paré; Gen Secr: Dr. J.H.V. Marchessault
Focus: Pediatrics
Periodical
News Bulletin (weekly) 02219

Canadian Paraplegic Association / Association Canadienne des Paraplégiques, 520 Sutherland Dr, Toronto, Ont. M4G 3V9
T: (416) 422-5640
Founded: 1945; Members: 2023
Focus: Pathology; Med 02220

Canadian Peace Research and Education Association / Association Canadienne de Recherhe et d'Education pour la Paix, c/o Prof. M.V. Naidu, Dept of Political Science, Brandon University, Brandon, Man. R7A 6A9
Focus: Prom Peace
Periodical
Peace Research 02221

Canadian Petroleum Law Foundation, 150 Ninth Av SW, Calgary, Alta. T2P 3H9
T: (403) 290-2941
Founded: 1963; Members: 203
Focus: Law 02222

Canadian Pharmaceutical Association (CPhA) / Association Pharmaceutique Canadienne, 1785 Alta Vista Dr, Ottawa, Ont. K1G 3Y6
T: (613) 523-7877
Founded: 1907; Members: 10500
Gen Secr: Leroy C. Fevang
Focus: Pharmacol
Periodical
Canadian Pharmaceutical Journal (weekly) 02223

Canadian Philosophical Association (CPA) / Association Canadienne de Philosophie, c/o Dept of Philosophy, University of Ottawa, Morisset Hall, Ottawa, Ont. K1N 6N5
T: (613) 238-2607
Founded: 1958; Members: 800
Gen Secr: Sylvie Aubin
Focus: Philos
Periodicals
Bulletin (weekly)
Dialogue (weekly) 02224

Canadian Physiological Society / Société Canadienne de Physiologie, c/o Dept of Physiology, Queens University, Kingston, Ont. K7L 3N6
T: (613) 545-2806; Fax: (613) 545-6880
Founded: 1936; Members: 526
Pres: Dr. L. Renaud; Gen Secr: Dr. F.J.R. Richmond
Focus: Physiology
Periodical
Physiology Canada (weekly) 02225

Canadian Phytopathological Society / Société Canadienne de Phytopathologie, c/o Agriculture Canada Research Station, POB 3000, Lethbridge, Alta. T1J 4B1
T: (403) 327-4561; Fax: (403) 382-3156
Founded: 1929; Members: 500
Pres: Dr. M. Caron; Gen Secr: Dr. D. Gaudet
Focus: Botany
Periodical
Canadian Journal of Plant Pathology (weekly)
. . . . 02226

Canadian Podiatry Association, 447 Sussec Dr, Ottawa, Ont. K1N 6Z4
T: (613) 236-0566
Focus: Orthopedics 02227

Canadian Polish Research Institute, 288 Roncesvalles Av, Toronto, Ont. M6R 2M4
T: (416) 535-6233
Founded: 1956
Pres: Rudolf K. Kogler
Focus: Hist; Sociology
Periodical
Studies (weekly) 02228

Canadian Political Science Association / Association Canadienne de Science Politique, 1 Stewart St, Ste 205, Ottawa, Ont. K1N 6H7
Founded: 1913
Pres: V. Seymour Wilson; Gen Secr: Joan F. Pond
Focus: Poli Sci
Periodicals
The Canadian Journal of Political Science: Revue canadienne de science politique
The Bulletin: Le Bulletin 02229

Canadian Psoriasis Foundation, 1565 Carling Av, Ste 400, Ottawa, Ont. K1Z 8R1
T: (613) 728-4000; Fax: (613) 725-9826
Founded: 1983
Focus: Derm
Periodical
Canadian Psoriasis Foundation Newsletter (weekly)
. . . . 02230

Canadian Psychiatric Association / Association des Psychiatres du Canada, 237 Argyle Av, Ste 200, Ottawa, Ont. K2P 1B8
T: (613) 234-2815; Fax: (613) 234-9857
Founded: 1951; Members: 2400
Pres: Dr. Jacques Drouin; Gen Secr: Alex Saunders
Focus: Psychiatry
Periodical
The Canadian Journal of Psychiatry (8 times annually) 02231

Canadian Psychiatric Research Foundation, 220 Yonge St, POB 607, Toronto, Ont. M5B 2H1
T: (416) 591-9289
Founded: 1980
Gen Secr: Leslie Webber
Focus: Psychiatry 02232

Canadian Psychoanalytic Society / Société Canadienne de Psychoanalyse, 7000 Côte des Neiges, Montréal, Qué. M3S 2C1
T: (514) 738-6105
Founded: 1952; Members: 248
Focus: Psychoan 02233

Canadian Psychological Association / Société Canadienne de Psychologie, Vincent Rd, Old Chelsea, Qué J0X 2N0
T: (819) 827-3927; Fax: (819) 827-4639
Founded: 1939; Members: 4000
Pres: Dr. Luc Granger; Gen Secr: Dr. Jean Pettifor
Focus: Psych
Periodicals
Canadian Journal of Behavioural Sience (weekly)
Canadian Journal of Psychology (weekly)
Canadian Psychology (weekly)
Highlights (weekly) 02234

Canadian Public Health Association, 1335 Carling Av, Ste 210, Ottawa, Ont. K1Z 8N8
T: (613) 725-3769
Founded: 1910; Members: 3500
Pres: Dr. Franklin M.M. White; Gen Secr: Gerald H. Dafoe
Focus: Public Health
Periodical
Canadian Journal of Public Health (weekly)
. . . . 02235

Canadian Public Relations Society / Société Canadienne des Relations Publiques,, 220 Laurier Av W, Ste 720, Ottawa, Ont. K1P 5Z9
T: (613) 232-1222; Tx: 053-3838
Founded: 1948; Members: 1650
Focus: Advert 02236

Canadian Public Relations Society (Nova Scotia), c/o I.W. Killam Hospital for Children, POB 3070, Halifax, N.S. B3J 3G9
T: (902) 424-3094
Focus: Advert 02237

Canadian Public Relations Society (Ottawa) / Société Canadienne des Relations Publiques (Ottawa), 220 Laurier Av W, Ste 720, Ottawa, Ont. K1P 5Z9
T: (613)) 232-7014; Tx: 0533838
Founded: 1957; Members: 160
Focus: Advert 02238

Canadian Quaternary Association / Association Canadienne pour l'Etude du Quaternaire, c/o Dept of Earth Sciences, University of Waterloo, Waterloo, Ont. N21 3G1
T: (519) 885-1211; Fax: (519) 746-2543
Founded: 1979
Gen Secr: A.V. Morgan
Focus: Geology; Geography; Bio; Archeol
Periodicals
CANQA Newsletter/Bulletin (weekly)
Géographie Physique et Quaternaire (3 times annually) 02239

Canadian Radiation Protection Association, 318 Lyon St, Ottawa, Ont. K1R 5W6
T: (613) 230-4883
Focus: Safety; Radiology
Periodical
Canadian Radiation Protection Association Bulletin (weekly) 02240

Canadian Railroad Historical Association / Association Canadienne d'Histoire Ferroviaire, POB 148, Saint Constant, Qué. J0L 1X0
T: (514) 632-2410
Founded: 1932; Members: 1700
Focus: Hist; Cultur Hist 02241

Canadian Research Institute for the Advancement of Women / Institut Canadien de Recherches pour l'Avancement de la Femme, 151 Slater St, Ste 415, Ottawa, Ont. K1P 5H3
T: (613) 563-0681/82
Founded: 1986; Members: 535
Focus: Sociology 02242

Canadian Research Management Association / Association Canadienne de la Gestion de Recherche, c/o Ontario Research Foundation, Sheridan Park, Mississauga, Ont. L5K 1B3
T: (416) 822-4111; Tx: 06982311
Founded: 1963; Members: 176
Focus: Business Admin 02243

Canadian Research Society for Children's Literature / Société Canadienne des Recherches pour Littérature Enfantine, c/o School of Library & Information Science, University of Western Ontario, London, Ont. N6G 1H1
T: (519) 679-3542
Founded: 1979
Focus: Lit 02244

Canadian Rodeo Historical Association, POB 1, Cochrane, Alta. T0L 0W0
T: (403) 932-4994
Focus: Cultur Hist 02245

Canadian Schizophrenia Foundation, 7375 Kingsway, Burnaby, B.C. V3N 3B5
T: (604) 521-1728
Founded: 1969; Members: 2000
Gen Secr: Steven Carter
Focus: Psychiatry
Periodicals
Health & Nutrition Update (weekly)
The Journal of Orthomolecular Medicine (weekly)
. . . . 02246

Canadian Science and Technology Historical Association, c/o Dept of Science Studies, Atkinson College York University, North York, Ont. M3J 1P3
Focus: Eng; Cultur Hist
Periodical
Scientia Canadensis (weekly) 02247

Canadian Semiotic Association / Association Canadienne de Sémiotique, c/o Dept of French, Queen's University, Kingston, Ont. K7L 3N6
T: (613) 547-5857
Founded: 1981; Members: 300
Focus: Ling 02248

Canadian Sheet Steel Building Institute / Institut Canadien de la Tole d'Acier pour le Bâtiment, 201 Consumers Rd, Ste 305, Willowdale, Ont. M2J 4G8
T: (416) 493-8780
Founded: 1961
Focus: Civil Eng 02249

Canadian Sickle Cell Society / Société de l'Anémie Calciforme du Canada, 1076 Bathurst St, Ste 305, Toronto, Ont. M5R 3G9
T: (416) 537-3475
Founded: 1976; Members: 978
Focus: Hematology 02250

Canadian Society for Cellular and Molecular Biology, c/o Centre de Recherche, Hôtel Dieu de Québec, 11 Côte du Palais, Québec, Qué. G1R 2J6
Founded: 1966; Members: 400
Pres: N. Marceau; Gen Secr: C. Cass
Focus: Cell Biol & Cancer Res
Periodical
Bulletin (3 times annually) 02251

Canadian Society for Civil Engineering / Société Canadienne de Génie Civil, 280 Albert St, Ste 202, Ottawa, Ont. K1P 5G8
T: (613) 232-4211; Fax: (613) 232-0390
Founded: 1972; Members: 6500
Focus: Civil Eng
Periodical
The Canadian Journal of Civil Engineering (weekly) 02252

Canadian Society for Clinical Investigation / Société Canadienne de Recherche Clinique, c/o Laboratory of Molecular Endocrinology, Royal Victoria Hospital, 687 Pine Av W, Montréal, Qué. H3A 1A1
T: (514) 842-8989
Founded: 1961; Members: 800
Focus: Med 02253

Canadian Society for Color / Société Canadienne pour la Couleur, Div of Physics, National Research Council, Ottawa, Ont. K1A 0R6
T: (613) 993-2504
Founded: 1972; Members: 130
Focus: Chem 02254

Canadian Society for Education through Art / Société Canadienne de l'Education par l'Art, 3186 Newbound Ct, Malton, Ont. L4T 1R9
T: (416) 677-7478
Founded: 1955; Members: 600
Focus: Educ
Periodicals
Canadian Review of Art Education Research
Newsletter
The Journal 02255

Canadian Society for Electrical and Computer Engineering, 280 Albert St, Ste 202, Ottawa, Ont. K1P 5G8
T: (613) 232-4211; Fax: (613) 232-0390
Founded: 1973
Focus: Electric Eng
Periodical
Canadian Electrical Engineering Journal (weekly) 02256

Canadian Society for Endocrinology and Metabolism / Société Canadienne d'Endocrinologie et Metabolisme, c/o McGill University, Allan Research Bldg, 1033 Pine Av, Rm 117, Montréal, Qué. H3A 1A1
T: (514) 842-1231
Founded: 1973
Focus: Endocrinology 02257

Canadian Society for Horticultural Science / Société Canadienne de Science Horticole, c/o Dept of Plant Science Research Station, Kentville, N.S. B4N 1J5
T: (902) 678-2171
Founded: 1956
Focus: Hort 02258

Canadian Society for Immunology (CSI) / Société Canadienne d'Immunologie, c/o Dept of Medicine, Saint Michaels Hospital, University of Toronto, Toronto, Ont. M5S 1A8
Founded: 1966; Members: 814
Pres: Dr. J. Bienenstock; Gen Secr: Dr. C.A. Ottaway
Focus: Immunology
Periodical
Bulletin (weekly) 02259

Canadian Society for International Health / Société Canadienne pour la Santé Internationale, 1565 Carling Av, Ste 400, Ottawa, Ont. K1Z 8R1
Founded: 1977; Members: 400
Focus: Public Health
Periodical
Synergy: Newsletter (weekly) 02260

Canadian Society for Italian Studies / Société Canadienne pour les Etudes Italiennes, c/o Dept of Modern Languages, University of Ottawa, Ottawa, Ont. K1N 6N5
T: (613) 231-5471
Founded: 1972; Members: 250
Focus: Hist
Periodical
Quaderni d'italianistica (weekly) 02261

Canadian Society for Mechanical Engineering (CSME) / Société Canadienne de Génie Mécanique, 280 Albert St, Ste 202, Ottawa, Ont. K1P 5G8
T: (613) 232-4211; Fax: (613) 232-0390
Founded: 2970; Members: 3780
Focus: Eng
Periodical
CSME Transactions 02262

Canadian Society for Nondestructive Testing, 135 Fennell Av W, Hamilton, Ont. L8N 3T2
T: (416) 387-1666
Gen Secr: N.G. Harding
Focus: Materials Sci 02263

Canadian Society for Nutritional Sciences, c/o Mount Saint Vincent University, Halifax, N.S. B3M 2J6
Founded: 1957; Members: 340
Pres: Dr. D. Fitzpatrick; Gen Secr: Dr. T. Glanville
Focus: Nutrition
Periodical
Nutrition Forum (weekly) 02264

Canadian Society for Psychomotor Learning and Sport Psychology, 21 Lakeland Point Dr, Kingston, Ont. K7M 4E8
T: (613) 389-8878
Focus: Psych 02265

Canadian Society for the Computational Study of Intelligence / Société Canadienne pour Etudes d'Intelligence par Ordinateur, 243 College St, Toronto, Ont. M5T 2Y1
T: (416) 593-4040
Founded: 1973
Focus: Computer & Info Sci 02266

Canadian Society for the History and Philosophy of Science, POB 6070, Station A, Montréal, Qué. H3C 3G1
T: (514) 923-924
Focus: Cultur Hist; Philos 02267

Canadian Society for the Study of Education (CSSE) / Société Canadienne pour l'Etude de l'Education, 1 Stewart St, Ste 205, Ottawa, Ont. K1N 6H7
T: (613) 230-3532; Fax: (613) 230-2746
Founded: 1972; Members: 1000
Gen Secr: Tim Howara
Focus: Educ
Periodicals
Canadian Journal of Education: Revue canadienne de l'éducation (weekly)
CSSE News: Nouvelles SCEE (8 times annually) 02268

Canadian Society for the Study of Higher Education / Société Canadienne pour l'Etude de l'Enseignement Supérieur, 151 Slater St, Ste 1001, Ottawa, Ont. K1P 5N1
T: (613) 563-1236
Founded: 1970; Members: 650
Focus: Educ
Periodical
Canadian Journal of Higher Education (3 times annually) 02269

Canadian Society for the Weizmann Institute of Science / Société Canadienne l'Institut Weizmann des Sciences, 345 Wilson Av, Ste 403, Downsview, Ont. M3H 5W1
T: (416) 630-0360
Founded: 1965
Focus: Sci 02270

Canadian Society for Training and Development / Société Canadienne pour la Formation et le Développement, POB 6754, Ottawa, Ont. K2A 3Z4
Founded: 1979; Members: 160
Focus: Econ; Educ 02271

Canadian Society of Agricultural Engineering / Société Canadienne de Génie Rural, 151 Slater St, Ste 907, Ottawa, Ont. K1P 5H4
T: (613) 232-9459; Tx: 0534306
Focus: Agri
Periodical
Canadian Agricultural Engineering Journal (weekly) 02272

Canadian Society of Agronomy / Société Canadienne d'Agronomie, 151 Slater St, Ste 907, Ottawa, Ont. K1P 5H4
T: (613) 232-9459
Founded: 1954; Members: 300
Focus: Agri 02273

Canadian Society of Animal Science (CSAS), c/o Agricultural Institute of Canada, 151 Slater St, Ste 907, Ottawa, Ont. K1P 5H4
T: (613) 232-9459; Fax: (613) 594-5190
Founded: 1925; Members: 550
Pres: P. Thacker; Gen Secr: V. Stevens
Focus: Zoology
Periodicals
Canadian Journal of Animal Science (weekly)
CSAS Newsletter (weekly) 02274

Canadian Society of Biblical Studies / Société Canadienne des Etudes Bibliques, c/o Dept of Religious Studies, University of Calgary, Calgary, Alta. T2N 1N4
T: (613) 231-3863
Founded: 1933; Members: 280
Pres: D. Jobling; Gen Secr: W.O. McCready
Focus: Rel & Theol
Periodicals
Society Bulletin (weekly)
Ugaritic Newsletter (weekly) 02275

Canadian Society of Cardiology Technologists, POB 3121, Winnipeg, Man. R3C 4E6
T: (204) 889-1593
Founded: 1970
Focus: Cardiol 02276

Canadian Society of Children's Authors, Illustrators and Performers (CANSCAIP) / Société Canadienne des Auteurs, Illustrateurs et Artistes pour Enfants, POB 280, Station L, Toronto, Ont. M6E 4Z2
T: (416) 922-7736
Founded: 1977
Focus: Graphic & Dec Arts, Design; Perf Arts; Lit
Periodical
CANSCAIP News (weekly) 02277

Canadian Society of Church History, c/o Dalhousie University, Halifax, N.S.
Founded: 1950
Gen Secr: T. Sinclair-Faulkner
Focus: Rel & Theol; Hist 02278

Canadian Society of Cinematographers, 1589 The Queensway, Unit 14, Toronto, Ont. M8Z 5W9
T: (416) 251-2211
Founded: 1959; Members: 270
Pres: Robert Rouveroy
Focus: Cinema 02279

Canadian Society of Clinical Neurophysiologists, c/o Dept of Medicine, Ottawa General Hospital, 501 Smyth Rd, Ottawa, Ont. K1H 8L6
Focus: Neurology 02280

Canadian Society of Cytology, c/o Saint Joseph's Hospital, Hamilton, Ont. L8N 4A6
T: (416) 521-6019
Gen Secr: Dr. Vicky Chen
Focus: Cell Biol & Cancer Res
Periodical
Canadian Society of Cytology Bulletin (weekly) 02281

Canadian Society of Electroencephalographers, Electromyographers and Clinical Neurophysiologists, c/o Centre Hospitalier Universitaire, Sherbrooke, Qué. J1H 5N4
T: (819) 563-5555
Founded: 1958; Members: 180
Pres: Dr. Jean Reiher
Focus: Neurology; Pathology; Med . . . 02282

Canadian Society of Engineering Management, 280 Albert St, Ste 202, Ottawa, K1P 5G8
T: (613) 232-4211; Fax: (613) 232-0390
Founded: 1972
Focus: Business Admin; Eng 02283

Canadian Society of Environmental Biologists, POB 962, Station F, Toronto, Ont. M4Y 2N9
Pres: Brent Lister
Focus: Bio; Ecology 02284

Canadian Society of Landscape Architects (CSLA) / Association des Architectes Paysagistes du Canada, POB 870, Station B, Ottawa, Ont. K1P 5P9
T: (613) 232-6342; Fax: (613) 763-9573
Founded: 1934; Members: 300
Focus: Archit
Periodicals
CSLA Bulletin (weekly)
Landscape Architectural Review (5 times annually) 02285

Canadian Society of Microbiologists (CSM) / Société Canadienne des Microbiologistes, 1200E Prince of Wales Dr, Ottawa, Ont. K2C 1M9
T: (613) 723-7233
Founded: 1951; Members: 814
Pres: Dr. P. Bourgaux; Gen Secr: P. Peterkin
Focus: Microbio
Periodicals
CSM Newsletter (3 times annually)
Programme & Abstracts (weekly) 02286

Canadian Society of Military Medals and Insignia, 15 Greenhill Dr, Thorold, Ont. L2V 1W4
T: (416) 227-0357
Founded: 1963; Members: 500
Focus: Hist; Military Sci
Periodicals
Journal (weekly)
Newsletter (weekly) 02287

Canadian Society of Orthopaedic Technologists / Société Canadienne des Technologistes en Orthopédie, 4433 Sheppard Av E, Ste 200, Agincourt, Ont. M1S 1V3
T: (416) 292-0687
Founded: 1973; Members: 240
Focus: Orthopedics
Periodicals
BodyCast (weekly)
NewsCast (weekly) 02288

Canadian Society of Otolaryngology, Head and Neck Surgery, 4953 Dundas St W, Ste 103, Islington, Ont. M9A 1B6
T: (416) 233-6034; Fax: (416) 239-8220
Members: 600
Pres: Dr. Melvin Schloss; Gen Secr: Dr. Howard Lampe
Focus: Otorhinolaryngology; Surgery
Periodical
Journal of Otolaryngology (weekly) . . . 02289

Canadian Society of Painters in Watercolour (CSPWC), c/o Visual Arts Ontario, 439 Wellington St W, Toronto, Ont. M5V 1E7
Founded: 1925; Members: 100
Pres: Rudolf Stussi
Focus: Fine Arts
Periodical
CSPWC Newsletter (3 times a year) . . 02290

Canadian Society of Petroleum Geologists, 206 Seventh Av SW, Calgary, Alta. T2P 0W7
T: (403) 264-5610
Founded: 1927; Members: 3700
Pres: Alice Payne; Gen Secr: Mike Morrison
Focus: Geology
Periodical
Bulletin of Canadian Petroleum Geology (weekly) 02291

Canadian Society of Plant Physiologists / Société Canadienne de Physiologie Végétale, c/o Dept of Botany, University of Guelph, Guelph, Ont. N1G 2W1
T: (519) 824-4120 ext 6000
Founded: 1958; Members: 640
Pres: Prof. J. Derek Bewley
Focus: Botany 02292

Canadian Society of Plastic Surgeons, 1333 Sheppard Av E, Ste 336, Willowdale, Ont. M2J 1V1
Focus: Surgery 02293

Canadian Society of Pulmonary and Cardiovascular Technology, POB 7175, Station A, Toronto, Ont. M5W 1X8
T: (705) 254-5181
Founded: 1972
Pres: Sauli Ahvenniemi
Focus: Pulmon Dis; Cardiol 02294

Canadian Society of Safety Engineering / Société Canadienne de la Santé et de la Sécurité au Travail, 6519B Mississauga Rd, Mississauga, Ont. L5N 1A6
T: (416) 567-7192; Fax: (416) 567-7191
Founded: 1949; Members: 2000
Focus: Safety 02295

Canadian Society of Soil Science / Société Canadienne de la Science du Sol, 151 Slater St, Ste 907, Ottawa, Ont. K1P 5H4
T: (613) 232-9459; Tx: 0534306
Founded: 1955; Members: 300
Focus: Agri 02296

Canadian Society of Zoologists, c/o Dept of Biology, York University, Toronto, Ont. M3J 1P3
T: (416) 667-2319
Focus: Zoology 02297

Canadian Sociology and Anthropology Association / Société Canadienne de Sociologie et d'Anthropologie, 1455 Blvd de Maisonneuve Ouest, Ste N-317-1, Montréal, Qué. H3G 1M8
T: (514) 848-8280
Founded: 1966; Members: 950
Focus: Anthro; Sociology
Periodical
Canadian Review of Sociology and Anthropology 02298

Canadian Speech and Hearing Society, 10230 111 Av, Edmonton, Alta. T5G 0B7
Members: 600
Gen Secr: Anita Yates
Focus: Logopedy
Periodical
Human Communication 02299

Canadian Spice Association, c/o Halford-Lewis Ltd, 465 Saint Jean St, Ste 606, Montréal, Qué. H2Y 2R6
T: (514) 842-7857
Founded: 1942; Members: 39
Focus: Food 02300

Canadian Standards Association / Association Canadienne de Normalisation, 178 Rexdale Blvd, Rexdale, Ont. M9W 1R3
T: (416) 747-4000
Founded: 1919; Members: 7500
Focus: Standards
Periodicals
CSA + Le Consommateur (weekly)
CSA + The Consumer (weekly)
Information Update
Standards Canada (weekly) 02301

Canadian Steel Environmental Association, 1 Yonge St, Ste 1400, Toronto, Ont. M5E 1J9
T: (416) 363-7261
Gen Secr: Alex C. Dick
Focus: Metallurgy 02302

Canadian Steel Industry Research Association / Association Canadienne pour la Recherche dans l'Industrie Sidérurgique, 1 Yonge St, Ste 1400, Toronto, Ont. M5E 1J9
T: (416) 363-7261
Founded: 1978
Focus: Metallurgy 02303

Canadian Stroke Recovery Association, 170 The Donway W, Ste 120, Don Mills, Ont. M3C 2G3
T: (416) 446-1580
Focus: Cardiol 02304

Canadian Teachers' Federation / Fédération Canadienne des Enseignants, 110 Argyle Av, Ottawa, Ont. K2P 1B4
T: (613) 232-1505; Tx: 0636700906; Fax: (613) 232-1886
Founded: 1920; Members: 15
Focus: Educ 02305

Canadian Technion Society / Société Canadienne Technion, 2085 Union Av, Ste 1480, Montréal, Qué. H3A 2C3
T: (514) 288-0682
Founded: 1950; Members: 1500
Focus: Eng 02306

Canadian Theatre Critics Association / Association des Critiques de Théâtre du Canada, 62 Wellesley St W, Ste 1806, Toronto, Ont. M5S 2X3
T: (416) 967-9446
Founded: 1981; Members: 65
Focus: Journalism; Perf Arts 02307

Canadian Thoracic Society (CTS), 75 Albert St, Ste 908, Ottawa, Ont. K1P 5E7
T: (613) 237-1208; Fax: (613) 563-3362
Founded: 1971; Members: 450
Pres: Dr. Morley Lertzma
Focus: Pulmon Dis 02308

Canadian Tinplate Recycling Council, POB 460, Hamilton, Ont. L8N 3J5
T: (416) 545-7478
Pres: R.A. Flemington
Focus: Ecology; Materials Sci 02309

Canadian Universities and Colleges Conference Officers Association, c/o University of Western Ontario, London, Ont. N6A 3K7
T: (519) 679-3991
Gen Secr: James McArthur
Focus: Educ 02310

Canadian University and College Counselling Association / Association du Counselling des Collèges et Universités du Canada, POB 1700, Victoria, B.C. V8P 4X8
T: (604) 592-1281
Founded: 1973
Focus: Public Admin; Educ 02311

Canadian University Music Society / Société de Musique des Universités Canadiennes, c/o School of Music, University of Windsor, Windsor, Ont. N9B 3PA
Fax: (519) 973-7050
Founded: 1964; Members: 350
Focus: Music
Periodical
Canadian University Music Review: Revue de musique des universités canadiennes (biennial) 02312

Canadian Urban Transit Association / Association Canadienne du Transport Urbain, 55 York St, Ste 901, Toronto, Ont. M5J 1R7
T: (416) 365-9800; Fax: (416) 365-1295
Founded: 1904; Members: 236
Focus: Transport
Periodicals
Annual Report (weekly)
Transit Fact Book & Membership Directory (weekly)
Transit Topics (weekly) 02313

Canadian Urological Association, c/o Victoria General Hospital, Halifax, N.S. B3H 2X9
T: (416) 595-3207
Founded: 1944; Members: 450
Focus: Urology 02314

Canadian Veterinary Medical Association / Association Canadienne des Vétérinaires, 339 Booth St, Ottawa, Ont. K1R 7K1
T: (613) 236-1162; Fax: (613) 236-9681
Founded: 1948; Members: 4000
Focus: Vet Med
Periodicals
Canadian Journal of Comparative Medicine (weekly)
Canadian Veterinary Journal (weekly) . . 02315

Canadian Water Quality Association / Association Canadienne pour le Qualité de l'Eau, 472 Lee Av, Waterloo, Ont. N2K 1X9
T: (519) 885-3854
Founded: 1960; Members: 180
Gen Secr: Lou J. Smith
Focus: Water Res; Ecology 02316

Canadian Welding Bureau / Bureau Canadien de Soudage, 7250 W Credit Av, Mississauga, Ont. L5N 5N1
T: (416) 542-1312; Tx: 06218506; Fax: (416) 542-1318
Founded: 1947; Members: 2000
Focus: Standards 02317

Canadian Wind Engineering Association / Association Canadienne du Génie Eolien, c/o Boundary Layer Wind Tunnel, University of Western Ontario, London, Ont. N6A 5B9
T: (519) 679-6662
Founded: 1976; Members: 300
Focus: Energy
Periodical
CWEA Newsletter (weekly) 02318

Canadian Writers' Foundation, 90 Kenilworth St, Ottawa, Ont. K1Y 3Y7
T: (613) 728-0602
Focus: Lit 02319

Cancer Research Society / Société de Recherche sur le Cancer, 19 Esterel, Pl Bonaventure, POB 183, Montréal, Qué. H5A 1H8
T: (514) 861-9227
Founded: 1946; Members: 4000
Focus: Cell Biol & Cancer Res
Periodical
The Recorder (biennial) 02320

Cardiology Technicians Associations of British Columbia, POB 76722, Stn S, Vancouver, B.C. V5R 5S7
T: (604) 526-2831
Founded: 1971; Members: 91
Pres: Bernice Osborne
Focus: Cardiol 02321

Cardiology Technologists Association of Ontario, 6 Parent Av, Downsview, Ont. M3M 1Z6
T: (416) 241-7887
Members: 500
Pres: Evelyn Wiggings
Focus: Cardiol
Periodical
Heartbeat (weekly) 02322

Catholic Public Schools Inter-Society Committee, POB 7000, Prince George, B.C. V2N 3Z2
T: (604) 964-7266
Pres: Mike Van Adrichem
Focus: Educ 02323

Centrale de l'Enseignement du Québec (CEQ), 1415 Jarry Est, Ste 300, Montréal, Qué. H2E 1A7
T: (514) 374-6660; Tx: 05131547
Founded: 1946; Members: 86000
Gen Secr: Michel Agnaieff
Focus: Educ
Periodical
Nouvelles CEQ (weekly) 02324

Central Ontario Industrial Relations Institute, 85 Richmond St W, Ste 200, Toronto, Ont. M5H 2C9
T: (416) 368-2364
Focus: Econ 02325

Centre d'Animation, de Développement et de Recherche en Education, 1940 Blvd Henri Bourassa Est, Montréal, Qué. H2B 1S2
T: (514) 381-8891; Fax: (514) 381-4086
Founded: 1968
Focus: Educ
Periodical
Prospectives (3 times annually) 02326

Centre de Bioéthique / Center for Bioethics, 110 Av des Pins Ouest, Montréal, Qué. H2W 1R7
T: (514) 842-1481; Tx: 05560398; Fax: (514) 843-5600
Founded: 1976
Focus: Med
Periodicals
Journal of Palliative Care (weekly)
Synapse: A Canadian News Service for Biomedical Ethics (weekly)
Synapse: Un service canadien d'information en éthique biomédicale (weekly) 02327

Centre for Monarchial Studies, 3050 Yonge St, Ste 206, Toronto, Ont. M4N 2K4
T: (416) 482-4157
Focus: Hist 02328

Centre for Research on Latin America and the Caribbean (CERLAC), c/o Founders College, York University, 4700 Keele St, Toronto, Ont. M3J 1P3
T: (416) 736-5237
Founded: 1978
Gen Secr: Dr. L.L. North
Focus: Soc Sci; Poli Sci; Econ 02329

Centre Franco-Ontarien de Resources Pédagogiques, 339 Rue Wilbrod, Ottawa, Ont. K1N 6M4
T: (613) 238-7957
Gen Secr: Gisèle Lalonde
Focus: Educ 02330

Ceramics and Stone Accident Prevention Association, 2 Bloor St E, Toronto, Ont. M4W 3C2
T: (416) 965-8888
Gen Secr: J.V. Findlay
Focus: Safety 02331

Cercles des Jeunes Naturalistes (CJN), 4101 Rue Sherbrooke Est, Ste 124, Montréal, Qué. H1X 2B2
T: (514) 252-3023; Tx: 05829647
Founded: 1931; Members: 3000
Pres: Gertrude de Champlain; Gen Secr: Elise Tousignant
Focus: Nat Sci
Periodicals
Feuillets du Naturaliste
Tracts CJN 02332

Cerebral Palsy Association in Alberta / Associtation de la Paralysie Cérébrale en Alberta, c/o Saint Gregory School, 5340 26 Av SW, Calgary, Alta. T3E 4Z4
T: (403) 246-3111
Members: 350
Gen Secr: 1976
Focus: Neurology
Periodical
CP Reporter (weekly) 02333

Cerebral Palsy Association of Manitoba, 825 Sherbrooke St, Winnipeg, Man. R3A 1M5
Focus: Neurology 02334

Cerebral Palsy Association of Newfoundland, POB 1700, St. John's, Nfld. A1C 2B8
T: (709) 753-5235, 739-1003
Founded: 1961; Members: 50
Pres: Andrea Crosbie
Focus: Neurology 02335

Cerebral Palsy Association of Nova Scotia, 3106 Barrington St, Halifax, N.S. B3K 2X6
T: (902) 425-3986
Focus: Neurology 02336

Cerebral Palsy Association of Prince Edward Island, POB 2702, Charlottetown, P.E.I. C1A 8C3
T: (902) 894-9488
Members: 50
Pres: Vernon Richards
Focus: Neurology 02337

Cerebral Palsy Association of Prince George and District, 1687 Strathcona Av, Prince George, B.C. V2L 4E7
T: (604) 563-7168
Founded: 1968
Gen Secr: Pat Bunn
Focus: Neurology 02338

Champlain Society (CS), Royal York Hotel, 100 Front Street W, Toronto, Ont. M5J IE3
T: (416) 363-8310
Founded: 1905; Members: 1250
Pres: Prof. Frederick H. Armstrong
Focus: Hist 02339

Charles H. Ivey Foundation, 201 Consumers Rd, Ste 105, Willowdale, Ont. M2J 4G8
Founded: 1958; Members: 10
Pres: C.R. Ivey
Focus: Sci 02340

Chemical Industries Accident Prevention Association, 2 Bloor St E, Toronto, Ont. M4W 3C2
T: (416) 965-8888
Gen Secr: J.V. Findlay
Focus: Safety; Chem 02341

Chemical Institute of Canada, 130 Slater St, Ste 550, Ottawa, Ont. K1P 6E2
T: (613) 232-6252; Fax: (613) 232-5862
Founded: 1945; Members: 10000
Gen Secr: Dr. Anne E. Alper
Focus: Chem
Periodicals
Canadian Chemical News (11 times annually)
The Canadian Journal of Chemical Engineering (weekly) 02342

Child Development Centre Society of Fort Saint John and District, 10408 105 Av, Fort Saint John, B.C. V1J 2M7
T: (604) 785-3200
Focus: Educ 02343

Children's Broadcast Institute / Institut de Radiotélédiffusion pour Enfants, 234 Eglinton Av E, Ste 405, Toronto, Ont. M4P 1G3
T: (416) 482-0321; Fax: (416) 482-0560
Founded: 1974; Members: 50
Focus: Mass Media
Periodicals
The Canadian Children's Film and Video Directory (biennial)
Children's Television (weekly)
Your Child and TV (weekly) 02344

Children's Oncology Care of Ontario, 356 Dundas St W, Toronto, Ont. M5T 1G5
T: (416) 977-0458
Founded: 1979
Pres: C.D. Chalmers
Focus: Cell Biol & Cancer Res; Pediatrics 02345

Children's Rehabilitation and Cerebral Palsy Association, 1195 W Eighth Av, Vancouver, B.C. V6H 1C5
T: (604) 734-2221; Fax: (604) 734-3533
Founded: 1945; Members: 140
Gen Secr: D.A. Ewen
Focus: Neurology; Pediatrics; Rehabil . . 02346

Christian Medical Society, 83 Princess Anne Cr, Islington, Ont. M9A 2P6
Focus: Med 02347

Church Library Association, 41 Moore Av, Aylmer, Ont. N5H 2Z6
T: (519) 773-5929
Founded: 1969; Members: 220
Pres: G. Friesen
Focus: Libraries & Bk Sci 02348

Cinémathèque Québécoise, 335 Blvd de Maisonneuve Est, Montréal, Qué. H2X 1K1
T: (514) 842-9763; Fax: (514) 842-1816
Focus: Cinema
Periodical
La Revue de la Cinémathèque (5 times annually) 02349

Civil Aviation Medical Association (CAMA), c/o Ministry of Transport, 330 Sparks St., Ottawa, Ont. K1A 0N5
T: (613) 990-2309; Tx: 0533336; Fax: (613) 957-2400
Pres: Dr. Steve Blizzard
Focus: Med 02350

Clinical Research Society of Toronto, c/o University of Toronto, 6353 Medical Sciences Bldg, Toronto, Ont. M5S 1A8
Focus: Med 02351

College of Dental Surgeons of British Columbia, 1125 Eighth Av W, Vancouver, B.C. V6H 3N4
T: (604) 736-3621
Founded: 1908; Members: 1800
Focus: Dent 02352

College of Dental Surgeons of Saskatchewan, 105 21 St E, Ste 811, Saskatoon, Sask. S7K 0B3
T: (306) 244-5072
Founded: 1905; Members: 325
Gen Secr: G.H. Peacocke
Focus: Dent 02353

College of Family Physicians of Canada / Collège des Médecins de Famille du Canada, 4000 Leslie St, Willowdale, Ont. M2K 2R9
T: (416) 493-7513
Founded: 1954; Members: 7000
Focus: Med
Periodicals
Canadian Family Physician (weekly)
Family Medicine Research Update (weekly) 02354

College of Physicians and Surgeons of Alberta, 9901 108 St, Edmonton, Ata. T5K 1G8
T: (403) 423-4764
Focus: Physiology; Surgery 02355

College of Physicians and Surgeons of British Columbia, 1807 Tenth Av W, Vancouver, B.C. V6J 2A9
T: (604) 736-5551
Founded: 1886; Members: 6800
Focus: Physiology; Surgery 02356

College of Physicians and Surgeons of Manitoba, 155 Carlton St, Ste 1410, Winnipeg, Man. R3C 3H8
T: (204) 947-1694
Founded: 1871; Members: 1850
Focus: Physiology; Surgery 02357

College of Physicians and Surgeons of New Brunswick, 400 Main St, Ste 1078, Saint John, N.B. E2K 4N5
T: (506) 658-0959
Focus: Med; Surgery 02358

College of Physicians and Surgeons of Ontario, 80 College St, Toronto, Ont. M5G 2E2
T: (416) 961-1711
Focus: Physiology; Surgery 02359

College of Physicians and Surgeons of Saskatchewan, 211 Fourth Av S, Saskatoon, Sask. S7K 1N1
T: (306) 244-7355
Members: 1526
Focus: Physiology; Surgery 02360

College of Psychologists of New Brunswick / Collège des Psychologues du Nouveau-Brunswick, POB 1194, Station A, Fredericton, N.B. E3B 5C8
T: (506) 854-7700
Founded: 1980; Members: 150
Focus: Physiology 02361

Collegium Internationale Allergologicum (CIA), c/o Faculty of Health Sciences, MacMaster University, 1200 Main St W, Hamilton, Ont. L8N 3Z5
T: (416) 525-9140; Fax: (416) 546-0800
Founded: 1954
Gen Secr: Dr. John Bienenstock
Focus: Med 02362

Committee on Atlantic Studies (CAS), c/o Dept of Political Science, Carleton University, Ottawa Ont. L1S 5B6
T: (613) 788-2780; Fax: (613) 788-4467
Founded: 1964; Members: 50
Gen Secr: Prof. Robert J. Jackson
Focus: Poli Sci 02363

Commonwealth Association for Education in Journalism and Communication (CAEJC), c/o Faculty of Law, University of Western Ontario, London, Ont. N6A 3K7
T: (519) 661-3348; Tx: 0647134; Fax: (519) 661-3790
Founded: 1985
Gen Secr: Prof. Robert Martin
Focus: Comm Sci; Educ
Periodical
CAEJC Newsletter (weekly) 02364

Commonwealth Association of Legislative Counsel (CALC), c/o Dept of the Attorney General, 9833 109 St, Edmonton, Alta. T5K 2E8
Founded: 1983
Gen Secr: Peter Pagano
Focus: Law
Periodical
The Loophole 02365

Commonwealth Association of Museums (CAM), c/o The Glenbow Museum, 130 Ninth Av SE, Calgary, Alta T2G 0P3
T: (403) 264-8300; Tx: 8955822; Fax: (403) 265-9769
Founded: 1974
Gen Secr: L. Irvine
Focus: Arts
Periodical
CAM Newsletter 02366

Commonwealth Association of Planners (CAP), 404-126 York St, Ottawa, Ont. K1N 5T5
T: (613) 233-2105; Fax: (613) 233-1984
Founded: 1971
Gen Secr: David H. Sherwood
Focus: Urban Plan
Periodical
Planners Newsletter (weekly) 02367

Commonwealth Geographical Bureau (CGB), c/o Dept of Geography, McGill University, 805 Rue Sherbrooke Ouest, Montréal, Qué. H3A 2K6
T: (514) 3984940
Founded: 1968
Gen Secr: Prof. John T. Parry
Focus: Geography
Periodical
CGB Newsletter (weekly) 02368

Commonwealth of Learning (COL), 1700-777 Dunsmuir St, POB 10428, Pacific Centre, Vancouver, B.C. V7Y 1K1
T: (604) 660-4675; Tx: 04507508; Fax: (604) 660-7472
Founded: 1988
Pres: Dr. James Maraj
Focus: Educ 02369

Community Planning Association of Canada / Association Canadienne d'Urbanisme, 2837 Dewdney, Regina, Sask. S4T 0X8
T: (306) 569-1611
Founded: 1949
Focus: Urban Plan 02370

Community Planning Association of Nova Scotia, POB 1259, Dartmouth, N.S. B2Y 4B9
T: (902) 469-7809
Founded: 1946; Members: 300
Gen Secr: Joanne Lamey
Focus: Urban Plan 02371

Comparative Federation and Federalism Research Committee, c/o Dept of Political Science, University of Windsor, Ont. 0N N9B, Windsor
T: (519) 253-4232; Fax: (519) 253-4232
Founded: 1984
Pres: Dr. C. Lloyd Brown-John
Focus: Poli Sci
Periodical
Newsletter (weekly) 02372

Composers, Authors and Publishers Association of Canada (CAPAC) / Association des Compositeurs, Auteurs et Editeurs du Canada, 1240 Bay St, Toronto, Ont. M5R 2C2
T: (416) 924-4427
Founded: 1925; Members: 15000
Focus: Lit; Music
Periodical
The Canadian Composer (10 times annually) 02373

Computer Communications Institute, 96 Peckham Av, Willowdale, Ont. M2R 2T5
T: (416) 222-3145
Focus: Computer & Info Sci 02374

Computer Law Association, Toronto Dominion Centre, POB 48, Toronto, Ont. M5K 1E6
T: (416) 362-1812
Gen Secr: Daniel G. Cooper
Focus: Law 02375

Confederation of Alberta Faculty Associations, c/o University of Alberta, 748 Education S, Edmonton, Alta. T6G 2G5
T: (403) 432-5630, 432-3098
Founded: 1971; Members: 12
Focus: Sci 02376

Conference Board of Canada, 25 McArthur Av, Ste 100, Ottawa, Ont. K1L 6R3
T: (613) 746-1261
Members: 800
Focus: Econ; Business Admin 02377

Conference of Alberta School Superintendents, 315 Sharon Av SW, Calgary, Alta. T3C 2G8
T: (403) 244-6825
Founded: 1964; Members: 208
Gen Secr: J.V. Van Tighem
Focus: Educ 02378

Conference of Defence Associations Institute / Institut du Congrès des Associations de Défense, 501-100 Gloucester St, Ottawa, Ont. K2P 0A4
T: (613) 563-1387; Fax: (613) 563-8508
Founded: 1987; Members: 600
Pres: B. Shapiro; Gen Secr: M.G. Paine
Focus: Military Sci 02379

Conférence Religieuse Canadienne / Canadian Religious Conference, 324 Laurier Av E, Ottawa, Ont. K1N 6P6
T: (613) 236-0824
Founded: 1954; Members: 400
Gen Secr: Albert Dumont
Focus: Rel & Theol 02380

Congregational Libraries Association of British Columbia, 1860 San Juan Av, Victoria, B.C. V8N 2J2
T: (604) 477-6303
Founded: 1971; Members: 135
Pres: Fran Rose
Focus: Libraries & Bk Sci 02381

Conseil des Recherches en Pêche et en Agro-Alimentaire du Québec, 200A Chemin Sainte-Foy, Québec, Qué. G1R 4X6
T: (418) 643-1229
Pres: André Vézina
Focus: Agri; Fisheries 02382

Conseil des Universités du Québec / Council of Universities of Quebec, 2700 Blvd Laurier, Sainte Foy, Qué. G1V 2L8
T: (418) 643-8592
Pres: Jacques L'Ecuyer
Focus: Sci 02383

Conseil International d'Education Mésologique (CIEM), c/o Ecole des Gradués, Bureau 2549, Université de Laval, Québec, Qué. G1K 7P4
T: (418) 656-3248; Tx: 05131621; Fax: (418) 656-3691
Founded: 1977
Gen Secr: Prof. Michel Maldague
Focus: Educ 02384

Conseil Québecoise pour l'Enfance et la Jeunesse, 3700 Rue Berri, Bureau 425, Montréal, Qué. H2L 4G9
T: (514) 842-5485
Founded: 1963; Members: 325
Gen Secr: Marguerite Wolfe
Focus: Educ
Periodical
Apprentissage et Socialisation en Piste (weekly) 02385

Conseil Supérieur de l'Education, 2050 Blvd Saint-Cyrille Ouest, Sainte Foy, Qué. G1V 2K8
T: (418) 643-3850
Founded: 1964; Members: 111
Pres: Robert Bisaillon; Gen Secr: Jean Proulx; Alain Durand
Focus: Educ
Periodical
Conseil-Education: Edu Council (15 times annually) 02386

Conservation and Development Association of Saskatchewan, POB 1076, Canora, Sask. S0A 0L0
T: (306) 563-5042
Founded: 1967; Members: 81
Gen Secr: Fred Petroff
Focus: Preserv Hist Monuments 02387

Conservation Council of New Brunswick / Conseil de la Conservation du Nouveau-Brunswick, 180 Saint John St, Fredericton, N.B. E3B 4A9
T: (506) 458-8747
Founded: 1969; Members: 450
Gen Secr: Mark Lutes
Focus: Preserv Hist Monuments
Periodical
Eco Allert (weekly) 02388

Conservation Council of Ontario, 489 College St, Ste 506, Toronto, Ont. M6G 1A5
Founded: 1952; Members: 120
Pres: David H. Hardy
Focus: Preserv Hist Monuments
Periodicals
Ontario Conservation News (11 times annually)
Worldwatch Papers 02389

Consortium-Distance Education Network, c/o CREAD, 3460 Rue de la Pérade, Sainte-Foy, Qué. G1X 3Y5
T: (418) 644-6910; Tx: 05131506; Fax: (418) 646-3039
Founded: 1990
Gen Secr: Armando Villarroel
Focus: Educ 02390

Consortium of Ontario Public Alternative Schools, 65 Culnan Av, Toronto, Ont. M8Z 5B1
T: (416) 259-4657
Founded: 1975; Members: 110
Pres: E. Bergey
Focus: Educ 02391

Construction Management Institute, 5799 Yonge St, Ste 901, Willowdale, Ont. M2M 3V3
Fax: (416) 224-1511
Focus: Civil Eng; Business Admin . . . 02392

Construction Safety Association of Ontaria, 74 Victoria St, Toronto, Ont. M5C 2A5
T: (416) 366-1501; Fax: (416) 366-0232
Founded: 1929; Members: 35000
Gen Secr: Leonard Sylvester
Focus: Safety
Periodical
The Counsellor (weekly) 02393

Corporation Professionnelle des Ergothèrapeutes du Québec, 1500 Blvd de Maisoneuve Est, Ste 400, Montréal, Qué. H2L 2B1
T: (514) 527-9281
Founded: 1974; Members: 600
Focus: Therapeutics 02394

Corporation Professionnelle des Médecins du Québec / Professional Corporation of Physicians of Quebec, 1440 Saint Catherine St W, Ste 914, Montréal, Qué. H3G 1S5
T: (514) 878-4441
Founded: 1847; Members: 14500
Focus: Physiology
Periodical
Bulletin (5 times annually) 02395

Corporation Professionnelle des Médecins Vétérinaires du Québec, 795 Rue du Palais, Ste 200, Saint Hyacinthe, Qué. J2S 5C6
T: (514) 774-1427; Fax: (514) 774-7635
Founded: 1902; Members: 1350
Pres: Paul Desrosiers
Focus: Vet Med
Periodicals
Le Médecin Vétérinaire du Québec (weekly)
Le Vétérinarius (weekly) 02396

Corporation Professionnelle des Orthophonistes et Audiologistes du Québec / Professional Corporation of Speech Therapists and Audiologists of Quebec, 4770 Rue de Salaberry, Montréal, Qué. H4J 1H6
Gen Secr: Marianne Berger
Focus: Logopedy; Therapeutics 02397

Corporation Professionnelle des Psychologues du Québec, 1575 Blvd Henri Bourassa Ouest, Ste 510, Montréal, Qué. H3M 3A9
T: (514) 337-3360
Founded: 1962; Members: 2380
Gen Secr: Yves Murray
Focus: Psych
Periodical
Les Cahiers du Psychologue Québecois (8 times annually) 02398

Corporation Professionnelle des Travailleurs Sociaux du Québec / Corporation of Professional Social Workers of the Province of Quebec, 5757 Decelles, Ste 335, Montréal, Qué. H3S 2C3
T: (514) 731-2749
Founded: 1960; Members: 1900
Gen Secr: Pierre Landry
Focus: Sociology 02399

Council for International Congresses of Dipterology, c/o Div of Entomology, University of Alberta, Edmonton, Alta. T6G 2E3
T: (403) 922-3221
Founded: 1986
Pres: Dr. Graham C.D. Griffiths
Focus: Entomology 02400

Council of Canadian Law Deans / Conseil des Doyens des Facultes de Droit du Canada, 57 Louis Pasteur, Ottawa, Ont. K1N 6N5
T: (613) 564-4060
Focus: Law 02401

Council of Canadian University Chemistry Chairmen / Conseil de Directeurs de Département de Chimie d'Universités Canadiennes, Dept of Chemistry, Simon Fraser University, Burnaby, B.C. V5A 1S6
Founded: 1970
Focus: Chem 02402

Council of Ontario Universities / Conseil des Universités de l'Ontario, 130 Saint George St, Ste 8039, Toronto, Ont. M5S 2T4
T: (416) 979-2165; Fax: (416) 979-8635
Founded: 1962
Focus: Sci 02403

Council of Parent Participation Pre-Schools in British Columbia, 7673 Sixth St, Burnaby, B.C. V3N 3M8
T: (604) 522-0834
Founded: 1945; Members: 2500
Focus: Educ 02404

Council of Prairie University Libraries, c/o University of Calgary Library, 2500 University Dr, Calgary, Alta. T2N 1N4
T: (403) 284-5954
Founded: 1968
Pres: Alan H. MacDonald
Focus: Libraries & Bk Sci 02405

Council of Private Technical Schools / Conseil des Ecoles Privées, 3035 Saint Antoine, Montréal, Qué. H3Z 1W8
T: (514) 935-2525
Founded: 1962
Focus: Educ 02406

Council of Regents for Colleges of Applied Arts and Technology, Queen's Park, Mowat Block, Toronto, Ont. M7A 1L2
T: (416) 965-4234
Members: 16
Gen Secr: D. Murdoch
Focus: Eng; Fine Arts 02407

Council of Western Canadian University Presidents / Conseil des Recteurs des Universités de l'Ouest Canadien, c/o Office of the President, University of Manitoba, Winnipeg, Man. R3T 2N2
T: (202) 474-9666
Focus: Sci 02408

Council on Homosexuality and Religion / Conseil de l'Homosexualité et la Religion, POB 1912, Winnipeg, Man. R3C 3R2
Founded: 1976; Members: 127
Focus: Rel & Theol 02409

County of York Law Association, Court House, University Av, Toronto, Ont. M5G 1T3
T: (416) 965-7489; Fax: (416) 947-9148
Founded: 1885
Focus: Law 02410

Cumberland County Family Planning Association, POB 661, Amherst, N.S. B4H 1X7
T: (902) 667-7500
Founded: 1981; Members: 44
Focus: Family Plan 02411

Cystic Fibrosis Foundation of Alberta, 10136 100 St, Ste 812, Edmonton Alta. T5J 0N8
T: (403) 426-4445
Focus: Physiology 02412

Defence Medical Association of Canada / Association Médicale de la Défense du Canada, POB 3207, Station D, Ottawa, Ont. K1P 6H7
Founded: 1892; Members: 200
Focus: Med 02413

Dental Association of Prince Edward Island, 184 Belvedere Av, Charlottetown, P.E.I. C1A 2Z1
T: (902) 894-4022
Members: 43
Focus: Dent 02414

Denturist Society of Alberta, 2 Athabascan Av, Sherwood Park, Alta. T8A 4E3
T: (403) 464-0300; Fax: (403) 467-9263
Founded: 1981; Members: 202
Pres: Shaun H. Yandt; Gen Secr: John G. Ashton
Focus: Dent
Periodical
Wild Rose Denturist (weekly) 02415

Denturist Society of Nova Scotia, 3627 Joseph Howe Dr, Halifax, N.S. B3L 4H8
T: (902) 455-8390
Focus: Dent 02416

Departments of Education Correspondence Schools Association (Canada), c/o Alberta Correspondence School, POB 4000, Barrhead, Alta. T0G 2P0
T: (403) 674-5333
Pres: Garry Popowich
Focus: Educ 02417

Donner Canadian Foundation / Fondation Canadienne Donner, Toronto-Dominion Centre, POB 122, Toronto, Ont. M5K 1H1
T: (416) 869-1091
Founded: 1950; Members: 22
Focus: Sci 02418

Dystonia Medical Research Foundation, 777 Hornby St, Ste 1800, Vancouver, B.C. V6Z 1S4
Focus: Med 02419

E.A. Baker Foundation for Prevention of Blindness / Fondation E.A. Baker pour la Prévention de la Cécité, 1931 Bayview Av, Toronto, Ont. M4G 4C8
Focus: Public Health; Ophthal 02420

Eastern Counties Lung Association, 1011 Leonard Av, POB 1521, Cornwall, Ont. K6J 1M1
T: (613) 932-4999
Gen Secr: M. Alexander
Focus: Pulmon Dis 02421

Eastern Newfoundland Engineering Society, POB 8414, St. John's, Nfld. A1B 3N7
Pres: W.W. Quinton
Focus: Eng 02422

Eastern Québec Teachers Association, 2559 Lalonde Av, Sainte Foy, Qué. G1W 1M8
Focus: Educ 02423

Eastern Townships Association of Teachers, POB 94, Cookshire, Qué. J0B 1M0
Focus: Educ 02424

Economic Council of Canada, POB 527, Ottawa, Ont. K1P 5V6
T: (613) 952-1711; Fax: (613) 952-2171
Founded: 1963
Pres: Judith Maxwell
Focus: Econ
Periodical
Au Courant (weekly) 02425

Economics Society of Alberta, c/o Esso Ressources Canada, 237 Fourth Av SW, Calgary, Alta. T2P 0H6
T: (403) 237-3437
Focus: Econ 02426

Education Relations Commission of Ontario, 111 Avenue Rd, Ste 400, Toronto, Ont. M5R 3J8
T: (416) 922-7679
Founded: 1975
Gen Secr: Robert H. Field
Focus: Educ 02427

Electric Vehicle Association of Canada (EVAC), 275 Slater St, Ste 500, Ottawa, Ont. K1P 5H9
T: (613) 236-2497
Founded: 1980; Members: 250
Gen Secr: Graham Donaldson
Focus: Auto Eng
Periodical
EVAC Newsletter (weekly) 02428

Electrolysis Association of Canada, 3319 Bloor St W, Toronto, Ont. M8X 1E7
T: (416) 231-3234
Focus: Physics 02429

Elizabeth Fry Society of Alberta (Edmonton), 10136 100 St, Ste 702, Edmonton, Alta. T5J 0P1
T: (403) 421-1175
Founded: 1979; Members: 90
Gen Secr: Shirley Lewis
Focus: Adult Educ; Sociology 02430

Elizabeth Fry Society of Manitoba, 51 Osborne St S, Winnipeg, Man. R3L 1Y2
T: (204) 474-2469
Focus: Adult Educ; Sociology 02431

Elizabeth Fry Society of New Brunswick, 18 Botsford St, Moncton, N.B. E1C 4W7
T: (506) 384-2756
Gen Secr: J.C. Cunningham
Focus: Adult Educ; Sociology 02432

Elizabeth Fry Society of Saskatchewan, 219 22 St E, Ste 301, Saskatoon, Sask. S7K 0G3
T: (306) 934-4606
Focus: Adult Educ; Sociology 02433

Energy Probe Research Foundation, 100 College St, Toronto, Ont. M5G 1L5
T: (416) 978-7014
Founded: 1974; Members: 13000
Pres: Anetta Turner
Focus: Energy 02434

Engineering Institute of Canada, 280 Albert St, Ste 202, Ottawa, Ont. K1P 5G8
T: (613) 232-4211; Fax: (613) 232-0390
Founded: 1887; Members: 16000
Pres: Colin Campbell; Gen Secr: Roger Blais
Focus: Eng

Periodicals
Canadian Electrical and Computer Engineering Journal (weekly)
Canadian Geotechnical Journal (weekly)
Canadian Journal of Civil Engineering (weekly)
Transactions of the Canadian Society for Mechanical Engineering (weekly) 02435

Entomological Society of British Columbia, c/o Biology Dept, University of Victoria, Victoria, B.C. V8W 2Y2
T: (604) 721-7102; Fax: (604) 721-7102
Founded: 1901; Members: 165
Focus: Entomology
Periodicals
Economic Entomology (weekly)
General Entomology (weekly)
Systematic Entomology (weekly) 02436

Entomological Society of Canada / Société Entomologique du Canada, 393 Winston Av, Ottawa, Ont. K2A 1Y8
T: (613) 725-2619; Fax: (613) 725-9349
Founded: 1863; Members: 800
Pres: Richard Ring; Gen Secr: R.J. West
Focus: Entomology
Periodicals
Bulletin (weekly)
Canadian Entomologist (weekly)
Memoirs (weekly) 02437

Entomological Society of Manitoba, c/o Agriculture Canada Research Station, 195 Dafoe Rd, Winnipeg, Man. R3T 2M9
Founded: 1944; Members: 150
Gen Secr: N.D.G. White
Focus: Entomology
Periodical
Proceedings of the Society (weekly) . . 02438

Entomological Society of Ontario, c/o Dept of Environmental Biology, University of Guelph, Guelph, Ont. N1G 2W1
T: (519) 824-4120
Founded: 1863; Members: 298
Pres: Dr. H. Goulet
Focus: Entomology
Periodical
Proceedings of the Entomological Society of Ontario (weekly) 02439

Entomological Society of Saskatchewan, c/o P.F.R.A. Tree Nursery, Indian Head, Sask. S0G 2K0
Founded: 1952; Members: 44
Gen Secr: Bruce Neill
Focus: Entomology
Periodicals
Annual Proceedings
Newsletter (weekly) 02440

Epilepsy Prince Edward Island, R.R. 1, Richmond, P.E.I. C0B 1Y0
Gen Secr: Dorothy Farish
Focus: Pathology 02441

Estonian Jurists' Association, 619 Bob O. Link Rd, Mississauga, Ont. L5J 2P5
T: (416) 823-2536
Pres: E.F. Tikenberg
Focus: Law 02442

Estonian Teachers' League, 105 Plateau Cr, Don Mills, Ont. M3C 1M9
T: (416) 447-7610
Founded: 1951; Members: 52
Focus: Educ; Ling 02443

E.W. Bickle Foundation, First Canadian Pl, POB 90, Toronto, Ont. M5X 1C5
Tx: (416) 864-3328
Founded: 1959; Members: 5
Gen Secr: E.J. Williams
Focus: Sci 02444

Expanded Polystyrene Association of Canada, 1262 Don Mills Rd, Don Mills, Ont. M3B 2W7
T: (416) 449-3444
Focus: Materials Sci 02445

Experimental Aircraft Association of Canada, POB 18, Mount Albert, Ont. L0G 1M0
Members: 1200
Pres: Lawrence Shaw
Focus: Aero 02446

Family Planning Resource Team, POB 1598, Sydney, N.S. B1P 6R8
T: (902) 539-5158
Focus: Family Plan 02447

Fédération des Affaires Sociales / Social Affairs Federation, 1601 Av De Lorimier, Montréal, Qué. H2K 4M5
T: (514) 598-2210
Members: 70000
Focus: Soc Sci; Sociology 02448

Fédération des Associations de Professeurs des Universités du Québec, 2715 Chemin de la Côte Sainte-Cathérine, Montréal, Qué. H3T 1B6
T: (514) 735-3654
Founded: 1969; Members: 5000
Gen Secr: Hubert Stéphenne
Focus: Sci 02449

Fédération des Collèges d'Enseignement Général et Professionel, 500 Blvd Crémazie Est, Montréal, Qué. H2P 1E7
T: (514) 381-8631; Fax: (514) 381-2263
Founded: 1969; Members: 88
Pres: Yvon Beaulieu; Gen Secr: Gaetan Boucher
Focus: Educ
Periodicals
Annuaire des Cégeps (weekly)
Cégepropos (weekly) 02450

Fédération des Commissions Scolaires Catholiques du Québec, 1001 Av Bégon, CP 490, Sainte Foy, Qué. G1V 4C7
T: (418) 651-3220
Gen Secr: Jacques Audy
Focus: Educ; Public Admin 02451

Fédération des Professionnelles et Professionnels de Cégeps et de Collèges, 8225 Blvd Saint-Laurent, Montréal, Qué. H2P 2M1
T: (514) 382-7670
Founded: 1974; Members: 440
Pres: Carmen Bourque
Focus: Sci 02452

Fédération du Québec pour le Planning des Naissances / Quebec Family Planning Federation, 3826 Rue Saint-Hubert, Montréal, Qué. H2L 4A5
T: (514) 842-9501
Founded: 1972
Focus: Family Plan 02453

Fédération Nationale des Enseignants et des Enseignantes du Québec, 1601 Rue De Lorimier, Montréal, Qué. H2K 4M5
T: (514) 598-2241
Founded: 1972; Members: 12000
Gen Secr: Denis Choinière
Focus: Educ 02454

Federation of English Speaking Catholic Teachers, 1857 de Maisonneuve W, Montréal, Qué. H3H 1J9
T: (514) 935-8571
Founded: 1943
Gen Secr: Leo Fernandes
Focus: Educ 02455

Federation of Independent School Associations in British Columbia, 150 Robson St, Vancouver, B.C. V6B 2A7
T: (604) 684-6023, 684-7846
Founded: 1966; Members: 4
Gen Secr: Garry Ensing
Focus: Educ 02456

Federation of Independent Schools in Canada / Fédération des Ecoles Indépendantes du Canada, 10766 97 St, Edmonton, Alta. T5H 2M1
T: (403) 424-7273
Founded: 1979
Focus: Educ 02457

Federation of Medical Women of Canada / Fédération des Femmes Médecins du Canada, 1815 Alta Vista Dr, Ste 201A, Ottawa, Ont. K1G 3Y6
T: (613) 731-1026
Founded: 1924; Members: 600
Focus: Med 02458

Federation of New Brunswick Faculty Associations / Fédération des Associations de Professeurs d'Université du Nouveau-Brunswick, 65 Brunswick, Ste 202, Fredericton, N.B. E3B 1G5
T: (506) 458-8977
Founded: 1969
Focus: Sci 02459

Federation of Women Teachers' Associations of Ontario, 1260 Bay St, Toronto, Ont. M5R 2B8
T: (416) 964-1232
Founded: 1918; Members: 30000
Focus: Educ 02460

Fédération Québécoise des Directeurs et Directrices d'Ecole, 7855 L.-H. Lafontaine, Ste 100, Anjou, Qué. H1K 4E4
T: (514) 353-7511; Fax: (514) 353-2064
Founded: 1961; Members: 3525
Pres: Jean-Marc Mathieu; Gen Secr: Micheline Lepage
Focus: Educ; Public Admin
Periodical
Information (5 times annually) 02461

Fire Fighters Burn Treatment Society, 11305 95 St, Edmonton, Alta. T5G 1L2
T: (403) 471-1220
Focus: Safety 02462

Fire Prevention Canada Association / Association Canadienne de Prévention des Incendies, 7 Liverpool Ct, Ste 1590, Ottawa, Ont. K1B 4L2
T: (613) 749-8200
Founded: 1956; Members: 1000
Focus: Safety 02463

Fish and Seafood Association of Ontario, 85 The E Mall, Ste 208, Toronto, Ont. M8Z 5W4
T: (416) 252-7185
Focus: Fisheries; Food 02464

Fisheries Association of British Columbia, 100 Pender St W, Ste 400, Vancouver, B.C. V6B 1R8
T: (604) 684-6454; Tx: 04508441
Focus: Fisheries 02465

Fondation du Québec des Maladies du Coeur / Quebec Heart Foundation, 440 Blvd Dorchester Ouest, Ste 1401, Montréal, Qué. H2Z 1V7
T: (514) 871-1551
Founded: 1955
Gen Secr: Jacqueline T. Ostiguy
Focus: Cardiol 02466

Fondation J. Armand Bombardier / J. Armand Bombardier Foundation, 1000 Rue J.A. Bombardier, CP 370, Valcourt, Qué. J0E 2L0
T: (514) 532-2258
Founded: 1965
Gen Secr: Camille Rouillard
Focus: Sci 02467

Fondation Justine Lacoste-Beaubien, 3175 Chemin de la Côte Sainte-Catherine, Montréal, Qué. H3T 1C5
T: (514) 731-4931
Focus: Sci 02468

Fondation Lionel Groulx, 257-261 Av Bloomfield, Outremont, Qué. H2V 3R6
T: (514) 271-4759
Founded: 1956
Focus: Sci 02469

Food Products Accident Prevention Association, 2 Bloor St E, Toronto, Ont. M4W 3C2
T: (416) 965-8888
Gen Secr: J.V. Findlay
Focus: Safety; Food 02470

Forest Products Accident Prevention Association / Association de Prévention des Accidents dans l'Industrie Forestière, POB 270, North Bay, Ont. P1B 8H2
T: (705) 472-4120; Fax: (705) 412-0207
Founded: 1915; Members: 1800
Focus: Forestry; Safety 02471

Forestry Canada – Ontario Region, c/o Great Lakes Forestry Centre, POB 490, Sault Sainte Marie, Ont. P6A 5M7
T: (705) 949-9461; Fax: (705) 759-5700
Founded: 1948
Focus: Forestry
Periodicals
Forestry Newsletter (weekly)
Program Review (weekly)
Survey Bulletin (3 times annually) . . . 02472

Foundation for Independent Research on Technology and Health, R.R. 1, Goodwood, Ont. L0C 1A0
T: (416) 294-3531
Founded: 1983; Members: 2
Gen Secr: Bruce M. Small
Focus: Eng; Public Health 02473

Foundation of the Canadian College of Health Service Executives, 17 York St, Ste 201, Ottawa, Ont. K1N 5S7
T: (613) 235-7218
Founded: 1972
Pres: Harvey H. Fox; Gen Secr: Richard G. Stock
Focus: Public Health
Periodical
Healthcare Management: Gestion des Soins de Santé (weekly) 02474

Fraser Institute, 626 Bute St, Vancouver B.C. V6E 3M1
T: (604) 688-0221; Fax: (604) 688-8539
Founded: 1974; Members: 1200
Gen Secr: Michael Walker
Focus: Sci
Periodicals
Fraser Forum (weekly)
On Balance (weekly) 02475

G.A. Frecker Association on Gerontology, c/o Memorial University of Newfoundland, St. John's, Nfld. A1B 3X9
T: (709) 737-7694
Founded: 1980; Members: 33
Pres: A. Kozma
Focus: Geriatrics 02476

Gairdner Foundation, 255 Yorkland Blvd, Ste 220, Willowdale, Ont. M2J 1S3
T: (416) 493-3101; CA: GAIRFOUND; Fax: (416) 493-8158
Founded: 1957
Gen Secr: Sally-Anne Hrica
Focus: Sci 02477

Garrod Association of Canada, c/o I.W. Killam Hospital for Children, 5850 University Av, POB 3070, Halifax, N.S. B3J 3G9
Focus: Pediatrics 02478

Genealogical Association of Nova Scotia, POB 3144, South Postal Station, Halifax, N.S. B3J 3G6
Founded: 1982; Members: 1135
Pres: Allan E. Marble
Focus: Genealogy
Periodical
The Nova Scotia Genealogist (3 times annually) 02479

Genealogical Institute of the Maritimes, POB 3142, South Postal Station, Halifax, N.S. B3H 3J6
T: (902) 424-6065
Founded: 1983; Members: 97
Pres: W.M. Glen; Gen Secr: G.D. Judge
Focus: Genealogy
Periodical
Annual Report (weekly) 02480

Genetics Society of Canada / Société de Génétique du Canada, 151 Slater St, Ste 907, Ottawa, Ont. K1P 5H4
T: (613) 232-9459; Fax: (613) 594-5190
Founded: 1956; Members: 565
Pres: Dr. John Heddle; Gen Secr: Dr. Graham Scoles
Focus: Genetics
Periodical
Bulletin (weekly) 02481

Geological Association of Canada / Association Géologique du Canada, c/o Dept of Earth Sciences, Memorial University of Newfoundland, Saint John's, Nfld. A1B 3X5
T: (709) 737-7660; Fax: (709) 737-2589
Founded: 1947; Members: 3000
Pres: Ian McHeath; Gen Secr: Elliott Burden
Focus: Geology
Periodical
Geoscience Canada (weekly) 02482

German-Canadian Historical Association, 4850 Côte-des-Neiges, Ste 1612, Montréal, Qué. H3V 1G5
T: (514) 737-5515
Founded: 1975; Members: 100
Gen Secr: Karin R. Gurttler
Focus: Hist
Periodical
Canadiana Germanica (weekly) 02483

Gerontology Association of Nova Scotia, 50 Pleasant St, POB 1312, Wolfville, N.S. B0P 1X0
Focus: Geriatrics 02484

G.K. Chesterton Society, 1437 College Dr, Saskatoon, Sask. S7N 0W6
T: (306) 966-9862
Focus: Lit 02485

Glenbow Alberta Institute, 130 Ninth Av SE, Calgary, Alta. T2G oP3
T: (403) 264-8300
Founded: 1955
Focus: Hist; Arts; Ethnology
Periodical
Glenbow (weekly) 02486

Goodwin's Foundation, POB 1043, Station B, Ottawa, Ont. K1P 5R1
T: (613) 237-1590
Founded: 1982; Members: 100
Focus: Sci 02487

Grain, Feed and Fertilizer Accident Prevention Association, 2 Bloor St E, Toronto, Ont. M4W 3C2
T: (416) 965-8888
Gen Secr: J.V. Findlay
Focus: Safety; Nutrition 02488

Grape and Wine Institute of British Columbia, POB 879, Station A, Kelowna, B.C. V1Y 7P5
T: (604) 642-1228
Pres: Fred Quance
Focus: Wines 02489

Greek Teachers Association, 30 Thorncliffe Park Dr, Toronto, Ont. M4H 1H8
T: (416) 421-5491
Focus: Educ 02490

Hamber Foundation, c/o Personal Trust Dept, The Canada Trust Company, POB 49390, Vancouver, B.C. V7X 1P3
Founded: 1965
Focus: Sci 02491

Hamilton Academy of Medicine, 286 N Victoria Av, Hamilton, Ont. L8L 5G4
T: (416) 528-1611
Founded: 1931
Focus: Med 02492

Hamilton Foundation, 120 King St W, Ste 205, Hamilton, Ont. L8P 4V2
T: (416) 523-5600
Founded: 1954; Members: 12
Gen Secr: Judith McCulloch
Focus: Sci 02493

Hamilton-Wentworth Lung Association, 570 Upper James St, Hamilton, Ont. L9C 2Y6
T: (416) 383-1608
Gen Secr: C. Pengelly
Focus: Pulmon Dis 02494

Harold Crabtree Foundation, 130 Albert St, Ste 2005, Ottawa, Ont. K1P 5G4
Focus: Med; Educ 02495

Health Labour Relations Association of British Columbia, 1212 Broadway W, Ste 500, Vancouver, B.C. V6H 3V1
T: (604) 736-8221; Tx: 0454261
Founded: 1975; Members: 140
Focus: Public Health 02496

Health Libraries Association of British Columbia (HLABC), c/o British Columbia Medical Service Library, 1807 Tenth Av W, Vancouver, B.C. V6J 1A9
T: (604) 736-5551
Founded: 1978
Focus: Libraries & Bk Sci; Public Health
Periodical
HLABC Forum (weekly) 02497

Heart and Stroke Foundation of Alberta, 1825 Park Rd SE, Calgary, Alta. T2G 3Y6
T: (403) 264-5549; Fax: (403) 237-0803
Founded: 1956; Members: 150
Gen Secr: John L. Paquet
Focus: Cardiol 02498

Heraldry Society of Canada / Société Héraldique du Canada, POB 8467, Station T, Ottawa, Ont. K1G 3H9
T: (416) 731-0867
Founded: 1966; Members: 600
Pres: Ian Campbell; Gen Secr: J.M. Matheson
Focus: Genealogy
Periodical
Heraldry in Canada (5 times annually) . 02499

H.G. Bertram Foundation, c/o Royal Trust Co, POB 980, Hamilton, Ont. L8N 3R2
T: (416) 522-8692
Focus: Sci 02500

Historical and Scientific Society of Manitoba, 190 Rupert Av, Winnipeg, Man. R3B 0N2
T: (204) 947-0559
Founded: 1879; Members: 750
Gen Secr: M.A. Jones
Focus: Hist; Sci 02501

Historical Society of Alberta, POB 4035, Station C, Calgary, Alta. T2T 5M9
T: (403) 264-8300
Founded: 1907; Members: 2450
Pres: Donna Coulter
Focus: Hist 02502

Historical Society of Mecklenburg Upper Canada, POB 406, Station K, Toronto, Ont. M4P 2G7
Focus: Hist 02503

Historical Society of the Gatineau, c/o Sheila Strang, Old Chelsea, Qué. J0X 2NO
Focus: Hist
Periodical
Up the Gatineau! (weekly) 02504

Hospital for Sick Children Foundation, 555 University Av, Ste 1218, Toronto, Ont. M5G 1X8
T: (416) 598-6166
Founded: 1972; Members: 330
Pres: Claus A Wirsig
Focus: Pediatrics 02505

Hospital Medical Records Institute, 250 Ferrand Dr, POB 3900, Don Mills, Ont. M3C 2T9
T: (416) 429-0464
Founded: 1963; Members: 650
Pres: R. Yungblut
Focus: Public Health 02506

Housing and Urban Development Association of Nova Scotia, c/o Veinotte Construction Ltd, Sunnybrae Av, Halifax, N.S. B3N 2E8
T: (902) 443-4214
Gen Secr: Edgar Veinotte
Focus: Urban Plan; Civil Eng 02507

Human Factors Association of Canada / Association Canadienne d'Ergonomie, 6519B Mississauga Rd, Mississauga, Ont. L5N 1A6
T: (416) 567-7193; Fax: (416) 567-7191
Founded: 1968; Members: 400
Gen Secr: Peter Fletcher
Focus: Physiology; Eng; Psych
Periodicals
Annual Conference Proceedings (weekly)
Communiqué (weekly) 02508

Human Nutrition Research Council of Ontario, POB 38, Stittsville, Ont. K0A 3G0
T: (613) 836-1125
Founded: 1980; Members: 12
Gen Secr: Dr. T. Keith Murray
Focus: Nutrition 02509

Hungarian Canadian Engineers' Association / Association des Ingénieurs Hongrois Canadiens, 31 Greengate Rd, Don Mills, Ont. M3B 1E7
T: (416) 447-5414; Fax: (416) 941-2807
Founded: 1953; Members: 400
Focus: Eng 02510

Huntington Society of Canada / Société Huntington du Canada, POB 333, Cambridge, Ont. N1R 5T8
T: (519) 622-1002
Founded: 1973; Members: 4500
Gen Secr: Roy H. Walker
Focus: Geography 02511

ICOM Museums/Musées Canada, 280 Metcalfe St, Ste 202, Ottawa, Ont. K2P 1R7
T: (613) 233-5653
Founded: 1975; Members: 275
Pres: Marie Couturier
Focus: Arts 02512

Independent Petroleum Association of Canada (IPAC), 707 Seventh Av SW, Ste 700, Calgary, Alta. T2P 0Z2
T: (403) 290-1530; Fax: (403) 290-1680
Founded: 1961; Members: 350
Gen Secr: Bob Reid
Focus: Petrochem
Periodical
IPAC Quarterly (weekly) 02513

Independent Schools Association of British Columbia, c/o Crofton House School, 3200 41 Av W, Vancouver, B.C. V6N 3E1
T: (604) 263-3255
Founded: 1930
Focus: Educ 02514

Indexing and Abstracting Society of Canada (IASC) / Société Canadienne pour l'Analyse de Documents, POB 744, Station F, Toronto, Ont. M4Y 2N6
Founded: 1977; Members: 200
Focus: Libraries & Bk Sci; Standards
Periodical
IASC Bulletin: Bulletin de la SCAD (weekly) 02515

Industrial Accident Prevention Association, 2 Bloor St W, Toronto, Ont. M4W 3N8
T: (416) 965-8888; Fax: (416) 963-1189
Gen Secr: R.H. Ramsay
Focus: Metallurgy
Periodical
Accident Prevention (10 times annually) . 02516

Information Resource Management Association of Canada (IRMAC), POB 5639, Station A, Toronto, Ont. M5W 1N8
Founded: 1971; Members: 250
Focus: Computer & Info Sci 02517

Innovation Management Institute of Canada / Institut Canadien de Gestion de l'Innovation, POB 6291, Station J, Ottawa, Ont. K2A 1T4
T: (613) 224-2940
Founded: 1975; Members: 90
Focus: Business Admin 02518

Institut Canadien d'Education des Adultes, 506 Rue Sainte-Catherine Est, Ste 800, Montréal, Qué. H2L 2C7
T: (514) 842-2766
Founded: 1956
Focus: Adult Educ 02519

Institut Canadien de Québec, 350 Saint-Joseph Est, Québec, Qué. G1K 3B2
T: (418) 529-0924
Founded: 1848; Members: 42
Gen Secr: Philippe Sauvageau
Focus: Sci 02520

Institut de Recherches Psychologiques (IRP) / Institute of Psychological Research, 34 Rue Fleury Ouest, Montréal, Qué. H3L 1S9
T: (514) 382-3000; Tx: 05826564
Founded: 1958
Focus: Psych
Periodical
Bulletin IRP (weekly) 02521

Institut d'Histoire de l'Amérique Française, 261 Bloomfield Av, Montréal, Qué. H2V 3R6
T: (514) 278-2232
Founded: 1947; Members: 1230
Pres: Jean Roy; Gen Secr: Sylvie Dépatie
Focus: Hist
Periodical
Revue d'Histoire de l'Amérique Française (weekly) 02522

Institut d'Horlogerie du Canada / Watchmaking Institute of Canada, 1012 Mount-Royal Est, Ste 107, Montréal, Qué. H2J 1X6
T: (514) 523-7623
Founded: 1951
Focus: Astronomy; Eng 02523

Institute for Aerospace Studies, 4925 Dufferin St, Downsview, Ont. M3H 5T6
Founded: 1949
Gen Secr: Prof. R.C. Tennyson
Focus: Aero
Periodicals
UTIAS Annual Report (weekly)
UTIAS Reports (weekly)
UTIAS Reviews (weekly)
UTIAS Technical Notes (weekly) 02524

Institute for Research on Public Policy / Institut de Recherches Politiques, POB 3670 South, Halifax, N.S. B3J 3K6
T: (902) 424-3801
Founded: 1972; Members: 71
Focus: Poli Sci
Periodicals
The Institute
Policy Options (weekly) 02525

Institute for the Encyclopedia of U.R.A.M., c/o Regis College, 15 Saint Mary St, Toronto, Ont. M4Y 2R5
T: (416) 922-2476
Founded: 1970; Members: 500
Gen Secr: Prof. Tibor Horvath
Focus: Humanities
Periodicals
Newsletter of the Institute for URAM (weekly)
Ultimate Reality and Meaning: Interdisciplinary Studies in the Philosophy (weekly) . . . 02526

Institute of Canadian Advertising / Institut de la Publicité Canadienne, 8 King St E, Ste 401, Toronto, Ont. M5C 1B5
T: (416) 368-2981
Founded: 1923; Members: 64
Focus: Advert 02527

Institute of Chartered Accountants of Ontario / Institut des Comptables Agréés de l'Ontario, 69 E Bloor St E, Toronto, Ont. M4W 1B3
T: (416) 962-1841
Founded: 1879; Members: 23000
Focus: Finance
Periodical
Check Mark 02528

Institute of Electrical and Electronics Engineers, 7061 Yonge St, Thornhill, Ont. L3T 2A6
T: (416) 881-1930; Fax: (416) 881-2057
Pres: Pamela E. Woodrow
Focus: Electric Eng; Electronic Eng . . 02529

Institute of Professional Libraries of Ontario, 36B Prince Arthur Av, Toronto, Ont. M5R 1A9
Focus: Libraries & Bk Sci 02530

Institute of Public Administration of Canada (IPAC) / Institut d'Administration Publique du Canada, 897 Bay St, Toronto, Ont. M5S 1Z7
T: (416) 923-7319; Fax: (416) 923-8994
Founded: 1947; Members: 3300
Focus: Public Admin
Periodical
IPAC Bulletin (weekly) 02531

Institute of Textile Science, c/o Dominion Textile Inc, POB 6250, Montréal, Qué. H3C 3L1
T: (514) 989-6415
Founded: 1956; Members: 150
Gen Secr: Arthur S. Klimes
Focus: Textiles
Periodical
Canadian Textile Journal (weekly) . . . 02532

Institut Militaire de Québec, CP 843, Haute-Ville, Québec, Qué. G1R 3R2
T: (418) 694-3667
Founded: 1928; Members: 250
Gen Secr: Georges H. Cloutier
Focus: Military Sci 02533

Insurers' Advisory Organization (IAO), 180 Dundas St W, Toronto, Ont. M5G 1Z9
T: (416) 597-1200; Fax: (416) 597-2180
Founded: 1883; Members: 65
Focus: Insurance
Periodicals
The IAO Inspector (weekly)
The Quarterly Report (weekly) 02534

Inter-American Commercial Arbitration Commission (Canadian Section) / Commission Interaméricaine d'Arbitrage Commercial (Section Canadienne), c/o Institut de Coopération Internationale, Université d'Ottawa, Ottawa, Ont. K1N 6N5
T: (613) 2321476; Tx: 0533338
Founded: 1934; Members: 35
Focus: Law 02535

Interior Forest Labour Relations Association, 203-2350 Hunter Rd, Kelowna, B.C. V1X 6C1
T: (604) 860-3592; Tx: 0485273
Founded: 1958
Pres: J.C. Todman
Focus: Forestry 02536

International Agricultural Exchange (Saskatchewan), Site 4, R.R. 1, POB 5, Maidstone, Sask. S0M 1M0
T: (306) 893-2764
Founded: 1963
Gen Secr: Anne Lundquist
Focus: Agri 02537

International Arthurian Society – North American Branch (IAS), c/o Dept of French, Dalhousie University, Halifax, N.S. B3H 3J5
T: (902) 424-2430
Founded: 1948; Members: 1200
Focus: Lit 02538

International Association for Cross-Cultural Psychology (IACCP), Psychology Dept, Queen's University, Kingston, Ont. K7L 3N6
T: (613) 547-2697
Founded: 1972; Members: 435
Pres: J.W. Berry
Focus: Psych
Periodicals
Bulletin (weekly)
Journal of Cross-Cultural Psychology (weekly) 02539

International Association of Botanical Gardens (IABG), c/o University of British Columbia, 6501 NW Marine Dr, Edinburgh, B.C. V6T IW5
T: (604) 228-3928
Founded: 1954; Members: 209
Focus: Botany 02540

International Association of Group Psychotherapy (IAGP) / Association Internationale de Psychothérapie de Groupe, c/o Dr. Fern Cramer-Azima, McGill University, 3666 McTavish St, Montréal, Qué. H3A 1Y2
Pres: Dr. Fern Cramer-Azima
Focus: Psych; Therapeutics 02541

International Association of Master Penmen and Teachers of Handwriting (IAMPTH), 34 Broadway Av, Ottawa, Ont. K1S 2V6
T: (613) 232-3014
Founded: 1950; Members: 200
Gen Secr: Fred D. Richardson
Focus: Educ
Periodical
Penmens News Letter (weekly) 02542

International Center for Comparative Criminology (ICCC) / Centre International de Criminologie Comparée, 3150 Jean-Brillant, Pavillon Lionel-Groulx, CP 6128, Succ A, Montréal, Qué. H3C 3J7
T: (514) 343-7065; CA: CRIMCOMPA; Fax: (514) 343-2269
Founded: 1969; Members: 20
Gen Secr: Jean-Paul Brodeur
Focus: Criminology
Periodical
Criminologie (weekly) 02543

International Center for Research on Bilingualism / Centre International de Recherche sur le Bilinguisme, Université Laval, Cité Universitaire, Sainte Foy, Qué. G1K 7P4
T: (418) 656-3232
Founded: 1967; Members: 160
Gen Secr: Lorne Laforge
Focus: Ling 02544

International Childbirth Education Association / Association Internationale d'Education Périnatale, 8207 Tenth St SW, Calgary, Alta. T2V 1M7
T: (403) 252-4683
Members: 1000
Gen Secr: Elaine Montgomery
Focus: Family Plan 02545

International Congress of University Adult Education (ICUAE), c/o Dr. John F. Morris, Dept of Extension and Summer Session, University of New Brunswick, POB 4400, Fredericton, N.B. E3B 5A3
T: (506) 453-4646; Fax: (506) 453-4599
Founded: 1960; Members: 300
Pres: Dr. Rex Nettele Ford; Gen Secr: Dr. John F. Morris
Focus: Adult Educ
Periodical
International Journal of University Adult Education (3 times annually) 02546

International Cooperation in History of Technology Committee (ICOHTEC), 120 Glen Rd, Toronto, Ont. M4W 2W2
Founded: 1968; Members: 116
Pres: Prof. C. Maccagni
Focus: Eng; Cultur Hist 02547

International Council for Adult Education / Conseil International d'Education des Adultes, 720 Bathurst Av, Ste 500, Toronto, Ont. M5S 2RY
T: (416) 588-1211; Tx: 06986766; Fax: (416) 588-5725
Founded: 1973
Focus: Adult Educ
Periodical
News from the Secretariat (weekly) . . 02548

International Council for Distance Education (ICDE), c/o Concordia University, 1455 de Maisonneuve W, Montréal, Qué. H3G IM8S
Founded: 1938; Members: 500
Focus: Educ
Periodical
Bulletin of the ICDE (3 times annually) . 02549

International Development Education Committee of Ontario / Comité du Développement de l'Éducation Internationale de l'Ontario, 252 Bloor St W, Ste 6298, Toronto, Ont. M5S 1V6
T: (416) 923-6641
Founded: 1975
Focus: Educ 02550

International Development Education Resources Association, 2524 Cypress St, Vancouver, B.C.
T: (604) 732-1496
Founded: 1974; Members: 350
Gen Secr: Pat Dalgleish
Focus: Educ
Periodicals
IDERA Calendor of Events (weekly)
IDERA Clipping Service (weekly) 02551

International Development Research Centre / Centre de Recherches pour le Développement International, 250 Albert St, POB 8500, Ottawa, Ont. K1G 3H9
T: (613) 236-6163; Tx: 0533753
Founded: 1970
Focus: Food; Agri; Nutrition; Public Health; Soc Sci; Computer & Info Sci
Periodicals
Annual Report (weekly)
Busqueda (weekly)
Quête d'avenirs (weekly)
Searching (weekly) 02552

International Federation for Medical and Biological Engineering, c/o John A. Hopps, DSC-Medical Engineering Section, National Research Council, Bldg M-50, Ste 164, Ottawa, Ont. K1A OR8
T: (613) 939-0287
Members: 13000
Gen Secr: Dr. J.A. Hopps
Focus: Eng; Med; Bio
Periodical
Journal (weekly) 02553

International Federation on Ageing, 380 Saint Antoine St W, Ste 3200, Montréal, Qué. M24 3X7
T: (514) 987-8191; Fax: (514) 987-1948
Focus: Geriatrics 02554

International Organization for Septuagint and Cognate Studies (IOSCS), c/o Dept of Near Eastern Studies, University of Toronto, Toronto, Ont. M5S 1A1
T: (416) 978-6599
Founded: 1968; Members: 228
Pres: A. Pietersma; Gen Secr: L. Greenspoon
Focus: Rel & Theol
Periodicals
Bulletin (weekly)
Septuagint and Cognate Studies (SCS) (weekly) 02555

International Peat Society (Canadian National Committee) / Société Internationale de la Tourbe (Le Comité National Canadien), c/o National Research Council Canada, 1411 Oxford St, Halifax, N.S. B3H 3Z1
T: (902) 426-7532
Founded: 1968; Members: 130
Focus: Materials Sci 02556

International Society for Heart Research (ISHR), c/o Dept of Physiology, University of Manitoba, 770 Bannatyne Av, Winnipeg, Man. R3E OW3
T: (204) 786-4336
Founded: 1967; Members: 1500
Gen Secr: Dr. N.S. Dhalla
Focus: Cardiol 02557

International Society for Research in Palmistry, 351 Victoria Av, Westmount, Qué. H3Z 2N1
Focus: Parapsych 02558

International Society of Biometeorology, c/o Prof. N.N. Barthakur, Dépt des Resources Renouvelables, Université McGill, Campus MacDonald, 21111 Lakeshore Rd, Sainte Anne de Bellevue, Qué. H9X
Founded: 1956; Members: 420
Pres: Prof. R.J. Reiter; Gen Secr: Prof. N.N. Barthakur
Focus: Geophys
Periodical
International Journal of Biometeorology (weekly) 02559

International Society of Electrophysiological Kinesiology (ISEK), Dept of Psychology, Concordia University, 7141 Sherbrooke St W, Montréal, Qué. H4B 1R6
T: (514) 482-0320
Founded: 1965; Members: 500
Pres: Dr. H.W. Ladd
Focus: Physiology
Periodical
Society Newsletter (weekly) 02560

International Theatre Institute, c/o Professional Association of Canadian Theatres, 64 Charles St E, Toronto, Ont. M4Y 1T1
T: (416) 968-3033
Focus: Perf Arts 02561

International Union of Pure and Applied Chemistry (Canadian National Committee), c/o Chemistry Dept, University of Alberta, Edmonton, Alta. T6G 2G2
Focus: Chem 02562

International Union of Pure and Applied Physics (IUPAP) / Union Internationale de Physique Pure et Appliquée, c/o Pavillon Vachon, Université Laval, Québec, Qué. G1K 7P4
T: (418) 656-2528
Founded: 1922
Focus: Physics
Periodical
News-Bulletin (weekly) 02563

Inter-University Council on Academic Exchanges with the USSR and Eastern Europe / Conseil Interuniversitaire sur les Echanges Académiques avec l'URSS et l'Europe de l'Est, c/o Dept of Germanic and Slavic Languages and Literatures, University of Waterloo, Waterloo, Ont. N2L 3G1
T: (519) 885-1211 ext 2259
Focus: Sci 02564

Island Association of Rehabilitation Workshops, c/o Tremploy, 190 Fitzroy St, Charlottetown, P.E.I. C1A 1S4
T: (902) 892-5338
Focus: Rehabil 02565

Jack Miner Migratory Bird Foundation, POB 39, Kingsville, Ont. N9Y 2E8
T: (519) 733-4034
Focus: Ornithology 02566

Jane Austen Society of North America (JASNA), c/o Eileen Sutherland, 4169 Lions Av, North Vancouver, B.C. V7R 3S2
T: (604) 988-0479
Founded: 1979; Members: 2500
Pres: Eileen Sutherland
Focus: Lit 02567

Jane Austen Society of North America, 4169 Lions Av, North Vancouver, B.C. V7R 3S2
T: (604) 988-0479
Founded: 1979; Members: 2000
Focus: Lit
Periodical
Persuasions (weekly) 02568

Jeunes Biologistes du Québec, 337 Rte 138, Saint-Fidèle, Qué. G0T 1T0
T: (418) 434-2209
Founded: 1956
Gen Secr: Rosaire Corbin
Focus: Bio 02569

J.P. Bickell Foundation, c/o National Trust Co, 21 King St E, Toronto, Ont. M5C 1B3
Tx: (416) 361-4096; Fax: (416) 361-4097
Founded: 1951; Members: 5
Gen Secr: D.R. Windeyer
Focus: Sci 02570

Juvenile Diabetes Foundation, 4632 Yonge St, Ste 201, Willowdale, Ont. M2N 5M1
T: (416) 224-2633
Founded: 1970
Pres: Annette Oelbaum
Focus: Diabetes 02571

J.W. MacConnell Foundation / Fondation J.W. McConnell, 1130 Sherbrooke St W, Ste 510, Montréal, Qué. H3A 2T1
T: (514) 288-2133
Founded: 1937
Gen Secr: Jacqueline Prevost
Focus: Sci 02572

Kinsmen Rehabilitation Foundation of British Columbia, 2256 12 Av W, Ste 100, Vancouver, B.C. V6K 2N5
T: (604) 736-8841
Founded: 1952
Focus: Rehabil 02573

Lakeshore Teachers Association / Association des Enseignants du Lakeshore, 84 Brunswick Blvd, Dollard des Ormeaux, Qué. H9B 2C5
T: (514) 683-9330
Founded: 1964; Members: 700
Focus: Educ 02574

Languages of Instruction Commission of Ontario, 25 Grosvenor St, Toronto, Ont. M4Y 1A9
T: (416) 965-3155
Gen Secr: G.C. Filion
Focus: Computer & Info Sci 02575

Lapidary, Rock and Mineral Society of British Columbia, 4705 47 Av, Delta, B.C. V4K 1P5
Pres: Wayne Belzer
Focus: Mineralogy 02576

Laubach Literacy of Canada (Prince Edward Island), 65 Ambrose St, Charlottetown, P.E.I. C1A 3P8
T: (902) 894-4312
Pres: Mabel MacSwain
Focus: Lit 02577

Law Foundation of Nova Scotia, POB 325, Halifax, N.S. B3J 2N7
T: (902) 422-8335
Founded: 1976; Members: 9
Gen Secr: Mary Helleiner
Focus: Law 02578

Law Society of Manitoba, 1400-155 Carlton St, Winnipeg, Man. R3C 3H8
T: (204) 942-5571
Founded: 1877
Gen Secr: Graeme Garson
Focus: Law
Periodical
Law Society of Manitoba Communiqué (10 times annually) 02579

Law Society of Newfoundland, POB 1028, Baird's Cove, Saint John's, Nfld. A1C 5M3
T: (709) 722-4740; Fax: (709) 722-8902
Founded: 1834
Focus: Law 02580

Law Society of Saskatchewan, 201-2212 Scarth St, Regina, Sask. S4P 2J6
T: (306) 569-8242
Founded: 1907
Gen Secr: Iain A. Mentiplay
Focus: Law
Periodicals
Index of Periodicals (weekly)
Journal (3 times annually)
This Week's Law (weekly) 02581

Law Society of the Northwest Territories, POB 2594, Yellowknife, N.W.T. X0E 1H0
T: (403) 873-5716
Founded: 1978; Members: 275
Gen Secr: Gordon R. Carter
Focus: Law 02582

Law Society of Upper Canada / Société du Barreau du Haut Canada, Osgoode Hall, 130 Queen St W, Toronto, Ont. M5H 2N6
T: (416) 947-3300; Fax: (416) 947-9070
Founded: 1797; Members: 21300
Focus: Law
Periodicals
Communique (weekly)
Gazette 02583

League of Canadian Poets, 24 Ryerson Av, Toronto, Ont. M5T 2P3
T: (416) 363-5047; Fax: (416) 860-0826
Founded: 1968; Members: 310
Gen Secr: Angela Rebeiro
Focus: Lit
Periodical
Newsletter (weekly) 02584

League of Education Administrators, Directors and Superintendents, c/o Saskatchewan Valley School, POB 809, Warnen, Sask. S0K 4S0
T: (306) 933-4414
Focus: Educ
Periodical
The Leader (weekly) 02585

Learning Disabilities Association of Canada / Troubles d'Apprentissage – Association Canadienne, 323 Chapel St, Ste 200, Ottawa, Ont. K1N 7Z2
T: (613) 238-5721
Founded: 1963; Members: 10000
Focus: Educ Handic
Periodical
National (weekly) 02586

Learning Resources Council, c/o Alberta Teacher's Association, 11010 142 St, Edmonton, Alta. T5N 1H9
T: (403) 453-2411
Founded: 1975; Members: 495
Pres: Diane Oberg
Focus: Educ 02587

Leon and Thea Koerner Foundation, c/o Personal Trust Dept, The Canada Trust Company, POB 49390, Vancouver, B.C. V7X 1P3
Founded: 1955
Focus: Sci 02588

Library Association of Alberta (L.A.A.), POB 1357, Main Post Office, Edmonton, Alta. T5J 2N2
Focus: Libraries & Bk Sci
Periodical
Letter of the L.A.A. (weekly) 02589

Library Association of Ottawa-Hull / Association des Bibliothèques d'Ottawa-Hull, POB 924, Station B, Ottawa, Ont. K1P 5P9
Members: 150
Focus: Libraries & Bk Sci 02590

Library Science Alumni Association, 47 Willcocks St, Toronto, Ont. M5S 1A1
Focus: Libraries & Bk Sci 02591

Library Technicians Association of British Columbia, POB 67515, Vancouver, B.C. V5W 3T9
T: (604) 434-7554
Founded: 1974; Members: 96
Pres: Marleen Caswell
Focus: Libraries & Bk Sci 02592

Ligue de Sécurite du Québec / Quebec Safety League, 6785 Rue Saint-Jacques Ouest, Montréal, Qué. H4B 1V3
T: (514) 482-9110; Fax: (514) 482-3398
Founded: 1923; Members: 600
Focus: Safety
Periodical
Famille Avertie (weekly) 02593

The Lithuanian Folk Art Institute / Lietuviu Tautodailes Institutas, 1573 Bloor St W, Toronto, Ont. M6P 1A6
Focus: Folklore 02594

Lung Association / Association Pulmonaire, 75 Albert St, Ste 908, Ottawa, Ont. K1P 5E7
T: (613) 237-1208; Fax: (613) 563-3362
Founded: 1900
Gen Secr: Bob Mech
Focus: Pulmon Dis 02595

Lung Association, Kawartha Pine Ridge Region, 299 Aylmer St N, Peterborough, Ont. K9J 7M4
T: (705) 742-6637
Pres: Barbara J. Hutchinson
Focus: Pulmon Dis 02596

MacBride Museum Society (MMS), POB 4037, Whitehorse, Y.T. Y1A 3S9
T: (403) 667-2709
Founded: 1950; Members: 200
Pres: Robin Armour; Gen Secr: Joanne Meehan
Focus: Hist 02597

The Malraux Society (MMM), c/o Dept of Romance Languages, University of Alberta, Edmonton, Alta. T6G 2E6
T: (403) 766-4177
Founded: 1969; Members: 300
Gen Secr: Prof. Robert S. Thornberry
Focus: Lit 02598

Manitoba Archaeological Society, POB 1171, Winnipeg, Man. R3C 2Y4
T: (204) 942-1243
Founded: 1961; Members: 300
Pres: Peter O. Walker
Focus: Archeol
Periodical
Manitoba Archaeological Quarterly (weekly) 02599

Manitoba Arts Council, 93 Lombard Av, Ste 100, Winnipeg, Man. R3B 3B1
T: (204) 944-2978
Focus: Arts 02600

Manitoba Association for Art Education, c/o Pinkham School, 765 Pacific Av, Winnipeg, Man. R3E 1G1
T: (204) 786-5749
Focus: Arts; Educ 02601

Manitoba Association for Multicultural Education, 1355 Main St, Winnipeg, Man. R2W 3T7
T: (204) 586-8591
Founded: 1983
Focus: Educ 02602

Manitoba Association for the Promotion of Ancestral Languages, 167 Lombard Av, Ste 1044, Winnipeg, Man. R3B 0V3
T: (204) 943-2222
Founded: 1983
Gen Secr: Helen Fernandes
Focus: Ling
Periodical
Manitoba Heritage Review (weekly) . . . 02603

Manitoba Association of Cardiology Technologists, 540 Mountain Av, Winnipeg, Man. R2W 1K9
T: (204) 582-1110
Founded: 1963; Members: 200
Pres: Elfriede Martens
Focus: Cardiol 02604

Manitoba Association of Library Technicians, POB 1872, Winnipeg, Man. R3C 1R1
T: (204) 669-4173
Focus: Libraries & Bk Sci 02605

Manitoba Association of Mathematics Teachers, c/o Victor Wyatt School, 485 Meadowood Dr, Winnipeg, Man. R2M 5C1
T: (204) 255-7880
Focus: Educ; Math 02606

Manitoba Association of Medical Radiation Technologists, 294 Duffield St, Winnipeg, Man. R3J 2J9
T: (204) 888-6190
Gen Secr: M.A. Lunn
Focus: Nucl Med; Radiology 02607

Manitoba Association of School Trustees / Association des Commissaires d'Ecoles du Manitoba, 191 Provencher Blvd, Winnipeg, Man. R2H 0G4
T: (204) 233-1595; Fax: (204) 231-1356
Founded: 1965
Focus: Educ
Periodicals
MAST (weekly)
Newsletter (10 times annually) 02608

Manitoba Association of Social Workers, 1767 Portage Av, Ste 5, Winnipeg, Man. R3J 0E7
T: (204) 888-9477
Founded: 1961; Members: 250
Pres: Leonard E. Carlson
Focus: Sociology
Periodical
The Manitoba Social Worker (8 times annually) 02609

Manitoba Association on Gerontology, 320 Sherbrooke St, Winnipeg, Man. R3B 2W6
T: (204) 775-5754
Focus: Geriatrics 02610

Manitoba Cancer Treatment and Research Foundation, 100 Olivia St, Winnipeg, Man. R3E 0V9
T: (204) 787-2197
Gen Secr: Dr. L.G. Israels
Focus: Cell Biol & Cancer Res 02611

Manitoba Cerebral Palsy Association, 354 Greenwood Av, Winnipeg, Man. R2M 4J6
Focus: Neurology 02612

Manitoba Committee on Children and Youth, 22 Middle Gate, Winnipeg, Man. R3C 2C4
Focus: Sociology; Educ 02613

Manitoba Dental Association, 296 Kennedy St, Ste 300, Winnipeg, Man. R3B 2M6
T: (204) 942-5211
Founded: 1883; Members: 475
Gen Secr: Ross H. McIntyre
Focus: Dent 02614

Manitoba Environmental Council, 500 Portage Av, Winnipeg, Man. R3C 3X1
T: (204) 945-7031
Founded: 1972; Members: 60
Focus: Ecology
Periodical
Report (weekly) 02615

Manitoba Epilepsy Association, 825 Sherbrooke St, Winnipeg, Man. R3A 1M5
T: (204) 783-0466
Pres: David Meighen
Focus: Pathology 02616

Manitoba Genealogical Society, POB 2066, Winnipeg, Man. R3C 3R4
T: (204) 9441153
Founded: 1976; Members: 460
Pres: Ruth M. Breckman
Focus: Genealogy
Periodicals
Generations (weekly)
Newsletter (weekly) 02617

Manitoba Health Libraries Association, c/o Misericordia General Hospital, 99 Cornish Av, Winnipeg, Man. R3C 1A2
T: (204) 786-8109
Pres: Sharon Allentuck
Focus: Libraries & Bk Sci; Public Health 02618

Manitoba Health Records Association, 961 Dudley Av, Winnipeg, Man. R3M 1R3
Members: 161
Pres: Evelyn Fondse
Focus: Public Health 02619

Manitoba Heart Foundation, 352 Donald St, Ste 301, Winnipeg, Man. R3B 2H8
T: (204) 942-0195
Founded: 1957; Members: 200
Gen Secr: Douglas I. Watson
Focus: Cardiol 02620

Manitoba Historical Society, 190 Rupert Av, Ste 211, Winnipeg, Man. R3B 0N2
T: (204) 947-0559
Focus: Hist 02621

Manitoba Indian Education Association, 294 Portage Av, Ste 301, Winnipeg, Man. R3C 0B9
T: (204) 947-0421
Focus: Educ 02622

Manitoba Institute of Agrologists, 1770 Ness Av, Winnipeg, Man. R3J 0Y1
T: (204) 889-7225
Founded: 1950; Members: 602
Focus: Agri 02623

Manitoba Institute of Registered Social Workers, 476 Clifton St, Winnipeg, Man. R3G 2X2
T: (204) 772-6605
Focus: Sociology 02624

Manitoba Library Association, c/o Winnipeg Centennial Library, 251 Donald St, Winnipeg, Man. R3C 3P5
Founded: 1936; Members: 322
Pres: Don Mills
Focus: Libraries & Bk Sci 02625

Manitoba Library Trustee Association, POB 1330, Neepawa, Man. R0J 1H0
T: (204) 269-7437
Founded: 1970
Pres: Christine Heywood
Focus: Libraries & Bk Sci 02626

Manitoba Lung Association, 629 McDermot Av, Winnipeg, Man. R3A 5P6
T: (204) 774-5501
Members: 51000
Gen Secr: R.F. Marks
Focus: Pulmon Dis 02627

Manitoba Medical Association, 125 Sherbrooke St, Winnipeg, Man. R3C 2B5
T: (204) 786-7565
Founded: 1908; Members: 1600
Gen Secr: John A. Laplume
Focus: Med 02628

Manitoba Naturalists Society (MNS), 302-128 James Av, Winnipeg, Man. R3B 0N8
T: (204) 943-9029
Founded: 1920; Members: 2400
Pres: Alison Elliott; Gen Secr: Elaine Taylor Turchyn
Focus: Nat Sci
Periodical
Bulletin (weekly) 02629

Manitoba Paramedical Association, 216-819 Sargent Av, Winnipeg, Man. R3E 0B9
T: (204) 775-2139
Founded: 1970; Members: 800
Gen Secr: Ron Wally
Focus: Public Health 02630

Manitoba Safety Council, 213 Notre Dame Av, Ste 202, Winnipeg, Man. R3B 1N3
T: (204) 949-1085
Founded: 1972; Members: 204
Gen Secr: D.A.J. Herbert
Focus: Safety 02631

Manitoba School Library Audio-Visual Association, c/o Manitoba Teachers Society, 191 Harcourt St, Winnipeg, Man. R3J 3H2
T: (204) 888-7961
Focus: Libraries & Bk Sci; Educ . . . 02632

Manitoba Social Science Teachers' Association, 1016 Buchanan Blvd, Ste 5, Winnipeg, Man. R2Y 1N5
T: (204) 888-3435
Members: 550
Pres: Dave Osborne
Focus: Educ; Soc Sci 02633

Manitoba Society for Training and Development, 155 Carlton St, POB 29, Winnipeg, Man. R3C 3H8
Members: 200
Focus: Educ 02634

Manitoba Society of Criminology / Société de Criminologie du Manitoba, POB 397, Winnipeg, Man. R3C 2H&
T: (204) 986-6391
Founded: 1969; Members: 210
Pres: Don Peters
Focus: Criminology 02635

Manitoba Society of Occupational Therapists (MSOT), 320 Sherbrooke St, Winnipeg, Man. R3B 2W6
T: (204) 775-5754; Fax: (204) 783-1183
Founded: 1964; Members: 200
Pres: Laurie Ringaert; Gen Secr: Karen Burns
Focus: Therapeutics
Periodical
Update (10 times annually) 02636

Manitoba Speech and Hearing Association, 285 Pembina Hwy, Ste 313, Winnipeg, Man. R3L 2E1
T: (204) 453-4539
Founded: 1961
Gen Secr: Betty Cote
Focus: Logopedy
Periodical
Hearsay (5 times annually) 02637

Manitoba Teachers' Society, 191 Harcourt St, Winnipeg, Man. R3J 3H2
T: (204) 888-7961
Founded: 1919; Members: 12500
Gen Secr: W.J. Pindera
Focus: Educ 02638

Manitoba Veterinary Medical Association, AC Services Complex, 545 University Cr, Winnipeg, Man. R3T 2N2
T: (204) 269-0625
Founded: 1890; Members: 235
Pres: Paul Schneider
Focus: Vet Med 02639

Maritime Provinces Higher Education Commission / Commission de l'Enseignement Supérieur des Provinces Maritimes, 450 Kings Pl, POB 6000, Fredericton, N.B. E3B 5H1
T: (506) 453-2844; Fax: (506) 453-2106
Founded: 1974; Members: 19
Focus: Educ
Periodicals
Annual Report (weekly)
Financial Plan (weekly)
Post-Secondary Programme Profile (weekly)
. . . . 02640

Max Bell Foundation, Toronto Dominion Centre, POB 122, Toronto, Ont. M5K 1H1
T: (416) 364-2814
Founded: 1965; Members: 6
Pres: George Gardiner
Focus: Sci 02641

Mechanics' Institute of Montréal, 1200 Atwater Av, Montréal, Qué. H3Z 1X4
T: (514) 935-7344
Founded: 1840; Members: 1200
Pres: Ralph Leavitt; Gen Secr: Sarah Jefferson
Focus: Eng
Periodical
Annual Report (weekly) 02642

Medical Council of Canada / Conseil Médical du Canada, POB 8234, Ottawa, Ont. K1G 3H7
T: (613) 521-6012; Fax: (613) 521-9417
Founded: 1912; Members: 39
Focus: Med 02643

Medical Council of New Brunswick, 10 Prince Edward St, Saint John, N.B. E2L 4M5
Pres: Dr. E.D. McCartney
Focus: Med 02644

Medical Council of Prince Edward Island, 206 Spring Park Rd, Charlottetown, P.E.I. C1A 3Y9
T: (902) 894-5316
Focus: Med 02645

Medical Research Council of Canada (MRC) / Conseil de Recherches Médicales du Canada, Jeanne Mance Bldg, Tunney's Pasture, Ottawa, Ont. K1A OW9
T: (613) 954-1814
Founded: 1960
Focus: Med
Periodicals
MRC Newsletter (weekly)
Report of the President (weekly) . . . 02646

Medical Society of Nova Scotia, 6080 Young St, Ste 305, Halifax, N.S. B3K 5L2
T: (902) 453-0205
Founded: 1861; Members: 1500
Gen Secr: D.D. Peacocke
Focus: Med 02647

Medical Society of Prince Edward Island, 279 Richmond St, Charlottetown, P.E.I. C1A 1J7
T: (902) 892-7527
Founded: 1855; Members: 159
Gen Secr: Marilyn Lowther
Focus: Med 02648

Medical Staff Council, 700 William Av, Winnipeg, Man. R3E 0Z3
Pres: Dr. W. Bebchuk
Focus: Med 02649

Medical Symposium Association of Canada, POB 3593, Station D, Edmonton, Alta. T5L 4J6
Pres: M.L. Jeppesen
Focus: Med 02650

Metropolitan Toronto Zoological Society, POB 280, West Hill, Ont. M1E 4R5
T: (416) 392-5900; Fax: (416) 392-5934
Focus: Zoology 02651

Micmac Association of Cultural Studies, POB 961, Sydney, N.S. B1P 6J4
T: (902) 539-8037, 539-8039; Tx: 019-35215
Founded: 1974; Members: 6000
Gen Secr: Peter J. Christmas
Focus: Arts; Cultur Hist 02652

Microscopical Society of Canada / Société de Microscopie du Canada, 150 College St, Rm 79, Toronto, Ont. M5S 1A8
T: (416) 978-8896
Founded: 1972; Members: 500
Focus: Optics 02653

Migraine Foundation, 390 Brunswick Av, Toronto, Ont. M5R 2Z4
T: (416) 920-4916
Founded: 1974; Members: 16000
Gen Secr: Rosemary Dudley
Focus: Sociology
Periodical
Migraine Report (weekly) 02654

Mineralogical Association of Canada, c/o Geological Society of Canada, 601 Booth St, Ottawa, Ont. K1A 0E8
Founded: 1955; Members: 2200
Focus: Mineralogy; Petrochem
Periodicals
The Canadian Mineralogist (weekly)
Newsletter (weekly) 02655

Mining Association of British Columbia, 1066 Hastings St W, Ste 480, POB 12540, Vancouver, B.C. V6E 3X1
T: (604) 681-4321; Tx: 04507784
Founded: 1921; Members: 54
Focus: Mining 02656

Mining Association of Manitoba, 155 Carlton St, Ste 1730, Winnipeg, Man. R3C 3H8
T: (204) 942-2789
Founded: 1942; Members: 12
Gen Secr: Henry Bloy
Focus: Mining 02657

Mining Society of Nova Scotia, 125 Main St, Glace Bay, N.S. B1A 4Y5
T: (902) 849-8656
Gen Secr: J. Coady Marsh
Focus: Mining 02658

Miramichi Historical Society, c/o Old Manse Library, 318 Jane St, Newcastle, N.B. E1V 2T5
T: (516) 622-3119
Founded: 1959; Members: 30
Focus: Hist 02659

Molson Family Foundation / Fondation de la Famille Molson, Pl d'Armes, CP 1600, Montréal, Qué. H2Y 3L3
T: (514) 521-1786
Founded: 1958; Members: 5
Focus: Sci 02660

Motor Vehicle Safety Association, c/o Richards Delivery Service, 7300 Kimbal St, Mississauga, Ont. L4T 3S8
T: (416) 676-1315
Founded: 1926; Members: 40
Gen Secr: Fred Ahrens
Focus: Safety 02661

M.S.I. Foundation, 10405 Jasper Av, Ste 1780, Edmonton, Alta. T5J 3N4
T: (403) 421-7532
Founded: 1970
Pres: R.W. Chapman
Focus: Sci 02662

Multicultural History Society of Ontario, 43 Queen's Park Cr E, Toronto, Ont. M5S 2C3
T: (416) 979-2973
Founded: 1976
Focus: Hist 02663

Multiple Dwelling Standards Association, 163 Beechwood Av, Willowdale, Ont. M2L 1J9
T: (416) 449-7700
Pres: Jan Schwartz
Focus: Standards 02664

Museums Association of Saskatchewan, 1808 Smith St, Regina, Sask. S4P 2N3
T: (306) 780-9279; Tx: 0712476
Founded: 1967; Members: 350
Gen Secr: Gayl Hipperson
Focus: Archives
Periodical
Liaison: Saskatchewan's Heritage Review (weekly)
. . . . 02665

Myasthenia Gravis Foundation of British Columbia, 1195 Eighth Av W, Vancouver, B.C. V6H 1C5
T: (604) 734-2221; Fax: (604) 734-3533
Founded: 1968; Members: 135
Focus: Neurology 02666

Nanaimo Neurological and Cerebral Palsy Association, 1685 Grant St, Nanaimo, B.C. V9S 2K4
T: (604) 753-0251
Gen Secr: Kay Rice
Focus: Neurology 02667

National Association for Photographic Art / Association Nationale d'Art Photographique, 22 Abbeville Rd, Scarborough, Ont. M1H 1Y3
T: (416) 438-0252
Founded: 1968; Members: 2800
Focus: Photo 02668

National Association of Women in Construction, 11508 119 St, Edmonton, Alta. T5G 2X7
T: (403) 454-6585
Focus: Eng 02669

National Audio Visual Association of Canada, 77 Mowat Av, Ste 209, Toronto, Ont. M6K 3E3
T: (416) 535-9799; Fax: (416) 536-1009
Founded: 1957; Members: 30
Gen Secr: Peter Dixon
Focus: Mass Media 02670

National Dental Examining Board of Canada / Bureau National d'Examen Dentaire du Canada, 100 Bronson Av, Ste 807, Ottawa, Ont. K1R 6G8
Focus: Dent 02671

National Design Council / Conseil National de Design, 235 Queen St, Ottawa, Ont. K1A 0H5
T: (613) 992-5004
Founded: 1961; Members: 17
Focus: Arts 02672

National Research Council of Canada (NRC) / Conseil National de Recherche, Montréal Rd, Bldg M-22, Ottawa, Ont. K1A 0R6
T: (613) 993-9101
Founded: 1916; Members: 3000
Pres: Larkin Kerwin
Focus: Sci 02673

Natural History Society of Prince Edward Island, POB 2346, Charlottetown, P.E.I. C1A 8C1
T: (902) 894-9595
Founded: 1969; Members: 155
Focus: Nat Sci
Periodical
Island Naturalist (5-6 times annually) . . 02674

Natural Sciences and Engineering Research Council of Canada (NSERC) / Conseil de Recherches en Sciences Naturelles et en Génie du Canada, 200 Kent St, Ottawa, Ont. K1A 1H5
T: (613) 995-5992
Founded: 1978; Members: 22
Focus: Nat Sci; Eng
Periodicals
Contact
NSERC Awards Guide 02675

Nature Conservancy of Canada / Société pour la Conservation des Sites Naturels, 2200 Yonge St, Ste 1710, Toronto, Ont. M4S 2C6
T: (416) 486-1011
Founded: 1962
Focus: Ecology 02676

New Brunswick Association of Pathologists, c/o Saint John Regional Hospital, POB 2100, Saint John, N.B. E2L 4L2
T: (506) 648-6501
Members: 18
Pres: John S. Mackay
Focus: Pathology 02677

New Brunswick Dental Society, 79 Neck Rd, Rothesay, N.B. E0G 2W0
T: (506) 847-4243
Founded: 1890; Members: 200
Focus: Dent 02678

New Brunswick Dietetic Association, POB 944, Saint John, N.B. E2L 4E3
Focus: Food 02679

New Brunswick Federation of Naturalists / Fédération des Naturalistes du Nouveau-Brunswick, c/o New Brunswick Museum, 277 Douglas Av, Saint John, N.B. E2K 1E5
T: (506) 693-1196
Founded: 1972; Members: 350
Focus: Nat Sci
Periodical
New Brunswick Naturalist: Le Naturaliste du New Brunswick (weekly) 02680

New Brunswick Genealogical Society / Société Généalogique du Nouveau-Brunswick, POB 3235, Station B, Fredericton, N.B. E3A 5G9
Founded: 1978; Members: 500
Pres: Bing Geldart
Focus: Genealogy 02681

New Brunswick Gerontology Association, 130 University Av, Saint John, N.B. E2K 4K3
Focus: Geriatrics 02682

New Brunswick Health Records Association, c/o Medical Record Dept, Saint John Regional Hospital, POB 2100, Saint John, N.B. E2L 4L2
T: (506) 648-6000
Founded: 1963; Members: 35
Focus: Public Health 02683

New Brunswick Heart Foundation / Fondation du Nouveau-Brunswick des Maladies du Coeur, 28 Germain St, Saint John, N.B. E2L 2E5
T: (506) 657-2062
Founded: 1962
Focus: Cardiol 02684

New Brunswick Historical Society, 120 Union St, Saint John, N.B. E2L 1A3
T: (506) 652-3590
Founded: 1874; Members: 325
Gen Secr: Willard F. Merritt
Focus: Hist 02685

New Brunswick Institute of Agrologists / Institut des Agronomes du Nouveau-Brunswick, POB 3479, Postal Station B, Fredericton, N.B. E3A 5H2
T: (506) 459-7541
Founded: 1960; Members: 160
Gen Secr: R.E.M. Routledge
Focus: Agri
Periodical
Newsletter (weekly) 02686

New Brunswick Lung Association / Association Pulmonaire du Nouveau-Brunswick, 65 Brunswick St, Fredericton, N.B. E3B 1G5
T: (506) 455-8961
Gen Secr: Margaret M. Murdoch
Focus: Pulmon Dis 02687

New Brunswick Medical Society / Société Médicale du Nouveau-Brunswick, 565 Priestman Centre, Ste 209, Fredericton, N.B. E3B 5X8
T: (506) 454-7745
Founded: 1867; Members: 880
Focus: Med 02688

New Brunswick Music Educators' Association, c/o Marshview Middle School, 21 McCarron Dr, Quispamsis, N.B. E0G 2W0
T: (506) 849-3615
Founded: 1971; Members: 120
Pres: Gerry Stephenson
Focus: Educ; Music 02689

New Brunswick Psychiatric Association / Association des Psychiatres du Nouveau-Brunswick, c/o Centracare, POB 3220, Saint John, N.B. E2M 4H7
T: (506) 672-7550
Founded: 1957; Members: 45
Gen Secr: D.B. Hade
Focus: Psychiatry 02690

New Brunswick Research and Productivity Council / Conseil de Recherche et de Productivité du Nouveau-Brunswick, POB 20000, Fredericton, N.B. E3B 6C2
T: (506) 452-8994; Tx: 01446115; Fax: (506) 452-1395
Founded: 1962
Focus: Econ
Periodicals
Alert (weekly)
Annual Report (weekly) 02691

New Brunswick Safety Council / Conseil de Securité du Nouveau-Brunswick, 151 Brunswick St, Fredericton, N.B. E3B 1G7
T: (506) 454-3555
Founded: 1967; Members: 125
Pres: Gladys E. Beattie
Focus: Safety 02692

New Brunswick School Superintendents Association / Association des Directeurs Généraux d'Ecoles du Nouveau-Brunswick, POB 239, Rexton, N.B. E0A 3L0
T: (506) 523-9114
Founded: 1967; Members: 44
Focus: Educ 02693

New Brunswick Society of Occupational Therapists / Société des Ergothérapeutes du Nouveau-Brunswick, c/o Dr. Everett Chalmers Hospital, POB 9000, Fredericton, N.B. E3B 5N5
T: (506) 452-5253
Founded: 1967; Members: 57
Pres: Alison J. Phalen
Focus: Therapeutics 02694

New Brunswick Speech and Hearing Association / Association d'Orthophonie et d'Audiologie du Nouveau-Brunswick, 180 Woodbridge St, Fredricton, N.B. E3B 4R3
Members: 40
Focus: Educ Handic 02695

New Brunswick Teachers' Federation / Fédération des Enseignants du Nouveau-Brunswick, POB 1535, Fredericton, N.B. E3B 5G2
T: (506) 452-8921; Fax: (506) 453-9795
Members: 8350
Focus: Educ 02696

New Brunswick Veterinary Medical Association, POB 20280, Fredericton, N.B. E3B 4Z7
T: (506) 455-9331
Founded: 1919
Gen Secr: L.B. Donald
Focus: Vet Med 02697

Newfoundland and Labrador Association of Occupational Therapists, POB 5423, Saint John's, Nfld. A1C 5W2
Founded: 1970; Members: 30
Focus: Therapeutics 02698

Newfoundland and Labrador Drama Society, 9 Davidson Pl, Saint John's, Nfld. A1E 1A4
Focus: Perf Arts 02699

Newfoundland and Labrador Veterinary Medical Association, POB 818, Mount Pearl, Nfld. A1N 3C8
T: (709) 368-8521
Members: 33
Pres: Andrew Fraser; Gen Secr: Neil E.B. Moss
Focus: Vet Med 02700

Newfoundland Association of Medical Radiation Technologists, c/o Radiology Dept, Health Science Center, Saint John's, Nfld. A1C 5S7
T: (709) 737-7161
Members: 300
Focus: Radiology; Nucl Med 02701

Newfoundland Association of Social Workers, POB 5244, East End PO, Saint John's, Nfld. A1C 5W1
Founded: 1950; Members: 110
Focus: Sociology 02702

Newfoundland Dietetic Association, POB 1756, Saint John's, Nfld. A1C 5P5
Founded: 1957
Focus: Nutrition 02703

Newfoundland Forest Protection Association, c/o Bowater Newfoundland Ltd, Corner Brook, Nfld. A2H 6J4
T: (709) 637-3342
Focus: Forestry 02704

Newfoundland Historical Society, Colonial Bldg, Military Rd, Rm 15, Saint John's, Nfld. A1C 2C7
T: (709) 722-3191
Founded: 1897; Members: 150
Focus: Hist 02705

Newfoundland Lung Association, c/o King George V Institute, POB 5250, Saint John's, Nfld. A1C 5W1
T: (709) 726-4664
Founded: 1944; Members: 10000
Focus: Pulmon Dis 02706

Newfoundland Medical Association, O'Mara Martin Bldg, Rawlins Cross, Saint John's, Nfld. A1C 2E4
T: (709) 726-7424
Founded: 1923; Members: 950
Focus: Med 02707

Newfoundland Medical Board, 15 Rowan St, Saint John's, Nfld. A1B 2X2
T: (709) 726-8546
Founded: 1893; Members: 12
Focus: Med 02708

Newfoundland Medical Council, 188 Waterford Bridge Rd, Saint John's, Nfld. A1E 1E2
Focus: Med 02709

Newfoundland Music Educators' Association, 19 Mount Saint Bernard Av, Corner Brook, Nfld. A2H 6E3
T: (709) 634-2787
Focus: Educ; Music 02710

Newfoundland Speech and Hearing Association, POB 8201, Station A, Saint John's, Nfld. A1G 3N4
Founded: 1978; Members: 22
Focus: Logopedy 02711

Newfoundland Teachers' Association, 3 Kenmount Rd, Saint John's, Nfld. A1B 1W1
T: (709) 726-3223
Founded: 1890; Members: 8450
Focus: Educ 02712

Newfoundland Thoracic Society, 47 Saint Clare Av, Saint John's, Nfld. A1C 2J9
T: (709) 579-2681
Founded: 1968; Members: 140
Focus: Pulmon Dis 02713

Newspaper Research Centre, 42 Charles St E, Ste 501, Toronto, Ont. M4Y 1T4
T: (416) 960-5030
Focus: Journalism 02714

North Island/Laurentian Teachers Union, 4335 Notre Dame Blvd, Ste 3, Chomedey-Laval, Qué. H7W 1T3
Focus: Educ 02715

North Okanagan Neurological Association, 2802 34 St, Vernon, B.C. V1T 5X1
T: (604) 549-1281
Founded: 1975
Gen Secr: Cherie Annand
Focus: Neurology 02716

Northwestern Ontario Water and Waste Conference, c/o City Engineer's Office, City Hall, 950 Memorial Av, Thunder Bay, Ont. P7B 4A2
T: (807) 625-2444
Founded: 1962; Members: 70
Gen Secr: Arthur G. McKay
Focus: Water Res 02717

Northwest Territories Law Foundation, POB 2594, Yellowknife, N.W.T. X1A 2P9
T: (403) 873-8275
Founded: 1981
Gen Secr: Gordon R. Carter
Focus: Law 02718

Northwest Territories Library Association, c/o Yellowknife Public Library, POB 694, Yellowknife, N.W.T. X0E 1H0
Focus: Libraries & Bk Sci 02719

Northwest Territories Music Educators' Association, c/o JBT Elementary School, POB 815, Fort Smith, N.W.T. X0E 0P0
T: (403) 872-8463
Founded: 1975; Members: 23
Focus: Educ; Music 02720

Northwest Territories Teachers' Association, POB 2340, Yellowknife, N.W.T. X1A 2P7
T: (403) 873-8501
Founded: 1953; Members: 756
Gen Secr: Blake W. Lyons
Focus: Educ 02721

Northwest Territories Teachers Association Home Economics Council, POB 2627, Inuvik, N.W.T. X0E 0T0
Pres: S. Nora Dixon
Focus: Educ; Home Econ 02722

Nova Scotia Association of Optometrists, POB 2718, Station M, Halifax, N.S. B3J 3P7
T: (902) 835-2020
Members: 59
Pres: Dr. Henry Smit; Gen Secr: Dr. Toby Mandelman
Focus: Ophthal; Optics
Periodical
Newsletter (weekly) 02723

Nova Scotia Association of Social Workers, 16 Grand View Dr, Dartmouth, N.S. B2W 1X5
T: (902) 434-0909
Founded: 1963; Members: 225
Gen Secr: Ruth Fisher
Focus: Soc Sci 02724

Nova Scotia Barristers' Society, 1815 Upper Water St, Halifax, N.S. B3J 1S7
T: (902) 422-1491
Founded: 1797
Focus: Law
Periodical
Nova Scotia Law News (weekly) . . . 02725

Nova Scotia Cardiology Technicians Association, 1935 Oxford St, Halifax, N.S. B3H 4A1
T: (902) 429-4083, 428-3526
Founded: 1967
Pres: Kathleen MacDonald
Focus: Cardiol 02726

Nova Scotia Chiropractic Association, 50 Hawthorne St, Antigonish, N.S. B2G 1A4
T: (902) 863-6924
Founded: 1953; Members: 21
Focus: Orthopedics 02727

Nova Scotia College Conference, Site 1, POB 23, Kingston, N.S. B0P 1R0
T: (902) 765-4887
Pres: Clary Laing
Focus: Educ 02728

Nova Scotia Confederation of University Faculty Associations, 1541 Barrington St, Ste 204, Halifax, N.S. B3J 1Z5
T: (902) 422-1204
Members: 1700
Focus: Sci 02729

Nova Scotia Criminology and Corrections Association, c/o CSC (Parole), POB 1473 North, Halifax, N.S. B3K 5H7
T: (902) 424-5473
Pres: Cheryl Picard
Focus: Criminology 02730

Nova Scotia Dental Association, 5991 Spring Garden Rd, Ste 604, Halifax, N.S. B3H 1Y6
T: (902) 420-0088
Founded: 1891; Members: 530
Focus: Dent
Periodical
Nova Scotia Dentist (weekly) 02731

Nova Scotia Dietetic Association, POB 8841, Station A, Halifax, N.S. B3K 5M5
Founded: 1958; Members: 132
Pres: Diane Fitzpatrick
Focus: Nutrition 02732

Nova Scotia Heart Foundation, 1657 Barrington St, Ste 321, Halifax, N.S. B3J 2Y3
T: (902) 423-7530
Gen Secr: Joan E. Fraser
Focus: Cardiol 02733

Nova Scotia Institute of Agrologists, POB 550, Truro, N.S. B2N 5E3
T: (902) 895-1571
Founded: 1953; Members: 230
Focus: Agri 02734

Nova Scotia Library Association, c/o Halifax City Regional Library, 5381 Spring Garden Rd, Halifax, N.S. B3J 1E9
T: (902) 426-6981
Focus: Libraries & Bk Sci
Periodical
Newsletter (weekly) 02735

Nova Scotia Lung Association, 17 Alma Crescent, Halifax, N.S. B3N 2C4
T: (902) 443-8141
Founded: 1926; Members: 150
Focus: Pulmon Dis 02736

Nova Scotia Medical Council, 7193 Quinpool Rd, Halifax, N.S. B3L 4K9
Pres: Dr. M.R. Banks
Focus: Med 02737

Nova Scotia Mineral and Gem Society, 116 Joffre St, Dartmouth, N.S. B2Y 3C9
T: (902) 469-1465
Pres: K.W. Hamilton
Focus: Mineralogy 02738

Nova Scotia Music Educators' Association, c/o Nova Scotia Teachers Union, POB 1060, Armdale, N.S. B3L 4L7
T: (902) 477-5621
Founded: 1959; Members: 300
Pres: Ninette Babineau
Focus: Educ; Music 02739

Nova Scotian Institute of Science (NSIS), c/o Killam Library, Dalhousie University, Halifax, N.S. B3H 4H8
T: (902) 494-2384
Founded: 1862; Members: 300
Pres: Dr. Jim Stewart; Gen Secr: Dr. David Goble
Focus: Sci
Periodical
Proceedings (weekly) 02740

Nova Scotia Pharmaceutical Society, 1526 Dresden Row, Halifax, N.S. B3J 2K2
Gen Secr: G.B. Locke
Focus: Pharmacol 02741

Nova Scotia Research Foundation Corporation, 100 Fenwick St, POB 790, Dartmouth, N.S. B2Y 3Z7
Founded: 1946; Members: 125
Pres: T.B. Nickerson; Gen Secr: R.F. MacNeill
Focus: Sci
Periodical
Annual Report (weekly) 02742

Nova Scotia Safety Council, 3627 Joseph Howe Dr, Halifax, N.S. B3L 4H8
T: (902) 454-9621
Founded: 1958; Members: 150
Gen Secr: George M. Currie
Focus: Safety 02743

Nova Scotia School Boards Association / Association des Conseils Scolaires de la Nouvelle-Ecosse, POB 605, Station M, Halifax, N.S. B3J 2R7
T: (902) 420-9191
Focus: Educ; Public Admin
Periodical
Newsletter (10 times annually) 02744

Nova Scotia Society of Medical Radiation Technologists, R.R. 1, POB 232, Armdale, N.S. B3L 4J1
T: (902) 424-6111
Founded: 1963; Members: 425
Gen Secr: Maria A. Nelson
Focus: Radiology; Nucl Med 02745

Nova Scotia Society of Occupational Therapists, POB 3381, Halifax, N.S. B3J 3J1
Founded: 1951; Members: 60
Focus: Therapeutics
Periodical
Info (weekly) 02746

Nova Scotia Teachers Union, POB 1060, Armdale, N.S. B3L 4L7
T: (902) 477-5621
Members: 11742
Gen Secr: Emmet Currie
Focus: Educ 02747

Nova Scotia Thoracic Society, c/o Nova Scotia Lung Association, 17 Alma Crescent, Halifax, N.S. B3N 3C4
T: (902) 443-8141
Founded: 1926
Focus: Pulmon Dis 02748

Nova Scotia Veterinary Medical Association, c/o Agricultural Centre, Kentville, N.S. B4N 1J5
T: (902) 678-2171 ext 264
Founded: 1913; Members: 150
Focus: Vet Med 02749

Numismatic Educational Services Association, POB 226, Barrie, Ont. L4M 4T2
T: (705) 737-0845
Founded: 1979; Members: 15
Gen Secr: K.B. Prophet
Focus: Numismatics 02750

Occupational Health and Safety Council of Prince Edward Island, 42 Great George St, Charlottetown, P.E.I. C1A 7N5
T: (902) 892-0941
Gen Secr: Phyllis MacDonald
Focus: Public Health; Safety 02751

Occupational Medical Association of Canada / Association des Médicins du Travail du Canada, c/o Medical Services, 290 Yonge St, Toronto, Ont. M5B 1C8
T: (416) 591-4242
Founded: 1972; Members: 250
Focus: Med 02752

The Ontario Archaeological Society, 126 Willowdale Av, Willowdale, Ont. M2N 4Y2
T: (416) 730-0797
Members: 800
Pres: Christine L. Caroppo; Gen Secr: Charles Garrad
Focus: Archeol; Anthro
Periodicals
Arch Notes (weekly)
Ontario Archaeology (weekly) 02753

Ontario Association for Continuing Education, 175 Saint Clair Av W, Toronto, Ont. M4V 1P7
Founded: 1971; Members: 400
Focus: Educ; Adult Educ 02754

Ontario Association for Curriculum Development, POB 931, Station B, London, Ont. N6A 5K1
T: (519) 685-0228
Founded: 1951; Members: 700
Gen Secr: Dr. Robert Howe
Focus: Educ
Periodical
Conference Report (weekly) 02755

Ontario Association of Alternative and Independent Schools / Association des Ecoles Alternatives et Indépendantes de l'Ontario, 2020 Bathurst St, Toronto, Ont. M5P 3L1
T: (416) 781-3591
Founded: 1975; Members: 510
Focus: Educ 02756

Ontario Association of Archivists, 168 Irving Av, Ottawa, Ont. K1Y 1Z5
T: (613) 996-8507
Focus: Archives 02757

Ontario Association of Education Administrative Officials / Association Ontarienne des Agents de l'Administration Scolaire, 252 Bloor St W, Ste N1201, Toronto, Ont. M5S 1V5
T: (416) 925-3276
Founded: 1971; Members: 560
Focus: Public Admin; Educ 02758

Ontario Association of Library Technician Instructors / Association des Professeurs des Bibliotechniciens d'Ontario, c/o Niagara College Library Technician Program, Woodlawn Rd, POB 1005, Welland, Ont. L3B 5S2
T: (416) 735-2211
Founded: 1973; Members: 20
Focus: Libraries & Bk Sci 02759

Ontario Association of Library Technicians, POB 682, Oakville, Ont. L6J 5C1
T: (613) 733-7421
Founded: 1974; Members: 500
Focus: Libraries & Bk Sci 02760

Ontario Association of Property Standards Officers, 200 Broadway St, Tillsonburg, Ont. N4G 5A7
Focus: Standards 02761

Ontario Athletic Therapists Association, 5 Frith Rd, Downsview, Ont. M3N 2L5
T: (416) 743-5012
Focus: Therapeutics 02762

Ontario Black History Society, 352 Sheppard Av E, Willowdale, Ont. M2N 3B4
T: (416) 225-1176
Founded: 1978
Focus: Hist 02763

Ontario Cancer Treatment and Research Foundation, 7 Overlea Blvd, Toronto, Ont. M4H 1A8
T: (416) 423-4240
Founded: 1943
Gen Secr: Dr. J.W. Meakin
Focus: Cell Biol & Cancer Res
Periodicals
Cancer in Ontario
Proceedings of the Clinical Conferences . 02764

Ontario Chiropractic Association (OCA), 1900 Bayview Av, Toronto, Ont. M4G 3E6
T: (416) 482-2707
Founded: 1929; Members: 810
Gen Secr: Shirley Stitt
Focus: Orthopedics
Periodical
OCA News (weekly) 02765

Ontario College of Percussion, 1656 Bayview Av, Toronto, Ont. M4T 3C2
T: (416) 483-9117
Focus: Music 02766

Ontario Community Development Society, c/o School of Agricultural Economics and Extension Education, University of Guelph, Guelph, Ont. N1G 2W1
Focus: Urban Plan 02767

Ontario Co-operative Education Association, c/o Hamilton Board of Education, 100 Main St, POB 558, Hamilton, Ont. L8N 3L1
T: (416) 527-5092
Founded: 1975; Members: 500
Pres: Jim O'Connor
Focus: Educ 02768

Ontario Council for Leadership in Educational Administration, 252 Bloor St W, Ste 6298, Toronto, Ont. M5S 1V5
T: (416) 923-6641
Founded: 1973
Gen Secr: Peter E. Angelini
Focus: Educ 02769

Ontario Council of Health, 700 Bay St, Toronto, Ont. M5G 1Z6
T: (416) 965-5031
Focus: Public Health 02770

Ontario Council on University Affairs / Conseil Ontarien des Affaires Universitaires, 700 Bay St, Toronto, Ont. M5G 1Z6
T: (416) 965-5233
Founded: 1974
Focus: Sci 02771

Ontario Deafness Research Foundation, 60 Bedford Rd, Toronto, Ont. M5R 2K2
T: (416) 964-9595
Gen Secr: Sandra Fox
Focus: Educ Handic 02772

Ontario Dental Association, 234 Saint George St, Toronto, Ont. M5R 2P1
T: (416) 922-3900
Members: 4380
Gen Secr: J.C. Gillies
Focus: Dent
Periodical
Ontario Dentist (weekly) 02773

Ontario Educational Research Council / Conseil Ontarien de Recherche Pédagogiques, 979 Finlay Av, Ajax, Ont. L1S 3V5
T: (416) 478-6621
Founded: 1959; Members: 20
Focus: Econ 02774

Ontario Electric Railway Historical Association, POB 121, Station A, Scarborough, Ont. M1K 5B9
T: (519) 856-9802
Founded: 1953; Members: 170
Focus: Cultur Hist 02775

Ontario Family Studies Home Economics Educators Association, 33 Claremont St, Thorold, Ont. L2V 1R5
T: (416) 227-1469
Pres: Betty Tieche
Focus: Educ; Home Econ 02776

Ontario Farm Drainage Association, POB 117, Chatham, Ont. N7M 5K1
T: (519) 351-5512
Focus: Agri 02777

The Ontario Genealogical Society, 40 Orchard View Blvd, Ste 251, Toronto, Ont. M4R 1B9
T: (416) 489-0734
Members: 5000
Gen Secr: Edna Hudson
Focus: Genealogy
Periodicals
Families (weekly)
Newsleaf (weekly) 02778

Ontario Gerontology Association / Association Ontarienne de Gérontologie, c/o Dept of Statistics, University of Waterloo, Waterloo, Ont. N2L 3G1
T: (519) 885-1211 ext 3468
Founded: 1982; Members: 300
Focus: Geriatrics
Periodical
Newsletter (weekly) 02779

Ontario Group Psychotherapy Association, 27 Renova Dr, Etobicoke, Ont. M9C 3E8
T: (416) 621-8413
Focus: Psych; Therapeutics 02780

Ontario Health Record Association, R.R. 1, Harrow, Ont. N0R 1G0
T: (519) 524-2944
Founded: 1936; Members: 850
Gen Secr: Pauline Baldwin
Focus: Public Health 02781

Ontario Historical Society, 5151 Yonge St, Willowdale, Ont. M2N 5P5
T: (416) 226-9011; Fax: (416) 226-2740
Founded: 1888; Members: 3300
Gen Secr: Dorothy Duncan
Focus: Hist
Periodicals
Approaching Ontario's Past (weekly)
Bulletin (weekly)
Ontario History 02782

Ontario Institute of Agrologists, c/o University of Guelph, 451 University Centre, Guelph, Ont. N1G 2W1
T: (519) 837-2820
Focus: Agri 02783

Ontario Library Association Literacy Guild, c/o North York Public Library, 5145 Yonge St, Toronto, Ont. M2N 5P4
T: (416) 226-9437
Pres: Annette Vafa
Focus: Libraries & Bk Sci; Lit 02784

Ontario Lung Association, 573 King St E, Toronto, Ont. M5A 1M5
T: (416) 864-1112; Fax: (416) 864-9916
Founded: 1945; Members: 900
Gen Secr: R. Ross Reid
Focus: Pulmon Dis
Periodicals
Lung Line (3 times annually)
Ontario Respiratory Care Society Update (3 times annually)
Ontario Thoracic Reviews (weekly) . . . 02785

Ontario Music Educators' Association, c/o Frontenac City Board of Education, 566 Grandview Av, London, Ont. N6K 3G5
T: (519) 472-4938
Founded: 1940; Members: 1350
Pres: Ronald Gilbert
Focus: Educ; Music 02786

Ontario Osteopathic Association, 45 Richmond St W, Ste 401, Toronto, Ont. M5H 1Z2
T: (416) 368-2006
Members: 21
Gen Secr: D.R. Firth
Focus: Pathology 02787

Ontario Podiatry Association, 37 Alvin Av, Toronto, Ont. M4T 2A2
T: (416) 927-9111
Founded: 1944
Pres: Dr. David Greenberg
Focus: Orthopedics 02788

Ontario Psychogeriatric Association, 83 Lakeshore Blvd, Kingston, Ont. K7M 6R4
T: (613) 546-1101
Founded: 1974; Members: 600
Focus: Geriatrics
Periodicals
Newsletter (5 times annually)
Proceeding (weekly) 02789

Ontario Public Interest Research Group, 85 Hastey Av, Ste 344, Ottawa, Ont. K1N 8Z4
T: (613) 230-3076
Focus: Sociology 02790

Ontario Public School Boards' Association, 200 McIntyre St E, POB 3110, North Bay, Ont. P1B 8H1
T: (705) 472-8170
Gen Secr: Wayne E. Perry
Focus: Educ 02791

Ontario Public School Teachers' Federation, 1260 Bay St, Toronto, Ont. M5R 2B7
T: (416) 928-1128; Fax: (416) 928-0179
Founded: 1921
Focus: Educ 02792

Ontario Registered Music Teachers Association, 48 Malvern Av, Toronto, Ont. M4E 3E3
T: (416) 694-0296
Focus: Educ; Music 02793

Ontario School Counsellors' Association / Association des Conseillers en Orientation de l'Ontario, 19 Treadgold Crescent, Don Mills, Ont. M3A 1X1
T: (416) 449-9321
Founded: 1964; Members: 900
Focus: Educ
Periodical
Ontario School Counsellor Association Reports (weekly) 02794

Ontario Secondary School Teachers' Federation, 60 Mobile Dr, Toronto, Ont. M4A 2P3
T: (416) 751-8300
Founded: 1919; Members: 36000
Gen Secr: L. Morris Richardson
Focus: Educ 02795

Ontario Society for Cable Television Engineering, 4560 Fieldgate Dr, Mississauga, Ont. L4W 3W6
T: (416) 629-1111
Pres: Karl Poirier
Focus: Eng 02796

Ontario Society for Industrial Archaeology, c/o Institute for History and Philosophy of Science and Technology, Victoria College, 73 Queen's Park Crescent E, Toronto, Ont. M5S 1K7
Gen Secr: Phyllis Rose
Focus: Archeol 02797

Ontario Society of Clinical Hypnosis, 200 Saint Clair Av W, Toronto, Ont. M4V 1R1
T: (416) 922-8300
Founded: 1972; Members: 400
Focus: Anesthetics 02798

Ontario Society of Occupational Therapists, 801 Eglinton Av W, Ste 400, Toronto, Ont. M5N 1E3
T: (416) 781-8290
Founded: 1923; Members: 680
Focus: Therapeutics 02799

The Ontario Speech and Hearing Association, 181 University Av, Ste 1202, Toronto, Ont. M5H 3M7
T: (416) 368-8132
Founded: 1958; Members: 500
Pres: D. Wolf
Focus: Logopedy 02800

Ontario Teachers' Federation / Fédération des Enseignants de l'Ontario, 1260 Bay St, Ste 700, Toronto, Ont. M5R 2B5
T: (416) 966-3424
Founded: 1944
Focus: Educ 02801

Ontario Teachers' Superannuation Commission, 5650 Yonge St, North York, Ont. M2M 4H5
Fax: (416) 730-5349
Members: 178839
Focus: Soc Sci
Periodicals
Annual Report (weekly)
Employer News (weekly)
Exchange (weekly) 02802

Ontario Thoracic Society, 573 King St E, Toronto, Ont. M5A 1M5
T: (416) 864-1112
Members: 325
Focus: Pulmon Dis
Periodical
The Ontario Thoracic Review (biennial) . 02803

Ontario Veterinary Association, 340 Woodlawn Rd W, Stes 24-25, Guelph, Ont. N1H 2X1
T: (519) 824-5600; Fax: (519) 824-6497
Focus: Vet Med
Periodical
Update (weekly) 02804

Ontario Vocational Educational Association, Baptist Church Rd, R.R. 2, Caledonia, Ont. N0A 1A0
T: (519) 752-4414
Focus: Educ 02805

Operation Eyesight Universal, POB 123, Station M, Calgary, Alta. T2P 2H6
T: (403) 283-6323
Founded: 1963
Pres: Ross C. McDonald
Focus: Ophthal
Periodical
Newsletter (weekly) 02806

Organization of Military Museums of Canada / Organisation des Musées Militaires du Canada, 330 Sussex Dr, Ottawa, Ont. K1A 0M8
T: (613) 996-1420; Fax: (613) 954-1016
Founded: 1968; Members: 250
Focus: Military Sci; Archives 02807

Osteoporosis Society of Canada / Société de l'Ostéoporose du Canada, 76 Saint Clair Av W, Ste 502, Toronto, Ont. M4V 1N2
T: (416) 922-1358
Founded: 1982
Pres: William C. Sturtridge; Gen Secr: Kathryn D. Robins
Focus: Pathology
Periodical
Osteoporosis: Bulletin for Physicians (weekly) 02808

Ottawa-Carleton Safety Council / Conseil de Securité d'Ottawa-Carleton, 190 Somerset St W, Ste 208, Ottawa, Ont. K2P 0J4
T: (613) 238-1513
Founded: 1956; Members: 100
Focus: Safety 02809

Pacific Coast Family Therapy Training Association, 1545 62 Av W, Vancouver, B.C. V6P 2E8
T: (604) 266-3141
Focus: Therapeutics 02810

Parkinson Foundation of Canada / Fondation Canadienne du Parkinson, 55 Bloor St W, Ste 232, Toronto, Ont. M4W 1A6
T: (416) 964-1155
Founded: 1967; Members: 4000
Focus: Neurology
Periodical
Network Newsletter (5 times annually) . 02811

PEN International, Centre Québécois, 615 Rue Belmont, Bureau 212, Montréal, Qué. H3B 2L8
T: (514) 398-0946
Founded: 1982; Members: 200
Pres: Jean Ethier-Blais; Gen Secr: René Le Clère
Focus: Lit 02812

Pharmacological Society of Canada, c/o Dept of Pharmacology and Therapeutics, Faculty of Medicine, University of British Columbia, Vancouver, B.C. V6T 1Z3
T: (604) 822-2039; Fax: (604) 822-6012
Founded: 1956; Members: 350
Pres: Dr. Frank La Bella; Gen Secr: Dr. C. Pang
Focus: Pharmacol
Periodical
Canadian Journal of Physiology and Pharmacology 02813

Pharmacy Association of Nova Scotia, POB 3214, Halifax, N.S. B3J 3H5
T: (902) 422-9583
Founded: 1979; Members: 500
Focus: Pharmacol 02814

Photo Electric Arts Foundation, POB 7109, Station A, Toronto, Ont. M5W 1X8
T: (416) 691-0630
Founded: 1978; Members: 20
Pres: Richard Hill
Focus: Arts 02815

The Photographic Historical Society of Canada, POB 115, Station S, Toronto, Ont. M1J 2Y6
T: (416) 247-3885
Founded: 1974; Members: 268
Pres: William B. Beller
Focus: Photo
Periodical
Photographic Canadiana (weekly) 02816

Physical Education Council of the New Brunswick Teachers Association, 805 Frampton Ln, Moncton, N.B. E1G 1S1
T: (506) 384-5011
Founded: 1969; Members: 195
Focus: Sports 02817

Physiotherapy Association of British Columbia (PABC), 222-4585 Canada Way, Burnaby, B.C. V5G 4L6
T: (604) 294-1664; Fax: (604) 294-9191
Focus: Physical Therapy
Periodical
PABC Report (weekly) 02818

Physiotherapy Foundation of Canada, c/o Physical Therapy Program, University of Western Ontario, London, Ont. N6A 5B8
Pres: Dr. Malcolm Peat
Focus: Physical Therapy 02819

Planning Institute of British Columbia, POB 24835, Station C, Vancouver, B.C. V5T 4E9
T: (604) 596-0391
Founded: 1958; Members: 340
Focus: Urban Plan 02820

Polish Canadian Librarians Association, 103 Avenue Rd, Apt 703, Toronto, Ont. M5R 2G9
Founded: 1977; Members: 64
Pres: Grace B. Kopec
Focus: Libraries & Bk Sci 02821

Polish Teachers' Association in Canada / Association des Instituteurs Polonais au Canada, 288 Roncesvalles Av, Toronto, Ont. M6R 2M4
T: (416) 532-2876
Founded: 1962; Members: 225
Focus: Educ; Ling 02822

Pollution Control Association of Ontario, 67 Poplar Crescent, Aurora, Ont. L4G 3M4
Founded: 1971; Members: 750
Gen Secr: S. Davey
Focus: Ecology 02823

Pollution Probe Foundation, 12 Madison Av, Toronto, Ont. M5R 2S1
T: (416) 978-7152
Founded: 1969; Members: 3000
Gen Secr: Colin Isaacs
Focus: Ecology 02824

Postal History Society of Canada, POB 3461, Station C, Ottawa, Ont. K1Y 4J6
T: (613) 728-2907
Founded: 1972; Members: 600
Focus: Cultur Hist 02825

Potash and Phosphate Institute of Canada / Institut Potasse et Phosphate du Canada, 555 Burnhamthorpe Rd, Etobicoke, Ont. M9C 2Y3
T: (416) 626-5023
Founded: 1970
Focus: Agri 02826

Prince Edward Island Association of Community Schools, Flat River, RR 3, Belle River, P.E.I. C0A 1B0
T: (902) 659-2811
Focus: Educ 02827

Prince Edward Island Association of Medical Radiation Technologists, POB 995, Montague, P.E.I. C0A 1R0
T: (902) 566-6292
Focus: Radiology; Nucl Med 02828

Prince Edward Island Association of Social Workers, POB 1888, Charlottetown, P.E.I. C1A 7N5
Members: 45
Pres: Sandy Simpson
Focus: Sociology
Periodical
Newsletter (weekly) 02829

Prince Edward Island Cerebral Palsy Association, POB 2702, Charlottetown, P.E.I. C1A 8C3
T: (902) 894-5176
Founded: 1953; Members: 150
Pres: Sophie MacCallum
Focus: Neurology 02830

Prince Edward Island Criminology and Corrections Association, 11 Yorkshire Dr, Charlottetown, P.E.I. C1A 6N7
T: (902) 894-816
Founded: 1974; Members: 20
Focus: Criminology 02831

Prince Edward Island Dietetic Association, POB 2575, Charlottetown, P.E.I. C1A 8C2
T: (902) 436-4760
Pres: Julia Donahoe
Focus: Nutrition 02832

Prince Edward Island Institute of Agrologists, POB 1600, Charlottetown, P.E.I. C1A 7N3
T: (902) 892-5465
Focus: Agri 02833

Prince Edward Island Lung Association, c/o Provincial Sanatorium, Charlottetown, P.E.I. C1A 2K1
T: (902) 894-7331
Gen Secr: Dr. W.R. Stewart
Focus: Pulmon Dis 02834

Prince Edward Island Medical Society, 279 Richmond St, Charlottetown, P.E.I. C1A 1J7
T: (902) 892-7527
Founded: 1855; Members: 150
Gen Secr: Marilyn Lowther
Focus: Med 02835

Prince Edward Island Museum and Heritage Foundation, 2 Kent St, Charlottetown, P.E.I. C1A 1M6
T: (902) 892-9127
Gen Secr: David Webber
Focus: Archives 02836

Prince Edward Island Physical Education Association, c/o Dept of Education, POB 2000, Charlottetown, P.E.I. C1A 7N8
T: (902) 368-4672
Founded: 1961; Members: 45
Gen Secr: Ken Goudet
Focus: Sports; Educ 02837

Prince Edward Island Psychiatric Association, c/o Health and Social Services, POB 2000, Charlottetown, P.E.I. C1A 7N8
T: (902) 892-5471
Focus: Psychiatry 02838

Prince Edward Island Society of Occupational Therapists, POB 2248, Charlottetown, P.E.I. C1A 8B9
Members: 17
Pres: Mari Basiletti
Focus: Therapeutics 02839

Prince Edward Island Speech and Hearing Association, POB 2000, Charlottetown, P.E.I. C1A 7N8
T: (902) 892-5471 ext 255
Founded: 1979; Members: 12
Pres: Dimphy Elsinga
Focus: Logopedy 02840

Prince Edward Island Teachers' Federation, POB 6000, Charlottetown, P.E.I. C1A 8B4
T: (902) 569-4157
Founded: 1880; Members: 1460
Gen Secr: James L. Blanchard
Focus: Educ 02841

Prince Edward Island Veterinary Medical Association, POB 1206, Charlottetown, P.E.I. C1A 7M8
T: (902) 566-7621
Founded: 1949; Members: 32
Gen Secr: Dr. S.C. Bellamy
Focus: Vet Med 02842

Print and Drawing Council of Canada, c/o Visual Arts Ontario, 439 Wellington St W, Toronto, Ont. M5V 1E7
Founded: 1977; Members: 350
Pres: Alison Brannen
Focus: Arts 02843

Printing Trades Accident Prevention Association, 2 Bloor St E, Toronto, Ont. M4W 3C2
T: (416) 965-8888
Gen Secr: J.V. Findlay
Focus: Safety 02844

Professional Development Institute / Institut Supérieur de Perfectionnement des Cadres I.S.P.C., POB 1181, Station B, Ottawa, Ont. K1P 5R2
T: (613) 523-3333; Tx: 0533159
Founded: 1973; Members: 431
Focus: Econ 02845

Professional Institute of the Public Service of Canada / Institut Professionnel de la Fonction Publique du Canada, 786 Bronson Av, Ottawa, Ont. K1S 4G4
T: (613) 237-6310; Tx: 0534718
Founded: 1920; Members: 18 000
Focus: Public Admin
Periodicals
Communications (weekly)
Pro Forum (weekly) 02846

Professional Marketing Research Society / Société Professionnelle de Recherche en Marketing, POB 5155, Toronto, Ont. M5W 1N5
T: (416) 487-4893
Founded: 1959; Members: 500
Focus: Marketing 02847

Provincial Association of Catholic Teachers (Québec), 2005 Saint Marc, Montréal, Qué. H3H 2G8
T: (514) 935-8612
Founded: 1969; Members: 3500
Gen Secr: Robert R. Dobie
Focus: Educ 02848

Provincial Association of Protestant Teachers of Québec / Association Provinciale des Enseignants Protestants du Québec, 84-J Brunswick Av, Dollard des Ormeaux, Qué. H9B 2C5
T: (514) 683-9330; Fax: (514) 683-1971
Founded: 1864; Members: 6000
Focus: Educ 02849

Provincial Medical Board of Nova Scotia, 1515 South Park St, Ste 3050, Halifax, N.S. B3J 2L2
T: (902) 422-5823
Pres: Dr. M.R. Macdonald
Focus: Med 02850

Psychological Society of Saskatchewan, POB 4528, Regina, Sask. S4P 3W7
T: (306) 586-5967
Founded: 1966; Members: 204
Pres: Dr. T. Greenough
Focus: Psych 02851

Public Affairs Council for Education, c/o Dir Public Relations and Information, University of Québec, 2875 Laurier Blvd, Sainte-Foy, Qué. G1V 2M3
Gen Secr: Kenneth Clements
Focus: Educ 02852

Pulp and Paper Research Institute of Canada, 570 Saint John's Blvd, Pointe Claire, Qué. H9R 3J9
T: (514) 630-4100
Founded: 1925
Pres: Peter Wrist
Focus: Materials Sci; Chem; Bio; Eng; Physics
Periodicals
Annual Report (weekly)
Trend (1-2 times annually) 02853

Quebec Association of Independent Schools / Association des Ecoles Privées du Québec, 484 Beaconsfield Blvd, Ste 203, Beaconsfield, Qué. H9W 4C4
T: (514) 694-5050
Founded: 1964; Members: 22
Pres: M. Fripp; Gen Secr: J. Gordon Pollock
Focus: Educ 02854

Recycling Council of Ontario, POB 310, Station P, Toronto, Ont. M5S 2S8
T: (416) 960-1025
Founded: 1978
Gen Secr: Dr. John Hanson
Focus: Ecology
Periodical
Ontario Recycling Update (weekly) . . . 02855

Registered Music Teachers' Association, 665 Campbell St, Winnipeg, Man. R3N 1C4
Focus: Educ; Music 02856

Reinforcing Steel Institute of Ontario, 1 Sparks Av, Ste T-2, Willowdale, Ont. M2H 2W1
T: (416) 499-4000; Fax: (416) 497-4143
Founded: 1973
Gen Secr: George Knapton
Focus: Metallurgy 02857

Reinsurance Research Council / Conseil de Recherche en Réassurance, First Canadian Pl, POB 100, Toronto, Ont. M5X 1B2
Focus: Insurance 02858

Research and Productivity Council, POB 6000, Fredericton, N.B. E3B 5H1
T: (506) 455-8994
Founded: 1962
Gen Secr: Dr. R.S. Boorman
Focus: Eng 02859

Research, Education and Assistance for Canadians with Herpes, POB 70, Station G, Toronto, Ont. M4M 3E8
T: (416) 698-6225
Founded: 1980; Members: 400
Pres: Andrea Howard
Focus: Derm 02860

Richard Ivey Foundation, 618 Richmond St, London, Ont. N6A 5J9
T: (519) 673-1280; Tx: 02499
Founded: 1947
Pres: R.M. Ivey
Focus: Sci 02861

Royal Architectural Institute of Canada / Institut Royal d'Architecture du Canada, 55 Murray St, Ste 330, Ottawa, Ont. K1N 5M3
Founded: 1908; Members: 3500
Gen Secr: William Shields
Focus: Archit
Periodicals
RAIC Bulletin (weekly)
RAIC Practice Notes (weekly) 02862

Royal Astronomical Society of Canada / Société Royale d'Astronomie du Canada, 136 Dupont St, Toronto, Ont. M5R 1V2
T: (416) 924-7973
Founded: 1890; Members: 3500
Gen Secr: Dr. David A. Tindall
Focus: Astronomy
Periodicals
Journal (weekly)
Observer's Handbook (weekly) 02863

Royal Canadian Academy of Arts / Académie Royale des Arts du Canada, 8 Adelaide St E, Toronto, Ont. M5C 1H5
T: (416) 363-9612; Fax: (416) 363-0920
Founded: 1880
Pres: Macy Dubois; Gen Secr: H.P.D. Van Ginkel
Focus: Arts 02864

Royal Canadian Geographical Society / Société Géographique Royale du Canada, 39 McArthur Av, Vanier, Ont. K1L 8L7
T: (613) 745-4629; Fax: (613) 744-0947
Founded: 1929; Members: 260000
Pres: T. Davidson
Focus: Geography 02865

Royal Canadian Institute, 720 Spadina Av, Ste 312, Toronto, Ont. M5S 2T9
T: (416) 928-2096
Founded: 1849; Members: 712
Pres: Dr. Bernhard Cinader
Focus: Sci 02866

Royal Canadian Military Institute, 426 University Av, Toronto, Ont. M5G 1S9
T: (416) 597-0286
Founded: 1890; Members: 3250
Gen Secr: Dario Vaccari
Focus: Military Sci 02867

Royal College of Dental Surgeons of Canada / Collège Royal des Chirurgiens Dentistes du Canada, 230 Saint George St, Toronto, Ont. M5R 2N5
T: (416) 961-6555
Founded: 1868; Members: 5000
Focus: Dent 02868

Royal College of Physicians and Surgeons of Canada (RCPSC) / Collège Royal des Médecins et Chirurgiens du Canada, 774 Echo Dr, Ottawa, Ont. K1S 1P3
T: (613) 746-8177; Fax: (613) 746-8833
Founded: 1929; Members: 24000
Pres: John Burgess; Gen Secr: Dr. Gilles D. Hurteau
Focus: Surgery; Physiology
Periodical
Annals RCPSC (weekly) 02869

Royal Nova Scotia Historical Society (RNSHS), POB 895, Armdale, Halifax, N.S. B3L 4K5
T: (902) 477-7422
Founded: 1878; Members: 350
Pres: Donald F. MacLean; Gen Secr: R. Barbour
Focus: Hist; Genealogy
Periodicals
Collections (weekly)
The Nova Scotia Genealogist (3 times annually) 02870

Royal Society of Canada / Société Royale du Canada, POB 9734, Ottawa, Ont. K1G 5J4
T: (613) 991-6990; Fax: (613) 991-6996
Founded: 1882; Members: 1150
Pres: John Meisel; Gen Secr: Pierre Hurtubise
Focus: Sci
Periodicals
Calendar: Annuaire
Presentation (weekly)
Proceedings and Transaction: Délibérations et Memoirs (weekly) 02871

RP Eye Research Foundation / Fondation RP pour la Recherche sur les Yeux, 185 Spadina Av, Ste 411, Toronto, Ont. M5T 2C6
T: (416) 598-4951
Founded: 1974
Focus: Ophthal
Periodical
Newsletter (weekly) 02872

R. Samuel McLaughlin Foundation, c/o National Trust Co., 21 King St E, Toronto, Ont. M5C 1B3
T: (416) 361-4096; Fax: (416) 361-4097
Founded: 1951; Members: 5
Gen Secr: D.R. Windeyer
Focus: Sci 02873

Rural Development Council of Prince Edward Island, c/o Sociology Dept, University of Prince Edward Island, Charlottetown, P.E.I. C1A 4P3
T: (902) 892-4121
Focus: Sociology 02874

Rural Education and Development Association, 14815 119 Av NW, Edmonton, Alta. T5L 2N9
T: (403) 451-5959
Focus: Educ 02875

Sales Research Club, c/o Commercial Travellers Association, 365 Blon St E, Ste 1002, Toronto, M4W 3L4
T: (416) 924-7724
Founded: 1933; Members: 80
Gen Secr: Terry Ruffell
Focus: Econ
Periodical
The Sales Researcher 02876

Sandford Fleming Foundation, POB 816, Waterloo, Ont. N2J 4C2
T: (519) 885-0910
Founded: 1976; Members: 46
Gen Secr: J.D. Weller
Focus: Sci 02877

Saskatchewan Amyotrophic Lateral Sclerosis Society Foundation, POB 546, Radville, Sask. S0C 2G0
Focus: Med; Pathology 02878

Saskatchewan Anti-Tuberculosis League, Fort San, Sansk. S0G 1T0
T: (306) 332-5984
Founded: 1911; Members: 120
Gen Secr: Ken More
Focus: Pulmon Dis 02879

Saskatchewan Association of Library Technicians, 342 Anderson Crescent, Saskatoon, Sask. S7H 4A3
Focus: Libraries & Bk Sci 02880

Saskatchewan Association of Medical Radiation Technologists, 2124 Saint Cecelia Av, Ste 401, Saskatoon, Sask. S7M 0P2
T: (306) 244-8645
Founded: 1956; Members: 400
Focus: Radiology; Nucl Med 02881

Saskatchewan Association of Pathologists, c/o Dept of Pathology, Regina General Hospital, Regina, Sask. S4P 0W5
T: (306) 359-4466
Founded: 1955; Members: 44
Pres: Dr. J. Vetters
Focus: Pathology 02882

Saskatchewan Association of Social Workers, 3304 Dewdney Av, Regina, Sask. S4T 0Y8
T: (306) 522-1511
Founded: 1967
Pres: Beth Predy
Focus: Sociology 02883

Saskatchewan Association of Teachers of German, 2317 Arlington Av, Saskatoon, Sask. S7J 2H8
T: (306) 373-1660
Founded: 1969; Members: 43
Focus: Educ; Ling 02884

Saskatchewan Association on Gerontology, POB 8546, Saskatoon, Sask. S7K 6K6
T: (306) 934-2112
Founded: 1980; Members: 180
Pres: Susan McDonald Wilson
Focus: Geriatrics 02885

Saskatchewan Cardiology Technicians Association, 214 Lindsay Pl, Ste 2, Saskatoon, Sask. S7N 3E6
Founded: 1968; Members: 140
Pres: Lynn Zinger
Focus: Cardiol 02886

Saskatchewan Dietetic Association (SDA), POB 3894, Regina, Sask. S4P 3R8
Founded: 1958; Members: 160
Pres: Sonja Matt
Focus: Nutrition
Periodical
SDA Newsletter (weekly) 02887

Saskatchewan Genealogical Society, POB 1894, Regina, Sask. S4P 3E1
Founded: 1967; Members: 530
Pres: Robert Pittendrigh
Focus: Genealogy
Periodical
Bulletin (weekly) 02888

Saskatchewan Geological Society, POB 234, Regina, Sask. S4P 2Z6
Founded: 1952; Members: 130
Gen Secr: J. Arthur
Focus: Geology 02889

Saskatchewan High School Principals' Group, POB 1108, Saskatoon, Sask. S7K 3N3
T: (306) 373-1660
Founded: 1960; Members: 160
Pres: A. Ledingham
Focus: Educ 02890

Saskatchewan Horticultural Association, POB 152, Balcarres, Sask. S0G 0C0
T: (306) 334-2201
Founded: 1927
Gen Secr: Kathryna Robb
Focus: Hort 02891

Saskatchewan Institute of Agrologists, 219 Robin Crescent, Ste 103, Saskatoon, Sask. S7L 6M8
T: (306) 242-2606
Founded: 1946; Members: 830
Focus: Agri
Periodical
Newsletter (weekly) 02892

Saskatchewan Library Association, POB 3388, Regina, Sask. S4P 3H1
T: (306) 780-9413
Founded: 1942; Members: 200
Pres: P. Resch
Focus: Libraries & Bk Sci
Periodical
Forum (5 times annually) 02893

Saskatchewan Lung Association, Fort San, Sask. S0G 1T0
T: (306) 332-5984
Focus: Pulmon Dis 02894

Saskatchewan Medical Association, 211 Fourth Av S, Saskatoon, Sask. S7K 1N1
T: (306) 244-2196
Founded: 1906; Members: 1150
Gen Secr: Dr. E.H. Baergen
Focus: Med 02895

Saskatchewan Pharmaceutical Association, 2631 28 Av, Ste 301, Regina, Sask. S4S 6X3
T: (306) 584-2292
Founded: 1911; Members: 1300
Focus: Pharmacol 02896

Saskatchewan Psychiatric Association, 1904 14 St E, Saskatoon, Sask. S7H 0B1
T: (306) 374-7822
Members: 40
Focus: Psychiatry 02897

Saskatchewan Registered Music Teachers' Association, 438 Egbert Av, Saskatoon, Sask. S7N 1X3
T: (306) 373-1209
Founded: 1932; Members: 250
Focus: Educ; Music 02898

Saskatchewan Research Council, 15 Innovation Blvd, Saskatoon, Sask. S7N 2X8
Tx: 0742484; Fax: (306) 933-7446
Founded: 1947
Pres: J. Hutch
Focus: Sci 02899

Saskatchewan Safety Council, 348 Victoria Av, Regina, Sask. S4N 0P6
T: (306) 527-3197
Founded: 1956; Members: 300
Gen Secr: R.M. Rowley
Focus: Safety 02900

Saskatchewan Society of Medical Laboratory Technologists, 614 Wardlow Rd, Saskatoon, Sask. S7M 4C9
Focus: Med 02901

Saskatchewan Society of Occupational Therapists, c/o O.T. Dept, Wascana Hospital, 23 & Av G, Regina, Sask. S4S 0A5
T: (306) 359-9255
Founded: 1903; Members: 5400
Pres: Jack MacKinnon
Focus: Therapeutics
Periodical
NBTA News (14 times annually) 02902

Saskatchewan Society of Osteopathic Physicians, 2228 Albert St, Regina, Sask. S4S 6X6
T: (306) 352-3167
Founded: 1914
Pres: Dr. Doris M. Tanner
Focus: Pathology; Physiology 02903

Saskatchewan Speech and Hearing Association, 70 Rawlinson Crescent, Regina, Sask. S4S 6B7
Gen Secr: Theresa Norlander
Focus: Logopedy 02904

Saskatchewan Teachers' Federation, POB 1108, Saskatoon, Sask. S7K 3N3
T: (306) 373-1660; Fax: (306) 374-1122
Founded: 1933; Members: 11700
Gen Secr: Fred Herron
Focus: Educ
Periodical
Saskatchewan Bulletin (15 times annually) 02905

Saskatchewan Veterinary Medical Association, 1025 Boychuk Dr, Ste 11, Saskatoon, Sask. S7H 5B2
T: (306) 955-7862
Focus: Vet Med 02906

Science Council of Canada / Conseil des Sciences du Canada, 100 Melcalfe St, Ottawa, Ont. K1P 5M1
T: (613) 996-2822
Founded: 1966
Pres: Dr. G.A. Kenney-Wallace; Gen Secr: E.V. Nyborg
Focus: Eng; Sci 02907

Science for Peace / Science et Paix, c/o University College, University of Toronto, Toronto, Ont. M5S 1A1
T: (416) 978-3606; Fax: 978-5848
Founded: 1981; Members: 1000
Pres: David Parnas; Gen Secr: Ericn Fawcett
Focus: Prom Peace
Periodicals
Canadian Papers in Peace Studies (weekly)
Science for Peace Bulletin (weekly) . . 02908

Sculptors' Society of Canada / Société des Sculpteurs du Canada, 40 Armadale Av, Toronto, Ont. M6S 3W8
Founded: 1928; Members: 110
Pres: May Marx
Focus: Arts 02909

Serena British Columbia, 6212 142 St, Surrey, B.C. V3W 5M3
T: (604) 596-8880
Founded: 1975; Members: 400
Pres: Joe Haddock; Janice Haddock
Focus: Family Plan
Periodical
Serena West Newsletter (weekly) . . . 02910

Serena Canada, 151 Holland, Ottawa, Ont. K1Y 0Y2
T: (613) 728-6536
Members: 931
Pres: Frank O'Connor; Marian O'Connor;
Gen Secr: Marie-Paule Doyle
Focus: Family Plan
Periodical
Serena Canada (weekly) 02911

Sex Information and Education Council of Canada (SIECCAN) / Conseil du Canada d'Informations et Education Sexuelles, 150 Laird Dr, Toronto, Ont. M4G 3V7
T: (416) 466-5304; Fax: (416) 467-8417
Founded: 1964; Members: 850
Focus: Educ
Periodicals
SIECCAN Journal (weekly)
SIECCAN Newsletter (2-3 times annually) 02912

Sir Joseph Flavelle Foundation, c/o National Trust Co., 21 King St E, Toronto, Ont. M5C 1B3
T: (416) 361-4096; Fax: (416) 361-4097
Founded: 1945; Members: 30
Gen Secr: D.R. Windeyer
Focus: Sci 02913

Social Science Federation of Canada / Fédération Canadienne des Sciences Sociales, 151 Slater St, Ste 415, Ottawa, Ont. K1P 5H3
T: (613) 238-6112; Fax: (613) 238-6114
Founded: 1941
Focus: Sociology 02914

Social Sciences and Humanities Research Council of Canada (SSHRC), 255 Albert St, POB 1610, Ottawa, Ont. K1P 6G4
T: (613) 995-5488
Founded: 1978; Members: 36
Pres: P. Leduc; Gen Secr: Louise Dandurand
Focus: Soc Sci; Humanities
Periodical
Annual Report 02915

Société Canadienne des Relations Publiques (Québec) / Canadian Public Relations Society (Québec), 1500 Rue Stanley, Ste 315, Montréal, Qué. H3A 1R3
T: (514) 842-7637
Founded: 1948; Members: 430
Focus: Advert
Periodical
Bulletin (weekly) 02916

Société d'Animation du Jardin et de l'Institut Botaniques (SAJIB), 4101 Sherbrooke St E, Ste 125, Montréal, Qué. H1X 2B2
T: (514) 252-1179
Founded: 1975; Members: 1000
Focus: Botany; Ecology; Hort
Periodical
Bulletin de la SAJIB (weekly) 02917

Société d'Archéologie et de Numismatique de Montréal / Antiquarian and Numismatic Society of Montréal, 280 Notre Dame St E, Montréal, Qué. H2Y 1C5
T: (514) 861-7182; Fax: (514) 861-8317
Founded: 1862; Members: 300
Pres: Pierre Brouillard; Gen Secr: Suzanne Lalumière
Focus: Archeol; Numismatics 02918

Société de Criminologie du Québec, 425 Viger Ouest, Local 620, Montréal, Qué. H2Z 1X2
T: (514) 873-4239
Founded: 1960; Members: 500
Gen Secr: Samir Rizkalla
Focus: Criminology
Periodical
Ressources et Vous (weekly) 02919

Société de Génealogie de Québec, CP 2234, Québec, Qué. G1K 7N8
T: (418) 683-5330
Founded: 1961; Members: 800
Pres: Jaqueline F. Asselin
Focus: Genealogy 02920

Societe d'Entomologie du Quebec / Entomological Society of Quebec, c/o Complexe Scientifique du Québec, Laboratoire d'Entomologie Forestière, 2700 Rue Einstein, Sainte-Foy, Qué. G1P 3W8
T: (418) 643-9679
Founded: 1873; Members: 250
Focus: Entomology 02921

Société de Protection des Plantes du Québec, c/o Station de Recherches Agricoles, CP 480, Saint-Hyacinthe, Qué. J2S 7B8
Founded: 1908; Members: 225
Pres: Richard Belanger; Gen Secr: L. Tartier
Focus: Ecology
Periodicals
Echos Phytosanitaires (weekly)
Phyto-Protection (weekly) 02922

Société des Auteurs, Eecherchistes, Documentalistes et Compositeurs, 1229 Rue Panet, Montréal, Qué. H2L 2Y6
T: (514) 526-9196; Fax: (514) 526-4124
Founded: 1949; Members: 550
Gen Secr: Yves Légaré
Focus: Lit; Doc; Music 02923

Société des Ecrivains Canadiens, 473 Rue Dolbeau, Québec, Qué. G1R 2R6
T: (418) 527-2940
Members: 80
Pres: Michel Champagne
Focus: Lit 02924

Société des Etudes Socialistes, c/o University College, University of Manitoba, Winnipeg, Man. R3T 2M8
T: (204) 474-9119
Founded: 1967; Members: 300
Focus: Socialism 02925

Société des Relations d'Affaires Hautes Etudes Commerciales, 5255 Av Decelles, Ste 2025C, Montréal, Qué. H3T 1V6
T: (514) 343-4544
Founded: 1963; Members: 200
Focus: Commerce 02926

Société de Thoracologie du Quebec / Quebec Thoracic Society, 3440 Av de l'Hôtel-de-Ville, Montréal, Qué. H2X 3B4
T: (514) 845-3129
Founded: 1956; Members: 83
Focus: Pulmon Dis 02927

Société Généalogique Canadienne-Française, CP 335, Station Place d'Armes, Montréal, Qué. H2Y 3H1
T: (514) 729-8366
Founded: 1943; Members: 3000
Pres: Normand Robert
Focus: Genealogy
Periodical
Mémoires (weekly) 02928

Société Historique Acadienne (SHA), POB 2363, Station A, Moncton, N.B. E1C 8J3
T: (506) 532-1280
Founded: 1960; Members: 700
Pres: Alonzo Le Blanc
Focus: Hist
Periodical
Cahiers de la Société Historique Acadienne (weekly) 02929

Société Historique de la Saskatchewan, c/o College West, Université de Régina, Regina, Sask. S4S 0A2
T: (306) 585-4177
Founded: 1978; Members: 30
Focus: Hist 02930

Société Historique de Québec, 43 Côte de la Fabrique, Québec, Qué. G1R 5M1
T: (418) 694-9740
Members: 550
Pres: Monique Duval
Focus: Hist
Periodical
Québecensia (weekly) 02931

Société Historique du Saguenay, CP 456, Chicoutimi, Qué. G7H 5C8
T: (418) 549-2805
Founded: 1934
Focus: Hist 02932

Société Linnéenne de Québec, c/o Aquarium de Québec, 1675 Av du Parc, Sainte Foy, Qué. G1W 4S3
T: (418) 653-8186
Founded: 1929; Members: 800
Pres: Robert Joly
Focus: Bio
Periodical
Le Linnéen (weekly) 02933

Société Nationale de Diffusion Educative et Culturelle / National Society for Educational and Cultural Advancement, 1755 René-Levesque Est, Montréal, Qué. H2K 4P6
T: (514) 598-1222; Fax: (514) 598-7872
Founded: 1966; Members: 21
Gen Secr: Joel Bévillard
Focus: Educ 02934

Sociéte pour vaincre la Pollution, 445 Rue Saint-François-Xavier, Montréal, Qué. H2Y 2T1
T: (514) 844-5477
Founded: 1970; Members: 400
Focus: Ecology 02935

Société Québécoise d'Assainissement des Eaux, 1055 Blvd Dorchester Est, Montréal, Qué. H2L 4S5
T: (514) 873-7411
Focus: Ophthal 02936

Société Zoologique de Québec, 8191 Av du Zoo, Charlesbourg, Qué. G1G 4G4
T: (418) 627-3072
Founded: 1932; Members: 400
Gen Secr: Céline Poulin
Focus: Zoology 02937

Society for Cable Television Engineering, 4560 Fieldgate Dr, Mississauga, Ont. L4W 3W6
T: (416) 629-1111
Founded: 1972; Members: 250
Pres: Karl Poirier
Focus: Eng 02938

Society for Indian and Northern Education, c/o College of Education, University of Saskatchewan, Saskatoon, Sask. S7N 0W0
Focus: Educ 02939

Society for the Study of Architecture in Canada / Société pour l'Etude de l'Architecture au Canada, POB 2935, Station D, Ottawa, Ont. K1P 5W9
Founded: 1974; Members: 300
Focus: Archit
Periodical
Nouvelle: News (weekly) 02940

Society of Chemical Industry (Canadian Section), c/o D.W.W. Kirkwood, Linde Canada Inc, 1 City Centre Dr, Ste 1200, Mississauga, Ont. L5B 1M2
T: (416) 803-1703; Fax: (416) 803-1687
Founded: 1902; Members: 200
Pres: Wilson D.C.
Focus: Chem 02941

Society of Christian Schools in British Columbia, 10638 132A St, Surrey, B.C. V3T 3X7
T: (604) 581-4444
Founded: 1962
Gen Secr: Harro van Brummelen
Focus: Educ 02942

Society of Commercial Seed Technologists (SCST), POB 1712, Brandon, Man. R7A 6S3
T: (204) 328-5313
Founded: 1922; Members: 210
Gen Secr: Marie T. Greeniaus
Focus: Agri
Periodical
Seed Technologist News (3 times annually) 02943

Society of Composers, Authors and Music Publishers of Canada, 41 Valleybrook Dr, Don Mills, Ont. M3B 2S6
T: (416) 445-8700; Fax: (416) 445-7108
Founded: 1990; Members: 41000
Gen Secr: Michael Rock
Focus: Music; Lit
Periodicals
The Canadian Composer (weekly)
Probe (weekly) 02944

Society of Management Accountants of Canada, 154 E Main St, POB 176, Hamilton, Ont. L8N 3C3
T: (416) 525-4100
Founded: 1940
Focus: Business Admin
Periodicals
Cost & Management (weekly)
RIA Digest (weekly) 02945

Society of Obstetricians and Gynaecologists of Canada / Société des Obstétriciens et Gynécologues du Canada, 14 Prince Arthur Av, Ste 210, Toronto, Ont. M5R 1A9
T: (416) 923-3513
Founded: 1945; Members: 1250
Gen Secr: Ruth Stockdale

Focus: Gynecology
Periodical
Bulletin (weekly) 02946

Society of Philosophy and Social Science, POB 401, Kingston, Ont. K7M 2Y7
T: (613) 544-3715
Founded: 1976; Members: 50
Pres: Edward R. Grenda
Focus: Philos; Soc Sci 02947

Society of Ukrainian Engineers and Associates in Canada, 2199 Bloor St W, Toronto, Ont. M6S 1N4
Founded: 1975
Pres: Dr. J.G. Kurys
Focus: Eng 02948

Spectroscopy Society of Canada, c/o T.R. Churchill, POB 332, Station A, Ottawa, Ont. K1N 8V3
Founded: 1967; Members: 500
Pres: Dr. D.C. Grégoire
Focus: Physics
Periodical
Canadian Journal of Spectroscopy . . . 02949

Speech and Hearing Association of Alberta, POB 86, Main P.O., Edmonton, Alta. T5J 2G9
Founded: 1964; Members: 210
Pres: Sara McClain
Focus: Logopedy 02950

Speech and Hearing Association of Nova Scotia, c/o Nova Scotia Hearing and Speech Clinic, 5919 South St, Halifax, N.S. B3J 3G6
Founded: 1978; Members: 45
Gen Secr: Maureen Merchant
Focus: Logopedy 02951

Speech and Hearing Association of Nova Scotia, POB 775, Station M, Halifax, N.S. B3J 2V2
Founded: 1978; Members: 61
Pres: Diane Roper
Focus: Logopedy 02952

Speech Foundation of Ontario, 93 Grenville St, Toronto, Ont. M5S 1B4
T: (416) 979-2437
Focus: Logopedy 02953

Spina Bifida and Hydrocephalus Association of Ontario, 55 Queen St, Ste 300, Toronto, Ont. M5C 1R6
T: (416) 364-1871; Fax: (416) 364-5437
Founded: 1971; Members: 1500
Pres: Robert Johnston
Focus: Pathology
Periodical
Current: Pamphlet (weekly) 02954

Spina Bifida Association of British Columbia, 9460 140 St, Surrey, B.C. V3V 5Z4
T: (604) 584-1361
Founded: 1976; Members: 50
Pres: Ellie Gelhorn
Focus: Pathology 02955

Spina Bifida Association of Canada, 633 Wellington Crescent, Winnipeg, Man. R3M 0A8
T: (204) 452-7580
Focus: Pathology 02956

Sport Medicine Council of Canada / Conseil Canadien de la Médecine Sportive, 333 River Rd, Tower C, Vanier, Ont. K1L 8H9
T: (613) 745-4866
Founded: 1978
Focus: Med; Sports 02957

Standards Council of Canada / Conseil Canadien des Normes, 350 Sparks St, Ste 1203, Ottawa, Ont. K1R 7S8
T: (613) 238-3222; Tx: 0534403
Founded: 1970; Members: 57
Focus: Standards 02958

Standards Engineering Society (Canadian Region), c/o Northern Telecom Electronics Ltd, 301 Moodie Dr, Nepean, Ont. K2H 9C4
T: (613) 726-4468
Founded: 1947; Members: 100
Gen Secr: Lorne K. Wagner
Focus: Standards 02959

Statistical Society of Canada (SSC) / Société Statistique du Canada, 1280 Main St W, Hamilton, Ont. L8S 4K1
T: (416) 525-9140
Founded: 1972; Members: 650
Focus: Stats
Periodicals
Canadian Journal of Statistics (weekly)
SSC Newsletter (weekly) 02960

Steel Castings Institute of Canada / Institut Canadien des Moulages d'Acier, 151 Slater St, Ste 307, Ottawa, Ont. K1P 5H3
T: (613) 232-2645; Fax: (613) 230-9607
Founded: 1917; Members: 11
Focus: Metallurgy 02961

Sugar Industry Technologists (SIT), POB 632, Sainte-Thérèse de Blainville, Qué. J7E 4K3
T: (514) 621-3524; Fax: (514) 965-1121
Founded: 1941; Members: 600
Gen Secr: Bruce A. Foster
Focus: Food
Periodical
Annual Technical Meeting Proceedings (weekly) 02962

Thyroid Foundation of Canada / Fondation du Canada pour les Maladies Thyroidiennes, POB 1643, Kingston, Ont. K7L 5C8
T: (613) 546-1576, 542-8330
Founded: 1980; Members: 1400
Pres: George Wright
Focus: Endocrinology 02963

Tiger Hills Arts Association, POB 58, Holland, Man. R0G 0X0
T: (204) 526-2063
Founded: 1978; Members: 147
Pres: Susan Dickens
Focus: Arts 02964

Toronto Area Archivists Group, POB 97, Station F, Toronto, Ont. M4Y 2L4
Founded: 1973; Members: 155
Pres: Kim Moir
Focus: Archives 02965

Toronto Psychoanalytic Society, 33 Price St, Toronto, Ont. M4W 1Z2
T: (416) 922-7770
Founded: 1970; Members: 112
Pres: H. Anderson
Focus: Psychoan 02966

Toronto Society of Financial Analysts, 390 Bay St, Ste 2000, Toronto, Ont. M5H 2Y2
T: (416) 366-5755
Founded: 1936; Members: 1070
Focus: Finance 02967

Traffic Injury Research Foundation of Canada / Fondation de Recherches sur les Blessures de la Route au Canada, 171 Nepean St, Ottawa, Ont. K2P 0B4
T: (613) 238-5235
Founded: 1964; Members: 250
Focus: Transport 02968

Transport 2000 Canada, 22 Metcalfe St, Ste 404, POB 858, Station B, Ottawa, Ont. K1P 5P9
T: (613) 235-2865
Founded: 1976; Members: 1000
Gen Secr: Nicholas G.L. Vincent
Focus: Transport 02969

Truss Plate Institute of Canada, POB 119, Concord, Ont. L4K 1B2
T: (416) 475-6670
Founded: 1971
Pres: Allan Gandell
Focus: Eng 02970

Turner's Syndrome Society / Association du Syndrome de Turner, Administrative Studies Bldg, Room 006, York University, Downsview, Ont. M3J 1P3
T: (416) 736-5023
Founded: 1981; Members: 150
Gen Secr: Sandi Hofbauer
Focus: Endocrinology 02971

Ultralight Aircraft Association of Canada, POB 625, Ottawa, Ont. K1P 3S7
T: (613) 236-4912
Members: 700
Pres: R. Morris
Focus: Aero 02972

Universities Art Association of Canada / Association d'Art des Universités du Canada, c/o Nova Scotia College of Art and Design, 5163 Duke St, Halifax, N.S. B3J 3J6
T: (403) 237-0373
Members: 200
Focus: Arts 02973

Upper Canada Society for the History and Philosophy of Science and Technology, c/o Dept of History, McMaster University, Hamilton, Ont. L8S 4L9
T: (416) 525-9140
Pres: George Grinelle
Focus: Cultur Hist; Philos 02974

Urban Development Institute of Ontario, 60 Bloor St W, Ste 1203, POB 12, Toronto, Ont. M4W 3B8
T: (416) 928-0491
Focus: Urban Plan 02975

Vancouver Museum Association, 1100 Chestnut St, Vancouver, B.C. V6J 3J9
T: (604) 736-4431; Fax: (604) 736-5417
Founded: 1968; Members: 2020
Pres: Rodney Ward; Gen Secr: David Hemphill
Focus: Hist; Anthro 02976

Vancouver Natural History Society, POB 3021, Vancouver, B.C. V6B 3X5
T: (604) 738-3177
Founded: 1918; Members: 1660
Pres: Val Schaefer
Focus: Nat Sci
Periodical
Discovery (weekly) 02977

Vancouver Safety Council, 96 Broadway E, Ste 204, Vancouver, B.C. V5T 1V6
T: (604) 877-1221
Focus: Safety 02978

Victoria Foundation, 877 Island Rd, Victoria, B.C. V8S 2V1
T: (604) 592-2132
Founded: 1936
Gen Secr: E.E. Chamberlin
Focus: Sci 02979

Victoria-Haliburton Lung Association, 12 King St, Lindsay, Ont. K9V 4R8
T: (705) 324-6497
Gen Secr: Lynn Kimmett
Focus: Pulmon Dis 02980

Victorian Studies Association of Western Canada, c/o Dept of English, University of Alberta Alberta, Edmonton, Alta. T6G 2E5
Focus: Hist
Periodical
Victorian Review (weekly) 02981

Visual Arts Ontario, 439 Wellington St W, Toronto, Ont. M5V 1E7
Founded: 1973; Members: 3000
Gen Secr: Hennie L. Wolff
Focus: Arts
Periodicals
Agenda (weekly)
Artviews (weekly) 02982

Waldorf School Association of Alberta, 9850 154 St, Edmonton, Alta. T5P 2G6
T: (902) 489-0919
Focus: Educ 02983

Waterloo Historical Society, POB 552, Station C, Kitchener, Ont. N2G 4A2
Founded: 1912
Gen Secr: Frances McIntosh
Focus: Hist
Periodical
Waterloo Historical Society (weekly) . . 02984

Welding Institute of Canada, 391 Burnhamthorpe Rd E, Oakville, Ont. L6J 6C9
T: (416) 845-9881
Founded: 1979; Members: 2500
Pres: N.F. Eaton
Focus: Eng 02985

West Coast Environmental Law Association, 207 Hastings St W, Ste 1001, Vancouver, B.C. V6B 1H7
T: (604) 684-7378; Fax: (604) 684-1312
Founded: 1974; Members: 300
Focus: Law 02986

West Coast Environmental Law Research Foundation, 207 Hastings St W, Ste 1001, Vancouver, B.C. V6B 1H7
T: (604) 684-7378
Founded: 1974; Members: 300
Focus: Law 02987

West Coast Library Association, c/o Western Region Libraries, POB 2007, Corner Brook, Nfld. A2H 6V7
T: (709) 634-7333
Focus: Libraries & Bk Sci 02988

Western Board of Music, c/o University of Alberta, Edmonton, Alta. T6G 2E6
T: (403) 492-3264
Founded: 1934
Gen Secr: Marguerite Abbott
Focus: Music 02989

Western Canada Water and Wastewater Association, POB 6168, Station A, Calgary, Alta. T2H 2L4
T: (403) 259-4041
Founded: 1948; Members: 2000
Focus: Water Res
Periodical
Bulletin (weekly) 02990

Western Canadian Society for Horticulture, c/o Alberta Horticultural Research Center, Brooks, Alta. T0J 0J0
T: (403) 362-3391
Founded: 1944; Members: 224
Focus: Hort 02991

Western Cleft Lip and Palate Association, 5953 Charles St, Burnaby, B.C. V5B 2G6
T: (604) 299-3588
Members: 150
Pres: Elizabeth Efting
Focus: Pathology 02992

Western Québec Teachers Association, 155 Mont Bleu, Ste 306, Hull, Qué. J8Z 1K4
Focus: Educ 02993

W. Garfield Weston Foundation, 22 Saint Clair Av E, Ste 2001, Toronto, Ont. M4T 2S3
T: (416) 922-4697
Founded: 1957
Focus: Sci 02994

Winnipeg Foundation, 167 Lombard Av, Ste 759, Winnipeg, Man. R3B 0V3
T: (204) 944-9474
Founded: 1921
Gen Secr: Alan G. Howison
Focus: Sci 02995

Winnipeg Society of Financial Analysts, c/o Richardson Greenshields of Canada Ltd, 1 Lombard Pl, Winnipeg, Man. R3B 0Y2
T: (204) 988-5629
Founded: 1952; Members: 85
Pres: Timothy E. Burt
Focus: Finance 02996

Woodworkers Accident Prevention Association, 2 Bloor St E, Toronto, Ont. M4W 3C2
T: (416) 965-8888
Focus: Safety; Forestry 02997

World Federation for Mental Health / Fédération Mondiale pour la Santé Mentale, c/o University of British Columbia, 107-2352 Health Sciences Mall, Vancouver, B.C. V6T 1W5
T: (604) 228-2332; CA: MENSANTE
Founded: 1948
Gen Secr: Roberta L. Beiser
Focus: Educ Handic 02998

Writers Development Trust, 24 Ryerson Av, Ste 207, Toronto, Ont. M5T 2P3
T: (416) 861-8222
Focus: Lit 02999

York Technology Association, 7305 Woodbine Av, Ste 132, Markham, Ont. L3R 3V7
T: (416) 477-1727; Fax: (416) 475-8104
Founded: 1982; Members: 120
Focus: Eng
Periodicals
Yorktech Directory (weekly)
Yorktech Newsletter (weekly) 03000

York-Toronto Lung Association, 573 King St E, Ste 201, Toronto, Ont. M5A 4L3
T: (416) 864-1112
Gen Secr: Robert Olsen
Focus: Pulmon Dis 03001

Youth Science Foundation / Fondation Sciences Jeunesse, 151 Slater St, Ste 904, Ottawa, Ont. K1P 5H3
T: (613) 238-1671
Founded: 1966; Members: 230
Focus: Sociology 03002

Yukon Historical and Museums Association, POB 4357, Whitehorse, Y.T. Y1A 3T5
T: (403) 667-4704
Focus: Hist; Archives 03003

Yukon Teachers' Association, 107 Main St, Ste 200, Whitehorse, Y.T. Y1A 2A7
T: (403) 668-6777
Members: 315
Pres: Ken Taylor
Focus: Educ 03004

Yukon Tuberculosis and Health Association, POB 4754, Whitehorse, Y.T. Y1A 4N6
T: (403) 667-7462
Founded: 1948
Pres: Corinne Cyr
Focus: Pulmon Dis; Public Health . . . 03005

Zoological Society of Manitoba, 190 Rupert Av, Winnipeg, Man. R3B 0N2
T: (204) 956-2830
Members: 100
Focus: Zoology 03006

Chile

Academia Chilena de Bellas Artes, Clasificador 1349, Correo Central, Santiago
T: (02) 331902
Founded: 1964
Pres: Carlos Riesco Grez; Gen Secr: Alejandro Sieve-King
Focus: Arts
Periodical
Boletín 03007

Academia Chilena de Ciencias, Clasificador 1349, Correo Central, Santiago
T: (02) 382847
Founded: 1964; Members: 57
Pres: Dr. Luis Vergas Fernández; Gen Secr: Dr. José Corvalán Díaz
Focus: Sci
Periodical
Boletín (weekly) 03008

Academia Chilena de Ciencias Sociales, Politicas y Morales, Clasificador 1349, Correo Central, Santiago
T: (02) 6331902; Fax: (02) 6326649
Founded: 1964; Members: 36
Pres: Juan de Dios Vial Larrain; Gen Secr: H. Godoy
Focus: Poli Sci; Sociology; Educ
Periodicals
Anales/Societas (weekly)
Boletín (3 times annually)
Folletos 03009

Academia Chilena de la Historia, Almirante Montt 453, Santiago
Founded: 1933; Members: 36
Pres: Fernando Campos Harriet
Focus: Hist
Periodicals
Boletín de la Academia
Historia de Chile 03010

Academia Chilena de la Lengua, Clasificador 1349, Correo Central, Santiago
T: (02) 382847
Founded: 1885; Members: 36
Pres: Roque E. Scarpa Straboni; Gen Secr: Ernesto Livacic Gazzano
Focus: Ling
Periodical
Boletín de la Academia Chilena de la Lengua 03011

Academia Chilena de Medicina, Clasificador 1349, Correo Central, Santiago
T: (02) 331902
Founded: 1964; Members: 93
Pres: Dr. Armando Roa; Gen Secr: Dr. Jaime Perez-Olea
Focus: Med
Periodicals
Boletín (weekly)
Proceedings of the Chilean History of Medicine . . . 03012

Appropriate Technology Study Centre for Latin America, Casilla 197-V, Valparaiso
T: (032) 213360; Tx: 330506; Fax: (032) 250383
Founded: 1979
Focus: Eng . . . 03013

Asociación Chilena de Microbiología (A.Ch.M.), Casilla 1075, Santiago 22
Founded: 1964; Members: 183
Pres: Dr. Manuel Rodriguez Leiva; Gen Secr: Dr. Ines Calderon Rivera
Focus: Microbio
Periodical
Anales de Microbiologia (weekly) . . . 03014

Asociación Chilena de Sismología e Ingeniería Antisísmica / Chilean Association of Seismology and Anti-Seismic Engineering, Casilla 2796, Santiago
Founded: 1963
Pres: Dr. Patricio Ruiz
Focus: Geophys . . . 03015

Asociación Chilena para la Investigación y Desarrollo del Hormigón Estructural, Marcoleta 250, Santiago
T: (02) 226501
Founded: 1980; Members: 45
Focus: Civil Eng . . . 03016

Asociación de Bibliotecarios de Chile, Casilla 3741, Santiago
Focus: Libraries & Bk Sci . . . 03017

Asociación de Informatica y Computación en Educación (ACHICE), Lo Barnechea 1772, Santiago
T: (02) 2167602; Fax: (02) 2167662
Founded: 1990; Members: 875
Pres: Fidel Oteiza; Gen Secr: Maria A. Palavicino
Focus: Computer & Info Sci . . . 03018

Asociación de Lingüística y Filología de América Latina (ALFAL), La Verbena 3882, Santiago
Founded: 1962
Focus: Ling
Periodical
Cuadernos de Lingüística (weekly) . . . 03019

Asociación Latinoamericana de Biomatemática / Latin American Association of Biomathematics, c/o Instituto Profesional de Santiago, Dieciocho 161, Santiago
Focus: Math . . . 03020

Caribbean Technical Cooperation Network on Upper Watershed Management, c/o FAO Regional Office, Av Santa Maria 6700, Casilla 10095, Santiago
T: (02) 2288056; Tx: 340279; Fax: (02) 484312
Founded: 1983
Gen Secr: Kyran Thelen
Focus: Eng
Periodical
Network Circular Letter . . . 03021

Centro de Investigación Minera y Metalúrgica, Av Parque Institucional 6500, Casilla 170, Santiago
T: (02) 2289544
Founded: 1972
Focus: Mineralogy; Metallurgy; Mining . . . 03022

Centro de Perfeccionamiento, Experimentación e Investigaciones Pedagógicas (CPEIP), c/o Curriculum Development and Research Center, Ministerio de Educación, Casilla 16162, Correo 9, Santiago
T: (02) 2167602; CA: CENTPERF; Fax: (02) 2167662
Founded: 1968; Members: 80
Pres: Gabriel de Pujadas Hermosilla; Gen Secr: R. Reyes Soto
Focus: Educ
Periodicals
English Teaching Newsletter (weekly)
Inred (weekly)
Revista Chilena de Educatión Quimica (10 times annually)
Revista de Educación (weekly)
Revista de Estudios (weekly)
Revista El Niñno Limitado (weekly) . . . 03023

Centro Interamericano de Enseñanza de Estadistica (CIENES), Calle Triana 820, Casilla 10015, Santiago
T: (02) 742526
Founded: 1962
Focus: Stats . . . 03024

Centro Interuniversitario de Desarollo (CINDA), Europa 2048, Santiago
T: (02) 2339869; Fax: (02) 2341117
Founded: 1971; Members: 16
Gen Secr: Ivan Lavados Montes
Focus: Develop Areas
Periodical
Boletín Informativo CINDA . . . 03025

Centro Latinoamericano de Demografía (CELADE) / Latin American Demographic Centre, Av Dag Hamarskjold s/n, Edificio Naciones Unidas, Casilla 91, Santiago
Fax: (02) 2080252
Founded: 1957
Focus: Stats
Periodicals
Boletín Demográfico (weekly)
DOCPAL Resúmentes sobre Población en América Latina (weekly)
Notas de Población (3 times annually) . 03026

Centro Nacional de Información y Documentación (CENID), Calle Canadá 308, Casilla 297V, Correo 21, Santiago
Founded: 1962
Focus: Doc
Periodicals
Seria Bibliográfica (weekly)
Serie Directorios (weekly)
Serie Estudios (weekly)
Serie Información y Documentación (weekly) . . . 03027

Colegio de Abogados, Palacio de los Tribunales de Justicia, Concepción
T: 23902
Focus: Law . . . 03028

Colegio de Architectos de Chile, Av Libertador Bernardo O'Higgins 115, Casilla 13377, Santiago
Fax: (02) 6399869
Founded: 1942; Members: 4500
Pres: H. Montecinos Barrientos; Gen Secr: Sergio Gonzalez Tapia
Focus: Archit; Eng; Energy; Civil Eng
Periodicals
Bienal de Arquitectura (biennial)
Boletín (weekly)
Revista CA (weekly) . . . 03029

Colegio de Bibliotecarios de Chile / Chilean Librarianship Association, Diagonal Paraguay 383, Depto 122, Casilla 3741, Santiago
T: (02) 2225652
Founded: 1969; Members: 1500
Focus: Libraries & Bk Sci
Periodical
Micronoticias (weekly) . . . 03030

Colegio de Dentistas de Chile, Tucapal 452, Concepción
T: (041) 25897
Focus: Dent . . . 03031

Colegio de Ingenieros de Chile, A. Pinto 372, Concepción
T: (041) 25615
Focus: Eng . . . 03032

Colegio de Quimico-Farmacéuticos de Chile, Casilla 1136, Santiago
Founded: 1942; Members: 2500
Focus: Chem; Pharmacol . . . 03033

Colegio de Químicos Farmacéuticos, Tucapal 340, Concepción
T: (041) 25502
Focus: Pharmacol . . . 03034

Colegio de Técnicos, Av Libertador Bernardo O'Higgins 650, Of 502, Concepción
T: (041) 21492
Focus: Eng . . . 03035

Colegio Farmacéutico de Chile, Casilla 265, Concepción
Founded: 1943
Focus: Pharmacol . . . 03036

Comisión Chilena de Energía Nuclear, Amunategui 95, Casilla 188D, Santiago
T: (02) 6990070; Tx: 340468; CA: NUCLEOTECNICA; Fax: (02) 6991618
Founded: 1965
Pres: Eduardo Bobadilla López; Gen Secr: Germán Piderit Alvear
Focus: Nucl Res
Periodical
Nucleotecnica (weekly) . . . 03037

Comisión Económica para América Latina y el Caribe / Economic Commission for Latin America and the Caribbean, Av Dag Hammarskjold s/n, Casilla 179D, Santiago
Focus: Econ; Stats
Periodicals
Anuario Estadistico de America Latina y el Caribe: Statistical Yearbook for Latin America and the Caribbean (weekly)
Boletín Demográfico (weekly)
Cepal Review (weekly)
Cuadernos de la Cepal (weekly)
Cuadernos del Ilpes (weekly)
DOCPAL Resúmenes sobre Población en América Latina (weekly)
Economic Survey of Latin America and the Caribbean (weekly)
Estudio Económico de América Latina y el Caribe (weekly)
Estudios e Informes de la Cepal (weekly)
Revista de la Cepal (3 times annually) . 03038

Comisíon Nacional de Investigación Científica y Tecnológica (CONTICYT), Calle Canadá 308, Casilla 297V, Santiago
T: (02) 744537; Fax: (02) 2096729
Founded: 1969
Pres: Enrique d' Etigny
Focus: Eng
Periodical
Panorama Cientifico (weekly) . . . 03039

Comité Chileno Veterinario de Zootecnia, Casilla 16202, Correo 9, Santiago
Founded: 1951
Focus: Vet Med; Zoology . . . 03040

Comité Nacional de Geografía, Geodesía y Geofísica, Nueva Santa Isabel 1640, Santiago
Fax: (02) 441677
Founded: 1935; Members: 119
Pres: Manuel Garate Meneses; Gen Secr: P. Gran Lopez
Focus: Surveying; Geography . . . 03041

Comité Nacional Pro Defensa de la Fauna y Flora (CODEFF) / National Committee for the Defence of Fauna and Flora, Sazi3 1885, Casilla 3675, Santiago
T: (02) 6968562
Founded: 1968; Members: 300
Pres: Pedro Fernández Bitterlich; Gen Secr: Gabriel Sanhueza Suárez
Focus: Ecology
Periodicals
CODEFF Actual (weekly)
CODEFF Informa (weekly) . . . 03042

Comité Oceanográfico Nacional, Casilla 324, Valparaíso
Founded: 1971
Focus: Oceanography . . . 03043

Commission on Livestock Development in Latin America and the Caribbean, c/o FAO Office for Latin America and the Caribbean, Av Santa María 6700, Casilla 10095, Santiago
T: (02) 2288056; Tx: 340279; Fax: (02) 4848056
Founded: 1987
Focus: Animal Husb . . . 03044

Confederación de Educadores de América Latina, c/o Ministerio de Educación, Av Libertador B. O'Higgins 1371, Santiago
Focus: Educ . . . 03045

Confederación Latinoamericano de Fisioterapía y Kinesiología (CLAFIC), c/o Colegio de Kinesiología, J. Diaz Garces 90, Casilla 9317, Correo Central, Santiago
Gen Secr: P. Mancilla
Focus: Physical Therapy . . . 03046

Consejo de Rectores de Universidades Chilenas (CRUCH), Moneda 673, Casilla 14798, Santiago
T: (02) 383137
Founded: 1954
Focus: Sci . . . 03047

Corporación de Fomento de la Producción (CORFO), Ramón Nieto 920, Santiago
Fax: (02) 6711058
Focus: Business Admin; Econ
Periodicals
Annual Report (weekly)
Chile Economic Report (weekly) . . . 03048

Corporación para el Desarrollo de la Ciencia (CODECI), Marcoleta 250, Casilla 3387, Santiago
T: (02) 2226501
Founded: 1978; Members: 120
Focus: Anthro; Archeol; Geology
Periodical
Revista CODECI (weekly) . . . 03049

Corporación Toesca para el Desarrollo de la Arquitectura, Marcoleta 250, Santiago
T: (02) 226501
Founded: 1980; Members: 50
Focus: Archit . . . 03050

Federación Latinoamericana de Parasitología, c/o Departamento der Parasitología, Universidad de Chile, Casilla 9183, Santiago
Focus: Microbio . . . 03051

Fundación Gildemeister, Augustinas Esquina Amunátegui 178, Casilla 99D, Santiago
Founded: 1947
Focus: Surgery; Cardiol . . . 03052

Instituto de Chile, Almirante Montt 453, Clasificador 1349, Correo Central, Santiago
T: (02) 6382847; Fax: (02) 6326649
Founded: 1964
Pres: Juan de Dios Vial Larraín; Gen Secr: Prof. Carlos Riesco
Focus: Sci
Periodical
Boletín . . . 03053

Instituto de Estudios y Publicaciones Juan Molina, Casilla de Correo 2974, Santiago
T: (02) 8441374
Founded: 1976; Members: 28
Focus: Nat Sci . . . 03054

Instituto de Ingenieros de Chile, San Martín 352, Casilla 487, Santiago
T: (02) 6984028
Founded: 1888; Members: 800
Focus: Eng
Periodicals
Anales
Revista Chilena de Ingeniería . . . 03055

Instituto de Investigaciones Agropecuarias, c/o Biblioteca Central, Casilla 439, Correo 3, Santiago
T: (02) 5586061
Founded: 1964
Focus: Agri
Periodicals
Agricultura Técnica (weekly)
Bibliografía Agrícola Chilena (weekly)
Investigación y Progreso Agropecuario: Carillanca (weekly)
Investigación y Progreso Agropecuario: La Platina (weekly)
Investigación y Progreso Agropecuario: Quilamapu (weekly)
Memoria Anual (weekly) . . . 03056

Instituto de Salud Publica de Chile, Av Marathon 1000, Casilla 48, Santiago
T: (02) 2391105; CA: BACTECHILE
Founded: 1929; Members: 720
Focus: Microbio
Periodical
Boletín (weekly) . . . 03057

Instituto Latinoamericano de Planificación Económica y Social, c/o United Nations, Av Dag Hammarskjöld, Casilla 1567, Santiago
Founded: 1962
Focus: Sociology; Econ . . . 03058

Instituto Oceanográfico de Valparaíso, Av Errázuriz 471, Casilla 117-V, Valparaíso
Founded: 1945
Focus: Oceanography . . . 03059

Latin American Association of Physiological Sciences, Canada 308, Santiago
T: (02) 2224516 ext 2850; Tx: 240395; Fax: (02) 2258427
Founded: 1957
Pres: Dr. José M. Anias Calderón; Gen Secr: Dr. Patricio Zapata
Focus: Physiology
Periodical
Acta Physiologica, Pharmacologica et Therapeutica Latino-Americana (weekly) . . . 03060

Liga Marítima de Chile, Av Errázuriz 471, Casilla 117V, Valparaíso
Founded: 1914; Members: 1350
Focus: Navig . . . 03061

Oficina Regional de Educación de la UNESCO para América Latina y el Caribe, Enrique Depiano 2058, Casilla 3187, Santiago
T: (02) 2049032; Tx: 340258; Fax: (02) 2091875
Founded: 1969
Focus: Educ
Periodicals
Boletin Bibliografico (weekly)
Boletin de Educación (3 times ammually)
Contacto (weekly)
FBEDOC Informa (weekly) . . . 03062

Servicio Nacional de Geología y Minería (SERNAGEOMIN) / National Service of Geology and Mining, Av Santa María 0104, Casilla 1347, Santiago
T: (02) 7375050; Fax: (02) 7372026
Founded: 1981; Members: 307
Pres: Hernán Danús Vásquez
Focus: Geology; Mining
Periodicals
Anuario de la Minería (weekly)
Boletín (weekly)
Carta Geológica de Chile (weekly)
Carta Hidrogeológica de Chile (weekly)
Carta Magnética de Chile (weekly)
Revista Geográfica de Chile
Revista Geológica de Chile (weekly) . . . 03063

Sistema Nacional de Información en Educación (SINIE) / National Educational Information System, c/o Ministerio de Educación Pública, Centro de Perfeccionamiento, Experimentación e Investigaciones Pedagogicas, Nido de Aguilas s/n, Casilla 16162, Correo 9, Santiago
T: (02) 2167602; Fax: (02) 2167662
Founded: 1983
Pres: Maria A. Palavicino
Focus: Educ
Periodical
INRED: Indices y Resumenes en Educación (weekly) . . . 03064

Sociedad Agronómica de Chile, MacIver 120, Of 36, Santiago
T: 384881
Founded: 1910; Members: 1900
Pres: Dr. L.A. Lizana; Gen Secr: Hector E. Núñez
Focus: Agri . . . 03065

Sociedad Arqueológica de la Serena, Casilla 125, La Serena
Founded: 1944
Focus: Archeol . . . 03066

Sociedad Chilena de Cancerología, c/o Fundación Arturo López Pérez, Rancagua 878, Santiago
T: 250063
Founded: 1964; Members: 64
Pres: Dr. Luis Orlandi; Gen Secr: Dr. Nelson Romero
Focus: Cell Biol & Cancer Res . . . 03067

Sociedad Chilena de Cardiología y Cirugía Cardiovascular / Chilean Society of Cardiology and Cardiovascular Surgery, Los Conquistadores 2251, Santiago
Founded: 1948; Members: 353
Pres: Oscar Gómez; Gen Secr: Fernando Florenzano
Focus: Cardiol 03068

Sociedad Chilena de Cirugía Plástica y Reparadora, Av Santa María 410, Santiago
Founded: 1944
Focus: Surgery 03069

Sociedad Chilena de Entomología, c/o Sección Entomología, Museo Nacional de Historia Natural, Interior Quinta Normal s/n, Casilla 21132, Santiago
Founded: 1923; Members: 175
Pres: Daniel Frias; Gen Secr: Marta Lewin
Focus: Entomology
Periodical
Revista Chilena de Entomología (weekly) 03070

Sociedad Chilena de Física, Casilla 5487, Santiago
Founded: 1960; Members: 254
Focus: Physics 03071

Sociedad Chilena de Gastroenterología, Esmeralda 678, Santiago
Focus: Gastroenter 03072

Sociedad Chilena de Gerontología, Av Bulnes 377, Dep 605, Santiago
T: (02) 67176
Founded: 1961; Members: 40
Focus: Geriatrics 03073

Sociedad Chilena de Hematología, MacIver 721, Casilla 639, Santiago
Founded: 1943; Members: 80
Pres: M. Bravo; Gen Secr: Myriam Campbell
Focus: Hematology 03074

Sociedad Chilena de Historia Natural, Casilla 787, Santiago
Founded: 1926
Pres: Dr. Jürgen Rotmann
Focus: Nat Sci 03075

Sociedad Chilena de Historia y Geografía, Casilla 1386, Santiago
Founded: 1911; Members: 311
Pres: Guillermo Donoso Vergara; Gen Secr: Alejandro Pizarro Soto
Focus: Geography; Hist
Periodical
Revista Chilena de Historia y Geografía 03076

Sociedad Chilena de Lingüística (SOCHIL), La Verbena 3882, Santiago
Founded: 1971; Members: 100
Pres: Ambrosio Rabanales; Gen Secr: Mario Bernales
Focus: Ling 03077

Sociedad Chilena de Obstetricia y Ginecología / Chilean Society of Obstetrics and Gynecology, Av Presidente Riesco 6007, Las Condes, Santiago
T: (02) 2112774; Fax: (02) 2112774
Founded: 1935; Members: 248
Pres: Dr. Patricio Vela P.; Gen Secr: Dr. Jorge Tisne T.
Focus: Gynecology
Periodical
Revista Chilena de Obstetricia y Ginecología (weekly) 03078

Sociedad Chilena de Parasitología, Condell 303, Casilla 50470, Santiago
Fax: (02) 5416840
Founded: 1964; Members: 160
Pres: Texia Gorman G.; Gen Secr: Myriam Lorca H.
Focus: Microbio
Periodical
Parasitología al Día (weekly) 03079

Sociedad Chilena de Patología de la Adaptación y del Mesenquima, c/o Hospital J.J. Aguirre, Santos Dumont 999, Santiago
Founded: 1954
Focus: Pathology 03080

Sociedad Chilena de Pediatría, Av Eliodoro Yañez 1984, Santiago
T: (02) 2254393; Fax: (02) 2232351
Founded: 1922; Members: 850
Pres: Dr. Gusravo Solar V.; Gen Secr: Dr. José Infante L.
Focus: Pediatrics
Periodical
Revista Chilena de Pediatria (weekly) . . 03081

Sociedad Chilena de Química, Casilla 2613, Concepción
Founded: 1945; Members: 1000
Focus: Chem
Periodical
Boletín de la Sociedad Chilena de Química (weekly) 03082

Sociedad Chilena de Sanidad, c/o Ministerio de Salud Pública, Huérfanos 1273, Santiago
Founded: 1947
Focus: Public Health 03083

Sociedad Chilena de Tisiología y Enfermedades Broncopulmonares, c/o Hospital del Tórax, J.M. Infante 717, Santiago
Founded: 1931
Focus: Pulmon Dis 03084

Sociedad Científica Chilena Claudio Gay, Casilla 2974, Santiago
Founded: 1946; Members: 143
Focus: Sci 03085

Sociedad Científica de Chile, Rosa Eguiguren 813, Casilla 696, Santiago
Founded: 1891; Members: 250
Pres: Rodrigo Miranda; Gen Secr: Héctor Cathalifaud
Focus: Sci 03086

Sociedad de Anatomía Normal y Patológica de Chile, c/o Museo Histórico, Santiago
Founded: 1938
Focus: Anat; Pathology 03087

Sociedad de Anestesiología de Chile, Casilla 4259, Santiago
Founded: 1946
Focus: Anesthetics 03088

Sociedad de Biología de Chile, Casilla 16164, Santiago
Founded: 1928
Focus: Bio
Periodicals
Archivos de Biología y Medicina Experimentales
Revista Chilena de Historia Natural . . 03089

Sociedad de Biología de Concepción, c/o Escuela de Química y Farmacía y Bioquímica, Casilla 29, Concepción
Founded: 1927
Focus: Bio 03090

Sociedad de Bioquímica de Concepción, c/o Escuela de Química y Farmacía y Bioquímica, Casilla 237, Concepción
Founded: 1957
Focus: Biochem 03091

Sociedad de Cirujanos de Chile, Providencia 1077, Dept 24, Casilla 2843, Santiago
T: (02) 2514442
Founded: 1949
Focus: Surgery
Periodicals
Informaciones de Secretaría (weekly)
Revista Chilena de Cirugía (weekly) . . 03092

Sociedad de Genética de Chile, Casilla 114-D, Santiago
T: (02) 2224516 ext 2835; Fax: (02) 2225515
Founded: 1964; Members: 100
Pres: Dr. Daniel Frias; Gen Secr: Dr. Manuel Santos
Focus: Genetics
Periodicals
Biological Research (weekly)
Revista Chilena de Historia Natural . . 03093

Sociedad de Medicina Veterinaria de Chile, Estado 337, Of 729, Santiago
T: (02) 397660
Founded: 1926; Members: 520
Focus: Vet Med 03094

Sociedad de Otorinolaringología de Valparaíso, Blas Cuenas 964, Valparaíso
Founded: 1945
Focus: Otorhinolaryngology 03095

Sociedad de Pediatría de Valparaíso, Av Brasil 1689, Valparaíso
Founded: 1932; Members: 45
Focus: Pediatrics 03096

Sociedad Geológica de Chile, Valentin Letelier 20, Dept 401, Casilla 13667, Correo 21, Santiago
T: (02) 6980481
Founded: 1962; Members: 400
Focus: Geology 03097

Sociedad Médica de Concepción, Casilla 60C, Concepción
Founded: 1886
Focus: Med 03098

Sociedad Médica de Santiago, Av Presidente Riesco 6007, Clasificador 1, Correo 27, Santiago
T: (02) 2205812
Founded: 1869; Members: 1800
Focus: Med
Periodical
Revista Médica de Chile (weekly) . . . 03099

Sociedad Médica de Valparaíso, Av Brasil 1689, Valparaíso
Founded: 1913; Members: 271
Focus: Med 03100

Sociedad Nacional de Agricultura / Federación Gremial, Tenderini 187, Casilla 40D, Santiago
T: (02) 396710; Tx: 240760; CA: SOCNADEA
Founded: 1838
Focus: Agri
Periodicals
Boletín de Mercado (weekly)
Boletín Económico (weekly)
Revista El Campesino 03101

Sociedad Nacional de Minería, Teatinos 20, Of 33, Casilla 1807, Santiago
Founded: 1883
Pres: Manuel Filiú Justiniano; Gen Secr: Julio Ascuí Latorre
Focus: Mining
Periodical
Boletín Minero 03102

Sociedad Odontológica de Concepción (SOC), Casilla 2107, Concepción
T: (041) 258997
Founded: 1924; Members: 350
Focus: Dent 03103

Unión de Federaciones Universitárias Chilenas (UFUCH), Bernardo O'Higgins 1058, Santiago
Focus: Educ 03104

China, People's Republic

Academy of Architectural Engineering, c/o Ministry of Building, Beijing
Founded: 1958
Focus: Civil Eng 03105

Academy of Building Materials, Hsi-chiao Pai-wan-chuang, Beijing
Focus: Materials Sci 03106

Academy of Cement Research, Beijing
Founded: 1954
Focus: Materials Sci 03107

Academy of Chemical Engineering, Beijing
Focus: Eng; Chem 03108

Academy of Highway Sciences, c/o Ministry of Communications, Beijing
Founded: 1952
Focus: Transport 03109

Academy of Hydrotechnology, Hsi-chiao Ching-wang-mu, Beijing
Founded: 1958
Focus: Eng; Hydrology 03110

Academy of Non-Ferrous Metallurgical Design, Huang-heng-tzu, Beijing
Focus: Metallurgy 03111

Academy of Petroleum Research, Beijing
Founded: 1958
Focus: Geology 03112

Acoustical Society of China, 17 Zhongguancun St, Beijing 100080
T: (01) 2533765
Founded: 1985; Members: 3035
Pres: Dinghua Guan; Gen Secr: Tong Chen
Focus: Acoustics
Periodicals
Acta Acustica (weekly)
Applied Acoustics (weekly)
Chinese Journal of Acoustics (weekly)
Noise and Vibration Control (weekly) . . 03113

Aerodynamics Research Society, POB 2425, Beijing
Pres: Genggan Zhuang
Focus: Eng 03114

The Architectural Society of China, Baiwanzhuang, West District, Beijing
T: (01) 893868; Fax: (01) 222302
Founded: 1953
Focus: Archit
Periodicals
Architectural Journal (weekly)
Architectural Knowledge
Journal of Building Structures (weekly) . 03115

Asian Packaging Federation (APF), c/o China National Export Commodities Packaging Research Institute, 28 Dong Hou Xiang, An Ding Men Wai, Beijing
Founded: 1966
Gen Secr: Zhongyin Wang
Focus: Materials Sci
Periodical
APF Bulletin (weekly) 03116

Beijing Academy of Agricultural Sciences, Beijing
Founded: 1958
Focus: Agri 03117

Beijing Academy of Coal Mine Design, Beijing
Founded: 1953
Focus: Mining 03118

Beijing Academy of Hydroelectrical Engineering Design, Beijing
Focus: Electric Eng 03119

Biophysical Society of China, POB 349, Beijing
T: (01) 2566757
Founded: 1980; Members: 1500
Pres: Prof. Shurong Wang; Gen Secr: Prof. Junxian Shen
Focus: Biophys 03120

Botanical Society of China, 141 Xizhimenwai, Beijing
Founded: 1933
Pres: Changku Hu
Focus: Botany
Periodicals
Acta Botanica Sinica (weekly)
Acta Mycologica Sinica (weekly)
Acta Phytoecologica et Geobotanica Sinica (weekly)
Acta Phytotaxonimica Sinica (weekly)
Bulletin of Biology (weekly)
Bulletin of Botany (weekly)
Plants 03121

Central South Academy of Industrial Architecture, Wu-han, Hupei
Focus: Archit 03122

Changchun Academy of Electrical Power Engineering Design, Ch'ang-ch'un, Chilin
Founded: 1957
Focus: Electric Eng 03123

Chekiang Academy of Agricultural Sciences, Hang-chou, Chechiang
Focus: Agri 03124

Chemical Industry and Engineering Society of China, POB 911, Beijing
T: (01) 466025; Tx: 22492
Founded: 1922; Members: 40000
Pres: Guangqi Yang; Gen Secr: Delin Yin
Focus: Eng 03125

China Academy of Railway Sciences, Xi Zhi Men Wai, Beijing 100081
T: (01) 2256050; CA: 6993; Fax: (01) 2255989
Founded: 1950; Members: 4000
Pres: Yizu Yi
Focus: Eng
Periodical
China Railway Sciences (weekly) . . . 03126

China Academy of Traditional Chinese Medicine, 18 Beixincang, Dongzhimennai, Beijing
T: (01) 446661; Tx: 210340
Founded: 1955
Pres: Shaowu Chen
Focus: Med
Periodicals
Chinese Acupuncture and Moxibustion (weekly)
Journal of Traditional Chinese Medicine . 03127

China Association for Science and Technology, 54 Sanlihe Rd, Beijing 100863
T: (01) 8321924; Tx: 20035; Fax: (01) 8321914
Founded: 1958; Members: 2000000
Pres: Guangya Zhu
Focus: Sci; Eng
Periodicals
Knowledge is Power (weekly)
Science and Technology Review (weekly) 03128

China Association of Traditional Chinese Medicine (CATCM), A4 Yinghualu, Hepingli Dongjie, Beijing 100029
T: (01) 4215576; Fax: (01) 4220867
Founded: 1979
Pres: Yuli Cui
Focus: Med
Periodical
China Journal of TCM (weekly) 03129

China Civil Engineering Society (CCES), Bai Wan Zhuang, POB 2500, Beijing
T: (01) 8311313; CA: CCES; Fax: (01) 8313669
Founded: 1953
Pres: Guohao Li; Gen Secr: Chenggang Li
Focus: Civil Eng
Periodical
China Civil Engineering Journal (weekly) 03130

China Coal Society, Hepingli, Beijing
T: (01) 4214931; Tx: 22504; Fax: (01) 4219234
Founded: 1962; Members: 48000
Pres: Weitang Fan; Gen Secr: Huizheng Pan
Focus: Mining
Periodical
Journal of the China Coal Society (weekly) 03131

China Computer Federation, POB 2704, Beijing
T: (01) 2562503; Tx: 222638; Fax: (01) 2567724
Founded: 1985; Members: 35700
Pres: Xiao Xiang Zhang; Gen Secr: Shukai Chen
Focus: Computer & Info Sci 03132

China Law Society, Bldg 22, 23 Fu Xing Lu, Beijing
T: (01) 8317529; Tx: 22505
Founded: 1982; Members: 4029
Pres: Zhongfang Wang; Gen Secr: Shutao Song
Focus: Law 03133

China Society of Fisheries, 31 Minfeng Lane, Beijing
T: (01) 660794; Tx: 22079
Members: 10000
Pres: Kejia Huang
Focus: Fisheries 03134

Chinese Abacus Association, Xizhimenwal, Shidaojkou, Beijing
T: (01) 896275
Founded: 1979; Members: 30000
Pres: Ming-Yuan Jiang; Gen Secr: Xin Li
Focus: Med 03135

Chinese Academy of Agricultural Sciences, Beijing
Founded: 1957
Pres: Liangshu Lu
Focus: Agri 03136

Chinese Academy of Coal Mining Sciences, Beijing
Pres: Weitang Fan
Focus: Mining 03137

Chinese Academy of Forestry, Wan Shou Shan, Beijing
Founded: 1978
Pres: Shu Huang
Focus: Forestry 03138

Chinese Academy of Geological Sciences (CAGS), Rd 26, Fuchengmenwai, Beijing 100037
T: (01) 8310893; Tx: 222721; Fax: (01) 8310894
Founded: 1959; Members: 5000
Pres: Yuchuan Chen
Focus: Geology
Periodical
Bulletin (weekly) 03139

Chinese Academy of Medical Sciences, Beijing
Founded: 1956
Pres: Fangzhou Gu
Focus: Med 03140

Chinese Academy of Meteorological Sciences, Beijing
T: (01) 8312277; Tx: 22094
Focus: Physics 03141

Chinese Academy of Sciences, 52 San Li He Rd, Beijing
T: (01) 868361; Tx: 22474
Founded: 1949
Pres: Guangzhao Zhou; Gen Secr: Qiheng Hu
Focus: Sci 03142

Chinese Academy of Sciences – Anhwei Branch, Ho-fei, Anhui
Founded: 1959
Focus: Sci 03143

Chinese Academy of Sciences – Central South Branch, Kuang-chou, Kuangtung
Founded: 1961
Focus: Sci 03144

Chinese Academy of Sciences – Chekiang Branch, Hang-chou, Chechiang
Founded: 1958
Focus: Sci 03145

Chinese Academy of Sciences – East China Branch, Yüeh-yang Lu, Shanghai
Founded: 1963
Focus: Sci 03146

Chinese Academy of Sciences – Fukien Branch, Hsia-men, Fuchien
Founded: 1960
Focus: Sci 03147

Chinese Academy of Sciences – Hopeh Branch, Pao-ting, Hopei
Founded: 1958
Focus: Sci 03148

Chinese Academy of Sciences – Kiangsu Branch, Nan-ching, Chiangsu
Founded: 1958
Focus: Sci 03149

Chinese Academy of Sciences – Kirin Branch, Ch'ang-ch'un, Chilin
Founded: 1958
Focus: Sci 03150

Chinese Academy of Sciences – Northwestern Branch, Hsi-an, Shanhsi
Founded: 1954
Focus: Sci 03151

Chinese Academy of Sciences – Shansi Branch, T'ai-yüan, Shanhsi
Founded: 1959
Focus: Sci 03152

Chinese Academy of Sciences – Shantung Branch, Chi-nan, Shantung
Founded: 1958
Focus: Sci 03153

Chinese Academy of Sciences – Sinkiang Branch, Wu-lu-mu-ch'i, Hsinchiang
Founded: 1959
Focus: Sci 03154

Chinese Academy of Social Sciences, 5 Jianguomen nei Da Jie, Beijing 100732
T: (01) 5137744; Tx: 22061; Fax: (01) 5138154
Founded: 1977; Members: 4865
Pres: Sheng Hu; Gen Secr: Jongshu Long
Focus: Soc Sci
Periodicals
Archaeology
Literary Review
Philosophical Research
Social Sciences in China (weekly)
Studies in Law
Studies on Chinese History
Studies on Economics
World Economy 03155

Chinese Academy of Space Technology, Beijing
Fax: (01) 8378237
Pres: Faren Qi
Focus: Aero 03156

Chinese Anti-Cancer Association, 87 Dagu Rd, Heping District, Tian Jing
Pres: Yuanxing Wu
Focus: Cell Biol & Cancer Res 03157

Chinese Anti-Tuberculosis Society, 42 Dung-si-xi-da St, Beijing
T: (01) 553685
Pres: Dingchen Huang
Focus: Pulmon Dis 03158

Chinese Archives Society, 21 Fengshen Hutong, Beijing
Founded: 1981; Members: 2200
Pres: Tong Pei
Focus: Archives 03159

Chinese Association of Agricultural Science Societies, c/o Ministry of Agruculture and Fisheries, Beijing
Pres: Liang-Shu Lu
Focus: Agri 03160

Chinese Association of Animal Science and Veterinary Medicine, Beijing Agricultural Exhibition Centre, Beijing
Pres: Lingfen Chen
Focus: Vet Med 03161

Chinese Association of Automation, POB 2728, Beijing
T: (01) 2544415; Fax: (01) 2545229
Founded: 1957; Members: 30000
Pres: Qiheng Hu
Focus: Eng
Periodicals
Acta Automatica Sinica
Automation Panorama
Information and Control
Pattern Recognition and Artificial Intelligence
Robot and Automation 03162

The Chinese Association of Fire Protection, 14 Dong Chang An St, Beijing
T: (01) 5122702
Founded: 1984; Members: 12000
Pres: Lei Ju; Gen Secr: Qi-Hong Wu
Focus: Eng 03163

Chinese Association of Integrated Traditional and Western Medicine, 18 Beixincang, Dongzhimennei, Beijing
T: (01) 4466613026
Founded: 1981; Members: 20000
Pres: Yueli Cui; Gen Secr: Weibo Lu
Focus: Med 03164

Chinese Astronomical Society (C.A.S.), 2 Beijing Xi Rd, Nanjing 210008
T: (025) 301096, 303921; Tx: 34144
Founded: 1922; Members: 1662
Pres: Qi-ben Li; Gen Secr: He-Qi Zhang
Focus: Astronomy
Periodicals
Acta Astronomica Sinica (weekly)
Acta Astrophysica Sinica (weekly)
Astronomical Circular (weekly)
Progress in Astronomy (weekly) 03165

Chinese Biochemical Society, c/o Shanghai Institute of Biochemistry, 320 Yue-Yang Rd, Shanghai
T: (021) 4374430; Fax: (021) 4338357
Founded: 1979; Members: 1500
Pres: L.X. Zhang; Gen Secr: G.R. Qi
Focus: Biochem
Periodicals
Chemistry of Life (weekly)
Chinese Biochemical Journal (weekly) . . 03166

Chinese Chemical Society, POB 2709, Beijing 100080
T: (01) 2568157, 2564020; Tx: 20035; CA: 9422, Beijing; Fax: (01) 2565854
Founded: 1932; Members: 25000
Pres: Prof. Guangxian Xu; Gen Secr: Prof. Mengyuan Cui
Focus: Chem
Periodicals
Acta Chimica Sinica (weekly)
Chinese Journal of Polymer Science (weekly)
Daxue Huaxue (weekly)
Fenxi Huaxue (weekly)
Gaofenzi Tongbao (weekly)
Gaofenzi Xuebao (weekly)
Huaxue Jiaoyu (weekly)
Huaxue Tongbao (weekly)
Huaxue Tongxun (weekly)
Huaxue Xuebao (weekly)
Journal of Molecular Sciences (weekly)
Sepu (weekly)
Wuji Huaxue (weekly)
Wuli Huaxue (weekly)
Yingyong Huaxue (weekly)
Youji Huaxue (weekly) 03167

Chinese Electrotechnical Society, 46 Sanlihe Rd, POB 2133, Beijing
T: (01) 8011242; Tx: 22341; CA: 0102; Fax: (01) 8011242
Founded: 1981; Members: 20000
Pres: Jingde Gao; Gen Secr: Zongrui Yu
Focus: Electric Eng 03168

Chinese Engineering Graphics Society, c/o Hua Zhong Institute of Technology, Wuhan
Pres: Fuxi Zhu
Focus: Eng 03169

Chinese Geological Society (G.S.C.), Baiwanzhuang, Fuchengmenwai, Beijng
T: (01) 893826
Founded: 1922; Members: 15000
Pres: Yuqi Cheng; Gen Secr: Zejiu Wang
Focus: Geology
Periodicals
Acta Geologia (weekly)
Geological Review (weekly) 03170

Chinese Geophysical Socitey (CGS), Qieng Hua Dong Lu, Beijing
Founded: 1949
Focus: Geophys
Periodicals
Acta Geophysica Sinica (weekly)
Bulletin of the Chinese Geophysical Society (weekly) 03171

Chinese Hydraulic Engineering Society, Bai Guang Rd, POB 2905, Beijing
Tx: 22466
Pres: Kai Yan
Focus: Mach Eng 03172

Chinese Information Processing Society, POB 2704, Beijing
Pres: Weichang Qian
Focus: Computer & Info Sci 03173

Chinese Light Industry Society, 3 Fuchenglu, Beijing
Pres: Long Ji
Focus: Eng 03174

Chinese Mathematical Society, Zong Guan Cun, Beijing
T: (01) 281022
Founded: 1935
Pres: Yuan Wang; Gen Secr: Zhong Li
Focus: Math 03175

Chinese Mechanical Engineering Society (CMES), Sanlihe Rd, Beijing 100823
T: (01) 8695319; Tx: 22341; Fax: (01) 8033613
Founded: 1936; Members: 180000
Pres: Yansun Lu; Gen Secr: Ruiquan Cheng
Focus: Eng
Periodical
Journal of Mechanical Engineering (weekly) 03176

Chinese Mechanics Society, c/o Institute of Mechanics, Zhong Guan Cun, Beijing
Pres: Lingxi Qian
Focus: Eng 03177

Chinese Medical Society, 42 Dongsi Xidajie, Beijing
Founded: 1914
Pres: Xiqing Bai
Focus: Med
Periodicals
Chinese Medical Journal (weekly)
National Medical Journal of China (weekly) 03178

Chinese Society for EC Studies, c/o Institute of World Economy, Fudan University, 220 Handad Rd, Shanghai
T: (021) 5492222 ext 3131; Tx: 33317; Fax: (021) 5491875
Founded: 1984; Members: 100
Pres: Kaixiang Yu; Gen Secr: Bingran Dai
Focus: Econ
Periodical
European Community Studies (weekly) . 03179

Chinese Society for Future Studies, A3 Wan Shou Si, Beijing
T: (01) 890861753
Founded: 1979; Members: 1000
Gen Secr: Dagong Du
Focus: Futurology 03180

Chinese Society for Oceanology and Limnology (C.S.O.L.), 7 Nanhai Rd, Quingdao 266071
T: (05327) 279062319; Tx: 32222; Fax: (05327) 270882
Founded: 1951; Members: 5900
Pres: Yunshan Qin; Gen Secr: Jinhai Dong
Focus: Hydrology; Oceanography
Periodicals
Chinese Journal of Oceanology and Limnology (weekly)
Oceanologia et Limnologia (weekly) . . . 03181

Chinese Society of Aeronautics and Astronautics (CSAA), 9 Xiaoguandongjie, Andingmenwai, Beijing
T: (01) 4916314; Tx: 22505; Fax: (01) 4914643
Founded: 1964; Members: 30000
Pres: Yu-Li Zhu; Gen Secr: Chu Xie
Focus: Aero 03182

Chinese Society of Agricultural Machinery, 1 Beishatan, Deshengmen Wai, Beijing
T: (01) 2017131 ext 2399; Tx: 222483; Fax: (01) 2017326
Founded: 1963; Members: 26000
Pres: Guangyuan He; Gen Secr: Wanjun Wang
Focus: Agri 03183

Chinese Society of Anatomy, 42 Dongsi Xidajie, Beijing
T: (01) 546231 ext 247
Founded: 1947; Members: 5000
Pres: Shepu Xue; Gen Secr: C. Ben
Focus: Anat 03184

Chinese Society of Astronautics, POB 838, Beijing
T: (01) 894602; Tx: 20026
Founded: 1979; Members: 5600
Pres: Xin-Min Ren; Gen Secr: Xian Wang
Focus: Aero 03185

Chinese Society of Chemical Engineering, Tung-ssu Erh-t'iao Pei-k'ou 1, Beijing
Founded: 1956
Focus: Eng; Chem 03186

Chinese Society of Electronics, Tung-ssu Liu T'iao, Beijing
Founded: 1956
Focus: Electronic Eng 03187

Chinese Society of Engineering Thermophysics, Zhong Guan Cun, POB 2706, Beijing
Founded: 1978; Members: 5000
Pres: Prof. Shoupan Liang; Prof. Shaoxi Shi; Prof. Xuejun, Wang, Prof. Buxuan Chen;
Gen Secr: Prof. Tianzhong Xu
Focus: Eng
Periodicals
Journal of Engineering Thermophysics (weekly)
Proceedings of Engineering Thermophysics Conference (weekly) 03188

Chinese Society of Environmental Sciences, Baiwanzhuang, Beijing
Tx: 22477
Founded: 1979; Members: 22000
Pres: Jingzhao Li; Gen Secr: Geping Qu
Focus: Ecology 03189

Chinese Society of Forestry, Wanshoushan, Beijing
Pres: Zhonglun Wu
Focus: Forestry 03190

Chinese Society of Geodesy, Photogrammetry and Cartography, Baiwanzhuang, Beijing
T: (01) 8322024; Tx: 222743; Fax: (01) 8311564
Founded: 1959; Members: 3000
Pres: Prof. Junyong Chen; Gen Secr: Prof. Xiaorong Zhung
Focus: Cart; Surveying
Periodical
Acta Geodetica et Cartographica Sinica . 03191

Chinese Society of High Energy Physics, POB 918, Beijing
T: (01) 811982; Tx: 22082
Founded: 1981; Members: 650
Pres: Minghan Ye; Gen Secr: Xueying Wang
Focus: Energy 03192

Chinese Society of History of Science, 1 Gong Yuan Xi Jie, Dongcheng District, Bejing
Pres: Jun Ke
Focus: Hist 03193

Chinese Society of Library Science, 7 Wen Jin St, Beijing
T: (01) 666331
Founded: 1981
Focus: Libraries & Bk Sci 03194

The Chinese Society of Metals, 46 Dongsixi Dajie, Beijing 100711
Fax: (01) 5124122
Founded: 1956
Focus: Metallurgy 03195

Crop Science Society of China, c/o Institute of Crop Science, Chinese Academy, Beijing
Pres: Jingxong Li
Focus: Agri 03196

Ecological Society of China, 19 Zhongguancun Lu, Beijing
T: (01) 2565694
Founded: 1979; Members: 2400
Pres: Prof. Ruyong Sun; Gen Secr: Prof. Dianmo Li
Focus: Ecology 03197

Entomological Society of China, 19 Zhongguancun Lu, Haidian, Beijing 100080
T: (01) 282266; Fax: (01) 2565689
Members: 7000
Pres: H.F. Chu; Gen Secr: M.Y. Liu
Focus: Entomology
Periodicals
Acta Entomologica (weekly)
Acta Zootaxonomia Sinica (weekly)
Kunchong zhishi: Entomological Knowledge (weekly) 03198

Genetics Society of China, Bldg 917, Datun Rd, Andingmenwai, Beijing
T: (01) 4914896; Fax: (01) 4914896
Founded: 1978; Members: 6500
Pres: Zhengsheng Li; Gen Secr: Shouyi Chen
Focus: Genetics
Periodical
Acta Genetica Sinica 03199

Geographical Society of China, 917 Bldg, Datun Rd, Beijing 100101
T: (01) 4911104; Fax: (01) 4911844
Founded: 1909; Members: 10000
Pres: Chuanjun Wu; Gen Secr: Jiazhen Zhang
Focus: Geography
Periodical
Acta Gepographica Sinica (weekly) . . . 03200

Heilongjiang Academy of Forestry Sciences (H.A.F.S.), Ha Ping Rd, Harbin 150040
T: 63961/65
Founded: 1960; Members: 1329
Pres: Zheng Zhou
Focus: Forestry
Periodicals
External Forestry (weekly)
Forestry Science and Technology (weekly) 03201

Honan Academy of Agricultural Sciences, Honan
Focus: Agri 03202

Hopeh Academy of Agricultural Sciences, Hopei
Focus: Agri 03203

Hunan Academy of Agricultural Sciences, Changsha 410125
T: (0731) 26145; CA: 4496
Founded: 1938; Members: 2600
Pres: Ren Qian
Focus: Agri
Periodical
Hunan Agricultural Sciences (weekly) . . 03204

Inner Mongolian Academy of Agricultural Sciences and Animal Husbandry, Hu-ho-hao-t'e, Neimengku
Focus: Agri 03205

Inner Mongolian Academy of Social Sciences, Huhehot
Founded: 1978
Focus: Econ; Philos; Sociology; Hist; Ling; Lit; Ethnology; Computer & Info Sci
Periodicals
Inner Mongolian Social Science: Chinese Edition (weekly)
Inner Mongolian Social Science: Mongolian Edition (weekly)
Mongolia Language and Literature (weekly) 03206

Jilin Academy of Agricultural Sciences, 5 West Xinghua St, Gongzhuling 136100
T: (04441) 5179; Tx: 83134
Founded: 1948
Focus: Agri 03207

Kiangsu Academy of Agricultural Sciences, Chiangsu
Focus: Agri 03208

Kwangtung Academy of Agricultural Sciences, Kuangtung
Focus: Agri 03209

Liaoning Academy of Agricultural Sciences, Shen-yang, Liaoning
Focus: Agri 03210

Mukden Academy of Chemical Engineering, Shen-yang, Liaoning
Founded: 1949
Focus: Eng; Chem 03211

Northwestern Academy of Industrial Architecture, Hsi-an, Shanhsi
Focus: Archit 03212

Palaeontological Society of China (PSC), c/o Nanjing Institute of Geology and Palaeontology, 1 Chi-Ming-Ssu, Nanjing 210008
T: (025) 6637395; Fax: (025) 712207
Founded: 1929; Members: 1200
Pres: Miman Zhang; Gen Secr: Xinan Mu
Focus: Paleontology
Periodical
Acta Palaeontologica Sinica (weekly) . . 03213

Shanghai Academy of Agricultural Sciences, Shanghai
Focus: Agri 03214

Shanghai Academy of Hydroelectrical Engineering Design, Shanghai
Focus: Electric Eng 03215

Shanghai Textile Engineering Society, 197 North Wulumuqi Rd, Shanghai
CA: 5737
Founded: 1930
Focus: Textiles 03216

Shansi Academy of Agricultural Sciences, Shanhsi
Focus: Agri 03217

Shantung Academy of Agricultural Sciences, Chi-nan, Shantung
Focus: Agri 03218

Shensi Academy of Agricultural Sciences, Wu-kung, Shenhsi
Founded: 1958
Focus: Agri 03219

Sian Academy of Electric Power Design, Hsi-an, Shanhsi
Founded: 1957
Focus: Electric Eng 03220

Society of Autmotive Engineering of China, POB 2728, Beijing
T: (01) 366544; Tx: 22656
Pres: Qiheng Hu
Focus: Eng 03221

Textile Academy, Ying Jia Fen, Beijing
CA: 9444
Founded: 1956
Focus: Textiles 03222

Wuhsi Academy of Textile Engineering, liande Rd, Wuhsi 214004
Founded: 1957
Focus: Textiles 03223

Xinjiang Academy of Agricultural Sciences, 38 Nanchang Rd, Urumqi, Xinjiang
Focus: Agri
Periodical
Xinjiang Agricultural Sciences (weekly) . 03224

Yünnan Academy of Agricultural Sciences, Yünnan
Focus: Agri 03225

China, Republic

Academia Historica, 225 Pei Yi Rd, Section 3, Taipei
Focus: Hist 03226

Academia Historica, 225 Sec 3, Pei Yi Rd, Hsintien, Taipei
Pres: Chi-Lu Huang; Gen Secr: Prof. Yun-Han Li
Focus: Hist 03227

Academia Sinica, Nankang, Taipei
T: (02) 7823142
Founded: 1928; Members: 105
Pres: Dr. Ta-You Wu; Gen Secr: Dr. Hong-Pang Wu
Focus: Sci 03228

Actuarial Institute of the Republic of China, 2-6 Alley 4, Lane 217, Chunghsiao East Rd, Section 3, Taipei
Founded: 1969
Focus: Insurance 03229

Agricultural Chemistry Society of China, c/o Dept of Agricultural Chemistry, College of Agriculture, National Taiwan University, Roosevelt Rd, Taipei
Founded: 1963
Focus: Chem 03230

Agricultural Extension Association of China, 14 Wenchou St, Taipei
Founded: 1955
Focus: Agri 03231

Agriculture Association of China, 14 Wenchow St, Taipei
Founded: 1917; Members: 2865
Pres: Chao-Chen Chen
Focus: Agri
Periodical
Journal (weekly) 03232

Asia and Oceania Federation of Nuclear Medicine and Biology, c/o Dept of Nuclear Medicine, Veterans General Hospital, Taipei 112
Founded: 1976
Gen Secr: Shin-Hwa Yeh
Focus: Bio; Nucl Med 03233

Asia and Pacific Association for the Control of Tobacco, 57 Fuxing Beilu, 12 Bldg, Unit 3, Taipei
Gen Secr: Dr. David Yen
Focus: Agri 03234

The Asia Foundation, 42 Chien Kuo North Rd, Taipei
Focus: Int'l Relat 03235

Asian-Australian Association of Animal Production Societies (AAAP), c/o Pig Research Institute, POB 23, Chunan, Miaoli 350
Gen Secr: L.C. Hsia
Focus: Animal Husb
Periodical
Asian-Australian Journal of Animal Science (weekly) 03236

Asian Crystallographic Association (ASCA), c/o Dept of Chemistry, National Taiwan University, Taipei 10764
Gen Secr: Prof. Yu Wang
Focus: Chem 03237

Asian Ecological Society, c/o Tunghai University, POB 843, Taichung 40704
Founded: 1977
Pres: Yun-Li Lin
Focus: Ecology
Periodical
Journal of Asian Ecology (weekly) . . . 03238

Asian Food Council, c/o CACCI, 122 Tunhua North Rd, Taipei 10590
T: (02) 7163016; Tx: 11144; Fax: (02) 7183683
Gen Secr: M. Chang
Focus: Food 03239

Asian Vegetable Research and Development Center (AVRDC), POB 42, Shanhua, Tainan 74199
T: (06) 5837801; Tx: 73560; Fax: (06) 5830009
Founded: 1971; Members: 9
Gen Secr: Emil Q. Javier
Focus: Hort
Periodicals
Centerpoint Newsmagazine (weekly)
Soybean Rust Newsletter (weekly) . . . 03240

Asia-Pacific Office Automation Council (APOAC), c/o CACCI, 122 Tunhua North Rd, Taipei 10590
T: (02) 7163016, 7164250; Tx: 11144; Fax: (02) 7163683
Gen Secr: C. Yen Johnson
Focus: Business Admin 03241

Association for Education through Art, Republic of China, 808 Chungcheng Rd, Wufeng Hsiang, Taichung Hsien
Founded: 1968
Focus: Arts; Educ 03242

Association for Physics and Chemistry Education of the Republic of China, 88 Roosevelt Rd, Section 5, Taipei
Founded: 1958
Focus: Chem; Educ; Physics 03243

Association for Socio-Economic Development in China, 301 Peita Rd, Hsinchu
Founded: 1963
Focus: Sociology; Econ 03244

Association of Animal Husbandry and Veterinary Medicine of Taiwan, 8 Kuanghua Rd, Nantou Hsien
Focus: Zoology; Vet Med 03245

Association of Child Education of the Republic of China, 94 Hoping East Rd, Section 3, Taipei
Founded: 1930
Focus: Educ 03246

The Association of Obstetrics and Gynecology of the Republic of China, c/o Dept of Obstetrics and Gynecology, Veterans General Hospital, 201 Shihpai Rd, Section 2, Taipei
Fax: (02) 8739512
Founded: 1961
Focus: Gynecology 03247

Astronautical Society of the Republic of China, c/o National Taiwan Science Hall, 41 Nanhai Rd, Taipei
Founded: 1958
Focus: Aero 03248

Astronomical Society of the Republic of China, c/o Taipei Observatory, Yuan Shan, Taipei 104
Founded: 1958; Members: 200
Pres: Dr. C.S. Shen; Gen Secr: Chang-Hsien Tsai
Focus: Astronomy 03249

Atomic Energy Council, 67 Lane 144, Keelung Rd, Section 4, Taipei 107
T: (02) 3924180; Tx: 26554
Founded: 1955; Members: 15
Pres: Chen-Hsing Yen; Gen Secr: Kuang-Chi Liu
Focus: Nucl Res
Periodicals
Chinese AEC Bulletin (weekly)
Nuclear Science Journal (weekly) . . . 03250

Biological Society of China, c/o Dept of Biology, National Taiwan Normal University, 162 Hoping East Rd, Section 1, Taipei
Founded: 1959
Focus: Bio 03251

China Academy, Hwa Kang, Yang Ming Shan
Founded: 1966
Pres: Chi-Yun Chang; Gen Secr: Wei-Ho Pan
Focus: Ethnology
Periodicals
Beautiful China Pictorial Monthly (weekly)
Chinese Culture (weekly)
Renaissance Monthly (weekly)
Sino-American Relations (weekly)
Sinological Monthly (weekly)
Sinological Quarterly (weekly) 03252

China Association of the Five Principles of Administrative Authority, 26 Ningpo West St, Taipei
Founded: 1962
Focus: Business Admin 03253

China Education Society, c/o National Taiwan Normal University, Hoping East Rd, Taipei
Founded: 1933
Focus: Educ 03254

China International Education Research Association, 173 Sinyi Rd, Section 2, Taipei
Founded: 1959
Focus: Educ 03255

China National Association of Literature and the Arts, 4 Lane 22, Nuigpo West St, Taipei
Focus: Lit; Arts 03256

China Social Education Society, 5 Chungshan South Rd, Taipei
Founded: 1931
Focus: Educ; Sociology 03257

The China Society, 7 Lane 52, Wenchow St, Taipei
Founded: 1960; Members: 100
Pres: Dr. Chi-Lu Chen; Gen Secr: Dr. Herbert Ma
Focus: Ethnology
Periodical
Journal (weekly) 03258

China Spiritual Therapy Study Association, 5 Alley 11, Lane 131, Olung St, Taipei
Founded: 1926
Focus: Therapeutics 03259

Chinese Association for Folklore, 422 Fulin Rd, Shihlin, POB 681292, Taipei
Founded: 1932
Pres: Prof. Lou-Kuang Lou; Gen Secr: Amy Lou
Focus: Ethnology 03260

Chinese Association for the Advancement of Science, 5 Chungshan South Rd, Taipei
Founded: 1917
Pres: Tien-Fong Cheng
Focus: Sci
Periodical
Science Education 03261

Chinese Association of Psychological Testing, c/o Psychological Laboratory, National Taiwan Normal Univeristy, Hoping East Rd, Taipei
Founded: 1935
Focus: Psych 03262

Chinese Buddhist Association, 23 Chungsiao East Rd, Section 1, Taipei
Founded: 1912
Focus: Rel & Theol 03263

Chinese Center, International PEN, 33 Lane 180, Kwang-Fu South Rd, Taipei 105
Founded: 1935
Focus: Lit
Periodical
The Chinese PEN (weekly) 03264

Chinese Chemical Society, POB 609, Taipei
Founded: 1932; Members: 6200
Pres: Y.S. Chen
Focus: Chem
Periodicals
Chemistry (weekly)
Journal 03265

Chinese Classical Music Association, 1 Lane 3, Linyi St, Taipei 100
Founded: 1951; Members: 1100
Focus: Music 03266

Chinese Film Critic's Association of China, 18 Hsining South Rd, Taipei
Founded: 1964
Focus: Cinema 03267

Chinese Forestry Association (C.F.A.), 2 Hang Chow-South Rd , Section 1, Taipei 100
T: (02) 3515441
Founded: 1915; Members: 1545
Pres: Chi-You Hsu
Focus: Forestry
Periodicals
Journal of Chinese Forestry (weekly)
Taiwan's Forestry Monthly (weekly) . . . 03268

Chinese Foundrymen Association, 1001 Kaonan Hwy, Nantzu, Kaohsiung 811
T: (07) 3534791; Fax: (07) 3524989
Founded: 1966; Members: 2389
Focus: Metallurgy
Periodicals
Casting (weekly)
Journal of the Chinese Foundrymen's Association (weekly) 03269

Chinese Historical Association, c/o College of Liberal Arts, National Taiwan University, Roosevelt Rd, Taipei
Founded: 1954
Focus: Hist 03270

Chinese Home Education Promotion Association, 7-1 Alley 6, Lane 238, Tun Hua North Rd, Taipei 104
T: (02) 7520944
Founded: 1960; Members: 5767
Focus: Home Econ; Educ 03271

Chinese Institute of Civil and Hydraulic Engineering, POB 499, Taipei
Founded: 1973; Members: 3700
Pres: Chi-Chang Yen
Focus: Civil Eng
Periodical
Civil and Hydraulic Engineering (weekly) . 03272

The Chinese Institute of Engineers, 4F, 1 Jen. Ai Rd, Section 2, Taipei
Founded: 1912; Members: 14116
Pres: Su-Ming Chang
Focus: Eng
Periodicals
CIE Newsletter (weekly)
Engineering Journal (weekly)
Journal of the Chinese Institute of Engineers (weekly) 03273

Chinese Institute of Mining and Metallurgical Engineers (CIMME), 2F, 38-1 Chin Nan Rd, Section 2, Taipei 100
T: (02) 3517046, 3960202; Fax: (02) 3975377
Founded: 1926
Pres: C.P. Lin
Focus: Mining; Metallurgy
Periodicals
The Metallurgy of Iron and Steel
Symposium on Steel Production Technology 03274

Chinese Language Society, c/o Taiwan Normal University, Hoping East Rd, Taipei
Founded: 1953
Gen Secr: Tzu-Shui Mao
Focus: Ling
Periodical
Chinese Language Monthly (weekly) . . 03275

Chinese Mathematical Society, c/o National Taiwan University, Taipei
Members: 550
Pres: Simon C. Shieh
Focus: Math 03276

Chinese Medical Association (CMA), 201 Shih-Pai Rd, Section 11, POB 3043, Taipei
T: (02) 8712121
Founded: 1915; Members: 1142
Pres: Dr. Chi-Shuen Tsou; Gen Secr: Dr. Tao-Chang Hsu
Focus: Med
Periodical
Chinese Medical Journal (weekly) . . . 03277

Chinese Medical History Association, 6 Lane 120, Hsinsheng Rd, Section 1, Taipei
Founded: 1963
Focus: Hist 03278

Chinese Medical Woman's Association, 21 Lane 504, Chung Hwa Rd, Section 4, Taipei
Founded: 1954
Focus: Med 03279

Chinese National Association for Mental Hygiene, c/o National Taiwan University Hospital, 1 Changteh St, Taipei
Founded: 1936
Focus: Hygiene 03280

Chinese National Foreign Relations Association, 94 Nanchang St, Section 1, Taipei
Pres: Kuo-Shu Huang
Focus: Poli Sci 03281

Chinese Physiological Society, 1 Jenai Rd, Section 1, Taipei
Founded: 1928
Focus: Physiology 03282

Chinese Psychological Association (CPA), c/o Dept of Psychology, National Taiwan University, Taipei 107
T: (02) 3630231 ext 2374; Fax: (02) 3629909
Founded: 1962; Members: 193
Focus: Psych
Periodical
Chinese Journal of Psychology (weekly) . 03283

Chinese Society for Electronic Data Processing, 66 Nanchang St, Section 1, Taipei
Founded: 1966
Focus: Computer & Info Sci 03284

Chinese Society for Materials Science, 195 Chung Hsing Rd, Section 4, Chutung, Hsinchu 31015
Founded: 1966; Members: 1281
Pres: O.C.C. Lin; Gen Secr: K.H. Liu
Focus: Materials Sci
Periodical
Chinese Journal of Materials Science (weekly) 03285

Chinese Society of Budgetary Management, 69-2 Lane 189, Antung St, Taipei
Founded: 1965
Focus: Finance 03286

Chinese Society of International Law, 187 Kinhwa St, Taipei
Founded: 1958
Focus: Law 03287

Chinese Statistical Association, 1 Nan Chung Rd, Section 1, Taipei
Founded: 1941; Members: 1082
Pres: C.C. Lee
Focus: Stats
Periodical
Chinese Statistical Journal 03288

The Chinese Taipei Pediatric Association, 11 Ching-Tao West, 4F-4, Taipei 100
T: (02) 3314917; Fax: (02) 3142184
Founded: 1960; Members: 2200
Pres: Prof. Hue-Hsiung Hsieh; Gen Secr: Prof. Tso-Ren Wang
Focus: Pediatrics
Periodical
Acta Pediatrica Sinica (weekly) 03289

Chinese Women Writer's Association, 16-5 Lane 61, Linyi St, Taipei
Founded: 1969
Focus: Lit 03290

Chinese Youth Academic Research Association, 219 Sungchiang Rd, Taipei
Founded: 1958
Focus: Sci 03291

Confucius-Mencius Society of the Republic of China, 45 Nanhai Rd, Taipei
Founded: 1960; Members: 3900
Pres: Dr. Li-Fu Chen; Gen Secr: Chung-Lin Hua
Focus: Philos
Periodicals
Confucius-Mencius Monthly (weekly)
Journal of the Confucius-Mencius Society 03292

Cooperative League of the Republic of China, 11-2 Fu Chow St, Taipei
Founded: 1928
Pres: Cheng-Kang Ku; Gen Secr: Yen Sung Chen
Focus: Business Admin; Econ 03293

Council of Agriculture, 37 Nanhai Rd, Taipei
T: (02) 3812991; Tx: 3310341
Founded: 1984
Pres: Paul M.H. Sun; Gen Secr: Ching-Lung Lee
Focus: Agri
Periodical
General Report (weekly) 03294

Early Childhood Education Society of the Republic of China, 321 Lane 124, Hu-lin St, Taipei
Founded: 1959
Focus: Educ 03295

Ethnological Society of China, c/o Institute of Ethnology, Academia Sinica, Nankang, Taipei
Founded: 1934
Focus: Ethnology 03296

Finance Association of China, 63-9 Hangchow South Rd, Section 1, Taipei
Founded: 1941
Focus: Finance 03297

The Geographical Society of China, 162 Hoping East Rd, Section 1, Taipei 106
T: (02) 3930874
Founded: 1934; Members: 800
Focus: Geography
Periodical
Bulletin (weekly) 03298

Geological Society of China, POB 3317, Taipei
Focus: Geology
Periodicals
Memoir
Proceedings 03299

Graphic Arts Association of China, 71 Paochao Rd, Hsientien, Taipei
Founded: 1956
Focus: Arts 03300

Historical Research Commission of Taiwan, 111 Yenping South Rd, Taipei
Founded: 1949
Focus: Hist 03301

International Academy of Chest Physicians and Surgeons of the American College of Chest Physicians, Republic of China Chapter, c/o National Taiwan University Hospital, 1 Changteh St, Taipei
T: (02) 3123456 ext 2527
Founded: 1969; Members: 166
Focus: Cardiol 03302

International Education Association of China, 173 Hsinyi Rd, Section 2, Taipei
Founded: 1959
Focus: Educ 03303

International House Association, Taipei Chapter, 18 Hsin Yi Rd , Section 3, Taipei
Pres: George Y.L. Wu
Focus: Urban Plan 03304

The Library Association of China (LAC), c/o National Central Library, 20 Chungshan South Rd, Taipei
T: (02) 3619132
Founded: 1953; Members: 2312
Gen Secr: Teresa Wang Chang
Focus: Libraries & Bk Sci
Periodical
Bulletin (weekly) 03305

Malacological Society of China, 2 Siangyang Rd, Taipei
Founded: 1970; Members: 400
Focus: Zoology
Periodical
Bulletin of Malacology (weekly) 03306

The Meteorological Society of the Republic of China, 64 Kung-Yuan Rd, Taipei
T: (02) 3713181 ext 230
Founded: 1958; Members: 400
Focus: Geophys
Periodicals
Atmospheric Science (weekly)
Bulletin of the Meteorological Society of the Republic of China (weekly)
Papers in Meteorological Research (weekly) 03307

Modern Fine Arts Association of Southern Taiwan, 17 Shulin St, Tainan
Founded: 1967
Focus: Arts 03308

Museum Association of China, 49 Nanhai Rd, Taipei
Founded: 1964
Focus: Arts 03309

National Audio-Visual Education Association of China, 162 Hoping East Rd, Section 1, POB 2225, Taipei 106
T: (02) 3511472
Founded: 1959; Members: 1500
Pres: Dr. Mason M.S. Chen; Gen Secr: Prof. Sheau-Ting Chang
Focus: Educ
Periodical
AV Newsletter (weekly) 03310

National Bar Association, 124 Chungking South Rd , Section 1, Taipei
Focus: Law 03311

National Institute for Compilation and Translation, 247 Choushan Rd, Taipei
Fax: (02) 3629256
Founded: 1932
Pres: Li-Yun Nancy Chao; Gen Secr: Chiu-Mei Chen
Focus: Ling
Periodicals
News Bulletin (weekly)
The Institute Periodical (weekly) 03312

National Music Council of China, 162 Hoping East Rd, Section 1, Taipei
Founded: 1957
Focus: Music 03313

National Science Council, 2 Canton St, Taipei
Founded: 1959
Pres: Li-An Chen
Focus: Sci
Periodicals
Abstracts of Research Papers (weekly)
National Science (weekly)
The NSC Review (weekly)
NSC Special Publication (weekly)
NSC Symposium Series (weekly)
Proceedings (weekly)
Science Bulletin (weekly) 03314

National Tax Research Association of China, 63-9 Hangchow South Rd, Section 1, Taipei
Founded: 1951
Focus: Law 03315

National Young Writers Association of China, 51 Hanchung St, Taipei
Founded: 1953
Focus: Lit 03316

Ophthalmological Society of the Republic of China, c/o Dept of Ophthalmology, National Taiwan University Hospital, 1 Chang-Teh St, Taipei 100
Fax: (02) 3832310
Founded: 1959
Focus: Ophthal
Periodical
Transactions of the Ophthalmological Society of the Republic of China (weekly) 03317

The Otolaryngological Society of the Republic of China (OSROC), 1 Changteh St, Taipei 10016
T: (02) 3819083
Founded: 1965; Members: 1107
Pres: Hsieh Ti; Gen Secr: Chuan-Jen Hsu
Focus: Otorhinolaryngology
Periodical
The Journal of the Otolaryngological Society of the Republic of China (weekly) 03318

Physical Society of China, POB 2330, Taipei
Focus: Physics
Periodical
Chinese Journal of Physics (weekly) . . 03319

Playwriters Association of the Republic of China, 218-1 Roosevelt Rd, Section 3, Taipei
Founded: 1970
Focus: Lit 03320

Population Association of China, 107 Roosevelt Rd, Section 4, Taipei
Founded: 1956
Focus: Sociology 03321

Public Administration Society of China, c/o Dept of Public Administration, College of Law and Commerce, National Chung-Hsing University, 53 Ho-Chiang St, Taipei
Founded: 1954
Focus: Business Admin 03322

The Radiological Society of the Republic of China, c/o Dept of Radiology, Veterans General Hospital, Shih-Pai, Taipei 11217
Fax: (02) 8733643
Founded: 1951
Gen Secr: Dr. Cheng Yen Chang
Focus: Radiology
Periodical
Chinese Journal of Radiology (weekly) . 03323

Republic of China Society of Cardiology (ROCSOC), c/o Veterans General Hospital, 201 Shih-Pai Rd, Taipei 112
T: (02) 8737628; Fax: (02) 8737629
Founded: 1960; Members: 730
Focus: Cardiol
Periodical
Chinese Journal of Cardiology: Acta Cardiologica Sinica (weekly) 03324

School Health Association of the Republic of China, 162 Hoping East Rd, Section 1, Taipei
Founded: 1962
Focus: Public Health 03325

The Society of Anaesthesiologist of the Republic of China (SAROC), 271 Roosevelt Rd, Section 3, Taipei 10764
T: (02) 3633912; Fax: (02) 3633912
Founded: 1956; Members: 1800
Pres: Prof. Peter P.C. Tan; Gen Secr: Eng-Chye Chuah
Focus: Anesthetics
Periodical
Acta Anaesthesiologica Sinica (5 times annually) 03326

Society of Chinese Acupuncture and Cauterizing, Chung Ching Bldg. 66-1, Chung Ching South Rd, Section 1, Taipei
Founded: 1955; Members: 3500
Focus: Anesthetics 03327

Society of Chinese Constitutional Law, 170 Keelung Rd, Section 2, Taipei
Founded: 1951
Focus: Law 03328

Society of the Chinese Borders History and Languages, c/o National Taiwan University Library, Taipei
Founded: 1958
Focus: Hist; Ling 03329

Special Education Association of the Republic of China, 320 Chungching North Rd, Section 3, Taipei
Founded: 1967
Focus: Educ 03330

Surgical Association of the Republic of China, c/o Dept of Surgery, National Taiwan University Hospital, Taipei
Founded: 1967
Focus: Surgery
Periodical
Journal (weekly) 03331

Television Academy of Arts and Sciences of the Republic of China, 10 Pateh Rd, Section 3, Taipei
Founded: 1969
Focus: Arts; Sci 03332

World Wide Ethical Society (WES), 264 Chien Kang Rd, Taipei
T: (02) 7670401
Founded: 1920; Members: 52000
Focus: Philos
Periodical
Morality Quarterly (weekly) 03333

Colombia

Academia Antioqueña de Historia, Carrera 43, No 53-37, Apdo Aéreo 7175, Medellín
T: 395576
Founded: 1903; Members: 60
Pres: Jaime Sierra Garcia; Gen Secr: Alicia Giraldo Gómez
Focus: Hist
Periodicals
Bolsilibros
Repertorio Histórico (3 times annually) . 03334

Academia Boyacense de Historia, Casa del Fundador, Tunja
T: 3441
Founded: 1905; Members: 30
Pres: Javier Ocampo Lopez; Gen Secr: Ramón Correa
Focus: Hist
Periodical
Repertorio Boyacense (weekly) 03335

Academia Colombiana de Ciencias Exactas, Físicas y Naturales, Carrera 3A, No 17-34, Apdo Aéreo 44763, Bogotá
T: 2414805
Founded: 1933; Members: 112
Pres: Luis Eduardo Mora-Osejo; Gen Secr: Santiago Díaz-Piedrahita
Focus: Nat Sci; Physics
Periodical
Revista 03336

Academia Colombiana de Historia, Calle 10, No 8-95, Apdo Aéreo 14428, Bogotá
Founded: 1902; Members: 40
Pres: Dr. Germán Arciniegas; Gen Secr: Roberto Velandia
Focus: Hist
Periodical
Boletín de Historia y Antigüedades . . . 03337

Academia Colombiana de Jurisprudencia, Calle 17, No 4-95, Of 210, Bogotá
T: 2414716
Founded: 1894; Members: 50
Pres: Hernando Morales M.
Focus: Law
Periodicals
Anuario
Revista (weekly) 03338

Academia Colombiana de la Lengua, Apdo Aéreo 13922, Bogotá
Founded: 1871; Members: 79
Pres: Eduardo Guzmán Esponda; Gen Secr: Horacio Bejarano Díaz
Focus: Ling
Periodicals
Anuario
Boletín 03339

Academia de Historia de Cartagena de Indias, Palacio de la Inquisición, Cartagena
T: 40036
Founded: 1912; Members: 190
Pres: Donaldo Bosa Herazo; Gen Secr: Carlos Villalba Bustillo
Focus: Hist
Periodical
Boletín Historial (weekly) 03340

Academia de Historia del Norte de Santander, Palacio Nacional, Cúcuta
T: 27796
Focus: Hist 03341

Academia Nacional de Medicina, Calle 60A, No 5-29, Apdo Aéreo 23224, Bogotá
Founded: 1890; Members: 40
Pres: Jorge Cavelier; Gen Secr: Juan Mendoza
Focus: Med
Periodicals
Medicina (3 times annually)
Temas Médicos (weekly) 03342

Asociación Colombiana de Bibliotecarios (ASCOLBI), Calle 10, No 3-16, Apdo Aéreo 30883, Bogotá
T: 2694219
Founded: 1942; Members: 1200
Pres: Saul Sanchez Toro; Gen Secr: B.N. Cardona de Gil
Focus: Libraries & Bk Sci
Periodical
Boletín (weekly) 03343

Asociación Colombiana de Facultades de Medicina, Calle 45A, No 9-77, Bogotá
T: 2322170
Founded: 1959; Members: 20
Pres: Dr. Luis A. Velez; Gen Secr: Abel Dueñas Padrón
Focus: Med
Periodical
Boletín de Medicamentos y Terapéutica: Boletín Informativo de Ascofame (weekly) . . . 03344

Asociación Colombiana de Fisioterapía, c/o Hospital Militar, Transversal 5, No 49-00, Bogotá
Founded: 1953; Members: 90
Focus: Physical Therapy 03345

Asociación Colombiana de Museos, Institutos y Casas de Cultura (ACOM), Calle 16, No 5-41, Bogotá
Focus: Arts 03346

Asociación Colombiana de Sociedades Científicas, Apdo Aéreo 6572, Bogotá
Founded: 1957; Members: 19
Focus: Sci 03347

Asociación Colombiana de Universidades (ASCUN), Carrera 10, No 88-45, Apdo Aéreo 012300, Bogotá
T: 2361424, 2186914; Fax: 2185098
Founded: 1958; Members: 61
Pres: Dr. Gabriel Jaime Cardora; Gen Secr: Dr. Jorge Rivadeneira Vargas
Focus: Sci
Periodical
Mundo Universitario (weekly) 03348

Asociación de Egresados de la Escuela Interamericana de Bibliotecología de la Universidad de Antioquia (ASEIBI), Apdo Aéreo 49983, Medellín
Focus: Libraries & Bk Sci
Periodical
Boletin ASEIBI (weekly) 03349

Asociación de Universidades Confiadas a la Compañía de Jesús en América Latina (AUSJAL) / Organization of Latin American Jesuit Universities and Faculties, Carrera 7, No 40-76, Of 501, Apdo 56710, Bogotá
T: 2875584, 2881511 ext 204; Fax: 2857524
Founded: 1985
Gen Secr: Jorge Hoyos
Focus: Educ
Periodical
Carta de AUSJAL (weekly) 03350

Asociación Latinoamericana de Física Médica / Latin American Association for Medical Physics, c/o Dep de Radioterapia, Hospital Universitario de Valle, Cali
Gen Secr: Herman Ramirez Contreras
Focus: Radiology 03351

Asociación Latinoamericana para Microbiologa, c/o Instituto Nacional de Salud, Apdo Aéreo 3495, Bogotá
Focus: Microbio 03352

Asociación Lationoamericana de Facultades de Odontologia (ALAFO), c/o Facultad de Odontologia, Universidad de Antioquia, Medellín
Founded: 1960
Focus: Dent 03353

Bibliotecarios Agríolas Colombiano, c/o Biblioteca de Tibaitatá, Apdo Aéreo 7984, Bogotá
Focus: Agri 03354

Centro de Educación en Administración de Salud (CEADS), Av 1, No 10-01, Apdo Aéreo 28504, Bogotá
T: 462723
Focus: Public Health 03355

Centro de Estudios sobre Desarrollo Económico, c/o Facultad de Economía, Universidad de los Andes, Carrera 1E, No 18A-10, Apdo Aéreo 4976, Bogotá
T: 2348156
Founded: 1958
Focus: Econ 03356

Centro de Investigación y Acción Social (CIAS), Carrera 5, No 33-A-08, Apdo Aéreo 25916, Bogotá
T: 324440
Focus: Sociology 03357

Centro Interamericano de Fotointerpretación (CIAF), Carrera 30, No 47A-57, Apdo Aéreo 53754, Bogotá
T: 2680300
Focus: Surveying
Periodical
Revista CIAF (weekly) 03358

Centro Interamericano de Vivienda y Planeamiento, Apdo Aéreo 6209, Bogotá
Founded: 1952
Focus: Sociology; Urban Plan 03359

Centro Internacional de Agricultura Tropical (CIAT), Apdo Aéreo 6713, Cali
T: 680111; Tx: 05769; CA: CINATROP
Focus: Agri; Develop Areas
Periodicals
Abstracts on Cassava (3 times annually)
Abstracts on Phaseolus Vulgaris (3 times annually)
Boletín Informativo de Pastos Tropicales (3 times annually)
Cassava Newsletter (weekly)
CIAT International (3 times annually)
CIAT Report (weekly)
Hojas de Frijol (3 times annually)
Jojas de Arroz (3 times annually)
Resumenes Pastos Tropicales (3 times annually)
. . . . 03360

Centro Nacional de Investigaciones de Café (CENICAFE), Chinchina
Founded: 1949
Focus: Agri
Periodicals
Anuario Meteorologico (weekly)
Avances Técnicos (weekly)
Boletín Técnico (weekly)
CENICAFE (3 times annually)
Resumenes de Cafe (5 times annually) . 03361

Centro Regional para el Fomento del Libro en América Latina (CERLAL), Calle 70, No 9-52, Apdo Aéreo 17438, Bogotá
T: 554594
Focus: Libraries & Bk Sci 03362

Colegio Colombiano de Cirujanos, c/o Hospital Militar Central, Transversal 5, No 49-00, Bogotá
Founded: 1954
Focus: Surgery 03363

Colegio de Bibliotecarios Colombianos, Apdo Aéreo 3272, Bogotá
Founded: 1963
Focus: Libraries & Bk Sci 03364

Comisión Episcopal Latino Americano (CELAM), Apdo Aéreo 5278, Bogotá
Focus: Rel & Theol 03365

Comité Nacional del Consejo Internacional de Museos (ICOM), c/o División de Museos y Restauración, Calle 24, No 5-60, Of 211, Bogotá
Focus: Arts; Ethnology 03366

Comité Nacional de Lucha contra el Cáncer (CIEC), Av 1, No 9-85, Bogotá
Founded: 1945
Focus: Cell Biol & Cancer Res 03367

Confederación Interamericana de Educación Católica, Apdo Aéreo 7478, Bogotá
Founded: 1945
Focus: Educ 03368

Confederacion Latinoamericana de Sociedades de Anestesiología (CLASA), Calle 118, No 20-08, Bogotá
T: 2138174
Founded: 1962; Members: 7000
Focus: Anesthetics 03369

Consejo Nacional de Archivos, Calle 24, No 5-60, Bogotá
Founded: 1961
Focus: Libraries & Bk Sci; Archives . . 03370

Consejo Nacional de Política Económica Planeación, c/o Biblioteca y Archivo, Departamento Nacional de Planeación, Carrera 26, No 13-19, Bogotá
Focus: Econ
Periodical
Revista de Planeación y Desarrollo (3 times annually) 03371

Corporación Latinoamericana de Investigación para el Desarrollo del Sector Rural y Zona Costeros, Apdo 102279, Bogotá
Focus: Develop Areas 03372

Departamento Administrativo Nacional de Estadística, c/o Centro Administrativo Nacional, Via El Dorado, Apdo Aéreo 80043, Bogotá
Founded: 1953
Focus: Stats 03373

Fondo Colombiano de Investigaciones Cientificas y Proyectos Especiales Francisco José de Caldas (COLCIENCIAS), Transversal 9-A, No 133-28, Apdo Aéreo 051580, Bogotá
T: 740660, 2169800; Tx: 44305; Fax: 2744460
Founded: 1968
Pres: Dr. Pedro José Amaya Pulido; Gen Secr: Isabel Forero de Moreno
Focus: Sci
Periodicals
Anales de Invemar (weekly)
Carta de Colciencias (weekly)
Ciencia, Tecnología y Desarrollo (weekly)
Colombia, Ciencia y Tecnología (weekly) 03374

Fundación OFA para el Avance de las Ciencias Biomédicas, Carrera 11A, No 93A-62, Apdo Aéreo 0061, Bogotá
T: 2366178, 2367119
Founded: 1979
Focus: Bio; Cell Biol & Cancer Res; Neurology
. . . . 03375

Grupo de Biblioteca y Documentación, Av El Dorado, Apdo Aéreo 8475, Bogotá
Focus: Libraries & Bk Sci; Doc 03376

Instituto Caro y Cuervo, Carrera 11, No 64-37, Apdo Aéreo 51502, Bogotá
T: 2557753
Founded: 1942; Members: 130
Focus: Ling; Lit; Hist
Periodicals
Anuario Bibliográfico Colombiano
Noticias Culturales (weekly)
Thesaurus: Boletín Cuadrimestral 03377

Instituto Central de Medicina Legal, Carrera 13, No 7-30, Bogotá
Founded: 1948
Focus: Med 03378

Instituto Colombiano Agropecuario, Calle 37, No 8-43, Apdo Aéreo 7984, Bogotá
Founded: 1962
Focus: Agri 03379

Instituto Colombiano de Antropología (ICAN), Carrera 7, No 28-66, Apdo Aéreo 407, Bogotá
T: 2836647, 2454843
Founded: 1941; Members: 32
Pres: Dr. Myriam Jimeno Santoyo
Focus: Anthro
Periodical
Revista Colombiana de Antropologia: Informes Antropologicos (weekly) 03380

Instituto Colombiano de Bienestar Familiar (ICBF), Av 68, Calle 64, Apdo Aéreo 18116, Bogotá
T: 314556
Focus: Sociology 03381

Instituto Colombiano de Crédito Educativo y Estudios Técnicos en el Exterior (ICETEX), Carrera 3, No 18-24, Apdo Aéreo 5735, Bogotá
Founded: 1950
Focus: Eng; Educ
Periodicals
Boletín desde Colombia (weekly)
Carta Informativa (weekly) 03382

Instituto Colombiano de Cultura, Carrera 3, No 18-24, Bogotá
Founded: 1968
Focus: Lit; Arts 03383

Instituto Colombiano de Cultura Hispánica, Calle 12, No 2-41, Bogotá
Founded: 1951
Focus: Lit; Hist 03384

Instituto Colombiano de la Reforma Agraria (INCORA), Av El Dorado, Apdo Aéreo 151046, Bogotá
Focus: Agri 03385

Instituto Colombiano de Normas Técnicas (ICONTEC), Carrera 37, No 52-95, Apdo Aéreo 14237, Bogotá
Fax: 2221435
Founded: 1963; Members: 950
Pres: Javier Henao Londoño; Gen Secr: Guillermo Higuera A.
Focus: Standards; Eng
Periodicals
Boletín Bibliográfico (weekly)
Boletín Informativo (weekly)
Normas y Calidades (weekly)
Standards (weekly) 03386

Instituto Colombiano de Oncologia Pediatrica / Colombian Institute of Pediatric Oncology, Calle 120, No 8-23, Bogotá
T: 2133067
Founded: 1982; Members: 6
Pres: Juan Manuel Zea; Gen Secr: Sonja Plazas de Perdomo
Focus: Cell Biol & Cancer Res; Pediatrics 03387

Instituto Colombiano de Pedagogía (ICOLPE), Carrera 7, No 27-52, Apdo Aéreo 52976, Bogotá
T: 822270
Focus: Educ 03388

Instituto Colombiano para el Fomento de la Educación Superior (ICFES) / Colombian Institute for the Development of Higher Education, Calle 17, No 3-40, Apdo Aéreo 6319, Bogotá
T: 2819311; Fax: 2868045
Founded: 1968; Members: 520
Pres: Dr. Marco Palacios Rozo; Gen Secr: Dr. Gustavo Sandoval
Focus: Educ; Adult Educ
Periodicals
Directorio de la Educación Superior en Colombio (bi-annually)
Educación Superior y Desarrollo (weekly) 03389

Instituto Colombinao Agropecuario (ICA), Apdo Aéreo 151123, Mosquera
T: 813080
Focus: Agri 03390

Instituto de Ciencias Naturales (ICN-MHN), Apdo Aéreo 7495, Bogotá
T: 2442387
Founded: 1936; Members: 57
Focus: Nat Sci; Botany; Geology; Zoology
Periodicals
Biblioteca José Jerónimo Triana (weekly)
Caldasía: Boletín (weekly)
Catalogo Ilustrado de las Plantas de Cundinamarca (weekly)
Flora de Colombia (weekly)
Lozanía: Acta Zoológica Colombiana (weekly)
Mutisía: Acta Botánica Colombiana (weekly)
. . . . 03391

Instituto Nacional de Cancerología, Calle 1, No 9-85, Apdo Aéreo 17158, Bogotá
T: 2464386
Founded: 1934
Focus: Cell Biol & Cancer Res; Radiology; Nucl Med; Intern Med; Surgery; Pathology . . 03392

Instituto Nacional de Investigaciones Geológico-Mineras (INGENOMINAS), Diagonal 53, No 34-53, Apdo Aéreo 4865, Bogotá
T: 2125400, 2121811
Founded: 1968; Members: 480
Focus: Geology; Mining 03393

Instituto Nacional de Medicina Legal, Carrera 13, No 7-30 y 7-46, Bogotá
Founded: 1948
Focus: Law; Med 03394

Instituto Nacional de Salud (INPES), Av el Dorado, Carrera 50, Apdo Aéreo 80334, Bogotá
Founded: 1968
Focus: Public Health 03395

IUS Federación Universitaria Nacional, Apdo Nacional 7503, Bogotá
Focus: Sci 03396

Junta Nacional de Folclore, c/o Instituto Caro y Cuervo, Apdo Aéreo 51502, Bogotá
Focus: Ethnology 03397

Liga Colombiana de Lucha Contra el Cancer, Calle 80, No. 5-77, Bogotá
T: 2116981, 2118894, 2118828
Founded: 1961
Focus: Cell Biol & Cancer Res 03398

PEN Club de Colombia, Apdo Aéreo 51748, Bogotá
Founded: 1983; Members: 50
Pres: David Mejía Velilla; Gen Secr: Maruja Vieira
Focus: Lit
Periodical
Noticias del PEN (weekly) 03399

Servicio Interamericana de Geodesía, c/o Instituto Geográfico Augustín Codazzi, Carrera 30, No 48-51, Bogotá
Focus: Surveying 03400

Servicio Nacional de Aprendizaje (SENA), Av Caracas, No 13-88, Apdo Aéreo 9801, Bogotá
T: 832965
Focus: Educ 03401

Sociedad Americana de Oftalmologia y Optometria, Apdo 091019, Bogotá
Focus: Ophthal 03402

Sociedad Antioqueña de Ingenieros y Architectos, Carrera 50, No 58-14, Apdo Aéreo 4754, Medellín
Founded: 1913; Members: 950
Focus: Archit; Eng 03403

Sociedad Colombiana de Biologí, Carrera 13, No 48-26, Bogotá
Founded: 1942; Members: 36
Pres: Dr. Gonzalo Montes; Gen Secr: Margaret Ordóñez-Smith
Focus: Bio 03404

Sociedad Colombiana de Cancerología, c/o Hospital Militar, Transversal 5, No 49-00, Bogotá
Founded: 1964
Focus: Cell Biol & Cancer Res 03405

Sociedad Colombiana de Cirugía, Calle 97A, No 10-65, Apdo 503, Bogotá
T: 2563007
Founded: 1973; Members: 300
Pres: Dr. E. Bozon Martinez; Gen Secr: Dr. Jaime Escallón
Focus: Surgery
Periodical
Revista 03406

Sociedad Colombiana de Economistas, Carrera 20, No 36-41, Apdo Aéreo 8429, Bogotá
T: 2459637
Founded: 1957; Members: 5500
Pres: Jorge Valencia Jaramillo; Gen Secr: Luis Alberto Avila
Focus: Econ
Periodical
Revista (weekly) 03407

Sociedad Colombiana de la Ciencia del Suelo (SCCS), Carrera II, No 66-34, Of 204, Apdo Aéreo 51791, Bogotá
Founded: 1955; Members: 400
Focus: Agri 03408

Sociedad Colombiana de Matemáticas (SCM), Edificio Camilo Torres, Bloque C, Mod 5, Of 302, Apdo Aéreo 2521, Bogotá 1
T: 2216829
Founded: 1955; Members: 800
Pres: Myriam Muñoz de Özak
Focus: Math
Periodicals
Lecturas Matematicas (3 times annually)
Revista Colombiana de Matemáticas (weekly)
. . . . 03409

Sociedad Colombiana de Obstetricia y Ginecologia (SCOG) / Colombian Society of Obstetrics and Gynecology, Carrera 9, No 20-13, Of 602, Bogotá
T: 2432511
Founded: 1943; Members: 130
Focus: Gynecology
Periodical
Revista Colombina de Obstetricia y Ginecologia (weekly) 03410

Sociedad Colombiana de Patología, c/o Dpto de Patología, Universidad del Valle, Cali
Founded: 1955; Members: 155
Pres: Dr. Edgar Duque; Gen Secr: Dr. José A. Dorado
Focus: Pathology 03411

Sociedad Colombiana de Pediatría, c/o Hospital Militar Central, Entrepiso 1, Bogotá
T: 2883985
Founded: 1917; Members: 150
Pres: Dr. Gabriel Lamus; Gen Secr: Enrique Gutierrez Saravia
Focus: Pediatrics
Periodical
Pediatría (weekly) 03412

Sociedad Colombiana de Psiquiatría, Carrera 18, No 84-87, Of 203, Bogotá 8
Founded: 1961; Members: 198
Pres: Dr. Gustavo Alvarez Cardona; Gen Secr: Dr. Germán Puerta Baptiste
Focus: Psychiatry
Periodical
Revista Colombiana de Psiquiatría (weekly) 03413

Sociedad Colombiana de Químicos e Ingenieros Químicos / Columbian Society of Chemists and Chemical Engineers, Av Jimenez, No 8-74, Of 501-513, Apdo Aéreo 10968, Bogotá
T: 2411480
Founded: 1941
Focus: Chem; Eng
Periodical
Química e Industria (weekly) 03414

Sociedad Colombina de Cardiología, c/o Hospital Militar, Transversal 5, No 49-00, Bogotá
Founded: 1944; Members: 101
Focus: Cardiol 03415

Sociedad de Antropología de Antioquía, c/o Universidad de Antioquia, Medellín
Founded: 1946
Focus: Anthro 03416

Sociedad de Ciencias Naturales Caldas, Apdo Aéreo 1180, Medellín
Founded: 1938
Focus: Nat Sci 03417

Sociedad de Pediatría y Puericultura del Atlántico, c/o Hospital Infantil San Francisco de Paula, Barranquilla
Founded: 1955
Focus: Pediatrics 03418

Sociedad Geográfica de Colombia, c/o Observatorio Astronómico Nacional, Apdo Nacional 2584, Bogotá
Founded: 1903; Members: 40
Focus: Geography 03419

Sociedad Jurídica de la Universidad Nacional, c/o Universidad Nacional, Apdo Aéreo 14490, Bogotá
Founded: 1908
Focus: Law 03420

Unión Parroquial del Sur, Calle 6, Sur 14-A, Bogotá
Focus: Rel & Theol 03421

Congo

Association for Health Information and Libraries in Africa (AHILA), c/o WHO/AFRO, POB 6, Brazzaville
T: 813860; Tx: 5217
Founded: 1980
Gen Secr: A. Ikama-Obambi
Focus: Public Health 03422

Bureau des Archives et Bibliothèques de Brazzaville, BP 2025, Brazzaville
Focus: Archives; Libraries & Bk Sci . . 03423

Centre Technique Forestier Tropical, BP 764, Pointe-Noire
Founded: 1958
Gen Secr: J.-C. Delwaulle
Focus: Forestry 03424

Conseil National de la Recherche Scientifique et Technique, Brazzaville
Founded: 1966
Focus: Sci; Eng 03425

Institut Africain, Mouyondzi
Focus: Ling 03426

Costa Rica

Academia Costarricense de la Lengua, Paseo de los Estudiantes, Apdo 157-1002, San José
Founded: 1923; Members: 18
Pres: Arturo Aguero Chaves; Gen Secr: Virginia Sandoval de Fonseca
Focus: Ling
Periodical
Boletín 03427

Academia Costarricense de Periodoncia, Apdo 1435, San José
Founded: 1965
Focus: Dent 03428

Academia de Geografía e Historia de Costa Rica, Apdo 4499, San José
Members: 31
Pres: Carlos Meléndez; Gen Secr: Oscar Aguilar
Focus: Hist
Periodical
Anales 03429

Asociación Costarricense de Bibliotecarios, Apdo 3308, San José
Founded: 1949
Focus: Libraries & Bk Sci 03430

Asociación Costarricense de Cirurgíca, Apdo 548-1000, San José
Founded: 1954; Members: 62
Pres: Dr. Eduardo Flores Montero; Gen Secr: Dr. Rodrigo Campos Rojas
Focus: Surgery 03431

Asociación Costarricense de Pediatría / Costarrican's Pediatric Association, Apdo 1654, San José
Founded: 1953; Members: 130
Focus: Pediatrics
Periodical
Carta Pediátrica (3 times annually) . . . 03432

Asociación de Cardiología, c/o Hospital San Juan de Dios, San José
Focus: Cardiol 03433

Asociación de Medicina Interna, c/o Hospital San Juan de Dios, San José
Focus: Intern Med 03434

Asociación de Obstetricia y Ginecología, c/o Hospital San Juan de Dios, San José
Focus: Gynecology 03435

Central American Institute for Business Administration, Apdo 960, Alajuela
T: 412255; Tx: 7040
Founded: 1961
Pres: Dr. Melvyn Copen
Focus: Business Admin
Periodical
Revista INCAE (weekly) 03436

Central American Institute of Public Administration, Edificio Schyfter, Av Central y Calle 2, Apdo 10025, San José
T: 223133; Tx: 2180
Founded: 1954
Pres: Carmen Maria Romero
Focus: Public Admin
Periodical
Revista Centroamericana de Administración Pública (weekly) 03437

Central American Institute of Social Studies, Coronado 2200, Apdo 75, San José
T: 290152; Fax: 293077
Founded: 1964
Pres: E. Delgado Vargas
Focus: Soc Sci 03438

Central American, Mexican and Caribbean Network for Bean Research, c/o PROFRIJOL, Coronado 2200, Apdo 55, San José
T: 294741; Tx: 2144
Founded: 1978
Focus: Crop Husb 03439

Central American University Confederation, Ciudad Universitaria Rodrigo Facio, Apdo 37, San José
T: 252744; Tx: 3011; Fax: 220478
Founded: 1948; Members: 7
Gen Secr: Rodrigo Fernández Vásquez
Focus: Educ
Periodicals
Carta Informativa de la Secretaría General (weekly)
Cuadernos de Investigación (weekly)
Estadisticas Universitarias (weekly) . . . 03440

Centro de Estudios Médicos Ricardo Moreno Cañas, Apdo 10151, San José
Founded: 1942; Members: 400
Focus: Geology 03441

Comisión de Energía Atómica de Costa Rica, Apdo 6681, San José
Founded: 1967
Focus: Nucl Res 03442

Confederación Centroamericana de Medicina del Deporte / Central American Federation of Sports Medicine, Apdo 172, Heredía 3000
Founded: 1986; Members: 7
Pres: Dr. Rafael Brenes Rojas
Focus: Med 03443

Consejo Superior Universitario Centroemericano (CSUCA), Ciudad Universitaria Rodrigo Facio, Apdo 37, San José
T: 252744
Founded: 1948
Focus: Adult Educ; Sci 03444

Instituto Centroamericano de Administración Pública (ICAP), Apdo 10025, San José
T: 223133; Tx: 2180
Founded: 1954
Focus: Poli Sci; Public Admin; Develop Areas
Periodical
Central American Journal of Public Administration (weekly) 03445

Instituto Centroamericano de Extensión de la Cultura (ICECU), Apdo 2948, 1000 San José
Founded: 1963
Focus: Educ
Periodical
Libro Almanaque Escuela para Todos (weekly) 03446

Instituto Interamericano de Cooperación para la Agricultura (IICA) / Inter-American Institute for Cooperation on Agriculture, Coronado 2200, Apdo 55, San José
T: 290222; Tx: 2144; CA: IICA San José
Founded: 1942
Focus: Agri
Periodicals
Desarrollo Rural en las Américas (weekly)
Turrialba (weekly) 03447

Instituto Latinoamericano de las Naciones Unidas para la Prevención del Delito y Tratamiento del Delincuente, Apdo 10338, San José
Founded: 1975
Focus: Law 03448

Inter-American Association of Agricultural Librarians and Documentalists, c/o Centro de Enseñanza e Investigación, Catie 7170, Turrialba
Founded: 1953
Focus: Libraries & Bk Sci; Doc
Periodicals
AIBDA Actualidades (weekly)
Boletín Especial (weekly)
Boletín Informativo de AIBDA (weekly)
Páginas de Contenido: Ciencias de la Información (weekly)
Revista AIBDA 03449

Junta Nacional de Planeamiento Económico, c/o Consejo Nacional de Producción, San José
Focus: Econ 03450

Organización de Estudios Tropicales, Apdo 16, San José
Focus: Geography 03451

Tropical Science Center, Apdo 83870, San José
T: 226241; Fax: 222759
Founded: 1962; Members: 45
Pres: J.A. Tosi; Gen Secr: Vicente Watson
Focus: Forestry; Ecology
Periodicals
Facsimile Series (weekly)
Notas Técnicas y Economicas (weekly)
Tropical Science Center Occassional Papers (weekly) 03452

Croatia

Croatiam Geographic Society, Marulićev trg 19, 41000 Zagreb
T: (041) 446728
Founded: 1897; Members: 600
Pres: Dr. M. Brazda
Focus: Geography 03453

Croatian Medical Association, Šubićeva ul 9, 41000 Zagreb
T: (041) 416820; Fax: (041) 416820
Founded: 1874; Members: 8500
Pres: Prof. Dr. M. Gjurašin; Gen Secr: Dr. J. Gjurović
Focus: Med 03454

Croatian Pharmaceutical Society, Masarykova 2, 41000 Zagreb
T: (041) 427944; Fax: (041) 431301
Founded: 1945; Members: 3000
Pres: M. Frleta-Mazzi
Focus: Pharmacol 03455

Društvo Ekonomista Hrvatske / Economists' Society of Croatia, Berislavićeva 6, 41000 Zagreb
T: (041) 445206
Focus: Econ 03456

Drustvo za Proučavanje i Unapredenje Pomorstva / Society for Research and Promotion of Maritime Sciences, Rade Končara 44, POB 391, Rijeka
Founded: 1962; Members: 323
Pres: B. Pecotić; Gen Secr: Dr. D. Crnković
Focus: Navig; Oceanography
Periodical
Pomorski Zbornik: Maritime Annals (weekly) 03457

European Association for Earthquake Engineering (EAEE), c/o Institute of Civil Engineering, POB 283, Zagreb
T: (041) 636444; Tx: 22275; Fax: (041) 533889
Founded: 1964
Gen Secr: Prof. Drazen Anicic
Focus: Eng 03458

Hrvatska Akademija Znanosti i Umjetnosti / Croatian Academy of Sciences and Arts, Zrinski trg 11, 41000 Zagreb
T: (041) 433661; Fax: (041) 433383
Founded: 1866; Members: 60
Pres: Dr. I. Supek; Gen Secr: Dr. M. Moguš
Focus: Sci; Arts
Periodicals
Bulletin International Starine
Ljetopis (weekly)
Rad 03459

Hrvatsko Bibliotekarsko Društvo / Croatian Library Association, c/o Nacionalna i Sveučilišna Biblioteka, Marulićev trg 21, 41000 Zagreb
T: (041) 446322; Tx: 22206; Fax: (041) 426676
Founded: 1948; Members: 1400
Pres: Dr. A. Horvat; Gen Secr: Tinka Katić
Focus: Libraries & Bk Sci
Periodical
Vjesnik Bibliotekara Hrvatske (weekly) . . 03460

Hrvatsko Numizmatičko Društvo (HND) / Croatian Numismatic Society, Habdeliceva 2, POB 181, 41000 Zagreb
T: (041) 431426
Founded: 1928; Members: 500
Pres: E. Fabry; Gen Secr: Prof. B. Prister; B. Kopac
Focus: Numismatics
Periodicals
Numizmatičke vijesti
Numizmatika
Obol 03461

Hrvatsko Prirodoslovno Društvo / Croatian Society of Natural Sciences, Ilica 16, 41000 Zagreb
T: (041) 425288
Founded: 1885; Members: 30000
Pres: Prof. Dr. D. Sunko; Gen Secr: Dr. H. Vančik
Focus: Nat Sci
Periodicals
Periodicum Biologorum (weekly)
Priroda (weekly) 03462

League of Jurists' Associations of Croatia, Zrinski trg 3, 41000 Zagreb
Focus: Law 03463

Nakladni Zavod Matice Hrvatske, Matice Hrvatske ul 2, POB 515, 41000 Zagreb
T: (041) 275522
Founded: 1842
Pres: B. Kreber
Focus: Ling; Lit 03464

Pedagoško-Književni zbor, Savez Pedagoških Društava Hrvatske, Trg Maršala Tita 4, 41000 Zagreb
T: (041) 420601
Founded: 1871; Members: 3000
Pres: H. Vrgoč; Gen Secr: A. Bikić
Focus: Educ 03465

Savez Muzejskih Društava Hrvatske / Association of Museums Societies of Croatia, Mesnička 5, 41000 Zagreb
T: (041) 426534
Founded: 1945; Members: 385
Pres: Branko Kirigin; Gen Secr: Jadranka Vinterhalter
Focus: Arts
Periodical
Vijesti muzealaca i konzervatora Hrvatske (weekly) 03466

Union of the Societies of Engineers and Technicians of Croatia, Berislavićeva 6, 4100 Zagreb
Founded: 1970
Pres: Prof. Dr. E. Nonveiller
Focus: Psychiatry 03467

Cuba

Academia Cubana de la Lengua, Av 19, No. 502, Vedado, La Habana 4
Founded: 1951
Pres: Dulce M. Loynaz; Gen Secr: Delio Carreras
Focus: Ling 03468

Academia de Ciencias de Cuba, Capitolio Nacional, La Habana 2
Founded: 1962
Pres: Prof. Dr. Rosa Elena Simeón Negrin; Gen Secr: Dr. Carlos Gómez Gutiérrez
Focus: Sci 03469

Asociación Cubana de Bibliotecarios, c/o Biblioteca Nacional, Plaza de la Revolución, La Habana
Focus: Libraries & Bk Sci 03470

Asociación Latinoamericana de Micología, c/o Jardín Botánica Nacional, Carretesa del Roci0 km 3 1/2, Calabazar, Habana
Founded: 1990
Gen Secr: Dr. Miguel Rodríguez
Focus: Botany 03471

Centro Asturiano, San Rafael 3, entre Zulueta y Monserrate, La Habana
Founded: 1928
Focus: Hist; Lit 03472

Centro de Estudios de Historia y Organización de la Ciencia Carlos J. Finlay, Cuba 460, Apdo 70, La Habana 1
Founded: 1861
Gen Secr: Dr. Tirso W. Sáenz
Focus: Hist; Nat Sci; Sci
Periodical
Conferencias y Estudios de Historia y Organización de la Ciencia (weekly) . . 03473

Centro de Información y Divulgación Agropecuario, Calle 11, No. 1057, Vedado, La Habana 4
Focus: Agri

Periodicals
Agrotecnia de Cuba (weekly)
Boletín de Reseñas
Ciencia y Técnica en la Agricultura
Informaciòn Express
Revista Cubana de Ciencias Veterinarias (weekly)
Revista Cubana de Reproducciòn Animal (weekly)
Revista Plantas Medicinales (weekly) . . 03474

Centro de Investigaciones Pesqueras, Av 1 y 26, Miramar, Marianao, La Habana
Founded: 1959
Focus: Agri 03475

Centro de Investigación Forestal / Forest Research Center, Calle 174, No. 1723, 17-B y 17-C, Siboney, La Habana
Founded: 1969; Members: 418
Focus: Forestry
Periodical
Revista Foresta Baracoa (weekly) . . . 03476

Centro Nacional de Información de Ciencias Médicas, Calle 23, No. 177, entre N y O, Vedado, La Habana
Focus: Med 03477

Centro Nacional de Investigaciones Científicas (CNIC), Av 25, Calle 158, Reparto Cubanacán, Apdo 6990, La Habana
Tx: CENIC
Founded: 1964
Focus: Bio; Chem; Physics
Periodicals
Revista de Ciencias Biológicas (weekly)
Revista de Ciencias Químicas (weekly) . 03478

Consejo Nacional de Tuberculosis Paulina Aldina, Calzada de Colombia y Octava, Marianao, La Habana
Founded: 1946
Focus: Pulmon Dis 03479

Grupo Nacional de Radiología, c/o Ministerio de Salud Pública, 23 y N, Vedado, La Habana 4
Founded: 1968
Focus: Radiology 03480

Sociedad Cubana de Historia de la Ciencia y de la Técnica, Cuba 460, La Habana
Founded: 1967
Focus: Nat Sci; Eng 03481

Sociedad Cubana de Historia de la Medicina, Cuba 460, La Habana 1
Focus: Med; Cultur Hist 03482

Sociedad Cubana de Ingenieros, Av de Bélgica 258, La Habana
Founded: 1908; Members: 500
Focus: Eng 03483

Sociedad de Radiología de La Habana, 23 y N, Vedado, La Habana, 4
Founded: 1968
Pres: Prof. Dr. Rafael Martorell; Gen Secr: Prof. Dr. Alfonso Meneses
Focus: Radiology 03484

Sociedad Económica de Amigos del País, Av Salvador Allende 710, La Habana
Founded: 1928
Focus: Econ 03485

Unión de Escritores y Artistas de Cuba, Calle 17 No. 351, La Habana, Vedado
Founded: 1961; Members: 66
Focus: Arts; Lit 03486

Cyprus

Association of Sports Medicine of the Balkan (ASMB), POB 5137, Nicosia
T: (02) 455762; Fax: (02) 464355
Founded: 1967
Gen Secr: Dr. N. Maroudias
Focus: Med 03487

Cyprus Astronautical Society, POB 3332, Limassol
Focus: Aero 03488

Cyprus Civil Engineers and Architects Association, Zena de Tyras Palace, POB 1825, Nicosia
T: (02) 41221
Founded: 1950; Members: 400
Focus: Archit 03489

Cyprus Economic Association, POB 1632, Nicosia
Focus: Econ 03490

Cyprus Geographical Association (CGA), POB 3656, Nicosia
Founded: 1968; Members: 200
Pres: G. Kadis; Gen Secr: Andreas Christodoulou
Focus: Geography
Periodical
The Geographical Chronicles (weekly) . . 03491

Cyprus Joint Technical Council, POB 1825, Nicosia
Focus: Eng 03492

Cyprus Numismatic Society (CNS), POB 3703, Nicosia
Founded: 1970; Members: 250
Focus: Numismatics
Periodical
Numismatic Report (weekly) 03493

Cyprus Ophthalmological Society, POB 406, Larnaca
T: (04) 631011; Fax: (04) 631370
Founded: 1970; Members: 45
Pres: A.P. Vorkas
Focus: Ophthal 03494

Cyprus Photogrammetric Society, G. Christoferou 9, Nicosia 112
Focus: Surveying 03495

Cyprus Research Centre / Kentron Epistemonikon Erevnon, POB 1952, Nicosia
Focus: Hist; Sociology
Periodicals
Epeteris (weekly)
Publications
Texts and Studies of the History of Cyprus 03496

Library Association of Cyprus, POB 1039, Nicosia
Founded: 1962
Pres: Costas D. Stephanou; Gen Secr: P.G. Rossos
Focus: Libraries & Bk Sci 03497

Pancyprian Medical Association, Zena de Tyras Palace, Nicosia
Focus: Med 03498

Society of Cypriot Studies, POB 1436, Nicosia
T: (02) 463205
Members: 200
Pres: Dr. Kypros Chrysanthis
Focus: Archeol; Hist; Ethnology; Ling
Periodical
Deltion: Bulletin (weekly) 03499

Czech Republic

Česká Akademie Věd (ČSAV) / Czech Academy of Sciences, Národní tř 3, 111 42 Praha 1
T: (02) 2358065; Tx: 121040
Founded: 1952; Members: 318
Pres: Josef Říman; Gen Secr: Karel Martinek
Focus: Sci
Periodicals
Acta Entomologica Bohemoslovaca (weekly)
Acta Technica (weekly)
Acta Virologica (weekly)
Aplikace matematiky (weekly)
Archeologické rozhledy (weekly)
Archiv orientální (weekly)
Biologia Plantarum (weekly)
Biologické listy (weekly)
Bulletin of the Czech Astronomical Institute (weekly)
Byzantinoslavica (weekly)
Časopis pro mineralogii a geologii (weekly)
Časopis pro pěestování matematiky (weekly)
Česká fysiologie (weekly)
Česká literatura (weekly)
Česká mykologie (weekly)
Česká psychologie (weekly)
Česká rusistika (5 times annually)
Český časopis historický (weekly)
Český časopis pro fyziku (weekly)
Český lid (weekly)
Chemické listy (weekly)
Collection of Czech Chemical Publications (weekly)
Czech Journal of Physics (weekly)
Czech Mathematical Journal (weekly)
Dějiny věd a techniky (weekly)
Ekonomicko-matematicky obzor (weekly)
Estetika (weekly)
Filozofický časopis (weekly)
Folia Biologica (weekly)
Folia Geobotanica e Phytotaxonomica (weekly)
Folia Microbiologica (weekly)
Folia Morphologica (weekly)
Folia Parasitologica (weekly)
Folia Zoologica (weekly)
Hudební věda (weekly)
Kybernetika (weekly)
Lidé a země (weekly)
Listy filologické (weekly)
Naše řeč (5 times annually)
Nový orient (10 times annually)
Památky archeologické (weekly)
Pedagogika (weekly)
Philologica Pragensia (weekly)
Photosynthetica (weekly)
Physiologia Bohemoslovaca (weekly)
Pokroky matematiky, fyziky a astronomie (weekly)
Politická ekonomie (weekly)
Právník (weekly)
Preslia (weekly) 03500

Česká Akademie Zemědělských Věd, Těšnov 65, 117 05 Praha 1
T: (02) 2320582; Tx: 121041; Fax: (02) 2328898
Founded: 1974; Members: 130
Pres: M. Spelina
Focus: Agri
Periodicals
Agricultural Literature (weekly)
Lesnictví (weekly)
Metodiky pro Zavádění Výzkumu do Praxe (weekly)
Přehled Lesnické a Myslivecké literatury (weekly)
Rostlinná Výroba (weekly)
Scientia Agriculturae Bohemoslovaca (weekly)
Zemědělská Ekonomika (weekly)
Zemědělská Informatika (weekly)
Zemědělská Technika (weekly)
Živočišná Výroba (weekly) 03501

Česká Archeologická Společnost (CAS) / Czech Archaeological Society, Malá Strana, Letenská 4, 118 01 Praha 1
T: (02) 539351
Founded: 1919; Members: 620
Pres: Karel Sklenář; Gen Secr: M. Lutovský
Focus: Archeol
Periodical
Zprávy ČAS: ČAS News (weekly) . . . 03502

Česká Biologicka Společnost, Joštova 10, 662 43 Brno
T: (05) 42126268; Fax: (05) 42213990
Founded: 1922
Pres: Prof. O. Nečas; Gen Secr: Prof. R. Janisch
Focus: Bio 03503

Česká Botanická Společnost, Benátská 2, 128 01 Praha 2
T: (02) 297941
Founded: 1912; Members: 1040
Pres: Dr. J. Holub; Gen Secr: Dr. L. Hrouda
Focus: Botany
Periodicals
Preslia (weekly)
Zprávy ČSBS (3 times annually) 03504

Česká Farmaceutická Společnost / Czech Pharmaceutical Association, Jungmannova 6, 110 00 Praha 1
T: (02) 2388157
Focus: Pharmacol
Periodical
Česko-Slovenská Farmacie (weekly) . . . 03505

Česká Geografická Společnost, Oldřichova 19, 128 00 Praha 2
Founded: 1894; Members: 1100
Pres: Prof. Dr. V. Gardavský; Gen Secr: Dr. D. Drbohlav
Focus: Geography
Periodical
Sborník: Journal (weekly) 03506

Česká Geologická Společnost / Czech Geological Society, Malostr nám 19, 1118 21 Praha 1
T: (02) 533641; Fax: (02) 533564
Founded: 1990; Members: 1200
Pres: Dr. Miroslav Štemprok
Focus: Mineralogy; Geology
Periodical
Journal of the Czech Geological Society (weekly) 03507

Česká Historická Společnost / Czech Historical Society, Husova 8, POB 63, 110 01 Praha 1
T: (02) 423338
Founded: 1934; Members: 650
Pres: Prof. Dr. J. Pátek; Gen Secr: Dr. E. Kubu
Focus: Hist
Periodical
Časopis Matice Moravské (weekly) . . . 03508

Česká Lékařská Společnost J.E. Purkyně / Czech Medical Association J.E. Purkyně, Sokolská 31, POB 88, 120 26 Praha 2
T: (02) 24915191; Fax: (02) 24216836
Founded: 1949; Members: 35000
Pres: Prof. Jaroslav Blahoš; Gen Secr: Prof. Karel Trnavský
Focus: Med
Periodicals
Anesteziologie a neodkladná péče: Anaesthesiology and Emergency Care (weekly)
Časopis lékařu českých: Czech Medical Journal (weekly)
Central European Journal of Public Health (weekly)
Česká revmatologie: Czech Rheumatology (weekly)
Česko-slovenská dermatologie: Czecho-Slovak Dermatology (weekly)
Česko-slovenská epidemiologie, mikrobiologie, immunologie: Czecho-Slovak Epidemiology, Microbiology, Immunology (weekly)
Česko-slovenská farmacie: Czecho-Slovak Pharmacy (weekly)
Československá fyziologie: Czecho-Slovak Physiology (weekly)
Česko-slovenská gastroenterologie a výživa: Czecho-Slovak Gastroenterology and Nutrition (weekly)
Česko-Slovenská gynekologie: Czecho-Slovak Gynaecology (weekly)
Česko-Slovenská hygiena: Czecho-Slovak Hygiene (weekly)
Česko-Slovenská neurologie a neurochirurgie: Czecho-Slovak Neurology and Neurosurgery (weekly)
Česko-Slovenská oftalmologie: Czecho-Slovak Ophthalmology (weekly)
Česko-Slovenská otolaryngologie a foniatrie: Czecho-Slovak Otorhinolaryngology and Phoniatrics (weekly)
Česko-Slovenská patologie s přílohou Soudní lékařství: Czecho-Slovak Pathology with Supplement on Forensic Medicine (weekly)
Česko-Slovenská pediatrie: Czecho-Slovak Paediatrics (weekly)
Česko-Slovenská psychiatrie: Czecho-Slovak Psychiatry (weekly)
Česko-Slovenská radiologie: Czecho-Slovak Radiology (weekly)
Česko-Slovenská stomatologie: Czecho-Slovak Stomatology (weekly)
Klinická biochemie a metabolismus: Clinical Biochemistry and Metabolism (weekly)
Klinická onkologie: Clinical Oncology (weekly)
Lékař a technika: Biomedical Engineering (weekly)
Pracovní lékařství: Occupational Medicine (weekly)
Praktické zubní lékařství: General Dentistry (weekly)
Praktický lékař: General Practitioner (weekly)
Rozhledy v chirurgii: Surgical Review (8 times annually)
Vnitřní lékařství: Internal Medicine (weekly) 03509

Česká Meteorologická Společnost / Czech Meterological Society, Na Šabatce 17, 143 06 Praha 4
T: (02) 85762548
Founded: 1959; Members: 270
Pres: Jan Bednář; Gen Secr: Dr. Karel Dubec
Focus: Geophys
Periodical
Bulletin for Members (weekly) 03510

Česká Národopisná Spolecnost / Czech Ethnographical Society, Národní třída 3, 111 42 Praha 1
T: (02) 227691
Founded: 1893; Members: 470
Pres: Dr. J. Kandert; Gen Secr: Dr. Lubomír Procházka
Focus: Ethnology 03511

Česká Oftalmologická Společnost (ČOS) / Czech Ophthalmological Society, c/o Eye Clinic, University Hospital, Šrobárova 50, 100 34 Praha 10
T: (02) 67162491; Fax: (02) 67172491
Founded: 1926; Members: 600
Pres: Prof. Pavel Kuchynka; Gen Secr: M. Janek
Focus: Ophthal
Periodical
Československá oftalmologie (weekly) . . 03512

Česká Orientalistická Společnost / Czech Society for Eastern Studies, Lázenšská 4, 118 37 Praha 1
T: (02) 532529
Founded: 1958
Pres: J. Patučkova; Gen Secr: Dr. L. Bareš
Focus: Ethnology 03513

Česká Pediatrická Společnost, Sokolska 31, 120 26 Praha 2
Focus: Pediatrics
Periodical
Československá pediatrie (weekly) . . . 03514

Česká Psychologická Společnost / Czech Psychological Association, U háje 12/268, 147 00 Praha 4
T: (02) 462568
Founded: 1927; Members: 2700
Pres: Prof. Dr. K. Pavlica; Gen Secr: Prof. Dr. E. Bedrnová
Focus: Psych 03515

Česká Radiologická Společnost / Czech Radiological Society, c/o Clinic for Diagnostic Radiology, Budínova 2, 180 81 Praha 8
T: (02) 822431
Founded: 1934; Members: 100
Pres: Prof. Dr. Jaromír Kolář; Gen Secr: Prof. Dr. Lúboš Vyhnánek
Focus: Radiology
Periodical
Československá radiologie (weekly) . . . 03516

Česká Společnost Anesteziologie a Resuscitace / Czech Society of Anaesthesiology and Resuscitation, c/o Dept of Intensive and Critical Care EMS, Šermířská 5, 169 00 Praha 6
Members: 1050
Focus: Anesthetics
Periodicals
Anesteziologie a neodkladná péče: Anaesthesiology and Emergency Medicine (weekly)
Information Bulletin (weekly)
Referátový výběr z anesteziologie a resuscitace: Reader's Digest – Anaesthesiology and Resuscitation (weekly) 03517

Česká Společnost Biochemická / Czech Biochemical Society, Národni 3, 111 42 Praha 1
Founded: 1993
Gen Secr: Prof. Václav Pačes
Focus: Biochem
Periodical
Bulletin (weekly) 03518

Česká Společnost Bioklimatologická (ČSBKS) / Czech Society of Bioclimatology, Šrobárova 48, 100 42 Praha 10
T: (02) 368116
Founded: 1965; Members: 300
Pres: Dr. V. Krečmer; Gen Secr: Dr. V. Pasák
Focus: Bio
Periodical
Zypravodaj (weekly) 03519

Česká Společnost Chemická / Czech Chemical Society, Hradčanské nám 12, 118 29 Praha 1
T: (02) 539074
Founded: 1866; Members: 2948
Pres: Prof. Dr. J. Horák; Gen Secr: Dr. F. Kaštánek
Focus: Chem
Periodical
Chemické Listy (weekly) 03520

Česká Společnost Ekonomická / Czech Economic Association, Politických Věznu 7, 111 73 Praha 1
T: (02) 261056
Founded: 1962
Focus: Econ 03521

Česká Společnost Entomologická / Czech Entomological Society, Viničná 7, 128 00 Praha 2
T: (02) 298726
Founded: 1904; Members: 1200
Focus: Entomology
Periodical
European Journal pf Entomology: Klapalekiana (weekly) 03522

Česká Společnost Fyziologie a Patologie Dýcháni / Czech Society of Respiratory Physiology and Pathology, c/o Česká Lékarská Společnost J.E. Purkyně, Sokolská 31, 120 26 Praha 2
T: (02) 294141
Founded: 1969; Members: 360
Pres: Prof. Jiří Kandus; Gen Secr: Prof. Martin Vízek
Focus: Pulmon Dis
Periodical
Bulletin (weekly) 03523

Česká Společnost Histo- a Cytochemnická / Czech Society for Histochemistry and Cytochemistry, Studničkova 2, 120 00 Praha 2
T: (02) 291751
Founded: 1962; Members: 210
Pres: Prof. Dr. Z. Lojda; Gen Secr: Dr. I. Juliš
Focus: Chem
Periodical
Proceedings (weekly) 03524

Česká Společnost Kybernetiku a Informatiku / Czech Society for Cybernetics and Information Sciences, Pod Vodárenskou Věží 4, 182 08 Praha 8
T: (02) 66052815; Fax: (02) 66414303
Founded: 1966; Members: 500
Pres: Jaroslav Vlček; Gen Secr: Radim Jiroušek
Focus: Cybernetics
Periodical
Kybernetika (weekly) 03525

Česká Spolecnost Mikrobiologická (CSM) / Czech Society for Microbiology, Vídeňská 1083, 142 20 Praha 4
T: (02) 4713221; Fax: (02) 4713221
Founded: 1928; Members: 1100
Pres: Prof. Libor Ebringer; Gen Secr: Dr. F. Kunc
Focus: Microbio
Periodicals
Bulletin České Společnosti Mikrobiologicke (weekly)
Folia Microbiologica (weekly) 03526

Česká Společnost Parasitologická / Czech Society of Parasitology, Viničná 7, Praha 2
Founded: 1993; Members: 300
Pres: Prof. Jaroslav Kulda; Gen Secr: Petr Horák
Focus: Biochem
Periodical
Zprávy České Společnosti Parasitologické (weekly) 03527

Česká Společnost pro Mechaniku / Czech Society for Mechanics, Dolejškova 5, 182 00 Praha 8
T: (02) 8587784
Founded: 1966; Members: 790
Pres: Prof. Dr. Ladislav Frýba; Gen Secr: Dr. Miloslav Okrouhlík
Focus: Mach Eng
Periodical
Bulletin (3 times annually) 03528

Česká Společnost pro Politické Vědy / Czech Society for Political Sciences, Národní 18, 116 91 Praha 1
T: (02) 203802
Founded: 1967
Pres: Prof. F. Šamalík; Gen Secr: L. Křížkovsky
Focus: Poli Sci 03529

Česká Společnost pro Vědy Zemědělské, Lesnické, Veterinárni a Potravinářské / Czech Society for Agriculture, Veterinary Sciences and Food Technology, Suchdol VŠZ, 160 21 Praha 6
T: (02) 3213640
Founded: 1968; Members: 750
Pres: Prof. F. Hron; Gen Secr: Dr. M. Valla
Focus: Agri
Periodical
Informační Zpravodaj (weekly) 03530

Česká Společnost Zoologická / Czech Zoological Society, Viničná 7, 128 44 Praha 2
T: (02) 294606
Founded: 1927; Members: 500
Pres: Č. Folk; Gen Secr: Dr. P. Roth
Focus: Zoology
Periodical
Věstník České Společnosti Zoologické (weekly) 03531

Česká Stomatologická Společnost / Czech Society of Stomatology, Karlovo nám 32, 121 11 Praha 2
T: (02) 293676
Founded: 1952; Members: 2000
Focus: Stomatology
Periodicals
Česká stomatologie (weekly)
Praktické zubní lékařství (10 times annually) 03532

Česká Vědecká Společnost pro Mykologii / Czech Scientific Society for Mycology, c/o Národní Muzeum, Václavske nám 68, 115 79 Praha 1
T: (02) 269451 ext 259
Founded: 1923
Focus: Botany, Specific 03533

Česká Vědecká Společnost pro Mykologii / Czech Society for Mycology, POB 106, 111 21 Praha 1
T: (02) 264405
Founded: 1946; Members: 310
Pres: Prof. Dr. Z. Urban; Gen Secr: Dr. S. Pouzar
Focus: Botany, Specific
Periodicals
Česká Mykologie (weekly)
Česká Mykologie: Czech Mycology (weekly)
Mykologické listy (weekly) 03534

Česká Vědecko-Technická Společnost pro Geodezii a Kartografii, Siroka 5, 110 01 Praha
Focus: Surveying 03535

Česká Vědeckothechnická Společnost / Czech Society for Sciences and Technology, Široká 5, 110 01 Praha 1
T: (02) 2321251; Tx: 122015
Founded: 1955; Members: 550000
Gen Secr: S. Zlatohlavý
Focus: Sci; Eng 03536

Czech Anatomical Society, U Nemocnice 3, 128 00 Praha 2
T: (02) 291951; Fax: (02) 297692
Founded: 1954; Members: 127
Pres: Prof. Dr. J. Slípka; Gen Secr: Prof. Dr. M. Doskočil
Focus: Anat
Periodical
Functional and Developmental Morphology (weekly) 03537

Czech Dermatological Society, Tř. Vítězného února 31, 120 26 Praha 2
T: (02) 24915195
Founded: 1925; Members: 600
Focus: Derm
Periodical
Česká dermatologie (weekly) 03538

Czech Oncologic Society, Sokolska 31, 120 26 Praha 2
Focus: Cell Biol & Cancer Res 03539

Czech Otolaryngological Society, Vítězného února 31, 120 26 Praha 2
T: (02) 299689
Founded: 1921; Members: 540
Focus: Otorhinolaryngology
Periodicals
Česká otolaryngologie (weekly)
Proceedings (weekly) 03540

Czech Society of Cardiology, c/o Dept of Medicine II, IKEM, Budejovicka 800, 146 22 Praha 4
Focus: Cardiol 03541

Czech Surgical Society, Jecna 17, 120 00 Praha 2
Focus: Surgery 03542

European Centre for Leisure and Education (ECLE), Jilská 1, 110 00 Praha 1
T: (02) 262078
Founded: 1968
Gen Secr: Ivan Savicky
Focus: Educ 03543

Gastroenterologická Společnost, c/o IV Medical Clinic, Universita Karlova, Ul Nemocnice 2, 128 08 Praha 2
Members: 70
Focus: Gastroenter 03544

International Association of Planetology (IAP), Geologická 3, Barandov, 152 00 Praha 5
Founded: 1968; Members: 130
Pres: Dr. R. Dejaiffe; Gen Secr: Dr. Nadezda Stovickova
Focus: Geology 03545

International Association on the Genesis of Ore Deposits (IAGOD), c/o Geological Survey, POB 65, 790 01 Jeseník
T: (0645) 4294; Tx: 62360; Fax: (02) 533564
Founded: 1964
Gen Secr: J. Aichler
Focus: Geology 03546

Jednota Českých Matematikû a Fysikû / Union of Czech Mathematicians and Physicists, Žitná 25, 115 67 Praha 1
Founded: 1862; Members: 3900
Pres: F. Nožička; Gen Secr: Š. Zajac
Focus: Math; Physics
Periodical
Pokroky matematiky, fyziky a astronomie (weekly) 03547

Kruh Moderních Filologu Při / Circle of Modern Philologists, U háje 12, 142 00 Praha 4
T: (02) 462568
Founded: 1911
Focus: Educ 03548

Literárněvědná Společnost / Literary Society, Nové Zámecké Schody 4, 118 00 Praha 1
T: (02) 530291
Founded: 1934; Members: 285
Pres: Dr. S. Wollman; Gen Secr: J. Vlášek
Focus: Lit 03549

Masarykova Česká Sociologická Společnost / Masaryk's Czech Sociological Association, Husova 4, 110 00 Praha 1
T: (02) 24220993
Founded: 1964; Members: 863
Focus: Sociology 03550

Matice Moravská / Samostatná sekce Historického Klubu při ČSAV, Gorkého 14, 602 00 Brno
T: (05) 41321258
Founded: 1849; Members: 560
Pres: Prof. Dr. Jan Janák
Focus: Hist; Lit
Periodical
Časopis Matice moravské (weekly) . . . 03551

Obec Architektů / The Society of Czech Architects, Letenská 5, 118 45 Praha 1
T: (02) 539743; Fax: (02) 3114926
Founded: 1990
Pres: Jan Kozel; Gen Secr: Karel Dušek
Focus: Archit 03552

Spektroskopická Společnost Jana Marca Marci / Jana Marca Marci Spectroscopic Society, Thákurova 7, 166 29 Praha 6
Fax: (02) 3112343
Founded: 1967; Members: 928
Focus: Physics
Periodical
Bulletin Spektroskopické Jana Marca Marci 03553

Společnost Antonína Dvořáka (SAD) / Antonín Dvořák Society, c/o Museum Antonína Dvořáka, Ke Karlova 20, 120 00 Praha 2
T: (02) 298214
Founded: 1931; Members: 875
Pres: Jarmil Burghauser
Focus: Music
Periodical
Zprávy SAD (weekly) 03554

Védecká Společnost pro Nauku o Kovech / Society for Metal Science, Žižkova 22, 616 62 Brno
Fax: (05) 412121301
Founded: 1966; Members: 255
Focus: Metallurgy 03555

Denmark

Akademiet for de skønne Kunster, Kongens Nytorv 1, 1050 København K
T: 33126860
Founded: 1754
Focus: Arts 03556

Akademiet for de tekniske Videnskaber (ATV) / Danish Academy of Technical Sciences, Lundtoftevej 266, 2800 Lyngby
T: 42881311; Fax: 45931377
Founded: 1937; Members: 625
Pres: Erik B. Rasmussen; Gen Secr: Vibeke Q. Zeuthen
Focus: Eng
Periodicals
Report, DK-edition (weekly)
Report, UK-edition (weekly) 03557

Akademikernes Centralorganisation (AC) / The Danish Confederation of Professional Associations, Nørre Voldgade 29, 1358 København K
T: 33128540; Fax: 33938540
Founded: 1972; Members: 177000
Pres: Alex Nielsen
Focus: Sci 03558

Arkivforeningen, c/o Rigsarkivet, Rigsdagsgården 9, 1218 København K
Founded: 1917; Members: 100
Focus: Archives
Periodical
Arkivforeningens Seminarrapporter (weekly) 03559

Arktisk Institut / Arctic Institute, Strandgade 100, 1401 København K
T: 32880150
Founded: 1954
Pres: C.C. Resting-Jeppesen; Gen Secr: Grethe Knutzen
Focus: Geography
Periodical
Acta Arctica 03560

Association for World Education (AWE), Skyum Bjerge, 7752 Snedsted
Founded: 1970; Members: 660
Focus: Educ
Periodical
Journal for World Education (weekly) . . 03561

Association Internationale du Théâtre pour l'Enfance et la Jeunesse (ASSITEJ), Frederiksborggade 20, 1350 København K
T: 33156900; Fax: 33131439
Founded: 1965; Members: 53
Pres: Michael Fitzgerald; Gen Secr: Michael Ramløse
Focus: Perf Arts 03562

Association of European Cancer Leagues (ECL), c/o Danish Cancer Society, Rosenvaengets Hovedvej 35, 2100 København
T: 31268866; Fax: 31264560
Founded: 1980
Pres: Ole Bang
Focus: Intern Med 03563

Astronomisk Selskab, c/o Observatoriet, Østervoldgade 3, 1350 København K
T: 35323999; Fax: 35323989
Founded: 1916; Members: 700
Focus: Astronomy
Periodical
Astronomisk Tidsskrift (weekly) 03564

Audiologopaedisk Forening (ALF) / Danish Speeck and Hearing Association, Graesager 207, 2980 Kokkedal
Founded: 1923; Members: 670
Focus: Logopedy
Periodicals
Dansk Audiologopaedi: Journal of the Danish Speech and Hearing Association (weekly)
Scandinavian Journal of Logopedics and Phoniatrics (weekly) 03565

Biokemisk Forening, Øster Farimagsgade 2a, 1353 København K
Founded: 1943; Members: 1000
Focus: Biochem 03566

Biologisk Selskab / Danish Biological Society, c/o Dept of Yeast Genetics, Carlsberg Laboratory, Gamle Carlsberg Vej 10, 2500 København
T: 33275331; Fax: 33274766
Founded: 1896; Members: 670
Pres: Thue W. Schwartz; Gen Secr: Morten C. Kielland-Brandt
Focus: Bio 03567

Bird Strike Committee Europe (BSCE), c/o Civil Aviation Administration, Ellebjergvej 50, POB 744, 2450 København
T: 36444848; Tx: 27096; Fax: 36440303
Founded: 1966; Members: 29
Pres: Hans Dahl
Focus: Ornithology 03568

Brandteknisk Selskab, c/o Dansk Ingeniorforening, Vester Farimagsgade 31, 1606 København K
Focus: Eng; Safety 03569

Bureau Permanent des Congrès Internationaux des Sciences Généalogique et Héraldique, c/o Rigsarkivet, Rigsdagsgården 9, 1218 København
T: 33923310
Founded: 1958
Gen Secr: Nils G. Bartholdy
Focus: Genealogy 03570

Byggecentrum / Building Centre, Dr. Neergaards Vej 15, 2970 Hørsholm
T: 45767373; Fax: 45767669
Founded: 1956
Gen Secr: Jørn Vibe Andreasen
Focus: Urban Plan; Archit 03571

Chakoten Dansk Militaershistorisk Selskab, Faelledvej 9, 2200 København N
T: 31310692
Founded: 1944; Members: 190
Focus: Military Sci; Hist
Periodical
Chakoten (weekly) 03572

Collegium Palynologicum Scandinavicum (CPS), c/o Geologisk Institute, Aarhus Universitet, Universitetsparken, 8000 Aarhus
Founded: 1969
Gen Secr: Dr. K. Raunsgaard Pedersen
Focus: Geology
Periodicals
Grana (weekly)
Palynott-nytt (3 times annually) 03573

Committee of the Acta-Endocrinologica Countries (CAEC), c/o Dept of Clinical Chemistry, Glostrup Hospital, 2600 Glostrup
Focus: Endocrinology 03574

Danatom, Tingvej 7, 4690 Haslev
Founded: 1956; Members: 35
Focus: Nucl Res 03575

Danmarks Biblioteksforening (DB) / Danish Library Association, Telegrafvej 5, 2750 Ballerup
T: 44681466; Fax: 44681103
Founded: 1905; Members: 2117
Pres: Søren Møller; Gen Secr: Flemming Ettrup
Focus: Libraries & Bk Sci
Periodicals
Biblioteksvejviser: Library Directory (weekly)
Bogens Verden: Periodical on Culture and Librarianship (10 times annually)
Kort Sagt!: Newsletter from D.L.A. . . . 03576

Danmarks Farmaceutiske Selskab / The Pharmaceutical Society of Denmark, Universitetsparken 2, 2100 København Ø
T: 31370850
Founded: 1912; Members: 700
Pres: Prof. V.N. Handlos; Gen Secr: Karin B. Jensen
Focus: Pharmacol 03577

Danmarks Forskningsbiblioteksforening / Danish Research Library Association, c/o Danmarks Tekniske Bibliotek, Anker Engelundsvej 1, 2800 Lyngby
Fax: 45939979
Pres: Lars Bjørnshauge; Gen Secr: Grethe Lillelund
Focus: Libraries & Bk Sci
Periodical
DF-Revy 03578

Danmarks Handelsskoleforening, c/o Handelsskolernes Laererforening, Godthåbsvej 106, 2000 København F
Founded: 1910; Members: 3300
Focus: Adult Educ 03579

Danmarks Jurist- og Økonomforbund, Gothersgade 133, 1123 København K
Founded: 1972; Members: 19000
Gen Secr: Niels Amfred
Focus: Law; Econ 03580

Danmarks Laererforening (DLF), Vandkunsten 12, 1467 København K
Founded: 1874; Members: 74000
Focus: Educ
Periodical
Folkeskoten (weekly) 03581

Danmarks Mikrobiologiske Selskab, c/o Klinisk Mikrobiologisk Afdeling, Rigshospitalet, Tagensvej 18, 2200 København N
Founded: 1958
Focus: Microbio 03582

Danmarks Naturfredningsforening (DN) / Danish Society for the Conservation of Nature, Noerregade 2, 1165 København F
T: 33322021; Fax: 33322202
Founded: 1911; Members: 265000
Pres: Prof. Svend Bichel; Gen Secr: David Rehling
Focus: Ecology
Periodical
Tidsskrift Natur og Miljø (weekly) . . . 03583

Danmarks Naturvidenskabelige Samfund, c/o Riso National Laboratory, Dept of Physics, 4000 Roskilde
Founded: 1911
Pres: A.R. Mackintosh; Gen Secr: Jens-Peter Lynov
Focus: Nat Sci 03584

Danmarks Realskoleforening, c/o Realskole, Vesterbrogade 6, 6000 Kolding
Founded: 1891
Focus: Educ 03585

Danmarks Skolebibliotekarforening / The Danish Association of School Librarians, Gøgevej 1, 4130 Viby
T: 42393440
Founded: 1953; Members: 2300
Pres: Gert Larsen
Focus: Libraries & Bk Sci
Periodical
Skolebiblioteket (9 times annually) . . . 03586

Danmarks Skolebiblioteksforening (DSF), Vesterbrogade 20, 1620 København V
T: 33253222; Fax: 33253223
Founded: 1933; Members: 250
Pres: Bente Frost; Gen Secr: Niels Jacobsen
Focus: Libraries & Bk Sci
Periodicals
Børn & Bøger (8 times annually)
Skolebiblioteksårbog (weekly) 03587

Dansk Agronomforening (DA), Strandvejen 863, 2930 Klampenborg
T: 31631166; Fax: 31638811
Founded: 1896; Members: 3550
Pres: Svend D. Vesterlund; Gen Secr: Anders Højgaard
Focus: Agri
Periodical
Ugeskrift for Jordbrug (weekly) 03588

Dansk Akustisk Selskab (DAS) / Danish Acoustical Society, c/o Danmarks Tekniske Højskole, Lundtoftevej 100, 2800 Lyngby
Founded: 1955; Members: 320
Focus: Acoustics 03589

Dansk Anaestesiologisk Selskab (DAS), c/o Arhus Amtssygehus, 8000 Arhus C
Founded: 1949; Members: 260
Focus: Anesthetics 03590

Dansk Astronautisk Forening, Postboks 31, 1002 København K
Founded: 1949; Members: 40
Focus: Eng 03591

Dansk Automationsselskab (DAu), c/o Symbion Science Park, Fruebjergvej 3, 2100 København
T: 39179999; Fax: 31205521
Founded: 1962; Members: 116
Pres: P. Martin Larsen; Gen Secr: Jette Balslev
Focus: Eng
Periodical
Medlemsnyt (weekly) 03592

Dansk Betonforening, c/o Dansk Ingeniørforening, Vester Farimagsgade 31, 1606 København V
T: 33156565
Founded: 1947; Members: 1030
Focus: Materials Sci 03593

Dansk Biologisk Selskab, c/o Tumorbiologisk Institut, Frederik den V's Vey 11, 2100 København Ø
Founded: 1896; Members: 660
Focus: Bio 03594

Dansk Botanisk Forening, Sølvgade 83, 1307 København K
T: 33141703
Founded: 1840; Members: 1250
Focus: Botany
Periodical
URT (weekly) 03595

Dansk Byplanlaboratorium / Danish Town Planning Institution, Peder Skrams Gade 2b, 1054 København
T: 33137281
Founded: 1921
Focus: Urban Plan
Periodical
Byplan (weekly) 03596

Dansk Cardiologisk Selskab (DCS) / Danish Society of Cardiology, c/o Cardiological Dept, University Hospital, Herlev Hospital, 2730 Herlev
T: 86126866; Fax: 86193075
Founded: 1960; Members: 550
Pres: Kristian Thygesen; Gen Secr: Anne Thomassen
Focus: Cardiol
Periodical
Newsletter (weekly) 03597

Dansk Cerealforening (DCF), c/o Dept of Biotechnology, DTH, Block 223, 2800 Lyngby
Founded: 1932; Members: 84
Focus: Agri 03598

Dansk Dataforening, Kronprinsensgade 14, 1114 København K
Founded: 1958; Members: 3500
Focus: Computer & Info Sci
Periodical
DATA posten (weekly) 03599

Dansk Dendrologisk Forening, Kirkegårdsvej 3a, 2970 Hørsholm
T: 42860641; Fax: 42860774
Founded: 1949
Focus: Forestry
Periodical
Dansk Dendrologisk Årsskrift 03600

Dansk Dermatologisk Selskab (DDS), c/o Dermatological Dept, Rigshospitalet, Blegdamsvej 9, 2100 København Ø
T: 35431554
Founded: 1899; Members: 210
Focus: Derm 03601

Danske Arkitekters Landsforbund (DAL) / Akademisk Arkitektforening / The Federation of Danish Architects, Bredgade 66, Postboks 1163, 1010 København K
T: 33131290; Fax: 33931203
Founded: 1879; Members: 6000
Pres: Viggo Grunnet; Gen Secr: B. Budholm
Focus: Archit
Periodicals
Arkitekten (18 times annually)
Arkitektur DK (weekly)
Artur (18 times annually) 03602

Danske Dermato-Venerologers Organisation, c/o Dr. Thais Hattel, Vesterbro 99, 9000 Alborg
Founded: 1970
Focus: Derm; Venereology 03603

Danske Fødsels- og Kvindelaegers Organisation, c/o Dr. Mogens Devantier, Søborg Hovedgade 199, 2860 Søborg
Founded: 1916
Focus: Gynecology 03604

Danske Forstkandidaters Forening (DFF) / Association of Danish Graduates in Forestry, Strandvejen 863, 2930 Klampenborg
T: 31631166; Fax: 31631790
Founded: 1897; Members: 595
Pres: J.C. Briand Petersen
Focus: Forestry
Periodical
Jord og Viden (26 times annually) . . . 03605

Danske Fysioterapeuter (D.F.) / Association of Danish Physiotherapists, Nørre Voldgade 90, 1358 København K
T: 33138211; Fax: 33938214
Founded: 1918; Members: 6015
Pres: Inger Brøndsted; Gen Secr: Elisabeth Haase
Focus: Therapeutics
Periodical
Danske Fysioterapeuter (weekly) 03606

Danske Interne Medicineres Organisation (DIMO), c/o Langtidsmedicinsk Afdeling Z, Bispebjerg Hospital, Bispebjerg Bakke, 2400 København NV
Founded: 1947; Members: 185
Focus: Med 03607

Dansk Komité for Historikernes Internationale Samarbejde / Danish Committee for International Historical Cooperation, c/o Institute of History, University, Njalsgade 102, 2300 København
Founded: 1926; Members: 45
Focus: Hist; Int'l Relat 03608

Danske Kunsthåndvårkeres Landssammenslutning (D.K.L.) / Danish Arts and Crafts Association, Linnesgade 20, 1361 København K
Founded: 1976; Members: 380
Gen Secr: Nina Linde
Focus: Arts; Fine Arts
Periodical
Dansk Kunsthåndverk (weekly) 03609

Danske Lunglaegers Organisation, c/o Jacob Winslow, Ringstedvej 6b, 4300 Holbaek
Members: 50
Focus: Pulmon Dis 03610

Danske Nervelaegers Organisation (DNO), c/o Odense Sygehus, 5000 Odense
Founded: 1915
Focus: Neurology 03611

Dansk Epilepsiforening / Danish Epilepsy Society, c/o University Clinic of Neurology, Hvidovre Hospital, Kettegård Alle 30, 2650 Hvidovre
T: 36323632; Fax: 31473941
Founded: 1970; Members: 39
Focus: Neurology 03612

Danske Radiologers Organisation (DRO) / Danish Association of Medical Imaging, c/o Dept of Radiology, Sønderborg Sygehus, 6400 Sønderborg
T: 74430311
Founded: 1921; Members: 220
Pres: Andreas Johannsen
Focus: X-Ray Tech 03613

Danske Veterinårhygiejnikeres Organisation (DVO), c/o Veterinårkontrollen, Volden 11, 4200 Slagense
Founded: 1898; Members: 510
Focus: Vet Med; Food; Hygiene 03614

Dansk Exlibris Selskab, Postboks 1519, 2700 Brønshøj
Founded: 1941
Focus: Libraries & Bk Sci 03615

Dansk Farmaceutforening / Danish Pharmacists Association, Toldbodgade 36, 1253 København K
Founded: 1873; Members: 2200
Focus: Pharmacol
Periodical
Farmaceutisk Tidende (weekly) 03616

Dansk Farmacihistorisk Selskab, c/o Danmarks Farmaceutiske Højskole, Universitetsparken 2, 2100 København Ø
Founded: 1953; Members: 430
Focus: Hist; Pharmacol
Periodical
Theriaca (weekly) 03617

Dansk Føderation for Informationsbehandling og Virksomhedsstyring (DANFIP), Kronprinsensgade 14, 1114 København K
Founded: 1968; Members: 4
Focus: Computer & Info Sci 03618

Dansk Forening for Europaret (DFE), c/o Plesner & Lunoe, Esplanaden 34, 1263 København K
Fax: 33120014
Founded: 1973; Members: 240
Focus: Law 03619

Dansk Forening for Jordbundsvidenskab / Danish Society of Soil Science, c/o Chemistry Dept, Royal Veterinary and Agricultural University, Thorvaldsensvej 40, 1871 Frederiksberg C
T: 35282419; Fax: 35282089
Founded: 1928; Members: 60
Pres: O.K. Borggaard; Gen Secr: B.T. Christensen
Focus: Agri 03620

Dansk Forening for Retssociologi, c/o Institute of Organisation, Howitzvej 60, 2000 København F
Founded: 1967; Members: 115
Focus: Law; Sociology 03621

Dansk Forfatterforening / Danish Writers' Association, Strandgade 6, 1401 København K
Founded: 1894
Focus: Lit
Periodical
Forfatteren 03622

Dansk Gastroenterologisk Selskab (DGS) / Danish Society of Gastroenterology, c/o Peter Funck-Jensen, Hvidovre Hospital, Kettegård Allé 30, 2650 Hvidovre
T: 36322259; Fax: 31473311
Founded: 1970; Members: 600
Pres: Peter Funck-Jensen; Gen Secr: Niels Qvist
Focus: Gastroenter 03623

Dansk Geofysisk Forening (DGF), Østerroldgade 10, 1350 København K
Founded: 1936; Members: 102
Focus: Geophys 03624

Dansk Geologisk Forening (DGF), Øster Voldgade 5-7, 1350 København K
T: 33135001; Fax: 33114637
Founded: 1893; Members: 750
Focus: Geology
Periodical
Bulletin of the Geological Society of Denmark (2-3 times annually) 03625

Dansk Geoteknisk Forening (DGF), Maglebjergvej 1, 2800 Lyngby
Founded: 1950; Members: 240
Focus: Civil Eng 03626

Dansk Gerontologisk Selskab / Danish Gerontological Society, c/o Langtidsmedicinsk Afdeling, Centralsygehuset, 4700 Naestved
Founded: 1957; Members: 175
Focus: Geriatrics 03627

Dansk Grafologisk Selskab (DGS), Mothsvej 49, 2840 Holte
Founded: 1952; Members: 59
Focus: Graphology 03628

Dansk Haematologisk Selskab (DHS), c/o Dept of Medicine and Haematology, Gentofte Hospital, 2900 Hellerup
Founded: 1972; Members: 62
Focus: Hematology 03629

Dansk Historielaererforening, c/o Jørgen Riskaer-Jørgensen, Bakkebøllevej 12, 4760 Vordingborg
T: 53776616
Founded: 1960; Members: 3000
Pres: Jørgen Riskaer-Jørgensen
Focus: Hist; Educ
Periodical
Historie & Samtid (weekly) 03630

Dansk Hortonomforening (DH) / Association of Danish Graduates in Horticulture, Strandvejen 863, 2930 Klampenborg
T: 31631166; Fax: 31631790
Founded: 1915; Members: 581
Pres: Bente Mortensen
Focus: Hort
Periodical
Ugeskrift for Jordbrug (weekly) 03631

Dansk Husflidsselskab / Danish Society of Domestic Crafts, Tyrebakken 11, 5300 Kerteminde
T: 65322096; Fax: 65325611
Founded: 1873; Members: 5000
Focus: Home Econ
Periodical
Husflid (weekly) 03632

Dansk Idraetslaererforening, Egemvaenget 143, 5800 Nyborg
Members: 5000
Focus: Sports 03633

Dansk Industrimedicinsk Selskab, Bindesbøllsvej 23, 2920 Charlottenlund
T: 31642220
Founded: 1950; Members: 342
Focus: Med; Hygiene 03634

Dansk Ingeniørforening, Vester Farimagsgade 29, 1780 København V
Founded: 1892; Members: 15000
Gen Secr: Helge S. Henriksen
Focus: Energy
Periodical
Ingeniøren (weekly) 03635

Dansk Kerneteknisk Selskab, c/o Dansk Ingeniørforening, Vester Farimagsgade 31, 1606 København V
Focus: Nucl Res 03636

Dansk Kirurgisk Selskab, c/o Kirurgisk Afdeling, Amtssygehuset, 2970 Hørsholm
Founded: 1908
Focus: Surgery 03637

Dansk Køleforening, c/o Refrigeration Laboratory, Danmarks Tekniske Højskole, Bygning 402, 2800 Lyngby
T: 42404590; Fax: 42404207
Founded: 1911; Members: 612
Focus: Electric Eng
Periodical
Scandinavian Refrigeration (weekly) . . . 03638

Dansk Komponistforening / Danish Composers' Society, Graebroedtorv 16, 1154 København K
T: 33135405; Fax: 33143219
Founded: 1913; Members: 153
Pres: Holm Mogens Winkel; Gen Secr: Kirsten Wrem
Focus: Music
Periodical
Danske Komponister af i Dag: Danish Composers of Today – Dänische Komponisten von heute 03639

Dansk Kriminalistforening, Gammel Torv 18, 1457 København K
Founded: 1899; Members: 450
Gen Secr: E.L.J. Høgh
Focus: Criminology 03640

Dansk Kulturhistorisk Museumsforening (DKM), Nr. Madsbadvej 6, 7884 Fur
Fax: 97593163
Founded: 1929; Members: 167
Pres: Michael Lauenborg; Gen Secr: Kirsten Rex-Andersen
Focus: Arts
Periodical
Arv og Eje 03641

Dansk Kunstmuseumsforening, c/o Statens Museum for Kunst, Sølvgade, 1307 København K
Founded: 1913
Focus: Arts 03642

Dansk Laererforeningen, Rosenørns Allé 18, 1970 København V
Founded: 1885; Members: 5213
Focus: Ling; Educ 03643

Dansk Lokalhistorisk Forening, c/o Ebbe Fels, Bondehavevej 146, 2880 Bagsvaerd
Founded: 1967; Members: 142
Focus: Hist 03644

Dansk Mathematisk Forening (DMF), Universitetsparken 5, 2100 København Ø
Founded: 1873; Members: 350
Focus: Math
Periodicals
Mathematica Scandinavia (weekly)
MAT-NYT (weekly)
normat (weekly) 03645

Dansk Medicinsk-Historisk Selskab, c/o Medicinsk-Historisk Museum, Bredgade 62, 1260 København K
Founded: 1917; Members: 350
Focus: Med; Cultur Hist
Periodical
Danish Medical Historical Yearbook (weekly) 03646

Dansk Medicinsk Selskab (DMS) / Danish Medical Society, c/o Domus Medica, Trondjemsgade 9, 2100 København
Founded: 1919; Members: 13864
Pres: Bent Harvald; Gen Secr: Steen Walter
Focus: Med
Periodical
Danish Medical Bulletin (weekly) 03647

Dansk Medikoteknisk Selskab, Bakkefaldet 44, 2840 Holte
Founded: 1973; Members: 370
Focus: Eng; Med 03648

Dansk Mejeringeniør Forening / Dairy Engineerings Association, Cikorievej 8, 5220 Odense SØ
Founded: 1887; Members: 1700
Focus: Dairy Sci; Eng
Periodicals
Danish Dairy and Food Industry
Maelkeritidende (weekly) 03649

Dansk Metallurgisk Selskab (DMS), c/o Danmarks Tekniske Højskole, Bygning 204, 2800 Lyngby
T: 42884022
Founded: 1933; Members: 300
Pres: Prof. P.N. Hansen; Gen Secr: K.G. Soerensen
Focus: Metallurgy
Periodical
Dansk Metallurgisk Selskabs Arbog (weekly) 03650

Dansk Musikbiblioteksforening / Association of Danish Music Libraries, c/o Dansk Musik Informations Center, Graabrødretorv 16, 1154 København K
T: 33112016; Fax: 33322016
Founded: 1961; Members: 150
Pres: Kirsten Voss-Eliasson
Focus: Libraries & Bk Sci; Music
Periodical
MUS'en (weekly) 03651

Dansk Musikpaedagogisk Forening (DMpF), Nørrebrogade 45, 2200 København N
Founded: 1898; Members: 1300
Focus: Music; Educ
Periodical
Opus: Music Magazine (weekly) 03652

Dansk Naturhistorisk Forening, Universitetsparken 15, 2100 København Ø
T: 35321000; Fax: 35321010
Founded: 1833; Members: 860
Pres: Ole E. Heie; Gen Secr: Thomas B. Sørensen
Focus: Hist; Nat Sci
Periodical
Dansk Naturhistorisk Forening Arsskrift (weekly) 03653

Dansk Nefrologisk Selskab, c/o Prof. Steen Olsen, University Institute of Pathology, Arhus Kommunehospital, 8000 Arhus C
Focus: Intern Med 03654

Dansk Neurologisk Selskab (DNS), c/o Per Hübbe, Nordkrog 16, 2900 Hellerup
Members: 259
Focus: Neurology 03655

Dansk Numismatisk Forening, c/o Preben Nielsen, Hallandsparken 134, 2630 Taastrup
T: 42521918
Founded: 1885
Focus: Numismatics
Periodical
numismatisk rapport (3 times annually) . 03656

Dansk Odontologisk Selskab, c/o Fritz Bundgård-Jørgensen, Standvejen 296b, 2930 Klampenborg
Founded: 1923
Focus: Dent 03657

Dansk Oftalmologisk Selskab / Danish Ophthalmological Society, c/o Rigshospitalets Øjenafdeling, Blegdamsvej 9, 2100 København
Founded: 1901; Members: 250
Focus: Ophthal
Periodicals
Acta Ophthalmalogica Scandinavica Transactions 03658

Dansk Økologisk Forening Oikos, c/o Institutet for Økologisk Botanik, Øster Farimagsgade 2d, 1353 København K
Focus: Ecology 03659

Dansk Ornithologisk Forening (DOF), Vesterbrogade 140, 1620 København V
T: 31314404; Fax: 31312435
Founded: 1906; Members: 10200
Pres: Christian Hjorth; Gen Secr: Arne Jensen
Focus: Ornithology
Periodicals
Dansk Ornitologisk Forenings Tidsskrift (weekly)
Fugle (5 times annually) 03660

Dansk Ortopaedisk Selskab (DOS), c/o Rigshospitalet, Blegdamsvej, 2100 København Ø
Founded: 1946; Members: 172
Focus: Orthopedics 03661

Dansk Oto-laryngologisk Selskab (DOS), c/o University ENT-Clinic, Gentofte Hospital, 2900 Hellerup
Founded: 1899; Members: 292
Focus: Otorhinolaryngology 03662

Dansk Pediatrisk Selskab (DPS), c/o Dr. Flemming Hart Hansen, Abrinken 155, 2830 Virum
Founded: 1908; Members: 200
Focus: Pediatrics 03663

Dansk Pneumologisk Selskab, c/o Dr. Finn V. Rasmussen, Sandholmgårdsvej 16B, 3450 Allrød
Founded: 1921; Members: 120
Focus: Pulmon Dis 03664

Dansk Pressehistorisk Selskab, c/o Berlingske Tidende, Pilestraede 34, 1147 København K
T: 33157575; Fax: 33131012
Members: 390
Focus: Hist; Journalism 03665

Dansk Psykoanalytisk Selskab, c/o Bispebjerg Hospital, Psychiatric Dept E, 2400 København NV
Members: 25
Focus: Psych; Psychoan
Periodical
The Scandinavian Psychoanalytic Review (weekly) 03666

Dansk Psykolog Forening (DP), Bjerregards Sidevej 4, 2500 Valby
T: 31163555; Fax: 36440855
Founded: 1947; Members: 3400
Pres: J. Petersen; Gen Secr: Preben Føltved
Focus: Psych
Periodical
Psykolog Nyt (23 times annually) . . . 03667

Dansk Radiologisk Selskab (DRS), c/o Radiological Dept, Herlev Hospital, 2730 Herlev
T: 02942733
Founded: 1921; Members: 500
Focus: Radiology 03668

Dansk Reumatologisk Selskab (D.S.F.), c/o Dept of Physical Medicine and Rheumatology, Bispebjerg Hospital, 2400 København N
Founded: 1921; Members: 353
Focus: Rheuma; Rehabil 03669

Dansk Selskab for Akupunktur, c/o Dr. Ole Dahl, H.C. Andersens Blvd 49, 1553 København V
Founded: 1974; Members: 382
Focus: Pathology 03670

Dansk Selskab for Allergologi og Immunologi, c/o University Hospital, Blegdamsvej 9, 2100 København Ø
Founded: 1946; Members: 161
Focus: Immunology 03671

Dansk Selskab for Almen Medicin (DSAM), Engrøjel 90, Postboks 123, 2670 Greve
T: 42903673
Founded: 1966; Members: 2025
Focus: Med
Periodical
Practicus (weekly) 03672

Dansk Selskab for Bygningsstatik (DSBy), c/o Danmarks Tekniske Højskole, Bygning 118, 2800 Lyngby
Founded: 1928; Members: 674
Focus: Civil Eng
Periodical
Bygningsstatische Meddelelser (weekly) . 03673

Dansk Selskab for Cancerforskning, c/o Rigshospitalet, Blegdamsvej 9, 2100 København Ø
Fax: 31356906
Founded: 1964
Pres: Hans Skovgaard Poulsen
Focus: Cell Biol & Cancer Res 03674

Dansk Selskab for Fotogrammetri og Landmåling (DSFL) / Danish Society for Photogrammetry and Surveying, Ålborg Universitet, Fibigerstråde 11, 9220 Ålborg
Fax: 98156541
Founded: 1934; Members: 475
Focus: Surveying
Periodical
DSFL Meddelelser (weekly) 03675

Dansk Selskab for Intern Medicin (DSIM), c/o Steno Memorial Hospital, 2820 Gentofte
Founded: 1916; Members: 754
Focus: Intern Med 03676

Dansk Selskab for Logopaedi og Foniatri / Danish Association of Logopedics and Phoniatrics, c/o Svend Prytz, Kongsdalvej 43, 2720 København
Focus: Logopedy; Ling 03677

Dansk Selskab for Materialprøvning og -forskning, c/o Dansk Ingeniørforening, Vester Farimagsgade 31, 1606 København V
Focus: Materials Sci 03678

Dansk Selskab for Musikforskning, c/o Musikhistorisk Museum, Abenra 30, 1124 København K
Founded: 1954
Focus: Music
Periodical
Danish Yearbook of Musicology (weekly) 03679

Dansk Selskab for Obstetrik og Gynaekologi, c/o Domus Medica, Kristianiagade 12a, 2100 København Ø
Founded: 1898; Members: 441
Focus: Gynecology 03680

Dansk Selskab for Oldtids- og Middelalderforskning (DSOM), c/o Nationalmuseet, Frederiksholms Kanal 12, 1220 København K
Founded: 1934; Members: 120
Pres: B. Jørgensen; Gen Secr: J. Lund
Focus: Hist 03681

Dansk Selskab for Oligofreniforskning, Svanevanget 32, 2100 København
Focus: Psychiatry 03682

Dansk Selskab for Operationsanalyse (DORS), Postbox 9, 3100 Hornbaek
Founded: 1962; Members: 250
Focus: Business Admin 03683

Dansk Selskab for Optometri (DSO), c/o Jens Rahlff, Gongesletten 5, 2950 Vedbaek
Founded: 1954; Members: 300
Focus: Optics 03684

Dansk Selskab for Opvarmnings- og Ventilationsteknik (DSOV), c/o Dansk Ingeniørforening, Vester Farimagsgade 29, 1606 København V
Founded: 1945
Focus: Air Cond; Eng 03685

Dansk Selskab for Patologi / Danish Society of Pathology, c/o Bartholin Instituttet, Øster Farimagsgade 5, 1399 København K
Focus: Pathology; Microbio 03686

Dansk Selskab for Social Medicin, c/o Afdeling for Scial Medicin, Odense Universitet, J.B. Winsløwsvej 17, 5000 Odense C
Founded: 1967; Members: 215
Focus: Hygiene 03687

Dansk Selskab for Teoretisk Statistik (DSTS), c/o Institut for Matematisk Statistik, Universitetsparken 5, 2100 København Ø
Founded: 1971; Members: 250
Focus: Stats 03688

Dansk Skattevidenskabelig Forening (DSF), Palaegade 4, 1261 København K
Founded: 1938; Members: 300
Focus: Law 03689

Dansk Skovforening / Danish Forestry Society, Amalievej 20, 1875 Frederiksberg
Founded: 1888
Pres: Gustav Berner; Gen Secr: Jens Thomsen
Focus: Forestry
Periodical
Dansk Skovforenings Tidsskrift and Skoven 03690

Dansk Socialradgiverforening (DSF) / Danish Association of Social Workers, Toldbodgade 19a, 1253 København K
T: 33913033; Fax: 33913069
Founded: 1938; Members: 7000
Pres: Anna Marie Møller
Focus: Sociology
Periodical
Socialradgiveren (weekly) 03691

Dansk Sociologisk Selskab, Howitzvej 60, 2000 København F
Focus: Sociology 03692

Dansk Sprogvaern, c/o B.T.Persson, Weirsoevej 10, 3500 Vaerløse
Founded: 1955
Focus: Ling 03693

Dansk Svejseteknisk Landsforening, Park Allé 345, 2605 Broendby
T: 43968800; Fax: 43962636
Founded: 1939
Focus: Eng
Periodical
Svejsning: Welding (weekly) 03694

Dansk Tandlaegeforening (DTF), Amaliegade 17, Postboks 143, 1004 København K
Founded: 1873; Members: 5633
Focus: Dent
Periodical
Tandlågebladet (weekly) 03695

Dansk Teknisk Laererforening (DTL), Rosenvångets Hovedvej 14, 2100 København Ø
Founded: 1907; Members: 2300
Focus: Educ
Periodical
DTL-Nyt (weekly) 03696

Dansk Vandteknisk Forening / Danish Water Supply Association, Vilhelm Becks Vej 60, 8260 Viby
T: 86112333
Founded: 1926; Members: 550
Pres: Poul Friis; Gen Secr: Anders Baekgaard
Focus: Water Res
Periodical
Vandteknik (10 times annually) 03697

Dansk Veterinårhistorisk Samfund / Danish Veterinary History Society, Vinkelgårdsvej 4, 6040 Egtved
Founded: 1934; Members: 500
Pres: J. Kristiansen
Focus: Vet Med
Periodical
Dansk Veterinårhistorisk Årbog (weekly) . 03698

danvak VVS Teknisk Forening / Danish Society of Heating, Ventilating and Airconditioning Engineers, Orholmvej 40b, 2800 Lyngby
T: 45877611; Fax: 45877677
Founded: 1964
Focus: Eng
Periodical
VVS danvak 03699

Den Danske Aktuarforening, c/o Tryg Forsikring, Parallelvej, Postboks 163, 2800 Lyngby
T: 42878811
Founded: 1901; Members: 150
Focus: Insurance 03700

Den Danske Dyrlaegeforening (DdD), Rosenlunds Allé 8, 2720 Vanlöse
T: 31710888; Fax: 31710322
Founded: 1849; Members: 2336
Pres: Anton Rosenbom; Gen Secr: Børge Jensen
Focus: Vet Med
Periodical
Dansk Veterinaertidsskrift (Danish Veterinary Journal) (weekly) 03701

Den Danske Historiske Forening, Njalsgade 102, 2300 København S
T: 31542211
Founded: 1839
Pres: Prof. Dr. E. Ladewig Petersen; Gen Secr: Carsten Due-Nielsen; Anders Monrad Møller
Focus: Hist
Periodical
Historisk Tidsskrift (weekly) 03702

Det Danske Afrika Selskab, c/o J. Konnild, Fragariavej 9, 2900 Hellerup
Focus: Ethnology 03703

Det Danske Bibelselskab (DBS), Frederiksborggade 50, 1360 København K
T: 33127835; Fax: 33932150
Founded: 1814; Members: 3500
Pres: Erik Norman Svendsen; Gen Secr: Morten Aagaard
Focus: Rel & Theol
Periodicals
Det Danske Bibelselskabs Årbog (weekly)
Nyt fra Bibelselskabet (weekly) 03704

Det Danske Hedeselskab (DDH) / Danish Land Development Service, Klostermarken 12, 8800 Viborg
T: 86676111; Tx: 66228; Fax: 86675101
Founded: 1866; Members: 10000
Pres: Olaf von Lowzow; Gen Secr: Anders Pedersen
Focus: Agri; Forestry; Ecology
Periodical
Vaekst: Hedeselskabets Tidsskrift (weekly) 03705

Det Danske Orgelselskab, c/o Erik Poppe, Dønnerupvej 2, 2720 Vanløse
Founded: 1970; Members: 600
Focus: Music
Periodical
Orglet (weekly) 03706

Det Danske Shakespeare Selskab, Rosenvaengets Allé 39, 2100 København Ø
Founded: 1961
Focus: Lit 03707

Det Danske Sprog- og Litteraturselskab (DSL), Frederiksholms Kanal 18a, 1220 København K
T: 33130660; Fax: 33140608
Founded: 1911; Members: 70
Pres: Prof. John Kousgård Sørensen; Gen Secr: Iver Kjär
Focus: Ling; Lit; Hist
Periodical
Pråsentationshefter (weekly) 03708

Det Grønlandske Selskab (DGS), L.E. Bruuns Vej 10, 2920 Charlottenlund
Founded: 1905; Members: 1300
Pres: Helge Schultz-Lorentzen; Gen Secr: Jørn Würtz
Focus: Geography
Periodicals
Grønland (10 times annually)
Saerskrifter (weekly) 03709

Det Kongelige Danske Videnskabernes Selskab (KDVS) / The Royal Danish Academy of Sciences and Letters, H. C. Andersens Blvd 35, 1553 København V
T: 33113240; Fax: 33910736
Founded: 1742; Members: 471
Pres: Dr. Erik Dal; Gen Secr: Prof. Dr. Thor A. Bak
Focus: Sci
Periodicals
Biologiske Skrifter (weekly)
Historisk-filosofiske Meddelelser (weekly)
Historisk-filosofiske Skrifter (weekly)
Matematisk-fysiske Meddelelser (weekly)
Oversigt over Selskabets Virksomhed (weekly) . . . 03710

Det Kongelige Nordiske Oldskriftselskab, Prinsens Palais, Frederiksholms Kanal 12, 1220 København K
Founded: 1825; Members: 1200
Pres: Queen Margrethe II; Gen Secr: Peter V. Petersen
Focus: Hist; Archeol
Periodicals
Aarbøger for Nordisk Oldkyndighed og Historie (weekly)
Nordiske Fortidsminder, Serie B – in quarto (weekly) . . . 03711

Det Krigsvidenskabelige Selskab, c/o Haerens Officersskole, Frederiksberg Slot, Postboks 214, 2000 Frederiksberg
T: 31162244 ext 243; Fax: 36440160
Founded: 1871; Members: 1100
Focus: Military Sci
Periodical
Militaert tidsskrift (8 times annually) . . . 03712

Det Laerde Selskab i Arhus, c/o Universitet, 8000 Arhus C
Founded: 1945; Members: 150
Focus: Sci . . . 03713

Det Medicinske Selskab i København / The Medical Society of Copenhagen, Trondhjemsgade 9, 2100 København Ø
T: 31380984
Founded: 1772; Members: 2416
Pres: Anders Tuxen; Gen Secr: Marianne Pontoppidan
Focus: Med . . . 03714

Det Udenrigspolitiske Selskab, Amaliegade 40a, 1256 København K
T: 33148886
Founded: 1946; Members: 800
Pres: Uffe Ellemann-Jensen; Gen Secr: Klaus Carsten Pedersen
Focus: Poli Sci
Periodicals
Udenrigs (weekly)
Udenrigspolitiske Skrifter (weekly) . . . 03715

DTL – Dansk Forening for Information og Dokumentation (DTL), c/o DTB, Anker Engelunds Vej 1, 2800 Lyngby
T: 42883088
Founded: 1959; Members: 344
Focus: Lit
Periodical
Skriftserie (weekly) . . . 03716

Elektroteknisk Forening, Ehlersvej 11, 2900 Hellerup
T: 49201819; Fax: 49201839
Founded: 1903; Members: 3000
Pres: Jakob B. Lyngsø; Gen Secr: Aage Hansen
Focus: Electric Eng . . . 03717

Entomologisk Forening, c/o Zoological Museum, Universitetsparken 15, 2100 København Ø
Founded: 1868; Members: 450
Pres: Niels Peder Kristensen; Gen Secr: Michael Hansen
Focus: Entomology
Periodical
Entomologiske Meddelelser (3 times annually) . . . 03718

Euroenviron, Aaboulevard 13, 1635 København V
T: 31394344; Fax: 31354344
Founded: 1989; Members: 13
Gen Secr: Lars Abel
Focus: Ecology . . . 03719

European Academy of Allergology and Clinical Immunology, c/o Allergy-Munksgaard, Norre Sogade 35, Postboks 2148, 1016 København K
Fax: 33129387
Focus: Immunology
Periodical
Allergy: European Journal of Allergy and Clinical Immunology (8 times annually) . . . 03720

European Advisory Committee on Health Research (EACHR), c/o WHO, Scherfigsvej 8, 2100 København
T: (01) 290111; Tx: 15348; Fax: (01) 181120
Founded: 1975; Members: 31
Gen Secr: Dr. H. Vuori
Focus: Public Health . . . 03721

European Association for Haematopathology (EAHP), c/o Laboratory of Immunohistology, Finsengade 12, 8000 Aarhus C
T: 86125555 ext 3841; Fax: 86182386
Members: 300
Gen Secr: Gorm Pallesen
Focus: Pathology . . . 03722

European College of Neuropsychopharmacology (ECNP), c/o Dept of Psychiatry, Frederiksborg General Hospital, 3400 Hillerød
T: 42261500; Fax: 86600244
Founded: 1985
Gen Secr: Dr. Per Bech
Focus: Pharmacol . . . 03723

European Council of National Associations of Independent Schools (ECNAIS), Langes Gaard 12, 4200 Slagelse
T: 53531845
Founded: 1988
Gen Secr: Kjeld Peter A. Olesen
Focus: Educ . . . 03724

Filologisk-Historiske Samfund, c/o Institute of Greek and Latin, University of Copenhagen, Nialsgade 92, 2300 København S
Founded: 1854; Members: 450
Pres: Dr. Minna S. Jensen
Focus: Ling; Lit
Periodical
Studier fra Sprog- og Oldtidsforskning . 03725

FORCE Institutterne / FORCE Institutes, Park Allé 345, 2600 Glostrup
T: 43969800; Tx: 33388; Fax: 43962636
Pres: Knud Rimmer; Gen Secr: Ernst Tiedemann
Focus: Eng . . . 03726

Foreningen af Danske Biologer. / The Association of Danish Biologists, c/o Kristian Lauridsen, Lindevangsvej 16B, 3460 Birkerød
Founded: 1976
Focus: Educ; Bio
Periodical
Biofag (weekly) . . . 03727

Foreningen af Danske Civiløkonomer (FDC), Börsen, Slotsholmsgade, 1216 København K
Founded: 1954; Members: 5300
Focus: Business Admin; Econ . . . 03728

Foreningen af Danske Kunstmuseer, Nr. Madsbadvej 6, 7884 Fur
Fax: 97593163
Founded: 1978; Members: 50
Pres: Christian Gether; Gen Secr: Kirsten Rex Andersen
Focus: Arts . . . 03729

Foreningen af Danske Museumsmaend, c/o Zoologisk Museum, Universitetsparken 15, 2100 København Ø
Founded: 1928; Members: 250
Focus: Arts . . . 03730

Foreningen af Geografilaerere ved de Gymnasiale Uddannelser, Blokvej 2e, 6760 Ribe
Focus: Geography; Educ
Periodical
Geo-Nyt (weekly) . . . 03731

Foreningen af Licentiater (FL), Gammeltorv 22, 1017 København K
Founded: 1937; Members: 150
Pres: Ejvind Hansen; Gen Secr: Karsten Raae
Focus: Agri . . . 03732

Foreningen af Medarbejdere ved Danmarks Forskningsbiblioteker, c/o Danmarks Tekniske Bibliotek, Anker Engelundsvej 1, 2800 Lyngby
T: 45939979
Founded: 1934
Pres: Lars Bjørnshauge; Gen Secr: Grethe Lillelund
Focus: Libraries & Bk Sci
Periodical
DF-Revy (10 times annually) . . . 03733

Foreningen af Speciallåger (FAS) / The Danish Association of Medical Specialists, Trondhjemsgade 9, 2100 København Ø
T: 31385500; Fax: 31386807
Founded: 1935; Members: 5233
Pres: Dr. Jan Søtoft; Gen Secr: Martin Teilmann
Focus: Med . . . 03734

Foreningen for Danske Lanbrugsskoler, c/o Grevevej 20, 2670 Greve
Founded: 1903; Members: 300
Focus: Adult Educ; Agri . . . 03735

Foreningen for National Kunst / Society for National Art, Landemarkt 3, 1000 København K
Founded: 1900
Pres: P.C.J. Stagsted; Gen Secr: O. Haase
Focus: Arts . . . 03736

Foreningen til Svampekundskabens Fremme / Danish Mycological Society, Postbox 102, 2860 Søborg
T: 31670649; Fax: 31670649
Founded: 1905; Members: 1800
Pres: Jørgen Albertsen
Focus: Botany, Specific
Periodical
SVAMPE (weekly) . . . 03737

Frankslaererforeningen, c/o Jørgen Lykke Petersen, Fuglebo 4, 2000 København F
Members: 500
Focus: Ling; Educ . . . 03738

Fysisk Forening, Blegdamsvej 17, 2100 København Ø
Founded: 1908; Members: 220
Focus: Physics . . . 03739

Gymnasieskolernes Laererforening, Lyngbyvej 32, 2100 København Ø
T: 31209591; Fax: 31207447
Founded: 1890; Members: 10000
Pres: Hans C. Thomsen; Gen Secr: Per Hersby
Focus: Educ
Periodical
Gymnasieskolen (weekly) . . . 03740

Gymnasieskolernes Tysklaererforening, c/o Nanna Bjargum, Egebjergvej 263, 4500 Nykøbing
T: 63428544
Founded: 1933; Members: 800
Focus: Ling; Educ
Periodical
Meddelelser FRA Gymnasieskolernes Tysklaererforening (weekly) . . . 03741

Historisk Samfund for Sønderjylland, Haderslevvej 45, 6200 Abenrå
T: 74624683
Focus: Hist
Periodicals
Sønderjyske Aarbøger (weekly)
Sønderjysk Maanedsskrift (weekly) . . . 03742

Historisk-Topografisk Selskab, Solbjergvej 25, 2000 Frederiksberg
Founded: 1944; Members: 500
Focus: Hist . . . 03743

International Association of Physical Education and Sports for Girls and Women (IAPESGW), Vestermøllevej 11, 8800 Viborg
Founded: 1953; Members: 400
Gen Secr: Mette Winkler
Focus: Sports
Periodical
Bulletin of IAPESGW . . . 03744

International Banking Research Institute, c/o Privatbanken A/S, Torvegade 2, 1249 København K
T: 31111111
Founded: 1951; Members: 55
Focus: Finance . . . 03745

International Council for the Exploration of the Sea (ICES), Palaegade 2-4, 1261 København K
T: 33154225; Fax: 33934215
Founded: 1902; Members: 18
Pres: Jakob Jakobsson; Gen Secr: Dr. Emory D. Anderson
Focus: Oceanography
Periodicals
Bulletin Statistique (weekly)
Cooperative Research Report (6-8 times annually)
Fiches d'Identification du Plancton (weekly)
ICES Oceanographic Data Lists and Inventories (3-4 times annually)
Identification Leaflets for Diseases and Parasites of Fish and Shellfish (weekly)
Journal du Conseil (2-3 times annually)
Rapports et Procès-Verbaux des Réunions (weekly)
Techniques in Marine Environmental Sciences (weekly) . . . 03746

International Law Association: Danish Branch, c/o A. Kaufmann, Skoubogade 1, 1158 København K
Founded: 1925
Focus: Law . . . 03747

International Society for Prosthetics and Orthotics (ISPO), Borgervaenget 5, 2100 København Ø
T: 31207260; Fax: 31181669
Founded: 1970; Members: 2600
Gen Secr: Norman A. Jacobs
Focus: Dent
Periodical
Prosthetics and Orthotics International (3 times annually) . . . 03748

International Society of Libraries and Museums of the Performing Arts, c/o Museum of Decorative Art, Bredgade 68, 1260 København
Founded: 1954; Members: 320
Focus: Libraries & Bk Sci; Perf Arts . . 03749

International Work Group for Indigenous Affairs (IWGJA), Fiolstraede 10, 1171 København K
T: 33124724
Founded: 1968; Members: 1400
Focus: Anthro
Periodicals
Documents (weekly)
Newsletter (weekly)
Yearbook (weekly) . . . 03750

Islandske Litteratursamfund i København, Helsevej 21, 3400 Hillerød
Founded: 1912
Gen Secr: P. Jonasson
Focus: Lit . . . 03751

Juridisk Forening, Esplanaden 34, 1263 København K
Founded: 1881; Members: 460
Gen Secr: H. Holm-Nielsen
Focus: Law . . . 03752

Jydsk Selskab for Fysik og Kemi (JSFK), c/o Dept of Chemistry, University, Langelandsgade 140, 8000 Arhus C
Founded: 1934; Members: 200
Focus: Chem; Physics . . . 03753

Jysk Arkaeologisk Selskab (JAS), Moesgård, 8270 Hœjbjerg
Founded: 1951; Members: 1500
Pres: Poul Kjaerum; Gen Secr: Hans Jørgen Madsen
Focus: Archeol
Periodical
Kuml: Journal of the JAS . . . 03754

Jysk Forening for Naturvidenskab, c/o Poul Hansen, Naturhistorisk Museum, 8000 Arhus C
Founded: 1903
Focus: Nat Sci . . . 03755

Jysk Selskab for Historie, Vester Allé 12, 8000 Arhus C
Fax: 86128560
Founded: 1866; Members: 1100
Pres: Prof. Dr. Henning Poulsen; Gen Secr: Jørgen Fink
Focus: Hist
Periodical
Nyt fra Historien (weekly) . . . 03756

Kemisk Forening, c/o H.C. Ørsted Institutet, Universitetsparken 5, 2100 København Ø
Founded: 1879; Members: 1000
Pres: Lars Carlsen; Gen Secr: Michael Gajhede
Focus: Chem . . . 03757

Kirkeligt Centrum, Vinstrups Allé 64, 2900 Hellerup
Founded: 1899
Focus: Rel & Theol . . . 03758

Kongelige Danske Geografiske Selskab, Øster Voldgade 10, 1350 København K
Founded: 1876; Members: 450
Gen Secr: Prof. S. Christiansen
Focus: Geography
Periodical
Geografisk Tidsskrift . . . 03759

Kongelige Danske Landhusholdningsselskab / Royal Danish Agricultural Society, Mariendalsvej 27, 2000 Frederiksberg
Founded: 1769
Pres: P. Skak Olufsen; A. Neimann-Sørensen; J. Krabbe; Gen Secr: Jens Wulff
Focus: Agri
Periodical
Tidsskrift for landøkonomi (weekly) . . . 03760

Kunstforeningen, Gammel Strand 48, 1202 København K
T: 33132964
Founded: 1825; Members: 3000
Pres: Jørgen Nørgaard; Gen Secr: Annette Stabell
Focus: Arts . . . 03761

Kunstnerforeningen af 18. November / Artists' Association of the 18th of November, Frederiksgade 8, København K
Founded: 1842
Focus: Arts . . . 03762

Landsforeningen af Fodterapeuter, Bjelkes Allé 43, 2200 København N
Founded: 1932; Members: 700
Focus: Therapeutics
Periodical
Fodterapeuten (10 times annually) . . . 03763

Landsforeningen for Polio-, Trafik- og Ulykkesskadede (PTU) / National Society of Polio and Accident Victims, Tuborgvej 5, 2900 Hellerup
T: 31629000; Fax: 31625439
Founded: 1953
Pres: Holger Kallehauge
Focus: Med; Rehabil
Periodical
PTU-Nyt (weekly) . . . 03764

Landsforeningen for Sukkersyge, Skt. Anne Plads 4, 5000 Odense C
Focus: Diabetes
Periodical
Tidsskrift for Sukkersyge (weekly) . . . 03765

Landsforeningen til Kraeftens Bekaempelse, Strandboulevarden 49, 2100 København Ø
T: 35268866; Fax: 35264560
Founded: 1928; Members: 366509
Focus: Cell Biol & Cancer Res . . . 03766

Lepidopterologisk Forening, Skovskellet 35a, 2840 Holte
Founded: 1941; Members: 500
Focus: Zoology
Periodical
Lepidoptera (weekly) . . . 03767

Levnedsmiddelselskabet (LEVS) / Danish Society of Food Science and Technology, c/o Dansk Ingeniørforening, Vester Farimagsgade 29, 1780 København V
T: 33156565; Fax: 33937171
Founded: 1959; Members: 1800
Pres: Prof. Jette Nielsen
Focus: Food . . . 03768

Lysteknisk Selskab (LTS), Byvej 12, Postbox 28 Stenløse
Founded: 1948; Members: 725
Focus: Electric Eng
Periodical
Lys (weekly) . . . 03769

Marinehistorisk Selskab, c/o Orlogsmuseet, Overgade 58, 1415 København K
Founded: 1951; Members: 450
Focus: Military Sci; Hist; Transport
Periodical
Marinehistorisk Tidsskrift (weekly) 03770

Matematiklaererforeningen, Pilegårdsparken 93, 3460 Birkerœd
Founded: 1931; Members: 1048
Focus: Math; Educ 03771

Mellemfolkeligt Samvirke / Danish Association for International Cooperation, Borgergade 10-14, 1300 København K
T: 33326244; Tx: 15928; CA: Mellemsam; Fax: 33156243
Founded: 1944; Members: 5500
Pres: Knud Vilby; Gen Secr: Poul Grosen
Focus: Int'l Relat
Periodicals
Kontakt (8 times annually)
MS Biblioteksnyt (18 times annually)
MS Revy (8 times annually)
ZAPP (weekly) 03772

Militaerteknisk Forening (MtF) / Danish Military Technical Society, c/o Dansk Ingeniørforening, Vester Farimagsgade 29, 1780 København V
T: 33156565; Fax: 33937171
Members: 625
Focus: Eng; Military Sci 03773

Nationaløkonomisk Forening, c/o Danmarks Nationalbank, Havnegade 5, 1093 København K
T: 31141411; Fax: 33325460
Founded: 1873; Members: 1500
Pres: Niels Christian Nielsen; Gen Secr: Kim Abildgren
Focus: Econ
Periodical
Nationaløkonomisk Tidsskrift (3 times annually) 03774

Naturhistorisk Forening for Jylland, c/o Natural History Museum, 8000 Arhus C
T: 86129777
Founded: 1911
Focus: Hist; Nat Sci
Periodical
Flora og Fauna (weekly) 03775

Naturhistorisk Forening for Nordsjaelland, Hyttebakken 4, 3400 Hillerød
Founded: 1967
Focus: Hist; Nat Sci 03776

Nordforsk, H. C. Andersens Blvd 18, 1553 København V
Focus: Sci 03777

Nordic Federation for Medical Education, c/o Rigshospitalet, Tagensvej 18, 2200 København N
T: 35375252; Fax: 46307850
Founded: 1966
Gen Secr: Jørgen Nystrup
Focus: Adult Educ; Med; Educ 03778

Nordic Society for Cell Biology, Bülowsvej 13, 1870 København
Founded: 1960; Members: 350
Focus: Cell Biol & Cancer Res 03779

Nordisk Akustik Selskab (NAS) / Acoustical Society of Scandinavia, c/o Technical University of Denmark, Lundtoftevej 100, 2800 Lyngby
Founded: 1956; Members: 800
Focus: Acoustics
Periodical
NAS-Proceedings 03780

Nordisk Byggedag (NBD) / Nordic Building Conference, c/o Architectfirm Karsten Palsson, Svanevej 26, 2400 København NV
Founded: 1927; Members: 222
Gen Secr: Karsten Palsson
Focus: Urban Plan
Periodical
Congress Literature 03781

Nordisk Cerealist Foreningen, c/o Afdeling for Teknisk Biokemi, Danmarks Tekniske Højskole, 2800 Lyngby
Founded: 1931; Members: 300
Focus: Agri 03782

Nordisk Forening for Celleforskning, c/o Biological Institute of the Carlsberg Foundation, Tagensvej 16, 2200 København N
Focus: Cell Biol & Cancer Res 03783

Nordisk Forening for Rettssociologi, c/o Copenhagen School of Economics, Howitzvej 60, 2000 København F
Founded: 1966
Focus: Law; Sociology
Periodical
Newsletter (weekly) 03784

Nordisk Institut for Asienstudier (NIAS) / Nordic Institute of Asian Studies, Njalsgade 84, 2300 København kSK
T: 31548844; Fax: 32962530
Founded: 1967
Gen Secr: Thommy Svensson
Focus: Humanities; Soc Sci
Periodicals
NIAS-nytt: Nordic Newsletter of Asian Studies (weekly)
NIAS Report (8 times annually)
SIAS Monograph Series (weekly)
SIAS Occasional Papers (weekly)
Studies on Asian Topics (weekly) . . . 03785

Nordisk Institut for Teoretisk Atomfysik (NORDITA), Blegdamsvej 17, 2100 København
T: 35325500; Tx: 15216; CA: NORDITA København; Fax: 31389157
Founded: 1957; Members: 5
Focus: Nucl Res; Physics
Periodicals
NORDITA Preprint (weekly)
NORDITA Virksomhedsberetning: NORDITA Report (weekly) 03786

Nordisk Kollegium for Fysisk Oceanografi, c/o Institut for Fysisk Oceanografi, Haraldsgade 6, 2200 København N
Fax: 1822565
Founded: 1965
Focus: Oceanography 03787

Nordisk Laederforskningsråd (NLFR), Gregersensvej, 2630 Taastrup
T: 43710663
Founded: 1960; Members: 13
Focus: Materials Sci 03788

Nordisk Neurokirurgisk Forening (NNF), c/o Neurosurgical Dept, Arhus Kommunehospital, Nørrebrogade 44, 8000 Arhus C
T: 86125555; Fax: 86139597
Founded: 1945; Members: 95
Pres: S. Valtonen; Gen Secr: Lennart Rabov
Focus: Surgery 03789

Nordisk Numismatisk Union, c/o Den Kgl. Mønt- og Medaillesamling, 1220 København K
Fax: 33155521
Focus: Numismatics
Periodical
Nordisk Numismatisk Unions Medlemsblad (8 times annually) 03790

Nordisk Statistisk Sekretariat, Sejrøgade 11, 2100 København Ø
Fax: 1184801
Founded: 1969
Focus: Stats
Periodicals
Nordisk Statistik Arsbok: Yearbook of Nordic Statistics (weekly)
Nordisk Statistik Skriftserie (weekly)
Tekniske Rapporter (weekly) 03791

Orientalsk Samfund, c/o University, Njalsgade 80, 1200 København S
Founded: 1915; Members: 50
Pres: Prof. S. Egerod; Gen Secr: Prof. J.P. Asmussen
Focus: Ethnology
Periodical
Acta Orientalia 03792

Paedagogisk Forening, c/o Hanne Thorsen, Gernersgade 1, 1319 København K
Founded: 1940; Members: 700
Focus: Sociology; Educ
Periodical
Det Paedagogiske Selskab: Dansk Paedagogisk Tidsskrift (9 times annually) 03793

Polymerteknisk Selskab (PTS), c/o Dansk Ingeniørforening, Vester Farimagsgade 31, 1606 København V
Founded: 1957; Members: 330
Focus: Eng 03794

Polyteknisk Forening, c/o Danmarks Tekniske Højskole, Anker Engelunds Vej 1, 2800 Lyngby
T: 42881104; Fax: 42881353
Founded: 1845; Members: 5493
Focus: Eng
Periodicals
Polygraphus (weekly)
Rusbogen (weekly) 03795

Rektorkollegiet / Conference of Danish Rectors, c/o Undervisningsministeriet, Frederiksholms Kanal 26, 1220 København K
T: 33925300; Tx: 16243; Fax: 33925075
Founded: 1967
Pres: Ove Nathan; Gen Secr: Ellen Hansen
Focus: Educ 03796

Samfundet for Dansk Genealogi og Personalhistorie, Rudersdalsvej 52, 2840 Holte
Founded: 1879; Members: 1100
Pres: Hans H. Worsøe; Gen Secr: Poul Steen
Focus: Genealogy
Periodical
Personalhistorisk Tidsskrift (weekly) . . . 03797

Samfundet til Udgivelse af Dansk Musik, Gråbrørestråde 18, 1156 København K
Founded: 1871; Members: 150
Pres: Klaus Ib Jørgensen
Focus: Music
Periodical
Bulletin (bi-annually) 03798

Samfund til Udgivelse af Gammel Nordisk Litteratur, c/o Prof. J. Helgason, Kjaerstrupvej 33, 2500 Valby
Founded: 1879
Focus: Lit 03799

Sammenslutningen af Danmarks Forskningsbiblioteker (SDF), c/o Danmarks Tekniske Bibliotek, Anker Engelundsvej 1, 2800 Lyngby
T: 45939979
Founded: 1949; Members: 135
Pres: Niels Mark; Gen Secr: Kaare Pedersen
Focus: Libraries & Bk Sci
Periodical
DF-Revy (10 times annually) 03800

Sammenslutningen af Danske Kunstforeninger / National Committee of Danish Art Societies, Lillebo Niverød, 2990 Niva
Founded: 1942; Members: 15000
Pres: Bjørn Harder; Gen Secr: Edith Dam
Focus: Arts 03801

Sammenslutningen af Lokalarkiver, Postbox 235, 7100 Vejle
T: 75840898; Fax: 75831801
Founded: 1949; Members: 400
Focus: Archives
Periodical
OmSLAget (weekly) 03802

Sammenslutningen af Medieforskere i Danmark, c/o Institut for Uddannelse og Socialisering, Alborg Universitetscenter, 9100 Alborg
Members: 71
Focus: Journalism 03803

Sammenslutningen af Praktiserende Dyrlaeger (SAPD), Rosenlunds Allé 8, 2720 Vanløse
T: 31710888
Founded: 1966; Members: 875
Focus: Vet Med 03804

Scandinavian Library Center (SLC), Telegrafvej 5, 2750 Ballerup
Founded: 1968
Focus: Libraries & Bk Sci 03805

Scandinavian Society for Economic and Social History, c/o University of Odense, Campusvej 55, 5230 Odense M
Fax: 65932974
Pres: Prof. Ulf Olsson
Focus: Hist; Econ; Sociology
Periodical
The Scandinavian Economic History Review (3 times annually) 03806

Scandinavian Sociological Association, c/o Munksgrad, Nørre Sogade 35, 1370 København K
Focus: Sociology 03807

Sektionen for Klinisk Neurofysiologi / Danish Society of EEG and Clinical Neurophysiology, c/o A. Fuglsang-Frederiksen, Brogaardsvej 26, 2680 Gentofte
Founded: 1946; Members: 39
Focus: Physiology 03808

Selskabet for Danmarks Kirkehistorie, c/o Institut for Kirkehistorie, Købmagergade 44-46, 1150 København K
T: 33152811
Founded: 1849; Members: 980
Focus: Hist; Rel & Theol
Periodical
Kirkehistoriske Samlinger (weekly) . . . 03809

Selskabet for Dansk Kulturhistorie, c/o Botanisk Centralbibliotek, 1307 København
Founded: 1936; Members: 30
Focus: Hist
Periodical
Kulturminder (weekly) 03810

Selskabet for Dansk Skolehistorie, c/o Dr. Vagn Skovgaard-Petersen, Taffelbays Allé 15, 2900 Hellerup
Founded: 1966
Focus: Hist; Educ
Periodical
Årbog for Dansk Skolehistorie: Yearbook on History of Danish Schools (Education) (weekly) 03811

Selskabet for Dansk Teaterhistorie, c/o B. Lemvigh-Müller, Gruntvigsvej 27b, 1864 København V
Founded: 1911
Focus: Hist; Perf Arts 03812

Selskabet for Filosofi og Psykologi, Kobmagergade 50, 1150 København K
Founded: 1926
Focus: Philos; Psych 03813

Selskabet for Historie og Samfundsøkonomi, c/o Institutet for Fremtidsforskning, Løngangstraede 25, 1468 København K
Founded: 1960
Pres: Prof. Dr. Niels Thomsen; Gen Secr: Steen Thomsen
Focus: Hist; Econ
Periodical
Økonomi og Politik (weekly) 03814

Selskabet for Københavns Historie, c/o Stadsarkivet, Radhuset, 1599 København V
T: 33662370; Fax: 33159212
Founded: 1935; Members: 1000
Focus: Hist 03815

Selskabet for Naturlaerens Udbredelse, c/o UNI-C, Vermundsgade 5, 2100 København Ø
Founded: 1824
Focus: Nat Sci
Periodical
Kwant (weekly) 03816

Selskabet for Tekniske Uddanelsespørgsmal, c/o Dansk Ingeniørforening, Vester Farimagsgade 31, 1606 København V
Focus: Educ; Eng 03817

Selskabet til Udgivelse af Danske Mindesmaerker, c/o Nivaagaards Malerisamling, Gammel Strandvej 2, 2990 Nivaa
Founded: 1919; Members: 200
Focus: Hist; Arts 03818

Selskab for Arbejdsmiljø, c/o Dansk Ingeniørforening, Vester Farimagsgade 31, 1606 København V
T: 33156565
Focus: Sociology
Periodical
Loke (weekly) 03819

Selskab for Nordisk Filologi, c/o Institut for Navneforskning, Københavns Universitet Amager, Njalsgade 80, 2300 København S
T: 31542211; Fax: 31544170
Founded: 1912; Members: 150
Focus: Ling; Lit
Periodical
Selskab for Nordisk Filologi Arsberetning (bi-annually) 03820

Skandinavisk Museumsforbund, Danish Section, Algade 48, 9000 Aalborg
T: 98124522
Founded: 1915; Members: 200
Pres: Brigitte Kjaer; Gen Secr: Viggo Petersen
Focus: Arts; Cultur Hist 03821

Socialpaedagogernes Landsforbund, Brolaeggerstraede 9, 1211 København K
T: 33140058; Fax: 33930604
Focus: Educ; Sociology
Periodical
Socialpaedagogen (weekly) 03822

Societas Heraldica Scandinavica / Heraldisk Selskab, c/o Dr. Ole Rostock, Sigmundsvej 8, 2880 Bagsvaerd
Founded: 1959; Members: 700
Focus: Genealogy
Periodical
Heraldisk Tidsskrift (weekly) 03823

Société Internationale pour l'Enseignement Commercial, Hunderupvej 122A, 5230 Odense M
T: 66121966; Fax: 66145794
Founded: 1901; Members: 2400
Gen Secr: Prof. Erik Lange
Focus: Adult Educ
Periodical
Revue Internationale pour l'Enseignment Commercial (weekly) 03824

Sønderjyllands Amatørarkaeologer, c/o Johan Jessen, Lindsnakkevej 7, 6200 Abenrå
Founded: 1966
Focus: Archeol 03825

Teknisk Skoleforening, c/o Ejnar Bo Pedersen, Munkehatten 28, 5220 Odense Ø
T: 66158725; Fax: 66154555
Founded: 1891; Members: 63
Focus: Educ; Eng 03826

Teologisk Forening ved Københavns Universitet, Valkendorfsgade 23, 1151 København K
Founded: 1905; Members: 227
Focus: Rel & Theol 03827

Vej- og Byplanforeningen, c/o Dansk Ingenıørforening, Vester Garimagsgade, 1606 København V
Focus: Civil Eng 03828

Dominican Republic

Academia Dominicana de la Historia, Calle Mercedes 50, Santo Domingo
Founded: 1931; Members: 67
Pres: E. Rodríguez Demorizi; Gen Secr: Dr. Frank Moya Pons
Focus: Hist
Periodical
Clio (weekly) 03829

Academia Dominicana de la Lengua, Av Tiradentes 66, Ensanche La Fe, Santo Domingo
Members: 14
Pres: Mariano Lebrón Saviñón; Gen Secr: Mariano Lebrón Saviñón
Focus: Ling 03830

Asociación Dominicana de Bibliotecarios (ASODOBI), c/o Biblioteca Nacional, Plaza de la Cultura, César Nicolás Penson 91, Santo Domingo
T: 6884086
Founded: 1974
Pres: P.J. Mella Chavier; Gen Secr: V. Regús
Focus: Libraries & Bk Sci
Periodical
El Papiro (weekly) 03831

Asociación Dominicana pro Bienestar de la Familia, Apdo 1053, Santo Domingo
Focus: Family Plan
Periodicals
Población y Desarrollo
Resúmenes sobre Población Dominicana 03832

Asociación Médica de Santiago, Apdo 445, Santiago de los Caballeros
Founded: 1941
Focus: Med 03833

Asociación Médica Dominicana, Apdo 1237, Santo Domingo
Founded: 1941; Members: 1551
Focus: Med 03834

Ateneo de Macoris, San Pedro de Macoris
Founded: 1890
Focus: Arts 03835

Grupo Bibliográfico Nacional de la República Dominicana, c/o Archivo General de la Nactión, Calle Chiclana de la Frontera, Santo Domingo
Focus: Libraries & Bk Sci 03836

Servicio de Documentación y Biblioteca, Palacio de Educación, Santo Domingo
Founded: 1958
Focus: Libraries & Bk Sci; Doc 03837

Sociedad Amantes de la Luz, España Esquina Av Central, Santiago de los Caballeros
Founded: 1874
Focus: Arts 03838

Ecuador

Academia Ecuatoriana de la Lengua, Apdo 3460, Quito
Founded: 1875
Pres: Galo René Pérez; Gen Secr: Piedad Larrea Borja
Focus: Lit
Periodicals
Memorias (weekly)
Obras de la Literatura Ecuatoriana (weekly) 03839

Academia Ecuatoriana de Medicina, c/o Casa de la Cultura Ecuatoriana, Apdo 67, Quito
Founded: 1958
Focus: Med 03840

Agence Latinoaméricaine d'Information (ALAI), Pasaje San Luis 104, Of 304, Casilla 596-A, Suc 3, Quito
T: (02) 572689; Fax: (02) 580835
Founded: 1976
Focus: Computer & Info Sci
Periodical
ALAI Servicio Mensual de Información y Documentación (weekly) 03841

Andean Instiute of Popular Arts, Calle Diego de Atienza y Av América, Quito
T: (02) 563096
Founded: 1977; Members: 8
Focus: Arts
Periodicals
Bocina de los Andes (weekly)
Revista IADAP (weekly) 03842

Asociación Ecuatoriana de Bibliotecarios (AEB), c/o Casa de la Cultura, Apdo 87, Quito
T: 528840
Focus: Libraries & Bk Sci 03843

Asociación Ecuatoriana de Museos (ASEM) / Ecuadorian Association of Museums, Caldas 340, Apdo 175B, Quito
Founded: 1977; Members: 60
Focus: Archives; Anthro; Nat Sci; Educ
Periodical
Boletín (weekly) 03844

Asociación Latinoamericana de Educación Radiofónica (ALER) / Latin American Association for Education by Radio, Valladolid 479, Casilla 4639-A, Quito
T: (02) 524358; Fax: (02) 503996
Founded: 1972; Members: 54
Gen Secr: Humberto Vandenbulcke
Focus: Educ
Periodical
Boletín Informativo ALER (weekly) . . . 03845

Asociación Latinoamericana de Escuelas y Facultades de Enfermería (ALADEFE) / Latin American Association of Nursing Science Schools and Departments, Casilla 1395-A, Suc 3, Quito
T: (02) 521053, 230764; Fax: (02) 521053
Founded: 1986
Gen Secr: Maria de Lourdes Velasco
Focus: Educ
Periodicals
Boletín Informativo ALADEFE (weekly)
Revista ALADEFE 03846

Casa de la Cultura Ecuatoriana, Av 6 de Diciembre 332, Apdo 67, Quito
Founded: 1944; Members: 25
Focus: Lit 03847

Centro Andino de Tecnología Rural (CATER), Av Pío Jamarillo Avarado y Pedro Vaca de la Cadena, Casilla 399, Loja
T: 961329; Tx: 4535
Founded: 1980
Focus: Eng 03848

Centro de Investigaciones Históricas, Apdo 7110, Guayaquil
Founded: 1930; Members: 15
Focus: Hist 03849

Centro Médico Federal del Azuay, Apdo 233, Cuenca
Founded: 1944
Focus: Med 03850

Comisión Ecuatoriana de Energía Atómica, Calle Cordero 779 y Av 6 de Diciembre, Apdo 2517, Quito
Founded: 1958; Members: 40
Focus: Nucl Res 03851

Dirección General de Geología y Minas, Carrión 1016, Quito
Founded: 1964; Members: 150
Focus: Geology; Mining 03852

Federación Nacional de Médicos del Ecuador, Av de los Estadios e Iñaquito, Quito
Founded: 1942; Members: 1435
Pres: Dr. Leonardo Malo Borrero; Gen Secr: Dr. Alfredo Pérez Rueda
Focus: Med 03853

Instituto Latinoamericano de Investigaciones Sociales (ILDIS) / Latin American Social Science Research Institute, Av Colón 1346, Apdo 367A, Quito
T: 543000; Tx: 2539
Founded: 1974
Focus: Sociology; Econ; Poli Sci . . . 03854

Sociedad Ecuatoriana de Alergía y Ciencias Afinas (SEACA), Av 12 de Octubre 2206, Apdo 2339, Quito
T: 235632
Founded: 1960; Members: 25
Focus: Immunology 03855

Sociedad Ecuatoriana de Astronomía, Apdo 165, Quito
Founded: 1956
Focus: Astronomy 03856

Sociedad Ecuatoriana de Pediatría, Apdo 5865, Guayaquil
Founded: 1945
Focus: Pediatrics 03857

Sociedad Latinoamericana de Farmacología, Apdo 8884-S7, Quito
T: 235632
Founded: 1960; Members: 1450
Focus: Pharmacol 03858

Egypt

Academy of Scientific Research and Technology, 101 Sharia Kasr El-Aini, Cairo
T: 551047; Tx: 93069
Founded: 1971
Pres: Prof. M.A. Hadi
Focus: Sci; Eng 03859

Academy of the Arabic Language, 15 Aziz Abaza St, Zamalek, Cairo
T: 3405026
Founded: 1932; Members: 40
Pres: Prof. Dr. Ibrahim Madkour; Gen Secr: Prof. Abdul-Salam Haroun
Focus: Ling
Periodicals
Councils and Conferences Proceedings (weekly)
Review of the Academy (weekly) . . . 03860

The African Society, 5 Ahmed Hishmat St, Zamalik, Cairo
T: (02) 3407658
Founded: 1972
Pres: Prof. Dr. Mahmod Mahfoz; Gen Secr: M. Fouad El Bidewy
Focus: Int'l Relat
Periodicals
Africa Newsletter
African Studies (weekly) 03861

African Soil Science Association (ASSS), c/o Dept of Soil Science, Faculty of Agriculture, Ain Shams University, Shobra El-Khema, Cairo
T: (02) 2201296; Tx: 94070; Fax: (02) 2913059
Founded: 1986; Members: 66
Pres: Prof. A.M. Elgala
Focus: Agri 03862

Afro-Asian Philosophy Association (AAPA), c/o Faculty of Education, Heliopolis, Cairo
T: (02) 2460531
Founded: 1978
Gen Secr: Dr. Mona Abousenna
Focus: Philos 03863

Afro-Asian Writers' Permanent Bureau, 104 Sharia Kasr El-Aini, Cairo
T: 29989, 26519
Founded: 1958
Focus: Lit 03864

Agricultural Guidance Research Institute, Nadi El Said St, Dokki
Founded: 1977
Focus: Agri 03865

Alexandria Medical Association, 4 Sharia G. Carducci, Alexandria
Founded: 1921; Members: 1200
Pres: Prof. H.S. El-Badawi; Gen Secr: Prof. Dr. T. Aboul-Azi
Focus: Med
Periodical
The Alexandria Medical Journal (weekly) 03866

Arab Aerospace Educational Organization, 11 Emad Al-Din St, Cairo
T: (02) 918561
Founded: 1986; Members: 11
Pres: Ali O. Ziko
Focus: Aero 03867

Arab Center for Energy Studies (ACES), c/o OAPEC, POB 108, Cairo
T: (02) 3542660; Tx: 21158; Fax: (02) 3542601
Founded: 1983
Focus: Develop Areas 03868

Arab Commission for International Law, c/o LAS, POB 11642, Cairo
T: (02) 750511; Fax: (02) 740331
Focus: Law 03869

Arab Council for Childhood and Development (ACCD), POB 15, Oman, Cairo
T: (02) 3484826; Tx: 20783; Fax: (02) 3803148
Founded: 1987
Gen Secr: Prof. Mamdouh Gabr
Focus: Educ
Periodicals
El-Mokhtar (weekly)
The State of the Child in the Arab World (weekly)
Towards a Better Life for the Arab Child (weekly) 03870

Arab Higher Committee for Pharmacological Affairs, c/o LAS, POB 11642, Cairo
T: (02) 750511; Fax: (02) 740331
Focus: Pharmacol 03871

Arab Institute of Navigation (AIN), POB 1029, Alexandria
T: (03) 5467221; Tx: 54160; Fax: (03) 4311882
Founded: 1978; Members: 276
Gen Secr: Dr. M. Abdel Aziz
Focus: Transport
Periodical
AIN News Bulletin (weekly) 03872

Arab Management Society (AMS), 23 Wadi El-Nil, Mahandesseen, Cairo
T: (02) 3446729; Tx: 92437; Fax: (02) 3445729
Founded: 1989
Pres: Talal Abu-Ghazaleh
Focus: Business Admin 03873

Arab Maritime Transport Academy, POB 1092, Alexandria
Founded: 1972
Focus: Transport
Periodicals
Bulletin of Maritime Transport Information Analysis (weekly)
College of Maritime Studies and Technology Newsletter (weekly)
Current Awareness Bulletin (weekly)
Journal of the Arab Maritime Transport Academy (weekly)
Maritime Research Bulletin (weekly)
News Bulletin (weekly) 03874

Arab Organization for Standardization and Metrology, 27 Sharia Dokki, Cairo
T: 818580
Founded: 1965
Focus: Standards 03875

Arab Scientific Advisory Committee for Blood Transfer, c/o LAS, Social Affairs Dept, POB 11642, Cairo
T: (02) 750511; Fax: (02) 740331
Focus: Med 03876

Arab States Regional Centre for Functional Literacy in Rural Areas (ASFEC), Sirs El-Layyan, Menoufia
Founded: 1952
Gen Secr: A.E.F. Galal
Focus: Adult Educ 03877

Armenian Artistic Union (A.A.U.), 3 Sharia Soliman, El-Halaby, POB 1060, Cairo
T: 742282
Founded: 1920; Members: 300
Pres: V. Depoyan
Focus: Arts 03878

Association for Medical Education in the Middle East, POB 1517, Alexandria
T: (03) 4930090; Fax: (03) 4838916
Pres: Prof. Bashir Hamad
Focus: Educ; Med 03879

Association of African Maritime Training Institutes (AAMTI) / Association des Instituts Africains de Formation Maritime, Gamal Abdel Nasser St, POB 1029, Alexandria
T: (03) 5497882; Tx: 54160; Fax: (03) 5467221
Founded: 1985; Members: 18
Gen Secr: A. Sadek
Focus: Navig
Periodical
AAMTI Newsletter (weekly) 03880

Association of Medical Schools in the Middle East (AMSME), c/o Faculty of Medicine, University, 22 Sharia Al-Gueish, Shatby, Alexandria
Focus: Med; Educ 03881

Atelier / Groupement d'Artistes et d'Ecrivains, 6 Victor Bassili St, Alexandria
T: (03) 4820526
Founded: 1934; Members: 350
Pres: Prof. Dr. Naima El-Shishiny; Gen Secr: Prof. Dr. Farouk Wahba
Focus: Arts; Lit; Photo; Music 03882

Cairo Demographic Centre (CDC), 78 St No 4, Hadhaba-El-Olya-Mokattam, Cairo 11571
T: (02) 929797; Tx: 92034
Founded: 1963
Pres: Dr. Sobhi Abdel-Hakim
Focus: Stats
Periodicals
CDC Acquisitions Lists (weekly)
CDC Newsletter (weekly) 03883

Cairo Odontological Society, 39 Kasr El-Nil, Cairo
Pres: Dr. A.M. Abdel-Azim; Gen Secr: Dr. J. Alcée
Focus: Dent 03884

Centre for Social Research and Documentation for the Arab Region (ARCSS), Zamalek PO, Cairo
T: (02) 3470019; Fax: (02) 4370019
Founded: 1978; Members: 5
Gen Secr: Dr. Ahmad M. Khalifa
Focus: Soc Sci
Periodical
ARCSS Newsletter (3 times annually) . . 03885

Eastern Mediterranean Hand Society, 13 Messaha St, Dokki, Cairo
T: (02) 3484483
Founded: 1980
Gen Secr: Prof. Dr. Wael M. Fahmy
Focus: Surgery 03886

Egyptian Agricultural Organization, Exhibition Grounds, Gezira, POB 63, Cairo
Founded: 1898
Focus: Agri 03887

Egyptian Association for Psychological Studies, Tager Bldg, 1 Osiris St, Garden City, Cairo
Founded: 1948; Members: 1100
Pres: Dr. Fouad A.-L.H. Abou-Hatab
Focus: Psych
Periodical
Yearbook of Psychology (weekly) . . . 03888

Egyptian Association of Archives, Librarianship and Information Science (EALJS), c/o Dept of Library and Information Science, Faculty of Arts, University, Cairo
T: 22167
Founded: 1978
Pres: Dr. Saad M. Hagrassy; Gen Secr: Dr. Hishmat A. Qasim
Focus: Archives; Libraries & Bk Sci; Computer & Info Sci 03889

Egyptian Botanical Society, 1 Ozoris St, Garden City, Cairo
Founded: 1956; Members: 230
Pres: Prof. Dr. A.M. Salama; Gen Secr: Dr. Mohamed Fawzy
Focus: Botany
Periodical
Egyptian Journal of Botany (3 times annually) 03890

Egyptian Dental Federation, 42 Sharia Kasr El-Aini, Cairo
Focus: Dent
Periodical
Egyptian Dental Journal (weekly) 03891

Egyptian Horticultural Society, POB 46, Cairo
Founded: 1915
Focus: Hort 03892

Egyptian Medical Association, 42 Sharia Kasr El-Aini, Cairo
Founded: 1919; Members: 2142
Pres: Prof. Dr. A. El-Kater; Gen Secr: Prof. Dr. A.H. Shaaban
Focus: Med
Periodical
Journal (weekly) 03893

Egyptian Organization for Biological Products and Vaccines, 51 Sharia Wezarat El Zeraa, Agouza, Giza
Focus: Bio 03894

Egyptian School Library Association (ESLA), 35 Sharia Alhalaa, Cairo
Focus: Libraries & Bk Sci 03895

The Egyptian Society for the Dissemination of Universal Culture and Knowledge (ESDUCK), 1081 Corniche El-Nil St, POB 24, Cairo
T: 20295, 25079
Focus: Ethnology; Cultur Hist 03896

Egyptian Society of Dairy Science, 1 Ouziris St, Garden City, Cairo
Founded: 1972
Pres: Dr. Ismael Yousry
Focus: Dairy Sci
Periodical
Egyptian Journal of Dairy Science . . . 03897

Egyptian Society of Engineers, 28 Sharia Ramses, Cairo
Founded: 1920
Pres: Prof. Dr. Ibrahim Adham El-Demirdash; Gen Secr: Dr. Mohamed M. El-Hashimy
Focus: Eng 03898

Egyptian Society of International Law, 16 Sharia Ramses, Cairo
T: (02) 5743152
Founded: 1945; Members: 410
Pres: Dr. F. Ried; Gen Secr: Dr. M. Chehab
Focus: Law
Periodical
Revue Egyptienne de Droit International (weekly) 03899

Egyptian Society of Medicine and Tropical Hygiene, 2 Sharia Fouad I, Alexandria
Founded: 1927
Pres: Dr. I. Abdel-Sayed; Gen Secr: Dr. J. Khouri
Focus: Hygiene; Trop Med 03900

Egyptian Society of Political Economy, Statistics and Legislation, 16 Sharia Ramses, POB 732, Cairo
Founded: 1909; Members: 1520
Pres: Dr. Gamal El-Oteifi; Gen Secr: Dr. I.A. Saleh
Focus: Law; Stats; Poli Sci; Econ
Periodical
L'Egypte Contemporaine (weekly) 03901

General Organization for Housing, Building and Planning Research, POB 1770, Cairo
T: 711564; Tx: 94025
Pres: A. Hamido
Focus: Urban Plan; Civil Eng 03902

Hellenic Society of Ptolemaic Egypt, 20 Sharia Fouad I, Alexandria
Founded: 1908
Pres: Dr. G. Partheniadis; Gen Secr: Costa A. Sandi
Focus: Hist 03903

High Council of Arts and Literature, 9 Sharia Hassan Sabri, Zamalek, Cairo
Founded: 1956
Focus: Arts; Lit 03904

High Council of Culture, 9 Sharia Hassan Sabri, Zamalek, Cairo
T: 818910, 814899
Founded: 1980
Focus: Arts; Lit 03905

Institut d'Egypte, 13 Sharia Sheikh Rihane, Cairo
Founded: 1798; Members: 160
Pres: Dr. Sileman Hazien; Gen Secr: P. Ghalioungui
Focus: Int'l Relat; Ethnology
Periodical
Bulletin (weekly) 03906

Institute of Arab Music, 2 Sharia Tewfik, Alexandria
Pres: Ahmed Bey Hassan; Gen Secr: A. Saad
Focus: Music 03907

Institute of Arab Music, 22 Sharia Ramses, Cairo
Focus: Music 03908

National Centre for Educational Research, Central Ministry of Education, 33 Sharia Falaky, Cairo
Founded: 1972
Gen Secr: Dr. Y.K. Youssef
Focus: Educ
Periodicals
Contemporary Trends in Education (weekly)
Educational Information Bulletin (weekly) . 03909

National Information and Documentation Centre (NIDOC), Sharia Al-Tahrir, Dokki, Cairo
T: 701696
Founded: 1955
Gen Secr: Dr. A.M. Gad
Focus: Doc; Computer & Info Sci
Periodical
Arab Science Abstracts 03910

National Research Centre, Al-Tahrir St, Dokki, Cairo
T: 701010; Tx: 94022
Founded: 1956
Gen Secr: Prof. Dr. M.B.E. Fayez
Focus: Sci
Periodical
Bulletin 03911

Ophthalmological Society of Egypt, Dar El Hekma, 42 Sharia Kasr El-Aini, Cairo
Founded: 1902; Members: 480
Pres: Prof. Dr. El-Said Khalil Abou Shousa; Gen Secr: Dr. Ahmad Ez El-Din Naim
Focus: Ophthal
Periodical
Annual Bulletin (weekly) 03912

Société Archéologique d'Alexandrie, 6 Sharia Mahmoud Moukhtar, Alexandria
Founded: 1893; Members: 100
Pres: Dr. A. Sadek; Gen Secr: D.A. Daoud
Focus: Archeol
Periodical
Bulletin 03913

Société Entomologique d'Egypte, 14 Sharia Ramses, POB 430, Cairo
Founded: 1907; Members: 502
Pres: Prof. M. Hafez; Gen Secr: Dr. A.H.M. Kamel
Focus: Entomology 03914

Society for Coptic Archaeology, 222 Sharia Ramses, Cairo
Founded: 1934; Members: 360
Pres: W.B. Ghali; Gen Secr: Dr. A. Khater
Focus: Archeol
Periodical
Bulletin (weekly) 03915

El Salvador

Academia Salvadoreña, Calle Poniente 13, San Salvador
Members: 22
Pres: Alfredo Martínez Moreno; Gen Secr: C.H. Ibarra
Focus: Sci 03916

Academia Salvadoreña de la Historia, Km 10 Planes de Renderoz, Col. Los Angeles, Villa Lilia 13, San Salvador
Founded: 1925
Pres: Jorge Lardé y Larín; Gen Secr: Pedro Escalante Mena
Focus: Hist
Periodical
Boletín (weekly) 03917

Asociación Centroamericana de Anatomía, c/o Escuela de Medicina, Arce, San Salvador
Founded: 1964
Focus: Anat 03918

Asociación de Bibliotecarios de El Salvador, c/o Biblioteca Nacional, Av Norte y Calle Delgado 8a, San Salvador
T: 216312
Focus: Libraries & Bk Sci 03919

Asociación de Radiólogos de América Central y Panamá, c/o Dr. Raúl Argüello, 5a Av Norte 48, San Salvador
Focus: Radiology 03920

Ateneo de El Salvador, Av España 225, Edificio Quán, Apdo 13a, San Salvador
Founded: 1912; Members: 132
Focus: Arts 03921

Central American Dermatological Society, 95 Av Norte 626, San Salvador 01117
Founded: 1957
Gen Secr: Dr. Enrique Hernandez Pérez
Focus: Derm 03922

Central American Paediatric Society, c/o Sociedad de Pediatría de El Salvador, Arce 1403, San Salvador
Focus: Pediatrics 03923

Central American Public Health Council, c/o ODECA, San Salvador
Focus: Public Health 03924

Centro de Estudios e Investigaciones Geotécnicas, Apdo 109, San Salvador
Founded: 1964
Focus: Geology 03925

Centro Nacional de Tecnología Agropecuario (CENTA), Final 1a Av Norte, Santa Tecla
Founded: 1942
Focus: Agri 03926

Colegio Médico de El Salvador, Final Pasaje 10, Col. Miramonte, San Salvador
Founded: 1943
Focus: Med 03927

Dirección General de Estadística y Censos, Arce 953, San Salvador
Founded: 1881
Focus: Stats 03928

Dirección General de Investigaciones Agronómicas, c/o Centro Nacional de Agronomía, Santa Tecla
Focus: Agri 03929

Sociedad de Anestiología de El Salvador, Gustavo Guerrero 640, San Salvador
Founded: 1958
Focus: Anesthetics 03930

Sociedad de Ginecología y Obstetricia de El Salvador, c/o Colegio Médico, Final Pasaje 10, Col. Miramonte, San Salvador
Founded: 1947; Members: 50
Focus: Gynecology 03931

Sociedad Médica de Salud Pública, c/o Colegio Médico, Final Pasaje 10, Col. Miramonte, San Salvador
Founded: 1960; Members: 50
Focus: Public Health 03932

Estonia

Academy of Sciences, Kohtu 6, 0106 Tallinn
T: 442129; Fax: 442149
Founded: 1938; Members: 52
Pres: A. Kõõrna; Gen Secr: U. Margna
Focus: Sci
Periodicals
Eesti Loodus: Nature of Estonia
Gorjutsie Slantsó: Oil Shale
Keel ja Kirjandus: Language and Literature
Toimetised: Proceedings
Uralica Linguistica 03933

Ethiopia

African Association for Public Administration and Management (AAPAM), POB 60087, Addis Ababa
T: 447200 ext 197; CA: APPAM
Founded: 1971; Members: 200
Focus: Public Admin
Periodical
AAPAM Newsletter (weekly) 03934

African Association of Science Editors (AASE), POB 2302, Addis Ababa
T: (01) 159813
Founded: 1985
Focus: Lit
Periodical
AASE Newsletter 03935

African Forum for Mathematical Ecology (AFME), c/o Asmara University, POB 32896, Asmara
Founded: 1990
Gen Secr: Kiflemariam Melake
Focus: Math; Ecology 03936

African Mountain Association (AMA), c/o University of Asmara, POB 1220, Asmara
T: (04) 113600; Tx: 42091
Founded: 1986
Gen Secr: Tewolde Berhan G. Egziabher
Focus: Geology
Periodical
African Mountains and Highlands Newsletter (weekly) 03937

African Small Ruminant Research Network, POB 5689, Addis Ababa
T: (01) 613215; Tx: 21207; Fax: (01) 611892
Focus: Agri
Periodical
Newsletter (weekly) 03938

African Training and Research Centre for Women (ATRCW), POB 3001, Addis Ababa
T: (01) 517200, 517000; Tx: 21029
Founded: 1975; Members: 51
Focus: Sociology
Periodical
ATRCW Update (weekly) 03939

All Africa Leprosy and Rehabilitation Training Center (ALERT), POB 165, Addis Ababa
T: (01) 711110; Tx: 21821; Fax: (01) 711199
Founded: 1965; Members: 18
Gen Secr: Neil Alldred
Focus: Rehabil
Periodical
ALERT (weekly) 03940

Association for Social Work Education in Africa (ASWEA) / Association pour l'Enseignement Social en Afrique, c/o Addis Ababa University, POB 1176, Addis Ababa
T: 126827
Founded: 1971; Members: 193
Focus: Adult Educ; Sociology
Periodicals
Journal for Social Work Education in Africa (1-2 times annually)
Professional Documents on Social Development (weekly) 03941

Association for the Advancement of Agricultural Sciences in Africa (AAASA), POB 30087, Addis Ababa
T: 443536
Founded: 1968
Gen Secr: Prof. M. El-Fouly
Focus: Agri
Periodicals
AAASA Newsletter (weekly)
African Journal of Agricultural Sciences (weekly) 03942

Ethiopian Library Association (ELA), POB 30530, Addis Ababa
T: 110844
Founded: 1961; Members: 150
Pres: A. Mengsteab; Gen Secr: B. Debela
Focus: Libraries & Bk Sci
Periodicals
Bulletin (weekly)
Directory of Ethiopian Libraries 03943

Ethiopian Medical Association, POB 2179, Addis Ababa
Founded: 1961
Pres: Dr. Abebe Haregewoin
Focus: Med
Periodical
Ethiopian Medical Journal (weekly) . . . 03944

International Livestock Centre for Africa (ILCA), POB 5689, Addis Ababa
T: 183215; Tx: 21207; Fax: 188191
Founded: 1974; Members: 11
Gen Secr: Dr. John Walsh
Focus: Agri
Periodicals
Annual Report (weekly)
Bulletin
Systems Studies Monographs 03945

Fiji

The Fiji Law Society, POB 144, Suva
Pres: K.N. Govind; Gen Secr: G.P. Shankar
Focus: Law 03946

Fiji Society, POB 1205, Suva
Founded: 1936; Members: 50
Pres: Ivan Williams
Focus: Sci
Periodical
Transactions (weekly) 03947

Finland

Äidinkielen opettajain liitto (ÄOL), Vilppulantie 2, 00700 Helsinki
T: (0) 3511763
Founded: 1948; Members: 2400
Focus: Ling; Educ; Lit
Periodicals
Virke (weekly)
Yearbook (weekly) 03948

Agronomiliitto / Finnish Association of Academic Agronomists, P. Makasiinikatu 6 A 8, 00130 Helsinki
T: (0) 171201
Founded: 1897; Members: 3300
Focus: Agri; Nutrition; Home Econ . . . 03949

Arkistoyhdistys / Archival Association, Rauhankatu 17, 00170 Helskini
T: (0) 176911
Focus: Archives
Periodical
Arkisto (weekly) 03950

Association of European Paediatric Cardiologists, c/o Children's Hospital, Stenbäcksg 11, 00290 Helsinki
T: (0) 4711; Fax: (0) 4714702
Founded: 1964
Gen Secr: Dr. Eric I. Wallgren
Focus: Cardiol
Periodical
Pediatric Cardiology (weekly) 03951

Baltic Marine Environment Protection Commission (HELCOM), Mannerheimintie 12A, 00100 Helsinki
T: (0) 602366; Tx: 125105; Fax: (0) 644577
Founded: 1974; Members: 6
Gen Secr: Fleming Otzen
Focus: Oceanography; Ecology 03952

Committee for Mapping the Flora of Europe, c/o Botanical Museum, University of Helsinki, Unioninkatu 44, 00170 Helsinki
Founded: 1965
Focus: Bio 03953

Ekonomiska Samfundet i Finland / Economic Society of Finland, c/o Svenska Handelshögskole, Arkadiagatan 22, 00100 Helsinki
Founded: 1894; Members: 900
Pres: Marianne Stenius; Gen Secr: Jan-Erik Krusberg
Focus: Econ
Periodical
Tidskrift (weekly) 03954

European Association for Environmental History, c/o Dept of Botany, University of Helsinki, Unioninkatu 44, 00170 Helsinki
T: (0) 1918610
Gen Secr: Prof. Yrjö Vasari
Focus: Nat Sci; Hist 03955

European Association for Research on Learning and Instruction (EARLI), PL 114, 20520 Turku
Gen Secr: Marja Vauras
Focus: Educ 03956

European Consortium for Mathematics in Industry (ECMI), c/o Rolf Nevanlinna Institute, Kirkkokatu 16, 15140 Lahti
Gen Secr: Sinikka Vaskelainen
Focus: Math 03957

Finlands Svenska Författareförening (FSF), Runebergsgatan 32 C 27, 00100 Helsinki
T: (0) 446871
Founded: 1919
Pres: Ingmar Svedberg; Gen Secr: Mette Jensen
Focus: Lit 03958

Finlands Svenska Lärarförbund (FSL), Järnvägsmannagatan 6, 00520 Helsinki
T: (0) 15021; Fax: (0) 142748
Founded: 1978; Members: 2271
Focus: Educ 03959

Finska Kemistsamfundet / Chemical Society of Finland, Hietaniemenkatu 2, 00100 Helsinki
Founded: 1891; Members: 574
Pres: Christel Lamberg-Allardt; Gen Secr: Mariann Holmberg
Focus: Chem 03960

Finska Läkaresällskapet / Medical Society of Finland, PL 316, 00171 Helsinki
Fax: (090) 1356463
Founded: 1835; Members: 950
Pres: Carl-Gustaf Standertskjöld-Nordenstam; Gen Secr: F. Pekonen
Focus: Med
Periodical
Finska Läkaresällskapets Handlingar (weekly) 03961

Geofysiikan Seura / Geophysical Society of Finland, c/o Geological Survey of Finland, Betonimichenkuja 4, 02150 Espoo
T: (0) 46931; Fax: (0) 462205
Founded: 1926; Members: 300
Pres: Lauri J. Pesonen; Gen Secr: Satu Mertanen
Focus: Physics; Geophys; Oceanography
Periodicals
Geofysiikan päivät
Geophysica (weekly) 03962

Geologian Tutkimuskeskus / Geological Survey of Finland, Betonimiehenkuja 4, 02150 Espoo
T: (0) 46931; Tx: 123185; Fax: (0) 462205
Founded: 1886; Members: 790
Focus: Geology
Periodicals
Geologian tutkimuskeskus, Opas (weekly)
Geologian tutkimuskeskus, Toimintakertomus
Geologian tutkimuskeskus, Tutkimusraportti
Geological Survey of Finland, Bulletin
Geological Survey of Finland, Special Paper
Maaperäkartta
Suomen geologinen kartta
Suomen geologinen yleiskartta 03963

Geologiliitto, Rautatieläisenkatu 6, 00520 Helsinki
Focus: Geology 03964

Historian Ystäväin Liitto / Society of the Friends of History, Kaupinkatu 22 A 5, 33500 Tampere
Founded: 1926; Members: 1500
Gen Secr: M. Linna
Focus: Hist
Periodicals
Historiallinen Aikakauskirja: Suuomen Historiallinen (weekly)
Historiallinen Kirjasto (weekly)
Historian Aitta (weekly) 03965

International Council for Laboratory Animal Science (ICLAS), c/o Dept of Physiology, University of Kuopio, PL 1627, 70211 Kuopio
T: (071) 163080; Tx: 42218; Fax: (071) 163410
Founded: 1956
Focus: Vet Med
Periodical
ICLAS-News (weekly) 03966

International Council of Sport Science and Physical Education (ICSSPE), c/o University of Jyväskylä, PL 35, 40351 Jyväskylä
T: (041) 603160; Fax: (041) 603161
Founded: 1958
Pres: Prof. Dr. Paavo Komi; Gen Secr: André Chaker
Focus: Sports
Periodicals
ICSSPE-Bulletin (weekly)
ICSSPE Sport Science Review (weekly) . 03967

International Law Association: Finnish Branch, Hallituskatu 11, PL 536, 00100 Helsinki
Founded: 1946; Members: 93
Focus: Law; Int'l Relat 03968

International Peat Society (IPS), Kuokkalantie 4, 40420 Jyskä
T: (041) 674042; Fax: (041) 677405
Founded: 1968
Gen Secr: Raimo Sopo
Focus: Geology
Periodicals
Bulletin of the IPS (weekly)
International Peat Journal (weekly)
Proceedings of International Peat Congresses (every fourth year) 03969

International Society for Folk-Narrative Research (ISFNR), Nordic Institute of Folklore, Henrikinkatu 3, 20500 Turku
Founded: 1960; Members: 439
Focus: Ethnology; Lit 03970

Juridiska Föreningen i Finland / Law Society of Finland, c/o Asianajotoim. H. Snellman, Aleksanterinkatu 48 A, 00100 Helsinki
T: (0) 177570
Founded: 1862; Members: 841
Focus: Law
Periodical
Tidskrift utgiven av Juridiska Föreningen i Finland 03971

Kalevalaseura, Hallituskatu 1, 00170 Helsinki
T: (0) 131231; Fax: (0) 13123220
Focus: Folklore; Ethnology; Ling; Hist
Periodical
Kalevalaseuran vuosikirja (weekly) . . . 03972

Kansantaloudellinen Yhdistys / Finnish Economic Association, c/o Timo Jalamo, KOP, PL 10, 00101 Helsinki
Founded: 1884; Members: 950
Pres: Heikki Koivisto; Gen Secr: Timo Jalamo
Focus: Econ
Periodical
Kansantaloudellinen Aikakauskirja: The Finnish Economic Journal (weekly) 03973

Kansantaloustieten Professorien ja Dosenttien Yhidistys, c/o Yhteiskuntatieteellinen Tiedekunta, Jyväsekylän Yliopisto, 40100 Jyväskylä 10
Focus: Econ; Educ; Adult Educ 03974

Kansanvalistusseura / The Society for Popular Culture, Museokatu 18 A 2, 00100 Helsinki
T: (0) 406632; Fax: (0) 441345
Founded: 1874
Focus: Adult Educ
Periodical
Adult Education in Finland (weekly) . . 03975

Kirjallisuudentutkijain Seura / The Literary Research Society, c/o Institution of Finnish Literature, University, Helsinki
Founded: 1927; Members: 250
Pres: Prof. Kai Laitinen; Gen Secr: P. Karttunen
Focus: Lit
Periodical
Kirjallisuudentutkijain Sauran Vuosikirja: The Yearbook of the Literary Research Society (weekly) 03976

Kirjastonhoitajaliitto, Rautatieläisenkatu 6, 00520 Helsinki
T: (0) 15021; Fax: (0) 147242
Founded: 1945; Members: 1700
Focus: Libraries & Bk Sci 03977

Klassillis-Filologinen Yhdistys / Klassisk-filologiska Föreningen / Classical Association of Finland, Hallituskatu 11-13, 00014 Helsingin Yliopisto
T: (0) 1912682; Fax: (0) 1912161
Founded: 1882; Members: 140
Pres: Prof. Dr. Heikki Solin; Gen Secr: Uta-Maria Liertz
Focus: Lit; Hist; Archeol
Periodical
Arctos: Acta philologica Fennica (weekly) 03978

Kotikielen Seura, Castrenianum, PL 3, 00014 Helsingin Yliopisto
Fax: (0) 1913329
Founded: 1876; Members: 400
Focus: Ling; Lit
Periodical
Virittäjä (weekly) 03979

Lääketieteellisen Fysiikan ja Tekniikan Yhdistys / Finnish Society for Medical Physics and Medical Engineering, PL 22, 33721 Tampere
Founded: 1968; Members: 250
Focus: Med; Physiology 03980

Lastentarhanopettajalütto, Rautatielaisenkatu 6, 00520 Helsinki
T: (0) 15021
Founded: 1919; Members: 7500
Focus: Educ
Periodical
Lastentarha (19 times annually) 03981

Maataloustuottajain Keskusliitto (MTK), Simonkatu 6, 00100 Helsinki
T: (0) 131151; Tx: 122474; Fax: (0) 13115425
Founded: 1917
Focus: Agri 03982

Museovirasto / National Board of Antiquities, PL 913, 00101 Helsinki
Founded: 1884
Focus: Hist; Ethnology; Preserv Hist Monuments; Numismatics; Archeol
Periodicals
Arkeologia Suomessa: Archaeology in Finland (weekly)
Museoviraston arkeologian julkaisuja
Museoviraston rakennushistorian osaston raportteja: Annual Report (weekly)
Museoviraston työväenkulttuurijulkaisuja
Nautica Fennica (weekly)
Rakennushistorian osaston julkaisuja
Suomen kirkot: Finland's Churches . . . 03983

Nordic Council for Scientific Information, c/o Helsinki University of Technology Library, Otaniementie 9, 02150 Espoo
T: (0) 4552633; Tx: 121591; Fax: (0) 4552576
Founded: 1976; Members: 9
Pres: Bendik Rugaas; Gen Secr: Harriet Rinne Mendes
Focus: Computer & Info Sci; Libraries & Bk Sci
Periodicals
Nordinfo (weekly)
Nordinfo – Nytt (weekly) 03984

Nordisk Kollegium for Fysisk Oceanografi (NKFO), c/o Institut of Marine Research, Asiakkaankatu 3, 00930 Helsinki
T: (0) 331044; Tx: 125731
Founded: 1966; Members: 8
Focus: Oceanography 03985

Nordisk Odontologisk Förening, c/o Institute of Dentistry, University of Helsinki, Mannerheimintie 172, 00300 Helsinki
T: (0) 47351
Founded: 1917
Focus: Dent
Periodical
Scandinavian Journal of Dental Research 03986

Opetusalan Ammattijarjestö (OAJ), Rautatieläisenkatu 6, 00520 Helsinki
Members: 36000
Focus: Educ 03987

Oulun Luonnonystäväin Yhdistys / Societas Amicorum Naturae Ouluensis, c/o Zoological Museum, University of Oulu, Linnanmaa, 90570 Oulu
T: (081) 5531251; Tx: 32375; Fax: (081) 5561278
Founded: 1925; Members: 430
Pres: Dr. Eino Erkinaro; Gen Secr: Pirkko Viro
Focus: Bio
Periodical
Aquilo (weekly) 03988

Säteilyturvakeskus (STUK) / Finnish Centre for Radiation and Nuclear Safety, PL 268, 00101 Helsinki
T: (0) 70821; Tx: 122691; Fax: (0) 7082210
Founded: 1958
Pres: Antti Vuorinen; Gen Secr: Antti Niittylä
Focus: Radiology; Safety
Periodicals
ST-guides
STUK-A-series
YVL-guides 03989

Scandinavian Dental Association, c/o Institute of Dentistry, University of Turku, Lemminkäisenkatu 2, 20520 Turku
Fax: (021) 6338356
Founded: 1866; Members: 1500
Focus: Dent 03990

Skandinaviska Simuleringssällskapet (SIMS) / Scandinavian Simulation Society, c/o VTT/Säh, Otakaari 7 B, 02150 Espoo
T: (0) 4566422; Fax: (0) 4550115
Founded: 1958; Members: 250
Pres: Erik Mosekilde; Gen Secr: Kaj Juslin
Focus: Eng 03991

Societas Biochemica, Biophysica et Microbiologica Fenniae, c/o Dept of Genetics, University of Helsinki, Arkadiankatu 7, PL 17, 00014 Helsingin Yliopisto
Founded: 1945; Members: 1100
Pres: Prof. Timo Korhonen; Gen Secr: Dr. Jarkko Hantula
Focus: Biophys; Biochem; Microbio . . . 03992

Societas Biologica Fennica Vanamo, Unioninkatu 44, PL 7, 00014 Helsingin Yliopisto
T: (0) 1918640
Founded: 1896; Members: 1600
Pres: Prof. Pertti Uotila; Gen Secr: Kurt Fagerstedt
Focus: Bio
Periodicals
Acta Botanica Fennica (weekly)
Acta Zoologica Fennica (weekly)
Annales Botanici Fennici (weekly)
Annales Zoologici Fennici (weekly)
Luonnon Tutkija: The Naturalist (5 times annually) 03993

Societas Medicinae Physicalis et Rehabilitationis Fenniae, c/o HYKS Dept of Physical Medicine, Haartmaninkatu 4, 00290 Helsinki
Founded: 1956; Members: 120
Focus: Rehabil; Physiology; Physical Therapy 03994

Societas pro Fauna et Flora Fennica (SFFF), c/o Botanical Museum, PL 47, 00014 Helsingin Yliopisto
T: (0) 7084787; Fax: (0) 7084830
Founded: 1821; Members: 870
Pres: Dr. K.-G. Widen; Gen Secr: R. Skyten
Focus: Botany; Zoology
Periodicals
Acta Botanica Fennica (weekly)
Acta Zoologica Fennica (weekly)
Annales Botanici Fennici (weekly)
Annales Zoologici Fennici (weekly)
Memoranda Societatis pro Fauna et Flora Fennica (weekly) 03995

Societas Scientiarum Fennica (SSF) / The Finnish Society of Sciences and Letters, Mariankatu 5, 00170 Helsinki
T: (0) 633005
Founded: 1838; Members: 170
Focus: Sci 03996

Suomalainen Lääkäriseura Duodecim / The Finnish Medical Society Duodecim, Kalevankatu 11 A, 00100 Helsinki
T: (0) 611050; Fax: (0) 611004
Founded: 1881; Members: 14300
Focus: Med
Periodicals
Annals of Medicine (weekly)
Lääketieteellinen Aikakauskirja Duodecim (weekly) 03997

Suomalainen Lakimiesyhdistys, Annankatu 16 B, 00120 Helsinki
T: (0) 603567
Founded: 1898; Members: 3000
Focus: Law 03998

Suomalainen Teologinen Kirjallisuusseura (STKS), Neitsytpolku 1 B, 00140 Helsinki
T: (0) 1913978
Founded: 1891; Members: 800
Pres: Dr. Eeva Martikainen; Gen Secr: Dr. Simo Peura
Focus: Rel & Theol
Periodical
Suomalainen Teologisen Kirjallisuusseuran julkaisuja 03999

Suomalainen Tiedeakatemia / Finnish Academy of Science and Letters, Mariankatu 5, 00170 Helsinki
T: (0) 636800; Fax: (0) 660117
Founded: 1908; Members: 418
Pres: Jouko Paunio; Gen Secr: Aarne Nyyssönen
Focus: Lit
Periodicals
Annales Academiae Scientiarum Fennicae
Folklore Fellows' Communications
Year Book (weekly) 04000

Suomalaisen Kirjallisuuden Seura (SKS) / Finnish Literature Society, Hallituskatu 1, PL 259, 00171 Helsinki
T: (0) 131231
Founded: 1831; Members: 2500
Pres: Prof. Dr. Heikki Ylikangas; Gen Secr: Urpo Vento
Focus: Lit; Folklore; Ling
Periodicals
Kirjallisuudentutkijain Seuran vuosikirja: Yearbook of the Literary Research Society (weekly)
Review of Finnish Linguistics and Ethnology
Studia Fennica (weekly) 04001

Suomalaisten Kemistien Seura (SKS), Hietaniemenkatu 2, 00100 Helsinki
T: (0) 408022
Founded: 1919; Members: 3000
Focus: Chem 04002

Suomalais-Ugrilainen Seura (SUS) / Société Finno-Ougrienne, Mariankatu 5-7, PL 320, 00171 Helsinki
T: (0) 1912888
Founded: 1883; Members: 650
Pres: Seppo Suhonen; Gen Secr: Dr. Juha Janhunen
Focus: Ling
Periodicals
Hilfsmittel für das Studium der finnisch-ugrischen Sprachen (weekly)
Journal de la Société Finno-Ougrienne (weekly)
Lexica Societatis Fenno-Ugricae (weekly)
Mémoires de la Société Finno-Ougrienne (weekly) 04003

Suomen Akatemia (SA), PL 57, 00551 Helsinki
T: (0) 77581; Tx: 123416; Fax: (0) 7758299
Founded: 1947
Pres: Prof. Erik Allardt; Gen Secr: Heikki Kallio
Focus: Sci 04004

Suomen Aktuaariyhdistys / The Actuarial Society of Finland, c/o Markku Paakkanen, Kansa Corporation Ltd, Hämeentie 33, PL 45, 00501 Helsinki
T: (0) 73168581; Fax: (0) 73168968
Founded: 1922; Members: 261
Pres: Matti Ruohonen
Focus: Insurance
Periodical
The Actuarial Society of Finland Working Papers (weekly) 04005

Suomen Allergologi- ja Immunologiyhdistys / Finnish Association of Allergology and Immunology, c/o Dept of Allergic Diseases, University Central Hospital, Meilahdentie 2, 00250 Helsinki
T: (0) 4716344; Fax: (0) 4716500
Founded: 1946; Members: 278
Pres: Tari Haahtela; Gen Secr: Ritva Sorva
Focus: Immunology 04006

Suomen Atomiteknillinen Seura (ATS), c/o Technical Research Centre, Nuclear Engineering Laboratory, Lönnrotinkatu 37, 00180 Helsinki
T: (0) 6160459; Tx: 124608
Founded: 1966; Members: 505
Focus: Nucl Res
Periodical
DiA-kunta 04007

Suomen Autoteknillinen Liitto (SATL), Köydenpunojankatu 8, 00180 Helsinki
T: (0) 6944724; Fax: (0) 6944027
Founded: 1934; Members: 5500
Focus: Auto Eng
Periodical
Suomen Autolehti (10 times annually) . . 04008

Suomen Avaruustutkimusseura, PL 507, 00101 Helsinki
Founded: 1959; Members: 100
Focus: Eng
Periodical
Avaruusluotain (weekly) 04009

Suomen Betoniyhdistys / Finska Betongföreningen, Mikonkatu 18B, 00100 Helsinki
T: (0) 651411; Fax: (0) 651145
Members: 730
Focus: Materials Sci
Periodicals
Betoni (weekly)
Nordic Concrete Research (weekly) . . . 04010

Suomen Egyptologinen Seura (SES) / Finnish Egyptological Society, c/o Raija Mattila, Fredrikinkatu 79 C 44, 00100 Helsinki
T: (0) 1913098
Founded: 1969; Members: 850
Focus: Archeol; Preserv Hist Monuments
Periodical
SES-ESF tiedotuslehti (weekly) 04011

Suomen Eksegeettinen Seura, Neitsytpolku 1 B, 00140 Helsinki
Founded: 1938; Members: 605
Focus: Rel & Theol 04012

Suomen Eläinlääkäriliitto, Rautatieläisenkatu 6, 00520 Helsinki
Fax: (0) 142214
Founded: 1892; Members: 1208
Focus: Vet Med

Periodical
Suomen Eläinlääkärilehti: Finsk Veterinärtidskrift (weekly) *04013*

Suomen Farmaseuttinen Yhdistys, c/o Dept of Pharmacy, University of Helsinki, PL 15, 00014 Helsingin Yliopisto
T: (0) 1912783; Fax: (0) 1912786
Founded: 1887; Members: 390
Pres: Martti Marvola; Gen Secr: Sirkku Saarela
Focus: Pharmacol *04014*

Suomen Farmasialiitto (SFL) / Finnish Pharmacists' Association, Rautatieläisenkatu 6, 00520 Helsinki
T: (0) 15021
Founded: 1917; Members: 7000
Focus: Pharmacol
Periodical
Semina (weekly) *04015*

Suomen Filosofinen Yhdistys (SFY) / The Philosophical Society of Finland, PL 24, 00014 Helsingin Yliopisto
T: (0) 1911; Fax: (0) 1917627
Founded: 1873; Members: 580
Pres: Ilkka Niiniluoto; Gen Secr: Ilpo Halonen
Focus: Philos
Periodicals
Acta Philosophica Fennica (weekly)
Ajatus-yearbook (weekly) *04016*

Suomen Fysiologiyhdistys / The Finnish Physiological Society, c/o Dept of Physiology, University, 90220 Oulu
T: (081) 332133
Founded: 1961; Members: 350
Focus: Physiology *04017*

Suomen Fysioterapeuttiliitto / Finlands Fysioterapeutförbund / The Finnish Association of Physiotherapists, Asemamiehenkatu 4, 00520 Helsinki
T: (0) 146817; Fax: (0) 1483054
Founded: 1943; Members: 6500
Pres: Outi Töytäri; Gen Secr: Anna-Maija Laitinen
Focus: Therapeutics
Periodical
Fysioterapia (8 times annually) *04018*

Suomen Fyysikkoseura / Finnish Physical Society, Siltavuorenpenger 20 M, PL 9, 00014 Helsingin Yliopisto
T: (0) 1918375; Fax: (0) 1918378
Founded: 1947; Members: 1100
Pres: Paul Hoyer; Gen Secr: Sari Nokkamen
Focus: Physics
Periodicals
Arkhimedes (weekly)
Fysiikka Tänään (weekly) *04019*

Suomen Gastroenterologiayhdistys / Finnish Society of Gastroenterology, c/o Duodecim, PL 713, 00101 Helsinki
T: (0) 611523; Fax: (0) 611004
Founded: 1956; Members: 482
Pres: Pekka Pikkarainen; Gen Secr: Anna-Liisa Karvonen
Focus: Gastroenter *04020*

Suomen Gemmologinen Seura / Gemmological Society of Finland, Vuorikatu 3 A 10, 00100 Helsinki
T: (0) 660562
Founded: 1960; Members: 160
Pres: Alf Larsson; Gen Secr: Anne Kangas
Focus: Mineralogy
Periodical
Gemmologian Työsaralta (weekly) . . . *04021*

Suomen Geologinen Seura, Kivimiehentie 1, 02150 Espoo
T: (0) 46931; Tx: 123185; Fax: (0) 462205
Founded: 1886; Members: 1000
Pres: Anneli Uutela; Gen Secr: Pekka Salonsaari
Focus: Geology
Periodical
Bulletin (1-2 times annually) *04022*

Suomen Geoteknillinen Yhdistys / Finnish Geotechnical Society, c/o VTT, PL 108, 02151 Espoo
T: (0) 4564681; Fax: (0) 467927
Founded: 1951; Members: 550
Focus: Mining; Geomorph *04023*

Suomen Hammaslääkäriliitto, Rautatieläisenkatu 6, 00520 Helsinki
T: (0) 15021
Founded: 1924; Members: 4148
Focus: Dent *04024*

Suomen Hammaslääkäriseura (SHS) / Finnish Dental Society, Rautatieläisenkatu 6, 00520 Helsinki
T: (0) 15021; Fax: (0) 146317
Founded: 1891; Members: 5400
Pres: Inkeri Rytomaa; Gen Secr: Leena Sandholm
Focus: Dent
Periodical
Proceedings of the Finnish Dental Society (weekly) *04025*

Suomen Heraldinen Seura, Kivelänkatu 1 C 14, 00260 Helsinki
Founded: 1957; Members: 77
Focus: Genealogy *04026*

Suomen Historiallinen Seura, Arkadiankatu 16 B 28, 00100 Helsinki
T: (0) 440369; Fax: (0) 441468
Founded: 1875; Members: 850
Pres: Mavjatta Hietala; Gen Secr: Rauno Enden
Focus: Hist
Periodicals
Historiallinen Arkisto (2-3 times annually)
Historiallisia Tutkimuksia (5-8 times annually)
Käsikirjoja (weekly)
Studia Historica (3-5 times annually)
Suomen Historian Lähteitä (15-20 times annually) *04027*

Suomen Hitsausteknillinen Yhdistys, Mäkelänkatu 36 A, 00510 Helsinki
T: (0) 7732199; Fax: (0) 7732661
Members: 3500
Focus: Eng
Periodical
Hitsaustekniikka-Svetsteknik (weekly) . . . *04028*

Suomen Hyönteistieteellinen Seura / The Entomological Society of Finland, P. Rautatiekatu 13, 00100 Helsinki
T: (0) 1911
Founded: 1935; Members: 350
Focus: Entomology
Periodicals
Acta Zoologica Fennica
Annales Entomologici Fennici (weekly)
Notulae Entomologicae (weekly) *04029*

Suomen Ilmailuliitto (SIL), c/o Malmin Lentoasema, 00700 Helsinki
Members: 12130
Focus: Eng
Periodical
Ilmailu (weekly) *04030*

Suomen Itämainen Seura / Finnish Oriental Society, c/o Dept of Asian and African Studies, University of Helsinki, PL 13, 00014 Helsingin Yliopisto
Fax: (0) 1912094
Founded: 1917; Members: 150
Focus: Ethnology
Periodical
Studia Orientalia (1-2 times annually) . . *04031*

Suomen Kansanopistoyhdistys / The Finnish Folk High School Association, P. Rautatiekatu 15 B 12, 00100 Helsinki
T: (0) 444413; Fax: (0) 445619
Founded: 1907; Members: 789
Focus: Adult Educ *04032*

Suomen Kardiologinen Seura, c/o Children's Hospital, University, Stenbäckinkatu 11, 00350 Helsinki
Founded: 1968; Members: 151
Focus: Cardiol *04033*

Suomen Kasvatusopillinen Yhdistys, Liisankatu 13, 00170 Helsinki
Founded: 1863; Members: 30
Pres: Olli Kohonen
Focus: Educ *04034*

Suomen Kasvatustieteellinen Seura / The Finnish Society for Educational Research, c/o Oulun Yliopisto Kasvatustieteiden Tiedekunta, PL 222, 90571 Oulu
T: (081) 5531011; Fax: (081) 5533600
Founded: 1967; Members: 275
Pres: Prof. Dr. Leena Syrjälä; Gen Secr: Saila Anttonen
Focus: Educ
Periodical
The Finnish Journal of Education . . . *04035*

Suomen Kemian Seura / Association of Finnish Chemical Societies, Hietaniemenkatu 2, 00100 Helsinki
T: (0) 408022; Fax: (0) 408780
Founded: 1970; Members: 4700
Pres: Prof. Jorma Sundquist; Gen Secr: Anneli Varmavuori
Focus: Chem *04036*

Suomen Kielen Seura / Finnish Language Society, c/o Humanistinen Tiedekunta, Turun Yliopisto, 20500 Turku
Fax: (021) 6336360
Focus: Ling
Periodical
Sananjalka (weekly) *04037*

Suomen Kirjailijaliitto / Association of Finnish Authors, Runeberginkatu 32 C, 00100 Helsinki
Founded: 1897; Members: 520
Pres: Jarkko Laine; Gen Secr: Päivi Liedes
Focus: Lit
Periodicals
Suomalaiset kertojat
Suomen Runotar *04038*

Suomen Kirjastoseura / The Finnish Library Association, Museokatu 18 A 5, 00100 Helsinki
T: (0) 492632; Fax: (0) 441345
Founded: 1910; Members: 2600
Pres: Mirja Ryynänen; Gen Secr: Tuula Haavisto
Focus: Libraries & Bk Sci
Periodical
Kirjastolehti (weekly) *04039*

Suomen Kirkkohistoriallinen Seura (SKHS) / Finnish Society for Church History, Fabianinkatu 7, PL 48, 00014 Helsingin Yliopisto
T: (0) 1913040; Fax: (0) 1913033
Founded: 1891; Members: 840
Pres: Prof. Simo Heininen; Gen Secr: Dr. Hannu Mustakallio
Focus: Hist; Rel & Theol
Periodicals
Suomen Kirkkohistoriallisen Seuran Toimituksia (3-4 times annually)
Suomen Kirkkohistoriallisen Seuran Vuosikirja: Yearbook (3-6 times annually) *04040*

Suomen Kirurgiyhdistys, Mäkelänkatu 2 A, 00500 Helsinki
T: (0) 3930768; Fax: (0) 3930801
Founded: 1925; Members: 1000
Gen Secr: Karl von Smitten
Focus: Surgery
Periodical
Annales Chirurgiae & Gynaecologiae (weekly) *04041*

Suomen Kliinisen Neurofysiologian Yhdistys, c/o Dept of Clinical Neurophysiology, University Central Hospital of Helsinki, Haartmaninkatu 4, 00290 Helsinki
Fax: (0) 4714009
Founded: 1972; Members: 95
Focus: Physiology *04042*

Suomen Korroosioyhdists Sky / Corrosion Society of Finland, PL 952, 00100 Helsinki
Founded: 1978; Members: 315
Focus: Metallurgy *04043*

Suomen Lainopillinen Yhdistys, Alexandersgatan 48 A, 00100 Helsinki
T: (0) 630302
Founded: 1862
Focus: Law *04044*

Suomen Lakimiesliitto / The Union of Finnish Lawyers, Uudenmaankatu 4-6 B, 00120 Helsinki
T: (0) 649201
Founded: 1944; Members: 14200
Focus: Law
Periodicals
Finnish Law I: Common Law
Finnish Law II: Public law
The Finnish Legal System: The English Language Publication *04045*

Suomen Lintutieteellinen Yhdistys, P. Rautatiekatu 13, 00100 Helsinki
Fax: (0) 1917443
Founded: 1924; Members: 645
Pres: Dr. Juha Tiainen; Gen Secr: Markku Mikkola
Focus: Ornithology
Periodical
Ornis Fennica (weekly) *04046*

Suomen Maantieteellinen Seura / Geographical Society of Finland, PL 4, 00014 Helsingin Yliopisto
T: (0) 1912434; Fax: (0) 1912641
Founded: 1888; Members: 1300
Pres: Leo Koutaniemi; Gen Secr: Prof. Paavo Talman
Focus: Geography
Periodicals
Fennia (weekly)
Terra (weekly) *04047*

Suomen Maataloustieteellinen Seura / Scientific Agricultural Society of Finland, c/o Dept of Horticulture, Agricultural Centre, 21500 Piikkio
T: (021) 727806; Fax: (021) 899999
Founded: 1909; Members: 600
Focus: Agri; Hort
Periodical
Journal (5-6 times annually) *04048*

Suomen Matemaattinen Yhdistys / Finnish Mathematical Society, c/o Dept of Mathematics, PL 4, 00014 Helsingin Yliopisto
T: (0) 1913206; Fax: (0) 1913213
Founded: 1868; Members: 360
Pres: Mika Seppälä; Gen Secr: Hans-Olav Tylli
Focus: Math
Periodical
Arkhimedes (weekly) *04049*

Suomen Metsätieteellinen Seura / The Society of Forestry in Finland, Unioninkatu 40 B, 00170 Helsinki
T: (0) 658707; Tx: 125181; Fax: (0) 1917619
Founded: 1909; Members: 560
Pres: Prof. Seppo Kellomäki; Gen Secr: Dr. E. Korpilahti
Focus: Forestry
Periodicals
Acta Forestalia Fennica (5-8 times annually)
Silva Fennica (weekly) *04050*

Suomen Muinaismuistoyhdistys (SMY), Nervanderinkatu 13, PL 913, 00101 Helsinki
T: (0) 662672
Founded: 1870; Members: 550
Pres: Dr. Torsten Edgren; Gen Secr: Marianne Schauman-Lönnqvist
Focus: Archeol; Ethnology; Hist
Periodicals
Aikakauskirja
Finskt Museum (weekly)
Iskos
Kansatieteellinen Arkisto
Suomen Museo (weekly) *04051*

Suomen Museoliitto (SML) / Finnish Museums Association, Annankatu 16 B 50, 00120 Helsinki
T: (0) 649001; Fax: (0) 608330
Founded: 1923; Members: 246
Pres: Matti Pelttari; Gen Secr: Anja-Tuulikki Huovinen
Focus: Arts
Periodicals
Museo (8 times annually)
Suomen museoliiton julkaisuja (weekly) . *04052*

Suomen Musiikinopettajain Liitto (SMOL), Fredrikinkatu 61 A 3, 00100 Helsinki
T: (0) 6943207
Founded: 1925; Members: 2700
Focus: Music; Educ *04053*

Suomen Näytelmäkirjailijaliitto / Dramatists' League of Finland, Vironkatu 12 B, 00170 Helsinki
Founded: 1921
Pres: Anto Seppälä; Gen Secr: Pirjo Westman
Focus: Perf Arts *04054*

Suomen Palontorjuntaliitto / Finlands Brandvärnsförbund / Finnish Fire Protection Association, Iso Roobertinkatu 7 A 4, 00120 Helsinki
T: (0) 649233
Founded: 1968; Members: 26
Focus: Safety *04055*

Suomen Psykologiliitto, Rautatieläisenkatu 6, 00520 Helsinki
T: (0) 150021; Fax: (0) 141716
Founded: 1957; Members: 3300
Focus: Psych
Periodical
Psykologiumtiset (weekly) *04056*

Suomen Radiologiyhdistys, PL 963, 00101 Helsinki
Founded: 1924; Members: 703
Focus: Radiology *04057*

Suomen Säveltäjät / Society of Finnish Composers, Runeberginkatu 15 A 11, 00100 Helsinki
T: (0) 445589; Fax: (0) 440181
Founded: 1945; Members: 101
Pres: Mikko Heiniö; Gen Secr: Maarit Anderzén
Focus: Music *04058*

Suomen Standardisoimisliitto / Finnish Standards Association, Maistraatinportti 2, PL 116, 00240 Helsinki
T: (0) 1499331; Tx: 122303; CA: Finnstandard
Founded: 1924
Gen Secr: Kari Kaartma
Focus: Standards
Periodicals
SFS Catalogue (weekly)
SFS-tiedotus (weekly) *04059*

Suomen Sukututkimusseura (SSS) / Genealogiska Samfundet i Finland / Genealogical Society of Finland, Mariankatu 7, 00170 Helsinki
T: (0) 179189
Founded: 1917; Members: 2300
Pres: Dr. Veli-Matti Autio; Gen Secr: Pertti Vuorinen
Focus: Genealogy
Periodicals
Genos: Suomen Sukututkimusseuran aikakauskirja (weekly)
Suomen Sukututkimusseuran julkaisuja: Skrifter (weekly)
Suomen Sukututkimusseuran vuosikirja: Årsskrift (weekly) *04060*

Suomen Syöpäyhdistys / Cancer Society of Finland, Liisankatu 21 B, 00170 Helsinki
T: (0) 1351011; Fax: (0) 1351093
Founded: 1936; Members: 130000
Pres: Prof. Jussi Huttunen; Gen Secr: Liisa Elovainio
Focus: Cell Biol & Cancer Res
Periodical
Syöpä-Cancer *04061*

Suomen Taideyhdistys / Finska Konstföreningen, Nervanderinkatu 3, 00100 Helsinki
T: (0) 494656; Fax: (0) 448794
Founded: 1846; Members: 3000
Pres: Jaakko Iloniemi; Gen Secr: Seppo Niinivaara
Focus: Arts *04062*

Suomen Tekstiiliteknillinen Liitto, Vallerinkatu 13-15 B, 33270 Tampere
T: (031) 440431; Fax: (031) 441808
Founded: 1936; Members: 984
Focus: Eng *04063*

Suomen Tieteellinen Kirjastoseura / Finnish Research Library Association, PL 217, 00171 Helsinki
Founded: 1929; Members: 600
Pres: Hellevi Urjölä; Gen Secr: Sirkka Kannisto
Focus: Libraries & Bk Sci
Periodicals
Guide to Research Libraries and Information Services in Finland (weekly)
Signum (8 times annually) *04064*

Suomen Tilastoseura / Statistiska Samfundet i Finland / The Finnish Statistical Society, c/o Statistics Finland, PL 504, 00101 Helsinki
T: (0) 17341; Fax: (0) 17343554
Founded: 1920; Members: 500
Pres: Gunnar Rosenqvist; Gen Secr: Ritva Panula
Focus: Stats
Periodicals
Scandinavian Journal of Statistics
Vuosikirja: Yearbook *04065*

Suomen Väestötieteen Yhdistys / Finnish Demographic Society, c/o Central Statistical Office of Finland, PL 770, 00101 Helsinki
Founded: 1973; Members: 169
Gen Secr: Timo Nikander
Focus: Stats 04066

Svenska Litteratursällskapet i Finland (SLS) / The Society of Swedish Literature in Finland, Mariegatan 8, 00170 Helsinki
T: (0) 177133; Fax: (0) 632820
Founded: 1885; Members: 1100
Pres: Prof. Johan Wrede; Gen Secr: Helena Solstrand-Pipping
Focus: Lit
Periodical
Skrifter utgivna av Svenska Litteratursällskapet i Finland (5-6 times annually) 04067

Svenska Tekniska Vetenskapsakademien i Finland (STV) / The Swedish Academy of Engineering Sciences in Finland, Folkskolegt 10 A, 00100 Helsinki
T: (0) 6944260; Fax: (0) 6945041
Founded: 1921; Members: 148
Pres: Prof. Tor-Magnus Enari; Gen Secr: Prof. Kenneth Holmberg
Focus: Eng
Periodicals
Förhandlingar Proceedings
Meddelanden-Reports (weekly) 04068

Tähtitieteellinen Yhdistys Ursa (URSA) / URSA Astronomical Association, Laivanvarustajankatu 9 C 54, 00140 Helsinki
T: (0) 174048; Fax: (0) 657728
Founded: 1921; Members: 5900
Pres: Prof. Raimo Lehti; Gen Secr: Seppo Linnaluoto
Focus: Astronomy
Periodicals
Tähdet: Stars (weekly)
Tähdet ja Avaruus: Stars and Space (weekly)
Ursa minor: Ursan jaostojentiedotuslehti (weekly) 04069

Taloushistoriallinen Yhdistys, Käsälä 6 E, 40250 Jyväskylä
T: (041) 271827
Founded: 1955; Members: 90
Focus: Econ; Hist 04070

Taloustieteellinen Seura / The Finnish Society for Economic Research, c/o Helsinki School of Economics, Runeberginkatu 22-24, 00100 Helsinki
T: (0) 4313493; Fax: (0) 4313738
Founded: 1936; Members: 700
Pres: Marianne Stenius; Gen Secr: Leena Kerkelä
Focus: Econ
Periodical
Finnish Economic Papers (biennial) . . . 04071

Tekniikan Akateemisten Liitto / The Finnish Association of Graduate Engineers, Ratavartijankatu 2, 00520 Helsinki
T: (0) 15901
Founded: 1896; Members: 35800
Gen Secr: Alari Kujala
Focus: Eng 04072

Teknillisten Oppilaitosten Opettajainliitto (TOOL), Rautatieläisenkatu 6, 00520 Helsinki
Founded: 1946; Members: 1065
Focus: Adult Educ 04073

Teknillisten Tieteiden Akatemia (TTA) / The Finnish Academy of Technology, Kansakoulukatu 10 A, 00100 Helsinki
T: (0) 6944260; Fax: (0) 6945041
Founded: 1957; Members: 337
Pres: Dr. Matti Kankaanpää; Gen Secr: Prof. Eino Tunkelo
Focus: Eng
Periodical
Acta Polytechnica Scandinavica (weekly) . 04074

Tekniska Föreningen i Finland, Banvaktsg 2, 00520 Helsinki
T: (0) 15901
Founded: 1880; Members: 2650
Pres: Göran Nyman; Gen Secr: Frej Gustafsson
Focus: Eng
Periodical
Forum för Ekonomi och Teknik 04075

Tietohuollon Neuvottelukunta / Finnish Council Council for Information Provision, PL 293, 00171 Helsinki
Tx: 122079; Fax: (0) 656765
Founded: 1991; Members: 10
Gen Secr: Annu Jylha-Pyykönen
Focus: Sci; Libraries & Bk Sci 04076

Tietopalveluseura / Finnish Society for Information Services, Harakantie 2, 02600 Espoo
T: (0) 518138; Fax: (0) 518167
Founded: 1947; Members: 752
Pres: Tuula Salo; Gen Secr: Gunilla Heikkinen
Focus: Computer & Info Sci
Periodical
Tietopaluelu (weekly) 04077

Turun Historiallinen Yhdistys (THY) / Turku Historical Society, c/o Institute of History, University of Turku, Henrikinkatu 2, 20500 Turku
T: (021) 6335365; Fax: (021) 6336585
Founded: 1923; Members: 604
Pres: Timo Soikkanen; Gen Secr: Tapani Kunttu
Focus: Hist
Periodical
Turun Historiallinen Arkisto 04078

Turun Soitannollinen Seura / Turku Music Society, c/o Sibelius Museum, Piispankatu 17, 20500 Turku
Founded: 1790; Members: 540
Pres: Prof. Dr. Arne Rousi; Gen Secr: Ilpo Tolvas
Focus: Music 04079

Työtehoseura / Work Efficiency Institute, Melkonkatu 16 A, 00210 Helsinki
T: (0) 6922445; Fax: (0) 6922084
Founded: 1300; Members: 1200
Gen Secr: Dr. Tarmo Hioma
Focus: Business Admin
Periodical
TEHO (10 times annually) 04080

Uusfilologinen Yhdistys (UFY) / Modern Language Society, Hallituskatu 11, PL 4, 00014 Helsingin Yliopisto
T: (0) 1912504
Founded: 1887; Members: 400
Pres: Prof. Marjatta Wis; Gen Secr: Minna Palander-Collin
Focus: Ling; Lit
Periodicals
Mémoires de la Société Néophilologique: Monograph Series (weekly)
Neuphilologische Mitteilungen (weekly) . . 04081

Valtiotieteellinen Yhdistys / Finnish Political Science Association, c/o Faculty of Political Science, University of Helsinki, PL 4, 00014 Helsingin Yliopisto
Founded: 1935; Members: 400
Pres: Prof. Dag Anckar; Gen Secr: Minna Pitkänen
Focus: Poli Sci
Periodical
Politiikka Géorges (weekly) 04082

France

Academia Ophthalmologica Internationalis, 6 Rue Sainte-Claire, 81000 Albi
T: 63541509; Fax: 63541509
Founded: 1975; Members: 49
Gen Secr: Dr. Pierre Amalric
Focus: Ophthal 04083

Académie Commerciale Internationale, 43 Rue de Tocqueville, 75017 Paris
T: (1) 47665134
Focus: Econ 04084

Académie d'Agriculture de France, 18 Rue de Bellechasse, 75007 Paris
T: (1) 47051037; Fax: (1) 45550978
Founded: 1761; Members: 370
Pres: Pierre Zert; Gen Secr: André Cauderon
Focus: Agri
Periodical
Comptes rendus de l'Académie d'Agriculture de France 04085

Académie d'Architecture, 9 Pl des Vosges, 75004 Paris
T: (1) 8878310, 8878574
Founded: 1840; Members: 100
Pres: Paul La Mache; Gen Secr: Emmanuel Besnard-Bernadac
Focus: Archit 04086

Académie d'Arles, c/o Musée Arlaten, Rue de la République, 13200 Arles
Focus: Sci 04087

Académie de Chirurgie, 26 Blvd Raspail, 75007 Paris
T: (1) 45482254
Founded: 1935; Members: 120
Gen Secr: Georges Cerbonnet
Focus: Surgery
Periodical
Chirurgie (weekly) 04088

Académie de la Réunion, c/o DRAC, 31 Rue Amiral Lacaze, 97400 Saint-Denis
T: 219171
Founded: 1913; Members: 25
Pres: Serge Ycard; Gen Secr: Y. Drouhet
Focus: Sci
Periodical
Bulletin 04089

Académie de Marine, 3 Av Octave Gréard, 75007 Paris
Founded: 1752; Members: 60
Pres: Paul Bastard; Gen Secr: Jacques Chatelle
Focus: Navig
Periodical
Communications et Mémoires (3 times annually) 04090

Académie de Nîmes, 16 Rue Dorée, 30000 Nîmes
Founded: 1682; Members: 60
Gen Secr: Jean Ménard
Focus: Sci
Periodicals
Bulletin Trimestriel (3 times annually)
Mémoires (weekly) 04091

Académie de Pharmacie de Paris, 4 Av de l'Observatoire, 75270 Paris Cédex 06
T: (1) 43255449; Fax: (1) 43290592
Founded: 1803; Members: 345
Pres: Paul Lechat; Gen Secr: Paul Rossignol
Focus: Pharmacol 04092

Académie des Beaux-Arts, 23 Quai de Conti, 75006 Paris
Founded: 1803; Members: 65
Pres: Henry Bernard; Gen Secr: Marcel Landowski
Focus: Arts
Periodical
Annuaire (weekly) 04093

Académie des Belles-Lettres, Sciences et Arts de La Rochelle, 30 Rue Gargoulleau, 17000 La Rochelle
Founded: 1732; Members: 50
Focus: Lit; Hist 04094

Académie des Inscriptions et Belles-Lettres, 23 Quai de Conti, 75006 Paris
T: (1) 3269282
Founded: 1663; Members: 145
Gen Secr: Jean Leclant
Focus: Lit
Periodicals
Journal des Savants
Mémoires 04095

Académie des Jeux Floraux, Hôtel d'Assézat et Clémence Isaure, Pl d'Assézat, 31000 Toulouse
T: 61212285
Founded: 1324; Members: 65
Gen Secr: Jean Sermet
Focus: Sci; Lit
Periodical
Recueil de l'Académie des Jeux Floraux (weekly) 04096

Académie des Lettres, Sciences et Arts d'Amiens, 60 Blvd de Saint-Quentin, 80090 Amiens
Founded: 1750
Focus: Sci; Lit; Arts 04097

Académie des Sciences, 23 Quai de Conti, 75006 Paris
T: (1) 44414367
Founded: 1666; Members: 370
Focus: Sci
Periodical
Comptes-Rendus de l'Académie des Sciences 04098

Académie des Sciences, Agriculture, Arts et Belles-Lettres d'Aix, c/o Musée P. Arbaud, 2a Rue du 4 Septembre, 13100 Aix-en-Provence
T: 42383895
Founded: 1829; Members: 150
Pres: Maurice Wolkonitsch; Gen Secr: Georges Souville
Focus: Arts; Agri; Sci; Lit; Hist
Periodical
Bulletin (weekly) 04099

Académie des Sciences, Arts et Belles-Lettres de Dijon, 5 Rue de l'Ecole de Droit, 21000 Dijon
Founded: 1740; Members: 550
Pres: Pierre Rat; Gen Secr: Martine Chauney
Focus: Arts; Sci; Lit; Hist
Periodical
Mémoires (bi-annually) 04100

Académie des Sciences, Belles-Lettres et Arts de Clermont, 19 Rue Bardoux, 63000 Clermont-Ferrand
Founded: 1747; Members: 542
Focus: Sci; Lit; Arts 04101

Académie des Sciences, Belles-Lettres et Arts de Lyon, Palais Saint-Jean, 4 Av Adolphe Max, 69005 Lyon
T: 78382654
Founded: 1700; Members: 52
Focus: Sci; Lit; Arts
Periodical
Mémoires de l'Académie (weekly) . . . 04102

Académie des Sciences, Belles-Lettres et Arts de Rouen, 190 Rue Beauvoisine, 76000 Rouen
Founded: 1744; Members: 50
Focus: Sci; Lit; Arts 04103

Académie des Sciences, Belles-Lettres et Arts de Savoie, Château, 73000 Chambéry
Focus: Sci; Lit; Arts 04104

Académie des Sciences d'Outre-Mer (ASOM), 15 Rue La Pérouse, 75116 Paris
T: (1) 47208793
Founded: 1922; Members: 275
Pres: Pierre Gentil; Gen Secr: Gilbert Mangin
Focus: Geography; Hist; Law
Periodicals
Hommes et Destins
Mondes et Cultures (weekly) 04105

Académie des Sciences et Lettres de Montpellier, c/o Bibliothèque Interuniversitaire, 4 Rue Ecole Mage, 34000 Montpellier
Founded: 1706
Gen Secr: Henri Vidal
Focus: Sci; Lit
Periodical
Bulletin de l'Académie des Sciences et Lettres de Montpellier (weekly) 04106

Académie des Sciences, Lettres et Arts d'Arras, c/o Archives Départementales, 12 Pl de la Préfecture, 62000 Arras
Focus: Sci; Lit; Arts
Periodical
Mémoires (weekly) 04107

Académie des Sciences, Lettres et Arts de Marseille, c/o Castel France, Saint-Loup, 13010 Marseille
Founded: 1715; Members: 40
Focus: Sci; Lit; Arts 04108

Académie des Sciences Morales et Politiques, 23 Quai de Conti, 75006 Paris
T: (1) 44414326; Fax: (1) 44414327
Founded: 1832; Members: 122
Pres: Jean Imbert; Gen Secr: Bernard Chenot
Focus: Philos; Poli Sci; Law; Econ
Periodical
Revue des Sciences Morales et Politiques: Nouvelle Série (weekly) 04109

Académie Diplomatique Internationale (A.D.I.), 4bis Av Hoche, 75008 Paris
T: (1) 42276618
Founded: 1926; Members: 586
Focus: Poli Sci 04110

Académie Française, 23 Quai de Conti, 75006 Paris
T: (1) 44414300; Fax: (1) 43294745
Founded: 1635; Members: 40
Focus: Ling; Lit
Periodicals
Annuaire de l'Académie Française (weekly)
Discours de réception des académiciens 04111

Académie Goncourt, c/o Drouant, Pl Gaillon, 75002 Paris
Founded: 1896; Members: 10
Pres: Hervé Bazin; Gen Secr: F. Nourissier
Focus: Lit 04112

Académie Internationale de Science Politique et d'Histoire Constitutionnelle, c/o Université de Paris I, 12 Pl du Panthéon, 75000 Paris
Founded: 1936
Focus: Poli Sci; Hist 04113

Académie Mallarmé, 38 Rue du Faubourg Saint-Jacques, 75014 Paris
Members: 30
Pres: Eugène Guillevic; Gen Secr: Jean Orizet
Focus: Lit 04114

Académie Montaigne, Le Doyenné, 72140 Sillé-le-Guillaume
Founded: 1924; Members: 20
Gen Secr: Constant Hubert
Focus: Lit 04115

Académie Nationale de Chirurgie Dentaire, 22 Rue Emile Ménier, 75116 Paris
T: (1) 47046540
Members: 190
Focus: Surgery
Periodical
Bulletin (weekly) 04116

Académie Nationale de Danse, 1 Rue Keller, 75011 Paris
T: (1) 48053262
Focus: Music 04117

Académie Nationale de Médecine, 16 Rue Bonaparte, 75272 Paris Cédex 06
T: (1) 43269680
Founded: 1820; Members: 130
Pres: René Küss; Gen Secr: Prof. André Lemaire
Focus: Med
Periodical
Bulletin (9 times annually) 04118

Académie Nationale de Metz, 20 en Nexirue, 57000 Metz
Focus: Sci; Hist
Periodicals
Bibliographie Lorraine (weekly)
Mémoires de l'Académie Nationale de Metz (weekly) 04119

Académie Nationale des Sciences, Belles-Lettres et Arts de Bordeaux, 1 Pl Bardineau, 33000 Bordeaux
Founded: 1712; Members: 40
Focus: Sci; Lit; Arts 04120

Académie Polonaise des Sciences, 74 Rue Lauriston, 75116 Paris
T: (1) 45536691
Focus: Sci
Periodical
Conférences (weekly) 04121

Académie Vétérinaire de France, 60 Blvd Latour-Maubourg, 75007 Paris
Founded: 1844; Members: 44
Gen Secr: M. Catsaras
Focus: Vet Med
Periodical
Bulletin (weekly) 04122

Administration Universitaire Francophone et Européenne en Médecine et Odontologie (AUFEMO) / Association of French-Speaking and European University Administrations in Medicine and Odontology, c/o Faculté de Médecine Cochin-Port Royal, 24 Rue du Faubourg Saint-Jacques, 75674 Paris Cédex 14
T: 1982
Pres: G. Vicente
Focus: Med; Dent
Periodicals
Arcanes (weekly)
La Chronique de l'AUFEMO (8 times annually)
Report (weekly) 04123

Advisory Group for Aerospace Research and Development (AGARD), 7 Rue Ancelle, 92200 Neuilly-sur-Seine
T: (1) 47385700; Tx: 610176; Fax: (1) 47385799
Founded: 1952; Members: 16
Focus: Aero 04124

African Technical Association (ATA), 23 Rue du Rocher, 75008 Paris
T: (1) 42942275; Tx: 282110; Fax: (1) 42942782
Founded: 1981
Gen Secr: Pierre Henri Bousez
Focus: Eng 04125

Agence Internationale de l'Energie (AIE) / International Energy Agency, 2 Rue André Pascal, 75775 Paris Cédex 16
Founded: 1974; Members: 21
Gen Secr: Helga Steeg
Focus: Energy
Periodicals
Monthly Oil Market Report (weekly)
Quarterly Oil and Gas Statistics (weekly) 04126

Agence Spatiale Européenne (ASE) / European Space Agency, 8-10 Rue Mario Nikis, 75738 Paris Cédex 15
T: (1) 42737654; Tx: 202746; CA: Spaceurop, Paris
Founded: 1966
Focus: Astronomy; Aero
Periodicals
Bulletin ESA (weekly)
Journal ESA (weekly)
Rapport Annuel (weekly) 04127

Agronomes sans Frontières, 43 Rue de la Glacière, 75013 Paris
T: (1) 43360362
Founded: 1985
Pres: Marcel Mazoyer
Focus: Agri 04128

American European Dietetic Association (AEDA), 26 Rue Charles Constantin, 78360 Montesson
Founded: 1978
Focus: Nutrition
Periodical
AEDA News (weekly) 04129

Amicale des Directeurs de Bibliothèques Universitaires (ADBU), 5 Rue Auguste Vacquerie, 75116 Paris
T: (1) 47230012
Founded: 1970; Members: 60
Focus: Libraries & Bk Sci
Periodical
Lettre de l'ADBU (weekly) 04130

Amicale Internationale de Phytosociologie, c/o Centre de Phytosociologie, Hameau de Haendries, 59270 Bailleul
T: 28490083
Founded: 1982; Members: 29
Gen Secr: Prof. Jean-Marie Genu
Focus: Botany 04131

Amici Thomae Mori (ATM), BP 808, 49008 Angers Cédex 01
Fax: 41887442
Founded: 1962; Members: 900
Focus: Lit; Hist; Rel & Theol
Periodicals
Gazette (weekly)
Moreana (weekly) 04132

Amis de Guy de Maupassant, 60 Rue Vaneau, 75007 Paris
Focus: Lit 04133

Amis de Rimbaud, c/o Pierre Petitfils, 13 Rue P.-L. Courier, 75007 Paris
T: (1) 45483420
Founded: 1927; Members: 350
Focus: Lit 04134

Arab Music Rostrum, c/o IMC, 1 Rue Miollis, 75732 Paris Cédex 15
T: (1) 45682550; Tx: 204461; Fax: (1) 43068798
Founded: 1990
Focus: Music 04135

Arab World Institute, 23 Quai Saint-Bernard, 75005 Paris
T: (1) 40513838
Founded: 1980; Members: 21
Gen Secr: Bassem El-Jisr
Focus: Cultur Hist
Periodical
Al-Moukhtarat (3 times annually) 04136

Asbestos International Association (AIA), 10 Rue de la Pépinière, 75008 Paris
T: (1) 45221400; Tx: 281133; Fax: (1) 42949886
Founded: 1976
Gen Secr: Daniel Bouige
Focus: Materials Sci
Periodical
AIA Newsletter (weekly) 04137

Asian Music Rostrum (ASMR), c/o UNESCO, 1 Rue Miollis, 75732 Paris Cédex 15
T: (1) 45682550; Tx: 204461; Fax: (1) 43068798
Founded: 1969
Focus: Music 04138

Asociación de Historadores Latinoamericanos y del Caribe (ADHILAC) / Association of Historians of Latin America and the Caribbean, c/o Université de Paris III, SHEAL, 28 Rue Saint-Guillaume, 75007 Paris
T: (1) 42223593
Founded: 1974
Gen Secr: F. Mauro
Focus: Hist
Periodicals
ADHILAC Informa
Bolletin ADHILAC 04139

Asociata Internationalà de Studi Romåne / International Association for Romanian Studies, 146 Blvd de Magenta, 75010 Paris
Gen Secr: Catherine Durandin
Focus: Ethnology 04140

Asocio de Studato Internacia pri Spiritaj kaj Teologiaj Instruoj (ASISTI), 7 Rue Parmentier, 18000 Bourges
Founded: 1989
Gen Secr: Christian Lavarenne
Focus: Rel & Theol
Periodical
Asistilo (weekly) 04141

Associated European Human Biological Centres, 21 Quai Montebello, 75005 Paris
T: (1) 46334034
Founded: 1963
Gen Secr: Dr. Georges Jaeger
Focus: Bio 04142

Associated Schools Project in Education for International Cooperation and Peace (ASP), c/o UNESCO, 7 Pl de Fontenoy, 75700 Paris
T: (1) 46681000; Tx: 204461; Fax: (1) 45680816
Founded: 1953
Focus: Educ
Periodicals
A Glimpse of the Associated Schools Project (weekly)
International Understanding at School (weekly)
. . . . 04143

Association Aéronautique et Astronautique de France (AAAF), 6 Rue Galilée, 75782 Paris Cédex 16
Founded: 1972; Members: 1800
Focus: Aero
Periodical
L'Aéronautique et l'Astronautique (weekly) 04144

Association Anthroposophique en France, 68 Rue de Caumartin, 75009 Paris
T: (1) 42810470
Focus: Philos 04145

Association Catéchétique Nationale pour l'Audio Visuel, 3 Rue Amyot, 75005 Paris
T: (1) 45872611
Focus: Educ
Periodical
Auvimages (5 times annually) 04146

Association Centrale des Vétérinaires (A.C.V.), 10 Pl Léon Blum, 75011 Paris
T: (1) 43562102
Founded: 1889; Members: 2500
Pres: Dr. J.-P. Marty; Gen Secr: Dr. R.A. Moal
Focus: Vet Med 04147

Association de Documentation pour l'Industrie Nationale, 28 Rue Notre-Dame de Nazareth, 75003 Paris
T: (1) 42785332; Tx: (1) 42741213
Founded: 1911; Members: 380
Gen Secr: Serge Bouchval
Focus: Doc 04148

Association de Gérontologie du 13e, 49 Rue Bobillot, 75013 Paris
T: (1) 45888814
Focus: Geriatrics 04149

Association de l'Economie des Institutions, 107 Av Henri Martin, 75016 Paris
T: (1) 45045755
Focus: Econ 04150

Association de l'Education Morale de la Jeunesse, 8 Rue Saint-Simon, 75007 Paris
T: (1) 45482097
Focus: Educ 04151

Association de Médecine Rurale (AMR), 60 Blvd Latour-Maubourg, 75007 Paris
T: (1) 40896104
Founded: 1950; Members: 1800
Focus: Med 04152

Association de Médecine Urbaine (AMU), 37 Rue de Bellefond, 75009 Paris
T: (1) 42850536
Founded: 1960
Focus: Med 04153

Association d'Enseignement Féminin Professionnel et Ménager, 64 Rue d'Assas, 75006 Paris
T: (1) 45489125
Focus: Adult Educ; Home Econ 04154

Association Dentaire Française (ADF), 92 Av de Wagran, 75017 Paris
T: (1) 42278900
Founded: 1970; Members: 15
Focus: Dent 04155

Association de Psychologie du Travail de Langue Française / Association of French-Speaking Psychologists of the Working Environment, 3 Rue Antoine Coypel, 78000 Versailles
T: 39512770
Founded: 1980
Pres: Prof. Pierre Goguelin
Focus: Psych 04156

Association de Psychologie Scientifique de Langue Française, c/o Institut de Psychologie, 28 Rue Serpente, 75006 Paris
Founded: 1960
Focus: Psych 04157

Association de Réadaption Psychopédagogique et Scolaire, 13 Rue de la Grange Batelière, 75009 Paris
T: (1) 47704050
Focus: Educ 04158

Association de Recherche et d'Etudes Catéchétiques, 172 Rue Raymond Losserand, 75014 Paris
T: (1) 45431497
Focus: Rel & Theol 04159

Association de Recherche et d'Expression dans l'Art, 36bis Rue Ballu, 75009 Paris
T: (1) 42850475
Focus: Arts 04160

Association de Recherches Universitaires Géographiques et Cartographiques (AUREG), 191 Rue Saint-Jacques, 75005 Paris
T: (1) 46340130
Founded: 1962; Members: 18
Focus: Geography; Cart 04161

Association des Actuaires Diplomés de l'Institut de Science Financière et d'Assurances, 243 Rue Saint-Honoré, 75001 Paris
Focus: Insurance 04162

Association des Amis d'Alfred de Vigny, 6 Av Constant Coquelin, 75007 Paris
T: (1) 42731286
Pres: Christiane Lefranc
Focus: Lit
Periodical
Bulletin (weekly) 04163

Association des Amis de Miguel Angel Asturias / Association of the Friends of Miguel Angel Asturias, 5 Rue Chabanais, 75002 Paris
T: (1) 47259234 ext 661
Founded: 1973
Focus: Lit
Periodicals
AAEA Journal (weekly)
AAEA Newsletter 04164

Association des Anatomistes (CNRS), BP 184, 54505 Vandoeuvre-les-Nancy
T: 83446065
Founded: 1899; Members: 1005
Gen Secr: Prof. G. Grignon
Focus: Anat
Periodical
Bulletin (weekly) 04165

Association des Archivistes Français, 60 Rue des Francs-Bourgeois, 75141 Paris Cédex 03
T: (1) 40276000
Founded: 1904; Members: 550
Pres: Michel Maréchal; Gen Secr: M.P. Arnauld; E. Gautier-Desvaux
Focus: Archives
Periodical
La Gazette des Archives (weekly) . . . 04166

Association des Artistes Sculpteurs, Architectes, Graveurs et Dessinateurs, 1 Rue La Bruyère, 75009 Paris
T: (1) 48748524
Focus: Archit; Arts 04167

Association des Auteurs de Films, 5 Rue Ballu, 75009 Paris
T: (1) 48745755
Focus: Cinema 04168

Association des Bibliothécaires Français (ABF), 7 Rue des Lions Saint-Paul, 75004 Paris
T: (1) 48879787; Fax: (1) 48879713
Founded: 1906; Members: 2500
Pres: Françoise Danset; Gen Secr: Cathéerine Schmitt
Focus: Libraries & Bk Sci
Periodical
Bulletin d'Informations (weekly) 04169

Association des Bibliothèques de Judaïca et Hébraïca en Europe / Association of Libraries of Judaica and Hebraica in Europe, c/o Bibliothèque de l'Alliance Israélite Universelle, 45 Rue La Bruyère, 75009 Paris
Founded: 1955; Members: 20
Focus: Libraries & Bk Sci 04170

Association des Bibliothèques Ecclésiastiques de France (ABEF), 6 Rue de Regard, 75006 Paris
Founded: 1963; Members: 190
Focus: Libraries & Bk Sci
Periodical
Bulletin (weekly) 04171

Association des Centres Médicaux de la Seine, 132 Rue du Faubourg Saint-Denis, 75010 Paris
T: (1) 46073187
Focus: Med 04172

Association des Chimistes, Ingénieurs et Cadres des Industries Agricoles et Alimentaires, 2 Rue Oratoire, 75001 Paris
T: (1) 42974138; Fax: (1) 42601198
Founded: 1883; Members: 2000
Pres: S. Touruère
Focus: Chem
Periodical
Revue IAA (10 times annually) 04173

Association des Conservateurs de Bibliothèques (ACB), 16 Rue Claude Bernard, 75231 Paris Cédex 05
T: (1) 44081786; Fax: (1) 44081784
Founded: 1992; Members: 600
Pres: Philippe Dupont; Gen Secr: Marie-Hélène Koenig
Focus: Libraries & Bk Sci
Periodicals
ABC Infos (weekly)
Annuaire des Associations de Bibliothécaires et Documentalistes (bi-annually) 04174

Association des Dermatologistes et Syphiligraphes de Langue Française / Association of French-Speaking Dermatologists and Syphilligraphers, c/o Service de Dermatologie, Hôpital Henri Mondor, Av du Mal de Lattre de Tassigny, 94010 Créteil
Founded: 1923
Gen Secr: Prof. J. Revuz
Focus: Derm; Venereology
Periodical
Annales de Dermatologie 04175

Association des Diplômés de l'Ecole de Bibliothécaires-Documentalistes, c/o Bibliothèque du Saulchoir, 43bis Rue de la Glacière, 75013 Paris
T: (1) 45870533
Founded: 1938
Focus: Libraries & Bk Sci; Doc
Periodical
Bulletin d'Information (weekly) 04176

Association des Directeurs des Centres Universitaires d'Administration des Entreprises, 162 Rue Saint-Charles, 75015 Paris
T: (1) 45583963
Focus: Business Admin 04177

Association des Ecoles de Service Social de la Région Parisienne, 250 Blvd Raspail, 75014 Paris
T: (1) 40336866
Focus: Educ; Sociology 04178

Association des Ecrivains Combattants, 8 Rue Roquépine, 75008 Paris
T: (1) 42650430
Founded: 1919; Members: 470
Focus: Lit 04179

Association des Ecrivains de Langue Française (A.D.E.L.F.), 14 Rue Broussais, 75014 Paris
T: (1) 43219599; Fax: (1) 43201222
Founded: 1926; Members: 2000
Pres: Edmond Jouve; Gen Secr: Simone Dreyfus
Focus: Lit
Periodical
Lettres et Cultures de Langue Française (weekly)
. . . . 04180

Association des Epidémiologists de Langue Française (ADELF), c/o INSERM U 292, Hôpital de Bicêtre, 78 Rue du General Leclerc, 94275 Le Kremlin-Bicêtre
T: (1) 45212251; Fax: (1) 45212075
Founded: 1976
Gen Secr: Dr. Nadine Job-Spira
Focus: Med 04181

Association des Etudes Tsiganes / Association for Gypsy Studies, 2 Rue d'Hautpoul, 75019 Paris
T: (1) 40400905
Founded: 1949
Focus: Ethnology
Periodical
Etudes Tsiganes (weekly) 04182

Association des Françaises Diplômées des Universités, 4 Rue de Chevreuse, 75006 Paris
T: (1) 43200132
Focus: Sci 04183

Association des Géographes de l'Est, 23 Blvd Albert Ier, BP 3397, 54015 Nancy Cédex
T: 83961614
Founded: 1961
Pres: Henri Nonn; Gen Secr: Claude Seyer
Focus: Geography
Periodical
Revue Géographique de l'Est (weekly) . 04184

Association des Géographes Français (AGF), 191 Rue Saint-Jacques, 75005 Paris
T: (1) 43290147
Founded: 1920; Members: 480
Pres: B. Bomer; Gen Secr: A. Metton
Focus: Geography
Periodical
Bulletin de l'Association de Géographes Français (5 times annually) 04185

Association des Hautes Etudes Hospitalières, 83-87 Av d'Italie, 75013 Paris
T: (1) 45885847
Founded: 1962
Focus: Public Admin; Public Health
Periodical
Magazine (weekly) 04186

Association des Industriels de France contre les Accidents du Travail (AIF), 10 Rue de Calais, 75441 Paris Cédex 09
T: (1) 48740220
Founded: 1883; Members: 6000
Focus: Safety 04187

Association des Informaticiens de Langue Française / Association of French-Speaking Computer Professionals and Users, 61 Rue de Vaugirard, 75006 Paris
Founded: 1981
Focus: Computer & Info Sci 04188

Association des Internes et Anciens Internes en Pharmacie des Hôpitaux de Nancy, 7 Rue Albert Lebrun, BP 403, 54001 Nancy Cédex
Founded: 1948; Members: 130
Focus: Pharmacol 04189

Association des Langues et Civilisations, Rue de Lille, 75007 Paris
T: (1) 42606705
Focus: Ling; Ethnology 04190

Association des Léprologues de Langue Française / Association of French-Language Leprologists, c/o Hôpital Saint-Louis, 2 Pl du Dr. A. Fournier, 75475 Paris Cédex 10
T: (1) 42494949
Founded: 1959
Gen Secr: Dr. Daniel Wallach
Focus: Derm
Periodical
Acta Leprologica (weekly) 04191

Association des Palynologues de Langue Française (APLF), c/o Institut de Géologie, 1 Rue de Blessig, 67084 Strasbourg
Founded: 1967
Focus: Geology 04192

Association des Pédiatres de Langue Française, c/o Service de Pédiatrie Néonatale, Hôpital A. Béclère, 92140 Clamart
Focus: Pediatrics 04193

Association des Physiologistes, c/o Prof. P. Delost, Laboratoire de Physiologie Animale, Ensemble Scientifique des Cézeaux, Université de Clermont, BP 45, 63170 Aubière
Founded: 1926; Members: 900
Focus: Physiology 04194

Association des Professeurs de Langues Vivantes de l'Enseignement Public (APLV), c/o Institut National de Recherche et de Documentation Pédagoqiques, 29 Rue d'Ulm, 75230 Paris Cédex 05
Members: 5500
Focus: Educ; Ling 04195

Association des Professeurs de Mathématiques de l'Enseignement Publique (APMEP), 13 Rue du Jura, 75013 Paris Cédex 05
Founded: 1910; Members: 13000
Focus: Math; Educ 04196

Association des Professeurs d'Histoire et de Géographie de l'Enseignement Public (A.P.H.G.), BP 831, 91001 Evry Cédex
Founded: 1910; Members: 12500
Focus: Educ; Hist; Geography
Periodical
Historiens et Gèographes (5-6 times annually) 04197

Association des Professionnels de l'Information et de la Documentation, 25 Rue Claude Tiller, 75012 Paris
T: (1) 43722525; Fax: (1) 43723041
Founded: 1963; Members: 5000
Pres: Jean Michel; Gen Secr: Olivier Thiebeauld
Focus: Libraries & Bk Sci; Doc
Periodical
Documentaliste – Sciences de l'Information (5 times annually) 04198

Association de Transfusion Sanguine, 6 Rue Alexandre Cabanel, 75015 Paris
T: (1) 45676905
Focus: Hematology 04199

Association d'Etudes pour l'Expansion de la Recherche Scientifique, 29 Rue d'Ulm, 75005 Paris
Focus: Sci 04200

Association d'Etudes Techniques des Industries de l'Estampage et de la Forge (ADETIEF TECHNIQUE), 9 Rue Pierre le Grand, 75008 Paris
T: (1) 47664761
Focus: Eng 04201

Association du Sport Scolaire et Universitaire, 13 Rue Saint-Lazare, 75009 Paris
T: (1) 48783423
Focus: Sports 04202

Association Episcopale Catéchistique (A.E.C.), 4 Av Vavin, 75006 Paris
T: (1) 43201343
Focus: Rel & Theol 04203

Association Européenne de Recherches et d'Echanges Pédagogiques (ASEREP), 28 Rue de l'Aude, 31500 Toulouse
T: 61341042; Fax: 341373
Founded: 1989
Gen Secr: Christian Raynal
Focus: Educ 04204

Association Européenne des Conservatoires, Académies de Musique et Musikhochschulen, c/o Conservatoire National de Région d'Angers, 26 Av Montaigne, 49100 Angers
T: 41886659; Fax: 41875281
Founded: 1953; Members: 111
Pres: Sir John Manduell; Gen Secr: John Richard Lowry
Focus: Music 04205

Association Européenne des Enseignants, Section Française (AEDE), BP 24507, 75327 Paris Cédex
Founded: 1956
Pres: Prof. Y.-H. Nouailhat
Focus: Educ
Periodical
Education européenne (5 times annually) 04206

Association Européenne de Thermographie (AET) / European Thermographic Association, c/o Laboratoire d'Electroradiologie, Faculté de Médecine, 11 Rue Humann, 67085 Strasbourg Cédex
T: 88360691
Founded: 1972
Gen Secr: Dr. Michel Gautherie
Focus: Physical Therapy
Periodical
Bulletin de l'AET (weekly) 04207

Association Européenne pour l'Administration de la Recherche Industrielle (EIRMA) / European Industrial Research Management Association, 38 Cours Albert Ier, 75008 Paris
T: (1) 42256044
Founded: 1966; Members: 155
Focus: Eng
Periodicals
Conference Papers (weekly)
Working Group Reports (weekly) 04208

Association Européenne pour l'Etude de l'Alimentation et du Développement de l'Enfant (ADE), c/o Hôpital Trousseau, 26 Av Dr. Arnold Netter, 75012 Paris
T: (1) 43461390 ext 3739; Fax: (1) 43467800
Founded: 1976; Members: 100
Pres: Dr. Z.L. Ostrowski
Focus: Educ; Nutrition
Periodical
ADE Bulletin 04209

Association for the Expansion of International Roles of the Languages of Continental Europe / Association pour l'Expansion du Rôle International des Langues d'Europe Continentale, 25 Blvd Jules Sandeau, 75116 Paris
T: (1) 45031003
Founded: 1987
Gen Secr: Alexandre Rousset
Focus: Ling
Periodical
Europe en Péril? (weekly) 04210

Association Française d'Acupuncture, 23 Rue Clapeyron, 75008 Paris
T: (1) 47024260
Focus: Anesthetics 04211

Association Française d'Astronomie (AFA), Observatoire du Parc Montsouris, 17 Rue Emile Deutsch de la Meurthe, 75014 Paris
T: (1) 45898144
Founded: 1946; Members: 12000
Focus: Astronomy
Periodical
Ciel et Espace (weekly) 04212

Association Française de Chirurgie (AFC), 26 Blvd Raspail, 75007 Paris
Founded: 1884; Members: 2000
Focus: Surgery 04213

Association Française de Formation, 63 Blvd des Batignolles, 75017 Paris
T: (1) 42935281
Focus: Adult Educ 04214

Association Française de Formation, Coopération, Promotion et Animation d'Entreprises, 12 Av Marceau, 75008 Paris
T: (1) 42253762
Focus: Adult Educ 04215

Association Française de Gemmologie (AFG), 14 Rue Cadet, 75009 Paris
Fax: (1) 42465016
Founded: 1964; Members: 1100
Focus: Mineralogy
Periodical
Revue de Gemmologie (weekly) 04216

Association Française de Génie Rural (AFGR), 30 Rue Las Cases, 75340 Paris Cédex 07
T: (1) 45559550
Founded: 1947; Members: 164
Focus: Agri
Periodical
Génie Rural: Column (weekly) 04217

Association Française de l'Eclairage (AFE), 52 Blvd Malesherbes, 75008 Paris
T: (1) 43872121; Fax: (1) 43871698
Founded: 1930; Members: 2000
Focus: Electric Eng
Periodicals
Lux: La Revue de l'Eclairage
Recommandations Relatives à l' Eclairage (weekly) 04218

Association Française de l'Ecole Paysanne, 11 Rue de Clichy, 75009 Paris
T: (1) 48746848
Focus: Educ 04219

Association Française de Management (CNOF), 119 Rue de Lille, 75007 Paris
T: (1) 45443880
Founded: 1926; Members: 1150
Focus: Business Admin; Econ
Periodical
Management France (weekly) 04220

Association Française de Normalisation (AFNOR) / French Association for Standardization, Tour Europe, 92080 Paris La Défence Cedex 07
T: (1) 42915555; Fax: (1) 42915656
Founded: 1926; Members: 5500
Pres: Philippe Boulin; Gen Secr: Bernard Vaucelle
Focus: Standards
Periodical
Enjeux (weekly) 04221

Association Française de Prévention des Accidents de Travail et Incendie, 17 Rue Salneuve, 75017 Paris
T: (1) 47665151
Focus: Safety 04222

Association Française de Psychiatrie, 23 Rue Pradier, 92410 Ville d'Avray
T: (1) 47091177; Fax: (1) 47506275
Founded: 1979; Members: 2200
Focus: Psychiatry
Periodicals
La Lettre de Psychiatrie Française (weekly)
Psychiatrie Française 04223

Association Française de Recherches d'Essais sur les Matériaux et les Constructions (AFREM), 12 Rue Brancion, 75015 Paris
T: (1) 45392233
Founded: 1948; Members: 65
Focus: Materials Sci; Civil Eng 04224

Association Française des Arabisants, 40 Av d'Iena, 75116 Paris
Founded: 1973; Members: 300
Focus: Ethnology 04225

Association Française de Science Economique (A.F.S.E.), c/o Université des Scienas Sociales, Pl Anatole France, 31042 Toulouse Cédex
Founded: 1952; Members: 700
Focus: Econ 04226

Association Française de Science Politique (AFSP), 224 Blvd Saint-Germain, 75007 Paris
T: (1) 45499221; Fax: (1) 45442027
Founded: 1949; Members: 985
Focus: Poli Sci
Periodical
Revue Française de Science Politique (weekly) 04227

Association Française des Enseignants de Français, 19 Rue des Martyrs, 75009 Paris
T: (1) 45264141; Fax: (1) 40169338
Founded: 1967; Members: 4000
Focus: Educ; Ling
Periodical
Le Francais Aujourd'hui: Le Supplément au Français Aujourd'hui (weekly) 04228

Association Française des Femmes Médecins, 123 Rue de Lille, 75007 Paris
T: (1) 45512929
Focus: Med 04229

Association Française des Hémophiles, 6 Rue Alexandre Cabanel, 75015 Paris
T: (1) 45677767; Fax: (1) 45678544
Pres: Dr. Patrick Wallet
Focus: Hematology
Periodical
L'Hémophile (3 times annually) 04230

Association Française des Historiens Economistes, 54 Blvd Raspail, 75006 Paris
Founded: 1965; Members: 150
Focus: Hist; Econ 04231

Association Française des Ingénieurs, Chimistes et Techniciens des Industries du Cuir (AFICTIC), BP 7003, 69342 Lyon Cédex 07
Founded: 1939; Members: 300
Gen Secr: Michel Aloy
Focus: Chem; Eng
Periodical
Revue Technique des Industries du Cuir 04232

Association Française des Instituts de Recherche sur le Développement (AFIRD), c/o Institut d'Etude du Développement Economique et Social, 58 Blvd Arago, 75013 Paris
T: (1) 43362355
Founded: 1975; Members: 45
Focus: Econ 04233

Association Française des Professeurs de Langues Vivantes, 19 Rue de la Glacière, 75013 Paris
T: (1) 47079482
Founded: 1902; Members: 4000
Pres: François Monnanteuil; Gen Secr: Sylvestre Vanuxem
Focus: Ling
Periodical
Les Langues Modernes (weekly) 04234

Association Française des Sociétés d'Etudes et de Conseils Exportatrices, 3 Rue Leon Bonnat, 75016 Paris
T: (1) 45244353
Focus: Econ 04235

Association Française d'Etude des Relations Professionnelles (C.N.A.M.), 33 Rue du Regard, 94260 Fresnes
Founded: 1969
Pres: Prof. J.C. Javillier
Focus: Sociology 04236

Association Française d'Etudes Américaines (AFEA), 1 Pl de l'Odéon, 75006 Paris
T: (1) 46330248
Founded: 1967; Members: 300
Focus: Ethnology 04237

Association Française d'Observateurs d'Etoiles Variables, c/o Observatoire Astronomique, 11 Rue de l'Université, 67000 Strasbourg
T: 88843711; Tx: 890506
Founded: 1921; Members: 110
Pres: E. Schweitzer; Gen Secr: D. Proust; M. Verdenet
Focus: Astronomy
Periodical
Bulletin (weekly) 04238

Association Française du Froid (AFF), 17 Rue Guillaume Appolinaire, 75006 Paris
Fax: (1) 62220042
Founded: 1908
Pres: Roland Violot; Gen Secr: Henri Minault
Focus: Air Cond; Eng
Periodical
Revue Générale du Froid (weekly) . . . 04239

Association Française Inter Médicale, 2 Rue de Clichy, 75009 Paris
T: (1) 48747329
Focus: Med 04240

Association Française pour la Protection des Eaux (AFPE), 195 Rue Saint-Jacques, 75005 Paris
T: (1) 43267053
Founded: 1960; Members: 1100
Focus: Ecology 04241

Association Française pour la Recherche et la Création Musicales, 9 Rue Chaptal, 75009 Paris
Founded: 1971
Pres: F. Bayle; Gen Secr: F. Dela
Focus: Music 04242

Association Française pour le Développement de la Stomatologie, c/o Institut de Stomatologie, Faculté de Médecine, 47 Blvd de l'Hôpital, 75013 Paris
T: (1) 43364927
Focus: Stomatology 04243

Association Française pour l'Etude des Eaux (AFEE), c/o Office International de l'Eau, Rue Edouard Chamberland, 87065 Limoges Cédex
T: 55114780; Fax: 55777224
Founded: 1949; Members: 450
Focus: Water Res
Periodical
Bulletin (11 times annually) 04244

Association Française pour l'Etude du Cancer (A.F.E.C.), 26 Rue d'Ulm, 75231 Paris Cédex 05
T: (1) 43291242
Founded: 1907; Members: 300
Focus: Cell Biol & Cancer Hes
Periodical
Bulletin du Cancer (5 times annually) . 04245

Association Française pour l'Etude du Quaternaire (A.F.E.Q.), c/o M. Clet, Centre Géomorphologie, CNRS, Rue des Tilleuls, 14000 Caen
T: 31455724; Fax: 31455757
Founded: 1962; Members: 650
Pres: J.P. Lautridou; Gen Secr: M. Clet
Focus: Geology
Periodical
Quaternaire: Bulletin de l'Association Française pour l'Etude du Quaternaire (weekly) . . 04246

Association Française pour l'Etude du Sol (A.F.E.S.), 4 Rue Redon, 78370 Plaisir
Founded: 1934; Members: 950
Pres: J.C. Remy; Gen Secr: M.C. Girard
Focus: Agri; Geomorph
Periodicals
Etude et Gestion des Sols (weekly)
European Journal of Soil Science (weekly) 04247

Association Française pour l'Information en Economie Ménagère, 11 Rue Toussaint Feron, 75013 Paris
T: (1) 40334551
Focus: Home Econ 04248

Association Francophone d'Education Comparée (AFEC), 1 Av Léon Journault, 92310 Sèvres
T: (1) 45327527
Founded: 1973
Focus: Educ
Periodicals
Education Comparée (weekly)
Etudes d'Education Comparée (weekly) . 04249

Association Francophone de Spectrométrie des Masses Solides, c/o Centre National d'Etudes des Télécommunications, 22301 Lannion
Founded: 1975
Focus: Physics 04250

Association Francophone Internationale des Groupes d'Animation de la Paraplégie (AFIGAP), c/o Centre de Rééducation Neurologique, Rte de Liverdy, Coubert, 77170 Brie-Comte-Robert
T: 64255810
Founded: 1979
Gen Secr: Dr. Marc Maury
Focus: Rehabil 04251

Association Générale des Conservateurs des Collections Publiques de France (AGCCPFI), 6 Rue des Pyramides, 75041 Paris Cédex 01
T: (1) 40153649; Fax: (1) 40153640
Founded: 1922; Members: 920
Pres: Geneviève Becquart; Gen Secr: Jean-Marcel Humbert
Focus: Arts
Periodical
Musées et Collections Publiques de France (weekly) 04252

Association Générale des Hygiénistes et Techniciens Municipaux (AGHTM), 9 Rue de Phalsbourg, 75017 Paris
T: (1) 44151550
Founded: 1905; Members: 1350
Focus: Hygiene; Eng
Periodical
Techniques, Sciences, Methodes (weekly) 04253

Association Générale des Médecins de France (A.G.M.F.), 60 Blvd de Latour-Maubourg, 75007 Paris Cédex 07
T: (1) 47054528
Pres: P. Baudouin; Gen Secr: Dr. A. Tochard
Focus: Med
Periodical
Bulletin 04254

Association Guillaume Budé (CSSF), 95 Blvd Raspail, 75006 Paris
T: (1) 42226915
Founded: 1917; Members: 3200
Focus: Lit 04255

Association Internationale de Droit Pénal (AIDP), BP 1146, 64011 Pau Cédex
Fax: 59807559
Founded: 1924
Gen Secr: Prof. R. Ottenhof
Focus: Law
Periodicals
Nouvelles Etudes Pénales (11 times annually)
Revue Internationale de Droit Pénal (weekly) 04256

Association Internationale de Géodésie / International Association of Geodesy, 140 Rue de Grenelle, 75700 Paris
Founded: 1922; Members: 78
Pres: Prof. J.J. Mueller; Gen Secr: M. Louis
Focus: Surveying 04257

Association Internationale de l'Inspection du Travail (AIIT) / International Association of Labour Inspections, c/o Direction Régionale du Travail, 66 Rue de Mouzaia, 75935 Paris Cédex 19
T: (1) 42003300; Tx: 220254
Founded: 1972
Focus: Business Admin 04258

Association Internationale de Littérature Comparée / International Comparative Literature Association, c/o Université de Paris-Sorbonne, 1 Rue Victor-Cousin, 75230 Paris Cédex 05
T: (1) 32911213
Founded: 1955; Members: 2800
Pres: Prof. E. Miner; Gen Secr: José Lambert; Prof. Lore Metzger
Focus: Lit 04259

Association Internationale de Médecine et de Biologie de l'Environnement (AIMBE), 115 Rue de la Pompe, 75116 Paris
T: (1) 45534504; Tx: 614584; CA: ECOMEBIO-Paris; Fax: (1) 45534175
Founded: 1971
Focus: Med; Bio; Ecology 04260

Association Internationale de Pédiatrie / Asociación Internacional de Pediatría / International Pediatric Association, Château de Longchamp, Bois de Boulogne, 75016 Paris
T: (1) 45271590; Tx: 648379; CA: Internaped Paris; Fax: (1) 45257367
Founded: 1912; Members: 110
Pres: Prof. Perla D. Santos-Ocampo; Gen Secr: Prof. Ihsan Dogramaci
Focus: Pediatrics
Periodical
International Child Health (weekly) . . . 04261

Association Internationale d'Epigraphie Grecque et Latine (A.I.E.G.L.), c/o Bibliothèque de la Sorbonne, 47 Rue des Ecoles, 75230 Paris Cédex 05
Founded: 1963
Gen Secr: Olivier Masson
Focus: Ling
Periodical
Nouvelles de l'A.I.E.G.L. 04262

Association Internationale de Recherche en Informatique Toxicologique (AIRIT), c/o CAP, Hôpital Fernand Widal, 200 Rue du Faubourg Saint-Denis, 75475 Paris
Founded: 1973
Pres: Prof. A. Efthymiou
Focus: Computer & Info Sci; Toxicology 04263

Association Internationale des Amis de Vasile Stanciu, 49 Blvd de Port Royal, 75013 Paris
T: (1) 60841151
Founded: 1987
Focus: Criminology 04264

Association Internationale des Arts Plastiques (AIAP), c/o UNESCO, 1 Rue Miollis, 75732 Paris Cédex 15
T: (1) 45682655
Founded: 1952; Members: 80
Gen Secr: C. Rubalcava
Focus: Arts
Periodical
Art-Journal of the Professional Artist (weekly) 04265

Association Internationale des Critiques de Théâtre (AICT), 6 Rue de Braque, 75003 Paris
Founded: 1956
Focus: Perf Arts
Periodical
Prospectus (3 times annually) 04266

Association Internationale des Critiques Littéraires (AICL), 58 Rue Claude-Bernard, 75005 Paris
Fax: (1) 43549299
Founded: 1969; Members: 700
Pres: Robert André
Focus: Lit
Periodical
Revue de l'AICL (weekly) 04267

Association Internationale des Démographes de Langue Française (AIDELF) / International Association of French-Language Demographers, 27 Rue du Commandeur, 75675 Paris Cédex 14
T: (1) 43201345; Fax: (1) 43277240
Founded: 1977
Gen Secr: Alain Parant
Focus: Soc Sci 04268

Association Internationale des Docteurs en Economie du Tourisme, c/o CHET, 8 Perspectives Mozart, Parc Mozart, 13100 Aix-en-Provence
Gen Secr: Prof. Dr. René Baretje
Focus: Business Admin 04269

Association Internationale des Ecoles des Sciences de l'Information (AIESI), c/o Institut d'Etudes Politiques, Cycle Information et Documentation, 27 Rue Saint-Guillaume, 75341 Paris Cédex 07
T: (1) 45495155
Founded: 1977; Members: 23
Pres: Martine Prévot-Hubert
Focus: Educ; Computer & Info Sci . . . 04270

Association Internationale des Educateurs de Jeunes Inadaptés (AIEJI), 3 Rue Pierre Brossolette, 64000 Pau
T: 59025962; Fax: 59026283
Founded: 1951; Members: 92
Focus: Educ 04271

Association Internationale des Etudes de l'Asie du Sud-Est (AIEAS), 22 Av du Président Wilson, 75116 Paris
T: (1) 45537301
Founded: 1973
Gen Secr: Prof. P.B. Lafont
Focus: Ethnology
Periodical
Lettre d'Information (weekly) 04272

Association Internationale des Etudes Françaises (AIEF), 11 Pl Marcelin Berthelot, 75005 Paris
T: (1) 43291211
Founded: 1949
Gen Secr: Prof. Robert Garapon
Focus: Ling; Lit
Periodical
Cahiers (weekly) 04273

Association Internationale des Professeurs de Langue et Littérature Russes / International Association of Teachers of Russian Language and Literature, c/o Société Française des Professeurs de Russe, 9 Rue Michelet, 75006 Paris
Founded: 1967
Focus: Ling; Lit; Educ 04274

Association Internationale des Professeurs et Maîtres de Conférences des Universités, 21 Rue Isabey, 54000 Nancy
T: 83281221
Founded: 1945
Gen Secr: L.-P. Laprévote
Focus: Sci 04275

Association Internationale des Sciences Juridiques (AISJ) / International Association of Legal Science, c/o CISS-UNESCO, 1 Rue Miollis, 75015 Paris
T: (1) 45682558; Fax: (1) 43068798
Founded: 1950
Gen Secr: M. Leker
Focus: Law 04276

Association Internationale des Sociologues de Langue Française (AISLF), c/o Université de Toulouse-le-Mirail, 5 Allée Antonio Machado, 31058 Toulouse Cédex
T: 61504374
Founded: 1956; Members: 1040
Gen Secr: Christiane Rondi
Focus: Sociology
Periodicals
Annuaire de l'AISLF
Bulletin de l'AISLF (weekly) 04277

Association Internationale des Statisticiens d'Enquêtes (AISE) / International Association of Survey Statisticians, 18 Blvd Adolphe Pinard, Bureau 1059, 75675 Paris Cédex 14
T: (1) 41175300
Founded: 1973; Members: 1435
Gen Secr: A. Charraud
Focus: Stats
Periodicals
International Statistical Information
International Statistical Review
Survey Statistician (weekly) 04278

Association Internationale des Techniciens Biologistes de Langue Française (ASSITEB), 109 Av Gabriel Peri, 94170 Le Perreux
T: 48711451
Founded: 1982
Pres: Nelly Marchal
Focus: Eng
Periodicals
Le Courier de l'ASSITEB
L'Information du Technicien Biologiste . . 04279

Association Internationale des Travaux en Souterrain (AITES) / International Tunnelling Association, 109 Av S. Allende, 69674 Bron Cédex
T: 78260455; Tx: 370008; Fax: 72372406
Founded: 1974
Gen Secr: C. Benenguier
Focus: Civil Eng
Periodical
Tunnelling and Underground Space Technology (weekly) 04280

Association Internationale des Universités (AIU) / International Association of Universities, 1 Rue Miollis, 75732 Paris Cédex 15
T: (1) 45682545; Fax: (1) 47347605
Founded: 1950; Members: 650
Pres: Dr. Walter Kamba; Gen Secr: Dr. Franz Eberhard
Focus: Educ
Periodicals
Bulletin (weekly)
Higher Education Policy (weekly) 04281

Association Internationale d'Histoire Economique / International Economic History Association, c/o Centre de Recherches Historiques, Ecole des Hautes Etudes en Sciences Sociales, 54 Blvd Raspail, 75270 Paris Cédex 06
T: (1) 49542440; CA: GENSECRAIHE Paris
Founded: 1965
Gen Secr: Prof. Joseph Goy
Focus: Hist; Econ 04282

Association Internationale d'Information Scolaire, Universitaire et Professionnelle (AIISUP), 20 Rue de l'Estrapade, 75005 Paris
T: (1) 43541027
Founded: 1956
Focus: Computer & Info Sci
Periodicals
Bibliographical Bulletin
Informations Universitaires et Professionnelles Internationales (weekly) 04283

Association Internationale d'Irradiation Industrielle (AIII) / Association of International Industrial Irradiation, 59 Rte de Paris, 69260 Charbonnières-les-Bains
T: 78871165; Tx: 370015; Fax: 78878831
Founded: 1970
Pres: Pierre E. Vidal
Focus: Eng
Periodical
Newsletter (weekly) 04284

Association Internationale d'Océanographie Médicale / International Association of Medical Oceanography, c/o CERBOM, Parc de la Côte, Av Jean Lorrain, 06000 Nice
Founded: 1975
Focus: Oceanography 04285

Association Internationale Données pour le Développement (DPD) / Data for Development International Association, 122 Av de Hambourg, 13008 Marseille
T: 91739018; Tx: 430258; Fax: (1) 91730138
Founded: 1975; Members: 198
Focus: Computer & Info Sci
Periodical
Data for Development Newletters (weekly) 04286

Association Internationale Francophone de Recherche Odontologique (AIFRO), c/o Faculté de Chirurgie Dentaire, 2 Pl Pasteur, 35000 Rennes
Founded: 1977
Gen Secr: E. Sinard-Savoie
Focus: Dent 04287

Association Internationale pour le Développement de l'Odonto-Stomatologie Tropicale / International Association for the Development of Tropical Odonto-Stomatology, c/o UER d'Odontologie, 146 Rue Léo Saignat, 33076 Bordeaux Cédex
T: 56909124
Founded: 1988
Gen Secr: Christine Canet
Focus: Stomatology; Dent
Periodical
Revue d'Odonto-Stomatologie Tropicale . 04288

Association Internationale pour le Développement des Gommes Naturelles / Association of International Development of Natural Gums, 4 Rue Frédéric Passy, 92200 Neuilly-sur-Seine
T: 47471850; Fax: 47471891
Founded: 1974
Focus: Materials Sci 04289

Association Internationale pour le Développement des Universités Internationales et Mondiales (AIDUM) / International Association for the Development of International and World Universities, 31 Av de Mun, 93600 Aulnay-sous-Bois
T: 48666698
Founded: 1953
Gen Secr: Marc Bullio
Focus: Sci
Periodical
Memoranda 04290

Association Internationale pour le Management du Sport, BP 149, 91240 Saint-Michel
T: 45391237
Founded: 1987
Pres: Robert Trottein
Focus: Business Admin 04291

Association Internationale pour l'Etude de la Paléontologie Humaine, c/o Institut de Paléontologie Humaine, 1 Rue René Panhard, 75013 Paris
T: (1) 43316291
Founded: 1982
Gen Secr: Prof. Henry de Lumley
Focus: Paleontology 04292

Association Internationale pour l'Etude du Comportement des Conuducteurs / International Drivers Behaviour Research Association, 6 Av Hoche, 75008 Paris
T: (1) 42679717; Tx: 643058; Fax: (1) 42279803
Founded: 1970
Gen Secr: T.E.A. Benjamin
Focus: Psych 04293

Association Interprofessionnelle de France (AINF), Rue Marcel Dassault, Zone Industrielle, BP 259, 59472 Seclin Cédex
T: 20329900; Fax: 20325166
Founded: 1894; Members: 3500
Pres: C. Bloch
Focus: Safety 04294

Association Interprofessionnelle des Centres Médicaux et Sociaux de la Région Parisienne (A.C.M.S.), 132 Rue du Faubourg Saint-Denis, 75010 Paris
T: (1) 42091800
Focus: Med; Sociology 04295

Association Interprofessionnelle pour la Formation Permanente dans le Commerce Textile, 69 Rue de Richelieu, 75001 Paris
T: (1) 2975493
Focus: Adult Educ; Textiles 04296

Association Les Amis de Gustave Courbet, c/o Musée Maison Natale Gustave Courbet, Pl Robert Fernier, 25290 Ornans
Fax: (1) 45244700
Focus: Arts
Periodical
Bulletin des Amis de Gustave Courbet (weekly) 04297

Association Linguistique Franco-Européenne (A.L.F.E.), 117 Rue de Rennes, 75006 Paris
T: (1) 45484563
Focus: Ling 04298

Association Littéraire et Artistique Internationale (ALAI), 55 Rue des Mathurins, 75008 Paris
T: (1) 42650445
Founded: 1878
Gen Secr: André Françon
Focus: Arts; Lit
Periodical
Bulletin de l'ALAI 04299

Association Lyonnaise de Criminologie et Anthropologie Sociale, 12 Av Rockefeller, 69008 Lyon
T: 72344858; Fax: 72120075
Founded: 1967
Focus: Criminology; Anthro 04300

Association Marc Bloch, 54 Blvd Raspail, 75006 Paris
Founded: 1949
Focus: Hist 04301

Association Maria Montessori, 51 Av Bugeaud, 75016 Paris
T: (1) 47235230
Focus: Educ 04302

Association Médico-Sociale Protestante de Langue Française, 68 Rue Marcel Bourdarias, 94140 Alfortville
T: 48992783, 43760666; Fax: 48980334
Founded: 1948
Gen Secr: Pierre-Yves Le Renard
Focus: Soc Sci; Med
Periodical
Ouvertures (weekly) 04303

Association Mondiale des Vétérinaires Microbiologistes, Immunologistes et Spécialistes des Maladies Infectieuses (A.M.V.M.I.), c/o Service de Microbiologie, Ecole Nationale Vétérinaire d'Alfort, 7 Av du Général de Gaulle, 94704 Maisons Alfort Cédex
T: (1) 43759211
Founded: 1967; Members: 240
Pres: Prof. Pilet
Focus: Vet Med
Periodical
Comparative Immunology Microbiology and infection disease (weekly) 04304

Association Nationale de la Recherche Technique (A.N.R.T), 101 Av Raymond Poincaré, 75116 Paris
T: (1) 45017227; Fax: (1) 47042520
Founded: 1953; Members: 264
Pres: G. Worms
Focus: Eng
Periodical
La Lettre Européenne du Progrés Technique (10 times annually) 04305

Association Nationale des Assistants de Service Social (ANAS), 15 Rue de Bruxelles, 75009 Paris
T: (1) 45220698
Focus: Public Admin
Periodical
Revue Française de Service Social (weekly) 04306

Association Nationale des Bibliothécaires, c/o Mlle Pintaparis, Cité Administrative, 17 Blvd Morland, 75004 Paris
Focus: Libraries & Bk Sci 04307

Association Nationale des Cours Professionnels pour les Préparateurs en Pharmacie, 41 Blvd de Magenta, 75010 Paris
T: (1) 42065900
Focus: Adult Educ; Pharmacol 04308

Association Nationale des Docteurs en Droit, 38bis Rue Fabert, 75007 Paris
T: (1) 47051165
Members: 4000
Focus: Law
Periodical
Droit et Economie (3 times annually) . . 04309

Association Nationale des Educateurs de Jeunes Inadaptés (ANEJ), 27 Rue de Maubeuge, 75009 Paris
T: (1) 48783917
Founded: 1947; Members: 2000
Focus: Educ 04310

Association Nationale des Professeurs en Economie Sociale et Familiale, 28 Pl Saint-Georges, 75009 Paris
T: (1) 42800782
Focus: Sociology; Educ 04311

Association Nationale pour la Formation et la Promotion Professionnelle dans l'Industrie et le Commerce de la Chaussure et des Cuirs et Peaux (A.F.P.I.C.), 5 Rue Joseph Sansboeuf, 75008 Paris
T: (1) 45222888
Focus: Adult Educ 04312

Association Nationale pour la Protection des Eaux, 195 Rue Saint-Jacques, 75005 Paris
T: (1) 43267053
Founded: 1960; Members: 800
Focus: Ecology 04313

Association Nationale pour la Protection des Villes d'Art, 39 Av de La Motte-Picquet, 75007 Paris
T: (1) 47053771
Founded: 1963
Pres: J. de Sacy
Focus: Preserv Hist Monuments 04314

Association Nationale pour la Réhabilitation Professionnelle par le Travail Protégé, 59 Blvd de Belleville, 75011 Paris
T: (1) 43571365
Focus: Rehabil 04315

Association Nationale pour l'Etude de la Neige et des Avalanches (ANENA), 15 Rue Ernest Calvat, 38000 Grenoble
T: 76513939
Founded: 1971; Members: 670
Pres: Jean-Guy Cupillard; Gen Secr: J. Lafeuille; F. Valla
Focus: Geology
Periodical
Neige et Avalanches (weekly) 04316

Association of African Geological Surveys / Association des Services Géologiques Africains, c/o CIFEG, Av de Concyr, BP 6517, 45065 Orléans
T: 38643657; Tx: 780258; Fax: 38643518
Founded: 1929
Focus: Geology
Periodical
African Geology 04317

Association of European Schools and Colleges of Optometry / Association Européenne des Ecoles et Collèges d'Optométrie, 134 Rte de Chartres, 91440 Bures-sur-Yvette
T: 69076737; Fax: 69287806
Founded: 1979; Members: 58
Gen Secr: Jean-Paul Roosen
Focus: Ophthal
Periodical
Communications (5 times annually) . . . 04318

Association of French-Speaking Planetariums, c/o Planetarium de Strasbourg, Université Louis Pasteur, Rue de l'Observatoire, 67000 Strasbourg
T: 88361251
Gen Secr: Agnes Acker
Focus: Astronomy 04319

Association Philotechnique, 47 Rue Saint-André des Arts, 75006 Paris
T: (1) 43264828
Focus: Ling 04320

Association pour la Fondation Internationale du Cinéma et de la Communication Audiovisuelle (AFICCA), 1 Av du Roi Albert, 06400 Cannes
T: 93940777; Fax: 93438895
Founded: 1987
Gen Secr: Antoine Virenque
Focus: Comm Sci; Cinema
Periodical
Script 04321

Association pour la Formation aux Professions Immobilières, 2 Impasse Mont Tonnerre, 75015 Paris
T: (1) 45672082
Focus: Adult Educ 04322

Association pour la Formation des Cadres de l'Industrie et du Commerce, 31 Av Pierre Ier de Serbie, 75784 Paris Cédex 16
T: (1) 47236158
Focus: Adult Educ 04323

Association pour la Formation Professionnelle dans les Industries Céréalières (A.F.P.I.C), 66 Rue La Boétie, 75008 Paris
Focus: Adult Educ 04324

Association pour la Formation Professionnelle dans les Industries de l'Ameublement (A.F.P.I.A.), 28 Av Daumesnil, 75012 Paris
T: (1) 43075907
Focus: Adult Educ 04325

Association pour la Médiathèque Public (AMP), c/o Bibliothèque Municipale, 37 Rue Saint-Georges, 59400 Cambrai
T: 27813520
Focus: Libraries & Bk Sci; Doc
Periodical
Médiathèques Publiques (weekly) 04326

Association pour la Prévention de la Pollution Atmosphérique (APPA), 58 Rue du Rocher, 75008 Paris
T: (1) 42936930
Founded: 1958; Members: 1000
Focus: Ecology 04327

Association pour la Promotion de la Pédagogie Nouvelle, 11 Rue Coetlogon, 75006 Paris
T: (1) 45445576
Focus: Educ 04328

Association pour la Recherche et le Développement en Informatique Chimique, 25 Rue Jussieu, 75005 Paris
T: (1) 43362525
Focus: Chem 04329

Association pour la Recherche Interculturelle (ARIC), c/o UFR de Psychologie, Université de Paris VIII, 2 Rue de la Liberté, 93526 Saint-Denis Cédex
Founded: 1984
Focus: Cultur Hist 04330

Association pour la Rééducation de la Parole et du Langage Oral et Ecrit et de la Voix (A.R.P.L.O.E.V.), 10 Rue de l'Arrivée, 75015 Paris
T: (1) 45444885
Founded: 1956; Members: 600
Focus: Logopedy
Periodical
Rééducation Orthophonique (weekly) . . 04331

Association pour le Développement de la Formation Professionnelle Continue dans les Industries Lourdes du Bois, 30 Av Marceau, 75008 Paris
T: (1) 42561732, 42561739
Focus: Adult Educ 04332

Association pour le Développement de la Formation Professionnelle dans les Transports, 46 Av de Villiers, 75017 Paris
T: (1) 47660360
Focus: Adult Educ 04333

Association pour le Développement de la Recherche en Toxicologie Expérimentale, 4 Av de l'Observatoire, 75006 Paris
T: (1) 43267122
Focus: Toxicology 04334

Association pour le Développement de la Stomatologie, 47 Blvd de l'Hôpital, 75013 Paris
T: (1) 43364927
Focus: Stomatology 04335

Association pour le Développement de la Traduction Automatique et de Linguistique Appliqué, 11 Rue de Lille, 75007 Paris
Focus: Ling 04336

Association pour le Développement de l'Enseignement et des Recherches Scientifiques auprès des Universités de la Région Parisienne (A.D.E.R.P.), 31 Av Pierre Ier de Serbie, 75116 Paris
T: (1) 47236158
Focus: Sci 04337

Association pour le Développement des Etudes Biologiques en Psychiatrie, 1 Rue Cabanis, 75014 Paris
T: (1) 45814557
Focus: Bio; Psychiatry 04338

Association pour le Développement des Relations Médicales entre la France et les Pays Etrangers, 16 Rue Bonaparte, 75272 Paris Cédex 06
Focus: Med 04339

Association pour le Développement des Techniques de Transport, d'Environnement et de Circulation (ATEC), 38 Av Emile Zola, 75015 Paris
T: (1) 45755611
Founded: 1973; Members: 1330
Focus: Transport
Periodical
T.E.C. (weekly) 04340

Association pour le Développement du Droit Mondial, 28 Rue Saint-Guillaume, 75007 Paris
Focus: Law 04341

Association pour l'Enseignement de l'Assurance, 8 Rue Chaptal, 75009 Paris
T: (1) 48747539
Focus: Insurance; Educ
Periodical
La Lettre d'Information (weekly) 04342

Association pour l'Innovation Scientifique, 41 Rue de Vaugirard, 75006 Paris
T: (1) 45480789
Focus: Sci 04343

Association Psychoanalytique de France, 24 Pl Dauphine, 75001 Paris
T: (1) 43298511; Fax: (1) 43261346
Members: 40
Focus: Psychoan 04344

Association Régionale d'Education Permanente, 22 Rue de Varenne, 75007 Paris
T: (1) 42220025
Focus: Educ 04345

Association Régionale des Oeuvres Educatives et des Vacances de l'Education Nationale (A.R.O.E.V.E.N.), 18 Passage Turquetil, 75011 Paris
T: (1) 43704501
Focus: Educ 04346

Association Régionale d'Informations Sociales, 30 Rue Gramont, 75002 Paris
T: (1) 42962336
Focus: Sociology 04347

Association Scientifique des Médecins Acupuncteurs de France (A.S.M.A.F.), 2 Rue du Général de Larminat, 75015 Paris
T: (1) 42733726
Founded: 1945; Members: 1500
Pres: Dr. Georges Cantoni; Gen Secr: Dr. H. Olivo
Focus: Anesthetics
Periodical
Méridiens (weekly) 04348

Association Stomatologique Internationale (ASI) / International Stomatological Association, 66 Rue La Boétie, 75008 Paris
Founded: 1907
Gen Secr: Prof. Jean Lakermance
Focus: Stomatology 04349

Association Technique de la Fonderie (ATF), 2 Rue de Bassano, 75116 Paris
T: (1) 47235550
Founded: 1911; Members: 2800
Focus: Metallurgy
Periodical
Hommes et Fondérie (weekly) 04350

Association Technique de la Réfrigération et de l'Equipement Ménager (A.T.R.E.M.), 28 Av Hoche, 75008 Paris
T: (1) 47662134, 5630200
Members: 4
Focus: Eng; Air Cond 04351

Association Technique de la Sidérurgie Française (ATS), Elysées la Défense, 19 Le Parvis, 92072 Paris la Défense Cédex 35
T: (1) 47678588; Tx: 611672; Fax: (1) 47678577
Founded: 1946; Members: 1100
Focus: Metallurgy 04352

Association Technique de l'Industrie du Gaz en France (ATG), 62 Rue de Courcelles, 75008 Paris
T: (1) 47660351
Founded: 1874; Members: 2150
Focus: Energy
Periodicals
Gaz d'Aujourd'hui (10 times annually)
Médiagaz (11 times annually) 04353

Association Technique de l'Industrie Papetière (ATIP), 154 Blvd Haussmann, 75008 Paris
T: (1) 42271191; Fax: (1) 45635309
Founded: 1947; Members: 1320
Focus: Eng
Periodical
Revue ATIP (6-8 times annually) . . . 04354

Association Technique Maritime et Aéronautique (ATMA), 47 Rue Monceau, 75008 Paris
T: (1) 45619911
Founded: 1889; Members: 1050
Focus: Eng; Aero
Periodical
Bulletin ATMA (weekly) 04355

Association Technique pour l'Etude de la Gestion des Institutions Publiques et des Entreprises Privées (ATEGIPE), 2 Av de la Bourbonnais, 75007 Paris
Founded: 1950; Members: 220
Focus: Business Admin 04356

Association Universitaire pour la Diffusion Internationale de la Recherche, 27 Rue Saint-Guillaume, 75341 Paris Cédex 07
Founded: 1972; Members: 25
Focus: Sci 04357

Association Universitaire pour le Développement de l'Enseignement et de la Culture en Afrique et à Madagascar (AUDECAM), 100 Rue de l'Université, 75007 Paris
T: (1) 45555638
Focus: Adult Educ 04358

Astronomical Data Centre, c/o Observatoire de Strasbourg, 11 Rue de l'Université, 67000 Strasbourg
T: 88358218; Tx: 890506; Fax: 88250160
Founded: 1972
Pres: Dr. Michel Crézé
Focus: Astronomy
Periodical
Bulletin d'Information du CDS (weekly) . 04359

Atlantic Gas Research Rxchange (AGRE), c/o Gaz de France, 361 Av du Président Wilson, BP 33, 93211 La Plaine-Saint-Denis
Founded: 1978
Gen Secr: Dr. Jean Pottier
Focus: Energy 04360

Bureau International de Documentation (BILD), 13 Av de l'Opéra, 75001 Paris
T: (1) 42603200
Focus: Doc 04361

Bureau International de Liaison et de Documentation (BILD), 50 Rue de Laborde, 75008 Paris
Founded: 1945
Pres: Joseph Rovan; Gen Secr: Horst Reinert
Focus: Int'l Relat
Periodicals
Documents
Revue des Questions Allemandes . . . 04362

Bureau International des Poids et Mesures (BIPM), Pavillon de Breteuil, 92310 Sèvres
T: (1) 45077070; Tx: 631351; Fax: (1) 45342021
Founded: 1875; Members: 47
Pres: Dr. D. Kind; Gen Secr: Dr. T.J. Quinn
Focus: Standards
Periodicals
Procès-Verbaux des séances du CIPM (weekly)
Sessions des Comités Consultatifs (8): Electricité, Photométrie et Radiométrie, Thermométrie, Définition du Mètre, Définition de la Seconde, Etalons des Mesure des Reyonnement Ionisants, Unités, Masse et grandeurs apparentées (weekly) 04363

Center for Studies, Research and Training in International Understanding and Cooperation (CERFCI), c/o FMACU-WFUCA, 1 Rue Miollis, 75015 Paris
T: (1) 45682818
Focus: Soc Sci; Econ 04364

Centre de Coopération Internationale en Recherche Agronomique pour le Développement (CIRAD), 42 Rue Scheffer, 75116 Paris
T: (1) 47043215; Tx: 648729; Fax: (1) 47551530
Founded: 1984
Focus: Agri
Periodical
Agritrop Tropical and Subtropical (weekly) 04365

Centre de Coopération pour les Recherches Scientifiques Relatives au Tabac (CORESTA) / Cooperation Centre for Scientific Research Relative to Tobacco, 53 Quai d'Orsay, 75347 Paris Cédex 07
T: (1) 45566019
Founded: 1956; Members: 130
Pres: Günther Hayn; Gen Secr: F. Jacob
Focus: Agri; Chem; Standards; Tobacco
Periodical
Bulletin du CORESTA (weekly) *04366*

Centre de Perfectionnement des Industries Textiles Rhône-Alpes (CEPITRA), 55 Montée de Choulans, 69323 Lyon Cédex 1
T: (1) 4285165
Focus: Eng; Textiles *04367*

Centre de Rencontres et d'Etudes des Dirigeants des Administrations Fiscales (CREDAF), 51 Rue de Rome, 75008 Paris
T: (1) 43874433; Tx: 280220; Fax: (1) 43874433
Founded: 1982
Gen Secr: Jacques Carion
Focus: Business Admin *04368*

Centre d'Etudes, de Documentation, d'Information et d'Action Sociales (CEDIAS), 5 Rue Las Cases, 75007 Paris
T: (1) 45516610, 47059246
Founded: 1963
Pres: Dr. M. Montalembert; Gen Secr: Brigitte Bouquet
Focus: Sociology; Doc; Poli Sci; Hist
Periodicals
Manuel de Placement (weekly)
Vie Sociale (weekly) *04369*

Centre d'Histoire Militaire et d'Etudes de Défense Nationale (CHMM), c/o Université Paul Valery, BP 5043, 34032 Montpellier Cédex
Founded: 1938
Focus: Hist; Military Sci
Periodical
Les Cahiers de Montpellier: Forces Armées et Politiques de Défense (weekly) *04370*

Centre Européen de Formation des Statisticiens Economistes des Pays en Voie de Développement (CESD), 3 Av Pierre Larousse, 92241 Malakoff Cédex
T: (1) 45401007; Tx: 204924; Fax: (1) 40921191
Founded: 1962
Gen Secr: L. Diop
Focus: Stats *04371*

Centre Européen de Recherches sur les Congrégations et Ordres Religieux (CERCOR), c/o Maison Rhône-Alpes des Sciences de l'Homme, 35 Rue du 11 Novembre, 42023 Saint-Etienne
T: 77389667
Founded: 1981; Members: 1400
Pres: Prof. Marcel Pacaut
Focus: Rel & Theol *04372*

Centre for the Advancement and Study of the European Currency, 16 Av Berthelot, 69007 Lyon
T: 72732820; Tx: 330949; Fax: 72734604
Founded: 1982; Members: 39
Pres: Michel Coste
Focus: Finance *04373*

Centre Français de Droit Comparé, 28 Rue Saint-Guillaume, 75007 Paris
T: (1) 42223593
Pres: Prof. Jacques Robert; Gen Secr: Prof. Xavier Blanc-Jouvan
Focus: Law
Periodical
Lettre du Centre Français de Droit Comparé (3 times annually) *04374*

Centre International de Cyto-Cybernétique (C.I.C. CYB.), 9 Av Niél, 75017 Paris
Founded: 1966; Members: 1280
Pres: Prof. Dr. Ernest Huant
Focus: Cybernetics *04375*

Centre International de l'Enfance, Château de Longchamp, Bois de Boulogne, 75016 Paris
T: (1) 45207992; Tx: 648379; Fax: (1) 45257367
Founded: 1949
Pres: Prof. Claude Griscelli; Gen Secr: Michèle Puybasset
Focus: Educ
Periodicals
Bulletins bibliographiques
L'Enfant en Milieu Tropical: Children's in the Tropics (weekly)
El Niño en el Tropico *04376*

Centre International d'Etude des Textiles Anciens (CIETA), 34 Rue de la Charité, 69002 Lyon
Founded: 1954; Members: 235
Focus: Hist; Textiles
Periodical
Bulletin du CIETA (weekly) *04377*

Centre International d'Etudes Latines (C.I.E.L.), 25 Rue du Mont Thabor, 75001 Paris
T: (1) 42608484
Focus: Ling *04378*

Centre International d'Etudes Romanes (CIER), 43 Rue Boissonade, 75014 Paris
T: (1) 43542476
Founded: 1952; Members: 500
Focus: Archeol *04379*

Centre International du Film pour l'Enfance et la Jeunesse (CIFEJ) / International Centre of Films for Children and Young People, 9 Rue Bargue, 75015 Paris
Founded: 1957
Pres: P. Golubovic; Gen Secr: L. Galandrin
Focus: Cinema
Periodicals
News from ICFCYP
Young Cinema International (weekly) . . *04380*

Centre International Scolaire de Correspondance Sonore (CISCS), Chanteloup, 10300 Sainte-Savine
Focus: Educ *04381*

Centre National de la Recherche Scientifique (CNRS), 20-22 Rue Saint-Armand, 75015 Paris
T: (1) 45331600; Tx: 200356; Fax: (1) 45339213
Founded: 1939
Focus: Sociology; Humanities
Periodicals
Annuaire de l'Afrique du Nord (weekly)
Annuaire des Pays de l'Océan Indien (weekly)
Annuaire Français de Droit International (weekly)
Antiquités Africaines (weekly)
Archaeonautica (weekly)
Bibliographie Annuelle de l'Histoire de France (weekly)
Cahiers de Micropaléontologie (weekly)
Cahiers du Seminaire d'Econométrie (weekly)
Etudes Celtiques (weekly)
Gallia (weekly)
Hermes (weekly)
Paleorient (weekly)
Paroisses et Communes de France (weekly)
Patrimoine au Présent (weekly)
Revue Archéologie Médiévale (weekly)
Revue Archéologique de l'Est et du Centre-Est (weekly)
Revue Archéologique de Narbonnaise (weekly)
Les Voies de la Création Théatrale (weekly) *04382*

Centre National d'Etudes Spatiales, 2 Pl Maurice Quentin, 75001 Paris
Founded: 1961
Pres: J.-L. Lions; Gen Secr: F. d' Allest
Focus: Aero *04383*

Centre Naturopa, c/o Conseil de l'Europe, 67075 Strasbourg Cédex
T: 88412000; Tx: 870943; Fax: 88412715
Focus: Ecology
Periodicals
Environment Features (weekly)
Library Bulletin (weekly)
Naturopa (3 times annually)
Naturopa-Newsletter (weekly) *04384*

Centre Technique et de Promotion des Laitiers Sidérurgiques (C.T.P.L.), 4 Pl de la Pyramide, 92070 Paris La Défense Cédex 33
T: (1) 49005800; Fax: (1) 49005858
Founded: 1948; Members: 3
Focus: Dairy Sci
Periodical
Revue Laitiers Sidérurgiques (2-3 times annually) *04385*

Cercle d'Etudes Architecturales, 38 Blvd Raspail, 75006 Paris
T: (1) 45483144
Focus: Archit *04386*

Cercle d'Etudes Pédiatriques (CEP), 153 Rue de Saussure, 75017 Paris
T: (1) 42671215
Founded: 1967; Members: 85
Pres: Dr. Jean Feigelson; Gen Secr: Dr. Marie F. Merklen
Focus: Pediatrics *04387*

Chambre Syndicale des Sociétés d'Etudes et de Conseils, 3 Rue Léon Bonnat, 75016 Paris
T: (1) 45244353
Focus: Sci
Periodicals
Lettre de l'Ingénieur (weekly)
Lettre de Syntec-Informatique (weekly) . *04388*

Club Européen d'Histoire de la Neurologie, c/o Service de Neurologie, Hôpital de l'Antiquaille, 1 Rue de l'Antiquaille, 69321 Lyon Cédex 05
T: 78258090
Gen Secr: M. Boucher
Focus: Neurology *04389*

Club of Rome (COR), 34 Rue d'Eylau, 75116 Paris
T: (1) 47044525; Fax: (1) 47044523
Founded: 1968; Members: 100
Gen Secr: Bertrand Schneider
Focus: Econ; Poli Sci; Soc Sci *04390*

Collège International pour l'Etude Scientifique des Techniques de Production Mécanique (CIRP), 10 Rue Mansart, 75009 Paris
T: (1) 45262180; Tx: 281029
Founded: 1950; Members: 433
Gen Secr: M. Véron
Focus: Eng
Periodical
CIRP Annals (weekly) *04391*

Comité d'Education Sanitaire et Sociale de la Pharmacie Française, 4 Av Ruysdael, 75008 Paris
T: (1) 46225428
Focus: Adult Educ; Pharmacol *04392*

Comité d'Entente des Ecoles de Formation en Economie Sociale Familiale, 28 Pl Saint-Georges, 75009 Paris
T: (1) 42800883
Focus: Educ; Sociology *04393*

Comité d'Etudes et de Liaison des Amendements Calcaires (C.E.L.A.C.), 30 Av de Messine, 75008 Paris
T: (1) 45630266; Fax: (1) 53750213
Focus: Agri *04394*

Comité d'Etudes et de Liaison Interprofessionnel de la Haute-Marne (C.E.L.I.H.M.), Résidence Gigny Val d'Ornel, BP 86, 52103 Saint-Dizier
Focus: Adult Educ *04395*

Comité d'Etudes et de Liaison Interprofessionnel du Département de l'Aisne (C.E.L.I.D.A.), 12 Blvd Roosevelt, 02100 Saint-Quentin
Focus: Adult Educ *04396*

Comité d'Etudes Fiscales et Contentieuses, 5 Rue La Boétie, 75008 Paris
T: (1) 42660850
Focus: Law *04397*

Comité d'Etudes pour un Nouveau Contrat Social, 17 Blvd Raspail, 75007 Paris
T: (1) 42221265
Focus: Sociology *04398*

Comité du Film Ethnographique (C.I.F.H.), c/o Musée de l'Homme, Pl du Trocadéro, 75016 Paris
T: (1) 47043820; Fax: (1) 45535282
Founded: 1952; Members: 20
Focus: Cinema *04399*

Comité Européen Permanent de Recherches pour la Protection des Populations contre les Risques d'Intoxication à Long Terme, c/o Faculté des Sciences Pharmaceutiques et Biologiques, Université René Descartes, 4 Av de l'Observatoire, 75006 Paris
T: (1) 43267122
Founded: 1957
Gen Secr: Prof. René Truhaut
Focus: Toxicology *04400*

Comité Européen pour l'Education des Enfants et Adolescents Précoces, Doués, Talentueux (EUROTALENT), c/o ALREP, 33 Av Franklin Roosevelt, 30000 Nîmes
T: 66648251
Focus: Educ; Adult Educ *04401*

Comité Européen pour les Problèmes Criminels (CEPC) / European Committee on Crime Problems, c/o Conseil de l'Europe, Div des Problèmes Criminels, 67075 Strasbourg Cédex
T: 88412000; Tx: 870943; Fax: 88412764
Founded: 1957
Pres: J. Schutte
Focus: Criminology; Law
Periodicals
International Exchange of Information on Current Criminological Research Projects in Member States (weekly)
Prison Information Bulletin (weekly) . . . *04402*

Comité Français d'Education et d'Assistance de l'Enfance Déficiente, 185 Rue de Charonne, 75011 Paris
T: (1) 43712570
Focus: Educ *04403*

Comité Français de l'Electricité, Tour Atlantique, 92080 Paris-La Défense Cédex 6
T: (1) 47781341; Fax: (1) 47739553
Pres: Guy Dallery; Gen Secr: Roger Le Goff
Focus: Electric Eng
Periodical
Les Cahiers Français de l'Electricité . . *04404*

Comité Historique du Centre-Est, 86 Rue Pasteut, 69007 Lyon
Focus: Hist *04405*

Comité International de l'AISS pour la Prévention des Risques Professionnels du Bâtiment et des Travaux Publics, c/o O.P.B.T.P., Tour Amboise, 204 Rond Point du Pont de Sèvres, 92516 Boulogne-Billancourt Cédex
T: (1) 46092654; Fax: (1) 46092740
Founded: 1968; Members: 40
Gen Secr: Pierre Serin
Focus: Safety *04406*

Comité International de Paléographie Latine (CIPL), c/o IRHT, 40 Av d'léna, 75116 Paris
T: (1) 47238939
Founded: 1953
Gen Secr: Denis Muzerelle
Focus: Ling *04407*

Comité International des Sciences Historiques (CISH), c/o Institut d'Histoire du Temps Présent, 44 Rue de l'Amiral Mouchez, 75014 Paris
T: (1) 42221213 ext 3270
Founded: 1926; Members: 80
Pres: Prof. T.C. Barker; Gen Secr: Prof. François Bédarida
Focus: Hist
Periodical
Bulletin d'Information (weekly) *04408*

Comité International pour l'Information et la Documentation en Sciences Sociales (CIDSS), 27 Rue Saint-Guillaume, 75341 Paris Cédex 07
T: (1) 44495050; Tx: 201002
Founded: 1950
Gen Secr: Prof. Jean Meyriat
Focus: Doc; Sociology
Periodicals
Current Sociology (3 times annually)
International Bibliography of the Social Sciences: Sociology, Political Science, Economics, Social and Cultural Anthropology (weekly)
International Political Science Abstracts (weekly) *04409*

Comité International Rom (CIR) / International Gypsy Committee, 5 Villa du Sud, 93380 Pierrefitte
Founded: 1954
Gen Secr: I.R. Djuric
Focus: Ethnology
Periodical
O Glaso Romano *04410*

Comité Mondial pour la Recherche Spatiale / Committee on Space Research, 51 Blvd de Montmorency, 75016 Paris
T: (1) 45250679; Fax: (1) 42889431
Founded: 1958; Members: 4057
Pres: Prof. W.I. Axford; Gen Secr: S. Grzedzielski
Focus: Astronomy
Periodicals
Advances in Space Research (weekly)
COSPAR Directory of Organization and Members (bi-annually)
COSPAR Information Bulletin (3 times annually) *04411*

Comité National contre les Maladies Respiratoires et la Tuberculose (CNTMR), 66 Blvd Saint-Michel, 75006 Paris
T: (1) 46345880
Founded: 1916; Members: 500
Pres: C. Molina; Gen Secr: M. Zerapha
Focus: Pulmon Dis
Periodical
Respirer (weekly) *04412*

Comité National de l'Enseignement Libre, 277 Rue Saint-Jacques, 75005 Paris
T: (1) 40331459
Focus: Educ *04413*

Comité National des Conseillers de l'Enseignement Technique, 39 Rue de la Roquette, 75011 Paris
T: (1) 48050608
Focus: Adult Educ; Eng *04414*

Comité National des Ecoles Françaises de Service Social, 9 Rue de l'Isly, 75008 Paris
T: (1) 42936322; Tx: 698805
Gen Secr: M.F. Marques
Focus: Educ; Sociology *04415*

Comité National Français de Géodesie et Géophysique, 136bis Rue de Grenelle, 75007 Paris
T: (1) 45554345
Members: 470
Focus: Surveying; Geophys *04416*

Comité National Français de Géographie, 191 Rue Saint-Jacques, 75005 Paris
Members: 500
Pres: J.R. Pitte; Gen Secr: A. Metton
Focus: Geography *04417*

Comité National Français de Mathématiciens, c/o Département de Mathématiques et d'Informatique, Ecole Normale Supérieure, 45 Rue d'Ulm, 75230 Paris Cedex 05
T: (1) 43291225
Founded: 1950
Pres: C. Houzel; Gen Secr: Mireille Deschamps
Focus: Math *04418*

Comité National Français des Recherches Antarctiques (CNFRA), 138bis Rue de Grenelle, 75700 Paris
Members: 150
Pres: C. Lorius; Gen Secr: G. Pillet
Focus: Geography
Periodical
Report (weekly) *04419*

Comité pour les Données Scientifiques et Technologiques (CODATA) / Committee on Data for Science and Technology, 51 Blvd de Montmorency, 75016 Paris
T: (1) 5250496
Founded: 1966
Focus: Computer & Info Sci
Periodicals
CODATA Bulletin (weekly)
CODATA Newsletter (weekly)
Conference Proceedings (weekly) *04420*

Comité Universitaire d'Information Pédagogique, 29 Rue d'Ulm, 75005 Paris
Focus: Educ; Adult Educ *04421*

Commission de la Carte Géologique du Monde / Commission for the Geological Map of the World, 77 Rue Claude Bernard, 75005 Paris
T: (1) 47072284; Tx: 206411; Fax: (1) 43367655
Founded: 1881; Members: 150
Pres: Jean Dercourt; Gen Secr: Philippe Bouysse

Focus: Cart
Periodical
CGMW Bulletin (weekly) 04422

Commission de l'Enseignement Supérieur en Biologie, 1 Rue Victor Cousin, 75005 Paris
Focus: Educ; Bio 04423

Commission Internationale de Bibliographie / International Bibliographic Commission, c/o Institut de Recherche et d'Histoire des Textes, 40 Av d'Iéna, 75016 Paris
Gen Secr: Michel Keul
Focus: Libraries & Bk Sci 04424

Commission Internationale des Grands Barrages (CIGB) / International Commission on Large Dams, 151 Blvd Haussmann, 75008 Paris
T: (1) 40426824; Tx: 641320
Founded: 1928
Gen Secr: J. Cotillon
Focus: Civil Eng
Periodical
ICOLD Technical Bulletin (3-4 times annually) 04425

Commission Internationale d'Etudes Historiques Latinoaméricaines et des Caraïbes (CIOHL), 28 Rue Saint-Guillaume, 75007 Paris
T: (1) 42223593
Founded: 1982
Gen Secr: Prof. Frédéric Mauro
Focus: Cultur Hist
Periodical
Newsletter (weekly) 04426

Commission Internationale d'Histoire des Mouvements Sociaux et des Structures Sociales, 9 Rue de Valence, 75005 Paris
T: (1) 47072864
Founded: 1953; Members: 26
Gen Secr: Denise Fauvel Rouif
Focus: Hist; Sociology 04427

Commission Océanographique Intergouvernementale (COI) / Intergovernmental Oceanographic Commission, c/o UNESCO, Pl de Fontenoy, 75700 Paris
T: (1) 45681000
Founded: 1960; Members: 117
Gen Secr: Dr. Gunnar Kullenberg
Focus: Oceanography 04428

Committee for European Marine Biology Symposia (EMBS), c/o Laboratoire d'Océanographie Biologique, Université de Bretagne Occidentale, 6 Av Le Gorgeu, 29287 Brest Cédex
Founded: 1966
Pres: Prof. M. Glémarec
Focus: Oceanography 04429

Committee for International Cooperation in National Research in Demography (CICRED), 27 Rue du Commandeur, 75675 Paris Cédex 14
T: (1) 43201345; Fax: (1) 43277240
Founded: 1972; Members: 300
Pres: Jean Bourgeois-Pichat
Focus: Soc Sci
Periodicals
CICRED Bulletin (weekly)
Directory of Demographic Research Centres (weekly) 04430

Committee on Data for Science and Technology (CODATA), 51 Blvd de Montmorency, 75016 Paris
T: (1) 45250496; Tx: 630553; Fax: (1) 42889431
Founded: 1966
Gen Secr: Phyllis Glaeser
Focus: Computer & Info Sci 04431

Community of European Management Schools (CEMS), 1 Rue de la Libération, 78350 Jouy-en-Josas
T: 39677457; Tx: 697942; Fax: 39677440
Members: 10
Focus: Educ; Business Admin 04432

Compagnie des Experts Architectes près la Cour d'Appel de Paris, 179 Rue Courcelles, 75116 Paris
Founded: 1928
Pres: J.-P. Sevaistre; Gen Secr: Bernard Guechot
Focus: Archit 04433

Confédération des Sociétés Scientifiques Françaises (CSSF), 11 Rue Pierre et Marie Curie, 75005 Paris
Founded: 1919
Focus: Sci 04434

Confédération des Syndicats Médicaux Français (CSMF), 60 Blvd de Latour-Maubourg, 75327 Paris Cedex 07
T: (1) 47055972
Founded: 1930; Members: 30000
Pres: Dr. Fedi; Gen Secr: Dr. J. Beaupère
Focus: Med
Periodical
Le Médecin de France (weekly) 04435

Confédération Européenne pour la Thérapie Physique (CETP), 24 Rue des Petits-Hôtels, 75010 Paris
T: (1) 42468007
Founded: 1938
Gen Secr: F. Chambon
Focus: Therapeutics 04436

Confédération Internationale des Associations de Médecines Alternatives Naturelles (CIAMAN), Av Becquerel, BP 37, 26700 Pierrelatte Cédex
T: (1) 75040336
Founded: 1974
Gen Secr: M. Monneret
Focus: Med
Periodical
Revue de Médecine Biotique 04437

Confédération Internationale des Sociétés d'Auteurs et Compositeurs (CISAC), 11 Rue Keppler, 75116 Paris
T: (1) 45535937
Founded: 1926; Members: 99
Gen Secr: J.A. Ziegler
Focus: Music; Lit
Periodical
Interauteurs (weekly) 04438

Confédération Nationale des Syndicats Dentaires (CNSD), 22 Av de Villiers, 75017 Paris
T: (1) 47660232
Founded: 1935; Members: 18000
Focus: Dent
Periodical
Le Chirurgien-Dentiste de France (weekly) 04439

Conférence des Présidents d'Universités, 12 Rue de l'Ecole de Médecine, 75270 Paris Cédex 06
T: (1) 43545049; Fax: (1) 43251651
Founded: 1971; Members: 86
Focus: Adult Educ 04440

Conférence Internationale des Doyens des Facultés de Médecine d'Expression Française (CIDMEF), c/o Faculté de Médecine, Université François Rabelais, BP 3223, 37032 Tours Cédex
T: 47376673; Tx: 752364; Fax: 47366099
Founded: 1981
Focus: Educ; Med
Periodical
Revue d'Education Médicale 04441

Conférence Internationale des Facultés de Droit ayant en Commun l'Usage du Français (CIFDUF), c/o Université de Droit, d'Economie et des Sciences, 3 Av Robert Schuman, 13628 Aix-en-Provence
T: 42201905
Founded: 1991
Pres: Louis Favoreu
Focus: Law; Educ 04442

Conférence Internationale des Facultés, Instituts et Ecoles de Pharmacie d'Expression Française (CIFPEF), 4 Av de l'Observatoire, 75270 Paris Cédex 06
T: (1) 46613325
Founded: 1987
Pres: Charles Souleau
Focus: Educ; Pharmacol 04443

Conférence Internationale des Formations d'Ingénieurs et Techniciens d'Expression Française (CITEF), c/o INSA, 20 Av Albert Einstein, 89821 Villeurbanne Cédex
T: 78948112
Founded: 1986
Gen Secr: N. Mongereau
Focus: Educ; Eng 04444

Conférence Internationale des Grands Réseaux Electriques à Haute Tension (CIGRE) / International Conference on Large High Voltage Electric Systems, 3 Rue de Metz, 75010 Paris
T: (1) 42465085; Fax: (1) 42465827
Founded: 1921; Members: 5000
Gen Secr: Y. Porcheron
Focus: Electric Eng
Periodicals
Electra (weekly)
Proceedings of the biennial Sessions . . 04445

Conférence Internationale des Responsables des Universités et Instituts à Dominante Scientifique et Technique d'Expression Française (CIRUISEF), c/o Université de Bordeaux I, 351 Cours de la Libération, 33405 Talence Cédex
T: 56846040
Founded: 1988
Pres: Jean Lascombe
Focus: Sci; Educ; Eng 04446

Conference on the Development and the Planning of Urban Transport in Developing Countries, c/o CODATU Association, Grande Arche, 92055 La Défense Cédex 04
T: (1) 40812304; Tx: 610835; Fax: (1) 40812393
Founded: 1980; Members: 9
Gen Secr: Christian Curé
Focus: Transport 04447

Congrès de Psychiatrie et de Neurologie de Langue Française, c/o Hôpital Universitaire Dupuytren, 2 Av Alexis Carrel, 87042 Limoges Cédex
Founded: 1890
Gen Secr: Prof. Jean-Marie Leger
Focus: Neurology; Psychiatry
Periodical
Annual Report (weekly) 04448

Congrès des Psychoanalystes de Langues Romanes, 187 Rue Saint-Jacques, 75005 Paris
Focus: Psychoan 04449

Congrès International de Médecine Légale et de Médecine Sociale de Langue Française / International French-Language Congresses of Forensic and Social Medicine, 2 Pl Mazas, 75012 Paris
T: (1) 43434254
Founded: 1911
Gen Secr: Prof. A.J. Chaumont
Focus: Hygiene 04450

Conité Européen Lex Informatica Mercatoriaque (CELIM), 868 Rte de Ganges, 34090 Montpellier
T: 67606586; Fax: 67604231
Founded: 1985
Gen Secr: Prof. Michel Vivnt
Focus: Computer & Info Sci 04451

Conseil International de la Danse (CIDD) / International Dance Council, c/o UNESCO, 1 Rue Miollis, 75732 Paris Cédex 15
T: (1) 45669422
Founded: 1973
Gen Secr: Nicole Luc-Marèchal
Focus: Music; Perf Arts 04452

Conseil International de la Langue Française, 11 Rue de Navarin, 75009 Paris
T: (1) 48787395; Fax: (1) 48784928
Founded: 1967; Members: 525
Focus: Ling
Periodicals
La Banque des Mots
Le Français Moderne 04453

Conseil International de la Musique (CIM), 1 Rue Miollis, 75732 Paris Cédex 15
T: (1) 45682550; Fax: (1) 43068798
Founded: 1949
Focus: Music
Periodical
Resonance (3 times annually) 04454

Conseil International de la Philosophie et des Sciences Humaines (CIPSH), c/o UNESCO, 1 Rue Miollis, 75732 Paris Cédex 15
T: (1) 45682685; Fax: (1) 40659480
Founded: 1949; Members: 13
Pres: Jean d' Ormesson; Gen Secr: Annelise Gaborieau
Focus: Humanities; Philos
Periodical
Diogenes (weekly) 04455

Conseil International des Monuments et des Sites / International Council on Monuments and Sites, 75 Rue de Temple, 75003 Paris
T: (1) 42773576; Fax: (1) 42775742
Founded: 1965; Members: 4500
Pres: Roland Silva; Gen Secr: H. Stovel
Focus: Preserv Hist Monuments
Periodical
Icomos News (weekly) 04456

Conseil International des Moyens du Film d'Enseignement (CIME) / International Council for Educational Media, c/o Institut Pédagogique National, 29 Rue d'Ulm, 75005 Paris
T: (1) 43261490
Founded: 1950
Gen Secr: R. Lefranc
Focus: Cinema 04457

Conseil International des Musées (ICOM), 1 Rue Miollis, 75015 Paris
T: (1) 47340500; Tx: 270602; Fax: (1) 43067862
Founded: 1946; Members: 10000
Pres: S. Ghose; Gen Secr: Elisabeth Des Portes
Focus: Arts
Periodical
ICOM News (weekly) 04458

Conseil International des Sciences Sociales (CISS), 1 Rue Miollis, 75015 Paris
T: (1) 45682558
Founded: 1952; Members: 14
Gen Secr: Prof. Luis I. Ramallo
Focus: Econ; Law; Poli Sci; Psych; Anthro; Geography
Periodicals
ISSC Newsletter
Social Science Information (weekly) . . . 04459

Conseil International des Unions Scientifiques (CIUS) / International Council of Scientific Unions, 51 Blvd de Montmorency, 75016 Paris
T: (1) 45250329; Tx: 645554; Fax: (1) 42889431
Founded: 1919; Members: 104
Gen Secr: J. Marton-Lefèvre
Focus: Sci
Periodical
ICSU Yearbook (weekly) 04460

Conseil International pour l'Information Scientifique et Technique / International Council for Scientific and Technical Information, 51 Blvd de Montmorency, 75016 Paris
T: (1) 45256592; Fax: (1) 42151262
Founded: 1984; Members: 51
Gen Secr: Daniel Confland
Focus: Computer & Info Sci
Periodical
Forum (weekly) 04461

Conseil Mondial d'Ethique des Droits de l'Animal (COMEDA), 10 Rue Gallieni, 92600 Asnières
Pres: Prof. Georges Heuse
Focus: Ecology 04462

Demeure Historique, 57 Quai de la Tournelle, 75005 Paris
T: (1) 43290286; Fax: (1) 43293644
Founded: 1924; Members: 2000
Pres: Marquis de Breteuil
Focus: Hist
Periodical
La Demeure Historique (weekly) 04463

Dialogue et Coopération, 140 Av Daumesnil, 75012 Paris
T: (1) 43440506
Founded: 1965
Pres: Jacqueline Cretté
Focus: Educ 04464

Dickens Fellowship, 29 Blvd Mariette, 62200 Boulogne-sur-Mer
Focus: Lit 04465

Division de Chimie Physique de la Société Française de Chimie, 10 Rue Vauquelin, 75005 Paris
T: (1) 47075448; Fax: (1) 43314222
Founded: 1908; Members: 1400
Focus: Chem
Periodical
Journal de Chimie physique et de physico-chimie biologique (weekly) 04466

Ecole Internationale de Bordeaux (EIB), 43 Rue Pierre Noailles, 33405 Talence Cédex
T: 56375059; Tx: 571741; Fax: 56044201
Founded: 1972
Focus: Educ 04467

Ecole Internationale d'Informatique de l'AFCET, 156 Blvd Péreire, 75017 Paris
T: (1) 47662419
Focus: Educ; Computer & Info Sci . . . 04468

Education for All Network, c/o NGO/UNESCO, 1 Rue Miollis, 75015 Paris
T: (1) 45683268; Fax: (1) 45671690
Founded: 1991
Gen Secr: Odile Moreau
Focus: Educ 04469

Equipe de Recherche Associée au C.N.R.S.-Laboratoire de Génétique, 351 Cours de la Libération, 33405 Talence
Focus: Genetics 04470

Esperanto Academy / Académie d'Espéranto, 5 Rue Léon Cogniet, 75017 Paris
T: (1) 42278916
Founded: 1905; Members: 45
Gen Secr: Jean Thierry
Focus: Ling 04471

Etudes Préhistoriques, 34 Av Limburg, 69110 Sainte-Foy-lès-Lyon
Founded: 1967; Members: 975
Focus: Hist
Periodical
Etudes Préhistoriques (weekly) 04472

Euro-African Asscociation for the Anthropology of Social Change and Development, BP 5045, 34032 Montpellier Cédex
T: 67617400; Fax: 67547800
Founded: 1991
Gen Secr: Jean-Pierre Chauveau
Focus: Anthro 04473

Eurocoast, c/o BRGM, Domaine de Luminy, 13009 Marseille
T: 91412446; Tx: 401585; Fax: 91411510
Founded: 1988
Gen Secr: Dr. L. Galtier
Focus: Geology; Eng 04474

EUROLAT – European Network for Studies on Laterites and Tropical Environment, c/o Institut de Géologie, Université Louis Pasteur, 1 Rue Blessig, 67084 Strasbourg
T: 88358564; Tx: 870260; Fax: 88607550
Founded: 1984; Members: 20
Pres: Prof. Yves Tardy
Focus: Geology 04475

European Academic and Research Network (EARN), c/o CIRCE, BP 167, 91403 Orsay Cédex
T: 69823973; Tx: 602166; Fax: 69285273
Founded: 1985
Gen Secr: H. Decers
Focus: Sci 04476

European Academy of Anaesthesiology / Académie Européenne d'Anesthésiologie, c/o Service d'Anesthésie et de Réanimation, Hôpital de Hautepierre, Av Molière, 67098 Strasbourg
T: 88289000; Fax: 88285464
Founded: 1976
Gen Secr: Prof. J.C. Otteni
Focus: Anesthetics 04477

European Aggregate Association (UEPG), 3 Rue Alfred Roll, 75849 Paris Cédex 17
Focus: Mach Eng 04478

European and Mediterranean Plant Protection Organization (EPPO), 1 Rue Le Nôtre, 75016 Paris
T: (1) 45207794; Fax: (1) 42248943
Founded: 1951; Members: 35
Pres: Dr. J. Thiault; Gen Secr: Dr. I.M. Smith
Focus: Ecology
Periodicals
Data Sheets on Quarantine Organismes (weekly)
EPPO-Bulletin (weekly)
Guidelines for the Biological Evaluation of Pesticides (weekly)
Reporting Service: Special Sheets for Quick Delivery of Urgent Information 04479

European Association for American Studies (EAAS), c/o Dépt d'Anglais, Faculté des Langues, Université Lyon II, 86 Rue Pasteur, 69007 Lyon
T: 78695601
Founded: 1954
Focus: Hist; Arts; Lit; Soc Sci
Periodical
European Contributions to American Studies 04480

European Association for Chinese Studies (EACS) / Association Européenne d'Etudes Chinoises, c/o Ecole des Hautes Etudes en Sciences Sociales, 54 Blvd Raspail, 75006 Paris
T: (1) 49542368; Tx: 203104; Fax: (1) 45449311
Founded: 1975
Gen Secr: Dr. Viviane Alleton
Focus: Cultur Hist
Periodical
AEDEC Newsletter (weekly) 04481

European Association for the International Space Year (EUREG), 16bis Av Bosuet, 75007 Paris
T: (1) 47051779; Fax: (1) 45519923
Gen Secr: J. Gomérieux
Focus: Aero 04482

European Association for the Study of Dreans (EASD) / Association Européenne pour l'Etude du Rêve, c/o Association Française pour l'Etude du Rêve, BP 30, 93451 Ile-Saint-Denis Cédex
Founded: 1987
Gen Secr: Henri Rojouan
Focus: Parapsych 04483

European Association of Contract Research Organizations (EACRO) / Association Européenne des Organisations de Recherche sous Contrat, c/o Bertin, BP 3, 78373 Plaisir Cédex
T: (1) 34818581; Tx: 696231
Founded: 1989; Members: 53
Pres: Georges Mordchelles-Regnier
Focus: Eng
Periodical
Eacronews (weekly) 04484

European Association of Establishments for Veterinary Education (EAEVE) / Association Européenne des Etablissements d'Enseignement Vétérinaire, 7 Av Général de Gaulle, 94700 Maisons-Alfort
T: (1) 43687334; Fax: (1) 43967131
Founded: 1988
Pres: Thomas Bernard
Focus: Educ; Vet Med 04485

European Association of Geochemistry, c/o Institut Physique du Globe, 4 Pl Jussieu, Tour 14-15, 3e Etage, 75252 Paris
Gen Secr: Dr. R.H.A. Wessels
Focus: Chem 04486

European Association of Marine Sciences and Techniques / Association Européenne des Sciences et Techniques de la Mer, c/o IGBA, 351 Cours de la Libération, 33405 Talence Cédex
T: 56806050; Fax: 56370774
Founded: 1985
Focus: Oceanography; Eng
Periodical
AESTM Newsletter (3 times annually) . . 04487

European Association of Museums of the History of Medical Sciences / Association Européenne des Musées de l'Histoire des Sciences Médicales, c/o Musée Claude Bernard, Saint-Julien en Beaujolais, 69440 Denicé
Founded: 1983
Focus: Med; Hist 04488

European Association of Plastic Surgeons (EURAPS) / Association Européenne des Chirurgiens Plastiques, 130 Rue de la Pompe, 75116 Paris
T: (1) 47274431; Fax: (1) 47276515
Founded: 1989; Members: 120
Gen Secr: Dr. Daniel Marchac
Focus: Surgery
Periodical
European Journal of Plastic Surgery . . 04489

European Association of Urology / Association Européenne d'Urologie, c/o Hôpital Cochin, 27 Rue du Faubourg Saint-Jacques, 75014 Paris
Gen Secr: Prof. A. Steg
Focus: Urology 04490

European Atomic Energy Society (EAES) / Société Européenne d'Energie Atomique, c/o Commissariat à l'Energie Atomique, 31-33 Rue de la Fédération, 75752 Paris Cédex 15
T: (1) 40561413
Founded: 1954; Members: 15
Gen Secr: P. Felten
Focus: Nucl Res 04491

European Audio Phonological Centers Association (EAPCA) / Association Européenne des Centres d'Audiologie, 23 Av de la République, 42000 Saint-Etienne
T: 77414720
Founded: 1967
Pres: Dr. Michel Courtoy; Gen Secr: Dr. P. Guichard-Duthel
Focus: Otorhinolaryngology 04492

European Brachytherapy Group, c/o Institut Gustave Roussy, 53 Rue Camille Desmoulins, 94805 Villejuif
T: 45594566; Fax: 45594727
Gen Secr: A. Gerbaulet
Focus: Therapeutics 04493

European Centre for Advanced Studies in Thermodynamics (ECAST), c/o LUAP Paris VII, Tour 33-2e, 52 Pl Jussieu, 75251 Paris Cédex 05
T: (1) 44277908
Founded: 1989
Pres: Prof. J. Chanu
Focus: Physics 04494

European Centre for Regional Development, 20 Pl des Halles, 67000 Strasbourg
T: 88223883; Tx: 870912; Fax: 88226482
Founded: 1985; Members: 71
Focus: Develop Areas 04495

European Committee for the Study of Salt (ECSS), 17 Rue Daru, 75008 Paris
T: (1) 47665290; Fax: (1) 47665266
Founded: 1958
Gen Secr: Bernard Moinier
Focus: Mineralogy 04496

European Committee on Ocean and Polar Sciences (ECOPS), c/o ESF, 1 Quai Lezay Marnésia, 67080 Strasbourg
T: 88767100; Tx: 890440; Fax: 88370532
Founded: 1990
Focus: Oceanography; Geography . . . 04497

European Consortium for Ocean Drilling (ECOD), c/o European Science Foundation, 1 Quai Lezay Marnésia, 67080 Strasbourg Cédex
T: 88767100; Tx: 890440; Fax: 88370532
Gen Secr: M. Fratta
Focus: Eng 04498

European Council on African Studies (ECAS), c/o Centre de Géographie Appliquée, 3 Rue de l'Argonne, 67060 Strasbourg Cédex
T: 88358247; Tx: 870260
Founded: 1983
Pres: A. Coupez; Gen Secr: J.-P. Blanck
Focus: Geography 04499

European Council on Chiropractic Education, 8 Chemin de Tucson, 86000 Poitiers
T: 49415825
Pres: Pierre Gruny
Focus: Educ 04500

European Council on Environmental Law, c/o Université des Sciences Juridiques, Politiques et Sociales, Pl d'Athènes, 67084 Strasbourg Cédex
Founded: 1974
Focus: Law 04501

European Documentation and Information System for Education (EUDISED), c/o Conseil de l'Europe, Direction de l'Enseignement, de la Culture et du Sport, BP 431R6, 67006 Strasbourg
T: 88412000; Tx: 870943; Fax: 88412788
Founded: 1968
Gen Secr: Dr. Michael Vorbeck
Focus: Educ 04502

European Dyslexia Association – International Organization for Specific Learning Disabilities (EDA), 46 Av de Port Royal des Champs, 78320 Le Mesnil-Saint-Denis
T: 34619252
Founded: 1987
Pres: Marcel Seynave; Gen Secr: Anne-Marie Montarnal
Focus: Educ 04503

European Federation of AIDS Research (EFAR), 51 Rue Liancourt, 75014 Paris
T: (1) 43216975; Fax: (1) 43214231
Founded: 1988
Pres: Prof. Luc Montagnier
Focus: Derm 04504

Expéditions Polaires Françaises, 47 Av Maréchal Fayolle, 75116 Paris
T: (1) 45041771; Fax: (1) 45031487
Focus: Geography 04505

Fédération Aéronautique Internationale (FAI), 10-12 Rue du Capitaine Ménard, 75015 Paris
T: (1) 45792477; Tx: 201327; Fax: (1) 45797315
Founded: 1905
Gen Secr: Max Bishop
Focus: Aero
Periodical
Air Sports International (3 times annually) 04506

Fédération Autonome de l'Education Nationale, 13 Av de Taillebourg, 75011 Paris
T: (1) 43732136; Fax: (1) 43700847
Focus: Educ
Periodical
Bulletin (weekly) 04507

Fédération Caribe de Santé Mentale, c/o Charles Saint-Cyr, Ravine Vilaine, Fort de France
Focus: Public Health; Psychiatry 04508

Fédération d'Associations et Groupements pour les Etudes Corses (FAGEC), BP 85, 20291 Bastia
Founded: 1970; Members: 16
Focus: Hist 04509

Fédération de l'Education Nationale (FEN), 48 Rue La Bruyére, 75009 Paris
T: (1) 42857101
Founded: 1946; Members: 500110
Focus: Educ
Periodicals
L'Enseignement Public (weekly)
F.E.N. HEBDO (weekly) 04510

Fédération des Amicales des Documentalistes et Bibliothcaires de l'Education Nationale, 29 Rue d'Ulm, 75007 Paris
Focus: Libraries & Bk Sci 04511

Fédération des Chambres Syndicales des Chirurgiens Dentistes de la Région de Paris, 32 Rue de la Victoire, 75009 Paris
T: (1) 48785573
Focus: Surgery; Dent 04512

Fédération des Gynécologues et Obstétriciens de Langue Française (FGOLF) / Federation of French-Language Gynecologists and Obstetricians, c/o Clinique Universitaire Baudelocque, 123 Blvd de Port-Royal, 75674 Paris Cédex 14
T: (1) 42341140
Founded: 1950; Members: 1700
Gen Secr: Prof. Jean René Zorn
Focus: Gynecology
Periodical
Journal de Gynécologie, Obstétrique et Biologie de la Réproduction (8 times annually) . 04513

Fédération des Médecins de France (FMF), 60 Rue Laugier, 75017 Paris
T: (1) 47634052
Founded: 1968
Focus: Med
Periodical
France Médecine (weekly) 04514

Fédération des Sociétés Francaises de Généalogie, d'Héraldique et de Sigillographie, 11 Blvd Pershing, 78000 Versailles
T: 39548516
Founded: 1969; Members: 9500
Focus: Genealogy
Periodical
Héraldique et Généalogie (weekly) . . . 04515

Fédération des Sociétés Historiques et Archéologiques de Paris et de l'Ile de France, 24 Rue Pavée, 75004 Paris
T: (1) 42744444
Founded: 1949
Focus: Hist; Archeol
Periodical
Paris et Ile-de-France: Mémoires publiés par la Fédération des Sociétés Historiques et Archéologiques de Paris et de l'Ile de France (weekly) 04516

Féderation des Sociétés Savantes de la Charente-Maritime, c/o Archives Départementales, 17000 La Rochelle
Founded: 1975
Gen Secr: Pascal Even
Focus: Hist; Archeol
Periodical
Revue de la Saintonge et de l'Aunis (weekly) 04517

Fédération des Syndicats Dentaires Libéraux (F.S.D.L.), 32 Rue de la Victoire, 75009 Paris
T: (1) 48785573; Fax: (1) 48782204
Focus: Dent
Periodicals
Flash Info
Le Libéral de Chaine
Le Libéral Dentaire 04518

Fédération des Syndicats Pharmaceutiques de France, 13 Rue Ballu, 75009 Paris
T: (1) 45261250
Members: 98
Focus: Pharmacol 04519

Fédération Française d'Education Physique et de Gymnastique Volontaire (FFEPGV), 2 Rue de Valois, 75001 Paris
T: (1) 42613844
Founded: 1888; Members: 76400
Focus: Sports
Periodical
Loisirs-Santé (5 times annually) 04520

Fédération Française de Sociétés des Sciences Naturelles, 57 Rue Cuvier, 75231 Paris Cédex 05
T: (1) 43382084
Founded: 1919; Members: 140
Pres: J. Lorenz; Gen Secr: R.F. Pujol
Focus: Nat Sci
Periodicals
L'Année Biologique (3 times annually)
Bulletin (weekly) 04521

Fédération Française de Spéléologie (FFS), 130 Rue Saint-Maur, 75011 Paris
T: (1) 43575654
Founded: 1963; Members: 6000
Focus: Speleology 04522

Fédération Française des Sociétés d'Amis de Musées (FFSAM), Place Henry de Montherlant, Hôtel d'Orsay, 75116 Paris
T: (1) 47236520
Focus: Arts
Periodical
Bulletin (weekly) 04523

Fédération Française des Sociétés de Protection de la Nature (FFSPN), 57 Rue Cuvier, 75231 Paris Cédex 05
T: (1) 43367995
Founded: 1968; Members: 400000
Focus: Ecology 04524

Fédération Française des Sociétés de Sciences Naturelles, 57 Rue Cuvier, 75231 Paris Cédex 05
T: (1) 43382084
Founded: 1919; Members: 140
Pres: J. Lorenz; Gen Secr: R.F. Pujol
Focus: Nat Sci
Periodicals
L'Année Biologique (3 times annually)
Bulletin (weekly) 04525

Fédération Française d'Etudes et de Sports Sous Marins, 24 Quai de Rive Neuve, 13007 Marseille
T: 91339931
Focus: Oceanography
Periodical
Etudes et Sports Sous Marins (weekly) . 04526

Fédération Générale des Syndicats de Biologistes, 133 Blvd du Montparnasse, 75006 Paris
T: (1) 43203910
Focus: Bio 04527

Fédération Historique de Provence, c/o Archives Départementales, 66b Rue Saint-Sébastien, 13259 Marseille Cédex 6
Founded: 1950; Members: 1050
Focus: Hist
Periodical
Provence Historique 04528

Fédération Hospitalière de France, 33 Av d'Italie, 75013 Paris
T: (1) 45843250
Focus: Med
Periodical
Revue Hospitalière de France (weekly) . 04529

Fédération Internationale Catholique d'Education Physique et Sportive (FICEP), 22 Rue Oberkampf, 75011 Paris
T: (1) 43385057
Founded: 1911; Members: 2500000
Focus: Sports 04530

Fédération Internationale des Associations d'Instituteurs, 3 Rue La Rochefoucauld, 75009 Paris
T: (1) 48745844; Fax: (1) 42852736
Focus: Educ 04531

Fédération Internationale des Centres d'Entraînement aux Méthodes d'Education Active (FICEMEA) / International Federation of Training Centres for the Promotion of Progressive Education, 76 Blvd de la Villette, 75940 Paris Cédex 19
Founded: 1954
Gen Secr: Claude Vercoutère
Focus: Educ
Periodical
FICEMEA Bulletin 04532

Fédération Internationale des Mouvements d'Ecole Moderne (FIMEN), 189 Av Francis Tonner, BP 109, 06322 Cannes-La Bocca Cédex
Founded: 1957
Focus: Educ
Periodicals
Art Enfantin (weekly)
Bibliothèque de travail (weekly)
L'Educateur (weekly) 04533

Fédération Internationale des Professeurs de Français (FIPF), 1 Av Léon Journault, 92310 Sèvres
T: (1) 46265316; Fax: (1) 46268169
Founded: 1969; Members: 125
Pres: Raymond Le Loch; Gen Secr: Jean Souillat
Focus: Ling; Educ
Periodicals
Dialogues et Cultures (weekly)
Une Lettre de la FIPF (weekly) 04534

Fédération Internationale des Professeurs de l'Enseignement Secondaire Officiel (F.I.P.E.S.O.) / International Federation of Secondary Teachers, 7 Rue de Villersexel, 75007 Paris
T: (1) 40632935; Fax: (1) 40632936
Founded: 1912; Members: 800000

Gen Secr: Louis Weber
Focus: Educ
Periodicals
Bulletin International de la F.I.P.E.S.O. (weekly)
Les Nouvelles de la F.I.P.E.S.O.: F.I.P.E.S.O. Newsletter (8 times annually) 04535

Fédération Internationale des Universités Catholiques (FIUC) / International Federation of Catholic Universities, 51 Rue Orfila, 75341 Paris
T: (1) 47972660; Fax: (1) 47972942
Founded: 1948; Members: 176
Pres: Prof. Julio Teran Dutari; Gen Secr: Marc Caudron
Focus: Sci
Periodical
News in Brief (weekly) 04536

Fédération Internationale pour l'Economie Familiale (FIEF) / International Federation for Home Economics, 5 Av de la Porte Brancion, 75015 Paris
T: (1) 48423474
Founded: 1908
Gen Secr: Odette Goncet
Focus: Home Econ
Periodical
Economie Familiale-Home Economics-Hauswirtschaft (weekly) 04537

Fédération Internationale pour l'Education des Parents (F.I.E.P.), 1 Av Léon Journault, 92311 Sèvres Cédex
T: (1) 45072164; Fax: (1) 46266927
Founded: 1964; Members: 120
Pres: Jean Auba; Gen Secr: Micheline Ducray
Focus: Adult Educ
Periodical
Lettre d'Information (weekly) 04538

Fédération Métallurgique Française, 6 Av de Massine, 75008 Paris
T: (1) 45221062
Focus: Metallurgy 04539

Fédération Nationale Aéronautique, 52 Rue Galilée, 75008 Paris
T: (1) 47203975; Fax: (1) 27203032
Focus: Aero
Periodical
Info Pilote (weekly) 04540

Fédération Nationale des Associations de Chimie de France, 28 Rue Saint-Dominique, 75007 Paris
Members: 12000
Focus: Chem 04541

Fédération Nationale des Syndicats Départementaux de Médecins Electro-Radiologistes Qualifiés, 60 Blvd Latour-Maubourg, 75327 Paris Cedex 07
T: (1) 45517784
Founded: 1907; Members: 3000
Pres: Dr. J. Moinard
Focus: Radiology
Periodical
Bulletin 04542

Fondation Nationale des Sciences Politiques, 27 Rue Saint-Guillaume, 75341 Paris Cédex 07
Tx: 201002
Founded: 1945
Pres: René Rémond; Gen Secr: S. Hurtig
Focus: Poli Sci
Periodicals
Bulletin Analytique de Documentation (weekly)
Mots (weekly)
Observations et Diagnostics Economiques (weekly)
Revue Economique (weekly)
Revue Française de Science Politique (weekly)
Vingtième Siècle: Revue d'Histoire (weekly) 04543

Fondation pour la Recherche Sociale, 14 Rue Saint-Benoît, 75006 Paris
Founded: 1965; Members: 15
Focus: Sociology; Econ
Periodical
Recherche Sociale (weekly) 04544

Fondation pour la Science – Centre International de Synthèse, 12 Rue Colbert, 75002 Paris
T: (1) 42975068; Fax: (1) 42974645
Founded: 1924; Members: 36
Focus: Sci; Philos
Periodicals
Revue de Synthèse
Revue d'Histoire des Sciences (weekly) . 04545

Fondation Saint-John Perse, Espace Méjanes, 8-10 Rue des Allumettes, 13098 Aix-en-Provence Cédex 2
Founded: 1975; Members: 400
Focus: Lit
Periodical
Cahiers Saint-John Perse (weekly) . . . 04546

France Intec, 43 Rue Decamps, 75016 Paris
Founded: 1895; Members: 10000
Focus: Eng 04547

Groupe de Recherche et pour l'Education et la Prospective (G.R.E.P.), 13-15 Rue des Petites Ecuries, 75010 Paris
T: (1) 48245036; Fax: (1) 48240054
Focus: Educ
Periodical
Pour 04548

Groupe de Recherche Génétique Epidémiologique, Château de Longchamp, Bois de Boulogne, 75016 Paris
T: (1) 47727791
Focus: Genetics; Immunology 04549

Groupe des Méthodes Pluridisciplinaires Contribuant à l'Archéologie (GMPCA), c/o L. Langouet, Laboratoire d'Archéométrie, Université de Rennes, 35042 Rennes Cédex
Tx: 243233
Founded: 1901
Pres: A. Tabbagh
Focus: Geophys
Periodical
Revue d'Archéométrie (weekly) 04550

Groupe d'Etude et de Synthèse des Microstructures, c/o Ecole Supérieure de Physique et Chimie Industrielle de la Ville de Paris, 10 Rue Vauquelin, 75231 Paris Cédex
Founded: 1970
Focus: Chem; Physics 04551

Groupe Leibniz, c/o Groupe d'Etude pour la Traduction Automatique, Université de Grenoble, 38000 Grenoble Cédex 53
Founded: 1974
Focus: Ling 04552

Groupement des Associations Dentaires Francophones (GADEF), 22 Av de Villiers, 75017 Paris
T: (1) 47660232
Founded: 1971
Pres: Dr. M. Pirard
Focus: Dent
Periodical
Bulletin du GADEF 04553

Groupement des Bureaux Médicaux, 8 Rue Botzaris, 75019 Paris
T: (1) 42056068
Focus: Med 04554

Groupement d'Etudes et de Réalisations Médicales, 154 Rue du Faubourg Saint-Denis, 75010 Paris
T: (1) 42088255
Focus: Med 04555

Groupement Industriel Européen d'Etudes Spatiales, 16 Av Bosquet, 75007 Paris
T: (1) 45558353
Founded: 1961; Members: 74
Focus: Aero
Periodical
Eurospace Bulletin (weekly) 04556

Groupement Médical d'Etudes sur l'Alcoolisme (GMEA), 6 Allée Dugay-Trouin, 44000 Nantes
Founded: 1942; Members: 950
Focus: Hygiene 04557

Groupement Médical Saint-Augustin, 43 Blvd Malesherbes, 75008 Paris
T: (1) 42654810
Focus: Med 04558

Groupement National d'Etude des Médecins du Bâtiment et des Travaux Publics, 6 Rue Paul Valery, 75016 Paris
T: (1) 47201020
Focus: Orthopedics; Traumatology . . . 04559

Groupement Professionnel National de l'Informatique, 43 Rue de Trevise, 75009 Paris
T: (1) 48246650
Focus: Computer & Info Sci 04560

Groupe Phonétique de Paris, c/o Institut de Phonétique, Université de Paris III, 19 Rue des Bernardins, 75005 Paris
Focus: Ling
Periodical
Report (weekly) 04561

Groupe pour l'Avancement des Sciences Analytiques, 88 Blvd Malesherbes, 75008 Paris
T: (1) 45639304; Fax: (1) 49530434
Gen Secr: J. Tranchant; D. Sandino
Focus: Physics; Chem
Periodical
Bulletin d'Information GAMS (weekly) . . 04562

Groupe Rhône-Alpes de Recherche et d'Etudes en Gestion, c/o Université Jean Moulin, BP 155, 69224 Lyon Cédex 1
Focus: Geography 04563

Institut de France, c/o Académie des Sciences, 23 Quai de Conti, 75006 Paris
T: (1) 44414367
Founded: 1795; Members: 160
Focus: Sci
Periodicals
Les comptes rendus de l'Académie des Sciences
Les nouvelles de l'Académie des Sciences
La vie des sciences 04564

Institut des Actuaires Français, 243 Rue Saint-Honoré, 75001 Paris
T: (1) 42601694
Founded: 1890; Members: 700
Pres: Victor Volcouve; Gen Secr: Jean Berthon
Focus: Insurance; Finance
Periodical
Bulletin de l'Institut des Actuaires Français (3 times annually) 04565

Institut des Sciences Historiques / Société Archéologique de France, 45 Rue Rémy Dumoncel, 75014 Paris
T: (1) 43275045
Founded: 1816; Members: 150
Pres: P. Montillet; Gen Secr: F. Gandrille
Focus: Hist
Periodical
La Science Historique (weekly) 04566

Institut Français d'Analyse de Groupe et de Psychodrame, 12 Rue Emile Deutsch de la Meurthe, 75014 Paris
T: (1) 5882322; Fax: (1) 45882322
Focus: Sociology; Psych; Therapeutics . 04567

Institut Français de l'Energie (I.F.E.), 3 Rue Henri Heine, 75016 Paris
T: (1) 45244614; Tx: 620232; Fax: (1) 40500754
Founded: 1951; Members: 46
Pres: Albert Robin; Gen Secr: Yves Chainet
Focus: Energy
Periodical
Actualité Combustibles Energie (weekly) . 04568

Institut Français d'Histoire Sociale, c/o Centre de Documentation et de Recherche, Archives Nationales, 60 Rue des Francs-Bourgeois, 75141 Paris Cédex 03
T: (1) 47072864
Founded: 1948; Members: 200
Pres: Jean-Pierre Chaline; Gen Secr: Denise Fauvel-Rouif
Focus: Hist 04569

Institut Géographique National, 136bis Rue de Grenelle, 75700 Paris
T: (1) 43988000; Fax: (1) 43988400
Founded: 1940; Members: 2000
Focus: Geography
Periodical
Bulletin d'Information de l'Institut Géographique National (weekly) 04570

Institut International d'Administration Publique, 2 Av de l'Observatoire, 75006 Paris
Fax: (1) 46332638
Focus: Public Admin
Periodical
Revue Française d'Administration Publique (weekly) 04571

Institut International de Droit d'Expression et d'Inspiration Françaises (I.D.E.F.), 27 Rue Oudinot, 75007 Paris
T: (1) 47831738; Fax: (1) 47831736
Founded: 1964
Pres: Raymond Barre; Gen Secr: Pierre Decheix
Focus: Law
Periodicals
Bulletin de l'I.D.E.F. (weekly)
Indépendance et Coopération (weekly) . 04572

Institut International de Planification de l'Education (IIPE) / International Institute for Educational Planning, 7 Rue Eugène Delacroix, 75116 Paris
T: (1) 45xk037700042822; Tx: 620074; CA: EDUPLAN Paris; Fax: (1) 40728366
Founded: 1963
Pres: Victor Urquidi; Gen Secr: Jacques Hallak
Focus: Educ
Periodical
IIEP Newsletter (weekly) 04573

Institut International des Droits de l'Homme / International Institute of Human Rights, 1 Quai Lezay-Marnésia, 67000 Strasbourg
Fax: 88363855
Founded: 1969
Pres: Denise Bindschedler-Robert; Gen Secr: Jean-Bernard Marie
Focus: Law 04574

Institut International du Froid (IIF) / International Institute of Refrigeration, 177 Blvd Malesherbes, 75017 Paris
T: (1) 42273235; Tx: 643269; Fax: (1) 47631798
Founded: 1908; Members: 800
Pres: L. Lucas
Focus: Air Cond
Periodicals
Bulletin (weekly)
International Journal of Refrigeration (weekly) 04575

Institut International du Théâtre (IIT), 1 Rue Miollis, 75732 Paris Cédex 15
T: (1) 45682650; Fax: (1) 43068798
Founded: 1948; Members: 88
Gen Secr: André-Louis Perinetti
Focus: Perf Arts 04576

Intergovernmental Council for the International Hydrological Programme (IHP), c/o UNESCO, Pl de Fontenoy, 75700 Paris
Founded: 1975; Members: 30
Focus: Water Res
Periodicals
Studies and Reports in Hydrology
Technical Documents in Hydrology
Technical Papers in Hydrology 04577

International Academy of the History of Science (IAHS), 12 Rue Colbert, 75002 Paris
Founded: 1929; Members: 260
Focus: Hist; Sci
Periodical
Archives Internationales d'Histoire des Sciences (weekly) 04578

International Association Futuribles (AIF), 55 Rue de Varenne, 75434 Paris Cédex 07
T: (1) 42226310; Fax: (1) 42226554
Founded: 1960
Gen Secr: Hugues de Jouvenel
Focus: Futurology
Periodicals
Actualités prospectives (weekly)
Revue Futuribles (weekly)
Vigie info (weekly) 04579

International Association of Agricultural Information Specialists, BP 5035, 34032 Montpellier Cédex 1
Fax: 67615820
Founded: 1955; Members: 660
Gen Secr: Dr. J. Van der Burg
Focus: Doc; Agri
Periodical
Bulletin (weekly) 04580

International Association of Engineering Geology (I.A.E.G.), c/o Laboratoire Central des Ponts et Chaussées, 58 Blvd Lefèbvre, 75732 Paris Cédex 15
T: (1) 40435243
Founded: 1964; Members: 5000
Pres: Prof. Ricardo Oliveira; Gen Secr: Dr. L. Primel
Focus: Geology
Periodical
Newsletter of the IAEG (weekly) 04581

International Association of Sanskrit Studies, 52 Rue du Cardinal Lemaire, 75231 Paris Cédex 05
Founded: 1977
Gen Secr: S. Lienhard
Focus: Lit; Ling; Cultur Hist; Arts; Archeol
Periodicals
Indologica Taurinensia
Newsletter 04582

International Astronautical Association (IAA), 3-5 Rue Mario Nikis, 75015 Paris
T: (1) 45674260; Tx: 205917; Fax: (1) 42737537
Founded: 1950; Members: 125
Gen Secr: Michelle Claudin
Focus: Aero
Periodical
Journal of the IAA (weekly) 04583

International Commission for Plant-Bee Relationships (ICPBR), c/o Laboratoire de Zoologie, INRA, 86600 Lusignan
T: 49556000; Fax: 49556088
Founded: 1951; Members: 190
Gen Secr: J.N. Tasei
Focus: Botany
Periodical
Reports of Meetings 04584

International Council on Archives (ICA), 60 Rue des Francs-Bourgeois, 75003 Paris
T: (1) 40276306; Fax: (1) 40276625
Founded: 1948; Members: 1300
Pres: Prof. Jean-Pierre Wallot; Gen Secr: Dr. Charles Kecskemeti
Focus: Archives
Periodicals
Archivum: International Review on Archives (weekly)
Bulletin CIA (weekly)
Janus (weekly) 04585

International Economic Association (I.E.A.), 23 Rue Campagne Première, 75014 Paris
T: (1) 43279144; Tx: 264918; Fax: (1) 42799216
Founded: 1950; Members: 50
Gen Secr: Jean-Paul Fitoussi
Focus: Econ
Periodical
Newsletter (1-2 times annually) 04586

International Research Council on Biokinetics of Impacts (IRCOBI), 109 Av Salvador Allende, 69500 Bron
T: 72362420; Fax: 72362437
Founded: 1973
Focus: Traumatology
Periodical
Proceedings (weekly) 04587

International Union for Health Education, 15 Rue de l'Ecole de Médecine, 75270 Paris Cédex 06
Founded: 1951; Members: 1200
Pres: Dr. M. Rajala
Focus: Educ
Periodical
HYGIE: International Journal of Health Education (weekly) 04588

International Union of Microbiological Societies (IUMS), c/o Institut de Biologie Moléculaire et Cellulaire du CNRS, 15 Rue Descartes, 67084 Strasbourg
T: 88417022; Fax: 88610680
Founded: 1927; Members: 68
Pres: Rita Colwell; Gen Secr: Marc H.V. Van Regenmortel
Focus: Microbio
Periodicals
Archives of Virology (weekly)
International Journal of Food Microbiology (weekly)
International Journal of Systematic Bacteriology (weekly)
Journal of Medical and Veterinary Mycology (weekly)

World Journal of Microbiology and Biotechnology (weekly) 04589

Jeunesses Littéraires de France, 117 Blvd Saint-Germain, 75279 Paris Cédex 06
Focus: Lit 04590

Jeunesses Musicales de France, 20 Rue Geoffroy l'Asnier, 75004 Paris
T: (1) 42781954; Tx: 215901
Founded: 1944; Members: 3800
Pres: J.M. Tournier; Gen Secr: Robert Berthier
Focus: Music 04591

Joint Committee on Climatic Changes and the Ocean, c/o UNESCO, 7 Pl de Fontenoy, 75700 Paris
T: (1) 45681000; Tx: 204461; CA: UNESCO, Paris; Fax: (1) 43061122
Focus: Oceanography; Physics 04592

Ligue Française contre la Sclérose en Plaques (L.F.S.E.P.), 17 Blvd Auguste Blanqui, 75013 Paris
T: (1) 42360182; Fax: (1) 42367158
Focus: Pathology
Periodical
Courier de la SEP (3 times annually) . 04593

Ligue Française de l'Enseignement et de l'Education Permanente, 3 Rue Récamier, 75341 Paris Cédex 07
T: (1) 45443871
Founded: 1866; Members: 3400000
Focus: Educ; Adult Educ 04594

Ligue Française de l'Enseignement Oroleis de Paris, 23 Rue Dagorno, 75012 Paris
T: (1) 43075930
Focus: Educ 04595

Ligue Française d'Hygiène Mentale (L.F.H.M.), 11 Rue Tronchet, 75008 Paris
T: (1) 42662070; Fax: (1) 42664489
Members: 100
Pres: Dr. Claude Leroy; Gen Secr: Natalie Alessandrini
Focus: Hygiene 04596

Ligue Internationale de l'Enseignement, de l'Education et de la Culture Populaire (LIEEP), 3 Rue Récamier, 75007 Paris Cédex 07
T: (1) 43589730
Founded: 1947; Members: 48
Gen Secr: F. Coursin
Focus: Educ
Periodical
Informations Internationales (weekly) . . 04597

Ligue Nationale contre le Cancer, 1 Av Stéphen Pichon, 75013 Paris
T: (1) 44068080; Fax: (1) 45865678
Founded: 1920; Members: 700000
Focus: Cell Biol & Cancer Res
Periodical
Vivre (weekly) 04598

Ligue pour la Protection des Oiseaux (LPO), La Corderie Royale, BP 263, 17305 Rochefort Cédex 05
T: 46821250
Founded: 1912; Members: 17000
Pres: Allan B. Dubourg; Gen Secr: Michel Terrasse
Focus: Ornithology
Periodical
L'Oiseau (weekly) 04599

Médecins sans Frontières, 8 Rue Saint-Sabin, 75011 Paris
T: (1) 40212929; Tx: 214360; Fax: (1) 48066868
Founded: 1971; Members: 2600
Pres: Dr. Rony Brauman; Gen Secr: Dr. Bernard Pécoul
Focus: Public Health; Med
Periodical
MSF: Bulletin (weekly) 04600

Mouvement Français pour la Qualité (MFQ), 5 Esplanade Charles de Gaulle, 92733 Nanterre Cédex
T: (1) 47290929; Fax: (1) 47253221
Founded: 1991; Members: 3500
Focus: Econ
Periodicals
Newsletter (11 times annually)
Newsletter: Qualité en Mouvement (5 times annually)
Regard sur la Qualité (weekly) 04601

Mouvement Universel de la Responsabilité Scientifique (MURS) / Universal Movement for Scientific Responsibility, 127 Blvd Saint-Michel, 75005 Paris
T: (1) 43264398; Fax: (1) 43543550
Founded: 1974
Pres: Jean Dausset; Gen Secr: Michel Barrault
Focus: Sci
Periodicals
Cahiers du MURS (3-4 times annually)
Lettre aux Générations 2000 04602

Naturalistes Parisiens, 45 Rue de Buffon, 75005 Paris
Founded: 1904; Members: 450
Pres: Adrien Roudier; Gen Secr: Claude Dupuis
Focus: Nat Sci
Periodical
Cahiers des Naturalistes: Bulletin des Naturalistes Parisiens (3 times annually) 04603

Office Général du Bâtiment et des Travaux Publics (O.G.B.T.P.), 55 Av Kléber, 75784 Paris Cédex 16
T: (1) 47201020
Founded: 1919
Pres: Michel Marconnet; Gen Secr: J. Ansaloni
Focus: Civil Eng 04604

Office National d'Etudes et des Recherches Aérospatiales, 29 Av de la Division Leclerc, 92320 Chatillon
Focus: Aero 04605

Office National d'Information sur les Enseignements et les Professions (O.N.I.S.E.P.), 48-50 Rue Albert, 75013 Paris
T: (1) 40776000; Fax: (1) 202962
Founded: 1970; Members: 576
Pres: Michel Praderie; Gen Secr: Pierre Mondon
Focus: Adult Educ; Educ
Periodicals
Avenirs (weekly)
Les Cahiers 04606

Ordre des Architectes, 140 Av Victor Hugo, 75016 Paris
T: (1) 47205856, 7205864
Founded: 1940
Focus: Archit
Periodical
Architectes (weekly) 04607

Ordre des Chirurgiens Dentistes, 174 Rue de Rivoli, 75001 Paris
T: (1) 42604973
Focus: Surgery; Dent 04608

Ordre des Géomètres-Experts (O.G.E.), 40 Av Hoche, 75008 Paris
T: (1) 45632426
Founded: 1946; Members: 2000
Pres: Jean Lamaison; Gen Secr: Marie Hélène Vellieux
Focus: Surveying
Periodical
Géomètre (weekly) 04609

Ordre National des Chirurgiens Dentistes, 7bis Rue Merimée, 75016 Paris
T: (1) 47235062
Focus: Surgery; Dent 04610

Ordre National des Médecins, 60 Blvd Latour-Maubourg, 75007 Paris
T: (1) 45513879
Focus: Med
Periodical
Bulletin de l'Ordre des Médecins (weekly) 04611

Organisation de la Jeunesse Esperantiste Française / French Youth Esperanto Organization, 4bis Rue de la Cerisaie, 75004 Paris
T: (1) 42786886
Founded: 1970; Members: 250
Pres: Philippe Berizzi; Gen Secr: Laurent Vignaud
Focus: Ling
Periodicals
Jefo Informa (weekly)
Koncize (weekly) 04612

Organisation Européenne pour l'Equipement de l'Aviation Civile, 17 Rue Hamelin, 75783 Paris Cédex 16
T: (1) 45057188; Fax: (1) 45530393
Founded: 1963; Members: 63
Pres: Sir John Charnley; Gen Secr: B. Perret
Focus: Electronic Eng 04613

Organisation Internationale contre le Trachome, c/o Hôpital de Créteil, Université Paris XII, 40 Av de Verdun, 94010 Créteil
Fax: (1) 48987787
Founded: 1929; Members: 600
Pres: Prof. Gabriel Coscas
Focus: Ophthal
Periodical
International Review of Trachoma (weekly) 04614

Organisation Internationale de Métrologie Légale (OIML), 11 Rue Turgot, 75009 Paris
T: (1) 48781282, 42852711; Tx: 215463; Fax: (1) 42821727
Founded: 1955; Members: 50
Focus: Standards
Periodicals
Bulletin de l'Organisation Internationale de Métrologie Légale (weekly)
OIML International Recommendations and International Documents 04615

Organisation Internationale de Recherche sur la Cellule, c/o UNESCO, Div of Scientific Research and Higher Education, 2 Pl de Fontenoy, 75007 Paris
T: (1) 45688378
Founded: 1962; Members: 395
Gen Secr: Dr. G.N. Cohen
Focus: Cell Biol & Cancer Res 04616

Organisation Scientifique des Industries du Bâtiment, 53 Blvd Lannes, 75116 Paris
T: (1) 45047089
Focus: Civil Eng 04617

Paris et son Histoire, 82 Rue Taitbout, 75009 Paris
T: (1) 45262677
Founded: 1952; Members: 5000
Focus: Hist
Periodical
Bulletin Paris et son Histoire (weekly) . 04618

Le Pays Bas-Normand, 21 Rue de la Planchette, 61100 Flers
T: 33643387
Founded: 1908; Members: 420
Pres: Dr. Jean Fournée; Gen Secr: Yves Letortu
Focus: Hist; Arts
Periodical
Revue (weekly) 04619

P.E.N. Maison Internationale, 6 Rue François Miron, 75004 Paris
T: (1) 42773787
Founded: 1921; Members: 600
Pres: René Tavernier; Gen Secr: Jean Orizet
Focus: Lit
Periodical
Informations (weekly) 04620

Program on Man and the Biosphere (MAB), c/o UNESCO, 7 Pl de Fontenoy, 75352 Paris Cédex 07
T: (1) 45684068; CA: 204461; Fax: (1) 40659897
Founded: 1971; Members: 112
Pres: Tomas Azcarate y Bang; Gen Secr: Pierre Lasserre
Focus: Ecology
Periodical
InfoMAB (bi-annually) 04621

Réunion Internationale des Laboratoires d'Essais et de Recherche sur les Matériaux et les Constructions (RILEM), 61 Av du Président Wilson, 94235 Cachan Cédex
T: (1) 47402397; Fax: (1) 47402074
Founded: 1947
Pres: Prof. Torben C. Hansen; Gen Secr: Michel Brusin
Focus: Civil Eng; Materials Sci 04622

Scientific Committee on Problems of the Environment (SCOPE), 51 Blvd de Montmorency, 75016 Paris
Fax: (1) 42881466
Founded: 1969; Members: 55
Pres: Prof. J.W.B. Stewart; Gen Secr: Prof. P. Bourdeau
Focus: Ecology
Periodicals
Newsletter (3 times annually)
SCOPE Series 04623

Societas Oto-Rhino-Laryngologia Latina, c/o Clinique ORL, 34000 Montpellier
Focus: Otorhinolaryngology 04624

Société Académique des Arts Libéraux de Paris, 3 Av Chanzy, BP 49, 94210 La Varenne Saint-Hilaire
T: (1) 42833603
Focus: Arts 04625

Société Africaine de Culture (SAC), 18 Rue des Ecoles, 75005 Paris
T: (1) 43297572
Founded: 1956
Focus: Ethnology 04626

Société Anatole France, 15 Rue Gustave Courbet, 75116 Paris
T: (1) 45537093
Founded: 1933; Members: 350
Focus: Lit 04627

Société Anatomique de Paris, 45 Rue des Saints-Péres, 75270 Paris Cédex 06
Founded: 1802
Focus: Anat
Periodical
Bulletins (weekly) 04628

Société Archéologique de Touraine (SAT), c/o Bibliothèque Municipale, 37042 Tours Cédex
Founded: 1840; Members: 1300
Focus: Archeol
Periodicals
Bulletin (weekly)
Mémoires (weekly) 04629

Société Archéologique et Historique du Limousin, 54 Rue Bourneville, 87032 Limoges Cédex
Founded: 1845
Focus: Archeol; Hist
Periodical
Bulletin de la Société Archéologique et Historique du Limousin (weekly) 04630

Société Asiatique de Paris (S.A.), 3 Rue Mazarine, 75006 Paris
T: (1) 44414314; Fax: (1) 44414316
Founded: 1822; Members: 650
Pres: André Caquot
Focus: Ethnology
Periodical
Journal Asiatique (bi-annually) 04631

Société Astronomique de Bordeaux, 71 Rue du Loup, 33000 Bordeaux
Founded: 1909; Members: 75
Focus: Astronomy 04632

Société Astronomique de France (S.A.F.), 3 Rue Beethoven, 75016 Paris
T: (1) 42241374; Fax: (1) 42307547
Founded: 1887; Members: 3000
Pres: Jean-Claude Ribes; Gen Secr: Joel Minois
Focus: Astronomy
Periodicals
L'Astronomie (weekly)
Observations et Travaux (weekly) . . . 04633

Société Astronomique de Lyon, c/o Observatoire de Lyon, 69230 Saint-Genis-Laval
Founded: 1901; Members: 160
Focus: Astronomy 04634

Société Botanique de France (S.B.F.), c/o Pharmaceutical Faculty, Rue J.B. Clement, 92296 Châtenay-Malabry Cédex
T: (1) 46835520; Fax: (1) 46831303
Founded: 1854; Members: 500
Pres: G. Ducreux; Gen Secr: G.G. Aymonin
Focus: Botany
Periodical
Acta Botanica Gallica (3-4 times annually) 04635

Société Cartographique de France, 48 Rue de Charenton, 75012 Paris
T: (1) 43072533
Focus: Cart 04636

Société Centrale d'Apiculture, 41 Rue Pernety, 75014 Paris
T: (1) 43062908
Focus: Bio 04637

Société d'Agriculture, Sciences, Belles-Lettres et Arts d'Orléans, 5 Rue Antoine Petit, 45000 Orléans
Founded: 1809; Members: 73
Focus: Agri; Sci; Lit; Arts 04638

Société d'Anthropologie de Paris (S.A.P.), 1 Rue René Panhard, 75013 Paris
Founded: 1859; Members: 550
Pres: Marie-Claude Chamla; Gen Secr: D. Ferembach
Focus: Anthro
Periodical
Bulletins et Mémoires (weekly) 04639

Société d'Archéologie et d'Histoire de la Manche, BP 600, 50010 Saint-Lo Cédex
Focus: Archeol; Hist 04640

Société d'Archéologie et d'Histoire de l'Aunis, Muséum Lafaille, 26 Rue Albert Ier, 17000 La Rochelle
Founded: 1970; Members: 250
Pres: Jean Floret
Focus: Hist; Archeol 04641

Société de Biogéographie, 57 Rue Cuvier, 75231 Paris Cédex 05
Founded: 1924; Members: 350
Gen Secr: Dr. C. Sastre
Focus: Geography
Periodicals
Compte Rendu des Séances
Mémoires 04642

Société de Biologie, c/o Collège de France, 11 Pl M. Berthelot, 75231 Paris Cédex 05
T: (1) 44271340
Founded: 1848; Members: 510
Pres: Prof. Yves Cohen; Gen Secr: Prof. Jacques Pulonovski
Focus: Bio
Periodical
Compte rendus de la Société de Biologie (weekly) 04643

Société de Biométrie Humaine, 41 Rue Gay Lussac, 75005 Paris
T: (1) 46330596
Founded: 1932; Members: 300
Gen Secr: J. Delmas
Focus: Bio; Anthro
Periodical
Cahiers d'Anthropologie et Biométrie Humaine (weekly) 04644

Société de Chimie Biologique, 4 Av de l'Observatoire, 75270 Paris Cédex 06
T: (1) 43251260
Founded: 1914; Members: 1129
Pres: P. Louisot; Gen Secr: J. Agneray; J.P. Ebel
Focus: Biochem
Periodicals
Biochimie (weekly)
Regard sur la Biochimie (weekly) . . . 04645

Société de Chimie Industrielle, 28 Rue Saint-Dominique, 75007 Paris
T: (1) 45556946; Fax: (1) 45559862
Founded: 1917; Members: 4000
Focus: Chem
Periodical
L'Actualité Chimique et Analysis (weekly) 04646

Société de Chimie Thérapeutique (SCT), 3 Rue Jean-Baptiste Clement, 92290 Châtenay Malabry
Founded: 1966; Members: 750
Focus: Chem 04647

Société de Chirurgie de Marseille, Hôtel Dieu, 13000 Marseille
Focus: Surgery 04648

Société de Chirurgie de Toulouse, 18 Rue de Languedoc, 31000 Toulouse
Focus: Surgery 04649

Société de Chirurgie Thoracique et Cardio-Vasculaire de Langue Française, c/o Hôpital Marie Lannelongue, 133 Av de la Résistance, 92350 Le Plessis Robinson
T: 46302133
Founded: 1948; Members: 420
Gen Secr: Dr. P. Levasseur
Focus: Surgery

Periodical
Annales de chirurgie (weekly) 04650

Société d'Economie et de Science Sociale, 80 Rue Vaneau, 75007 Paris
Founded: 1856; Members: 300
Pres: Edouard Secretan
Focus: Sociology; Econ
Periodical
Les Etudes Sociales (weekly) 04651

Société d'Economie Politique, c/o Librairie Sirey, 22 Rue Soufflot, 75005 Paris
Pres: D. Villey
Focus: Econ; Poli Sci
Periodical
Annales d'Economie Politique 04652

Société de Démographie Historique, 54 Blvd Raspail, 75006 Paris
T: (1) 49542556
Founded: 1963
Pres: Jean-Pierre Bardet; Gen Secr: Fauve-Chamoux
Focus: Hist
Periodicals
Annales de démographie historique (weekly)
Bulletin dh (3 times annually) 04653

Société de Géographie, 184 Blvd Saint-Germain, 75006 Paris
T: (1) 45485462
Founded: 1821; Members: 900
Focus: Geography
Periodical
Acta Geographica (weekly) 04654

Société de Géographie Commerciale de Bordeaux, 71 Rue du Loup, 33000 Bordeaux
Focus: Geography 04655

Société de Géographie Commerciale de Paris, 8 Rue Roquépine, 75008 Paris
T: (1) 42664626
Founded: 1873; Members: 490
Pres: Jacques Augarde
Focus: Geography
Periodical
Revue Economique Française (weekly) . 04656

Société de Géographie de Lille, 77 Rue Nationale, 59800 Lille
T: 20572745
Founded: 1880; Members: 1500
Focus: Geography
Periodical
Hommes et Terres du Nord (weekly) . . 04657

Société de Géographie de Lyon, 74 Rue Pasteur, 69007 Lyon
Founded: 1873
Focus: Geography
Periodical
Revue de Géographie de Lyon (weekly) 04658

Société de Géographie de Toulouse, Pl d'Assézat, 31000 Toulouse
Focus: Geography 04659

Société de Législation Comparée, 28 Rue Saint-Guillaume, 75007 Paris
T: (1) 45444467; Fax: (1) 45494165
Founded: 1869; Members: 1300
Pres: Georges Flécheux; Gen Secr: Prof. Xavier Blanc-Jouvan
Focus: Law
Periodicals
Journées de la Société de Legislation Comparée (weekly)
Revue Internationale de Droit Comparé (weekly)
. . . . 04660

Société de l'Histoire de France, c/o Ecole des Chartes, 19 Rue de la Sorbonne, 75005 Paris
Founded: 1834; Members: 400
Pres: Pierre Chaunu; Gen Secr: André Vernet
Focus: Hist
Periodicals
Annuaire-Bulletin (weekly)
Mémoires 04661

Société de l'Histoire de l'Art Français, Palais du Louvre, Pavillon de Marsan, 107 Rue de Rivoli, 75001 Paris
Founded: 1872; Members: 1000
Pres: Jean-Pierre Samoyault; Gen Secr: Bruno Foucart
Focus: Hist; Arts; Cultur Hist
Periodicals
Archives de l'Art Français (weekly)
Bulletin de la Société de l'Histoire de l'Art Français (weekly) 04662

Société de l'Histoire du Protestantisme Français (S.H.P.F.), 54 Rue des Saints-Pères, 75007 Paris
T: (1) 45486207
Founded: 1852
Pres: Roger Zuber; Gen Secr: Jean-Hugues Carbonnier
Focus: Hist; Rel & Theol
Periodicals
Bulletin Historique et Littéraire (weekly)
Cahiers de Généalogie Protestante (weekly)
. . . . 04663

Société de Linguistique de Paris, c/o Ecole Pratique des Hautes Etudes, 47 Rue des Ecoles, 75005 Paris
Founded: 1864; Members: 750
Pres: Culioli
Focus: Ling
Periodicals
Bulletin de la Société de Linguistique de Paris (weekly)
Collection linguistique (weekly) 04664

Société de Linguistique Romane, 8 Quai Rouget de Lisle, 67000 Strasbourg
T: 88350084
Founded: 1925; Members: 1000
Pres: Prof. Dr. M. Pfister; Gen Secr: Prof. Dr. G. Straka
Focus: Ling 04665

Société de Médecine, Chirurgie et Pharmacie de Toulouse, Pl d'Assézat, 31000 Toulouse
Focus: Surgery; Med; Pharmacol 04666

Société de Médecine de Strasbourg, c/o Faculté de Médecine, 4 Rue Kirschleger, 67085 Strasbourg Cédex
Founded: 1919; Members: 450
Pres: Prof. J.L. Schlienger
Focus: Med
Periodical
Bulletin (weekly) 04667

Société de Médecine et de Chirurgie de Bordeaux, 15 Rue du Professeur Demons, 33000 Bordeaux
Focus: Surgery; Med 04668

Société de Médecine et d'Hygiène du Travail, 15 Rue de l'Ecole de Médecine, 75006 Paris
T: (1) 6331188
Members: 250
Focus: Hygiene 04669

Société de Médecine Légale et de Criminologie de France, 2 Pl Mazas, 75012 Paris
T: (1) 43434254
Founded: 1868; Members: 400
Pres: A. Marin; Gen Secr: C. Piva
Focus: Criminology; Med; Forensic Med; Toxicology
Periodicals
Droit Médical
Médecine Légale (weekly) 04670

Société d'Emulation Historique et Littéraire d'Abbéville, 81 Pl du Général de Gaulle, 80000 Abbéville
Founded: 1797; Members: 300
Focus: Hist; Lit
Periodical
Bulletin (weekly) 04671

Société de Mythologie Française, 3 Rue Saint-Laurent, 75010 Paris
Founded: 1950
Focus: Hist
Periodicals
Bulletin (3 times annually)
Bulletin de la Société de Mythologie Française
. . . . 04672

Société d'Encouragement pour l'Industrie Nationale (S.E.I.N.), 4 Pl Saint-Germain-des-Prés, 75006 Paris
T: (1) 45485561
Founded: 1801
Gen Secr: Christian Pascault
Focus: Econ
Periodical
L'Industrie Nationale 04673

Société de Neurochirurgie de Langue Française, c/o Hôpital Neurologique, 59 Blvd Pinel, 69394 Lyon Cédex 3
T: 8540077
Founded: 1950; Members: 523
Pres: F. Cohadon; Gen Secr: G. Fischer
Focus: Surgery
Periodical
Neurochirurgie (weekly) 04674

Société de Neurophysiologie Clinique de Langue Française, c/o Dr. Plouin, Hôpital Saint-Vincent de Paul, 75674 Paris Cédex 14
Founded: 1949; Members: 800
Focus: Med; Physiology; Neurology
Periodical
Neurophysiologie Clinique 04675

Société de Neuropsychologie de Langue Française, 47 Blvd de l'Hôpital, 75013 Paris
T: (1) 61498926; Fax: (1) 61499524
Founded: 1977; Members: 450
Gen Secr: Alain Agniel
Focus: Psych
Periodical
Revue de Neuropsychologie (weekly) . . 04676

Société de Nutrition et de Diététique de Langue Française, Hôtel Dieu, 1 Pl du Parvis Notre Dame, 75004 Paris
T: (1) 42348452; Fax: (1) 40510057
Founded: 1963
Focus: Nutrition
Periodical
Cahiers de Nutrition et de Diététique (weekly)
. . . . 04677

Société de Pathologie Comparée, 4 Rue Théodule Ribot, 75017 Paris
T: (1) 46225319
Founded: 1901
Focus: Pathology 04678

Société de Pathologie Exotique, 25 Rue du Docteur Roux, 75015 Paris
T: (1) 45668869
Founded: 1908; Members: 446
Focus: Pathology; Trop Med
Periodical
Bulletin de la Societé de Pathologie Exotique (5 times annually) 04679

Société de Pharmacie de Bordeaux, c/o Faculté de Pharmacie, 3 Pl de la Victoire, 33076 Bordeaux Cédex
T: 56913424
Founded: 1834; Members: 300
Focus: Pharmacol
Periodical
Bulletin (weekly) 04680

Société de Pharmacie de Lyon, 8 Av Rockefeller, 69373 Lyon Cédex 08
Focus: Pharmacol
Periodical
Bulletin des Travaux de la Société de Pharmacie de Lyon (weekly) 04681

Société de Pharmacie de Marseille, 92 Rue Blanqui, 13001 Marseille
Focus: Pharmacol 04682

Société de Philosophie de Toulouse, 4 Rue Albert Lantmann, 31000 Toulouse
Focus: Philos 04683

Société de Psychologie Médicale de Langue Française, 28 Rue du Ranelagh, 75016 Paris
Founded: 1960; Members: 200
Focus: Psych
Periodical
Psychologie Médicale (10 times annually) 04684

Société de Réanimation de Langue Française (S.R.L.F.) / French-Language Society for Reanimation, c/o C.H.V. Côte de Nacre, Av de la Côte de Nacre, 14040 Caen
T: 31064708; Tx: 31475157
Founded: 1971; Members: 1000
Gen Secr: Prof. P. Charbonneau
Focus: Med
Periodical
Bulletin S.R.L.F. 04685

Société de Recherches et d'Etudes Historiques Corses, c/o Musée Fesch, 20000 Ajaccio
Focus: Hist 04686

Société de Recherches Géophysiques, 81 Rue Laugier, 75017 Paris
T: (1) 47544391
Focus: Geophys 04687

Société de Recherches Pharmaceutiques et Cientifiques, 6 Rue Lincoln, 75008 Paris
T: (1) 42252265
Focus: Pharmacol 04688

Société de Recherches Psychothérapiques de Langue Française, c/o Dr. J.C. Benoit, 1bis Rue Deroisin, 78000 Versailles
T: 39514868
Founded: 1961; Members: 250
Focus: Therapeutics
Periodical
Annales de Psychothérapie (weekly) . . 04689

Société d'Ergonomie de Langue Française (SELF), c/o Laboratoire de Physiologie, Faculté de Médecine de Caen, CHU-Côte de Nâcre, 14032 Caen Cédex
T: 31448112
Founded: 1963; Members: 308
Gen Secr: Prof. Michel Pottier
Focus: Physiology; Eng; Psych
Periodicals
Bulletin de Liaison de la SELF (weekly)
Le Travail Humain (weekly) 04690

Société des Africanistes, c/o Musée de l'Homme, Pl du Trocadéro, 75116 Paris
T: (1) 47277255
Founded: 1931; Members: 350
Gen Secr: A. Deluz
Focus: Ethnology
Periodical
Journal des Africanistes (weekly) 04691

Société des Agriculteurs de France (SAF), 8 Rue d'Athènes, 75009 Paris
T: (1) 42857227
Focus: Agri 04692

Société des Américanistes, c/o Musée de l'Homme, 17 Pl du Trocadéro, 75116 Paris
T: (1) 47046311
Founded: 1895; Members: 500
Focus: Ethnology; Archeol
Periodical
Journal (weekly) 04693

Société des Amis de la Revue de Géographie de Lyon, 74 Rue Pasteur, 69007 Lyon
Founded: 1923
Pres: J. Bethemont
Focus: Geography
Periodical
Revue de Géographie de Lyon (weekly) 04694

Société des Amis de Marcel Proust et des Amis d'Illiers-Combray, 49 Rue Vineuse, 75016 Paris
Focus: Lit 04695

Société des Amis d'Eugène Delacroix, 6 Pl Furstenberg, 75006 Paris
T: (1) 45622934
Founded: 1928; Members: 245
Focus: Arts 04696

Société des Amis du Louvre, 34 Quai du Louvre, 75001 Paris
Fax: (1) 40205344
Founded: 1897; Members: 37000
Pres: François Puaux; Gen Secr: Louis-Antoine Prat
Focus: Fine Arts
Periodical
Chronique 04697

Société des Amis du Musée de l'Homme, c/o Musée de l'Homme, Pl du Trocadéro, 75116 Paris
T: (1) 47046210
Focus: Ethnology; Anthro; Hist 04698

Société des Anciens Textes Français (S.A.T.F), c/o Ecole des Chartes, 19 Rue de la Sorbonne, 75005 Paris
T: (1) 46334182
Founded: 1875
Pres: Prof. A. Micha; Gen Secr: Prof. J. Monfrin
Focus: Lit 04699

Société des Auteurs et Compositeurs Dramatiques, 11bis Rue Ballu, 75009 Paris
T: (1) 40234444; Fax: (1) 43267428
Founded: 1777; Members: 28000
Pres: Claude Brulé
Focus: Music; Lit; Perf Arts
Periodical
Revue 04700

Sociéte des Bibliophiles de Guyenne, c/o Bibliothèque Municipale, 3 Rue Mably, 33075 Bordeaux
Focus: Libraries & Bk Sci
Periodical
Revue Française d'Histoire du Livre (weekly)
. . . . 04701

Société des Ecoles du Dimanche, 152 Rue Léon Maurice Nordmann, 75013 Paris
T: (1) 43318659
Focus: Educ 04702

Société des Etudes Latines (CSSF), 1 Rue Victor-Cousin, 75230 Paris Cédex 05
Founded: 1923
Pres: Alain Michel
Focus: Lit
Periodicals
Collection d'Etudes Latines
Revue des Etudes Latines 04703

Société des Etudes Renaniennes, 167 Blvd Malesherbes, 75017 Paris
Focus: Rel & Theol 04704

Société des Experts-Chimistes de France, 39bis Rue de Dantzig, 75015 Paris
Founded: 1912; Members: 1200
Pres: Jean-Guy Faugère; Gen Secr: Colette Courcelles
Focus: Chem
Periodical
Annales des Falsifications de l'Expertise Chimique et Toxicologique 04705

Société des Explorateurs et des Voyageurs Français, 9bis Av de Montespan, 75116 Paris
T: (1) 47042272
Focus: Geography 04706

Société des Gens de Lettres de France, 38 Rue du Faubourg Saint-Jacques, 75014 Paris
T: (1) 43541866
Founded: 1838
Pres: Didier Decoin; Gen Secr: Jacques Bens
Focus: Lit 04707

Société des Lépidopteristes Français, c/o Muséum d'Histoire Naturelle, 45bis Rue Buffon, 75005 Paris
Focus: Entomology 04708

Société des Lettres, Sciences et Arts de la Haute-Auvergne, c/o Archives Départementales, Rue du 139. R.I., 15012 Aurillac Cédex
Founded: 1899; Members: 650
Pres: Marcel Delzons; Gen Secr: Christian Marchi
Focus: Geology; Botany; Hist; Lit; Archeol 04709

Société des Médecins-Chefs des Compagnies Européennes d'Aviation, c/o Service Médical d'Air France, 1 Sq Max Hymans, 75015 Paris
Focus: Med 04710

Société des Océanistes, c/o Musée de l'Hommes, Pl du Trocadéro, 75116 Paris
T: (1) 47046340
Founded: 1938; Members: 560
Pres: J. Garanger; Gen Secr: Michel Panoff
Focus: Ethnology
Periodicals
Journal (weekly)
Publications 04711

Société des Poètes Français, 38 Rue du Faubourg Saint-Jacques, 75014 Paris
Founded: 1902; Members: 1600
Pres: Dr. Pierre Osenat; Gen Secr: Paule Le Milbeau
Focus: Lit
Periodical
Bulletin Trimestriel (3 times annually) . . *04712*

Société des Professeurs de Dessin et d'Arts Plastiques de l'Enseignement Secondaire, c/o Lycée Buffon, 16 Blvd Pasteur, 75015 Paris
Founded: 1901; Members: 1050
Focus: Educ; Arts *04713*

Société des Sciences, Arts et Belles-Lettres de Bayeux (S.A.B.L.), Hôtel de Ville, 14400 Bayeux
Founded: 1891; Members: 100
Focus: Arts; Lit; Sci *04714*

Société des Sciences et Arts, 22 Rue La Bourdonnais, 97400 Saint-Denis
Focus: Sci; Arts *04715*

Société des Sciences Historiques et Naturelles de la Corse, c/o Lycée Marbeuf, 20200 Bastia
Focus: Hist; Nat Sci *04716*

Société des Sciences Naturelles de Bourgogne, c/o Mus. Hist. Nat., 1 Av Albert Ier, 21000 Dijon
Founded: 1913
Pres: Prof. Jean Pages; Gen Secr: Gérard Ferrière
Focus: Nat Sci
Periodical
Bulletin Scientifique de Bourgogne (weekly)
. . . . *04717*

Société des Sciences Physiques et Naturelles de Bordeaux, c/o Université de Bordeaux, 351 Cours de la Libération, 33405 Talence
Founded: 1850; Members: 300
Focus: Nat Sci; Physics *04718*

Société de Statistique de Paris, c/o Institut Henri Poincaré, 11 Rue Pierre et Marie Curie, 75005 Paris
Founded: 1860; Members: 1040
Pres: Georges La Calue; Gen Secr: Annie Morin
Focus: Stats
Periodical
Journal de la Société de Statistique de Paris (3 times annually) *04719*

Société de Statistique, d'Histoire et d'Archéologie de Marseille et de Provence, Palais de la Bourse, 13231 Marseille
Founded: 1827; Members: 600
Focus: Archeol; Hist; Stats *04720*

Société de Stomatologie de France, 20 Passage Dauphine, 75006 Paris
Gen Secr: Dr. R. Bataille
Focus: Stomatology; Surgery *04721*

Société d'Ethnographie de Paris, 6 Rue Champfleury, 75007 Paris
Founded: 1859; Members: 210
Focus: Ethnology; Soc Sci
Periodical
L'Ethnographie (weekly) *04722*

Société d'Ethnologie Française (SEF), 6 Av du Mahatma Gandhi, 75116 Paris
T: (1) 44176000; Fax: (1) 44176060
Founded: 1947; Members: 430
Pres: Marc Abélès; Gen Secr: Jacques Cheyronnaud
Focus: Ethnology
Periodical
Ethnologie Française (weekly) *04723*

Société d'Ethnozoologie et d'Ethnobotanique (SEZEB), c/o Laboratoire d'Ethnobotanique, Muséum National d'Histoire Naturelle, 57 Rue Cuvier, 75005 Paris
T: (1) 43316957
Members: 120
Focus: Zoology; Botany *04724*

Société de Transplantation, 2 Pl du Docteur Fournier, 75010 Paris
Founded: 1966
Focus: Surgery *04725*

Société d'Etude de Psychodrame Pratique et Théorique (SEPT), 10 Rue des Lions, 75004 Paris
Founded: 1964; Members: 150
Focus: Psych
Periodical
Psychodrame (3 times annually) *04726*

Société d'Etude du Dix-Septième Siècle, c/o Collège de France, 11 Pl Marcelin Bethelot, 75231 Paris Cédex 05
Founded: 1949; Members: 1100
Pres: Jacques Truchet; Gen Secr: Daniel Aris
Focus: Hist; Lit; Philos; Arts
Periodical
XVIIe Siècle (weekly) *04727*

Société d'Etudes Ardennaises (SEA), c/o Archives Départementales des Ardennes, BP 831, 08011 Charleville-Mézières
Founded: 1955; Members: 550
Focus: Hist
Periodical
Revue Historique Ardennaise (weekly) . . *04728*

Société d'Etudes Dantesques, c/o Centre Universitaire Méditerranéen, 65 Promenade des Anglais, 06000 Nice
Founded: 1935; Members: 21
Gen Secr: Simon Lorenzi
Focus: Lit
Periodical
Bulletin *04729*

Société d'Etudes Economiques et Comptables, 62 Rue Jouffroy, 75017 Paris
T: (1) 9241353
Focus: Econ; Law *04730*

Société d'Etudes et de Contrôles Juridiques, 12 Rue de la Paix, 75002 Paris
T: (1) 42616558
Focus: Law *04731*

Société d'Etudes et de Documentation Economiques, Industrielles et Sociales (SEDEIS), 34 Av Charles de Gaulle, 92200 Neuilly-sur-Seine
T: 47228561
Founded: 1948
Pres: Jacques Plassard
Focus: Doc; Sociology; Econ
Periodicals
Analyses de la SEDEIS (weekly)
Chroniques d'Actualité (weekly) *04732*

Société d'Etudes et de Recherches Biologiques (S.E.R.B.), 53 Rue Villiers de l'Isle Adam, 75020 Paris
T: (1) 46368853
Focus: Bio *04733*

Société d'Etudes et de Recherches en Sciences Sociales (S.E.R.E.S.), 10 Rue Richer, 75009 Paris
T: (1) 47706471
Focus: Sociology *04734*

Société d'Etudes et de Recherches pour la Connaissance de l'Homme (S.E.R.C.H.), 5 Rue du Commandant Marchand, 75016 Paris
T: (1) 45003727
Focus: Ethnology *04735*

Société d'Etudes et de Soins pour les Enfants Paralysés (SESEP), Château de Longchamp, Bois de Boulogne, 75016 Paris
T: (1) 45275979
Founded: 1947; Members: 550
Focus: Educ Handic; Pediatrics *04736*

Société d'Etudes Ferroviaires, 13 Rue Chardin, 75016 Paris
T: (1) 48706282
Focus: Transport *04737*

Société d'Etudes Financières et Meunières (S.E.F.I.M.), 66 Rue La Boétie, 75008 Paris
T: (1) 43594580
Focus: Finance; Eng *04738*

Société d'Etudes Folkloriques du Centre-Ouest (SEFCO), Maison de Jeannette, Les Granges, 17400 Saint-Jean d'Angely
Founded: 1962; Members: 2950
Focus: Ethnology; Ling
Periodicals
Aguiaine (weekly)
Le Subiet (weekly) *04739*

Société d'Etudes Hispaniques et de Diffusion de la Culture Française à l'Etranger, 65 Rue Solférino, 24000 Périgueux
Focus: Ethnology *04740*

Société d'Etudes Historiques, c/o Ecole des Hautes Etudes, La Sorbonne, 47 Rue des Ecoles, 75005 Paris
Founded: 1833
Focus: Hist *04741*

Société d'Etudes Italiennes, Grand Palais, Perron Alexandre III, Cours La Reine, 75008 Paris
Focus: Ethnology *04742*

Société d'Etudes Jaurésiennes, 131 Rue de l'Abbé Groult, 75015 Paris
Founded: 1959
Gen Secr: Gilles Candar
Focus: Poli Sci
Periodical
Bulletin (weekly) *04743*

Société d'Etudes Juives, 19 Rue de Téhéran, 75008 Paris
T: (1) 45631728
Founded: 1880; Members: 245
Focus: Rel & Theol; Hist; Lit; Sociology; Arts
Periodical
Revue des Etudes Juives *04744*

Société d'Etudes Juridiques, Economiques et Fiscales (S.E.J.E.F.), 191 Rue Saint-Honoré, 75001 Paris
T: (1) 42606880
Focus: Law; Econ *04745*

Société d'Etudes Linguistiques et Anthropologiques de France (SELAF), 5 Rue de Marseille, 75010 Paris
T: (1) 42088393
Founded: 1964
Pres: J.M.C. Thomas
Focus: Ling; Anthro; Ethnology
Periodicals
Acquisition du language et pathologie
Applications et Transferts
Arctique
Bibliothèque de la SELAF
Description de langues et monographies ethnolinguistiques
Ethnomusicologie
Ethnosciences
Etudes ethnolinguistiques Maghreb-Sahara
Europe de tradition orale
LACITO-Documents
Langues et cultures africaines
Langues et cultures du Pacifique
Langues et sociétés de l'Amérique traditionnelle
Linguistique générale
Numéros spéciaux
Oralité – Documents et Etudes
Sociolinguistique: systèmes de langues et interaction sociales et culturelles
Tradition Orale *04746*

Société d'Etudes Médiévales, 24 Rue de la Chaine, 86022 Poitiers
T: 49410386; Fax: 49018537
Founded: 1958; Members: 60
Focus: Hist
Periodical
Cahiers de Civilisation Médiévale (3 times annually) *04747*

Société d'Etudes Minières, Industrielles et Financières, 7 Blvd de la Madeleine, 75001 Paris
T: (1) 42611337
Focus: Mining; Eng; Finance *04748*

Société d'Etudes Ornithologiques, c/o Laboratoire d'Ecologie, Muséum National d'Histoire Naturelle, 4 Av du Petit Château, 91800 Brunoy
Founded: 1929; Members: 1100
Pres: Camille Ferry; Gen Secr: Jacques Perrin de Brichambaut
Focus: Ornithology *04749*

Société d'Etudes pour le Développement Economique et Social (S.E.D.E.S), 105 Rue de Lille, 75007 Paris
T: (1) 45558759
Founded: 1958; Members: 285
Focus: Sociology; Econ *04750*

Société d'Etudes Psychiques, 2 Rue des Fabriques, 54000 Nancy
Focus: Psychiatry *04751*

Société d'Etudes Robespierristes, c/o Faculté de Lettres, Université de Paris I, 17 Rue de la Sorbonne, 75231 Paris Cédex 05
Focus: Hist
Periodical
Annales Historiques de la Révolution Française (weekly) *04752*

Société d'Etudes Romantiques, c/o Prof. Dr. Jacques Saint-Gérand, 29 Blvd Gergovia, 63037 Clermont-Ferrand Cédex
T: 73346502
Focus: Hist; Lit; Philos
Periodical
Romantique (weekly) *04753*

Société d'Etudes Scientifiques et de Recherches (SEURI), 10ter Chemin du Parc, 95220 Herblay
T: (1) 34500007; Tx: 250303; Fax: (1) 39970799
Focus: Sci *04754*

Société d'Etudes Techniques, 13 Blvd de Strasbourg, 75010 Paris
T: (1) 47705251
Focus: Eng *04755*

Société d'Histoire de Bordeaux, 71 Rue du Loup, 33000 Bordeaux
Focus: Hist *04756*

Société d'Histoire de la Médecine Hébraïque, c/o Dr. I. Simon, 177 Blvd Malesherbes, 75017 Paris
Focus: Hist
Periodicals
Revue d'Histoire de la Médecine Hébraique
Revue Trimestrielle (3 times annually) . *04757*

Société d'Histoire de la Pharmacie (S.H.P.), c/o Faculté de Pharmacie, 4 Av de l'Observatoire, 75270 Paris Cédex 06
T: (1) 43258315
Founded: 1913; Members: 1300
Pres: Jean Flahaut; Gen Secr: Christian Warolin
Focus: Hist
Periodical
Revue d'Histoire de la Pharmacie (weekly)
. . . . *04758*

Société d'Histoire du Droit (S.H.D.), 158 Rue Saint-Jacques, 75005 Paris
Founded: 1913; Members: 500
Pres: Prof. Gérard Sautel; Gen Secr: Prof. Olivier Guillot
Focus: Hist; Law *04759*

Société d'Histoire du Droit Normand, c/o Faculté de Droit, Université, 14000 Caen
Focus: Hist; Law *04760*

Société d'Histoire du Théâtre, 98 Blvd Kellermann, 75013 Paris
T: (1) 45884655
Founded: 1948
Pres: François Périer; Gen Secr: Rose Marie Moudoues
Focus: Hist; Perf Arts
Periodical
Revue d'Histoire du Théâtre (weekly) . . *04761*

Société d'Histoire et d'Archéologie de Bretagne, 20 Av Jules Ferry, 35700 Rennes
T: 99380370; Fax: 99871074
Founded: 1919; Members: 750
Focus: Hist; Archeol
Periodical
Archives Historiques (weekly) *04762*

Société d'Histoire et d'Archéologie de la Lorraine (SHAL), c/o Archives Départementales de la Lorraine, 1 Allée du Château, 57070 Saint Julien les Metz
Founded: 1888; Members: 1200
Focus: Hist; Archeol
Periodical
Les Cahiers Lorrains (3 times annually) . *04763*

Société d'Histoire et d'Archéologie Le Vieux Montmartre, c/o Musée de Montmartre, 12 Rue Cortot, 75018 Paris
T: (1) 45066111; Fax: (1) 45063075
Founded: 1886; Members: 800
Pres: Jean-Claude Gouvernon
Focus: Arts; Lit *04764*

Société d'Histoire Générale et d'Histoire Diplomatique, 13 Rue Soufflot, 75005 Paris
Founded: 1887; Members: 400
Focus: Hist; Poli Sci
Periodical
Revue d'Histoire Diplomatique *04765*

Société d'Histoire Moderne, 47 Blvd Bessières, 75017 Paris
T: (1) 46277074
Founded: 1901; Members: 1100
Pres: Serge Berstein; Gen Secr: Guy Boquet
Focus: Hist *04766*

Société d'Histoire Religieuse de la France, 28 Rue d'Assas, 75006 Paris
Founded: 1914; Members: 900
Pres: Michel Mollat du Jourdin; Gen Secr: Bernard Barbiche
Focus: Hist
Periodical
Revue d'Histoire de l'Eglise de France (weekly)
. . . . *04767*

Société d'Hygiène International, 5 Sq Henri Delormel, 75014 Paris
T: (1) 5429590
Focus: Hygiene *04768*

Société d'Obstétrique et de Gynécologie de Marseille, c/o Hôpital de la Conception, 13001 Marseille
Focus: Gynecology *04769*

Société d'Obstétrique et de Gynécologie de Toulouse (S.O.G.T.), c/o Hôpital de la Grave, 31052 Toulouse
Fax: 61777934
Founded: 1908; Members: 58
Focus: Gynecology
Periodical
Journal de Gynécologie Obstétrique et Biologie de la Reproduction *04770*

Société d'Océanographie de France, 195 Rue Saint-Jacques, 75005 Paris
Founded: 1897
Focus: Oceanography *04771*

Société d'Ophtalmologie de l'Est de la France, 133 Rue Saint-Dizier, 54000 Nancy
Focus: Ophthal *04772*

Société d'Ophtalmologie de Lyon, c/o Hôpital Edouard Herriot, 69001 Lyon
Focus: Ophthal *04773*

Société d'Ophtalmologie de Paris (SOP), 108 Rue du Bac, 75007 Paris
Founded: 1888; Members: 200
Gen Secr: Dr. Jean-Paul Boissin
Focus: Ophthal
Periodical
Bulletin (weekly) *04774*

Société d'Oto-Neuro-Ophthalmologie du Sud-Est de la France, c/o Hôpital de la Timone, 13001 Marseille
Focus: Otorhinolaryngology; Neurology; Ophthal
. . . . *04775*

Société du Salon d'Automne, Grand Palais, Porte H, 75008 Paris
Founded: 1903
Pres: M. Avoy
Focus: Arts *04776*

Société Entomologique de France, c/o Muséum d'Histoire Naturelle, 45 Rue Buffon, 75005 Paris
T: (1) 60793384
Founded: 1832; Members: 600
Gen Secr: J. Péricart
Focus: Entomology
Periodicals
Annales (weekly)
Bulletin (5 times annually) *04777*

Société Européenne de Radiologie Cardio-Vasculaire et de Radiologie d'Intervention / European Society of Cardio-Vascular Radiology and Interventional Radiology, c/o Dépt de Radiologie, Hôpital Cardio-Vasculaire, 69393 Lyon Cédex 03
Founded: 1975; Members: 200
Gen Secr: Prof. F. Pindt
Focus: Radiology
Periodical
Annales de Radiologie 04778

Société Européenne de Radiologie Pédiatrique, c/o Hôpital des Enfants Malades, 149 Rue de Sèvres, 75730 Paris Cédex 15
Focus: Radiology 04779

Société Financière Européenne / European Financial Society, 20 Rue de la Paix, 75002 Paris
T: (1) 42615747
Focus: Finance 04780

Société Française d'Acoustique, 33 Rue Croulebarbe, 75013 Paris
T: (1) 45355400; Fax: (1) 43317426
Founded: 1948; Members: 1200
Focus: Physics; Acoustics
Periodical
Acta Acustica (weekly) 04781

Société Française d'Allergologie, 1 Rue du Val-de-Grâce, 75005 Paris
T: (1) 40333448
Founded: 1947; Members: 825
Gen Secr: Prof. P. Gervais
Focus: Immunology
Periodical
Revue Française d'Allergologie et d'Immunologie Clinique (5 times annually) 04782

Société Française d'Anesthésie et de Réanimation, 74 Rue Raynouard, 75016 Paris
T: (1) 45258525; Fax: (1) 40503522
Founded: 1960; Members: 2900
Pres: Prof. P. Scherpereel; Gen Secr: Prof. F. d' Athis
Focus: Anesthetics
Periodical
Annales Francaises d'Anesthésie et de Réanimation (weekly) 04783

Société Française d'Angéiologie, 3 Rue Jacques Dulud, 92200 Neuilly-sur-Seine
T: (1) 46370754
Founded: 1947; Members: 350
Pres: Prof. Sénac; Gen Secr: du Arfi de Laigue
Focus: Hematology
Periodical
Angéiologie (8 times annually) 04784

Société Française d'Archéocivilisation et de Folklore, c/o Ecole des Hautes Etudes en Sciences Sociales, 54 Blvd Raspail, 75006 Paris
T: (1) 45443979
Founded: 1928
Focus: Archeol 04785

Société Française d'Archéologie (S.F.A.), c/o Musée des Monuments Français, Palais de Chaillot, 1 Pl du Trocadéro, 75116 Paris
T: (1) 47047896; Fax: (1) 44053425
Founded: 1834; Members: 2850
Pres: Alain Brandenburg; Gen Secr: Philippe Dubost
Focus: Archeol
Periodicals
Bulletin Monumental (weekly)
Congrès Archéologique de France (weekly) 04786

Société Française d'Art Contemporain, 28bis Blvd de Sebastopol, 75004 Paris
T: (1) 42773846
Focus: Arts 04787

Société Française de Biologie Clinique, c/o Laboratoire de Chimie Clinique et Biologie Moléculaire, 15 Rue de l'Ecole de Médecine, 75270 Paris Cédex 06
Gen Secr: Prof. L. Hartmann
Focus: Biochem; Biophys 04788

Société Française de Cardiologie, 15 Rue de Madrid, 75008 Paris
T: (1) 43879514; Fax: (1) 43871714
Founded: 1937; Members: 2200
Focus: Cardiol
Periodical
Archives des Maladies du Coeur et des Vaisseaux (weekly) 04789

Société Française de Céramique (SFC), 23 Rue de Cronstadt, 75015 Paris
T: (1) 40432300; Fax: (1) 45315804
Founded: 1945
Focus: Arts 04790

Société Française de Chimie, 250 Rue Saint-Jacques, 75005 Paris
T: (1) 43252078
Founded: 1984; Members: 5000
Pres: J.B. Donnet; Gen Secr: R. Hamelin
Focus: Chem
Periodicals
L'Actualité Chimique
Bulletin de la Société Chimique de France
Journal de Chimie Physique 04791

Société Française de Chirurgie Orthopédique et Traumatologique (SOFCOT), c/o Hôpital Cochin, 27 Rue du Faubourg Saint-Jacques, 75674 Paris Cédex 13
T: (1) 46336479
Members: 1800
Pres: B. Glorion
Focus: Surgery; Orthopedics; Traumatology
Periodical
Revue de Chirurgie Orthopédique . . . 04792

Société Française de Chirurgie Pédiatrique (SPCP), c/o Hôpital Debrousse, 29 Rue Soeur Bouvier, 69322 Lyon Cédex 05
Founded: 1959
Pres: M. Lacheretz; Gen Secr: G. Montfort
Focus: Surgery
Periodical
Chirurgie Pédiatrique (weekly) 04793

Société Française de Chirurgie Plastique et Reconstructive (SFCPR), 40 Rue Bichat, 75010 Paris
T: (1) 42066244
Founded: 1953; Members: 360
Pres: Dr. A. de Coninck; Gen Secr: Prof. P. Banzet
Focus: Surgery 04794

Société Française de Chronométrie et de Microtechnique (SFCM), 39 Av de l'Observatoire, 25003 Besançon Cédex
Founded: 1931; Members: 350
Focus: Eng 04795

Société Française d'Ecologie, c/o Muséum National d'Histoire Naturelle, 4 Av du Petit Château, 91800 Brunoy
Focus: Ecology 04796

Société Française de Composition, 24 Rue de la Banque, 75002 Paris
T: (1) 42612332
Focus: Music 04797

Société Française d'Economie Rurale (SFER), 16 Rue Claude Bernard, 75341 Paris Cédex 05
Founded: 1949; Members: 400
Pres: Jean M. Boussard; Gen Secr: Lucien Bourgeois
Focus: Agri
Periodical
Economie Rurale (weekly) 04798

Société Française de Dermatologie et de Syphilographie, 37 Rue Galilée, 75016 Paris
Pres: M. Gastinel; Gen Secr: M. Degos
Focus: Derm; Venereology 04799

Société Française de Droit Aérien et Spatial (SFDAS), 17 Av de Lamballe, 75016 Paris
T: (1) 45244650
Founded: 1954; Members: 321
Focus: Law
Periodical
Revue Française de Droit Aérien et Spatial (3 times annually) 04800

Société Française de Génétique (SFG) / French Society of Genetics, c/o R. Motta, C.S.E.A.L.-C.N.R.S., 3b Rue de la Férollerie, 45071 Orléans Cédex 2
T: 38515438
Founded: 1987; Members: 533
Pres: Ethel Moustacchi; Gen Secr: Roland Motta
Focus: Genetics
Periodical
Bulletin de la SFG (3 times annually) . 04801

Société Française de Géographie Economique, 148 Blvd Malesherbes, 75017 Paris
T: (1) 46223457
Founded: 1936
Focus: Geography 04802

Société Française de Graphologie, 5 Rue Las Cases, 75007 Paris
T: (1) 45554694
Focus: Graphology
Periodical
La Graphologie (3 times annually) . . . 04803

Société Française de Gynécologie (SFG), 20 Rue Clément Marot, 75008 Paris
T: (1) 47208781
Founded: 1931; Members: 582
Pres: J.P. Wolff; Gen Secr: A. Gorins
Focus: Gynecology
Periodical
Gynécologie (weekly) 04804

Société Française d'Egyptologie (SFE), Collège de France, 11 Pl Marcelin-Berthelot, 75231 Paris Cédex 05
T: (1) 43256211
Founded: 1923; Members: 850
Pres: J. Vercoutter; Gen Secr: V. Laurent
Focus: Archeol; Hist
Periodicals
Bulletin
Revue d'Egyptologie 04805

Société Française de Malacologie (SFM), c/o Laboratoire de Biologie des Invertébrés Marins et de Malacologie, Muséum National d'Histoire Naturelle, 55 Rue de Buffon, 75005 Paris
T: (1) 40793091
Founded: 1969; Members: 350
Focus: Zoology
Periodical
Haliotis (weekly) 04806

Société Française de Médecine Aérospatiale, c/o Laboratoire de Médecine Aérospatiale, Centre d'Essais au Vol, 91228 Bretigny Cédex
Founded: 1960; Members: 700
Pres: Prof. F. Crance; Gen Secr: Prof. Henri Marotte
Focus: Med; Physiology
Periodical
Médecine Aéronautique et Spatiale (weekly) 04807

Société Française de Médecine du Sport (SFMS), 1 Rue Lacretelle, 75015 Paris
T: (1) 48285562
Founded: 1921; Members: 3000
Focus: Med; Physical Therapy 04808

Société Française de Médecine du Trafic (SFMT), 21 Rue de l'Ecole de Médecine, 75006 Paris
T: (1) 46035373
Focus: Med 04809

Société Française de Médecine Esthétique (SFME), 154 Rue A. Silvestre, 92400 Courbevoie
T: (1) 45063259; Fax: (1) 45063911
Founded: 1973; Members: 200
Pres: J.J. Legrand
Focus: Surgery; Derm
Periodical
Journal de Médecine Esthétique et de Chirurgie Dermatologique (weekly) 04810

Société Française de Médecine Générale (SFMG), 29 Av du Général Leclerc, 75014 Paris
T: (1) 43208593
Founded: 1973; Members: 500
Focus: Med
Periodical
Documents de Recherches en Médecine Générale (weekly) 04811

Société Française de Médecine Orthopédique et Thérapeutique Manuelle, 6 Rue Jean Richepin, 75116 Paris
T: (1) 45041048
Founded: 1965; Members: 240
Focus: Orthopedics; Therapeutics
Periodical
Revue de Médecine Orthopédique (weekly) 04812

Société Française de Médecine Préventive et Sociale, 60 Blvd de Latour-Maubourg, 75340 Paris Cédex 07
T: (1) 42605438
Founded: 1953; Members: 700
Focus: Hygiene
Periodical
Bulletin de Liaison (weekly) 04813

Société Française de Médecine Psychosomatique, 22 Rue Legendre, 75017 Paris
Focus: Psychiatry
Periodical
Revue de Médecine Psychosomatique (weekly) 04814

Société Française de Mesothérapie (S.F.M.), 87 Blvd Suchet, 75016 Paris
Founded: 1964
Focus: Therapeutics
Periodical
S.F.M. Bulletin (3 times annually) . . . 04815

Société Française de Métallurgie et de Matériaux (SFMM), 1 Rue Paul Cézanne, 75008 Paris
T: (1) 49537237; Fax: (1) 49537100
Founded: 1945; Members: 1500
Gen Secr: Yves Franchot
Focus: Metallurgy; Materials Sci 04816

Société Française de Microbiologie, c/o Institut Pasteur, 28 Rue du Docteur Roux, 75724 Paris Cédex 15
T: (1) 45665800
Founded: 1937; Members: 1030
Focus: Microbio 04817

Société Française de Microscopie Electronique (S.F.M.E.), 67 Rue Maurice Günsbourg, 94200 Ivry-sur-Seine
Founded: 1959; Members: 2300
Pres: Prof. Alain Bourret
Focus: Electronic Eng
Periodicals
Biology of the Cell (9 times annually)
Journal de Microscopie et de Spectroscopie Electronique (weekly) 04818

Société Française de Minéralogie et de Cristallographie, 4 Pl Jussieu, Casier 115, Pour 16/26, 75252 Paris Cédex 05
Fax: (1) 44276024
Founded: 1878; Members: 600
Gen Secr: H. Suquet
Focus: Mineralogy
Periodical
European Journal of Mineralogy (weekly) 04819

Société Française de Musicologie, 2 Rue Louvois, 75002 Paris
Founded: 1917; Members: 550
Focus: Music
Periodical
Revue de Musicologie (weekly) 04820

Société Française de Mycologie Médicale, c/o Institut Pasteur, 25 Rue du Docteur Roux, 75724 Paris Cédex 15
T: (1) 45688254; Tx: 250609
Founded: 1956; Members: 500
Pres: J.Y. Rastide
Focus: Botany, Specific
Periodical
Journal de Mycologie Médicale (weekly) . 04821

Société Française d'Endocrinologie, c/o Masson Editeur, 120 Blvd Saint-Germain, 75280 Paris Cédex 06
Founded: 1939; Members: 550
Focus: Endocrinology
Periodical
Annales d'Endocrinologie (weekly) . . . 04822

Société Française de Néonatologie, c/o Hôpital Port Royal, 123 Blvd de Port Royal, 75674 Paris Cédex 14
T: (1) 42341260
Members: 1200
Focus: Pediatrics; Physiology; Pathology
Periodical
Progrès en Néonatologie (weekly) . . . 04823

Société Française d'Energie Nucléaire (S.F.E.N.), 48 Rue de la Procession, 75015 Paris
T: (1) 45670770
Focus: Energy; Nucl Res
Periodicals
Bulletin de Liaison (weekly)
Revue Générale Nucléaire (weekly) . . . 04824

Société Française de Neurologie, 120 Blvd Saint-Germain, 75006 Paris
Fax: (1) 44245247
Founded: 1899
Focus: Neurology
Periodical
Revue Neurologique (10 times annually) . 04825

Société Française de Numismatique (SFN), c/o Bibliothèque Nationale, Cabinet des Médailles, 58 Rue de Richelieu, 75084 Paris Cédex 02
T: (1) 47038344
Founded: 1865; Members: 700
Focus: Numismatics
Periodicals
Bulletin (10 times annually)
Revue Numismatique (weekly) 04826

Société Française de Pathologie Respiratoire, 66 Blvd Saint-Michel, 75005 Paris
Founded: 1945
Focus: Pathology 04827

Société Française de Pédagogie (SFP), 6 Rue du Champ de l'Alouette, 75013 Paris
Founded: 1902; Members: 10000
Pres: M. Bonissel; Gen Secr: M. Gevrey
Focus: Educ
Periodical
Bulletin Trimestriel (3 times annually) . . 04828

Société Française de Pédiatrie, c/o Hôpital des Enfants Malades, 149 Rue de Sèvres, 75743 Paris Cédex 15
Founded: 1929; Members: 750
Pres: Prof. F. Larbre; Gen Secr: Prof. Jean Frezal
Focus: Pediatrics
Periodical
Archives Françaises de Pédiatrie (weekly) 04829

Société Française de Philosophie, c/o Centre International de Synthèse, 12 Rue Colbert, 75002 Paris
Founded: 1901; Members: 180
Pres: Bernard Bourgeois; Gen Secr: Jean-Marie Beyssade
Focus: Philos
Periodicals
Bulletin de la Société Française de Philosophie (3 times annually)
Revue de Metaphysique et de Morale (3 times annually) 04830

Société Française de Phlébologie, 46 Rue Saint-Lambert, 75015 Paris
T: (1) 45330271; Fax: (1) 42507518
Founded: 1947; Members: 1700
Pres: Pierre Wallois; Gen Secr: F. Vin
Focus: Med
Periodical
Phlébologie (weekly) 04831

Société Française de Phoniatrie, c/o Clinique ORL, Hôpital Lariboisière, Rue Ambroise Paré, 75010 Paris
Founded: 1934; Members: 320
Focus: Otorhinolaryngology 04832

Société Française de Photogrammétrie et de Télédétection, 2 Av Pasteur, 94160 Saint-Mandé
T: (1) 43741215
Founded: 1959; Members: 615
Pres: Colette M. Girard; Gen Secr: Alain Baudoin
Focus: Surveying
Periodical
Bulletin (weekly) 04833

Société Française de Photographie et Cinématographie, 9 Rue Montalembert, 75007 Paris
T: (1) 42223717
Founded: 1854; Members: 1200
Gen Secr: Norbert Le Roy
Focus: Photo 04834

Société Française de Physiologie Végétale, 4 Pl Jussieu, 75230 Paris Cédex 05
Founded: 1955; Members: 600
Pres: P. Mazliak; Gen Secr: B. Camara
Focus: Botany
Periodical
Physiologie Végétale 04835

Société Française de Physique (SFP), 33 Rue Croulebarbe, 75013 Paris
T: (1) 47073292
Founded: 1874; Members: 3200
Focus: Physics
Periodicals
Bulletin (weekly)
Journal de Physique (weekly)
Lettres et Journal de Physique (weekly)
Revue de Physique Appliquée (weekly) . 04836

Société Française de Phytiatrie et de Phytopharmacie, c/o C.N.R.A., Rte de Saint-Cyr, 78000 Versailles
T: 49507522
Founded: 1951; Members: 1000
Focus: Pharmacol; Botany 04837

Société Française de Psychologie (S.F.P.), 28 Rue Serpente, 75006 Paris
T: (1) 43267967
Founded: 1920; Members: 1500
Gen Secr: Jean-Marc Monteil
Focus: Psych
Periodical
Psychologie Française (weekly) 04838

Société Française de Radiologie Médicale, c/o Hôpital Lariboisière, 2 Rue Ambroise Paré, 75010 Paris
Founded: 1909; Members: 1800
Gen Secr: M. Blery
Focus: Radiology
Periodicals
Diagnostic and Interventional Radiology
Revue d'Imagerie Médicale 04839

Société Française de Santé Publique (S.F.S.P.), c/o Centre de Médecine Préventive, 2 Av du Doyen J. Parisot, 54501 Vandoeuvre lès Nancy Cédex
T: 83443917; Fax: 83443776
Members: 400
Pres: Denis Zmirou; Gen Secr: Jean F. Collin
Focus: Public Health
Periodicals
Santé et Société (weekly)
Santé Publique (weekly) 04840

Société Française de Sciences et Techniques Pharmaceutiques, 9 Rue de la Montagne Sainte-Geneviève, 75005 Paris
T: (1) 43268137; Fax: (1) 43298252
Members: 1200
Focus: Pharmacol
Periodical
Sciences et Techniques Pharmaceutiques (10 times annually) 04841

Société Française de Sexologie Clinique / French Society for Clinical Sexology, 75 Blvd de Courcelles, 75008 Paris
T: (1) 47638129
Founded: 1974; Members: 300
Pres: Charles Gellman; Gen Secr: Gilbert Tordjman
Focus: Public Health
Periodical
Cahiers de Sexologie Clinique (weekly) . 04842

Société Française de Sociologie (S.F.S.), 59-61 Rue Pouchet, 75849 Paris Cédex 17
T: (1) 40251099
Founded: 1962; Members: 545
Pres: Claudine Herzlich; Gen Secr: Sabine Erbes-Seguin
Focus: Sociology 04843

Societé Française des Physiciens d'Hôpital (S.F.P.H.), c/o Institut Curie, 26 Rue d'Ulm, 75231 Paris Cedex 05
Founded: 1972; Members: 340
Pres: Alain Noel; Gen Secr: Pascal Louisot
Focus: Physics
Periodicals
Bulletin du Cancer
Journal Européen de Radiothérapie (weekly)
. . . . 04844

Société Française des Professeurs de Russe (S.P.R.), 9 Rue Michelet, 75006 Paris
Focus: Educ; Ling
Periodicals
L'Enseignement du Russe (weekly)
S.P.R. Informations (weekly) 04845

Société Française des Thermiciens (SFT), 3 Rue Henri Heine, 75016 Paris
T: (1) 42245935; Fax: (1) 40500754
Members: 4000
Focus: Eng
Periodical
Revue Générale de Thermique (weekly) . 04846

Société Française des Urbanistes (S.F.U.), 38 Rue Eugène Oudiné, 75013 Paris
T: (1) 45850220
Focus: Urban Plan 04847

Société Française de Thérapeutique et de Pharmacologe Clinique, c/o Doin Editeur, 8 Pl de l'Odéon, 75006 Paris
Founded: 1866; Members: 1000
Pres: Prof. A. Tillement; Gen Secr: Prof. F.C. Hugues
Focus: Therapeutics
Periodical
Thérapie (weekly) 04848

Société Française de Toxicologie, 4 Av de l'Observatoire, 75006 Paris
Founded: 1974; Members: 90
Focus: Toxicology 04849

Société Française d'Etude du Dix-Huitième Siècle, c/o R. Granderoute, 12 Av du Stade Nautique, 64000 Pau
T: 59029139
Focus: Hist; Lit
Periodical
Dix Huitième Siècle (weekly) 04850

Société Française d'Etudes des Phénomènes Psychiques, 1 Rue des Gâtines, 75020 Paris
T: (1) 43493080
Founded: 1893; Members: 1000
Pres: A. Croonenberghs
Focus: Psych
Periodical
La Tribune Psychique (weekly) 04851

Société Française d'Etudes et de Réalisations Cartographiques, 5 Rue Papillon, 75009 Paris
T: (1) 45230491
Focus: Cart 04852

Société Française d'Etudes Juridiques, 101 Av Raymond Poincaré, 75016 Paris
T: (1) 47238768
Focus: Law 04853

Société Française d'Hématologie, 96 Rue Didot, 75014 Paris
Focus: Hematology 04854

Société Française d'Héraldique et de Sigillographie, 60 Rue des Francs-Bourgeois, 75141 Paris Cédex 03
Founded: 1937; Members: 200
Pres: Robert-Henri Bautier; Gen Secr: Jean-Luc Chassel
Focus: Genealogy
Periodical
Revue française d'héraldique et de sigillographie (weekly) 04855

Société Française d'Histoire de la Médecine (S.F.H.M), 38bis Rue de Courlancy, 51100 Reims
T: 26483260; Fax: 26483271
Founded: 1902; Members: 500
Pres: P. Lefebvre; Gen Secr: Prof. Alain Ségal
Focus: Med; Cultur Hist
Periodical
Histoire des Sciences Médicales (weekly) 04856

Société Française d'Histoire d'Outre-Mer (SFHOM), 9 Rue Robert de Flers, 75015 Paris
T: (1) 40584873
Founded: 1913; Members: 500
Pres: Charles-Robert Ageron; Gen Secr: Roger Pasquier
Focus: Hist
Periodical
Revue Française d'Histoire d'Outre Mer (weekly)
. . . . 04857

Société Française d'Hydrologie et de Climatologie Médicales, 1 Rue Monticelli, 75014 Paris
Founded: 1853; Members: 413
Pres: Dr. R. Capoduro; Gen Secr: Dr. G. Girault
Focus: Water Res; Geophys
Periodical
La Presse Thermale et Climatique (weekly)
. . . . 04858

Société Française d'Hygiène, de Médecine Sociale et de Génie Sanitaire, c/o Département de Santé Publique, Faculté de Médecine de Nancy, BP 184, 54505 Vandoeuvre-lès-Nancy Cédex
T: 83514415
Founded: 1877; Members: 400
Gen Secr: J.P. Deschamps
Focus: Public Health; Hygiene; Med . . 04859

Société Française d'Ichtyologie (S.F.I.), c/o Muséum National d'Histoire Naturelle, 43 Rue Cuvier, 75231 Paris Cédex 05
T: (1) 40793755; Fax: (1) 40793771
Founded: 1976; Members: 300
Pres: R. Galzin; Gen Secr: G. Duhamel
Focus: Zoology; Ecology; Fisheries
Periodical
Cybium: Annuaire des Ichtyologistes Français (weekly) 04860

Société Française d'Ophthalmologie (SFO), 9 Rue Mathurin Régnier, 75015 Paris
T: (1) 47342021
Founded: 1883; Members: 6800
Gen Secr: Prof. Hubert Bourgeois
Focus: Ophthal
Periodicals
Ophthalmologie
Rapport (weekly) 04861

Société Française d'Optique Physiologique, c/o Laboratoire de Physique du Muséum, 43 Rue Cuvier, 75231 Paris Cédex 05
T: (1) 45873898
Founded: 1966
Focus: Optics; Physiology 04862

Société Française d'Orthopédie, 7 Rue de Duras, 75008 Paris
T: (1) 42652899
Focus: Orthopedics 04863

Société Française d'Oto-Rhino-Laryngologie et de Pathologie Cervico-Faciale, 9 Rue Villebois-Mareuil, 75017 Paris
Founded: 1880; Members: 1500
Pres: Dr. R. Batisse; Gen Secr: Prof. Charles Freche
Focus: Otorhinolaryngology
Periodicals
Comptes Rendus
Rapports Discutés au Congrès 04864

Société Française d'Urologie, 6 Av Constant Coquelin, 75007 Paris
Founded: 1919; Members: 50
Pres: Dr. Boissonnat
Focus: Urology
Periodical
Journal d'Urologie 04865

Société Française du Vide (S.F.V.), 19 Rue du Renard, 75004 Paris
T: (1) 42781582; Fax: (1) 42786320
Founded: 1945; Members: 2000
Focus: Eng
Periodical
Le Vide, les Couches Minces (weekly) . 04866

Société Française pour le Droit International (SFDI), c/o Centre d'Etudes Internationales et Européennes, Faculté de Droit, Pl d'Athènes, 67084 Strasbourg Cédex
Founded: 1967; Members: 350
Focus: Law 04867

Société Géologique de France (SGF), 77 Rue Claude Bernard, 75005 Paris
T: (1) 43317735
Founded: 1830; Members: 2500
Focus: Geology
Periodicals
Bulletin de la Société Géologique de France (weekly)
Géochronique (weekly)
Mémoires de la Société Géologique de France (weekly) 04868

Société Géologique et Minéralogique de Bretagne (SGMB), c/o Institut de Géologie, Faculté des Sciences, Av du Général Leclerc, 35042 Rennes Cédex
T: 99364815
Founded: 1920; Members: 230
Focus: Geology; Mineralogy
Periodicals
Bulletin (weekly)
Mémoires (weekly) 04869

Société Historique, Archéologique et Littéraire de Lyon, c/o Archives Municipales de Lyon, 4 Av Adolphe Max, 69005 Lyon
Founded: 1807; Members: 100
Pres: Dr. Maurice Boucher; Gen Secr: Paul Fenga
Focus: Archeol; Hist; Lit
Periodical
Bulletin de la Société Historique, Archéologique et Littéraire de Lyon (weekly) 04870

Société Historique de la Province de Maine, 26 Rue des Chanoines, 72000 Le Mans
Focus: Hist
Periodical
La Province du Maine (weekly) 04871

Société Historique du Bas-Limousin, c/o Robert Joudoux, 13 Pl Municipale, 19000 Tulle
Focus: Hist 04872

Société Historique et Archéologique du Périgord, 18 Rue du Plantier, 24000 Périgueux
Founded: 1874
Focus: Hist; Archeol
Periodical
Bulletin de la Société Historique et Archéologique du Périgord (weekly) 04873

Société Hydrotechnique de France, 199 Rue de Grenelle, 75007 Paris
T: (1) 47051337; Fax: (1) 45569746
Founded: 1912; Members: 800
Pres: Georges Maurin; Gen Secr: Max Perrin
Focus: Eng; Hydrology
Periodical
La Houille Blanche: Revue Internationale de l'Eau (8 times annually) 04874

Société Internationale de Bibliographie Classique (SIBC) / International Society of Classical Bibliography, 11 Av René Coty, 75014 Paris
T: (1) 43276790
Founded: 1923
Gen Secr: Pierre Paul Corsetti
Focus: Libraries & Bk Sci
Periodical
L'Année Philologique: Bibliographie Critique et Analytique de l'Antiquité Gréco-Latine (weekly)
. . . . 04875

Société Internationale de Biologie Mathématique, 11bis Av de la Providence, 92160 Antony
Founded: 1962
Pres: Dr. Collot; Gen Secr: Maréchal
Focus: Bio; Math
Periodical
Revue de Bio-Mathématique (weekly) . . 04876

Société Internationale de Criminologie (SIC) / International Society of Criminology, 4 Rue de Mondovi, 75001 Paris
T: (1) 42618022 ext 5825
Founded: 1934; Members: 1200
Gen Secr: Georges Picca
Focus: Criminology
Periodical
Newsletter 04877

Société Internationale de Podologie Médico-Chirurgicale, 93 Av du Docteur Picaud, 06150 Cannes-la-Bocca
Focus: Surgery 04878

Société Internationale de Psychopathologie de l'Expression (SIPE) / International Society of Art and Psychopathology, c/o Clinique des Maladies Mentales et de l'Encephale, 100 Rue de la Santé, 75014 Paris
T: (1) 45895521
Founded: 1959; Members: 600
Pres: Prof. Robert Volmat
Focus: Psychiatry
Periodical
Newsletter 04879

Société Internationale de Psycho-Prophylaxie Obstétricale, 31 Rue Saint-Guillaume, 75007 Paris
T: (1) 45481513
Founded: 1958
Gen Secr: Dr. Pierre Vellay
Focus: Gynecology
Periodical
Bulletin (weekly) 04880

Société Internationale des Amis de Montaigne, BP 913, 75073 Paris Cédex 02
Founded: 1912; Members: 780
Focus: Lit
Periodical
Bulletin (weekly) 04881

Société Internationale de Transfusion Sanguine (SITS) / International Society of Blood Transfusion, c/o CNTS, BP 100, 91943 Les Ulis Cédex
T: (1) 69072040
Founded: 1937; Members: 1780
Pres: G. Archer; Gen Secr: G. Garretta
Focus: Hematology
Periodical
Transfusion Today (weekly) 04882

Société Internationale d'Urologie (SIU), 9 Blvd du Temple, 75003 Paris
T: (1) 42784009
Founded: 1910; Members: 1500
Gen Secr: Prof. Dr. A. Jardin
Focus: Urology 04883

Société Internationale pour la Lutte contre le Cancer du Sein / International Society Against Breast Cancer, 26 Rue de la Faisanderie, 75116 Paris
Founded: 1973
Pres: Prof. K.H. Hollmann
Focus: Cell Biol & Cancer Res 04884

Société Italienne des Auteurs et Editeurs, 65 Rue La Boétie, 75008 Paris
T: (1) 42250147
Focus: Lit 04885

Société J. S. Bach, 95 Rue Vaugirard, 75006 Paris
Focus: Music 04886

Société Juridique et Fiscale de France, 2bis Rue de Villiers, 92309 Levallois Perret Cédex
Focus: Law 04887

Société Linnéenne de Provence, c/o Lycée Thiers, Pl du Lycée, 13001 Marseille
Founded: 1909; Members: 350
Focus: Nat Sci 04888

Société Longuédocienne de Géographie, c/o Université P. Valéry, BP 5043, 34032 Montpellier Cédex
T: 67142183
Focus: Geography 04889

Société Mathématique de France (S.M.F.), c/o Institut H. Poincaré, 11 Rue Pierre et Marie Curie, 75005 Paris Cédex 05
T: (1) 44276796
Founded: 1872; Members: 1650
Pres: Daniel Barlet
Focus: Math
Periodicals
Astérisque (weekly)
Bulletin et Mémoires de la S.M.F. (weekly)
Gazette des Mathématiciens (weekly)
Officiel des Mathématiques (weekly) . . 04890

Société Médicale des Hôpitaux de Paris, 45 Quai de la Tournelle, 75005 Paris
Gen Secr: Charles Haas
Focus: Med
Periodical
Annales de Médecine Interne 04891

Société Médicale d'Imagerie, Enseignement et Recherche (SMIER), c/o Hôpital Saint-Antoine, 75571 Paris Cédex 12
T: (1) 49282460; Fax: (1) 49282687
Founded: 1955
Pres: F. Vicari; Gen Secr: C. Florent
Focus: Radiology
Periodical
Acta Endoscopica (weekly) 04892

Société Médico-Chirurgicale des Hôpitaux et Formations Sanitaires des Armées, 277bis Rue Saint-Jacques, 75005 Paris
Founded: 1969
Focus: Surgery 04893

Société Médico-Chirurgicale des Hôpitaux Libres, 1 Pl d'Iéna, 75016 Paris
Pres: L. Michelet; Gen Secr: J.A. Huet; A.D. Herschberg; J. Valletta
Focus: Surgery 04894

Société Médico-Psychologique, c/o J. Parant, 14-16 Av R. Schuman, 92100 Boulogne
Founded: 1852; Members: 675
Focus: Med; Psych; Psychiatry
Periodical
Annales Médico-Psychologiques (10 times annually) 04895

Société Météorologique de France (SMF), 73-77 Rue de Sèvres, 92100 Boulogne-Billancourt
Founded: 1852; Members: 950
Gen Secr: A. Villevieille
Focus: Geophys
Periodical
La Météorologie 04896

Société Mycologique de France, 18 Rue de l'Ermitage, 75020 Paris
T: (1) 43663540
Founded: 1884; Members: 2000
Pres: J. Perreau; Gen Secr: Henri Romagnesi
Focus: Botany, Specific
Periodical
Bulletin Trimestriel (3 times annually) . . 04897

Société Nationale Académique de Cherbourg, 21 Rue Bonhomme, 50100 Cherbourg
T: 33532806
Founded: 1755; Members: 40
Pres: André Poirier; Gen Secr: Claude Dréno
Focus: Sci
Periodical
Mémoires (every 4-5 years) 04898

Société Nationale de Protection de la Nature (SNPN), 57 Rue Cuvier, BP 405, 75221 Paris Cédex 05
T: (1) 47073195; Fax: (1) 47070716
Founded: 1854; Members: 8500
Focus: Ecology
Periodicals
Le Courier de la Nature (weekly)
La Terre et la Via (weekly) 04899

Société Nationale des Architectes de France, 8 Rue Albert Samain, 75017 Paris
Founded: 1872; Members: 500
Pres: J.-F. Bellat; Gen Secr: Roger Laine
Focus: Archit
Periodical
Le Monitor des Architectes 04900

Société Nationale des Beaux-Arts (S.N.B.A.), 11 Rue Berryer, 75008 Paris
Founded: 1890; Members: 800
Pres: F. Baboulet; Gen Secr: E. Audfray
Focus: Arts 04901

Société Nationale des Sciences Naturelles et Mathématiques, 21 Rue Bonhomme, 50100 Cherbourg
T: 33532806
Focus: Math; Nat Sci; Chem; Physics
Periodical
Mémoires 04902

Société Nationale de Transfusion Sanguine (SNTS), 6 Rue A. Cabanel, 75739 Paris Cédex 15
T: (1) 43067000
Founded: 1955; Members: 510
Focus: Hematology
Periodicals
Blood Transfusion and Immunhematology (weekly)
Revue Française de Transfusion et Hémobiologie 04903

Société Nationale d'Horticulture de France (SNHF), 84 Rue de Grenelle, 75007 Paris
T: (1) 45488100
Founded: 1827; Members: 13000
Pres: Michel Cointat; Gen Secr: André Genin
Focus: Hort
Periodical
Jardins de France (10 times annually) . 04904

Société Nationale Française de Gastro-Entérologie (SNFGE), 33 Blvd Picpus, 75571 Paris Cédex 12
Founded: 1911; Members: 570
Focus: Gastroenter 04905

Société Nationale Française de Rééducation et Réadaption Fonctionnelles, 13 Blvd Raspail, 75007 Paris
Focus: Rehabil 04906

Société Odontologique de Paris (S.O.P.), 239 Rue du Faubourg Saint-Martin, 75010 Paris
T: (1) 42030505
Members: 2500
Pres: Dr. M. Fitoussi; Gen Secr: Dr. Y. Bismuth
Focus: Dent
Periodical
Revue d'Odonto-Stomatologie (weekly) . . 04907

Société Ornithologique de France (SOF), c/o Muséum National d'Histoire Naturelle, 55 Rue de Buffon, 75005 Paris
T: (1) 43310249
Founded: 1909; Members: 910
Focus: Ornithology 04908

Société Parisienne d'Etudes et de Recherches Foncières (SOPEREF), 88bis Rue Jouffroy, 75017 Paris
T: (1) 42670023
Focus: Geology 04909

Société Parisienne d'Etudes Spéciales, 10 Rue Castagnary, 75015 Paris
T: (1) 48420939
Focus: Sci 04910

Société Phycologique de France, c/o Laboratoire de Biologie Végétale Marine, Université Paris VI, 7 Quai Saint-Bernard, 75230 Paris Cédex 05
Founded: 1950; Members: 150
Focus: Botany 04911

Société pour la Protection des Paysages, Sites et Monuments, 39 Av de La Motte-Picquet, 75007 Paris
T: (1) 47053771
Founded: 1901; Members: 7000
Pres: J. de Sacy
Focus: Ecology
Periodical
Sites et Monuments (weekly) 04912

Société Provençale de Pédiatrie, c/o Hôpital de la Conception, 13001 Marseille
Focus: Pediatrics 04913

Société Psychoanalytique de Paris, 187 Rue Saint-Jacques, 75005 Paris
T: (1) 46333290
Founded: 1926; Members: 285
Focus: Psychoan 04914

Société Racinienne, 52 Rue J. Dulud, 92200 Neuilly-sur-Seine
Focus: Lit 04915

Société Saint-Simon, 11d Allée d'Honneur, 92330 Sceaux
T: (1) 45788270
Pres: Philippe Hourcade
Focus: Hist 04916

Société Savoisienne d'Histoire et d'Archéologie, c/o Musée Savoisien, Sq de Lannoy de Bissy, 73000 Chambéry
T: 79334448
Focus: Hist; Archeol 04917

Société Scientifique de Bretagne, c/o Faculté des Sciences, 35000 Rennes
Founded: 1924; Members: 280
Focus: Sci 04918

Société Scientifique d'Hygiène Alimentaire (SSHA), 16 Rue de l'Estrapade, 75005 Paris
T: (1) 43251185
Founded: 1904; Members: 1182
Pres: Dr. Guy Ebrard
Focus: Hygiene 04919

Société Technique d'Etudes Mécaniques et d'Outillage (S.T.E.M.O.), 22 Passage Dumas, 75011 Paris
T: (1) 43713513
Focus: Eng 04920

Société Théosophique de France, 4 Sq Rapp, 75007 Paris
T: (1) 45513179
Focus: Rel & Theol 04921

Société Vétérinaire Pratique de France (S.V.P.F.), 10 Pl Léon Blum, 75011 Paris
Founded: 1879; Members: 1730
Pres: Dr. Christian Rondeau; Gen Secr: Dr. Lucien Pigoury
Focus: Vet Med
Periodical
Bulletin de la Société Vétérinaire Pratique de France (weekly) 04922

Société Zoologique de France, 195 Rue Saint-Jacques, 75005 Paris
T: (1) 40793110
Founded: 1876; Members: 750
Pres: Prof. B. Condé; Gen Secr: Prof. J.-L. d' Hondt
Focus: Zoology
Periodicals
Bulletin (weekly)
Mémoires (weekly) 04923

Syndicat des Chirurgiens Dentistes de Paris, 4 Rue La Vrillière, 75001 Paris
T: (1) 42964339
Focus: Surgery; Dent
Periodical
Le Chirurgien-Dentiste de Paris (weekly) 04924

Syndicat des Ecrivains, 38 Rue du Faubourg Saint-Jacques, 75014 Paris
T: (1) 43220647
Focus: Lit 04925

Syndicat des Enseignants, 209 Blvd Saint-Germain, 75007 Paris
T: (1) 45443842; Fax: (1) 45440119
Focus: Educ 04926

Syndicat des Psychiatres Français (SPF), 23 Rue Pradier, 92410 Ville d'Avray
T: (1) 47091177; Fax: (1) 47506275
Founded: 1967; Members: 1500
Focus: Psychiatry
Periodicals
Annuaire des Psychiatres Français (bi-annually)
La Lettre de Psychiatrie Française (weekly)
Psychiatrie Française (weekly) 04927

Syndicat des Spécialistes Français en Orthopédie Dento-Faciale (SSFODF), 92 Av de Wagram, 75017 Paris
T: (1) 42278900
Founded: 1961; Members: 600
Focus: Orthopedics 04928

Syndicat des Vétérinaires de la Région Parisienne, 8 Rue Pierre Guerin, 75016 Paris
T: (1) 42886799
Focus: Vet Med 04929

Syndicat Général de l'Education Nationale, 5 Rue des Feuillantines, 75005 Paris
T: (1) 43266243
Focus: Educ 04930

Syndicat Général des Personnels de l'Education Nationale, 55 Rue Pixerecourt, 75020 Paris
T: (1) 46367693
Focus: Educ 04931

Syndicat National de l'Enseignement Supérieur, 78 Rue du Faubourg Saint-Denis, 75010 Paris
T: (1) 47709035
Founded: 1937; Members: 10780
Focus: Educ 04932

Syndicat National de l'Enseignement Technique, 74 Rue de la Fédération, 75015 Paris
T: (1) 47836130; Fax: (1) 47832669
Members: 11000
Focus: Adult Educ 04933

Syndicat National de l'Intendance de l'Education Nationale, 22bis Rue de Paradis, 75010 Paris
T: (1) 45231991
Focus: Educ 04934

Syndicat National de l'Orthopédie Française, 18 Blvd des Filles du Calvaire, 75011 Paris
T: (1) 43556220
Focus: Orthopedics 04935

Syndicat National des Allergologistes Français, 19 Rue Eupatoria, 37000 Tours
T: 47543276
Founded: 1959; Members: 400
Pres: Dr. Paul Fleury; Gen Secr: Dr. F. Duguet
Focus: Immunology 04936

Syndicat National des Anesthésistes-Réanimateurs Français, 185 Rue Saint-Maur, 75010 Paris
T: (1) 42380868
Pres: Dr. Frayssinhes; Gen Secr: Dr. Chapus
Focus: Anesthetics 04937

Syndicat National des Auteurs et Compositeurs de Musique (SNAC), 80 Rue Taitbout, 75442 Paris Cédex 09
T: (1) 48749630
Founded: 1946
Pres: Antoine Duhamel; Gen Secr: Emmanuel Rengervé
Focus: Music; Lit
Periodical
Le Bulletin des Auteurs (3 times annually) 04938

Syndicat National des Chefs d'Etablissements d'Enseignement Libre, 277 Rue Saint-Jacques, 75005 Paris
T: (1) 43296550; Fax: (1) 46340884
Focus: Educ
Periodical
Fiches Syndicales (weekly) 04939

Syndicat National des Chirurgiens Français (SNCF), c/o Confédération des Syndicats Médicaux Français, 60 Blvd de Latour-Maubourg, 75007 Paris
T: (1) 47053751
Members: 1200
Focus: Surgery 04940

Syndicat National des Chirurgiens Plasticiens, 32 Blvd de Courcelles, 75017 Paris
T: (1) 49246154
Founded: 1968; Members: 125
Focus: Surgery 04941

Syndicat National des Collèges de la Région Parisienne, 20 Rue Neuve des Boulets, 75011 Paris
T: (1) 43714043
Focus: Educ 04942

Syndicat National des Collèges et des Lycées, 13 Av Taillebourg, 75011 Paris
T: (1) 43732136; Fax: (1) 43700847
Focus: Educ
Periodical
Bulletin (weekly) 04943

Syndicat National des Enseignements de Second Degré (S.N.E.S.), 1 Rue de Courty, 75007 Paris
T: (1) 40632900; Fax: (1) 40632909
Founded: 1928
Focus: Educ
Periodical
L'Université Syndicaliste (weekly) 04944

Syndicat National des Instituteurs et Professeurs de Collège, 5 Rue Paul Louis Courier, 75007 Paris
T: (1) 42221021
Founded: 1920; Members: 320000
Focus: Educ 04945

Syndicat National des Lycées et Collèges (SNALC), 4 Rue de Trévise, 75009 Paris
T: (1) 45230514; Fax: (1) 42462660
Founded: 1905; Members: 15000
Pres: Françoise Angoulvant
Focus: Educ
Periodicals
Quinzaine Universitaire (weekly)
SNALC-Info S. 1 (weekly) 04946

Syndicat National des Médecins Acupuncteurs de France, 14 Blvd de Courcelles, 75017 Paris
T: (1) 46222975
Founded: 1946; Members: 1500
Focus: Anesthetics 04947

Syndicat National des Médecins Anatomo-Cyto-Pathologistes Français, c/o Faculte X Bichat, 16 Rue Henri Huchard, 75018 Paris
T: (1) 42633534
Pres: Dr. Marie Claire Imbert; Gen Secr: Dr. Jean Paul Donzel; Jacques Hassoun
Focus: Anat; Pathology 04948

Syndicat National des Médecins Biologistes, 133 Blvd de Montparnasse, 75006 Paris
T: (1) 43203838
Founded: 1927
Pres: Dr. Gérard Gallez; Gen Secr: Dr. Jean Gayraud
Focus: Med; Bio 04949

Syndicat National des Médecins de Groupe, 26 Rue de Clichy, 75009 Paris
T: (1) 42800607
Focus: Med
Periodicals
Lettre Hebdomadaire des Médecins de Groupe (weekly)
Médecins de Groupe (weekly) 04950

Syndicat National des Médecins des Hôpitaux Publics, 60 Blvd Latour-Maubourg, 75007 Paris
T: (1) 4556018
Focus: Med
Periodical
Médecine Hospitaliére (weekly) 04951

Syndicat National des Médecins du Sport (S.N.M.S.), 60 Blvd de Latour-Maubourg, 75340 Paris Cédex 07
T: (1) 47053751
Founded: 1948
Focus: Med
Periodicals
Cinésiologie: La Revue Internationale des Médecins du Sport (weekly)
Lettre Syndicale (weekly) 04952

Syndicat National des Médecins Electro-Radiologistes Qualifiés, 60 Blvd Latour-Moubourg, 75007 Paris
T: (1) 45517784
Focus: Radiology 04953

Syndicat National des Médecins Français Spécialistes des Maladies du Coeur et des Vaisseaux, 147 Blvd Brune, 75014 Paris
T: (1) 45430810
Founded: 1948; Members: 10
Pres: Dr. Jacques Vadot; Gen Secr: Dr. Gérard Rousselet
Focus: Cardiol 04954

Syndicat National des Médecins Homéopathes Français, c/o Hôpital Saint-Jacques, 37 Rue des Volontaires, 75730 Paris Cédex 15
Focus: Homeopathy 04955

Syndicat National des Médecins Ostéothérapeutes Français (SNMOF), 67 Rue Raymond Poincaré, 54000 Nancy
T: 83406061; Fax: 83273460
Founded: 1953; Members: 600
Pres: Dr. Jean-Louis Garcia; Gen Secr: Dr. Daniel Le Corgne
Focus: Therapeutics 04956

Syndicat National des Médecins Phlébologues Français, c/o Dr. Sapin, 55 Rue de Varenne, 75007 Paris
Focus: Physiology 04957

Syndicat National des Médecins Rhumatologues, 19 Blvd Pierre Brossolette, 92160 Antony
T: 42373888
Pres: Dr. Yves Lac de Fougères; Gen Secr: Dr. Raymond Cohen
Focus: Rheuma 04958

Syndicat National des Médecins Spécialisés en Phoniatrie, 16 Rue des Ursulines, 93200 Saint-Denis
Founded: 1966
Focus: Otorhinolaryngology 04959

Syndicat National des Médecins Spécialistes de l'Endocrinologie et de la Nutrition, 71 Av des Ternes, 75017 Paris
T: (1) 45741919
Founded: 1960; Members: 200
Pres: Dr. Nicolas Gueritée; Gen Secr: Dr. Jean Michel Daninos
Focus: Endocrinology; Nutrition 04960

Syndicat National des Médecins Spécialistes en Stomatologie et Chirurgie Maxillo-Faciale, c/o Dr. Etienne Alexandre, 3 Rue de Rivoli, 75004 Paris
T: (1) 42728960
Pres: Dr. Etienne Alexandre; Gen Secr: Dr. Francis Dujarric
Focus: Stomatology 04961

Syndicat National des Oto-Rhino-Laryngologistes Français, 12 Rue de Logelbach, 75017 Paris
Founded: 1907; Members: 1250
Focus: Otorhinolaryngology 04962

Syndicat National des Pédiatres Français (SNPF), c/o Dr. J.L. Boy, 28 Rue August Audollent, 63000 Clermont-Ferrand
Founded: 1952; Members: 800
Focus: Pediatrics 04963

Syndicat National des Personnels de Direction de l'Enseignement Secondaire (SNPDES), c/o Lycée Jacquard, 2 Rue Bouret, 75019 Paris
Members: 3400
Focus: Educ 04964

Syndicat National des Professeurs d'Arts Martiaux (S.N.P.A.M.), 68 Rue Castagnary, 75015 Paris
T: (1) 45314129
Founded: 1957; Members: 250
Pres: Luc Levannier
Focus: Sports 04965

Syndicat National des Professeurs des Ecoles Normales d'Instituteurs (SNPEN), 48 Rue La Bruyère, 75009 Paris
Founded: 1946; Members: 1600
Focus: Educ 04966

Syndicat National des Psychiatres des Hôpitaux (SNPH), 1 Rue Cabanis, 75014 Paris
T: (1) 43319950
Founded: 1945; Members: 900
Focus: Psychiatry 04967

Syndicat National des Vétérinaires Français, 10 Pl Léon Blum, 75011 Paris
T: (1) 43791152
Focus: Vet Med
Periodicals
La Dépèche Technique (weekly)
La Dépèche Vétérinaire (weekly) 04968

Syndicat National Français des Dermatologistes et Vénéréologistes, 60 Blvd de Latour-Maubourg, 75007 Paris
T: (1) 47053751
Founded: 1928
Focus: Derm; Venereology
Periodicals
La Lettre du Président (weekly)
L'Officiel des Dermatologistes et Vénéréologistes (3 times annually) 04969

Syndicat National Professionnel des Médecins du Travail (SNPMT), 12 Impasse Mas, 31000 Toulouse
T: 61992077; Fax: 61627566
Founded: 1949
Focus: Hygiene
Periodical
Revue Médecine et Travail 04970

Union Astronomique Internationale (UAI) / International Astronomical Union, 98bis Blvd Arago, 75014 Paris
T: (1) 43258358; Tx: 205671
Founded: 1919; Members: 7700
Gen Secr: D. McNally
Focus: Astronomy
Periodical
IAU/UAI Information Bulletin (weekly) . . 04971

Union Centrale des Arts Décoratifs (U.C.A.D.), 107 Rue de Rivoli, 75001 Paris
T: (1) 42603214
Founded: 1864; Members: 2200
Pres: Antoine Riboud; Gen Secr: Pierre Lambertin; Thierry Bondoux
Focus: Arts 04972

Union des Biologistes de France, 4 Rue Pasquier, 75008 Paris
T: (1) 42651597; Fax: (1) 42655805
Pres: Adrien Bedossa; Gen Secr: Michèle Hanne
Focus: Bio; Med
Periodical
Le Nouveau Biologiste (10 times annually) 04973

Union des Océanographes de France, 195 Rue Saint-Jacques, 75005 Paris
T: (1) 43256310; Fax: (1) 40517316
Founded: 1901
Focus: Oceanography
Periodical
Journal de Recherche Océanographique . 04974

Union des Physiciens, 44 Blvd Saint-Michel, 75270 Paris Cédex 06
T: (1) 43297308
Founded: 1906; Members: 11000
Pres: Prof. M.A. Touren
Focus: Physics; Chem; Educ
Periodical
Bulletin (weekly) 04975

Union des Professeurs de Spéciales (Mathématiques et Sciences Physiques) (U.P.S.), c/o Musée Pédagogique, 29 Rue d'Ulm, 75230 Paris Cédex 05
Founded: 1928; Members: 1000
Pres: A. Simon; Gen Secr: Lucien Sellier
Focus: Educ 04976

Union des Travailleurs Espérantistes des Pays de Langue Française / SAT-AMIKARO, 67 Av Gambetta, 75020 Paris
T: (1) 47978705
Focus: Ling 04977

Union Fédérative des Sociétés d'Education Physique et de Préparation Militaire, 23 Rue La Sourdière, 75001 Paris
T: (1) 42612774
Focus: Sports; Military Sci 04978

Union Française des Organismes de Documentation (U.F.O.D.), 16 Rue Jules Claretie, 75016 Paris
T: (1) 45040771
Founded: 1932
Focus: Doc 04979

Union Française pour l'Espéranto, 4bis Rue de la Cerisaie, 75004 Paris
T: (1) 42786886
Focus: Ling
Periodical
Franca Esperantisto: Revue Française d'Espéranto (8 times annually) 04980

Union Générale des Auteurs et Musiciens Professionnels, 71 Rue du Faubourg Saint-Martin, 75010 Paris
T: (1) 42082290
Focus: Lit; Music 04981

Union Géodésique et Géophysique Internationale (UGGI), c/o Bureau Gravimétrique International, 18 Av Edouard Belin, 31055 Toulouse Cédex
T: 61332989; Tx: 530776; Fax: 61253098
Founded: 1919
Gen Secr: Dr. Georges Balmino
Focus: Geology; Geophys
Periodical
Chronique UGGI (weekly) 04982

Union Internationale d'Electrothermie (UIE) / International Union for Electroheat, Tour Atlantique, 92080 Paris-La Défense Cédex 06
T: (1) 47789934; Fax: (1) 49060373
Founded: 1953
Pres: J. Finet; Gen Secr: G. Vanderschueren
Focus: Electric Eng 04983

Union Internationale de Phlébologie, 106 Av de Suffren, 75015 Paris
T: (1) 43069909; Fax: (1) 43069057
Founded: 1960; Members: 16
Pres: André Davy; Gen Secr: Dr. Pierre Wallois
Focus: Med
Periodical
Phlébologie (weekly) 04984

Union Internationale des Architectes (UIA) / International Union of Architects, 51 Rue Raymonard, 75016 Paris
T: (1) 45243688; Tx: 614855; CA: UNIARCH; Fax: (1) 45240278
Founded: 1948; Members: 900000
Gen Secr: Vassilis Sgoutas
Focus: Archit
Periodical
Bulletin d'Information (weekly) 04985

Union Internationale des Etudes Orientales et Asiatiques (UIEOA) / International Union for Oriental and Asian Studies, 77 Quai du Port-au-Fouarre, 94100 Saint-Maur
Founded: 1951
Pres: Prof. R.N. Dandekar; Gen Secr: Prof. Louis Bazin
Focus: Ethnology 04986

Union Internationale des Femmes Architectes (UIFA) / International Union of Women Architects, 14 Rue Dumont d'Urville, 75116 Paris
T: (1) 47208882
Members: 1000
Pres: S. d' Herbez de la Tour
Focus: Archit 04987

Union Internationale des Sciences Biologiques (UIBS) / International Union of Biological Sciences, 51 Blvd de Montmorency, 75016 Paris
T: (1) 45250009
Founded: 1922
Gen Secr: Dr. Talal Younès
Focus: Bio
Periodical
Biology International (weekly) 04988

Union Internationale des Sociétés d'Aide à la Santé Mentale, 39 Rue Charles Monselet, 33000 Bordeaux
T: 56816005
Founded: 1972
Focus: Psychiatry 04989

Union Internationale pour la Liberté d'Enseignement (UILE) / International Union for the Liberty of Education, 17 Rue Monceau, 75006 Paris
Founded: 1950
Pres: R. Despretz; Gen Secr: Edouard Lizop
Focus: Educ
Periodical
Congress Report 04990

Union Internationale Thérapeutique, 8 Pl de l'Odéon, 75006 Paris
Founded: 1934
Focus: Therapeutics 04991

Union Nationale des Médecins Spécialistes Confédérés, 60 Blvd de Latour-Maubourg, 75340 Paris Cédex 07
T: (1) 47053751; Fax: (1) 45415270
Members: 9000
Focus: Med 04992

Union Nationale des Techniciens Biologistes (UNATEB), 109 Av Gabriel Péri, 94170 Le Perreux
T: (1) 48714344; Fax: (1) 48711431
Founded: 1963; Members: 3000
Focus: Bio
Periodicals
Le Biotechnologiste (weekly)
UNATEB Actualités (weekly) 04993

Union Nationale Patronale des Prothésistes Dentaire, 27 Av Stephen Pichon, 75013 Paris
T: (1) 45870104
Focus: Dent 04994

Union Professionnelle des Professeurs, Cadres et Techniciens du Secrétariat et de la Comptabilité, 21 Rue Croulebarbe, 75013 Paris
T: (1) 43312248
Focus: Adult Educ 04995

Union Syndicale Nationale des Angiologues, 102 Av du Général Leclerc, 75014 Paris
T: (1) 45436458
Pres: Dr. Daniel Paitel; Gen Secr: Dr. Roger Elkrieff
Focus: Hematology 04996

Union Technique de l'Automobile, du Motocycle et du Cycle (U.T.A.C.), 157 Rue Lecourbe, 75015 Paris
T: (1) 48425390
Focus: Auto Eng
Periodicals
Album des Normes de l'Automobile
Bulletin Mensuel de Documentation (weekly) 04997

Union Technique de l'Electricité (UTE), Immeuble L'Avoisier, 92052 Paris-La Défense Cédex 64
T: (1) 47685020
Founded: 1907; Members: 7
Focus: Electric Eng
Periodicals
Bulletin (3 times annually)
Catalogue (weekly) 04998

Universités Unies pour l'Environnement, c/o AIMBE, 115 Rue de la Pompe, 75116 Paris
T: (1) 45534504, 45530970; Tx: 614584; CA: ECOMEBIO-Paris; Fax: (1) 45534175
Focus: Ecology 04999

Vieilles Maisons Françaises (VMF), 93 Rue de l'Université, 75007 Paris
T: (1) 40526171; Fax: (1) 45511225
Founded: 1962; Members: 16000
Pres: Baron de Grandmaison
Focus: Archit
Periodical
Vieilles Maisons Françaises 05000

World Federation of Scientific Workers, BP 404, 93514 Montreuil Cédex
T: (1) 45818003
Founded: 1946; Members: 400000
Pres: Dr. C. Russell; Gen Secr: M.A. Jaeglé
Focus: Sci
Periodical
Scientific World (weekly) 05001

World Medical Association (WMA), 28 Av des Alpes, BP 6363, 01212 Ferney-Voltaire Cédex
Fax: 50405937
Founded: 1947; Members: 59
Gen Secr: Dr. André Wynen
Focus: Med
Periodical
World Medical Journal (weekly) 05002

French Polynesia

Société des Etudes Océaniennes (S.E.O.), BP 110, Papeete, Tahiti
T: 439887
Founded: 1917; Members: 600
Pres: Dr. P. Moortgat
Focus: Ethnology; Archeol
Periodical
Bulletin (weekly) 05003

Gabon

African Women Jurists Federation, BP 4347, Libreville
Founded: 1978
Gen Secr: Pauline Nyngone
Focus: Law
Periodical
Le Message (weekly) 05004

Gambia

African Centre for Democracy and Human Rights Studies, Kairiba Av, Kombo St, Mary Division, Banjul
T: 94525; Tx: 94962
Founded: 1989
Gen Secr: Raymond C. Sock
Focus: Poli Sci; Law
Periodical
African Centre Human Rights Newsletter (weekly) 05005

Georgia

Amateur Society of Basque Language, Rustaveli 52, 380008 Tbilisi
Pres: S.V. Dzidziguri
Focus: Ling 05006

Commission for the Study of Production Forces and Natural Resources, Paliashvili 87, 380030 Tbilisi
T: 223216
Pres: A.A. Dzidzguri
Focus: Econ 05007

Commission on Biosphere and Ecology Research, Rustaveli 52, 380008 Tbilisi
Pres: A.N. Tavkhelidze
Focus: Biophys; Ecology 05008

Council on Co-ordinating Scientific Studies of the Georgian Language, Dzerzhinski 8, 380007 Tbilisi
Pres: K.B. Lomtatidze
Focus: Ling 05009

Council on the History of Natural Sciences and Technology, Rustaveli 52, 380008 Tbilisi
Pres: V.D. Parkadze
Focus: Hist; Sci; Eng 05010

Georgian Academy of Sciences, Pr Rustaveli 52, 380008 Tbilisi
T: 998961
Members: 137
Pres: A.N. Tavkhelidze; Gen Secr: L.K. Sanadze
Focus: Sci
Periodicals
Bulletin (weekly)
Matsne: Herald (weekly)
Metsnierba da Technika (weekly)
Yearbook of Iberian-Caucasian Linguistics (weekly) 05011

Georgian Bio-Mecico-Technical Society, Telavi 51, 380003 Tbilisi
Pres: K.S. Nadreishvili
Focus: Med 05012

Georgian Botanical Society, Kodzhorskoe Shoss, 380007 Tbilisi
Pres: G.S. Nakhutsrishvili
Focus: Botany, Specific 05013

Georgian Commission on Archaeology, Rustaveli 52, 380008 Tbilisi
Pres: A.M. Apakidze
Focus: Archeol 05014

Georgian Geographical Society, Ketskhoveli 11, 380007 Tbilisi
Pres: V.S. Dzhaoshvili
Focus: Geography 05015

Georgian Geological Society, Rustaveli 52, 380008 Tbilisi
Pres: L.K. Gabunia
Focus: Geology 05016

Georgian History Society, Rustaveli 52, 380008 Tbilisi
Pres: A.M. Apakidze
Focus: Hist 05017

Georgian National Committee on UNESCO Long-Term Programme „Man and the Biosphere", Rustaveli 52, 380008 Tbilisi
Pres: L.K. Gabunia
Focus: Biophys 05018

Georgian National Speleological Society, Rukhadze 1, 380093 Tbilisi
Focus: Geology 05019

Georgian Philosophy Centre, Rustaveli 29, 380008 Tbilisi
Pres: N.Z. Chavchavadze
Focus: Philos 05020

Georgian Society of Biochemistry, University 2, 380043 Tbilisi
Pres: N.G. Aleksidze
Focus: Biochem 05021

Georgian Society of Genetics and Selectionists, Gotua 3, 380060 Tbilisi
Pres: T.G. Chanishvili
Focus: Genetics 05022

Georgian Society of Helminthologists, Chavchavadze, Prof. B.E., 380030 Tbilisi
Pres: Prof. B.E. Kurashvili
Focus: Zoology 05023

Georgian Society of Patho-Anatomists, V. Pshavela 27b, 380077 Tbilisi
Pres: T.I. Dekanosidze
Focus: Anat 05024

Georgian Society of Physiologists, Gudamakari 2, 380092 Tbilisi
Pres: V.M. Okudzjava
Focus: Physiology 05025

Georgian Society of Psychologists, Jashvili 22, 380007 Tbilisi
Pres: N.Z. Nadirashvili
Focus: Psych 05026

Scientific-Technical Council on Computer Technology, Mathematical Modelling, Automation of Scientific Research and Instrument Making, Rustaveli 52, 380008 Tbilisi
Pres: N.S. Amaglobeli
Focus: Computer & Info Sci; Math; Eng 05027

Germany

Aachener Geschichtsverein, Fischmarkt 3, 52062 Aachen
Focus: Hist
Periodicals
Aachener Beiträge für Baugeschichte und Heimatkunst (weekly)
Zeitschrift des Aachener Geschichtsvereins (weekly) 05028

Abwassertechnische Vereinigung e.V. (ATV), Markt 71, 53757 Sankt Augustin
T: (02241) 2320; Fax: (02241) 23235
Founded: 1948; Members: 7290
Focus: Water Res
Periodical
Korrespondenz Abwasser (weekly) . . . 05029

Ärztekammer Berlin, Flottenstr 28-42, 13407 Berlin
T: (030) 408060
Pres: Dr. Ellis E. Huber; Gen Secr: Josef Kloppenborg
Focus: Med
Periodical
Berliner Ärzte (weekly) 05030

Ärztekammer Bremen, Schwachhauser Heerstr 26-28, 28209 Bremen
T: (0421) 340051
Focus: Med 05031

Ärztekammer des Saarlandes, Faktoreistr 4, 66111 Saarbrücken
T: (0681) 40030
Pres: Prof. Dr. Franz Carl Loch; Gen Secr: Heinz Jürgen Lander
Focus: Med
Periodical
Saarländisches Ärzteblatt (weekly) . . . 05032

Ärztekammer des Saarlandes, Abteilung Zahnärzte, Puccinistr 2, 66119 Saarbrücken
T: (0681) 586080
Pres: Dr. Werner Röhrig
Focus: Dent 05033

Ärztekammer Frankfurt, Georg-Voigt-Str 15, 60325 Frankfurt
T: (069) 79201
Focus: Med 05034

Ärztekammer Hamburg, Humboldtstr 56, 22083 Hamburg
T: (040) 228020
Pres: Dr. Rolf Bialas; Gen Secr: Dr. Klaus-Heinrich Damm
Focus: Med
Periodical
Hamburger Ärzteblatt (weekly) 05035

Ärztekammer Land Brandenburg, Thiemstr 41, 03050 Cottbus
T: (0355) 422012
Pres: Dr. Roger Kirchner; Gen Secr: Dr. Reinhard Heiber
Focus: Med 05036

Ärztekammer Mecklenburg-Vorpommern, Humboldtstr 6, 18055 Rostock
T: (0381) 22265
Pres: Dr. Andreas Crusius
Focus: Med 05037

Ärztekammer Niedersachsen, Berliner Allee 20, 30175 Hannover
T: (0511) 38002; Tx: 922969; Fax: (0511) 3802240
Pres: Prof. Dr. H. Eckel; Gen Secr: Dr. Ulrich Kirchhoff
Focus: Med
Periodical
Niedersächsisches Ärzteblatt (weekly) . . 05038

Ärztekammer Nordrhein (ÄKNo), Tersteegenstr 31, 40474 Düsseldorf
T: (0211) 43020; Fax: (0211) 4302200
Founded: 1945; Members: 38000
Pres: Dr. Jörg D. Hoppe; Gen Secr: Dr. Wolfgang Klitzsch; Dr. Robert D. Schäfer
Focus: Med
Periodical
Rheinisches Ärzteblatt (weekly) 05039

Ärztekammer Sachsen-Anhalt, Zollstr 12, 39114 Magdeburg
T: (0391) 33861
Pres: Prof. Dr. Walter Brandstätter; Gen Secr: Dr. Peter Eichelmann
Focus: Med 05040

Ärztekammer Schleswig-Holstein, Bismarckallee 8-12, 23795 Bad Segeberg
T: (04551) 8030
Pres: Dr. Ingeborg Retzlaff; Gen Secr: Dr. Karl-Werner Ratschko
Focus: Med
Periodical
Schleswig-Holsteinisches Ärzteblatt (weekly) 05041

Ärztekammer Westfalen-Lippe, Kaiser-Wilhelm-Ring 4-6, 48145 Münster
T: (0251) 37500; Tx: 892612; Fax: (0251) 3750450
Focus: Med
Periodical
Westfälisches Ärzteblatt (weekly) 05042

Ärztliche Gesellschaft für Physiotherapie, Kneippärztebund e.V., Seb.-Kneipp-Promenade 28-30, 53902 Bad Münstereifel
T: (02253) 911
Founded: 1894; Members: 700
Focus: Therapeutics 05043

Agrarsoziale Gesellschaft e.V. (ASG), Kurze Geismarstr 33-35, 37073 Göttingen
T: (0551) 497090
Founded: 1947; Members: 580
Pres: Manfred Merforth
Focus: Sociology
Periodicals
Geschäfts- und Arbeitsbericht (weekly)
Kleine Reihe (weekly)
Ländlicher Raum: Rundbrief der ASG (weekly)
Materialsammlung (weekly)
Schriftenreihe für Ländliche Sozialfragen (weekly) 05044

AID Auswertungs- und Informationsdienst für Ernährung, Landwirtschaft und Forsten e.V., Konstantinstr 124, 53179 Bonn
T: 0228) 84990
Pres: Dr. A. Padberg; Gen Secr: Dr. Horst Schanze
Focus: Agri; Forestry; Nutrition
Periodicals
AID Verbraucherdienst: Zeitschrift für Fach-, Lehr- und Beratungskräfte im Bereich Ernährung (weekly)
Ausbildung und Beratung in Land- und Hauswirtschaft: Monatsschrift für Lehr- und Beratungskräfte (weekly) 05045

Akademie der Arbeit in der Universität, Mertonstr 30, 60325 Frankfurt
T: (069) 772021; Fax: (069) 7073469
Pres: Prof. Dr. Diether Döring; Dr. Otto Ernst Kempen; Prof. Dr. Renate Neubäumer
Focus: Sociology; Law; Econ
Periodical
Mitteilungen (weekly) 05046

Akademie der Diözese Rottenburg-Stuttgart, Im Schellenkönig 61, 70184 Stuttgart
T: (0711) 16406; Fax: (0711) 1640777
Gen Secr: Dr. Gebhard Fürst
Focus: Rel & Theol 05047

Akademie der Künste / Academy of Arts, Hanseatenweg 10, 10557 Berlin
T: (030) 390070; Fax: (030) 3900771
Founded: 1696; Members: 256
Pres: Walter Jens
Focus: Arts; Fine Arts 05048

Akademie der Wissenschaften in Göttingen, Theaterstr 7, 37073 Göttingen
T: (0551) 395362; Fax: (0551) 395365
Founded: 1751; Members: 250
Pres: Prof. Dr. Ulrich Mölk; Gen Secr: Prof. Dr. Heinz Georg Wagner
Focus: Sci
Periodicals
Abhandlungen: I. Philologisch-Historische Klasse, II. Mathematisch-Physikalische Klasse (weekly)
Göttingische Gelehrte Anzeigen (weekly)
Jahrbuch
Nachrichten: I. Philologisch-Historische Klasse, II. Mathematisch-Physikalische Klasse (weekly) 05049

Akademie der Wissenschaften und der Literatur zu Mainz, Geschwister-Scholl-Str 2, 55131 Mainz
T: (06131) 5770
Founded: 1949; Members: 220
Pres: Prof. Dr. Clemens Zintzen; Gen Secr: Dr. Wulf Thommel
Focus: Sci
Periodicals
Abhandlungen: Geistes- und Sozialwissenschaftliche Klasse (weekly)
Abhandlungen: Klasse der Literatur (weekly)
Abhandlungen: Mathematisch-Naturwissenschaftliche Klasse (weekly)
Altern und Entwicklung: Aging and Development (weekly)
Basic Aspects of Glaucoma Research (weekly)
Biona Report (weekly)
Erdwissenschaftliche Forschung: Kommission für Erdwissenschaftliche Forschung (weekly)
Forschungen zur Antiken Sklaverei: Kommission für Geschichte des Altertums (weekly)
Forschungen zur neueren Medizin- und Biologiegeschichte (weekly)
Fortschritte der Zoologie (weekly)
Funktionsanalyse Biologischer Systeme (weekly)
Historische Forschungen (weekly)
Hydronymia Europaea (weekly)
Hydronymia Germaniae (weekly)
Information Processing in Animals (weekly)
Jahrbuch (weekly)
Karl-August-Forster-Lectures: Mathematisch-Naturwissenschaftliche Klasse (weekly)
Kiniogenases (weekly)
Mainzer Reihe (weekly)
Medizinhistorisches Journal: Kommission für Geschichte der Medizin und der Naturwissenschaften (weekly)
Medizinische Forschung (weekly)
Microfauna Marina: Mathematisch-Naturwissenschaftliche Klasse (weekly)
Musikalische Denkmäler (weekly)
Neue Studien zur Musikwissenschaft (weekly)
Ökosystemanalyse und Umweltforschung (weekly)
Paläoklimaforschung (weekly)
Research in Molecular Biology: Mathematisch-Naturwissenschaftliche Klasse (weekly)
Soemmering-Forschungen (weekly)
Studien zu den Bogazköy-Texten (weekly)
Studien zu den Fundmünzen der Antike (weekly)
Tropische und Subtropische Pflanzenwelt (weekly)
Untersuchungen zur Sprach- und Literaturgeschichte der Romanischen Völker (weekly)
Veröffentlichungen: Orientalische Kommission (weekly)
Verschollene und Vergessene (weekly) . 05050

Akademie der Wissenschaften zu Berlin / Academy of Sciences and Technology in Berlin, Griegstr. 5-7, 14193 Berlin
T: (030) 8209050
Founded: 1987; Members: 31
Pres: Prof. Dr. Horst Albach; Gen Secr: Dr. Wolfgang Holl
Focus: Sci 05051

Akademie für Arbeit und Sozialwesen, Sophienstr 6-8, 66111 Saarbrücken
T: (0681) 40051
Focus: Sociology 05052

Akademie für das Grafische Gewerbe, Pranckhstr 2, 80335 München
T: (089) 555761; Fax: (089) 524718
Founded: 1927
Pres: Dietmar Leischner
Focus: Eng 05053

Akademie für Fernstudium und Weiterbildung Bad Harzburg, Golfstr 11, 38667 Bad Harzburg
T: (05322) 54091; Fax: (05322) 2014
Founded: 1990; Members: 3
Gen Secr: Horst K. Bülow
Focus: Educ 05054

Akademie für Führungskräfte der Wirtschaft e.V. (AFW), Postf 1116, 38667 Bad Harzburg
T: (05322) 730; Tx: 957623
Founded: 1956
Pres: Prof. Dr. Fritz Raidt; Gen Secr: Prof.Dr. Reinhard Höhn
Focus: Business Admin
Periodical
Management heute (weekly) 05055

Akademie für öffentliches Gesundheitswesen in Düsseldorf (AföG), Auf'm Hennekamp 70, 40225 Düsseldorf
T: (0211) 310960; Fax: (0211) 310669
Founded: 1971; Members: 7
Focus: Public Health
Periodical
Blickpunkt (weekly) 05056

Akademie für Organisation, Kaiserstr 3, 53113 Bonn
T: (0228) 210021
Gen Secr: Dr. Reiner Chrobok
Focus: Business Admin 05057

Akademie für Politische Bildung, Buchensee 1, 82327 Tutzing
T: (08158) 2560
Founded: 1957
Focus: Poli Sci 05058

Akademie für Publizistik in Hamburg e.V., Magdalenenstr 64a, 20148 Hamburg
T: (040) 447142, 447644
Pres: Eberhard Maseberg; Gen Secr: Dr. W. Teichert
Focus: Journalism 05059

Akademie für Raumforschung und Landesplanung (ARL), Hohenzollernstr 11, 30161 Hannover
T: (0511) 348420; Fax: (0511) 3484241
Founded: 1946; Members: 450
Pres: Dr. Gottfried Schmitz; Gen Secr: Prof. Dr. Werner Schramm
Focus: Urban Plan
Periodical
Raumforschung und Raumordnung: Spatial Research and Spatial Management (weekly)
. . . . 05060

Akademie gemeinnütziger Wissenschaften zu Erfurt e.V., Anger 57, 99084 Erfurt
T: (0361) 26155
Focus: Soc Sci 05061

Akademie Klausenhof, Klausenhofstr 100, 46499 Hamminkeln
T: (02852) 890; Fax: (02852) 89300
Gen Secr: Dr. Alois Becker
Focus: Educ 05062

Akademie Kontakte der Kontinente, Langenbachstr 1, 53113 Bonn
Focus: Poli Sci 05063

Akademie Remscheid für musische Bildung und Medienerziehung e.V. (ARS), Küppelstein 34, 42857 Remscheid
T: (02191) 7940
Founded: 1958
Pres: Dr. Max Fuchs
Focus: Arts; Educ
Periodicals
kulturarbeit aktuell (weekly)
RAT – Remscheider Arbeitshilfen und Texte (weekly) 05064

Akademischer Verein Hütte, Carmerstr 12, 10623 Berlin
Founded: 1846
Focus: Metallurgy 05065

Aktion Bildungsinformation e.V. (AB), Alte Poststr 5, 70173 Stuttgart
T: (0711) 299335
Pres: Eberhard Kleinmann; Gen Secr: Werner Kinzinger
Focus: Educ 05066

Aktion Psychisch Kranke / Vereinigung zur Reform der Versorgung psychisch Kranker e.V., Gaurheindorfer Str 15, 53111 Bonn
T: (0228) 631545
Pres: Bernhard Jagoda; Gen Secr: Christine Przytulla
Focus: Psychiatry 05067

Aktionsgemeinschaft Natur- und Umweltschutz Baden-Württemberg, Olgastr 19, 70182 Stuttgart
T: (0711) 241460; Fax: (0711) 2360556
Focus: Ecology 05068

Aktionsgemeinschaft Soziale Marktwirtschaft e.V. (ASM), Mohlstr 26, 72074 Tübingen
T: (07071) 550600; Fax: (07071) 550601
Founded: 1953
Pres: Prof. Dr. Joachim Starbatty
Focus: Econ
Periodical
ASM-Bulletin (weekly) 05069

Albertus-Magnus-Institut, Adenauerallee 19, 53111 Bonn
T: (0228) 263849
Founded: 1931; Members: 6
Pres: Prof. Dr. Wilhelm Kübel
Focus: Rel & Theol; Philos 05070

Allgemeine Ärztliche Gesellschaft für Psychotherapie (AÄGP), Friedrich-Lau-Str 7, 40474 Düsseldorf
T: (0211) 450741; Fax: (0211) 450741
Founded: 1926; Members: 1583
Pres: Prof. Dr. Henning H. Studt; Gen Secr: Dr. Reinhard Hirsch
Focus: Therapeutics 05071

Allgemeine Gesellschaft für Philosophie in Deutschland e.V., c/o Zentrum für Philosophie, Universität, Otto-Behaghel-Str 10, 35394 Giessen
Fax: (0641) 7022504
Focus: Philos 05072

Allgemeiner Cäcilien-Verband für Deutschland (ACV), Andreasstr 9, 93059 Regensburg
T: (0941) 84339; Fax: (0941) 84339
Founded: 1868; Members: 318000
Pres: Prof. Dr. Wolfgang Bretschneider
Focus: Music
Periodicals
Kirchenmusikalisches Jahrbuch (weekly)
Musica Sacra (weekly) 05073

Anatomische Gesellschaft, c/o Institut für Anatomie, Medizinische Universität, Ratzeburger Allee 160, 23538 Lübeck
T: (0451) 5004030; Fax: (0451) 5004034
Founded: 1886; Members: 1025
Gen Secr: Prof. Dr. Wolfgang Kühnel
Focus: Anat
Periodical
Anatomischer Anzeiger: Annals of Anatomy (weekly) 05074

Anthropos Institut, Arnold-Janssen-Str 20, 53754 Sankt Augustin
Focus: Anthro
Periodical
Anthropos: International Review of Ethnology and Linguistics Internationale Zeitschrift für Völker- und Sprachenkunde (weekly) 05075

Anthroposophische Gesellschaft in Deutschland, Zur Uhlandshöhe 10, 70188 Stuttgart
T: (0711) 1643121; Fax: (0711) 1643130
Gen Secr: Friedhelm Dörmann
Focus: Humanities; Educ; Philos
Periodical
die Drei: Zeitschrift für Anthroposophie (weekly) . . . 05076

Anwenderverband Deutscher Informationsverarbeiter e.V. (adi), Skipperweg 3, 24159 Kiel
T: (0431) 37790
Focus: Computer & Info Sci
Periodical
Online (weekly) . . . 05077

Arbeitsausschuß Wälzlager im DIN Deutsches Institut für Normung e.V., Kamekestr 2-8, 50672 Köln
T: (0221) 57131
Pres: W. Plzack; Gen Secr: H. Tepper
Focus: Standards; Eng . . . 05078

Arbeitsgemeinschaft Allensbach e.V., Kappeler-Berg-Str 54, 78476 Allensbach
Focus: Stats
Periodical
Allensbacher Almanach . . . 05079

Arbeitsgemeinschaft Allergiekrankes Kind, Hauptstr 29, 35745 Herborn
T: (02772) 41237
Founded: 1977; Members: 5842
Pres: Marianne Stock
Focus: Immunology
Periodical
Infoblatt (weekly) . . . 05080

Arbeitsgemeinschaft beruflicher und ehrenamtlicher Naturschutz (ABN), Konstantinstr 110, 53179 Bonn
Founded: 1947; Members: 700
Pres: Prof. Dr. Guenter Preuss; Gen Secr: Prof. Dr. Wolfgang Erz
Focus: Ecology
Periodical
Jahrbuch für Naturschutz und Landschaftspflege (weekly) . . . 05081

Arbeitsgemeinschaft Bremer Schule e.V., Konsul-Cassel-Str 17, 28357 Bremen
Focus: Educ . . . 05082

Arbeitsgemeinschaft der Archive und Bibliotheken in der evangelischen Kirche, Veilhofstr 28, 90489 Nürnberg
Fax: (0911) 5819683
Focus: Archives; Libraries & Bk Sci . . . 05083

Arbeitsgemeinschaft der Deutschen Werkkunstschulen, Am Wandrahm 23, 28195 Bremen
Focus: Arts . . . 05084

Arbeitsgemeinschaft der Direktoren der Institute für Leibesübungen an Universitäten und Hochschulen der Bundesrepublik Deutschland, Kaiserstr 12, 76131 Karlsruhe
Focus: Sports . . . 05085

Arbeitsgemeinschaft der Grossforschungs-Einrichtungen (AGF), Ahrstr 45, 53175 Bonn
T: (0228) 376741
Pres: Prof. Dr. Joachim Treusch; Gen Secr: Dr. Klaus Fleischmann
Focus: Sci . . . 05086

Arbeitsgemeinschaft der kirchlichen Büchereiverbände, Wittelsbacherring 9, 53115 Bonn
T: (0228) 72580
Gen Secr: Erich Hodick
Focus: Libraries & Bk Sci . . . 05087

Arbeitsgemeinschaft der Ordenshochschulen (AGO), Pallottistr 3, 56179 Vallendar
T: (0261) 64021
Focus: Rel & Theol; Educ . . . 05088

Arbeitsgemeinschaft der Parlaments- und Behördenbibliotheken (APBB), c/o Bibliothek des Deutschen Patentamts, 80297 München
T: (089) 21952606; Tx: 523534; Fax: (089) 21952221
Members: 500
Pres: Hubert Rothe; Gen Secr: Johann Fuchs
Focus: Libraries & Bk Sci
Periodicals
Arbeitshefte der Arbeitsgemeinschaft der Parlaments- und Behördenbibliotheken (weekly)
Mitteilungen der Arbeitsgemeinschaft der Parlaments- und Behördenbibliotheken (weekly) . . . 05089

Arbeitsgemeinschaft der Regionalbibliotheken / Joint Association of Regional Libraries, c/o Niedersächsische Landesbibliothek, Waterloostr 8, 30169 Hannover
Fax: (0511) 15785
Pres: Wolfgang Dittrich
Focus: Libraries & Bk Sci . . . 05090

Arbeitsgemeinschaft der Seminarlehrer im Saarländischen Philologenverband, Stennweilerstr 35, 66564 Ottweiler
Focus: Adult Educ . . . 05091

Arbeitsgemeinschaft der Sozialdemokraten im Gesundheitswesen (ASG), Ollenhauerstr 1, 53113 Bonn
T: (0228) 532456
Pres: Horst Peter
Focus: Public Health . . . 05092

Arbeitsgemeinschaft der Spezialbibliotheken e.V. (ASpB), c/o Kekulé-Bibliothek, Bayer AG, Bayerwerk, 51373 Leverkusen
T: (0214) 307819; Fax: (0214) 307407
Founded: 1946; Members: 671
Pres: Prof. Dr. Wolfrudolf Laux
Focus: Libraries & Bk Sci
Periodical
Report . . . 05093

Arbeitsgemeinschaft der Verbände Gemeinnütziger Privatschulen in der Bundesrepublik, Lortzingstr 50a, 50931 Köln
Focus: Educ . . . 05094

Arbeitsgemeinschaft der wissenschaftlichen Institute des Handwerks der EG-Länder / Working Committee of the Scientific Institutes for Crafts in the EEC Countries, c/o Institut für Handwerkswirtschaft, Max-Joseph-Str 4, 80333 München
T: (089) 593671; Fax: (089) 553453
Founded: 1961
Gen Secr: Dr. Klaus Laub
Focus: Econ . . . 05095

Arbeitsgemeinschaft Deutsche Lateinamerika-Forschung (ADLAF), c/o Zentralinstitut für Lateinamerika-Studien, Katholische Universität Eichstätt, Ostenstr 26-28, 85071 Eichstätt
T: (08421) 20403; Fax: (08421) 20599
Founded: 1965; Members: 260
Pres: Prof. Dr. Karl Kohut
Focus: Sci
Periodical
ADLAF-Info (weekly) . . . 05096

Arbeitsgemeinschaft deutscher wirtschaftswissenschaftlicher Forschungsinstitute e.V., Poschingerstr 5, 81679 München
Founded: 1949; Members: 10
Pres: Prof. Dr. K.H. Oppenländer
Focus: Econ
Periodical
Gemeinschaftsdiagnose: Die Lage der Weltwirtschaft und der westdeutschen Wirtschaft im Frühjahr und im Herbst eines jeden Jahres (weekly) . . . 05097

Arbeitsgemeinschaft Fernwärme e.V. (AGFW), Stresemannallee 30, 60596 Frankfurt
T: (069) 63041; Tx: 411284; Fax: (069) 6304289
Founded: 1971; Members: 381
Pres: Karl Otto Abt; Gen Secr: Hans Neuffer
Focus: Air Cond
Periodical
Fernwärme International: District Heating/Chauffage Urbain (weekly) . . . 05098

Arbeitsgemeinschaft Freier Schulen, Vereinigungen und Verbände gemeinnütziger Schulen in freier Trägerschaft, Am Schlachtensee 2, 14163 Berlin
T: (030) 8012079
Focus: Educ . . . 05099

Arbeitsgemeinschaft für Abfallwirtschaft, Lindenallee 13-17, 50968 Köln
T: (0221) 37711
Focus: Ecology . . . 05100

Arbeitsgemeinschaft für angewandte Sozialforschung (AgaS), Blutenburgstr 93, 80634 München
T: (089) 132005
Focus: Sociology . . . 05101

Arbeitsgemeinschaft für betriebliche Altersversorgung e.V. (ABA), Postf 101208, 69002 Heidelberg
T: (06221) 21422, 20619; Fax: (06221) 24210
Founded: 1938; Members: 1200
Pres: Dr. B.-Jürgen Andresen; Gen Secr: Michael Lubnow
Focus: Soc Sci; Poli Sci
Periodical
Betriebliche Altersversorgung (8 times annually) . . . 05102

Arbeitsgemeinschaft für Elektronenoptik e.V., Karl-Friedrich-Str 17, 76133 Karlsruhe
T: (0721) 30805
Members: 31
Focus: Optics . . . 05103

Arbeitsgemeinschaft für Jugendhilfe, Haager Weg 44, 53127 Bonn
T: (0228) 910240; Fax: (0228) 9102466
Focus: Poli Sci
Periodical
Forum Jugendhilfe (weekly) . . . 05104

Arbeitsgemeinschaft für juristisches Bibliotheks- und Dokumentationswesen (AJBD), c/o Teilbibliothek Recht der Universitätsbibliothek, Eichleitnerstr 30, 86159 Augsburg
T: (0821) 598335
Founded: 1971; Members: 150
Focus: Libraries & Bk Sci; Doc; Law
Periodical
Recht, Bibliothek, Dokumentation (3 times annually) . . . 05105

Arbeitsgemeinschaft für Kieferchirurgie innerhalb der Deutschen Gesellschaft für Zahn-, Mund- und Kieferheilkunde, Arnold-Heller-Str 16, 24105 Kiel
T: (0431) 5972833
Founded: 1956; Members: 300
Pres: Prof. Dr. P. Reichart; Gen Secr: Prof. Dr. Dr. B. Hoffmeister
Focus: Surgery
Periodical
Deutsche zahnärztliche Zeitschrift (weekly) 05106

Arbeitsgemeinschaft für Krebsbekämpfung des Landes Niedersachsen e.V. / Niedersächsische Krebsgesellschaft, Ellernstr 36, 30175 Hannover
T: (0511) 815091
Focus: Cell Biol & Cancer Res . . . 05107

Arbeitsgemeinschaft für Landschaftsentwicklung (AGL), Godesberger Allee 142-148, 53175 Bonn
T: (0228) 8100239
Founded: 1970; Members: 9
Focus: Ecology . . . 05108

Arbeitsgemeinschaft für medizinisches Bibliothekswesen / Joint Association for Medical Librarianship, c/o Boehringer Mannheim GmbH, Zentralbibliothek, Sandhofer Str 116, 68305 Mannheim
T: (0621) 7592376; Fax: (0621) 7594419
Gen Secr: Peter Stadler
Focus: Libraries & Bk Sci . . . 05109

Arbeitsgemeinschaft für Osteuropaforschung, Wilhelmstr 36, 72074 Tübingen
Focus: Geography . . . 05110

Arbeitsgemeinschaft für sparsamen und umweltfreundlichen Energieverbrauch e.V. (ASUE), Heidenkampsweg 101, 20097 Hamburg
T: (040) 234509
Pres: Ulrich Hartmann; Gen Secr: Ulrich Ingenillem; Bernhard Vogt
Focus: Energy . . . 05111

Arbeitsgemeinschaft für Umweltfragen e.V. (AGU), Matthias-Grünewald-Str 1-3, 53175 Bonn
T: (0228) 375005; Fax: (0228) 375515
Pres: Prof. Dr. Kurt Oeser; Gen Secr: Arnim Schmülling
Focus: Ecology
Periodical
Das Umweltgespräch: Schriftenreihe . . . 05112

Arbeitsgemeinschaft für wirtschaftliche Verwaltung e.V. (AWV), Düsseldorfer Str 40, 65760 Eschborn
T: (06196) 495388; Fax: (06196) 496351
Pres: Werner Strohmayr; Gen Secr: Gert Abram
Focus: Business Admin
Periodical
AWV-Informationen (weekly) . . . 05113

Arbeitsgemeinschaft für zeitgemässes Bauen e.V., Walkerdamm 17, 24103 Kiel
T: (0431) 63064
Pres: Klaus Petersen; Gen Secr: Dieter Selk
Focus: Civil Eng
Periodical
Mitteilungsblätter (3-4 times annually) . . . 05114

Arbeitsgemeinschaft Getreideforschung e.V. (AGF) / Association of Cereal Research, Schützenberg 10, 32756 Detmold
T: (05231) 25530; Fax: (05231) 20505
Pres: K. Moorahrend; Gen Secr: K. Niebuhr
Focus: Agri
Periodical
Getreide, Mehl und Brot (weekly) . . . 05115

Arbeitsgemeinschaft Grünland und Futterbau in der Gesellschaft für Pflanzenbauwissenschaften, Ludwigstr 23, 35390 Giessen
Founded: 1957; Members: 65
Pres: Prof. Dr. Wilhelm Opitz von Boberfeld
Focus: Agri . . . 05116

Arbeitsgemeinschaft Hauswirtschaft e.V., Poppelsdorfer Allee 15, 53115 Bonn
T: (0228) 224063; Fax: (0228) 210827
Pres: Siglinde Porsch
Focus: Home Econ . . . 05117

Arbeitsgemeinschaft Historischer Kommissionen und Landesgeschichtlicher Institute e.V. / Association of Historic Councils and Regional History Institutes, c/o Johann-Gottfried-Herder-Institut, Gisonenweg 5-7, 35037 Marburg
T: (06421) 1840; Fax: (06421) 184139
Founded: 1898; Members: 55
Pres: Prof. Dr. Roderich Schmidt; Gen Secr: Dr. Winfried Irgang
Focus: Hist
Periodical
Mitteilungsblatt (weekly) . . . 05118

Arbeitsgemeinschaft Industriebau (AGI), Ebertpl 1, 50668 Köln
T: (0221) 729693
Founded: 1959; Members: 100
Focus: Civil Eng
Periodical
Industriebau (weekly) . . . 05119

Arbeitsgemeinschaft industrieller Forschungsvereinigungen „Otto von Guericke" e.V., Bayenthalgürtel 23, 50968 Köln
T: (0221) 376800; Fax: (0221) 3768027
Founded: 1954; Members: 104
Pres: Hans Wohlfart; Gen Secr: Dr. Hans Klein
Focus: Eng
Periodical
Akzente, Profile, Innovationen (weekly) . 05120

Arbeitsgemeinschaft Kartoffelforschung e.V., Am Schützenberg 10, 32756 Detmold
T: (05231) 25530; Fax: (05231) 20505
Pres: Werner Engel; Gen Secr: Klaus Niebuhr
Focus: Agri . . . 05121

Arbeitsgemeinschaft katholisch-theologischer Bibliotheken (AKThB), Leostr 21, 33098 Paderborn
T: (05251) 290480; Fax: (05251) 282575
Founded: 1947; Members: 139
Pres: Hermann-Josef Schmalor
Focus: Libraries & Bk Sci
Periodical
Mitteilungsblatt (weekly) . . . 05122

Arbeitsgemeinschaft Korrosion e.V., Theodor-Heuss-Allee 25, 60486 Frankfurt
T: (069) 7564209; Fax: (069) 7564201
Members: 26
Focus: Metallurgy
Periodical
Werkstoffe und Korrosion . . . 05123

Arbeitsgemeinschaft Media-Analyse e.V. (AG.MA), Wolfsgangstr 92, 60322 Frankfurt
T: (069) 550391
Pres: Alfred Müller; Gen Secr: Hans-Erdmann Scheler
Focus: Mass Media
Periodical
Media-Analyse (MA) (weekly) . . . 05124

Arbeitsgemeinschaft Sozialwissenschaftlicher Institute e.V. (ASI), Lennéstr 30, 53113 Bonn
T: (0228) 22810; Fax: (0228) 2281120
Founded: 1953; Members: 102
Pres: Dr. Joachim Scharioth; Gen Secr: Dr. Hans-Christoph Hobohm
Focus: Soc Sci
Periodical
Soziale Welt: Zeitschrift für sozialwissenschaftliche Forschung und Praxis (weekly) . . . 05125

Arbeitsgemeinschaft Spina bifida und Hydrocephalus e.V. (ASbH), Münsterstr 13, 44145 Dortmund
T: (0231) 834777; Fax: (0231) 833911
Focus: Physiology
Periodical
ASbH-Brief (weekly) . . . 05126

Arbeitsgemeinschaft Verstärkte Kunststoffe e. V., Am Hauptbahnhof 10, 60329 Frankfurt
T: (069) 250920; Fax: (069) 250919
Focus: Materials Sci . . . 05127

Arbeitsgemeinschaft Versuchsreaktor (AVR), Auf der Lausward, 40221 Düsseldorf
Focus: Nucl Res . . . 05128

Arbeitsgemeinschaft Währungsethik, Goldammerweg 181, 50858 Köln
T: (0221) 583510
Focus: Econ . . . 05129

Arbeitsgemeinschaft wildbiologischer und jagdkundlicher Forschungsstätten, c/o Forschungsstelle für Jagdkunde und Wildschadenverhütung des Landes Nordrhein-Westfalen, Forsthaus Hardt, Pützchens Chaussee 228, 53229 Bonn
T: (0228) 482115
Focus: Forestry; Bio . . . 05130

Arbeitsgemeinschaft wissenschaftliche Literatur e.V., Grosser Hirschgraben 17-21, 60311 Frankfurt
T: (069) 287242
Focus: Lit . . . 05131

Arbeitsgemeinschaft zur Verbesserung der Agrarstruktur in Hessen (AVA), Alexanderstr 2, 65187 Wiesbaden
Focus: Agri . . . 05132

Arbeitsgruppe für empirische Bildungsforschung e.V. (AfeB), Werderstr 38, 69120 Heidelberg
T: (06221) 49128
Focus: Educ
Periodical
Übersicht über die bisherigen Arbeiten (weekly) . . . 05133

Arbeitsgruppe für strukturelle Molekularbiologie in der Max-Planck-Gesellschaft, Notkerstr 85, 22603 Hamburg
T: (040) 89982801; Fax: (040) 891314
Gen Secr: Dr. Eckhard Mandelkow
Focus: Bio . . . 05134

Arbeitskreis Bildung und Politik Rheinland, Hauptstr 77, 53424 Remagen
T: (02642) 644
Focus: Educ; Poli Sci . . . 05135

Arbeitskreis Chemische Industrie, Palmstr 17, 50672 Köln
Focus: Chem; Ecology . . . 05136

Arbeitskreis deutscher Bildungsstätten e.V., Haager Weg 44, 53127 Bonn
T: (0228) 285065
Pres: Moritz von Engelhardt; Gen Secr: Mechthild Merfeld
Focus: Educ; Adult Educ
Periodical
Außerschulische Bildung (weekly) 05137

Arbeitskreis Ethnomedizin, Curschmannstr 33, 20251 Hamburg
Focus: Med; Ethnology
Periodical
Ethnomedizin (weekly) 05138

Arbeitskreis für Betriebsführung München (ABM), Waldhaus, 86943 Thaining
T: (08172) 249
Focus: Business Admin 05139

Arbeitskreis für Hochschuldidaktik, Schlüterstr 28, 20146 Hamburg
Focus: Adult Educ 05140

Arbeitskreis für Medizinische Geographie im Zentralverband der deutschen Geographen, c/o Geomedizinische Forschungsstelle der Heidelberger Akademie der Wissenschaften, Karlstr 4, 69117 Heidelberg
T: (06221) 543272
Founded: 1972; Members: 60
Focus: Geography 05141

Arbeitskreis für Ost-West-Fragen, Weserstr 37, 32602 Vlotho
Focus: Poli Sci 05142

Arbeitskreis für Schulmusik und allgemeine Musikpädagogik e.V., Winterleitenweg 65, 97082 Würzburg
T: (0931) 71315; Fax: (0931) 71315
Founded: 1953; Members: 1500
Pres: Prof. Dr. Volker Schütz; Gen Secr: Winfried Noack
Focus: Music; Educ
Periodical
Neue Musikzeitung (weekly) 05143

Arbeitskreis für Wehrforschung (AfW), Charlottenstr 44, 70182 Stuttgart
T: (0711) 244726
Founded: 1954
Focus: Military Sci 05144

Arbeitskreis Gesundheitskunde e.V., Feldbergstr 11, 78112 Sankt Georgen
T: (07724) 6148
Focus: Public Health 05145

Arbeitskreis katholischer Schulen in freier Trägerschaft in der Bundesrepublik Deutschland, Kaiserstr 163, 53113 Bonn
T: (0228) 103246
Pres: Prof. Dr. Bernhard Krautter; Gen Secr: Nikolaus Kircher
Focus: Educ; Rel & Theol
Periodical
Engagement: Zeitschrift für Erziehung und Schule (weekly) 05146

Arbeitskreis Rhetorik in Wirtschaft, Politik und Verwaltung, Auf dem Steinchen 8, 53127 Bonn
T: (0228) 281900, 281989; Fax: (0228) 281989
Pres: Prof. Dr. Gerhard Lange
Focus: Ling 05147

Arbeitswissenschaft im Landbau e.V., Fruwirthstr 48, 70539 Stuttgart
T: (0711) 4592816
Founded: 1941; Members: 100
Pres: Prof. Dr. K. Landau; Gen Secr: Dr. Frisch
Focus: Agri; Eng; Standards 05148

Arbeit und Leben / Arbeitskreis für die Bundesrepublik Deutschland e.V., Tersteegenstr 61-63, 40474 Düsseldorf
T: (0211) 434686; Fax: (0211) 459593
Pres: Dr. Dieter Eich; Gen Secr: Theo W. Länge
Focus: Sociology; Poli Sci; Educ . . . 05149

Arbeo-Gesellschaft e.V., 85777 Bachenhausen
Focus: Hist; Lit 05150

Archäologische Gesellschaft zu Berlin, Podbielskiallee 69-71, 14195 Berlin
T: (030) 830080
Founded: 1842; Members: 240
Pres: Prof. Dr. Adolf H. Borbein; Gen Secr: Dr. Antje Krug
Focus: Archeol
Periodical
Berliner Winckelmannsprogramm (weekly) 05151

Arnold-Bergstraesser-Institut für kulturwissenschaftliche Forschung e.V. / Forschungsinstitut zu Politik und Gesellschaft überseeischer Länder, Windausstr 16, 79110 Freiburg
T: (0761) 85091; Fax: (0761) 892967
Pres: Prof. Dr. Theodor Hanf; Prof. Dr. Dieter Oberndörfer; Gen Secr: Dr. Heribert Weiland
Focus: Poli Sci; Ethnology; Sociology; Econ 05152

ASB Management-Zentrum-Heidelberg e.V., Gaisbergstr 11-13, 69115 Heidelberg
T: (06221) 9888; Fax: (06221) 988682
Pres: Hanns Schmidt; Gen Secr: Heinz-Jürgen Kochann
Focus: Business Admin
Periodical
ASB-Aktuell (weekly) 05153

Association Européenne pour l'Enseignement de l'Architecture (AEEA) / European Association for the Teaching of Architecture, c/o Fachbereich Architektur, Gesamthochschule, Henschelstr 2, 34127 Kassel
Pres: Prof. A. Jockusch
Focus: Educ; Archit 05154

Association Internationale des Professeurs de Philosophie (AIPPHi) / International Association of Teachers of Philosophy, Am Schirrhof 11, 32427 Minden
T: (0571) 23474
Founded: 1964
Gen Secr: Luise Dreyer
Focus: Philos
Periodical
Bulletin d'Information AIPPHi 05155

Association of European Federations of Agro-Engineers (AEFA), Auf der Helte 23, 53604 Bad Honnef
T: (02224) 6124
Founded: 1974
Gen Secr: Caspar von der Crone
Focus: Agri 05156

Astronomische Gesellschaft e.V. (A.G.), c/o Landessternwarte, Königstuhl, 69117 Heidelberg
Founded: 1863; Members: 630
Pres: Prof. Dr. W. Hillebrandt; Gen Secr: Dr. Gerhard Kklare
Focus: Astronomy
Periodical
Mitteilungen (weekly) 05157

Ausschuss Normenpraxis im DIN Deutsches Institut für Normung e.V. (ANP), Burggrafenstr 6, 10787 Berlin
T: (030) 26012337
Members: 700
Pres: Norbert Zimmermann; Gen Secr: Gerhard Senk
Focus: Standards 05158

AW produktplanung / Arbeitsgemeinschaft der Wirtschaft für Produktdesign und Produktplanung e.V., Holteyer Str 6, 45289 Essen
T: (0201) 570294
Focus: Graphic & Dec Arts, Design . . 05159

AWV-Fachausschuss Mikrofilm/Optische Informationssysteme, Düsseldorfer Str 40, 65760 Eschborn
T: (06196) 495374
Focus: Doc
Periodical
AWV-Informationen (weekly) 05160

Bach-Verein Köln e.V., Minoritenstr 7, 50667 Köln
T: (0221) 2216058
Founded: 1931; Members: 1050
Focus: Music 05161

Baden-Württembergischer Sportärzteverband, Bezirksgruppe Süd-Baden, Hugstetter Str 55, 79106 Freiburg
T: (0761) 2011
Focus: Med 05162

Badischer Landesverein für Naturkunde und Naturschutz e.V. (BLNN), Gerberau 32, 79098 Freiburg
T: (0761) 2163325; Fax: (0761) 21636717
Founded: 1882; Members: 600
Pres: Dr. H. Koerner
Focus: Ecology
Periodical
Mitteilungen des Badischen Landesvereins für Naturkunde und Naturschutz (weekly) . . 05163

Baltische Gesellschaft in Deutschland e. V., Gunezrainerstr 6, 80802 München
Focus: Hist
Periodical
Mitteilungen aus Baltischem Leben (weekly) 05164

Bankakademie e.V., Postf 100341, 60318 Frankfurt
T: (069) 1540080; Fax: (069) 551461
Founded: 1957; Members: 13
Gen Secr: Dr. Helmut Reinboth
Focus: Finance 05165

Battelle-Institut e.V., Am Römerhof 35, 60486 Frankfurt
T: (069) 79080; Tx: 411966; Fax: (069) 790880
Founded: 1952
Gen Secr: Dr. G.H. Hewig
Focus: Nat Sci; Eng
Periodical
Battelle-Information 05166

Bayerische Akademie der Schönen Künste, Max-Joseph-Pl 3, 80539 München
T: (089) 294622; Fax: (089) 2285885
Founded: 1948; Members: 200
Pres: Prof. Dr. Heinz Friedrich; Gen Secr: Dr. Oswald Georg Bauer
Focus: Arts 05167

Bayerische Akademie der Werbung, Briennerstr 47, 80333 München
T: (089) 593450
Focus: Advert 05168

Bayerische Akademie der Wissenschaften, Marstallpl 8, 80539 München
T: (089) 230310; Fax: (089) 23031100
Founded: 1759; Members: 286
Pres: Prof. Dr. Dr. Horst Fuhrmann; Gen Secr: Monika Stoermer
Focus: Sci
Periodicals
Abhandlungen und Sitzungsbericht: Mathematisch-Naturwissenschaftliche Klasse (weekly)
Abhandlungen und Sitzungsbericht: Philosophisch-Historische Klasse (5-10 times annually)
Jahrbuch (weekly) 05169

Bayerische Akademie für Arbeits- und Sozialmedizin, Pfarrstr 3, 80538 München
T: (089) 2984260
Founded: 1968
Focus: Med; Hygiene 05170

Bayerische Botanische Gesellschaft (BBG), Menzinger Str 67, 80638 München
T: (089) 17861264; Fax: (089) 172638
Founded: 1890; Members: 900
Pres: Dr. W. Lippert
Focus: Botany
Periodical
Berichte (weekly) 05171

Bayerische Kommission für die Internationale Erdmessung, Marstallpl 8, 80539 München
T: (089) 23031113; Tx: 5213550; Fax: (089) 23031100
Founded: 1868
Focus: Surveying 05172

Bayerische Krebsgesellschaft e.V., Tumblinger Str 4, 80337 München
T: (089) 531175, 539524
Founded: 1925
Pres: Prof. Dr. Hans Erhart; Gen Secr: Dr. Ludwig Lutz
Focus: Cell Biol & Cancer Res
Periodical
Rundschreiben 05173

Bayerische Landesärztekammer, Mühlbaurstr 16, 81677 München
T: (089) 41471
Pres: Dr. Hans Hege; Gen Secr: Dr. Enzo Amarotico; Dr. Horst Frenzel
Focus: Med
Periodical
Bayerisches Ärzteblatt (weekly) 05174

Bayerische Landeszahnärztekammer (BLZK), Fallstr 34, 81369 München
T: (089) 724010
Members: 11000
Pres: Dr. Dr. Joseph Kastenbauer; Gen Secr: Dr. F. Schmeller
Focus: Dent
Periodical
Bayerisches Zahnärzteblatt (BZB) (weekly) 05175

Bayerischer Holzwirtschaftsrat, Prannerstr 9, 80333 München
T: (089) 294561; Tx: 524067
Focus: Forestry 05176

Bayerischer Landesverein für Familienkunde e.V., Ludwigstr 14, 80539 München
Focus: Genealogy
Periodical
Blätter (3 times annually) 05177

Bayerischer Lehrer- und Lehrerinnenverband (BLLV), Bavariaring 37, 80336 München
T: (089) 778026
Pres: Albin Dannhäuser; Gen Secr: Dr. Dieter Reithmeier
Focus: Educ
Periodical
Bayerische Schule (weekly) 05178

Bayerische Röntgengesellschaft, c/o Städt. Krankenanstalten, Jakob-Henle-Str 1, 90766 Fürth
Focus: X-Ray Tech 05179

Bayerischer Sportärzteverband e.V., Nymphenburger Str 81, 80636 München
Pres: Dr. Eugen Gossner
Focus: Med
Periodical
Sportärztliche Mitteilungen (weekly) . . . 05180

Bayerischer Volkshochschulverband e.V., Fäustlestr 5, 80339 München
T: (089) 510800
Pres: Josef Deimer; Gen Secr: Hermann Kumpfmüller
Focus: Adult Educ
Periodical
Das Forum: Zeitschrift der Volkshochschulen in Bayern (weekly) 05181

Beilstein-Institut für Literatur der Organischen Chemie, Varrentrappstr 40-42, 60486 Frankfurt
T: (069) 7917251; Fax: (069) 7917669
Focus: Chem
Periodical
Beilstein: Handbuch der Organischen Chemie (17 times annually) 05182

Bergbau-Forschung, Franz-Fischer-Weg 61, 45307 Essen
T: (0201) 1051
Focus: Mining 05183

Bergischer Geschichtsverein e. V., Friedrich-Engels-Allee 89-91, 42285 Wuppertal
Focus: Hist
Periodicals
Romerike Berge (weekly)
Zeitschrift des Bergischen Geschichtsvereins (weekly) 05184

Berlin-Brandenburgische Akademie der Wissenschaften, Jägerstr. 22-23, 10117 Berlin
T: (030) 20370-345; Fax: (030) 20370500
Founded: 1992; Members: 75
Pres: Prof. Dr. Hubert Markl; Gen Secr: Diepold salvini-Plawen
Focus: Sci 05185

Berliner Arbeitskreis Information, c/o Universitätsbibliothek der TU Berlin, Str des 17. Juni 135, 10623 Berlin
Gen Secr: Kurt Penke
Focus: Computer & Info Sci; Doc
Periodical
Mitteilungen (3-4 times annually) 05186

Berliner Gesellschaft für Anthropologie, Ethnologie und Urgeschichte / Berlin Society for Anthropology, Ethnology and Prehistory, Schloß Charlottenburg, Langhansbau, 14059 Berlin
T: (030) 32091282
Founded: 1869; Members: 285
Pres: Prof. Dr. Georg Pfeffer; Gen Secr: Dr. Claudius Müller
Focus: Anthro; Hist; Ethnology
Periodical
Mitteilungen (weekly) 05187

Berliner Mathematische Gesellschaft e.V., Str des 17. Juni 136, 10623 Berlin
Founded: 1899
Pres: Prof. Dr. E.J. Thiele
Focus: Math
Periodical
Sitzungsberichte 05188

Berliner Medizinische Gesellschaft, Fregestr 73, 12159 Berlin
Pres: Prof. Dr. Hans Herken; Gen Secr: Prof. Dr. K.-O. Habermehl
Focus: Med 05189

Berliner Orthopädische Gesellschaft e.V., Clayallee 229-233, 14195 Berlin
T: (030) 3311660; Fax: (030) 810041
Pres: Dr. Martin Talke; Gen Secr: Dr. Martin Käding
Focus: Orthopedics 05190

Berliner Sportärztebund e.V., Forckenbeckstr 20, 14199 Berlin
T: (030) 8232056; Fax: (030) 8238870
Founded: 1950; Members: 510
Pres: Dr. Folker Boldt
Focus: Med 05191

Berufsverband der Ärzte für Orthopädie e.V., Stephanienstr 88, 76133 Karlsruhe
T: (0721) 25820
Members: 2300
Focus: Orthopedics
Periodical
Informationen (weekly) 05192

Berufsverband der Augenärzte Deutschlands e.V. (BVA), Wildenbruchstr 21, 40545 Düsseldorf
T: (0211) 570310; Fax: (0211) 579912
Founded: 1950
Pres: Dr. G. Kraffel; Gen Secr: Dr. M. Freigang
Focus: Ophthal
Periodical
Der Augenarzt (weekly) 05193

Berufsverband der Berliner Hals-, Nasen-, Ohren-Ärzte, Hohenzollerndamm 91, 14199 Berlin
T: (030) 8262021
Focus: Otorhinolaryngology 05194

Berufsverband der Deutschen Chirurgen (BDC), Wendemuthstr 5, 22041 Hamburg
T: (040) 682059; Fax: (040) 684821
Founded: 1960; Members: 10000
Pres: Prof. Dr. Karl Hempel
Focus: Surgery
Periodical
Der Chirurg BDC: Informationen des Berufsverbandes der Deutschen Chirurgen (weekly) 05195

Berufsverband der Deutschen Dermatologen e.V., Westliche Karl-Friedrich-Str 32, 75172 Pforzheim
T: (07231) 5802
Focus: Derm
Periodical
Der Deutsche Dermatologe (weekly) . . 05196

Berufsverband der Deutschen Fachärzte für Urologie, Julius-Leber-Str 10, 22765 Hamburg
Founded: 1953
Focus: Urology 05197

Berufsverband der Deutschen Radiologen und Nuklearmediziner e.V., Sonnenstr 3, 80331 München
T: (089) 592690
Pres: Dr. Klaus Wallnöfer
Focus: Radiology; Nucl Med
Periodicals
Mitglieder-Info (10 times annually)
Röntgenpraxis (weekly) 05198

Berufsverband der Deutschen Urologen e.V., Erdingerstr 17, 84405 Dorfen
T: (08081) 41313
Pres: Dr. Klaus Schalkhäuser
Focus: Urology
Periodical
Der Urologe (weekly) 05199

Berufsverband der Kinderärzte Deutschlands e.V., Mielenforster Str 2, 51069 Köln
T: (0221) 6804064; Fax: (0221) 683204
Members: 9600
Pres: Dr. Wolfgang Meinrenken; Gen Secr: J. Radbruch
Focus: Pediatrics
Periodical
der kinderarzt: Zeitschrift für Kinderheilkunde und Jugendmedizin (weekly) 05200

Berufsverband der Praktischen Ärzte und Ärzte für Allgemeinmedizin Deutschlands e.V. (BPA), Theodor-Heuss-Ring 14, 50668 Köln
T: (0221) 160670
Pres: Dr. Klaus-Dieter Kossow; Gen Secr: Dieter Robert Adam
Focus: Med
Periodical
Der Praktische Arzt (20 times annually) . 05201

Berufsverband Deutscher Hörgeschädigtenpädagogen / Berufsverband der Lehrer an Gehörlosen- und Schwerhörigenschulen, Hammer Str 124, 22043 Hamburg
T: (040) 6828730; Fax: (040) 68287340
Pres: Christiane Hartmann-Börner
Focus: Educ Handic
Periodicals
Das bunte Blatt (weekly)
Hörgeschädigtenpädagogik (weekly) . . . 05202

Berufsverband Deutscher Nervenärzte e.V. (BVDN), Goethestr 21, 60313 Frankfurt
T: (069) 285974
Members: 1600
Focus: Neurology 05203

Berufsverband Deutscher Psychologen e.V. (BDP), Heilsbachstr 22, 53123 Bonn
T: (0228) 641054/56
Members: 6000
Focus: Psych
Periodical
Report Psychologie (weekly) 05204

Berufsverband Deutscher Soziologen e.V. (BDS), Feilenstr 2, 33602 Bielefeld
T: (0521) 170733
Founded: 1976; Members: 700
Focus: Sociology
Periodical
Sozialwissenschaften und Berufspraxis (weekly) 05205

Berufsverband Geprüfter Graphologen/Psychologen e.V. (BGG/P), Cimbernstr 70c, 81377 München
T: (089) 7145880
Founded: 1952; Members: 85
Focus: Graphology; Psych
Periodical
Angewandte Graphologie und Charakterkunde (3 times annually) 05206

Beta Beta Delta, c/o Fachhochschule für Bibliothekswesen, Wolframstr 32, 70191 Stuttgart
Founded: 1983
Pres: Margareta Payer
Focus: Libraries & Bk Sci 05207

Beton-Verein Berlin e.V., Bundesallee 23, 10717 Berlin
T: (030) 860301
Focus: Materials Sci 05208

Betriebswirtschafts-Akademie e.V. (BWA), Taunusstr 54, 65183 Wiesbaden
T: (06121) 5341
Founded: 1961
Focus: Business Admin 05209

Bezirksärztekammer Koblenz, Emil-Schüller-Str 45-47, 56068 Koblenz
T: (0261) 390010
Pres: Dr. Hans Jöckel
Focus: Med 05210

Bezirksärztekammer Nordwürttemberg (BÄK NW), Jahnstr 32, 70597 Stuttgart
T: (0711) 769810; Fax: (0711) 7698159
Founded: 1955; Members: 12500
Pres: Dr. Karl-Heinz Kamp; Gen Secr: Dr. Helmut Paris
Focus: Med
Periodical
Ärzteblatt Baden-Württemberg (weekly) . 05211

Bezirksärztekammer Pfalz, Maximilianstr 22, 67433 Neustadt
T: (06321) 8930
Focus: Med 05212

Bezirksärztekammer Trier, Balduinstr 10-14, 54290 Trier
T: (0651) 45011
Focus: Med 05213

Bezirkszahnärztekammer Koblenz, Bahnhofstr 32, 56068 Koblenz
T: (0261) 36681
Pres: Dr. Richard Pickel
Focus: Med 05214

Bezirkszahnärztekammer Pfalz, Brunhildenstr 1, 67059 Ludwigshafen
T: (0621) 519111; Fax: (0621) 622972
Focus: Med 05215

Bezirkszahnärztekammer Rheinhessen, Eppichmauergasse 1, 55116 Mainz
T: (06131) 287760; Fax: (06131) 225706
Pres: Dr. Bernd Stern; Gen Secr: Maxim Hasselwander
Focus: Med 05216

Bezirkszahnärztekammer Trier, Saarstr 52, 54290 Trier
T: (0651) 45150
Pres: Dr. Karl-Josef Wilbertz; Gen Secr: Elfriede Jungen
Focus: Med 05217

Bildungswerk der Bayerischen Wirtschaft e.V. (bbw), Brienner Str 7, 80333 München
T: (089) 2900260; Fax: (089) 29002655
Pres: Dr. Manfred Scholz; Gen Secr: Karl-Georg Nickel
Focus: Adult Educ 05218

Bildungswerk der Konrad-Adenauer-Stiftung, Politische Akademie der Konrad-Adenauer-Stiftung, Urfelder Str 221, 50389 Wesseling
T: (02236) 7071
Focus: Poli Sci
Periodical
Eichholzbrief (weekly) 05219

Bildungswerk der Nordrhein-Westfälischen Wirtschaft e.V., Brunnenstr 28, 58332 Schwelm
T: (02336) 7018
Focus: Adult Educ 05220

Braunschweigische Wissenschaftliche Gesellschaft (BWG), Fallersleber-Tor-Wall 16, 38100 Braunschweig
T: (0531) 3914596
Founded: 1944; Members: 110
Focus: Sci 05221

Bremer Ausschuss für Wirtschaftsforschung (BAW), Schlachte 10-11, 28195 Bremen
T: (0421) 3978804; Tx: 244804; Fax: (0421) 3978810
Founded: 1947
Pres: Dr. Wolfram Elsner
Focus: Econ
Periodicals
BAW-Monatsbericht (weekly)
Bremer Zeitschrift für Wirtschaftspolitik (weekly)
Regionalwirtschaftliche Studien (weekly) . 05222

Bremer Gesellschaft für Wirtschaftsforschung e.V., Am Brill 21-23, 28195 Bremen
T: (0421) 170807
Pres: Prof. Dr. Rolf Stuchtey; Gen Secr: Prof. Dr. Alfons Lemper
Focus: Econ 05223

Bremer Sportärztebund, Horner Str 70, W-2800 Bremen
T: (0421) 44925144
Focus: Med 05224

Bund der Freien Waldorfschulen e.V., Heidehofstr 32, 70184 Stuttgart
T: (0711) 210420; Fax: (0711) 2104219
Focus: Educ
Periodical
Erziehungskunst (weekly) 05225

Bund Deutscher Architekten (BDA), Ippendorfer Allee 14b, 53127 Bonn
T: (0228) 285011; Fax: (0228) 285465
Pres: Erhard Tränkner; Gen Secr: Carl Steckeweh
Focus: Archit
Periodical
Der Architekt (weekly) 05226

Bund Deutscher Baumeister, Architekten und Ingenieure (BDB), Kennedeyallee 11, 53175 Bonn
T: (0228) 376784; Fax: (0228) 376057
Pres: Hermannjosef Beu; Gen Secr: Dr. Hans Rudolf Sangenstedt
Focus: Archit
Periodical
Deutsche Bauzeitung (weekly) 05227

Bund Deutscher Kunsterzieher e.V. (BDK), Jakobistr 40, 30163 Hannover
T: (0511) 662229
Pres: Wilfried Schlosser
Focus: Arts; Educ
Periodical
BDK-Mitteilungen (weekly) 05228

Bundesärztekammer (BÄK), Herbert-Lewin-Str 1, 50931 Köln
T: (0221) 40040
Founded: 1947
Pres: Dr. Karsten Vilmar; Gen Secr: Prof. Dr. Christoph Fuchs
Focus: Med
Periodicals
Deutsches Ärzteblatt: Publikationsorgan der deutschen Ärzteschaft (weekly)
Medizin Heute: Zeitschrift für Patienten (weekly) 05229

Bundesakademie für musikalische Jugendbildung, Postf 1158, 78635 Trossingen
T: (07425) 94930; Fax: (07425) 949321
Pres: Prof. Dr. Hans-Walter Berg
Focus: Music; Educ 05230

Bundesakademie für öffentliche Verwaltung, Deutschhermstr 93, 53177 Bonn
T: (0228) 331005
Focus: Adult Educ 05231

Bundes-Arbeitsgemeinschaft Akademischer Räte in der Bundesrepublik, c/o Institut für Datenverarbeitung, Arcisstr 21, 80333 München
Members: 10000
Focus: Public Admin 05232

Bundesarbeitsgemeinschaft der katholisch-kirchlichen Büchereiarbeit, Wittelsbacherring 9, 53115 Bonn
T: (0228) 72580; Fax: (0228) 630389
Gen Secr: Norbert Brockmann
Focus: Libraries & Bk Sci 05233

Bundesarbeitsgemeinschaft für Rehabilitation, Walter-Kolb-Str 9-11, 60594 Frankfurt
T: (069) 6050180; Tx: 412536
Pres: Dr. Rolf Wuthows; Gen Secr: Bernd Steinke
Focus: Rehabil
Periodicals
BAR-REHA-INFO (weekly)
Die Rehabilitation (weekly) 05234

Bundesarbeitsgemeinschaft katholischer Familienbildungsstätten, Prinz-Georg-Str 44, 40477 Düsseldorf
T: (0211) 4499245; Fax: (0211) 4499259
Founded: 1956; Members: 120
Gen Secr: Gislinde Fischer-Köhler
Focus: Adult Educ
Periodical
BAG-Magazin (3 times annually) 05235

Bundesarbeitsgemeinschaft Schule-Wirtschaft, Gustav-Heinemann-Ufer 84-88, 50968 Köln
T: (0221) 3708303
Pres: Dr. Roland Delbos; Gen Secr: Hans-Jürgen Brackmann
Focus: Educ
Periodical
Schule-Wirtschaft (weekly) 05236

Bundesarbeitsgemeinschaft zur Förderung haltungsgefährdeter Kinder und Jugendlicher e.V., Fischtorpl 17, 55116 Mainz
T: (06131) 227440
Focus: Orthopedics; Rehabil
Periodical
Haltung und Bewegung (weekly) 05237

Bundesarchitektenkammer, Königswinterer Str 709, 53227 Bonn
T: (0228) 441051; Fax: (0228) 442760
Pres: Prof. Gerhart Laage; Gen Secr: Dr. Henning Hillmann
Focus: Archit 05238

Bundesausschuß Betriebswirtschaft (BBW), Düsseldorfer Str 40, 65760 Eschborn
T: (06196) 495257; Tx: 418362; Fax: (06196) 495303
Pres: Dr. Hannspeter Neubert; Gen Secr: Erich John
Focus: Business Admin 05239

Bundesausschuss für Wissenschaft und Bildung des Deutschen Sportbundes, Otto-Fleck-Schneise 12, 60528 Frankfurt
T: (069) 67000; Fax: (069) 674906
Pres: Hans Hansen
Focus: Sports
Periodical
Sportwissenschaft (weekly) 05240

Bundesverband Bildender Künstler Bundesrepublik Deutschland e.V. (BBK), Meckenheimer Allee 85, 53115 Bonn
T: (0228) 630406; Fax: (0228) 696994
Members: 15000
Pres: Dierk Engelken; Gen Secr: Dr. Ursula Cramer
Focus: Fine Arts
Periodical
Kulturpolitik (weekly) 05241

Bundesverband der beamteten Tierärzte, Geilenkirchener Str 39, 52525 Heinsberg
T: (02452) 2875
Focus: Vet Med 05242

Bundesverband der freiberuflichen und unabhängigen Sachverständigen für das Kraftfahrzeugwesen e.V. (BVSK), Kantering 57, 53639 Königswinter
T: (02223) 22000; Tx: 885233
Focus: Auto Eng 05243

Bundesverband der Friedrich-Bödecker-Kreise e.V., Fischtorpl 23, 55116 Mainz
T: (06131) 2889018, 2889023; Fax: (06131) 230333
Founded: 1981
Focus: Lit
Periodical
Autoren lesen vor Schülern – Autoren sprechen mit Schülern (weekly) 05244

Bundesverband der öffentlichen angestellten und vereidigten Chemiker e.V., Grosse Bleichen 34, 20354 Hamburg
T: (040) 343435, 345210
Focus: Chem 05245

Bundesverband der Pneumologen, Schloßstr 22, 45468 Mülheim
T: (0208) 474608
Focus: Pulmon Dis 05246

Bundesverband der Vertrauens- und Rentenversicherungsärzte, Ubbo-Emmius-Str 2, 26603 Aurich
T: (04941) 2933
Focus: Med 05247

Bundesverband Deutscher Ärzte für Mund-Kiefer-Gesichtschirurgie e.V., Harburger Rathausstr 41, 21073 Hamburg
T: (040) 777070; Fax: (040) 779606
Pres: Dr. Dr. W.K. Busch; Gen Secr: Prof. Dr. Dr. W. Steinhilber
Focus: Surgery 05248

Bundesverband Deutscher Ärzte für Naturheilverfahren e.V., Sophienstr 2, 80333 München
T: (089) 557578; Fax: (089) 5502535
Pres: Dr. Michael Probst; Dr. Christian Hentschel
Focus: Med
Periodical
Heilkunst: Forum anerkannter Naturheilverfahren (weekly) 05249

Bundesverband Deutscher Leibeserzieher, Lenzhalde 66, 70192 Stuttgart
Focus: Sports 05250

Bundesverband Deutscher Privatschulen, Darmstädter Landstr 85a, 60598 Frankfurt
T: (069) 614058; Fax: (069) 626763
Founded: 1946; Members: 480
Pres: Joachim Böttcher; Gen Secr: Christian Lucas
Focus: Educ
Periodical
Freie Bildung und Erziehung (weekly) . . 05251

Bundesverband für den Selbstschutz, Deutschherrenstr 93-95, 53177 Bonn
T: (0228) 8401
Focus: Eng
Periodical
Zivilschutz Magazin: Fachzeitschrift für Zivilschutz, Katastrophenschutz und Selbstschutz (weekly) 05252

Bundesverband Hilfe für das autistische Kind e.V., Bebelallee 141, 22297 Hamburg
T: (040) 5115604; Fax: (040) 5110813
Focus: Educ Handic; Therapeutics
Periodical
Autismus (weekly) 05253

Bundesverband Jugend und Film e.V. (BJF), Schweizer Str 6, 60594 Frankfurt
T: (069) 610439; Fax: (069) 6032185
Pres: Bernt Lindner; Gen Secr: Reinhold T. Schöffel
Focus: Cinema
Periodical
BJF-magazin (weekly) 05254

Bundesverband Katholischer Ingenieure und Wirtschaftler Deutschlands, Venusbergweg 1, 53115 Bonn
T: (0228) 217942; Fax: (0228) 218452
Founded: 1961; Members: 170
Pres: Prof. Dr. Peter Treier; Gen Secr: Urban Zinser
Focus: Eng; Econ
Periodical
tum: technik und mensch (weekly) . . . 05255

Bundesverband Legasthenie e.V., Gneisenaustr 2, 30175 Hannover
T: (0511) 853465; Fax: (0511) 858065
Focus: Psych; Educ Handic
Periodical
LRS (weekly) 05256

Bundesvereinigung für Gesundheit e.V., Heilsbachstr 30, 53123 Bonn
T: (0228) 987270; Fax: (0228) 6420024
Founded: 1954; Members: 260
Pres: Dr. Hans-Peter Voigt; Gen Secr: Gottfried Neuhaus
Focus: Public Health
Periodical
Gesundheits-Informations-Dienst (weekly) . 05257

Bundesvereinigung Kulturelle Jugendbildung e.V., Küppelstein 34, 42857 Remscheid
T: (02191) 794390; Fax: (02191) 794389
Focus: Educ 05258

Bundesvereinigung Lebenshilfe für geistig Behinderte e.V., Raiffeisenstr 18, 35043 Marburg
T: (06421) 43007/09
Focus: Educ Handic
Periodical
Die Lebenshilfe-Zeitung (weekly) 05259

Bundesvereinigung Logistik e.V. (BVL), Contrescarpe 45, 28195 Bremen
T: (0421) 335680; Fax: (0421) 320369
Members: 2000
Pres: Dr. Hanspeter Stabenau
Focus: Econ; Marketing
Periodical
Logistik Heute (10 times annually) . . . 05260

Bundeszahnärztekammer (BDZ), Universitätsstr 71, 50931 Köln
T: (0221) 40010; Fax: (0221) 404035
Founded: 1953; Members: 17
Pres: Dr. Fritz-Josef Willmes; Gen Secr: Dr. Detlef Schulze-Wilk
Focus: Dent
Periodical
Zahnärztliche Mitteilungen 05261

Bund Freiheit der Wissenschaft e.V., Elisabethstr 3, 53177 Bonn
T: (0228) 352083; Fax: (0228) 358658
Pres: Clemens Christians; Prof. Dr. Klaus W. Hempfer; Prof. Dr. A. Woll; Gen Secr: Birgit Harz
Focus: Sci
Periodical
Freiheit der Wissenschaft (weekly) . . . 05262

Bund für Deutsche Schrift und Sprache, Postf 1110, 26189 Ahlhorn
T: (04435) 1313; Fax: (04435) 3623
Founded: 1918
Focus: Lit; Ling
Periodical
Die Deutsche Schrift: Vierteljahreshefte zur Förderung der deutschen Sprache und Schrift (weekly) . . . 05263

Bund für freie und angewandte Kunst e.V., Herdweg 29, 64285 Darmstadt
T: (06151) 64371
Focus: Arts . . . 05264

Bund für Lebensmittelrecht und Lebensmittelkunde e.V. (BLL), Godesberger Allee 157, 53175 Bonn
T: (0228) 819930; Fax: (0228) 375069
Founded: 1955; Members: 500
Pres: Dr. Karl Schneider; Gen Secr: Dr. Matthias Horst
Focus: Food; Law
Periodical
BLL-Schriftenreihe (weekly) . . . 05265

Bund für Umwelt und Naturschutz Deutschland e.V. (BUND), Im Rheingarten 7, 53225 Bonn
T: (0228) 400970; Fax: (0228) 4009740
Founded: 1975; Members: 220000
Pres: Hubert Weinzierl; Gen Secr: P. Onno
Focus: Ecology
Periodical
Natur & Umwelt (weekly) . . . 05266

Bund für Umwelt und Naturschutz Deutschland, Landesverband Hessen e.V., Kelsterbacher Str 28, 64546 Mörfelden-Walldorf
T: (06105) 44041; Fax: (06105) 44691
Focus: Ecology
Periodical
Natur und Umwelt (weekly) . . . 05267

Bund katholischer Erzieher Deutschlands (BKED), Hedwig-Dransfeld-Pl 4, 45143 Essen
T: (0201) 623029
Focus: Educ . . . 05268

Bund Naturschutz in Bayern e.V., Kirchenstr 88, 81675 München
T: (089) 4599180; Fax: (089) 485866
Founded: 1913
Pres: Hubert Weinzierl; Gen Secr: Helmut Steininger
Focus: Ecology
Periodical
Natur & Umwelt (weekly) . . . 05269

Bund Technischer Experten e.V., Busestr 42, 28213 Bremen
T: (0421) 211014/15
Founded: 1924
Focus: Eng . . . 05270

Carl-Cranz-Gesellschaft e.V. / Gesellschaft für technisch-wissenschaftliche Weiterbildung, Postf 1112, 82234 Weßling
T: (08153) 28413; Tx: 0526419; Fax: (08153) 281345
Founded: 1961
Pres: Prof.Dr. H.H. Homuth; Gen Secr: H.G. Apelt
Focus: Eng; Adult Educ . . . 05271

Carl Duisberg Gesellschaft e.V., Hohenstaufenring 30-32, 50674 Köln
T: (0221) 20980; Fax: (0221) 2098111
Founded: 1949; Members: 1000
Gen Secr: Dr. Norbert Schneider
Focus: Sci
Periodicals
Carl Duisberg Forum (weekly)
Echo aus Deutschland (weekly) . . . 05272

Catholic Media Council / Publizistische Medienplanung für Entwicklungsländer e.V., Anton-Kurze-Allee 2, 52074 Aachen
T: (0241) 73081; Fax: (0241) 73462
Focus: Journalism; Mass Media; Develop Areas
Periodical
Information Bulletin (weekly) . . . 05273

CENELEC Electronic Components Committee (CECC), Gartenstr 179, 60596 Frankfurt
T: (069) 639171; Tx: 4032175; Fax: (069) 639427
Founded: 1970
Gen Secr: C. Weaver
Focus: Electronic Eng . . . 05274

Center for International Research on Economic Tendency Surveys (CIRET), Poschingerstr 5, 81679 München
T: (089) 92241; Tx: 522269; Fax: (089) 985369
Founded: 1969
Gen Secr: Prof. Dr. K.H. Oppenländer
Focus: Econ . . . 05275

Commission Internationale pour la Protection du Lac de Constance / Internationale Gewässerschutzkommission für den Bodensee, c/o Landesamt für Wasserwirtschaft, Lazarettstr 67, 80636 München
T: (089) 1259348; Fax: (089) 1259435
Founded: 1960
Gen Secr: B. Engstle
Focus: Ecology . . . 05276

Confederation of European Laryngectomees, Luisenstr 20, 36179 Bebra
Pres: Hans Friedrich Nemnich
Focus: Otorhinolaryngology . . . 05277

Conference of Baltic Oceanographers (CBO), c/o Institut für Meereskunde, Universität Kiel, Düsternbrooker Weg 20, 24105 Kiel
T: (0431) 5971; Tx: 0292619
Founded: 1957
Gen Secr: Dr. J.C. Duinker
Focus: Oceanography . . . 05278

Conseil International des Associations de Bibliothèques de Théologie, Postf 250104, 50517 Köln
T: (0221) 3382110; Fax: (0221) 3382103
Founded: 1961; Members: 9
Pres: Dr. A. Geuns; Gen Secr: Dr. Isolde Dunke
Focus: Libraries & Bk Sci . . . 05279

Cusanus-Institut, Domfreihof 3, 54210 Trier
T: (0651) 44503
Founded: 1960; Members: 230
Pres: Prof. Dr. Klaus Kremer; Prof. Dr. Klaus Reinhardt
Focus: Philos; Law; Rel & Theol
Periodicals
Kleine Schriften der Cusanus-Gesellschaft
Mitteilungen und Forschungsbeiträge der Cusanus-Gesellschaft . . . 05280

Dachverband Psychosozialer Hilfsvereinigungen e.V., Graurheindorfer Str 15, 53111 Bonn
T: (0228) 631548
Focus: Rehabil; Psych; Therapeutics . . . 05281

Dachverband wissenschaftlicher Gesellschaften der Agrar-, Forst-, Ernährungs-, Veterinär- und Umweltforschung e.V., Eschborner Landstr 122, 60489 Frankfurt
T: (069) 24788104; Tx: 413185; Fax: (069) 24788110
Founded: 1973; Members: 28
Pres: Prof. Dr. C. Thoroe; Gen Secr: Dr. R. Dörre
Focus: Ecology; Vet Med; Agri; Nutrition; Forestry
Periodicals
Agrarspectrum (weekly)
Agrarspectrum: Schriftenreihe . . . 05282

Design Zentrum Nordrhein-Westfalen e.V., Hindenburgstr 25-27, 45127 Essen
T: (0201) 820210; Fax: (0201) 231903
Founded: 1954
Pres: Klaus J. Maack; Gen Secr: Dr. Peter Zec
Focus: Graphic & Dec Arts, Design . . . 05283

Deutsche Akademie der Darstellenden Künste e.V., Bockenheimer Landstr 104, 60323 Frankfurt
T: (069) 7410049
Pres: Dr. Günther Ruhle; Gen Secr: Cathrin Ehrlich
Focus: Arts; Perf Arts; Cinema . . . 05284

Deutsche Akademie der Naturforscher Leopoldina, August-Bebel-Str 50a, 06108 Halle
T: (0345) 25014; Fax: (0345) 21727
Founded: 1652; Members: 1000
Pres: Prof. Dr. Benno Parthier
Focus: Nat Sci; Med
Periodicals
Acta Historica Leopoldina: Abhandlungen aus dem Archiv für Geschichte der Naturforschung und Medizin der Deutschen Akademie der Naturforscher Leopoldina (weekly)
Informationen (weekly)
Leopoldina: Mitteilungen der Deutschen Akademie der Naturforscher Leopoldina, Reihe 3 (weekly)
Nova Acta Leopoldina: Abhandlungen der Deutschen Akademie der Naturforscher Leopoldina (weekly) . . . 05285

Deutsche Akademie für Kinder- und Jugendliteratur e.V., Hauptstr 42, 97332 Würzburg
T: (0931) 4355
Founded: 1976; Members: 30
Pres: Prof. Dr. Heinrich Pleticha
Focus: Lit
Periodical
Volkacher Bote: Mitteilungsblatt . . . 05286

Deutsche Akademie für medizinische Fortbildung, Schöne Aussicht 2, 34117 Kassel
Focus: Med; Adult Educ . . . 05287

Deutsche Akademie für Nuklearmedizin, Karl-Wiechert-Allee 9, 30625 Hannover
Founded: 1968; Members: 80
Focus: Nucl Med . . . 05288

Deutsche Akademie für Sprache und Dichtung e.V., Alexandraweg 23, 64287 Darmstadt
T: (06151) 40920; Fax: (06151) 409299
Founded: 1949; Members: 134
Pres: Prof. Dr. Herbert Heckmann; Gen Secr: Dr. Gerhard Dette
Focus: Ling; Lit
Periodical
Jahrbuch der Deutschen Akademie (weekly) . . . 05289

Deutsche Akademie für Städtebau und Landesplanung e.V., Kaiserpl 4, 80803 München
T: (089) 332077
Founded: 1922; Members: 450
Pres: Prof. Dr. G. Albers
Focus: Urban Plan
Periodical
Mitteilungen (weekly) . . . 05290

Deutsche Akademie für Verkehrswissenschaft e.V., Agnesstr 53, 22301 Hamburg
T: (040) 475341
Founded: 1963; Members: 467
Focus: Transport; Law . . . 05291

Deutsche Arbeitsgemeinschaft für Paradontologie, Bäckerstr 102, 38640 Goslar
Focus: Dent . . . 05292

Deutsche Arbeitsgemeinschaft genealogischer Verbände e.V., c/o Nordrhein-Westfälisches Personenstandsarchiv Rheinland, Schloßstr 12, 50321 Brühl
T: (02232) 42948; Fax: (02232) 42948
Pres: Dr. Jörg Füchtner
Focus: Genealogy
Periodical
Genealogie (weekly) . . . 05293

Deutsche Arbeitsgemeinschaft Vakuum (DAGV), Postf 1913, 52428 Jülich
T: (02461) 615925
Founded: 1963; Members: 400
Focus: Eng; Physics . . . 05294

Deutsche Botanische Gesellschaft e.V., Untere Karspüle 2, 37073 Göttingen
T: (0551) 33347
Founded: 1882
Pres: Prof. Dr. W. Nultsch; Gen Secr: Dr. Wulf Koch
Focus: Botany
Periodical
Botanica Acta: Berichte der Deutschen Botanischen Gesellschaft (3 times annually) . . . 05295

Deutsche Bunsen-Gesellschaft für Physikalische Chemie e.V. (DBG), Varrentrappstr 40-42, 60486 Frankfurt
T: (069) 7917201
Founded: 1894; Members: 1500
Pres: Prof. Dr. W. Gruenbein; Gen Secr: Dr. Heinz Behret
Focus: Chem
Periodical
Berichte der Bunsen-Gesellschaft für Physikalische Chemie: An International Journal of Physical Chemistry (weekly) . . . 05296

Deutsche Dendrologische Gesellschaft, Hawstr 28, 54290 Trier
T: (0651) 33061; Fax: (0651) 33062
Founded: 1892
Focus: Forestry
Periodical
Mitteilungen (weekly) . . . 05297

Deutsche Dermatologische Gesellschaft, Schittenhelmstr 7, 24105 Kiel
Founded: 1888
Pres: Prof. Dr. Enno Christophers; Gen Secr: Prof. Dr. Gernot Rassner
Focus: Derm . . . 05298

Deutsche Diabetes-Gesellschaft, c/o Diabetesklinik, Wielandstr 23, 32445 Bad Oeynhausen
T: (05731) 9702
Members: 685
Focus: Diabetes
Periodical
Diabetologie-Informationen (weekly) . . . 05299

Deutsche EEG-Gesellschaft, Wissmannstr 1a, 14193 Berlin
Founded: 1950; Members: 1200
Pres: Prof. Dr. M. Stöhr; Gen Secr: Prof. Dr. S. Kubicki
Focus: Med; Neurology . . . 05300

Deutsche Elektrotechnische Kommission im DIN und VDE (DKE), Stresemannallee 15, 60596 Frankfurt
T: (069) 63080; Tx: 412871; Fax: (069) 6312925
Founded: 1970
Pres: Prof. Dr. Karl-Heinz Schneider; Gen Secr: Karl-Ludwig Orth
Focus: Electric Eng; Standards
Periodicals
DIN-Mitteilungen + elektronorm (weekly)
Elektrotechnische Zeitschrift (weekly) . . . 05301

Deutsche Exlibris Gesellschaft e.V., Ringstr 109, 78465 Konstanz
Founded: 1891
Pres: Dr. Gernot Blum; Gen Secr: Manfred Neureiter
Focus: Libraries & Bk Sci
Periodicals
Jahrbuch für Exlibriskunst und Graphik
Mitteilungen (weekly) . . . 05302

Deutsche farbwissenschaftliche Gesellschaft e.V. (DfwG), Unter den Eichen 87, 12203 Berlin
T: (030) 81045400, 81045409
Founded: 1974; Members: 200
Focus: Materials Sci . . . 05303

Deutsche Film- und Fernsehakademie Berlin, Pommernallee 1, 14052 Berlin
T: (030) 303071
Focus: Cinema . . . 05304

Deutsche Forschungsgemeinschaft (DFG), Kennedyallee 40, 53175 Bonn
T: (0228) 8851; Fax: (0228) 8852221
Founded: 1951; Members: 79
Pres: Prof. Dr. Wolfgang Frühwald; Gen Secr: Burkhart Müller
Focus: Sci
Periodicals
Forschung (weekly)
German Research (3 times annually) . . . 05305

Deutsche Forschungsgesellschaft für Oberflächenbehandlung e.V., Adersstr 94, 40215 Düsseldorf
T: (0211) 370457; Fax: (0211) 370459
Founded: 1949; Members: 170
Pres: Artur Goldschmidt; Gen Secr: Helmut Vesper
Focus: Eng . . . 05306

Deutsche Forschungs- und Versuchsanstalt für Luft- und Raumfahrt e.V. / German Aerospace Research Establishment, Linder Höhe, 51147 Köln
T: (02203) 6010; Tx: 88100; Fax: (02203) 67310
Founded: 1969; Members: 130
Pres: Prof. Dr. Walter Kröll
Focus: Aero
Periodicals
DFVLR-Forschungsberichte (weekly)
DFVLR-Mitteilungen (weekly)
DFVLR-Nachrichten (weekly)
Wissenschaftliche Berichte der Forschungsbereiche
Zeitschrift für Flugwissenschaften und Weltraumforschung (weekly) . . . 05307

Deutsche Gartenbauwissenschaftliche Gesellschaft e.V., Herrenhäuser Str 2, 30419 Hannover
T: (0511) 7622638
Pres: Prof. Dr. J. Meyer
Focus: Hort . . . 05308

Deutsche Gemmologische Gesellschaft e.V. / Deutsche Gesellschaft für Edelsteinkunde / Gemmological Association of Germany, Professor-Schlossmacher-Str 1, 55743 Idar-Oberstein
T: (06781) 43011; Tx: 426269; Fax: (06781) 41616
Founded: 1932
Pres: Prof.Dr. Hermann Bank; Gen Secr: Dr. Ulrich Henn
Focus: Mineralogy
Periodical
Zeitschrift der Deutschen Gemmologischen Gesellschaft (weekly) . . . 05309

Deutsche Geodätische Kommission (DGK), Marstallpl 8, 80539 München
T: (089) 23031113; Tx: 5213550; Fax: (089) 23031100
Founded: 1952; Members: 34
Pres: Prof. Dr. Dietrich Möller; Gen Secr: Prof. Dr. Klaus Schnädelbach
Focus: Surveying; Cart . . . 05310

Deutsche Geologische Gesellschaft (D.G.G.), Postf 510153, 30631 Hannover
T: (0511) 6432507; Fax: (0511) 6432304
Founded: 1848; Members: 3100
Pres: Prof. Dr. Dierk Henningsen
Focus: Geology
Periodicals
Nachrichten Deutsche Geologische Gesellschaft (weekly)
Zeitschrift der Deutschen Geologischen Gesellschaft (weekly) . . . 05311

Deutsche Geophysikalische Gesellschaft e.V. (DGG), Corrensstr 24, 48149 Münster
T: (0251) 834733; Fax: (0251) 838397
Founded: 1922; Members: 880
Pres: Prof. Dr. R. Hänel; Gen Secr: Dr. H. Jödicke
Focus: Geophys
Periodicals
Geophysical Journal International (weekly)
Mitteilungen (weekly) . . . 05312

Deutsche Gesellschaft für Aesthetische Medizin, Augustenburger Pl 1, 13353 Berlin
Focus: Med . . . 05313

Deutsche Gesellschaft für Agrarrecht / Vereinigung für Agrar- und Umweltrecht e.V., Postf 1969, 53009 Bonn
T: (0228) 703140; Fax: (0228) 703498
Members: 450
Pres: Prof. Dr. Ekkehard Pabsch; Gen Secr: Marianne May
Focus: Law; Agri
Periodical
Agrarrecht: Zeitschrift für das gesamte Recht der Landwirtschaft, der Agrarmärkte und des ländlichen Raumes (weekly) . . . 05314

Deutsche Gesellschaft für Allergieforschung, Liebermeisterstr 8, 72076 Tübingen
Focus: Immunology . . . 05315

Deutsche Gesellschaft für Allergie- und Immunitätsforschung, c/o BG-Kliniken Bergmannsheil, Gilsingstr 14, 44789 Bochum
Founded: 1951
Pres: Prof. Dr. Dr. J. Ring; Gen Secr: Prof. Dr. G. Schultze-Werninghaus

Focus: Immunology
Periodical
Allergo Journal (weekly) 05316

Deutsche Gesellschaft für allgemeine und angewandte Entomologie e.V. (DGaaE), c/o Biologische Bundesanstalt, Institut für Pflanzenschutz im Obstbau, Schwabenheimerstr 101, 69221 Dossenheim
Founded: 1976; Members: 550
Pres: Prof. Dr. E. Dickler; Gen Secr: Dr. Heidrun Vogt
Focus: Entomology
Periodical
Mitteilungen der Deutschen Gesellschaft für allgemeine und angewandte Entomologie (weekly) 05317

Deutsche Gesellschaft für Amerikastudien e.V. (DGfA), c/o Wirtschafts- und Sozialwissenschaftliche Fakultät, Universität Erlangen-Nürnberg, Lange Gasse 20, 90403 Nürnberg
T: (0911) 5302296
Founded: 1953; Members: 504
Focus: Ethnology
Periodical
Amerikastudien/American Studies: Vierteljahresschrift (weekly) 05318

Deutsche Gesellschaft für Anästhesiologie und Intensivmedizin (DGAI) / German Society of Anaesthesiology and Intensive Medicine, Roritzerstr 27, 90419 Nürnberg
T: (0911) 933780; Fax: (0911) 3938195
Founded: 1953; Members: 6100
Pres: Prof. Dr. G. Benad; Gen Secr: Holger Sorgatz
Focus: Anesthetics; Med
Periodical
Anästhesiologie und Intensivmedizin (weekly) 05319

Deutsche Gesellschaft für Analytische Psychologie e.V. (DGAP), Wedellstr 16-18, 12247 Berlin
T: (030) 7745561
Founded: 1961; Members: 350
Pres: Dr. Rudolf Müller
Focus: Psych; Psychoan
Periodical
Analytische Psychologie (weekly) 05320

Deutsche Gesellschaft für angewandte Optik e.V. (DGaO), c/o Carl Zeiss Jena GmbH, Tatzendpromenade 1a, 07745 Jena
T: (03641) 5882454; Tx: 331545; Fax: (03641) 5882023
Founded: 1923; Members: 580
Pres: Prof. Dr. T. Tschudi; Gen Secr: Dr. F. Merkle
Focus: Optics
Periodical
Optik (weekly) 05321

Deutsche Gesellschaft für Angiologie, c/o Medizinische Klinik, Hirschlandstr 97, 73730 Esslingen
T: (07365) 31031
Founded: 1971; Members: 400
Focus: Hematology 05322

Deutsche Gesellschaft für Anthropologie, Albertstr 11, 79104 Freiburg
Focus: Anthro 05323

Deutsche Gesellschaft für Arbeitsmedizin e.V., Pfarrstr 3, 80538 München
T: (089) 2184259/60
Members: 720
Focus: Hygiene 05324

Deutsche Gesellschaft für Asienkunde e.V. / German Association for Asian Studies, Rothenbaumchaussee 32, 20148 Hamburg
T: (040) 445891
Founded: 1967; Members: 735
Pres: Hans Klein
Focus: Ethnology; Poli Sci; Econ; Arts
Periodical
Asien: Deutsche Zeitschrift für Politik, Wirtschaft und Kultur (weekly) 05325

Deutsche Gesellschaft für Auswärtige Politik e.V. (DGAP), Adenauerallee 131, 53113 Bonn
T: (0228) 26750; Fax: (0228) 2675173
Founded: 1955; Members: 1500
Pres: C. Peter Henle
Focus: Poli Sci
Periodicals
Europa-Archiv: Zeitschrift für internationale Politik (weekly)
Die Internationale Politik (weekly) . . . 05326

Deutsche Gesellschaft für Bauingenieurwesen e.V., Barbarossapl 2, 76137 Karlsruhe
Founded: 1946; Members: 490
Pres: Prof. Dr. Wilhelm Strickler; Gen Secr: Gerhart Bochmann
Focus: Civil Eng 05327

Deutsche Gesellschaft für Baukybernetik e.V (DGBK), Riemenschneiderstr 9, 37603 Holzminden
T: (05531) 4040
Focus: Civil Eng; Cybernetics
Periodical
Bauen mit Kopf (bi-annually) 05328

Deutsche Gesellschaft für Baurecht e.V., Schumannstr 53, 60325 Frankfurt
T: (069) 748893; Fax: (069) 747083
Members: 550
Focus: Law 05329

Deutsche Gesellschaft für Bevölkerungswissenschaft e.V., Gustav-Stresemann-Ring 6, 65180 Wiesbaden
T: (0611) 752599; Fax: (0611) 39544
Founded: 1952
Pres: Prof. Dr. Karl Schwarz; Gen Secr: Dr. Johannes Otto
Focus: Sociology; Stats
Periodical
Acta Demographica (weekly) 05330

Deutsche Gesellschaft für Biomedizinische Technik e.V., Markgrafenstr 11, 10969 Berlin
T: (030) 2516029
Founded: 1961; Members: 512
Focus: Eng; Med; Bio
Periodical
Biomedizinische Technik: Biomedical Engineering (10 times annually) 05331

Deutsche Gesellschaft für Biophysik, c/o Fakultät für Physik, Technische Universität München, James-Franck-Str, 85748 Garching
T: (089) 32092552
Founded: 1961
Pres: Prof. Dr. H. Rüterjans; Gen Secr: Prof. Dr. F. Parak
Focus: Biophys 05332

Deutsche Gesellschaft für Bluttransfusion und Immunhämatologie e.V., Sandhofstr 1, 60528 Frankfurt
T: (069) 6782283, 6782204
Focus: Hematology
Periodical
Blut (weekly) 05333

Deutsche Gesellschaft für Chemisches Apparatewesen, Chemische Technik und Biotechnologie e.V. (DECHEMA), Theodor-Heuss-Allee 25, 60486 Frankfurt
T: (069) 75640; Tx: 412490; CA: Dechema Frankfurtmain; Fax: (069) 7564201
Founded: 1926; Members: 2700
Pres: Prof. Dr. Heinz-Gerhard Franck; Gen Secr: Prof. Dr. Dieter Behrens
Focus: Eng
Periodicals
Biotechnologie – Verfahren, Anlagen, Apparate (weekly)
Chemie-Ingenieur-Technik (weekly)
Literaturkurzberichte Chemische Technik und Biotechnologie (weekly)
Materialwissenschaft und Werkstofftechnik: Journal of Materials Technology and Testing (weekly)
Werkstoffe und Korrosion: Materials and Corrosion (weekly) 05334

Deutsche Gesellschaft für Chirurgie, Venusbergweg 1, 53115 Bonn
Founded: 1872; Members: 3444
Pres: Prof.Dr. Leo Koslowski; Gen Secr: Prof. Dr. Dr. F. Stelzner
Focus: Surgery
Periodical
Langenbecks Archiv für Chirurgie . . . 05335

Deutsche Gesellschaft für Christliche Kunst e.V., Wittelsbacherpl 2, 80333 München
T: (089) 288645
Founded: 1893; Members: 900
Pres: Prof. Dr. P. Urban Rapp
Focus: Arts
Periodical
Hefte der Deutschen Gesellschaft für christliche Kunst (weekly) 05336

Deutsche Gesellschaft für Chronometrie e.V. (DGC), Postf 101013, 70009 Stuttgart
T: (0711) 242502; Fax: (0711) 608116
Founded: 1949; Members: 750
Pres: Prof. Dr. Friedrich Assmus
Focus: Eng; Electronic Eng; Astronomy
Periodical
DGC-Mitteilungen (weekly) 05337

Deutsche Gesellschaft für das Badewesen e.V. (DGfdB), Postf 100910, 45009 Essen
T: (0201) 233030; Fax: (0201) 221310
Founded: 1899
Pres: Felix Zimmermann; Gen Secr: Friedrich R. Kunze
Focus: Public Health
Periodical
Archiv des Badewesens (weekly) . . . 05338

Deutsche Gesellschaft für die Bekämpfung der Muskelkrankheiten e.V., Rennerstr 4, 79106 Freiburg
T: (0761) 277932, 278024; Fax: (0761) 281043
Focus: Pathology
Periodical
Muskelreport (weekly) 05339

Deutsche Gesellschaft für die Vereinten Nationen e.V., Poppelsdorfer Allee 55, 53115 Bonn
T: (0228) 213646; Fax: (0228) 217492
Pres: Dr. Helga Timm; Gen Secr: Joachim Krause
Focus: Poli Sci
Periodical
Vereinte Nationen (weekly) 05340

Deutsche Gesellschaft für Dokumentation e.V. (DGD), Hanauer Landstr 126-128, 60314 Frankfurt
T: (069) 430313; Fax: (069) 4909096
Founded: 1948; Members: 1800
Pres: A. de Kemp; Gen Secr: Hans Nerlich
Focus: Doc
Periodicals
DGD-Newsletter (weekly)
Nachrichten für Dokumentation (weekly)
OLBG-INFO (weekly) 05341

Deutsche Gesellschaft für Dynamische Psychiatrie (DGDP), Geiselgasteigstr 203, 81545 München
T: (089) 644017
Founded: 1980
Focus: Psychiatry 05342

Deutsche Gesellschaft für Elektronenmikroskopie e.V. (DGE), c/o Fritz-Haber-Institut der Max-Planck-Gesellschaft, Faradayweg 4-6, 14195 Berlin
Founded: 1949; Members: 750
Pres: Dr. K. Zierold; Gen Secr: Dr. B. Tesche
Focus: Nat Sci
Periodical
Elektronenmikroskopie (weekly) 05343

Deutsche Gesellschaft für Endokrinologie, Postfach 650311, 14129 Berlin
T: (030) 4685802; Tx: 182030; Fax: (030) 46916614
Founded: 1953; Members: 1200
Pres: Prof. Dr. E. Nieschlag; Gen Secr: Dr. Ursula-F. Habenicht
Focus: Endocrinology
Periodical
Endokrinologie-Informationen (weekly) . . 05344

Deutsche Gesellschaft für Erd- und Grundbau e.V. (DGEG), Kronprinzenstr 35A, 45128 Essen
T: (0201) 227677
Founded: 1951; Members: 950
Focus: Civil Eng
Periodicals
Dokumentation für Bodenmechanik, Grundbau, Felsmechanik, Ingenieurgeologie
Geotechnical Abstracts (weekly)
Geotechnik (weekly) 05345

Deutsche Gesellschaft für Ernährung e.V. (DGE), Feldbergstr 28, 60323 Frankfurt
T: (069) 9714060; Fax: (069) 97140699
Founded: 1953; Members: 2000
Pres: Prof. Dr. Volker Pudel; Gen Secr: Dr. Helmut Oberritter
Focus: Nutrition
Periodical
Ernährungs-Umschau: Organ der DGE (weekly) 05346

Deutsche Gesellschaft für Erziehungswissenschaft (DGfE), c/o Institut für Allgemeine Pädagogik, Humboldt-Universität, Unter den Linden 6, 10099 Berlin
T: (030) 20932587; Fax: (030) 20932345
Founded: 1963; Members: 1400
Pres: Prof. Dr. Dietrich Benner
Focus: Educ
Periodicals
Arbeitsberichte (bi-annually)
Tagungsberichte 05347

Deutsche Gesellschaft für Fettwissenschaft e.V., Soester Str 13, 48155 Münster
T: (0251) 64745
Founded: 1935
Pres: Dr. W. Umbach; Gen Secr: Dr. Heinrich Brüning
Focus: Materials Sci
Periodical
Fett-Wissenschaft, Technologie (14 times annually) 05348

Deutsche Gesellschaft für Filmdokumentation, Traubenstr 9, 65207 Wiesbaden
Focus: Cinema; Doc 05349

Deutsche Gesellschaft für Film- und Fernsehforschung, Willroiderstr 6, 81545 München
T: (089) 646948
Focus: Cinema 05350

Deutsche Gesellschaft für Forschung im Graphischen Gewerbe, Brunnerstr 2, 80804 München
Founded: 1951
Focus: Eng 05351

Deutsche Gesellschaft für Galvano- und Oberflächentechnik e.V. (DGO), Horionpl 6, 40213 Düsseldorf
T: (0211) 132381; Fax: (0211) 327199
Founded: 1961; Members: 1800
Pres: Willi Metzger; Gen Secr: Dr. K.-W. Kramer
Focus: Eng 05352

Deutsche Gesellschaft für Gartenkunst und Landschaftspflege e.V. (DGGL), Markgrafenstr 14, 76131 Karlsruhe
T: (0721) 1332950
Focus: Hort; Ecology
Periodical
Garten und Landschaft (weekly) 05353

Deutsche Gesellschaft für Gerontologie (DGG), c/o Medizinische Universität, Ratzeburger Allee 160, 23538 Lübeck
T: (0451) 5002400; Fax: (0451) 5006518
Founded: 1967; Members: 730
Pres: Prof. Dr. Rudolf-M. Schütz
Focus: Geriatrics 05354

Deutsche Gesellschaft für Geschichte der Medizin, Naturwissenschaft und Technik e.V., c/o Astronomisch-Physikalisches Kabinett, Brüder-Grimm-Pl 1, 80336 München
T: (089) 21803252
Founded: 1901; Members: 701
Pres: Prof. Dr. M. Folkerts; Gen Secr: Prof. Dr. Uta Lindgren
Focus: Cultur Hist; Med; Nat Sci; Eng
Periodical
Nachrichtenblatt (3 times annually) . . . 05355

Deutsche Gesellschaft für Gesundheitsvorsorge e.V. (DGGV), Driescher Hecke 19, 51375 Leverkusen
T: (0214) 56744
Founded: 1968; Members: 800
Focus: Public Health 05356

Deutsche Gesellschaft für Gynäkologie und Geburtshilfe, c/o Städtisches Marienkrankenhaus, Mariahilfbergweg 7, 92224 Amberg
Founded: 1885
Gen Secr: Prof. Dr. D. Berg
Focus: Gynecology 05357

Deutsche Gesellschaft für Hämatologie und Onkologie e.V., c/o Medizinische Klinik III, Klinikum Großhadern, Marchioninistr 15, 81377 München
T: (089) 70956403
Members: 580
Pres: Prof. Dr. H. Heimpel; Gen Secr: Prof. Dr. K. Possinger
Focus: Hematology; Cell Biol & Cancer Res
Periodicals
Ann. Haemat. (weekly)
Onkologie (weekly) 05358

Deutsche Gesellschaft für Hals-Nasen-Ohren-Heilkunde, Kopf- und Hals-Chirurgie, Hittorfstr 17, 53129 Bonn
T: (0228) 231770; Fax: (0228) 239325
Founded: 1922; Members: 2660
Pres: Prof. Dr. Bernd Freigang; Gen Secr: Prof. Dr. Harald Feldmann
Focus: Surgery; Otorhinolaryngology
Periodical
HNO-Informationen (weekly) 05359

Deutsche Gesellschaft für Heereskunde e.V. (DGfHK), Augustin-Wibbelt-Str 8, 59269 Beckum
Founded: 1898; Members: 1100
Focus: Military Sci
Periodical
Zeitschrift für Heereskunde: Wissenschaftliches Organ für die Kulturgeschichte der Streitkräfte, ihrer Bekleidung, Bewaffnung und Ausrüstung, für heeresmuseale Nachrichten und Sammler-Mitteilungen (weekly) 05360

Deutsche Gesellschaft für Herpetologie und Terrarienkunde e.V. (DGHT), Postf 1421, 53351 Rheinbach
T: (02255) 6086; Fax: (02255) 1726
Founded: 1964; Members: 5100
Pres: Ingo Pauler
Focus: Zoology
Periodicals
elaphe (weekly)
Salamandra: Zeitschrift für Herpetologie und Terrarienkunde (weekly) 05361

Deutsche Gesellschaft für Herz- und Kreislaufforschung, c/o Institut für Experimentelle Chirurgie, Heinrich-Heine-Universität, Moorenstr 5, 40225 Düsseldorf
T: (0211) 3115255; Fax: (0211) 3113550
Founded: 1928; Members: 2800
Pres: Prof. Dr. H. Kreuzer; Gen Secr: Prof. Dr. G. Arnold
Focus: Cardiol
Periodical
Zeitschrift für Kardiologie: Verhandlungsbericht (weekly) 05362

Deutsche Gesellschaft für Holzforschung e.V. (DGfH), Schwanthalerstr 79, 80336 München
T: (089) 5389057; Fax: (089) 531657
Founded: 1942; Members: 600
Pres: Karl-Josef Hüttemann; Gen Secr: Joachim Tebbe
Focus: Forestry
Periodicals
DGfH-Nachrichten
Informationsdienste Holz der Entwicklungsgemeinschaft Holzbau in der DGfH
Merkhefte
Mitteilungshefte 05363

Deutsche Gesellschaft für Hopfenforschung e.V. (DGfH), Hüll 5, 85283 Wolnzach
T: (08442) 3597
Founded: 1926; Members: 150
Pres: Dr. Georg Beer; Gen Secr: Hermann Schlicker
Focus: Agri
Periodical
Forschungsberichte (weekly) 05364

Deutsche Gesellschaft für Hydrokultur e.V., Kurt-Schumacher-Str 36, 45699 Herten
Focus: Hydrology 05365

Deutsche Gesellschaft für Hygiene und Mikrobiologie e.V. (DGHM), c/o Hygiene-Institut, Universität, Im Neuenheimer Feld 324, 69120 Heidelberg
T: (06221) 568310; Fax: (06221) 565857
Founded: 1906; Members: 1600
Pres: Prof. Dr. D. Bitter-Suermann; Gen Secr: Prof. Dr. Hans-G. Sonntag
Focus: Microbio; Hygiene 05366

Deutsche Gesellschaft für Innere Medizin, Humboldtstr 14, 65189 Wiesbaden
T: (0611) 307946; Fax: (0611) 378260
Founded: 1882; Members: 4000
Pres: Prof. Dr. Dr. M. Classen; Gen Secr: Prof. Dr. Dr. H.G. Lasch
Focus: Intern Med
Periodical
Supplement der Medizinischen Klinik (weekly) 05367

Deutsche Gesellschaft für Internistische Intensivmedizin, Rübenkamp 148, 22307 Hamburg
Members: 950
Focus: Intern Med 05368

Deutsche Gesellschaft für Kartographie e.V. (D.G.f.K.), Arno-Holz-Str 12, 12165 Berlin
Founded: 1950; Members: 2300
Pres: Prof. Dr. Ulrich Freitag; Gen Secr: Jürgen Bosserhoff
Focus: Cart
Periodical
Kartographische Nachrichten (weekly) . . 05369

Deutsche Gesellschaft für Kieferorthopädie e.V., Schlesierstr 21, 97078 Würzburg
T: (0931) 31479
Founded: 1907; Members: 1500
Focus: Dent
Periodical
Fortschritte der Kieferorthopädie (weekly) 05370

Deutsche Gesellschaft für Kinderheilkunde, c/o Kinderklinik, Medizinische Hochschule, Konstanty-Gutschow-Str 8, 30625 Hannover
T: (0511) 5323212
Founded: 1883; Members: 6000
Pres: Prof. Dr. Johannes Brodehl
Focus: Pediatrics
Periodical
Monatsschrift für Kinderheilkunde (weekly) 05371

Deutsche Gesellschaft für Kinder- und Jugendpsychiatrie e.V., Hans-Sachs-Str 6, 35039 Marburg
T: (06421) 285334; Fax: (06421) 285667
Founded: 1950; Members: 650
Pres: Prof. Dr. Gerd Lehmkuhl; Gen Secr: Prof. Dr. Dr. Helmut Remschmidt
Focus: Psychiatry
Periodical
Zeitschrift für Kinder- und Jugendpsychiatrie 05372

Deutsche Gesellschaft für Kommunikationsforschung, Willroider Str 6, 81545 München
T: (089) 646948
Focus: Comm Sci 05373

Deutsche Gesellschaft für Laboratoriumsmedizin e.V., Witzelstr 63, 40225 Düsseldorf
T: (0211) 340556; Fax: (0211) 341930
Members: 600
Pres: Prof. Dr. Jürgen D. Kruse-Jarres; Gen Secr: Hartmut Reineck
Focus: Med 05374

Deutsche Gesellschaft für Lichtforschung, Hochstädter Landstr 23, 63454 Hanau
Founded: 1927
Focus: Physics 05375

Deutsche Gesellschaft für Logistik e.V. (DGFL), Heinrich-Hertz-Str 4, 44227 Dortmund
T: (0231) 75443220
Founded: 1973; Members: 1000
Pres: H. Stübig; Gen Secr: Dr. Karin Bockelmann
Focus: Marketing; Econ
Periodical
Logistik Spektrum: Logistik in Industrie, Handel und Dienstleistung (weekly) 05376

Deutsche Gesellschaft für Luft- und Raumfahrt e.V. (DGLR), Godesberger Allee 70, 53175 Bonn
T: (0228) 376726; Fax: (0228) 374755
Founded: 1968; Members: 3500
Pres: Hans E.W. Hoffmann; Gen Secr: Heinz Schwaebisch
Focus: Aero
Periodical
Luft- und Raumfahrt (weekly) 05377

Deutsche Gesellschaft für Luft- und Raumfahrtmedizin e.V. (DGLRM), c/o Abt. Neurologie, Bundeswehrkrankenhaus, Oberer Eselsberg, 89081 Ulm
T: (0731) 1712060
Founded: 1961
Focus: Med 05378

Deutsche Gesellschaft für Lungenkrankheiten und Tuberkulose, Hugstetter Str 55, 79106 Freiburg
Focus: Pulmon Dis 05379

Deutsche Gesellschaft für Manuelle Medizin e.V., Heerstr 162, 56154 Boppard
T: (06742) 5917
Founded: 1953; Members: 1300
Pres: Dr. Alfred Möhrle; Gen Secr: Dr. Gerhard Schneider
Focus: Med 05380

Deutsche Gesellschaft für Materialkunde e.V. (DGM), Adenauerallee 21, 61440 Oberursel
T: (06171) 4081; Fax: (06171) 52554
Founded: 1919; Members: 3200
Pres: Prof. Dr. H. Mecking; Gen Secr: Dr. P.P. Schepp
Focus: Metallurgy
Periodical
Zeitschrift für Metallkunde 05381

Deutsche Gesellschaft für Medizinische Informatik, Biometrie und Epidemiologie (GMDS), Herbert-Lewin-Str 1, 50931 Köln
T: (0221) 4004256
Founded: 1955; Members: 600
Pres: Prof. Dr. Jörg Michaelis; Gen Secr: Franz F. Stobrawa
Focus: Doc; Med; Stats
Periodical
Mitteilungsblatt der GMDS (3 times annually) 05382

Deutsche Gesellschaft für Medizinische Soziologie, Am Hochsträß 8, 89081 Ulm
T: (0731) 1762927; Fax: (0731) 1762038
Focus: Sociology
Periodical
Medizinsoziologie (weekly) 05383

Deutsche Gesellschaft für Missionswissenschaft (DGMW), Eckenerstr 1, 69121 Heidelberg
T: (06221) 480935
Founded: 1918; Members: 128
Focus: Rel & Theol 05384

Deutsche Gesellschaft für Moor- und Torfkunde / German Society for Bog and Peat Research, Alfred-Bentz-Haus, Stilleweg 2, 30655 Hannover
Founded: 1970; Members: 150
Pres: Prof. Dr. J.-D. Becker-Platen
Focus: Geology
Periodical
TELMA 05385

Deutsche Gesellschaft für Mund-, Kiefer- und Gesichtschirurgie (DGMKG), c/o Klinik und Poliklinik für Mund-, Kiefer- und Gesichtschirurgie, Welschnonnenstr 17, 53111 Bonn
T: (0228) 2872452; Fax: (0228) 2872604
Founded: 1951; Members: 600
Pres: Prof. Dr. Dr. J.-Erich Hausamen; Gen Secr: Prof. Dr. Dr. Rudolf H. Reich
Focus: Surgery
Periodical
Zeitschrift für Mund-, Kiefer- und Gesichtschirurgie (weekly) 05386

Deutsche Gesellschaft für Neurochirurgie, Hufelandstr 55, 45147 Essen
Founded: 1950; Members: 244
Pres: Prof. Dr. W. Grote; Gen Secr: Prof. Dr. H. Dietz
Focus: Surgery 05387

Deutsche Gesellschaft für Neurologie, Josef-Schneider-Str 11, 97080 Würzburg
Founded: 1906
Focus: Neurology 05388

Deutsche Gesellschaft für Neuropathologie und Neuroanatomie e.V., c/o Institut für Neuropathologie der Universität, Thalkirchner Str 36, 80337 München
Focus: Neurology 05389

Deutsche Gesellschaft für Neuroradiologie, Josef-Schneider-Str. 11, 97080 Würzburg
Focus: Radiology 05390

Deutsche Gesellschaft für Nuklearmedizin, Sigmund-Freud-Str 25, 53127 Bonn
Members: 750
Pres: Prof. Dr. H.J. Biersack
Focus: Nucl Med
Periodical
Nuklearmedizin (weekly) 05391

Deutsche Gesellschaft für Orthopädie und Traumatologie e.V. (DGOT), Allmeygang 6, 65929 Frankfurt
T: (069) 318561; Fax: (069) 318561
Founded: 1901
Focus: Orthopedics
Periodical
Mitteilungsblatt der Deutschen Gesellschaft für Orthopädie und Traumatologie (weekly) . 05392

Deutsche Gesellschaft für Ortung und Navigation e.V. (DGON), Pempelforter Str 47, 40211 Düsseldorf
T: (0211) 369909; Fax: (0211) 351645
Founded: 1951; Members: 165
Pres: Dr. Manfred Böhm; Gen Secr: Peter Becker
Focus: Navig
Periodical
Ortung und Navigation (3 times annually) 05393

Deutsche Gesellschaft für Osteuropakunde e.V. (DGO), Schaperstr 30, 10719 Berlin
T: (030) 2184172
Founded: 1913; Members: 800
Pres: Dr. Otto W. von Amerongen; Gen Secr: Dr. Ernst von Eicke
Focus: Ethnology
Periodicals
Osteuropa (weekly)
Osteuropa-Recht (weekly)
Osteuropa-Wirtschaft (weekly) 05394

Deutsche Gesellschaft für Parasitologie e.V. (DGP), c/o Behringwerke AG, Postf 1140, 35001 Marburg
T: (06421) 392606; Fax: (06421) 394757
Founded: 1961; Members: 628
Pres: Prof. Dr. H.J. Bürger; Gen Secr: Dr. B. Enders
Focus: Med; Vet Med; Zoology 05395

Deutsche Gesellschaft für Parodontologie (DGP), Sierichstr 60, 22301 Hamburg
T: (040) 2793335
Founded: 1925; Members: 1980
Focus: Dent
Periodical
DGP-Nachrichten (weekly) 05396

Deutsche Gesellschaft für Pathologie, c/o Senckenbergisches Zentrum der Pathologie der Universität, Theodor-Stern-Kai 7, 60596 Frankfurt
T: (069) 63015364
Founded: 1897; Members: 950
Gen Secr: Prof. Dr. Klaus Hübner
Focus: Pathology 05397

Deutsche Gesellschaft für Perinatale Medizin, c/o Frauenklinik, Pulsstr 4, 14059 Berlin
T: (030) 6253081
Focus: Gynecology
Periodical
Perinatal-Medizin (weekly) 05398

Deutsche Gesellschaft für Personalführung e.V., Niederkasseler Lohweg 16, 40547 Düsseldorf
T: (0211) 59780; Fax: (0211) 5978199
Pres: Hans Kauth; Gen Secr: Dr. D. Walz
Focus: Business Admin 05399

Deutsche Gesellschaft für Pharmakologie und Toxikologie, Postf 4119, 64271 Darmstadt
T: (06151) 722592; Tx: 4193280; Fax: (06151) 713314
Members: 2200
Pres: Prof. Dr. H. Greim; Gen Secr: Prof. Dr. H.P. Wolf
Focus: Pharmacol; Toxicology
Periodical
Naunyn-Schmiedeberg's Archive of Pharmacology (weekly) 05400

Deutsche Gesellschaft für Phlebologie, c/o Allergie- und Hautklinik, Lippestr 9-11, 26548 Norderney
T: (04932) 805404; Fax: (04932) 805200
Founded: 1953; Members: 1200
Pres: Prof. Dr. W. Hach; Gen Secr: Prof. Dr. W. Lechner
Focus: Physiology; Pathology; Gastroenter
Periodical
Phlebologie und Proktologie (3 times annually) 05401

Deutsche Gesellschaft für Photogrammetrie und Fernerkundung, c/o Institut für Photogrammetrie und Kartographie, Universität der Bundeswehr, Werner-Heisenberg-Weg 39, 85579 Neubiberg
T: (089) 60043448; Tx: 05215800; Fax: (089) 60044090
Founded: 1909; Members: 850
Pres: Prof. Dr. Egon Dorrer
Focus: Surveying
Periodical
Bildmessung und Luftbildwesen (weekly) . 05402

Deutsche Gesellschaft für Photographie e.V. (DGPh), Rheingasse 8-12, 50676 Köln
T: (0221) 2402037; Fax: (0221) 2402035
Founded: 1951; Members: 700
Pres: Dr. Hans Friderichs; Gen Secr: Gert Koshofer
Focus: Photo
Periodical
DGPh Intern (weekly) 05403

Deutsche Gesellschaft für Physikalische Medizin und Rehabilitation, K.-Gutschow-Str 8, 30625 Hannover
Pres: Prof. Dr. E. Conradi; Gen Secr: Prof. Dr. A. Gehrke
Focus: Rehabil; Physiology; Physical Therapy
Periodical
Zeitschrift Physikalische Medizin, Rehabilitationsmedizin, Kurortmedizin (weekly) 05404

Deutsche Gesellschaft für Pilzkunde, Kaiserstr 2, 76131 Karlsruhe
Founded: 1921
Focus: Botany, Specific 05405

Deutsche Gesellschaft für Plastische und Wiederherstellende Chirurgie e.V., c/o Diakoniekrankenhaus, Elise-Averdieck-Str 17, 27342 Rotenburg
T: (04261) 772127; Fax: (04261) 772128
Founded: 1961; Members: 427
Gen Secr: Dr. Hans Rudolph
Focus: Surgery 05406

Deutsche Gesellschaft für Pneumologie (DGP), c/o Klinik für Lungen- und Bronchialerkrankungen, Waldhof Elgershausen, 35753 Greifenstein
T: (06449) 77261
Pres: Prof. Dr. R. Loddenkemper; Gen Secr: Prof. Dr. H. Morr
Focus: Pulmon Dis
Periodical
Pneumologie (weekly) 05407

Deutsche Gesellschaft für Poesie- und Bibliotherapie e.V. (DGPB), Alteburger Str 7, 50678 Köln
Pres: Traute Pape
Focus: Therapeutics 05408

Deutsche Gesellschaft für Polarforschung (DeGePo), Columbusstr, 27568 Bremerhaven
T: (0471) 4831210
Founded: 1936; Members: 370
Pres: Prof. Dr. D. Möller; Gen Secr: Prof. Dr. H. Miller
Focus: Geography; Geology; Geophys; Bio
Periodical
Polarforschung (weekly) 05409

Deutsche Gesellschaft für Psychiatrie und Nervenheilkunde e.V. (DGPN), Joseph-Stelzmann-Str 9, 50931 Köln
T: (0221) 4786357; Fax: (0221) 4786398
Founded: 1842; Members: 1100
Pres: Prof. Dr. Uwe Henrik Peters; Gen Secr: Prof. Dr. Reinhold Schüttler
Focus: Neurology; Psychiatry
Periodicals
Nervenarzt
Spektrum 05410

Deutsche Gesellschaft für Psychologie e.V. (DGfPs), c/o Psychologisches Institut IV, Fliednerstr 21, 48149 Münster
T: (0251) 834153; Fax: (0251) 838319
Founded: 1904; Members: 1500
Pres: Prof. Dr. Urs Baumann; Gen Secr: Prof. Dr. Amélie Mummendey
Focus: Psych
Periodical
Psychologische Rundschau (weekly) . . 05411

Deutsche Gesellschaft für Psychosomatische Medizin e.V. (DGPM), Geiselgasteigstr 203, 81545 München
T: (089) 641016
Focus: Psychiatry; Neurology; Psych . . 05412

Deutsche Gesellschaft für Publizistik- und Kommunikationswissenschaft e.V., c/o Lehrstuhl für Journalistik I, Katholische Universität Eichstätt, Ostenstr 26, 85071 Eichstätt
T: (08421) 20564; Fax: (08421) 20553
Founded: 1963; Members: 400
Pres: Prof. Dr. Walter Hömberg
Focus: Journalism; Sociology; Comm Sci
Periodical
Aviso: Informationsdienst der Deutschen Gesellschaft für Publizistik- und Kommunikationswissenschaft (3 times annually) 05413

Deutsche Gesellschaft für Qualität e.V. (DGQ), Kurhessenstr 95, 60431 Frankfurt
T: (069) 520128
Focus: Materials Sci
Periodical
Qualität und Zuverlässigkeit: Zeitschrift für industrielle Qualitätssicherung (weekly) . . 05414

Deutsche Gesellschaft für Qualitätsforschung (Pflanzliche Nahrungsmittel) e.V. (DGQ), c/o Lehrstuhl für Gemüsebau, TU Weihenstephan, 85354 Freising
T: (08161) 713723; Fax: (08161) 714491
Founded: 1955
Pres: Prof. Dr. J. Leichmann; Gen Secr: Dr. J. Habben
Focus: Botany
Periodical
Proceedings der Jahreskongresse (weekly) 05415

Deutsche Gesellschaft für Rechtsmedizin, Melatengürtel 60-62, 50823 Köln
T: (0221) 4784280; Fax: (0221) 4784261
Founded: 1904; Members: 400
Pres: Prof. Dr. Michael Staak
Focus: Forensic Med
Periodical
Zeitschrift für Rechtsmedizin: Journal of Legal Medicine (weekly) 05416

Deutsche Gesellschaft für Rheumatologie e.V., c/o Rheumaklinik, 24576 Bad Bramstedt
Founded: 1927; Members: 657
Gen Secr: Dr. H.J. Albrecht
Focus: Rheuma
Periodical
Zeitschrift für Rheumatologie (weekly) . . 05417

Deutsche Gesellschaft für Säugetierkunde e.V., c/o Zoologisches Institut, Auf der Morgenstelle 28, 72076 Tübingen
T: (07071) 292958; Fax: (07071) 294634
Founded: 1926; Members: 600
Pres: Prof. Dr. Uwe Schmidt; Gen Secr: Prof. Dr. Hans Erkert
Focus: Zoology
Periodical
Zeitschrift für Säugetierkunde: International Journal of Mammalian Biology (weekly) 05418

Deutsche Gesellschaft für Sexualforschung e.V. (DGfS), c/o Sexualberatungsstelle der Abteilung für Sexualforschung, Universität, Poppenhusenstr 12, 22305 Hamburg
Founded: 1950; Members: 220
Pres: Prof. M. Hauch; Gen Secr: Sonja Düring
Focus: Psych
Periodical
Beiträge zur Sexualforschung 05419

Deutsche Gesellschaft für Sonnenenergie e.V. (DGS), Augustenstr 79, 80333 München
T: (089) 524071
Members: 6000
Focus: Energy
Periodical
Sonnenenergie (weekly) 05420

Deutsche Gesellschaft für Soziale Psychiatrie e.V. (DGSP), Stuppstr 14, 50823 Köln
Founded: 1970; Members: 2700
Pres: Holger Vulturius; Gen Secr: Richard Suhre
Focus: Psychiatry
Periodical
Soziale Psychiatrie (5 times annually) . . 05421

Deutsche Gesellschaft für Sozialmedizin und Prävention e.V., Overbergstr 17, 44801 Bochum
T: (0234) 7004868; Fax: (0234) 7007922
Founded: 1964; Members: 400
Pres: Prof. Dr. J. Gostomzyk
Focus: Hygiene
Periodical
Sozial- und Präventivmedizin (weekly) . . 05422

Deutsche Gesellschaft für Sozialpädiatrie e.V. (DGSP), Heiglhofstr 63, 81377 München
T: (089) 71009312
Founded: 1909; Members: 750
Pres: Prof. Dr. Dr. Theodor Hellbrügge; Gen Secr: Prof. Dr. Hubertus von Voss
Focus: Pediatrics 05423

Deutsche Gesellschaft für Soziologie, c/o Universität, Schloss, 68163 Mannheim
Pres: Prof. Wolfgang Zapf
Focus: Sociology
Periodical
Soziologie (weekly) 05424

Deutsche Gesellschaft für Sprachheilpädagogik e.V. (DGS), Leonberger Ring 1, 12349 Berlin
T: (030) 6057965
Founded: 1927; Members: 3000
Focus: Logopedy
Periodical
Die Sprachheilarbeit (weekly) 05425

Deutsche Gesellschaft für Sprach- und Stimmheilkunde, Kardinal-von-Galen-Ring 10, 48149 Münster
Founded: 1925
Focus: Logopedy 05426

Deutsche Gesellschaft für Sprachwissenschaft / German Society for Linguistics, c/o Philosophische Fakultät, Universität Passau, Innstr 40, 94030 Passau
Founded: 1978; Members: 800
Pres: Prof. Dr. Rudi Keller; Gen Secr: Prof. Dr. Rudolf Emons
Focus: Ling
Periodicals
Bulletin (weekly)
Zeitschrift für Sprachwissenschaft (weekly) 05427

Deutsche Gesellschaft für Sprechwissenschaft und Sprecherziehung e.V. (DGSS), Dompl 23, 48143 Münster
T: (0251) 834429
Founded: 1920; Members: 360
Focus: Ling; Educ 05428

Deutsche Gesellschaft für Suchtforschung und Suchttherapie e.V., Westring 2, 59065 Hamm
T: (02381) 90150; Fax: (02381) 15331
Pres: Dr. Hans Watzl; Gen Secr: E. Göcke
Focus: Public Health; Educ; Sociology; Psych; Med
Periodical
Sucht: Zeitschrift für Wissenschaft und Praxis 05429

Deutsche Gesellschaft für Technische Zusammenarbeit (GTZ), Dag-Hammarskjöld-Weg 1-5, 65726 Eschborn
T: (06196) 790
Founded: 1975
Focus: Eng
Periodicals
Akzente (weekly)
Gate (weekly) 05430

Deutsche Gesellschaft für Thorax-, Herz- und Gefässchirurgie, c/o Kerckhoff-Institut, Parkstr 1, 61231 Bad Nauheim
Founded: 1971; Members: 800
Pres: Prof. Dr. A. Hehrlein; Gen Secr: Prof. Dr. M. Leitz
Focus: Surgery
Periodical
The Thoracic and Cardiovascular Surgeon (weekly) 05431

Deutsche Gesellschaft für Unfallheilkunde e.V., Theodor-Stern-Kai 7, 60596 Frankfurt
Focus: Surgery
Periodical
Unfallheilkunde: Traumatology (weekly) . 05432

Deutsche Gesellschaft für Urologie, Humboldtstr 5, 30169 Hannover
Focus: Urology 05433

Deutsche Gesellschaft für Verdauungs- und Stoffwechselkrankheiten, c/o Zentrum der Inneren Medizin, Theodor-Stern-Kai 7, 60596 Frankfurt
T: (069) 63011
Founded: 1913
Focus: Intern Med
Periodical
Zeitschrift für Gastroenterologie (weekly) . 05434

Deutsche Gesellschaft für Verhaltenstherapie e.V. (DGVT), Postf 1343, 72003 Tübingen
T: (07071) 41211
Founded: 1968
Pres: Gerhard Brückner; Hertha Collin; Cornelia Paulus; Peter Schulz; Bernd Stefanides
Focus: Therapeutics
Periodical
Verhaltenstherapie und Psychosoziale Praxis (weekly) 05435

Deutsche Gesellschaft für Versicherungsmathematik e.V., Walter-Flex-Str 3, 53113 Bonn
T: (0228) 9162210; Fax: (0228) 9162211
Founded: 1948; Members: 838
Pres: Horst Becker; Gen Secr: Dr. Martin Balleer
Focus: Insurance
Periodical
Blätter (weekly) 05436

Deutsche Gesellschaft für Völkerkunde e.V. (DGV), Werderring 10, 79098 Freiburg
T: (0761) 2034461
Founded: 1929; Members: 350
Focus: Ethnology
Periodical
Zeitschrift für Ethnologie 05437

Deutsche Gesellschaft für Völkerrecht, c/o Institut für Völkerrecht, Adenauerallee 24-42, 53113 Bonn
Founded: 1917
Pres: Prof. Dr. Christian Tomuschat
Focus: Law 05438

Deutsche Gesellschaft für Volkskunde e.V. (DGV), Friedländer Weg 2, 37085 Göttingen
Founded: 1904; Members: 1000
Pres: Prof. Dr. Rolf Wilhelm Brednich
Focus: Ethnology
Periodicals
dgv informationen (weekly)
Zeitschrift für Volkskunde (weekly) . . . 05439

Deutsche Gesellschaft für Wehrtechnik e.V. (DWT), Deutschherrenstr 157, 53179 Bonn
T: (0228) 330007
Founded: 1957; Members: 853
Pres: Dr. Ernst Grosch; Gen Secr: Hans-Heinz Feldhoff
Focus: Military Sci
Periodical
Wehrtechnik (weekly) 05440

Deutsche Gesellschaft für Wirtschaftliche Fertigung und Sicherheitstechnik e.V. (DGW), Gilbachweg 18, 41564 Kaarst
T: (02101) 68212
Focus: Safety; Mach Eng 05441

Deutsche Gesellschaft für Wohnmedizin und Bauhygiene e.V., Nelkenweg 5, 76291 Spöck-Stutensee
T: (07249) 6932
Pres: Prof. Dr. J. Beckert; Gen Secr: Prof. Dr. H.-G. Sonntag
Focus: Hygiene
Periodical
Wohnmedizin 05442

Deutsche Gesellschaft für Zahnärztliche Prothetik und Werkstoffkunde e.V. (DGZPW), Konstanty-Gutschow-Str 8, 30625 Hannover
T: (0511) 5324773; Fax: (0511) 5324790
Pres: Prof. Dr. A. Roßbach; Gen Secr: Dr. H. Tschernitschek
Focus: Dent 05443

Deutsche Gesellschaft für Zahnerhaltung, Prinzregentenstr 1, 10717 Berlin
Focus: Dent 05444

Deutsche Gesellschaft für Zahn-, Mund- und Kieferheilkunde, Lindemannstr 96, 40237 Düsseldorf
T: (0211) 682296
Founded: 1859; Members: 9500
Focus: Dent
Periodical
Deutsche Zahnärztliche Zeitschrift: DZZ (weekly) 05445

Deutsche Gesellschaft für Zerstörungsfreie Prüfung e.V. (DGZfP), Unter den Eichen 87, 12205 Berlin
T: (030) 8114001; Tx: 181788; Fax: (030) 8114003
Founded: 1933; Members: 1100
Pres: Prof. Dr. Dierk Schnitger; Gen Secr: Wolfgang Bock
Focus: Eng
Periodical
Materialprüfung 05446

Deutsche Gesellschaft für Züchtungskunde e.V. (DGfZ), Adenauerallee 174, 53113 Bonn
T: (0228) 213411; Fax: (0228) 223497
Founded: 1905; Members: 1000
Gen Secr: Dr. Klaus Wemken
Focus: Zoology
Periodical
Züchtungskunde (weekly) 05447

Deutsche Gesellschaft zur Förderung der Gehörlosen und Schwerhörigen e.V., Veit-Stoß-Str 14, 80687 München
T: (089) 588848
Founded: 1962; Members: 12
Pres: Peter Donath
Focus: Rehabil; Educ Handic
Periodical
hörgeschädigte kinder (weekly) 05448

Deutsche Glastechnische Gesellschaft e.V. (DGG), Mendelssohnstr 75-77, 60325 Frankfurt
T: (069) 749088
Founded: 1922; Members: 1400
Pres: Manfred Werner; Gen Secr: Prof. Dr. Helmut A. Schaeffer
Focus: Eng
Periodical
Glastechnische Berichte (weekly) 05449

Deutsche Graphologische Vereinigung e.V. (DGV) / Berufsverband Deutscher Graphologen, Postf 4031, 58239 Schwerte
T: (02304) 73024
Focus: Graphology 05450

Deutsche Gruppenpsychotherapeutische Gesellschaft e.V. (DGG), Kantstr 120-121, 10625 Berlin
T: (030) 3132698; Fax: (030) 3136959
Pres: Dr. Dorothee Doldinger; Gen Secr: Dr. Alfred Doldinger
Focus: Therapeutics
Periodical
Dynamische Psychiatrie: Internationale Zeitschrift für Psychiatrie und Psychoanalyse (weekly) 05451

Deutsche Hämophilieberatung / Verein zur Beratung bei Blutungskrankheiten e.V., Lessingstr 61, 45772 Marl-Hüls
T: (02365) 21503
Pres: Ingeborg Köster
Focus: Hematology 05452

Deutsche Hämophiliegesellschaft zur Bekämpfung von Blutungskrankheiten e.V., Halenseering 3, 22149 Hamburg
T: (040) 6722970; Fax: (040) 6724944
Founded: 1956
Focus: Hematology
Periodical
Hämophilie-Blätter (weekly) 05453

Deutsche Hauptstelle gegen die Suchtgefahren e.V. (DHS), Postf 1369, 59003 Hamm
T: (02381) 90150; Fax: (02381) 15331
Pres: Knut Lehmann; Gen Secr: Rolf Hillinghorst
Focus: Public Health; Educ; Sociology; Psych; Med
Periodical
Sucht: Zeitschrift für Wissenschaft und Praxis (6 times annually) 05454

Deutsche Ileostomie-Kolostomie-Urostomie-Vereinigung e.V., Kepserstr 50, Postf 1265, 85312 Freising
T: (08161) 84911; Fax: (08161) 85521
Focus: Rehabil
Periodical
ILCO-Praxis: Organ der Deutschen Ileostomie-Colostomie-Urostomie-Vereinigung (weekly) 05455

Deutsche Jazz-Föderation e.V., Kleine Bockenheimer Str 12, 60313 Frankfurt
Focus: Music 05456

Deutsche Kakteen-Gesellschaft e.V., Nordstr 30, 26939 Ovelgönne
T: (04480) 1408; Fax: (04480) 1564
Founded: 1892; Members: 10000
Pres: Prof. Dr. Wilhelm Barthlott; Gen Secr: Horst Berk
Focus: Hort
Periodical
Kakteen und andere Sukkulenten (weekly) 05457

Deutsche Kautschuk-Gesellschaft e.V. (DKG), Zeppelinallee 69, 60487 Frankfurt
T: (069) 7936153; Fax: (069) 7936155
Founded: 1926; Members: 1550
Pres: Dr. Burkhard Meister; Gen Secr: Fritz Katzensteiner
Focus: Materials Sci 05458

Deutsche Keramische Gesellschaft e.V. (DKG), Frankfurter Str 196, 51147 Köln
T: (02203) 69069; Fax: (02203) 69301
Founded: 1919; Members: 1400
Pres: Günther Schmidt; Gen Secr: Dr. Markus Blumenberg
Focus: Materials Sci
Periodicals
Ceramic Forum International: Berichte (10 times annually)
Fachausschussberichte (weekly)
Fortschrittsberichte (weekly) 05459

Deutsche Kommission für Ingenieurausbildung, Graf-Recke-Str 84, 40239 Düsseldorf
T: (0211) 6214277; Fax: (0211) 6214575
Pres: Prof. Dr. H. Weinerth; Gen Secr: Dr. Franz-Josef Schlösser
Focus: Adult Educ; Eng 05460

Deutsche Krebsgesellschaft e.V., Paul-Ehrlich-Str 41, 60596 Frankfurt
T: (069) 6300960; Fax: (069) 639130
Founded: 1970; Members: 2000
Pres: Prof. Dr. H. Maas; Gen Secr: Prof. Dr. P. Drings
Focus: Cell Biol & Cancer Res
Periodicals
Journal of Cancer Research and Clinical Oncology (weekly)
Mitteilungen (weekly) 05461

Deutsche Landjugend-Akademie Fredeburg (DLA), Johannes-Hummel-Weg 1, 57392 Schmallenberg
T: (02972) 7007
Founded: 1948; Members: 6
Focus: Agri
Periodical
Fredeburger Hefte (weekly) 05462

Deutsche Landwirtschafts-Gesellschaft e.V. (DLG), Eschborner Landstr 122, 60489 Frankfurt
T: (069) 247880; Tx: 413185; Fax: (069) 7241554
Founded: 1885; Members: 12000
Pres: Günter Flessner; Gen Secr: Dr. Dietrich Rieger
Focus: Agri
Periodical
DLG-Mitteilungen (weekly) 05463

Deutsche Lichttechnische Gesellschaft e.V., Burggrafenstr 6, 10787 Berlin
T: (030) 26012439; Fax: (030) 26011723
Founded: 1912; Members: 600
Pres: H. Richter; Gen Secr: Dr. Michael Seidl
Focus: Electric Eng 05464

Deutsche Malakozoologische Gesellschaft (DMG), c/o Forschungsinstitut Senckenberg, Senckenberganlage 25, 60325 Frankfurt
T: (069) 75421; Fax: (069) 746238
Founded: 1868; Members: 270
Pres: Dr. Vollrath Wiese; Gen Secr: Dr. Ronald Janssen
Focus: Zoology
Periodical
Mitteilungen der Deutschen Malakozoologischen Gesellschaft (weekly) 05465

Deutsche Mathematiker-Vereinigung e.V. (DMV), Albertstr 24, 79104 Freiburg
T: (0761) 278020; Fax: (0761) 272698
Founded: 1890; Members: 2600
Pres: Prof. Dr. M. Grötschel; Gen Secr: Prof. Dr. D. Ferus
Focus: Math
Periodical
Jahresbericht (weekly) 05466

Deutsche Medizinische Arbeitsgemeinschaft für Herd- und Regulationsforschung e.V. (DAH), Scharnhorststr 21, 52351 Düren
T: (02421) 34842
Founded: 1950
Focus: Intern Med 05467

Deutsche MERU Gesellschaft, Am Berg 2, 49143 Bissendorf
T: (05402) 8833; Fax: (05402) 7149
Founded: 1973; Members: 500
Pres: Dr. Klaus Volkamer
Focus: Sci
Periodical
Mitteilungsblätter der Deutschen MERU-Gesellschaft (weekly) 05468

Deutsche Meteorologische Gesellschaft e.V. (DMG), Mont Royal, 56841 Traben-Trarbach
T: (06541) 18205
Founded: 1883; Members: 1733
Pres: Prof. Dr. K. Labitzke; Gen Secr: G. Zehnpfund
Focus: Geophys; Oceanography
Periodicals
Beiträge zur Physik der Atmosphäre (weekly)
Meteorologische Zeitschrift (weekly) . . . 05469

Deutsche Meteorologische Gesellschaft e.V., Zweigverein Hamburg (DMG), Bernhard-Nocht-Str 76, 20359 Hamburg
T: (040) 31908824; Fax: (040) 31908803
Founded: 1883; Members: 438
Pres: Hans-Joachim Heinemann; Gen Secr: Reinhard Zöllner
Focus: Oceanography; Geophys 05470

Deutsche Mineralogische Gesellschaft e.V. (DMG), c/o Institut für Mineralogie, WWU, 48149 Münster
Fax: (0251) 838397
Founded: 1908; Members: 1800
Gen Secr: Prof. Dr. H. Kroll
Focus: Mineralogy
Periodical
European Journal of Mineralogy (weekly) 05471

Deutsche Montessori Gesellschaft, Postf 5461, 97004 Würzburg
Focus: Educ
Periodical
Das Kind: Halbjahresschrift für Montessori-Pädagogik (weekly) 05472

Deutsche Morgenländische Gesellschaft e.V., c/o Südasien-Institut, Im Neuenheimer Feld 330, 69120 Heidelberg
Founded: 1845; Members: 698
Gen Secr: Manfred Hake
Focus: Ling; Lit; Hist; Ethnology
Periodicals
Journal of the Nepal Research Centre (weekly)
ZDMG (weekly) 05473

Deutsche Mozart-Gesellschaft e.V. (DMG), Karlstr 6, 86150 Augsburg
T: (0821) 518588; Fax: (0821) 157228
Founded: 1951; Members: 3500
Pres: Dr. Friedhelm Brusniak; Gen Secr: Brigitte Löder
Focus: Music
Periodical
Acta Mozartiana (3 times annually) . . . 05474

Deutsche Multiple Sklerose Gesellschaft Bundesverband e.V. (DMSG), Vahrenswalderstr 205-207, 30165 Hannover
T: (0511) 633023; Fax: (0511) 633887
Founded: 1952; Members: 33000
Pres: Prof. Dr. Dr. Hermann Hoffmann; Gen Secr: Dorothee Pitschnau
Focus: Med; Pathology
Periodical
Aktiv (weekly) 05475

Deutsche Neurovegetative Gesellschaft, Rosspfad 19, 40489 Düsseldorf
T: (0211) 402236
Founded: 1960
Focus: Neurology 05476

Deutsche Numismatische Gesellschaft / Verband der Deutschen Münzvereine e.V., Hans-Purrmann-Allee 26, 67346 Speyer
T: (06232) 92458
Pres: Dr. Rainer Albert; Gen Secr: Eugen Zepp
Focus: Numismatics
Periodical
Numismatisches Nachrichtenblatt (weekly) 05477

Deutsche Ophthalmologische Gesellschaft Heidelberg (DOG), Im Neuenheimer Feld 400, 69120 Heidelberg
T: (06221) 411787; Fax: (06221) 484616
Founded: 1857; Members: 2800
Pres: Prof. Dr. Hans-Jürgen Thiel; Gen Secr: Prof. Dr. Hans Eberhard Völcker
Focus: Ophthal
Periodical
Der Ophthalmologe (weekly) 05478

Deutsche Orchideen-Gesellschaft, Arndtstr 8, 27367 Sottrum
T: (04264) 9017
Focus: Botany
Periodical
Die Orchidee (weekly) 05479

Deutsche Orient-Gesellschaft e.V. (DOG), c/o Altorientalisches Seminar der FU, Bitterstr 8-12, 14195 Berlin
T: (030) 8383347
Founded: 1898; Members: 700
Pres: Prof. Dr. Johannes Renger; Gen Secr: Dr. Volkmar Fritz
Focus: Archeol; Hist
Periodical
Mitteilungen der Deutschen Orient-Gesellschaft (weekly) 05480

Deutsche Ornithologen-Gesellschaft e.V., Möggingen, 78315 Radolfzell
Founded: 1850; Members: 2200
Pres: Prof. Dr. Peter Berthold
Focus: Ornithology
Periodical
Journal für Ornithologie (weekly) 05481

Deutsche Parlamentarische Gesellschaft e.V., Dahlmannstr 7, 53113 Bonn
T: (0228) 212654
Pres: Reinhard Freiherr von Schorlemer; Gen Secr: Ingrid von Hagen
Focus: Poli Sci 05482

Deutsche Paul-Tillich-Gesellschaft e.V., David-Hilbert-Str 15, 37085 Göttingen
T: (0551) 46480
Founded: 1960; Members: 239
Focus: Rel & Theol 05483

Deutsche Pharmakologische Gesellschaft e.V., Postf 101709, 42017 Wuppertal
T: (0202) 368327
Members: 1540
Focus: Pharmacol
Periodical
Naunyn-Schmiedeberg's Archives of Pharmacology 05484

Deutsche Pharmazeutische Gesellschaft e.V. (DPhG), Ginnheimer Str 26, 65760 Eschborn
T: (06196) 928274
Founded: 1890; Members: 4000
Pres: Prof. Dr. Hans-Hartwig Otto; Gen Secr: Dr. Hartmut Morck
Focus: Pharmacol
Periodicals
Archiv der Pharmazie (weekly)
Pharmazie in unserer Zeit (weekly) . . . 05485

Deutsche Phono-Akademie e.V., Grelckstr 36, 22529 Hamburg
T: (040) 581935; Fax: (040) 582842
Pres: Gerd Gebhardt; Gen Secr: Prof. Werner Hay
Focus: Music 05486

Deutsche Physikalische Gesellschaft e.V. (DPG), Hauptstr 5, 53604 Bad Honnef
T: (02224) 92320; Fax: (02224) 923250
Founded: 1845; Members: 25000
Pres: Prof. Dr. H. Schopper; Gen Secr: Dr. W. Heinicke
Focus: Physics
Periodicals
Physikalische Berichte (weekly)
Physikalische Blätter (weekly)
Verhandlungen der DPG (weekly) . . . 05487

Deutsche Physiologische Gesellschaft e.V., Im Neuenheimer Feld 326, 69120 Heidelberg
T: (06221) 564033; Fax: (06221) 564561
Founded: 1904; Members: 820
Pres: Prof. Dr. Ulrich Zwiener; Gen Secr: Prof. Dr. Wolfgang Kuschinsky
Focus: Physiology 05488

Deutsche Phytomedizinische Gesellschaft e.V. (DPG), Essenheimerstr 144, 55128 Mainz
T: (06131) 993080
Founded: 1949; Members: 1722
Pres: Dr. P. Kraus; Gen Secr: Dr. R. Gessner
Focus: Crop Husb 05489

Deutsche Planungsgesellschaft EG Bonn (dp), Rheinweg 31, 53113 Bonn
T: (0228) 233604
Focus: Econ 05490

Deutsche Psychoanalytische Gesellschaft e.V. (DPG), Nussbaumstr 7, 80336 München
Founded: 1910; Members: 300
Pres: Prof. Dr. Michael Ermann; Gen Secr: Dr. Paul Bernhard
Focus: Psychoan
Periodicals
Praxis der Kinderpsychologie und Kinderpsychiatrie
Zeitschrift für Psychosomatische Medizin und Psychoanalyse 05491

Deutsche Psychoanalytische Vereinigung e.V. (DPV), Sulzaer Str 3, 14199 Berlin
T: (030) 8264547; Fax: (030) 8266090
Founded: 1950; Members: 712
Pres: Dr. Carl Nedelmann; Gen Secr: Dr. Ludwig Haesler
Focus: Psych 05492

Deutsche Quartärvereinigung (Deuqua), Postf 510153, 30631 Hannover
T: (0511) 6432487; Tx: 0923730; Fax: (0511) 6432304
Founded: 1950; Members: 650
Pres: Prof. Dr. H. Hagedorn
Focus: Geology
Periodical
Eiszeitalter und Gegenwart (weekly) . . 05493

Deutscher Ärztinnenbund e.V., Herbert-Lewin-Str 5, 50931 Köln
T: (0221) 4004282; Fax: (0221) 4004287
Founded: 1946; Members: 2000
Pres: Dr. Ingeborg Retzlaff
Focus: Med
Periodical
Ärztin: Mitteilungsblatt des Deutschen Ärztinnenbundes (weekly) 05494

Deutscher Akademikerinnenbund e.V. (DAB), Weiltinger Str 8, 90449 Nürnberg
T: (0911) 673128; Fax: (0911) 6880327
Members: 2000
Pres: Luise Joppe
Focus: Sci
Periodical
Mitteilungsblatt des Deutschen Akademikerinnenbundes e.V. (weekly) . . 05495

Deutscher Akademischer Austauschdienst (DAAD), Kennedyallee 50, 53175 Bonn
T: (0228) 8820; Tx: 885515
Pres: Prof. Dr. Theodor Berchem; Gen Secr: Dr. Christian Bade
Focus: Sci
Periodical
Hochschule und Ausland (weekly) . . . 05496

Deutscher Altphilologen-Verband, Mitterlängstr 13, 82178 Puchheim
T: (089) 803814
Members: 6000
Pres: Prof. Dr. Friedrich Maier
Focus: Ling; Lit
Periodical
Mitteilungsblatt des Deutschen Altphilologenverbandes (weekly) 05497

Deutscher Arbeitsgerichtsverband e.V., Blumenthalstr 33, 50670 Köln
T: (0221) 7740347
Pres: Prof. Dr. Peter Hanau; Gen Secr: Dr. Heinz-Jürgen Kalb
Focus: Law
Periodical
Mitteilungen (weekly) 05498

Deutscher Arbeitsring für Lärmbekämpfung e.V. (DAL), Frankenstr 25, 40476 Düsseldorf
T: (0211) 488499
Focus: Public Health
Periodical
Zeitschrift für Lärmbekämpfung (weekly) . 05499

Deutscher Ausschuss für Stahlbau (DASt), Ebertpl 1, 50668 Köln
T: (0221) 77310
Focus: Civil Eng 05500

Deutscher Autoren-Verband e.V. (DAV), Künstlerhaus, Sophienstr 2, 30159 Hannover
T: (0511) 322068
Founded: 1946; Members: 150
Focus: Lit 05501

Deutscher Bäderverband e.V., Schumannstr 111, 53113 Bonn
T: (0228) 262040
Founded: 1892; Members: 6
Pres: Dr. Christoph Kirschner; Gen Secr: Antonius Weber
Focus: Public Health
Periodical
Heilbad und Kurort: Zeitschrift für das gesamte Bäderwesen (weekly) 05502

Deutscher Berufsverband der Hals-Nasen-Ohrenärzte e.V., Mühlenhof 2-4, 24534 Neumünster
T: (04321) 45035; Fax: (04321) 44348
Founded: 1951; Members: 3500
Pres: Prof. Dr. Klaus Seifert; Dr. Klaus Otto; Gen Secr: Renate Burghard
Focus: Otorhinolaryngology
Periodical
HNO-Mitteilungen (weekly) 05503

Deutscher Berufsverband der Sozialarbeiter und Sozialpädagogen e.V. (DBS), Schützenbahn 17, 45127 Essen
T: (0201) 239666; Fax: (0201) 200259
Founded: 1916; Members: 3500
Pres: G.-Rolf. Hille
Focus: Educ; Sociology; Adult Educ
Periodicals
DBS Pressedienst (weekly)
Der Sozialarbeiter (weekly) 05504

Deutscher Beton-Verein e.V. (DBV), Bahnhofstr 61, 65185 Wiesbaden
T: (0611) 14030
Founded: 1898; Members: 640
Pres: Dr. Hans Luber; Gen Secr: Dr. Manfred Stiller
Focus: Materials Sci; Graphic & Dec Arts, Design; Standards
Periodical
Vorträge Betontag 05505

Deutscher Bibliotheksverband e.V. (DBV), Bundesallee 184-185, 10717 Berlin
T: (030) 8505274; Fax: (030) 8542240
Founded: 1949; Members: 2300
Pres: Jochen Dieckmann; Gen Secr: Prof. Günter Beyersdorff
Focus: Libraries & Bk Sci
Periodical
DBV-INFO (weekly) 05506

Deutscher Dampfkesselausschuss (DDA), Kurfürstenstr 56, 45138 Essen
T: (0201) 284881
Focus: Eng 05507

Deutscher Diabetiker-Verband e.V., Hahnbrunner Str 46, 67659 Kaiserslautern
T: (0631) 76488
Pres: Dr. Heinz Bürger-Büsing
Focus: Diabetes
Periodical
Diabetes aktuell / Hallo Du auch (weekly) 05508

Deutsche Religionsgeschichtliche Studiengesellschaft, Denkmalstr 5, 66119 Saarbrücken
Founded: 1970
Pres: Prof. Dr. A. Rupp
Focus: Rel & Theol 05509

Deutscher Erfinderring e.V., Schlegelstr 17, 90491 Nürnberg
Focus: Eng
Periodical
Erfinder und Neuheitendienst (weekly) . . 05510

Deutscher Forstverein e.V. (DFV), Stresemannallee 61, 60596 Frankfurt
T: (069) 638674; Fax: (069) 6312981
Pres: Dr. Wolfgang Dertz; Gen Secr: Christian Freiherr von Bethmann
Focus: Forestry
Periodical
Tagungsbericht (every second year) . . 05511

Deutscher Forstwirtschaftsrat e.V. (DFWR), Münstereifeler Str 19, 53349 Rheinbach
T: (02226) 2350; Fax: (02226) 5792
Founded: 1950; Members: 64
Pres: Erich Naujack; Gen Secr: Horst Womelsdorf
Focus: Forestry
Periodical
DFWR-Dreijahresbericht (every 3 years) . 05512

Deutscher Germanistenverband, c/o Germanistisches Institut, Rheinisch-Westfälische Technische Hochschule, Eilfschornsteinstr 15, 52056 Aachen
T: (0241) 806076
Pres: Prof. Dr. Ludwig Jäger
Focus: Ling; Lit
Periodical
Mitteilungen des Deutschen Germanistenverbandes (weekly) 05513

Deutsche Rheologische Gesellschaft e.V. (DRG), Unter den Eichen 87, 12200 Berlin
T: (030) 81041529
Founded: 1951; Members: 280
Pres: Prof. Dr. M.H. Wagner; Gen Secr: Prof. Dr. K.-H. Habig
Focus: Materials Sci
Periodical
Rheologica Acta 05514

Deutsche Rheuma-Liga Bundesverband e.V., Rheinallee 69, 53173 Bonn
T: (0228) 355425
Pres: Gudrun Schaich-Walch; Gen Secr: Silvia Wollersheim
Focus: Rheuma
Periodical
Mobil: Das Rheuma-Magazin (weekly) . . 05515

Deutscher Hochschulverband, Rheinallee 18, 53173 Bonn
T: (0228) 364002; Fax: (0228) 353403
Founded: 1950; Members: 15000
Pres: Prof.Dr. Hartmut Schiedermair; Gen Secr: Dr. Michael Hartmer
Focus: Educ
Periodical
Mitteilungen des Hochschulverbandes (weekly) 05516

Deutscher Holzwirtschaftsrat, Mainzer Str 64, 65185 Wiesbaden
Focus: Forestry 05517

Deutsche Richterakademie, Berliner Allee 7, 54295 Trier
T: (0651) 33021; Fax: (0651) 300210
Pres: Werner Jastroch
Focus: Law
Periodical
Justiz und Recht 05518

Deutscher Juristen-Fakultätentag, Domerschulstr 16, 97070 Würzburg
T: (0931) 31899; Fax: (0931) 57047
Members: 40
Pres: Prof. Dr. Franz-Ludwig Knemeyer
Focus: Law 05519

Deutscher Juristentag e.V., Oxfordstr 21, 53111 Bonn
T: (0228) 983910
Founded: 1860; Members: 8200
Gen Secr: Felix Busse
Focus: Law 05520

Deutscher Kälte- und Klimatechnischer Verein e.V. (DKV), Pfaffenwaldring 10, 70569 Stuttgart
T: (0711) 6853200; Fax: (0711) 6853242
Founded: 1909; Members: 1200
Pres: Eckart Brandner; Gen Secr: Irene Reichert
Focus: Air Cond 05521

Deutscher Kassenarztverband e.V., Darmstädter Str 29, 64521 Gross-Gerau
T: (06152) 54648; Fax: (06152) 52131
Founded: 1969; Members: 10300
Pres: Dr. Helmut Walther
Focus: Med 05522

Deutscher Kommunikationsverband BDW e.V., Königswinterer Str 552, 53227 Bonn
T: (0228) 444560; Fax: (0228) 444503
Pres: Werner D. Ludwig
Focus: Advert
Periodical
Insight Kommunikation (weekly) 05523

Deutscher Komponisten-Verband e.V., Kadettenweg 80b, 12205 Berlin
T: (030) 8334121; Fax: (030) 8330713
Founded: 1954; Members: 1250
Pres: Karl Heinz Wahren; Gen Secr: Marianne Augustin
Focus: Music
Periodical
Informationen (2-3 times annually) . . . 05524

Deutscher Künstlerbund e.V., Zeughofstr 20, 10997 Berlin
T: (030) 6189191; Fax: (030) 6117082
Pres: Prof. Verena Vernunft; Gen Secr: Ursula Binder
Focus: Arts 05525

Deutscher Lehrerverband Niedersachsen, Grosse Packhofstr 28, 30159 Hannover
Focus: Educ 05526

Deutscher Markscheider-Verein e.V. (DMV), Postf 101809, 44621 Herne
T: (02323) 154311; Fax: (02323) 152380
Founded: 1879; Members: 500
Pres: Manfred Böhmer; Gen Secr: Hartmut Heinke
Focus: Mining
Periodical
Das Markscheidewesen (weekly) 05527

Deutscher Medizinischer Informationsdienst e.V. (DMI), Siesmayerstr 15, 60323 Frankfurt
T: (069) 745882; Tx: 0414950
Focus: Public Health; Educ 05528

Deutscher Museumsbund e.V. (DMB), Erbprinzenstr 13, 76133 Karlsruhe
T: (0721) 175161
Founded: 1917; Members: 1700
Pres: Prof. Dr. Siegfried Rietschel
Focus: Archives; Arts
Periodical
Museumskunde (3 times annually) . . . 05529

Deutscher Musikrat e.V., Nationalkomitee der Bundesrepublik Deutschland im Internationalen Musikrat (DMR), Am Michaelshof 4a, 53177 Bonn
T: (0228) 83080; Tx: 886868; Fax: (0228) 352650
Founded: 1953; Members: 86
Pres: Prof. Dr. Franz Müller-Heuser; Gen Secr: Prof. Dr. Andreas Eckhardt
Focus: Music
Periodical
Musikforum: Referate und Informationen des Deutschen Musikrates (weekly) 05530

Deutscher Naturheilbund e.V., Am Wiesenbach 42, 74564 Crailsheim
T: (07951) 42432
Founded: 1889; Members: 15000
Pres: Dr. Johann Abele; Gen Secr: Alfred Adis
Focus: Med
Periodical
Der Naturarzt (weekly) 05531

Deutscher Naturkundeverein e.V., Zavelsteinstr 38b, 70469 Stuttgart
T: (0711) 855896
Focus: Nat Sci 05532

Deutscher Naturschutzring e.V., Bundesverband für Umweltschutz (DNR), Am Michaelshof 8-10, 53177 Bonn
T: (0228) 359005; Fax: (0228) 359096
Pres: Prof. Dr. Wolfgang Engelhardt; Helmut Röscheisen
Focus: Ecology
Periodical
DNR-Kurier (weekly) 05533

Deutscher Nautischer Verein von 1868 e.V., Elbchaussee 277, 22605 Hamburg
T: (040) 823670
Founded: 1868; Members: 4270
Pres: Gerd Trulsen; Gen Secr: Garrit Leemrijze
Focus: Navig 05534

Deutsche Röntgengesellschaft (DRG), Frankfurter Str 231, 63263 Neu-Isenburg
T: (06102) 4032
Founded: 1905; Members: 1950
Pres: Prof. Dr. Horst Sack
Focus: Bio; Radiology; X-Ray Tech; Nucl Med 05535

Deutscher Philologen-Verband e.V. (DPhV), Bahnhofsweg 8, 82008 Unterhaching
T: (089) 6251619; Fax: (089) 6251818
Founded: 1904
Pres: Heinz Durner; Gen Secr: Gabriele Lipp
Focus: Ling; Lit
Periodical
Die Höhere Schule (weekly) 05536

Deutscher Politologen-Verband e.V. (dp), Peter-Schwingen-Str 11, 53177 Bonn
T: (0228) 321000
Founded: 1965; Members: 600
Pres: Heinz J.H. Fleischhauer
Focus: Poli Sci 05537

Deutscher Rat für Landespflege (DRL), Konstantinstr 110, 53179 Bonn
T: (0228) 331097; CA: (0228) 334727
Founded: 1962; Members: 22
Pres: Prof. Dr. Dr. Wolfgang Haber; Gen Secr: Prof. Dr. Klaus Borchard
Focus: Ecology
Periodical
Schriftenreihe des DRL (weekly) 05538

Deutscher Sportärztebund (DSÄB) / Deutsche Gesellschaft für Sportmedizin e.V., Handschuhsheimer Landstr 55, 69121 Heidelberg
T: (06221) 470880, 480919
Founded: 1950; Members: 5500
Focus: Med; Physical Therapy
Periodicals
Abdruck von durchzuführenden Weiterbildungs- und Fortbildungskursen (weekly)
Deutsche Zeitschrift für Sportmedizin (weekly) 05539

Deutscher Sportlehrerverband e.V. (DSLV), Am Rasselberg 16, 35578 Wetzlar
T: (06441) 28444; Fax: (06441) 26697
Founded: 1949; Members: 12000
Pres: Hansjörg Kofink; Gen Secr: Irmgard Rau
Focus: Sports 05540

Deutscher Stahlbau-Verband (DSTV), Ebertpl 1, 50668 Köln
T: (0221) 77310
Members: 163
Focus: Civil Eng
Periodical
Stahlbau-Nachrichten (weekly) 05541

Deutscher Stenografielehrerverband e.V. / Verband der Lehrer für Bürowirtschaft, Kurzschrift und Maschinenschreiben, Carsten-Meyn-Weg 6, 22399 Hamburg
T: (040) 6028135
Focus: Adult Educ 05542

Deutscher Tonkünstlerverband e.V. (DTKV), Linprunstr 16, 80335 München
T: (089) 5234054
Founded: 1844
Pres: Prof. Dr. Inka Stumpfl
Focus: Music; Educ
Periodicals
Musik und Bildung (weekly)
Neue Musikzeitung (weekly) 05543

Deutscher Verband Evangelischer Büchereien e.V. / Zentralstelle der Büchereiarbeit in der Evangelischen Kirche in Deutschland, Bürgerstr 2, 37073 Göttingen
T: (0551) 74917
Pres: Horstdieter Wildner; Gen Secr: Gabriele Kassenbrock
Focus: Libraries & Bk Sci
Periodicals
Buchauswahl (weekly)
Der Evangelische Buchberater (weekly) . 05544

Deutscher Verband Farbe (DVF), Unter den Eichen 87, 12203 Berlin
T: (030) 81045400
Founded: 1976; Members: 500
Focus: Materials Sci 05545

Deutscher Verband Forstlicher Forschungsanstalten (DVFFA), Schloss, 67705 Trippstadt
T: (06306) 8311; Fax: (06306) 2821
Founded: 1872; Members: 72
Pres: Prof. Dr. Axel Roeder
Focus: Forestry 05546

Deutscher Verband für Angewandte Geographie e.V. (DVAG), Elbchaussee 24, 22765 Hamburg
T: (040) 3905059
Pres: Margret Mergen; Gen Secr: Kornelia Gretsch
Focus: Geography
Periodical
Standort (3 times annually) 05547

Deutscher Verband für das Skilehrwesen e.V. (DVS), Postf 1449, 87554 Oberstdorf
T: (08322) 80126; Fax: (08322) 80226
Founded: 1951
Pres: Dr. Harald Kiedaisch
Focus: Sports 05548

Deutscher Verband für Materialforschung und -prüfung e.V. (DVM), Unter den Eichen 87, 12205 Berlin
T: (030) 8113066; Fax: (030) 8119359
Founded: 1896; Members: 350
Pres: Dr. Manfred Wilhelm; Gen Secr: Ingrid Maslinski
Focus: Materials Sci
Periodical
Materialprüfung – Materials Testing – Matériaux Essais et Recherches (weekly) 05549

Deutscher Verband für Physiotherapie / Zentralverband der Krankengymnasten/Physiotherapeuten e.V., Postf 210280, 50528 Köln
T: (0221) 884031
Pres: Eckhardt Böhle; Gen Secr: Heinz Christian Esser
Focus: Therapeutics 05550

Deutscher Verband für Schweisstechnik e.V. (DVS), Postf 101965, 40010 Düsseldorf
T: (0211) 15910; Fax: (0211) 1575950
Founded: 1947; Members: 20000
Pres: Heinz F. Landré; Gen Secr: Dr. Detlef von Hofe
Focus: Eng
Periodicals
Der Praktiker (weekly)
Schweissen und Schneiden (weekly)
Verbindungstechnik in der Elektronik (weekly) 05551

Deutscher Verband für Wasserwirtschaft und Kulturbau e.V. (DVWK) / German Association for Water Resources and Land Improvement, Gluckstr 2, 53115 Bonn
T: (0228) 983870; Fax: (0228) 9838733
Members: 2400
Pres: Karl Hans Heil; Gen Secr: Dr. Wolfram Dirksen
Focus: Hydrology; Water Res; Agri
Periodicals
Wasser und Boden (weekly)
Wasserwirtschaft (weekly) 05552

Deutscher Verband für Wohnungswesen, Städtebau und Raumplanung e.V., Suebenstr 1, 53175 Bonn
T: (0228) 376951; Fax: (0228) 376953
Founded: 1946; Members: 700
Pres: Karl Ravens; Gen Secr: Dr. Hans-Ludwig Oberbeckmann
Focus: Urban Plan
Periodical
Stadtbau-Informationen (weekly) 05553

Deutscher Verband Technischer Assistenten in der Medizin e.V., Spaldingstr 110b, 20097 Hamburg
T: (040) 231436
Pres: Heidi Schramm; Gen Secr: Almuth Never
Focus: Eng; Med
Periodical
mta praxis (weekly) 05554

Deutscher Verband technisch-wissenschaftlicher Vereine (DVT), Graf-Recke-Str 84, 40239 Düsseldorf
T: (0211) 6214499
Founded: 1916; Members: 103
Pres: Dr. H. Gassert; Gen Secr: Dr. Jörg Debelius
Focus: Eng 05555

Deutscher Verein des Gas- und Wasserfaches e.V. (DVGW), Hauptstr 71-79, 65727 Eschborn
T: (06196) 70170; Tx: 4072874; Fax: (06196) 481152
Founded: 1859; Members: 6010
Pres: Rolf Beyer; Gen Secr: Dr. W. Feind; Dr. W. Merkel
Focus: Water Res; Energy
Periodical
gwf – Das Gas- und Wasserfach: Ausgaben Gas/Erdgas und Wasser/Abwasser (weekly) 05556

Deutscher Verein für Internationales Seerecht e.V. (DVIS), Esplanade 6, 20354 Hamburg
T: (040) 350970
Founded: 1898; Members: 330
Pres: Dr. Hans-Christian Albrecht
Focus: Law 05557

Deutscher Verein für Kunstwissenschaft e.V., Jebensstr 1, 10623 Berlin
T: (030) 3139932
Founded: 1908; Members: 1150
Pres: Prof. Dr. Henning Bock; Gen Secr: Dr. Brigitte Hüfler
Focus: Arts
Periodical
Zeitschrift des Deutschen Vereins für Kunstwissenschaft (3 times annually) . . 05558

Deutscher Verein für Vermessungswesen e.V. (DVW), Akademiestr 3, 69117 Heidelberg
Founded: 1871; Members: 3300
Pres: Dr. H. Röhrs
Focus: Surveying
Periodical
Zeitschrift für Vermessungswesen (weekly) 05559

Deutscher Verein für Versicherungswissenschaft e.V. (DVfVW), Johannisberger Str 31, 14187 Berlin
T: (030) 8212031; Fax: (030) 8222875
Founded: 1899; Members: 1670
Pres: Prof. Dr. Dr. Robert Schwebler; Gen Secr: Dr. Ulrich Schlie
Focus: Insurance
Periodical
Zeitschrift für die gesamte Versicherungswissenschaft (weekly) . . . 05560

Deutscher Verkehrssicherheitsrat e.V. (DVR), Herbert-Rabius-Str 24, 53222 Bonn
T: (0228) 400010; Fax: (0228) 4000167
Members: 260
Pres: Dr. Gerhard Schork; Gen Secr: Herbert Warnke
Focus: Transport; Safety
Periodical
DVR-report (weekly) 05561

Deutscher Volkshochschul-Verband e.V. (DVV), Obere Wilhelmstr 32, 53225 Bonn
T: (0228) 975690; Tx: 228321; CA: dvvbon; Fax: (0228) 8209550
Founded: 1953; Members: 16
Focus: Adult Educ
Periodicals
Adult Education and Development (weekly)
Agenda (weekly)
Volkshochschule (weekly) 05562

Deutscher Werkbund e.V., Weißadlergasse 4, 60311 Frankfurt
T: (069) 290658; Fax: (069) 2979991
Pres: Prof. Dr. Hermann Glaser; Gen Secr: Dr. Regine Halter
Focus: Educ
Periodical
Werk und Zeit (weekly) 05563

Deutscher Wissenschaftler Verband, Höhlenweg 31, 38642 Goslar
Focus: Humanities
Periodical
DWV-Mitteilungen (weekly) 05564

Deutscher Zentralausschuss für Chemie (DZfCh), Varrentrappstr 40-42, 60486 Frankfurt
T: (069) 7917323; Fax: (069) 7917322
Pres: Prof. Dr. Dr. Dr. H.A. Staab; Gen Secr: Prof. Dr. Heindirk tom Diek
Focus: Chem 05565

Deutscher Zentralverein Homöopathischer Ärzte e.V., Münsterstr 10, 53111 Bonn
Founded: 1829; Members: 2500
Pres: Dr. Anton Drähne; Gen Secr: Gernot Baur
Focus: Homeopathy
Periodicals
AHZ: Allgemeine homöopathische Zeitung (weekly)
KHZ: Klassische hom. Zeitung (weekly) . 05566

Deutsches Anwaltsinstitut e.V., Brüderstr 2, 44787 Bochum
T: (0234) 12566
Focus: Law 05567

Deutsches Archäologisches Institut, Podbielskiallee 69-71, 14195 Berlin
T: (030) 830080
Founded: 1829
Pres: Prof. Dr. Helmut Kyrieleis
Focus: Archeol
Periodicals
Archäologische Berichte aus dem Yemen
Archäologische Berichte aus Iran
Archäologische Bibliographie
Archäologischer Anzeiger
Athenische Mitteilungen
Baghdader Mitteilungen
Beiträge zur Allgemeinen und Vergleichenden Archäologie
Berichte der Römisch-Germanischen-Kommission
Chiron
Damaszener Mitteilungen
Germania
Istanbuler Mitteilungen
Jahrbuch
Madrider Mitteilungen
Mitteilungen des Deutschen Archäologischen Instituts, Abteilung Kairo
Römische Mitteilungen 05568

Deutsches Atomforum e.V. (DAtF), Heussallee 10, 53113 Bonn
T: (0228) 5070
Founded: 1959
Pres: Dr. Claus Berke
Focus: Nucl Res
Periodical
Atom-Informationen (weekly) 05569

Deutsches Bibliotheksinstitut (dbi) / German Library Institute, Bundesallee 184-185, 10717 Berlin
T: (030) 85050; Tx: 308512; Fax: (030) 8505100
Pres: Prof. Günter Beyersdorff
Focus: Libraries & Bk Sci
Periodicals
Bibliotheksdienst (weekly)
Bibliotheksinfo (weekly)
Forum Musikbibliothek (weekly)
Schulbibliothek Aktuell (weekly) 05570

Deutsche Schillergesellschaft e.V. (DSG), Schillerhöhe 8-10, 71672 Marbach
T: (07144) 8480
Founded: 1895; Members: 2700
Pres: Prof. Dr. Eberhard Lämmert; Gen Secr: Dr. Ulrich Ott
Focus: Lit
Periodicals
Jahrbuch der Deutschen Schiller-Gesellschaft (weekly)
Marbacher Arbeitskreis für Geschichte der Germanistik, Mitteilungen (2-3 times annually) 05571

Deutsche Sekretärinnen-Akademie, Ackerstr 90, 40233 Düsseldorf
T: (0211) 662766
Focus: Adult Educ 05572

Deutsche Sektion der Internationalen Liga gegen Epilepsie, Herforder Str 5-7, 33602 Bielefeld
T: (0521) 124192
Founded: 1957; Members: 1400
Pres: Prof. Dr. A. Stefan; Gen Secr: Dr. P. Zahner
Focus: Pathology
Periodical
Rundbrief (3 times annually) 05573

Deutsche Sektion des Internationalen Instituts für Verwaltungswissenschaften, Gaurheindorferstr 198, 53117 Bonn
Pres: Prof. Dr. Dr. Heinrich Siedentopf; Gen Secr: Dr. Dietmar Seiler
Focus: Public Admin
Periodicals
Schriften der Deutschen Sektion (weekly)
Verwaltungswissenschaftliche Informationen (3 times annually) 05574

Deutsches Elektronen-Synchrotron (DESY), Notkestr 85, 22603 Hamburg
T: (040) 89980; Tx: 215127; Fax: (040) 89983282
Founded: 1959
Focus: Physics
Periodicals
DESY Journal
Forschungsberichte
High Energy Physics Index (weekly)
Pressespiegel DESY 05575

Deutsches Forum für Entwicklungspolitik, Herwarthstr 16, 53115 Bonn
Focus: Poli Sci 05576

Deutsche Shakespeare-Gesellschaft, Markt 13, 99423 Weimar
T: (03643) 64076; Fax: (03643) 64076
Founded: 1864; Members: 2000
Focus: Lit
Periodical
Shakespeare-Jahrbuch (weekly) 05577

Deutsche Shakespeare-Gesellschaft West e.V., Rathaus, 44777 Bochum
T: (0234) 311842
Founded: 1963; Members: 1950
Pres: Prof. Dr. Ulrich Suerbaum; Gen Secr: Prof. Dr. Raimund Borgmeier
Focus: Lit
Periodical
Jahrbuch (weekly) 05578

Deutsches Handwerksinstitut e.V. (DHI), Max-Joseph-Str 4, 80333 München
T: (089) 593671, 594132; Fax: (089) 553453
Focus: Eng; Educ; Law; Econ 05579

Deutsches High-Fidelity Institut e.V. (DHFI), Karlstr 19-21, 60329 Frankfurt
T: (069) 2556409; Tx: 411372; Fax: (069) 236521
Focus: Acoustics 05580

Deutsches Institut für Ärztliche Mission, Paul-Lechler-Str 24, 72076 Tübingen
Fax: (07071) 27125
Focus: Med
Periodical
Nachrichten aus Ärztlicher Mission (weekly) 05581

Deutsches Institut für angewandte Kommunikation und Projektförderung e.V., Leibnizstr 69, 53177 Bonn
Focus: Comm Sci 05582

Deutsches Institut für Betriebswirtschaft e.V. (DIB), Börsenstr 4, 60313 Frankfurt
T: (069) 2197245
Founded: 1943
Gen Secr: Wolfgang Werner
Focus: Business Admin
Periodicals
Betriebliches Vorschlagswesen
Fachzeitschrift für die Praxis in Wirtschaft und Verwaltung (weekly) 05583

Deutsches Institut für Filmkunde e.V. (DIF), Schaumainkai 41, 60596 Frankfurt
T: (069) 617045; Fax: (069) 620060
Pres: Dr. Gerd Albrecht; Gen Secr: Peter Franz
Focus: Cinema
Periodical
Die Information (weekly) 05584

Deutsches Institut für Internationale Pädagogische Forschung (DIPF), Schloßstr 29, 60486 Frankfurt
T: (069) 247080; Fax: 24708444
Founded: 1952; Members: 100
Focus: Educ
Periodical
Mitteilungen und Nachrichten 05585

Deutsches Institut für medizinische Dokumentation und Information (DIMDI) / German Institute for Medical Documentation and Information, Weißhausstr 27, 50939 Köln
T: (0221) 47241; Tx: 8881364; Fax: (0221) 411429
Founded: 1969
Pres: Dr. Rolf Fritz
Focus: Doc; Med 05586

Deutsches Institut für Urbanistik / German Institute for Urban Affairs, Strasse des 17. Juni 112, 10623 Berlin
T: (030) 390010; Fax: (030) 39001100
Focus: Urban Plan
Periodicals
Archiv für Kommunalwissenschaften (weekly)
Difv-Materialien (5-10 times annually)
Graue Literatur zur Orts-, Regional- und Landesplanung (weekly)
Informationen zur modernen Stadtgeschichte (weekly)
Kommunalwissenschaftliche Dissertationen (weekly) 05587

Deutsches Institut für Vormundschaftswesen, Postfach 102020, 69010 Heidelberg
Fax: (06221) 981828
Focus: Family Plan
Periodical
Der Amtsvormund (weekly) 05588

Deutsches Institut für Wirtschaftsforschung (DIW), Königin-Luise-Str 5, 14195 Berlin
T: (030) 829910; Fax: (030) 82991200
Founded: 1925
Pres: Prof. Dr. Lutz Hoffmann
Focus: Econ
Periodicals
Vierteljahrshefte zur Wirtschaftsforschung (weekly)
Wochenbericht des DIW (weekly) 05589

Deutsches Jugendinstitut e.V. (DJI), Freibadstr 30, 81543 München
T: (089) 623060; Fax: (089) 62306162
Founded: 1961; Members: 140
Pres: Prof. Dr. Ingo Richter
Focus: Lit
Periodicals
Diskurs: Studien zu Kindheit, Jugend, Familie und Gesellschaft (weekly)
Dokumentation Bibliographie Jugendhilfe (weekly) 05590

Deutsches Komitee Instandhaltung e.V. (DKIN), Brehmstr 78, 40239 Düsseldorf
T: (0211) 623240; Fax: (0211) 625204
Founded: 1970; Members: 19
Pres: Rolf B. Neurath; Gen Secr: Katharina Schlosser
Focus: Eng
Periodical
DKIN-Empfehlungen (weekly) 05591

Deutsches Krebsforschungszentrum (DKFZ), Postf 101949, 69009 Heidelberg
T: (06221) 422850; Fax: (06221) 422840
Founded: 1964
Pres: Prof. Dr. Dr. Harald zur Hausen; Gen Secr: Dr. Reinhard Grunwald
Focus: Cell Biol & Cancer Res
Periodical
einblick: Zeitschrift des Deutschen Krebsforschungszentrums (3-4 times annually) 05592

Deutsches Kunststoff-Institut, Schloßgartenstr 6, 64289 Darmstadt
T: (06151) 162106
Focus: Materials Sci
Periodical
Mitteilungen aus dem Deutschen Kunststoff-Institut (weekly) 05593

Deutsches Kupfer-Institut e.V., Knesebeckstr 96, 10623 Berlin
T: (030) 310271; Tx: 184643; Fax: (030) 3128826
Founded: 1927; Members: 65
Focus: Materials Sci 05594

Deutsches Nationales Komitee des Weltenergierats (DNK), Graf-Recke-Str 84, 40239 Düsseldorf
T: (0211) 6214499
Pres: Dr. K. Barthelt; Gen Secr: Dr. J. Debelius
Focus: Energy 05595

Deutsches Nationalkomitee für Denkmalschutz, c/o Bundesministerium des Innern, Postf 170290, 53108 Bonn
T: (0228) 6815563
Pres: Hans Zehetmair; Gen Secr: Hans Günter Kowalski
Focus: Preserv Hist Monuments
Periodical
Denkmalschutz-Informationen (weekly) 05596

Deutsches Optisches Komitee, c/o Institut für medizinische Optik der Universität München, Theresienstr 37, 80333 München
T: (089) 23941
Focus: Optics 05597

Deutsches Orient-Institut, Mittelweg 150, 20148 Hamburg
T: (040) 441481
Founded: 1960
Pres: Prof. Dr. Udo Steinbach; Gen Secr: Dr. Thomas Koszinowski
Focus: Ethnology; Hist; Poli Sci
Periodicals
Mitteilungen (weekly)
Orient (weekly) 05598

Deutsche Statistische Gesellschaft (DStG), c/o Universität, 78464 Konstanz
T: (07531) 883758
Founded: 1911; Members: 800
Pres: Prof. Dr. S. Heiler; Gen Secr: R. Jeske
Focus: Stats
Periodical
Allgemeines Statistisches Archiv (weekly) 05599

Deutsches Textilforschungszentrum Nord-West e.V., Frankenring 2, 47798 Krefeld
Founded: 1990; Members: 100
Pres: Prof. Dr. E. Schollmeyer
Focus: Textiles 05600

Deutsche Stiftung für internationale Entwicklung, Hans-Böckler-Str 5, 53225 Bonn
T: (0228) 40010
Focus: Int'l Relat; Econ
Periodical
Entwicklung und Zusammenarbeit (weekly) 05601

Deutsche Straßenliga e.V. / Vereinigung zur Förderung des Straßen- und Verkehrswesens, Herderstr 56, 53173 Bonn
T: (0228) 956830
Members: 99
Pres: Prof. Dr. Hans J. Kayser; Gen Secr: Dr. Wolfgang Neumann
Focus: Transport
Periodicals
Straße und Wirtschaft (weekly)
Straße-Verkehr-Wirtschaft: (SVW)-Infodienst (weekly) 05602

Deutsche Studiengesellschaft für Publizistik (Stupu), Königstr 1a, 70173 Stuttgart
T: (0711) 293165, 294353
Founded: 1956; Members: 47
Focus: Journalism 05603

Deutsches Übersee-Institut, Neuer Jungfernstieg 21, 20354 Hamburg
T: (040) 3562593; Fax: (040) 3562547
Pres: Dr. Werner Draguhn
Focus: Marketing; Int'l Relat
Periodicals
Jahrbuch Dritte Welt: Daten, Übersichten, Analysen (weekly)
Nord-Süd aktuell (weekly) 05604

Deutsches wissenschaftliches Steuerinstitut der Steuerberater und Steuerbevollmächtigten e.V., Dechenstr 14, 53115 Bonn
T: (0228) 637822
Pres: Dr. Wilfried Dann; Gen Secr: Katrin L. Schulz
Focus: Finance 05605

Deutsches Wollforschungsinstitut (DWI), Veltmanpl 8, 52062 Aachen
T: (0241) 44690; Tx: 832829; Fax: (0241) 4469100
Founded: 1952; Members: 106
Pres: Hans Wohlfart; Gen Secr: Prof. Dr. Hartwig Höcker
Focus: Textiles
Periodical
DWI-Report (weekly) 05606

Deutsches Zentralinstitut für soziale Fragen, Bernadottestr 94, 14195 Berlin
Fax: (030) 8314750
Focus: Soc Sci
Periodical
Soziale Arbeit: Deutsche Zeitschrift für soziale und sozialverwandte Gebiete (weekly) 05607

Deutsches Zentralkomitee zur Bekämpfung der Tuberkulose (DZK), c/o Abteilung für Pneumologie, Universität, Langenbeckstr 1, 55101 Mainz
Founded: 1893; Members: 32
Pres: Prof. Dr. R. Loddenkemper; Gen Secr: Prof. Dr. R. Ferlinz
Focus: Pulmon Dis
Periodical
Pneumologie (weekly) 05608

Deutsches Zentrum für Altersfragen e.V., Manfred-von-Richthofen-Str 2, 12101 Berlin
T: (030) 7866071; Fax: (030) 7854350
Founded: 1973
Focus: Geriatrics; Soc Sci
Periodical
Altenhilfe: Beispiele, Informationen, Meinungen (weekly) 05609

Deutsche Tierärzteschaft e.V. (DT), Oxfordstr 10, 53111 Bonn
T: (0228) 655760; Fax: (0228) 692767
Members: 25
Pres: Prof. Dr. Günter Pschorn; Gen Secr: Eberhardt Rösener
Focus: Vet Med
Periodical
Deutsches Tierärzteblatt (weekly) 05610

Deutsche Tropenmedizinische Gesellschaft e.V. (DTG), Postf 800248, 65902 Frankfurt
Founded: 1962; Members: 650
Pres: Prof. Dr. D. Mehlitz; Gen Secr: Dr. R. Snethlage
Focus: Trop Med 05611

Deutsche Vereinigung für die Rehabilitation Behinderter e.V., Friedrich-Ebert Anlage 9, 69117 Heidelberg
T: (06221) 25485; Fax: (06221) 166009
Pres: Dr. Wolfgang Blumenthal; Gen Secr: Gerhard André
Focus: Rehabil
Periodical
Die Rehabilitation: Zeitschrift für alle Fragen der medizinischen, schulisch-beruflichen und sozialen Eingliederung (weekly) 05612

Deutsche Vereinigung für Finanzanalyse und Anlageberatung e.V. (DVFA), Fichtestr 18, 63303 Dreieich
T: (06103) 68185; Fax: (06103) 68194
Founded: 1960; Members: 600
Gen Secr: Ulrike Diehl
Focus: Finance
Periodical
Beiträge zur Wertpapieranalyse (weekly) . 05613

Deutsche Vereinigung für gewerblichen Rechtsschutz und Urheberrecht e.V., Theodor-Heuss-Ring 19-21, 50668 Köln
T: (0221) 77160; Fax: (0221) 7716110
Members: 2700
Pres: Karlheinz Quack; Gen Secr: Dr. Ralf Vieregge
Focus: Law
Periodical
Gewerblicher Rechtsschutz und Urheberrecht (GRUR) (weekly) 05614

Deutsche Vereinigung für Internationales Steuerrecht, Gustav-Heinemann-Ufer 84-88, 50968 Köln
T: (0221) 3708586
Founded: 1852; Members: 500
Focus: Finance 05615

Deutsche Vereinigung für internationales Steuerrecht im Verband der Fiscal Association, Bayerische Sektion e.V., Galeriestr 6a, 80539 München
T: (089) 292244
Focus: Finance 05616

Deutsche Vereinigung für Parlamentsfragen e.V., Bundeshaus, 53113 Bonn
T: (0228) 162442, 162250; Fax: (0228) 162671
Pres: Dr. Jürgen Anton Rüttgers; Gen Secr: Gunter Gabrysch
Focus: Poli Sci
Periodical
Zeitschrift für Parlamentsfragen (weekly) . 05617

Deutsche Vereinigung für Politische Wissenschaft (DVPW), Residenzschloss, 64283 Darmstadt
T: (06151) 163197; Fax: (06151) 162397
Founded: 1951; Members: 1200
Pres: Prof. Dr. Gerhard Lehmbruch; Gen Secr: Karin Bogusch
Focus: Poli Sci
Periodicals
Politische Vierteljahresschrift (weekly)
PVS-Literatur (weekly) 05618

Deutsche Vereinigung für Religionsgeschichte (DVRG), c/o Seminar für Religionswissenschaft, Wilhelm-Busch-Str 22, 30167 Hannover
Founded: 1950; Members: 154
Pres: Prof. Dr. Dr. Peter Antes
Focus: Rel & Theol 05619

Deutsche Vereinigung für Sportwissenschaft, c/o Universität Hamburg, Von-Melle-Park 8, 20146 Hamburg
T: (040) 458109; Fax: (040) 453745
Founded: 1976
Pres: Prof. Dr. Karlheinz Scherler; Gen Secr: Frederik Borkenhagen
Focus: Sports
Periodical
dvs-Informationen (weekly) 05620

Deutsche Vereinigung für Verbrennungsforschung e.V., c/o VGB, Postf 103932, 45039 Essen
T: (0201) 8128216
Pres: Prof. Dr. H.-D. Schilling; Gen Secr: Prof. Dr. J. Jacobs
Focus: Eng 05621

Deutsche Vereinigung zur Bekämpfung der Viruskrankheiten e.V. (DVV), Pettenkoferstr 9a, 80336 München
T: (089) 533401
Focus: Immunology; Microbio 05622

Deutsche Verkehrswissenschaftliche Gesellschaft e.V. (DVWG), Brüderstr 53, 51427 Bergisch Gladbach
T: (02204) 60027; Fax: (02204) 67743
Founded: 1908; Members: 3100
Pres: Dr. Horst Weigelt; Gen Secr: Sigurd Rielke
Focus: Transport
Periodical
Internationales Verkehrswesen (weekly) . 05623

Deutsche Veterinärmedizinische Gesellschaft e.V. (DVG), Frankfurter Str 89, 35392 Giessen
T: (0641) 24466
Founded: 1952; Members: 4000
Pres: Prof. Dr. Dr. Eberhard Grunert; Gen Secr: Prof. Dr. H. Baljev
Focus: Vet Med 05624

Deutsche Werbewissenschaftliche Gesellschaft e.V. (DWG), Königswinterer Str 552, 53227 Bonn
T: (0228) 444560; Fax: (0228) 444503
Founded: 1919
Pres: Prof. Dr. Dr. Arnold Hermanns; Gen Secr: Lutz E. Weidner
Focus: Advert
Periodical
Werbeforschung und Praxis (weekly) . . 05625

Deutsche Wissenschaftliche Kommission für Meeresforschung, Palmaille 9, 22767 Hamburg
T: (040) 389050; Fax: (040) 38905129
Pres: Prof. Dr. Jens Meincke; Gen Secr: Prof. Dr. Alfred Post
Focus: Oceanography
Periodical
Archive of Fishery and Marine Research: Archiv für Fischerei und Meeresforschung (2-3 times annually) 05626

Deutsche Zeitungswissenschaftliche Vereinigung e.V. (ZWV), Karolinenpl 3, 80333 München
Focus: Journalism 05627

Deutsche Zentrale für Volksgesundheitspflege e.V. (DZV), Münchener Str 48, 60329 Frankfurt
T: (069) 235761/62
Founded: 1955; Members: 70
Focus: Public Health 05628

Deutsche Zoologische Gesellschaft e.V. (DZG), Poppelsdorfer Schloss, 53115 Bonn
Founded: 1890; Members: 1600
Pres: Prof. Dr. W. Rathmayer; Gen Secr: Dr. Helga Eichelberg
Focus: Zoology
Periodical
Verhandlungen (weekly) 05629

Deutsch-Pazifische Gesellschaft e.V. (DPG), c/o Dr. F. Steinbauer, Feichtmayrstr 25, 80992 München
T: (089) 151158; Fax: (089) 151833
Founded: 1974; Members: 550
Pres: Dr. F. Steinbauer
Focus: Int'l Relat
Periodical
Bulletins: Informationshefte (weekly) . . . 05630

Deutschsprachige Arbeitsgemeinschaft für Handchirurgie, c/o Abteilung für Handchirurgie und Plastische Chirurgie, Bergedorfer Str 10, 21033 Hamburg
T: (031) 643534; Fax: (031) 262419
Founded: 1960; Members: 78
Gen Secr: Prof. Dr. Dieter Buck-Gramcko
Focus: Surgery 05631

DIN Deutsches Institut für Normung e.V., Burggrafenstr 6, 10787 Berlin
T: (030) 26010; Tx: 184273; Fax: (030) 2601231
Founded: 1917; Members: 5400
Pres: Eberhard Möllmann; Gen Secr: Prof. Dr. Helmut Reihlen
Focus: Standards
Periodical
DIN-Mitteilungen + elektronorm (weekly) . 05632

Dokumentationsring Elektrotechnik, c/o Siemens AG, Werner-von-Siemens-Str 50, 91052 Erlangen
Focus: Electric Eng 05633

Dokumentationsring Pädagogik (DOPAED) / Leitstelle des DOPAED / DOPAED Coordination office, c/o Deutsches Institut für Internationale Pädagogische Forschung, Schloßstr 29, 60486 Frankfurt
T: (069) 247080; Fax: (069) 2470844
Founded: 1964; Members: 30
Pres: Hartmut Müller
Focus: Educ
Periodical
Bibliographie Pädagogik: Bibliographie Pédagogique (1-2 times annually) 05634

Dramatiker-Union e.V. (DU), Bismarckstr 107, 10625 Berlin
T: (030) 317676; Fax: (030) 317677
Founded: 1871; Members: 450
Pres: Prof. Curth Flatow; Gen Secr: Eckhard Schulz
Focus: Perf Arts 05635

EAAP-FAO Global Data Bank for Animal Genetic Resources, c/o Institut für Tierzucht, Bünteweg 17, 30559 Hannover
Founded: 1988
Gen Secr: D. Simon
Focus: Animal Husb 05636

Eichendorff-Gesellschaft e.V., Bahnhofstr 71, 40883 Ratingen
T: (02102) 9650
Founded: 1931; Members: 520
Pres: Prof. Dr. Peter H. Neumann; Gen Secr: Dr. Volkmar Stein
Focus: Lit
Periodicals
Aurora: Jahrbuch der Eichendorff-Gesellschaft (weekly)
Aurora-Buchreihe (weekly) 05637

Electronic Components Quality Assurance Committee (ECQAC), c/o CECC, Gartenstr 179, 60596 Frankfurt
T: (069) 639171; Tx: 4032175; Fax: (069) 639427
Focus: Electronic Eng 05638

Elektrotechnischer Verein Berlin e.V., Bismarckstr 33, 10625 Berlin
T: (030) 3414566
Members: 1200
Focus: Electronic Eng
Periodical
ETV-Mitteilungen (weekly) 05639

Energietechnische Gesellschaft im VDE, Stresemannallee 15, 60596 Frankfurt
T: (069) 6308346; Tx: 412871
Members: 6000
Pres: Prof. E.F. Peschke; Gen Secr: Dr. G. Jesse
Focus: Energy
Periodicals
Elektrotechnische Zeitschrift (weekly)
European Transactions on Electrical Power Engineering (weekly) 05640

Energie- und Umweltzentrum am Deister e.V. (E.u.U.Z.), 31832 Springe-Eldagsen
T: (05044) 380, 1880; Fax: (05044) 4640
Focus: Energy; Ecology
Periodical
Rundbrief des E.u.U.Z. (weekly) 05641

Ernst Barlach Gesellschaft e.V., Kippingstr 2, 20144 Hamburg
T: (040) 459160
Pres: Dr. Ekkehard W. Nümann
Focus: Arts; Fine Arts; Lit 05642

Ernst-Mach-Institut, Eckerstr 4, 79104 Freiburg
Focus: Physics
Periodical
Bericht (weekly) 05643

E.T.A. Hoffmann-Gesellschaft e.V., Nonnenbrücke 1, 96047 Bamberg
T: 28173
Pres: Dr. Reinhard Heinritz
Focus: Lit
Periodical
Mitteilungen (weekly) 05644

Euro-Handelsinstitut e.V., Spichernstr 55, 50672 Köln
T: (0221) 579930; Fax: (0221) 5799345
Pres: H.J. Zellekens; Gen Secr: Dr. Bernd Hullier
Focus: Business Admin
Periodical
dynamik im handel 05645

Europäische Akademie Bayern e.V., Leuchtenbergring 3, 81677 München
T: (089) 4708186; Fax: (089) 4705709
Pres: Hans Zehetmair; Gen Secr: Michael Jörger
Focus: Educ 05646

Europäische Akademie Berlin e.V. (EAB), Bismarckallee 46-48, 14193 Berlin
T: (030) 8262095; Fax: (030) 8266410
Founded: 1964; Members: 17
Pres: Dr. Giuseppe Vita; Gen Secr: Dr. Eckart D. Stratenschulte
Focus: Educ 05647

Europäische Akademie Otzenhausen e.V. (EAO), Europahausstr, 66620 Nonnweiler
T: (06873) 306, 406; Fax: (06873) 1452
Founded: 1954; Members: 30
Focus: Poli Sci
Periodical
Dokumente und Schriften der Europäischen Akademie Otzenhausen (weekly) 05648

Europäische Bildungs- und Aktionsgemeinschaft (EBAG), Bertha-von-Suttner-Pl 13, 53111 Bonn
T: (0228) 657702, 654014; Fax: (0228) 657685
Pres: Dr. Otto Schmuck; Gen Secr: Richard Heiser
Focus: Educ 05649

Europäische Bildungs- und Begegnungszentren (EBZ), Taubenberg 84, 65510 Idstein
T: (06126) 53423; Fax: (06126) 54630
Founded: 1977; Members: 10
Pres: Walter Buschmann
Focus: Educ 05650

Europäische Föderation Biotechnologie (EFB) / European Federation of Biotechnology, c/o Dechema, Postf 150104, 60061 Frankfurt
T: (069) 7564209; Fax: (069) 7564201
Founded: 1978
Pres: Prof. Dr. G. Kreysa
Focus: Biophys; Biochem
Periodical
EFB-Newsletter (weekly) 05651

Europäische Föderation für Chemie-Ingenieur-Wesen (EFCIW) / European Federation of Chemical Engineering, Theodor-Heuss-Allee 25, 60486 Frankfurt
T: (069) 7564235; Fax: (069) 7564201
Founded: 1953; Members: 73
Gen Secr: Prof. Dr. G. Kreysa
Focus: Eng; Chem
Periodical
EFCE Newsletter (weekly) 05652

Europäische Föderation Korrosion (EFK) / European Federation of Corrosion, c/o DECHEMA, Theodor-Heuss-Allee 25, 60486 Frankfurt
T: (069) 7564209; Fax: (069) 7564201
Founded: 1955; Members: 75
Gen Secr: Prof. Dr. G. Kreysa
Focus: Metallurgy
Periodical
EFC Newsletter (weekly) 05653

Europäische Märchengesellschaft e.V. (EMG), Schloß Bentlage, 48432 Rheine
T: (05971) 12117; Fax: (05971) 53046
Pres: Dr. Ursula Heindrichs; Gen Secr: Thomas Bücksteeg
Focus: Lit; Folklore 05654

Europäischer Erzieherbund e.V., Sektion Deutschland (EEB), Julius-Brecht-Str 16, 65824 Schwalbach
T: (06196) 85515; Fax: (06196) 85515
Founded: 1956
Pres: Prof. Dr. Wolfgang Mickel; Gen Secr: Jürgen Kummetat
Focus: Educ 05655

Europäischer Verband für Produktivitätsförderung, c/o Dr. Helms, Elbchaussee 352, 22609 Hamburg
T: (040) 823011; Fax: (040) 826594
Focus: Business Admin; Eng
Periodical
News Release (weekly) 05656

Europäische Staatsbürger-Akademie e.V., Adenauerallee 59, 46399 Bocholt
T: (02871) 3430; Fax: (02871) 343101
Founded: 1961; Members: 26
Pres: Hans Peters; Gen Secr: Gerhard Eickhorn
Focus: Adult Educ 05657

Europäische Vereinigung für Eigentumsbildung, Hermann-Lindrath-Gesellschaft e.V., Am Südbahnhof 2, 30171 Hannover
T: (0511) 810906
Founded: 1960; Members: 250
Pres: Carl Doehring
Focus: Sociology; Finance
Periodical
Pressedienst (3-4 times annually) . . . 05658

European Aluminium Association (EAA), Königsallee 30, 40212 Düsseldorf
T: (0211) 80871; Tx: 8586407; Fax: (0211) 324098
Founded: 1981
Gen Secr: Dr. Hansgeorg Seebauer
Focus: Metallurgy
Periodical
World Aluminium Abstracts (weekly) . . 05659

European Aluminium Foil Association (EAFA), Schumannstr 46, 60325 Frankfurt
T: (069) 748024
Founded: 1972; Members: 15
Focus: Metallurgy 05660

European Association for Chinese Law (EACL), Karl-Marx-Allee 199, 52066 Aachen
T: (0241) 602599
Founded: 1984
Focus: Law
Periodical
EACL Information Bulletin (weekly) . . . 05661

European Association for Research and Development in Higher Education (EARDHE), c/o Unit for Staff and Development Research into Higher Education, Free University Berlin, Habelschwerdter Allee 34a, 14195 Berlin
Founded: 1972
Gen Secr: Ilona Yenal
Focus: Educ 05662

European Association for Special Education (EASE), Reutlinger Str 31, 70597 Stuttgart
Founded: 1970
Pres: Klaus Wenz; Gen Secr: Arie J. Breur
Focus: Educ
Periodical
EASE Circular Letter (weekly) 05663

European Association for the Development of Databases in Education and Training / Europäische Vereinigung der Datenbanken in der Aus- und Weiterbildung, c/o Data-Print GmbH, Tauentzienstr 4, 10789 Berlin
T: (030) 2199960; Fax: (030) 21999678
Founded: 1988
Gen Secr: B. Böttcher
Focus: Computer & Info Sci 05664

European Association for Theoretical Computer Science (EATCS), c/o Fachbereich 17, Universität Paderborn, Warburger Str 100, 33098 Paderborn
Founded: 1972
Gen Secr: Prof. Dr. B. Monien
Focus: Computer & Info Sci
Periodicals
Bulletin of the EATCS (3 times annually)
Theoretical Computer Science 05665

European Association for the Study of Diabetes (EASD) / Europäische Gesellschaft für Diabetologie, Auf'm Hennekamp 32, 40225 Düsseldorf
T: (0211) 316738; Fax: (0211) 3190987
Founded: 1965; Members: 3600
Gen Secr: Dr. Viktor Jörgens
Focus: Diabetes 05666

European Association for the Study of Diabetes (EASD), Auf'm Hennekamp 32, 40225 Düsseldorf
T: (0211) 316738; Fax: (0211) 3190987
Founded: 1964; Members: 5000
Gen Secr: Dr. V. Jörgens
Focus: Diabetes
Periodical
Diabetologia (weekly) 05667

European Association of Business and Management Teachers (EBM), c/o Fachbereich Wirtschaft, Fachhochschule Dortmund, Postf 335, 44227 Dortmund
T: (0231) 39010
Founded: 1985
Pres: Alan Hale; Gen Secr: Hans Hantke
Focus: Educ; Business Admin
Periodical
EBM Review (3 times annually) 05668

European Association of Neurosurgical Societies (EANS), c/o Universitätsklinikum Steglitz, Hindenburgdamm 30, 12203 Berlin
Founded: 1972; Members: 21
Pres: Prof. Dr. Mario Brock
Focus: Surgery 05669

European Astronaut Centre (EAC) / Europäisches Astronautenzentrum, c/o ESA-EAC, Linder Höhe, 51147 Köln
T: (02203) 60010; Fax: (02203) 600166
Founded: 1990; Members: 13
Pres: Andres Ripoll
Focus: Aero 05670

European Authors' Association Die Kogge / Europäische Autorenvereinigung Die Kogge, Ziegeleiweg 6, 32429 Minden
T: (0571) 53568
Founded: 1924
Focus: Lit 05671

European Baptist Theological Teachers' Conference, c/o Theologisches Seminar Hamburg, Rennbahnstr 115, 22111 Hamburg
T: (040) 655850; Fax: (040) 65585134
Founded: 1954
Focus: Educ 05672

European Centre for the Development of Vocational Training / Europäisches Zentrum für die Förderung der Berufsbildung, Bundesallee 22, 10717 Berlin
T: (030) 884120; Tx: 184163; Fax: (030) 88412222
Founded: 1975; Members: 12
Focus: Educ 05673

European Collaboration on Measurement Standards (EUROMET), c/o PTB, Bundesallee 100, 38116 Braunschweig
T: (0531) 5928103; Fax: (0531) 5924006
Founded: 1987
Gen Secr: Dr. P. Drath
Focus: Standards 05674

European Committee for Future Accelerators (ECFA), c/o III. Physikalisches Institut, Rheinisch-Westfälische Hochschule, Sommerfeldstr 26-28, 52074 Aachen
Founded: 1962
Pres: Prof. Günter Flügge
Focus: Eng 05675

European Council of Management (CECIOS), c/o RKW, Düsseldorfer Str 40, 65760 Eschborn
T: (06196) 495366; Tx: 4072755; Fax: (06196) 495303
Gen Secr: Dr. H. Müller
Focus: Business Admin 05676

European Fast Reactor Association, c/o Kernforschungszentrum Karlsruhe, Leopoldhafen 2, 76344 Eggenstein
Gen Secr: Dr. W. Marth
Focus: Nucl Res 05677

European Federation of Corrosion (EFC), c/o DECHEMA, Theodor-Heuss-Allee 25, 60486 Frankfurt
T: (069) 7564209; Tx: 0412490; Fax: (069) 7564201
Founded: 1955
Gen Secr: Prof. Dr. G. Kreysa
Focus: Metallurgy
Periodical
EFC Newsletter 05678

European Federation of Cytology Societies (EFCS), c/o Zentrum für Pathologie, Institut für Cytopathologie, Universität, Moorenstr 5, 40225 Düsseldorf
T: (0221) 3112524; Tx: 8587348; Fax: (0221) 342229
Founded: 1969
Gen Secr: Dr. Peter Pfitzer
Focus: Cell Biol & Cancer Res 05679

European Society of Neuroradiology, c/o Institut für Neuroradiology, Universität des Saarlandes, 66421 Homburg
Fax: (06841) 164310
Pres: Dr. J.P. Braun; Gen Secr: Prof. A. Valavanis
Focus: Radiology
Periodical
Neuroradiology 05680

European Weed Research Society (EWRS), c/o Dr. Birgit Krauskopf, Bayer AG, Bayerwerk, Postf E-Reg., 51368 Leverkusen
T: (02173) 384928; Tx: 85103275; Fax: (02173) 383593
Founded: 1959; Members: 590
Focus: Botany
Periodical
Weed Research (weekly) 05681

Evangelische Akademie Bad Boll, 73087 Bad Boll
T: (07164) 791
Founded: 1945
Focus: Adult Educ
Periodicals
Aktuelle Gespräche (weekly)
Arbeitshilfen (weekly)
Dokumente, Texte und Tendenzen (weekly) 05682

Evangelische Akademie Baden, Postf 2269, 76137 Karlsruhe
T: (0721) 9349297; Fax: (0721) 8349349
Founded: 1947
Gen Secr: Klaus Nagorni
Focus: Adult Educ
Periodical
Diskussionen: Zeitschrift für Akademiearbeit und Erwachsenenbildung 05683

Evangelische Akademie Hofgeismar, Schlösschen Schönburg, Gesundbrunnen, 34369 Hofgeismar
T: (05671) 8810
Focus: Adult Educ
Periodical
Anstösse (weekly) 05684

Evangelische Akademie Loccum, Münchehäger Str 6, 31575 Rehburg-Loccum
T: (05766) 810; Fax: (05766) 81188
Founded: 1952
Pres: Hans May; Gen Secr: Ernst Bohnenkamp
Focus: Adult Educ
Periodicals
Forum Loccum (weekly)
Loccumer Protokolle 05685

Evangelische Akademien in Deutschland e.V., Akademieweg 11, 73087 Bad Boll
T: (07164) 791
Founded: 1947; Members: 31
Pres: Hans May; Gen Secr: Dr. Fritz Erich Anhelm
Focus: Adult Educ 05686

Evangelische Akademie Tutzing, Schloßstr 2-4, 82327 Tutzing
T: (08158) 2510; Fax: (08158) 251133
Founded: 1947
Pres: Dr. Friedemann Greiner; Gen Secr: Martin Kurz
Focus: Adult Educ
Periodicals
Tutzinger Blätter (weekly)
Tutzinger Materialien (weekly) 05687

Evangelische Akademikerschaft in Deutschland, Kniebisstr 29, 70188 Stuttgart
T: (0711) 282015/16
Focus: Sci
Periodical
Radius (weekly) 05688

Evangelischer Erziehungs-Verband e.V. (EREV) / Bundesverband evangelischer Einrichtungen und Dienste, Lister Meile 87, 30161 Hannover
T: (0511) 660266; Fax: (0511) 660222
Founded: 1920; Members: 400
Pres: Klaus Kinkel; Gen Secr: Hans Bauer
Focus: Educ

Periodicals
EREV-Schriftenreihe (weekly)
Evangelische Jugendhilfe (5 times annually) . . . 05689

Evangelische Sozialakademie Friedewald, Schlossstr 1, 57520 Friedewald
T: (02743) 2091
Founded: 1949; Members: 10
Focus: Sociology; Rel & Theol; Econ . . 05690

Evangelische Studiengemeinschaft e.V. / Protestant Institute for Interdisciplinary Studies, Schmeilweg 5, 69118 Heidelberg
T: (06221) 14061
Founded: 1957; Members: 20
Focus: Rel & Theol; Philos; Nat Sci
Periodicals
Forschungen und Berichte
Texte und Materialien . . . 05691

Fachinformationszentrum Technik e.V., Ostbahnhofstr 13, 60314 Frankfurt
T: (069) 4308241; Fax: (069) 4308200
Focus: Textiles
Periodical
Informationsdienst Textiltechnik Bekleidungstechnik Textilmaschinenbau (weekly) . . . 05692

Fachschaft Berliner Chirurgen e.V., Technowpromenade 65, 13437 Berlin
T: (030) 4112870
Focus: Surgery . . . 05693

Fachverband Deutscher Heilpraktiker e.V., Moorweg 10, 53123 Bonn
T: (0228) 611049; Fax: (0228) 641084
Gen Secr: Klaus Schwarzbach
Focus: Med
Periodical
Der Heilpraktiker (weekly) . . . 05694

Fachverband Moderne Fremdsprachen (FMF), Marconistr 30b, 86179 Augsburg
T: (0821) 85237
Founded: 1880; Members: 5500
Pres: Prof. Dr. Konrad Schröder
Focus: Ling
Periodical
Neusprachliche Mitteilungen aus Wissenschaft und Praxis (weekly) . . . 05695

Fachverband Pulvermetallurgie (Fpm), Postf 921, 58009 Hagen
T: (02331) 958817; Fax: (02331) 51046
Founded: 1948; Members: 43
Pres: Dr. Lothar Albano-Müller; Gen Secr: Dr. Hermann Hassel
Focus: Metallurgy
Periodicals
powder metallurgy international (weekly)
Pulvermetallurgie in Wissenschaft und Praxis (weekly) . . . 05696

Fachverband Textilunterricht e.V., Münzstr 6, 48143 Münster
T: (0251) 42595
Founded: 1975; Members: 500
Pres: Dr. Ruth Blechwenn
Focus: Educ; Textiles
Periodical
Textil-Info für Unterricht & Bildung (weekly) . . . 05697

Faust-Gesellschaft, Nachtigallenstr 52, 75417 Mühlacker
Focus: Hist . . . 05698

Ferdinand Tönnies-Gesellschaft e.V., Freiligrathstr 11, 24116 Kiel
T: (0431) 551107; Fax: (0431) 552993
Pres: Lars Clausen
Focus: Sociology
Periodicals
Skizze (weekly)
Tönnies-Forum (weekly) . . . 05699

Filmbewertungsstelle Wiesbaden (FBW), Postf 120240, 65080 Wiesbaden
T: (0611) 9660040; Tx: 04186691
Pres: Steffen Wolf
Focus: Cinema . . . 05700

Filmkritiker Kooperative, Kreitmayerstr 3, 80335 München
Focus: Cinema
Periodical
Filmkritik (weekly) . . . 05701

Födergemeinschaft für Absatz- und Werbeforschung, Friedensstr 11, 60311 Frankfurt
Focus: Marketing; Advert . . . 05702

Fördergemeinschaft für das Süddeutsche Kunststoff-Zentrum e.V. (FSKZ), Frankfurter Str 15-17, 97082 Würzburg
T: (0931) 41040; Tx: 68448; Fax: (0931) 4194177
Pres: Dr. Volker Hülck; Gen Secr: Dr. Otto Schwarz
Focus: Materials Sci; Adult Educ; Eng . 05703

Förderungsgemeinschaft der Kartoffelwirtschaft e.V., Dethlingen, 29633 Munster
T: (05192) 2282
Founded: 1949
Focus: Nutrition; Food; Agri
Periodical
Der Kartoffelbau: Fachzeitschrift über Züchtung, Vermehrung, Produktionstechnik, Verwertung und Ökonomik (weekly) . . . 05704

FOGRA Forschungsgesellschaft Druck e.V., Postf 800469, 81604 München
T: (089) 431820
Pres: Jan te Neues; Gen Secr: Dr. Hans-Joachim Falge
Focus: Eng
Periodicals
FOGRA Aktuell (weekly)
Fogra-Literaturdienst (weekly)
Fogra-Literatur-Profil (weekly)
Fogra-Mitteilungen (weekly)
Fogra-Patentschau (weekly) . . . 05705

Forschungsgemeinschaft Angewandte Geophysik e.V., Postf 3266, 30032 Hannover
T: (0511) 326331; Fax: (0511) 328501
Founded: 1954; Members: 16
Pres: Prof. Dr. M. Kürsten; Gen Secr: Otto Lenz
Focus: Geophys . . . 05706

Forschungsgemeinschaft Eisenhüttenschlacken, Bliersheimer Str 62, 47229 Duisburg
T: (02065) 47087; Fax: (02065) 47529
Focus: Metallurgy; Ecology; Eng . . . 05707

Forschungsgemeinschaft Feuerfest e.V., An der Elisabethkirche 27, 53113 Bonn
T: (0228) 915080; Fax: (0228) 9150866
Focus: Materials Sci
Periodical
Literaturbericht (weekly) . . . 05708

Forschungsgemeinschaft für Hochspannungs- und Hochstromtechnik e.V. (FGH), Hallenweg, 68219 Mannheim
T: (0621) 89970
Founded: 1921; Members: 77
Focus: Electric Eng . . . 05709

Forschungsgemeinschaft für technisches Glas e.V. (FtG), Bronnbach 28, 97877 Wertheim
T: (09342) 92120; Fax: (09342) 921292
Founded: 1950; Members: 41
Pres: P. Hahmann; Gen Secr: Dr. H.-H. Fahrenkrog; K.J. Hermann
Focus: Materials Sci . . . 05710

Forschungsgemeinschaft für Verpackungs- und Lebensmitteltechnik e.V., Schragenhofstr 35, 80992 München
T: (089) 1411091
Focus: Food . . . 05711

Forschungsgemeinschaft Industrieofenbau e.V., Lyoner Str 18, 60528 Frankfurt
T: (069) 66030; Tx: 411321, 413152; Fax: (069) 6603692
Pres: Prof. Dr. Bernhard Wielage; Gen Secr: Dr. G. Habig
Focus: Eng . . . 05712

Forschungsgemeinschaft Kalk und Mörtel e.V., Annastr 67-71, 50969 Köln
T: (0221) 376920
Focus: Materials Sci
Periodical
Forschungsberichte der Forschungsgemeinschaft Kalk und Mörtel e.V. (weekly) . . . 05713

Forschungsgemeinschaft Kraftpapiere und Papiersäcke, Nerotal 4, 65193 Wiesbaden
T: (0611) 524041/42
Focus: Materials Sci . . . 05714

Forschungsgemeinschaft Naturstein-Industrie e.V., Buschstr 22, 53113 Bonn
T: (0228) 213234
Members: 12
Pres: Dr. L. Bäumler; Gen Secr: K.-H. Plock
Focus: Materials Sci
Periodical
Die Naturstein-Industrie (weekly) . . . 05715

Forschungsgemeinschaft Werkzeuge und Werkstoffe e.V. (FGW), Postf 100320, 42803 Remscheid
T: (02191) 900300; Fax: (02191) 900320
Focus: Eng . . . 05716

Forschungsgemeinschaft Zink e.V., Friedrich-Ebert-Str 37-39, 40210 Düsseldorf
T: (0211) 350867; Fax: (0211) 350869
Focus: Materials Sci . . . 05717

Forschungsgesellschaft Druckmaschinen e.V. (FGD), Postf 710864, 60498 Frankfurt
T: (069) 6603451; Fax: (069) 6603675
Founded: 1955; Members: 24
Pres: Hermamm Thomas
Focus: Mach Eng . . . 05718

Forschungsgesellschaft für Agrarpolitik und Agrarsoziologie e.V., Meckenheimer Allee 125, 53115 Bonn
T: (0228) 634781, 634788
Founded: 1953; Members: 61
Pres: Prof. Dr. Wilhelm Henrichsmeyer; Gen Secr: Dr. Richard Struff
Focus: Poli Sci; Econ; Sociology . . . 05719

Forschungsgesellschaft für Strassen- und Verkehrswesen e.V., Postf 501362, 50973 Köln
T: (0221) 397035; Fax: (0221) 393747
Focus: Civil Eng; Transport
Periodicals
Arbeitsgruppe Asphaltstrassen: Schriftenreihe (weekly)
Arbeitsgruppe Betonstrassen: Schriftenreihe (weekly)
Arbeitsgruppe Erd- und Grundbau (weekly)
Arbeitsgruppe Mineralstoffe im Strassenbau (weekly)
Dokumentation Straße: Kurzauszüge aus dem Schrifttum über das Straßenwesen (weekly)
Forschung im Straßenwesen (weekly)
Forschungsarbeiten aus dem Strassenwesen (weekly)
Strassenverkehrstechnik (weekly)
Strasse und Autobahn (weekly) . . . 05720

Forschungsgesellschaft Kunststoffe e.V., Schlossgartenstr 6, 64289 Darmstadt
T: (06151) 291150; Fax: (06151) 292855
Founded: 1953; Members: 168
Pres: Dr. Werner Wagner; Gen Secr: Klaus Darlapp
Focus: Materials Sci
Periodical
Literaturschnelldienst: Kunststoffe, Kautschuk, Fasern (weekly) . . . 05721

Forschungsgesellschaft Landschaftsentwicklung Landschaftsbau e.V., Colmantstr 32, 53115 Bonn
T: (0228) 691810; Fax: (0228) 650098
Focus: Ecology . . . 05722

Forschungsgesellschaft Stahlverformung e.V., Postf 4009, 58040 Hagen
T: (02331) 958841; Fax: (02331) 51046
Focus: Eng . . . 05723

Forschungsgesellschaft Steinzeugindustrie e.V., Max-Planck-Str 6, 50858 Köln 40
Focus: Materials Sci . . . 05724

Forschungs-Gesellschaft Verfahrenstechnik e.V., Graf-Recke-Str 84, 40239 Düsseldorf
T: (0211) 6214552; Fax: (0211) 6214159
Founded: 1951
Pres: Dr. Henning Bode; Gen Secr: Dr. Hanns-Dieter Butzmann
Focus: Eng . . . 05725

Forschungsgruppe für Anthropologie und Religionsgeschichte e.V., Denkmalstr 5, 66119 Saarbrücken
Founded: 1970
Pres: Prof. Dr. A. Rupp
Focus: Anthro; Rel & Theol
Periodicals
Forschungen zur Anthropologie und Religionsgeschichte
Jahrbuch für Anthropologie und Religionsgeschichte (weekly) . . . 05726

Forschungsgruppe Köln, Barbarossapl 2, 50674 Köln
T: (0221) 235177/79
Focus: Sci . . . 05727

Forschungsinstitut für Arbeiterbildung e.V. (FIAB), Kirchpl 2, 45657 Recklinghausen
T: (02361) 57034; Fax: (02361) 183362
Founded: 1980
Focus: Adult Educ
Periodical
Jahrbuch Arbeit, Bildung, Kultur . . . 05728

Forschungsinstitut für Gesellschaftspolitik und beratende Sozialwissenschaft e.V., Nikolausberger Weg 11, 37073 Göttingen
T: (00551) 394798, 56309
Focus: Poli Sci; Sociology
Periodical
Mitteilungsblatt (weekly) . . . 05729

Forschungsinstitut für Pigmente und Lacke e.V., Allmandring 37, 70569 Stuttgart
T: (0711) 687800; Fax: (0711) 6878079
Founded: 1951; Members: 37
Pres: Prof. Dr. Lothar Dulog; Gen Secr: Dr. Joachim Engemann
Focus: Materials Sci . . . 05730

Forschungsinstitut für Rationalisierung e.V. / FIR, Pontdriesch 14-16, 52062 Aachen
T: (0241) 477050; Fax: (0241) 402401
Focus: Business Admin
Periodical
FIR+IAW-Mitteilungen (3 times annually) . 05731

Forschungsinstitut für Wärmeschutz e.V. (FIW), Postf 1525, 82157 Gräfelfing
T: (089) 858000; Fax: (089) 8580040
Founded: 1918; Members: 140
Pres: Peter Hefter; Gen Secr: Joachim Achtziger; Horst Zehendner
Focus: Civil Eng
Periodicals
FIW-Informationen
Mitteilungen aus dem Forschungsinstitut für Wärmeschutz . . . 05732

Forschungskuratorium Gesamttextil / Ständiger Ausschuss des Gesamtverbandes der Textilindustrie in der Bundesrepublik Deutschland -Gesamttextil- e.V., Frankfurter Str 10-14, 65760 Eschborn
T: (06196) 9660
Pres: Mathias Schek; Gen Secr: Ralf Lawaczeck
Focus: Textiles
Periodical
Textilforschung: Berichte des Forschungskuratoriums Gesamttextil . . . 05733

Forschungskuratorium Maschinenbau e.V., Lyoner Str 18, 60528 Frankfurt
T: (069) 66031
Pres: Prof. Dr. Hubertus Christ; Gen Secr: Hartmut Geisendorf
Focus: Mach Eng . . . 05734

Forschungsrat Kältetechnik e.V., Lyoner Str 18, 60528 Frankfurt
T: (069) 6603277; Fax: (069) 6603218
Pres: Prof. Dr. Helmut Lotz; Gen Secr: Dr. Karin Jahn
Focus: Air Cond . . . 05735

Forschungsstelle des Bundesverbandes der Deutschen Ziegelindustrie e.V., Schaumburg-Lippe-Str 4, 53103 Bonn
T: (0228) 914930; Tx: 886884; Fax: (0228) 9149327
Founded: 1953; Members: 100
Pres: K.-H. Brakemeier
Focus: Materials Sci
Periodical
ZI – Ziegelindustrie International (weekly) 05736

Forschungsstelle für Acetylen, Marsbruchstr 186, 44287 Dortmund
T: (0231) 451194
Focus: Chem . . . 05737

Forschungsstelle für den Handel Berlin e.V. (FfH), Fehrbelliner Pl 3, 10707 Berlin
T: (030) 8630940; Fax: (030) 86309444
Pres: Prof. Dr. Volker Trommsdorff; Gen Secr: Dr. Helmut Bunge
Focus: Commerce
Periodical
mitteilungen aus der FfH (weekly) . . . 05738

Forschungsstelle für internationale Agrar- und Wirtschaftsentwicklung e.V., Ringstr 19, 69115 Heidelberg
T: (06221) 193056; Fax: (06221) 167482
Pres: Prof. Dr. O. Gans; Gen Secr: Dr. O.C. Kirsch
Focus: Agri . . . 05739

Forschungsverband für den Handelsvertreter- und Handelsmaklerberuf, Geleniusstr 1, 50931 Köln
T: (0221) 514043
Pres: Eberhard Runge; Gen Secr: Dr. Andreas Paffhausen
Focus: Advert; Commerce . . . 05740

Forschungsvereinigung Antriebstechnik e.V., Lyoner Str 18, 60528 Frankfurt
T: (069) 66030
Pres: Dr. Manfred Hirt; Gen Secr: Dr. Heinz Muno
Focus: Eng . . . 05741

Forschungsvereinigung Automobiltechnik e.V. (FAT), Westendstr 61, 60325 Frankfurt
T: (069) 75701
Pres: Dr. Erika Emmerich
Focus: Auto Eng . . . 05742

Forschungsvereinigung der Deutschen Asphaltindustrie e.V., Geleitstr 105, 63067 Offenbach
T: (069) 883305
Focus: Materials Sci . . . 05743

Forschungsvereinigung der Gipsindustrie e.V., Birkenweg 13, 64295 Darmstadt
T: (06151) 314310; Fax: (06151) 316549
Focus: Materials Sci . . . 05744

Forschungsvereinigung der Rheinischen Bimsindustrie e.V., Sandkauler Weg 1, 56564 Neuwied
T: (02631) 22227/28
Focus: Materials Sci . . . 05745

Forschungsvereinigung Elektrotechnik beim ZVEI e.V., Stresemannallee 19, 60596 Frankfurt
T: (069) 6302277
Focus: Electric Eng . . . 05746

Forschungsvereinigung Feinmechanik und Optik e.V., Pipinstr 16, 50667 Köln
Founded: 1963
Pres: Dr. Ernst Pohlen
Focus: Eng; Optics . . . 05747

Forschungsvereinigung für angewandte Schloß-, Beschlag- und präventive Sicherheitstechnik e.V., Offerstr 12, 42551 Velbert
T: (02051) 95060; Fax: (02051) 950620
Focus: Eng . . . 05748

Forschungsvereinigung für Luft- und Trocknungstechnik e.V. (FLT), Lyoner Str 18, 60528 Frankfurt
T: (069) 660289; Tx: 411321, 413152;
CA: Maschinenverein Frankfurtmain
Founded: 1964; Members: 40
Pres: E. Offergeld; Gen Secr: F.A. Kitzerow
Focus: Air Cond
Periodical
Forschungsberichte aus dem Gebiet der Luft- und Trocknungstechnik (weekly) . . . 05749

Forschungsvereinigung Kalk-Sand e.V., Entenfangweg 15, 30419 Hannover
T: (0511) 793077; Fax: (0511) 750333
Focus: Materials Sci
Periodical
Forschungsberichte (weekly) . . . 05750

Forschungsvereinigung Porenbeton e.V., Postf 1826, 65008 Wiesbaden
T: (0611) 85086; Fax: (0611) 809707
Focus: Materials Sci . . . 05751

Forschungsvereinigung Programmiersprachen für Fertigungseinrichtungen e.V., Peterstr 17, 52062 Aachen
T: (0241) 25607
Pres: Prof. Dr. W. Maßberg; Gen Secr: Dr. Wolfgang Budde
Focus: Computer & Info Sci 05752

Forschungsvereinigung Schweißen und Schneiden e.V., Aachener Str 172, 40223 Düsseldorf
T: (0211) 15910; Tx: 8582583; Fax: (0211) 1591200
Focus: Metallurgy
Periodicals
Der Praktiker (weekly)
Schweißen und Schneiden: Welding and Cutting (weekly)
Verbindungstechnik in der Elektronik (weekly)
. . . . 05753

Forschungsvereinigung Verbrennungskraftmaschinen e.V. (FVV), Postf 710864, 60498 Frankfurt
T: (069) 66030
Founded: 1956; Members: 52
Pres: Prof. Dr. Ulrich Seiffert; Gen Secr: hartmut Geisendorf
Focus: Auto Eng 05754

Forschungsvereinigung Ziegelindustrie, Schaumburg-Lippe-Str 4, 53113 Bonn
T: (0228) 914930; Tx: 886884; Fax: (0228) 9149327
Focus: Materials Sci 05755

Forschungszentrum des Deutschen Schiffbaues e.V., An der Alster 1, 20099 Hamburg
T: (040) 245505
Focus: Navig 05756

Fränkische Geographische Gesellschaft e.V., Kochstr 4, 91054 Erlangen
T: (09131) 852645; Fax: (09131) 852634
Founded: 1954; Members: 1050
Pres: Prof. Dr. Horst Kopp; Gen Secr: Prof. Dr. Winfried Killisch
Focus: Geography
Periodical
Mitteilungen der Fränkischen Geographischen Gesellschaft 05757

Frankfurter Geographische Gesellschaft e.V. (FGG), Senckenberganlage 36, 60325 Frankfurt
T: (069) 7982913; Fax: (069) 7988382
Founded: 1836; Members: 300
Pres: Prof. Dr. K. Lutz; Prof. Dr. G. Nagel; Dr. H. Piltz; Gen Secr: Dr. F. Fuchs
Focus: Geography
Periodical
Frankfurter Geographische Hefte (weekly) 05758

Fraunhofer-Gesellschaft zur Förderung der angewandten Forschung e.V., Leonrodstr 54, 80636 München
T: (089) 120501; Fax: (089) 1205317
Founded: 1949
Pres: Prof. Dr. Dr. Hans-Jürgen Warnecke
Focus: Nat Sci; Eng; Electronic Eng; Computer & Info Sci; Materials Sci; Civil Eng; Energy; Ecology; Public Health; Econ
Periodicals
Forschungsplan (weekly)
Der Fraunhofer (weekly)
Jahresbericht/Annual Report (weekly)
Mediendienst/Research News (weekly)
Medienspiegel (weekly) 05759

Freier Deutscher Autorenverband e.V. (FDA), Karolinenpl 3, 80333 München
T: (089) 224452
Pres: Prof. Dr. Nikolaus Lobkowicz; Gen Secr: Reinhard Hauschild
Focus: Lit 05760

Freier Verband Deutscher Zahnärzte e.V., Mallwitzstr 16, 53177 Bonn
T: (0228) 85570; Tx: 0885706; Fax: (0228) 347967
Founded: 1955
Pres: Dr. Ralph Gutmann; Gen Secr: Manfred Gilles
Focus: Dent
Periodicals
Der Freie Zahnarzt (weekly)
Zahnarzt aktuell: Informationsdienst für Mitglieder des Freien Verbandes Deutscher Zahnärzte e.V. (weekly) 05761

Freie Vereinigung von Fachleuten öffentlicher Verkehrsbetriebe (FV), Bochumer Str 4, 45879 Gelsenkirchen
T: (0209) 1584230
Members: 270
Pres: Fred Büsing
Focus: Transport 05762

Freundeskreis Deutscher Auslandsschulen e.V., Adenauerallee 148, 53113 Bonn
T: (0228) 104426; Tx: 886805; Fax: (0228) 104158
Pres: Hans-Peter Stihl; Gen Secr: Rolf Raddatz
Focus: Educ 05763

Friedrich-Ebert-Stiftung e.V. / The Friedrich Ebert Foundation, Godesberger Allee 149, 53175 Bonn
T: (0228) 8830; Tx: 885479; Fax: (0228) 883396
Founded: 1925
Pres: Holger Börner; Gen Secr: Klaus-Peter Schneider
Focus: Adult Educ; Hist
Periodicals
Archiv für Sozialgeschichte
Schriftenreihe des Forschungsinstituts der Friedrich-Ebert-Stiftung
Vierteljahresberichte des Forschungsinstituts der Friedrich-Ebert-Stiftung 05764

Frobenius-Gesellschaft e.V., Liebigstr 41, 60323 Frankfurt
T: (069) 721012; Fax: (069) 173725, 722538
Founded: 1938; Members: 170
Pres: Werner Busch
Focus: Ethnology; Hist; Cultur Hist
Periodicals
Ergebnisse der Frobenius-Expeditionen
Paideuma-Mitteilungen zur Kulturkunde (weekly)
Sonderschriften des Frobenius-Instituts
Studien zur Kulturkunde 05765

Frontinus-Gesellschaft e.V., Brüderstr 53, 51427 Bergisch Gladbach
T: (02204) 43724
Pres: Dr. Fritz Gläser
Focus: Eng
Periodicals
Frontinus-Schriftenreihe
Geschichte der Wasserversorgung . . . 05766

Fuldaer Geschichtsverein e.V., Stadtschloss, 36010 Fulda
Focus: Hist
Periodical
Fuldaer Geschichtsblätter (weekly) . . . 05767

Gaswärme-Institut e.V., Hafenstr 101, 45356 Essen
T: (0201) 341023/26
Focus: Air Cond 05768

Gemeinnütziger Verein zur Förderung von Philosophie und Theologie e.V., Rheinsdorfer Burgweg 9, 53332 Bornheim
Pres: Prof. Dr. Karl Kutsch; Gen Secr: Kurt Theodor Fröhlich
Focus: Philos; Rel & Theol 05769

Gemeinschaft katholischer Studierender und Akademiker, August-Bebel-Str 42, 68519 Viernheim
T: (06204) 4801
Focus: Sci 05770

Gemeinschaftsausschuss Kaltformgebung e.V., Kaiserswerther Str 137, 40474 Düsseldorf
Focus: Metallurgy 05771

Gemeinschaftswerk der Evangelischen Publizistik e.V., Emil-von-Behring-Str 3, 60394 Frankfurt
T: (069) 580980; Tx: 176997347; Fax: (069) 58098100
Pres: Dr. A. Haarbeck; Gen Secr: Hans Norbert Janowski
Focus: Journalism; Rel & Theol
Periodicals
epd Film (weekly)
Evangelische Information (weekly)
medienpraktisch (weekly)
medium (weekly) 05772

Gemeinschaft zur Förderung der privaten deutschen Pflanzenzüchtung e.V. (GFP), Kaufmannstr 71, 53115 Bonn
T: (0228) 965410
Focus: Crop Husb 05773

Gemologisches Institut (GIE) / Gilde Internationaler Edelsteinexperten, Kaiserfeld 2, 55743 Idar-Oberstein
T: (06781) 36161; Fax: (06781) 36162
Pres: Peter O. Reiter
Focus: Mineralogy 05774

Genealogische Gesellschaft Hamburg e.V., Postf 302042, 20307 Hamburg
Focus: Genealogy
Periodical
Familiengeschichte in Norddeutschland (weekly)
. . . . 05775

Geographische Gesellschaft, Heinrich-Vogl-Str 7, 81479 München
Focus: Geography
Periodical
Mitteilungen (weekly) 05776

Geographische Gesellschaft Bremen, c/o Übersee-Museum, Bahnhofspl 13, 28195 Bremen
T: (0421) 3619744
Founded: 1877
Gen Secr: Dr. Dieter Heintze
Focus: Geography; Ethnology 05777

Geographische Gesellschaft in Hamburg e.V., Bundesstr 55, 20146 Hamburg
Founded: 1873
Focus: Geography
Periodical
Mitteilungen der Geographischen Gesellschaft in Hamburg (1-2 times annually) 05778

Geographische Gesellschaft zu Hannover (GGH), Schneiderberg 50, 30167 Hannover
T: (0511) 7622233
Founded: 1878; Members: 600
Pres: Prof. Dr. H. Buchholz; Gen Secr: Prof. Dr. A. Arnold
Focus: Geography
Periodical
Hannoversche Geographische Arbeiten (1-2 times annually) 05779

GEO-KART / Wirtschaftsverband für Geodäsie und Kartographie e.V., c/o Carl Zeiss, Postf 1380, 73444 Oberkochen
T: (07364) 203722
Pres: Dr. Dierk Hobbie; Gen Secr: Prof. Dr. Dr. J. Nittinger
Focus: Surveying; Cart 05780

Geologische Vereinigung e.V., Vulkanstr 23, 56743 Mendig
T: (02652) 1508
Founded: 1910; Members: 2900
Pres: Prof. Dr. Wolfgang Schlager; Gen Secr: Dr. Carl-Detlef Cornelius
Focus: Geology
Periodical
Geologische Rundschau (3 times annually) 05781

Georg-Agricola-Gesellschaft zur Förderung der Geschichte der Naturwissenschaften und der Technik e.V., Postf 105463, 40045 Düsseldorf
T: (0211) 4547116; Tx: 08584721; Fax: (0211) 4547111
Founded: 1926; Members: 284
Pres: Prof. Dr. W. Dettmering; Gen Secr: R. Gabrisch
Focus: Nat Sci; Eng 05782

Georg-Friedrich-Händel-Gesellschaft e.V., Große Nikolaistr 3-5, 06108 Halle
T: (045) 24606
Founded: 1955
Pres: Prof. Dr. Bernd Baselt; Gen Secr: Dr. Karin Zauft
Focus: Music
Periodical
Händel-Jahrbuch (weekly) 05783

Georg-von-Vollmar-Akademie e.V., Landwehrstr 37, 80336 München
T: (089) 595223; Fax: (089) 5232549
Pres: Dr. Helmuth Rothemund
Focus: Poli Sci
Periodical
Kochel-Brief (weekly) 05784

Germana Esperanto Asocio r.a. / Deutscher Esperanto-Bund e.V., Rheinweg 15, 53113 Bonn
T: (0228) 235898
Founded: 1906; Members: 3000
Pres: Dr. Wolfgang Schwanzer
Focus: Ling
Periodical
Esperanto aktuell (8 times annually) . . 05785

Germania Judaica / Kölner Bibliothek zur Geschichte des deutschen Judentums e.V., Josef-Haubrich-Hof 1, 50676 Köln
T: (0221) 232349
Founded: 1959
Gen Secr: Dr. Monika Richarz
Focus: Hist 05786

Gesamtverein der deutschen Geschichts- und Altertumsvereine, Severinstr 222-228, 50676 Köln
T: (0221) 2212327
Founded: 1852
Pres: Prof. Dr. Hugo Stehkämper; Gen Secr: Prof. Dr. Hans-Eugen Specker
Focus: Hist
Periodical
Blätter für Deutsche Landesgeschichte (weekly)
. . . . 05787

Gesellschaft anthroposophischer Ärzte, Trossinger Str 53, 70619 Stuttgart
Focus: Med
Periodical
Der Merkurstab: Beiträge zu einer Entwicklung der Heilkunst 05788

Gesellschaft der Ärzte für Erfahrungsheilkunde e.V., Fritz-Frey-Str 21, 69121 Heidelberg
T: (06221) 406222; Fax: (06221) 400727
Founded: 1966; Members: 1350
Gen Secr: Dr. Ewald Fischer
Focus: Med
Periodical
Erfahrungsheilkunde (weekly) 05789

Gesellschaft der Bibliophilen e.V., Am Pfad 1d, 97204 Höchberg
Focus: Libraries & Bk Sci
Periodical
Wandelhalle der Bücherfreunde (weekly) . 05790

Gesellschaft der Musik- und Kunstfreunde Heidelberg e.V., Köpfelweg 25, 69118 Heidelberg
T: (06223) 40695; Fax: (06223) 47135
Founded: 1945; Members: 600
Pres: Franz Günther Büscher
Focus: Music; Arts
Periodical
Die Gesellschaft (3 times annually) . . . 05791

Gesellschaft des Bauwesens e.V. (GdB), RKW-Haus, Düsseldorfer Str 40, 65760 Eschborn
T: (06196) 43143
Founded: 1962; Members: 1000
Pres: Wolfram Zeller; Gen Secr: Horst Wetzel
Focus: Civil Eng 05792

Gesellschaft Deutscher Chemiker e.V. (GDCh), Varrentrappstr 40-42, 60486 Frankfurt
T: (069) 7917320; Tx: 4170497; Fax: (069) 7917322
Founded: 1867; Members: 27000
Pres: Prof. Dr. Dr. H. Noeth; Gen Secr: Prof. Dr. H. tom Dieck
Focus: Chem
Periodicals
Angewandte Chemie (weekly)
Angewandte Chemie International Edition (weekly)
Chemie-Ingenieur-Technik (weekly)
Chemie in unserer Zeit (weekly)
Chemische Berichte (weekly)
Liebigs Annalen der Chemie (weekly)
Nachrichten aus Chemie, Technik und Laboratorium (weekly) 05793

Gesellschaft Deutscher Metallhütten- und Bergleute e.V. (GDMB), Postf 1054, 38668 Clausthal-Zellerfeld
T: (05323) 3438; Fax: (05323) 78804
Founded: 1912; Members: 2000
Pres: Dr. Rolfroderich Nemitz; Gen Secr: D. Dornbusch
Focus: Mining; Metallurgy
Periodicals
Erzmetall (weekly)
Schriftenreihe der GDMB (weekly) . . . 05794

Gesellschaft Deutscher Naturforscher und Ärzte e.V. (GDNÄ), Postf 120190, 51349 Leverkusen
T: (0214) 49990; Fax: (0241) 3071640
Founded: 1822; Members: 6500
Pres: Prof. Dr. Hubert Markl; Gen Secr: Dr. Ernst Truscheit
Focus: Med; Nat Sci 05795

Gesellschaft für Agrargeschichte, Postf 700562, 70574 Stuttgart
T: (0711) 4592618
Founded: 1953; Members: 340
Pres: Josef Hubert Graf von Neippberg; Gen Secr: Prof. Dr. Harald Winkel
Focus: Hist; Agri
Periodical
Zeitschrift für Agrargeschichte und Agrarsoziologie (weekly) 05796

Gesellschaft für angewandte Informatik (GAI), Frankenstr 2, 61352 Bad Homburg
T: (06172) 47262
Focus: Computer & Info Sci 05797

Gesellschaft für Angewandte Mathematik und Mechanik (GAMM), c/o Meerestechnik II, Technische Universität, Eissendorfer Str 42, 21071 Hamburg
Founded: 1922; Members: 1700
Pres: Prof. Dr. R. Mennicken; Gen Secr: Prof. Dr. E. Kreuzer
Focus: Math; Eng
Periodical
GAMM-Mitteilungen (weekly) 05798

Gesellschaft für Anlagen- und Reaktorsicherheit (GRS), Schwertnergasse 1, 50667 Köln
T: (0221) 20680; Tx: 2214123; Fax: (0221) 2068442
Gen Secr: Prof. Dr. Adolf Birkhofer; Gerald Hennenhöfer
Focus: Safety; Nucl Res
Periodicals
GRS-Bericht (weekly)
GRS-Spektrum (weekly) 05799

Gesellschaft für Anthropologie und Humangenetik, c/o Abt Humanbiologie/Anthropologie, Fachbereich Biologie, Universität, 28359 Bremen
Pres: Prof. Dr. H. Walter
Focus: Anthro; Genetics 05800

Gesellschaft für Arbeitswissenschaft e.V. (GfA), Ardeystr 67, 44139 Dortmund
T: (0231) 124243; Fax: (0231) 1084308
Founded: 1953; Members: 644
Pres: Prof. Dr. H. Luczak; Gen Secr: Prof. Dr. W. Laurig
Focus: Business Admin
Periodical
Zeitschrift für Arbeitswissenschaft (weekly) 05801

Gesellschaft für Arzneipflanzenforschung e.V. (GfA) / Society for Medicinal Plant Research, Am Grundbach 5, 97271 Kleinrinderfeld
T: (0931) 800271; Fax: (0931) 800275
Founded: 1953; Members: 1280
Pres: Prof. Dr. G. Franz; Gen Secr: Dr. B. Frank
Focus: Pharmacol
Periodical
Planta Medica (weekly) 05802

Gesellschaft für bedrohte Völker e.V., Postf 2024, 37010 Göttingen
T: (0551) 499060; Fax: (0551) 58028
Focus: Sociology; Poli Sci; Ethnology
Periodical
Pogrom: Zeitschrift für bedrohte Völker . 05803

Gesellschaft für Bibliothekswesen und Dokumentation des Landbaues (GBDL), Engesserstr 20, 76131 Karlsruhe
T: (0721) 6625148; Fax: (0721) 6625111
Founded: 1958; Members: 150
Pres: Prof. Dr. W. Laux; Gen Secr: Dr. Thomas Storck

Focus: Libraries & Bk Sci; Doc; Agri
Periodical
Mitteilungen der Gesellschaft für Bibliothekswesen und Dokumentation des Landbaues (weekly) 05804

Gesellschaft für Biologische Chemie e.V. (GBCh), Bahnhofstr 9-15, 82327 Tutzing
Founded: 1947; Members: 4500
Pres: Prof. Dr. G. Maass; Gen Secr: Dr. K. Beaucamp
Focus: Biochem
Periodical
Hoppe-Seyler's Zeitschrift für Physiologische Chemie (3 times annually) 05805

Gesellschaft für Datenschutz und Datensicherung e.V. (GDD), Irmintrudisstr 1a, 53111 Bonn
T: (0228) 694313; Fax: (0228) 695638
Founded: 1976; Members: 1000
Pres: Bernd Hentschel; Gen Secr: Thomas Muthlein
Focus: Law; Computer & Info Sci
Periodical
Recht der Datenverarbeitung (5 times annually) 05806

Gesellschaft für Deutsche Postgeschichte, Schaumainkai 53, 60552 Frankfurt 70
Focus: Hist; Public Admin
Periodical
Archiv für deutsche Postgeschichte (weekly) 05807

Gesellschaft für deutsche Sprache e.V. (GfdS), Taunusstr 11, 65183 Wiesbaden
T: (0611) 520031; Fax: (0611) 51313
Founded: 1947; Members: 1950
Pres: Prof. Dr. Günther Pflug
Focus: Ling
Periodicals
Muttersprache: Zeitschrift zur Pflege und Erforschung der deutschen Sprache (weekly)
Der Sprachdienst (weekly) 05808

Gesellschaft für Deutschlandforschung e.V., Stresemannstr 90, 10963 Berlin
T: (030) 2614350; Fax: (030) 2651248
Founded: 1978; Members: 700
Pres: Prof. Dr. Jens Hacker
Focus: Law; Educ; Econ; Poli Sci . . . 05809

Gesellschaft für Dezentralisierte Energiewirtschaft e.V., Stuttgarter Str 95, 71638 Ludwigsburg
T: (07141) 20777
Focus: Energy; Econ 05810

Gesellschaft für die Geschichte und Bibliographie des Brauwesens e.V., Seestr 13, 13353 Berlin
T: (030) 4509234, 4509264; Tx: 186043; CA: Amylum Berlin; Fax: (030) 4536069
Founded: 1913; Members: 250
Pres: Dr. Uwe Paulsen; Gen Secr: Dr. Hans Günter Schultze-Berndt
Focus: Eng
Periodical
Jahrbuch (weekly) 05811

Gesellschaft für Elektrische Hochleistungsprüfungen (PEHLA), Theodor-Stern-Kai 1, Hochhaus Süd, 60596 Frankfurt
T: (069) 66418148; Tx: 411076
Focus: Electric Eng 05812

Gesellschaft für empirische soziologische Forschung e.V., Marienstr 2, 90402 Nürnberg
T: (0911) 224333; Fax: (0911) 225685
Founded: 1949; Members: 13
Pres: Dr. Herbert Rische; Gen Secr: Dr. Rainer Wasilewski
Focus: Sociology 05813

Gesellschaft für Epilepsieforschung e.V., Königsweg 3, 33617 Bielefeld
T: (0521) 1443568
Founded: 1955
Pres: P. Johannes Busch; Gen Secr: Eberhard Schmidt
Focus: Med 05814

Gesellschaft für Erdkunde zu Berlin (GFE), Arno-Holz-Str 14, 12165 Berlin
T: (030) 7919001; Fax: (030) 7933249
Founded: 1828; Members: 500
Focus: Geography
Periodical
Die Erde (weekly) 05815

Gesellschaft für Erdkunde zu Köln e.V., c/o Geographisches Institut, Universität, Albertus-Magnus-Pl, 50931 Köln
Focus: Geography 05816

Gesellschaft für Erd- und Völkerkunde, Meckenheimer Allee 166, 53115 Bonn
T: (0228) 737506
Founded: 1910; Members: 300
Pres: Prof. Dr. Eberhard Mayer; Gen Secr: Dr. Günter Menz
Focus: Geography; Ethnology 05817

Gesellschaft für Erd- und Völkerkunde zu Stuttgart e.V., Hegelpl 1, 70174 Stuttgart
T: (0711) 20503222
Founded: 1882; Members: 1100
Pres: Prof. Dr. Christoph Borcherdt
Focus: Geography; Ethnology 05818

Gesellschaft für Ernährungsbiologie e.V., Veterinärstr 13, 80539 München
Members: 200
Pres: Prof. Dr. G. Wolfram; Gen Secr: Dr. Ingrid Ascher
Focus: Nutrition; Bio 05819

Gesellschaft für Ernährungsphysiologie, Eschborner Landstr 122, 60489 Frankfurt
T: (069) 24788308; Fax: (069) 24788114
Founded: 1953; Members: 106
Pres: Prof. Dr. D. Giesecke; Gen Secr: Dr. Hans-Hermann Freese
Focus: Vet Med
Periodical
Proceedings of the Society of Nutrition Physiology (1-2 times annually) 05820

Gesellschaft für Familienkunde in Franken e.V., Archivstr 17, 90408 Nürnberg
Focus: Genealogy
Periodical
Blätter für Fränkische Familienkunde (weekly) 05821

Gesellschaft für Finanzwirtschaft in der Unternehmensführung e.V. (GEFIU), c/o Metallgesellschaft AG, Reuterweg 14, 60323 Frankfurt
T: (069) 1593376; Tx: 41225; Fax: (069) 1592125
Focus: Finance; Business Admin 05822

Gesellschaft für Geistesgeschichte e.V., Drosselweg 8, 91056 Erlangen
T: (09131) 41187
Founded: 1958; Members: 150
Pres: Prof. Dr. Julius H. Schoeps; Gen Secr: Joachim Fricke
Focus: Hist; Humanities
Periodical
Zeitschrift für Religions- und Geistesgeschichte (weekly) 05823

Gesellschaft für Geographie und Geologie Bochum, Tippelspfad 3, 44803 Bochum
Focus: Geography; Geology 05824

Gesellschaft für Geschichte und Kultur e.V., Postf 2513, 53015 Bonn
T: (0228) 611106; Fax: (0228) 611106
Pres: Peter Wegener; Gen Secr: Kurt Wittkowski
Focus: Hist; Cultur Hist
Periodical
Archiv für Geschichte und Kultur . . . 05825

Gesellschaft für Goldschmiedekunst, Altstädter Markt 6, 63450 Hanau
T: (06181) 256556; Fax: (06181) 256554
Pres: Torsten Bröhan; Gen Secr: Dr. C. Weber
Focus: Fine Arts 05826

Gesellschaft für Historische Waffen- und Kostümkunde, c/o Historisches Museum, Pferdestr 6, 30159 Hannover
T: (0511) 1683965; Fax: (0511) 1685003
Founded: 1896; Members: 236
Pres: Dr. Alheidis von Rohr; Gen Secr: Gerhard Grosse-Löscher
Focus: Hist; Military Sci
Periodical
Zeitschrift für Historische Waffen- und Kostümkunde (weekly) 05827

Gesellschaft für Immunologie e.V., c/o Dr. Friedrich R. Seiler, Behringwerke AG, Emil-von-Behring-Str 76, 35041 Marburg
T: (06421) 392338; Fax: (06421) 394663
Founded: 1968; Members: 1350
Pres: Prof. Dr. Fritz Melchers; Gen Secr: Dr. Friedrich R. Seiler
Focus: Immunology 05828

Gesellschaft für Informatik e.V. (GI), Godesberger Allee 99, 53175 Bonn
T: (0228) 376751; Fax: (0228) 378178
Founded: 1970; Members: 18000
Pres: Prof. Dr. Roland Vollmar; Gen Secr: Dr. Hermann Rampacher
Focus: Computer & Info Sci
Periodical
Informatik-Spektrum (weekly) 05829

Gesellschaft für Informationsverarbeitung in der Landwirtschaft (GIL), Birkheckenstr 100a, 70599 Stuttgart
Founded: 1980; Members: 180
Gen Secr: Prof. Dr. Hans Geidel
Focus: Computer & Info Sci; Agri
Periodical
Infoagrar (weekly) 05830

Gesellschaft für Informationsvermittlung und Technologieberatung, Blumenstr 1, 80331 München
T: (089) 263060; Fax: (089) 2608471
Focus: Business Admin; Econ; Electronic Eng; Electric Eng; Energy; Chem; Med; Bio . 05831

Gesellschaft für Interkulturelle Germanistik e.V. (GIG), c/o Institut für Internationale Kommunikation und Auswärtige Kulturarbeit, Jahnstr 8-10, 95444 Bayreuth
T: (0921) 515345; Fax: (0921) 511207
Focus: Ling; Lit 05832

Gesellschaft für internationale Geldgeschichte (GIG), Rotlintstr 66, 60318 Frankfurt
T: (069) 498921; Fax: (06105) 71356
Pres: Christian Stoess; Gen Secr: Ilse Wagner
Focus: Finance; Econ
Periodical
Geldgeschichtliche Nachrichten (weekly) . 05833

Gesellschaft für internationale Sprache e.V., Schaumanns Kamp 126, 21465 Reinbek
Focus: Ling
Periodical
Interlinguistika Informa Servo (weekly) . . 05834

Gesellschaft für Klassifikation e.V. (GfKl), c/o Institut für Statistik, RWTH, Wüllnerstr 3, 52056 Aachen
T: (0241) 804573; Fax: (0241) 804776
Founded: 1977; Members: 280
Pres: Prof. Dr. H.H. Bock
Focus: Libraries & Bk Sci 05835

Gesellschaft für Konsum-, Markt- und Absatzforschung (GfK), Nordwestring 101, 90319 Nürnberg
T: (0911) 3950; Fax: (0911) 3952209
Founded: 1934; Members: 840
Pres: Klaus Hehl
Focus: Marketing 05836

Gesellschaft für Literatur in Nordrhein-Westfalen e.V., Wilmergasse 12, 48143 Münster
T: (0251) 46877
Founded: 1977
Pres: Wolfram Dorn; Gen Secr: M. Damwerth
Focus: Lit 05837

Gesellschaft für Lungen- und Atmungsforschung e.V., Gilsingstr 14, 44789 Bochum
T: (0234) 3026444; Fax: (0234) 3026420
Founded: 1962; Members: 400
Gen Secr: Prof. Dr. Gerhard Schultze-Werninghaus
Focus: Pulmon Dis
Periodical
Atemwegs- und Lungenkrankheiten (weekly) 05838

Gesellschaft für Mathematik und Datenverarbeitung, Postf 1316, 53731 Sankt Augustin
T: (02241) 141; Fax: (02241) 142619
Focus: Math; Computer & Info Sci . . . 05839

Gesellschaft für mathematische Forschung e.V., Lorenzenhof, Walke, 77709 Oberwolfach
T: (07834) 9790; Fax: (07834) 97938
Founded: 1959; Members: 53
Gen Secr: Prof. Dr. Martin Barner
Focus: Math
Periodical
Archiv der Mathematik (weekly) 05840

Gesellschaft für Mittelrheinische Kirchengeschichte, Karmeliterstr 1-3, 56068 Koblenz
Focus: Hist; Rel & Theol
Periodical
Archiv für Mittelrheinische Kirchengeschichte (weekly) 05841

Gesellschaft für Musikforschung e.V. (GfM), Heinrich-Schütz-Allee 35, 34131 Kassel
T: (0561) 31050
Founded: 1947; Members: 1700
Pres: Prof. Dr. Klaus W. Niemöller; Gen Secr: Barbara Schumann
Focus: Music
Periodical
Die Musikforschung (weekly) 05842

Gesellschaft für Naturkunde in Württemberg e.V., Rosenstein 1, 70191 Stuttgart
T: (0711) 89360
Founded: 1844; Members: 850
Pres: Prof. Dr. Ulrich Kull; Gen Secr: Dr. Horst Janus
Focus: Ecology
Periodical
Jahreshefte 05843

Gesellschaft für Natur- und Völkerkunde Ostasiens e.V., c/o Japanisches Seminar, Von-Melle-Park 6, 20146 Hamburg
Focus: Ethnology
Periodicals
Kagami: Japanischer Zeitschriftenspiegel (3 times annually)
Nachrichten der Gesellschaft für Natur- und Völkerkunde Ostasiens e.V.: Zeitschrift für Kultur und Geschichte Ost- und Südostasiens (weekly) 05844

Gesellschaft für Neue Musik e.V., Ernst-Schwendler-Str 2-4, 60320 Frankfurt
T: (069) 446030; Fax: (069) 441063
Members: 400
Pres: Prof. Friedrich Goldmann; Gen Secr: Ulf Werner
Focus: Music 05845

Gesellschaft für öffentliche Wirtschaft e.V. (GöWG), Sarrazinstr 11-15, 12159 Berlin
T: (030) 8521045/46
Founded: 1951; Members: 100
Pres: Felix Zimmermann; Gen Secr: Wolf Leetz
Focus: Econ
Periodical
Zeitschrift für öffentliche und gemeinwirtschaftliche Unternehmen (weekly) 05846

Gesellschaft für Organisation e.V. (GfürO), Gutenbergstr 23, 35390 Giessen
T: (0641) 36033
Founded: 1922
Pres: Peter Quirin; Gen Secr: Dr. Reiner Chrobok
Focus: Public Admin; Business Admin
Periodical
Zeitschrift Führung + Organisation (8 times annually) 05847

Gesellschaft für Pädagogik und Information e.V., Rathenaustr 16, 33102 Paderborn
T: (05251) 34024
Founded: 1964; Members: 400
Pres: Prof. Dr. Dr. G.E. Ortner; Prof. Dr. U. Lehnert
Focus: Educ; Computer & Info Sci
Periodical
Pädagogik und Information 05848

Gesellschaft für Pädiatrische Radiologie, c/o Röntgenabt. der Universitäts-Kinderklinik, Josef-Schneider-Str, 97080 Würzburg
Focus: Radiology 05849

Gesellschaft für prä- und postoperative Tumortherapie e.V., Sackgasse 8, 55278 Undenheim
T: (06737) 1298
Founded: 1969; Members: 150
Focus: Therapeutics 05850

Gesellschaft für praktische Energiekunde e.V. (GFPE), Am Blütenanger 71, 80995 München
T: (089) 1581210
Founded: 1949; Members: 177
Focus: Energy 05851

Gesellschaft für Programmierte Instruktion und Mediendidaktik e.V., c/o Seminar Anglistik/Didaktik, Justus-Liebig-Universität, Otto-Behaghel-Str 10, 35394 Giessen
Founded: 1964; Members: 600
Focus: Mass Media; Computer & Info Sci; Educ
Periodical
Neue Unterrichtspraxis 05852

Gesellschaft für Psychotherapie, Psychosomatik und Medizinische Psychologie e.V., Karl-Tauchnitz-Str 25, 04107 Lipzig
T: (041) 328503; Tx: (041) 051350
Founded: 1960; Members: 1800
Pres: Prof. Dr. M. Geyer
Focus: Psych; Therapeutics 05853

Gesellschaft für publizistische Bildungsarbeit e.V., Haus Busch, Helfe, 58099 Hagen
T: (02331) 64555; Fax: (02331) 65587
Focus: Journalism; Educ 05854

Gesellschaft für rationale Verkehrspolitik e.V. (GRV), Bromberger Str 5, 40599 Düsseldorf
T: (0211) 741507
Founded: 1970; Members: 242
Pres: Dr. Alfons Thoma; Gen Secr: Werner Kammer
Focus: Poli Sci
Periodical
GRV-Nachrichten (weekly) 05855

Gesellschaft für Rationelle Energieverwendung e.V., Theodor-Heuss-Pl 7, 14062 Berlin
T: (030) 3016090
Focus: Energy
Periodical
Merkblätter (weekly) 05856

Gesellschaft für Rechtsvergleichung e.V., Humboldtallee 15, 37073 Göttingen
T: (0551) 59035
Founded: 1950; Members: 1100
Pres: Prof. Dr. Peter Schlechtriem
Focus: Law
Periodicals
Arbeiten zur Rechtsvergleichung
Ausländische Aktiengesetze
Mitteilungen (weekly) 05857

Gesellschaft für Regionalforschung e.V. (GfR), Kardinal-Gahlen-Weg 10, 53175 Bonn
T: (0228) 310563
Pres: Prof. Uwe Schubert; Gen Secr: Brigitte Kirschner
Focus: Hist; Sociology
Periodical
Seminarberichte (weekly) 05858

Gesellschaft für Sicherheitswissenschaft e.V. (GfS), c/o Prof. Dr. Peter C. Compes, Bergische Universität, Gaußstr 20, 42097 Wuppertal
T: (0202) 4392060
Members: 250
Pres: Prof. Dr. Peter C. Compes
Focus: Safety
Periodicals
Berichte über GfS-Sommer-Symposien
Sicherheitswissenschaftliche Monographien 05859

Gesellschaft für Sozial- und Wirtschaftsgeschichte, c/o Institut für Sozial- und Wirtschaftsgeschichte, Universität Heidelberg, Grabengasse 14, 69117 Heidelberg
T: (06221) 542933
Founded: 1961; Members: 220
Pres: Prof. Dr. E. Schremmer; Gen Secr: Prof. Dr. J. Schneider
Focus: Hist; Soc Sci; Econ 05860

Gesellschaft für sozialwissenschaftliche Forschung in der Medizin, Werderring 16, 79098 Freiburg
T: (0761) 36349
Gen Secr: Klaus Riemann
Focus: Med; Sociology 05861

Gesellschaft für Strahlen- und Umweltforschung (GSF), Ingolstädter Landstr 1, 91465 Ergersheim
T: (09847) 31870
Founded: 1964
Focus: Ecology; Nat Sci
Periodical
GSF-Bericht 05862

Gesellschaft für Technologiefolgenforschung e.V. (GTF), Hohenzollerndamm 91, 14199 Berlin
T: (030) 8255070
Focus: Eng; Sociology; Ecology 05863

Gesellschaft für Tribologie e.V. (GfT), Ernststr 12, 47443 Moers
T: (02841) 54213
Founded: 1959; Members: 410
Focus: Eng 05864

Gesellschaft für Übernationale Zusammenarbeit e.V., Bachstr 32, 53115 Bonn
T: (0228) 7290080
Founded: 1945
Pres: Dr. Franz Schoser; Prof. Joseph Rovan; Gen Secr: Horst Reinert
Focus: Poli Sci
Periodicals
Dokumente: Zeitschrift für den deutsch-französischen Dialog
Revue des questions allemandes – Documents 05865

Gesellschaft für Unternehmensgeschichte e.V., Bonner Str 211, 50968 Köln
T: (0221) 387737; Fax: (0221) 342760
Pres: Dr. Reinhart Freudenberg; Gen Secr: Prof. Manfred Pohl
Focus: Hist; Econ
Periodicals
Anno: Magazin für Unternehmensgeschichte (weekly)
Zeitschrift für Unternehmensgeschichte (weekly) 05866

Gesellschaft für Ursachenforschung bei Verkehrsunfällen e.V. (GUVU), Hermeskeiler Str 17, 50935 Köln
T: (0221) 4303373
Founded: 1959; Members: 280
Pres: Alfred Zerban; Gen Secr: Prof. Dr. Klaus Engels
Focus: Transport
Periodicals
sicher aktuell
VS aktuell (weekly) 05867

Gesellschaft für Versicherungswissenschaft und -gestaltung e.V. (GVG), Prälat-Otto-Müller-Pl 2, 50670 Köln
T: (0221) 726965; Fax: (0221) 730373
Founded: 1947
Pres: Dr. Herbert Rische; Gen Secr: Dr. Volker Leienbach
Focus: Insurance
Periodicals
Informationsdienst (10 times annually)
Schriftenreihe (2-3 times annually) . . . 05868

Gesellschaft für Wirbelsäulenforschung, c/o BG-Unfallklinik, Friedberger Landstr 430, 60389 Frankfurt
Focus: Anat 05869

Gesellschaft für Wirtschaftskunde e.V., Albertus-Magnus-Str 2, 57072 Siegen
T: (0271) 52799
Pres: Prof. Dr. Gerhard Merk
Focus: Econ 05870

Gesellschaft für Wirtschafts- und Sozialwissenschaften, Seminargebäude A5, 68159 Mannheim
T: (0621) 2923457; Fax: (0621) 2925007
Founded: 1872; Members: 1700
Pres: Prof. Dr. Heinz König
Focus: Sociology; Econ
Periodical
Zeitschrift für Wirtschafts- und Sozialwissenschaften (weekly) 05871

Gesellschaft für wissenschaftliche Gesprächspsychotherapie e.V. (GwG), Richard-Wagner-Str 12, 50674 Köln
T: (0221) 237917
Members: 7300
Focus: Educ Handic
Periodicals
GwG-Zeitschrift (weekly)
Informationsblätter der GwG (weekly) . . 05872

Gesellschaft für Wissenschaft und Leben im Rheinisch-Westfälischen Industriegebiet e.V. (GWL), Hollestr 1, 45127 Essen
Founded: 1919
Pres: Prof. Dr. D. Weis; Gen Secr: Prof. Dr. E. Steinmetz
Focus: Sci; Sociology 05873

Gesellschaft für Wohnungsrecht und Wohnungswirtschaft Köln e.V., Gyrhofstr 2, 50931 Köln
T: (0221) 4702311
Focus: Urban Plan; Finance; Law . . . 05874

Gesellschaft Information Bildung e.V. (GIB), Schloßstr 29, 60486 Frankfurt
T: (069) 247080; Fax: (069) 24708444
Founded: 1992; Members: 28
Pres: Hartmut Müller
Focus: Doc; Educ 05875

Gesellschaft Rheinischer Ornithologen e.V., Schlesische Str 80, 40231 Düsseldorf
T: (0211) 214885
Focus: Ornithology
Periodical
Charadrius: Zeitschrift für Vogelkunde, Vogelschutz und Naturschutz in Nordrhein-Westfalen (weekly) 05876

Gesellschaft Sozialwissenschaftlicher Infrastruktureinrichtungen e.V. (GESIS), B2, 1, 68159 Mannheim
Founded: 1986; Members: 3
Focus: Soc Sci 05877

Gesellschaft zum Studium strukturpolitischer Fragen e.V., Postf 120333, 53045 Bonn
T: (0228) 691316; Fax: (0228) 697327
Pres: Dr. Ludolf von Wartenberg; Gen Secr: Hans-Hermann Lutzke
Focus: Poli Sci
Periodical
Die strukturpolitische Information 05878

Gesellschaft zur Erforschung des Markenwesens e.V., Schöne Aussicht 59, 65193 Wiesbaden
T: (0611) 58670; Fax: (0611) 586727
Founded: 1954
Pres: Walter Eggers; Gen Secr: Barbara von Moeller
Focus: Econ 05879

Gesellschaft zur Förderung der Erforschung der Zuckerkrankheit e.V., Auf'm Hennekamp 65, 40225 Düsseldorf
T: (0211) 33821
Focus: Hematology 05880

Gesellschaft zur Förderung der finanzwissenschaftlichen Forschung e.V., Zülpicher Str 182, 50937 Köln
T: (0221) 426979; Fax: (0221) 422352
Founded: 1949; Members: 150
Pres: Helmut Loehr; Gen Secr: Dr. Dieter Ewringmann
Focus: Finance 05881

Gesellschaft zur Förderung der Lufthygiene und Silikoseforschung e.V., Auf'm Hennekamp 50, 40225 Düsseldorf
T: (0211) 33890; Tx: 8582164; Fax: (0211) 3190910
Founded: 1962; Members: 9
Pres: Prof. Dr. Hans Schadewaldt; Gen Secr: Prof. Dr. Hans-Werner Schlipkoter
Focus: Pulmon Dis; Toxicology; Immunology; Hygiene
Periodicals
Jahresberichte
Umwelthygiene 05882

Gesellschaft zur Förderung der Segelflugforschung e.V., Herrenstr 14-16, 79098 Freiburg
T: (0761) 52719
Focus: Aero 05883

Gesellschaft zur Förderung der Spektrochemie und angewandten Spektroskopie e.V., Bunsen-Kirchhoff-Str 11, 44139 Dortmund
T: (0231) 13920; Fax: (0231) 1392120
Founded: 1952; Members: 45
Pres: Dr. Günther Breil
Focus: Chem; Physics 05884

Gesellschaft zur Förderung der Wissenschaftlichen Forschung über das Spar- und Girowesen e.V., Simrockstr 4, 53113 Bonn
T: (0228) 204241; Fax: (0228) 204705
Founded: 1955; Members: 643
Pres: Dr. H. Rehm; Gen Secr: Dr. E. Ketzel
Focus: Finance 05885

Gesellschaft zur Förderung der wissenschaftlichen Zusammenarbeit mit der Universität Tel-Aviv, Lucas-Cranach-Str 17, 53175 Bonn
T: (0228) 374005
Focus: Sci 05886

Gesellschaft zur Förderung des Unternehmernachwuchses e.V. (GFU), Lichtentaler Str 92, 76530 Baden-Baden
T: (07221) 700402; Fax: (07221) 700422
Founded: 1955; Members: 105
Pres: Dr. Wilfried Guth; Gen Secr: Dr. Peter Zürn
Focus: Adult Educ 05887

Gesellschaft zur Förderung Frankfurter Malerei des 19. und 20. Jahnhunderts e.V., Atzelbergstr 9, 60329 Frankfurt
T: (069) 479920
Focus: Fine Arts 05888

Gesellschaft zur Förderung Pädagogischer Forschung e.V. (GFPF), Schloßstr 29, 60486 Frankfurt
T: (069) 69770245; Tx: 4170331; Fax: (069) 69708228
Founded: 1950; Members: 500
Pres: Hans Krollmann; Gen Secr: P. Doebrich
Focus: Educ
Periodical
Zeitschrift für erziehungs- und sozialwissenschaftliche Forschung (weekly) 05889

Gesellschaft zur Herausgabe des Corpus Catholicorum, Johannisstr 8-10, 48143 Münster
Founded: 1917
Focus: Rel & Theol 05890

GES-Gesellschaft für elektronische Systemforschung, 78476 Allensbach
Focus: Electronic Eng 05891

Gesundheitspolitische Gesellschaft e.V., Weimarer Str 8, 24106 Kiel
T: (0431) 338550
Pres: Prof. Dr. Fritz Beske
Focus: Poli Sci 05892

Gesundheitstechnische Gesellschaft e.V. (GG), Alt-Marienfelde 12d, 12277 Berlin
T: (030) 726046
Founded: 1949
Pres: Prof. Dr. H. Protz; Gen Secr: M. Samp
Focus: Eng; Public Health
Periodical
GG-Nachrichten (10 times annually) . . 05893

GFM-GETAS Gesellschaft für Marketing-, Kommunikations- und Sozialforschung, Langelohstr 134, 22545 Hamburg
Founded: 1945
Focus: Marketing; Soc Sci 05894

GKSS-Forschungszentrum Geesthacht (GKSS), Max-Planck-Str, 21502 Geesthacht
Founded: 1956; Members: 700
Focus: Eng
Periodical
GKSS-Information (weekly) 05895

Görres-Gesellschaft zur Pflege der Wissenschaft, Postf 101618, 50456 Köln
T: (0221) 738317
Founded: 1876; Members: 3000
Pres: Prof. Dr. Dr. Paul Mikat; Gen Secr: Prof. Dr. Rudolf Schieffer
Focus: Sci
Periodicals
Historisches Jahrbuch
Jahrbuch für Volkskunde
Kirchenmusikalisches Jahrbuch
Literaturwissenschaftliches Jahrbuch
Oriens Christianus
Philosophisches Jahrbuch
Portugiesische Forschungen
Römische Quartalschrift (weekly)
Spanische Forschungen
Vierteljahresschrift für wissenschaftliche Pädagogik
Zeitschrift für Klinische Psychologie, Psychopathologie und Psychotherapie . . 05896

Goethe-Gesellschaft in Weimar e.V., Burgpl 4, 99403 Weimar
T: (03643) 202050; Fax: (03643) 202050
Founded: 1885; Members: 5000
Pres: Prof. Dr. Werner Keller; Gen Secr: Dr. Gunter Rentzsch
Focus: Lit 05897

Göttinger Arbeitskreis, Calsowstr 54, 37085 Göttingen
T: (0551) 55848; Fax: (0551) 486203
Pres: Prof. Dr. Boris Meissner; Gen Secr: Dr. Alfred Eisfeld
Focus: Hist
Periodical
Deutsche in der ehemaligen Sowjetunion (weekly) 05898

Gottfried-Wilhelm-Leibniz-Gesellschaft e.V., c/o Niedersächsische Landesbibliothek, Waterloostr 8, 30169 Hannover
T: (0511) 1267331; Fax: (0511) 1267202
Founded: 1966; Members: 370
Pres: Prof. Dr. E.G. Mahrenholz; Gen Secr: Prof. Dr. W. Totok
Focus: Philos
Periodicals
Studia Leibnitiana
Studia Leibnitiana Sonderhefte
Studia Leibnitiana Supplementa 05899

Gruppe Ökologie, c/o Institut für ökologische Forschung und Bildung e.V., Kleine Düvelstr 21, 30171 Hannover
T: (0511) 853055; Fax: (0511) 853062
Focus: Ecology 05900

Gustav Freytag Gesellschaft e.V., Bahnhofstr 71, 40883 Ratingen
T: (02102) 65203
Pres: Dr. Jürgen Matoni
Focus: Lit
Periodical
Gustav-Freytag-Blätter (weekly) 05901

Gutenberg-Gesellschaft (GG) / Internationale Vereinigung für Geschichte und Gegenwart der Druckkunst, Liebfrauenpl 5, 55116 Mainz
T: (06131) 226420; Fax: (06131) 123488
Founded: 1901; Members: 1650
Pres: Hermann-Hartmut Weyel; Gen Secr: Gertraude Benöhr
Focus: Hist; Eng 05902

GVC/VDI-Gesellschaft Verfahrenstechnik und Chemieingenieurwesen / Prozess- und Umwelttechnik, Graf-Recke-Str 84, 40239 Düsseldorf
T: (0211) 6214256; Tx: 8586525; Fax: (0211) 6214575
Founded: 1934; Members: 8000
Pres: Prof. Dr. M. Bohnet; Gen Secr: Prof. Dr. H. Cremer
Focus: Eng
Periodicals
Chemical Engineering and Technology (weekly)
Chemie-Ingenieur-Technik (weekly) . . . 05903

Hafenbautechnische Gesellschaft e.V. (HTG), Dalmannstr 1, 20457 Hamburg
T: (040) 32851
Founded: 1914; Members: 1400
Pres: Dr. Jochen Müller; Gen Secr: Karlheinz Pöpping
Focus: Civil Eng
Periodicals
Binnenschiffahrt (weekly)
Hansa (weekly) 05904

Hahn-Schickard-Gesellschaft für angewandte Forschung e.V., Breitscheidstr 2b, 70174 Stuttgart
T: (0711) 293174; Fax: (0711) 2268304
Founded: 1954; Members: 80
Pres: Dr. M. Berner; Gen Secr: K. Schuler
Focus: Eng 05905

Hamburger Autorenvereinigung e.V., Jürgensallee 13, 22609 Hamburg
T: (040) 876758
Founded: 1977
Pres: Rosemarie Fiedler-Winter; Gen Secr: G. Laub
Focus: Lit 05906

Hamburger Gesellschaft für Völkerrecht und Auswärtige Politik e.V., Rothenbaumchaussee 21-23, 20148 Hamburg
T: (040) 41234607
Gen Secr: Dr. Karl-Andreas Hernekamp
Focus: Poli Sci; Law
Periodical
Verfassung und Recht in Übersee: Law and Politics in Africa, Asia and Latin America (weekly) 05907

Hannoversches Forschungsinstitut für Fertigungsfragen e.V. (HFF), Welfengarten 1a, 30167 Hannover
T: (0511) 7622264
Focus: Standards; Eng 05908

Hartmannbund / Verband der Ärzte Deutschlands e.V., Godesberger Allee 54, 53175 Bonn
T: (0228) 81040; Fax: (0228) 18104155
Founded: 1900; Members: 41553
Pres: Dr. Hans-Jürgen Thomas; Gen Secr: Klaus Nöldner
Focus: Public Health
Periodical
Hartmannbund-Magazin (weekly) 05909

Harzverein für Geschichte und Altertumskunde, Zehntstr 24, 38640 Goslar
Focus: Hist; Archeol
Periodical
Harz-Zeitschrift (weekly) 05910

Heidelberger Akademie der Wissenschaften, Karlstr 4, 69117 Heidelberg
T: (06221) 543265; Fax: (06221) 543355
Founded: 1909
Pres: Prof. Dr. Albrecht Dihle; Gen Secr: Gunther Jost
Focus: Sci
Periodical
Zentralblatt für Mathematik und Grenzgebiete (weekly) 05911

Hessische Akademie für Bürowirtschaft e.V. (HAB), Kandelstr 7, 60528 Frankfurt
T: (069) 673857
Founded: 1965; Members: 420
Focus: Adult Educ
Periodical
HAB-Journal (weekly) 05912

Hessische Krebsgesellschaft e.V., Heinrich-Heine-Str 44, 35039 Marburg
T: (06421) 63324; Fax: (06421) 600711
Founded: 1952; Members: 130
Pres: Prof. Dr. W.-D. Hirschmann; Gen Secr: Dr. Margret Schrader
Focus: Cell Biol & Cancer Res 05913

Hessischer Philologen-Verband, Hellmundstr 5, 65183 Wiesbaden
Focus: Ling
Periodical
blickpunkt schule (8 times annually) . . 05914

Historische Kommision der Deutschen Gesellschaft für Erziehungswissenschaft, Lüerstr 3, 30175 Hannover
T: (0511) 7629412/13; Fax: (0511) 7623456
Founded: 1967; Members: 300
Focus: Hist; Educ
Periodical
Information zur erziehungs- und bildungshistorischen Forschung (IZEBF) (2-3 times annually) 05915

Historische Kommission des Börsenvereins des Deutschen Buchhandels, Grosser Hirschgraben 17-21, 60311 Frankfurt
T: (069) 1306287; Fax: (069) 1306382
Founded: 1953; Members: 27
Pres: Dr. Wulf D. von Lucius; Gen Secr: Dr. Monika Estermann
Focus: Hist; Libraries & Bk Sci
Periodical
Archiv für Geschichte des Buchwesens (weekly) 05916

Historische Kommission zu Berlin, Kirchweg 33, 14129 Berlin
T: (030) 8160010
Focus: Hist
Periodical
I.W.K.: Internationale wissenschaftliche Korrespondenz zur Geschichte der deutschen Arbeiterbewegung (weekly) 05917

Historische Kommission zur Erforschung des Pietismus an der Universität Münster, Georgskommende 7, 48143 Münster
Focus: Hist; Rel & Theol
Periodicals
Arbeiten zur Geschichte des Pietismus
Pietismus und Neuzeit: Ein Jahrbuch zur Geschichte des neueren Protestantismus (weekly)
Texte zur Geschichte des Pietismus (weekly)
. . . . 05918

Historischer Verein Bamberg, Untere Sandstr 46, 46049 Bamberg
T: (0951) 28936
Founded: 1830; Members: 1001
Pres: Prof. Dr. Gerd Zimmermann
Focus: Hist
Periodical
Bericht des Historischen Vereins Bamberg (weekly)
. . . . 05919

Historischer Verein der Pfalz e.V., Postf 1429, 67324 Speyer
Focus: Hist
Periodical
Mitteilungen (weekly) 05920

Historischer Verein Dillingen an der Donau, Örtelstr 10, 89407 Dillingen
Focus: Hist
Periodical
Jahrbuch (weekly) 05921

Historischer Verein für die Saargegend e.V., c/o Stadtarchiv, Postf 439, 66104 Saarbrücken
T: (0681) 9051546
Founded: 1839; Members: 850
Pres: Dr. Dieter Staerk; Gen Secr: Dr. Fritz Jacoby
Focus: Hist
Periodical
Zeitschrift für die Geschichte der Saargegend (weekly) 05922

Historischer Verein für Hessen, c/o Staatsarchiv, Karolinenpl 3, 64289 Darmstadt
T: (06151) 165900; Fax: (06151) 165901
Focus: Hist
Periodicals
Archiv für Hessische Geschichte und Altertumskunde (weekly)
Darmstädter Archivschriften (weekly)
Hessische Beiträge zur Geschichte der Arbeiterbewegung (weekly) 05923

Historischer Verein für Oberfranken e.V., Ludwigstr 21, Postf 110263, 95421 Bayreuth
T: (0921) 65307
Founded: 1827; Members: 770
Pres: Wolfgang Winkler; Gen Secr: Norbert Hübsch
Focus: Hist
Periodical
Archiv für Geschichte von Oberfranken (weekly)
. . . . 05924

Historischer Verein für Schwaben, Schaezlerstr 25, 86152 Augsburg
Gen Secr: Rainer Frank
Focus: Hist
Periodical
Zeitschrift des Historischen Vereins für Schwaben (weekly) 05925

Historischer Verein für Württembergisch Franken, c/o Stadtarchiv, Postf 100180, 74501 Schwäbisch Hall
Focus: Hist
Periodical
Württembergisch Franken (weekly) . . . 05926

Historischer Verein Rupertiwinkel e.V., Postf 1108, 83405 Laufen
Focus: Hist
Periodical
Das Salzfaß (weekly) 05927

Hochschullehrerbund e.V. (HLB), Rüngsdorfer Str 4c, 53173 Bonn
T: (0228) 352271; Fax: (0228) 354512
Founded: 1973
Pres: Prof. Werner Kuntze; Gen Secr: Dr. Hubert Mücke
Focus: Educ
Periodical
Die Neue Hochschule (weekly) 05928

Hochschulrektorenkonferenz, Ahrstr 39, 53175 Bonn
T: (0228) 8870; Fax: (0228) 887110
Founded: 1949; Members: 233
Pres: Prof. Dr. Hans-Uwe Erichsen
Focus: Adult Educ
Periodical
Bibliographie der Hochschulrektorenkonferenz (20 times annually) 05929

Hölderlin-Gesellschaft, Hölderlinhaus, Bursagasse 6, 72070 Tübingen
Founded: 1943; Members: 1500
Pres: Prof. Dr. Gerhard Kurz; Gen Secr: Valérie Lawitschka
Focus: Lit
Periodical
Hölderlin-Jahrbuch (bi-annually) 05930

Hohenzollerischer Geschichtsverein, Postf 526, 72482 Sigmaringen
T: (07571) 101558; Fax: (07571) 552
Focus: Hist
Periodicals
Hohenzollerische Heimat (weekly)
Zeitschrift für Hohenzollerische Geschichte (weekly)
. . . . 05931

Hüttentechnische Vereinigung der Deutschen Glasindustrie e.V. (HVG), Mendelssohnstr 75-77, 60325 Frankfurt
T: (069) 749088; Fax: (069) 749719
Founded: 1920; Members: 33
Pres: Erich Schuster; Gen Secr: Prof. Dr. Dr. Helmut A. Schaeffer
Focus: Eng; Materials Sci
Periodical
HVG-Mitteilungen (3 times annually) . . 05932

Humanistische Union e.V., Bräuhausstr 2, 80331 München
T: (089) 226441/42
Focus: Poli Sci
Periodical
Vorgänge: Zeitschrift für Bürgerrechte und Gesellschaftspolitik 05933

Humboldt-Gesellschaft für Wissenschaft, Kunst und Bildung e.V., Riedlach 12, 68307 Mannheim
T: (0621) 771236
Founded: 1962; Members: 580
Pres: Prof. Dr. Herbert Kessler; Gen Secr: Prof. Dr. Gudrun Höhl
Focus: Sci; Arts
Periodical
Mitteilungen (1-2 times annually) 05934

HWWA – Institut für Wirtschaftsforschung – Hamburg, Neuer Jungfernstieg 21, 20354 Hamburg
T: (040) 35620; Tx: 211458
Focus: Econ; Commerce; Marketing; Int'l Relat; Develop Areas
Periodicals
Bibliographie der Wirtschaftspresse: Documentation of selected articles from periodicals
Intereconomics: Review of International Trade and Development
Konjunktur von Morgen: Brief fortnightly Survey of German and World Business Cycles and of the World's Commodity Marktets (weekly)
Wirtschaftsdienst (weekly)
Wirtschaftsdienst: A Monthly Magazine on Economic Policies (weekly) 05935

Ifo Institut für Wirtschaftsforschung e.V. / Ifo Institute for Economic Research, Poschingerstr 5, 81679 München
T: (089) 92240; Fax: (089) 985369
Founded: 1949
Pres: Prof. Dr. K. H. Oppenländer
Focus: Econ
Periodicals
Ifo-Digest (weekly)
Ifo-Schnelldienst (3 times monthly)
Ifo-Studien (weekly)
Wirtschaftskonjunktur (weekly) 05936

Immuno / Medizinisch-wissenschaftliche Information, Slevogtstr 3-5, 69126 Heidelberg
Focus: Doc; Med; Immunology 05937

INCA-FIEJ Research Association (IFRA), Washingtonpl 1, 64287 Darmstadt
T: (06151) 70050; Tx: 419273; Fax: (06151) 784542
Founded: 1961; Members: 975
Gen Secr: Dr. Friedrich W. Burkhardt
Focus: Journalism
Periodicals
Newspaper Techniques (weekly)
technique de presse (weekly)
Zeitungstechnik (weekly) 05938

Industrie-Gemeinschaft Aerosole e.V. (IGA), Karlstr 21, 60329 Frankfurt
T: (069) 25561508; Fax: (069) 25561471
Members: 89
Pres: Dr. Helmut Rudolff; Gen Secr: Dr. Walter Schütz
Focus: Eng 05939

INFODAS / Gesellschaft für Systementwicklung und Informationsverarbeitung, Rhonestr 2, 50765 Köln
T: (0221) 709120; Fax: (0221) 7091255
Founded: 1974
Gen Secr: Hans-Jürgen Klinge
Focus: Computer & Info Sci 05940

Informatica / Gesellschaft für die EDV-Ausbildung, Ahastr 5, 64285 Darmstadt
T: (06151) 662646
Focus: Computer & Info Sci 05941

Informationstechnische Gesellschaft im VDE (ITG), Stresemannallee 15, 60596 Frankfurt
T: (069) 6308360; Tx: 412871; Fax: (069) 6308273
Founded: 1954; Members: 10000
Pres: Dr. Hans Schüßler; Gen Secr: Dr. Volker Schanz
Focus: Electric Eng; Comm
Periodical
Nachrichtentechnische Zeitschrift: Communications Journal (weekly) 05942

Institut der deutschen Wirtschaft e.V., Gustav-Heinemann-Ufer 84-88, 50968 Köln
T: (0221) 370801; Tx: 8882768; CA: Deutstitut; Fax: (0221) 3708192
Founded: 1951; Members: 116
Pres: Dr. Manfred Lennings; Gen Secr: Prof. Dr. Gerhard Fels
Focus: Econ; Sociology
Periodical
iwd: Informationsdienst des Instituts der deutschen Wirtschaft (weekly) 05943

Institut der Hessischen Volkshochschulen, Winterbachstr 38, 60320 Frankfurt
T: (069) 5600080; Fax: (069) 56000810
Pres: Lothar Arabin; Gen Secr: Dr. Enno Knobel
Focus: Adult Educ
Periodical
Hessische Blätter für Volksbildung (weekly) 05944

Institut Finanzen und Steuern e.V. / Finance and Taxation Institute, Markt 10, 53111 Bonn
T: (0228) 654246; Fax: (0228) 691258
Founded: 1949
Gen Secr: Dr. Adalbert Uelner
Focus: Finance
Periodical
Grüne Briefe (weekly) 05945

Institut für angewandte Arbeitswissenschaft e.V. (IfaA), Marienburger Str 7, 50968 Köln
T: (0221) 376030
Pres: Prof. Dr. Heinrich Kürpick
Focus: Sci
Periodical
Angewandte Arbeitswissenschaft (weekly) . 05946

Institut für Angewandte Geodäsie, Richard-Strauss-Allee 11, 60598 Frankfurt
T: (069) 63331; Tx: 413592; Fax: (069) 6333425
Founded: 1952
Pres: Pof. Dr. Hermann Seeger
Focus: Surveying
Periodicals
Mitteilungen (weekly)
Nachrichten aus dem Karten- und Vermessungswesen (weekly) 05947

Institut für Angewandte Pädagogische Forschung e.V. (IfAPF), Heinrich-Beensen-Str 8, 30926 Seelze
T: (0511) 492671
Focus: Educ
Periodical
Informationen des IfAPF (weekly) . . . 05948

Institut für angewandte Verbraucherforschung e.V., Aachener Str 89, 50931 Köln
T: (02234) 71910; Fax: (02234) 71060
Focus: Sociology; Ecology 05949

Institut für Angewandte Wirtschaftsforschung (IAW), Ob dem Himmelreich 1, 72074 Tübingen
T: (07071) 98960; Fax: (07071) 989699
Pres: Prof. Dr. Adolf Wagner
Focus: Econ; Marketing
Periodical
Mitteilungen des Instituts für Angewandte Wirtschaftsforschung (weekly) 05950

Institut für angewandte Wirtschaftsforschung im Mittelstand (IWM), Hirschburgweg 5, 40629 Düsseldorf
T: (0211) 686631
Focus: Marketing; Econ 05951

Institut für Auslandsbeziehungen, Charlottenpl 17, 70173 Stuttgart
T: (0711) 22250; Tx: 723772; Fax: (0711) 2264346
Founded: 1917
Pres: Paul Harro Piazolo; Gen Secr: Klaus Daweke
Focus: Int'l Relat
Periodicals
Materialien zum Internationalen Kulturaustausch (weekly)
Schriftenreihe Dokumentation (weekly)
Zeitschrift für Kulturaustausch (weekly) . 05952

Institut für bankhistorische Forschung e.V., Goethepl 9, 60313 Frankfurt
T: (069) 287739
Pres: Dr. Ernst-Moritz Lipp; Gen Secr: Gabriele Jachmich
Focus: Finance; Hist
Periodical
Bankhistorisches Archiv (weekly) 05953

Institut für Bauforschung e.V., An der Markuskirche 1, 30163 Hannover
T: (0511) 661096/97; Fax: (0511) 661098
Members: 30
Pres: Prof. Dr. Joachim Arlt
Focus: Civil Eng 05954

Institut für Chemiefasern, Körschtalstr 26, 73770 Denkendorf
T: (0711) 3408101
Focus: Textiles 05955

Institut für den Wissenschaftlichen Film, Nonnenstieg 72, 37075 Göttingen
T: (0551) 2020; Tx: 96691; Fax: (0551) 202200
Focus: Cinema; Sci
Periodical
Publikationen zu wissenschaftlichen Filmen (weekly) 05956

Institut für Deutsches und Internationales Baurecht e.V., Schumannstr 53, 60325 Frankfurt
T: (069) 748893; Fax: (069) 7411775
Pres: Prof. Wolfgang Heiermann; Gen Secr: Dr. Hans-Georg Watzke
Focus: Law 05957

Institut für Energie- und Umweltforschung Heidelberg e.V. (IFEU), Wilhelm-Blum-Str 12-14, 69120 Heidelberg
T: (06221) 47670
Focus: Energy; Ecology 05958

Institut für Europäische Umweltpolitik, Aloys-Schulte-Str 6, 53129 Bonn
T: (0228) 213810; Fax: (0228) 221982
Focus: Ecology; Eng 05959

Institut für Film und Bild in Wissenschaft und Unterricht, Postf 260, 82026 Grünwald
T: (089) 64971; Fax: (089) 6497300
Pres: Dieter Kamm; Gen Secr: Manfred Gaibinger
Focus: Cinema
Periodical
FWU-Magazin (weekly) 05960

Institut für Gesellschaftswissenschaften Walberberg e.V., Simrockstr 19, 53113 Bonn
Focus: Sociology
Periodical
Die Neue Ordnung (weekly) 05961

Institut für gewerbliche Wasserwirtschaft und Luftreinhaltung e.V. / Institute of Water Resources Management and Air Pollution Control, Wankelstr 33, 50996 Köln
Founded: 1956; Members: 2500
Pres: Dr. S. Schopka
Focus: Water Res; Ecology
Periodicals
IWL-Forum (weekly)
IWL-Umweltbrief (weekly) 05962

Institut für Handwerkswirtschaft München, Max-Joseph-Str 4, 80333 München
T: (089) 593671, 594132; Fax: (089) 553453
Focus: Econ 05963

Institut für Länderkunde e.V., Beethovenstr 4, 04107 Leipzig
T: (0341) 328003; Fax: (0341) 294872
Focus: Geography 05964

Institut für nationale und internationale Fleisch- und Ernährungswirtschaft, Rombachweg 11, 69118 Heidelberg
T: (06221) 80528, 80529; Tx: 461834; Fax: (06221) 809029
Founded: 1977
Focus: Nutrition; Food; Develop Areas; Doc
. . . . 05965

Institut für Neue Musik und Musikerziehung e.V., Grafenstr 35, 64283 Darmstadt
T: (06151) 23062
Founded: 1948; Members: 400
Focus: Music; Educ
Periodical
Veröffentlichungen des INMM (weekly) . 05966

Institut für Neue Technische Form, Eugen-Bracht-Weg 6, 64287 Darmstadt
T: (06151) 48008
Focus: Standards 05967

Institut für ökologische Forschung und Bildung e.V., Kettelerstr 15, 48147 Münster
T: (0251) 26091
Focus: Ecology 05968

Institut für Schadenverhütung und Schadenforschung der öffentlich-rechtlichen Versicherer e.V. (IfS), Preetzer Str 75, 24143 Kiel
T: (0431) 775780
Pres: Reinhard Schäfer
Focus: Insurance
Periodical
IfS-Information (weekly) 05969

Institut für Sozialarbeit und Sozialpädagogik, Am Stockborn 5-7, 60439 Frankfurt
T: (069) 582025; Fax: (069) 582029
Focus: Educ; Sociology; Adult Educ
Periodical
Informationsdienst zur Ausländerarbeit (weekly)
. . . . 05970

Institut für Sozialforschung und Sozialwirtschaft e.V., Trillerweg 68, 66117 Saarbrücken
T: (0681) 954240; Fax: (0681) 9542427
Founded: 1969
Pres: Peter Ochs; Gen Secr: Dr. Hermann Kotthoff
Focus: Sociology; Econ 05971

Institut für Städtebau, Wohnungswirtschaft und Bausparwesen e.V., Neefestr 2a, 53115 Bonn
T: (0228) 259950; Fax: (0228) 259919
Pres: Walter Englert; Gen Secr: Peter Rohland
Focus: Urban Plan; Finance 05972

Institut für Technik der Betriebsführung im Handwerk, Karl-Friedrich-Str 17, 76133 Karlsruhe
T: (0721) 931030; Fax: (0721) 9310350
Gen Secr: Dr. Gerold Hantsch
Focus: Business Admin 05973

Institut für technische Weiterbildung Berlin e.V. (ITW), Luxemburger Str 10, 13353 Berlin
T: (030) 456010; Fax: (030) 4539039
Founded: 1967; Members: 26
Focus: Eng; Adult Educ 05974

Institut für Textil- und Faserforschung Stuttgart, Körschtalstr 26, 73770 Denkendorf
T: (0711) 34080
Pres: F. Gerber
Focus: Textiles 05975

Institut für Urheber- und Medienrecht e.V., Wiedenmayerstr 32, 80538 München
T: (089) 2913474; Fax: (089) 221528
Pres: Prof. Dr. Manfred Rehbinder
Focus: Law; Mass Media
Periodicals
UFITA (weekly)
ZUM: Zeitschrift für Urheber- und Medienrecht (weekly) 05976

Institut für Wirtschaft und Gesellschaft Bonn e.V. (IWG Bonn) / Bonn Institute for Economic and Social Research, Ahrstr 45, 53175 Bonn
T: (0228) 372044; Tx: 885420; Fax: (0228) 375869
Founded: 1977
Pres: Prof. Dr. Meinhard Miegel; Prof. Dr. Kurt H. Biedenkopf; Thorwald Risler
Focus: Econ; Poli Sci
Periodicals
IWG-Berichte
IWG-Impulse
IWG-Mitteilungen 05977

Institut für Wissenschaftliche Zusammenarbeit, Landhausstr 18, 72074 Tübingen
Focus: Sci 05978

Institut für Zeitungsforschung, Wißstr 4, 44122 Dortmund
T: (0231) 5023221
Pres: Prof. Dr. Hans Bohrmann
Focus: Journalism 05979

Institut für Ziegelforschung Essen e.V., Am Zehnthof 197-203, 45307 Essen
T: (0201) 5921301; Fax: (0201) 5921320
Focus: Materials Sci 05980

Institut Neue Wirtschaft e.V., Kurze Mühren 2, 20095 Hamburg
T: (040) 308010; Fax: (040) 30801107
Focus: Econ 05981

Interessengemeinschaft deutschsprachiger Autoren, Vosspl 2, 23701 Eutin
Focus: Lit
Periodical
Das Literarische Wort (weekly) 05982

Interessengemeinschaft für Lederforschung und Häuteschädenbekämpfung im Verband der deutschen Lederindustrie e.V., Leverkuser Str 20, 65929 Frankfurt
Focus: Materials Sci 05983

International Academy of Cytology (IAC) / Académie Internationale de Cytologie, Hugstettenstr 55, 79106 Freiburg
T: (0761) 2703012; Fax: (0761) 2703122
Founded: 1957; Members: 1950
Gen Secr: Manuel Hilgarth
Focus: Cell Biol & Cancer Res
Periodical
Acta Cytologica (weekly) 05984

International Commission for Uniform Methods of Sugar Analysis (ICUMSA), c/o Institut für landwirtschaftliche Technologie und Zuckerindustrie, Langer Kamp 5, 38106 Braunschweig
T: (0531) 380090; Tx: 952359; Fax: (0531) 3800988
Focus: Nutrition 05985

International Commission of Sugar Technology, Donauwörther Str 50, 86641 Rain
T: (09002) 71210; Fax: (09002) 71346
Founded: 1949; Members: 200
Focus: Eng
Periodical
Proceedings of General Assemblies . . 05986

International Copyright Society, Rosenheimer Str 11, 81667 München
T: (089) 4800300; Tx: 522306; Fax: (089) 48003408
Founded: 1954; Members: 402
Focus: Law
Periodical
Yearbook (weekly) 05987

Internationale Akademie für Pathologie, Deutsche Abteilung e.V., Röttgener Str 101, 53127 Bonn
T: (0228) 282404; Fax: (0228) 284796
Founded: 1964; Members: 1355
Pres: Prof. Dr. G. Delling; Gen Secr: Prof. Dr. V. Totovic
Focus: Pathology 05988

Internationale Arbeitsgemeinschaft der Archiv-, Bibliotheks- und Graphikrestauratoren (IADA), Wehrdaer Str 135, 35041 Marburg
T: (06421) 81758; Fax: (06421) 82506
Founded: 1957; Members: 530
Pres: M. Koch; Gen Secr: Ludwig Ritterpusch
Focus: Archives; Libraries & Bk Sci; Arts
Periodical
Restauro (weekly) 05989

Internationale Biometrische Gesellschaft, Deutsche Region (IBS), Goethestr 23, 52064 Aachen
T: (0241) 8089796
Founded: 1953
Pres: Prof. Dr. H. Thöni; Gen Secr: Prof. Dr. R. Repges
Focus: Bio 05990

Internationale Gesellschaft für Geschichte der Pharmazie, Graf-Moltke-Str 46, 28211 Bremen
T: (0421) 345525
Founded: 1926; Members: 1300
Gen Secr: Dr. Gerald Schröder
Focus: Hist; Pharmacol 05991

Internationale Heinrich Schütz-Gesellschaft e.V. (ISG), Heinrich-Schütz-Allee 35, 34131 Kassel
T: (0561) 31050
Founded: 1930; Members: 1500
Pres: Prof. Dr. Arno Forchert; Gen Secr: Sieglinde Fröhlich
Focus: Music
Periodicals
Acta Sagittariana (weekly)
Schütz-Jahrbuch (weekly) 05992

Internationale Kommission zum Schutze des Rheins gegen Verunreinigung / International Commission for the Protection of the Rhine against Pollution, Kaiserin-Augusta-Anlagen 15, 56068 Koblenz
T: (0261) 12495; Fax: (0261) 36572
Founded: 1963; Members: 5
Gen Secr: D. Hogervost
Focus: Ecology
Periodical
Zahlentafeln der Physikalisch-Chemischen Untersuchungen des Rheinwassers (weekly) 05993

Internationaler Arbeitskreis für Musik e.V. (IAM), Heinrich-Schütz-Allee 29, 34131 Kassel
T: (0561) 37927
Founded: 1951; Members: 5000
Pres: Prof. Diether de la Motte; Gen Secr: Adolf Lang
Focus: Music
Periodical
IAM-Journal 05994

Internationaler Arbeitskreis Sonnenberg, Bankpl 8, 38100 Braunschweig
T: (0531) 49242; Fax: (0531) 42512
Gen Secr: Fritz Eitzel
Focus: Sci
Periodicals
Internationale Briefe / International Journal / Revue Internationale (weekly)
Sonnenberg-News (weekly) 05995

Internationaler Kunstkritikerverband, Sektion der Bundesrepublik Deutschland e.V. (AICA), Colmantstr 15, 53115 Bonn
T: (0228) 631591
Pres: Dr. Horst Richter; Gen Secr: Bernhard Rohe
Focus: Arts 05996

Internationaler Museumsrat, Deutsches Nationalkomitee, c/o Deutsches Museum, Museumsinsel 1, 80538 München
T: (089) 179251
Founded: 1950; Members: 500
Focus: Arts 05997

Internationaler Rat für Umweltrecht / International Council of Environmental Law, Adenauerallee 214, 53113 Bonn
T: (0228) 2692240; Fax: (0228) 2692250
Founded: 1969; Members: 256
Focus: Law; Ecology
Periodical
Environmental Policy and Law (8 times annually) 05998

Internationales Dokumentations- und Studienzentrum für Jugendkonflikte, Gaußstr 20, 42119 Wuppertal
T: (0202) 4392308
Focus: Law
Periodical
Cahier (weekly) 05999

Internationales Institut für Öffentliche Finanzen, c/o Universität des Saarlandes, 66041 Saarbrücken
T: (0681) 3023653; Fax: (0681) 3024369
Founded: 1937; Members: 850
Pres: Dr. Vito Tanzi; Gen Secr: Birgit Schneider
Focus: Finance
Periodical
Proceedings (weekly) 06000

Internationales Institut für Traditionelle Musik e.V., Winklerstr 20, 12623 Berlin
T: (030) 8262853, 8261889; Tx: 182875; Fax: (030) 8259991
Founded: 1963; Members: 31
Pres: Max Peter Baumann; Gen Secr: Reinhard Weichmann
Focus: Music; Doc
Periodical
The World of Music: Intercultural Music Studies (3 times annually) 06001

Internationales Zentralinstitut für das Jugend- und Bildungsfernsehen, Rundfunkpl 1, 80335 München
T: (089) 59002140; Tx: 521070; Fax: (089) 59002379
Founded: 1965
Gen Secr: Paul Löhr
Focus: Mass Media; Educ
Periodical
Televizion (weekly) 06002

Internationale Vereinigung der Musikbibliotheken, Musikarchive und Musikdokumentationszentren, Gruppe Bundesrepublik Deutschland (AIBM), c/o Deutsches Musikarchiv der Deutschen Bibliothek, Postf 450229, 12172 Berlin
T: (030) 770020
Founded: 1951; Members: 200
Pres: Dr. Joachim Jaenecke; Gen Secr: Dr. Bettina von Seyfried
Focus: Libraries & Bk Sci
Periodical
Forum Musikbibliothek (weekly) 06003

Internationale Vereinigung für Rechts- und Sozialphilosophie e.V., c/o Juristisches Seminar, Georg-August-Universität, Pl der Göttinger Sieben 6, 37073 Göttingen
T: (0551) 397384; Fax: (0551) 394872
Founded: 1909
Pres: Prof. Dr. Ralf Dreier; Gen Secr: Dr. D. Buchwald
Focus: Law; Sociology; Philos 06004

Internationale Vereinigung für Vegetationskunde (IVV), Wilhelm-Weber-Str 2, 37073 Göttingen
T: (0551) 395700; Fax: (0551) 398449
Founded: 1937; Members: 1200
Pres: Prof. Dr. Sandro Pignatti; Gen Secr: Prof. Dr. Hartmut Dierschke
Focus: Botany
Periodical
Journal of Vegetation Science (5 times annually) 06005

Internationale Vereinigung von Versicherungsjuristen (A.I.D.A.), Deutsche Landesgruppe, c/o Deutscher Verein für Versicherungswissenschaft e.V., Johannisberger Str 31, 14197 Berlin
T: (030) 8212031; Fax: (030) 8222875
Founded: 1960
Pres: Prof. Dr. Ulrich Hübner
Focus: Law
Periodical
Zeitschrift für die gesamte Versicherungswissenschaft (weekly) . . . 06006

International Geographical Union (IGU), c/o Geographisches Institut, Universität, Meckenheimer Allee 166, 53115 Bonn
T: (0228) 739287; Fax: (0228) 739272
Founded: 1923; Members: 73
Pres: Prof. H.T. Verstappen; Gen Secr: Prof. Dr. Eckart Ehlers
Focus: Geography
Periodical
Bulletins (weekly) 06007

International Institute of Public Finance, c/o Universität, Postf 1150, 66041 Saarbrücken
Focus: Finance
Periodical
Papers and Proceedings (weekly) . . . 06008

International Mineralogical Association (IMA), c/o Institut für Mineralogie, Universität, Lahnberge, 35043 Marburg
T: (6421) 285617
Founded: 1958
Gen Secr: Prof. Stefan Hafner
Focus: Mineralogy
Periodical
IMA News 06009

International Society for Group Activity in Education (ISGE), Schlittweg 34, 69198 Schriesheim
T: (06203) 62717
Founded: 1972
Pres: Prof. Dr. Ernst Meyer; Prof. Dr. Wincenty Okon; Gen Secr: Göte Rudvall
Focus: Educ
Periodical
Forum Pädagogik: Zeitschrift für pädagogische Modelle und soziale Probleme (weekly) . 06010

International Society of Developmental Biologists (ISDS), c/o Max-Planck-Institut für Biophysikalische Chemie, Postf 2841, 37018 Göttingen
Founded: 1950; Members: 800
Pres: Prof. P. Gruss
Focus: Bio
Periodical
ISDS Newsletter (weekly) 06011

Inter Nationes e.V., Kennedyallee 91-103, 53175 Bonn
T: (0228) 8801; Tx: 17228308; Fax: (0228) 880457
Pres: Dr. Dieter W. Benecke
Focus: Int'l Relat
Periodicals
Bildung und Wissenschaft (weekly)
Fikrun wa fan (Geist und Leben) (weekly)
hand in hand (3 times annually)
Humboldt (3 times annually)
Kulturchronik (weekly) 06012

Interparlamentarische Arbeitsgemeinschaft, Postf 120110, 53043 Bonn
T: (0228) 2692212, 2692228; Fax: (0228) 2692251
Pres: Herrmann Leeb; Gen Secr: Dr. Wolfgang Burhenne
Focus: Poli Sci 06013

Intitut für Baustoffprüfung und Fußbodenforschung, Industriestr 19, 53842 Troisdorf
T: (02241) 42042; Fax: (02241) 404295
Focus: Civil Eng; Materials Sci 06014

Intitut für Bildungsmedien e.V., Zeppelinallee 33, 60325 Frankfurt
T: (069) 709046; Tx: 416213; Fax: (069) 7071870
Founded: 1971
Pres: Andreas Baer
Focus: Mass Media; Educ 06015

Joachim Jungius-Gesellschaft der Wissenschaften e.V. (JJG), Edmund-Siemers-Allee 1, 20146 Hamburg
T: (040) 417444
Founded: 1947; Members: 119
Pres: Prof. Dr. Gerhard Seifert
Focus: Humanities; Nat Sci 06016

Johannes-Althusius-Gesellschaft e.V., Universitätsstr 14-16, 48143 Münster
T: (0251) 4901
Pres: Prof. Dr. Dieter Wyduckel; Gen Secr: Prof. Dr. Hans Ulrich Scupin
Focus: Law; Hist 06017

Johann Gottfried Herder-Forschungsrat e.V., Gisonenweg 7, 35037 Marburg
T: (06421) 1840
Founded: 1950
Pres: Prof. Dr. Hans Lemberg; Gen Secr: Dr. Hugo Weczerka
Focus: Hist 06018

Jung-Stilling-Gesellschaft e.V., Postf 100433, 57004 Siegen
T: (0271) 52799
Pres: Prof. Dr. Gerhard Merk
Focus: Hist 06019

Kant-Gesellschaft, Am alten Forsthaus 16, 53125 Bonn
Founded: 1904
Focus: Philos 06020

Kassenärztliche Bundesvereinigung (KBV), Herbert-Levin-Str 3, 50931 Köln
T: (0221) 40050
Founded: 1955
Gen Secr: Dr. Reiner Hess
Focus: Med 06021

Kassenärztliche Vereinigung Bayerns, Mühlbaurstr 16, 81677 München
T: (089) 41471; Fax: (089) 4147324
Pres: Dr. Lothar Wittek; Gen Secr: Robert Schmitt
Focus: Med 06022

Kassenärztliche Vereinigung Berlin, Bismarckstr 95-96, 10625 Berlin
T: (030) 310031272; Fax: (030) 31003210
Pres: Dr. Roderich Nehls; Gen Secr: Hans Werner Krapf
Focus: Med
Periodical
Mitteilungsblatt (weekly) 06023

Kassenärztliche Vereinigung Bremen, Schwachhauser Heerstr 26-28, 28209 Bremen
T: (0421) 340051
Focus: Med 06024

Kassenärztliche Vereinigung Hamburg, Humboldtstr 56, 22083 Hamburg
T: (040) 228021
Focus: Med 06025

Kassenärztliche Vereinigung Hessen, Georg-Voigt-Str 15, 60325 Frankfurt
T: (069) 79201
Focus: Med 06026

Kassenärztliche Vereinigung Koblenz, Emil-Schüller-Str 14-16, 56073 Koblenz
T: (0261) 12552
Focus: Med 06027

Kassenärztliche Vereinigung Niedersachsen, Postf 3167, 30175 Hannover
T: (0511) 38003
Pres: Dr. Kirsten Strahl; Gen Secr: Dr. Lothar Feige
Focus: Med 06028

Kassenärztliche Vereinigung Nordbaden, Postf 3806, 76023 Karlsruhe
T: (0721) 5961151; Fax: (0721) 5961188
Pres: Hans-Jürgen Schmidt; Gen Secr: Günter Abendschön
Focus: Med 06029

Kassenärztliche Vereinigung Nordrhein, Emanuel-Leutze-Str 8, 40547 Düsseldorf
T: (0211) 59701
Focus: Med 06030

Kassenärztliche Vereinigung Nord-Württemberg, Albstadtweg 11, 70567 Stuttgart
T: (0711) 78750
Pres: Dr. Wolfgang Mohr; Gen Secr: Dr. Thomas Zalewski
Focus: Med

Periodical
Kassenarzt-aktuell (weekly) 06031

Kassenärztliche Vereinigung Pfalz, Maximilianstr 22, 67433 Neustadt
T: (06321) 8930
Pres: Dr. Gudrun Blaul; Gen Secr: Martin Schöning
Focus: Med 06032

Kassenärztliche Vereinigung Rheinhessen, Hindenburgstr 32, 55118 Mainz
T: (06131) 63020
Pres: Dr. Günter Gerhardt; Gen Secr: Gerhard Roth
Focus: Med 06033

Kassenärztliche Vereinigung Saarland, Faktoreistr 4, 66111 Saarbrücken
T: (0681) 40031
Focus: Med 06034

Kassenärztliche Vereinigung Schleswig-Holstein, Bismarckallee 1-3, 23795 Bad Segeberg
T: 890
Pres: Dr. Eckhard Weisner; Gen Secr: Dr. Bodo Kosanke
Focus: Med
Periodical
Schleswig-Holsteinisches Ärzteblatt . . . 06035

Kassenärztliche Vereinigung Südbaden, Sundgauallee 27, 79114 Freiburg
T: (0761) 8840
Focus: Med 06036

Kassenärztliche Vereinigung Südwürttemberg, Wächterstr 76, 72074 Tübingen
T: (07071) 2080
Pres: Prof. Dr. Wolfgang Brech; Gen Secr: Burckhard Szidat
Focus: Med 06037

Kassenärztliche Vereinigung Trier, Balduinstr 10-12, 54290 Trier
T: (0651) 45011
Focus: Med 06038

Kassenärztliche Vereinigung Westfalen-Lippe, Westfalendamm 45, 44141 Dortmund
T: (0231) 41071
Focus: Med 06039

Kassenzahnärztliche Bundesvereinigung (KZBV), Universitätsstr 73, 50931 Köln
T: (0221) 40010; Fax: (0221) 404035
Founded: 1955; Members: 17
Pres: Wilfied Schad; Gen Secr: Dr. Burkhard Tiemann
Focus: Dent
Periodical
Zahnärztliche Mitteilungen: ZM (weekly) . 06040

Kassenzahnärztliche Vereinigung Berlin, Georg-Wilhelm-Str 16, 10711 Berlin
T: (030) 890040; Fax: (030) 89004102
Pres: Dr. Klaus Degner; Gen Secr: Helmut Depke
Focus: Dent
Periodical
Mitteilungsblatt der Berliner Zahnärzte (weekly) 06041

Kassenzahnärztliche Vereinigung für den Regierungsbezirk Freiburg, Schönauer Str 4, 79115 Freiburg
T: (0761) 490410
Focus: Dent 06042

Kassenzahnärztliche Vereinigung für den Regierungsbezirk Karlsruhe, Joseph-Meyer-Str 8-10, 68167 Mannheim
T: (0621) 23335
Focus: Dent 06043

Kassenzahnärztliche Vereinigung für den Regierungsbezirk Stuttgart, Heinrich-Baumann-Str 1-3, 70190 Stuttgart
T: (0711) 283243
Focus: Dent 06044

Kassenzahnärztliche Vereinigung für den Regierungsbezirk Tübingen, Wilhelmstr 133, 72074 Tübingen
T: (07071) 212611/13
Focus: Dent 06045

Kassenzahnärztliche Vereinigung Hamburg, Katharinenbrücke 1, 20457 Hamburg
T: (040) 363011
Focus: Dent 06046

Kassenzahnärztliche Vereinigung Hessen, Lyoner Str 21, 60528 Frankfurt
T: (069) 66071
Focus: Dent
Periodical
Der hessische Zahnarzt: DHZ (weekly) . 06047

Kassenzahnärztliche Vereinigung im Lande Bremen, Emmastr 220, 28213 Bremen
T: (0421) 211035/39
Focus: Dent 06048

Kassenzahnärztliche Vereinigung Koblenz-Trier, Poststr 4-8, 56068 Koblenz
T: (0261) 38047
Focus: Dent 06049

Kassenzahnärztliche Vereinigung Niedersachsen, Berliner Allee 14, 30175 Hannover
T: (0511) 34931
Focus: Dent
Periodical
Niedersächsisches Zahnärzteblatt (weekly) 06050

Kassenzahnärztliche Vereinigung Nordrhein, Lindemannstr 36-42, 40237 Düsseldorf
T: (0211) 68850; Fax: (0211) 6885333
Focus: Dent
Periodical
Rheinisches Zahnärzteblatt (weekly) . . . 06051

Kassenzahnärztliche Vereinigung Pfalz, Brunhildenstr 1, 67059 Ludwigshafen
T: (0621) 519111; Fax: (0621) 622972
Focus: Dent 06052

Kassenzahnärztliche Vereinigung Rheinhessen, Eppichmauergasse 1, 55116 Mainz
T: (06131) 287760; Fax: (06131) 225706
Pres: Dr. K.-P. Sitte; Gen Secr: M. Hasselwander
Focus: Dent 06053

Kassenzahnärztliche Vereinigung Saarland, Puccinistr 2, 66119 Saarbrücken
T: (0681) 586080
Focus: Dent 06054

Kassenzahnärztliche Vereinigung Schleswig-Holstein, Westring 498, 24106 Kiel
T: (0431) 38970; Fax: (0431) 389710
Focus: Dent 06055

Kassenzahnärztliche Vereinigung Westfalen-Lippe, Auf der Horst 25, 48147 Münster
T: (0251) 5070; Fax: (0251) 507117
Focus: Dent
Periodical
Zahnärzteblatt Westfalen-Lippe (weekly) . 06056

Katalyse-Umweltgruppe Köln e.V., Friesenstr 84, 50670 Köln
T: (0221) 122166
Focus: Ecology 06057

Katholische Ärztearbeit Deutschlands, Venusbergweg 1, 53115 Bonn
T: (0228) 217942; Fax: (0228) 218452
Founded: 1958; Members: 670
Pres: Dr. Ursula Brandenburg; Gen Secr: Urban Zinser
Focus: Med
Periodical
Renovatio: Zeitschrift für das interdisziplinäre Gespräch (weekly) 06058

Katholische Akademie in Bayern, Mandlstr 23, 80802 München
T: (089) 381020
Founded: 1957
Focus: Adult Educ
Periodicals
Schriften der Katholischen Akademie in Bayern
zur debatte: Themen der Katholischen Akademie in Bayern (weekly) 06059

Katholische Akademie Trier, Hinter dem Dom 1, 54290 Trier
T: (0651) 86055
Gen Secr: Gerhard Schwetje
Focus: Adult Educ 06060

Katholische Akademikerarbeit Deutschlands, Venusbergweg 1, 53115 Bonn
T: (0228) 217942; Fax: (0228) 218452
Founded: 1979; Members: 920000
Pres: Bernhard Mihm; Gen Secr: Urban Zinser
Focus: Sci
Periodical
Schriftenreihe der Katholischen Akademikerarbeit Deutschlands (weekly) 06061

Katholische Bundesarbeitsgemeinschaft für Erwachsenenbildung (KBE), René-Schickele-Str 10, 53123 Bonn
T: (0228) 643081; Fax: (0228) 643083
Founded: 1957
Pres: Erwin Müller-Ruckwitt; Gen Secr: Bernhard Nacke
Focus: Adult Educ
Periodical
Erwachsenenbildung: Adult Education (weekly) 06062

Katholische Erwachsenenbildung im Lande Niedersachsen e.V., Hohenzollernstr 22, 30161 Hannover
T: (0511) 348500; Fax: (0511) 3485033
Founded: 1956; Members: 101
Pres: Hermann Bringmann; Gen Secr: Hubert Stuntebeck
Focus: Adult Educ
Periodical
KEB-NordWest: Mitteilungsblatt für kath. Erwachsenenbildung (5 times annually) . 06063

Katholische Erziehergemeinschaft in Bayern (KEG), Herzogspitalstr 13, 80331 München
T: (089) 267041; Fax: (089) 2606387
Pres: Prof. Dr. Konrad Macht; Gen Secr: Martin Gerbert
Focus: Educ; Sociology
Periodicals
Christ und Bildung (weekly)
paed (weekly)
Treffpunkt Kindergarten/Forum Sozialpädagogik (weekly) 06064

Katholische Juristenarbeit Deutschlands, Venusbergweg 1, 53115 Bonn
T: (0228) 217942; Fax: (0228) 218452
Founded: 1968; Members: 115
Pres: Dr. Horst-Harald Lewandowski; Gen Secr: Urban Zinser
Focus: Law 06065

Katholische Pädagogenarbeit Deutschlands, Venusbergweg 1, 53115 Bonn
T: (0228) 217942; Fax: (0228) 218452
Founded: 1978; Members: 100
Pres: Dr. Gabriele Peus; Gen Secr: Urban Zinser
Focus: Educ 06066

Katholischer Akademikerverband (KAV), Venusbergweg 1, 53115 Bonn
T: (0228) 217942; Fax: (0228) 218452
Founded: 1913
Pres: Norbert Darga; Gen Secr: Urban Zinser
Focus: Sci
Periodical
Renovatio: Zeitschrift für das interdisziplinäre Gespräch (weekly) 06067

Katholischer Akademischer Ausländer-Dienst, Hausdorffstr 151, 53129 Bonn
T: (0228) 230007; Fax: (0228) 230009
Founded: 1956
Pres: Prof. Dr. Peter Hünermann; Gen Secr: Dr. Hermann Weber
Focus: Sci
Periodical
KAAD-Korrespondenz (1-2 times annually) 06068

Katholisches Institut für Medieninformation e.V., Am Hof 28, 50667 Köln
T: (0221) 9254630; Tx: 881155; Fax: (0221) 9254636
Focus: Mass Media
Periodicals
Fernseh-Dienst (weekly)
film-dienst (weekly)
Funk-Korrespondenz (weekly) 06069

Kestner-Gesellschaft, Warmbüchenstr 16, 30159 Hannover
T: (0511) 327081; Fax: (0511) 3681699
Founded: 1916; Members: 3000
Pres: Dr. Wolfgang Wagner; Gen Secr: Dr. Carl Haenlein
Focus: Arts 06070

Klinische Forschungsgruppe für Multiple Sklerose, c/o Neurologische Klinik der Universität, Josef-Schneider-Str 11, 97080 Würzburg
T: (0931) 2012201; Fax: (0931) 2012697
Gen Secr: Prof. Dr. K.V. Toyka
Focus: Med 06071

Klinische Forschungsgruppe für Reproduktionsmedizin, c/o Frauenklinik der Universität, Steinfurter Str 107, 48149 Münster
T: (0251) 836096
Focus: Med 06072

Kneipp-Bund e.V. / Bundesverband für Gesundheitsförderung, Adolf-Scholz-Allee 6, 86825 Bad Wörishofen
T: (08247) 30020; Fax: (08247) 300299
Pres: Engelbert Memminger
Focus: Public Health 06073

Kollegium der Medizinjournalisten, Brünnsteinstr 13, 83080 Oberaudorf
T: (08033) 2327
Focus: Med; Journalism 06074

Kommission für Alte Geschichte und Epigraphik des Deutschen Archäologischen Instituts, Amalienstr 73, 80799 München
T: (089) 281045; Fax: (089) 2805161
Focus: Archeol
Periodical
Chiron: Mitteilungen der Kommission für Alte Geschichte und Epigraphik des Deutschen Archäologischen Instituts (weekly) . . . 06075

Kommission für Erforschung der Agrar- und Wirtschaftsverhältnisse des Europäischen Ostens e.V., Otto-Behaghel-Str 10, 35394 Giessen
T: (0641) 7022835; Fax: (0641) 7022837
Founded: 1957; Members: 8
Gen Secr: Prof. Dr. Eberhard Schinke
Focus: Agri; Hist
Periodicals
Giessener Abhandlungen zur Agrar- und Wirtschaftsforschung des europäischen Ostens (8-10 times annually)
Osteuropastudien der Hochschulen des Landes Hessen: Reihe I (8-10 times annually) . 06076

Kommission für Geschichte des Parlamentarismus und der politischen Parteien e.V., Colmantstr 39, 53115 Bonn
Fax: (0228) 695810
Founded: 1951; Members: 18
Pres: Prof. Dr. Rudolf Morsey; Gen Secr: Dr. Martin Schumacher
Focus: Hist; Poli Sci 06077

Kommission für geschichtliche Landeskunde in Baden-Württemberg, Eugenstr 7, 70182 Stuttgart
T: (0711) 2125262; Fax: (0711) 2125283
Focus: Hist
Periodicals
Zeitschrift für die Geschichte des Oberrheins (weekly)
Zeitschrift für Württembergische Landesgeschichte (weekly) 06078

Kommission Reinhaltung der Luft im VDI und DIN, Postf 101139, 40002 Düsseldorf
T: (0211) 6214532; Fax: (0211) 6214157
Pres: Dr. Herbert Gassert; Gen Secr: Dr. Klaus Grefen
Focus: Ecology
Periodical
Staub-Reinhaltung der Luft (weekly) . . 06079

Kommunalpolitische Vereinigung der CDU und CSU Deutschland, Friedrich-Ebert-Allee 73, 53113 Bonn
T: (0228) 544246
Pres: Dr. Horst Waffenschmidt; Gen Secr: Dr. Joseph-Theodor Blank
Focus: Poli Sci
Periodical
Kommunalpolitische Blätter (weekly) . . . 06080

Konferenz der deutschen Akademien der Wissenschaften, Geschwister-Scholl-Str 2, 55131 Mainz
T: (06131) 57728; Fax: (06131) 57740
Founded: 1973; Members: 6
Gen Secr: Dr. Günter Brenner
Focus: Sci 06081

Konferenz der Landesfilmdienste in der Bundesrepublik Deutschland e.V., Rheinallee 59, 53173 Bonn
T: (0228) 355002; Fax: (0228) 358269
Pres: Günter Dach; Gen Secr: Heinz-Joachim Herrmann
Focus: Cinema 06082

Kongreßgesellschaft für ärztliche Fortbildung e.V., Klingsorstr 21, 12167 Berlin
T: (030) 7913091; Fax: (030) 7913994
Founded: 1952; Members: 13
Pres: Prof. Dr. Rudolf Häring; Gen Secr: Prof. Dr. R. Gotzen
Focus: Med; Adult Educ 06083

Kulturpolitische Gesellschaft e.V., Stirnband 10, 58093 Hagen
T: (02331) 586553; Fax: (02331) 56824
Pres: Dr. Olaf Schwencke; Gen Secr: Dr. Norbert Sievers
Focus: Poli Sci 06084

Kuratorium der Deutschen Wirtschaft für Berufsbildung, Buschstr 83, 53113 Bonn
T: (0228) 212076/77; Fax: (0228) 212079
Focus: Adult Educ; Econ 06085

Kuratorium für Forschung und Technik der Zellstoff- und Papierindustrie, Adenauerallee 55, 53113 Bonn
Fax: (0228) 2670562
Focus: Eng 06086

Kuratorium für Kulturbauwesen, Welfengarten 1, 30167 Hannover
Founded: 1949
Focus: Civil Eng 06087

Kuratorium für Technik und Bauwesen in der Landwirtschaft e.V. (KTBL), Bartningstr 49, 64289 Darmstadt
T: (06151) 70010; Fax: (06151) 700123
Pres: Prof. Dr. Hans Schön; Gen Secr: Harald Kühner
Focus: Civil Eng; Eng; Agri
Periodicals
Landtechnik: Fachzeitschrift für Agrartechnik und ländliches Bauen (weekly)
Schriften (weekly) 06088

Kuratorium für Waldarbeit und Forsttechnik e.V. (KWF) / Federal Centre of Forest Operations and Techniques, Spremberger Str 1, 64823 Groß-Umstadt
T: (06078) 2017
Founded: 1962; Members: 1200
Focus: Forestry
Periodical
Forsttechnische Informationen: Mitteilungsblatt des KWF (weekly) 06089

Ländliche Erwachsenenbildung in Niedersachsen e.V. (LEB), Marienstr 9-11, 30171 Hannover
T: (0511) 304110; Tx: 922424; Fax: (0511) 3631615
Founded: 1951; Members: 80
Pres: Werner Kollenrott; Gen Secr: Dr. G. Lippert
Focus: Adult Educ 06090

Landesärztekammer Baden-Württemberg, Postf 700361, 70573 Stuttgart
T: (0711) 769890
Pres: Prof. Dr. Friedrich-Wilhelm Kolkmann; Gen Secr: Gerd Eggstein
Focus: Public Health
Periodical
Ärzteblatt Baden-Württemberg, (ÄBW): Offizielles Organ der Landesärztekammer Baden-Württemberg (weekly) 06091

Landesärztekammer Hessen, Brossstr 6, 60487 Frankfurt
T: (069) 79480; Fax: (069) 7948128
Pres: Dr. Alfred Möhrle; Gen Secr: Dr. Michael Popovic
Focus: Med
Periodical
Hessisches Ärzteblatt (weekly) 06092

Landesärztekammer Rheinland-Pfalz, Deutschhauspl 3, 55116 Mainz
T: (06131) 288220; Fax: (06131) 2882288
Pres: Dr. Hans Engelhard; Gen Secr: Dr. Jochen Wimmenauer
Focus: Med
Periodical
Ärzteblatt Rheinland-Pfalz (weekly) . . . 06093

Landesärztekammer Thüringen, Stoystr 2, 07743 Jena
T: (03641) 25541
Pres: Dr. E. Beleites; Gen Secr: Dr. Andreas Braunsdorf; Helmut Heck; Dr. Wolfgang Thöle
Focus: Med . . . *06094*

Landesarbeitsgemeinschaft Jugend und Literatur NRW e.V., c/o Brigitte Müller-Beyreiss, Von-Werth-Str 159, 50259 Pulheim-Brauweiler
T: (02234) 84286; Fax: (02234) 89724
Focus: Lit; Educ . . . *06095*

Landesverband der Volkshochschulen Niedersachsens e.V., Bödekerstr 16, 30161 Hannover
T: (0511) 348410; Fax: (0511) 3484125
Pres: Horst Milde; Gen Secr: Bernd Rebens
Focus: Adult Educ . . . *06096*

Landesverband der Volkshochschulen Schleswig-Holsteins e.V., Theodor-Heuss-Ring 132, 24143 Kiel
T: (0431) 76018
Members: 170
Focus: Adult Educ . . . *06097*

Landesverband der Volkshochschulen von Nordrhein-Westfalen e.V., Reinoldistr 8, 44135 Dortmund
T: (0231) 9520580; Fax: (0231) 95205823
Members: 143
Pres: Kurt Krüger; Gen Secr: Reiner Hammelrath
Focus: Adult Educ
Periodical
Volkshochschule: Zeitschrift des Deutschen Volkshochschul-Verbandes (weekly) . . . *06098*

Landeszahnärztekammer Baden-Württemberg, Herdweg 50, 70174 Stuttgart
T: (0711) 227160; Fax: (0711) 295144
Pres: Dr. Rüdiger Engel; Gen Secr: Johann Glück
Focus: Dent
Periodical
Zahnärzteblatt Baden-Württemberg (weekly) *06099*

Landeszahnärztekammer Hessen, Lyoner Str 21, 60528 Frankfurt
T: (069) 66070; Fax: (069) 6666945
Pres: Dr. P. Witzel; Gen Secr: Hans-Jürgen Schroeder
Focus: Dent
Periodical
Der Hessische Zahnarzt (weekly) . . . *06100*

Landeszahnärztekammer Rheinland-Pfalz, Frauenlobpl 2, 55118 Mainz
T: (06131) 618061; Fax: (06131) 672985
Pres: Dr. Rüdiger Krebs; Gen Secr: Renate Hehr-Mahler
Focus: Dent
Periodical
Zahnärztliche Informationen (weekly) . . *06101*

Leiterkreis der Katholischen Akademien, c/o Katholische Akademie Schwerte, Postf 1429, 58209 Schwerte
T: (02304) 4770; Fax: (02304) 47724
Founded: 1958; Members: 24
Pres: Gerhard Krems
Focus: Adult Educ; Poli Sci; Cultur Hist; Rel & Theol; Econ . . . *06102*

Lernen Fördern – Bundesverband zur Förderung Lernbehinderter e.V., Rolandstr 61, 50677 Köln
T: (0221) 380666
Pres: Margarethe Boomers; Gen Secr: Rudolf C. Zelfel
Focus: Rehabil; Educ Handic
Periodical
Lernen Fördern (weekly) . . . *06103*

List Gesellschaft e.V., c/o Ruhr-Universität, GC 3/159, 44780 Bochum
T: (0234) 705151; Fax: (0234) 7094144
Founded: 1954; Members: 950
Pres: Prof. Dr. Dieter Spethmann
Focus: Econ
Periodical
List Forum für Wirtschafts- und Finanzpolitik (weekly) . . . *06104*

Literarischer Verein in Stuttgart e.V. / Stuttgart Literary Society, Rosenbergstr 113, 70193 Stuttgart
T: (0711) 638264/65; Fax: (0711) 6369010
Founded: 1839
Focus: Lit
Periodical
Bibliothek: Editionsreihe (weekly) . . . *06105*

Literarisches Colloquium, Am Sandwerder 5, 14109 Berlin
Focus: Lit
Periodical
Sprache im Technischen Zeitalter (weekly) *06106*

Mainzer Altertumsverein, Rheinallee 3b, 55116 Mainz
Focus: Hist; Arts; Archeol
Periodical
Mainzer Zeitschrift (weekly) . . . *06107*

Management Akademie (MA), Rolandstr 9, 45128 Essen
T: (0201) 8100450
Focus: Adult Educ
Periodicals
Handeln (weekly)
MA Spezial (weekly) . . . *06108*

Marburger Bund (MB) / Verband der angestellten und beamteten Ärzte Deutschlands e.V., Riehler Str 6, 50668 Köln
T: (0221) 733173, 724624; Fax: (0221) 733697
Founded: 1947; Members: 33000
Pres: Dr. Frank Ulrich Montgomery; Gen Secr: Dr. Dieter Boeck
Focus: Med
Periodical
Der Arzt im Krankenhaus und im Gesundheitswesen: Monatsschrift des Marburger Bundes (weekly) . . . *06109*

Margarine-Institut für gesunde Ernährung, Friesenweg 1, 22763 Hamburg
T: (040) 882091; Fax: (040) 882093
Pres: Prof. Karl-Friedrich Gander; Gen Secr: Dr. Ingo Witte
Focus: Nutrition . . . *06110*

Mathematische Gesellschaft in Hamburg (MGH), Bundesstr 55, 20146 Hamburg
Founded: 1690; Members: 400
Focus: Math
Periodical
Mitteilungen der Mathematischen Gesellschaft in Hamburg (weekly) . . . *06111*

Mathematisch-Naturwissenschaftlicher Fakultätentag, Institut für Physikalische Chemie, Bundesstr 45, 20146 Hamburg
T: (040) 41233458/59
Focus: Math; Nat Sci; Adult Educ . . . *06112*

Max-Eyth-Gesellschaft für Agrartechnik e.V., Barningstr 49, 64289 Darmstadt
T: (06151) 7001124
Pres: Prof. Dr. Arno Gego; Gen Secr: Dr. Jürgen Frisch
Focus: Eng; Agri . . . *06113*

Max-Planck-Gesellschaft zur Förderung der Wissenschaften e.V. (MPG), Hofgartenstr 2, 80539 München
T: (089) 21081; Fax: (089) 229850
Founded: 1911
Pres: Prof. Dr. Hans F. Zacher; Gen Secr: Dr. Wolfgang Hasenclever
Focus: Sci
Periodical
MPG-Spiegel (weekly) . . . *06114*

MEDICA, Deutsche Gesellschaft zur Förderung der Medizinischen Diagnostik e.V., Postf 700149, 70571 Stuttgart
T: (0711) 761454, 763443; Fax: (0711) 766992
Focus: Med
Periodical
Medica (weekly) . . . *06115*

Medical Women's International Association (MWIA), Herbert-Lewin-Str 1, 50931 Köln
T: (0221) 4004558; Fax: (0221) 4004557
Founded: 1919; Members: 19138
Focus: Med
Periodical
Congress Report (every 2-3 years) . . . *06116*

Medizinischer Fakultätentag der Bundesrepublik Deutschland (MFT) / Association of all Medical Faculties in the Federal Republic of Germany, Domagkstr 12, 48129 Münster
T: (0251) 835504; Fax: (0251) 835524
Founded: 1913
Pres: Prof. Dr. Fritz H. Kemper
Focus: Med . . . *06117*

Medizinisch Pharmazeutische Studiengesellschaft e.V., Dreizehnmorgenweg 44, 53175 Bonn
T: (0228) 819990; Fax: (0228) 8199999
Founded: 1961
Focus: Pharmacol . . . *06118*

Mommsen-Gesellschaft / Verband der deutschen Forscher auf dem Gebiete des griechisch-römischen Altertums, c/o Institut für Klassische Altertumskunde, Olshausenstr 40, 24098 Kiel
T: (0431) 8802660; Fax: (0431) 8802286
Founded: 1949; Members: 500
Pres: Prof. Dr. Ernst-Richard Schwinge; Gen Secr: Dr. Bardo Maria Gauly
Focus: Lit; Hist; Archeol . . . *06119*

Monumenta Germaniae Historica, Postf 340223, 80099 München
Founded: 1819
Pres: Prof. Dr. Dr. Horst Fuhrmann; Gen Secr: Dr. Wolfram Setz
Focus: Hist
Periodical
Deutsches Archiv für Erforschung des Mittelalters . . . *06120*

Moses Mendelssohn Zentrum für europäisch-jüdische Studien, Am Neuen Palais 10, 14469 Potsdam
T: (0331) 972163; Fax: (0331) 972163
Focus: Cultur Hist . . . *06121*

Münchener Tierärztliche Gesellschaft (MTG), Veterinärstr 13, 80539 München
T: (089) 21802900
Founded: 1873; Members: 180
Focus: Vet Med . . . *06122*

Münchner Dermatologische Gesellschaft, c/o Dermatologische Universitäts-Klinik, Frauenlobstr 9-11, 80337 München
Focus: Derm
Periodical
Der Hautarzt (weekly) . . . *06123*

Münchner Entomologische Gesellschaft e.V., Münchhausenstr 21, 81247 München
Founded: 1905; Members: 670
Pres: Dr. Wolfgang Dierl
Focus: Entomology
Periodicals
Mitteilungen der Münchner Entomologischen Gesellschaft (weekly)
Nachrichtenblatt der Bayerischen Entomologen (weekly) . . . *06124*

Münchner Kreis / Übernationale Vereinigung für Kommunikationsforschung e.V., Tal 16, 80331 München
T: (089) 223238; Fax: (089) 225407
Pres: Prof. Dr. Eberhard Witte; Gen Secr: Walter Lämmle
Focus: Comm Sci
Periodical
Telecommunications (weekly) . . . *06125*

Nah- und Mittelost-Verein e.V., Mittelweg 151, 20148 Hamburg
T: (040) 440251
Founded: 1950; Members: 500
Pres: Dr. Rolf-E. Breuer; Gen Secr: Dr. Otto Plassmann
Focus: Ethnology . . . *06126*

Naturforschende Gesellschaft Bamberg e.V., Bergstr 14, 96191 Viereth-Trunstadt
Founded: 1834; Members: 250
Pres: Dr. Ernst Unger; Dr. K. Garleff
Focus: Nat Sci
Periodical
Berichte der Naturforschenden Gesellschaft Bamberg e.V. (weekly) . . . *06127*

Naturforschende Gesellschaft Freiburg, Albertstr 23b, 79104 Freiburg
Founded: 1821
Focus: Nat Sci
Periodical
Berichte der Naturforschenden Gesellschaft Freiburg (weekly) . . . *06128*

Naturhistorische Gesellschaft Hannover (NHG), Postf 510153, 30631 Hannover
T: (0511) 24702456
Founded: 1797; Members: 428
Pres: Dr. J.D. Becker-Platen
Focus: Nat Sci
Periodicals
Beihefte zu den Berichten der Naturhistorischen Gesellschaft Hannover (weekly)
Berichte der Naturhistorischen Gesellschaft Hannover (weekly) . . . *06129*

Naturhistorische Gesellschaft Nürnberg e.V. (NHG), Gewerbemuseumspl 4, 90403 Nürnberg
T: (0911) 227970
Founded: 1801; Members: 2100
Pres: Rainer Ott; Gen Secr: Kathrin Göbel
Focus: Hist; Nat Sci; Ethnology; Anthro
Periodical
Natur und Mensch (weekly) . . . *06130*

Naturhistorischer Verein der Rheinlande und Westfalens, Nussallee 15a, 53115 Bonn
T: (0228) 692377
Focus: Nat Sci
Periodicals
Decheniana (weekly)
Decheniana-Beihefte (weekly) . . . *06131*

Naturschutzbund Deutschland e.V., Herbert-Rabius-Str 26, 53225 Bonn
T: (0228) 975610; Fax: (0228) 9756190
Founded: 1899; Members: 200000
Pres: Jochen Flasbarth; Gen Secr: Uwe Hüser
Focus: Ecology
Periodical
Naturschutz heute (weekly) . . . *06132*

Naturwissenschaftlicher und Historischer Verein für das Land Lippe e.V., Willi-Hofmann-Str 2, 32756 Detmold
T: (05231) 766110; Fax: (05231) 766124
Focus: Nat Sci; Hist
Periodical
Lippische Mitteilungen aus Geschichte und Landeskunde (weekly) . . . *06133*

Naturwissenschaftlicher Verein für das Fürstentum Lüneburg von 1851 e.V., Salzstr 26, 21335 Lüneburg
T: (04131) 403883
Founded: 1851; Members: 470
Pres: Henry Makowski; Gen Secr: Prof. Dr. Kurt Horst
Focus: Nat Sci . . . *06134*

Naturwissenschaftlicher Verein in Hamburg, Martin-Luther-King-Pl 3, 20146 Hamburg
Fax: (040) 41233937
Founded: 1837; Members: 600
Focus: Nat Sci
Periodicals
Abhandlungen (weekly)
Verhandlungen (weekly) . . . *06135*

Naturwissenschaftlicher Verein zu Bremen, c/o Übersee-Museum, Bahnhofspl 13, 28195 Bremen
Fax: (0421) 3619291
Founded: 1864
Focus: Nat Sci
Periodical
Abhandlungen des Naturwissenschaftlichen Vereins zu Bremen (weekly) . . . *06136*

NAV-Virchowbund (NAV) / Verband der niedergelassenen Ärzte Deutschlands e.V., Belfortstr 9, 50668 Köln
Founded: 1949; Members: 17870
Pres: Dr. Erwin Hirschmann; Gen Secr: Hartwig Lange
Focus: Med
Periodical
Der Niedergelassene Arzt (weekly) . . . *06137*

Nephrologischer Arbeitskreis Saar-Pfalz-Mosel e.V., c/o Medizinische Klinik III, Städtisches Krankenhaus, 67653 Kaiserslautern
T: (0631) 2031256
Pres: Prof. Dr. F.W. Albert
Focus: Med . . . *06138*

Neue Bachgesellschaft e.V., Thomaskirchhof 16, 04109 Leipzig
T: (0341) 275308; Fax: (0341) 275308
Founded: 1900
Pres: Prof. Dr. Dr. Helmuth Rilling
Focus: Music
Periodical
Bach-Jahrbuch (weekly) . . . *06139*

Neue Kriminologische Gesellschaft / Wissenschaftliche Vereinigung deutscher, österreichischer und schweizerischer Kriminologen, c/o Institut für Kriminologie der Universität, Corrensstr 34, 72076 Tübingen
T: (07071) 292931; Fax: (07071) 292041
Founded: 1988; Members: 300
Pres: Prof. Dr. Dr. Günther Kaiser; Gen Secr: Dr. Werner Maschke
Focus: Criminology . . . *06140*

Niedersächsischer Bund für freie Erwachsenenbildung e.V., Marienstr 9-11, 30171 Hannover
T: (0511) 364910; Fax: (0511) 322925
Founded: 1954; Members: 6
Pres: Klaus Schaede; Gen Secr: Jürgen Castendyk
Focus: Adult Educ
Periodical
Erwachsenenbildung: Berichte & Informationen der Erwachsenenbildung in Niedersachsen (weekly) . . . *06141*

Niedersächsischer Landesverband der Heimvolkshochschulen e.V., Marienstr 11, 30171 Hannover
T: (0511) 326962; Fax: (0511) 329738
Founded: 1961; Members: 24
Pres: Bernd Bensch; Gen Secr: Ingrid Ellinghaus
Focus: Adult Educ . . . *06142*

Nordrhein-Westfälische Gesellschaft für Urologie, c/o Städtische Krankenanstalten, Caprivistr 1, 49076 Osnabrück
Focus: Urology . . . *06143*

Nordwestdeutsche Gesellschaft für ärztliche Fortbildung e.V., Todendorfer Str 14, 22964 Steinburg
T: (04534) 8202
Focus: Med; Adult Educ . . . *06144*

Nordwestdeutsche Gesellschaft für innere Medizin, Blankeneser Landstr 68, 22587 Hamburg
T: (040) 860720, 862024
Founded: 1924; Members: 10
Focus: Intern Med
Periodical
Kongressberichte der Tagungen der NWDGIM (weekly) . . . *06145*

Nordwestdeutsche Vereinigung der Hals-Nasen-Ohrenärzte, c/o Universitäts-HNO-Klinik, Hospitalstr 20, 24105 Kiel
Focus: Otorhinolaryngology . . . *06146*

Normenausschuss Akustik, Lärmminderung und Schwingungstechnik im DIN Deutsches Institut für Normung e.V., Burggrafenstr 6, 10787 Berlin
T: (030) 26012367; Fax: (030) 26011231
Pres: Prof. Dr. Ludwig Schreiber; Gen Secr: Hans-Peter Groda
Focus: Civil Eng; Standards . . . *06147*

Normenausschuss Anstrichstoffe und ähnliche Beschichtungsstoffe im DIN Deutsches Institut für Normung e.V., Burggrafenstr 6, 10787 Berlin
T: (030) 26011
Pres: Dr. Günther Duve; Gen Secr: Ekkehard Fritzsche
Focus: Materials Sci; Standards . . . *06148*

Normenausschuss Armaturen im DIN Deutsches Institut für Normung e.V. (NAA), Kamekestr 2-8, 50672 Köln
T: (0221) 57131
Pres: Otto Seppelfricke; Gen Secr: Dr. Ingo Richter
Focus: Standards; Eng . . . *06149*

Normenausschuss Bauwesen im DIN Deutsches Institut für Normung e.V., Burggrafenstr 6, 10787 Berlin
T: (030) 26012501; Tx: 184273; Fax: (030) 26011180
Pres: Dr. Lothar Mayer; Gen Secr: Eckhard Vogel
Focus: Civil Eng; Standards
Periodical
Mitteilungen aus der Baunormung (weekly) *06150*

Normenausschuss Bergbau im DIN Deutsches Institut für Normung e.V., Franz-Fischer-Weg 61, 45307 Essen
T: (0201) 1721558; Fax: (0201) 1721462
Founded: 1922
Pres: Walter Ostermann; Gen Secr: Horst Michaely
Focus: Mining; Standards
Periodical
Bergbau-Verzeichnis (weekly) 06151

Normenausschuss Bibliotheks- und Dokumentationswesen im DIN Deutsches Institut für Normung e.V. (NABD), Burggrafenstr 6, 10787 Berlin
T: (030) 26012791; Tx: 184273; Fax: (030) 26011231
Pres: Dr. W. Neubauer; Gen Secr: Edith Lechner
Focus: Libraries & Bk Sci; Doc; Standards
Periodical
NABD-Mitteilungen (3 times annually) . . 06152

Normenausschuss Bild und Film im DIN Deutsches Institut für Normung e.V. (photokinonorm), Burggrafenstr 6, 10787 Berlin
T: (030) 26012433; Fax: (030) 26011723
Pres: Dr. Klaus Nieswandt; Gen Secr: Dr. Michael Seidl
Focus: Standards; Photo 06153

Normenausschuss Bühnentechnik in Theatern und Mehrzweckhallen im DIN Deutsches Institut für Normung e.V. (FNTh), Burggrafenstr 6, 10787 Berlin
T: (030) 26012694
Pres: Michael Schumacher; Gen Secr: Dr. Michael Seidl
Focus: Standards; Arts 06154

Normenausschuss Bürowesen im DIN Deutsches Institut für Normung e.V., Burggrafenstr. 4-6, 10787 Berlin
T: (030) 26011
Focus: Standards; Business Admin . . . 06155

Normenausschuss Chemischer Apparatebau im DIN Deutsches Institut für Normung e.V. (FNCA), Kamekestr 8, 50672 Köln
T: (0221) 5713522; Fax: (0221) 5713414
Founded: 1950
Pres: W. Becker; Gen Secr: Dr. I. Richter
Focus: Standards; Eng
Periodical
DIN-Mitteilungen (weekly) 06156

Normenausschuss Dental im DIN Deutsches Institut für Normung e.V., Westliche 56, 75172 Pforzheim
T: (07231) 357058
Founded: 1969
Pres: Dr. J. Eberlein; Gen Secr: Dr. H.-P. Keller
Focus: Standards; Dent 06157

Normenausschuss Dichtungen im DIN Deutsches Institut für Normung e.V., Kamekestr 2-8, 50672 Köln
T: (0221) 57131
Focus: Standards; Eng 06158

Normenausschuss Druckgasanlagen im DIN Deutsches Institut für Normung e.V., Burggrafenstr 4-10, 10787 Berlin
T: (030) 2602351
Focus: Standards; Eng 06159

Normenausschuss Druck- und Reproduktionstechnik im DIN Deutsches Institut für Normung e.V. (NDR), Burggrafenstr 4-6, 10787 Berlin
T: (030) 26011
Focus: Eng; Standards 06160

Normenausschuss Einheiten und Formelgrössen im DIN Deutsches Institut für Normung e.V. (AEF), Burggrafenstr 6, 10787 Berlin
T: (030) 26012498; Fax: (030) 26011231
Founded: 1907; Members: 180
Pres: Dr. Günther Garlichs; Gen Secr: B. Brinkmann
Focus: Standards 06161

Normenausschuss Eisen-, Blech- und Metallwaren im DIN Deutsches Institut für Normung e.V. (NA EBM), Kaiserswerther Str 135, 40474 Düsseldorf
T: (0211) 454930; Fax: (0211) 4549369
Pres: Dr. Fricke; Gen Secr: Horst Junker
Focus: Eng; Standards 06162

Normenausschuß Eisen und Stahl im DIN Deutsches Institut für Normung e.V., Sohnstr 65, 40237 Düsseldorf
T: (0211) 88941
Founded: 1947
Focus: Metallurgy; Standards 06163

Normenausschuss Erdöl- und Erdgasgewinnung im DIN Deutsches Institut für Normung e.V. (NÖG), Kamekestr 8, 50672 Köln
T: (0221) 5713522
Founded: 1975; Members: 25
Pres: F. Schlemm; Gen Secr: Dr. I. Richter
Focus: Eng; Standards; Mining 06164

Normenausschuss Ergonomie im DIN Deutsches Institut für Normung e.V., Burggrafenstr 6, 10787 Berlin
T: (030) 26011; Fax: (030) 2601231
Focus: Standards 06165

Normenausschuss Fahrräder im DIN Deutsches Institut für Normung e.V. (NAFA), Kamekestr 2-8, 50672 Köln
T: (0221) 57131
Focus: Eng; Standards 06166

Normenausschuss Farbe im DIN Deutsches Institut für Normung e.V. (FNF), Burggrafenstr 6, 10787 Berlin
T: (030) 26012433
Members: 150
Pres: Prof. Dr. Heinz Terstiege; Gen Secr: Dr. Michael Seidl
Focus: Materials Sci; Standards 06167

Normenausschuss Feinmechanik und Optik im DIN Deutsches Institut für Normung e.V. (NAFuO), Westliche 56, 75172 Pforzheim
T: (07231) 91880; Fax: (07231) 356973
Founded: 1941
Pres: Dr. L. Brauer; Gen Secr: Dr. K. Gindele
Focus: Standards; Optics; Eng 06168

Normenausschuss Feuerwehrwesen im DIN Deutsches Institut für Normung e.V. (FNFW), Burggrafenstr 6, 10787 Berlin
T: (030) 2601340/41
Focus: Eng; Standards; Safety 06169

Normenausschuss Gastechnik im DIN Deutsches Institut für Normung e.V., Hauptstr 71-76, 65760 Eschborn
T: (06196) 70170
Pres: R. Günnewig; Gen Secr: J. Neun
Focus: Eng; Standards
Periodicals
gwf/Gas-Erdgas (weekly)
gwf/Wasser-Abwasser (weekly) 06170

Normenausschuss Giessereiwesen im DIN Deutsches Institut für Normung e.V., Kamekestr 8, 50672 Köln
T: (0221) 57130; Fax: (0221) 5713414
Pres: Dr. J. von Hirsch; Gen Secr: Heinz Mohr
Focus: Standards; Metallurgy 06171

Normenausschuss Gleitlager im DIN Deutsches Institut für Normung e.V. (NGL), Kamekestr 2-8, 50672 Köln
T: (0221) 57131
Pres: W. Hilgers; Gen Secr: H. Tepper
Focus: Standards; Eng 06172

Normenausschuss Graphische Symbole im DIN Deutsches Institut für Normung e.V. (NGS), Burggrafenstr 6, 10787 Berlin
T: (030) 26011463
Focus: Standards; Eng 06173

Normenausschuss Grundlagen der Normung im DIN Deutsches Institut für Normung e.V. (NG), Burggrafenstr 6, 10787 Berlin
T: (030) 26011; Tx: 184273
Focus: Standards
Periodical
DIN-Mitteilungen + electronorm: Zentralorgan der deutschen Normung (weekly) 06174

Normenausschuss Hauswirtschaft im DIN Deutsches Institut für Normung e.V. (NHW), Burggrafenstr 4-10, 10787 Berlin
T: (030) 26011
Focus: Standards; Home Econ 06175

Normenausschuss Heiz-, Koch- und Wärmegeräte im DIN Deutsches Institut für Normung e.V. (NH), Am Hauptbahnhof 10, 60329 Frankfurt
T: (069) 234157
Focus: Electric Eng; Standards 06176

Normenausschuss Heiz- und Raumlufttechnik im DIN Deutsches Institut für Normung e.V. (NHR), Burggrafenstr 6, 10787 Berlin
T: (030) 26012351; Fax: (030) 26011231
Founded: 1929; Members: 800
Pres: Jürgen Diehl; Gen Secr: Uwe Rechentin
Focus: Eng; Standards 06177

Normenausschuss Holzwirtschaft und Möbel im DIN Deutsches Institut für Normung e.V., Kamekestr 2-8, 50672 Köln
T: (0221) 57130
Founded: 1949; Members: 480
Pres: Prof. Dr. D. Noack; Gen Secr: Holger Lorentzen
Focus: Econ; Standards
Periodical
NHM Info (weekly) 06178

Normenausschuß Informationsverarbeitungssysteme im DIN Deutsches Institut für Normung e.V. (NI), Burggrafenstr 6, 10772 Berlin
T: (030) 26012465
Pres: Werner Brodbeck; Gen Secr: Dr. Ingo Wende
Focus: Computer & Info Sci; Standards
Periodical
DIN-Mitteilungen (weekly) 06179

Normenausschuss Instandhaltung im DIN Deutsches Institut für Normung e.V. (NIN), Kamekestr 2-8, 50672 Köln
T: (0221) 5713307
Pres: Dieter Renkes; Gen Secr: Dr. G. Hellwig
Focus: Eng; Standards 06180

Normenausschuss Kältetechnik im DIN Deutsches Institut für Normung e.V. (FNKä), Kamekestr. 2-8, 50672 Köln
T: (0221) 5713514
Pres: Prof. Dr. H. Lotz; Gen Secr: Dr. I. Richter
Focus: Eng; Standards
Periodical
DIN-Mitteilungen (weekly) 06181

Normenausschuss Kautschuktechnik im DIN Deutsches Institut für Normung e.V., Zeppelinallee 69, 60487 Frankfurt
T: (069) 79360; Fax: (069) 7936150
Focus: Eng; Standards 06182

Normenausschuss Kerntechnik im DIN Deutsches Institut für Normung e.V. (NKe), Burggrafenstr 6, 10787 Berlin
T: (030) 26012701
Members: 450
Pres: Dr. H. Bilger; Gen Secr: Prof. Dr. K. Becker
Focus: Nucl Res; Standards
Periodical
IKN Informationen Kerntechnische Normung (weekly) 06183

Normenausschuss Kommunale Technik im DIN Deutsches Institut für Normung e.V. (NKT), Burggrafenstr 6, 10787 Berlin
T: (030) 2601340/41
Focus: Eng; Standards 06184

Normenausschuss Kraftfahrzeuge im DIN Deutsches Institut für Normung e.V., Westendstr 61, 60325 Frankfurt
T: (069) 75701
Pres: Prof. Gunter Zimmermeyer; Gen Secr: Walter Sicks
Focus: Eng; Standards; Auto Eng . . . 06185

Normenausschuss Kunststoffe im DIN Deutsches Institut für Normung e.V., Burggrafenstr 6, 10787 Berlin
T: (030) 26012352
Founded: 1946; Members: 600
Pres: M. Hawerkamp; Gen Secr: Dr. D. Hayer
Focus: Materials Sci; Standards 06186

Normenausschuss Laborgeräte und Laboreinrichtungen im DIN Deutsches Institut für Normung e.V., Theodor-Heuss-Allee 25, 60486 Frankfurt
T: (069) 7564255
Founded: 1926; Members: 230
Focus: Standards; Eng 06187

Normenausschuss Lebensmittel und Landwirtschaftliche Produkte im DIN Deutsches Institut für Normung e.V. (NAL), Burggrafenstr 6, 10787 Berlin
T: (030) 26012445; Fax: (030) 26011186
Pres: Prof. Dr. K. Paulus; Gen Secr: Dr. Ulrike Bohnsack
Focus: Food; Standards 06188

Normenausschuss Lichttechnik im DIN Deutsches Institut für Normung e.V. (FNL), Burggrafenstr 6, 10787 Berlin
T: (030) 26012433
Members: 400
Pres: Dr. Werner Kebschull; Gen Secr: Dr. Michael Seidl
Focus: Electric Eng; Standards 06189

Normenausschuss Maschinenbau im DIN Deutsches Institut für Normung e.V. (NAM), Lyoner Str 18, 60528 Frankfurt
T: (069) 6603341; Fax: (069) 6603557
Founded: 1949
Pres: Dr. Wolfgang Hansen; Gen Secr: Wilhelm Dey
Focus: Mach Eng; Standards 06190

Normenausschuss Materialprüfung im DIN Deutsches Institut für Normung e.V. (NMP), Burggrafenstr 6, 10787 Berlin
T: (030) 26012712
Pres: Prof. Dr. Horst Czichos; Gen Secr: Dr. W. Rauls
Focus: Materials Sci; Standards 06191

Normenausschuss Mechanische Verbindungselemente im DIN Deutsches Institut für Normung e.V., Kamekestr 8, 50672 Köln
T: (0221) 5713307
Pres: K. Keyser; Gen Secr: Dr. G. Hellwig
Focus: Eng; Standards 06192

Normenausschuss Medizin im DIN Deutsches Institut für Normung e.V. (NAMed), Burggrafenstr 6, 10772 Berlin
T: (030) 26012413; Tx: 184273;
CA: Deutschnormen Berlin; Fax: (030) 26011231
Founded: 1967
Pres: Prof. Dr. W. Thefeld; Gen Secr: G. Herfurth
Focus: Standards; Med
Periodical
DIN-Mitteilungen + elektronorm: Zentralorgan der deutschen Normung (weekly) 06193

Normenausschuss Nichteisenmetalle im DIN Deutsches Institut für Normung e.V., Kamekestr 8, 50672 Köln
T: (0221) 57130; Fax: (0221) 5713414
Pres: Dr. A.W. Baukloh; Gen Secr: Heinz Mohr
Focus: Standards; Metallurgy 06194

Normenausschuss Papier und Pappe im DIN Deutsches Institut für Normung e.V., Burggrafenstr 6, 10787 Berlin
T: (030) 26012685
Pres: Dr. H.L. Baumgarten; Gen Secr: Manfred Krause
Focus: Standards; Materials Sci 06195

Normenausschuss Persönliche Schutzausrüstung und Sicherheitskennzeichnung im DIN Deutsches Insitut für Normung e.V. (NPS), Burggrafenstr 4-10, 10787 Berlin
T: (030) 26011
Focus: Standards; Safety 06196

Normenausschuss Pigmente und Füllstoffe im DIN Deutsches Institut für Normung e.V. (NPF), Burggrafenstr 6, 10787 Berlin
T: (030) 26011
Pres: Dr. Klaus Udo Meckenstock; Gen Secr: Ekkehard Fritzsche
Focus: Materials Sci; Standards 06197

Normenausschuss Pulvermetallurgie im DIN Deutsches Institut für Normung e.V. (NPu), Kamekestr 2-8, 50672 Köln
T: (0221) 57131
Pres: W. Löhmer; Gen Secr: H. Tepper
Focus: Standards; Metallurgy 06198

Normenausschuss Rohre, Rohrverbindungen und Rohrleitungen im DIN Deutsches Institut für Normung e.V., Kamekestr 8, 50672 Köln
T: (0221) 57130; Fax: (0221) 5713414
Pres: A. Dahlinger; Gen Secr: P. Richter
Focus: Standards; Eng
Periodical
DIN-Mitteilungen (weekly) 06199

Normenausschuss Rundstahlketten im DIN Deutsches Institut für Normung e.V. (NRK), Kamekestr 2-8, 50672 Köln
T: (0221) 57131
Focus: Standards; Eng 06200

Normenausschuss Schienenfahrzeuge im DIN Deutsches Institut für Normung e.V., Panoramaweg 1, 34131 Kassel
T: (0561) 35056; Fax: (0561) 315709
Founded: 1958; Members: 75
Pres: A. Atzorn; Gen Secr: L. Gregel
Focus: Eng; Standards 06201

Normenausschuss Schmiedetechnik im DIN Deutsches Institut für Normung e.V., Goldene Pforte 1, 58093 Hagen
T: (02331) 958835; Tx: 823806
Pres: Dr. Kaspar Vieregge; Gen Secr: Werner Decker
Focus: Standards; Eng 06202

Normenausschuss Schmuck im DIN Deutsches Insitut für Normung e.V., Westliche Karl-Friedrich Str 56, 75171 Pforzheim
T: (07231) 375058
Founded: 1973
Focus: Standards 06203

Normenausschuss Schweisstechnik im DIN Deutsches Institut für Normung e.V. (NAS), Burggrafenstr 6, 10787 Berlin
T: (030) 26012342; Fax: (030) 2601231
Founded: 1925
Pres: Prof. Dr. H.J. Krause; Gen Secr: F. Zentner
Focus: Standards; Eng 06204

Normenausschuss Siebböden und Kornmessung im DIN Deutsches Institut für Normung e.V. (NASK), Burggrafenstr 6, 10787 Berlin
T: (030) 26012536; Fax: (030) 26011180
Pres: Prof. Dr. Kurt Leschonski; Gen Secr: Peter Fröhlich
Focus: Standards; Eng 06205

Normenausschuss Sport- und Freizeitgerät im DIN Deutsches Institut für Normung e.V., Kamekestr 8, 50672 Köln
T: (0221) 5713512
Pres: A. Nagel; Gen Secr: H. Lorentzen
Focus: Standards; Eng; Sports
Periodical
NA Sport Info (weekly) 06206

Normenausschuss Stahldraht und Stahldrahterzeugnisse im DIN Deutsches Institut für Normung e.V., Kamekestr 2-8, 50672 Köln
T: (0221) 5713307
Gen Secr: Dr. G. Hellwig
Focus: Standards; Eng 06207

Normenausschuss Terminologie im DIN Deutsches Institut für Normung e.V., Burggrafenstr 6, 10787 Berlin
T: (030) 26012318
Pres: Dr. H.-R. Spiegel; Gen Secr: E.M. Baxmann-Krafft
Focus: Standards 06208

Normenausschuss Textil und Textilmaschinen im DIN Deutsches Institut für Normung e.V. (Textilnorm), Burggrafenstr 6, 10787 Berlin
T: (030) 2601432
Focus: Standards; Mach Eng; Textiles
Periodical
Textilnorm Mitteilungen (3 times annually) 06209

Normenausschuss Transportkette im DIN Deutsches Institut für Normung e.V. (NTK), Burggrafenstr 6, 10787 Berlin
T: (030) 2601497; Tx: 184273;
CA: Deutschnormen Berlin; Fax: (030) 2601231
Focus: Standards; Eng; Transport . . . 06210

Normenausschuss Überwachungsbedürftige Anlagen im DIN Deutsches Institut für Normung e.V., Kamekestr 8, 50672 Köln
T: (0221) 5713522; Fax: (0221) 5713414
Gen Secr: Dr. I. Richter
Focus: Standards; Eng
Periodical
DIN-Mitteilungen (weekly) *06211*

Normenausschuss Uhren und Schmuck im DIN Deutsches Institut für Normung e.V., Westliche Karl-Friedrich-Str 56, 75172 Pforzheim
T: (07231) 91880; Fax: (07231) 356993
Founded: 1941
Pres: W. Duckwitz; Gen Secr: Dr. K. Gindele
Focus: Eng; Standards *06212*

Normenausschuss Vakuumtechnik im DIN Deutsches Institut für Normung e.V. (NAV), Kamekestr 2-8, 50672 Köln
T: (0221) 57131
Pres: Dr. H. Henning; Gen Secr: Dr. Ingo Richter
Focus: Standards; Eng *06213*

Normenausschuss Verpackungswesen im DIN Deutsches Institut für Normung e.V. (NAVp), Burggrafenstr 6, 10787 Berlin
T: (030) 26011; Tx: 184273; CA: Deutschnormen Berlin; Fax: (030) 2601231
Founded: 1948; Members: 5901
Focus: Standards; Materials Sci *06214*

Normenausschuss Waagenbau im DIN Deutsches Institut für Normung e.V. (NWB), Burggrafenstr 6, 10787 Berlin
T: (030) 26012367; Fax: (030) 26011231
Pres: D. Buer; Gen Secr: Hans-Peter Grode
Focus: Standards; Eng *06215*

Normenausschuss Wärmebehandlungstechnik metallischer Werkstoffe im DIN Deutsches Institut für Normung e.V. (NWT), Kamekestr 2-8, 50672 Köln
T: (0221) 57131
Pres: Dr. R. Liedtke; Gen Secr: H. Tepper
Focus: Standards; Eng; Metallurgy . . . *06216*

Normenausschuss Wasserwesen im DIN Deutsches Institut für Normung e.V. (NAW), Burggrafenstr 4-10, 10787 Berlin
T: (030) 2601421
Founded: 1952; Members: 750
Focus: Standards; Water Res *06217*

Normenausschuss Werkzeuge und Spannzeuge im DIN Deutsches Institut für Normung e.V., Kamekestr 8, 50672 Köln
T: (0221) 57130; Fax: (0221) 5713414
Pres: Wolfgang Kelch; Gen Secr: E. Barthel
Focus: Standards; Eng *06218*

Normenausschuss Werkzeugmaschinen im DIN Deutsches Institut für Normung e.V. (NWM), Corneliusstr 4, 60325 Frankfurt
T: (069) 7560810; Fax: (069) 7568111
Pres: Dr. I. Faulstich; Gen Secr: H.-P. Leonhardt
Focus: Eng; Standards; Mach Eng . . . *06219*

Normenausschuss Zeichnungswesen im DIN Deutsches Institut für Normung e.V. (NZ), Burggrafenstr 6, 10787 Berlin
T: (030) 26012349; Fax: (030) 26011163
Pres: H.F. Wagner; Gen Secr: H.W. Geschke
Focus: Standards; Econ
Periodical
DIN-Mitteilungen + elektronorm: Zentralorgan der deutschen Normung (weekly) *06220*

Numismatische Kommission der Länder in der Bundesrepublik Deutschland, c/o Staatliche Museen zu Berlin, Münzkabinett, Bodestr 1-3, 10178 Berlin
T: (030) 20355510; Fax: (030) 2004950
Pres: Dr. Bernd Kluge; Gen Secr: Prof. Dr. N. Klüßendorf
Focus: Numismatics *06221*

Oberrheinische Gesellschaft für Geburtshilfe und Gynäkologie, c/o Universitäts-Frauenklinik, Schleichstr 4, 72076 Tübingen
T: (07071) 222955
Founded: 1905; Members: 400
Focus: Gynecology *06222*

Öko-Institut, Institut für angewandte Ökologie e.V., Postf 6226, 79038 Freiburg
T: (0761) 473031; Fax: (0761) 475437
Founded: 1977; Members: 5500
Gen Secr: Christiane Friedrich
Focus: Ecology
Periodical
Öko-Mitteilungen (weekly) *06223*

Orchester-Akademie des Berliner Philharmonischen Orchesters e.V., Matthäikirchstr 1, 10785 Berlin
T: (030) 2628604
Focus: Music *06224*

Ost-Akademie e.V., Herderstr 1-11, 21335 Lüneburg
T: (04131) 42094; Fax: (04131) 405084
Members: 50
Pres: Prof. Dr. Helmut de Rudder; Gen Secr: Dr. Bernhard Schalhorn
Focus: Hist; Poli Sci; Cultur Hist
Periodical
Deutsche Studien (weekly) *06225*

Osteuropa-Institut München, Scheinerstr 11, 81679 München
T: (089) 983821, 987341; Fax: (089) 9210110
Founded: 1952
Focus: Hist; Econ; Ethnology
Periodical
Economic Systems (weekly) *06226*

Otto A. Friedrich-Kuratorium für Grundlagenforschung zur Eigentumspolitik, Gustav-Heinemann-Ufer 72, 50968 Köln
T: (0221) 37950, 384085
Focus: Poli Sci *06227*

Outward Bound – Deutsche Gesellschaft für Europäische Erziehung e.V., Nymphenburger Str 42, 80335 München
T: (089) 181058; Fax: (089) 183933
Founded: 1950
Pres: Jochen von Bredow; Gen Secr: Rainer Güttler
Focus: Educ *06228*

Pädagogische Arbeitsstelle des Deutschen Volkshochschul-Verbandes e.V., Holzhausenstr 21, 60322 Frankfurt
T: (069) 1540050
Focus: Adult Educ *06229*

Pädagogisches Zentrum, Uhlandstr 96-97, 10717 Berlin
T: (030) 86871
Focus: Educ
Periodicals
Informationen für den Biologie-Unterricht (3 times annually)
Informationen für den Chemieunterricht (weekly)
Informationen zur DDR-Pädagogik (3 times annually) *06230*

Paläontologische Gesellschaft, c/o Dr. Werner, Senckenberg-Institut, Senckenberg-Anlage 25, 60325 Frankfurt
T: (069) 7940046
Founded: 1912; Members: 938
Pres: Prof. Dr. J. Remane; Gen Secr: Dr. R. Werner
Focus: Paleontology
Periodical
Paläontologische Zeitschrift (weekly) . . . *06231*

PEN Zentrum Bundesrepublik Deutschland, Sandstr 10, 64283 Darmstadt
T: (06151) 23120; Fax: (06151) 293414
Founded: 1951; Members: 488
Pres: Carl Amery; Gen Secr: Hanns Werner Schwarze
Focus: Lit *06232*

Peter-Schwingen-Gesellschaft e.V., Muffendorfer Hauptstr 62, 53177 Bonn
T: (0228) 325998
Pres: Dr. Horst Heidemann; Gen Secr: Dr. Pia Heckes
Focus: Fine Arts *06233*

Pharma-Dokumentationsring e.V. (PDR), Pharma Research Centre, 42096 Wuppertal
T: (0202) 368495; Fax: (0202) 364200
Founded: 1958; Members: 27
Pres: Dr. A. Mullen
Focus: Pharmacol *06234*

Philosophischer Fakultätentag, c/o Fachbereich Evangelische Theologie, Bau 8/308, Im Stadtwald, 66123 Saarbrücken
T: (0681) 3022949; Fax: (0681) 3024234
Pres: Prof. Dr. Gert Hummel
Focus: Educ; Philos *06235*

Physikalisch-Medizinische Sozietät zu Erlangen, Universitätsstr 40, 91054 Erlangen
Focus: Med; Physics
Periodical
Sitzungsberichte der Physikalisch-Medizinischen Sozietät zu Erlangen (1-2 times annually) *06236*

Politische Akademie Biggesee, 57489 Attendorn-Neulisternohl
T: (02722) 7390
Founded: 1951; Members: 110
Pres: Franz Becker; Gen Secr: Alfred Hagedorn
Focus: Poli Sci
Periodical
Meinungen, Informationen, Nachrichten (weekly) *06237*

POLLICHIA e.V., Saarlandstr 13, 76855 Annweiler
T: (06346) 7353; Fax: (06346) 7245
Founded: 1840; Members: 3800
Pres: Prof. Dr. Günter Preuß; Gen Secr: Prof. Dr. Norbert Hailer
Focus: Nat Sci; Ecology
Periodicals
Mitteilungen der POLLICHIA (weekly)
Pfälzer Heimat (weekly) *06238*

Postakademie, Schloss, 63924 Kleinheubach
Focus: Adult Educ *06239*

Pro Familia / Deutsche Gesellschaft für Sexualpädagogik und Sexualberatung e.V., Stresemannallee 3, 60596 Frankfurt
T: (069) 639001; Fax: (069) 639852
Founded: 1952
Pres: Prof. Dr. Uta Meier; Gen Secr: Elke Thoß
Focus: Sociology
Periodical
pro familia magazin: Sexualpädagogik und Familienplanung (weekly) *06240*

Prüf- und Forschungsinstitut für die Schuhherstellung e.V., Hans-Sachs-Str 2, 66955 Pirmasens
T: (06331) 74017; Tx: 452406
Focus: Materials Sci *06241*

Psychobiologische Gesellschaft, Freundhofweg 5, 45479 Mülheim
T: (0208) 485041
Founded: 1953; Members: 130
Focus: Bio; Psych *06242*

Rabanus-Maurus-Akademie, Eschenheimer Anlage 21, 60318 Frankfurt
T: (069) 554538
Focus: Rel & Theol *06243*

Rat für Formgebung, Postf 150311, 60063 Frankfurt
T: (069) 747919; Fax: (069) 7410911
Pres: Prof. Dr. Dieter Rams; Gen Secr: Dr. Hans Höger
Focus: Graphic & Dec Arts, Design
Periodical
Design Report (weekly) *06244*

Rationalisierungs-Gemeinschaft Bauwesen im RKW, Düsseldorfer Str 40, 65760 Eschborn
T: (06196) 495312
Founded: 1952
Pres: Gabriele Jany; Gen Secr: Horst Wetzel
Focus: Business Admin; Civil Eng
Periodical
Informationen Bau-Rationalisierung ibp (6-8 times annually) *06245*

Rationalisierungs-Gemeinschaft Verpackung im RKW, Düsseldorfer Str 40, 65760 Eschborn
T: (06196) 495200; Fax: (06196) 495303
Focus: Business Admin
Periodical
Informationsdienst Verpackung IV (weekly) *06246*

Rationalisierungs-Kuratorium der Deutschen Wirtschaft e.V. (RKW), Düsseldorfer Str 40, 65760 Eschborn
T: (06196) 4951; Tx: 4072755
Founded: 1921; Members: 8000
Pres: Dr. Otmar Franz; Gen Secr: Hubert Borns; Dr. Herbert Müller
Focus: Business Admin
Periodical
Wirtschaft & Produktivität (weekly) . . . *06247*

Rationalisierungs-Kuratorium der Deutschen Wirtschaft e.V., Landesgruppe Baden-Württemberg, Königstr 49, 70173 Stuttgart
T: (0711) 229980; Fax: (0711) 2299810
Pres: Richard Hirschmann; Gen Secr: Dr. Albrecht Fridrich
Focus: Business Admin
Periodical
RKW-Mitteilungen (weekly) *06248*

Rationalisierungs-Kuratorium der Deutschen Wirtschaft e.V., Landesgruppe Berlin, Rankestr 5-6, 10789 Berlin
T: (030) 8844800; Fax: (030) 88448025
Pres: Ernst-Henning Graf von Hardenberg; Gen Secr: Dr. Gernot Schneider
Focus: Business Admin *06249*

Rationalisierungs-Kuratorium der Deutschen Wirtschaft e.V., Landesgruppe Bremen, Balgebrückstr 3-5, 28195 Bremen
T: (0421) 323316; Fax: (0421) 326218
Pres: Peter Kloess; Gen Secr: Dr. Dieter Porschen
Focus: Business Admin *06250*

Rationalisierungs-Kuratorium der Deutschen Wirtschaft e.V., Landesgruppe Hamburg, Heilwigstr 33, 20249 Hamburg
T: (040) 4602087; Fax: (040) 482032
Pres: Ralf Bacia; Gen Secr: Hans-Jürgen Rabe
Focus: Business Admin *06251*

Rationalisierungs-Kuratorium der Deutschen Wirtschaft e.V., Landesgruppe Hessen, Düsseldorfer Str 40, 65760 Eschborn
T: (06196) 495358; Fax: (06196) 495368
Pres: Dr. Richard Gehrunger; Gen Secr: Bernd Siebenhaar
Focus: Business Admin *06252*

Rationalisierungs-Kuratorium der Deutschen Wirtschaft e.V., Landesgruppe Niedersachsen, Friesenstr 14, 30161 Hannover
T: (0511) 338030
Pres: Günter Schwank; Gen Secr: Alois Vilgis
Focus: Business Admin
Periodical
wir produktiv (weekly) *06253*

Rationalisierungs-Kuratorium der Deutschen Wirtschaft e.V., Landesgruppe Nord-Ost, Holtenauer Str 94, 24105 Kiel
T: (0431) 563075
Pres: Adolf-F. Stein; Gen Secr: Erwin Roloff
Focus: Business Admin
Periodical
Wirtschaft & Produktivität (weekly) . . . *06254*

Rationalisierungs-Kuratorium der Deutschen Wirtschaft e.V., Landesgruppe Nordrhein-Westfalen, Sohnstr 70, 40237 Düsseldorf
T: (0211) 666196
Focus: Business Admin *06255*

Rationalisierungs-Kuratorium der Deutschen Wirtschaft e.V., Landesgruppe Rheinland-Pfalz, Schillerstr 26-28, 55116 Mainz
T: (06131) 286610; Fax: (06131) 286619
Pres: Anton Malburg; Gen Secr: Dieter Ibielski
Focus: Business Admin
Periodical
RKW Kompass (weekly) *06256*

Rationalisierungs-Kuratorium für Landwirtschaft e.V., Am Kamp 13, 24783 Osterrönfeld
T: (04331) 847940; Fax: (04331) 847950
Pres: Eberhard Herweg; Gen Secr: Dr. Hardwin Traulsen
Focus: Agri *06257*

Rat von Sachverständigen für Umweltfragen, Gustav-Stresemann-Ring 11, 65180 Wiesbaden
T: (0611) 7632210; Fax: (0611) 7311269
Pres: Prof. Dr. H.W. Thoenes; Gen Secr: Dr. G. Halbritter
Focus: Ecology *06258*

Rechts- und Staatswissenschaftliche Gesellschaft, Königswall 26, 45657 Recklinghausen
T: (02361) 26819, 23630
Focus: Law; Poli Sci *06259*

Rechts- und Staatswissenschaftliche Vereinigung Düsseldorf e.V., Cecilienallee 3, 40474 Düsseldorf
T: (0211) 4971515
Founded: 1949; Members: 360
Pres: Dr. Heinrich Wiesen
Focus: Poli Sci; Law *06260*

REFA-Verband für Arbeitsstudien und Betriebsorganisation e.V., Wittichstr 2, 64295 Darmstadt
T: (06151) 88010; Fax: (06151) 8801109
Founded: 1924; Members: 45600
Pres: Josef Schwartmann; Gen Secr: Dr. Edgar Theis
Focus: Business Admin
Periodicals
Fortschrittliche Betriebsführung und Industrial Engineering (weekly)
REFA-Nachrichten (weekly) *06261*

Regionale Organisation der FDI für Europa (ERO), Universitätsstr 71, 50861 Köln
T: (0221) 4001204; Fax: (0221) 404035
Founded: 1965; Members: 27
Pres: A. Schneider; Gen Secr: Dr. J. Bjornvad
Focus: Dent *06262*

Rheinische Naturforschende Gesellschaft, Reichklarastr 1, 55116 Mainz
Focus: Nat Sci
Periodicals
Mainzer Naturwissenschaftliches Archiv (weekly)
Mitteilungen der Rheinischen Naturforschenden Gesellschaft (weekly)
Museumsführer (weekly) *06263*

Rheinische Vereinigung für Volkskunde, Am Hofgarten 22, 53113 Bonn
T: (0228) 737618
Founded: 1947; Members: 200
Pres: Prof. Dr. H.L. Cox
Focus: Ethnology
Periodicals
Rheinisches Jahrbuch für Volkskunde
Rheinisch-Westfälische Zeitschrift für Volkskunde *06264*

Rheinisch-Westfälische Akademie der Wissenschaften, Palmenstr 16, 40217 Düsseldorf
T: (0211) 342051; Fax: (0211) 341475
Founded: 1950
Pres: Prof. Dr. Hans Schadewaldt
Focus: Humanities; Eng; Econ; Philos
Periodicals
Abhandlungen (weekly)
Jahrbuch (weekly)
Jahresprogramm (weekly)
Sitzungsberichte (weekly) *06265*

Rheinisch-Westfälische Auslandsgesellschaft e.V., Postf 103334, 44033 Dortmund
T: (0231) 838000; Tx: 822158; Fax: (0231) 8380055
Pres: Horst Schiffmann; Gen Secr: Günter Löb
Focus: Int'l Relat
Periodical
Brücken (weekly) *06266*

Rheinisch-Westfälisches Institut für Wirtschaftsforschung, Hohenzollernstr 1-3, 45128 Essen
T: (0201) 81490; Fax: (0201) 8149209
Pres: Prof. Dr. Paul Klemmer; Dr. Ullrich Heilemann
Focus: Econ; Marketing
Periodicals
RWI-Handwerksberichte (weekly)
RWI-Konjunkturberichte (weekly)
RWI-Konjunkturbriefe (weekly)
RWI-Mitteilungen (weekly)
RWI-Papiere (weekly)
Schriftenreihe des RWI (weekly) *06267*

Rheinisch-Westfälische Vereinigung für Lungen- und Bronchialheilkunde, Tüschener Weg 40, 45239 Essen
T: (0201) 4309201; Fax: (0201) 4309498
Focus: Pulmon Dis *06268*

Richard-Wagner-Verband Bayreuth e.V., Weberhof 4, 95448 Bayreuth
T: (0921) 21512
Pres: Paul Götz
Focus: Music 06269

Saarländische Gesellschaft für zahnärztliche Fortbildung, Puccinistr 2, 66119 Saarbrücken
Pres: Prof. Dr. Dr. J. Dumbach
Focus: Dent 06270

Saarländischer Gymnasiallehrerverband e.V., Losheimer Str 16, 66679 Losheim
T: 2216
Focus: Educ 06271

Saarländisch-Pfälzische Internistengesellschaft e.V., c/o Medizinische Klinik C, Bremserstr 79, 67063 Ludwigshafen
T: (0621) 5084100
Pres: Prof. Dr. F.W. Albert; Gen Secr: Prof. Dr. E. Börner
Focus: Intern Med 06272

Sächsische Akademie der Wissenschaften zu Leipzig (SAW), Goethestr 3-5, 04109 Leipzig
T: (0341) 281081
Founded: 1846; Members: 140
Pres: Prof. Dr. Günter Hause
Focus: Sci
Periodicals
Abhandlungen (weekly)
Sitzungsberichte (weekly) 06273

Sächsische Landesärztekammer, Kaitzer Str 2, 01069 Dresden
T: (0351) 4678220; Fax: (0351) 4678237
Pres: Prof. Dr. Heinrich Diettrich; Gen Secr: Dr. Verena Diefenbach
Focus: Med 06274

Schiffbautechnische Gesellschaft e.V. (STG), Lämmersieth 72, 22305 Hamburg
T: (040) 6904910; Fax: (040) 6900341
Founded: 1899; Members: 2000
Pres: Prof. Dr. Harald Keil; Gen Secr: Dr. Hans-Joachim Dreier
Focus: Eng
Periodical
Jahrbuch der Schiffbautechnischen Gesellschaft (weekly) 06275

Schleswig-Holsteinische Geselllschaft für Zahn-, Mund- und Kieferheilkunde, Ratzeburger Allee 160, 23562 Lübeck
Focus: Otorhinolaryngology; Dent 06276

Schmalenbach-Gesellschaft / Deutsche Gesellschaft für Betriebswirtschaft e.V., Nonnendammallee 101, 13629 Berlin
T: (030) 3827024
Founded: 1978; Members: 1500
Pres: Dr. Joachim Funk; Gen Secr: Dr. Gertrud Fuchs-Wegner
Focus: Business Admin
Periodical
Schmalenbachs Zeitschrift für betriebswirtschaftliche Forschung (weekly) 06277

Schopenhauer-Gesellschaft e.V., c/o Philologisches Institut, Universität Bonn, Am Hof 1a, 53113 Bonn
Founded: 1911; Members: 915
Pres: Prof. Dr. M. Ingenkamp
Focus: Philos
Periodical
Schopenhauer-Jahrbuch (weekly) 06278

Schutzgemeinschaft Alt Bamberg e.V., Schillerpl 9, 96047 Bamberg
T: (0951) 202521
Founded: 1968; Members: 300
Pres: Rainer Hartmann; Gen Secr: Werner Hottelmann
Focus: Preserv Hist Monuments
Periodical
Informationsheft (weekly) 06279

Senckenbergische Naturforschende Gesellschaft (SNG), Senckenberganlage 25, 60325 Frankfurt
T: (069) 75420; Fax: (069) 746238
Founded: 1817; Members: 5000
Pres: Dr. Hanns C. Schroeder-Hohenwarth; Gen Secr: Prof. Dr. Willi Ziegler
Focus: Nat Sci 06280

SNV Studiengesellschaft Verkehr, Lokstedter Weg 24, 20251 Hamburg
T: (040) 460680; Tx: 2173596; Fax: (040) 4608155
Founded: 1971
Gen Secr: Dr. Alexander Flechtner
Focus: Transport 06281

Sorbisches Institut e.V., Ernst-Thälmann-Str 6, 02625 Bautzen
T: (03591) 44303; Fax: (03591) 44340
Focus: Ethnology 06282

Sozialakademie Dortmund, Hohe Str 141, 44139 Dortmund
T: (0231) 126059
Founded: 1947
Pres: Prof. Dr. P. Warneke; Gen Secr: Ernst-Ottmar Nölle
Focus: Sociology 06283

Sportärztebund Hamburg, Eppendorfer Baum 8, 20249 Hamburg
T: (040) 483880
Focus: Med 06284

Sportärztebund Hessen, Otto-Fleck-Schneise 10, 60528 Frankfurt
T: (069) 67800923; Fax: (069) 6708505
Focus: Med
Periodical
Sportärzteverband Hessen – aktuell (weekly) 06285

Sportärztebund Niedersachsen, Sprangerweg 2, 37075 Göttingen
T: (0551) 22414
Focus: Med 06286

Sportärztebund Nordrhein, Bernhard-Letterhaus-Str 17, 51377 Leverkusen
T: (0214) 51804
Focus: Med 06287

Sportärztebund Rheinland-Pfalz, Roonstr 10, 67655 Kaiserslautern
T: (0631) 16079; Fax: (0631) 25021
Focus: Med 06288

Sportärzteverband Schleswig-Holstein, Olshausenstr 40-60, 24098 Kiel
T: (0431) 8803775
Focus: Med 06289

Sprachverband Deutsch für ausländische Arbeitnehmer e.V., Raimundistr 2, 55118 Mainz
T: (06131) 964440; Fax: (06131) 9644444
Pres: Rüdiger Vogt; Gen Secr: Gerhard Fiedler
Focus: Ling
Periodicals
Bildungsarbeit in der Zweitsprache Deutsch: Konzepte und Materialien (3 times annually)
Deutsch lernen: Zeitschrift für den Sprachunterricht mit ausländischen Arbeitnehmern (weekly) 06290

Staats- und Handelspolitische Gesellschaft, Königswall 26, 45657 Recklinghausen
T: (02361) 26819, 23630
Focus: Commerce; Poli Sci 06291

Staats- und Wirtschaftspolitische Gesellschaft e.V., Parkallee 84, 20144 Hamburg
T: (040) 446541/42
Focus: Econ; Poli Sci
Periodical
Deutschland-Journal (weekly) 06292

Ständiger Arbeitsausschuss für die Tagungen der Nobelpreisträger in Lindau, Postf 1325, 88103 Lindau
Focus: Sci 06293

Ständiger Ausschuss für Geographische Namen (StAGN), Richard-Strauss-Allee 11, 60578 Frankfurt
T: (069) 6333316
Focus: Ling 06294

Stifterverband für die Deutsche Wissenschaft, Postf 164460, 45224 Essen
T: (0201) 84010; Fax: (0201) 8401301
Founded: 1920; Members: 4800
Pres: Dr. Klaus Liesen; Gen Secr: Dr. H. Niemeyer
Focus: Humanities; Sci
Periodicals
Arbeitsschriften Forschung und Entwicklung in der Wirtschaft
Jahresberichte über die Tätigkeit des Stifterverbands
Materialien aus dem Stiftungszentrum
Materialien zur Bildungspolitik
Wirtschaft und Wissenschaft 06295

Stifterverband Metalle / Gesellschaft zur Förderung der Metallforschung, Tersteegenstr 28, 40474 Düsseldorf
T: (0211) 454710; Tx: 08584721; Fax: (0211) 4647111
Focus: Metallurgy 06296

Studiengemeinschaft für Fertigbau e.V., Panoramaweg 11, 65191 Wiesbaden
T: (06121) 562191; Fax: (06121) 564699
Focus: Civil Eng 06297

Studiengemeinschaft Holzleimbau e.V., Füllenbachstr 6, 40474 Düsseldorf
T: (0211) 434635; Fax: (0211) 452314
Focus: Civil Eng 06298

Studiengesellschaft für den kombinierten Verkehr e.V. (SGKV), Börsenpl 1, 60313 Frankfurt
T: (069) 283571; Fax: (069) 285920
Founded: 1928; Members: 154
Pres: Ferdinand von Peter; Gen Secr: Dr. Christoph Seidelmann
Focus: Transport 06299

Studiengesellschaft für Holzschwellenoberbau e.V., Mainzer Str 64, 65185 Wiesbaden
T: (06121) 300020; Fax: (06121) 309175
Focus: Materials Sci
Periodical
Die Holzschwelle (weekly) 06300

Studiengesellschaft für Stahlleitplanken e.V., Spandauer Str 25, 57072 Siegen
T: (0271) 53039
Focus: Materials Sci 06301

Studiengesellschaft für unterirdische Verkehrsanlagen e.V. (STUVA), Mathias-Brüggen-Str 41, 50827 Köln
T: (0221) 597950; Fax: (0221) 5979550
Founded: 1960; Members: 230
Pres: Prof. Dr. Günter Girnau; Gen Secr: Dr. Alfred Haack
Focus: Transport
Periodical
Tunnel (weekly) 06302

Studiengesellschaft Stahlanwendung e.V., Breite Str 69, 40213 Düsseldorf
T: (0211) 829382
Focus: Eng
Periodical
Forschungsberichte (weekly) 06303

Studiengruppe Entwicklung Technischer Hilfsmittel für Behinderte, Bergheimer Str 143, 69115 Heidelberg
T: (06221) 24511
Members: 12
Focus: Eng 06304

Studiengruppe für Sozialforschung e.V., 83250 Marquartstein
T: (08641) 7130
Focus: Sociology 06305

Studiengruppe Unternehmer in der Gesellschaft, Ölmühlweg 37b, 61462 Königstein
T: (06174) 7348
Focus: Econ 06306

Studienkreis für Presserecht und Pressefreiheit, Königstr 1a, 70173 Stuttgart
T: (0711) 294353, 293165
Founded: 1956
Focus: Law 06307

Süddeutsches Kunststoff-Zentrum (SKZ) / Institut für Kunststoffverarbeitung, -anwendung und -prüfung, Frankfurter Str 15-17, 97082 Würzburg
T: (0931) 41040; Tx: 068448; Fax: (0931) 4194177
Pres: Dr. O. Schwarz
Focus: Materials Sci; Adult Educ . . . 06308

Südosteuropa-Gesellschaft e.V. (SOG), Widenmayerstr 49, 80538 München
T: (089) 2285291; Fax: (089) 2289469
Founded: 1952
Pres: Dr. Walter Althammer; Gen Secr: Dr. Roland Schönfeld
Focus: Ethnology; Econ; Int'l Relat; Hist; Arts; Ling
Periodical
Südosteuropa-Mitteilungen (weekly) . . . 06309

Technische Akademie Wuppertal e.V. (TAW) / Akademie für Fort- und Weiterbildung von Fach- und Führungskräften, Hubertusallee 18, 42117 Wuppertal
T: (0202) 74950
Founded: 1948
Pres: Erich Giese
Focus: Eng 06310

Technische Fördergemeinschaft Holzsilo (TFS), Malterer Str 18f, 79102 Freiburg
T: (0761) 32321
Focus: Materials Sci 06311

Technische Vereinigung der Großkraftwerksbetreiber e.V. (VGB), Klinkestr 27-31, 45136 Essen
Pres: Prof. Dr. Werner Hlubek; Gen Secr: Prof. Dr. Hans-Dieter Schilling
Focus: Energy
Periodical
VGB Kraftwerkstechnik (weekly) 06312

Thomas-Morus-Akademie Bensberg, Overather Str 51, 51429 Bergisch Gladbach
T: (02204) 54781/82
Focus: Rel & Theol; Adult Educ; Soc Sci
Periodicals
Bensberger Manuskripte
Bensberger Protokolle 06313

Tübinger Förderkreis zur Erforschung der Troas – Freunde von Troia, c/o Prof. Dr. Manfred Korfmann, Institut für Ur- und Frühgeschichte, Schloß Hohentübingen, 72074 Tübingen
Focus: Archeol 06314

Unabhängiger Ärzteverband Deutschlands e.V., Hohenstaufenring 39, 50674 Köln
T: (0221) 217659
Members: 1250
Focus: Med
Periodical
Der freie Arzt (weekly) 06315

VDD-Berufsverband Dokumentation, Information, Kommunikation, Postf 2509, 53111 Bonn
T: (040) 330261
Founded: 1961; Members: 400
Pres: Dr. Winfried Schmitz-Esser; Gen Secr: Dr. Hans Peter Jäger
Focus: Doc
Periodical
VDD-Schriftenreihe (weekly) 06316

VDEh-Gesellschaft zur Förderung der Eisenforschung, Breite Str 27, 40213 Düsseldorf
T: (0211) 88941
Founded: 1966; Members: 50
Focus: Metallurgy 06317

VDE/VDI-Gesellschaft Mikroelektronik, Stresemannallee 15, 60596 Frankfurt
T: (069) 6308330; Tx: 412871; CA: Elektrobund; Fax: (069) 6308273
Founded: 1987; Members: 3300
Focus: Electronic Eng
Periodical
mikroelektronik: Entwicklung und Produktion – Technik und Wirtschaft (weekly) 06318

VDI-Gesellschaft Agrartechnik, Graf-Recke-Str 84, 40239 Düsseldorf
T: (0211) 6214264; Fax: (0211) 6214575
Focus: Eng; Agri
Periodical
Mitglieder-Informationen (weekly) 06319

VDI-Gesellschaft Bautechnik / VDI-Society for Civil Engineering, Graf-Recke-Str 84, 40239 Düsseldorf
T: (0211) 6214575; Fax: (0211) 8586525
Gen Secr: Reinhold Jesorsky
Focus: Civil Eng
Periodical
Bauingenieur 06320

VDI-Gesellschaft Energietechnik, Graf-Recke-Str 84, 40239 Düsseldorf
T: (0211) 6214416; Fax: (0211) 6214575
Pres: Prof. Dr. Dr. Helmut Schaefer; Gen Secr: Dr. Erich Sauer
Focus: Energy 06321

VDI-Gesellschaft Entwicklung Konstruktion Vertrieb (VDI-EKW), Graf-Recke-Str 84, 40239 Düsseldorf
T: (0211) 6214239; Tx: 8586525; Fax: (0211) 6214575
Founded: 1973; Members: 16000
Focus: Eng
Periodicals
Absatzwirtschaft: Zeitschrift für Marketing (weekly)
Konstruktion: Zeitschrift für Konstruktion und Entwicklung im Maschinen-, Apparate- und Gerätebau (weekly) 06322

VDI-Gesellschaft Fahrzeugtechnik, Graf-Recke-Str 84, 40239 Düsseldorf
T: (0211) 6214264; Fax: (0211) 6214264
Focus: Auto Eng
Periodical
Mitglieder-Informationen (weekly) 06323

VDI-Gesellschaft Fördertechnik Materialfluss Logistik, Graf-Recke-Str 84, 40239 Düsseldorf
T: (0211) 6214258; Tx: 586525; Fax: (0211) 6214155
Pres: H. Schulte; Gen Secr: Dr. H. Redeker
Focus: Eng 06324

VDI-Gesellschaft Kunststofftechnik, Graf-Recke-Str 84, 40239 Düsseldorf
T: (0211) 62141; Tx: 8586525
Members: 6500
Pres: Dr. Kurt Weirauch; Gen Secr: Ludwig Vollrath
Focus: Eng 06325

VDI-Gesellschaft Produktionstechnik, Graf-Recke-Str 84, 40239 Düsseldorf
T: (0211) 6214231/32; Fax: (0211) 6214575
Focus: Eng
Periodical
WT-Werkstattechnik (weekly) 06326

VDI-Gesellschaft Technische Gebäudeausrüstung (VDI-TGA) / VDI-Society for Technical Building Services, Graf-Recke-Str 84, 40239 Düsseldorf
T: (0211) 6214251; Tx: 8586525; Fax: (0211) 6214575
Founded: 1935; Members: 7250
Pres: Prof. Dr. Heinz Bach; Gen Secr: Undine Stricker-Berghoff
Focus: Eng
Periodical
HLH Heizung Lüftung/Klima Haustechnik (weekly) 06327

VDI-Gesellschaft Werkstofftechnik, Graf-Recke-Str 84, 40239 Düsseldorf
T: (0211) 62140
Pres: Prof. Dr. H.W. Grünling; Gen Secr: Dr. Ludwig Vollrath
Focus: Materials Sci
Periodicals
Ingenieur-Werkstoffe (weekly)
Materialprüfung (weekly) 06328

VDI-Kommission Lärmminderung, Graf-Recke-Str 84, 40239 Düsseldorf
T: (0211) 62141
Focus: Eng; Ecology 06329

VDI/VDE-Gesellschaft Mess- und Automatisierungstechnik (GMA), Postf 101139, 40002 Düsseldorf
T: (0211) 6214224; Tx: 08586525; Fax: (0211) 6214575
Members: 15500
Pres: Prof. Dr. Martin Polke; Gen Secr: Herbert Wiefels
Focus: Eng 06330

VDI/VDE-Gesellschaft Mikro- und Feinwerktechnik, Postf 101139, 40002 Düsseldorf
T: (0211) 6214230; Fax: (0211) 6214575
Gen Secr: Dr. Helmut Lauruschkat
Focus: Eng
Periodical
F & M: Feinwerktechnik & Meßtechnik (8 times annually) 06331

Verband Bildung und Erziehung e.V. (VBE), Dreizehnmorgenweg 36, 53175 Bonn
T: (0228) 959930; Fax: (0228) 378934
Members: 130000
Pres: Dr. Wilhelm Ebert; Gen Secr: Michael Zimmermann
Focus: Educ; Adult Educ

Periodical
Forum E (weekly) 06332

Verband der Betriebs- und Werksärzte e.V. (VDBW) / Berufsverband deutscher Arbeitsmediziner, Marie-Alexandra-Str 36, 76135 Karlsruhe
T: (0721) 33660; Fax: (0721) 30245
Pres: Dr. Friedrich Helbing; Gen Secr: Götz Busse
Focus: Med 06333

Verband der Bibliotheken des Landes Nordrhein-Westfalen e.V., Königsholz 2, 58453 Witten
T: (02302) 83704; Fax: (02302) 83704
Founded: 1949; Members: 320
Pres: Bernhard Adams; Gen Secr: Richard Grigoleit
Focus: Libraries & Bk Sci
Periodical
Mitteilungsblatt (weekly) 06334

Verband der deutschen Höhlen- und Karstforscher e.V., Keplerstr 1, 70771 Leinfelden-Echterdingen
Founded: 1955; Members: 603
Focus: Geology; Speleology
Periodicals
Die Höhle: Zeitschrift für Karst- und Höhlenkunde (weekly)
Karst und Höhle (weekly)
Mitteilungen des Verbandes der deutschen Höhlen- und Karstforscher e.V. (weekly) 06335

Verband der Dozenten an Deutschen Ingenieurschulen, Lorscher Str 13, 55129 Mainz
Focus: Adult Educ; Eng 06336

Verband der Gemeinde-Tierärzte Baden-Württembergs, c/o Städtischer Schlachthof, Bahnackerstr 14, 76532 Baden-Baden
Focus: Vet Med 06337

Verband der Gemeinschaften der Künstlerinnen und Kunstfreunde e.V. (GEDOK), Einern 29, 42279 Wuppertal
T: (0202) 524642; Fax: (0202) 522539
Members: 4500
Pres: Dr. Renate Massmann; Gen Secr: Jacobina Reuschenberg
Focus: Fine Arts 06338

Verband der Geschichtslehrer Deutschlands, Redderkamp 24, 25335 Bokholt-Hanredder
T: (04123) 2384
Founded: 1949; Members: 5000
Focus: Hist; Educ
Periodicals
Geschichte in Wissenschaft und Unterricht (weekly)
Geschichte, Politik und ihre Didaktik (weekly)
Geschichte und Politik in der Schule (weekly)
Informationen für den Geschichts- und Gemeinschaftskundelehrer (weekly) . . . 06339

Verband der Historiker Deutschlands, c/o Max-Planck-Institut für Geschichte, Hermann-Föge-Weg 11, 49560 Göttingen
T: (0551) 49560; Fax: (0551) 495670
Founded: 1893; Members: 1800
Pres: Prof. Dr. Lothar Gall; Gen Secr: Prof. Dr. Otto Gerhard Oexle
Focus: Hist 06340

Verband der leitenden Krankenhausärzte Deutschlands e.V., Tersteegenstr 9, 40474 Düsseldorf
T: (0211) 434033
Focus: Med
Periodical
Arzt und Krankenhaus (weekly) 06341

Verband der Materialprüfungsämter e.V. (VMPA), Beethovenstr 52, 38106 Braunschweig
T: (0531) 3915499; Fax: (0531) 3914573
Founded: 1948; Members: 48
Pres: Prof. Dr. H. Falkner; Gen Secr: J. Günther
Focus: Materials Sci 06342

Verband der Technischen Überwachungs-Vereine e.V. (VdTÜV), Kurfürstenstr 56, 45138 Essen
T: (0201) 89870; Fax: (0201) 8987120
Founded: 1949; Members: 18
Pres: Dr. Ernst Schadow; Gen Secr: Dr. Lutz K. Wessely
Focus: Eng; Safety
Periodical
TÜ-Technische Überwachung (weekly) . . 06343

Verband der Tierheilpraktiker e.V., Am Postbichl 29, 86405 Meitingen
T: (08271) 1322
Focus: Vet Med 06344

Verband der Volkshochschulen des Landes Bremen, Schwachhauser Heerstr 67, 28211 Bremen
T: (0421) 4963666
Focus: Adult Educ 06345

Verband der Volkshochschulen des Saarlandes e.V., Bahnhofstr 47-49, 66111 Saarbrücken
T: (0681) 36680; Fax: (0681) 36610
Founded: 1959; Members: 16
Pres: Dr. Brunhilde Peter; Gen Secr: Dr. Detlef Oppermann
Focus: Adult Educ 06346

Verband der Volkshochschulen von Rheinland-Pfalz e.V., Postf 4069, 55030 Mainz
T: (06131) 234567
Founded: 1947; Members: 77
Pres: Kurt Beck; Gen Secr: Lothar Bentin
Focus: Adult Educ 06347

Verband Deutscher Agrarjournalisten e.V. (VDAJ), Godesberger Allee 142-148, 53175 Bonn
T: (0228) 731171; Fax: (0228) 373260
Members: 663
Pres: Hans-Heinrich Matthiesen; Gen Secr: Dr. Christiane Volkinsfeld
Focus: Journalism; Agri 06348

Verband Deutscher Badeärzte e.V., Elisabethstr 7, 32545 Bad Oeynhausen
T: (05731) 21203; Fax: (05731) 260880
Pres: Dr. Wolfram Enders; Gen Secr: Dr. Hartwig Raeder
Focus: Med
Periodical
Physikalische Medizin, Rehabilitationsmedizin, Kurortmedizin (weekly) 06349

Verband Deutscher Biologen e.V. (VDBiol) / German Association of Biologists, c/o Institut 9, Fachbereich 06, Universität Hamburg, Von-Melle-Park 8, 20146 Hamburg
Pres: Prof. Dr. G. Schaefer; Gen Secr: D. Schetat
Focus: Bio
Periodical
Biologen in unserer Zeit (weekly) . . . 06350

Verband Deutscher Elektrotechniker e.V. (VDE), Stresemannallee 15, 60596 Frankfurt
T: (069) 63080
Members: 36000
Pres: Arno Treptow; Gen Secr: Dr. Friedrich Dankward Althoff
Focus: Electric Eng
Periodicals
dialog (10 times annually)
etz-Archiv (weekly)
etz Elektrotechnische Zeitschrift (weekly)
me mikroelektronik (weekly)
ntz-Archiv (weekly)
ntz Nachrichtentechnische Zeitschrift (weekly)
. . . . 06351

Verband Deutscher Hochschullehrer der Geographie, c/o Geographisches Institut, Universität, Im Neuenheimer Feld 348, 69120 Heidelberg
T: (6221) 564570, 564590
Focus: Geography 06352

Verband Deutscher Kunsthistoriker e.V., c/o Institut für Kunstgeschichte, Technische Hochschule, Petersenstr 15, 64287 Darmstadt
T: (06151) 162130, 163230
Founded: 1948
Pres: Prof. Dr. H. Röttgen; Gen Secr: Dr. Michael Groblewski
Focus: Hist; Arts; Cultur Hist 06353

Verband Deutscher Landwirtschaftlicher Untersuchungs- und Forschungsanstalten (VDLUFA), Bismarckstr 41a, 64293 Darmstadt
T: (06151) 26485; Fax: (06151) 293370
Founded: 1888
Pres: Dr. Dietrich Heller; Gen Secr: Helmut Zarges
Focus: Agri
Periodicals
Landwirtschaftliche Forschung (weekly)
VDLUFA-Schriftenreihe (weekly) 06354

Verband Deutscher Lehrer im Ausland e.V., Pellwormer Str 13, 25813 Husum
T: (04841) 3126
Pres: Dieter Forster; Gen Secr: Wolfgang Baier
Focus: Educ 06355

Verband deutscher Musikschulen e.V., Plittersdorfer Str 93, 53173 Bonn
T: (0228) 957060; Fax: (0228) 9570633
Pres: Reinhart von Gutzeit; Gen Secr: Rainer Mehlig
Focus: Music; Educ 06356

Verband Deutscher Physikalischer Gesellschaften, Gänsheidestr 15a, 70184 Stuttgart
Focus: Physics 06357

Verband Deutscher Realschullehrer im Deutschen Beamtenbund (VDR), Viersener Str 57b, 41751 Viersen
T: (02162) 52246
Focus: Educ 06358

Verband Deutscher Schiffahrts-Sachverständiger e.V. (V.D.S.S.), Steinhöft 11, 20459 Hamburg
T: (040) 373062; Fax: (040) 373474
Pres: Gerd Weselmann
Focus: Navig 06359

Verband Deutscher Schulgeographen e.V., Breslauer Str 26, 30938 Burgwedel
T: (05139) 1205
Founded: 1912; Members: 4900
Pres: Dr. Dieter Richter
Focus: Geography; Educ 06360

Verband Deutscher Schulmusiker e.V. / Interessenverband der Musiklehrer an allgemeinbildenden Schulen und der diese Berufsgruppe ausbildenden Hochschullehrer, Weihergarten 5, 55116 Mainz
T: (06131) 234049
Pres: Prof. Dr. Dieter Zimmerschied; Gen Secr: Brigitte Franken
Focus: Music; Educ
Periodical
Musik und Bildung: Praxis Musikerziehung (11 times annually) 06361

Verband deutscher Waldvogelpfleger und Vogelschützer, Zukunftstr 25, 55130 Mainz
T: (06131) 87179
Focus: Ornithology 06362

Verband deutscher Werkbibliotheken e.V., c/o Bayer AG, Nobelstr 33b, 51373 Leverkusen
T: (0214) 41097; Fax: (0214) 44386
Pres: Heinrich-Joachim Möller
Focus: Libraries & Bk Sci 06363

Verband Deutscher Zoodirektoren e.V., Tiergartenstr 3, 69045 Heidelberg
T: (06221) 411161; Fax: (06221) 49242
Pres: Dr. Dieter Poley
Focus: Zoology
Periodical
Der Zoologische Garten (weekly) 06364

Verband Hochschule und Wissenschaft im Deutschen Beamtenbund (VHW), Dreizehnmorgenweg 36, 53175 Bonn
T: (0228) 378331
Founded: 1973; Members: 2500
Pres: Prof. Dr. Reinhard Kuhnert; Gen Secr: Volker Klinkhardt
Focus: Adult Educ; Sci
Periodical
VHW-Mitteilungen (weekly) 06365

Verband Katholischer Landvolkshochschulen Deutschlands, Drachenfelsstr 4, 53604 Bad Honnef
T: (02224) 93800; Fax: (02224) 938080
Founded: 1948; Members: 21
Pres: Jochen Deutschenbauer; Gen Secr: Franz Gunkel
Focus: Adult Educ 06366

Verband Physikalische Therapie / Vereinigung für die physiotherapeutischen Berufe e.V., Hofweg 15, 22085 Hamburg
T: (040) 2201236; Fax: (040) 2205537
Pres: Bruno Blum; Gen Secr: Peter Bröcker
Focus: Therapeutics
Periodical
Physikalische Therapie in Theorie und Praxis (weekly) 06367

Verein der Bibliothekare an Öffentlichen Bibliotheken e.V. (VBB), Postf 1324, 72703 Reutlingen
T: (07121) 36999; Fax: (07121) 300433
Founded: 1949; Members: 4500
Pres: Konrad Umlauf; Gen Secr: Katharina Boulanger
Focus: Libraries & Bk Sci
Periodical
Buch und Bibliothek (BuB) (10 times annually)
. . . . 06368

Verein der Diplom-Bibliothekare an wissenschaftlichen Bibliotheken e.V. (VdDB), c/o Universitätsbibliothek, Universitätsstr 31, 93053 Regensburg
T: (0941) 9433952
Founded: 1948; Members: 2600
Pres: Marianne Saule; Gen Secr: Marianne Groß
Focus: Libraries & Bk Sci
Periodical
Rundschreiben (weekly) 06369

Verein der Textilchemiker und Coloristen e.V. (VTCC), Rohrbacher Str 76, 69115 Heidelberg
T: (06221) 21865
Founded: 1948; Members: 1450
Pres: Dr. Wolfgang Schwindt; Gen Secr: Käthe Ellerkamm
Focus: Chem 06370

Verein der Zellstoff- und Papier-Chemiker und -Ingenieure e.V., Berliner Allee 56, 64295 Darmstadt
Fax: (06151) 311076
Founded: 1905; Members: 2000
Pres: B. Steinbeis
Focus: Materials Sci
Periodical
Das Papier (weekly) 06371

Verein Deutscher Archivare (VDA), c/o Generaldirektion der Staatlichen Archivare Bayerns, Postf 221152, 80501 München
T: (089) 28638484; Fax: (089) 28638615
Founded: 1947; Members: 1800
Pres: Dr. Hermann Rumschöttel
Focus: Archives
Periodicals
Der Archivar (weekly)
Archive und Archivare in der Bundesrepublik Deutschland, Österreich und der Schweiz (weekly)
. . . . 06372

Verein Deutscher Bibliothekare e.V. (VDB), c/o Universitätsbibliothek Mainz, Jakob-Welter-Weg 6, 55128 Mainz
T: (06131) 392644; Fax: (06131) 394159
Founded: 1900; Members: 1350
Pres: Dr. Andreas Anderhub; Gen Secr: Dr. Monika Hagenmaier-Farnbauer
Focus: Libraries & Bk Sci
Periodicals
Jahrbuch der Deutschen Bibliotheken (weekly)
Zeitschrift für Bibliothekswesen und Bibliographie: Offizielle Nachrichten 06373

Verein Deutscher Eisenhüttenleute (VDEh), Sohnstr 65, 40237 Düsseldorf
T: (0211) 67070; Tx: 8582512; Fax: (0211) 6707310
Founded: 1860; Members: 10000
Pres: Dr. K.A. Zimmermann; Gen Secr: Dr. D. Springorum
Focus: Metallurgy
Periodicals
Literaturschau Stahl und Eisen (weekly)
MPT Metallurgical Plant and Technology International (weekly)
Stahl und Eisen (weekly)
Steel Research (weekly) 06374

Verein Deutscher Emailfachleute e.V., Zehlendorfer Str 24, 58097 Hagen
T: (02331) 10880; Fax: (02331) 108833
Pres: Klaus Wendel; Gen Secr: Horst Völker
Focus: Eng
Periodical
Mitteilungen (weekly) 06375

Verein Deutscher Giessereifachleute e.V. (VDG), Sohnstr 70, 40237 Düsseldorf
T: (0211) 68710; Tx: 8586885; Fax: (0211) 6871333
Founded: 1909; Members: 3900
Pres: Wilhelm Kuhlgatz; Gen Secr: Dr. Niels Ketscher
Focus: Metallurgy
Periodicals
Casting Plant and Technology: CP+T International (weekly)
Giesserei (26 times annually)
Giessereiforschung (weekly)
Giesserei-Literaturschau (weekly) 06376

Verein Deutscher Ingenieure (VDI) / Association of German Engineers, Graf-Recke-Str 84, 40239 Düsseldorf
T: (0211) 62140; Fax: (0211) 6214575
Founded: 1856
Gen Secr: Dr. Peter Gerber
Focus: Eng
Periodicals
Forschung im Ingenieurwesen (weekly)
Umwelt (weekly)
VDI-Nachrichten (weekly)
VDI-Zeitschrift (weekly) 06377

Verein Deutscher Zuckertechniker (VDZ), c/o Nordharzer Zucker AG, 38315 Schladen
Focus: Food 06378

Verein für bayerische Kirchengeschichte, Veilhofstr 28, 90489 Nürnberg
Fax: (0911) 5819683
Focus: Hist; Rel & Theol
Periodical
Zeitschrift für bayerische Kirchengeschichte (weekly) 06379

Verein für Binnenschiffahrt und Wasserstraßen e.V., Dammstr 15-17, 47118 Duisburg
T: (0203) 800060; Tx: 855692; Fax: (0203) 8000621
Pres: Dr. H. Zünkler; Gen Secr: Dr. H.U. Pabst
Focus: Transport
Periodical
Binnenschiffahrt: Zeitschrift für Binnenschiffahrt und Wasserstraßen (ZfB) (weekly) 06380

Verein für das Forschungsinstitut für Edelmetalle und Metallchemie e.V., Katharinenstr 17, 73525 Schwäbisch Gmünd
T: (07171) 10060; Fax: (07171) 100654
Founded: 1922
Gen Secr: Prof. Dr. Ch. J. Raub
Focus: Metallurgy 06381

Verein für Familienforschung in Ost- und Westpreussen e.V., In de Krümm 10, 21147 Hamburg
Focus: Genealogy
Periodical
Altpreussische Geschlechterkunde 06382

Verein für Familien- und Wappenkunde in Württemberg und Baden e.V., Postf 105441, 70047 Stuttgart
Focus: Genealogy
Periodical
Südwestdeutsche Blätter für Familien- und Wappenkunde (weekly) 06383

Verein für Forstliche Standortskunde und Forstpflanzenzüchtung e.V., Wonnhaldestr 4, 79100 Freiburg
T: (0761) 4018283
Founded: 1951; Members: 1050
Pres: Prof. Dr. Hans-Ulrich Moosmayer; Gen Secr: Gerhard Mühlhäußer
Focus: Forestry
Periodical
Mitteilungen des Vereins für Forstliche Standortskunde und Forstpflanzenzüchtung (weekly)
. . . . 06384

Verein für Gerberei-Chemie und -Technik e.V. (VGCT), Camerloherstr 83, 80689 München
T: (089) 565711
Focus: Chem; Eng
Periodical
Das Leder (weekly) 06385

Verein für Geschichte des Hegaus e.V., August-Ruf-Str 7, 78224 Singen
T: (07731) 85463; Tx: 793726; Fax: (07731) 69154
Focus: Hist
Periodical
Hegau (weekly) 06386

Verein für Geschichte und Landeskunde von Osnabrück, Schloßstr 29, 49074 Osnabrück
T: (0541) 28577
Focus: Hist
Periodicals
Heimatkunde des Osnabrücker Landes in Einzelbeispielen (weekly)
Osnabrücker Geschichtsquellen und Forschungen (weekly)
Osnabrücker Mitteilungen (weekly)
Osnabrücker Urkundenbuch 06387

Verein für Kommunalwirtschaft und Kommunalpolitik e.V., Niederrheinstr 10, 40474 Düsseldorf
T: (0211) 431670, 431335
Members: 600
Focus: Poli Sci; Econ
Periodical
Kommunalwirtschaft (weekly) 06388

Verein für Kommunalwissenschaften e.V., Strasse des 17. Juni 112, 10623 Berlin
T: (030) 390010
Pres: Prof. Dr. Heinrich Mäding; Gen Secr: Dr. Rolf-Peter Löhr
Focus: Poli Sci
Periodicals
Archiv für Kommunalwissenschaften (weekly)
Informationen zur modernen Stadtgeschichte 06389

Verein für technische Holzfragen e.V., Bienroder Weg 54e, 38108 Braunschweig
T: (0531) 39090
Focus: Materials Sci 06390

Verein für Versicherungs-Wissenschaft und -Praxis Nordhessen e.V., Kölnische Str 108, 34119 Kassel
T: (0561) 7881131
Focus: Insurance 06391

Verein für Wasser-, Boden- und Lufthygiene e.V., Postf 311420, 10644 Berlin
T: (030) 27065746; Fax: (030) 4145800
Gen Secr: H. Nobis-Wicherding
Focus: Hygiene
Periodicals
Literaturberichte über Wasser, Abwasser und feste Abfallstoffe (weekly)
Schriftenreihe (weekly) 06392

Verein für Westfälische Kirchengeschichte, Altstädter Kirchpl 1-3, 33602 Bielefeld
Focus: Hist; Rel & Theol
Periodical
Jahrbuch für Westfälische Kirchengeschichte (weekly) 06393

Vereinigung Bayerischer Augenärzte, c/o Augenklinik des Klinikums rechts der Isar, Ismaninger Str 22, 81675 München
Focus: Ophthal 06394

Vereinigung der Ärzte der Medizinaluntersuchungsämter, Alte Poststr 11, 49074 Osnabrück
Focus: Med 06395

Vereinigung der Bayerischen Chirurgen e.V., c/o Kreiskrankenhaus Alt/Neuötting, Vinzenz-von-Paul-Str 10, 84503 Altötting
T: (08671) 509211; Fax: (08671) 509290
Founded: 1911; Members: 780
Gen Secr: Prof. Dr. Hartwig Bauer
Focus: Surgery 06396

Vereinigung der Freunde der Mineralogie und Geologie e.V. (VFMG), Blumenthalstr 40, 69120 Heidelberg
T: (06221) 413411
Pres: Bolko Cruse
Focus: Mineralogy; Geology
Periodical
Der Aufschluss (weekly) 06397

Vereinigung der Hochschullehrer für Zahn-, Mund- und Kieferheilkunde, c/o Zahnärztliches Institut, Universität, Ludwig-Rehn-Str 14, 60596 Frankfurt
Focus: Adult Educ; Dent 06398

Vereinigung der Landesdenkmalpfleger in der Bundesrepublik Deutschland, Scharnhorststr 1, 30175 Hannover
T: (0511) 1085263; Fax: (0511) 1085328
Founded: 1949; Members: 354
Focus: Preserv Hist Monuments
Periodical
Deutsche Kunst und Denkmalpflege (weekly) 06399

Vereinigung der Praktischen und Allgemeinärzte Bayerns e.V., Ludmillastr 13, 81543 München
T: (089) 655505
Pres: Dr. Klaus Meyer-Lutterloh
Focus: Med
Periodical
Hausarzt Bayern (weekly) 06400

Vereinigung der unabhängigen freiberuflichen Versicherungs- und Wirtschaftsmathematiker in der Bundesrepublik Deutschland e.V. (I.A.C.A.) / Deutsche Sektion der International Association of Consulting Actuaries, Nördliche Münchner Str 5, 82031 Grünwald
T: (089) 6416040; Tx: 5213897; Fax: (089) 64160420
Founded: 1968; Members: 20
Pres: Dr. Karl-Josef Bode
Focus: Math; Insurance; Econ 06401

Vereinigung der Versicherungs-Betriebswirte e.V. (VVB), Landgrafenstr 1, 50931 Köln
T: (0221) 404398
Founded: 1951; Members: 1250
Focus: Insurance; Business Admin
Periodical
Versicherungs-Betriebswirt (6 times annually) 06402

Vereinigung deutscher Ärzte, Eiserne Hand 3, 60318 Frankfurt
T: (069) 598059
Focus: Med 06403

Vereinigung Deutscher Gewässerschutz e.V. (VDG), Matthias-Grünewald-Str 1-3, 53175 Bonn
T: (0228) 375007; Fax: (0228) 375515
Pres: Prof. Dr. Dieter Flinspach; Gen Secr: Arnim Schmülling
Focus: Ecology 06404

Vereinigung deutscher Landerziehungsheime, Am Schlachtensee 2, 14163 Berlin
T: (030) 8012079
Founded: 1924; Members: 15
Focus: Educ 06405

Vereinigung Deutscher Neuropathologen und Neuroanatomen, c/o Prof. Dr. W. Schachenmayr, Institut für Neuropathologie, Arndtstr 16, 35392 Giessen
Focus: Anat; Pathology; Neurology . . . 06406

Vereinigung Freischaffender Architekten Deutschlands e.V. (VFA), Poppelsdorfer Allee 48, 53115 Bonn
T: (0228) 631568
Focus: Archit 06407

Vereinigung für angewandte Botanik, Grisebachstr 6, 37077 Göttingen
T: (0551) 393748; Fax: (0551) 393759
Founded: 1902; Members: 370
Pres: Prof. Dr. Michael Runge; Gen Secr: Prof. Dr. Hans-Jürgen Jäger
Focus: Botany
Periodical
Angewandte Botanik (3 times annually) . 06408

Vereinigung für Bankbetriebsorganisation e.V., Schaumainkai 69, 60596 Frankfurt
T: (069) 625011; Fax: (069) 628631
Gen Secr: Manfred Gogolin
Focus: Finance
Periodical
vbo-informationen (weekly) 06409

Vereinigung Getreide-, Markt- und Ernährungsforschung e.V. (GMF), Kronprinzenstr 51, 53173 Bonn
T: (0228) 355019, 355010; Tx: 885584; Fax: (0228) 356972
Pres: Hermann Friedenburg; Gen Secr: Prof. Dr. Werner Steller
Focus: Marketing; Nutrition 06410

Vereinigung Stadt-, Regional- und Landesplanung e.V. (SRL), Weg am Kötterberg 3, 44807 Bochum
T: (0234) 501514; Fax: (0234) 501243
Pres: Dr. Hille von Seggern; Gen Secr: Eitel-Friedrich Beyer
Focus: Urban Plan
Periodical
SRL-Information (weekly) 06411

Vereinigung Süddeutscher Orthopäden e.V., Maria-Viktoria-Str 9, 76530 Baden-Baden
Focus: Orthopedics
Periodical
Orthopädische Praxis (weekly) 06412

Vereinigung Südwestdeutscher Dermatologen, c/o Universitäts-Hautklinik, Hauptstr 7, 79104 Freiburg
Focus: Derm 06413

Vereinigung Südwestdeutscher HNO-Ärzte, c/o Prof. Dr. Matzker, Neufelder Str 32, 51067 Köln
Focus: Otorhinolaryngology 06414

Vereinigung Südwestdeutscher Radiologen und Nuklearmediziner, c/o Dr. S. Bosnjakovic-Büscher, Radiologische Abteilung, Städt. Krankenhaus Sindelfingen, 71065 Sindelfingen
Focus: Med; X-Ray Tech; Nucl Med . . 06415

Vereinigung Westdeutscher Hals-, Nasen- und Ohrenärzte, c/o HNO-Klinik, Städtische Krankenanstalten, Neufelder Str 32, 51067 Köln
Founded: 1897; Members: 1370
Gen Secr: Prof. Dr. T. Brusis
Focus: Otorhinolaryngology 06416

Vereinigung zur Erforschung der Neueren Geschichte e.V., Argelanderstr 59, 53115 Bonn
T: (0228) 216205
Founded: 1957
Pres: Prof. Dr. Konrad Repgen; Gen Secr: Dr. K. Abmeier
Focus: Hist 06417

Vereinigung zur Förderung der technischen Optik e.V., Ernst-Leitz-Str 30, 35578 Wetzlar
Focus: Optics 06418

Vereinigung zur Förderung des Deutschen Brandschutzes e.V. (VFDB), Postf 1231, 48338 Altenberge
T: (02505) 2617
Founded: 1950; Members: 1540
Pres: Hans Jochen Blätte; Gen Secr: Hanns-Helmuth Spohn
Focus: Safety
Periodicals
vfdb-Zeitschrift: Forschung und Technik im Brandschutz (weekly)
vfdb-Zeitschrift: Forschung und Technik im Brandschutz (weekly) 06419

Vereinigung zur Förderung des Instituts für Kunststoffverarbeitung in Industrie und Handwerk an der Rhein.-Westf. Technischen Hochschule Aachen e.V., Pontstr 49, 52062 Aachen
T: (0241) 803806; Fax: (0241) 404551
Focus: Materials Sci 06420

Verein katholischer deutscher Lehrerinnen e.V. (VkdL), Hedwig-Dransfeld-Pl 4, 45143 Essen
T: (0201) 623029; Fax: (0201) 621587
Founded: 1885; Members: 10000
Pres: Nelly Friedrich
Focus: Educ
Periodical
Katholische Bildung (weekly) 06421

Verein Naturschutzpark e.V., 29646 Niederhaverbeck
T: (05198) 408; Fax: (05198) 668
Founded: 1909; Members: 5000
Gen Secr: Dr. Eberhard Jüttner
Focus: Ecology
Periodical
Naturschutz- und Naturparke (weekly) . . 06422

Verein Nordfriesisches Institut e.V., Süderstr 30, 25821 Bredstedt
T: (04671) 2081; Fax: (04671) 1333
Focus: Hist; Lit; Ling
Periodicals
Der Maueranker (weekly)
Das nordfriesische Jahrbuch (weekly)
Nordfriesland (weekly) 06423

Verein von Altertumsfreunden im Rheinlande, Colmantstr 14-16, 53115 Bonn
T: (0228) 72940
Founded: 1848; Members: 1500
Pres: Prof. Dr. Gerhard Wirth
Focus: Hist; Cultur Hist
Periodical
Bonner Jahrbücher (weekly) 06424

Verein zur Förderung der deutschen Tanz- und Unterhaltungsmusik e.V., Friedrich-Wilhelm-Str 31, 53113 Bonn
Founded: 1966
Focus: Music 06425

Verein zur Förderung der Gießerei-Industrie e.V., Sohnstr 70, 40237 Düsseldorf
T: (0211) 68711
Focus: Metallurgy 06426

Verein zur Förderung der Versicherungswissenschaft in Hamburg e.V., Überseering 45, 22297 Hamburg
T: (040) 63761
Pres: Klemens Wesselkork
Focus: Insurance 06427

Verein zur Förderung der Versicherungswissenschaft in München e.V., Schackstr 3, 80539 München
T: (089) 390460; Fax: (089) 390836
Focus: Insurance
Periodical
Schriftenreihe des Vereins zur Förderung der Versicherungswissenschaft in München e.V. (3-6 times annually) 06428

Versicherungswissenschaftlicher und Versicherungswirtschaftlicher Fördererverein e.V., Rathenaupl 16-18, 90489 Nürnberg
T: (0911) 5312470
Members: 50
Focus: Insurance 06429

Versicherungswissenschaftlicher Verein in Hamburg e.V., Schlüterstr 28, 20146 Hamburg
T: (040) 41232629; Fax: (040) 41236252
Members: 723
Pres: Prof. Dr. Gerrit Winter
Focus: Insurance 06430

Versuchsgrubengesellschaft, Tremoniastr 13, 44137 Dortmund
T: (0231) 16884
Focus: Mining
Periodical
Grubensicherheitliche Kurzberichte (weekly) 06431

VGB-Forschungsstiftung / Forschungsstiftung der VGB Technische Vereinigung der Großkraftwerksbetreiber e.V., Klinkestr 27-31, 45136 Essen
T: (0201) 8128216; Fax: (0201) 8128286
Focus: Energy
Periodical
VGB-Kraftwerkstechnik (weekly) 06432

Volkshochschulverband Baden-Württemberg e.V., Raiffeisenstr 14, 70771 Leinfelden-Oberaichen
T: (0711) 759000
Pres: Prof. Dr. Günther Dohmen; Gen Secr: Renate Krausnick-Horst
Focus: Adult Educ 06433

Volkskundliche Kommission für Westfalen / Landschaftsverband Westfalen-Lippe, Dompl 23, 48143 Münster
T: (0251) 834404; Fax: (0251) 838393
Founded: 1928; Members: 40
Pres: Prof. Dr. Dr. Günter Wiegelmann; Gen Secr: Prof. Dr. Dietmar Sauermann
Focus: Ethnology
Periodical
Rheinisch-Westfälische Zeitschrift für Volkskunde (weekly) 06434

Volks- und Betriebswirtschaftliche Vereinigung im Rheinisch-Westfälischen Industriegebiet e.V., Mercatorstr 22-24, 47051 Duisburg
T: (0203) 2821288
Founded: 1920; Members: 100
Pres: Dr. Theodor Pieper; Gen Secr: Theodor Friedhoff
Focus: Business Admin; Econ 06435

Walter Eucken Institut e.V., Goethestr 10, 79100 Freiburg
T: (0761) 78088
Focus: Int'l Relat 06436

Weltbund für Erneuerung der Erziehung, Deutschsprachige Sektion (WEE), Keplerstr 87, 69120 Heidelberg
T: (06221) 477502
Founded: 1920
Pres: Prof. Dr. Horst Hörner; Gen Secr: Prof. Dr. Hans Christoph Berg
Focus: Educ
Periodical
Forum Pädagogik (weekly) 06437

Westdeutsche Gesellschaft für Familienkunde e.V. (WGfF), Wallstr 96, 51063 Köln
T: (0221) 628512
Founded: 1913; Members: 800
Focus: Genealogy
Periodical
Mitteilungen (weekly) 06438

Westdeutscher Medizinischer Fakultätentag (WMFT), Schillerstr 25, 91054 Erlangen
Founded: 1913; Members: 29
Focus: Med; Adult Educ 06439

Westfälische Gesellschaft für Zahn-, Mund- und Kieferheilkunde, Waldeyerstr 30, 48149 Münster
Pres: Prof. Dr. R. Marxkors
Focus: Dent 06440

West- und Süddeutscher Verband für Altertumsforschung, Schillerstr 11, 55116 Mainz
T: (06131) 392667; Fax: (06131) 393227
Founded: 1900; Members: 140
Pres: Prof. Dr. H. Ament
Focus: Archeol 06441

Wilhelm-Busch-Gesellschaft e.V., Georgengarten 1, 30167 Hannover
T: (0511) 714076/77; Fax: (0511) 7011222
Founded: 1930; Members: 3600
Pres: Klaus Schaede; Gen Secr: Gisela Vetter
Focus: Lit; Fine Arts
Periodical
Mitteilungen der Wilhelm-Busch-Gesellschaft 06442

Wirtschaftsakademie für Lehrer e.V., Hindenburgring 12a, 38667 Bad Harzburg
T: (05322) 730; Tx: 957623
Founded: 1959
Pres: Prof. Dr. Reinhard Höhn
Focus: Econ 06443

Wirtschaftspolitische Gesellschaft von 1947, Klüberstr 15, 60325 Frankfurt
T: (069) 722995
Founded: 1947
Pres: Ulrich von Puffendorf
Focus: Poli Sci
Periodical
Offene Welt (weekly) 06444

Wirtschaftspolitischer Club Bonn e.V., Walter-Flex-Str 2, 53113 Bonn
T: (0228) 917720; Fax: (0228) 239521
Members: 200
Gen Secr: Peter Sattler
Focus: Poli Sci; Econ 06445

Wissenschaftliche Gesellschaft an der Johann Wolfgang Goethe Universität Frankfurt am Main, Postf 111932, 60054 Frankfurt
T: (069) 7982139
Focus: Sci
Periodical
Sitzungsberichte, Frankfurter Wissenschaftliche Beiträge (weekly) 06446

Wissenschaftliche Gesellschaft für Europarecht, c/o Universität Trier, Postf 3825, 54296 Trier
T: (0651) 2012512
Founded: 1962; Members: 480
Pres: Prof. Dr. Jürgen Schwarze; Gen Secr: Prof. Dr. Peter-Christian Müller-Graff
Focus: Law

Periodical
Europarecht (weekly) 06447

Wissenschaftliche Gesellschaft für Theologie e.V., Liebermeisterstr 12, 72076 Tübingen
T: (07071) 292888
Founded: 1973; Members: 590
Pres: Prof. Dr. J. Mehlhausen
Focus: Rel & Theol 06448

Wissenschaftlicher Verein für Verkehrswesen e.V. (WVV), Postf 500500, 44221 Dortmund
T: (0231) 7552270; Fax: (0231) 751532
Founded: 1971; Members: 165
Pres: Dr. Günter Spiekermann
Focus: Transport 06449

Wissenschaftliche Vereinigung für Augenoptik und Optometrie e.V. (WVAO), Adam-Karrillon-Str 32, 55118 Mainz
T: (06131) 613061
Pres: Malte Volz; Gen Secr: Hartmut Glaser
Focus: Ophthal; Optics
Periodical
Optometrie (weekly) 06450

Wissenschaftsrat, Brohlerstr 11, 50968 Köln
T: (0221) 37760
Pres: Prof. Dr. Gerhard Neuweiler; Gen Secr: Dr. Winfried Benz
Focus: Sci
Periodical
Empfehlungen und Stellungnahmen (weekly)
. . . . 06451

Wissenschaftszentrum Berlin, Jägerstr 22-23, 10117 Berlin
T: (030) 20370366; Fax: (030) 20370680
Focus: Sci 06452

Wittheit zu Bremen e.V., Baumwollbörse, Zimmer 334, 28195 Bremen
T: (0421) 323347; Fax: (0421) 327019
Founded: 1924; Members: 150
Pres: Prof. Dr. Dr. Christian Marzahn
Focus: Sci 06453

Württembergische Bibliotheksgesellschaft, Vereinigung der Freunde der Landesbibliothek, Postf 105441, 70047 Stuttgart
T: (0711) 2124428; Fax: (0711) 2125422
Founded: 1946; Members: 500
Pres: Dr. Wulf von Lucius
Focus: Libraries & Bk Sci 06454

Wuppertaler Kreis e.V. / Deutsche Vereinigung zur Förderung der Weiterbildung von Führungskräften, Schönhauser Str 64, 50968 Köln
T: (0221) 372018; Fax: (0221) 385952
Pres: Albrecht Bendziula; Gen Secr: Carsten R. Löwe
Focus: Adult Educ; Business Admin . . 06455

Zahnärztekammer Berlin, Georg-Wilhelm-Str 14-16, 10711 Berlin
T: (030) 3028002/03
Focus: Dent 06456

Zahnärztekammer Bremen, Emmastr 220, 28213 Bremen
T: (0421) 211035/39
Focus: Dent 06457

Zahnärztekammer Hamburg, Möllner Landstr 31, 22111 Hamburg
T: (040) 7334050; Fax: (040) 7325828
Focus: Dent
Periodical
Hamburger Zahnärzteblatt (weekly) . . . 06458

Zahnärztekammer Niedersachsen, Hildesheimer Str 35, 30169 Hannover
T: (0511) 853085
Focus: Dent
Periodical
Niedersächsisches Zahnärzteblatt (weekly) 06459

Zahnärztekammer Nordrhein, Emanuel-Leutze-Str 8, 40547 Düsseldorf
T: (0211) 526050
Pres: Dr. Joachim Schulz-Bongert; Gen Secr: Klaus Merse
Focus: Dent
Periodical
Rheinisches Zahnärzteblatt (weekly) . . . 06460

Zahnärztekammer Schleswig-Holstein, Westring 498, 24106 Kiel
T: (0431) 3897200; Fax: (0431) 3897210
Pres: Dr. Rüdiger Schultz; Gen Secr: Dr. Thomas Ruff
Focus: Dent
Periodical
Mitteilungsblatt (8 times annually) . . . 06461

Zahnärztekammer Westfalen-Lippe, Auf der Horst 29, 48147 Münster
T: (0251) 490902; Fax: (0251) 4909117
Founded: 1954
Pres: Dr. Dr. Jürgen Weitkamp; Gen Secr: Dr. Jochen Neumann-Wedekindt
Focus: Dent
Periodical
Zahnärzteblatt Westfalen-Lippe (weekly) . 06462

Zentralausschuss für Deutsche Landeskunde e.V., Postf 111932, 60054 Frankfurt
T: (069) 7982404
Founded: 1882
Pres: Prof. Dr. Klaus Wolf; Gen Secr: Dr. Franz Schymik
Focus: Geography
Periodicals
Berichte zur deutschen Landeskunde (weekly)
Deutsche Landschaften: Landeskundliche Erläuterungen zur Topographischen Karte 1:50000 (weekly)
Forschungen zur deutschen Landeskunde (1-2 times annually)
Schrifttumsberichte zur deutschen Landeskunde (weekly) 06463

Zentrale für Fallstudien e.V., Schloss Gracht, 50374 Erftstadt
T: (02235) 4060
Focus: Educ 06464

Zentralinstitut für Kunstgeschichte, Meiserstr 10, 80333 München
Fax: (089) 5236752
Focus: Cultur Hist
Periodicals
Bibliographie zur Kunstgeschichtlichen Literatur in Ost-, Mittelost- und Südosteuropäischen Zeitschriften (weekly)
Kunstchronik (weekly) 06465

Zentralstelle für Agrardokumentation und -information (ZADI), Villichgasse 17, 53177 Bonn
T: (0228) 95480; Fax: (0228) 9548149
Founded: 1969
Gen Secr: Dr. Anton Mangstl
Focus: Agri; Doc; Ecology 06466

Zentralstelle für Pilzforschung und Pilzverwertung, Breslauer Str 3, 68753 Waghäusel
T: (07254) 74271
Founded: 1951
Pres: Dr Dorita Bötticher; Gen Secr: Dr. Bernhard Bötticher
Focus: Botany, Specific 06467

Zentralverband der Deutschen Geographen, c/o Geographisches Institut, Universität, Im Neuenheimer Feld 348, 69120 Heidelberg
T: (06221) 564570; Fax: (06221) 564996
Pres: Prof. Dr. Dietrich Barsch; Gen Secr: Prof. Dr. Heinz Karrasch
Focus: Geography 06468

Ghana

Accra Regional Maritime Academy, POB 1115, Accra
T: 712343, 712775
Pres: S.A. Ibrahim
Focus: Oceanography 06469

African Commission on Agricultural Statistics, c/o FAO, Regional Office for Africa, POB 1628, Accra
T: 666851; Tx: 2139
Founded: 1961; Members: 43
Focus: Agri; Stats 06470

African Forestry and Wildlife Commission (AFWC), c/o FAO, Regional Office for Africa, POB 1628, Accra
T: 666851; Tx: 2139; Fax: 668427
Founded: 1959; Members: 41
Focus: Forestry; Ecology 06471

African Gerontological Society, c/o Dept of Sociology, University of Ghana, POB 25, Legon
Focus: Geriatrics 06472

African Network for the Development of Ecological Agriculture (ANDEA), POB 444, Mamprobi, Accra
Focus: Agri 06473

African Union for Scientific Development (AUSD), POB 365, Accra
Gen Secr: Victor Cornelius Anson Yevu
Focus: Sci 06474

All Africa Teachers' Organization (AATO), POB 7431, Accra
T: (021) 221515; Tx: 2269
Founded: 1974
Gen Secr: Thomas A. Bediako
Focus: Educ
Periodical
All Africa Teachers' Organization Newsletter (weekly) 06475

Arts Council of Ghana, POB 2738, Accra
T: 64099
Founded: 1958
Pres: Prof. J. H. Nketia; Gen Secr: Charles E. Phillips
Focus: Arts 06476

Association of African Universities (AAU) / Association des Universités Africaines, POB 5744, Accra
T: (021) 663281; Tx: 2284; Fax: (021) 664293
Founded: 1987; Members: 95
Gen Secr: Prof. Donald E.U. Ekong
Focus: Educ
Periodical
AAU Newsletter (3 times annually) . . . 06477

Classical Association of Ghana, c/o University, POB 25, Legon, Accra
Founded: 1952; Members: 17
Focus: Hist 06478

Council for Scientific and Industrial Research (CSIR), POB M32, Accra
T: 77651
Founded: 1958
Gen Secr: Dr. R.G.J. Butler
Focus: Sci; Eng
Periodicals
Annual Report (weekly)
CSIR Handbook
Ghana Journal of Agricultural Science
Ghana Journal of Science 06479

Economic Society of Ghana, POB 22, Legon, Accra
Founded: 1957; Members: 500
Focus: Econ
Periodicals
Economic Bulletin of Ghana
Social and Economic Affairs (weekly) . . 06480

Geological Survey of Ghana, POB M80, Accra
Founded: 1913
Gen Secr: G.O. Kesse
Focus: Geology
Periodical
Annual Report (weekly) 06481

Ghana Academy of Arts and Sciences, POB M32, Accra
T: 777651
Founded: 1959
Pres: Dr. E. Evans-Anfom; Gen Secr: Prof. E. Laing
Focus: Arts; Sci
Periodicals
J.B. Danquah Memorial Lectures (weekly)
Proceedings (weekly) 06482

Ghana Association of Writers, POB 4414, Accra
Founded: 1957
Pres: A. Okai; Gen Secr: J.E. Allotey-Pappoe
Focus: Lit
Periodicals
Angla (weekly)
Takra (weekly) 06483

Ghana Bar Association, POB 4150, Accra
Pres: Peter A. Adjetey; Gen Secr: N. Kuenyehia
Focus: Law 06484

Ghana Geographical Association (G.G.A.), c/o Dept of Geography, University, Accra
Founded: 1955
Pres: Prof. E.V.T. Engmann; Gen Secr: Dr. J.L. Gyamfi-Fenteng
Focus: Geography
Periodical
Bulletin (weekly) 06485

Ghana Library Association, POB 4105, Accra
T: 76591
Founded: 1962
Pres: J.A. Villars; Gen Secr: D.B. Addo
Focus: Libraries & Bk Sci
Periodical
Ghana Library Journal (weekly) 06486

Ghana Library Board, POB 663, Accra
Founded: 1950
Focus: Libraries & Bk Sci
Periodical
Ghana National Bibliography (weekly) . . 06487

Ghana Meteorological Services Department, POB 87, Legon, Accra
Founded: 1937; Members: 409
Gen Secr: S.E. Tandoh
Focus: Astrophys 06488

Ghana Science Association, POB 7, Legon, Accra
Founded: 1959
Pres: Prof. A.N. de Heer-Amissah; Gen Secr: Dr. E.C. Quaye
Focus: Sci
Periodical
The Ghana Journal of Science 06489

Ghana Sociological Association, c/o Dept of Sociology, University of Ghana, Accra
Founded: 1961; Members: 215
Pres: J.N. Amaa; Gen Secr: Dr. C.K. Brown
Focus: Sociology
Periodical
Ghana Journal of Sociology 06490

Historical Society of Ghana, POB 12, Legon, Accra
Founded: 1952; Members: 600
Pres: T.A. Osae; Gen Secr: J.G.K. Tengey
Focus: Hist
Periodicals
Transactions (weekly)
West African Journal for History Teachers (weekly) 06491

Pharmaceutical Society of Ghana, POB 2133, Accra
Founded: 1935; Members: 600
Pres: K.A. Ohene-Manu; Gen Secr: M. Appiah
Focus: Pharmacol 06492

West African Examinations Council (WAEC), POB 125, Accra
T: (021) 221511; CA: REXCIL ACCRA
Founded: 1952
Focus: Educ
Periodicals
Annual Report (weekly)
WAEC News (weekly) 06493

West African Science Association, c/o Botany Dept, University of Ghana, POB 7, Legon, Accra
Founded: 1953; Members: 7
Pres: Prof. A. Hontoundji; Gen Secr: Dr. J.K.B.A. Ata
Focus: Sci
Periodical
Journal (weekly) 06494

Gibraltar

Gibraltar Ornithological and Natural History Society, c/o Gibraltar Natural History Field Centre, Jew's Gate, Upper Rock Nature Reserve, POB 843, Gibraltar
T: 72639, 74022; Fax: 74022
Founded: 1978; Members: 231
Gen Secr: J.E. Cortes
Focus: Ornithology; Botany
Periodicals
Alectoris (weekly)
Gibraltar Nature News (weekly)
Strak of Gibraltar Bird Observation Report (weekly) 06495

The Gibraltar Society, John Mackintosh Hall, Gibraltar
Founded: 1929; Members: 301
Pres: E.A.J. Canessa; Gen Secr: E.F.I. Garcia
Focus: Hist 06496

Gibraltar Teachers' Association (GTA), 40 Town Range, Gibraltar
Founded: 1962; Members: 300
Pres: J.J. Cortes; Gen Secr: J. Wright
Focus: Educ 06497

Greece

Akadimia Athinon, El. Venizelou 28, 10679 Athinai
Founded: 1926; Members: 100
Focus: Sci
Periodicals
Mnimeia Ellinikis Historias: Documents of Greek History
Pragmateiai: Papers
Praktika tes Akademias Athinon: Proceedings (weekly) 06498

Archaeologiki Hetairia, Panepistimiou 22, 10672 Athinai
T: (01) 3644996
Founded: 1837; Members: 526
Focus: Archeol
Periodicals
Archaeologiki Ephimeris (weekly)
Ergon (weekly)
O Mentor (weekly)
Praktika (weekly) 06499

Association of Arts and Letters, Mitropoleos 38, 10563 Athinai
T: (021) 3233033
Founded: 1938; Members: 1500
Pres: A. Evangeliatos; Gen Secr: G. Saranti
Focus: Ling; Lit; Arts 06500

Athens Centre of Ekistics, Stratiotikou Syndesmou 24, Athinai
T: (01) 3623216; Tx: 215227
Founded: 1963
Pres: Panayotis C. Psomopoulos
Focus: Soc Sci 06501

Balkan Physical Union, c/o Dept of Physics, University of Thessaloniki, 54006 Thessaloniki
T: (031) 992715
Gen Secr: E.K. Polychroniadis
Focus: Physics 06502

Elliniki Anaisthisiologiki Etaireia, Ionos Dragoumi 34, Athinai 162
T: (01) 7242566
Members: 205
Focus: Anesthetics
Periodical
Acta Anaesthisiologica Hellenica (weekly) 06503

Elliniki Astronautiki Etaireia, Voulis 14, Athinai 126
T: (01) 3227666
Founded: 1957; Members: 75
Focus: Aero 06504

Elliniki Cheirourgiki Etaireia, Papadiamantopoulou 150, Athinai 623
T: (01) 7701813
Members: 300
Focus: Surgery 06505

Elliniki Geografiki Etaireia / Hellenic Geographical Society, Voucourestiou 11, Athinai 10671
T: (01) 3631112
Founded: 1919; Members: 137
Pres: Dimitrios Dimitriades; Gen Secr: George S. Ivantchos
Focus: Geography
Periodical
Bulletin 06506

Elliniki Kardiologiki Etairia, Sisini 17, 11528 Athinai
T: (01) 7221633; Tx: 222408; Fax: (01) 7226139
Focus: Cardiol
Periodical
Hellenic Cardiological Review (weekly) . 06507

Elliniki Ktiniatriki Eteria / Hellenic Veterinary Medical Society, Kentrikon Tachydromeion, POB 3546, Athinai
Fax: (01) 5241189
Founded: 1924; Members: 262
Pres: Dr. A.T. Rantsios
Focus: Vet Med
Periodical
Bulletin of the Hellenic Veterinary Medical Society (weekly) 06508

Elliniki Laographiki Etaireia, Didotou 12, Athinai
T: (01) 633110
Focus: Ethnology 06509

Elliniki Mathimatiki Eteria (H.M.E.), Panepistimiou 34, 10679 Athinai
T: (01) 3616532; Fax: (01) 3641025
Founded: 1918; Members: 13000
Pres: Prof. Elias Lipitakis; Gen Secr: Djiachristos Vaggelis
Focus: Math
Periodicals
Bulletin
Enimerossi (weekly)
Euclides (5 times annually)
Mathematical Review 06510

Elliniki Microbiologiki Etaireia, c/o Dept of Microbiology, Univeristy, POB 1540, Athinai
T: (01) 7785638
Founded: 1947; Members: 1330
Focus: Microbio 06511

Elliniki Nomismatiki Etaireia / Hellenic Numismatic Society, Didotou 45, 10680 Athinai
T: (01) 3615585; Fax: (01) 3634296
Founded: 1970; Members: 300
Pres: Anastasios P. Tzamalis; Gen Secr: Panos Tazedakis
Focus: Numismatics
Periodical
Numismatika Chronika (weekly) 06512

Elliniki Paidiatriki Etairia, Hippokratous 65, POB 1519, Athinai 144
T: (01) 3638602
Focus: Pediatrics 06513

Elliniki Pharmakeutiki Etaireia, Emm. Benakis 30, 10678 Athinai
T: (01) 3609108
Founded: 1932; Members: 550
Focus: Pharmacol
Periodical
Archeia tis Pharmakeftikis (weekly) . . . 06514

Elliniki Spilaiologiki Etaireia, Spilaiologiki Etaireia, Athinai 135
T: (01) 617824
Focus: Speleology 06515

Ellinikon Kentron Paragogikotitos (Elpeka), Kapodistriou 28, Athinai 147
T: (01) 3600411
Founded: 1953
Focus: Business Admin 06516

Ellinikos Organismos Tupopoiesseos (ELOT) / Hellenic Organization for Standardization, Acharnon 313, 11145 Athinai
T: (01) 2015025
Founded: 1976
Focus: Standards
Periodical
Catalogue of Hellenic Standards (weekly) 06517

Enosis Ellinon Bibliothekarion / Greek Library Association, Skoulenion 4, 10561 Athinai
T: (01) 3226625; Fax: (01) 3226625
Founded: 1968; Members: 1000
Pres: K. Hadjopoulos; Gen Secr: A. Salomon
Focus: Libraries & Bk Sci
Periodical
Bibliothikes kai Plicophocis: Bibliothèques et information (weekly) 06518

Enosis Ellinon Chimikon / Association of Greek Chemists, Kaningos 27, 10682 Athinai
T: (01) 3621524
Founded: 1924; Members: 5000
Pres: P. Hamakiotis; Gen Secr: D. Psomas
Focus: Chem
Periodical
Chimika Chronika 06519

Enosis Ellinon Physikon, Grivaion 6, Athinai 144
T: (01) 3635701
Founded: 1930; Members: 1500
Focus: Physics 06520

Enosis Hellinon Mousourgon / League of Greek Composers, Odos Mitropoleos 38, Athinai
T: (01) 3223302; Fax: (01) 3223302
Founded: 1931; Members: 105
Pres: Theodoros Antoniou; Gen Secr: Georgios Koucoupos
Focus: Music 06521

Epimelitirion Technikon tis Ellados, Karageorgi Servias 4, Athinai 125
T: (01) 3226511
Focus: Eng 06522

Etaireia Byzantinon kai Metabyzantinon Meleton / Society for Byzantine and Post-byzantine Studies, c/o Dept of Byzantine Literature, School of Philosophy, University of Athens, Ilissia, 15771 Athinai
T: (01) 7291488, 7291490
Founded: 1977; Members: 45
Pres: A.D. Kominis; Gen Secr: Helen Papailiopoulou
Focus: Hist; Lit; Arts
Periodicals
Diptycha
Diptychon Parafylla 06523

Etaireia Byzantinon Spoudon / Society for Byzantine Studies, Odos Aristeidou 8, 10559 Athinai
Founded: 1919; Members: 250
Pres: N.B. Tomadakis; Gen Secr: P.G. Nikolopoulos
Focus: Cultur Hist
Periodical
Epetiris Etairias Byzantinin Spoudon: EEBS (weekly) 06524

Etaireia Makedonikon Spoudon (EMS) / Society for Macedonian Studies, Ethnikis Amynis 4, 54621 Thessaloniki
T: (031) 271195, 270343; Fax: (031) 271501
Founded: 1939; Members: 508
Pres: Prof. Constantinos Vavouskos
Focus: Hist
Periodicals
Ellinika: Classical studies (weekly)
Makedonika: History, Folklore, Archaeology of North Greece (weekly) 06525

Etaireia Odontostomatologikis Ereunis, M. Asias 70, 11527 Athinai
T: (01) 7780671
Founded: 1948; Members: 70
Focus: Dent; Stomatology
Periodical
Odontostomatological Progress (weekly) . 06526

European Institute of Environmental Cybernetics, Kerasundos 4, 16232 Athinai
T: (01) 7669466
Founded: 1970
Focus: Cybernetics 06527

Greek National Committee for Astronomy, c/o Akademia Athinon, Odos Anagnostopoulou 14, 10673 Athinai
Fax: (01) 3631606
Focus: Astronomy 06528

Greek National Committee for Space Research, c/o Akademia Athinon, Odos Anagnostopoulon 14, 10673 Athinai
Fax: (01) 3631606
Focus: Astronomy 06529

Greek National Committee for the Quiet Sun International Years, c/o Akademia Athinou, Odos Panepistimiou, Athinai
Focus: Astronomy 06530

Hellenic Association of University Women, Voulis 44a, 10557 Athinai
Pres: Irene Dilari; Gen Secr: Flora Kamari
Focus: Adult Educ 06531

Hellenic Institute of International and Foreign Law, Solonos 73, 10679 Athinai
Fax: (01) 3619777
Founded: 1939
Focus: Int'l Relat; Law
Periodical
Revue Hellénique de Droit International (weekly) 06532

Helliniki Epitropi Atomikis Energhias / Greek Atomic Energy Commission, Aghia Paraskevi, POB 60228, 15310 Athinai
T: (01) 6513111; Tx: 216199
Founded: 1954
Pres: M. Antonopoulos-Domis; Gen Secr: Prof. N. Antoniou
Focus: Nucl Res 06533

Hetaireia Hellenon Philologon / Society of Greek Philologists, Euripidou 12, POB 3373, 10559 Athinai
T: (01) 3213363
Founded: 1948; Members: 1900
Pres: P. Georgountzos; Gen Secr: G. Korres
Focus: Ling; Lit; Hist
Periodical
Platon: Deltion tes Hetaireias Hellenon Philologon (weekly) 06534

Hetairia Hellinon Logotechnon / Society of Greek Men of Letters, Gennadiou 8, 10678 Athinai
Founded: 1934; Members: 700
Pres: Ilias Simopoulos; Gen Secr: Panos Panagiotounis
Focus: Lit 06535

Hetairia Hellinon Thetricon Syngrapheon / Greek Playwrights' Association, Asklipiou 33, 10680 Athinai
Fax: (01) 3614219
Founded: 1908; Members: 500
Pres: Lakis Michailidis; Gen Secr: George Christofilakis
Focus: Lit 06536

Istoriki kai Ethnologiki Etaireia tis Ellados / Historical and Ethnological Society of Greece, Old Parliament, Stadiou, Athinai
T: (01) 3226370
Founded: 1882
Pres: D.A. Gezoutas; Gen Secr: V.C. Mazarakis-Aenian
Focus: Ethnology; Hist
Periodical
Bulletin (weekly) 06537

Kallitechnikon Epimelitirion Ellados (K.E.E.), Mitropoleos 38, Athinai 126
T: (01) 3231230
Founded: 1944; Members: 1100
Focus: Arts 06538

Omospondia Didaskaliki Ellados, Xenofontos 15a, Athinai 118
T: (01) 3236547
Members: 15000
Focus: Educ 06539

Omospondia Panellinios Syndesmon Dasoponon, Veranzerou 42, Athinai
T: (01) 5237512
Focus: Forestry 06540

Panellinios Enosis Technikon (PET), Vranzerou 34, Athinai
T: (01) 5233756
Founded: 1957; Members: 6500
Focus: Eng 06541

Panellinios Omospondia Syndesmon Geoponon, L. Katsoni 12, Athinai
T: (01) 6467516
Focus: Agri 06542

PEN Centre, Skoufa 60a, Athinai 144
Focus: Lit 06543

Syllogos Architktonon Diplomatouchon Anotaton Scholon (SADAS), Ipitou 3, Athinai 118
T: (01) 3236431
Founded: 1922; Members: 2100
Focus: Archit 06544

Syllogos Iatrikos Panellinios, Semitelou 2, Athinai 611
Focus: Med 06545

Syllogos pros Diadosin ton Hellenikon Grammaton / Society for the Promotion of Greek Education, Odos Pindarou 15, Athinai 136
T: (01) 3612370
Founded: 1869; Members: 9
Pres: Philip Dragoumis; Gen Secr: A. Panayotis
Focus: Educ 06546

Union of Middle East Mediterranean Paediatric Societies (UMEMPS), Karneadou 18, Athinai 139
T: (01) 720057
Founded: 1966; Members: 15
Focus: Pediatrics 06547

Women's Literary Society, Evrou 4, Athinai 611
Focus: Lit 06548

World Society for Ekistics (W.S.E.), Strat. Syndesmou 24, 10673 Athinai
Fax: (01) 3633395
Gen Secr: P. Psomopoulos
Focus: Anthro; Sociology; Urban Plan; Archit 06549

Grenada

Caribbean Agricultural and Rural Development, Advisory and Training Service (CARDATS), POB 270, Saint George's
Founded: 1973; Members: 7
Gen Secr: Hugh A. Saul
Focus: Develop Areas; Agri
Periodical
Project Progress Report (weekly) . . . 06550

Caribbean Association of Catholic Teachers (CACT), Morne Jaloux, Saint George's
Founded: 1967
Gen Secr: Anthony Parke
Focus: Educ 06551

Guam

Association of South Pacific Environmental Institutions (ASPEI), c/o College of Arts and Sciences, University of Guam, UOG Station, Mangilao
T: 7342921 ext 287; Tx: 7216275; Fax: 7343118
Founded: 1986
Pres: Prof. H.I. Manner
Focus: Ecology 06552

Guatemala

Academia de Ciencias Médicas, Físicas y Naturales de Guatemala, 13 Calle 1-25, Zona 1, Apdo 569, Guatemala City
Founded: 1945; Members: 69
Pres: Dr. Carlos Cossich; Gen Secr: Rubén Mayorga
Focus: Med; Nat Sci; Physics
Periodical
Annals (weekly) 06553

Academia de Geografía e Historia de Guatemala (AGHG), 3a Av 8-35, Zona 1, Guatemala City
T: 535141, 23544
Founded: 1923; Members: 50
Pres: Jorge Luján Muñoz; Gen Secr: Ana M. Urruela de Quezada
Focus: Geography; Hist
Periodical
Anales de la Academia de Geografía e Historia (weekly) 06554

Academia de la Lengua Maya Quiché, 7a Calle 11-27, Zona 1, Quezaltenango
Founded: 1959; Members: 20
Pres: Prof. Adrián Ines Chávez; Gen Secr: Prof. Víctor Salvador de León Toledo
Focus: Ling
Periodical
El Idioma Quiché y su Grafía 06555

Academia Guatemalteca de la Lengua, 12 Calle 6-60, Zona 9, Guatemala City
T: 322824
Founded: 1887
Pres: David Vela; Gen Secr: Mario Alberto Carrera
Focus: Sci
Periodical
Boletín 06556

Asociación Bibliotecológica Guatemalteca, c/o Biblioteca Nacional, 5a Av 7-26, Zona 1, Guatemala City
Founded: 1969
Focus: Libraries & Bk Sci 06557

Asociación Centroamericana de Historia Natural (ACAHN), Apdo 1120, Guatemala City
Focus: Nat Sci 06558

Asociación de Ortodoncistas de Guatemala, 13 Calle 3-43, Zona 10, Guatemala City
Founded: 1946
Focus: Dent 06559

Asociacion Latinoamericana de Escuelas de Cirurgía Dental, Av las Américas 21-69, Zona 10, Guatemala City
Founded: 1946
Focus: Dent 06560

Asociación Pediátrica de Guatemala, 12 Av 12-72, Zona 1, Guatemala City
Founded: 1945
Focus: Pediatrics 06561

Casa de la Cultura de Occidente, 7a Calle 11-35, Zona 1, Quezaltenango
Founded: 1958
Focus: Hist 06562

Center for Meso American Studies on Appropriate Technology, 1a Av 32-21, Zona 12, Apdo 1160, Guatemala City
T: (02) 762018; CA: (02) 762355
Founded: 1976
Gen Secr: Dr. Edgardo Cáceres
Focus: Eng
Periodical
Boletín RED (weekly) 06563

Central American Research Institute for Industry, Av La Reforma 4-47, Zona 10, Apdo 1552, Guatemala City
T: (02) 310631; Tx: 5312; Fax: (02) 317466
Founded: 1956; Members: 6
Pres: Ludwig Ingram
Focus: Econ
Periodical
Revista ICAITI 06564

Colegio de Ingenieros de Guatemala, 7A Av 39-60, Zona 8, Guatemala City
Founded: 1947; Members: 1342
Focus: Eng 06565

Corporación Centroamericana de Servicios de Navegación Aéreal (COCESNA), c/o SIECA, Apdo 1237, Guatemala City
Focus: Transport 06566

Escuela Regional de Ingeniería Sanitaria (ERIS), c/o Universidad de San Carlos, Ciudad Universitaria, Zona 12, Guatemala City
T: 760424; Tx: 5950
Gen Secr: Arturo Pazos-Sosa
Focus: Eng 06567

Federación de Universidades Privadas de América Central (FUPAC), c/o Universidad Rafael Landivar, Apdo 1273, Guatemala City
Founded: 1966
Focus: Sci 06568

Instituto Centroamericano de Investigación y Tecnología (ICAITI), Av La Reforma 4-47, Zona 10, Apdo 1552, Guatemala City
Tx: 5312; CA: ICAITI
Founded: 1956
Focus: Eng 06569

Instituto de Nutrición de Centro América y Panamá (INCAP) / Institute of Nutrition of Central America and Panama, Apdo 1188, 01901 Guatemala City
T: 723762/67; Tx: 5696; CA: INCAPGU;
Fax: 715658
Founded: 1949
Gen Secr: Luis Octavio Angel
Focus: Food

Periodicals
ASI & PROPAG Bulletin: Informative Pamphlets (weekly)
Report (weekly) 06570

Instituto Nacional de Sismología, Vulcanología, Meteorología e Hidrología (INSIVUMEH), 7a Av 14-57, Zona 13, Guatemala
T: 314967, 314986, 319183; CA: INSIVUMEH; Fax: 363944
Founded: 1925; Members: 512
Gen Secr: Estuardo Velásquez Vásquez
Focus: Geology; Geophys; Water Res
Periodicals
Climatic Bulletin (weekly)
Hydrologic Bulletin (weekly)
Meteorologic Bulletin (weekly)
Seismologic Bulletin (weekly)
Underground Water Resources Publication (weekly) 06571

Secretaria Permanente del Tratado General de Integración Económica Centroamericana, Apdo 1237, Guatemala City
Focus: Econ 06572

Sociedad Centroamericana de Dermatologia, 4a Av 1-56, Zona 1, Guatemala City
Founded: 1957
Focus: Derm 06573

Sociedad Pro-Arte Musical, 12 Calle 2-09, Zona 3, Apdo 980, Guatemala City
Founded: 1945; Members: 200
Pres: Lulú C. de Herrarte; Gen Secr: Dora G. de Mendizábal
Focus: Music 06574

Guyana

Adult Education Association of Guyana, POB 101111, Georgetown
T: 52639
Founded: 1957
Pres: Rudolph Davidson; Gen Secr: Newton L. Profitt
Focus: Adult Educ 06575

Guyana Institute of International Affairs, 189 Charlotte St, Lacytown, POB 812, Georgetown
Founded: 1965; Members: 175
Pres: Donald A.B. Trotman
Focus: Poli Sci
Periodical
Annual Journal of International Affairs (weekly) 06576

Guyana Library Association (GLA), c/o National Library, 76-77 Main St, POB 10240, Georgetown
T: 62690
Founded: 1968; Members: 42
Pres: Wenda Stephenson; Gen Secr: Alethea John
Focus: Libraries & Bk Sci
Periodical
Guyana Library Association Bulletin (weekly) 06577

Guyana Society, Company Path, Georgetown
Focus: Agri; Econ 06578

Haiti

Conseil National des Recherches Scientifiques, c/o Département de la Santé Publique et de la Population, Port-au-Prince
Founded: 1963
Focus: Nat Sci 06579

Honduras

Academia Hondureña, Apdo 38, Tegucigalpa
Founded: 1949; Members: 28
Pres: Miguel R. Ortega; Gen Secr: Jorge Fidel Durón
Focus: Sci
Periodical
Boletín 06580

Academia Hondureña de Geografía e Historia, Apdo 619, Tegucigalpa
Founded: 1968
Pres: Dr. Ramón E. Cruz; Gen Secr: Fernando Ferrari Bustillo
Focus: Geography; Hist
Periodical
Revista 06581

Asociación de Bibliotecarios y Archiveros de Honduras (ABAH), 11a Calle, 1a y 2a Av 105, Comayagu#E5ela, Tegucigalpa
Founded: 1951; Members: 53
Focus: Libraries & Bk Sci; Archives . . 06582

Asociación de Facultades de Medicina (ACAFAM), c/o Facultad de Ciencias Médicas, Universidad de Honduras, Tegucigalpa
T: 323975
Focus: Med 06583

Instituto Hondureño de Cultura Interamericana, 2a Av 520 5a y 6a Calle, Comayaguela, Apdo 201, Tegucigalpa
Founded: 1939; Members: 900
Pres: Vicente Diaz Reyes
Focus: Sci; Arts; Hist 06584

Sociedad Centroamericana de Cardiología, c/o Clínicas Unidas, Tegucigalpa
Focus: Cardiol 06585

Sociedad de Geografía e Historia de Honduras, Av de la Policía Central, Tegucigalpa
Founded: 1927
Focus: Geography; Hist 06586

Hong Kong

Asia and Oceania Society for Comparative Endocrinology (AOSCE), c/o Dept of Zoology, University of Hong Kong, Hong Kong
Founded: 1987
Gen Secr: Prof. D.K.O. Chan
Focus: Endocrinology 06587

Asian Association of Management Organizations (AAMO), c/o Hong Kong Management Association, 14/F Fairmont House, 8 Cotton Tree Dr, Central, Hong Kong
Founded: 1960
Gen Secr: John Hung
Focus: Business Admin
Periodical
AAMO Newsletter 06588

Asian Dermatological Association, c/o Federation of Medical Societies of Hong Kong, 4/F Duke of Windsor Bldg, 15 Hennesy Rd, Hong Kong
T: 5278898; Fax: 8650345
Gen Secr: Dr. Stephen Ngai
Focus: Derm 06589

Asian Foundation for the Prevention of Blindness, 247 Nam Cheong St, Kowloon, Hong Kong
T: 9958660 ext 300; Fax: 7880040
Founded: 1981
Gen Secr: Grace Chan
Focus: Ophthal 06590

Asian Pacific Federation of Human Resource Management (APFHRM), c/o Hong Kong Institute of Personnel Management, Hong Kong Jewellery Bldg, 178-180 Queen's Rd, GPOB 4404, Hong Kong
T: 440236, 422787; Fax: 8543885
Founded: 1968
Focus: Business Admin
Periodical
APFHRM Newsletter 06591

Asian Surgical Association, c/o Dept of Surgery, Queen Mary Hospital, University of Hong Kong, Hong Kong
T: 8192610; Fax: 8551897
Founded: 1976
Gen Secr: Dr. John Wong
Focus: Surgery 06592

Hong Kong Library Association (HKLA), GPOB 10095, Hong Kong
T: 468161
Founded: 1958; Members: 300
Pres: Elaine Morgan; Gen Secr: Edith Wu
Focus: Libraries & Bk Sci
Periodicals
HKLA Journal (weekly)
Newsletter (8 times annually) 06593

International Council of Associations for Science Education (ICASE), c/o Dept of Curriculum Studies, University of Hong Kong, Hong Kong
T: (5) 8170837; Tx: (5) 479907; Fax: (5) 8550543
Founded: 1973; Members: 75
Gen Secr: Dr. Jack Holbrook
Focus: Sci; Educ
Periodical
ICASE Newsletter (weekly) 06594

Royal Asiatic Society, Hong Kong Branch, POB 3864, Hong Kong
Founded: 1847; Members: 560
Pres: David Gilkes; Gen Secr: David Shiel
Focus: Ethnology; Hist; Cultur Hist
Periodical
Journal (weekly) 06595

Hungary

Bör-, Cipö-, és Börfeldolgozóipari Tudományos Egyesület / Scientific Society of the Leather, Shoe and Allied Industries, Kossuth Lajos utca 6, 1053 Budapest
Founded: 1930
Pres: Dr. T. Karnitscher
Focus: Materials Sci
Periodical
Bör és Cipötechnika: Leather and Shoemaking 06596

Bolyai János Matematikai Társulat / János Bolyai Mathematical Society, Anker-köz 1, 1061 Budapest
T: (1) 427741
Founded: 1891
Focus: Math
Periodicals
A Matematika Tanitása
Combinatorica
Középiskolai Matematikai Lapok
Matematikai Lapok
Periodica Mathematica Hungarica 06597

Commission of the History of Historiography, c/o Institut d'Histoire de l'Académie des Sciences de Hongrie, POB 9, 1250 Budapest
Founded: 1980; Members: 95
Gen Secr: Dr. Ferenc Glatz
Focus: Hist
Periodical
Storia della Storiografia (weekly) 06598

Council of Directors of Institutes of Tropical Medicine in Europe (TROPMEDEUROPE), c/o Hungarian Tropical Health Institute, Gyáli út 5-7, 1097 Budapest
T: (01) 1142412; Tx: 225104; Fax: (01) 1330960
Founded: 1983
Gen Secr: Prof. F. Várnai
Focus: Trop Med 06599

Energiagazdálkodási Tudományos Egyesület (ETE), Kossuth Lajos tér 6-8, 1055 Budapest
T: (1) 532751
Founded: 1949; Members: 4200
Pres: Gyula Czipper; Gen Secr: Győző Wiegand
Focus: Energy
Periodicals
Energia és Atomtechnika (weekly)
Energiagazdálkodás (weekly) 06600

Eötvös Loránd Fizikai Társulat / Roland Eötvös Physical Society, Fö utca 68, 1371 Budapest
T: (1) 159065; Tx: 225792
Founded: 1891; Members: 2500
Pres: Prof. Dezsö Kiss; Gen Secr: Prof. Peter Richter
Focus: Physics
Periodical
Fizikai Szemle (weekly) 06601

Epitéstudományi Egyesület / Scientific Society for Building, Fö utca 68, 1027 Budapest
T: (01) 2018416; Tx: 224343; Fax: (01) 1561215
Founded: 1949; Members: 5213
Pres: Dr. Miklós Márkus; Gen Secr: Gábor Farkas
Focus: Civil Eng
Periodicals
Magyar Epitöipar: Hungarian Building Industry (weekly)
Magyar Epületgépészet: Sanitary Engeneering (weekly) 06602

European Distance Education Network (EDEN), c/o TEMPUS Hungarian Office, Ajtósi Dürer sor 19-21, 1440 Budapest
T: (1) 2515641; Fax: (1) 1534991
Founded: 1991
Focus: Educ 06603

Faipari Tudományos Egyesület / Scientific Society of the Timber Industry, Anker-köz 1, 1061 Budapest
T: (1) 227861
Founded: 1950; Members: 3300
Pres: Tibor Kara; Gen Secr: Dr. G. Dalocsa
Focus: Eng
Periodical
Faipar 06604

Geodéziai és Kartográfiai Egyesület (GKE) / Society for Geodesy and Cartography, Fö utca 68, 1027 Budapest
T: (1) 158641
Founded: 1956; Members: 2500
Pres: Dr. I. Joó; Gen Secr: Dr. Ferenc Karsay
Focus: Surveying; Cart
Periodical
Geodézia és Kartográfia 06605

Gépipari Tudományos Egyesület (GTE) / Scientific Society of Mechanical Engineers, Fö utca 68, 1027 Budapest
T: (1) 2020656; Tx: 224343; Fax: (1) 2020252
Founded: 1949; Members: 12000
Pres: Prof. Dr. János Ginszler; Gen Secr: Dr. János Rittinger
Focus: Eng
Periodicals
Gép: Machine
Gépgyártástechnológia: Manufacturing Processes
Gépipar: Machine Industry
Jármüvek, Epitöipari és Mezögazdasági Gépek: Vehicles, Building and Agricultural Machines
Múanyag és Gumi: Plastics and Rubber 06606

Hiradástechnikai Tudományos Egyesület / Scientific Society for Telecommunication, Kossuth Lajos tér 6-8, 1055 Budapest
Fax: (01) 561215
Founded: 1949; Members: 4500
Pres: Dr. Géza Gordos; Gen Secr: Gábor Halhi
Focus: Electric Eng
Periodicals
Híradástechnika: Telecommunication (weekly)
Hírlevél: Newsletter (weekly) 06607

International Measurement Confederation (IMEKO), Kossuth Lajos tér 6-8, 1055 Budapest
T: (1) 1531562; Fax: (1) 1531406
Founded: 1958; Members: 33
Gen Secr: Prof. Dr. Tamas Kemény
Focus: Eng
Periodical
Measurement (weekly) 06608

Közlekedéstudományi Egyesület (KTE) / Scientific Society for Transport, Kossuth Lajos tér 6-8, 1055 Budapest
T: (1) 530562
Founded: 1949; Members: 13500
Pres: Prof. Endre Kerkápoly; Gen Secr: Dr. J. Zahumenszky
Focus: Transport 06609

Korányi Sandor Társaság, c/o I. Dept of Obstetrics and Gynecology, Bavoss utca 27, 1088 Budapest
T: (1) 1140461
Focus: Med 06610

Liszt Ferenc Társaság / F. Liszt Society, Vörösmarty utca 35, 1064 Budapest
T: (1) 421573, 229804
Founded: 1893; Members: 941
Focus: Music 06611

Magyar Agrártudományi Egyesület (MAE) / Hungarian Society of Agricultural Sciences, Kossuth Lajos tér 6-8, 1055 Budapest
T: (1) 1530651; Fax: (1) 1531950
Founded: 1951; Members: 12600
Pres: Dr. P. Horn; Gen Secr: Dr. V. Marillai
Focus: Agri
Periodical
Agrárvilág 06612

Magyar Allergologiai és Klinikai Immunológiai Társaság (MAKIT), c/o Dr. Willibald Wiltner, 3221 Kékestető
Founded: 1967; Members: 250
Focus: Immunology 06613

Magyar Altalános Orvosok Tudományos Egyesülete (MAOTE), Bessenyei utca 27, 1133 Budapest
Founded: 1967; Members: 2185
Focus: Med
Periodical
Medicus Universalis (weekly) 06614

Magyar Anatómusok, Histologusok és Embryologusok Társasága / Hungarian Society of Anatomists, Histologists and Embryologists, c/o Dept of Anatomy, Semmelweis University Medical School, Türoltó utca 58, POB 95, 1094 Budapest
T: (1) 335158
Founded: 1966; Members: 171
Pres: Prof. G. Székely; Gen Secr: Prof. T. Donáth
Focus: Anat; Bio 06615

Magyar Angiologiai Társaság, POB 88, 1450 Budapest
Founded: 1961; Members: 160
Focus: Hematology 06616

Magyar Asztonautikai Egyesület / Hungarian Astronautical Society, Fö utca 68, 1027 Budapest
T: (1) 159813
Founded: 1986; Members: 450
Pres: Dr. I. Almár; Gen Secr: Dr. G. Major
Focus: Eng 06617

Magyar Balneológiai Egyesület, c/o Dr. Lóránt Fröhlich, Frankel Leo utca 25, 1023 Budapest
T: (1) 154280
Founded: 1967
Focus: Physical Therapy 06618

Magyar Biofizikai Társaság (MBFT) / Hungarian Biophysical Society, Fö utca 68, POB 433, 1371 Budapest
T: (1) 2021216; Fax: (1) 2021216
Founded: 1961; Members: 500
Pres: Prof. Lajos Keszthelyi; Gen Secr: Dr. Sándor Györgyi
Focus: Physics
Periodicals
Bulletin of the Hungarian Biophysical Society
Magyar Biofizikai Társaság Ertesitöje (every 3 years)
Newsletter (weekly)
Newsletter (weekly) 06619

Magyar Biokemiai Társaság / Hungarian Biochemical Society, Anker-köz 1, 1061 Budapest
T: (1) 229446; Tx: 225369
Founded: 1942
Pres: Dr. G. Dénes; Gen Secr: Dr. Egon Hidvégi
Focus: Biochem
Periodical
Biokémia 06620

Magyar Biológiai Társaság (MBT) / Hungarian Biological Society, Fö utca 68, 1027 Budapest
T: (1) 354360
Founded: 1952; Members: 750
Focus: Bio 06621

Magyar Dermatologiai Társulat, Maria utca 41, 1085 Budapest
T: (1) 135400
Founded: 1928; Members: 434
Focus: Derm 06622

Magyar Diabetes Társaság / Hungarian Diabetes Association, Pihenö utca 1, POB 1, 1529 Budapest
Fax: (1) 1763521
Founded: 1970
Gen Secr: Laszlo Koranyi
Focus: Diabetes
Periodical
Diabetologia Hungarica (weekly) 06623

Magyar Elektrotechnikai Egyesület (MEE) / Hungarian Electrotechnical Association, Kossuth Lajos tér 6-8, 1055 Budapest
T: (1) 1120662; Tx: 225792; Fax: (1) 1534069
Founded: 1900; Members: 6500
Pres: Dr. T. Horváth; Gen Secr: Dr. P.I. Timár

Focus: Electric Eng
Periodical
Elektrotechnika (weekly) 06624

Magyar Elemezésipari Tudományos Egyesület (METE) / Hungarian Scientific Society for the Food Industry, Akadémia utca 1-3, 1361 Budapest
T: (1) 1122859; Fax: (1) 1310288
Founded: 1949; Members: 7000
Pres: Prof. P. Biacs; Gen Secr: Dr. Zoltán Hernádi
Focus: Food
Periodical
Élelmezési Ipar (weekly) 06625

Magyar Elettani Társaság / Hungarian Physiological Society, Nagyvárad tér 4, POB 370, 1445 Budapest
T: (1) 137017
Founded: 1931; Members: 600
Pres: Prof. L. Hársing; Gen Secr: Prof. E. Monos
Focus: Pathology; Physiology
Periodical
Kisérletes Orvostudomány (weekly) . . . 06626

Magyar Farmakológiai Társaság (MFT), POB 370, 1445 Budapest
T: (1) 1137015; Fax: (1) 1138090
Members: 490
Focus: Pharmacol 06627

Magyar Fögorvosok Egyesülete (MFE), Szentdirályi utca 40, 1088 Budapest
T: (01) 330970
Founded: 1878; Members: 100
Focus: Dent 06628

Magyar Földrajzi Társaság (MFT), Népköztarsaság utca 62, 1062 Budapest
T: (1) 117688
Founded: 1872; Members: 1150
Focus: Geography
Periodical
Földrajzi Közlemenyek: Geographical Review (weekly) 06629

Magyar Gastroenterologiai Társaság, c/o Prof. Dr. I. Wittman, János Kórház, Diósárk utca 1, 1125 Budapest
Members: 820
Focus: Gastroenter 06630

Magyar Geofizikusok Egyesülete (MGE), Fö utca 68, POB 433, 1371 Budapest
T: (1) 2019815; Fax: (1) 2019815
Founded: 1954; Members: 700
Focus: Geophys
Periodical
Magyar Geofizika (weekly) 06631

Magyar Gerontológiai Társaság / Hungarian Association of Gerontology, c/o Gerontology Center, Semmelweis Medical University, Rökk Szilárd utca 13, POB 45, 1428 Budapest
T: (1) 135411; Fax: (1) 1141830
Founded: 1967; Members: 240
Pres: Dr. E. Beregi
Focus: Geriatrics; Bio; Soc Sci 06632

Magyar Gyógyszerészeti Társaság / Hungarian Pharmaceutical Society, Zrinyi utca 3, 1051 Budapest
T: (1) 181573
Founded: 1924; Members: 3000
Pres: Dr. G. Szász; Gen Secr: Dr. Z. Vincze
Focus: Pharmacol
Periodicals
Acta Pharmaceutica Hungarica (weekly)
Gyógyszerészet (weekly) 06633

Magyar Hidrológiai Társaság / Hungarian Hydrological Society, Fö utca 68, 1027 Budapest
T: (1) 150063
Founded: 1917; Members: 5600
Focus: Water Res; Hydrology
Periodicals
Hidrológiai Közlöny (weekly)
Hidrológiai Tájékoztató (weekly) 06634

Magyarhoni Földtani Társulat / Hungarian Geological Society, Fö utca 68, 1027 Budapest
T: (1) 2019129; Tx: 224343; Fax: (1) 1561215
Founded: 1848; Members: 1400
Pres: Dr. Tibor Kecskeméti; Gen Secr: DSr. János Halmai
Focus: Geology
Periodicals
Általános Földtani Szemle: General Geological Review (weekly)
Annals of the History of Hungarian Geology: Földtani Tudománytörtáneti Évkönyv (weekly)
Discussiones Palaeontologicae: Öslénytani Viták (weekly)
Engineering Geological Review: Mérnökgeológiai-Környezetföldtani Szemle (weekly)
Földtani Közlöny (weekly) 06635

Magyar Immunológiai Társaság, c/o Országos Reuma- és Fizioterápiás Intézet, POB 114, 1502 Budapest
T: (1) 665822
Founded: 1972
Focus: Immunology
Periodical
Acta Microbiologica et Immunologica Hungarica 06636

Magyar Iparjogvédelmi Egyesület (MIE) / Hungarian Association for the Protection of Industrial Property, Kossuth Lajos tér 6-8, 1055 Budapest
T: (1) 313380, 321544; Tx: 225792
Founded: 1962; Members: 2 010
Pres: Dr. István Bihari; Gen Secr: Dr. Endre Lontai
Focus: Eng
Periodical
MIE Közleményi 06637

Magyar Irodalomtörténeti Társaság / Society of Hungarian Literary History, Pesti Barnabás utca 1, 1052 Budapest
Founded: 1912
Pres: Dezsö Keresztury; Gen Secr: Sándor Iván Kovács
Focus: Lit
Periodical
Irodalomtörténet: Literary History (weekly) 06638

Magyar Irók Szövetsége / Association of Hungarian Writers, Bajza utca 18, 1062 Budapest
Founded: 1945
Pres: József Tornei
Focus: Lit
Periodicals
Kortárs (weekly)
Magyar Napló (weekly) 06639

Magyar Jogász Egylet (MJSZ) / Hungarian Lawyers Association, Szemere utca 10, 1054 Budapest
T: (1) 1114880; Fax: (1) 1114013
Founded: 1949; Members: 9000
Focus: Law
Periodicals
Jogásnapló (weekly)
Jogász Szövetségi Értekezések (weekly)
Magyar Jog (weekly) 06640

Magyar Kardiologusok Társasága / Hungarian Society of Cardiology, POB 88, 1450 Budapest
T: (1) 2151220; Tx: 227888; Fax: (1) 2157067
Founded: 1957; Members: 1050
Pres: Prof. Dr. Gyula Papp; Gen Secr: Dr. Joseph Borbola
Focus: Cardiol
Periodical
Cardiologia Hungarica (weekly) 06641

Magyar Karszt- és Barlangkutató Társulat, Anker-köz 1-3, 1061 Budapest
T: (1) 217293
Founded: 1910; Members: 850
Focus: Speleology
Periodicals
Beszámoló (weekly)
Karszt és Barlang (weekly) 06642

Magyar Kémikusok Egyesülete (MKE) / Hungarian Chemical Society, Fö utca 68, 1027 Budapest
T: (1) 2016883; Tx: 224343; Fax: (1) 2018056
Founded: 1907; Members: 5000
Pres: Dr. G. Náray-Szabó; Gen Secr: Dr. L. Harsányi
Focus: Chem
Periodicals
Magyar Kémiai Folyóirat: Hungarian Journal of Chemistry (weekly)
Magyar Kémikusok Lapja: Hungarian Chemical Journal (weekly) 06643

Magyar Könyvtárosok Egyesülete (MKE) / Association of Hungarian Librarians, Szabó Ervin tér 1, 1088 Budapest
T: (1) 1185815; Fax: (1) 1182050
Founded: 1935; Members: 2300
Pres: Dr. Tibor Horváth; Gen Secr: István Papp
Focus: Libraries & Bk Sci
Periodical
Hirlevél Magyar Könyvtárosok Egyesülete tagjaihoz (weekly) 06644

Magyar Meteorológiai Társaság / Hungarian Meteorological Society, Fö utca 68, 1027 Budapest
T: (1) 2017525
Founded: 1925; Members: 450
Pres: Dr. Pál Ambrózy; Gen Secr: Dr. Iván Mersich
Focus: Geophys 06645

Magyar Mikrobiológiai Társaság (MMT) / Hungarian Society of Microbiology, POB 64, 1966 Budapest
T: (1) 137652
Founded: 1951; Members: 358
Focus: Microbio 06646

Magyar Néprajzi Társasag / Hungarian Ethnographical Society, Kossuth Lajos tér 12, 1055 Budapest
T: (1) 326340
Founded: 1889; Members: 942
Pres: I. Balassa; Gen Secr: Miklos Szilagys
Focus: Ethnology
Periodicals
Ethnographia (weekly)
Néprajzi Hirek 06647

Magyar Numizmatikai Társulat, Csepreghu utca 4, 1088 Budapest
T: (1) 131058
Founded: 1901; Members: 507
Pres: Dr. Jenö Fitz; Gen Secr: Dr. Gyula Rádóczy
Focus: Numismatics
Periodicals
Evkönyv (weekly)
Numizmatikai Közlöny (weekly) 06648

Magyar Nyelvtudományi Társaság (MNyT) / Hungarian Linguistic Society, Pesti Barnabás utca 1, 1052 Budapest
T: (1) 1376819
Founded: 1904; Members: 691
Pres: Prof. Dr. Loránd Benkö; Gen Secr: J. Kiss
Focus: Ling
Periodicals
A Magyar Nyelvtudományi Társaság Kiadványai (weekly)
Magyar Nyelv (weekly) 06649

Magyar Orvostudományi Társaságok és Egyesületek Szövetsége / Federation of Hungarian Medical Societies, Nádor utca 36, 1051 Budapest
T: (1) 1123807; Fax: (1) 1837918
Members: 69
Focus: Med
Periodicals
Intermed (weekly)
MOTESZ Magazin 06650

Magyar Pathológusok Társasága, c/o Dept of Pathology, Haynal Imre University of Health Sciences, Szabolos utca 35, 1389 Budapest
T: (1) 1403892
Members: 400
Focus: Pathology
Periodicals
Morphologia és Igazságügyi Orvosi Szemle (weekly)
Pathologia (weekly) 06651

Magyar PEN Club / Hungarian PEN Club, Vörösmarty tér 1, 1051 Budapest
T: (1) 1184143
Founded: 1926; Members: 329
Pres: Miklós Hubay; Gen Secr: Imre Szász
Focus: Lit
Periodicals
The Hungarian PEN
Le PEN Hongrois (weekly) 06652

Magyar Pszichológiai Társaság / Hungarian Psychological Association, Meredek utca 1, POB 220, 1536 Budapest
T: (1) 260663
Founded: 1928; Members: 1 054
Pres: Lajos Bartha; Gen Secr: Judit Ungárné-Komoly
Focus: Psych
Periodical
Pszichológiai Szemle: Psychological Review 06653

Magyar Radiológusok Társasága / Societas Radiologorum Hungarorum / Society of the Hungarian Radiologists, Mohai utca 2, 1115 Budapest
T: (1) 1813674; Fax: (1) 1853652
Founded: 1922; Members: 500
Pres: Dr. Gábor Vadon; Gen Secr: Dr. Béla Fornet
Focus: Radiology
Periodical
Magyar Radiologia (weekly) 06654

Magyar Régészeti és Müvészettörténeti Társulat / Hungarian Society of Archaeology and History of Fine Arts, Múzeum krt 14, Budapest 8
Focus: Fine Arts; Archeol
Periodicals
Archaeológiai Ertesítö
Müvészettörténeti Értesítö 06655

Magyar Rovartani Társaság, Barross utca 13, 1088 Budapest
T: (1) 130035
Founded: 1910; Members: 300
Pres: Dr. Z. Mészáros; Gen Secr: A. Podlusány
Focus: Entomology
Periodical
Folia Entomologica Hungarica 06656

Magyar Sebész Társaság / Societas Chirurgica Hungarica, Ra1th Gy utca 7-9, 1122 Budapest
T: (1) 550330; Fax: (1) 1562402
Founded: 1906; Members: 1345
Focus: Surgery
Periodical
Magyar Sebészet (weekly) 06657

Magyar Szinházi Intézet / Hungarian Theatre Institute, Krisztina krt 57, 1016 Budapest
Founded: 1957; Members: 35
Focus: Perf Arts
Periodicals
Hungarian Theatre Hungarian Drama
Magyar Szinházi Hirek
Szinházelméleti Füzetek
Szinháztörténeti Könyvtár 06658

Magyar Történelmi Társulat / Hungarian Historical Society, I. Uri utca 51-53, 1250 Budapest
T: (1) 160160
Founded: 1867
Pres: István Diószegi; Gen Secr: Klára Hegyi
Focus: Hist
Periodicals
Magyarország Ujabbkori Forrásai
Századok 06659

Magyar Traumatológus Társaság / Hungarian Society for Trauma Surgery, Fiumei utca 17, POB 21, 1430 Budapest
T: (1) 1337599; Fax: (1) 1330966
Founded: 1967; Members: 488
Focus: Traumatology
Periodical
Magyar Traumatologia, Orthopaedia, Kézsebézet Plasztikai Sebészet (weekly) 06660

Magyar Tudományos Akadémia / Hungarian Academy of Sciences, Roosevelt tér 9, 1051 Budapest
T: (1) 382344; Tx: 224139; Fax: (1) 1128483
Founded: 1825; Members: 217
Pres: Domokos Kosáry; Gen Secr: László Keviczky
Focus: Sci
Periodicals
Acta Agronomica
Acta Alimentaria
Acta Antiqua
Acta Archaeologica
Acta Biochimica et Biphysica
Acta Biologica
Acta Botanica
Acta Chimica
Acta Chirurgica
Acta Ethnographica
Acta Geodaetica, Geophysica et Montanistica
Acta Geologica
Acta Historiae Artium
Acta Historica
Acta Juridica
Acta Linguistica
Acta Litteraria
Acta Mathematica
Acta Medica
Acta Microbiologica
Acta Morphologica
Acta Oeconomica
Acta Orientalia
Acta Paediatrica
Acta Physica
Acta Physiologica
Acta Phytopathologica
Acta Technica
Acta Veterinaria
Acta Zoologica
Studia Musicologica
Studia Scientiarium Mathematicarum
Studia Slavica 06661

Magyar Urbanisztikai Társaság (MUT) / Society for Human Settlements, Rákóczi utca 7, 1088 Budapest
T: (1) 339959
Founded: 1966; Members: 698
Pres: Dr. J. Szabó; Gen Secr: G. Barna
Focus: Urban Plan
Periodical
Városépites (weekly) 06662

Magyar Zenei Tanács / Hungarian Music Council, V. Vörösmarty tér 1, POB 47, 1364 Budapest
T: (1) 1184267; Fax: (1) 1178267
Founded: 1990; Members: 82
Gen Secr: Katalin Forrai
Focus: Music
Periodicals
Magyar Zene (weekly)
Polifónia (weekly) 06663

Méréstechnikai és Automatizálási Tudományos Egyesület (MATE) / Scientific Society of Measurement and Automation, Kossuth Lajos tér 6-8, POB 451, 1055 Budapest
T: (1) 531406
Founded: 1952; Members: 3000
Pres: Prof. Dr. I. Martos; Gen Secr: P. Reiniger
Focus: Eng
Periodical
Mérés és Automatika: Measurement and Automation (weekly) 06664

Müszaki és Természettudományi Egyesületek Szövetségi Kamarája / Federal Chamber of Technical and Scientific Societies, Kossuth L. tér 6-8, 1055 Budapest
Founded: 1948
Focus: Eng; Nat Sci
Periodical
Müszaki Magazin 06665

Neumann János Számitógéptudományi Társaság / John von Neumann Computer Society, Báthory utca 16, 1054 Budapest
T: (1) 1329349; Tx: 225369; Fax: (1) 1318140
Founded: 1968; Members: 5500
Pres: M. Havass; Gen Secr: Maria Föth
Focus: Computer & Info Sci 06666

Optikai, Akusztikai és Filmtechnikai Egyesület (OPAKF) / Optical, Acoustical and Filmtechnical Society, Fö utca 68, 1027 Budapest
T: (1) 354943; Tx: 224343; Fax: (1) 561215
Founded: 1933; Members: 1700
Pres: Gábor Heckenast; Gen Secr: Olivér Petrik
Focus: Cinema; Optics 06667

Országos Állategészségügyi Intézet / Central Veterinary Institute, Tábornok utca 2, 1149 Budapest
Fax: (1) 2525177
Founded: 1928; Members: 118
Focus: Vet Med 06668

Országos Erdészeti Egyesület / Hungarian Forestry Association, Fő utca 68, 1027 Budapest
T: (1) 2016283; Fax: (1) 2017737
Founded: 1866; Members: 4500
Pres: András Schmotzer; Gen Secr: Gábor Barátossy
Focus: Forestry
Periodical
Erdészeti Lapok: Forestry Bulletin (weekly) 06669

Országos Magyar Bányászati és Kohászati Egyesület (OMBKE) / Hungarian Mining and Metallurgical Society, Fő utca 68, 1027 Budapest
T: (1) 2017337; Tx: 224343
Founded: 1892; Members: 9500
Pres: Dr. István Tóth; Gen Secr: Dr. Pál Tardy
Focus: Mining; Metallurgy
Periodicals
Bányászat: Mining (weekly)
Bohászat: Metallurgy (weekly)
Kőolaj és Földgáz: Oil and Gas
Öntöde: Foundry 06670

Országos Magyar Cecília Társulat (OMCE) / National Hungarian Cecilia Society, Fehérvári utca 82, 1119 Budapest
Founded: 1897; Members: 2800
Pres: G. Szakos; Gen Secr: László Tardy
Focus: Music 06671

Papir- és Nyomdaipari Müszaki Egyesület / Technical Association of the Paper and Printing Industry, Fő utca 68, 1027 Budapest
T: (01) 321748, 113250
Founded: 1948; Members: 3800
Focus: Eng
Periodicals
Magyar Grafika (weekly)
Papiripar (weekly) 06672

Szervezési és Vezetési Tudományos Társaság / Society for Organization and Management Science, II. Fő utca 68, POB 433, 1371 Budapest
T: (1) 154090; CA: 61225792; Fax: (1) 561215
Founded: 1970; Members: 14 500
Pres: Dr. Ferenc Trethon; Gen Secr: Dr. János Polonkai
Focus: Business Admin; Public Admin
Periodical
Ipar-Gazdaság: Industry-Economy (weekly) 06673

Szilikátipari Tudományos Egyesület, Anker-köz 1, POB 240, 1368 Budapest
T: (1) 226497
Founded: 1949; Members: 1500
Focus: Eng
Periodical
Epitőanyog: Building Materials (weekly) . 06674

Textilipari Müszaki és Tudományos Egyesület (TMTE) / Hungarian Society of Textile Technology and Science, Fő utca 68, 1027 Budapest
T: (1) 2018782; Tx: 224343; Fax: (1) 1561215
Founded: 1948; Members: 7200
Pres: Dr. Frigyes Geleji; Gen Secr: Gábor Kelényi
Focus: Eng; Textiles
Periodicals
Cotton Industry
Flax, Hemp, Synthetic Fibre Industry Bulletin
Haberdashery Technology
Hungarian Textile Engineering
Knitwear Industry Review
Textile Cleaning
Textile Economy
Wool Industry Review 06675

Tudományos Ismeretterjesztő Társulat / Society for Dissemination of Sciences, Bródy Sándor utca 16, 1088 Budapest
T: (1) 343380
Founded: 1841; Members: 27600
Pres: Adám György; Gen Secr: Ferenc Rottler
Focus: Nat Sci
Periodicals
Békési Élet: Békés Life (weekly)
Borsodi Szemle: Borsodi Review (weekly)
Egészég: Health
Élet és Tudomány: Life and Science (weekly)
Föld és Ég: Earth and Sky (weekly)
Müemlékvédelem: Care of Ancient Monuments (weekly)
Természet Világa: World of Nature (weekly)
Valóság: Reality (weekly) 06676

Iceland

Arkitektafélag Islands / Icelandic Architects' Association, 101 Reykjavik
Members: 260
Pres: Ormar P. Gubmundsson
Focus: Arts
Periodical
Arkitídindi (3 times annually) 06677

Bókavardafélag Islands (BVFI), POB 1497, 121 Reykjavik
Founded: 1960; Members: 120
Focus: Libraries & Bk Sci 06678

Búnadarfélag Islands / Agricultural Society of Iceland, Baendahöllinni, Hagatorgi, POB 7080, 127 Reykjavik
T: (1) 19200; Fax: (1) 628290
Founded: 1899; Members: 4000
Pres: Hjörtur E. Thérarinsson; Gen Secr: Jónas Jónsson
Focus: Agri
Periodicals
Bùnadarrit (weekly)
Freyr (weekly)
Handbök Bånoa (weekly)
Hrossaraektin (weekly)
Nautgriparaektin (weekly)
Saudfjárraektin (weekly) 06679

Danske Selskab i Reykjavik / Danish Society, POB 222, Reykjavik
Founded: 1923
Pres: H. Fenger
Focus: Int'l Relat 06680

Félag Bókavarda í Rannsókarbókasöfnum / Islandic Research Librarians Association, POB 5382, 125 Reykjavik
T: (1) 13080
Gen Secr: O. Benediktsdóttir
Focus: Libraries & Bk Sci 06681

Félag Enskukennara á Islandi (FEKI), POB 7122, 127 Reykjavik
T: (1) 35042
Focus: Ling; Educ
Periodical
Málfridur (weekly) 06682

Félag Háls-, Nef- og Eyrnalaekna, c/o Prof. Stefan Olafsson, Hringbraut 34, Reykjavik
Focus: Otorhinolaryngology 06683

Félag Islenskra Röntgenlaekna / Icelandic Society of Radiology, c/o Röntgendeild, Landakotsspítalinn, Reykjavik
Fax: (1) 627314
Founded: 1957; Members: 35
Pres: Gudmundur J. Eliasson; Gen Secr: Einar Steingrimsson
Focus: X-Ray Tech 06684

Félag Islenskra Tryggingastaerdfraedinga (FIT) / The Society of Icelandic Actuaries, Sudarlandsbraut 6, 105 Reykjavik
Founded: 1967; Members: 10
Focus: Insurance; Stats 06685

Félag Menntaskólakennara (FM), c/o Bandalag Haskólamanna, Studentaheimilinu vid Hringbraut, Reykjavik
Members: 250
Focus: Educ 06686

Gedverndarfélag Islands, Hafnarstraeti 5, POB 467, Reykjavik
T: (1) 12139
Founded: 1949; Members: 893
Focus: Psychiatry 06687

Hid Islenska Fornleifafélag, c/o National Museum of Iceland, POB 1489, 121 Reykjavik
T: (1) 28888
Founded: 1880; Members: 660
Pres: Hördur Agustsson; Gen Secr: Thorhallur Vilmundarson
Focus: Archeol
Periodical
Arbók (weekly) 06688

Hid Islenska Nattúrufraedifélag, POB 846, 121 Reykjavik
Founded: 1889; Members: 1800
Focus: Bio
Periodical
Náttúrufrådingurinn (weekly) 06689

International PEN Centre, Fifuhvammsvegi 19, Reykjavik
Gen Secr: Gisli Astthorsson
Focus: Lit 06690

Islenzka bókmennatafélag / The Islandic Literary Society, Thingholtst 3, POB 1252, 121 Reykjavik
Founded: 1816; Members: 1450
Pres: Sigurdur Líndal
Focus: Lit
Periodicals
Annual Journal (weekly)
Skirnir 06691

Islenzka nátturúfrådifélag, Hid / The Icelandic Natural History Society, POB 846, Reykjavik
Founded: 1889; Members: 1900
Focus: Hist
Periodical
Náttúrufraedingurinn (weekly) 06692

Kennarasamband Islands (K.I.) / The Icelandic Teachers Association, Laufásvegur 81, 101 Reykjavik
T: (1) 624080; Fax: (1) 623470
Founded: 1980; Members: 3800
Pres: Svanhildur Kaaber; Gen Secr: Valgeir Gestsson
Focus: Educ
Periodicals
Félagsblad K.I. (weekly)
Nýmenntamál (weekly) 06693

Loeknafélag Islands (L.I.) / Icelandic Medical Association, Domus Medica, Egilsgötu 3, 101 Reykjavik
T: (1) 18331, 18660; Fax: (1) 624452
Founded: 1918; Members: 850
Pres: H. Thordarson; Gen Secr: P. Thordarson
Focus: Med
Periodicals
Fréttabréf Laekna: Physicians News (10 times annually)
Laeknabladid: Icelandic Medical Journal (19 times annually) 06694

Menntamálarád / Arts Council, POB 1398, Reykjavik
Founded: 1928
Gen Secr: Einar Laxness
Focus: Arts
Periodicals
Almanak
Andvari
Studia Islandica 06695

Rannsóknarád Ríkisins / The National Research Council, Laugavegur 13, 101 Reykjavik
T: (1) 621320; Fax: (1) 29814
Founded: 1987; Members: 7
Pres: Petur Stefansson; Gen Secr: Vilhjálmur Lúdviksson
Focus: Sci
Periodicals
Newsletter in Icelandic (weekly)
Report on Research in Iceland (weekly) . 06696

Rannsóknastofnun Fiskidnadarins, Skúlagata 4, 101 Reykjavik
Founded: 1934
Focus: Food 06697

Skógraektarfélag Islands / The Forest Association of Iceland, Ránargötu 18, Reykjavik
T: (1) 18150; Fax: (1) 627131
Founded: 1930; Members: 7000
Pres: Hulda Valtysdóttir; Gen Secr: Brynjótlur Jónsson
Focus: Forestry
Periodical
Yearbook (weekly) 06698

Skurdlaeknafélag Islands, c/o St. Joseph's Hospital, Reykjavik
Focus: Surgery 06699

Sögufélag / The Historical Society, POB 1078, 121 Reykjavik
T: (1) 14620
Founded: 1902; Members: 1250
Pres: Heimir Thorleifsson; Gen Secr: Margrét Gudmundsdóttir
Focus: Hist
Periodical
Saga: Ný saga (weekly) 06700

Surtseyjarfélagid / Surtsey Research Society, POB 352, Reykjavik
Founded: 1965; Members: 65
Focus: Nat Sci
Periodical
Surtsey Research Progress Reports . . 06701

Svaefingalaeknafélag Islands / Icelandic Society of Anesthesiologists, c/o Dept of Anaesthesia, Landspitalinn, Reykjavik
T: (1) 29000
Founded: 1960; Members: 29
Pres: Thorsteinn Sv. Stefansson
Focus: Anesthetics 06702

Tannlaeknafélag Islands (T.F.I.) / Icelandic Dental Associatio, Sídumúli 35, POB 8596, 105 Reykjavik
T: (1) 34646; Fax: (1) 33562
Founded: 1927; Members: 240
Pres: Jon Asgeir Eyjolfsson; Gen Secr: Sigridur Dagbjartsdóttir
Focus: Dent
Periodical
Tannlaeknabladid (weekly) 06703

Tónlistarfélagid / Music Society, Gardastrati 17, 101 Reykjavik
Founded: 1930
Pres: Baldvin Tryggvason; Gen Secr: Rut Magnússon
Focus: Music 06704

Tónskáldafélag Islands / Icelandic Composers' Society, Laufásvegi 40, 101 Reykjavik
T: (1) 24972
Founded: 1945; Members: 35
Pres: John A. Speight; Gen Secr: Thorunn Gudmundsdóttir
Focus: Cinema 06705

Verkfraedingafélag Islands / Association of Chartered Engineers in Iceland, Engjateigur 9, 109 Reykjavik
T: (1) 688511; Fax: (1) 689703
Founded: 1912; Members: 1076
Pres: Gudmundur Thorarinsson; Gen Secr: Arnbjörg Edda Gudbjörnsdóttir
Focus: Eng
Periodicals
Arbók (weekly)
Fréttabréf (weekly) 06706

Vísindafélag Islands, c/o Háskóla Islands, 101 Reykjavik
Founded: 1918; Members: 75
Focus: Sci
Periodical
RIT (weekly) 06707

India

The Academy of Zoology, Khandari Rd, Agra 2
Founded: 1954; Members: 1000
Pres: Prof. Beni Charan Mahendra; Gen Secr: Dr. D.P.S. Bhati
Focus: Zoology 06708

Aeronautical Society of India, 13-B, Indraprastha Estate, New Delhi 110002
Founded: 1948; Members: 2000
Pres: I.M. Chopra; Gen Secr: Dr. S.A. Hussainy
Focus: Aero 06709

Agri-Horticultural Society of India, Alipur Rd, Calcutta 700027
Founded: 1820; Members: 1200
Pres: R.S. Chitlangia; Gen Secr: B.N. Mazumder
Focus: Agri; Hort
Periodicals
Horticultural Bulletin (weekly)
Monthly Garden News Sheet (weekly) . 06710

Agri-Horticultural Society of Madras, Cathedral PO, Madras 600086
Founded: 1835; Members: 1200
Pres: R.S. Chitlangia; Gen Secr: B.N. Mazumder
Focus: Hort; Agri
Periodicals
Horticultural Bulletin (weekly)
Monthly Garden News Sheet (weekly) . 06711

Ahmedabad Textile Industry's Research Association (ATIRA), P.O. Polytechnic, Ahmadabad 380015
T: 442671
Founded: 1947; Members: 350
Focus: Textiles
Periodical
ATIRA Communications on Textiles (weekly) 06712

Allahabad Mathematical Society, 10 C.S.P. Singh Marg, Allahabad 211001
Founded: 1958; Members: 250
Gen Secr: Dr. P. Srivastava
Focus: Math
Periodicals
Bulletin
Indian Journal of Mathematics (3 times annually) 06713

All India Fine Arts and Crafts Society, Old Mill Rd, New Delhi 110001
Founded: 1928; Members: 510
Pres: Dr. M.S. Randhawa; Gen Secr: S.S. Bhagat
Focus: Arts
Periodicals
Arts News (weekly)
Roopa Lekha (weekly) 06714

Andhra Historical Research Society, Godavari Bund Rd, Rajahmundry
Founded: 1922
Focus: Hist 06715

Anthropological Society of Bombay, 209 Dr. Dadabhaj Naoroji Rd, Fort, Bombay 400001
Founded: 1886
Pres: Dr. J.F. Bulsara; Gen Secr: Sapur F. Desai
Focus: Anthro 06716

Anthropological Survey of India, 27 Jawaharlal Nehru Rd, Calcutta 700016
T: (033) 239726; CA: Anthropos, Calcutta
Founded: 1945
Gen Secr: A.K. Danda
Focus: Anthro
Periodicals
Annual Report (weekly)
Folklore Series (weekly)
Human Science
Linguistic Series (weekly)
Memoir (weekly)
Newsletter (weekly)
Occasional Publication (weekly) 06717

Art Society of India, Sandhurst House, 524 S.V.P. Rd, Bombay 400004
T: (022) 3888550
Founded: 1918; Members: 425
Pres: Harish B. Talim; Gen Secr: R.D. Pareek
Focus: Arts 06718

Asian-African Legal Consultative Committee (AALCC), 27 Ring Rd, New Dehli 110024
T: (011) 624161
Founded: 1956; Members: 39
Focus: Law 06719

Asian and Pacific Centre for Transfer of Technology (APCTT), 49 Palace Rd, POB 115, Bangalore 560812
T: (0812) 266931; Tx: 845719; Fax: (0812) 263105
Founded: 1977
Gen Secr: O.C. Bugge
Focus: Eng
Periodical
Asia-Pacific Tech Monitor (weekly) . . . 06720

Asian Center for Organization, Research and Development (ACORD), C-125 Greater Kailash I, New Delhi 110048
T: (011) 6410616
Founded: 1981
Gen Secr: Prof. Brij Kapur
Focus: Business Admin 06721

Asian Environmental Society (AES), U-112A Vidhata House, Vikas Marg, New Delhi 110092
T: (011) 7222279
Founded: 1972; Members: 60
Gen Secr: Dr. Desh Bandhu
Focus: Ecology 06722

Asian Grain Legumes Network, c/o ICRISAT, Patancheru 502324
T: (0842) 224016; Tx: 422203; Fax: (0842) 241239
Founded: 1986
Gen Secr: Donald G. Faris
Focus: Hort 06723

Asian Network of Human Resource Development Planning Institutes, c/o ILO/ARTEP, POB 643, New Delhi 110001
T: (011) 344853; Tx: 66498
Founded: 1986
Focus: Soc Sci
Periodicals
HRD Documentation Bulletin (weekly)
HRD Newsletter (3 times annually) . . . 06724

Asian Regional Coordinating Committee on Hydrology (ARCCOH), c/o National Institute of Hydrology, Roorkee 247667
T: (01332) 2106; Tx: (0597) 205; Fax: (01332) 2123
Founded: 1980; Members: 28
Focus: Hydrology
Periodical
ARCCOH Newsletter (weekly) 06725

Asia-Oceania Association of Otolaryngological Societies, c/o Association of Otolaryngologists of India, Patil House, H.M. Patil Marg, Dadar, Bombay 400028
T: (022) 451320, 4306229
Gen Secr: Dr. V.K. Sukhtankar
Focus: Otorhinolaryngology 06726

Asia Theological Association (ATA), POB 3432, Bangalore 560034
T: (0812) 531154, 532516; Fax: (0812) 569387
Founded: 1970
Gen Secr: Dr. Ken R. Gnanakan
Focus: Rel & Theol
Periodical
ATA News (weekly) 06727

Asiatic Society, 1 Park St, Calcutta 700016
T: (033) 290779, 297250; Tx: 0215238; Fax: (033) 290355
Founded: 1784; Members: 1292
Pres: Dr. M. Chakrabarty; Gen Secr: Dr. Chandan Raychoudhury
Focus: Ethnology
Periodicals
Bibliotheka Indica
Journal
Monthly Bulletin (weekly)
Year Book (weekly) 06728

Asiatic Society of Bombay, Town Hall, Bombay 400023
T: (022) 2860956
Founded: 1804; Members: 1970
Pres: J.S.H. Singh; Gen Secr: Ramesh Jamindar
Focus: Ethnology
Periodical
Journal 06729

Association for Engineering Education in South and Central Asia, c/o Indian Society for Technical Education, SJCE Campus, Mysore 570006
T: 35191, 37664
Gen Secr: Prof. M.V. Ranganath
Focus: Educ; Eng 06730

Association for Medical Education in South-East Asia, Yojana Bhavan, New Delhi 110001
T: (011) 660170; Tx: 3165656
Gen Secr: Prof. J.S. Bajaj
Focus: Educ; Med 06731

Association of Asian Social Science Research Councils (AASSREC) / Association des Conseils Asiatiques pour la Recherche en Sciences Sociales, c/o Indian Council of Social Science Research, 35 Ferozehah Rd, New Delhi 110001
T: (011) 384734; Tx: 61083
Founded: 1973
Gen Secr: Prof. D.N. Dhanagare
Focus: Soc Sci
Periodicals
AASSREC Panorama Newsletter (weekly)
Asian Social Scientists 06732

Association of Indian Universities, AIU House, 16 Kotla Marg, New Delhi 110002
T: (011) 3315105, 3310059, 3312429; Tx: 3166180; CA: ASINDU
Founded: 1925; Members: 175
Focus: Sci 06733

Association of Management Development Institutions in Asia, Bella Vista, Hyderabad 500049
T: (0842) 36952
Focus: Business Admin 06734

Association of Surgeons of India, 18 Swamy Sivananda Salai, Adams Rd, Chepauk, Madras 600005
Founded: 1938; Members: 6000
Pres: Dr. T.E. Udwadia; Gen Secr: Dr. K. Chockalingam
Focus: Surgery
Periodical
Indian Journal of Surgery (weekly) . . . 06735

Astronomical Society of India, c/o Dept of Astronomy, Osmania University, Hyderabad 500007
Founded: 1973; Members: 500
Focus: Astronomy
Periodical
Bulletin of the Astronomical Society of India (weekly) 06736

Bal Bhavan Society, c/o Dept of Biochemstry, University, 35 Ballygunge Circular Rd, Calcutta 700019
Founded: 1958
Focus: Educ 06737

Bal Bhavan Society, Kotla Rd, New Delhi 110001
Founded: 1958
Pres: R.D. Barakataki; Gen Secr: Shanta Gandhi
Focus: Educ 06738

Bar Association of India, Chamber 93, Supreme Court Bldg, New Dehli 110001
Focus: Law
Periodical
The Indian Advocate (weekly) 06739

Bengal Natural History Society, c/o Natural History Museum, Darjeeling
Founded: 1923; Members: 100
Pres: T.S. Broca; Gen Secr: N. Pal
Focus: Hist; Nat Sci
Periodical
Journal (weekly) 06740

Bharata Ganita Parisad, c/o Dept of Mathematics and Astronomy, University, Lucknow
T: 75944
Founded: 1950; Members: 400
Pres: Prof. H.C. Khare; Gen Secr: Dr. D. Singh
Focus: Math
Periodical
Ganita 06741

Bharata Itihasa Samshodhaka Mandala, 1321 Sadashiva Peth, Poona 411030
Founded: 1910; Members: 750
Pres: R.S. Walimbe; Gen Secr: Dr. Usha Ranade
Focus: Hist
Periodical
Journal (weekly) 06742

Bihar Research Society, Museum Bldgs, Patna 800001
Founded: 1915; Members: 180
Pres: Dr. J.C. Jha; Gen Secr: M.S. Pandey
Focus: Hist
Periodical
Journal 06743

Bombay Art Society (BAS), c/o Jehangir Art Gallery, Bombay 400023
T: (022) 2044058
Founded: 1888; Members: 850
Pres: Dr. Saryu Doshi; Gen Secr: G.S. Adivrekar
Focus: Arts
Periodical
Art Journal 06744

Bombay Historical Society, c/o Prince of Wales Museum, Bombay 400001
Founded: 1925
Focus: Hist 06745

Bombay Medical Union, Blavatsky Lodge Bldg, Chowpatty, Bombay 400007
Founded: 1883; Members: 250
Pres: Dr. U.N. Bastodkar; Gen Secr: Dr. S.M.K. Thacker
Focus: Med 06746

Bombay Natural History Society, Hornbill House, Shaheed Bhagat Singh Rd, Bombay 400023
T: (022) 225155, 243869; CA: Hornbill
Founded: 1883; Members: 2500
Focus: Nat Sci
Periodicals
Hornbill (weekly)
Journal of the Bombay Natural History Society (3 times annually) 06747

Bombay Textile Research Association (BTRA), L.B. Shastri Marg, Ghatkopar, Bombay 400086
T: (022) 5152651; CA: MILLITRA Bombay; Fax: (022) 5149759
Founded: 1954
Focus: Textiles
Periodicals
BTRA Bulletin (weekly)
BTRA Cleanings (weekly)
BTRA Current Textile Literature (weekly)
BTRA Scan (weekly) 06748

Botanical Survey of India, P/8 Brabourne Rd, Calcutta 700001
Founded: 1890
Gen Secr: Dr. M.P. Nayar
Focus: Botany
Periodicals
Bulletin (weekly)
Indian Floras
Records and Reports (weekly) 06749

Calcutta Mathematical Society (C.M.S.), 92 Acharya Prafulla Chandra Rd, Calcutta 700009
Founded: 1908; Members: 520
Pres: Prof. L. Debnath; Gen Secr: Dr. B.N. Mandal
Focus: Math
Periodicals
The Bulletin of the Calcutta Mathematical Society (weekly)
News Bulletin (10 times annually) . . . 06750

Calcutta Statistical Association, c/o Calcutta University, New Sience Bldg, 35 B.C. Rd, Calcutta 700019
Founded: 1945
Focus: Stats
Periodical
Calcutta Statistical Association Bulletin (weekly) 06751

Committee on Science and Technology in Developing Countries (COSTED), 21 Gandhi Mandap Rd, Guindy, Madras 600025
T: (044) 419466; Tx: 412014; Fax: (044) 414543
Founded: 1966
Gen Secr: R.R. Daniel
Focus: Sci; Eng
Periodical
COSTED News (3 times annually) . . . 06752

Council of Scientific and Industrial Research, Rafi Marg, New Delhi
T: (011) 3710512; Tx: 65202; Fax: (011) 3710018
Founded: 1942
Gen Secr: Dr. S.K. Joshi
Focus: Sci; Eng
Periodicals
CSIR News (weekly)
Current Literature on Science of Science (weekly)
Indian Journal of Biochemistry & Biophysics (weekly)
Indian Journal of Chemistry, Section A (weekly)
Indian Journal of Chemistry, Section B (weekly)
Indian Journal of Experimental Biology (weekly)
Indian Journal of Fibre and Textile Research (weekly)
Indian Journal of Marine Sciences (weekly)
Indian Journal of Pure and Applied Physics (weekly)
Indian Journal of Radio and Space Physics (weekly)
Indian Journal of Technology (Including Engineering) (weekly)
Journal of Scientific and Industrial Research (weekly)
Medicinal & Aromatic Plants Abstracts (weekly)
Research & Industry (weekly)
Science-Ki-Duniya (weekly)
Science Reporter (weekly)
Vigyan Pragati (weekly) 06753

Crafts Council of Western India, 59 L. Jagmohandas Marg, Bombay 40006
Founded: 1972; Members: 155
Pres: Roshan Kalapesi; Gen Secr: Jane Verma; Jamini Ahluwalia
Focus: Arts 06754

Delhi Library Association, c/o Hardinge Public Library, Queen's Garden, POB 1270, Dehli 110006
Focus: Libraries & Bk Sci 06755

Eastern Regional Organization for Planning and Housing (EAROPH), 4-A Ring Rd, Indraprashta Estate, New Delhi 110002
T: (011) 274809
Founded: 1958; Members: 231
Gen Secr: Prof. C.S. Chandrasekhara
Focus: Urban Plan
Periodical
EAROPH News and Notes (weekly) . . 06756

The Electrochemical Society of India, c/o Indian Institute of Science, Bangalore 560012
T: (0812) 340977; Tx: 08468349; CA: Science, Bangalore
Founded: 1964
Pres: Dr. R.P. Dambal; Gen Secr: Prof. D.K. Padma
Focus: Electrochem
Periodical
The Journal of the Electrochemical Society of India (weekly) 06757

Ethnographic and Folk Culture Society, Ram Krishna Marg, Faizabad Rd, POB 209, Lucknow 226007
Founded: 1945; Members: 386
Focus: Ethnology
Periodicals
The Esatern Anthropologist (weekly)
Manav (weekly) 06758

Federation of Indian Library Associations, Misri Bazar, Patiala, Punjab
Focus: Libraries & Bk Sci 06759

Geographical Society of India, c/o Dept of Geography, University of Calcutta, 35 Ballygunge Circular Rd, Calcutta 700019
T: (033) 473681
Founded: 1933; Members: 650
Pres: Prof. Sunil K. Munsi; Gen Secr: Dr. R. Bhattacharya
Focus: Geography
Periodical
Geographical Review of India (weekly) . 06760

Geological, Mining and Metallurgical Society of India, c/o Geology Dept, University of Calcutta, 35 B.C. Rd, Calcutta 700019
Founded: 1924; Members: 206
Pres: Prof. Saurindranath Sen; Gen Secr: Dr. B.K. Samanta; Dr. A. Mitra
Focus: Mining; Geology; Metallurgy
Periodicals
Bulletin (weekly)
Journal (weekly) 06761

Geological Survey of India, 27 Jawaharlal Nehru Rd, Calcutta 700013
Founded: 1851
Gen Secr: D.P. Dhoundial
Focus: Geology
Periodicals
Bulletin
Catalogue Series
Geological Survey of India News
Indian Minerals
Manual Series
Memoirs
Miscellaneous Publications
Palaeontologica Indica
Records
Special Publications 06762

Gujarat Research Society, Samshodhan Sadan, Ramkrishna Mission Marg, Khar, Bombay 400052
Founded: 1936
Pres: Dr. M.R. Shah
Focus: Hist
Periodical
Journal (weekly) 06763

Helminthological Society of India, c/o Dept of Parasitology, U.P. College of Veterinary Science and Animal Husbandry, Mathura
Pres: Prof. S.N. Singh; Gen Secr: Prof. B.P. Pande
Focus: Vet Med
Periodical
Indian Journal of Helminthology (weekly) 06764

Hyderabad Educational Conference, 19 Bachelors' Quarters, Jawaharlal Nehru Rd, Hyderabad
Founded: 1913
Focus: Educ 06765

Hyderabad Educational Conference, 19 Bachelor's Quarters, Jawaharlal Nehru Rd, Hyderabad
Founded: 1913
Pres: Syed Masood Ali; Gen Secr: Gouse Mohiuddin
Focus: Educ 06766

Indian Academy of Sciences, C.V. Raman Av, Sadashivanagar, POB 8005, Bangalore 560080
T: (0812) 342546; Tx: 08452178; CA: Academy Bangalore; Fax: (0812) 346094
Founded: 1934; Members: 700
Pres: Prof. R. Narasimha; Gen Secr: Prof. N. Viswanadham; Prof. V. Krishnan
Focus: Sci
Periodicals
Bulletin of Materials Science (weekly)
Current Science (weekly)
Journal of Astrophysics & Astronomy (weekly)
Journal of Biosciences (weekly)
Journal of Genetics (weekly)
Journal of Genetics (weekly)
Pramana: Journal of Physics (weekly)
Proceedings
Proceedings (Chemical Sciences) (weekly)
Proceedings (Earth and Planetary Sciences) (weekly)
Proceedings (Mathematical Sciences) (weekly)
Sadhana: Engineering Sciences (weekly) . 06767

Indian Adult Education Association (I.A.E.A.), 17-B Indraprastha Marg, New Delhi 110002
T: (011) 3319282, 3722206; CA: ADEDASSO
Founded: 1939; Members: 1200
Pres: B.S. Garg; Gen Secr: K.C. Choudhary
Focus: Adult Educ
Periodicals
I.A.E.A. Newsletter (weekly)
Indian Journal of Adult Education (weekly)
Jagoo aur Jagao (weekly)
Proudh Shiksha (weekly) 06768

Indian Anthropological Association, c/o Dept of Anthropology, University of Delhi, Delhi 110007
Founded: 1964; Members: 300
Pres: Prof. Dr. L.P. Vidyarth; Gen Secr: Dr. P.K. Datta
Focus: Anthro
Periodicals
Anthropologists in India (every 2-3 years)
Indian Anthropologist (weekly)
News Bulletin (weekly) 06769

Indian Association for the Cultivation of Science (I.A.C.S.), Jadavpur, Calcutta 700032
Founded: 1876; Members: 1600
Pres: Prof. S.K. Mukherjee; Gen Secr: Prof. A.K. Barua
Focus: Physics; Chem; Math
Periodical
Indian Journal of Physics (weekly) . . . 06770

Indian Association of Academic Librarians, c/o Jawaharlal Nehru University Library, New Mehrauli Rd, New Delhi 110067
T: (011) 650005
Focus: Libraries & Bk Sci
Periodical
Newsletter 06771

Indian Association of Biological Sciences, c/o Life Science Centre, University of Calcutta, Calcutta 700019
Pres: Prof. B.K. Bachhawat; Gen Secr: Prof. T.M. Das
Focus: Bio *06772*

Indian Association of Geohydrologists, c/o Geological Survey of India, 4 Chowringhee Lane, Calcutta 700016
Founded: 1964; Members: 440
Pres: V. Subramanyam; Gen Secr: A.K. Roy
Focus: Water Res
Periodical
Indian Geohydrology *06773*

Indian Association of Parasitologists, 110 Chittaranjan Av, Calcutta 700012
Pres: Dr. H.N. Ray; Gen Secr: Dr. A.B. Chaudhury
Focus: Microbio *06774*

Indian Association of Special Libraries and Information Centres (IASLIC), P-291 CIT Scheme 6-M Kankurgachi, Calcutta 700054
T: (033) 359651
Founded: 1955; Members: 855
Pres: Dr. Sankar Sen; Gen Secr: Dr. S.M. Ganguly
Focus: Libraries & Bk Sci; Computer & Info Sci
Periodicals
IASLIC Bulletin (weekly)
Indian Library Science Abstracts (weekly)
Marketing of Library & Information Services in India (weekly) *06775*

Indian Association of Systematic Zoologists, c/o Zoological Survey of India, 34 Chitteranjan Av, Calcutta 700012
Founded: 1947
Pres: Dr. A.P. Kapur
Focus: Zoology *06776*

Indian Association of Teachers of Library Science (IATLIS), c/o Dept of Library and Information Science, Nagpur University, Nagpur 5
Founded: 1971
Focus: Adult Educ; Libraries & Bk Sci
Periodicals
Bulletin of Classification Society of India
IATLIS Communication (weekly) *06777*

Indian Biophysical Society, c/o Saha Institute of Nuclear Physics, 92 Acharya Prafulla Chandra Rd, Calcutta 700009
T: (033) 354281
Founded: 1965; Members: 200
Pres: Prof. N.N. Saha; Gen Secr: Prof. D.P. Burma
Focus: Biophys
Periodical
Prodeedings (weekly) *06778*

Indian Botanical Society, c/o Dept of Botany, University of Madras, Chepauk, Madras 600005
Pres: K.S. Thind; Gen Secr: Prof. K.S. Bhargava
Focus: Botany
Periodical
Journal *06779*

Indian Brain Research Association, c/o Dept of Biochemstry, Calcutta University, 35 Ballygunge Circular Rd, Calcutta 700019
Founded: 1964; Members: 300
Pres: Prof. J.J. Ghosh
Focus: Neurology
Periodical
Brain News (weekly) *06780*

Indian Cancer Society, c/o Lady Ratan Tata Medical & Research Centre, Cooperage, M. Karve Rd., Bombay 400021
T: (022) 2029941; CA: CANCERHIND
Founded: 1951
Gen Secr: Dr. D.J. Jussawalla
Focus: Cell Biol & Cancer Res
Periodical
Indian Journal of Cancer (weekly) . . . *06781*

Indian Ceramic Society, c/o Central Glass and Ceramics Research Institute, Calcutta 700032
Founded: 1928
Pres: A.A. Ganpule; Gen Secr: Dr. K.P. Srivastava
Focus: Materials Sci
Periodical
Transactions (weekly) *06782*

Indian Chemical Society, 92 Acharya Prafulla Chandra Rd, Calcutta 700009
T: (033) 353478
Founded: 1924; Members: 1800
Pres: Prof. K.K. Rohatgi-Mukherjee; Gen Secr: Prof. S.K. Talapatra
Focus: Chem
Periodical
Journal (weekly) *06783*

Indian College Library Association, 66 Ranjan Colony, Hyderabad 500253
T: 525202
Pres: A.P. Jain
Focus: Libraries & Bk Sci *06784*

Indian Council of Agricultural Research, Krishi Anusandhan Bhavan, Dr. Rajendra Prasad Rd, New Dehli 110012
T: (011) 388991
Founded: 1929
Pres: R.B. Singh; Gen Secr: Dr. N.S. Randhawa
Focus: Agri *06785*

Indian Council of Historical Research, 35 Ferozeshah Rd, New Delhi 110001
T: (011) 384347
Pres: Prof. Ravinder Kumar; Gen Secr: Dr. T.R. Sareen
Focus: Hist
Periodical
The Indian Historical Review (weekly) . . *06786*

Indian Council of Medical Research (I.C.M.R.), Medical Enclave, Ansari Nagar, POB 4508, New Dehli 110029
T: (011) 653980; Tx: 63067
Founded: 1911
Gen Secr: Dr. A.S. Paintal
Focus: Med
Periodicals
Annual Report (weekly)
Journal of Medical Research (weekly) . . *06787*

Indian Council of Social Science Research (ICSSR), 35 Ferozeshah Rd, New Dehli 110001
T: (011) 386208; Tx: 3160183; CA: Icsores, New Delhi
Founded: 1969
Pres: Prof. S. Chakravarty
Focus: Sociology
Periodicals
ICSSR Journal of Abstracts and Reviews: Economics
ICSSR Journal of Abstracts and Reviews: Geography
ICSSR Journal of Abstracts and Reviews: Political Science
ICSSR Journal of Abstracts and Reviews: Sociology and Social Anthropology
ICSSR Newsletter (weekly)
ICSSR Research Abstracts (weekly)
Indian Dissertation Abstracts
Indian Journal of Social Science (weekly)
Indian Psychological Abstracts *06788*

Indian Council of World Affairs (I.C.W.A.), Sapru House, Barakhamba Rd, New Dehli 110001
T: (011) 3317248, 3319055
Founded: 1943; Members: 2900
Pres: Harcharan Singh Josh; Gen Secr: S.C. Parasher
Focus: Poli Sci
Periodicals
Foreign Affairs Report (weekly)
India Quarterly (weekly) *06789*

Indian Dairy Association (IDA), IDA House, Sector IV, R.K. Puram, New Dehli 110022
Members: 2000
Focus: Agri
Periodicals
Dairying-in-India (weekly)
Indian Dairyman (weekly)
Indian Journal of Dairy Science (weekly) *06790*

Indian Economic Association, c/o Delhi School of Economics, Dehli 110009
Founded: 1918
Pres: Prof. V.M. Dandekar; Gen Secr: Prof. K.A. Naqvi
Focus: Econ
Periodical
Indian Economic Journal *06791*

Indian Folklore Society, 3 Abdul Hamid St, Calcutta, 700069
T: (033) 236334
Focus: Folklore
Periodical
Folklore (weekly) *06792*

Indian Geographical Society, c/o Dept of Geography, Madras University, Madras 600005
Fax: (044) 566693
Focus: Geography *06793*

Indian Geologists' Association, c/o Dept of Geology, Panjab University, Chandigarh 160014
T: (0172) 22740; Tx: 395464; Fax: (0172) 54549
Founded: 1968; Members: 250
Pres: Prof. R.S. Chaudhri; Gen Secr: Dr. Naresh Kochhar
Focus: Geology
Periodical
Bulletin of the Indian Geologists' Association (weekly) *06794*

Indian Institute of Architects, Prospect Chambers Annexe, Dr. D.N. Rd, Bombay 40001
Founded: 1928; Members: 4200
Pres: M.G. Deobhakta; Gen Secr: A.V. Ogale; S.A. Deshpande
Focus: Archit *06795*

Indian Institute of Metals (IIM), 33A Chowringhee Rd, Calcutta 700071
T: (033) 294648
Founded: 1946; Members: 5300
Pres: Dr. J.J. Irani; Gen Secr: Dr. L. R. Vaidyanath
Focus: Metallurgy
Periodicals
Journal of Alloy Phase Diagrams (3 times annually)
Metal News (weekly) *06796*

Indian Jute Industries' Research Association, 17 Taratola Rd, POB 12, Calcutta 700053
Founded: 1966
Focus: Eng *06797*

Indian Law Institute, Opp. Supreme Court, Bhagwandas Rd, New Delhi 110001
T: (011) 387526
Founded: 1956
Gen Secr: Prof. P.M. Bakshi
Focus: Law
Periodicals
Index to Indian Legal Periodicals (weekly)
Journal of the Indian Law Institute (weekly) *06798*

Indian Library Association (ILA), A/40-41, Flat 201, Ansal Bldgs, Dr. Mukherjee Nagar, Delhi 110009
T: (011) 7117743
Founded: 1933; Members: 2780
Pres: C.P. Vashishth; Gen Secr: A.P. Gakhar
Focus: Libraries & Bk Sci
Periodical
Bulletin (weekly) *06799*

Indian Mathematical Society, c/o Dr. S.P. Arya, Maitreyi College, Bapu Dham Complex, Chanakyapuri, New Delhi 110021
Founded: 1907; Members: 1000
Pres: Prof. M.K. Singal; Gen Secr: Dr. S.P. Arya
Focus: Math
Periodical
Journal of the Indian Mathematical Society: Mathematics Student (weekly) *06800*

The Indian Medical Association (IMA), IMA House, Indraprastha Marg, New Dehli 110002
T: (011) 3318680; CA: INMEDICI
Founded: 1928; Members: 70000
Pres: Dr. N. Satyanarayana; Gen Secr: Dr. N.K. Grover
Focus: Med
Periodicals
Annals of IMA Academy of Medical Specialities (weekly)
Bulletin on Continuing Medical Education (weekly)
Journal of the Indian Medical Association (weekly) *06801*

Indian Musicological Society, Jambu Bet, Dandia Bazar, Baroda 390001
T: 555388
Founded: 1979
Gen Secr: Prof. R.C. Mehta
Focus: Music
Periodical
Journal of the Indian Musicological Society (weekly) *06802*

Indian National Academy of Engineering, c/o Institution of Electronics and Telecommunication Engineers, 2 Institutional Area, Lodi Rd, New Dehli 110003
Founded: 1987
Pres: J. Krishna; Gen Secr: S.N. Mitra
Focus: Eng *06803*

Indian National Science Academy, Bahadur Shah Zafar Marg, New Delhi 110002
T: (011) 3313153
Founded: 1935; Members: 686
Pres: Dr. A.S. Paintal; Gen Secr: Prof. H.Y.M. Ram; Dr. P.K. Dass
Focus: Sci
Periodicals
Indian Journal of History of Science
Indian Journal of Pure and Applied Mathematics
Proceedings: Pt.A: Physical Sciences
Proceedings: Pt.B: Biological Sciences
Yearbook *06804*

Indian Optometric Association, POB 2812, New Delhi 110060
Focus: Ophthal
Periodical
Optometry (weekly) *06805*

Indian Pharmaceutical Association (I.P.A.), Kalina, Santacruz East, Bombay 400098
T: (022) 6122401
Founded: 1935; Members: 5500
Pres: Dr. P.M. Naik; Gen Secr: Dr. J.K. Lalla
Focus: Pharmacol
Periodicals
Indian Journal of Pharmaceutical Sciences (weekly)
Pharmatimes (weekly) *06806*

Indian Phytopathological Society (I.P.S.), c/o Indian Agricultural Research Institute, New Dehli 110012
Fax: (011) 5752006
Founded: 1947; Members: 1720
Pres: Dr. A.K. Sarbhoy; Gen Secr: Dr. Dinesh Kumar
Focus: Botany
Periodical
Indian Phytopathology (weekly) *06807*

Indian Political Science Association, c/o Centre for Anna Studies, University of Madras, Madras 600005
T: 568778; Tx: 416376; CA: Madras University
Founded: 1937; Members: 3000
Pres: Prof. L.S. Rathore; Gen Secr: Prof. K.P. Singh
Focus: Poli Sci *06808*

Indian Psychoanalytical Society, 14 Parsibagan Lane, Calcutta 700009
Focus: Psychoan *06809*

Indian Psychometric and Educational Research Association, c/o Dept of Education, Patna Training College, Patna 800004
Founded: 1969; Members: 330
Pres: Dr. A.K.P. Sinha; Gen Secr: Dr. R.P. Singh
Focus: Educ
Periodical
Indian Journal of Psychometry and Education *06810*

Indian Public Health Association, 110 Chittaranjan Av, Calcutta 700073
Focus: Public Health
Periodical
Indian Journal of Public Health (weekly) *06811*

Indian Rubber Manufacturers Research Association (I.R.M.R.A.), Plot B-88, Rd U 2, Wagle Industrial Estate, Thane 400604
T: 593910
Founded: 1959; Members: 48
Gen Secr: Dr. W. Millns
Focus: Materials Sci *06812*

Indian Science Congress Association (ISCA), 14 Dr. Biresh Guha St, Calcutta 700017
T: (033) 2474530
Founded: 1914; Members: 10000
Pres: Prof. P.N. Srivastava; Gen Secr: Prof. D.P. Chrkraborty
Focus: Sci; Eng; Med
Periodicals
Everyman's Science (weekly)
Proceedings (weekly) *06813*

Indian Society for Nuclear Techniques in Agriculture and Biology, c/o Nuclear Research Laboratory, Indian Agricultural Research Institute, New Delhi 110012
Pres: Dr. M.S. Chatrath; Gen Secr: Dr. M.S. Sachdev
Focus: Nucl Res; Agri; Bio *06814*

The Indian Society of Agricultural Economics (ISAE), 46-48 Esplanade Mansions, Mahatma Gandhi Rd, Bombay 400001
T: (022) 242542
Founded: 1939; Members: 1000
Pres: Prof. V.M. Dandekar; Gen Secr: Dr. Tara Shukla
Focus: Agri
Periodical
Indian Journal of Agricultural Economics (weekly) *06815*

Indian Society of Criminology, c/o University of Madras, Madras 600005
Founded: 1970; Members: 600
Pres: Justice S. Mohan; Gen Secr: Prof. Dr. K.V. Kaliappan
Focus: Criminology
Periodical
Indian Journal of Criminology (weekly) . *06816*

Indian Society of Earthquake Technology, Roorkee, Uttar Pradesh
Focus: Geophys *06817*

Indian Society of Genetics and Plant Breeding, c/o Genetics Div, Indian Agricultural Research Institute, New Dehli 110012
Founded: 1941; Members: 1200
Pres: Dr. R.S. Rana; Gen Secr: Dr. V.L. Chopra
Focus: Botany; Genetics
Periodical
Journal *06818*

Indian Society of Oriental Art (I.S.O.A.), 15 Park St, Calcutta 700016
Founded: 1907; Members: 362
Pres: Nirmal Kumar Biswas; Gen Secr: Indira Nag Chaudhuri
Focus: Arts
Periodical
The Journal of the Indian Society of Oriental Art (weekly) *06819*

Indian Society of Soil Science, c/o Div of Soil Science and Agricultural Chemistry, Indian Agricultural Research Institute, New Delhi 110012
T: (011) 5720991
Founded: 1934; Members: 1875
Pres: Dr. N.N. Goswami; Gen Secr: Dr. G. Narayanasamy
Focus: Agri
Periodicals
Bulletin of the Indian Society of Soil Science (weekly)
Journal of the Indian Society of Soil Science (weekly) *06820*

Indian Space Research Organization (ISRO), c/o Dept of Space, Antariksh Bhavan, New Bel Rd, Bangalore 560094
T: (0812) 334474; Fax: (0812) 334229
Founded: 1969
Pres: Prof. U.R. Rao; Gen Secr: R.K. Verma
Focus: Astronomy *06821*

Indian Standards Institution (ISI), 9 Bahadur Shah Zafar Marg, Manak Bhavan, New Dehli 110002
T: (011) 270131, 272166
Founded: 1947
Focus: Standards
Periodicals
Current Published information on Standardization (weekly)
ISI Annual Report (weekly)

ISI Bulletin (weekly)
ISI Handbook (weekly)
Manakdoot (weekly)
Standards Worldover (weekly) 06822

Indo-British Historical Society, 21 Rajaram Mehta Av, Madras 600029
T: (044) 422404
Founded: 1968; Members: 675
Pres: Dharma Vira; Gen Secr: G.T. Verghese
Focus: Hist
Periodical
Indo-British Review (weekly) 06823

Inland Fisheries Society of India, c/o Central Inland Capture Fisheries Research Institute, Barrackpore 743101
T: 561191; CA: Fishsearch Barrackpore India
Founded: 1969; Members: 800
Pres: Dr. Arun G. Jhingran; Gen Secr: Apurba Ghosh
Focus: Fisheries 06824

Institute of Chartered Accountants of India, Indraprastha Marg, New Delhi 110002
Founded: 1949; Members: 32000
Pres: P.A. Nair; Gen Secr: R.L. Choppa
Focus: Business Admin 06825

International Academy of Indian Culture, J-22 Hauz Khas Enclave, New Delhi 110016
T: (011) 665800
Founded: 1935
Pres: Dr. L. Chandra
Focus: Ethnology
Periodical
Satapitaka Series (weekly) 06826

International Society of Plant Morphologists, c/o Dept of Botany, University of Delhi, Dehli 110007
T: (011) 2918983
Founded: 1951; Members: 500
Pres: Prof. H.Y. Mohan Ram; Gen Secr: Prof. S.P. Bhatnagar
Focus: Botany
Periodical
Phytomorphology (weekly) 06827

International Tamil League, c/o Thenmozhi, 5 Arunachala St, Chepauk, Madras 600005
Founded: 1968; Members: 10000
Pres: G.D. Appavanar
Focus: Ling
Periodicals
Tamilchittu
Tamil Nilam (weekly)
Thenmozhi (weekly) 06828

Islamic Research Association, 8 Shepherd Rd, Bombay 400008
Founded: 1933
Focus: Rel & Theol 06829

Jammu and Kashmir Academy of Art, Culture and Languages, Srinagar
T: 73425, 74558
Founded: 1958
Pres: Dr. Abdullah Farooq; Gen Secr: M.Y. Taing
Focus: Arts; Ethnology; Ling; Lit
Periodical
Sheeraza (weekly) 06830

Kamarupa Anusandhan Samiti / Assam Research Society, Gauhati, Assam
Founded: 1912; Members: 200
Pres: Biswanarayan Shastri; Gen Secr: Dr. Dharmeswar Chutia; Dr. R.D. Choudhury
Focus: Hist; Archeol
Periodical
Journal of Assam Research Society (weekly)
. . . . 06831

Karnatak Historical Research Society, Diwan Bahadur Rodda Rd, Dharwar 1
Founded: 1914; Members: 150
Pres: N. Shrinivasrao; Gen Secr: U.B. Bidi; Dr. P.R. Panchamukhi
Focus: Hist
Periodical
Karnatak Historical Review (weekly) . . 06832

Linguistic Society of India, c/o Deccan College, Poona 411006
T: 27231
Founded: 1928; Members: 691
Pres: Dr. R.N. Srivastava; Gen Secr: Dr. S.R. Sharma
Focus: Ling
Periodical
Indian Linguistics (weekly) 06833

Madras Literary Society and Auxiliary of the Royal Asiatic Society, College Rd, Madras 600006
Pres: J.S. Suryamurthy; Gen Secr: S.V.B. Row
Focus: Lit 06834

Medical Council of India, Temple Lane, Kotla Rd, POB 337, New Dehli 110002
Founded: 1934
Pres: Dr. A.K.B. Sinha; Gen Secr: Dr. P.S. Jain
Focus: Med
Periodical
Indian Medical Register 06835

Mineralogical Society of India, Manasa Gangotri, Mysore 6
Founded: 1959; Members: 400
Pres: Dr. A. Viswanathiah; Gen Secr: Dr. P.N. Satish
Focus: Mineralogy
Periodical
The Indian Mineralogist (weekly) 06836

Museums Association of India (MAI), c/o National Museum of Natural History, Barakhmba Rd, New Dehli 110001
T: (011) 387073
Founded: 1944; Members: 550
Pres: Dr. M.L. Nigam; Gen Secr: P.G. Gupte
Focus: Arts
Periodicals
Journal of Indian Museums
Museums Newsletter 06837

Mysore Horticultural Society, Lalbaugh, Bangalore 560004
Founded: 1912
Focus: Hort 06838

Mythic Society, 2 Nrupathunga Rd, Bangalore 560002
T: (0812) 29034
Founded: 1909; Members: 373
Pres: Prof. K.T. Pandurangi; Gen Secr: Dr. M.K.L.N. Sastry
Focus: Philos
Periodical
Journal (weekly) 06839

National Academy of Art, Rabindra Bhavan, New Dehli 110001
T: (011) 387241
Founded: 1934
Pres: Prof. S. Chaudhuri; Gen Secr: M. Rajaram
Focus: Arts
Periodical
Lalit Kala (weekly) 06840

National Academy of Sciences, 5 Lajpatrai Rd, Allahabad 211 002
T: 55224; CA: NATACADEMY
Founded: 1930; Members: 2957
Pres: Dr. M.S. Swaminathan; Gen Secr: Prof. H.C. Khare; Dr. Manju Sharma
Focus: Sci
Periodicals
Annual Number (weekly)
National Academy of Sciences Letters (weekly)
Proceedings Sec. A: Physical Sciences (weekly)
Proceedings Sec. B: Biological Sciences (weekly)
. . . . 06841

National Council of Applied Economic Research (NCAER), Parisila Bhavan , 11 Indraprastha Estate, New Dehli 110002
T: (011) 3317860; Tx: 3165880; CA: Arthsandan;
Fax: (011) 3327164
Founded: 1956
Focus: Econ 06842

National Council of Educational Research and Training, Sri Aurobindi Marg, New Dehli 110016
Founded: 1961
Gen Secr: Dr. P.L. Malhotra
Focus: Educ
Periodicals
Indian Educational Review (weekly)
Journal of Indian Education (weekly)
School Science (weekly) 06843

National Geographical Society of India, c/o Dept of Geography, Banaras Hindu University, Varanasi 221005
Fax: (0542) 312059
Founded: 1948; Members: 502
Pres: Dr. Hari Narain; Gen Secr: Prof. R.L. Singh; Dr. Rana P.B. Singh
Focus: Geography
Periodical
National Geographical Journal of India (weekly)
. . . . 06844

National Institute of Design (NID), Paldi, Ahmedabad 380007
T: (0272) 79692; Tx: 121322; CA: Institute;
Fax: (0272) 77236
Founded: 1961
Pres: Vikas Satwalekar
Focus: Graphic & Dec Arts, Design
Periodicals
Design Concerns: Student Colloquium Papers
Design Folio (weekly) 06845

National Productivity Council, 5-6 Institutional Area, Lodi Rd, New Delhi 110003
Fax: (011) 4615002
Founded: 1958
Gen Secr: Siladitya Ghosh
Focus: Business Admin
Periodicals
Energy Management (weekly)
Maintenance (weekly)
Productivity (weekly)
Productivity News (weekly) 06846

Operational Research Society of India (ORSI), c/o CSIR, Rafi Marg, New Dehli 110001
Focus: Computer & Info Sci 06847

Optical Society of India (OSI), c/o Applied Physics Dept, University of Calcutta, 92 Acharya Prafulla Chandra Rd, Calcutta 700009
T: (033) 3508386
Founded: 1965; Members: 450
Pres: Prof. K. Singh; Gen Secr: Prof. A.K. Chakraborty
Focus: Optics
Periodical
Journal of Optics (weekly) 06848

Palynological Society of India, c/o Environmental Resources Research Centre, POB 1230, Peroorkada P.O., Thiruvananltapuram 695005
T: (047) 432159
Founded: 1965; Members: 285
Pres: Dr. P.K.K. Nair; Gen Secr: Prof. P.M. Mathew
Focus: Botany
Periodical
Journal of Palynology (weekly) 06849

PEN All-India Centre, 40 New Marine Lines, Bombay 400020
Founded: 1933; Members: 700
Pres: U. Joshi; Gen Secr: N. Ezekiel
Focus: Lit
Periodical
The Indian PEN (weekly) 06850

Pharmacy Council of India, Combined Councils' Bldg, Temple Lane, Kotla Rd, POB 7020, New Dehli 110002
Founded: 1949
Pres: V.C. Sane; Gen Secr: D.K. Jain
Focus: Pharmacol 06851

Rajasthan Academy of Science, c/o Birla College, Pilani 333031
Founded: 1951
Pres: Prof. V.L. Narayanan
Focus: Sci
Periodical
Proceedings 06852

Research Designs and Standards Organization, c/o Ministry of Railways, Manak Nagar, Lucknow 226011
Founded: 1957
Gen Secr: R.M. Sambamoorthi
Focus: Eng; Standards
Periodicals
Documentation Notes (weekly)
Indian Railway: Technical Bulletin (weekly) 06853

Sahitya Akademi / National Academy of Letters, Rabindra Bhavan, 35 Ferozshah Rd, New Dehli 110001
Founded: 1954
Pres: Prof. U.R. Anantha Murthy; Gen Secr: Prof. I.N. Choudhuri
Focus: Lit
Periodicals
Indian Literature (weekly)
Samakaleena Bharateeya Sahitya (weekly)
Samskrita Pratibha (weekly) 06854

Sangeet Natak Akademi, Rabindra Bhavan, 35 Feroze Shah Rd, New Delhi 110001
T: (011) 387246; Tx: 65466
Founded: 1953
Pres: G. Karnad; Gen Secr: Malik Usha
Focus: Music
Periodical
Sangeet Natak (weekly) 06855

Sanskrit Academy, Sanskrit College Bldgs, Mylapore, Madras 600004
Focus: Ling 06856

Silk and Art Silk Mills' Research Association (SASMIRA), Sasmira Marg, Worli, Bombay 400025
T: (022) 4935351
Founded: 1950; Members: 151
Pres: D.N. Shroff; Gen Secr: E.K. Raghavan
Focus: Eng
Periodicals
Man-made Textiles India (weekly)
Sasmira Bulletin (weekly)
Sasmira Technical Digest (weekly) . . . 06857

Society of Biological Chemists, India, c/o Indian Institute of Science, Bangalore 560012
Founded: 1930; Members: 1600
Pres: Dr. P.M. Bhargava
Focus: Biochem; Bio
Periodicals
Biochemical Review (weekly)
Proceedings: Abstract of Papers Presented at the Annual Meeting 06858

Society of Young Scientists (S.Y.S.), c/o All India Institute of Medical Sciences, 111 Ashwini Block, Ansaringar, New Dehli 110029
Founded: 1974; Members: 351
Focus: Sci 06859

South Indian Horticultural Association, Lawley Rd, Post, Coimbatore 641003
Focus: Hort 06860

South India Textile Research Association, POB 3205 Coimbatore 641014
T: (0422) 574367; Fax: (0422) 571896
Focus: Textiles 06861

Sri Aurobindo Centre, Adhchini, Junction of Sri Aurobindo Marg and Qutab Hotel Rd, New Delhi 110017
Pres: Dharma Vira; Gen Secr: Dr. K.M. Agarwala
Focus: Cultur Hist
Periodical
AHANA (weekly) 06862

Survey of India / Map Record and Issue Office, Hathibarkala, Dehra Dun 248001
Founded: 1767
Focus: Geography 06863

Tamil Association, Karanthai Tamil Sangam, Thanjavur 2
Founded: 1911; Members: 216
Pres: S. Ramanathan; Gen Secr: T.K.P. Govindasami
Focus: Ling
Periodical
Tamil Pozhil (weekly) 06864

The Theosophical Society, c/o Theosophy Science Centre, Adyar, Madras 600020
T: (044) 417198
Founded: 1990; Members: 500
Gen Secr: Dr. A. Kannan
Focus: Philos; Parapsych; Sociology; Eng; Rel & Theol
Periodical
Holistic Science (weekly) 06865

Tripura Library Association, Agartala
Focus: Libraries & Bk Sci 06866

United Lodge of Theosophists, 40 New Marine Lines, Bombay 400020
CA: Avyahara, Bombay
Founded: 1929
Focus: Rel & Theol; Philos
Periodical
The Theosophical Movement: A Magazine devoted to the Living of the Higher Life (weekly) 06867

United Schools International (USI), USO House, 6 Special Institutional Area, New Dehli 110057
T: (011) 661103, 663998
Founded: 1961
Focus: Educ
Periodical
World Informo (weekly) 06868

Uttar Pradesh (India) Library Association, c/o Amir-ud-Daula Public Library, Kaiserbagh, Lucknow 226001
T: 43473
Founded: 1956
Focus: Libraries & Bk Sci
Periodical
Lucknow Librarian (weekly) 06869

Zoological Society of Calcutta, 35 Ballygunge Circular Rd, Calcutta 700019
Focus: Zoology 06870

Zoological Society of India, c/o Zoological Survey of India, 34 Chittaran Av, Calcutta 700012
Founded: 1916
Focus: Zoology 06871

Indonesia

Akademi Teknologi Kulit, Djalan Diponegoro 101, Jogjakarta
Gen Secr: P. Sukarbowo
Focus: Eng 06872

ASEAN Council of Teachers (ACT), c/o Persatuan Guru Republik Indonesia, Jalan Tanah Abang Tiga 24, Jakarta
Founded: 1978
Focus: Educ 06873

ASEAN Energy Management and Research Training Centre (AEEMTRC), c/o ASEAN, Jalan Sisingamangaraja 70A, POB 2072, Jakarta
T: (021) 716451, 712991, 711988; Tx: 47213
Focus: Business Admin 06874

ASEAN Federation of Endocrine Societies (AFES), c/o Div of Endocrinology and Metabolism, Dept of Medicine, University of Indonesia, Salemba 6, Jakarta 10430
Pres: Dr. Slamet Suyono
Focus: Endocrinology
Periodical
Journal of AFES 06875

ASEAN Law Association (ALA), c/o MKK, Wisma Metropolitan II Lantai 14, Jalan Jenderal Sudirman, Jakarta
Founded: 1979
Gen Secr: Prof. Dr. Komar Kantatmaadja
Focus: Law
Periodicals
ALA Newsletter
ASEAN Law Journal 06876

ASEAN Population Coordination Unit (APCU), POB 186, Jakarta 10002
T: (02) 8195251, 8191308; Tx: 48181
Founded: 1979; Members: 6
Focus: Soc Sci 06877

ASEAN Population Information Network (ASEANPOPIN), c/o APCU, Jalan Lt Jen M.T. Harijono 9-11, POB 186, Jakarta 10002
T: (02) 8195251, 8191308; Tx: 48181
Focus: Soc Sci 06878

The Asia Foundation, Djalan Darmawangsa 50, Kebayoran Baru, Jakarta
Focus: Int'l Relat 06879

Asian Confederation of Physical Therapy (ACPT), c/o Indonesian Physiotherapy Association, Jalan Puskesmas 20, Otista, Jakarta
Founded: 1980
Gen Secr: M. Hardjono
Focus: Physical Therapy 06880

Asian Physics Education Network (ASPEN), c/o UNESCO/ROTSEA, Jalan M.H. Thamrin 14, Tromolpos, Jakarta 10012
T: (021) 321308; Tx: 61464; Fax: (021) 334498
Founded: 1981
Gen Secr: U.S. Kuruppu
Focus: Educ; Physics
Periodicals
Asia-Pacific Journal of Physics Education
Asia-Pacific Physics News
ASPEN Newsletter (weekly) *06881*

Asian Society of Oto-Rhino-Laryngology, Jalan Proklamasi 42c, Jakarta 10320
Focus: Otorhinolaryngology *06882*

Asia Pacific League against Rheumatism (APLAR), c/o Seroja Arthritis Centre, Jalan Seroja Dalam 7, Semarang 50241
Founded: 1963; Members: 17
Gen Secr: John Darmawan
Focus: Rheuma *06883*

Asia Soil Conservation Network for the Humid Tropics (ASOCON), POB 133, Jakarta 10270
Fax: (02) 5700263
Gen Secr: W.J. Godert
Focus: Geology *06884*

Asosiasi Perpustakaan, Arsip dan Dokumentasi Indonesia (APADI), Medan Merdeka Selatan 11, Jakarta
Focus: Libraries & Bk Sci; Doc; Archives *06885*

Astronomical Association of Indonesia, c/o Jakarta Planetarium, Cikini Raya 73, Jakarta
Founded: 1920
Pres: Prof. Dr. B. Hidayat; Gen Secr: Dr. S. Darsa
Focus: Astronomy *06886*

Cooperative Program in Technological Research and Higher Education in Southeast Asia and the Pacific, c/o ROSTSEA, Jalan Thamrin 14, Jakarta 10002
T: (021) 321308; Tx: 44178
Focus: Educ; Eng
Periodical
Technology for Development in Southeast Asia and the Pacific (weekly) *06887*

Himpunam Pustakawan Chusus Indonesia (HPCI), Djalan Raden Saleh 43, Jakarta
Focus: Libraries & Bk Sci *06888*

Ikatan Dokter Indonesia, Djalan Dr. Samratulangi 29, Jakarta 10350
Fax: (021) 3900473
Founded: 1950; Members: 30000
Focus: Med *06889*

Ikatan Pustakawan Indonesia (IPI) / Indonesian Library Association, Djalan Medan Merdeka Selatan 11, POB 3624, Jakarta 10110
T: (021) 375718; Fax: (021) 3103554
Founded: 1954
Pres: Soekarman Kartosedono; Gen Secr: Ipon S. Purawijaya
Focus: Libraries & Bk Sci
Periodical
Majalah Ikatan Pustakawan Indonesia . . *06890*

Jajasan Dana Normalisasi Indonesia, Djalan Braga 38, Bandung
Founded: 1920
Focus: Standards *06891*

Perkumpulan Penggemar Alam di Indonesia, c/o Herbarium Bogoriense, Bogor
Founded: 1911
Focus: Hist; Nat Sci *06892*

Persatuan Insinyur Indonesia / Indonesian Institute of Engineers, Djalan Teuku Umar 30B, Jakarta Pusat
Pres: I. Sumantri; Gen Secr: I.J. Madjid
Focus: Eng
Periodical
Majalah Insinyur Indonesia (weekly) . . *06893*

Pusat Pembinaan dan Pengembangan Bahasa / National Centre for Language Development, Djalan Daksinapati Barat IV, Rawamangun, POB 2625, Jakarta 13220
T: (021) 4896558, 4894564; Fax: (021) 4880407
Founded: 1975; Members: 200
Pres: Alwi Hasan; Gen Secr: Dini Hasjmi
Focus: Ling; Lit
Periodicals
Bahasa dan Sastra (weekly)
Informasi Pustaka Kebahasaan (weekly)
Lembar Komunikasi (weekly) *06894*

Yayasan Dana Normalisasi Indonesia / Foundation for Indonesian Standardisation, Djalan Braga 40, Bandung
T: (022) 59220
Founded: 1920
Focus: Standards *06895*

Iran

The Ancient Iran Cultural Society, Jomhorie Eslamie Av, Shahrokh St, Teheran
Founded: 1961
Pres: A. Quoreshi
Focus: Hist *06896*

Association of Ophthalmists, c/o Faculty of Medicine, University, Teheran
Focus: Ophthal *06897*

British Institute of Persian Studies, Khiaban Dr. Ali Shariati, Kucheh Alvand, Gholhak, POB 11365-6844, Teheran 19396
Founded: 1961; Members: 850
Gen Secr: S.J. Whitwell
Focus: Hist; Archeol
Periodical
Iran *06898*

Iranian Society of Microbiology, c/o Dept of Microbiology and Immunology, Faculty of Medicine, University, Teheran
Founded: 1940; Members: 185
Focus: Microbio *06899*

Medical Nomenclature Society of Iran, c/o University College of Medicine and Pharmacy, Isfahan
Focus: Med; Standards *06900*

National Cartographic Centre, Azadi Sq, POB 1844, Teheran
Focus: Cart *06901*

Philosophy and Humanities Society, c/o Faculty of Arts, University, Teheran
Focus: Humanities; Philos *06902*

Sazemane Pachuhesh va Barnamerizi Amuzeshi / Organization of Research and Educational Planning, c/o Ministry of Education, Iranshahr Av, Martyr Musavi Bldg, POB 14367, Teheran
T: (021) 839262/64
Founded: 1972; Members: 250
Pres: G.A. Haddad Adel
Focus: Educ; Public Admin
Periodical
Roshd: Educational Journal for Students and Teachers (weekly) *06903*

Society of Iranian Clinicians, c/o Faculty of Medicine, University, Enghelab Av, Teheran
Pres: Prof. Y. Adle
Focus: Med *06904*

Iraq

Arab Academy of Music, POB 6150, Baghdad
T: (01) 5373891
Founded: 1971; Members: 20
Gen Secr: Munir Bashir
Focus: Music
Periodical
Arab Music Magazine (weekly) *06905*

Arab Bureau for Prevention of Crime, POB 28025, Baghdad
Gen Secr: Hamadi Asi Al-Falahi
Focus: Criminology *06906*

Arab Dental Federation, 38 Al-Sabragh St, Baghdad
Founded: 1969
Gen Secr: Prof. A. Al-Sibahi Al-Mansour
Focus: Dent *06907*

Arab Federation for Technical Education (AFTE), POB 718, Baghdad
T: (01) 7189801; Tx: 213541
Founded: 1981
Gen Secr: Ahmed B. Alnaib
Focus: Educ
Periodicals
Arab Journal for Technical Education (3 times annually)
Arab Technical Education Newsletter (weekly) *06908*

Arab Gulf States Information Documentation Centre (AGSIDC), POB 5063, Baghdad
T: (01) 5555962; Tx: 213267; Fax: (01) 5567629
Founded: 1980; Members: 7
Gen Secr: Dr. Jasim Mohammed Jirjees
Focus: Computer & Info Sci
Periodical
Journal – Information Documentation (weekly) *06909*

Arab Historians Association (AHA), POB 4085, Baghdad
T: (01) 4438868; Fax: (01) 740331
Founded: 1974
Gen Secr: Prof. Dr. Mustafa A. Al-Najjar
Focus: Hist
Periodicals
Arab Historian (weekly)
The Union News (weekly) *06910*

Arab Industrial Development and Mining Organization (AIDMO), POB 3156, Sadoon, Baghdad
T: (01) 7748546; Tx: 2828; Fax: (01) 7184658
Focus: Econ; Mining *06911*

Arab Institute for Training and Research in Statistics, POB 588, Baghdad
Focus: Stats *06912*

Arab Literacy and Adult Education Organization (ARLO), 113 Abu Nawas St, POB 3217, Al-Saadoon, Baghdad
T: (01) 8874850; Tx: 2564
Founded: 1988; Members: 21
Gen Secr: Prof. Dr. Musari Al-Rawi
Focus: Adult Educ
Periodical
Education of the Masses (3 times annually) *06913*

Arab Petroleum Training Institute (APTI), POB 6037, Al-Tajeyat, Baghdad
T: (01) 5234100; Tx: 2728
Founded: 1978; Members: 11
Focus: Petrochem
Periodical
Arabic Training (weekly) *06914*

Arab Union of Veterinary Surgeons, POB 27098, Baghdad
Focus: Vet Med *06915*

Association of Arab Geologists / Association des Géologues Arabes, c/o Iraq National Oil Company, Al-Khulani Sq, Baghdad
Tx: 2204
Gen Secr: Dr. Sammi Sherif
Focus: Geology *06916*

Iraq Academy, Waziriyah, Baghdad
Founded: 1947
Pres: Dr. S.A. Al-Ali; Gen Secr: Dr. N.H. Al-Qissi
Focus: Sci
Periodicals
Arabs before Islam
Bulletin (weekly)
Majallat al Majimma'al Ilmi al Iraqi (weekly) *06917*

Iraqi Medical Society, Maari St, Al Mansoor, Baghdad
Founded: 1920; Members: 871
Pres: F.H. Ghali; Gen Secr: A.K. Al-Khateer
Focus: Med *06918*

Society of Iraqi Artists, Damascus St, Baghdad
Founded: 1956
Pres: Noori Al-Rawi; Gen Secr: Amer Alubidi
Focus: Arts *06919*

Teachers' Society, Sharia Al-Askari, Baghdad
Founded: 1942
Focus: Educ *06920*

Ireland

Agricultural Science Association, 21 Upper Mount St, Dublin 2
T: (01) 760781
Focus: Agri *06921*

Aosdana, 70 Merrion Sq, Dublin 2
T: (01) 6611840
Founded: 1981; Members: 150
Focus: Arts *06922*

Apothecaries' Hall, 95 Merrion Sq, Dublin 2
Founded: 1791
Pres: Dr. F.J. O'Donnell
Focus: Lit *06923*

Architectural Association of Ireland (AAI), 8 Merrion Sq, Dublin 2
T: (01) 6761703
Founded: 1896; Members: 450
Pres: Sean Manon
Focus: Archit
Periodicals
AAI Green Book (weekly)
AAI Newsletter (weekly) *06924*

Arts Council, 70 Merrion Sq, Dublin 2
T: (01) 6611840
Founded: 1951
Gen Secr: Adrian Munnelly
Focus: Arts
Periodicals
Annual Report (weekly)
Art Matters (weekly) *06925*

Association of Consulting Engineers of Ireland, 63 Haddington Rd, Dublin 4
T: (01) 600374
Focus: Eng *06926*

Association of Irish Headmistresses, Glengara Park, Dun Laoghaire, Co. Dublin
Focus: Educ *06927*

Association of Irish Jurists, c/o Law Library, Four Courts, Dublin 7
T: (01) 8720622; Fax: (01) 8720455
Focus: Law *06928*

Association of Irish Musical Societies, 5 Rathgar Av, Dublin 6
T: (01) 970247
Focus: Music *06929*

Association of Ophthalmic Opticians of Ireland, 11 Harrington St, Dublin 8
T: (01) 753498
Focus: Ophthal; Optics *06930*

Association of Secondary Teachers, Ireland, ASTI House, Winetavern St, Dublin 8
T: (01) 719144; Fax: (01) 719280
Gen Secr: C.M. Lennon
Focus: Educ
Periodicals
ASTIR (9 times annually)
The Secondary Teacher (weekly) . . . *06931*

Authors' Guild of Ireland, 130 Furry Park Rd, Dublin 5
T: (01) 331189
Focus: Lit *06932*

Automobile, General Engineering and Mechanical Operations Union, 22 North Frederick St, Dublin 1
T: (01) 744233
Focus: Auto Eng; Eng *06933*

The Bram Stoker Society (BSS), Regent House, Trinity College, Dublin 2
Fax: (01) 6772694
Founded: 1980; Members: 200
Pres: Leslie Shepard; Gen Secr: David Lass
Focus: Lit *06934*

Building Materials Federation, Confederation House, Kildare St, Dublin 2
T: (01) 779801
Focus: Materials Sci; Civil Eng *06935*

Church Education Society for Ireland, 28 Bachelor's Walk, Dublin 1
Founded: 1839
Gen Secr: A. Wilson
Focus: Rel & Theol; Educ
Periodical
Annual Report (weekly) *06936*

Conradh na Gaeilge / Gaelic League, 6 Sráid Fhearchair, Dublin 2
Founded: 1893; Members: 200
Pres: Ite Ni Chionnaith; Gen Secr: Sean MacMathuna
Focus: Ling
Periodicals
An t Ultach (weekly)
Rosc (weekly) *06937*

Cork Historical and Archaeological Society (CHAS), Ballysheehy Lodge, Clogheen, Cork Co. Cork
Founded: 1891; Members: 500
Pres: G. O'Crualaoich; Gen Secr: Patrick Holohan
Focus: Archeol; Hist *06938*

County Louth Archaeological and Historical Society (C.L.A.H.S.), 5 Oliver Plunkett Park, Dundalk, Co. Louth
Founded: 1903; Members: 650
Focus: Archeol; Hist
Periodical
County Louth Archaeological and Historical Journal (weekly) *06939*

Cystic Fibrosis Association of Ireland, 24 Lower Rathmines Rd, Dublin 6
T: (01) 962433
Focus: Physiology
Periodical
Newsletter (weekly) *06940*

The Dental Council, 57 Merrion Sq, Dublin 2
T: (01) 6762069
Founded: 1928; Members: 19
Pres: Dr. Daniel I. Keane; Gen Secr: Thomas Farren
Focus: Dent
Periodical
Register of Dentists (weekly) *06941*

Dublin University Biological Association, c/o Trinity College, Dublin
Founded: 1874; Members: 400
Pres: Prof. Ian Temperley; Gen Secr: Emer Loughrey
Focus: Bio *06942*

Electro-Technical Council of Ireland, Ballymun Rd, Dublin 9
T: (01) 376773
Focus: Electric Eng *06943*

Engineering and Scientific Association of Ireland, 13 Mather Rd, Mount Merrion, Co. Dublin
Founded: 1903; Members: 100
Pres: T.A. McInerney; Gen Secr: M.J. Higgins
Focus: Eng; Sci
Periodical
Annual Report (weekly) *06944*

Espace Vidéo Européen (EVE), c/o Irish Film Institute, 6 Eustace St, Dublin 2
T: (01) 6795744; Fax: (01) 6799657
Founded: 1990
Gen Secr: Norma Cairns
Focus: Cinema *06945*

European Alliance of Muscular Dystrophy Associations (EAMDA), c/o MDI, Carmichael House, North Brunowick St, Dublin 7
Founded: 1972; Members: 20
Gen Secr: Judy Windle
Focus: Med
Periodicals
EAMDA Bulletin
EAMDA Newsletter *06946*

European Healthcare Management Association (EHMA), Vergemount Hall, Clonskeagh, Dublin 6
T: (01) 2839299; Fax: (01) 2838653
Founded: 1966; Members: 130
Pres: Dr. J. Manuel Freire; Gen Secr: Philip C. Berman
Focus: Public Health
Periodical
Newsletter (weekly) *06947*

Federation of Irish Film Societies, 65 Harcourt St, Dublin 2
T: (01) 712982
Founded: 1977; Members: 4000
Focus: Cinema *06948*

Folklore of Ireland Society, c/o Dept of Irish Folklore, University College, Belfield, Dublin 4
T: (01) 2693244
Founded: 1927
Gen Secr: M. Ross
Focus: Ethnology; Hist
Periodical
Béaloideas: Journal (weekly) *06949*

Friends of Medieval Dublin (FMD), c/o Dept of Medieval History, University College, Belfield, Dublin 4
Fax: (01) 2837022
Founded: 1976
Pres: Dr. H.B. Clarke
Focus: Hist; Archeol *06950*

The Friends of the National Collections of Ireland, c/o Hugh Lane Municipal Gallery of Modern Art, Charlemont House, Parnell Sq, Dublin 1
Founded: 1924
Gen Secr: C. Magan
Focus: Arts *06951*

Genealogical Office, c/o State Heraldic Museum, Kildare St, Dublin
Fax: (01) 6766690
Founded: 1552
Pres: D.F. Begley
Focus: Genealogy *06952*

The Geographical Society of Ireland (G.S.I.), c/o Dept of Geography, University College, Galway, Co. Galway
T: (091) 24411
Founded: 1934; Members: 200
Pres: Dr. Anngret Simms; Gen Secr: Dr. Mary Cawley
Focus: Geography
Periodicals
Geonews (bi-annually)
Irish Geography (bi-annually) *06953*

Health Research Board, 73 Lower Baggot St, Dublin 2
T: (01) 6761176; Fax: (01) 6611856
Founded: 1987
Gen Secr: John O'Gorman
Focus: Public Health
Periodical
Annual Report (weekly) *06954*

Honorable Society of King's Inns, Henrietta St, Dublin 1
T: (01) 8744840; Fax: (01) 8726048
Founded: 1542; Members: 900
Focus: Law *06955*

Incorporated Law Society of Ireland, Solicitors' Buildings, Blackhall Pl, Dublin 7
T: (01) 710711; Tx: 31219; Fax: (01) 710704
Founded: 1841; Members: 3300
Gen Secr: James J. Ivers
Focus: Law
Periodicals
Gazette (weekly)
Law Directory (weekly) *06956*

Institiuid Teangeolaiochta Eireann / Linguistics Institute of Ireland, 31 Fitzwilliam Pl, Dublin 2
T: (01) 6765489; Fax: (01) 6610004
Founded: 1972; Members: 23
Pres: Prof. C.R. O'Cleirigh; Gen Secr: Eoghan MacAogain
Focus: Ling
Periodicals
Annual Report (weekly)
Teangeolas (weekly) *06957*

Institute of Advertising Practitioners in Ireland (IAPI), 35 Upper Fitzwilliam St, Dublin 2
T: (01) 6765991, 6764876; Fax: (01) 6614589
Founded: 1964
Focus: Advert *06958*

Institute of Architectural and Associated Technology, 8 Merrion Sq, Dublin 2
Founded: 1966; Members: 185
Pres: James F. Kirwan; Gen Secr: Robert McKee
Focus: Archit; Civil Eng
Periodical
Newsletter (weekly) *06959*

Institute of Chartered Accountants in Ireland, 87-89 Pembroke Rd, Ballsbridge, Dublin 4
Founded: 1888; Members: 5880
Pres: R.F. Hussey; Gen Secr: R.L. Donovan
Focus: Stats *06960*

The Institute of Chemistry of Ireland / Institiúid Ceimice Na hEireann, c/o Royal Dublin Society, Science Section, Ballsbridge, Dublin 4
Founded: 1950; Members: 600
Pres: Conor Murphy; Gen Secr: Dr. James P. Ryan
Focus: Chem
Periodicals
Irish Chemical News (3 times annually)
Irish Chemical Newsletter (bi-annually) . *06961*

Institute of Industrial Engineers (IIE), 35-39 Shelbourne Rd, Dublin 4
T: (01) 686244
Focus: Eng *06962*

Institute of Management Consultants in Ireland, Harcourt House, Harcourt St, Dublin 2
T: (01) 757971
Focus: Business Admin *06963*

Institute of Public Administration (IPA), Vergemount Hall, Clonskea, Dublin 6
T: (01) 697011
Founded: 1957
Focus: Advert
Periodicals
Administration: Journal (weekly)
Young Citizen: Magazine (7 times annually) *06964*

Institute of Taxation in Ireland, 15 Fitzwilliam Sq, Dublin 2
T: (01) 688181
Focus: Finance *06965*

Institution of Civil Engineers (Republic of Ireland), 56 Springhill Park, Killiney, Co. Dublin
T: (01) 853389
Founded: 1818; Members: 538
Pres: Vincent Curley; Gen Secr: Patrick Dullaghan
Focus: Eng
Periodicals
Municipal Engineer (weekly)
New Civil Engineer (weekly) *06966*

Institution of Electrical Engineers (Irish Branch), 6 Tivoli Close, Dun Laoghaire, Co. Dublin
Pres: Dr. T.J. Gallagher; Gen Secr: C.J. Bruce
Focus: Electric Eng *06967*

Institution of Engineers of Ireland (IEI), 22 Clyde Rd, Ballsbridge, Dublin 4
T: (01) 684341
Founded: 1835; Members: 5900
Pres: P. Pigott; Gen Secr: M.B. O'Donovan
Focus: Eng
Periodicals
Monthly Journal (weekly)
Transactions *06968*

Insurance Institute of Ireland, 32 Nassau St, Dublin 2
T: (01) 6772753; Fax: (01) 6772621
Focus: Insurance *06969*

International Union of Food Science and Technology, c/o National Food Centre, Dunsinea, Castleknock, Dublin 15
T: (01) 383222; Tx: 31947
Founded: 1970; Members: 43
Pres: Prof. E. von Sydow; Gen Secr: Dr. D.E. Hood
Focus: Food
Periodical
IUFost Newsletter (weekly) *06970*

Irish Academy of Letters, c/o School of Irish Studies, Thomas Prior House, Merrion Rd, Dublin 4
Founded: 1932
Focus: Lit *06971*

Irish Association for Economic Geology, c/o Geological Survey of Ireland, Haddington Rd, Dublin 4
T: (01) 6715233; Fax: (01) 6691782
Founded: 1973
Pres: Jim Geraghty
Focus: Geology
Periodical
IAEG Annual Review (weekly) *06972*

Irish Association of Civil Liberty, 8 Dawson St, Dublin 2
Focus: Poli Sci *06973*

Irish Association of Curriculum Development, c/o E. Sides, 5 Ailesbury Gardens, Ballsbridge, Dublin 4
T: (01) 693811
Focus: Educ
Periodical
Compass (bi-annually) *06974*

Irish Association of Professional Archaeologists (IAPA), POB 2252, Dublin
Founded: 1973
Gen Secr: Celie Orahilly
Focus: Archeol *06975*

Irish Association of Social Workers, 114-116 Pearse St, Dublin
T: (01) 771930, 773253, 771749
Focus: Soc Sci
Periodical
The Irish Social Worker (weekly) . . . *06976*

Irish Astronomical Society, POB 2547, Dublin 15
T: (01) 8202135
Founded: 1937; Members: 305
Pres: Robin Moore; Gen Secr: James J. Lynch
Focus: Astronomy
Periodical
Orbit (weekly) *06977*

Irish Cancer Society, 22 Earlsfoot Terrace, Dublin 2
T: (01) 757048
Founded: 1963
Focus: Cell Biol & Cancer Res *06978*

Irish Commercial Horticultural Association, c/o Irish Farm Centre, Naas Rd, Dublin 12
T: (01) 501166
Focus: Hort *06979*

Irish Computer Society, 17 Earlsfort Terrace, Dublin 2
Fax: (01) 6620788
Founded: 1972
Focus: Computer & Info Sci *06980*

Irish Dental Association (IDA), 10 Richview Office Park, Clonskeagh Rd, Dublin 14
T: (01) 2830499
Founded: 1922
Focus: Dent
Periodical
Journal of the Irish Dental Association (weekly) *06981*

Irish Epilepsy Association, 249 Crumlin Rd, Dublin 12
T: (01) 557500; Fax: (01) 557013
Focus: Pathology
Periodical
Epilepsy News (3 times annually) . . . *06982*

Irish Family Planning Association (IFPA), 36-37 Lower Ormond Quay, Dublin 1
T: (01) 8725033; Fax: (01) 8726639
Focus: Family Plan
Periodicals
Annual Report (weekly)
Newsletter (weekly) *06983*

Irish Federation of University Teachers (IFUT), 11 Merrion Sq, Dublin 2
T: (01) 6610910; Fax: (01) 6610909
Focus: Educ
Periodical
IFUT News *06984*

Irish Film Society, c/o 9 Annaville Park, Dublin 14
T: (01) 773591
Founded: 1936
Focus: Cinema *06985*

Irish Georgian Society (IGS), 42 Merrion Sq, Dublin 2
T: (01) 6767053
Founded: 1958
Focus: Hist; Archit
Periodical
Bulletin (weekly) *06986*

Irish Grassland and Animal Production Association, Belclare, Tuam, Co. Galway
T: (091) 98140; Fax: (093) 55430
Focus: Animal Husb; Agri *06987*

Irish Institute of Purchasing and Materials Management, 90 Saint Stephen's Green, Dublin 2
T: (01) 752552
Focus: Materials Sci; Commerce *06988*

Irish Management Institute (IMI), Sandyford Rd, Dublin 14
T: (01) 2956911
Founded: 1952
Focus: Business Admin *06989*

Irish Manuscripts Commission (I.M.C.), 73 Merrion Sq, Dublin 2
Founded: 1928; Members: 18
Pres: Brian Trainor
Focus: Hist
Periodical
Analecta Hibernica *06990*

Irish Maritime Law Association, Merrion Hall, Strand Rd, Dublin 4
T: (01) 695522
Focus: Law *06991*

Irish Medical Association (IMA), 10 Fitzwilliam Pl, Dublin 2
Founded: 1936; Members: 5000
Gen Secr: George McNeice
Focus: Med
Periodical
Irish Medical Journal (weekly) *06992*

Irish Mining and Quarrying Society, 87-89 Waterloo Rd, Dublin 4
T: (01) 685193
Focus: Mining *06993*

Irish National Teachers' Organisation (I.N.T.O.), 35 Parnell Sq, Dublin 1
T: (01) 8722533; Fax: (01) 8422462
Founded: 1868; Members: 24000
Pres: M. McGorry; Gen Secr: Joe O'Toole
Focus: Educ
Periodicals
Education Today (3 times annually)
Tuarascail (10 times annually) *06994*

Irish PEN, 26 Rosslyn Killarney Rd, Bray, Co. Wicklow
Founded: 1921; Members: 60
Pres: F.J.A. Gaughan; Gen Secr: Arthur Flynn
Focus: Lit *06995*

Irish Productivity Centre (IPC), 35-39 Shelbourne Rd, Dublin 4
T: (01) 6686244; Fax: (01) 6686525
Founded: 1963
Focus: Business Admin; Econ *06996*

Irish Psychoanalytical Association, 2 Belgrave Terrace, Monkstown, Dublin
Founded: 1942; Members: 8
Focus: Psychoan *06997*

Irish Quality Control Association, 3 Saint Stephen's Green, Dublin 2
T: (01) 781755
Focus: Materials Sci *06998*

Irish Science Teachers Association, c/o Dept of Chemistry, University College, Bellfield, Dublin 4
Focus: Nat Sci; Educ *06999*

The Irish Society for Design and Craftwork, 112 Ranelagh, Dublin 6
Founded: 1894
Pres: D. O'Murcada; Gen Secr: Angela O'Brien
Focus: Arts *07000*

Irish Society of Arts and Commerce, 55 Fairview Strand, Dublin
Founded: 1911
Pres: Edward Kissane; Gen Secr: A. von Muntz
Focus: Arts *07001*

Irish Textiles Federation, Confederation House, Kildare St, Dublin 2
T: (01) 779801
Focus: Textiles *07002*

Irish Timber Growers Association, 31 Pembroke Rd, Dublin 4
T: (01) 689018
Focus: Forestry *07003*

Irish Veterinary Association, 53 Lansdowne Rd, Ballsbridge, Dublin 4
T: (01) 6685263; Fax: (01) 6604345
Founded: 1888
Focus: Vet Med
Periodical
Irish Veterinary Journal (weekly) *07004*

Irish Welding Association, Fitzwilton House, Wilton Pl, Dublin 2
T: (01) 760306, 682222; Tx: 24258;
CA: Colybrand Dublin
Focus: Eng *07005*

Irish Wildbird Conservancy, Ruttledge House, 8 Longford Pl, Monkstown, Co. Dublin
T: (01) 804322
Founded: 1969
Focus: Ornithology
Periodicals
Irish Birds (weekly)
Irish Wildbird Conservancy News (weekly) *07006*

Library Association of Ireland, 53 Upper Mount St, Dublin 2
Founded: 1928
Focus: Libraries & Bk Sci
Periodical
An Leabharlann: The Irish Library (weekly) *07007*

Maritime Institute of Ireland, Haigh Terrace, Dun Laoghaire, Co. Dublin
T: (01) 800969
Focus: Navig *07008*

The Marketing Society, 19-22-Upper Pembroke St, Dublin 2
T: (01) 761196
Founded: 1971; Members: 360
Focus: Advert *07009*

The Medical Council, 8 Lower Hatch St, Dublin 2
Founded: 1978
Focus: Med
Periodical
General Register of Medical Practitioneers (every 5 years) *07010*

Medical Research Council of Ireland (M.R.C.I.), 9 Clyde Rd, Ballsbridge, Dublin 4
Founded: 1937; Members: 9
Focus: Med
Periodical
Annual Report (weekly) *07011*

Medical Union, 51 Harcourt St, Dublin 2
T: (01) 781562
Focus: Med *07012*

Mental Health Association of Ireland, 2 Herbert Av, Dublin 4
T: (01) 695096
Focus: Psychiatry *07013*

The Military History Society of Ireland, c/o University College Dublin, Newman House, 86 Saint Stephen's Green, Dublin 2
Fax: (01) 962094
Founded: 1949; Members: 700
Gen Secr: Dr. K.P. Ferguson; J.P. Coyle
Focus: Hist; Military Sci
Periodical
The Irish Sword (weekly) *07014*

The Music Association of Ireland, 5 North Frederick St, Dublin 1
Founded: 1948; Members: 850
Pres: E.N. Chathalriabhaigh; Gen Secr: Margot Doherty
Focus: Music
Periodicals
Annual Report (weekly)
Soundpost (weekly) *07015*

National Development Association, 3 Saint Stephen's Green, Dublin 2
Founded: 1967; Members: 100
Focus: Agri *07016*

National Safety Council, 4 Northbrook Rd, Ranelagh, Dublin 6
T: (01) 963422; Fax: (01) 963306
Founded: 1988
Focus: Safety *07017*

Old Dublin Society, c/o City Assembly House, 58 South William St, Dublin 2
Founded: 1934; Members: 15000
Pres: Anthony P. Behan; Gen Secr: S. Smith
Focus: Hist

Periodical
Dublin Historical Record (weekly) . . . 07018

Pharmaceutical Society of Ireland, 37 Northumberland Rd, Dublin 4
T: (01) 6600699, 5600551; Fax: (01) 6681461
Founded: 1875; Members: 2053
Pres: Timothy Lawleh; Gen Secr: Eugenie Canavan
Focus: Pharmacol
Periodicals
Calendar (weekly)
The Irish Pharmacy Journal (weekly) . . 07019

Photographic Society of Ireland (PSI), 38-39 Parnell Sq, Dublin 1
Founded: 1854; Members: 600
Focus: Arts 07020

Royal Academy of Medicine, 6 Kildare St, Dublin 2
Founded: 1882; Members: 1500
Pres: J.W. Dundee
Focus: Med
Periodical
Irish Journal of Medical Science (weekly) 07021

Royal College of Physicians of Ireland, 6 Kildare St, Dublin 2
T: (01) 6616677
Founded: 1654
Pres: Dr. J.S. Doyle; Gen Secr: J.W. Bailey
Focus: Med 07022

Royal College of Surgeons in Ireland (RCSI), 123 Saint Stephen's Green, Dublin 2
T: (01) 4780200; Fax: (01) 4782100
Founded: 1784
Focus: Surgery
Periodicals
Journal of the Irish College of Physicians and Surgeons (weekly)
Nursing Review (weekly) 07023

Royal Dublin Society (R.D.S.), Ballsbridge, Dublin 4
T: (01) 6680866; Fax: (01) 6604014
Founded: 1731; Members: 12000
Gen Secr: Bridgeen McCloskey
Focus: Sci
Periodical
Royal Dublin Society Seminar Proceedings (weekly) 07024

Royal Hibernian Academy of Arts, 15 Ely Pl, Dublin 2
T: (01) 612558
Founded: 1823; Members: 54
Pres: Thomas Ryan; Gen Secr: John Coyle
Focus: Archit; Arts 07025

Royal Institute of Architects of Ireland (RIAI), 8 Merrion Sq, Dublin 2
Fax: (01) 610948
Founded: 1839; Members: 1350
Gen Secr: John Graby
Focus: Archit
Periodicals
Irish Architect (weekly)
RIAI Yearbook (weekly) 07026

Royal Institution of Chartered Surveyors, 8 Merrion Sq, Dublin 2
Founded: 1868
Focus: Surveying 07027

Royal Irish Academy (R.I.A.), 19 Dawson St, Dublin 2
T: (01) 6762570, 6764222; Fax: (01) 6762346
Founded: 1785; Members: 250
Pres: Prof. J.D. Scanlan; Gen Secr: Patrick Buckley
Focus: Sci; Humanities; Soc Sci
Periodicals
Erin (weekly)
Irish Journal of Earth Sciences (weekly)
Irish Studies in International Affairs
Proceedings of the Royal Irish Academy (weekly)
. . . . 07028

The Royal Irish Academy of Music (RIAM), 36-38 Westland Row, Dublin 2
T: (01) 6764412/13
Founded: 1856; Members: 135
Focus: Music 07029

The Royal Society of Antiquaries of Ireland (R.S.A.I.), 63 Merrion Sq, Dublin 2
T: (01) 6761749
Founded: 1849; Members: 1050
Focus: Archeol
Periodical
Journal of The Royal Society of Antiquaries of Ireland (weekly) 07030

Royal Zoological Society of Ireland (R.Z.S.I.), Phoenix Park, Dublin 8
T: (01) 6771424; Fax: (01) 6771660
Founded: 1830; Members: 8396
Pres: Prof. David McConnell; Gen Secr: M. Sinanan; D. Kilroy
Focus: Zoology
Periodical
Annual Report (weekly) 07031

Society of Designers in Ireland, 8 Merrion Sq, Dublin 2
T: (01) 807646
Founded: 1972; Members: 181
Focus: Graphic & Dec Arts, Design . . 07032

Society of Irish Foresters, c/o Royal Dublin Society, Ballbridge, Dublin 4
Founded: 1942; Members: 630
Pres: B. Wright
Focus: Forestry
Periodical
Irish Forestry 07033

Society of the Irish Motor Industry, 5 Upper Pembroke St, Dublin 2
T: (01) 6761690; Fax: (01) 6619213
Founded: 1968
Focus: Auto Eng
Periodical
Irish Motor Industry (weekly) 07034

Statistical and Social Inquiry Society of Ireland (SSISI), c/o Allied Irish Banks, Bankcentre, Ballsbridge, Dublin 4
T: (01) 6600311
Founded: 1847; Members: 420
Pres: Dr. Padraig McGowan; Gen Secr: Dr. D. de Buitleir
Focus: Sociology; Stats
Periodical
Journal of the Statistical and Social Inquiry Society of Ireland (weekly) 07035

Theosophical Society in Ireland, 31 Pembroke Rd, Dublin 4
Founded: 1919
Gen Secr: M. Finnegan
Focus: Rel & Theol 07036

University Philosophical Society (U.P.S.), c/o Trinity College, Dublin 2
T: (01) 712127; Tx: 25442
Founded: 1684; Members: 1150
Pres: Alan Doherty; Gen Secr: Christine O'Rourke
Focus: Philos
Periodical
Laws 07037

Veterinary Council, 53 Lansdowne Rd, Dublin 4
T: (01) 6684402; Fax: (01) 6684402
Founded: 1931; Members: 1823
Focus: Vet Med 07038

Israel

Academic Circle of Tel Aviv, POB 2425, Tel Aviv
Founded: 1956
Focus: Sci 07039

Academy of the Hebrew Language, POB 3449, Jerusalem 91034
T: (02) 632242; Fax: (02) 617065
Founded: 1953; Members: 50
Pres: Prof. M. Bar-Asher; Gen Secr: G. Birkenbaum
Focus: Ling
Periodicals
Lamed Lěšoněkha (weekly)
Lěšonénu: A Quarterly for the Study of the Hebrew Language and Cognate Subjects (weekly)
Lěšonénu Laam (weekly) 07040

Agricultural Research Organization, c/o The Volcani Center, POB 6, Bet Dagan
Fax: (03) 993998
Founded: 1921
Focus: Agri 07041

Architectural Association of Israel, POB 2425, Tel Aviv
Founded: 1952
Focus: Archit 07042

Arthur Rubinstein International Music Society, 9 Vilna St, POB 6108, Tel Aviv
T: (03) 5239449; Fax: (03) 5239793
Founded: 1980
Pres: Jan Jacob Bistritzky
Focus: Music
Periodical
Bulletin (weekly) 07043

Association for the Advancement of Science in Israel, c/o Dept of Physics, Bar-Ilan University, Ramat-Gan 52100
T: (03) 5318433; Tx: 361311; Fax: (03) 5353298
Founded: 1953; Members: 5000
Pres: Prof. M. Jammer
Focus: Sci
Periodical
Proceedings of Congress of Scientific Societies
. . . . 07044

Association Internationale des Etudes Arméniennes (AIEA), POB 16174, Jerusalem 91161
T: (02) 412906
Founded: 1982
Pres: Prof. Dr. Michael Edward Stone
Focus: Ethnology 07045

Association of Engineers and Architects in Israel, 200 Dizengoff Rd, POB 3082, Tel Aviv 63462
T: (03) 5240274; Tx: 371690; Fax: (03) 5235993
Founded: 1922
Focus: Archit
Periodicals
Bulletin (weekly)
Handassa 07046

Association of Religious Writers, POB 7440, Jerusalem
Founded: 1963; Members: 100
Pres: J.E. Chen
Focus: Lit
Periodical
Mabua (weekly) 07047

Biochemical Society of Israel, c/o Dept of Chemical Immunology, Weizmann Institute, Rehovot
Members: 450
Pres: Prof. I. Pecht
Focus: Biochem 07048

Botanical Society of Israel, c/o Dept of Biology, Ben-Gurion University, Beer Sheva
T: (057) 661373
Founded: 1936; Members: 200
Pres: Prof. Chanan Itai; Gen Secr: Dr. Nurit Roth-Bejerano
Focus: Botany 07049

Clinical Paediatric Club of Israel, c/o Paediatric Dept, Hospital, Zerifin 70300
Fax: (04) 49502
Founded: 1953
Focus: Pediatrics 07050

Coordination Office of Paediatric Endocrine Societies (COPES), c/o Institute of Paediatric and Adolescent Endocrinology, Beilinson Medical Centre, Petah Tikva 49100
T: (03) 9225108; Fax: (03) 9229685
Gen Secr: Prof. Z. Laron
Focus: Endocrinology; Pediatrics 07051

Development Study Center (DSC), POB 2355, Rehovot 76120
T: (08) 474111; Tx: 381378; Fax: (08) 475884
Founded: 1963
Focus: Develop Areas 07052

Ecumenical Institute for Theological Research, POB 19556, Jerusalem
T: (02) 713451
Founded: 1971
Focus: Rel & Theol 07053

The Genetics Society of Israel, c/o Agricultural Research Organization, Regional Experiment Station, Neve Ya'ar
Founded: 1960; Members: 200
Focus: Genetics 07054

Hebrew Writers Association in Israel, POB 7111, Tel Aviv
T: (03) 6953256; Fax: (03) 6919681
Founded: 1921; Members: 425
Focus: Lit
Periodical
Moznayim (weekly) 07055

Historical Society of Israel, POB 4179, Jerusalem 91041
T: (02) 669464; Fax: (02) 662135
Founded: 1936; Members: 1000
Pres: Prof. Yosef Kaplan; Gen Secr: Zvi Yekutiel
Focus: Hist
Periodical
Zion (weekly) 07056

Industrial Medical Association, c/o Dr. P. Berstein, 43 Ha-Nasi Blvd, Haifa
Founded: 1958
Focus: Hygiene 07057

International Society of Computerized and Quantitative EMG, POB 9117, Jerusalem 91090
T: (02) 731050; Fax: (02) 713086
Founded: 1981
Pres: Dr. Arieh N. Gilai; Gen Secr: Prof. Joe F. Jabre
Focus: Physiology; Med; Neurology . . . 07058

Isarel Plastics and Rubber Center, Technion City, Nveh Sha'anan, Haifa 32000
Fax: (04) 227582
Founded: 1986; Members: 45
Gen Secr: S. Keniy
Focus: Materials Sci 07059

The Israel Academy of Sciences and Humanities, 43 Jabotinsky Rd, Einstein Sq, POB 4040, Jerusalem 91040
T: (02) 636211; Fax: (02) 66609
Founded: 1959; Members: 50
Pres: Prof. Joshua Jortner; Gen Secr: Dr. Meir Zadok
Focus: Humanities; Sci 07060

Israel Association for Applied Animal Genetics, c/o Ministry of Agriculture, Haifa
Founded: 1957
Focus: Zoology 07061

Israel Association for Asian Studies, c/o Institute of Asian and African Studies, Hebrew University, Jerusalem
Founded: 1972
Focus: Ethnology 07062

Israel Association for Physical Medicine and Rheumatology, c/o Dr. G. Levy-Zackes, 102 Rothschild Blvd, Tel Aviv
Founded: 1951
Focus: Rheuma; Physiology 07063

Israel Association of Archaeologists, POB 586, Jerusalem
Founded: 1955
Gen Secr: Dr. M.W. Prausnitz
Focus: Archeol 07064

Israel Association of General Practitioners, c/o Dr. W. Mainzer, Kerem Maharal, Hof Ha-Karmel
Founded: 1958
Focus: Med 07065

Israel Association of Plastic Surgeons, c/o Tel-Hashomer Hospital, Ramat-Gan
Founded: 1956
Focus: Surgery 07066

Israel Atomic Energy Commission, 26 Rh. Hauniversita, Ramat Aviv, POB 17120, Tel Aviv 61070
T: (03) 422922; Tx: 33450; Fax: (03) 422974
Founded: 1952
Gen Secr: Dr. S.Y. Ettinger
Focus: Nucl Res 07067

Israel Bar Association, 95 Eben Gvirol St, POB 14152, Tel Aviv 64047
Members: 7500
Pres: Prof. David Libai
Focus: Law
Periodical
Od-Maida (weekly) 07068

Israel Crystallographic Society (ICS), c/o Dept of Chemistry, Israel Institute of Technology, Haifa 32000
T: (04) 293716; Fax: (04) 233735
Founded: 1952; Members: 50
Focus: Mineralogy 07069

Israel Dermatological Society, c/o Dept of Dermatology, Beilinson Medical School, Tel Aviv University Sackler School of Medicine, Petah Tikva
Fax: (03) 9219685
Founded: 1927; Members: 105
Focus: Derm 07070

Israel Exploration Society (I.E.S.), 5 Avida St, POB 7041, Jerusalem 91070
T: (02) 227991; Fax: (02) 247772
Founded: 1913; Members: 5500
Pres: Prof. Avraham Biran; Gen Secr: Joseph Aviram
Focus: Archeol
Periodicals
Eretz-Israel (weekly)
Israel Exploration Journal (weekly)
Qadmoniot (weekly) 07071

Israel Geographical Society (IGS), c/o Dept of Geography, The Hebrew University, Jerusalem
T: (02) 883017
Founded: 1962; Members: 300
Pres: Prof. Dr. Amiram Gonen; Gen Secr: Gabriel Lipshitz
Focus: Geography 07072

Israel Geological Society, POB 1239, Jerusalem
Members: 400
Pres: Dr. Yaacov Arkin; Gen Secr: Dr. Rivka Amit
Focus: Geology
Periodical
Abstracts of Annual Meeting (weekly) . . 07073

Israel Gerontological Society (IGS), POB 11243, Tel Aviv 11243
T: (03) 232725
Founded: 1956; Members: 450
Pres: Dr. Hayim Har-Paz
Focus: Bio; Soc Sci
Periodicals
Gerontology (weekly)
Yedyion Haaguda (weekly) 07074

Israel Heart Society, c/o Dr. S. Rogel, Mayer de Rothschild Hadassah University Hospital, Jerusalem
Founded: 1950
Focus: Cardiol 07075

Israeli Dental Association (I.D.A.), 49 Bar Kochba St, Tel Aviv
Founded: 1920; Members: 1600
Focus: Dent 07076

Israeli Neurological Association, c/o Dept of Neurology, The Chaim Sheba Medical Center, Tel Hashomer
Founded: 1970; Members: 100
Focus: Neurology 07077

The Israel Institute of Productivity (I.I.P.), 4 Henrietta Szold St, POB 33010, Tel Aviv 61330
T: (03) 430231
Founded: 1951; Members: 2500
Gen Secr: Israel Meidan
Focus: Business Admin
Periodical
Ha Mif'al: The Enterprise (weekly) . . . 07078

Israeli Urological Association, c/o Dr. M. Sassoon, 26 Helsinki St, Tel Aviv 62996
Founded: 1935
Focus: Urology 07079

Israel Library Association (ILA), POB 303, Tel Aviv
T: 61002
Founded: 1952
Pres: A. Vilner; Gen Secr: R. Eindelstein
Focus: Libraries & Bk Sci
Periodicals
Meida La Sefran
Yad-La-Kore 07080

Israel Mathematical Union, c/o Dept of Mathematics and Computer Science, Ben-Gurion University, Beer Sheva 84105
T: (057) 461609; Fax: (057) 281340
Founded: 1953; Members: 210
Pres: Prof. M. Cohen; Gen Secr: Prof. D. Berend
Focus: Math
Periodical
Proceedings of the Annual Meeting of the Israel Mathematical Union (weekly) *07081*

Israel Medical Association (I.M.A.), 39 Shaul Hamelech Blvd, POB 33003, Tel Aviv 61330
T: (03) 696639
Founded: 1912; Members: 20000
Pres: Dr. Miriam Zangen
Focus: Med
Periodicals
Harefuha (weekly)
Mikhtav Lekhaver (weekly) *07082*

Israel Music Institute, POB 3004, Tel Aviv 61030
Tx: 341118; Fax: (03) 236926
Founded: 1961
Pres: Paul Landau
Focus: Music *07083*

Israel Oriental Society (I.O.S.), c/o Hebrew University, Jerusalem 91905
T: (02) 883633
Founded: 1949; Members: 2500
Pres: T. Kollek; Gen Secr: Oded Aron
Focus: Ethnology
Periodical
Hamizrah Hehadash: The New East . . *07084*

Israel Pediatric Association, c/o Hasharon Hospital, Petach Tikva 49375
Fax: (03) 9372343
Founded: 1935
Focus: Pediatrics *07085*

Israel Physical Society (IPS), c/o Dept of Physics, Israel Institute of Technology, Haifa 32000
T: (04) 293020
Founded: 1954; Members: 250
Pres: Prof. G. Shaviv; Gen Secr: Dr. D. Levine
Focus: Physics
Periodical
IPS Bulletin (weekly) *07086*

Israel Political Science Association, c/o Dept of Political Science, University of Haifa, Haifa
Founded: 1921
Pres: Prof. Gabriel Bendor; Gen Secr: Y. Zolmarovitch
Focus: Poli Sci *07087*

Israel Prehistoric Society, POB 1502, Jerusalem 91014
Founded: 1958
Gen Secr: Dr. A. Gopher
Focus: Anthro
Periodical
Mitekufat Haeven: Journal of the Israel Prehistoric Society (weekly) *07088*

Israel Radiological Society, POB 8833, Haifa 31087
T: (04) 530880
Founded: 1927; Members: 200
Pres: Prof. A. Rosenberger; Gen Secr: Dr. A. Prober
Focus: Radiology *07089*

Israel Society for Biblical Research, c/o World Jewish Bible Center, 2 Ha Askan, Jerusalem
T: (02) 759152
Pres: S.J. Kreutner
Focus: Rel & Theol
Periodical
Beth Mikra (weekly) *07090*

Israel Society for Experimental Biology and Medicine, c/o Weizmann Institute of Science, Rehovoth
Founded: 1962
Focus: Bio; Med *07091*

Israel Society for Gastroenterology, c/o Israel Medical Association, 1 Heftman St, Tel Aviv
Founded: 1953
Focus: Gastroenter *07092*

Israel Society for Hematology and Blood Transfusion, c/o Dr. A. Eldor, Dept of Hematology, POB 12000, Jerusalem 91120
Founded: 1957; Members: 162
Focus: Hematology *07093*

Israel Society for Microbiology (ISM), POB 12206, Jerusalem 91120
Fax: (02) 346198
Founded: 1932; Members: 600
Pres: Prof. Itzhak Kanane; Gen Secr: Prof. David Merzbach
Focus: Microbio
Periodical
Newsletter (5 times annually) *07094*

Israel Society of Aeronautics and Astronautics (I.S.A.A.), POB 2956, Tel Aviv 61028
Founded: 1951; Members: 400
Focus: Aero
Periodical
Bi'af (weekly) *07095*

Israel Society of Allergology, 23 Balfour St, Tel Aviv
Founded: 1949; Members: 30
Pres: Dr. N. Lass
Focus: Immunology *07096*

Israel Society of Anesthesiologists (I.S.A.), c/o Dr. G. Gurman, Haifa City Medical Center (Rothschild), Haifa
T: 671671
Founded: 1952
Focus: Anesthetics
Periodical
I.S.A. Newsletter (weekly) *07097*

Israel Society of Clinical Neurophysiology (ISCN), POB 9117, Jerusalem 91090
T: (02) 2412251; Fax: (02) 2713086
Founded: 1960
Pres: Dr. Arieh N. Gilai; Gen Secr: Prof. Pablo Solzi
Focus: Med; Physiology; Neurology . . . *07098*

Israel Society of Criminology, POB 1260, Jerusalem
Founded: 1955; Members: 350
Pres: Justice H. Cohen; Gen Secr: Dr. M. Horowitz
Focus: Criminology *07099*

Israel Society of Logic and Philosophy of Science, c/o Hebrew University, Jerusalem
Founded: 1959
Focus: Philos *07100*

Israel Society of Pathologists, c/o Prof. Elena Kessler, Hasharon Hospital, Petan Tiqva
Founded: 1950; Members: 100
Pres: Prof. Elena Kessler
Focus: Pathology *07101*

Israel Society of Soil Mechanics and Foundation Engineering, c/o Association of Engineers and Architects in Israel, 200 Dizengoff Rd, Tel Aviv
Founded: 1948
Focus: Agri *07102*

Israel Society of Soil Science, c/o National and University Institute of Agriculture, Rehovoth
Founded: 1951
Focus: Agri *07103*

Israel Society of Special Libraries and Information Centres (ISLIC), Kyriat Atidim, POB 43074, Tel Aviv 61430
T: (03) 492064
Founded: 1966; Members: 850
Pres: L. Frenkiel
Focus: Libraries & Bk Sci; Computer & Info Sci
Periodical
Bulletin (weekly) *07104*

Israel Veterinary Medical Association, POB 1871, Tel Aviv
Founded: 1922
Focus: Vet Med *07105*

Jerusalem Philosophical Society, c/o Hebrew University, Jerusalem
T: (02) 883747; Fax: (02) 322545
Founded: 1943; Members: 130
Focus: Philos
Periodical
IYYUN: The Jerusalem Philosophical Quarterly (weekly) *07106*

Museums Association of Israel, POB 33288, Tel Aviv 61332
Founded: 1966; Members: 51
Pres: Marc Scheps
Focus: Arts *07107*

National Council for Research and Development (NCRD), POB 18195, Jerusalem 91181
T: (02) 277060; Tx: 26188
Founded: 1950; Members: 40
Gen Secr: Y. Saphir
Focus: Sci
Periodical
Scientific Research in Israel (every 2-3 years) *07108*

The Natural Resources Research Organization, 38 Keren Hayesod St, Jerusalem
Focus: Geology *07109*

Oto-Laryngological Society of Israel, c/o Israel Medical Association, 1 Heftman St, Tel Aviv
Founded: 1925
Focus: Otorhinolaryngology *07110*

Society for the Protection of Nature in Israel (SPNI), 4 Hashfela St, Tel Aviv 66183
T: (03) 335063; Tx: 33488; Fax: (03) 377695
Founded: 1953; Members: 45000
Pres: Yossi Leshem; Gen Secr: Yoav Sagi
Focus: Ecology
Periodicals
Eretz Magazine (weekly)
Pashosh (weekly)
Teva Ve'Aretz (weekly) *07111*

Society of Authors, Composers and Music Publishers in Israel (ACUM), 118 Rothschild Blvd, POB 14220, Tel Aviv 65271
T: (03) 5620115; Tx: 35770; CA: ACUMEMIS, Tel-Aviv; Fax: (03) 5620119
Founded: 1936; Members: 1700
Pres: Shlomo Tanny; Gen Secr: Ran Kedar
Focus: Lit; Music
Periodical
Chadshot ACUM (3 times annually) . . *07112*

Society of Orthopedic Surgeons of the Israel Medical Association, 1 Heftman St, Tel Aviv
Founded: 1934
Focus: Surgery; Orthopedics *07113*

Technion Research and Development Foundation Ltd, Senate House, Technion City, Haifa 32000
Founded: 1952
Gen Secr: Prof. U. Shamir
Focus: Eng; Materials Sci
Periodicals
Hamatechet
In the Field of Building
Research Report *07114*

Tel Aviv Astronomical Association, 13 de Haas St, Tel Aviv
Founded: 1961
Focus: Astronomy *07115*

Verband deutschsprachiger Schriftsteller in Israel, POB 1356, Tel Aviv 61013
Focus: Lit
Periodical
Rundschreiben (5-6 times annually) . . . *07116*

Wilfrid Israel House for Oriental Art and Studies, Kibbutz Hazorea, Post Hazorea 30060
Founded: 1951
Gen Secr: Dr. M. Meron
Focus: Arts; Fine Arts; Archeol *07117*

Yad Izhak Ben-Zvi, POB 7660, Jerusalem
Fax: (02) 638310
Founded: 1964
Focus: Hist; Cultur Hist
Periodicals
Cathedra: History of Palestine Studies (weekly)
Pe'amim: Studies in the Cultural Heritage of Oriental Jewry (weekly) *07118*

Yeshivat Dvar Yerushalayim, 18 Blau, POB 5454, Jerusalem
T: (02) 817647
Founded: 1970
Pres: B. Horovitz
Focus: Rel & Theol; Hist
Periodical
Jewish Studies Magazine (weekly) . . . *07119*

Zoological Society of Israel, c/o Dept of Zoology, Hebrew University, Jerusalem 91904
T: (02) 585971
Founded: 1940; Members: 300
Pres: Prof. A. Borut
Focus: Zoology *07120*

Italy

Academia Belgica, Via Omero 8, 00197 Roma
T: (06) 3201889; Fax: (06) 3208361
Founded: 1939
Pres: Prof. J. Hanesse
Focus: Hist
Periodical
Bulletin de l'Institut Historique Belge de Rome (weekly) *07121*

Academia Cardinalis Bessarionis / Cultus et Lectura Patrum, c/o Convento di S. Giacomo, Lungotevere Farnesina 12, 00165 Roma
T: (06) 655758
Founded: 1975; Members: 14
Focus: Rel & Theol
Periodical
Bessarione: Atti Convegni sul Paleocristiano *07122*

Academia Gentium Pro Pace (A.GEN.P.P.), CP 6326, 00195 Roma
T: (06) 3335449
Founded: 1967; Members: 2750
Pres: Prof. Bruno Marchese Di Magny Rigon; Gen Secr: M.G. Da Feltre
Focus: Rel & Theol; Philos; Physics; Med; Lit; Fine Arts
Periodicals
AGENPP Press: Agenzia di Stampa Quotidiana
Annali dell'Athenaeum
Annuario *07123*

Academia Petrarca di Lettere, Arti e Scienze, Via dell'Orto, 52100 Arezzo
T: (0575) 24700
Founded: 1810; Members: 413
Pres: Prof. Alberto Fatucchi; Gen Secr: Dr. Guido Goti
Focus: Lit; Arts; Sci
Periodicals
Atti e Memorie
Studi Petrarcheschi *07124*

Accademia Agraria, Via Mazza 9, 61100 Pesaro
Founded: 1828
Pres: Prof. Bruno Bruni
Focus: Agri
Periodical
Esercitazioni della Accademia Agraria di Pesaro *07125*

Accademia Albertina di Belle Arti e Liceo Artistico, Via Accademia Albertina 6, 10123 Torino
Founded: 1652
Focus: Arts *07126*

Accademia Ambrosiana Medici Umanisti e Scrittori (A.A.M.U.S.), Viale Lunigiana 5, 20125 Milano
T: (02) 603715
Focus: Humanities *07127*

Accademia Americana, Via Angelo Masina 5, 00153 Roma
T: (06) 588653
Founded: 1894
Pres: Sophie Consagra; Gen Secr: Joseph Connors
Focus: Sci *07128*

Accademia Anatomico-Chirurgica, c/o Biblioteca della Facoltà di Medicina e Chirurgia, Università degli Studi, Policlinico, XIV Settembre 1860, Via Enrico dal Pozzo, CP 72 Succ 3, 06100 Perugia
T: (075) 5733923
Founded: 1802
Focus: Surgery; Anat
Periodical
Annali della Facoltà di Medicina e Chirurgia dell'Università degli Studi di Perugia e Atti dell'Accademia Anatomico-Chirurgica (weekly) *07129*

Accademia Artistica Internazionale Pinocchio d'Oro (A.A.I.P.D.), Via Foria 26, 80137 Napoli
T: (081) 445780
Founded: 1964
Focus: Perf Arts
Periodical
Il Malcontento (weekly) *07130*

Accademia Biella Cultura, c/o Circolo Sociale Biellese, CP 383, 13051 Biella
Focus: Arts
Periodical
Premio Biella Poesia Junior: Biella Poetry Junior Price (weekly) *07131*

Accademia Corale Stefano Tempia, Via del Carmine 28, 10122 Torino
T: (011) 547372
Founded: 1875
Focus: Music *07132*

Accademia Cosentina, Piazza XV Marzo 7, 87100 Cosenza
T: (0984) 25007
Founded: 1507
Focus: Humanities
Periodical
Atti (weekly) *07133*

Accademia Culturale d'Europa, Viale IV Novembre, Villa Silvera, 01030 Bassano Romano
T: 634115
Focus: Humanities
Periodical
Il Torchio Artistico e Letterario (weekly) . *07134*

Accademia Culturale di Rapallo, c/o Ufficio Stampa del Comune, 16035 Rapallo
T: (0185) 50201
Founded: 1978
Focus: Humanities; Adult Educ *07135*

Accademia degli Abruzzi per le Scienze e le Arti (A.A.S.A.), Via Saline 18, 66010 Chieti
T: (0871) 684708
Founded: 1965
Focus: Sci; Adult Educ
Periodical
Quaderni del Sapere Scientifico (weekly) *07136*

Accademia degli Euteleti, Loggiati di San Domenico 3, CP 30, 56027 San Miniato
Founded: 1644
Focus: Lit; Hist; Sci; Arts
Periodical
Bollettino dell'Accademia degli Euteleti (weekly) *07137*

Accademia degli Incamminati, Via del Cappuccini, 47015 Modigliana
T: 91131
Founded: 1650
Focus: Sci; Lit; Arts; Agri; Business Admin *07138*

Accademia degli Incolti, c/o Collegio Nazareno, Largo Nazareno 25, 00187 Roma
T: (06) 6787700
Founded: 1658
Focus: Educ *07139*

Accademia degli Ottimi, Via Bezzecca 1b, 00185 Roma
T: (06) 480251
Focus: Educ *07140*

Accademia degli Sbalzati, Via XX Settembre 46, 52037 Sansepolcro
T: 76909
Founded: 1964
Focus: Arts; Perf Arts *07141*

Accademia dei Filedoni, Piazza Italia 2, 06100 Perugia
Founded: 1816
Focus: Sci *07142*

Accademia dei Filodrammatici, Piazza Paolo Ferrari 6, 20121 Milano
T: (02) 872564
Founded: 1796
Focus: Arts; Perf Arts 07143

Accademia dei Filopatridi (Rubiconia), Piazza Borghesi 11, 47039 Savignano sul Rubicone
T: (0541) 945107
Founded: 1651; Members: 239
Pres: Prof. Lorenzo Cappelli; Gen Secr: Fermo Fellini
Focus: Hist; Lit; Arts; Sci
Periodical
Quaderni della Rubiconia Accademia dei Filopatridi (weekly) 07144

Accademia dei Gelati, Piazza San Rocco, 67038 Scanno
Focus: Hist; Arts 07145

Accademia dei Sepolti, Via Buomparenti 7, 56048 Volterra
T: (0588) 86558
Founded: 1597
Focus: Sci
Periodical
Rassegna Volterrana (weekly) 07146

Accademia della Crusca, Via di Castello 46, 50141 Firenze
T: (055) 454277; Fax: (055) 454279
Founded: 1583; Members: 15
Pres: Prof. Giovanni Nencioni
Focus: Sci; Ling
Periodicals
Studi di Filologia Italiana: Bollettino dell'Accademia della Crusca (weekly)
Studi di Grammatica Italiana (weekly)
Studi di Lessicografia Italiana (weekly) . 07147

Accademia delle Scienze dell'Istituto di Bologna, Via Zamboni 31, 40126 Bologna
T: (051) 222596
Founded: 1711; Members: 260
Pres: Prof. Dario Graffi; Gen Secr: Prof. Silvano Leghissa
Focus: Sci 07148

Accademia delle Scienze di Ferrara, Via Romei 3, 44100 Ferrara
T: (0532) 205209
Founded: 1823; Members: 271
Pres: Prof. Gabriele Battaglia; Gen Secr: Vincenzo Caputo
Focus: Sci
Periodical
Atti dell'Accademia e Supplementi (weekly) 07149

Accademia delle Scienze di Torino, Via Maria Vittoria 3, 10123 Torino
T: (011) 510047
Founded: 1757; Members: 240
Pres: Prof. Silvio Romano
Focus: Sci
Periodicals
Atti dell'Accademia delle Scienze di Torino: Classe di Scienze fisiche (weekly)
Atti dell'Accademia delle Scienze di Torino: Classe di Scienze morali (weekly)
Memorie dell'Accademia delle Scienze di Torino (weekly)
Memorie dell'Accademia delle Scienze di Torino: Classe di Scienze morali (weekly) . . . 07150

Accademia delle Scienze e delle Arti degli Ardenti di Viterbo, Largo Cesare Battisti 2, 01100 Viterbo
T: (0761) 30220
Founded: 1480
Focus: Cultur Hist 07151

Accademia delle Scienze Mediche di Palermo, c/o Clinica Chirurgica B, Via Liborio Giuffrè 5, 90127 Palermo
T: (091) 230808
Founded: 1621
Pres: Prof. P. di Voti; Gen Secr: Prof. Pietro Bazan
Focus: Med
Periodical
Atti (weekly) 07152

Accademia di Agricoltura di Torino, Via Andrea Doria 10, 10123 Torino
T: (011) 8127470; Fax: (011) 8127470
Founded: 1785; Members: 150
Pres: Prof. Pier Luigi Ghisleni; Gen Secr: Dr. Attilio Salsotto
Focus: Agri
Periodical
Annali dell'Accademia di Agricoltura di Torino (weekly) 07153

Accademia di Agricoltura Scienze e Lettere, Palazzo Erbisti, Via Leoncino 6, 37121 Verona
T: (045) 8003668
Focus: Sci; Agri; Ling
Periodical
Atti e Memorie (weekly) 07154

Accademia di Belle Arti, Via Ricasoli 66, 50122 Firenze
Founded: 1801
Focus: Arts 07155

Accademia di Belle Arti, Via Brera 28, 20121 Milano
Founded: 1776
Focus: Arts 07156

Accademia di Belle Arti, Piazza S. Francesco 5, 06100 Perugia
Founded: 1546
Focus: Arts 07157

Accademia di Belle Arti, Via di Roma 13, 48100 Ravenna
T: (0544) 213641; Fax: (0544) 30378
Founded: 1827
Focus: Arts 07158

Accademia di Belle Arti e Liceo Artistico, Via Belle Arti 54, 40126 Bologna
Focus: Arts 07159

Accademia di Belle Arti e Liceo Artistico, Piazza dell' Accademia 1, 54033 Carrara
Focus: Arts 07160

Accademia di Belle Arti e Liceo Artistico, Via Lombardia 7, 73100 Lecce
Focus: Arts 07161

Accademia di Belle Arti e Liceo Artistico, Via Bellini 36, 80135 Napoli
Founded: 1838
Focus: Arts 07162

Accademia di Belle Arti e Liceo Artistico, Via Papireto 20, 90134 Palermo
Focus: Arts 07163

Accademia di Belle Arti e Liceo Artistico, Via Ripetta 222, 00186 Roma
Focus: Arts 07164

Accademia di Belle Arti e Liceo Artistico, Dorsoduro 1050, 30123 Venezia
Founded: 1750
Focus: Arts 07165

Accademia di Costume e di Moda, Libero Istituto di Studi Superiori di Belle Arti, Piazza Farnese 144, 00186 Roma
T: (06) 6568169, 6564132
Founded: 1964
Focus: Graphic & Dec Arts, Design . . 07166

Accademia di Danimarca, Via Omero 18, 00197 Roma
T: (06) 3200951; Fax: (06) 3222717
Founded: 1956
Pres: Prof. Dr. Otto Steen Due; Gen Secr: Dr. Karen Ascani
Focus: Sci
Periodical
Analecta Romana Instituti Danici (weekly) 07167

Accademia di Francia, Viale Trinità dei Monti 1, 00187 Roma
T: (06) 6797142
Founded: 1666
Pres: Jean-Marie Drot; Gen Secr: André Haize
Focus: Sci 07168

Accademia di Medicina di Torino, Via Po 18, 10123 Torino
Founded: 1846
Pres: Prof. Angelo Carbonara; Gen Secr: Prof. Ettore Masenti
Focus: Med
Periodical
Giornale dell'Accademia di Medicina di Torino (weekly) 07169

Accademia di Paestum Eremo Italico, Via Trieste 9, 84085 Mercato San Severino
T: 879191
Founded: 1949
Focus: Lit; Sci; Arts; Archeol; Journalism
Periodical
Fiorisce un Cenacolo (weekly) 07170

Accademia di Relazioni Pubbliche (A.R.P.), Via XX Settembre 26, 00187 Roma
T: (06) 4950350
Founded: 1970
Focus: Comm Sci 07171

Accademia di Romania, Piazza Josè de San Martin 1, Valle Giulia, 00197 Roma
T: (06) 3601898, 3601594
Founded: 1920
Focus: Ethnology; Sci
Periodicals
Bollettino Bibliografico
Notiziario dell'Accademia di Romania . . 07172

Accademia di Scienze, Lettere e Arti, Palazzo del Liceo Classico, 55100 Lucca
Founded: 1819
Focus: Cultur Hist; Hist 07173

Accademia di Scienze, Lettere e Arti, Via Beato Odorico da Pordenone 9, 33100 Udine
Founded: 1606
Pres: Prof. Piercarlo Caracci; Gen Secr: Giuseppe Fornasir
Focus: Lit; Sci; Sociology; Arts
Periodical
Atti dell'Accademia di Scienze, Lettere e Arti (weekly) 07174

Accademia di Scienze, Lettere e Belle Arti degli Zelanti e dei Dafnici, Piazza Duomo 1, 95024 Acireale
T: (095) 604557
Founded: 1934
Focus: Sci; Lit; Arts
Periodical
Memorie e Rendiconti (weekly) 07175

Accademia di Scienze, Lettere ed Arti, Piazza Indipendenza 17, 90129 Palermo
T: (091) 420862
Focus: Arts; Lit; Sci
Periodical
Atti dell'Accademia di Scienze, Lettere ed Arti di Palermo (weekly) 07176

Accademia Economico Agraria dei Georgofili, Loggiato degli Uffizi, 50122 Firenze
T: (055) 213360
Founded: 1753
Focus: Econ; Agri
Periodicals
Agraria dei Georgofili
I Georgofili: Atti della Accademia Economico
Rivista di Storia dell'Agricoltura 07177

Accademia Etrusca, Palazzo Casali, Piazza Signorelli, 52044 Cortona-Arezzo
Pres: Dr. Guglielmo Maetzke
Focus: Sci
Periodicals
Fonti e Testi (weekly)
L'Annuario (weekly)
Note e Documenti (weekly) 07178

Accademia Euro-Afro-Asiatica del Turismo, c/o Espomanifestotur, CP 170, 95100 Catania
T: (095) 7273122; Fax: (095) 7273122
Focus: Travel
Periodical
Euroturismo Progetto 2000 07179

Accademia Europea Dentisti Implantologi (A.E.D.I.), Piazza Bertarelli 4, 20122 Milano
T: (02) 879298
Founded: 1978
Focus: Dent
Periodical
Rivista Europea di Implantologia (weekly) 07180

Accademia Filarmonica di Bologna (Reale), Via F. Guerrazzi 13, 40125 Bologna
T: (051) 222997
Founded: 1666
Pres: Ettore Campogalliani; Gen Secr: Dr. Luciano Lanzarini
Focus: Music 07181

Accademia Filarmonica di Verona, Via dei Mutilati 4l, 37100 Verona
Founded: 1543
Focus: Music 07182

Accademia Filarmonica Romana, Via Flaminia 118, 00196 Roma
T: (06) 3201752; Fax: (06) 3210410
Founded: 1821; Members: 1000
Pres: Adriana Panni
Focus: Music
Periodical
Il Giornale della Filarmonica (weekly) . . 07183

Accademia Fulginia di Arti, Lettere, Scienze, Via Tasso 6, 06034 Foligno
T: (0742) 21469
Founded: 1961
Focus: Arts; Lit; Sci
Periodical
Bollettino Storico della Città di Foligno . 07184

Accademia Georgica, Piazza della Repubblica 9, 62010 Treia
T: 515138
Founded: 1430
Focus: Agri 07185

Accademia Gioenia di Scienze Naturali, Corso Italia 55, 95129 Catania
T: (095) 240601
Founded: 1824; Members: 136
Pres: Prof. A. Arcoria; Gen Secr: Prof. F. Furnari
Focus: Nat Sci
Periodicals
Atti
Bolletino delle Sedute 07186

Accademia Gli Amici Dei Sacri Lari, Terrazza Nunzienza, Viale Enrico Fermi 4, 24100 Bergamo
T: (035) 237458
Founded: 1961
Focus: Rel & Theol; Lit 07187

Accademia Il Tetradramma, Via IV Novembre 152, 00187 Roma
T: (06) 6784964, 6784991
Focus: Arts; Lit; Sci 07188

Accademia Internazionale d'Arte Moderna (A.I.A.M.), Via Giulio Sacchetti 10, 00167 Roma
T: (06) 6373303
Founded: 1975
Focus: Arts; Educ
Periodical
Il Notiziario dell'Accademia Internazionale d'Arte Moderna (weekly) 07189

Accademia Internazionale della Tavola Rotonda, Via Zante 21, 20138 Milano
T: (02) 7490506
Founded: 1957
Focus: Arts; Sci; Hist; Educ
Periodical
La Tavola Rotonda (weekly) 07190

Accademia Internazionale di Medicina Legale e di Medicina Sociale, Viale Regina Elena 336, 00161 Roma
Founded: 1938
Focus: Med; Hygiene
Periodical
Acta Medicinae Legalis et Medicinae Socialis 07191

Accademia Internazionale per le Scienze Economiche, Sociali e Sanitarie (A.I.S.E.S.S.), Via Francesco d'Ovidio 135, 00137 Roma
T: (06) 8280261
Focus: Econ; Soc Sci; Hygiene 07192

Accademia Italiana di Medicina Omeopatica Hahnemanniana (AIMOH), Piazza Navona 49, 00186 Roma
T: (06) 659030
Focus: Homeopathy
Periodical
Rassegna Italiana di Medicina Omeopatica (weekly) 07193

Accademia Italiana di Scienze Forestali, Piazza Edison 11, 50133 Firenze
T: (055) 570348; Fax: (055) 575724
Founded: 1951; Members: 203
Pres: Prof. Fiorenzo Mancini
Focus: Forestry
Periodicals
Annali dell'Accademia Italiana di Scienze Forestali (weekly)
Bollettino della Bibliografia Forestale Italiana (weekly)
L'Italia Forestale e Montana (weekly) . . 07194

Accademia Italiana di Stenografia e di Dattilografia Giuseppe Aliprandi, Via Ricasoli 9, 50122 Firenze
T: (055) 2398641; Fax: (055) 289719
Focus: Educ
Periodical
Specializzazione (weekly) 07195

Accademia Italiana di Storia della Farmacia (A.I.S.F.), Via G. Pardo Roquez 1, 56100 Pisa
T: (050) 27397
Focus: Pharmacol 07196

Accademia Italiana di Studi Filatelici e Numismatici, Piazza Battisti 1, CP 102, 42100 Reggio Emilia
Founded: 1975
Focus: Numismatics 07197

Accademia Lancisiana di Roma, Borgo S. Spirito 3, 00193 Roma
T: (06) 68308539
Founded: 1715
Focus: Humanities
Periodical
Atti dell'Accademia Lancisiana di Roma . 07198

Accademia Letteraria Italiana, Arcadia, c/o Biblioteca Angelica, Piazza S. Agostino 8, 00186 Roma
T: (06) 655874
Founded: 1690
Focus: Lit
Periodicals
Atti e Memorie: Serie Terza
Quaderni dell'Accademia Arcadia 07199

Accademia Ligure di Scienze e Lettere, Piazza G. Matteotti 5, 16123 Genova
T: (010) 565570, 566080; Fax: (010) 566080
Founded: 1890; Members: 180
Pres: Prof. L. Brian; Gen Secr: Prof. A.F. Bellezza
Focus: Sci; Lit
Periodical
Atti della Accademia Ligure di Scienze e Lettere (weekly) 07200

Accademia Ligustica de Belle Arti, Piazza Raffaele de Ferrari 5, 16121 Genova
T: (010) 581957
Focus: Arts; Ling 07201

Accademia Lunigianese di Scienze Giovanni Capellini, Via XX Settembre 148, 19100 La Spezia
Focus: Sci; Geography 07202

Accademia Medica, Via Benedetto XV 6, 16132 Genova
Founded: 1885
Focus: Med 07203

Accademia Medica di Roma, c/o Policlinico Umberto I, Viale del Policlinico, 00161 Roma
T: (06) 4957818
Founded: 1876; Members: 400
Pres: Prof. Giorgio Monticelli; Gen Secr: Prof. L. Travia
Focus: Med
Periodical
Bollettino ed Atti dell'Accademia Medica di Roma 07204

Accademia Medica Lombarda, c/o Ospedale Policlinico, Padiglione Monteggia, Via Francesco Sforza 35, 20122 Milano
T: (02) 598941
Founded: 1912
Pres: Prof. Walter Montorsi
Focus: Med; Bio
Periodical
Atti dell'Accademia Medica Lombarda (weekly) 07205

Accademia Medica Pistoiese Filippo Pacini, c/o Ospedale Civile, Piazza Giovanni XXIII, 51100 Pistoia
T: (0573) 372209
Founded: 1928
Focus: Med; Surgery
Periodical
Bollettino (weekly) 07206

Accademia Medico-Chirurgica del Piceno, c/o Ospedale Civile Umberto 1, 60100 Ancona
Founded: 1932
Focus: Med; Surgery
Periodical
Rassegna Medico-Chirurgica del Piceno . 07207

Accademia Medico-Fisica Fiorentina, c/o Istituto di Radiologia, Policlinico Universitario di Careggi, Viale Morgagni, 50134 Firenze
T: (055) 410084
Focus: Med; Physiology 07208

Accademia Musicale Chigiana, Via di Città 89, 53100 Siena
Founded: 1932
Gen Secr: Guido Turchi
Focus: Music
Periodical
Chigiana: Rivista annuale di studi musicologici (weekly) 07209

Accademia Musicale Ottorino Respighi (A.M.O.R.), Via di Villa Maggiorani 20, 00168 Roma
T: (06) 336261
Founded: 1978
Focus: Music
Periodicals
Atti dei Convegni Musicologici Annuali
Libro dei Programmi Annuali 07210

Accademia Nazionale dei Lincei, Via della Lungara 10, 00165 Roma
T: (06) 6838831; Fax: (06) 6893616
Founded: 1603; Members: 332
Pres: Prof. Giorgio Salvini; Gen Secr: Dr. Cesare F. Golisano
Focus: Sci
Periodicals
Adunanze straordinarie per il conferimento dei premi A. Feltrinelli (weekly)
Annuario (weekly)
Atti dei Convegni Lincei (weekly)
Bollettino dei Classici (weekly)
Celebrazioni lincee (weekly)
Contributi del Centro Linceo Interdisciplinare B. Segre (weekly)
Indici e sussidi bibliografici della Biblioteca (weekly)
Memorie della classe di scienze morali, storiche e filologiche (weekly)
Memorie lincee, matematica e applicazioni (weekly)
Monumenti Antichi (weekly)
Notizie degli scavi di antichità (weekly)
Problemi attuali di scienza e di cultura (weekly)
Rendiconti della classe di scienze morali, storiche e filologiche (weekly)
Rendiconti delle Adunanze Solenni (weekly)
Rendiconti lincei, scienze fisiche e naturali (weekly) 07211

Accademia Nazionale dei Sartori, Largo dei Lombardi 21, 00186 Roma
T: (06) 6794041
Founded: 1947
Focus: Sci
Periodical
Il Maestro Sarto (weekly) 07212

Accademia Nazionale delle Scienze, detta dei XL, Palazzo Civiltá del Lavoro, 00144 Roma
T: (06) 5925557
Founded: 1782; Members: 52
Pres: Prof. G.B. Marini Bettolo; Gen Secr: Prof. Alessandro Ballio
Focus: Nat Sci
Periodicals
Annali
Annuario dell'Accademia Nazionale dei XL
Rendiconti dell'Accademia Nazionale dei XL (weekly)
Studi e Documenti 07213

Accademia Nazionale di Agricoltura (ANA), Via Castiglione 11, 40124 Bologna
T: (051) 268809
Founded: 1807; Members: 220
Pres: Prof. Giuseppe Medici; Gen Secr: Dr. Tullio Romualdi
Focus: Agri
Periodical
Annali (weekly) 07214

Accademia Nazionale di Arte Drammatica Silvio d'Amico, Via Vincenzo Bellini 16, 00198 Roma
T: (06) 853680
Founded: 1935
Focus: Perf Arts 07215

Accademia Nazionale di Belle Arti di Parma, Viale Paolo Toschi 1, 43100 Parma
T: (0521) 22270
Founded: 1752
Focus: Arts 07216

Accademia Nazionale di Danza, Largo Arrigo VII 5, 00153 Roma
Founded: 1948
Focus: Perf Arts 07217

Accademia Nazionale di Entomologia, Via Romana, 50125 Firenze
T: (055) 220531
Focus: Entomology 07218

Accademia Nazionale di Marina Mercantile, Via Garibaldi 4, 16124 Genova
Founded: 1945; Members: 50
Focus: Navig
Periodical
Atti della Accademia Nazionale di Marina Mercantile 07219

Accademia Nazionale di San Luca, Piazza dell'Accademia di San Luca 77, 00187 Roma
T: (06) 6790324; Fax: (06) 6790324
Members: 55
Focus: Arts
Periodical
Annuario dell'Accademia Nazionale di San Luca 07220

Accademia Nazionale di Santa Cecilia, Via Vittoria 6, 00187 Roma
T: (06) 6790389, 6783996; Tx: 614150; CA: Concerti Roma; Fax: (06) 6782796
Founded: 1566; Members: 100
Pres: Francesco Siciliani
Focus: Music
Periodicals
Annuario dell'Accademia Nazionale di Santa Cecilia
Studi Musicali 07221

Accademia Nazionale di Scienze, Lettere e Arti, Corso Vittorio Emanuele II 59, 41100 Modena
T: (059) 225566; Fax: (059) 225566
Founded: 1683; Members: 130
Pres: Prof. Gustavo Vignocchi
Focus: Arts; Lit; Sci
Periodical
Atti e Memorie 07222

Accademia Nazionale Italiana di Entomologia, c/o Istituto Sperimentale per la Zoologia Agraria, Via Romana 15-17, 50125 Firenze
T: (055) 220531
Pres: Prof. Rodolfo Zocchi; Gen Secr: Prof. Antonello Crovetti
Focus: Entomology 07223

Accademia Nazionale Virgiliana di Scienze, Lettere ed Arti di Mantova, Via Accademia 47, 46100 Mantova
T: (0376) 320314; Fax: (0376) 222774
Founded: 1562; Members: 95
Pres: Prof. Claudio Gallico; Gen Secr: Ciro Ferrari
Focus: Arts; Lit; Sci
Periodical
Atti e Memorie (weekly) 07224

Accademia Olimpica, Largo Goethe 3, 36100 Vicenza
T: (0444) 324376, 320396; Fax: (0444) 321875
Founded: 1555
Pres: Prof. Alessandro Faedo; Gen Secr: Virgilio Marzot
Focus: Nat Sci; Econ; Philos; Arts
Periodical
Odeo Olimpico 07225

Accademia Polacca delle Scienze / Polska Akademia Nauk, Palazzo Doria, Vicolo Doria 2, 00187 Roma
T: (06) 6792170; Fax: (06) 6794087
Founded: 1927
Gen Secr: Prof. Krzysztof Zaboklicki
Focus: Sci; Ethnology; Humanities; Soc Sci
Periodical
Conferenze (5 times annually) 07226

Accademia Pomposiana, Via Roma 31, 44021 Codigoro
T: 93484
Founded: 1972
Focus: Arts; Lit
Periodical
Seriarte (weekly) 07227

Accademia Pontaniana, Via Mezzocannone 8, 80134 Napoli
T: (081) 207075
Founded: 1443
Pres: Prof. Giuseppe Martano; Gen Secr: Prof. Guido Guerra
Focus: Sci; Lit; Arts
Periodicals
Atti dell'Accademia Pontaniana
Fonti Aragenesi
Quaderni
Registri della Cancelleria Angioina . . . 07228

Accademia Pratese di Medicina e Scienze, c/o Ospedale Civile, Piazza Ospedale 5, 50047 Prato
T: (0574) 49001
Founded: 1958
Focus: Med; Sci 07229

Accademia Prenestina del Cimento di Musica, Lettere, Scienze, Arti Visive e Figurative, Via Orazio Marucchi 5, 00036 Palestrina
T: (06) 9558935
Founded: 1977
Focus: Music; Lit; Comm Sci; Arts . . . 07230

Accademia Raffaello, Via Raffaello 57, 61029 Urbino
T: (0722) 4735
Founded: 1869
Focus: Cultur Hist; Fine Arts 07231

Accademia Romana di Cultura, Via Vittorio Fiorini 15, 00179 Roma
T: (06) 723543
Founded: 1949
Focus: Arts; Lit; Sci 07232

Accademia Romana di Scienze Mediche e Biologiche, Via IV Novembre 152, 00187 Roma
T: (06) 6784964
Focus: Med; Bio
Periodicals
Messaggio Medico: Il Corriere di Roma (weekly)
Quaderni Scientifico-Tecnici 07233

Accademia Roveretana degli Agiati, Via Camestrini 1, 38068 Rovereto
T: (0464) 436663
Founded: 1750
Pres: Prof. Danilo Vettori; Gen Secr: Dr. Gianfranco Zandonati
Focus: Sci; Lit
Periodical
Atti (weekly) 07234

Accademia Salentina di Lettere ed Arti, Via Idomeneo 77a, 73100 Lecce
T: (0832) 41127
Founded: 1949
Focus: Lit; Arts 07235

Accademia Scientifica, Letteraria, Artistica del Frignano Lo Scoltenna, Palazzo del Credito, Via Cesare Costa 27, 41027 Pievepelago
T: (0536) 71470
Founded: 1902
Pres: Antonio Galli; Gen Secr: Angelo Pasquesi
Focus: Lit; Arts
Periodical
Rassegna Frignanese: Rivista di Cultura e di Studi Regionali (weekly) 07236

Accademia Senese degli Intronati, Palazzo Patrizi Piccolomini, Via di Città 75, 53100 Siena
T: (0577) 284073
Founded: 1528
Focus: Lit; Arts; Hist
Periodical
Bullettino Senese di Storia Patria (weekly) 07237

Accademia Simba, Via XX Settembre 49, 00187 Roma
T: (06) 483572, 4755592
Focus: Econ; Soc Sci; Poli Sci
Periodical
Corriere Africano: Mensile di Relazioni Africa-Europa-Medio Oriente (weekly) 07238

Accademia Spagnola di Belle Arti, Piazza S. Pietro in Montorio 3, 00153 Roma
T: (06) 5816013, 5818607, 582806
Founded: 1873
Focus: Arts 07239

Accademia Spoletina, Palazzo Mauri, Via Brignone, 06049 Spoleto
T: (0743) 24191
Founded: 1477; Members: 205
Pres: Filippo De Marchis; Gen Secr: Prof. Romano Cordella
Focus: Hist; Cultur Hist; Arts
Periodical
Spoletium (weekly) 07240

Accademia Tedesca Villa Massimo / Deutsche Akademie Villa Massimo, Largo di Villa Massimo 1-2, 00161 Roma
T: (06) 44236394; Fax: (06) 44290771
Founded: 1913
Pres: Dr. Jürgen Schilling
Focus: Arts; Lit; Music 07241

Accademia Tiberina, Via del Vantaggio 22, 00186 Roma
Founded: 1813; Members: 2200
Pres: Prof. Dr. Igor Istomin-Duranti; Gen Secr: Silvia Raómini
Focus: Sci 07242

Accademia Toscana di Scienze e Lettere La Colombaria, Via S. Egidio 23, 50122 Firenze
T: (055) 2396628
Founded: 1735
Pres: Prof. Francesco Adorno; Gen Secr: Prof. Francesco Mazzoni
Focus: Sci; Lit
Periodical
Atti e Memorie (weekly) 07243

Accademia Universale Citta' Eterna, Via Vincenzo Brunacci 15, 00146 Roma
T: (06) 5576604, 5577188
Founded: 1970
Focus: Lit; Arts; Sociology
Periodicals
Città Eterna (weekly)
VIP Gran Premio 07244

Accademia Universale Guglielmo Marconi, Via Ugo Fleres 27, 00137 Roma
T: (06) 825848
Founded: 1978
Focus: Arts; Lit; Sci
Periodical
Teleuropa: Mensile di Cultura, Arte e Attualità (weekly) 07245

Accademia Valdarnese del Poggio, Via Poggio Bracciolini 40, 52025 Montevarchi
T: 981227
Founded: 1450
Focus: Geology; Paleontology
Periodical
Memorie Valdarnesi 07246

Agricultural Libraries Network (AGLINET), c/o FAO, Via delle Terme di Caracalla, 00100 Roma
T: (06) 57971; Tx: 610181
Founded: 1971; Members: 27
Focus: Libraries & Bk Sci 07247

Apimondia, Corso Vittorio Emanuele 101, 00186 Roma
T: (06) 6512286; Tx: 612533; Fax: (06) 6548578
Founded: 1949; Members: 77
Gen Secr: Dr. Silvestro Cannamela
Focus: Entomology
Periodicals
Agrindex (weekly)
Apiacta (weekly) 07248

Aquatic Sciences and Fisheries Information System (ASFIS), c/o FAO, Fisheries Dept, Fishery Information, Data and Statistics Service, Via delle Terme di Caracalla, 00100 Roma
T: (06) 57971; Tx: 610181; Fax: (06) 5146172
Founded: 1972; Members: 12
Gen Secr: Dr. Richard Pepe
Focus: Fisheries 07249

Arbeitsgemeinschaft der Kunstbibliotheken, c/o Deutsches Archäologisches Institut Rom, Via Sardegna 79, 00187 Roma
T: (06) 4817812; Fax: (06) 4884973
Pres: Dr. Horst Blanck
Focus: Libraries & Bk Sci 07250

Association of Advisers on Education in International Religious Congregations (EDUC International), c/o International School Sisters of Notre-Dame, Via della Stazione Aurelia 95, 00165 Roma
T: (06) 6808065
Founded: 1967
Gen Secr: Dr. Janet Marie Abbacchi
Focus: Educ 07251

Association of Agricultural Research Institutions in the Near East and North Africa (AARINENA), c/o FAO Regional Office for the Near East, Via delle Terme di Caracalla, 00100 Roma
T: (06) 5793804; Tx: 610181
Founded: 1985; Members: 11
Gen Secr: Dr. A.W. El-Moursi
Focus: Agri 07252

Association of European Operational Research Societies (EURO), c/o Dip di Elettronica, Informatica e Sistemistica, Università di Bologna, Viale Risorgimento 2, 40136 Bologna
T: (051) 6443028
Founded: 1975
Gen Secr: Prof. Paolo Toth
Focus: Computer & Info Sci
Periodicals
EURO Bulletin
European Journal of Operational Research (weekly) 07253

Associazione Alessandro Scarlatti, Piazza dei Martiri 58, 80121 Napoli
T: (081) 406011
Founded: 1919
Focus: Music 07254

Associazione Anestesisti Rianimatori Ospedalieri Italiani (AAROI), Via Massimo Stanzione 15, 80129 Napoli
T: (081) 378724
Founded: 1959; Members: 1400
Focus: Anesthetics 07255

Associazione Archaeologica Romana, Vicolo del Governo Vecchio 8, 00186 Roma
Founded: 1902; Members: 500
Focus: Archeol 07256

Associazione Archeologica Allumiere Adolfo Klitsche de la Grange, c/o Museo Civico, 00051 Allumiere
Founded: 1954
Focus: Archeol
Periodical
Notiziario del Museo Civico (weekly) . . 07257

Associazione Archeologica Centumcellae, Piazza Leandro 5, 00053 Civitavecchia
T: (0766) 500783
Founded: 1911
Pres: Dr. Odoardo Toti; Gen Secr: Francesco Nastasi
Focus: Archeol
Periodical
Bollettino d'Informazioni 07258

Associazione Archeologica Romana, Vicolo del Governo Vecchio 8, 00186 Roma
T: (06) 64565647
Founded: 1902
Focus: Archeol
Periodical
Romana Gens (weekly) 07259

Associazione Archivistica Ecclesiastica, Piazza S. Calisto 14, 00153 Roma
T: (06) 67015228
Pres: Prof. Vincenzo Monachino; Gen Secr: Prof. Emanuele Boaga
Focus: Archives
Periodicals
Archiva Ecclesiae
Notiziario: Organo di Collegamento (weekly) 07260

Associazione Astrofili Bolognesi (A.A.B.), CP 313, 40100 Bologna
T: (051) 517800
Founded: 1967; Members: 250
Pres: Franco Tulipani; Gen Secr: Gianmarco Passerini
Focus: Astronomy
Periodical
Giornale dell' A.A.B. (weekly) 07261

Associazione Bresciana di Ricerche Economiche (A.B.R.E.), c/o Camera di Commercio, Via Luigi Einaudi 23, 25100 Brescia
T: (030) 45061
Founded: 1960
Focus: Econ 07262

Associazione Campana degli Insegnanti di Scienze Naturali (ACISN), c/o Università degli Studi, Via Mezzocannone 8, 80134 Napoli
T: (081) 612849
Founded: 1969
Focus: Nat Sci
Periodical
ACISN: Bollettino Periodico delle Attività . 07263

Associazione Centri di Orientamento Scolastico Professionale e Sociale (COSPES), Piazza Ateneo Salesiano 1, 00139 Roma
T: (06) 8184641
Focus: Educ 07264

Associazione Criogenica Italiana (A.Cr.I.), c/o Istituto di Scienze Fisiche, Facoltà di Scienze, Università degli Studi, Via Dodecaneso 33, 16146 Genova
T: (010) 59931
Founded: 1980
Focus: Eng; Air Cond 07265

Associazione degli Africanisti Italiani, c/o Dipartimento di Studi Politici e Sociali, Sezione di Studi Afro-Asiatici, Via Strada Nuova 65, 27100 Pavia
T: (0382) 37358
Founded: 1967; Members: 70
Focus: Ethnology
Periodical
Bollettino (weekly) 07266

Associazione degli Statistici, Via Roma 12, 33100 Udine
T: (0432) 204198
Founded: 1976
Focus: Stats
Periodical
Statistica (weekly) 07267

Associazione dei Biologi delle Facoltá di Farmacia (A.B.F.), c/o Istituto di Farmacologia, Universitá, Largo E. Meneghetti 2, 35100 Padova
T: (049) 20110, 23857
Founded: 1973
Focus: Pharmacol 07268

Associazione dei Critici Letterari Italiani, Via Bu Meliana 12, 00195 Roma
T: (06) 3565003
Founded: 1969
Focus: Lit 07269

Associazione dei Geografi Italiani (A.GE.I.), c/o Istituto dell'Enciclopedia Italiana, Piazza Paganica 4, 00186 Roma
T: (06) 650881
Founded: 1977
Pres: Prof. Marcello Zunica; Gen Secr: Prof. Pietro Mura
Focus: Geography 07270

Associazione di Cultura Lao Silesu, Piazza Quintino Sella 34, 09016 Iglesias
T: (0781) 41902
Founded: 1963
Focus: Lit; Arts; Folklore
Periodical
La Questione Sarda-Autonomia (weekly) . 07271

Associazione di Cultura Romana Te Roma Sequor, Via Gregoriana 25, 00187 Roma
Founded: 1925
Focus: Hist
Periodical
Te Roma Sequor (weekly) 07272

Associazione Dietetica Italiana (ADI), Via dei Penitenzieri 13, 00193 Roma
Focus: Hematology 07273

Associazione di Ricerca e Interventi Psicosociali e Psicoterapeutici (ARIPS), Via Brescia 6, 25080 Molinetto di Mazzano
T: (030) 2620589, 2791407
Founded: 1978
Focus: Sociology; Psych
Periodical
Notizie ARIPS (weekly) 07274

Associazione Educatrice Italiana (AEI), Via Trinita dei Pelligrini 16, 00186 Roma
T: (06) 6561555
Founded: 1927
Focus: Educ 07275

Associazione Forense Italiana, Palazzo Giustizia, Piazza Cavour, 00193 Roma
Focus: Law 07276

Associazione Forestale Italiana, Via Guido d'Arezzo 16, 00198 Roma
Pres: Dr. Alfonso Froncillo
Focus: Forestry 07277

Associazione Genetica Italiana (AGI), c/o Prof. G.A. Danieli, Dipartimento di Biologia, Via Trieste 75, 35100 Padova
T: (049) 8071978
Founded: 1953; Members: 450
Gen Secr: Prof. G.A. Danieli
Focus: Genetics
Periodical
Atti dell'Associazione Genetica Italiana (weekly) 07278

Associazione Geo-Archeologica Italiana, c/o Instituto di Archeologia Università di Roma, Piazzale Aldo Moro 5, 00185 Roma
T: (06) 857540
Founded: 1968
Focus: Geology; Archeol
Periodical
Geo-Archeologia (weekly) 07279

Associazione Geofisica Italiana (AGI), c/o C.N.R.-I.F.A., Piazzale Luigi Sturzo 31, 00144 Roma
T: (06) 5910941; Fax: (06) 5915790
Founded: 1952; Members: 200
Pres: Marcello Pagliari; Gen Secr: Dr. Mauro Basili
Focus: Geophys
Periodical
Bollettino Geofisico (weekly) 07280

Associazione Geotecnica Italiana, Via Bormida 2, 00198 Roma
T: (06) 8416120; Fax: (06) 8842265
Founded: 1947
Pres: Sandro Martinetti; Gen Secr: Sergio Di Maio
Focus: Mining
Periodicals
Atti dei Convegni Nazionale di Geotecnica
Rivista Italiana di Geotecnica (weekly) . 07281

Associazione Giacomo Boni per la Difesa dei Monumenti di Roma Antica, c/o Direzione del Palatino e del Foro Romano, Piazza S. Maria Nova 53, 00186 Roma
T: (06) 6790333
Founded: 1973
Focus: Preserv Hist Monuments 07282

Associazione Grafologica Italiana (A.G.I.), Scale San Francesco 8, 60100 Ancona
T: (071) 201759
Focus: Graphology
Periodical
Attualità Grafologica (weekly) 07283

Associazione Idrotecnica Italiana (AII), Viale Regina Margherita 239, 00198 Roma
T: (06) 4404493
Founded: 1923
Pres: Umberto Ucelli; Gen Secr: Pasquale Penta
Focus: Water Res
Periodical
Idrotecnica (weekly) 07284

Associazione Internazionale Centro Studi di Storia e Documentazione delle Regioni, Stradone Vescovado 3, 42100 Reggio Emilia
T: (0522) 35621
Focus: Doc; Hist 07285

Associazione Internazionale di Archeologia Classica (AIAC), Piazza S. Marco 49, 00186 Roma
T: (06) 6798798
Founded: 1945; Members: 450
Gen Secr: Prof. Maria F. Squarciapino
Focus: Archeol
Periodicals
Annuario
Fasti Archaeologici (weekly) 07286

Associazione Internazionale di Diritto Nucleare, Via S. Ilaria 2, 00199 Roma
Founded: 1970; Members: 150
Focus: Law 07287

Associazione Internazionale di Poesia, Via Angelo Poliziano 69, 00184 Roma
Focus: Lit
Periodical
Il Giornale dei Poeti 07288

Associazione Internazionale Giuristi Italia-USA, Via Castelfranco Veneto 90, 00191 Roma
T: (06) 3277781
Founded: 1979
Focus: Law 07289

Associazione Internazionale per lo Studio del Diritto Canonico, c/o Istituto di Diritto Canonico, Facoltà di Giurisprudenza, Università degli Studi, Città Universitaria, Piazzale delle Scienze, 00185 Roma
T: (06) 491319
Focus: Rel & Theol 07290

Associazione Italiana Biblioteche (AIB), c/o Biblioteca Nazionale Centrale, Viale Castro Pretorio 105, 00185 Roma
T: (06) 4463532
Founded: 1930; Members: 2000
Pres: Tommaso Giordano; Gen Secr: Luca Bellingeri
Focus: Libraries & Bk Sci
Periodicals
AIB Notizie (weekly)
Bolletino AIB (weekly) 07291

Associazione Italiana Condizionamento dell'Aria, Riscaldamento e Refrigerazione, Via Sardegna 32, 20146 Milano
Founded: 1960
Pres: Mario Costantino
Focus: Air Cond
Periodical
Condizionamento dell'Aria (weekly) . . . 07292

Associazione Italiana degli Insegnanti di Geografia (AIIG), c/o G. Valussi, Via P. Valussi 2, 34141 Trieste
Members: 2080
Focus: Educ; Geography
Periodical
La Geografia nelle Scuole (weekly) . . 07293

Associazione Italiana degli Slavisti (A.I.S.), c/o Istituto di Filologia Slava, Facoltà di Lettere e Filosofia, Università di Pisa, Via del Collegio Ricci 10, 56100 Pisa
T: (050) 28466
Founded: 1971
Pres: Prof. Giuseppe Dell'Agata; Gen Secr: Prof. Giovanna Brogi Bercoff
Focus: Ling; Lit
Periodicals
Europa Orientalis: Studi e Ricerche sulle Culture dell'Est Europeo
Ricerche Slavistiche 07294

Associazione Italiana dei Chimici del Cuoio, c/o Prof. Giuseppe De Simone, Via Salerno 37, 10152 Torino
Focus: Chem
Periodical
Cuoio, Pelli e Materie Concianti 07295

Associazione Italiana dei Giuristi Europei, Via Nomentana 76, 00161 Roma
Focus: Law 07296

Associazione Italiana di Aeronautica e Astronautica (AIDAA), Via Po 50, 00198 Roma
T: (06) 8445894
Members: 494
Focus: Aero
Periodical
L'Aerotecnica-Missili e Spazio (weekly) . 07297

Associazione Italiana di Anestesia Odontostomatologica, CP 1630, 40100 Bologna
T: (051) 247784
Founded: 1971
Pres: Prof. Giovanni Manani; Gen Secr: Dr. Luigi Baldinelli
Focus: Anesthetics; Dent
Periodical
Giornale di Anestesia Stomatologica (weekly) 07298

Associazione Italiana di Anglistica (A.I.A.), Via della Faggiola 7, 56126 Pisa
Focus: Ling
Periodical
Textus (weekly) 07299

Associazione Italiana di Cardiostimolazione (A.I.C.), Viale Gramsci 14, 56125 Pisa
T: (050) 21051; Fax: (050) 21051
Founded: 1978; Members: 508
Pres: Enrico Adornato; Gen Secr: Maria Grazia Bongiorni
Focus: Cardiol
Periodical
Cardiostimolazione 07300

Associazione Italiana di Cartografia (AIC), c/o Istituto di Geografia dell'Università, Largo S. Marcellino 10, 80138 Napoli
Founded: 1963
Focus: Cart
Periodical
Bollettino dell'Associazione Italiana di Cartografia 07301

Associazione Italiana di Chimica Tessile e Coloristica (AICTC), Viale Sarca 223, 20126 Milano
Focus: Chem; Textiles
Periodical
Bollettino dell'Associazione Italiana di Chimica Tessile e Coloristica (weekly) 07302

Associazione Italiana di Cinematografia Scientifica, Via Alfonso Borelli 50, 00161 Roma
T: (06) 490820
Focus: Cinema 07303

Associazione Italiana di Cultura Classica, c/o Istituto di Filologia Classica, Università, Piazza Brunelleschi 4, 50121 Firenze
Focus: Ethnology 07304

Associazione Italiana di Dietetica e Nutrizione Clinica (ADI), Via dei Penitenzieri 13, 00193 Roma
T: (06) 6564096
Members: 155
Focus: Hematology; Food 07305

Associazione Italiana di Diritto Marittimo, Via Po 1, 00198 Roma
Focus: Law 07306

Associazione Italiana di Documentazione e di Informazione (AIDI), Via Vittoria Colonna 39, 00193 Roma
T: (06) 3604841
Founded: 1966
Focus: Doc; Computer & Info Sci . . . 07307

Associazione Italiana di Fisica Sanitaria e di Protezione contro le Radiazioni (AIFSPR), c/o Centro Studi Fisico-Biologici, Corso Polonia 14, 10126 Torino
T: (011) 693358
Focus: Public Health; Ecology 07308

Associazione Italiana di Genio Rurale (AIGR), Via Gradenigo 6, 35131 Padova
T: (049) 8071065; Fax: (049) 8070615
Founded: 1959; Members: 373
Focus: Civil Eng; Agri
Periodical
Rivista di Ingegneria Agraria (weekly) . . 07309

Associazione Italiana di Immuno-Oncologia Clinico-Pratica, c/o Istituti Ospedalieri, Viale Albertoni, 46100 Mantova
T: (0376) 329261
Founded: 1978
Focus: Immunology 07310

Associazione Italiana di Ingegneria Chimica, Corso Venezia 16, 20121 Milano
T: (02) 791175
Founded: 1958
Focus: Eng; Chem
Periodical
Rivista I.C.P. 07311

Associazione Italiana di Ingegneria Medica e Biologica (A.I.I.M.B.), c/o Facolta di Ingegneria, Via Claudio 21, 80125 Napoli
T: (081) 620522; Tx: 710333
Founded: 1966; Members: 160
Focus: Med; Bio; Eng 07312

Associazione Italiana di Medicina Aeronautica e Spaziale (AIMAS), Via Piero Gobetti 2a, 00185 Roma
T: (06) 463538
Founded: 1952
Focus: Med 07313

Associazione Italiana di Medicina dell'Assicurazione Vita, Viale Regina Elena 336, 00161 Roma
Focus: Med
Periodical
Atti del Convegno Nazionale di Medicina dell'Assicurazione Vita 07314

Associazione Italiana di Metallurgia (A.I.M.), Piazzale Rodolfo Morandi 2, 20121 Milano
T: (02) 76020551, 76021132; Tx: 323831; CA: ASSOMETAL Milano; Fax: (02) 784236
Founded: 1946
Pres: Dr. Giuseppe Orlando; Gen Secr: Aurelio Ciccocioppo
Focus: Metallurgy
Periodical
La Metallurgia Italiana (weekly) 07315

Associazione Italiana di Microbiologia Applicata (SIMA), Via Novara 89, 20153 Milano
T: (02) 4047941; Fax: (02) 40090010
Founded: 1975; Members: 250
Pres: Livio Leali; Gen Secr: Roberto Ligugnana
Focus: Microbio 07316

Associazione Italiana di Oncologia Medica (A.I.O.M.), Via G. Venezian 1, 20133 Milano
T: (02) 2664352, 2390676
Founded: 1973
Pres: Prof. Maurizio Tonato; Gen Secr: Dr. Alberto Scanni
Focus: Cell Biol & Cancer Res
Periodicals
Abstracts Annual Meeting
Association News
Update in Medical Oncology 07317

Associazione Italiana di Protezione contro le Radiazione (A.I.R.P.), c/o Enea, Viale G.B. Ercolani 8, 40138 Bologna
T: (051) 498259; Fax: (051) 498131
Founded: 1958
Pres: Dr. Gilberto Busuoli; Gen Secr: Dr. Paolo Vecchia
Focus: Radiology
Periodical
Bollettino A.I.R.P. (weekly) 07318

Associazione Italiana di Psicologia dello Sport (A.I.P.S.), Via Aldighieri 7, 40100 Ferrara
T: (0532) 34401
Founded: 1974
Focus: Psych; Sports 07319

Associazione Italiana di Radiobiologia Medica (AIRBM), c/o Istituto di Radiologia, Università, Piazza S. Mazia di Gesu, 95124 Catania
Founded: 1959; Members: 140
Focus: Bio; Radiology 07320

Associazione Italiana di Radiologia e Medicina Nucleare (A.I.R.M.N.), c/o Istituto di Radiologia, Università degli Studi La Sapienza, Policlinico Umberto Primo, 00161 Roma
Founded: 1959
Pres: Prof. Carissimo Biagini; Gen Secr: Prof. Vincenzo Cavallo
Focus: Radiology; Nucl Med
Periodicals
La Radiologia Medica (weekly)
Il Radiologo (weekly) 07321

Associazione Italiana di Ricerca Operativa (AIRO), c/o ILVA, Via Ilva 3, 16128 Genova
T: (010) 314415
Founded: 1961; Members: 500
Pres: Paolo Toth; Gen Secr: Agnese Bagnasco
Focus: Business Admin
Periodical
Ricerca Operativa (3 times annually) . . 07322

Associazione Italiana di Scienze Politiche e Sociali (AISPS), Viale Bruno Buozzi 105, 00197 Roma
Focus: Poli Sci; Sociology 07323

Associazione Italiana di Sociologia (AIS), Via del Quadraro 14, 00174 Roma
T: (06) 7672209
Founded: 1972
Focus: Sociology 07324

Associazione Italiana di Strumentisti, Via Giulio Carcano 24, 20141 Milano
T: (02) 8435844
Focus: Eng 07325

Associazione Italiana di Studio delle Relazioni Industriali, Via Brescia 29, 00198 Roma
Focus: Business Admin 07326

Associazione Italiana di Studi Semiotici (A.I.S.S.), Strade Comunale di Pecetto 244, 10131 Torino
T: (011) 658052
Focus: Ling 07327

Associazione Italiana di Tecnica Navale (ATENA), Salita S. Caterina 10, 16123 Genova
T: (010) 542254
Founded: 1946
Focus: Navig 07328

Associazione Italiana di Tecnologia Alimentare (A.I.T.A.), c/o Istituto di Tecnologie Alimentari, Via Celoria 2, 20133 Milano
T: (02) 70631971; Fax: (02) 70638625
Founded: 1979
Focus: Nutrition
Periodical
Tecnologie Alimentari: Sistemi per Produrre 07329

Associazione Italiana di Terapia Occupazionale (A.I.T.O.), Via Peralba 9, 00141 Roma
Founded: 1977
Focus: Psychiatry
Periodical
Il Bagatto (weekly) 07330

Associazione Italiana di Terapie Psicologiche (AITP), Corso XXII Marzo 57, 20129 Milano
T: (02) 726489, 7388427
Founded: 1978
Focus: Psychiatry 07331

Associazione Italiana Giuristi Democratici (A.I.G.D.), Viale Carso 51, 00195 Roma
T: (06) 315664
Founded: 1946
Focus: Law
Periodical
Democrazia e Diritto: Trimestrale di Diritto e Giurisprudenza (weekly) 07332

Associazione Italiana per gli Studi di Marketing (AISM), Via Olmetto 3, 20123 Milano
T: (02) 863293
Founded: 1954; Members: 750
Focus: Marketing
Periodical
Giornale di Marketing (weekly) 07333

Associazione Italiana per la Difesa degli Interessi di Diabetici, Via del Scofa 14, 00186 Roma
T: (06) 6543784
Focus: Diabetes 07334

Associazione Italiana per l'Analisi delle Sollecitazioni (A.I.A.S.), c/o Dipartimento di Energetica, Università, Via Valerio 10, 34127 Trieste
T: (040) 6763430; Fax: (040) 568469
Founded: 1971; Members: 400
Pres: Prof. Alessandro Freddi; Gen Secr: Prof. Fulvio Di Marino
Focus: Eng
Periodical
Notiziario A.I.A.S. (weekly) 07335

Associazione Italiana per la Promozione degli Studi e delle Ricerche per l'Edifizia (AIRE) / Italian Association for the Promotion of Building Research and Studies, Via B. Angelico 3, 20133 Milano
T: (02) 716281; Fax: (02) 716295
Founded: 1964; Members: 5
Pres: Enzo Collio; Gen Secr: Angela Astori
Focus: Civil Eng 07336

Associazione Italiana per la Psicologia Umanistica e Transpersonale (I.A.H.P.), Via Adolfo Ravà 61, 00142 Roma
T: (06) 5402291
Focus: Psych 07337

Associazione Italiana per la Qualità (AICQ), Piazza Armando Diaz 2, 20123 Milano
T: (02) 8052285
Founded: 1955; Members: 2000
Focus: Materials Sci
Periodicals
Lettera Circolare Qualità
Qualità (weekly) 07338

Associazione Italiana per la Ricerca Industriale (A.I.R.I.), Viale Gorizia 25, 00198 Roma
T: (06) 8848871; Fax: (06) 8552949
Founded: 1974
Focus: Eng; Econ
Periodical
Notizie A.I.R.I. (weekly) 07339

Associazione Italiana per la Ricerca nell'Impiego degli Elastomeri (AIRIEL), Via San Vittore 36, 20123 Milano
T: (02) 4988168; Fax: (02) 435432
Founded: 1968; Members: 44
Pres: Arrigo Pini; Gen Secr: Ettore Lauretti
Focus: Materials Sci
Periodical
Notiziario dell'AIRIEL L'Industria della Gomma/Elastica (weekly) 07340

Associazione Italiana per L'Educazione Sanitaria (A.I.E.S.), c/o Centro Sperimentale per l'Educazione Sanitaria, Via del Giochetto, 06100 Perugia
T: (075) 28377
Founded: 1967
Focus: Hygiene
Periodical
Notiziario in La Salute Umana (weekly) . 07341

Associazione Italiana per le Ricerche di Storia del Cinema, Via Tuscolana 1522, 00173 Roma
Founded: 1964
Focus: Cinema
Periodical
Bollettino dell'Associazione (weekly) . . . 07342

Associazione Italiana per l'Informatica et il Calcolo Automatico, Piazzale Rodolfo Morandi 2, 20121 Milano
T: (02) 784970, 874607; Fax: (02) 784236
Focus: Computer & Info Sci
Periodical
Informatica (weekly) 07343

Associazione Italiana per lo Studio del Dolore (A.I.S.D.), c/o Clinica Medica, Università degli Studi, Viale G.B. Morgagni 85, 50134 Firenze
T: (055) 412063
Founded: 1976
Focus: Pathology
Periodical
Bollettino dell'Associazione Italiana per lo Studio del Dolore (weekly) 07344

Associazione Italiana per lo Studio della Psicologia Analitica (A.I.P.A.), Via Cola di Rienzo 28, 00192 Roma
T: (06) 388210
Founded: 1961
Focus: Psych; Psychoan
Periodical
Rivista di Psicologia Analitica 07345

Associazione Italiana per lo Sviluppo Internazionale (AISI), Via Luigi Lilio 19, 00143 Roma
T: (06) 5146089
Founded: 1977
Focus: Int'l Relat
Periodicals
Note Informative AISI (weekly)
Quaderni AISI (weekly) 07346

Associazione Italiana Santa Cecilia per la Musica Sacra (AISC), Piazza S. Calisto 16, 00153 Roma
Founded: 1877; Members: 3000
Focus: Music
Periodical
Bollettino Ceciliano (weekly) 07347

Associazione Italiana Scientifica di Metapsichica (A.I.S.M.), Via S. Vittore 19, 20123 Milano
Focus: Parapsych
Periodical
Metapsichica (weekly) 07348

Associazione Italiana Sclerosi Multipla (AISM), Via della Magliana 279, 00146 Roma
T: (06) 5267923
Focus: Pathology
Periodical
Notiziario AISM (weekly) 07349

Associazione Italiana Socioanalisi Individuale (A.I.S.I.), Via del Don 6, 20123 Milano
T: (02) 8358328
Founded: 1976
Pres: Dr. Giorgio Castelletti
Focus: Psych; Sociology 07350

Associazione Italiana Studi Americanisti (A.I.S.A.), Corso Solferino 29, 16122 Genova
T: (010) 814737
Founded: 1964
Focus: Ethnology; Hist; Archeol; Anthro
Periodical
Terra America (weekly) 07351

Associazione Italiana Tecnico-Economica del Cemento (AITEC), Via di S. Teresa 23, 00193 Roma
T: (06) 8548505; Fax: (06) 8416176
Founded: 1960; Members: 45
Focus: Civil Eng
Periodicals
Cemento (weekly)
L'Industria Italiana del Cemento (weekly) 07352

Associazione Italiana tra Foniatri e Logopedisti, c/o Centro Medico di Foniatria, Via Bergamo 10, 35100 Padova
Focus: Logopedy 07353

Associazione La Nostra Famiglia, 22037 Ponte Lambro
T: (031) 625111; Fax: (031) 625275
Founded: 1948
Pres: Zaira Spreafico
Focus: Educ Handic
Periodicals
Notiziario d'Informazione del Gruppo Amici de La Nostra Famiglia
Saggi: Rivista di Neuropsicologia Infantile, Psicopedagogia, Riabilitazione 07354

Associazione Medica Italiana di Idroclimatologia, Talassologia e Terapia Fisica, Via Roverto 11, 00198 Roma
T: (06) 863505
Founded: 1896
Focus: Physical Therapy; Therapeutics
Periodical
La Clinica Termale (weekly) 07355

Associazione Medica Italiana per lo Studio della Ipnosi (A.M.I.S.I.), Via Paisiello 28, 20131 Milano
T: (02) 2365493; Fax: (02) 2365493
Founded: 1959
Focus: Psychoan
Periodical
Rivista Italiana di Ipnosi Clinica e Sperimentale 07356

Associazione Medici Dentisti Italiani (AMDI), Via Savoia 78, 00198 Roma
T: (06) 8540535; Fax: (06) 8414133
Members: 4000
Focus: Dent; Stomatology
Periodicals
Bollettino AMDI: Annuario degli Atti Ufficiali dell'Associazione (weekly)
Fronte Stomatologico (weekly)
Rivista Italiana di Stomatologia (weekly) . 07357

Associazione Micologica ed Ecologica Romana (A.M.E.R.), c/o Cattedra di Micologia Orto Botanico, Università degli Studi, Largo Cristina di Svezia 24, 00165 Roma
Founded: 1973
Focus: Ecology; Botany, Specific 07358

Associazione Nazionale Archivistica Italiana (ANAI), Via di Ponziano 15, 00152 Roma
T: (06) 585067
Founded: 1948
Focus: Archives
Periodical
Archivi e Cultura (weekly) 07359

Associazione Nazionale Assistenti Sociali (AssNAS), c/o CISS, Via Duilio, 00192 Roma
T: (06) 318696
Focus: Sociology 07360

Associazione Nazionale degli Urbanisti (ASSURBANISTI), CP 348, 31100 Treviso
Founded: 1977; Members: 250
Focus: Urban Plan
Periodical
Il Giornale degli Urbanisti: Organo Ufficiale (weekly) 07361

Associazione Nazionale dei Musei di Enti Locali e Istituzionale, c/o Museo Civico, Piazza del Santo 10, 35100 Padova
T: (049) 23106
Focus: Music 07362

Associazione Nazionale dei Musei Italiani, Piazza S. Marco 49, 00186 Roma
T: (06) 6791343
Founded: 1955; Members: 615
Focus: Arts
Periodical
Musei e Gallerie d'Italia 07363

Associazione Nazionale dei Periti Grafici a Base Psicologica (ANPGP), Corso XXII Marzo 57, 20129 Milano
T: (02) 7388427; Fax: (02) 7491051
Founded: 1969; Members: 98
Pres: Prof. Rolando Marchesan; Gen Secr: Prof. Leila Mazzi Biraschi
Focus: Psych; Graphology 07364

Associazione Nazionale del Libero Pensiero Giordano Bruno, Casella Postale 6089, Roma-Prati
T: (06) 3274807
Founded: 1903
Focus: Philos
Periodical
La Ragione (weekly) 07365

Associazione Nazionale di Ingegneria Nucleare, Piazza Sallustio 24, 00187 Roma
T: (06) 486415
Focus: Nucl Res 07366

Associazione Nazionale di Ingegneria Sanitaria (ANDIS), Piazza Sallustio 24, 00187 Roma
T: (06) 487397
Founded: 1953
Focus: Eng
Periodical
Ingegneria Sanitaria (weekly) 07367

Associazione Nazionale di Meccanica (ASMECCANICA), c/o F.A.S.T., Piazzale Rodolfo Morandi 2, 20121 Milano
T: (02) 784991
Founded: 1970
Focus: Eng
Periodical
La Meccanica Italiana (weekly) 07368

Associazione Nazionale Disegno di Macchine (A.D.M.), Viale delle Scienze, 90128 Palermo
T: (091) 427238
Founded: 1974
Focus: Mach Eng
Periodical
Disegno di Macchine 07369

Associazione Nazionale Esercenti Teatri (ANET), Via Villa Patrizi 10, 00161 Roma
Focus: Perf Arts 07370

Associazione Nazionale Famiglie di Fanciulli e Adulti Subnormali (ANFFAS), Borgo Sant'Angelo 19, 00193 Roma
T: (06) 6547454
Focus: Educ Handic 07371

Associazione Nazionale Filosofia Arti Scienze (FAS), Via Oberdan 15, 40126 Bologna
T: (051) 232506
Founded: 1958
Focus: Philos; Sci
Periodical
Quaderni FAS 07372

Associazione Nazionale Industria Meccanica Varia ed Affine (A.N.I.M.A.), Piazza Armando Diaz 2, 20123 Milano
T: (02) 809006
Founded: 1945
Focus: Eng
Periodical
L'Industria Meccanica (weekly) 07373

Associazione Nazionale Ingegneri ed Architetti Italiani (ANIAI), Piazza Sallustio 24, 00187 Roma
T: (06) 486415, 4744397
Focus: Archit; Eng
Periodicals
Bollettino di Informazione ANIAI (weekly)
Il Giornale dell'Ingegnere (weekly)
L'Ingegnere: Rivista Tecnica di Ingegneria e di Architettura (weekly)
Ingegneria Nucleare (weekly)
Ingegneria Sanitaria (weekly)
Quaderni ANIAI
Tecnica Ospedaliera (weekly) 07374

Associazione Nazionale Ingegneri Minerari (A.N.I.M.), Piazza Sallustio 24, 00187 Roma
T: (06) 4744397
Founded: 1962
Focus: Mineralogy 07375

Associazione Nazionale Insegnanti di Disegno (ANID), c/o Prof. Luigi Varone, Salita Arenella 13a, 80129 Napoli
T: (081) 378927
Focus: Adult Educ 07376

Associazione Nazionale Italiana Esperti Scientifici del Turismo (ANIEST), Via C. Federici 2, 00147 Roma
T: (06) 5134973
Founded: 1964; Members: 125
Focus: Econ
Periodical
Rassegna di Studi Turistici (weekly) . . 07377

Associazione Nazionale Italiana per l'Automazione (ANIPLA), Piazzale Rodolfo Morandi 2, 20121 Milano
T: (02) 702311
Members: 800
Focus: Eng 07378

Associazione Nazionale per i Centri Storico-Artistici, Palazzo dei Consoli, 06024 Gubbio
Founded: 1961
Focus: Cultur Hist
Periodical
Bollettino dell'Associazione (weekly) . . . 07379

Associazione Nazionale per il Progresso della Scuola Italiana, Palazzo Odescalchi, Piazza SS. Apostoli 80, 00186 Roma
T: (06) 6780004
Founded: 1970
Focus: Educ
Periodical
Politica della Scuola (weekly) 07380

Associazione Nazionale per la Scuola Italiana (ANSI), Via Paolo Emilio 57, 00192 Roma
T: (06) 383632
Founded: 1945
Focus: Educ
Periodical
Rinnovare la Scuola (weekly) 07381

Associazione Nazionale per lo Studio dei Problemi del Credito, Via Lisbona 9, 00198 Roma
T: (06) 863036
Focus: Finance 07382

Associazione Nazionale Professori Universitari di Ruolo (ANPUR), c/o Prof. Dr. V. Castellano, Via Ippolerate 79, 00161 Roma
Focus: Adult Educ 07383

Associazione Otologica Ospedaliera Italiano (AOOI), c/o Prof. Lucio Coppo, Via Boezio 14, 00192 Roma
T: (06) 353264
Founded: 1947; Members: 1126
Focus: Otorhinolaryngology
Periodical
Acta Otorhinolaryngologica Italica (weekly) 07384

Associazione Ottica Italiana (AOI), Largo Enrico Fermi 1, 50125 Firenze
T: (055) 221163
Founded: 1926
Focus: Optics
Periodical
Luce e Immagini (weekly) 07385

Associazione Pedagogica Italiana, c/o S.S. Macchietti, Via Cassia Aurelia II 32, 53043 Chiusi Città
T: (0578) 20451; Fax: (0578) 20248
Founded: 1951; Members: 40
Gen Secr: Prof. Aldo D'Alfonso
Focus: Educ
Periodicals
Bollettino della As.Pe.I.
Prospettiva E.P. (weekly) 07386

Associazione per il Diabete Infantile e Giovanile (A.D.I.G.), c/o Servizio di Diabetologia, Istituto di Clinica Pediatrica, Università degli Studi, Policlinico Umberto 1, Viale Regina Elena 324, 00161 Roma
T: (06) 490962
Founded: 1976
Focus: Diabetes; Pediatrics
Periodical
Scritti in Favore dei Bambini Diabetici (weekly)
. . . . 07387

Associazione per Imola Storico-Artistica (I.S.A.), Via Emilia 125, 40026 Imola
T: (0542) 20908; Fax: (0542) 55082
Founded: 1938
Pres: Dr. Gian Franco Fontana
Focus: Preserv Hist Monuments; Arts; Hist
. . . . 07388

Associazione per la Conservazione delle Tradizioni Popolari, Palazzo Fatta, Piazza Marina 19, 90133 Palermo
T: (091) 328060
Founded: 1965
Focus: Ethnology; Folklore; Perf Arts
Periodical
Studi e Materiali per la Storia della Cultura Popolare (weekly) 07389

Associazione per l'Agricoltura Biodinamica, Via Privata Vasto 4, 20121 Milano
T: (02) 29002544; Fax: (02) 29002544
Founded: 1947
Focus: Agri
Periodical
Terra Biodinamica (weekly) 07390

Associazione per le Previsioni Econometriche (PROMETEIA), Via Santa 12, 40125 Bologna
T: (051) 226789
Focus: Econ 07391

Associazione per lo Studio del Problema Mondiale dei Rifugiati, Sezione Italiana / Association for the Study of the World Refugee Problem, A.W.R., Piazzale di Porta Pia 121, 00198 Roma
T: (06) 8445514
Founded: 1960
Focus: Poli Sci; Sociology 07392

Associazione per lo Sviluppo degli Studi di Banca e Borsa (ASSBB), c/o Banca Popolare Commercio e Industria, Via Moscova 33, 20121 Milano
Founded: 1973
Focus: Finance 07393

Associazione per lo Sviluppo delle Scienze Religiose in Italia, Via San Vitale 114, 40125 Bologna
T: (051) 239532; Fax: (051) 230658
Pres: Dr. Enzo Bianchi; Gen Secr: Dr. Massimo Toschi
Focus: Rel & Theol; Hist 07394

Associazione per lo Sviluppo dell'Istruzione e della Formazione Professionale (A.S.I.P.), Via Federico Cesi 30, 00193 Roma
T: (06) 310429
Founded: 1971
Focus: Develop Areas
Periodical
Sviluppo, Formazione, Economia, Cooperazione Internazionale (weekly) 07395

Associazione per lo Sviluppo di Studi e Ricerche nell'Industria Tessile Laniera Oreste Rivetti, Piazza Lamarmora 5, 13051 Biella
T: (015) 20490, 21655
Focus: Textiles 07396

Associazione Piemontese di Studi Filosofici, Via Torino 55, 13051 Biella
T: (015) 22756
Focus: Philos 07397

Associazione Professionale Italiana Medici Oculisti (APIMO), Via degli Scialoia 18, 00196 Roma
T: (06) 3610333
Focus: Ophthal 07398

Associazione Psicanalitica Italiana, Via Crivelli 15, 20122 Milano
T: (02) 588804
Founded: 1976
Focus: Psychoan
Periodicals
Clinica: Rivista Internazionale di Psichiatria
Vel: Collana Periodica di Psicanalisi (weekly)
. . . . 07399

Associazione Romana di Entomologia (ARDE), c/o Museo Civico di Zoologia, Via Ulisse Aldrovandi 18, 00197 Roma
T: (06) 3216586
Founded: 1945; Members: 340
Pres: Prof. Mario Pinzari; Gen Secr: Dr. Emanuele Piattella
Focus: Entomology
Periodical
Bollettino dell'Associazione Romana di Entomologia (weekly) 07400

Associazione Scientifica di Produzione Animale (ASPA), c/o Istituto Zootecnico Università, Via F. Delpino 1, 80137 Napoli
T: (081) 441273
Founded: 1973; Members: 130
Focus: Zoology
Periodical
Zootecnica e Nutrizione Animale (weekly) 07401

Associazione Siciliana per le Lettere e le Arti (A.S.L.A.), Via XX Settembre 68, 90141 Palermo
T: (091) 211410
Focus: Lit; Arts 07402

Associazione Sociologi Lucani, Via Ammiraglio Ruggiero 71, 85044 Lauria Inferiore
Focus: Sociology 07403

Associazione Sole Italico, Corso Ovidio 191, 67039 Sulmona
T: 51606
Founded: 1971
Focus: Ethnology; Arts
Periodical
Bollettino Informativo (weekly) 07404

Associazione Studi sull'Informazione (A.S.I.), Viale Ferdinando di Savoia 5, 20124 Milano
Founded: 1975
Focus: Comm
Periodical
Problemi dell'Informazione (weekly) . . . 07405

Associazione Tecnica Italiana per la Cellulosa e la Carta (ATICLCA), Via Sandro Botticelli 19, 20133 Milano
T: (02) 2664141; Fax: (02) 2664141
Founded: 1967; Members: 1160
Focus: Eng
Periodical
Industria della Carta (weekly) 07406

Associazione Tecnica Italiana per la Cinematografia (A.T.I.C.), Viale Regina Margherita 286, 00198 Roma
T: (06) 44231480; Fax: (06) 4404128
Founded: 1947
Focus: Cinema
Periodicals
Note di Tecnica Cinematografica (weekly)
Notiziario (weekly) 07407

Associazione Teologica Italiana per lo Studio della Morale (A.T.I.S.M.), c/o Seminario Vescovile, Via Monte S. Gabriele 60, 28100 Novara
T: (0321) 21783
Founded: 1965
Focus: Philos
Periodical
Rivista di Teologia Morale 07408

Associazione Termotecnica Italiana (ATI), c/o Istituto di Fisica Tecnica e Impianti Nucleari, Facoltà di Ingegneria, Politecnico, Corso Duca degli Abruzzi 24, 10129 Torino
T: (011) 537353, 553636
Founded: 1947
Focus: Air Cond
Periodicals
Annuario ATI
Atti dei Congressi Nazionali
La Termotecnica (weekly) 07409

Associazione Urania / Associazione Ligure per lo Studio e la Divulgazione dell'Astronomia e dell'Astronautica, c/o Museo Civico di Storia Naturale Giacomo Doria, Via Brigata Liguria 9, 16121 Genova
T: (010) 564567
Founded: 1951; Members: 103
Pres: Adriano Cravero; Gen Secr: Diego Guido Torrisi
Focus: Astronomy 07410

Automazione Energia Informazione (AEI), Viale Monza 259, 20126 Milano
T: (02) 257791; Fax: (02) 2570512
Founded: 1896; Members: 11600
Pres: Salvatore Randi; Gen Secr: Giacinto Spegiorin
Focus: Electronic Eng
Periodicals
Alta Frequenza: Rivista di Elettronica (weekly)
L'Elettrotecnica (weekly)
L'Energia Elettrica (weekly)
ETT-European Transactions on Telecommunications and Related Technologies (weekly) . . . 07411

Bioelectrochemical Society (BES), c/o Dipartimento di Biologia, Via Imerio 42, 40126 Bologna
T: (051) 351293; Fax: (051) 242576
Gen Secr: Prof. Dr. B.A. Melandri
Focus: Biochem 07412

Centre for Coordination of Research of the International Federation of Catholic Universities, Piazza della Pilotta 4, 00187 Roma
T: (06) 6786253; Fax: (06) 6786253
Founded: 1975
Pres: A.A. Roest Crollius
Focus: Educ 07413

Centre for Human Evolution Studies, Via Antonio Bertoloni 28b, 00197 Roma
T: (06) 873420; Fax: (06) 808594
Founded: 1978
Gen Secr: Prof. Francesco Cigliano
Focus: Bio
Periodicals
The Brain and the Integration of Sciences (weekly)
Cultura e Natura (weekly) 07414

Centro Appenninico del Terminillo Carlo Jucci (C.A.T.), Via Comunali 43, 02100 Rieti
T: (0746) 40291
Founded: 1949
Focus: Food 07415

Centro Bibliografica Francescano, c/o Convento Le Grazie, Viale San Lorenzo, 82100 Benevento
T: (0824) 21678
Founded: 1958
Focus: Libraries & Bk Sci
Periodical
Voce Francescana (weekly) 07416

Centro Camuno di Studi Preistorici ed Etnologici (CCSP), 25044 Capo di Ponte
T: (0364) 42091; Fax: (0364) 42572
Founded: 1964
Pres: Claudio Beretta; Gen Secr: Emmanuel Anati
Focus: Ethnology; Arts; Archeol; Anthro; Hist
Periodicals
B.C. Notizie
Bollettino del Centro Camuno di Studi Preistorici
. . . . 07417

Centro d'Arte e di Cultura, Via Castiglione 8, 40124 Bologna
T: (051) 234044
Pres: Federico Masè Dari; Gen Secr: Guglielmo Franchi
Focus: Arts; Lit 07418

Centro di Demodossalogia / Centro Studi e Indagini sull'Opinione Pubblica, Via R. Zampieri 49, 00159 Roma
T: (06) 4390253
Founded: 1939
Focus: Psych; Journalism; Travel
Periodical
Il Mezzo: Notiziario 07419

Centro di Documentazione, Via degli Orafi 29, CP 347, 51100 Pistoia
T: (0573) 367144
Founded: 1977
Focus: Doc
Periodicals
Fogli di Informazione (weekly)
Notiziario del Centro (weekly) 07420

Centro di Documentazione Economica per Giornalisti, Via Cicerone 28, 00193 Roma
T: (06) 3605994, 674132
Founded: 1964
Focus: Econ; Stats; Doc 07421

Centro di Documentazione e di Iniziativa Politica (C.I.D.I.P.), Piazza SS. Apostoli 80, 00186 Roma
T: (06) 6780004
Focus: Poli Sci 07422

Centro di Documentazione e Promozione Archeologica, Arco de Banchi 8, 00186 Roma
T: (06) 655838
Founded: 1970
Focus: Archeol
Periodical
Agenda dell'Archeologia Italiana 07423

Centro di Documentazione Giornalistica, Piazza di Pietra 26, 00186 Roma
T: (06) 6791496; Fax: (06) 6797492
Pres: Dr. Achille Cardini; Gen Secr: Dr. Marcella Cardini
Focus: Journalism; Doc
Periodical
L'Agenda del Giornalista 07424

Centro di Documentazione Statistica Internazionale, c/o Istituto di Statistica, Facoltà di Economia e Commercio, Università degli Studi, Via Belle Arti 41, 40126 Bologna
T: (051) 265945, 225892
Founded: 1970
Focus: Stats
Periodical
Statistica 07425

Centro di Documentazione Storica per l'Alto Adige (C.D.S.-A.A.), c/o Dr. Ferruccio Bravi, Piazza Mazzini 20, 39100 Bolzano
Founded: 1968
Pres: Dr. Ferruccio Bravi
Focus: Hist 07426

Centro di Documentazione, Studi e Ricerche Jacques Maritain, Via Pagliarani 4, 47037 Rimini
T: (0541) 50194
Founded: 1978
Focus: Hist
Periodicals
Discorsi e Immagini
Ricerca e Presenza: Mensile di Notizie ed Informazione Bibliografica (weekly) . . . 07427

Centro di Formazione e Studi per il Mezzogiorno (FORMEZ), Via Salaria 229, 00199 Roma
T: (06) 841101
Founded: 1965
Focus: Econ 07428

Centro di Informazioni Sterilizzazione e Aborto (C.I.S.A.), Corso di Porta Vigentina 15a, 20122 Milano
T: (02) 581203
Founded: 1973
Focus: Law; Med; Family Plan 07429

Centro di Musicologia Walter Stauffer, c/o Palazzo Raimondi, Corso Garibaldi 20, 26100 Cremona
Focus: Music 07430

Centro di Ricerca Applicata e Documentazione (C.R.A.D.), Via Pradamano 2a, 33100 Udine
T: (0432) 520543; Fax: (0432) 522755
Founded: 1971
Focus: Ecology; Radiology
Periodical
Technical Reports (weekly) 07431

Centro di Ricerca e di Studio sul Movimento dei Disciplinati, c/o Oratorio di S. Agostino, Piazza Domenico Lupattelli 1c, 06100 Perugia
T: (075) 64458
Founded: 1963
Focus: Rel & Theol; Hist
Periodical
Quaderni del Centro 07432

Centro di Ricerca Pergamene Medievali e Protocolli Notarili, Via di Ponziano 15, 00162 Roma
T: (06) 585067
Founded: 1965
Gen Secr: Prof. Maria Luisa Lombardo
Focus: Archives; Hist
Periodical
Archivi e Cultura: Organo Culturale dell Centro di Ricerca Pergamene Medievali e Protocolli Notarili (weekly) 07433

Centro di Ricerca per il Teatro (C.R.T.), Via Ulisse Dini 7, 20142 Milano
T: (02) 8439878; Fax: (02) 8466592
Founded: 1974
Pres: Prof. Sisto Dalla Palma; Gen Secr: Dr. Giorgio Zorcù
Focus: Perf Arts 07434

Centro di Ricerche Biopsichiche di Padova (C.R.B.), Via Dante 60, 35100 Padova
T: (049) 657996
Founded: 1957
Pres: Prof. Conte Giorgio Foresti; Gen Secr: Dr. Carla Berlanda
Focus: Psych; Therapeutics
Periodical
Centro Ricerche Biopsichiche 07435

Centro di Studi Aziendali e Amministrativi, Via XI Febbraio 78, 26100 Cremona
Founded: 1979
Focus: Econ; Law 07436

Centro di Studi Bonaventuriani, Viale Fratelli Agosti, 01022 Bagnoregio
Founded: 1953
Pres: Prof. Pietro Prini; Gen Secr: Anna Petrangeli Papini
Focus: Philos; Rel & Theol
Periodical
Bollettino del Centro di Studi Bonaventuriani, Doctor Seraphicus 07437

Centro di Studi Chimico-Fisici di Macromolecole Sintetiche e Naturali, c/o Istituto di Chimica Industriale, Università di Genova, Corso Europa 30, 16132 Genova
T: (010) 510262
Focus: Chem 07438

Centro di Studi e Applicazioni di Organizzazione Aziendale della Produzione e dei Trasporti (C.S.A.O.), c/o Istituto Trasporti e Organizzazione Industriale, Politecnico, Corso Duca degli Abruzzi 24, 10129 Torino
T: (011) 512763
Founded: 1960
Focus: Business Admin; Transport
Periodical
Sicurezza Notizie (weekly) 07439

Centro di Studi e di Ricerche per la Psicoterapia della Coppia e della Famiglia, Viale Regina Margherita 37, 00198 Roma
T: (06) 852006, 852130
Focus: Psychiatry; Family Plan 07440

Centro di Studi e Documentazione delle Ricerche sulla Didattica dell'Educazione Fisica e dello Sport, Mostra Oltremare, 80100 Napoli
T: (081) 610264
Focus: Sports 07441

Centro di Studi e Ricerche di Museologia Agraria, c/o Museo Lombardo di Storia dell'Agricoltura, CP 908, 20101 Milano
Founded: 1975
Pres: Prof. Elio Baldacci
Focus: Cultur Hist; Agri
Periodical
Acta Museorum Italicorum Agricultare: Rivista di Storia dell'Agricoltura (weekly) 07442

Centro di Studi e Ricerche sulla Nutrizione e Sugli Alimenti, c/o Università degli Studi, Via del Taglio, Zona Cornocchio, 43100 Parma
T: (0521) 96224
Founded: 1971
Focus: Nutrition 07443

Centro di Studi Etruschi, Via Maitiani 6, 05018 Orvieto
T: 5216
Focus: Hist 07444

Centro di Studi Filologici e Linguistici Siciliani, c/o Istituto di Filologiae Linguistica, Facoltà di Lettere, Viale delle Scienze, 90128 Palermo
Founded: 1951
Pres: Prof. Mario Fasino; Gen Secr: Prof. Giovanni Ruffino
Focus: Ling
Periodical
Bollettino del Centro di Studi Filologici e Linguistici Siciliani 07445

Centro di Studi Filosofici di Gallarate, Via Donatello 24, 35100 Padova
T: (049) 651444
Founded: 1945
Focus: Philos 07446

Centro di Studi Grafici, Via Silvio Pellico 6, 20121 Milano
T: (02) 862988
Focus: Graphic & Dec Arts, Design . . 07447

Centro di Studi Metodologici, c/o Facoltà di Magistero, Università degli Studi, Via Sant'Ottavio 20, 10124 Torino
T: (011) 876611
Founded: 1948
Focus: Doc; Humanities 07448

Centro di Studio e Documentazione sul Vietnam e il Terzo Mondo (CE.S.VIET.), Via Petitti 19, 20149 Milano
T: (02) 321682
Founded: 1967
Focus: Geography 07449

Centro di Studi per la Storia dell'Architettura, Casa dei Crescenzi, Via del Teatro di Marcello 54, 0186 Roma
T: (06) 461540, 6793017
Founded: 1952
Focus: Archit; Cultur Hist
Periodical
Bollettino (weekly) 07450

Centro di Studi per l'Educazione Fisica (C.S.E.F.), Via Guerrazzi 13, 40125 Bologna
T: (051) 225951
Founded: 1956
Focus: Sports
Periodical
Educazione Fisica e Sport nella Scuola (weekly) 07451

Centro di Studi Pratici di Agricoltura, Fondazione Fratelli Gustavo e Severino Navarra, Via Cortevecchia 3, 44100 Ferrara
T: (0532) 34765
Founded: 1923
Focus: Agri 07452

Centro di Studi Preistorici e Archeologici, c/o Musei Civici di Villa Mirabello, 21100 Varese
T: (0332) 281590
Founded: 1953
Pres: Prof. Mario Mirabella Roberti; Gen Secr: Constantino Storti
Focus: Anthro; Hist; Archeol
Periodical
Sibrium: Collana di Studi e Documentazioni 07453

Centro di Studi Sociali e Politici Lorenzo Milani, Via Pietro Aretini 16, 52100 Arezzo
T: (0575) 23353
Focus: Sociology; Poli Sci 07454

Centro di Studi Storici ed Etnografici del Piceno, Lungo Castellano Sisto V 36, 63100 Ascoli Piceno
T: (0736) 53458
Founded: 1976
Focus: Ethnology; Hist
Periodical
Piceno (weekly) 07455

Centro di Studi Storici Maceratesi, CP 49, 62100 Macerata
Founded: 1965
Focus: Hist
Periodical
Studi Maceratesi: Collana 07456

Centro di Studi sul Teatro Medioevale e Rinascimentale, Piazza S. Pellegrino, Palazzo degli Alessandri, 01100 Viterbo
T: (0761) 39131
Founded: 1975
Focus: Perf Arts 07457

Centro di Vita Europea, Via Baldo degli Ubaldi 272, 00167 Roma
T: (06) 6374703
Founded: 1966
Focus: Econ; Poli Sci
Periodicals
Anni Nuovi (weekly)
Vita Europea (weekly) 07458

Centro Emilia-Romagna per la Storia del Giornalismo (C.E.R.S.G.), c/o Dipartimento di Discipline Storiche, Largo Trombetti 1, 40126 Bologna
T: (051) 236230, 267273
Founded: 1975; Members: 80
Focus: Journalism
Periodical
Giornalismo Emiliano-Romagnolo (weekly) 07459

Centro Esperantista Contro l'Imperialismo Linguistico (C.E.C.I.L.), Via Monte Nero 22, 20135 Milano
T: (02) 597386
Founded: 1969
Focus: Ling 07460

Centro Europeo per il Progresso della Scuola, Palazzo Odescalchi, Piazza SS. Apostoli 80, 00187 Roma
T: (06) 6780004
Founded: 1970
Focus: Educ 07461

Centro Europeo per il Progresso Economico e Sociale (C.E.P.E.S.), Via Palestro 34, 00185 Roma
T: (06) 462259
Focus: Econ; Sociology 07462

Centro Europeo per la Diffusione della Cultura (C.E.D.C.) / Centre Européen de Diffusion de la Culture, Via Filippi 63, Roma
Founded: 1927
Focus: Lit; Arts
Periodicals
La Capitale (weekly)
Scarabée: Belgio (weekly) 07463

Centro Informazione Farine e Pane, Via del Viminale 43, 00184 Roma
T: (06) 480641/42
Founded: 1973
Focus: Food 07464

Centro Informazioni e Studi sulla Comunita' Europea (CISMEC), Via Gaetano Donizetti 53, 20122 Milano
T: (02) 705068, 793448, 796282, 796586
Founded: 1959
Focus: Econ; Poli Sci
Periodical
Dimensione Europa: Periodico d'Informazioni e Studi sulla Comunità Europea 07465

Centro Internazionale della Pace (C.I.P.), Corso Re Umberto 38, 10128 Torino
Founded: 1967
Pres: Monsignore Giuliano Gennaro Pierino; Gen Secr: Dr. Giovanni Tempo
Focus: Prom Peace
Periodical
Civitas Pacis (weekly) 07466

Centro Internazionale di Documentazione e Comunicazione (IDOC) / International Documentation and Communication Center, Via S. Maria dell'Anima 30, 00186 Roma
T: (06) 6868332; Fax: (06) 6832766
Founded: 1965
Pres: Michael Traber; Gen Secr: Heinrich Hunke
Focus: Doc; Comm Sci
Periodical
IDOC Internazionale (weekly) 07467

Centro Internazionale di Ipnosi Medica e Psicologica / International Center of Medical and Psychological Hypnosis, c/o Istituto di Indagini Psicologiche, Corso XXII Marzo 57, 20129 Milano
T: (02) 7388427; Fax: (02) 7491051
Founded: 1969
Pres: Prof. Rolando Marchesan; Gen Secr: Dr. Aurora Zavertanik
Focus: Med; Psych; Depth Psych . . . 07468

Centro Internazionale di Ricerche sulle Strutture Ambiente Pio Manzù, 47040 Verucchio
Focus: Ecology 07469

Centro Internazionale di Scienze Meccaniche (CISM) / International Centre for Mechanical Sciences, Piazza Garibaldi 18, 33100 Udine
T: (0432) 294989, 22523
Founded: 1968
Focus: Eng
Periodical
Mechanics Research Communications . . 07470

Centro Internazionale di Studi Archeologici Maiuri, Via Quattro Orologi, 80056 Ercolano
T: 32417
Focus: Archeol 07471

Centro Internazionale di Studi di Architettura Andrea Palladio, Domus Comestabilis, Basilica Palladiana, CP 835, 36100 Vicenza
T: (0444) 323014; Fax: (0444) 322869
Founded: 1959
Pres: Danilo Longhi
Focus: Archit
Periodical
Annali di Architettura (weekly) 07472

Centro Internazionale di Studi e Documentazione sulle Comunità Europee (C.I.S.D.C.E.), Corso Magenta 61, 20123 Milano
T: (02) 48009072; Fax: (02) 48009067
Founded: 1958
Pres: Prof. G.M. Ubertazzi; Gen Secr: Ester Friz
Focus: Econ
Periodical
Euroinformazioni (weekly) 07473

Centro Internazionale di Studi Rosminiani, 28049 Stresa
T: (0323) 30091; Fax: (0323) 31623
Founded: 1966
Focus: Rel & Theol; Poli Sci; Philos
Periodicals
Bollettino Charitas: Di Spiritualità
Rivista Rosminiana: Di Cultura e Filosofia 07474

Centro Internazionale di Studi Sardi (C.I.S.S.), c/o Dipartimento di Biologia Sperimentale, Facoltà di Scienze Matematiche, Fisiche e Naturali, Università degli Studi, Via G.T. Porcell 2, 09100 Cagliari
T: (070) 659294
Founded: 1955
Pres: Prof. Carlo Maxia; Gen Secr: Prof. Giovanni Floris
Focus: Ethnology; Folklore
Periodical
Atti Congresso Internazionale di Studi Sardi 07475

Centro Internazionale di Studi Umanistici, c/o Facoltà di Filosofia, Università, Via Nomentana 118, 00161 Roma
Founded: 1978
Focus: Humanities; Philos
Periodical
Archivio di Filosofia 07476

Centro Internazionale Magistrati Luigi Severini, c/o Palazzo di Giustizia, 06100 Perugia
T: (075) 28214, 24254
Founded: 1954
Focus: Law 07477

Centro Internazionale per gli Studi sulla Irrigazione, Piazza Pradaval 16, 37100 Verona
T: (045) 24682
Founded: 1953
Focus: Water Res
Periodical
L'Irrigazione (weekly) 07478

Centro Internazionale per l'Avanzamento della Ricerca e dell'Educazione, Via Rivetti 61, 13069 Vigliano Biellese
Founded: 1974
Focus: Educ
Periodical
Yoga & Ayurveda (weekly) 07479

Centro Internazionale per le Communicazione Sociali (CISCOM), Via della Pisana 1111, CP 9092, 00163 Roma
T: (06) 6592915; Fax: (06) 6592929
Gen Secr: Antonio Martinelli
Focus: Soc Sci 07480

Centro Internazionale per l'Educazione Artistica della Fondazione Giorgio Cini (C.I.ED.ART.), Isola di San Giorgio Maggiore, 30124 Venezia
T: (041) 8990
Focus: Educ; Arts 07481

Centro Internazionale per l'Iniziativa Giuridica (C.I.I.G.), Via Foggia 6, 00161 Roma
T: (06) 5242540/06
Founded: 1973
Focus: Law
Periodical
Iniziativa Giuridica 07482

Centro Internazionale per lo Studio dei Papiri Ercolanesi (C.I.S.P.E.), c/o Istituto di Papirologia, Università degli Studi, Via Porta di Massa 1, 80133 Napoli
Founded: 1969
Pres: Prof. Giovanni Pugliese Carratelli; Gen Secr: Prof. Marcello Gigante
Focus: Materials Sci
Periodical
Cronache Ercolanesi (weekly) 07483

Centro Internazionale Ricerche sulle Strutture Ambientali Pio Manzu (C.I.R.S.A.) / International Research Centre on the Habitat Pio Manzu, 47040 Verucchio
T: (0541) 678139; Fax: (0541) 670172
Founded: 1969
Focus: Develop Areas
Periodical
Strutture Ambientali (weekly) 07484

Centro Internazionale Sonnenberg per l'Italia, c/o Luciano Mosso, Corso Turati 11, 10128 Torino
T: (011) 5683598; Fax: (011) 5683598
Founded: 1964
Pres: Dr. Lorenzo Cavallo; Gen Secr: Luciano Mosso
Focus: Econ; Educ
Periodical
Sonnenberg Internationale Briefe, International Journal, Revue Internationale: Rivista Trilingue (Tedesco, Inglese, Francese) (weekly) . . 07485

Centro Internazionale Studi Famiglia (C.I.S.F.), Via Duccio di Boninsegna 10, 20145 Milano
T: (02) 48012040; Fax: (02) 48009938
Founded: 1973
Focus: Family Plan 07486

Centro Internazionale Studi Musicali (C.I.S.M.), Largo del Nazareno 8, 00187 Roma
T: (06) 6790360
Founded: 1962
Focus: Music 07487

Centro Internazionale Studi Umanistici Scientifici Psicologici, Via Rosa Raimondi Garibaldi 42, 00145 Roma
T: (06) 5134891
Focus: Psych 07488

Centro Isec, Iniziative per Studi e Convegni, Piazza Cinque Giornate 2, 00192 Roma
T: (06) 3216657; Fax: (06) 3215143
Founded: 1969
Focus: Educ 07489

Centro Italiano di Antropologia Culturale / Italian Center of Culture Anthropology, c/o Facoltà di Magistero, Dipartimento di Sociologica, Via Torino 95, 00184 Roma
T: (06) 4758400, 851216
Founded: 1964
Focus: Anthro
Periodical
Antropologia Culturale: Proceedings of Seminars (weekly) 07490

Centro Italiano di Biostatistica (BIOS), Via Filippo Nicolai 49, 00136 Roma
T: (06) 3451503
Founded: 1950
Pres: Prof. Stefano Somogyi
Focus: Stats; Bio 07491

Centro Italiano di Musica Antica (C.I.M.A.), c/o Chiesa Valdese, Via Flaminia 808, 00191 Roma
T: (06) 3277073
Founded: 1979
Focus: Music 07492

Centro Italiano di Parapsicologia (C.I.P.), Viale Calascione 5a, 80132 Napoli
T: (081) 647343
Founded: 1963
Focus: Parapsych
Periodical
Informazioni di Parapsicologia (weekly) . 07493

Centro Italiano di Ricerche e d'Informazione sull'Economia delle Imprese Pubbliche e di Pubblico Interesse (C.I.R.I.E.C.), Via Fratelli Gabba 6, 20121 Milano
T: (02) 86460848; Fax: (02) 865198
Founded: 1956
Focus: Econ
Periodical
Economia Pubblica (weekly) 07494

Centro Italiano di Sessuologia (C.I.S.), Via della Pigna 13, 00186 Roma
T: (06) 6789407
Founded: 1959
Focus: Med; Hygiene; Psych

Periodical
Rivista di Sessuologia (weekly) 07495

Centro Italiano di Solidarieta' (Ce.I.S.), Piazza Benedetto Cairoli 118, 00186 Roma
T: (06) 659469, 657905
Founded: 1971
Focus: Toxicology
Periodical
Il Delfino (weekly) 07496

Centro Italiano di Studi Aziendali (C.I.S.A.), Via S. Maria Fulcorina 17, 20123 Milano
T: (02) 804996
Founded: 1952
Focus: Business Admin
Periodical
L'Ufficio Moderno: Notiziario (weekly) . . 07497

Centro Italiano di Studi di Diritto dell'Energia (CISDEN), Via Paisiello 28, 00198 Roma
T: (06) 8442587
Founded: 1969
Focus: Nucl Res; Law
Periodical
Bollettino di Informazioni 07498

Centro Italiano di Studi e Programmazioni per la Pesca (C.I.S.P.P.), Via Cassia 6, 00191 Roma
T: (06) 399690, 390606
Founded: 1963
Focus: Fisheries 07499

Centro Italiano di Studi Europei (C.I.S.E.), Via del Corso 267, 00186 Roma
T: (06) 6780004
Founded: 1960
Focus: Poli Sci
Periodical
L'Italia e l'Europa: Rassegna di Diritto, Economia, Politica, Società (weekly) 07500

Centro Italiano di Studi Finanziari, Lungotevere Anguillara 9, 00153 Roma
T: (06) 5110027
Focus: Finance 07501

Centro Italiano di Studi Sociali Economici e Giuridici (C.I.S.S.E.G.), Via De Carolis 73, 00136 Roma
T: (06) 3497297
Focus: Sociology; Law; Econ 07502

Centro Italiano per lo Studio e lo Sviluppo dell'Agopuntura Moderna e dell'Altra Medicina (C.I.S.S.A.M.), Via Canove 17, 36100 Vicenza
T: (0444) 39208
Founded: 1973
Focus: Med; Physical Therapy
Periodical
Agopuntura Moderna, Scienze dell'Uomo Totale: Rivista di Bioterapia e Psicobiofisica (weekly)
. . . . 07503

Centro Italiano per lo Studio e lo Sviluppo della Psicoterapia e dell'Autogenes Training (C.I.S.S.P.A.T.), Piazza de Gasperi 41, 35100 Padova
T: (049) 650861
Founded: 1970
Focus: Psychiatry
Periodicals
Psicoterapie, Metodi e Tecniche (weekly)
Psyche (weekly) 07504

Centro Italiano Ricerca e Informazione Economica (CIRIEC), Via Flavia 47, 00187 Roma
T: (06) 482890
Focus: Econ 07505

Centro Italiano Ricerche e Studi Assicurativi (C.I.R.S.A.), Via del Corso 184, 00168 Roma
T: (06) 6789127, 6798847
Focus: Insurance 07506

Centro Italiano Ricerche e Studi Trasporto Aereo (CIRSTA), Via Toscani 78, 00152 Roma
T: (06) 5313952
Founded: 1978
Focus: Aero; Transport 07507

Centro Italiano Studi Containers (CISCo), Via Garibaldi 4, 16124 Genova
Founded: 1967; Members: 140
Focus: Transport
Periodicals
CISCo-News: Notiziario (weekly)
Quaderni CISCo 07508

Centro Italiano Studi Politici Economici Sociali (C.I.S.P.E.S.), Via del Tritone 62, 00187 Roma
T: (06) 6797672, 6796539
Founded: 1963
Focus: Econ 07509

Centro Italiano Studi sull'Arte dello Spettacolo (C.I.S.A.S.), Via Angelo Poliziano 51, 00184 Roma
T: (06) 733092
Founded: 1965
Pres: Gabriella Maria Montefoschi; Gen Secr: Dr. Marela Caputo
Focus: Cinema; Perf Arts
Periodicals
Agenzia Giornalistica SAFJ-Press
Corriere di Roma 07510

Centro Italiano Sviluppo Impieghi Acciaio (C.I.S.I.A.), Piazza Velasca 8, 20122 Milano
T: (02) 865840, 8059692, 8057046
Founded: 1954
Focus: Materials Sci
Periodical
Acciaio (weekly) 07511

Centro Italo-Nipponico di Studi Economici (C.I.N.S.E.), Piazza Armando Diaz 7, 20123 Milano
T: (02) 802896
Founded: 1973
Focus: Econ
Periodical
Nuovo Giappone (weekly) 07512

Centro Lattiero Caseario di Assistenza e Sperimentazione Antonio Bizzozero, Via Pomponio Torelli 17, 43100 Parma
T: (0521) 44916
Founded: 1956
Focus: Dairy Sci
Periodical
Bollettino Tecnico (weekly) 07513

Centro Ligure di Storia Sociale, Piazza Campetto 8a, 16123 Genova
T: (010) 297408
Founded: 1955
Pres: Dr. Luca Borzani
Focus: Hist; Sociology; Socialism
Periodical
Movimento Operaio e Socialista (weekly) 07514

Centro Luigi Lavazza per gli Studi e Ricerche sul Caffe', Corso Novara 59, 10154 Torino
T: (011) 2398
Focus: Coffee, Tea, Cocoa 07515

Centro Nazionale di Dietobiologia ed Igiene della Alimentazione, Via Mario Pagano 31a, 20145 Milano
T: (02) 496303
Focus: Nutrition 07516

Centro Nazionale di Studi Urbanistici, Via Antonio Bertoloni 31, 00197 Roma
T: (06) 805101, 805103
Founded: 1964
Focus: Urban Plan 07517

Centro per gli Studi di Tecnica Navale (CE.TE.NA.), Via Al Molo Giano, 16126 Genova
T: (010) 590521, 589349, 542481
Founded: 1962
Pres: Vittorio Marulli
Focus: Navig 07518

Centro per gli Studi e le Applicazioni delle Risorse Energetiche (C.S.A.R.E.), Corso del Popolo 261, 45100 Rovigo
T: (0425) 26478
Founded: 1977
Focus: Energy
Periodical
Bollettino del C.S.A.R.E. (weekly) . . . 07519

Centro per gli Studi sui Mercati Esteri (CEME), Via Giulio Alberoni 8a, 00198 Roma
T: (06) 854144
Founded: 1956
Focus: Econ; Marketing; Int'l Relat
Periodical
Mondo Aperto 07520

Centro per gli Studi sui Sistemi Distributivi e il Turismo (CESDIT), Corso Venezia 49, 20121 Milano
T: (02) 7750
Founded: 1977
Focus: Commerce; Travel 07521

Centro per la Diffusione del Libro Lucano, Via Pretoria 210, 85100 Potenza
T: (0971) 24570
Focus: Hist 07522

Centro per la Documentazione Automatica (C.D.A.), Via Cusani 10, 20121 Milano
T: (02) 870444
Founded: 1962
Focus: Doc; Law 07523

Centro per la Riforma del Diritto di Famiglia, c/o Avv. Giuliana Fuà, Via Enrico Dandolo 4, 20122 Milano
T: (02) 705136
Founded: 1967
Focus: Family Plan 07524

Centro per la Statistica Aziendale, Via A. Baldesi 20, 50131 Firenze
T: (055) 50713
Founded: 1935
Focus: Econ
Periodicals
Circolare (weekly)
La Congiuntura in Toscana (weekly)
Index (weekly)
Lettere d'Affari (weekly)
Previsioni a Breve Termine (weekly)
Prontuario Economico del Turista: Spesa Giornaliera del Viaggiatore in 42 Paesi (weekly)
. . . . 07525

Centro per la Storia della Tradizione Aristotelica Nel Veneto, Via Accademia 5, 35100 Padova
T: (049) 24034
Focus: Cultur Hist; Philos 07526

Centro per lo Studio dei Dialetti Veneti dell'Istria, c/o Istituto di Glottologia, Università degli Studi, Via dell'Università 1, 34100 Trieste
T: (040) 722274
Founded: 1970
Focus: Ling
Periodical
Bollettino del Centro per lo Studio dei Dialetti Veneti dell'Istria (weekly) 07527

Centro Piombinese di Studi Storici, c/o Biblioteca Comunale, Via Cavour 152, 57025 Piombino
T: 33110
Focus: Hist 07528

Centro Polesano di Studi Storici, Archeologici ed Etnografici (C.P.S.S.A.E.), c/o Museo delle Civiltà in Polesine, Piazzale S. Bortolo 18, CP 106, 45100 Rovigo
T: (0425) 21021
Founded: 1964
Focus: Ethnology; Hist; Archeol
Periodicals
Padusa
Padusa Notiziario (weekly) 07529

Centro Provinciale Impiego Combinato Tecniche Agricole (Centro I.C.T.A.), c/o Camera di Commercio, Piazza della Costituzione 8, 40128 Bologna
T: (051) 519061
Founded: 1960
Focus: Agri; Eng 07530

Centro Psico-Pedagogico Didattico (Ce.Psi.Pe.Di.), Via Galeotti 5, 40127 Bologna
T: (051) 502842
Founded: 1978
Focus: Educ; Psych
Periodical
Problemi d'Oggi: Bimestrale di Scienze Umane e Formazione Sociale (weekly) 07531

Centro Regionale di Studi Sociali V.G. Galati, Piazza Stocco 21, Lamezia Terme
T: 23889
Focus: Sociology 07532

Centro Ricerche Applicazione Bioritmo (C.R.A.B.), Viale dell'Università 25, 00185 Roma
T: (06) 4951759
Focus: Med; Genetics 07533

Centro Ricerche Archeologiche e Scavi di Torino per il Medio Oriente e l'Asia, Via Gaudenzio Ferrari 1, 10124 Torino
Founded: 1963
Focus: Archeol
Periodical
Mesopotamia 07534

Centro Ricerche Cosmetologiche, Via Montoggio 49, 00168 Roma
T: (06) 6378788
Founded: 1972
Focus: Pharmacol
Periodical
Journal of Applied Cosmetology (weekly) 07535

Centro Ricerche Didattiche Ugo Morin (C.R.D.M.), Via S. Giacomo 4, 31010 Paderno del Grappa
T: (0423) 330441; Fax: (0423) 539098
Founded: 1976
Focus: Educ
Periodical
L'Insegnamento della Matematica e delle Scienze Integrate (weekly) 07536

Centro Ricerche di Storia e Arte Bitontina (C.R.S.A.B.), Viale Giovanni XXIII 129, 70032 Bitonto
Founded: 1968
Focus: Cultur Hist; Ethnology; Archeol; Music
Periodicals
Studi Bitontini
La tua Città 07537

Centro Ricerche Economiche ed Operative della Cooperazione (C.R.E.O.C.), Piazza S. Maria degli Angeli a Pizzofalcone 1, 80132 Napoli
T: (081) 400711
Founded: 1972
Focus: Econ 07538

Centro Ricerche Economiche Sociologiche e di Mercato nell'Edilizia (C.R.E.S.M.E.), Via Sebenico 2, 00198 Roma
T: (06) 850158, 865952, 867227
Founded: 1962
Focus: Econ; Sociology; Marketing
Periodicals
Bulletins (weekly)
Cahiers (weekly)
Volumes (weekly) 07539

Centro Ricerche Metapsichiche e Psicofoniche (Ce.Ri.Me.Ps.), Via Mancini 3, 63023 Fermo
T: (0734) 24231
Founded: 1977
Focus: Parapsych 07540

Centro Ricerche Socio-Religiose, Largo Donnaregina 22, 80138 Napoli
T: (081) 449589
Founded: 1966
Focus: Rel & Theol; Sociology
Periodical
Annuario dell'Archidiocesi di Napoli (weekly)
. . . . 07541

Centro Ricerche Urbanistiche e di Progettazione, Via Beato Angelico 23, 20133 Milano
T: (02) 7387917
Founded: 1975
Focus: Urban Plan 07542

Centro Romano per lo Studio dei Problemi di Attualita' Sociale, Via dei Savorelli 38, 00165 Roma
T: (06) 6221865
Focus: Sociology; Poli Sci 07543

Centro Rossiniano di Studi, c/o Fondazione Gioacchino Rossini, Piazza Olivieri 5, 61100 Pesaro
T: (0721) 30053
Founded: 1869
Focus: Music
Periodical
Bollettino del Centro Rossiniano di Studi 07544

Centro Sperimentale di Cinematografia (C.S.C.), Via Tuscolana 1524, 00173 Roma
T: (06) 740046; Fax: (06) 7111619
Founded: 1935
Focus: Cinema
Periodical
Bianco e Nero 07545

Centro Sperimentale Italiano di Giornalismo, Viale Caldara 13, 20122 Milano
T: (02) 584223
Focus: Journalism 07546

Centro Sperimentale Metallurgico, Via di Castel Romano, 00129 Roma
T: (06) 64951
Founded: 1963
Focus: Metallurgy 07547

Centro Studi Archaeologici di Boscoreale, Boscotrecase e Trecase, c/o Biblioteca Comunale Francesco Cangemi, Piazza Pace 36, 80041 Boscoreale
Founded: 1974
Pres: Angelo Bianco; Gen Secr: Dr. Angelandrea Casale
Focus: Archeol
Periodicals
Bollettino del Centro Studi Archaeologici (weekly)
Rivista sull'Arte, la Storia e il Folklore di Somma Vesuviana (weekly) 07548

Centro Studi Assicurativi Piero Sacerdoti, Corso Venezia 8, 20121 Milano
T: (02) 794235
Founded: 1951
Focus: Insurance
Periodical
Diritto e Pratica nell'Assicurazione (weekly) 07549

Centro Studi Cinematografici (C.S.C.), Via Gregorio VII 6, 00165 Roma
T: (06) 6382605
Pres: Baldo Vallero; Gen Secr: Franco Scarmigna
Focus: Cinema
Periodical
Notiziario del C.S.C. (weekly) 07550

Centro Studi della Cooperazione nel Veneto, Piazzale del Mutilato 6, 36100 Vicenza
T: (0444) 24226
Founded: 1961
Pres: Dr. Giancarlo De Biasio; Gen Secr: M. Carolina Zanetti
Focus: Econ; Agri 07551

Centro Studi di Diritto del Lavoro, c/o Palazzo di Giustizia, Corso Porta Vittoria, 20122 Milano
Focus: Law 07552

Centro Studi di Diritto Fluviale e della Navigazione Interna, S. Marco, Calle Avvocati 392, 30124 Venezia
T: (041) 5220762; Fax: (041) 5236357
Pres: Marcello Olivi; Gen Secr: Renata Farnea Radich
Focus: Navig; Law 07553

Centro Studi di Diritto Sportivo (CE.S.DI.S.), c/o Studio Legale Dal Lago, Contrà Canove Nuove 3, 36100 Vicenza
T: (0444) 44044
Founded: 1972
Focus: Law; Sports 07554

Centro Studi di Economia Applicata all'Ingegneria (C.S.E.I.), c/o Facoltà di Ingegneria, Università degli Studi, Viale Japigia 182, 70126 Bari
T: (080) 580639
Founded: 1975
Pres: Prof. Umberto Ruggiero
Focus: Econ; Eng 07555

Centro Studi di Economia Applicata all'Ingegneria (C.S.E.I.), c/o Facoltà d'Ingegneria, Università degli Studi, Via Claudio 21, 80125 Napoli
T: (081) 614641
Founded: 1968
Focus: Econ; Eng
Periodical
Notizie C.S.E.I. (weekly) 07556

Centro Studi di Estimo e di Economia Territoriale (Ce.S.E.T.), c/o Istituto di Economia e Politica Agraria, Facoltà di Agraria, Università degli Studi, Piazzale delle Cascine 18, 50144 Firenze
T: (055) 352051
Focus: Econ 07557

Centro Studi di Poesia e di Storia delle Poetiche, Via Bu Meliana 12, 00195 Roma
Focus: Lit 07558

Centro Studi di Politica Economica (CEEP), Via San Francesco da Paola 17, 10123 Torino
T: (011) 8170449
Founded: 1973
Focus: Econ; Poli Sci
Periodical
Ambiente (weekly) 07559

Centro Studi di Psicologia e Sociologia Applicate ad Indirizzo Adleriano (C.E.P.S.A.), Via Michele Amari 47, 00179 Roma
T: (06) 7807897, 7857216
Founded: 1969
Pres: Dr. Giovanna Pisano; Gen Secr: Prof. Francesco Maria Scala
Focus: Psych; Econ
Periodical
Cepsa-Informa (3 times annually) . . . 07560

Centro Studi di Psicoterapia e Psicologia Clinica, Via Antonio Cecchi 3, 16129 Genova
T: (010) 541092
Founded: 1971
Pres: Dr. Giandomenico Montinari; Gen Secr: Dr. Giovanni Pesenti
Focus: Psychiatry
Periodical
Aggiornamenti di Psicoterapia e Psicologia Clinica (weekly) 07561

Centro Studi Diritto Comunitario, Via in Lucina 10, 00186 Roma
T: (06) 6789560, 6783144
Focus: Law 07562

Centro Studi di Storia Locale, Via Dante 37, 54100 Massa
T: (0585) 41720
Founded: 1974
Focus: Hist 07563

Centro Studi e Applicazioni in Tecnologie Avanzate (C.S.A.T.A.), c/o Istituto di Fisica, Università degli Studi, Via G. Amendola 173, 70126 Bari
T: (080) 583388
Founded: 1969
Focus: Eng 07564

Centro Studi e Archivio della Comunicazione, Università di Parma, Piazzale della Pace 7a, 43100 Parma
T: (0521) 270847; Fax: (0521) 207125
Focus: Photo 07565

Centro Studi Economici (CE.S.E.), Piazza Colonna 355, 00187 Roma
T: (06) 6794065
Focus: Econ 07566

Centro Studi Economici e Sociali Giuseppe Toniolo, Piazza Giuseppe Toniolo 2, 56100 Pisa
T: (050) 571181
Founded: 1966
Pres: Prof. Silvio Trucco
Focus: Econ
Periodical
Studi Economici e Sociali (3 times annually) 07567

Centro Studi Economici per l'Alta Italia, Via Alessandro Manzoni 31, 20121 Milano
T: (02) 639634
Focus: Econ 07568

Centro Studi ed Esperienze Scout Baden Powell, Via Achille Papa 17, 00195 Roma
T: (06) 3603618
Founded: 1974
Pres: Dr. Guido Palombi
Focus: Educ
Periodical
Esperienze e Progetti (weekly) 07569

Centro Studi e di Educazione Civica Enrico Mattei, Via Giovanni Cravero 45, 10154 Torino
T: (011) 267880
Founded: 1963
Focus: Educ
Periodical
Opinioni Libere (weekly) 07570

Centro Studi e Documentazione della Cultura Armena (CSDCA), Via Melzi d'Eril 6, 20154 Milano
T: (02) 342718
Founded: 1967
Focus: Hist; Arts
Periodical
Documents of American Architecture (weekly) 07571

Centro Studi e Iniziative Pier Santi Mattarella (C.S.I.P.M.), Piazza Malta 27-29, 9110 Trapani
T: (0923) 22241, 22881
Founded: 1981
Focus: Educ
Periodical
Faro (weekly) 07572

Centro Studi e Ricerche per la Conoscenza della Liguria Attraverso le Testimonianze dei Viaggiatori Stranieri / Research Institute for Foreign travellers in Italy, Via F. Mignone 46, 16133 Genova
T: (010) 3450119
Focus: Hist; Geography; Travel
Periodicals
Architettura Navale agli Inizi del '600 (weekly)
Il Porto di Genova agli Inizi del '600 (weekly)
Viaggiare in Liguria (weekly)
Viaggiatori Stranieri in Liguria (weekly)
Viaggiatori Tedeschi del '700 (weekly) . 07573

Centro Studi e Ricerche sui Rapporti Umani, Via dei Lucchesi 21a, 00187 Roma
T: (06) 67015283
Focus: Soc Sci 07574

Centro Studi Filippo e Marta Larizza per la Formazione Permanente degli Educatori e per la Prevenzione del Disadattamento Giovanile, c/o Collegio Vescovile Graziani, Via Ca' Rezzonico 6, 36061 Bassano del Grappa
T: (0424) 524369
Founded: 1979
Pres: Prof. Luigi Secco
Focus: Educ
Periodical
Larizza Informazioni (3 times annually) . 07575

Centro Studi Mario Mazza, Via Fassolo 29, 16126 Genova
T: (010) 267155; Fax: (010) 267155
Founded: 1962
Pres: Giancarlo Volpato; Gen Secr: Giancarlo Coppa
Focus: Educ
Periodical
Strade Aperte (weekly) 07576

Centro Studi Mutualistici Emancipazione e Partecipazione (S.M.S.), Via Francesco d'Ovidio 135, 00137 Roma
T: (06) 8280261
Founded: 1974
Focus: Educ; Hist; Sociology 07577

Centro Studi Nord e Sud, Via Chiatamone 7, 80121 Napoli
T: (081) 418347
Focus: Geography; Sociology
Periodical
Nord e Sud (weekly) 07578

Centro Studi Parlamentari, Via della Rosetta 5, 00186 Roma
T: (06) 6547838
Focus: Poli Sci 07579

Centro Studi per il Mezzogiorno A. Ajon, Palazzo Grasso, Viale de Gasperi, 95024 Acireale
T: (095) 605705
Focus: Poli Sci; Hist; Econ 07580

Centro Studi per il Progresso della Educazione Sanitaria e del Diritto Sanitario (STUDES), Via Azuni 9, 00196 Roma
T: (06) 3600093
Founded: 1962
Focus: Public Health; Educ
Periodical
Rassegna Amministrativa della Sanità (weekly) 07581

Centro Studi per la Programmazione Economica e Sociale (C.S.P.E.S.), Via della Vite 27, 00187 Roma
Focus: Econ; Sociology 07582

Centro Studi per la Valorizzazione delle Risorse del Mezzogiorno, Corso Umberto I 22, 80138 Napoli
T: (081) 5527744
Founded: 1990
Pres: Nicola Squitieri; Gen Secr: Antonio Pisanti
Focus: Econ; Poli Sci
Periodicals
Incontri di Studio: Collana di Quaderni Meridionalistici (weekly)
Politica Meridionalista: Rivista Mensile di Cultura Economia Attualità (weekly) 07583

Centro Studi Piero Gobetti, Via Antonio Fabro 6, 10122 Torino
T: (011) 531429
Founded: 1961
Focus: Hist
Periodical
Mezzosecolo (weekly) 07584

Centro Studi Pietro Mancini, Corso Telesio 53, 87100 Cosenza
T: (0984) 29983
Focus: Perf Arts; Cinema; Journalism . . 07585

Centro Studi Politici e Sociali Alcide de Gasperi, Via Nicola Serra 123, 87100 Cosenza
T: (0984) 32865
Founded: 1964
Focus: Poli Sci; Educ; Sociology
Periodical
Prospettive 07586

Centro Studi Politico-Sociali Achille Grandi, Piazza Sant'Ambrogio 15, 20123 Milano
T: (02) 893838, 867148, 871110
Founded: 1961
Focus: Poli Sci; Sociology 07587

Centro Studi Problemi Medici (C.S.P.M.), Corso Sempione 77, 20149 Milano
T: (02) 3189587
Focus: Med 07588

Centro Studi Ricerche e Documentazioni per l'Agricoltura Siciliana, Via Milano 58, 00184 Roma
T: (06) 461893
Focus: Agri 07589

Centro Studi Ricerche Ligabue, S. Croce 499, 30135 Venezia
T: (041) 5225127; Fax: (041) 791661
Founded: 1978
Pres: Giancarlo Ligabue
Focus: Archeol; Paleontology; Anthro; Nat Sci 07590

Centro Studi Russia Cristiana (R.C.), Via Martinengo 16, 20139 Milano
T: (02) 564145
Founded: 1973
Focus: Rel & Theol; Hist; Lit
Periodical
Russia Cristiana (weekly) 07591

Centro Studi Santa Veronica Giuliani, Via del Marchese Paolo 13, 06012 Città di Castello
Founded: 1978; Members: 350
Pres: Pellegrino Tomaso Ronchi
Focus: Rel & Theol 07592

Centro Studi Storici di Mestre (C.S.S.), c/o Bilbioteca Civica, Via Piave 5, 30170 Mestre
T: (041) 952010
Founded: 1961
Focus: Cultur Hist 07593

Centro Studi Storici Sociali, Via Castiglione 6, 40124 Bologna
T: (051) 223979
Focus: Hist 07594

Centro Studi sulla Resistenza, Piazza della Repubblica 3, 61029 Urbino
Focus: Poli Sci 07595

Centro Studi Wilhelm Reich, Salita Cupa Caiafa 36, 80122 Napoli
T: (081) 664389
Focus: Psych 07596

Centro Studi Zingari, Via dei Barbieri 22, 00186 Roma
T: (06) 6833181; Fax: (06) 6868760
Founded: 1970
Pres: Bruno Nicolini
Focus: Ethnology
Periodical
Lacio Drom: Bimestrale di Studi Zingari (weekly) 07597

Centro Superiore di Logica e Scienze Comparate (C.S.L.S.C.), Via Belmeloro 3, 40126 Bologna
T: (051) 235471
Focus: Logic 07598

Centro Sviluppo Impiego Diesel (CE.S.I.D.), Via Italo Svevo 1, 34144 Trieste
Founded: 1980
Focus: Auto Eng 07599

Centro Thomas Mann, Via Zanardelli 36, 00186 Roma
T: (06) 659766
Focus: Lit 07600

Cineteca Italiana Archivio Storico del Film, Villa Comunale, Via Palestro 16, 20121 Milano
T: (02) 799224
Founded: 1947
Focus: Cinema 07601

Circolo Filosofico di Studi Tomistici, Largo Angelicum 1a, 00184 Roma
T: (06) 681261
Founded: 1912
Focus: Rel & Theol; Philos
Periodical
Idea: Rivista di Cultura e di Vita Sociale (weekly) 07602

Circolo Giuridico Italiano, Via della Conciliazione 15, 00193 Roma
T: (06) 6569941
Focus: Law 07603

Circolo Speleologico Romano (CSR), Via Ulisse Aldrovandi 18, 00197 Roma
T: (06) 3216223
Founded: 1904; Members: 150
Focus: Speleology
Periodicals
Notiziario del Circolo Speleologico Romano (weekly)
Quaderni di Speleologia 07604

Club Turati, Via Brera 18, 20121 Milano
T: (02) 877903, 877873
Focus: Poli Sci 07605

Codex Coordinating Committee for Africa, Via Terme di Caracalla, 00100 Roma
T: (06) 57971; Tx: 610181; Fax: (06) 57973152
Founded: 1974
Focus: Food 07606

Collegio degli Ingegneri Ferroviari Italiani (C.I.F.I.), Via Giovanni Giolitti 34, 00185 Roma
T: (06) 4882129; Fax: (06) 4742987
Founded: 1899
Pres: Silvio Rizzotti; Gen Secr: Bruno Cirillo
Focus: Transport
Periodicals
L'Ingegneria Ferroviaria
La Tecnica Professionale 07607

Collegio dei Tecnici dell' Acciaio (CTA), Piazzale Rodolfo Morandi 2, 20121 Milano
T: (02) 784711
Founded: 1966; Members: 600
Focus: Eng; Metallurgy
Periodical
Costruzioni Metalliche: Organo Ufficiale del CTA 07608

Collegium Biologicum Europa, Via Agrigento 6, 00161 Roma
Focus: Bio 07609

Collegium Internationale Neuro-Psychopharmacologicum (CINP), c/o Centre of Neuropharmacology, University of Milano, Via Balzaretti 9, 20133 Milano
T: (02) 20488331
Founded: 1957
Gen Secr: Giorgio Racagni
Focus: Pharmacol 07610

Collegium Musicum di Latina (C.M.L.), Via Antonio Rosmini 10, 04100 Latina
Founded: 1968
Focus: Music 07611

Comitato dei Geografi Italiani, Via Giorgio Baglivi 3, 00161 Roma
T: (06) 866070
Founded: 1965; Members: 105
Focus: Geography 07612

Comitato Elettrotecnico Italiano (C.E.I.), Viale Monza 259, 20126 Milano
Focus: Electric Eng 07613

Comitato Glaciologico Italiano (CGI), Via Accademia delle Scienze 5, 10123 Torino
T: (011) 553525
Founded: 1914; Members: 40
Focus: Geology
Periodical
Geografia Fisica e Dinamica Quaternaria: Bollettino del Comitato Claciologico Italiano (weekly) 07614

Comitato Italiano per l'Educazione Sanitaria (C.I.E.S.), c/o Centro Sperimentale per l'Educazione Sanitaria, Via del Giochetto, 06100 Perugia
T: (075) 28377
Focus: Public Health; Educ
Periodicals
Educazione Sanitaria e Medicina Preventiva (weekly)
La Salute Umana (weekly) 07615

Comitato Italiano per lo Studio dei Problemi della Popolazione (CISP), Via Nomentana 41, 00161 Roma
T: (06) 859555
Founded: 1928
Pres: Prof. N. Federici; Gen Secr: Dr. E. Brighenti
Focus: Sociology
Periodical
Genus: Rivista fondata da Corrado Gini edita sotto il Patrocinio del Consiglio Nazionale delle Ricerche semestrale 07616

Comitato Italiano per lo Studio del Dolore in Oncologia (C.I.S.D.O.), c/o Istituto Regina Elena, Viale Regina Elena 291, 00161 Roma
T: (06) 497931
Focus: Cell Biol & Cancer Res 07617

Comitato Nazionale Italiana per l'Organizzazione Scientifica (ENIOS), Palazzo della Civiltá del Lavoro, 00144 Roma
T: (06) 595147
Focus: Business Admin 07618

Comitato per Bologna Storica-Artistica, Strada Maggiore 71, 40125 Bologna
T: (051) 347764
Focus: Hist; Arts 07619

Comitato Termotecnico Italiano (C.T.I.), c/o Istituto di Fisica Tecnica, Politecnico, 10100 Torino
Founded: 1950
Focus: Air Cond 07620

Commissione di Studio dei Fenomeni di Corrosione Elettrolitica, Viale Risorgimento 2, Bologna
Focus: Metallurgy 07621

Commission for Controlling the Desert Locust in the Eastern Region of its Distribution Area in South West Asia (DL/SWA), c/o Migrant Pest Group, FAO, Via delle Terme di Caracalla, 00100 Roma
T: (06) 5794021; Tx: 610181; Fax: (06) 5146172
Founded: 1964; Members: 4
Gen Secr: Laurence McCulloch
Focus: Agri 07622

Committee for the European Development of Science and Technology, c/o ENEA, Viale Regina Margherita 125, 00198 Roma
T: (06) 85281; Tx: 610183
Founded: 1982
Pres: Dr. Umberto Colombo
Focus: Sci; Eng 07623

Committee on Forest Development in the Tropics, c/o FAO, Via delle Terme di Caracalla, 00100 Roma
Tx: 610181
Founded: 1967; Members: 41
Focus: Forestry *07624*

Community of Mediterranean Universities (CMU), c/o Università degli Studi, Piazza Umberto I 1, 70121 Bari
T: (080) 360786; Tx: 810598; Fax: (080) 369108
Founded: 1983; Members: 100
Focus: Educ
Periodical
CMU Bulletin (weekly) *07625*

Consiglio Nazionale degli Architetti, Corso Rinascimento 11, 00186 Roma
T: (06) 6561374
Focus: Archit *07626*

Consiglio Nazionale dei Chimici, c/o Comprofessionist, Via Sicilia 5, 00187 Roma
T: (06) 485641
Focus: Chem *07627*

Consiglio Nazionale delle Ricerche (CNR), Piazzale Aldo Moro 7, 00185 Roma
T: (06) 49931; Tx: 610076; CA: CORICERCHE
Founded: 1923
Pres: Prof. Luigi Rossi Bernardi; Gen Secr: Dr. Bruno Colle
Focus: Sci
Periodicals
Bollettino Ufficiale del CNR
Monografie Scientifiche del CNR *07628*

Consiglio Nazionale Forense, Via Arenula 71, 00186 Roma
T: (06) 6542689
Focus: Forensic Med *07629*

Consociatio Internationalis Musicae Sacrae (CIMS), Via di Rossa 21, 00165 Roma
T: (06) 6638792
Founded: 1963; Members: 663
Gen Secr: Vincent Alvares
Focus: Music *07630*

Current Agricultural Research Information System (CARIS), c/o FAO-GII, Via delle Terme di Caracalla, 00100 Roma
T: (06) 57974391; Tx: 610181; Fax: (06) 5146172
Founded: 1979; Members: 152
Pres: A.I. Lebowitz
Focus: Agri *07631*

Deputazione di Storia Patria per gli Abruzzi, Corso Umberto 19, 67100 L'Aquila
T: (0862) 22581
Founded: 1888
Focus: Hist
Periodical
Bollettino della Deputazione Abruzzese di Storia Patria (weekly) *07632*

Deputazione di Storia Patria per il Friuli, Palazzo Mantica, Via Manin 18, CP 319, 33100 Udine
T: (0432) 21924
Founded: 1918
Focus: Hist
Periodical
Memorie Storiche Forogiuliesi *07633*

Deputazione di Storia Patria per la Calabria (D.S.P.C.), c/o Museo Nazionale, Piazza G. de Nava 26, 89100 Reggio Calabria
T: (0985) 21949
Founded: 1957
Pres: Prof. Maria Mariotti; Gen Secr: Prof. Mirella Mafrici
Focus: Hist
Periodical
Rivista Storica Calabrese (Nuova Serie) (weekly) *07634*

Deputazione di Storia Patria per la Lucania, Piazza Vittorio Emanuele 14, 85100 Potenza
Founded: 1957
Pres: Prof. Vincenzo Verrastro; Gen Secr: Prof. Tommaso Pedio
Focus: Hist
Periodical
Bollettino Storico della Basilicata *07635*

Deputazione di Storia Patria per la Sardegna, Via Cadello 9bis, 09100 Cagliari
T: (070) 4092764; Fax: (070) 502521
Founded: 1955
Focus: Hist *07636*

Deputazione di Storia Patria per la Toscana, Piazza dei Giudici 1, 50122 Firenze
T: (055) 23251
Founded: 1862
Pres: Prof. Arnaldo D'Addario; Gen Secr: Dr. Giulio Prunai
Focus: Hist
Periodical
Archivio Storico Italiano (weekly) *07637*

Deputazione di Storia Patria per le Antiche Province Modenesi, c/o Aedes Muratoriana, Via Pomposa 1, 41100 Modena
T: (059) 241104
Founded: 1860
Pres: Prof. Giorgio Boccolari; Gen Secr: Prof. Lidia Righi
Focus: Hist
Periodical
Atti e Memorie (weekly) *07638*

Deputazione di Storia Patria per le Marche, Piazza B. Stracca 3, 60100 Ancona
T: (071) 29411
Founded: 1863
Focus: Hist
Periodical
Atti e Memorie della Deputazione di Storia Patria per le Marche (weekly) *07639*

Deputazione di Storia Patria per le Province di Romagna, c/o Istituto di Discipline Storiche e Giuridiche, Facoltà di Magistero, Largo Trombetti 1, 40126 Bologna
T: (051) 236210
Focus: Hist
Periodical
Atti e Memorie della Deputazione di Storia Patria per le Province di Romagna (weekly) . *07640*

Deputazione di Storia Patria per le Province Parmensi, Borgo Schizzati 3, 43100 Parma
T: (0521) 238661
Founded: 1860
Focus: Hist
Periodical
Archivio Storico per le Province Parmensi *07641*

Deputazione di Storia Patria per le Venezie, S. Croce, Calle del Tintor 1583, 30135 Venezia
T: (041) 524487
Founded: 1871
Focus: Hist
Periodical
Archivio Veneto *07642*

Deputazione di Storia Patria per l'Umbria, Palazzo della Penna, Via Podiani 11, 06100 Perugia
T: (075) 5727057
Founded: 1896
Pres: Dr. Giovanni Antonelli; Gen Secr: Prof. Paola Pimpinelli
Focus: Hist
Periodical
Bollettino della Deputazione di Storia Patria per l'Umbria (weekly) *07643*

Deputazione Provinciale Ferrarese di Storia Patria, Via Cairoli 13, 44100 Ferrara
Founded: 1884
Focus: Hist
Periodical
Atti e Memorie della Deputazione Provinciale Ferrarese di Storia Patria *07644*

Deputazione Reggiana di Storia Patria, Via I. Pindemonte 12, 42100 Reggio Emilia
Focus: Hist
Periodical
Bollettino Storico Reggiano (weekly) . . *07645*

Deputazione Subalpina di Storia Patria, Via Principe Amedeo 5, 10123 Torino
T: (011) 537226
Founded: 1833
Focus: Hist
Periodical
Bollettino Storico Bibliografico Subalpino (weekly) *07646*

Documentation and Research Centre, Via dei Verbiti 1, CP 5080, 00100 Roma
T: (06) 5741350
Founded: 1964; Members: 69
Gen Secr: W. Jenkinson
Focus: Doc; Rel & Theol
Periodical
SEDOS Bulletin (weekly) *07647*

Earthnet Programme Office (EPO), c/o ESRIN, Via Galileo Galilei, CP 65, 00044 Frascati
T: (06) 941801; Tx: 610637; Fax: (06) 94180361
Founded: 1978; Members: 13
Gen Secr: Livio Marelli
Focus: Geology
Periodical
Earth Observation Quarterly (weekly) . . *07648*

Ente Autonomo La Biennale de Venezia, S. Marco 1364a, 30100 Venezia
T: (041) 700311
Focus: Arts; Archit; Perf Arts; Music; Cinema *07649*

Ente di Unificazione Navale (UNAV) / Marine Standardization Office, Via al Nolo Giano, Calata Grazie, 16126 Genova
T: (010) 5995795; Fax: (010) 5995790
Founded: 1938
Pres: Mario Carlo Ramacciotti
Focus: Navig
Periodicals
Note di Informazioni Trimestrali e Note su Attività Tecnica
Tabelle Italiane Navali *07650*

Ente Eugenio e Claudio Faina per l'Istruzione Professionale Agraria, Via Torino 45, 00184 Roma
T: (06) 4751735
Founded: 1912
Focus: Agri; Educ *07651*

Ente Fauna Siciliana, CP 76, 96017 Noto
Fax: (0931) 813273
Founded: 1973
Pres: Prof. Marcello La Greca; Gen Secr: Bruno Ragonese
Focus: Ecology; Botany
Periodical
Atti del Convegno Siciliano di Ecologia . *07652*

Ente Friulano di Economia Montana, c/o Palazzo della Provincia, Piazza Patriarcato 3, 33100 Udine
T: (0432) 22804, 61233
Focus: Econ *07653*

Ente Istruzione Professionale Artigiana (E.I.P.A.), Piazzetta Pattari 4, 20122 Milano
T: (02) 802851/54
Focus: Adult Educ *07654*

Ente Nazionale Francesco Petrarca, c/o Università degli Studi, Palazzo del Bo, Via VIII Febbraio 2, 35122 Padova
T: (049) 8803763
Founded: 1971
Pres: Prof. Luigi Gui; Gen Secr: Prof. Francesco Piovan
Focus: Lit
Periodicals
Censimento dei Codici Petrarcheschi
Lectura Petrarce
Studi sul Petrarca *07655*

Ente Nazionale Italiano di Unificazione (UNI), Piazza Diaz 2, 20123 Milano
T: (02) 876914; Tx: 312481; CA: Unificazione Milano
Founded: 1921
Focus: Standards
Periodical
L'Unificazione (weekly) *07656*

Ente per la Valorizzazione dei Vini Astigiani (E.V.V.A.), c/o Camera di Commercio, Piazza Medici 8, 14100 Asti
T: (0141) 53011
Founded: 1967
Focus: Wines *07657*

Ente per le Nuove Tecnologie, l'Energia e l'Ambiente, Viale Regina Margherita 125, 00198 Roma
T: (06) 85281; Fax: (06) 85282591
Founded: 1960
Pres: Nicola Cabibbo
Focus: Nucl Res *07658*

Ente Studi Economici per la Calabria (E.S.E.C.), Via A. Arabia 7, 87100 Cosenza
Founded: 1950
Focus: Econ *07659*

ESRIN, Via Galileo Galilei, CP 64, 00044 Frascati
T: (06) 941801; Tx: 610637; Fax: (06) 94180361
Founded: 1966; Members: 13
Focus: Aero *07660*

Europa Club, Sezione Italiana (EKIS) / Associazione per il Superamento delle Barriere Linguistiche in Europa, Via Rovereto 14, 37126 Verona
T: (045) 915837
Focus: Ling *07661*

European Academy of Allergology and Clinical Immunology (EAACI), c/o Clinica Medica I, Policlinico Umberto I, 00161 Roma
T: (06) 4466172; Fax: (06) 4940594
Founded: 1956
Gen Secr: Prof. Sergio Bonini
Focus: Immunology *07662*

European Association for Bioeconomic Studies (EABS), Via Larga, 20122 Milano
Founded: 1991
Focus: Econ *07663*

European Association for the Visual Studies of Man (EAVSoM), c/o Instituto di Antropologia, Via di Proconsolo 12, 50122 Firenze
T: (055) 214049
Gen Secr: Prof. Paolo Chiozzi
Focus: Anthro *07664*

European Association of Remote Sensing Laboratories (EARSeL), c/o Istituto di Gasdinamica, Facoltà di Ingegneria, Università di Napoli, Piazzale Tecchio 80, 80125 Napoli
T: (081) 7682159; Tx: 722392; Fax: (081) 7682160
Founded: 1977; Members: 230
Pres: Prof. S. Vetrella
Focus: Geology
Periodical
EARSeL News (weekly) *07665*

European Association of Senior Hospital Physicians, c/o Primario Divisione Medicina d'Urgenza, 33100 Udine
Founded: 1965
Pres: Dr. Franco Perraro
Focus: Med *07666*

European Association of Social Medicine, Via Sacchi 24, 10128 Torino
T: (011) 517017
Founded: 1955
Gen Secr: Prof. Enrico Belli
Focus: Med *07667*

European Cell Biology Organization (ECBO), c/o Dept of Pharmacology, University of Milan, Via Vanvitelli 32, 20129 Milano
T: (02) 70146254; Fax: (02) 7490937
Founded: 1969
Gen Secr: Francesco Clementi
Focus: Cell Biol & Cancer Res *07668*

European Centre for Research and Development in Primary Health Care (EuroCentre-PHC), Via Marzia 16, 06100 Perugia
T: (075) 6962487; Fax: (075) 6963904
Founded: 1991
Gen Secr: Marlène Läubli
Focus: Public Health *07669*

European Centre for the Validation of Alternative Testing Methods (ECVAM), c/o Joint Research Centre, Environment Institute, 21020 Ispra
T: (0332) 789834; Tx: 380042; Fax: (0332) 789222
Founded: 1991
Gen Secr: Dr. Friedrich Geiss
Focus: Materials Sci *07670*

European Centre for Training Craftsmen in the Conservation of the Architectural Heritage, Isola di San Servolo, CP 676, 30100 Venezia
T: (041) 5268546; Tx: 420362; Fax: (041) 2760211
Founded: 1977
Gen Secr: Wolfdietrich Elbert
Focus: Preserv Hist Monuments *07671*

European Centre of Environmental Studies (ECES), Via Po 14, 10123 Torino
T: (011) 8127167; Fax: (011) 832870
Founded: 1990
Pres: Carlo Savini
Focus: Ecology *07672*

European Commission for the Control of Foot-and-Mouth Disease (EUFMD), c/o FAO, Viale delle Terme di Caracalla, 00153 Roma
T: (06) 51971; Tx: 610181; Fax: (06) 57975749
Founded: 1954; Members: 29
Focus: Vet Med *07673*

European Communities Biologists Association (ECBA), c/o Ordine Nazionale dei Biologi, Via di San Anselmo 11, 00153 Roma
T: (06) 5780553; Fax: (06) 5740682
Founded: 1975
Gen Secr: Prof. Dr. Stefano Dumontet
Focus: Bio *07674*

European Consortium for Church and State Research, c/o Sezione Relazioni tra Stato e Confessioni Religiose, Università degli Studi di Firenze, Via Laura 48, 50121 Firenze
Focus: Poli Sci *07675*

European Coordinating Committee for Artificial Intelligence (ECCAI), c/o Istituto per la Ricerca Scientifica e Tecnologica, Loc Panté di Povo, 38050 Povo
T: (0461) 814444; Tx: 400874; Fax: (0461) 810851
Founded: 1982; Members: 25
Pres: Dr. Oliviero Stock; Gen Secr: Susan Struthers
Focus: Computer & Info Sci *07676*

European Council for Social Research on Latin America, Via Arcangelo Corelli 10, 00198 Roma
T: (06) 79237364; Fax: (06) 79200083
Founded: 1971
Gen Secr: Prof. Pierangelo Catalano
Focus: Soc Sci *07677*

European Dialysis and Transplant Association – European Renal Association (EDTA/ERA), CP 474, Parma Sud, 43100 Parma
T: (0521) 290343; Fax: (0521) 291777
Founded: 1964
Gen Secr: Prof. Vincenzo Cambi
Focus: Surgery; Hematology *07678*

European Federation for the Advancement of Anaesthesia in Dentistry, CP 1630, 40100 Bologna
T: (051) 247784
Founded: 1973
Pres: Dr. Luigi Baldinelli
Focus: Anesthetics *07679*

European Network of Scientific Information Referral Centres (EUSIREF), CP 64, 00044 Frascati
Focus: Sci *07680*

European Safeguards Research and Development Association (ESARDA), c/o Commission of the European Communities Joint Research Centre, 21020 Ispra
T: (0332) 789372; Tx: 380042, 380058; Fax: (0332) 789509
Founded: 1969; Members: 10
Pres: G. Dean; Gen Secr: C. Foggi
Focus: Eng
Periodical
ESARDA Bulletin (weekly) *07681*

European Society for Clinical Respiratory Physiology, c/o Fondazione Menarini, Piazza del Carmine 4, 20121 Milano
T: (02) 874932
Founded: 1966
Gen Secr: L. Allegra
Focus: Physiology
Periodical
SEPCR Newsletter (weekly) *07682*

European Society for the Study of Ultrasonics, c/o Centro Minerva Medica, Via L. Spallanzani, 00161 Roma
Focus: Physics *07683*

European Society of Hypnosis in Psychotherapy and Psychosomatic Medicine-Italian Constituent Society, c/o Centro Studi di Ipnosi Clinica H. Bernheim, Via Valverde 65, 37100 Verona
T: (045) 30795
Founded: 1964
Pres: Prof. Gualtiero Guantieri; Gen Secr: Dr. Piero Roncaroli
Focus: Therapeutics
Periodical
International Journal of Clinical and Experimental Hypnosis 07684

Famiglia e Libertà / Comitato Tecnico Permanente per lo Studio della Legislazione e dei Rimedi, Via della Conciliazione 15, 00193 Roma
T: (06) 6544841
Founded: 1965
Focus: Family Plan; Sociology 07685

Federazione delle Associazioni Scientifiche e Tecniche (FAST), Piazzale Rodolfo Morandi 2, 20121 Milano
T: (02) 783051; Fax: (02) 782485
Founded: 1961; Members: 33
Pres: Francesco Sponzilli; Gen Secr: Dr. Alberto Pieri
Focus: Sci; Eng
Periodicals
Alta Frequenza
Automazione e Strumentazione
L'Elettrotecnica
La Fonderia Italiana
La Meccanica Italiana
Medicina dello Sport: Mensile di fisiopatologia dello sport (weekly)
La Metallurgia Italiana
Notiziario della FAST
La Termotecnica 07686

Federazione delle Istituzioni Antropologiche Italiane (F.I.A.I.), c/o Istituto di Antropologia Fisica, Università degli Studi, Via Balbi 4, 16126 Genova
T: (010) 204654
Founded: 1977
Focus: Anthro 07687

Federazione Europea di Zootecnia (F.E.Z.) / Fédération Européenne de Zootechnie / European Association for Animal Production, Via A. Tortonia 15a, 00161 Roma
T: (06) 8840785; Fax: (06) 8441733
Founded: 1949
Pres: Prof. A. Nardone; Gen Secr: N. Frydlender
Focus: Zoology
Periodical
Livestock Production Science (weekly) . . 07688

Federazione Italiana contro la Tubercolosi e le Malattie Polmonari Sociali, Via G. da Procida 7, 00162 Roma
T: (06) 44240682; Fax: (06) 44240682
Focus: Pulmon Dis
Periodical
Lotta contro la Tubercolosi e le Malattie Polmonari Sociali (weekly) 07689

Federazione Italiana dei Cineclub (FEDIC), Via de Villa Patrizi 10, 00161 Roma
T: (06) 421901
Founded: 1949
Focus: Cinema
Periodical
Giornale dello Spettacolo: Notiziario FEDIC 07690

Federazione Italiana delle Scienze e delle Attività Motorie (F.I.S.A.M.), Via Federico Cesi 30, 00193 Roma
T: (06) 315577
Founded: 1965
Focus: Auto Eng
Periodical
Società Domani, Dibattiti 07691

Federazione Italiana Dottori in Agraria e Forestali (FEDERAGRONOMI), Via Livenza 6, 00198 Roma
T: (06) 8416036; Fax: (06) 8555961
Founded: 1944; Members: 74
Focus: Agri; Forestry
Periodical
Il Dottore in Scienze Agrarie e Forestali (weekly) 07692

Federazione Medico-Sportiva Italiana (FMSI), Viale Tiziano 70, 00196 Roma
T: (06) 394670, 393221; Fax: (06) 36858206
Pres: Prof. Gustavo Tuccimei; Gen Secr: Dr. Emilio Gasbarrone
Focus: Med; Sports
Periodical
Medicina dello Sport (weekly) 07693

Federazione Nazionale Collegi Tecnici di Radiologia Medica, c/o G.C. Luisetti, Viale S. Marco 56, 30173 Mestre
Focus: Radiology 07694

Federazione Nazionale degli Ordini dei Medici Chirurghi e Odontoiatri, Piazza Cola di Rienzo 80a, 00192 Roma
T: (06) 362031
Founded: 1946; Members: 34
Focus: Med
Periodicals
Federazione Medica (weekly)
Il Medico d'Italia 07695

Federazione Nazionale degli Ordini dei Veterinari Italiani (FNOVI), Via del Tritone 125, 00186 Roma
T: (06) 485923
Focus: Vet Med 07696

Federazione Nazionale Insegnanti Educazione Fisica (FNIEF), Via Pasquale Revoltella 41, 00152 Roma
T: (06) 533417
Focus: Sports 07697

Federazione Nazionale pro Natura, c/o Istituto per lo Sviluppo Economico dell'Appennino, Via Marchesana 12, 40124 Bologna
T: (051) 231999
Founded: 1959
Focus: Ecology
Periodical
Natura Società (weekly) 07698

Fondazione Centro di Documentazione Ebraica Contemporanea (F.C.D.E.C.), Via Eupili 8, 20145 Milano
T: (02) 316338, 316092
Founded: 1955
Pres: Dr. Luisella Mortara Ottolenghi; Gen Secr: N.N.
Focus: Hist 07699

Forum Italiano dell'Energia Nucleare (FIEN), Via Paisello 26-28, 00198 Roma
T: (06) 868291
Founded: 1958; Members: 233
Focus: Nucl Res 07700

Gruppi Archeologici d'Italia (G.A.I.), Via Tacito 41, 00193 Roma
T: (06) 6874028
Founded: 1960; Members: 3000
Focus: Archeol
Periodical
Archeologia (weekly) 07701

Gruppo Italiano di Linguistica Applicata (GILA), c/o Istituto del Consiglio Nazionale delle Richerche, Via S. Maria 36, 56100 Pisa
Focus: Ling 07702

International Association for the Study of Canon Law, c/o Universitá degli Studi, Cittá Universitaria, 00100 Roma
Founded: 1973
Focus: Rel & Theol; Law 07703

International Association for Veterinary Homeopathy, Via Soresina 16, 20144 Milano
Focus: Vet Med
Periodical
International Journal for Veterinary Homeopathy (weekly) 07704

International Association for Water Law (IAWL), Via Montevideo 5, 00198 Roma
T: (06) 8548932
Founded: 1967; Members: 93
Focus: Law
Periodicals
Annales Juris Aquarum (weekly)
Aquaforum (weekly) 07705

International Association of Biblicists and Orientalists, Piazza Duomo 4, 48100 Ravenna
T: (0544) 23432
Founded: 1967
Pres: Dr. Angelo Duranti
Focus: Rel & Theol; Ethnology
Periodical
Biblia Revuo (weekly) 07706

International Association of Engineering Geology, Sezione Italiana, c/o Istituto di Geologia Applicata e Geotecnica, Facoltà di Ingegneria, Università degli Studi, Via Re David 200, 70125 Bari
T: (080) 228369
Founded: 1972
Focus: Eng; Geology
Periodical
Bulletin of the International Association of Engineering Geology 07707

International Board for Plant Genetic Resources (IBPGR), Via delle Sette Chiese 142, 00145 Roma
T: (06) 518921; Tx: 4900005322; Fax: (06) 5750309
Founded: 1974
Focus: Genetics
Periodical
Plant Genetic Resources Newsletter (weekly) 07708

International Centre for Advanced Technical and Vocational Training / Centre International de Perfectionnement Professionnel et Technique, Corso Unità d'Italia 125, 10127 Torino
T: (011) 633733; Tx: 221449; CA: INTERLAB Torino; Fax: (011) 638842
Founded: 1963
Focus: Adult Educ; Eng; Sociology; Econ; Marketing; Finance; Travel; Energy; Educ; Develop Areas; Animal Husb; Agri; Fisheries; Doc
Periodical
Bulletin (3 times annually) 07709

International Centre for Mechanical Sciences, Piazza Garibaldi 18, 33100 Udine
T: (0432) 294989
Founded: 1968
Gen Secr: Prof. Giovanni Bianchi
Focus: Eng
Periodical
Mechanics Research Communications (weekly) 07710

International Centre for Theoretical Physics (ICTP), CP 586, 34136 Trieste
T: (040) 22401; Tx: 460392; CA: CENTRATOM; Fax: (040) 224163
Founded: 1964
Focus: Physics
Periodicals
Annual Report of Scientific Activities (weekly)
Directory of Mathematicians from Dev. Countries (weekly)
Directory of Physicists from Dev. Countries (weekly)
News from ICTP (weekly) 07711

International Committee for Recording the Productivity of Milk Animals (ICRPMA), Via A. Torlonia 15a, 00161 Roma
T: (06) 8840785; Fax: (06) 8441733
Founded: 1951
Pres: Dr. K. Meyn; Gen Secr: N Frydlender
Focus: Agri 07712

International Council of Museums (I.C.O.M.) / Comitato Nazionale Italiano, c/o Museo Nazionale della Scienza e della Tecnica, Via S. Vittore 19, 20123 Milano
Tx: 334147
Founded: 1976
Focus: Archives
Periodical
Notiziario (weekly) 07713

International Juridical Organization for Environment and Development (IJO), Via Barberini 3, 00187 Roma
T: (06) 4742117; Tx: 614046; CA: Juricountries Rome; Fax: (06) 4745779
Founded: 1964
Pres: Mario Guttieres; Gen Secr: M.E. Sikabonyi
Focus: Law; Energy; Ecology
Periodical
IJO Newsletter (weekly) 07714

International Law Association, Sezione Italiana (I.L.A.), Via delle Quattro Fontane 15, 00184 Roma
T: (06) 463979, 460981; Tx: 616313; Fax: (06) 4820686
Pres: Prof. Francesco Capotorti; Gen Secr: A. Giuseppe Guerreri
Focus: Law 07715

International League of Esperantist Teachers, Via Palestro 36, 54100 Massa
T: (0585) 41756
Founded: 1949; Members: 2700
Gen Secr: Catina Dazzini
Focus: Ling; Educ
Periodical
IPORO Internacia Pedagogia Revuo (3 times annually) 07716

International Poplar Commission (IPC), c/o FAO, Viale delle Terme di Caracalla, 00153 Roma
T: (06) 57971; Fax: (06) 5795137
Founded: 1947; Members: 35
Gen Secr: J.B. Ball
Focus: Forestry
Periodical
Session Report 07717

International Society for Medical and Psychological Hypnosis (ISMPH), c/o Istituto di Indagini Psicologiche, Corso XXII Marzo 57, 20129 Milano
T: (02) 7388427; Fax: (02) 7491051
Founded: 1985; Members: 112
Pres: Prof. Rolando Marchesan; Dr. Milton V. Kline; Gen Secr: Dr. Aurora Zavertanik
Focus: Med; Psych 07718

International Society for the Study of Infectious and Parasitic Diseases, c/o Ospedale Amadeo di Savoia, Corso Svizzera 164, 10149 Torino
Focus: Microbio; Immunology 07719

International Society for Twin Studies (ISTS), c/o Mendel Institute, Piazza Galeno 5, 00162 Roma
Founded: 1974; Members: 227
Pres: Dr. Luigi Gedda; Gen Secr: Dr. Paolo Parisi
Focus: Behav Sci; Genetics; Med
Periodicals
Acta Geneticae Medicae et Gemellologiae/Twin Research (weekly)
Twins (weekly) 07720

International Society of Theoretical and Experimental Hypnosis (ISTEH), c/o Istituto di Indagini Psicologiche, Corso XXII Marzo 57, 20129 Milano
T: (02) 7388427; Fax: (02) 7401051
Founded: 1988; Members: 65
Pres: Prof. Rolando Marchesan
Focus: Psychoan 07721

International Solar Energy Society, Sezione Italiana (I.S.E.S.), Via Francesco Crispi 72, 80121 Napoli
T: (081) 666684
Founded: 1964
Focus: Energy
Periodicals
Energie Alternative: Bimestrale della Sezione Italiana (weekly)
ISES News (weekly)
Solar Energy
Sun World 07722

International Study Group for Steroid Hormones (ISGSH), c/o Clinica Medica, Policlinico Umberto I, 00161 Roma
T: (06) 4940568; Fax: (06) 490530
Founded: 1961
Pres: Prof. Francesco Sciarra
Focus: Pathology
Periodical
Proceedings of the Meetings 07723

Istituto Agronomico per l'Oltremare (I.A.O.), Via Antonio Cocchi 4, 50131 Firenze
T: (055) 573201; Tx: 574549; Fax: (055) 580314
Founded: 1904; Members: 61
Pres: Gian Luigi Curotti
Focus: Agri
Periodical
Rivista di Agricoltura Subtropicale e Tropicale (weekly) 07724

Istituto Cooperativo per l'Innovazione (ICIE) / Istituto Nazionale per la Ricerca Applicata ed il Trasferimento Tecnologico della Lega Nazionale delle Cooperative e Mutue / Cooperative Institute for Innovation, Via Nomentana 133, 00161 Roma
T: (06) 8845848; Fax: (06) 8550250
Founded: 1972; Members: 30
Pres: Pietro Andreotti
Focus: Archit
Periodical
Innovazione: Trimestrale di Informazione Tecnico-Scientifica (weekly) 07725

Istituto Cooperazione Economica Internazionale (I.C.E.I.), Via Tommaso Salvini 3, 20122 Milano
T: (02) 799144, 784723
Founded: 1978
Pres: Michele Achilli; Gen Secr: Lucia Lanzanova
Focus: Econ; Int'l Relat 07726

Istituto di Sociologia Internazionale di Gorizia (I.S.I.G.) / Institute of International Sociology, Gorizia, Via Mazzini 13, 34170 Gorizia
T: (0481) 32580; Fax: (0481) 532094
Founded: 1968
Pres: Dr. Mario Brancati; Gen Secr: Prof. Alberto Gasparini
Focus: Sociology; Int'l Relat; Poli Sci
Periodicals
Futuribili (weekly)
ISIG (weekly) 07727

Istituto Elettrotecnico Nazionale Galileo Ferraris (IEN), Corso Massimo d'Azeglio 42, 10125 Torino
T: (011) 3488933; Tx: 211553; Fax: (011) 6507611
Founded: 1935; Members: 150
Pres: Prof. Gian Federico Micheletti; Gen Secr: Dr. Carlo Manacorda
Focus: Electric Eng
Periodical
Pubblicazioni IEN (weekly) 07728

Istituto Geografico Militare (IGMI), Via Cesare Battisti 10-12, 50100 Firenze
T: (055) 27751
Founded: 1872; Members: 404
Pres: Enrico Borgenni
Focus: Geography
Periodicals
Bollettino di Geodesia e Scienze Affini (weekly)
L'Universo (weekly) 07729

Istituto Internazionale delle Comunicazioni (I.I.C.), Villa Piaggio, Via Pertinace, 16125 Genova
T: (010) 294683/84; Fax: (010) 200883
Founded: 1962
Pres: Prof. Francesco Carassa
Focus: Comm
Periodical
Notiziario I.I.C.: Mensile d'Informazione (weekly) 07730

Istituto Internazionale di Diritto Umanitario (IIDU), Corso Cavallotti 115, 18038 San Remo
T: (0184) 541848; Fax: (0184) 541600
Founded: 1970; Members: 119
Focus: Law 07731

Istituto Internazionale di Studi Liguri (IISL), Via Romana 39bis, 18012 Bordighera
T: (0184) 263601; Fax: (0184) 263601
Founded: 1947; Members: 2000
Focus: Hist
Periodicals
Cahiers Ligures de Préhistoire et de Protohistoire (weekly)
Giornale Storico della Lunigiana e del Territorio Lucense (weekly)
Rivista di Studi Liguri (weekly)
Rivista Ingauna e Intemelia
Studi Genuensi (weekly) 07732

Istituto Internazionale di Vulcanologia, Viale Regina Margherita 6, 95123 Catania
Focus: Geology 07733

Istituto Internazionale per l'Unificazione del Diritto Privato, Via Panisperna 28, 00184 Roma
T: (06) 6783189
Founded: 1926; Members: 51
Pres: Prof. Riccardo Monaco
Focus: Law
Periodicals
NEWS Bulletin (3 times annually)
Uniform Law Review (weekly) 07734

Istituto Italiano degli Attuari, Via del Corea 3, 00186 Roma
T: (06) 3226051
Founded: 1930; Members: 355
Focus: Insurance
Periodical
Giornale (weekly) 07735

Istituto Italiano della Saldatura, Lungobisagno Istria 15, 16141 Genova
T: (010) 83411; Fax: (010) 867780
Founded: 1948; Members: 800
Focus: Eng
Periodical
Rivista Italiana della Saldatura (weekly) . 07736

Istituto Italiano di Numismatica (IIN), Palazzo Barberini, Via Quattro Fontane 13, 00184 Roma
T: (06) 4743603
Founded: 1912
Focus: Numismatics
Periodical
Annali dell'Istituto Italiano di Numismatica (weekly)
. . . . 07737

Istituto Italiano di Paleontologia Umana (I.I.P.U.), Piazza Mincio 2, 00198 Roma
T: (06) 8557598
Founded: 1927; Members: 235
Focus: Paleontology
Periodicals
Memorie
Quaternaria Nova: Atti – pubblicati nella rivista
. . . . 07738

Istituto Italiano di Pubblicismo (I.I.P.), c/o Facoltà di Scienze Statistiche, Università degli Studi, Piazzale delle Scienze 2, 00185 Roma
Founded: 1947
Focus: Journalism; Comm Sci
Periodicals
Bibliografia Italiana sull'Informazione (weekly)
Saggi e Studi di Pubblicistica (weekly) . 07739

Istituto Italiano di Storia della Chimica, Via G. B. Morgagni 32, 00161 Roma
Focus: Chem; Cultur Hist 07740

Istituto Italiano di Studi Germanici, Via Calandrelli 25, 00153 Roma
T: (06) 582465, 5897577; Fax: (06) 5835929
Founded: 1932
Gen Secr: Prof. Dr. Paolo Chiarini
Focus: Ling; Lit
Periodical
Studi Germanici (3 times annually) . . . 07741

Istituto Italiano per il Medio ed Estremo Oriente (ISMEO), Via Merulana 248, 00185 Roma
T: (06) 732741; Tx: 624163
Founded: 1933
Pres: Prof. Gherardo Gnoli
Focus: Ethnology
Periodicals
Cina (weekly)
East and West (weekly)
Il Giappone (weekly) 07742

Istituto Italiano per la Storia Antica, Via Milano 76, 00184 Roma
T: (06) 460597
Founded: 1935; Members: 5
Focus: Hist 07743

Istituto Italiano per la Storia della Musica (IISM), Via Vittoria 6, 00187 Roma
T: (06) 6798259
Founded: 1938
Pres: Prof. Raffaello Monterosso
Focus: Hist; Music 07744

Istituto Italiano Studi di Ipnosi Clinica e Psicoterapia, Via Valverde 65, 37100 Verona
T: (045) 30795
Pres: Prof. Gualtiero Guantieri; Gen Secr: Dr. Andrea Gambacciani
Focus: Therapeutics
Periodical
Bernheim Newsletter (3 times annually) . 07745

Istituto Italo-Africano, Via U. Aldrovandi 16, 00197 Roma
Focus: Ethnology; Hist; Econ; Poli Sci . 07746

Istituto Nazionale delle Assicurazioni (I.N.A.), Via Sallustiana 51, 00187 Roma
T: (06) 4882497; Fax: (06) 47224595
Founded: 1912
Pres: Lorenzo Pallesi
Focus: Insurance
Periodical
Assicurazioni: Bimestrale di Diritto, Economia e Finanza delle Assicurazioni Private (weekly)
. . . . 07747

Istituto Nazionale di Ottica (INO), Largo Enrico Fermi 6, 50125 Firenze
T: (055) 23081; Fax: (055) 2337755
Founded: 1927; Members: 32
Pres: Prof. Fortunato Tito Arecchi; Gen Secr: Chiara Sergardi Burlamacchi
Focus: Optics 07748

Istituto Nazionale di Urbanistica (INU), Via S. Caterina da Siena 46, 00186 Roma
T: (06) 6793559
Founded: 1929; Members: 2488
Focus: Urban Plan
Periodicals
Urbanistica (weekly)
Urbanistica Informazioni (weekly) 07749

Istituto per gli Studi di Politica Internazionale (ISPI), Via Clerici 5, 20121 Milano
T: (02) 878266/68
Founded: 1933; Members: 80
Pres: Egidio Ortona; Gen Secr: Prof. Giuliano Urbani
Focus: Poli Sci
Periodicals
Quaderni-Papers (10-12 times annually)
Relazioni Internazionali (weekly) 07750

Istituto per la Cooperazione Economica Internazionale e i Problemi di Sviluppo (I.C.E.P.S.) / Institute for International Economic Cooperation and Development, Via Cola di Rienzo 11, 00192 Roma
T: (06) 3215095; Fax: (06) 3214690
Founded: 1966; Members: 55
Pres: Angelo Maria Sanza; Gen Secr: Giuseppe Bonanno di Linguaglossa
Focus: Econ; Int'l Relat; Develop Areas
Periodical
Booklets and Documents (weekly) . . . 07751

Istituto per la Cooperazione Universitaria (I.C.U.), Viale Grossini 26, 00198 Roma
T: (06) 85300782; Tx: 621148; Fax: (06) 8443204
Founded: 1966
Pres: Prof. Raffaello Cortesini; Gen Secr: Prof. Umberto Farri
Focus: Adult Educ
Periodical
SIPE – Servizio Stampa Educazione e Sviluppo: Notiziario d'informazione (weekly) 07752

Istituto per la Storia del Risorgimento Italiano (ISTOR), Piazza Venezia, Vittoriano, 00186 Roma
T: (06) 6793526, 6793589
Founded: 1935; Members: 3500
Pres: Prof. Emilia Morelli; Gen Secr: Prof. Romano Ugocini
Focus: Hist
Periodical
Rassegna Storica del Risorgimento (weekly)
. . . . 07753

Istituto per l'Oriente (I.P.O.), Via Alberto Caroncini 19, 00197 Roma
T: (06) 804106
Founded: 1921; Members: 200
Focus: Ethnology
Periodicals
Oriens Antiquus
Oriente Moderno
Rassegna di Studi Etiopici 07754

Istituto Storico Italiano per l'Età moderna e contemporanea, Via Michelangelo Caetani 32, 00186 Roma
T: (06) 68806922; Fax: (06) 6875127
Founded: 1934
Pres: Prof. Luigi Lotti
Focus: Hist
Periodical
Annuario dell'Istituto Storico Italiano per l'Età moderna e contemporanea (bi-annually) . 07755

ITALCONSULT / Società Generale per Progettazioni, Consulenze e Partecipazioni, Via Giorgione 163, 00147 Roma
T: (06) 54671
Founded: 1957
Focus: Eng 07756

Italia Nostra – Associazione Nazionale per la Tutela del Patrimonio Storico Artistico e Naturale della Nazione, Corso Vittorio Emanuele 287, 00186 Roma
T: (06) 6565751
Founded: 1955; Members: 15000
Focus: Preserv Hist Monuments
Periodical
Italia Nostra 07757

ITALSIEL / Società Italiana Sistemi Informativi Elettronici, Via Isonzo 21b, 00198 Roma
T: (06) 841351
Founded: 1969
Focus: Electronic Eng; Computer & Info Sci
. . . . 07758

Keats-Shelley Memorial Association (KSMA), Piazza di Spagna 26, 00187 Roma
T: (06) 6784235; Fax: (06) 6784167
Founded: 1906
Pres: Kenneth Prichard-Jones; Gen Secr: Warren Obluck
Focus: Lit
Periodicals
Keats-Shelley Bulletin (weekly)
Keats-Shelley Journal (weekly) 07759

Latin-Mediterranean Society of Pharmacy, Via Belmeloro 6, 40126 Bologna
Founded: 1953
Focus: Pharmacol 07760

Lega Italiana per la Lotta contro i Tumori, Via Alessandro Torlonia 15, 00161 Roma
T: (06) 867382, 8845024; Fax: (06) 8450362
Focus: Cell Biol & Cancer Res 07761

Ordine Nazionale degli Attuari, Via del Corea 3, 00186 Roma
T: (06) 6794014
Focus: Insurance 07762

P.E.N. International Centre, Via Muzio Clementi 64, 00193 Roma
Pres: Maria Bellonci; Gen Secr: Rosario Assunto
Focus: Lit 07763

Sezione di Training Autogeno del Centro Internazionale de Ipnosi, c/o Istituto di Indagini Psicologiche, Corso XXII Marzo 57, 20129 Milano
T: (02) 7388427; Fax: (02) 7491051
Members: 72
Pres: Dr. Carlo Lunardi; Dr. Santi Gatto
Focus: Psych 07764

Sindacato Autonomo Scuola Media Italiana (SASMI), Viale Trastevere 60, 00153 Roma
T: (06) 5803367
Focus: Educ 07765

Sindacato Nazionale Autori Drammatici (SNAD), Via dei Baullari 4, 00186 Roma
Founded: 1948
Focus: Lit 07766

Sindacato Nazionale Istruzione Artistica (S.N.I.A.), Via Antonino Pio 40, 00145 Roma
Focus: Arts; Educ
Periodical
Mensile di Informazione Sindacale sull'Istruzione Artistica (weekly) 07767

Sindacato Nazionale Scrittori (SNS), Via Basento 52d, 00198 Roma
T: (06) 8440837
Founded: 1945; Members: 1400
Focus: Lit 07768

Società Adriatica di Scienze (SAS) / Adriatic Society of Sciences, Via Orsera 3, CP 1029, 34145 Trieste
Founded: 1874; Members: 200
Pres: Prof. Dulio Lausi; Gen Secr: Dr. Tiziana C. Velari
Focus: Sci
Periodical
Bollettino della Società Adriatica di Scienze (weekly) 07769

Società Archeologica Comense, c/o Civico Museo, Giovio, Piazzale delle Medaglie d'Oro Comasche 1, 22100 Como
T: (031) 211343, 269022
Founded: 1902
Pres: Cesare Piovan; Gen Secr: Alberto Pozzi
Focus: Archeol
Periodical
Rivista Archeologica della Antica Provincia e Diocesi di Como (weekly) 07770

Società Archeologica Viterbese Pro Ferento, Via S. Pietro 80, 01100 Viterbo
T: 1927
Focus: Archeol 07771

Società Astronomica Italiana, Via Brera 28, 20121 Milano
T: (02) 874444
Founded: 1920; Members: 550
Focus: Astronomy
Periodicals
Giornale di Astronomica (weekly)
Memorie della Società Astronomica Italiana (weekly) 07772

Società Botanica Italiana (S.B.I.), c/o Dipartimento di Biologia Vegetale, Facoltà di Scienze Matematiche, Fisiche e Naturali, Università degli Studi, Via Giorgio La Pira, 50121 Firenze
T: (055) 2757379; Fax: (055) 2757379
Founded: 1888; Members: 900
Focus: Botany
Periodicals
Giornale Botanico Italiano (weekly)
Informatore Botanico Italiano (weekly) . . 07773

Società Chimica Italiana (SCI), Viale Liegi 48, 00198 Roma
T: (06) 8549691; Fax: (06) 8548734
Founded: 1909; Members: 5000
Pres: Prof. Ivano Bertini; Gen Secr: Dr. M.A. Berardi
Focus: Chem
Periodicals
Annali di Chimica (weekly)
La Chimica e l'Industria: Quaderni dell'Ingegnere Chimico Italiano (weekly)
Chimica nella Scuola (weekly)
Il Farmaco (weekly)
Gazzetta Chimica Italiana (weekly) . . . 07774

Società Dalmata di Storia Patria, c/o Società Dante Alighieri, Palazzo Firenze, Piazza Firenze 27, 00186 Roma
T: (06) 686992
Focus: Hist
Periodical
Atti e Memorie 07775

Società Dante Alighieri, Piazza Firenze 27, 00186 Roma
T: (06) 6873694
Founded: 1889
Pres: Prof. Massimo Pallottino; Gen Secr: Dr. Giuseppe Cota
Focus: Lit
Periodical
Pagine della Dante (3 times annually) . 07776

Società Dantesca Italiana (S.D.I), Via dell'Arte della Lana 1, 50123 Firenze
T: (055) 294580
Founded: 1888; Members: 260
Pres: Prof. Francesco Mazzoni
Focus: Lit
Periodicals
Quaderni degli Studi Danteschi
Quaderni del Centro di Documentazione Dantesca e Medievale
Studi Danteschi (weekly) 07777

Società Dauna di Cultura, c/o Palazzetto della Cultura e dell'Arte, Via Ferdinando Galiani 1, 71100 Foggia
T: (0881) 23042
Founded: 1947
Focus: Lit; Arts; Eng
Periodical
Rassegna di Studi Dauni (weekly) . . . 07778

Società degli Amici del Museo Civico di Storia Naturale Giacomo Doria, Via Brigata Liguria 9, 16121 Genova
T: (010) 585753
Founded: 1927
Pres: Dr. Enzo Coddé; Gen Secr: Dr. Anna Maria Cominetti
Focus: Mineralogy; Paleontology; Botany; Zoology
. . . . 07779

Società degli Ingegneri e degli Architetti in Torino, Corso Massimo d'Azeglio 42, 10125 Torino
T: (011) 6508511; Fax: (011) 6508168
Founded: 1866
Pres: Prof. Giorgio De Ferrari; Gen Secr: Beatrice Coda Negozio
Focus: Archit; Eng
Periodical
Atti e Rassegna Tecnica della Società degli Ingegneri e degli Architetti in Torino (weekly)
. . . . 07780

Società dei Naturalisti in Napoli, Via Mezzocannone 8, 08134 Napoli
T: (081) 207922
Founded: 1881; Members: 350
Pres: Prof. Aldo Napoletamo; Gen Secr: Prof. Teresa De Cunzo
Focus: Nat Sci
Periodicals
Bollettino della Società dei Naturalisti in Napoli
Memorie: Supplemento al Bollettino (weekly)
. . . . 07781

Società di Fotogrammetria e Topografia (SIFET), Piazzale Rodolfo Morandi 2, 20121 Milano
Founded: 1951
Focus: Surveying 07782

Società di Letture e Conversazioni Scientifiche, Piazza Fontane Marose 6, 16123 Genova
Founded: 1866
Focus: Lit; Sci 07783

Società di Linguistica Italiana (S.L.I.), Via M. Caetani 32, 00186 Roma
T: (06) 651613; Fax: (06) 66987044
Founded: 1966; Members: 1000
Pres: Prof. Lorenzo Renzi; Gen Secr: Prof. Emanuele Banfi
Focus: Ling
Periodical
Bollettino della S.L.I. (weekly) 07784

Società di Medicina Legale e delle Assicurazioni, c/o Istituto di Medicina Legale e delle Assicurazione, Università degli Studi, Viale Regina Elena 336, 00161 Roma
T: (06) 4952941
Founded: 1907
Focus: Public Health
Periodical
Zacchia (weekly) 07785

Società di Minerva, c/o Biblioteca Civica, Piazza A. Hortis 4, 34123 Trieste
Founded: 1810; Members: 171
Pres: Gino Pavan
Focus: Hist
Periodical
Archeografo Triestino 07786

Società di Ortopedia e Traumatologia dell'Istituto Meridionale ed Insulare (S.O.T.I.M.I.), c/o Clinica Ortopedica I, Facoltà di Medicina, Università degli Studi, Via S. Andrea delle Dame 4, 80138 Napoli
T: (081) 459824
Founded: 1956
Focus: Traumatology
Periodical
Atti e Memorie della S.O.T.I.M.I. (weekly) 07787

Società di Scienze Farmacologiche Applicate (S.S.F.A.), Via G. Jan 18, 20129 Milano
Fax: (02) 29520179
Pres: Dr. Domenico Criscuolo; Gen Secr: Dr. Paolo Baroldi
Focus: Pharmacol
Periodical
Bolletino di Informazione (weekly) . . . 07788

Società di Scienze Naturali del Trentino, c/o Museo Tridentino di Scienze Naturali, Via Calepina 14, 38100 Trento
T: (0461) 234760
Founded: 1948
Focus: Nat Sci
Periodical
Natura Alpina (weekly) 07789

Società di Storia Patria di Terra di Lavoro, Palazzo Reale, 81100 Caserta
T: (0823) 326037
Founded: 1952
Focus: Hist; Sociology
Periodical
Archivio Storico di Terra di Lavoro (weekly) 07790

Società di Storia Patria per la Puglia, Piazza Umberto 2, 70121 Bari
T: (080) 237538
Pres: Prof. Francesco Maria De Robertis; Gen Secr: Prof. Mauro Spagnoletti
Focus: Hist
Periodical
Archivio Storico Pugliese (weekly) . . . 07791

Società di Storia Patria per la Sicilia Orientale, Piazza Stesicoro 29, 95124 Catania
T: (095) 275920
Founded: 1903
Focus: Hist
Periodicals
Archivio Storico per la Sicilia Orientale
Quaderni di Filologia e Letteratura Siciliana 07792

Società di Studi Celestiniani (S.S.C.), c/o Biblioteca Comunale, 86170 Isernia
T: 2372
Founded: 1919
Focus: Hist; Rel & Theol 07793

Società di Studi Geografici (S.S.G.), Via S. Gallo 10, 50129 Firenze
T: (055) 2757956
Founded: 1895; Members: 550
Pres: Paolo Roberto Federici; Gen Secr: Laura Cassi
Focus: Geography
Periodical
Rivista Geografica Italiana (weekly) . . . 07794

Società di Studi Romagnoli (S.S.R.), c/o Biblioteca Malatestiana, Piazza Bufalini 1, 47023 Cesena
T: (0547) 21297
Founded: 1949
Focus: Hist; Geography
Periodical
Studi Romagnoli (weekly) 07795

Società di Studi Trentini di Scienze Storiche, Via Petrarca 36, 38100 Trento
T: (0461) 983383
Focus: Hist 07796

Società di Studi Valdesi, Via Beckwith 3, 10066 Torre Pellice
Founded: 1881; Members: 650
Focus: Hist
Periodical
Bollettino (weekly) 07797

Società Ecologica Friulana (S.E.F.), Viale delle Rose 60, 33030 Campoformido
T: (0432) 69716
Founded: 1973
Focus: Hist; Lit; Poli Sci; Ecology
Periodical
Corriere del Friuli 07798

Società Economica di Chiavari, Via Ravaschieri 15, 16043 Chiavari
T: 309941
Founded: 1791
Focus: Econ
Periodicals
Atti della Società Economica di Chiavari
Enciclopedia Araldica 07799

Società Emiliana Pro Montibus et Silvis, c/o Accademia Nazionale di Agricoltura, Via Castiglione 11, 40124 Bologna
T: (051) 268809
Pres: Prof. Gabriele Goidanich; Gen Secr: Dr. Giorgio Monti
Focus: Ecology
Periodical
Natura & Montagna (weekly) 07800

Società Entomologica Italiana, Via Brigata Liguria 9, 16121 Genova
Founded: 1869; Members: 980
Pres: Prof. Cesare Conci; Gen Secr: Dr. Roberto Poggi
Focus: Entomology
Periodicals
Bollettino
Memorie: Supplemento al Bollettino . . . 07801

Società Europea di Cultura / Société Européenne de Culture, S. Marco 2516, 30124 Venezia
T: (041) 5230210
Founded: 1950
Pres: Prof. Vincenzo Cappelletti; Gen Secr: Michelle Campagnolo-Bouvier
Focus: Arts; Int'l Relat
Periodical
Comprendre (weekly) 07802

Società Europea di Patologia / European Society of Pathology, c/o Istituto di Anatomia e Istologia Patologica, Via Francesco Sforza 38, 20122 Milano
T: (02) 874214
Focus: Pathology 07803

Società Farmaceutica del Mediterraneo Latino, c/o Istituto di Farmacognosia, Facoltà di Farmacio dell'Universita, Villaggio Annunziata, 98100 Messina
T: (090) 1632
Founded: 1953
Focus: Pharmacol
Periodical
Pharmacia Mediterranea 07804

Società Filarmonica di Trento, Via Verdi 30, 38100 Trento
T: (0461) 21830
Founded: 1795
Focus: Music 07805

Società Filologica Friulana G.I. Ascoli (S.F.F.), Via Manin 18, 33100 Udine
T: (0432) 501598; Fax: (0432) 511766
Founded: 1919
Pres: Prof. Manlio Michelutti; Gen Secr: Prof. Gianfranco Ellero
Focus: Ling; Lit
Periodicals
Ce Fastu? (weekly)
Sot la Nape (weekly)
Strolic (weekly) 07806

Società Filologica Romana, Piazzale Aldo Moro, Città Universitaria, 00185 Roma
T: (06) 491919
Founded: 1901
Pres: Prof. Aurelio Roncaglia
Focus: Ling; Lit
Periodical
Studi Romanzi 07807

Società Filosofica Calabrese (S.F.C.), Via C. Battisti 12, 89015 Palmi
T: 22523
Founded: 1948
Focus: Philos 07808

Società Filosofica Italiana (S.F.I.), Corso Magenta 83, 20123 Milano
T: (02) 463386
Founded: 1906; Members: 1900
Pres: Prof. Armando Rigobello; Gen Secr: Prof. Luciana Vigone
Focus: Philos
Periodical
Bollettino della Società Filosofica Italiana (weekly) 07809

Società Filosofica Romana, c/o Istituto di Filosofia, Università, Città Universitaria, 00185 Roma
T: 4991
Focus: Philos 07810

Società Gallaratese per gli Studi Patri, Chiostro di S. Francesco, Via Borgo Antico 2, 2013 Gallarate
T: 795092
Founded: 1896
Focus: Hist
Periodical
Rassegna Gallaratese di Storia e d'Arte (weekly) 07811

Società Geografica Italiana (S.G.I), Via della Navicella 12, 00184 Roma
T: (06) 7008279; Fax: (06) 7004677
Founded: 1867; Members: 1300
Pres: Prof. Gaetano Ferro; Gen Secr: Prof. Franco Salvatori
Focus: Geography
Periodicals
Bibliografia Geografica della Regione Italiana (weekly)
Bollettino della Società Geografica Italiana (weekly)
Memorie della S.G.I.: Monografie scientifiche 07812

Società Geologica Italiana (S.G.I.), Città Universitaria, Piazzale delle Scienze 2, 00100 Roma
T: (06) 4959390
Founded: 1881; Members: 2036
Focus: Geology
Periodicals
Bollettino (weekly)
Memorie (weekly) 07813

Società Incoraggiamento Arti e Mestieri, Via S. Maria 18, 20123 Milano
Founded: 1838
Focus: Arts 07814

Società Internazionale di Diritto Penale Militare e di Diritto della Guerra, Gruppo Italiano / Société Internationale de Droit Penal Militaire et de Droit de la Guerre; Groupe Italiennne, Viale delle Milizie 5c, 00192 Roma
T: (06) 386052
Founded: 1956
Focus: Law; Int'l Relat
Periodical
Rivista di Diritto Penale e di Diritto della Guerra 07815

Società Internazionale di Psicologia della Scrittura (S.I.P.S.) / International Society for the Psychology of Writing, Corso XXII Marzo 57, 20129 Milano
T: (02) 7426489, 7388427
Founded: 1961; Members: 187
Pres: Prof. Marco Marchesan; Gen Secr: Prof. Rolando Marchesan
Focus: Lit; Psych; Graphology
Periodical
Rivista Internazionale di Psicologia e Ipnosi (weekly) 07816

Società Internazionale di Studi Francescani, Piazza del Comune 27, 06081 Assisi
T: (075) 813210
Founded: 1972
Focus: Rel & Theol; Hist 07817

Società Internazionale di Studi Gemellari / International Society for Twin Studies, c/o Istituto Gregorio Mendel, Piazza Galeno 5, 00162 Roma
T: (06) 862055
Pres: Prof. Luigi Gedda; Gen Secr: Prof. Paolo Parisi
Focus: Gynecology; Bio; Genetics
Periodicals
Acta Genetica Medicae et Gemellologiae: Twin Research (weekly)
Twins (weekly) 07818

Società Internazionale di Tecnica Idrotermale (S.I.T.H.), Via de Gasperi 144, 80053 Castellamare di Stabia
T: 8715322
Focus: Hydrology; Water Res 07819

Società Istriana di Archeologia e Storia Patria (S.I.A.S.P.), c/o Archivio di Stato, Via La Marmora 17, 34139 Trieste
Founded: 1884
Focus: Hist; Archeol; Folklore; Geography; Ling; Numismatics
Periodical
Atti e Memorie (weekly) 07820

Società Italiana Amici dei Fiori, Via dei Tavolini 8, 50123 Firenze
Founded: 1931
Focus: Hort; Botany
Periodical
Il Giardino Fiorito (weekly) 07821

Società Italiana Attività Nervosa Superiore (SIANS), c/o Clinica Psychiatrica Romeo Vuoli, Via G.F. Besta 1, 20161 Milano
T: (02) 584100
Focus: Neurology 07822

Società Italiana Calcolo Ricerca Economica Operativa (S.I.C.R.E.O.), Via Polistena 10, 00173 Roma
T: (06) 7485435, 7485476, 7485480
Founded: 1960
Focus: Econ 07823

Società Italiana degli Economisti, Via Garribaldi 4, 16124 Genova
Founded: 1950; Members: 170
Focus: Econ
Periodical
Bollettino (weekly) 07824

Societá Italiana della Continenza, c/o Cattedra I'Patologia Ostetrica e Ginecologica, Università di Roma, Policlinico Umberto I, 00161 Roma
T: (06) 4959341
Focus: Pathology; Gynecology 07825

Società Italiana della Scienza del Suolo (S.I.S.S.), c/o Istituto Sperimentale per lo Studio e la Difesa del Suolo, Piazza M. D'Azeglio 30, 50121 Firenze
T: (055) 2477242; Fax: (055) 241485
Members: 250
Pres: Prof. Giovanni Fierotti; Gen Secr: Prof. Giulio Ronchetti
Focus: Geology
Periodical
Bollettino della Società Italiana della Scienza del Suolo: Nuova Serie (weekly) 07826

Società Italiana della Trasfusione del Sangue, Corso Principe Oddone 18, 10153 Torino
T: (011) 471515
Founded: 1936; Members: 150
Focus: Hematology
Periodical
Rivista di Emoterapia ed Immunoematologia 07827

Società Italiana delle Scienze Veterinarie (S.I.S.Vet), Via Antonio Bianchi 1, 25100 Brescia
T: (030) 52516
Founded: 1947; Members: 1100
Focus: Vet Med
Periodical
Atti (weekly) 07828

Società Italiana di Agopuntura, Corso Principe Oddone 18, 10153 Torino
T: (011) 471515
Founded: 1968
Focus: Anesthetics
Periodical
Rivista Italiana di Agopuntura 07829

Società Italiana di Agronomia (SIA) / Italian Society of Agronomy, c/o Istituto di Agronomia, Facoltà di Agraria, Università degli Studi, Via Filippo Re 6-8, 40126 Bologna
T: (051) 351510; Fax: (051) 351511
Founded: 1966
Pres: Prof. Luigi Cavazza
Focus: Agri
Periodical
Rivista di Agronomia (weekly) 07830

Società Italiana di Allergologia e Immunologia Clinica, c/o Policlinico Umberto I, Viale del Policlinico, 00100 Roma
T: (06) 4950068
Founded: 1953
Focus: Immunology
Periodical
Folia Allergologica et Immunologica Clinica (weekly) 07831

Società Italiana di Anatomia, c/o Istituto di Anatomia Umana Normale, Policlinico di Careggi, 50134 Firenze
T: (055) 410084
Founded: 1929; Members: 300
Focus: Anat 07832

Società Italiana di Anatomia Patologica, Via P. Castelli, 98100 Messina
T: (090) 44895
Focus: Anat; Pathology 07833

Società Italiana di Andrologia (S.I.A.) / Society of Andrology, c/o Centro di Andrologia, 1. Clinica Medica, Università degli Studi, Policlinico S. Chiara, 56100 Pisa
T: (050) 26135
Founded: 1976; Members: 400
Focus: Pathology
Periodicals
Andrologia
Archives of Andrology
Fertility and Sterility
Infertility
International Journal of Andrology
Journal of Andrology
Journal of Endocrinological Investigation . 07834

Società Italiana di Anestesia, Analgesia, Rianimazione e Terapia Intensiva, c/o Istituto di Anestesiologia e Rianimazione, Università degli Studi di Firenze, Viale Morgagni 85, 50134 Firenze
Fax: (055) 430393
Members: 4000
Focus: Anesthetics
Periodicals
Acta Anesthesiologica Italia (weekly)
Minerva Anesthesiologica (weekly) . . . 07835

Società Italiana di Anestesiologia Rianimazione e Terapia del Dolore (S.I.A.R.), c/o Istituto di Anestesiologia e Rianimazione, Facoltà di Medicina e Chirurgia, Università degli Studi, Viale Ennio, 70124 Bari
T: (080) 250395
Founded: 1934
Focus: Anesthetics 07836

Società Italiana di Angiologia, c/o Athenaeum Angiologicum Santorianum, Domus Camiliana, Via San Ranierino 9, 56100 Pisa
T: (050) 26358
Founded: 1953
Focus: Hematology 07837

Società Italiana di Antropologia ed Etnologia, Via del Proconsolo 12, 50122 Firenze
T: (055) 214049
Founded: 1871
Focus: Anthro; Ethnology
Periodical
Archivio per l'Antropologia e l'Etnologia (weekly) 07838

Società Italiana di Audiologia e Foniatria (SIA), c/o Clinica ORL, Università, Policlinico Umberto I, Viale del Policlinico, 00161 Roma
T: (06) 490054
Founded: 1967; Members: 300
Focus: Otorhinolaryngology
Periodical
Bollettino di Audiologia e Foniatria (weekly) 07839

Societá Italiana di Aziendologia, Viale dei Mille 38, 20129 Milano
T: (02) 7490247
Focus: Business Admin
Periodical
Tecniche Direzionali (weekly) 07840

Società Italiana di Biochimica, c/o Istituto di Chimica Biologica, Università, Via del Giochetto, CP 3, 06100 Perugia
Founded: 1951
Focus: Chem; Biochem 07841

Società Italiana di Biogeografia (S.I.B.), c/o Istituto di Biologia Generale, Università degli Studi, Via T. Pendola 62, 53100 Siena
T: (0577) 287006; Fax: (0577) 44474
Founded: 1954
Pres: Prof. B. Baccetti
Focus: Botany; Paleontology
Periodical
Biogeographia 07842

Società Italiana di Biologia Marina (S.I.B.M.), c/o Acquario Comunale, Piazzale Mascagni 1, 57100 Livorno
T: (0586) 805504
Founded: 1969
Focus: Bio; Oceanography 07843

Società Italiana di Biologia Sperimentale (S.I.B.S.), c/o Istituto di Fisiologia Umana, Facoltà di Medicina e Chirurgia II, Università degli Studi, Via Sergio Pansini 5, 80131 Napoli
T: (081) 5453022; Fax: (081) 5453045
Focus: Bio
Periodical
Bollettino Società Italiana di Biologia Sperimentale (weekly) 07844

Società Italiana di Biometria (SIB), c/o Dipartimento di Statistica e Matematica, Via Curtatone 1, 50123 Firenze
T: (055) 284845
Founded: 1963
Focus: Bio 07845

Società Italiana di Buiatria, c/o Istituto di Patologia Speciale e Clinica Medica Veterinaria, Facoltà di Medicina Veterinaria, Università degli Studi, Via Nizza 52, 10126 Torino
T: (011) 651828, 688769
Focus: Vet Med
Periodicals
Atti della Società Italiana di Buiatria (weekly)
Notiziario Buiatrico (weekly) 07846

Società Italiana di Cancerologia, Via Giacomo Venezian 1, 20133 Milano
Fax: (02) 2664342
Focus: Cell Biol & Cancer Res 07847

Società Italiana di Cardiologia, Corso di Francia 197, 00191 Roma
T: (06) 36309819
Focus: Cardiol
Periodical
Cardiologia (weekly) 07848

Società Italiana di Chemioterapia (SIC), c/o Istituto C. Forlanini, Viale Taramelli 5, 27100 Pavia
T: (0382) 23707
Founded: 1953; Members: 210
Focus: Therapeutics
Periodical
Giornale Italiano di Chemioterapia (weekly) 07849

Società Italiana di Chirurgia, c/o Clinica Chirurgica B, Policlnico Umberto I, 00161 Roma
T: (06) 4956239
Founded: 1882
Focus: Surgery
Periodicals
Archivio ed Atti (weekly)
Bollettino (weekly) 07850

Società Italiana di Chirurgia Cardiaca e Vascolare, c/o Istituto di Chirurgia del Cuore e Grossi Vasi, Policlinico Umberto I, Lungotevere Flaminio 30, 00196 Roma
T: (06) 3230227; Fax: (06) 3220744
Founded: 1963; Members: 720
Pres: Prof. Domenico Bertini; Gen Secr: Prof. Paolo Micozzi
Focus: Surgery
Periodical
Archivio di Chirurgia Toracica e Cardiovascolare (weekly) 07851

Società Italiana di Chirurgia Clinica, c/o Policlinico, Clinica Chirurgia B, 00161 Roma
T: (06) 4956239
Founded: 1882; Members: 220
Focus: Surgery 07852

Società Italiana di Chirurgia della Mano, Largo Piero Palagi 1, 50139 Firenze
T: (055) 427811
Focus: Surgery
Periodical
Rivista Italiana di Chirurgia della Mano . 07853

Società Italiana di Chirurgia d'Urgenza, di Pronto Soccorso e di Terapia Intensiva Chirurgica, c/o Istituto di Chirurgia d'Urgenza, Facoltà di Medicina e Chirurgia, Università degli Studi, Via Sforza 35, 20122 Milano
T: (02) 581635
Focus: Surgery 07854

Società Italiana di Chirurgia Estetica, Via della Camilluccia 643, 00135 Roma
T: (06) 36304792; Fax: (06) 36303204
Founded: 1970
Pres: Giorgio Fischer
Focus: Surgery 07855

Società Italiana di Chirurgia Pediatrica, c/o Ospedale Bambino Gesu, Piazza S. Onofrio 4, 00165 Roma
T: (06) 68591
Founded: 1963; Members: 379
Focus: Surgery; Pediatrics
Periodical
Rassegna Italiana di Chirurgia Pediatrica (weekly) 07856

Società Italiana di Chirurgia Plastica, c/o Div Chirurgia Plastica, Istituto Ospitalieri, 37100 Verona
T: (045) 912600
Focus: Surgery 07857

Società Italiana di Chirurgia Toracica, Via Augusto Murri 4, 00161 Roma
T: (06) 4958300
Founded: 194
Focus: Surgery 07858

Società Italiana di Citologia Clinica e Sociale, c/o Istituto Regina Elena, Viale Regina Elena 291, 00161 Roma
T: (06) 497931
Founded: 1961
Focus: Cell Biol & Cancer Res
Periodical
Citologia Informazioni 07859

Società Italiana di Criminologia, c/o Istituto di Antropologia Criminale, Via Bartolo Longo 72, 00156 Roma
T: (06) 416671
Founded: 1957
Focus: Criminology
Periodical
Rassegna di Criminologia (weekly) . . . 07860

Società Italiana di Dermatologia e Sifilografia (SIDES), c/o Istituto S. Gallicano, Via S. Gallicano 25a, 00153 Roma
T: (06) 5813741, 5892390
Founded: 1883; Members: 330
Focus: Derm; Venereology 07861

Società Italiana di Diabetologia, c/o Ospedale S. Giovanni Battista e della Città di Torino, Istituto di Endocrinologia, Corso Polonia 14, 10126 Torino
Focus: Diabetes 07862

Società Italiana di Diabetologia e Endocrinologia Pediatrica, c/o Istituto di Clinica Pediatrica III dell'Università, Via G.B. Grassi 74, 20157 Milano
T: (02) 3556241
Focus: Diabetes 07863

Società Italiana di Economia Agraria (S.I.D.E.A.), c/o Istituto di Estimo Rurale, Facoltà di Agraria, Università degli Studi, Via Filippo Re 10, 40126 Bologna
T: (051) 243268; Fax: (051) 252187
Founded: 1962; Members: 300
Focus: Agri
Periodical
Rivista di Economia Agraria: Atti del Convegno di Studi (weekly) 07864

Società Italiana di Economia Demografia e Statistica (SIEDS), CP 12003, Belsito, 00136 Roma
T: (06) 3451503
Founded: 1938; Members: 600
Pres: Prof. Stefano Somogyi; Gen Secr: Prof. M. Natale
Focus: Econ; Stats
Periodicals
Atti Sociali
Collano di Studi e Monografie
Rivista Italiana di Economia, Demografia e Statistica (weekly) 07865

Società Italiana di Elettroencefalografia e Neurofisiologia, c/o Clinica Neurologica, Facoltàtà di Medicina, Università degli Studi, Via Ugo Foscolo 7, 40123 Bologna
T: (051) 585158
Founded: 1949; Members: 270
Focus: Neurology; Physiology 07866

Società Italiana di Ematologia, c/o Istituto di Clinica Medica Generale e Terapia Medica I, Padiglione Granelli, Policlinico, Via S. Sforza 35, 20122 Milano
T: (02) 573166
Focus: Hematology
Periodical
Haematologica (weekly) 07867

Società Italiana di Endocrinologia (SIE), Via Cosenza 8, 00161 Roma
T: (06) 44231346; Fax: (06) 4404294
Founded: 1951; Members: 851
Pres: Prof. Mario Andreoli
Focus: Endocrinology
Periodical
Journal Endocrinological Investigation (weekly) 07868

Società Italiana di Ergonomia (S.I.E.), Via S. Barnaba 8, 20122 Milano
T: (02) 57992613; Fax: (02) 55187172
Founded: 1961; Members: 170
Pres: Prof. Antonio Grieco; Gen Secr: Dr. Marco Fregoso
Focus: Sociology
Periodical
Notiziario S.I.E. (weekly) 07869

Società Italiana di Ergonomia Stomatologica (S.I.E.S.), Via Commenda 10, 20122 Milano
T: (02) 593593
Founded: 1968
Focus: Stomatology; Physiology 07870

Società Italiana di Farmacologia, Via Balzaretti 9, 20133 Milano
T: (02) 29404672; Fax: (02) 29404961
Founded: 1946; Members: 1099
Pres: Prof. Rodolfo Paoletti; Gen Secr: Prof. Giampaolo Velo
Focus: Pharmacol
Periodical
Pharmacological Research (weekly) . . . 07871

Società Italiana di Farmacologia Clinica, c/o Istituto di Farmacologia, Facoltà di Medicina e Chirurgia, Università degli Studi, Via Roma 55, 56100 Pisa
T: (050) 22312, 46009; Tx: 590035
Focus: Pharmacol
Periodical
Drugs under Experimental and Clinical Research (weekly) 07872

Società Italiana di Filosofia Giuridica e Politica, c/o Facoltà di Giurisprudenza, Università, 00185 Roma
T: (06) 490489; Fax: (06) 49910951
Founded: 1936
Pres: Prof. Dino Fiorot; Gen Secr: Dr. Maurizio Basciu
Focus: Philos; Law; Poli Sci
Periodicals
Proceedings of the Società Italiana di Filosofia Giuridica e Politica (bi-annually)
Rivista Internazionale di Filosofia del Diritto (weekly) 07873

Società Italiana di Fisica, Via Castiglione 101, 40136 Bologna
T: (051) 331554
Founded: 1897; Members: 1023
Pres: Prof. R.A. Ricci
Focus: Physics
Periodicals
Europhysics Letters (3 times per month)
Giornale di Fisica (weekly)
Il Nuovo Cimento A (weekly)
Il Nuovo Cimento B (weekly)
Il Nuovo Cimento C (weekly)
Il Nuovo Cimento D (weekly)
Il Nuovo Saggiatore (weekly)
Rivista del Nuovo Cimento (weekly) . . 07874

Società Italiana di Fisiologia, c/o Istituto di Fisiologia Umana, Viale Margagni 65, 50134 Firenze
Focus: Physiology 07875

Società Italiana di Foniatria, c/o Clinica ORL, Ospedale Garibaldi, Università, 95100 Catania
Focus: Otorhinolaryngology 07876

Società Italiana di Fotobiologia, c/o Laboratorio di Radiobiologia Animale del C.N.E.N., Località Casaccia, 00060 S. Maria di Galeria
Focus: Radiology; Bio 07877

Società Italiana di Fotogrammetria e Topografia (S.I.F.E.T.), Piazzale Morandi 2, 20121 Milano
T: (02) 296621
Founded: 1951
Focus: Surveying
Periodical
Bollettino della Società Italiana di Fotogrammetria e Topografia (weekly) 07878

Società Italiana di Gastroenterologia (SIGE), Clinica Medica, Università S. Orsola, 40100 Bologna
Founded: 1937
Focus: Gastroenter 07879

Società Italiana di Genetica Agraria (SIGA) / Italian Society of Agricultural Genetics, c/o Istituto Sperimentale per la Cerealicoltura, S.S. 16 km. 675, 71100 Foggia
T: (0881) 42972; Fax: (0881) 693150
Founded: 1954; Members: 400
Pres: Prof. Franco Lorenzetti; Gen Secr: Dr. Natale Di Fonzo
Focus: Genetics 07880

Società Italiana di Geofisica e Meteorologia, c/o Università degli Studi, Via Balbi 30, 16126 Genova
Focus: Geophys
Periodical
Rivista Italiana di Geofisica e Scienze Affini 07881

Società Italiana di Gerontologia e Geriatria, Via G.C. Vanini 5, 50129 Firenze
T: (055) 474330; Fax: (055) 461217
Members: 2500
Focus: Geriatrics
Periodicals
Giornale della Arteriosclerosi
Giornale di Gerontologia (weekly) . . . 07882

Società Italiana di Ginecologia Pediatrica, Via L. Spallazani 11, 00161 Roma
Founded: 1978
Focus: Pediatrics; Gynecology 07883

Società Italiana di Ginnastica Medica, Medicina Fisica e Riabilitazione (SIGM), Via Crivelli 20, 20122 Milano
T: (02) 5453328
Founded: 1952
Focus: Rehabil; Physical Therapy
Periodical
La Ginnastica Medica (weekly) 07884

Società Italiana di Glottologia (S.I.G.), c/o Istituto di Glottologia, Via S. Maria 36, 56100 Pisa
Founded: 1970
Focus: Ling 07885

Società Italiana di Immunoematologia e Trasfusione del Sangue (S.I.I.T.S. – A.I.C.T.), Via Rocca Sinibalda 71, 00199 Roma
T: (06) 86203724; Fax: (06) 8600740
Founded: 1954
Focus: Hematology; Immunology
Periodical
La Trasfusione del Sangue (weekly) . . 07886

Società Italiana di Immunologia e di Immunopatologia (S.I.I.I.), c/o Istituto di Clinica Medica a, Università degli Studi, Policlinico, 70124 Bari
T: (080) 360713, 369728
Founded: 1968
Focus: Immunology; Pathology
Periodical
Proceedings 07887

Societá Italiana di Ingegneria, Aerofotogrammetria e Topografia (S.I.A.T.), Via Giuseppe Armellini 35, 00143 Roma
T: (06) 5920946
Founded: 1961
Focus: Surgery 07888

Società Italiana di Ippologia, c/o Istituto di Clinica Chirurgica Veterinaria, Facoltà di Veternaria, Università degli Studi, Strada Comunale del Cornocchio, 43100 Roma
T: (06) 96287
Focus: Vet Med 07889

Società Italiana di Laringologia, Otologia, Rinologia e Patologia Cervico-Facciale, c/o Istituto di Clinica Otorinolaringolatrica I, Facoltà di Medicina e Chirurgia, Università degli Studi, Viale del Policlinico, 00161 Roma
T: (06) 490051
Focus: Otorhinolaryngology
Periodical
Acta O.R.L. Italica (weekly) 07890

Società Italiana di Liposcultura, Via della Camilluccia 643, 00135 Roma
T: (06) 36304792; Fax: (06) 36303204
Founded: 1986
Pres: Giorgio Fischer
Focus: Med 07891

Società Italiana di Malacologia (S.I.M.), c/o Acquario Civico, Viale Gadio 2, 20121 Milano
T: (02) 872867; Fax: (02) 325721
Founded: 1965
Pres: Piero Piani
Focus: Zoology
Periodicals
Bollettino Malacologico (weekly)
Lavori S.I.M. (weekly)
Notiziario S.I.M. (weekly) 07892

Società Italiana di Medicina del Lavoro e di Igiene Industriale, Via Severino Boezio 24, 27100 Pavia
T: (0382) 301221
Founded: 1906; Members: 1000
Focus: Med; Hygiene
Periodical
Bollettino Lavoro 07893

Società Italiana di Medicina del Traffico (SIMT), c/o Clinica Ortopedica e Traumatologica dell'Università, Piazzale delle Scienze, 00185 Roma
T: (06) 491672
Founded: 1958; Members: 700
Focus: Traumatology; Orthopedics; Surgery
Periodical
Rassegna di Medicina del Traffico (weekly) 07894

Società Italiana di Medicina e Igiene della Scuola, Via Vincenzo Monti 57, 20145 Milano
T: (02) 4817404, 48007052
Founded: 1951
Pres: Dr. Marcello Cantoni; Gen Secr: Dr. Ada Azria
Focus: Med; Hygiene
Periodical
Rivista Italiana di Medicina e Igiene della Scuola (weekly) 07895

Società Italiana di Medicina Estetica, Viale Mazzini 13, 00195 Roma
T: (06) 351514, 385377
Founded: 1975
Focus: Physiology
Periodical
La Medicina Estetica (weekly) 07896

Società Italiana di Medicina Fisica e Riabilitazione (S.I.M.F.E.R.), c/o Centro F. Cornaglia, Corso Savona 25, 10024 Moncalieri
T: (011) 645596
Focus: Physiology; Rehabil 07897

Società Italiana di Medicina Interna (SIMI), Via Savoia 37, 00198 Roma
T: (06) 8554399; Fax: (06) 8559067
Founded: 1888
Pres: Prof. Gian Gastone Neri Serneri; Gen Secr: Prof. Franco Dammacco
Focus: Intern Med

Periodical
Rivista Ufficiale Annali Italiani di Medicina Interna (weekly) 07898

Società Italiana di Medicina Legale e delle Assicurazioni (S.I.M.L.A.), Viale Regina Elena 336, 00161 Roma
T: (06) 4952941
Focus: Med
Periodical
Minerva Medico-Legale (weekly) 07899

Società Italiana di Medicina Preventiva e Sociale (S.I.M.P.S.), Via Caffaro 19, 16124 Genova
Founded: 1966
Focus: Hygiene; Public Health 07900

Società Italiana di Medicina Psicosomatica (SIMP), Via della Camilluccia 195, 00135 Roma
T: (06) 3420230
Founded: 1966; Members: 1000
Pres: Prof. Ferruccio Antonelli; Gen Secr: Dr. Mario Cimica
Focus: Psychiatry
Periodical
Medicina Psicosomatica (weekly) 07901

Società Italiana di Medicina Sociale (SIMS), Corso Bramante 83, 10126 Torino
T: (011) 687272
Founded: 1947; Members: 500
Pres: Prof. Ferdinando Antoniotti; Gen Secr: Prof. Enrico Belli
Focus: Hygiene
Periodical
Rivista Italiana di Medicina Sociale e Preventiva (weekly) 07902

Società Italiana di Medicina Subacquea ed Iperbarica (S.I.M.S.I.), Via Schipa 68, 80122 Napoli
T: (081) 664462
Focus: Therapeutics 07903

Società Italiana di Mesoterapia, Viale Mazzini 13, 00195 Roma
T: (06) 351514
Focus: Therapeutics 07904

Società Italiana di Meteorologia Applicata (S.I.M.A.), Via Flavia 104, 00187 Roma
T: (06) 486415, 4744397
Founded: 1979
Pres: Prof. Guiseppe Cena; Gen Secr: Achille Somma
Focus: Geophys 07905

Società Italiana di Microangiologia e Microcorcolazione, c/o Athenaeum Angiologicum Santorianum, Domus Camilliana, Via S. Ranierino 9, 56100 Pisa
T: (050) 26358
Focus: Microbio 07906

Società Italiana di Microbiologia, c/o Istituto di Microbiologia, Via Androne 81, 95124 Catania
T: (095) 312633; Fax: (095) 312633
Founded: 1962
Focus: Bio; Microbio 07907

Società Italiana di Mineralogia e Petrologia (SIMP), c/o Museo di Storia Naturale, Corso Venezia 55, 20121 Milano
T: (02) 702018
Founded: 1941; Members: 530
Focus: Mineralogy; Petrochem
Periodical
Rendiconti (weekly) 07908

Società Italiana di Musicologia (SIDM), Via Galliera 3, 40121 Bologna
Founded: 1964; Members: 630
Pres: Prof. Agostino Ziino
Focus: Music
Periodicals
Monumenti Musicali Italiani
Quaderni della Rivista Italiana di Musicologia (weekly)
Rivista Italiana di Musicologia 07909

Società Italiana di Neurochirurgia (SIN), c/o Ospedale Generale Regionale, 60100 Ancona
T: (071) 56636
Founded: 1948; Members: 220
Focus: Surgery
Periodical
Journal of Neurosurgical Sciences . . . 07910

Società Italiana di Neurologia, c/o Istituto di Clinica Neurologica, Università Cattolica del S. Cuore, Via Pineta Sacchetti 526, 00168 Roma
Founded: 1905; Members: 785
Focus: Neurology 07911

Società Italiana di Neuropediatria, c/o Istituto di Clinica Pediatrica, Facoltà di Medicina e Chirurgia, Università degli Studi, Via Mattioli 10, 53100 Siena
T: (0577) 283284
Focus: Pediatrics; Neurology 07912

Società Italiana di Neuropsichiatria Infantile (SINPI), c/o Cattedra di Neuropsichiatria Infantile, Via Pansini 5, 80131 Napoli
Focus: Psychiatry 07913

Società Italiana di Neuroradiologia, Via Generale Orsini 40, 80132 Napoli
T: (081) 393483
Focus: Radiology 07914

Società Italiana di Neurosonologia (S.I.N.S.), Via Casaglia 65, 40135 Bologna
T: (051) 589092
Founded: 1970
Focus: Neurology 07915

Società Italiana di Nipiologia, c/o Istituto di Clinica Pediatrica II, Viale Regina Elena 324, 00161 Roma
T: (06) 4951738, 490962
Founded: 1915
Focus: Pediatrics
Periodical
Minerva Nipiologica (weekly) 07916

Società Italiana di Nutrizione Umana, c/o Cattedra di Scienza dell'Alimentazione, Viale Forlanini 1, 00161 Roma
T: (06) 858276, 860584
Focus: Food; Nutrition 07917

Società Italiana di Odontoiatria Infantile (SIOI), Viale Shakespeare 47, 00144 Roma
T: (06) 596443
Founded: 1956; Members: 250
Focus: Dent
Periodical
Bollettino della SIOI (weekly) 07918

Società Italiana di Odontostomatologia e Chirurgia Maxillo-Facclale (S.I.O.C.M.F.), c/o Clinica Odontostomatologica, Via Verdi 28, 67100 L'Aquila
T: (0862) 411176; Fax: (0862) 411176
Founded: 1957; Members: 1200
Pres: Prof. Sergio Tartaro; Gen Secr: Prof. Franco Marci
Focus: Dent; Stomatology; Surgery
Periodical
Minerva Stomatologica (weekly) 07919

Società Italiana di Oncologia Ginecologica, Via O. Tommasini 1, 00162 Roma
Founded: 1977
Focus: Cell Biol & Cancer Res; Gynecology
. . . . 07920

Società Italiana di Ortopedia e Traumatologia (S.I.O.T.), c/o Clinica Ortopedicae Traumatologica, Università degli Studi, Piazzale delle Scienze 5, 00185 Roma
T: (06) 491672
Founded: 1906; Members: 2300
Pres: Prof. Lombardo Perugia; Gen Secr: Prof. Marcello Pizzetti
Focus: Orthopedics; Traumatology . . . 07921

Società Italiana di Ostetricia e Ginecologia (S.I.O.G.), c/o Policlinico Umberto I, 00161 Roma
T: (06) 4954292
Founded: 1892; Members: 1050
Focus: Gynecology
Periodical
Notiziario e Bollettino 07922

Società Italiana di Oto-Neuro-Oftalmologia, c/o Clinica Oculista, Università degli Studi, 40138 Bologna
T: (051) 392948
Founded: 1923
Focus: Otorhinolaryngology; Neurology; Ophthal
. . . . 07923

Società Italiana di Otorinolaringologia e Chirurgia Cervico Facciale (S.I.O.e Ch.C.-F.), c/o II. Clinica Otorinolaringolatrica, Policlinico Umberto I, Viale del Policlinico, 00161 Roma
T: (06) 490051
Founded: 1893; Members: 850
Focus: Otorhinolaryngology
Periodical
Acta O.R.L. Italica (weekly) 07924

Società Italiana di Otorinolaringologia Pediatrica (S.I.O.P.), c/o Ospedale Bambino Gesù, Piazza S. Onofrio 4, 00165 Roma
T: (06) 68592498; Fax: (06) 68592522
Founded: 1978; Members: 391
Focus: Otorhinolaryngology; Pediatrics
Periodical
L'Otorinolaringologia Pediatrica (3 times annually)
. . . . 07925

Società Italiana di Parassitologia, c/o Istituto di Parassitologia, Università degli Studi di Roma, Piazzale Aldo Moro 5, 00185 Roma
T: (06) 49914932; Fax: (06) 49914644
Founded: 1959; Members: 325
Pres: Prof. Mario Coluzzi; Gen Secr: Dr. Michele Maroli
Focus: Med
Periodical
Parassitologia (3 times annually) 07926

Società Italiana di Patologia, c/o Istituto di Anatomia e Istologia Patologica, Via Francesco Sforza 38, 20122 Milano
T: (02) 874200, 874214
Founded: 1948; Members: 400
Focus: Pathology
Periodicals
Atti
Bollettino 07927

Società Italiana di Patologia Aviare (S.I.P.A.), c/o Istituto di Patologia Generale e Anatomia Patologica, Via S. Costanzo 4, 06100 Perugia
T: (075) 5854534; Fax: (075) 5854532
Founded: 1960; Members: 152
Pres: Prof. G. Adrubali; Gen Secr: Prof. D. Gallazzi
Focus: Vet Med
Periodical
Zootecnica International: Atti del Convegno Annuale (weekly) 07928

Società Italiana di Patologia ed Allevamento dei Suini / Italian Society of Swine Pathology and Breeding, c/o Istituto Zooprofilattico, Via del Mercati 13a, 43100 Parma
T: (0521) 293538; Fax: (0521) 291314
Founded: 1974; Members: 500
Pres: Prof. Franco Scatozza; Gen Secr: Dr. Giuseppe Barigazzi
Focus: Vet Med
Periodical
Atti (weekly) 07929

Società Italiana di Patologia e di Allevamento degli Ovini e dei Caprini (S.I.P.A.O.C.), c/o Sezione Diagnostica dell'Istituto Zooprofilattico, Via Passo Gravina 193, 95125 Catania
T: (095) 338585
Founded: 1974
Focus: Vet Med 07930

Società Italiana di Patologia Vascolare (S.I.P.V.), Via Nazionale 230, 00184 Roma
T: (06) 4757192
Focus: Pathology; Physiology; Biochem . 07931

Società Italiana di Pediatria (S.I.P.), c/o Clinica Pediatrica, Università, Viale Regina Elena 324, 00100 Roma
T: (06) 4951738, 5870442
Founded: 1936; Members: 5500
Focus: Pediatrics
Periodical
Rivista Italiana di Pediatria 07932

Società Italiana di Pneumologia, c/o Edizioni Minerva Medica, Corso Bramante 83, 10126 Torino
T: (011) 678282; Fax: (011) 3121736
Focus: Pulmon Dis
Periodical
Minerva Pneumologica (weekly) 07933

Società Italiana di Psichiatria (SIP), c/o Istituto di Clinica Psichiatrica, Via Francesco Sforza 35, 20100 Milano
T: (02) 5483075
Focus: Psychiatry
Periodical
Bollettino 07934

Società Italiana di Psichiatria Biologica, c/o Clinica Psichiatrica, Largo Madonna delle Grazie, 80138 Napoli
T: (081) 5666502
Pres: Prof. Mario Maj
Focus: Psychiatry; Bio
Periodical
Bollettino di Psichiatria Biologica (weekly) 07935

Società Italiana di Psicologia (S.I.Ps.), Via Due Macelli 66, 00187 Roma
Focus: Psych
Periodical
Psicologia Italiana (weekly) 07936

Società Italiana di Psicologia Individuale (S.I.P.I.), Via Giason del Maino 19a, 20146 Milano
T: (02) 464355, 4592614
Founded: 1969
Focus: Psych
Periodical
Rivista di Psicologia Individuale (weekly) 07937

Società Italiana di Psicologia Scientifica (SIPS), c/o Istituto di Psicologia, Università Policlinico Filciuzza, 90121 Palermo
Founded: 1971
Focus: Psych 07938

Società Italiana di Psicosintesi Terapeutica (S.I.P.T.), Via San Domenico 16, 50133 Firenze
T: (055) 570140
Founded: 1973; Members: 11
Focus: Therapeutics
Periodical
Psicosintesi Clinica 07939

Società Italiana di Psicoterapia Analitica Immaginativa (S.I.P.A.I.), Via Mantova 5, 26100 Cremona
T: (0372) 430334
Founded: 1975
Focus: Therapeutics 07940

Società Italiana di Radiologia Medica e di Medicina Nucleare, c/o Prof. Aldo Perussia, Via Comelico 2, 20135 Milano
T: (02) 540766
Founded: 1913
Focus: Radiology; Nucl Med 07941

Società Italiana di Reologia (S.R.I.), c/o Istituto Principi Ingegneria Chimica, Università, Piazza V. Tecchio, 80125 Napoli
T: (081) 611800, 610966
Founded: 1970; Members: 87
Focus: Physics 07942

Società Italiana di Reumatologia, c/o Cattedra di Reumatologia, Università degli Studi, Policlinico, 70124 Bari
T: (080) 369933
Founded: 1050
Focus: Rheuma
Periodical
Reumatismo 07943

Società Italiana di Scienze Farmaceutiche (SISF), Via Giorgio Jan 18, 20129 Milano
T: (02) 29513303; Fax: (02) 29520179
Founded: 1953; Members: 500
Focus: Pharmacol
Periodical
Cronache Farmaceutiche (weekly) . . . 07944

Società Italiana di Scienze Fisiche e Matematiche Mathesis, c/o Istituto Matematico, Facoltà di Scienze Matematiche, Fisiche e Naturali, Università degli Studi, Via Vicenza 23, 00185 Roma
T: (06) 4954641
Focus: Math; Physics 07945

Società Italiana di Scienze Naturali, c/o Museo Civico di Storia Naturale, Corso Venezia 55, 20121 Milano
T: (02) 62085405
Founded: 1857; Members: 900
Pres: Dr. Luigi Cagnolaro; Gen Secr: Dr. Bona Bianchi Potenza
Focus: Nat Sci
Periodicals
Atti della Società Italiana di Scienze Naturali e del Museo Civico di Storia Naturale di Milano
Memorie della Società Italiana di Scienze Naturali e del Museo Civico di Storia Naturale di Milano
Natura: Rivista di Scienze Naturali (weekly)
Paleontologia Lombarda: Nuova Serie
Rivista Italiana di Ornitologia 07946

Società Italiana di Senologia, Via Lazzaro Spallanzani 11, 00161 Roma
T: (06) 862289
Focus: Med
Periodical
Giornale Italiano di Senologia (weekly) . 07947

Società Italiana di Sessuologia Clinica, Via Luigi Rizzo 50, 00136 Roma
T: (06) 311384
Focus: Venereology 07948

Società Italiana di Sessuologia Medica, Via di Tor Fiorenza 13, 00199 Roma
T: (06) 8391557
Focus: Venereology 07949

Società Italiana di Sociologà, Piazza delle Scienze 5, 00185 Roma
Founded: 1937
Focus: Sociology 07950

Società Italiana di Statistica (SIS), Salita de Crescenzi 26, 00186 Roma
T: (06) 6869845; Fax: (06) 68806742
Founded: 1939; Members: 1000
Pres: Prof. Alfredo Rizzi; Gen Secr: Prof. Dionisia Maffioli
Focus: Stats
Periodicals
SIS-Bollettino (weekly)
SIS-Informazioni (weekly) 07951

Società Italiana di Storia della Medicina (S.I.S.MED.), Viale Oriani 42, 40137 Bologna
T: (051) 344750
Founded: 1956
Focus: Med; Cultur Hist
Periodical
Rivista di Storia della Medicina 07952

Società Italiana di Studi sul Secolo XVIII, c/o Accademia Letteraria dell'Arcadia, Piazza S. Agostino 8, 00186 Roma
T: (06) 655874
Founded: 1978
Focus: Hist
Periodical
Bollettino 07953

Società Italiana di Terapia Familiare, Via Reno 60, 00198 Roma
T: (06) 864261
Founded: 1976
Focus: Therapeutics
Periodical
Terapia Familiare (weekly) 07954

Società Italiana di Tossicologia, c/o Istituto di Farmacologia, Facoltà di Medicina e Chirurgia, Università Cattolica del S. Cuore, Via Pineta Sacchetti 644, 00168 Roma
T: (06) 33054253
Founded: 1967
Focus: Toxicology
Periodical
Rivista di Tossicologia Sperimentale e Clinica
. . . . 07955

Società Italiana di Urodinamica, c/o Clinica Urologica, Policlinico S. Orsola, Via Massarenti 9, 40138 Bologna
T: (051) 397931
Founded: 1977
Focus: Urology 07956

Società Italiana di Urologia (S.I.U.) / Italian Urological Association, Viale Cortina d'Ampezzo 49, 00135 Roma
T: (06) 3273328
Founded: 1922; Members: 1500
Pres: Prof. Carlo Corbi; Gen Secr: Prof. Lucio Miano
Focus: Urology

Periodicals
Acta Urologica Italica (weekly)
Aula Medica (3 times annually)
Journal of Nephrology (3 times annually)
La Settimana degli Ospedali (weekly) . . *07957*

Società Italiana Medica del Training Autogeno (S.I.M.T.A.), c/o Istituto di Psicologia, Facoltà di Medicina e Chirurgia, Università degli Studi, Viale Berti Pichat 5, 40127 Bologna
T: (051) 276610
Focus: Therapeutics *07958*

Società Italiana Medici e Operatori Geriatrici, Via Ippolito Nievo 16, 50129 Firenze
T: (055) 474510
Founded: 1965; Members: 800
Pres: Prof. Ippolito Scardigli
Focus: Geriatrics
Periodical
Medicina Geriatrica (weekly) *07959*

Società Italiana Medico-Chirurgica di Pronto Soccorso, c/o Pronto Soccorso del Policlinico S. Orsola, Via Massarenti 9, 40136 Bologna
T: (051) 342416
Founded: 1973
Focus: Surgery *07960*

Società Italiana Organi Artificiali, c/o Patologia Chirugica, Policlinico Umberto I, Università degli Studi, 00161 Roma
T: (06) 4950741/43
Focus: Surgery *07961*

Società Italiana per gli Archivi Sanitari Ospedalieri (S.I.A.S.O.), c/o Ospedale Versilia, Direzione Sanitaria, Via A. Fratti, 55049 Viareggio
T: (0584) 31438, 47283
Founded: 1975
Focus: Doc; Med *07962*

Società Italiana per il Progresso della Zootecnia, Via Monte Ortigara 35, 20137 Milano
T: (02) 576527
Focus: Zoology *07963*

Società Italiana per l'Antropologia e la Etnologia (SIAE), Via del Proconsolo 12, 50122 Firenze
T: (055) 2396449
Members: 160
Focus: Anthro; Ethnology
Periodical
Archivio per l'Antropologia e la Etnologia (weekly) *07964*

Società Italiana per l'Archeologia e la Storia delle Arti (S.I.A.S.A.), c/o Facoltà di Architettura, Università degli Studi, Via Monteoliveto 3, 80134 Napoli
T: (081) 320878
Focus: Archeol; Arts; Cultur Hist *07965*

Società Italiana per la Robotica Industriale (S.I.R.I.), c/o Federazione delle Associazioni Scientifiche e Tecniche, Piazzale Rodolfo Morandi 2, 20121 Milano
T: (02) 783051
Focus: Electronic Eng *07966*

Società Italiana per l'Educazione Musicale (S.I.E.M.), Via Clerici 10, 20121 Milano
Founded: 1969
Pres: Prof. Johannella Tafuri; Gen Secr: Dr. Romeo Della Bella
Focus: Music; Educ
Periodical
Musica Domani: Trimestrale di Pedagogia Musicale (weekly) *07967*

Società Italiana per le Scienze Ambientali: Biometeorologia, Bioclimatologia ed Ecologia (S.I.S.A.), c/o Cattedra di Idrologia e Climatologia Medica, Università degli Studi, Via Luigi Vanvitelli 32, 20129 Milano
T: (02) 7386914
Founded: 1970
Focus: Ecology; Geophys *07968*

Società Italiana per l'Organizzazione Internazionale (S.I.O.I.), Piazza di S. Marco 51, 00186 Roma
T: (06) 6793566, 6793949; Fax: (06) 6789102
Founded: 1944
Pres: Umberto La Rocca; Gen Secr: Prof. Luigi Ferrari Bravo
Focus: Business Admin
Periodical
La Comunità Internazionale (weekly) . . *07969*

Società Italiana per lo Studio della Cancerogenesi Ambientale ed Epidemiologia dei Tumori (S.I.S.C.A.), c/o Istituto di Oncologia, Università degli Studi, Viale Benedetto XV 10, 16132 Genova
T: (010) 302754, 300767
Founded: 1977
Focus: Public Health; Cell Biol & Cancer Res *07970*

Società Italiana per lo Studio della Fertilità e della Sterilità, c/o Clinica Ostetrico-Ginecologico, 00168 Roma
T: (06) 33054979
Focus: Venereology; Gynecology *07971*

Società Italiana per lo Studio dell'Arteriosclerosi (S.I.S.A.), c/o Istituto di Clinica Medica la Università degli Studi, Policlinico, 70124 Bari
T: (080) 369728
Founded: 1964
Focus: Pathology *07972*

Società Italiana per lo Studio delle Sostanze Grasse, Via del Lauro 3, 20121 Milano
T: (02) 897280
Focus: Materials Sci *07973*

Società Italiana per lo Studio e l'Applicazione del Pirodiserbo, c/o Istituto di Patologia Vegetale, Facoltà di Agraria, Università Cattolica del S. Cuore, 29100 Piacenza
T: (0523) 62600
Founded: 1968
Focus: Ecology *07974*

Società Italiana pro Deontologia Sanitaria (S.I.De.S.), Piazza S. Agostino 24, 20123 Milano
T: (02) 490887
Founded: 1958
Focus: Hygiene
Periodical
Medicina Sociale: Organo della S.I.De.S. e della Società Italiana di Medicina Sociale . . *07975*

Società Jonico-Salentina de Medicina e Chirurgia, c/o Ospedale Regionale, Via Bruno, 74100 Taranto
Focus: Med; Surgery
Periodical
Bollettino della Società Jonico Salentina di Medicina e Chirurgia *07976*

Società Laziale Abruzzese di Medicina del Lavoro, Viale Regina Elena 336, 00162 Roma
T: (06) 492515
Focus: Hygiene; Pathology *07977*

Società Laziale – Abruzzese Marchigiana Molisana di Ostetricia e Ginecologia (L.A.M.), c/o Istituto di Clinica Ostetrica e Ginecologica, Policlinico Umberto I, Viale Regina Elena 326, 00181 Roma
T: (06) 4955260
Founded: 1900
Focus: Gynecology
Periodical
Aggiornamenti in Ostetricia e Ginecologia (weekly) *07978*

Società Letteraria, Piazzetta Scalette Rubiani 1, 37121 Verona
T: (045) 8030641; Fax: (045) 595949
Founded: 1808
Pres: Dr. Giambattista Ruffo; Gen Secr: Gloria Rivolta
Focus: Lit
Periodical
Bollettino della Società Letteraria di Verona (weekly) *07979*

Società Ligure di Storia Patria, Via Albaro 11, 16145 Genova
T: (010) 308683
Founded: 1857
Focus: Hist
Periodical
Atti (weekly) *07980*

Società Lombarda di Criminologia, Via Carlo Crivelli 20, 20122 Milano
T: (02) 76820
Founded: 1956
Focus: Criminology *07981*

Società Mazziniana Pensiero e Azione, Via Angelo Brunetti 60, 00186 Roma
T: (06) 687410
Founded: 1921
Focus: Hist; Poli Sci *07982*

Società Medica Chirurgica di Bologna, c/o Archiginnasio, Piazza Galvani 1, 40124 Bologna
T: (051) 231488
Founded: 1823; Members: 300
Focus: Surgery
Periodical
Bollettino delle Scienze Mediche (weekly) *07983*

Società Medico-Chirurgica, Facoltà di Medicina e Chirurgia, Università degli Studi, Via Crisanzio 3, 70122 Bari
T: (080) 218849
Focus: Surgery *07984*

Società Medico-Chirurgica di Ferrara, c/o Sez. di Chirurgia Pediatrica, Arcispedale S. Anna, Corso Giovecca 203, 44100 Ferrara
T: (0532) 395580
Founded: 1956
Pres: Prof. Paolo Georgacopulo; Gen Secr: Dr. Andrea Frabchella
Focus: Surgery *07985*

Società Medico-Chirurgica di Modena, c/o Clinica Chirurgica Generale dell'Università degli Studi, Policlinico, Via del Pozzo, 41100 Modena
Founded: 1873
Focus: Surgery
Periodical
Bollettino (weekly) *07986*

Società Messinese di Storia Patria, c/o Palazzo dell'Università, 98100 Messina
T: 1900
Focus: Hist
Periodical
Archivio Storico Messinese *07987*

Società Napoletana di Chirurgia, c/o Istituto di Chirurgia d'Urgenza. I Facoltà di Medicina e Chirurgia, Università degli Studi, Piazza Miraglia, 80138 Napoli
T: (081) 446765
Founded: 1945
Focus: Surgery *07988*

Società Napoletana di Storia Patria, Piazza Municipio Maschio Angioino, 80133 Napoli
T: (081) 5510353; Fax: (081) 5529238
Founded: 1876; Members: 650
Pres: Prof. Giuseppe Galasso; Gen Secr: Sabatino Santangelo
Focus: Hist
Periodical
Archivio Storico per le Province Napoletane (weekly) *07989*

Società Naturalisti Veronesi F. Zorzi, c/o Museo Civico di Storia Naturale, Lungadige Porta Vittoria 9, 37129 Verona
T: (045) 8001987
Pres: Dr. Lorenzo Sorbini; Gen Secr: Giuseppina De Mori
Focus: Nat Sci *07990*

Società Nazionale di Informatica delle Camere di Commercio per la Gestione dei Centri Elettronici Reteconnessi Valutazione Elaborazione Dati (CERVED), Piazza Sallustio 21, 00187 Roma
T: (06) 4741268, 486403
Founded: 1974
Focus: Electronic Eng; Computer & Info Sci *07991*

Società Nazionale di Scienze, Lettere ed Arti, Ex Società Reale, Via Mezzacannone 8, 80134 Napoli
T: (081) 324634
Focus: Arts; Lit; Sci *07992*

Società Nucleare Italiana (S.N.I.), CP 414, 50100 Firenze
T: (055) 4392854
Founded: 1975
Focus: Nucl Res *07993*

Società Oftalmologica Italiana, c/o Istituto di Clinica Oculistica, Facoltà di Medicina, Università degli Studi, Via Massarenti 5, 40138 Bologna
T: (051) 392848
Founded: 1942
Focus: Ophthal *07994*

Società Ornitologica Italiana (S.O.I.), c/o Museo Ornitologico e di Scienze Naturali Loggetta Lombardesca, 48100 Ravenna
T: (0544) 35625
Pres: Azelio Ortali; Gen Secr: Antonio Caterini
Focus: Ornithology
Periodical
Gli Uccelli d'Italia (weekly) *07995*

Società Ornitologica Reggiana (S.O.R.), c/o Ferdinando Messori, Via del Cristo 6, 42100 Reggio Emilia
T: (0522) 37865
Founded: 1938
Focus: Ornithology
Periodicals
Il Giornale degli Uccelli (weekly)
Italia Ornitologica: Organo Ufficiale della Federazione Ornicoltori Italiani (weekly)
Uccelli (weekly) *07996*

Società Orticola Italiana (S.O.I.), Piazza G. Puccini 4, 50144 Firenze
T: (055) 361688
Founded: 1953
Focus: Forestry; Ecology; Hort
Periodical
Notiziario di Ortoflorofrutticoltura (weekly) *07997*

Società Pavese di Storia Patria, Via Belli 9, 27100 Pavia
T: (0382) 22154
Focus: Hist *07998*

Società per gli Studi Storici, Archeologici ed Artistici della Provincia di Cuneo (SSSAA), c/o Biblioteca Civica, Via Cacciatori delle Alpi 9, 12100 Cuneo
T: (0171) 634367
Founded: 1929
Focus: Hist; Archeol; Arts
Periodical
Bollettino della SSSAA (weekly) *07999*

Società per la Formazione la Ricerca e l'Addestramento per le Aziende e le Organizzazioni (FORRAD), Via Fabio Filzi 25a, 20124 Milano
T: (02) 6570941
Founded: 1968
Focus: Econ; Business Admin
Periodical
Forrad Informazioni *08000*

Società per la Matematica e l'Economia Applicate (SOMEA), Piazza del Collegio Romano 2, 00186 Roma
T: (06) 6798443, 6798487
Founded: 1967
Focus: Math; Econ *08001*

Società per le Belle Arti ed Esposizione Permanente, c/o Palazzo Sociale, Via Filippo Turati 34, 20121 Milano
T: (02) 639803, 661445
Founded: 1884
Focus: Fine Arts
Periodical
Società per le Belle Arti ed Esposizione Permanente (weekly) *08002*

Società Piemontese, Ligure, Lombarda di Ortopedia e Traumatologia, c/o Istituto di Clinica Ortopedica, Facoltà di Medicina e Chirurgia, Università degli Studi, Ospedale S. Martino, 16132 Genova
T: (010) 502739
Focus: Orthopedics; Traumatology . . . *08003*

Società Pistoiese di Storia Patria, Vicolo della Sapienza 12, CP 339, 51100 Pistoia
T: (0573) 23712
Founded: 1898
Pres: Prof. Enrico Coturri; Gen Secr: Vanna Vignali
Focus: Hist
Periodical
Bullettino Storico Pistoiese (weekly) . . . *08004*

Società Promotrice di Belle Arti, Casina Pompeiana nella Villa Comunale, 80121 Napoli
T: (081) 425188
Founded: 1861
Focus: Fine Arts *08005*

Società Reggiana d'Archeologia, Via Lero 6, 42100 Reggio Emilia
T: (0522) 433029
Founded: 1968
Focus: Archeol *08006*

Società Reggiana di Studi Storici, Corso Cairoli 6, 42100 Reggio Emilia
T: (0522) 434316
Founded: 1978
Pres: Prof. Davide Dazzi; Gen Secr: Prof. Aurelia Fresta
Focus: Hist
Periodical
Reggio Storia (3 times annually) *08007*

Società Ricerche Impianti Nucleari (SORIN), 13100 Vercelli
Focus: Nucl Res *08008*

Società Romana di Chirurgia, c/o Patologia Speciale Chirurgica, Università degli Studi, Policlinico Umberto I, 00161 Roma
T: (06) 4954876
Founded: 1939
Focus: Surgery *08009*

Società Romana di Storia Patria, Piazza della Chiesa Nuova 18, 00186 Roma
T: (06) 68307513
Founded: 1876; Members: 100
Pres: Prof. Letizia Ermini Pani; Gen Secr: Prof. Jean Coste
Focus: Hist
Periodical
Archivio della Società Romana di Storia Patria (weekly) *08010*

Società Sassarese per le Scienze Giuridiche, c/o Isituto Giuridico dell'Università, Piazza Università, 07100 Sassari
Focus: Law *08011*

Società Savonese di Storia Patria (S.S.S.P.), Piazza della Maddalena 14, CP 358, 17100 Savona
T: (019) 811960
Founded: 1885; Members: 274
Pres: Prof. Almerino Lunardon
Focus: Hist; Archeol
Periodicals
Atti e Memorie (weekly)
Sabazia (weekly) *08012*

Società Siciliana per la Storia Patria, Piazza S. Domenico 1, 90133 Palermo
T: (091) 582774
Founded: 1872
Focus: Hist
Periodical
Archivio Storico Siciliano (weekly) . . . *08013*

Società Siracusana di Storia Patria, c/o Villa Reimann, Via Necropoli Grotticelle 14, 96100 Siracusa
Focus: Hist
Periodical
Archivio Storico Siracusano (weekly) . . *08014*

Società Speleologica Italiana (SSI), c/o Museo Civico di Storia Naturale, Corso Venezia 55, 20121 Milano
T: (02) 702018, 795381
Founded: 1950; Members: 559
Focus: Speleology
Periodicals
Le Grotte d'Italia
Speleologia *08015*

Società Storica Catanese, Via Etnea 248, 95131 Catania
T: (095) 434782
Founded: 1955
Focus: Hist *08016*

Società Storica di Terra d'Otranto, Palazzo Adorni, Via Umberto I 32, 73100 Lecce
T: (0832) 41938
Founded: 1966
Focus: Hist
Periodical
Rivista Storica del Mezzogiorno (weekly) *08017*

Società Storica Lombarda, Via Morone 1, 20121 Milano
T: (02) 860118
Founded: 1874; Members: 450
Focus: Hist
Periodical
Archivio Storico Lombardo *08018*

Società Storica Novarese, Via S. Gaudenzio 15, 28100 Novara
Fax: (0321) 34339
Founded: 1907
Pres: Roberto Di Tieri
Focus: Hist
Periodical
Bollettino Storico per la Provincia di Novara (weekly) *08019*

Società Storica Pisana, c/o Dipartimento di Medievistica, Facoltà di Lettere, Università degli Studi, Via Derna 1, 56126 Pisa
T: (050) 29475; Fax: (050) 40949
Focus: Hist
Periodical
Bollettino Storico Pisano *08020*

Società Tarquiniense d'Arte e Storia, Via delle Torri 29-33, 01016 Tarquinia
T: (0766) 858194
Founded: 1917
Pres: Bruno Blasi; Gen Secr: Romano Andreaus
Focus: Arts; Hist; Cultur Hist
Periodical
Bolletino (weekly) *08021*

Società Teosofica in Italia (S.T.I.), Via Enrico Toti 3, 34131 Trieste
Founded: 1901
Focus: Rel & Theol
Periodical
Rivista Italiana di Teosofia (weekly) . . *08022*

Società Tiburtina di Storia e d'Arte, Villa d'Este, 00019 Tivoli
T: (0774) 22187
Founded: 1919
Pres: Prof. Camillo Pieraltini; Gen Secr: Dr. Vincenzo Giovanni Pacifici
Focus: Hist; Arts; Cultur Hist
Periodical
Atti e Memorie della Società Tiburtina di Storia e d'Arte (weekly) *08023*

Società Torricelliana di Scienze e Lettere, Corso Garibaldi 2, CP 179, 48018 Faenza
Founded: 1947
Focus: Lit; Sci
Periodical
Torricelliana (weekly) *08024*

Società Toscana di Orticoltura, Via delle Terme 4, 50123 Firenze
T: 1852
Focus: Hort
Periodical
Rivista della Ortoflorofrutticoltura Italiana . *08025*

Società Toscana di Scienze Naturali (STSN), Via S. Maria 53, 56100 Pisa
T: (050) 22413
Founded: 1847
Focus: Nat Sci
Periodical
Atti della Società Toscana di Scienze Naturali *08026*

Société de la Flore Valdôtaine (S.F.V.), Piazza Chanoux 8, 11100 Aosta
Founded: 1858
Focus: Geophys
Periodicals
Bulletin (weekly)
Revue Valdôtaine d'Histoire Naturelle (weekly) *08027*

Society for International Development (SID), Palazzo Civiltà del Lavoro, 00144 Roma
T: (06) 5917897; Fax: (06) 5919836
Founded: 1957; Members: 9000
Pres: Maurice Williams
Focus: Int'l Relat
Periodicals
Compass (weekly)
Development (weekly)
Development Hotline (weekly)
Meridian (weekly) *08028*

Studio Teologico Accademico Bolognese (S.T.A.B.), Piazza S. Domenico 13, 40124 Bologna
T: (051) 265756, 232238
Founded: 1978
Focus: Rel & Theol
Periodical
Sacra Doctrina *08029*

SVP Italia, Via Piccinni 3, 20131 Milano
T: (02) 2043451; Tx: 335649; Fax: (02) 29531000
Founded: 1962
Pres: Iginio Lagioni
Focus: Doc; Stats
Periodical
Notiziario SVP (weekly) *08030*

Tecnagro, Via Tommaso Grossi 6, 00184 Roma
T: (06) 7008896; Fax: (06) 7009380
Founded: 1976
Focus: Agri *08031*

Tecneco, Via Papiria, 61032 Fano
T: (0721) 86741
Founded: 1971
Focus: Energy; Water Res *08032*

Tecnocasa, Viale Lombardia 43, 67100 L'Aquila
T: (0862) 61223
Founded: 1972
Focus: Civil Eng *08033*

Tecnocentro Italiano (T.I.), c/o Dr. Fasola, Via Donizetti 45, 20122 Milano
T: (02) 780848
Founded: 1965
Pres: Dr. Giuseppe Fasola
Focus: Law; Poli Sci *08034*

Tecnofarmaci / Società Consortile per Azioni per lo Sviluppo della Ricerca Farmaceutica, Piazza Indipendenza 24, 00040 Pomezia
T: (06) 9111961; Tx: 611522; Fax: (06) 9111956
Founded: 1974
Pres: Prof. Rodolfo Paoletti
Focus: Pharmacol *08035*

Tecnomare, S. Marco 3584, 30124 Venezia
T: (041) 796711; Tx: 410484; Fax: (041) 5200209
Founded: 1971
Pres: Giuseppe Muscarella
Focus: Water Res; Oceanography . . . *08036*

Tecnotessile / Centro di Ricerche, Via Valentini 14, 50047 Prato
T: (0574) 35741/42
Founded: 1972
Focus: Textiles *08037*

Unione Accademica Nazionale (U.A.N.), Via della Lungara 230, 00165 Roma
T: (06) 6875024; Fax: (06) 6869066
Founded: 1923
Focus: Hist; Archeol *08038*

Unione Antropologica Italiana (U.A.I.), c/o Istituto di Antropologia, Via Selmi 1, 40126 Bologna
T: (051) 243272
Founded: 1972
Pres: Enzo Lucchetti
Focus: Anthro
Periodical
Antropologia Contemporanea (weekly) . . *08039*

Unione Associazioni Regionali (UN.A.R.), Corso Vittorio Emanuele 24, 00186 Roma
T: (06) 6797491, 6797717, 6782779, 8393783
Founded: 1977
Focus: Sociology
Periodical
Presenze Regionali (weekly) *08040*

Unione Bolognese Naturalisti (U.B.N.), c/o Dipartimento di Biologia, Università degli Studi, Via S. Giacomo 9, 40126 Bologna
Founded: 1950
Focus: Nat Sci
Periodicals
Natura e Montagna (weekly)
Notiziario (weekly) *08041*

Unione Cattolica Italiana Insegnanti Medi (UCIIM), Via Crescenzio 25, 00193 Roma
T: (06) 6875584, 6542701
Founded: 1944; Members: 20000
Pres: Prof. Cesarina Checcacci; Gen Secr: Prof. Eugenio Boldoni
Focus: Educ
Periodical
La Scuola l'Uomo (weekly) *08042*

Unione Consultori Italiani Prematrimoniali e Matrimoniali (U.C.I.P.E.M.), Via Giuseppe Garibaldi 3, 40124 Bologna
T: (051) 224531
Founded: 1968
Focus: Family Plan *08043*

Unione della Legion d'Oro (U.L.D.O.), Via del Vantaggio 22, 00186 Roma
T: (06) 6788658
Founded: 1954
Focus: Sci
Periodical
La Voce dell'Unione: Atti Annuali (weekly) *08044*

Unione Erpetologica Italiana, Via Ulisse Aldrovandi 18, 00197 Roma
T: (06) 873586
Focus: Zoology *08045*

Unione Giuristi Cattolici Italiani (U.G.C.I.), Via della Conciliazione 1, 00193 Roma
T: (06) 6564865
Founded: 1948
Focus: Rel & Theol; Law
Periodicals
Iustitia (weekly)
Quaderni di Justitia (weekly)
Vita dell'Unione: Bollettino Mensile (weekly) *08046*

Unione Italiana Lotta alla Distrofia Muscolare (U.I.L.D.M.), Via P.P. Vergerio 17, 35126 Padova
T: (049) 757361; Fax: (049) 757033
Founded: 1961
Focus: Physical Therapy; Psych
Periodical
Distrofia Muscolare (weekly) *08047*

Unione Matematica Italiana (UMI), Piazza Porta San Donato 5, 40127 Bologna
T: (051) 229909
Founded: 1922; Members: 1855
Focus: Math
Periodicals
Bollettino (weekly)
Notiziario (weekly) *08048*

Unione Micologica Italiana (U.M.I.), c/o Centro di Micologia, Via Filippo Re 8, 40126 Bologna
T: (051) 351401
Founded: 1969
Pres: Prof. Gabriele Goidànich; Gen Secr: Prof. Gilberto Govi
Focus: Botany, Specific
Periodical
Micologia Italiana (weekly) *08049*

Unione Tecnica Italiana Farmacisti (UTIFar), Via Casaregis 52, 16129 Genova
T: (010) 587345; Fax: (010) 587013
Founded: 1959
Pres: Dr. Luigi Casanova; Gen Secr: Dr. Pietro Dettori
Focus: Pharmacol
Periodical
Collegamento (weekly) *08050*

Union Generela di Ladins dla Dolomites, Val Gardena-Cêsa di Ladins-Str Rezia 83, 39046 Ortisei-Urtijëi
Founded: 1949
Focus: Ethnology
Periodical
La Usc di Ladins (weekly) *08051*

Union Ladins Val Badia (ULVB), S. Leonardo Scola Vedla, Pedraces
Founded: 1967
Focus: Ling; Ethnology; Lit
Periodical
La Usc di Ladins (weekly) *08052*

Union Mondiale des Enseignants Catholiques (UMEC) / World Union of Catholic Teachers, Piazza S. Calisto 16, 00153 Roma
T: (06) 69887286
Founded: 1950; Members: 45
Focus: Educ
Periodicals
Nouvelles UMEC (weekly)
Proceedings of the World Congress (every 3 years)
WUCT Newsletter (weekly) *08053*

World Association for Animal Production, Via A. Torlonia 15a, 00161 Roma
Gen Secr: N. Frydlender
Focus: Animal Husb *08054*

World Future Studies Federation (WFSF), CP 6203, 00100 Roma
T: (06) 872529; Tx: 613529
Founded: 1973; Members: 500
Pres: Dr. Eleonora Masini
Focus: Futurology
Periodical
WFSF Newsletter (weekly) *08055*

Ivory Coast

Académie Régionale des Sciences et Techniques de la Mer d'Abidjan / Regional Maritime Academy, Abidjan, BP V 158, Abidjan
T: 452830; Tx: 23399
Members: 15
Gen Secr: Sogodogo Souleimane
Focus: Oceanography *08056*

African Institute for Economic and Social Development, BP 08 BP8, Abidjan
T: 441594; Fax: 440641
Founded: 1962
Gen Secr: René Roi
Focus: Develop Areas; Soc Sci; Econ . *08057*

African Music Rostrum (AFMR), c/o Comité Ivoirien de la Musique, BP 48 08, Abidjan
T: 440457
Founded: 1970
Gen Secr: Aclepo Yapo
Focus: Music *08058*

African Oil Palm Development Association (AFOPDA), BP 15 341, Abidjan
T: 251518
Founded: 1985
Gen Secr: B.H. Sounou
Focus: Botany
Periodical
Africa Palm (weekly) *08059*

African Union of Sports Medicine, c/o Centre National de Médecine du Sport, BP 61, El-Biar, Algiers
T: (02) 782863; Tx: 52390
Founded: 1982; Members: 9
Pres: Prof. Constant Roux
Focus: Med *08060*

Centre de Recherches Océanographiques (CRO), BP V18, Abidjan
Focus: Oceanography
Periodicals
Archives scientifiques (weekly)
Documents scientifiques du CRO (weekly) *08061*

Centre de Recherches Zootechniques, BP 449, Bouaké
Focus: Zoology *08062*

Centre des Sciences Humaines, BP 1600, Abidjan
Founded: 1960
Focus: Ethnology; Sociology *08063*

Centre Technique Forestier Tropical, BP 8033, Abidjan 08
Founded: 1962
Focus: Forestry *08064*

Institut Africain pour le Développement Economique et Social (INADES), 08 BP 8, Abidjan 08
T: 441594
Focus: Econ; Sociology
Periodicals
Bibliographies commented (weekly)
Fichier Afrique (weekly)
Listes d'acquisition (weekly)
Rapport d'activité (weekly) *08065*

Société pour le Développement Minier de la Côte d'Ivoire (SODEMI), BP 2816, Abidjan
T: 442994/96
Founded: 1962; Members: 200
Focus: Mining *08066*

Jamaica

Association for Commonwealth Language and Literature Studies (ACLALS), c/o Dept of English, University of the West Indies, Mona, Kingston 7
Tx: 2123
Founded: 1965
Pres: Prof. Edward Baugh
Focus: Ling; Lit
Periodical
ACLALS Newsletter *08067*

Association of Caribbean Economists (ACE), POB 735, Kingston 8
T: (0809) 9278283; Tx: 3607; Fax: (0809) 92724096
Focus: Econ *08068*

British Caribbean Veterinary Association, Hope, Kingston
Focus: Vet Med *08069*

Caribbean Council of Engineering Organizations (CCEO), 2-4 Ruthven Rd, 122, Kingston 10
Focus: Eng *08070*

Caribbean Council of Legal Education (CLE), Mona Campus, Kingston 7
Founded: 1971
Focus: Law; Educ *08071*

Caribbean Energy Information System (CEIS), c/o Scientific Research Council, POB 350, Kingston 6
Gen Secr: Mona Whyte
Focus: Energy *08072*

Caribbean Food and Nutrition Institute (CFNI), c/o University of the West Indies, Mona, POB 140, Kingston 7
T: 9271540/41, 9271927; CA: Cajanus, Kingston
Founded: 1967
Pres: Dr. W.P. Patterson
Focus: Food; Nutrition
Periodical
Cajanus (weekly) *08073*

Caribbean Institute of Mass Communications (CARIMAC), c/o University of the West Indies, Mona
Founded: 1974
Gen Secr: A. Brown
Focus: Comm Sci
Periodicals
CARIMAC Newsletter
Crossover (3 times annually) *08074*

Caribbean Regional Drug Testing Laboratory (CRDTL), c/o Government Chemist Dept, Hope Gardens, Kingston 6
Founded: 1975; Members: 14
Focus: Pharmacol *08075*

Commonwealth Association of Planners (CAP), The Professional Centre, 2-4 Ruthven Rd, Kingston 10
T: 92988805; Fax: 9299242
Founded: 1971; Members: 11500
Focus: Urban Plan
Periodical
CAP News (weekly) *08076*

Commonwealth Caribbean Medical Research Council (CCMRC), c/o Tropical Metabolism Research Unit, University of the West Indies, Mona, Kingston
Founded: 1972; Members: 20
Gen Secr: Molly McGann
Focus: Med *08077*

Commonwealth Library Association (COMLA), POB 40, Mandeville
T: 9620703; CA: COMLA; Fax: 9622770
Founded: 1972; Members: 52
Pres: S. Ferguson; Gen Secr: Joan Swaby
Focus: Libraries & Bk Sci
Periodical
COMLA Newsletter (weekly) *08078*

Geological Society of Jamaica, c/o Dept of Geology, University of the West Indies, Mona, Kinston 7
T: 9272728; Fax: 9271640
Founded: 1958; Members: 80
Pres: Dr. Trevor Jackson; Gen Secr: Janette Scott
Focus: Geology
Periodical
The Journal of the Geological Society of Jamaica (weekly) 08079

Institute of Jamaica, 12-16 East St, Kingston
Founded: 1879
Focus: Anthro; Ethnology; Hist 08080

International P.E.N. Club, c/o Institute of Jamaica, 12-16 East St, Kingston
Founded: 1947; Members: 50
Focus: Lit 08081

Jamaica Historical Society, 14-16 East St, Kingston
Focus: Hist 08082

Jamaica Library Association (JLA), POB 58, Kingston 5
Founded: 1949; Members: 211
Pres: Albertina Jefferson
Focus: Libraries & Bk Sci
Periodicals
Jamaica Library Association Annual Report (weekly)
Jamaica Library Association Bulletin (weekly)
JLA News (weekly) 08083

Jamaican Association of Sugar Technologists, c/o Sugar Research Dept, Mandeville
Founded: 1937; Members: 326
Pres: C.A. Gordon; Gen Secr: H.M. Thompson
Focus: Food 08084

Jamaica National Trust Commission, Hope Gardens, POB 473, Kingston 6
Founded: 1959
Focus: Urban Plan 08085

Jamaican Geographical Society, c/o Geography Dept, University of the West Indies, Mona, Kingston 7
Focus: Geography
Periodical
Newsletter (weekly) 08086

Medical Association of Jamaica (MAJ), 19 Ruthven Rd, Kingston 10
T: 65451
Founded: 1966; Members: 300
Focus: Med
Periodical
Newsletter (weekly) 08087

Medical Research Council Laboratories (MRC LABS), c/o University of the West Indies, Mona, Kingston 7
T: 9272471
Founded: 1962; Members: 22
Pres: Dr. D.A. Rees; Gen Secr: Graham R. Serjeant
Focus: Hematology 08088

Scientific Research Council, POB 350, Kingston 6
Founded: 1960; Members: 16
Focus: Sci 08089

Japan

Africa Kyokai / Africa Society of Japan, 1-11-2 Toranomon, Minato-ku, Tokyo 105
T: (03) 5011878
Focus: Int'l Relat 08090

Ajia Chosakai / Asian Affairs Research Council, c/o The Mainichi Newspapers, 1-1 Hitotsubashi, Chiyoda-ku, Tokyo 100
T: (03) 2111616
Focus: Int'l Relat; Poli Sci
Periodicals
Ajia Jiho: Monthly Asia Review (weekly)
Asia Quarterly (weekly) 08091

Ajiakurabu / Asian Club, World Trade Center, Ste 2405, 2-4-1 Hamamatsu-cho, Minato-ku, Tokyo 105
T: (03) 4356071
Focus: Int'l Relat 08092

Ajia Seikei Gakkai / Society for Asian Political and Economic Studies, c/o Prof. T. Yamada, Dept of Political Science, Keio University, 2-15-45 Mita, Minato-ku, Tokyo 108
Founded: 1953; Members: 600
Pres: T. Okabe
Focus: Poli Sci; Econ
Periodical
Ajia Kenkyu (weekly) 08093

Asia Crime Prevention Foundation (ACPF), 60 Sunshine Bldg, 1-1 Higashi-Ikebukuro 3-chome, Toshima-ku, Tokyo 170
T: (03) 9871444; Fax: (03) 9869549
Founded: 1982
Pres: Sugiichiro Watari
Focus: Criminology
Periodicals
Asian Journal of Crime Prevention and Criminal Justice
Newsletter of the Asia Crime Prevention Foundation 08094

Asian and Oceanian Thyroid Association (AOTA), c/o First Dept of Internal Medicine, School of Medicine, Nagasaki University, 7-1 Sakamoto-machi, Nagasaki 852
Founded: 1975
Pres: Prof. Shigenobo Nagataki
Focus: Intern Med 08095

Asian and Pacific Federation of Organizations for Cancer Research and Control (APFOCC), c/o Institute of Preventive Oncology, 1-2 Ichigaya-Sadoharacho, Shinjuku-ku, Tokyo 162
T: (03) 32672556
Founded: 1973; Members: 18
Gen Secr: Dr. T. Hirayama
Focus: Intern Med 08096

Asian Association on Remote Sensing (AARS), c/o Institute of Industrial Science, University of Tokyo, 7-22 Roppongi, Minato-ku, Tokyo 106
Tx: 2427317; Fax: (03) 34792762
Gen Secr: Prof. Shunji Murai
Focus: Surveying 08097

Asian Council of Securities Analysts (ASAC), c/o Security Analysts Association of Japan, Daini Shokenkaikan Bldg, 2-1-1 Nihonbashi-Kayabacho, Chuo-ku, Tokyo
T: (03) 3661515; Fax: (03) 6665843
Founded: 1978
Gen Secr: Yukio Miyasaka
Focus: Finance 08098

Asian Fluid Mechanics Committee (AFMC), c/o Institute of Flow Research, 6-10-205 Akasaka, Minato-ku, Tokyo
Founded: 1982
Pres: Prof. H. Sato
Focus: Eng 08099

Asian Health Institute (AHI), 987 Minamiyama, Nisshin, Aichi-Gun, Aichi 470
T: (05617) 31950; Fax: (05617) 31990
Founded: 1980
Gen Secr: Dr. Hiromi Kawahara
Focus: Public Health
Periodical
AHI News (weekly) 08100

Asian-Pacific Organization for Cell Biology (APOCB), c/o Shigei Medical Research Institute, 2117 Yamada, Okayama 701
T: (0862) 823113; Tx: 823115
Founded: 1988
Gen Secr: Dr. Tohru Okigaki
Focus: Cell Biol & Cancer Res 08101

Asian Pacific Society of Respirology, c/o Japan Society of Chest Diseases, 4-1-11 Hongo, Bunkyo-ku, Tokyo 113
Focus: Intern Med 08102

Asian Pacific University Presidents Conference, c/o Tokai University, 2-28-4 Tomigaya, Shibuya-ku, Tokyo 151
T: (03) 34672211; Tx: 2423402; Fax: (03) 34604515
Founded: 1987; Members: 56
Pres: Shigeyoshi Matsumae
Focus: Educ 08103

Asian Parasite Control Organization (APCO), c/o Hokenkaikan Bekkan, 1-2 Sadohara-cho, Ichigaya, Shinjuku-ku, Tokyo 162
T: (03) 32681800; Fax: (03) 32668767
Founded: 1974
Gen Secr: Chojiro Kunii
Focus: Med 08104

Asian Patent Attorneys Association (APAA), c/o Asamura Patent Office, New Ohtemachi Bldg, 2-1 Ohtemachi 2-chome, Chiyoda-ku, Tokyo 100
T: (03) 32113651; Tx: 22979; Fax: (03) 32113651
Founded: 1969
Focus: Law
Periodical
APAA News 08105

Asian Population and Development Association (APDA), Nagatacho TBR Bldg, Room 710, 10-2 Nagatacho 2-chome, Chiyoda-ku, Tokyo 100
T: (03) 5817796; Tx: 2222034
Founded: 1982
Focus: Sociology 08106

Asian Productivity Organization (APO), 4-14 Akasaka 8-chome, Minato-ku, Tokyo 107
T: (03) 34087221; Tx: 26477; Fax: (03) 34087220
Founded: 1961
Gen Secr: Nagao Yoshida
Focus: Business Admin 08107

Asian Rural Institute – Rural Leaders Training Center, 442-1 Tzukinokizawa, Nishinasuno, Tochigi 329-27
T: (0287) 363111; Fax: (0287) 375833
Founded: 1973
Gen Secr: Toshihiro Takami
Focus: Agri
Periodical
ART Network Bulletin 08108

Asian Society for Adapted Physical Education and Exercise (ASAPE), c/o Institute of Special Education, University of Tsukuba, 1-1-1 Tennoudai, Tsukuba-shi, Ibaraki-ken, Tsukuba 305
T: (0298) 536748; Fax: (0298) 536504
Founded: 1986; Members: 204
Gen Secr: Hideo Nakata
Focus: Sports
Periodical
ASAPE Newsletter (weekly) 08109

Asian Society for Sport Psychology, c/o Dept of Physical Education, Nihon University, 3-25-40 Sakurajousui, Setagaya-ku, Tokyo 156
Founded: 1988
Gen Secr: Dr. A. Fujita
Focus: Psych 08110

Asia Pacific Academy of Ophthalmology (APAO), c/o Dept of Ophthalmology, School of Medicine, Juntendo University, 3-1-3 Hongom, Bunkyo-ku, Tokyo 113
T: (03) 38133111
Founded: 1957; Members: 21
Gen Secr: Dr. Akira Nakajima
Focus: Ophthal 08111

Association of Asian-Pacific Operational Research Societies (APORS), c/o Operations Research Society of Japan, Gakkai Center Bldg, 2-4-16 Yayoi, Bunkyo-ku, Tokyo 113
T: (03) 38153352
Founded: 1985; Members: 10
Gen Secr: Prof. Kunihiro Wakamaya
Focus: Business Admin
Periodical
Asia-Pacific Journal of Operational Research (weekly) 08112

Association of Medical Doctors for Asia (AMDA), c/o Suganami Hospital, 1/310 Narazu, Okayana 701
T: (0862) 847676
Founded: 1984; Members: 200
Pres: Dr. Shigeru Suganami
Focus: Med 08113

Bigaku-Kai / The Japanese Society for Aesthetics, c/o Faculty of Letters, University of Tokyo, 7-3-1 Hongo, Bunkyo-ku, Tokyo 113
Founded: 1950; Members: 1366
Pres: Prof. Dr. Masao Yamamoto
Focus: Philos
Periodicals
Aesthetics (bi-annually)
Bigaku (weekly) 08114

Bijutsu-shi Gakkai / Japanese Art History Society, c/o Tokyo National Research Institute of Cultural Properties, 13-27 Ueno Park, Taito-ku, Tokyo 110
Founded: 1949; Members: 665
Focus: Hist; Arts
Periodical
Journal (weekly) 08115

Chigaku Dantai Kenkyu-Kai / The Association for the Geological Collaboration in Japan, Aizu Tenpo Bldg, 1-8-7, Minami-Ikebukuro, Toshima-ku, Tokyo 171
T: (03) 9833378
Founded: 1947; Members: 2800
Pres: Takayasu Karsumi
Focus: Geology
Periodical
Chikyu-kagaku: Earth Science (weekly) . 08116

Chusei Tetsugakkai, c/o Institute of Philosphy, Keio University, 45-15-2, Minato-ku, Mita, Tokyo 108
T: (03) 4534511
Founded: 1952; Members: 380
Pres: Inagaki Ryosuke; Gen Secr: Yauchi Yoshiaki
Focus: Philos 08117

Daini-Tokyo Bengoshikai Toshokan / Daini-Tokyo Bar Association, 1-1-4 Kasumigaseki, Chiyoda-ku, Tokyo 100
T: (03) 5812255
Focus: Law 08118

Denki Kagaku Kyokai / Institute of Electrical Engineers of Japan, Shin-Yurakucho Bldg, 1-12-1, Yuraku-cho, Chiyoda-ku, Tokyo 100
Founded: 1933
Focus: Chem
Periodical
Denki Kagaku 08119

Denki Tsushin Kyokai, Shin-Yuraku-cho Kaikan, 1-11 Yuraku-cho, Chiyoda-ku, Tokyo 100
Founded: 1938
Focus: Electric Eng 08120

Denshi Joho Tsushin Gakkai / Institute of Electronics, Information and Communication Engineers of Japan, Kikai-Shinko-Kaikan Bldg, 3-5-8 Shibakoen, Minato-ku, Tokyo 105
Founded: 1917; Members: 35000
Pres: K. Maeda
Focus: Electronic Eng; Comm Sci
Periodicals
Journal, Transactions (weekly)
Original Contributions in English and Abstracts in English from the Transactions (weekly) . 08121

Doboku-Gakkai / Japan Society of Civil Engineers, 1 Yotsuya, Shinjuku-ku, Tokyo 160
T: (03) 3553441
Founded: 1914; Members: 30219
Pres: Prof. Dr. Kiyoshi Horikawa; Gen Secr: Yoshitaka Okamoto
Focus: Civil Eng
Periodicals
Civil Engineering in Japan
Coastal Engineering in Japan (weekly)
Journal (weekly)
Proceedings (weekly) 08122

Doshitsu Kogakkai / Japanese Society of Soil Mechanics and Foundation Engineering, Sugayama Bldg, 4F Kanda Awaji-cho 2-23, Chiyoda-ku, Tokyo 101
T: (03) 2517661
Founded: 1949; Members: 15000
Pres: Kikuo Kotoda; Gen Secr: Masao Yamakawa
Focus: Civil Eng
Periodicals
Soils and Foundations (weekly)
Tsuchi to Kiso (weekly) 08123

East and Southeast Asia Federation of Soil Science Societies (ESAFS), c/o JSSSPN, 202 26-10, Hongo 6-chome, Bunkyo-ku, Tokyo 113
Founded: 1990
Gen Secr: Dr. Naoko Nishizawa
Focus: Geology 08124

East Asian Bird Protection Society, c/o Wild Bird Society of Japan, Aoyama Flower Bldg, 1-1-4 Shibuya, Shibuya-ku, Tokyo 150
Focus: Ornithology 08125

East-West Sign Language Association (EWSLA), c/o Dept of Linguistics, International Christian University, 10-2, 3 Chome, Mitaka, Tokyo 181
Founded: 1980; Members: 200
Focus: Ling
Periodical
Language Sciences (weekly) 08126

Engei Gakkai / Japanese Society for Horticultural Science, c/o Faculty of Agriculture, Kyoto University, Sakyo-ku, Kyoto
Founded: 1923; Members: 2855
Pres: T. Asahira; Gen Secr: Y. Takeda
Focus: Hort
Periodical
Journal (weekly) 08127

Gakujutsu Bunken Fukyu-Kai / Association for Science Documents Information, c/o Tokyo Institute of Technology, 2-15-1, O-okayama, Meguro-ku, Tokyo 152
Founded: 1933
Pres: S. Kanbara
Focus: Doc
Periodical
Reports on Progress in Polymer Physics in Japan 08128

Handotai Kenkyu Shinkokai Handotai Kenkyusho Toshoshitsu / Semiconductor Research Foundation, Kawauchi, Sendai 980
T: (0222) 237287
Focus: Electric Eng
Periodicals
Semiconductor Science and Technology (weekly)
SRI Bulletin (weekly) 08129

Hanshin Doitsubungakukai / Der Japanische Verein für Germanistik, c/o Kobe Daigaku Bungakubu, Rokkodac-cho, Nada-ku, Kobe 657
Founded: 1952
Focus: Ling; Lit
Periodical
Doitsu-Bungaku-Ronko: Forschungsberichte zur Germanistik (weekly) 08130

Hikaku-ho Gakkai / Japan Society of Comparative Law, c/o Faculty of Law, Tokyo University, Hongo, Bunkyo-ku, Tokyo 113
Founded: 1950; Members: 780
Pres: H. Tanaka
Focus: Law
Periodical
Hikakuhô Konkyû: Comparative Law Journal (weekly) 08131

Hiroshima Shikagu Kenkyukai, c/o Hiroshima Daigaku Bungakubu, Higashi Sendamachi, Hiroshima 730
Founded: 1929
Focus: Hist
Periodical
Shigaku-Kenkyu: Review of Historical Studies (weekly) 08132

Hiroshima Tetsugakkai / Hiroshima Philosphical Society, c/o Hiroshima Daigaku Bungakubu, 1-1-89 Higashisenda-machi, Hiroshima-shi, Hiroshima-ken 730
T: (0822) 411221
Focus: Philos 08133

Hogaku Kyokai / Jurisprudence Association, c/o Faculty of Law, University of Tokyo, 7-3-1 Hongo, Bunkyo-ku, Tokyo 113
T: (03) 38122111; Fax: (03) 38167375
Founded: 1884; Members: 600
Pres: M. Nishio
Focus: Law
Periodical
Hogaku Kyokai Zassi (weekly) 08134

Hokkaido Keizai Rengokai / Hokkaido Economic Federation, Nihon Seimei Bldg, 6F, 4 Nishi, Kita 3-jo, Chuo-ku, Sapporo 060
T: (011) 2216166
Focus: Econ 08135

Hosei-shi Gakkai / Legal History Association, c/o University of Tokyo, Hongo, Bunkyo-ku, Tokyo 113
Founded: 1949; Members: 400
Pres: T. Sera
Focus: Hist; Law
Periodical
Legal History Review (weekly) 08136

Hosokai / Lawyers' Association, 1-1 Kasumigaseki, Chiyoda-ku, Tokyo
Founded: 1891; Members: 20000
Pres: Kazuto Ishida; Gen Secr: K. Nagai
Focus: Law
Periodical
Hoso Jiho 08137

Hyomen Gijutsu Kyokai / Surface Finishing Society of Japan, Kyodo Bldg, 2, Kanda Iwamoto-cho, Chiyoda-ku, Tokyo 101
T: (03) 32523286; Fax: (03) 32523288
Focus: Fisheries
Periodical
Hyomen Gijutsu (weekly) 08138

Ikomasan Tenmon Kyokai / Ikomasan Astronomical Society, Ikoma-Sanzyo, Ikoma-gun, Nara Ken
Founded: 1942; Members: 705
Pres: Joe Ueta; Gen Secr: H. Hamane
Focus: Astronomy
Periodical
Tenmon Kyositu: Astronomical Class (weekly) 08139

Information Processing Society of Japan (IPSJ), STEC Joho Bldg, 1-24-1, Nishi-shinjuku, Shinjuku-ku, Tokyo 160
T: (03) 53223535
Founded: 1960; Members: 32000
Pres: Dr. Y. Mizuno
Focus: Computer & Info Sci
Periodicals
Joho Shori (weekly)
Joho Shori Gakkai Ronbunshi: Transactions of the Information Processing Society of Japan (weekly) 08140

International Association for Religion and Parapsychology (IARP), 4-11-7 Inokashira, Mitaka-Shi, Tokyo 181
T: (0422) 483535; Fax: (0422) 483548
Founded: 1972; Members: 2000
Pres: Hiroshi Motoyama
Focus: Psych; Parapsych 08141

International Association of Traffic and Safety Sciences (IATSS), 2-6-20 Yaesu, Chuo-ku, Tokyo 104
Founded: 1974; Members: 34
Gen Secr: K. Mikami
Focus: Transport
Periodicals
IATSS Research (biannually)
IATSS Review (weekly)
Statistic of Road Traffic Accidents in Japan (weekly)
White Paper on Transportation Safety (weekly) 08142

International Congress on Fracture (ICF), c/o Prof. T. Yorobori, Institute for Fracture and Safety, 802, 17-18, 1-chome, Kamisugi, Sendai
Founded: 1965; Members: 30
Pres: D. Francois; Gen Secr: Prof. T. Yokobori
Focus: Materials Sci
Periodical
Proceedings 08143

International Palaeontological Association (IPA), c/o Dept of Geology and Mineralogy, Faculty of Sciences, Hokkaido University, Sapporo 060
T: (011) 7162111; Fax: (011) 7199394
Founded: 1933; Members: 1550
Gen Secr: Dr. Makoto Kato
Focus: Paleontology
Periodical
Lathaia (weekly) 08144

Itaria Gakkai, c/o Kyoto Dagaku Bungakubu, Yoshida Honmachi, Sakyo-ku, Kyoto 606
Founded: 1940
Focus: Ethnology 08145

Japan Techno-Economics Society (JATES), Masuda Bldg, 2-4-5 Iidabashi, Chiyoda-ku, Tokyo 102
T: (03) 32635501; CA: Supratechno; Fax: (03) 32635504
Founded: 1966; Members: 2814
Pres: Koji Kobayashi; Gen Secr: Yoshio Ishikawa
Focus: Business Admin
Periodical
Technology and Economy (weekly) . . . 08146

Japan Weather Association, Kaiji Center Bldg, 4-5 Kojimachi, Chiyoda-ku, Tokyo
Founded: 1950
Pres: Naoshi Machida
Focus: Geophys
Periodicals
Geophysical Magazine
Journal of Meteorological Research (weekly)
Oceanographical Magazine (weekly) . . . 08147

Jinbun Chiri Gakkai / The Human Geographical Society of Japan, Kinki-chiho Hatsumei Center Godo Bldg, 14, Yoshida Kawara-cho, Sakyo-ku, Kyoto 606
T: (075) 7517687; Fax: (075) 7517687
Founded: 1948; Members: 1780
Focus: Geography 08148

Jishin Gakkai, c/o Dagaku Jishin Kenkyusho, 1-1-1 Yayoi, Bunkyo-ku, Tokyo 113
Founded: 1929
Focus: Geology 08149

Kagaku Kisoron Gakkai / Japan Association for Philosophy of Science, c/o Otyanamizu University, 2-1-1, Doruka, Bunkyo-ku, Tokyo 112
T: (03) 39433151 ext 310; Fax: (03) 39439637
Founded: 1954; Members: 550
Pres: Natuhiko Yoshida; Gen Secr: Murakami Yoichiro
Focus: Philos
Periodicals
Annals of the Japan Association for Philosophy of Science (weekly)
Kagaku Kisoron Kenkyu (weekly) . . . 08150

Kaigai Nogyo Kaihatsu Kyokai / Overseas Agricultural Development Association, Ajia Kaikan, 8-10-32 Akasaka, Minato-ku, Tokyo 107
T: (03) 4783508
Focus: Agri 08151

Kajima Heiwa Kenkyujo / Kajima Institute of International Peace, 6-5-30 Akasaka, Minato-ku, Tokyo
T: (03) 55612027
Gen Secr: Kowaguchi Seiji
Focus: Int'l Relat 08152

Kami Parupu Gijutsu Kyokai / The Japanese Technical Association of the Pulp and Paper Industry, c/o Kami-Parupu Kaikan, 9-11, 3-chome, Ginza, Chuo-ku, Tokyo 104
Founded: 1947
Focus: Eng
Periodical
Japan Tappi (weekly) 08153

Kansai Keizai Rengokai / Kansai Economic Federation, Nakanoshima Center Bldg, 6-2-27 Nakanoshima, Kita-ku, Osaka 530
T: (06) 4410101
Focus: Econ; Hist 08154

Kansai Zosen Kyokai / The Kansai Society of Naval Architects, c/o Dept of Naval Architecture, Osaka University, 2-1 Yamada-oka, Suita, Osaka 565
T: (06) 8775111; Fax: (06) 8785364
Founded: 1912; Members: 2400
Pres: I. Tanaka
Focus: Eng
Periodicals
Bulletin of the Kansai Society of Naval Architects (weekly)
Journal of The Kansai Society of Naval Architects (weekly) 08155

Keikinzoku Gakkai / Japanese Institute of Light Metals, Nihonbashi Asahiseimei Bldg, 1-3, Nihonbashi 2-chome, Chuo-ku, Tokyo 103
T: (03) 32733041; Fax: (03) 32132918
Founded: 1951; Members: 2020
Pres: K. Hirano
Focus: Metallurgy
Periodical
Journal (weekly) 08156

Keizai Chiri Gakkai / Japan Association of Economic Geographers, c/o Institute of Economic Geography, Hitotsubashi University, 2-1 Naka, Kunitachi, Tokyo 186
T: (0425) 721101 ext 5374; Fax: (0425) 711893
Founded: 1954; Members: 720
Pres: M. Ishii
Focus: Geography
Periodical
Annals of the Japan Association of Economic Geographers (weekly) 08157

Keizai Chosa Kai Shiryoshitsu / Economic Research Assocation, Yaesukogyo Ginza Bldg, 3-15-10 Ginza, Chuo-ku, Tokyo 104
T: (03) 5931171
Focus: Econ 08158

Keizaigaku-shi Gakkai / Japan Society for the History of Social and Economic Thought, c/o Faculty of Economics, Konan University, 8-9-1 Okamoto, Higashinada-ku, Kobe
Founded: 1949; Members: 790
Pres: M. Tanaka
Focus: Hist; Econ; Sociology
Periodical
Annual Bulletin (weekly) 08159

Keizai-ho Gakkai / Association of Economic Jurisprudence, c/o Hitotsubashi University, Kunitachi, Tokyo 186
Founded: 1951; Members: 280
Focus: Law
Periodical
Journal (weekly) 08160

Keizai Tokei Kenkyukai, c/o Tokyo Kyoiku Daigaku, Tennodai 1-1-1, Sakura-mura, Niihari-gun, Ibaraki-ken, Tokyo 300
Founded: 1955
Focus: Stats; Econ 08161

Keizu Kyokai Toshokan / Genealogical Society, 5-10-30, Minami-Azubu, Minato-ku, Tokyo 106
T: (03) 34402764; Fax: (03) 34402774
Focus: Genealogy 08162

Kekkaku Yobokai Kekkaku Kenkyusho Toshoshitsu / Japan Anti-Tuberculosis Assocation, 3-2-24 Kiyose, Tokyo 204
T: (0424) 935711; Fax: (0424) 924600
Founded: 1939; Members: 55
Pres: Mazakazu Aoki
Focus: Pulmon Dis
Periodicals
Red Double-Barred Cross (weekly)
Review of Tuberculosis for Public Health Nurses (weekly) 08163

Ken-i Kai Ganka Toshokan / Ken-Ikai Foundation, 5-3-15 Ginza, Chuo-ku, Tokyo 104
T: (03) 5710294
Focus: Ophthal 08164

Kikai Shinko Kyokai Keizai Kenkyo-sho Shiryoshitsu / Japan Society for the Promotion of Machine Industry, Economic Research Institute, Kishinkyo Bldg, 3-5-8 Shiba-koen, Minato-ku, Tokyo 105
T: (03) 4348255
Gen Secr: Hidetaka Sudo
Focus: Econ; Mach Eng
Periodical
KSK Scanner (weekly) 08165

Kobunshi Gakkai / Society of Polymer Science, Nagaoka Bldg, 2-4-7 Tsukiji, Chuo-ku, Tokyo 104
T: (03) 35433765; Fax: (03) 35458560
Founded: 1951; Members: 12800
Pres: A. Abe; Gen Secr: T. Takahiko
Focus: Physics; Chem
Periodicals
Kobunshi (weekly)
Polymer Journal (weekly)
Polymer Preprints, Japan (weekly) . . . 08166

Koeki Jigyo Gakkai, c/o Nihon Denki Kyokai, 1-3 Yuraku-cho, Chiyoda-ku, Tokyo 100
Founded: 1948
Focus: Military Sci 08167

Kogyo Kayaku Kyokai / The Industrial Explosives Society of Japan, Gumma Bldg, 2-3-21 Nihonbashi, Chuo-ku, Tokyo 103
T: (03) 80716715; Fax: (03) 32717592
Founded: 1939; Members: 971
Pres: I. Fukuyama
Focus: Eng
Periodical
Journal (weekly) 08168

Kokka Gakkai, c/o Tokyo Dagaku Hogakubu, 7-3-1 Hongo, Bunkyo-ku, Tokyo 113
Founded: 1877
Focus: Poli Sci; Sociology 08169

Koko Eisei Gakkai, c/o Nihon Shika Dagaku, 1-9-20 Fujimi-cho, Chiiyoda-ku, Tokyo 102
Founded: 1952
Focus: Dent 08170

Kokugo Gakkai / Society for the Study of Japanese Language, c/o Faculty of Letters, University of Tokyo, Hongo, Bunkyo-ku, Tokyo 113
Founded: 1944; Members: 1500
Pres: E. Iwabuchi
Focus: Ling
Periodical
Studies in the Japanese Language (weekly) 08171

Kokusaiho Gakkai / Association of International Law, c/o Faculty of Law, University of Tokyo, Hongo, Bunkyo-ku, Tokyo 113
T: (03) 38122111
Founded: 1897; Members: 804
Pres: S. Kozai
Focus: Law
Periodical
Journal of International Law and Diplomacy (weekly) 08172

Kokusai Keizai Gakkai, Seiko Biru, 7-2-1 Minami-Aoyama, Minato-ku, Tokyo 107
Founded: 1950
Focus: Econ 08173

Kuki-Chowa Eisei Kogakkai / Society of Heating, Airconditioning and Sanitary Engineers of Japan, 1-8-1 Kitashinjuku, Shinjuku-ku, Tokyo
Founded: 1917; Members: 18000
Pres: T. Shinohara
Focus: Eng; Air Cond
Periodicals
Journal (weekly)
Transactions (3 times annually) 08174

Kyoikushi Gakkai, c/o Kyoiku Dagaku, Tennodai 1-1-1, Sakura-mura, Niihari-gun, Ibaraki-ken, Tokyo 300
Founded: 1956
Focus: Educ 08175

Kyoiku Tetsugakkai / Society of Educational Philosophy, c/o Sophia University, Kioi-cho, Chiyoda-ku, Tokyo 102
Founded: 1957; Members: 510
Pres: T. Oura
Focus: Educ; Philos
Periodical
Studies in the Philosophy of Education . 08176

Kyosei Kyokai Kyosei Toshokan / Japanese Correctional Assocation, 3-37-2 Arai, Nakano-ku, Tokyo 165
T: (03) 3874451
Focus: Criminology 08177

Kyoto Daigaku Keizai Gakkai, c/o Kyoto Dagaku Keizai, Yoshida-Honmachi, Sakyo-ku, Kyoto 606
Founded: 1919
Focus: Econ
Periodical
Keizai Ronso: Economic Review (weekly) 08178

Kyoto Tetsugakkai, c/o Kyoto Daigaku Bungakubu, Yoshida Honmachi, Sakyo-ku, Kyoto 606
Founded: 1915
Focus: Philos
Periodical
Tesugaku-Kenkyu: The Journal of Philosophical Studies (weekly) 08179

Manyo Gakkai / Society for Manyo Studies, c/o Kansai University, Senriyama Suita-shi, Osaka 564
Founded: 1951; Members: 810
Focus: Sci
Periodical
Manyo (weekly) 08180

Minji Soshoho Gakkai / Japan Association of Civil Procedure Law, c/o Hitotsubashi Daigaku Hogakubu, Naka 2-1, Kunitachi-shi, Tokyo 186
Founded: 1949
Focus: Law
Periodical
Journal of Civil Procedure (weekly) . . . 08181

Minzokugaku Shinkokai / Japanese Society of Ethnology, 3-1-17 Higashi-cho, Hoya-shi, Tokyo 188
T: (0424) 215003
Focus: Folklore 08182

Moralogy Kenkyusho / Institute of Moralogy, 2-1-1 Hikarigaoka, Kashiwa-shi, Chiba 277
Fax: (0471) 761177
Founded: 1926
Pres: M. Hiroike
Focus: Philos
Periodical
Studies in Moralogy 08183

National Cancer Center, 5-1-1 Tsukiji, Chuo-ku, Tokyo 104
T: (03) 5422511; Fax: (03) 5453567
Founded: 1962; Members: 800
Pres: T. Sugimura; Gen Secr: H. Ichikawa; S. Takayama
Focus: Cell Biol & Cancer Res
Periodicals
Annual Report (weekly)
Bone Tumor Registration in Japan (weekly)
Clinical Staging of Lung Cancer
Collected Papers (weekly)
Japanese Journal of Clinical Oncology (weekly)
Registration and Clinical Statistics of Stomach Cancer in Japan
Report of Hematologic Neoplasmas Registration in Japan (weekly) 08184

Nichibei Hogakkai / Japanese American Society for Legal Studies, c/o Faculty of Law, University of Tokyo, Hongo, Bunkyo-ku, Tokyo 113
Founded: 1964; Members: 950
Gen Secr: T. Kinoshita
Focus: Law
Periodical
Amerika Ho: Law in the United States (weekly) 08185

Nihon Aisotopu Kyokai / Japan Radioisotope Association, 28-45 Hon-Komagome 2-chome, Bunkyo-ku, Tokyo 113
T: (03) 39467111; CA: Japanisotope Tokyo; Fax: (03) 39462640
Founded: 1951; Members: 7500
Pres: Kiku Nakao; Gen Secr: Susumu Suzuki
Focus: Eng; Chem; Radiology
Periodicals
Isotope News (weekly)
Radioisotopes (weekly) 08186

Nihon Arerugi Gakkai / Japanese Society of Allergology, 33-19-802, Hakusan-1, Bunkyo-ku, Tokyo 113
T: (03) 8160280
Founded: 1952; Members: 4300
Gen Secr: Yoshitami Kimura
Focus: Med
Periodical
Japanese Journal of Allergology (weekly) 08187

Nihon Bearingu Kogyokai / Japan Bearing Industrial Assocation, Kikaishinkokaikan, 3-5-8 Shibakoen, Minato-ku, Tokyo 105
T: (03) 4330927
Focus: Mach Eng 08188

Nihon Bunseki Kagaku-Kai / Japan Society for Analytical Chemistry, Gotanda Sanhaitsu, 26-2, Nishigotanda 1-chome, Shinagawa-ku, Tokyo 141
T: (03) 4903351; Fax: (03) 34903572
Founded: 1952; Members: 9103
Pres: M. Tanaka; Gen Secr: Dr. T. Fujinuki
Focus: Chem
Periodicals
Analytical Sciences (weekly)
Bunseki
Bunseki Kagaku (weekly) 08189

Nihon Chugoku Gakkai, c/o Yushima Seido, 1-4-25 Yushima, Bunkyo-ku, Tokyo 113
Founded: 1949
Focus: Ling; Lit 08190

Nihon Denpun Gakkai / Japanese Society of Starch Science, c/o National Food Research Institute, Ministry of Agriculture, Forestry and Fisheries, 2-1-2 Kannondai, Tsukuba, Ibaraki 305
Founded: 1952; Members: 1200
Pres: H. Fuwa
Focus: Food

Periodical
Denpun Kagaku: Journal for Starch and Its Related Carbohydrates and Enzymes . . 08191

Nihon Denshi Kogyo Shinko Kyokai / Japan Electronic Industry Developent Assocation, Kikai Shinko Kaikan, 3-5-8 Shibakoen, Minato-ku, Tokyo 105
T: (03) 4348211
Focus: Electronic Eng 08192

Nihon Dobutsu Shinri Gakkai / The Japanese Society for Animal Psychology, c/o Institute of Psychology, University of Tsukuba, 1-1-1 Ten'noudai, Tsukuba-shi, Ibaraki 305
T: (0298) 534720
Founded: 1933; Members: 410
Pres: Fujita Osamu; Gen Secr: Tomihara Kazuya
Focus: Vet Med
Periodical
Dobutsu-shinrigaku-kenkyu: The Japanese Journal of Animal Psychology (weekly) 08193

Nihon Dokubun Gakkai / Japanese Society of German Literature, c/o Ikubundo, Hongo 5-30-21, Bunkyo-ku, Tokyo 113
Founded: 1947; Members: 2640
Pres: S. Shibata
Focus: Lit
Periodical
Doitsu Bungaku: German Literature (weekly) 08194

Nihon Do Senta / Japan Copper Development Association, Konwa Bldg, 1-12-22 Tsukiji, Chuo-ku, Tokyo 104
T: (03) 5426631
Focus: Metallurgy 08195

Nihon Eibungakkai / English Literary Society of Japan, 601 Kenkyusha Bldg, 9 Surugadai 2-chome, Kanda, Chiyoda-ku, Tokyo 101
T: (03) 2937528
Founded: 1917; Members: 3600
Pres: Y. Takamatsu
Focus: Lit; Ling
Periodical
Studies in English Literature (3 times annually) 08196

Nihon Eisei Gakkai / Japanese Society for Hygiene, c/o Dept of Hygiene and Preventive Medicine, Faculty of Medicine, University of Tokyo, 7-3-1 Hongo, Bunkyo-ku, Tokyo 113
T: (03) 38122111 ext 3437; Fax: (03) 56842297
Founded: 1902; Members: 2757
Gen Secr: Prof. O. Wada
Focus: Hygiene
Periodical
Japanese Journal of Hygiene (weekly) . 08197

Nihon Esperanto Gakkai / Japana Esperanto-Instituto / Japan Esperanto Institute, Waseda-mati 12-3, Sinzyuku-ku, Tokyo 162
T: (03) 32034581; Fax: (03) 32034582
Founded: 1919; Members: 1400
Pres: Sadaziro Kubo; Gen Secr: Yosio Isino
Focus: Ling
Periodical
La Revuo Orienta: Esperanto (weekly) . 08198

Nihon Furansu-go Furansu-bun-gaku-kai / Japanese Society of French Language and Literature, c/o Nichi-Futsu Kaikan, 2-3, Kanda-Surugadai, Chiyoda-ku, Tokyo
Founded: 1946; Members: 1800
Pres: H. Tajima
Focus: Ling; Lit
Periodicals
Bulletin (weekly)
Studies in French Language and Literature (weekly) 08199

Nihon Gakujutsu Shinko-kai / Japan Society for the Promotion of Science, 5-3-1 Kojimachi, Chiyoda-ku, Tokyo 102
T: (03) 2631721; Tx: 32281
Founded: 1967
Pres: T. Sawada; Gen Secr: F. Sakai
Focus: Nat Sci
Periodicals
Annual Report (weekly)
Japanese Scientific Monthly (weekly) . . 08200

Nihon Gas Kyokai Chosabu Chosaka / Japan Gas Association, 15-12, Toranomon 1-chome, Minato-ku, Tokyo 105
T: (03) 5020111
Founded: 1947; Members: 248
Focus: Energy
Periodical
Journal (weekly) 08201

Nihon Gengogakkai / Linguistic Society of Japan, c/o Sanseido Co. Ltd, 2-22-14 Misaki-cho, Chiyoda-ku, Tokyo 101
Founded: 1938; Members: 1700
Pres: Katsumi Matsumoto
Focus: Ling
Periodical
Gengo Kenkyu: Journal of the Linguistic Society of Japan (weekly) 08202

Nihon Genshiryoku Gakkai / Atomic Energy Society of Japan, 1-1-13, Shimbashi, Minato-ku, Tokyo 105
T: (03) 35081261; Fax: (03) 35816128
Founded: 1959; Members: 6630
Gen Secr: Minoru Masamoto
Focus: Nucl Res
Periodicals
Journal of Nuclear Science and Technology (weekly)
Journal of the Atomic Energy Society of Japan (weekly) 08203

Nihon Gomu Kyokai, 5-26 Motoakasaka, 1-chome, Minato-ku, Tokyo 107
T: (03) 4012957
Founded: 1928; Members: 3500
Focus: Materials Sci 08204

Nihon Hakubutsukan Kyokai / Japanese Association of Museums, Shoyu-Kaikan 3-3-1, Chiyoda-ku, Tokyo
Focus: Archives
Periodical
Museum Studies (weekly) 08205

Nihon Hoken Gakkai, Shinkokusai Bldg, 4-1, Marunouchi 3-chome, Chiyoda-ku, Tokyo 100
T: (03) 2130661
Founded: 1940; Members: 850
Focus: Insurance 08206

Nihon Hoken Igakkai / Association of Life Insurance Medicine of Japan, c/o Life Insurance Association of Japan, Shi-Kokusai Bldg, 3-4-1 Marunouchi, Chiyoda-ku, Tokyo 100
T: (03) 2862735
Founded: 1901; Members: 982
Pres: Hiroshi Tsukamoto; Gen Secr: Hiroo Iwata
Focus: Med 08207

Nihon Hozon Shika Gakkai, c/o Nihon Shika Daigaku, 1-9-20 Fujimi-cho, Chiyoda-ku, Tokyo 102
Founded: 1955
Focus: Dent 08208

Nihon Iden Gakkai / Genetics Society of Japan, c/o National Institute of Genetics, 1, 111 Yata, Mishima, Shizuoka 411
Fax: (0559) 756240
Founded: 1920
Pres: K. Moriwaki
Focus: Genetics
Periodical
Japanese Journal of Genetics (weekly) . 08209

Nihon Ikushu Gakkai / Japanese Society of Breeding, c/o Faculty of Agriculture, University of Tokyo, 1-1-1 Yayoi, Bunkyo-ku, Tokyo 113
Founded: 1951; Members: 1800
Pres: Yuzo Fiutsuhara; Gen Secr: Seishi Ninomiya
Focus: Bio
Periodical
Japanese Journal of Breeding (weekly) . 08210

Nihon Indogaku Bukkyogaku Kai / Japanese Association of Indian and Buddhist Studies, c/o Dept of Indian Philosophy and Buddhist Studies, University of Tokyo, 7-3-1 Hongo, Bunkyo-ku, Tokyo 113
T: (03) 38122111
Founded: 1951; Members: 2340
Pres: S. Mayeda; Gen Secr: Y. Ejima
Focus: Philos; Rel & Theol
Periodical
Journal of Indian and Buddhist Studies (weekly) 08211

Nihon Ishi-Kai / The Japan Medical Association, 2-28-16, Honkomagome, Bunkyo-ku, Tokyo 113
Founded: 1947
Focus: Med
Periodicals
Asian Medical Journal (weekly)
Journal of The Japan Medical Association (weekly) 08212

Nihon Ishinkin Gakkai, c/o Juntendo Daigaku, 2-1-1 Hongo, Bunkyo-ku, Tokyo 113
Founded: 1956
Focus: Botany, Specific
Periodical
Japanese Journal of Medical Mycology (weekly) 08213

Nihon Jui Gakkai / Japanese Society of Veterinary Science, Rakunokaikan Bldg, 1-37-20 Yoyogi, Shibuya-ku, Tokyo 151
T: (03) 33790636; Fax: (03) 33790636
Founded: 1885; Members: 4530
Pres: Prof. Dr. Tomotari Mitsuoka; Gen Secr: Prof. Dr. Michio Takahashi
Focus: Vet Med
Periodical
The Journal of Veterinary Medical Science (weekly) 08214

Nihon Junkanki Gakkai / Japanese Circulation Society, c/o Kinki Invention Center, 14 Yoshida Kawahara-cho, Sakyo-ku, Kyoto 606
T: (075) 7518643; CA: Circul Japan; Fax: (075) 7713060
Founded: 1935; Members: 16854
Pres: C. Kawai
Focus: Cardiol
Periodical
Japanese Circulation Journal (weekly) . . 08215

Nihon Kagaku Gijutsu Joho Sentah (JICST) / Japan Information Center of Science and Technology, POB 1478, Tokyo
T: (03) 35816411; Fax: (03) 35933375
Founded: 1957; Members: 325
Pres: M. Nakamura
Focus: Doc
Periodicals
Current Bibliography on Science and Technology: Kagaku Gijutsu Bunken Sokuho (weekly)
Journal of Information Processing and Management: Joho Kanri
Proceedings of the Annual Meeting on Information Science and Technology 08216

Nihon Kairui Gakkai / Malacological Society of Japan, c/o National Science Museum, 3-23-1 Hyakunin-cho, Shinjuku-ku, Tokyo 169
T: (03) 33642311; Fax: (03) 33642316
Founded: 1928; Members: 900
Pres: T. Habe
Focus: Zoology
Periodicals
Chiribotan (weekly)
Venus (weekly) 08217

Nihon Kakuigakukai / Japanese Society of Nuclear Medicine, c/o Japan Radioisotope Association, 28-45 Hon-Komagome 2-chome, Bunkyo-ku, Tokyo
Fax: (03) 39472535
Founded: 1963; Members: 3500
Pres: M. Furudate
Focus: Med
Periodical
Japanese Journal of Nuclear Medicine (weekly) 08218

Nihon Kandenchi Kogyokai / Japan Dry Battery Assocation, Dai 9 Mori Bldg, 1-2-2 Atago, Minato-ku, Tokyo 105
T: (03) 4362471
Focus: Electrochem 08219

Nihon Kasai Gakkai / Japanese Association for Fire Science and Engineering, c/o Business Center for Academic Societies, 2-chome 4-16 Yayoi, Bunkyo-ku, Tokyo 113
Founded: 1951; Members: 2000
Pres: H. Saito
Focus: Econ
Periodicals
Bulletin (weekly)
Kasai (weekly) 08220

Nihon Kensetsu Kikaika Kyokai, Kikai Shinko Bldg, 5-8, 3-chome, Shiba Park, Minato-ku, Tokyo 105
T: (03) 4331501
Founded: 1949; Members: 1527
Focus: Civil Eng 08221

Nihon Kikai Gakkai / Japan Society of Mechanical Engineers, Sanshin Hokusei Bldg, 2-4-9 Yoyogi, Shibuya-ku, Tokyo
T: (03) 33796781; Fax: (03) 33790936
Founded: 1897; Members: 45000
Pres: K. Tsuchiya
Focus: Eng
Periodicals
International Journal (weekly)
Journal
Transactions (weekly)
Yearbook (weekly) 08222

Nihon Kikaku Kyokai Gaikoku Kikaku Raiburari / Nihon Kikaku Kyokai Gaikoku Kikaku Raiburari / Japanese Standards Assocation, 4-1-24 Akasaka, Minato-ku, Tokyo 107
T: (03) 5838001
Focus: Standards 08223

Nihon Kin Gakkai / Mycological Society of Japan, c/o Business Center for Academic Societies, 4-1 Yayoi 2-chome, Bunkyo-ku, Tokyo 113
T: (03) 8151903
Founded: 1956; Members: 1100
Pres: K. Tubaki
Focus: Botany, Specific
Periodicals
News (weekly)
Transactions (weekly) 08224

Nihon Kirisutokyo Gakkai / The Japan Society of Christian Culture, c/o Research Institute of Christian Culture, Sophia University, 7-1, Kioichou, Chiyoda-ku, Tokyo 102
T: (03) 32383190; Fax: (03) 32384145
Founded: 1952; Members: 669
Pres: Peter Nemeshegyi; Gen Secr: Shunichi Takayanagi
Focus: Rel & Theol
Periodical
Nihon no Shingaku: Theological Studies in Japan (weekly) 08225

Nihon Kobutsu Gakkai, c/o Tokyo Daigaku Kyoyobu, 3-8-1 Komaba, Meguro-ku, Tokyo 153
Founded: 1952
Focus: Mineralogy 08226

Nihon Koho Gakkai / Japan Public Law Association, c/o Faculty of Law, University of Tokyo, Hongo, Bunkyo-ku, Tokyo 113
Founded: 1948; Members: 800
Pres: J. Tanaka
Focus: Law
Periodical
Koho-Kenkyu (weekly) 08227

Nihon Kokogaku Kyokai / Japanese Archaeologists Association, 5-15-5 Hirai, Edogawa-ku, Tokyo 132
T: (03) 36186608
Founded: 1948
Focus: Archeol
Periodical
Archaeologia Japonica (weekly) 08228

Nihon Koko Geka Gakkai, Tokyo Joshi Ika Daigaku, 10 Kawada-cho, Shinjuku-ku, Tokyo
Founded: 1952; Members: 2000
Gen Secr: M. Murase
Focus: Surgery; Med
Periodical
Japanese Journal of Oral Surgery (weekly) 08229

Nihon Koku Eisei Gakkai / Japanese Society for Dental Health, c/o Koku Hoken Kyokai, 44-2 Komagome, 1-chome, Toshima-ku, Tokyo 170
Founded: 1952; Members: 2000
Pres: Y. Sakakibara
Focus: Dent
Periodical
Journal (weekly) 08230

Nihon Kokukai Gakkai / Japanese Stomatological Society, Dept of Oral Surgery, School of Medicine, University of Tokyo, 7-3-1, Hongo, Bunkyo-ku, Tokyo
T: (03) 8155411
Founded: 1947; Members: 3600
Gen Secr: I. Yamashita
Focus: Stomatology
Periodical
Journal of the Japanese Stomatological Society (weekly) 08231

Nihon Kokusai Kyoiku Kyokai / Association of International Education, 4-5-29 Komaba, Meguro-ku, Tokyo 153
T: (03) 4673521
Focus: Int'l Relat; Educ
Periodical
ABC's of Study in Japan (weekly) . . . 08232

Nihon Kotsu Igakkai / Japanese Association of Transportation Medicine, c/o Chuo Tetsudo Byouin, 2-1-3 Yoyogi, Shibuya-ku, Tokyo 151
T: (03) 3791111
Founded: 1947; Members: 5000
Focus: Med
Periodical
Journal of Transportation Medicine (weekly) 08233

Nihon Kotsu Kyokai Toshokan / Japan Transportation Association, Shin-Kokusai Bldg, 3-4 Marunouchi, Chiyoda-ku, Tokyo 100
T: (03) 216-4082
Focus: Transport 08234

Nihon Kuho Gakkai, c/o Chuo Daigaku, 3-9 Surugadai, Kanda, Chiyoda-ku, Tokyo 101
Founded: 1955
Focus: Law 08235

Nihon Kyobu Geka Gakkai, c/o Tokyo Daigaku Igakubu, 7-3-1 Hongo, Bunkyo-ku, Tokyo 113
Founded: 1948
Focus: Surgery 08236

Nihon Kyoiku Gakkai / Japanese Society for the Study of Education, 5-26-5-901 Hongo, Bunkyo-ku, Tokyo 113
T: (03) 8182505; Fax: (03) 8166898
Founded: 1941; Members: 3000
Pres: T. Ohta
Focus: Educ
Periodical
Japanese Journal of Educational Research (weekly) 08237

Nihon Kyoiku-shakai Gakkai / Japan Society for the Study of Educational Sociology, c/o Faculty of Education, University of Tokyo, 7-3-1 Hongo, Bunkyo-ku, Tokyo 113
Founded: 1949; Members: 800
Pres: M. Aso
Focus: Educ; Sociology
Periodical
Journal of Educational Sociology (weekly) 08238

Nihon Kyoiku Shinri Gakkai, c/o Tokyo Daigaku Kyoiku Gakubu, 7-3-1 Hongo, Bunkyo-ku, Tokyo 113
Founded: 1952
Focus: Educ; Psych
Periodicals
The Annual Report of Educational Psychology in Japan (weekly)
The Japanese Journal of Educational Psychology (weekly) 08239

Nihon Kyosei Shikagakkai / Japan Orthodontic Society, c/o Koku Hoken Kyokai, 1-44-2 Komagome, Toshima-ku, Tokyo 170
T: (03) 39478891; Fax: (03) 39478341
Founded: 1932; Members: 3200
Pres: Tetsuo Iizuka; Gen Secr: Michiyasu Sato
Focus: Dent
Periodical
The Journal of the Japan Orthodontic Society (weekly) 08240

Nihon Masui Gakkai / Japan Society of Anesthesiology, TY Bldg, 3-18-11 Hongo, Bunkyo-ku, Tokyo 113
Founded: 1954; Members: 3711
Pres: K. Amaha
Focus: Anesthetics
Periodical
Journal of Anesthesia (weekly) 08241

Nihon Myakkan Gakkai, c/o Tokyo Daigaku Igakubu, 7-3-1 Hongo, Bunkyo-ku, Tokyo 113
Founded: 1960
Focus: Sci 08242

Nihon Naika Gakkai / Japanese Society of Internal Medicine, Hongo Daiichi Bldg, 34-3, Hongo 3-chome, Bunkyo-ku, Tokyo 113
T: (03) 38135991
Founded: 1903; Members: 63000
Pres: A. Shibata
Focus: Med
Periodicals
Internal Medicine (weekly)
Journal (weekly) 08243

Nihon Nettai Igakkai / Japanese Society of Tropical Medicine, c/o Institute of Tropical Medicine, Nagasaki University, 1-12-4 Sakamoto, Nagasaki 852
T: (0958) 472111; Fax: (0958) 476607
Founded: 1959
Focus: Med
Periodical
Japanese Journal of Tropical Medicine and Hygiene (weekly) 08244

Nihon Noshinkei Geka Gakkai, c/o Tokyo Daigaku Igakubu, 7-3-1 Hongo, Bunkyo-ku, Tokyo 113
Founded: 1948
Focus: Surgery 08245

Nihon Oyo Shinri-gakkai / Japan Association of Applied Psychology, c/o Dept of Psychology, College of Humanities and Sciences, Nihon University, 3-25-40 Sakurajosui, Setagaya-ku, Tokyo
T: (03) 3291151
Founded: 1931
Pres: H. Yamamoto; Gen Secr: K. Mural
Focus: Psych 08246

Nihon PEN Kurabu / Japan PEN Club, Room 265, Syuwa Residential Hotel, 9-1-7, Akasaka, Minato-ku, Tokyo 107
T: (03) 4021171; Fax: (03) 44025951
Founded: 1935; Members: 1000
Pres: S. Endo; Gen Secr: H. Ozaki
Focus: Lit
Periodical
Japanese Literature Today (weekly) . . 08247

Nihon Reito Kyokai / Japanese Association of Refrigeration, Sanei Bldg, 8 San-ei-cho, Shinguku-ku, Tokyo 160
T: (03) 33595231; Fax: (03) 33595233
Founded: 1925; Members: 6000
Pres: Hidehiko Takahashi; Gen Secr: Shigeru Mori
Focus: Electric Eng
Periodicals
Nihon Reito Kyokai Ronbunshu: JAR Transactions (3 times annually)
Reito: Refrigeration (weekly) 08248

Nihon Rinrigakukai / Japanese Society for Ethics, c/o Dept of Ethics, Faculty of Letters, University of Tokyo, Bunkyo-ku, Tokyo 113
Founded: 1950; Members: 800
Pres: Y. Yazima
Focus: Philos
Periodical
Rinrigakukaironshu (weekly) 08249

Nihon Rinsho Byori Gakkai / Japan Society of Clinical Pathology, Well-Stone Heights 203, 2-11-10, Hongo, Bunkyo-ku, Tokyo 113
Fax: (03) 8138150
Founded: 1951; Members: 3500
Pres: Tadashi Kawai; Gen Secr: Kinya Kawano
Focus: Pathology
Periodical
The Japanese Journal of Clinical Pathology (weekly) 08250

Nihon Ronen Igakukai / Japan Geriatrics Society, Kyorin Bldg 702, 4-2-1 Yushima, Bunkyo-ku, Tokyo 113
Founded: 1959; Members: 4500
Pres: Prof. H. Orimo
Focus: Geriatrics
Periodical
Japanese Journal of Geriatrics (weekly) . 08251

Nihon Sakumotsu Gakkai / Crop Science Society of Japan, c/o Faculty of Agriculture, University of Tokyo, Bunkyo-ku, Tokyo 113
T: (03) 8122111
Founded: 1927; Members: 1900
Pres: Y. Tuno; Gen Secr: N. Inoue
Focus: Crop Husb; Agri
Periodical
Japanese Journal of Crop Science (weekly) 08252

Nihon Sangyo Eisei Gakkai, c/o Koei Biru, 1-29-8 Shinjuku, Tokyo 160
Fax: (03) 33561536
Founded: 1929
Focus: Sci
Periodical
Japanese Journal of Industrial Health (weekly) 08253

Nihon Sanshi Gakkai / Japanese Society of Sericultural Science, c/o National Institute of Sericultural and Entomological Science, Owashi 1-2, Tsubuka, Ibaraki 305
T: 66070
Founded: 1929; Members: 1313
Pres: Dr. Akinori Shimazaki; Gen Secr: Dr. Hiroaki Yanagawa
Focus: Agri
Periodical
The Journal of Sericultural Science of Japan (weekly) 08254

Nihon Seibutsu Butsuri Gakkai / Biophysical Society of Japan, c/o Dept of Biophysical Engineering, Faculty of Engineering Science, Osaka University, Machikaneyama, Toyonaka, Osaka 560
T: (06) 8440943; Fax: (06) 8439354
Founded: 1960; Members: 3000
Pres: Kosai Michiki
Focus: Biophys
Periodical
Seibutsu Butsuri: Biophysics (weekly) . . 08255

Nihon Seibutsu Kankyo Chosetsu Kenkyukai / Japanese Society of Environment Control in Biology, c/o Dept of Agricultural Engineering, University of Tokyo, Tokyo 113
T: (03) 8122111 ext 5356
Founded: 1969; Members: 752
Pres: Prof. Dr. Jiro Sugi; Gen Secr: Prof. Dr. Tadashi Takakura
Focus: Bio
Periodical
Environment Control in Biology (weekly) . 08256

Nihon Seishin Shinkei Gakkai / Japanese Society of Psychiatry and Neurology, Hongo Sky Bldg 38-11, 3-chome, Hongo, Bunkyo-ku, Tokyo 113
Founded: 1902; Members: 6500
Pres: Y. Kasahara
Focus: Neurology; Psychiatry
Periodical
Seishin Shinkeigaku Zasshi (weekly) . . 08257

Nihon Sen'i Kikai Gakkai / Textile Machinery Society of Japan, c/o Osaka Kagaku Gijutsu Senta, 1-8-4, Utsubo-Honmachi, Nishi-ku, Osaka 550
Fax: (06) 4434694
Founded: 1948
Focus: Mach Eng
Periodicals
Journal: English Edition (weekly)
Journal: Japanese Edition (weekly) . . . 08258

Nihon Setchakuzai Kogyokai Toshoshitsu / Japan Adhesive Industry Association, Fukushima Bldg, 1-15-10 Uchikanda, Chiyoda-ku, Tokyo 101
T: (03) 2913303
Focus: Materials Sci 08259

Nihon Shakai Shinri Gakkai / The Japanese Society of Social Psychology, c/o Dept of Sociology, Faculty of Letters, Keio University, 2-15-45, Minato-ku, Tokyo 108
Founded: 1960; Members: 1209
Pres: Kimiyoshi Hirora
Focus: Psych
Periodical
Research in Social Psychology: Bulletin of the Japanese Society of Social Psychology (3 times a year) 08260

Nihon Shinku Kyokai, Kikai Shinko Kaikan, 21-1-5 Shiba-koen, Minato-ku, Tokyo 105
Founded: 1958
Focus: Physics 08261

Nihon Shinrigakkai / Japanese Psychological Association, 2-40-14-902, Hongo, Bunkyo-ku, Tokyo 113
Founded: 1927; Members: 4800
Pres: Yoshinori Matsuyama
Focus: Psych
Periodicals
Japanese Journal of Psychology (weekly)
Japanese Psychological Research (weekly) 08262

Nihon Shokaki Naishikyo Gakkai / Japan Gastroenterological Endoscopy Society, 3-22, Ogawa-machi, Kanda, Chiyoda-ku, Tokyo 101
Founded: 1961; Members: 22950
Pres: Prof. Takao Sakita
Focus: Gastroenter
Periodical
Gastroenterological Endoscopy (weekly) . 08263

Nihon Shoni Geka Gakkai / Japanese Society of Pediatric Surgeons, c/o Business Center for Academic Societies, 2-4-16 Yayoi, Bunkyo-ku, Tokyo 113
Founded: 1964
Focus: Pediatrics
Periodical
Journal (7 times annually) 08264

Nihon Shonika Gakkai / Societas Paediatrica Japonica / Japan Pediatric Society, 4 Daiichi Magami Bldg, 1-1-5, Koraku, Bunkyo-ku, Tokyo 112
Founded: 1896; Members: 13000
Pres: M. Okuni
Focus: Pediatrics
Periodical
Acta Paediatrica Japonica (weekly) . . . 08265

Nihon Shukyo Gakkai / Japanese Association for Religious Studies, c/o Dept of Religious Studies, University of Tokyo, 7-3-1 Hongo, Bunkyo-ku, Tokyo 113
T: (03) 38122111; Fax: (03) 38159989
Founded: 1930; Members: 1900
Focus: Rel & Theol
Periodical
Journal of Religious Studies (weekly) . . 08266

Nihon Tairyoku Igakkai, c/o Tokyo Jieikai Ika Dagaku, Minato-ku, Tokyo 105
Founded: 1949
Focus: Med 08267

Nihon Tenmon Gakkai / Astronomical Society of Japan, c/o National Astronomical Observatory, Osawa Mitaka, Tokyo
Fax: (0422) 311359
Founded: 1908; Members: 2028
Pres: U. Yutaka
Focus: Astronomy
Periodicals
The Astronomical Herold (weekly)
Publications (weekly) 08268

Nihon Tetsugakkai, c/o Toyo Daigaku Bungakubu, 5-28-20 Hakusan, Bunkyo-ku, Tokyo 112
Founded: 1949
Focus: Philos 08269

Nihon Tokei Gakkai / Japan Statistical Society, c/o Institute of Statistical Mathematics, 4-6-7 Minami-azabu, Minato-ku, Tokyo 106
T: (03) 4425801
Founded: 1931; Members: 1164
Pres: T. Nakamura
Focus: Stats
Periodical
Journal (weekly) 08270

Nihon Toshokan Kyokai / Japan Library Association, 1-10, 1-chome, Taishido, Setagaya-ku, Tokyo 154
T: (03) 34106411; Fax: (03) 34217588
Founded: 1892; Members: 8300
Gen Secr: Reiko Sakagawa
Focus: Libraries & Bk Sci
Periodicals
Gendai no Toshokan (weekly)
Nihon no Sankotosho Shikiban (weekly)
Toshokan Zasshi (weekly) 08271

Nihon Uirusu Gakkai / Society of Japanese Virologists, c/o Business Centre for Academic Societies, 4-16 Yayoi 2-chome, Bunkyo-ku, Tokyo 113
Founded: 1958; Members: 2141
Pres: Dr. N. Ishida
Focus: Med
Periodicals
Microbiology and Immunology
Virus (weekly) 08272

Nihon Yakuri Gakkai / Japanese Pharmacological Society, 2-4-16 Yayoi, Bunkyo-ku, Tokyo 113
T: (03) 38144828
Founded: 1927; Members: 5600
Pres: M. Endo
Focus: Pharmacol
Periodicals
Folia Pharmacologica Japonica (weekly)
The Japanese Journal of Pharmacology (weekly) 08273

Nihon Yosetsu Kyokai, Sampo Sakuma Bldg 1-11, Kanda Sakuma-cho, Chiyoda-ku, Tokyo 101
T: (03) 2530581
Founded: 1949; Members: 750
Focus: Eng 08274

Nihon Zairyo Gakkai, 1-101 Yoshida-Izumidono-machi, Sakyo-ku, Kyoto 606
Founded: 1952
Focus: Materials Sci 08275

Nihon Zairyo Kyodo Gakkai, c/o Research Institute for Stenght and Facture of Materials, Tohoku University, Armaki-Aoba, Sendai 980
Founded: 1966; Members: 1068
Focus: Materials Sci 08276

Nihon Zaisei Gakkai / Japanese Association of Fiscal Studies, c/o Hitotsubashi University, Kunitachi, Tokyo 186
Founded: 1940; Members: 195
Focus: Finance 08277

Nihon Zeiho Gakkai, c/o Ichiro Nakagawa, Iwakura-Hase-machi 1051-3, Sakyo-ku, Kyoto 606
Founded: 1951
Focus: Law 08278

Nihon Zosen Gakkai / Society of Naval Architects of Japan, Sempaku-Shinko Bldg, 15-16 Toranomon, 1-chome, Minato-ku, Tokyo 105
Founded: 1897; Members: 5300
Pres: Y. Fujita
Focus: Eng
Periodicals
Bulletin (weekly)
Journal (weekly)
Naval Architecture and Ocean Engineering (weekly) 08279

Nippon Afurika Gakkai, c/o Tokyo Daigaku Rigakubu, 7-31 Hongo, Bunkyo-ku, Tokyo 113
Founded: 1964
Focus: Ethnology 08280

Nippon Bitamin Gakkai / Vitamin Society of Japan, c/o Nippon Italy Kyoto-Kaikan, 4-Ushinomiya-cho, Yoshida, Sakyo-ku, Kyoto 606
T: (075) 7510314
Founded: 1947; Members: 2000
Pres: Dr. K. Okuda
Focus: Nutrition; Food; Pharmacol; Med; Agri; Chem
Periodicals
Journal of Nutritional Science and Vitaminology (weekly)
Vitamins (weekly) 08281

Nippon Bungaku Kyokai / Japanese Literature Association, 2-17-10 Minami-otsuka, Toyoshima-ku, Tokyo 170
Founded: 1946; Members: 1500
Pres: T. Hirosue
Focus: Lit
Periodical
Japanese Literature (weekly) 08282

Nippon Bunko Gakkai / Spectroscopial Society of Japan, Clean Bldg, 1-13 Kanda-Awaji-cho, Chiyoda-ku, Tokyo 101
Founded: 1951; Members: 1200
Pres: K. Shibata; Gen Secr: Y. Nihei
Focus: Physics; Chem
Periodical
Journal (weekly) 08283

Nippon Butsuri Gakkai / The Physical Society of Japan, Room 211, Kikai-Shinko Bldg, 3-5-8 Shiba-Koen, Minato-ku, Tokyo 105
T: (03) 34342671; Fax: (03) 34320997
Founded: 1977; Members: 17062
Pres: Tetsuro Kobayashi
Focus: Physics
Periodicals
Butsuri (weekly)
Japanese Journal of Applied Physics (weekly)
Journal of the Physical Society of Japan (weekly)
Progress of Theoretical Physics (weekly) 08284

Nippon Butsuri-Kagaku Kenkyukai / Physico-Chemical Society of Japan, c/o Research Institute for Production Development, 15 Simogamo Morimoto-cho, Sakyo-ku, Kyoto 606
T: (075) 7811107
Founded: 1926; Members: 360
Pres: Jono Wasaburo; Gen Secr: Tadashi Makita
Focus: Physics; Chem 08285

Nippon Byori Gakkai / Japanese Society of Pathology, c/o Dept of Pathology, Faculty of Medicine, University of Tokyo, Bunkyo-ku, Tokyo 113
Fax: (03) 38158379
Founded: 1911; Members: 4000
Gen Secr: R. Machinami
Focus: Pathology
Periodicals
Annual of Pathological Autopsy Cases in Japan
Pathologica Japonica (weekly)
Transactions 08286

Nippon Chikusan Gakkai / Japanese Society of Zootechnical Science, 201 Nagatani Corporas, Ikenohata 2-9-4, Taito-ku, Tokyo 110
Founded: 1924; Members: 2000
Pres: Y. Shoda
Focus: Zoology
Periodical
Japanese Journal of Zootechnical Science (weekly) 08287

Nippon Chikyu Denki Ziki Gakkai / Society of Terrestrial Magnetism and Electricity of Japan, c/o Japan Academic Societies Business Centre, 2-4-16 Yayoi, Bunkyo-ku, Tokyo 113
T: (03) 8151903
Founded: 1947; Members: 750
Gen Secr: Prof. T. Oguti
Focus: Physics
Periodical
Journal of Geomagnetism and Geo-Electricity (weekly) 08288

Nippon Chiri Gakkai / Assocation of Japanese Geographers, c/o Bldg of Academic Societies Centre, 2-4-16, Yayoi, Bunkyo-ku, Tokyo 113
T: (03) 38151912; Fax: (03) 38151672
Founded: 1925; Members: 3000
Pres: Tsuneyoshi Ukita
Focus: Geography
Periodical
Geographical Review of Japan: Chirigaku Hyoron (weekly) 08289

Nippon Chishitsu Gakkai / Geological Society of Japan, Maruishi Bldg, 1-10-4 Kajicho, Chiyoda-ku, Tokyo 101
T: (03) 2527242
Founded: 1893; Members: 5081
Pres: Y. Kuroda
Focus: Geology
Periodical
Journal (weekly) 08290

Nippon Chô Gakkai / Ornithological Society of Japan, c/o Dept of Zoology, National Science Museum, 3-23-1 Hyakunin-cho, Shinjuku-ku, Tokyo 160
Founded: 1912; Members: 800
Pres: N. Kuroda
Focus: Ornithology
Periodical
Japanese Journal of Ornithology (weekly) 08291

Nippon Dai-Yonki Gakkai / Japan Association for Quaternary Research, c/o Business Centre for Academic Societies, Honkomagome 5-16-9, Bunkyo-ku, Tokyo 113
Fax: (03) 58145820
Founded: 1956; Members: 1800
Pres: K. Sohma
Focus: Geology

Periodical
The Quaternary Research (weekly) . . . *08292*

Nippon Denki Kyokai Chosabu Chosaka / The Japan Electric Association, Danki Bldg, 1-7-1 Yuraku-cho, Chiyoda-ku, Tokyo 100
T: (03) 2160551
Focus: Energy *08293*

Nippon Denshi Kenbikyo Gakkai / Japanese Society of Electron Microscopy, c/o Business Centre for Academic Societies, 4-16 Yayoi 2-chome, Bunkyo-ku, Tokyo 113
Founded: 1949; Members: 2690
Pres: K. Ogawa
Focus: Physics
Periodical
Journal (weekly) *08294*

Nippon Dobutsu Gakkai / The Zoological Society of Japan, Toshin Bldg, 2-27-2 Hongo, Bunkyo-ku, Tokyo 113
Founded: 1878; Members: 2420
Pres: Hideo Mohri
Focus: Zoology
Periodical
Zoological Science (weekly) *08295*

Nippon Dokumenesyon Kyokai / Japan Documentation Society, 5-7, Koisikaw 2, Bunkyo-ku, Tokyo 112
T: (03) 8133791
Founded: 1950; Members: 1200
Focus: Doc
Periodicals
Informant: Microfiche-Editon (weekly)
Journal (weekly) *08296*

Nippon Dokyo Gakkai / Japan Society of Taoistic Research, c/o Dept of Oriental Philosophy, School of Literature, Waseda University, 1-24-1, Toyama, Shinjuku-ku, Tokyo 162
T: (03) 2034111
Founded: 1950; Members: 500
Pres: H. Kusuyama
Focus: Ethnology; Rel & Theol
Periodical
Journal of Eastern Religions (weekly) . . *08297*

Nippon Engeki Gakkai / Japanese Society for Theatre Research, c/o Waseda University, 1-6-1 Nishi-Waseda, Shinjuku-ku, Tokyo
Founded: 1949
Pres: S. Endo
Focus: Perf Arts
Periodical
Kiyo (weekly) *08298*

Nippon Gan Gakkai / Japanese Cancer Association, c/o Cancer Institute, Kami-Ikebukuro 1-37-1, Toshima-ku, Tokyo 170
T: (03) 39180111
Founded: 1907; Members: 14000
Pres: Dr. Iro Nobuyuki
Focus: Cell Biol & Cancer Res
Periodicals
Gann-Monograph on Cancer Research (weekly)
Japanese Journal of Cancer Research (weekly)
. . . . *08299*

Nippon Ganseki Kobutsu Kosho Gakkai / Japanese Association of Mineralogists, Petrologists and Economic Geologists, c/o Faculty of Science, Tohoku University, Aoba, Sendai 980
T: (022) 2243852; Fax: (022) 2243852
Founded: 1928; Members: 1000
Pres: Banno Shohei; Gen Secr: Kanisawa Satoshi
Focus: Geology; Mineralogy; Petrochem
Periodical
Journal of Mineralogy, Petrology and Economic Geology (weekly) *08300*

Nippon Geka Gakkai / Japan Surgical Society, Nippon Ishi Kaikan, 2-5 Kanda Surugadai, Chiyoda-ku, Tokyo 101
Founded: 1899; Members: 12750
Pres: N. Shimada
Focus: Surgery
Periodical
Journal (weekly) *08301*

Nippon Gyosei Gakkai / Japanese Society for Public Administration, c/o Meiji University, Kanda-Surugadai 1-1, Chiyoda-ku, Tokyo
Founded: 1945; Members: 400
Pres: A. Sato
Focus: Business Admin
Periodical
Nenpo (weekly) *08302*

Nippon Hifuka Gakkai, 3-14-10 Hongo, Bunkyo-ku, Tokyo 113
Founded: 1901
Focus: Derm
Periodicals
The Journal of Dermatology (weekly)
Nihon Hifuka Gakkai Zasshi (14 times annually)
. . . . *08303*

Nippon Hikaku Bungakukai / Comparative Literature Society of Japan, c/o Aoyamagakiun University, Shibuya-ku, Tokyo 150
Founded: 1948; Members: 400
Pres: K. Nakajima
Focus: Lit
Periodicals
Bulletin (weekly)
Journal (weekly) *08304*

Nippon Hinyoki-ka Gakkai / The Japanese Urological Association, Taisei Bldg, Hongo 3-14-10, Bunkyo-ku, Tokyo
T: (03) 38147921; Fax: (03) 38144117
Founded: 1912; Members: 5780
Pres: Osamu Yoshida
Focus: Urology
Periodical
The Japanese Journal of Urology (13 times annually) *08305*

Nippon Hoi Gakkai / Medico-Legal Society of Japan, c/o Dept of Legal Medicine, Faculty of Medicine, University of Tokyo, 7-3-1 Hongo, Bunkyo-ku, Tokyo 113
T: (03) 8122111
Founded: 1914; Members: 964
Pres: I. Shikata
Focus: Med; Law
Periodical
Journal (weekly) *08306*

Nippon Hoshakai Gakkai / Japan Association of Sociology of Law, c/o University of Tokyo, Hongo, Bunkyo-ku, Tokyo 113
Founded: 1947; Members: 990
Pres: N. Toshitani; Gen Secr: Kahei Rokumoto
Focus: Law; Sociology
Periodical
Sociology of Law (weekly) *08307*

Nippon Hoshasen Eikyo Gakkai / Japan Radiation Research Society, c/o National Institute of Radiological Sciences, 4-9-1 Anagawa, Inage-ku, Chiba-shi 263
T: (0472) 2512111; Fax: (0472) 2569616
Founded: 1954; Members: 981
Pres: Akihiro Shima; Gen Secr: Koki Sato
Focus: Radiology
Periodical
Journal of Radiation Research (weekly) . *08308*

Nippon Hotetsu Gakkai / Japanese Association for Legal Philosophy, c/o Faculty of Law, Kyoto University, Sakyo-ku, Kyoto
Founded: 1948; Members: 343
Pres: S. Kato
Focus: Philos
Periodical
Annual (weekly) *08309*

Nippon Hotetsu Shika Gakkai / Japan Prosthodontic Society, c/o Koku Hoken Kyokai, 1-44-2 Komagome, Toshima-ku, Tokyo
Founded: 1931; Members: 3700
Pres: K. Hiranuma
Focus: Dent
Periodical
Journal of the Japan Prosthodontic Society (weekly) *08310*

Nippon Igaku Hoshasen Gakkai / Japan Radiological Society, Room 301 Akamon Habitation, 5-29-13, Hongo, Bunkyo-ku, Tokyo
T: (03) 8143077
Founded: 1923; Members: 4500
Pres: M. Iio
Focus: Radiology
Periodical
Nippon Acta Radiologca (weekly) . . . *08311*

Nippon Imono Kyokai, 8-12-13 Ginza, Chuo-ku, Tokyo 104
T: (03) 5412758
Founded: 1932; Members: 7
Focus: Metallurgy *08312*

Nippon Jibi-Inkoka Gakkai / Oto-Rhino-Laryngological Society of Japan, c/o Chateau-Takanawa, 3-23-14 Takanawa, Minato-ku, Tokyo
T: (03) 34433085; Fax: (03) 34433037
Founded: 1893; Members: 9600
Pres: Nomura Yasuya
Focus: Otorhinolaryngology
Periodical
Journal of Otolaryngology of Japan (weekly)
. . . . *08313*

Nippon Jidoseigyo Kyokai / Japan Association of Automatic Control Engineers, 14, Kawahara-cho, Yoshida, Sakyo-ku, Kyoto
Founded: 1957; Members: 1964
Pres: M. Kuwahara
Focus: Eng
Periodical
Systems and Control (weekly) *08314*

Nippon Jinruigaku Kai / Anthropological Society of Nippon, c/o Business Center for Academic Societies, 4-16 Yayoi 2-chome, Bunkyo-ku, Tokyo 113
Fax: (03) 8175800
Founded: 1884; Members: 900
Pres: Bin Yamaguchi
Focus: Anthro
Periodical
Journal of the Anthropological Society of Nippon (weekly) *08315*

Nippon Jinrui Iden Gakkai / The Japan Society of Human Genetics, c/o Dept of Human Genetics, Tokyo Medical and Dental University, 1-5-45 Yushima, Bunkyo-ku, Tokyo 113
Founded: 1957; Members: 900
Pres: Dr. Ei Matsunaga; Gen Secr: Dr. Takashi Imamura
Focus: Genetics
Periodical
The Japanese Journal of Human Genetics (weekly) *08316*

Nippon Junkatsu Gakkai / Japan Society of Lubrication Engineers, Kikai Shinko Kaikan 407-2, 3-5-8 Shiba Koen, Minato-ku, Tokyo 105
T: (03) 4341926
Founded: 1956; Members: 2700
Pres: H. Mori
Focus: Med
Periodical
Journal (weekly) *08317*

Nippon Kagakukai / Chemical Society of Japan, 1-5 Kanda-Surugadai, Chiyoda-ku, Tokyo
Founded: 1878; Members: 35000
Pres: Tamaru Kenzi
Focus: Chem
Periodicals
Bulletin (weekly)
Chemistry Letters
Kagaku to Kogyo
Nippon Kagaku Kaishi *08318*

Nippon Kagaku Kyoiku Gakukai / Japan Society for Science Education, c/o National Institute for Educational Research, 6-5-22, Shimomeguro, Meguro-ku, Tokyo 153
Founded: 1977; Members: 1050
Pres: Dr. M. Oki
Focus: Educ
Periodicals
Journal (weekly)
Letter (weekly) *08319*

Nippon Kaibo Gakkai / Japanese Association of Anatomiats, c/o Business Center for Academic Societies, 4-16, Yayoi 2-chome, Bunkyo-ku, Tokyo 113
T: (03) 8151903
Founded: 1893; Members: 2300
Pres: K. Yasuda
Focus: Anat
Periodical
Acta Anatomica Nipponica (weekly) . . . *08320*

Nippon Kaiho Gakkai / Maritime Law Association of Japan, c/o Chuo University, Higashinakano, Hachioji-shi, Tokyo
Founded: 1950; Members: 165
Pres: T. Ishii
Focus: Law
Periodical
Report *08321*

Nippon Kaisui Gakkai / Society of Sea Water Science, Japan, c/o Japan Salt Industry Association, 7-15-14, Roppongi, Minato-ku, Tokyo 106
T: (03) 4026414
Founded: 1950; Members: 485
Pres: T. Yamare
Focus: Water Res *08322*

Nippon Kaiyo Gakkai / Oceanographical Society of Japan, MK Bldg 202, 1-6-14 Minamidai, Nakano-ku, Tokyo 164
Founded: 1941; Members: 1650
Pres: T. Nannichi
Focus: Oceanography
Periodical
Journal (weekly) *08323*

Nippon Kango Kyokai / Japanese Nursing Assocation, 1-2-3 Umezono, Kiyoseshi, Tokyo 204
T: (0424) 927466
Founded: 1946
Pres: Mito Takako
Focus: Med *08324*

Nippon Kazan Gakkai / Volcanological Society of Japan, Earthquake Research Institute, University of Tokyo, 1-1-1, Yayoi-cho, Bunkyo-ku, Tokyo 113
T: (03) 8137421
Founded: 1932; Members: 950
Pres: S. Aramaki
Focus: Geophys
Periodicals
Bulletin (weekly)
Bulletin of Volcanic Eruptions (weekly) . *08325*

Nippon Keiei Gakkai / Japan Society for the Study of Business Administration, Institute of Business Research, Hitotsubashi University, Naka 2-1, Kunitachi, Tokyo 186
T: (0425) 721101
Founded: 1926; Members: 1930
Pres: Moriyuki Tajima
Focus: Business Admin
Periodical
Keieigaku Ronshu (weekly) *08326*

Nippon Keiho Gakkai / Criminal Law Society of Japan, c/o University of Tokyo, Hongo, Bunkyo-ku, Tokyo
Founded: 1949; Members: 727
Pres: K. Matsuo
Focus: Criminology
Periodical
Journal (weekly) *08327*

Nippon Keizai Seisaku Gakkai / Japan Economic Policy Association, c/o Keio University, Mita, Minato-ku, Tokyo 108
Founded: 1940; Members: 862
Pres: T. Yamanaka
Focus: Poli Sci
Periodical
Annals *08328*

Nippon Kekkaku-byo Gakkai / The Japanese Society for Tuberculosis, 3-1-24 Matsuyama, Kiyose-shi, Tokyo 204
Fax: (0424) 918315
Founded: 1923; Members: 3000
Pres: K. Hara
Focus: Pulmon Dis
Periodical
Kekkaku (weekly) *08329*

Nippon Kessho Gakkai / Crystallographic Society of Japan, c/o Business Center for Academic Societies, 2-4-16 Yayoi, Bunkyo-ku, Tokyo 113
Founded: 1950; Members: 1000
Pres: Y. Iidaka
Focus: Mineralogy
Periodical
Journal (weekly) *08330*

Nippon Ketsueki Gakkai / Japan Haematological Society, c/o Kyoto Daigaku Igakubu Fuzoku Byoin, Kawahara, Shogoin, Sakyo-ku, Kyoto 606
Fax: (075) 7520761
Founded: 1937; Members: 5518
Pres: H. Uchino
Focus: Hematology
Periodical
Acta Haematologica Japonica: International Journal of Hematology (8 times annually) . . . *08331*

Nippon Kikan-Shokudo-ka Gakkai / Japan Bronco-Esophagological Society, Uno Bldg, 2-3-11, Koraku, Bunkyo-ku, Tokyo
Founded: 1949; Members: 3600
Pres: Dr. K. Togawa
Focus: Pulmon Dis
Periodical
Journal (weekly) *08332*

Nippon Kinzoku Gakkai / Japan Institute of Metals, Aoba Aramaki, Sendai 980
T: (022) 2233685; Tx: 852250
Founded: 1937; Members: 11000
Pres: O. Saeki
Focus: Metallurgy
Periodicals
Bulletin (weekly)
Journal (weekly)
Transactions (weekly) *08333*

Nippon Kisei-chu Gakkai / Japanese Society of Parasitology, c/o Institute of Medical Science, University of Tokyo, Minato-ku, Tokyo 108
Founded: 1926; Members: 930
Gen Secr: Prof. H. Tanaka
Focus: Microbio
Periodical
Japanese Journal of Parasitology (weekly) *08334*

Nippon Kisho Gakkai / Meteorological Society of Japan, c/o Japan Meteorological Agency, 1-3-4 Ote-machi, Chiyoda-ku, Tokyo 100
Founded: 1882; Members: 4240
Pres: T. Asai
Focus: Geophys
Periodicals
Journal of the Meteorological Society of Japan (weekly)
Papers in Meteorology and Geophysics (weekly)
Tenki (weekly) *08335*

Nippon Kogakukai / Japan Federation of Engineering Societies, Nihon-kogyo-Kaikan nai, Nogizaka Bldg, 6-41, Akasaka 9-chome, Minato-ku, Tokyo 107
Founded: 1879; Members: 64
Focus: Eng
Periodical
Journal (weekly) *08336*

Nippon Kokai Gakkai / Japan Institute of Navigation, c/o Tokyo University of Mercantile Marine, 2-1-6 Etchujima, Koto-ku, Tokyo 135
T: (03) 6411171
Founded: 1948; Members: 940
Pres: Prof. K. Oikawa
Focus: Navig
Periodicals
Journal (weekly)
Navigation (weekly) *08337*

Nippon Koku Gakkai / Japan Society of Aeronautical and Space Science, 1-18-1 Shinbashi, Minato-ku, Tokyo
Founded: 1934; Members: 1800
Focus: Mach Eng
Periodicals
Journal (weekly)
Transactions (weekly) *08338*

Nippon Kokusai Seiji Gakkai / Japan Association of International Relations, c/o Keio Daigaku, 2-15-45 Mita, Minato-ku, Tokyo 108
Founded: 1956; Members: 1400
Focus: Poli Sci
Periodical
International Relations (3 times annually) *08339*

Nippon Koku Uchu Gakkai / Japan Society for Aeronautical and Space Sciences, 18-2 Shinbashi 1-chome, Minato-ku, Tokyo 105
T: (03) 5010463
Founded: 1934; Members: 3100
Pres: Prof. M. Sunakawa
Focus: Eng
Periodicals
Journal (weekly)
Transactions (weekly) *08340*

Nippon Kontyu Gakkai / Entomological Society of Japan, c/o Dept of Zoology, National Science Museum (Natural History), 3-23-1 Hyakunin-cho, Shinjuku, Tokyo 160
T: (03) 3642311
Founded: 1917; Members: 1200
Pres: T. Shirozu
Focus: Entomology
Periodical
Kontyu (weekly) 08341

Nippon Koseibutsu Gakkai / Palaeontological Society of Japan, c/o Business Center for Academic Societies, Yayoi 2-4-16, Bunky-ku, Tokyo 113
Founded: 1935; Members: 600
Pres: Y. Takayanagi
Focus: Paleontology
Periodicals
Fossils (weekly)
Special Papers (weekly)
Transactions and Proceedings (weekly) . 08342

Nippon Koshu-Eisei Kyokai / Japan Public Health Association, Koei Bldg, 29-8, Shinjuku 1-chome, Shinjuku-ku, Tokyo
Founded: 1883; Members: 5000
Pres: M. Seijo
Focus: Public Health
Periodicals
Japanese Journal of Public Health (weekly)
Public Health Information (weekly) . . . 08343

Nippon Mokuzai Gakkai / Japan Wood Research Society, 21-4-407, Hongo 6-chome, Bunkyo-ku, Tokyo 113
T: (03) 38160396; Fax: (03) 38186568
Founded: 1955; Members: 2240
Pres: Dr. Mikiaki Ohkuma; Gen Secr: Dr. Shigenori Kuga
Focus: Forestry
Periodical
Mokuzai Gakkaishi: Journal of the Japan Wood Reseach Society (weekly) 08344

Nippon Nensho Gakkai / Combustion Society of Japan, c/o Dept of Reaction Chemistry, Faculty of Engineering, University of Tokyo, 7 Hongo, Bunkyo-ku, Tokyo 113
Fax: (03) 356842509
Founded: 1953; Members: 580
Pres: T. Hirano
Focus: Chem; Eng
Periodical
Nensho Kenkyu (weekly) 08345

Nippon Nensho Kenkyukai / The Japanese Section of the Combustion Institute / Combustion Society of Japan, c/o Kyoto Daigaku Kogakubu, Yoshida, Sakyo-ku, Kyoto 606
Founded: 1954; Members: 400
Pres: Prof. T. Hirano; Gen Secr: Prof. H. Jinno
Focus: Eng
Periodical
Nensho-Kenkyu (3 times annually) . . . 08346

Nippon Netsushori Gijutsu Kyokai / The Japan Society for Heat Treatment, Shinsen Bldg 4F, 8-2 Shinsen, Shibuya-ku, Tokyo 150
T: (03) 4617116
Focus: Energy
Periodical
Netsu Shori: Journal of the Japan Society for Heat Treatment (weekly) 08347

Nippon Nogei Kagaku Kai / Japan Society for Bioscience, Biotechnology and Agrochemistry, 2-4-16 Yayoi, Bunkyo-ku, Tokyo 113
Founded: 1924; Members: 15162
Pres: Hideaki Yamada
Focus: Agri; Chem; Bio
Periodicals
Bioscience, Biotechnology and Biochemistry (weekly)
Journal of the Japan Society for Bioscience, Biotechnology and Agrochemistry (weekly) 08348

Nippon Nogyo-Kisho Gakkai / Society of Agricultural Meteorology of Japan, c/o Dept of Agricultural Engineering, Faculty of Agriculture, University of Tokyo, Yayoi 1-1-1, Bunkyo-ku, Tokyo 113
T: (03) 38122111 ext 5355
Founded: 1942; Members: 1169
Focus: Agri; Geophys
Periodical
Nogyo-Kisho: Journal of Agricultural Meteorology (weekly) 08349

Nippon Noritsu Kyokai Shiryoshitsu / Japan Management Association, Kyoritsu Bldg, 3-1-22 Shiba-koen, Minato-ku, Tokyo 105
T: (03) 4346211
Focus: Business Admin 08350

Nippon No-Shinkei Gek Gakkai / Japan Neurosurgical Society, c/o Dept of Neurosurgery, Faculty of Medicine, University of Tokyo, 7-3-1 Hongo, Bunkyo-ku, Tokyo
Fax: (03) 38122090
Founded: 1948; Members: 5905
Pres: S. Joya
Focus: Neurology; Surgery
Periodicals
Neurologia Medico-Chirurgica I (weekly)
Neurologia Medico-Chirurgica II (weekly) . 08351

Nippon Ongaku Gakkai / The Musicological Society of Japan, c/o Tokyo Geijutsu Daigaku, Ueno-Park, Taito-ku, Tokyo 110
T: (03) 38286111
Founded: 1952; Members: 1300
Focus: Music
Periodical
Ongaku Gaku: Journal of the Musicological Society of Japan (weekly) 08352

Nippon Onkyo Gakkai / Acoustical Society of Japan, Ikeda Bldg, 2-7-7 Yoyogi, Shibuya-ku, Tokyo 151
T: (03) 33791200; Fax: (03) 33791456
Founded: 1936; Members: 4100
Pres: T. Sone
Focus: Acoustics
Periodicals
Journal (weekly)
The Journal (E): English Edition (weekly)
Reports of Spring and Autumn Meetings (weekly) 08353

Nippon Orient Gakkai / Society for Near Eastern Studies in Japan, c/o Tokyo Tenrikyokan, 1-9 Nishiki-cho, Kanda, Chiyoda-ku, Tokyo 101
Founded: 1954
Focus: Ethnology
Periodical
Bulletin of the Society for Near Eastern Studies in Japan 08354

Nippon Oyo-Dobutsu-Konchu Gakkai / Japanese Society of Applied Entomology and Zoology, c/o Japan Plant Protection Association, 11-43-1 Komagome, Toshima-ku, Tokyo 170
Founded: 1957; Members: 1900
Pres: Mitsuhashi Jun
Focus: Entomology; Zoology
Periodicals
Applied Entomology and Zoology (weekly)
Japanese Journal of Applied Entomology and Zoology (weekly) 08355

Nippon Rai Gakkai / Japanese Leprosy Association, 4-2-1, Aoba-cho, Higashimurayama-shi, Tokyo 189
Founded: 1927; Members: 355
Pres: S. Yamamoto
Focus: Derm
Periodical
La Lepro (weekly) 08356

Nippon Rikusui Gakkai / Japanese Society of Limnology, c/o Dept of Chemistry, Faculty of Science, Tokyo Metropolitan University, 1-1 Minamiohsawa, Hachioji, Tokyo
T: (0426) 772531
Founded: 1931; Members: 1215
Pres: Yasuhiko Tezuka; Gen Secr: Prof. Norio Ogura
Focus: Hydrology; Bio; Chem; Physics
Periodical
Rikusui-gaku Zasshi: Japanese Journal of Limnology (weekly) 08357

Nippon Ringakukai / Japanese Forestry Society, c/o Japan Forest Technical Association, 7 Rokuban-cho, Chiyoda-ku, Tokyo
T: (03) 32612766; Fax: (03) 32612766
Founded: 1914; Members: 2800
Pres: S. Sasaki
Focus: Forestry
Periodicals
Journal (weekly)
Transactions of the Annual Meeting (weekly) 08358

Nippon Rodo-ho Gakkai / Japanese Labour Law Association, c/o Waseda University, Nishi Waseda, Shinjuku-ku, Tokyo
Founded: 1950; Members: 550
Pres: K. Hokao
Focus: Sci
Periodical
Journal (weekly) 08359

Nippon Rosiya Bungakkai / Russian Literary Society in Japan, c/o Baba-ken, Tokyo Institute of Technology, 2-12-1, O-okayama, Meguro-ku, Tokyo 152
Members: 380
Pres: Masanobu Togo; Gen Secr: T. Egawa
Focus: Ling; Lit
Periodical
Bulletin 08360

Nippon Saikingakkai / Japanese Society for Bacteriology, c/o Business Centre for Academic Societies, 2-4-16 Yayoi, Bunkyo-ku, Tokyo 113
Founded: 1927; Members: 3000
Pres: S. Sasaki
Focus: Sci
Periodical
Japanese Journal of Bacteriology, Microbiology and Immunology 08361

Nippon Sanka-Fujinka Gakkai / Japan Society of Obstetrics and Gynaecology, c/o Hoken Kaikan Bldg, 1-1 Sadohara-cho, Ichigaya, Shinjuku-ku, Tokyo 162
Founded: 1949
Pres: Prof. Yoshiteru Terashima
Focus: Gynecology
Periodical
Acta Obstetrica et Gynaecologica Japonica (weekly) 08362

Nippon Seikei Geka Gakkai / Japanese Orthopaedic Association, Fuse Bldg, 30-10, Hongo 3-chome, Bunkyo-ku, Tokyo 113
Fax: (03) 38182337
Founded: 1926; Members: 16545
Pres: Y. Yamauchi
Focus: Orthopedics
Periodical
Journal (weekly) 08363

Nippon Seiri Gakkai / Physiological Society of Japan, Fuse Bldg, Hongo 3-30-10, Bunkyo-ku, Tokyo 113
Founded: 1922; Members: 3500
Pres: Masao Ito
Focus: Physiology
Periodical
Journal 08364

Nippon Seitai Gakkai / The Ecological Society of Japan, c/o Biological Institute, Faculty of Science, Tohoku University, Aoba-ku, Sendai 980
T: (022) 2250636; Fax: (022) 2250636
Founded: 1954; Members: 2600
Pres: Oshima Yasuyuki; Gen Secr: Nishihira Moritaka
Focus: Ecology
Periodicals
Ecological Research (3 times annually)
Japanese Journal of Ecology (3 times annually) 08365

Nippon Seiyo Koten Gakkai / The Classical Society of Japan, Dept of Classics, Faculty of Letters, Kyoto University, Kyoto
Founded: 1950; Members: 550
Pres: Norio Fujisawa
Focus: Lit; Hist; Philos
Periodical
Journal of Classical Studies (weekly) . . 08366

Nippon Seiyoshi Gakkai / The Japanese Society of Western History, c/o Osaka University, Toyonaka-shi, 1-1 Machikaneyama-cho, Osaka 560
Founded: 1948; Members: 900
Pres: T. Horii
Focus: Hist
Periodical
Studies in Western History (weekly) . . 08367

Nippon Seizi Gakkai / Japanese Political Science Association, c/o Faculty of Law, Rikkyo University, 3-34-1, Nishi-Ikebukuro, Toshima-ku, Tokyo 171
Members: 820
Pres: Jiro Kamishima
Focus: Poli Sci 08368

Nippon Seramikkusu Kyokai / The Ceramic Society of Japan, 22-17, 2-chome, Hyakunin-cho, Shinjuku-ku, Tokyo 169
Fax: (03) 33625714
Founded: 1891; Members: 7500
Pres: H. Yanagida; Gen Secr: Y. Suzuki
Focus: Materials Sci
Periodicals
Bulletin of the Ceramic Society of Japan (weekly)
Journal of the Ceramic Society of Japan (weekly) 08369

Nippon Shakai Gakkai / Japanese Sociological Society, c/o Dept of Sociology, Faculty of Letters, University of Tokyo, 7-3-1 Hongo, Bunkyo-ku, Tokyo
Founded: 1923; Members: 2500
Pres: Joji Watanuki
Focus: Sociology
Periodicals
International Journal of Japanese Sociology (weekly)
Shakaigaku Hyóron: Japanese Sociological Review (weekly) 08370

Nippon Shashin Gakkai / Society of Photographic Science and Technology of Japan, c/o Tokyo Polytechnic Institute, 2-9-5 Hon-cho, Nakano-ku, Tokyo 164
Founded: 1925; Members: 1550
Pres: T. Wakabayashi
Focus: Eng
Periodical
Journal (weekly) 08371

Nippon Shashin Sokuryo Gakkai / Japan Society of Photogrammetry and Remote Sensing, Daichi Honan Bldg 502, 2-8-17 Minami Ikebukoro, Toshima-ku, Tokyo 171
Fax: (03) 39847402
Founded: 1962; Members: 1038
Pres: Kazuhiko Otake; Gen Secr: Ryutaru Tateishi
Focus: Surveying
Periodical
Journal (weekly) 08372

Nippon Shiho Gakkai / Japan Association of Private Law, c/o University of Tokyo, Hongo, Bunkyo-ku, Tokyo 113
Founded: 1948; Members: 688
Pres: T. Suzuki
Focus: Law
Periodical
Journal (weekly) 08373

Nippon Shika Hoshasen Gakkai / Japanese Society of Dental Radiology, c/o Tokyo Dental College, 2-9-18, Misakicho, Chiyoda-ku, Tokyo
Founded: 1951; Members: 319
Pres: S. Ando
Focus: Radiology
Periodical
Dental Radiology (weekly) 08374

Nippon Shika Hozon Gakkai / Japanese Society of Conservative Dentistry, c/o Tokyo Dental University, 1 Misaki-cho Kanda, Chiyoda-ku, Tokyo
Founded: 1955; Members: 826
Pres: E. Sekine
Focus: Dent
Periodical
Journal (weekly) 08375

Nippon Shika Igakkai / Japanese Association for Dental Science, c/o Japan Dental Association, 4-1-20 Kudan-kita, Chiyoda-ku, Tokyo
T: (03) 32629214; Fax: (03) 32629885
Founded: 1949; Members: 80000
Pres: Prof. H. Sekine
Focus: Dent
Periodicals
Dentistry in Japan (weekly)
Journal (weekly) 08376

Nippon Shimbun Gakkai / Japan Society for Studies in Journalism and Mass Communication, c/o Institute of Journalism and Communication Studies, University of Tokyo, 7 Hongo, Bunkyo-ku, Tokyo 113
T: (03) 8122111 ext 5921
Founded: 1951; Members: 666
Pres: T. Sato
Focus: Journalism
Periodical
Japanese Journalism Review (weekly) . . 08377

Nippon Shinkei Gakkai / Japanese Society of Neurology, Ichimaru Bldg, 2-31-21 Yushima, Bunkyo-ku, Tokyo 113
T: (03) 8151080; Fax: 815-1931
Founded: 1960; Members: 6216
Pres: Dr. Toru Mannen
Focus: Neurology
Periodical
Rinsho Shinkei: Clinical Neurology (weekly) 08378

Nippon Shinkeikagaku Gakkai / Japan Neuroscience Society, c/o Business Center for Academic Societies, 5-16-9 Honkomagome, Bunkyo-ku, Tokyo 113
Founded: 1974; Members: 2100
Pres: Masao Ito
Focus: Neurology
Periodicals
Neuroscience News (weekly)
Neuroscience Research (weekly) 08379

Nippon Shogyo Gakkai / Japan Society of Commercial Sciences, c/o Meiji University, 1-1 Surugadai, Kanda, Chiyoda-ku, Tokyo 101
Founded: 1951; Members: 429
Pres: K. Fukuda
Focus: Business Admin 08380

Nippon Shokaki-byo Gakkai / The Japanese Society of Gastroenterology, Ginza Orient Bldg, Ginza 8-9-13, Chuo-ku, Tokyo 104
T: (03) 35734297; Fax: (03) 32892359
Founded: 1899; Members: 123000
Focus: Gastroenter
Periodicals
Gastroenterologia Japonica (weekly)
Nihon Shokaki-byo Gakkai Zasshi (weekly) 08381

Nippon Shokubutsu-Byori Gakkai / Phytopathological Society of Japan, c/o Japan Plant Protection Association, 1-43-11, Komagome, Toshima-ku, Tokyo 170
T: (03) 39436021; Fax: (03) 39436021
Founded: 1916; Members: 1924
Pres: T. Tsuchizaki; Gen Secr: T. Hibi
Focus: Crop Husb
Periodical
Annals of the Phytopathological Society of Japan (weekly) 08382

Nippon Shokubutsu Gakkai / Botanical Society of Japan, Toshin Bldg, 2-chome, 27-2 Hongo, Bunkyo-ku, Tokyo
T: (03) 38145675; Fax: (03) 38145461
Founded: 1882; Members: 2300
Pres: K. Iwatsuki
Focus: Botany
Periodical
Journal of Plant Research (weekly) . . 08383

Nippon Shokubutsu Seiri Gakkai / Japanese Society of Plant Physiologists, Shimotachiuri Ogawa Higashi, Kamikyo-ku, Kyoto 602
T: (075) 4413157
Founded: 1959; Members: 2792
Pres: K. Asada
Focus: Botany
Periodical
Plant and Cell Physiology (8 times annually) 08384

Nippon Sokuchi Gakkai / Geodetic Society of Japan, c/o Geogaphical Survey Institute, 1 Kitazato, Tsukuba-shi, Ibaraki 305
T: 641111
Founded: 1954; Members: 600
Pres: I. Nakagawa
Focus: Surveying
Periodical
Journal of the Geodetic Society of Japan (weekly) 08385

Nippon Sugaku Kai / Mathematical Society of Japan, 25-9-203 Hongo 4-chome, Bunkyo-ku, Tokyo 113
Founded: 1877; Members: 4700
Pres: Akio Hattori
Focus: Math
Periodicals
Japanese Journal of Mathematics (weekly)
Journal of the Mathematical Society of Japan (weekly)
Publications of the Mathematical Society of Japan (weekly)
Sugaku (weekly) 08386

Nippon Sugaku Kyoiku Gakkai / Japan Society of Mathematical Education, Koishikawa, POB 18, Tokyo 112
Founded: 1919; Members: 4732
Pres: Prof. Y. Matsuo
Focus: Math; Educ
Periodicals
Journal (weekly)
Report on Mathematical Education (weekly)
Supplementary Issue 08387

Nippon Suisan Gakkai / Japanese Society of Scientific Fisheries, c/o Tokyo University of Fisheries, 5-7, Konan-4, Minato-ku, Tokyo 108
T: (03) 4712165; Fax: (03) 4712165
Founded: 1932; Members: 4589
Pres: Dr. Kanehisa Hashimoto; Gen Secr: Dr. Haruyuki Kanehiro
Focus: Fisheries
Periodical
Bulletin: Nippon Suisan Gakkaishi (weekly) 08388

Nippon Syoyakugakkai / Japanese Society of Pharmacognosy, c/o Business Centre for Academic Societies, 4-16 Yayoi 2-chome, Bunkyo-ku, Tokyo 113
T: (03) 8151903
Founded: 1946; Members: 1027
Pres: M. Konoshima
Focus: Pharmacol
Periodical
Japanese Journal of Pharmacognosy (weekly)
. . . . 08389

Nippon Taiiku Gakkai / Japanese Society of Physical Education, c/o School of Physical Education, University of Tokyo, Hongo 7-3-1, Bunkyo-ku, Tokyo 113
Founded: 1950; Members: 2315
Pres: M. Matsui
Focus: Sports
Periodicals
Journal of Health and Physical Education (weekly)
Research Journal (weekly) 08390

Nippon Teiinoshujutsu Kenkyukai / Japanese Society for Stereotactic Functional Neurosurgery, c/o Dept of Neurosurgery, Nihon University, 30-1 Ohyaguchi Kamimachi, Itabashi-ku, Tokyo
Founded: 1963; Members: 450
Gen Secr: Prof. T. Tsubokawa
Focus: Neurology
Periodical
Summary of Annual Meeting (weekly) . . 08391

Nippon Tekko Kyokai / The Iron and Steel Institute of Japan, Keidanren Kaikan, 1-9-4 Otemachi, Chiyoda-ku, Tokyo 100
T: (03) 32796021; Tx: 02228153;
CA: Nippontekkokyo, Tokyo; Fax: (03) 32451355
Founded: 1915; Members: 10160
Pres: Shunkichi Miyoshi; Gen Secr: Jim Shimada
Focus: Metallurgy
Periodicals
ISIJ International (weekly)
Tetsu-to-Hagane (weekly) 08392

Nippon Tonyo-byo Gakkai / Japan Diabetes Society, Hongo Sky Bldg 403, 3-38-11 Hongo, Bunkyo-ku, Tokyo 113
Founded: 1938
Focus: Diabetes
Periodical
The Journal of the Japan Diabetes Society
. . . . 08393

Nippon Toshokan Gakkai / Japan Society of Library Science, c/o Faculty of Sociology, Toyo University, 28-20 Hakusan 5-chome, Bunkyo-ku, Tokyo 112
Founded: 1953; Members: 570
Focus: Libraries & Bk Sci
Periodicals
Annals (weekly)
Bibliography of Library Science (weekly) . 08394

Nippon Yakugaku-Kai / The Pharmaceutical Society of Japan, 12-15, Shibuya 2-chome, Shibuya-ku, Tokyo 150
Founded: 1881; Members: 18875
Pres: Nojima Syoshichi; Gen Secr: Kanaoka Yuichi
Focus: Pharmacol
Periodicals
Biological Pharmaceutical Bulletin (weekly)
Chemical Pharmaceutical Bulletin (weekly)
Eiseikagaku (weekly)
Farumashia (weekly)
Yakugaku Zasshi (weekly) 08395

Nippon Yukagaku Kyokai / Japan Oil Chemists' Society, Yushi Kogyo Kaikan, 3-13-11 Nihonbashi, Chuo-ku, Tokyo 103
T: (03) 32717463; Fax: (03) 32717464
Founded: 1951; Members: 2600
Pres: Tetsutaro Hashimoto
Focus: Petrochem
Periodical
Journal (weekly) 08396

Nippon Yuketsu Gakkai / Japan Society of Blood Transfusion, Nisseki Chuo-Ketsueki Center, 1-31-4 Hiroo, Shibuya-ku, Tokyo
Founded: 1954; Members: 1260
Pres: S. Murakami
Focus: Hematology
Periodical
Journal (weekly) 08397

Nogyo-Doboku Gakkai / Japanese Society of Irrigatin, Drainage and Reclamation Engineering, Nogyo Doboku-Kaikan, 34-4 Shinbashi 5-chome, Tokyo 105
Founded: 1929; Members: 13000
Pres: T. Sawada
Focus: Sci
Periodicals
Journal (weekly)
Transactions (weekly) 08398

Nogyo-Ho Gakkai / Japan Agricultural Law Association, c/o Faculty of Law, University of Nihon, Chiyoda-ku, Tokyo 101
Founded: 1956; Members: 200
Pres: T. Ogura
Focus: Agri
Periodical
Nogyo-ho Kenkyu (weekly) 08399

Nogyokikai Gakkai / Society of Agricultural Machinery, c/o Dept of Agricultural Engineering, Faculty of Agriculture, University of Tokyo, Bunkyo-ku, Tokyo 113
Founded: 1937; Members: 1650
Pres: A. Hosokawa
Focus: Eng
Periodical
Journal (weekly) 08400

Okinawa Kyokai Chosa Engoka, Gloria Bldg, 3-6-15 Kasumigaseki, Chiyoda-ku, Tokyo 100
T: (03) 5800641
Focus: Poli Sci 08401

Oyo-buturi Gakkai (JJAP) / Japan Society of Applied Physics, Kikai-Shinko Bldg, 3-5-8 Shiba-Koen, Minato-ku, Tokyo 105
Founded: 1932; Members: 9000
Pres: T. Okada
Focus: Physics
Periodicals
JJAP (weekly)
Oyo-buturi 08402

Raten Amerika Kyokai Kenkyubu / Latin-American Association, Dai-2 Jingumae Bldg, 2-6-14 Jingumae, Shibuya-ku, Tokyo 150
T: (03) 4032661
Focus: Econ; Law 08403

Riron Keiryo Keizai Gakkai / Japan Association of Economics and Econometrics, c/o Institute of Statistical Research, Shinbashi 1-18-16, Minato-ku, Tokyo 105
Founded: 1934; Members: 147
Pres: Yasuo Uekawa
Focus: Econ 08404

Rissho Koseikai / Rissho Koseikai Fuzoku Kosei Toshokan, Gyogakuen Bldg, 1-2-8 Wada, Suginami-ku, Tokyo 166
T: (03) 3840111
Focus: Rel & Theol 08405

San-yo Gijutsu Shinkokai / Sanyo Association for Advancement of Science and Technology, 16-1-1 Hon-machi, Kurashiki-shi, Okayama-ken 710
T: (0846) 226655
Focus: Sci; Eng 08406

Sapporo Norin Gakkai / Sapporo Society of Agriculture and Fisheries, c/o Faculty of Agriculture, Hokkaido University, Kita-ku, Sapporo 160
Founded: 1908
Pres: T. Ui
Focus: Forestry; Agri
Periodical
Journal (weekly) 08407

Seisan Gijutsu Kenkyusho / Institute of Industrial Science, c/o University of Tokyo, 7-22-1 Roppongi, Minato-ku, Tokyo 106
T: (03) 34026231; Tx: (0242) 3216; Fax: (03) 34025078
Founded: 1949
Pres: Prof. Dr. M. Harashima
Focus: Business Admin
Periodical
Report (6-8 times annually) 08408

Seito Kogyokai Jimubu Chosaka / Japan Sugar Refiners' Assocation, 5-7, Samban-cho, Chiyoda-ku, Tokyo 102
T: (03) 2620176
Focus: Agri; Econ 08409

Sekai Keizai Chosakai Shiryoshitsu, Seiko Bldg, 7-2-1 Minami-aoyama, Minato-ku, Tokyo 107
T: (03) 4001671
Focus: Econ 08410

Sekiyu Gijutsu Kyokai, Keidanren Kaikan, 1-9-4 Otemachi, Chiyoda-ku, Tokyo 100
Founded: 1933
Focus: Eng 08411

Sekiyu Remmei Kohobu Shiryoka / Petroleum Association of Japan, Keidanren Bldg, 1-9-4 Otemachi, Chiyoda-ku, Tokyo 100
T: (03) 2793811
Focus: Petrochem
Periodical
Japan Oil Statistics Today (weekly) . . 08412

Sen-i Gakkai / Society of Fibre Science and Technology, 3-3-9-208 Kamiosaki, Shinagawa-ku, Tokyo 141
Founded: 1943; Members: 3000
Pres: H. Inagaki
Focus: Eng
Periodical
Journal (weekly) 08413

Senshokutai Gakkai, c/o International Christian University, 1500 Ozawa, Mitaka-shi, Tokyo 181
Founded: 1949
Focus: Bio 08414

Shakai Keizaishi Gakkai, c/o Dept of Economics, Sophia University, Kioi-cho, Chiyoda-ku, Tokyo 102
T: (03) 32383090; Fax: (03) 32383090
Founded: 1930; Members: 1179
Focus: Hist; Econ
Periodical
Shakai-Keizai-Shigaku (weekly) 08415

Shakai Seisaku Gakkai, c/o Hosei Daigaku Ohara Shakai Mondai Kenkyusho, 2-17-1 Fujimi-cho, Chiyoda-ku, Tokyo 102
Founded: 1950
Focus: Poli Sci 08416

Shigaku-kai / Historical Society of Japan, c/o Faculty of Letters, University of Tokyo, Hongo, Bunkyo-ku, Tokyo 113
Founded: 1889; Members: 2470
Pres: O. Naruse
Focus: Hist
Periodical
Shigaku-Zasshi 08417

Shigaku Kenkyukai, c/o Kyoto Daigaku Bungakubu, Yoshida Honmachi, Sakyo-ku, Kyoto 606
T: (075) 7532787
Founded: 1908; Members: 1460
Pres: Yoshihiro Kawachi; Gen Secr: Sadako Ishida
Focus: Hist; Archeol; Geography
Periodical
The Shirin or the Journal of History (weekly)
. . . . 08418

Shika Kiso Igakkai, c/o Tokyo Shika Daigaku, 2-9-18 Misaki-cho, Chiyoda-ku, Tokyo 101
Founded: 1967
Focus: Otorhinolaryngology 08419

Shinto Gakkai, c/o Izumo Taisha Bunshi, 7-18-5 Rappongi, Minato-ku, Tokyo 105
Founded: 1938
Focus: Rel & Theol 08420

Shinto Shukyo Gakkai, c/o Kokugakuin Daigaku, 4-10-28 Higashi, Shibuya-ku, Tokyo 150
Founded: 1947
Focus: Rel & Theol 08421

Shokubai Gakkai / Catalysis Society of Japan, Shin-ikedayama Mansions 302, 5-21-13 Higashi-gotana, Sinagawa-ku, Tokyo 141
T: (03) 34442126; Fax: (03) 34448794
Founded: 1958; Members: 1415
Pres: Y. Izumi; Gen Secr: K. Yoshino
Focus: Chem
Periodical
Shokubai: Catalyst (weekly) 08422

Shokubutsu Bunrui Chiri Gakkai / Phytogeographical Society, c/o Dept of Botany, Kyoto University, Kyoto 606
Founded: 1932; Members: 460
Pres: S. Kitamura
Focus: Botany
Periodical
Acta Phytotaxonomica et Geobotanica (weekly)
. . . . 08423

Shomei Gakkai / Illuminating Engineering Institute of Japan, 7-1, Yurakucho 1-chome, Chiyoda-ku, Tokyo 100
T: (03) 2010645
Founded: 1916; Members: 4891
Pres: K. Kobayashi
Focus: Electric Eng
Periodical
Journal of Light and Visual Environment 08424

The Society of Fermentation and Bioengineering, Japan, c/o Faculty of Engineering, Osaka University, 2-1 Yamadaoka, Suita, Osaka 565
T: (06) 8762731; Fax: (06) 8762773
Founded: 1923; Members: 3100
Pres: Prof. Shiro Nagai
Focus: Food
Periodicals
Journal of Fermentation and Bioengineering (weekly)
Seibutsu-kogaku Kaishi (weekly) 08425

Society of Japanese Historical Research, c/o Kokugakuin University, 4-10-28 Higashi, Shibuya-ku, Tokyo 150
Founded: 1910
Gen Secr: H.R. Tsubaki
Focus: Hist
Periodical
Kokushigaku 08426

Suifumeitokukai Shokokan, 1215-1, Migawa-machi, Mito-shi, Ibaraki-ken 310
T: (0292) 412721
Focus: Cultur Hist 08427

Tensor Society, c/o Kawaguchi Institute of Mathematical Sciences, 2-7-15 Matsu-ga-oka, Chigasaki 253
Founded: 1937; Members: 550
Pres: Prof. Dr. A. Kawaguchi; Gen Secr: Prof. T. Kawaguchi; Prof. H. Kawaguchi
Focus: Math
Periodical
Tensor 08428

Tetsugaku-kai / Philosophical Society, c/o Faculty of Letters, University of Tokyo, 7-3-1 Hongo, Bunkyo-ku, Tokyo 113
Founded: 1884; Members: 500
Pres: J. Katsura
Focus: Philos
Periodical
Tetsugaku-zasshi (weekly) 08429

Toa Igaku Kyokai, 2-20 Shin-Ogawa, Shinjuku-ku, Tokyo 162
Founded: 1954
Focus: Med 08430

Toa Kumo Gakkai, c/o Oitemon Gakuin Daigaku, Ibaraki-shi, Osaka 567
Founded: 1936
Focus: Sci 08431

Tochi Seidoshi Gakkai, c/o Institute of Social Science, University of Tokyo, 3-1, Hongo 7-chome, Bunkyo-ku, Tokyo 113
Founded: 1948; Members: 950
Focus: Hist; Econ; Agri
Periodical
Tochi Seido Shigaku: The Journal of Agrarian History (weekly) 08432

Todai Chugoku Gakkai / The Sinological Society, c/o Faculty of Letters, The University of Tokyo, Bunkyo-ku, Tokyo
T: (03) 8122111, 8123746
Founded: 1985; Members: 800
Pres: Ito Toramaru; Gen Secr: Akira Takeda
Focus: Ethnology
Periodical
Hugoku-Shakai to Bunka: China-Society and Culture 08433

Toho Gakkai / Institute of Eastern Culture, 4-1, Nishi-Kanda 2-chome, Chiyoda-ku, Tokyo 101
T: (03) 32611061, 32627221; Fax: (03) 32627227
Founded: 1947; Members: 1469
Pres: Dr. Jikido Takasaki; Gen Secr: Hiroshi Yanase
Focus: Hist; Lit; Ling; Philos
Periodicals
Acta Asiatica (weekly)
Books and Articles on Oriental Subjects published in Japan (weekly)
Tohogaku (weekly)
Transactions of the International Conference of Orientalists in Japan (weekly) 08434

Tokei Kagaku Kenkyukai / Research Association of Statistical Sciences, c/o Kyushu University 33, Fukuoka 812
Founded: 1941
Focus: Stats; Cybernetics
Periodical
Bulletin of Informatics and Cybernetics (weekly)
. . . . 08435

Tokyo Bengoshikai Toshokan / Tokyo Bar Association, 1-1-4 Kasumigaseki, Chiyoda-ku, Tokyo 100
T: (03) 35812201; Fax: (03) 35810865
Founded: 1911; Members: 3340
Focus: Law
Periodicals
Libra (weekly)
Toben Shinbun (weekly) 08436

Tokyo Chigaku Kyokai / Tokyo Geographical Society, Chigaku-Kaikan, 12-2 Nibancho, Chiyoda-ku, Tokyo 102
Fax: (03) 32630257
Founded: 1879; Members: 695
Pres: Dr. Isamu Kobayashi; Gen Secr: Dr. Tadao Arita
Focus: Geology; Geography; Geophys
Periodical
Journal of Geography (7 times annually) 08437

Tokyo Daigaku Keizai Gakkai / Society of Economics, c/o Faculty of Economics, University of Tokyo, Bunkyo-ku, Tokyo 113
Founded: 1922; Members: 500
Pres: M. Nakamura
Focus: Econ
Periodical
Journal of Economics (weekly) 08438

Toshi Kaihatsu Kyokai Joho Sabisu Senta Shiryoshitsu / Urban Developers' Association of Japan, Akasaka Tokyu Bldg, 2-14-3 Nagat-cho, Chiyoda-ku, Tokyo 100
T: (03) 5803871
Focus: Urban Plan 08439

Toyo Gakujutsu Kyôkai, c/o Toyo Bunko, 2-28-21 Honkomagome, Bunkyo-ku, Tokyo 113
Founded: 1907; Members: 456
Focus: Ethnology 08440

Toyo Ongaku Gakkai / The Society for Research in Asiatic Music, c/o Dept of Musicology, Faculty of Music, Tokyo University of Fine Art and Music, 12-8 Ueno Park, Taito-ku, Tokyo 110
Founded: 1936; Members: 750
Pres: Mario Yokomichi
Focus: Music
Periodical
Journal of the Society for Research in Asiatic Music (weekly) *08441*

Toyoshi Kenkyukai / Society of Oriental Research, c/o Kyoto Daigaku Bungakubu, Yoshida-honmachi, Sakyo-ku, Kyoto 606
Founded: 1935
Focus: Hist
Periodical
The Toyoshi-Kenkyu: The Journal of Oriental History (weekly) *08442*

Un-yu Chosakyoku Johobu Toshoshitsu / Institute of Transportation Economics, 7-1-1, Ueno, Taito-ku, Tokyo 110
T: (03) 8414101; Fax: (03) 8414030
Gen Secr: Tsunetake Oota
Focus: Transport
Periodical
Unyu-To-Keizai: Transportation & Economy (weekly) *08443*

Waka Bungakkai, c/o Showa Joshi Daigaku Kenkyukan, 1-7-57 Taishido, Setagaya-ku, Tokyo 154
Focus: Lit *08444*

World Association of Societies of Pathology (Anatomic and Clinical) (WASP), c/o Japan Clinical Pathology Foundation for International Exchange, Sakura-Sugamo Bldg 7F, Sugamo 2-11-1, Toshima-ku, Tokyo 170
T: (03) 39188161; Fax: (03) 39496168
Founded: 1947; Members: 50
Pres: Prof. Dr. Tadashi Kawai; Gen Secr: George W. Pennington
Focus: Pathology
Periodical
News Bulletin (2-3 times annually) . . . *08445*

Yosetsu Gakkai / Japan Welding Society, 1-11 Sakuma-cho, Kanda, Chiyoda-ku, Tokyo 101
T: (03) 2530488; Fax: (03) 2533059
Founded: 1925; Members: 5000
Pres: Prof. Dr. K. Nishiguchi
Focus: Eng
Periodicals
Journal (weekly)
Transactions (weekly) *08446*

Zenkoku Nogyokozo Kaizen Kyobai Chosabu / National Assocation of Agrarian Structure Improvement, Zenkoku Choson Kaikan, 1-11-35 Nagata-cho, Chiyoda-ku, Tokyo 100
T: (03) 5813960
Focus: Agri *08447*

Zisin Gakkai / Seismological Society of Japan, c/o Earthquake Research Institute, University of Tokyo, 1-1-1 Yayoi, Bunkyo-ku, Tokyo
Founded: 1929; Members: 1300
Pres: K. Tsumuya
Focus: Geophys
Periodicals
Journal of Physics of the Earth (weekly)
Zisin: Journal (weekly) *08448*

Jordan

Arab Administrative Development Organization (ARADO), POB 17159, Amman
T: (06) 814118, 811394; Tx: 21594; Fax: (06) 816972
Founded: 1961; Members: 19
Gen Secr: Dr. Ahmed Sakr Ashour
Focus: Public Admin
Periodical
Arab Journal of Administration (weekly) . *08449*

Arab Biosciences Network (ARABN), c/o Higher Council for Science and Technology, POB 560, Jubaiha
T: (06) 840401; Tx: 23019
Founded: 1983
Gen Secr: Prof. Victor Billeh
Focus: Bio
Periodical
Arab Biosciences Newsletter *08450*

Association of Arab Universities (AARU) / Association des Universités Arabes, POB 401, Jubeyha, Amman
T: (06) 845131; Tx: 23855; Fax: (06) 832994
Founded: 1964; Members: 70
Focus: Educ
Periodical
AARU Bulletin (weekly) *08451*

Jordan Library Association, POB 6289, Amman
T: 629412
Founded: 1963; Members: 450
Pres: Farouk Mo'az; Gen Secr: Medhat Mar'ei
Focus: Libraries & Bk Sci
Periodicals
The Jordanian National Bibliography (weekly)
Rissalat al-Maktaba: Message of the Library (weekly) *08452*

Jordan Research Council, POB 6070, Amman
Founded: 1964; Members: 53
Gen Secr: Issam Khairy
Focus: Sci *08453*

Royal Scientific Society (R.S.S.), POB 925819, Amman
T: 844701; Tx: 21276
Founded: 1970; Members: 350
Pres: Dr. Jawad A. Anani
Focus: Sci
Periodicals
Current List of Periodical Holdings
Monthly Accession List (weekly) *08454*

Kazakhstan

Council for Study of Productive Forces, Ul Kurmangazy 29, 480021 Alma-Ata
T: 695051
Pres: U.B. Baimuratov
Focus: Econ *08455*

Kazakh Academy of Sciences, Ul Shevchenko 28, 480021 Alma-Ata
Members: 125
Pres: U.M. Sultangazin; Gen Secr: V.N. Okolovich
Focus: Sci
Periodicals
Izvestiya: Bulletin
Vestnik: Herald *08456*

Kenya

ACCE Institute for Communication Development and Research (ACCE/ICDR), POB 47495, Nairobi
T: (02) 27043, 334244 ext 2068; Tx: 22175
Founded: 1983
Focus: Comm Sci *08457*

African Academy of Sciences (AAS), POB 14798, Nairobi
T: (02) 802176, 802182; Tx: 25446; Fax: (02) 802185
Founded: 1985; Members: 80
Pres: Prof. Thomas R. Odhiambo
Focus: Nat Sci
Periodicals
Discovery and Innovations (weekly)
Whydah (weekly) *08458*

African Association for Biological Nitrogen Fixation (AABNF), c/o Dept of Soil Science, University of Nairobi, POB 30197, Nairobi
Founded: 1984
Pres: Prof. Alaa El Din; Gen Secr: Prof. S.O. Keya
Focus: Agri *08459*

African Association for Correspondence Education (AACE), c/o Correspondence Course Unit, POB 30688, Nairobi
Founded: 1973
Focus: Educ *08460*

African Association for Literacy and Adult Education (AALAE), Finance House, Loita St, Nairobi
T: (02) 22391, 331512; Tx: 22096; Fax: (02) 340849
Founded: 1984
Focus: Adult Educ *08461*

African Association of Dermatology, POB 46471, Nairobi
Focus: Derm *08462*

African Association of Insect Scientists (AAIS), POB 59862, Nairobi
T: (02) 802501, 602509; Tx: 22053; Fax: (02) 803360
Founded: 1978; Members: 2000
Gen Secr: Dr. J.P.R. Ochieng-Odero
Focus: Entomology
Periodicals
Bulletin of African Insect Science
Insect Science and its Application . . . *08463*

African Centre for Technology Studies (ACTS), POB 45917, Nairobi
T: (02) 744047; Fax: (02) 743995
Founded: 1982
Focus: Eng *08464*

African Commission on Mathematics Education, c/o Bureau of Educational Research, POB 43844, Nairobi
T: (02) 817356
Focus: Educ; Math *08465*

African Council for Communication Education (ACCE), POB 47495, Nairobi
T: (02) 27043; Tx: 25148
Founded: 1976; Members: 333
Pres: Kwame Boafo
Focus: Educ; Comm Sci
Periodicals
Africa Media Review (3 times annually)
AFRICOM Newsletter (3 times annually) . *08466*

African Economic Research Consortium (AERC), POB 47543, Nairobi
T: (02) 225234, 228057; Tx: 22480; Fax: (02) 218840
Founded: 1988
Gen Secr: Jeffrey C. Fine
Focus: Econ
Periodical
AERC Newsletter (weekly) *08467*

African Elephant and Rhino Specialist Group (AERSG), c/o Wildlife Conservation International, POB 62844, Nairobi
T: (02) 221699; Tx: 22165; Fax: (02) 215969
Founded: 1982
Pres: Dr. David Western
Focus: Zoology
Periodical
Pachyderm *08468*

African Institute for Higher Technical Training and Research (AIHTTR), c/o Kenya Regional Training Institution, POB 53763, Nairobi
T: (02) 335661, 220060; Tx: 22529
Founded: 1979; Members: 15
Gen Secr: J.K. Kioko
Focus: Eng *08469*

African Medical and Research Foundation (AMREF), POB 30125, Nairobi
T: (02) 501301, 500508; Tx: 23254; Fax: (02) 506112
Founded: 1957; Members: 12
Gen Secr: Dr. Michael S. Gerber
Focus: Med *08470*

African Network of Scientific and Technological Institutions (ANSTI), c/o ROSTA, POB 30592, Nairobi
T: (02) 333930; Tx: 22275; Fax: (02) 521045
Founded: 1980
Focus: Nat Sci; Eng
Periodicals
African Journal of Science and Technology
ANSTI Newsletter *08471*

African Regional Organization for Standardization (ARSO), City Hall Annexe, Muindi Mbingu St., POB 57363, Nairobi
T: (02) 24561, 330882; Tx: 22097; Fax: (02) 729228
Founded: 1977; Members: 23
Gen Secr: Zawdu Felleke
Focus: Standards
Periodical
ARSO Newsletter (weekly) *08472*

African Society for Environmental Studies Programme (ASESP), POB 44777, Nairobi
T: (02) 747960, 740817
Founded: 1968; Members: 17
Gen Secr: Dr. Peter Muyanda-Mutebi
Focus: Ecology
Periodical
African Social Studies Forum (weekly) . *08473*

African Training Centre for Literacy and Adult Education, c/o AALAE, Finance House, Loita St, POB 50768, Nairobi
T: (02) 22391, 331512; Tx: 22096; Fax: (02) 340849
Founded: 1991
Focus: Adult Educ; Lit *08474*

African Trypanotolerant Livestock Network,' c/o ILCA, POB 46847, Nairobi
T: (02) 292013, 592093
Gen Secr: G. d' Ieteren
Focus: Vet Med *08475*

African Water Network, c/o KWAHO, POB 61470, Nairobi
Founded: 1988
Focus: Water Res *08476*

Agricultural Society of Kenya, POB 30176, Nairobi
T: (02) 566655
Founded: 1901; Members: 12000
Pres: Dr. E.J. Mrabu; Gen Secr: H.R. Were
Focus: Agri
Periodical
The Kenya Farmer (weekly) *08477*

Association for Teacher Education in Africa, POB 45869, Nairobi
Focus: Educ *08478*

Association of Faculties of Science in African Universities (AFSAU), c/o University of Nairobi, POB 30197, Nairobi
T: (02) 43181
Founded: 1977
Gen Secr: Prof. Raphael Munavu
Focus: Nat Sci *08479*

Association of International Schools in Africa (AISA), c/o International School of Kenya, POB 14103, Nairobi
T: (02) 582578; Tx: 22370; Fax: (02) 582451
Members: 61
Gen Secr: Connie Buford
Focus: Educ
Periodical
AISA Newsletter (weekly) *08480*

Association of Surgeons of East Africa, POB 30726, Nairobi
T: (02) 340930
Founded: 1950; Members: 500
Pres: M.Y.D. Kodwavwala; Gen Secr: K.C. Rankin
Focus: Surgery
Periodical
Proceedings *08481*

Association of Theological Institutions of Eastern Africa (ATIEA), POB 50784, Nairobi
Focus: Rel & Theol *08482*

Basic Education Resource Centre (BERC), c/o Kenyatta University, POB 43844, Nairobi
T: (02) 810901
Founded: 1975
Pres: Dr. John O. Shiundu
Focus: Educ *08483*

Centre for African Family Studies (CAFS), POB 60054, Nairobi
T: (02) 747144; Tx: 22792; Fax: (02) 747160
Founded: 1975
Focus: Family Plan
Periodical
CAFS Newsletter (weekly) *08484*

Church Organization Research and Advisory Trust of Africa (CARAT AFRICA), POB 42493, Nairobi
T: (02) 331698; Tx: 25782
Founded: 1975
Focus: Business Admin *08485*

Climate Network Africa, c/o KC, POB 21136, Nairobi
T: (02) 226028; Fax: (02) 336742
Founded: 1991
Focus: Geophys *08486*

Committee of International Development Institutions on the Environment (CIDIE), c/o UNEP, POB 30552, Nairobi
T: (02) 230800; Tx: 22068; Fax: (02) 226886
Founded: 1980
Gen Secr: H. Abaza
Focus: Develop Areas; Ecology *08487*

Commonwealth Association of Polytechnics in Africa (CAPA), POB 52428, Nairobi
T: (02) 338156; Tx: 22529
Founded: 1978; Members: 110
Gen Secr: Dr. Pius Igharo
Focus: Educ
Periodicals
CAPA Journal of Technical Education and Training (weekly)
CAPA Newsletter (weekly) *08488*

Desert Locust Control Organization for Eastern Africa (DLCOEA), POB 30023, Nairobi
T: (02) 501704; Tx: 25510; Fax: (02) 611648
Founded: 1962
Focus: Geology *08489*

Desert Locust Control Organization for Eastern Africa, POB 4255, Nairobi
Founded: 1962
Focus: Ecology
Periodical
Desert Locus Situation Report (weekly) . *08490*

Earthwatch, c/o UNEP, POB 30552, Nairobi
T: (02) 333930; Tx: 22068; Fax: (02) 520711
Founded: 1972
Focus: Ecology *08491*

East African Academy, POB 47288, Nairobi
T: 22976
Focus: Sci *08492*

East Africa Natural History Society (EANHS), POB 44486, Nairobi
T: (02) 742131
Founded: 1909; Members: 800
Focus: Nat Sci; Hist *08493*

East African Engineering Consultants (EAEC), Madison Insurance House, Upper Hill Rd, POB 30707, Nairobi
T: (02) 711415; Fax: (02) 569389
Founded: 1967
Focus: Eng *08494*

East African Industrial Research Organization, POB 30650, Nairobi
Founded: 1948
Focus: Eng *08495*

East African Library Association (EALA), POB 46031, Nairobi
Founded: 1956
Focus: Libraries & Bk Sci *08496*

East African Natural History Society Nairobi (E.A.N.H.S.), POB 44486, Nairobi
T: (02) 742131
Founded: 1909; Members: 1250
Pres: Prof. S.G. Njuguna
Focus: Hist; Nat Sci
Periodicals
EANHS Bulletin (weekly)
The Journal of the East African Natural History Society (weekly) *08497*

East African Wild Life Society (E.A.W.L.S.), Nairobi Hilton Bldg, Mama Ngina St, POB 20110, Nairobi
T: (02) 337422, 27047, 331888
Founded: 1956; Members: 10000
Pres: Dr. Richard E. Leakey; Gen Secr: K. Nehemiah
Focus: Ecology
Periodicals
The African Journal of Ecology (weekly)
Swara Magazine: Wildlife News (weekly) *08498*

East Asian Seas Action Plan, c/o UNEP, POB 30552, Nairobi
T: (02) 333930; Tx: 22068; Fax: (02) 520711
Founded: 1981
Focus: Oceanography *08499*

Eastern Africa Environment Network (EAEN), Museum Hill Centre, Museum Hill Rd, POB 20110, Nairobi
T: (02) 748170; Fax: (02) 746868
Founded: 1001
Gen Secr: Charles J. Cara
Focus: Ecology *08500*

Eastern and Southern African Regional Branch of the International Council on Archives (ESARBICA), c/o Kenya National Archives, POB 49210, Nairobi
T: (02) 28979
Founded: 1969; Members: 13
Gen Secr: Musila Musembi
Focus: Archives
Periodical
ESARBICA Journal (weekly) *08501*

Historical Association of Kenya, POB 31028, Nairobi
Founded: 1966
Pres: Prof. B.A. Ogot
Focus: Hist
Periodicals
Hadith Series (weekly)
Kenya Historical Review (weekly) . . . *08502*

Kenya Dental Association (KDA), POB 20059, Nairobi
Founded: 1943; Members: 110
Focus: Dent *08503*

Kenya Library Association, POB 46031, Nairobi
Founded: 1956; Members: 200
Pres: M.J. Ongany; Gen Secr: Lily Nyariki
Focus: Libraries & Bk Sci
Periodicals
Kelias News (weekly)
Maktaba: Official Journal (weekly) . . . *08504*

Kenya National Academy of Sciences, POB 39450, Nairobi
Founded: 1977; Members: 50
Focus: Arts; Sci
Periodicals
The Kenya Journal of Sciences: Series A (weekly)
The Kenya Journal of Sciences: Series B (weekly)
The Kenya Journal of Sciences: Series C (weekly) *08505*

Ophthalmological Society of East Africa (OSEA), c/o Kenyatta National Hospital, POB 20723, Nairobi
Focus: Ophthal *08506*

Standing Conference of African University Libraries (SCAUL), c/o University, POB 30197, Nairobi
T: (02) 334244
Founded: 1964
Focus: Libraries & Bk Sci *08507*

Korea, Democratic People's Republic

Academy of Agricultural Science, Ryongsong Dictrict, Pyongyang
Founded: 1948
Pres: Yong Kyun Ri
Focus: Agri *08508*

Academy of Fisheries, Sinpo City, South Hamgyong Province
Founded: 1969
Gen Secr: Yun Il Kim
Focus: Zoology *08509*

Academy of Forestry Science, Taesong District, Pyongyang
Founded: 1948
Pres: Rok Jae Im
Focus: Forestry *08510*

Academy of Light Industry Science, Songyo District, Pyongyang
Founded: 1954
Gen Secr: Kye Sok Kim
Focus: Eng; Econ *08511*

Academy of Medical Sciences, Pyongyang
Pres: Chun Hyop Paik
Focus: Med *08512*

Academy of Railway Sciences, Hyongjaesan District, Pyongyang
Gen Secr: Hyon Ju Kim
Focus: Eng *08513*

Academy of Sciences, Ryonmot-dong, Jangsan St, Sosong District, Pyongyang
T: 51956
Founded: 1952
Pres: Kyong Pong Kim
Focus: Sci
Periodical
Journal (weekly) *08514*

Academy of Social Sciences, Central District, Pyongyang
Founded: 1952
Pres: Hyong Sop Yang
Focus: Sociology *08515*

Library Association of the Democratic People's Republic of Korea, c/o State Central Library, Pyongyang
Founded: 1953
Focus: Libraries & Bk Sci *08516*

Korea, Republic

Asia Foundation, Gwang Wha Moon, POB 738, Seoul 110-607
Founded: 1954
Gen Secr: Ben Kremenak
Focus: Int'l Relat *08517*

Asian Association of Open Universities (AAOU), c/o Korea Air and Correspondance University, 169 Tongsung-dong, Chongro-gu, Seoul
Founded: 1987
Focus: Educ *08518*

Asian Federation of Catholic Medical Associations (AFCMA), c/o Catholic University Graduate School, 505 Banpo-dong, Socho-ku, Seoul 137701
T: (02) 5910598; Fax: (02) 5910598
Founded: 1960
Pres: Dr. Yong Whee Bahk
Focus: Med *08519*

Asian Federation of Societies for Ultrasound in Medicine and Biology, c/o Dept of Radiology, Seoul National University Hospital, 28 Yongong-dong, Chongno-gu, Seoul 110744
Fax: (02) 7660950
Focus: Radiology *08520*

Asian Pacific Association for the Study of the Liver, c/o Dept of Internal Medicine, Catholic University Medical College, 62 Youido-gu, Seoul 150010
Gen Secr: Prof. W.K. Chung
Focus: Intern Med *08521*

Asian-Pacific Society of Cardiology (APSC), c/o Seoul National University Hospital, Room 9326, Seoul 110744
Fax: (02) 7442819
Founded: 1956
Gen Secr: Young Woo Lee
Focus: Cardiol
Periodical
APSC Newsletter (weekly) *08522*

Asian Regional Association for Home Economics (ARAHE), c/o College of Home Economics, Yonsei University, 134 Shinchon-dong, Sudaemun-ku, Seoul 120749
Gen Secr: Ki Yull Lee
Focus: Home Econ *08523*

Asian Society of Agricultural Economists, c/o Korea Rural Economics Institute, 4-102 Hong-dong, Dongdaemoon-ku, Seoul 131
T: (02) 9627311; Fax: (02) 9656959
Focus: Agri *08524*

Asia-Pacific Lawyers Association, Korea Re-Insurance Bldg, 80 Soosong-Dong, Chongro-ku, Seoul
T: (02) 7355661; Tx: 23250; Fax: (02) 7355206
Founded: 1984
Pres: Byong Ho Lee
Focus: Law *08525*

International Cultural Society of Korea (ICSK), Daewoo Foundation Bldg 526, 5-ga, Namdaemunno, Chung-gu, CPO 2147, Seoul 100-095
Fax: (02) 7572049
Founded: 1972
Pres: Seong-jin Kim
Focus: Int'l Relat
Periodicals
Le Courrier de la Corée (weekly)
Koreana: English Edition (weekly)
Koreana: Japanese Edition (weekly)
Korea Newsreview (weekly) *08526*

Korea Branch of the Royal Asiatic Society, CPOB 255, Seoul 100
T: (02) 7639483
Founded: 1900; Members: 1600
Pres: H.G. Underwood
Focus: Ethnology
Periodical
RAS Korea Branch Transactions (weekly) *08527*

Korean Association for the Biological Sciences, c/o Sung Kyun Kwan University, Seoul
Focus: Bio *08528*

Korean Association of Sinology, c/o Asiatic Research Center, Korea University, Anam-dong, Seoul
Founded: 1955; Members: 100
Pres: Jun-Yop Kim
Focus: Ethnology
Periodical
Journal of Chinese Studies *08529*

Korean Chemical Society, 35, 5-Ka, Anam-Dong, Seongbuk-Ku, Seoul 136-075
T: (02) 9265457; Fax: (02) 9235589
Founded: 1946; Members: 2595
Focus: Chem
Periodicals
Bulletin (weekly)
Chemical Education (weekly)
ChemWorld (weekly)
Journal of the Korean Chemical Society (weekly)
The Korean Journal of Medicinal Chemistry (weekly) *08530*

Korean Economic Association, 45, 4-ga Namdae-mun-ro, Chung-gu, Seoul
Founded: 1952; Members: 300
Pres: Hochin Choi; Gen Secr: Pil Woo Kee
Focus: Econ
Periodical
Korean Economic Review (weekly) . . . *08531*

Korean Forestry Society, c/o College of Agriculture, Seoul National University, Suwon, Kyonggido 440-744
Fax: (0331) 2915830
Founded: 1960; Members: 634
Focus: Forestry
Periodical
Journal of the Korean Forestry Society (weekly) *08532*

Korean Geographical Society, c/o Dept of Geography, College of Social Sciences, Seoul National University, Seoul 151-742
T: (02) 8860101 ext 2430; Tx: 29664; Fax: (02) 8855272
Founded: 1945; Members: 585
Pres: Bo-Woong Chang; Gen Secr: Sam Ock Park
Focus: Geography
Periodical
Chirihak (weekly) *08533*

Korean Historical Association, 2-5, Myong-nuyun-dong 3-ga, Chongno-ku, Seoul
Founded: 1952; Members: 400
Gen Secr: W.S. Lee; H.S. Cha
Focus: Hist
Periodical
Yoksa Hakbo (weekly) *08534*

Korean Library and Information Science Society (KOLIS), c/o Dept of Library and Information Science, Yonsei University, 134 Shinchon-dong, Sodamun-ku, Seoul 120-749
Founded: 1971; Members: 500
Pres: Prof. Byung-Mock Rhee; Gen Secr: Prof. Sang-Wan Han
Focus: Libraries & Bk Sci; Computer & Info Sci
Periodical
Hankuk Munhunjungbohakhoeji: Journal of the Korean Library and Information Science Society (weekly) *08535*

Korean Library Association, c/o Central National Library, 100-177, 1-ga, Hoehyun-dong, Chung-gu, Seoul
T: (02) 7524864, 7525613
Founded: 1955; Members: 970
Pres: Hyo-Soon Song; Gen Secr: Dae-Kwon
Focus: Libraries & Bk Sci
Periodical
KLA Bulletin (weekly) *08536*

Korean Medical Association (KMA), CPOB 2062, Seoul
Fax: (02) 7921296
Founded: 1908; Members: 33628
Pres: Jae Joum Kim; Gen Secr: Jwa Ryong Yang
Focus: Med
Periodicals
Journal (weekly)
Newspaper (weekly) *08537*

Korean Micro-Library Association, c/o Library Association, Central National Library Bldg, 6 Sogong-dong, 6 Chung-ku, Seoul
Focus: Libraries & Bk Sci *08538*

Korean Psychological Association, c/o Dept of Psychology, Seoul National University, Shinrim 2-dong, Kwanak-gu, Seoul
T: (02) 8770101
Founded: 1946; Members: 297
Pres: Whan-Chang Tong; Gen Secr: Woo-Rheem Chang
Focus: Psych
Periodicals
Korean Journal of Clinical Psychology (weekly)
Korean Journal of Industrial Psychology (weekly)
Korean Journal of Psychology (weekly)
Korean Journal of Social Psychology (weekly) *08539*

The Korean Society of Pharmacology, c/o Dept of Pharmacology, Yonsei University College of Medicine, 134 Shin-Chon-Dong, Seo-Dae-Moon-ku, Seoul 120-752
Founded: 1947; Members: 335
Pres: Chang Koh Kye; Gen Secr: Soo Ahn Young
Focus: Pharmacol
Periodical
The Korean Journal of Pharmacology . . *08540*

Music Association of Korea, Room 303, FACO Bldg, 81-6, Sejongro, Chongno-gu, Seoul
Founded: 1961; Members: 700
Pres: Dr. Tai Joon Park; Gen Secr: Dae Yup Sohn
Focus: Music *08541*

National Academy of Arts, 1 Sejongro, Chongno-gu, Seoul 50
T: (02) 7203902
Founded: 1954
Pres: Kim Dong Ni; Gen Secr: Kwon Jong Buk
Focus: Arts
Periodicals
Bibliography
Bulletin
Journal
Survey of Korean Arts *08542*

National Academy of Sciences, San 94, Panpo-dong, Seocho-gu, Seoul 137-044
T: (02) 5340737; Fax: (02) 5373183
Founded: 1954; Members: 150
Pres: E.-Hyock Kwon; Gen Secr: Soo-Jung Lo
Focus: Sci
Periodicals
Bibliography
Bulletin
Journal
Report on International Symposium . . . *08543*

Kuwait

Arab Center for Medical Literature (ACML), POB 5225, Safat, Kuwait
T: 2416915, 2419086; Tx: 44675; Fax: 2416931
Founded: 1980; Members: 21
Gen Secr: Dr. Abdel Rahman Al-Awadi
Focus: Lit *08544*

Arab Medical Information Network (AMIN), c/o ACML, POB 5225, Safat, Kuwait
T: 2416915, 2419086; Tx: 44675; Fax: 2416931
Founded: 1963
Focus: Med *08545*

Arab Planning Institute (API), POB 5834, Safat, Kuwait
T: 843130, 844061; Tx: 22996; Fax: 4846891
Founded: 1972; Members: 15
Gen Secr: Abdulla M. Ali
Focus: Econ *08546*

Educational Innovation Programme for Development in the Arab States (EIPDAS), Shuwaikh B-70451, Kuwait, POB 64085
T: 4838691; Tx: 23178; Fax: 4836742
Founded: 1979
Gen Secr: A. El-Atrash
Focus: Educ *08547*

Kyrgyzstan

Kyrgyz Academy of Sciences, Leninskii Pr 265a, 720071 Bishkek
T: 264541
Members: 79
Pres: I.T. Aitmatov; Gen Secr: V.P. Zhivogliadov
Focus: Sci
Periodical
Izvestiya: Bulletin *08548*

Kyrgyz Commission on Earthquake Forecasting, Asanbay 52-54, 720060 Bishkek
T: 418137
Pres: E.M. Mamyrov
Focus: Geology *08549*

Kyrgyz Genetics and Selection Society, Leninskii pr 265a, 720071 Bishkek
T: 243994
Pres: M.M. Tokobaev
Focus: Genetics *08550*

Kyrgyz Geographical Society, Dzerzhinskogo 30, 720081 Bishkek
T: 264721
Pres: Umurzakov
Focus: Geography *08551*

Laos

Lao Buddhist Fellowship, Maha Kudy, That Luang, Vientiane
Founded: 1964
Pres: Thong Khoune Anantasunthone; Gen Secr: Siho Sihavong
Focus: Arts *08552*

Laos-China Association, c/o National Plan Office, POB 46, Vientiane
Founded: 1976
Pres: Ma Khaikhamphi Thoune
Focus: Int'l Relat *08553*

Laos-Soviet Association, c/o Ministry of Agriculture, Forestry and Irrigation, Nongbone Rd, Vientiane
Founded: 1976
Pres: Khamsouk Saisompheng
Focus: Int'l Relat *08554*

Laos-Viet Nam Association, c/o Ministry of Public Health, Fa Ngum Rd, Vientiane
Founded: 1976
Pres: Chao Souk Vongsak
Focus: Int'l Relat *08555*

Latvia

Council on Libraries, Turgeneva iela 19, 226524 Riga
Pres: E. Silins
Focus: Libraries & Bk Sci *08556*

Latvian Academy of Medicine, Dzirciema iela 16, 226007 Riga
T: 459243; Tx: 161172; Fax: 225039
Founded: 1951
Pres: Prof. V. Korzans
Focus: Med *08557*

Latvian Academy of Sciences, Turgeneva iela 19, 226524 Riga
T: 225361; Fax: 228784
Members: 137
Pres: J.J. Lielpeteris; Gen Secr: A. Silins
Focus: Sci
Periodicals
Khimiya Geterotsiklicheskikh Soedinenii: Chemistry of Heterocyclic Compounds
Latvijas Fizikas un Tehnisko Zinatnu Zurnals: Latvian Journal of Physics and Technical Sciences
Latvijas Kimijas Zurnals: Latvian Chemical Journal
Magnitnaya Gidrodinamika: Magnetic Hydrodynamics
Mekhanika Kompozitnykh Materialov: Mechanics of Composite Materials
Vestis: Proceedings 08558

Terminological Commission, Turgenev iela 19, 226524 Riga
Pres: A.Y. Blinkena
Focus: Sci 08559

Lebanon

Association Libanaise des Sciences Juridiques, c/o Faculté de Droit et des Sciences Economiques, Université Saint Joseph, BP 293, Beirut
T: 326636
Founded: 1963; Members: 40
Pres: P. Gannagé
Focus: Law
Periodicals
Etudes Juridiques
Proche-Orient 08560

Association of Theological Institutes in the Middle East (ATIME), c/o Dept of Theological Concerns, Immeuble DEEB, Rue Makhoul, BP 5376, Beirut
T: (01) 3448994; Tx: 22662
Founded: 1967; Members: 11
Focus: Rel & Theol 08561

Lebanese Library Association, c/o University Library, American University of Beirut, Beirut
T: 865250 ext 28354; Tx: 20801
Founded: 1960
Pres: Leila Hanhan; Gen Secr: Linda Sadaka
Focus: Libraries & Bk Sci 08562

Liberia

Geological, Mining and Metallurgical Society of Liberia, POB 902, Monrovia
Founded: 1964; Members: 78
Pres: C.S. Wotorson; Gen Secr: Dr. M.-H. Neufville
Focus: Geology; Mining; Metallurgy
Periodical
Bulletin (weekly) 08563

Liberia Arts and Crafts Association, POB 885, Monrovia
Founded: 1964; Members: 14
Pres: R.V. Richards
Focus: Arts 08564

Society of Liberian Authors, POB 2468, Monrovia
Founded: 1959
Focus: Lit
Periodical
Kaafa (weekly) 08565

Libya

African Centre for Applied Research and Training in Social Development (ACARTSOD), POB 80606, Tripoli
T: (021) 833640; Tx: 20803
Founded: 1977; Members: 26
Gen Secr: Dr. Mohammed Duri
Focus: Soc Sci
Periodical
ACARTSOD Newsletter (3 times annually) 08566

Arab Federation of Construction, Wood and Building Materials, Ousama St, POB 10828, Tripoli
Gen Secr: Abou Al-Qassem Taleb
Focus: Materials Sci 08567

Intellectual Society of Libya, 136 Shar'a Baladia, POB 1017, Tripoli
Founded: 1959
Focus: Sci 08568

Liechtenstein

Historischer Verein, 9490 Vaduz
T: (075) 2366343; Fax: (075) 3921747
Founded: 1901; Members: 800
Pres: Dr. Alois Ospelt; Gen Secr: Dr. Rupert Quaderer
Focus: Hist
Periodical
Jahrbuch des Historischen Vereins für das Fürstentum Liechtenstein 08569

Internationale Alpenschutzkommission / International Commission for the Protection of Alps, Heiligkreuz 52, 9490 Vaduz
Founded: 1952
Pres: Josef Biedermann; Gen Secr: Ulf Tödter
Focus: Ecology
Periodical
Info-Bulletin (weekly) 08570

Liechtensteinische Gesellschaft für Umweltschutz (LGU), Heiligkreuz 52, 9490 Vaduz
T: (075) 2325262
Founded: 1973; Members: 720
Pres: Barbara Rheinberger; Gen Secr: Wilfried Marxer-Schädler
Focus: Ecology
Periodicals
LGU-Mitteilungen (weekly)
Liechtensteiner Umweltbericht (2-3 times annually) 08571

Liechtensteinischer Ärzteverein, Postf 464, 9490 Vaduz
Focus: Med 08572

Lithuania

Lithuanian Academy of Sciences, Gedimino pr 3, 2600 Vilnius
Tx: 261141
Founded: 1941; Members: 130
Pres: J.K. Pozela; Gen Secr: E.J. Vilkas
Focus: Sci
Periodicals
Astronomical Bulletin
Lithuanian Mathematical Journal
Lithuanistika: Lithuanian Studies
Mokslas ir Tekhnika: Science and Technology
Trudy Akademii Nauk Litvy: Bulletin of the Lithuanian Academy 08573

PEN Centre of Lithuania, Sirvydo 6, 2600 Vilnius
T: 628643
Founded: 1989; Members: 36
Pres: Kornelijus Platelis; Gen Secr: Galina Bauzyte-Cepinskiene
Focus: Lit 08574

Luxembourg

Association des Médecins et Médecins-Dentistes du Grand-Duché de Luxembourg, 29 Rue de Vianden, 2680 Luxembourg
T: 444033; Fax: 458349
Founded: 1904; Members: 820
Focus: Med; Dent
Periodical
Le Corps Médical (weekly) 08575

Association des Professeurs de l'Enseignement Secondaire et Supérieur du Grand-Duché de Luxembourg (APESS), 26 Rue J. P. Beicht, Luxembourg
Founded: 1912; Members: 800
Focus: Educ 08576

Association Européenne de Méthodes Médicales Nouvelles, 33 Rue du Fort Neipperg, 2230 Luxembourg
Founded: 1987
Gen Secr: Robert Van Sinay
Focus: Med 08577

Association Internationale d'Orientation Scolaire et Professionelle (AIOSP), 257 Rte d'Arlon, Strassen
Founded: 1953
Focus: Adult Educ; Educ 08578

Association Luxembourgeoise des Kinésithérapeutes Diplomés (A.L.K.D.), BP 645, 2016 Luxembourg
Founded: 1967; Members: 105
Focus: Therapeutics
Periodical
Kiné Info (weekly) 08579

Association of European Psychiatrists / Association Européenne de Psychiatrie, c/o Centre Hospitalier de Luxembourg, 4 Rue Barblé, 1210 Luxembourg
T: 112706
Founded: 1983
Gen Secr: Prof. C. Pull
Focus: Psychiatry 08580

Commission Grand-Ducale d'Instruction, 29 Rue Aldringen, 2926 Luxembourg
T: 4785211; Fax: 466815
Founded: 1843; Members: 10
Pres: Jean-Pierre Kraemer; Gen Secr: Gérard Gretsch
Focus: Educ 08581

European Association for the Transfer of Technologies, Innovations and Industrial Information (TII) / Association Européenne pour le Transfert des Technologies, de l'Innovation et de l'Information Industrielle, 3 Rue des Capucins, 1313 Luxembourg
T: 463035; Tx: 6105111; Fax: 462185
Founded: 1984; Members: 470
Gen Secr: Christopher John Hull
Focus: Eng 08582

European Association of Information Services / Association Européenne des Centres de Dissémination des Informations Scientifiques, BP 1416, 1014 Luxembourg
T: 422474; Fax: 422474
Founded: 1970; Members: 200
Gen Secr: D. Mahon
Focus: Computer & Info Sci 08583

European Business Associates On-line (EBA), 22 Rue Demier Sol, 2543 Luxembourg
Founded: 1982
Pres: Mary Clark
Focus: Business Admin 08584

European Centre for Parliamentary Research and Documentation (ECPRD), c/o Parlement Européen, 2929 Luxembourg
T: 43001 ext 2118; Tx: 3494; Fax: 434071
Founded: 1977; Members: 33
Gen Secr: Klaus Pöhle
Focus: Poli Sci 08585

Ligue Luxembourgeoise pour la Protection de la Nature et des Oiseaux (LLPNO), BP 709, 2017 Luxembourg
T: 472312
Founded: 1920; Members: 4500
Focus: Ornithology
Periodical
Regulus (weekly) 08586

Société des Naturalistes Luxembourgeois (SNL), CP 327, Luxembourg
Founded: 1890; Members: 480
Pres: Claude Meisch; Gen Secr: M. Molitor
Focus: Nat Sci
Periodical
Bulletin de la Société des Naturalistes Luxembourgeois (weekly) 08587

Société des Sciences Médicales du Grand-Duché de Luxembourg, 3 Rue Conrad, Luxembourg
T: 440437
Founded: 1861; Members: 800
Focus: Med
Periodical
Bulletin (2-3 times annually) 08588

Société Luxembourgeoise de Radiologie, 29 Rue de Vianden, Luxembourg
T: 475023
Founded: 1968; Members: 14
Focus: Radiology 08589

Syndicat National des Enseignants (S.N.E.), BP 2437, Luxembourg
T: 481118
Founded: 1948; Members: 2300
Focus: Educ
Periodical
Ecole et Vie: Bulletin dınformation syndical, pédagogique et culturel (8 times annually) 08590

Macao

Circulo de Cultura Musical / Circle of Musical Culture, Largo de Santo Agostinho 3, Macao 68
Founded: 1952
Gen Secr: F.X.F. Garcia
Focus: Music 08591

European Committee of Construction Economists (CEEC), 234 Rue du Faubourg Saint-Honoré, 75008 Paris
T: (1) 42890805; Tx: 650464; Fax: (1) 42890806
Founded: 1979
Gen Secr: Gordon F. Wheatley 08592

Macedonia

Društvo na Istoričarite na Umetnosta od Makedonija / Society of Art Historians of Macedonia, c/o Arheološki Muzej na Makedonija, Curčiska bb, 91000 Skopje
Founded: 1970; Members: 130
Pres: Milanka Boškovska; Gen Secr: Mate Boškovski
Focus: Arts; Cultur Hist
Periodical
Likovna Umetnost: Plastic Arts 08593

Društvo na Kompozitorite na Makedonija / Society of Macedonian Composers, Maksim Gorki 18, 91000 Skopje
Founded: 1950; Members: 40
Pres: Vlastimir Nikolovski; Gen Secr: Sotir Golaboski
Focus: Music
Periodical
Bilten: Bulletin 08594

Društvo na Likovnite Umetnici na Makedonija / Society of Plastic Arts of Macedonia, c/o Umetnička Galerija, Kruševska 1a, 91000 Skopje, (041) 10110
Founded: 1944; Members: 150
Pres: Boro Mitrikeski; Gen Secr: Naso Bekaroski
Focus: Fine Arts 08595

Društvo na Literaturnite Preveduvači na Makedonija / Society of Literary Translators of Macedonia, POB 3, 91000 Skopje
Founded: 1955; Members: 102
Pres: Prof. Dr. Božidar Nastev; Gen Secr: Taško Širilov
Focus: Lit 08596

Društvo na Muzejskite Rabotnici na Makedonija / Museum Society of Macedonia, c/o Muzei na Grad Skopje, Mito Hadži-Vasilev-Jasmin bb, 91000 Skopje
Founded: 1951; Members: 100
Pres: Kuzman Geogrievski; Gen Secr: Galena Kuculovska
Focus: Archives 08597

Društvo na Pisatelite na Makedonija / Society of Writers of Macedonia, Maksim Gorki 18, 91000 Skopje
T: (091) 236205
Founded: 1951; Members: 194
Pres: T. Calovski; Gen Secr: T. Petrovski; R. Siljan
Focus: Lit 08598

Društvo za Filozofija, Sociologija i Politikologija na Makedonija / Society for Philosophy, Sociology and Politics of Macedonia, c/o Institut za Sociološki i Političko-Pravni Istraživanja, Bul Partizanski Odredi bb, 91000 Skopje
Founded: 1960; Members: 170
Pres: Dr. Dragan Taškovski; Gen Secr: Sveta Škarić
Focus: Philos; Sociology; Poli Sci
Periodical
Zbornik: Collected Papers 08599

Društvo za Nauka i Umetnost / Association of Sciences and Art, POB 145, 97000 Bitola
T: (097) 22683
Founded: 1960
Pres: S. Cvetanovski; Gen Secr: T. Ognenovski
Focus: Psych 08600

Farmaceutsko Društvo na Makedonija / Pharmacological Society of Macedonia, Ivo Ribar Lola 6, 91000 Skopje
Pres: Lazar Tolov; Gen Secr: Galaba Srbinovska
Focus: Pharmacol
Periodical
Bilten: Bulletin 08601

Geografsko Društvo na Makedonija / Geographical Society of Macedonia, c/o Geografski Institut pri Prirodomatematički Fakultet, POB 146, 91000 Skopje
Founded: 1949; Members: 600
Pres: Prof. V. Gramatnikovski; Gen Secr: N. Panov
Focus: Geography
Periodicals
Geografski Razgledi: Geographical Surveys
Geografski Vidik: Geographical Look . . 08602

Makedonska Akademija na Naukite i Umetnostite (MANU) / Macedonian Academy of Sciences and Art, Bul Krste Misirkov bb, POB 428, 91000 Skopje
T: (091) 235506
Founded: 1967; Members: 47
Pres: K. Bogoev; Gen Secr: A. Andreevski
Focus: Sci; Arts
Periodicals
Letopis (weekly)
Prilozi na Oddelenieto za Bioloski i Medicinski Nauki (weekly)
Prilozi na Oddelenieto za Matematicko-Tehnicki Nauki
Prilozi na Oddelenieto za Opstestveni Nauki (weekly) 08603

Makedonsko Geološko Društvo / Macedonian Geological Society, c/o Geološki Zavod, POB 28, 91001 Skopje
T: (091) 230873; Tx: 51871
Founded: 1954; Members: 300
Pres: Nikola Tudžarov; Gen Secr: Roza Petrovska
Focus: Geology 08604

Makedonsko Lekarsko Društvo / Medical Society of Macedonia, Gradski zid blok 11/6, 91000 Skopje
Founded: 1946; Members: 2000
Focus: Med
Periodical
Makedonski Medicinski Pregled: Macedonian Medical Review 08605

Republički Zavod za Zaštita na Spomenicite na Kulturata / Institute for the Protection of Cultural Monuments of Macedonia, Maršal Tito bb, Gorče Petrov, 91000 Skopje
Founded: 1949
Gen Secr: Dimitar Kornakov
Focus: Preserv Hist Monuments
Periodical
Kulturno-Istorisko Nasledstvo na Makedonija: The Cultural and Historical Heritage of Macedonia 08606

Sojuz na Društvata na Arhivskite Rabotnici na Makedonija / Union of Societies of Archivists of Macedonia, Ul Gligor Prličev br 3, 91000 Skopje
T: (091) 237211; Fax: (091) 234461
Founded: 1954; Members: 340
Focus: Archives
Periodical
Makedonski Arhivist (weekly) 08607

Sojuz na Društvata na Bibliotekarite na Makedonija / Union of Librarians' Associations of Macedonia, c/o Narodna i Univerzitetska Biblioteka "Kliment Ohridski", Bul Goce Delčev br 6, 91000 Skopje
Founded: 1949; Members: 550

Pres: T. Pikov; Gen Secr: P. Matkovska
Focus: Libraries & Bk Sci
Periodical
Bibliotekarska iskra (weekly) 08608

Sojuz na Društvata na Istoričarite na Makedonija / Union of Societies of Historians of Macedonia, c/o Institut za Nacionalna Istorija, Ul Gligor Prličev br 3, POB 591, 91000 Skopje
T: (091) 239036
Founded: 1952
Pres: Dr. A. Trajanovski; Gen Secr: T. Cepreganov
Focus: Hist
Periodical
Istorija: History 08609

Sojuz na Društvata na Matematičarite i Informatičarite na Makedonija / Society of Mathematicians and Computerists of Macedonia, POB 162, 91000 Skopje
T: (091) 261330; Fax: (091) 228141
Founded: 1950
Gen Secr: Prof. Dr. N. Celakoski; Prof. Dr. D. Karčicka
Focus: Math; Computer & Info Sci
Periodical
Matematički Bilten: Mathematical Bulletin . 08610

Sojuz na Društvata na Veterinarnite Lekari i Tehničari na Makedonija / Union of Associations of Veterinary Surgeons and Technicians of Macedonia, c/o Veterinaren Institut, Lazar Pop-Trajkov 5, POB 95, 91000 Skopje
Founded: 1950; Members: 450
Pres: Siljan Zaharievski; Gen Secr: A. Blaže
Focus: Vet Med
Periodical
Makednonski Veterinaren Pregled: Macedonian Veterinary Review 08611

Sojuz na Društvata za Makedonski Jazik i Literatura / Union of Associations for Macedonian Language and Literature, c/o Filološki Fakultet, Bul Krste Misirkov bb, 91000 Skopje
Founded: 1954; Members: 700
Pres: Kiril Koneski; Gen Secr: V. Tuševski
Focus: Lit
Periodical
Literaturen Zbor: Literary Word 08612

Sojuz na Ekonomistite na Makedonija / Union of Economists of Macedonia, c/o Ekonomiski Fakultet, K. Misirkov bb, 91000 Skopje
T: (091) 224311; Fax: (091) 224973
Founded: 1950; Members: 3000
Pres: Prof. Dr. T. Fiti; Gen Secr: A. Spasovski
Focus: Econ
Periodical
Stopanski pregled: Economic Review . . 08613

Sojuz na Inženeri i Tehničari na Makedonija / Society of Engineers and Technicians of Macedonia, Nikola Vapcarov bb, 91000 Skopje
Founded: 1945; Members: 27000
Pres: Prof. Dr. Dime Lazarov; Gen Secr: Boro Ravnjanski
Focus: Eng 08614

Sojuz na Inženeri i Tehničari po Sumarstvo i Industrija za Prerabotka na Drvo na Makedonija / Union of Forestry Engineers and Technicians of Macedonia, c/o Šumarski Institut, Engelsova 2, 91000 Skopje
Founded: 1952; Members: 500
Pres: Zivko Minčev; Gen Secr: Mile Stamenkov
Focus: Forestry
Periodical
Sumarski Pregled: Forester's Review . . 08615

Sojuz na Združenijata na Pravnicite na Makedonija / Union of Associations of Jurists of Macedonia, XII Udarna Brigada 2, 91000 Skopje
Founded: 1946; Members: 4000
Pres: Boro Dogandžiski; Gen Secr: P. Golubovski
Focus: Law
Periodical
Pravna misla 08616

Sojuz na Zemjodelskite Inženeri i Tehničari na SR Makedonija / Union of Agricultural Engineers and Technicians of Macedonia, c/o Zemjodelskošumarski Fakultet, 91000 Skopje
Founded: 1945; Members: 5000
Pres: Prof. Dr. Risto Lozanovski; Gen Secr: Dr. Risto Ilkovski
Focus: Agri
Periodical
Socijalističko Zemjodelstvo: Socialist Agriculture 08617

Združenie na Arheolozite na Makedonija / Archaeological Society of Macedonia, c/o Muzej na Makedonija, 91000 Skopje
T: (091) 220222
Founded: 1970; Members: 150
Pres: D. Mitrevski; Gen Secr: M. Ivanovski
Focus: Archeol
Periodical
Macedoniae Acta Archaeologica 08618

Združenie na Folkloristite na Makedonija / Association of Folklorists of Macedonia, c/o Institut za Folklor, Ruzveltova 3, 91000 Skopje
T: (091) 233876; Fax: (091) 233319
Founded: 1952; Members: 60
Pres: G. Smokvarski; Gen Secr: E. Lafazanovsky
Focus: Folklore 08619

Madagascar

Académie Malgache, Tsimbazaza, BP 6217, Antananarivo
T: 21084
Founded: 1902; Members: 140
Pres: Dr. C. Rabenoro
Focus: Sci
Periodicals
Bulletin
Mémoires 08620

Malawi

Association for the Taxonomic Study of the Flora of Tropical Africa, c/o National Herbarium and Botanic Gardens, POB 528, Zomba
T: 523145; Tx: 45252; Fax: 522108
Founded: 1950; Members: 800
Gen Secr: Dr. J. Seyani
Periodicals
AETFAT Bulletin (weekly)
AETFAT Index (weekly) 08621

Geological Survey of Malawi, Liwonde Rd, POB 27, Zomba
Founded: 1921
Focus: Surveying
Periodicals
Bulletin
Memoirs
Records 08622

Malawi Library Association, POB 429, Zomba
T: 522222
Founded: 1976; Members: 130
Pres: J.J. Uta; Gen Secr: G.B. Shaba
Focus: Libraries & Bk Sci
Periodical
Bulletin 08623

Society of Malawi, POB 125, Blantyre
Founded: 1948; Members: 550
Pres: A. Schwarz; Gen Secr: B.C. Lamport-Stokes
Focus: Hist; Sci
Periodical
Journal (weekly) 08624

Tea Research Foundation (Central Africa), POB 51, Mulanje
T: (462) 261, 271, 255; Tx: 44458;
CA: Teasearch Mulanje
Founded: 1966
Focus: Coffee, Tea, Cocoa
Periodicals
Newsletter (weekly)
Report (weekly) 08625

Malaysia

Action for Rational Drugs in Asia (ARDA), c/o IOCU, POB 1045, Penang
T: (04) 371396; Tx: 40164; Fax: (04) 366506
Founded: 1986; Members: 16
Gen Secr: Dr. K. Balasubramaniam
Focus: Pharmacol 08626

ASEAN Federation for Psychiatric and Mental Health (AFPMH), c/o Dept of Psychological Medicine, Faculty of Medicine, University of Malaya, 59100 Kuala Lumpur
T: (03) 7502068, 7556477; Tx: 39854; Fax: (03) 75736611
Founded: 1981
Gen Secr: Dr. S. Jeyarajah
Focus: Psychiatry
Periodical
ASEAN Mental Health Bulletin 08627

ASEAN Food Handling Bureau (AFHB), Level 3, G14 and G15, Damansara Town Centre, 50490 Kuala Lumpur
T: (03) 2551088, 2544199; Tx: 31555; Fax: (03) 2552787
Founded: 1981
Focus: Food
Periodicals
ASEAN Food Handling Newsletter (weekly)
ASEAN Food Handling Project Review
ASEAN Food Journal (weekly) 08628

ASEAN Institute for Physics (ASEANIP), c/o Dept of Physics, University of Malaya, 59100 Kuala Lumpur
T: (03) 7555466 ext 386
Founded: 1989
Pres: Prof. B.C. Tan
Focus: Physics 08629

ASEAN Institute of Forest Management (AFM), IGB Plaza, Ste 903, Jalan Kampar 6, 50400 Kuala Lumpur
T: (03) 4429251
Founded: 1986
Focus: Forestry; Business Admin . . . 08630

ASEAN Plant Quarantine Centre and Training Institute (PLANTI), PMB 209, 43400 Serdang
T: (03) 9486010, 9486016; Tx: 21302; Fax: (03) 9486023
Founded: 1980
Gen Secr: M. Sivanaser
Focus: Botany, Specific
Periodical
Plantinews (3 times annually) 08631

Asia Foundation, 197-7 Jalan Ampang, Kuala Lumpur
Founded: 1954
Gen Secr: L.T. Forman
Focus: Sociology; Econ; Educ 08632

Asian Association of National Languages (ASANAL), c/o Language Center, University of Malaya, 59100 Kuala Lumpur
T: (03) 7560181
Focus: Ling 08633

Asian Institute for Development Communication (AIDCOM), APDC Bldg, Persiaran Duta, 50480 Kuala Lumpur
T: (03) 2542558; Tx: 31533; Fax: (03) 2543785
Founded: 1986
Gen Secr: Zel Leong
Focus: Comm Sci
Periodicals
AIDCOM Information
Journal of Development Communication . 08634

Asian Wetland Bureau (AWB), c/o Institute of Advanced Studies, University of Malaya, Lembah Pantai, 59100 Kuala Lumpur
T: (03) 7572176; Tx: 39845; Fax: (03) 7571225
Founded: 1983
Focus: Geology 08635

Asia Pacific Forum on Women, Law and Development (APWLD), APWLD Bldg, Persiaran Duta, POB 12224, 50770 Kuala Lumpur
T: (03) 2550648; Tx: 31655; Fax: (03) 2541371
Founded: 1986
Gen Secr: Nimalka Fernando
Focus: Soc Sci; Law 08636

Asia-Pacific People's Environment Network (APPEN), c/o Sahabat Alam Malaysia, 43 Salween Rd, 10050 Penang
T: (04) 376930; Tx: 40989; Fax: (04) 375705
Founded: 1983
Gen Secr: S.M. Mond Idris
Focus: Ecology 08637

Association for Medical Education in the Western Pacific Region (AMEWPR), c/o Dept of Medical Education, Faculty of Medicine, University, Jalan Raja Muda Abdul Aziz, 50300 Kuala Lumpur
T: (03) 2923066; Tx: 31496; Fax: (03) 2912659
Founded: 1988
Gen Secr: Prof. Sharifah H. Shahabudin
Focus: Educ; Med 08638

Association of Development Research and Training Institutes of Asia and the Pacific (ADIPA), c/o APDC, Pesiaran Duta, POB 12224, 50770 Kuala Lumpur
T: (03) 2548088; Tx: 30676; Fax: (03) 2550316
Founded: 1973; Members: 100
Gen Secr: Prof. Suk Bum Yoon
Focus: Develop Areas
Periodical
ADIPA Newsletter 08639

Association of South-East Asian Nation Countries' Union of Polymer Science (ASEANCUPS), c/o Plastics and Rubber Institute, 47000 Sungai Ehsan
Gen Secr: Dr. Chan Boon Lye
Focus: Materials Sci 08640

Commonwealth Association of Scientific Agricultural Societies (CASAS), c/o FPM, Menara Boustead, Jalan Raja Chulan, 50200 Kuala Lumpur
Founded: 1978
Pres: Alladin Hashim
Focus: Agri
Periodical
CASAS Newsletter (weekly) 08641

Geological Survey of Malaysia, Scrivenor Rd, Ipoh, Perak
Founded: 1903; Members: 792
Gen Secr: E.H. Yin
Focus: Surveying; Geology
Periodicals
Annual Report (weekly)
District Memoirs
Economic Bulletins
Map Bulletins
Professional Papers (West)
Regional Memoirs
Reports and Bulletins (East) 08642

Geological Survey of Malaysia, POB 560, 93712 Kuching, Sarawak
T: (082) 240152
Founded: 1949; Members: 117
Focus: Geology
Periodicals
Annual Report (weekly)
Bulletin
Geological Papers
Memoirs
Report 08643

Malayian Nature Society, POB 10750, 50724 Kuala Lumpur
Founded: 1940; Members: 2650
Pres: Prof. Dr. Salleh Mohd Nor; Gen Secr: Prof. P.N. Avadhani
Focus: Hist; Nat Sci
Periodical
The Malayan Nature Journal (weekly) . . 08644

Malaysian Biochemical Society, c/o Biochemistry Dept, University of Malaya, Kuala Lumpur
T: (03) 7502884; Tx: 39845; CA: Unisel;
Fax: (03) 7573661
Founded: 1973; Members: 110
Pres: Ramasamy Perumal; Gen Secr: Mahmud Rahman
Focus: Biochem
Periodical
Proceedings of Annual Conference (weekly)
. . . . 08645

Malaysian Branch of the Royal Asiatic Society, 130M Jalan Thamby Abdullah, off Jalan Tun Sambanthan, Brickfields, 50470 Kuala Lumpur
Founded: 1877; Members: 846
Focus: Ethnology
Periodical
Journal of the Malaysian Branch of the Royal Asiatic Society (weekly) 08646

Malaysian Historical Society, 958 Jalan Hose, Kuala Lumpur
T: (03) 481469
Founded: 1953; Members: 160
Pres: Dato Musa Hitam
Focus: Hist
Periodicals
Malaysia Dari Segi Sejarah (weekly)
Malaysia in History 08647

Malaysian Institute of Architects, 4-6 Jalan Tangsi, POB 10855, Kuala Lumpur
T: (03) 928733
Founded: 1967; Members: 994
Pres: Haji Hajeedar Bin Haji Abdul Majid
Focus: Archit
Periodicals
Berita Akitek (weekly)
Majallah Akitek (weekly)
Panduan Akitek (weekly) 08648

Malaysian Library Association, POB 12545, 50782 Kuala Lumpur
Founded: 1955
Pres: R.A. Rashid; Gen Secr: Ahmad Ridzuan Bin Wan Chik
Focus: Libraries & Bk Sci
Periodicals
Berita PPM (weekly)
Majalah PPM (weekly) 08649

Malaysian Medical Association (MMA), MMA House, 124 Jalan Pahang, POB S-20, 53000 Kuala Lumpur
T: (03) 4420617
Founded: 1959; Members: 4915
Pres: Dr. S. Mahmood; Gen Secr: A.N. Khor
Focus: Med
Periodicals
Medical Journal of Malaysia (weekly)
MMA Newsletter (weekly) 08650

Malaysian Rubber Research and Development Board, 148 Jalan Ampang, POB 10508, 50716 Kuala Lumpur
Founded: 1959
Pres: Datok Ahmad Farouk Bin Haji S.M. Ishak
Focus: Materials Sci
Periodicals
Annual Report (weekly)
Getah Asli (weekly)
Malaysian Rubber Review (weekly)
Natural Rubber Technology (weekly)
Rubber Developments (weekly) 08651

Malaysian Scientific Association, POB 911, Kuala Lumpur
Founded: 1954; Members: 170
Pres: Dr. M.K. Rajakumar; Gen Secr: Prof. Dr. J.I. Furtado
Focus: Sci 08652

Malaysian Society of Anaesthesiologists, MMA House, 124 Jalan Pahang, Kuala Lumpur
T: (03) 980617
Members: 100
Focus: Anesthetics 08653

Malaysian Zoological Society, 301 Lee Yan Lian Bldg, Jalan Tun Perak, 50050 Kuala Lumpur
Pres: Y.B. Hutson; Gen Secr: Yuen Tang
Focus: Zoology 08654

Tamil Language Society, c/o Dept of Indian Studies, University of Malaysia, Kuala Lumpur
Founded: 1957; Members: 350
Pres: M. Jaykumar; Gen Secr: L. Krishnan
Focus: Ling
Periodical
Tamil Olu (weekly) 08655

Mali

International African Migratory Locust Organisation, BP 136, Bamako
Founded: 1955; Members: 17
Focus: Entomology

Periodicals
Annual Report (weekly)
Locusta (weekly)
Monthly Information Bulletin (weekly) . . 08656

Malta

Agrarian Society, Palazzo de la Salle, 219 Republic St, Valletta
T: 26345
Founded: 1844; Members: 200
Pres: Prof. Joseph A. Micallef; Gen Secr: Joseph Borg
Focus: Agri 08657

Chamber of Architects and Civil Engineers, 1 Wilga St, Paceville
Founded: 1910
Focus: Archit 08658

Dental Association of Malta, c/o Centre of Professional Bodies, Medisle Village, Saint Andrews
Founded: 1946; Members: 64
Focus: Dent 08659

The Economic Society of Malta, c/o Malta Federation, Pacevile
Founded: 1971; Members: 31
Focus: Econ 08660

Euro-Mediterranean Centre on Marine Contamination Hazards, c/o Foundation for International Studies, University of Malta, Saint Paul St, Valletta
T: 224067; Tx: 1673; Fax: 230551
Founded: 1986
Gen Secr: Anthony Micallef
Focus: Oceanography; Ecology 08661

Ghaqda Bibljotekarji (Gh.B.), c/o Public Library, Floriana
Founded: 1969; Members: 70
Pres: Joseph Grima; Gen Secr: Joseph Debattista
Focus: Libraries & Bk Sci
Periodicals
Ghaqda Bibljotekarji: Library Association Newsletter (weekly)
Occasional Papers Series (weekly) . . . 08662

Ghaqda Tal Folklor / Folklore Society, 78 Saint Trophimus St, Sliema
Founded: 1964; Members: 150
Focus: Ethnology
Periodical
L-Imnara (weekly) 08663

Graduate Teachers' Association, c/o Malta Federation, Paceville
T: 38851
Founded: 1969; Members: 220
Focus: Educ 08664

Malta Geographical Society, c/o E. Sambrook, 3 Saint Pancras Flats, Nazju Ellul St, Gzira
T: 511886
Founded: 1943; Members: 220
Focus: Geography 08665

Malta Historical Society (MHS), Kosi Kot, Triq Guze Ellul Mercer, Qormi
Founded: 1950; Members: 350
Focus: Hist
Periodicals
Melita Historica (weekly)
Proceedings of History Week (weekly) . 08666

Malta Union of Teachers (MUT), 213 Republic St, Valletta
T: 27815, 22663
Founded: 1919; Members: 3400
Focus: Educ 08667

Medical Association of Malta (MAM), c/o Malta Federation of Professional Bodies, 1 Wilga St, Paceville
T: 38851
Members: 320
Focus: Med 08668

The Ornithological Society (MOS), POB 498, Valletta
T: 40278, 48023
Founded: 1962; Members: 500
Focus: Ornithology
Periodicals
Bird's Eye View (weekly)
Il-Merill (weekly)
L-Ghasfur U L-Ambjent Naturali: The Bird and the Natural Habitat (weekly) 08669

Professional Librarians of Malta Association (PLMA), c/o Malta Federation of Professional Bodies, 1 Wilga St, Paceville
T: 38851
Founded: 1971
Focus: Libraries & Bk Sci 08670

Society for the Study and Conservation of Nature, POB 459, Valletta
T: 248558
Founded: 1962; Members: 1500
Pres: Vince R. Attard; Gen Secr: Annick Bonnello
Focus: Ecology; Botany; Paleontology
Periodicals
Central Mediterranean Naturalist (weekly)
Il-Ballotra (weekly)
In-Natura (weekly)
Potamon (weekly) 08671

Mauritania

Conseil Panafricain pour la Protection de l'Environnement et le Développement (CPPED), BP 994, Nouakchott
T: 53077
Founded: 1982
Gen Secr: Dr. W. Nsanga
Focus: Ecology 08672

Mauritius

Centre for Documentation, Research and Training on the Islands of the South West Indian Ocean, POB 91, Rose Hill
T: 2126821; Tx: 4277; Fax: 2080076
Founded: 1981
Pres: Pynee Chellapermal
Focus: Geography
Periodical
Bulletin IBION (3 times annually) . . . 08673

Royal Society of Arts and Sciences of Mauritius, c/o Sugar Industry Research Institute, Réduit
T: 541061; Tx: 4899; Fax: 541971
Founded: 1847; Members: 173
Pres: J.M. Paturau; Gen Secr: J.C. Autrey
Focus: Arts; Sci
Periodical
Proceedings 08674

Société de l'Histoire de l'Ile Maurice, c/o Service Bureau, Pl Foch, Port Louis
Founded: 1938; Members: 810
Gen Secr: G. Ramet
Focus: Hist
Periodical
Bulletin 08675

Société de Technologie Agricole et Sucrière de l'Ile Maurice, c/o Mauritius Sugar Industry Research Institute, Réduit
Fax: 541971
Founded: 1910; Members: 350
Focus: Agri; Eng
Periodical
La Revue Agricole et Sucrière de l'Ile Maurice (3 times annually) 08676

Mexico

Academia de Arte Dramática, c/o Universidad, Hermosillo, Sonora
Focus: Perf Arts 08677

Academia de Artes Plásticas, c/o Universidad, Guanajuato
Focus: Arts 08678

Academia de Artes Plásticas, c/o Universidad, Hermosillo, Sonora
Focus: Arts 08679

Academia de Ciencias Históricas de Monterrey, Apdo 389, Monterrey, Nuevo León
Founded: 1947
Focus: Hist 08680

Academia de Dramática, c/o Universidad, Guanajuato
Focus: Perf Arts 08681

Academia de la Investigación Científica / Mexcian Academy of Scientific Research, Av San Jerónimo 260, Col. Jardines de Pedregal, 04500 México, D.F.
T: (5) 5504000; Fax: (5) 5501143
Founded: 1959; Members: 755
Pres: Antonio Peñz; Gen Secr: Mario Campesino
Focus: Sci
Periodical
Ciencia (weekly) 08682

Academia de Música, c/o Universidad, Guanajuato
Focus: Music 08683

Academia de Música, c/o Universidad, Hermosillo, Sonora
Focus: Music 08684

Academia Mexicana de Jurisprudencia y Legislación, 5 de Mayo 32, México, D.F.
Founded: 1889
Focus: Law 08685

Academia Mexicana de la Historia, Plaza de Carlos Pacheco 21, Cuauhtémoc, 06070 México, D.F.
Founded: 1919
Pres: Prof. Luis González y González
Focus: Hist
Periodical
Memorias (weekly) 08686

Academia Mexicana de la Lengua, Donceles 66, 06010 México, D.F.
Founded: 1957; Members: 23
Pres: José Luis Martínez; Gen Secr: Manuel Alcalá Anaya
Focus: Ling 08687

Academia Nacional de Ciencias, Apdo M-77-98, México 1, D.F.
Founded: 1884; Members: 24
Pres: Dr. Manuel Velasco Suárez; Gen Secr: Dr. Antonio Pompa y Pompa
Focus: Sci
Periodicals
Memorias
Revista 08688

Academia Nacional de Historia y Geografía, Londres 60, México 6, D.F.
Founded: 1925; Members: 179
Pres: Antonio Fernández del Castillo
Focus: Geography; Hist
Periodical
Revista 08689

Academia Nacional de Medicina, c/o Centro Médico Nacional, Bloque B, Av Cuauhtémoc 330, Apdo M 8075, México, D.F.
T: (5) 7617044 ext 139; Fax: (5) 5784271
Founded: 1864; Members: 302
Pres: Dr. Victor M. Espinosa de los Reyes; Gen Secr: Dr. Miguel Tanimoto Weki
Focus: Med
Periodical
Gaceta Médica de México (weekly) . . 08690

Anuarios de Filosofía y Letras, c/o Facultad de Filosofía y Letras, Ciudad Universitaria, México 20, D.F.
Focus: Lit; Philos 08691

Asociación de Ingenieros y Arquitectos de México, Calle del Puente de Alvarado 58, México, D.F.
Founded: 1868; Members: 560
Focus: Archit; Eng 08692

Asociación Interamericana de Gastroenteología (AIGA), Pizarras 171, México 20, D.F.
Founded: 1948
Focus: Gastroenter 08693

Asociación Latinoamericana de Ecodesarrollo / Latin American Association for Ecological Development, Apdo 1-110, 21000 Mexicali
Founded: 1978
Focus: Ecology 08694

Asociación Latinoamericana de Sociología (ALAS), c/o Instituto de Investigaciones Sociales, Torre de Humanidades, Universidad, México 20, D.F.
Focus: Sociology 08695

Asociación Médica del Hospital Beistegui, Regina 7, México, D.F.
Founded: 1963
Focus: Med 08696

Asociación Médica Franco-Mexicana, Dr. Balmis 148, México 7, D.F.
Founded: 1928; Members: 600
Focus: Med 08697

Asociación Mexicana de Administración Científica, Durango 167, México, D.F.
Focus: Law 08698

Asociación Mexicana de Bibliotecarios, AC (AMBAC) / Mexican Librarians Association, Angel Urraza 817a, Col. del Valle, 03100 México, D.F.
T: (5) 5751135; Fax: (5) 5751135
Founded: 1954; Members: 800
Pres: Surya Peniche Sánchez Macgregor; Gen Secr: Elsa Ramírez Leyva
Focus: Libraries & Bk Sci
Periodicals
Memorias (weekly)
Noticiero de la AMBAC (weekly) 08699

Asociación Mexicana de Facultades y Escuelas de Medicina, Av V. Carranza 870, Apdo 836, San Luis de Potosí
Founded: 1957
Focus: Med 08700

Asociación Mexicana de Géologos Petroleros, Ciprés 176, México 4, D.F.
Founded: 1949; Members: 600
Focus: Geology 08701

Asociación Mexicana de Ginecologia y Obstetricia, Av Baja California 311, Cuauhtémoc, 06100 México
Founded: 1945; Members: 350
Focus: Gynecology
Periodical
Ginecologia y Obstetricia de México (weekly) 08702

Asociación Mexicana de Orquideologia, Apdo 53123, 11320 México, D.F.
T: (5) 2031909, 2942862; Tx: 1763374; Fax: (5) 5314349
Focus: Botany
Periodical
Orquidea 08703

Asociación Mexicana de Profesores de Microbiología y Parasitología en Escuelas de Medicina, Tolsa 238, Guadalajara, Jalisco
Founded: 1962
Focus: Microbio 08704

Asociación Nacional de Universidades e Institutos de Enseñanza Superior, Insurgentes Sur 2133, 01000 México D.F.
Founded: 1950
Focus: Educ; Sci
Periodical
Revista de la Educación Superior (weekly) 08705

Asociación Panamericana de Cirurgía Pediátrica (APCP), Calzada de Tlalpan 4515, México 22, D.F.
T: 5730094
Founded: 1966; Members: 600
Focus: Surgery 08706

Ateneo de Ciencias y Artes de Chiapas, 3a Oriente 28, Tuxtla Gutiérrez
Founded: 1942; Members: 76
Focus: Arts 08707

Barra Mexicana-Colegio de Abogados, Varsovia 1, 06600 México D.F.
Founded: 1923; Members: 1040
Pres: Miguel I. Estrada Sámano; Gen Secr: Alejandro Manterola Martínez
Focus: Law
Periodical
El Foro (3 times annually) 08708

Center for Economic and Social Studies in the Third World (CESSTEM), Porfirio Diaz 50, San Jerónimo Lidice, 10200 México, D.F.
T: (05) 5952088
Founded: 1976
Gen Secr: Jorge Nuno
Focus: Soc Sci; Econ 08709

Center of Coordination and Diffusion of Latin American Studies / Centro Coordinador y Difusor de los Estudios Latinoamericanos, Torre I de Humanidades, Ciudad Univ, 04510 México, D.F.
T: 5505745
Founded: 1978
Gen Secr: Dr. Leopoldo Zea
Focus: Soc Sci; Geography; Hist
Periodical
Carta del CCYDEL (3 times annually) . 08710

Center of Regional Cooperation in Adult Education for Latin America, Quinta Eréndira s/n, 61600 Patzcuaro
T: (0454) 21475; Fax: (0454) 20092
Founded: 1951
Pres: Prof. Mario Aguilerra
Focus: Adult Educ
Periodicals
Cuadernos del CREFAL (weekly)
Revista Interamericana de Educación de Adultos 08711

Centre for Latin American Monetary Studies, Durango 54, Cuauhtemoc, Colonia Roma, 06700 México D.F.
T: (05) 5330300; Tx: 1771229; Fax: (05) 5146554
Founded: 1949
Gen Secr: Ramón Lacuona
Focus: Finance
Periodicals
CEMLA Bulletin (weekly)
Money Affairs (weekly)
Montaria (weekly) 08712

Centro Científico y Técnico Francés en México, Liverpool 67, Col. Juarez, 06600 México, D.F.
Founded: 1960
Focus: Eng; Nat Sci
Periodical
Interface (3 times annually) 08713

Centro de Estudios Económicos y Sociales del Tercer Mundo, Coronel Porfirio Díaz 50, Col. San Jeronimo Lidice, 10200 México, D.F.
T: 5955950
Focus: Econ; Sociology 08714

Centro de Estudios Educativos, Av Revolución 1291, México 20, D.F.
Founded: 1963
Focus: Educ 08715

Centro de Estudios Monetarios Latinoamericanos (CEMLA), Durango 54, México 7, D.F.
Founded: 1949
Focus: Finance 08716

Centro de Información Científica y Humanística, c/o Universidad Autónoma de México, Ciudad Universitaria, Apdo 70-392, 04510 México 20, D.F.
Founded: 1971
Gen Secr: Margarita Almada de Ascencio
Focus: Humanities
Periodicals
Bibliografía Latinoamericana (weekly)
Clase: Citas Latinoamericanas en Sociología en Ciencias y Humanidades (weekly)
Inforum (weekly)
Periodica: Indice de Revistas Latinoamericanas en Ciencias (weekly) 08717

Centro de Investigación y de Estudios Avanzados del Instituto Politécnico Nacional, Av Instituto Politécnico Nacional 2508, Apdo 14-740, 07360 México, D.F.
T: 5862770, 5862849; Tx: 72826; CA: POLINVEST; Fax: 7548707
Founded: 1961; Members: 470
Pres: Dr. Héctor O. Nava Jaimes; Gen Secr: Romeo Campesino
Focus: Biochem; Physics; Electronic Eng; Math; Chem; Energy; Cell Biol & Cancer Res
Periodical
Avance y Perspectiva (weekly) 08718

Centro Internacional de Mejoramiento de Maíz y Trigo (CIMMYT), Lisboa 27, Apdo 6-641, 06600 México, D.F.
Founded: 1966
Focus: Agri
Periodicals
Barley Yellow Dwarf Newsletter
Boletín de Información Científica
CIMMYT Aujourd'hui
CIMMYT Economics Paper
CIMMYT Economics Working Paper
CIMMYT EN
CIMMYT hechos y tendencias mundiales relacinadas con trigo
CIMMYT hechos y tendencias mundiales relacionadas con maíz
CIMMYT Hoy
CIMMYT IN
CIMMYT Réalité et tendances-le mais dans le monde: le potentiel maisicole de l'Afrique subsaharienne
CIMMYT Research Report
CIMMYT Today
CIMMYT Wheat Special Report
CIMMYT World Maze Facts and Trends
CIMMYT World Wheat Facts and Trends
Documento de trabajo de Programa de Economía del CIMMYT
Farming System Bulletin
Monografías de Economía del CIMMYT
Scientific Information Bulletin 08719

Centro Nacional de Cálcula, c/o Unidad Profesional Zacatenco IPN-Lindavista, México 14, D.F.
Founded: 1963
Focus: Math 08720

Centro Nacional de Ciencias y Tecnologías Marinas, Apdo 512, Veracruz
Founded: 1957; Members: 23
Focus: Eng 08721

Centro Regional de Educación de Adultos y Alfabetización Funcional para América Latina (CREFAL), Quinta Eréndira, 61600 Pátzcuaro
Fax: 20092
Founded: 1951
Pres: Luis Benavides Ilizaliturri
Focus: Educ; Adult Educ
Periodicals
Cuadernos del CREFAL (3 times annually)
Retablo de Papel (3 times annually)
Revista Interamericana de Educación de Adultos (weekly) 08722

Centro Vasco, Madero 6, México, D.F.
Founded: 1800
Focus: Lit; Hist 08723

El Colegio Nacional, Luis González Obregón 23, 06020 México, D.F.
Fax: (5) 7021779
Founded: 1943; Members: 40
Focus: Educ
Periodical
Memoria de El Colegio Nacional (weekly) 08724

Comisión Latinoamericana de Investigadores en Sorgo (CLAIS), Apdo 6641, 06000 México D.F.
T: (05) 7613311; Fax: (05) 41069
Gen Secr: Dr. Compton L. Paul
Focus: Crop Husb 08725

Commission of Studies for Latin American Church History, Celaya 21-402, Colonia Hipódromo, 06100 México, D.F.
T: (05) 5745661
Founded: 1971
Pres: Dr. Enrique Dussel
Focus: Rel & Theol
Periodical
Boletín CEHILA (weekly) 08726

Confederación de Educadores Americanos, Venezuela 38, México 1, D.F.
Focus: Educ 08727

Consejo Interamericano de Archiveros (CITA), c/o Archivo Nacional de México, México, D.F.
Focus: Libraries & Bk Sci 08728

Consejo Nacional de Ciencia y Tecnologia (CONACYT), Insurgentes Sur 1677, México 20, D.F.
Founded: 1970
Focus: Nat Sci; Eng 08729

Departamento de Antropología e Historia de Nayarit, Av México 91, Tepic, Nayarit
Founded: 1946
Focus: Anthro; Hist 08730

Departamento de Educación Audiovisual, c/o Instituto Politécnico Nacional, Prolongación de Carpio 475, México, D.F.
Focus: Educ 08731

Environmental Training Network for Latin America and the Caribbean, c/o Naciones Unidas, Oficina Regional para América Latina, Av Presidente Masaryk 29, Colonia Polanco, 11570 México, D.F.
T: (02) 2034975; Tx: 71055; Fax: (02) 2034465
Founded: 1980
Gen Secr: Enrique Leff
Focus: Ecology 08732

Equipe 7 des Arts Visuels, 2 Lerdo 35, Azcapotzalco, México 16, D.F.
T: 5611271
Focus: Arts 08733

Federación Latinoamericana de Parasitología (FLAP), c/o Departamento de Parasitología, Universidad, Apdo 20372, México 20, D.F.
Founded: 1963
Focus: Microbio 08734

Instituto de Ecología / Ecological Institute, Km 2,5 Carretera Antigua a Coatepec, Apdo 63, 91000 Xalapa
T: 86000, 86110, 86209, 86310, 86409; Tx: 015542; Fax: 86510
Founded: 1974; Members: 150
Pres: Dr. Gonzalo Halffter
Focus: Ecology
Periodicals
Acta Botanica (3 times annually)
Acta Zoologica (3 times annually) . . . 08735

Instituto Indigenista Interamericano / Inter-American Indian Institute, Nubes 232, Col. Pedregal de San Angel, 01900 México, D.F.
T: (5) 5680819; Fax: (5) 6521274
Founded: 1940
Pres: Dr. José Matos Mar
Focus: Ethnology
Periodicals
América Indígena (weekly)
Anuario Indigenista (weekly) 08736

Instituto Nacional de Bellas Artes, Paseo de la Reforma y Campo Marte s/n, 11580 México
Tx: 1761347
Founded: 1947
Pres: Victor Sandoval de León
Focus: Arts 08737

Instituto Nacional de Higiene, Calzada, Mariano Escobedo 20, México, D.F.
Fax: (5) 276693
Founded: 1904; Members: 153
Focus: Hygiene 08738

Instituto Nacional de Investigaciones Agrícolas, Apdo 6882, México 6, D.F.
Founded: 1960
Focus: Agri 08739

Instituto Panamericano de Geografía e Historia (IPGH) / Pan American Institute of Geography and History, Ex-Arzobispado 29, 11860 México, D.F.
Founded: 1928; Members: 21
Gen Secr: Leopoldo Rodriguez
Focus: Geography; Hist
Periodicals
Boletín de Antropología Americana (weekly)
Folklore Americano (weekly)
Revista Cartográfica (weekly)
Revista de Historia de Amercia (weekly)
Revista Geofísica (weekly)
Revista Geográfica (weekly) 08740

International Association of Gerontology (IAG), c/o Dr. Samuel Bravo, Jojutla 91, Tlalpan, México, D.F.
T: (5) 5735056; Tx: 1760240; Fax: (5) 6795842
Founded: 1950; Members: 53
Pres: Dr. Samuel Bravo; Gen Secr: Dr. Joaquin Gonzalez
Focus: Geriatrics
Periodicals
Gerontology: International Journal of Experimental and Clinical Gerontology (weekly)
IAG Newsletter (weekly) 08741

International Communication Agency (ICA), Libertad 1492, Guadalajara, Jalisco
Founded: 1949
Focus: Lit 08742

International Communication Agency (ICA), Londres 16, México 6, D.F.
Founded: 1942
Focus: Lit 08743

International Communication Agency (ICA), Av Constitución 411, Apdo 152, Monterrey
Founded: 1949
Focus: Lit 08744

Sociedad Agronómica Mexicana, López 23, México, D.F.
Focus: Agri 08745

Sociedad Astronómica de México, Jardín Felipe Xicotencath, Col. Alamos 13, México 13, D.F.
Founded: 1902
Focus: Astronomy 08746

Sociedad de Educación, c/o Sección Educacional, Edificio del Banco de Londres y México, México, D.F.
Focus: Educ 08747

Sociedad de Estudios Biológicos, Balderas 94, México, D.F.
Focus: Bio 08748

Sociedad de Oftalmología del Hospital de Oftalmológico de Nuestra Señora de la Luz, c/o Escuela de Medicina, Ezequiel Montes 135, México, D.F.
Founded: 1893
Focus: Ophthal 08749

Sociedad Geológica Mexicana, Jaime Torres Bodet 176, Col. Santa Maria La Ribera, 06400 México
Founded: 1904; Members: 760
Pres: Bernardo Martell Andrade; Gen Secr: Luis Velazquez Aguirre
Focus: Geology
Periodical
Boletín de la Sociedad Geológica Mexicana (weekly) 08750

Sociedad Interamericana de Cardiología (IASC), Juan Badiano 1, México 22, D.F.
T: 5732911
Founded: 1946; Members: 4000
Focus: Cardiol 08751

Sociedad Latinoamericana de Alergología, Dr. Márquez 162, México, D.F.
Founded: 1964; Members: 485
Focus: Immunology 08752

Sociedad Matemática Mexicana (SMM), c/o Instituto de Matemáticas, UNAM, Apdo 70-450, 04510 México, D.F.
T: (5) 6224520; Fax: (5) 5489499
Founded: 1943; Members: 800
Pres: Dr. José Carlos Gómez L.
Focus: Math
Periodical
Miscelanea Matematica: Boletín de la Sociedad Matematica Mexicana and Aportaciones Matematicas 08753

Sociedad Médica del Hospital General, Dr. Balmis y Dr. Pasteur, México, D.F.
Founded: 1939
Focus: Med 08754

Sociedad Médica del Hospital Oftalmológico de Nuestra Señora de la Luz, Ezequiel Montes 135, México, D.F.
Founded: 1966
Focus: Ophthal 08755

Sociedad Mexicana de Antropología (S.M.A.), Apdo 105-100, 11581 México, D.F.
Founded: 1937; Members: 300
Gen Secr: Dr. Lourdes Arizpe; Dr. Alejandro Martínez
Focus: Anthro
Periodical
Revista Mexicana de Estudios Antropológicos (bianual) 08756

Sociedad Mexicana de Bibliografía, c/o Hemeroteca Nacional, Carmen y San Ildefonso, México, D.F.
Founded: 1945; Members: 91
Focus: Libraries & Bk Sci 08757

Sociedad Mexicana de Cardiología (SMC) / Mexican Society of Cardiology, Juan Badiano 1, Tlalpan, 14080 México, D.F.
T: (5) 5732911 ext 295; Fax: (5) 5732111
Founded: 1935; Members: 310
Focus: Cardiol
Periodical
Quehacer cardiológico 08758

Sociedad Mexicana de Entomología, Apdo 63, 56230 Chapingo
Founded: 1952; Members: 900
Pres: Jorge Vera Graciano; Gen Secr: Sergio Ibáñez Bernal
Focus: Entomology
Periodical
Folia Entomológica Mexicana (weekly) . . 08759

Sociedad Mexicana de Estudios Psico-Pedagógicos, Nayarit 86, México, D.F.
Focus: Educ; Psych 08760

Sociedad Mexicana de Eugenesia, Acapulco 44, México, D.F.
Founded: 1931; Members: 150
Focus: Genetics 08761

Sociedad Mexicana de Fitogenética (SOMEFI) / Mexican Society for Plant Breeding and Genetics, c/o Colegio de Postgraduados y Universidad Autonoma Chapingo, Apdo 21, 56230 Chapingo
T: 42200 ext 5795
Founded: 1965
Pres: Dr. Victor A. González Hernández; Gen Secr: Dr. Manuel Livera Muñoz
Focus: Botany
Periodical
Fitotecnia (weekly) 08762

Sociedad Mexicana de Fitopatología (SMF), c/o Centro de Fitopatología, Colegio de Postgraduados, 56230 Montecillo
T: 45211; Fax: 45077
Founded: 1960; Members: 400
Pres: Dr. Ignacio Cid del Prado Vera; Gen Secr: M.C. Pilar Rodriguez
Focus: Botany
Periodical
Revista Mexicana de Fitopatología: El Vector (weekly) 08763

Sociedad Mexicana de Geografía y Estadística, Justo Sierra 19, Apdo 10-739, 06020 México, D.F.
Founded: 1833; Members: 975
Focus: Geography; Stats
Periodical
Boletín (3 times annually) 08764

Sociedad Mexicana de Historia de la Ciencia y la Tecnología, Av Dr. Vertiz 724, México 12, D.F.
Focus: Cultur Hist
Periodical
Anales (weekly) 08765

Sociedad Mexicana de Historia Natural, Av Dr. Vertiz 724, México 12, D.F.
Founded: 1868; Members: 400
Focus: Nat Sci 08766

Sociedad Mexicana de Historia y Filosofia de la Medicina, Administración Urbana 107, Cuauhtémoc, Apdo 107-035, 06760 México, D.F.
Founded: 1957; Members: 260
Pres: Juan Somolinos Palencia
Focus: Hist
Periodical
Boletín (weekly) 08767

Sociedad Mexicana de Neurologia y Psiquiatria, Tlalpan, 14410 México, D.F.
T: 5732822 ext 29
Founded: 1947; Members: 400
Focus: Neurology; Psychiatry
Periodical
Revista de Neurologia, Neurocirugia y Psiquiatria (weekly) 08768

Sociedad Mexicana de Nutrición y Endocrinología, Viaducto Tlalpan y Av San Fernando, México 22, D.F.
Founded: 1960; Members: 235
Focus: Endocrinology; Food 08769

Sociedad Mexicana de Pediatría, Tehuantepec 86-503, Col. Roma Sur, 06760 México, D.F.
T: (5) 5648371; Fax: (5) 5647739
Founded: 1930; Members: 450
Pres: Dr. Eduardo Aparicio Frías
Focus: Pediatrics
Periodical
Revista Mexicana de Pediatría (weekly) . 08770

Sociedad Mexicana de Salud Pública, Leibnitz 32, México 5, D.F.
Founded: 1944; Members: 750
Focus: Public Health 08771

Sociedad Mexicana de Seguridad Radiologica (SMSR) / Radiological Protection Mexican Association, Insurgentes Sur 1806, 01030 México, D.F.
T: 5349402, 5219045; Fax: 5341405, 5213792
Founded: 1975; Members: 120
Pres: Víctor Manuel Tovar; Gen Secr: Raúl Ortiz Magaña
Focus: Radiology
Periodical
Boletín SMSR (weekly) 08772

Sociedad Mexicana de Tisiología, Av Coyoacán 1707, México 12, D.F.
Founded: 1932
Focus: Pulmon Dis 08773

Sociedad Nuclear Mexicana / Mexican Nuclear Society, Sierra Mojada 447, Col. Lomas de Barrilaco, 11010 México, D.F.
T: 5219402, 5406567; Tx: 1773824; Fax: 5213798
Founded: 1988; Members: 250
Pres: Dr. Carlos Vélez Ocón; Gen Secr: Carlos Enrique Cervantes de Gortari
Focus: Nucl Res 08774

Sociedad Nuevoleonesa de Historia, Geografía y Estadística, c/o Biblioteca Universitaria Alfonso Reyes, Apdo 1575, Monterrey
Founded: 1942; Members: 80
Focus: Geography; Hist; Stats 08775

Sociedad Química de México, Ciprés 176, México 4, D.F.
T: 5470790
Founded: 1956; Members: 2000
Focus: Chem 08776

Unión de Universidades de América Latina, Ciudad Universitaria, Delegación Alvaro Obregon, Apdo 70232, 04510 México, D.F.
T: 5489786, 5480269; Tx: 1760180; CA: UDUAL-México
Founded: 1949; Members: 121
Focus: Sci
Periodicals
Gaceta UDUAL (weekly)
Universidades (weekly) 08777

Moldova, Republic of

Entomological Society of Moldova, Str Academiei 1, 277028 Chişinău
T: 739896
Pres: B.V. Verescighin
Focus: Zoology 08778

Geographic Society of Moldova, Bul Stefan cel Mare 1, 277012 Chişinău
T: 228428
Pres: A.M. Capcelea
Focus: Geography 08779

Hydrobiological Society of Moldova, Str Academiei 1, 277028 Chişinău
T: 739809
Pres: F.P. Cioric
Focus: Bio 08780

Mendeleev Chemical Society, Str Academiei 3, 277028 Chişinău
T: 739755
Pres: K.I. Turta
Focus: Chem 08781

Microbiological Society of Moldova, Str Academiei 1, 277028 Chişinău
T: 739916
Pres: G.V. Mereniuc
Focus: Microbio 08782

Moldovan Academy of Sciences, Bul Stefan cel Mare 1, 277001 Chişinău
T: 261478; Fax: 262091
Founded: 1961; Members: 61
Pres: A.M. Andries; Gen Secr: G.V. Siscanu
Focus: Sci
Periodicals
Elektronnaya Obrabotka Materialov: Electronic Processing of Materials (weekly)
Izvestiya: Bulletin
Revista de Filozofie si Drept: Journal of Philosophy and Law (weekly)
Revista de Istorie a Moldovei: Moldovan Historical Journal (weekly)
Revista de Lingvistica si Stiinta Literara: Journal of Linguistics and Study of Literature (weekly)
. . . . 08783

Moldovan Society of Animal Protection, Str Academiei 1, 277028 Chişinău
T: 739821
Pres: P.I. Nesterov
Focus: Ecology 08784

Moldovan Sociological Association, Str Puşkin 38, 279000 Beltsi
T: 23222
Pres: N.V. Turcanu
Focus: Sociology 08785

Ornithological Society of Moldova, Str Academiei 1, 277028 Chişinău
T: 737509
Pres: I.M. Ganea
Focus: Ornithology 08786

PEN Centre of Moldova, Bul Miron Costin 21, 277068 Chişinău
T: 443540
Pres: Andrei Burac
Focus: Lit 08787

Physical Society of Moldova, Str Academiei 5, 277028 Chişinău
T: 725887
Pres: T.I. Malinovschil
Focus: Physics 08788

Protozoological Society of Moldova, Str Academiei 1, 277028 Chişinău
T: 737511
Pres: C.A. Andriuta
Focus: Zoology 08789

Society of Botanists of Moldova, Str Padurilor 18, 277018 Chişinău
T: 523896
Pres: A.G. Negru
Focus: Botany 08790

Society of Genetics of Moldova, Str Gribova 44, 277049 Chişinău
T: 432308
Pres: V.D. Siminel
Focus: Genetics 08791

Society of Plant Physiologists of Moldova, Str Padurilor 22, 277028 Chişinău
T: 555514
Pres: S.I. Toma
Focus: Botany 08792

Teriological Society of Moldova, Str Academiei 1, 277028 Chişinău
T: 725566
Pres: A.I. Munteanu
Focus: Nat Sci 08793

Monaco

Association Monégasque de Préhistoire (A.P.S.M.), c/o Musée d'Anthropologie Préhistorique de Monaco, Blvd du Jardin Exotique, 98000 Monaco
T: 93158006
Founded: 1952; Members: 56
Focus: Speleology; Hist
Periodical
Bulletin du Musée d'Anthropologie Préhistorique de Monaco (weekly) 08794

Comité Arctique International (CAI), c/o Centre Scientifique de Monaco, 16 Blvd de Suisse, 98030 Monte Carlo
Founded: 1979
Pres: C. Solamito
Focus: Geography 08795

Commission Internationale pour l'Exploration Scientifique de la Mer Méditerranée (CIESM), 16 Blvd de Suisse, 98000 Monaco
T: 93303879; Fax: 92161195
Founded: 1910; Members: 2521
Pres: Prince Rainier III de Monaco; Gen Secr: Prof. F. Doumenge
Focus: Oceanography
Periodical
Proceedings of CIESM Congress . . . 08796

Organisation Hydrographique Internationale (OHI) / International Hydrographic Organization, 7 Av Président J. F. Kennedy, Monte Carlo
T: 506587; Tx: 479164; CA: Burhydint Monaco; Fax: 252003
Founded: 1921; Members: 57
Pres: Christian-R. Andreasen
Focus: Water Res
Periodicals
IHO Yearbook (weekly)
International Hydrographic Bulletin (weekly)
International Hydrographic Review (weekly) 08797

Mongolia

Academy of Sciences, Ulan Bator
Founded: 1921
Pres: N. Sodnom; Gen Secr: L. Dorj
Focus: Sci
Periodicals
Journal of the MPR Academy of Sciences (weekly)
Studia Archaeologica
Studia Ethnographica
Studia Folklorica
Studia Historica
Studia Mongolica
Studia Museologica 08798

Morocco

African Network of Administrative Information (ANAI), BP 310, Tangiers
T: (09) 36601; Tx: 33664
Founded: 1981
Gen Secr: T.H. Temsamani
Focus: Computer & Info Sci
Periodical
ANAI Index (weekly) 08799

African Training and Research Centre in Administration for Development, Av Mohamed V, BP 310, Tangiers
T: (09) 36601; Tx: 33864; Fax: (09) 943572
Founded: 1964; Members: 27
Gen Secr: Mamadou Thiam
Focus: Public Admin 08800

Association des Amateurs de la Musique Andalouse, 133 Av Ziraoui, Casablanca
Founded: 1956
Gen Secr: H.D. Benjelloun
Focus: Music 08801

Association Nationale des Géographes Marocains, c/o Faculté des Lettres, Université Mohamed V, Rabat
Founded: 1916; Members: 500
Gen Secr: M. Laghaout
Focus: Geography
Periodical
Revue de Géographie du Maroc (weekly) 08802

Association of African Faculties of Agriculture / Association des Facultés d'Agriculture d'Afrique, c/o Institut Agronomique et Vétérinaire Hassan II, BP 6280, Rabat
T: (07) 774702; Tx: 31873; Fax: (07) 774702
Founded: 1973
Gen Secr: Prof. Dr. A.O. Tantawy
Focus: Educ; Agri
Periodical
AFAA Newsletter (weekly) 08803

Centre d'Etudes, de Documentations et d'Informations Economiques et Sociales (C.E.D.I.E.S.), 23 Blvd Mohamed Abdouh, Casablanca
T: 52696
Focus: Sociology; Econ
Periodical
C.E.D.I.E.S. Informations 08804

Centre National de Documentation, BP 826, Rabat
Founded: 1968
Focus: Econ; Sociology; Doc
Periodicals
Index de la Documentation Économique, Scientifique et Technique (weekly)
Index Rétrospectifs Specialisés (weekly) . 08805

Comité National de Géographie du Maroc, c/o Institut Universitaire de la Recherche Scientifique, Chari Ma Al Ainine, BP 6287, Rabat
Founded: 1947
Focus: Geography 08806

Conférence Internationale des Directeurs et Doyens des Etablissements d'Enseignement Supérieur et Facultés d'Expression Française des Sciences de l'Agriculture et de l'Alimentation (CIDEAFA), c/o Institut Agronomique et Vétérinaire Hassan II, BP 6214, Rabat
T: (07) 770935; Fax: (07) 778110
Founded: 1990
Pres: Dr. Mohammed Sedrati
Focus: Educ; Agri; Food 08807

Société des Sciences Naturelles et Physiques du Maroc, c/o Institut Scientifique Chérifien, Av Moulay Chérif, Rabat
Founded: 1920; Members: 420
Pres: H. Faraj; Gen Secr: A. Sasson
Focus: Nat Sci; Physics
Periodicals
Bulletin
Bulletin de la Section des Naturalistes Enseignants 08808

Société d'Etudes Economiques, Sociales et Statistiques du Maroc, BP 535, Chellah, Rabat
Founded: 1933; Members: 20
Gen Secr: N. El Fassi
Focus: Sociology; Stats; Econ
Periodical
Signes du Présent (weekly) 08809

Société d'Horticulture et d'Acclimatation du Maroc, BP 13854, Casablanca
Founded: 1914; Members: 260
Pres: J. Duplat; Gen Secr: R. Tripotin
Focus: Hort 08810

Myanmar

Burma Medical Research Council, 5 Zafar Shah Rd, Rangoon
Focus: Med 08811

Burma Research Society, c/o Universities' Central Library, University Post Office, Rangoon
Founded: 1910; Members: 1040
Pres: U Htin Gyi; Gen Secr: Dr. U. Shein
Focus: Sci
Periodical
Journal (weekly) 08812

Central Research Organization, Kanbe, Yankin Post Office, Rangoon
Gen Secr: Dr. Mehm Thet San
Focus: Sci 08813

Namibia

Arts Association of Namibia, POB 994, Windhoek
T: (061) 31160
Founded: 1947; Members: 540
Gen Secr: A.H. Eins
Focus: Arts
Periodical
Newsletter (weekly) 08814

Institute of South West African Architects, POB 1478, Windhoek 9000
T: (061) 35119
Founded: 1952; Members: 80
Pres: C.M. Gay
Focus: Archit 08815

South West Africa Scientific Society, Corner John Meinert-Leutwein St, POB 67, Windhoek 9000
T: (061) 225372
Founded: 1925; Members: 1200
Pres: Dr. K.F.R. Budack; Gen Secr: A. Henrichsen
Focus: Sci
Periodicals
Dinteria
Journal SWA Scientific Society
Namibiana 08816

Nepal

Royal Nepal Academy, Kamaladi, Kathmandu
T: 211283
Founded: 1957; Members: 178
Gen Secr: V.B. Malla
Focus: Sci
Periodicals
Kabita
Prajna 08817

Netherlands

Actuarieel Genootschap (AG), Postbus 259, 1000 AG Amsterdam
T: (020) 5942685; Fax: (020) 6937968
Founded: 1888; Members: 750
Pres: J. van der Starre; Gen Secr: R.H. Sprenkels
Focus: Insurance
Periodical
De Actuaris (weekly) 08818

Algemene Bond ter Bevordering van Beroepsonderwijs, Bezuidenhoutseweg 211, 2594 AK 's-Gravenhage
T: (070) 854505
Focus: Adult Educ 08819

Algemene Bond van Onderwijzend Personeel (ABOP), Herengracht 54, 1015 BN Amsterdam
T: (020) 264633
Founded: 1966; Members: 43000
Focus: Educ 08820

Algemene Nederlandse Vereniging voor Wijsbegeerte (A.N.V.W.), Prinses Christinalaan 55, 1421 BE Uithoorn
Founded: 1933; Members: 159
Focus: Philos 08821

Architektenraad, Keizersgracht 321, Postbus 19611, 1016 EE Amsterdam
Founded: 1946; Members: 2700
Pres: H.B.A. Verhagen; Gen Secr: J. Höweler
Focus: Archit 08822

Association for Europeam Astronauts (AEA), c/o ESTEC, Keplerlaan 1, 2201 AZ Noordwijk
Founded: 1984
Pres: W. Ockels
Focus: Aero 08823

Association Internationale d'Etudes Occitanes (AIEO), President Kennedylaan 220, 2343 GX Oegstgeest
Founded: 1981
Pres: Prof. Dr. Q.I.M. Mok; Gen Secr: Gérard Gouiran
Focus: Ethnology
Periodical
Bulletin de l'AIEO 08824

Association Montessori Internationale / International Montessori Association, Koninginneweg 161, 1075 CN Amsterdam
T: (020) 6798932; Fax: (020) 6767341
Founded: 1929
Gen Secr: Fahmida Malik
Focus: Educ
Periodical
Communications (weekly) 08825

Association of European Correspondence Schools (AECS), c/o Koninklijke PBNA, Postbus 9053, 6800 GS Arnhem
T: (085) 575740; Fax: (085) 516102
Founded: 1985; Members: 77
Focus: Educ
Periodical
Epistolodidaktika (weekly) 08826

Association of European Schools of Planning (AESOP), c/o Vakgroep Planologie, Katholieke Universiteit Nijmegen, Thomas Van Aquinostr 5, 6500 KD Nijmegen
T: (080) 615471; Fax: (080) 616220
Founded: 1987; Members: 73
Gen Secr: Dr. Myriam Jansen-Verbeke
Focus: Educ
Periodicals
AESOP News (weekly)
AESOP Papers (weekly) 08827

Association of Roman Ceramic Archaeologists, Etudestr 62, 6544 RT Nijmegen
Founded: 1957; Members: 300
Focus: Archeol
Periodicals
Acta Supplementa
Kongressacta 08828

Association of Schools of Public Health in the European Region (ASPHER), c/o Dept of Health Sciences, School of Public Health, University of Limburg, Postbus 616, 6200 MD Maastricht
T: (043) 881235; Fax: (043) 670932
Founded: 1966
Focus: Educ; Public Health 08829

Auteursunie, Vondelstr 90, 1054 GN Amsterdam
T: (020) 186324
Founded: 1967; Members: 4
Focus: Lit 08830

Bataafsch Genootschap der Proefondervindelijke Wijsbegeerte, Postbus 597, 3000 AN Rotterdam
Founded: 1769; Members: 375
Pres: Prof. Dr. H.G. van Eijk; Gen Secr: Dr. J.R. ter Molen
Focus: Philos; Sci
Periodical
Nieuwe Verhandelingen (weekly) 08831

Benelux Society for Microcirculation, c/o Dept of Biophysics, University of Limburg, Postbus 616, 6200 MD Maastricht
Founded: 1986; Members: 40
Gen Secr: D.W. Slaaf
Focus: Biophys 08832

Bernoulli Society for Mathematical Statistics and Probability, Prinses Beatrixlaan 428, Postbus 950, 2270 AZ Voorburg
T: (070) 3375737; CA: Statist Voorburg; Fax: (070) 3860025
Founded: 1975; Members: 1470
Pres: O.E. Barndorff-Nielsen; Gen Secr: Z.E. Kenessey
Focus: Math; Stats 08833

Bond Heemschut, Nieuwezijds Kolk 28, 1012 PV Amsterdam
T: (020) 6225292; Fax: (020) 6240571
Founded: 1911; Members: 10500
Focus: Archit; Ecology
Periodical
Heemschut (weekly) 08834

Bond van Nederlandse Stedebouwkundigen (BNS), Waterlooplein 219, Postbus 19126, 1000 GC Amsterdam
T: (020) 6276820
Founded: 1935; Members: 425
Pres: J.C. Vogelij; Gen Secr: Dr. J.H. Crum-Haverman
Focus: Urban Plan
Periodical
Blauwekamer/Profiel (weekly) 08835

Caribbean Institute of Perinatology (CIP), c/o Dept of Obstetrics and Gynaecology, State University Hospital, Oostersingel 59, 9712 EZ Groningen
T: (050) 613173; Fax: (050) 613143
Founded: 1988
Pres: Dr. S. Leon; Gen Secr: Prof. Dr. E. Rudy Boersma
Focus: Gynecology 08836

Center for Research and Documentation on the World Language Problem, c/o CED, Nieuwe Binnenweg 176, 3015 BJ Rotterdam
T: (010) 4361044; Tx: 23721; Fax: (010) 4361751
Founded: 1952
Gen Secr: Prof. H. Tonkin
Focus: Ling 08837

Centraal Bureau vóor Genealogie / Central Bureau for Genealogy, Prins Willem Alexanderhof 22, Postbus 11755, 2502 AT 's-Gravenhage
T: (070) 3814651; Fax: (070) 3478394
Founded: 1945
Focus: Genealogy
Periodicals
Jaarboek (weekly)
Mededelingen (weekly)
Nederland's Adelsboek (weekly)
Nederland's Patriciaat (weekly) 08838

Centrum voor de Studie van het Onderwijs in Ontwikkelingslanden / Centre for the Study of Education in Developing Countries, Kortenaerkade 11, Postbus 29777, 2502 LT 's-Gravenhage
T: (070) 4260291; Tx: 35361; Fax: (070) 4260299
Founded: 1963; Members: 15
Pres: Prof. Dr. L.F.B. Dubbeldam
Focus: Educ
Periodical
Verhandelingen 08839

Centrum voor Plantenveredelings- en Reproduktieonderzoek (CPRO) / Centre for Plant Breeding and Reproduction Research, Postbus 16, 6700 AA Wageningen
Focus: Agri; Crop Husb
Periodical
Annual Report (weekly) 08840

Centrum voor Wiskunde en Informatica / Centre for Mathematics and Computer Science, Kruislaan 413, 1098 SJ Amsterdam
T: (020) 5929333; Tx: 12571; Fax: (020) 5924199
Founded: 1946; Members: 150
Focus: Math; Computer & Info Sci
Periodicals
CWI Monographs
CWI Syllabus
CWI Tracts
Jaarverslagen: Annual Report (weekly)
Report Series (weekly) 08841

Comité International d'Histoire de l'Art, Kromme Nieuwegracht 29, 3512 HD Utrecht
T: (030) 314886; Fax: (030) 314886
Pres: Prof. Dr. Thomas Gaehtgens; Gen Secr: Prof. Dr. W. Reinink
Focus: Hist; Arts; Cultur Hist 08842

Commission of Socialist Teachers of the European Community (CSTEC), Kanonsdijk 128, 7205 AE Zutphen
T: (05750) 12439
Gen Secr: B. Prenger
Focus: Educ 08843

Committee on Water Research (COWAR), c/o TNO Committee on Hydrological Research, Postbus 6067, 2600 JA Delft
T: (015) 697262; Fax: (015) 564801
Founded: 1964
Gen Secr: H.J. Colenbrander
Focus: Hydrology
Periodical
COWAR Bulletin (weekly) 08844

Conference of European Computer User Associations (CECUA), c/o BSG, Postbus 432, 1400 AK Bussum
Founded: 1978
Focus: Computer & Info Sci 08845

Coordinating Committee for Intercontinental Research Networking (CCIRN), c/o RARE, Singel 466-468, 1012 WP Amsterdam
Founded: 1987
Focus: Sci 08846

Esperanto Writers' Association, Volkerakstr 38, 1078 XT Amsterdam
T: (020) 6712664
Founded: 1957; Members: 83
Pres: William Auld
Focus: Lit 08847

Europa Nostra/IBI, Lange Voorhout 35, 2514 EC 's-Gravenhage
T: (070) 3560333; Fax: (070) 3617865
Founded: 1991
Focus: Preserv Hist Monuments 08848

Europe 2000, Ardennenlaan 11, 5601 JN Son en Breugel
Founded: 1989
Gen Secr: T. Sleeswijk Visser
Focus: Poli Sci 08849

European Asphalt Pavement Association (EAPA), Straatweg 68, Postbus 175, 3620 AD Breukelen
T: (03462) 66868; Fax: (03462) 63505
Founded: 1973
Gen Secr: Max von Devivere
Focus: Civil Eng
Periodical
European Asphalt Magazine (weekly) . . 08850

European Association for Cancer Education (EACE), c/o BOOG, Bloemsingel 1, 9713 BZ Groningen
T: (050) 632888; Fax: (050) 632883
Founded: 1987
Gen Secr: Dr. Wim Bender
Focus: Educ; Cell Biol & Cancer Res
Periodical
Journal of Cancer Education 08851

European Association for Cognitive Ergonomics (EACE), c/o Dept of Mathematics and Computer Science, Vrije Universiteit, De Boelelaan 1081A, 1081 HV Amsterdam
T: (020) 5485591; Fax: (020) 6426275
Founded: 1987
Focus: Computer & Info Sci
Periodical
EACE Newsletter (weekly) 08852

European Association for Gastroenterology and Endoscopy (EAGE), c/o Academic Medical Center, Melbergdreef 9, 1105 AZ Amsterdam
Founded: 1970
Pres: Prof. Guido N.J. Tytgat
Focus: Gastroenter
Periodical
European Journal for Gastroenterology and Hepatology 08853

European Association for Grey Literature Exploitation (EAGLE), Postbus 40407, 2209 LK 's-Gravenhage
T: (070) 314028; Fax: (070) 3140493
Founded: 1981
Gen Secr: Dr. R.H.A. Wessels
Focus: Lit 08854

European Association for Gynaecologists and Obstetricians, c/o Dept of Obstetrics and Gynecology, University Hospital, Postbus 9101, 6500 HB Nijmegen
T: (080) 514725; Tx: 48232; Fax: (080) 540576
Founded: 1985; Members: 100
Gen Secr: Prof. Dr. T. Eskes
Focus: Gynecology
Periodical
European Journal of Obstetrics, Gynecology and Reproductive Biology 08855

European Association for International Education (EAIE), Van Diemenstr 344, 1013 CR Amsterdam
T: (020) 6252727; Fax: (020) 6209406
Founded: 1989
Focus: Educ
Periodical
EAIE Newsletter (3 times annually) . . . 08856

European Association for Japanese Studies (EAJS), c/o Vakgroup van Japan en Korea, Rijksuniversiteit Leiden, Postbus 9515, 2300 RA Leiden
T: (071) 272549; Fax: (071) 272615
Pres: Prof. Dr. Adriana Boscaro
Focus: Cultur Hist 08857

European Association for Microprocessing and Microprogramming (EUROMICRO), Postbus 2346, 7301 EA Apeldoorn
T: (055) 557372; Fax: (055) 557393
Founded: 1973
Focus: Computer & Info Sci 08858

European Association for Population Studies (EAPS), Postbus 11676, 2502 AR 's-Gravenhage
T: (070) 3469482; Tx: 31138; Fax: (070) 3647187
Founded: 1983
Gen Secr: Guillaume J. Wunsch
Focus: Soc Sci
Periodicals
EAPS Newsletter (2-3 times annually)
European Journal of Population (weekly) 08859

European Association for Potato Research (EAPR), Postbus 20, 6700 AA Wageningen
T: (08370) 19112
Founded: 1956; Members: 550
Gen Secr: C.D. van Loon
Focus: Food 08860

European Association for Research on Plant Breeding (EUCARPIA), c/o Breeding Station Wiersum, Rendierweg 10, Postbus 94, 8250 AB Dronten
T: (08370) 19112
Founded: 1956; Members: 1345
Gen Secr: I.M. Mesken
Focus: Crop Husb 08861

European Association for Research on Plant Breeding (EUCARPIA), Postbus 315, 6700 AH Wageningen
T: (08370) 82838; Fax: (08370) 83457
Founded: 1956; Members: 1000
Pres: Dr. P.W.A. Tigerstedt; Gen Secr: Dr. M.J. de Jeu
Focus: Botany; Crop Husb; Genetics
Periodicals
EUCARPIA Bulletin (weekly)
Proceedings of Section Meetings (weekly) 08862

European Association for the Study of the Liver (EASL), c/o Div of Hepatology and Gastroenterology, Academic Medical Center, Meibergdreef 9, 1105 AZ Amsterdam
T: (020) 5663632; Fax: (020) 6917033
Founded: 1966
Gen Secr: Dr. Peter L.M. Jansen
Focus: Intern Med
Periodical
Journal of Hepatology (weekly) 08863

European Association of Diabetes Educators (EADE), Postbus 3023, 2301 DA Leiden
T: (071) 210905
Founded: 1986
Pres: Marianne Wayenberg
Focus: Educ; Diabetes 08864

European Association of Distance Teaching Universities (EADTU), c/o Open Universiteit, Valkenburgerweg 167, Postbus 2960, 6401 DL Heerlen
T: (045) 762272; Tx: 4556559; Fax: (045) 741473
Founded: 1987
Focus: Educ
Periodical
EADTU-News (weekly) 08865

European Association of Experimental Social Psychology (EAESP), c/o Vakgroep Sociale Psychologie, Universiteit van Amsterdam, Roetersstr 15, 1018 WB Amsterdam
T: (020) 5256790; Fax: (020) 5256710
Founded: 1964; Members: 680
Pres: Prof. Dr. G. Semin; Gen Secr: Prof. Dr. A.S.R. Manstead
Focus: Psych
Periodical
European Journal of Experimental Social Psychology (weekly) 08866

European Association of Exploration Geophysicists (E.A.E.G.), Laan van Vollenhove 3039, Postbus 298, 3700 AG Zeist
T: (03404) 62655; Fax: (03404) 62640
Founded: 1951; Members: 4000
Pres: I. Gausland; Gen Secr: J.-C. Grosset
Focus: Geophys
Periodicals
First Break (weekly)
Geophysical Prospecting (8 times annually)
Tidal Gravity Corrections (weekly) . . . 08867

European Association of Law and Economics, c/o University of Maastricht, Postbus 616, 6200 MD Maastricht
T: (043) 887060; Tx: 56726; Fax: (043) 256538
Founded: 1984; Members: 280
Gen Secr: Michael G. Faure
Focus: Econ; Law 08868

European Association of Petroleum Geoscientists and Engineers (EAPG), Utrechtseweg 62, Postbus 298, 3700 AG Zeist
T: (03404) 62655; Fax: (03404) 62640
Founded: 1989
Focus: Geology 08869

European Association of State Veterinary Officers (EASVO), c/o Veterinaire Inspectie van de Volksgezondheid, Wijnhaven 78, 3011 WT Rotterdam
Founded: 1980
Gen Secr: Dr. K. Minderhoud
Focus: Vet Med 08870

European Association of Teachers, Koningsholster 64, 6573 VV Beek-Ubbergen
T: (08895) 42854
Founded: 1956; Members: 25000
Gen Secr: Dr. Guus N.M. Wijngaards
Focus: Educ 08871

European Association of Veterinary Pharmacology and Toxicology (EAVPT), c/o Dept of Veterinary Pharmacology and Toxicology, Faculty of Veterinary Medicine, University of Utrecht, Postbus 80176, 3508 TD Utrecht
Focus: Vet Med 08872

European Bureau of Adult Education (EBAE), Troelstralaan 2, Postbus 367, 3800 AJ Amersfoort
T: (033) 654116
Founded: 1953; Members: 160
Pres: P. Federighi; Gen Secr: W. Bax
Focus: Adult Educ
Periodicals
Conference Reports (weekly)
Newsletter (weekly) 08873

European Calcified Tissue Society, c/o Dept of Oral Cell Biology, Van der Boechorststr 7, 1080 BT Amsterdam
Founded: 1981
Gen Secr: Prof. Dr. E.M. Burger
Focus: Cell Biol & Cancer Res 08874

European Cartographic Institute, Laan Copes Van Cattenburch 79, 2585 EW 's-Gravenhage
T: (02) 2161545; Fax: (02) 2163026
Pres: M. Plaizier
Focus: Cart 08875

European Centre for Development Policy Management (ECDPM), Onze Lieve Vrouweplein 21, 6211 HE Maastricht
T: (043) 255121; Tx: 56493; Fax: (043) 253636
Founded: 1986
Gen Secr: Dr. L. de la Rive
Focus: Poli Sci 08876

European Centre for Work and Society (ECWS), Hoogbrugstr 43, Postbus 3073, 6202 NB Maastricht
T: (02) 216724; Fax: (02) 255712
Founded: 1979
Gen Secr: J. Hillenius
Focus: Soc Sci 08877

European College of Gerodontology (ECG), c/o Dept of Oral Function, Dental School Nijmegen, Postbus 9101, 6500 HB Nijmegen
Founded: 1990
Gen Secr: Dr. C. de Baat
Focus: Dent 08878

European College of Obstetrics and Gynaecology (ECOG), c/o Office of Post Graduate Medical Education, Erasmus University Medical School, Postbus 1738, 3000 DR Rotterdam
T: (010) 4087880
Founded: 1992
Gen Secr: M. Wenckebach
Focus: Gynecology 08879

European Committee for the Advancement of Thermal Sciences and Heat Transfer (EUROTHERM), c/o Delft University of Technology, Postbus 5046, 2600 GA Delft
T: (015) 784104; Tx: 38151; Fax: (015) 622814
Founded: 1986
Pres: Prof. C.J. Hoogendoorn; Gen Secr: Prof. Keith Cornwell
Focus: Eng 08880

European Committee on Radiopharmaceuticals, c/o Dr. D. den Hoed Kliniek, Postbus 5201, 3008 AE Rotterdam
Founded: 1978
Pres: Prof. Dr. P.H. Cox
Focus: Pharmacol 08881

European Company Lawyers Association (ECLA), Prinses Beatrixlaan 5, Postbus 93093, 2509 AB 's-Gravenhage
T: (070) 3497397; Fax: (070) 3819508
Founded: 1990
Gen Secr: Anne B. Scheltema Beduin
Focus: Law 08882

European Development Education Curriculum Network (EDECN), Nassauplein 8, 1815 GM Alkmaar
T: (072) 118502; Fax: (072) 151221
Founded: 1982; Members: 48
Gen Secr: C.B. Zwaga
Focus: Educ 08883

European Endangered Species Programme (EEP), c/o Amsterdam Zoo, Postbus 20164, 1000 HD Amsterdam
T: (020) 6207476; Fax: (020) 6253931
Founded: 1985
Focus: Zoology 08884

European Environmental Research Organization (EERO), General Foulkesweg 70, Postbus 191, 6700 AD Wageningen
T: (08370) 84817; Fax: (08370) 84818
Founded: 1987; Members: 50
Gen Secr: Dr. John V. Lake
Focus: Ecology 08885

European Federation of Associations of Market Research Organizations (EFAMRO), c/o ESOMAR, J.J. Viottastr 29, 1071 JP Amsterdam
T: (020) 6642141; Tx: 18535; Fax: (020) 6642922
Founded: 1992; Members: 6
Gen Secr: Bryan A. Bates
Focus: Marketing 08886

European Federation of Branches of the World's Poultry Science Association, c/o Spelderholt Institute for Poultry Research, Beekbergen
T: 1808; Tx: 30757
Founded: 1960; Members: 3900
Gen Secr: D.A. Ehlhardt
Focus: Zoology
Periodical
World's Poultry Science Journal (3 times annually) 08887

European Federation of the Associations of Dieticians (EFAD), c/o Nederlandse Vereniging van Dietisten, Boterstr 1a, Postbus 341, 5340 AH Oss
T: (04120) 24543; Fax: (04120) 37736
Founded: 1978
Gen Secr: H.M. Van Oosten
Focus: Nutrition 08888

European Grassland Federation, c/o Netherlands Fertilizer Institute, Badhuisweg 139, 2597 JN 's-Gravenhage
T: (070) 525071
Founded: 1963; Members: 20
Focus: Agri
Periodical
Proceedings (bi-annually) 08889

European Packaging Federation (EPF), c/o Nederlands Verpakkingscentrum, Postbus 164, 2800 AD Gouda
Founded: 1953; Members: 14
Gen Secr: P.F.H. Janssen
Focus: Materials Sci 08890

European Photochemistry Association (EPA), c/o Laboratory of Organic Chemistry, University of Amsterdam, Nieuwe Achtergracht 129, 1018 WS Amsterdam
T: (020) 5255676; Fax: (020) 5255670
Founded: 1968; Members: 980
Pres: Prof. Dr. Jan W. Verhoeven
Focus: Chem
Periodical
EPA-Newsletter (weekly) 08891

European Society for Opinion and Marketing Research (ESOMAR) / Association Européenne pour les Etudes d'Opinion et de Marketing, J.J. Viottastr 29, 1071 JP Amsterdam
T: (020) 6642141; Tx: 18535; Fax: (020) 6642922
Founded: 1948
Pres: Mary Goodyear; Gen Secr: Bryan Bates
Focus: Marketing
Periodicals
Annual Congress Book of Papers (weekly)
The ESOMAR Directory (weekly)
Marketing and Research Today (weekly)
Newsbrief (weekly)
Seminar Books of Papers (8-10 times annually)
The Monograph Series (weekly) 08892

European Society of Cardiology (E.S.C.) / Société Européenne de Cardiologie, Postbus 23410, 3001 KK Rotterdam
T: (010) 186014145; Fax: (010) 186014258
Founded: 1950
Pres: M.E. Bertrand; Gen Secr: G. Breithardt
Focus: Cardiol
Periodical
European Heart Journal (weekly) 08893

Federatie van Organisaties van Bibliotheek-, Informatie-, Dokumentatiewezen (FOBID) / Federation of Library Information and Documentation Organisations, Taco Scheltemastr 5, Postbus 93166, 2509 AD 's-Gravenhage
Founded: 1975
Focus: Doc; Libraries & Bk Sci 08894

Fédération Internationale d'Information et de Documentation (FID) / International Federation for Information and Documentation, Prins Willem-Alexanderhof 5, 2595 BE 's-Gravenhage
T: (070) 3140671; Tx: 34402; Fax: (070) 3140667
Founded: 1895; Members: 425
Pres: Ritva Launo; Gen Secr: Ben G. Goedegebuure
Focus: Doc
Periodicals
Education and Training Programmes for Information Personnel: ET Newsletter (weekly)
FID Directory (biennial)
FID News Bulletin (weekly)
International Forum of Information and Documentation (weekly) 08895

FOM-Instituut voor Atoom- en Molecuulfysica / FOM Institute of Atomic and Molecular Physics, Kruislaan 407, Postbus 41883, 1009 DB Amsterdam
T: (020) 946711; Tx: 18109; Fax: (020) 6684106
Founded: 1953; Members: 200
Pres: Dr. K.H. Chang; Gen Secr: Prof. Dr. F.W. Saris
Focus: Nucl Res; Physics
Periodical
Annual Report (weekly) 08896

FOM-Instituut voor Plasmafysica Rijnhuizen / Institute of Plasmaphysics Rijnhuizen, Edisonbaan 14, Postbus 1207, 3430 BE Nieuwegein
Fax: (03402) 31204
Founded: 1959; Members: 120
Focus: Nucl Res
Periodical
Annual Status Report (weekly) 08897

Fries Genootschap van Geschied-, Oudheid- en Taalkunde, c/o Fries Museum, Turfmarkt 24, 8911 KT Leeuwarden
Founded: 1827; Members: 1800
Focus: Archeol; Hist
Periodicals
Fries Museum Bulletin (weekly)
De Vrije Fries (weekly) 08898

Fryske Akademy (FA) / Frisian Academy, Doelestrijtte 8, 8911 DX Leeuwarden
T: (058) 131414; Fax: (058) 131409
Founded: 1938; Members: 400
Focus: Ling; Hist; Soc Sci
Periodicals
It Beaken (weekly)
Ut de Smidte (weekly) 08899

Gelre / Vereniging tot Beoefening van Gelderse Geschiedenis, Oudheidkunde en Recht, Markt 1, 6811 CG Arnhem
T: (085) 599335
Founded: 1897; Members: 800
Focus: Hist
Periodical
Bijdragen en Mededelingen (weekly) . . 08900

Genootschap Amstelodamum, Singel 436h, 1017 AV Amsterdam
T: (020) 246740
Focus: Hist 08901

Genootschap Architectura et Amicitia, Waterlooplein 67, 1011 PB Amsterdam
T: (020) 220188
Founded: 1855; Members: 300
Focus: Archit 08902

Genootschap ter Bevordering van Melkkunde, Postbus 5806, 2280 HV Rijswijk
Founded: 1908; Members: 273
Focus: Dairy Sci
Periodical
Netherlands Milk and Dairy Journal (weekly) 08903

Genootschap ter Bevordering von Natuur-, Genees- en Heelkunde, Plantage Muidergracht 12, 1018 TV Amsterdam
T: (020) 22125
Founded: 1790; Members: 900
Focus: Med; Public Health 08904

Genootschap voor Wetenschappelijke Filosofie, c/o Dr. G. Corver, Fazantweg 12, 9765 JM Paterswolde
Founded: 1933; Members: 100
Focus: Philos 08905

Geschiedkundige Vereniging Die Haghe, Galvanistr 95, 2517 RB 's-Gravenhage
T: (070) 624562
Founded: 1890; Members: 1100
Focus: Hist
Periodical
Yearbook (weekly) 08906

Hague Academy of International Law (HAIL), Carnegieplein 2, 2517 KJ 's-Gravenhage
T: (070) 3469680
Founded: 1923
Focus: Law 08907

Historisch Genootschap De Maze, c/o Gemeentearchief, Robert Fruinstr 52, 3021 XE Rotterdam
T: (010) 775166
Founded: 1931; Members: 90
Focus: Hist 08908

Historisch Genootschap Roterodamum, 's-Landswerf 177, 3063 GE (010) Rotterdam
T: 127869
Focus: Hist 08909

Hollandsche Maatschappij der Wetenschappen (H.M.W.) / Dutch Society of Sciences, Postbus 9698, 2003 LR Haarlem
T: (023) 321773
Founded: 1752; Members: 450
Pres: M. Enschedé; Gen Secr: Prof. Dr. D.M. Schenkeveld; Prof. Dr. A.A. Verrijn Stuart
Focus: Sci
Periodical
Haarlemse Voordrachten (weekly) . . . 08910

Institute for Esperanto in Commerce and Industry (IECI), Postbus 72, Valkenswaard
Founded: 1963
Focus: Ling 08911

Instituut voor Mechanisatie, Arbeid en Gebouwen (IMAG) / Institute of Agricultural Engineering, Mansholtlaan 10-12, 6708 PA Wageningen
T: (08370) 76300; Fax: (08370) 25670
Founded: 1974; Members: 155
Focus: Agri
Periodicals
Annual Report (weekly)
IMAG-publikaties (8 times annually)
IMAG-Research Reports 08912

Instituut voor Onderwijskundige Dienstverlening, Postbus 9104, 6500 HE Nijmegen
T: (080) 612470; Fax: (080) 615551
Focus: Educ 08913

Instituut voor Perceptie Onderzoek (IPO) / Institute for Perception Research, Postbus 513, 5600 MB Eindhoven
Fax: (040) 773876
Founded: 1957; Members: 900
Focus: Biophys
Periodical
IPO Annual Progress Report (weekly) . . 08914

Instituut voor Toegepaste Sociale Wetenschapen (ITS) / Institute for Applied Social Sciences, Toernooiveld 5, Postbus 9048, 6500 KJ Nijmegen
T: (080) 653500; Fax: (080) 653599
Founded: 1965; Members: 140
Pres: A.J. Mens; Gen Secr: A.J. Buster
Focus: Soc Sci
Periodical
ITS-publikaties 08915

Internationaal Instituut voor Sociale Geschiedenis (Stichting) (IISG) / International Institute of Social History, Cruquiusweg 31, 1019 AT Amsterdam
T: (020) 6685866; Fax: (020) 6654181
Founded: 1935
Pres: H.M. van de Kar
Focus: Hist
Periodicals
Annual Report (weekly)
International Review of Social History (3 times annually) 08916

International Academy of Management (IAM), c/o CIOS, Van Alkemadelaan 700, Postbus 90730, 2509 LS 's-Gravenhage
Founded: 1958; Members: 180
Focus: Business Admin 08917

International Association for Hydraulic Research (IAHR), Rotterdamseweg 185, Postbus 177, 2600 MH Delft
T: (015) 569353; Tx: 38176; Fax: (015) 619674
Founded: 1935; Members: 2360
Gen Secr: H.J. Overbeek
Focus: Eng
Periodicals
IAHR Bulletin (5-6 times annually)
Journal of Hydraulic Research (5-6 times annually)
Proceedings of Biennial Congresses (bi-annually) 08918

International Association for Media in Science (IAMS), Postbus 80125, 3508 TC Utrecht
T: (030) 534500; Fax: (030) 515463
Pres: Dr. Jan T. Goldschmeding; Gen Secr: Werner Grosse
Focus: Mass Media
Periodical
Newsletter 08919

International Association for Official Statistics, Prinses Beatrixlaan 428, Postbus 950, 2270 AZ Voorburg
T: (070) 3375737; Fax: (070) 3860025
Founded: 1985; Members: 540
Pres: M. Snorrason; Gen Secr: Z.E. Kenessey
Focus: Stats 08920

International Association for Statistical Computing (IASC), Prinses Beatrixlaan 428, Postbus 950, 2270 AZ Voorburg
T: (070) 3357737; CA: Statist Voorburg;
Fax: (070) 3860025
Founded: 1977; Members: 775
Pres: N. Carlo Lauro; Gen Secr: Z.E. Kenessey
Focus: Stats; Computer & Info Sci
Periodical
CSOA: Computational Statistics and Data Analysis (3 times annually) 08921

The International Association for the Evaluation of Educational Achievement, Sweelinckplein 14, 2517 GK 's-Gravenhage
T: (070) 3469679; Fax: (070) 3609951
Founded: 1959
Pres: Tjeerd Plomp
Focus: Educ
Periodical
IEA Newsletter (weekly) 08922

International Association for the History of Glass, c/o Piet C. Ritsema van Eck, Rijksmuseum, Postbus 74888, 1070 DN Amsterdam
Founded: 1958; Members: 200
Focus: Hist 08923

International Association of Applied Psychology, Montessorilaan 3, 6525 HR Nijmegen
Founded: 1920; Members: 2122
Gen Secr: C.J. de Wolff
Focus: Psych
Periodicals
International Review of Applied Psychology (weekly)
Newsletter (bi-annually) 08924

International Association of Wood Anatomists (IAWA), c/o Institute of Systematic Botany, Heidelberglaan 2, 3584 CS Utrecht
T: (030) 532643; Fax: (030) 518061
Founded: 1931; Members: 500
Gen Secr: Ben J.H. ter Welle
Focus: Forestry
Periodical
IAWA Journal (weekly) 08925

International Bureau of Fiscal Documentation (IBFD), Sarphatistr 600, Postbus 20237, 1000 HE Amsterdam
T: (020) 6267726; Tx: 13217; CA: Forintax;
Fax: (020) 6228658
Founded: 1938
Pres: J.F. Avery Jones; Gen Secr: H.M.A.L. Hamaekers
Focus: Law; Doc
Periodicals
Bulletin for International Fiscal Documentation (weekly)
European Taxation (weekly)
International VAT Monitor (weekly) . . . 08926

International Colour Association, c/o Philips Lighting, Postbus 80020, 5600 JM Eindhoven
Founded: 1967; Members: 28
Gen Secr: Dr. C. van Triqt
Focus: Chem 08927

International Commission for the Nomenclature of Cultivated Plants (ICNCP), c/o Government Institute for Research on Varieties of Cultivated Plants, Dept for Horticultural Botany, Postbus 32, 6700 AA Wageningen
Founded: 1955; Members: 24
Gen Secr: F. Schneider
Focus: Botany 08928

International Commission for the Nomenclature of Cultivated Plants, c/o RIVRO, Postbus 32, 6700 AA Wageningen
Members: 24
Gen Secr: W.A. Brandenburg
Focus: Botany 08929

International Commitee for Histochemistry and Cytochemistry (ICHC), Wassenaarseweg 62, 2333 AL Leiden
Founded: 1960
Focus: Chem 08930

International Council for Children's Play (ICCP), c/o Institute for Special Education, State University Groningen, Oude Boteringestr 1, 9712 GA Groningen
T: (050) 114813
Founded: 1959; Members: 325
Pres: André Michelet; Gen Secr: Dr. Rimmert van der Kooij
Focus: Educ 08931

International Federation for Housing and Planning (IFHP), Wassenaarseweg 43, 2596 CG 's-Gravenhage
T: (070) 3244557; Fax: (070) 3282085
Founded: 1913; Members: 1200
Focus: Civil Eng; Urban Plan
Periodicals
Congress Papers Proceedings (weekly)
IFHP News Sheet (7 times annually) . . 08932

International Federation of Free Teachers' Unions (FFTU), Herengracht 54-56, 1015 BN Amsterdam
T: (020) 249072, 264633; Tx: 17118; Fax: (020) 274205
Founded: 1951; Members: 7100000
Gen Secr: Fred van Leeuwen
Focus: Educ
Periodicals
Information Bulletin
Newsletter (weekly)
TRI (weekly) 08933

International Federation of Library Associations and Institutions (IFLA), c/o Koninklijke Bibliotheek, Prins Willem Alexanderhof 5, Postbus 95312, 2509 CH 's-Gravenhage
T: (070) 3140884; Tx: 34402; Fax: (070) 3834827
Founded: 1927; Members: 1100
Pres: Robert Wedgeworth; Gen Secr: Leo Voogt
Focus: Libraries & Bk Sci
Periodicals
IFLA Annual (weekly)
IFLA Directory (biennial)
IFLA Journal (weekly)
IFLA Trends (bi-annually) 08934

International Federation of Ophthalmological Societies (IFOS), c/o Sint Radboudziekenhuis, Universiteit van Nijmegen, Postbus 9101, 6500 HB Nijmegen
T: (080) 613138; Fax: (080) 540522
Founded: 1962; Members: 60
Pres: Prof. A. Nakajima; Gen Secr: Prof. A.F. Deutman
Focus: Ophthal
Periodical
Proceedings (every 4 years) 08935

International Humanist and Ethical Union (IHEU), Nieuwegracht 69a, 3512 LG Utrecht
T: (030) 312155; Fax: (030) 364169
Founded: 1952
Focus: Philos
Periodical
International Humanist (weekly) 08936

International Huntington Association (IHA), c/o Gerrit Dommerholt, Callunahof 8, 7217 ST Harfsen
T: (05733) 1595; Fax: (070) 3500050
Founded: 1979; Members: 32
Pres: Gerrit Dommerholt; Gen Secr: Robyn Kapp
Focus: Med
Periodical
Newsletter (weekly) 08937

International Organisation for the Study of the Endurance of Wire Ropes, c/o Dept of Transport Technology, University of Technology Delft, Postbus 5034, 2600 GA Delft
T: (015) 782011; Tx: 38151; Fax: (015) 781397
Founded: 1963
Pres: Dr. L. Wiek
Focus: Materials Sci
Periodical
Bulletin (weekly) 08938

International Radiation Protection Association (IPRA), Postbus 662, 5600 MB Eindhoven
T: (040) 473355; Fax: (040) 435020
Founded: 1966; Members: 13000
Pres: Charles B. Meinhold; Gen Secr: Chris J. Huyskens
Focus: Radiology
Periodical
IRPA-Bulletin (weekly) 08939

International Society for Horticultural Science (I.S.H.S.), De Dreijen 6, 6703 BC Wageningen
T: (08370) 21747; Tx: 45760

Founded: 1959; Members: 3000
Focus: Hort
Periodicals
Acta Horticulturae
Chronica Horticulturae (weekly) 08940

International Society for Soilless Culture (ISOSC), Postbus 52, 6700 AB Wageningen
T: (08370) 13809; Fax: (08370) 23457
Founded: 1955; Members: 405
Pres: Rick S. Donnan; Gen Secr: Abram A. Steiner
Focus: Agri 08941

International Society of City and Regional Planners (ISoCaRP), Mauritskade 23, 2514 HD 's-Gravenhage
T: (070) 3462654; Fax: (070) 3617909
Founded: 1965; Members: 432
Gen Secr: Dr. Erica C. Poventrud-Boeke
Focus: Urban Plan
Periodical
The News Bulletin (weekly) 08942

International Society of Paediatric Oncology, Postbus 3283, 5203 DG 's-Hertogenbosch
T: (073) 429285; Fax: (073) 414766
Founded: 1969; Members: 560
Gen Secr: Prof. A.N. Cratt
Focus: Pediatrics; Cell Biol & Cancer Res
Periodical
Medical and Paediatric Oncology (8 times annually) 08943

International Statistical Institute (ISI), Prinses Beatrixlaan 428, Postbus 950, 2270 AZ Voorburg
T: (070) 3375737; CA: Statist Voorburg; Fax: (070) 3860025
Founded: 1885; Members: 1800
Pres: Prof. J.K. Ghosh; Gen Secr: Dr. Z.E. Kenessey
Focus: Stats
Periodicals
International Statistical Review (3 times annually)
Short Book Reviews (3 times annually)
Statistical Theory and Method Abstracts (weekly) 08944

International Technical and Scientific Organization for Soaring Flight (OSTIV), c/o DFVLR, Van Halewijnplein 37, 8031 Wessling
Founded: 1948
Focus: Eng; Sci 08945

Katholiek Documentatie Centrum (KDC), Erasmuslaan 36, 6521 MG Nijmegen
Fax: (080) 615944
Focus: Hist; Sociology; Rel & Theol
Periodicals
Bronnen en Studies (weekly)
Jaarboeken (weekly)
Scripta (weekly)
Sleutels (weekly) 08946

Koninklijke Maatschappij tot Bevordering der Bouwkunst, Bond van Nederlandse Architekten (B.N.A.), Keizersgracht 321, 1016 EE Amsterdam
T: (020) 5553666; Fax: (020) 5953699
Founded: 1842; Members: 2700
Pres: Prof. C. Weeber; Gen Secr: D. van der Veer
Focus: Archit
Periodical
Architectuur/Bouwen (weekly) 08947

Koninklijke Maatschappij Tuinbouw en Plantkunde (K.M.T.P.), Kwekerijweg 2, 2597 JK 's-Gravenhage
T: (070) 3514551; Fax: (070) 3522579
Founded: 1872; Members: 60000
Focus: Hort; Botany
Periodical
Groei & Bloei (weekly) 08948

Koninklijke Maatschappij voor Natuurkunde onder de Zinspreuk Diligentia, Lange Voorhout 5, 2514 EA 's-Gravenhage
Founded: 1793; Members: 530
Focus: Nat Sci
Periodical
Natuurkundige Voordrachten, Nieuwe Reeks (weekly) 08949

Koninklijke Nederlandsche Heidemaatschappij, Lovinklaan 6, 6021 HX Arnhem
T: (085) 455146; Fax: (085) 515235
Founded: 1888; Members: 2500
Pres: Dr. C.J. Rijnvos; Gen Secr: M.A.E. Chorus-Menken
Focus: Ecology
Periodical
Heidemijtijdschrift (weekly) 08950

Koninklijke Nederlandsche Maatschappij tot Bevordering der Geneeskunst (KNMG), Lomanlaan 103, 3526 XD Utrecht
T: (030) 885411
Founded: 1849; Members: 19000
Focus: Med 08951

Koninklijke Nederlandse Akademie van Wetenschappen / Royal Netherlands Academy of Arts and Sciences, Kloveniersburgwal 29, Postbus 19121, 1000 GC Amsterdam
T: (020) 5510700; Fax: (020) 6204941
Founded: 1808
Focus: Sci
Periodicals
Akademie Nieuws (5 times annually)
Medelingen: Nieuwe Reeks, Letterkunde
Proceedings: Series A: Mathematical Sciences; Series B: Palaeontology, Geology, Chemistry, Physics and Anthropology; Series C: Biological and Medical Sciences (weekly)
Verhandelingen: Nieuwe Reeks, Letterkunde
Verhandelingen, Series I, II: Transactions 08952

Koninklijke Nederlandse Bosbouw Vereniging, Postbus 139, 6800 AC Arnhem
T: (085) 454141
Founded: 1910; Members: 320
Focus: Forestry
Periodical
Nederlands Bosbouw Tijdschrift (weekly) . 08953

Koninklijke Nederlandse Botanische Vereniging (K.N.B.V.), Toernooiveld, 6525 ED Nijmegen
T: (080) 653380; Fax: (080) 553450
Founded: 1845; Members: 800
Focus: Botany
Periodicals
Acta Botanica Neerlandica (weekly)
Bionieuws (weekly) 08954

Koninklijke Nederlandse Chemische Vereniging (KNCV), Burnierstr 1, 2596 HV 's-Gravenhage
T: (070) 469406
Founded: 1903; Members: 8000
Focus: Chem 08955

Koninklijke Nederlandse Maatschappij ter Bevordering der Pharmacie / Royal Netherlands Association for Advancement of Pharmacy, Alexanderstr 11, 2514 JL 's-Gravenhage
T: (070) 624111
Founded: 1842; Members: 3100
Pres: Prof. Dr. T.F.J. Tromp; Gen Secr: A. van Zijl
Focus: Pharmacol
Periodical
Pharmaceutisch Weekblad 08956

Koninklijke Nederlandse Maatschappij voor Diergeneeskunde (K.N.M.v.D.), Julianalaan 10, 3581 NT Utrecht
T: (030) 510111; Fax: (030) 511707
Founded: 1862; Members: 3420
Pres: Prof. Dr. E.H. Kampelmacher; Gen Secr: Dr. T. Jorna
Focus: Vet Med
Periodicals
Tijdschrift voor Diergeneeskunde (23 times annually)
Veehouder en Dieren Arts (weekly)
Veterinary Quarterly (weekly) 08957

Koninklijke Nederlandse Natuurhistorische Vereniging (KNNV) / Vereniging voor Veldbiologie / Royal Dutch Society for Natural History, Oudegracht 237, 3511 NK Utrecht
T: (030) 314797
Founded: 1901; Members: 9500
Pres: J.W.G. Pfeiffer; Gen Secr: H.N. van der Voort
Focus: Bio; Nat Sci
Periodicals
Natura (11 times annually)
Wetenschappelijke Mededelingen (10 times annually) 08958

Koninklijke Nederlandse Toonkunstenaars-vereniging / Royal Netherlands Association of Musicians, Van Miereveldstraat 13, Amsterdam
Founded: 1875; Members: 2000
Gen Secr: Theo van Eijk
Focus: Music
Periodical
Muziek en Dans 08959

Koninklijke Nederlandse Vereniging voor Luchtvaart (KNVL), Jozef Israelsplein 8, 2596 AS 's-Gravenhage
T: (070) 245457
Founded: 1907; Members: 14000
Focus: Eng 08960

Koninklijke Vereniging van Leraren en Onderwijzers in de Lichamelijke Opvoeding, Zinzendorflaan 9, Postbus 398, 3700 AJ Zeist
T: (03404) 20847; Fax: (03404) 12810
Founded: 1862; Members: 8000
Focus: Sports
Periodical
Lichamelijke Opvoeding (18 times annually) 08961

Koninklijk Genootschap voor Landbouwwetenschap (KGvL) / Royal Netherlands Society for Agricultural Science, Postbus 79, 6700 AB Wageningen
T: (08370) 83537
Founded: 1886; Members: 6200
Pres: A.A. Jongebreur; Gen Secr: P.C. Kuijpers
Focus: Agri
Periodicals
Adressenlijst: Address-List of all Dutch Agricultural Alumni (weekly)
LT-Journal (weekly)
Netherlands Journal of Agricultural Science (weekly) 08962

Koninklijk Instituut van Ingenieurs (KIvI) / Royal Institution of Engineers in the Netherlands, Prinsessegracht 23, Postbus 30424, 2514 AP 's-Gravenhage
T: (070) 3919900; Tx: 33641; Fax: (070) 3919840
Founded: 1847; Members: 18000
Pres: J.M. Ossewaarde; Gen Secr: J.N.P. Haarsma
Focus: Eng
Periodicals
De Ingenieur (weekly)
De Ingenieurskrant (weekly)
Ingenieursnieuws (weekly)
Journal A: Automatic Control
Yearbook (bi-annually) 08963

Koninklijk Instituut voor Taal-, Land- en Volkenkunde (KITLV) / Royal Institute of Linguistics and Anthropology, Reuvensplaats 2, Postbus 9515, 2300 RA Leiden
T: (071) 272295; Fax: (071) 272638
Founded: 1851; Members: 2000
Gen Secr: Prof. Dr. P. Boomgaard
Focus: Anthro; Ling; Ethnology
Periodicals
Bijdragen tot de Taal-, Land- en Volkenkunde (3 times annually)
European Newsletter on Southeast Asian Studies
Excerpta Indonesica
New West Indian Guide 08964

Koninklijk Nederlands Aardrijkskundige Genootschap (K.N.A.G.) / Royal Dutch Geographical Society, Weteringschans 12, 1017 SG Amsterdam
T: (020) 277716
Founded: 1874; Members: 3100
Pres: Dr. H.B. Eenhoorn; Gen Secr: Dr. H. van Steijn
Focus: Geography
Periodicals
Geografisch Tijdschrift (5 times annually)
De Nieuwe Geografenkrant (10 times annually)
De Nieuwe Geografenkrant (10 times annually)
Tijdschrift voor Economische en Sociale Geografie: Journal of Economic and Social Geography (5 times annually) 08965

Koninklijk Nederlandsch Genootschap voor Geslacht- en Wapenkunde, Postbus 11755, 2502 AT 's-Gravenhage
T: (070) 814651
Members: 1400
Focus: Genealogy
Periodicals
De Nederlandsche Leeuw (weekly)
Nederlandse Genealogieen (weekly)
Series Publicaties (weekly)
Series Werken (weekly) 08966

Koninklijk Nederlands Geologisch Mijnbouwkundig Genootschap (KNGMG) / Royal Geological and Mining Society of the Netherlands, Postbus 157, 2000 AD Haarlem
Fax: (023) 312328
Founded: 1912; Members: 1650
Pres: Prof. Dr. S.B. Kroonenberg; Gen Secr: Dr. L.J. Witte
Focus: Mining; Geology
Periodicals
Geologie en Mijnbouw: International Journal of the Royal Geological and Mining Society of the Netherlands (weekly)
Nieuwsbrief (10 times annually)
Verhandelingen van het KNGMG (weekly) 08967

Koninklijk Oudheidkundig Genootschap Amsterdam (K.O.G.), Postbus 74888, 1070 DN Amsterdam
Founded: 1858; Members: 450
Pres: Prof. Dr. D.P. Blok
Focus: Arts; Cultur Hist
Periodical
Annual Reports (every 4-5 years) . . . 08968

Kristaina Esperantista Ligo Internacia (KELI) / International Christian Esperanto Association, Koningsmantel 4, 2403 HZ Alphen
Founded: 1911; Members: 1300
Pres: A. Burkhardt; Gen Secr: E.A. van Dijk
Focus: Ling 08969

Landelijke Huisartsen Vereniging, Lomanlaan 103, 3526 XD Utrecht
T: (030) 885411
Focus: Med 08970

Landelijke Specialisten Vereniging (LSV), Lomanlaan 103, Postbus 20057, 3502 LB Utrecht
T: (030) 823301
Founded: 1946; Members: 6200
Focus: Med 08971

Limburgs Geschied- en Oudheidkundig Genootschap (LGOG), Postbus 83, 6200 AB Maastricht
T: (043) 212586; Fax: (043) 218572
Founded: 1863; Members: 2900
Focus: Hist
Periodicals
Archeologie in Limburg
Limb. Tijdschrift voor Genealogie (weekly)
De Maasgouw (weekly)
Werken LGOG (weekly) 08972

Maatschappij Arti et Amicitiae, Rokin 112, 1012 LB Amsterdam
T: (020) 233508
Founded: 1839; Members: 1200
Focus: Arts 08973

Maatschappij der Nederlandse Letterkunde, c/o Universiteitsbibliotheek, Witte Singel 27, POB 9501, 2300 RA Leiden
Founded: 1766; Members: 1100
Focus: Lit; Ling; Hist
Periodical
Jaarboek van de Maatschappij der Nederlandse Letterkunde te Leiden (weekly) 08974

Maatschappij tot Bevordering der Toonkunst, Jacob van Campenstr 59, 1072 BD Amsterdam
T: (020) 713091
Founded: 1829; Members: 6000
Focus: Music 08975

Mijnbouwkundige Vereeniging, Mijnbouwstr 120, 2628 RX Delft
T: (015) 786039
Founded: 1892; Members: 1000
Focus: Mining
Periodical
Jaarboek 08976

Monumentenraad / Monuments and Historic Buildings Council, Broederplein 41, 3703 CD Zeist
Founded: 1961; Members: 40
Pres: Y.P.W. van der Werff; Gen Secr: F.J.L. van Dulm
Focus: Preserv Hist Monuments
Periodical
Kunstreisboek voor Nederland: Annual Report 08977

Nationale Vereniging voor Economisch Onderwijs, Kapelweidtje 9, 1861 JH Bergen
Focus: Educ; Econ 08978

Natuurhistorisch Genootschap in Limburg (NHG) / Society of Natural History in Limburg, Postbus 882, 6200 AW Maastricht
T: (043) 213671
Founded: 1910; Members: 1132
Pres: A.J.W. Lenders; Gen Secr: Dr. H. Schmitz
Focus: Hist; Nat Sci
Periodicals
Limburgse Vogels (weekly)
Natuurhistorisch Maandblad (weekly)
SOK-medelingen (weekly) 08979

Nederlands Bibliotheek en Lektuur Centrum / Dutch Centre for Public Libraries and Literature, Taco Scheltemastraat 5, Postbus 93054, 2509 AB 's-Gravenhage
T: (070) 3141500; Fax: (070) 3141600
Founded: 1972
Focus: Lit; Libraries & Bk Sci
Periodical
Bibliotheek en Samenleving 08980

Nederlands Bureau voor Bibliotheekwezen en Informatieverzorging (NBBI) / Netherlands Organization for Libraries and Information Services, Burg. Van Karnebeeklaan 19, 2585 BA 's-Gravenhage
T: (070) 3607833; Fax: (070) 3615011
Founded: 1987
Pres: Dr. T. van Kooten
Focus: Libraries & Bk Sci; Computer & Info Sci 08981

Nederlandsche Internisten Vereeniging (NIV), c/o Domus Medica, Lomanlaan 103, 3526 XD Utrecht
T: (030) 823380229; Fax: (030) 802290
Founded: 1931; Members: 1700
Focus: Intern Med
Periodicals
The Netherlands Journal of Medicine (weekly)
The Newsletter (weekly) 08982

Nederlandsche Maatschappij tot Bevordering der Tandheelkunde / Dutch Dental Organization, Postbus 2000, 3430 CA Nieuwegein
T: (03402) 76276; Tx: 26401; Fax: (03402) 48994
Founded: 1914; Members: 7200
Focus: Dent
Periodical
Nederlands Tandartsenblad: Dutch Dental Journal (weekly) 08983

Nederlandsche Vereniging voor Druk- en Boekkunst / Netherlands Society for the Art of Printing and Bookproduction, Bestevaerstr 10, 2014 AL Haarlem
Founded: 1938; Members: 300
Pres: Dr. C.J. Keyser; Gen Secr: F. Mayer
Focus: Graphic & Dec Arts, Design; Libraries & Bk Sci
Periodical
Mededelingen (weekly) 08984

Nederlandsche Vereniging voor Levensverzekeringgeneeskunde, c/o Nationale Nederlanden, Schiekade 130, 3032 AL Rotterdam
T: (010) 652700
Focus: Med 08985

Nederlandse Anatomen Vereniging, c/o Laboratorium voor Anatomie en Embryologie, Vrije Universiteit, Postbus 7161, 1007 MC Amsterdam
Focus: Anat 08986

Nederlandse Astronomenclub, c/o Dr. J.B.G.M. Bloemen, Leiden Observatory, Postbus, 2300 RA Leiden
T: (071) 275818
Focus: Astronomy 08987

Nederlandse Bond voor Natuurgeneeswijze, Frans van Mierisstr 57, 1071 RL Amsterdam
T: (020) 723552
Focus: Med 08988

Nederlands Economisch Instituut / Netherlands Economic Institute, K.P. van der Mandelelaan 11, Postbus 4175, 3006 AD Rotterdam 3016
T: (010) 4538800; Fax: (010) 4530768
Focus: Econ
Periodical
Economisch Statistische Berichten (weekly) 08989

Nederlandse Dendrologische Vereniging, Vogellaan 6, 2771 JX Boskoop
T: 4266
Focus: Forestry 08990

Nederlandse Dierkundige Vereniging (NDV), c/o Vakgroep Experimentale Diermorfologie en Celbiologie, Marijkeweg 40, 6709 PG Wageningen
Fax: (08370) 83962
Founded: 1872; Members: 800
Focus: Zoology
Periodical
Netherlands Journal of Zoology (weekly) . 08991

Nederlandse Entomologische Vereniging (NEV), Plantage Middenlaan 64, 1018 DH Amsterdam
Founded: 1845; Members: 600
Focus: Entomology
Periodicals
Entomologia Experimentalis et Applicata (weekly)
Entomologische Berichte (weekly)
Monografieën van de Nederlandse Entomologische Vereniging (weekly)
Tijdschrift voor Entomologie (weekly) . . 08992

Nederlandse Genealogische Vereniging, Postbus 976, 1000 AZ Amsterdam
T: (020) 766780
Focus: Genealogy
Periodical
Gens Nostra (weekly) 08993

Nederlandse Genetische Vereniging, c/o Clusius Laboratorium, Wassenaarseweg 64, 2333 AL Leiden
Founded: 1915; Members: 250
Focus: Genetics 08994

Nederlandse Museumvereniging (NMV), Wierdijk 18, 1601 LA Enkhuizen
T: (02280) 5692
Founded: 1926; Members: 700
Focus: Arts 08995

Nederlandse Mycologische Vereniging (N.M.V.), c/o Biologisch Station, Kampsweg 27, 9418 PD Wyster
T: (05936) 2441
Founded: 1908; Members: 500
Pres: Dr. E.J.M. Arnolds; Gen Secr: M.T. Veerkamp
Focus: Botany, Specific
Periodical
Coolia: Journal (weekly) 08996

Nederlandse Organisatie voor Internationale Samenwerking in het Hoger Onderwijs (NUFFIG) / Netherlands Organisation for International Cooperation in Higher Education, Kortenaerkade 11, Postbus 29777, 2502 LT 's-Gravenhage
T: (070) 4260260; Tx: 33565; CA: NUFFIC The Hague; Fax: (070) 510513
Founded: 1952
Focus: Educ 08997

Nederlandse Organisatie voor Wetenschappelijk Onderzoek / Netherlands Organization for Scientific Research, Laan van Nieuw Oost Indië 131, Postbus 93138, 2509 AC 's-Gravenhage
Gen Secr: W. Hutter
Focus: Humanities; Soc Sci; Bio; Eng; Med; Nat Sci
Periodical
Jaarboek: Annual Report (weekly) . . . 08998

Nederlandse Ornithologische Unie (N.O.U.), Boomgaardweg 44, 3984 KK Odijk
Founded: 1901; Members: 950
Focus: Ornithology
Periodicals
Ardea (weekly)
Limosa (weekly) 08999

Nederlandse Patholoog Anatomen Vereniging (NPAV), c/o Prof. Dr. J.W. Arends, Afd Pathologie, Academisch Ziekenhuis, Postbus 1910, 6201 BX Maastricht
Founded: 1920; Members: 200
Focus: Anat; Pathology 09000

Nederlandse Sint-Gregorius Vereniging ter Bevordering van Liturgische Muziek, Oudwijk 21, 3581 TG Utrecht
T: (030) 513739
Focus: Music
Periodical
Gregoriusblad (weekly) 09001

Nederlandse Toonkunstenaarsraad / Council of Organisations of Musicians in the Netherlands, Valeriusplein 20, Amsterdam
Founded: 1948
Focus: Music
Periodical
A Musical Guide for Holland 09002

Nederlandse Tuinbouwraad, Schiefbaanstr 29, 2596 RC 's-Gravenhage
T: (070) 450600; Tx: 32185; Fax: (070) 453902
Founded: 1908; Members: 9
Focus: Hort 09003

Nederlandse Vereniging van Artsen voor Revalidatie en Physische Geneeskunde (VRA), Van Swietenlaan 4, 9700 RM Groningen
T: (050) 266810
Founded: 1955; Members: 230
Pres: Dr. H. Berghauser; Gen Secr: Dr. A. Coster
Focus: Rehabil; Physical Therapy; Physiology 09004

Nederlandse Vereniging van Bibliothecarissen, Documentalisten en Literatuuronderzoekers, Nolweg 13d, 4209 AW Schelluinen
Founded: 1912
Pres: Dr. G.A.J.S. van Marle; Gen Secr: H. J. Krikke-Scholten
Focus: Libraries & Bk Sci; Doc; Lit . . 09005

Nederlandse Vereniging van Gieterijtechnici (NVvGT), c/o AVNEG, Bredewater 20, Postbus 190, 2700 AD Zoetermeer
T: 214516
Founded: 1927; Members: 260
Focus: Metallurgy
Periodical
Gietwerk Perspektief (weekly) 09006

Nederlandse Vereniging van Laboratoriumartsen, c/o Laboratorium voor de Volksgezondheid in Friesland, Postbus 21020, 8900 JA Leeuwarden
Gen Secr: Dr. J.E. Degener
Focus: Med 09007

Nederlandse Vereniging van Neurochirurgen (NVvN), c/o Neurochirurgische Afdeling, Academisch Ziekenhuis Rotterdam, Dr. Molewaterplein 40, 3015 GD Rotterdam
Fax: (010) 4633735
Founded: 1952; Members: 87
Focus: Surgery
Periodical
Clinical Neurology and Neurosurgery: CNN (weekly) 09008

Nederlandse Vereniging van Pedagogen, Onderwijskundigen en Andragologen / Dutch Society of Educational Psychologists, Korte Elisabethstr 11, 3511 JG Utrecht
Founded: 1962
Focus: Educ
Periodical
Nederlands Tijdschrift voor Opvoeding, Vorming en Onderwijs 09009

Nederlandse Vereniging van Radiologische Laboranten (NVRL), Catharijnesingel 73, Utrecht
T: (030) 318842
Founded: 1950; Members: 2400
Focus: Radiology; Nucl Med
Periodical
Gamma (weekly) 09010

Nederlandse Vereniging van Specialisten in den Dento-Maxillaire Orthopaedie, Weezenhof 14-16, Nijmegen
Founded: 1963
Focus: Orthopedics; Dent 09011

Nederlandse Vereniging van Tandartsen (NVvT), Langegracht 39, 3601 AJ Maarssen
T: (03465) 64511
Founded: 1904; Members: 2350
Focus: Dent 09012

Nederlandse Vereniging van Wiskundeleraren (NVvW), Traviatastr 132, 2555 VJ 's-Gravenhage
T: (070) 3687998
Founded: 1925; Members: 3400
Pres: Dr. J. van Lint; Gen Secr: Dr. J.W. Maassen
Focus: Educ; Math
Periodical
Euclides (10 times annually) 09013

Nederlandse Vereniging voor Afvalwaterbehandeling en Waterkwaliteitsbeheer, Postbus 70, 2280 AB Rijswijk
T: (070) 902720
Founded: 1958; Members: 1900
Focus: Water Res
Periodical
H20-Journal (weekly) 09014

Nederlandse Vereniging voor Algemene Gezondheidszorg (NVAG), Postbus 70032, 3000 LP Rotterdam
T: (010) 4339615
Founded: 1980; Members: 300
Focus: Public Health; Hygiene
Periodical
NVAG-Bulletin (weekly) 09015

Nederlandse Vereniging voor Anesthesiologie (NVA), Lomanlaan 103, Postbus 20063, 3502 HB Utrecht
T: (030) 823387; Fax: (030) 101053
Founded: 1948; Members: 702
Pres: Prof. Dr. J.W. van Kleef; Gen Secr: J.M. Bongertman-Diek
Focus: Anesthetics
Periodical
Nederlands Tijdschrift voor Anesthesiologie MTVA (weekly) 09016

Nederlandse Vereniging voor Cardiologie, c/o Dept of Cardiology, Sint Antonius Hospital, Koekoekslaan 1, 3435 CM Nieuwegein
T: (03402) 47669; Fax: (03402) 34420
Focus: Cardiol
Periodical
Netherlands Journal of Cardiology (weekly) 09017

Nederlandse Vereniging voor Gastro-Enterologie (NVGE), Postbus 657, 2003 RR Haarlem
Founded: 1913; Members: 725
Focus: Gastroenter 09018

Nederlandse Vereniging voor Geodesie (NVG), Waltersingel 1, 7314 NK Apeldoorn
T: (055) 285111
Founded: 1970; Members: 500
Focus: Surveying
Periodical
NGT Geodesia (weekly) 09019

Nederlandse Vereniging voor Heelkunde, Lomanlaan 103, 3526 XD Utrecht
T: (030) 823327; Fax: (030) 823329
Focus: Surgery
Periodical
The European Journal of Surgery . . . 09020

Nederlandse Vereniging voor Internationaal Recht (NVIR) / Netherlands Branch of the International Law Association, Alexanderstr 20-22, Postbus 30461, 2500 GL 's-Gravenhage
T: (070) 3420300; Fax: (070) 3420359
Founded: 1910; Members: 570
Pres: A.H.A. Soons; Gen Secr: E.N. Frohn
Focus: Law
Periodical
Mededelingen (weekly) 09021

Nederlandse Vereniging voor Kindergeneeskunde, c/o Pediatric Dept, Free University Hospital, De Boelelan 1117, 1081 HV Amsterdam
T: (020) 5482395
Focus: Pediatrics 09022

Nederlandse Vereniging voor Logica en Wijsbegeerte der Exacte Wetenschappen, c/o Dept of Philosophy, Heidelberglaan 2, 3508 TC Utrecht
T: (030) 532173
Founded: 1947; Members: 175
Focus: Philos 09023

Nederlandse Vereniging voor Luchtvaarttechniek (NVvL), Anthony Fokkerweg 2, 1059 CM Amsterdam
T: (020) 5113113; Fax: (020) 178024
Founded: 1941; Members: 330
Gen Secr: F.J. Sterk
Focus: Aero
Periodical
NVvL Jaarboek (weekly) 09024

Nederlandse Vereniging voor Management (NIVE), Neuhuyskade 40, Postbus 90730, 2509 LS 's-Gravenhage
T: (070) 180180; Tx: 32626
Founded: 1925; Members: 2515
Focus: Business Admin 09025

Nederlandse Vereniging voor Medisch Onderwijs / Dutch Association for Medical Education, Universiteitsweg 100, Postbus 80030, 3508 TA Utrecht
T: (030) 538344
Focus: Med; Educ
Periodical
Bulletin: Officieel Blad (3 times annually) 09026

Nederlandse Vereniging voor Microbiologie (NVvM) / Netherlands Society for Microbiology, c/o R.I.V.M., Postbus 1, 3720 BA Bilthoven
T: (030) 7433068; Fax: (030) 293651
Founded: 1911; Members: 1100
Pres: Prof. Dr. J. Verhoef; Gen Secr: Dr. A.M. Breure
Focus: Microbio
Periodical
Bionieuws (20 times annually) 09027

Nederlandse Vereniging voor Mondziekten en Kaakchirurgie, Sportlaan 600, 2566 TP 's-Gravenhage
Founded: 1956; Members: 300
Focus: Otorhinolaryngology; Surgery . . 09028

Nederlandse Vereniging voor Neurologie, Postbus 20050, 3502 LB Utrecht
T: (030) 823343
Founded: 1895; Members: 750
Pres: Dr. J.J. van der Sande; Gen Secr: G. Meeuwissen-Schrijver
Focus: Neurology
Periodical
Clinical Neurology and Neurosurgery (weekly) 09029

Nederlandse Vereniging voor Obstetrie en Gynaecologie, Lomanlaan 103, 3526 XD Utrecht
T: (030) 823328; Fax: (030) 823329
Focus: Gynecology
Periodical
Nederlands Tijdschrift voor Obstetrie en Gynaecologie 09030

Nederlandse Vereniging voor Orthodontische Studie (NVOS), Schelluinsevliet 5, 4203 NB Gorinchem
Fax: (01830) 33177
Members: 750
Focus: Dent 09031

Nederlandse Vereniging voor Parasitologie, c/o Dept of Medical Parasitology, Postbus, 6500 HB Nijmegen
Founded: 1962; Members: 150
Focus: Microbio 09032

Nederlandse Vereniging voor Pathologie, c/o Dr. R.W.M. Giard, Afd Pathologie, St. Clara Ziekenhuis, Olympiaweg 350, 3078 HT Rotterdam
Focus: Pathology 09033

Nederlandse Vereniging voor Personeelbeleid (NVP), Catharijnesingel 53, Postbus 19124, 3501 DC Utrecht
T: (030) 367101; Fax: (030) 343991
Focus: Business Admin
Periodical
Personeelbeleid (weekly) 09034

Nederlandse Vereniging voor Produktieleiding (NPL), Van Alkemadelaan 700, 90730, 2509 LS 's-Gravenhage
T: (070) 264341
Founded: 1970
Focus: Business Admin
Periodical
Bedrijfsvoering (weekly) 09035

Nederlandse Vereniging voor Psychiatrie, Lomanlaan 103, Postbus 20062, 3502 LB Utrecht
T: (030) 823303; Fax: (030) 888400
Founded: 1871; Members: 1714
Pres: Prof. Dr. F.G. Zitman; Gen Secr: S. Maurer
Focus: Psychiatry 09036

Nederlandse Vereniging voor Psychotherapie, Koningslaan 12, 3583 GC Utrecht
Fax: (030) 5228660508
Founded: 1930
Pres: Prof. Dr. R.E. Abraham
Focus: Psych; Therapeutics 09037

Nederlandse Vereniging voor Thoraxchirurgie, c/o Dept of Cardiothoracic Surgery, University Hospital Groningen, Postbus 30001, 9700 RB Groningen
Focus: Surgery 09038

Nederlandse Vereniging voor Toegepaste Taalwetenschap, c/o Instituut voor Toegepaste Taalkunde, Rijksuniversiteit, Wilhelminapark 11, 3581 NC Utrecht
Founded: 1972; Members: 400
Focus: Ling
Periodical
Toegepaste Taalwetenschap: Publications in Applied Linguistics (weekly) 09039

Nederlandse Vereniging voor Tropische Geneeskunde (N.V.T.G.), Postbus 96984, 2509 JJ 's-Gravenhage
T: (070) 3241860; Fax: (070) 3241860
Founded: 1907; Members: 919
Focus: Trop Med
Periodical
Medicus Tropicus 09040

Nederlandse Vereniging voor Urologie, Lomanlaan 103, 3526 XD Utrecht
T: (030) 823328; Fax: (030) 823329
Founded: 1908
Focus: Urology 09041

Nederlandse Vereniging voor Veiligheidskunde (NVVK) / Netherlands Society on Safety Engineering, c/o GAK, Afd AB, Postbus 8300, 1005 CA Amsterdam
T: (020) 455351
Founded: 1947; Members: 950
Pres: P.J.A. van Buchem; Gen Secr: E. van Ree
Focus: Safety
Periodical
NVVK-Nieuws (5 times annually) 09042

Nederlandse Vereniging voor Zeegeschiedenis, c/o Dr. J.P. Sigmond, Leeuerikstr 14, 2333 VZ Leiden
Founded: 1961; Members: 487
Focus: Hist; Transport 09043

Nederlandse Vereniging vor Dermatologie en Venereologie, Nieuwe Ginnekenstr 14, 4811 NR Breda
T: (076) 137596
Founded: 1896; Members: 450
Focus: Derm; Venereology 09044

Nederlandse Werkgroep van Praktizijns in de Natuurlijke Geneeskunst (NWP), Ruige Velddreef 133, 3831 PG Leusden
T: (033) 953133; Fax: (033) 953133
Focus: Med 09045

Nederlandse Zootechnische Vereniging / Dutch Association for Animal Production, Deylerweg 99, 2241 AC Wassenaar
T: (01751) 12617
Founded: 1931; Members: 250
Pres: D. Oostendorp; Gen Secr: J.J. Bakker
Focus: Zoology 09046

Nederlands Filosofisch Genootschap, P.J. Oudlaan 2, 3705 VP Zeist
T: (03404) 15987
Focus: Philos 09047

Nederlands Genootschap van Leraren (NGL), Postbus 407, 3300 AK Dordrecht
T: (078) 131611; Fax: (070) 130178
Founded: 1972; Members: 20000
Focus: Educ
Periodical
NGL blad (weekly) 09048

Nederlands Genootschap voor Anthropologie, Linnaeusstr 2a, Amsterdam
Focus: Anthro 09049

Nederlands Genootschap voor Fysiotherapie (N.G.v.F.), Postbus 240, 3800 AE Amersfoort
T: (033) 622400; Fax: (033) 616462
Founded: 1889; Members: 14000
Focus: Therapeutics
Periodicals
Fysiovisie (11 times annually)
Nederlands Tijdschrift voor Fysiotherapie (11 times annually)
Nederlands Tijdschrift voor Manuele Therapie (weekly) 09050

Nederlands Historisch Genootschap (N.H.G.), Postbus 90406, 2509 LK 's-Gravenhage
Founded: 1845; Members: 1700
Pres: Prof. Dr. C. Fassem; Gen Secr: Dr. G.N. van der Plaat
Focus: Hist
Periodicals
Bijdragen en Mededelingen betreffende de Geschiedenis der Nederlanden
Nederlandse Historische Bronnen
Werken NHG 09051

Nederlands Huisartsen Genootschap (N.H.G.), Lomanlaan 103, 3526 XD Utrecht
T: (030) 881700; Fax: (030) 870668
Founded: 1956; Members: 4600
Pres: F.J.M. König; Gen Secr: J.P.M. van der Voort
Focus: Med
Periodicals
Huisarts en Wetenschap (13 times annually)
NHG-Standards for GP's (8 times annually)
. . . . 09052

Nederlands Instituut voor Internationale Betrekkingen Clingendael / Netherlands Institute of International Relations Clingendael, Clingendael 7, Postbus 93080, 2597 VH 's-Gravenhage
T: (070) 3245384; Fax: (070) 3282002
Founded: 1983; Members: 900
Focus: Int'l Relat
Periodical
Internationale Spectator (weekly) 09053

Nederlands Instituut voor Marketing (NIMA) / Netherlands Institute of Marketing, Hogehilweg 8, Postbus 7352, 1007 JJ Amsterdam
T: (020) 6974821; Fax: (020) 6913971
Founded: 1966; Members: 3200
Focus: Marketing
Periodical
Tijdschrift voor Marketing (weekly) . . . 09054

Nederlands Oogheelkundig Gezelschap (NOG), c/o Prof. Dr. F. Hendrikse, Postbus 1041, 6201 BA Maastricht
T: (04406) 42808
Founded: 1892; Members: 70 50
Focus: Ophthal 09055

Nederlands Psychoanalytisch Genootschap (N.P.G.), Maliestr 1a, 3581 SH Utrecht
T: (030) 333300
Founded: 1947; Members: 170
Pres: J.N. Voorhoeve; Gen Secr: J.H. Scheffer
Focus: Psychoan
Periodical
Psychoanalytisch Forum (weekly) 09056

Nederlands-Zuidafrikaanse Vereniging / Netherlands South African Society, Keizersgracht 141, 1015 CK Amsterdam
Fax: (020) 6382596
Founded: 1881; Members: 1000
Pres: Prof. Dr. G.J. Schulte; Gen Secr: F.H.H. Pols
Focus: Int'l Relat
Periodicals
Annual Report (weekly)
Zuid-Afrika (weekly) 09057

Nederlanedse Orthopaedische Vereniging (N.O.V.), Postbus 9011, 6500 GM Nijmegen
T: (080) 659911
Founded: 1898
Focus: Orthopedics 09058

Netherlands Centre of the International PEN, Rogneurdonk 17, 6218 HG Maastricht
T: (043) 433498; Fax: (043) 433498
Founded: 1923; Members: 305
Pres: Dr. Rudolf Geel; Gen Secr: Hans van de Waarsenburg
Focus: Lit
Periodical
PEN-Kwartaal (weekly) 09059

Permanent International Committee of Linguists, c/o Instituut voor Nederlandse Lexicologie, Witte Singel, Matthias de Vrieshof 2-3, 2311 BZ Leiden
Founded: 1928; Members: 50
Pres: Prof. Dr. R.H. Robins; Gen Secr: Prof. Dr. P.G.J. van Sterkenburg
Focus: Ling
Periodical
Linguistique Bibliographie 09060

Pharma-Dokumentationsring e.V., c/o Organon International B.V., POB 20, 5340 BH Oss
T: (04120) 62409
Focus: Pharmacol 09061

Protestants-Christelijke Onderwijsvakorganisatie (PCO), Postbus 87868, 2508 DG 's-Gravenhage
T: (070) 3522541; Fax: (070) 3522841
Focus: Educ
Periodical
Magazine (weekly) 09062

Raad voor de Kunst, R.J. Schimmelpennincklaan 3, 2517 JN 's-Gravenhage
T: (070) 3469619; Fax: (070) 3614727
Founded: 1947; Members: 60
Pres: J. de Ruiter; Gen Secr: A. Nicolai
Focus: Arts
Periodicals
Knipselkrant (weekly)
Recommendations (weekly) 09063

Raad voor het Cultuurbeheer / Cultural Heritage Council, Sir Winston Churchillaan 362, Postbus 5406, 2280 HK Rijswijk
Fax: (070) 3406237
Founded: 1961; Members: 40
Pres: C.G. Goekoop; Gen Secr: Dr. C.W.M. Hendriks
Focus: Preserv Hist Monuments 09064

Research Group for European Migration Problems (REMP), Pauwenlaan 17, 2566 TA 's-Gravenhage
T: (070) 647784
Founded: 1952; Members: 178
Focus: Sociology 09065

Rijksbureau voor Kunsthistorische Documentatie (R.K.D.) / Netherlands Institute for Art History, Prins Willem-Alexanderhof 5, Postbus 90418, 2509 LK 's-Gravenhage
T: (070) 3471514; Fax: (070) 3475005
Founded: 1932
Pres: R.E.O. Ekkart; Gen Secr: S.M.L. Vader
Focus: Arts; Hist; Doc
Periodical
Periodical Oud Holland (weekly) 09066

Samenwerkingsverband van de Universiteitsbibliotheken, de Koninklijke Bibliotheek en de Bibliotheek van de Koninklijke Nederlandse Akademie van Wetenschappen (UKB), c/o Bibliotheek der Vrije Universiteit, De Boelelaan 1103, 1081 HV Amsterdam
T: (020) 5483690
Founded: 1977; Members: 15
Pres: N.P. van den Berg; Gen Secr: J.H. de Swart
Focus: Libraries & Bk Sci 09067

Sectie Operationele Research (SOR), De Lairessestr 111-115, 1075 HH Amsterdam
Founded: 1959; Members: 464
Focus: Computer & Info Sci 09068

Société Universitaire Européenne de Recherches Financières (SUERF), Warandelaan 2, 5037 AB Tilburg
T: (013) 662435
Founded: 1963
Pres: Christian de Boissieu; Gen Secr: Jacques Sijben
Focus: Finance
Periodicals
SUERF Papers on Monetary Policy and Financial Systems
SUERF Reprints
SUERF Series
SUERF Translations 09069

Stichting Centrale Raad voor de Academies van Bouwkunst / Central Council of the Academies for Architecture, Keizersgracht 321, 1016 EE Amsterdam
Focus: Archit; Urban Plan 09070

Stichting Economisch Instituut voor de Bouwnijverheid / Economic Institute for the Building Industry, De Cuserstr 89, 1081 CN Amsterdam
Founded: 1956
Focus: Civil Eng
Periodical
Bouw Werk: De Bouw in Feiten, Cijfers en Analyses (weekly) 09071

Stichting Koninklijk Zoölogisch Genootschap Natura Artis Magistra / Royal Zoological Society, Plantage Kerklaan 40, Postbus 20164, 1000 HD Amsterdam
Fax: (020) 5233409
Founded: 1838
Focus: Zoology
Periodical
Dieren 09072

Stichting Natuur en Milieu / The Netherlands Society for Nature and Environment, Donkerstr 17, 3511 KB Utrecht
T: (030) 331328; Fax: (030) 331311
Founded: 1972; Members: 7000
Focus: Ecology
Periodical
Natuur en Milieu (weekly) 09073

Stichting Nederlands Agronomisch-Historisch Instituut / Netherlands Agronomic-Historical Foundation, Oude Kijk in 't Jatstr 26, 9712 EK Groningen
Founded: 1949
Focus: Agri; Hist 09074

Stichting Nederlands Agronomisch-Historisch Instituut / Netherlands Agronomic-Historical Foundation, Vismarkt 40, Groningen
Founded: 1949
Pres: R.J. Clevering; Gen Secr: A.F. Stroink
Focus: Agri; Hist
Periodicals
Historia Agriculturae
Historia Agriculturae 09075

Stichting Verenigd Nederlands Filminstituut / Netherlands Film and TV Academy, Steynlaan 8, Postbus 515, Hilversum
Founded: 1948
Focus: Cinema
Periodicals
Groepsmedia (weekly)
Skoop (weekly) 09076

Stichting voor de Technische Wetenschappen (STW) / Technology Foundation, Van Vollenhovenlaan 661, Postbus 3021, 3502 GA Utrecht
T: (030) 923211; Tx: 26401; Fax: (030) 961536
Founded: 1981
Pres: Prof. Dr. A. Verruijt; Gen Secr: Dr. C. le Pair
Focus: Eng
Periodical
Annual Report (weekly) 09077

Stichting voor Fundamenteel Onderzoek der Materie (FOM) / Foundation for Fundamental Research on Matter, Van Vollenhovenlaan 659, Postbus 3021, 3502 GA Utrecht
T: (030) 923211; Tx: 40445; Fax: (030) 946099
Founded: 1946
Pres: Prof. Dr. J.J.M. Beenakker; Gen Secr: Dr. K.H. Chang
Focus: Physics 09078

Stichting voor Wetenschappelijk Onderzoek van de Tropen (WOTRO) / Foundation for the Advancement of Tropical Research, Laan van Nieuw Oost Indië 131, Postbus 93138, 2509 AC 's-Gravenhage
Founded: 1964
Gen Secr: R.R. van Kessel-Hagesteyn
Focus: Develop Areas
Periodical
Annual Report (weekly) 09079

Studievereniging voor Psychical Research (SPR), Eemwijkplein 16, 2271 RA Voorburg
T: (070) 863732
Founded: 1919; Members: 600
Focus: Psych 09080

Theater Instituut Nederland, Herengracht 168, 1016 BP Amsterdam
T: (020) 6235104
Founded: 1978
Focus: Perf Arts
Periodicals
Dansjaarboek (weekly)
Theaterjaarboek (weekly)
Toneel Theatraal 09081

Unitas Malacologica, c/o Nationaal Natuurhistorisch Museum, Postbus 9517, 2300 RA Leiden
T: (071) 133344
Founded: 1962; Members: 400
Focus: Zoology 09082

Universal Esperanto Association (UEA), Nieuwe Binnenweg 176, 3015 BJ Rotterdam
T: (010) 4361044; Fax: (010) 4361751
Founded: 1908; Members: 26700
Pres: Prof. John C. Wells; Gen Secr: Ian Jackson
Focus: Ling
Periodicals
Esperanto (weekly)
Kontakto (weekly) 09083

Vereniging Het Nederlandsch Economisch-Historisch Archief, Cruquiusweg 31, 1019 AT Amsterdam
T: (020) 6685866; Fax: (020) 6654181
Founded: 1914
Focus: Hist; Econ
Periodicals
Economisch- en sociaal-historisch Jaarboek (weekly)
NEHA-Bulletin (weekly) 09084

Vereniging Het Nederlands Kanker Instituut / Netherlands Cancer Institute, Plesmanlaan 121, 1066 CX Amsterdam
T: (020) 5129111; Fax: (020) 172625
Founded: 1913
Pres: J.D. Hooglandt; Gen Secr: P. den Tex
Focus: Cell Biol & Cancer Res; Biochem; Genetics; Immunology
Periodical
Annual Report (weekly) 09085

Vereniging Natuurmonumenten, Noordereinde 60, 1243 JJ 's-Graveland
T: (035) 62004
Founded: 1905; Members: 250000
Pres: Dr. P. Winsemius; Gen Secr: P.J. van Herwerden
Focus: Ecology
Periodical
Natuurbehoud (weekly) 09086

Vereniging tot Bevordering der Homoeopathie in Nederland, Postbus 82027, 2500 NA 's-Gravenhage
Focus: Homeopathy 09087

Vereniging van Archivarissen in Nederland (VAN), Postbus 897, 8901 Leeuwarden
Founded: 1891; Members: 750
Focus: Archives
Periodicals
Nederlands Archievenblad (weekly)
Nieuws van Archieven (weekly) 09088

Vereniging van Docenten in Geschiedenis en Staatsrichting in Nederland (VGN), Patrijslaan 12a, 2261 ED Leidschendam
T: (070) 3277923; Fax: (070) 3853669
Founded: 1958; Members: 2500
Focus: Hist; Adult Educ
Periodical
Kleio (10 times annually) 09089

Vereniging van Homoeopathische Artsen in Nederland, Eykmanlaan 8, 3571 JS Utrecht
T: (030) 711606
Focus: Homeopathy 09090

Vereniging van Katholieke Leraren Sint-Bonaventura (V.K.L.), Reviusstr 68, 7552 GL Hengelo
Founded: 1918; Members: 5000
Focus: Educ 09091

Vereniging van Medische Analisten (V.v.M.A.), Wilhelminapark 52, 3581 NM Utrecht
T: (030) 523792
Founded: 1946; Members: 4500
Focus: Med
Periodical
Analyse (weekly) 09092

Vereniging van Nederlandse Kunsthistorici (VNK), Oude Boteringestr 81, 9712 GG Groningen
Members: 481
Focus: Arts; Hist 09093

Vereniging voor Agrarisch Recht (VAR) / Agricultural Law Association, Hollandseweg 1, 6706 KN Wageningen
Fax: (08370) 4763
Founded: 1959; Members: 400
Focus: Law 09094

Vereniging voor Arbeidsrecht, Postbus 132, 3440 AC Woerden
T: (03480) 21752; Fax: (03480) 18460
Founded: 1946; Members: 650
Focus: Law
Periodical
Geschriften van de Vereniging voor Arbeidsrecht (weekly) 09095

Vereniging voor Bouwrecht, Wassenaarseweg 23, 2596 CE 's-Gravenhage
T: (070) 3245544
Focus: Law 09096

Vereniging voor Calvinistische Wijsbegeerte (V.v.C.W.), c/o Centrale Interfaculteit, Vrije Universiteit, Amsterdam
Founded: 1935; Members: 600
Focus: Philos
Periodicals
Beweging (weekly)
Philosophia Reformata (weekly) 09097

Vereniging voor de Staathuishoudkunde, Achterom 98, 2611 PS Delft
Focus: Poli Sci 09098

Vereniging voor Filosofie-Onderwijs, César Franckrode 10, 2717 BD Zoetermeer
T: 215359
Members: 250
Focus: Philos; Educ 09099

Vereniging voor het Theologisch Bibliothecariaat (VTB), Postbus 289, 6500 AG Nijmegen
T: (080) 515478, 512162, 228467
Founded: 1947; Members: 120
Focus: Libraries & Bk Sci
Periodicals
Bibliografie Doctorale Scripties Theologie (weekly)
Mededelingen van de VTB (weekly) . . 09100

Vereniging voor Hoger Beroepsonderwijs, Europaboulevard 23, 1079 PC Amsterdam
T: (020) 429333
Focus: Adult Educ 09101

Vereniging voor Nederlandse Muziekgeschiedenis (VNM), Postbus 1514, 3500 BM Utrecht
T: (030) 735004
Founded: 1868; Members: 570
Focus: Hist; Music
Periodical
Tijdschrift van de Vereniging voor Nederlandse Muziekgeschiedenis (weekly) 09102

Vereniging voor Penningkunst, Van der Meystr 1, 1815 GP Alkmaar
T: (072) 120041
Founded: 1925; Members: 700
Focus: Numismatics
Periodical
De Beeldenaar (weekly) 09103

Vereniging voor Statistiek (V.V.S.), Postbus 282, 1850 AG Heiloo
T: (072) 330973; Fax: (072) 333372
Founded: 1945; Members: 1260
Pres: Prof. Dr. J. Wessels; Gen Secr: Dr. A.W. Ambergen
Focus: Stats
Periodicals
Kwantitatieve Methoden (3 times annually)
Statistica Neerlandica (weekly)
V.V.S.-Bulletin (10 times annually) . . . 09104

Vereniging voor Wijsbegeerte te s'-Gravenhage, c/o C. Meeuwstra, Leuvensestr 23, 2587 GA 's-Gravenhage
Founded: 1907; Members: 100
Focus: Philos 09105

Vereniging voor Wijsbegeerte van het Recht / Netherlands Association for the Philosophy of Law, G. van Amstellaan 17, 1181 EJ Amstelveen
T: (020) 6901333
Founded: 1918; Members: 400
Focus: Law; Philos
Periodicals
Journal for Legal Philosophy and Jurisprudence (3 times annually)
Rechtsfilosofie en rechtstheorie (3 times annually) 09106

Vereniging voor Zuivelindustrie en Melkhygiëne (VVZM), Laan van Meerdervoort 18-20, 2517 AK 's-Gravenhage
T: (070) 634936
Focus: Food 09107

Volkenrechtelijk Instituut / International Law Institute, Utrecht Univèrsity, Janskerkhof 3, Utrecht
Founded: 1955
Focus: Law
Periodical
Nova et Vetera Iuris Gentium 09108

Wagnervereeniging / Wagner Society, Gabriël Metsustr 32, Amsterdam
Founded: 1883
Focus: Music 09109

Wiskundig Genootschap (W.G.), Postbus 4079, 1009 AB Amsterdam
Founded: 1778; Members: 1250
Pres: Prof. Dr. E. Thomas; Gen Secr: Dr. R.W. Goldbach
Focus: Math
Periodicals
Mededelingen van het Wiskundig Genootschap (weekly)
Nieuw Archief voor Wiskunde (weekly) . 09110

World Federation of Neurosurgical Societies (WFNS), Bergweg 12, 6523 MD Nijmegen
T: (080) 231146
Founded: 1955; Members: 52
Pres: Dr. K. Clark; Gen Secr: Prof. Dr. H.A.D. Walder
Focus: Surgery 09111

Netherlands Antilles

Caribbean Association on Mental Retardation and other Developmental Disabilities (CAMRODD), POB 3688, Jong Bloed 171
T: (09) 78824; Fax: (09) 676540
Founded: 1972
Pres: Aminta Costa Gomez
Focus: Rehabil
Periodical
CAMRODD Bulletin (weekly) 09112

New Caledonia

Council of Pacific Arts, c/o SPC, Anse Vata, BP D5, Nouméa Cédex
T: 262000; Tx: 3139; Fax: 263818
Focus: Arts 09113

Société des Etudes Mélanésiennes, Nouméa
T: 272342
Founded: 1938
Focus: Ethnology
Periodical
Etudes Mélanésiennes (weekly) 09114

New Zealand

Agronomy Society of New Zealand, c/o Plant Science Dept, Lincoln University, POB 84, Canterbury
Members: 295
Gen Secr: Dr. B.A. McKenzie
Focus: Agri 09115

Art Galleries and Museums Association of New Zealand (AGMANZ), c/o Museum of New Zealand, POB 467, Wellington
T: (04) 3859609; Fax: (04) 3857157
Founded: 1947; Members: 400
Pres: Sherry Reynolds
Focus: Arts; Archives
Periodicals
AGMANZ News (weekly)
Newsletter (weekly)
New Zealand Museum Journal (weekly) . 09116

Asian Peace Research Association (APRA), c/o Dept of Sociology, University of Canterbury, Christchurch
T: (03) 642982; Fax: (03) 642999
Founded: 1980
Gen Secr: Dr. Fevin Clements
Focus: Prom Peace 09117

Association for Engineering Education in Southeast Asia and the Pacific (AEESEAP), c/o Faculty of Engineering, University of Canterbury, Private Bag, Christchurch
T: (03) 667001; Fax: (03) 642705
Founded: 1973; Members: 210
Focus: Educ; Eng
Periodicals
AEESEAP Journal of Engineering Education (weekly)
AEESEAP Newsletter (weekly) 09118

Auckland Medical Research Foundation, POB 7151, Auckland
T: (09) 372886
Founded: 1956
Pres: Sir Henry Cooper; Gen Secr: Leslie Corkery
Focus: Med
Periodical
Annual Report (weekly) 09119

Australian Association of Clinical Biochemists – New Zealand Branch, c/o Hamilton Medical Laboratory, POB 52, Hamilton
Tx: 8380594
Founded: 1967; Members: 150
Gen Secr: G. Scheurich
Focus: Biochem
Periodical
Newsletter (weekly) 09120

Canterbury Medical Research Foundation, POB 2682, Christchurch
Gen Secr: G.R. Wood
Focus: Med 09121

Clean Air Society of Australia and New Zealand (N.Z. Branch), POB 27116, Wellington
Members: 86
Gen Secr: R.C. Pilgrim
Focus: Ecology
Periodical
Clean Air (weekly) 09122

Commonwealth Heraldry Board, POB 23-056, Papatoetoe, Auckland
T: 2787415
Founded: 1969
Pres: Dr. A.E. Tonson
Focus: Cultur Hist
Periodical
Commonwealth Heraldry Bulletin (weekly) 09123

Entomological Society of New Zealand, c/o S. Millar, 8 Maymorn Rd, Te Marua, Upper Hutt
Founded: 1951; Members: 262
Gen Secr: S. Millar
Focus: Entomology
Periodicals
New Zealand Entomologist (weekly)
Weta (weekly) 09124

Foundation for Research, Science and Technology, POB 12240, Wellington
T: (04) 731862; Fax: (04) 731841
Founded: 1989
Focus: Sci
Periodical
Annual Report (weekly) 09125

Geological Society of New Zealand (GSNZ), c/o New Zealand Geological Survey, POB 30368, Lower Hutt
T: (04) 699059; Fax: (04) 691479
Founded: 1955; Members: 820
Pres: R.B. Stewart; Gen Secr: J. Palmer
Focus: Geology
Periodical
Newsletter (weekly) 09126

Hawke's Bay Medical Research Foundation, POB 596, Napier
Focus: Med 09127

Institute of Energy (New Zealand Section), c/o Dr. E.R. Palmer, 7 Ngahere St, Stokes Valley, Wellington
Members: 96
Focus: Energy 09128

Institution of Professional Engineers New Zealand, 101 Molesworth St, POB 12241, Wellington
Founded: 1914; Members: 6000
Pres: Dr. R.J. Aspen; Gen Secr: A.T. Mitchell
Focus: Eng
Periodical
New Zealand Engineering (weekly) . . . 09129

Medical Research Council of New Zealand (MRCNZ), Wellesley St, POB 5541, Auckland
T: (09) 798227
Founded: 1950; Members: 14
Pres: B.H. Smith; Gen Secr: Dr. J.V. Hodge
Focus: Med
Periodical
Research Review (weekly) 09130

Meteorological Society of New Zealand, POB 6523, Wellington
T: (04) 4729379
Founded: 1979; Members: 410
Gen Secr: Dr. M.J. Salinger
Focus: Geophys
Periodical
Newsletter (weekly) 09131

New Zealand Academy of Fine Arts, Buckle St, Private Bag, Wellington
T: (04) 859267, 844911
Founded: 1882; Members: 2000
Focus: Arts 09132

New Zealand Archaeological Association (N.Z.A.A.), POB 6337, Dunedin North
Founded: 1957; Members: 500
Pres: R. Hooker; Gen Secr: M. White
Focus: Archeol
Periodicals
Archaeology in New Zealand (weekly)
New Zealand Journal of Archaeology (weekly) 09133

New Zealand Association of Clinical Biochemists, c/o Wellington Hospital, POB 47, Wellington
Members: 150
Pres: Dr. C.J. Lovell-Smith; Gen Secr: Dr. P.L. Hurst
Focus: Biochem
Periodical
Newsletter (weekly) 09134

New Zealand Association of Scientists, POB 1874, Wellington
Founded: 1940
Pres: David Penny; Gen Secr: Pat Kelly
Focus: Sci
Periodical
New Zealand Science Review (weekly) . 09135

New Zealand Association of Soil Conservators, POB 204, Blenheim
Members: 262
Gen Secr: C.G. Tozer
Focus: Ecology 09136

New Zealand Biochemical Society, c/o Dept of Biochemistry and Microbiology, Lincoln University, POB 84, Canterbury
Fax: (03) 3253851
Members: 200
Pres: Prof. C.C. Winterbourn; Gen Secr: Dr. M.F. Barnes
Focus: Biochem 09137

New Zealand Book Council, Book House, Boulcott St, POB 11377, Wellington
T: (04) 4991569
Founded: 1972; Members: 1180
Pres: Fiona Kidman; Gen Secr: Philippa Christmas
Focus: Libraries & Bk Sci 09138

New Zealand Cartographic Society, POB 12454, Thorndon
Founded: 1971; Members: 220
Pres: R.B. Phillips; Gen Secr: D. Harvey
Focus: Cart
Periodicals
Cartogram Newsletter (weekly)
New Zealand Cartographic Journal (weekly) 09139

New Zealand Computer Society, POB 10044, Wellington
T: (04) 4731043; Fax: (04) 4731025
Founded: 1960; Members: 2000
Gen Secr: R.A. Henry
Focus: Electronic Eng
Periodical
Journal of Computing (weekly) 09140

New Zealand Council for Educational Research, POB 3237, Wellington
Founded: 1933; Members: 40
Focus: Educ
Periodicals
Annual Report (weekly)
Newsletter (weekly)
Set: Research Information for Teachers (weekly) 09141

New Zealand Dairy Technology Society, c/o Dairy Science Section, Anchor Products Ltd, Waitoa
T: (07) 8893989
Members: 250
Gen Secr: H. Singh
Focus: Eng 09142

New Zealand Dietetic Association, POB 5065, Wellington
Members: 372
Gen Secr: N. Goodman
Focus: Nutrition
Periodical
Journal (weekly) 09143

New Zealand Ecological Society, POB 25178, Christchurch
Members: 474
Gen Secr: Caroline Mason
Focus: Ecology 09144

New Zealand Electronics Institute (NZEI), POB 755, Auckland
Founded: 1946; Members: 419
Gen Secr: D.R. Muir
Focus: Electronic Eng
Periodicals
Newelectronics Journal (weekly)
NZEI Ralph Slade Memorial Lecture (weekly) 09145

New Zealand Fertiliser Manufacturers Research Association (NZFMRA), 61 Otara Rd, POB 23637, Auckland
T: (09) 2747184
Founded: 1947; Members: 27
Focus: Chem
Periodicals
NZFMRA Annual Report (weekly)
Proceedings, NZFMRA Symposia (weekly)
Proceedings, NZFMRA Technical Conferences (weekly) 09146

New Zealand Genetical Society, c/o Crop Research Div, DSIR, Private Bag, Christchurch
T: (03) 252511; Tx: 4703; Fax: (03) 252074
Founded: 1949; Members: 131
Pres: Dr. D.W.R. White; Gen Secr: Dr. A.J. Conner
Focus: Genetics 09147

New Zealand Geographical Society, c/o Dept of Geography, University of Canterbury, Private Bag, Christchurch
Founded: 1944; Members: 1200
Pres: Dr. Mary Watson; Gen Secr: I.F. Owens
Focus: Geography
Periodicals
New Zealand Geographer (weekly)
New Zealand Journal of Geography (weekly)
Proceedings (weekly) 09148

New Zealand Geophysical Society, POB 1320, Wellington
T: (04) 4738208; Fax: (04) 4710977
Founded: 1980; Members: 140
Pres: D.J. Darby; Gen Secr: T.H. Webb
Focus: Geophys
Periodical
Newsletter (3 times annually) 09149

New Zealand Historical Association, c/o History Dept, Canterbury University, Christchurch
T: (03) 677001; Tx: 61109; Fax: (03) 642999
Founded: 1979; Members: 300
Pres: Ann Parsonson; Gen Secr: Luke Trainor
Focus: Hist
Periodical
Newsletter (weekly) 09150

New Zealand Historic Places Trust, POB 2629, Wellington
T: (04) 4724341; Fax: (04) 4990669
Founded: 1954; Members: 23500
Pres: Dr. Tim Beaglehde; Geoffrey Whitehead
Focus: Hist 09151

New Zealand Hydrological Society (NZHS), POB 12300, Wellington
T: (04) 3517099; Fax: (04) 3517091
Founded: 1961; Members: 500
Pres: M.P. Mosley; Gen Secr: L. Rowe
Focus: Water Res
Periodical
Journal of Hydrology (New Zealand) (weekly) 09152

New Zealand Institute of Agricultural Science (NZIAS), POB 19560, Christchurch
T: (03) 3842432; Fax: (03) 3842432
Founded: 1954; Members: 1007
Gen Secr: R. Sheppard
Focus: Agri
Periodicals
Bulletin of Agricultural and Horticultural Sciences (weekly)
New Zealand Agricultural Science (weekly) 09153

New Zealand Institute of Architects, Greenock House, 102-112 Lambton Quay, POB 438, Wellington
T: (04) 4735346
Founded: 1905
Pres: J.W. Harrison; Gen Secr: A.K. Purdie
Focus: Archit 09154

New Zealand Institute of Chemistry, POB 12347, Wellington
T: (04) 4739444; Fax: (04) 4732324
Founded: 1931; Members: 1500
Pres: D.S. Winter; Gen Secr: A.A. Turner
Focus: Chem
Periodical
Chemistry in New Zealand (weekly) . . 09155

New Zealand Institute of Food Science and Technology (N.Z.I.F.S.T.), POB 15052, Christchurch
T: (03) 3889269
Founded: 1965; Members: 700
Pres: Dennis Thomas
Focus: Eng; Food
Periodical
Food Technologist (weekly) 09156

New Zealand Institute of Forestry, POB 19840, Christchurch
T: (03) 3942432; Fax: (03) 3942432
Founded: 1926; Members: 800
Pres: P.F. Olsen; Gen Secr: R. Van Rossen
Focus: Forestry
Periodical
New Zealand Forestry (weekly) 09157

New Zealand Institute of International Affairs (N.Z.I.I.A.), c/o Victoria University of Wellington, POB 600, Wellington
T: (04) 727430; Fax: (04) 731261
Founded: 1934; Members: 600
Pres: Sir Wallace Rowling; Gen Secr: D.G. Holborow
Focus: Poli Sci
Periodical
New Zealand International Review (weekly) . . . 09158

The New Zealand Institute of Management (NZIM), 101 Molesworth St, Wellington
Founded: 1945; Members: 5980
Focus: Business Admin . . . 09159

New Zealand Institute of Physics (NZIP), c/o Dept of Physics, University of Auckland, Private Bag, Auckland
Founded: 1982; Members: 360
Gen Secr: Dr. G.D. Putt
Focus: Physics . . . 09160

New Zealand Institute of Surveyors, POB 831, Wellington
T: (04) 4711774; Fax: (04) 4711907
Founded: 1888; Members: 1300
Pres: A.G. Blaikie
Focus: Eng
Periodical
Journal (weekly) . . . 09161

New Zealand Law Society, 26 Waring Taylor St, POB 5041, Wellington
T: (04) 4727837; Fax: (04) 4737909
Founded: 1869; Members: 6069
Gen Secr: A.D. Ritchie
Focus: Law
Periodical
Law Talk (weekly) . . . 09162

New Zealand Leather and Shoe Research Association (LASRA), Dairy Farm Rd, Private Bag, Palmerston North
T: (063) 82108; Fax: (063) 81185
Founded: 1928; Members: 162
Pres: N. Dobson; Gen Secr: A. Passman
Focus: Materials Sci; Microbio
Periodicals
Annual Conference of Fellmongers and Hide Processors
Annual Conference of Leather Technicians
Annual Report (weekly) . . . 09163

New Zealand Library Association (NZLA), 20 Brandon St, POB 12212, Wellington
T: (04) 735834; Fax: (04) 4991480
Founded: 1910; Members: 1700
Pres: S. Sutherland
Focus: Libraries & Bk Sci
Periodicals
Library Life (weekly)
New Zealand Libraries (weekly) . . . 09164

New Zealand Limnological Society, POB 6016, Rotorua
Founded: 1968; Members: 200
Gen Secr: Dr. K.W. Rowe
Focus: Hydrology
Periodical
Newsletter (weekly) . . . 09165

New Zealand Maori Arts and Crafts Institute, POB 334, Rotorua
T: (073) 489047; Fax: (073) 489045
Founded: 1963
Pres: Dr. J.E.H. Marsh
Focus: Arts
Periodical
Whaka . . . 09166

New Zealand Marine Sciences Society, POB 297, Wellington
T: (04) 3861029; Fax: (04) 3861299
Founded: 1960; Members: 317
Gen Secr: Dr. S. Hanchet
Focus: Oceanography
Periodical
Newsletter (weekly) . . . 09167

New Zealand Mathematical Society, c/o Dept of Mathematics and Statistics, Otago University, POB 56, Dunedin
T: (04) 4797758; Fax: (04) 4798427
Founded: 1972; Members: 200
Pres: Prof. D.A. Holton; Gen Secr: Dr. R. Aldred
Focus: Math
Periodical
Newsletter (3 times annually) . . . 09168

New Zealand Medical Association, POB 156, Wellington
T: (04) 4724741; Fax: (04) 4710838
Founded: 1887; Members: 4500
Pres: Dr. A. Scott; Gen Secr: P. Faulkner
Focus: Med
Periodical
The New Zealand Medical Journal (weekly) . . . 09169

New Zealand Microbiological Society, c/o Dept of Oral Biology and Oral Pathology, University of Otago, POB 647, Dunedin
T: (03) 4797076; Fax: (03) 4790673
Founded: 1956; Members: 460
Pres: Dr. A.J. Robinson; Gen Secr: Dr. H.F. Jenkinson
Focus: Microbio
Periodical
Newsletter (weekly) . . . 09170

The New Zealand National Society for Earthquake Engineering, POB 17268, Wellington
T: (04) 4766866; Fax: (04) 4766866
Founded: 1968; Members: 700
Pres: P.J. North; Gen Secr: M.D. Brice
Focus: Civil Eng
Periodical
Bulletin (weekly) . . . 09171

New Zealand Pottery and Ceramics Research Association (PACRA), 2 Bell Rd, Gracefield, Lower Hutt
T: 666919
Founded: 1947; Members: 7
Focus: Materials Sci . . . 09172

New Zealand Psychological Society, POB 4092, Wellington
Founded: 1962; Members: 800
Gen Secr: L. Crowther
Focus: Psych
Periodicals
Bulletin (weekly)
NZ Journal of Psychology (weekly) . . . 09173

New Zealand Society for Electron Microscopy, c/o Dept of Anatomy, School of Medicine, University of Auckland, Private Bag, Auckland
Members: 80
Gen Secr: H. Holloway
Focus: Optics; Electronic Eng
Periodical
E.M. News (weekly) . . . 09174

New Zealand Society for Horticultural Science, POB 8264, Havelock North
Members: 201
Gen Secr: J.A. Geelen
Focus: Hort . . . 09175

New Zealand Society for Parasitology, c/o Ruakura Agricultural Centre, Private Bag, Hamilton
T: (07) 8385558
Founded: 1972; Members: 100
Pres: Dr. G.W. Yeates; Gen Secr: B.C. Hosking
Focus: Microbio; Zoology; Immunology; Entomology
Periodical
Proceedings (weekly) . . . 09176

New Zealand Society of Animal Production, c/o Ruakura Agricultural Centre, Private Bag, Hamilton
T: (071) 8562839; Fax: (071) 385658
Members: 746
Pres: D. Wallace; Gen Secr: P. Kilgarriff
Focus: Agri
Periodical
Proceedings of the New Zealand Society of Animal Production (weekly) . . . 09177

New Zealand Society of Dairy Science and Technology (NZSDST), c/o Technical Services, Bay Milk Products Ltd, Private Bag, Edgecumbe
T: (076) 49011
Founded: 1963; Members: 285
Gen Secr: W. French
Focus: Food; Eng . . . 09178

New Zealand Society of Plant Physiologists, c/o National Institute of Horticultural Products, Private Bag, Palmerston North
Members: 90
Gen Secr: D.S. Bertaud
Focus: Botany . . . 09179

New Zealand Society of Soil Science (NZSSS), c/o Dept of Soil Science, Lincoln University, POB 84, Canterbury
T: (03) 252811; Fax: (03) 252994
Founded: 1952; Members: 300
Pres: H.K.J. Powell; Gen Secr: R.G. McLaren
Focus: Agri
Periodical
Soil News (weekly) . . . 09180

New Zealand Statistical Association, POB 1731, Wellington
T: (04) 4954685
Founded: 1950; Members: 290
Pres: C.J. Thompson; Gen Secr: A.G. Gray
Focus: Stats
Periodicals
Newsletter (weekly)
Statistician (2-3 times annually) . . . 09181

New Zealand Veterinary Association (N.Z.V.A.), POB 27499, Wellington
T: (04) 843632
Founded: 1923; Members: 1150
Pres: Jim Edwards; Gen Secr: Bob Duckworth
Focus: Vet Med
Periodical
New Zealand Veterinary Journal (weekly) 09182

Nutrition Society of New Zealand, c/o Dept of Human Nutrition, University of Otago, POB 56, Dunedin
Members: 200
Gen Secr: Christine Thomson
Focus: Food
Periodical
Proceedings of thr Nutrition Society of New Zealand (weekly) . . . 09183

Operational Research Society of New Zealand, POB 904, Wellington
T: (04) 121855
Founded: 1964; Members: 170
Gen Secr: G. Eng
Focus: Computer & Info Sci
Periodical
New Zealand Operational Research (weekly) . . . 09184

Ornithological Society of New Zealand (OSNZ), POB 12397, Wellington
Founded: 1939; Members: 1150
Pres: Brian D. Bell; Gen Secr: R.A. Empson
Focus: Ornithology
Periodical
OSNZ News (weekly) . . . 09185

Palmerston North Medical Research Foundation, POB 648, Palmerston North
Founded: 1959
Gen Secr: John Forsythe
Focus: Med
Periodical
Annual Report (weekly) . . . 09186

P.E.N. New Zealand, 631 Birkenhead, POB 34, Auckland
Founded: 1934; Members: 650
Focus: Lit
Periodical
The New Zealand Author (weekly) . . . 09187

Physiological Society of New Zealand, c/o Dept of Physiology and Anatomy, Massey University, Palmerston North
T: (063) 69079
Founded: 1972; Members: 174
Gen Secr: Dr. R.J. Pack
Focus: Physiology
Periodical
Proceedings of the Physiological Society of New Zealand (weekly) . . . 09188

Polynesian Society, c/o Anthropology Dept, University of Auckland, Private Bag, Auckland
T: (09) 737999; Tx: 21480; Fax: (09) 3033429
Founded: 1892; Members: 1100
Pres: Prof. Bruce Biggs; Gen Secr: Prof. J. Huntsman
Focus: Ethnology; Anthro; Hist; Antique
Periodicals
Journal of Polynesian Society
Memoirs (weekly) . . . 09189

Population Association of New Zealand, POB 225, Wellington
T: (04) 4716146; Fax: (04) 4714412
Founded: 1974; Members: 132
Gen Secr: N.J. Pole
Focus: Soc Sci
Periodical
New Zealand Population Review (weekly) 09190

Queen Elizabeth II Arts Council of New Zealand, POB 3806, Wellington
T: (04) 4730880; Fax: (04) 4712865
Founded: 1963; Members: 13
Pres: Jenny Pattrick; Gen Secr: Peter Quin
Focus: Arts
Periodical
Arts Fines (weekly) . . . 09191

Ross Dependency Research Committee, c/o Dept of Scientific and Industrial Research, Private Bag, Wellington 1
T: (04) 729979
Pres: Dr. T. Hatherton; Gen Secr: P. Coulson
Focus: Geography . . . 09192

Royal Aeronautical Society, New Zealand Division, c/o BP Oil Nz Ltd, POB 892, Wellington
Founded: 1945; Members: 230
Focus: Aero . . . 09193

Royal Agricultural Society of New Zealand, POB 3095, Wellington
T: (04) 4724190; Fax: (04) 4712278
Founded: 1924
Pres: W.C. Scott; Gen Secr: H.R. Holbrook
Focus: Agri
Periodical
Bulletin (3 times annually) . . . 09194

Royal Astronomical Society of New Zealand (R.A.S.N.Z.), POB 3181, Wellington
Founded: 1920; Members: 347
Pres: Dr. P. Cottrell; Gen Secr: G. Whiteford
Focus: Astronomy
Periodicals
Newsletter (weekly)
Southern Stars (weekly) . . . 09195

Royal Society of New Zealand (RSNZ), POB 598, Wellington
T: (04) 4727421; CA: Royalsoc; Fax: (04) 4731841
Founded: 1867; Members: 17000
Pres: Prof. J.N. Dodd; Gen Secr: V.R. Moore
Focus: Sci
Periodicals
Bulletin (weekly)
Journal (weekly)
Proceedings (weekly) . . . 09196

Systematics Association of New Zealand, c/o Dept of Botany, University of Otago, Dunedin
Members: 180
Gen Secr: G.S. Ridley
Focus: Botany . . . 09197

Veterinary Services Council, POB 417, Wellington
Founded: 1946
Focus: Vet Med . . . 09198

Waikato Geological and Lapidary Society, POB 62, Hamilton
Founded: 1966
Pres: G.T. Matthews; Gen Secr: R. Ray
Focus: Geology . . . 09199

Wellington Medical Research Foundation, c/o The Secretary, POB 14240, Wellington
T: (04) 888179
Focus: Med . . . 09200

Nicaragua

Academia Nacional de Filosofía, Managua
Founded: 1964
Focus: Philos . . . 09201

Academia Nicaragüense de la Lengua, Apdo 2711, Managua
Founded: 1928; Members: 13
Pres: Pablo Antonio Cuadra; Gen Secr: Julio Ycaza Tigerino
Focus: Ling . . . 09202

Asociación de Bibliotecas Universitarias y Especializadas de Nicaragua (ABUEN), Apdo 68, León
Focus: Libraries & Bk Sci . . . 09203

Asociación Nicaragüense de Bibliotecarios (ASNIBI), c/o Biblioteca Nacional, Calle del Triunfo 302, Apdo 101, Managua
Focus: Libraries & Bk Sci . . . 09204

Sociedad de Oftalmología Nicaragüense, c/o Clínica Especializada, Managua
Founded: 1949
Focus: Ophthal . . . 09205

Sociedad Nicaragüense de Psiquiatría y Psicología, c/o Centro México, Managua
Founded: 1962
Focus: Psychiatry; Psych . . . 09206

Niger

Centre for Linguistic and Historical Studies by Oral Tradition, BP 878, Niamey
T: 735414; Tx: 5422
Founded: 1968
Gen Secr: D. Laya
Focus: Ling; Hist . . . 09207

Centre Régional de Formation et d'Application en Agrométéorologie et Hydrologie Opérationnelle, BP 11011, Niamey
T: 733987; Tx: 5448; Fax: 732435
Founded: 1974; Members: 9
Focus: Geophys; Hydrology . . . 09208

Nigeria

Accrediting Council for Theological Education in Africa (ACTEA), PMB 2049, Kaduna
Founded: 1976
Pres: Dr. Tite Tienou; Gen Secr: George Foxall
Focus: Educ
Periodicals
ACTEA Bulletin
ACTEA Tools and Studies
ASTIN Bulletin (weekly)
TEE Newsletter . . . 09209

Africa Genetics Association, POB 10123, Ugbowo, Benin City
Founded: 1982
Pres: Prof. O.S.A. Aromose
Focus: Genetics . . . 09210

Africa Leadership Forum, POB 2286, Abeokuta, Ogun State
Tx: 246680
Founded: 1988
Pres: Olusegun Obasanjo
Focus: Poli Sci . . . 09211

African Association of Political Science (AAPS), c/o Dept of Social Sciences, Lagos State University, Badagry Express-Way Ojo, Lagos
T: (01) 884209; Tx: 22638
Founded: 1973
Gen Secr: L. Adele Jinadu
Focus: Poli Sci
Periodicals
AAPS Newsletter
African Journal of Political Economy
African Review . . . 09212

African Centre for Fertilizer Development (ACFD), c/o OAU, 26-28 Marina, Lagos
Gen Secr: Prof. A. Williams
Focus: Agri . . . 09213

African Council of Food and Nutrition Sciences (AFRONUS), POB 5160, Harare
T: 728991; Tx: 6221
Founded: 1988
Gen Secr: Dr. T.N. Maletnlema
Focus: Nutrition; Food . . . 09214

African Curriculum Organization (ACO), c/o Institute of Education, University of Ibadan, Ibadan
T: (022) 62550
Members: 20
Gen Secr: Prof. P.A.I. Obanya
Focus: Soc Sci 09215

African Feed Resources Research Network, c/o ILCA Liaison Office, POB 3211, Harare
Founded: 1991
Gen Secr: Prof. Jackson A. Kategile
Focus: Agri 09216

African Mathematical Union (AMU), c/o Dept of Mathematics, University of Ibadan, Ibadan
Pres: Prof. Aderemi O. Kuku
Focus: Math
Periodical
Afrika Mathematika 09217

African Peace Research Institute (APRI), POB 51757, Falomo Ikoyi, Lagos
T: (01) 633437
Founded: 1985
Gen Secr: Dr. Peter Bushel Okoh
Focus: Prom Peace
Periodical
APRI Journal (weekly) 09218

African Regional Centre for Engineering Design and Manufacturing (ARCEDEM), PMB 19, UII Post Office, Ibadan
T: (022) 710180; Tx: 31167
Founded: 1979; Members: 25
Gen Secr: Dr. M.F. Abdel-Rahman
Focus: Eng
Periodical
ARCEDEM Bulletin (weekly) 09219

African Regional Network for Microbiology (ARNM), c/o School of Biological Sciences, Imo State University, PMN 2000, Okigwi
T: (083) 232214
Founded: 1977
Pres: Prof. S.O. Emejuaiwe
Focus: Microbio
Periodical
Newsletter 09220

African Statistical Association (AFSA), c/o Federal Office of Statistics, 36-38 Broad St, PMB 12528, Lagos
Focus: Stats 09221

Africa Regional Centre for Information Science (ARCIS), c/o University of Ibadan, 6 Benue Rd, Ibadan
T: (022) 400550 ext 2654; Tx: 31433, 31233
Founded: 1990
Gen Secr: Prof. W. Aiyepeku
Focus: Computer & Info Sci 09222

Alley Farming Network for Tropical Africa (AFNETA), c/o ITA, Oyo Rd, PMB 5320, Ibadan
T: (022) 400300, 400314; Tx: 20311
Founded: 1988
Gen Secr: A.N. Atta-Krah
Focus: Agri
Periodical
Afnetan (weekly) 09223

Ecological Society of Nigeria, c/o Dept of Biological Sciences, University of Lagos, Lagos
Founded: 1973; Members: 100
Pres: Dr. A.W.A. Edwards; Gen Secr: S.L.O. Malaka
Focus: Ecology 09224

Ecumenical Association of Third World Theologians (EATWOT), POB 499, Port Harcourt
Founded: 1976
Gen Secr: Teresa Okure
Focus: Rel & Theol
Periodical
Voices from the Third World (weekly) . 09225

Entomological Society of Nigeria, c/o Dept of Crop Protection, Ahmadu Bello University, Zaria
Founded: 1965; Members: 170
Focus: Entomology
Periodicals
The Nigerian Entomologists' Magazine (weekly)
Nigerian Journal of Entomology (weekly)
Occasional Publications of the Entomological Society of Nigeria (weekly) 09226

Fisheries Society of Nigeria, PMB 12529, Lagos
Founded: 1976; Members: 500
Pres: B.F. Dada; Gen Secr: B.S. Moses
Focus: Zoology
Periodicals
Advisory Notes
Fishery Bulletin
Nigerian Journal of Fisheries and Hydrobiology
Proceedings 09227

Forestry Association of Nigeria, POB 4185, Ibadan
Founded: 1970; Members: 400
Focus: Forestry
Periodicals
Nigerian Journal of Forestry (weekly)
Proceedings of Annual Conferences (weekly)
. . . . 09228

Genetics Society of Nigeria, c/o International Institut of Tropical Agriculture, Oyo Rd, PMB 5320, Ibadan
Founded: 1972; Members: 75
Pres: Dr. O.A. Ojomo; Gen Secr: Dr. A.O. Abifarin
Focus: Genetics
Periodical
Proceedings 09229

Geological Survey of Nigeria, PMB 2007, Kaduna South
T: 212003
Founded: 1919
Gen Secr: J.I. Nehikhare
Focus: Geology
Periodicals
Annual Report (weekly)
Bulletin
Occasional Papers
Records 09230

Historical Society of Nigeria, c/o Dept of History, University of Lagos, Lagos
Founded: 1955
Pres: Prof. O. Ikhime; Gen Secr: Prof. A.I. Asiwaju
Focus: Hist
Periodicals
Bulletin of News (weekly)
Journal
Tarikh (weekly) 09231

The Medical and Dental Council of Nigeria, 25 Ahmed Onibudo St, Victoria Island, PMB 12611, Lagos
T: (01) 613323; Tx: Medcouncil
Founded: 1963; Members: 50
Pres: Dr. M.P. Otolorin; Gen Secr: Dr. C.O. Ezeani
Focus: Med; Stomatology
Periodical
Medical and Dental Register (weekly) . . 09232

Nigeria Educational Research Council, POB 8058, Lagos
Founded: 1965; Members: 30
Pres: Prof. S.N. Nwosu; Gen Secr: J.M. Akintola
Focus: Educ 09233

Nigerian Academy of Science, c/o Dept of Computer Science, University of Lagos, PMB 1004, University of Lagos Post Office, Akoka
Founded: 1977; Members: 45
Focus: Sci
Periodicals
The Discourses of the Academy (weekly)
The Proceedings of the Academy (weekly) 09234

Nigerian Bar Association, 25 Odion Rd, POB 403, Warri
Founded: 1962
Pres: Dr. M. Odje; Gen Secr: D. Akande
Focus: Law 09235

Nigerian Economic Society, c/o Dept of Economics, University of Ibadan, Ibadan
Founded: 1957; Members: 1000
Pres: James O. Osakwe; Gen Secr: Dr. E.O. Ojameruaye
Focus: Econ
Periodicals
Nigerian Journal of Economic and Social Studies (3 times annually)
Proceedings of Annual Conferences . . 09236

Nigerian Geographical Association, c/o Dept of Geography, University of Ibadan, Ibadan
Founded: 1955; Members: 500
Pres: Prof. R.K. Udo; Gen Secr: Dr. I. Adalemo
Focus: Geography
Periodical
Nigerian Geographical Journal 09237

Nigerian Institute of International Affairs, Kofo Abayomi Rd, GPOB 1727, Lagos
T: (01) 615606; Tx: 22638; CA: Internations, Lagos
Founded: 1961
Pres: Prof. G.O. Niia
Focus: Poli Sci
Periodicals
Bulletin (weekly)
Dialogue Series (weekly)
Lecture Series (weekly)
Monograph Series (weekly)
Nigerian Forum (weekly)
Nigerian Journal of International Affairs (weekly)
. . . . 09238

Nigerian Institute of Management (NIM), Plot 22, Idowu Taylor St, Victoria Island, POB 2557, Lagos
T: (01) 615105, 616203
Founded: 1961; Members: 20000
Gen Secr: Prof. O. Iyanda
Focus: Business Admin
Periodical
Management in Nigeria (weekly) 09239

Nigerian Library Association (NLA), c/o National Library of Nigeria, PMB 12626, Lagos
Founded: 1962; Members: 1000
Pres: J.A. Dosunmu; Gen Secr: L.I. Ehigiator
Focus: Libraries & Bk Sci
Periodicals
Nigerian Libraries (3 times annually)
NLA Newsletter (weekly) 09240

Nigerian Society for Microbiology, c/o Dept of Medical Microbiology, University College Hospital, Ibadan
Founded: 1973; Members: 130
Pres: Prof. E.O. Ogunba; Gen Secr: Prof. K.O. Alausa
Focus: Microbio
Periodical
Nigerian Journal for Microbiology (weekly) 09241

Nigerian Veterinary Medical Association, c/o Nigerian Veterinary Research Institute, POB 38, Vom, Plateau State
Founded: 1963; Members: 2155
Pres: Dr. A.A. Fabunmi; Gen Secr: Dr. O.G. Oguntoyinbo
Focus: Vet Med
Periodical
Nigerian Veterinary Journal (weekly) . . 09242

Nutrition Society of Nigeria, c/o Dept of Food Science and Technology, Obafemi Amolowo University, Ile-Ife
Founded: 1966; Members: 350
Pres: Prof. O. Bassir; Gen Secr: Dr. J.B. Fashakin
Focus: Food
Periodical
Nigerian Nutrition Newsletter 09243

Organization of African Unity – Scientific, Technical and Research Commission, PMB 2359, Lagos
T: (01) 633430; 633289; 633359; Tx: 22199; CA: Tecnafrica
Founded: 1965; Members: 51
Focus: Sci; Eng
Periodicals
African Soils (3 times annually)
Bulletin des Epizooties en Afrique (weekly)
Bulletin Interafricain d'Informations Phytosanitaires (weekly)
Bulletin of Epizootic Diseases of Africa (weekly)
Inter-African Phytosanitary Bulletin (weekly)
Journal of African Meldicinal Plants (weekly)
. . . . 09244

Pan African Association of Neurological Sciences, c/o Prof. T.O. Dada, 346 Herbert Macaulay St, POB 457, Yaba, Lagos
Focus: Neurology 09245

West African Association of Agricultural Economists, c/o Dept of Agricultural Economics, University of Ibadan, Ibadan
Fax: (087) 222872
Founded: 1972; Members: 275
Pres: Prof. A.E. Ikpi; Gen Secr: Dr. T. Eponou
Focus: Agri
Periodical
West African Journal of Agricultural Economics
. . . . 09246

Norway

Arkivarforeningen, Folke Bernadottes vei 21, Oslo
Founded: 1936; Members: 53
Focus: Archives
Periodical
Norsk arkivforum 09247

Association for Nordic Dialysis and Transplant Personnel, Nordslettvn 10c, 7038 Trondheim
T: (07) 965044
Founded: 1972
Pres: Ingrid Lian
Focus: Hematology
Periodical
Nordiatrans Newspaper (weekly) 09248

Committee on Conceptual and Terminological Analysis (COCTA), c/o Institute of Political Science, Postboks 1097, Blindern, Oslo 3
T: (02) 455050
Founded: 1970; Members: 150
Pres: Prof. Jan-Erik Lane
Focus: Poli Sci
Periodicals
COCTA News (weekly)
Journal of Theoretical Politics (weekly) . 09249

Den Geofysiske Kommisjon / Geophysical Commission, c/o Det Norske Meteorologiske Institutt, Niels Henrik Abels vei 40, Postboks 320, Blindern, 0313 Oslo
Founded: 1917; Members: 5
Focus: Geophys
Periodical
Geofysiske Publikasjoner 09250

Den Norske Aktuarforening, Hansteens gt 2, Postboks 2429, Solli, 0202 Oslo
T: 2555000; Fax: 2434456
Founded: 1904; Members: 215
Focus: Insurance; Stats
Periodical
Scandinavian Actuarial Journal: Published in cooperation with the societies of actuaries in Denmark, Finland and Sweden (weekly) . 09251

Den Norske Historiske Forening, c/o Avd for Historie, Postboks 1008, Blindern, 0315 Oslo
T: 22856759; Fax: 22854828
Founded: 1869; Members: 1500
Focus: Hist 09252

Den Norske Laegeforening (DNLF), Fjellvn 5, 1324 Lysaker
T: 67124600; Fax: 67124620
Founded: 1886; Members: 11700
Pres: Dr. Knut Eldjarn; Gen Secr: Dr. Harry Martin Svabø
Focus: Med 09253

Den Norske Mikrobionomforening (DNM), c/o Statens Institutt for Folkehelse, Postuttak, Oslo
T: 2356020
Founded: 1973; Members: 294
Focus: Microbio 09254

Den Norske Tannlegeforening (NTF) / The Norwegian Dental Association, Frederik Stangsgt 20, Postboks 3063, Elisenberg, 0207 Oslo
T: 22551350; Fax: 22551109
Founded: 1884; Members: 4500
Focus: Dent
Periodical
Den Norske Tannlegeforenings Tidende (17 times annually) 09255

Den Norske Veterinaerforening (DNV), Sognsvn 4, 0451 Oslo
T: 22567650; Fax: 22690450
Founded: 1888; Members: 1600
Focus: Vet Med
Periodical
Norsk Veterinaertidsskrift (weekly) . . . 09256

Den Polytekniske Forening, Rosenkrantzgt 7, 0155 Oslo
T: 22426870; Fax: 22425887
Founded: 1815; Members: 8000
Pres: Rolf Skår; Gen Secr: Nils C. Tommeraas
Focus: Eng
Periodical
Teknisk Ukeblad (weekly) 09257

Det Kongelige Norske Videnskabers Selskab (D.K.N.V.S.) / The Royal Norwegian Society of Sciences and Letters, Erling Skakkes gt 47b, 7013 Trondheim
T: (07) 3592167; Fax: (07) 3595895
Founded: 1760; Members: 347
Pres: Prof. Haakon Olsen; Gen Secr: Prof. Nils Søvik
Focus: Sci
Periodical
Skrifter: Scientific Papers (weekly) . . . 09258

Det Norske Geografiske Selskap, c/o Avd for Samfunnsgeografi, Iks, Universitetet i Oslo, Postboks 1056, Blindern, 0316 Oslo
T: 22856943; Fax: 22854828
Founded: 1889; Members: 500
Focus: Geography
Periodical
Norsk Geografisk Tidsskrift (weekly) . . 09259

Det Norske Hageselskap, Motzfeldtsgt 1, Postboks 9008, Grønland, 0133 Oslo
T: 22173360; Fax: 22172319
Founded: 1884
Focus: Hort 09260

Det Norske Medicinske Selskab, Drammensvn 44, 0271 Oslo
T: 2440644
Founded: 1833; Members: 600
Focus: Med 09261

Det Norske Samlaget, Trondheimsvegen 15, Postboks 4672, Sofienberg, 0506 Oslo
T: 22687600; Fax: 22687502
Founded: 1868; Members: 3500
Focus: Lit
Periodicals
Maal og Minne (weekly)
Norsk Litterår Årbok (weekly)
Syn og Segn (weekly) 09262

Det Norske Skogselskap, Wergelandsvn 23b, 0167 Oslo
T: 22469857
Founded: 1898; Members: 12500
Focus: Forestry
Periodical
Norsk Skogbruk (weekly) 09263

Det Norske Videnskaps-Akademi / The Norwegian Academy of Science and Letters, Drammensvn 78, Postboks 7585, Skillebekk, 0205 Oslo
T: 444296; Fax: 562656
Founded: 1857; Members: 558
Focus: Sci 09264

European Brain and Behaviour Society (EBBS), c/o Institute of Preclinical Medicine, Dept of Neurophysiology, University of Oslo, Postboks 1104, Blindern, 0317 Oslo 3
Fax: (02) 454949
Founded: 1969; Members: 430
Gen Secr: Dr. Terje Sagvolden
Focus: Neurology
Periodical
EBBS Newsletter 09265

Foreningen til Norske Fortidsminnesmerkers Bevaring / Society for the Preservation of Ancient Monuments in Norway, Dronningengsgt 11, 0152 Oslo
Founded: 1844; Members: 8000
Gen Secr: Seppo Heinonen
Focus: Preserv Hist Monuments

Periodicals
Årbok for Foreningen til norske Fortidsminnesmerkers Bevaring (weekly)
Fortidsvern (weekly) 09266

Fysikkforeningen, c/o Fysisk Institut, Universitetet, Postboks 1048, Blindern, Oslo
Founded: 1938; Members: 110
Focus: Physics 09267

Hovedkomiteen for Norsk Forskning / Central Committee for Norwegian Research, Postboks 8031, Dep, Oslo
Founded: 1965
Focus: Sci 09268

Høyokoleutdannedes Forbund, Postboks 9200, Grønland, 0134 Oslo
Founded: 1956; Members: 900
Focus: Business Admin
Periodical
Etcetera (weekly) 09269

International Federation of Social Workers (IFSW), Postboks 4649, Sofienberg, 0506 Oslo
T: (02) 2031152; Fax: (02) 2031114
Founded: 1928; Members: 300000
Gen Secr: Tom Johannesen
Focus: Soc Sci
Periodicals
IFSW Newsletter (3 times annually)
International Social Work (weekly) . . . 09270

International Political Science Association (IPSA), c/o Institute of Political Science, University of Oslo, Box 1097, Blindern, 0317 Oslo
Pres: Carole Pateman; Gen Secr: Francesco Kjellberg
Focus: Poli Sci 09271

Joint Committee of the Nordic Natural Science Research Councils / The Research Council of Norway, Postboks 2700, St. Hanshaugen, 0131 Oslo
T: 22037000; Fax: 22037001
Founded: 1967; Members: 5
Gen Secr: Signe D. Urbye
Focus: Nat Sci 09272

Kirkehistorisk Samfunn / Church History Society, Hamana terrasse 81, 1300 Sandvika
Founded: 49; Members: 49
Focus: Rel & Theol; Hist
Periodical
Norvegia Sacra 09273

Kommunale Bibliotekarbeiderers Forening / Municipal Librarians' Association, c/o Kari Hjelde, Oppegårdbibliotekene, 1410 Kolbotn
Focus: Libraries & Bk Sci
Periodical
Kontakten (weekly) 09274

Landslaget for Lokalhistorie, c/o Historisk Institutt, 7055 Dragvoll
T: 7596433
Founded: 1920; Members: 240
Focus: Hist
Periodical
Heimen (weekly) 09275

Landslaget for Språklig Samling (LSS), Postboks 636, Sentrum, 0106 Oslo
Founded: 1959
Focus: Ling
Periodical
Språklig Samling (weekly) 09276

Landslaget Musikk i Skolen, Toftesgt 69, 0552 Oslo
T: 22714646; Fax: 22375511
Founded: 1956
Focus: Educ; Music 09277

Medicinske Selskap i Bergen / Medical Society of Bergen, c/o Med. Dept B, Haukeland Hospital, 5016 Bergen
Founded: 1831; Members: 220
Pres: Dr. Leif Utne; Gen Secr: Dr. A. Schreiner
Focus: Med
Periodical
Medicinsk Revue (weekly) 09278

Nordic Road Safety Council, c/o Samferdselsdepartementet, Möllergatan 1-3, Oslo 3
T: 2348150; Fax: 2349570
Founded: 1971; Members: 5
Focus: Transport
Periodical
Rapporte (weekly) 09279

Nordisk Anaesthesiologisk Forening (NAF) / The Scandinavian Society of Anaesthesiologists, c/o Dept of Anaesthesia, Ullevaal Hospital, Oslo
Founded: 1950; Members: 900
Focus: Anesthetics
Periodical
Acta Anaesthesiologica Scandinavica (weekly) 09280

Norges Almenvitenskapelige Forskningsråd (NAVF), Munthesgt 29, Oslo 2
T: 2565290
Founded: 1949; Members: 101
Focus: Sci 09281

Norges Fiskeriforskningsråd (NFFR) / The Norwegian Fisheries Research Council, Postboks 1853, 7001 Trondheim
T: (07) 915580
Founded: 1971; Members: 33
Focus: Fisheries 09282

Norges Geologiske Undersökelse (NGU) / Geological Survey of Norway, Postboks 3006, Lade, 7002 Trondheim
T: (07) 7904011; Fax: (07) 7921620
Founded: 1858
Pres: Knut S. Heier
Focus: Geology
Periodicals
Bulletin
Skrifter 09283

Norges Geotekniske Institutt / Norwegian Geotechnical Institute, Postboks 3930, Ullevaal Hageby, 0806 Oslo
T: 22230388; Fax: 22230448
Founded: 1953; Members: 130
Focus: Geophys 09284

Norges Kunstnerråd / Norwegian Artists' Council, Kronprinsensgt 1, Postboks 1341, Vika, 0113 Oslo
T: 22837332; Fax: 22837326
Founded: 1940
Pres: Ivar Eskeland; Gen Secr: Carl H. Iversen
Focus: Arts 09285

Norges Landbruksvitenskapelige Forskningsråd / The Agricultural Research Council of Norway, Ökernvn 145, Postboks 8154, Dep, 0033 Oslo
Founded: 1949; Members: 30
Focus: Agri 09286

Norges Standardiseringsforbund (NSF) / Norwegian Standards Association, Postboks 7020, Homansbyen, 0306 Oslo
T: 2466094; Tx: 19050; CA: Standardisering; Fax: 2464457
Founded: 1923; Members: 180
Focus: Standards
Periodicals
Catalogue of Norwegian Standards (weekly)
Standardisering (weekly) 09287

Norges Tekniske Vitenskapsakademi, 7000 Trondheim
Founded: 1955
Focus: Eng 09288

Norges Teknisk-Naturvitenskapelige Forskningsråd (NTNF) / Royal Norwegian Council for Scientific and Industrial Research, Sognsvn 72, Tåsen, Oslo
T: 237685; Tx: 76951
Founded: 1946; Members: 40
Focus: Nat Sci; Eng 09289

Norsk Anestesiologisk Forening, c/o Gjövik Fylkessykehus, 2800 Gjövik
Founded: 1949; Members: 160
Focus: Anesthetics 09290

Norsk Arkeologisk Selskap, Frederiksgt 2, Oslo
Founded: 1936; Members: 1000
Focus: Archeol 09291

Norsk Astronautisk Forening / Norwegian Astronautical Society, Postboks 52, Blindern, 0313 Oslo
T: 22210702
Founded: 1951; Members: 400
Pres: Cathrine Malmby; Gen Secr: Johannes Fossen
Focus: Eng; Aero
Periodical
Nytt om Romfart (weekly) 09292

Norsk Bibliotekforening (NBF) / Norwegian Library Association, Malerhaugvn 20, 0661 Oslo
T: 22688550; Fax: 22672368
Founded: 1913; Members: 3600
Pres: T. Minken; Gen Secr: B. Aaker
Focus: Libraries & Bk Sci
Periodical
Internkontakt (11 times annually) 09293

Norsk Botanisk Forening (NBF) / Norwegian Botanical Association, c/o Botanisk Museum, Trandheimsvn 23b, 0562 Oslo
Founded: 1935; Members: 1250
Focus: Botany
Periodical
Blyttia (weekly) 09294

Norske Akademi for Sprog og Litteratur, Stortinget, Oslo
Founded: 1953
Focus: Ling; Lit 09295

Norske Arkitekters Landsforbund (NAL), Josefinesgt 34, 0351 Oslo
T: 22602290; Fax: 22695948
Founded: 1911; Members: 2150
Focus: Archit
Periodicals
Arkitektnytt (20 times annually)
Byggekunst (8 times annually) 09296

Norske Billedkunstnere / The Association of Norwegian Visual Artists, Kongensgt 3, 0153 Oslo
T: 22421357
Founded: 1888; Members: 1900
Pres: Siri Kvitvik; Gen Secr: Jon Øien
Focus: Fine Arts
Periodical
Billedkunstneren (10 times annually) . . 09297

Norske Fagbibliotek Forening (NFF), Malerhaugvn 20, 0661 Oslo
T: 22688576
Founded: 1947; Members: 680
Focus: Libraries & Bk Sci 09298

Norske Forfatterforening / Norwegian Authors' Society, Rådhusgata 7, Oslo
Founded: 1893; Members: 400
Focus: Lit 09299

Norske Fysioterapeuters Forbund (NFF) / Norwegian Physiotherapist Association, Pilestredet 56, Postboks 7009, Homansbyen, 0306 Oslo
T: 22697800; Fax: 22565825
Founded: 1936; Members: 5800
Pres: Toril B. Buene; Gen Secr: Kari Haug
Focus: Therapeutics
Periodical
Fysioterapeuten (16 times annually) . . 09300

Norske Havforskeres Forening (NHF), c/o Svein Sundby Havforskningsinst., Postboks 1870, 5011 Nordnes
Founded: 1949; Members: 260
Focus: Oceanography 09301

Norske Kunst- og Kulturhistoriske Museer, Ullevålsvn 11, 0165 Oslo
T: 22201402; Fax: 22112337
Founded: 1918; Members: 331
Focus: Arts; Hist
Periodical
Museumsnytt (weekly) 09302

Norske Meierifolks Landsforening, Nedre Slottsgt 23, Postboks 398, 0103 Oslo
T: 22422520; Fax: 22413801
Founded: 1914; Members: 1344
Gen Secr: Steinar Hnsby
Focus: Dairy Sci
Periodical
Meieriposten (weekly) 09303

Norske Musikklaereres Landsforbund (N.M.L.L.), Östre Strandgt 17a, 4600 Kristiansand
T: (042) 25989
Founded: 1914; Members: 750
Focus: Music; Educ 09304

Norske Naturhistoriske Museers Landsforbund (NNML), c/o Tromsø Museum, 9000 Tromsø
T: (083) 86080
Founded: 1938; Members: 165
Focus: Hist; Nat Sci
Periodical
Museumsnytt (weekly) 09305

Norske Sivilingeniørers Forening / Norwegian Society of Chartered Engineers, Postboks 2312, Solli, 0201 Oslo
T: 22947500; Fax: 22947501
Founded: 1874; Members: 30000
Pres: Berit Kvaeren; Gen Secr: Terje Norddal
Focus: Civil Eng
Periodicals
Bygg
Elektro (weekly)
Sivilingeniören (weekly)
Teknisk Ukeblad (weekly) 09306

Norske Siviløkonomers Forening (NSF), Kai Munks vei 41, Postboks 23, Tåsen, 0801 Oslo
T: 22234369; Fax: 22180151
Founded: 1939; Members: 9035
Gen Secr: Sverre Bjønnes
Focus: Business Admin 09307

Norsk Faglaererlag (NF), Waldemar Thranesgt 1a, 0171 Oslo
T: 2465658; Fax: 2467467
Founded: 1914; Members: 7400
Focus: Adult Educ
Periodical
Yrke (10 times annually) 09308

Norsk Forening for Internasjonal Rett (NFIR) / Norsk Avdeling av International Law Association, Fridtjof Nansens plass 4, Postboks 1600, Vika, Oslo
T: 2411834
Founded: 1925; Members: 102
Focus: Law 09309

Norsk Forening for Mikrobiologi, c/o Dept of Microbiology, Agricultural University, Postboks 40, 1432 Aas
T: 941060
Founded: 1959; Members: 220
Focus: Microbio 09310

Norsk Forening mot Støy, Kingosgt 22, Oslo
T: 2352302
Founded: 1963
Focus: Ecology 09311

Norsk Forsikringsjuridisk Forening, c/o Sparebanken Oslo Akershus, Grensen 3, Oslo
Founded: 1934; Members: 900
Focus: Law
Periodical
Average (weekly) 09312

Norsk Fysiologisk Forening, c/o Institutt for ernåringsforskning, Universitetet i Oslo, Postboks 1046, Blindern, Oslo
Focus: Physiology 09313

Norsk Fysisk Selskap (NFS), c/o Institutt for Energiteknikk, Postboks 40, 2007 Kjeller
T: 63806075
Founded: 1951; Members: 710
Pres: Thormod Henriksen; Gen Secr: Gerd Jarrett
Focus: Physics
Periodical
Fra Fysikkens Verden (weekly) 09314

Norsk Gastroenterologisk Selskap, c/o Dept of Surgery, Aker Hospital, Oslo
Members: 30
Focus: Gastroenter 09315

Norsk Geofysisk Forening (NGF), c/o River and Harbour Laboratory, Klaebuvn 153, 7000 Trondheim
T: (07) 92300
Founded: 1918; Members: 167
Focus: Geophys 09316

Norsk Geologisk Forening, Postboks 3006, Lade, 7002 Trondheim
Fax: (07) 921620
Founded: 1905; Members: 900
Focus: Geology
Periodical
Norsk Geologisk Tidsskrift (weekly) . . . 09317

Norsk Geoteknisk Forening (NGF) / Norwegian Geotechnical Society, Sognsvn 72, Postboks 3930, Ullevål Hageby, 0806 Oslo
T: 22230388; Tx: 19787; Fax: 22230448
Founded: 1950; Members: 338
Focus: Mining 09318

Norsk Heraldisk Forening, Postboks 313, Sentrum, 0103 Oslo
Founded: 1969
Focus: Genealogy 09319

Norsk Homøopatisk Pasientforening, Postboks 412, 7001 Trondheim
T: (07) 3522307; Fax: (07) 3522307
Founded: 1951; Members: 1200
Pres: Thore Aalberg; Gen Secr: Aslak Steinsbekk
Focus: Homeopathy
Periodical
Homøopatisk Tidsskrift (weekly) 09320

Norsk Institutt for By- og Regionforskning (NIBR) / Norwegian Institute for Urban and Regional Research, Gaustadalléen 21, Postboks 44, Blindern, 0313 Oslo
T: 22958800; Fax: 22607774
Founded: 1965; Members: 80
Focus: Urban Plan
Periodicals
NIBR Notater
NIBR Rapport 09321

Norsk Kirurgisk Forening / Norwegian Surgical Society, Fjellvn 5, 1324 Lysaker
Founded: 1911; Members: 730
Pres: Hans Olav Beisland
Focus: Surgery
Periodical
Vitenskapelige Forhandlinger: Scientific Proceedings (weekly) 09322

Norsk Kjemisk Selskap (NKS) / Norwegian Chemical Society, Postboks 1107, Blindern, 0317 Oslo
T: 22855531; Fax: 22855441
Founded: 1893; Members: 2200
Pres: Prof. Leiv K. Sydnes; Gen Secr: Ove Kjølberg
Focus: Chem
Periodical
KJEMI (10 times annually) 09323

Norsk Korrosjonsteknisk Forening, Rosenkrantzgt 7, Oslo
Members: 150
Focus: Metallurgy 09324

Norsk Laererlag / Norwegian Union of Teachers, Rosenkrantzgt 15, Oslo
T: 2415875
Founded: 1892; Members: 59000
Pres: Per Woeien; Gen Secr: Magne Askeland
Focus: Educ
Periodical
Norsk Skoleblad (weekly) 09325

Norsk Logopedlag (NLL) / Norwegian Association of Logopedists, 7041 Trondheim
T: (07) 509192
Founded: 1948; Members: 1000
Pres: Hanna Jensen
Focus: Logopedy
Periodical
Norsk Tidsskrift for Logopedi (weekly) . 09326

Norsk Lokalhistorisk Institutt (NLI) / Norwegian Institute of Local History, Folke Bernadottes vei 21, Kringsja, Oslo
T: 237480
Founded: 1955
Focus: Hist 09327

Norsk Matematisk Forening, c/o Matematisk Institutt, Universitetet i Oslo, Postboks 1053, Blindern, 0316 Oslo
Fax: 22854349
Founded: 1918
Focus: Math
Periodical
Nordisk Matematisk Tidskrift (weekly) . . 09328

Norsk Metallurgisk Selskap, Rosenkrantzgt 7, Oslo
T: 2330741
Founded: 1936; Members: 440
Focus: Metallurgy 09329

Norsk Meteorologforening (NMF) / The Norwegian Association of Meteorologists, c/o Det Norske Meteorologiske Institutt, Postboks 43, Blindern, 0313 Oslo
T: 22963000; Fax: 22963050

Founded: 1961; Members: 110
Pres: K. Gislefoss
Focus: Geophys 09330

Norsk Musikkinformasjon / Norwegian Music Information Centre, Toftesgt 69, 0552 Oslo
T: 22370909; Fax: 22356938
Founded: 1978
Gen Secr: Jostein Simble
Focus: Music
Periodical
Listen to Norway (3 times annually) . . 09331

Norsk Naturforvalterforbund / Norwegian Association of Agriculture Graduates, Parkvn 37, Oslo
T: 447877; Fax: 551630
Founded: 1970; Members: 4098
Focus: Agri
Periodical
NaFo-nytt (23 times annually) 09332

Norsk Operasjonsanalyseforening, c/o Norwegian State Railways Executive Offices, Storgt 33, Oslo
T: 209550
Founded: 1959; Members: 190
Focus: Computer & Info Sci 09333

Norsk P.E.N., Urtegt 50, 0187 Oslo
T: 22571220; Tx: 71230
Pres: Toril Brekke
Focus: Lit 09334

Norsk Radiologisk Forening (NRF), c/o Røntgenavd, Ullevål Sykehus, 0407 Oslo
Founded: 1920; Members: 280
Focus: Radiology
Periodical
Noraforum (weekly) 09335

Norsk Regnesentral / Norwegian Computing Centre, Gaustadalléen 23, Postboks 114, Blindern, 0314 Oslo
T: 453500; Tx: 76518; Fax: 697660
Founded: 1952
Focus: Computer & Info Sci 09336

Norsk Samfunnsgeografisk Forening (NSGF), c/o Geografisk Institutt, Universitetet i Trondheim, 7055 Dragvoll
Founded: 1974; Members: 200
Focus: Geography
Periodicals
NSGF-Meldingsblad: Newsletter (5-6 times annually)
Skrifter fra NSGF: Conference Proceedings (weekly) 09337

Norsk Senter for Samferdselsforskning / Norwegian Center for Transport Research, Grensesvingen 7, Postboks 6110, Etterstad, 0602 Oslo
T: 22573800; Fax: 22570290
Founded: 1964
Focus: Transport
Periodical
Samferdsel: Communication 09338

Norsk Slektshistorisk Forening, Postboks 9562, Egertorvet, 0128 Oslo
Founded: 1926; Members: 1300
Focus: Genealogy 09339

Norsk Tekstil Teknisk Forbund (NTTF), Lars Hillesgt 34, 5000 Bergen
T: (05) 327240
Founded: 1941; Members: 749
Focus: Textiles 09340

Norwegian Council of Cultural Affairs, Grev Wedels plass 1, 0151 Oslo
T: 22334042; Fax: 22428919
Founded: 1965
Focus: Arts
Periodical
The Norwegian Cultural Fund 09341

Organization of Nordic Teachers Associations, c/o Norsk Laererlag, Rosenkrantzgt 15, 0160 Oslo
Founded: 1968
Focus: Educ 09342

Papirindustriens Forskningsinstitutt / The Norwegian Pulp and Paper Research Institute, Postboks 24, Blindern, 0319 Oslo
Tx: 78171; Fax: 22468014
Founded: 1923; Members: 60
Focus: Materials Sci 09343

Riksbibliotektjenesten / National Office for Research and Special Libraries, Bygdøy Allé 21, Postboks 2439, Solli, 0201 Oslo
T: 22430880; Fax: 22560981
Focus: Libraries & Bk Sci
Periodicals
Annual Report (weekly)
Skrifter (weekly)
Synopsis (weekly) 09344

Scandinavian Society for Electron Microscopy, c/o National Institute of Public Health, Oslo
T: 2356020
Founded: 1948; Members: 600
Focus: Nat Sci 09345

Selskapet for Lyskultur, Blommenholmvn 1, Postboks 18, 1301 Sandvika
T: 67547930, 67547948
Founded: 1936; Members: 405
Gen Secr: Johan Ihme
Focus: Eng
Periodical
News from Lightening Culture (weekly) . 09346

Selskapet til Vitenskapenes Fremme / Society for the Advancement of Science, c/o Fysisk Institutt, Universitetet i Bergen, Allegt 55, 5007 Bergen
T: (05) 318334
Founded: 1927; Members: 277
Focus: Sci 09347

Skogbrukets og Skogindustrienes Forskningsforening (SSFF) / The Norwegian Forestry and Forest Industry Association, Hoegskolevn 12, 1432 Aas
T: 64949000; Fax: 64942980
Founded: 1947
Focus: Forestry 09348

Sosialøkonomenes Forening (SF), Youngstorget, Postboks 8872, 0028 Oslo
T: 22170035; Fax: 22173155
Founded: 1908; Members: 946
Focus: Econ
Periodicals
Norsk Økonomisk Tidsskrift
Sosialøkonomen 09349

Standardiseringsforeningen, c/o Norges Standardiseringsforbund, Haakon Vii's gt 2, Oslo
T: 2416820
Founded: 1974; Members: 110
Focus: Standards 09350

Statistics Norway, Skippergt 15, Postboks 8131, Dep, 0033 Oslo
T: 22864500; Fax: 22864973
Founded: 1876; Members: 800
Focus: Stats
Periodicals
Norway's Official Statistics (NOS)
Reports from the Central Bureau of Statistics (REP)
Social Economic Studies (SES)
Statistical Analyses (SA) 09351

Statsøkonomisk Forening, Dronningensgt 16, Oslo
T: 2413820
Founded: 1883; Members: 400
Focus: Econ
Periodical
Statsøkonomisk Tidsskrift (weekly) . . . 09352

Stiftelsen for Industriell og Teknisk Forskning ved Norges Tekniske Høgskole (SINTEF) / The Foundation for Scientific and Industrial Research at the Norwegian Institute of Technology, 7034 Trondheim NTH
Founded: 1950
Focus: Eng
Periodical
Reports 09353

The Teachers' Union, Wergelandsvn 15, 0167 Oslo
T: 22030000; Fax: 22110542
Founded: 1993; Members: 33500
Pres: Anders Folkestad; Gen Secr: Helle Hogner
Focus: Educ
Periodical
Skoleforum (weekly) 09354

Oman

Historical Association of Oman, POB 3941, Ruwi
Founded: 1971; Members: 145
Focus: Hist
Periodical
Bulletin (weekly) 09355

Pakistan

Academy of Letters, 36 48 St, Islamabad
Founded: 1979
Focus: Lit 09356

All-Pakistan Educational Conference, Conference Hall, 1-J45/10, Syed Altaf Ali Brelvi Rd, Karachi
T: (021) 611196
Founded: 1951; Members: 118
Pres: Dr. F.U. Bagai; Gen Secr: S.M.A. Brelvi
Focus: Educ
Periodical
Al-Ilm (weekly) 09357

Architects Regional Council Asia (ARCASIA), 404 Noor Estate, Shara El-Faisal, Karachi 75350
Founded: 1979; Members: 12
Pres: Syeed Akeel Bilgrami
Focus: Archit 09358

Arts Council of Pakistan, M.R. Kayani Rd, Karachi
T: (021) 515108
Founded: 1956; Members: 31140
Pres: S.A. Siddiqi; Gen Secr: Q. Akbar
Focus: Arts 09359

Asia Foundation, POB 1165, Islamabad
T: (051) 820507
Gen Secr: D. Roen Repp
Focus: Int'l Relat; Lit 09360

Baluchi Academy, Mekran House, Sariab Rd, Quetta
T: 22248
Founded: 1961; Members: 37
Pres: B.A. Baluch; Gen Secr: A. Baluch
Focus: Ling; Lit 09361

Defence Housing Society, c/o College of Physicians and Surgeons, 7 Central St, Karachi 46
T: (021) 541172, 542210
Founded: 1962; Members: 528
Focus: Surgery; Physiology 09362

Federal Library Association, 169 Sawar Rd, Rawalpindi
Focus: Libraries & Bk Sci 09363

Hamdard Foundation, c/o Hamdard Centre, Nazimabad, Karachi 74600
T: (021) 616001; Tx: 24529
Founded: 1953
Pres: Hakim Mohammed Said
Focus: Soc Sci
Periodicals
Akhbar-ut-Tib (weekly)
Hamdard-i-Sehat
Hamdard Islamicus
Hamdard Medicus (weekly)
Hamdard Naunehal (weekly) 09364

Idarah-i-Yadgar-i-Ghalib, Nazimabad 2, POB 2268, Karachi 18
Founded: 1968
Pres: A.M. Malik; Gen Secr: M. Zaman
Focus: Lit
Periodical
Ghalib 09365

Institute of Cost and Management Accountants of Pakistan, Soldier Bazaar, POB 7284, Karachi 3
T: (021) 719907; Tx: 2733
Founded: 1951; Members: 825
Gen Secr: Z.H. Subzwari
Focus: Econ; Business Admin
Periodicals
Industrial Accountant (weekly)
Professional Information Bulletin (weekly)
Students' Handbook (weekly) 09366

Institute of Islamic Culture, Club Rd, Lahore 3
T: (042) 53908; CA: ICULT
Founded: 1950
Pres: Prof. M. Saeed Sheikh; Gen Secr: M.A. Dar
Focus: Lit; Ethnology
Periodical
Al-Ma'arif (weekly) 09367

Institution of Electrical Engineers, 4 Lawrence Rd, Lahore
Founded: 1969
Pres: Javid Akhtar
Focus: Electric Eng
Periodicals
Journal (weekly)
Newsletter (weekly) 09368

Institution of Engineers, Pakistan (I.E.P.), I.E.P. Round About, Engineering Centre, Gulberg III, Lahore
T: (042) 872927
Founded: 1948; Members: 51207
Gen Secr: Qaisar Zaman
Focus: Eng
Periodical
The Pakistan Engineer (weekly) 09369

Iqbal Academy, GPOB 1308, Lahore
T: (042) 57214
Founded: 1951
Gen Secr: Prof. M. Munawwar
Focus: Ling; Lit
Periodicals
Iqbaliat (weekly)
Iqbaliat Farsi (weekly)
Iqbal Review (weekly) 09370

Jamiyat-ul-Falah, Akbar Rd, Saddar, POB 7141, Karachi 3
T: (021) 721394
Founded: 1950; Members: 250
Gen Secr: Prof. Syed Lutfullah
Focus: Rel & Theol
Periodical
Voice of Islam (weekly) 09371

Karachi Theosophical Society, Jamshed Memorial Hall, M.A. Jinnah Rd, Karachi 1
T: (021) 721275
Founded: 1896; Members: 124
Pres: D.F. Mirza; Gen Secr: G.K. Minwalla
Focus: Rel & Theol
Periodical
Theosophy in Karachi (weekly) 09372

Lok Virsa / National Institute of Folk and Traditional Heritage, Garden Av, Shakatparian Hills, Box 1184, Islamabad
T: (051) 813756; Tx: 54468; Fax: (051) 813756
Founded: 1974
Gen Secr: Uxi Mufti
Focus: Cultur Hist; Ethnology; Folklore . 09373

Mehran Library Association, POB 126, Hyderabad
Focus: Libraries & Bk Sci
Periodical
Newsletter 09374

National Book Council of Pakistan, Block 14-D, Al-Markaz F/8, Islamabad
T: (051) 850892
Founded: 1981
Gen Secr: R. Ahmad
Focus: Libraries & Bk Sci
Periodical
Kitab (weekly) 09375

National Science Council of Pakistan, 63 School Rd, Islamabad
Founded: 1961
Focus: Sci 09376

Pakistam Anti Tuberculosis Association, Rm 8, Block 55, Karachi 72400
T: (021) 5688011
Pres: Mian Fazl-i-Ahmed; Gen Secr: Muhammad Nawaz
Focus: Intern Med
Periodical
The Challenge (weekly) 09377

Pakistan Academy of Letters, H-8/1, Islamabad
T: (051) 855447
Founded: 1976
Pres: G.R.A. Agro; Gen Secr: Arif Iftikhar
Focus: Lit
Periodicals
Academy (weekly)
Adbiyat (weekly) 09378

Pakistan Academy of Sciences, 3 Constitution Av, G-3, Islamabad
Founded: 1953; Members: 53
Pres: Dr. Amir Muhammed; Gen Secr: Dr. M.D. Shami
Focus: Sci
Periodicals
Monographs (weekly)
Proceedings (weekly)
Proceedings of Symposia (weekly) . . . 09379

Pakistan Atomic Energy Commission, POB 1114, Islamabad
Fax: (051) 819031 ext 2550
Founded: 1956
Pres: Dr. Ishfaq Ahmad
Focus: Nucl Res
Periodical
The Nucleus (weekly) 09380

Pakistan Board for Advancement of Literature, Narsing Das Garden, Club Rd, Lahore
Focus: Lit 09381

Pakistan Concrete Institute, 11 Bambino Chambers, Garden Rd, Karachi 0310
Pres: Umar Munshi; Gen Secr: Rehman Akhtar
Focus: Eng; Civil Eng 09382

Pakistan Council of Architects and Town Planners, c/o Defence Housing Authority, E-6 Fourth Gizri St, Karachi 46
T: (021) 537416
Founded: 1983
Pres: Y. Lari
Focus: Archit; Urban Plan 09383

Pakistan Council of Scientific and Industrial Research (PCSIR), c/o Press Centre, Shahrah-e-Kemal Ataturk, Karachi 1
T: (021) 212256; Fax: (021) 2636704
Founded: 1953
Pres: Dr. A.Q. Ansari; Gen Secr: Dr. S.M.A. Hai
Focus: Sci; Business Admin
Periodicals
Karawan-e-Science (weekly)
Pakistan Journal of Scientific and Industrial Research (weekly)
Science Chronicle (weekly)
Technology Digest (weekly) 09384

Pakistan Historical Society, c/o Co-operative Housing Society, 30 New Karachi, Karachi 5
T: (021) 410847
Founded: 1951
Pres: H.M. Said; Gen Secr: Dr. S.M. Haq
Focus: Hist
Periodical
Journal (weekly) 09385

Pakistan Institute of International Affairs, Aiwan-e-Sadar Rd, Karachi 1
T: (021) 512891
Founded: 1947; Members: 650
Gen Secr: R.A. Siddiqui
Focus: Poli Sci
Periodical
Pakistan Horizon (weekly) 09386

Pakistan Library Association, c/o University of Karachi, POB 8455, Karachi
Founded: 1957; Members: 800
Pres: Dr. J. Jalibi; Gen Secr: S.A. Khan
Focus: Libraries & Bk Sci
Periodicals
Conference Proceedings (weekly)
Journal (weekly)
Newsletter (weekly) 09387

Pakistan Medical Association (PMA), PMA House, Garden Rd, POB 7267, Karachi 3
T: (021) 714632
Pres: I. Ahmad; Gen Secr: Dr. H. Ahmed
Focus: Med 09388

Pakistan Medical Research Council (PMRC), 162/0/III, Minhas House, PECHS, Karachi 29
T: (021) 416522
Founded: 1953; Members: 19
Pres: M.A.Z. Mohydin; Gen Secr: Dr. M.A. Pervaiz
Focus: Med
Periodical
Pakistan Journal of Medical Research (weekly) . . . 09389

Pakistan Museum Association, c/o National Museum, Burns Garden, Karachi
T: (021) 211341
Founded: 1949
Pres: Dr. F.A. Khan; Gen Secr: M.A. Haleem
Focus: Arts
Periodicals
Museums Journal of Pakistan (weekly)
Museum Studies . . . 09390

Pakistan Philosophical Congress, c/o Dept of Philosophy, University of the Punjab, New Campus, Lahore 20
T: (042) 85134
Founded: 1954
Pres: Dr. W.A. Farooqi; Gen Secr: G. Irfan
Focus: Philos
Periodicals
Annual Proceedings (weekly)
Pakistan Philosophical Journal (weekly) . 09391

Pashto Academy, c/o University of Peshawar, Peshawar
Founded: 1955
Focus: Ling; Lit
Periodical
Pakhto (weekly) . . . 09392

Punjab Bureau of Education, 15A Mahmud Ghaznavi Rd, Lahore 54000
Founded: 1958
Focus: Educ
Periodicals
Educational Statistics Higher Education (weekly)
Educational Statistics (Schools) (weekly)
Sanvi Taleem (weekly)
Taleem-e-Tadrees (weekly) . . . 09393

Punjabi Adabi Academy, 12-G, Model Town, Lahore 14
Founded: 1957
Focus: Ling; Lit . . . 09394

Punjab Text Board, 21 E-11, Gulberg III, Lahore
Focus: Lit . . . 09395

Punjab University Historical Society, University Hall, Lahore
Founded: 1911
Focus: Hist . . . 09396

Quaid-i-Azam Academy, 297 M.A. Jinnah Rd, POB 894, Karachi 5
T: (021) 718184; CA: Qaidacademy
Founded: 1976
Gen Secr: Sharif Al Mujahid
Focus: Sci
Periodical
Journal of Pakistan Studies . . . 09397

Research Society of Pakistan, c/o University of the Punjab, 2 Narsingdas Garden, Club Rd, Lahore 3
T: (042) 65907
Founded: 1963
Focus: Hist; Cultur Hist
Periodical
Journal (weekly) . . . 09398

Scientific Society of Pakistan, c/o University, University Campus, Karachi 32
T: (021) 463144
Founded: 1954; Members: 3500
Pres: Dr. S.I. Ali; Gen Secr: A. Hasan
Focus: Sci
Periodicals
Jadeed Science (weekly)
Proceedings of Annual Science Conferences
Science Bachchon Key Liye (weekly)
Science Name (weekly) . . . 09399

Shah Waliullah Academy, Hyderabad
T: (0221) 24154
Founded: 1963
Gen Secr: A.G.M. Qasmi
Focus: Philos
Periodical
Alwali (weekly) . . . 09400

Sindhi Adabi Board, Sindh University Campus, Tamshoro, Sindh
Founded: 1951
Pres: Makhdum Muhammad Zaman Talib-ul-Maula; Gen Secr: Habibullah Siddiqui Patai
Focus: Lit
Periodicals
Gul Phul (weekly)
Mehran (weekly) . . . 09401

Sind Library Association, POB 126, Hyderabad
Founded: 1966
Focus: Libraries & Bk Sci . . . 09402

Society for the Promotion and Improvement of Libraries, c/o Hamdard Library, Al-Majeed Centre, Nazimabad, Karachi 18
Focus: Libraries & Bk Sci . . . 09403

Urdu Academy, 33C Model Town A, Bahawalpur
T: 2381
Founded: 1959
Gen Secr: M.H. Shihab
Focus: Ling; Lit
Periodical
Az-Zubair (weekly) . . . 09404

Panama

Academia Nacional de Ciencias de Panamá, Apdo 4570, Panamá City
Founded: 1942; Members: 24
Focus: Sci . . . 09405

Academia Panameña de la Historia, Apdo 973, Zona 1, Panamá City
Founded: 1921
Pres: Miguel A. Martín; Gen Secr: Rogelio Alfaro
Focus: Hist
Periodical
Boletín . . . 09406

Academia Panameña de la Lengua, Apdo 1748, Zona 1, Panamá City
Members: 12
Pres: Ismael García S.; Gen Secr: T. Díaz Blaitry
Focus: Ling
Periodical
Boletín (weekly) . . . 09407

Asociación Centroamericana de Sociología (ACAS) / Central American Association of Sociology, Apdo 63093, El Dorado, Panamá City
T: 2320028
Founded: 1974; Members: 150
Gen Secr: Marco A. Gandásegui
Focus: Sociology . . . 09408

Asociación de Bibliotecarios Graduados del Istmo de Panamá, c/o Biblioteca de la Universidad de Panamá, Panamá City 3
T: 238786
Focus: Libraries & Bk Sci . . . 09409

Asociación Panameña de Bibliotecarios, Apdo 3435, Panamá City
Focus: Libraries & Bk Sci . . . 09410

Asociación Panamericana de Oftalmología, Apdo 1189, Panamá City 1
Founded: 1940; Members: 4000
Focus: Ophthal . . . 09411

Association for Cooperation in Banana Research in the Caribbean and Tropical America, c/o UPEB, Apdo 4273, Panamá City 5
T: 636310; Tx: 2568
Founded: 1965
Focus: Agri . . . 09412

Central American Society of Pharmacology, c/o Facultad de Medicina, Universidad de Panamá, Panamá City
T: 643701; Fax: 635622
Founded: 1975
Gen Secr: Ceferino Sanchez
Focus: Pharmacol . . . 09413

Centro para el Desarrollo de la Capacidad Nacional de Investigación, c/o Estafeta Universitaria, Universidad de Panamá, Panamá City
Founded: 1976
Pres: Dr. Alfred Soler B.
Focus: Sci
Periodicals
Carta Informativa
Revista Scientia . . . 09414

Comisión Nacional de Arqueología y Monumentos Históricos (CONAMOH), Apdo 662, Panamá City
T: 628130
Founded: 1953
Pres: M. Isaac; Gen Secr: J. Zulma de Luca
Focus: Archeol; Preserv Hist Monuments 09415

Consejo de Economía Nacional, Av 3a, Panamá City
Focus: Econ . . . 09416

Consejo Nacional de Ciencia, Apdo 3277, Panamá City
Founded: 1963
Focus: Sci . . . 09417

Educational and Development Foundation of the Latin American Confederation of Credit Unions, c/o FECOLAC, Apdo 3280, Panamá City 3
T: (0507) 273322; Tx: 3592; Fax: (0507) 273768
Founded: 1979; Members: 2518
Gen Secr: Dr. Ramiro Valderrama
Focus: Educ . . . 09418

Federación Odontológica de Centro América y Panamá, Apdo 4115, Panamá City 5
Founded: 1953
Focus: Dent . . . 09419

Inter-American Statistical Institute (IASI), Apdo 5139, Panamá City 5
T: (507) 641349, 64367; Fax: (507) 644601
Founded: 1940
Gen Secr: Evelio O. Fabbroni
Focus: Stats
Periodicals
Estadística (weekly)
Newsletter (2-3 times annually) . . . 09420

Papua New Guinea

Papua New Guinea Library Association (PNGLA), POB 5368, Boroko
Founded: 1973; Members: 200
Pres: Margaret J. Obi; Gen Secr: Jenny Wal
Focus: Libraries & Bk Sci
Periodicals
Directory of Libraries in Papua New Guinea
PNGLA Librarian's Calendar (weekly)
PNGLA Nius (weekly)
Toktok bilong haus buk (weekly) . . . 09421

Papua New Guinea Scientific Society, c/o National Museum, POB 5560, Boroko
Founded: 1949; Members: 203
Pres: H. Sakulas
Focus: Sci
Periodical
Proceedings (weekly) . . . 09422

Papua New Guinea Teachers' Association, POB 6546, Boroko
Founded: 1971; Members: 10500
Focus: Educ . . . 09423

Paraguay

Academia de la Lengua y Cultura Guaraní, Calle España y Mompox, Asunción
Founded: 1975
Pres: Dr. Rufino Arevalo Paris; Gen Secr: Antonio E. Gonzáles
Focus: Ling; Lit; Hist
Periodical
Revista . . . 09424

Asciación Indeginista del Paraguay, Calle España y Mompox, Casilla 1838, Asunción
Founded: 1942
Focus: Ethnology . . . 09425

Asociación de Bibliotecarios del Paraguay, Casilla 1505, Asunción
Focus: Libraries & Bk Sci . . . 09426

Asociación de Bibliotecarios Universitarios del Paraguay, c/o Escuela de Bibliotecología, Universidad Nacional, Av España 1098, Asunción
Focus: Libraries & Bk Sci . . . 09427

Centro Paraguayo de Estudios de Desarrollo Económico y Social, Casilla 1189, Asunción
Focus: Sociology; Econ . . . 09428

Comisión Paraguaya de Documentación e Información, c/o Instituto de Ciencias, Universidad Nacional, Av España 1098, Asunción
Focus: Doc . . . 09429

Federación Universitaria del Paraguay, c/o Universidad, Colón 63, Asunción
Focus: Adult Educ . . . 09430

Instituto Nacional de Parasitología, c/o Instituto de Microbiología, Facultad de Medicina, Casilla 1102, Asunción
Founded: 1963; Members: 5
Focus: Microbio . . . 09431

Servicio Cooperativo Interamericano de Salud Pública (SCISP), Av Pettirossi y Brasil, Asunción
Focus: Public Health . . . 09432

Servicio Técnico Interamericano de Cooperación Agrícola, Casilla 819, Asunción
Founded: 1943
Focus: Agri; Eng . . . 09433

Sociedad Científica del Paraguay, Av España 505, Asunción
Founded: 1921; Members: 80
Pres: Dr. Andrés Barbero; Gen Secr: G. Tell Bertoni
Focus: Sci
Periodical
Revista . . . 09434

Sociedad de Pediatría y Puericultura del Paraguay, 25 de Mayo y Tacuaí, Asunción
Founded: 1928; Members: 28
Focus: Pediatrics . . . 09435

Unión Sudamericana de Asociaciones de Ingenieros (USAI), Casilla 336, Asunción
Founded: 1935
Focus: Eng . . . 09436

Peru

Academia de Estomatología del Perú, Apdo 2467, Lima
Founded: 1929
Pres: Dr. Hernán Villena Martinez; Gen Secr: Dr. Carlos Vélez Vargas
Focus: Stomatology
Periodicals
Actualidades Académicas
Boletín . . . 09437

Academia Nacional de Ciencias Exactas, Físicas y Naturales de Lima, Casilla 1979, Lima
Founded: 1939
Focus: Nat Sci; Physics
Periodical
Actas . . . 09438

Academia Nacional de Medicina, Camaná 773, Apdo 987, Lima
Founded: 1884; Members: 106
Gen Secr: Dr. Jorge Voto Bernales
Focus: Med . . . 09439

Academia Peruana de Cirurgía, Camaná 773, Lima
Founded: 1940; Members: 100
Pres: Dr. Luis Gurmendi
Focus: Surgery
Periodical
Revista . . . 09440

Academia Peruana de la Lengua, Jr. Conde de Superunda 298, Lima 1
Members: 30
Pres: Dr. Augusto Tamayo Vargas; Gen Secr: Dr. E. Núñez
Focus: Hist
Periodical
Boletín (weekly) . . . 09441

Agrupación de Bibliotecas para la Información Socio-Económica (ABIISE), Apdo 2874, Lima 100
T: 351760
Focus: Libraries & Bk Sci . . . 09442

Andean Commission of Jurists, Los Sauces 285, San Isidro, Lima 27
T: (014) 407907, 428094; Fax: (014) 426468
Founded: 1980
Gen Secr: Dr. Diego Garcia Sayán
Focus: Law
Periodicals
Andean Newsletter (weekly)
Boletín de la Comisión Andina de Juristas (weekly) . . . 09443

Andean Institute for Population Studies and Development, Lola Pardo Vargas 325, Urbanización Aurora Miraflores, Lima 18
Focus: Soc Sci . . . 09444

Andean Institute of Social Studies, Av Arequipa 2064, Lima 1
T: (014) 289929; Tx: 21127
Focus: Soc Sci . . . 09445

Andean Technological Information System, c/o JUNAC, Paseo de la República 3895, Casilla 181177, Lima 27
T: (014) 414212; Tx: 20104; Fax: (014) 420911
Founded: 1980
Focus: Computer & Info Sci
Periodical
Boletín Informativo SAIT (weekly) . . . 09446

Asociación de Artistas Aficionados, Ica 323, Lima
Founded: 1938; Members: 254
Focus: Arts; Perf Arts . . . 09447

Asociación de Bibliotecarios y Documentalistas Agrícolas del Perú (ABYDAP) / Peruvian Association of Agricultural Librarians and Documentalists, c/o Biblioteca Agrícola Nacional, Universidad Nacional Agraria, Apdo 456, Lima
Founded: 1976
Focus: Agri
Periodicals
ABYDAP Informa (weekly)
Boletín Técnico (weekly) . . . 09448

Asociación de Ingenieros Civiles del Perú, Nicolás de Piérola 788, Lima
Focus: Civil Eng . . . 09449

Asociación Electrotécnica Peruana, Av República de Chile 284, Lima
Founded: 1943
Focus: Electric Eng . . . 09450

Asociación Latinoamericana Científico de Plantas, c/o Universidad Agraria, Apdo 456, Lima
Focus: Botany . . . 09451

Asociación Médica Peruana Daniel A. Carrión, Jirón Ucayali 218, Lima
Founded: 1920; Members: 1500
Focus: Med . . . 09452

Asociación Nacional de Escritores y Artistas (ANEA), Puno 421, Lima 1
Founded: 1938; Members: 954
Focus: Lit . . . 09453

Asociación Peruana de Archiveros, c/o Archivo Nacional, Palacio de la Nación Justicia, Apdo 3124, Lima
Focus: Archives . . . 09454

Asociación Peruana de Astronomía (A.P.A.), Enrique Palacios 374, Chorrillos, Lima 9
Founded: 1946; Members: 400
Focus: Astronomy . . . 09455

Asociación Peruana de Bibliotecarios, Apdo 3760, Lima
Focus: Libraries & Bk Sci . . . 09456

Center for Andean Regional Studies Bartolomé de las Casas, Pampa de la Allianza 465, Apdo 477, Cusco
T: (084) 232544
Founded: 1974
Gen Secr: Dr. Guido Delran
Focus: Hist . . . 09457

Centro de Estudios Histórico-Militares del Perú, Paseo Colón 190, Lima 1
Focus: Hist; Military Sci . . . 09458

Centro de Investigación y Restauración de Bienes Monumentales del Instituto Nacional de Cultura, Casilla 5247, Lima
Focus: Preserv Hist Monuments 09459

Centro del PEN Internacional, Apdo 1161, Lima
Founded: 1940; Members: 25
Pres: Dr. José Gálvez; Gen Secr: Fernando Romero
Focus: Lit 09460

Centro Latinoamericana de Estudios y Difusión de la Construcción en Tierra (CLEDTIERRA), Apdo 5603, Correo Central, Lima
T: (014) 406027; Tx: 25201
Gen Secr: Silvia Matuk
Focus: Eng 09461

Centro Nacional de Patologia Animal, Apdo 1128, Lima
Focus: Vet Med 09462

CIAT Andean Zone Network for Bean Research, Apdo 14-0185, Lima 14
T: 411711; Fax: 411711
Founded: 1988
Gen Secr: Guillermo E. Gálvez
Focus: Crop Husb 09463

Colegio de Arquitectos del Perú, Av San Felipe 999, Lima 11
Founded: 1962; Members: 1611
Focus: Archit
Periodical
Habitar 09464

Consejo Andino de Cienccia y Tecnología (CACYT), c/o Depto de Tecnología, JUNAC, Casilla 548, Lima 18
T: 414212; Tx: 20104; Fax: 420911
Founded: 1983
Gen Secr: Prof. Carlos Aguirre
Focus: Sci; Eng 09465

Consejo Nacional de la Universidad Peruana, Calle Aldabas 3, Surco, Apdo 4664, Lima 33
Founded: 1969; Members: 33
Focus: Sci 09466

Federación de Sociedades Latinoamericanas del Cáncer, Apdo 4135, Lima
Focus: Cell Biol & Cancer Res 09467

Federación Médica Peruana, Apdo 4439, Lima
Founded: 1942; Members: 1230
Pres: Dr. Vicente Ubillús; Gen Secr: Dr. Enrique Fernández
Focus: Med
Periodical
Boletín de la Federación Médica Peruana 09468

Instituto de Zoonosis e Investigación Pecuaria (IZIP), Camilo Carrillo 402, Apdo 1128, Lima
T: 248530
Founded: 1940; Members: 150
Focus: Zoology 09469

Instituto Geográfico Nacional, Apdo 2038, Lima
Founded: 1921; Members: 260
Focus: Geography
Periodical
Boletín Informativo 09470

Instituto Peruano de Ingenieros Mecánicos, Av República de Chile 284, Lima
Focus: Eng 09471

Junta de Control de Energía Atómica, Espinar 250, San Miguel, Apdo 914, Lima
Focus: Nucl Res 09472

Liga Nacional de Higiene y Profilaxia Social, Apdo 2563, Lima
Founded: 1923
Focus: Hygiene 09473

Servicio de Investigación y Promoción Agraria, c/o Estación Experimental de la Molina, Apdo 2791, Lima
Founded: 1927
Focus: Agri 09474

Sociedad de Ingenieros del Perú, Av N. de Piérola 788, Apdo 1314, Lima
Focus: Eng 09475

Sociedad Entomológica del Perú, Apdo 4796, Lima
Founded: 1956; Members: 600
Pres: Oswaldo Gamero; Gen Secr: Pedro G. Aguilar
Focus: Entomology
Periodical
Revista Peruana de Entomología . . . 09476

Sociedad Geográfica de Lima, Jirón Puno 456, Apdo 100-1176, Lima 100
Founded: 1888
Pres: Dr. Santiago E. Antunez de Mayolo
Focus: Geography
Periodicals
Anuario Geográfico del Perú (weekly)
Boletín de la Sociedad Geográfica de Lima (weekly)
Diccionario Geográfico del Perú (weekly) 09477

Sociedad Geológica del Perú (SGP), Arnaldo Marquez 2277, Lima
T: 623948
Founded: 1924; Members: 700
Focus: Geology 09478

Sociedad Latinoamericana de Investigación Pediátrica, Jirón Hancayo 190, Lima
Focus: Pediatrics 09479

Sociedad Nacional Agraria, A. Miró Quesada 327-341, Apdo 350, Lima
Founded: 1824
Focus: Agri 09480

Sociedad Nacional de Minería, Pl San Martín 917, Lima
Focus: Mining 09481

Sociedad Peruana de Espeleología, Porta 540, Miraflores, Lima
Founded: 1965
Focus: Speleology 09482

Sociedad Peruana de Eugenesia, Apdo 2563, Lima
Founded: 1943
Focus: Genetics 09483

Sociedad Peruana de Historia de la Medicina, Apdo 987, Lima
Founded: 1939; Members: 110
Pres: Dr. Carlos E. Paz Soldán; Gen Secr: Dr. J.B. Lastres
Focus: Med; Cultur Hist
Periodical
Anales 09484

Sociedad Peruana de Ortopedia y Traumatología, Villalta 218, Lima
Focus: Orthopedics; Traumatology . . . 09485

Sociedad Peruana de Tisiología y Enfermedades Respiratorias, Domingo Casanova 116, Lince, Lima
Founded: 1935
Focus: Pulmon Dis 09486

Sociedad Química del Perú, Apdo 891, Lima, 100
Founded: 1933
Pres: Dr. Gastón Pons Muzzo; Gen Secr: Dr. Juan de Dios Guevara
Focus: Chem
Periodical
Boletín (weekly) 09487

Philippines

Academia Filipina, 1746 Singalong, Paco 2801, Manila
Members: 21
Pres: R.P.A. Hidalgo; Gen Secr: Francisco Zaragoza
Focus: Sci 09488

Agricultural Economics Society of South East Asia (AESSEA), c/o Land Bank of the Philippines, Manila
Founded: 1975
Pres: Pablito M. Villegas
Focus: Agri 09489

ASEAN Association for Planning and Housing (AAPH), Strata Bldg, Emerald Av, Pasig, Manila
T: (02) 6313462
Founded: 1979
Gen Secr: Teresita Flor Ruiz
Focus: Urban Plan
Periodicals
AAPH Bulletin (weekly)
AAPH Data Resouces (weekly)
Interlink (weekly) 09490

ASEAN Neurological Society, Room 205, PCS Bldg, 992 EDSA, Quezon City, Manila
T: (02) 984973
Focus: Neurology 09491

ASEAN Training Centre for Preventive Drug Education, c/o Dept of Health Education, College of Education, University of the Philippines, Diliman, Quezon City, Manila 3004
T: (02) 999756; Fax: (02) 992863
Founded: 1980; Members: 6
Gen Secr: Evelina A. Mejilano
Focus: Educ 09492

Asian Alliance of Appropriate Technology Practitioners (APPROTECH ASIA), c/o Philippine Social Development Center, Magallanes corner Real Sts, Intramuros, Manila
T: (02) 479918, 497041; Tx: 40404; Fax: (02) 8189720
Founded: 1980; Members: 17
Gen Secr: Lilia O. Ramos
Focus: Eng 09493

Asian Association for Biology Education (AABE), c/o ISMED, University of the Philippines, Diliman, Quezon City, Manila
T: (02) 984276
Founded: 1965
Gen Secr: Lucille C. Gregorio
Focus: Educ; Bio 09494

Asian Association of Agricultural Colleges and Universities (AAACU), c/o SEARCA College, Laguna 4031
T: (0942) 2576, 2290; Tx: 40904
Gen Secr: Dr. S.A. Srinilta
Focus: Educ; Agri 09495

Asian Confederation of Teachers (ACT), POB 163, Manila
Focus: Educ 09496

Asian Fisheries Society (AFS), c/o MOC, POB 1501, Makati, Manila
T: (02) 8180466; Tx: 45658
Founded: 1984; Members: 1000
Pres: Dr. Chua Tia Eng
Focus: Fisheries
Periodical
Asian Fisheries Science (weekly) . . . 09497

Asian Institute of Management (AIM), 123 Paseo de Roxas, Makati, Manila 3117
T: (02) 863260, 874011; Tx: 63778; Fax: (02) 8179240
Founded: 1968
Focus: Business Admin
Periodical
The Asian Manager (weekly) 09498

Asian Institute of Tourism (AIT), c/o University of the Philippines, Dilman, Quezon City, Manila 3004
T: (02) 969071
Founded: 1976
Gen Secr: Evangeline M. Ortiz
Focus: Econ
Periodical
AIT Newsbrief 09499

Asian NGO Coalition for Agrarian Reform and Rural Development (ANGOC), POB 870, Makati, Manila 1200
T: (02) 8163033, 8151198; Tx: 23136; Fax: (02) 8151198
Founded: 1979
Focus: Agri
Periodicals
Information Notes (weekly)
Lok Niti (weekly) 09500

Asian-Pacific Weed Science Society (APWSS), C-214 Bioscience Bldg, U.P. Los Banos College, Laguna 4031
Founded: 1967; Members: 300
Gen Secr: Dr. Aurora M. Baltazar
Focus: Agri; Botany; Hort
Periodicals
APWSS Newsletter (2-3 times annually)
Proceedings of Conferences 09501

Asian Recycling Association (ARA), Prosamapi Cooperative, Isarog Farms, Palestina Pili, Camarines Sur
Founded: 1978
Gen Secr: Portia A. Nayve
Focus: Ecology
Periodical
ARA Newsletter (weekly) 09502

Asian Regional Training and Development Organization (ARTDO), V.V. Soliven Bldg, Ste 2039, EDA Greenhills, San Juan, Manila 1500
T: (02) 707742
Founded: 1974
Gen Secr: Bernardo F. Ople
Focus: Educ
Periodicals
ARTDO Journal (weekly)
ARTDO Report (weekly) 09503

Asian Rice Farming Network, c/o IRRI, POB 933, Manila
Founded: 1974
Focus: Agri 09504

Asian Women's Research and Action Network (AWRAN), c/o Philippina, 12 Pasajede La Paz, Project 4, Quezon City, Manila
Founded: 1982
Focus: Soc Sci 09505

Asia Pacific Grouping of Consulting Engineers (ASPAC), Emerald Av, Pasig, POB 12019, Ortigas Central Post Office, Manila
T: (02) 6312782
Gen Secr: P. Hein
Focus: Eng 09506

Asia Pacific Physics Teachers and Educators Association (APPTEA), c/o ISMED, University of the Philippines, Diliman, Quezon City, Manila
Founded: 1986
Gen Secr: Estela J. Rodriguez
Focus: Educ; Physics 09507

Asia Pacific Quality Control Organization (APQCO), Philippine Centre for Economic Development, Ste 217, POB 116, UP Campus, Diliman
T: (02) 961449 ext 217; Tx: 40404; Fax: (02) 5217225
Founded: 1981; Members: 13
Gen Secr: Miflora M. Gatchalian
Focus: Materials Sci 09508

Association of Deans of Southeast Asian Graduate Schools of Management (ADSGM), Paseo de Roxas 123, Legaspi Village, Makati, Manila
T: (02) 874011; Tx: 63778; Fax: (02) 8179240
Pres: Dr. Gaston Z. Ortigas
Focus: Educ; Business Admin 09509

Association of Pediatric Societies of the Southeast Asian Region (APSSEAR), c/o Medical Center of Manila, 1122 General Luna St, Ste 326, POB EA 100, Manila
T: (02) 507874; Fax: (02) 7216569
Founded: 1974
Gen Secr: Dr. Perla D. Santos Ocampo
Focus: Pediatrics
Periodical
Bulletin of the Association of Pediatric Societies of the Southeast Asian Region (weekly) . 09510

Association of Southeast Asian Marine Scientists (ASEAMS), c/o UP, POB 1, Diliman, Manila
T: (02) 982471; Tx: 2231
Founded: 1986
Pres: Edgardo D. Gomez
Focus: Oceanography 09511

Association of Special Libraries of the Philippines (ASLP), c/o College of Public Administration Library, University of the Philippines, POB 474, Manila
Founded: 1954; Members: 300
Pres: F.C. Mercado; Gen Secr: Edna P. Ortiz
Focus: Libraries & Bk Sci
Periodicals
ASLP Bulletin (weekly)
ASLP Newsletter (weekly) 09512

Colombo Plan Staff College for Technician Education (CPSC), Bldg Block C, UI Complex, Meralco Av, Pasig, Manila
T: (02) 6730886; Tx: 43175; Fax: (02) 6730891
Founded: 1974; Members: 21
Gen Secr: Isaac Goodine
Focus: Educ
Periodical
Colombo Plan Staff College for Technician Education Newsletter (weekly) 09513

Confederation of Medical Associations in Asia and Oceania (CMAAO), 862 Guillermo Masangkay St, Binondo, Manila 1006
T: (02) 216405
Founded: 1956
Gen Secr: Dr. Primitivo D. Chua
Focus: Med 09514

Conference of Asian-Pacific Pastoral Institutes (CAPPI), c/o EAPI, POB 221, UP Campus, Manila
T: (02) 983182
Founded: 1986
Gen Secr: Geoffrey King
Focus: Rel & Theol
Periodical
CAPPI Newsletter (weekly) 09515

Crop Science Society of the Philippines, c/o Institute of Plant Breeding, UPLB, Laguna
Founded: 1970; Members: 2500
Pres: William D. Dar; Gen Secr: Manuel M. Lantin
Focus: Agri
Periodical
Philippine Journal of Crop Science (3 times annually) 09516

East Asian Pastoral Institute (EAPI), c/o Ateneo de Manila University, Quezon City, Manila
T: (03) 983182; Fax: (03) 9217534
Founded: 1954
Gen Secr: Geoffrey King
Focus: Rel & Theol
Periodicals
The Bridge Newsletter
East Asian Pastoral Review (weekly) . . 09517

Eastern Regional Organization for Public Administration (EROPA), POB 464, Manila
T: (02) 993861
Founded: 1958; Members: 91
Gen Secr: Raul P. de Guzman
Focus: Public Admin
Periodicals
Asian Review of Public Administration
EROPA Bulletin (weekly) 09518

ESCAP/WMO Typhoon Committee (TC), 1300 Domestic Rd, Pasay, POB 7285, Domestic Airport, Manila
T: (02) 9228055; Tx: 22250
Founded: 1968; Members: 10
Gen Secr: Roman L. Kintanar
Focus: Geophys
Periodical
ESCAP/WMO Newsletter 09519

International Association of Historians of Asia (IAHA), National Archives, T.M. Kalaw St, Manila
Founded: 1961
Focus: Hist 09520

Los Baños Biological Club / Baños Biological Club, c/o College, Laguna
Founded: 1923
Focus: Bio 09521

National Research Council of the Philippines (NRCP), Bicutan Taguig, Manila
Founded: 1933
Pres: Dr. Ernesto P. Sonido
Focus: Sci
Periodicals
NRCP Research Bulletin (weekly)
NRCP Technical Bulletin (weekly) . . . 09522

Philippine Association of Nutrition, c/o Food and Nutrition Research Institute, Corner Taft Av and Pedro Gil St, Ermita, Manila
Founded: 1947; Members: 1100
Pres: Dr. Segundo C. Serrano

Focus: Food
Periodical
Philippine Journal of Nutrition (weekly) . 09523

Philippine Atomic Energy Commission, Don Mariano Marcos Av, Diliman, Quezon City
Founded: 1958
Focus: Nucl Res 09524

Philippine Council of Chemists, 2227 Severino Reyes St, Santa Cruz, POB 1202, Manila 2805
Founded: 1958; Members: 200
Pres: Miguel G. Ampil; Gen Secr: P.B. Carbonell
Focus: Chem 09525

Philippine Historical Association (PHA), c/o University of the East, Sampaloc, Manila
Founded: 1955; Members: 500
Pres: Prof. D.G. Capino; Gen Secr: Prof. J.L. Revilla
Focus: Hist
Periodical
PHA Bulletin (weekly) 09526

Philippine Institute of Architects, POB 350, Manila
Founded: 1933; Members: 405
Pres: Manuel T. Mañosa; Gen Secr: M.P. Angeles
Focus: Archit 09527

Philippine Library Association, c/o National Library, T.M. Kalaw St, Room 301, Manila
T: (02) 590177; Tx: 40726
Founded: 1923; Members: 1025
Pres: Cora M. Nera; Gen Secr: C.A. Santacruz
Focus: Libraries & Bk Sci
Periodicals
Bulletin (weekly)
Newsletter 09528

Philippine Medical Association (P.M.A.), P.M.A. House, North Av, Quezon City
T: (02) 973514, 974974; Fax: (02) 974974
Founded: 1903; Members: 140
Pres: Dr. P. Chua; Gen Secr: A. Carino
Focus: Med
Periodical
P.M.A. Newsletter 09529

Philippine Paediatric Society, POB 3527, Manila
Founded: 1947; Members: 620
Pres: E.M. Rigor; Gen Secr: Miguel Noche
Focus: Pediatrics
Periodical
Philippine Journal of Paediatrics (weekly) 09530

Philippine Pharmaceutical Association, Cardinal Bldg, Corner Herran and F. Agoncillo Sts, Ermita, Manila
Founded: 1920; Members: 4300
Pres: Lourdes Talag Echauz; Gen Secr: Dr. Elvira F. Silva
Focus: Pharmacol
Periodical
Journal 09531

Philippine Society of Parasitology, c/o College of Public Health, University of the Philippines, 625 Pedro Gil St, POB EA-460, Ermita, Manila 1000
T: (02) 596808; Fax: (02) 5211394
Founded: 1930; Members: 450
Pres: W.U. Tiu
Focus: Med 09532

Philippine Veterinary Medical Association (PVMA), c/o College of Veterinary Medicine, University of the Philippines, Diliman, Quezon City 1101
T: (02) 995436, 8161107
Founded: 1907; Members: 2800
Pres: Dr. Grace D. de Ocampo; Gen Secr: Dr. Elizabeth A. Rubio-Lazaro
Focus: Vet Med 09533

Philosophical Association of the Philippines, POB 3797, Manila
Founded: 1973; Members: 200
Pres: Dr. R.C. Reyes; Gen Secr: Prof. J.L. Revilla
Focus: Philos
Periodical
Philippine Journal of Philosophy (weekly) 09534

Radioisotope Society of the Philippines (RSP), c/o Philippine Nuclear Research Institute, Don Marino Marcos Av, Diliman, Quezon City
Founded: 1961; Members: 2500
Pres: Alumanda M. Dela Rosa; Gen Secr: Maria F.O. Medina
Focus: Nucl Res
Periodical
The Nucleus (weekly) 09535

Society for the Advancement of Research, c/o Los Baños College, University of the Philippines, Laguna
Founded: 1930; Members: 530
Pres: Dr. Felix Librero; Gen Secr: Dr. B.R. Sumayo
Focus: Agri 09536

Society for the Advancement of the Vegetable Industry (SAVI), c/o Los Baños College, University of the Philippines, Laguna
Founded: 1967
Pres: E.T. Rasco; Gen Secr: Maria Malixi-Paje
Focus: Food
Periodical
Proceedings of Annual Seminar-Workshop 09537

United Technological Organizations of the Philippines, 512-516 Samanillo Bldg, Escolta, Manila
Founded: 1946
Focus: Eng 09538

Poland

Białostockie Towarzystwo Naukowe / Societas Scientiarum Bialostocensis, c/o Dom Technika, Ul Skłodowskiej-Curie 2, 15-097 Bialystok
T: 21227
Focus: Sci
Periodical
Białostoczyzna (weekly) 09539

Bydgoskie Towarzystwo Naukowe (BTN), Ul Jezuicka 4, 85-102 Bydgoszcz
T: 222268
Founded: 1959; Members: 359
Focus: Sci 09540

Central European Mass Communication Research Documentation Centre (CECOM), Rynek Glowny 23, 31-008 Krakow
Founded: 1974
Gen Secr: Walery Pisarek
Focus: Mass Media 09541

Copernicus Astronomical Centre, Ul Bartycka 18, 00-716 Warszawa
Founded: 1957
Focus: Astronomy 09542

Gdánskie Towarzystwo Naukowe (GTN), Ul Grodzka 12, 80-841 Gdánsk
T: (058) 312124
Founded: 1922; Members: 498
Pres: Prof. W. Prosnak; Gen Secr: Dr. A. Zbierski
Focus: Sci
Periodicals
Acta Biologica (weekly)
Acta Technica Gedanensia (weekly)
Gdańskie Studia Jezykoznawcze (weekly)
Gdańsk Wczesnośredniowieczny (weekly)
Peribalticum (weekly)
Pomorskie Monografie Toponomastyczne (weekly)
Pomorze Gdańskie (weekly)
Rocznik Gdański (weekly)
Seria Zródel (weekly) 09543

International Commission on Trichinellosis (ICT), Ul Przybyszewskiego 49, 60-355 Poznan
Founded: 1958; Members: 42
Focus: Vet Med
Periodical
Proceedings of the International Commission on Trichinellosis (weekly) 09544

Lódzkie Towarzystwo Naukowe (LTN) / Société des Sciences et Lettres de Łódź, Piotrkowska 179, 90-447 Łódź
T: (042) 361026, 361995; Fax: (042) 362415
Founded: 1936; Members: 250
Pres: Prof. Dr. Stanisław Diszewski; Gen Secr: Prof. Dr. Julian Lawrynowicz
Focus: Sci
Periodicals
Acta Archaeologica Lodziensia (weekly)
Acta Geographica Lodziensia
Biuletyn Peryglacjalny (weekly)
Bulletin de la Société des Sciences et des Lettres de Łódź
Prace Polonistyczne (weekly)
Przeglad Socjologiczny (weekly)
Rozprawy Komisji Jeżykowej (weekly)
Sprawozdania z Czynności i Posiedzeń Naukowych /various/
Studia Prawno-Ekonomiczne (weekly)
Sylwetki Lódzkich Uczonych (weekly)
Szlakami nauki (weekly)
Zagadnienia Rodzajów Literackich (weekly) 09545

Lubelskie Towarzystwo Naukowe, Plac Litewski 5, Lublin
T: (081) 26236
Founded: 1958; Members: 200
Focus: Sci 09546

Lubuskie Towarzystwo Naukowe, Ul Gwardii Ludowej 14, 65-536 Zielona Góra
T: (068) 63375
Focus: Sci
Periodicals
Przeglad Lubuski
Rocznik Lubuski 09547

Opolskie Towarzystwo Przyjaciót Nauk (OTPN), Ul Ozimska 46, 45-058 Opole
T: (077) 32112
Founded: 1955; Members: 327
Focus: Sci
Periodicals
Wydział Języka i Literatury
Wydział Nauk Historyczno-Społecznych
Wydział Nauk Medycznych
Wydział Nauk Przyrodniczych 09548

Polska Akademia Nauk (PAN), Palac Kultury i Nauki, POB 24, 00-901 Warszawa
T: (022) 200211; Tx: 813929
Founded: 1952; Members: 417
Pres: Jan Kostrzewski
Focus: Sci
Periodical
Bulletin 09549

Polskie Lekarskie Towarzystwo Radiologiczne (PLTR), Barnacha 1a, 02-097 Warszawa
T: (022) 223005; Fax: (022) 223005
Founded: 1925; Members: 1280
Focus: Radiology
Periodical
Polski Przeglag Radiologii (weekly) . . . 09550

Polskie Towarzystwo Anatomiczne, Chalubinskiego 5, 02-004 Warszawa
T: (022) 281041
Founded: 1923; Members: 400
Focus: Anat 09551

Polskie Towarzystwo Antropologiczne, Ul Marymoncka 34, 01-813 Warszawa
T: (022) 340431
Founded: 1925; Members: 500
Focus: Anthro 09552

Polskie Towarzystwo Archeologiczne i Numizmatyczne (PTAiN), Jezuicka 6, 00-958 Warszawa
T: (022) 313928
Founded: 1953; Members: 7000
Pres: Leszek Kokocinski; Gen Secr: Kazimierz Madej
Focus: Archeol; Numismatics
Periodicals
Biuletyn Numizmatyczny (weekly)
Wiadomości Numizmatyczne (weekly)
Z Otchłani Wieków (weekly) 09553

Polskie Towarzystwo Astronautyczne (PTA), c/o Military Institute of Aviation Medicine, Ul Krasińskiego 54, 01-755 Warszawa
T: (022) 250441 ext 62684
Founded: 1954; Members: 420
Pres: Prof. Dr. Piotr Wolański; Gen Secr: Zbigniew Sarol
Focus: Aero
Periodicals
Astronautyka: Astronautics (weekly)
Postepy Astronautyki: Progress in Astronautics (weekly) 09554

Polskie Towarzystwo Astronomiczne (PTA), Bartycka 18, 00-716 Warszawa
T: (022) 410041
Founded: 1923; Members: 240
Focus: Astronomy 09555

Polskie Towarzystwo Balneologii, Bioklimatologii i Medycyny Fizykalnej, Ul Slowackiego 8-10, 60-823 Poznan
T: (061) 46338, 46547
Founded: 1954; Members: 500
Focus: Physiology; Physical Therapy . . 09556

Polskie Towarzystwo Biochemiczne / Polish Biochemical Society, Ul Freta 16, 00-227 Warszawa
T: (022) 311304
Founded: 1958; Members: 1320
Focus: Chem; Biochem
Periodicals
Monografie Biochemiczne: Biochemical Monographs (weekly)
Postopy Biochemii: Advances in Biochemistry (weekly) 09557

Polskie Towarzystwo Botaniczne, Ul Rekowiecka 26-30, 02-528 Warszawa
T: (022) 499340
Founded: 1922; Members: 1200
Focus: Botany
Periodicals
Acta Agrobotanica (weekly)
Acta Myologica (weekly)
Acta Societatis Botanicorum Poloniae (weekly)
Monographiae Botanicae (weekly)
Rocznik Dendrologiczny (weekly)
Wiadomości Botaniczne (weekly) 09558

Polskie Towarzystwo Chemiczne (PTCh), Ul Freta 16, 00-227 Warszawa
T: (022) 311304
Founded: 1918; Members: 2900
Focus: Chem
Periodical
Wiadomości Chemiczne (weekly) 09559

Polskie Towarzystwo Dermatologiczne / Polish Dermatological Society, Ul Przybyszewskiego 49, 60-355 Poznań
T: (061) 676841 ext 285
Founded: 1922; Members: 1200
Pres: Prof. Jerzy Bowszyc; Gen Secr: Dr. W. Silny
Focus: Derm; Venereology
Periodical
Przeglad Dermatologiczny (weekly) . . . 09560

Polskie Towarzystwo Dermatologiczne (PTD), c/o Dept of Dermatology, Warsaw School of Medicine, Ul Koszykowa 82a, 02-008 Warszawa
T: (022) 215180
Founded: 1922; Members: 1300
Focus: Derm
Periodical
Przeglad Dermatologiczny: Polish Journal of Dermatology (weekly) 09561

Polskie Towarzystwo Ekonomiczne (PTE) / Polish Economic Society, Nowy Swiat 49, 00-042 Warszawa
T: (022) 274857
Founded: 1918; Members: 11500
Pres: Prof. Zdzislaw Sadowski; Gen Secr: Prof. Urszula Plowiec
Focus: Econ
Periodical
Ekonomista (weekly) 09562

Polskie Towarzystwo Endokrynologiczne, Plac Starynkiewicza 3, 02-015 Warszawa
T: (022) 281159
Founded: 1951; Members: 400
Focus: Endocrinology
Periodical
Endokrynologia Polska (weekly) 09563

Polskie Towarzystwo Entomologiczne (PTE), Nowy Swiat 72, 00-330 Warszawa
T: (022) 269617
Founded: 1920; Members: 800
Focus: Entomology 09564

Polskie Towarzystwo Farmaceutyczne, Dluga 16, 00-238 Warszawa
T: (022) 311542
Founded: 1947; Members: 3000
Focus: Pharmacol
Periodicals
Acta Poloniae Pharmaceutica (weekly)
Bromatologia i Chemia Toksykologiczna (weekly)
Farmacja Polska (weekly) 09565

Polskie Towarzystwo Filologiczne (PTF), Krakowskie Przedmiéscie 1, 00-047 Warszawa
Founded: 1893; Members: 500
Focus: Ling; Lit 09566

Polskie Towarzystwo Filozoficzne (PTF), Palac Staszica, 00-330 Warszawa
Founded: 1904; Members: 738
Focus: Philos
Periodical
Ruch Filozoficzny: Philosophical Movement (weekly) 09567

Polskie Towarzystwo Fizjologiczne, Jaczewskiego 8, 20-090 Lublin
T: (081) 26510
Founded: 1936; Members: 669
Focus: Physiology 09568

Polskie Towarzystwo Fizyczne (PTF) / Polish Physical Society, Hoza 69, 00-681 Warszawa
T: (02) 6212668; Fax: (02) 6212668
Founded: 1919; Members: 1840
Pres: Henryk Szymczak; Gen Secr: Ireneusz Strzalkowski
Focus: Physics
Periodicals
Acta Physica Polonica ser. A: Europhysics Journal (weekly)
Acta Physica Polonica ser. B: Europhysics Journal (weekly)
Delta (weekly)
Postepy Fizyki: Advances in Physics (weekly)
Reports on Mathematical Physics (weekly) 09569

Polskie Towarzystwo Fizyki Medycznej, Wawelska 15, 02-034 Warszawa
T: (022) 224431
Founded: 1965; Members: 380
Focus: Physiology
Periodical
Postępy Fizyki Medycznej: Progress in Medical Physics (weekly) 09570

Polskie Towarzystwo Geograficzne (PTG) / Polish Geographical Society, Krakowskie Przedmiéscie 30, 00-325 Warszawa
T: (022) 261794
Founded: 1918; Members: 2300
Pres: Prof. Dr. Wojciech Stankowski; Gen Secr: Dr. Maria M. Wilczyńska-Woloszyn
Focus: Geography
Periodicals
Czasopismo Geograficzne (weekly)
Fotointerpretacja w Geografii (weekly)
Polski Przeglad Kartograficzny (weekly)
Poznaj Świat (weekly) 09571

Polskie Towarzystwo Geologiczne (PTG) / Geological Society of Poland, Oleandry 2a, 30-063 Kraków
T: (012) 332041
Founded: 1921; Members: 1450
Pres: Prof. Dr. Andrzej Slaczka; Gen Secr: Dr. Janusz Magiera
Focus: Geology
Periodical
Annales Societatis Geologorum Poloniae: Rocznik Polskiego Towarzystwa Geologicznego . . 09572

Polskie Towarzystwo Ginekologiczne, Ul Karowa 2, 00-315 Warszawa
T: (022) 261754
Founded: 1922; Members: 2025
Focus: Gynecology 09573

Polskie Towarzystwo Gleboznawcze (P.T.G.), Ul Wisniowa 61, 02-520 Warszawa
T: (022) 494816
Founded: 1937; Members: 1070
Focus: Agri 09574

Polskie Towarzystwo Hematologiczne, Ul Marcinkowskiego 1, 50-368 Wroclaw
T: (071) 224959
Founded: 1950; Members: 280
Focus: Hematology 09575

Polskie Towarzystwo Higieny Psychicznej, Ul Targowa 59, 03-729 Warszawa
T: (022) 186599
Founded: 1958; Members: 450
Focus: Hygiene

Periodicals
Zagadnienia Wychowawcze a Zdrowie Psychiczne: Upbringing Problems and Mental (weekly)
Zdrowie Psychiczne: Mental Health (weekly)
. . . . 09576

Polskie Towarzystwo Historii Medycyny i Farmacji / Polish Society of History of Medicine and Pharmacy, Ul Długa 16, 00-238 Warszawa
T: (022) 310241
Founded: 1957; Members: 400
Focus: Med; Pharmacol; Cultur Hist
Periodical
Archiwum Historii Medycyny (weekly) . . 09577

Polskie Towarzystwo Historyczne (PTH), Rynek Starego Miasta 29-31, Warszawa
T: (022) 316341
Founded: 1886; Members: 3000
Focus: Hist 09578

Polskie Towarzystwo Immunologiczne (PTI) / Polish Society for Immunology, c/o Dept of Pathophysiology and Immunology, Institute of Basic Medical Sciences, Military Medical Academy, Pl Hallera 1, 90-647 Łódź
T: (042) 324273; Fax: (042) 324273
Founded: 1969; Members: 439
Focus: Immunology 09579

Polskie Towarzystwo Jezykoznawcze (PTJ), Al A. Mickiewicza 9-11, 31-120 Kraków
Founded: 1925; Members: 763
Pres: Prof. Irena Bajerowa; Gen Secr: Dr. W. Sedzik
Focus: Ling
Periodical
Biuletyn Polskiego Towarzystwa Jezykoznawczego: Bulletin de la Société Polonaise de Linguistique (weekly) 09580

Polskie Towarzystwo Kardiologiczne (P.T.K.), Nowogrodzka 59, 02-006 Warszawa
T: (022) 283507
Founded: 1962; Members: 1236
Focus: Cardiol 09581

Polskie Towarzystwo Lekarski (PTL), Al Ujazdowskie 24, 00-478 Warszawa
T: (022) 288699
Founded: 1951; Members: 20000
Focus: Med 09582

Polskie Towarzystwo Lesne (PTL), Ul Bitwy Warszawskiej 1920 R.3, 02-362 Warszawa
T: (022) 221470; Fax: (022) 224935
Founded: 1882; Members: 3229
Pres: Prof. Dr. Andrzej Szujecki; Gen Secr: Dr. G. Wetuszewski
Focus: Forestry
Periodical
Sylwan (weekly) 09583

Polskie Towarzystwo Ludoznawcze (PTL) / Polish Ethnological Society, Szewska 36, 50-139 Wroclaw
T: (071) 443832, 444613
Founded: 1895; Members: 1154
Pres: Dr. Zygmunt Klodnicki
Focus: Ethnology
Periodicals
Archiwum Ethnograficzne: Ethnographic Archives (weekly)
Atlas Polskich Strojów Ludowych: Atlas of Polish Folk Costumes (weekly)
Biblioteka Populamonaukowa: Library of Popular Science (weekly)
Literatura Ludowa: Folk Literature (weekly)
Łódzkie Studia Etnograficzne: Ethnographic Works and Materials (weekly)
Lud: The People (weekly)
Prace Ethnologiczne: Ethnological Works (weekly)
Prace i Materiały Etnograficzne: Ethnographic Works and Materials (weekly) 09584

Polskie Towarzystwo Matematyczne (PTM) / Polish Mathematical Society, Ul Sniadeckich 8, 00-950 Warszawa
T: (022) 299592
Founded: 1919; Members: 2115
Pres: Julian Musielak; Gen Secr: Jan Butkiewicz
Focus: Math
Periodicals
Commentationes Mathematicae (weekly)
Dydaktyka Matematyki (weekly)
Fundamenta Informaticae (weekly)
Matematyka Stosowana (weekly)
Wiadomości Matematyczne (weekly) . . . 09585

Polskie Towarzystwo Mechaniki Teoretycznej i Stosowanej / Polish Society of Theoretical and Applied Mechanics, c/o Wydzial Inzynierii Ladowej Politechniki Warszawskiej, Al Armii Ludowej 16, Pok 706, 00-632 Warszawa
T: (022) 257180
Founded: 1958; Members: 1192
Focus: Eng; Physics
Periodical
Mechanika Teoretyczna i Stosowana: Journal of Theoretical and Applied Mechanics (weekly)
. . . . 09586

Polskie Towarzystwo Medycyny Pracy, Ul B. Bieruta 20, 41-200 Sosnowiec
T: 660640
Founded: 1969; Members: 2590
Focus: Med 09587

Polskie Towarzystwo Mikrobiologów / Polish Society of Microbiology, Ul Chocimska 24, 00-791 Warszawa
Founded: 1927; Members: 600
Focus: Microbio
Periodicals
Acta Microbiologica Polonica (weekly)
Medycyna Doswiadczalna i Mikrobiologia (weekly)
Postepy Mikrobiologii (weekly) 09588

Polskie Towarzystwo Milosników Astronomii, Ul Solskiego 30, 31-027 Kraków
T: (012) 23892
Founded: 1921; Members: 2500
Focus: Astronomy 09589

Polskie Towarzystwo Mineralogiczne, Al Mickiewicza 30, 30-059 Kraków
T: (012) 34330
Founded: 1969; Members: 150
Focus: Mineralogy 09590

Polskie Towarzystwo Nauk Weterynaryjnych (PTNW) / Polish Society of Veterinary Sciences, Ul Grochowska 272, 03-849 Warszawa
T: (022) 103397
Founded: 1953; Members: 2620
Pres: Prof. Dr. E. Prost; Gen Secr: Dr. T. Frymus
Focus: Vet Med
Periodical
Medycyna Weterynaryjna (weekly) . . . 09591

Polskie Towarzystwo Nautologiczne (PTN) / Polish Nautological Society, Sienkiewicza 3, 81-374 Gdynia
T: (058) 204975
Founded: 1958
Pres: Prof. Dr. Daniel Duda; Gen Secr: Aleksander Gosk
Focus: Navig
Periodical
Nautologia (weekly) 09592

Polskie Towarzystwo Neofilologiczne (PTN), Ul Mielczynskiego 27-29, 61-725 Poznan
T: (061) 532682
Founded: 1929; Members: 2120
Focus: Ling; Lit
Periodical
Neofilolog (weekly) 09593

Polskie Towarzystwo Neurologiczne (P.T.N.) / Polish Neurological Society, c/o Neurological Rehabilitation Hospital, Ul Ujejskiego 37, 05-510 Konstancin-Jeziorna
T: 564041; Fax: 564972
Founded: 1934; Members: 1400
Focus: Neurology
Periodical
Neurologia i Neurochirurgia Polska (weekly)
. . . . 09594

Polskie Towarzystwo Orientalistyczne (PTO) / Polish Oriental Society, Ul Sniadeckich 8, 00-656 Warszawa
T: (022) 217332
Founded: 1922; Members: 300
Focus: Ethnology
Periodical
Przegląd Orientalistyczny: Oriental Review (weekly)
. . . . 09595

Polskie Towarzystwo Ortopedyczne i Traumatologiczne / Polish Orthopedic and Traumatologic Society, Ul Lindleys, 02-005 Warszawa
T: (058) 418652
Founded: 1928; Members: 1625
Pres: Prof. Dr. J. Szczekot; Gen Secr: Prof. Dr. Stanislaw Mazurkiewicz
Focus: Orthopedics; Traumatology
Periodical
Chirurgia Narządów Ruchu i Ortopedia Polska (weekly) 09596

Polskie Towarzystwo Otolaryngologiczne (PTOL) / Polish Otolaryngological Society, Ul Banacha 1A, 02-097 Warszawa
T: (022) 235975
Founded: 1889; Members: 1340
Pres: Prof. Dr. S. Betlejewski
Focus: Otorhinolaryngology
Periodical
Otolaryngologia Polska: Polish Journal of Otolaryngology (weekly) 09597

Polskie Towarzystwo Parazytologiczne / Polish Parasitological Society, Ul C. Norwida 29, 50-375 Wroclaw
T: (071) 226661
Founded: 1948; Members: 410
Focus: Microbio
Periodicals
Katalog Fauny Pasozytniczej Polski (weekly)
Monografie parazytologiczne (weekly)
Wiadomości Parazytologiczne (weekly) . . 09598

Polskie Towarzystwo Patologów / Polish Society of Pathologists, Ul S. Zeromskiego 113, 90-549 Łódź
T: (042) 25145
Founded: 1958; Members: 700
Pres: Prof. Andrzej Kulig; Gen Secr: Dr. Krzysztof W. Zielinski
Focus: Pathology
Periodical
Patologia Polska (weekly) 09599

Polskie Towarzystwo Pediatryczne (PTP), Ul Karowa 31, 00-324 Warszawa
T: (022) 267147
Founded: 1908; Members: 5600
Focus: Pediatrics
Periodicals
Pediatria Polska (weekly)
Przeglad Pediatryczny (weekly) 09600

Polskie Towarzystwo Przyrodników im. Kopernika, Ul Rakowiecka 36, 02-532 Warszawa
T: (022) 490171
Founded: 1875; Members: 2300
Focus: Nat Sci
Periodicals
Kosmos (weekly)
Wszechświat (weekly) 09601

Polskie Towarzystwo Psychiatryczne (P.T.P.) / The Polish Psychiatric Association, Sobieskiego 1-9, 02-957 Warszawa
T: (022) 215959 ext 33; Fax: (022) 215695
Founded: 1920; Members: 1000
Pres: Prof. Jacek Bomba; Gen Secr: Janusz Heitzman
Focus: Psychiatry
Periodicals
Psychiatria Polska (weekly)
Psychoterapia (weekly) 09602

Polskie Towarzystwo Psychologiczne (PTP) / Polish Psychological Association, Ul Stawki 5-7, 00-183 Warszawa
T: (022) 311368; Fax: (022) 311368
Founded: 1950; Members: 4300
Pres: Dr. M. Toeplitz-Winiewska; Gen Secr: Krzysztof Broclawik
Focus: Psych
Periodicals
Nowiny Psychologiczne (10 times annually)
Przeglad Psychologiczny (weekly) . . . 09603

Polskie Towarzystwo Semiotyczne (PTS), Palac Kultury i Nauki, 00-901 Warszawa
Founded: 1968; Members: 115
Pres: Prof. Jerzy Pelc
Focus: Ling
Periodical
Studia Semiotyczne: Semiotic Studies (weekly)
. . . . 09604

Polskie Towarzystwo Socjologiczne (PTS) / Polish Sociological Association, Nowy Swiat 72, 00-330 Warszawa
T: (022) 267737; Fax: (022) 267737
Founded: 1957; Members: 1040
Pres: Prof. A. Kloskowska; Gen Secr: Dr. E. Tarkowska
Focus: Sociology
Periodical
Polish Sociological Review (weekly) . . 09605

Polskie Towarzystwo Stomatologiczne (PTS) / Polish Dental Association, Ul Nowotki 21, 90-202 Łódź
T: (042) 321735
Founded: 1951; Members: 9017
Pres: Prof. Dr. Włodzimierz Józefowicz
Focus: Stomatology
Periodicals
Czasopismo Stomatologiczne (weekly)
Protetyka Stomatologiczna (weekly) . . . 09606

Polskie Towarzystwo Urologiczne (PTU), Ul Warszawska 52, Katowice
T: (032) 30978
Founded: 1949; Members: 280
Focus: Urology 09607

Polskie Towarzystwo Zoologiczne, Sienkiewicza 21, 50-335 Wroclaw
T: (071) 225041, 222817; Fax: (071) 282817
Founded: 1937; Members: 1100
Focus: Zoology
Periodicals
Notatki Ornitologiczne (weekly)
Przeglad Zoologiczny (weekly)
The Ring (weekly)
Zoologica Poloniae (weekly) 09608

Polskie Towarzystwo Zootechniczne (PTZ) / Polish Society of Animal Production, Ul Kaliska 9, 02-316 Warszawa
T: (022) 221723
Founded: 1922; Members: 1290
Pres: Dr. J. Luchowiec; Gen Secr: Dr. Anna Otwinowska
Focus: Zoology
Periodicals
Animal Production Review (weekly)
Animal Production Review Applied Sciences Report (weekly) 09609

Polski Klub Literacki P.E.N. / Polish P.E.N. Centre, Pałac Kultury i Nauki, 00-901 Warszawa
T: (022) 263948
Founded: 1925; Members: 270
Pres: Juliusz Zulawski; Gen Secr: Anna Trzeciakowska
Focus: Lit 09610

Polski Komitet Pomiarów Automatyki, c/o Naczelna Organizacja Techniczna w Polsce, UL Czackiego 3-5, 00-043 Warszawa
Focus: Eng 09611

Poznánskie Towarzystwo Przyjaciól Nauk (PTPN), Ul Sew. Mielzyńskiego 27-29, 61-725 Poznan
T: (061) 527441
Founded: 1857; Members: 1022
Focus: Sci
Periodicals
Badania Fizjograficzne: Seria A, B, C (weekly)
Badania z Dziejów Spolecznych i Gospodarczych (weekly)
Bulletin: Série D: Sciences Biologiques (weekly)
Fizyka Dielektryków i Radiospektroskopia (weekly)
Lingua Posnaniensis (weekly)
Prace Komisji Archeologicznej (weekly)
Prace Komisji Automatyki (weekly)
Prace Komisji Biologicznej (weekly)
Prace Komisji Etnograficznej (weekly)
Prace Komisji Filologicznej (weekly)
Prace Komisji Filozoficznej (weekly)
Prace Komisji Geologiczno-Geograficznej (weekly)
Prace Komisji Historii Sztuki (weekly)
Prace Komisji Historycznej (weekly)
Prace Komisji Jezykoznawczej (weekly)
Prace Komisji Nauk Ekonomicznych (weekly)
Prace Komisji Nauk Rolniczych i Komisji Nauk Leśnych (weekly)
Prace Komisji Nauk Spolecznych (weekly)
Prace Monograficzne nad Przyroda Wielkopolskiego Parku Narodowego pod Poznaniem (weekly)
Roczniki Dziejów Spolecznych i Gospodarczych
Roczniki Historyczne (weekly)
Slavia Occidentalis (weekly)
Sprawozdania (weekly)
Studia nad Historia Prawa Polskeigo (weekly)
Wydawnictwa Źródłowe Komisji Historycznej (weekly) 09612

Stowarzyszenie Archiwistow Polskich, Nowy Swiat 72, Warszawa
Focus: Archives 09613

Stowarzyszenie Autorów ZAIKS / Society of Authors ZAIKS, Hipoteczna 2, 00-092 Warszawa
T: (022) 276061; Fax: (022) 6351347
Founded: 1918; Members: 4000
Pres: Tadeusz Wojciech Maklakiewicz; Gen Secr: Witold Kolodziejski
Focus: Lit 09614

Stowarzyszenie Bibliotekarzy Polskich / Polish Librarians Association, Konopczyńskiego 5-7, 00-953 Warszawa
Founded: 1917; Members: 13500
Pres: Stanisław Czajka; Gen Secr: Jan Waluszewski
Focus: Libraries & Bk Sci
Periodicals
Bibliotekarz (weekly)
Poradnik Bibliotekarza (weekly)
Przeglad Biblioteczny (weekly) 09615

Stowarzyszenie Historyków Sztuki (SHS) / Association of Art Historians in Poland, Rynek Starego Miasta 27, 00-272 Warszawa
T: (022) 313773
Founded: 1934; Members: 1602
Pres: Prof. Dr. Tadeusz Chrzanowski; Gen Secr: Prof. Dr. Maria Poprzecka
Focus: Arts; Cultur Hist; Preserv Hist Monuments
Periodical
Kronika Stowarzyszenia Historyków Sztuki (weekly)
. . . . 09616

Szczecinskie Towarzystwo Naukowe, Rycerska 3, 70-537 Szczecin
Founded: 1956
Focus: Sci 09617

Towarzystwo Anestezjologów Polskich (TAP), Ul Dluga 1-2, 61-848 Poznan
T: (061) 51021
Founded: 1958; Members: 1350
Focus: Anesthetics 09618

Towarzystwo Chirurgów Polskich (TChP), Ul Banacha 1a, 00-957 Warszawa
T: (022) 236411 ext 467
Founded: 1889; Members: 2100
Pres: Prof. Dr. Z. Paplinski; Gen Secr: Prof. Dr. Jan Nielubowicz
Focus: Surgery 09619

Towarzystwo imienia Fryderyka Chopina (TiFC) / Frederick Chopin Society, Okólnik 1, 00-368 Warszawa
T: (022) 275471; Fax: (022) 279599
Founded: 1934; Members: 750
Pres: Tadeusz Chmielewski; Gen Secr: Albert Grudzinski
Focus: Music
Periodical
Rocznik Chopinowski: Chopin Studies . . 09620

Towarzystwo Internistów Polskich (TIP), Pasteura 4, 50-367 Wroclaw
T: (071) 210765
Founded: 1906; Members: 3000
Focus: Intern Med 09621

Towarzystwo Literackie im. Adama Mickiewicza, Nowy Swiat 72, 00-330 Warszawa
T: (022) 265231 ext 79
Founded: 1886; Members: 1500
Focus: Lit
Periodical
Rocznik Towarzystwa Literackiego im. Adama Mickiewicza (weekly) 09622

Towarzystwo Milosników Historii i Zabytków Krakowa (TMHiZK), Swietego Jana 12, 31-018 Kraków
T: (012) 25398
Founded: 1820; Members: 405
Focus: Preserv Hist Monuments; Hist . . 09623

Towarzystwo Milosnikow Jezyka Polskiego (TMJP) / Society of Friends of Polish Language, Straszewskiego 27, 31-113 Kraków
T: (012) 222699
Founded: 1920; Members: 800
Pres: Prof. Dr. Stanislaw Urbaíczyk
Focus: Ling
Periodical
Jezyk Polski (5 times annually) 09624

Towarzystwo Naukowe Organizacji i Kierownictwa / Scientific Society of Organization and Management, Koszykowa 6, 00-564 Warszawa
T: (022) 292127; Tx: 813649; Fax: (022) 292127
Founded: 1925; Members: 32000
Pres: Prof. Dr. Henryk Sadownik
Focus: Business Admin
Periodicals
Organizational Problems (weekly)
Organization Review (weekly) 09625

Towarzystwo Naukowe Płockie / Societas Scientiarum Plocensis / The Scientific Society of Płock, Plac Narutowicza 8, 09-402 Płock
T: 22604, 29477
Founded: 1820; Members: 607
Focus: Sci
Periodical
Sprawozdanie reczne, Notatki Płockie: Yearbook (weekly) 09626

Towarzystwo Naukowe w Toruniu (TNT), Ul Wysoka 16, 87-100 Torun
T: (056) 23941
Founded: 1875; Members: 538
Pres: Prof. Dr. Marian Biskup; Gen Secr: Prof. Dr. Miroslaw Nestorowicz
Focus: Sci; Hist
Periodicals
Fontes (weekly)
Prace Archeologiczne (weekly)
Prace Popularnonaukowe (weekly)
Prace Wydziału Filologiczno-Filozoficznego (weekly)
Roczniki TNT (3 times annually)
Sprawozdania TNT (weekly)
Studia Iuridica (weekly)
Studia Societatis Scientiarum Torunensis: Sectio C – geographia et geologia (weekly)
Studia Societatis Scientiarum Torunensis: Sectio D – botanica (weekly)
Studia Societatis Scientiarum Torunensis: Sectio E – zoologia (weekly)
Studia Societatis Scientiarum Torunensis: Sectio F – astronomia (weekly)
Studia Societatis Scientiarum Torunensis: Sectio G – physiologia (weekly)
Studia Societatis Scientiarum Torunensis: Sectio H – medicina
Zapiski Historyczne TNT (weekly) . . . 09627

Towarzystwo Przyjaciól Nauk w Przymyslu (TPN), Rynek 4, 37-700 Przemysl
T: 5601
Founded: 1909; Members: 249
Pres: Dr. Zdzislaw Budzynski; Gen Secr: Tadeusz Burzyński
Focus: Sci
Periodicals
Biblioteka Przemyska
Rocznik Przemyski (weekly) 09628

Towarzystwo Urbanistów Polskich (T.U.P.), Plac Zamkowy 10, 00-277 Warszawa
T: (022) 310773, 312830
Founded: 1923; Members: 2900
Focus: Urban Plan
Periodical
Miasto: The Town (weekly) 09629

Wroclawskie Towarzystwo Naukowe (WTN), Rosenbergów 13, 51-616 Wroclaw
Founded: 1946; Members: 540
Pres: Prof. Dr. Mieczysław Klimowicz; Gen Secr: Prof. Dr. Jacek Kolbuszewski
Focus: Sci
Periodicals
Annales Silesiae (weekly)
Litteraria (weekly)
Rozprawy Komisji Jezykowej (weekly)
Sprawozdania,: Reports A and B (weekly) 09630

Zrzeszenie Polskich Towarzystw Medycznych (ZPTM) / Federation of Polish Medical Societies, Ul Karowa 31, 00-324 Warszawa
T: (022) 266320
Founded: 1965; Members: 50
Pres: Prof. B. Górnicki; Prof. A. Rytowska; Gen Secr: Prof. C. Korczak
Focus: Med
Periodicals
Activities of the Federation of Polish Medical Societies
Problemy Medycyny i Farmacji (1-2 times annually)
Register of the Federation of Polish Medical Societies (2-3 times annually) 09631

Zydowski Instytut Historyczny w Polsce (ZIH), Ul Tlomackie 3-5, 00-090 Warszawa
T: (022) 279221; Fax: (022) 278372
Founded: 1947; Members: 120
Pres: Prof. O. Grinberg; Gen Secr: M. Friconan
Focus: Hist
Periodicals
Biuletyn ZIH (weekly)
Bleter far Geszichte (weekly) 09632

Portugal

Academia das Ciências de Lisboa, Rua da Academia das Ciências 19, 1200 Lisboa
T: (01) 3463866; Fax: (01) 3420395
Founded: 1779; Members: 163
Pres: Prof. José Pinto Peixoto; Gen Secr: José Manuel Toscano-Rico
Focus: Sci
Periodicals
Anuário Académico
Boletim 09633

Academia Nacional de Belas-Artes, Largo da Academia Nacional de Belas-Artes, Lisboa
Founded: 1932; Members: 20
Pres: A. de Carvalho; Gen Secr: Dr. Fernando de Pamplona
Focus: Arts
Periodicals
Belas Artes: Revista-Boletim
Boletim 09634

Academia Portuguesa da História, Largo da Rosa 5, Lisboa
T: (01) 868997
Founded: 1720; Members: 40
Pres: Prof. J.V. Serrão; Gen Secr: Prof. I. da Rosa Pereira
Focus: Hist
Periodicals
Anais
Boletim
Documentos Medievais Portugueses . . 09635

Associação Central de Agricultura Portuguesa, Rua D. Dinis 2, Lisboa 2
T: (01) 682462
Focus: Agri 09636

Associação das Universidades de Lingua Portuguesa (AULP) / Assoiation of Portuguese-Speaking Universities, Alameda de S. António dos Capuchos 1, 1100 Lisboa
T: (01) 545434
Founded: 1986
Gen Secr: Dr. Manuel J. Coelho da Silva
Focus: Educ 09637

Associação dos Arqueólogos Portugueses, c/o Museu Arqueológico, Largo do Carmo, Lisboa
T: (01) 3460473
Founded: 1863; Members: 600
Pres: Dr. E. da Cunha Serrão; Gen Secr: Dr. E. Cabral
Focus: Archeol
Periodical
Arqueólogia e História (weekly) 09638

Associação dos Técnicos e Auxiliares de Radiologia de Portugal, Rua Dona Estefania 47, Lisboa 1
T: (01) 555867
Focus: Radiology 09639

Associação Portuguesa de Bibliotecários, Arquivistas e Documentalistas, Campo Grande 83, 1751 Lisboa Codex
T: (01) 767862
Founded: 1973; Members: 1150
Pres: Luis F. Abreu Nunes; Gen Secr: A.P. Gordo
Focus: Libraries & Bk Sci; Archives; Doc
Periodicals
Cadernos de Biblioteconomia, Arquivistica e Documentagão (weekly)
Noticia (weekly) 09640

Associação Portuguesa de Economistas, Rua da Estrela 8, 1200 Lisboa
T: (01) 661584
Focus: Econ 09641

Associação Portuguesa de Escritores (APE), Rua de S. Domingos à Lapa 17, 1200 Lisboa
Founded: 1973; Members: 600
Pres: O. Lopes; Gen Secr: C.E. da Costa
Focus: Lit
Periodical
Loreto 13 (weekly) 09642

Associação Portuguesa de Estudos Clássicos, c/o Faculdade de Letras, 3000 Coimbra
Fax: (039) 36733
Focus: Hist 09643

Associação Portuguesa de Fisioterapeutas, Av Pedro Alvares Cabral 1a, Lisboa 2
T: (01) 658656
Focus: Therapeutics 09644

Associação Portuguesa de Fotogrametria, c/o Instituto Geográfico e Cadastral, Praca de Estrela, 1200 Lisboa
T: (01) 666023
Focus: Surveying 09645

Associação Portuguesa de Fundição (APF), Rua do Campo Alegre 672, Porto
T: (02) 690675, 6000764
Founded: 1964; Members: 200
Pres: José Costa Silva; Gen Secr: Manuel Botelho Chaves
Focus: Metallurgy
Periodicals
Anuario (weekly)
Fundigao (weekly) 09646

Associação Portuguesa de Management (APM), Rua Rodrigo da Fonseca 182, 1000 Lisboa
Fax: (01) 3883266
Focus: Business Admin 09647

Associação Portuguesa de Odontologia, Br. Sabrosa 91, Lisboa 1
T: (01) 841633
Focus: Dent 09648

Associação Portuguesa para a Qualidade Industrial, Praça das Indústrias, 1300 Lisboa
T: (01) 636443; Tx: 44719; CA: INDUSTRIPORT
Focus: Materials Sci
Periodical
Qualidade (3 times annually) 09649

Associação Protectora dos Diabéticos de Portugal (A.P.D.P.), Rua do Salitre 118-120, 1200 Lisboa
T: (01) 3880041; Fax: (01) 659371
Founded: 1926; Members: 767
Pres: José P. Cardoso de Oliveira
Focus: Diabetes
Periodical
Bulletin A.P.D.P. (every 3 years) . . . 09650

Associação Técnica da Indústria do Cimento (ATIC), Av 5 Outubro 54, 1000 Lisboa
T: (01) 547538; Fax: (01) 3525099
Founded: 1965; Members: 5
Focus: Eng
Periodical
ATIC Magazine (weekly) 09651

Association of Paediatric Education in Europe (APEE), c/o Serv Pediatri-HSM, Av Prof. Egas Moniz, 1699 Lisboa Codex
Fax: (01) 764059
Founded: 1970; Members: 70
Gen Secr: Prof. Lincoln J. Silva
Focus: Educ
Periodical
APEE Bulletin 09652

Associção dos Arquitectos Portugueses (AAP) / Portuguese Architects' Association, Av 24 de Julho 52, 1200 Lisboa
T: (01) 3951401
Founded: 1978; Members: 5020
Pres: Pedro Brandão; Gen Secr: Jorge Silva
Focus: Archit
Periodical
Journal Arquitectos (10 times annually) . 09653

Centro de Estudos de História e Cartografia Antiga, Rua Jau 54, 1300 Lisboa
T: (01) 645321
Founded: 1958; Members: 9
Pres: E. Madeira Santos
Focus: Hist; Cart 09654

Colégio Ibero-Latino-Americano de Dermatologia (C.I.L.A.D.), Av da Liberdade 90, 1298 Lisboa Codex
T: (01) 322540
Founded: 1948; Members: 1500
Focus: Derm 09655

Conselho de Reitores das Universidades Portuguesas, Campo dos Mártires da Pátria 2, 1100 Lisboa
T: (01) 549170; Tx: 44733
Pres: Prof. E. Pina; Gen Secr: Dr. M. Marchante
Focus: Educ 09656

Fundação Calouste Gulbenkian, Berna 45, 1093 Lisboa Codex
T: (01) 735131; Tx: 12345
Pres: Dr. José Azeredo Perdigąo
Focus: Sci 09657

Instituto Açoriano de Cultura, Apdo 67, 9701 Angra do Heroismo Codex
Founded: 1956
Pres: Dr. Jorge Paulus Banno; Gen Secr: Dr. A. Fraga Barcelos
Focus: Educ 09658

Instituto de Coímbra, Rua da Ilha, Coímbra
Founded: 1851; Members: 599
Pres: Prof. Luis . Mendonça de Albuquerque; Gen Secr: Armando Carneida Silva
Focus: Educ 09659

Instituto dos Actuarios Portugueses (IAP), Rua Rodrigo da Fonseca 76, Lisboa
T: (01) 557830
Founded: 1945; Members: 150
Focus: Insurance 09660

Instituto Portugués de Arqueologia, História e Etnografia / Portuguese Archaeological, Historical and Ethnographical Institute, Praça do Imperio, Edificio dos Jerónimos, Belém, Lisboa
Gen Secr: Dr. João L. Saavedra Machado
Focus: Hist
Periodical
Ethnos 09661

International Society for Rock Mechanics (ISRM), c/o Laboratório Nacional de Engenharia Civil, Av do Brasil 101, 1799 Lisboa Codex
T: (01) 882131
Founded: 1962; Members: 5000
Pres: Charles Fairhurst; Gen Secr: José Delgado Rodrigues
Focus: Geology
Periodical
Journal (weekly) 09662

Junta Nacional de Investigação Científica e Tecnológica (JNICT/CDCT) / Centro de Documentação Cinetífica e Técnica, Av Prof. Gama Pinto 2, 1699 Lisboa
T: (01) 772886, 731300, 731350; Fax: (01) 7965622
Founded: 1936
Focus: Doc 09663

Ordem dos Engenheiros / Association of Engineers, Av de António Augusto de Aguiar 3-D, 1097 Lisboa Codex
T: (01) 562438; Tx: 62340
Founded: 1936
Pres: J.A. Simões Cortez; Gen Secr: H.P. Pereira
Focus: Eng
Periodical
Review Ingenium (weekly) 09664

Ordem dos Médicos / Medical Association, Av Almirante Reis 242, 1000 Lisboa
T: (01) 805492; Tx: 42904
Founded: 1938; Members: 24851
Pres: A. Gentil Martins
Focus: Med
Periodical
Revista (weekly) 09665

Prevenção Rodoviária Portuguesa (PRP) / Portuguese Road Safety Association, Rua Rosa Araújo 12, 1200 Lisboa
T: (01) 554982, 554753, 560418, 574160; Tx: 15570; Fax: (01) 570745
Founded: 1965; Members: 237
Pres: António Jervis Pereira; Gen Secr: José Miguel Trigoso
Focus: Transport
Periodicals
Boletim Informativo (weekly)
PRP Activities Report (weekly) 09666

Real Instituto Arqueológico de Portugal, Alameida das Linhas de Torres 97, 1700 Lisboa
T: (01) 7591109
Founded: 1868
Pres: Dr. José A. Falcąo; Gen Secr: Dr. Jorge M. Rodrigues Ferreira
Focus: Archeol 09667

Real Sociedade Arqueológica Lusitana, 7540 Santiago de Cacém
T: 826380
Founded: 1849; Members: 163
Pres: Dr. José António Falcąo; Gen Secr: Dr. Lilia Ribeiro da Silva
Focus: Archeol
Periodicals
Anais da Real Sociedade Arqueológica Lusitana (weekly)
Memórias da Real Sociedade Arqueológica Lusitana (weekly)
Repertorium Fontium Artis Historiae Portugaliae Instaurandum
Trabalhos da Real Sociedade Arqueológica Lusitana (weekly) 09668

Sindicato Nacional dos Odontologistas Portuguesas, Arroios 179, Lisboa 1
T: (01) 546389
Focus: Dent 09669

Sindicato Nacional dos Professionais de Serviço Social, Rua Luciano Cordeiro 18, Lisboa
T: (01) 537963
Focus: Sociology 09670

Sindicato Nacional dos Professores, Conde Redondo 22, Lisboa 1
T: (01) 573014
Founded: 1940; Members: 6000
Focus: Educ 09671

Sindicato Nacional dos Protésicos Dentários, Palmira 66, Lisboa 1
T: (01) 840075
Focus: Dent 09672

Sociedade Anatómica Luso-Hispano-Americana / Portuguese-Spanish-Latin American Anatomical Society, Av Egas Moniz, Lisboa 4
T: 1935
Pres: Prof. Dr. Armando dos Santos Ferreira
Focus: Anat
Periodicals
Actas dos Congressos
Anuário Estatistico: Continente e Ilhas Adjacentes
Arquivo de Anatomia e Antropologia
Boletim Mensal de Estatistica
Comércio Externo (weekly)
Estatisticas Agricolas
Estatisticas da Educação
Estatisticas Demográficas
Estatisticas Industriais
Estatisticas Monetárias e Financeiras
Folha Mensal do Estado das Culturas e Previsão de Colheitas 09673

Sociedade Anatómica Portuguesa (SAP), c/o Lab de Anatomia Normal, Faculdade de Medicina de Coímbra, 3049 Coímbra Codex
Founded: 1930; Members: 246
Pres: Prof. Dr. R. Teixeira
Focus: Anat 09674

Sociedade Astronómica de Portugal, Edificio Faculdade de Ciencias, Rua da Escola Politecnica, Lisboa 2
T: (01) 661521
Focus: Astronomy 09675

Sociedade Broteriana, c/o Dept de Botânica, Universidade, 3049 Coimbra
T: (039) 22897; Fax: (039) 20780
Founded: 1880; Members: 300
Pres: Prof. Dr. José Firmino Moreira Mesquita; Gen Secr: Jorge Paiva
Focus: Botany
Periodicals
Anuário (weekly)
Boletim (weekly)
Memórias (weekly) 09676

Sociedade Científica da Universidade Católica Portuguesa, c/o Universidade Católica Portuguesa, Palma de Cima, 1600 Lisboa
T: (01) 7265817
Founded: 1980; Members: 240
Pres: José da Cruz Policarpo; Gen Secr: Prof. J. Bacelar e Oliveira
Focus: Sci 09677

Sociedade das Ciências Médicas de Lisboa, Av da República 34, 1000 Lisboa
T: (01) 772730
Founded: 1822; Members: 2000
Focus: Med 09678

Sociedade de Ciências Agrárias de Portugal, Rua da Junqueira 299, 1300 Lisboa
T: (01) 3633719
Focus: Agri
Periodical
Revista de Ciencias Agrarias (weekly) . 09679

Sociedade de Estudos Açoreanos Afonso Chaves / Society for Research on the Azores, Ponta Delgada
Founded: 1932; Members: 100
Focus: Nat Sci; Geophys
Periodical
Açoreana (weekly) 09680

Sociedade de Estudos Técnicos SARL-SETEC, Rua Joaquin António de Aguiar 73, Lisboa 1
Members: 10000
Pres: Armando Lencastre; Gen Secr: Jorge Ramiro
Focus: Eng
Periodical
Boletim Informativo Nacional (weekly) . . 09681

Sociedade de Geografia de Lisboa, Rua das Portas de Santo Antão 100, 1100 Lisboa
T: (01) 3425401; Fax: (01) 3464553
Founded: 1875
Pres: António E. Sousa Leitão; Gen Secr: Nuno Pedro da Silva
Focus: Geography
Periodical
Boletim (weekly) 09682

Sociedade de Lingua Portuguesa, Rua São José 41, Lisboa 2
T: (01) 363949
Founded: 1949; Members: 8000
Focus: Ling 09683

Sociedade Farmacêutica Lusitana / Portuguese Pharmaceutical Society, Rua da Sociedade Farmacêutica 18, Lisboa
T: (01) 41433
Founded: 1835; Members: 4500
Pres: Prof. Dr. Alfredo Albuquerque
Focus: Pharmacol
Periodical
Revista Portuguesa de Farmcacia . . . 09684

Sociedade Geológica de Portugal, c/o Escola Politécnica, Faculty of Science, University, 1294 Lisboa Codex
Founded: 1940; Members: 600
Pres: Prof. R. Rocha; Gen Secr: Prof. M.T. Azevedo
Focus: Geology
Periodicals
Boletim (weekly)
Maleo 09685

Sociedade Histórica da Independência de Portugal, Palacio da Independencia, Largo de São Domingo 11, 1200 Lisboa
T: (01) 3428987; Fax: (01) 3460754
Focus: Hist
Periodicals
Independência: Society's Review
Report (weekly) 09686

Sociedade Martins Sarmento / Martins Sarmento Society, Rua de Paio Galvão, Guimarães
Founded: 1882; Members: 425
Pres: G. Alves
Focus: Lit
Periodical
Revista de Guimarães 09687

Sociedade Nacional de Belas Artes / National Society of Fine Arts, c/o Palacio das Belas Artes, Rua Barata Salgueiro 36, Lisboa
T: (01) 521293
Founded: 1901
Pres: Fernando de Azevedo
Focus: Arts 09688

Sociedade Portuguesa de Alergologia e Imunologia Clínica (SPAIC) / Portuguese Society of Allergology and Clinical Immunology, c/o Faculdade de Medicina, Av Egas Moniz, 1699 Lisboa
T: (01) 764673
Members: 140
Focus: Immunology
Periodical
Allergologia e Imunologia Clinica: Boletim (weekly) 09689

Sociedade Portuguesa de Anestesiologia, Av da República 34, Lisboa 1
Founded: 1955; Members: 145
Focus: Anesthetics 09690

Sociedade Portuguesa de Antropologia e Etnologia, Praça Gomes Teixeira, 4000 Porto
T: (02) 310290 ext 276
Founded: 1918; Members: 138
Pres: Prof. Dr. Susana Oliveira Jorge; Gen Secr: Crisanda Urbano Unsworth
Focus: Anthro; Ethnology
Periodical
Trabalhos de Antropologia e Etnologia (weekly) 09691

Sociedade Portuguesa de Autores, Av Duque de Loulé 31, 1098 Lisboa Codex
T: (01) 578320; Tx: 42563
Founded: 1925; Members: 8671
Pres: Dr. Luiz Francisco Rebello
Focus: Lit
Periodical
Autores (weekly) 09692

Sociedade Portuguesa de Bioquímica, c/o Instituto de Química Fisiológica, Faculdade de Medicina, Lisboa
Members: 230
Focus: Biochem 09693

Sociedade Portuguesa de Cardiologia, Campo Grande 28, 1700 Lisboa
T: (01) 7970685; Fax: (01) 7931095
Focus: Cardiol
Periodicals
Boletim da Sociedade Portuguesa de Cardiologia
Revista Portuguesa de Cardiologia: Portuguese Journal of Cardiology 09694

Sociedade Portuguesa de Ciências Naturais, c/o Faculdade de Ciências, Rua da Escola Politécnica, 1294 Lisboa Codex
T: (01) 661521
Founded: 1907
Focus: Nat Sci
Periodicals
Boletim
Natura
Naturalia 09695

Sociedade Portuguesa de Ciências Veterinárias, Rua D. Dinis 2a, 1200 Lisboa
T: (01) 680188
Founded: 1902; Members: 625
Focus: Vet Med
Periodical
Revista Portuguesa de Ciências Veterinárias (weekly) 09696

Sociedade Portuguesa de Dermatologia e Venereologia, c/o Hospital do Desterro, Lisboa 2
Founded: 1942; Members: 105
Focus: Derm; Venereology 09697

Sociedade Portuguesa de Educação Médica, Av da República 34, 1000 Lisboa
Fax: (01) 7977578
Members: 130
Focus: Educ
Periodical
Boletim da Sociedade Portuguesa de Educação Médica (3 times annually) 09698

Sociedade Portuguesa de Especialistas de Pequenos Animales / Portuguese Society of Specialists in Small Animals, Rua de D. Dinis 2-A, 1200 Lisboa
Pres: Dr. Antonio Marques de Almeida
Focus: Zoology 09699

Sociedade Portuguesa de Espeleologia, Rua Saraiva Carvalho 233, 1300 Lisboa
T: (01) 666291
Founded: 1948; Members: 150
Focus: Speleology
Periodical
Algar (weekly) 09700

Sociedade Portuguesa de Gastroenterologia, c/o Dr. J.M.C. Ribeiro, Av Infante Santo 55, Lisboa 3
Members: 145
Focus: Gastroenter 09701

Sociedade Portuguesa de Hemorreologia e Microcirculaçã0, Apdo 4098, 1502 Lisboa Codex
Fax: (01) 7977578
Members: 242
Focus: Med
Periodical
Boletim da Sociedade Portuguesa de Hemorreologia e Microcirculação (weekly) 09702

Sociedade Portuguesa de Higiene Alimentar, Rua de D. Dinis 2a, 1200 Lisboa
Pres: A. Martins Mendes
Focus: Nutrition 09703

Sociedade Portuguesa de Medicina Fisica e Reabilitação (S.P.M.F.R.) / Portuguese Society of Physical Medicine and Rehabilitation, Av Almirante Gago Coutinho 151, 1700 Lisboa
T: (01) 8470654; Fax: (01) 8471215
Members: 300
Pres: Dr. Maria Lidia Ramalho; Gen Secr: João Pinheiro
Focus: Rehabil
Periodical
Medicina Fisica e de Reabilitação (weekly) 09704

Sociedade Portuguesa de Neurologia e Psiquiatria, c/o Hospital Miguel Bombarda, Rua da Cruz Carreira, Lisboa
T: (01) 49849
Focus: Neurology; Psychiatry 09705

Sociedade Portuguesa de Numismática (SPN), Rua Costa Cabral 664, Porto
T: (02) 496029
Founded: 1952; Members: 1600
Focus: Numismatics 09706

Sociedade Portuguesa de Nutrição e Alimentação Animal, Rua de D. Dinis 2a, 1200 Lisboa
Pres: Dr. J. Portugal
Focus: Agri 09707

Sociedade Portuguesa de Oftalmologia, Av Almirante Gago Countinho 151, 1700 Lisboa
T: (01) 8470654; Fax: (01) 8471215
Founded: 1939; Members: 660
Focus: Ophthal
Periodical
Revista da Sociedade Portuguesa de Oftalmologia (3 times annually) 09708

Sociedade Portuguesa de Ortopedia e Traumatologia, Av Cons. Barjona Freitas 5, Lisboa 4
T: (01) 784541
Focus: Orthopedics; Traumatology . . . 09709

Sociedade Portuguesa de Otorinolaringologia e Bronco-Esofagologia, Av da Liberdade 65, 1200 Lisboa
Members: 200
Focus: Otorhinolaryngology
Periodical
Boletim (weekly) 09710

Sociedade Portuguesa de Pediatria, c/o Hospital Santa Maria, Av 28 de Maio, Lisboa 4
Focus: Pediatrics 09711

Sociedade Portuguesa de Química (SPQ), Av da República 37, 1000 Lisboa
T: (01) 734637
Founded: 1974; Members: 3000
Pres: Prof. Alberto Romão Dias; Gen Secr: Prof. Carlos Castro
Focus: Chem
Periodicals
Boletim da Sociedade Portuguesa de Química (weekly)
Revista Portuguesa de Qúimica (weekly) 09712

Sociedade Portuguesa de Reprodução Animal, Rua de D. Dinis 2a, 1200 Lisboa
Pres: Dr. A.J.B. Cristina Alves
Focus: Agri 09713

Sociedade Portuguesa de Reumatologia (SPR) / Portugese Society of Rheumatology, Rua de Dona Estefânia 187-189, 1000 Lisboa
T: (01) 572326
Founded: 1972; Members: 133
Focus: Rheuma
Periodicals
Acta Reumatológica Portuguesa (weekly)
Boletim Informativo (weekly) 09714

Sociedade Portuguesa Veterinária de Anatomia Comparativa / Portuguese Veterinary Society of Comparative Anatomy, Rua de D. Dinis 2a, 1200 Lisboa
T: (01) 680188
Founded: 1974
Pres: Dr. Paulo Marques
Focus: Vet Med; Anat 09715

Sociedade Portuguesa Veterinária de Estudos Sociológicos / Portuguese Society of Sociological Veterinary Studies, Rua de D. Dinis 2a, 1200 Lisboa
Pres: Prof. Paulo Marques
Focus: Vet Med; Sociology 09716

Sociedade Portuguese de Estomatologia (S.P.E.), Av Rainha D. Amélia 36, 1600 Lisboa
T: (01) 793948
Members: 533
Focus: Stomatology 09717

Puerto Rico

Academia Puertorriqueña de la Historia, Av Wilson 1308, Santurce
Founded: 1932; Members: 40
Pres: Aurelio Tió
Focus: Hist 09718

Academia Puertorriqueña de la Lengua Española, Apdo 4008, San Juan, PR 00936
Founded: 1955; Members: 22
Pres: Salvador Tió; Gen Secr: Segundo Cardona
Focus: Ling
Periodical
Boletín (weekly) 09719

Asociación de Maestros de Puerto Rico, Ponce de León 33, Hato Rey, PR 00917
Focus: Educ 09720

Asociación de Psicólogos de Puerto Rico, Apdo 21816, Rio Piedras, PR 00928
Focus: Psych 09721

Association of Caribbean Universities and Research Institutes (UNICA), POB 11532, Caparra Heights Station, San Juan, PR 00922
T: 7894388
Founded: 1968; Members: 51
Focus: Sci
Periodicals
Caribbean Educational Bulletin (3 times annually)
Newsletter (weekly) 09722

Association of Caribbean University, Research and Institutional Libraries (ACURIL), Apdo 23317, Estación de la Universidad, San Juan, PR 00931
T: (809) 7640000 ext 2359, 7908054
Founded: 1969
Gen Secr: Oneida R. Ortiz
Focus: Libraries & Bk Sci
Periodical
ACURIL Newsletter 09723

Association of Marine Laboratories of the Caribbean (AIMLC), c/o Dept of Marine Sciences, University of Puerto Rico, POB 5000, Mayaguez, PR 00709-5000
T: (809) 899-2048; Tx: 3452024; Fax: (809) 265-2880
Founded: 1957; Members: 350
Gen Secr: Dr. Ernest H. Williams
Focus: Oceanography
Periodicals
Newsletter (weekly)
Proceedings of the Association of Marine Laboratories of the Caribbean (weekly) . 09724

Ateneo de Ponce, Apdo 1923, Ponce, PR 00731
Founded: 1956
Pres: Hilda Chavier; Gen Secr: Vicente Ruiz
Focus: Arts; Lit 09725

Ateneo Puertorriqueño, Edificio Ateneo, Av Ponce de León, San Juan, PR 00901
Founded: 1876; Members: 550
Pres: Eduardo Morales Coll; Gen Secr: Prof. Roberto Ramos Perea
Focus: Arts; Lit 09726

Business Education Research of America (BERA), c/o Enrique Piniero, POB 304, Hato Rey, PR 00919
T: (809) 763-1010
Founded: 1901; Members: 25
Pres: Enrique Piniero
Focus: Adult Educ
Periodicals
A Touch of Class (weekly)
Business Today (3 times annually) . . . 09727

Caribbean Hospitality Training Centre (CHTI), 18 Marseilles St, Santurce, PR 00907-1672
T: (809) 724-6023; Tx: 3252361; Fax: (809) 725-9108
Founded: 1980
Gen Secr: Eva L. Diaz
Focus: Educ 09728

Caribbean Studies Association (CSA), POB X, UPR Station, Rio Piedras, PR 00931
T: (809) 763-0812
Founded: 1974; Members: 650
Gen Secr: Angel Calderon-Cruz
Focus: Ethnology 09729

Congreso de Poesía de Puerto Rico, c/o Colegio de Agricultura y Artes Mecánicas, Mayaguez, PR 00708
Focus: Lit 09730

Sociedad Mayaguezana por Bellas Artes, Mayaguez, PR 00708
Focus: Arts 09731

Sociedad Puertorriqueña de Escritores, Apdo 4692, San Juan
Founded: 1937; Members: 81
Pres: Ernesto Juan Fonfrías
Focus: Lit 09732

Réunion

Association Historique Internationale de l'Océan Indien, c/o Archives Départementales de la Réunion, Le Chaudron, 97490 Sainte-Clotilde
Founded: 1960
Gen Secr: M. Chabin
Focus: Hist 09733

Romania

Academia de Stiinte Agricole si Silvice (ASAS) / Academy of Agricultural and Forest Sciences, Bd Màràsti 61, Bucuresti
T: (01) 180699; Tx: 11394
Founded: 1969; Members: 100
Focus: Forestry; Agri
Periodicals
Buletinul informativ al Academiei de Stiinte Agricole si Silvice (weekly)
Bulletin de l'Académie des Sciences Agricoles et Forestières (weekly) *09734*

Academia de Stiinte Medicale (A.S.M.) / Academy of Medical Sciences, Bd 1 Mai 11, 79173 Bucuresti
T: (01) 502393
Founded: 1934; Members: 175
Pres: Prof. Stefan M. Milcu
Focus: Med
Periodical
Buletinul (weekly) *09735*

Academia de Stiinte Sociale si Politice (A.S.S.P.) / Academy of Social and Political Sciences, Str Onesti 11, 70119 Bucuresti
T: (01) 157620
Founded: 1970; Members: 183
Focus: Poli Sci; Sociology
Periodicals
Dacia: Revue d'Archéologie et d'Histoire Ancienne (weekly)
Revista de filozofie: Journal of Philosophy (weekly)
Revista de istorie: Journal of History (weekly)
Revista de istorie şi teorie literară: Journal of Literary History and Theory (weekly)
Revista de psihologie: Journal of Psychology (weekly)
Revue des Etudes Sud-Est Européennes (weekly)
Revue Roumaine des Sciences Sociales: Série des Sociologie/Série des Sciences Economiques (weekly)
Revue Roumaine des Sciences Sociales: Série de Psychologie (weekly)
Revue Roumaine des Scienes Sociales: Série de Sciences Juridiques (weekly)
Revue Roumaine des Scienes Sociales: Série des Philosophie et Logique (weekly)
Revue Roumaine d'Etudes Internationales (weekly)
Revue Roumaine d'Histoire (weekly)
Revue Roumaine d'Histoire de l'Art: Série Théâtre, Musique, Cinéma (weekly)
Revue Roumaine d'Histoire de l'Art/Serie Beaux-Arts (weekly)
Studii şi cercetări de istoria artei, seria Artă plastică: Studies and Researches in Art History/Fine Arts Series (weekly)
Studii şi cercetări de istoria artei, seria Teatru, muzică, cinematografie: Studies and Researches in Art History/Theatre, Music, Cinematography Series (weekly)
Studii şi cercetări de istorie veche şi arheologie: Studies and Researches in Ancient History and Archeology (weekly)
Studii şi cercetări juridice: Juridical Studies and Researches (weekly)
Synthesis: Bulletin du Comité National de Littérature Comparée (weekly)
Thraco-dacica (weekly)
Viiturul social: The Social Future, Journal of Sociology and Political Science (weekly) . *09736*

Academia Romana, Calea Victoriei 125, 71102 Bucuresti
Founded: 1948; Members: 195
Pres: Mihai Draganescu; Gen Secr: Nicolae N. Constantinescu
Focus: Sci
Periodicals
L'Analyse Numérique et la Théorie de l'Approximation
Annuaire Roumain d'Anthropologie
Cahiers de Linguistique Théoretique et Appliquée
Cellulose Chemistry and Technology
Mathematica-Revue d'Analyse Numérique et de Théorie del 'Approximation
Revue Romaine de Physique
Revue Roumaine de Biochimie
Revue Roumaine de Biologie: Série de Biologie Animale
Revue Roumaine de Biologie: Série de Biologie Végétale
Revue Roumaine de Chimie
Revue Roumaine de Géologie, Géophysique et Géographie
Revue Roumaine de Linguistique
Revue Roumaine de Mathématiques Pures et Appliquées
Revue Roumaine des Sciences Techniques Série de Mécanique Appliquée
Studii si cercetari matematice: Studies and Research in Mathematics *09737*

Asociatia Bibliotecarilor din Romania, c/o Biblioteca Centrala de Stat, Str Jop Ghica 4, 70018 Bucuresti
T: (01) 503765
Founded: 1956
Focus: Libraries & Bk Sci *09738*

Asociaţia Cineaştilor din Romania / Cinema Workers' Association of Romania, Str Mendeleev 28-30, Sector 1, 70169 Bucuresti
T: (01) 505741
Founded: 1963; Members: 1200
Focus: Cinema *09739*

Asociatia de Drept International si Relatii Internationale din Romania (A.D.I.R.I), Soseana Kiseleff 47, 71268 Bucuresti
T: (01) 185462
Founded: 1966; Members: 600
Pres: Vlad-Andrei Moga
Focus: Law; Int'l Relat; Poli Sci
Periodical
Revue Roumaine d'Études Internationales (weekly) *09740*

Asociaţia Generala a Economiştilor din Romania (A.G.E.R.) / Romanian General Association of Economists, Calea Griviţei 21, 78101 Bucuresti
T: (01) 6507820; Fax: (01) 3129717
Founded: 1990; Members: 20000
Pres: Prof. Nicolae N. Constantinescu; Gen Secr: Prof. Aurel Işfănescu
Focus: Econ
Periodical
Economistul (weekly) *09741*

Asociatia Juristilor din Romania, Bd Gral Magheru 22, 70158 Bucuresti
T: (01) 593440
Founded: 1949; Members: 15000
Focus: Law
Periodical
Revista Romana de Drept (weekly) . . *09742*

Asociatia Oamenilor de Stiinta din Romania (AOS), Str Gabriel Péri 1, 70148 Bucuresti
T: (01) 136234
Founded: 1956
Focus: Sci *09743*

Asociaţia Psihiatricâ Română (A.P.R.) / Romanian Psychiatric Association, M. Kogâlniceanu av 95a, 70603 Bucuresti
T: (01) 3113471
Founded: 1918
Pres: Prof. Dr. George Ionescu; Gen Secr: Dr. Floria Tudose
Focus: Psychiatry
Periodical
The Romanian Review of Psychiatry, Child Psychiatry and Clinical Psychiatry (weekly) *09744*

Asociatia Psihologilor din Romania, Str Onesti 11, 70119 Bucuresti
Founded: 1964
Focus: Psych
Periodicals
Revista de Psihologie
Revue des Sciences Sociales, Série de Psychologie *09745*

Asociatia Romana de Ştiinte Politice / Romanian Association of Political Sciences, Sos. Kiseleff 47, 71268 Bucuresti
Founded: 1968; Members: 357
Pres: G. Macovescu; Gen Secr: J. Ceterchi
Focus: Poli Sci
Periodical
Viitorul social *09746*

Balkan Medical Union / Union Médicale Balkanique, Str Gabriel Péri 1, CP 149, 70148 Bucuresti
T: (01) 167846
Founded: 1932; Members: 2000
Focus: Med
Periodicals
Annuaire
Archives (weekly)
Bulletin (weekly) *09747*

Comitetul National al Geologilor din Romania (C.N.G.R.S.R.), Str Mendeleev 36, 70169 Bucuresti
Founded: 1962
Focus: Geology *09748*

Entente Médicale Méditerranéenne (EMM), c/o Union Médicale Balkanique, Strada Gabriel Péri 1, CP 149, 70148 Bucuresti
T: (0) 137857
Founded: 1980
Focus: Med *09749*

European Centre for Higher Education, c/o CEPES-UNESCO, Palatul Kretulescu, Strada Stirbei Voda 39, Bucuresti
T: (01) 6130839; Tx: 11658; Fax: (01) 6415025
Founded: 1972; Members: 44
Gen Secr: Carin Berg
Focus: Educ
Periodical
Higher Education in Europe (weekly) . . *09750*

European Committee for Scientific and Cultural Relations with Romania, Strada Politechicii, 77213 Bucuresti
Focus: Sci *09751*

International Association of South-East European Studies, Str Ion Frimu 9, Bucuresti
T: (01) 507410, 507470
Founded: 1963
Focus: Ethnology
Periodicals
Bulletin d'Archéologie Sud-Est Européenne
Bulletin de l'AIESEE (weekly) *09752*

Latin Language Mathematicians' Group, c/o Institut de Mathématiques, Académie des Sciences de la RSR, Calea Grivitei 21, Bucuresti
Founded: 1955
Focus: Math *09753*

Romanian Medical Association, Sos. Berceni 10, 75622 Bucuresti
T: (01) 6827570 ext 160; Fax: (01) 3129867
Members: 90000
Focus: Med
Periodical
Viaţa Medicalá (weekly) *09754*

Societatea de Anestezie si Terapie Intensiva, Str Progresului 10, 70754 Bucuresti
Founded: 1973; Members: 46
Pres: Dr. Z. Filipescu; Gen Secr: Dr. Radu Simonescu
Focus: Anesthetics; Therapeutics *09755*

Societatea de Balneologie, c/o USSM, Str Progresului 10, 70754 Bucuresti
Focus: Physical Therapy *09756*

Societatea de Cardiologie, c/o USSM, Str Progresului 10, 70754 Bucuresti
Founded: 1947
Pres: Prof. M. Anton; Gen Secr: Prof. P. Popescu
Focus: Cardiol *09757*

Societatea de Chirurgie, c/o USSM, Str Progresului 10, 70754 Bucuresti
Founded: 1898
Pres: Prof. Ion Juvara
Focus: Surgery
Periodical
Chirurgie (weekly) *09758*

Societatea de Dermato-Venerologie, c/o USSM, Str Progresului 10, 70754 Bucuresti
Founded: 1928
Pres: Dr. A. Dimitrescu; Gen Secr: I. Forsea
Focus: Derm; Venereology
Periodical
Dermato-Venerologia (weekly) *09759*

Societatea de Endocrinologie, c/o USSM, Str Progresului 10, 70754 Bucuresti
Founded: 1918
Pres: S. Milcu; Gen Secr: Dr. C. Tasca
Focus: Endocrinology *09760*

Societatea de Farmacie, c/o USSM, Str Progresului 10, 70754 Bucuresti
Founded: 1880; Members: 1200
Pres: Prof. D. Dobrescu; Gen Secr: Dr. O. Contz
Focus: Pharmacol
Periodical
Farmacia (weekly) *09761*

Societatea de Fiziologie, c/o USSM, Str Progresului 10, 70754 Bucuresti
Founded: 1949
Pres: Prof. P. Groza; Gen Secr: Prof. Dr. Elvira Miulescu
Focus: Physiology *09762*

Societatea de Ftiziologie, c/o USSM, Str Progresului 10, 70754 Bucuresti
Founded: 1930
Focus: Pathology *09763*

Societatea de Gastroenterologie, c/o USSM, Str Progresului 10, 70754 Bucuresti
Founded: 1959; Members: 60
Pres: A. Oproiu; Gen Secr: Bendedict Gheorghescu
Focus: Gastroenter *09764*

Societatea de Gerontologie, c/o USSM, Str Progresului 10, 70754 Bucuresti
Founded: 1956
Pres: Prof. A. Aslan; Gen Secr: Dr. Lidia Hartia
Focus: Geriatrics *09765*

Societatea de Histochimie si Citochimie, c/o USSM, Str Progresului 10, 70754 Bucuresti
Founded: 1964
Pres: Prof. I. Diculescu; Gen Secr: Prof. Doina Onicescu
Focus: Chem *09766*

Societatea de Igiena si Sànàtate Publica, c/o USSM, Str Progresului 10, 70754 Bucuresti
Founded: 1949
Pres: Prof. Dr. P. Manu; Gen Secr: Dr. T. Niculescu
Focus: Public Health; Hygiene
Periodical
Igiena *09767*

Societatea de Istorie Medicinei, c/o USSM, Str Progresului 10, 70754 Bucuresti
Founded: 1929
Pres: Prof. B. Dutescu; Gen Secr: N. Marcu
Focus: Hist *09768*

Societatea de Medicinà Generalà, c/o USSM, Str Progresului 10, 70754 Bucuresti
Founded: 1961
Pres: Prof. G. Panaitescu; Gen Secr: Dr. M. Radulescu; Dr. Mircea Angelescu
Focus: Med *09769*

Societatea de Medicinà Internà, c/o USSM, Str Progresului 10, 70754 Bucuresti
Founded: 1919; Members: 50
Pres: Prof. Rudolf Geib
Focus: Intern Med
Periodical
Medicina Interna (weekly) *09770*

Societatea de Medicinà Sportiva, c/o AMR, Str Progresului 10, 70754 Bucuresti
Founded: 1932
Pres: Prof. Dr. Ioan Dragan; Gen Secr: Dr. Marta Baroga
Focus: Physical Therapy; Sports *09771*

Societatea de Medici si Naturalisti / The Society of Physicians and Natural Scientists, Independentei Blvd 16, POB 25, 66000 Iasi
T: (098) 142980; Fax: (098) 142980
Founded: 1830
Pres: Prof. L. Haulica; Gen Secr: Traian Mihaescu
Focus: Med
Periodical
Revista Medico-Chirurgicala: Medical Surgical Journal (weekly) *09772*

Societatea de Microbiologie, c/o Institutul Cantacuzino, Bucuresti
Focus: Microbio
Periodical
Bacteriologia, Virusologia, Parazitologia si Epidemiologia (weekly) *09773*

Societatea de Obstetricà si Ginecologie, c/o USSM, Str Progresului 10, 70754 Bucuresti
Founded: 1900
Pres: Prof. Dr. Dan Alessandrescu; Gen Secr: Dr. G. Teodoru
Focus: Gynecology
Periodical
Obstetrica si Ginecologia (weekly) . . . *09774*

Societatea de Oftalmologie, c/o USSM, Str Progresului 10, 70754 Bucuresti
Founded: 1922
Pres: Prof. Dr. M. David; Gen Secr: Prof. M. Olteanu
Focus: Ophthal
Periodical
Oftalmologia (weekly) *09775*

Societatea de Oto-Rino-Laringologie, c/o USSM, Str Progresului 10, 70754 Bucuresti
Founded: 1908; Members: 100
Pres: Prof. D. Hociota; Gen Secr: R. Calarasu
Focus: Otorhinolaryngology
Periodical
Oto-Rino-Laringologie (weekly) *09776*

Societatea de Patologie Infectioasà, c/o USSM, Str Progresului 10, 70754 Bucuresti
Founded: 1958
Focus: Pathology *09777*

Societatea de Pediatrie, c/o USSM, Str Progresului 10, 70754 Bucuresti
Founded: 1925; Members: 100
Pres: Prof. Rasvan Priscu; Gen Secr: Dr. T. Popescu
Focus: Pediatrics
Periodical
Pediatrie (weekly) *09778*

Societatea de Radiologie, c/o USSM, Str Progresului 10, 70754 Bucuresti
Founded: 1924
Gen Secr: Dr. I. Tudosio
Focus: Radiology
Periodical
Radiologie (weekly) *09779*

Societatea de Stiinte Biologice din Romania, Alleea Portocalilor 1-3, 76258 Bucuresti
T: (01) 496602
Founded: 1949; Members: 9000
Focus: Bio *09780*

Societatea de Stiinte Farmaceutice / Society of Pharmaceutical Sciences, c/o Asociatia Medicala Romana, Str Progresului 10, 79754 Bucuresti
T: (01) 6141071
Founded: 1880; Members: 320
Pres: Prof. Dr. D. Dobrescu; Gen Secr: Angela Grasu
Focus: Pharmacol
Periodical
Farmacia (weekly) *09781*

Societatea de Stiinte Filologice din Romania (S.S.F.), Bd Schitu Magureanu 1, 70626 Bucuresti
T: (01) 151792
Founded: 1949; Members: 12000
Focus: Ling; Lit
Periodicals
Buletinul SSF (weekly)
Limba si literatura romana (weekly)
Limbile moderne in scoala (weekly)
Studii de literatura universala (weekly) . *09782*

Societatea de Stiinte Fizice si Chimice din Romania, Bd Schitu Magureanu 1, 70626 Bucuresti
T: (01) 147508
Founded: 1964; Members: 8000
Focus: Chem; Physics
Periodicals
Buletin de fizica si chimie (weekly)
Revista de Fizica si Chimie (weekly) . . *09783*

Societatea de Stiinte Geografice din Romania, Bd Nicolae Bălcescu 1, 70111 Bucuresti
T: (01) 149350
Founded: 1875; Members: 5000
Focus: Geography
Periodicals
Buletinul S.S.G.: Bulletin de la Société Roumaine de Géographie
In ajutorul profesorului de geografie: A l'Aide du Professeur de Géographie
Lecturi geografice: Des Lectures Géographiques
Terra (weekly) *09784*

Societatea de Stiinte Geologice din Romania, Str Berzei 46, Bucuresti
Founded: 1930; Members: 850
Focus: Geology 09785

Societatea de Stiinte Matematice din Romania (SSM), Str Academiei 14, 79547 Bucuresti
T: (01) 144653
Founded: 1895; Members: 14000
Focus: Math
Periodicals
Bulletin Mathématique (weekly)
Gazeta Mathematica (weekly)
Mathematikai Lapok (weekly) 09786

Societatea de Stomatologie, c/o USSM, Str Progresului 10, 70754 Bucuresti
Founded: 1938
Pres: Prof. Lucian Ene; Gen Secr: Prof. C. Burlibasa
Focus: Stomatology
Periodical
Stomatologie (weekly) 09787

Societatea de Studii Clasice din Romania, Str Spiru Haret 12, 70738 Bucuresti
Focus: Archeol
Periodical
Studii clasice (weekly) 09788

Societatea Nationala de Medicina Generala din Romania (SNMG) / National Society of General Practice of Romania, Str Progresului 10, 70754 Bucuresti
T: (01) 6141062; Fax: (01) 3121357
Founded: 1961; Members: 6000
Pres: Adrian Restian; Gen Secr: Grigore Rusoi
Focus: Med
Periodical
Revista Medicala Romana: Romanian Medical Journal (weekly) 09789

Societatea Nationala Romana pentru Stiinta Solului (S.N.R.S.S.) / National Romanian Society for Soil Science, Bd Màràsti 61, 71331 Bucuresti
T: (01) 172180; Tx: 11394
Founded: 1961; Members: 600
Focus: Agri
Periodical
Stiința Solului (3 times annually) 09790

Societatea Numismatica Romana (SNR), Str I.C. Frimu 11, Sect 1, 71119 Bucuresti
T: (01) 503410
Founded: 1903; Members: 1810
Focus: Numismatics
Periodical
Buletinul (weekly) 09791

Societatea Romana de Linguistica, Bd Republicii 13, 70031 Bucuresti
T: (01) 141717
Founded: 1970
Focus: Ling 09792

Societatea Romana de Neurochirurgie / Romanian Society of Neurosurgery, c/o AMR, Sos. Berceni 10, 75622 Bucuresti
T: (01) 6827570 ext 160; Fax: (01) 3129867
Founded: 1982; Members: 115
Pres: Prof. A. Constantinovici; Gen Secr: Prof. A.V. Ciurea
Focus: Surgery; Neurology
Periodical
Romanian Neurosurgery (weekly) 09793

Societatea Română de Ortopedie și Traumatologie (SOROT) / Romanian Society for Orthopaedics and Traumatology, Str Progresului 10, 70754 Bucuresti
T: (01) 6141062
Founded: 1935; Members: 312
Pres: Dr. Ion Dinulescu; Gen Secr: Dr. Dinu Antonescu
Focus: Orthopedics; Traumatology
Periodical
Revista de Ortopedie si Traumatologie (weekly) 09794

Societatea Romana de Sprijia a Virstnicilor Suferinzi de Afectiuni de tip Alzheimer (R.S.S.P.S.A.D.) / Romanian Support Society for People Suffering of Alzheimer Type Diseases, Str Gh. Marinescu 3, Bucuresti
T: (01) 3113471; Fax: (01) 3212268
Founded: 1992; Members: 171
Pres: Dr. Catalina Tudose
Focus: Neurology
Periodical
Romanian Alzheimer Newsletter 09795

Uniunea Arhitectilor din Romania, Str Academiei 18-20, 79182 Bucuresti
T: (01) 140743
Founded: 1952; Members: 3000
Focus: Archit
Periodical
Arhitectura (weekly) 09796

Uniunea Artistilor Plastici din Romania, Str Nicolae Iorga 21, Bucuresti
T: (01) 507380
Founded: 1950; Members: 1650
Focus: Arts
Periodical
Arta (weekly) 09797

Uniunea Scriitorilor din Romania / Writers' Union of the SRR, Calea Victoriei 115, Bucuresti
T: (01) 507245
Founded: 1877
Focus: Lit
Periodicals
Convorbiri literare
Igaz Szó
Knijevni Jivot
Luceafaírul
Orizont
România Literaraí
Secolul 20
Steaua
Utunk
Vatra
Viata Româneascaí 09798

Russia

Agro-Industrial Society, Ul Kirova 13, 101000 Moskva
T: 9243809
Pres: V.I. Fisinin
Focus: Agri; Econ 09799

Aircraft Building Society, Leningradskii Pr 24a, 125040 Moskva
T: 2142288
Pres: A.M. Batkov
Focus: Aero 09800

Association of Economic Scientific Institutions, Krasikova 27, 117218 Moskva
T: 1290427
Focus: Econ 09801

Association of International Law, Ul Frunze 10, 119841 Moskva
Pres: Prof. G.I. Tunkin
Focus: Law 09802

Association of Orientalists, Ul Rozhdestvenka 12, 103753 Moskva
T: 9285764
Pres: M.S. Kapitsa
Focus: Ethnology 09803

Association of Political Sciences, Ul Frunze 10, 118941 Moskva
Pres: Dr. G.K. Shakhnazarov
Focus: Poli Sci 09804

Association of Sinologists, Ul Krasikova 27, 117848 Moskva
Pres: M.L. Titarenko
Focus: Ethnology 09805

Astronomical and Geodesical Society, Sadovo-Kudrinskaya ul 24, 103001 Moskva
Pres: Y.D. Bulanzhe
Focus: Astronomy 09806

Automobile and Road Building Society, B. Ovchinnikovsky per 12, 113184 Moskva
T: 2314813
Pres: A.K. Vasilev
Focus: Auto Eng; Civil Eng 09807

Biochemical Society, Ul Vavilova 34, 117991 Moskva
T: 1359779
Pres: S.E. Severin
Focus: Biochem 09808

Biological Engineering Society, Ul Lesteva 18, 113809 Moskva
T: 2366075
Pres: V.E. Matveyev
Focus: Eng 09809

Civil Engineering Society, Podsosensky per 25, 103062 Moskva
T: 2978799
Pres: I.I. Ishenko
Focus: Eng 09810

CMEA Coordination Centre for Study of World Oceans Development of Techniques for Exploration and Utilization of Resources, c/o Institute of Oceanologie, Ul Krasikova 23, 117218 Moskva
T: 1245979; Tx: 411968
Founded: 1971
Focus: Oceanography
Periodical
Information Bulletin of the Coordinating Centre 09811

Commission for a Linguistic Atlas of Europe, Ul Semashko 1, 103009 Moskva
T: 2903152
Pres: V.V. Ivanov
Focus: Ling; Cart 09812

Commission for Atomic Energy, Leninskii Pr 14, 117901 Moskva
T: 19692101
Pres: A.P. Aleksandrov
Focus: Nucl Res 09813

Commission for Computer Technology, Ul Vavilova 40, 117967 Moskva
T: 1352489
Pres: A.A. Dorodnitsyn
Focus: Computer & Info Sci 09814

Commission for Computer Technology, Pr Akademika Lavrenteva 17, 630090 Novosibirsk
T: 355650
Pres: A.S. Alekseev
Focus: Electronic Eng 09815

Commission for Co-ordination of Research in State Nature Reserves, Leninskii Pr 33, 117071 Moskva
T: 2322088
Pres: V.E. Sokolov
Focus: Nat Sci 09816

Commission for Ecology, Leninskii Pr 14, 117071 Moskva
T: 2343506
Pres: G.I. Marchuk
Focus: Ecology 09817

Commission for International Scientific Links, Leninskii Pr 14, 117071 Moskva
T: 2344485; Tx: 411095
Pres: I.M. Makarov
Focus: Sci 09818

Commission for International Tectonic Maps, Pyzhevsky per 7, 109017 Moskva
T: 2308157
Pres: V.E. Khain
Focus: Cart 09819

Commission for Links in Social Science with the American Council of Learned Societies, Khlebnii per 2, 121814 Moskva
T: 2030673
Focus: Soc Sci 09820

Commission for Links with US Research Establishments in the Use of Technology and New Communications Technologies in Education, Yaroslavskaya ul 13, 129366 Moskva
T: 2374532
Pres: E.P. Velikhov
Focus: Educ 09821

Commission for Management of the Development of Cities, Pr 60-letiya Oktyabrya 9, 117312 Moskva
T: 1357575
Pres: D.M. Gvishiani
Focus: Urban Plan 09822

Commission for Mechanics and Physics of Polymers, Ul Vavilova 28, 117813 Moskva
T: 1356384
Pres: G.L. Slonimsky
Focus: Physics 09823

Commission for Meteorites and Space Dust, Universitetsky Pr 3, 630090 Novosibirsk
T: 353654
Pres: Y.A. Dolgov
Focus: Astronomy 09824

Commission for Nuclear Physics, Leninskii Pr 14, 117901 Moskva
T: 2342286
Pres: M.A. Markov
Focus: Physics 09825

Commission for Philology and Phonetics, Volkhonka 18, 121019 Moskva
T: 2026621
Pres: T.V. Gamkrelidze
Focus: Ling; Lit 09826

Commission for Scientific and Technical Co-operation of the Academy of Sciences and Organizations of Moscow Oblast, Ul Vavilova 34, 117995 Moskva
T: 2229393
Pres: V.N. Melkishev
Focus: Eng; Sci 09827

Commission for Socio-ecological Research, Pr 60-letiya Oktyabrya 9, 117312 Moskva
T: 1354332
Pres: D.M. Gvishiani
Focus: Ecology 09828

Commission for Space Toponomy, Pyatnitskaya ul 48, 109017 Moskva
Pres: M.Y. Marov
Focus: Aero 09829

Commission for Synchroton Irradiation, Ul Kosygina 2, 117973 Moskva
T: 1969206
Pres: S.T. Belyaev
Focus: Eng 09830

Commission for the Co-ordination of Co-operation of Humanities Institutions of the Russian Academy of Sciences with UNESCO, Ul Volkhonka 18, 121019 Moskva
T: 2374421
Pres: E.P. Chelyshev
Focus: Humanities 09831

Commission for the Development of Scientific Co-operation with Great Britain, Leninskii Pr 59, 117333 Moskva
T: 1356541
Pres: B.K. Vainshtein
Focus: Sci 09832

Commission for the Effective Use of Shales in the Russian Economy, Izhorskaya ul 13-19, 127412 Moskva
T: 4859663
Pres: A.E. Sheindlin
Focus: Econ 09833

Commission for the European (Vienna) Centre for Co-ordination of Research and Documentation in Social Sciences, Ul Krasikova 28, 117418 Moskva
T: 1286851
Pres: V.A. Vinogradov
Focus: Soc Sci 09834

Commission for the History of Geological Knowledge and Geological Study of the Russian Federation, Pyzhevsky per 7, 109017 Moskva
T: 1357866
Pres: V.V. Tikhomirov
Focus: Geology 09835

Commission for the History of Philology, Volkhonka 18, 121019 Moskva
T: 9393248
Pres: P.A. Nikolaev
Focus: Cultur Hist; Ling; Lit 09836

Commission for the Processing of the Scientific Legacy of Academician V.I. Vernadsky, Bankovsky per 2, 101000 Moskva
T: 9237482
Pres: A.L. Yanshin
Focus: Sci 09837

Commission for the Prospects of Development of Science in the Russian Federation, Ul Vavilova 44, 117333 Moskva
T: 9301195
Pres: G.A. Mesyats
Focus: Sci 09838

Commission for the Protection of Natural Waters, Ul Vavilova 44, 117333 Moskva
T: 3147885
Pres: B.N. Laskorin
Focus: Ecology 09839

Commission for the Study of Productive Forces and Natural Resources, Maronovsky per 26, 117049 Moskva
T: 2382112
Pres: A.G. Aganbegyan
Focus: Econ 09840

Commission for the Study of the Arctic, Ul Vavilova 7, 117822 Moskva
T: 2342968
Pres: N.P. Laverov
Focus: Geography 09841

Commission for the Use of Computers and for Raising the Qualifications of Computer Users, Ul Vavilova 44, 117333 Moskva
T: 2374532
Pres: E.P. Velikhov
Focus: Computer & Info Sci 09842

Commission for the World's Oceans, Ul Vavilova 44, 117333 Moskva
T: 1351568
Pres: L.M. Brekhovskikh
Focus: Oceanography 09843

Commission for Work with Young People, Ul Kosygina 4, 117977 Moskva
T: 2374532
Pres: E.P. Velikhov
Focus: Educ 09844

Commission on Terminology, Leninskii pr 265a, 720071 Bishkek
Pres: S.T. Tabysahaliev 09845

Committee for Russian Scientists in Defence of Peace and against Nuclear War, Profsoyuznaya 84, 117810 Moskva
T: 3335288
Pres: R.Z. Sagdeev
Focus: Prom Peace 09846

Committee for Systems Analysis, Ul Krasikova 32, 117418 Moskva
T: 2663308
Pres: V.S. Mikhalevich
Focus: Computer & Info Sci 09847

Committee for the Coordination of Construction of Scientific Apparatus and Antomation of Research Work, Leninskii Pr 48, 117829 Moskva
Founded: 1978
Gen Secr: Prof. W.L. Talrose
Focus: Eng 09848

Committee for UNESCO Programme „Man and Biosphere", Ul Fersmana 13, 117312 Moskva
T: 2322088
Pres: V.E. Sokolov
Focus: Soc Sci 09849

Committee of the International Programme „Litosphere", Staromonetnii 22, 109180 Moskva
T: 2312944
Pres: A.L. Yanshin
Focus: Geophys 09850

Committee of the International Programme of Geological Correlation, Pyzhevsky per 7, 109017 Moskva
T: 2317500
Pres: V.A. Zharikov
Focus: Geology 09851

Committee on Meteorites, Ul M. Ulyanovoi 3, 117313 Moskva
T: 1385789
Pres: Y.A. Shukulyukov
Focus: Astrophys 09852

Committee on Petrography, Staromonetny per 35, 109017 Moskva
T: 2331635
Pres: O.A. Bogatikov
Focus: Geology 09853

Consultative Working Group for the Preparation of New Questions of Longterm Prospects of Development of Energy, Ul Vavilova 44, 117333 Moskva
T: 2324872
Pres: M.A. Styrikovich
Focus: Energy 09854

Co-ordination Committee for Computer Technology, Leninskii Pr 14, 117901 Moskva
T: 2342549
Pres: G.I. Marchuk
Focus: Computer & Info Sci 09855

Co-ordination Council for Information on Achievements, Leninskii Pr 61, 117333 Moskva
T: 2344485
Pres: I.M. Makarov
Focus: Sci 09856

Co-ordination Council for Scientific Problems Linked with Ecological Consequences of the Use of New Technological Systems, Leninskii Pr 14, 117901 Moskva
Pres: A.P. Aleksandrov
Focus: Ecology 09857

Council for International Co-operation in Social Sciences, Ul Dimitrova 35, 113095 Moskva
T: 2387600
Pres: P.N. Fedoseev
Focus: Soc Sci 09858

Council for Links between the Academy of Sciences and Higher Education, Leninskii Pr 14, 117901 Moskva
T: 2325801
Pres: A.A. Logunov
Focus: Educ 09859

Council for Metrological Provision and Standardization, Bolshoi Pr 612, 199178 St. Petersburg
T: 2178602
Pres: N.S. Solomenko
Focus: Standards 09860

Council for the Study of Productive Forces, Ul Vavilova 7, 117822 Moskva
T: 1356358
Pres: L.A. Kozlov
Focus: Business Admin 09861

Council of Scientific Medical Societies, Rakhmanovskii per 3, 101431 Moskva
Pres: G.N. Serdyukovskaya; Gen Secr: S.S. Yarmonenko
Focus: Med 09862

Council on International Cooperation in Research and Uses of Outer Space (INTERCOSMOS), Leninskii Pr 14, 117901 Moskva
T: (095) 2343828; Tx: 7564
Founded: 1970
Pres: G. Kharitonov
Focus: Aero 09863

D.I. Mendeleev Chemical Society, Krivokolenniy per 12, 101907 Moskva
T: 9257285
Pres: A.V. Fokin
Focus: Chem 09864

Economics Society, B. Cheremushkinskaya ul 34, 117259 Moskva
T: 1201321
Pres: V.S. Pavlov
Focus: Econ 09865

Entomological Society, Universitetskaya Nab. 1, 199034 St. Petersburg
T: 2181212
Pres: G.S. Medvedev
Focus: Zoology 09866

Federation of Anaesthesiologists and Reanimatologists, Botkin Hospital kv 3, 125101 Moskva
T: 9459725; Fax: 9459725
Founded: 1991
Pres: E.A. Damir; Gen Secr: I. Molchanov
Focus: Anesthetics 09867

Ferrous Metallurgy Society, Baumanskaya ul 9, 107865 Moskva
T: 2670988
Pres: N.I. Drozdov
Focus: Metallurgy 09868

Flour, Fodder and Grain Storage Society, Chistoprudnyi bul 12a, 101859 Moskva
T: 9286733
Pres: M.L. Timoshyshin
Focus: Agri 09869

Geological Society, 2-Roshinskaya ul 10, 113191 Moskva
T: 2373333
Pres: V.F. Rogov
Focus: Geology 09870

Geomorphological Commission, Staromonetnii 29, 109017 Moskva
T: 2380360
Pres: D.A. Timofeev
Focus: Geology 09871

Group of the Far Eastern Division, Ul Vavilova 44, 117333 Moskva
T: 1359019
Pres: V.S. Astashov
Focus: Sci 09872

Hydrobiological Society, Ul Gorkogo 27, 103050 Moskva
Pres: L.M. Suschenya
Focus: Bio 09873

I.I. Polzunov Science Production Association for Research and Design of Power Equipment, Politekhnicheskaya ul 24, 194021 St. Petersburg
T: 2779213
Pres: V.K. Ryzhkov
Focus: Mach Eng 09874

Institute for Standardization, Pr Kalinina 56, 121205 Moskva
T: 2908789; Tx: 411141
Founded: 1962
Gen Secr: M.A. Dovbenko
Focus: Standards
Periodical
Aspects of Standardization (weekly) . . 09875

International Confederation of Theatre, Tverskaya ul 12, 103009 Moskva
T: 2092436 app 228; Fax: 2095249
Founded: 1992
Pres: K.Y. Lavrov
Focus: Perf Arts 09876

International Society for Pathophysiology, Ul Baltiiskaya 8, 124315 Moskva
T: 1528655; Fax: 1518540
Members: 1200
Pres: Prof. G.N. Kryzhanovsky; Gen Secr: V. Shinkarenko
Focus: Physiology 09877

I.P. Pavlov Physiological Society, Pogodinskaya ul 6, 119121 Moskva
Pres: O.G. Gazenko
Focus: Physiology 09878

Mapping and Prospecting Engineering Society, Ul Krzhizhanovskogo 14, 117801 Moskva
T: 1243560
Pres: A.A. Drazhnyuk
Focus: Eng 09879

Mational Scientific Medical Society of Forensic Medical Officers, Ul Sadovaya-Kudrinskaya 32, 123242 Moskva
Pres: A.B. Kapustin; Gen Secr: G.N. Nazarov
Focus: Forensic Med 09880

Mechanical Engineering Society, Ul Chkalova 64, 109004 Moskva
T: 9258332
Pres: B.N. Sokolov
Focus: Eng 09881

Medical Engineering Society, Ul Kasatkina 3, 129301 Moskva
T: 2839784
Pres: B.I. Leonov
Focus: Eng 09882

Microbiological Society, Pr 60-letiya Oktyabrya 7, 117811 Moskva
Focus: Microbio 09883

Mining Engineering Society, Karetnyi ryad 10, 103006 Moskva
T: 2998815
Pres: A.P. Fisun
Focus: Mining 09884

Moscow Science-Production Association, Ul Ulsacheva 35, 119048 Moskva
T: 2455656
Pres: V.V. Klyuev
Focus: Sci 09885

Moscow Society of Naturalists, Ul Gertsena 6, 103009 Moskva
T: 2036704
Founded: 1805; Members: 2700
Pres: A.L. Yanshin
Focus: Nat Sci
Periodical
Byulleten Moskovskogo Obshchestva Ispytatelei Prirody (weekly) 09886

Municipal Economy Soiety, Trekhprudny per 11-13, 103001 Moskva
T: 2998300
Pres: A.F. Poryadin
Focus: Econ 09887

Museum Council, Ul Dm. Ulyanova 19, 117036 Moskva
T: 1251121
Pres: B.A. Rybakov
Focus: Arts 09888

National Committee for the Collection and Assessment of Numerical Data in Science and Technology, Leninskii Pr 14, 117901 Moskva
T: 2324205
Pres: V.V. Sychev
Focus: Computer & Info Sci 09889

National Committee for Thermal Analysis, Leninskii Pr 312, 117907 Moskva
T: 2323420
Pres: V.B. Iazarev
Focus: Physics 09890

National Committee of Biochemists, Leninskii Pr 33, 117071 Moskva
Pres: S.E. Severin
Focus: Biochem 09891

National Committee of Biologists, Leninskii Pr 33, 117071 Moskva
Pres: V.E. Sokolov
Focus: Bio 09892

National Committee of Chemists, Ul A.N. Kosygina 2b, Moskva
Focus: Chem 09893

National Committee of Finno-Ugric Philologists, Ul Semashko 1, 103009 Moskva
Pres: T.-R. O. Viitso
Focus: Ling; Lit 09894

National Committee of Geologists, Pyzhevsky per 7, 109017 Moskva
T: 2308151
Pres: N.P. Laverov
Focus: Geology 09895

National Committee of History and Philosophy of Natural Science and Technology, Volkhonka 14, 119842 Moskva
T: 2039320
Pres: I.T. Frolov
Focus: Hist; Philos; Nat Sci; Eng . . . 09896

National Committee of Mathematicians, Ul Vavilova 42, 117966 Moskva
Focus: Math 09897

National Committee of Russian Historians, Ul Dm. Ulyanova 19, 117036 Moskva
T: 1260529
Pres: S.L. Tikhvinsky
Focus: Hist 09898

National Committee of Slavonic Philologists, Volkhonka 18, 121019 Moskva
Pres: N.I. Tolstoi
Focus: Ling; Lit 09899

National Committee of the International Council of Scientific Unions, Leninskii Pr 14, 117901 Moskva
Pres: V.A. Ambartsumyan
Focus: Sci 09900

National Committee of the International Scientific Radio Union, Pr Marksa 18, 103907 Moskva
Pres: V.V. Migulin
Focus: Mass Media 09901

National Committee of the Pacific Ocean Research Association, Ulyanovskaya ul 51, 109004 Moskva
T: 2724786
Pres: N.A. Shilo
Focus: Oceanography 09902

National Committee of the Scientific Committee for Problems of the Environment, Ul Vavilova 40, 117697 Moskva
T: 1352489
Pres: A.A. Dorodnitsyn
Focus: Ecology 09903

National Committee of Turkish Philologists, Ul Semashko 1, 103009 Moskva
Pres: E.R. Tenishev
Focus: Ling; Lit 09904

National Committee on Automatic Control, Profsoyuznaya 65, Moskva
Pres: V.A. Trapeznikov
Focus: Mach Eng 09905

National Committee on the International Biological Programme, Leninskii Pr 14, 117901 Moskva
Focus: Bio 09906

National Committee on Theoretical and Applied Mechanics, Pr Vernadskogo 101, 177526 Moskva
Focus: Mach Eng 09907

National Immunological Society, Kashirskoe shosse 24, 115478 Moskva
T: 1118333; Fax: 1171027
Founded: 1983; Members: 600
Pres: R.V. Petrov; Gen Secr: S.Y. Sidorovich
Focus: Immunology 09908

National Medical and Technical Scientific Society, Ul Kasatkina 3, 129301 Moskva
T: 1879723; Tx: 412209; Fax: 1873734
Founded: 1968
Pres: B.I. Leonov; Gen Secr: B.E. Belousov
Focus: Med; Eng
Periodical
Biomedical Engineering (weekly) 09909

National Ophthalmological Society, Ul Sadovo-Chernogryazskaya 14, 103064 Moskva
Pres: E.S. Avetisov; Gen Secr: T.I. Forofonofa
Focus: Ophthal 09910

National Pharmaceutical Society, Ul Krasikova 34, 117418 Moskva
Pres: M.T. Alyushin; Gen Secr: R.S. Skulkova
Focus: Pharmacol 09911

National Scientific Medical Society of Anatomists, Histologists and Embryologists, Ul Ostrovityanova 1, 117869 Moskva
Pres: V.V. Kupriyanov; Gen Secr: V.V. Korolev
Focus: Anat; Physiology 09912

National Scientific Medical Society of Endocrinologists, Ul Dm. Ulyanova 11, 117036 Moskva
Pres: V.G. Baranov; Gen Secr: N.T. Starkova
Focus: Endocrinology 09913

National Scientific Medical Society of Gastroenterologists, Ul Pogodinskaya 5, 119435 Moskva
Pres: V.C. Vasilenko; Gen Secr: M.A. Vinogradova
Focus: Gastroenter 09914

National Scientific Medical Society of Haematologists and Transfusiologists, Novozykovskii Pr 4a, 125167 Moskva
Pres: V.N. Shabalin; Gen Secr: M.P. Khokhlova
Focus: Hematology 09915

National Scientific Medical Society of History of Medicine, Petrovirigskii per 6-8, 101838 Moskva
Pres: Y.P. Lisitsyn; Gen Secr: I.V. Vengrova
Focus: Hist; Med 09916

National Scientific Medical Society of Hygienists, Mechnikova per 5, 103064 Moskva
Pres: G.N. Serdyukovskaya; Gen Secr: A.G. Sukharev
Focus: Hygiene 09917

National Scientific Medical Society of Infectionists, Botkinskii per 3, 125284 Moskva
Pres: V.N. Nikiforov; Gen Secr: N.M. Belyaeva
Focus: Med 09918

National Scientific Medical Society of Medical Geneticists, Ul Moskvoreche 1, 115478 Moskva
T: 1118594; Fax: 3240702
Founded: 1978; Members: 1500
Pres: E.K. Ginter; Gen Secr: N.P. Kuleshov
Focus: Genetics 09919

National Scientific Medical Society of Nephrologists, Ul Rossolimo 11a, 119021 Moskva
T: 2485333
Founded: 1969
Pres: N.A. Mukhin; Gen Secr: S.O. Androsova
Focus: Med 09920

National Scientific Medical Society of Neuropathologists and Psychiatrists, Ul Kropotkinskii per 23, 119034 Moskva
Pres: G.V. Morozov; Gen Secr: G.Y. Lukacher
Focus: Pathology; Psychiatry 09921

National Scientific Medical Society of Neurosurgeons, Ul Fadeeva 5, 125047 Moskva
Pres: A.N. Konovalov; Gen Secr: F.A. Serbienko
Focus: Surgery 09922

National Scientific Medical Society of Obstetricians and Gynaecologists, Ul Shabolovka 57, 113163 Moskva
Pres: G.M. Saveleva; Gen Secr: T.V. Chervakova
Focus: Gynecology 09923

National Scientific Medical Society of Oncologists, Ul St. Petersburgskaya 68, 188646 St. Petersburg
Pres: N.P. Napalkov; Gen Secr: E.S. Kiseleva
Focus: Cell Biol & Cancer Res 09924

National Scientific Medical Society of Oto-Rhino-Laryngologists, Bolshaya Pirogovskaya 6, 119435 Moskva
Pres: N.A. Preobrazhenskii; Gen Secr: N.P. Konstantinova
Focus: Otorhinolaryngology 09925

National Scientific Medical Society of Paediatriciams, Lomonosovskii Pr 2, 117963 Moskva
Pres: M.Y. Studenikin; Gen Secr: G.V. Yatsyk
Focus: Pediatrics 09926

National Scientific Medical Society of Phtisiologists, Ul 6 kilometr Severnoi Zheleznoi Dorogi, 107564 Moskva
Pres: A.G. Khomenko; Gen Secr: V.V. Erokhin
Focus: Med 09927

National Scientific Medical Society of Physical Therapists and Health-Resort Physicians, Kalinina Pr 50, 121099 Moskva
Pres: A.N. Obrosov; Gen Secr: V.D. Grigoreva
Focus: Physical Therapy; Med 09928

National Scientific Medical Society of Physicians-Analysts, Ul Sadovaya-Kudrinskaya 3, 123242 Moskva
Pres: B.F. Korovkin; Gen Secr: R.L. Martsishevskaya
Focus: Med 09929

National Scientific Medical Society of Physicians in Curative Physical Culture and Sports Medicine, Lomonovskii Pr 2, 117963 Moskva
Pres: S.V. Kruzshev; Gen Secr: A.V. Sokova
Focus: Med 09930

National Scientific Medical Society of Rheumatologists, Kashirskoye shosse 34a, 115522 Moskva
T: 1144490
Founded: 1928; Members: 1418
Pres: V.A. Nassonova; Gen Secr: L.N. Denissov
Focus: Rheuma
Periodical
Revmatologhia (weekly) 09931

National Scientific Medical Society of Roentgenologists and Radiologists, Ul Profsoyuznaya 86, 117837 Moskva
Pres: A.S. Pavlov; Gen Secr: V.Z. Agranat
Focus: X-Ray Tech; Radiology 09932

National Scientific Medical Society of Stomatologists, Ul Pogodinskaya 5, 119435 Moskva
Pres: N.N. Bazhanov; Gen Secr: V.M. Bezrukov
Focus: Stomatology 09933

National Scientific Medical Society of Surgeons, Abrikosovskii per 2, 119874 Moskva
Pres: B.V. Petrovskii; Gen Secr: M.I. Perelman
Focus: Surgery 09934

National Scientific Medical Society of Toxicologists, Ul Bekhtereva 1, 193019 St. Petersburg
Pres: S.N. Golikov; Gen Secr: L.A. Timofeevskaya
Focus: Toxicology 09935

National Scientific Medical Society of Traumatic Surgeons and Orthopaedists, Ul Priorova 10, 125299 Moskva
Pres: Y.G. Shaposhnikov; Gen Secr: S.M. Zhuravlev
Focus: Surgery; Orthopedics 09936

National Scientific Medical Society of Venereologists and Dermatologists, Ul Korolenko 3, 107076 Moskva
Pres: O.K. Shaposhnikov; Gen Secr: V.N. Mordovtsev
Focus: Venereology; Derm 09937

National Scientific Medical Soiety of Therapists, Cherepkovskaya 15, 121500 Moskva
Pres: A.S. Smetnev; Gen Secr: B.A. Sidorenko
Focus: Therapeutics 09938

National Scientific Society of Urological Surgeons, 3-ya Parkovaya 51, 105483 Moskva
T: 1649652
Founded: 1925; Members: 5500
Pres: N.A. Lopatkin; Gen Secr: A.F. Darenkov
Focus: Urology; Surgery
Periodical
Urology and Nephrology (weekly) . . . 09939

Paper and Wood-Working Society, Ul 25 Oktyabrya 8, 103012 Moskva
T: 9244728
Pres: Y.A. Guskov
Focus: Forestry 09940

Permanent Commission of Gosplan, Ul Vavilova 44, 117333 Moskva
T: 1969331
Pres: A.P. Alksandrov
Focus: Energy 09941

Petroleum and Gas Society, Leninskii Pr 63, 117876 Moskva
T: 1358866
Pres: S.T. Toplov
Focus: Energy 09942

Philosophy Society, Smolensky bul 20, 121002 Moskva
T: 2012402
Pres: I.T. Frolov
Focus: Philos 09943

Power and Electrical Power Engineering Society, Stremyannaya ul 10, 191025 St. Petersburg
T: 3113277
Pres: N.N. Tikhodeyev
Focus: Electric Eng; Energy 09944

Press and Publishing Engineering Society, Volkov per 7-9, 123376 Moskva
T: 2521431
Pres: B.A. Kuzmin
Focus: Eng 09945

Programme Committee Surface Physics, Chemistry and Mechanics, Chenogolovka, 142342 Moskva
T: 2374532
Pres: E.P. Velikhov
Focus: Chem; Physics; Eng 09946

Pushkin Commission, Nab. Makarova 4, 199034 St. Petersburg
T: 2181601
Pres: D.S. Likhachev
Focus: Lit 09947

Radio Engineering, Electronics and Telecommunications Society, Kuznetskii Most 20, 103897 Moskva
T: 9217108
Pres: Y.V. Gulyaev
Focus: Electric Eng; Electronic Eng . . 09948

Rersearch Council for Applied Mathematics, Ul Vavilova 44, 117333 Moskva
Pres: V.P. Maslov
Focus: Math 09949

Russian Academy of Agricultural Sciences, Bolshoi Kkaritonevsky per, 107814 Moskva
T: 2073942
Founded: 1929
Pres: A.A. Nikonov
Focus: Agri
Periodicals
Doklady: Proceedings
Mekhanizatsiya i Elektrificatsiya Selskogo Khozyaistva: Mechanization and Electrification of Agriculture
Selektsiya i Semenovodstvo: Selection and Seed Science
Selskokhozyaisstvennaya Biologiya: Agricultural Biology
Vestnik Selskokhozyaistvennoi Nauki: Agricultural Science Journal 09950

Russian Academy of Arts, Kropotkinskaya ul 21, 119034 Moskva
T: 2013971; Fax: 2902088
Founded: 1757
Pres: N.A. Ponomarev; Gen Secr: V.A. Lenjashin
Focus: Arts 09951

Russian Academy of Medical Sciences, Ul Solyanka 14, 109801 Moskva
T: 2970504
Founded: 1944
Pres: V.I. Pokrovskii; Gen Secr: D.S. Sarkisov
Focus: Med
Periodicals
Arkhiv Anatomii, Gistologii i Embriologii: Anatomy, Histology and Embryology Archive
Arkhiv Patologii i Meditsiny: Pathology and Medicine Archive
Byulleten Eksperimentalnoi Biologii i Meditsiny: Bulletin of Experimental Biology and Medicine
Byulleten Sibirskogo Otdeleniya Rossiiskoi AMN: Bulletin of the Siberian Division of the Russian Academy of Medical Sciences
Immunologiya: Immunology
Meditsinskaya Radiologiya: Medical Radiology
Patologicheskaya Fiziologiya i Eksperimentalnaya Terapiya: Pathological Physiology and Experimental Therapy
Pharmakologiya i Toksikologiya: Pharmacology and Toxicology
Vestnik Rossiiskoi Akademii Meditsinskikh Nauk: Journal of the Russian Academy of Medical Sciences
Voprosy Meditsinskoi Khimii: Problems of Medical Chemistry
Voprosy Virusologii Khimii: Problems of Virology
. . . . 09952

Russian Academy of Pedagogical Sciences, Pogodinskaya ul 8, 119905 Moskva
T: 2451641
Founded: 1943
Pres: Prof. V.G. Kostomarov; Gen Secr: Prof. N.D. Nikandrov
Focus: Educ
Periodicals
Defektologiya: Defectology (weekly)
Kavnt: Quantum (weekly)
Pedagogika: Pedagogics (weekly)
Russkii Yazyk i Natsionalnoi Shkole: Russian in the National School
Semya i Shkola: Family and School (weekly)
Voprosy Psikhologii: Problems of Psychology (weekly) 09953

Russian Academy of Sciences, Leninskii Pr 14, 117901 Moskva
T: 2342153; Tx: 411964
Founded: 1725
Pres: Y.S. Osipov; Gen Secr: I.M. Makarov
Focus: Sci
Periodicals
Agrokhimiya: Agrochemistry (weekly)
Akusticheskii Zhurnal: Acoustics Journal (weekly)
Algebra i Analiz: Algebra and Analysis (weekly)
Astronomicheskii Vestnik: Astronomical Herald (weekly)
Astronomicheskii Zhurnal: Astronomy Journal (weekly)
Atomnaya Energiya: Atomic Energy (weekly)
Avtomekhanika i Telemekhanika: Automation and Telemechanics (weekly)
Avtometriya: Autometry (weekly)
Aziya i Afrika Segodnya: Asia and Africa Today (weekly)
Biofizika: Biophysics (weekly)
Biokhimiya: Biochemistry (weekly)
Biologicheski Membrany: Biological Membranes (weekly)
Biologiya Morya: Biology of the Sea (weekly)
Bioorganicheskaya Khimiya: Bioorganic Chemistry (weekly)
Botanicheskii Zhurnal: Journal of Botany (weekly)
Chelovek: Man
Defektokopiya: Defectoscopy (weekly)
Diskretnaya Matematika: Discrete Mathematics
Doklady Akademii Nauk Rossiiskoi: Proceedings of the Russian Academy of Sciences (3 times monthly)
Ekologiya: Ecology (weekly)
Ekonomika i Matematicheskie Metody: Economics and Mathematical Methods (weekly)
Ekonomika i Organizatsiya Promyshlennogo Proizvodstva: Economics amd the Organisation of Industrial Production (weekly)
Elektrichestvo: Electricity (weekly)
Elektrokhimiya: Electrochemistry (weekly)
Elektronnoe Modelirovanie: Electronic Modelling (weekly)
Energiya: Energy (weekly)
Entomologicheskoe Obozrenie: Entomological Survey (weekly)
Fizika Goreniya i Vzryva: Physics of Combustion and Explosion (weekly)
Fizika i Khimiya Obrabotki Materialov: Physics and Chemistry of Materials Processing (weekly)
Fizika i Khimiya Stekla: Physics and Chemistry of Glass (weekly)
Fizika i Tekhnika Poluprovodnikov: Semiconductor Physics and Technology
Fizika Metallov i Metallovedenie: Physics of Metals and Metal Science (weekly)
Fizika Plazmy: Plasma Physics (weekly)
Fizika Tverdogo Tela: Solid State Physics (weekly)
Fiziko-Tekhnicheskie Problemy Razrabotki Poleznykh Iskopaemykh: Physical and Technical Problems of Mineral Exploitation (weekly)
Fiziologicheskii Zhurnal: Physiological Journal (weekly)
Fiziologiya Cheloveka: Human Physiology (weekly)
Fiziologiya Rastenii: Plant Physiology (weekly)
Funktsionalnyi Analiz i Ego Prilozhenie: Functional Analysis and its Application (weekly)
Genetika: Genetics (weekly)
Geografiya i Prirodnya Resursy: Geography and Natural Resources (weekly)
Geokhimiya: Geochemistry (weekly)
Geologiya i Geofizika: Geology and Geophysics (weekly)
Geologiya Rudnykh Mestorozhdenii: Geology of Ore Deposits (weekly)
Geomagnetizm i Aeronomiya: Geomagnetism and Aeronomy (weekly)
Geomorfologiya: Geomorphology (weekly)
Geotekhtonika: Geotectonics (weekly)
Inzhnernaya Geologiya: Engineering Geology (weekly)
Issledovanie Zemli iz Kosmosa: Investigation of the Earth from Space (weekly)
Izvestiya Rossiiskoi Akademii Nauk: Bulletin of the Russian Academy of Sciences
Izvestiya Sibirskogo Otdeleniya Rossiiskoi Akademii Nauk: Bulletin of the Siberian Branch of the Russian Academy of Sciences
Khimicheskaya Fizika: Chemical Physics (weekly)
Khimiya i Tekhnologiya Topliv i Masel: Chemistry and Technology of Fuels and Oils (weekly)
Khimiya i Tekhnologiya Vody: Water Chemistry and Technology (weekly)
Khimiya i Zhizn: Chemistry and Life (weekly)
Khimiya Prirodnykh Soedinenii: Chemistry of Natural Compounds (weekly)
Khimiya Tverdogo Topliva: Solid Fuel Chemistry (weekly)
Khimiya Vysokikh Energii: High Energy Chemistry (weekly)
Kinetika i Kataliz: Kinetics and Catalysis (weekly)
Kolloidny Zhurnal: Colloids Journal (weekly)
Kompleksnoe Ispolzovanie Mineralnogo Syrya: Comprehensive Utilisation of Mineral Raw Materials (weekly)
Koordinatsionnaya Khimiya: Coordination Chemistry (weekly)
Kosmicheskie Issledovaniya: Space Research (weekly)
Kristallografiya: Crystallography (weekly)
Kvant: Quantum (weekly)
Kvantovaya Elektronika: Quantum Electronics (weekly)
Latinskaya Amerika: Latin America (weekly)
Lesovedenie: Forestry Studies (weekly)
Litologiya i Poleznye Iskopaemye: Lithology and Minerals
Magnitnyi Rezonans i ego Primenenie: Magnetic Resonance and its Application
Matematicheskii Sbornik: Mathematical Collection (weekly)
Matematicheskii Zametki: Mathematical Notes (weekly)
Mikrobiologiya: Microbiology (weekly)
Mikroelektronika: Microelectronics (weekly)
Mineralogicheskii Zhurnal: Mineralogical Journal (weekly)
Mirovaya Ekonomika i Mezhdunarodnye Otnosheniya: World Economics and International Relations (weekly)
Molekulyarnaya Biologiya: Molecular Biology (weekly)
Nauchnoe Proborostroenie: Scientific Instrumentation (weekly)
Nauka v Rossii: Science in Russia (weekly)
Neftekhimiya: Petrochemistry
Neirofiziologiya: Neurophysiology (weekly)
Neirokhimiya: Neurochemistry (weekly)
Novaya i Noveishaya Istoriya: Modern and Contemporary History (weekly)
Obshchestvennye Nauki: Social Sciences (weekly)
Okeanologiya: Oceanology (weekly)
Ontogenez: Ontogenesis (weekly)
Optika Atmosfery: Optics of Atmosphere (weekly)
Optika i Spektroskopiya: Optics and Spectroscopy (weekly)
Otechestvennaya Istoriya: The Nation's History (weekly)
Otechestvennye Arkhivy: National Archives (weekly)
Paleontologicheskii Zhurnal: Palaeontological Journal (weekly)
Parazitologiya: Parasitology (weekly)
Pochvovedenie: Soil Science (weekly)
Poverkhnost: Fizika, Khimiya, Mekhanika: Surface: Physics, Chemistry, Mechanics (weekly)
Pribory i Tekhnika Eksperimenta: Instruments and Equipment for Experiments (weekly)
Prikladnaya Biokhimiya i Mikrobiologiya: Applied Biochemistry and Microbiology (weekly)
Prikladnaya Matematika i Mekhanika: Applied Mathematics and Mechanics (weekly)
Priroda: Nature (weekly)
Problemy Dalnego Vostoka: Problems of the Far East (weekly)
Problemy Mashinostroeniya i Nadezhnosti Mashin: Problems of Engineering and Machine Reliability (weekly)
Problemy Peredachi Informatsii: Problems of Information Transmission (weekly)
Programmirovanie: Programming (weekly)
Psikhologicheskii Zhurnal: Psychological Journal (weekly)
Radiobiologiya: Radiobiology (weekly)
Radiokhimiya: Radiochemistry (weekly)
Rastitelnye Resursy: Plant Resources (weekly)
Rossiiskaya Arkheologiya: Russian Archaeology (weekly)
Rossiiskii Musei: Russian Museums (weekly)
Russkaya Literatura: Russian Literature (weekly)
Russkaya Rech: Russian Speech (weekly)
Sibirskii Matematicheskii Zhurnal: Siberian Mathematical Journal (weekly)
Slavyanovedenie: Slavonic Studies (weekly)
Sotsiologicheskie Issledovaniya: Sociological Research (weekly)
Sverkhverdye Materialy: Superhard Materials (weekly)
Teoreticheskaya i Matematicheskaya Fizika: Theoretical and Mathematical Physics (weekly)
Teoreticheskie Osnovy Khimicheskoi Tekhnologii: Theoretical Foundations of Chemical Technology (weekly)
Teoriya Veroyatnostei i ee Primenenie: Probability Theory and its Application (weekly)
Teploenergetika: Heat and Power Engineering (weekly)
Teplofizika Vysokikhk Temperatur: High Temperature Thermal Physics (weekly)
Tikhookeanskaya Geologiya: Pacific Ocean Geology (weekly)
Trenie i Iznos: Friction and Wear (weekly)
Tsitologiya: Cytology (weekly)
Uspekhi Fizicheskikh Nauk: Progress of Physics (weekly)
Uspekhi Fiziologicheskikh Nauk: Progress of Physiology (weekly)
Uspekhi Khimii: Progress in Chemistry (weekly)
Uspekhi Matematicheskikh Nauk: Progress in Mathematics (weekly)
Uspekhi Sovremennoi Biologii: Progress in Modern Biology (weekly)
Vestnik Dalnevostochnogo Otdeleniya Rossiiskoi Akademii Nauk: Journal of the Far Eastern Division of the Russian Academy of Sciences
Vestnik Drevnei Istorii: Journal of Ancient History (weekly)
Vestnik Rossiiskoi Akademii Nauk: Journal of the Russian Academy of Sciences (weekly)
Vodnye Resursy: Water Resources (weekly)
Voprosy Ekonomiki: Economic Questions (weekly)
Voprosy Filosofii: Questions of Philosophy (weekly)
Voprosy Ikhtiologii: Questions of Ichthyology (weekly)
Voprosy Literatury: Questions of Literature (weekly)
Voprosy Yazykoznaniya: Questions of Linguistics (weekly)
Vorposy Istorii: Questions of History (weekly)
Vorposy Istorii Estestvoznaniya i Tekhniki: History of Natural Sciences and Technology (weekly)
Vostok: East (weekly)
Vulkanologiya i Seismologiya: Vulcanology and Seismology (weekly)
Vysokomolekulyarnye Soedineniya: High Molecular Compounds (weekly)
Yadernaya Fizika: Nuclear Physics (weekly)
Zapiski Mineralogo Obshchestva: Notes of the Mineralogical Society (weekly)
Zaschita Metallov: Protection of Metals (weekly)
Zemlya i Vselennaya: Earth and Universe (weekly)
Zhurnal Analiticheskoi Khimii: Journal of Analytical Chemistry (weekly)
Zhurnal Eksperimentalnoi i Teoreticheskoi Fiziki: Journal of Experimental and Theoretical Physics (weekly)
Zhurnal Evolyutsionnoi Biokhimii i Fiziologii: Journal of Evolutionary Biochemistry and Physiology (weekly)
Zhurnal Fizicheskoi Khimii: Journal of Physical Chemistry (weekly)
Zhurnal Nauchnoi i Prikladnoi Fotografii i Kinematografii: Journal of Scientific and Applied Photography and Cinematography (weekly)
Zhurnal Neorganicheskoi Khimii: Journal of Inorganic Chemistry (weekly)
Zhurnal Obshchei Biologii: Journal of General Biology (weekly)
Zhurnal Obshchei Khimii: Journal of General Chemistry (weekly)
Zhurnal Organicheskoi Khimii: Journal of Organic Chemistry (weekly)
Zhurnal Prikladnoi Khimii: Journal of Applied Chemistry (weekly)
Zhurnal Prikladnoi Mekhaniki i Tekhnicheskoi Fiziki: Journal of Applied Mechanics and Technical Physics (weekly)
Zhurnal Strukturnoi Khimii: Journal of Structural Chemistry (weekly)
Zhurnal Tekhnicheskoi Fiziki: Journal of Technical Physics (weekly)
Zhurnal Vychislityelnoi Matematiki i Matematicheskoy Fiziki: Journal of Computational Mathematics and Mathematical Physics (weekly)
Zhurnal Vysshei Nervnoi Deyatelnosti: Journal of Higher Nervous Activity (weekly)
Zoologicheskii Zhurnal: Zoological Journal (weekly)
. . . . 09954

Russian Association for Comparative Literature, Ul Vorovskogo 25a, 121069 Moskva
T: 2901709
Pres: Y.B. Vipper
Focus: Lit 09955

Russian Botanical Society, Ul Prof. Popova 2, 197022 St. Petersburg
T: 2340092
Pres: A.L. Takhtadzan
Focus: Botany 09956

Russian Geographical Society, Per Grivtsova 10, 190000 St. Petersburg
T: 3156312
Pres: A.F. Treshnikov
Focus: Geography 09957

Russian Linguistics Society, Ul Semashko 1, 103009 Moskva
Pres: T.V. Gamkrelidze
Focus: Ling 09958

Russian Mineralogical Society, V.O. Liniya 2, 199026 St. Petersburg
T: 2188640
Pres: D.V. Rundkvist
Focus: Mineralogy 09959

Russian Palaeontological Society, Srednii Pr 74, 199026 St. Petersburg
T: 2189121
Pres: B.S. Sokolov
Focus: Paleontology 09960

Russian Palestine Society, Volkhonka 14, 119842 Moskva
T: 2039398
Pres: O.G. Peresypkin
Focus: Hist 09961

Russian PEN Centre, Ul Neglinnaya 18, 103031 Moskva
T: 2094589; Fax: 2000293
Founded: 1989; Members: 121
Pres: A. Bitov; Gen Secr: V. Stabnikov
Focus: Lit 09962

Russian Pharmacological Society, Ul Baltiiskaya 8, 125315 Moskva
T: 1511881; Fax: 1511261
Founded: 1958; Members: 615
Pres: D.A. Kharkevich; Gen Secr: S.A. Borisenko
Focus: Pharmacol
Periodical
Farmakologia i Toksikologia: Pharmacology and Toxicology (weekly) 09963

Russian Pugwash Committee, Leninskii Pr 14, 117901 Moskva
T: 1373545
Pres: V.I. Goldansky
Focus: Eng 09964

Russian Society of Genetics and Breeders, Ul Fersmana 11, 117312 Moskva
Pres: V.A. Strunnikov
Focus: Genetics; Animal Husb 09965

Russian Union of Composers, Ul Nezhdanovoi 8-10, Moskva
T: 2295218; Tx: 411702; Fax: 2004273
Founded: 1960; Members: 1500
Pres: V. Kazenin
Focus: Music
Periodicals
Musical Life
Musical Review
Muzikalnaya Akademia 09966

Scientific Medical Society of Anatomists-Pathologists, Bolshaya Serpuhovskaya 27, 109801 Moskva
Pres: D.S. Sarkisov; Gen Secr: V.P. Tumanov
Focus: Anat; Pathology 09967

Scientific-Technical Association, Pr Ogorodnikova 26, 198103 St. Petersburg
T: 2512850
Pres: M.L. Aleksandrov
Focus: Eng 09968

Shipbuilding Engineering Society, Nevskii Pr 44, 191011 St. Petersburg
T: 3155027
Pres: I.V. Gorynin
Focus: Eng 09969

Society for Railway Transport, Ul K. Marksa 11, 107262 Moskva
T: 2626180
Pres: G.M. Korenko
Focus: Transport 09970

Society for Trade and Commerce, Gogolevskii bul 9, 121019 Moskva
T: 2032246
Pres: Y.K. Tvildiani
Focus: Commerce 09971

Society of Cardiology, Petroverigskii ul 10, 101953 Moskva
T: 9238636
Founded: 1958
Pres: Dr. R. Oganov; Gen Secr: Dr. N. Perova
Focus: Cardiol
Periodical
Kardiologiya: Cardiology (weekly) 09972

Society of Herminthologists, B. Cheremushkinskaya 28, 117259 Moskva
Pres: A.S. Bessonov
Focus: Bio 09973

Society of Light Industry, Ul Vavilova 69, 117846 Moskva
T: 1347009
Pres: R.A. Chayanov
Focus: Mach Eng 09974

Society of Mammalogists, Ul Vavilova 44, 117033 Moskva
T: 2322088
Pres: V.E. Sokolov
Focus: Zoology 09975

Society of Non-Ferrous Metallurgy, Per Pechatnikova 7, 103045 Moskva
T: 2084838
Pres: V.S. Lobanov
Focus: Metallurgy 09976

Society of Ornithologists, I.Y. Kotelnichesky per 10, 109240 Moskva
T: 2974496
Pres: V.D. Ilichev
Focus: Ornithology 09977

Society of Protozoologists, Tikhoretskii per 4, 194064 St. Petersburg
T: 2474496
Pres: Y.I. Polyanskii
Focus: Zoology 09978

Society of Psychologists, Yarovslavskaya ul 13, 129366 Moskva
T: 2827212
Pres: E.V. Shorokhova
Focus: Psych 09979

Society of the Food Industry, Kuznetskii most 19, 103031 Moskva
T: 9252611
Pres: A.N. Bogatyrev
Focus: Food 09980

Society of the Instrument Building Industry, Pr K. Marksa 17, 121019 Moskva
T: 2033503
Pres: G.I. Kavalerov
Focus: Mach Eng 09981

Society of the Timber and Forestry Industry, Ul Chernyshevskogo 29, 103062 Moskva
T: 9239570
Pres: Y.A. Yagodnikov
Focus: Forestry 09982

Sociological Association, Ul Krzhizhanovskogo 24, 117418 Moskva
T: 1208257
Pres: T.I. Zavlavskaya
Focus: Sociology 09983

Soil Science Society, Pyzhevskii per 7, 109017 Moskva
Pres: G.V. Dobrovolsky
Focus: Geology 09984

Theatre Union of the Russian Federation, Ul Gorkogo 16, 103009 Moskva
T: 2299152; Tx: 411030
Members: 30124
Pres: M.A. Ulyanov
Focus: Perf Arts
Periodicals
Information from the Secretariat (weekly)
Problems of Contemporary Theatre . . . 09985

Tropical Committee, Leninskii Pr 33, 117071 Moskva
T: 2341209
Pres: V.E. Sokolov
Focus: Geography 09986

Union of Russian Architects, Ul Shchuseva 3, 103889 Moskva
T: 2902579
Founded: 1932; Members: 20000
Gen Secr: Y.P. Platonov
Focus: Archit 09987

Union of Russian Filmmakers, Vasilevskaya 13, 123825 Moskva
T: 2515370
Focus: Cinema 09988

Union of Russian Writers, Ul Vorovskogo 52, Moskva
T: 2916350
Founded: 1935; Members: 9500
Pres: G.M. Markov
Focus: Lit
Periodicals
Inostrannaya Literatura (weekly)
Literaturnaya Gazeta (weekly)
Novyi Mir (weekly) 09989

Union of Scientific and Learned Societies, Kursovoi per 17, 119034 Moskva
T: 2030928
Pres: A.Y. Ishlinskii
Focus: Sci 09990

Water Management Society, Staropansky per 3, 103012 Moskva
T: 9252446
Focus: Water Res 09991

Znanie, Proezd Servova 4, 101813 Moskva
T: 9213381
Founded: 1947; Members: 3300000
Pres: N.G. Basov
Focus: Sci
Periodical
Science and Life (weekly) 09992

Rwanda

African and Mauritian Institute of Statistics and Applied Economics, BP 1109, Kigali
T: (07) 84989
Founded: 1975; Members: 9
Gen Secr: Idrissa Guira
Focus: Stats; Econ 09993

St. Lucia

Caribbean Association for Rehabilitation Therapists (CARDA), c/o CART, POB 1068, Castries
Founded: 1979
Focus: Rehabil; Therapeutics 09994

Caribbean Environmental Health Institute (CEHI), POB 1111, Castries
T: (045) 4521412; Tx: 6248; Fax: (045) 4532721
Members: 16
Gen Secr: Dr. Naresh C. Singh
Focus: Ecology 09995

San Marino

European Centre for Disaster Medicine, c/o Ospedale di Stato, Via Toscana, 47031 San Marino
T: (0541) 994535, 903706; Tx: 893
Founded: 1986
Focus: Med 09996

Saudi Arabia

Arab Bureau of Education for the Gulf States (ABEGS), POB 3908, Riyadh 11481
T: (01) 4789889; Tx: 401441; CA: Tarbia;
Fax: (01) 4783165
Founded: 1975; Members: 6
Pres: Dr. Ali M. Al-Towagry
Focus: Educ
Periodicals
Arab Gulf Journal of Scientific Research (weekly)
Risalat Ul-Khalee Al-Arabi (weekly) . . . 09997

Arab Satellite Communications Organization (ARABSAT), POB 1038, Riyadh
T: (01) 4646666; Tx: 401400
Founded: 1976; Members: 21
Gen Secr: Dr. Abdel-Quader Al-Bairi
Focus: Comm Sci 09998

Arab Security Studies and Training Center (ASSTC), POB 6830, Riyadh 11452
T: (01) 246344; Tx: 400949; Fax: (01) 2464713
Pres: Dr. Farouk A. Al-Rahman Murad
Focus: Criminology 09999

Arab Urban Development Institute (AUDI), POB 6892, Riyadh
T: (01) 4419158, 4419876, 4418100; Tx: 403566;
Fax: (01) 4418235
Founded: 1980
Gen Secr: Dr. M.A. Al-Hammad
Focus: Urban Plan 10000

Center for Research in Islamic Education, POB 1034, Mecca
Founded: 1980; Members: 15
Focus: Educ 10001

Commission for Controlling the Desert Locust in the Near East (CCDLNE), POB 327, Jeddah
T: (02) 6445941; Fax: (02) 6799563
Founded: 1965; Members: 13
Gen Secr: A. Khasawneh
Focus: Agri 10002

Higher Council for Promotion of Arts and Letters, Riyadh
Gen Secr: A.B.A. Idris
Focus: Arts; Lit 10003

King Abdul Aziz Research Centre, POB 2945, Riyadh 11461
Founded: 1972
Gen Secr: Dr. Tami bin Hudaif Al-Buqmi
Focus: Hist; Geography; Lit; Arts
Periodical
Addarah (weekly) 10004

Saudi Biological Society, c/o Botany Department, Faculty of Science, King Saud University, POB 2455, Riyadh 11451
T: (01) 4675826; Tx: 201019
Founded: 1975; Members: 341
Pres: Dr. Abdul Aziz Hamid Abu-Zinada;
Gen Secr: Dr. I. Nader
Focus: Bio
Periodicals
Abstract and Programme of Annual Conference
Proceedings of the Saudi Biological Society (weekly) 10005

Society of Esaff Alkhairia, Mecca
Founded: 1946
Pres: Muhammad Sarour Al-Sabban; Gen Secr: Ahmed Sibai
Focus: Hist 10006

Senegal

Africa AIDS Research Network, BP 7318, Dakar
Focus: Immunology 10007

African Biosciences Network (ABN), c/o UNESCO-BREDA, BP 3311, Dakar
T: 235083, 234614; Tx: 21735, 51410
Founded: 1981
Gen Secr: Dr. Amadou T. Bâ
Focus: Bio
Periodical
ABN Newsletter (weekly) 10008

African Centre for Monetary Studies (ACMS), 15 Blvd Franklin Roosevelt, Dakar
T: 233821; Tx: 61256; Fax: 217760
Founded: 1978; Members: 34
Gen Secr: Andrew K. Mullei
Focus: Finance
Periodicals
ACMS Research Report
Financial Journal
Financial News Analysis 10009

African Council for Social and Human Sciences, c/o BREDA, BP 3311, Dakar
Founded: 1990
Pres: Prof. Lloyd Sachikonye
Focus: Soc Sci 10010

African Institute for Economic Development and Planning, BP 3186, Dakar
T: 231020, 214831; Tx: 51579; Fax: 212158
Founded: 1962; Members: 51
Gen Secr: Dr. Jeggan C. Senghor
Focus: Develop Areas; Econ 10011

African Network for Integrated Development (ANID), BP 12085, Dakar
T: 224495; Tx: 1304; Fax: 221544
Founded: 1985
Gen Secr: Abdou El Mazide Ndiaye
Focus: Develop Areas
Periodicals
Impact (weekly)
Jef-Jel (weekly) 10012

African Regional Centre for Technology (ARCT), BP 2435, Dakar
T: 227712; Tx: 61282; Fax: 257712
Founded: 1977
Gen Secr: Dr. Ousmane Kane
Focus: Eng 10013

African Standing Conference on Bibliographic Control (ASCOBIC), Immeuble Administratif, Av Roume, Dakar
T: 21507223
Members: 13
Gen Secr: S. Mbaye
Focus: Libraries & Bk Sci 10014

Agency for the Safety of Aerial Navigation in Africa, BP 8132, Dakar
T: 201080, 231041; Tx: 31519; Fax: 200600
Founded: 1959
Focus: Aero 10015

Association de Radiologie d'Afrique Francophone, 41bis Rue Carnot, BP 3566, Dakar
Gen Secr: Dr. A. N'Doye
Focus: Radiology 10016

Association des Bibliothèques de l'Enseignement Supérieure de l'Afrique de l'Ouest Francophone (ABESAO) / Association of Higher Education Libraries in French-Speaking West Africa, c/o Bibliothèque Universitaire, BP 2006, Dakar
T: 246981; Fax: 242379
Founded: 1986
Gen Secr: Henri Sene
Focus: Libraries & Bk Sci 10017

Association des Etablissements d'Enseignement Vétérinaire Totalement ou Partiellement de Langue Française (AEEVTPLF), c/o EISVM, BP 5077, Dakar
T: 230545, 256692; Tx: 51403; Fax: 254283
Founded: 1982
Focus: Vet Med 10018

Association des Facultés ou Etablissements de Lettres et Sciences Humaines des Universités d'Expression Française (AFELSH), c/o Faculté des Lettres et Sciences Humaines, Université Cheikh Anta Diop, Fann, Dakar
T: (0221) 213158
Founded: 1988
Pres: Aloyse-Raymond Ndiaye
Focus: Humanities 10019

Association des Professionnelles Africaines de la Communication (APAC), BP 4234, Dakar
Founded: 1984; Members: 300
Gen Secr: Fatoumata Sow
Focus: Comm Sci
Periodical
La Satellite (weekly) 10020

Association of African Women for Research and Development (AAWORD) / Association des Femmes Africaines pour la Recherche sur le Développement, BP 3304, Dakar
T: 252572, 230211; Tx: 61339; Fax: 241289
Founded: 1977; Members: 400
Focus: Develop Areas
Periodical
Flash: African Trade (weekly) 10021

Association of Maize Researchers in Africa (AMRA), BP 6236, Dakar
Founded: 1985
Gen Secr: Dr. A. Camara
Focus: Agri 10022

Centre Africain d'Etudes Supérieures en Gestion (CESAG), BP 3802, Dakar
T: 219254; Tx: 1423; Fax: 213215
Founded: 1983; Members: 7
Gen Secr: Julien Keita
Focus: Business Admin 10023

Comité Scientifique Inter-Africain Post-Récolte / Inter-African Scientific Committee Post-Harvest, c/o Bureau Africain de l'AUPELF, BP 10017, Liberté, Dakar
T: 212927; Tx: 906267
Founded: 1980
Focus: Food 10024

Conférence des Facultés et Ecoles de Médecine d'Afrique d'Expression Française, c/o Faculté de Médecine et de Pharmacie, Fann, Dakar
Pres: Prof. R. Ndoye
Focus: Educ; Med 10025

Council for the Development of Economic and Social Research in Africa (CODESRIA), BP 3304, Dakar
T: 230211; Tx: 61339; Fax: 241289
Founded: 1973
Gen Secr: Thandika Mkandawire
Focus: Sociology; Econ
Periodicals
Africa Development (weekly)
CODESRIA Bulletin (weekly) 10026

Ecole de Bibliothécaires, Archivistes et Documentalistes (EBAD), c/o University Cheikh Anta Diop de Dakar, BP 3252, Dakar
T: 257660; Tx: 51262; Fax: 252883
Founded: 1963
Gen Secr: Ousmane Sane
Focus: Doc; Libraries & Bk Sci; Archives 10027

Environmental Development in the Third World, 4-5 Rue Kléber, BP 3370, Dakar
T: 224229; Fax: 222695
Founded: 1972
Gen Secr: Jacques Bugnicourt
Focus: Ecology 10028

Sierra Leone

African Paediatric Club, c/o Dr. M.A.S. Jalloh, Government Hospital, Kenema
Founded: 1973
Focus: Pediatrics 10029

Historical Society of Sierra Leone (HSSL), c/o Dept of History, Fourah Bay College, University of Sierra Leone, Freetown
T: 281
Founded: 1975; Members: 30
Pres: G. S. Anthony; Gen Secr: Dr. A.J.G. Wyse
Focus: Hist
Periodical
Journal (weekly) 10030

Sierra Leone Association of Archivists, Librarians and Information Scientists, c/o Sierra Leone Library Board, POB 326, Freetown
T: 23848
Founded: 1970; Members: 50
Pres: A. Thomas; Gen Secr: P.K. Kargbo
Focus: Libraries & Bk Sci
Periodical
Bulletin (weekly) 10031

Sierra Leone Science Association, c/o Dept of Physics, Fourah Bay College, University of Sierra Leone, Freetown
Pres: Dr. E.R.T. Awunor-Renner; Gen Secr: Dr. H.G. Morgan
Focus: Sci 10032

Singapore

Academy of Medicine, Singapore, College of Medicine Bldg, 16 College Rd, Singapore 0316
T: 2238968; Tx: 40173; Fax: 2255155
Founded: 1957; Members: 708
Pres: Dr. Ngoh Chuan Tan; Gen Secr: Y.L. Lam
Focus: Med
Periodical
Annals of the Academy (weekly) . . . 10033

Asia and Oceania Federation of Obstetrics and Gynecology (AOFOG), c/o Dept of Obstetrics and Gynecology, National University Hospital, Lower Kent Ridge Rd, Singapore 0511
Founded: 1957; Members: 21
Gen Secr: Prof. S.S. Ratnam
Focus: Gynecology
Periodical
Asia-Oceania Journal of Obstetrics and Gynecology (weekly) 10034

Asian Mass Communication Research and Information Centre (AMIC), 39 Newton Rd, Singapore 11
T: 2515106; Tx: 55524; Fax: 2534525
Founded: 1971
Gen Secr: Vijay Menon
Focus: Comm Sci
Periodicals
Asian Journal of Communication (weekly)
Asian Mass Communication Bulletin . . 10035

Asian Network for Biological Sciences (ANBS), c/o Dept of Botany, National University of Singapore, Lower Kent Ridge Rd, Singapore 0511
T: 7722711; Tx: 33943
Founded: 1978
Gen Secr: Prof. A.N. Rao
Focus: Bio 10036

Asian Network for Industrial Technology Information and Extension (TECHNONET ASIA), 291 Serangoon Rd, Singapore 0821
T: 2912372; Tx: 55002; Fax: 2922372
Founded: 1973; Members: 15
Gen Secr: Dr. C. Anton Balasuriya
Focus: Eng
Periodicals
Newsletter (weekly)
Technonet Digest (weekly) 10037

Asian-Pacific Association for Laser Medicine and Surgery (APALMS), 3 Mount Elizabeth, 06-03 Mount Elizabeth Medical Centre, Singapore 0922
Gen Secr: Dr. Fong Poh Him
Focus: Surgery 10038

Asian Pacific Dental Federation (APDF), 841 Mountbatten Rd, Singapore 1543
T: 3453125; Tx: 34189; Fax: 3442116
Founded: 1955
Gen Secr: Dr. Oliver Hennedige
Focus: Dent
Periodical
APDF Newsletter (weekly) 10039

Asian-Pacific Political Science Association (APPSA), c/o Dept of Political Science, University of Singapore, 10 Kent Ridge Crescent, Singapore 0511
T: 7723970; Tx: 51111; Fax: 7770751
Founded: 1984
Gen Secr: Prof. Chan Heng Chee
Focus: Poli Sci
Periodical
APPSA Newsletter (3 times annually) . . 10040

Asian Pacific Section of the International Confederation for Plastic and Reconstructive Surgery, c/o Dept of Plastic and Reconstructive Surgery, Singapore General Hospital, Outram Rd, Singapore 0316
Founded: 1970
Gen Secr: Dr. S.T. Lee
Focus: Surgery
Periodical
Asian Pacific Section of IPRS Newsletter 10041

Asian-Pacific Tax and Investment Research Centre (APTIRC), 2 Nassim Rd, Singapore 1025
T: 2351954; Tx: 50257; Fax: 7331540
Founded: 1982
Gen Secr: Judy Lee-Chong
Focus: Finance
Periodical
Asian-Pacific Tax and Investment Bulletin (weekly) 10042

Association for Theological Education in South East Asia (ATESEA), 324 Onan Rd, Singapore 1542
T: 3447316; Fax: 3447316
Founded: 1957
Gen Secr: Dr. Yeow Choo Lak
Focus: Educ
Periodical
Asia Journal of Theology (weekly) . . . 10043

The China Society, Maxwell Rd, POB 3738, Singapore 9057
Founded: 1948; Members: 148
Pres: Simon H. Kek-Chay
Focus: Ethnology
Periodical
Journal 10044

Chinese Language and Research Centre, Jurong Campus, Upper Jurong Rd, Singapore 2264
Focus: Ling 10045

Economic Research Centre, Kent Ridge, Singapore 0511
Focus: Econ 10046

Indian Fine Arts Society, St. Michael's Mansion, St. Michael's Rd, POB 2812, Singapore 12
Founded: 1949
Focus: Arts 10047

Institute of Physics, Singapore, c/o Dept of Physics, National University of Singapore, Kent Ridge, Singapore 0511
Fax: 7774279
Founded: 1973; Members: 153
Pres: Dr. Yung Kuo Lim; Gen Secr: Dr. Meng Hau Kuok
Focus: Physics
Periodicals
Bulletin (weekly)
Physics Update (bi-annually)
Singapore Journal of Physics 10048

Institute of Southeast Asian Studies, Heng Mui Keng Terrace, Pasir Panjang, Singapore 0511
Fax: 7781735
Founded: 1968
Gen Secr: K.S. Sandhu
Focus: Sociology; Poli Sci; Econ
Periodicals
ASEAN Economic Bulletin (3 times annually)
Contemporary Southeast Asia (weekly)
SOJOURN: Social Issues in Southeast Asia (weekly)
Southeast Asian Affairs (weekly) 10049

International Commission on Occupational Health, c/o Dept of COFM, National University Hospital, Lower Kent Ridge Rd, Singapore 0511
T: 7724290; Fax: 7791489
Founded: 1906; Members: 1900
Focus: Public Health
Periodical
Newsletter (weekly) 10050

Library Association of Singapore (LAS), c/o NBDCS Secretariat, Bukit Merah Branch Library, Bukit Merah Central, Singapore 0315
Founded: 1955; Members: 414
Pres: R. Ramachandran; Gen Secr: Amayeet Kaur Gill
Focus: Libraries & Bk Sci
Periodicals
LAS Newsletter (weekly)
Singapore Libraries (weekly) 10051

Singapore Association for the Advancement of Science, c/o Singapore Science Centre, Science Centre Rd, Singapore 2260
T: 5603316; Fax: 5659533
Founded: 1976
Pres: Dr. Ang Kok Peng; Gen Secr: Dr. Chia Woon Kim
Focus: Sci 10052

Singapore Institute of Architects (SIA), 02-393, Block 23, Outram Park, Singapore 0316
T: 2203456; Tx: 22652; CA: INSTOFARCH
Founded: 1923; Members: 853
Pres: Gan Eng Oon; Gen Secr: Chia Kok Leong
Focus: Archit
Periodicals
SIA Journal (weekly)
Yearbook (weekly) 10053

Singapore Institute of International Affairs (SIIA), 6 Nassim Rd, Singapore 1025
T: 7349600; Fax: 7336217
Members: 80
Pres: Dr. Teik Soon Lau; Gen Secr: M. Rajretnam
Focus: Int'l Relat 10054

Singapore Medical Association (SMA), Alumni Medical Centre, 2 College Rd, Singapore 0316
T: 2231264; Fax: 2247827
Founded: 1959; Members: 2200
Pres: Dr. Kok Soo Tan; Gen Secr: Dr. Lip Kee Yap
Focus: Med
Periodicals
Singapore Medical Journal (weekly)
SMA Newsletter (weekly) 10055

Singapore National Academy of Science (SNAS), c/o Singapore Science Centre, Science Centre Rd, Singapore 2260
T: 5603316; Fax: 5659533
Founded: 1971; Members: 1000
Pres: Prof. Leo Tau Wee Hin; Gen Secr: Dr. Chia Woon Kim
Focus: Sci
Periodical
Journal of the Singapore National Academy of Science (weekly) 10056

Slovakia

Jednota Slovenských Matematikov a Fyzikov / Association of Slovak Mathematicians and Physicists, Gorkého 5, 811 01 Bratislava
T: (07) 59094
Founded: 1862
Pres: Dr. M. Greguš; Gen Secr: Dr. E. Adlerová
Focus: Math; Physics 10057

Kružok Moderných Filológov / Union for Modern Philology, Gondova 2, 818 01 Bratislava
T: (07) 56471
Founded: 1956; Members: 205
Pres: Prof. Dr. E. Terray; Gen Secr: Dr. J. Kerdo
Focus: Ling; Lit 10058

Organizačné Stredisko Vedeckých Spoločností pri SAV / Organization Centre of the Scientific Societies of the SAS, Gorkého 13, 811 01 Bratislava
T: (07) 330084
Founded: 1973
Gen Secr: A. Urland
Focus: Sci 10059

Slovenská Akadémia Vied (SAV) / Slovak Academy of Sciences, Štefánikova 49, 814 38 Bratislava
T: (07) 492751; Tx: 93261; Fax: (07) 496849
Founded: 1953; Members: 60
Pres: L. Macho; Gen Secr: Dr. A. Hajduk
Focus: Sci
Periodicals
Acta Physica Slovaca
Acta Virologica
Slavica Slovaca
Sociológia 10060

Slovenská Antropologická Spoločnost / Slovak Anthropological Society, Mlynská dolina 2, 842 15 Bratislava
T: (07) 324706
Founded: 1965; Members: 137
Pres: Dr. M. Pospíšil; Gen Secr: Dr. I. Drobný
Focus: Anthro 10061

Slovenská Archeologická Spoločnost / Slovak Archaeological Society, 949 21 Nitra-Hrad
T: (087) 28481; Fax: (087) 414329
Founded: 1956; Members: 330
Pres: Dr. D. Bialeková; Gen Secr: Dr. K. Kuzmová
Focus: Archeol 10062

Slovenská Astronomická Spoločnost / Slovak Astronomical Society, 059 60 Tatranská Lomnica
T: 967866; Tx: 78277; Fax: 967656
Founded: 1959; Members: 353
Pres: Dr. V. Rušin; Gen Secr: Dr. R. Komžík
Focus: Astronomy 10063

Slovenská Biochemiká Spoločnost / Slovak Biochemical Society, Kalinčiakova 8, 832 32 Bratislava
T: (07) 65356
Founded: 1959; Members: 450
Pres: Prof. Dr. V. Mézeš
Focus: Biochem 10064

Slovenská Bioklimatologická Spoločnost / Slovak Bioclimatology Society, Mickiewiczova 13, 812 54 Bratislava
T: (07) 54430
Founded: 1966; Members: 180
Pres: D. Zachar; Gen Secr: Dr. Z. Cabojová
Focus: Biophys 10065

Slovenská Biologická Spoločnost / Slovak Biological Soiety, Sasinkova 4, 811 08 Bratislava
T: (07) 685415
Founded: 1967; Members: 180
Pres: Prof. G. Čatár; Gen Secr: Dr. I. Tomo
Focus: Bio 10066

Slovenská Botanická Spoločnost / Slovak Botanical Society, Dúbravská 14, 842 23 Bratislava
T: (07) 3782502
Founded: 1955; Members: 152
Pres: Dr. F. Hindák; Gen Secr: Dr. K. Goliašová
Focus: Botany 10067

Slovenská Chemická Společnost / Slovak Chemical Society, Radlinského 9, 812 37 Bratislava
T: (07) 495205
Founded: 1940; Members: 1400
Pres: Prof. Dr. E. Borsig; Gen Secr: Prof. M. Uher
Focus: Chem
Periodical
Chemické zvesti: Chemical Papers (weekly) 10068

Slovenská Demografická a Štatistická Spoločnost / Slovak Demographic and Statistical Society, Obrancov mieru 4, 812 86 Bratislava
T: (07) 671692
Founded: 1968; Members: 152
Pres: D. Vojtko
Focus: Stats 10069

Slovenská Ekonomická Spoločnost / Slovak Economics Society, Stefánikova 2, 813 74 Bratislava
T: (07) 335540
Founded: 1964; Members: 400
Pres: Prof. L. Andrášik; Gen Secr: Dr. I. Milko
Focus: Econ 10070

Slovenská Entonologická Spoločnost / Slovak Entomology Society, Sienkiewiczova 1, 814 34 Bratislava
T: (07) 332985
Founded: 1957; Members: 218
Pres: Dr. A. Blahutiak
Focus: Zoology 10071

Slovenská Geografická Spoločnost / Slovak Geographical Society, Stefánikova 49, 814 73 Bratislava
T: (07) 492751
Founded: 1945; Members: 480
Pres: J. Drdoš; Gen Secr: Dr. D. Kollár
Focus: Geography 10072

Slovenská Geologická Spoločnost / Slovak Geological Society, Mlynská dolina 1, 817 04 Bratislava
T: (07) 3705111; Tx: 82446; Fax: (07) 371940
Founded: 1965; Members: 1023
Pres: Dr. O. Samuel; Gen Secr: Dr. O. Franko
Focus: Geology 10073

Slovenská Historická Spoločnost / Slovak Historical Society, Klemensova 19, 813 64 Bratislava
T: (07) 56321
Founded: 1946; Members: 785
Pres: R. Marsina; Gen Secr: Dr. J. Lukačka
Focus: Hist 10074

Slovenská Jazykovedná Spoločnost / Slovak Linguistics Society, Panská 26, 813 64 Bratislava
T: (07) 331763
Founded: 1967; Members: 260
Pres: Prof. Dr. K. Buzássyová; Gen Secr: Dr. M. Nábělková
Focus: Ling 10075

Slovenská Jednota Klassických Filológov / Slovak Association of Classical Philologists, Gondova 2, 818 01 Bratislava
T: (07) 58041
Founded: 1969
Pres: Dr. P. Kuklica; Gen Secr: Dr. D. Škoviera
Focus: Ling; Lit 10076

Slovenská Lekárska Spoločnost (SLS) / Slovak Medical Society, Mickiewiczova 18, 813 22 Bratislava
T: (07) 50354
Founded: 1969; Members: 33061
Focus: Med 10077

Slovenská Literárnovedná Spoločnost / Slovak Literary Society, Konventná 13, 813 64 Bratislava
T: (07) 313391
Founded: 1958; Members: 320
Pres: Dr. F. Štraus; Gen Secr: Dr. M. Bátorová
Focus: Lit 10078

Slovenská Meteorologická Spoločnost / Slovak Meteorological Society, Jeséniova 17, 833 15 Bratislava
T: (07) 42030; Tx: 93265
Founded: 1960; Members: 170
Pres: Dr. I. Panenka; Gen Secr: Dr. F. Hesek
Focus: Astrophys 10079

Slovenská Národopisná Spoločnost / Slovak Ethnography Society, Jakubovo nám 12, 813 64 Bratislava
T: (07) 334956
Founded: 1958; Members: 400
Pres: O. Danglová; Gen Secr: Dr. H. Hlošková
Focus: Ethnology 10080

Slovenská Orientalistická Spoločnost / Slovak Society for Oriental Studies, Klemensova 19, 813 64 Bratislava
T: (07) 56321
Founded: 1960; Members: 42
Pres: Dr. J. Múčka; Gen Secr: Dr. J. Chmiel
Focus: Ethnology 10081

Slovenská Parazitologická Spoločnost / Slovak Society for Parasitology, Hlinkova 3, 400 01 Košice
T: 4234455; Fax: 4231414
Founded: 1992; Members: 82
Pres: Prof. P. Dubinský; Gen Secr: M. Sabová
Focus: Med
Periodical
Správy Slovenskej Parazitologickej Spoločnosti (weekly) 10082

Slovenská Pedagogická Spoločnost / Slovak Education Society, c/o Filozofická Fakulta UPJS, 080 01 Prešov
T: (0791) 32869
Founded: 1965; Members: 365
Pres: Prof. Dr. J. Ríčalka; Gen Secr: Dr. J. Gallo
Focus: Educ 10083

Slovenská Pneumologická a Ftizeologická Spoločnost / Slovak Society of Pneumology and Phtisiology, Krajinska 93-103, 825 56 Bratislava
T: (07) 2401398; Fax: (07) 243622
Founded: 1968; Members: 504
Pres: Prof. Peter Krištúfek; Gen Secr: Prof. Stefan Urban
Focus: Pulmon Dis 10084

Slovenská Psychologická Spoločnost / Slovak Psychological Society, Gondova 2, 818 01 Bratislava
T: (07) 56471
Founded: 1959; Members: 1400
Pres: Dr. T. Kollárik; Gen Secr: Dr. E. Kollárová
Focus: Psych 10085

Slovenská Sociologická Spoločnost / Slovak Sociological Society, Hviezdoslavovo nám 10, 811 02 Bratislava
T: (07) 331480
Founded: 1964; Members: 356
Pres: Dr. Z. Kusá; Gen Secr: Dr. M. Piscová
Focus: Sociology 10086

Slovenská Spoločnost Medzinárodné Právo / Slovak Society for International Law, Klemensova 19, 813 64 Bratislava
T: (07) 56321
Founded: 1969; Members: 63
Pres: Prof. Dr. J. Azud; Gen Secr: Dr. C. Kandráčová
Focus: Law 10087

Slovenská Spoločnost pre Dejiny Vied a Techniky pri SAV / Slovak Society for History of Science and Technology, Klemensova 19, 813 64 Bratislava
T: (07) 57645
Founded: 1965; Members: 300
Pres: Dr. M. Skladaný; Gen Secr: Dr. O. Pöss
Focus: Cultur Hist 10088

Slovenská Spoločnost pre Kybernetiku a Informatiku / Slovak Society for Cybernetics and Informatics, Dúbravská cesta 9, 842 37 Bratislava
T: (07) 373271; Tx: 93355; Fax: (07) 376045
Founded: 1966; Members: 270
Pres: Prof. J. Mikleš; Gen Secr: B. Hrúz
Focus: Cybernetics; Computer & Info Sci 10089

Slovenská Spoločnost pre Mechaniku / Slovak Society for Mechanics, Dúbravská cesta 9, 842 20 Bratislava
T: (07) 3782530
Founded: 1967; Members: 225
Pres: Prof. Dr. J. Brilla; Gen Secr: Dr. V. Sládek
Focus: Physics 10090

Slovenská Spoločnost pre Vedy Polnohospodarske, Lesnické a Potravinárské / Slovak Society for Agriculture, Forestry and Food, 900 28 Ivanka pri Dunaji
T: 943456
Founded: 2968; Members: 300
Pres: J. Královič; Gen Secr: A. Dandár
Focus: Agri; Forestry; Food 10091

Slovenská Zoologická Spoločnost / Slovak Zoological Society, Obrancov mieru 3, 814 34 Bratislava
T: (07) 335435
Founded: 1956; Members: 300
Pres: L. Weismann; Gen Secr: Dr. E. Kalivodová
Focus: Zoology 10092

Slovenské Filozofické Združenie (SFZ) / The Slovak Philosophical Association, Klemensova 19, 813 64 Bratislava
T: (07) 321215; Fax: (07) 321215
Founded: 1990; Members: 280
Pres: Dr. Jana Balážová; Gen Secr: Dr. Tibor Pichler
Focus: Philos 10093

Spolok Architektov Slovenska / Slovak Architects Society, Panská 15, 811 01 Bratislava
T: (07) 335711; Fax: (07) 335744
Founded: 1956; Members: 1400
Pres: Š. Šlachta; Gen Secr: P. Guldan
Focus: Archit 10094

Zväz Slovenských Knihovníkov a Informatikov / Association of the Slovak Librarians and Information Workers, Michalská 1, 814 17 Bratislava
T: (07) 330557; Tx: 93255
Founded: 1969; Members: 2907
Focus: Libraries & Bk Sci
Periodicals
Ročenka knihovníckej sekcie ZKI (weekly)
Zborník INFOS (weekly)
Zväzový bulletin (weekly) 10095

Slovenia

Arhivsko Društvo Slovenije (ADS), Zvezdarska 1, 61000 Ljubljana
T: (061) 20002
Founded: 1954; Members: 131
Focus: Archives 10096

Association of Engineers and Technicians of Slovenia, Erjavčeva 15, 61000 Ljubljana
Founded: 1953
Focus: Eng
Periodical
Nova Proizvodnja (weekly) 10097

Društvo Matematikov, Fizikov i Astronomov Slovenije / Association of Mathematicians, Physicists and Astronomers of Slovenia, POB 64, 61111 Ljubljana
Founded: 1949; Members: 1020
Pres: Dr. P. Petek
Focus: Math; Physics; Astronomy
Periodicals
Kujizuica Sipma (weekly)
Obzornik za matematiko in fiziko (weekly)
Presek: list za mlade matematike, fizike in astronome (weekly) 10098

Društvo Slovenskih Skladateljev (DSS) / Society of Slovene Composers, Trg Francoske Revolucije 6, 61000 Ljubljana
T: (061) 213487; Fax: (061) 213487
Founded: 1945; Members: 106
Pres: P. Mihelčič; Gen Secr: M. Strmčnik
Focus: Music
Periodical
Edicije DSS 10099

Društvo za Medicinsku i Biološku Tehniku, c/o Faculty of Electrical Engineering, 61000 Ljubljana
T: (061) 61342
Members: 57
Focus: Eng 10100

Jamarska zveza Slovenije (JZS) / Speleological Association of Slovenia, POB 44, 61109 Ljubljana
T: (061) 315666
Founded: 1889; Members: 1000
Pres: B. Urbar; Gen Secr: T. Bukovec
Focus: Speleology
Periodical
Naše jame (weekly) 10101

Prirodoslovno Društvo Slovenije v Ljubljani, Novi trg 4, 61000 Ljubljana
T: (061) 22786
Founded: 1934; Members: 3000
Focus: Sci 10102

Raziskovalna Skupnost Slovenije / Research Community of Slovenia, Tržaška 42, 61115 Ljubljana
Fax: (061) 261125
Founded: 1953
Pres: Savin Jogan; Gen Secr: V. Kaučič
Focus: Eng; Nat Sci
Periodical
Raziskovalec: Researcher (weekly) . . . 10103

Slavistično Društvo Slovenije / Society for Slavonic Studies in Slovenia, Aškerčeva 12, 61000 Ljubljana
T: (061) 332611 ext 274
Founded: 1935; Members: 1500
Pres: Dr. M. Orožen
Focus: Ling; Lit
Periodicals
Jezik in Slovstvo (8 times annually)
Slavistična Revija (weekly) 10104

Slovenska Akademija Znanosti in Umetnosti (SAZU) / Slovenian Academy of Sciences, Novi Trg 3, 61000 Ljubljana
T: (061) 156068; Fax: (061) 155232
Members: 86
Pres: F. Bernik; Gen Secr: M. Drovenik
Focus: Sci
Periodicals
Arheološki vestnik: Acta Archaeologica (weekly)
Biblioteka (weekly)
Dela: Opera (weekly)
Geografski zbornik: Acta Geographica (weekly)
Krasoslovni zbornik: Acta Carsologica (weekly)
Letopis-Yearbook (weekly)
Razprave: Dissertations (weekly)
Slovenski biografski leksikon (weekly)
Traditiones (weekly) 10105

Slovenska Matica / Slovenian Society, Trg Osvoboditve 7, 61000 Ljubljana
T: (061) 214200
Founded: 1864; Members: 2800
Pres: Prof. Dr. P. Simoniti; Gen Secr: Drago Jančar
Focus: Ling; Lit 10106

Slovensko Umetnostno Zgodovinsko Društvo / Slovenian Society of Historians of Art, Aškerčeva 12, 61000 Ljubljana
T: (061) 150001 ext 247; Fax: (061) 159337
Founded: 1921; Members: 299
Pres: Nace Šumi
Focus: Cultur Hist
Periodical
Archives d'Histoire de l'Art: Zbornik za Umetnostno Zgodovino 10107

Society for Natural Sciences of Slovenia, Novi trg 4, 61000 Ljubljana
Founded: 1934; Members: 3000
Pres: R. Kavčič; Gen Secr: T. Wraber
Focus: Nat Sci
Periodical
Proteus (10 times annually) 10108

Society of Jurists of Slovenia, Dalmati-nova 4, 61000 Ljubljana
Founded: 1947; Members: 1073
Pres: Jože Pavličič
Focus: Law
Periodical
Jurist 10109

Union of Arts of the Russian Federation, Ul Pokrovka 37, 103062 Moskva
Focus: Arts 10110

Zveza Bibliotekarskih Društev Slovenije (DBS) / Library Association of Slovenia, Turjaška 1, 61000 Ljubljana
T: (061) 150131
Founded: 1947; Members: 919
Pres: I. Kanič; Gen Secr: T. Kobe
Focus: Libraries & Bk Sci
Periodical
Knjižnica (weekly) 10111

Zveza Društev Arhitektov Slovenije (ZDAS), Erjagavčeva 15, 61000 Ljubljana
T: (061) 21608
Founded: 1951; Members: 622
Focus: Archit 10112

Zveza Društev za Varilno Tehniko Slovenije, Erjavćeva 15, 61000 Ljubljana
T: (061) 221631
Focus: Eng
Periodical
Varilna Tehnika (weekly) 10113

Zveza Ekonomistov Slovenije, Trubarjeva 3, 61000 Ljubljana
T: (061) 224433
Focus: Econ 10114

Zveza Geografskih Društev Slovenije / Association of the Geographical Societies of Slovenia, Aškerčeva 12, 61000 Ljubljana
T: (061) 150001; Fax: (061) 159337
Founded: 1922; Members: 700
Pres: Prof. Dr. M. Jeršič; Gen Secr: I. Jurinčič; M. Krevs
Focus: Geography
Periodicals
Geografski obzornik (weekly)
Geografski vestnik (weekly)
Zborniki geografskih zborovanj: Congress Proceedings 10115

Zveza Pedagoskih Društev Slovenije / Pedagocical Society of Slovenia, Gosposka ulica 3, 61000 Ljubljana
Founded: 1920
Pres: M. Vidmar
Focus: Educ
Periodical
Sodobna pedagogika (weekly) 10116

Zveza Zgodovinskih Društev Slovenije / Slovenian Historical Association, Aškerčeva 12, 61000 Ljubljana
T: (061) 150001 ext 210; Fax: (061) 159337
Founded: 1839; Members: 1395
Pres: Dr. D. Mihelič
Focus: Hist 10117

South Africa

Aerial Survey, Photogrammetric and Remote Sensing Research Group, c/o Dept of Land Surveying, University of Natal, King George V Av, Durban 4001
T: (031) 352461
Focus: Surveying 10118

Aeronautical Society of South Africa / Lugvaartkundige Vereniging van Suid-Afrika, POB 130774, Bryanston 2021
T: (011) 7063763
Founded: 1911; Members: 621
Pres: G. Eckermann; Gen Secr: D.P. du Plooy
Focus: Aero
Periodical
Journal (weekly) 10119

Africa Institute of South Africa, Corner Hamilton and Belvedere Sts, Arcadia, POB 630, Pretoria 0001
T: (012) 286970; Fax: (012) 3238163
Founded: 1960
Gen Secr: Dr. S.F. Coetzee
Focus: Ethnology; Econ; Geography; Poli Sci; Int'l Relat
Periodical
Africa Insight (weekly) 10120

African Association for the Study of Liver Diseases, c/o Dept of Medicine, Liver Research Centre, University of Cape Town, Cape Town
T: (021) 471250; Fax: (021) 4486815
Gen Secr: R.E. Kirsch
Focus: Intern Med 10121

African Library Association of South Africa (ALASA), c/o University of the North Library, Private Bag X5090, Pietersburg 0700
T: Sovenga 33
Founded: 1964
Focus: Libraries & Bk Sci 10122

The Associated Scientific and Technical Societies of South Africa (ASTS), c/o Observatory, 18a Gill St, Marshalltown 2198
T: (011) 4871512; Fax: (011) 6481876
Founded: 1920; Members: 78000
Pres: Dr. R.P. Viljoen; Gen Secr: J.A. Nel
Focus: Sci; Eng
Periodical
Annual Proceedings (weekly) 10123

Association of Surgeons of South Africa / Chirurgiese Vereniging van Suid-Africa, POB 52027, Saxonwold, Johannesburg 2132
T: (011) 8371011
Founded: 1943; Members: 250
Pres: P. Perdikis; Gen Secr: V.E. Sorour
Focus: Surgery
Periodical
South African Journal of Surgery . . . 10124

Astronomical Society of Southern Africa, c/o South African Astronomical Observatory, POB 9, Observatory 7935
Founded: 1922; Members: 500
Pres: C.R.G. Turk; Gen Secr: H.E. Krumm
Focus: Astronomy
Periodicals
Handbook (weekly)
Notes (weekly) 10125

The Botanical Society of South Africa, Kirstenbosch, Claremont 7735
T: (021) 7972090; Fax: (021) 7972376
Founded: 1913; Members: 20000
Pres: Prof. O.A.M. Lewis; Gen Secr: D.B. Barends
Focus: Botany; Ecology; Hort
Periodicals
S.A. Wild Flower Guide Series (weekly)
Veld & Flora (weekly) 10126

Cancer Association of South Africa, 139 Smit St, Braamfontein 2017
T: (011) 4032825; Fax: (011) 4031946
Founded: 1931
Focus: Cell Biol & Cancer Res
Periodical
S.A.-Cancer Bulletin (weekly) 10127

Carbohydrate and Lipid Metabolism Research Group, c/o Dept of Medicine and Chemical Pathology, Medical School, University of the Witwatersrand, Hospital Hill, Johannesburg 2001
T: (011) 7241561
Focus: Med; Pathology; Cell Biol & Cancer Res 10128

Chemical Engineering Research Group, c/o Dept of Chemical Engineering, University of Natal, King George V Av, Durban 4001
T: (031) 253411
Focus: Chem; Eng 10129

Classical Association of South Africa, c/o Dept of Latin, University of Stellenbosch, Stellenbosch 7599
T: (021) 8083573, 8083142; Fax: (021) 8084336
Founded: 1956; Members: 400
Pres: Prof. L. Cilliers; Gen Secr: L.F. van Ryneveld
Focus: Ling; Lit
Periodicals
Acta Classica (weekly)
Akroterion (weekly) 10130

Climatology Research Group, c/o Dept of Geography and Environmental Studies, University of the Witwatersrand, Johannesburg 2001
T: (011) 7161111
Focus: Geography 10131

The College of Medicine of South Africa, 17 Milner Rd, Rondebosch 7700
T: (021) 6899533; Fax: (021) 6853766
Members: 4500
Focus: Med
Periodical
Transactions of the College of Medicine of South Africa (weekly) 10132

CSIR, POB 395, Pretoria 0001
T: (012) 8412911; Fax: (012) 8413789
Founded: 1945; Members: 3100
Pres: Dr. J.B. Clark
Focus: Sci; Eng
Periodicals
Chemdata (weekly)
CSIR Annual Report (weekly)
Technobrief (weekly)
Watertek Newsletter (weekly) 10133

Dental Association of South Africa (DASA) / Die Tandheelkundige Vereniging van Suid-Afrika, Private Bag 1, Houghton 2041
T: (011) 6424687; Fax: (011) 6425718
Founded: 1922; Members: 2900
Pres: Prof. R. Lurie; Gen Secr: Dr. Helmut Heydt
Focus: Dent
Periodical
Journal of the Dental Association of South Africa 10134

Diffuse Obstructive Pulmonary Syndrome Research Group, c/o Dept of Internal Medicine, Medical School, University of Stellenbosch, POB 63, Tygerberg 7505
Focus: Intern Med 10135

Division of Energy Technology (Enertek), c/o Scientia, POB 395, Pretoria 0001
T: (012) 8414946; Fax: (012) 8412135
Focus: Chem; Eng 10136

Economic Society of South Africa, POB 929, Pretoria 0001
Founded: 1925; Members: 570
Focus: Econ
Periodical
The South African Journal of Economics (weekly) 10137

The Electron Microscopy Society of Southern Africa, c/o Electron Microscope Unit, Medical School, POB 17039, Congella, Durban 4013
Founded: 1962; Members: 335
Focus: Optics
Periodical
Proceedings (weekly) 10138

English Academy of Southern Africa, POB 124, Wits 2050
Founded: 1961
Focus: Ling
Periodical
English Academy Review (weekly) . . . 10139

Entomological Society of Southern Africa, POB 103, Pretoria 0001
Founded: 1937; Members: 450
Focus: Entomology
Periodicals
African Entomology (weekly)
Memoirs (weekly) 10140

Federasie van Afrikaanse Kultuurverenigings (F.A.K.), POB 91050, Auckland Park 2006
T: (011) 7267134; Fax: (011) 7262073
Founded: 1929; Members: 4000
Pres: Prof. J.A. Heyns; Gen Secr: H. Cronjé
Focus: Hist
Periodical
Handhaaf Newsletter (weekly) 10141

Federasie van Rapportryerskorpse, POB 91001, Auckland Park 2006
Founded: 1961; Members: 14500
Pres: G.J. Erasmus; Gen Secr: J.G. du Plessis
Focus: Ling; Lit
Periodical
Die Rapportryer (weekly) 10142

Genealogical Society of South Africa, POB 1344, Kelvin 2054
Founded: 1963; Members: 1350
Pres: C. Roeloffze; Gen Secr: A.J. Smith
Focus: Genealogy
Periodical
Familia (weekly) 10143

The Geological Society of South Africa (GSSA) / Die Geologiese Vereniging van Suid-Afrika, POB 44283, Linden 2104
T: (011) 8882288; Fax: (011) 8882181
Founded: 1895; Members: 1900
Pres: C. Frick; Gen Secr: Dr. J.J. Cilliers
Focus: Geology
Periodical
South African Journal of Geology . . . 10144

Heraldry Society of Southern Africa, POB 44245, Claremont 7735
Founded: 1953; Members: 150
Pres: Dr. C. Pama; Gen Secr: Michael Purcell
Focus: Genealogy
Periodical
Arma (weekly) 10145

Herpetological Association of Africa (H.A.A.), c/o Port Elizabeth Museum, POB 13147, Port Elizabeth 6013
T: (041) 561051; Fax: (041) 562175
Members: 300
Gen Secr: Dr. W. R. Branch
Focus: Zoology
Periodicals
African Herp News
Journal of the Herpetological Association of Africa 10146

Human Sciences Research Council / Raad vir Geesteswetenskaplike Navorsing, 134 Pretorius St, Private Bag X41, Pretoria 0001
T: (012) 2029111; Tx: 321710; CA: RAGEN, Pretoria; Fax: (012) 3265362
Pres: Dr. R.H. Stumpf
Focus: Humanities; Sociology; Arts
Periodicals
Afford Ability (weekly)
Africa 2001 Dialogue with the Future (weekly)
CSD/SWO Bulletin
HSRC Centre for Constitution Analysis
HSRC/RGN In Focus (weekly)
Information Update (weekly)
Prodder Newsletter (weekly)
Thambodala 10147

Institute of Bankers in South Africa, POB 10335, Johannesburg 2000
Founded: 1904; Members: 20000
Gen Secr: P. Kraak
Focus: Finance
Periodical
The South African Banker (weekly) . . . 10148

Institute of South African Architects, 9 Gordon Hill, Parktown, Houghton 2041
Founded: 1927; Members: 3500
Focus: Archit
Periodical
Architecture (SA) (weekly) 10149

Joint Council of Scientific Societies, POB 13480, Yeoville 2143
Founded: 1968
Focus: Sci 10150

The Medical Association of South Africa (M.A.S.A.) / Die Mediese Vereniging van Suid-Afrika, 428 King's Hwy, Lynnwood 0081
T: (012) 476101; Fax: (012) 471815
Founded: 1927; Members: 11000
Pres: S. Kay; Gen Secr: B.B. Mandell
Focus: Med
Periodicals
Journal for Continuing Medical Education (weekly)
South African Medical Journal (weekly)
South African Medical News (weekly) . . 10151

Medical Graduates' Association of the University of the Witwatersrand, c/o Medical School, Hospital St, Johannesburg 2000
Founded: 1934; Members: 1100
Focus: Med 10152

Medical Research Council, c/o School of Dentistry, University of the Witwatersrand, Private Bag 3, PO Wits, Johannesburg 2050
T: (011) 7164163; Tx: 427125; CA: Uniwits; Fax: (011) 7168030
Focus: Dent 10153

Mintek, 200 Hans Strijdom Dr, Private Bag X3015, Randburg 2125
T: (011) 7094111; Tx: 424867; CA: Minteksa, Johannesburg
Pres: Dr. A. Edwards
Focus: Mineralogy
Periodicals
Annual Report (weekly)
Mintek Bulletin: Mintek Research Digest (weekly) 10154

National Association for Clean Air (NACA), POB 5777, Johannesburg 2000
T: (011) 6462210
Founded: 1969; Members: 450
Focus: Ecology
Periodical
Clean Air Journal (weekly) 10155

The National Association of Scientists, Lynnwood, POB 11346, Pretoria 0001
Founded: 1968; Members: 1180
Focus: Sci 10156

Natural Products Chemistry Research Group, c/o Dept of Chemistry, University of the Witwatersrand, Johannesburg 2001
T: (011) 394011
Focus: Chem 10157

The Nutrition Society of Southern Africa, c/o Dept of Human Nutrition, POB 177, Medunsa 0204
T: (012) 5294499; Fax: (012) 582323
Founded: 1957; Members: 305
Pres: Prof. J.M. Pettifor; Gen Secr: P.M.N. Kuzwayo
Focus: Food
Periodical
SA Journal of Food Science and Nutrition (weekly) 10158

Oral and Dental Research Institute, POB X1, Tygerberg 7505
T: (021) 9312246; Fax: (021) 9312287
Focus: Dent
Periodicals
Archives of Environmental Health
Clinical Preventive Dentistry
Community Dentistry and Oral Epidemiology
International Journal of Pediatric Otorhinolaryngology
Journal of Clinical Orthodontics
Journal of Oral Pathology and Medicine
Journal of the Dental Association of South Africa
South African Medical Journal 10159

Organic Soil Association of South Africa, POB 47100, Parklands 2121
Founded: 1948; Members: 999
Focus: Agri; Hort; Nutrition; Public Health
Periodical
Soil News (weekly) 10160

Physiological and Biochemical Society of Southern Africa, c/o Dept of Physiology, University, Pretoria 0002
Founded: 1971; Members: 76
Focus: Physiology; Biochem 10161

Pollution Research Group, c/o Dept of Chemical Engineering, King George V Av, Durban 4001
T: (031) 8163375; Tx: 612231; Fax: (031) 8163131
Founded: 1970
Pres: Prof. C.A. Buckley
Focus: Chem; Eng; Ecology 10162

Preclinical Diagnostic Chemistry Research Group of the South African Medical Research Council, c/o Dept of Chemical Pathology, University of Natal, 719 Umbilo Rd, Congella 4013
T: 254211
Focus: Chem; Pathology 10163

Pretoria Horticultural Society, POB 1186, Pretoria 0001
Founded: 1916; Members: 700
Focus: Hort 10164

Primate Behaviour Research Group, c/o University of Witwatersrand, PO Wits, Johannesburg 2050
Focus: Behav Sci 10165

Prosthodontics Society of South Africa, Private Bag X1, Houghton 2041
Founded: 1970; Members: 250
Focus: Dent 10166

Psychological Association of South Africa, POB 2729, Pretoria 0001
T: (012) 3261981; Fax: (012) 3261981
Founded: 1982; Members: 1600
Focus: Psych
Periodical
South African Journal of Psychology (weekly) 10167

Research Group for Lung Metabolism, c/o Dept of Medical Physiology and Biochemistry, Faculty of Medicine, POB 63, Tygerberg 7505
T: (021) 933131
Focus: Pulmon Dis 10168

Research Group for Neurochemistry, c/o Dept of Chemical Pathology, Medical School, University of Stellenbosch, POB 63, Tygerberg 7505
T: (021) 9310111
Founded: 1977
Focus: Neurology; Chem 10169

Royal Aeronautical Society, Southern Africa Division, POB 130774, Bryanston 2021
Founded: 1945; Members: 130
Pres: B.H. Prescott; Gen Secr: N.A.H. Barraud
Focus: Aero 10170

Royal Society of South Africa, P.D. Hahn Bldg, POB 594, Cape Town 8000
T: (021) 6502543; Fax: (021) 6503726
Founded: 1877; Members: 400
Pres: Prof. G.F.R. Ellis; Gen Secr: Prof. M.A. Cluver
Focus: Sci
Periodical
Transactions (weekly) 10171

The Science Writers' Association of South Africa, POB 686, Johannesburg 2000
Founded: 1959; Members: 30
Focus: Lit 10172

Society for Endocrinology, Metabolism and Diabetes of Southern Africa, POB 783155, Sandton 2146
T: (011) 7837275
Founded: 1960; Members: 87
Pres: Prof. W. P. U. Jackson; Gen Secr: Dr. L.A. Distiller
Focus: Diabetes; Endocrinology 10173

Society for Experimental Biology, c/o South African Institute for Medical Research, Hospital St, Johannesburg 2001
Founded: 1951; Members: 44
Focus: Bio 10174

Society of Physiologists, Biochemists and Pharmacologists, c/o Dept of Biochemistry, University, Pretoria 0002
Members: 50
Focus: Pharmacol; Physiology; Biochem . 10175

Soil Science Society of Southern Africa (S.S.S.S.A.), POB 1821, Pretoria 0001
Founded: 1953; Members: 200
Focus: Agri 10176

South African Academy of Science and Arts / Suid-Afrikaanse Akademie vir Wetenskap en Kuns, Engelenburghuis, Hamilton St, POB 538, Pretoria 0001
T: (012) 285082
Founded: 1909; Members: 1610
Pres: Dr. C.F. Garbeis; Gen Secr: Dr. D.J.C. Geldenhuys
Focus: Sci; Arts
Periodicals
Jaarverslag (weekly)
Nuusbrief
SA Tydskrif vir Natuurwetenskappe en Tegnologie
Tydskrif vir Geesteswetenskappe 10177

South African Archaeological Society (S.A.A.S.), POB 15700, Vlaeberg 8012
T: 243330
Founded: 1945; Members: 1250
Pres: Dr. T.M.O. Maggs; Gen Secr: Dr. J. Sealy
Focus: Archeol
Periodicals
The Digging Stick (3 times annually)
Goodwin Series (weekly)
Monograph Series (weekly)
S.A. Archaeological Bulletin (weekly) . . 10178

The South African Association for Food Science and Technology (SAAFoST), POB 2140, Edenvale 1610
T: (011) 6096322; Fax: (011) 4524928
Founded: 1961; Members: 1200
Pres: A.N. Starke; Gen Secr: P. McDonald
Focus: Food; Med
Periodicals
Food Review (weekly)
The SA Journal of Food Science & Nutrition (weekly) 10179

South African Association for Technical and Vocational Education (S.A.A.T.V.E), POB 40684, Arcadia 0007
T: (012) 3236851; Fax: (012) 3230185
Founded: 1895; Members: 4800
Pres: P.P. Peach; Gen Secr: J. Le Roux
Focus: Adult Educ; Educ
Periodical
Journal of Technical and Vocational Education in South Africa (weekly) 10180

South African Association of Arts, POB 6188, Pretoria 0001
T: (012) 3222601
Founded: 1945; Members: 5000
Pres: Marilyn Martin
Focus: Arts
Periodical
South Africa Arts Calendar (weekly) . . 10181

The South African Association of Botanists, c/o Botany Dept, University of Fort Hare, Alice 5700
Founded: 1968; Members: 230
Focus: Botany 10182

The South African Association of Physicists in Medicine and Biology, c/o Div of Production Technology, POB 395, Pretoria 0001
Founded: 1960; Members: 61
Focus: Bio; Med; Physiology
Periodical
Congress Brochure (weekly) 10183

South African Biological Society / Suid-Afrikaanse Biologiese Vereniging, POB 820, Pretoria 0001
Founded: 1907; Members: 144
Pres: Dr. N. van Rooyen; Gen Secr: E.A. Boomker
Focus: Bio
Periodical
Journal (weekly) 10184

South African Ceramic Society, POB 13702, Northmead 1511
Founded: 1967
Focus: Materials Sci
Periodicals
Conference Proceedings (weekly)
Keramicos (weekly) 10185

The South African Chemical Institute (S.A.C.I.), POB 93480, Yeoville 2143
T: (011) 4871543; Fax: (011) 6481876
Founded: 1912; Members: 1700
Pres: Dr. M.D. Booth; Gen Secr: Prof. C.J. Rademeyer
Focus: Chem
Periodicals
Chemical Processing SA (weekly)
SA Journal of Chemistry (weekly) . . . 10186

South African Council for Automation and Computation (S.A.C.A.C.), POB 395, Pretoria 0001
Founded: 1961; Members: 60
Focus: Electric Eng; Computer & Info Sci 10187

South African Crystallographic Society, c/o Prof. G.J. Kruger, Chemistry Dept, POB 524, Johannesburg 2000
Fax: (011) 4892363
Founded: 1958
Focus: Mineralogy 10188

South African Dietetics and Home Economics Association, POB 3046, Stellenbosch 7600
Founded: 1953; Members: 300
Focus: Hematology; Home Econ 10189

South African Filtration Society (SAFJL), 159 Duxbury Rd, Hillcrest 0181
T: 747346
Founded: 1978; Members: 453
Focus: Eng 10190

South African Genetic Society, c/o Dept of Genetics, University, Stellenbosch 7600
Founded: 1956
Focus: Genetics 10191

South African Geographical Society (S.A.G.S.), POB 128, Wits 2050
T: 3391951
Founded: 1917; Members: 475
Pres: Prof. J.J. McCarthy; Gen Secr: Dr. J. Fairhurst
Focus: Geography
Periodicals
South African Geographical Journal (weekly)
South African Landscape Series (weekly) 10192

South African Institute for Librarianship and Information Science (SAJBJ), POB 36575, Menlo Park, Pretoria 0102
Founded: 1930; Members: 2851
Pres: Prof. S.P. Manaka; Gen Secr: Prof. A. Louw
Focus: Libraries & Bk Sci; Computer & Info Sci
Periodicals
Newsletter (weekly)
South African Journal for Librarianship and Information Science (weekly) 10193

The South African Institute for Medical Research, Hospital St, POB 1038, Johannesburg 2000
T: (011) 7250511; CA: BACTERIA, Johannesburg
Focus: Med 10194

South African Institute of Aeronautical Engineers (SAIAeE), POB 27335, Sunnyside 0132
T: (011) 9273162
Founded: 1977; Members: 140
Pres: P. Marwick
Focus: Aero; Eng
Periodical
Aeronautica Meridiana (weekly) 10195

South African Institute of Assayers and Analysts, Kelvin House, 2 Hollard St, POB 61019, Mashalltown 2107
Founded: 1919; Members: 181
Pres: J.W. Barnett
Focus: Eng
Periodical
Bulletin (weekly) 10196

The South African Institute of International Affairs, Jan Smuts House, University of the Witwatersrand, POB 31596, Braamfontein 2017
T: (011) 3392021; Tx: 427291; Fax: (011) 3392154
Founded: 1934; Members: 3500
Pres: C.B. Strauss; Gen Secr: Prof. John Barratt
Focus: Poli Sci
Periodicals
Bibliographical Series (weekly)
Bradlow Series (weekly)
South African Journal of International Affairs (weekly)
Special Studies (weekly) 10197

The South African Institute of Mining and Metallurgy, Cape Towers, 11-13 MacLaren St, POB 61127, Marshalltown 2107
T: (011) 8341273; Fax: (011) 8385923
Founded: 1884; Members: 2186
Pres: J.P. Hoffman; Gen Secr: Doris Gardner
Focus: Mining; Metallurgy; Physics
Periodical
Journal of the South African Institute of Mining and Metallurgy (weekly) 10198

The South African Institute of Organization and Methods, POB 693, Pretoria 0001
Founded: 1960
Focus: Business Admin
Periodical
Organization & Methods Newsletter (weekly) 10199

The South African Institute of Physics (S.A.I.P.), POB 72, Faure 7131
Founded: 1955; Members: 600
Pres: Prof. F.J.W. Hahne; Gen Secr: Prof. B. Spoelstra
Focus: Physics 10200

South African Institute of Printing (S.A.I.P.), Pearl Assurance House, Heerengracht, Cape Town 8001
T: (021) 254210; Fax: (021) 215485
Founded: 1969; Members: 137
Pres: W. Judge; Gen Secr: L. Coetzee
Focus: Eng; Graphic & Dec Arts, Design 10201

South African Institute of Race Relations, POB 31044, Braamfontein 2017
Founded: 1929; Members: 2200
Focus: Sociology; Poli Sci; Educ
Periodicals
Race Relations News (weekly)
Race Relations Survey (weekly) 10202

The South African Institution of Civil Engineers, POB 93495, Yeoville 2143
T: (011) 6481184; Fax: (011) 6487427
Founded: 1903; Members: 7000
Pres: B. Bruce; Gen Secr: D.B. Botha
Focus: Civil Eng
Periodicals
Civil Engineering (weekly)
Journal of the South African Institution of Civil Engineers 10203

South African Market Research Association (SAMRA), POB 9858, Johannesburg 2000
Founded: 1963; Members: 494
Focus: Marketing
Periodical
Report (weekly) 10204

The South African Mathematical Society (SAMS), POB 395, Pretoria 0001
T: (012) 869211
Founded: 1957; Members: 300
Focus: Math
Periodical
Quaestiones Mathematicae (weekly) . . . 10205

The South African Medical Research Council, Francie van Zijl Av, Parowvallei, POB 19070, Tygerberg 7505
T: (021) 9380911; CA: MEDRES, Cape Town; Fax: (021) 9380200
Founded: 1969
Focus: Med
Periodicals
AIDS Bulletin (weekly)
CAPFSA Reporter (weekly)
Continuum (weekly)
Documentum (weekly)
MRC Annual Report (weekly)
MRC News (weekly)
Salute (weekly)
Urbanisation and Health Newsletter (weekly) 10206

The South African National Committee on Illumination (SANCI), POB 395, Pretoria 0001
T: (012) 869211
Founded: 1954; Members: 200
Focus: Electric Eng 10207

South African National Group of the International Society for Rock Mechanics (S.A.N.G.O.R.M.), POB 61809, Marshalltown 2107
T: (011) 7263020; Tx: 426070; CA: Bullion, Johannesburg; Fax: (011) 7265405
Founded: 1969; Members: 417
Pres: Dr. T.R. Stacey; Gen Secr: S.A. Thorpe
Focus: Mining; Civil Eng 10208

South African National Multiple Sclerosis Society, 295 Villiers Rd, Walmer 6070
Fax: (041) 512900
Founded: 1963
Focus: Pathology; Med 10209

South African National Tuberculosis Association (SANTA), 621 Leisk House, Corner Bree and Rissik Sts, Johannesburg 2001
T: (011) 299636
Founded: 1947; Members: 230
Pres: P.H. Anderson; Gen Secr: J. Hylton Smith
Focus: Pulmon Dis
Periodicals
SANTA Health Magazine (weekly)
SANTA News (weekly) 10210

The South African Numismatic Society, POB 1689, Cape Town 8000
T: (021) 244412
Founded: 1941; Members: 340
Pres: Dr. W.D.F. Malherbe; Gen Secr: Rita Miller
Focus: Numismatics
Periodical
Bulletin (weekly) 10211

South African Optometric Association, POB 3966, Pretoria 0001
Founded: 1924; Members: 714
Focus: Optics
Periodical
South African Optometrist (weekly) . . . 10212

South African Orthopaedic Association, 1203b Durdoc Centre, 460 Smith St, Durban 4001
Founded: 1942; Members: 216
Pres: J.J. van Niekerk
Focus: Orthopedics
Periodical
Journal of Bone and Joint Surgery . . 10213

South African PEN Centre, 2 Scott Rd, Claremont 7735
Founded: 1060; Members: 100
Gen Secr: A. Naudé
Focus: Lit
Periodical
Newsletter 10214

The South African Rheumatism and Arthritis Association, c/o Brenthurst Clinic, 4 Park Lane, Parktown, Johannesburg 2001
Focus: Rheuma 10215

South African Society for Photogrammetry, Remote Sensing and Cartography, POB 69, Newlands 7725
Founded: 1958; Members: 300
Focus: Cart
Periodical
South African Journal of Photogrammetry, Remote Sensing and Cartography (weekly) . . . 10216

The South African Society of Anaesthesists (SASA), Suite 101, Cargo corner Sevemth Av, Rosebank 2196
Focus: Anesthetics 10217

South African Society of Otorhinolaryngology, 532 Louis Leipoldt Medical Centre, Bellville 7530
Members: 100
Focus: Otorhinolaryngology 10218

The South African Society of Physiotherapy, POB 11151, Johannesburg 2000
Founded: 1925; Members: 1400
Focus: Therapeutics
Periodical
The South African Journal of Physiotherpy: Die Soid-Afrikaanse Tydskrif Fisioterapie (weekly) 10219

South African Speech and Hearing Association / Suid-Afrikaanse Vereniging vir Spraak- en Gehoorheelkunde, POB 31782, Braamfontein 2017
T: (011) 394011 ext 660
Founded: 1946; Members: 600
Focus: Logopedy
Periodical
The South African Journal of Communication Disorders (weekly) 10220

The South African Statistical Association, POB 27321, Sunnyside 0132
T: (011)4981623
Founded: 1953; Members: 400
Pres: Prof. C.F. Smit; Gen Secr: L. Verhoef
Focus: Stats
Periodical
South African Statistical Journal (weekly) 10221

South African Veterinary Association, POB 25033, Monument Park 0105
T: (012) 3461150; Fax: (012) 3462929
Founded: 1903; Members: 1364
Pres: Prof. G.F. Bath; Gen Secr: V. da Silva
Focus: Vet Med
Periodical
Journal of the South African Veterinary Association (weekly) 10222

Southern Africa Cardiac Society, c/o Chelmsford Medical Centre, Saint Augustine's Hospital, 107 Chelmsford, Durban 4001
Members: 300
Focus: Cardiol 10223

Southern African Association for the Advancement of Science, Jakaranda 301, Beckettstr 304, Arcadia 0083
T: (012) 4202169; Fax: (012) 433254
Founded: 1902; Members: 124
Focus: Sci 10224

Southern African Institute of Forestry (SAIF), POB 1022, Pretoria 0001
T: (012) 473479
Founded: 1967; Members: 390
Pres: J.H. Steyn; Gen Secr: C. Viljoen
Focus: Forestry
Periodical
South African Forestry Journal (weekly) . 10225

The Southern African Museums Association (SAMA), POB 29294, Sunnyside 0132
T: (012) 3411320
Founded: 1936; Members: 675
Pres: B. Wilmot; Gen Secr: T. van der Merwe
Focus: Arts
Periodicals
SAMAB: Southern African Museums Association Bulletin (weekly)
Samantix (weekly) 10226

Southern African Ornithological Society (S.A.O.S.), POB 84394, Greenside, Johannesburg 2034
T: (011) 8884147; Fax: (011) 7827013
Founded: 1930; Members: 5000
Focus: Ornithology
Periodicals
Birding in Southern Africa (weekly)
Ostrich (weekly) 10227

Southern African Society of Aquatic Scientists, c/o Freshwater Research Unit, Zoology Dept, University of Cape Town, Rondebosch 7700
T: (021) 6503635; Tx: 522208; Fax: (021) 6503726
Founded: 1963; Members: 350
Pres: Dr. J.A. Day; Gen Secr: Dr. B. Gale
Focus: Hydrology
Periodical
Journal (weekly) 10228

Spectroscopic Society of South Africa, c/o Dept of Chemistry, University of Pretoria, Pretoria 0002
T: (012) 4202511; Fax: (012) 432963
Founded: 1969; Members: 400
Gen Secr: Prof. C.J. Rademeyer
Focus: Chem
Periodical
Newsletter (weekly) 10229

Succulent Society of South Africa, POB 1193, Pretoria 0001
Founded: 1962; Members: 2500
Focus: Botany
Periodical
Aloe: Journal of the Succulent Society of South Africa 10230

Transvaal Underwater Research Group, POB 5180, Johannesburg 2000
Founded: 1952; Members: 150
Focus: Oceanography 10231

The Tree Society of Southern Africa, POB 4116, Johannesburg 2000
Founded: 1948; Members: 500
Pres: Prof. A.B.A. Brink
Focus: Forestry
Periodicals
Newsletter (weekly)
Trees in South Africa (weekly) 10232

Van Riebeeck Society (VRS), c/o South African Library, POB 496, Cape Town 8000
T: (021) 238424
Founded: 1918; Members: 1500
Pres: Dr. F.R. Bradlow; Gen Secr: Dr. C. Pama
Focus: Hist 10233

Water Research Commission, 491 18 Av, Rietfontein, Pretoria 0001
T: (012) 3300340; CA: WATERKOM, Pretoria; Fax: (012) 3312565
Focus: Hydrology
Periodicals
SA Water Bulletin
Water SA 10234

Water Systems Research Group, c/o Dept of Civil Engineering, University of the Witwatersrand, 1 Jan Smuts Av, PO Wits, Johannesburg 2050
Focus: Hydrology; Water Res 10235

Wildlife Society of Southern Africa, POB 44189, Linden 2104
Founded: 1902; Members: 26000
Pres: Dr. E.A. Zaloumis
Focus: Ecology
Periodicals
African Wildlife (weekly)
Toktokki (weekly) 10236

Zoological Society of Southern Africa (ZSSA), c/o Dept of Zoology, University, Port Elizabeth 6000
T: (041) 5311335
Members: 500
Focus: Zoology
Periodical
South African Journal of Zoology (weekly) 10237

Spain

Academia de Buenas Letras de Barcelona, Calle Obispo Cassador 3, Barcelona 22
T: (03) 3150010
Founded: 1729
Pres: M. de Riquer; Gen Secr: José Alsina Clota
Focus: Lit
Periodicals
Boletín
Memorias 10238

Academia de Ciencias Exactas, Físicas, Químicas y Naturales de Zaragoza, c/o Facultad de Ciencias, Ciudad Universitaria, Zaragoza
Founded: 1916; Members: 47
Pres: Juan Sancho San Roman; Gen Secr: Enrique Melendez Andreu
Focus: Chem; Nat Sci; Physics; Math
Periodical
Revista (weekly) 10239

Academia de Ciencias Médicas de Bilbao, Lersundi 9, 48009 Bilbao
T: (04) 4233768
Founded: 1895; Members: 1300
Pres: Dr. Isaac Fernandez Martin-Granizo; Gen Secr: Dr. Aurelio Aparicio Ferrero
Focus: Med
Periodical
Gaceta Médica de Bilbao (weekly) . . . 10240

Acadèmia de Ciencies Mèdiques de Catalunya i de Baleares, Paseo de la Bonanova 47, 08017 Barcelona
T: (03) 2123895
Founded: 1898; Members: 10000
Pres: M. Foz i Sala; Gen Secr: P. Pardo i Peret
Focus: Med
Periodical
Monografies Mèdiques 10241

Academia Española de Dermatología y Sifilografía, Sandoval 7, Madrid 10
Founded: 1909; Members: 489
Pres: Prof. D. José Cabre Piera; Gen Secr: R. Morán López
Focus: Derm; Venereology
Periodical
Actas Dermosifiliográficas 10242

Academia Iberoamericana y Filipina de Historia Postal, c/o Dirección General de Correos y Telecomunicación, Palacio de Comunicaciones, Madrid 14
Founded: 1930; Members: 89
Pres: A. Martín García; Gen Secr: José Jusdado Martín
Focus: Hist
Periodical
Boletín (weekly) 10243

Academia Médico-Quirúrgica Española, Villanueva 11, Madrid 1
Founded: 1891; Members: 492
Pres: Prof. Eduardo Arias Vallejo; Gen Secr: Dr. Julio Múñiz González
Focus: Surgery; Med
Periodical
Anales 10244

Agrupación Astronáutica Española (AAE), Rosellon 134, Barcelona 36
Founded: 1953; Members: 85
Focus: Aero 10245

Amics dels Museus de Catalunya / Friends of the Museums Association, Palacio de la Virreina, La Rambla 99, 08002 Barcelona
T: (03) 3014379; Fax: (03) 3189421
Founded: 1933; Members: 1000
Pres: J.M. Garrut; Gen Secr: Fauusto Serra de Dalmases
Focus: Archives
Periodical
Historiales 10246

Asociación de Escritores y Artistas Españoles / Writers' and Artists' Association, Calle de Leganitos 10, 28013 Madrid
Founded: 1872; Members: 645
Pres: Luis Cervera Vera; Gen Secr: José Gerardo Manrique de Lara
Focus: Lit 10247

Asociación de Investigación Técnica de la Industria Papelera Española, Carretera de La Coruna Km 7, 28040 Madrid
T: (01) 3070977; Fax: (01) 3572828
Founded: 1963; Members: 373
Focus: Eng
Periodicals
Información Técnico-Económica (weekly)
Investigación y Técnica del Papel (weekly)
. . . . 10248

Asociación del Cuerpo Nacional Veterinario, Carranza 3, Madrid 10
T: (01) 4465725
Founded: 1954; Members: 250
Focus: Vet Med 10249

Asociación de Medicina Aeronáutica y Espacial, Paseo de la Bonanova 47, Barcelona 6
Founded: 1960; Members: 50
Focus: Med 10250

Asociación de Peritos Agrícolas del Estado, Av de José Antonio 29, Madrid
T: (01) 2223354
Focus: Agri 10251

Asociación de Personal Investigador del CSIC (API), Serrano 113, 28006 Madrid
Fax: (01) 5627518
Founded: 1956; Members: 1149
Pres: Dr. Ernesto Garcia Lopez; Gen Secr: Dr. Monique Novaes-Ledieu
Focus: Sci 10252

Asociación Electrotécnica Española, Valencia 169, Barcelona
T: (03) 2541449
Focus: Electric Eng 10253

Asociación Española Astronáutica, Paseo del Pinto Rosales 34, Madrid 8
T: (01) 2479800
Focus: Aero 10254

Asociación Española contra el Cancer, Marco Aurelio 14, Barcelona
T: (03) 2002099
Focus: Cell Biol & Cancer Res 10255

Asociación Española de Amigos de los Castillos (AEAC), Bárbara de Braganza 8, Madrid 4
T: (01) 4191829
Founded: 1952; Members: 2608
Focus: Preserv Hist Monuments 10256

Asociación Española de Bibliotecarios, Archiveros, Museologos y Documentalistas (ANABAD), Recoletos 5, Apdo 14281, 28001 Madrid
T: (01) 2751727
Founded: 1949; Members: 1300
Pres: Alonso Vicenta Cortés; Gen Secr: Roca Maria Rosario Fernández
Focus: Libraries & Bk Sci; Archives; Archeol
Periodical
Boletín de la ANABAD (weekly) 10257

Asociación Española de Biopatología Clínica, Sandoval 7, Madrid 10
Founded: 1946; Members: 800
Focus: Pathology 10258

Asociación Española de Cine Científico (ASECIC), Serrano 150, Madrid 6
Founded: 1966; Members: 305
Focus: Cinema 10259

Asociación Española de Cirujanos, Santa Isabel 51, Madrid 12
Founded: 1934; Members: 1000
Focus: Surgery 10260

Asociación Española de Gemología, Paseo de Gracia 64, 08007 Barcelona
T: (03) 2151398; Fax: (03) 2154312
Focus: Mineralogy
Periodical
Gemologia (weekly) 10261

Asociación Española de la Lucha contra la Poliomielitis (ALPE) / Spanish Polio Association, Casarrubuelos 5, 28015 Madrid
T: (01) 4480864; Fax: (01) 5942338
Gen Secr: Nieves Sanchiz Pons
Focus: Med 10262

Asociación Española de Logopedia, Foniatria y Audiología (AELFA), Provenza 319, 08037 Barcelona
T: (03) 2577818
Founded: 1960; Members: 975
Focus: Logopedy; Otorhinolaryngology
Periodical
Revista de Logopedia (weekly) 10263

Asociación Española de Marketing (AEM), Paseo de la Castellana 122, 28046 Madrid
T: (01) 2627205; Tx: 48660
Founded: 1966
Focus: Advert 10264

Asociación Española de Orientalistas (AEO), c/o Rectorado, Universidad Autónoma, 28049 Madrid
Focus: Ethnology
Periodical
Boletin de la Asociación Española de Orientalistas (weekly) 10265

Asociación Española de Pediatría, Villanueva 11, Madrid 1
Focus: Pediatrics 10266

Asociación Española de Psicoterapia Analítica, Serrano Jover 6, Madrid
T: (01) 2480328
Focus: Therapeutics 10267

Asociación Española de Químicos y Coloristas Textiles (AEQCT), c/o Joaquín Vidal Artes, Paseo General Mola 35, Barcelona 9
Focus: Chem; Textiles 10268

Asociación Española de Técnicos de Cerveza y Malta, Juan de la Cierva 3, Madrid 6
T: (01) 2622900
Members: 510
Focus: Eng 10269

Asociación Española de Técnicos de Radiología (AETR), Victor Tradera 12, Madrid 8
Founded: 1973
Focus: Radiology 10270

Asociación Española para el Estudio Científico del Retraso Mental, General Oraa 19, Madrid 6
Focus: Psychiatry 10271

Asociación Española para el Progreso de las Ciencias (A.E.P.C.), Valverde 24, Madrid 13
T: (01) 2212529
Founded: 1908; Members: 450
Pres: Prof. Angel González-Alvarez; Gen Secr: Prof. José M. Torroja
Focus: Sci
Periodical
Las Ciencias (weekly) 10272

Asociación Iberoamericana de Educación Superior a Distancia, c/o UNED, Apdo 50487, 28080 Madrid
Founded: 1981; Members: 53
Pres: Dr. Mariano Artés Gómez; Gen Secr: Dr. D. Luis Tejero Escribano
Focus: Educ
Periodical
Revista Iberoamericana de Educación Superior a Distancia (weekly) 10273

Asociación Internacional Veterinaria de Producción Animal (AJVPA), c/o Asociación Mundial Veterinaria, Isabel la Católica 12, 28013 Madrid
T: (01) 2471838
Founded: 1951; Members: 2000
Pres: Prof. Dr. A. de Vuyst; Gen Secr: Prof. Dr. Carlos L. de Cuenca
Focus: Vet Med
Periodical
Zootechnia (weekly) 10274

Asociación Latinoamericana de Derecho Aeronáutico y Espacial (ALADA) / Latin American Association for Aeronautical and Space Law, Av Reyes Católicos 4, 28040 Madrid
T: (01) 2440600 ext 211
Gen Secr: Dr. Enrique Mapelli y López
Focus: Law 10275

Asociación Nacional de Químicos de España / National Association of Chemists, Lagasca 83, Madrid 6
Founded: 1945; Members: 9000
Pres: José Luis Negro López; Gen Secr: Joaquín Corado López
Focus: Chem
Periodical
Química e Industria (weekly) 10276

Asociación Nacional de Veterinarios Titulares, Av de José Antonio 88, Madrid 13
T: (01) 2412667
Focus: Vet Med 10277

Asociación Numismática Española (ANE), Gran Via de les Corts Catalanes 627, 08010 Barcelona
T: (03) 3188245; Fax: (03) 3189062
Founded: 1955; Members: 2465
Focus: Numismatics
Periodical
Gaceta Numismatica 10278

Asociación para el Progreso de la Dirección, Montalba 3, Madrid
T: (01) 2324529
Focus: Business Admin 10279

Asociación para la Defensa de la Naturaleza (ADENA) / World Wildlife Fund, Santa Engracia 6, Madrid 10
T: (01) 4102401/02
Focus: Ecology
Periodical
Panda (weekly) 10280

Asociación Química Española de la Industria del Cuero, c/o Enrique Roig, Av de José Antonio 608, Barcelona 7
Focus: Chem 10281

Association for Dental Education in Europe (ADEE), c/o Faculdad de Odontologia, Universidad, Madrid
Founded: 1975
Gen Secr: Dr. Mariano Danz
Focus: Educ; Dent 10282

Ateneo Barcelonés / Barcelona Athenaeum, Calle Canuda 6, 08002 Barcelona
Founded: 1860; Members: 4200
Pres: Heribert Barrera Costa; Gen Secr: Xavier Renau Nanen
Focus: Arts; Lit 10283

Ateneo Científico, Literario y Artístico, Calle del Prado 21, Madrid 14
T: (01) 4296251
Founded: 1820; Members: 6500
Pres: José Prat García; Gen Secr: Fernando Mansilla Izquierdo
Focus: Arts; Lit 10284

Ateneo Científico, Literario y Artístico / Scientific, Literary and Artistic Athenaeum, Calle Cifuentes 25, Mahón, Minorca
Founded: 1905; Members: 630
Pres: F. Tutzó Bennasar; Gen Secr: Gabriel Pons Olives
Focus: Lit; Arts
Periodical
Revista de Menorca (weekly) 10285

Center of Information and Research SIIS, Reina Regente 5, 20003 San Sebastián
T: (043) 423656; Fax: (043) 293007
Founded: 1972
Focus: Rehabil 10286

Centro de Estudios Constitucionales / Centre for Constitutional Studies, Pl de la Marina Española 9, 28013 Madrid
Fax: (01) 5478549
Founded: 1977
Pres: Francisco J. Laporta San Miguel;
Gen Secr: Daniel Villagra Blanco
Focus: Poli Sci; Law
Periodicals
Derecho Privado y Constitución (3 times annually)
Revista de Administración Pública
Revista de Economía Política
Revista de Estudios Internacionales (weekly)
Revista de Estudios Políticos (weekly)
Revista de Instituciones Europeas
Revista del Centro de Estudios Constitucionales (3 times annually)
Revista de Política Social (weekly)
Revista Española de Derecho Constitucional
. . . . 10287

Colegio de Abogados de Barcelona / Barcelona College of Lawyers, Calle Mallorca 283, Barcelona 37
T: (03) 2154612
Founded: 1832; Members: 6200
Pres: Antonio Plascencia Monleón; Gen Secr: José Gasch Riudor
Focus: Law
Periodicals
Anuario de Sociologia y Psicologia Juridicas
Revista Juridica de Cataluña 10288

Colegio de Arquitectos de Cataluña / College of Architects of Catalonia, Pl Nueva 5, 08002 Barcelona
Founded: 1931; Members: 3247
Gen Secr: Pedro Serra Amengual
Focus: Archit
Periodical
Cuadernos de Arquitectura y Urbanismo . 10289

Colegio Notarial / College of Notaries, Calle Notariado 4, Barcelona 1
T: (03) 3174800
Members: 346
Pres: V. Font Boix
Focus: Law 10290

Colegio Oficial de Ingenieros Industriales de Cataluña, Via Layetana 39, 08003 Barcelona
Pres: C. Ponsa Ballart; Gen Secr: J. Cornet Colom
Focus: Eng 10291

Comité International Permanent des Etudes Mycéniennes, c/o Universidad del Pais Vasco, 01006 Vitoria
T: (045) 139811; Fax: (045) 138227
Founded: 1956
Gen Secr: Prof. José L. Melena
Focus: Cultur Hist 10292

Comité Nacional Español del Consejo Internacional de la Música / National Committee of the International Music Council, Av Reina Victoria 58, Madrid 3
Focus: Music 10293

Conferencia de Rectores de las Universidades del Estado, c/o Escuela de Enfermería, Menéndez Pidal s/n, 14004 Córdoba
T: (057) 201903; Tx: 76561
Founded: 1978; Members: 31
Pres: Prof. Dr. Rafael Portaencasa Baeza;
Gen Secr: Prof. Dr. Manuel Peláez del Rosal
Focus: Educ 10294

Consejo de Universidades, Cuidad Universitaria s/n, 28040 Madrid
T: (01) 4497437
Founded: 1983
Focus: Educ 10295

Consejo General de Colegios Oficiales de Diplomados en Trabajo Social, Campomanes 10, 28013 Madrid
T: (01) 5415776; Fax: (01) 5590277
Founded: 1967; Members: 36
Pres: Maria Jesús Utrilla Moya
Focus: Sociology 10296

Consejo General de Colegios Oficiales de Doctores y Licenciados en Filosofía y Letras y en Ciencias / General Council of Official Colleges of Doctors and Licenciates in Philosophy, Letters and Science, Bolsa 11, 28012 Madrid
Founded: 1944; Members: 55800
Pres: José Luis Negro Fernández; Gen Secr: Roberto Salmerón Sanz
Focus: Lit; Philos 10297

Consejo General de Colegios Oficiales de Farmacéuticos, Villanueva 11, Madrid
T: (01) 2251020
Founded: 1942; Members: 13500
Pres: Ernesto Marco Cañizares
Focus: Pharmacol
Periodical
Boletín de Información 10298

Consejo General de Colegios Oficiales de Odontólogos y Estomatólogos de España, Villanueva 11, 28001 Madrid
T: (01) 5770638; Fax: (01) 5770639
Members: 11914
Pres: Dr. José Maria Lara Sanz; Gen Secr: Dr. José Antonio Zafra Anta
Focus: Dent; Stomatology
Periodical
Revista de Actualidad Odontoestomatológica Española (weekly) 10299

Consejo General de Colegios Veterinarios de España, Villanueva 11, Madrid 1
T: (01) 2767330
Founded: 1923; Members: 7400
Focus: Vet Med 10300

Consejo General de Oficiales de Ingenieros Técnicos Agrícolas de España, Plaza de Santo Domingo 13, Madrid
Founded: 1949; Members: 3668
Pres: José Luis Linaza de la Cruz; Gen Secr: Constantino Lagaron Yañez
Focus: Agri 10301

Consejo Superior de los Colegios de Arquitectos de España, Paseo de la Castellana 12, Apdo 688, Madrid 1
T: (01) 4352200; Tx: 46004
Members: 18
Focus: Archit
Periodical
CSC Arquitectos-Q (weekly) 10302

Departamento de Historia del Arte Diego Velázquez, c/o Centro de Estudios Históricos, Centro Superior de Investigaciones Cientificas, Duque de Medinaceli 6, 28014 Madrid
Fax: (01) 5854878
Founded: 1939; Members: 9
Gen Secr: Enrique Arias Angles
Focus: Arts
Periodical
Archivo Español de Arte (3 times annually)
. . . . 10303

European Association of Coleopterology, c/o Dept de Biologia Animal, Facultat de Biologia, Universitat de Barcelona, Diagonal 645, 08028 Barcelona
T: (03) 3308851 ext 165; Fax: (03) 4110887
Pres: Dr. Marina Blas
Focus: Vet Med 10304

European Association of Football Teams Physicians, Diagonal 648, 08017 Barcelona
T: (03) 2052213; Tx: 51902
Gen Secr: Dr. Carlos Bestit
Focus: Med 10305

European Bioelectromagnetics Association, c/o Hospital Ramon y Cajal, Carretera de Comenar km 9, 28934 Madrid
T: (01) 3581275
Pres: Dr. Jocelyne Leal
Focus: Physics *10306*

European Centre for Professional Training in Environment and Tourism, Viriato 21, 28010 Madrid
T: (01) 5930831; Fax: (01) 5930980
Focus: Educ *10307*

European Cooperative Research Network on Olives, c/o Centro de Investigación y Desarrollo Agrario, DGIEA-INIA, Alameda del Obispo, 14080 Cordoba
Founded: 1974
Gen Secr: Juan M. Caballero
Focus: Crop Husb
Periodical
Olea *10308*

Federación de Urbanismo y de la Vivienda, Pl del Cordon 1, Madrid
Focus: Urban Plan *10309*

Federación Española de Religiosos de Enseñanza (FERE) / Spanish Federation of Religious Teaching, Conde de Peñalver 45, Apdo 53052, 28006 Madrid
Founded: 1957
Pres: R.P. Aureliano Laguna Vegas; Gen Secr: Santiago Martín Jiménez
Focus: Rel & Theol; Educ
Periodicals
Boletín de la FERE (weekly)
Educadores (weekly) *10310*

Fundación Juan March, Castelló 77, 28006 Madrid
T: (01) 4354240; Tx: 45406; Fax: (01) 5763420
Founded: 1955
Pres: Juan March Delgado; Gen Secr: José Luis Yuste Grijalba
Focus: Arts; Music
Periodicals
Anales
Boletín Informativo (weekly)
Saber Leer (weekly) *10311*

Institución Fernando el Católico, Palacio Provincial, Plaza de España, 50004 Zaragoza
Founded: 1943; Members: 150
Pres: José Marco Berges; Gen Secr: José Barranco López
Focus: Arts; Ling; Lit *10312*

Institut Agrícola Català de Sant Isidre / Catalan Agricultural Institute, Pl Sant Josep Oriol 4, Barcelona 2
T: (03) 3011636
Founded: 1851; Members: 10356
Pres: Ignacio de Puig i Girona; Gen Secr: Fernando de Múller i de Dalmases
Focus: Agri
Periodicals
Boletín (weekly)
Calendari del Pagés (weekly)
Revista (weekly) *10313*

Institut del Teatre / Theatrical Institute, Sant Pere Més Boix 7, 08003 Barcelona
T: (03) 3173974/78; Fax: (03) 3017943, 3173779
Founded: 1913
Pres: Jordi Coca; Gen Secr: Joan Castells
Focus: Perf Arts
Periodical
Estudis escènics (3 times annually) . . *10314*

Institut d'Estudis Catalans, Carrer de París 150, Apdo 1146, Barcelona
Founded: 1907
Pres: J. Ainaud de Lasarte; Gen Secr: R. Aramon i Serra
Focus: Hist *10315*

Instituto Amatller de Arte Hispánico / Institute of Hispanic Art, Paseo de Gracia 41, 08007 Barcelona
T: (03) 2160175; Fax: (03) 4875827
Founded: 1941; Members: 6
Gen Secr: Santiago Alcolea
Focus: Arts; Fine Arts *10316*

Instituto Aula de Mediterráneo, c/o Universidad de Valencia, Valencia
Founded: 1942; Members: 545
Pres: Dr. F. Sánchez Castañer
Focus: Lit *10317*

Instituto de España, San Bernardo 49, 28008 Madrid
T: (01) 2224885
Founded: 1938
Pres: Fernando Chueca Goitia
Focus: Sci
Periodical
Anuario (weekly) *10318*

Instituto de Estudios Africanos, Castellana 5, Madrid 1
Founded: 1945; Members: 20
Pres: Eduardo Blanco Rodríguez; Gen Secr: J. Ventura Bañares
Focus: Ethnology *10319*

Instituto de Estudios Asturianos, Plaza de Porlier 5, Oviedo
T: (085) 211760
Founded: 1946; Members: 50
Gen Secr: Jesús Evaristo Casariego
Focus: Hist *10320*

Instituto de Farmacología Española, Mendez Alvaro 57, 28045 Madrid
T: (01) 2273020; Tx: 45448; Fax: (01) 4676431
Focus: Pharmacol *10321*

Instituto de Filologia, Duque de Medinaceli 6, 28014 Madrid
T: (01) 4292017
Founded: 1939; Members: 15
Pres: Natalio Fernández Marcos; Gen Secr: Leonor Carracedo
Focus: Ling; Lit
Periodicals
Al-Qantara
Anales Cervantinos (weekly)
Emerita (weekly)
Revista de Dialectologia y Tradiciones Populares (weekly)
Revista de Filología Española
Revista de Literatura
Sefarad *10322*

Instituto de Historia y Cultura Naval, Juan de Mena 1, 28014 Madrid
T: (01) 5327151
Founded: 1942; Members: 10
Pres: Ricardo Cerezo Martinez
Focus: Hist; Transport *10323*

Instituto de la Engeniería de España, General Arrando 38, 28010 Madrid
Founded: 1905; Members: 20000
Gen Secr: Jaime Tornos
Focus: Eng *10324*

Instituto Geográfico Nacional, Calle del General Ibáñez de Ibero 3, 28003 Madrid
T: (01) 5333800; Tx: 23465; Fax: (01) 5546743
Founded: 1870; Members: 1200
Gen Secr: Angel Arévalo Barroso
Focus: Geography *10325*

Instituto Nacional de Estadística (INE), Paseo de la Castellana 183, 28046 Madrid
T: (01) 2799300; Tx: 22224
Founded: 1945
Focus: Stats
Periodicals
Anuario Estadística de España (weekly)
Boletín Mensual de Estadística (weekly)
Revista Estadística Española *10326*

Instituto Nacional de Meteorología, Ciudad Universitaria, Apdo 285, Madrid
Founded: 1887; Members: 5011
Pres: M. Bautista Pérez
Focus: Physics
Periodical
Boletín Meteorológico *10327*

International Association for Shell and Spatial Structures (IASS), Alfonso XII 3, 28014 Madrid
Founded: 1959; Members: 700
Gen Secr: A. Casas
Focus: Civil Eng
Periodical
IASS Bulletin (3 times annually) *10328*

International Congress of Carboniferous Stratigraphy and Geology (ICC), c/o Instituto Geológico y Minero de España, Ríos Rosas 23, Madrid 3
Founded: 1927
Gen Secr: G. Ortuño Aznar
Focus: Geology
Periodical
Congress Proceedings *10329*

International Sociological Association (ISA) / Association Internationale de Sociologie, Pinar 25, 28223 Madrid
T: (01) 3527650; Fax: (01) 3524945
Founded: 1949; Members: 3750
Pres: T.K. Oommen; Gen Secr: I. Barlinska
Focus: Sociology
Periodicals
Current Sociology/Sociologie Contemporaine (3 times annually)
International Sociology (weekly)
ISA Bulletin (weekly)
Sage Studies in International Sociology (weekly) *10330*

International Towing Tank Conference, Canal de Experiencias Hidrodinamicas, El Pardo, 28048 Madrid
Founded: 1932; Members: 70
Gen Secr: Mariano Perez-Sobrino
Focus: Eng *10331*

Oficina Internacional de Información y Observación del Español, Av de los Reyes Católicos 4, Madrid 3
T: (01) 2440600
Founded: 1963
Gen Secr: Manuel Alvar
Focus: Ling; Doc
Periodicals
Español Actual
Lingüistica Española Actual *10332*

Organización de Estados Iberoamericanos para la Educación, la Ciencia y la Cultura (OEI), Bravo Murillo 38, 28015 Madrid
T: (01) 5944382; Fax: (01) 5943286
Founded: 1949; Members: 20
Gen Secr: José Torreblanca Prieto
Focus: Educ; Sci
Periodical
Revista Iberoamericana de Educación (weekly) *10333*

Organización Médica Colegial-Consejo General de Colegios Médicos de Españna, Villanueva 11, Madrid 1
T: (01) 2258410
Founded: 1930; Members: 21
Pres: Ricardo Ferré
Focus: Med
Periodicals
Boletín Formativo e Informativo (weekly)
Medicina de España *10334*

Patronato de Biología Animal, Embajadores 68, Madrid 12
Founded: 1933
Pres: Dr. Angel Campano López; Gen Secr: Dr. Rafael Diaz Montilla
Focus: Zoology
Periodical
Revista (weekly) *10335*

PEN Club, c/o Libréria Turner, Génova 3, Madrid
Focus: Lit *10336*

Real Academia de Bellas Artes de la Purísima Concepción, Rastro, Casa de Cervantes, 47001 Valladolid
Founded: 1746; Members: 30
Pres: Dr. Nicomedes Sanz y Ruiz de la Peña; Gen Secr: Dr. Nemesio Montero Pérez
Focus: Arts *10337*

Real Academia de Bellas Artes de San Fernando, Alcalá 13, 28014 Madrid
T: (01) 5221546
Founded: 1752
Pres: Federico Sopeña; Gen Secr: Enrique Pardo Canalis
Focus: Arts
Periodical
Boletín *10338*

Real Academia de Bellas Artes de Santa Isabel de Hungría, Abades 12-14, 41004 Sevilla
T: (05)4221198
Founded: 1660; Members: 36
Pres: Antonio de la Banda y Vargas; Gen Secr: Ramón Corzo Sánchez
Focus: Arts
Periodical
Boletín de Bellas Artes y Temas de Estetica y Arte (weekly) *10339*

Real Academia de Bellas Artes de San Telmo, Málaga
Founded: 1849; Members: 28
Pres: José Luis Estrada Segalerva; Gen Secr: B. Peña Hinojosa; Luis B. Hernández de Santa-Olalla
Focus: Arts *10340*

Real Academia de Bellas Artes y Ciencias Históricas de Toledo, Esteban Illán 9, Toledo
Founded: 1916; Members: 33
Pres: Dr. Julio Porres Martín-Cleto; Gen Secr: E. Pedraza Ruiz
Focus: Hist; Arts
Periodical
Toletum *10341*

Real Academia de Ciencias Exactas, Físicas y Naturales, Valverde 22-24, 28004 Madrid
T: (01) 5212529; Fax: (01) 5325716
Founded: 1847; Members: 42
Pres: Angel Martín Municio; Gen Secr: José Javier Etayo Miqueo
Focus: Nat Sci; Physics; Sci
Periodicals
Anuario (weekly)
Memoria
Revista *10342*

Real Academia de Ciencias Morales y Políticas (R.A.C.M.Y.P.), Pl de la Villa 2, 28005 Madrid
T: (01) 2481330
Founded: 1857; Members: 36
Pres: Luis Díez del Corral; Gen Secr: S. del Campo Urbano
Focus: Philos; Poli Sci
Periodicals
Anales (weekly)
Catalogo (weekly) *10343*

Real Academia de Ciencias y Artes de Barcelona (RACAB), Rambla de los Estudios 115, 08002 Barcelona
T: (03) 3170536
Founded: 1770; Members: 95
Pres: Prof. Enric Freixa; Gen Secr: Prof. Dr. Manuel Puigcerver
Focus: Sci; Arts
Periodical
Memorias de la Real Academia de Ciencias y Artes de Barcelona (4-6 times annually) *10344*

Real Academia de Córdoba de Ciencias, Bellas Letras y Nobles Artes, Ambrosio de Morales 9, 14003 Córdoba
T: (057) 413168
Founded: 1810; Members: 35
Pres: Angel Aroca Lara; Gen Secr: Joaquin Criado Costa
Focus: Arts; Lit
Periodical
Boletín (weekly) *10345*

Real Academia de Farmacia, Farmacia 11, 28004 Madrid
T: (01) 5310307
Founded: 1589; Members: 40
Pres: Angel Santos-Ruíz; Gen Secr: Manuel Ortega Mata
Focus: Pharmacol
Periodical
Anales (weekly) *10346*

Real Academia de Jurisprudencia y Legislación, Marqués de Cubas 13, 28014 Madrid
T: (01) 2222069
Founded: 1730; Members: 40
Pres: Antonio Hernández Gil; Gen Secr: Raimundo Fernández-Cuesta y Merelo
Focus: Law *10347*

Real Academia de la Historia, León 21, 28014 Madrid
T: (01) 4290611
Founded: 1738; Members: 36
Pres: Emilio García Gómez; Gen Secr: Eloy Benito Ruano
Focus: Hist
Periodical
Boletín *10348*

Real Academia de la Lengua Vasca, Plaza Barria 15, 48005 Bilbao
Founded: 1919; Members: 24
Pres: Jean Haritschelhar; Gen Secr: Endrike Knörr
Focus: Ling
Periodicals
003-Euskararen Lekukoak
Euskera (weekly)
Iker
Jagon
Onomasticon Vasconiae *10349*

Real Academia de Medicina de Sevilla, Pl de España, Sevilla
Founded: 1697; Members: 150
Pres: Dr. Gabriel Sánchez de la Cuesta y Gutiérrez; Gen Secr: Dr. Lucas Bermudo Fernández
Focus: Med *10350*

Real Academia de Medicina y Cirugía de Palma de Mallorca, Morey 20, Palma de Mallorca
Members: 19
Pres: José Sampol Vidal; Gen Secr: Santiago Forteza
Focus: Surgery; Med *10351*

Real Academia de Nobles y Bellas Artes de San Luis, Pl de Los Sitios 6, Zaragoza
Founded: 1792; Members: 140
Pres: Angel Canellas López; Gen Secr: Adolfo Castillo Genzor
Focus: Arts
Periodical
Boletín (weekly) *10352*

Real Academia Española, Felipe IV 4, 28014 Madrid
T: (01) 2394605
Founded: 1713; Members: 115
Pres: Rafael Lapesa Helgar; Gen Secr: Alonso Zamora Vicente
Focus: Ling; Lit *10353*

Real Academia Gallega, Palacio Municipal, La Coruña
Founded: 1905; Members: 40
Pres: Domingo García-Sabell; Gen Secr: Francisco Vales Villamarín
Focus: Sci
Periodical
Boletín *10354*

Real Academia Hispano-Americana, Pl de San Francisco 3, Apdo 16, Cádiz
Founded: 1910; Members: 29
Pres: José M. Pemán y Pemartín; Gen Secr: Manuel Antonio Rendón y Gómez
Focus: Sci
Periodical
Boletín *10355*

Real Academia Nacional de Medicina, Arrieta 12, 28013 Madrid
T: (01) 2470318
Founded: 1732; Members: 190
Pres: José Botella Llusiá; Gen Secr: Dr. V. Matilla Gómez
Focus: Med
Periodicals
Anales
Biblioteca Clásica de la Medicina Española *10356*

Real Academia Sevillana de Buenas Letras, Abades 14, 41004 Sevilla
Founded: 1751; Members: 135
Pres: Prof. Francisco Morales Padrón; Gen Secr: Dr. Enrique de la Vega Viguera
Focus: Lit
Periodical
Boletín (weekly) *10357*

Real Sociedad Arqueológica Tarraconense (R.S.A.T.), Calle Mayor 35, Apdo 573, Tarragona
T: (077) 233789
Founded: 1844; Members: 750
Pres: E. Olive i Martínez; Gen Secr: Joan Menchon i Bes
Focus: Archeol

Periodical
Butlleti Arqueologic (weekly) 10358

Real Sociedad Bascongada de los Amigos del Pais (R.S.B.A.P.), Ramón Ma. de Lilí 6, 20002 San Sebastian
T: (043) 285577
Founded: 1765; Members: 1200
Pres: José M. Aycart Orbegozo
Focus: Hist
Periodicals
Boletín de la Real Sociedad Bascongada de los Amigos del Pais (weekly)
Orria (weekly) 10359

Real Sociedad Económica de Amigos del País de Tenerife / Royal Economic Society of Friends of Tenerife, c/o La Laguna de Tenerife, San Agustín 23, Tenerife
Founded: 1777; Members: 490
Pres: Marqués de Villanueva del Pardo; Gen Secr: Manuel D. Quintana
Focus: Ethnology; Econ 10360

Real Sociedad Española de Física, c/o Facultades de Física y Química, Ciudad Universitaria, 28040 Madrid
T: (01) 3944359; Fax: (01) 5433879
Founded: 1903; Members: 1300
Pres: J.M. Savirón de Cidón; Gen Secr: J.M. Pastor Benavides
Focus: Chem; Physics
Periodical
Anales de Física (weekly) 10361

Real Sociedad Española de Historia Natural, c/o Facultades de Biología y Geología, Ciudad Universitaria, 28040 Madrid
T: (01) 3945000
Founded: 1871; Members: 800
Focus: Hist; Nat Sci
Periodicals
Actas (weekly)
Boletín: Sección Biológica (weekly)
Boletín: Sección Geológica (weekly) . . 10362

Real Sociedad Española de Química, c/o Facultad de Ciencias Químicas, Ciudad Universitaria, Madrid
Founded: 1903
Pres: J. Antonio Rodríguez Renuncio; Gen Secr: Carlos Seoane Prado
Focus: Chem
Periodical
Anales de Quimica 10363

Real Sociedad Fotográfica (RSF), Príncipe 16, 28012 Madrid
T: (01) 5224300
Founded: 1899; Members: 750
Pres: Maria Teresa Gutierrez Barranco; Gen Secr: Juan Carlos Martin
Focus: Arts
Periodical
Boletín (weekly) 10364

Real Sociedad Geográfica, Valverde 22, Madrid 13
T: (01) 5323831
Founded: 1876; Members: 494
Pres: Dr. José M. Torroja Menéndez
Focus: Geography
Periodicals
Boletín (weekly)
Hoja Informativa (weekly) 10365

Real Sociedad Matemática Española (R.S.M.E.), Serrano 123, 28006 Madrid
T: (01) 2619800
Founded: 1911; Members: 600
Pres: Pedro Luis García Pérez; Gen Secr: Juan Llovet Verdugo
Focus: Math
Periodical
Revista Matemática Iberoamericana . . . 10366

Real Sociedad Vascongada de los Amigos del País, c/o Museo de San Telmo, Pl Ignacio Zuloaga, San Sebastián 3
Founded: 1764; Members: 24
Pres: Ignacio M. Barriola Irigoyen
Focus: Hist
Periodicals
Anuario de Eusko-Folklore Aranzadiana Orria
Boletín (weekly)
Boletín de Estudios Históricos sobre San Sebastián (weekly)
Boletín de la Cofradía Vasca de Gastronomía Egan
Munibe 10367

Seminario de Filología Vasca Julio de Urquijo / Julio de Urquijo Seminary of Basque Philology, Palacio de la Diputación de Guipúzcoa, San Sebastián
Founded: 1953
Pres: Luis Michelena
Focus: Ling; Lit
Periodical
Anuario 10368

Servicio de Investigación Prehistórica y Museo de Prehistoria (S.I.P.), Calle de la Corona 36, 46003 Valencia
T: (06) 3917164
Founded: 1927; Members: 9
Pres: Bernardo Marti Oliver
Focus: Hist
Periodicals
Archivo de Prehistoria Levantina (bi-annually)
Serie de Trabajos Varios: Monografias (weekly)
. . . . 10369

Sociedad Anatómica Española, c/o Facultad de Medicina, Granada
Founded: 1949; Members: 250
Focus: Anat 10370

Sociedad de Ciencias Aranzadi, c/o Museo San Telmo, Pl I. Zuloaga, 20003 San Sebastián
T: (043) 422945; Fax: (043) 421316
Founded: 1947; Members: 2000
Pres: Jesus Altuna; Gen Secr: Francisco Etxeberria
Focus: Nat Sci; Anthro; Archeol
Periodicals
Aranzadiana (weekly)
Boletín de Astronomia (weekly)
Munibe (Antropologia-Arkeologia) (weekly)
Munibe (Ciencias Naturales) (weekly) . . 10371

Sociedad de Ciencias, Letras y Artes El Museo Canario / Scientific, Literary and Art Society, Dr. Chil 33, Las Palmas
Founded: 1879
Pres: José M. Alzola
Focus: Lit; Arts; Ethnology
Periodical
El Museo Canario (weekly) 10372

Sociedad de Estadística e Investigación Operativa (S.E.I.O.), Serrano 123, 28006 Madrid
T: (01) 2619800
Founded: 1962; Members: 400
Pres: Marco A. Lopez Cerda; Gen Secr: José Antonio Romero Cordero
Focus: Computer & Info Sci; Stats
Periodicals
Trabajos de Estadística (3 times annually)
Trabajos de Investigación Operativa . . 10373

Sociedad de Pediatría de Madrid y Castilla La Mancha, Villanueva 11, Madrid 1
Founded: 1913; Members: 750
Pres: A. Nogales; Gen Secr: J.I. Sánchez Díaz
Focus: Pediatrics 10374

Sociedad de Pediatría de Madrid y Región Centro / Paediatrics Society of Madrid and the Central Region, Villanueva 11, Madrid 1
Founded: 1913; Members: 750
Focus: Pediatrics
Periodical
Boletín (weekly) 10375

Sociedad Española de Acústica, Serrano 144, Madrid
Focus: Acoustics 10376

Sociedad Española de Anestesiología y Reanimación (SEDAR), Villanueva 11, Madrid 1
Founded: 1953; Members: 1470
Focus: Anesthetics 10377

Sociedad Española de Bioquímica (SEB), Velazquez 144, Madrid 6
Founded: 1960; Members: 1100
Focus: Biochem 10378

Sociedad Española de Bromatología, c/o Facultad de Farmacia, Ciudad Universitaria, 28040 Madrid
Focus: Food
Periodicals
Anales de Bromatologia (weekly)
Semestral (weekly) 10379

Sociedad Española de Cerámica y Vidrio / Spanish Society of Ceramic and Glass, Carretera de Valencia km 24.300, Arganda del Rey, Madrid
T: (01) 8700550; Fax: (01) 5590575
Founded: 1960; Members: 700
Gen Secr: Dr. José R. Jurado Egea
Focus: Materials Sci
Periodical
Boletín (weekly) 10380

Sociedad Española de Ciencias Fisiológicas, c/o Facultad de Medicina, Ciudad Universitaria, Madrid 3
Focus: Physiology 10381

Sociedad Española de Cirugía Oral y Maxillofacial, Villanueva 11, Madrid 1
Focus: Surgery 10382

Sociedad Española de Cirugía Ortopédica y Traumatología, Biblioteca Ayala 20, Madrid
T: (01) 2769966
Focus: Surgery 10383

Sociedad Española de Cirugía Plástica, Reparadora y Estética, Santa Isabel 51, 28012 Madrid
Members: 466
Focus: Surgery
Periodical
Cirugía Plástica Iberolatinoamericana (weekly)
. . . . 10384

Sociedad Española de Diabetes, Santa Isabel 51, Madrid 12
Founded: 1954; Members: 3470
Focus: Diabetes 10385

Sociedad Española de Estudios Clásicos, Hortaleza 104, 28004 Madrid
Focus: Hist
Periodical
Estudios Clasicos (weekly) 10386

Sociedad Española de Filosofía, Serrano 127, Madrid 6
Focus: Philos 10387

Sociedad Española de Horticultura (S.E.H.) / Spanish Horticultural Society, Arrieta 7, Madrid
T: (01) 5483124; Fax: (01) 5475556
Founded: 1880; Members: 550
Pres: Francisco Lorenzo Baeza; Gen Secr: Alberto Juanco Simon
Focus: Hort
Periodical
S.E.H. (weekly) 10388

Sociedad Española de Mecánica del Suelo y Cimentaciones, Alfonso XII 3, 28014 Madrid
Fax: (01) 5277442
Focus: Civil Eng
Periodical
Boletin de la Sociedad Española de Mecanica del Suelo y Cimentaciones (weekly) 10389

Sociedad Española de Medicina Psicosomática y Psicoterapía, Paseo Bonanova 47, Barcelona 17
Founded: 1955; Members: 195
Focus: Psychiatry; Therapeutics 10390

Sociedad Española de Microbiología (SEM) / Spanish Society for Microbiology, Hortaleza 104, 28004 Madrid
T: (01) 3092322; Tx: (01) 3082511
Founded: 1946; Members: 2000
Pres: Prof. F. Ruiz Berraquero; Gen Secr: Dr. J. Antonio Leal
Focus: Microbio
Periodical
Microbiologia SEM (weekly) 10391

Sociedad Española de Optica (SEDO) / Spanish Optical Society, Serrano 121, 28006 Madrid
T: (01) 5616070; Fax: (01) 5645557
Founded: 1968; Members: 320
Pres: Prof. M.J. Yzvel; Gen Secr: Dr. P. Artal
Focus: Optics 10392

Sociedad Española de Otorinolaringología y Patología Cévico-Facial, Villanueva 11, Madrid 1
T: (01) 4312692
Founded: 1966; Members: 1100
Focus: Otorhinolaryngology
Periodical
Acta Otorinolaringologica Española (weekly)
. . . . 10393

Sociedad Española de Patología Digestiva y de la Nutrición, Almagro 38, Madrid
Founded: 1933; Members: 800
Pres: Dr. H.G. Mogena
Focus: Pathology; Nutrition
Periodical
Revista Española de las Enfermedades del Aparato Digestivo y de la Nutrición . . 10394

Sociedad Española de Psicoanàlisis / Spanish Psychoanalytical Society, Alicante 27, 08022 Barcelona
T: (03) 2125839; Fax: (03) 2125839
Founded: 1959; Members: 55
Focus: Psychoan
Periodical
Revista Catalana de Psicoanàlisi (weekly) 10395

Sociedad Española de Radiología Medica (S.E.R.M.), Goya 38, 28001 Madrid
Founded: 1946; Members: 1209
Pres: Dr. José Manrique
Focus: Radiology; Nucl Med
Periodical
Radiología (weekly) 10396

Sociedad Española de Rehabilitación y Medicina Fisica (S.E.R.), Villanueva 11, 28001 Madrid
Focus: Rehabil
Periodical
Rehabilitación (weekly) 10397

Sociedad General de Autores de España (SGAE) / General Society of Spanish Authors, Fernando VI 4, Apdo 484, 28004 Madrid
Pres: Juan José Alonso Millán; Gen Secr: Javier Moscoso del Prado
Focus: Lit
Periodical
Boletín (weekly) 10398

Sociedad Ibero-Americana de Estudios Numismáticos, c/o Museo de la Fabrica Nacional de Moneda & Timbre, Jorge Juan 106, 28009 Madrid
Focus: Numismatics 10399

Sociedad Luso-Española de Neurocirugía, c/o S. Obrador, Eduardo Dato 23, Madrid 10
Members: 114
Focus: Surgery 10400

Sociedad Oftalmológica Hispano-Americano, c/o Clínica N. S. de la Concepción, Av Reyes Católicos 2, Madrid
Focus: Ophthal 10401

Sociedad Veterinaria de Zootecnica de España / Spanish Veterinary Society of Zootechnics, Isabel la Católica 12, Madrid
T: (01) 2471838
Founded: 1945
Pres: Prof. Dr. C.L. de Cuenca; Gen Secr: Prof. Dr. J.M. Cid Diaz
Focus: Zoology; Animal Husb
Periodical
Zootecnia (weekly) 10402

Societat Arqueològica Lul-liana (S.A.L.), Monti-Sion 9, 07010 Palma de Mallorca
Founded: 1881; Members: 3400
Pres: Maria Barceló Crespí; Gen Secr: Gabriel Ensenyat Pujol
Focus: Archeol
Periodical
Butlleti de la Societat Arqueològica Lul-liana (weekly) 10403

World Federation of Associations of Pediatric Surgeons, c/o Dept de Cirugia Pediatrica, Hospital Universitario Materno-Infantil, Vall d'Hebron 119-129, 08035 Barcelona
T: (03) 4272000 ext 2312; Fax: (03) 4282171
Founded: 1974
Gen Secr: Prof. José Boix-Ochoa
Focus: Surgery 10404

World Veterinary Association (WVA) / Asociación Mundial de Veterinaria, Isabel la Católica 12, 28013 Madrid
T: (01) 2471838; Fax: (01) 2488301
Founded: 1959; Members: 58
Pres: Prof. Dr. J.F. Figueroa; Gen Secr: Prof. Dr. C.L. de Cuenca
Focus: Vet Med
Periodical
Bulletin of the WVA (bi-annually) . . . 10405

Sri Lanka

Asian Packaging Information Centre (APIC), c/o Aitken Spence Printing, 315 Vauxhall St, Colombo 2
Founded: 1974
Pres: Stanley Wickremeratne
Focus: Materials Sci
Periodical
APIC Journal (weekly) 10406

Asian-South Pacific Bureau of Adult Education (ASPBAE), 30/63A Longden Pl, Colombo 7
T: (01) 589844; Tx: 21537; Fax: (01) 580721
Founded: 1964
Gen Secr: Dr. W.M.K. Wijetunga
Focus: Adult Educ
Periodicals
Asian Journal of Surgery (weekly)
ASPBAE Courier Service (3 times annually)
ASPBAE News (3 times annually) . . . 10407

Buddhist Academy of Ceylon, 109 Rosmead Pl, Colombo
Focus: Rel & Theol 10408

Ceylon Gemmologists Association, 61 Abdul Caffoor Mawatha, Colombo 3
Pres: Prof. K. Kularatnam
Focus: Mineralogy 10409

Ceylon Geographical Society, 61 Abdul Caffoor Mawatha, Colombo 3
Founded: 1938; Members: 100
Pres: Prof. K. Kularatnam; Gen Secr: Dr. K.U. Sirinanda
Focus: Geography
Periodical
The Ceylon Geographer (weekly) . . . 10410

Ceylon Humanist Society, Rutnam Institute Bldg, University Lane, Jaffna 7
Pres: J.T. Rutnam; Gen Secr: O.M. de Alwis
Focus: Philos
Periodical
Journal (weekly) 10411

Ceylon Institute of World Affairs, c/o M. de Silva, 82B Ward Pl, Colombo 7
Founded: 1957
Pres: Anton Muttukumaru
Focus: Poli Sci 10412

Ceylon Palaeological Society, 61 Abdul Caffoor Mawatha, Colombo 3
Pres: Prof. K. Kularatnam
Focus: Paleontology 10413

Ceylon Society of Arts, c/o Art Gallery, Ananda Coomarassamy Mawatha, Colombo 7
T: (01) 93067
Founded: 1887
Pres: Dr. A.R. Abeyasinghe; Gen Secr: D. Jayasuriya
Focus: Arts 10414

The Classical Association of Ceylon, 8 Selbourne Rd, Colombo 3
Founded: 1935; Members: 146
Pres: E.C.S. Perera; Gen Secr: N.L. Rajasingham
Focus: Hist 10415

Maha Bodhi Society of Ceylon, 130 Maligakande Rd, Maradana, Colombo 10
Founded: 1891; Members: 12000
Pres: G.N. Jayasuriya; Gen Secr: N. Wijenaike
Focus: Rel & Theol
Periodical
Sinhala Bauddhaya (weekly) 10416

National Education Society of Ceylon, c/o Dept of Humanities Education, University of Colombo, POB 1490, Colombo 3
Focus: Educ 10417

Royal Asiatic Society of Sri Lanka, c/o Mahaveli Centre, 86 Ananda Coomaraswanuy Mawatha, Colombo 7
T: (01) 699249
Founded: 1845
Pres: Dr. C.G. Uragoda; Gen Secr: Dr. L.S. Dewaraja; M.H. Sirisoma; K.D. Pardnavitana
Focus: Ethnology; Hist; Ling; Lit; Arts; Soc Sci
Periodical
Journal (weekly) 10418

Sri Lanka Association for the Advancement of Science, 120/10 Wijerama Mawatha, Colombo 7
T: (01) 91681
Founded: 1944; Members: 2000
Pres: Prof. D. Attygalle; Gen Secr: Dr. W.A. Amarasir
Focus: Sci
Periodicals
Proceedings
Vidya Viyapathi 10419

Sri Lanka Library Association, c/o The Professional Centre, 175/75 Bauddhaloka Mawatha, Colombo 7
T: (01) 589103
Founded: 1960; Members: 113
Pres: W.B. Dorakumbara; Gen Secr: W. Ranasinghe
Focus: Libraries & Bk Sci
Periodical
Ceylon Library Review (weekly) 10420

Sri Lanka Medical Association, 6 Wijerama Mawatha, Colombo 7
Founded: 1885
Focus: Med
Periodical
The Ceylon Medical Journal (weekly) . . 10421

Theosophical Society of Ceylon, 49 Peterson Lane, Colombo 6
Focus: Rel & Theol 10422

Sudan

Agricultural Research Corporation, POB 126, Wad Medani
Founded: 1918; Members: 160
Gen Secr: O.I. Gameel
Focus: Agri 10423

Arab Centre for Agricultural Documentation and Information (ACADI), c/o AOAD, POB 474, Khartoum
T: (011) 78760; Tx: 22554
Founded: 1981
Focus: Agri 10424

Association des Professeurs de Français en Afrique (APFA) / Association of French Teachers in Africa, c/o Dept of French, Faculty of Arts, University of Khartoum, POB 321, Khartoum
T: (011) 78201
Founded: 1981; Members: 20
Pres: Younis Elamin
Focus: Educ; Ling
Periodical
JAFTA Journal (weekly) 10425

Commission for Archaeology, POB 178, Khartoum
Focus: Archeol 10426

National Council for Research, POB 2404, Khartoum
Founded: 1970
Pres: Prof. H.E.W. Habashi; Gen Secr: Prof. M.O. Khidir
Focus: Sci 10427

Philosophical Society, POB 526, Khartoum
Founded: 1946
Focus: Philos
Periodicals
Proceedings of Annual Conferences
Sudan Notes and Records 10428

Sudan Library Association, POB 32, Khartoum North
Focus: Libraries & Bk Sci 10429

Suriname

Centre for Agricultural Research in Suriname, POB 1914, Paramaribo
T: 60244
Founded: 1965
Focus: Agri 10430

Geologisch Mijnbouwkundige Dienst, Kleine Waterstr 2-6, Paramaribo
Focus: Mining; Geology 10431

Stichting Cultureel Centrum Suriname, POB 1241, Paramaribo
Fax: 10555
Founded: 1947
Pres: Dr. R. Venetiaan
Focus: Arts; Hist; Lit 10432

Stichting voor Wetenschappelijk Ondersoek van de Tropen (WOTRO) / Netherlands Foundation for the Advancement of Tropical Research (WOTRO), POB 1914, Paramaribo
Focus: Sci 10433

Swaziland

Swaziland Art Society, POB 812, Mbabane
Founded: 1970; Members: 60
Pres: F. Berrangé
Focus: Arts 10434

Swaziland Library Association, POB 2309, Mbabane
T: 42633; Tx: 2270
Founded: 1984
Pres: B.J.K. Kingsley; Gen Secr: M. Gyinah
Focus: Libraries & Bk Sci 10435

Sweden

Asociación Nórdica de Intercambio en Educación Popular con América Latina (FONOLA-SOL), Box 1001, 442 25 Kungälv
T: (0303) 11485, 14192
Focus: Educ 10436

Association of Nordic Paper Historians, St Eriksgt 130, 113 43 Stockholm
T: (08) 338369
Founded: 1968
Pres: Jan Olof Ruden
Focus: Hist 10437

Baltic Marine Biologists (BMB), c/o Institute of Marine Research, Box 4, 453 21 Lysekil
T: (0523) 14180; Fax: (0523) 13977
Founded: 1968; Members: 400
Gen Secr: Dr. Bernt I. Dybern
Focus: Bio; Oceanography
Periodicals
The Baltic Marine Biologists Publications (weekly)
Proceedings of the Baltic Marine Biological Symposia (bi-annually) 10438

Bergianska Stiftelsen / Bergius Foundation, Box 50017, 104 05 Stockholm
Founded: 1976
Focus: Hort; Botany 10439

Brandförsvarsföreningen (SBF), 115 87 Stockholm
T: (08) 7837000
Founded: 1919; Members: 2200
Focus: Safety
Periodical
Brandförsvar (10 times annually) 10440

Cancerfonden – Riksföreningen mot Cancer / Swedish Cancer Society, Box 17096, 104 62 Stockholm
T: (08) 7722800; Fax: (08) 7202208
Founded: 1951
Pres: Arne Ljungquist; Gen Secr: Bo S. Oscarsson
Focus: Cell Biol & Cancer Res
Periodicals
Forskning Nu (weekly)
Rädda Livet (weekly) 10441

Carl Johans Förbundet, Box 135, 751 04 Uppsala
T: (018) 152170
Founded: 1848; Members: 950
Focus: Hist 10442

Centre for the Study of International Relations, Döbelnsgt 81, Box 19112, 104 32 Stockholm
Tx: 12453; Fax: (08) 6120592
Founded: 1971
Pres: Cläes Palme; Gen Secr: Prof. Jacob W. F. Sundberg; Prof. Dr. Lars Hjerner; Prof. Dr. Richard K. T. Hsieh
Focus: Int'l Relat
Periodical
Review 10443

Chemical Societies of the Nordic Countries, c/o Svenska Kemistsamfundet, Wallingt 26b, 111 24 Stockholm
T: (08) 115260
Focus: Chem 10444

Coalition Clean Baltic (CCB), c/o SNF, Box 4510, 102 65 Stockholm
Founded: 1990; Members: 14
Gen Secr: Bertil Hagerhall
Focus: Ecology 10445

Committee on Family Research (CFR), c/o Sociologiska Institutionen, Uppsala Universitet, Thunbergsv 3, Box 821, 751 08 Uppsala
Founded: 1950; Members: 210
Pres: Dr. Jan Trost; Gen Secr: Dr. Barbara James
Focus: Sociology
Periodical
CFR-Gazette (3 times annually) 10446

Conference of the Baltic Oceanographers, c/o Hans Dahlin, Swedish Meteorological and Hydrological Institute, 601 76 Norrköping
Focus: Oceanography 10447

Council of Nordic Teachers' Associations, c/o Sveriges Lärarförbund, Box 12229, 102 26 Stockholm
T: (08) 7376500; Tx: 14283; Fax: (08) 569415
Founded: 1968
Gen Secr: Jan Krylbom
Focus: Educ 10448

Dag Hammarskjöld Foundation, Dag Hammarskjöld Centre, Övre Slottsgt 2, 753 10 Uppsala
T: (018) 128872; Tx: 76234; Fax: (018) 122072
Founded: 1962
Gen Secr: Sven Hamrell; Olle Nordberg
Focus: Develop Areas
Periodical
Development Dialogue (weekly) 10449

Esperantist Ornithologists' Association, Telegrafgt 5, 149 00 Nynäshamn
Founded: 1961
Focus: Ornithology 10450

European Association of Labour Economists (EALE), c/o Swedish Institute for Social Research, University of Stockholm, 106 91 Stockholm
T: (08) 8163448; Fax: (08) 8154670
Founded: 1989
Gen Secr: Prof. Eskil Wadensjö
Focus: Soc Sci
Periodical
EALE Newsletter 10451

European Association of Perinatal Medicine, c/o Unit of Pediatric Physiology, University Hospital, 751 85 Uppsala
T: (018) 167515
Founded: 1968
Gen Secr: Dr. L.E. Bratteby
Focus: Pediatrics 10452

European Drosophila Centre, c/o Dept of Genetics, Umea University, 901 87 Umea
T: (090) 165000
Members: 10
Focus: Genetics 10453

European Ecological Federation, c/o University of Lund, Ecology Bldg, 223 62 Lund
T: (046) 148188
Gen Secr: Dr. P.H. Enckell
Focus: Ecology 10454

European Federation of Productivity Services (EFPS), c/o Sveriges Rationaliseringsförbund, Box 4324, 102 67 Stockholm
T: (08) 249225; Fax: (08) 6433061
Founded: 1961
Focus: Business Admin 10455

European Incoherent Scatter Scientific Association (EISCAT), Box 812, 981 28 Kiruna
T: (0980) 79161; Tx: 8778
Founded: 1975
Focus: Sci
Periodical
Annual Report (weekly) 10456

Flygtekniska Föreningen (FTF) / The Swedish Society of Aeronautics and Astronautics, c/o Swedish Space Corporation, Box 4107, 171 04 Stockholm
T: (08) 6276298; Fax: (08) 987069
Founded: 1933; Members: 1867
Focus: Aero 10457

Förbundet Sveriges Arbetsterapeuter (FSA) / Swedish Association of Occupational Therapists, Box 760, 131 24 Nacka
T: (08) 7162850; Fax: (08) 7185392
Founded: 1979; Members: 6600
Pres: Inga-Britt Lindström
Focus: Therapeutics
Periodical
Arbetsterapeuten (16 times annually) . . 10458

Föreningen för Vattenhygien (FVH), Box 519, 162 15 Vällingby 5
T: 228580, 316097
Founded: 1944; Members: 1200
Focus: Hygiene
Periodical
Vatten (weekly) 10459

Föreningen Svenska Tonsättare (F.S.T.) / Society of Swedish Composers, Sandhamsgt 79, Box 27327, 102 54 Stockholm
T: (08) 7838842; Tx: 15591
Founded: 1918; Members: 133
Pres: Sten Hanson
Focus: Music 10460

Företagsekonomiska Föreningen, c/o Gunnar Bergström, Dalagt 36, 113 24 Stockholm
T: (08) 313192
Founded: 1936
Focus: Business Admin 10461

Forskningsrådsnämnden (FRN) / Swedish Council for Planning and Coordination of Research, Sveavägen 166, Box 6710, 113 85 Stockholm
T: (08) 6100600; Fax: (08) 6100633
Founded: 1977
Focus: Sci
Periodical
Källa 10462

Fylkingen / Society of Contemporary Music and Intermedia Art, Münchenbryggeriet, Söder Mälarstrand 27, 102 66 Stockholm
T: (08) 845443; Fax: (08) 6693868
Founded: 1933; Members: 130
Pres: Tankred Kent; Gen Secr: Erik Mikael Karlsson
Focus: Music
Periodical
Hz: The Fylkingen Bulletin (weekly) . . 10463

Geografilärarnas Riksförening, c/o Gegrafiska Institutionen, Sölvegt 13, 223 62 Lund
T: (046) 108414; Fax: (046) 108401
Founded: 1933; Members: 1150
Focus: Geography; Educ
Periodical
Geografiska Notiser (weekly) 10464

Geografiska Förbundet (GF), c/o Dept of Physical Geography, 106 91 Stockholm
T: (08) 340860
Members: 360
Focus: Geography 10465

Geologiska Föreningen / Geological Society, c/o SGU, Box 670, 751 28 Uppsala
Founded: 1871; Members: 550
Pres: David G. Gee; Gen Secr: Monica Beckholmen
Focus: Geology
Periodical
Geologiska Föreningens i Stockholm Förhandlingar: GFF (weekly) 10466

Göteborgs Läkarförening, Storgt 35, 411 38 Göteborg
T: (031) 119654
Focus: Med 10467

Humanistisk-Samhällsvetenskapliga Forskningsradet (HSFR) / Swedish Council for Research in the Humanities and Social Sciences, Box 6712, 113 85 Stockholm
T: (08) 6100660; Fax: (08) 6100680
Founded: 1977; Members: 11
Pres: Dr. Birgit Antonsson; Gen Secr: Prof. Bo Särlvik
Focus: Humanities; Soc Sci
Periodicals
Brytpunkt
Tvärsnitt (weekly) 10468

Ingenjörsvetenskapsakademien (IVA), Grev Turegt 14, Box 5073, 102 42 Stockholm
T: (08) 7912900; Fax: (08) 6115623
Founded: 1919; Members: 733
Pres: Hans G. Forsberg; Gen Secr: Bengt Thulin
Focus: Eng
Periodicals
IVA-Newsletter (weekly)
IVA-Nytt (4-5 times annually) 10469

International Association for the Evaluation of Educational Achievement (IEA), c/o Institutionen för Internationell Pedagogik, Universitet, 106 91 Stockholm
T: 154000
Founded: 1959; Members: 36
Pres: Alan C. Purves
Focus: Educ
Periodical
IEA Newsletter (weekly) 10470

International Association of Medical Laboratory Technologists (IAMLT), Östermalmsg 19, 114 26 Stockholm
T: (08) 103031; Fax: (08) 109061
Founded: 1954; Members: 200000
Pres: Ulla-Britt Lindholm; Gen Secr: Margareta Haag
Focus: Med
Periodical
Med Tel International (weekly) 10471

International Association of Music Libraries, Archives and Documentation Centres, c/o Swedish Music History Archive, Nybrokajen 13, Box 16326, 103 26 Stockholm
T: (08) 6664562; Fax: (08) 6664565
Founded: 1951; Members: 1900
Focus: Libraries & Bk Sci; Music; Archives; Doc
Periodical
Fontes Artis Musicae (weekly) 10472

International Association of Sound Archives (IASA), c/o ALB, Box 27890, 115 93 Stockholm
Members: 370
Pres: James McCarty; Gen Secr: Sven Allerstrand
Focus: Archives
Periodical
IASA Journal 10473

International Association on Mechanization of Field Experiments (IAMFE), Box 7033, 750 07 Uppsala
T: (018) 671000; Fax: (018) 673529
Founded: 1964
Pres: Egil Øyjord; Gen Secr: Kenneth Alness
Focus: Agri
Periodical
Proceeding of IAMFE Conferences . . . 10474

International Law Association, Swedish Branch, Box 16050, 103 22 Stockholm
T: (08) 231200; Tx: 15638
Founded: 1922
Pres: Lars Hjerner; Gen Secr: Ulf Franke
Focus: Law; Int'l Relat 10475

International Research Group on Wood Preservation (IRG), c/o IRG Secretariat, Box 5607, 114 86 Stockholm
T: (08) 101453; Fax: (08) 108081
Founded: 1969; Members: 300
Pres: Dr. Anthony F. Bravery; Gen Secr: Jörán Jermer
Focus: Forestry
Periodicals
Annual Report (weekly)
IRG Documents (more than 100 times annually) 10476

International Society for Fat Research (ISF), c/o AB Karlshamns Oljefabriker, 292 00 Karlshamn
T: (0454) 82000
Founded: 1954; Members: 1500
Gen Secr: Dr. Ragnar Ohlson
Focus: Food 10477

Karolinska Förbundet, c/o Ulla Johanson, Kungsgt 77, 112 27 Stockholm
Founded: 1910; Members: 1000
Focus: Hist
Periodical
Karolinska Förbundets årsbok (weekly) . 10478

Kartografiska Sällskapet, c/o Lantmäteriverket, 801 82 Gävle
T: (026) 153000
Founded: 1908; Members: 1300
Focus: Cart
Periodical
Kartbladet (weekly) 10479

Kommittén för Petrokemisk Forskning och Utveckling, Box 5073, 161 05 Stockholm
Focus: Petrochem 10480

Konsthistoriska Sällskapet, c/o Institutet för Konstvetenskap, Universitet, Box, 106 91 Stockholm 50
T: (08) 150160
Founded: 1914; Members: 550
Focus: Hist; Arts 10481

Kungliga Akademien för de fria Konsterna, Fredsgt 12, 103 26 Stockholm
T: (08) 232945; Fax: (08) 7905924
Founded: 1735; Members: 115
Pres: Henry Montgomery; Gen Secr: Bo Sylvan
Focus: Arts
Periodical
Exhibition Catalogues (weekly) 10482

Kungliga Fysiografiska Sällskapet i Lund, Stortorget 6, 222 23 Lund
T: (046) 132528; Tx: 33533; Fax: (046) 104439
Founded: 1772; Members: 385
Pres: Prof. Nils Johnson; Gen Secr: Prof. Sten Ahrland
Focus: Sci; Med; Eng
Periodical
Arsbok (weekly) 10483

Kungliga Gustav Adolfs Akademien (KGAA), Klostergt 2, 753 21 Uppsala
T: (018) 548783
Founded: 1932; Members: 205
Pres: Dr. Sven G. Svenson
Focus: Educ
Periodicals
Ethnologia Scandinavica
Namn och Bygd (weekly) 10484

Kungliga Humanistiska Vetenskaps-Samfundet i Uppsala / Royal Society of the Humanities at Uppsala, Ihres väg 9, 752 63 Uppsala
Founded: 1889
Pres: Prof. Anita Jacobson-Widding; Gen Secr: Prof. J.-O. Tjäder
Focus: Humanities
Periodicals
Arsbok
Skrifter: Acta
Yearbook 10485

Kungliga Krigsvetenskapsakademien / Royal Swedish Academy of Military Sciences, 107 87 Stockholm
Founded: 1796; Members: 360
Pres: Carl-Olof Ternryd; Gen Secr: Helge Gard
Focus: Military Sci
Periodical
Handlingar och Tidskrift (weekly) 10486

Kungliga Musikaliska Akademien (KMA), Blasieholmstorg 8, 111 48 Stockholm
T: (08) 6115720; Fax: (08) 6118718
Founded: 1771; Members: 218
Pres: Anders R. Öhman; Gen Secr: Bengt Holmstrand
Focus: Music
Periodicals
Årsskrift (weekly)
Musica Sveciae 10487

Kungliga Samfundet för Utgivande av Handskrifter rörande Skandinaviens Historia, c/o National Archives, Box 12541, 102 29 Stockholm
Founded: 1821; Members: 100
Focus: Hist
Periodical
Kunglig Samfundets Handlinger (weekly) . 10488

Kungliga Skogs- och Lantbruksakademien (KSLA), Box 6806, 113 86 Stockholm
Founded: 1811; Members: 340
Focus: Forestry; Agri
Periodicals
Acta Agriculturae Scandinavica (weekly)
Kungl. Skogs- och Lantbruksakademiens Tidskrift (weekly)
Scandinavian Journal of Forest Research (weekly) 10489

Kungliga Vetenskapsakademien (KVA) / The Royal Swedish Academy of Sciences, Lilla Frescativ 4, Box 50005, 104 05 Stockholm
T: (08) 6739500; Tx: 17073; Fax: (08) 155670
Founded: 1739; Members: 385
Pres: Prof. Torvard C. Laurent; Gen Secr: Prof. Carl-Olof Jacobson
Focus: Sci
Periodicals
Acta Mathematica (weekly)
Acta Zoologica (weekly)
Ambio (8 times annually)
Arkiv för Matematik (weekly)
Chemica Scripta (weekly)
Physica Scripta (weekly)
Zoologica Scripta (weekly) 10490

Kungliga Vetenskaps- och Vitterhets-Samhället i Göteborg / Royal Society of Arts and Science of Göteborg, Box 5096, 402 22 Göteborg
Founded: 1778; Members: 207
Gen Secr: Prof. Jan Hult
Focus: Arts; Sci
Periodicals
Årsbok (weekly)
Botanica (weekly)
Geophysica (weekly)
Humaniora (weekly)
Interdisciplinaria (weekly)
Minnestal: Obituaries (weekly)
Zoologica (weekly) 10491

Kungliga Vetenskaps-Societeten i Uppsala, Larsgt 1, 753 10 Uppsala
T: (018) 131270
Founded: 1710; Members: 230
Pres: Prof. M. Carlsson; Gen Secr: Prof. Lars-Olof Sundelöf
Focus: Sci
Periodicals
Arsbok (weekly)
Nova Acta 10492

Kungliga Vitterhets Historie och Antikvitets Akademien (KVHAA), Villagt 3, Box 5622, 114 86 Stockholm
T: (08) 200936; Fax: (08) 200981
Founded: 1753; Members: 150
Pres: Prof. S. Strömholm; Gen Secr: Prof. S. Helmfrid
Focus: Archeol; Hist; Lit
Periodical
Fornvännen (weekly) 10493

Lärarförbundet, Segelbåtsvägen 15, Box 12229, 102 26 Stockholm
T: (08) 7376500; Fax: (08) 6569415
Founded: 1991; Members: 195000
Focus: Educ 10494

Lärarnas Riksförbund (LR), Sveavägen 50, Box 3529, 103 69 Stockholm
T: (08) 6132700; Fax: (08) 219136
Founded: 1884; Members: 55000
Pres: Ove Engman; Gen Secr: Gunnar Häggström
Focus: Educ
Periodical
Skolvärlden (32 times annually) 10495

Legitimerade Sjukgymnasters Riksförbund, Wallingt 5, Box 3196, 103 63 Stockholm
T: (08) 241490; Fax: (08) 217931
Founded: 1943; Members: 8200
Focus: Therapeutics
Periodical
Sjukgymnasten (weekly) 10496

Litteraturfrämjandet, Bellmansgt 30, 116 47 Stockholm
T: (08) 449175; Fax: (08) 701646
Focus: Lit
Periodical
En bok för alla 10497

Lunds Botaniska Förening (LBF) / Botanical Society of Lund, c/o The Botanical Museum, Ö. Vallgt 18, 223 61 Lund
Fax: (046) 104234
Founded: 1858; Members: 670
Focus: Botany
Periodical
Lungs Botaniska Förening Medlemsblad (1-2 times annually) 10498

Lunds Matematiska Sällskap, c/o Matematiska Institutionen, Box 118, 221 07 Lund
Founded: 1923; Members: 150
Focus: Math 10499

Matematiska Föreningen, c/o Dept of Mathematics, Uppsala University, Lägerhyddsvägen 2, Box 480, 751 06 Uppsala
T: (018) 183200; Fax: (018) 183201
Founded: 1853; Members: 50
Pres: Inger Sigstam; Gen Secr: Mats E. Andersson
Focus: Math 10500

Medicinska Forskningsrådet / Medical Research Council, Box 6713, 113 85 Stockholm
Fax: (08) 6100777
Founded: 1945
Pres: Gunnar Brodin; Gen Secr: Tore Scherstén
Focus: Med 10501

Musikaliska Konstföreningen / Musical Art Association, Blasieholmstorg 8, 111 48 Stockholm
T: (08) 116920
Founded: 1859; Members: 150
Pres: Ingmar Milveden; Gen Secr: Gösta Percy
Focus: Music 10502

Nationalekonomiska Föreningen / Swedish Economic Association, c/o Swedish Institute for Social Research, 106 91 Stockholm
T: (08) 162000
Founded: 1877; Members: 1200
Pres: Prof. Eskil Wadensjö; Gen Secr: Dr. Lena Schröder
Focus: Econ
Periodical
Ekonomisk Debatt (8 times annually) . . 10503

Naturvetenskapliga Forskningsrådet / Swedish Natural Science Research Council, Box 6711, 113 85 Stockholm
T: (08) 6100700; Tx: 13599
Founded: 1977
Pres: Kerstin Niblaeus; Gen Secr: Prof. Gunnar Öquist
Focus: Nat Sci
Periodicals
Annual Report (weekly)
Yearbook 10504

Nobelstiftelsen / Nobel Foundation, Nobel House, Sturegt 14, Box 5232, 102 45 Stockholm
T: (08) 6630920; Tx: 12382; Fax: (08) 6603847
Founded: 1900
Pres: Prof. Bengt Samuelsson; Gen Secr: Michael Sohlman
Focus: Sci 10505

Nordic Association of Applied Geophysics, University of Luleå, 951 87 Luleå
T: (0920) 91000; Tx: 80447
Founded: 1964; Members: 400
Gen Secr: Gustav Lindkvist
Focus: Geophys 10506

Nordiska Afrikainstitutet / Scandinavian Institute of African Studies, Box 1703, 751 47 Uppsala
T: (018) 155480; Fax: (018) 695629
Founded: 1962
Focus: Develop Areas
Periodicals
Africana
Annual Newsletter (weekly)
Research Report
Seminar Proceedings 10507

Nordiska Institutet för Samhällsplanering / Nordic Institute for Studies in Urban and Regional Planning, Box 1658, 111 86 Stockholm
Fax: (08) 6115105
Founded: 1968
Pres: Niels Østergård
Focus: Urban Plan
Periodicals
Arsrapport (weekly)
Grupparbeten
Rapporter 10508

Nordiska Samarbetskommittén för Internationell Politik, inklusive Konflikt- och Fredsforskning, Box 1253, 111 82 Stockholm
T: (08) 234060
Founded: 1966; Members: 5
Focus: Poli Sci; Military Sci
Periodicals
Cooperation and Conflict: Nordic Journal of International Politics
Newsletter: International Studies in the Nordic Countries 10509

Nordiska Samfundet för Latinamerika-forskning (NOSALF) / Scandinavian Association for Research on Latin America, c/o Latinamerika-Institutet, 106 91 Stockholm
T: (08) 162884
Focus: Int'l Relat
Periodicals
Nordic Journal of Latin American Studies
Studies on Latin America 10510

Nordisk Förening för Tillämpad Geofysik, c/o Geological Survey of Sweden, 104 05 Stockholm 50
Founded: 1964; Members: 174
Focus: Geophys 10511

Nordisk Genetikerforening, c/o Wallenberg Laboratory, Universitet, 104 05 Stockholm 50
Founded: 1960; Members: 282
Focus: Genetics 10512

Oikos, Ecology Bldg, University of Lund, 223 62 Lund
Fax: (046) 104716
Focus: Ecology
Periodicals
Ecography (weekly)
Oikos (9 times annually) 10513

Pennklubben, c/o Bonniers, Box 3159, 103 63 Stockholm
Founded: 1922; Members: 400
Pres: Agneta Plejel; Gen Secr: Hans Isaksson
Focus: Lit 10514

Riksföreningen för Lärarna i Moderna Språk (LMS), Box 41, 425 02 Hisings Kärra
T: (031) 571640; Fax: (031) 572643
Founded: 1938; Members: 5200
Focus: Ling; Educ
Periodicals
LMS-Lingua (weekly)
Moderna Språk (weekly) 10515

SACO Sveriges Akademikers Centralorganisation, Lilla Nygt 14, Box 2206, 103 15 Stockholm
T: (08) 225200; Fax: (08) 247701
Founded: 1975; Members: 355000
Pres: Jörgen Ullenhag
Focus: Sci
Periodical
SACO Tidningen Akademiker (weekly) . . 10516

Samfundet De Nio, c/o Anders R. Öhman, Smalandsgt 20, 111 46 Stockholm
Founded: 1913; Members: 9
Pres: I. Jonsson; Gen Secr: Anders R. Öhman
Focus: Lit 10517

Scandinavian Forum for Lipid Research and Technology, c/o SIK, Box 27022, 400 23 Göteborg
T: (031) 400120
Founded: 1969; Members: 400
Focus: Cell Biol & Cancer Res 10518

Scandinavian Orthopaedic Association, c/o Ortopedkliniken, Karolinska Sjukhuset, 104 01 Stockholm
Focus: Orthopedics 10519

Skogs- och Jordbrukets Forskningsråd (SJFR) / Council for Forestry and Agricultural Research, Box 6488, 113 82 Stockholm
T: (08) 7360910; Tx: 13433; Fax: (08) 332915
Pres: Hans Ekelund; Gen Secr: Prof. Jan-Erik Hällgren
Focus: Forestry; Agri
Periodical
SJFR Eltgottråd (5 times annually) . . . 10520

Skolledarförbundet (SLF), Vasagt 48, Box 3266, 103 65 Stockholm
T: (08) 6969780; Fax: (08) 249844
Founded: 1966; Members: 5700
Focus: Educ
Periodical
Skolledaren (10 times annually) 10521

Standardiseringskommissionen i Sverige (SIS) / Swedish Standards Institution, Tegnergt 11, Box 3295, 103 66 Stockholm
T: (08) 6135200; Tx: 17453; CA: standardis stockholm; Fax: (08) 4117035
Founded: 1922; Members: 24
Pres: Gunnar Engman; Gen Secr: Lars Wallin
Focus: Standards
Periodicals
Månadens Standard (weekly)
Nyhetsbrev Europa Standardisering (weekly)
Nytt om Niotusen (weekly)
Teknik & Standard (weekly) 10522

Statens Kulturråd / Swedish Council for Cultural Affairs, Box 7843, 103 98 Stockholm
T: (08) 6797260; Fax: (08) 6111349
Founded: 1974
Pres: Lars Bergquist; Gen Secr: Göran Löfdahl
Focus: Cultur Hist
Periodical
Kulturrådet 10523

Statens Råd för Byggnadsforskning (BFR) / Swedish Council for Building Research, Sankt Göransgt 66, 112 98 Stockholm
T: (08) 6177300; Fax: (08) 6537462
Founded: 1960
Focus: Archit; Energy
Periodicals
Documents: Research Reports in English (weekly)
Newsletter (3 times annually)
Rapporter: Research Reports in Swedish (150 times annually)
Synopses (9 times annually) 10524

Statistika Föreningen / Statistical Society, c/o Statistika Centralbyran (Statistics Sweden), Statsistika Centralbyrån, 115 81 Stockholm
Founded: 1901; Members: 600
Pres: Birgitta Hedman; Gen Secr: Anders Rickardsson
Focus: Stats 10525

Statsvetenskapliga Förbundet / Swedish Political Science Association, c/o Statsvetenskapliga Institutionen, Stockholms Universitet, Stockholm
T: (08) 162000
Founded: 1971; Members: 300
Focus: Poli Sci
Periodical
Politologen (weekly) 10526

Strindbergssällskapet / The Strindberg Society, c/o C.O. Johansson, Bomullsvägen 14, 196 38 Kungsängen
T: (08) 58171746; Fax: (08) 110141
Founded: 1945; Members: 850
Focus: Lit
Periodical
Medelanden fra från Strindbergssällskapet (weekly) 10527

Studieförbundet Närnigsliv och Samhälle (SNS), Sköldungagt 2, 114 27 Stockholm
T: (08) 232520
Founded: 1948
Focus: Econ 10528

Styrelsen för Teknisk Utveckling / Swedish National Board for Technical Development, Liljeholmsvägen 32, Box 43200, 100 72 Stockholm
T: (08) 7754000; Tx: 10840
Focus: Eng

Periodicals
Energiteknik (weekly)
Energy Technology (5 times annually)
New Swedish Technology (weekly)
Technik i Tiden (weekly) *10529*

Svenska Akademien, Box 2118, 103 13 Stockholm
T: (08) 106524; Fax: (08) 244225
Founded: 1786; Members: 18
Focus: Ling; Lit
Periodical
Svenska Akademiens Handlingar (weekly) *10530*

Svenska Akademiska Rektorskonferensen, c/o Uppsala University, Box 256, 751 05 Uppsala
T: (018) 182500; Tx: 76024; Fax: (018) 111853
Founded: 1966; Members: 24
Pres: Prof. Stig Strömholm; Gen Secr: Lennart Ståhle
Focus: Educ *10531*

Svenska Aktuarieföreningen / Swedish Insurance Federation, Box 1436, 111 84 Stockholm
T: (08) 7837150
Founded: 1904; Members: 227
Pres: Alf Guldberg; Gen Secr: Arne Sandström
Focus: Insurance
Periodical
Scandinavian Actuarial Journal (weekly) . *10532*

Svenska Akustiska Sällskapet (SAS) / Acoustical Society of Sweden, Box 276, 401 24 Göteborg
Founded: 1945; Members: 275
Gen Secr: L. Landström
Focus: Acoustics *10533*

Svenska Arkeologiska Samfundet / Swedish Archaeological Society, Box 5405, 114 84 Stockholm
T: (08) 7839000
Founded: 1947
Focus: Archeol
Periodical
Swedish Archaeology (every 5 years) . . *10534*

Svenska Arkitekters Riksförbund (SAR) / National Association of Swedish Architects, Norrlandsgt 18, 111 43 Stockholm
T: (08) 240230
Founded: 1936; Members: 3600
Pres: Ulf Gillberg; Gen Secr: M. Beckman
Focus: Archit
Periodical
Arkitekttidningen (16 times annually) . . *10535*

Svenska Arkivsamfundet / The Swedish Archival Association, c/o Riksarkivet, Fyrverkarbacken 13-17, Box 12541, 10229 Stockholm
T: (08) 7376350; Fax: (08) 7376474
Founded: 1952; Members: 650
Pres: Erik Norberg; Gen Secr: Ulf Söderberg
Focus: Archives
Periodical
Arkiv, samhälle och forskning: Svenska arkiv samfundets skriftserie (3 times annually) . *10536*

Svenska Astronomiska Sällskapet, c/o Stockholms Observatorium, 133 36 Saltsjöbaden
T: 164463; Fax: 7174719
Founded: 1919; Members: 1250
Focus: Astronomy
Periodical
Astronomisk Tidskrift (weekly) *10537*

Svenska Bergmannaföreningen (SBF), Box 5283, 102 46 Stockholm
T: (08) 6639905
Founded: 1941; Members: 1800
Focus: Mining
Periodical
Bergsmannen med JKA (weekly) *10538*

Svenska Bibliotekariesamfundet (SBS) / Swedish Association of University and Research Librarians, c/o Staffan Lööf, University of Borås, Box 874, 501 15 Borås
T: (033) 164000
Founded: 1921; Members: 1100
Pres: Tomas Lidman
Focus: Libraries & Bk Sci
Periodical
Bibliotekariesamfundet meddelar (weekly) . *10539*

Svenska Botaniska Föreningen, c/o Naturhistoriska Riksmuseet, Box 50007, 104 05 Stockholm
Founded: 1907; Members: 2800
Focus: Botany
Periodical
Svensk Botanisk Tidskrift (weekly) . . . *10540*

Svenska Dermatologiska Sällskapet (SDS) / Swedish Society of Dermatology and Venerology, c/o Svenska Läkaresällskapet, Klara Östra Kyrkogt 10, 101 27 Stockholm
T: (08) 243350
Members: 350
Focus: Derm; Venereology *10541*

Svenska Färgeritekniska Riksförbundet (SFR), c/o Arne Wärnegard, Valasgt 16, 412 74 Göteborg
Founded: 1924; Members: 363
Focus: Eng *10542*

Svenska Föreningen för Ljuskultur, Box 5512, 114 85 Stockholm
T: (08) 675834
Founded: 1926; Members: 39
Focus: Electric Eng *10543*

Svenska Föreningen för Medicinsk Teknik och Fysik / Swedish Society for Medical Engineering and Medical Physics, c/o Dept of Biomedical Engineering, University Hospital, 901 85 Umea
Fax: (090) 136717
Founded: 1956; Members: 932
Gen Secr: Heikki Teriö
Focus: Med; Physiology; Eng *10544*

Svenska Föreningen för Mikrobiologi / Swedish Society for Microbiology, c/o Dept of Microbiological Ecology, Lund University, Helgonavägen 5, 223 62 Lund
T: (046) 109614; Fax: (046) 104158
Founded: 1964; Members: 400
Pres: Dr. Bengt Söderström; Gen Secr: Dr. Hans-Börje Jansson
Focus: Microbio *10545*

Svenska Försäkringsföreningen / Swedish Insurance Society, Slöjdgt 9, 111 57 Stockholm
T: (08) 242860; Fax: (08) 241320
Founded: 1875; Members: 4000
Pres: Erland Strömbäck; Gen Secr: Anders Kleverman
Focus: Insurance
Periodicals
Scandinavian Insurance Quarterly (weekly)
Swedish Insurance Yearbook (weekly) . . *10546*

Svenska Folkbibliotekarie Förbundet (SFF) / Swedish Association of Public-Library Employees, c/o DIK-förbundet, Box 760, 131 24 Nacka
T: (08) 7162880; Fax: (08) 7165291
Founded: 1938; Members: 3200
Pres: Christina Persson
Focus: Libraries & Bk Sci
Periodical
DIK-forum (20 times annually) *10547*

Svenska Fornminnesföreningen, c/o Statens Historika Museum, Box 5405, 114 84 Stockholm
T: (08) 228900
Founded: 1871; Members: 350
Focus: Archeol; Archit *10548*

Svenska Fornskriftsällskapet (SFS), c/o Kungliga Biblioteket, Box 5039, 102 41 Stockholm
T: (08) 241040
Founded: 1843; Members: 230
Focus: Lit *10549*

Svenska Fysikersamfundet / Swedish Physical Society, c/o Svedberg Laboratory, Uppsala University, Box 533, 751 21 Uppsala
T: (018) 183112; Tx: 76088; Fax: (018) 183833
Founded: 1920; Members: 1000
Pres: Prof. Gunnar Tibell; Gen Secr: Dr. Curt Ekström
Focus: Physics
Periodicals
Fysik-Aktuellt (weekly)
Kosmos (weekly) *10550*

Svenska Geofysiska Föreningen (SGF), c/o Swedish Meteorological and Hydrological Institute, 601 76 Norrköping
T: (011) 108000
Founded: 1920; Members: 210
Pres: Sture Wickerts; Gen Secr: Lars E. Olsson
Focus: Geophys
Periodical
Tellus *10551*

Svenska Geotekniska Föreningen (SGF) / Swedish Geotechnical Society, 581 93 Linköping
T: (013) 115100; Tx: 50125; Fax: (013) 131696
Founded: 1950; Members: 700
Pres: Eskil Sellgren; Gen Secr: Bengt Rydell
Focus: Geology; Eng
Periodical
SGF Rapport (5 times annually) *10552*

Svenska Gymnastikläraresällskapet (SGS), Bjönrstigen 13, 162 40 Vällingby
T: (08) 899701
Founded: 1884; Members: 4000
Focus: Sports
Periodical
Tidskrift i Gymnastik & Idrott (10 times annually) *10553*

Svenska Historiska Föreningen, Box 5405, 114 84 Stockholm
T: (08) 7832502; Fax: (08) 7832515
Founded: 1881; Members: 2200
Pres: Carl Göran Andrae; Gen Secr: Lars Magnusson
Focus: Hist
Periodical
Historisk Tidskrift: Svensk Historisk Bibliografi (weekly) *10554*

Svenska Kemistsamfundet, Wallingt 24, 111 24 Stockholm
T: (08) 106678
Founded: 1883; Members: 4300
Focus: Chem
Periodicals
Acta Chemica Scandinavica (10 times annually)
Kemisk Tidskrift (weekly) *10555*

Svenska Klassikerförbundet, c/o Institutionen för Klassiska Språk, Stockholms Universitet, Universitetsvägen 10E, 106 91 Stockholm
T: (08) 163465; Fax: (08) 164307
Founded: 1935; Members: 490
Pres: Gunhild Vidén; Gen Secr: Anders Ohlsson
Focus: Hist; Ling *10556*

Svenska Kriminalistföreningen, c/o Ministry of Justice, Box, 103 30 Stockholm
T: (08) 7634625
Founded: 1911; Members: 450
Focus: Criminology
Periodical
Nordisk Tidskrift for Kriminalvidenskab (weekly) *10557*

Svenska Kyltekniska Föreningen / The Swedish Society of Refrigeration, Box 4113, 175 04 Järfälla
T: (08) 58026135; Fax: (08) 58081793
Founded: 1942; Members: 1300
Pres: H. Olson; Gen Secr: Urban Flyckt
Focus: Eng *10558*

Svenska Kyrkohistoriska Föreningen, c/o Teologiska Institutionen, Box 1604, 751 46 Uppsala
Founded: 1900; Members: 337
Focus: Hist; Rel & Theol *10559*

Svenska Läkaresällskapet (SLS), Klara Östra Kyrkogt 10, Box 738, 101 35 Stockholm
T: (08) 243350
Founded: 1807; Members: 13000
Pres: Prof. Kerstin Hagenfeldt
Focus: Med
Periodical
Svenska Läkaresällskapets Handlingar: Hygiea (weekly) *10560*

Svenska Linné-Sällskapet / Swedish Linnaeus Society, Box 1530, 751 45 Uppsala
T: (018) 136540
Founded: 1917; Members: 600
Focus: Nat Sci; Sci
Periodical
Yearbook of the Swedish Linnaeus Society (bi-annually) *10561*

Svenska Litteratursällskapet, c/o Litteraturvetenskapliga Institutionen, Box 1909, 751 49 Uppsala
T: (018) 182943
Founded: 1880; Members: 600
Pres: Prof. Lars Furuland
Focus: Lit
Periodical
Samlaren: Tidskrift för svensk litteraturvetenskaplig forskning (weekly) *10562*

Svenska Livsmedelstekniska Föreningen, Katarinavägen 20, 116 45 Stockholm
T: (08) 7145045; Fax: (08) 408045
Focus: Eng; Food
Periodical
Livsmedelsteknik (9 times annually) . . . *10563*

Svenska Matematikersamfundet, c/o Dept of Mathematics, University of Umea, 901 87 Umea
T: (090) 165000
Founded: 1950; Members: 540
Pres: Dr. Urban Cegrell; Gen Secr: Tord Sjödin
Focus: Math
Periodicals
Mathematica Scandinavia
Nordisk Matematisk Tidskrift *10564*

Svenska Museiföreningen / The Swedish Museums Association, Linnégt 89, Box 27151, 102 52 Stockholm
T: (08) 7832960; Fax: (08) 6606034
Founded: 1906; Members: 1350
Pres: Elisabeth Fleetwood; Gen Secr: Katarina Årre
Focus: Arts
Periodical
Svenska Museer: Debate and Information on Museum Policies and Museology (weekly) *10565*

Svenska Nationalkomittén för Geologi, c/o Prof. Dr. Tryggve Troedsson, Dept of Forest Soils, Box 7001, 750 07 Uppsala
Founded: 1959; Members: 12
Pres: J.O. Carlsson; Gen Secr: Prof. Tryggve Troedsson
Focus: Geology *10566*

Svenska Nationalkommittén för Kristallografi, Box 50005, 104 05 Stockholm
T: (08) 6739500; Fax: (08) 155670
Focus: Mineralogy *10567*

Svenska Naturskyddsföreningen (SNF) / The Swedish Society for Nature Conservation, Box 4625, 116 91 Stockholm
T: (08) 7026500; Fax: (08) 7020855
Founded: 1909; Members: 200000
Pres: Ulf von Sydow; Gen Secr: Gunnar Landborn
Focus: Ecology
Periodical
Sveriges Natur (7 times annually) . . . *10568*

Svenska Österbottens Litteraturförening / Swedish Österbottens Literary Assocoiation, c/o Olof Haegerstrand, Fasanvaegen 4, 775 00 Krylbo
Focus: Lit
Periodical
Horisont *10569*

Svenska Ogonläkareföreningen, c/o Dept of Ophthalmology, Karolinska Hospital, 104 01 Stockholm
Focus: Ophthal *10570*

Svenska Operationsanalysföreningen (SORA), Box 20, 104 50 Stockholm
Founded: 1959; Members: 600
Focus: Computer & Info Sci *10571*

Svenska Psykoanalytiska Föreningen, Västerlånggt 60, 111 29 Stockholm
T: (08) 108095
Founded: 1934; Members: 76
Focus: Graphic & Dec Arts, Design; Psych; Psychoan *10572*

Svenska Sällskapet för Antropologi och Geografi (SSAG), c/o Naturgeografiska Institutionen, Stockholms Universitet, 106 91 Stockholm
Fax: (08) 164818
Founded: 1880; Members: 1100
Pres: Göran Hoppe; Gen Secr: Gunhild Rosquist
Focus: Anthro; Geography
Periodicals
Atlas of Sweden
Geografiska Annaler
Ymer *10573*

Svenska Samfundet för Informationsbehandling (SSI), Box 22114, 104 22 Stockholm
Focus: Computer & Info Sci *10574*

Svenska Samfundet för Musikforskning / Swedish Society for Musicology, c/o Statens Musiksamlingar, Box 16326, 103 26 Stockholm
T: (08) 112058
Founded: 1919; Members: 550
Pres: Prof. Anna Johnson; Gen Secr: Karin Hallgren
Focus: Music
Periodicals
Monumenta Musicae Svecicae
Musik i Sverige
Svensk Tidskrift för Musikforskning (weekly) *10575*

Svenska Växtgeografiska Sällskapet, c/o Växtbiologiska Institutionen, Box 559, 751 22 Uppsala
T: (018) 139955
Founded: 1923; Members: 430
Focus: Botany; Bio
Periodicals
Acta Phytogeographica Suecica (weekly)
Växtekologiska Studier (weekly) *10576*

Svenska Yrketsutbildningsföreningen, Box 172, 771 01 Ludvika
Focus: Educ *10577*

Svensk Flyghistorisk Förening / Swedish Aviation Historical Society, Box 308, 101 26 Stockholm
Focus: Hist; Aero
Periodical
Svensk Flyghistorisk Tidskrift (weekly) . . *10578*

Svensk Förening för Anatomi, c/o Dept of Anatomy, University of Umeå, 901 87 Umeå
Fax: (090) 165480
Founded: 1969; Members: 50
Focus: Anat *10579*

Svensk Förening för Anestesi och Intensivvård (SAFI) / Swedish Society for Anaesthesia and Intensive Care, c/o Dept of Anaesthesiology, Linköping University Hospital, 581 85 Linköping
Fax: (013) 222836
Founded: 1946; Members: 1220
Gen Secr: Claes Lennmarken
Focus: Anesthetics *10580*

Svensk Förening för Gastroenterologi och Gastrointestinal Endoskopi (SFGGE), c/o Dept of Surgery, Helsingborg Hospital, 251 87 Helsingborg
Fax: (042) 242731
Founded: 1954; Members: 500
Gen Secr: Hans Graffner
Focus: Gastroenter *10581*

Svensk Förening for Radiofysik, c/o Dept of Medical Physics, Danderyd Hospital, 182 03 Danderyd
Focus: Physics *10582*

Svensk Kirurgisk Förening (SKF) / Swedish Surgical Society, c/o Kirurgiska Kliniken, Akademiska Sjukhuset, 751 85 Uppsala
T: (08) 7292000
Founded: 1905; Members: 1400
Pres: Ulf Haglund; Gen Secr: Bengt Gerdin
Focus: Surgery
Periodical
Svensk Kirurgi (weekly) *10583*

Svensk Otolaryngologisk Förening, c/o Sabbatsbergs Sjukhus, Box 6401, 113 82 Stockholm
Members: 350
Focus: Otorhinolaryngology *10584*

Sveriges Allmänna Biblioteksförening (SAB), Box 3127, 103 62 Stockholm
T: (08) 7230082
Founded: 1915; Members: 2500
Focus: Libraries & Bk Sci
Periodical
Biblioteksbladet *10585*

Sveriges Allmänna Konstförening (S.A.K.), Stora Nygt 5, Box 2151, 103 14 Stockholm
T: (08) 104676/77
Founded: 1832; Members: 22000
Pres: Stig Ramel; Gen Secr: Göran Nilsson
Focus: Arts
Periodical
Art publications (weekly) 10586

Sveriges Arkivtjänstemäns Förening, c/o DIK-Förbundet, Box 760, 131 24 Nacka
T: (08) 7162880; Fax: (08) 7165291
Founded: 1936; Members: 440
Pres: Per Matsson
Focus: Archives
Periodical
DIK-forum (20 times annually) 10587

Sveriges Civilingenjörsförbund (CF) / The Swedish Association of Graduate Engineers, Malmskillnadsgt 48, Box 1419, 111 84 Stockholm
T: (08) 142000
Founded: 1861; Members: 45000
Pres: Ingrid Bruce; Gen Secr: Sven Magnusson
Focus: Civil Eng
Periodicals
Civilingenjören (weekly)
Ny Teknik (weekly) 10588

Sveriges Författarförbund / Swedish Writers' Union, Drottninggt 88b, 111 36 Stockholm
T: (08) 7912280; Fax: (08) 7912285
Founded: 1893; Members: 2000
Pres: Peter Curman; Gen Secr: John Erik Forslund
Focus: Lit
Periodical
Författaren 10589

Sveriges Gemmologiska Riksförening, c/o Jonas Gevers, Box 19121, 104 32 Stockholm
Fax: (08) 8207317
Founded: 1947
Focus: Mineralogy 10590

Sveriges Geologiska Undersökning (SGU), Villavägen 18, 751 28 Uppsala
T: (018) 179000
Founded: 1858; Members: 267
Focus: Geology 10591

Sveriges Museimannaförbund / Swedish Association of Museum Curators, DIK-förbundet, Box 760, 131 24 Nacka
T: (08) 7162880; Fax: (08) 7165291
Members: 3220
Pres: Olof Stroh
Focus: Arts
Periodical
DIK-forum (20 times annually) 10592

Sveriges Praktiserande Läkares Förening, Villagt 5, Box 5610, 114 86 Stockholm
T: (08) 224740
Focus: Med 10593

Sveriges Rationaliseringsförbund (SRF), Box 4324, 102 67 Stockholm
T: (08) 249225; Fax: (08) 6433061
Founded: 1956; Members: 257
Focus: Business Admin
Periodical
SRF-Information (weekly) 10594

Sveriges Socianomer Riksförbund (SSR), Mariedalsv 4, 112 51 Stockholm
T: (08) 6174400
Founded: 1958; Members: 24103
Focus: Sociology
Periodicals
Socionomen (8 times annually)
SSR-tidningen (40 times annually)
Triss (Students) (weekly) 10595

Sveriges Tandläkarförbund (STF) / The Swedish Dental Association, Nybrogt 53, Box 5843, 102 48 Stockholm
T: (08) 6661500; Fax: (08) 6625842
Founded: 1908; Members: 11900
Pres: Prof. Göran Koch; Gen Secr: Gunnar Luthman
Focus: Dent
Periodicals
Swedish Dental Journal (weekly)
Tandläkartidningen: The Journal of the S.D.A. (every third week) 10596

Sveriges Vetenskapliga Specialbiblioteks Förening (SVSF), c/o Utrikesdepartementets Bibliotek, Box 16121, 103 23 Stockholm
Founded: 1945; Members: 200
Focus: Libraries & Bk Sci 10597

Sveriges Veterinärförbund (SVF), Kungsholms Hamnplan 7, Box 12709, 112 94 Stockholm
T: (08) 6542480; Fax: (08) 6517082
Founded: 1860; Members: 1860
Pres: Herbert Lundström; Gen Secr: Sven Oskarsson
Focus: Vet Med
Periodical
Svensk Veterinärtidning (15 times annually) 10598

Svetstekniska Föreningen (SVF), Box 5073, 102 42 Stockholm
T: (08) 220760; Fax: 808) 6799404
Founded: 1941; Members: 3200
Pres: Stig-Erik Erikson; Gen Secr: Joakim Hedegård
Focus: Eng
Periodical
Svetsen (weekly) 10599

Teknikkonsulterna (SKIF) / Swedish Association of Consulting Engineers, Kungsholmstorg 1, Box 22076, 104 22 Stockholm
Fax: (08) 6502972
Founded: 1910; Members: 200
Pres: Olof Hulthén
Focus: Eng
Periodical
Konsulttidningen (weekly) 10600

Tekniska Litteratursällskapet / Swedish Society for Technical Documentation, Box 5073, 102 42 Stockholm
Gen Secr: Birgitta Levin
Focus: Lit; Doc
Periodical
Tidskrift för Documentation (weekly) . . 10601

Tekniska Samfundet i Göteborg, Viktor Rydbergsgt 14, 411 32 Göteborg
Focus: Eng 10602

Utrikespolitiska Institutet (U.I.) / The Swedish Institute of International Affairs, Box 1253, 111 82 Stockholm
T: (08) 234060; Fax: (08) 201049
Founded: 1938
Pres: Leif Leifland; Gen Secr: R. Lindahl
Focus: Poli Sci
Periodicals
Internationella studier (5 times annually)
Världspolitikens dagsfrågor (weekly) . . . 10603

Vetenskapliga Bibliotekens Tjänstemannaförening (VBT) / Association of Research Library Employees, c/o DIK-förbundet, Box 760, 131 24 Nacka
T: (08) 7162880; Fax: (08) 7165291
Founded: 1959; Members: 740
Pres: C. Larsson
Focus: Sci
Periodical
DIK-forum (20 times annually) 10604

Wenner-Gren Center Foundation for Scientific Research (WGC), Sveavägen 166, 113 46 Stockholm
Founded: 1962
Pres: Dr. Jan Wallander; Gen Secr: Prof. Torvard Laurent
Focus: Nat Sci
Periodicals
The Wenner-Gren Center International Symposium Series
Wenner-Gren Center Svenska Symposier 10605

Switzerland

Académie Internationale de Céramique / Internationale Akademie für Keramik / International Academy of Ceramics, c/o Musée Ariana, 10 Av de la Paix, 1202 Genève
T: (022) 342950; Fax: (022) 7337011
Founded: 1953; Members: 390
Pres: Dr. Rudolf Schnyder
Focus: Arts
Periodical
Céramique 10606

Académie Internationale d'Héraldique, c/o Musée d'Art et d'Histoire, 2 Rue Charles Galland, 1206 Genève
Founded: 1949; Members: 90
Pres: Dr. Jean-Claude Loutsch; Gen Secr: Roger Harmignies
Focus: Genealogy 10607

Advisory Committee for the Coordination of Information Systems (ACCIS), Palais des Nations, 1211 Genève 10
T: (022) 7988591; Tx: 289696; Fax: (022) 339879
Founded: 1983; Members: 35
Focus: Computer & Info Sci
Periodical
ACCIS Newsletter (weekly) 10608

Ärztekommission für Rettungswesen SRK (AKOR SRK), Rainmattstr 10, Postf, 3001 Bern
T: (031) 3877111; Fax: (031) 3112793
Pres: Dr. R.P. Maeder; Gen Secr: Ulrich Schüle
Focus: Public Health 10609

Aktion 100, Verein zur Förderung der Sicherheit im Strassenverkehr, Funkstr 107, 3084 Wabern
T: (031) 540430
Focus: Transport
Periodical
Bulletin der Aktion (1-2 times annually) . 10610

Aktion Freiheitliche Bodenordnung / Action pour un Droit Foncier Libéral, Stampfenbachstr 69, 8035 Zürich
T: (01) 3632240
Focus: Law 10611

Ala – Schweizerische Gesellschaft für Vogelkunde und Vogelschutz, c/o Schweizerische Vogelwarte, 6204 Sempach
T: (041) 990022; Fax: (041) 994007
Founded: 1909
Pres: Dr. Luc Schifferli; Gen Secr: Renate Horváth
Focus: Ornithology
Periodical
Der Ornithologische Beobachter (weekly) . 10612

Allgemeine Anthroposophische Gesellschaft, c/o Goetheanum, 4143 Dornach
T: (061) 7014242
Founded: 1923
Pres: Manfred Schmidt-Brabant
Focus: Philos 10613

Allgemeine Geschichtforschende Gesellschaft der Schweiz (AGGS), Länggasstr 49, 3000 Bern
T: (031) 658093
Founded: 1841; Members: 1000
Pres: Prof. Dr. Beatrix Mesmer
Focus: Hist
Periodical
Schweizerische Zeitschrift für Geschichte (weekly) 10614

Alliance Graphique Internationale (AGI), Sonnhaldenstr 3, 8032 Zürich
Founded: 1952
Focus: Eng 10615

Antiquarische Gesellschaft in Zürich (AGZ), c/o Staatsarchiv, Postf, 8057 Zürich
Fax: (01) 3641815
Founded: 1832; Members: 620
Pres: Dr. Werner Widmer
Focus: Hist
Periodical
Mitteilungen der Antiquarischen Gesellschaft 10616

Aqua Viva, Schweizerische Aktionsgemeinschaft zur Erhaltung der Flüsse und Seen, Seilerstr 27, Postf 5242, 3001 Bern
T: (031) 3813252
Pres: Otto Zwygart; Gen Secr: Erika Winzeler
Focus: Ecology
Periodical
Natur und Mensch (6-8 times annually) . 10617

Arbeitsgemeinschaft für orale Implantologie, Bleicherweg 39, 8002 Zürich
T: (01) 2013613; Fax: (01) 2025459
Focus: Dent 10618

Association des Bibliothèques Internationales / Association of International Libraries, c/o WHO Library, 1211 Genève 27
T: (022) 7912071; Fax: (022) 7881836
Founded: 1963
Pres: Dr. D. Avriel
Focus: Libraries & Bk Sci 10619

Association des Ecoles Internationales / International Schools Association, CIC Case 20, 1211 Genève 20
T: (022) 7336717; CA: Interschools Genève
Founded: 1951; Members: 85
Focus: Educ
Periodical
ISA Bulletin (3 times annually) 10620

Association des Instituts d'Etudes Européennes (AIEE) / Association of Institutes of European Studies, c/o European Cultural Centre, 122 Rue de Lausanne, 1200 Genève
Founded: 1951
Gen Secr: Prof. Dusan Sidjanski
Focus: Cultur Hist 10621

Association des Sociétés Suisses des Professeurs de Langues Vivantes, Imfangring 10, 6005 Luzern
T: (041) 448771
Focus: Educ; Ling 10622

Association d'Histoire et de Science Politique (HISPO), Eichholzstr 96, 3084 Wabern
T: (031) 9610257
Founded: 1977; Members: 165
Pres: L. Altermatt
Focus: Hist; Poli Sci 10623

Association Européenne Francophone pour les Etudes Bahá'íes, Dufourstr 13, 3005 Bern
Focus: Rel & Theol 10624

Association for Computational Linguistics-Europe, c/o IDSIA, Corso Elvezia 36, 6900 Lugano
T: (091) 228881
Focus: Computer & Info Sci 10625

Association for the International Collective Management of Audiovisual Works (AGICOA), 25 Rte de Ferney, 1202 Genève
T: (022) 7344580; Tx: 414225; Fax: (022) 7344762
Founded: 1981; Members: 16
Gen Secr: Rodolphe Egli
Focus: Law 10626

Association Internationale de la Securité Sociale, 4 Rue des Morillons, CP 1, 1211 Genève 22
Founded: 1927; Members: 315
Focus: Sociology
Periodicals
Internationale Revue für soziale Sicherheit (weekly)
International Social Security Review (weekly)
Revista Internacional de Seguridad Social (weekly)
Revue Internationale de Sécurité Sociale (weekly)
Weltbibliographie der sozialen Sicherheit (weekly) 10627

Association Internationale des Directeurs d'Ecoles Hôtelières, c/o Ecole Hôtelière de Lausanne, 1000 Lausanne 25
T: (021) 7851111
Founded: 1955
Gen Secr: Jean-Louis Aeschlimann
Focus: Adult Educ
Periodical
EUHOFA International Journal (weekly) . 10628

Association Internationale pour l'Histoire Contemporaine de l'Europe / International Association for Contemporary History of Europe, 132 Rue de Lausanne, 1211 Genève
Founded: 1968
Gen Secr: Antoine Fleury-Juhei
Focus: Hist
Periodical
Bulletin de Liaison et d'Information . . . 10629

Association of International Consultants on Human Rights, CP 529, 1211 Genève
T: (022) 7364452; Fax: (022) 7364863
Founded: 1983
Pres: Daniel Prémont
Focus: Law 10630

Association Olympique Internationale pour la Recherche Médico-Sportive (AIORMS), c/o COI, Château de Vidy, 1007 Lausanne
T: (021) 253271; Tx: 454024; Fax: (021) 241552
Focus: Med 10631

Association Suisse de Droit Aérien et Spatial (ASDA), 18 Rue du Marché, 1204 Genève
T: (022) 286522
Founded: 1954; Members: 220
Focus: Law
Periodical
ASDA SVLR (3 times annually) 10632

Association Suisse de Science Politique / Swiss Political Science Association, c/o Département de Science Politique, Université de Genève, 102 Blvd Carl Vogt, 1205 Genève
T: (022) 7058379
Founded: 1959; Members: 800
Pres: Prof. Dr. Pierre Allan
Focus: Poli Sci
Periodicals
Année Politique Suisse (weekly)
Annuaire Suisse de Science Politique (weekly)
Bulletin de l'Association Suisse de Science Politique (weekly) 10633

Association Suisse des Ecoles Hôtelières, c/o Centre International de Glion, 1823 Glion-sur-Montreux
T: (021) 9634841; Tx: 453171; Fax: (021) 9631384
Focus: Adult Educ 10634

Bernische Botanische Gesellschaft, Altenbergrain 21, 3013 Bern
T: (031) 6314911
Founded: 1918; Members: 380
Pres: Dr. Klaus Ammann; Gen Secr: Christine Keller
Focus: Botany
Periodical
Sitzungsberichte (weekly) 10635

Budapest Union for the International Recognition of the Deposit of Microorganisms for the Purposes of Patent Procedure, c/o WIPO, 34 Chemin des Colombettes, CP 18, 1211 Genève 20
T: (022) 7309111; Tx: 412912; Fax: (022) 7335428
Founded: 1977
Focus: Law 10636

Bund für vereinfachte rechtschreibung, Pflugstr 18, 8006 Zürich
T: (01) 3628846
Founded: 1924; Members: 1500
Focus: Ling
Periodical
Rechtschreibung: Mitteilungen des Bundes für vereinfachte rechtschreibung (3 times annually) 10637

Bund Schweizer Architekten (BSA), Keltenstr 45, 8044 Zürich
T: (01) 2522852
Founded: 1908; Members: 545
Pres: J. Blumer
Focus: Archit
Periodical
Werk, Bauen und Wohnen (10 times annually) 10638

Bureau International d'Education (BIE) / International Bureau of Education, CP 199, 1211 Genève 20
T: (022) 7981455; Tx: 415771; CA: Intereduc Genève; Fax: (022) 7981486
Founded: 1925; Members: 174
Pres: Juan Carlos Tedesco
Focus: Educ
Periodicals
Bulletin du Bureau International d'Education: Bulletin of the International Bureau of Education (weekly)
Educational Innovation and Education (weekly) 10639

Bureau International de la Paix / International Peace Bureau, 41 Rue de Zürich, 1201 Genève
T: (022) 316429
Founded: 1891; Members: 79
Gen Secr: C. Archer
Focus: Poli Sci; Prom Peace
Periodicals
Geneva Monitor-Disarmement
IPB Geneva News 10640

Campagne d'Education Civique Européenne, c/o European Cultural Centre, 122 Rue de Lausanne, 1211 Genève 21
T: (022) 7322803
Founded: 1960
Focus: Educ 10641

Centre Européen de Réflexion et d'Etude en Thermodynamique (CERET), c/o Polytechnique de Lausanne, 1015 Lausanne
T: (021) 6933506
Focus: Physics 10642

Centre for the Independence of Judges and Lawyers (CIJL), 109 Rte de Chêne, BP 45, 1224 Genève
T: (022) 493545; Tx: 418531; Fax: (022) 493145
Founded: 1978
Gen Secr: Mona Rishmawi
Focus: Law
Periodical
CIJL Bulletin (weekly) 10643

Centre Suisse de Documentation en Matière d'Enseignement et d'Education (CESDOC) / Schweizerische Dokumentationsstelle für Schul- und Bildungsfragen, 15 Rte des Morillons, 1218 Grand-Saconnex
T: (022) 7984531; Fax: (022) 7883173
Founded: 1962
Gen Secr: Peter Gentinetta
Focus: Doc; Educ
Periodicals
Bibliographie Pédagogique Suisse (weekly)
Bulletin: Mitteilungen (weekly) 10644

Christlicher Lehrer- und Erzieherverein der Schweiz, Schiltmatthalde 15, 6048 Horw
T: 473729
Focus: Educ
Periodical
Schweizer Schule (weekly) 10645

CIP International Study and Research Center in Psychosynteresis, CP 937, 1001 Lausanne
T: (032) 936674; Fax: (032) 936740
Founded: 1984
Focus: Psychoan 10646

Collaborative International Pesticides Analytical Council (CIPAC), c/o Eidgenössische Forschungsanstalt für Obst, Wein und Gartenbau, 8820 Wädenswil
T: (01) 7836111
Founded: 1957
Pres: Dr. H.P. Bosshardt
Focus: Agri 10647

Collegium Romanicum (CR), c/o Section d'Italien, Faculté des Lettres, Université de Lausanne, 1015 Lausanne
Founded: 1947; Members: 167
Pres: Prof. Dr. Jean-Jacques Marchand; Gen Secr: Dr. Denis Fachard
Focus: Ling; Lit
Periodicals
Versants (weekly)
Vox Romanica (weekly) 10648

Commission Electrotechnique Internationale (CEI) / International Electrotechnical Commission, 3 Rue de Varembé, 1211 Genève 20
T: (022) 7340150; Tx: 414121; Fax: (022) 7333843
Founded: 1906
Gen Secr: A.M. Raeburn
Focus: Electric Eng
Periodicals
IEC Bulletin (weekly)
IEC Yearbook (weekly) 10649

Commission Internationale pour la Protection des Eaux du Léman contre la Pollution (CIPEL), 23 Av de Chailly, CP 80, 1000 Lausanne
T: (021) 73314
Focus: Ecology 10650

Commission Médicale Chrétienne (CMC) / Christliche Gesundheitskommission / Christian Medical Commission, 150 Rte de Ferney, CP 2100, 1211 Genève 2
T: (022) 916111; Tx: 415730; Fax: (022) 7910361
Founded: 1967
Pres: Margareta Sköld
Focus: Public Health
Periodical
CMC Contact Newsletter (weekly) . . . 10651

Commission pour l'Encouragement de la Recherche Scientifique, Monbijoustr 28, Postf 7023, 3001 Bern
T: (031) 612143
Founded: 1944; Members: 14
Pres: Prof. Dr. H. Sieber; Gen Secr: Dr. Peter Kuentz
Focus: Sci 10652

Conférence des Directeurs des Ecoles Techniques Supérieures de Suisse, Postf 805, 8401 Winterthur
T: (052) 2677171
Focus: Eng 10653

Conférence des Statisticiens Européens / Conference of European Statisticians, c/o Statistical Div, ECE, Palais des Nations, 1211 Genève 10
T: (022) 8171234; Tx: 289696
Founded: 1953; Members: 34
Gen Secr: John Kelly
Focus: Stats
Periodical
Statistical Journal (weekly) 10654

Conseil des Organisations Internationales des Sciences Médicales / Council for International Organizations of Medical Sciences, c/o World Health Organization, Via Appia, 1211 Genève 27
T: (022) 913406; Tx: 27821; Fax: (022) 910746
Founded: 1949; Members: 60
Pres: Dr. F. Vilardell; Gen Secr: Dr. Zbigniew Bankowski
Focus: Med
Periodicals
Calendar of Congresses of Medical Sciences (weekly)
CIOMS Organization and Activities/Directory of Members (bi-annually) 10655

Conseil International sur les Problèmes de l'Alcoolisme et des Toxicomanies (CIPAT) / International Council on Alcohol and Addictions, CP 189, 1001 Lausanne
T: (021) 209865; Fax: (021) 209817
Founded: 1907
Pres: Dr. Ibrahim Al-Awaji; Gen Secr: Dr. Eva Tongue
Focus: Toxicology
Periodical
ICAA News (weekly) 10656

Consultative Council for Postal Studies (CCPS), c/o UPU, Weltpoststr 4, 3000 Bern 15
T: (031) 432211; Tx: 912761; Fax: (031) 432210
Founded: 1957
Gen Secr: Adwaldo C. Botto de Barros
Focus: Eng 10657

Cooperative Program for Monitoring and Evaluation of the Long-range Transmission of Air Pollutants in Europe (EMEP), c/o CEE, Palais des Nations, 1211 Genève 10
T: (022) 346011; Tx: 412962
Founded: 1979
Focus: Ecology 10658

Dachverband Schweizerischer Lehrerinnen und Lehrer, Ringstr 54, Postf 189, 8057 Zürich
T: (01) 3118303; Fax: (01) 3118315
Founded: 1849; Members: 40000
Pres: Beat W. Zemp; Gen Secr: Urs Schildknecht
Focus: Educ
Periodical
Schweizerische Lehrerinnen- und Lehrer-Zeitung SLZ (weekly) 10659

Deutschschweizerischer Sprachverein (DSSV), Kutscherweg 3, 3047 Bremgarten
T: (031) 234019
Focus: Ling
Periodical
Sprachspiegel (weekly) 10660

Deutschschweizerisches PEN-Zentrum, Postf 6018, 3001 Bern
T: (031) 446677
Founded: 1932; Members: 200
Pres: Ernst Reinhardt; Gen Secr: Hans Erpf
Focus: Lit
Periodical
PEN-Brief (weekly) 10661

Development Innovations and Networks, 3 Rue de Varembé, 1211 Genève 20
T: (022) 7341716; Tx: 414148; Fax: (022) 7400011
Founded: 1980; Members: 159
Gen Secr: Fernand Vincent
Focus: Develop Areas
Periodical
IRED Forum (weekly) 10662

Digital Equipment Computer Users Society (Decus Europe), CP 510, Petit Lancy, 1213 Genève 1
Gen Secr: Pierre Antoine Riedi
Focus: Computer & Info Sci 10663

Ecological and Toxicological Association of the Dyestuffs Manufacturing Industry (ETAD), Clarastr 4, 4005 Basel
T: (061) 68122; Fax: (061) 6914278
Founded: 1974; Members: 32
Gen Secr: Dr. Rudolf Anliker
Focus: Ecology; Toxicology 10664

Ecumenical Institute Bossey, Château de Bossey, 1298 Céligny
T: (022) 7762531; Fax: (022) 7760169
Founded: 1946
Gen Secr: Prof. Dr. Jacques Nicole
Focus: Rel & Theol
Periodicals
Bossey Newsletter
Ministerial Formation 10665

Eidgenössischer Musikverband, 6252 Dagmersellen
T: (062) 861450; Fax: (062) 861445
Pres: Josef Meier; Gen Secr: Josef Zinner
Focus: Music
Periodical
Schweizerische Blasmusikzeitung (weekly) 10666

Euro-International Committee for Concrete / Comité Euro-International du Béton, c/o Dept de Génie Civil, Ecole Polytechnique de Lausanne, CP 88, 1015 Lausanne
T: (021) 6932747
Founded: 1953
Focus: Materials Sci 10667

Europäische Gesellschaft für Schriftpsychologie und Schriftexpertise / European Society of Handwriting Psychology, Postf 88, 8041 Zürich
Fax: (01) 4816288
Members: 835
Pres: Rudolf Känzig
Focus: Graphology
Periodical
EGS-Bulletin (weekly) 10668

Europäische Organisation für Qualität / European Organization for Quality, Postf 5032, 3001 Bern
T: (031) 216166; Fax: (031) 216951
Founded: 1957; Members: 25
Gen Secr: M. Conrad
Focus: Business Admin
Periodical
EOQ Quality (weekly) 10669

Europäische Rektorenkonferenz, 10 Rue du Conseil Général, 1211 Genève 4
T: (022) 3292644, 3292251; Tx: 428380; Fax: (022) 3292821
Founded: 1959; Members: 490
Pres: Prof. Dr. Hinrich Seidel; Gen Secr: Dr. A. Barblan
Focus: Educ
Periodical
CRE-action (weekly) 10670

Europäisches Zentrum für die Bildung im Versicherungswesen, Kirchlistr 2, 9010 Sankt Gallen
T: (071) 251515; Fax: (071) 251392
Focus: Adult Educ; Insurance 10671

European Association for Cardio-Thoracic Surgery (AECTS), c/o Clinic for Cardiovascular Surgery, University Hospital, 8091 Zürich
T: (01) 2553298
Founded: 1986
Gen Secr: Prof. Marko Turina
Focus: Surgery
Periodical
European Journal of Cardio-Thoracic Surgery (weekly) 10672

European Association for Computer Graphics / Association Européenne pour les Graphiques sur l'Ordinateur, CP 16, 1288 Aire-la-Ville
Founded: 1980
Focus: Computer & Info Sci 10673

European Association for Signal Processing (EURASIP), CP 134, 1000 Lausanne 13
T: (021) 6932626; Tx: 454062; Fax: (021) 6932603
Founded: 1978
Gen Secr: Prof. M. Kunt
Focus: Computer & Info Sci
Periodicals
EURASIP Newsletter (3 times annually)
Image Communication (weekly)
Speech Communication (weekly) 10674

European Association for the Advancement of Radiation Curing by UV, EB and Laser Beams, Pérolles 24, 1700 Fribourg
T: (037) 226465; Fax: (037) 226545
Founded: 1988; Members: 266
Gen Secr: R. Rijnaart
Focus: Radiology 10675

European Association for Transactional Analysis / Association Européenne d'Analyse Transactionnelle, 59 Case Grand-Pré, 1211 Genève 16
Founded: 1975; Members: 20
Focus: Psychoan 10676

European Association of Development Research and Training / Association Européenne des Instituts de Recherche et de Formation en Matière de Développement, 10 Rue Richemont, CP 272, 1211 Genève 21
T: (022) 7314648; Tx: 412584; Fax: (022) 7385797
Founded: 1975; Members: 200
Gen Secr: Claude Auroi
Focus: Sci
Periodicals
EADI Newsletter (2-3 times annually)
European Journal of Development Research (weekly) 10677

European Association of Music Conservatories, Academies and High Schools, c/o Konservatorium für Musik und Theater, Kramgasse 36, 3011 Bern
T: (031) 226221; Fax: (031) 212053
Founded: 1953; Members: 99
Gen Secr: Urs Frauchiger
Focus: Music 10678

European Association of Music Festivals (EAMF) / Association Européenne des Festivals de Musique, 122 Rue de Lausanne, 1202 Genève
T: (022) 7322803; Fax: (022) 7384012
Founded: 1951; Members: 53
Pres: F. de Ruiter
Focus: Music 10679

European Centre for Insurance Education and Training, Kirchlistr 2, 9010 Sankt Gallen
T: (071) 251515; Fax: (071) 251392
Founded: 1957
Focus: Insurance; Educ 10680

European Chemoreception Research Organization (ECRO), c/o Universität Zürich-Irchel, Winterthurerstr 190, 8057 Zürich
T: (01) 2575413; Fax: (01) 2574004
Founded: 1970
Gen Secr: Dr. D. Glaser
Focus: Chem 10681

European Council of Coloproctology, c/o Policlinique de Chirurgie, Hôpital Cantonal de Genève, 24 Rue Micheli-du-Crest, 1211 Genève 4
T: (022) 3727902
Gen Secr: Dr. Marc-Claude Marti
Focus: Surgery 10682

European Crystallographic Committee (ECC), c/o Laboratoire de Cristallographie, 24 Quai Ernest Ansermet, 1211 Genève 4
Founded: 1972
Gen Secr: Dr. H.D. Flack
Focus: Geology 10683

European Cytoskeletal Club, c/o Dépt de Pathologie, Faculté de Médecine CMU, Université de Genève, 1 Rue Michel Servet, 1211 Genève 4
Founded: 1980
Gen Secr: Prof. G. Gabbiani
Focus: Pathology 10684

European Nuclear Society (E.N.S.), Monbijoustr 5, Postf 5032, 3001 Bern
T: (031) 216111; Tx: 912110; Fax: (031) 229203
Founded: 1975; Members: 70
Focus: Nucl Res
Periodical
Nuclear Europe (weekly) 10685

European Society for Pediatric Nephrology (ESPN), c/o Universitäts-Kinderklinik, Steinwiesstr 75, 8032 Zürich
Founded: 1967; Members: 230
Focus: Pediatrics
Periodical
Abstracts of Communications 10686

Fachgruppe Carosserie- und Fahrzeugtechnik des STV, Kienbergweg 6, 4450 Sissach
T: (061) 983507
Focus: Eng; Auto Eng 10687

Fachgruppe für Arbeiten im Ausland des SIA (FAA), Selnaustr 16, Postf, 8039 Zürich
T: (01) 2831515; Fax: (01) 2016335
Pres: Willem E. Pleines
Focus: Eng; Archit 10688

Fachgruppe für Architektur des SIA, Selnaustr 16, 8039 Zürich
T: (01) 2831515; Fax: (01) 2016335
Members: 710
Pres: P. Giorgis
Focus: Archit 10689

Fachgruppe für Betriebstechnik des STV, Im Geissacker 24, 8404 Winterthur
T: (052) 814832
Focus: Eng 10690

Fachgruppe für Brückenbau und Hochbau des SIA (FBH), Selnaustr 16, 8039 Zürich
T: (01) 2831515; Fax: (01) 2016335
Members: 1000
Pres: M. Hartenbach
Focus: Civil Eng 10691

Fachgruppe für das Management im Bauwesen des SIA (FMB), Selnaustr 16, 8039 Zürich
T: (01) 2011570; Fax: (01) 2016335
Focus: Business Admin 10692

Fachgruppe für Elektronik des STV, CP 958, 1001 Lausanne
T: (021) 391151
Focus: Electronic Eng 10693

Fachgruppe für industrielles Bauen des SIA, Selnaustr 16, 8039 Zürich
T: (01) 2831515; Fax: (01) 2016335
Founded: 1970; Members: 203
Pres: Hermann Käser
Focus: Civil Eng 10694

Fachgruppe für Raumplanung und Umwelt des SIA, Selnaustr 16, 8039 Zürich
T: (01) 2831515; Fax: (01) 2016395
Members: 300
Pres: M. von Känel
Focus: Urban Plan; Ecology 10695

Fachgruppe für Untertagbau des SIA (FGU), Selnaustr 16, Postf, 8039 Zürich
T: (01) 2831515; Fax: (01) 2016335
Pres: Prof. Dr. Kalman Kovari
Focus: Civil Eng

Periodical
FGU-Dokumentationsbulletin (weekly) . . 10696

Fachgruppe für Verfahrenstechnik des SIA, c/o Hoffmann-La Roche, 4002 Basel
Focus: Eng 10697

Fachgruppe für Vermessung und Kulturtechnik (Deutsch-Schweiz) des STV, Weingartenstr, 8501 Weiningen
T: (054) 75765
Focus: Surveying 10698

Fédération des Ecoles Privées de la Suisse Romande, La Combe, 1180 Rolle
T: 752727
Focus: Educ 10699

Fédération des Sociétés d'Agriculture de la Suisse Romande (F.S.A.S.R.), 3 Av des Jordils, CP 186, 1000 Lausanne 6
T: (021) 6177477
Founded: 1881; Members: 28000
Gen Secr: P.-Y. Felley
Focus: Agri
Periodical
Revue Suisse de Viticulture, d'Arboriculture et d'Horticulture (weekly) 10700

Fédération Internationale des Associations d'Etudes Classiques (FIEC) / International Federation of the Societies of Classical Studies, 6 Chemin Aux-Folies, 1293 Bellevue
T: (022) 7742656
Founded: 1948; Members: 58
Pres: Prof. J. Irigoin; Gen Secr: Prof. Dr. François Paschoud
Focus: Hist; Ling; Lit; Archeol
Periodical
Année philologique (weekly) 10701

Fédération Internationale des Femmes Diplômées des Universités (FIFDU) / International Federation of University Women, 37 Quai Wilson, 1201 Genève
T: (022) 7312380; Fax: (022) 780440
Founded: 1919; Members: 230000
Gen Secr: Dorothy Davies
Focus: Sci
Periodical
IFUW Newsletter (weekly) 10702

Fédération Internationale des Sociétés de Philosophie (FISP), c/o Séminaire de Philosophie, Université, Miséricorde, 1700 Fribourg
T: (037) 242669; Tx: 36110; Fax: (037) 219703
Founded: 1948; Members: 86
Pres: Evandro Agazzi; Gen Secr: Ioanna Kucuradi
Focus: Philos 10703

Fédération Suisse des Ecoles Privées, 40 Rue des Vollandes, 1211 Genève 6
T: (022) 355706
Focus: Educ 10704

Fondation Hindemith, c/o M.M. Décombaz, 40 Rue du Simplon, 1800 Vevey
Founded: 1968
Pres: Dr. A. Briner
Focus: Music
Periodical
Hindemith-Jahrbuch (weekly) 10705

Forschungsgemeinschaft für Nationalökonomie, Dufourstr 48, 9000 Sankt Gallen
T: (071) 302300; Fax: (071) 302646
Focus: Econ 10706

Geographische Gesellschaft Bern, Hallerstr 12, 3012 Bern
Fax: (031) 658511
Founded: 1873; Members: 600
Pres: Dr. M. Hasler
Focus: Geography
Periodicals
Berner Geographische Mitteilungen (weekly)
Jahrbuch 10707

Geographisch-Ethnographische Gesellschaft Zürich (GEGZ), c/o Geographisches Institut, Universität Zürich-Irchel, Winterthurerstr 190, 8057 Zürich
T: (01) 2575121
Founded: 1889; Members: 560
Pres: Prof. Dr. G. Furrer; Gen Secr: Dr. S. Wyder
Focus: Geography
Periodical
Geographica Helvetica (weekly) 10708

Geographisch-Ethnologische Gesellschaft Basel (GEG), Klingelbergstr 16, 4056 Basel
T: (061) 2673660
Founded: 1923; Members: 570
Pres: Dr. Bernhard Gardi
Focus: Ethnology; Geography
Periodicals
Basler Beiträge zur Geographie (weekly)
Physiogeographica: Basler Beiträge zur Physiogeographie (weekly)
Regio Basiliensis: Basler Zeitschrift für Geographie/Revue de Géographie de Bâle (3 times annually) 10709

Gesellschaft der Freunde alter Musikinstrumente, Bündtenweg 62, 4102 Binningen
T: (061) 4218363
Founded: 1953; Members: 170
Pres: Georg F. Senn; Gen Secr: Andreas Schlegel
Focus: Music
Periodical
Glareana (weekly) 10710

Gesellschaft für Arzneipflanzenforschung e.V., c/o Pharmazeutisches Institut, ETH Zürich, ETH-Zentrum, 8092 Zürich
T: (01) 2563166
Focus: Pharmacol; Therapeutics 10711

Gesellschaft für das Schweizerische Landesmuseum, Museumstr 2, 8023 Zürich
T: (01) 2211010
Focus: Arts 10712

Gesellschaft für deutsche Sprache und Literatur in Zürich, c/o Deutsches Seminar der Universität, Rämistr 74-76, 8001 Zürich
T: (01) 2572561; Fax: (01) 2620250
Founded: 1894; Members: 200
Pres: Dr. Rudolf Schwarzenbach; Gen Secr: Dr. Claudia Brinker
Focus: Ling; Lit 10713

Gesellschaft für Forschung auf biophysikalischen Grenzgebieten (GFBG), Im Rehwechsel 19, 4102 Binningen
T: 475813
Focus: Biophys 10714

Gesellschaft für Führungspraxis und Personalentwicklung (GfP), Löwenstr 16, 8021 Zürich
T: (01) 2118158
Focus: Business Admin 10715

Gesellschaft für Schweizerische Kunstgeschichte (GSK), Pavillonweg 2, Postf 6321, 3001 Bern
T: (031) 3014281; Fax: (031) 3016991
Founded: 1880; Members: 9000
Pres: Prof. Dr. Johannes Anderegg; Gen Secr: Dr.0 N. Caviezel
Focus: Hist; Arts
Periodical
Unsere Kunstdenkmäler/Nos Monuments d'Art et d'Histoire/I nostri Monumenti Storici (weekly) 10716

Gesellschaft für Versuchstierkunde (GV), c/o Dr. U. Märki, BRL Ltd, Wölferstr 4, 4414 Füllinsdorf
Gen Secr: Dr. U. Märki
Focus: Zoology 10717

Gesellschaft Pro Vindonissa, c/o Vindonissa-Museum, 5200 Brugg
T: (056) 412184
Pres: Hugo Doppler
Focus: Archeol
Periodical
Jahresbericht der Gesellschaft Pro Vindonissa (weekly) 10718

Gesellschaft schweizerischer Amtsärzte, c/o Bundesamt für Gesundheitswesen, Postf, 3001 Bern
T: (031) 3229508
Pres: Dr. A.J. Seiler
Focus: Med 10719

Gesellschaft Schweizerischer Maler, Bildhauer und Architekten (GSMBA) / Société des Peintres, Sculpteurs et Architectes Suisses, Kirchpl 9, 4132 Muttenz
T: (061) 617480
Founded: 1865; Members: 1600
Pres: Pierre Casè; Gen Secr: Esther Brunner-Buchser
Focus: Archit
Periodical
Schweizer Kunst (weekly) 10720

Gesellschaft Schweizerischer Tierärzte (GST), Länggassstr 8, Postf 6324, 3001 Bern
T: (031) 3025500; Fax: (031) 3028841
Founded: 1813; Members: 2000
Pres: Dr. J.-P. Siegfried; Gen Secr: B. Josi
Focus: Vet Med
Periodicals
GST-Bulletin (10 times annually)
Schweizer Archiv für Tierheilkunde (weekly) 10721

Gesellschaft schweizerischer Zeichenlehrer, Kloster, 8840 Einsiedeln
T: (055) 534431
Focus: Educ; Arts 10722

Groupement Romand du Marketing, 2 Av Agassiz, 1001 Lausanne
T: (021) 3197111
Founded: 1945; Members: 135
Gen Secr: M. Louis Mayer
Focus: Marketing 10723

Gruppe Olten, Schweizer Autorengruppe, Hauptstr 87, 8274 Tägerwilen
T: (072) 692353; Fax: (072) 692983
Pres: Manfred Züfle; Gen Secr: Jochen Kelter
Focus: Lit
Periodical
Mitteilungsblatt (3-5 times annually) . . . 10724

Heimverband Schweiz, Seegartenstr 2, 8008 Zürich
T: (01) 3834948
Pres: Gämperle. Walter; Gen Secr: Werner Vonaesch
Focus: Sociology
Periodical
Fachzeitschrift Heim 10725

Historisch-Antiquarischer Verein Heiden, 9410 Heiden
Founded: 1874
Pres: Rudolf Rohner
Focus: Hist 10726

Historischer Verein des Kantons Bern, c/o Stadt- und Universitäts-Bibliothek, Münstergasse 61, 3000 Bern 7
Founded: 1847
Focus: Hist
Periodicals
Archiv des Historischen Vereins des Kantons Bern (weekly)
Berner Zeitschrift für Geschichte und Heimatkunde (weekly) 10727

Historische und Antiquarische Gesellschaft zu Basel (HAG), c/o Universitätsbibliothek, Schönbeinstr 18-20, 4056 Basel
T: (061) 2673111
Founded: 1836; Members: 600
Focus: Hist
Periodical
Basler Zeitschrift für Geschichte und Altertumskunde (weekly) 10728

Institut International de Psychologie et de Psychothérapie Charles Baudouin, 25 Chemin des Rannaux, 1296 Coppet
Founded: 1924
Pres: Claude Piron
Focus: Psych; Therapeutics 10729

Institut National Genevois, 1 Promenade du Pin, 1204 Genève
Founded: 1853; Members: 850
Pres: Peter Tschopp; Gen Secr: Monique Tanner
Focus: Nat Sci; Humanities; Sci
Periodicals
Bulletin
Mémoires 10730

Institut Romand de Recherches et de Documentation Pédagogiques (IRDP), 43-45 Faubourg de l'Hôpital, CP 54, 2000 Neuchâtel
T: (038) 244191; Fax: (038) 259947
Founded: 1969
Pres: J. Guinand; Gen Secr: J.-A. Tschoumy
Focus: Doc; Educ
Periodicals
Coordination
Report 10731

Institut Suisse de Recherches Expérimentales sur le Cancer / Schweizerisches Institut für Experimentelle Krebsforschung / Swiss Institute for Experimental Cancer Research, 155 Chemin des Boveresses, 1066 Epalinges
T: (021) 3165858; CA: ISRECANCER Epalinges; Fax: (021) 6526933
Founded: 1964; Members: 156
Pres: J.-J. Cevey; Gen Secr: Prof. B. Hirt
Focus: Cell Biol & Cancer Res
Periodical
International Journal of Cancer (weekly) . 10732

International Association for the Study of Insurance Economics, 18 Chemin Rieu, 1208 Genève
T: (022) 3470938; Tx: 23358; Fax: (022) 3472078
Founded: 1973
Pres: Prof. Orio Giarini
Focus: Insurance
Periodical
The Geneva Papers on Risk and Insurance (weekly) 10733

International Board on Books for Young People (IBBY), Nonnenweg 12, Postf, 4003 Basel
T: (061) 2722917; Fax: (061) 2722757
Founded: 1953
Gen Secr: Leena Maissen
Focus: Lit
Periodical
Bookbird: World of Children's Books (weekly) 10734

International Computing Centre (ICC), Palais des Nations, 1211 Genève 10
T: (022) 7913201; Fax: (022) 7919746
Founded: 1971
Gen Secr: Raymond Wee
Focus: Computer & Info Sci 10735

Internationale Architekten-Union, Sektion Schweiz, c/o SIA, Selnaustr 16, Postf, 8039 Zürich
T: (01) 2831515; Fax: (01) 2016335
Pres: Regina Gonthier
Focus: Archit 10736

Internationale Föderation der Vereine der Textilchemiker und Coloristen (IFVTCC) / International Federation of Associations of Textile Chemists and Colourists, Hollenweg 8a, 4153 Reinach
Founded: 1930; Members: 11
Gen Secr: Dr. Pierre Albrecht
Focus: Chem; Textiles 10737

Internationale Gesellschaft für Musikwissenschaft (IGMW) / Société Internationale de Musicologie / International Musicological Society, Postf 1561, 4001 Basel
Founded: 1927; Members: 1400
Gen Secr: Rudolf Häusler
Focus: Music
Periodical
Acta Musicologica (weekly) 10738

Internationales Kali-Institut / International Potash Institute, Schneidergasse 27, Postf 1609, 4001 Basel
T: (061) 2612922; Fax: (061) 2612925
Founded: 1952
Focus: Agri
Periodicals
International Fertilizer Correspondent/Corresponsal Internacional Agrícola (weekly)
Kali-Briefe/Potash Review/Revue de la Potassa/Revista de la Potasa (weekly) . 10739

Internationales Komitee Giessereitechnischer Vereinigungen, Konradstr 9, Postf 7190, 8023 Zürich
T: (01) 2719090; Fax: (01) 2719292
Founded: 1927
Gen Secr: Dr. J. Gerster
Focus: Metallurgy 10740

Internationale Union Demokratisch-Sozialistischer Erzieher (IUDSE), Amslergut, 5103 Wildegg
T: (064) 531562
Founded: 1951; Members: 18
Focus: Educ 10741

Internationale Vereinigung für Brückenbau und Hochbau (IVBH) (IABSE) / Association Internationale des Ponts et Charpentes (AIPC) / International Association for Bridge and Structural Engineering, c/o ETH-Hönggerberg, 8093 Zürich
T: (01) 6332647; CA: IABSE Zürich; Fax: (01) 3712131
Founded: 1929; Members: 3500
Gen Secr: Alain Golay
Focus: Civil Eng
Periodicals
IABSE Congress Report
IABSE Report
Structural Engineering Documents
Structural Engineering International (weekly) 10742

Internationale Vereinigung für gewerblichen Rechtsschutz (IVfgR) / Association Internationale pour la Protection de la Propriété Industrielle / International Association for the Protection of Industrial Property, Bleicherweg 58, 8027 Zürich
T: (01) 2041212; Tx: 815656; Fax: (01) 2027502
Founded: 1897; Members: 6980
Focus: Law
Periodical
AIPPI Newsletter (3-7 times annually) . . 10743

Internationale Vereinigung für Natürliche Wirtschaftsordnung (INWO) / Ligue Internationale pour l'Ordre Economique Naturel / International Association for a Natural Economic Order, Postf, 5001 Aarau
Pres: Werner Rosenberger
Focus: Econ
Periodical
Evolution (11 times annually) 10744

Internationale Vereinigung für Saatgutprüfung / Association Internationale d'Essais de Semences / International Seed Testing Association, Postf 412, 8046 Zürich
Founded: 1924; Members: 180
Pres: Don Scott; Gen Secr: Prof. A. Lovato
Focus: Agri
Periodicals
ISTA News Bulletin (weekly)
Seed Science and Technology (3 times annually) 10745

Internationale Vereinigung gegen den Lärm, Hirschenpl 7, 6004 Luzern
T: (041) 513013; Fax: (041) 529093
Founded: 1959
Pres: Karel Novotny; Gen Secr: Dr. W. Aecherli
Focus: Ecology
Periodical
Informationsbulletin (weekly) 10746

Internationale Vereinigung wissenschaftlicher Fremdenverkehrsexperten, Varnbüelstr 19, 9000 Sankt Gallen
T: (071) 302530; Fax: (071) 302536
Founded: 1949; Members: 380
Pres: Prof. Dr. C. Kaspar; Gen Secr: Dr. H.P. Schmidhauser
Focus: Econ; Ecology; Soc Sci
Periodical
Zeitschrift für Fremdenverkehr (weekly) . 10747

Internationale Veterinäranatomische Nomenklatur-Kommission (IVANK) / International Committee on Veterinary Anatomical Nomenclature, Winterthurerstr 260, 8057 Zürich
Founded: 1957; Members: 23
Gen Secr: Dr. Josef Frewein
Focus: Vet Med 10748

International Federation for Psychotherapy, c/o Prof. Dr. Edgar Heim, Murtenstr 21, 3010 Bern
Founded: 1934; Members: 4000
Pres: Prof. Dr. Edgar Heim; Gen Secr: Prof. Dr. Wolfgang Senf
Focus: Psych; Therapeutics 10749

International Industrial Relations Association (IIRA), c/o International Labour Office, 1211 Genève 22
Fax: (022) 7884709
Founded: 1966; Members: 1500

Gen Secr: William R. Simpson
Focus: Econ; Business Admin
Periodicals
IIRA Bulletin (weekly)
IIRA Membership Directory (bi-annually)
Proceedings of IIRA World Congresses . 10750

International Institute for Management Development (IMD), CP 915, 1001 Lausanne
Founded: 1946
Focus: Business Admin 10751

International Organization for Biological Control of Noxious Animals and Plants, c/o Swiss Federal Research Station for Agronomy, Reckenholzstr 191-211, 8046 Zürich
Fax: (01) 3777201
Founded: 1956
Gen Secr: Dr. F. Bigler
Focus: Bio
Periodical
Entomophaga (weekly) 10752

International Organization for Standardization, 1 Rue de Varembé, CP 56, 1211 Genève 20
T: (022) 7490111; Fax: (022) 7333430
Founded: 1947; Members: 92
Focus: Standards
Periodicals
ISO 9000 News (weekly)
ISO Bulletin (weekly) 10753

International Project Management Association (INTERNET), c/o FIDES/IBB, Tödistr 47, Postf 656, 8027 Zürich
T: (01) 2493198; Fax: (01) 2493064
Founded: 1971; Members: 4000
Gen Secr: Hans Peter Blaser
Focus: Business Admin
Periodical
International Journal of Project Management (weekly) 10754

International Public Relations Association (IPRA), CP 2100, 1211 Genève 2
T: (022) 7910550; Fax: (022) 7880336
Founded: 1955; Members: 1000
Focus: Advert
Periodicals
Gold Papers
IPRA Newsletter (weekly)
IPRA Review (weekly)
Members Manual 10755

International Society for General Relativity and Gravitation, c/o Institute of Theoretical Physics, Sidlerstr 5, 3012 Bern
Founded: 1974; Members: 350
Gen Secr: Dr. A. Held
Focus: Physics
Periodical
General Relativity and Gravitation (weekly) 10756

International Society of Internal Medicine (ISIM), c/o Regionalspital, 4900 Langenthal
Founded: 1948; Members: 3000
Gen Secr: Dr. Rolf A. Streuli
Focus: Intern Med
Periodical
ISIM-Bulletin (weekly) 10757

International Telecommunication Union (ITU), Pl des Nations, 1211 Genève 20
T: (022) 7305111; Tx: 421000; Fax: (022) 7337256
Founded: 1865; Members: 172
Gen Secr: Dr. Pekka Tarjanne
Focus: Comm 10758

International Union for Conservation of Nature and Natural Resources (IUCN), 28 Rue Mauverney, 1196 Gland
T: (022) 9990001; Tx: 419624; CA: IUCNATURE Gland; Fax: (022) 9990002
Founded: 1948; Members: 773
Pres: Shridath Ramphal; Gen Secr: Dr. M.W. Holdgate
Focus: Bio; Ecology
Periodicals
Interact (weekly)
IUCN Bulletin (weekly) 10759

Kommission Sportmed des Schweizerischen Landesverbandes für Sport, Laubeggstr 70, 3006 Bern
T: (031) 3597111; Tx: 911815; Fax: (031) 3523380
Pres: Dr. Peter Jenoure
Focus: Med 10760

Konferenz Schweizerischer Gymnasialrektoren, c/o Gymnase Cantonal de Nyon, CP 293, 1260 Nyon
T: (022) 3612437; Fax: (022) 3610485
Pres: Daniel Noverraz
Focus: Educ
Periodical
Colloquium (weekly) 10761

Konferenz Schweizerischer Handelsschulrektoren, 7 Via Vela, 6830 Chiasso
T: (091) 447320
Pres: Ettore Cavadini; Gen Secr: Josef Arnold
Focus: Adult Educ 10762

Kunstverein Sankt Gallen, Museumstr 32, 9000 Sankt Gallen
T: (071) 253355
Founded: 1827; Members: 2500
Pres: Dr. H.P. Müller; Gen Secr: C. Kalthoff
Focus: Arts 10763

Laboratoire Européen pour la Physique des Particules, Meyrin, 1211 Genève 23
T: (022) 7676111; Tx: 419000; Fax: (022) 7676555
Founded: 1954
Pres: Sir William Mitchell; Gen Secr: Carlo Rubbia
Focus: Nucl Res
Periodicals
CERN Courier/Courrier (7 times annually)
Courrier CERN / CERN Courier (10 times annually) 10764

Mouvement International des Intellectuels Catholiques/MIIC-Pax Romana / International Catholic Movement for Intellectual and Cultural Affairs/ICMICA-Pax Romana, 37-39 Rue de Vermont, CP 85, 1211 Genève 20
T: (022) 7336740; Fax: (022) 7336749
Pres: Mary J. Mwingira; Gen Secr: M. Mauricio Molina
Focus: Sci
Periodical
Convergence 10765

Naturforschende Gesellschaft des Kantons Glarus (NGG), Berglirain 12, 8750 Glarus
Founded: 1881; Members: 190
Focus: Nat Sci
Periodical
Mitteilungen der Naturforschenden Gesellschaft des Kantons Glarus (weekly) 10766

Naturforschende Gesellschaft in Basel, c/o Dr. Hubert Meindl, Kornfeldstr 77, 4125 Riehen
Founded: 1817; Members: 580
Focus: Nat Sci
Periodical
Verhandlungen der Naturforschenden Gesellschaft in Basel (weekly) 10767

Naturforschende Gesellschaft in Bern (NGB), c/o Stadt- und Universitätsbibliothek, Münstergasse 61, 3000 Bern 7
T: (031) 3203258
Founded: 1786; Members: 480
Pres: Dr. J. Zettel
Focus: Nat Sci
Periodical
Mitteilungen (weekly) 10768

Naturforschende Gesellschaft in Zürich (NGZ), Bundtacherstr 5, 8127 Forch
Founded: 1746; Members: 1200
Focus: Nat Sci
Periodicals
Neujahrsblatt
Vierteljahresschrift (weekly) 10769

Naturforschende Gesellschaft Luzern (NGL), Denkmalstr 4, 6006 Luzern
T: (041) 514340
Founded: 1855; Members: 680
Focus: Nat Sci
Periodical
Mitteilungen der Naturforschenden Gesellschaft Luzern (every 1-2 years) 10770

Naturforschende Gesellschaft Schaffhausen, Postf 432, 8201 Schaffhausen
Founded: 1822; Members: 520
Pres: Dr. J. Walter; Gen Secr: H. Lustenberger
Focus: Nat Sci
Periodicals
Mitteilungen
Neujahrsblätter 10771

Naturforschende Gesellschaft Solothurn, c/o Dr. Peter Berger, Hofmatt, 4582 Bruggelen
Founded: 1823; Members: 373
Focus: Nat Sci
Periodical
Mitteilungen der Naturforschenden Gesellschaft des Kantons Solothurn 10772

Naturwissenschaftliche Gesellschaft Winterthur, Frohbergstr 21, 8542 Wiesendangen
Founded: 1884; Members: 334
Pres: Dr. K.F. Kaiser
Focus: Nat Sci
Periodical
Mitteilungen (every 3 years) 10773

Neue Schweizerische Chemische Gesellschaft, Postf 313, 4010 Basel
T: (061) 2724950; Fax: (061) 2724089
Founded: 1901; Members: 2300
Pres: Dr. K. Heusler; Gen Secr: Dr. K. Gubler
Focus: Chem
Periodical
Helvetica Chimica Acta (9 times annually) 10774

Opera Svizzera dei Monumenti d'Arte (OSMA), Casa San Francesco, Via Cappuccini 8, 6600 Locarno
Founded: 1963; Members: 12000
Pres: R. Respini; Gen Secr: E. Ruesch
Focus: Hist; Arts
Periodical
I Monumenti d'Arte e di Storia del Canton Ticino 10775

Organisation Internationale de Protection Civile (OIPC), 10-12 Chemin de Surville, 1213 Petit-Lancy
T: (022) 7934433; Fax: (022) 7934428
Founded: 1931; Members: 43
Gen Secr: Znaidi Sadok
Focus: Eng
Periodical
Revue Internationale de Protection Civile 10776

Organisation Météorologique Mondiale (OMM) / World Meteorological Organization, 41 Av Giuseppe Motta, CP 2300, 1211 Genève 2
T: (022) 7308111; Fax: (022) 7342326
Founded: 1951; Members: 160
Pres: Zou Jingmeng; Gen Secr: Prof. G.O.P. Obasi
Focus: Geophys
Periodical
WMO Bulletin (weekly) 10777

PEN Club de Suisse Romande, 34 Rue de la Gabelle, 1227 Carouge
Founded: 1949; Members: 65
Pres: Jean-Pierre Moulin; Gen Secr: Brigitte Mantilleri
Focus: Lit
Periodical
Lettre de Information (1-2 times annually) 10778

Permanent Working Group of European Junior Hospital Doctors, c/o VSAO, Postf 229, 3000 Bern 6
Gen Secr: Dr. Hans-U. Würsten
Focus: Med 10779

Physikalische Gesellschaft Zürich (PGZ), c/o Physik-Institut der Kantonsschule Räwibühl, Rämistr 54, 8001 Zürich
T: (01) 2656382
Founded: 1887; Members: 590
Pres: Prof. Dr. H.R. Ott; Gen Secr: Prof. Dr. F. Meier
Focus: Physics 10780

Pro Helvetia, Hirschengraben 22, 8024 Zürich
T: (01) 2519600; Tx: 56969; Fax: (01) 2519606
Founded: 1939; Members: 35
Pres: R. Simmen; Gen Secr: Urs Frauchiger
Focus: Int'l Relat; Arts
Periodical
Passagen (weekly) 10781

Renaissance, Schweizerischer Verband katholischer Akademiker-Gesellschaften, c/o Rudolf Tuor, Oberseeburgrain 1, 6006 Luzern
Focus: Sci 10782

Rheinaubund, Schweizerische Arbeitsgemeinschaft für Natur und Heimat, Neustadt 29, Postf 584, 8201 Schaffhausen
T: (053) 252658
Focus: Ecology
Periodical
Natur und Mensch: Schweizerische Blätter für Natur und Heimatschutz (weekly) . . . 10783

Schweizer Heimatschutz, Postf, 8032 Zürich
T: (01) 2522660; Fax: (01) 2522870
Pres: Ronald Grisard
Focus: Ecology
Periodical
Heimatschutz/Sauvegarde (weekly) . . . 10784

Schweizerische Ärztegesellschaft für Manuelle Medizin, c/o Klinik Wilhelm Schulthess, Neumünsterallee 10, 8032 Zürich
T: (01) 3857431
Founded: 1959; Members: 1000
Focus: Med 10785

Schweizerische Afrika-Gesellschaft (SAG), Postf, 3000 Bern
T: (031) 447022
Founded: 1974; Members: 170
Pres: Dr. Claude Savary; Gen Secr: Lorenz Homberger
Focus: Ethnology
Periodical
Newsletter (weekly) 10786

Schweizerische Akademie der medizinischen Wissenschaften (SAMW), Peterspl 13, 4051 Basel
T: (061) 2614977; Fax: (061) 2614934
Founded: 1943; Members: 57
Pres: Prof. Dr. A.F. Muller; Gen Secr: Dr. J. Gelzer
Focus: Med
Periodical
Bulletin (weekly) 10787

Schweizerische Akademie der Naturwissenschaften (SNG), Bärenpl 2, 3011 Bern
T: (031) 223375; Fax: (031) 213291
Founded: 1815; Members: 25000
Pres: Prof. Dr. Paul Walter; Gen Secr: Dr. Peter Schindler
Focus: Nat Sci
Periodical
Jahrbuch der Schweizerischen Akademie der Naturwissenschaften (weekly) 10788

Schweizerische Akademie der Sozial- und Geisteswissenschaften, Hirschengraben 11, 3001 Bern
T: (031) 223311
Founded: 1946
Pres: Prof. Dr. Carl Pfaff; Gen Secr: Dr. Beat Sitter-Liver
Focus: Humanities; Soc Sci
Periodical
Bulletin (3-4 times annually) 10789

Schweizerische Akademie der Technischen Wissenschaften (SATW), Postf, 8039 Zürich
T: (01) 2831616; Fax: (01) 2831620
Founded: 1981; Members: 45000
Pres: Prof. Dr. J.-C. Badoux
Focus: Eng 10790

Schweizerische Akademische Gesellschaft der Anglisten (SAGA) / Société Suisse d'Etudes Anglaises, c/o Département d'Anglais, Faculté des Lettres, Université de Genève, 22 Blvd des Philosophes, 1205 Genève
T: (022) 7057034, 7057027; Fax: (022) 7810171
Founded: 1949; Members: 61
Pres: Prof. Dr. Richard Waswo; Gen Secr: Prof. Dr. Udo Fries
Focus: Ling; Lit
Periodicals
English Studies at Swiss Universities (weekly)
Swiss Papers in English Language and Literature (bi-annually) 10791

Schweizerische Arbeitsgemeinschaft für akademische Berufs- und Studienberatung (AGAB), Zentralstr 28, 6002 Luzern
T: (041) 245252
Founded: 1959
Pres: Pierre Kaech; Gen Secr: Pierre Kaech
Focus: Educ 10792

Schweizerische Arbeitsgemeinschaft für hauswirtschaftliche Bildungs- und Berufsfragen, Kürbergstr 33, 8049 Zürich
T: (01) 3421484
Pres: T. Huber; Gen Secr: Alice Bürki
Focus: Home Econ 10793

Schweizerische Arbeitsgemeinschaft für Jugendmusik und Musikerziehung (SAJM), Windenbodenstr, 6345 Neuheim
T: (042) 522829
Founded: 1956; Members: 1700
Focus: Music; Educ
Periodical
SAJM (weekly) 10794

Schweizerische Arbeitsgemeinschaft für Raumfahrt (SAFR), Postf 4215, 6002 Luzern
T: (041) 558567
Founded: 1959
Focus: Aero
Periodical
SAFR-Mitteilungen (weekly) 10795

Schweizerische Arbeitsgemeinschaft für Rehabilitation, c/o Dr. M. Maeder, Zum Burgfeldenhof 40, 4055 Basel
Pres: Dr. Heinz Christoph; Gen Secr: Dr. M. Maeder
Focus: Rehabil 10796

Schweizerische Arbeitsgemeinschaft für Schul- und Jugendzahnpflege, 8135 Langnau
T: 7133287
Focus: Dent 10797

Schweizerische Asiengesellschaft (SA), c/o Ostasiatisches Seminar der Universität, Zürichbergstr 4, 8032 Zürich
T: (01) 2573181
Founded: 1947; Members: 180
Pres: Prof. Dr. Robert H. Gassmann
Focus: Ethnology; Philos; Rel & Theol; Lit; Soc Sci
Periodicals
Asiatische Studien/Etudes Asiatiques (weekly)
Schweizer Asiatische Studien 10798

Schweizerische Bibliophilen-Gesellschaft, Voltastr 43, 8044 Zürich
Founded: 1921; Members: 700
Pres: Dr. Conrad Ulrich
Focus: Libraries & Bk Sci
Periodicals
Librarium (3 times annually)
Stultifera Navis 10799

Schweizerische Botanische Gesellschaft (SBG), c/o Geobotanisches Institut, Zollikerstr 107, 8008 Zürich
Founded: 1890
Pres: Dr. P. Geissler; Gen Secr: Dr. F. Jacquemoud
Focus: Botany
Periodical
Botanica Helvetica (2-4 times annually) . 10800

Schweizerische Diabetes-Gesellschaft (SDG), Forchstr 95, 8032 Zürich
T: (01) 3831315; Fax: (01) 4228912
Pres: Ernst Feldmann; Gen Secr: Dr. J. Schultheiss
Focus: Diabetes
Periodical
Journal des Diabétiques (5 times annually) 10801

Schweizerische Direktoren-Konferenz gewerblicher Berufs- und Fachschulen, Burgerstr 24, 6003 Luzern
T: (041) 232981
Focus: Adult Educ 10802

Schweizerische Energie-Stiftung, Sihlquai 67, 8005 Zürich
T: (01) 2715464; Fax: (01) 2730369
Founded: 1977
Pres: R. Bär; Gen Secr: Karl Wellinger

Focus: Energy
Periodical
Energie + Umwelt (weekly) 10803

Schweizerische Entomologische Gesellschaft (SEG/SES) / Société Entomologique Suisse, c/o Entomologisches Institut, ETH-Zentrum, 8092 Zürich
T: (01) 2563919
Founded: 1858; Members: 385
Pres: Dr. M. Brancucci; Gen Secr: Dr. C. Flückiger
Focus: Entomology
Periodicals
Insecta Helvetica (weekly)
Mitteilungen der Schweizerischen Entomologischen Gesellschaft/ Bulletin de la Société Entomologique Suisse (weekly) 10804

Schweizerische Esperanto-Gesellschaft, Zumhofstr 22, 6010 Kriens
T: (041) 222967
Pres: Claude Gacond; Gen Secr: Andy E. Kuenzli
Focus: Ling
Periodical
Svisa Esperanto-Societi Informas (9 times annually) 10805

Schweizerische Ethnologische Gesellschaft (SEG) / Société Suisse d'Ethnologie, Schwanengasse 7, 3011 Bern
T: (031) 658998
Focus: Ethnology
Periodicals
Ethnologica Helvetica (weekly)
SEG/SSE – Information (weekly) 10806

Schweizerische Franchising-Vereinigung (SFV), Schlossbergstr 22, Postf 125, 8702 Zollikon
T: 3914477
Focus: Econ 10807

Schweizerische Gesellschaft der Kernfachleute (SGK), c/o Schweizerische Vereinigung für Atomenergie, Monbijoustr 5, Postf 5032, 3001 Bern
Founded: 1958; Members: 250
Focus: Nucl Res 10808

Schweizerische Gesellschaft für Agrarwirtschaft, Sonneggstr 33, 8092 Zürich
T: (01) 25656391; Fax: (01) 2523410
Pres: Bernard Lehmann; Gen Secr: Michel Roux
Focus: Agri
Periodical
Schweizerische Zeitschrift für Agrarwirtschaft und Agrarsoziologie (weekly) 10809

Schweizerische Gesellschaft für Astrophysik und Astronomie, c/o Oberservatoire de Genève, 1290 Sauverny
T: (022) 552611
Founded: 1968; Members: 105
Focus: Astrophys; Astronomy 10810

Schweizerische Gesellschaft für Aussenpolitik (SGA), Stapferhaus, Schloss, 5600 Lenzburg
T: (064) 515751
Founded: 1968; Members: 600
Pres: Dr. Raymond Probst; Gen Secr: M. Kirchhofer
Focus: Poli Sci 10811

Schweizerische Gesellschaft für Automatik (SGA), Postgasse 17, 3011 Bern
T: (031) 3120512; Fax: (031) 3121250
Founded: 1956
Pres: Prof. Dr. A.H. Glattfelder
Focus: Eng
Periodical
SGA-ASSPA-Bulletin (weekly) 10812

Schweizerische Gesellschaft für Balneologie und Bioklimatologie, c/o Dr. R. Eberhard, Heilbadzentrum, 7500 Sankt Moritz
T: (082) 37171
Founded: 1902; Members: 126
Pres: Dr. O. Knüsel; Gen Secr: Dr. R. Eberhard
Focus: Physical Therapy
Periodical
Congress report 10813

Schweizerische Gesellschaft für Bildungs- und Erziehungsfragen, Höflistr 12, Postf 313, 8135 Langnau
T: (01) 7132141
Focus: Educ
Periodical
Die Presserundschau (weekly) 10814

Schweizerische Gesellschaft für cystische Fibrose, Bellevuestr 166, 3095 Spiegel
T: (031) 9722828; Fax: (031) 9722828
Pres: Jean Clivaz; Gen Secr: Regula Salm-Mueller
Focus: Cell Biol & Cancer Res; Pathology
Periodical
cf-Bulletin (weekly) 10815

Schweizerische Gesellschaft für Dermatologie und Venereologie, c/o Département de Dermatologie, Université, CHUV, 1011 Lausanne
T: (021) 414646
Pres: Prof. Dr. T. Rutli; Gen Secr: Prof. Dr. E. Frenk
Focus: Derm; Venereology 10816

Schweizerische Gesellschaft für die Rechte der Urheber musikalischer Werke (SUISA) / Swiss Society for Rights of Authors of Musical Works, Bellariastr 82, 8038 Zürich
T: (01) 4856666; Tx: 816576; Fax: (01) 4854333
Founded: 1942; Members: 10100
Pres: Hans Ulrich Lehmann
Focus: Music; Law 10817

Schweizerische Gesellschaft für Ernährungsforschung (SGE), c/o Geriatrische Universitätsklinik, Kantonsspital, 4031 Basel
T: (061) 2652954; Fax: (061) 2652670
Focus: Food 10818

Schweizerische Gesellschaft für Familienforschung (SGFF), c/o Dr. Hans B. Kälin, Steinbühlallee 189, 4054 Basel
T: (061) 3017455
Pres: Dr. Hans B. Kälin
Focus: Genealogy
Periodicals
Mitteilungen der SGFF (3 times annually)
SGFF-Jahrbuch (weekly) 10819

Schweizerische Gesellschaft für Feintechnik (SGFT), Kirchenweg 4, 8032 Zürich
T: (01) 478400
Focus: Eng 10820

Schweizerische Gesellschaft für Gerontologie (SGG), c/o Klinik Geriatrie-Rehabilitation, Zieglerspital, 3001 Bern
Founded: 1953; Members: 950
Focus: Geriatrics 10821

Schweizerische Gesellschaft für Geschichte der Medizin und der Naturwissenschaften / Société Suisse d'Histoire de la Médicine et des Sciences Naturelles, Bühlstr 26, 3000 Bern 9
Founded: 1921; Members: 280
Gen Secr: Prof. Dr. U. Boschung
Focus: Med; Cultur Hist
Periodical
Gesnerus (weekly) 10822

Schweizerische Gesellschaft für Gynäkologie und Geburtshilfe (SGGG), c/o CHUV, 1011 Lausanne
T: (021) 3142523; Fax: (021) 3142670
Founded: 1964; Members: 690
Pres: Prof. Dr. P. de Grandi; Gen Secr: Dr. Hansjörg Welti
Focus: Gynecology 10823

Schweizerische Gesellschaft für Innere Medizin, c/o Medizinische Abteilung, Regionalspital, 4900 Langenthal
Founded: 1932; Members: 1421
Pres: Prof. P. Burckhardt
Focus: Intern Med 10824

Schweizerische Gesellschaft für Kardiologie / Société Suisse de Cardiologie, c/o Inselspital, Freiburgstr 3010, 4031 Bern
T: (031) 632111
Focus: Cardiol 10825

Schweizerische Gesellschaft für Kartographie, c/o ETH, Hönggerberg, 8093 Zürich
Fax: (01) 3720438
Founded: 1969; Members: 256
Pres: Prof. Ernst Spiess
Focus: Cart
Periodicals
Kartographische Nachrichten (weekly)
Kartographische Schriftenreihe (weekly) . 10826

Schweizerische Gesellschaft für kaufmännisches Bildungswesen, 15 Rue Le Corbusier, 1208 Genève
T: (022) 3463131; Fax: (022) 7892732
Focus: Adult Educ
Periodical
Schweizerische Zeitschrift für kaufmännisches Bildungswesen 10827

Schweizerische Gesellschaft für Kommunikations- und Medienwissenschaft (SGKM), c/o SRG Forschungsdienst, Giacomettistr 1, Postf, 3000 Bern 15
T: (031) 439430; Fax: (031) 439436
Founded: 1974
Pres: Prof. Dr. Matthias Steinmann
Focus: Comm Sci
Periodical
SGKM-Bulletin (weekly) 10828

Schweizerische Gesellschaft für Lebensmittel-Wissenschaft und -Technologie (SGLWT), Postf 561, 8820 Wädenswil
Members: 530
Pres: Dr. R. Buri
Focus: Food; Eng
Periodicals
Lebensmittel-Technologie (weekly)
Lebensmittel-Wissenschaft und -Technologie: Food Science & Technology (weekly) 10829

Schweizerische Gesellschaft für Lehrerinnen- und Lehrerbildung (SGL), Muesmattstr 27, 3012 Bern
T: (031) 658316
Members: 500
Pres: Prof. Dr. Hans Badertscher
Focus: Educ; Psych
Periodical
Beiträge zur Lehrerbildung (3 times annually) 10830

Schweizerische Gesellschaft für Logik und Philosophie der Wissenschaften, c/o Philosophisches Seminar, Universität, Falkenpl 16, 3012 Bern
Founded: 1957; Members: 42
Focus: Philos 10831

Schweizerische Gesellschaft für Marketing, Bleicherweg 21, 8022 Zürich
T: (01) 2023425
Founded: 1941; Members: 320
Focus: Marketing 10832

Schweizerische Gesellschaft für Mikrobiologie, c/o Institut Microbion, CP, 1001 Lausanne
Focus: Microbio 10833

Schweizerische Gesellschaft für Mikrotechnik, c/o VSM, Kirchenweg 4, 8032 Zürich
T: (01) 3844844; Tx: 816519
Founded: 1962; Members: 108
Pres: P. Stauber; Gen Secr: K. Eckstein
Focus: Electronic Eng 10834

Schweizerische Gesellschaft für Muskelkranke (SGMK), Forchstr 136, 8032 Zürich
T: (01) 4221634; Fax: (01) 4225931
Pres: R.T. Fässler; Gen Secr: Sylvia Schaetzle
Focus: Physiology; Pathology
Periodical
Mitteilungsblatt (1-2 times annually) . . . 10835

Schweizerische Gesellschaft für Oberflächentechnik, Hänselsmatt 2, 2544 Bettlach
T: (065) 553574; Fax: (065) 553038
Founded: 1955; Members: 550
Focus: Eng
Periodical
Oberfläche/Surface (weekly) 10836

Schweizerische Gesellschaft für Onkologie, c/o ISREC, 1066 Epalinges
T: (021) 333061
Focus: Cell Biol & Cancer Res 10837

Schweizerische Gesellschaft für orientalische Altertumswissenschaft / Société Suisse pour l'Etude du Proche Orient Ancien / Swiss Society for the Study of the Ancient Near East, c/o Prof. Albert de Pury, 4 Blvd des Tranchées, 1205 Genève
T: (022) 478748
Focus: Hist
Periodical
Orbis Biblicus et Orientalis (weekly) . . 10838

Schweizerische Gesellschaft für Orthopädie (SGO) / Société Suisse d'Orthopédie, Postf, 3000 Bern 25
T: (031) 3329619; Fax: (031) 3329879
Pres: Dr. Andreas Burckhardt; Gen Secr: Prof. Dr. André Gächter
Focus: Orthopedics
Periodical
Nachrichten-Bulletin SGO (3 times annually) 10839

Schweizerische Gesellschaft für Oto-Rhino-Laryngologie, Hals- und Gesichtschirurgie / Société Suisse d'Oto-Rhino-Laryngologie et de Chirurgie Cervico-Faciale, c/o Universitäts-HNO-Klinik, Inselspital, 3010 Bern
T: (031) 642921
Founded: 1913; Members: 250
Focus: Surgery; Otorhinolaryngology
Periodical
ORL-Verhandlungsberichte der wissenschaftlichen Frühjahrsversammlung (weekly) 10840

Schweizerische Gesellschaft für Personalfragen (SGP), Löwenstr 20, 8001 Zürich
T: (01) 2119544
Founded: 1972; Members: 2200
Focus: Business Admin
Periodical
SGP Mitteilungen (weekly) 10841

Schweizerische Gesellschaft für Physikalische Medizin und Rehabilitation, c/o Schweizerische Rheumaliga, Renggerstr 71, 8038 Zürich
T: (01) 4825600
Founded: 1946
Focus: Physiology 10842

Schweizerische Gesellschaft für Pneumologie (SGP), Falkenpl 9, Postf, 3001 Bern
T: (031) 3020822; Fax: (031) 3028833
Pres: Prof. Dr. T.C. Medici; Gen Secr: A. von Allman
Focus: Pulmon Dis
Periodical
Tuberkulose und Lungenkrankheiten: Beilage zum Bulletin des Bundesamtes für Gesundheitswesen (8 times annually) 10843

Schweizerische Gesellschaft für Psychologie, c/o Institut de Psychologie, 2 BFSH, 1015 Lausanne
Founded: 1943; Members: 450
Pres: Prof. Dr. F. Gaillard
Focus: Psych
Periodicals
Psychologie
Schweizerische Zeitschrift für Psychologie (weekly) 10844

Schweizerische Gesellschaft für Rheumatologie, c/o Schweizerische Rheumaliga, Renggerstr 71, 8038 Zürich
T: (01) 4825600
Focus: Rheuma
Periodical
Zeitschrift für Rheumatologie 10845

Schweizerische Gesellschaft für Skandinavische Studien (SGSS), c/o Deutsches Seminar, Abt für Nordische Philologie, Universität Zürich, Rämistr 74, 8001 Zürich
T: (01) 2572511; Fax: (01) 2620250
Founded: 1961; Members: 230
Pres: Prof. Dr. Oskar Bandle; Gen Secr: Dr. Thomas Seiler
Focus: Ling; Lit
Periodical
Beiträge zur nordischen Philologie (weekly) 10846

Schweizerische Gesellschaft für Sozial- und Präventivmedizin / Société Suisse de Médecine Sociale et Préventive, Finkenhubelweg 11, 3012 Bern
T: (031) 648631
Members: 580
Focus: Hygiene
Periodical
Sozial- und Präventivmedizin: Médecine Sociale et Préventive (weekly) 10847

Schweizerische Gesellschaft für Soziologie / Société Suisse de Sociologie, Rämistr 69, 8001 Zürich
Founded: 1955; Members: 450
Pres: René Levy; Gen Secr: Peter Rusterholz
Focus: Sociology
Periodicals
Bulletin (weekly)
Schweizerische Zeitschrift für Soziologie: Revue Suisse de Sociologie (3 times annually) . 10848

Schweizerische Gesellschaft für Statistik und Volkswirtschaft / Société Suisse de Statistique et d'Economie Politique, c/o Bundesamt für Statistik, 3003 Bern
T: (031) 618699
Founded: 1864; Members: 1000
Pres: Prof. Claude Jeanrenaud; Gen Secr: Martine Jacot-Diserens
Focus: Stats; Econ
Periodical
Schweizerische Zeitschrift für Volkswirtschaft und Statistik (weekly) 10849

Schweizerische Gesellschaft für Theaterkultur (SGTK), Herrenholzweg 33, 8906 Bonstetten
T: (01) 7000357
Founded: 1927; Members: 550
Focus: Perf Arts
Periodicals
Mimos: Mitteilungen der SGIK (weekly)
Schriften der Schweizerischen Gesellschaft für Theaterkultur
Schweizer Theater-Jahrbuch (weekly)
Szene Schweiz/Scène suisse/Scena svizzera 10850

Schweizerische Gesellschaft für Tropenmedizin und Parasitologie, Raemistr 3, 8001 Zürich
Focus: Trop Med 10851

Schweizerische Gesellschaft für Umweltschutz (SGU), Merkurstr 45, 8032 Zürich
T: (01) 2512826
Focus: Ecology
Periodical
SGU Bulletin (weekly) 10852

Schweizerische Gesellschaft für Ur- und Frühgeschichte, Petersgraben 9-11, 4001 Basel
T: (0611) 253078
Members: 3000
Gen Secr: Dr. Urs Niffeler
Focus: Archeol
Periodicals
Archäologie der Schweiz, Archéologie Suisse, Archeologia Svizzera (weekly)
Jahrbuch SGUF, Annuaire SSPA, Annuario SSPA 10853

Schweizerische Gesellschaft für Vakuum-Physik und -Technik, c/o Institut für Technische Physik, ETH, Hönggerberg, 8093 Zürich
Focus: Physics; Eng 10854

Schweizerische Gesellschaft für Versicherungsrecht, Postf 4338, 8022 Zürich
T: (01) 2843260
Pres: Prof. Dr. Pierre Tercier; Gen Secr: Prof. Dr. Moritz Kuhn
Focus: Law 10855

Schweizerische Gesellschaft für Volkskunde, Augustinergasse 19, 4051 Basel
T: (061) 2619900
Founded: 1896; Members: 1900
Pres: Dr. Hans Schnyder
Focus: Ethnology
Periodicals
Folklore suisse / Folclore svizzero (3 times annually)
Schweizerisches Archiv für Volkskunde / Archives Suisses des Traditions Populaires (weekly)
Schweizer Volkskunde: Korrespondenzblatt (weekly) 10856

Schweizerische Gesellschaft Pro Technorama, Postf 3, 8404 Winterthur
T: (052) 2430505; Fax: (052) 2422967
Members: 5500
Pres: Othmar Hegi
Focus: Eng; Cultur Hist
Periodical
Pro Technorama (weekly) 10857

Schweizerische Graphologische Berufsvereinigung, Grundstr 6, 8703 Erlenbach
T: (033) 9153166
Focus: Graphology 10858

Schweizerische Hämophilie-Gesellschaft (SHG), Seestr 45, Postf 531, 8027 Zürich
T: (01) 2810855; Fax: (01) 2810855
Pres: Ursula Fries; Gen Secr: M. Angst
Focus: Hematology
Periodical
Bulletin (weekly) 10859

Schweizerische Heilpädagogische Gesellschaft, Brunnmattstr 38, Postf, 3000 Bern 14
T: (031) 262629; Fax: (031) 264521
Pres: Dr. A. Lüthy
Focus: Educ; Rehabil
Periodicals
Pages Romandes (weekly)
Schweizerische Heilpädagogische Rundschau (SHR) (weekly) 10860

Schweizerische Hochschulkonferenz (SHK), Wildhainweg 21, 3012 Bern
T: (031) 245533; Fax: (031) 241792
Founded: 1969; Members: 29
Pres: H.R. Striebel; Gen Secr: N. Ischi
Focus: Adult Educ
Periodicals
Jahresbericht (weekly)
Politique de le Science (4-6 times annually) 10861

Schweizerische Hochschulrektoren-Konferenz, Seidenweg 68, 3012 Bern
T: (031) 3028341; Fax: (031) 3026811
Pres: Prof. Dr. Luc Weber; Gen Secr: Dr. Rudolf Nägeli
Focus: Educ 10862

Schweizerische Informationskonferenz, Postf, 4003 Basel
T: (061) 257171
Focus: Computer & Info Sci 10863

Schweizerische katholische Arbeitsgemeinschaft für Ausländerfragen (SKAF), Neustadtstr 7, 6003 Luzern
T: (041) 230347; Fax: (041) 235846
Pres: Dr. Walter Gut; Gen Secr: Dr. Urs Köppel
Focus: Rel & Theol 10864

Schweizerische Konferenz der Rektoren der Berufsschulen kaufmännischer Richtung, Postf 6390, 8023 Zürich
T: (01) 2024710
Focus: Adult Educ 10865

Schweizerische Krebsliga, Monbijoustr 61, Postf 8219, 3001 Bern
T: (031) 3722767; Fax: (031) 3718301
Pres: Prof. W. Lehmann; Gen Secr: M. Wieser
Focus: Cell Biol & Cancer Res 10866

Schweizerische Kriminalistische Gesellschaft / Société Suisse de Droit Pénal, 6512 Giubiasco
Focus: Law
Periodical
Rechtsprechung in Strafsachen (weekly) . 10867

Schweizerische Lichttechnische Gesellschaft (SLG), Postgasse 17, 3011 Bern
T: (031) 3122251; Fax: (031) 3121250
Pres: Philippe Joye; Gen Secr: Dr. Charles Giroud
Focus: Eng; Energy
Periodical
Mitteilung an die Mitglieder (weekly) . . 10868

Schweizerische Liga gegen den Lärm, Hirschenpl 7, 6004 Luzern
T: (041) 513013
Pres: Prof. Dr. Meinrad Schär; Gen Secr: Dr. Willy Aecherli
Focus: Ecology 10869

Schweizerische Liga gegen Epilepsie (SLgE), Feldeggstr 71, Postf 129, 8032 Zürich
T: (01) 3830531
Focus: Pathology
Periodicals
Epilepsie: Informationsblatt (weekly)
Epilepsie-Selbsthilfe-Zeitung: Kontakte (weekly) 10870

Schweizerische Management-Gesellschaft (SMG) / Swiss Management Association, Bleicherweg 64a, Postf, 8027 Zürich
T: (01) 2022325; Fax: (01) 2022320
Founded: 1961; Members: 800
Pres: Fred Sutter; Gen Secr: Petra Kaiser
Focus: Business Admin 10871

Schweizerische Mathematische Gesellschaft (SMG), c/o Mathematisches Institut, Universität Zürich, Rämistr 74, 8001 Zürich
Founded: 1910; Members: 460
Pres: Prof. Harald Holmann; Gen Secr: Prof. P. Hess
Focus: Math
Periodicals
Commentarii Mathematici Helvetici (weekly)
Elemente der Mathematik (weekly) . . . 10872

Schweizerische Multiple Sklerose Gesellschaft (SMSG), Brinerstr 1, 8036 Zürich
T: (01) 4614600; Fax: (01) 4570939
Members: 9000
Pres: Dr. Paul Eisenring; Gen Secr: Walter Lerch
Focus: Pathology
Periodical
MS Aktuell/SP Actuel/Attualitá SM (weekly) 10873

Schweizerische Musikforschende Gesellschaft, Sonnenweg 23, 4020 Basel
T: (061) 3122973
Founded: 1916; Members: 540
Pres: Prof. Dr. Ernst Lichtenhahn; Gen Secr: Dr. Dorothea Baumann
Focus: Music
Periodicals
Jahrbuch (weekly)
Schweizer Musikdenkmäler 10874

Schweizerische Neurologische Gesellschaft (SNG), c/o Neurologische Klinik, Kantonsspital Aarau, 5000 Aarau
Founded: 1908; Members: 228
Pres: Prof. A.J. Steck; Gen Secr: Prof. K. Hess
Focus: Neurology
Periodical
Schweizer Archiv für Neurologie 10875

Schweizerische Normen-Vereinigung (SNV), Mühlebachstr 54, 8008 Zürich
T: (01) 2545454; Fax: (01) 2545474
Founded: 1919; Members: 500
Pres: G. Hongler; Gen Secr: Dr. H. Zürrer
Focus: Standards
Periodicals
SNV Bulletin (11 times annually)
switec Information 10876

Schweizerische Numismatische Gesellschaft / Société Suisse de Numismatique, c/o Stadt- und Universitätsbibliothek, Münstergasse 61, 3000 Bern 7
Members: 743
Pres: Giovanni-M. Staffieri; Gen Secr: Ruth Feller
Focus: Numismatics
Periodicals
Schweizerische Numismatische Rundschau: Revue Suisse de Numismatique (weekly)
Schweizerischer Münzkatalog (weekly)
Schweizer Münzblätter: Gazette Numismatique Suisse (weekly) 10877

Schweizerische Ophthalmologische Gesellschaft (S.O.G.), Zugerstr 30, 8810 Horgen
T: (01) 7252229; Fax: (01) 7252237
Founded: 1927; Members: 300
Focus: Ophthal 10878

Schweizerische Paläontologische Gesellschaft, c/o Musée d'Histoire Naturelle, 1 Rte de Malagnou, CP 6434, 1211 Genève 6
T: (022) 7353445; Fax: (022) 7359130
Founded: 1921; Members: 255
Gen Secr: Dr. D. Decrouez
Focus: Paleontology
Periodical
Bericht 10879

Schweizerische Philosophische Gesellschaft / Société Suisse de Philosophie, Dufourstr 31, 9000 Sankt Gallen
T: (071) 259311
Founded: 1940; Members: 675
Pres: Dr. Daniel Brühlmeier; Gen Secr: Dr. Eduard Marbach
Focus: Philos
Periodicals
Studia Philosophica (weekly)
Supplementa (weekly) 10880

Schweizerische Physikalische Gesellschaft / Société Suisse de Physique / Swiss Physical Society, 24 Quai Ernest Ansermet, 1211 Genève 4
T: (022) 7026253; Fax: (022) 7812192
Founded: 1925; Members: 1300
Focus: Physics
Periodical
Helvetica Physica Acta (8 times annually) 10881

Schweizerischer Altphilologen-Verband, Sommergasse 46, 4056 Basel
Focus: Ling; Lit
Periodical
Bulletin (weekly) 10882

Schweizerischer Anglistenverband, c/o Liceo Cantonale, Viale Cattaneo 4, 6900 Lugano
T: (091) 236561; Fax: (091) 237754
Focus: Ling; Lit 10883

Schweizerischer Anwaltsverband, Bollwerk 21, 3001 Bern
T: (031) 3122505; Tx: 815358; Fax: (031) 3123103
Founded: 1898; Members: 4800
Pres: Arnaldo Bolla; Gen Secr: Dr. Max P. Oesch
Focus: Law
Periodical
Der Schweizer Anwalt (weekly) 10884

Schweizerischer Apotheker-Verein / Société Suisse de Pharmacie, Stationsstr 12, 3097 Bern-Liebefeld
T: (031) 9715858; Fax: (031) 9721569
Founded: 1843; Members: 4500
Focus: Pharmacol
Periodicals
Index Nominum
Pharmaceutica Acta Helvetiae
Schweizerische Apothekerzeitung 10885

Schweizerischer Ballettlehrer-Verband, Holgass-Str 47, 8634 Hombrechtikon
T: (055) 424494
Focus: Educ; Perf Arts 10886

Schweizerischer Berufsverband diplomierter Sozialarbeiterinnen und Sozialpädagoginnen, Holligenstr 70, 3008 Bern 21
T: (031) 3822822; Fax: (031) 3821125
Focus: Sociology; Educ
Periodicals
Sozialarbeit (weekly)
Travail Social (weekly) 10887

Schweizerischer Berufsverband für Angewandte Psychologie (SBAP), Winkelweg 3, 8127 Forch
T: (01) 9803620; Fax: (01) 9803620
Founded: 1952; Members: 380
Pres: Urs Rüegsegger; Gen Secr: M. Hefti
Focus: Psych
Periodical
SBAP-Bulletin (weekly) 10888

Schweizerischer Bund für Elternbildung (SBE), Obergrundstr 65, 6003 Luzern
T: (041) 234874
Focus: Adult Educ
Periodical
Bulletin SBE (weekly) 10889

Schweizerischer Bund für Naturschutz, Wartenbergstr 22, Postf, 4020 Basel
T: (061) 3127442; Fax: (061) 3127447
Pres: Jacques Morier-Genoud; Gen Secr: Otto Sieber
Focus: Ecology
Periodicals
Protection de la Nature (8 times annually)
Schweizer Naturschutz (8 times annually) 10890

Schweizerischer Chemiker-Verband, Im Rehwechsel 19, 4102 Binningen
T: 329069
Focus: Chem
Periodical
Chimia (weekly) 10891

Schweizerischer Elektrotechnischer Verein (SEV), Postf, 8034 Zürich
T: (01) 3849222; Tx: 817431; Fax: (01) 4221426
Founded: 1889; Members: 6700
Pres: Jules Peter; Gen Secr: Dr. Johannes Heyner
Focus: Electric Eng 10892

Schweizerischer Fachverband der diplomierten medizinischen Laborantinnen und Laboranten, Waisenhausstr 14, 9000 Sankt Gallen
Fax: (071) 234073
Focus: Med; Eng
Periodical
Labor und Medizin (weekly) 10893

Schweizerischer Fachverband für Schweiss- und Schneidmaterial (SFAS), Birchstr 230, 8050 Zürich
T: (01) 3012121
Focus: Materials Sci 10894

Schweizerischer Fachverband Sozialdienst in Spitälern, c/o Käthi Lüthi, Zieglerspital, 3007 Bern
Focus: Sociology 10895

Schweizerischer Forstverein (SFV), c/o ETH-Zentrum, HG FO 21.1, 8092 Zürich
T: (01) 2565205
Founded: 1843; Members: 1100
Focus: Forestry
Periodical
Schweizerische Zeitschrift für Forstwesen (weekly) 10896

Schweizerischer Handelslehrerverein / Société Suisse des Professeurs de Sciences Commerciales, Arvenweg 12, 8840 Einsiedeln
Focus: Adult Educ
Periodical
Schweizerische Zeitschrift für kaufmännisches Bildungswesen: Revue Suisse pour l'enseignement commercial (weekly) 10897

Schweizerische Rheumaliga, Renggerstr 71, 8038 Zürich
T: (01) 4825600; Fax: (01) 4826439
Pres: Dr. G. Rezzonico; Gen Secr: G. Dalvit
Focus: Rheuma
Periodical
Forum R (weekly) 10898

Schweizerischer Ingenieur- und Architekten-Verein (SIA), Selnaustr 16, 8039 Zürich
Founded: 1837; Members: 7400
Pres: Dr. H.-H. Gasser; Gen Secr: C. Reinhart
Focus: Archit; Civil Eng
Periodicals
Ingénieurs et Architectes Suisses
Rivista Tecnica della Svizzera Italiana
Schweizer Ingenieur und Architekt . . . 10899

Schweizerischer Musikpädagogischer Verband / Société Suisse de Pédagogie Musicale, Forchstr 460, 8702 Zürich
T: (01) 531752
Founded: 1893; Members: 4500
Pres: J. Roman Widmer; Gen Secr: Elisabeth Eichert
Focus: Music; Educ
Periodicals
Agenda du Musicien: Musikerkalender
Bulletin Mensuel SSPM: Mitteilungsblatt
Feuillets Suisses de Pédagogie Musicale: Schweizer musikpädagogische Blätter . . 10900

Schweizerischer Notaren-Verband, Thunstr 73, 3000 Bern 7
T: (031) 3511451
Founded: 1920; Members: 1300
Pres: Pierre Natural; Gen Secr: Andreas B. Notter
Focus: Law 10901

Schweizerischer Physiotherapeuten-Verband (SPV), Stadtstr 30, Postf, 6204 Sempach-Stadt
T: (041) 993388; Fax: (041) 993381
Founded: 1920; Members: 3300
Focus: Therapeutics
Periodical
Der Physiotherapeut (weekly) 10902

Schweizerischer Romanistenverband / Association Suisse des Romanistes, c/o WBZ, Bruchstr 9a, Postf, 6000 Luzern 7
Fax: (041) 220079
Members: 520
Focus: Ling; Lit
Periodical
Lettre Circulaire (weekly) 10903

Schweizerischer Schriftstellerinnen- und Schriftsteller-Verband (SSV), Kirchgasse 25, Postf, 8001 Zürich 1
T: (01) 3020261; Fax: (01) 2613153
Founded: 1912; Members: 650
Pres: Janine Massard; Gen Secr: Lou Pflüger
Focus: Lit
Periodical
Forum der Schriftsteller: Forum des Ecrivains (weekly) 10904

Schweizerischer Technischer Verband (STV), Weinbergstr 41, 8006 Zürich
T: (01) 2613794
Founded: 1905; Members: 17000
Pres: Jean-Michel Stotzer; Gen Secr: A. Waltenspühl
Focus: Eng
Periodical
Schweizerische Technische Zeitschrift (weekly) 10905

Schweizerischer Verband der Akademikerinnen, Masanserstr 93, 7000 Chur
T: (081) 273391
Focus: Sci 10906

Schweizerischer Verband der Ingenieur-Agronomen und der Lebensmittel-Ingenieure / Association Suisse des Ingénieurs Agronomes et des Ingénieurs en Technologie Alimentaire, Länggasse 79, 3052 Zollikofen
T: (031) 570668; Fax: (031) 574925
Founded: 1901; Members: 2200
Pres: S. Guarneri; Gen Secr: O. Meyer
Focus: Agri; Nutrition
Periodical
Bulletin 10907

Schweizerischer Verband der Lehrer an kaufmännischen Berufsschulen, Laubacherweg 3, 6033 Buchrain
T: 333262
Focus: Adult Educ 10908

Schweizerischer Verband für Betriebsorganisation und Fertigungstechnik (SVBF), Postf 350, 8401 Winterthur
Founded: 1946; Members: 2900
Pres: Hans Betschart; Gen Secr: Friedy Frauenknecht
Focus: Business Admin
Periodical
P+P, Planung + Produktion: Fachzeitschrift für Organisation, Rationalisierung, Informationsverarbeitung, Produktionstechnik (weekly) 10909

Schweizerischer Verband für die Materialtechnik (SVMT), c/o Mat Search, 15 Chemin Jean Pavillard, 1009 Pully
T: (021) 7290154; Fax: (021) 7290156
Focus: Materials Sci
Periodical
Oberflächen-Werkstoffe (11 times annually) 10910

Schweizerischer Verband für die Wärmebehandlung der Werkstoffe (SVW), c/o Technikum Winterthur, Ingenieurschule, Postf, 8401 Winterthur
T: (052) 826349
Focus: Eng 10911

Schweizerischer Verband für Fernunterricht / Swiss Association for Education by Correspondence, Postf 866, 8280 Kreuzlingen
T: (072) 724444; Fax: (072) 725562
Founded: 1971
Focus: Educ 10912

Schweizerischer Verband für Frauenrechte, 8 Av de Georgette, 1003 Lausanne
T: (021) 207732
Pres: Simone Chapuis-Bischof
Focus: Law
Periodical
Contact (weekly) 10913

Schweizerischer Verband für Landtechnik (SVLT), Postf 53, 5223 Riniken
T: (056) 412022
Founded: 1924; Members: 38000
Focus: Eng; Agri
Periodical
Schweizer Landtechnik: Technique Agricole (15 times annually) 10914

Schweizerischer Verband für Sport in der Schule (SVSS), Neubrückstr 155, Postf, 3000 Bern 26
T: (031) 3028802; Fax: (031) 3028812
Members: 8000
Pres: Rose-Marie Repond
Focus: Sports; Educ
Periodical
Sporterziehung in der Schule (weekly) . 10915

Schweizerischer Verband kantonal approbierter Zahnärzte, Bäckerstr 15, 8004 Zürich
T: (01) 2424722
Focus: Dent 10916

Schweizerischer Verband von Fachleuten für Alkoholgefährdeten- und Suchtkrankenhilfe, Naglerwiesenstr 4, 8099 Zürich
Pres: Katharina Müller; Martin Rüther
Focus: Soc Sci
Periodical
Sozialarbeit & Suchtprobleme (10 times annually) 10917

Schweizerischer Verein der Chemiker-Coloristen (SVCC), Postf 41, 4665 Oftringen
T: 431111
Founded: 1922
Focus: Chem 10918

Schweizerischer Verein des Gas- und Wasserfaches (SVGW) / Swiss Gas and Water Industry Association, SGWA, Grütlistr 44, Postf 658, 8027 Zürich
T: (01) 2021633
Members: 1029
Focus: Water Res; Energy
Periodical
gwa: Gas Wasser Abwasser (weekly) . . 10919

Schweizerischer Verein für Schweisstechnik, St. Alban-Rheinweg 222, 4006 Basel
T: (061) 3178484; Fax: (061) 3178480
Pres: J. Imler; Gen Secr: Dr. P. Kunzmann
Focus: Eng
Periodical
Schweißtechnik als Bestandteil der Technica (weekly) 10920

Schweizerischer Verein für Vermessung und Kulturtechnik (SVVK), Postf 732, 4501 Solothurn
T: (065) 246503; Fax: (065) 224744
Founded: 1902; Members: 750
Pres: Paul Gfeller; Gen Secr: Silvia Steiner
Focus: Surveying; Eng
Periodical
Vermessung, Photogrammetrie, Kulturtechnik (weekly) 10921

Schweizerischer Verkehrssicherheitsrat, Schwanengasse 3, 3001 Bern
T: (031) 3123638; Fax: (031) 3121839
Pres: Hans Ulrich Büschi; Gen Secr: Dr. Jakob Schälchli
Focus: Transport 10922

Schweizerischer Werkbund (SWB), Limmatstr 118, 8031 Zürich
T: (01) 2727176
Founded: 1913; Members: 1000
Pres: Martin Heller
Focus: Arts; Ecology
Periodical
SWB-Information (weekly) 10923

Schweizerischer Wissenschaftsrat (SWR), Wildhainweg 9, Postf 5675, 3001 Bern
T: (031) 619666
Founded: 1965; Members: 24
Focus: Sci
Periodicals
Futura/Fer: Ergebnisse des Projektes Forschungspolitische Früherkennung des Schweizerischen Wissenschaftsrates (weekly)
Jahresbericht (weekly)
Wissenschaftspolitik: Mitteilungsblatt der schweizerischen wissenschaftlichen Instanzen 10924

Schweizerische Schillerstiftung / Fondation Schiller Suisse, Im Ring 2, 8126 Zumikon
Founded: 1905; Members: 450
Pres: Prof. Dr. Egon Wilhelm
Focus: Lit
Periodical
Jahresbericht (weekly) 10925

Schweizerisches Institut für das Gesundheitswesen, Pfrundweg 14, 5001 Aarau
T: (064) 247161; Fax: (064) 245138
Focus: Public Health 10926

Schweizerisches Institut für Kunstwissenschaft / Institut Suisse pour l'Etude de l'Art, Zollikerstr 32, 8008 Zürich
T: (01) 7100021; Fax: (01) 7100021
Founded: 1951; Members: 2000
Pres: Dr. Johannes Fulda; Gen Secr: Dr. Hans-Jörg Heusser
Focus: Arts
Periodicals
Jahrbuch (weekly)
Jahresbericht (weekly) 10927

Schweizerisches Institut für Volkskunde, Augustinergasse 19, 4051 Basel
T: (061) 2619900
Founded: 1937; Members: 1900
Gen Secr: Dr. Hans Schnyder; Ernst Huber
Focus: Ethnology 10928

Schweizerisches Nationalkomitee für geistige Gesundheit, Hohensteinweg 28, 8055 Zürich
T: (01) 4620344
Focus: Psychiatry
Periodical
Kriminologisches Bulletin (weekly) . . . 10929

Schweizerische Sprachwissenschaftliche Gesellschaft / Société Suisse de Linguistique, c/o Romanisches Seminar, Universität, Stapfelberg 7, 4051 Basel
Founded: 1947; Members: 190
Pres: Prof. Dr. Georges Lüdi; Gen Secr: Prof. Dr. A. Kristol
Focus: Ling
Periodical
Cahiers Ferdinand de Saussure (weekly) 10930

Schweizerische Stiftung für Alpine Forschungen / Swiss Foundation for Alpine Research, Binzstr 23, 8045 Zürich
T: (01) 4610147
Founded: 1939; Members: 11
Pres: Dr. Albert Eggler; Gen Secr: Dr. Eduard Leuthold
Focus: Geography
Periodical
The Mountain World 10931

Schweizerische Stiftung für Angewandte Psychologie / Fondation Suisse pour la Psychologie Appliquée, c/o Institut de Psychologie, Université, Rte des Fougères, 1701 Fribourg
Founded: 1927; Members: 150
Pres: Dr. H. Schmid; Gen Secr: Daniel Burri
Focus: Psych
Periodical
Psychologie für die Praxis (3 times annually) 10932

Schweizerische Studiengruppe für Konsumentenfragen, Hallerstr 58, 3012 Bern
Focus: Econ 10933

Schweizerische Theologische Gesellschaft (SThG), Postf 2323, 3001 Bern
T: (061) 9215884
Founded: 1965; Members: 240
Pres: Dr. Christoph Wassermann; Gen Secr: Prof. Dr. Christoph Barben
Focus: Rel & Theol 10934

Schweizerische Trachtenvereinigung / Fédération Nationale des Costumes Suisses, Mühlegasse 15, Postf, 3400 Burgdorf
T: (034) 222239
Founded: 1926; Members: 28000
Pres: Hansruedi Spichiger
Focus: Folklore
Periodicals
Costumes et Coutures
Heimatleben 10935

Schweizerische Vereinigung der Lack- und Farben-Chemiker (SVLFC), c/o Streit AG, Postf 2006, 8502 Frauenfeld
T: (054) 7201144; Fax: (054) 7201148
Founded: 1948; Members: 300
Pres: Hans Rufenacht; Gen Secr: J. Sarbach
Focus: Chem
Periodical
Farbe und Lack 10936

Schweizerische Vereinigung der Versicherungsmathematiker (VSVM), c/o Schweizerische Rückversichungs-Gesellschaft, Mythenquai 50-60, Postf, 8022 Zürich
T: (01) 2852681
Founded: 1905; Members: 800
Focus: Insurance
Periodical
Mitteilungen (3 times annually) 10937

Schweizerische Vereinigung diplomierter Chemiker HTL (SVCT) / Swiss Association of Certified Chemists HTL, Postf 46, 4007 Basel
T: (061) 3246561; Fax: (061) 3246301
Founded: 1946; Members: 1100
Gen Secr: Beatrice Halbeisen
Focus: Chem
Periodical
Mitglieder-Bulletin: à jour (weekly) . . . 10938

Schweizerische Vereinigung für Altertumswissenschaft / Association Suisse pour l'Etude de l'Antiquité, c/o Seminar für klassische Philologie, Nadelberg 6, 4051 Basel
T: (061) 2618065
Founded: 1943; Members: 138
Pres: Prof. Fritz Graf
Focus: Archeol; Hist
Periodical
Museum Helveticum: Schweizerische Zeitschrift für klassische Altertumswissenschaft (weekly) 10939

Schweizerische Vereinigung für Atomenergie (SVA), Monbijoustr 5, Postf 5032, 3001 Bern
T: (031) 3115882; Fax: (031) 3206831
Founded: 1958; Members: 820
Pres: Dr. Hans Jörg Huber; Gen Secr: Dr. Peter Hählen
Focus: Nucl Res
Periodicals
Bulletin (weekly)
Kernpunkte (weekly) 10940

Schweizerische Vereinigung für Datenverarbeitung, Postf 373, 8037 Zürich
T: (057) 333705; Fax: (057) 334100
Pres: Dr. Kurt Egg; Gen Secr: R. Zimmerli
Focus: Computer & Info Sci
Periodical
Output (weekly) 10941

Schweizerische Vereinigung für Dokumentation (SVD), Postf 200, 3605 Thun
T: (033) 282592; Fax: (033) 284247
Founded: 1939; Members: 500
Pres: Dr. E. Wyss; Gen Secr: Theo Brenzikofer
Focus: Doc
Periodical
ARBIDO (weekly) 10942

Schweizerische Vereinigung für Ernährung (SVE) / Swiss Association for Nutrition, Postf, 3052 Zollikofen
T: (031) 9112422; Fax: (031) 8112477
Founded: 1965; Members: 6000
Pres: Prof. Dr. D. Pometta; Gen Secr: Hansjörg Ryser
Focus: Food
Periodical
Wissenschaftliche Schriftenreihe 10943

Schweizerische Vereinigung für Erwachsenenbildung (SVEB) / Swiss Federation for Adult Education, Oerlikonerstr 38, 8057 Zürich
T: (01) 3116455; Fax: (01) 3116459
Founded: 1951
Pres: Dr. D. Witzig; Gen Secr: Dr. A. Schläfli
Focus: Adult Educ
Periodicals
Education permanente (weekly)
SVEB-Bulletin (5 times annually) 10944

Schweizerische Vereinigung für Filmkultur, Gerechtigkeitsgasse 22, 3011 Bern
T: (031) 224333
Focus: Cinema 10945

Schweizerische Vereinigung für Gesundheitstechnik (SVG), Walchestr 33, Postf 305, 8035 Zürich
Focus: Eng 10946

Schweizerische Vereinigung für Internationales Recht, Postf 690, 8027 Zürich
Founded: 1914; Members: 700
Pres: Prof. Dr. F. Knoepfler; Gen Secr: Dr. P.M. Gutzwiller
Focus: Law
Periodicals
Schweizerisches Jahrbuch für Internationales Recht (weekly)
Swiss Studies in International Law . . . 10947

Schweizerische Vereinigung für Kleintiermedizin (SVK), c/o Gesellschaft Schweizerischer Tierärzte, Postf 6324, 3001 Bern
Fax: (031) 3028841
Founded: 1969; Members: 530
Focus: Vet Med 10948

Schweizerische Vereinigung für Landesplanung (VLP), Schänzlihalde 21, 3013 Bern
T: (031) 3326444; Fax: (031) 3321428
Pres: Eduard Belser; Gen Secr: Rudolf Muggli
Focus: Urban Plan 10949

Schweizerische Vereinigung für Schiedsgerichtsbarkeit / Association Suisse de l'Arbitrage, c/o Rechtsanwälte Bär & Karrer, Seefeldstr 19, 8024 ZZürich
T: (01) 2615150; Fax: (01) 2513025
Pres: Dr. Marc Blessing; Gen Secr: Jean-Paul Chapuis
Focus: Law
Periodical
Bulletin (weekly) 10950

Schweizerische Vereinigung für Sozialpolitik / Association Suisse de Politique Sociale, Affolternstr 123, 8050 Zürich
T: (01) 3113727
Pres: Prof. Antonin Wagner; Gen Secr: Dr. Günther Latzel
Focus: Poli Sci
Periodical
Das sozialpolitische Forum (weekly) . . 10951

Schweizerische Vereinigung für Steuerrecht, c/o Schweizerische Bankgesellschaft, Abt Steuerdienst, Bahnhofstr 45, 8021 Zürich
T: (01) 2342361
Focus: Law 10952

Schweizerische Vereinigung für Urheber- und Medienrecht (SVUM), Frohburgstr 116, 8057 Zürich
T: (01) 3610702, 2573046; Fax: (01) 3631290
Founded: 1946
Pres: Prof. Dr. Manfred Rehbinder; Gen Secr: Dr. Reto M. Hilty
Focus: Law 10953

Schweizerische Vereinigung für Weltraumtechnik, Postf 2613, 3001 Bern
T: (031) 220382
Focus: Aero 10954

Schweizerische Vereinigung für Zukunftsforschung (SZF) / Swiss Society for Futures Research, Brunnenwiesli 7, Postf, 8810 Horgen
T: 7257810
Founded: 1970; Members: 440
Focus: Futurology
Periodical
Zukunftsforschung: Informationen über Zukunftsforschung, Planung und Zukunftsgestaltung (5 times annually) 10955

Schweizerische Vereinigung gegen Tuberkulose und Lungenkrankheiten (SVTL), Falkenpl 9, Postf, 3001 Bern
T: (031) 3020822; Fax: (031) 3028833
Pres: Dr. R. de Haller; Gen Secr: A. von Allmen
Focus: Pulmon Dis
Periodical
Tuberkulose und Lungenkrankheiten: Beilage zum Bulletin des Bundesamtes für Gesundheitswesen (8 times annually) 10956

Schweizerische Vereinigung zum Schutz und zur Förderung des Berggebietes (VSB), Neue Simplonstr 40, 3900 Brig
T: (028) 234212
Focus: Ecology
Periodical
VSB Bulletin (weekly) 10957

Schweizerische Verkehrswirtschaftliche Gesellschaft, c/o Institut für Verkehrswirtschaft, Hochschule Sankt Gallen, Vambüelstr 19, 9000 Sankt Gallen
T: (071) 302525; Fax: (071) 302536
Focus: Transport
Periodical
Jahrbuch der Schweizerischen Verkehrswirtschaft (weekly) 10958

Schweizerische Zahnärzte-Gesellschaft, Münzgraben 2, 3011 Bern
T: (031) 227628
Focus: Dent
Periodical
Schweizerische Monatsschrift für Zahnmedizin (13 times annually) 10959

Schweizerische Zentralstelle für Hochschulwesen, Seidenweg 68, 3012 Bern
T: (031) 3022350
Founded: 1920
Pres: Prof. Dr. Luc Weber; Gen Secr: Dr. Rudolf Nägeli
Focus: Educ 10960

Schweizerische Zoologische Gesellschaft (SZG), c/o Zoologisches Institut, Rheinsprung 9, 4051 Basel
T: (061) 252535
Founded: 1893; Members: 630
Focus: Zoology
Periodical
Revue Suisse de Zoologie: Annales d'Histoire Naturelle de Genève (weekly) 10961

Schweizer Musikrat, Bahnhofstr 78, 5000 Aarau
T: (064) 229423; Fax: (064) 244767
Founded: 1964
Pres: Willi Gohl; Gen Secr: Ursula Bally-Fahr
Focus: Music
Periodical
Musikalische Berufsstudien in der Schweiz: Etudes Musicales Professionnelles en Suisse . . 10962

Società Retorumantscha (SRR), Rohanstr 5, 7000 Chur
T: (081) 223435
Founded: 1885; Members: 660
Focus: Ling
Periodicals
Annalas da la Società Retorumantscha (weekly)
DRG, Dicziunari Rumantsch Grischun (3 times annually) 10963

Società Storica Locarnese (SSL), Via Barchee 1, 6616 Locarno
Founded: 1955; Members: 120
Pres: Augusto Rima; Gen Secr: Prof. Luisa Bolla
Focus: Hist
Periodical
Bollettino SSL (weekly) 10964

Societá Ticinese di Scienze Naturali, c/o Museo Cantonale di Storia Naturale, Viale Cattaneo 4, 6900 Lugano
Founded: 1903
Pres: Dr. G. Cotti; Gen Secr: B. Jann
Focus: Nat Sci
Periodical
Bollettino (weekly) 10965

Société Académique, c/o Université, 2000 Neuchâtel
Founded: 1889; Members: 308
Focus: Sci 10966

Société Astronomique de Suisse (SAG) / Schweizerische Astronomische Gesellschaft, 10 Ch du Marais-Long, 1217 Meyrin
T: (022) 7823228
Founded: 1938; Members: 3600
Pres: Dr. Heinz Strübin; Gen Secr: Paul-Emile Müller
Focus: Astronomy
Periodical
Orion (weekly) 10967

Société de Géographie de Genève, 2 Rue de l'Athénée, 1205 Genève
Founded: 1858; Members: 270
Pres: Georges Carrel
Focus: Geography
Periodical
Le Globe: Bulletin et Mémoires (weekly) 10968

Société de Physique et d'Histoire Naturelle de Genève, c/o Muséum d'Histoire Naturelle, Rte de Malagnou, CP 6434, 1211 Genève 6
Fax: (022) 7353445
Founded: 1790; Members: 215
Pres: J. Naef; Gen Secr: Louis de Roguin
Focus: Hist; Nat Sci; Physics
Periodical
Archives des Sciences (3 times annually) 10969

Société des Professeurs d'Allemand en Suisse Romande et Italienne, 22 Chemin de la Gradelle, 1224 Genève
T: (022) 483944
Focus: Educ; Ling 10970

Société d'Histoire de la Suisse Romande, c/o Bibliothèque Cantonale et Universitaire, Dorigny, 1015 Lausanne
T: (021) 6923225
Founded: 1837; Members: 500
Pres: Antoine Lugon
Focus: Hist
Periodical
Mémoires et Documents 10971

Société d'Histoire et d'Archéologie, c/o Bibliothèque Publique et Universitaire, Les Bastions, 1211 Genève 4
T: (022) 208266
Founded: 1838; Members: 530
Pres: Guy Le Comte; Gen Secr: Daniel Aquillon
Focus: Archeol; Hist
Periodicals
Bibliographie genevoise (weekly)
Bulletin (weekly)
Mémoires et Documents (weekly) . . . 10972

Société et Fédération Internationale de Cardiologie (SIC) / International Society and Federation of Cardiology, ISFC, 34 Rue de l'Athénée, CP 117, 1211 Genève 12
T: (022) 3476755; Fax: (022) 3471028
Founded: 1978; Members: 60
Pres: Dr. D. Kelly; Gen Secr: M.B. de Figueiredo
Focus: Cardiol
Periodical
Heartbeat (weekly) 10973

Société Fribourgeoise des Sciences Naturelles, c/o Faculté des Sciences, Université, Pérolles, 1700 Fribourg
T: (037) 826111
Founded: 1886; Members: 340
Focus: Nat Sci
Periodical
Bulletin de la Société Fribourgeoise des Sciences Naturelles (weekly) 10974

Société Géologique Suisse / Schweizerische Geologische Gesellschaft, c/o Prof. A. Strasser, Institut de Géologie, Pérolles, 1700 Fribourg
T: (037) 826384
Founded: 1882; Members: 1007
Pres: Prof. A. Strasser; Gen Secr: Prof. M. Burkhard
Focus: Geology
Periodical
Eclogae Geologicae Helvetiae (3 times annually) 10975

Société Internationale de Chirurgie (SIC) / International Society of Surgery, Hauptstr 63, Postf 411, 4153 Reinach
T: (061) 7117036; Fax: (061) 7117303
Founded: 1902; Members: 3500
Pres: Prof. Michael Trede; Gen Secr: Prof. Thomas Rüedi
Focus: Surgery
Periodical
The World Journal of Surgery (weekly) . 10976

Société Internationale de Médecine Interne / International Society of Internal Medicine, c/o Medizinische Abteilung, Regionalspital, 4900 Langenthal
Founded: 1948
Gen Secr: Prof. Rolf A. Streuli
Focus: Intern Med
Periodical
ISIM Bulletin (weekly) 10977

Société Jean-Jacques Rousseau, c/o Charles Wirz, 26 Rue Voltaire, 1201 Genève
T: (022) 3448050
Founded: 1904
Focus: Lit; Philos
Periodical
Annales de la Société Jean-Jacques Rousseau (weekly) 10978

Société Médicale de la Suisse Romande, 8 Rue de la Dent-Blanche, 1950 Sion
T: (027) 220156
Focus: Med 10979

Société Neuchâteloise des Sciences Naturelles, c/o Bibliothèque Publique et Universitaire, 3 Pl Numa-Droz, 2000 Neuchâtel
T: (038) 207300
Founded: 1832; Members: 400
Focus: Nat Sci
Periodical
Bulletin (weekly) 10980

Société Pédagogique de la Suisse Romande, CP 408, 1227 Carouge
Focus: Educ
Periodical
L'Educateur (29 times annually) 10981

Société Romande d'Audiophonologie et de Pathologie du Langage (SRAPL), 13 Pl de la Liberté, 2800 Delémont
Founded: 1968; Members: 500
Focus: Otorhinolaryngology 10982

Société Suisse de Chirurgie / Schweizerische Gesellschaft für Chirurgie, c/o Policlinique de Chirurgie, Hôpital Cantonal Universitaire, 1211 Genève 14
Founded: 1913; Members: 900
Pres: Prof. Dr. P. Matter; Gen Secr: Prof. Dr. M.C. Marti
Focus: Surgery
Periodical
Helvetica Chirurgica Acta 10983

Société Suisse de Juristes, 1 Av Tribunal Fédéral, 1002 Lausanne
T: (021) 202265
Founded: 1861; Members: 3615
Focus: Law 10984

Société Suisse des Américanistes, 65-67 Blvd Carl-Vogt, 1205 Genève
Founded: 1949; Members: 220
Pres: Louis Necker; Gen Secr: Gerhard Baer
Focus: Ethnology
Periodical
Bulletin (weekly) 10985

Société Suisse des Professeurs de Français, Chemin de Grande-Cour, 1249 Chancy
T: 561070
Focus: Educ; Ling 10986

Société Suisse des Professeurs de Musique de l'Enseignement Secondaire, Alpstr 34, 6020 Emmenbrücke
T: (041) 554252
Founded: 1958
Focus: Music; Educ 10987

Société Suisse d'Héraldique, 46 Blvd des Tranchées, 1206 Genève
T: (022) 473512
Founded: 1891; Members: 550
Focus: Genealogy
Periodical
Archives Héraldiques Suisses (weekly) . 10988

Société Suisse pour la Recherche en Education, 11 Rte de Drize, 1227 Carouge
T: (022) 209333
Pres: Dr. Edo Poglia
Focus: Educ
Periodical
Bildungsforschung und Bildungspraxis: Education et Recherche (3 times annually) 10989

Société Vaudoise des Sciences Naturelles, Palais de Rumine, 1005 Lausanne
T: (021) 3124334
Founded: 1819; Members: 560
Focus: Nat Sci
Periodicals
Bulletin de la Société Vaudoise des Sciences Naturelles (weekly)
Mémoires de la Société Vaudoise des Sciences Naturelles (weekly) 10990

Société Vaudoise d'Histoire et d'Archéologie (SVHA), c/o Archives Cantonales, 32 Rue de la Mouline, 1022 Chavannes-près-Renens
T: (021) 6923511
Founded: 1902; Members: 1170
Focus: Archeol; Hist
Periodical
Revue historique vaudoise (weekly) . . . 10991

Stiftung für Humanwissenschaftliche Grundlagenforschung, Postf 112, 8030 Zürich
T: (01) 3830922
Founded: 1970
Pres: Prof. Jules Angst; Gen Secr: Dr. Walter Bodmer
Focus: Humanities 10992

Union Internationale contre le Cancer (UICC) / International Union against Cancer, 3 Rue du Conseil-Général, 1205 Genève
T: (022) 3201811; Fax: (022) 3201810
Founded: 1934; Members: 240
Gen Secr: A.J. Turnbull
Focus: Cell Biol & Cancer Res
Periodicals
International Journal of Cancer (weekly)
UICC Calendar of International Meetings on Cancer (weekly)
UICC News (weekly) 10993

Union Internationale pour la Protection des Obtentions Végétales (UPOV) / International Union for the Protection of New Varieties of Plants, 34 Chemin des Colombettes, 1211 Genève 20
T: (022) 7309111; Tx: 412912; Fax: (022) 7335428
Founded: 1961; Members: 23
Gen Secr: Arpad Bogsch
Focus: Law; Agri
Periodical
Plant Variety Protection (weekly) 10994

Verband der Heilpädagogischen Ausbildungsinstitute der Schweiz (VHpA) / Union Suisse des Instituts de Formation en Pédagogie Curative (UIPC), Obergrundstr 61, 6003 Luzern
T: (041) 231883
Focus: Educ; Rehabil 10995

Verband der Kantonschemiker der Schweiz, Muesmattstr. 19, Postf, 3000 Bern 9
T: (031) 6331111
Focus: Chem 10996

Verband der Museen der Schweiz, Baselstr 7, 4500 Solothurn
T: (065) 236710; Fax: (065) 238583
Pres: Dr. Josef Brülisauer; Gen Secr: Verena von Sury Zumsteg
Focus: Arts
Periodical
Information (weekly) 10997

Verband der Schweizer Geographen / Association Suisse de Géographie, c/o Geographisches Institut, Universität Zürich, Winterthurerstr 190, 8057 Zürich
T: (01) 2575180
Founded: 1990; Members: 190
Pres: Prof. Hans Elsasser; Gen Secr: Dr. Gilbert Thélin
Focus: Geography
Periodical
Geographica Helvetica (weekly) 10998

Verband der Schweizerischen Volkshochschulen (AUPS) / Association des Universités Populaires Suisses, Hallerstr 58, 3000 Bern
T: (031) 3028209; Fax: (031) 3025646
Founded: 1943; Members: 85
Pres: Prof. Dr. Urs Hochstrasser; Gen Secr: M. Hofstetter
Focus: Adult Educ 10999

Verband deutschschweizerischer Ärzte-Gesellschaften (Vedag), Gässlistr 17, 8856 Tuggen
T: 512212
Focus: Med
Periodical
Sprechstunde: Das Magazin der Ärzte für Ihre Gesundheit (weekly) 11000

Verband freierwerbender Schweizer Architekten (FSAI), Zeughausstr 12, 8853 Lachen
T: (055) 633263
Focus: Archit 11001

Verband Jüdischer Lehrer und Kantoren der Schweiz, Brandschenkesteig 12, 8002 Zürich
T: (01) 2020025
Founded: 1926; Members: 62
Pres: Erich A. Hausmann; Gen Secr: Michel Bollag
Focus: Educ
Periodical
Bulletin 11002

Verband Musikschulen Schweiz, Postf 49, 4410 Liestal
T: (061) 9221300; Fax: (061) 9221302
Pres: Willi Renggli
Focus: Educ; Music
Periodical
Animato (weekly) 11003

Verband Schweizerischer Abwasserfachleute (VSA), Strassburgstr 10, Postf, 8026 Zürich
T: (01) 2412585
Founded: 1944; Members: 1300
Pres: Fritz Conradin; Gen Secr: Hanspeter Walser
Focus: Water Res; Eng; Ecology . . . 11004

Verband Schweizerischer Assistenz- und Oberärzte (VSAO) / Association Suisse des Médecins Assistants et Chefs de Clinique, Dahlhölzliweg 3, Postf 229, 3000 Bern 6
T: (031) 3511573; Fax: (031) 3529008
Founded: 1945
Gen Secr: Walter Krähenmann
Focus: Med
Periodical
Bulletin VSAO (weekly) 11005

Verband Schweizerischer Ergotherapeuten (VSE), Langstr 31, 8004 Zürich
T: (01) 2425464; Fax: (01) 2915440
Members: 1100
Pres: Fabienne Pauchard; Gen Secr: Erica Kuster
Focus: Therapeutics
Periodical
Ergotherapie (weekly) 11006

Verband Schweizerischer Marketing- und Sozialforscher (SMS), c/o IHA Institut für Marktanalysen AG, 6052 Hergiswil
T: (041) 952222
Focus: Marketing; Soc Sci
Periodical
Media-Trend-Journal (3 times annually) . 11007

Verband Schweizerischer Vermessungs-Techniker (VSVT), c/o Marja Balmer, Gyrischachenstr 61, 3400 Burgdorf
T: (034) 229804; Fax: (034) 229804
Founded: 1929; Members: 1450
Pres: Paul Richle
Focus: Surveying
Periodical
Vermessung-Photogrammetrie-Kulturtechnik (weekly) 11008

Verein der Freien Pädagogischen Akademie, Postf, 8908 Hedingen
T: (01) 7615235
Focus: Educ 11009

Vereinigung der Freunde antiker Kunst, c/o Archäologisches Seminar, Universität, Schönbeinstr 20, 4056 Basel
T: (061) 2673063
Founded: 1956; Members: 750
Pres: Dr. Gérard Seiterle; Gen Secr: Kurt Müller
Focus: Arts
Periodical
Antike Kunst (weekly) 11010

Vereinigung der Gymnastiklehrer, Hirschgartnerweg 31, 8057 Zürich
T: (01) 469267
Focus: Adult Educ; Sports 11011

Vereinigung der Lehrer für Maschinenschreiben und Bürotechnik im SKV (VLMB), Landschaustr 30, 6006 Luzern
T: (041) 362362
Focus: Adult Educ 11012

Vereinigung für Berufsbildung der Schweizerischen Versicherungswirtschaft (VBV), Bubenbergpl 10, Postf 8625, 3001 Bern
T: (031) 3111722; Fax: (031) 3117756
Pres: E. Plain; Gen Secr: M. Stettler
Focus: Insurance
Periodical
VBV/AFA Info-Bulletin (weekly) 11013

Vereinigung für Rechsstaat und Individualrechte, Benedikt-Hugi-Str 14, 4500 Solothurn
T: (065) 226795
Focus: Law 11014

Vereinigung für Walsertum (VfW), Bachhalteweg 19, 3900 Brig
T: (028) 233389
Focus: Hist
Periodical
Wir Walser (weekly) 11015

Vereinigung Schweizerischer Archivare (VSA), c/o Schweizerisches Bundesarchiv, Archivstr 24, 3003 Bern
T: (031) 618989
Founded: 1922; Members: 200
Pres: Dr. R. Aebersold
Focus: Archives
Periodicals
ARBIDO-Bulletin (8 times annually)
ARBIDO-Revue (weekly) 11016

Vereinigung Schweizerischer Bibliothekare (VSB) / Association des Bibliothécaires Suisses, Hallwylstr 15, 3003 Bern
T: (031) 618978; Fax: (031) 898463
Founded: 1895; Members: 1450
Focus: Libraries & Bk Sci
Periodical
ARBIDO-B; ARBIDO-R (8 times annually) 11017

Vereinigung Schweizerischer Hochschuldozenten, Altwiesenstr 1, 8116 Würenlos
T: (056) 742683
Founded: 1919; Members: 1500
Pres: Prof. Dr. Jean-Louis Duc; Gen Secr: Prof.Dr. Andreas Miller
Focus: Adult Educ
Periodical
Bulletin (weekly) 11018

Vereinigung Schweizerischer Kinder- und Jugendpsychologen (SKJP), Bielstr 9, 4500 Solothurn
T: (065) 212961
Founded: 1969; Members: 392
Pres: Werner Graf; Gen Secr: Paul Schmid
Focus: Psych
Periodical
Der Jugendpsychologe (weekly) 11019

Vereinigung Schweizerischer Naturwissenschaftslehrer (VSN), c/o Dr. Willy Bachmann, Heithigstr 37, 8173 Neerach
T: (01) 8581639
Members: 730
Pres: Janine Digout
Focus: Nat Sci; Educ
Periodical
c+b: Chemie und Biologie (weekly) . . . 11020

Vereinigung Schweizerischer Petroleumgeologen und -Ingenieure, Schützenmattweg 13, 4460 Gelterkinden
T: (061) 992605
Focus: Geology 11021

Vereinigung Umwelt und Bevölkerung (ECOPOP) / Association Ecologie et Population, Postf 313, 3052 Zollikofen
T: (031) 9113466
Pres: Margrit Annen; Gen Secr: Hansjörg Seiler
Focus: Sociology
Periodical
EWPOP-Bulletin (3 times annually) . . . 11022

Verein katholischer Lehrerinnen der Schweiz, Dammstr 11, 6280 Hochdorf
T: (041) 881236
Founded: 1891
Pres: Marlina Blum; Gen Secr: Iwana Höltschi
Focus: Educ
Periodicals
Lehreragenda (weekly)
Notenheft (weekly)
Unterrichtsheft (weekly)
Unterrichtsjournal (weekly) 11023

Verein Schweizerdeutsch, Birkenstr 3, 8853 Lachen
T: (055) 632280
Founded: 1938
Pres: Dr. Stefan Fuchs
Focus: Ling
Periodical
Mundart: Forum des Vereins Schweizerdeutsch (weekly) 11024

Verein Schweizerischer Geschichtslehrer, Steingrubenweg 239, 4125 Riehen
T: (061) 671883
Founded: 1913
Pres: Pierre Felder
Focus: Hist; Educ
Periodical
Bulletin (weekly) 11025

Verein zur Förderung der Augenoptik, Holbeinstr 20, 8008 Zürich
T: (01) 344241
Focus: Optics 11026

Verein zur Förderung der Wasser- und Lufthygiene (VFWL), Spanweidstr 3, 8006 Zürich
T: (01) 3634922
Focus: Hygiene
Periodical
Boden-Wasser-Luft 11027

Volksgesundheit Schweiz (VGS), Splügenstr 3, 8027 Zürich
T: (01) 2023433
Focus: Public Health
Periodical
VGS-Gesundheitsmagazin (weekly) . . . 11028

Waldwirtschaftsverband Schweiz, Rosenweg 14, 4501 Solothurn
T: (065) 231011
Members: 270
Focus: Forestry
Periodical
Wald + Holz, La Foret: Organe des Waldwirtschaftsverbandes Schweiz (weekly) 11029

Wissenschaftliche Vereinigung zur Pflege des Wirtschafts- und Konsumentenschutzrechts / Scientific Association for Commercial and Consumer Law, Löwenstr 55-57, 8023 Zürich
T: (01) 2118737; Tx: 814295; Fax: (01) 2120122
Founded: 1978; Members: 210
Pres: Prof. Dr. Dr. Hans Giger; Gen Secr: E. Stähli-Seidl
Focus: Law
Periodical
Schriftenreihe zum Konsumentenschutzrecht 11030

World Association for the School as an Instrument of Peace, 5-7 Rue du Simplon, 1207 Genève
T: (022) 7352422; Fax: (022) 7364853
Founded: 1967
Focus: Educ; Law
Periodical
Ecole e Paix (weekly) 11031

World Confederation of Organizations of the Teaching Profession, 5 Av du Moulin, 1110 Morges
T: (021) 8017467; Tx: 458219; Fax: (021) 8017469
Founded: 1952; Members: 191
Gen Secr: Robert Harris
Focus: Educ
Periodical
Echo (weekly) 11032

World Health Organization (WHO), 1211 Genève 27
T: (022) 7912111; Tx: 27821; CA: UNISANTE-Genève; Fax: (022) 7910746
Founded: 1946; Members: 180
Gen Secr: Dr. Hiroshi Nakajima
Focus: Public Health
Periodicals
Bulletin of the World Health Organization (weekly)
International Digest of Health Legislation (weekly)
Weekly Epidemiological Record (weekly)
WHO Drug Information (weekly)
World Health (weekly)
World Health Forum (weekly)
World Health Statistics Quarterly (weekly) 11033

World Intellectual Property Organization (WIPO), 34 Chemin des Colombettes, CP 18, 1211 Genève 20
T: (022) 7309111; Tx: 412912; Fax: (022) 7335428
Founded: 1967; Members: 140
Gen Secr: Dr. Arpad Bogsch
Focus: Law
Periodicals
Copyright (weekly)
Le Droit d'Auteur (weekly)
Industrial Property (weekly)
International Design & Bulletin (weekly)
Les Marques Internationales (weekly)
PCT Gazette: Gazette of International Patent Applications (weekly)
La Propriété Industrielle (weekly) 11034

World Psychiatric Association, c/o Hôpital Psychogériatrique, 1008 Prilly
T: (021) 6436111; Fax: (021) 6436238
Founded: 1961; Members: 64000
Focus: Psychiatry 11035

Syria

L'Académie Arabe de Damas / The Arab Academy of Damascus, POB 327, Damascus
Founded: 1919
Focus: Ling; Lit
Periodical
Revue de l'Académie Arabe de Damas (weekly) 11036

Arab Agronomists Union, POB 3800, Damascus
T: (011) 335852, 333017
Gen Secr: Dr. Yaha Bakour
Focus: Agri 11037

Arab Center for the Studies of Arid Zones and Dry Lands (ACSAD), POB 2440, Damascus
T: (011) 755713; Tx: 412697
Founded: 1971; Members: 17
Gen Secr: Dr. Mohamed El-Khash
Focus: Geology 11038

Arab Centre for Information Studies on Population, Development and Construction, 20 Av Yabroudi, POB 11542, Damascus
T: (011) 425303
Founded: 1974
Gen Secr: Zoubeir Seif El-Islam
Focus: Soc Sci; Computer & Info Sci . 11039

Arab Council for Medical Specialization, Elharsh St, El-Mazza, Damascus
T: (011) 248942; Tx: 419101
Gen Secr: Dr. Abou-Dajaja Awad
Focus: Med 11040

Arab Federation for the Organs of the Deaf (AFOD), POB 4230, Damascus
T: (011) 420652
Founded: 1972
Pres: Abdul-Aziz Shuaib
Focus: Med
Periodical
AFOD Newsletter (00) 11041

Arab Institute for Forestry and Ranges, POB 142, Bouga, Lattakia
T: 22000, 28459
Focus: Forestry 11042

Arab States Regional Broadcasting Training Centre, POB 5333, Damascus
T: (011) 661206, 214449; Tx: 419140
Gen Secr: Zouheir Al-Baradi
Focus: Educ 11043

Arab Union for Cement and Building Materials (AUCBM), POB 9015, Damascus
T: (011) 665070; Tx: 412602
Founded: 1977; Members: 70
Gen Secr: Ahmad Al-Roussan
Focus: Materials Sci
Periodicals
Al-Omran Al-Arabi (weekly)
AUCBM Journal (weekly) 11044

Tajikistan

Tajik Academy of Sciences, Pr Rudaki 33, 734025 Dushanbe
T: 225083
Founded: 1951; Members: 690
Pres: S.K. Negmatullaev; Gen Secr: G.H Salibaev
Focus: Sci
Periodicals
Doklady: Report
Izvestiya: Bulletin
Problemy Gastroenterologii: Problems of Gastroenterology 11045

Tanzania

African Participatory Research Network, POB 5627, Dar es Salaam
Focus: Law 11046

Association of Management Training Institutions of Eastern and Southern Africa (AMTIESA), POB 3030, Arusha
T: (057) 2881; Tx: 42076; Fax: (057) 7776
Founded: 1985; Members: 27
Gen Secr: Max Mwanahiba
Focus: Business Admin
Periodical
AMTIESA Bulletin (weekly) 11047

Confederation of African Medical Associations and Societies (CAMAS), c/o Medical Association of Tanzania, POB 701, Dar es Salaam
Tx: 41505; Fax: (051) 46229
Founded: 1982
Gen Secr: Prof. A. Shao
Focus: Med 11048

East Africa Association for Theological Education by Extension (EAATEE), POB 32, Njombe
Founded: 1982
Gen Secr: John Simalenga
Focus: Rel & Theol; Educ 11049

East Africa Medical Research Council (EAMRC), Town Centre, Arusha
T: (057) 2810
Founded: 1962
Focus: Med 11050

East African Agriculture and Forestry Research Organization (EAAFRO), c/o East African Community, POB 3181, Arusha
Founded: 1948
Focus: Forestry; Agri 11051

East African Literature Bureau, POB 1408, Dar es Salaam
Founded: 1948
Focus: Lit 11052

East, Central and Southern African College of Nursing (ECSACON), POB 1009, Arusha
T: (057) 2961; Tx: 42121; Fax: (057) 2714
Founded: 1988
Gen Secr: Rosalie Kakande
Focus: Educ; Med 11053

Eastern African Centre for Research on Oral Traditions and African National Languages (EACROTANAL), POB 600, Zanzibar
T: 30786
Founded: 1979
Gen Secr: Henri Rahaingoson
Focus: Ling
Periodical
EACROTANAL Newsletter 11054

Eastern Africa Statistical Training Centre (EASTC), c/o University of Dar es Salaam, POB 35193, Dar es Salaam
Founded: 1965; Members: 12
Focus: Stats 11055

Eastern and Southern African Management Institute (ESAMI), POB 3030, Arusha
T: (057) 2881; Tx: 42076; Fax: (057) 7776
Founded: 1980; Members: 11
Gen Secr: Dr. Kasuka S. Mutukwa
Focus: Business Admin
Periodicals
Africa Management Development Forum
ESAMI Newsletter (weekly) 11056

Eastern and Southern African Mineral Resources Development Centre (ESAMRDC), POB 1250, Dodoma
T: (0611) 20364; Tx: 53324
Founded: 1975; Members: 18
Gen Secr: J.E.N. Kagule-Magambo
Focus: Mineralogy
Periodical
ESAMRDC Newsletter (weekly) 11057

Eastern and Southern African Universities Research Programme (ESAURP), POB 35121, Dar es Salaam
T: (051) 73687; Tx: 41327; Fax: (051) 26921
Founded: 1977
Gen Secr: Prof. T.L. Maliyamkono
Focus: Educ 11058

Tanzania Commission for Science and Technology, POB 4302, Dar es Salaam
T: (051) 75311; Fax: (051) 75313
Founded: 1968
Focus: Sci; Eng
Periodicals
Annual Report (weekly)
S & T News (weekly) 11059

Tanzania Library Association (TLA), POB 2645, Dar es Salaam
Founded: 1965; Members: 200
Pres: Theophilus E. Mlaki; Gen Secr: M. Ngaiza
Focus: Libraries & Bk Sci
Periodicals
Matukio (weekly)
Someni: Tanzania Library Association Journal (weekly) 11060

Tanzania Medical Association, POB 9083, Dar es Salaam
Focus: Med 11061

The Tanzania Society, POB 511, Dar es Salaam
T: (051) 44311
Founded: 1936; Members: 1200
Focus: Ethnology; Hist
Periodical
Tanzania Notes and Records (weekly) . 11062

Tanzania Veterinary Association (TVA), POB 3174, Morogoro
T: (056) 3236, 4652; Tx: 55308; Fax: (056) 3718, 3177
Founded: 1968; Members: 380
Pres: Prof. P. Msolla; Gen Secr: Prof. A.D. Maeda-Machangu
Focus: Vet Med
Periodicals
Proceedings of the Tanzania Veterinary Association Scientific Conferences (weekly)
Tanzania Veterinary Journal (weekly) . . 11063

Thailand

Agricultural Information Development Scheme, c/o ESCAP, Agriculture and Rural Development Div, UN Bldg, Rajadammen Av, Bangkok 10200
T: (02) 2829161, 2829181; Tx: 2392
Founded: 1979
Focus: Computer & Info Sci
Periodical
Agricultural Information Development Bulletin (weekly) 11064

Agricultural Science Society of Thailand, c/o Kasetsart University, POB 1070, Bangkok 10903
Founded: 1967
Pres: Yookti Sarikaphuti; Gen Secr: Vichai Haruthaithansasan
Focus: Agri 11065

Applied Scientific Research Corporation of Thailand, 196 Phahonyothin Rd, Bangkok
Focus: Sci 11066

ASEAN Federation of Engineering Organizations (AFEO), c/o Engineering Institute of Thailand, Henri Dunant Rd, Bangkok 10500
Founded: 1980; Members: 6
Focus: Eng 11067

ASEAN Federation of Plastic Surgeons, c/o Dept of Surgery, Chulalongkorn Hospital, Sirindhorn Bldg, Bangkok 10500
Gen Secr: Dr. Thavorn Charoonsmith
Focus: Surgery 11068

ASEAN Institute for Health Development (AIHD), c/o Mahidol University, 25/5 Putthamonthon 4, Salaya, Nakhon Chaisri, Nakhon Pathom 73170
T: (02) 4419040; Tx: 84770; Fax: (02) 4419044
Founded: 1982
Gen Secr: Dr. Yawarat Porapakkham
Focus: Public Health 11069

ASEAN Otorhinolaryngological Head and Neck Federation, c/o Dept of Otolaryngology, Siriraj Hospital, Mahidol University, Bangkok 10700
T: (02) 4113254, 4114816
Founded: 1980
Gen Secr: Dr. Suchitra Prasansuk
Focus: Otorhinolaryngology 11070

ASEAN Sub Committee on Non-Conventional Energy Research (SCNCER), c/o King Mongkut's Institute of Technology, Thonburi, Bangkok 10140
T: (02) 4279242, 4275208; Tx: 72383; Fax: (02) 4278077
Founded: 1979; Members: 6
Focus: Energy
Periodical
SCNCER Newsletter 11071

Asia and Pacific Commission on Agricultural Statistics (APCAS), c/o FAO Regional Office, Maliwan Mansion, 39 Phra Atit Rd, Bangkok 10200
T: (02) 2817844; Tx: 82815; Fax: (02) 2800445
Founded: 1963
Gen Secr: H. Som
Focus: Stats; Agri 11072

Asia and Pacific Plant Protection Commission (APPPC), c/o FAO Regional Office, Maliwan Mansion, Phra Atit Rd, Bangkok 10200
T: (02) 2817844; Tx: 2815; Fax: (02) 2800445
Founded: 1955; Members: 25
Focus: Agri 11073

Asia and Pacific Programme of Educational Innovation for Development (APED), c/o UNESCO PROAP, 920 Sukhumvit Rd, POB 967, Prakanong Post Office, Bangkok 10110
T: (02) 3910879; Tx: 20591; Fax: (02) 3910866
Founded: 1973
Gen Secr: Ahmed Medayat
Focus: Educ 11074

Asia Foundation, POB 1910, Bangkok 5
Gen Secr: Allen C. Choate
Focus: Educ; Econ; Sociology 11075

Asian and Pacific Information Network on Medicinal and Aromatic Plants (APINMAP), c/o Ministry of University Affairs, 328 Sri Ayutthya Rd, Bangkok 10400
T: (02) 2589853, 2451157; Tx: 72145; Fax: (02) 2871443
Founded: 1987
Gen Secr: Prof. Kamchorn Manunapichu
Focus: Botany 11076

Asian and Pacific Project for Labour Administration (ARPLA), Labour Dept Bldg, Fuangnakorn Rd, Bangkok
Focus: Business Admin 11077

Asian Association of Occupational Health (AAOH), c/o Occupational Health Dept, Faculty of Public Health, Mahidol University, 420/1 Rajvidhi Rd, Bangkok 10400
T: (02) 2457793; Tx: 84770; Fax: (02) 2467765
Founded: 1964
Pres: Malinee Wongphanich
Focus: Public Health 11078

Asian Centre of Educational Innovation for Development (ACEID), c/o UNESCO PROAP, 920 Sukhumvit Rd, POB 967, Prakanong Post Office, Bangkok 10110
T: (02) 3910879; Tx: 20591
Founded: 1973
Pres: L. de la Cruz
Focus: Educ 11079

Asian Disaster Preparedness Center (ADPC), GPOB 2754, Bangkok 10501
T: (02) 5245353; Tx: 84276; Fax: (02) 5245360
Founded: 1986
Pres: B.A.O. Ward
Focus: Eng 11080

Asian Geotechnical Engineering Information Center (AGE) 10501, POB 2754, Bangkok
T: (02) 5245862; Fax: (02) 5162126
Founded: 1973
Gen Secr: H. Arthur Vespry
Focus: Eng
Periodicals
AGE Current Awareness Service (weekly)
AGE News (weekly) 11081

Asian Institute of Technology Library and Regional Documentation Centre, POB 2754, Bangkok
Focus: Eng; Doc
Periodicals
Abstracts of AIT Reports and Publications on Energy (weekly)
AGE Current Awareness Service (weekly)
AGE News (weekly)
AGE Refdex (weekly)
ENFO Newsletter (weekly)
Environmental Sanitation Abstracts (3 times annually)
Environmental Sanitation Reviews (weekly)
Ferrocement Abstracts (weekly)
IFIC Do it yourself Series (weekly)
IFIC Slide Presentation Series (weekly)
IFIC Specialized Bibliographies (weekly)
Journal of Ferrocement (weekly)
RERIC Holdings List (weekly)
RERIC International Energy Journal (weekly)
RERIC Membership Directory (weekly)
RERIC News (weekly) 11082

Asian-Pacific Federation of Therapeutic Communities (APFTC), 31/1 Eramai, 10 Ekamai Rd, Bangkok 10110
Founded: 1984
Pres: J. Amnuay Intuputi
Focus: Therapeutics 11083

Asian Society for Environmental Protection (ASEP), c/o SEAPO, Asian Institute of Technology, POB 2754, Bangkok 10501
T: (02) 516011029; Tx: 84276; Fax: (02) 5162126
Founded: 1984; Members: 200
Gen Secr: Guenter Tharun
Focus: Ecology
Periodical
ASEP Newsletter (weekly) 11084

Asia-Pacific Forestry Commission (APFC), c/o FAO Regional Office, Maliwan Mansion, Phra Atit Rd, Bangkok 10200
T: (02) 2817844; Tx: 82815; Fax: (02) 2800445
Founded: 1949
Focus: Forestry 11085

Asia-Pacific Information Network in Social Sciences (APINESS), c/o RUSHSAP, UNESCO Annex Office, 24/1 Sukhumvit Soi 59, POB 967, Bangkok 10110
T: (02) 3811347; Tx: 20591; Fax: (02) 3910866
Founded: 1986
Focus: Soc Sci
Periodicals
APINESS Newsletter
APPEN Features Service
Asia-Pacific Environment Newsletter (weekly)
Environmental News Digest (3 times annually)
. . . . 11086

Asia-Pacific Programme of Education for All (APPEAL), c/o UNESCO PROAP, 920 Sukhumvit Rd, POB 967, Prakanong Post Office, Bangkok 10110
T: (02) 3910879; Tx: 20591; Fax: (02) 3910866
Founded: 1986; Members: 29
Focus: Educ 11087

Association of Geoscientists for International Development (AGID), c/o Asian Institute of Technology, POB 2754, Bangkok 10501
T: (02) 5245514; Tx: 84276; Fax: (02) 5162126
Founded: 1974
Gen Secr: Dr. J.L. Rau
Focus: Geology
Periodicals
AGID News (weekly)
Bulletin des Géosciences en Afrique de l'Ouest (weekly)
Geociencias (weekly)
Middle East Geoscience News (weekly)
South and West Asian Geoscience Newsletter (3 times annually)
West African Geoscience Newsletter . . 11088

Association of Southeast Asian Institutions of Higher Learning (ASAIHL), c/o Chulalongkorn University, Ratasastra Bldg 2, Henri Dunant Rd., Bangkok 10330
T: (02) 256966; Tx: 72432; Fax: (02) 2554441
Founded: 1956; Members: 120
Gen Secr: Dr. Ninnat Olanvoravuth
Focus: Educ 11089

Committee for Co-ordination of Investigations of the Lower Mekong Basin, Pibultham Villa, Kasatsuk Bridge, Bangkok 10330
T: (02) 2250029; Tx: 21322; Fax: (02) 2252796
Founded: 1957
Focus: Water Res 11090

Environmental Sanitation Information Center (ENSIC), c/o AIT, POB 2754, Bangkok 10501
T: (02) 5160110; Tx: 84276; Fax: (02) 5162126
Founded: 1978
Focus: Ecology 11091

ESCAP/FAO/UNIDO Fertilizer, Development and Information Network for Asia and the Pacific, c/o FADINAP/ESCAP Agriculture and Rural Development Div, United Nations Bldg, Rajadamnern Av, Bangkok 10200
T: (02) 2829161; Tx: 82392; Fax: (02) 2812402
Founded: 1978; Members: 25
Focus: Agri 11092

Medical Association of Thailand, 3 Silom St, Bangkok
Founded: 1921; Members: 3057
Pres: Prof. Dr. S. Niyomsen; Gen Secr: Prof. Dr. S. Unakol
Focus: Med
Periodical
Journal 11093

National Culture Commission, Ratchadapisek Rd, Huay Khwang, Bangkok 10310
T: (02) 2470013
Founded: 1979
Gen Secr: Somkid Chotigavanit
Focus: Sci 11094

Office of the Atomic Energy Commission for Peace, 16 Thanon Vibhavadee Rangsit, Bangkhen, Bangkok 10900
T: (02) 5790138; Tx: 87161; CA: ATOMTHAI BANGKOK
Founded: 1962; Members: 300
Gen Secr: Athorn Patumasootra
Focus: Nucl Res 11095

Rectors' Conference, c/o Planning Div, Ministry of University Affairs, Si-Ayuthya Rd, Bangkok 10400
Founded: 1971; Members: 14
Pres: Dr. N. Bhamarapra; Gen Secr: Prof. Dr. U. Boonprasert
Focus: Educ 11096

Royal Institute, Thanon Na Phra Larn, Bangkok 10200
T: (02) 2214822
Founded: 1933; Members: 161
Pres: B. Chatamra
Focus: Arts; Sci
Periodicals
Journal (weekly)
Saranukrom Thai (weekly) 11097

Science Society of Thailand, c/o Faculty of Science, Chulalongkorn University, Phya Thai Rd, Bangkok 10500
T: (02) 252987; Tx: 20217; Fax: (02) 2554441
Founded: 1948; Members: 2000
Pres: Dr. K. Mongkolkul; Gen Secr: K. Bunyaklat
Focus: Sci
Periodicals
Journal (weekly)
Science (weekly) 11098

The Siam Society, 131 Soi Asoke, Sukhumvit 21, POB 65, Bangkok
T: (02) 2583491
Founded: 1904; Members: 1500
Pres: M.R. Patanachai Jayant; Gen Secr: V.M. di Croco
Focus: Ethnology
Periodicals
Journal (weekly)
Natural History Bulletin (weekly) 11099

Southeast Asian Society of Soil Engineering, POB 2754, Bangkok
Focus: Eng 11100

Thai-Bhara Cultural Lodge and Swami Satyananda Puri Foundation, 136/I Siriphongs Rd, Bangkok
Founded: 1940; Members: 450
Pres: Dr. O. Ketusinh; Gen Secr: P.S. Karess
Focus: Ling; Lit; Philos 11101

Thai Library Association (TLA), 273 Vibhavadee Rangsit Rd, Phyathai, Bangkok 10400
Founded: 1954; Members: 900
Pres: M. Chavalit; Gen Secr: K. Suckcharoen
Focus: Libraries & Bk Sci
Periodicals
TLA Bulletin (weekly)
The World of Books 11102

Togo

African Association of Education for Development (ASAFED), BP 3907, Lomé
T: 216316; Tx: 5131
Founded: 1978
Pres: Koffi Attignon
Focus: Educ
Periodical
Famille et Développement (weekly) . . . 11103

Association Togolaise pour le Développement de la Documentation, des Bibliothèques, Archives et Musées, c/o Bibliothèque de l'Université, BP 1515, Lomé
Founded: 1959; Members: 60
Pres: K. Attignon; Gen Secr: E. Amah
Focus: Doc; Archives; Arts 11104

Trinidad and Tobago

Agricultural Society of Trinidad and Tobago, 44 Pembroke St, Port of Spain
Founded: 1894; Members: 528
Pres: Ian F. McDonald; Gen Secr: Leo C. Nanton
Focus: Agri
Periodical
Journal (weekly) 11105

Caribbean Agricultural Research and Development Institute (CARDI), UWI Campus, Saint Augustine
Founded: 1975; Members: 12
Focus: Agri 11106

Caribbean Agro-Economic Society, c/o Dept of Agricultural Economics and Farm Management, University of the West Indies, UWI Campus, Saint Augustine
Founded: 1974
Pres: Dr. Carlisle A. Pemberton
Focus: Agri
Periodicals
Caribbean Agro-Economic Society Journal
Caribbean Agro-Economic Society Newsletter (weekly) 11107

Caribbean Association for Feminist Research and Action (CAFRA), PO Bag 442, Tunapuna Post Office, Tunapuna
Founded: 1985
Gen Secr: Catherine Shepherd
Focus: Soc Sci
Periodicals
CAFRA News (weekly)
Novedades CAFRA (weekly) 11108

Caribbean Association of Nutritionists and Dieticians (CANDI), c/o University of the West Indies, 16 Rowland Rd, West Moorings
Founded: 1972
Gen Secr: Eunice Warner
Focus: Nutrition 11109

Caribbean Council for Science and Technology (CCST), 22-24 Saint Vincent St, POB 1113, Port of Spain
Founded: 1980; Members: 13
Focus: Nat Sci; Eng
Periodical
CCST Newsletter (weekly) 11110

Caribbean Documentation Centre (CDC), 22-24 Saint Vincent St, POB 1113, Port of Spain
Members: 21
Focus: Doc
Periodical
Current Awareness Bulletin (weekly) . . 11111

Caribbean Federation for Mental Health, c/o Saint Ann's Hospital, POB 65, Port of Spain
Focus: Psychiatry 11112

Caribbean Industrial Research Institute (CARIRI), Tunapuna PO, Tunapuna
Founded: 1970
Gen Secr: Dr. Desmond A. Ali
Focus: Business Admin
Periodicals
Caribbean Biotechnology Newsletter (weekly)
CARIRI Technochat (weekly) 11113

Caribbean Information System for Economic and Social Planning (CARISPLAN), 22 Saint Vincent St, POB 1113, Port of Spain
Founded: 1979; Members: 21
Gen Secr: Wilma J. Primus
Focus: Soc Sci; Econ
Periodical
CARISPLAN Abstracts (weekly) 11114

Caribbean Information System for the Agricultural Sciences (CAGRIS), c/o Main Library, University of the West Indies, Saint Augustine
Focus: Agri
Periodicals
Cagrindex (weekly)
CAGRIS Newsletter 11115

Caribbean Meteorological Organization (CMO), POB 461, Port of Spain
Founded: 1973; Members: 16
Gen Secr: C.E. Berridge
Focus: Geophys 11116

Caribbean Plant Protection Commission (CPPC), c/o Ministry of Food Production, Marine Exploitation, Forestry and the Environment, Research Div, Central Experiment Station, Centeno, Arima
Founded: 1967; Members: 22
Gen Secr: Mona T. Jones
Focus: Ecology 11117

Caribbean Regional Branch of the International Council on Archives (CARBICA), 105 Saint Vincent St, POB 763, Port of Spain
Founded: 1965
Pres: Edwina Peters
Focus: Archives 11118

Caribbean Sports Medicine Association (CASMA), 77 Picton St, Newton-Hooslook, Port of Spain
Founded: 1986
Pres: Dr. Terry Ali
Focus: Med 11119

Eastern Caribbean Institute of Agriculture and Forestry (ECIAF), Centeno, Arima
Founded: 1954
Focus: Agri; Forestry 11120

Historical Society of Trinidad and Tobago, c/o Ministry of Culture, Eastern Main Rd, Laventille, Port of Spain
Pres: E.E. Williams; Gen Secr: E. Johnson
Focus: Hist 11121

Library Association of Trinidad and Tobago, POB 1275, Port of Spain
Founded: 1960
Pres: E. Mohammed; Gen Secr: E. Peters
Focus: Libraries & Bk Sci
Periodical
Bulletin 11122

Sugar Technologist's Association of Trinidad and Tobago, 80 Abercromby St, POB 230, Port of Spain
Founded: 1967
Focus: Eng 11123

Theosophical Society of Trinidad, Eastern Main Rd, Guaico
Gen Secr: L.R. Khillawan
Focus: Rel & Theol 11124

Tobago District Agricultural Society, Main St, Scarborough
Pres: R.H. Harrower; Gen Secr: S.A. Davies
Focus: Agri 11125

Trinidad Music Association, c/o Bishop Anstey High School, Abercromby St, Port of Spain
Founded: 1941; Members: 102
Pres: R. Johnstone; Gen Secr: Velma Jardine
Focus: Music 11126

Tunisia

Arab Atomic Energy Agency (AAEA), El-Manzah 5, POB 402, 1004 Tunis
T: (01) 766010; Tx: 14896
Founded: 1988
Gen Secr: Dr. Ali A. Abdullah
Focus: Nucl Res 11127

Arab Committee for Ottoman Studies (ACOS), BP 50, 1118 Zaghouan
T: (02) 76446
Founded: 1983
Pres: Abdeljehil Temimi
Focus: Hist 11128

Arab Federation of Sports Medicine, 6 Rue d'Annaba, Tunis
Founded: 1982
Focus: Med 11129

Arab Information Network for Terminology (ARABTERM), c/o INNOPRI, BP 23, Belvedère, 1012 Tunis
Founded: 1989
Focus: Computer & Info Sci 11130

Arab League Educational, Cultural and Scientific Organization, BP 1120, Tunis
T: (01) 784466; Tx: 13825; Fax: (01) 784965
Founded: 1964; Members: 20
Gen Secr: Dr. Musari Al-Rawi
Focus: Sci; Educ
Periodicals
ALECSO Newsletter
Al-Lissan Al-Arabi Magazine
Arab Culture Magazine
Arab Manuscripts Magazine
Bulletin of Arab Publications
Bulletin of Educational Statistics for the Arab States 11131

Arab Medical Union, c/o Union des Médecins Arabes, 16 Rue de Touraine, Belvedere, 1002 Tunis
T: (01) 792736
Founded: 1961; Members: 15
Gen Secr: Prof. Aziz El-Matri
Focus: Med
Periodical
Revue Médicale de l'Union des Médecins Arabes (weekly) 11132

Arab Regional Branch of the International Council on Archives (ARBICA), BP 50, Zaghouan, Tunis
T: (02) 76446
Founded: 1972; Members: 19
Focus: Archives
Periodical
Arab Archives Journal (weekly) 11133

Association Africaine de Microbiologie et d'Hygiène Alimentaire (AAMHA) / African Association of Microbiology and Food Hygiene, c/o Dept de Biochimie, Faculté de Médecine, Ibn El-Jazzar, 4000 Sousse
T: (03) 22600 ext 431; Tx: 30988
Founded: 1984
Pres: Prof. Mongi Jemmali
Focus: Microbio; Food
Periodical
Microbiologie et Hygiène Alimentaire (weekly) 11134

Association Internationale des Archives Francophones (AIAF), c/o Archives Nationales de Tunisie, Pl du Gouvernement, 1020 Tunis
T: (01) 260556; Tx: 14530
Pres: Moncef Fakhfath
Focus: Archives 11135

Association Maghrébine des Etudes de la Population, 9bis Rue Khaled Ibn El-Qualid, Mutuelleville, Tunis
T: (01) 204760
Gen Secr: Mongi B'Chir
Focus: Sociology 11136

Association of Arab Institutes and Centres for Economic and Social Research (AICARDES) / Association des Instituts et des Centres Arabes pour les Recherches Economiques et Sociales, c/o Institut d'Economie Quantitative, 27 Rue du Liban, Belvédère, 1002 Tunis
T: (01) 283214, 283216; Tx: 15117
Founded: 1977
Gen Secr: Hédi Gorbel
Focus: Soc Sci; Econ 11137

Association Tunisienne des Bibliothécaires, Documentalistes et Archivistes, BP 380, 1015 Tunis
Founded: 1966
Focus: Libraries & Bk Sci; Doc; Archives
Periodical
Rassid (3 times annually) 11138

Comité International d'Etudes Morisques, BP 50, 1118 Zaghouan
T: (02) 76446
Focus: Ethnology 11139

Comité National des Musées, c/o Musée National du Bardo, Tunis
Founded: 1961
Pres: N. Attya
Focus: Arts
Periodical
Les Musées de Tunisie 11140

Conseil International sur les Mathématiques dans les Pays en Voie de Développement (ICOMIDC), c/o Dépt de Mathématiques, Faculté des Sciences de Tunis, 1060 Campus Universitaire, Tunis
Gen Secr: Noureddine Boudriga
Focus: Math 11141

Institut National d'Archéologie et d'Art, 4 Pl du Château, 1008 Tunis
T: (01) 261622, 261259
Founded: 1957; Members: 80
Focus: Archeol; Ethnology
Periodicals
Africa
Bibliothèque Archéologique
Cahiers des Arts et Traditions Populaires
Etudes Hispano-Andalouses
Notes et Documents 11142

Union Nationale des Arts Plastiques, Musée du Belvédère, Tunis
Focus: Fine Arts 11143

Turkey

Danubian League against Thrombosis and Hemorrhagic Disorders, c/o Cerpahpasa Medical School, Anhul University, Levent Begonay Sok 6, 80620 Istanbul
Gen Secr: Prof. Orhan N. Ulutin
Focus: Med 11144

Fizik Tedavi ve Rehabilitasyon Derneği / The Society of Physical Medicine and Rehabilitation, c/o Fiziksel Tip ve Rehabilitasyon Anabilim Dali, Istanbul Tip Fakültesi, Capa, 34390 Istanbul
T: (01) 525014; Fax: (01) 631167
Founded: 1963; Members: 120
Focus: Therapeutics; Rehabil
Periodical
Fizik Tedavi Rehabilitasyon Dergisi (weekly) 11145

Istanbul Dişhekimleri Odasi, Cumhuriyet Cad, Safir Apt 361/3 Harbiye, Istanbul
T: (01) 2444442; Fax: (01) 2445909
Focus: Dent
Periodicals
Dişhekimliğinde Klinik (weekly)
Ido Dergi (weekly) 11146

Jeoloji Mühendisleri Odasi / Chamber of Geological Engineers of Turkay, PK 464, Kizilay, 06424 Ankara
T: (04) 343601
Founded: 1946; Members: 3200
Focus: Geology
Periodical
Türkiye Kurum Bülteni (weekly) 11147

Milletlerarasi Sark Tetkikleri Cemiyeti / International Society for Oriental Research, Türkiyat Enstitüsü, Bayezit, Istanbul
Founded: 1947
Pres: Prof. Fuad Köprülü
Focus: Ethnology 11148

P.E.N. Yazarlar Derneği / P.E.N., Turkish Centre, Cağaloğlu Yokusu 40, Istanbul
Founded: 1951; Members: 70
Pres: Yasar Nabi Nayir
Focus: Lit 11149

Türk Anesteziyoloji ve Reanimasyon Derneği (TARD), c/o Cerrahpasa Tip Fakültesi, Istanbul
T: (01) 5861526; Fax: (01) 5295600
Founded: 1956; Members: 600
Pres: Prof. Dr. Sadi Sun; Gen Secr: Prof. Dr. L. Telci
Focus: Anesthetics
Periodical
Türk Anesteziyoloji ve Reanimas yon Demégi Mecmuasi (weekly) 11150

Türk Biyoloji Derneği, PK 144, Sirkeci, Istanbul
Founded: 1949; Members: 200
Focus: Bio 11151

Türk Cerrahi Cemiyeti / Turkish Surgical Society, Valikonagi Cad 10, Harbiye, Istanbul
Founded: 1931; Members: 870
Focus: Surgery 11152

Türk Diabetler Cemiyeti, Meyva Sok 10, Hrabiye, Istanbul
Focus: Diabetes 11153

Türk Dil Kurumu (TDK) / Turkish Linguistic Society, Atatürk Bulvari 217, Ankara
Founded: 1932; Members: 40
Pres: Prof. Dr. H. Eren; Gen Secr: H. Selçuk
Focus: Ling
Periodical
Türk Dili (weekly) 11154

Türk Eczacilari Birliği / Turkish Pharmaceutical Association, 26 Ortaklar Han, Cağaloğlu, Istanbul
Focus: Pharmacol
Periodical
Türk Eczacilari Birligi Mecmuasi (weekly) 11155

Türk Fizik Derneği, c/o CNAEM, PK 1, Havaalani, Istanbul
T: (01) 737515
Founded: 1950; Members: 282
Focus: Physics 11156

Türk Gastroenterologii Derneği, c/o Dept of Gastroenterology, Gulhane Military Medical Academy, Ankara
Focus: Gastroenter 11157

Türk Halk Bilgisi Derneği / Turkish Folklore Society, PK 46, Aksary, Istanbul
Founded: 1946
Pres: S.Y. Ataman; Gen Secr: Ihsan Hincer
Focus: Ethnology 11158

Türk Hukuk Kurumu / Turkish Law Association, 2 Cad 55/6 Bahcelievler, Ankara
Founded: 1934
Focus: Law
Periodical
Türk Hukuk Lugati 11159

Türkiye Aktüerler Cemiyeti, Siraselviler Cad 87, Taksim, Istanbul
Focus: Insurance 11160

Türkiye Fitopatolji Derneği / Turkish Phytopathological Society, c/o Zirai Mücadele Araştirma Enstitüsü, Bornova, 35040 Izmir
T: (051) 880030; Fax: (051) 741653
Members: 250
Pres: Dr. Hikmet Saygili; Gen Secr: Dr. Emin Onan
Focus: Bio
Periodical
The Journal of Turkish Phytopathology (2-3 times annually) 11161

Türkiye Ziraat Odalari Birligi / Union of Turkish Chambers of Agriculture, Izmir Cad 24, Yenisehir, 06440 Ankara
T: (04) 4171274; Tx: 46380; CA: Ziraat Odalari; Fax: (04) 4173068
Founded: 1963; Members: 425
Pres: Erol Baraz; Gen Secr: Fikret Celenligil
Focus: Agri
Periodicals
Agricultural and Economic Report (weekly)
Farmer and Rural World (weekly) . . . 11162

Türk Kanser Arastirma ve Savas Kurumu, PK 1078, Yenisehir, Ankara
Focus: Cell Biol & Cancer Res
Periodical
Kanser: The Turkish Journal of Cancer (weekly) 11163

Türk Kardiyoloji Derneği / Turkish Society of Cardiology, Nisbetiye Cad 37, Etiler, 80630 Istanbul
T: (01) 2516335; Fax: (01) 2573787
Founded: 1963; Members: 350
Focus: Cardiol
Periodical
Archives of the TSC (weekly) 11164

Türk Kütüphaneciler Derneği (TKD) / Turkish Librarians' Association, Elgün Sok 8/8, Kizilay, 06440 Ankara
T: (041) 2301325; Fax: (041) 2301325
Founded: 1949; Members: 1341
Pres: Prof. Dr. Tülin Sgălamtunc; Gen Secr: Ayhan Kaygusuz
Focus: Libraries & Bk Sci
Periodical
Türk Kütüphaneciliği (weekly) 11165

Türk Mikrobiyoloji Cemiyeti / Turkish Microbiological Society, PK 57, Beyazit, 34492 Istanbul
Fax: (01) 2304409
Founded: 1931; Members: 258
Pres: Prof. Dr. Özdem Ang; Gen Secr: Prof. Dr. Candan Bozok
Focus: Microbio
Periodicals
Infeksion Dergisi
Türk Mikrobiyoloji Cemiyeti Dergisi (weekly) 11166

Türk Mühendis ve Mimar Odalari Birligi (TMMB) / Union of Chambers of Turkish Engineers and Architects, Konur Sok 4, Yenisehir, Ankara
T: (04) 41181275; Fax: (04) 4174824
Founded: 1954; Members: 170000
Focus: Archit; Eng
Periodicals
Elektrik Mühendisliği: Electrical Engineering
Endüstri Mühendisliği: Industrial Engineering
Harita ve Kadastro Mühendisliği: Surveying Rngineering
Jeofizik: Geophysics
Jeoloji Mühendisliği: Geological Engineering
Kimya Mühendisliği: Chemical Engineering
Madencilik: Mining
Metalurji: Metallurgy
Mimarlik: Architecture
Mühendis ve Makina: Engineers and Machines
Orman Mühendisliği: Forest Engineering
Tarim ve Mühendislik: Agriculture and Engineering
Tekstil ve Mühendis: Textile and Engineering
Tesisat Mühendisliği: Installation Engineering
Türkiye Mühendislik Haberleri: News of Engineering in Turkey 11167

Türk Nöro-Psikiyatri Derneĭ, c/o I.Ü. Istanbul Tip Fakültesi, Psikiyatri Kliniği, 34390 Capa, Istanbul
Founded: 1914; Members: 600
Pres: Prof. Dr. Selim Özaydin; Gen Secr: Prof. Dr. Imadettin Akkök
Focus: Neurology; Psychiatry
Periodical
Nöroppsikiyatri Arsivi: Archives of Neuro-Psychiatry (weekly) 11168

Türk Oftalmoloji Derneği, c/o Capa Göz Klinigi, Istanbul
Members: 300
Focus: Ophthal 11169

Türk Ortopedi ve Travmatoloji Dernegi / Turkish Association of Orthopaedics and Traumatology, c/o Istanbul Tip Fakültesi Ortopedi ve Travmatoloji Klinigi, Topkapi, 34390 Istanbul
T: (01) 5213687
Founded: 1939; Members: 510
Pres: Dr. Alp Göksan; Gen Secr: Prof. Dr. I. Remzi Tözün
Focus: Orthopedics; Traumatology
Periodical
Acta Orthopaedica et Traumatologica Turcica (5 times annually) 11170

Türk Otomatik Kontrol Kurumu (TOK), c/o Teknik Üniversitesi, Istanbul
T: (01) 433100
Founded: 1959
Focus: Eng 11171

Türk Oto-Rino-Laringoloji Cemiyeti, c/o Capa Kulak Bogaz, Burun Klinigi, Istanbul
Founded: 1930; Members: 150
Focus: Otorhinolaryngology 11172

Türk Sirfi ve Tatbiki Matematik Derneği / Turkish Society of Pure and Applied Mathematics, Dedeefendi Cad 8, Sehzadebasi, Istanbul
Founded: 1948; Members: 100
Pres: Prof. Cahit Arf
Focus: Math 11173

Türk Standartlari Enstitüsü (TSE), Necatibey Cad 112, Bakanliklar, Ankara
T: 187240
Founded: 1960
Focus: Standards
Periodical
Standard (weekly) 11174

Türk Tarih Kurumu (TTK) / Turkish Historial Society, Kizilay Sok 1, Ankara
T: (04) 121100
Founded: 1931; Members: 41
Pres: Prof. Dr. Y. Yücel; Gen Secr: Prof. Dr. Ekrem Akurgal
Focus: Hist
Periodicals
Belgeler (weekly)
Belleten (3 times annually) 11175

Türk Tüberküloz Cemiyeti / Turkish Tuberculosis Society, Selime Hatun, Sağlik Sok, Taksim, Istanbul
Founded: 1937
Focus: Immunology 11176

Türk Uroloji Dernegi / Turkish Urological Society, c/o Dr. Cafer Yildiran, Talimhane Lamartin Cad 46/4, Eren ap Taksim, Istanbul
Founded: 1933; Members: 180
Pres: Dr. Sedat Tellaloğlu; Gen Secr: Dr. Cafer Yildiran
Focus: Urology
Periodical
Türk Uroloji Dergisi: Turkish Journal of Urology (weekly) 11177

Türk Veteriner Hekimleri Dernegi / Turkish Veterinary Medicine Association, Saglik Sok 21-3, Yenisehir, Ankara
Founded: 1930
Focus: Vet Med 11178

Yeni Felsefe Cemiyeti / The New Philosophical Society, Isik Lisesi, Nisantasi, Istanbul
Founded: 1943
Focus: Philos 11179

Turkmenistan

Turkmen Academy of Sciences, Ul Gogolya 15, 744000 Ashkhabad
Members: 55
Pres: A.G. Babaev; Gen Secr: V.N. Nikolaev
Focus: Sci
Periodicals
Izvestiya: Bulletin
Problemy Osvoyeniya Pustyn: Problems of Desert Development 11180

Uganda

Association for Teacher Education in Africa, c/o Prof. J.C.B. Bigala, Makerere University, POB 7062, Kampala
Founded: 1970; Members: 50
Pres: Prof. M. Mohapeloa; Gen Secr: Prof. J.C.B. Bigala
Focus: Educ
Periodicals
Education in Eastern Africa
Journal of West African Education . . . 11181

Association of Eastern and Southern African Universities / Association des Universités de l'Afrique de l'Est et du Sud, c/o Inter-University Council, POB 7110, Kampala
T: (041) 256251; Tx: 62179
Founded: 1984; Members: 20
Pres: Prof. Tom Tlou
Focus: Educ 11182

Cotton Research Corporation, POB 7084, Kampala
Focus: Crop Husb 11183

East African Agricultural Economics Society, c/o Dept of Rural Economy and Extension, Makerere University, POB 7062, Kampala
Founded: 1967
Focus: Agri 11184

East African School of Librarianship (EASL), c/o Makerere University, POB 7062, Kampala
T: (041) 554342; Tx: 61351
Founded: 1963
Gen Secr: Dr. S.A.H. Abidi
Focus: Libraries & Bk Sci; Educ
Periodicals
EASL Bulletin
EASL Newsletter 11185

Inter-University Council for East Africa, POB 7110, Kampala
T: (041) 56251
Founded: 1980; Members: 700
Gen Secr: E.K. Kigozi
Focus: Adult Educ
Periodicals
Newsletter
Report (weekly) 11186

Uganda Library Association (ULA), POB 5894, Kampala
Founded: 1972; Members: 140
Pres: P. Birungi; Gen Secr: L.M. Ssengero
Focus: Libraries & Bk Sci
Periodical
Ugandan Libraries 11187

The Uganda Society, POB 4980, Kampala
Founded: 1933; Members: 600
Focus: Hist; Lit
Periodical
The Uganda Journal 11188

Ukraine

Gerontology and Geriatrics Society, Vyshgorodskaya ul 67, 252655 Kiev
T: 4304068; Fax: 4329956
Founded: 1963; Members: 1880
Pres: V.V. Bezrukov; Gen Secr: Y.K. Duplenko
Focus: Geriatrics
Periodical
Problems of Ageing and Longevity (weekly) . . . 11189

Scientific and Technical Societies National Headquarters, Ul Artema 21, 252053 Kiev
T: 2124234
Focus: Sci; Eng . . . 11190

Ukrainian Academy of Agrarian Sciences, Ul Suvorova 9, 252010 Kiev
T: 2901085; Tx: 131487
Focus: Agri . . . 11191

Ukrainian Academy of Medical Sciences, Ul Gamarnikova 42, 252075 Kiev
Focus: Med . . . 11192

Ukrainian Academy of Sciences, Vladimirskaya ul 54, 252601 Kiev
T: 2256366; Tx: 131376
Founded: 1919; Members: 356
Pres: B.E. Paton; Gen Secr: B.S. Stognii
Focus: Sci
Periodicals
Arkheologiya: Archaeology
Avtomaticheskaya Svarka: Automatic Welding
Avtomatika: Automation
Biopolimery i Kletki: Biopolymers and Cells
Dopovidi Akademii Nauk Ukrainy: Report of the Ukrainian Academy of Sciences
Ekonomika Ukrainy: Economy of the Ukraine
Eksperimentalnaya Onkologyia: Experimental Oncology
Electronnoe Modelirovanie: Electronic Modelling
Filosofska i Sotsiologichna Dumka: Philosophy and Sociological Thought
Fizika Nizkikh Temperatur: Low-Temperature Physics
Fiziko-khimicheskaya Mekhanika Materialov: Physical and Chemical Mechanics of Materials
Fiziologicheskii Zhurnal: Physiological Journal
Fiziologiya i Biokhimiya Kulturnykh Rastenii: Physiology and Biochemistry of Cultivated Plants
Geofizicheskii Zhurnal: Geophysical Journal
Geologicheskii Zhurnal: Geological Journal
Gidrobiologicheskii Zhurnal: Hydrobiological Journal
Khimicheskaya Tekhnologiya: Chemical Technology
Khimiya i Tekhnologiya Vody: Water Chemistry and Engineering
Kibernetika, Kinematika i Fizika Nebesnykh Tel: Cybernetics, Kinematics and Physics of Heavenly Bodies
Kriobiologiya: Cryobiology
Literaturoznavstvo: Literary Studies
Metallofizika: Physics of Metals
Morskoi Gidrofizicheskii Zhurnal: Marine Hydrophysical Journal
Movoznavstvo: Linguistics
Narodna Tvorchist ta Etnografiya: Folk Art and Ethnography
Nikrobiologicheskii Zhurnal: Microbiological Journal
Poroshkovaya Metallurgiya: Powder Metallurgy
Prikladnaya Mekhanika: Applied Mechanics
Problemy Prochnosti: Problems of Strength
Problemy Spetsialnoi Elektrometallurgii: Problems of Special Electrometallurgy
Promyshlennaya Teplofizika: Industrial Thermal Physics
Radyanske Pravo: Soviet Law
Sverkhtverdye Materialy: Superhard Materials
Tekhnicheskaya Diagnostika i Nerazrushayushchii Kontrol: Technical Diagnostics and Non-destructive Testing
Tekhnicheskaya Elektrodinamika: Technical Electrodynamics
Teoreticheskaya i Eksperimentalnaya Khimiya: Theoretical and Experimental Chemistry
Tsitologiya i Genetika: Cytology and Genetics
Ukrainskii Biokhimicheskii Zhurnal: Ukrainian Biochemical Journal
Ukrainskii Botanichnyi Zhurnal: Ukrainian Botanical Journal
Ukrainskii Fizicheskii Zhurnal: Ukrainiam Physics Journal
Ukrainskii Istorichnyi Zhurnal: Ukrainian Historical Journal
Ukrainskii Khimicheskii Zhurnal: Ukrainian Chemical Journal
Ukrainskii Matematicheskii Zhurnal: Ukrainian Mathematical Journal
Upravlyayusshchie Sistemy i Mashiny: Control Systems and Computers
Vestnik Zoologiya: Zoological Journal
Visnyk Akademii Nauk Ukrainy: Journal of the Ukrainian Academy of Sciences . . . 11193

United Kingdom

Aberdeen-Angus Cattle Society, 6 King's Pl, Perth PH2 8AD
T: (0738) 22477; Fax: (0738) 36436
Founded: 1879; Members: 1100
Focus: Animal Husb
Periodicals
Aberdeen-Angus Herdbook (weekly)
Aberdeen-Angus Review (weekly) . . . 11194

Abergavenny and Border Counties Agricultural Society, 5 Lion St, Abergavenny NP7 5PE
T: (0873) 3152
Focus: Agri . . . 11195

Abortion Law Reform Association (ALRA) / A Woman's Right to Choose Campaign, 27-35 Mortimer St, London W1N 7RJ
T: (071) 6377264
Founded: 1936
Focus: Law
Periodical
Breaking Chains (weekly) . . . 11196

Academia Europaea, 31 Old Burlington St, London W1
T: (071) 7345402; Fax: (071) 2875115
Founded: 1988; Members: 1000
Gen Secr: Dr. Craig Sinclair
Focus: Educ . . . 11197

Academi Gymreig / Welsh Academy, Mount Stuart House, Mount Stuart Sq, Cardiff CF1 6DQ
T: (0222) 492025
Founded: 1959; Members: 60
Pres: J.E.C. Williams; Gen Secr: Dafydd Rogers
Focus: Ling; Lit
Periodicals
Cyfres Cyfieith-iadau'r Academi
Taliesin . . . 11198

Accountants Study Group of the EEC, c/o Institute of Chartered Acoountants in England and Hall, Moorgate Pl, London EC2
Focus: Business Admin . . . 11199

Action Against Allergy (AAA), 43 The Downs, London SW20 8HG
Focus: Immunology . . . 11200

Action for Dysphasic Adults (ADA), 1 Royal St, London SE1 7LL
T: (071) 2619572
Founded: 1980
Focus: Logopedy
Periodical
Newsletter (weekly) . . . 11201

Action in International Medicine (AIM), 46 Cleveland St, London W1P 6DB
T: (071) 6363610; Fax: (071) 6363612
Gen Secr: Prof. A.P. Haines
Focus: Med . . . 11202

Acton Society Trust, 5 Addison Bridge Pl, London W14 8XP
Founded: 1948
Pres: Edward Goodman
Focus: Econ; Poli Sci . . . 11203

Acupuncture Scientific and Clinical Advisory Group, 19 Richmond Hill, Clifton, Bristol BS8 1BA
T: (0272) 739477
Focus: Anesthetics . . . 11204

Advertising Association (AA), Abford House, 15 Wilton Rd, London SW1V 1NJ
T: (071) 8282771
Focus: Advert
Periodical
International Journal of Advertising (weekly) . . . 11205

Advisory Centre for Education (ACE), 1B Aberdeen Studio, 22-24 Highbury Grove, London N5 2EA
T: (071) 3548321; Fax: (071) 3591962
Founded: 1960
Pres: Lord Young of Dartington; Gen Secr: Prof. Sally Tomlinson
Focus: Educ
Periodical
ACE-Bulletin (weekly) . . . 11206

Advisory Committee on Protection of the Sea (ACOPS), 57 Duke St, London W1M 5DH
T: (071) 4990704; Fax: (071) 4933092
Founded: 1952
Gen Secr: Dr. Viktor Sebek
Focus: Oceanography; Ecology
Periodicals
ACOPS News (weekly)
ACOPS Yearbook (weekly) . . . 11207

The Aeroplane Collection (TAC), 38 Saint Mark's Av, Oldfield Brow, Altrincham WA14 4JB
Members: 36
Gen Secr: John Davidson
Focus: Aero
Periodical
Control Column . . . 11208

The Aetherius Society, 757 Fulham Rd, London SW6 5UU
T: (071) 736418
Founded: 1956
Gen Secr: Dr. Richard Lawrence
Focus: Rel & Theol
Periodical
Cosmic Voice (weekly) . . . 11209

African Society of International and Comparative Law, 22 Highbury Grove, London N6 2EA
T: (071) 7040610
Gen Secr: Emile K.M. Yakpo
Focus: Law . . . 11210

African Studies Association of the United Kingdom (ASAUK), c/o Royal African Society, SOAS, Thornhaugh St, Russell Sq, London WC1H 0XG
T: (071) 3236253
Founded: 1963; Members: 600
Pres: Prof. Christopher Clapham
Focus: Ethnology
Periodical
African Affairs (weekly) . . . 11211

Agency for Cooperation and Research in Development, Francis House, Francis St, London SW1P 1DQ
T: (071) 8287611; Tx: 8954437; Fax: (071) 9766113
Founded: 1974
Gen Secr: Mark Sinclair
Focus: Develop Areas . . . 11212

Agricultural and Food Research Council (AFCR), Polaris House, North Star Av, Swindon SN2 1UH
T: (0793) 413200; Fax: (0793) 413201
Founded: 1931
Pres: Sir Alistair Grant; Gen Secr: Prof. T.L. Blundell
Focus: Agri; Food
Periodicals
AFCR News
Annual Report (weekly)
Corporate Plan
Handbook of Agricultural and Food Research . . . 11213

Agricultural Economics Society (AES), c/o Dept of Agricultural and Food Economics, Queens University of Belfast, Newforge Lane, Belfast BT9 5PX
T: (0232) 661166; Fax: (0232) 668384
Founded: 1926; Members: 760
Pres: Prof. D.R. Colman; Gen Secr: Dr. J. Davis
Focus: Agri
Periodical
Journal of Agricultural Economics (3 times annually) . . . 11214

Agricultural Education Association (AEA), c/o Askham Bryan College, York YO2 3PR
T: (0904) 706232
Founded: 1894; Members: 300
Focus: Adult Educ; Agri . . . 11215

Agricultural Research Institute of Northern Ireland, Large Park, Hillsborough Bt26 6DR
T: (0846) 682484; Fax: (0846) 689594
Focus: Agri
Periodical
Report on Agricultural Research (weekly) 11216

Aircraft Research Association (ARA), Manton Lane, Bedford MK41 7PF
T: (0234) 350681; Tx: 825056; Fax: (0234) 328584
Founded: 1952; Members: 4
Pres: Dr. J.E. Green; Gen Secr: M.P. Carr
Focus: Aero . . . 11217

Airfields Environment Federation (AEF), 5 High Timber St, London EC4V 3NS
T: (071) 3298139; Fax: (071) 3298160
Founded: 1975
Gen Secr: Peter Fairhurst
Focus: Aero
Periodical
Newsletter (weekly) . . . 11218

Airship Association, 20 Myddelton Gardens , Winchmore Hill, London N21
T: (081) 3601357
Founded: 1971
Focus: Aero . . . 11219

Alcuin Club, All Saints Vicarage, Highlands Rd, Runcorn WA7 4PS
T: (0928) 575666; Fax: (0928) 581868
Founded: 1897; Members: 650
Focus: Rel & Theol
Periodicals
Collections (weekly)
Joint Liturgical Studies (3 times annually) 11220

Alkan Society, 21 Heronswood, Salisbury SP2 8DH
T: (0722) 325771
Founded: 1977; Members: 140
Gen Secr: Peter Grove
Focus: Music
Periodical
Bulletin (3 times annually) . . . 11221

Alternative Society (AS), 9 Morton Av, Kidlington, Oxon
T: 3413
Founded: 1972
Focus: Sociology . . . 11222

Amateur Entomologists' Society (A.E.S.), 22 Salisbury Rd, Feltham TW13 5DP
Founded: 1935; Members: 2000
Pres: Duncan Reavey
Focus: Entomology
Periodical
Bulletin: Wants & Exchange List (weekly) 11223

American Civil War Round Table, United Kingdom (ACWRTUK), 98 Kew Green, Kew, Richmond TW9 3AP
T: (081) 9409950
Founded: 1953; Members: 105
Focus: Hist
Periodicals
Crossfire (weekly)
Newsletter (weekly) . . . 11224

American Dental Society of Europe, 40 Harley St, London W1N 1AB
Fax: (071) 4362015
Founded: 1873; Members: 145
Focus: Dent . . . 11225

Anaesthetic Research Society, c/o Dept of Anaesthesia, Royal Liverpool Hospital, Prescot St, POB 147, Liverpool
T: (051) 7064000; Fax: (051) 7065815
Members: 300
Gen Secr: Dr. J. Hunter
Focus: Anesthetics
Periodical
Proceedings . . . 11226

Anatomical Society of Great Britain and Ireland, c/o Dept of Anatomy and Experimental Pathology, Saint Salvator's College, Bute Medical Bldgs, Saint Andrews KY16 9TS
T: (0334) 76161 ext 7106
Founded: 1887; Members: 700
Pres: Dr. D. Brynmor Thomas; Gen Secr: Prof. J.A. Firth
Focus: Anat
Periodical
Journal of Anatomy (weekly) . . . 11227

Ancient Monuments Society (AMS), Saint Ann's Vestry Hall, 2 Church, London EC4V 5HB
T: (071) 2363934
Founded: 1924; Members: 2000
Focus: Preserv Hist Monuments
Periodical
Transactions (weekly) . . . 11228

An Comunn Gaidhealach (ACG) / The Highland Association, 109 Church St, Inverness IV1 1EU
T: (0463) 231226
Founded: 1891; Members: 2600
Gen Secr: Donald M. MacLean
Focus: Ethnology; Ling . . . 11229

Anglesey Antiquarian Society and Field Club (A.A.S.), 22 Lon Ganol, Menai Bridge LL59 5LU, Anglesey
T: (0248) 712566
Founded: 1911; Members: 1100
Focus: Archeol; Hist . . . 11230

Anglo-Continental Dental Society (ACDS), 75 Wimpole St, London W1
T: (071) 9353242
Founded: 1942
Focus: Dent . . . 11231

Anglo-European College of Chiropractic, 13 Parkwood Rd, Bournemouth BH5 2DF
T: (202) 431021; Fax: (202) 417352
Founded: 1965
Pres: Prof. Brian N. Kliger
Focus: Med . . . 11232

Anglo-Norman Text Society (ANTS), c/o Birkbeck College, Malet St, London WCIE 7HX
Founded: 1938; Members: 370
Focus: Lit
Periodical
ANTS Publications (weekly) . . . 11233

Animal Breeding Research Organization (ABRO), King's Bldgs, West Main Rd, Edinburgh EH9 3JQ
T: (031) 6676901
Founded: 1947
Focus: Animal Husb . . . 11234

Animal Diseases Research Association, c/o Moredun Research Institute, 408 Gilmerton Rd, Edinburgh EH17 7JH
T: (031) 6643262; Fax: (031) 6648001
Founded: 1920; Members: 2000
Pres: J. Stobo; Gen Secr: M.J. Mackenzie
Focus: Vet Med
Periodicals
Association Report and Accounts (weekly)
Moredun Research Institute: Scientific Report (bi-annually) . . . 11235

Animal Health Trust, POB 5, Newmarket CB8 7DW
T: (0638) 661111; Tx: 818418; Fax: (0638) 665789
Pres: A.J. Higgins; Gen Secr: Eve Fremantle
Focus: Vet Med
Periodical
Annual Report (weekly) . . . 11236

Anthroposophical Society in Great Britain, 35 Park Rd, London NW1 6XT
T: (071) 7234400
Founded: 1923; Members: 2000
Focus: Philos . . . 11237

Antiquarian Horological Society (A.H.S.), New House, High St, Ticehurst, Wadhurst TN5 7AL
T: (0580) 200155
Founded: 1953; Members: 2000
Pres: Prof. Arnold Wolfendale; Gen Secr: Maria Collins
Focus: Hist; Eng
Periodical
Antiquarian Horology (weekly) . . . 11238

Appropriate Health Resources and Technologies Action Group (AHRTAG), 1 London Bridge St, London SE1 8SG
T: (071) 3781403; Tx: 912881; Fax: (071) 4036003
Founded: 1977
Gen Secr: K. Atawell
Focus: Public Health
Periodicals
AIDS Action (weekly)
ARI News (weekly)
CBR News (3 times annually)
Dialogue on Diarrhoea (weekly) 11239

Arab Research Centre, 76-78 Notting Hill Gate, London W11 3HS
T: (071) 2212425; Fax: (071) 2215899
Founded: 1979; Members: 50
Pres: Abdel Majid Farid; Gen Secr: A. Amall
Focus: Ethnology
Periodicals
Arab Paper (10 times annually)
The Arab Researcher 11240

Arboricultural Association, Ampfield House, Romsey SO51 9PA
T: (0794) 68717; Fax: (0794) 68978
Founded: 1964; Members: 2000
Focus: Forestry
Periodical
Arboricultural Journal: The International Journal of Urban Forestry (weekly) 11241

Architects and Surveyors Institute (ASI), 15 Saint Mary St, Chippenham, Wilts
T: (0249) 444505; Fax: (0249) 443602
Founded: 1926; Members: 6000
Pres: M.V. Lelliott; Gen Secr: B.A. Hunt
Focus: Archit; Surveying
Periodical
ASI Journal (weekly) 11242

Architectural and Archaeological Society for the County of Buckingham, c/o County Museum, Church St, Aylesbury HP20 2QP
T: (0296) 20984
Founded: 1847; Members: 602
Focus: Archit; Archeol
Periodical
Records of Buckinghamshire (weekly) . . 11243

Architectural and Archaeological Society of Durham and Northumberland, c/o Dept of Archaeology, Saddler St, Durham City DH1 3NU
T: (091) 3743633
Founded: 1861; Members: 300
Focus: Archeol; Archit
Periodical
Durham Archaeological Journal (weekly) . 11244

Architectural Association (AA), 34-36 Bedford Sq, London WC1B 3ES
T: (071) 6360974
Founded: 1847; Members: 3000
Focus: Archit
Periodical
AA Files (weekly) 11245

Aristotelian Society, c/o Dept of Philosophy, Birkbeck College, Malet St, London WC1E 7HX
T: (071) 2551724
Founded: 1880; Members: 850
Gen Secr: Sebastian Gardner
Focus: Philos
Periodicals
Monographs
Proceedings (weekly) 11246

Armagh Field Naturalists Society, 14a Dobbins Grove, Armagh BT60 1BG
Founded: 1952; Members: 50
Focus: Nat Sci 11247

Arms and Armour Society of Great Britain, 1 Vicarage Lane, Laleham, Staines, Middx
Focus: Military Sci
Periodicals
Journal (weekly)
The Military Chest (weekly) 11248

Arnold Bennett Literary Society, Burslem Leisure Centre, Market Pl, Stoke-on-Trent ST6 3DS
T: (0782) 813363
Founded: 1954
Focus: Lit 11249

The Arthritis and Rheumatism Council (ARC), Copeman House, Saint Mary's Court, Chesterfield S41 7TD
T: (0246) 558033
Founded: 1937
Pres: Lord Dainton; Gen Secr: J. Norton
Focus: Rheuma
Periodical
Reports on Rheumatic Diseases (weekly) 11250

Art Libraries Society of the United Kingdom and Ireland (ARLIS UK & Ireland), 18 College Rd, Bromsgrove B60 2NE
T: (0527) 579298
Founded: 1969; Members: 750
Pres: Ian Monie; Gen Secr: Sonia French
Focus: Libraries & Bk Sci
Periodicals
Art Libraries Journal (weekly)
News-sheet (weekly) 11251

Arts Association, 227 Goldhawk Rd, London W12 8ER
T: (081) 7434378
Founded: 1953; Members: 8500
Focus: Arts 11252

Arts Council of Great Britain, 14 Great Peter St, London SW1P 3NQ
T: (071) 3330010
Founded: 1940; Members: 20
Pres: Lord Peter Palumbo; Gen Secr: Anthony Everitt
Focus: Arts
Periodical
Report (weekly) 11253

Arts Council of Northern Ireland, 181a Stranmillis Rd, Belfast BT9 5DU
Founded: 1943
Focus: Arts
Periodical
Annual Report (weekly) 11254

Ashmolean Natural History Society of Oxfordshire (ANHSO), Pickets Heath, The Ridgeway, Boars Hill, Oxford OX1 5EZ
Founded: 1828; Members: 300
Pres: Dr. A.W. McDonald; Gen Secr: Dr. C. Huxley-Lambrick
Focus: Hist; Nat Sci
Periodical
Fritillary (weekly) 11255

Aslib, The Association for Information Management, 20-24 Old St, London EC1N 9AP
T: (071) 2534488; Tx: 23667; Fax: (071) 4300514
Founded: 1924; Members: 2100
Pres: Lord Ezra; Gen Secr: Roger Bowes
Focus: Libraries & Bk Sci
Periodicals
Aslib Booklist (weekly)
Aslib Informations (weekly)
Aslib Proceedings (weekly)
Current Awareness Bulletin (weekly)
Forthcoming International Scientific and Technical Conferences (weekly)
Journal of Documentation (weekly)
Program (weekly) 11256

Association for Adult and Continuing Education (AACE), Hamilton House, Mabledon Pl, London WC1H 9BH
T: (071) 3801314
Founded: 1978; Members: 3000
Focus: Adult Educ 11257

Association for All Speech Impaired Children (AFASIC), 347 Central Markets, London EC1A 9NH
T: (071) 2363632
Founded: 1968; Members: 750
Focus: Logopedy
Periodicals
Annual Report (weekly)
Newsletter (weekly) 11258

Association for Applied Hypnosis (AAH), 33 Abbey Park Rd, Grimsby DN32 0HS
T: (0472) 47702
Founded: 1980; Members: 25
Focus: Depth Psych 11259

Association for Brain Damaged Children, Clifton House, Saint Pauls Rd, Coventry CV6 5DE
T: (0203) 665450
Founded: 1969; Members: 100
Focus: Pediatrics; Logopedy 11260

Association for British Music (ABM), 2 Union Pl, Boston PE21 6PS
T: (0205) 60541
Founded: 1977; Members: 600
Focus: Music 11261

Association for Child Psychology and Psychiatry (ACPP), 70 Borough High St, London SE1 1XF
T: (071) 4037458; Fax: (071) 4037081
Founded: 1957; Members: 2500
Gen Secr: R. Whithear
Focus: Psych; Psychiatry; Soc Sci; Educ 11262

Association for Educational and Training Technology (A.E.T.T.), Higher Millbrook, Beavor Lane, Axminster EX13 5EQ
Founded: 1965
Pres: Dr. Peter Clarke; Gen Secr: N.R. Winterburn
Focus: Educ
Periodicals
Aspects of Educational Technology (weekly)
Educational and Training Technology International (weekly)
International Yearbook of Educational and Training Technology (bi-annually) 11263

Association for Group and Individual Psychotherapy, 29 Saint Marles Crescent, London NW1
T: (071) 4859141
Focus: Psych; Therapeutics 11264

Association for Independent Disabled Self-Sufficiency (AIDS), 7 Alfred St, Bath BA1 2QU
T: (0225) 25197
Founded: 1977; Members: 200
Focus: Educ Handic
Periodical
Phoenix (weekly) 11265

Association for Industrial Archaeology (AIA), c/o The Wharfage, Ironbridge, Telford TF8 7AW
T: (0952) 432751
Founded: 1973; Members: 630
Gen Secr: A. Patrick
Focus: Archeol
Periodicals
AIA Bulletin (weekly)
Industrial Archaeology Review (weekly) . 11266

Association for Information Management (ASLIB), 20-24 Old St, London EC1V 9AP
T: (071) 2534488; Tx: 23667; Fax: (071) 4300514
Founded: 1924
Gen Secr: Roger Bowes
Focus: Business Admin
Periodicals
ASLIB Book List (weekly)
ASLIB Information (10 times annually)
ASLIB Proceedings (10 times annually)
Computing, Communications, Media and Socio-Technology-Trend Monitor (8 times annually)
Critique (10 times annually)
Current Awareness Bulletin (10 times annually)
Forthcoming International Scientific and Technical Conferences (weekly)
Index to Theses with Abstracts (weekly)
The Intelligent Enterprise (weekly)
IT Link incorporating Automation Notes (weekly)
Journal of Documentation (weekly)
Online Notes (weekly)
Program-Automated Library and Information Systems (weekly)
Records Management Journal (weekly) . 11267

Association for International Cancer Research, Moor House, Cameron-Markby
Founded: 1979; Members: 9
Focus: Cell Biol & Cancer Res 11268

Association for Language Learning, 16 Regent Pl, Rugby CV21 2PN
Fax: (0788) 546443
Founded: 1990; Members: 5600
Pres: Madeleine Bedford; Gen Secr: Christine Wilding
Focus: Ling; Educ
Periodicals
Francophonie (weekly)
German Teaching (weekly)
Language Learning Journal (weekly)
Language World Newsheet (weekly)
Rusistika (weekly)
Tuttitalia (weekly)
Vida Hispanica (weekly) 11269

Association for Latin Liturgy (ALL), Hillside, Combe Hill, Combe Saint Nicholas, Chard TA20 3NW
T: (0460) 65029
Founded: 1969; Members: 400
Focus: Rel & Theol 11270

Association for Literary and Linguistic Computing (ALLC), c/o Dept of English, University College of North Wales, Bangor LL57 2DG
T: (0248) 351151
Founded: 1973; Members: 420
Gen Secr: Thomas M. Corns
Focus: Computer & Info Sci
Periodical
Literary & Linguistic Computing (weekly) . 11271

Association for Medical Deans in Europe (AMDE), c/o Dept of Medicine, Institute of Clinical Science, Grosvenor Rd, Belfast BT12 6HJ
T: (0232) 240503; Fax: (0232) 240899
Founded: 1979
Pres: Prof. A.H. Gary Love
Focus: Med 11272

Association for Medical Education in Europe (AMEE), c/o Centre for Medical Education, Ninewells Hospital and Medical School, Dundee DD1 9SY
T: (0382) 60111; Fax: (0382) 645748
Founded: 1972
Gen Secr: Prof. R.M. Harden
Focus: Educ; Med
Periodical
Medical Education (weekly) 11273

Association for Medical Physics Technology (AMPT), 30 Burnham Av, Ickenham, Middx
Founded: 1952; Members: 242
Focus: Physics; Eng 11274

Association for Petroleum and Explosives Administration (APEA), c/o Trading Standards Dept, Hinchingbrooke Cottage, Brampton Rd, Huntington PE18 8NA
Founded: 1974; Members: 350
Focus: Safety
Periodical
The Bulletin (weekly) 11275

Association for Professionals in Services for Adolescents (APSA), 13 Bonaly Dr, Edinburgh EH13 0EJ
T: (031) 4411049
Founded: 1969; Members: 500
Focus: Psychiatry
Periodical
Journal of Adolescence (weekly) 11276

Association for Radiation Research (ARR), c/o Dr. K.M. Prise, Gray Laboratory, Mount Vernon Hospital, POB 100, Northwood HA6 2JR
Founded: 1958; Members: 330
Pres: Prof. P. Wardman; Gen Secr: Dr. K.M. Prise
Focus: Radiology 11277

Association for Recurrent Education (ARE), c/o Dept of Education Management, 36 Collegiate Crescent, Sheffield
T: 56101
Founded: 1975; Members: 65
Focus: Educ 11278

Association for Religious Education (ARE), 17 Clover Close, Cumnor Hill OX2 9JH
T: (08676) 3030
Founded: 1968
Focus: Rel & Theol; Educ 11279

Association for Research in the Voluntary and Community Sector (ARVAC), Unit 29, Wivenhoe Business Centre, Brook St, Wivenhoe CO7 9DP
T: (0206) 824281
Founded: 1978; Members: 250
Gen Secr: Pat Marsden
Focus: Public Admin
Periodical
ARVAC Bulletin (weekly) 11280

Association for Research into Restricted Growth (ARRG), 5 Teak Walk, Witham CM8 2SX
Founded: 1970; Members: 350
Focus: Pathology 11281

The Association for Science Education (A.S.E.), College Ln, Hatfield AL10 9AA
T: (07072) 67411
Founded: 1963; Members: 23500
Gen Secr: Dr. David S. Moore
Focus: Educ; Sci
Periodicals
ASE Primary Science (3 times annually)
Education in Science (5 times annually)
Primary Science Review (3 times annually)
The School Science Review (weekly) . . 11282

Association for Scottish Literary Studies (ASLS), c/o Dept of English, University of Aberdeen, Aberdeen AB9 2UB
T: (0224) 40241
Founded: 1970; Members: 750
Focus: Lit
Periodicals
New Writing Scotland (weekly)
Scottish Language (weekly)
Scottish Literary Journal (weekly)
Scottish Literary Journal Supplements (weekly) 11283

Association for Spina Bifida and Hydrocephalus (ASBAH), 42 Park Rd, Peterborough PE1 2UQ
T: (0733) 555988; Fax: (0733) 555985
Founded: 1966; Members: 10000
Focus: Med
Periodical
Link (weekly) 11284

Association for Studies in the Conservation of Historic Buildings (ASCHB), c/o Institute of Archaeology, 31-34 Gordon Sq, London WC1H 0PY
T: (071) 5809148; CA: (071) 5809639
Founded: 1968; Members: 450
Focus: Preserv Hist Monuments
Periodicals
Newsletter (weekly)
Transactions (weekly) 11285

Association for Teaching Psychology (ATPsych), c/o The British Psychological Society, 48 Princess Rd East, Leicester LE1 7DR
T: (0533) 549568
Founded: 1971; Members: 500
Focus: Psych; Educ
Periodical
Psychology Teaching (weekly) 11286

Association for the Education and Welfare of the Visually Handicapped (A.E.W.V.R.), Saint John's School House, Hadzor, Droitwich Spa WR9 7DR
Founded: 1979; Members: 328
Focus: Educ Handic
Periodical
British Journal of Visual Impairment (3 times annually) 11287

The Association for the Protection of Rural Scotland (A.P.R.S.), Gladstone's Land, 483 Lawnmarket, Edinburgh EH1 2NT
T: (031) 2256592
Founded: 1926; Members: 1110
Gen Secr: Elizabeth Garland
Focus: Ecology
Periodicals
Annual Report (weekly)
Newsletter (weekly) 11288

Association for the Reduction of Aircraft Noise (ARAN), 11 First St, London SW3
T: (071) 5841848
Founded: 1964
Focus: Ecology 11289

Association for the Reform of the Latin Teaching (A.R.L.T.), 33 Lane End Drive, Knaphell, Woking GU21 2QQ
T: (021) 4540895
Founded: 1911; Members: 1000
Focus: Ling; Educ

Periodical
Latin Teaching (weekly) 11290

Association for the Study of Animal Behaviour (ASAB), c/o Dept of Biology, Royal Holloway and Bedford New College, Egham Hill, Egham TW20 0EX
Founded: 1936; Members: 1030
Pres: Prof. Dr. N.R. Chalmers; Gen Secr: Dr. C.K. Catchpole
Focus: Behav Sci
Periodical
Animal Behaviour (weekly) 11291

Association for the Study of German Politics (ASGP), c/o Dept of Business Studies and Languages, Sheffield City Polytechnic, Sheffield S17 4AB
Founded: 1974
Gen Secr: Dr. Ian King
Focus: Poli Sci 11292

Association for the Study of Medical Education (ASME), c/o Ninewells Hospital and Medical School, Dundee DD1 9SY
Founded: 1957; Members: 800
Pres: Sir John Walton; Gen Secr: Prof. John Anderson
Focus: Adult Educ; Med
Periodicals
Annual Report (weekly)
Medical Education (weekly) 11293

Association for the Study of Obesity (ASO), 50 Ruby Rd, London E17 4RF
T: (081) 5210653
Founded: 1967; Members: 210
Gen Secr: Michael Enser
Focus: Physiology 11294

Association for the Teaching of the Social Sciences (ATSS), 19 Mandeville Gardens, Saint Albans Al1 40A
Founded: 1963; Members: 1081
Focus: Educ; Sociology
Periodicals
Briefings (3 times annually)
The Social Science Teacher (3 times annually)
. . . . 11295

Association for Veterinary Clinical Pharmacology and Therapeutics, Cedar Cottage, Sutton Pl, Abinger Hammer, Dorking, Surrey
Focus: Vet Med; Pharmacol 11296

Association in Scotland to Research into Astronautics (ASTRA), 720 Glasgow Rd, Craigneuk, Wishaw, Lanarks
Focus: Astronomy
Periodicals
Asgard (weekly)
Space Explorer (weekly)
Spacereport (weekly) 11297

Association of Advisers in Design and Technical Studies (AADTS), 33 Foxhill Crescent, Camberley GU15 1PR
T: (0276) 23846
Members: 120
Focus: Eng; Graphic & Dec Arts, Design 11298

Association of Agricultural Education Staffs (AAES), 43 Saint John's Rd, Mogerhanger MK44 3RJ
T: (0767) 40464
Founded: 1946; Members: 792
Focus: Adult Educ; Agri 11299

Association of Anaesthetists of Great Britain and Ireland, 9 Bedford Sq, London WC1B 3RA
T: (071) 6311650; Fax: (071) 6314352
Founded: 1932; Members: 6000
Pres: Dr. W.R. MacRae; Gen Secr: Dr. R.S. Vaughan
Focus: Anesthetics
Periodical
Anaesthesia (weekly) 11300

The Association of Applied Biologists (AAB), c/o Horticultural Research International, Worthing Rd, Littlehampton BN17 6LP
Founded: 1904; Members: 1200
Pres: Prof. N.L. Innes; Gen Secr: Dr. F.A. Langton
Focus: Bio
Periodical
Annals of Applied Biology (weekly) . . . 11301

Association of Art Institutions (AAI), 24 Widemarsh St, Hereford HR4 9EP
T: (0432) 66653
Founded: 1942; Members: 500
Focus: Arts 11302

Association of Arts Centres in Scotland (AACS), 50 Forest Rd, Aberdeen AB2 4BP
T: (0224) 37115
Founded: 1970; Members: 40
Focus: Arts 11303

Association of Basic Science Teachers in Dentistry, Dr. Josie A Beeley, c/o Oral Sciences Dept, Dental School, University of Glasgow,, 378 Sauchiehall St, Glasgow G23JZ
T: (041) 3327020 ext 205; Tx: 777070;
Fax: (041) 3531593
Founded: 1978; Members: 90
Pres: Prof. Colin Robinson; Gen Secr: Dr. Josie Beeley
Focus: Dent 11304

Association of Beauty Teachers (ABT), 38a Portsmouth Rd, Southampton SO2 9AD
T: (0703) 422695
Founded: 1967; Members: 300
Gen Secr: J.M. Palmer
Focus: Educ
Periodical
Health and Beauty World (weekly) . . . 11305

Association of Blind and Partially-Sights Teachers and Students (ABAPSTAS), POB 6727, London WC1N 3XX
Founded: 1970; Members: 200
Focus: Educ; Educ Handic; Adult Educ . 11306

Association of British Climatologists (ABC), c/o Dept of Environmental and Geographical Studies, Roehampton Institute, Wimbledon Parkside, London SW19 5NN
T: (081) 3923473; Tx: 54348
Founded: 1974; Members: 120
Focus: Geophys
Periodical
Directory of Climatologists 11307

Association of British Correspondence Colleges (A.B.C.C.), 6 Francis Grove, London SW19 4DT
T: (081) 5449559
Founded: 1955; Members: 20
Focus: Educ 11308

Association of British Dental Surgery Assistants, 29 London St, Fleetwood FY7 6JY
T: (0253) 778631; Fax: (0253) 777268
Founded: 1940
Pres: J. Robins; Gen Secr: P.A. Swain
Focus: Surgery; Dent
Periodical
The British Dental Surgery Assistant (weekly)
. . . . 11309

Association of British Dispensing Opticians, 6 Hurlingham Business Park, Sulivan Rd, London SW6 3DU
T: (071) 7365531; Fax: (071) 7315531
Focus: Optics 11310

Association of British Forensic Specialists, POB 389, Moseley, Birmingham
T: (021) 4497735
Focus: Forensic Med 11311

Association of British Geodesists (ABG), c/o Dept of Land Surveying, Polytechnic of East London, Longbridge Rd, Dagenham RM8 2AS
T: (081) 5907722
Founded: 1950; Members: 60
Focus: Surveying 11312

Association of British Neurologists, 9 Fitzroy Sq, London W1P 5AH
Founded: 1933; Members: 560
Pres: Dr. J.R. Heron; Gen Secr: Prof. C. Kennard
Focus: Neurology 11313

Association of British Paediatric Nurses (ABPN), c/o Central Nursing Office, Hospital for Sick Children, Great Ormond St, London WC1N 3JH
T: (071) 4059200
Founded: 1938; Members: 1600
Focus: Pediatrics
Periodicals
Newssheet (weekly)
Spotlight on Children (weekly) 11314

Association of British Pewter Craftsmen (ABPC), 136 Hagley Rd, Edgbaston, Birmingham B16 9PN
T: (021) 4544141; Fax: (021) 4544949
Members: 28
Gen Secr: P.W. Adams
Focus: Materials Sci
Periodical
Pewter Review (00) 11315

Association of British Science Writers (ABSW), c/o The British Association, 23 Saville Row, London W1X 1AB
T: (071) 4391205; Fax: (071) 7341658
Founded: 1947; Members: 450
Gen Secr: Dr. Peter Briggs
Focus: Lit; Sci
Periodical
The Science Reporter (weekly) 11316

Association of British Spectroscopists, c/o Dr. D.V. Bowen, Physical Sciences, Pfizer Central Research, Sandwich, Kent CT13 9NJ
T: (0304) 616438; Fax: (0304) 616726
Founded: 1969; Members: 50
Pres: Dr. D.V. Bowen
Focus: Physics; Chem; Bio 11317

Association of British Theological and Philosophical Libraries (ABTAPL), c/o Bible Society's Library, Cambridge University Library, West Rd, Cambridge CB3 9DR
T: (0223) 333000 ext 3075; Tx: 81395;
Fax: (0223) 333160
Founded: 1956; Members: 290
Pres: Judith Powles; Gen Secr: Alan F. Jesson
Focus: Libraries & Bk Sci
Periodical
Bulletin (3 times annually) 11318

Association of Building Engineers, Jubilee House, Billing Brook Rd, Weston Favell, Northampton NN3 4NW
T: (0604) 404121
Founded: 1925; Members: 4800
Pres: D.F. James; Gen Secr: W.A. Black
Focus: Civil Eng
Periodicals
Building Engineer (10 times annually)
Fire Surveyor (weekly) 11319

Association of Career Teachers (ACT), Hillsboro, Castledine St, Loughborough, Leics
T: (0509) 214617
Founded: 1975; Members: 1000
Focus: Adult Educ 11320

Association of Chief Architects of Scottish Local Authorities, c/o Lanark District Council, 57 High St, Lanark
T: (0786) 3111
Founded: 1952
Focus: Archit 11321

Association of Child Psychotherapists, Burgh House, New End Sq, London NW3
T: (071) 7948881; Fax: (081) 3616794
Members: 150
Focus: Therapeutics
Periodical
Journal of Child Psychology 11322

Association of Christian Teachers (ACT), 130 City Rd, London EC1V 2NJ
T: (071) 4862561
Founded: 1971; Members: 2800
Focus: Educ 11323

Association of Clinical Biochemists (ACB), Burlington House, Piccadilly, London W1V 0BN
T: (071) 4378656
Founded: 1953; Members: 2000
Pres: Prof. V. Marks; Gen Secr: Dr. A.M. Kelly
Focus: Biochem
Periodicals
Annals of Clinical Biochemistry (weekly)
Current Advances in Clinical Chemistry (weekly)
News Sheet (weekly) 11324

Association of Clinical Pathologists (A.C.P.), c/o School of Biological Sciences, Falmer, Brighton BN1 9QG
T: (0273) 606755
Founded: 1927; Members: 2350
Pres: Prof. F.V. Flynn; Gen Secr: Dr. W.R. Timperley
Focus: Pathology
Periodical
Journal of Clinical Pathology (weekly) . . 11325

Association of Colleges for Further and Higher Education, Swindon College, Regent Circus, Swindon SN1 1PT
T: (0793) 513193
Founded: 1893; Members: 362
Gen Secr: C. Brain
Focus: Educ; Adult Educ 11326

Association of Commonwealth Archivists and Records Managers, c/o Institute of Commonwealth Studies, 28 Russell Sq, London WC1B 5DS
Founded: 1984
Focus: Archives
Periodical
Association of Commonwealth Archivists and Records Managers Newsletter 11327

Association of Commonwealth Teachers, 42 Camborne Av, London W13
T: (081) 5673221
Focus: Educ 11328

Association of Commonwealth Universities (A.C.U.), 36 Gordon Sq, London WC1H 0PF
T: (071) 3878572; CA: Acumen London;
Fax: (071) 3872655
Founded: 1913; Members: 420
Pres: Prof. B.L. Clarkson; Gen Secr: Dr. A. Christodoulou
Focus: Educ
Periodicals
A.C.U. Bulletin of Current Documentation (5 times annually)
Commonwealth Universities Yearbook (weekly)
. . . . 11329

Association of Community Health Councils for England and Wales, 362 Euston Rd, London NW1 3BL
T: (071) 3884814
Focus: Public Health
Periodical
CHC News (10 times annually) 11330

Association of Community Workers (ACW), Grindon Lodge, Beech Grove Rd, Newcastle-upon-Tyne NE4 2RS
T: (091) 2724341
Founded: 1968; Members: 650
Gen Secr: L.A. Leach
Focus: Sociology
Periodicals
Community Work (10 times annually)
Talking Point (10 times annually) . . . 11331

Association of Consulting Actuaries (ACA), 39-45 Tottenham Court Rd, London W1P 0JP
T: (071) 636777
Founded: 1951; Members: 140
Focus: Insurance 11332

Association of County Public Health Officers (AssCPHO's), 1 Cloatley Rd, Hankerton, Malmesbury, Wilts
T: (06667) 441
Founded: 1946; Members: 70
Gen Secr: N.J. Durnford
Focus: Public Health 11333

Association of Crematorium Medical Referees, 59 Kings Rd, Westcliff-on-Sea, Essex
T: (0702) 77878
Founded: 1953
Focus: Med 11334

Association of Dental Hospitals of Great Britain and Northern Ireland, Saint Chad's, Queensway, Birmingham B4 6NN
T: (021) 2368611
Focus: Dent 11335

Association of Directors of Education in Scotland (A.D.E.S.), c/o Tayside Region Dept of Education, Tayside House, Dundee DD1 3RJ
T: (0382) 303654
Founded: 1920; Members: 150
Gen Secr: Sandy Watson
Focus: Educ 11336

Association of Directors of Social Services (ADSS), Town Hall, Stockport SK4 3XE
T: (061) 4747896; Fax: (061) 4747895
Founded: 1971; Members: 120
Gen Secr: R.J. Lewis
Focus: Sociology 11337

Association of Directors of Social Work (ADSW), c/o Social Work Dept, Newtown St, Boswells, Melrose TD6 0SA
T: (0835) 23366
Founded: 1969; Members: 68
Focus: Sociology 11338

Association of Disabled Professionals (ADP), 170 Benton Hill, Horbury WF4 5HW
T: (0924) 270335; Fax: (0924) 276498
Founded: 1970; Members: 507
Pres: S.J. Maynard
Focus: Rehabil
Periodical
House Bulletin (weekly) 11339

Association of District Medical Officers, c/o Royal United Hospital, Combe Park, Bath BA1 3NG
Founded: 1974
Focus: Med 11340

Association of Educational Psychologists (AEP), 3 Sunderland Rd, Durham DH1 2LH
T: (091) 3849512
Founded: 1962; Members: 450
Gen Secr: J.B. Harrison-Jennings
Focus: Educ; Psych
Periodical
Journal (weekly) 11341

Association of European Geological Societies (AEGS), c/o Geological Society, Burlington House, Piccadilly, London W1V 0JU
T: (071) 4349944
Founded: 1987
Gen Secr: R.M. Bateman
Focus: Geology 11342

Association of European Latin Americanist Historians, c/o Institute of Latin American Studies, University of Liverpool, POB 147, Liverpool L69 3BX
T: (051) 7943079
Founded: 1972
Gen Secr: Prof. John Fisher
Focus: Hist
Periodical
Historía Latinoamericana en Europa (weekly)
. . . . 11343

Association of European Open Air Museums (AEOM), c/o Weald and Downland Open Air Museum, Singleton, Chichester PO18 0EU
T: (024363) 348
Founded: 1966
Pres: Christopher Zeuner
Focus: Cultur Hist 11344

Association of Genealogists and Record Agents (AGRA), 29 Badgers Close, Horsham RH12 5RU
Founded: 1968; Members: 100
Gen Secr: D.R. Young
Focus: Genealogy; Hist 11345

Association of General Practitioner Hospitals, c/o Saint Chad Health Centre, The Dimbles, Lichfield, Staffs
Founded: 1969
Focus: Public Health 11346

Association of Governing Bodies of Girls Public Schools (GBGSA), 27 Church Rd, Steep, Petersfield GU32 2DW
T: (0730) 4823
Founded: 1942; Members: 233
Focus: Educ 11347

Association of Governing Bodies of Public Schools (Boys) (GBA), 27 Church Rd, Steep, Petersfield GU32 2DW
T: (0730) 4823
Founded: 1941; Members: 247
Focus: Educ 11348

Association of Hairdressing Teachers in Colleges of Further Education, 4 Samuel St, Balby, Doncaster
T: 851815
Founded: 1964; Members: 360
Focus: Adult Educ 11349

Association of Head and Neck Oncologists of Great Britain, 330 Gray's Inn Rd, London WC1X 8DA
T: (071) 8378855
Founded: 1968; Members: 150
Focus: Cell Biol & Cancer Res; Physiology 11350

Association of Headmistresses of Preparatory Schools (AHMPS), 138 Church St, London W8 4BN
T: (071) 7272316
Founded: 1929
Focus: Educ 11351

Association of Heads of Independent Schools (AHIS), c/o Eton College, 5 Gullivers, Windor SP8 4ER
T: (07535) 60902
Founded: 1924; Members: 120
Focus: Educ 11352

Association of Health Care Information and Medical Records Officers (AMRO), Whitecroft, Wilmore Hill Lane, Hopton ST18 0AW
Founded: 1948; Members: 1000
Gen Secr: D. Harrison
Focus: Med; Public Health 11353

Association of Health Service Treasurers (AHST), c/o South Warwickshire Health Authority, Westgate House, Market St, Warwick CV34 4DE
T: (0926) 493491
Members: 488
Focus: Public Health; Finance 11354

Association of Hispanists of Great Britain and Ireland (ABH), c/o Dept of Spanish, Queen Mary College, University of London, Mile End Rd, London E1 4NS
T: (081) 9804811
Founded: 1955; Members: 400
Focus: Lit 11355

Association of Independent Libraries, c/o Portico Library, 57 Mosley St, Manchester M2 3HY
Founded: 1989; Members: 16
Pres: Lord Quinton; Gen Secr: Janet Allan
Focus: Libraries & Bk Sci
Periodicals
Directory (weekly)
Newsletter (weekly) 11356

Association of Independent Research and Technology Organisations, POB 330, Cambridge CB5 8DU
T: (0223) 467831; Fax: (0223) 462051
Founded: 1975; Members: 35
Gen Secr: J.A. Bennett
Focus: Eng 11357

Association of Institute and School of Education In-Service Tutors (AJSEJT), c/o School of Education, Leazes Rd, Durham
Founded: 1970; Members: 36
Focus: Educ 11358

Association of International Accountants / Association des Experts-Comptables Internationaux, South Bank Bldg, Kingsway, Team Valley, Gateshead NE11 0JS
T: (091) 4824409; Fax: (091) 4825578
Founded: 1928
Gen Secr: J.R.A. Turnbull
Focus: Law 11359

The Association of Law Teachers (ALT), c/o Coalville Technical College, Bridge Rd, Coalville, Leicester LE6 2QR
T: (0533) 836136
Founded: 1965; Members: 700
Focus: Educ; Law 11360

Association of Lecturers in Accountancy (ALJA), 62 North Gyle Lane, Edinburgh EH12 8LD
T: (031) 3396144
Founded: 1964; Members: 450
Focus: Adult Educ 11361

Association of Lecturers in Colleges of Education in Scotland (ALCES), Beech Grove KA8 0SR
T: (0292) 60321
Members: 1190
Focus: Educ 11362

Association of London Chief Librarians (ALCL), c/o Central Library, Barking IG11 7NB
T: (081) 5178666; Fax: (081) 5941156
Founded: 1965; Members: 34
Pres: G. Allen; Gen Secr: A.J. Hill
Focus: Libraries & Bk Sci
Periodical
Directory of London Public Libraries (weekly) 11363

Association of Manufacturing Chemists, 1 Wardrobe Pl, Carter Lane, London EC4V 5AJ
T: (071) 2485971
Members: 250
Focus: Chem 11364

Association of Marine Engineering Schools (AMES), c/o Liverpool Polytechnic, Byrom St, Liverpool L3 3AF
T: (051) 2073581
Founded: 1951; Members: 22
Focus: Adult Educ; Eng 11365

Association of Meat Inspectors (AMJ), 10 Shaftesbury Av, Barnet EN5 5JA
T: (081) 4408712
Founded: 1965; Members: 620
Focus: Vet Med 11366

Association of Medical Advisers to the Pharmaceutical Industry, 41 Queen's Gate, London SW7 5HU
T: (071) 5899076
Focus: Pharmacol 11367

Association of Municipal Engineers, 1 Great George St, London SW1P 3AA
T: (071) 2227722; Fax: (071) 2227500
Founded: 1873; Members: 9972
Focus: Eng
Periodical
Municipal Engineer (weekly) 11368

Association of National European and Mediterranean Societies of Gastroenterology (ASNEMGE), c/o Gastroenterology Unit, Guy's Hospital, London SE1 9RT
T: (071) 9554564; Fax: (071) 4076689
Founded: 1947
Gen Secr: Prof. Dr. R.H. Dowling
Focus: Gastroenter 11369

Association of National Health Service Officers (ANHSO), 28 Worget Rd, Wareham BH20 4PN
T: (09295) 6712
Members: 5000
Focus: Public Health 11370

Association of National Park Officers (ANPO), c/o Lake District National Park Authority, Kendal LA9 4CH
T: (0539) 24555
Members: 34
Gen Secr: R. Foster
Focus: Forestry; Ecology 11371

Association of Noise Consultants (ANC), 6 Trap Rd, Guilden Morden SG8 0JE
T: (0763) 852958
Founded: 1973; Members: 40
Pres: Dr. A.J. Jones; Gen Secr: Dr. J. Williams
Focus: Eng 11372

Association of Northumberland Local History Societies, c/o Literary and Philosophical Society, Westgate Rd, Newcastle-upon-Tyne NE1 1SE
T: (091) 2320192
Founded: 1966; Members: 3500
Gen Secr: C.M. Fraser
Focus: Hist
Periodical
Tyne & Tweed (weekly) 11373

Association of Official Architects (AOA), 66 Portland Pl, London W1N 4AD
T: (071) 5805533
Founded: 1958; Members: 500
Focus: Archit 11374

Association of Painting Craft Teachers (A.P.C.T.), 5 Raven Court, Churchwood Dr, Saint Leonards-on-Sea TN38 9RL
Founded: 1921; Members: 380
Gen Secr: Michael A. Thomas
Focus: Adult Educ 11375

Association of Police Surgeons (APSGB), Creaton House, Northampton NN6 8ND
T: (060124) 722
Members: 1000
Pres: Dr. R.A.R. Lawrence; Gen Secr: Dr. Hugh de la Haye Davies
Focus: Forensic Med
Periodical
The Police Surgeon (weekly) 11376

Association of Polytechnic Teachers (APT), 27 Elphinstone Rd, Southsea PO5 3HP
T: (0705) 818625
Founded: 1973; Members: 3200
Focus: Educ; Eng
Periodical
Bulletin (8 times annually) 11377

Association of Principals of Colleges (APC), Turnford, Broxborn, Herts
Founded: 1921; Members: 600
Focus: Educ 11378

Association of Principals of Colleges (Northern Ireland Branch) (APC), c/o Technical College, Lisburn BT27 4SU
Members: 30
Focus: Educ; Eng; Arts; Sci; Econ; Business Admin 11379

Association of Professional Scientists and Technologists, 14 Harley St, London W1N 1AA
T: (071) 6367021
Focus: Sci; Eng 11380

Association of Psychiatrists in Training (APJT), c/o Royal Edinburgh Hospital, Edinburgh EH10 5HG
T: (031) 4472011
Founded: 1972; Members: 450
Focus: Psychiatry 11381

Association of Public Analysts (A.P.A.), Burlington House, Piccadilly, London W1V 0BN
Founded: 1953; Members: 100
Pres: A. Harrison; Gen Secr: H. Barnett
Focus: Food; Ecology
Periodical
Journal (weekly) 11382

Association of Recognised English Language Services (ARELS), 2 Pontypool Pl, London SE1 8QF
T: (071) 2423136; Fax: (071) 9289378
Founded: 1984; Members: 200
Gen Secr: Richard Walker-Arnott
Focus: Ling; Educ
Periodical
Learn English in Britain (weekly) . . . 11383

Association of Religious in Education, 41 Cromwell Rd, London SW7 2DH
T: (081) 9463788
Founded: 1969; Members: 1100
Pres: Josephine Egan; Gen Secr: Hilda Mitchell
Focus: Educ; Rel & Theol
Periodical
A.R.E. Bulletin (weekly) 11384

Association of Researchers in Medicine and Science (ARMS), c/o Clinical Science Laboratories, Guy's Hospital, London SE1 9RT
Founded: 1978; Members: 300
Gen Secr: Dr. J.R. Archer
Focus: Med
Periodical
Newsletter (weekly) 11385

Association of School Natural History Societies (ASNHS), c/o Lancing College, Lancing BN15 0RN
T: (07917) 2213
Founded: 1946; Members: 300
Focus: Hist; Nat Sci 11386

Association of Schools of Public Health in the European Region (ASPHER), c/o Dept of Community Health, University of Bristol, Canynge Hall, Whiteladies Rd, Bristol BS8 2PR
T: (0272) 24161 ext 1100
Founded: 1966
Focus: Educ; Public Health 11387

Association of Scientific, Technical and Managerial Staffs, 79 Camder Rd, London NW1 9ES
T: (071) 2674422; Tx: 25226
Focus: Sci; Eng; Business Admin
Periodicals
Industry News (weekly)
Medical World (weekly) 11388

Association of Scottish Local Health Councils (ASLHC), 29 Castle Terrace, Edinburgh EH1 2EL
T: (031) 2295782
Founded: 1977
Focus: Public Health
Periodical
Health Matters (weekly) 11389

Association of Show and Agricultural Organisations (ASAO), The Showground, Winthorpe, Newark NG24 2NY
T: (0636) 702627
Founded: 1924; Members: 190
Gen Secr: J.N. Armitage
Focus: Agri
Periodical
Official List of Shows and Sales (weekly) 11390

Association of Ski Schools in Great Britain (ASSGB), c/o White Mountain Ski School, 6 High St, Grantown-on-Spey
Founded: 1964
Focus: Sports 11391

Association of Social Anthropologists of the Commonwealth (ASA), 50 Fitzroy St., London W1P 5HS
T: (071) 3870455
Founded: 1946; Members: 440
Pres: Sir Raymond Firth; Gen Secr: Dr. H.L. Moore
Focus: Anthro; Sociology
Periodicals
Annals (weekly)
ASA Newsletter (weekly)
ASA Studies (weekly)
Monograph Series (weekly)
Research Methods Series (weekly) . . . 11392

Association of Social Research Organisation (ASRO), c/o Overseas Development Institute, Regent's College, Inner Circle, Regent's Park, London NW1 4NS
T: (071) 4877413
Founded: 1972; Members: 29
Gen Secr: T.M. Quirke
Focus: Sociology 11393

Association of Surgeons of Great Britain and Ireland, c/o Royal College of Surgeons, 35-43 Lincoln's Inn Fields, London WC2A 3PN
T: (071) 4056753
Founded: 1920; Members: 2000
Pres: Prof. T.G. Parks; Gen Secr: B.F. Ribeiro
Focus: Surgery 11394

Association of Swimming Therapy (AST), 4 Oak St, Shrewsbury SY3 7RH
T: (0743) 344393
Founded: 1952; Members: 30000
Gen Secr: Ted Cowen
Focus: Therapeutics
Periodical
Report (weekly) 11395

Association of Teachers of Management (ATM), c/o Polytechnic of Central London, 35 Marylebone Rd, London NW1 5LS
T: (071) 4865811
Founded: 1960
Focus: Business Admin; Educ 11396

Association of Teachers of Mathematics (A.T.M.), 7 Shaftesbury St, Derby DE23 8YB
Founded: 1952; Members: 4000
Pres: David Rooke; Gen Secr: Marjorie Gorman
Focus: Math; Educ
Periodicals
Mathematics Teaching (weekly)
Micromath
Termley 11397

Association of Teachers of Printing and Allied Subjects (ATPAS), 39 Childwall Av, Liverpool L16 0JE
T: (051) 7371660
Founded: 1930; Members: 450
Focus: Adult Educ; Eng 11398

Association of Track and Field Statisticians (ATFS), Poste Restante, Larkhill PO, Salisbury SP4 8PY
T: (0980) 33371 ext 5446
Founded: 1950
Gen Secr: David Martin
Focus: Stats
Periodical
ATFS Bulletin (3 times annually) . . . 11399

Association of Tutors (A.O.T.), 4 Hawthorn Way, Cambridge CB4 1AX
T: (0223) 313464; Fax: (0223) 355352
Founded: 1958; Members: 70
Focus: Adult Educ 11400

Association of Tutors in Adult Education (ATAE), 9 Northfolk Walk, Springfield Park, Sandiacre, Nottingham
Founded: 1912; Members: 450
Focus: Adult Educ 11401

Association of University Radiation Protection Officers (AURPO), c/o University Hospital of Wales, Heath Park, Cardiff CF6 4XW
T: (0222) 742003; Fax: (0222) 742012
Founded: 1962; Members: 122
Gen Secr: A.R. Richards
Focus: Radiology
Periodical
AURPO Newsletter (3 times annually) . 11402

Association of University Teachers (AUT), 1 Pembridge Rd, London W11 3HJ
T: (071) 2214370
Founded: 1919; Members: 27000
Focus: Adult Educ 11403

Association of University Teachers in Accounting (AUTA), c/o Div of Economic Studies, University, Sheffield S10 2TN
T: (0742) 78555
Founded: 1947; Members: 110
Focus: Business Admin 11404

Association of University Teachers (Scotland) (A.U.T.(S.)), c/o Dept of Statistics, University, Glasgow G12 8QW
T: (041) 3304047
Founded: 1922; Members: 4800
Pres: J. Duffy; Gen Secr: P. Breeze
Focus: Educ 11405

Association of Veterinary Anaesthetists of Great Britain and Ireland (AVA), Worlington House, Worlington, Bury Saint Edmunds, Suffolk
T: (0934) 852581
Founded: 1964
Focus: Vet Med 11406

Association of Veterinary Teachers and Research Workers (A.V.T.R.W.), c/o Moredun Research Institute, 408 Gilmerton Rd, Edinburgh EH17 7JH
T: (031) 6643262
Founded: 1948; Members: 800
Focus: Educ; Vet Med 11407

Association of Vice-Principals in Colleges (AVPC), c/o Windsor & Maidenhead College, Boyn Hill Ave, Maidenhead SL6 4EZ
T: (0628) 25221
Founded: 1965; Members: 150
Focus: Educ
Periodical
Conference Reports (weekly) 11408

Association of Voluntary Aided Secondary Schools (AVASS), 10 C Reddons Rd, Beckenham BR3 1LZ
T: (081) 7787270
Founded: 1968; Members: 125
Focus: Educ 11409

Association of Workers for Maladjusted Children (AWMC), c/o Sue Panter, 28 High St, Brigstock, Kettering NN14 3NA
T: (0536) 373337
Founded: 1951; Members: 1250
Focus: Educ Handic

Periodical
Therapeutic Care and Education (3 times annually) 11410

Association of Young Irish Archaeologists (AYJA), c/o Susan Mannion, Dept of Archaeology, The Queen's University, 17 University Sq, Belfast BT7 1NN
T: (0232) 245133
Founded: 1968; Members: 125
Focus: Archeol 11411

Association to Combat Huntington's Chorea / Combat, 34a Station Rd, Hinckley LE10 1AP
T: (0455) 615558
Founded: 1971; Members: 6600
Pres: Winifred Walmsley; Gen Secr: R.V. Bates
Focus: Med
Periodical
Newsletter (weekly) 11412

Assurance Medical Society (AMS), 11 Chandos St, London W1M 0EB
T: (071) 6066159
Founded: 1893; Members: 260
Focus: Med
Periodical
Transactions (weekly) 11413

Astronomical Society of Edinburgh (A.S.E.), City Observatory, Calton Hill, Edinburgh EH7 5AA
T: (031) 6644857
Founded: 1924; Members: 100
Focus: Astronomy 11414

Atlantis Research Centre (ARC), 14 Montpellier Villas, Brighton BN1 3DG
T: (0273) 25544
Founded: 1947; Members: 300
Focus: Archeol 11415

Audio Engineering Society, British Section (AES, UK), 8 Granville Rd, Sevenoaks
Founded: 1970; Members: 365
Focus: Eng
Periodical
Journal (10 times annually) 11416

The Avicultural Society (A.S.), Windsor Forest Stud, Mill Ride, Ascot SL5 8LT
T: (03447) 5444
Founded: 1894; Members: 1000
Focus: Ornithology 11417

Ayrshire Agricultural Association, 24 Beresford Terrace, Ayr KA7 2EL
T: (0292) 66168
Members: 1196
Focus: Agri 11418

Ayrshire Archaeological and Natural History Society (A.A.N.H.S.), 1 Portmark Av, Ayr KA7 4DD
T: (0292) 42077
Founded: 1947; Members: 250
Focus: Archeol; Hist; Nat Sci 11419

Balint Society, Tollgate Health Centre, 220 Tollgate Rd, London E6 4JS
T: (071) 4745656
Founded: 1970; Members: 150
Gen Secr: Dr. David Watt
Focus: Psych
Periodical
Journal of the Balint Society (weekly) . 11420

The Bantock Society, 101 Crouch Hill, London N8
T: (071) 7068055
Founded: 1946; Members: 100
Focus: Lit 11421

The Baptist Historical Society (B.H.S.), Baptist House, 129 Broadway, Oxford OXU 8RT
Founded: 1908; Members: 600
Pres: Dr. W.M.S. West; Gen Secr: R. Hayden
Focus: Hist
Periodical
Baptist Quarterly (weekly) 11422

Bar Association for Commerce, Finance and Industry, 63 Great Cumberland Pl, London W1H 7LJ
T: (071) 7239556
Founded: 1965; Members: 550
Focus: Law 11423

Barbirolli Society, 8 Tunnel Rd, Retford DN22 7TA
Founded: 1972; Members: 240
Focus: Music
Periodical
Newsletter (weekly) 11424

Barge and Canal Development Association (BCDA), 33 Walnut Crescent, Peacock Estate, Wakefield WF2 0EU
T: (0924) 366677
Founded: 1972
Focus: Transport
Periodical
Report (weekly) 11425

Basic English Foundation, 13 Beaumont St, London W1N 1FF
Founded: 1947
Focus: Ling 11426

Bath and Camerton Archaeological Society, 61 Pulteney St, Bath BA2 4DN
T: (0225) 66272
Founded: 1947
Focus: Archeol 11427

Battery Vehicle Society (BVS), 3 Steyning Court, Steyning Av, Peacehaven BN9 8LU
T: 4193
Founded: 1973; Members: 109
Focus: Electric Eng 11428

BC Society, 2 Ham Farm Cottages, Hurst Rd, Hassocks BN6 9NN
T: (07918) 5013
Founded: 1977; Members: 210
Focus: Archeol
Periodical
BC News (weekly) 11429

Bedfordshire Archaeological Council, 36 Saint Andrews Rd, Bedford
Founded: 1963
Focus: Archeol 11430

Bedfordshire Historical Record Society (BHRS), c/o County Record Office, County Hall, Bedford MK42 9AP
T: (0234) 228833
Founded: 1913; Members: 250
Pres: E. Whitbroad; Gen Secr: Betty Chambers
Focus: Hist
Periodical
BHRS Publications (weekly) 11431

Belfast Natural History and Philosophical Society (BNHPS), c/o Linen Hall Library, 17 Donegall Sq, Belfast
Founded: 1821; Members: 100
Gen Secr: J.C. Gray
Focus: Hist; Nat Sci; Philos
Periodical
Proceedings and Reports 11432

Benesh Institute of Choreology, 12 Lisson Grove, London NW1 6TS
T: (071) 2583041; Fax: (071) 7246434
Founded: 1962; Members: 100
Pres: Sir Peter Wright; Gen Secr: Andrew Ward
Focus: Music
Periodicals
The Choreologist (weekly)
Newsletter (weekly) 11433

Benslow Music Trust, Little Benslow Hills, Benslow Lane, Hitchin SG4 9RB
T: (0462) 459446; Fax: (0462) 440171
Founded: 1929; Members: 1500
Pres: Lady Evelyn Barbirolli; Gen Secr: Keith Stent
Focus: Music; Educ
Periodical
Course Brochure (weekly) 11434

Bertrand Russell Society, 9 Naseby Av, Higher Blackley, Manchester M9 2JJ
T: (061) 7950307
Founded: 1974
Focus: Philos 11435

Berwickshire Naturalists' Club (BNC), The Hill, Coldingham, Eyemouth TD14 5QB
T: (03903) 209
Founded: 1831; Members: 350
Focus: Nat Sci; Antique
Periodical
HBNC (weekly) 11436

BHRA Fluid Engineering, Cranfield, Bedford MK43 0AJ
T: (0234) 750422
Founded: 1947; Members: 200
Pres: Sir Ronald Mason; Gen Secr: Ian Cooper
Focus: Eng
Periodicals
Civil Engineering (weekly)
Computer Aided Process Control (weekly)
Condition Monitor (weekly)
Drilling News (weekly)
Fluid Flow Measurement (weekly)
Fluid Power (weekly)
Fluid Sealing (weekly)
Industrial Aerodynamics (weekly)
Industrial Corrosion (weekly)
Industrial Setting Report (weekly)
Mixing & Separation (weekly)
Multiphase Update (weekly)
Offshore Engineering (weekly)
Pipelines (weekly)
Pumps (weekly)
Solid-Liquid Flow (weekly)
Tribos (weekly)
World Ports & Harbours (weekly) . . . 11437

The Bibliographical Society, c/o British Library, Great Russell St, London W1B 3DG
T: (071) 3237567; Fax: (071) 3237566
Founded: 1892; Members: 1200
Pres: P.H. Davison; Gen Secr: M.M. Foot
Focus: Libraries & Bk Sci
Periodical
The Library (weekly) 11438

BIBRA Toxicology International, Woodmansterne Rd, Carshalton SM5 4DS
T: (081) 6434411; Fax: (081) 6617029
Founded: 1960; Members: 350
Pres: Lord Zuckerman; Gen Secr: Dr. S.E. Jaggers
Focus: Bio
Periodicals
BIBRA Bulletin (10 times annually)
BIBRA Toxicity Profiles (100 times annually)
Food and Chemical Toxicology (weekly)
Toxicology in Vitro (weekly) 11439

The Biochemical Society, 59 Portland Pl, London W1N 3AJ
T: (071) 5805530; Fax: (071) 6377626
Founded: 1911; Members: 9000
Pres: Prof. A.D.B. Malcolm; Gen Secr: G.D. Jones
Focus: Biochem
Periodicals
The Biochemical Journal (weekly)
Biochemical Society Symposia (weekly)
Biochemical Society Transactions (weekly)
The Biochemist (weekly)
Clinical Science (14 times annually)
Essays in Biochemistry (weekly) 11440

Biodeterioration Society, c/o International Mycological Institute, Ferry Ln, Kew TW9 3AF
T: (081) 9404086; Tx: 9312102252; Fax: (081) 3321171
Founded: 1969
Gen Secr: Dr. J. Kelley
Focus: Botany
Periodical
International Biodeterioration 11441

Bio-dynamic Agricultural Association (BDAA), Woodman Lane, Clent, Stourbridge DY9 9PX
T: (0562) 884933; Fax: (0562) 882619
Founded: 1929; Members: 650
Focus: Agri
Periodical
Star and Furrow (weekly) 11442

Biological Engineering Society (B.E.S.), c/o Royal College of Surgeons, Lincoln's Inn Fields, London WC2A 3PN
T: (071) 2427750
Founded: 1960; Members: 600
Gen Secr: Dr. R.E. Trotman
Focus: Eng
Periodicals
Journal of Biomedical Engineering (weekly)
Proceedings of Conferences and Symposia 11443

Biorhythmic Research Association, 22 Far Lane, Normanton-on-Soar, Loughborough, Leics
Focus: Bio 11444

Bird Life International, Wellsbrook Court, Girton Rd, Cambridge CB3 0NA
T: (0223) 277318; Fax: (0223) 277200
Founded: 1922
Focus: Ornithology
Periodicals
Annual Report (weekly)
World Birdwatch (weekly) 11445

Birmingham and Midland Institute (BMI), Margaret St, Birmingham B3 3BS
T: (021) 2363591
Founded: 1854
Focus: Arts; Lit
Periodical
B.M.I. Magazine (weekly) 11446

Birmingham and Warwickshire Archaeological Society (B.W.A.S.), c/o Birmingham and Midland Institute, Margaret St, Birmingham B3 3BS
T: (021) 2363591
Founded: 1870; Members: 250
Pres: Dr. T.R. Slater; Gen Secr: A.J. Wilson
Focus: Archeol
Periodical
Birmingham and Warwickshire Archaeological Society Transactions (weekly) 11447

Birmingham Bibliographical Society, c/o Main Library, University of Birmingham, Edgbaston, Birmingham B15 2TT
T: (021) 4145814; Fax: (021) 4714691
Founded: 1972; Members: 50
Focus: Libraries & Bk Sci 11448

The Birmingham Metallurgical Association, c/o Dept of Mechanical and Production Engineering, City of Birmingham Polytechnic, Franchise St, Birmingham B42 2SU
T: (021) 3566911
Founded: 1903; Members: 1500
Focus: Metallurgy 11449

Birmingham Natural History Society (BNHS), 55 Selby Close, Yardley, Birmingham B26 6AP
Founded: 1858; Members: 173
Focus: Hist; Nat Sci 11450

Birmingham Transport Historical Group (BTHG), 43 Shorncliffe Rd, Folkestone CT20 2UD
T: (0303) 59243
Founded: 1963; Members: 26
Focus: Transport; Hist 11451

Blackcountry Society (BCS), 15 Claydon Rd, Wallheath, Kingswinford DY6 0HR
T: (0384) 293656
Founded: 1967; Members: 2000
Focus: Archeol; Hist
Periodical
The Blackcountry Man Quarterly Magazine (weekly) 11452

Blair Bell Research Society (BBRS), c/o Dept of Obstetrics and Gynaecology, Clinical Science Bldg, Leicester Royal Infirmary, POB 65, Leicester LE2 2EG
T: (0533) 523164; Fax: (0533) 523107
Founded: 1962; Members: 532
Pres: Prof. William Dunlop; Gen Secr: Dr. James Drife
Focus: Gynecology 11453

Bliss Classification Association (BCA), c/o The Library, Fitzwilliam College, Cambridge CB3 0DG
Fax: (0223) 464162
Founded: 1967; Members: 130
Pres: Jack Mills; Gen Secr: Christina Panagiotidou
Focus: Libraries & Bk Sci
Periodical
Bliss Classification Bulletin (weekly) . . . 11454

BMT Cortec, c/o Wallsend Research Station, Wallsend NE28 6UY
T: (091) 2625242; Tx: 53476; Fax: (091) 2638754
Founded: 1944
Focus: Eng
Periodical
BMT Abstracts (weekly) 11455

BNF Metals Technology Centre (BNF), c/o Grove Laboratories, Denchworth Rd, Wantage OX12 9BJ
T: (02357) 2992; Tx: 837166
Founded: 1920; Members: 4
Focus: Eng; Metallurgy
Periodical
BNF Abstracts (weekly) 11456

Boarding Schools Association (BSA), 43 Raglan Rd, Reigate RH2 0DU
T: (0737) 226450; Fax: (0737) 226775
Founded: 1966; Members: 485
Focus: Educ
Periodicals
Boarding School (weekly)
Occasional Papers (weekly)
Research Topics (weekly) 11457

Bone and Tooth Society, c/o Bone Research Laboratory, Churchill Hospital, Headington, Oxford
T: (0865) 64811
Founded: 1950; Members: 248
Focus: Dent 11458

Bookplate Society, 9 Lyndale Av, London NW2 2QD
T: (071) 4351059
Members: 300
Focus: Graphic & Dec Arts, Design; Libraries & Bk Sci
Periodicals
Journal (weekly)
Newsletter (weekly) 11459

Book Trust, Book House, 45 East Hill, London SW18 2QZ
T: (081) 8709055
Founded: 1925; Members: 2000
Pres: Sir Simon Horney; Gen Secr: Beverley Anderson
Focus: Lit
Periodical
Reading Matters (weekly) 11460

Border Union Agricultural Society (BUAS), Showground Office, Springwood Park, Kelso TD5 8LS
T: (05732) 2188
Founded: 1813; Members: 1065
Focus: Agri 11461

Borthwick Institute of Historical Research, Saint Anthony's Hall, Peaseholme Green, York YO1 2PW
T: (0904) 59861
Focus: Hist 11462

The Botanical Society of Scotland (B.S.S.), c/o Royal Botanic Garden, Inverleith Row, Edinburgh EH3 5LR
T: (031) 5527171
Founded: 1836; Members: 570
Pres: Jackie Muscott; Gen Secr: R. Galt
Focus: Botany
Periodical
Botanical Society of Scotland (weekly) . 11463

Botanical Society of the British Isles (B.S.B.I), c/o Dept of Botany, Natural History Museum, Cromwell Rd, London SW7 5BD
Founded: 1836; Members: 2700
Pres: Dr. F. Perring; Gen Secr: M. Briggs
Focus: Botany
Periodicals
BSBI Abstracts (weekly)
Watsonia (weekly) 11464

Branch Line Society (BLS), 15 Springwood Hall Gardens, Gledholt, Huddersfield HD1 4HA
T: (0484) 25782
Founded: 1955; Members: 1000
Focus: Transport
Periodical
Branch Line News (weekly) 11465

Brewery History Society, 10 Ringstead Court, Ringstead Rd, Sutton, Surrey
Focus: Hist 11466

BRF International (BRF), Lyttel Hall, Nutfield RH1 4HY
Founded: 1948
Focus: Eng
Periodical
Industry Review (weekly) 11467

Bristol and Gloucestershire Archaeological Society, 22 Beaumont Rd, Gloucester GL2 0EJ
T: (0452) 302610
Founded: 1876; Members: 1000
Pres: K. Branigan; Gen Secr: D.J.H. Smith
Focus: Archeol; Hist
Periodicals
Gloucestershire Record Series (weekly)
Transactions (weekly) 11468

Bristol Industrial Archaeological Society (BIAS), c/o City Museum, Queens Rd, Bristol BS8 1RL
T: (0272) 299771
Founded: 1967; Members: 400
Pres: Owen H. Ward
Focus: Archeol
Periodicals
BIAS Bulletin (3 times annually)
BIAS Journal (weekly) 11469

British Academy (B.A.F.T.A.), 20-21 Cornwall Terrace, London NW1 4QP
T: (071) 4875966; Tx: 23194; Fax: (071) 2243807
Founded: 1901; Members: 523
Pres: Sir Keith Thomas; Gen Secr: P.W.H. Brown
Focus: Cinema
Periodicals
Annual Report (weekly)
Proceedings (weekly) 11470

British Academy of Film and Television Arts (BAFTA), 195 Piccadilly, London W1V 9LG
T: (071) 734 0022; Fax: (071) 7341792
Founded: 1976; Members: 2200
Focus: Cinema 11471

British Academy of Forensic Sciences (B.A.F.S.), c/o Anaesthetics Unit, The London Hospital Medical College, Turner St, London E1 2AD
T: (071) 3779201
Founded: 1959; Members: 500
Gen Secr: Dr. Patricia J. Flynn
Focus: Forensic Med
Periodical
Medicine, Science and the Law (weekly) 11472

British Acupuncture Association and Register (BAAR), 34 Alderney St, London SW1V 4EU
T: (071) 8341012
Founded: 1960; Members: 230
Pres: Dr. E.W. Johnson
Focus: Med; Physical Therapy
Periodical
The British Journal of Acupuncture (weekly)
. . . . 11473

British Agricultural History Society (BAHS), c/o Rural History Centre, Reading University, Whiteknights, POB 229, Reading RG6 2AG
T: (0734) 318660
Founded: 1953; Members: 820
Pres: M.A. Havinden; Gen Secr: Dr. R. Perren
Focus: Hist; Agri
Periodical
Agricultural History Review (weekly) . . 11474

British Agricultural Marketing Research Group (BAMRG), Cedar Cott, Upper Bolney, Henley-on-Thames
Founded: 1971
Focus: Marketing 11475

British American Scientific Research Association (BASRA), 59 Swiss Av, Watford WD1 7LL
Founded: 1962; Members: 55
Gen Secr: C.J. Cooksey
Focus: Sci
Periodical
BASRA (weekly) 11476

British and European Geranium Society (BEGS), 56 Shrigley Rd., Higher Poynton SK12 1TF
T: (0625) 873056
Founded: 1970; Members: 2000
Gen Secr: Doris P. Codling
Focus: Botany 11477

British and Foreign Bible Society (BFBS), Stonchill Green, Westlea, Swindon SN5 7DG
T: (0793) 513713; Tx: 44283; Fax: (0793) 512539
Founded: 1804
Gen Secr: Neil Crosbie
Focus: Rel & Theol
Periodical
Word in Action (weekly) 11478

British and Irish Association of Law Librarians (BIALL), c/o Pinseat & Co., 3 Colmore Circus, Birmingham B4 6BH
T: (021) 2001050; Fax: (021) 6261040
Founded: 1969; Members: 580
Focus: Libraries & Bk Sci
Periodicals
Law Librarian (weekly)
Newsletter (weekly) 11479

British Arachnological Society, Burns Farm, Cornhill, Banff AB45 2DL
T: (04666) 231
Founded: 1963; Members: 600
Gen Secr: M.J. Roberts
Focus: Zoology
Periodicals
Bulletin (3 times annually)
Newsletter (3 times annually) 11480

British Archaeological Association (BAA), c/o Conway Library, Courtauld Institute, Somerset House, Strand, Lomdon WC2 0RN
Founded: 1843; Members: 700
Pres: L. Keen
Focus: Arts; Cultur Hist
Periodicals
Conference Transactions (weekly)
Journal (weekly) 11481

British Artists in Glass (BAG), Prospect House, Farlam, Brampton CA8 1LA
T: (06976) 203
Founded: 1977; Members: 140
Focus: Fine Arts 11482

British Arts, 227 Goldhawk Rd, London W12 8ER
Founded: 1953; Members: 10800
Focus: Arts 11483

British Association for Accident and Emergency Medicine (BAEM), c/o The Royal College of Surgeons of England, 35-43 Lincoln's Inn Fields, London WC2A 3PN
T: (071) 8319405; Fax: (071) 4050318
Founded: 1967; Members: 712
Pres: Dr. Keith Little; Gen Secr: Stephen Miles
Focus: Surgery
Periodical
Archives of Emergency Medicine (weekly) 11484

British Association for American Studies (B.A.A.S.), c/o Dept of English Literature, The University, Newcastle-upon-Tyne NE1 7RU
T: (091) 2227755
Founded: 1955; Members: 500
Focus: Ethnology; Hist; Lit
Periodicals
BAAS Pamphlets in American Studies (3 times annually)
Journal of American Studies (3 times annually)
. . . . 11485

British Association for Applied Linguistics (BAAL), 8a Hill Rd, Clevedon BA21 7HH
T: (0272) 876519
Founded: 1968; Members: 500
Pres: P. Meara; Gen Secr: Jill Bourne
Focus: Ling
Periodicals
Applied Linguistics (3 times annually)
Newsletter 11486

British Association for Brazing and Soldering (BABS), c/o The Welding Institute, Abington House, Cambridge CB1 6AL
T: (0223) 891162; Tx: 81183; Fax: (0223) 892588
Founded: 1970
Gen Secr: J.B. Dunkerton
Focus: Eng; Metallurgy
Periodical
Welding and Metal Fabrication (10 times annually)
. . . . 11487

British Association for Canadian Studies (BACS), c/o Dept of History, Birkbeck College, Malet St, London WC1
T: (071) 5806622
Founded: 1975; Members: 120
Focus: Hist
Periodical
Bulletin of Canadian Studies (weekly) . . 11488

British Association for Cancer Research (BACR), c/o Institute of Biology, 20 Queensberry Pl, London SW7 2DZ
T: (071) 5818333; Fax: 8239409
Founded: 1960; Members: 1100
Focus: Cell Biol & Cancer Res 11489

British Association for Commercial and Industrial Education (BACIE), 35 Harbour Exchange Sq, London E14 9GE
T: (071) 9878989; Fax: (071) 9879898
Founded: 1919; Members: 600
Focus: Adult Educ
Periodical
BACIE Journal 11490

British Association for Counselling (BAC), 37a Sheep St, Rugby CV21 3BX
T: (0788) 78328/9
Founded: 1977; Members: 1800
Pres: Prof. Douglas Hooper; Gen Secr: Elisabeth Davies
Focus: Therapeutics
Periodicals
Counselling (weekly)
Newsletter (weekly) 11491

British Association for Immediate Care (BASICS), 7 Black Horse Lane, Ipswich ID1 2EF
T: (0473) 218407
Founded: 1977; Members: 2200
Gen Secr: R. Bailey
Focus: Med
Periodical
BASICS (3 times annually) 11492

British Association for Information and Libraries Education and Research, c/o Dept of Information Studies, University of Sheffield, Sheffield S1 4DP
Founded: 1952; Members: 18
Pres: Prof. T. Wilson
Focus: Libraries & Bk Sci; Computer & Info Sci
. . . . 11493

British Association for Psychopharmacology (BAP), c/o Dept of Psychiatry, Guy's Hospital Medical School, London SE1 9RT
T: (071) 4077600
Founded: 1974; Members: 315
Focus: Pharmacol 11494

British Association for the Advancement of Science (B.A.A.S.), Fortress House, 23 Savile Row, London W1X 1AB
T: (071) 4943326; Tx: 092091; Fax: (071) 7341658
Founded: 1831; Members: 2000
Pres: Dr. Anne McLaren; Gen Secr: Dr. P. Briggs
Focus: Sci
Periodicals
Science & the Public (weekly)
Scope (weekly) 11495

British Association for the History of Religions, c/o Open University in Wales, 24 Cathedral Rd, Cardiff CF1 9SA
Founded: 1954; Members: 170
Pres: Prof. C.G. Williams; Gen Secr: Dr. T. Thomas
Focus: Hist; Rel & Theol
Periodical
Bulletin (3 times annually) 11496

British Association for the Study of Religions, c/o Open University in Wales, 24 Cathedral Rd, Cardiff CF1 9SA
Founded: 1954; Members: 200
Pres: Prof. Ursula King; Gen Secr: Dr. T. Thomas
Focus: Rel & Theol
Periodical
Bulletin (3 times annually) 11497

British Association in Forensic Medicine (B.A.F.M.), c/o Dept of Forensic Medicine, Saint James' Hospital, Leeds LS9 7TF
T: (0532) 431897
Founded: 1950; Members: 180
Pres: P.J. Stevens; Gen Secr: M.A. Green
Focus: Forensic Med 11498

British Association of Academic Phoneticians, c/o Phonetics Laboratory, Dept of English Language, University of Glasgow, Glasgow G12 9QQ
T: (041) 3398855 ext 4596; Fax: (041) 3078030
Founded: 1984; Members: 75
Gen Secr: M.K.C. MacMahon
Focus: Ling 11499

British Association of Advisers and Lecturers in Pysical Education (B.A.A.L.P.E.), Nelson House, 6 The Beacon, Exmouth EX8 2AG
T: (0395) 263247; Fax: (0395) 276348
Founded: 1920; Members: 290
Pres: P. Whitlam; Gen Secr: G.M. Edmondson
Focus: Sports
Periodical
Bulletin of Physical Education: Safe Practice in Physical Education (3 times annually) . . 11500

British Association of Art Therapists (BAAT), 11a Richmond Rd, Brighton BN2 3RL
Founded: 1964; Members: 600
Pres: Diane E. Waller; Gen Secr: Geoffrey Hicks
Focus: Therapeutics
Periodical
Inscape (weekly) 11501

British Association of Blind Esperantists, 32 Benson St, Middlesborough T55 6JQ
Founded: 1935
Focus: Ling 11502

British Association of Clinical Anatomists (BACA), c/o Dept of Anatomy, University, Oxford Rd, Manchester M13 9PT
T: (061) 2738241 ext 155
Founded: 1977; Members: 200
Focus: Anat; Surgery
Periodicals
LM & Rules
Proceedings of Meetings (weekly) . . . 11503

British Association of Cosmetic Surgeons (BACS), 138 Harley St, London W1
T: (071) 9350554
Founded: 1980; Members: 15
Focus: Surgery 11504

British Association of Crystal Growth (BACG), c/o Allen Clark Research Centre, The Plessey Company Ltd, Caswell, Towcester, Northants
T: 50581
Founded: 1969; Members: 240
Focus: Mineralogy 11505

British Association of Friends of Museums (BAFM), 548 Wilbraham Rd, Manchester M21 1LB
T: (061) 8818640
Founded: 1973; Members: 230
Pres: Sir John Hale; Gen Secr: Rosemary Marsh
Focus: Arts; Sci
Periodicals
Newsletter (weekly)
Yearbook (weekly) 11506

British Association of Hair Transplant Surgeons, 125 Worlds End Lane, Quinton, Birmingham B32 1JX
T: (021) 4225282
Focus: Surgery 11507

British Association of Homoeopathic Pharmacists (BAHP), 19a Cavendish Sq, London W1M 9AD
T: (071) 6293204/05
Founded: 1980; Members: 100
Focus: Homeopathy 11508

British Association of Homoeopathic Veterinary Surgeons (BHAVS), Chinham House, Stanford-in-the-Vale SN7 8NQ
T: (0367) 710324; Fax: (0367) 718243
Founded: 1981; Members: 150
Gen Secr: Christopher Day
Focus: Homeopathy; Vet Med
Periodical
Newsletter (weekly) 11509

British Association of Landscape Industries (BALI), 9 Henry St, Keighley BD21 3DR
T: (0535) 606139; Fax: (0535) 610269
Founded: 1972; Members: 750
Gen Secr: John Topping
Focus: Hort
Periodical
Newsletter (weekly) 11510

British Association of Numismatic Societies (BANS), c/o Philip Memick, Bush Boake Allen, Blackhorse Lane, London E17 5QP
Founded: 1953
Pres: D. Sellwood
Focus: Numismatics 11511

British Association of Occupational Therapists (BAOT), 20 Rede Pl, London W2 4TU
T: (071) 2299738
Founded: 1944; Members: 6700
Focus: Therapeutics 11512

British Association of Oral and Maxillo-Facial Surgeons (BAOS), c/o Royal College of Surgeons, 35-43 Lincoln's Inn Fields, London WC2A 3PN
T: (071) 4058074
Founded: 1962; Members: 1150
Gen Secr: M.R. Bromige
Focus: Surgery; Otorhinolaryngology
Periodical
The British Journal of Oral & Maxillo-Facial Surgery (weekly) 11513

British Association of Orthodontists (BAO), 16 Castle Hill, Maidenhead SL6 4JJ
T: (0628) 23279
Founded: 1965; Members: 200
Focus: Dent 11514

British Association of Otolaryngologists (BAO), c/o Royal College of Surgeons of England, 35-43 Lincoln's Inn Fields, London WC2A 3PN
T: (071) 4048373; Fax: (071) 4050318
Founded: 1943; Members: 800
Pres: A.G. Kerr; Gen Secr: I.S. Mackam
Focus: Otorhinolaryngology 11515

British Association of Paediatric Nephrology, c/o East Birmingham Hospital, Bordesley Green East, Birmingham B9 5ST
T: (021) 7724311
Focus: Pediatrics 11516

British Association of Paediatric Surgeons (B.A.P.S.), c/o Royal College of Surgeons, Nicolson St, Edinburgh EH8 9DW
T: (031) 6683975; Fax: (031) 6671905
Founded: 1954; Members: 744
Focus: Surgery; Pediatrics 11517

British Association of Picture Libraries, c/o Winton, 10 Cheyne Row, London SW3
T: (071) 3525824
Focus: Libraries & Bk Sci 11518

British Association of Plastic Surgeons (BAPS), c/o Royal College of Surgeons, 35-43 Lincoln's Inn Fields, London WC2A 3PN
T: (071) 8315161; Fax: (071) 8314041
Founded: 1946; Members: 561
Pres: R. Sanders; Gen Secr: P.S. Muray
Focus: Surgery
Periodical
British Journal of Plastic Surgery (8 times annually) 11519

British Association of Psychotherapists, 37 Mapesbury Rd, London NW2 4HJ
T: (081) 4529823; Fax: (081) 4525182
Founded: 1951; Members: 450
Focus: Psych; Therapeutics 11520

British Association of Seed Analysts (BASA), 3 Whitehall Court, London SW1A 2EQ
T: (071) 9303611; Fax: (071) 9303952
Founded: 1961
Focus: Agri 11521

British Association of Social Psychiatry (BASP), 16 Lichfield Rd, Kew TW9 3JR
Founded: 1964; Members: 125
Focus: Psychiatry
Periodical
The International Journal of Social Psychiatry
. . . . 11522

British Association of Social Workers (BASW), 16 Kent St, Birmingham B5 6RD
T: (021) 6223911; Fax: (021) 6224860
Founded: 1970; Members: 9700
Focus: Sociology

Periodicals
BASW News (weekly)
British Journal of Social Work (weekly)
Practice (weekly)
Rostrum (weekly) 11523

British Association of Sport Medicine (BASM), 49 Blakes Lane, New Malden KT3 6N3, Surrey
T: 9490607
Members: 730
Focus: Med 11524

British Association of Surgical Oncology (BASO), c/o Royal College of Surgeons, 35-43 Lincoln's Inn Fields, London WC2A 3PN
T: (071) 4055612; CA: Collsurg London
Founded: 1973; Members: 411
Gen Secr: T.G. Allen-Mersh
Focus: Surgery; Cell Biol & Cancer Res
Periodical
European Journal of Surgical Oncology (weekly) 11525

British Association of Teachers of the Deaf (BATOD), The Rycroft Centre, Stanley Rd, Cheadle Hulme SK8 6RF
T: (061) 4375951
Founded: 1976
Focus: Educ Handic
Periodical
Journal: Teacher of the Deaf (weekly) . 11526

British Association of the Experiment in International Living (EIL), Otesaga, Upper Wyche, Malvern WR14 4EN
T: (06845) 62577/78
Founded: 1932; Members: 2000
Focus: Int'l Relat; Travel 11527

British Association of Urological Surgeons (BAUS), c/o Royal College of Surgeons, 35-43 Lincoln's Inn Fields, London WC2A 3PN
T: (071) 4051390
Founded: 1945; Members: 1100
Pres: R.T. Turner-Warwick; Gen Secr: M.G. Royle
Focus: Urology; Surgery 11528

British Association Representing Breeders (B.A.R.B.), 9 Portland St, King's Lynn, Norfolk PE30 1PB
T: (0553) 773094; Fax: (0553) 772804
Founded: 1973; Members: 41
Pres: Gareth Fryer; Gen Secr: D. Dealtrey
Focus: Botany, Specific
Periodical
BARB News and Views (weekly) . . . 11529

British Astronomical Association (BAA), Burlington House, Piccadilly, London W1V 9AG
T: (071) 7344145
Founded: 1890; Members: 4500
Pres: Dr. J.W. Mason; Gen Secr: Rosa Atwell
Focus: Astronomy
Periodicals
British Astronomical Association Handbook (weekly)
Journal of the British Astronomical Association (weekly) 11530

British Aviation Preservation Council (BAPC), Stonewharf, 2 Dale Rd, Coalbrookdale, Telford TF8 7DT
T: (0952) 433534
Founded: 1967; Members: 144
Pres: Sir Peter G. Masefield; Gen Secr: J.D. Storer
Focus: Aero
Periodical
Update (weekly) 11531

British Ballet Organization (B.B.O.), 39 Lonsdale Rd, London SW13 9JP
T: (081) 7481241; Fax: (081) 7481301
Founded: 1930; Members: 1000
Focus: Perf Arts
Periodical
The Dancer Magazine (weekly) 11532

British Balloon and Airship Club (BBAC), 47 Vaughan Way, Leicester LE1 4SG
T: (021) 6434050
Founded: 1966; Members: 2000
Focus: Aero
Periodical
The Aerostat (weekly) 11533

British Brush Manufacturers Research Association (BBMRA), c/o Dept of Textile Industries, University, 4 Southampton Row, London WC1B 4AB
Founded: 1946; Members: 77
Focus: Eng 11534

British Bryological Society (BBS), c/o National Museum of Wales, Cardiff CF1 3NP
T: (0222) 397951; Fax: (0222) 373219
Founded: 1896; Members: 520
Pres: Dr. R.E. Langton; Gen Secr: Dr. M.E. Newton
Focus: Botany
Periodicals
Bulletin of the British Bryological Society (weekly)
Journal of Bryology (weekly) 11535

British Butterfly Conservation Society (BBCS), POB 222, Dedham, Colchester CO7 6EY
T: (0206) 322342
Founded: 1968; Members: 10100
Focus: Ecology

Periodical
News Bulletin (3 times annually) 11536

British Cardiac Society, 7 Fitzroy Sq, London W1P 5AH
T: (071) 3833887; Fax: (071) 3880903
Founded: 1937
Gen Secr: Dr. D. Dymond
Focus: Cardiol
Periodicals
British Heart Journal (weekly)
Cardiovascolar Research (weekly) . . . 11537

British Cartographic Society (BCS), 13 Sheldrake Gardens, Hordle, Lymington SO4 10FJ
Founded: 1963; Members: 1006
Pres: Dr. C. Board; Gen Secr: C.E. Beattie
Focus: Cart
Periodicals
The Cartographic Journal (weekly)
Newsletter (weekly) 11538

British Caspian Trust (BCT), Colonsay, Hampton Lovett, Droitwich, Worcs
T: (029923) 495
Founded: 1975; Members: 138
Focus: Folklore
Periodical
British Caspian Trust News-Letter (weekly) 11539

British Cave Research Association (B.C.R.A.), 6 Worcester Terrace, Bristol BS8 3JW
T: (0272) 741848
Founded: 1973; Members: 800
Focus: Speleology
Periodicals
Caves and Caving (weekly)
Cave Science: The Transactions of the B.C.R.A. (weekly) 11540

British Cement Association (BCA), Century House, Telford Av, Crowthorne RG11 6YS
T: (0344) 762676; Fax: (0344) 761214
Founded: 1935
Focus: Materials Sci 11541

British Ceramic Society (B.C.S.), Shelton House, Stoke Rd, Stoke-on-Trent ST4 2DR
T: (0782) 23116
Founded: 1900; Members: 960
Focus: Materials Sci 11542

British Ceramic Tile Council, Federation House, Stoke-on-Trent ST4 2RU
T: (0782) 747147
Founded: 1960; Members: 5
Focus: Materials Sci 11543

British College of Optometrists, 10 Knaresborough Pl, London SW5 0TG
T: (071) 3737765; Fax: (071) 3731143
Founded: 1980; Members: 7312
Pres: R.W. Chappell; Gen Secr: P.D. Leigh
Focus: Optics
Periodical
Ophthalmic and Physiological Optics (weekly) 11544

British Committee for Standards in Haematology (B.C.S.H.), c/o Dept of Haematology, Middlesex Hospital, Riding House St, London W1P 7LD
T: (071) 6368333
Members: 6
Focus: Hematology 11545

British Compressed Air Society (BCAS), 8 Leicester St, London WC2H 7BN
T: (071) 4370678; Tx: 263536
Focus: Eng 11546

The British Computer Society (BCS), 13 Mansfield St, London W1M 0BP
T: (071) 6370471; Fax: (071) 6311049
Founded: 1957; Members: 34000
Gen Secr: J.R. Brookes
Focus: Computer & Info Sci
Periodicals
Computer Bulletin (weekly)
Computer Journal (weekly)
Formal Aspects of Computing (weekly)
Interacting with Computers (weekly)
IT in Nursing (weekly)
Software Engineering Journal (weekly)
What's on in Computing (weekly) . . . 11547

The British Cryogenics Council, c/o The Institution of Mechanical Engineers, 1 Birdcage Walk, London SW2H 9SS
T: (071) 2227899
Founded: 1967; Members: 12
Pres: Prof. R.G. Scurlock; Gen Secr: J.S. Harris
Focus: Physics; Eng
Periodical
British Cryogenics Council Newsletter (weekly) 11548

British Dental Association (BDA), 63-64 Wimpole St, London W1M 8AL
T: (071) 9350875; Fax: 4875232
Founded: 1880; Members: 15000
Gen Secr: J.M.G. Hunt
Focus: Dent
Periodical
British Dental Journal (weekly) 11549

British Dental Hygienists Association (B.D.H.A.), 64 Wimpole St, London W1M 8AL
T: (071) 9350875
Founded: 1949; Members: 900
Focus: Dent 11550

British Diabetic Association (B.D.A.), 10 Queen Anne St, London W1M 0BD
T: (071) 3231531; Fax: (071) 6373644
Founded: 1934; Members: 130000
Pres: Sir Harry Secombe; Gen Secr: Michael Cooper
Focus: Diabetes
Periodicals
Balance (weekly)
Diabetes Contents (weekly)
Diabetic Medicine (10 times annually) . . 11551

The British Dietetic Association (BDA), Elizabeth House, 22 Suffolk St, Queensway, Birmingham B1 1LS
T: (021) 6435483
Founded: 1936; Members: 2650
Pres: C.A. Middleton; Gen Secr: J.C.J. Grigg
Focus: Nutrition
Periodical
Journal of Human Nutrition and Dietetics (weekly) 11552

British Dyslexia Association (BDA), 98 London Rd, Reading RG1 5AU
Founded: 1972
Focus: Logopedy
Periodical
Dyslexia Contact (weekly) 11553

The British Ecological Society (B.E.S.), 26 Blades Court, Deodar Rd, Putney, London SW15 2NU
T: (081) 8719797; Fax: (081) 8719779
Founded: 1913; Members: 4700
Pres: Prof. R.M. May; Gen Secr: Dr. R. Hammond
Focus: Ecology
Periodicals
Functional Ecology (weekly)
Journal of Animal Ecology
Journal of Applied Ecology (3 times annually)
Journal of Ecology
Symposium (weekly) 11554

British Educational Management and Administration Society, 5 Hill View Rd, Osney, Oxford OX2 0DA
Founded: 1971
Pres: Christopher Price; Gen Secr: Lesley Anderson
Focus: Educ
Periodicals
Educational Management and Administration (3 times annually)
Management in Education 11555

British Electrophoresis Society, c/o Dept of Medical Microbiology, University of Aberdeen, Medical School Bldgs, Foresterhill, Aberdeen AB9 2ZD
T: (0224) 681818 ext 52505; Fax: (0224) 685604
Founded: 1982; Members: 225
Pres: Dr. Michael Dunn; Gen Secr: Dr. Phillip Cash
Focus: Chem 11556

British Endodontic Society (B.E.S.), 40 Wimpole St, London W1M 7AF
T: (071) 4863648
Founded: 1963
Focus: Dent 11557

British Epilepsy Association (BEA), Anstey House, 40 Hanover Sq, Leeds LS3 1BE
T: (0532) 439393; Fax: (0532) 428804
Founded: 1950; Members: 13500
Focus: Pathology
Periodical
Epilepsy Today (weekly) 11558

British Federation of Festivals, 198 Park Ln, Macclesfield SK11 6UD
T: (0625) 428297
Founded: 1921
Gen Secr: E. Whitehead
Focus: Music 11559

British Federation of Film Societies (BFFS), 81 Dean St, London W1V 6AA
T: (071) 4374355
Founded: 1945; Members: 650
Focus: Cinema
Periodicals
BFFS Handbook (3 times annually)
BFFS Register (weekly)
Film Magazine (weekly) 11560

British Federation of Women Graduates (BFWG), 4 Mandeville Courtyard, 142 Battersea Park Rd, London SW11 4NB
T: (071) 4988037
Founded: 1907; Members: 2000
Pres: Griselda Kenyon; Gen Secr: Annabel Stein
Focus: Sci
Periodical
BFWG News 11561

British Film Institute (BFI), 21 Stephen St, London W1P 1PL
T: (071) 2551444; Tx: 27624; Fax: (071) 4367950
Founded: 1933; Members: 34000
Gen Secr: W. Stevenson
Focus: Cinema
Periodicals
BFI Film and Television Handbook (weekly)
Directions (weekly)
Sight and Sound (weekly) 11562

British Fluoridation Society, Sandlebrook, Mill Lane, Alderley Edge SK9 7TY
T: (0565) 873936; Fax: (0565) 873936
Founded: 1968; Members: 550
Focus: Public Health 11563

British Food Manufacturing Industries Research Association (BFMIRA), Randalls Rd, Leatherhead KT22 7RY
T: 376761
Founded: 1919; Members: 675
Focus: Food
Periodical
Abstracts (weekly) 11564

British Geological Survey, Keyworth NG12 5GG
T: (0602) 363100; Tx: 378173
Founded: 1835
Focus: Geology 11565

British Geomorphological Research Group (BGRG), 1 Kensington Gore, London SW7 2AR
Founded: 1960; Members: 400
Focus: Geology 11566

British Geotechnical Society (BGS), c/o Institution of Civil Engineers, 1-7 Great George St, London SW1P 3AA
T: (071) 2227722
Founded: 1959; Members: 350
Focus: Eng 11567

British Geriatrics Society (B.G.S.), 1 Saint Andrews Pl, London NW1 4LB
T: (071) 9354004; Fax: (071) 2240454
Founded: 1947; Members: 1572
Pres: Dr. Michael Denham; Gen Secr: Dr. B.S.D. Sastry
Focus: Geriatrics
Periodical
Age and Ageing 11568

British Goat Society (BGS), 34-36 Fore St, Bovey Tracey TQ13 9AD
T: (0626) 833168
Founded: 1879; Members: 1500
Focus: Dairy Sci
Periodicals
The Herd Book (weekly)
The Monthly Journal (weekly)
The Year Book (weekly) 11569

British Grassland Society (BGS), c/o AFRC IGAP, Hurley, Maidenhead SL6 5LR
T: (062882) 3626
Founded: 1945; Members: 1350
Focus: Agri
Periodicals
Grass Farmer (3 times annually)
Grass & Forage Science (weekly) . . . 11570

British Herpetological Society (B.H.S.), c/o Zoological Society of London, Regent's Park, London NW1 4RY
Founded: 1948; Members: 860
Pres: Prof. S.L. Cloudsley-Thompson
Focus: Zoology
Periodicals
Bulletin (weekly)
Herpetological Journal (weekly) 11571

The British Homoeopathic Association (B.H.A.), 27a Devonshire St, London W1N 1RJ
T: (071) 9352163
Founded: 1902; Members: 3750
Gen Secr: Enid Segall
Focus: Homeopathy
Periodical
Homoeopathy (weekly) 11572

British Horological Institute (BHI), Upton Hall, Upton NG23 5TE
T: (0636) 813795; Fax: (0636) 812258
Founded: 1858; Members: 3594
Gen Secr: W.M.G. Evans
Focus: Eng
Periodical
Horological Journal (weekly) 11573

British Humanist Association (BHA), 14 Lamb's Conduit Passage, London WC1R 4RH
T: (071) 4300908; Fax: (071) 4301271
Founded: 1963
Focus: Philos
Periodical
Humanist News (8 times annually) . . . 11574

British Hydromechanics Research Association (BHRA), Cranfield, Bedford MK43 0AJ
T: (0234) 750422; Tx: 825059
Founded: 1947
Focus: Mach Eng
Periodicals
Civil Engineering Hydraulics Abstracts (weekly)
Fluid Flow Measurements Abstracts (weekly)
Fluid Power Abstracts (weekly)
Fluid Sealing Abstracts (weekly)
Industrial Aerodynamics Abstracts (weekly)
International Dredging Abstracts (weekly)
Pump Abstracts (weekly)
Solid Liquid Flow Abstracts (weekly)
Tribos (weekly) 11575

British Hypnosis Research Association (BHRA), 15 The Bank, Somersham PE17 3DJ
T: (048) 840915
Founded: 1979
Focus: Depth Psych 11576

British Hypnotherapy Association (BHA), 67 Upper Berkeley St, London W1H 7DH
T: (071) 7234443
Founded: 1958; Members: 310
Focus: Therapeutics 11577

British Ichthyological Society (BIS), 60 Newfields, Welwyn Garden City AL8 6YT
T: 29038
Founded: 1961; Members: 126
Focus: Zoology 11578

British Institute for Brain Injured Children (BIBIC), Knowle Hall, Knowle, Bridgwater TA7 8PJ
T: (0278) 684060; Fax: (0278) 685573
Founded: 1972
Focus: Educ Handic
Periodical
Newsletter (weekly) 11579

The British Institute of Cleaning Science (BICS), Whitworth Chambers, George Row, Northampton NN1 1DF
T: (0604) 230075
Founded: 1961; Members: 1975
Gen Secr: Peter Andrews
Focus: Eng 11580

British Institute of Human Rights (BIHR), Kings College, Strand, London WC2R 2LS
T: (071) 8732352; Fax: (071) 8361799
Founded: 1970
Pres: Lord Scarman
Focus: Poli Sci
Periodical
Human Rights Case Digest (weekly) . . 11581

British Institute of Industrial Therapy (BIIT), Exmoor House, Methuen St, Southampton SO2 0FQ
T: (0703) 635345
Founded: 1980; Members: 500
Pres: Dr. Mounir Ekdawi; Gen Secr: Dr. Alan Whitehead
Focus: Therapeutics
Periodical
Industrial Therapy (weekly) 11582

British Institute of International and Comparative Law (B.I.I.C.L.), 17 Russell Sq, London WC1B 5DR
T: (071) 6365802; Fax: (071) 3232016
Founded: 1958; Members: 2500
Pres: Lord Goff; Gen Secr: J.P. Gardner
Focus: Law
Periodicals
Bulletin of Legal Developments (weekly)
International and Comparative Law (weekly) 11583

British Institute of Learning Disabilities, Wolverhampton Rd, Kidderminster DY10 3PP
T: (0562) 850251; Fax: (0562) 851970
Founded: 1972; Members: 2000
Focus: Educ Handic
Periodicals
Current Awareness Service (weekly)
Mental Handicap (weekly)
Mental Handicap Bulletin (weekly)
Mental Handicap Research (weekly) . . 11584

British Institute of Musculoskeletal Medicine, 27 Green Lane, Northwood HA6 2PX
T: (0923) 825583
Founded: 1992; Members: 357
Pres: Dr. M. Hutson; Gen Secr: Dr. P.G. Skew
Focus: Med
Periodical
Journal of Orthopaedic Medicine (weekly) 11585

British Institute of Persian Studies, 38 Thomas More House, Barbican, London EC2Y 8BT
Founded: 1961; Members: 500
Focus: Archeol; Arts; Anthro; Archit; Ling; Lit
Periodical
Iran (weekly) 11586

British Institute of Practical Psychology (B.I.P.P.), 67 Highbury New Park, London N5 2EZ
T: (071) 2263569
Founded: 1933; Members: 2652
Focus: Psych 11587

British Institute of Professional Photography, Fox Talbot House, Amwelf End, Ware SG12 9HN
T: (0920) 464011
Founded: 1901; Members: 5500
Focus: Photo
Periodicals
The Photographer (weekly)
The Register of Members and Guide for Buyers of Photography (weekly) 11588

British Institute of Radiology (B.I.R.), 36 Portland Pl, London W1N 4AT
T: (071) 5804085; Fax: (071) 2553209
Founded: 1897; Members: 1800
Pres: Dr. W. Hately; Gen Secr: Mary-Anne Piggett
Focus: Radiology
Periodical
The British Journal of Radiology 11589

British Institute of Surgical Technologists (B.I.S.T.), 1 Webbs Court, Buckhurst Av, Sevenoaks TN13 1LZ
Founded: 1935; Members: 1350
Pres: L.B. Ward; Gen Secr: I.F. Sherwood
Focus: Surgery
Periodical
Journal (weekly) 11590

British Insurance Law Association (B.I.L.A.), 229-230 Shoreditch High St, London E1 6PJ
T: (071) 3750471
Founded: 1964; Members: 320
Pres: Prof. A.L. Diamond; Gen Secr: R. Hanson-James
Focus: Law
Periodical
BILA Bulletin (weekly) 11591

British Interlingua Society, 14 Ventnor Court, Wostenholm Rd, Sheffield S7 1LB
T: (0742) 582931
Founded: 1955; Members: 26
Focus: Ling
Periodical
Lingua e Vita (3 times annually) . . . 11592

British Internal Combustion Engine Research Institute (BJCERJ), 111 Buckingham Av, Slough SL1 4PH
T: 27371
Founded: 1966; Members: 55
Focus: Eng 11593

British International Studies Association, c/o Dept of Politics, University of Lancaster, Bailrigg LA1 4YF
Founded: 1975
Focus: Poli Sci 11594

British Interplanetary Society, 27-29 South Lambeth Rd, London SW8 1SZ
T: (071) 7353160
Focus: Astronomy
Periodicals
Journal of the British Interplanetary Society (weekly)
Spaceflight (weekly) 11595

British Iris Society (BIS), 43 Sea Lane, Goring-by-Sea, Worthing BN12 4QD
Members: 900
Focus: Botany, Specific
Periodical
The Iris Yearbook (weekly) 11596

British Jazz Society, 10 Southfield Gardens, Twickenham, Middx
T: 8920133
Founded: 1964
Focus: Music 11597

British Kidney Patient Association (BKPA), Bordon, Hants
T: 2022
Founded: 1975
Focus: Otorhinolaryngology
Periodical
Silver Lining Appeal Brochure (weekly) . 11598

British Kinematograph Sound and Television Society (BKSTS), Victoria House, Vernon Pl, London WC1B 4DF
T: (071) 2428400; Fax: (071) 4053560
Founded: 1931; Members: 2500
Focus: Cinema
Periodicals
BKSTS News (10 times annually)
Cinema Technology (weekly)
Image Technology (10 times annually) . 11599

British Laboratory Animals Veterinary Association, c/o Chemical Defense Establishment, Porton, Salisbury, Wilts
Focus: Vet Med 11600

British Leather Confederation, Leather Trade House, Kings Park Rd, Moulton Park, Northampton NN3 1JD
T: (0604) 494131; Tx: 317124; Fax: (0604) 648220
Founded: 1920
Gen Secr: R.P. Pearson
Focus: Materials Sci 11601

British Leprosy Relief Association (LEPRA), Fairfax House, Causton Rd, Colchester CO1 1PU
T: (0206) 562286; Fax: (0206) 762151
Founded: 1924
Focus: Derm
Periodical
Leprosy Review (weekly) 11602

British Library National Sound Archive, 29 Exhibition Rd, London SW7 2AS
T: (071) 5896603; Fax: (071) 8238970
Founded: 1955
Focus: Music
Periodical
POMPI: Popular Music Periodicals Index (weekly) 11603

British Lichen Society, c/o Dept of Botany, The Natural History Museum, London SW7 5BD
T: (071) 9388852
Founded: 1958; Members: 600
Pres: F.S. Dobson; Gen Secr: O.W. Purvis
Focus: Botany
Periodicals
British Lichen Society Bulletin (weekly)
The Lichenologist (3 times annually) . . 11604

British Marine Aquarist Association (BMAA), 139 Bradford Av, Hull HU9 4LZ
Founded: 1970; Members: 300
Focus: Oceanography
Periodical
Marinews (weekly) 11605

British Measures Group (BMG), 6 Park Rd, Teddington, Middx
Founded: 1973
Focus: Standards 11606

British Medical Association (BMA), BMA House, Tavistock Sq, London WC1H 9JP
T: (071) 3874499; Fax: (071) 3836148
Founded: 1832; Members: 74000
Gen Secr: E.M. Armstrong
Focus: Med
Periodical
British Medical Journal (weekly) 11607

British Microcirculation Society, c/o Dept of Physiology, Charing Cross Hospital Medical School, Fulham Palace Rd, London W6 8RF
T: (081) 7482040 ext 2850
Founded: 1963; Members: 140
Focus: Physiology; Intern Med 11608

British Migraine Association, 178a High Rd, West Byfeet KT14 7ED
T: (0932) 352468
Founded: 1958; Members: 12000
Focus: Med
Periodical
Newsletter (3 times annually) 11609

British Model Soldier Society (BMSS), 22 Lynwood Rd, Ealing, London, W5 1JJ
T: (081) 9985230
Founded: 1935; Members: 850
Gen Secr: David Pearce
Focus: Military Sci
Periodical
The Bulletin (weekly) 11610

British Morgan Horse Society (BMHS), George and Dragon Hall, Mary Pl, London W11
T: (071) 2298155
Founded: 1975
Pres: Tony Phillips; Gen Secr: Emma Scheck
Focus: Zoology
Periodical
Morgan Horse Magazine (weekly) . . . 11611

British Museum Society, c/o British Museum, Bloomsbury, London WC1B 3DG
T: (071) 3238605
Founded: 1968; Members: 9400
Focus: Arts
Periodical
British Museum Magazine (3 times annually) 11612

British Music Society (BMS), 7 Tudor Gardens, Upminster RM14 3DE
T: (0708) 224795
Founded: 1978; Members: 450
Gen Secr: Stephen Trowell
Focus: Music
Periodicals
Journal (weekly)
Newsletter (weekly) 11613

British Mycological Society (B.M.S.), c/o School of Biological Sciences, University of Portsmouth, King Henry I St, Portsmouth PO1 2DY
T: (0705) 842024; Fax: (0705) 842070
Founded: 1896; Members: 2000
Gen Secr: Dr. S.T. Moss
Focus: Botany, Specific
Periodicals
Mycological Research (weekly)
Mycologist (weekly) 11614

British National Committee on Space Research, c/o The Royal Society, 6 Carlton House Terrace, London SW1Y 5AG
Founded: 1959
Focus: Astronomy 11615

British Natural Hygiene Society (BNHS), Shalimar, Harold Grove, Frinton-on-Sea CO13 9BD
T: (0255) 672823
Founded: 1959; Members: 300
Focus: Hygiene
Periodical
The Hygienist (weekly) 11616

British Naturalists' Association (B.E.N.A.), 48 Russell Way, Higham Ferrers NN9 8EJ
T: (0933) 314672; Fax: (0933) 314672
Founded: 1905; Members: 2500
Pres: Lord Skelmersdale; Gen Secr: J.F. Pearton
Focus: Ecology
Periodicals
British Naturalist (weekly)
Country-Side (weekly) 11617

British Naturopathic and Osteopathic Association (BNOA), 6 Netherhall Gardens, London NW3 5RR
T: (071) 4358728
Founded: 1925; Members: 250
Focus: Med; Public Health
Periodical
British Naturopathic Journal (weekly) . . 11618

British Non-Ferrous Metals Research Association (BNFMRA), c/o Grove Laboratories, Denchworth Rd, Wantage
T: 2992
Founded: 1920
Focus: Metallurgy 11619

British Nuclear Energy Society (BNES), 1-7 Great George St, London SW1P 3AA
T: (071) 2227722
Founded: 1962; Members: 1189
Pres: Dr. D. Pooley; Gen Secr: P.A.F. Bacos
Focus: Nucl Res
Periodical
Nuclear Energy (weekly) 11620

British Numismatic Society (B.N.S.), c/o Royal Mint, Llantrisant, Pontyclun CF7 8YT
T: (0443) 222111; Fax: (0443) 228799
Founded: 1903; Members: 530
Focus: Numismatics
Periodical
British Numismatic Journal (weekly) . . . 11621

British Nutrition Foundation (BNF), High Holborn House, 52-54 High Holborn, London WC1V 6RQ
T: (071) 4046504; Fax: (071) 4046747
Founded: 1967
Gen Secr: O.M. Conning
Focus: Nutrition
Periodical
BNF Bulletin 11622

British Occupational Hygiene Society (B.O.H.S.), 1 Saint Andrews Pl, Regents Park, London NW1 4LB
T: (071) 4864860
Founded: 1953; Members: 1287
Focus: Hygiene
Periodicals
Annals of Occupational Hygiene (weekly)
Technical Guide Series (weekly)
Technical Handbook Series (weekly) . . 11623

British Origami Society (BOS), 11 Yarningale Rd, Kings Heath, Birmingham B14 6LT
T: (021) 4431330
Founded: 1967; Members: 250
Focus: Fine Arts
Periodical
British Origami (weekly) 11624

British Ornithologists' Club (BOC), 1 Uppingham Rd, Oakham LE15 6JB
Founded: 1892; Members: 600
Focus: Ornithology
Periodical
Bulletin of the British Ornithologists' Club (weekly) 11625

British Ornithologists' Union (B.O.U.), c/o The Natural History Museum, Tring HP23 6AP
T: (0442) 890080; Fax: (0442) 890693
Founded: 1858; Members: 2000
Pres: Prof. Janet Kear; Gen Secr: Dr. C.J. Feare
Focus: Ornithology
Periodical
Ibis (weekly) 11626

British Orthopaedic Association (BOA), c/o Royal College of Surgeons, 35-43 Lincoln's Inn Fields, London WC2A 3PN
T: (071) 4056507; Fax: (071) 8312676
Founded: 1918; Members: 2778
Pres: A.W.F. Lettin; Gen Secr: I.J. Leslie
Focus: Orthopedics 11627

British Orthoptic Society (B.O.S.), Tavistock House North, Tavistock Sq, London WC1H 9HX
T: (071) 3877992; Fax: (071) 3832584
Founded: 1937; Members: 1000
Focus: Ophthal
Periodical
British Orthoptic Journal (weekly) . . . 11628

British Paediatric Association (BPA), 5 Saint Andrew's Pl, London NW1 4LB
T: (071) 4866151
Founded: 1928; Members: 2028
Pres: Prof. Sir David Hull; Gen Secr: Dr. R. MacFaul
Focus: Pediatrics
Periodical
Disease in Childhood (weekly) 11629

British Pharmacological Society (B.P.S.), c/o Medical College, Charterhouse Sq, London EC1M 6BQ
T: (071) 9826170; Fax: (071) 9826173
Founded: 1931; Members: 2200
Gen Secr: Dr. J. Marlagan
Focus: Pharmacol
Periodicals
British Journal of Chemical Pharmacology
British Journal of Pharmacology 11630

British Photobiology Society (B.P.S.), c/o Dept of Optometry, City University, London EC1V 7DD
T: (071) 2534399 ext 4309; Fax: 8378068
Founded: 1955; Members: 350
Pres: Dr. A.R. Young; Gen Secr: Dr. R.A. Douglas
Focus: Bio
Periodical
Photobiology Bulletin (weekly) 11631

British Phycological Society (BPS), c/o Botany Dept, The Natural History Museum, London
T: (071) 9389352
Founded: 1952; Members: 700
Pres: Dr. M.J. Dring; Gen Secr: Dr. E.J. Cox
Focus: Botany
Periodical
European Journal of Phycology (weekly) . 11632

British Postgraduate Medical Federation (BPMF), 33 Millman St, London WC1N 3EJ
T: (071) 8316222
Founded: 1945
Focus: Med 11633

British Postmark Society (BPS), 21 Empress Way, Euxton, Chorley PR7 6QB
T: (0257) 269652
Founded: 1958; Members: 300
Focus: Cultur Hist
Periodical
Bulletin (weekly) 11634

British Psychoanalytical Society, 63 New Cavendish St, London W1M 7RD
T: (071) 5804952
Founded: 1913; Members: 421
Gen Secr: M. Garfield
Focus: Psychoan
Periodical
International Journal of Psycho-Analysis (weekly) 11635

British Psychological Society (BPS), 48 Princess Rd East, Leicester LE1 7DR
T: (0533) 549568; Fax: (0533) 470787
Founded: 1901; Members: 19000
Pres: Prof. G. Lindsay; Gen Secr: Dr. J. Greeger
Focus: Psych
Periodicals
British Journal of Clinical Psychology
British Journal of Developmental Psychology
British Journal of Educational Psychology
British Journal of Mathematical and Statistical Psychology
British Journal of Medical Psychology
British Journal of Psychology
British Journal of Social Psychology
Journal of Occupational and Organizational Psychology
Selection & Development Review . . . 11636

The British Puppet and Model Theatre Guild (BPMTG), 18 Maple Rd, Yeading UB4 9LP
Founded: 1925; Members: 400
Focus: Perf Arts
Periodicals
Magazine (weekly)
Newsletter (10 times annually) 11637

British Records Association (BRA), 18 Padbury Court, London E2 7EH
T: (071) 7291415
Founded: 1932; Members: 1000
Pres: Sir Thomas Bingham; Gen Secr: Tim Harris
Focus: Hist
Periodicals
Archives (weekly)
Archives and the User (weekly) 11638

British Record Society (BRS), c/o College of Arms, Queen Victoria St, London EC4V 4BT
T: (071) 2369612
Founded: 1887; Members: 250
Focus: Hist; Doc
Periodical
Indexes to Testamentary Records . . . 11639

British Retinitis Pigmentosa Society (BRPS), POB 350, Buckingham MK18 5EL
T: (02806) 363; Fax: (02806) 515
Founded: 1975; Members: 2500
Gen Secr: L. Cantor
Focus: Ophthal
Periodical
BRPS Newsletter (weekly) 11640

British Robot Association (BRA), Aston Science Park, Love Lane, Birmingham B7 4BJ
T: (021) 6281745
Founded: 1977; Members: 85
Gen Secr: Don Pitt
Focus: Computer & Info Sci 11641

British Safety Council (BSC), 62-64 Chancellor's Rd, London W6 9RS
T: (081) 7411231
Founded: 1957; Members: 18000
Focus: Safety 11642

British Schools Exploring Society (BSES), c/o Royal Geographical Society, 1 Kensington Gore, London SW7 2AR
T: (071) 5840710; Fax: (071) 5817995
Founded: 1932
Focus: Educ
Periodical
Expedition Report (weekly) 11643

British Science Fiction Association (BSFA), 27 Albemarle Dr, Grove, Wantage OX12 0NB
Founded: 1958; Members: 1200
Gen Secr: A. Cook
Focus: Lit
Periodicals
Focus: New Writers Forum (3 times annually)
Matrix (weekly)
Vector (weekly) 11644

British Shell Collectors Club (BSCC), 368 Kingston Rd, New Maldon KT3 3RX
T: (081) 5410110
Founded: 1973
Focus: Zoology
Periodical
BSCC Newsletter (weekly) 11645

British Small Animal Veterinary Association (BSAVA), Kingsley House, Church Lane, Shurdington, Cheltenham GL51 5TQ
T: (0242) 862994; Fax: (0242) 863009
Members: 3400
Focus: Vet Med
Periodical
Journal of Small Animal Practice (weekly) 11646

British Social Biology Council (B.S.B.C.), 69 Eccleston Sq, London SW1V 1PJ
T: (071) 8342091
Founded: 1935; Members: 15
Gen Secr: V. Box
Focus: Bio
Periodical
Social Biology and Human Affairs (weekly) 11647

British Society for Agricultural Labour Science (BSALS), c/o Work Science Laboratory, Dept of Agriculture and Horticulture, Reading University, Earley Gate, Reading RG6 2AT
T: (0734) 85123
Founded: 1969; Members: 140
Focus: Agri
Periodical
Agricultural Manpower (weekly) 11648

British Society for Allergy and Clinical Immunology (BSACI), c/o Wythenshawe Hospital, Manchester M13 9PT
Members: 500
Focus: Immunology
Periodical
Clinical Allergy (weekly) 11649

British Society for Antimicrobial Chemotherapy (BSAC), c/o Birmingham Heartlands Hospital, Birmingham B9 5JT
T: (021) 7731740; Fax: (021) 7668752
Founded: 1971; Members: 850
Pres: Prof. D.C.E. Speller; Gen Secr: Dr. M.J. Wood
Focus: Therapeutics
Periodical
Journal of Antimicrobial Chemotherapy (weekly) 11650

British Society for Cell Biology (B.S.C.B.), c/o Dept of Zoology, Downing St, Cambridge CB2 3EJ
T: (0223) 336631, 336600
Founded: 1959; Members: 1400
Focus: Cell Biol & Cancer Res
Periodical
Newsletter (weekly) 11651

British Society for Clinical Cytology (BSCC), c/o Cytology Laboratory, Saint Mary's Hospital, Whithworth Park, Manchester M13 0JH
T: (061) 2249633
Founded: 1962
Focus: Cell Biol & Cancer Res 11652

British Society for Developmental Biology (BSDB), c/o Dept of Developmental Biology, Marishal College, University of Aberdeen, Aberdeen AB9 1AS
T: (0224) 40241
Founded: 1964; Members: 330
Focus: Bio 11653

British Society for Eighteenth Century Studies (BSECS), c/o Dept of French, King's College, University of Aberdeen, Aberdeen AB9 2UB
T: (0224) 272148
Founded: 1971; Members: 450
Gen Secr: Dr. John Dunkley
Focus: Lit; Hist
Periodicals
Bulletin (weekly)
Journal (weekly) 11654

British Society for Electronic Music (BSEM), 49 Deodar Rd, London SW15
T: (081) 8742363
Founded: 1969; Members: 112
Focus: Music 11655

British Society for Haematology (B.S.H.), 2 Carlton House Terrace, London SW1Y 5AF
T: (071) 6437305; Fax: (071) 7700933
Founded: 1960; Members: 950
Focus: Hematology
Periodical
The British Journal of Haematology (weekly) 11656

The British Society for Immunology (BSI), POB 35, East Grinstead RH19 3UT
T: (0342) 312032
Founded: 1956; Members: 2500
Focus: Immunology
Periodicals
BSI Newsletter (weekly)
Clinical & Experimental Immunology (weekly)
Immunology (weekly) 11657

British Society for Middle Eastern Studies, c/o Faculty of Oriental Studies, Sidgwick Av, Cambridge CB3 9DA
T: (0223) 335127; Fax: (0223) 335110
Founded: 1973; Members: 700
Pres: Sir James Craig; Gen Secr: Dr. C. Holes
Focus: Sci
Periodicals
British Journal of Middle Eastern Studies (weekly)
Newsletter (3 times annually) 11658

British Society for Music Therapy (BSMT), 69 Avondale Av, East Barnet EN4 8NB
T: (081) 3688879
Founded: 1958; Members: 400
Focus: Therapeutics
Periodicals
BSMT Bulletin (3 times annually)
Journal of British Music Therapy (weekly) 11659

British Society for Oral Medicine, c/o Eastman Dental Hospital, 256 Gray's Inn Rd, London WC1X 8LD
T: (071) 9151172
Gen Secr: Dr. Joanna Zakrzewska
Focus: Dent 11660

British Society for Parasitology, 62 London Rd, Reading RG1 5AS
T: (0734) 861345
Founded: 1962
Focus: Microbio 11661

The British Society for Phenomenology (B.S.P.), 81 Plymouth Rd, Penarth, Cardiff CF6 2DE
Founded: 1968; Members: 90
Pres: Dr. A. Grieder; Gen Secr: Dr. Nicholas Davey
Focus: Philos
Periodical
Journal (3 times annually) 11662

British Society for Plant Pathology (BSPP), c/o Agricultural Development and Advisory Service, Lawnswood, Leeds LS16 5PY
T: (0532) 611222; Fax: (0532) 301141
Founded: 1981; Members: 680
Gen Secr: Dr. N.V. Hardwick
Focus: Botany; Agri; Microbio
Periodical
Plant Pathology (weekly) 11663

British Society for Research on Ageing (BSRA), c/o School of Biological Sciences, University of Manchester, Stopford Bldg, Oxford Rd, Manchester M13 9PT
T: (061) 2755252
Founded: 1945; Members: 230
Gen Secr: Dr. I. Davies
Focus: Bio
Periodical
Lifespan (weekly) 11664

British Society for Restorative Dentistry (B.S.R.D.), c/o Guy's Hospital Dental School, London Bridge, London SE1 9RT
T: (071) 4077600 Ext. 3013/4
Founded: 1968; Members: 350
Focus: Dent 11665

British Society for Rheumatology, 3 Saint Andrew's Pl, London NW1 4LB
T: (071) 2243739; Fax: (071) 2240156
Founded: 1983; Members: 1300
Pres: Dr. A.G. Mowat; Gen Secr: Kate Baillie
Focus: Rheuma
Periodical
British Journal of Rheumatology (weekly) 11666

British Society for Social Responsibility in Science (BSSRS), 25 Horsell Rd, London N5 1XL
T: (071) 6079615
Founded: 1969; Members: 600
Focus: Sociology
Periodical
Science for People (weekly) 11667

British Society for Strain Measurement (B.S.S.M.), Exchange Bldg, Quayside, Newcastle-upon-Tyne NE1 3BJ
T: (091) 2617971; Fax: (091) 2324472
Founded: 1964; Members: 900
Pres: G.F. Chalmers; Gen Secr: G. Buckley
Focus: Eng
Periodical
Strain (weekly) 11668

British Society for Surgery of the Hand, c/o Royal College of Surgeons, 35-43 Lincoln's Inn Fields, London WC2A 3PN
T: (071) 8315161; Fax: (071) 8314041
Founded: 1969; Members: 636
Pres: W. Souter; Gen Secr: R.E. Page
Focus: Surgery
Periodical
Journal of Hand Surgery (weekly) . . . 11669

British Society for the History of Mathematics, c/o Faculty of Mathematics, Open University, Walton Hall, Milton Keynes MK7 6AA
T: (0908) 652351; Tx: 633045; Fax: (0908) 652140
Pres: J. Fauvel; Gen Secr: Dr. J.E. Barrow-Green
Focus: Hist; Math
Periodical
The British Society for the History of Mathematics Newsletter (weekly) 11670

British Society for the History of Medicine, 149 Harley St, London W1N 1HG
T: (071) 9354444
Founded: 1965
Focus: Hist 11671

British Society for the History of Pharmacy (BSHP), c/o Royal Pharmaceutical Society of Great Britain, 36 York Pl, Edinburgh EH1 3HU
Founded: 1967; Members: 205
Focus: Hist; Pharmacol
Periodical
Pharmaceutical Historian (weekly) . . . 11672

The British Society for the History of Science (BSHS), 31 High St, Stanford in the Vale, Faringdon SN7 8LH
T: (0367) 718963; Fax: (0367) 718963
Founded: 1947; Members: 850
Pres: Prof. Geoffrey N. Cantor
Focus: Hist; Nat Sci
Periodicals
The British Journal for the History of Science
Monograph Series
Newsletter 11673

British Society for the Philosophy of Science (BSPS), c/o Dept of Physics, Kings College, London
Founded: 1948; Members: 520
Pres: Prof. M.L.G. Redhead; Gen Secr: George Ross
Focus: Philos
Periodical
The British Journal for the Philosophy of Science (weekly) 11674

British Society for the Promotion of Vegetable Research, c/o Institute of Horticultural Research, Wellesbourne CV35 9EF
T: (0789) 470382; Fax: (0789) 472552
Founded: 1949; Members: 220
Pres: Dr. H.G. Jones; Gen Secr: A.A. Dalby
Focus: Food
Periodical
Annual Report (weekly) 11675

British Society for the Study of Infection, c/o Eastern Hospital, Homerton Grove, London
Founded: 1974; Members: 452
Focus: Immunology 11676

British Society for the Study of Mental Subnormality (BSSMS), c/o Cranage Hospital, Cranage CW48 EG
T: (0477) 32021
Founded: 1952
Focus: Psychiatry; Psych; Educ Handic
Periodicals
British Journal of Mental Subnormality (weekly)
Newsletter (weekly) 11677

British Society for the Study of Orthodontics (BSSO), c/o Eastman Dental Hospital, 256 Grays Inn Rd, London WC1X 8LD
Founded: 1908; Members: 1140
Pres: R. Marx; Gen Secr: T.G. Bennett
Focus: Dent
Periodical
British Journal of Orthodontics (weekly) . 11678

British Society of Aesthetics (BSA), c/o Faculty of Art and Design, Nottingham Trent University, Burton St, Nottingham NG1 4BU
T: (0602) 418418; Tx: 377534; Fax: (0602) 486403
Founded: 1960; Members: 360
Pres: Prof. Richard Wollheim; Gen Secr: Richard Woodfield
Focus: Philos
Periodical
British Journal of Aesthetics (weekly) . . 11679

British Society of Animal Production (BSAP), POB 3, Penicuik EH26 0RZ
Fax: (031) 4455636
Founded: 1944; Members: 1350
Pres: Dr. J.J. Robinson; Gen Secr: M.A. Steele
Focus: Agri
Periodical
Animal Production (weekly) 11680

British Society of Audiology (B.S.A.), 80 Brighton Rd, Reading RG6 1PS
T: (0734) 660622; Fax: (0734) 351915
Founded: 1967; Members: 1500
Pres: B.C. Rowe; Gen Secr: Dr. G.J. Sutton
Focus: Otorhinolaryngology
Periodical
British Journal of Audiology (weekly) . . 11681

The British Society of Dowsers (B.S.D.), Sycamore Barn, Hastingleigh, Ashford, TN25 5HW
T: (0233) 750253
Founded: 1933; Members: 987
Gen Secr: M.D. Rust
Focus: Geology; Hydrology; Public Health
Periodical
Journal (weekly) 11682

British Society of Flavourists (BSF), c/o International Flavours & Fragrances IFF (GB) Ltd, Flavour Division, Duddery Hill, Haverhill CB9 8LG
T: (0440) 704488
Founded: 1971; Members: 393
Focus: Botany; Food
Periodical
Newsletter (weekly) 11683

British Society of Gastroenterology (B.S.G.), 3 Saint Andrew's Pl, London NW1 4LB
T: (071) 3873534; Fax: (071) 4873735
Members: 1200
Pres: Prof. M.S. Losowsky; Gen Secr: Prof. M.J.G. Farthing
Focus: Gastroenter 11684

British Society of Hypnotherapists (BSH), 51 Queen Anne St, London W1
T: (071) 9357075
Founded: 1950
Focus: Therapeutics 11685

British Society of Medical and Dental Hypnosis, 10 Chillerton Rd, London SW17
T: (081) 6723025
Focus: Med; Dent 11686

British Society of Painters in Oil, Pastels and Acrylic, c/o Leslie Simpson, 41 Lister St, Riverside Gardens, Ilkley LS29 9ET
T: (0943) 609075
Founded: 1986; Members: 50
Gen Secr: Leslie Simpson
Focus: Fine Arts 11687

British Society of Rheology, 27 Alexander Rd, Stotfold, Hitchin SG5 4NA
T: (0763) 244280; Fax: (0763) 244298
Founded: 1940; Members: 600
Pres: Dr. D. C.-H. Cheng; Gen Secr: C.A. Moules
Focus: Materials Sci 11688

British Society of Scientific Glassblowers (BSSG), 21 Grebe Av, Eccleston Park, Saint Helens WA10 3QL
T: (051) 7091438; Fax: (051) 7070631
Founded: 1960; Members: 300
Gen Secr: Peter Halliwell
Focus: Materials Sci
Periodical
Journal (weekly) 11689

British Society of Soil Science (BSSS), c/o Dept of Soil Science, University of Reading, London Rd, Reading RG1 5AQ
T: (0734) 318911
Founded: 1947; Members: 1000
Pres: Dr. T. Batey; Gen Secr: Dr. S. Nortcliff
Focus: Agri
Periodicals
Journal of Soil Science (weekly)
Soil Use and Management (weekly) . . 11690

The British Sociological Association (BSA), Unit 3G, Mountjoy Research Centre, Stockton Rd, Durham DH1 3UR
T: (091) 3830839; Fax: (091) 3830782
Founded: 1951; Members: 2400
Pres: Prof. M. Barrett; Gen Secr: Nicola Boyne
Focus: Sociology
Periodical
Sociology, Work, Employment and Society (weekly) 11691

British Standards Institution (BSI), 2 Park St, London W1A 2BS
T: (071) 6299000; Tx: 266933
Founded: 1901
Pres: Sir Philip Foreman; Gen Secr: Dr. I. Dunstan
Focus: Standards
Periodicals
Annual Report (weekly)
BSI News (weekly) 11692

British Theatre Association (B.T.A.), c/o Darwin Bldg, Regent's College, Regent's Park, London NW1
T: (071) 9352571
Founded: 1919; Members: 2500
Pres: Earl of Bessborough; Gen Secr: Sally Meades
Focus: Perf Arts
Periodical
Drama (weekly) 11693

British Theatre Institute (BTI), 30 Clareville St, London SW7
Focus: Perf Arts 11694

British Tinnitus Association (BTA), c/o Tinnitus Support Service, 105 Gower St, London WC1E 6AH
T: (071) 3878033, 3878079
Founded: 1979; Members: 6000
Focus: Otorhinolaryngology
Periodical
BTA Newsletter (weekly) 11695

British Trolleybus Society (BTS), 14 Ilkley Rd, Caversham, Reading RG4 7BD
Founded: 1961; Members: 365
Focus: Transport
Periodical
Trolleybus (weekly) 11696

British Trust for Ornithology (BTO), The Nunnery, Thetford IP24 2PU
T: (0842) 750050; Fax: (0842) 750030
Founded: 1933; Members: 9000
Pres: R.P. Howard; Gen Secr: Dr. J.J.D. Greenwood
Focus: Ornithology
Periodicals
Bird Study (3 times annually)
BTO News (weekly)
Ringing and Migration (3 times annually) 11697

British Tunnelling Society, c/o Institution of Civil Engineers, Great George St, London SW1
T: (071) 2227722
Founded: 1971; Members: 700
Focus: Civil Eng 11698

British Universities Association of Slavists (BUAS), c/o Dept of Russian, University of Reading, Reading RG6 2AA
T: (0734) 875123
Founded: 1956; Members: 300
Focus: Ling; Lit; Hist
Periodical
Research Work in Progress (weekly) . . 11699

British Universities Film and Video Council (BUFVC), 55 Greek St, London WIV 5LR
T: (071) 7343687; Fax: (071) 2873914
Founded: 1948; Members: 260
Focus: Cinema
Periodical
Viewfinder Magazine 11700

British Universities Industrial Relations Association (BUIRA), c/o Dept of Management Sciences, University of Manchester Institute of Science and Technology, POB 88, Manchester M6O 1QD
T: (061) 2363311
Founded: 1950; Members: 375
Focus: Stats 11701

British Urban and Regional Information System Association (BURISA), c/o County Council, Social Services Dept, The Castle, Winchester SO23 8UQ
T: (0962) 847257
Founded: 1972
Pres: D. Ward
Focus: Computer & Info Sci
Periodical
Newsletter (6 times annually) 11702

British Veterinary Association (BVA), 7 Mansfield St, London W1M0AT
T: (071) 6366541; Fax: (071) 4362970
Founded: 1882; Members: 8500
Pres: Robert Young; Gen Secr: James Baird
Focus: Vet Med
Periodicals
In Practice (weekly)
Research in Veterinary Science (weekly)
Veterinary Record (weekly) 11703

British Veterinary Radiology Association (BVRA), c/o Royal Veterinary College, Hawkshead Lane, North Mymms, Hatfield, Herts
Focus: Vet Med; Radiology
Periodical
BVRA Abstracts (weekly) 11704

British Veterinary Zoological Society, Chartley House, Swannells Wood, Studham, Beds
Focus: Vet Med; Zoology 11705

British Vexillological Society (BVS), 80 Copers Cope Rd, Beckenham, Herts
Founded: 1981; Members: 217
Focus: Cultur Hist
Periodical
Raising the Standard (weekly) 11706

British Watercolour Society, c/o Leslie Simpson, Ralston House, 41 Lister St, Riverside Gardens, Ilkley LS29 9ET
T: (0943) 609075; Fax: (0943) 609075
Focus: Arts 11707

British Wind Energy Association, c/o Dept of Technology (EM), Open University, Milton Keynes, Bucks
Focus: Energy 11708

British Wood Preserving and Damp-proofing Association (BWPA), 4 Romford Rd, Stratford, London E15 4EA
T: (081) 5192588; Fax: (081) 5193444
Founded: 1930; Members: 400
Gen Secr: J. David
Focus: Ecology
Periodicals
Information Leaflets
News Sheet (weekly) 11709

British Zeolite Association (BZA), c/o Dept of Chemistry, The City University, Northampton, London ECIV OHB
T: (071) 2534399; Tx: 263896
Founded: 1977; Members: 100
Focus: Mineralogy 11710

Brittle Bone Society (BBS), 112 City Rd, Dundee DD2 2PW
T: (0382) 817771; Fax: (0382) 816348
Founded: 1972; Members: 1500
Focus: Physiology
Periodical
Newsletter (weekly) 11711

The Brontë Society, Brontë Parsonage, Haworth, Keighley BD22 8DR
T: (0535) 42323; Fax: (0535) 647131
Founded: 1893; Members: 3500
Pres: Lord Briggs of Lewes; Gen Secr: Ruth M. Battye
Focus: Lit
Periodical
Transactions (bi-annually) 11712

Browning Society of London, 2 Sedlescombe Rd, London SW6
T: (071) 3855361
Founded: 1970; Members: 70
Focus: Lit 11713

Brunel Society, c/o Brunel Technical College, Ashley Down, Bristol BS7 9BU
T: (0272) 41241
Founded: 1968; Members: 303
Focus: Eng 11714

The Buddhist Society, 58 Eccleston Sq, London SW1V 1PH
T: (071) 8345858
Founded: 1924; Members: 3000
Focus: Rel & Theol
Periodical
The Middle Way (weekly) 11715

Building Services Research and Information Association (BSRIA), Old Bracknell Lane West, Bracknell RG12 7AH
T: (0344) 426511; Fax: (0344) 487575
Founded: 1959; Members: 760
Gen Secr: T. Wood
Focus: Civil Eng
Periodicals
Engineering Services Management (weekly)
International Building Services Abstracts (weekly)
Statistics Bulletin (weekly) 11716

The Burns Federation, c/o Dick Institute, Elmbank Av, Kilmarnock KA1 3BU
T: (0563) 26401; Fax: (0563) 29661
Founded: 1885; Members: 250
Focus: Lit 11717

Buteshire Natural History Society, Mecknoch Farm, Rothesay, Bute PA20 0QA
T: (0700) 502409
Founded: 1905; Members: 200
Pres: L. Cumming; Gen Secr: E.M. Johnston
Focus: Hist; Nat Sci
Periodical
Transactions of the Buteshire Natural History Society 11718

Byron Society, 6 Gertrude St, London SW10 0JN
T: (071) 3525112
Founded: 1971; Members: 958
Focus: Lit
Periodical
The Byron Society Journal (weekly) . . 11719

C.A.B. International, C.A.B. International Centre, Wallingford OX10 8DE
T: (0491) 832111; Tx: 847964; Fax: (0491) 833508
Founded: 1929; Members: 34
Gen Secr: D. Laing
Focus: Agri
Periodicals
Abstracts on Hygiene and Communicable Diseases (weekly)
Agbiotech News and Information (weekly)
Agricultural Engineering Abstracts (weekly)
Agroforestry Abstracts (weekly)
Aids Newsletter (weekly)
Animal Breeding Abstracts (weekly)
Bibliography of Systematic Mycology (weekly)
Biocontrol News and Information (weekly)
Biodeterioration Abstracts (weekly)
Bulletin of Entomological Research (weekly)
Crop Physiology Abstracts (weekly)
Current Aids Literature (weekly)
Dairy Science Abstracts (weekly)
Descriptions of Pathogenic Fungi & Bacteria (weekly)
Distribution Maps of Pests (weekly)
Distribution Maps of Plant Diseases (weekly)
Faba Bean Abstracts (weekly)
Field Crop Abstracts (weekly)
Forest Products Abstracts (weekly)
Forestry Abstracts (weekly)
Helminthological Abstracts (weekly)
Herbage Abstracts (weekly)
Horticultural Abstracts (weekly)
Index of Current Research on Pigs (weekly)
Index of Fungi (weekly)
Index Veterinarius (weekly)
Irrigation and Drainage Abstracts (weekly)
Leisure, Recreation and Tourism Abstracts (weekly)
Maize Abstracts (weekly)
Nematological Abstracts (weekly)
Nutrition Abstracts and Reviews: Human and Experimental (weekly)
Nutrition Abstracts and Reviews: Livestock Feeds and Feeding (weekly)
Ornamental Horticulture (weekly)
Pig News and Information (weekly)
Plant Breeding Abstracts (weekly)
Plant Growth Regulator Abstracts (weekly)
Post Harvest News and Information (weekly)
Potato Abstracts (weekly)
Poultry Abstracts (weekly)
Protozoological Abstracts (weekly)
Public Health News (weekly)
Review of Agricultural Entomology (weekly)
Review of Medical and Veterinary Entomology (weekly)
Review of Medical and Veterinary Mycology (weekly)
Review of Plant Pathology (weekly)
Rice Abstracts (weekly)
Rural Development Abstracts (weekly)
Seed Abstracts (weekly)
Soils and Fertilizers (weekly)
Sorghum and Millets Abstracts (weekly)
Soyabean Abstracts (weekly)
Sugar Industry Abstracts (weekly)
Tropical Diseases Bulletin (weekly)
Veterinary Bulletin (weekly)
Weed Abstracts (weekly)
Wheat, Barley and Triticale Abstracts (weekly)
World Agric. Econ. & Rural Sociology Abstract (weekly) 11720

CAB International Bureau of Crop Protection, c/o CAB International, Wallingford OX10 8DE
T: (0491) 32111; Tx: 847964; Fax: (0491) 33508
Founded: 1987
Gen Secr: Dr. P.R. Scott
Focus: Agri 11721

CAB International Bureau of Horticulture and Plantation Crops, c/o CAB International, Wallingford OX10 8DE
T: (0491) 32111; Tx: 847964; Fax: (0491) 33508
Founded: 1929
Gen Secr: S. Bhat
Focus: Hort; Agri 11722

C.A.B. International Bureau of Horticulture and Plantation Crops, Wallingford OX10 8DE
T: (0491) 832111; Tx: 847964; Fax: (0491) 833508
Focus: Hort; Botany
Periodicals
Horticultural Abstracts (weekly)
Ornamental Horticulture (weekly)
Postharvest News and Information (weekly)
Seed Abstracts (weekly) 11723

CAB International Bureau of Nutrition, c/o CAB International, Wallingford OX10 8DE
T: (0491) 32111; Tx: 847964; Fax: (0491) 33508
Founded: 1929; Members: 29
Gen Secr: E. Dodsworth
Focus: Nutrition 11724

CAB International Bureau of Plant Breeding and Genetics, c/o CAB International, Wallingford OX10 8DE
T: (0491) 32111; Tx: 847964; Fax: (0491) 33508
Founded: 1929; Members: 29
Gen Secr: Ray Watkins
Focus: Botany; Agri 11725

CAB International Bureau of Plant Breeding and Genetics, c/o CAB International, Wallingford OX10 8DE
T: (0491) 32111; Tx: 847964; Fax: (0491) 33508
Founded: 1929; Members: 29
Gen Secr: Ray Watkins
Focus: Botany; Agri 11726

C.A.B. International Department of Dairy Science and Technology, Wallingford OX10 8DE
T: (0491) 832111; Tx: 847964; Fax: (0491) 833508
Founded: 1938
Focus: Agri
Periodical
Dairy Science Abstracts (weekly) 11727

C.A.B. International Department of Soils, Wallingford OX10 8DE
T: (0491) 32111; Tx: 847964; Fax: (0491) 33508
Founded: 1929
Focus: Agri
Periodicals
Irrigation and Drainage Abstracts (weekly)
Soils and Fertilizers (weekly) 11728

CAB International Division of Animal Health and Medical Parasitology, c/o CAB International, Wallingford OX10 8DE
T: (0491) 32111; Tx: 847964; Fax: (0491) 33508
Founded: 1929; Members: 29
Gen Secr: G. Phillips
Focus: Vet Med 11729

C.A.B. International Division of Animal Production, Wallingford OX10 8DE
T: (0491) 83211; Tx: 847964; Fax: (0491) 833508
Founded: 1929
Focus: Genetics; Animal Husb
Periodicals
AgBiotech News and Information (weekly)
Animal Breeding Abstracts (weekly)
Poultry Abstracts (weekly) 11730

CAB International Forestry Bureau, c/o CAB International, Wallingford OX10 8DE
T: (0491) 32111; Tx: 847964; Fax: (0491) 33508
Founded: 1938; Members: 29
Gen Secr: Dr. K. Becker
Focus: Forestry 11731

C.A.B. International Institute of Entomology, 56 Queen's Gate, London SW7 5JR
T: (071) 5840067; Tx: 847964
Focus: Entomology
Periodicals
Bulletin of Entomological Research (weekly)
Distribution Maps of Pests (18 times annually)
Review of Applied Entomology, Series A (Agricultural), Series B (Medical and Veterinary) (weekly) 11732

Caernarvonshire Historical Society (CHS), Shire Hall St, Caernarfon LL55 1SH
T: (0286) 679088
Founded: 1938; Members: 600

Pres: Lady G. Roberts; Gen Secr: Bryn R. Parry
Focus: Hist
Periodical
Transactions (weekly) 11733

Cambrian Archaeological Association, The Laurels, Westfield Rd, Newport NP9 4ND
T: (0633) 262449
Founded: 1846; Members: 857
Pres: W. Gwyn Thomas; Gen Secr: Dr. J.M. Hughes
Focus: Archeol
Periodical
Archaeologia Cambrensis 11734

Cambridge Antiquarian Society (C.A.S.), 39 Highsett Hills Rd, Cambridge CB2 1NZ
T: (0223) 355515
Founded: 1840; Members: 495
Gen Secr: Dr. M. Hesse
Focus: Archeol; Hist
Periodicals
Conduit (weekly)
Proceedings (weekly) 11735

Cambridge Bibliographical Society, c/o Cambridge University Library, West Rd, Cambridge CB3 9DR
T: (0223) 333000; Tx: 81395
Founded: 1949; Members: 500
Pres: Dr. F.H. Stubbings; Gen Secr: N. Thwaite
Focus: Libraries & Bk Sci
Periodicals
Monographs (weekly)
Transactions (weekly) 11736

Cambridge Philosophical Society, c/o Scientific Periodicals Library, Arts School, Benet St, Cambridge CB2 3PY
T: (0223) 334743
Founded: 1819; Members: 1550
Pres: Prof. P.P.G. Bateson; Gen Secr: Judith M. Winton Thomas
Focus: Philos; Nat Sci
Periodicals
Biological Reviews (weekly)
Mathematical Proceedings (weekly) . . . 11737

Cambridge Refrigeration Technology (CRT) / Shipowners Refrigerated Cargo Research Association, 140 Newmarket Rd, Cambridge CB5 8HE
T: (0223) 65101; Tx: 81604; Fax: (0223) 461522
Founded: 1945; Members: 27
Focus: Transport
Periodical
Newsletter (weekly) 11738

Cambridge Society for Industrial Archaeology (CSIA), Engineers House, Riverside, Cambridge
Founded: 1968; Members: 150
Focus: Archeol 11739

Campaign for the Advancement of State Education (C.A.S.E.), 43 Littleheath Charlton, London SE7
T: (081) 3177213
Founded: 1960; Members: 2500
Focus: Educ 11740

Campaign for the Protection of Rural Wales (CPRW), 31 High St, Welshpool SY21 7JP
T: (0938) 552525, 556212; Fax: (0938) 552741
Founded: 1928; Members: 4000
Pres: Lord Williams of Elvel; Gen Secr: Dr. Neil Caldwell
Focus: Ecology
Periodical
Rural Wales Magazine (3 times annually) 11741

Campden Food and Drink Research Association (CFDRA), Chipping Campden, Glos
T: (0386) 840319; Fax: (0386) 841306
Founded: 1919; Members: 620
Pres: Lord King's Norton; Gen Secr: Prof. C. Dennis
Focus: Food 11742

Cancer Research Campaign (CRC), Cambridge House, 8-10 Cambridge Terrace, London NW1 4JL
T: (071) 2241333
Founded: 1923; Members: 104
Pres: Duke of Gloucester; Gen Secr: David de Peyer
Focus: Cell Biol & Cancer Res
Periodicals
Annual Report (weekly)
British Journal of Cancer (weekly) . . . 11743

Canterbury and York Society, 15 Cusack Close, Strawberry Hill, Twickenham TW1 4TB
Founded: 1902; Members: 200
Pres: Dr. D.M. Owen; Gen Secr: Dr. C. Harper-Bill
Focus: Hist 11744

Careers Research and Advisory Centre (CRAC), Sheraton House, Castle Park, Cambridge CB3 0AX
T: (0223) 260277
Founded: 1964
Focus: Adult Educ 11745

Carlyle Society of Edinburgh, 38 Grange Rd, Edinburgh EH9 1UL
Founded: 1929; Members: 65
Focus: Lit 11746

Castings Technology International, 7 East Bank Rd, Sheffield S2 3PT
T: (0742) 28647; Fax: (0742) 730852
Founded: 1968; Members: 210
Focus: Metallurgy
Periodical
Annual Conference Proceedings (weekly) 11747

Catholic Record Society (C.R.S.), 114 Mount St, London W1Y 6AH
Founded: 1904; Members: 800
Gen Secr: R. Rendel
Focus: Hist
Periodical
Recusant History (weekly) 11748

Central Bureau for Educational Visits and Exchanges, Seymour Mews House, Seymour Mews, London WIH 9PE
Fax: (071) 9355741
Founded: 1948
Pres: J.A. Carter; Gen Secr: A.H. Male
Focus: Educ
Periodicals
A Year Between
Home from Home
Study Holidays
Teach Abroad
Volunteer Work
Working Holidays (weekly) 11749

Central Council of Physical Recreation, Francis House, Francis St, London SW1P 1DE
T: (071) 8283163; Fax: (071) 6308820
Founded: 1935
Pres: Philip Duke of Edinburgh; Gen Secr: Peter Lawson
Focus: Sports 11750

Central Scotland Aviation Group (CSAG), Craigmount Av, Edinburgh EH12 8ED
Founded: 1965; Members: 662
Focus: Aero
Periodicals
Scottish Air News (weekly)
Scottish Fly-Over Supplement (weekly) . 11751

Central Sterilising Club, c/o Dept of Pathology, Kingston Hospital, Kingston-upon-Thames KT2 7BD
Founded: 1960; Members: 500
Focus: Physiology 11752

Centre for Alternative Technology, Llwyngwern Quarry, Machynlleth SY20 9AZ
T: (0654) 702400; Fax: (0654) 702782
Founded: 1974; Members: 300
Focus: Eng; Ecology
Periodical
Clean Slate (weekly) 11753

Centre for Iberian Studies (ISSA), c/o University of Keele, Keele ST5 5BG
T: (0782) 621111; Fax: (0782) 613847
Founded: 1968; Members: 300
Focus: Soc Sci
Periodical
Iberian Studies (weekly) 11754

CERAM Research, Queens Rd, Penkhull, Stoke-on-Trent ST4 7LQ
T: (0782) 45431; Tx: 36228; Fax: (0782) 412331
Founded: 1948
Pres: Prof. Sir Ronald Mason; Gen Secr: Dr. N.E. Sanderson
Focus: Materials Sci
Periodicals
CERAM Research News (weekly)
CERAM Research Progress (weekly)
World Ceramics Abstracts (weekly) . . . 11755

Challenger Society for Marine Science, c/o Institute of Oceanographic Science, Wormley, Godalming GU8 5UB
T: (0428) 684141
Founded: 1903; Members: 400
Pres: Dr. B.S. McCartney; Gen Secr: H.S.J. Roe
Focus: Oceanography
Periodical
Ocean Challenge (weekly) 11756

The Charles Lamb Society, 1a Royston Rd, Richmond TW10 6LT
T: (081) 9403837
Founded: 1935
Pres: Prof. John Beer; Gen Secr: M.R. Huxstep
Focus: Lit
Periodical
The Charles Lamb Society Bulletin (weekly)
. . . . 11757

Charles Rennie Mackintosh Society (CRM Society), Queens Cross, 870 Garscube Rd, Glasgow G20 7EL
T: (041) 9466600; Fax: (041) 9452321
Founded: 1973; Members: 1500
Focus: Archit; Fine Arts
Periodical
Newsletter (3 times annually) 11758

Chartered Association of Certified Accountants, 29 Lincoln's Inn Fields, London WC2A 3EE
T: (071) 2426855; Tx: 24381; Fax: (071) 8318054
Founded: 1904; Members: 40000
Pres: K. Duncan; Gen Secr: A.L. Rose
Focus: Law
Periodicals
Accountants' Guide
Certified Accountants (weekly)
List of Members
Students' Newsletter (weekly) 11759

The Chartered Institute of Bankers, 10 Lombard St, London EC3V 9AS
Founded: 1879; Members: 95000
Pres: Brian Pearse; Gen Secr: Eric Glover
Focus: Finance
Periodical
Banking World (weekly) 11760

The Chartered Institute of Building, Englemere, Kings Ride, Ascot SL5 8BJ
T: (0990) 23355
Founded: 1834; Members: 33500
Pres: Robert Wharton; Gen Secr: Keith Banbury
Focus: Civil Eng
Periodicals
Chartered Builder (26 times annually)
Construction Computing (3 times annually)
List of Building Courses
Yearbook & Directory of Members . . . 11761

Chartered Institute of Management Accountants, 63 Portland Pl, London W1N 4AB
T: (071) 6372311
Founded: 1919; Members: 34000
Gen Secr: Sir George Vallings
Focus: Finance
Periodicals
Management Accounting
Management Accounting Research . . . 11762

Chartered Institute of Marketing, Moor Hall, Cookham, Maidenhead SL6 9QH
T: (0628) 524922; Fax: (0628) 849462
Founded: 1911; Members: 24000
Focus: Marketing
Periodicals
Journal of Marketing Management (weekly)
Marketing Business (10 times annually)
Marketing Success (weekly)
State of Marketing (weekly) 11763

Chartered Institute of Patent Agents (CIPA), Staple Inn Bldgs, High Holborn, London WC1V 7PZ
T: (071) 4059450; Fax: (071) 4300471
Founded: 1882; Members: 2050
Gen Secr: M.C. Ralph
Focus: Law
Periodicals
CIPA (weekly)
Register of Patent Agents (weekly) . . . 11764

Chartered Institute of Public Finance and Accountancy (CIPFA), 3 Robert St, London WC2N 6BH
T: (071) 8958823; Fax: (071) 8958825
Founded: 1885; Members: 14500
Gen Secr: N.P. Hepworth
Focus: Business Admin; Finance
Periodicals
Public Finance and Accountancy (weekly)
Public Money and Management (weekly) 11765

The Chartered Institute of Transport (CIT), 80 Portland Pl, London W1N 4DP
T: (071) 6369952; CA: Transitute London; Fax: (071) 6370511
Founded: 1919; Members: 21000
Pres: HRH The Princess Royal; Gen Secr: R.P. Botwood
Focus: Transport
Periodicals
Proceedings (weekly)
Transport (weekly) 11766

Chartered Institution of Building Services Engineers, 222 Balham High Rd, London SW12 9BS
T: (081) 6755211
Founded: 1897; Members: 12500
Pres: David Arnold; Gen Secr: A.V. Ramsay
Focus: Civil Eng
Periodicals
Building Services (weekly)
Building Services Engineering Research and Technology (weekly)
Lighting Research and Technology (weekly)
. . . . 11767

Chartered Insurance Institute (C.I.I.), 20 Aldermanbury, London EC2V 7HY
T: (071) 6063835; Fax: (071) 7260131
Founded: 1897; Members: 68000
Pres: R. Hill; Gen Secr: D.E. Bland
Focus: Insurance
Periodicals
C.I.I. Journal (weekly)
Society of Fellows Journal (weekly) . . 11768

Chartered Society of Designers, 29 Bedford Sq, London WC1B 3EG
T: (071) 6311510; Fax: (071) 5802338
Founded: 1930; Members: 8500
Pres: Jeremy Rewse-Davies; Gen Secr: Brian Lymbery
Focus: Graphic & Dec Arts, Design
Periodicals
Design Review (weekly)
Newsletter (weekly) 11769

The Chartered Society of Physiotherapy (CSP), 14 Bedford Row, London WC1R 4ED
T: (071) 2421941; Fax: (071) 8314509
Founded: 1894; Members: 25000
Gen Secr: Toby Simon
Focus: Therapeutics
Periodical
Physiotherapy (weekly) 11770

Chemicals Notation Association (CNA), c/o Wellcome Foudation Ltd, Temple Hill, Dartford DA1 5AH
Founded: 1969; Members: 77
Focus: Chem 11771

Chemometrics Society, c/o Pfizer Central Research, Sandwich CT13 9NJ
T: (03046) 3511
Founded: 1976; Members: 20
Focus: Chem 11772

The Chester Archaeological Society (C.A.S.), 24 Nicholas St, Chester CH1 2NX
T: (0244) 317108
Founded: 1849; Members: 300
Gen Secr: B.E. Harris
Focus: Archeol
Periodical
Journal (weekly) 11773

Chester Society of Natural Science, Literatur and Art, c/o The Grosvenor Museum, Grosvenor St, Chester CH1 2DD
T: (0244) 21616
Focus: Arts; Lit; Nat Sci 11774

The Chest, Heart and Stroke Association, Tavistock House North , Tavistock Sq, London WC1H 9JE
T: (071) 3873012
Founded: 1899
Pres: Duke of Kent; Gen Secr: Sir David Atkinson
Focus: Pulmon Dis; Cardiol 11775

Chiltern Society, Silver How, Little Hollis, Great Missenden HP16 9HZ
T: 3524
Founded: 1965; Members: 3800
Focus: Preserv Hist Monuments; Ecology
Periodical
Chiltern News (weekly) 11776

China Policy Study Group (CPSG), 62 Parliament Hill, London NW3 2TJ
T: (071) 4353416
Founded: 1964
Focus: Poli Sci 11777

China Society, 31b Torrington Sq, London WC1
T: (071) 6367985
Founded: 1906; Members: 150
Focus: Ethnology 11778

Chippendale Society, Temple Newsam House, Leeds LS15 0AE
T: (0532) 647321
Founded: 1963; Members: 405
Focus: Arts 11779

Chiropractic Advancement Association (CAA), 38 The Island, Thames Ditton KT7 0SQ
T: (081) 3982098
Founded: 1965; Members: 2000
Focus: Orthopedics
Periodicals
The British Chiropractic Handbook
News Bulletin (bi-annually) 11780

Choir Schools Association, c/o Wells Cathedral School, Wells, Somerset
T: (0749) 672117
Founded: 1921; Members: 40
Focus: Educ; Music 11781

The Chopin Society, 42 Beechcroft Gardens, Wembley Park HA9 8EP
T: 9045962
Founded: 1971; Members: 60
Focus: Music
Periodical
Newsletter (weekly) 11782

Christian Education Movement (CEM), Royal Buildings, Victoria St, Derby DE1 1GW
T: (0332) 296655
Founded: 1965
Focus: Educ
Periodicals
The British Journal of Religions Education
RE Today 11783

Church Music Association (CMA), 15 Denbigh Rd, London W11 2SJ
T: (071) 7276387
Founded: 1955
Focus: Music 11784

Cinema Organ Society (COS), 38 Duxford Rd, Whittlesford, Cambridge CB2 4ND
Founded: 1953; Members: 2500
Focus: Music
Periodical
Cinema Organ (weekly) 11785

Cinema Theatre Association (CTA), 66 Harrowcleve Gardens, Teddington TW11 1DJ
Founded: 1967; Members: 400
Pres: Anthony Moss; Gen Secr: Adam Unger
Focus: Archit; Hist
Periodicals
Bulletin (weekly)
Picture House (3 times annually) . . . 11786

Circle of State Librarians (CSL), c/o Dept of Trade and Industry, Room 116, 123 Victoria St, London SW1E 6RB
T: (071) 2156697; Fax: (071) 2155665
Founded: 1946; Members: 600
Focus: Libraries & Bk Sci
Periodical
State Librarian (3 times annually) . . . 11787

Citizens Protection Society (CPS), 611 Collingwood House, Dolphin Sq, London SW1V 3NF
T: (071) 8340887
Founded: 1970; Members: 50000
Focus: Criminology 11788

City and Guilds of London Institute, 76 Portland Pl, London W1N 4AA
T: (071) 2782468; Fax: (071) 4367630
Founded: 1878
Gen Secr: John Barnes
Focus: Educ
Periodicals
Broadsheet (3 times annually)
Handbook (weekly)
Report and Accounts (weekly) 11789

City of Stoke on Trent Museum Archaeological Society (S.O.T.M.A.S.), c/o City Museum and Art Gallery, Bethesda St, Hanley, Stoke-on-Trent ST1 3DW
T: (0782) 202173
Founded: 1959; Members: 60
Pres: E.E. Royle; Gen Secr: R.H. Outram
Focus: Archeol
Periodicals
Arc' News (weekly)
Staffordshire Archaeological Studies (weekly)
. . . . 11790

The Civic Trust, 17 Carlton House Terrace, London SW1Y 5AW
T: (071) 9300914
Founded: 1957
Gen Secr: Martin Bradshaw
Focus: Ecology
Periodical
Urban Focus (weekly) 11791

The Classical Association, c/o Dr. M. Schofield, Saint John's College, Cambridge CB2 1TP
Founded: 1903; Members: 4000
Pres: Prof. F.G.B. Millar; Gen Secr: Dr. Malcolm Schofield
Focus: Lit
Periodicals
Classical Quarterly (weekly)
Classical Review (weekly)
Greece and Rome (weekly)
Proceedings of the Classical Association (weekly)
. . . . 11792

Cleveland Industrial Archaeology Society (CIAS), 8 Loweswater Crescent, Stockton-on-Tees, Cleveland TS18 4PY
Founded: 1965; Members: 73
Focus: Archeol 11793

Clinical Theology Association (CTA), Saint Mary's House, Church Westcote, Oxford OX7 6SF
T: (0993) 830209
Founded: 1962; Members: 1000
Pres: Richard Darby; Gen Secr: P.J. Van de Kasteele
Focus: Rel & Theol; Psych
Periodical
Contact (weekly) 11794

Collaborative International Pesticides Analytic Council (CIPAC), 61 Finchley Court, Ballards Lane, London N3
Founded: 1957
Focus: Agri
Periodical
CIPAC Proceedings 11795

College of Ophthalmologists, 17 Cornwall Terrace, London NW1 4QW
T: (071) 9350702; Fax: (071) 9359838
Founded: 1880; Members: 1734
Pres: P. Wright; Gen Secr: P.A. Hunter
Focus: Ophthal
Periodical
Eye (weekly) 11796

The College of Preceptors, Coppice Row, Theydon Bois, Epping CM16 7DN
T: (0992) 812727; Fax: (0992) 814690
Founded: 1846; Members: 3500
Pres: Lady Bridget Plowden; Gen Secr: T.F. Wheatley
Focus: Educ
Periodical
Education Today (3 times annually) . . 11797

The College of Radiographers, 14 Upper Wimpole St, London W1M 8BN
T: (071) 9355726/27
Founded: 1920; Members: 9500
Focus: Radiology
Periodicals
Radiography (weekly)
Radiography News (weekly) 11798

Colour Group (Great Britain), c/o Kodak Ltd, Research Division, Wealdstone, Harrow HA4 1TY
T: 4274380
Founded: 1960; Members: 300
Focus: Materials Sci 11799

Comité de Liaison des Géomètres-Experts Européens, 12 Great George St, London SW1P 3AD
T: (071) 2227000; Tx: 915443; Fax: (071) 2229430
Founded: 1972
Gen Secr: Alain Bourcy
Focus: Math 11800

Comité Dentaire de Liaison de la Communauté Européenne, c/o British Dental Association, 64 Wimpole St, London W1M 8AL
T: (01) 9350875; Fax: (01) 4875232
Focus: Dent 11801

Commemorative Collectors Society, 25 Farndale Close, Long Eaton, Nottingham NG10 3PA
T: (0602) 727666
Founded: 1972; Members: 5016
Gen Secr: S.N. Jackson
Focus: Graphic & Dec Arts, Design
Periodicals
Collecting Commemorabilia (weekly)
Newsletter (weekly) 11802

Commission Internationale de Marketing, c/o EMA Central Secretariat, 18 Saint Peters Steps, Brixham
Founded: 1971
Focus: Marketing
Periodical
Journal of International Marketing and Marketing Research 11803

Committee of Directors of Polytechnics, Kirkman House, 12-14 Whitfield St, London W1P 6AX
T: (071) 6379939; Fax: (071) 4364966
Founded: 1970; Members: 34
Pres: J.M. Stoddart; Gen Secr: Dr. R.J. Brown
Focus: Educ
Periodicals
First Destinations of Polytechnic Students (weekly)
Polytechnic Courses: A Guide to Full-Time and Sandwich Courses (weekly)
Polytechnic Courses Handbook (weekly) . 11804

Committee of Vice-Chancellors and Principals of the Universities of the United Kingdom (CVCP), 29 Tavistock Sq, London WC1H 9EZ
T: (071) 3879231; Fax: (071) 3888649
Founded: 1918
Pres: Dr. Kenneth Edwards; Gen Secr: Tom Burgner
Focus: Sci
Periodical
British Universities' Guide to Graduate Study (weekly) 11805

Commons, Open Spaces and Footpaths Preservation Society, 25A Bell St, Henley-on-Thames RG9 2BA
T: (0491) 573535
Founded: 1865; Members: 2500
Gen Secr: Kate Ashbrook
Focus: Ecology
Periodical
Open Space (3 times annually) 11806

Commonwealth Advisory Aeronautical Research Council (CAARC), Saint Christopher House, Southwark St, London SE1 0RD
Founded: 1946; Members: 7
Gen Secr: Diana Halliday
Focus: Aero 11807

Commonwealth Association for Development (CAD), 168 Towerbridge Rd, London SE1 3LS
T: (071) 3577017; Fax: (071) 3577113
Founded: 1988
Gen Secr: Dr. Charles I. Kejeh
Focus: Develop Areas
Periodicals
CAD Newsletter (3 times annually)
Journal of the Commonwealth Association for Development (weekly)
Research in the Commonwealth (weekly) 11808

Commonwealth Association for Mental Handicap and Developmental Disabilities (CAMHDD), 36a Osberton Pl, Sheffield S11 8XL
T: (0742) 682695
Founded: 1983
Gen Secr: Dr. V.R. Pandurangi
Focus: Rehabil
Periodical
CAMHDD Newsletter 11809

Commonwealth Association for Public Administration and Management (CAPAM), c/o COMSEC, Marlborough House, Pall Mall, London SW1Y 5HX
T: (071) 8393411; Tx: 27678; Fax: (071) 9300827
Focus: Public Admin; Business Admin . 11810

Commonwealth Association of Architects – Education Committee, 66 Portland Pl, London W1N 4AD
T: (071) 6368276; Tx: 22914; Fax: (071) 2551541
Founded: 1966
Gen Secr: George Wilson
Focus: Educ; Archit 11811

Commonwealth Association of Science and Mathematics Educators (CASME), c/o Commonwealth Secretariat, Marlborough House, Pall Mall, London SW1Y 5HX
T: (071) 8396039
Founded: 1974
Focus: Educ; Sci; Math 11812

Commonwealth Association on Surveying and Land Economy, 15 Greycoat Pl, London SW1P 1SB
T: (071) 2228961; Fax: (071) 9768304
Founded: 1969; Members: 50
Focus: Surveying; Agri
Periodical
Survey Review (weekly) 11813

Commonwealth Bureau of Nutrition (CAB), c/o Rowett Research Institute, Bucksburn, Aberdeen AB2 9SB
T: (0224) 712162
Founded: 1929
Focus: Agri
Periodicals
Nutrition Abstracts and Reviews: Livestock Feeds and Feeding (weekly)
Nutrition Abstructs an Reviews: Human and Experimental (weekly) 11814

Commonwealth Commercial Crime Unit (CCU), c/o Legal Div, Commonwealth Secretariat, Marlborough House, Pall Mall, London SW1Y 5HX
T: (071) 8393411; Tx: 27678; Fax: (071) 9300827
Founded: 1981
Focus: Law 11815

Commonwealth Consultative Group on Technology Management (CCGTM), c/o Economic Affairs Div, Commonwealth Secretariat, Marlborough House, Pall Mall, London SW1Y 5HX
T: (071) 8393411; Tx: 27678; Fax: (071) 9300827
Founded: 1990
Gen Secr: Michaela Y. Smith
Focus: Business Admin 11816

Commonwealth Defence Science Organisation (CDSO), c/o Ministry of Defence, Main Bldg, Whitehall, London SW1A 2HB
T: (071) 2183294
Founded: 1946; Members: 14
Focus: Military Sci 11817

Commonwealth Dental Association, 64 Wimpole St, London W1M 8AL
Founded: 1991
Gen Secr: Dr. Norman Whitehouse
Focus: Dent 11818

Commonwealth Education Liaison Committee (CELC), c/o Education Programme, Human Resource Development Group, Commonwealth Secretariat, Marlborough House, Pall Mall, London SW1Y 5HX
T: (071) 8393411; Tx: 27678; Fax: (071) 9300827
Founded: 1960
Focus: Educ 11819

Commonwealth Education Programme (CELC), Marlborough House, Pall Mall, London SW1Y 5HX
T: (071) 8393411; Fax: 9300827
Founded: 1960
Focus: Educ
Periodical
Commonwealth Education News (3 times annually)
. . . . 11820

Commonwealth Forestry Association (C.F.A.), c/o Oxford Forestry Institute, South Parks Rd, Oxford OX1 3RB
T: (0865) 275072; Fax: (0865) 275074
Founded: 1921; Members: 1600
Pres: Duke of Buccleuch and Queensberry; Gen Secr: P.J. Wood
Focus: Forestry
Periodical
Review (weekly) 11821

Commonwealth Independent Centre for the Study of the South African Economy and International Finance, c/o London School of Economics, Houghton St, London WC2A 2AE
Founded: 1990
Gen Secr: Dr. Jonathan Leape
Focus: Econ 11822

Commonwealth Industrial Training and Experience Programme (CITEP), c/o Fellowships and Training Programme, Commonwealth Secretariat, Marlborough House, Pall Mall, London SW1Y 5HX
T: (071) 8393411; Tx: 27678; Fax: (071) 9300827
Founded: 1986
Focus: Educ 11823

Commonwealth Institute, Kensington High St, London W8
T: (071) 6034535; Fax: (071) 6027374
Focus: Sci 11824

Commonwealth Lawyers' Association (CLA), c/o Law Society, 50 Chancery Ln, London WC2A 1SX
T: (071) 2421222; Tx: 261203; Fax: (071) 4059522
Founded: 1969
Gen Secr: H.C. Adamson
Focus: Law
Periodicals
CLANews
The Commonwealth Lawyer 11825

Commonwealth Legal Education Association (CLEA), c/o Commonwealth Secretariat, Marlborough House, Pall Mall, London SW1Y 5HX
T: (071) 8393411
Founded: 1971; Members: 189
Focus: Educ; Law
Periodical
Commonwealth Education (weekly) . . . 11826

Commonwealth Medical Association, BMA House, Tavistock Sq, London WC1H 9JP
T: (071) 3836351
Founded: 1962
Focus: Med 11827

Commonwealth Music Association, 1b Berry Pl, London EC1V 0JD
T: (071) 2530437; Fax: (071) 6082394
Founded: 1990
Gen Secr: Anne McClellan
Focus: Music 11828

Commonwealth Pharmaceutical Association (CPA), 1 Lambeth High St, London SE1 7JN
T: (071) 7359141; Tx: 9312131542; Fax: (071) 7357629
Founded: 1970
Gen Secr: Raymond Dickinson
Focus: Pharmacol
Periodical
CPA Newsletter (weekly) 11829

Commonwealth Secretariat, Education Programme, Human Resource, Development Group, Marlborough House, Pall Mall, London SW1Y 5HX
T: (071) 8393411
Gen Secr: P.R.C. Williams
Focus: Educ
Periodical
Commonwealth Education News (3 times annually)
. . . . 11830

Commonwealth Trust, 18 Northumberland Av, London WC2N 5BJ
T: (071) 9306733; Fax: (071) 9309705
Pres: Sir Oliver Forster; Gen Secr: Sir David Thorne
Focus: Sci
Periodical
Newsletter (3 times annually) 11831

The Company Chemists' Association (CCA), 1 Thane Rd West, Nottingham NG2 3AA
T: (0602) 506111
Founded: 1898; Members: 1800
Focus: Pharmacol 11832

The Composers' Guild of Great Britain, 34 Hanway St, London W1P 9DE
T: (071) 4360007
Founded: 1944; Members: 500
Pres: Sir Peter Maxwell Davies; Gen Secr: Heather Rosenblatt
Focus: Music
Periodical
Compass News (weekly) 11833

Computer Arts Society (CAS), 50-51 Russell Sq, London WC1B 4JX
T: (071) 6363783
Founded: 1969; Members: 480
Focus: Computer & Info Sci
Periodical
Page (weekly) 11834

Conchological Society of Great Britain and Ireland, 26 Courtland Av, Ilford IG1 3DW
Founded: 1876; Members: 700
Focus: Zoology
Periodical
Journal of Conchology (weekly) 11835

Confederate Historical Society (CHS), 19 Montague Av, Leigh-on-Sea SS9 3SL
T: 78075
Founded: 1962
Focus: Hist 11836

Confederation of British Industry, Centre Point, 103 New Oxford St, London WC1A 1DU
T: (071) 3797400; Tx: 21332; Fax: (071) 2401578
Pres: Sir Michael Angus; Gen Secr: Howard Davies
Focus: Eng
Periodicals
Economic Situation Report (weekly)
Financial Times Distributive Trades Survey (weekly)
Industrial Trends Survey (weekly) . . . 11837

Construction Industry Computing Association (CICA), Guildhall Pl, Cambridge CB2 3QQ
T: (0223) 311246
Founded: 1973
Focus: Computer & Info Sci
Periodicals
Computer Newssheet
Construction Industry Software Selector
Evaluation Reports 11838

Construction Industry Research and Information Association (CIRIA), 6 Storey's Gate, London SW1P 3AU
T: (071) 2228891; Fax: 2221708
Founded: 1967; Members: 600
Pres: Terrel Wyatt; Gen Secr: Dr. P. Bransby
Focus: Civil Eng 11839

Contemporary Art Society (C.A.S.), c/o Tate Gallery, 20 John Islip St, London SW1P 4LL
T: (071) 8215323
Founded: 1910; Members: 1500
Pres: David Gordon; Gen Secr: Gill Hedley
Focus: Arts 11840

Coordinating European Council for the Development of Performance Tests for Lubricants and Engine Fuels (CEC), 61 New Cavendish St, London W1M 8AR
T: (071) 6361004; Tx: 264380; Fax: (071) 2551472
Founded: 1963
Gen Secr: John Heritage
Focus: Chem 11841

Cornish Methodist Historical Association (CMHA), 17 Knight's Meadow, Carnon Downs TR15 3AD
Founded: 1960; Members: 350
Focus: Hist; Rel & Theol 11842

Cornish Mining Development Association (CMDA), 33 Wellington Rd, Camborne TR14 7LH
T: (02092) 714245
Founded: 1948; Members: 120
Focus: Mining 11843

Cornwall Archaeological Society (CAS), Trezeres, Harleigh Rd, Bodmin, Cornwall
T: 2700
Founded: 1961; Members: 600
Focus: Archeol
Periodical
Cornish Archaeology (weekly) 11844

COSPAS-SARSAT, c/o INMARSAT, 40 Melton St, London NW1 2EQ
T: (071) 7281391; Tx: 297201; Fax: (071) 3877480
Founded: 1979
Gen Secr: Daniel Levesque
Focus: Aero 11845

The Costume Society, c/o K. Smalley, 117 Princethorpe House, Warwick Estate, London W2 5SX
T: (071) 2869520
Founded: 1965; Members: 1000
Focus: Cultur Hist
Periodical
Costume (weekly) 11846

Council for British Archaeology (CBA), Bowes Morrell House, 111 Walmgate, York YO1 2UA
T: (0904) 671417
Founded: 1944; Members: 400
Pres: P. Addyman; Gen Secr: R. Morris
Focus: Archeol
Periodicals
Archaeology in Britain (weekly)
British Archaeological Bibliography (weekly)
British Archaeological News (10 times annually)
. . . . 11847

Council for British Geography, c/o Royal Geographical Society, 1 Kensington Gore, London SW7 2AR
T: (071) 5846371
Founded: 1988
Pres: Rex Walford; Gen Secr: Dr. D. Mottershead
Focus: Geography 11848

Council for Dance Education and Training (CDET), 5 Tavistock Pl, Room 301, London WC1H 9SS
Founded: 1978
Focus: Perf Arts 11849

Council for Education in World Citizenship, Seymour Mews House, Seymour Mews, London W1H 9PE
T: (071) 9351752; Fax: (071) 9355548
Founded: 1939
Gen Secr: Patricia Rogers
Focus: Educ
Periodicals
Broadsheet Digest
Current Affairs Broadsheet (weekly)
Newsletter 11850

Council for Environmental Conservation, 80 York Way, London N1 9AG
T: (071) 2784736
Founded: 1969
Pres: Duke of Wellington; Gen Secr: David Hughes
Focus: Ecology
Periodical
Habitat (10 times annually) 11851

Council for Environmental Education (CEE), c/o University of Reading, London Rd, Reading RG1 5AQ
T: (0734) 756061
Founded: 1968
Gen Secr: Ewan McLeish
Focus: Educ
Periodicals
Annual Review of Environmental Education (weekly)
Earthlines (weekly)
Newsheet (10 times annually) 11852

Council for the Accreditation of Correspondence Colleges (CACC), 27 Marylebone Rd, London NW1 5JS
T: (071) 9355391; Fax: (071) 9352540
Founded: 1969; Members: 44
Pres: T.B. Degenhardt
Focus: Educ
Periodical
Information Leaflet (weekly) 11853

Council for the Care of Churches (CCC), 83 London Wall, London EC2M 5NA
T: (071) 6380971
Founded: 1921
Pres: Christopher Campling; Gen Secr: Thomas Cocke
Focus: Preserv Hist Monuments
Periodical
Churchscape (weekly) 11854

Council for the Protection of Rural England, 25 Buckingham Palace Rd, London SW1W 0PP
T: (071) 9766433
Founded: 1926; Members: 45000
Focus: Ecology
Periodicals
Membership Magazine (3 times annually)
Report (weekly) 11855

Council of British Geography (COBRIG), c/o Royal Geographical Society, 1 Kensington Gore, London SW7 2AR
T: (071) 5846371
Founded: 1988
Pres: Eleanor Rawling; Gen Secr: Dr. Derek Mottershead
Focus: Geography 11856

Council of Legal Education (C.L.E.), c/o Inns of Court School of Law, 39 Eagle St, London WC1R 4AJ
T: (071) 4045787; Fax: (071) 8314188
Founded: 1852
Pres: Justice Phillips; Gen Secr: M.A. Phillips
Focus: Educ; Law
Periodical
A Range of Information Publications (weekly)
. . . . 11857

Council of Professors of Building, c/o Dept of Engineering and Construction, University of Aston, Birmingham, B4 7ET
Founded: 1978; Members: 10
Focus: Archit 11858

County, City and Borough Architects' Association, Civic Centre, POB 26, Harrow HA1 2ZX
T: 8635611
Founded: 1950; Members: 300
Focus: Archit 11859

County Education Officers' Society (CEOS), c/o Education Dept, The Castle, Winchester SO23 8UG
T: (0962) 54411
Founded: 1955; Members: 130
Focus: Educ 11860

County Emergency Planning Officers Society (CEPO), County Hall, Worlester WR5 2NP
T: (0905) 353366 ext 3020
Founded: 1966; Members: 62
Focus: Educ 11861

County Museum Society, c/o Hereford and Worcester County Museum, Hartlebury Castle, Hartlebury, Kidderminster DY11 7XZ
T: (0299) 250416
Founded: 1965
Focus: Arts; Hist 11862

County Planning Officers Society, County Hall, Glenfield, Leicester LE3 8RJ
T: (0533) 871313; Tx: 341478
Founded: 1973; Members: 110
Focus: Agri 11863

County Surveyors Society (CSS), c/o County Surveyors Dept, Northampton House, Northampton NN1 2HZ
T: (0604) 34833
Founded: 1885; Members: 120
Focus: Transport 11864

Coventry and District Archaeological Society (CADAS), 20 Harvey Close, Allesley, Coventry CV5 9FV
T: (0203) 402462
Focus: Archeol 11865

Cowethas Lacha Sten Cernow (CIC/SLS) / Stannary Law Society, Gün Hengesten, Gunnislake, Cornwall
Founded: 1975
Focus: Law
Periodical
Stannary Law Journal (weekly) 11866

Critics' Circle, 7 Lloyd Sq, London WC1X 9BA
T: (071) 8374379
Founded: 1913
Focus: Lit 11867

Critics' Guild, 9 Compayne Gardens, London NW6
T: (071) 3280809
Founded: 1950
Focus: Journalism 11868

Cross and Cockade, Great Britain Society of World Ware One Aero Historians, 23A Winchester St, Farnborough GU14 6AJ
T: (0252) 547435
Founded: 1968; Members: 1200
Focus: Hist
Periodical
Journal (weekly) 11869

Crown Imperial Society, 37 Wolsey Close, Southall UB2 4NQ
Founded: 1973
Focus: Hist 11870

Cumberland and Westmorland Antiquarian and Archaeological Society (C.W.A.A.S.), 2 High Tenterfell, Kendal LA9 4PG
T: (0539) 728288
Founded: 1866; Members: 875
Pres: B. Harbottle; Gen Secr: R. Hall
Focus: Archeol; Hist
Periodical
Transactions (weekly) 11871

Cutlery and Allied Trades Research Association (CATRA), Henry St, Sheffield S3 7EQ
T: (0742) 769736; Fax: (0742) 722151
Founded: 1960; Members: 100
Focus: Materials Sci 11872

Cymdeithas Cysyllyiadau Cyhoeddus Cymru (Welsh Public Relations Association), c/o British Steel Corporation, Welsh Division, Gabalfa, Cardiff
T: (0222) 62161; Tx: 49151
Founded: 1978
Focus: Advert 11873

Cymdeithas Hanes Sir Ddinbych / Denbighshire Historical Society, Gorffwysfa, Llansannan, Denbigh LL16 5HN
T: (074577) 386
Founded: 1951; Members: 700
Gen Secr: W.C. Wynne-Woodhouse
Focus: Hist
Periodical
Transactions (weekly) 11874

Cymdeithas yr Iaith Gymraeg, Pen Roc, Rhodfa'r Mor, Aberystwyth SY23 2AZ
T: (0970) 624501; Fax: (0970) 627122
Founded: 1962; Members: 5000
Focus: Ling 11875

Cystic Fibrosis Research Trust, Alexandra House, 5 Blyth Rd, Bromley BR1 3RS
T: (081) 4647211; Fax: (081) 3130472
Founded: 1964; Members: 20000
Pres: Sir John Batten; Gen Secr: G.J. Edkins
Focus: Physiology
Periodicals
CF News
Directory (weekly)
Information Leaflets 11876

Dartmoor Pony Society, Puddaven Farm, North Bovey, Newton Abbot TQ13 8RJ
T: (0647) 433203
Founded: 1946; Members: 550
Focus: Animal Husb
Periodical
Dartmoor Diary 11877

Dartmoor Preservation Association (DPA), 4 Oxford Gardens, Mannamead, Plymouth PL3 4SF
T: (0752) 21875
Founded: 1883; Members: 1500
Focus: Ecology 11878

Delius Society, 16 Slade Close, Waldenlade, Chatham, Kent
Founded: 1962; Members: 380
Focus: Music 11879

Delphinium Society, Birklands, Douglas Grove, Farnham GU10 3HP
Founded: 1928; Members: 1500
Focus: Zoology 11880

Dentists' Liaison Committee for the EEC, 64 Wimpole St, London W1M 8AL
T: (071) 9350876; Fax: (071) 4875232
Founded: 1960
Gen Secr: N. Whitehouse
Focus: Dent 11881

Derbyshire Archaeological Society (D.A.S.), 7 Ashleigh Dr, Uttoxeter ST14 7RC
T: 66568
Founded: 1878; Members: 675
Focus: Archeol
Periodical
Derbyshire Archaeological Journal (weekly) 11882

Design and Industries Association (DIA), 12 Carlton House Terrace, London SW1
T: (071) 9300540
Founded: 1915; Members: 1150
Focus: Eng 11883

Design Council, 28 Haymarket, London SW1Y 4SU
T: (071) 8398000; Tx: 8812963
Founded: 1944
Gen Secr: Ivor Owen
Focus: Graphic & Dec Arts, Design
Periodicals
Design (weekly)
Designing (3 times annually)
Engineering (weekly) 11884

The Design Council – Scotland, 45 Gordon St, Glasgow G1 3LZ
T: (041) 2216121; Fax: (041) 2218799
Founded: 1948
Gen Secr: F.H. Binnie
Focus: Graphic & Dec Arts, Design
Periodicals
The Big Paper (3 times annually)
Designing (3 times annually)
Engineering Design Education & Training (weekly)
. . . . 11885

Designers and Art Directors Association of London (DADA), 12 Carlton House Terrace, London SW1Y 5AH
T: (071) 8392964
Founded: 1962; Members: 157
Focus: Arts
Periodicals
Annual of the Best Advertising, Graphics, Television & Editorial Design (weekly)
Design & Art Direction (weekly) 11886

Design Research Centre for the Gold, Silver and Jewellery Industries, Saint Dunstan's House, Carey Lane, London EC2V 8AE
T: (071) 6067260
Founded: 1951; Members: 222
Focus: Arts 11887

Devon and Cornwall Record Society, c/o Devon and Exeter Institution, 7 The Close, Exeter EX1 1EZ
T: (0392) 74727
Founded: 1904; Members: 775
Pres: The Bishop of Exeter
Focus: Hist
Periodicals
Extra Series (weekly)
New Series (weekly) 11888

Devon Archaeological Society (DAS), c/o City Museum, Queen St, Exeter, Devon
T: (0392) 72340
Founded: 1929; Members: 900
Pres: H. Quinnell; Gen Secr: D. Butler
Focus: Archeol 11889

The Devonshire Association for the Advancement of Science, Literature and Art (D.A.), 7 Cathedral Close, Exeter EX1 1EZ
T: (0392) 52461; Fax: (0392) 52461
Founded: 1862; Members: 1830
Pres: Sir Jack Boles; Gen Secr: E.A.K. Patrick
Focus: Arts; Lit; Sci
Periodical
Report & Transactions (weekly) 11890

Dickens Fellowship, 48 Doughty St, London WC1N 2LF
T: (071) 4052127
Founded: 1902; Members: 5000
Pres: Prof. J. Lokin; Gen Secr: Edward G. Preston
Focus: Lit
Periodical
The Dickensian (3 times annually) . . . 11891

Diplomatic and Commonwealth Writers Association of Britain (DCWAB), 8 Plender Court, College Pl, London NW1 0DH
Founded: 1960
Focus: Lit 11892

Direct Investigation Group on Aerial Phenomena (DIGAP), 24 Bent Fold Dr, Unsworth, Bury BL9 8NG
T: (061) 7664560
Founded: 1953; Members: 20
Focus: Physics 11893

Disabled Living Foundation (DLF), 380-384 Harrow Rd, London W9 2HU
T: (071) 2896111; Fax: (071) 2662922
Founded: 1970
Pres: Lady Hamilton; Gen Secr: Charles Moore
Focus: Educ Handic
Periodicals
DLF Hamilton Index
Wheelchair Information Pack (weekly) . . 11894

Disinfected Mail Study Circle (DMSC), 25 Sinclair Grove, London NW11 9JH
T: (081) 4559190
Founded: 1974; Members: 130
Pres: V. Denis Vandervelde; Gen Secr: Hans Smith
Focus: Med; Cultur Hist; Public Health
Periodical
Pratique (weekly) 11895

Distributive Trades Education and Training Council, 56 Russell Sq, London WC1M 4HP
T: (071) 6369811
Focus: Adult Educ 11896

District Planning Officers Society (DPOS), c/o Wealden District Council, Council Offices, Pine Grove, Crowborough TN6 1DH
T: (0892) 602405
Founded: 1974; Members: 320
Gen Secr: A.J. Brown
Focus: Urban Plan
Periodical
Newsletter (weekly) 11897

Dolmetsch Foundation, Lavant Park Farm, West Lavant, Chichester PO18 9AH
T: (0243) 528612
Founded: 1928; Members: 600
Pres: Lord Bridges; Gen Secr: Peter D. Ritchie
Focus: Music
Periodicals
Bulletin Newsletter (3 times annually)
The Consort (weekly) 11898

Dolmetsch Historical Dance Society (DHDS), 28 Mead Crescent, Swaythling, Southampton, Hants
T: (0703) 553252
Founded: 1970; Members: 120
Focus: Perf Arts
Periodical
Historical Dance (weekly) 11899

The Donizetti Society, 56 Harbut Rd, London SW11 2RB
T: (071) 2284928
Founded: 1973; Members: 820
Focus: Music
Periodical
Journal (bi-annually) 11900

Dorothy L. Sayers Society, Rose Cottage, Malthouse Lane, Hurstpierpoint BN6 9JY
T: (0273) 833444
Founded: 1976; Members: 480
Gen Secr: Christopher J. Dean
Focus: Hist; Lit
Periodicals
Annual Seminar Proceedings
Bulletin (weekly) 11901

Dorset Natural History and Archaeological Society, 66 High West St, Dorchester DT1 1XA
T: (0305) 262735
Founded: 1875; Members: 2000
Focus: Archeol; Hist; Nat Sci
Periodicals
Dorset Monographs (weekly)
Proceedings of the Dorset Natural History and Archaeological Society (weekly) 11902

Dorset Record Society (DRS), c/o Dorset County Museum, High West St, Dorchester DT1 1XA
Founded: 1962; Members: 1900
Focus: Hist 11903

The Dozenal Society of Great Britain (DSGB), Walnut Bank, Underhill, Moulsford OX10 9JH
Founded: 1958; Members: 70
Gen Secr: Arthur F. Whillock
Focus: Standards
Periodical
The Dozenal Journal (weekly) 11904

Drama Association of Wales (DAW) / Cymdeithas Ddrama Cymru, c/o Library, Singleton Rd, Cardiff CF2 2ET
T: (0222) 452200; Fax: (0222) 452277
Founded: 1965; Members: 700
Gen Secr: Heidi T. Jones
Focus: Perf Arts
Periodical
DAWN: DAWN Plus 11905

Drama Board Association, Witches Broom, Yester Park, Chilehurst BR7 5DQ
Focus: Perf Arts 11906

Drop Forging Research Association (DFRA), Shepherd St, Sheffield S3 7BA
T: (0742) 27463
Founded: 1960; Members: 84
Focus: Metallurgy
Periodicals
Annual Report (weekly)
Newsletter (weekly) 11907

Dryland Professional Network, c/o IIED, 3 Endsleigh St, London WC1
Tx: 261681
Founded: 1987; Members: 56
Gen Secr: Dr. Camilla Toulmin
Focus: Geology 11908

Dry Stone Walling Association, 1 The Old School, Pant Glas, Oswestry SY10 7HS
T: (0691) 654019
Founded: 1968; Members: 300
Focus: Civil Eng 11909

The Dugdale Society, c/o The Shakespeare Centre, Stratford-upon-Avon CV37 6QW
T: (0789) 204016
Founded: 1920; Members: 317
Pres: Sir William Dugdale
Focus: Hist 11910

Dumfriesshire and Galloway Natural History and Antiquarian Society, 9 Gillbrae Court, Dumfries DG1 4DH
T: 5769
Founded: 1862; Members: 400
Focus: Archeol; Hist 11911

Dunedin Society, 104 Hill St, Glasgow G3
T: (041) 3320727
Founded: 1911
Focus: Arts 11912

Durham County Local History Society, c/o Durham County Record Office, County Hall, Durham DH1 5UL
T: (091) 3833575
Founded: 1964; Members: 210
Pres: H.J. Smith; Gen Secr: J. Gill
Focus: Hist 11913

Dyslexia Institute, 133 Gresham Rd, Staines TW18 2AJ
T: (0784) 463851; Fax: (0784) 460747
Founded: 1972; Members: 250
Focus: Educ Handic
Periodicals
Dyslexia Review (weekly)
Parents Guide-Special Needs-Special Provision (weekly) 11914

Early English Text Society (EETS), Christ Church, Oxford OX1 1DP
Founded: 1864; Members: 1090
Pres: Prof. John Burrow; Gen Secr: R.F.S. Hamer
Focus: Lit 11915

Eastern Arts Association, 8-9 Bridge St, Cambridge CB2 1UH
T: (0223) 357596
Founded: 1971
Focus: Arts 11916

East Hertfordshire Archaeological Society (EHAS), 1 Marsh Lane, Stanstead Abbots, Ware SG12 8HH
T: (0920) 870664
Founded: 1898; Members: 200
Focus: Archeol
Periodical
Hertfordshire Archaeology (bi-annually) . . 11917

East London History Society, 18 Hawkdene Rd, London E4 7PF
T: (081) 5244506
Founded: 1952; Members: 165
Focus: Hist 11918

East Lothian Antiquarian and Field Naturalists' Society, Inchgarth, East Links, Dunbar EH42 1LT
T: (0368) 63335
Founded: 1924; Members: 250
Pres: Prof. R. Mitchison; Gen Secr: Stephen A. Bunyan
Focus: Archeol; Hist; Nat Sci
Periodical
Transactions (bi-annually) 11919

East Midlands Arts Board, Mountfields House, Forest Rd, Loughborough LE11 3HU
T: (0509) 218292; Fax: (0509) 262214
Founded: 1969
Focus: Arts 11920

East of England Agricultural Society, East of England Showground, Peterborough PE2 0XE
T: (0733) 234451
Founded: 1797; Members: 7267
Focus: Agri
Periodical
East of England Journal (weekly) . . . 11921

East Riding Archaeological Society (ERAS), 7 Mill Rd, Swanland, Hull HU14 3PJ
T: (0482) 632244
Founded: 1960; Members: 230
Focus: Archeol 11922

Ecclesiastical History Society (E.H.S.), c/o Dept of Medieval History, University of Glasgow, Glasgow G12 8QQ
Founded: 1961; Members: 850
Pres: Dr. D.M. Thompson; Gen Secr: M.J. Kennedy
Focus: Hist; Rel & Theol
Periodical
Studies in Church History (weekly) . . . 11923

Ecclesiological Society, Saint Andrew-by-the-Wardrobe, Queen Victoria St, London EC4V 5DE
Founded: 1839; Members: 400
Pres: S.E. Dykes Bower; Gen Secr: Prof. K.H. Murta
Focus: Rel & Theol
Periodicals
Ecclesiology Today (3 times annually)
Papers (3 times annually) 11924

Ecological Physics Research Group, c/o Cranfield Institute of Technology, Cranfield, Bedford MK43 0AL
T: (0234) 750993
Focus: Ecology; Physics 11925

Economic and Social Science Research Association (ESSRA), 177 Vauxhall Bridge Rd, London SW1V 1ER
T: (071) 8344979
Founded: 1969
Focus: Econ; Sociology 11926

Economic History Society, c/o Dept of Economics and Related Studies, University of York, Heslington YO1 5DD
Founded: 1927; Members: 2000
Pres: Prof. B.B. Supple; Gen Secr: Dr. D.T. Jenkins
Focus: Hist; Econ
Periodical
The Economic History Review (weekly) . 11927

The Economics and Business Education Association, 1a Keymer Rd, Hassocks BN6 6AD
T: (0273) 846033; Fax: (0273) 844646
Founded: 1948; Members: 3200
Pres: Chris Marsden
Focus: Econ; Educ
Periodical
Economics and Business Education (weekly) 11928

Economic Study Association, 8 Rathbone Pl, London W1
T: (071) 6360467
Founded: 1966
Focus: Econ 11929

Economists Advisory Group, 54b Tottenham Court Rd, London W1
T: (071) 3234923
Focus: Econ 11930

Edinburgh Architectural Association (EAA), 15 Rutland Sq, Edinburgh EH1 2BE
T: (031) 2297205
Founded: 1858
Focus: Archit 11931

Edinburgh Bibliographical Society (E.B.S.), c/o New College Library, Mound Pl, Edinburgh EH1 2LU
T: (031) 2258400 ext 256; Fax: (031) 2200952
Founded: 1890; Members: 200
Pres: R.A. Paxton; Gen Secr: M.C.T. Simpson
Focus: Libraries & Bk Sci
Periodical
Transactions 11932

Edinburgh Festival Society, 21 Market St, Edinburgh EH1 1BW
T: (031) 2264001
Founded: 1947
Gen Secr: Frank Dunlop
Focus: Music
Periodicals
Festival Preliminary Brochure (weekly)
Festival Programme Brochure (weekly)
Souvenir Festival Guide (weekly) 11933

The Edinburgh Highland Reel and Strathspey Society, 78 Milton Rd West, Edinburgh EH15 1QY
T: (031) 6695927
Founded: 1881; Members: 120
Focus: Music 11934

Edinburgh Mathematical Society (E.M.S.), c/o Dept of Mathematics and Statistics, King's Bldgs, Mayfield Rd, Edinburgh EH9 3JZ
T: (031) 6505060; Fax: (031) 6506553
Founded: 1883; Members: 310
Pres: Prof. A.D.D. Craik; Gen Secr: Dr. C. J. Smyth
Focus: Math
Periodical
Proceedings (3 times annually) 11935

Edinburgh Medical Missionary Society (EMMS), 7 Washington Lane, Edinburgh EH11 2HA
T: (031) 3133828; Fax: (031) 3134662
Founded: 1841
Pres: Dr. Andrew Young; Gen Secr: Fred M. Aitken
Focus: Med
Periodical
Healing Hand (3 times annually) 11936

Edinburgh Obstetrical Society, c/o Simpson Memorial Maternity Pavilion, Royal Infirmary, Edinburgh EH3 9YW
T: (031) 2292477
Founded: 1840; Members: 100
Focus: Gynecology 11937

Edinburgh Royal Choral Union (ERCU), 43 Inverleith Gardens, Edinburgh EH3 5PR
T: (031) 5523874; Fax: (031) 4473398
Founded: 1858; Members: 100
Focus: Music 11938

Edinburgh Sir Walter Scott Club, 37 Queen St, Edinburgh
T: (031) 2252914
Founded: 1894; Members: 475
Focus: Lit 11939

Edmonton Hundred Historical Society, c/o Town Hall, Green Lanes, London N13
T: (081) 8866555 ext 15
Founded: 1936; Members: 356
Focus: Hist 11940

Educational Centres Association (ECA), c/o The Chequer Centre, Chequer St, London EC1
T: (071) 2514158
Founded: 1921; Members: 100
Focus: Educ 11941

Educational Development Association (EDA), The Castle, Wisbech, Cambs
T: 61515
Founded: 1888; Members: 4000
Focus: Educ 11942

Educational Drama Association (E.D.A.), c/o Vauxhall Gardens School, Barrack St, Birmingham B7 4HA
Founded: 1948; Members: 380
Focus: Educ; Perf Arts 11943

The Educational Institute of Design, Craft and Technology (EIDCT), c/o Fred R. Willmore, 34 Burton St, Melton Mowbray LE13 1AF
T: (0664) 69754
Founded: 1891; Members: 1200
Gen Secr: Fred R. Willmore
Focus: Educ; Graphic & Dec Arts, Design; Eng
Periodicals
CDT Yearbook and Directory (bi-annually)
Designing and Making (3 times annually) 11944

Educational Institute of Scotland (EIS), 46 Moray Pl, Edinburgh EH3 6BH
T: (031) 2256244; Fax: (031) 2203151
Founded: 1847; Members: 47500
Pres: Willie Hart; Gen Secr: Jim Martin
Focus: Educ
Periodical
Scottish Educational Journal (weekly) . . 11945

Educational Television Association (E.T.A.), The King's Manor, Exhibition Sq, York YO1 2EP
T: (0904) 433929; Fax: (0904) 433929
Founded: 1965; Members: 200
Pres: Robert McPherson; Gen Secr: Josie Key
Focus: Educ
Periodical
Journal of Educational Television (2-3 times annually) 11946

Edwardian Studies Association, 125 Markyate Rd, Dagenham, Essex
Founded: 1975
Focus: Hist 11947

The Egypt Exploration Society (E.E.S.), 3 Doughty Mews, London WC1N 2PG
T: (071) 2421880
Founded: 1882; Members: 2700
Gen Secr: Dr. Patricia Spencer
Focus: Ethnology
Periodicals
Graeco-Roman Memoirs (weekly)
Journal of Egyptian Archaeology (weekly) 11948

Eighteen Nineties Society, 3 Kemplay Rd, London NW3 1TA
T: (071) 7945030
Founded: 1963; Members: 80
Focus: Hist 11949

Electoral Reform Society of Great Britain and Ireland, 6 Chancel St, London SE1 0UU
T: (071) 9281622; Fax: (071) 9284366
Founded: 1884; Members: 2000
Pres: Baroness Seear; Gen Secr: E.M. Syddique
Focus: Law
Periodical
Representation (weekly) 11950

Electric Railway Society (ERS), 17 Catherine Dr, Sutton Coldfield B73 6AX
T: (021) 3548332
Founded: 1946; Members: 350
Gen Secr: Dr. I.D.O. Frew
Focus: Transport
Periodical
Journal (weekly) 11951

Electric Vehicle Development Group (EVDG), c/o The City University, Northampton, London ECIV 0HB
T: (071) 2532432; Tx: 262284
Founded: 1978; Members: 70
Focus: Electric Eng
Periodical
Electric Vehicle Developments (weekly) . 11952

Electroencephalographic Society (EEG Society), c/o Dept of Anatomy, Medical School, Birmingham University, Birmingham B15 2TJ
T: (021) 4721301 ext 3346-3367
Founded: 1943; Members: 170
Focus: Neurology
Periodical
The EEG Journal (weekly) 11953

Electro Physiological Technologists' Association (E.P.T.A.), c/o EEG Dept, Hospital for Sick Children, London WC1N 3JH
T: (071) 4059200 ext 5154
Founded: 1949; Members: 562
Pres: Dr. R. Cooper; Gen Secr: J. Chaloner
Focus: Physiology
Periodical
Journal of Electrophysiological Technology (weekly) 11954

Elgar Society, The Ridge, Pilgrim Pl, Woldingham CR3 7AG
T: 3252
Founded: 1951; Members: 730
Focus: Music 11955

Elgar Society (London), 7 Batchworth Lane, Northwood HA6 3AU
T: 22010
Founded: 1971; Members: 690
Focus: Music 11956

Engineering Committee on Oceanic Resources (ECOR), 76 Mark Ln, London EC3R 7JN
T: (071) 4810750; Fax: (071) 4814001
Founded: 1971
Focus: Eng
Periodical
ECOR Newsletter (weekly) 11957

The Engineering Council, 10 Maltravers St, London WC2R 3ER
T: (071) 2407891; Tx: 297177; Fax: (071) 2407517
Founded: 1981
Pres: Duke of Kent; Gen Secr: Lewis Chelton
Focus: Eng
Periodical
Newsletter (weekly) 11958

Engineering Equipment and Materials Users' Association (E.E.M.U.A.), 14-15 Belgrave Sq, London SW1X 8PS
T: (071) 2355316; Fax: (071) 2456937
Founded: 1950; Members: 18
Gen Secr: R.W. Snudden
Focus: Energy
Periodical
E.E.M.U.A. Publications (weekly) 11959

English Association, The Vicarage, Priory Gardens, London W4 1TT
T: (081) 9954236
Founded: 1906; Members: 1500
Pres: Prof. Martin Dodsworth; Gen Secr: Dr. Ruth Fairbanks-Joseph
Focus: Ling; Lit
Periodicals
English (3 times annually)
Essays and Studies (weekly)
The Year's Work in English Studies (weekly) 11960

The English Folk Dance and Song Society (EFDSS), 2 Regent's Park Rd, London NW1 7AY
T: (071) 4852206/08
Founded: 1932; Members: 8000
Pres: Princess Margaret; Gen Secr: Brenda Godrich
Focus: Music
Periodicals
English Dance and Song (3 times annually)
Folk Directory (weekly)
Folk Music Journal (weekly) 11961

English Goethe Society (EGS), c/o Dept of German, University College, Gower St, London WC1E 6BT
T: (071) 3877050
Founded: 1886
Focus: Lit
Periodical
Publications of the English Goethe Society (weekly) 11962

English Guernsey Cattle Society (EGCS), Bury Farm, Pednor Rd, Chesham HP5 2LA
Founded: 1884; Members: 800
Focus: Agri
Periodicals
EGCS Herdbook & Type & Production Register (weekly)
Guernsey Breeders' Journal (weekly) . . 11963

English Place-Name Society (EPNS), c/o School of English Studies, University of Nottingham, Nottingham NG7 2RD
T: (0602) 519119
Founded: 1923; Members: 650
Pres: Dr. Margaret Gelling
Focus: Ling
Periodical
Journal and Volume (weekly) 11964

English Speaking Board (International), 32 Norwood Av, Southport PR9 7EG
T: (0704) 231366
Founded: 1953
Pres: Arthur Ridings
Focus: Ling 11965

English-Speaking Board (International) (E.S.B.), 26a Princes St, Southport PR8 1EQ
T: (0704) 501730
Founded: 1953; Members: 1500
Pres: Christabel Burniston; Gen Secr: Arthur Ridings
Focus: Ling
Periodicals
Spoken English (3 times annually)
Spoken English (3 times annually) . . . 11966

English Spelling Association (ISA), 11 First St, London SW3
Founded: 1974; Members: 77
Focus: Ling 11967

English Westerners Society (EWS), 39a Kildare Terrace, London W2
Founded: 1954; Members: 300
Focus: Hist
Periodicals
Brand Book (1-2 times annually)
Tally Sheet (3 times annually) 11968

Environmental Council, 80 York Way, London N1 9AG
T: (071) 2784736; Fax: (071) 8379688
Founded: 1969
Gen Secr: Steve Robinson
Focus: Ecology
Periodical
Habitat (10 times annually) 11969

The Ephemera Society (Ephsoc), 84 Marylebone High St, London W1M 3DE
T: (071) 4874669; Fax: (071) 9357305
Founded: 1975; Members: 950
Focus: Cultur Hist
Periodical
The Ephemerist (weekly) 11970

ERA Technology (ERA), Cleeve Rd, Leatherhead KT22 7SA
T: (0372) 374151; Tx: 264045; Fax: (0372) 374496
Founded: 1920; Members: 390
Gen Secr: M.J. Withers
Focus: Electric Eng
Periodical
ERA Technology News (weekly) 11971

Ergonomics Society (ES), Devonshire House, Devonshire Sq, Loughborough LE11 3DW
T: (0509) 234904; Fax: (0509) 234904
Founded: 1949; Members: 1023
Pres: R. Sell; Gen Secr: T. Mayfield
Focus: Psych; Physiology; Anat; Eng
Periodicals
Applied Ergonomics (weekly)
Ergonomic Abstracts
Ergonomics (weekly)
The Ergonomist (weekly) 11972

Esperanto-Asocio de Britio (EAB), 140 Holland Park Av, London W11 4UF
T: (071) 7277821
Founded: 1976; Members: 1050
Focus: Ling
Periodical
La Brita Esperantisto (weekly) 11973

Esperanto Teachers Association (E.T.A.), 7 Fairacre Rd, Barwell, Leicester LE9 8HH
T: Earl Shilton 44403
Founded: 1939; Members: 160
Focus: Ling; Educ 11974

Essex Agricultural Society, The Showground, Great Leighs, Chelmsford, Essex
T: (024534) 259
Founded: 1858; Members: 1500
Focus: Agri
Periodical
Show Catalogue (weekly) 11975

Essex Archaeological and Historical Congress (EAHC), Low Hill House, Stratford Saint Mary CO7 6JX
T: (0206) 37239
Founded: 1964; Members: 70
Focus: Archeol; Hist
Periodicals
Essex Journal (3 times annually)
Newsletter (3 times annually) 11976

Essex Society for Archaeology and History (EAS), Hollytrees, Colchester CO1 1UG
Founded: 1852; Members: 466
Focus: Archeol; Hist
Periodicals
Essex Archaeological News (3 times annually)
Essex Archaeology and History (weekly) . 11977

Essex Wildlife Trust (ENT), c/o Fingringhoe Wick Conservation Centre, South Green Rd, Fingringhoe, Colchester CO5 7DN
T: (0206) 729678; Fax: (0206) 729298
Founded: 1959
Focus: Ecology
Periodical
Essex Wildlife Magazine (3 times annually) 11978

Estuarine and Brackish-Water Sciences Association (EBSA), c/o Dept of Zoology, University of Cambridge, Downing St, Cambridge CB2 3EJ
T: (0223) 58717
Founded: 1971; Members: 385
Focus: Water Res
Periodicals
E.B.S.A. Bulletin (weekly)
Estuarine, Coastal and Shelf Science (weekly) 11979

EUCLID, c/o Ministry of Defence, Room 2409, Main Bldg, Whitehall, London SW1A 2HB
T: (071) 2189000
Founded: 1990; Members: 13
Gen Secr: Alison Blake
Focus: Military Sci 11980

EUROMINERALS – Confederation of Learned/Engineering Societies in the Mineral Industry, c/o Institute of Mining and Metallurgy, 44 Portland Pl, London W1N 4BR
Founded: 1990
Gen Secr: Michael J. Jones
Focus: Geology 11981

Europeam Federation of Endocrine Societies, c/o Unit of Metabolic Medicine, Saint Mary's Hospital Medical School, London W2 1PG
T: (071) 7261353
Founded: 1984
Gen Secr: Prof. V.H.T. James
Focus: Endocrinology 11982

European Academy of Facial Surgery, 107 Harley St, London W1
T: (071) 9353171
Founded: 1977
Pres: T.R. Bull
Focus: Surgery
Periodical
Monographs in Facial Plastic Surgery (weekly) 11983

European Air Law Assciation (EALA), 66 Chartfield Av, London SW15 6HQ
T: (081) 7883513; Fax: (081) 7892467
Founded: 1988
Pres: Prof. P.D. Dagtoglou
Focus: Law 11984

European Airlines Electronic Committee (EAEC), c/o British Airways, S409 Heathrow Airport, London
Founded: 1953
Pres: Clive Baxter
Focus: Electronic Eng 11985

European Association for Aquatic Mammals (EAAM), Chartley House, Swannells Wood, Studham, Dunstable LU6 2QB
T: (0582) 872487; Fax: (0582) 872315
Founded: 1972
Gen Secr: V.J.A. Manton
Focus: Zoology
Periodical
EAAM Newsletter (weekly) 11986

European Association for Behaviour Therapy (EABT), c/o Dept of Psychology, Northwick Park Hospital, Watford Rd, Harrow
T: (081) 8692277; Fax: (081) 8793326
Founded: 1970
Gen Secr: Dr. R. Holland
Focus: Therapeutics
Periodical
European Behavioural Therapists (3 times annually) 11987

European Association for Cancer Research (EACR), c/o Cancer Research Campaign Laboratories, University of Nottingham, Nottingham NG7 2RD
Founded: 1968; Members: 580
Focus: Cell Biol & Cancer Res 11988

European Association for Cranio-Maxillo-Facial Surgery, c/o Dept of Oral and Maxillo-Facial Surgery, King's College Hospital, London SE5 8RX
T: (071) 2746222; Fax: (071) 3263185
Founded: 1970
Gen Secr: Prof. J.H. Sowray
Focus: Surgery
Periodical
Journal of Cranio-Maxillo-Facial Surgery . 11989

European Association for Evolutionary Political Economy (EAPE), c/o Dept of Economics and Government, Newcastle-upon-Tyne NE1 8ST
T: (091) 2326002 ext 3939; Fax: (091) 2358017
Founded: 1988
Gen Secr: G. Hodgson
Focus: Poli Sci; Econ 11990

European Association for Personnel Management (EAPM), c/o IPM, Camp Rd, Wimbledon, London SW19 4UX
T: (081) 9469100; Fax: (081) 9472570
Founded: 1962
Gen Secr: Brian Ward-Lilley
Focus: Business Admin
Periodical
EAPM Newsletter (weekly) 11991

European Association for the Conservation of Energy (EuroACE), 9 Sherlock Mews, London W1M 3RH
T: (071) 9351495; Fax: (071) 9358346
Founded: 1991
Gen Secr: Andrew Warren
Focus: Energy 11992

European Association for the Science of Air Pollution (EURASAP), c/o Air Pollution Group, Mechanical Engineering Dept, Imperial College, London SW7 2AZ
T: (071) 5895111; Tx: 929484; Fax: (071) 5847596
Founded: 1986
Focus: Ecology 11993

European Association of Cardiothoracic Anaesthesiologists (EACTA), c/o Papworth Hospital, Papworth Everard CB3 8RE
T: (0480) 830541; Fax: (0480) 831114
Founded: 1986
Gen Secr: Dr. R.D. Latimer
Focus: Anesthetics 11994

European Association of Fisheries Economists (EAFE), c/o CMRE, Portsmouth Polytechnic, Locksway Rd, Southsea PO4 8JF
T: (0705) 844085; Fax: (0705) 844037
Founded: 1989
Gen Secr: Arthur E. Neiland
Focus: Fisheries 11995

European Association of Internal Medicine, c/o Royal Sussex County Hospital, Eastern Rd, Brighton BN2 5BE
T: (0273) 696955; Fax: (0273) 684554
Founded: 1969
Gen Secr: Dr. Christopher Davidson
Focus: Intern Med 11996

European Association of Internal Medicine, c/o Royal Sussex County Hospital, Eastern Rd, Brighton BN2 5BE
T: (0273) 696955 ext 4088; Fax: (0273) 684554
Members: 612
Pres: Prof. Ugo Carcassi; Gen Secr: Dr. Christopher Davidson
Focus: Intern Med
Periodical
European Journal of Internal Medicine (weekly) 11997

European Association of Nuclear Medicine (EANM), c/o Institute of Nuclear Medicine, Middlesex Hospital Medical School, Mortimer St, London W1N
Founded: 1988
Gen Secr: Prof. Dr. P.J. Ell
Focus: Nucl Med
Periodical
EJNM Journal (weekly) 11998

European Association of Organic Geochemists, c/o Dept of Geology, University of Newcastle-upon-Tyne, Drummond Bldg, Newcastle-upon-Tyne NE1 7RU
T: (091) 2226000; Tx: 53654
Gen Secr: Prof. Stephen R. Larter
Focus: Chem 11999

European Association of Science Editors (EASE), 13 Wimpole St, London W1M 7AB
T: (071) 6363175
Members: 1982
Pres: Dr. John W. Glen; Gen Secr: Maeve O'Connor
Focus: Sci; Med
Periodical
European Science Editing 12000

European Association of South Asian Archaeologists, c/o Ancient India and Iran Trust, 23 Brooklamds Av, Cambridge CB2 2BG
T: (0223) 356841; Fax: (0223) 61125
Founded: 1970
Gen Secr: Dr. Bridget Allchin
Focus: Archeol 12001

European Atherosclerosis Society (EAS), c/o Dept of Internal Medicine, Medical School, Saint Bartholomew Hospital, London EC1A 7BE
T: (071) 6018431; Fax: (071) 6018042
Founded: 1964; Members: 82
Pres: Dr. David J. Galton
Focus: Intern Med 12002

The European-Atlantic Movement (TEAM), 7 Cathedral Close, Exeter EX1 1EZ
T: (0392) 74908
Founded: 1958; Members: 520
Gen Secr: V.J. Fitzgerald
Focus: Hist 12003

European Centre for Medium-Range Weather Forecasts, Shinfield Park, Reading RG2 9AX
T: (0734) 499000; Tx: 847908; Fax: (0734) 869450
Founded: 1973; Members: 18
Gen Secr: Dr. David Burridge
Focus: Geophys 12004

European Centre for Pollution Research, c/o Queen Mary and Westfield College, University of London, London E1 4NS
Gen Secr: B. Nath
Focus: Ecology 12005

European Centre for Traditional and Regional Cultures (ECTARC), Parade St, Llangollen LL20 8RB
T: (0978) 861514; Fax: (0978) 861804
Founded: 1988
Gen Secr: S. Davies
Focus: Cultur Hist 12006

European Cetacean Society (ECS), c/o Dept of Zoology, University of Oxford, South Parks Rd, Oxford OX1 3PS
T: (0865) 727984; Fax: (0865) 727984
Founded: 1987
Gen Secr: Dr. P.G.H. Evans
Focus: Zoology 12007

European Chemical Marketing Association (ECMRA), c/o FEMTA, Studio 38, Wimbledon Business Centre, Riverside Rd, London SW17 0BA
T: (081) 8790709; Fax: (081) 9472637
Founded: 1967
Focus: Marketing 12008

European Collection of Animal Cell Cultures (ECACC), c/o PHLS Centre for Applied Microbiology and Research, Porton Down, Salisbury SP4 0UG
T: (0980) 610391; Tx: 47683; Fax: (0980) 611096
Gen Secr: Dr. Alan Doyle
Focus: Microbio 12009

European College of Marketing and Marketing Research (ECMMR), 18 Saint Peters Steps, Brixham
Founded: 1969
Gen Secr: Dr. W. Newill
Focus: Marketing 12010

European Committee for Biological Effects of Carbon Black, 27 Chiltington Ray, Brighton BN2 8HB
Focus: Bio 12011

European Committee for Civil Engineers (ECCE), c/o ICE, Great George St, London SW1P 3AA
T: (071) 2227722; Tx: 935637; Fax: (071) 2227500
Founded: 1985
Gen Secr: G. Hornby
Focus: Eng 12012

European Communities Chemistry Council, c/o Royal Society of Chemistry, Burlington House, Piccadilly, London W1 0BN
T: (071) 4378656; Tx: 268001; Fax: (071) 4378883
Founded: 1973; Members: 19
Gen Secr: E.K. McEwan
Focus: Chem 12013

European Communities Clinical Chemistry Committee, Burlington House, Piccadilly, London W1 0BN
T: (071) 4378656; Tx: 268001; Fax: (071) 4378883
Founded: 1973; Members: 12
Focus: Chem 12014

European Confederation of Huntington Associations, c/o Dick Bates, 34a Station Rd, Hinckley LE10 1AP
T: (0455) 615558
Pres: R.V. Bates; Gen Secr: Carys Farmer-Little
Focus: Med 12015

European Consortium for Political Research (ECPR), c/o University of Essex, Wivenhoe Park, Colchester CO4 3SQ
T: (0206) 872501; Tx: 98440; Fax: (0206) 872500
Founded: 1970; Members: 163
Gen Secr: Prof. Kenneth Newton
Focus: Poli Sci 12016

European Consortium for Political Research (ECPR), c/o University of Essex, Wivenhoe Park, Colchester CO4 3SQ
T: (0206) 872501; Fax: (0206) 872500
Founded: 1970; Members: 175
Pres: Prof. Giorgio Freddi; Gen Secr: Prof. Ken Newton
Focus: Poli Sci
Periodicals
ECPR News (3 times annually)
European Journal of Political Research (8 times annually) 12017

European Contact Dermatitis Society (ECDS), c/o Dept of Dermatology, Royal Victoria Hospital, Grosvenor Rd, Belfast BT12 6BA
T: (0232) 240503
Gen Secr: Dr. Desmond Burrows
Focus: Derm 12018

European Contact Lens Society of Ophthalmologists (ECLSO), 143 Harley St, London W1N 1DJ
T: (071) 9350886
Founded: 1969
Pres: Dr. Jonathan Kersley
Focus: Ophthal 12019

European Council for Clinical and Laboratory Standardization (ECCLS), c/o School of Postgraduate Medical Education, University of Warwick, Coventry CV4 7AL
T: (0203) 523913; Fax: (0203) 461606
Founded: 1979
Gen Secr: Prof. N.K. Shinton
Focus: Standards 12020

European Council for Industrial Marketing, 18 Saint Peters Steps, Brixham, Devon
Founded: 1968
Gen Secr: Dr. A.J. Williamson
Focus: Business Admin 12021

European Council of International Schools (ECIS), 21b Lavant St, Petersfield GU32 3EL
T: (0730) 268244; Fax: (0730) 267914
Founded: 1965
Gen Secr: Michael Maybury
Focus: Educ 12022

European Council of Optics and Optometry (ECOO), 233-234 Blackfriars Rd, London SE1 8NW
Founded: 1960
Gen Secr: Ian Hunter
Focus: Optics 12023

European Council of Town Planners (ECTP), c/o Royal Town Planning Institute, 26 Portland Pl, London W1N 4BE
T: (071) 6369107; Fax: (071) 3231582
Founded: 1985
Gen Secr: David R. Fryer
Focus: Urban Plan 12024

European Dental Society, 10 Pike's End, Eastcote, Pinner HA5 2EX
T: (081) 8680837
Founded: 1943
Gen Secr: Dr. Caroline Shanbury
Focus: Dent 12025

European Design Education Network (EDEN), 120 Bothwell St, Glasgow G2 7JP
Gen Secr: A. Barron
Focus: Educ 12026

European Electrostatic Discharge Association (EESDA), c/o Marketplace PR, Technology House, Old Wokingham Rd, POB 90, Crowthorne RG11 6PX
T: (0344) 780022; Fax: (0344) 773231
Gen Secr: L. Ashburner
Focus: Materials Sci 12027

European Environmental Mutagen Society (EEMS), c/o Dept of Genetics, West Mains Rd, Edinburgh EH9 3JN
Focus: Genetics 12028

European Federation of Associations of Industrial Safety and Medical Officers, c/o Institution of Industrial Safety Officers, 222 Uppingham Rd, Leicester
T: (0533) 768424
Founded: 1952; Members: 18
Focus: Safety; Hygiene 12029

European Group for Organization Studies (EGOS), c/o Sociology Dept, Plymouth Polytechnic, Drake Circus, Plymouth PL4 8AA
T: (0752) 21312
Focus: Business Admin
Periodical
International Yearbook of Organization Studies (weekly) 12030

European Industrial Marketing Research Society, c/o EMA, 9 Aston Rd, Nuneaton, Warwicks
Focus: Marketing 12031

European Marketing Research Board International, 79-81 Uxbridge Rd, London W5 5SU
Focus: Marketing
Periodicals
AB TGI (weekly)
TGI (weekly)
Youth TGI (weekly) 12032

European Mechanics Council (Euromech), c/o Dept of Applied Mathematics and Theoretical Physics, University of Cambridge, Cambridge CB3 9EW
Fax: (0223) 312984
Founded: 1964
Pres: Prof. D.G. Crighton; Gen Secr: Prof. B. Lundberg
Focus: Physics 12033

The European Movement, 158 Buckingham Palace Rd, London SW1W 9TR
Fax: (071) 8248124
Founded: 1948
Gen Secr: S. Woodard
Focus: Econ
Periodicals
Enterprise Europe Bulletin (weekly)
Facts (weekly)
New Europe Papers (weekly) 12034

European Organization for Caries Research (ORCA), c/o Div of Oral Biology, Leeds Dental Institute, Clarendon Way, Leeds LS2 9LU
Fax: (0532) 336158
Founded: 1953; Members: 320
Focus: Dent
Periodical
Caries Research (weekly) 12035

European Orthodontic Society (E.O.S.), Flat 31, 49 Hallam St, London W1M 5LL
T: (071) 9352795; Fax: (071) 9352795
Founded: 1907; Members: 2000
Gen Secr: Prof. J.P. Moss
Focus: Dent
Periodical
European Journal of Orthodontics (weekly) 12036

European Psycho-Analytical Federation, 15 Elsworthy Rd, London NW3
T: (071) 7220338
Founded: 1967
Focus: Psych 12037

European Society for Clinical Investigation, c/o Guy's Hospital Medical School, Guy's Tower, London SE1 9RT
Founded: 1967; Members: 880
Focus: Med 12038

European Society for Comparative Endocrinology, c/o Dept of Physiology and Biochemistry, University of Reading, Reading
T: (0734) 85123
Founded: 1965
Focus: Endocrinology 12039

European Society of Nematologists, c/o Scottish Crop Research Institute, Invergowrie DD2 5DA
T: (08267) 731
Founded: 1955; Members: 510
Focus: Microbio 12040

European Space Association (ESA), 32 Carment Dr, Ardeer, Stevenston, Ayrs
T: 61740
Founded: 1977
Focus: Astronomy 12041

European Underseas Bio-Medical Society (EUBS), 6 Parkhill Av, Aberdeen AB2 0FP
Founded: 1971; Members: 300
Focus: Med 12042

Exeter Industrial Archaeology Group, c/o Dept of Economic History, Exeter University, Exeter EX4 4RJ
T: (0392) 77911
Founded: 1969; Members: 50
Focus: Archeol
Periodical
Bulletin (weekly) 12043

Experimental Psychology Society (EPS), c/o Dept of Psychology, Royal Holloway and Bedford New College, Egham TW20 0EX
Founded: 1947; Members: 480
Pres: Prof. P.M.A. Rabbitt; Gen Secr: Dr. E. Funnell
Focus: Psych
Periodical
Quarterly Journal of Experimental Psychology (weekly) 12044

Fabian Society, 11 Dartmouth St, London SW1H 9BN
T: (071) 2228877; Fax: (071) 9767153
Founded: 1884; Members: 4200
Pres: H.D. Hughes; Gen Secr: Simon Crine
Focus: Poli Sci
Periodicals
Fabian Discussion Papers (weekly)
Fabian Pamphlet (weekly)
Fabian Research Series (weekly)
Fabian Review (weekly)
Young Fabian Pamphlets (weekly) . . . 12045

Fabric Care Research Association (FCRA), c/o Forest House Laboratories, Knaresborough Rd, Harrogate HG2 7LZ
T: (0423) 885977; Fax: (0423) 880045
Founded: 1976; Members: 750
Pres: A.W. Jones; Gen Secr: C.J. Tebbs
Focus: Eng 12046

Faculty of Actuaries in Scotland, 23 Saint Andrew Sq, Edinburgh EH2 1AQ
T: (031) 5571575; Fax: (031) 5576702
Founded: 1856; Members: 1322
Gen Secr: W.W. Mair
Focus: Insurance
Periodicals
Transactions (weekly)
Year Book 12047

Faculty of Advocates, c/o Advocates Library, Parliament House, Edinburgh EH1 1RF
T: (031) 2265071
Founded: 1532
Pres: A.C.M. Johnston; Gen Secr: J.R. Doherty
Focus: Law 12048

Faculty of Astrological Studies, BM 7470, London WC1N 3XX
Fax: (071) 7006479
Founded: 1948
Focus: Parapsych 12049

Faculty of Building (FB), 12 High St, Elstree WD6 3EP
T: (081) 2070366
Founded: 1945; Members: 4850
Gen Secr: John E.G. Wilson
Focus: Archit; Civil Eng
Periodicals
Brunel Lectures (weekly)
Lord Bossom Lectures (weekly)
Royal Society Lectures (weekly) 12050

Faculty of Dental Surgery, c/o Royal College of Surgeons, 35-43 Lincoln's Inn Fields, London WC2A 3PN
T: (071) 4053474; Tx: 936573; Fax: (071) 8319438
Founded: 1947; Members: 6500
Pres: Kenneth Ray; Gen Secr: Albert de Looze
Focus: Surgery; Dent 12051

Faculty of Homoeopathy, c/o Royal London Homeopathic Hospital, Great Ormond St, London WC1N 3HR
T: (071) 8378833
Founded: 1950
Focus: Homeopathy
Periodical
The British Homeopathic Journal (weekly) 12052

Faculty of Royal Designers for Industry, c/o Royal Society of Arts, John Adam St, London WC2N 6EZ
T: (071) 9305115
Founded: 1936; Members: 120
Pres: Prof. Charles Handy; Gen Secr: Christopher Lucas
Focus: Graphic & Dec Arts, Design . . 12053

Faculty of Teachers in Commerce, 141 Bedford Rd, Sutton Coldfield B75 6DB
T: (021) 3781265
Founded: 1872; Members: 1450
Focus: Adult Educ
Periodical
Teacher in Commerce: The National Journal of the Commercial Teaching Profession (weekly) 12054

Fair Organ Preservation Society (FOPS), 47 Hawthorne Av, Hellesdon, Norwich, Norfolk
T: (0603) 412258
Founded: 1958; Members: 1600
Focus: Music 12055

The Family Planning Association (FPA), 27-35 Mortimer St, London W1N 7RJ
T: (071) 6367866; Fax: 4363288
Founded: 1930
Focus: Family Plan
Periodicals
Family Planning Today (weekly)
Latest Literature in Family Planning (weekly) 12056

Farm and Food Society (FAFS), 4 Willifield Way, London NW11 7XT
T: (081) 4550634
Founded: 1966
Focus: Agri; Food
Periodical
Farm and Food News (3 times annually) 12057

Farm Buildings Association (FBA), Roseleigh, Deddington, Oxford OX5 4SP
T: (0869) 38234
Founded: 1956; Members: 620
Focus: Agri; Archit 12058

Fauna and Flora Preservation Society (FFPS), 1 Kensington Gore, London SW7 2AR
T: (071) 8238899; Fax: (071) 8239690
Founded: 1903; Members: 4500
Pres: Sir Peter Scott
Focus: Ecology
Periodical
ORYX (weekly) 12059

Federal Trust for Education and Research, 158 Buckingham Palace Rd, London SW1W 9TR
T: (071) 2599990; Fax: (071) 2599505
Founded: 1945
Gen Secr: Andrew Duff
Focus: Educ 12060

Federation for Ulster Local Studies, c/o Dr. Bill Crawford, 8 Fitzwilliam St, Belfast BT9 6AW
T: (0232) 235254
Members: 191
Gen Secr: Kathleen Gormley
Focus: Hist
Periodical
Ulster Local Studies (weekly) 12061

Fédération Internationale d'Education Physique (FIEP), 4 Cleevecroft Av, Bishops Cleeve, Cheltenham GL52 4JZ
Founded: 1923
Pres: John C. Andrews
Focus: Sports
Periodical
FIEP Bulletin (weekly) 12062

Federation of British Artists, 17 Carlton House Terrace, London SW1Y 5BD
T: (071) 9306844; Fax: (071) 8397830
Founded: 1886; Members: 56
Focus: Fine Arts; Arts
Periodical
The Art Magazine (weekly) 12063

Federation of European Veterinaries in Industry and Research (FEVIR), Squibb House, Animal Health Div, 141-149 Staines Rd, Hounslow TW3 3JB
Founded: 1975
Focus: Vet Med 12064

Federation of Family History Societies (FFHS), c/o The Benson Room, Birmingham and Midland Institute, Margaret St, Birmingham B3 3BS
Founded: 1974; Members: 165
Focus: Genealogy
Periodical
Family History News & Digest (weekly) . 12065

Federation of London Area Dental Committees, Tavistock House North, Tavistock Sq, London
T: (071) 3871493
Focus: Dent 12066

Federation of Old Cornwall Societies (FOCS), Tremarsh, Launceston, Cornwall
T: (0566) 3509
Founded: 1924
Focus: Hist
Periodical
Old Cornwall (weekly) 12067

Federation of Veterinarians of the EEC, 32 Belgrave Sq, London SW1X 8AP
Founded: 1961
Focus: Vet Med 12068

The Federation of Zoological Gardens of Great Britain and Ireland, c/o Zoological Gardens, Regent's Park, London NW1 4RY
T: (071) 5860230; Fax: (071) 7224427
Founded: 1965; Members: 52
Pres: Prof. R.J. Wheater; Gen Secr: P.J. Olney
Focus: Zoology
Periodical
Zoo Federation News (3 times annually) 12069

The Fellowship for Freedom in Medicine (FFM), Stockburry House, Church St, Storrington RH20 4LD
T: (09066) 2679
Founded: 1948; Members: 400
Focus: Med 12070

Fellowship of British Christian Esperantists (FBCE), 5 Osborne Close, Hornchurch RM11 1HJ
Founded: 1953
Focus: Ling 12071

Fellowship of Christian Writers, 104 Evelyn St, London SE8 5DD
T: (081) 6925787
Founded: 1969; Members: 400
Focus: Lit
Periodical
Newsheet (weekly) 12072

Fellowship of Postgraduate Medicine, 6 Saint Andrews Pl, London NW1 4LB
T: (071) 9355556
Founded: 1919
Focus: Med
Periodical
Postgraduate Medical Journal (weekly) . 12073

Field Studies Council (FSC), Preston Montford, Montford Bridge, Shrewsburry SY4 1HW
T: (0743) 850674
Founded: 1943; Members: 6000
Pres: I. Mercer; Gen Secr: A.D. Thomas
Focus: Nat Sci
Periodicals
Annual Report (weekly)
Field Studies (weekly)
Programmes of Courses 12074

Filtration Society (FS), 7 Manor Close, Oadby, Leicester LE2 4FE
T: (0533) 720536; Tx: 34319
Founded: 1964; Members: 1300
Gen Secr: Peter Swift
Focus: Ecology
Periodical
Filtration & Separation Magazine (weekly) 12075

Fire Protection Association (FPA), Aldermary House, Queen St, London EC4N 1TJ
T: (071) 2485222
Founded: 1946; Members: 6300
Focus: Eng 12076

Fisheries Society of the British Isles (FSBI), 141 Newmarket Rd, Cambridge CB5 8HA
Founded: 1967; Members: 293
Focus: Zoology

Periodicals
F.S.B.I. Newsletter (3 times annually)
Journal of Fish Biology (weekly) 12077

Flag Institute, 10 Vicarage Rd, Chester CH2 3HZ
T: (0244) 351335; Fax: (0244) 341894
Founded: 1971; Members: 350
Pres: B.E. Nicolls; Gen Secr: W.G. Crampton
Focus: Hist
Periodical
Flagmaster (weekly) 12078

Flintshire Historical Society (FHS), 50 Hafod Park, Mold, Clwyd CH7 1QW
T: (0352) 752582
Founded: 1911; Members: 450
Gen Secr: I.M. Read
Focus: Hist; Archeol
Periodical
Journal 12079

Flora Europaea Organization (FEO), c/o Dept of Botany, Liverpool Museum, William Brown St, Liverpool L3 8EN
T: (051) 2070001; Fax: (051) 2981395
Founded: 1956
Gen Secr: Dr. J.R. Edmondson
Focus: Botany
Periodical
Notulae Systematicae ad Floram Europaeam Spectantes (weekly) 12080

Flour Miling and Baking Research Association (FMBRA), Chorleywood, Rickmansworth WD3 5SH
T: 4111
Founded: 1967; Members: 600
Focus: Home Econ 12081

Folklore Society, c/o University College, Gower St, London WC1E 6BT
T: (071) 3875894
Founded: 1878; Members: 1000
Pres: Dr. Jacqueline Simpson; Gen Secr: Thomas W. Brown
Focus: Ethnology
Periodical
Folklore (weekly) 12082

Food Education Society (FES), 160 Piccadilly, London W1 0NQ
T: (071) 5841001
Founded: 1908
Focus: Food 12083

Foreign Affairs Circle, Church House, Petersham, Richmond TW10 7AA
T: (0748) 9484833
Founded: 1962; Members: 300
Focus: Poli Sci 12084

The Forensic Science Society, 18a Mount Parade, Harrogate HG1 1BX
T: (0423) 506068; Fax: (0423) 530948
Founded: 1959
Focus: Law
Periodical
Journal (weekly) 12085

Fort Cumberland and Portsmouth Militaria Society (FC & PMS), 49 Lichfield Rd, Portsmouth PO3 6DD
T: (0705) 668981, 754003
Founded: 1966
Focus: Military Sci; Hist
Periodicals
Point News: Historical Newssheets (weekly)
Point Papers Series (weekly) 12086

Fortress Study Group (FSG), Blackwater Forge House, Blackwater, Newport PO30 3BJ
T: (0983) 526207
Founded: 1975; Members: 500
Gen Secr: D.W. Quarmby
Focus: Military Sci
Periodicals
Casemate (3 times annually)
Fort (weekly) 12087

Foundation for Business Responsibilities (FBR), 40 Doughty St, London WC1N 2LF
T: (071) 4055195
Founded: 1966
Focus: Econ
Periodicals
Sir Frederick Hooper Essay Award (weekly)
Sir George Earle Memorial Lecture (weekly)
. . . . 12088

Foundation for the Study of Infant Deaths / Cot Death Research and Support, 35 Belgrave Sq, London SW1X 8QB
T: (071) 2350965; Fax: (071) 8231986
Founded: 1971; Members: 150
Gen Secr: Joyce Epstein
Focus: Pediatrics
Periodical
Newsletter (3 times annually) 12089

The Francis Bacon Society, Canonbury Tower, Islington, London N1 2NQ
T: (071) 3701233
Founded: 1886
Pres: Francis Cowper; Gen Secr: Clive McNeir
Focus: Lit
Periodicals
Baconiana (weekly)
Jottings (weekly) 12090

Franco-British Society, Room 623, Linen Hall, 162-168 Regent St, London W1R 5TB
T: (071) 7340815
Founded: 1944
Pres: Marquis of Lansdowne; Gen Secr: Marian Clarke
Focus: Int'l Relat
Periodical
Newsletter (weekly) 12091

Freshwater Biological Association, Ferry House, Ambleside LA22 0LP
T: (05394) 42468; Fax: (05394) 46914
Founded: 1929; Members: 2000
Pres: J. Jeffery; Gen Secr: Prof. J.G. Jones
Focus: Bio; Hydrology
Periodicals
Freshwater Forum (3 times annually)
Occasional Publications (weekly)
Scientific Publications (weekly) 12092

Friends Historical Society, Friends House, Euston Rd, London NW1 2BJ
Founded: 1903; Members: 400
Focus: Hist
Periodical
Journal (weekly) 12093

Friends of The National Libraries (FNL), c/o The British Library, Great Russell St, London WC1B 3DG
T: (071) 3237559
Founded: 1931; Members: 850
Pres: Lord Egremont; Gen Secr: A. Payne
Focus: Libraries & Bk Sci
Periodical
Friends of the National Libraries (weekly) 12094

Furniture History Society (FHS), c/o Dept of Furniture and Woodwork, Victoria and Albert Museum, London SW7 2RL
Founded: 1964; Members: 1350
Focus: Hist
Periodical
Furniture History (weekly) 12095

Furniture Industry Research Association (FIRA), Maxwell Rd, Stevenage SG1 2EW
T: (0438) 313433; Fax: (0438) 727607
Founded: 1961; Members: 950
Pres: A.F. Robinson; Gen Secr: D.M. Heughan
Focus: Eng
Periodical
FIRA Bulletin (weekly) 12096

Gaelic League of Scotland, c/o Highlanders Institute, 34 Berkley St, Glasgow G3
T: (041) 2214543
Founded: 1912; Members: 300
Focus: Ling 12097

The Galpin Society for the Study of Musical Instruments, 38 Eastfield Rd, Western Park, Leicester LE3 6FE
T: (0533) 855136
Founded: 1946; Members: 1500
Focus: Music 12098

Galton Institute, 19 Northfields Prospect, Northfields, London SW18 1PE
T: (081) 8747257
Founded: 1907; Members: 400
Pres: Prof. G. Ainsworth Harrison; Gen Secr: L. Brooks
Focus: Genetics
Periodical
Symposium Proceedings (weekly) . . . 12099

The Game Conservancy Trust, Burgate Manor, Fordingbridge SP6 1EF
T: (0425) 652381; Fax: (0425) 655848
Founded: 1970; Members: 23000
Pres: Duke of Westminster; Gen Secr: Dr. G.R. Potts
Focus: Ecology
Periodicals
Game Conservancy Annual Review (weekly)
Game Conservancy Newsletter (3 times annually)
. . . . 12100

Gemmological Association and Gem Testing Laboratory of Great Britain, 27 Greville St, London EC1N 8SU
T: (071) 4043334; Fax: (071) 4048843
Founded: 1931; Members: 4000
Focus: Mineralogy
Periodical
Journal of Gemmology (weekly) 12101

General Council and Register of Osteopaths (G.C.R.O.), 56 London St., Reading RG1 4SQ
T: (0734) 576585; Fax: (0734) 566246
Founded: 1936; Members: 1693
Focus: Med; Pathology 12102

General Council of the Bar, 3 Bedford Row, London WC1R 4DB
T: (071) 2420082; Fax: (071) 8319217
Founded: 1883
Pres: John Rowe
Focus: Law
Periodicals
Code of Conduct
Counsel (weekly) 12103

General Dental Council (G.D.C.), 37 Wimpole St, London W1M 8DQ
T: (071) 4862171; Fax: (071) 2243294
Founded: 1956; Members: 50
Pres: Prof. D.K. Mason
Focus: Dent
Periodicals
The Dentists Register (weekly)
The Rolls of Dental Auxiliaries (weekly) . 12104

General Dental Practitioner's Association (GDPA), 49 Cromwell Grove, Levenshulme, Manchester M19 3QD
T: (061) 2247442
Founded: 1954; Members: 2000
Focus: Dent
Periodical
The Probe (weekly) 12105

General Medical Council, 44 Hallam St, London W1N 6AE
T: (071) 5807642
Founded: 1858; Members: 91
Pres: Prof. Sir Robert Kilpatrick
Focus: Med
Periodical
The Medical Register (weekly) 12106

General Studies Association (G.S.A.), Swallowhurst, Sinnington, York Y06 6SH
T: Kirkbymoorside 31962
Founded: 1962; Members: 900
Focus: Educ 12107

The General Teaching Council for Scotland, 5 Royal Terrace, Edinburgh EH7 5AF
T: (031) 5560072; Fax: (031) 5576773
Founded: 1965; Members: 49
Focus: Educ
Periodical
Link (3 times annually) 12108

The Genetical Society of Great Britain, c/o MRC Human Genetics Unit, Western General Hospital, Crewe Rd, Edinburgh EH4 2XU
T: (031) 3322471; Fax: (031) 3432620
Founded: 1919; Members: 1600
Pres: Prof. P. Nurse; Gen Secr: Prof. J.P.W. Young
Focus: Genetics
Periodicals
Genes and Development (weekly)
Heredity (weekly) 12109

The Geographical Association (G.A.), 343 Fulwood Rd, Sheffield S10 3BP
T: (0742) 670666; Fax: (0742) 670688
Founded: 1893; Members: 10500
Pres: Prof. A.S. Goudie; Gen Secr: F.M. Sear
Focus: Geography
Periodicals
Geography (weekly)
Primary Geographer (weekly)
Teaching Geography (weekly) 12110

The Geological Society, Burlington House, Piccadilly, London W1V 0JU
T: (071) 4349944; Fax: (071) 4398975
Founded: 1807; Members: 7700
Pres: Prof. C.D. Curtis; Gen Secr: R.M. Bateman
Focus: Geology
Periodicals
Journal of Marine and Petroleum Geology (weekly)
Journal of the Geological Society (weekly)
Quarterly Journal of Engineering Geology (weekly)
. . . . 12111

Geologists' Association (G.A.), Burlington House, Piccadilly, London W1V 9AG
Founded: 1858; Members: 2400
Pres: Dr. J.E. Robinson; Gen Secr: S.E. Stafford
Focus: Geology
Periodicals
Geologists' Association Circular (weekly)
Proceedings (weekly) 12112

George Eliot Fellowship, 71 Stepping Stones Rd, Coventry CV5 8JT
T: (0203) 592231
Founded: 1930; Members: 400
Focus: Lit
Periodical
Review (weekly) 12113

Gilbert and Sullivan Society of Edinburgh, 35 Paisley Av, Edinburgh EH8 7LG
T: (031) 6612528
Founded: 1924
Focus: Music 12114

Girl's Schools Association (GSA), c/o Headington School, Oxford OX3 7TD
T: (0865) 62711
Founded: 1973; Members: 207
Focus: Educ 12115

Glamorgan History Society (G.H.S.), c/o Glamorgan Record Office, Mid Glamorgan County Hall, Cathays Park, Cardiff CF1 3NE
Members: 350
Gen Secr: H.M. Thomas
Focus: Hist
Periodical
Morgannwg (weekly) 12116

Glasgow Agricultural Society, 4 Halloway Park, Ayr
T: (0292) 64295
Focus: Agri 12117

Glasgow Archaeological Society (G.A.S.), 8 Tavistock Dr, Glasgow G43 2SJ
T: (041) 3398855
Founded: 1856; Members: 250
Focus: Archeol
Periodicals
Bulletin (weekly)
Glasgow Archaeological Journal (weekly) . 12118

Glasgow Mathematical Association (GMA), c/o Dept of Mathematics, University of Glasgow, University Gardens, Glasgow G12 8QW
T: (041) 3398855
Founded: 1927; Members: 60
Pres: R. Bailey; Gen Secr: S. Rowan
Focus: Math 12119

Glasgow Obstetrical and Gynaecological Society, c/o Deparment of Obstetrics and Gynaecology, Royal Maternity Hospital, Glasgow G4 0NA
T: (041) 5522435
Founded: 1886; Members: 150
Focus: Gynecology 12120

Glenn Miller Society, 244 Edgware Rd, London W2 1DS
T: (071) 2624604
Focus: Music 12121

Gloucestershire Society for Industrial Archaeology, Oak House, Hamshill, Coaley, Dursley GL11 5EH
T: (0453) 860595
Founded: 1964; Members: 250
Pres: W.V. Awdry; Gen Secr: Dr. R. Wilson
Focus: Archeol
Periodicals
Journal (weekly)
Newsletter (weekly) 12122

Good Gardeners' Association (GGA), Pinetum Lodge, Churcham GL2 8AD
T: (0452) 750402
Focus: Hort
Periodical
Newsletter (weekly) 12123

Gooseberry Society, 14 Hillary Rise, Barnet EN5 5AZ
T: (081) 4413120
Founded: 1980
Focus: Nutrition 12124

Gower Society, c/o Royal Institution of South Wales, Victoria Rd, Swansea SA1 1SN
Founded: 1947; Members: 1600
Focus: Ecology
Periodical
Gower (weekly) 12125

Greater London Federation of Parent-Teacher Associations, 113 Kingston Rd, London SW19
T: (081) 5409299
Focus: Educ 12126

Great Western Society (GWS), Didcot OX11 7NJ
T: (0235) 817200
Founded: 1961; Members: 4200
Focus: Transport
Periodical
Great Western Echo (weekly) 12127

The Greek Institute, 34 Bush Hill Rd, London N21 2DS
T: (081) 3607968
Founded: 1969; Members: 350
Pres: Dr.Kypros Tofallis
Focus: Ling; Lit; Hist
Periodical
Greek Review (weekly) 12128

Group-Analytic Society, 1 Daleham Gardens, London NW3 5BY
T: (071) 4312693
Founded: 1952; Members: 310
Focus: Behav Sci
Periodical
Group Analysis (3 times annually) . . . 12129

Group and Association of County Medical Officers of Health of England and Wales, c/o Dr. G. Ramage, County Health Dept, Martin St, Stafford
Founded: 1902
Focus: Public Health 12130

Group for the Study of Irish Historic Settlement (GSIHS), c/o School of Earth Sciences, Northern Ireland Polytechnic, Jordanstown, Newtonabbey BT37 0QB
T: Whiteabbey 65131
Founded: 1969; Members: 250
Focus: Hist 12131

Guild of Catholic Doctors, White Lodge, Radnor Rd, Westbury-on-Trym, Bristol BS9 4DX
T: (0272) 624617
Founded: 1923; Members: 1700
Focus: Med 12132

Guild of Church Musicians, Hillbrow, Blechingley, Surrey
T: (088374) 3168
Founded: 1888
Gen Secr: John Ewington
Focus: Music
Periodicals
Laudate (2-3 times annually)
Year Book (weekly) 12133

Guild of Guide Lecturers (GGL), 52 Borough High St, London SE1 1XN
T: (071) 3781705; Fax: (071) 4031115
Founded: 1950
Focus: Adult Educ

Periodicals
Member's Bulletin (weekly)
Newsletter 12134

Guild of Health, 26 Queen Anne St, London W1M 9LB
T: (071) 5802492
Founded: 1904
Focus: Med
Periodical
Way of Life (weekly) 12135

Guild of Pastoral Psychology, POB 1107, London W3 6ZP
T: (081) 9938366
Founded: 1936; Members: 550
Focus: Psych
Periodicals
Bulletin (weekly)
Lectures (weekly) 12136

Guild of Travel Writers (GTW), 31 Riverside Court, Reading RG4 8A2
T: (0734) 481384
Members: 100
Focus: Lit 12137

Gwent Wildlife Trust (OWL), 16 White Swan Court, Monmouth NP5 3NY
T: (0600) 715501
Founded: 1964; Members: 1500
Focus: Ecology
Periodical
Wild about Gwent (3 times annually) . . 12138

Gypsy Lore Society (GLS), Coutts House, Sandon near Stafford
T: 265
Founded: 1888; Members: 700
Focus: Ethnology 12139

Haemophilia Society, 123 Westminster Bridge Rd, London SE1 7HR
T: (071) 9282020; Fax: (071) 6201416
Founded: 1950; Members: 4000
Gen Secr: David G. Watters
Focus: Hematology
Periodicals
Bulletin (weekly)
Update (weekly) 12140

The Hakluyt Society, c/o Map Library, British Library, Great Russell St, London WC1B 3DG
T: (0986) 86359; Fax: (0986) 868181
Founded: 1846; Members: 2400
Pres: Prof. P.E.H. Hair; Gen Secr: Sarah Tyacke; Dr. W.F. Ryan
Focus: Geography; Hist
Periodicals
Hakluyt Society – Extra Series (weekly)
Hakluyt Society – Second Series (weekly) 12141

Haldane Society, 35 Wellington St, London WC2
Founded: 1930; Members: 500
Focus: Sociology 12142

Halle Concerts Society, 30 Cross St, Manchester M2 7BA
T: (061) 8348363; Fax: (061) 8321669
Founded: 1858; Members: 4000
Pres: Sebastian de Ferranti; Gen Secr: Geoffrey Ogden
Focus: Music
Periodicals
Halle News (3 times anually)
Halle Year Book (weekly) 12143

Hampshire Field Club and Archaeological Society, c/o King Alfred's College, Winchester SO22 4NR
Founded: 1885; Members: 600
Pres: Adrian Rance
Focus: Archeol; Hist; Nat Sci
Periodicals
Newsletter (weekly)
Proceedings (weekly) 12144

Hansard Society for Parliamentary Government, Saint Philip's Bldg, Sheffield St, London WC2A 2EX
T: (071) 9557478; Fax: (071) 9557492
Founded: 1944; Members: 500
Pres: Lord Hayhoe; Gen Secr: David Harris
Focus: Poli Sci
Periodical
Parliamentary Affairs: A Journal of Comparative Politics (weekly) 12145

Hardy Plant Society (HPS), Garden Cottage, 214 Ruxley Lane, West Ewell KT19 9EZ
Founded: 1957; Members: 1900
Focus: Hort
Periodicals
Bulletin (weekly)
News Letter (3 times annually) 12146

The Harleian Society, c/o College of Arms, Queen Victoria St, London EC4V 4BT
Founded: 1869
Pres: J.P. Brooke-Little; Gen Secr: P.L. Gwynn-Jones
Focus: Genealogy 12147

The Harveian Society of London, 11 Chandos St, London W1M 0EB
T: (071) 5801043
Founded: 1831; Members: 448
Pres: Dr. Jenny Madagan; Gen Secr: M.C. Griffiths
Focus: Med 12148

HATRA, 7 Gregory Blvd, Nottingham NG7 6LD
T: (0602) 623311; Tx: 378230; Fax: (0602) 625450
Founded: 1949; Members: 200
Pres: P.D. Smith; Gen Secr: C.J. Moye
Focus: Eng
Periodical
Knitstats (weekly) 12149

Havergal Brian Society (HBS), 5 Eastbury Rd, Watford WD1 4PT
T: (0923) 224607
Founded: 1974; Members: 240
Gen Secr: Dr. A. Marshall
Focus: Music
Periodical
Newsletter (weekly) 12150

Hawick Archaeological Society (H.A.S.), Orrock-House, Stirches Rd, Hawick TD9 7HF
T: (0450) 75546
Founded: 1856; Members: 550
Pres: Hugh K. Mackay; Gen Secr: Ian W. Landles
Focus: Archeol
Periodical
Transactions (weekly) 12151

Hawk and Owl Trust, c/o Bird of Prey Section, Zoological Society of London, Regents Park, London NW1 4RY
T: (071) 6037756; Fax: (071) 6037756
Members: 475
Focus: Ornithology
Periodicals
Annual Report (weekly)
Newsletter (3 times annually) 12152

Head Teachers' Association of Scotland (HAS), c/o Royal Academy, Midhills Rd, Inverness IV2 3NG
T: (0463) 33424
Focus: Educ 12153

Health Service Social Worker Group, 112 Wood St, Barnet, Herts
Focus: Public Health; Sociology 12154

Hearing Concern (BAHOH), 7-11 Armstrong Rd, London W3 7JL
T: (081) 7431110; Fax: (081) 7429043
Founded: 1947; Members: 6000
Pres: Prof. I.G. Taylor; Gen Secr: Marwood Braund
Focus: Rehabil
Periodical
Hark (weekly) 12155

Heather Society, Denbeigh, All Saints Rd, Creeting Saint Mary, Ipswich IP6 8PJ
T: (0449) 711220
Founded: 1963; Members: 900
Focus: Botany, Specific
Periodicals
Bulletin
Yearbook (weekly) 12156

Henry Bradshaw Society (H.B.S.), c/o University of East Anglia, Norwich NR4 7TJ
T: (0603) 592454; Fax: (0603) 250454
Founded: 1890; Members: 350
Pres: Prof. Henry Chadwick; Gen Secr: David Chadd
Focus: Rel & Theol 12157

Henry Doubleday Research Association (HDRA), c/o National Centre for Organic Gardening, Ryton-on-Dunsmore, Coventry CV8 3LG
T: (0203) 303517; Fax: (0203) 639229
Founded: 1958; Members: 5600
Pres: Earl Kitchener; Gen Secr: Alan Gear
Focus: Hort; Agri
Periodical
Henry Doubleday Research Association Newsletter (weekly) 12158

Henry Williamson Society, Ryburn House, Ryburn House, Freshwater PO40 9HJ
T: (0983) 752207
Members: 310
Focus: Lit
Periodical
Journal (weekly) 12159

Henty Society, 60 Painswick Rd, Cheltenham GL50 2ER
T: (0242) 516578
Founded: 1976; Members: 139
Focus: Lit
Periodicals
Bulletin (weekly)
The Double Ressure (weekly) 12160

The Heraldry Society, 44-45 Museum St, London WC1A 1LH
T: (071) 4302172
Founded: 1947; Members: 1600
Pres: Duke of Norfolk; Gen Secr: M. Miles
Focus: Genealogy
Periodicals
The Coat of Arms (weekly)
The Heraldry Gazette (weekly) 12161

Heraldry Society of Scotland, c/o National Museum of Antiquities of Scotland, Queen St, Edinburgh EH2 1JD
T: (031) 5568921
Founded: 1977; Members: 180
Focus: Genealogy
Periodical
The Double Treasure (weekly) 12162

Hertfordshire Local History Council (HLHC), Lamb Cottage, Whitwell, Hitchin, Herts
Founded: 1951; Members: 200
Focus: Hist 12163

Hertfordshire Local History Council (HLHC), Lamb Cottage, Whitwell, Hitchin, Herts
Founded: 1951
Focus: Cultur Hist
Periodical
Hertfordshire's Past (weekly) 12164

Hertfordshire Natural History Society and Field Club, 9 Hill Rise, Potters Bar EN6 2RX
T: (0707) 57586
Founded: 1875; Members: 400
Focus: Geology; Ornithology
Periodicals
Bird NL (weekly)
Transactions (weekly) 12165

Hesketh Hubbard Art Society, 17 Carlton House Terrace, London SW1Y 5BD
T: (071) 9306844; Fax: (071) 8397830
Founded: 1930
Pres: Simon Whittle
Focus: Arts; Fine Arts
Periodical
Annual Catalogue (weekly) 12166

H.G. Wells Society, c/o English Dept, Nene College, Moulton Park NN2 7AL
T: (0604) 735500; Fax: (0604) 720636
Founded: 1960
Gen Secr: S. Hardy
Focus: Lit
Periodicals
Newsletter
Wellsian 12167

Higher Education Funding Council for England, Northavon House, Coldharbour Ln, Bristol BS16 1QD
T: (0272) 317317; Fax: (0272) 317173
Founded: 1992
Pres: Sir Ron Dearing; Gen Secr: Prof. Graeme Davies
Focus: Educ 12168

High Pressure Technology Association (H.P.T.A.), c/o Dept of Mechanical Engineering, Leeds University, Leeds LS2 9JT
T: (0532) 31751
Founded: 1967; Members: 200
Focus: Eng 12169

Hispanic and Luso-Brazilian Council, 2 Belgrave Sq, London SW1X 8PJ
T: (071) 2352303
Founded: 1943; Members: 600
Pres: Viscount Montgomery of Alamein; Gen Secr: Sir Kenneth James
Focus: Ethnology; Hist; Geography; Lit
Periodical
Bulletin (weekly) 12170

Historic Aircraft Association (HAA), c/o Civil Aviation Authority, Airworthiness Division, Brabazon House, Redhill, Surrey
Founded: 1979; Members: 70
Focus: Aero 12171

Historical Association (HA), 59a Kennington Park Rd, London SE11 4JH
T: (071) 7353901; Fax: (071) 5824989
Founded: 1906; Members: 7000
Gen Secr: M. Stiles
Focus: Hist
Periodicals
The Historian (3 times annually)
Teaching History (3 times annually) . . 12172

The Historical Metallurgy Society (HMS), 22 Easterfield Dr, Southgate, Swansea SA3 2DB
T: (0792) 233223
Founded: 1962; Members: 400
Gen Secr: Peter Hutchison
Focus: Hist
Periodical
The Journal of the Historical Metallurgy Society (weekly) 12173

Historical Newspaper Service, 8 Monks Av, New Barnet EN5 1DB
T: (081) 4403159
Focus: Lit; Journalism 12174

Historical Society of the Church in Wales, c/o Trinity College, Carmarthen SA31 3EP
T: 7971
Founded: 1946; Members: 430
Focus: Hist; Rel & Theol 12175

Historical Society of the Methodist Church in Wales, Llys Myfyr, Pwllheli, Caerns
T: 2608
Founded: 1944
Focus: Hist; Rel & Theol 12176

Historic Breechloading Smallarms Association, c/o Imperial War Museum, Lambeth Rd, London SE1 6HZ
Focus: Hist; Military Sci 12177

Historic Society of Lancashire and Cheshire (H.S.L.C.), Stand Park Rd, Liverpool LI6 9JD
Founded: 1848; Members: 500
Pres: J.I. Kermode; Gen Secr: J.E. Hollinshead
Focus: Hist
Periodical
Transactions (weekly) 12178

History of Education Society, c/o B.J. Starkey, 4 Marydene Dr, Leicester LE5 6HD
T: (0533) 416899
Founded: 1967; Members: 450
Focus: Hist; Educ
Periodical
History of Education Society Bulletin (weekly) 12179

Honourable Society of Cymmrodorion, 30 Eastcastle St, London W1N 7PD
T: (071) 6362033
Founded: 1751; Members: 1600
Pres: Dr. Benjamin G. Jones
Focus: Arts; Lit
Periodical
Transactions (weekly) 12180

Honours Graduate Teachers' Association (HGTA), 25 Montgomerie Terrace, Ayr KA7 1SL
T: 84964
Founded: 1964; Members: 200
Focus: Educ 12181

Horatian Society, 4 Breams Bldgs, London EC4A 1AQ
T: (071) 3535835
Founded: 1934
Focus: Lit 12182

Hospitals Consultants and Specialists Association (H.C.S.A.), 1 Kingslere Rd, Overton Rg25 3JP
T: (0256) 771777; Fax: (0256) 770999
Founded: 1947; Members: 3120
Pres: Dr. F. Foster-Thompson; Gen Secr: S.J. Charkham
Focus: Med
Periodical
The Consultant (weekly) 12183

Hotel, Catering and Institutional Management Associaton, 191 Trinity Rd, London SW17
T: (081) 6724251
Founded: 1960
Focus: Econ 12184

Housman Society (HS), c/o Area Library, 70 New Rd, Bromsgrove, Worcestershire
Founded: 1973; Members: 200
Focus: Lit
Periodical
Journal (weekly) 12185

Huguenot Society of Great Britain and Ireland, c/o The Huguenot Library, University College, Gower St, London WC1E 6BT
Founded: 1885; Members: 1432
Gen Secr: Mary Bayliss
Focus: Hist
Periodicals
Proceedings (weekly)
Quarto Series (weekly) 12186

Hunter Archaeological Society, 37 Chesterwood Dr, Sheffield S10 5DD
Founded: 1912; Members: 340
Focus: Archeol
Periodical
Transactions (weekly) 12187

Hunterian Society, Brampton House, 60 Grove End Rd, London NW8 9NH
Founded: 1819; Members: 450
Gen Secr: Dr. J. Rosenberg
Focus: Med
Periodical
Transactions (weekly) 12188

Huntingdonshire Local History Society, Old Red Lion House, Thurleigh, Beds
Founded: 1957; Members: 300
Focus: Hist
Periodical
Records (weekly) 12189

Hyper Active Children's Support Group (HACSG), 71 Whyke Lane, Chichester PO19 2LD
Founded: 1978; Members: 2520
Gen Secr: Sally Bunday
Focus: Public Health; Educ Handic
Periodical
Newsletter (3 times annually) 12190

Ileostomy Association of Great Britain and Ireland, Amblehurst House, Black Scotch Lane, Mansfield NG18 4PF
T: (0623) 28099; Fax: (0623) 28099
Founded: 1956; Members: 9200
Pres: Prof. N.S. Williams; Gen Secr: David S. Eades
Focus: Gastroenter
Periodical
ia Journal (weekly) 12191

Incorporated Association of Architects, Jubilee House, Billing Brook Rd, Weston Favell NN3 4NW
T: 0604) 404121
Founded: 1925; Members: 4800
Pres: E. Heading; Gen Secr: W.A. Black
Focus: Archit
Periodicals
Architect and Surveyor (weekly)
Fire Surveyor (weekly) 12192

Incorporated Association of Organists, c/o Richard Popple, 24 Hither Green Ln, Abbey Park, Redditch B98 9BW
T: (0527) 65555
Founded: 1913; Members: 7500
Pres: Dr. Roy Massey; Gen Secr: Richard Popple
Focus: Music
Periodical
The Organists' Review (weekly) 12193

Incorporated Association of Preparatory Schools (IAPS), 138 Kensington Church St, London W8 4BN
T: (071) 7272316
Founded: 1892; Members: 1150
Focus: Educ
Periodical
The Preparatory Schools Review (weekly) 12194

Incorporated Society of Musicians, 10 Stratford Pl, London W1N 9AE
T: (071) 6294413; Fax: (071) 4081538
Founded: 1882; Members: 4000
Pres: Gillian Weir; Gen Secr: Neil Hoyle
Focus: Music
Periodicals
Music Journal (weekly)
Register of Specialist Teachers (weekly)
Yearbook (weekly) 12195

Incorporated Society of Registered Naturopaths (ISRN), 291 Gilmerton Rd, Edinburgh EH16 5UQ
Founded: 1934
Focus: Med; Pathology 12196

Independent Association of Telecommunications Users (TUA), 3-7 Stamford St, London SE1 9NT
T: (081) 8837229
Founded: 1979; Members: 200
Focus: Comm 12197

Independent Schools Joint Committee (ISJC), 27 Church Rd, Steep, Petersfield GU32 2DW
Founded: 1974; Members: 10
Focus: Educ 12198

Industrial Fire Protection Association of Great Britain (IFPA), Aldermary House, Queen St, London EC4
T: (071) 2485222
Founded: 1941; Members: 800
Focus: Safety 12199

Industrial Locomotive Society (ILS), c/o Channings, Kettlewell Hill, Woking GU21 4JA
T: (0483) 761435
Founded: 1946; Members: 250
Focus: Mach Eng
Periodical
Industrial Locomotive (weekly) 12200

Industrial Marketing Research Association (IMRA), 11 Bird St, Lichfield WS13 6PW
T: (0543) 263448; Fax: (0543) 250929
Founded: 1963; Members: 790
Focus: Marketing
Periodicals
Annual Conference Papers (weekly)
European Guide to Industrial Marketing Consultancy (weekly)
Symposium Papers 12201

Industrial Society, 3 Carlton House Terrace, London SW1Y 5DG
T: (071) 8394300
Founded: 1918; Members: 15000
Focus: Comm Sci
Periodicals
Directory of Sources
Industrial Society (weekly) 12202

Industrial Unit of Tribology (IUT), c/o Dept of Mechanical Engineering, University of Leeds, Leeds 9JT
Founded: 1966; Members: 250
Focus: Eng 12203

Institute for Animal Health, Compton, Newbury RG16 0NN
T: (0635) 578411; Fax: (0635) 578844
Focus: Agri; Vet Med
Periodical
Annual Report (weekly) 12204

Institute for Animal Health Pirbright, c/o Pirbright Laboratory, Lash Rd, Pirbright Woking GU24 0NF
T: (0483) 232441; Tx: 859137; Fax: (0483) 232448
Focus: Immunology; Biochem; Genetics . 12205

Institute for Consumer Ergonomics (ICE), 75 Swingbridge Rd, Loughborough LE11 0JB
T: (0509) 236161; Tx: 34319; Fax: (0509) 610725
Founded: 1970
Gen Secr: I.A.R. Galer
Focus: Public Health 12206

Institute for Cultural Research (ICR), POB 13, Tunbridge Wells TN3 0HQ
T: (089286) 2045
Founded: 1968; Members: 450
Pres: Cecil C.M.G. Robert; Gen Secr: Henri Bortoft
Focus: Cultur Hist 12207

Institute for Fiscal Studies (IFS), 7 Ridgmount St, London WC1E 7AE
T: (071) 6363784; Fax: (071) 3234780
Founded: 1969; Members: 1000
Gen Secr: Andrew Dilnot
Focus: Econ
Periodical
Fiscal Studies (weekly) 12208

Institute for Scientific Information, 132 High St, Uxbridge, Middx
T: (0895) 30085
Focus: Sci 12209

Institute for Sex Education and Research (ISER), 40 School Rd, Birmingham B13
T: (021) 4490892
Founded: 1969
Focus: Behav Sci 12210

Institute of Actuaries, Staple Inn Hall, High Holborn, London WC1V 7QJ
T: (071) 2420106
Founded: 1848; Members: 8483
Pres: L.J. Martin; Gen Secr: A.G. Tait
Focus: Insurance
Periodical
Journal (3 times annually) 12211

Institute of Administrative Management (IAM), 40 Chatsworth Parade, Petts Wood, Orpington BR5 1RW
T: (0689) 75555; Fax: (0689) 8952569
Founded: 1915; Members: 10000
Gen Secr: M.J. Ainsworth
Focus: Business Admin
Periodicals
Office & Information Management International (10 times annually)
Office Job Evaluation Manual (weekly)
Office Salaries Analysis (weekly) 12212

Institute of Asphalt Technology (IAT), Unit 18, Central Trading Estate, Staines TW18 4XE
T: (0784) 465387
Founded: 1966; Members: 1480
Gen Secr: J.F. Hills
Focus: Materials Sci
Periodical
Asphalt Technology (weekly) 12213

Institute of Bankers in Scotland, 19-20 Rutland Sq, Edinburgh EH1 2DE
T: (031) 2299869
Founded: 1875; Members: 14500
Gen Secr: Dr. C.W. Munn
Focus: Adult Educ; Finance
Periodical
The Scottish Banker (weekly) 12214

Institute of Biology, 20-22 Queensberry Pl, London SW7 2DZ
T: (071) 5818333; Fax: (071) 8239409
Founded: 1950; Members: 15000
Pres: Prof. C.R.W. Spedding; Gen Secr: Dr. R. Priestley
Focus: Bio
Periodicals
Biologist (5 times annually)
Journal of Biological Education (weekly) . 12215

Institute of British Geographers, 1 Kensington Gore, London SW7 2AR
T: (071) 5846371
Founded: 1933; Members: 2000
Focus: Geography
Periodicals
Area (weekly)
Transactions (weekly) 12216

Institute of Business Administration, 25 Bridgeman Terrace, Wigan WN1 1TD
T: (0442) 43572; Fax: (0442) 829350
Founded: 1946
Focus: Business Admin
Periodical
Business Administrator (weekly) 12217

Institute of Ceramics, Shelton House, Stoke Rd, Shelton, Stoke-on-Trent ST4 2DR
T: (0782) 23116
Founded: 1900; Members: 2000
Gen Secr: S.B. Buchanan
Focus: Materials Sci
Periodicals
Proceedings
Transactions and Journal (weekly) . . . 12218

Institute of Chartered Accountants in England and Wales, Chartered Accountants' Hall, Moorgate Pl, London EC2P 2BJ
T: (071) 6287060; Tx: 884443
Founded: 1880; Members: 86000
Pres: M.A. Chamberlain; Gen Secr: A.J. Colquhoun
Focus: Econ
Periodicals
Accountancy (weekly)
Accountants Digest (weekly)
Accounting and Business Research (weekly) 12219

Institute of Chartered Accountants of Scotland, 27 Queen St, Edinburgh EH2 1LA
T: (031) 2255673; Fax: (031) 2253813
Founded: 1854; Members: 12900
Pres: J.A. Denholm; Gen Secr: P.W. Johnston
Focus: Econ
Periodicals
The Accountant's Magazine (weekly)
Annual Report (weekly)
Official Directory of Members (weekly) . 12220

Institute of Chartered Foresters (IFGB), 7a Saint Colme St, Edinburgh EH3 6AA
T: (031) 2252705; Fax: (031) 2206128
Founded: 1925; Members: 1400
Pres: D.W.G. Taylor; Gen Secr: Margaret W. Dick
Focus: Forestry
Periodical
Forestry (weekly) 12221

Institute of Chartered Secretaries and Administrators, 16 Park Crescent, London W1N 4AH
T: (071) 5804741
Founded: 1891; Members: 44000
Pres: J.H. Bingley; Gen Secr: John Ainsworth
Focus: Public Admin
Periodicals
Administrator (weekly)
Company Secretarial Practice 12222

Institute of Clay Technology (ICT), c/o Butterley Building Materials Ltd, Wellington St, Ripley DE5 3DZ
T: (0773) 3661
Founded: 1927; Members: 771
Focus: Materials Sci
Periodical
Euroclay (weekly) 12223

Institute of Community Studies, 18 Victoria Park Sq, London E2 9PF
T: (081) 9806263
Founded: 1954
Pres: M. Young
Focus: Sociology 12224

Institute of Concrete Technology (ICT), POB 255, Beaconsfield HP9 1JE
T: (0494) 674572
Founded: 1972; Members: 600
Gen Secr: Roy Jolly
Focus: Civil Eng
Periodical
Convention Symposium Papers (weekly) . 12225

Institute of Construction Management (ICM), 397 City Rd, London EC1V 1NE
T: (071) 2780471; Fax: (071) 8373194
Founded: 1974; Members: 1300
Gen Secr: J.W. Lawrence
Focus: Civil Eng
Periodicals
Construction Management-Focus (weekly)
ICM Reference Book and List of Members (weekly) 12226

Institute of Contemporary Arts, 12 Carlton House Terrace, London SW1V 5AH
T: (071) 9300493; Fax: (071) 8730051
Founded: 1948; Members: 7000
Pres: Mik Flood
Focus: Arts 12227

Institute of Contemporary History and Wiener Library, 4 Devonshire St, London W1N 2BH
T: (071) 6367247; Fax: (071) 4366428
Founded: 1933
Gen Secr: Dr. David Cesarani
Focus: Hist 12228

Institute of Corrosion, POB 253, Leighton Buzzard LU7 7WB
T: (0525) 851771; Fax: (0525) 376690
Founded: 1975; Members: 1600
Pres: B.S. Wyatt; Gen Secr: K.M. Vincent
Focus: Metallurgy
Periodicals
Annual Report (weekly)
Corrosion Science (weekly)
Industrial Corrosion (weekly) 12229

Institute of Credit Management (ICM), Easton House, Easton on the Hill, Stamford PE90 3NH
T: (0780) 56777
Founded: 1939; Members: 3400
Focus: Finance
Periodical
Credit Management (weekly) 12230

Institute of Data Processing Management (IDPM), 18 Henrietta St, London WC2E 8NU
T: (071) 2403304
Founded: 1978; Members: 7000
Focus: Computer & Info Sci
Periodicals
DP International (weekly)
DP International Quarterly Journal (weekly) 12231

Institute of Directors (IOD), 116 Pall Mall, London SW1Y 5ED
T: (071) 8391233; Fax: (071) 9301949
Founded: 1903; Members: 38000
Focus: Business Admin
Periodical
The Director (weekly) 12232

Institute of Economic Affairs (IEA), 2 Lord North St, London SW1P 3LB
T: (071) 7993745; Fax: (071) 7992137
Founded: 1955
Gen Secr: Lord Harris
Focus: Econ
Periodicals
Economic Affairs (weekly)
Hobart Papers
IEA Readings
Occasional Papers
Research Monographs 12233

Institute of Electrolysis (I of E), 251 Seymour Grove, Manchester M16 0DS
T: (061) 8815306
Founded: 1944; Members: 298
Focus: Physics 12234

Institute of Energy, 18 Devonshire St, London W1N 2AU
T: (071) 5807124; Fax: (071) 5804420
Founded: 1927; Members: 6000
Pres: Prof. J.S. Harrison; Gen Secr: J.E.H. Leach
Focus: Energy
Periodicals
Energy World (10 times annually)
Energy World Yearbook (weekly)
Fuel and Energy Abstracts (weekly)
Journal of the Institute of Energy (weekly) 12235

Institute of Engineers and Technicians (IE), 100 Grove Vale, London SE22 8DR
T: (081) 6931255; Fax: (0795) 535468
Founded: 1948; Members: 5500
Pres: Frank Shaw; Gen Secr: Dr. Tony Deeson
Focus: Eng
Periodical
Journal (weekly) 12236

Institute of Fisheries Management (IFM), Balmaha, Coldwells Rd, Holmer, Hereford
T: (0432) 276225
Founded: 1969; Members: 1200
Gen Secr: E. Staite
Focus: Fisheries
Periodical
Fish (weekly) 12237

Institute of Food Research / Biotechnology and Biological Sciences Research Council, Earley Gate, Whiteknights Rd, Reading RG6 2EF
T: (0734) 357000; Fax: (0734) 267917
Gen Secr: Prof. Douglas Georgala
Focus: Nutrition; Agri
Periodicals
Newsletter (weekly)
Report (weekly) 12238

Institute of Food Science and Technology of the United Kingdom (IFST), 5 Cambridge Court, 210 Shepherd's Bush Rd, London W6 7NL
T: (071) 6036316
Founded: 1964; Members: 3500
Pres: K.G. Anderson; Gen Secr: H.G. Wild
Focus: Food; Eng
Periodicals
Food Science and Technology Today (weekly)
International Journal of Food Science and Technology (weekly) 12239

The Institute of Group-Analysis, Group-Analysis Society (London), 1 Daleham Gardens, London NW3 5BY
T: (071) 4312693; Fax: (071) 4317276
Founded: 1971; Members: 170
Pres: Dr. Janet Boakes; Gen Secr: Brenda Ling
Focus: Behav Sci
Periodical
Group Analysis (3 times annually) . . . 12240

Institute of Health Education (IHE), 14 High Elm Rd, Hale Barns, Ches
T: (061) 9808276; Fax: (061) 9807446
Founded: 1962
Focus: Public Health
Periodical
Journal of the Institute of Health Education (weekly) 12241

Institute of Heraldic and Genealogical Studies (IHGS), Northgate, Canterbury CT1 1BA
T: (0227) 68664
Founded: 1961; Members: 124
Pres: Viscount Monckton of Brenchley
Focus: Genealogy
Periodical
Family History (weekly) 12242

Institute of Hospital Engineering, 20 Landport Terrace, Portsmouth PO1 2RG
T: (0705) 823186; Fax: (0705) 815927
Founded: 1943
Focus: Eng
Periodical
Institute Journal (10 times annually) . . 12243

Institute of Housing, Octavia House, Westwood Business Park, Westwood Way, Coventry CV4 8JP
T: (0203) 694433
Founded: 1965; Members: 5400
Focus: Home Econ
Periodicals
Housing (weekly)
Inside Housing (weekly) 12244

Institute of Hydrology, MacLean Bldg, Crowmarsh Gifford, Wallingford OX10 8BB
T: (0491) 38800; Tx: 849365
Focus: Hydrology 12245

Institute of Information Scientists, 44 Museum St, London WC1A 1LY
T: (071) 8318003; Fax: (071) 4301270
Founded: 1958; Members: 2750
Pres: Marino Saksida; Gen Secr: Sarah Carter
Focus: Computer & Info Sci
Periodicals
Inform (10 times annually)
IT Link (weekly)
Journal of Information Science 12246

Institute of Inventors, 19-21 Fosse Way, London W13 0BZ
Founded: 1964; Members: 700
Focus: Eng
Periodical
New Invention Lists (weekly) 12247

Institute of Investment Management and Research, 211-213 High St, Bromley BR1 1NY
T: (081) 4640811; Fax: (081) 3130587
Founded: 1956; Members: 3028
Focus: Finance
Periodical
The Professional Investor (10 times annually) 12248

Institute of Jewish Affairs (IJA), 79 Wimpole St, London W1M 7DD
T: (071) 9358266; Fax: (071) 9353252
Founded: 1941
Pres: Lord Rothschild; Gen Secr: A. Lerman
Focus: Hist
Periodicals
East European Jewish Affairs (weekly)
Patterns of Prejudice (weekly)
Research Reports (10 times annually) . 12249

Institute of Leisure and Amenity Management (ILAM), Lower Basildon, Reading RG8 9NE
T: (0491) 874222; Fax: (0491) 874059
Founded: 1983; Members: 6000
Gen Secr: Alan Smith
Focus: Travel
Periodical
The Leisure Manager (weekly) 12250

Institute of Linguists, 24a Highbury Grove, London N5 2EA
T: (071) 3597445
Founded: 1910; Members: 6700
Pres: John Drew; Gen Secr: Edda Ostarhild
Focus: Ling
Periodical
The Linguist (weekly) 12251

Institute of Logistics (ILDM), Douglas House, Queens Sq, Corby NN 17 1PL
T: (0536) 205500; Fax: (0536) 400979
Founded: 1993; Members: 11000
Focus: Physics; Logic
Periodical
Logistics Focus (weekly) 12252

Institute of Management, Management House, Cottingham Rd, Corby NN17 1TT
T: (0536) 204222; Fax: (0536) 201651
Founded: 1947; Members: 73500
Pres: James Watson; Gen Secr: Roger Young
Focus: Business Admin
Periodicals
Management Today (weekly)
Professional Manager (weekly) 12253

Institute of Management Services (IMS), 1 Cecil Court, London Rd, Enfield EN2 6DD
T: (081) 3637452; Fax: (081) 3678149
Founded: 1941; Members: 9000
Pres: Lord Chilver; Gen Secr: Frank O'Connolly
Focus: Business Admin
Periodical
Management Services (weekly) 12254

Institute of Marine Engineers, 76 Mark Lane, London EC3R 7JN
T: (071) 4818493; Tx: 886841; Fax: (071) 4881854
Founded: 1889; Members: 16500
Pres: G. Geddes; Gen Secr: J.E. Sloggett
Focus: Eng
Periodicals
Bulletin (weekly)
Marine Engineers Review (weekly)
Offshore Technology (weekly)
Technical Conference Reports
Technical Transactions 12255

Institute of Materials, 1 Carlton House Terrace, London SW1Y 5DB
T: (071) 8394071; Fax: (071) 8392078
Founded: 1985; Members: 11500
Pres: Sir John Collyear; Gen Secr: Dr. J.A. Catterall
Focus: Materials Sci
Periodicals
Historical Metallurgy (weekly)
Ironmaking and Steelmaking (weekly)
Materials Science and Technology (weekly)
Metals and Materials (weekly)
Surface Engineering (weekly) 12256

Institute of Materials Management (IMM), c/o Institute of Logistics, Douglas House, Queens Sq, Corby NN17 1PL
T: (0536) 205500
Founded: 1952; Members: 3500
Focus: Business Admin
Periodicals
IMM News (weekly)
Logistics Today (weekly)
Members' Reference Book & Buyers' Guide (weekly) 12257

Institute of Mathematics and its Applications (IMA), 16 Nelson St, Southend-on-Sea SS1 1EF
T: (0702) 354020; Fax: (0702) 354111
Founded: 1964; Members: 7500
Pres: Prof. J.C.R. Hunt
Focus: Math
Periodicals
Bulletin (weekly)
IMA Journal of Applied Mathematics (weekly)
IMA Journal of Mathematical Control and Information (weekly)
IMA Journal of Mathematics Applied in Business and Industry (weekly)
IMA Journal of Mathematics Applied in Medicine and Biology (weekly)
IMA Journal of Numerical Analysis (weekly)
Teaching Mathematics and its Applications (weekly) 12258

Institute of Measurement and Control, 87 Gower St, London WC1E 6AA
T: (071) 3874949; Fax: (071) 3888431
Founded: 1944; Members: 6000
Pres: C.J.A. Bosworth; Gen Secr: M.J. Yates
Focus: Eng
Periodicals
Instrument Engineer's Yearbook (weekly)
Measurement and Control (weekly)
Transactions (weekly) 12259

Institute of Media Executives (Inst M E), Ely House, Somerford, Willenhall, West Midlands
T: (0902) 61249
Founded: 1973; Members: 151
Focus: Mass Media 12260

Institute of Medical Laboratory Sciences (IMLS), 12 Queen Anne St, London W1M 0AU
T: (071) 6368192; Fax: (071) 4364946
Founded: 1912; Members: 14000
Focus: Med
Periodicals
British Journal of Biomedical Science (weekly)
IMLS Gazette (weekly) 12261

The Institute of Metals, 1 Carlton House Terrace, London SW1Y 5DB
T: (071) 8394071; Tx: 8814813; Fax: 8392289
Founded: 1985; Members: 12500
Pres: C.E.H. Morris; Gen Secr: Dr. J.A. Catterall
Focus: Metallurgy
Periodicals
British Corrosion Journal (weekly)
Historical Metallurgy (weekly)
International Materials Review (weekly)
Ironmaking and Steelmaking (weekly)
Materials Science and Technology (weekly)
Metals and Materials (weekly)
Powder Metallurgy (weekly)
Steel in the USSR (weekly)
Surface Engineering (weekly) 12262

Institute of Municipal Building Management (IMBM), 30 Rhiw Rd, Colwyn Bay LL29 7TP
T: (0492) 31348
Founded: 1951; Members: 1660
Focus: Urban Plan 12263

Institute of Occupational Medicine, 600 Roxburgh Pl, Edinburgh EH8 9SU
T: (031) 6675131
Founded: 1969
Focus: Hygiene 12264

Institute of Packaging (IOP), Sysonby Lodge, Nottingham Rd, Melton Mowbray LE13 0NU
T: (0664) 500055; Fax: (0664) 64164
Founded: 1947; Members: 4000
Pres: Victor Watson; Gen Secr: Jerry Berragan
Focus: Eng
Periodical
Panorama (weekly) 12265

Institute of Personnel Management (IPM), IPM House, Camp Rd, London SW19 4UX
T: (081) 9469100; Fax: (081) 9472570
Founded: 1913; Members: 50000
Focus: Business Admin
Periodicals
Personnel Management (weekly)
Personnel Management Plus 12266

Institute of Petroleum (IP), 61 New Cavendish St, London W1M 8AR
T: (071) 6361004; Fax: (071) 2551472
Founded: 1913; Members: 7000
Pres: Charles M. Smith; Gen Secr: Ian Ward
Focus: Petrochem
Periodicals
IP Statistics Service (weekly)
Petroleum Review (weekly) 12267

Institute of Physics, 47 Belgrave Sq, London SW1X 8QX
T: (071) 2356111; Fax: (071) 2596002
Founded: 1874; Members: 18000
Pres: C. Foxell; Gen Secr: Dr. A. Jones
Focus: Physics
Periodicals
Bioimaging (weekly)
Classical and Quantum Gravity (weekly)
Distributed Systems Engineering (weekly)
European Journal of Physics (weekly)
High Performance Polymers (weekly)
Inverse Problems (weekly)
Journal of Hard Materials (weekly)
Journal of Micromechanics and Microengineering (weekly)
Journal of Physics A: Mathematical and General (18 times annually)
Journal of Physics B: Atomic, Molecular and Optical Phyisics (weekly)
Journal of Physics C: Condensed Matter (weekly)
Journal of Physics D: Applied Physics (weekly)
Journal of Physics G: Nuclear Physics (weekly)
Journal of Radiological Protection (weekly)
Measurement Science and Technology (weekly)
Modelling and Simulation in Materials Science and Engineering (weekly)
Nano Technology (weekly)
Network (weekly)
Nonlinearity (weekly)
Opto and Laser Europe (weekly)
Physics Education (weekly)
Physics in Medicine and Biology (weekly)
Physics World (weekly)
Physiological Measurement (weekly)
Plasma Physics and Controlled Fusion (weekly)
Plasma Sources Science and Technology (weekly)
Public Understanding of Science (weekly)
Pure and Applied Optics (weekly)
Quantum Optics (weekly)
Reports on Progress in Physics (weekly)
Semiconductor Science and Technology (weekly)
Smart Materials and Structures (weekly)
Superconductor Science and Technology (weekly)
Waves in Random Media (weekly) . . . 12268

Institute of Population Registration (IPR), 96 Herongate Rd, Wanstead, London E12 5EQ
Founded: 1962; Members: 600
Focus: Stats
Periodicals
Annual Report (weekly)
Journal (weekly)
Newsletter (weekly) 12269

Institute of Printing (IOP), 8 Lonsdale Gardens, Tunbridge Wells TN1 1NU
T: (0892) 38118
Founded: 1961; Members: 2000
Focus: Eng 12270

Institute of Professional Investigators (IPI), 31a Wellington St, Blackburn BB1 8AF
T: (0254) 680072
Founded: 1976; Members: 400
Focus: Adult Educ
Periodical
The Professional Investigator (weekly) . . 12271

Institute of Psycho-Analysis, 63 New Cavendish St, London W1M 7RD
T: (071) 5804952
Founded: 1924; Members: 268
Focus: Psychoan
Periodicals
British Psycho-Analytical Society AR (weekly)
British Psycho-Analytical Society Bulletin (weekly)
British Psycho-Analytical Society Roster (weekly)
International Journal of Psycho-Analysis (weekly)
International Review of Psycho-Analysis (weekly) 12272

Institute of Psycho-Sexual Medicine, 11 Chandos St, London W1M 9DE
Focus: Med
Periodical
Institute of Psychosexual Medicine Journal (weekly) 12273

Institute of Purchasing and Supply (IPS), IPS House, High St, Ascot SL5 7HU
T: (0990) 23711
Founded: 1931; Members: 15000
Focus: Commerce
Periodicals
Procurement Weekly (weekly)
Purchasing & Supply Management (weekly) 12274

Institute of Quarrying, 7 Regent St, Nottingham NG1 5BY
T: (0602) 411315
Founded: 1917; Members: 5000
Gen Secr: J. Berridge
Focus: Mining
Periodical
Quarry Management (weekly) 12275

Institute of Race Relations, 2-6 Leeke St, London WC1X 9HS
T: (071) 8370041; Fax: (071) 2780623
Founded: 1958; Members: 490
Gen Secr: A. Sivanandan
Focus: Sociology
Periodical
Race and Class (weekly) 12276

Institute of Refrigeration, Kelvin House, 76 Mill Lane, Carshalton, Surrey
T: (081) 6177033; Fax: (081) 7730165
Founded: 1899; Members: 2500
Pres: A.H. Brown; Gen Secr: M.J. Horlick
Focus: Food; Eng
Periodical
Proceedings (weekly) 12277

Institute of Sales and Marketing Management (Inst SMM), 24 Warwick Rd, Royal Leamington Spa CV32 5JH
T: (0926) 37621/4; Tx: 311746
Founded: 1967; Members: 16000
Focus: Marketing
Periodical
Sales & Marketing Management 12278

Institute of Science Technology (IST), Mansell House, 22 Bole St, Lichfield WS13 6LP
T: (0543) 251346
Founded: 1954; Members: 1500
Pres: Lord Perry of Walton; Gen Secr: R. Dow
Focus: Eng
Periodical
Science Technology (weekly) 12279

Institute of Scientific and Technical Communicators (ISTC), 2-16 Goodge St, London W1P 1FF
T: (071) 4364425; Fax: (071) 5800747
Founded: 1972; Members: 1893
Pres: P. Greenfield; Gen Secr: A. Brobyn
Focus: Sci; Eng
Periodical
The Communicator (10 times annually) . 12280

Institute of Sheet Metal Engineering (ISME), Exeter House, 48 Holloway Mead, Birmingham B1 1NU
T: (021) 6222560; Fax: (021) 6666316
Founded: 1946; Members: 400
Pres: R. Jaf; Gen Secr: C. Bates
Focus: Eng
Periodicals
Oracle (weekly)
Sheet Metal Industries: Trade Journal (weekly) 12281

Institute of Small Business (ISB), 57-61 Mortimer St, London W1
T: (071) 6374383
Focus: Commerce
Periodical
Business Ideas Letter (weekly) 12282

Institute of Sports Medicine (ISM), 10 Nottingham Pl, London W1M 4AX
T: (071) 4861303
Founded: 1963
Focus: Med 12283

Institute of Statisticians, 43 Saint Peter's Sq, Preston PR1 7BX
T: (0772) 204237; Fax: (0772) 204476
Founded: 1948; Members: 2500
Pres: Sir John Boreham; Gen Secr: D.A. Holland
Focus: Stats
Periodicals
The Professional Statistician (weekly)
The Statistician (weekly) 12284

Institute of Supervisory Management (ISM), 22 Bore St, Lichfield WS13 6LP
T: (0543) 251346; Fax: (0543) 415804
Founded: 1947; Members: 20000
Focus: Commerce
Periodical
Modern Management (weekly) 12285

Institute of Technicians in Venereology, c/o Orsett Hospital, Orsett, Essex
Founded: 1951
Focus: Venereology 12286

Institute of Training and Development (ITD), Marlow House, Institute Rd, Marlow SL7 1BN
T: (0628) 890123; Fax: (0628) 890208
Founded: 1964; Members: 11000
Gen Secr: George B. Webster
Focus: Econ; Educ
Periodical
Training and Development (weekly) . . . 12287

Institute of Translation and Interpreting, 377 City Rd, London EC1V 1NA
T: (071) 7137600; Fax: (071) 7137650
Founded: 1986; Members: 950
Pres: R. Fletcher; Gen Secr: D. Castellano
Focus: Ling
Periodical
ITI Bulletin (weekly) 12288

Institute of Trichologists (IT), 228 Stockwell Rd, London SW9 9SU
T: (071) 7332056
Founded: 1902; Members: 230
Focus: Med; Immunology
Periodical
Trichologist 12289

Institute of Value Management (IVM), c/o Delta Executive Devt Centre Ltd, 22 Cavendish Av, Buxton SK17 9AE
T: (0298) 2284
Founded: 1966; Members: 80
Focus: Econ
Periodical
Value (weekly) 12290

Institute of Wood Science (IWSc), Hughenden Valley, High Wycombe HP14 4NU
T: (0494) 565374
Founded: 1955; Members: 1500
Gen Secr: M.W. Holloway
Focus: Forestry
Periodical
Journal (weekly) 12291

Institution of Agricultural Engineers (IAgrE), West End Rd, Silsoe, Bedford MK45 4DU
T: (0525) 861096; Fax: (0525) 861660
Founded: 1938; Members: 2400
Gen Secr: M.H. Hurst
Focus: Agri; Eng

Periodical
The Agricultural Engineer incorporating Soil & Water (weekly) 12292

The Institution of Chemical Engineers, 165-171 Railway Terrace, Rugby CV21 3HQ
T: (0788) 578214; Tx: 311780; Fax: (0788) 560833
Founded: 1922; Members: 13051
Gen Secr: Dr. T.J. Evans
Focus: Eng
Periodicals
Chemical Engineer (weekly)
Chemical Engineer Diary & Institution News (weekly)
Chemical Engineering Research & Design: Transactions of The Institution of Chemical Engineers (weekly)
Environmental Protection Bulletin (weekly)
Loss Prevention Bulletin (weekly) . . . 12293

Institution of Civil Engineers (ICE), 1-7 Great George St, London SW1P 3AA
T: (071) 2227722; Tx: 935637; Fax: (071) 2227500
Founded: 1818; Members: 76397
Gen Secr: Roger Dobson
Focus: Civil Eng
Periodicals
Advances in Cement Research (weekly)
The Concrete Yearbook (weekly)
Construction Today (weekly)
Geotechnique (weekly)
Ground Engineering (weekly)
Ground Engineering Yearbook (weekly)
Magazine of Concrete Research (weekly)
New Civil Engineer (weekly)
Nuclear Energy (weekly)
Offshore Engineer (weekly)
Offshore Engineer Yearbook (weekly)
Proceedings: civil engineering (weekly)
Proceedings: Municipal Engineer (weekly)
Proceedings: structures and buildings (weekly)
Proceedings: Transport (weekly)
Proceedings: Water, Maritime and Energy (weekly)
Steel Construction Yearbook (weekly)
Underground (weekly)
Underground Services Directory (weekly)
Waste, Recycling and Environmental Yearbook (weekly)
World Water (weekly) 12294

Institution of Corrosion Science and Technology, POB 253, Leighton Buzzard LU7 7WB
T: (0525) 851771; Fax: (0525) 376690
Founded: 1975; Members: 1500
Pres: Dr. J.L. Bown; Gen Secr: K.M. Vincent
Focus: Metallurgy
Periodicals
Corrosion Science (weekly)
Industrial Corrosion (8 times annually) . 12295

Institution of Electrical Engineers (IEE), Savoy Pl, London WC2R 0BL
T: (071) 2401871; Tx: 261176; Fax: 2407735
Founded: 1871; Members: 129000
Gen Secr: Dr. J.C. Williams
Focus: Electric Eng; Electronic Eng
Periodicals
Computers & Control Abstracts (weekly)
Computing Control Engineering (weekly)
Electrical & Electronics Abstracts (weekly)
Electronics Education (3 times annually)
Electronics Letters (weekly)
Engineering Management Journal (weekly)
Engineering Science and Education Journal (weekly)
IEE News (11 times annually)
IEE Proceedings (weekly)
IEE Review (weekly)
Intelligent Systems Engineering (weekly)
Manufacturing Engineer (weekly)
Medical & Biological Engineering & Computing (weekly)
Physics Abstracts (weekly)
Power Engineering Journal (weekly)
Software Engineering Journal (weekly) . 12296

Institution of Electronics, 659 Oldham Rd, Rochdale OL16 4PE
Founded: 1930; Members: 2500
Gen Secr: W. Birtwistle
Focus: Electronic Eng
Periodical
Proceedings (weekly) 12297

Institution of Electronics and Electrical Incorporated Engineers (IEEIE), Savoy Hill House, Savoy Hill, London WC2R 0BS
T: (071) 8363357; Fax: (071) 4979006
Founded: 1965; Members: 30000
Gen Secr: A.C. Gingell
Focus: Electric Eng; Electronic Eng
Periodical
Electrotechnology (weekly) 12298

Institution of Engineering Designers, Courtleigh, Westbury Leigh, Westbury BA13 3TA
T: (0373) 822801; Fax: (0373) 858085
Founded: 1945; Members: 6000
Gen Secr: M.J. Osborne
Focus: Graphic & Dec Arts, Design
Periodical
The Engineering Designer (weekly) . . . 12299

Institution of Engineers amd Shipbuilders in Scotland, 1 Atlantic Quay, Broomielaw, Glasgow G2 8JE
T: (041) 2483721
Founded: 1857; Members: 800
Pres: I.C. Broadley; Gen Secr: E.W. Bell
Focus: Eng
Periodicals
Transactions (weekly)
Year Book and List of Members (bi-annually) 12300

Institution of Environmental Health Officers (IEH), Chadwick House, 48 Rushworth St, London SE1 0QT
T: (071) 9286006
Founded: 1883; Members: 6000
Pres: E. Foskett; Gen Secr: A.M. Tanner
Focus: Ecology
Periodicals
Environmental Health Journal (weekly)
Environmental Health News (weekly) . . 12301

Institution of Environmental Sciences, 14 Princes Gate, Hyde Park, London SW7 1PU
T: (0252) 515511; Fax: (0252) 549682
Founded: 1971; Members: 740
Pres: Dr. M.J. Romeril; Gen Secr: Dr. J.F. Potter
Focus: Nat Sci
Periodicals
The Environmentalist (weekly)
I.E.S. News Sheet (weekly)
I.E.S. Proceedings (weekly) 12302

Institution of Fire Engineers (IFE), 148 New Walk, Leicester LE1 7QB
T: (0533) 553654; Fax: (0533) 471231
Founded: 1918; Members: 10000
Pres: D.E. Williams; Gen Secr: C.E. Mackwood
Focus: Eng
Periodicals
Dictionary of Fire Technology
Fire Technology – Calculations
Fire Technology – Chemistry & Combustion
How Did It Start?
Quarterly (weekly) 12303

Institution of Gas Engineers (IGE), 17 Grosvenor Crescent, London SW1X 7ES
T: (071) 2459811; Fax: (071) 2451229
Founded: 1863; Members: 6000
Gen Secr: Derek Chapman
Focus: Eng; Energy
Periodical
Gas Engineering & Management (10 times annually) 12304

Institution of Highways and Transportation, 3 Lygon Pl, Ebury St, London SW1W 0JS
T: (071) 7305245
Founded: 1930; Members: 8750
Focus: Civil Eng
Periodical
Highways and Transportation 12305

Institution of Lighting Engineers, 9 Lawford Rd, Rugby CV21 2DZ
T: (0788) 76492; Fax: (0788) 540145
Founded: 1924; Members: 1750
Pres: G.L.T. Pritchard; Gen Secr: D. Barnes
Focus: Electric Eng
Periodical
Lighting Journal (weekly) 12306

Institution of Mechanical Engineers (IMechE), 1 Birdcage Walk, London SW1H 9JJ
T: (071) 2227899; Tx: 917944; Fax: (071) 2224557
Founded: 1847; Members: 77000
Pres: Dr. A.A. Denton; Gen Secr: Dr. R.A. Pike
Focus: Eng
Periodicals
Automotive Engineering (weekly)
Engineering Materials (weekly)
Engineering News (weekly)
Green Engineering (weekly)
Proceedings (weekly) 12307

Institution of Mechanical Incorporated Engineers (IMIE), 3 Birdcage Walk, London SW1H 9JN
T: (071) 7991808; Fax: (071) 7992243
Founded: 1988; Members: 8000
Pres: Lord Gregson; Gen Secr: Ian Barnes
Focus: Eng
Periodical
Mechanical Incorporated Engineer (weekly) 12308

Institution of Mining and Metallurgy (IMM), 44 Portland Pl, London W1N 4BR
T: (071) 5803802; Fax: (071) 4365388
Founded: 1892; Members: 4850
Pres: Prof. N. Warner; Gen Secr: M.J. Jones
Focus: Mining; Metallurgy
Periodicals
IMM Abstracts (weekly)
New Minerals Industry International (weekly)
Transactions (3 times annually) 12309

Institution of Mining Engineers (IMinE), Danum House, South Parade, Doncaster DN1 2DY
T: (0302) 320486; Fax: (0302) 340554
Founded: 1889; Members: 3968
Gen Secr: W.J.W. Bourne
Focus: Eng; Mining
Periodical
The Mining Engineer (weekly) 12310

Institution of Nuclear Engineers (INucE), 1 Penerley Rd, London SE6 2LQ
T: (081) 6981500; Fax: (081) 6956409
Founded: 1959; Members: 1500
Gen Secr: Sheila Blackburn
Focus: Nucl Res
Periodical
The Nuclear Engineer (weekly) 12311

Institution of Occupational Safety and Health, 222 Uppingham Rd, Leicester LE5 0QG
T: (0533) 768424
Founded: 1947
Focus: Public Health
Periodical
Protection (weekly) 12312

Institution of Plant Engineers (IPE), 77 Great Peter St, London SW1P 2EZ
T: (071) 2332855; Fax: (071) 2332604
Founded: 1946; Members: 6750
Pres: J.B. Widdop; Gen Secr: R.S. Pratt
Focus: Eng; Botany
Periodical
The Plant Engineer (weekly) 12313

Institution of Polish Engineers in Great Britain (STP), 238-246 King St, London W6 0RF
T: (081) 7411940
Founded: 1940; Members: 425
Focus: Eng; Ethnology
Periodicals
Bulletin (weekly)
Technika i Nauka (weekly) 12314

Institution of Post Office Electrical Engineers (IPOEE), 2-12 Gresham St, London EC2V 7AG
T: (071) 2512655
Founded: 1906; Members: 13000
Focus: Electric Eng
Periodical
Journal (weekly) 12315

Institution of Railway Signal Engineers (IRSE), 1 Badlake Close, Badlake Hill, Dawlish EX7 9JA
T: (0626) 888096; Fax: (0626) 888571
Founded: 1912; Members: 2300
Gen Secr: R.L. Weedon
Focus: Eng
Periodical
Proceedings (weekly) 12316

Institution of Structural Engineers, 11 Upper Belgrave St, London SW1X 8BH
T: (071) 2354535; Fax: (071) 2354294
Founded: 1908; Members: 23000
Pres: Dr. H.P.J. Taylor; Gen Secr: D.J. Clark
Focus: Eng
Periodical
The Structural Engineer (weekly) 12317

Institution of Water and Environment Management, 15 John St, London WC1N 2EB
T: (071) 8313110; Fax: (071) 4054967
Founded: 1895
Focus: Ecology; Hydrology
Periodicals
Journal of the Institution of Water and Environment Management (weekly)
Newsletter (weekly)
Year Book (weekly) 12318

Intercontinental Church Society, 175 Tower Bridge Rd, London SE1 2AQ
T: (071) 4074588; Fax: (071) 3780541
Founded: 1823
Pres: Lord of Newick Brentford; Gen Secr: Dr. P.K.L. Schmiegelow
Focus: Rel & Theol
Periodical
Intercon (3 times annually) 12319

Intermediate Technology Development Group (ITDG), Myson House, Railway Terrace, Rugby CV21 3HT
T: (0788) 560631; Tx: 317466; Fax: (0788) 540270
Founded: 1965
Pres: Prince of Wales
Focus: Eng
Periodicals
Appropriate Technology (weekly)
Waterlines (weekly) 12320

International African Institute (IAI), Thornhaugh St, Russell Sq, London WC1H 0XG
T: (071) 8313068
Founded: 1926
Pres: Prof. William A. Shack; Gen Secr: Prof. David Parkin
Focus: Ethnology; Rel & Theol
Periodical
Africa and Africa Bibliography (weekly) . 12321

International Anatomical Nomenclature Committee, c/o Dept of Anatomy, Guy's Hospital Medical School, London SE1 9RT
Focus: Anat; Standards 12322

International Association for Esperanto in Libraries, 14 Elmdale Rd, London N13 4UL
T: (081) 8889411
Founded: 1971; Members: 100
Focus: Ling; Libraries & Bk Sci 12323

International Association for Mass Communication Research (IAMCR), c/o Centre for Mass Communication Research, University of Leicester, Leicester LE1 7LT
Founded: 1957; Members: 1000
Focus: Journalism 12324

International Association for Scandinavian Studies (IASS), c/o EUR, University of East Anglia, Norwich NR4 7TJ
T: (0603) 56161; Fax: (0603) 58553
Founded: 1962; Members: 145
Focus: Ethnology 12325

International Association of Agricultural Museums, c/o Rural History Centre, University of Reading, Reading RG6 2AG
T: (0734) 318660; Tx: 847813; Fax: (0734) 751264
Founded: 1968
Pres: Dr. Nitra Vontorcic
Focus: Agri
Periodical
Acta Museorum Agriculturae Pragae (weekly) 12326

International Association of Applied Linguistics (IAAL), c/o Dept of Linguistics, University of Edinburgh, Edinburgh EH8 9LN
Founded: 1964; Members: 22
Focus: Ling 12327

International Association of Biological Standardization (IABS), South Wind, Streat, Hassocks BN6 8RT
T: 890366
Founded: 1955; Members: 600
Focus: Standards; Bio
Periodical
Journal of Biological Standardization (weekly) 12328

International Association of Environmental Coordinators (IAEC), Corylus Burton Way, Chalfont Saint Gills
Founded: 1976; Members: 150
Focus: Ecology 12329

International Association of Institutes of Navigation, c/o Royal Insitute of Navigation, 1 Kensington Gore, London SW7 2AT
T: (071) 5895021; Fax: (071) 8238671
Founded: 1975
Focus: Navig 12330

International Association of Music Libraries, Archives and Documentation Centres (United Kingdom Branch), c/o University of Exeter Library, Stocker Rd, Exeter EX4 4PT
Gen Secr: J. Crawley
Focus: Libraries & Bk Sci
Periodical
BRIO (weekly) 12331

International Association of Paediatric Dentistry, c/o Dept of Child Dental Health, The London Hospital Medical College, Turner St, London E1 2AD
T: (071) 3777058 ext 2199; Fax: (071) 3777058
Founded: 1969
Focus: Dent
Periodical
Journal of the IAPD (weekly) 12332

International Association of Teachers of English as a Foreign Language (IATEFL), 3 Kingsdown Chambers, Kingsdown Park, Tankerton, Whitstable CT5 2DJ
Fax: (0227) 274415
Founded: 1967; Members: 3000
Pres: Prof. David Crystal
Focus: Ling; Educ
Periodical
Newsletter (5 times annually) 12333

International Association of Technological University Libraries (IATUL), c/o Michael Breaks, Heriot-Watt University Library, Edinburgh EH14 4AS
T: (031) 4495111; Fax: (031) 4513164
Founded: 1955; Members: 202
Pres: Dr. Gerard Van Marle; Gen Secr: Michael Breks
Focus: Libraries & Bk Sci
Periodical
IATUL Newsletter (weekly) 12334

International Association of University Professors of English (IAUPE), c/o Penwithian Higher Fore St, Marazion TR17 0BQ
Founded: 1951; Members: 660
Focus: Adult Educ; Ling
Periodical
Bulletin (bi-annually) 12335

International Association of Volcanology and Chemistry of the Earth's Interior JAVCEJ, c/o Dept of Earth Siences, University, Leeds LS2 9JT
Founded: 1919
Focus: Geology; Chem 12336

International Association on Water Quality (IAWQ), 1 Queen Anne's Gate, London SW1H 9BT
T: (071) 2223848; Tx: 918518; Fax: (071) 2331197
Founded: 1965; Members: 4600
Pres: Prof. P. Grau; Gen Secr: A. Milburn
Focus: Water Res; Hygiene; Ecology

Periodicals
Water Quality International (weekly)
Water Research (weekly)
Water Science and Technology (weekly) . 12337

International Bar Association (IBA), 2 Harewood Pl, London W1R 9HB
Fax: (071) 4090456
Founded: 1947; Members: 12500
Gen Secr: Madeleine May
Focus: Law
Periodicals
International Bar News (weekly)
International Business Lawyer (weekly)
International Legal Practioner (weekly) . . 12338

International Bee Research Association (IBRA), 18 North Rd, Cardiff CF1 3DY
T: (0222) 372409; Tx: 262433; Fax: (0222) 665522
Founded: 1949; Members: 1500
Pres: Prof. W.E. Kerr; Gen Secr: Andrew Matheson
Focus: Zoology
Periodicals
Apicultural Abstracts (weekly)
Bee World (weekly)
Journal of Apicultural Research (weekly) . 12339

International Cerebral Palsy Society (ICPS), 19 Saint Mary's Grove, Chiswick, London W4 3LL
T: (071) 7949761
Founded: 1969; Members: 300
Focus: Rehabil 12340

International Commission for Orders of Chivalry, 1 Darnaway St, Edinburgh EH3 6DW
T: 2251896
Founded: 1960
Focus: Hist 12341

International Commission for the History of Representative and Parliamentary Institutions, Arts Bldg, University of Sussex, Brighton BN1 9QN
T: (0273) 606755
Founded: 1936
Focus: Hist
Periodical
Parliaments, Estates and Representation (weekly)
. . . . 12342

International Commission on Physics Education, c/o Science Dept, Malvern, Worcs
Founded: 1960; Members: 14
Focus: Physics; Educ 12343

International Commission on Polar Meteorology (ICPM), c/o British Antarctic Survey, Atmospheric Sciences Division, Madingley Rd, Cambridge CB3 0ET
Focus: Geophys 12344

International Commission on Zoological Nomenclature (ICZN), c/o Natural History Museum, Cromwell Rd, London SW7 5BD
T: (071) 9389387
Founded: 1895; Members: 29
Focus: Zoology; Standards
Periodical
Bulletin of Zoological Nomenclature (weekly)
. . . . 12345

International Conference on Social Science and Medicine, c/o Centre for Social Research, University of Sussex, Falmer, Brighton BN1 9RF
T: (0273) 66755
Founded: 1968; Members: 150
Focus: Med; Hygiene 12346

International Council of Kinetography (ICKL), 250 Burges Rd, London E6 2E5
Founded: 1959; Members: 79
Focus: Perf Arts
Periodical
Conference Proceedings (bi-annually) . . 12347

International Dental Federation (IDF), 7 Carlisle St, London W1V 5RG
Founded: 1900; Members: 350000
Pres: Dr. Clive Ross; Gen Secr: Dr. P.-A. Zillin
Focus: Dent
Periodicals
FDI Dental World (weekly)
International Dental Journal (weekly) . . 12348

International Enamellers Institute, Ripley near Derby
Founded: 1957
Focus: Eng 12349

International Esperanto-Association of Jurists (IEAJ), 77 Grasmere Av, Wembley HA9 8TF
T: 9049081
Founded: 1957; Members: 305
Focus: Law
Periodical
Internacia Jura Revuo (weekly) 12350

International Federation for Modern Languages and Literatures, c/o The Queen's University of Belfast, Belfast BT7 1NN
Founded: 1928
Focus: Ling; Lit 12351

International Federation for Theatre Research (IFTR), c/o Dept of Theatre Studies, University of Lancaster, Bailrigg, Lancaster
Founded: 1955; Members: 164
Focus: Perf Arts 12352

International Federation of Airworthiness (IFA), 58 Whiteheath Av, Ruislip HA4 7PW
T: (0895) 672504; Tx: 8951771; Fax: (0895) 676656
Founded: 1975; Members: 110
Pres: Stewart John; Gen Secr: Jim Rainbow
Focus: Eng
Periodical
International Airworthiness News (weekly) 12353

International Federation of Anti-Leprosy Associations (ICEP), 234 Blythe Rd, London W14 0HJ
T: (071) 6026925; Tx: 894241
Founded: 1966; Members: 24
Focus: Derm
Periodicals
ICEP Coordinate Budget (3 times annually)
ICEP Flash (weekly) 12354

International Federation of Cell Biology (IFCB), c/o Imperial Cancer Research Fund, Lincoln's Inn Fields, London WC2A 3PX
Founded: 1947
Focus: Cell Biol & Cancer Res
Periodical
Cell Biology International Reports (weekly) 12355

International Federation of Gynecology and Obstetrics (FIGO), 27 Sussex Pl, London NW1 4RG
T: (071) 7232951; Fax: (071) 7247725
Founded: 1954; Members: 94
Pres: J.J. Sciarra; Gen Secr: Prof. D.V.I. Fairweather
Focus: Gynecology
Periodical
International Journal of Gynecology an Obstetrics (weekly) 12356

International Federation of Multiple Sclerosis Societies (I.F.M.S.S.), 10 Heddon St, London W1R 7LJ
T: (071) 7349120; Fax: (071) 2872587
Founded: 1967; Members: 34
Pres: William P. Benton; Gen Secr: Sylvia Lawry
Focus: Med
Periodicals
Annual Report (weekly)
MS Research in Progress (biennially)
MS Research Report (weekly)
Therapeutic Claims in MS (biennially)
Update (weekly) 12357

International Federation of Practitioners of Natural Therapeutics (IFPNT), 21 Bingham Pl, London W1M 3FH
T: (071) 9356933
Founded: 1963
Focus: Therapeutics 12358

International Federation of Societies of Cosmetic Chemists (IFSCC), 57 Guildford St, Luton LU1 2NL
T: (0582) 26661; Fax: (0582) 405217
Founded: 1959; Members: 32
Gen Secr: Lorna Weston
Focus: Chem 12359

International Federation of Sportive Medicine (IFSM), Farnham Fark Rehabilitation Centre, Farnham Royal, Slough SL2 3LR
Founded: 1928; Members: 20000
Focus: Med 12360

International Federation of Surgical Colleges, c/o Royal College of Surgeons, 18 Nicolson St, Edinburgh EH8 9DW
Founded: 1958; Members: 42
Focus: Educ; Surgery 12361

International Federation on Aging – European Office (IFA), Astral House, 1268 London Rd, London SW16 4EJ
T: (081) 6798000; Tx: 24667; Fax: (081) 6796069
Founded: 1973
Gen Secr: Sally Greengross
Focus: Geriatrics; Adult Educ
Periodical
Aging International (bi-annually) 12362

International Filariasis Association (IFA), c/o Dept of Helminthology, London School of Hygiene and Tropical Medicine, Kepped St, London WC1E 7HT
T: (01) 8255154
Founded: 1963; Members: 200
Focus: Med; Immunology 12363

International Food Information Service (IFIS), Lane End House, Shinfield, Reading RG2 9BB
T: (0734) 883895; Fax: (0734) 885065
Founded: 1968
Gen Secr: J.R. Metcalfe
Focus: Food
Periodicals
Food Science and Technology Abstracts (weekly)
Food Science Profiles (weekly) 12364

International Genetics Federation, c/o Plant Breeding Institute, Maris Lane, Trumpington, Cambridge CB2 2LQ
Founded: 1911
Focus: Genetics 12365

International Glaciological Society (IGS), c/o SPRI, Lensfield Rd, Cambridge CB2 1ER
T: (0223) 355974; Fax: (0223) 336543
Founded: 1936; Members: 900
Focus: Geology; Oceanography
Periodicals
Annals of Glaciology: Conference Proceedings (1-2 times annually)
ICE-News Bulletin (3 times annually)
Journal of Glaciology (3 times annually) . 12366

International Hydrofoil and Multihull Society, 34 Riverside, Martham, Great Yarmouth NR29 4RG
Founded: 1954
Focus: Eng
Periodical
Safe Multhull Cruising (weekly) 12367

International Hydrofoil Society (IHS), 17 Melcombe Court, Dorset Sq, London NW1 6EP
Founded: 1970
Focus: Eng; Oceanography 12368

International Institute for Conservation of Historic and Artistic Works (IIC), 6 Buckingham St, London WC2N 6BA
T: (071) 8395975; Fax: (071) 9761564
Founded: 1950; Members: 3800
Focus: Preserv Hist Monuments
Periodicals
Art & Archaeology Technical Abstracts (weekly)
Studies in Conservation (weekly) 12369

International Institute of Social Economics (IISE), Enholmes Hall, Patrington, Hull HU12 0PR
T: (0964) 630033; Tx: 51317
Founded: 1972; Members: 43
Focus: Econ
Periodical
International Journal of Social Economics (7 times annually) 12370

International Language Society of Great Britain (ILSGB), 18 Lane Head Rise, Staincross Common, Mapplewell, Barnsley S75 6NQ
Founded: 1908
Focus: Ling 12371

International Language Union, 14 Stray Towers, Victoria Rd, Harrogate HG2 O1J
Founded: 1908
Focus: Ling 12372

International Law Association (ILA), 17 Russell Sq, London WC1B 5DR
T: (071) 3232978; Fax: (071) 3233580
Founded: 1873; Members: 4000
Pres: Prof. M. El-Said El-Dakkak; Gen Secr: Prof. D.J.C. Wyld
Focus: Law
Periodical
Conference Reports (biennial) 12373

International Lead and Zinc Study Group (ILZSG), Metro House, 58 Saint James St, London SW1A 1LD
T: (071) 499-9373; Tx: 299819; Fax: (071) 4933725
Founded: 1959; Members: 30
Focus: Metallurgy; Mining
Periodical
Lead and Zinc Statistics (weekly) . . . 12374

International Management Centre from Buckingham (IMCB), 13 Castle St, Buckingham MK18 1BP
T: (0280) 817222; Fax: (0280) 813297
Founded: 1964; Members: 2800
Pres: Baroness Dr. Caroline Cox; Gen Secr: Dr. Bamie Pettman
Focus: Business Admin
Periodical
Management Decision (weekly) 12375

International Map Collector's Society (IMCS), Woodstock, Flyford Flavell WR7 4BS
T: (038682) 619
Founded: 1980; Members: 200
Focus: Cart 12376

International Medical Society of Paraplegia, c/o National Spinal Injuries Centre, Stoke Mandeville Hospital, Aylesbury HP21 8AL
T: (0296) 315866; Fax: (0296) 315268
Founded: 1961; Members: 1000
Focus: Pathology 12377

International Mycological Institute (IMI), Bakeham Ln, Egham TW20 9TV
T: (0784) 470111; Fax: 3321171
Focus: Botany, Specific
Periodicals
Bibliography of Systematic Mycology (weekly)
IMI Descriptions of Pathogenic Fungi and Bacteria (weekly)
IMI Distribution Maps of Plant Diseases (weekly)
Index of Fungi (weekly)
Mycopathologia
Systema Ascomycetum (weekly) 12378

International Numismatic Commission, c/o Dept of Coins and Medals, British Museum, London WC1B 3DG
T: (071) 3238227; Fax: (071) 3238171
Founded: 1936; Members: 132
Pres: Dr. Cécile Morrisson; Gen Secr: Dr. Andrew Burnett
Focus: Numismatics
Periodicals
Compte rendu (weekly)
International Numismatic Newsletter (weekly)
. . . . 12379

International Organization for the Study of the Old Testament (IOSOT), 34 Gough Way, Cambridge CB3 9LN
T: (0223) 63219
Founded: 1950
Focus: Rel & Theol 12380

International Organization of Palaeobotany (I.O.P.), c/o N.E. London Polytechnic, Romford Rd, London E15
T: (081) 5907722
Founded: 1950; Members: 1300
Focus: Botany 12381

International P.E.N. Writers Association, 9-10 Charterhouse Bldgs, Goswell Rd., London EC1M 7AT
T: (071) 2534308
Focus: Lit
Periodical
P.E.N. International (weekly) 12382

International Petroleum Industry Environmental Conservation Association (IPIECA), 110 Euston Rd, London NW1
Founded: 1974; Members: 19
Focus: Ecology 12383

International Phonetic Association (IPA), c/o Dept of Linguistics and Phonetics, University of Leeds, Leeds LS2 9JT
T: (0532) 333563
Founded: 1886; Members: 750
Focus: Ling
Periodical
Journal of the International Phonetic Association (bi-annually) 12384

International Psychoanalytical Association (IPA), Broomhills, Woodside Lane, London N12 8UD
T: (081) 4468324; Fax: 4454729
Founded: 1910; Members: 7200
Focus: Psychoan
Periodicals
Bulletin (weekly)
IPA Newsletter (weekly)
IPA Roster (weekly) 12385

International Rubber Study Group (IRSG), York House, Empire Way, Wembley HA9 0PA
T: (081) 9037707; Tx: 8951293; Fax: (081) 9032848
Founded: 1944; Members: 24
Focus: Materials Sci
Periodicals
International Rubber Digest (weekly)
Proceedings of Annual Assembly (weekly)
Proceedings of International Discussion Forum (weekly)
Rubber Statistical Bulletin (weekly) . . . 12386

International Seismological Centre (ISC), Pipers Lane, Thatcham, Newbury RG13 4NS
T: (0635) 861022; Tx: 849303; Fax: (0635) 872351
Founded: 1964
Focus: Geology
Periodicals
Bibliography of Seismology (weekly)
Bulletin of the International Seismological Centre (weekly)
Felt and Damaging Earthquakes (weekly)
Regional Catalogue of Earthquakes (weekly)
. . . . 12387

International Society for Applied Ethology, c/o Dr. McAppleby, IERM, University of Edinburgh, West Mains Rd, Edinburgh EH9 3JG
T: (031) 6671041
Founded: 1966; Members: 400
Focus: Vet Med
Periodical
Newsletter (Biannual) 12388

International Society for Microelectronics, Tapestries Coach House, Harbertonford, Devon
Focus: Electronic Eng 12389

International Society for Soil Mechanics and Foundation Engineering (ISSMFE), c/o Engineering Dept, University, Trumpington St, Cambridge CB2 1PZ
T: (0223) 355020; Tx: 81239; Fax: (0223) 359675
Founded: 1936
Focus: Eng
Periodical
ISSMFE News (weekly) 12390

International Society of Audiology, 330 Gray's Inn Rd, London WC1
Founded: 1952
Focus: Otorhinolaryngology 12391

International Society of Development Biologists (ISDB), Bassett Crescent East, Southampton SO9 3TU
Founded: 1911; Members: 800
Focus: Bio
Periodical
All Differentiation (weekly) 12392

International Society of Neuropathology, c/o National Hospital for Nervous Diseases, Queen Sq, London WC1N 3BG
Focus: Pathology 12393

International Society of Radiographers and Radiological Technicians (ISRRT), 18 Merthyr Rd, Whitchurch, Cardiff CF4 1DG
Founded: 1959; Members: 52
Focus: Radiology 12394

International Special Committee on Radio Interference (CISPR), c/o British Standards Institution, 2 Park St, London W1A 2BS
T: (071) 6299000; Fax: 6290506
Founded: 1933
Gen Secr: T.H. Chapman
Focus: Electric Eng 12395

International Study Group for Mathematics Learning (ISGML), c/o LVCE Research Institute, Ealing Technical College, Woodlands Av, Acton, London W3 9DN
T: (081) 9926944
Founded: 1961
Focus: Educ; Math 12396

International Tar Conference (ITC), c/o BTJA, 132-135, Sloane St, London SW1X 9BB
T: (071) 7305212
Founded: 1952
Focus: Materials Sci 12397

International Union Against the Venereal Diseases and the Treponematoses (IUVDT), c/o The Pread Street Clinic, Saint Mary's Hospital, London W2 1NY
T: (071) 2621123
Founded: 1923
Focus: Venereology 12398

International Union for Quaternary Research, c/o London University, Egham TW20 0EX
T: (0273) 748919; Fax: (0273) 748919
Founded: 1928; Members: 37
Focus: Geology
Periodical
Quaternary International (weekly) 12399

International Union of Air Pollution Prevention and Environmental Protection Associations, 136 North St, Brighton BN1 1RG
T: (0273) 326313; Fax: (0273) 735802
Founded: 1964; Members: 34
Pres: Dr. A. Estlander; Gen Secr: J. Langston
Focus: Ecology
Periodicals
Handbook (bi-annually)
Newsletter (weekly)
World Congress Proceedings 12400

International Union of Crystallography (IUCr), 5 Abbey Sq, Chester CH1 2HU
T: (0244) 42878
Founded: 1947; Members: 33
Pres: Prof. M. Nardelli; Gen Secr: Dr. J.N. King
Focus: Mineralogy
Periodicals
Acta Crystallographica, Section B (weekly)
Acta Crystallographica, Section C (weekly)
Acta Cystallographica, Section A (weekly)
Journal of Applied Crystallography (weekly)
Midemia Structures and Dimensions (weekly)
Structure Reports (weekly) 12401

International Union of Independent Laboratories / Union Internationale des Laboratoires Independants, c/o Harry Stanger Ltd, The Laboratories, Fortune Lane, Elstree WD6 3HQ
T: (081) 2073191; Fax: 2074706
Founded: 1960; Members: 780
Pres: H. Hochschwender; Gen Secr: D.H. Stanger
Focus: Sci 12402

International Union of Nutritional Sciences (IUNS), c/o Institute of Biology, 20 Queensberry Pl, London SW7 2DZ
Founded: 1946; Members: 50
Focus: Nutrition
Periodicals
IUNS Directory (weekly)
IUNS Newsletter (bi-annually) 12403

International Union of Pure and Applied Chemistry (IUPAC), Bank Court Chambers, 2-3 Pound Way, Templars Sq, Cowley, Oxford, OX4 3YF
T: (0865) 747744; Tx: 83147; Fax: (0865) 747510
Founded: 1919; Members: 45
Focus: Chem
Periodicals
Chemistry International (weekly)
Pure and Applied Chemistry (weekly) . . 12404

International Waterfowl and Wetlands Research Bureau (IWRB), Slimbridge GL2 7BX
T: (0453) 890624, 890634
Founded: 1954
Pres: Dr. M. Moser
Focus: Ornithology
Periodical
IWRB News (weekly) 12405

International Wool Secretariat (IWS), Wool House, 6-7 Carlton Gardens, London SW1Y 5AE
T: (071) 9307300; Tx: 263926
Founded: 1937; Members: 980
Focus: Marketing; Textiles 12406

International Wool Study Group (I.W.S.G.), 123 Victoria St, London SW1E 6RB
T: (071) 2156214
Founded: 1946; Members: 17
Gen Secr: M.T. Dunn
Focus: Materials Sci
Periodicals
Wool (weekly)
Wool Questionnaire (weekly) 12407

Interplanetary Space Travel Research Association (ISTRA), 21 Hargwyne St, London SW9
T: (071) 7334814
Founded: 1957; Members: 6000
Focus: Astronomy 12408

Intractable Pain Society of Great Britain and Ireland (IPS), c/o Dept of Anaesthetics, Derbyshire Royal Infirmary, London Rd, Derby DE1 2QJ
T: (0332) 47141
Founded: 1968; Members: 196
Focus: Anesthetics
Periodical
Forum (2-3 times annually) 12409

Irish Association for Cultural, Economic and Social Relations, 31 Castlehill Rd, Belfast
Focus: Econ; Sociology 12410

Irish Heritage, 32 The Grove, London N3 1QJ
T: (081) 3462726
Founded: 1973; Members: 201
Focus: Lit; Music; Perf Arts 12411

Isle of Man Natural History and Antiquarian Society, c/o Manx Museum, Douglas
Founded: 1879; Members: 850
Focus: Hist; Nat Sci; Archeol
Periodical
Proceedings 12412

Isle of Wight Natural History and Archaeological Society, Swiss Chalet, Rylstone Gardens, Shanklin PO37 6RG
Founded: 1919
Focus: Hist; Nat Sci; Archeol
Periodicals
Isle of Wight Birds (weekly)
Proceedings (weekly) 12413

Isotype Institute, 116 Haverstock Hill, London NW3
Founded: 1942
Focus: Educ 12414

Jane Austen Society, Carton House, Medstead, Alton GU34 5PE
T: (0420) 562469
Founded: 1940; Members: 1800
Focus: Lit
Periodical
Annual Report (weekly) 12415

Jazz Centre Society (JCS), 35 Great Russell St, London WC1
T: (071) 5808532
Founded: 1968
Focus: Music 12416

Jersey Association of the National Association of Schoolmasters, 3 Palace Close, Bagatelle, Saint Saviour's, Jersey
Focus: Educ 12417

Jewish Historical Society of England (JHSE), 33 Seymour Pl, London W1H 5AP
T: (071) 7234404
Founded: 1893; Members: 900
Pres: Israel Finesten; Gen Secr: Cyril Drukker
Focus: Hist
Periodical
Transactions (18 times annually) 12418

Johann Strauss Society of Great Britain, 12 Bishams Court, Church Hill, Caterham, CR3 6SE
Founded: 1964; Members: 500
Focus: Music
Periodical
Tritsch-Tratsch (weekly) 12419

Johnson Society, c/o Johnson Birthplace Museum, Breadmarket St, Lichfield WS13 6LG
T: (05432) 24972
Founded: 1910; Members: 590
Focus: Lit
Periodical
Transactions (weekly) 12420

Johnson Society of London, The Manse, Tower Rd, Hindhead GU26 6SU
T: 4167
Founded: 1928
Focus: Lit 12421

The Joint Association of Classical Teachers (JACT), 31-34 Gordon Sq, London WC1H 0PY
Founded: 1962; Members: 2200
Gen Secr: J. Cohen
Focus: Hist; Educ
Periodicals
JACT Bulletin
JACT Review (weekly)
Omnibus 12422

The Josephine Butler Society, 49 Hawkshead Lane, North Mymms, Hatfield AL9 7TD
T: (0707) 43150
Founded: 1870; Members: 160
Focus: Law
Periodical
News and Views (weekly) 12423

Junior Astronomical Society (JAS), 36 Fairway, Keyworth, Nottingham NG12 5DU
Founded: 1953; Members: 2100
Gen Secr: Guy N.H. Fennimore
Focus: Astronomy
Periodicals
Circular Newsletter (weekly)
Popular Astronomy (weekly) 12424

Jury System Reform Society (JSRS), 18 Faversham Rd, Beckenham BR3 3PN
Founded: 1979; Members: 1435
Focus: Law
Periodical
Justice for Whom? 12425

Keats-Shelley Memorial Association (K-SMA), 1 Lewis Rd, Radford Semele, Leamington Spa CV31 1UB
T: (071) 4352062
Founded: 1906; Members: 252
Focus: Lit
Periodical
The Keats-Shelley Review 12426

Kent and Sussex Poetry Society, c/o D.M. Hulse, Costens, Carpenters Lane, Hadlow TN11 0EY
T: (0732) 851404
Founded: 1946; Members: 50
Focus: Lit
Periodical
Poetry Folio (weekly) 12427

Kent Archaeological Society (K.A.S.), Three Elms, Woodlands Lane, Shorne, Gravesend DA12 3HH
T: (0474) 822280
Founded: 1857; Members: 1620
Focus: Archeol
Periodical
Archaeologia Cantiana (weekly) 12428

Kent County Agricultural Society, County Showground, Detling, Maidstone ME14 3JF
T: (0622) 630975; Fax: (0622) 630978
Founded: 1923; Members: 3000
Gen Secr: Frances Day
Focus: Agri
Periodical
Kent View (weekly) 12429

Kilvert Society, 91 Hallow Rd, Worcester
Focus: Sci 12430

The Kipling Society, POB 68, Haslemere GU27 2YR
T: (0428) 652709
Founded: 1927; Members: 1000
Pres: Dr. Michael Brock; Gen Secr: Norman Entract
Focus: Lit
Periodical
The Kipling Journal (weekly) 12431

Laboratory Animal Science Association (LASA), c/o Charing Cross Hospital Medical School, Saint Dunstans Rd, London LS2 9JT
T: (081) 7482040
Founded: 1963; Members: 360
Focus: Zoology 12432

Lace Research Association (LRA), 7 Gregory Blvd, Nottingham NG7 6LD
T: (0602) 623311
Founded: 1952; Members: 23
Focus: Materials Sci 12433

Lace Society, Belvedere, Bradley, Wrexham, Clwyd
T: Brymbo 8144
Founded: 1962; Members: 220
Focus: Materials Sci 12434

Lakeland Dialect Society (LDS), 8 Barras Close, Morton Park, Carlisle CA2 6PR
T: (0228) 20539
Founded: 1938; Members: 280
Focus: Ling 12435

Lambeg Industrial Research Association (LIRA), 14 Lambeg Rd, Lambeg, Lisburn BT27 4RJ
T: (0846) 662255
Founded: 1919; Members: 102
Pres: A. McIntosh; Gen Secr: Dr. W.W. Foster
Focus: Eng 12436

Lancashire and Cheshire Fauna Society, c/o Entomology Dept, Manchester Museum, The University, Manchester M13 9PL
T: (061) 2733333
Founded: 1914; Members: 200
Focus: Zoology 12437

Lancashire Dialect Society (L.D.S.), 30 Broadoak Rd., Bramhall, Stockport
T: (061) 4408272
Founded: 1951; Members: 300
Gen Secr: Peter Wright
Focus: Ling
Periodical
Journal (weekly) 12438

Lancashire Parish Register Society (LPRS), 65 Hillfield Rd, Hemel Hempstead HP2 4AB
Founded: 1897
Gen Secr: Neil Hudson
Focus: Hist 12439

The Landscape Institute (LI), 6-7 Barnard Mews, London SW11 1QU
T: (071) 7389166
Founded: 1929; Members: 3500
Focus: Hort
Periodicals
Landscape Design (10 times annually)
Register of Practices (weekly) 12440

Law Society, Law Society's Hall, 113 Chancery Ln, London WC2A 1PL
T: (071) 2421222; Tx: 261203; Fax: (071) 8310344
Founded: 1825; Members: 76000
Gen Secr: John Hayes
Focus: Law
Periodical
Gazette (weekly) 12441

Law Society of Scotland, 26 Drumsheugh Gardens, Edinburgh EH3 7YR
T: (031) 2267411; Fax: (031) 2252934
Founded: 1949; Members: 8268
Focus: Law
Periodicals
Journal (weekly)
Scottish Civil Law Reports (weekly)
Scottish Criminal Case Reports (weekly) . 12442

League for the Exchange of Commonwealth Teachers (LECT), 7 Lion Yard, Tremadoc Rd, Clapham, London SW4 7NF
T: (071) 4981101
Founded: 1901; Members: 3000
Focus: Educ
Periodical
Exchange Teacher: Annual Report (weekly)
. . . . 12443

Leeds Philosophical and Literary Society, c/o City Museum, Calverley St, Leeds
T: (0532) 452894
Founded: 1820; Members: 150
Pres: B. Mortimer; Gen Secr: Dr. A.C. Chadwick
Focus: Lit; Philos
Periodical
Proceedings (weekly) 12444

Leicestershire Archaeological and Historical Society (LAHS), Guildhall, Leicester LE1 5FQ
Founded: 1855; Members: 700
Focus: Archeol; Hist 12445

Leicestershire Local History Council, c/o Leicestershire Records Office, 57 New Walk, Leicester LE1 7JB
T: (0533) 544566
Founded: 1966
Focus: Hist
Periodical
The Leicestershire Historian (weekly) . . 12446

Leisure Studies Association (LSA), c/o Scottish Tourist Board, 23 Ravelston Terrace, Edinburgh EH4 3EY
T: (031) 3322433; Fax: (031) 72272
Founded: 1975; Members: 200
Pres: Les Haywood; Gen Secr: Brian Hay
Focus: Travel
Periodicals
Conference Reports (weekly)
Leisure Studies (3 times annually) . . . 12447

Lewis Carroll Society, 55 Heath Cottages, Chislehurst Common BR7 5ND
T: 4675734
Founded: 1969; Members: 235
Focus: Lit 12448

Liberal Industrial Relations Association (LIRA), Bury Fold Farm, Darwen BB5 2QQ
T: 74709
Founded: 1973; Members: 70
Focus: Econ 12449

Library Association (L.A.), 7 Ridgmount St, London WC1E 7AE
T: (071) 6367543; Tx: 21897; Fax: (073) 4367218
Founded: 1877; Members: 25000
Gen Secr: Ross Shimmon
Focus: Libraries & Bk Sci
Periodicals
Library Association Record (weekly)
Year Book (weekly) 12450

The Library Association – Personnel, Training and Education Group, c/o Perry Library, South Bank University, 10b Borough Rd, London SE1 0AA
T: (071) 8156644; Fax: (071) 8156699
Founded: 1968; Members: 1200
Focus: Libraries & Bk Sci
Periodical
Personnel, Training and Education Journal (3 times annually) 12451

Lincoln Record Society (L.R.S.), c/o Lincoln Cathedral Library, Lincoln LN2 1PZ
T: (0522) 544544
Founded: 1910; Members: 340
Pres: Prof. J.C. Holt; Gen Secr: N.H. Bennett
Focus: Hist 12452

Linguistics Association of Great Britain (L.A.G.B.), c/o Dept of English Studies, Glasgow G1 1XH
T: (041) 5524400
Founded: 1959; Members: 700
Pres: Prof. K. Brown; Gen Secr: Dr. N. Fabb
Focus: Ling
Periodicals
British Linguistic Newsletter (weekly)
Journal of Linguistics (weekly) 12453

Linnean Society of London, Burlington House, Piccadilly, London W1V 0LQ
T: (071) 4344479; Fax: (071) 2879364
Founded: 1788; Members: 1800
Pres: Prof. B.G. Gardiner; Gen Secr: Dr. J.C. Marsden

Focus: Hist; Nat Sci
Periodicals
Biological Journal (weekly)
Botanical Journal (weekly)
The Linnean (3 times annually)
Zoological Journal (weekly) 12454

The Liszt Society, 135 Stevenage Rd, Fulham, London SW6 6PB
T: (071) 3819751
Founded: 1950; Members: 260
Pres: Leslie Howard; Gen Secr: J. Audrey Ellison
Focus: Music
Periodicals
The Liszt Society Journal (weekly)
Newsletter (weekly) 12455

Literary and Philosophical Society of Liverpool, 13 Hilbre Av, Wallasey L44 5RR
T: (051) 6384309
Founded: 1812; Members: 50
Focus: Lit; Philos 12456

Literary and Philosophical Society of Newcastle upon Tyne, Westgate Rd, Newcastle-upon-Tyne NE1 1ST
T: (091) 2320192
Founded: 1793; Members: 1300
Focus: Lit; Philos; Hist 12457

Local Population Studies Society (LPSS), 10 Holmbush Rd, London SW15
T: (081) 7892571
Founded: 1973; Members: 370
Focus: Cultur Hist
Periodicals
Local Population Studies (weekly)
LPS Supplements (weekly) 12458

Loch Ness Phenomena Investigation Bureau (LNPIB), c/o Loch Ness Investigation, Drumnadrochit, Invernessshire
Founded: 1961
Focus: Zoology 12459

London and Middlesex Archaeological Society (LAMAS), c/o Museum of London, 150 London Wall, London EC2Y 5HN
T: (071) 6003699
Founded: 1855; Members: 850
Pres: Harvey Sheldon; Gen Secr: Malcolm Harden
Focus: Archeol; Hist
Periodicals
Newsletter (3 times annually)
Transactions (weekly) 12460

London Association of Science Teachers (LAST), 61 Crescent Lane, London SW4
Focus: Educ 12461

London Mathematical Society (L.M.S.), Burlington House, Piccadilly, London W1V ONL
T: (071) 4375377; Fax: (071) 4394629
Founded: 1865; Members: 1970
Pres: Prof. J.R. Ringrose; Gen Secr: Prof. R.Y. Sharp
Focus: Math
Periodicals
LMS Bulletin (weekly)
LMS Journal (weekly)
LMS Proceedings (weekly)
Nonlinearity (weekly) 12462

London Medieval Society (LMS), c/o Dept of German, Westfield College, Kidderpore Av, London NW3 7ST
T: (071) 4357141
Founded: 1945; Members: 150
Focus: Hist 12463

London Metallurgical Society, 43 Melbourne Rd, Wallington, Surrey
Founded: 1966
Focus: Metallurgy 12464

London Natural History Society (LNHS), 21 Green Way, Frinton-on-Sea CO13 9AL
T: (0255) 674678
Founded: 1858; Members: 1200
Pres: R. Day; Gen Secr: A.J. Barrett
Focus: Hist; Nat Sci
Periodicals
London Bird Report (weekly)
London Naturalist (weekly) 12465

London Orchestral Association, 13-14 Archer St, London W1V 7HG
T: (071) 4371588
Focus: Music 12466

London Record Society (L.R.S.), c/o Institute of Historical Research, Senate House, London WC1E 7HU
T: (071) 6360272
Founded: 1964; Members: 360
Pres: H.S. Cobb; Gen Secr: H.J. Creaton
Focus: Hist 12467

London Society, Senate House, Malet St, London WC1E 7HU
T: (071) 5805537
Founded: 1912; Members: 450
Pres: Duke of Gloucester; Gen Secr: Benita Jones
Focus: Urban Plan
Periodical
Journal (weekly) 12468

London Subterranean Survey Association (LSSA), c/o Faculty of Design and the Built Environment, University of East London, Duncan House, High St, London E15 2JB
T: (081) 5907722 ext 3256
Founded: 1968; Members: 25
Focus: Surveying
Periodical
London's Infrastructure (3 times annually) 12469

London Topographical Society (LTS), 36 Old Deer Park Gardens, Richmond TW9 2TL
T: (081) 9405419
Founded: 1880; Members: 900
Pres: Duke of Edinburgh; Gen Secr: Patrick Frazer
Focus: Hist
Periodical
London Topographical News (weekly) . . 12470

London Underground Railway Society (TLURS), 113 Wandle Rd, Morden SM4 6AD
Founded: 1961
Focus: Transport 12471

London Wargames Section (LWS), 48 East View, Barnet EN5 5TN
Founded: 1967
Focus: Military Sci 12472

London Welsh Association (LWA), 157 Gray's Inn Rd, London WC1
T: (071) 8373722
Focus: Ling 12473

Lute Society, 71 Priory Rd, Kew Gardens, Richmond TW9 3DH
T: (0748) 9407086
Founded: 1956; Members: 720
Focus: Music 12474

Macmillan Cancer Fund, 15-19 Britten St, London SW3 3TZ
T: (071) 3517811; Fax: 3768098
Founded: 1911; Members: 5600
Pres: Marchioness of Zetland; Gen Secr: Douglas Scott
Focus: Cell Biol & Cancer Res 12475

Maghreb Studies Association, c/o The Maghreb Bookshop, 45 Burton St, London WC1H 9AL
T: (071) 3881840
Founded: 1981
Pres: Prof. H. Bouraoui; Gen Secr: M. Ben-Madani
Focus: Ethnology
Periodical
The Maghreb Review 12476

Maha Bodhi Society of Sri Lanka (U.K.), 5 Heathfield Gardens, London W4 4JU
T: (081) 9959493; Fax: (081) 7423107
Founded: 1926; Members: 175
Focus: Rel & Theol
Periodical
Samadhi (weekly) 12477

The Malacological Society of London, c/o Canterbury Christ Church College, North Holmes Rd, Canterbury CT1 1QU
Founded: 1893; Members: 386
Pres: Dr. J.D. Taylor; Gen Secr: Dr. G.B.J. Dussart
Focus: Zoology
Periodical
The Journal of Molluscan Studies (3 times annually) 12478

Malaysian Rubber Producers' Research Association (MRPRA), Tun Abdul Razak Laboratory, Brickendonbury, Hertford SG13 8NL
T: (0992) 54966; Fax: (0992) 554837
Founded: 1938
Pres: Ishak Ahmad Farouk bin Haji; Gen Secr: Dr. C.S.L. Baker
Focus: Materials Sci
Periodical
Rubber Developments (weekly) 12479

Malone Society, c/o Reference Div, Oxford University Press, Walton St, Oxford OX2 6DP
T: (0865) 58229
Founded: 1906; Members: 750
Gen Secr: Prof. John Creaser
Focus: Lit
Periodicals
Annual Report (weekly)
Bulletin (1-2 times annually) 12480

Management Research Groups (MRG), c/o Institute of Management, Management House, Cottingham Rd, Corby NN17 1TT
T: (0536) 204222; Fax: (0536) 201651
Founded: 1926; Members: 280
Focus: Business Admin 12481

Management Systems Training Council (MSTC), 2235-2237 Coventry Rd, Sheldon, Birmingham B26 3NW
Founded: 1957; Members: 120
Focus: Adult Educ; Business Admin . . 12482

Manchester Geographical Society, 385 Com Exchange Bldgs, Manchester M4 3EY
T: (061) 8342965
Founded: 1884; Members: 120
Gen Secr: Dr. B.P. Hindle
Focus: Geography
Periodical
The Manchester Geographer (weekly) . . 12483

Manchester Literary and Philosophical Society, 14 Kennedy St, Manchester M2 4BY
T: (061) 2283638; Fax: (061) 2369482
Founded: 1781; Members: 720
Pres: E. Cass; Gen Secr: Dr. C.J.E. Kempster
Focus: Lit; Philos
Periodical
Memoirs and Proceedings (weekly) . . . 12484

Manchester Medical Society, c/o John Rylands University Library, Oxford Rd, Manchester M13 9PP
T: (061) 2736048; Fax: (061) 2728046
Founded: 1834; Members: 2250
Pres: Prof. I.E. Gillespie; Gen Secr: Barbara Lowndes
Focus: Med
Periodical
Annual Report 12485

Manchester Region Industrial Archaeology Society (M.R.I.A.S.), 5 Hilton Rd, Disley SK12 2JV
T: (0663) 763346
Founded: 1964
Focus: Archeol
Periodical
Newsletter (3 times annually) 12486

Manchester Statistical Society (MSS), c/o CIS Bldg, Miller St, Manchester M60 0AL
T: (061) 8328686
Founded: 1833; Members: 235
Pres: John D. Ilett; Gen Secr: P.K. Berry
Focus: Poli Sci; Econ
Periodical
Transactions of the Manchester Statistical Society (weekly) 12487

Manorial Society of Great Britain (MSGB), 65 Belmont Hill, London SE13 5AX
T: (081) 8520200
Founded: 1906; Members: 2300
Focus: Genealogy; Hist
Periodical
Bulletin (weekly) 12488

Manpower Society, c/o Shell Refining and Marketing Ltd, Shell Centre, London SE1
T: (071) 9344518
Founded: 1970; Members: 570
Focus: Business Admin 12489

Mansfield Law Club (MLC), c/o London Guildhall University, 84 Moorgate, London EC2M 6SQ
T: (071) 3201000; Fax: (071) 3201525
Founded: 1947; Members: 800
Focus: Law 12490

The Manx Gaelic Society, 3 Glencrutchery Rd, Douglas, Isle of Man
Founded: 1899; Members: 200
Focus: Ling 12491

Marine Biological Association of the United Kingdom (MBA), The Laboratory, Citadel Hill, Plymouth PL1 2PB
T: (0752) 222772; Fax: (0752) 670637
Founded: 1884; Members: 1350
Pres: Sir Crispin Tickell; Gen Secr: Dr. M. Whitfield
Focus: Bio; Oceanography
Periodicals
Journal of the Marine Biological Association of the United Kingdom (weekly)
MBA News (weekly) 12492

Marine Conservation Society (MCS), Candle Cottage, Kempley, Glos
Founded: 1979; Members: 580
Focus: Ecology
Periodical
Newsletter (weekly) 12493

Marine Librarians Association (MLA), c/o Marine Society, 202 Lambeth Rd, London SE1 7JW
Founded: 1972; Members: 145
Focus: Libraries & Bk Sci 12494

Maritime Trust, 2 Greenwich Church St, London SE10 9BG
T: (081) 8582698
Founded: 1969; Members: 800
Pres: Philip Duke of Edinburgh; Gen Secr: Janet Hales
Focus: Oceanography
Periodical
Newsletter of the Friends of the Maritime Trust (weekly) 12495

Marlowe Society, 8 Norhyrst Av, London SE26
T: (081) 6532809
Founded: 1956; Members: 260
Focus: Perf Arts 12496

Marquetry Society, 2a The Ridgeway, Saint Albans AL4 9AU
T: 68241
Founded: 1952; Members: 700
Focus: Fine Arts
Periodical
Marquetarian (weekly) 12497

Mathematical Association, 259 London Rd, Leicester LE2 3BE
T: (0533) 703877; Fax: (0533) 448508
Founded: 1871; Members: 6300
Focus: Math
Periodicals
The Mathematical Gazette (weekly)
Mathematics in School (5 times annually) 12498

Mathematical Instruction Subcommittee (MIS), c/o The Royal Society, 6 Carlton House Terrace, London SW1Y 5AG
Founded: 1984
Pres: Prof. N.L. Biggs
Focus: Educ; Math
Periodical
MIS Newsletter (weekly) 12499

Medau Society of Great Britain and Northern Ireland, 220 Balham High Rd, London SW12
T: (081) 6737333
Founded: 1952; Members: 2140
Focus: Educ 12500

Media Studies Association (MSA), c/o The School of Communication, Trinity and All Saints' College, Browenberrie Lane, Horsforth, Leeds LS18 5HD
T: (0532) 584341
Founded: 1977; Members: 130
Focus: Mass Media
Periodical
Media Reporter (weekly) 12501

Medical Contact Lens Association (MCLA), Elm Park Rd, London N21
Founded: 1965; Members: 80
Focus: Ophthal 12502

The Medical Defence Union (MDU), 3 Devonshire Pl, London W1N 2EA
T: (071) 4866181; Tx: 8955275; Fax: (071) 9355503
Founded: 1885; Members: 150000
Pres: Dr. Ian K. Fry; Gen Secr: Dr. John Wall
Focus: Med
Periodicals
Annual Report of The Medical Defence Union (weekly)
Healthcare Risk Management Bulletin (weekly)
International Journal of The Medical Defence Union (weekly)
Journal of The Medical Defence Union (weekly)
MDU Nurse (bi-annually) 12503

Medical Officers of Schools Association (M.O.S.A.), 11 Chandos St, London, W1M 0EB
T: (071) 5801043
Founded: 1884; Members: 437
Pres: Dr. J.H.D. Briscoe; Gen Secr: Dr. R.W.E. Harrington
Focus: Public Health
Periodical
Proceedings and Report (weekly) . . . 12504

Medical Practices Committee, 80-94 Newington Causeway, London SE1 6EF
T: (071) 9722930; Fax: (071) 4722985
Focus: Med 12505

Medical Practitioners' Union (MPU), 50 Southwark St, London SE1 1VN
T: (071) 3787255
Founded: 1914; Members: 5024
Pres: Dr. S.J. Watkins; Gen Secr: J. Fisher
Focus: Med
Periodical
Medical World (weekly) 12506

Medical Protection Society (MPS), 50 Hallam St, London W1N 6DE
T: (071) 6370541; Fax: (071) 6360690
Founded: 1892
Focus: Med 12507

Medical Research Council (MRC), 20 Park Crescent, London W1N 4AL
T: (071) 6365422; Tx: 24897; CA: Medresco London; Fax: (071) 4366179
Founded: 1920
Pres: Sir David Plastow; Gen Secr: Sir D. Rees
Focus: Med
Periodicals
Annual Report (weekly)
Corporate Plan (weekly)
Handbook (weekly)
MRC News (weekly)
Scientific Strategy (weekly) 12508

Medical Research Society (MRS), c/o Dept of Medicine, Royal Free Hospital, Pond St, London NW3 2QG
T: (071) 7940500
Focus: Med 12509

Medical Sciences Historical Society, c/o School of Dental Surgery, University of Liverpool, Pembroke Pl, Liverpool L69 3BX
Focus: Med; Cultur Hist 12510

Medical Society for the Study of Radiesthesia (MSSR), 24 Browning Av, Bournemouth BH5 1NN
T: (202) 35149
Members: 19
Focus: Radiology 12511

Medical Society for the Study of Venereal Diseases (MSSVD), c/o Dept of Genitourinary Medicine, King's College Hospital, 16-22 Caldecot Rd, London SE5 9RS
T: (071) 3263470; Fax: (071) 3263458
Founded: 1922; Members: 550
Pres: Dr. James Bingham; Gen Secr: Dr. T.J. McManus
Focus: Venereology

Periodical
Genito-Urinary Medicine (weekly) 12512

The Medical Society of London, 11 Chandos St, London W1M 0EB
T: (071) 5801043
Founded: 1773; Members: 558
Pres: N. Kirby; Gen Secr: Dr. P. Last; Dr. P. Mitchell-Heggs
Focus: Med 12513

Medical Women's Federation (MWF), Tavistock House North, Tavistock Sq, London WC1H 9HX
T: (071) 3877765
Founded: 1917; Members: 3000
Pres: Dr. Nuala Sterling; Gen Secr: Dr. Patricia Price
Focus: Med
Periodical
Medical Woman (3 times annually) . . . 12514

Medico-Legal Society, 15 Saint Botolph St, London EC3A 7NJ
T: (071) 2472277; Fax: (071) 7828507
Founded: 1901; Members: 650
Pres: Dr. John Wall; Gen Secr: E. Pygott
Focus: Med; Law
Periodical
Medico-Legal Journal (weekly) 12515

Medieval Combat Society, 20 Winterfold Close, London SW19
T: (081) 7899307
Focus: Hist 12516

Medieval Settlement Research Group, c/o Environment Dept, County Hall, Taunton TA1 4DY
T: (0823) 255426; Fax: (0823) 334346
Founded: 1986; Members: 500
Focus: Hist; Archeol
Periodical
Annual Report (weekly) 12517

Men of the Stones (MS), 25 Cromarty Rd, Stamford PE9 2TQ
T: (0780) 53527
Founded: 1947; Members: 600
Focus: Archit
Periodical
Yearbook & Directory (weekly) 12518

Mercia Cinema Society, 64 Somerton Dr, Erdington, Birmingham B23 5ST
Founded: 1980; Members: 129
Focus: Cinema
Periodical
Mercia Bioscope (weekly) 12519

Merioneth Agricultural Society, Tir Y Dail, Cader Rd, Dolgellau, Gwynedd
T: (0341) 422837
Founded: 1868; Members: 500
Focus: Agri
Periodical
Show Catalogue (weekly) 12520

Merseyside Archaeological Society (MAS), 14 Southwood Rd, Saint Michaels, Liverpool L17 7BQ
T: (051) 7288505
Founded: 1977; Members: 120
Focus: Archeol 12521

Merseyside Aviation Society (MAS), Room 14, Hangar No 2, Liverpool Airport, Liverpool L24 8QQ
Founded: 1956; Members: 475
Focus: Aero
Periodical
Flypast 12522

Mervyn Peake Society (MPS), c/o John Watney, Five Elm Park Gardens, Flat 36, London SW10 9QQ
Founded: 1975; Members: 123
Gen Secr: John Watney
Focus: Lit 12523

Metals Society (MS), 1 Carlton House Terrace, London SW1Y 5DB
T: (071) 8394071
Founded: 1974; Members: 5000
Focus: Metallurgy
Periodicals
British Corrosion Journal
Metals Abstracts (weekly)
Metals Abstracts Index (weekly)
Metals & Materials (weekly)
Metals Science (weekly)
Metals Technology (weekly)
Powder Metallurgy (weekly) 12524

Midlands Asthma and Allergy Research Association (MAARA), The Allergy and Asthma Treatment and Research Centre, 12 Vernon St, Derby DE1 1FT
T: (0332) 362461
Founded: 1968
Focus: Immunology 12525

Military Heraldry Society, 37 Wolsey Close, Southall, Middx
T: (081) 5744425
Focus: Military Sci
Periodical
Formation Sign (weekly) 12526

Military Historical Society (MHS), 30 Edgeborough Way, Bromley BR1 2UA
T: (081) 4607341
Founded: 1948
Pres: Lord Carver
Focus: Hist; Military Sci
Periodical
Bulletin (weekly) 12527

Mind Association, c/o Dept of Logic and Metaphysics, University of Saint Andrews, Saint Andrews KY16 9AL
Founded: 1900; Members: 873
Gen Secr: Peter Clark
Focus: Philos
Periodical
Mind (weekly) 12528

MIND – National Association for Mental Health, 22 Harley St, London W1N 2ED
T: (071) 6370741; Fax: (071) 3230061
Founded: 1946; Members: 1600
Pres: Timothy F. Durkin; Gen Secr: J. Clements
Focus: Psychiatry
Periodical
Open Mind (weekly) 12529

Mineralogical Society of Great Britain and Ireland, 41 Queen's Gate, London SW7 5HR
T: (071) 5847516; Fax: (071) 8288021
Founded: 1876; Members: 1150
Pres: Prof. C.M.B. Henderson; Gen Secr: Dr. G.M. Manby
Focus: Mineralogy
Periodicals
Clay Minerals (weekly)
Mineralogical Abstracts (weekly)
Mineralogical Magazine (weekly)
Monographs (weekly) 12530

Minerals Engineering Society (MES), 2 Alder Grove, Stapenhill, Burton-on-Trent DE15 9QR
T: 45510
Founded: 1958; Members: 730
Focus: Mining 12531

Mining Institute of Scotland, c/o National Coal Board, Green End, Edinburgh EH17 7PZ
T: (031) 6632811
Founded: 1878; Members: 390
Focus: Mining 12532

Minority Rights Group (MRG), 379 Brixton Rd, London SW9 7DE
T: (071) 9789498; Fax: (071) 7386265
Founded: 1967
Focus: Law
Periodical
MRG Reports (weekly) 12533

Modern Churchpeople's Union (MCU), Saint Martin's Vicarage, 25 Birch Grove, London W3 9SP
Founded: 1898; Members: 600
Pres: Peter Selby; Gen Secr: Nicholas Henderson
Focus: Rel & Theol
Periodical
Modern Believing (weekly) 12534

Modern Humanities Research Association (MHRA), c/o Birkbeck College, Malet St, London WC1E 7HX
T: (071) 6316103
Founded: 1918; Members: 1000
Pres: Prof. G. Price; Gen Secr: Prof. D.A. Wells
Focus: Lit
Periodicals
Annual Bibliography of English Language and Literature (weekly)
Modern Language Review (weekly)
Portuguese Studies (weekly)
Slavonic and East European Review (weekly)
Yearbook of English Studies (weekly)
The Year's Work in Modern Language Studies (weekly) 12535

The Montessori Society A.M.I. (UK), 26 Lyndhurst Gardens, London NW3 5NW
T: (071) 4357874
Founded: 1935
Focus: Educ
Periodical
Quarterly (weekly) 12536

Monumental Brass Society, 57 Leeside Crescent, London NW11 0HA
Founded: 1887; Members: 600
Focus: Preserv Hist Monuments
Periodicals
Bulletin (3 times annually)
Portfolio (weekly)
Transaction (weekly) 12537

Moray Society, 1 High St, Elgin IV30 1EQ
T: (0343) 543675
Founded: 1836
Focus: Hist; Archeol; Paleontology . . . 12538

The Morris Ring, 21 Eccles Rd, Ipswich IP2 9RG
T: (0473) 682540
Founded: 1934
Focus: Music 12539

Motor Industry Research Association (MIRA), Watling St, Nuneaton CV10 OTU
T: (0203) 348541
Founded: 1946; Members: 180
Pres: Sir John Collyear; Gen Secr: D. Swallow
Focus: Auto Eng
Periodicals
Agest for Japan SAE Revlon (3 times annually)
Agest for SAE Publications (weekly)
Automobile Abstracts (weekly)
Business News Index (weekly) 12540

The Multiple Sclerosis Society of Great Britain and Northern Ireland, 25 Effie Rd, London SW6 1EE
T: (071) 7366267
Founded: 1953; Members: 60000
Gen Secr: J. Walford
Focus: Med; Pathology
Periodical
M.S. News (3 times annually) 12541

Muscular Dystrophy Group of Great Britain & Northern Ireland, 7-11 Prescott Pl, London SW4 6BS
T: (071) 7208055; Fax: 4980670
Founded: 1959; Members: 425
Pres: Sir Richard Attenborough; Gen Secr: F. Logan
Focus: Med; Pathology
Periodicals
In Focus Newsletter (weekly)
The Search (3 times annually) 12542

Museum Attendants Association, c/o City Museum and Art Gallery, Museum Rd, Old Portsmouth PO1 2LJ
T: (705) 811527
Focus: Arts 12543

Museums and Galleries Commission, 16 Queen Anne's Gate, London SW1H 9AA
T: (071) 2334200; Fax: (071) 2333686
Founded: 1931; Members: 15
Pres: Graham Greene; Gen Secr: Peter Longman
Focus: Arts
Periodical
Annual Report (weekly) 12544

The Museums Association, 42 Clerkenwell Close, London EC1R 0PA
T: (071) 2501834
Founded: 1889; Members: 3500
Pres: Sue Pearce; Gen Secr: Mark Taylor
Focus: Arts
Periodicals
Museums Journal (weekly)
Museums Yearbook (weekly) 12545

Museum Training Institute, 55 Well St, Bradford BD1 5PS
T: (0274) 391056; Fax: (0274) 394890
Founded: 1989
Gen Secr: Simon Roodhouse
Focus: Arts; Educ
Periodical
News (weekly) 12546

Music Advisers' National Association (MANA), c/o Education Office, 22 Northgate St, Warwick CV34 4SR
T: (0926) 493431 ext 2149
Founded: 1947
Focus: Music; Educ 12547

Musical Box Society of Great Britain (MBSOGB), Landbeach, Cambridge CB4 4DT
T: (0223) 860332
Founded: 1962; Members: 1000
Focus: Music 12548

Music Masters' and Mistresses' Association (MMA), 71 Portsmouth Rd, Guildford GU2 5BS
Founded: 1903; Members: 536
Focus: Music; Educ 12549

Names Society, 32 Speer Rd, Thames Ditton KT7 OPW
T: 3980761
Founded: 1967; Members: 390
Focus: Ling 12550

Napoleonic Association (NA), 59 Grimbald Rd, Knaresborough HG5 8HD
T: (042386) 4857
Founded: 1975; Members: 500
Focus: Military Sci
Periodical
Directory of Wargame Section (weekly) . 12551

Napoleonic Society, 4 Boscombe Av, London E10 6HY
T: (081) 5393876
Founded: 1970; Members: 120
Focus: Hist 12552

Narrow Gauge Railway Society (NGRS), 47 Birchington Av, Birchencliffe, Huddersfield HD3 3RD
T: (0422) 74526
Founded: 1951; Members: 800
Focus: Transport 12553

National Adult School Organisation (N.A.S.O.), Norfolk House, Smallbrook Queensway, Birmingham B5 4LJ
T: (021) 6439297
Founded: 1798; Members: 1500
Pres: Leslie Hill; Gen Secr: William Scarle
Focus: Adult Educ
Periodicals
One and All Magazine (weekly)
Study Handbook (weekly) 12554

National Anti-Vivisection Society (NAVS), 51 Harley St, London W1N 1DD
T: (071) 5804034
Founded: 1875
Gen Secr: J. Creamer
Focus: Vet Med
Periodical
The Campaigner and Animals Defender (weekly) 12555

National Art-Collections Fund (NACF), 7 Cromwell Pl, South Kensington, London SW7 2JN
T: (071) 8210404
Founded: 1903; Members: 36000
Pres: Sir Nicholas Goodison; Gen Secr: David Barrie
Focus: Arts
Periodical
NACF Magazine (weekly) 12556

National Association for Environmental Education (NAEE), c/o Wolverhampton University, Walsall Campus, Gorway Rd, Walsall WS1 3BD
Founded: 1960
Focus: Educ
Periodical
Environmental Education (3 times annually) 12557

National Association for Film in Education (NAFE), 26 Freta Rd, Bexleyheath, Kent
T: 3049519
Founded: 1971
Focus: Cinema; Educ 12558

National Association for Multiracial Education (NAME), The Northbrook Centre, Penn Rd, Slough SL2 1PH
T: 23416
Founded: 1966; Members: 515
Focus: Educ 12559

National Association for Outdoor Education (NAOE), c/o Scout Dike Outdoor Centre, Penistone, Sheffield S30 6GF
T: Penistone 2285
Founded: 1969; Members: 450
Focus: Educ 12560

National Association for Road Safety Instruction in Schools, 16 Woodward Av, Hendon, London NW4 4NY
T: (081) 2025787
Founded: 1970
Focus: Educ; Transport 12561

National Association for Soviet and East European Studies (NASEES), c/o Faculty of Social Sciences, Open University, Milton Keynes MK7 6AA
T: (0908) 653257
Members: 300
Focus: Ethnology 12562

National Association for the Relief of Pagets Disease, 413 Middleton Rd, Rhodes, Manchester M24 4QZ
T: (061) 6431998
Founded: 1973; Members: 500
Focus: Cell Biol & Cancer Res 12563

National Association for the Teaching of English (N.A.T.E.), Broadfield Business Centre, 50 Broadfield Rd, Sheffield S8 0XJ
T: (0742) 555419; Fax: (0742) 555296
Founded: 1963; Members: 4500
Pres: Shona Walton; Gen Secr: Ann Malcolm
Focus: Ling; Educ
Periodicals
English in Education (3 times annually)
NATE News (3 times annually) 12564

National Association of Careers and Guidance Teachers (NACGT), 18 Broad Mead, Townbridge BA14 9BX
T: 63482
Founded: 1969; Members: 1600
Focus: Adult Educ 12565

National Association of Drama Advisers (NADA), c/o Havering Education Authority, Mercury House, Mercury Gardens, Romford, Essex
T: 66999
Founded: 1960; Members: 75
Focus: Perf Arts 12566

National Association of Head Teachers (NAHT), 1 Heath Sq, Boltro Rd, Haywards Heath RH16 1BJ
T: (0444) 458133
Founded: 1897; Members: 19700
Focus: Educ
Periodical
Head Teachers Review 12567

National Association of Language Advisers (NALA), c/o Association for Language Learning, 16 Regent Pl, Rugby CV21 2PN
T: (0788) 526443
Founded: 1983; Members: 140
Focus: Ling 12568

National Association of Principal Agricultural Education Officers (NAPAEO), c/o Bishop Burton College, Bishop Burton, Beverley HU17 8QG
T: (0964) 550481; Fax: (0964) 551190
Members: 50
Focus: Adult Educ; Agri; Hort; Forestry; Fisheries; Ecology 12569

National Association of Schoolmasters (NASUWT) / Union of Women Teachers, Hillscourt Education Centre, Rednal, Birmingham B45 8R5
T: (021) 4536150; Fax: (021) 4537224
Founded: 1923; Members: 125000
Gen Secr: Nigel de Grachy
Focus: Educ
Periodicals
Career Teacher (10 times annually)
Teaching Today (3 times annually) . . . 12570

National Association of Teachers in Further and Higher Education (NATFHE), Hamilton House, Mabledon Pl, London WC1H 9BH
T: (071) 3876806, 3882745
Founded: 1975; Members: 73000
Focus: Educ; Adult Educ
Periodical
NATFHE Journal (9 times annually) . . 12571

National Association of Teachers of the Mentally Handicapped (NATMH), 25 Grecian St, Aylesbury, Bucks
Founded: 1935
Focus: Educ Handic
Periodical
Teaching and Training (weekly) 12572

National Associations of Teachers of Home Economics and Technology (NATHE), Hamilton House, Mabledon Pl, London WC1H 9BJ
T: (071) 3871441; Fax: 3837230
Founded: 1896; Members: 5000
Pres: E. Grieve; Gen Secr: Geoffrey Thompson
Focus: Educ; Home Econ
Periodical
Modus (10 times annually) 12573

National Asthma Campaign, Providence House, Providence Pl, London N1 0NT
T: (071) 2262260
Founded: 1927
Pres: Duchess of Gloucester
Focus: Pulmon Dis
Periodical
Asthma News (weekly) 12574

National Autistic Society (NAS), 276 Willesden Lane, London NW2 5RB
Founded: 1963; Members: 3500
Focus: Rehabil
Periodicals
Communication (3 times annually)
Connection (3 times annually) 12575

National Back Pain Association (BPA), 31-33 Park Rd, Teddington TW11 0AB
T: (081) 9775474; Fax: (081) 9435318
Founded: 1968
Focus: Med
Periodical
Talkback (weekly) 12576

National Bible Society of Scotland (NBSS), 7 Hampton Terrace, Edinburgh EH12 5XU
T: (031) 3379701
Founded: 1861
Focus: Rel & Theol 12577

National Campaign for Nursery Education (NCNE), 23 Albert St, London NW1 7LU
Founded: 1966; Members: 180
Focus: Educ; Public Health 12578

National Caving Association, c/o Whernside Cave and Fell Centre, Dent near Sedburgh, Derby DE6 3FE
Focus: Speleology 12579

National Christian Education Council (NCEC), Robert Denholm House, Nutfield, Redhill, Surrey
T: (0737) 822411; Fax: (0737) 822116
Founded: 1803
Focus: Educ 12580

National Conference of University Professors, c/o Biology and Biochemistry Dept, Brunel University, Uxbridge UB8 3PH
T: (0895) 74000; Fax: (0895) 74348
Founded: 1989; Members: 570
Pres: Prof. T.B. Benjamin; Gen Secr: Prof. R.L. Willson
Focus: Educ 12581

National Council for Educational Technology, Sir William Lyons Rd, Science Park, Coventry CV4 7EZ
T: (0203) 416994; Fax: (0203) 411418
Founded: 1973; Members: 80
Pres: J. Richards; Gen Secr: Margaret Bell
Focus: Educ
Periodical
British Journal of Educational Technology (3 times annually) 12582

National Council of Corrosion Societies, 1 Carlton House Terrace, London SW1Y 5DB
Focus: Metallurgy 12583

National Council of Jewish Religious Day Schools, 5 The Bishop's Av, London N2
T: (081) 4445117
Focus: Rel & Theol; Educ 12584

National Council of Psycho-Therapists, 1 Clovelly Rd, London W5
T: (081) 5670262
Focus: Psych; Therapeutics 12585

National Eczema Society, 4 Tavistock Pl, London WC1H 9RA
T: (071) 3884097; Fax: (071) 7130733
Founded: 1976; Members: 10000
Focus: Derm
Periodical
Exchange (weekly) 12586

National Education Association (NEA), Highcroft, Tewin Close, Tewin Wood, Herts
T: 262
Founded: 1970; Members: 33000
Focus: Educ 12587

National Federation of Continuative Teachers' Associations (NFCTA), 110 North Cray Rd, Bexley DA5 3NA
T: 7038029
Founded: 1920
Focus: Adult Educ 12588

National Federation of Sailing Schools (NFSS), c/o Lymington Seamanship and Sailing Centre, 21 New St, Lymington SO4 9BH
T: (0590) 77601
Founded: 1964; Members: 48
Focus: Sports 12589

National Federation of Voluntary Literacy Schemes (NFVLS), 131 Camberwell Rd, London SE5 0HF
T: (071) 7038083
Founded: 1977
Focus: Adult Educ
Periodical
Wall Paper (weekly) 12590

National Foundation for Educational Research in England and Wales (NFER), The Mere, Upton Park, Slough SL1 2DQ
T: (075) 74123
Founded: 1947
Pres: Sir James Hamilton; Gen Secr: Dr. Clare Burstall
Focus: Educ; Adult Educ
Periodicals
Educational Research (3 times annually)
Educational Research News
Research Reports 12591

National Foundry and Engineering Training Association (NFETA), Fleming House, Renfrew St, Glasgow G3 6TG
T: (041) 3320826
Founded: 1966; Members: 300
Focus: Adult Educ; Eng; Metallurgy . . 12592

National Housing and Town Planning Council (NHTPC), 14-18 Old St, London EC1V 9AB
T: (071) 2512363
Founded: 1900; Members: 400
Pres: Ann Holmes
Focus: Urban Plan
Periodical
Housing and Planning Review (weekly) . 12593

National Institute for Medical Research, The Ridgeway, Mill Hill, London NW7 1AA
T: (081) 9593666
Founded: 1920
Focus: Med 12594

National Institute for Social Work (NISW), 5-7 Tavistock Pl, London WC1H 9SS
T: (071) 3879681; Fax: (071) 3877968
Founded: 1961
Pres: Sir Peter Barclay; Gen Secr: D. Statham
Focus: Sociology
Periodical
Social Care Update (weekly) 12595

National Institute of Adult Continuing Education (England and Wales) (NIACE), 19b De Montfort St, Leicester LE1 7GE
T: (0533) 551451; Fax: (0533) 554792
Founded: 1949
Gen Secr: Alan Tuckett
Focus: Adult Educ
Periodicals
Adults Learning (weekly)
Studies in the Education of Adults (weekly)
Time to Learn (weekly)
Year Book of Adult Continuing Education (weekly) 12596

National Institute of Agricultural Botany (NIAB), Huntingdon Rd, Cambridge CB3 0LE
T: (0223) 76381
Founded: 1919; Members: 5340
Focus: Botany 12597

National Institute of Economic and Social Research (NIESR), 2 Dean Trench St, London SW1P 3HE
T: (071) 2227665
Founded: 1938
Focus: Soc Sci; Econ
Periodicals
National Institute Briefing Notes (weekly)
National Institute Discussion Papers (weekly)
National Institute Economic Review (weekly)
National Institute Report Series (weekly) . 12598

National Institute of Industrial Psychology (NIIP), The Mere, Upton Park, Slough SL1 2DQ
Founded: 1921
Gen Secr: J.A. Fox
Focus: Psych 12599

National Institute of Medical Herbalists (NIMH), 9 Palace Gate, Exeter EX1 1JA
T: (0392) 426022
Founded: 1864; Members: 311
Focus: Med 12600

National Materials Handling Centre (NMHC), c/o Cranfield Institute of Technology, Cranfield, Bedford MK43 0AL
T: (0234) 750323; Tx: 825072; Fax: (0234) 750875
Members: 300
Focus: Materials Sci
Periodical
International Distribution & Handling Review (weekly) 12601

National Operatic and Dramatic Association (NODA), 1 Crestfield St, London WC1H 8AU
T: (071) 8375655; Fax: (071) 8330609
Founded: 1899; Members: 3000
Gen Secr: Mark Thorburn
Focus: Perf Arts
Periodical
NODA National News (weekly) 12602

National Pure Water Association (NPWA), Meridan, Cae Goody Lane, Ellesmere SY12 9DW
T: (0691) 623015
Founded: 1960
Pres: Dr. Peter Mansfield; Gen Secr: N. Brugge
Focus: Ecology 12603

National Schizophrenia Fellowship (NSF), 29 Victoria Rd, Surbiton KT6 4JT
T: 3903651
Founded: 1973; Members: 45
Focus: Psychiatry 12604

National Secular Society, 702 Holloway Rd, London N19 3NL
T: (071) 2721266
Founded: 1866
Focus: Philos 12605

National Society (Church of England), Church House, Great Smith St, London SW1P 3NZ
T: (071) 2221672; Fax: (071) 2332572
Founded: 1811
Gen Secr: Geoffrey Duncan
Focus: Rel & Theol; Educ
Periodicals
Crosscurrent (3 times annually)
Together (9 times annually) 12606

National Society for Clean Air (NSCA), 136 North St, Brighton BN1 1RG
T: (0273) 326313; Fax: (0273) 735082
Founded: 1899; Members: 1500
Pres: Baroness Platt; Gen Secr: Dr. Tom Crossett
Focus: Ecology
Periodicals
Clean Air (weekly)
NSCA Members' Handbook (weekly)
Proceedings of Annual Conferences and Seminars (weekly)
Workshop Proceedings (weekly) 12607

National Society for Education in Art and Design (NSEAD), The Gatehouse, Corsham Court, Corsham SN13 0BZ
T: (0249) 714825; Fax: (0249) 716138
Founded: 1888; Members: 2500
Pres: Michael Yeomans; Gen Secr: John Steers
Focus: Arts; Educ
Periodical
Journal of Art & Design Education: NSEAD Newsletter (3 times annually) 12608

National Society for Epilepsy (N.S.E.), Chalfont Centre for Epilepsy, Chalfont Saint Peter SL9 0RJ
T: Chalfont Saint Giles 3991
Founded: 1892
Focus: Pathology 12609

National Society for Research into Allergy (NSRA), POB 45, Hinckley LE10 1JY
T: (0455) 851546; Fax: (0455) 851546
Founded: 1980
Focus: Immunology
Periodical
Reaction (weekly) 12610

National Society for Transplant Surgery, 11 Alma Rd, Cardiff CF2 5BD
T: (0222) 40039
Founded: 1968
Focus: Surgery 12611

National Strict Baptist Sunday School Association (NSBSSA), 87 Pynchbek, Bishops Stortford, Essex
Founded: 1938; Members: 200
Focus: Rel & Theol; Educ
Periodical
Go Teach (weekly) 12612

National Trolleybus Association (NTA), 29A Hillfield Park, London N10 3QT
T: (081) 8833202
Founded: 1963; Members: 650
Focus: Transport
Periodical
Trolleybus Magazine (weekly) 12613

National Trust for Places of Historic Interest or Natural Beauty, 36 Queen Anne's Gate, London SW1H 9AS
T: (071) 2229251; Fax: (071) 2225097
Founded: 1895; Members: 2100000
Pres: Lord Chorley; Gen Secr: Angus Stirling
Focus: Preserv Hist Monuments; Ecology
Periodicals
Annual Report (weekly)
Members' and Visitors' Handbook
Newsletter
Properties of the National Trust 12614

National Trust for Scotland, 5 Charlotte Sq, Edinburgh EH2 4DU
T: (031) 2265922; Fax: (031) 2206266
Founded: 1931; Members: 218000
Gen Secr: Lester Borley
Focus: Preserv Hist Monuments; Ecology
Periodicals
The Educational Guide to NTS
Guide to Properties
Heritage Scotland (weekly)
Welcome 12615

National Union of Teachers (N.U.T.), Hamilton House, Mabledon Pl, London WC1H 9BD
T: (071) 3886191
Founded: 1870; Members: 180000
Focus: Educ
Periodicals
Annual Report (weekly)
Education Review (weekly)
The Teacher (weekly) 12616

National Waterways Transport Association (NWTA), Central House, 32-66 High St, Stratford, London E16 2PS
T: (081) 5194872
Founded: 1975
Gen Secr: F.P. Morris
Focus: Transport
Periodical
Newsletter (3 times annually) 12617

National Zoological Association of Great Britain, Stowlangtoft, Bury Saint Edmunds IP31 3JW
T: Pakenham 30623
Founded: 1972; Members: 61
Focus: Zoology 12618

Natural Environment Research Council (NERC), Polaris House, North Star Av, Swindon SN2 1EU
T: (0793) 411500; Tx: 444293; Fax: (0793) 411501
Founded: 1965
Pres: Prof. John Knill
Focus: Nat Sci
Periodicals
Annual Report (weekly)
NERC News (weekly) 12619

Natural History Society of Northumbria, c/o Hancock Museum, Newcastle-upon-Tyne NE2 4PT
T: (091) 2326386
Founded: 1829; Members: 850
Pres: Duke of Northumberland; Gen Secr: David Noble-Rollin
Focus: Hist; Nat Sci
Periodical
Transactions (weekly) 12620

The Nautical Institute, 202 Lambeth Rd, London SE1 7LQ
T: (071) 9281351
Founded: 1972; Members: 6000
Focus: Navig
Periodical
Seaways (weekly) 12621

Navy Records Society, Chatsworth House, 66-70 Saint Mary Axe, London EC3A 8BD
Fax: (071) 3825400
Founded: 1893; Members: 800
Pres: Lord Carrington; Gen Secr: A.J. McMillan
Focus: Hist
Periodical
Annual Report (weekly) 12622

Neonatal Society, c/o John Radcliffe Hospital, Oxford OX3 9DU
T: (0865) 221355; Fax: (0865) 221366
Founded: 1959; Members: 370
Gen Secr: Dr. Andrew Wilkinson
Focus: Physiology
Periodical
Early Human Development (3 times annually) 12623

Newark and Nottinghamshire Agricultural Society, The Showground, Winthorpe, Newark NG24 2NY
T: (0636) 702627
Founded: 1799; Members: 2500
Gen Secr: J.N. Armitage
Focus: Agri
Periodical
Catalogue (weekly) 12624

The Newcomen Society, c/o Science Museum, South Kensington, London SW7 2DD
T: (071) 5891793; Fax: (071) 5891793
Founded: 1920; Members: 1000
Gen Secr: Clive Ellam
Focus: Hist; Eng

Periodicals
Bulletin (3 times annually)
Transactions (weekly) 12625

New English Art Club, 17 Carlton House Terrace, London SW1Y 5BD
T: (071) 9306844; Fax: (071) 8397830
Founded: 1886; Members: 55
Gen Secr: William Bowyer
Focus: Arts
Periodical
Catalogue of Annual Exhibition (weekly) . 12626

Nicholas Roerich Society, 91 Fitzjohn's Av, London NW3 6NX
T: (071) 4355490
Focus: Lit 12627

NICOD, 31 Ulsterville Av, Belfast BT9 7AS
T: (0232) 666188; Fax: (0232) 682400
Founded: 1941; Members: 150
Focus: Orthopedics
Periodical
Annual Report (weekly) 12628

Norfolk and Norwich Archaeological Society, Garsett House, Saint Andrew's Hall Plain, Norwich NR3 1AT
Founded: 1846; Members: 609
Focus: Archeol
Periodical
Norfolk Archaelogy (weekly) 12629

Norfolk Record Society, 425 Unthank Rd, Norwich NR4 7QB
T: (0603) 52937; Tx: 97197
Founded: 1931; Members: 250
Focus: Hist
Periodical
Papers of Nathaniel Bacon (weekly) . . 12630

Northamptonshire Archaeological Society (NAS), 54 The Knoll, Earls Barton, Northants
Founded: 1974; Members: 180
Focus: Archeol 12631

Northamptonshire Natural History Society and Field Club (N.N.H.S.), Humfrey Rooms, Castilian Terrace, Northampton NN1 1LD
Founded: 1876; Members: 440
Focus: Hist; Nat Sci
Periodical
Journal (weekly) 12632

Northamptonshire Record Society, Wootton Hall Park, Northampton NN4 9BQ
T: (0604) 762297
Founded: 1920
Pres: Sir Hereward Wake
Focus: Hist
Periodical
Journal (weekly) 12633

Northern Horticultural Society, c/o Harlow Carr Botanical Gardens, Crag Lane, Harrogate HG3 1QB
T: (0423) 565418
Founded: 1946; Members: 10500
Focus: Hort
Periodicals
Index Seminum (weekly)
The Northern Gardener (weekly) 12634

Northern Ireland Association for Mental Health, 80 University St, Belfast BT7 1HE
T: (0232) 328474
Members: 570
Focus: Psychiatry
Periodical
Mental Health Matters (5 times annually) 12635

Northern Ireland Chest, Heart and Stroke Association, 21 Dublin Rd, Belfast BT2 7FJ
T: (0232) 320184
Focus: Cardiol; Pulmon Dis; Rehabil
Periodical
Masterstroke (weekly) 12636

Northern Ireland Civil Rights Association (NICRA), 2 Marquis St, Belfast BT1 1JJ
T: (0232) 23351
Founded: 1967; Members: 205
Focus: Law 12637

Northern Ireland Polio Fellowship (NIPF), 198 Belvoir Dr, Belvoir Park, Belfast BT8 4PJ
T: (0232) 643367
Founded: 1948
Focus: Med 12638

Northern Ireland Women Teacher's Association (NIWTA), 4 Ormiston Dr, Belfast BT4 3JS
Founded: 1923; Members: 200
Focus: Educ 12639

The North of England Zoological Society (N.E.Z.S.), c/o Zoological Garden, Upton, Chester CH2 1LH
T: (0244) 380280; Fax: (0244) 371273
Founded: 1934; Members: 3300
Pres: Duke Westminster; Gen Secr: Dr. M.R. Brambell
Focus: Zoology
Periodical
Chester Zoo Life (3 times annually) . . 12640

North Somerset Agricultural Society, 5 Thiery Rd, Brislington, Bristol BS4 2NX
T: (0272) 778463
Focus: Agri 12641

North Staffordshire Field Club, c/o School of Sciences, Staffordshire University, College Rd, Stoke-on-Trent ST4 2DE
T: (0782) 744531
Founded: 1865; Members: 260
Gen Secr: R.A. Tribbeck
Focus: Archeol; Hist; Nat Sci
Periodical
Transactions (weekly) 12642

North Tyne and Redesdale Agricultural Society, Stobswood House, Stobswood, Morpeth NE61 5QA
Focus: Agri 12643

North Wales Agricultural Society, Rhos-y-Wylfa, Nant-y-Garth, Portdinorwic LL56 4QB
T: (0248) 670531
Focus: Agri 12644

North Wales Arts Association, 10 Wellfield House, Bangor LL57 1ER
T: (0248) 353248; Fax: (0248) 351077
Founded: 1967
Gen Secr: L. Williams
Focus: Arts
Periodicals
Ben Bowen Thomas Lecture (weekly)
Report (weekly) 12645

North Western Naturalists' Union (NWNU), 59 Moss Lane, Bramhall, Stockport SK7 1EQ
T: (061) 4392899
Founded: 1929
Focus: Nat Sci
Periodicals
Newsletter (weekly)
North Western Naturalist (weekly) . . . 12646

North Western Society for Industrial Archaeology and History (NWSIAH), c/o Merseyside County Museum, William Brown St, Liverpool L3 8EN
T: (051) 2070001
Founded: 1964; Members: 120
Focus: Archeol; Hist 12647

North West Regional Association of Industrial Safety Groups, 1 Moor Av, Penwortham, Preston PR1 0ND
T: (0305) 42603
Founded: 1959
Focus: Safety 12648

Nottingham and Nottinghamshire Field Club, Nottingham Rd, Burton Joyce, Nottingham NG14 5BD
T: (0602) 863256
Founded: 1889; Members: 100
Focus: Nat Sci 12649

Nottinghamshire Local History Association, 110 Mansfield Rd, Nottingham NG1 4HL
T: (0602) 53681
Founded: 1953; Members: 140
Focus: Hist
Periodical
Nottinghamshire Historian (weekly) . . . 12650

Nutrition Society, 10 Cambridge Court, 210 Shepherds Bush Rd, London W6 7NJ
Founded: 1941; Members: 1350
Pres: Dr. R.G. Whitehead
Focus: Food
Periodicals
The British Journal of The Nutrition Society (weekly)
The Proceedings of The Nutrition Society (3 times annually) 12651

Odinic Rite (O.R.), BCM Runic, London WC1N 3XX
Founded: 1973
Pres: Geoffrey L.D. Holley-Heimgest
Focus: Rel & Theol
Periodicals
OR Briefing (weekly)
Rimstock (weekly) 12652

Oil and Colour Chemists' Association, 967 Harrow Rd, Wembley HA0 2SF
T: (081) 9081086; Fax: (081) 9081219
Founded: 1918; Members: 2500
Pres: J. Memmings; Gen Secr: C. Pacey-Day
Focus: Chem
Periodical
Surface Coating's International (weekly) . 12653

Omnibus Society (OS), 103a Streatham Hill, London SW2 4UD
T: (081) 6745280
Founded: 1929; Members: 1000
Focus: Transport 12654

Operational Research Society (ORS), Neville House, Waterloo St, Birmingham B2 5TX
T: (021) 6430236; Fax: (021) 6313485
Founded: 1954; Members: 3000
Focus: Computer & Info Sci
Periodicals
European Journal of Information Systems (weekly)
Insight (weekly)
Journal of the Operational Research Society (weekly)
Newsletter (weekly) 12655

Oral History Society, Dept of Sociology, University of Essex, Colchester, Essex
T: (0206) 873333
Members: 1000
Gen Secr: Robert Perks
Focus: Hist
Periodical
Oral History (weekly) 12656

Orders and Medals Research Society (OMRS), 123 Turnpike Link, Croydon CR0 5NU
T: (081) 6802701
Founded: 1942; Members: 2500
Focus: Military Sci; Hist
Periodical
Orders & Medals (weekly) 12657

Organic Federation, POB 8, Malvern WR14 2NQ
T: 4598
Founded: 1975; Members: 24
Focus: Public Health 12658

Organisation of Professional Users of Statistics (OPUS), Lancaster House, More Lane, Esher, Surrey
T: 63121
Members: 5
Focus: Stats 12659

Organisation of Teachers of Transport Studies (OTTS), c/o Southall College of Technology, Beaconsfield Rd, Southall UB1 1DP
T: (081) 5743448
Founded: 1971; Members: 102
Focus: Educ; Transport
Periodical
Seminar & Annual Report (weekly) . . . 12660

Oriental Ceramic Society (OCS), 31b Torrington Sq, London WC1E 7JL
T: (071) 6367985
Founded: 1921
Pres: Dr. Jessica Rawson; Gen Secr: Jean Martin
Focus: Arts
Periodical
Transactions of the Oriental Ceramic Society (weekly) 12661

Oriental Numismatic Society (ONS), 30 Warren Rd, Woodley, Reading RG5 3AR
T: (0734) 693528
Founded: 1970; Members: 535
Focus: Numismatics
Periodicals
ONS Newsletter (5-6 times annually)
ONS Occasional Papers (weekly) . . . 12662

Ornithological Society of the Middle East (OSME), The Lodge, Sandy SG19 2DL
T: (0767) 80551
Founded: 1967; Members: 850
Focus: Ornithology
Periodicals
Bulletin (weekly)
Sandgrouse (weekly) 12663

Osteopathic Association of Great Britain (OAGB), 1-4 Suffolk St, London SW1Y 4HG
T: (01) 292056
Founded: 1925; Members: 347
Focus: Pathology
Periodicals
British Osteopathic Journal
Osteopathic Newsletter (weekly) 12664

Osteopathic Medical Association, 114 Wigmore St, London
T: (071) 5806147
Focus: Med; Pathology 12665

Overseas Doctors Association in the UK (ODA), 28-32 Princess St, Manchester M1 4LB
T: (061) 236 5594
Founded: 1975; Members: 61283
Focus: Med
Periodical
News Review (weekly) 12666

Oxford Centre for Islamic Studies, c/o Saint Cross College, Oxford OX1 3TU
T: (0865) 725077; Tx: 83147; Fax: (0865) 248942
Founded: 1985
Gen Secr: Dr. Farhan Ahmad Nizami
Focus: Rel & Theol
Periodical
Journal of Islamic Studies (weekly) . . . 12667

Oxford Preservation Trust, 10 Turn Again Lane, Saint Ebbes, Oxford OX1 1QL
T: (0865) 242918
Founded: 1927; Members: 1000
Pres: Dr. D.C.M. Yardle; Gen Secr: Moyra Haynes
Focus: Preserv Hist Monuments; Ecology
Periodical
Annual Report (weekly) 12668

Oxfordshire Architectural and Historical Society (OAHS), 3 Wytham St, Oxford OXI 4SO
T: (0865) 245262
Founded: 1839; Members: 370
Focus: Archit; Hist; Archeol
Periodical
Oxoniensia (weekly) 12669

Oxfordshire Record Society (O.R.S.), c/o Bodleian Library, Oxford OX1 3BG
Founded: 1919; Members: 350
Focus: Hist 12670

Oxford University Archaeological Society (OUAS), c/o Ashmolean Museum, Oxford
Founded: 1911; Members: 300
Focus: Archeol 12671

Paediatric Welfare and Research Association, 22 Tytherington Dr, Tytherington, Macclesfield, Cheshire
T: (0625) 25228
Founded: 1968; Members: 102
Focus: Pediatrics 12672

Paint Research Association (PRA), 8 Waldegrave Rd, Teddington TW11 8LD
T: (081) 9774427; Fax: (081) 9434705
Founded: 1926; Members: 180
Pres: Eustache Bancken; Gen Secr: John A. Bernie
Focus: Materials Sci
Periodicals
Coating Regulations and Environmental Issues (weekly)
Comet (weekly)
Paint Titles (weekly)
PRA Newsletter (6 times monthly)
World Surface Coating Abstracts (weekly) 12673

Palaeontographical Society, c/o British Geological Survey, Keyworth, Nottingham NG12 5GG
Founded: 1847; Members: 600
Pres: M. House; Gen Secr: S.P. Tunnicliff
Focus: Paleontology
Periodical
Monographs (weekly) 12674

Palaeontological Association, c/o Dept of Geology, University of Wales College of Cardiff, POB 914, Cardiff CF1 3YE
T: (0222) 874338; Fax: (0222) 874326
Founded: 1957; Members: 1600
Focus: Paleontology
Periodicals
Palaeontology (weekly)
Special Papers in Palaeontology (weekly) 12675

Palestine Exploration Fund, 2 Hinde Mews, Marylebone Lane, London W1M 5RR
T: (071) 9355379
Founded: 1865; Members: 900
Pres: Archbishop of Canterbury; Gen Secr: Yolande Hodson
Focus: Geography; Archeol; Hist
Periodical
Palestine Exploration Quarterly (weekly) . 12676

Pali Text Society (PTS), 62 South Lodge, Circus Rd, London NW8 9ET
T: (071) 2864280
Founded: 1881
Focus: Lit 12677

Parkinson's Disease Society of the United Kingdom (PDS), 22 Upper Woburn Pl, London WC1H 0RA
T: (071) 3833513; Fax: (071) 3835754
Founded: 1969; Members: 40000
Gen Secr: Michael Whelan
Focus: Med; Neurology
Periodical
The Parkinson Newsletter (3-4 times annually) 12678

Parliamentary and Scientific Committee, 16 Great College St, London SW1P 3RX
T: (071) 2227085
Focus: Sci; Eng
Periodical
Science in Parliament (5 times annually) 12679

Parrot Society, 19A De Parys, Bedford
T: (0234) 58922
Founded: 1966; Members: 4000
Focus: Ornithology
Periodical
Magazine (weekly) 12680

Pastel Society, 17 Carlton House Terrace, London SW1Y 5BD
T: (071) 9306844; Fax: (071) 8397830
Members: 60
Pres: John Blockley
Focus: Arts; Fine Arts
Periodical
Annual Catalogue (weekly) 12681

Pathological Society of Great Britain and Ireland, 2 Carlton House Terrace, London SW1Y 5AF
T: (071) 9761260; Fax: (071) 9761267
Founded: 1906; Members: 1600
Gen Secr: Prof. F. Walker
Focus: Pathology
Periodicals
Journal of Medical Microbiology (weekly)
Journal of Pathology (weekly)
Reviews in Medical Microbiology (weekly) 12682

Peak District Mines Historical Society (PDMHS), 85 Peveril Rd, Beeston, Nottingham
T: (0602) 257945
Founded: 1959
Focus: Mining; Hist 12683

Peakland Archaeological Society, 12 The Crescent, Hayfield Rd, Chapel in le Frith, Stockport
Focus: Archeol 12684

Pembrokeshire Agricultural Society, Withybush Airfield, Haverfordwest, Dyfed
Focus: Agri 12685

Pembrokeshire Historical Society, Dolan Dwrbach, Scloddau Fishguard SA65 9RN
T: 3707
Founded: 1950; Members: 150
Focus: Hist
Periodical
Journal of the Pembrokeshire Historical Society (weekly) 12686

Pendragon Society, 22 Alma Rd, Clifton, Bristol BS8 2BS
T: (0272) 33032
Founded: 1959; Members: 150
Focus: Hist 12687

P.E.N. English Centre, 7 Dilke St, London SW3 4JE
T: (071) 3526303
Founded: 1921
Focus: Lit 12688

Perthshire Society of Natural Science (PSNS), c/o Museum, George St, Perth
T: (0738) 32488
Founded: 1867
Focus: Nat Sci
Periodical
The Journal of the Perthshire Society of Natural Science (bi-annually) 12689

Peter Warlock Society, 32a Chipperfield House, Cale St, London SW3 3SA
T: (071) 5899595
Founded: 1963; Members: 200
Focus: Music
Periodical
Newsletter 12690

Petroleum Exploration Society of Great Britain (PESGB), 6 Saint James's Sq, London SW1
Founded: 1965; Members: 960
Focus: Geology 12691

Pharmaceutical Society of Northern Ireland (PSNI), 73 University St, Belfast BT7 1HL
T: (0232) 326927; Fax: (0232) 439919
Founded: 1925; Members: 1400
Gen Secr: Derek J. Lawson
Focus: Pharmacol 12692

The Philological Society, c/o School of Oriental and African Studies, Thornhaugh St, Russell Sq, London WC1H 0XG
T: (071) 6372388; Tx: 291829; Fax: (071) 4363844
Founded: 1842; Members: 704
Pres: Prof. P.H. Matthews; Gen Secr: Prof. R.J. Hayward
Focus: Ling
Periodical
Transactions (weekly) 12693

The Philosophical Society of England, 78 Saint Andrew's Rd, Portslade, Brighton BN4 1DE
T: (0273) 414204
Founded: 1913
Pres: John Wilson; Gen Secr: Dr. P. Faunch
Focus: Philos
Periodical
The Philosopher 12694

Philosophy of Education Society of Great Britain, c/o Education Centre, New University of Ulster, Coleraine, Ulster
T: 4141
Founded: 1964; Members: 560
Focus: Philos; Educ 12695

The Photogrammetric Society, c/o Dept of Photogrammetry and Surveying, University College London, Gower St, London WC1E 6BT
T: (071) 3877050; Fax: (071) 3800453
Founded: 1952; Members: 600
Pres: R. Wood; Gen Secr: Dr. R.P. Kirby
Focus: Surveying
Periodical
The Photogrammetric Record (weekly) . 12696

Physical Education Association of Great Britain and Northern Ireland, Ling House, 5 Western Court, Bromley St, Digbeth, Birmingham B9 4AN
T: (021) 7530909
Founded: 1899
Pres: N. Armstrong; Gen Secr: P. Harrison
Focus: Sports
Periodical
British Journal of Physical Education (weekly) 12697

The Physiological Society, POB 506, Oxford OX1 3XE
T: (0865) 798498; Fax: (0865) 798092
Founded: 1876; Members: 1669
Gen Secr: Dr. C.A.R. Boyd
Focus: Physiology
Periodicals
Experimental Physiology (weekly)
The Journal of Physiology (14 times annually) 12698

Phytochemical Society of Europe (PSE), c/o Department of Biochemistry, University College of Swansea, Swansea SA2 8PP
T: (0792) 295376
Founded: 1957; Members: 500
Focus: Botany; Chem; Biochem
Periodical
PSE Symposia (weekly) 12699

Pipe Roll Society, c/o Public Record Office, Chancery Lane, London WC2A 1LR
T: (071) 4050741
Founded: 1883; Members: 300
Focus: Hist
Periodical
Annual Report (weekly) 12700

Pira International, Randalls Rd, Leatherhead KT22 7RU
T: (0372) 376161; Fax: (0372) 360104
Founded: 1965; Members: 1000
Focus: Eng
Periodicals
DTP Publishing Commentary (10 times annually)
International Packaging Abstracts (weekly)
Now-World Publishing Monitor
Paper and Board Abstracts (weekly)
Paper Technology (weekly)
Printing Abstracts (weekly)
UK Printing Industry Statistics (weekly) . 12701

Plainsong and Mediaeval Music Society, c/o Dr. Stephen Farmer, Magdalene College, Cambridge CB3 0AG
Founded: 1888
Pres: Prof. H. Chadwick; Gen Secr: Dr. S. Farmer
Focus: Music
Periodical
Plainsong and Mediaeval Music (weekly) 12702

The Plastics and Rubber Institute (PRI), 11 Hobart Pl, London SW1W 0HL
T: (071) 2459555; Tx: 915719; Fax: (071) 8231379
Founded: 1975; Members: 10000
Pres: S. Allen; Gen Secr: J. Hawkins
Focus: Materials Sci
Periodicals
Plastics and Rubber International (weekly)
Plastics and Rubber Processing and Applications (weekly)
Progress in Rubber and Plastic Technology (weekly) 12703

Play Matters – The National Association of Toy and Leisure Libraries, 68 Churchway, London NW1 1LT
T: (071) 3879592; Fax: (071) 3332714
Founded: 1976; Members: 950
Focus: Libraries & Bk Sci
Periodicals
ARK Newsletter (weekly)
Good Toy Guide (weekly) 12704

The Plymouth Athenaeum, Derry's Cross, Plymouth PL1 2SW
T: (0752) 266079
Founded: 1812; Members: 430
Gen Secr: W.E. Pope
Focus: Sci; Arts 12705

Poetry Society (PS), 22 Betterton St, London WC2H 9BU
T: (071) 2404810; Fax: (071) 2404818
Founded: 1909
Gen Secr: Chris Green
Focus: Lit
Periodical
Poetry Review (weekly) 12706

Poets' Theatre Guild, 44 Croham Park Av, South Croydon CR2 7HL
T: 6888803
Focus: Lit 12707

Polish Educational Society, 238 King St, London W6
T: (081) 7411993
Focus: Educ
Periodical
Hychowanie Ojczyste (weekly) 12708

Polish Medical Association, 14 Collingham Gardens, London SW5
T: (071) 3731087
Focus: Med 12709

Polish Underground Movement (1939-1945) Study Trust (PUMST), 11 Leopold Rd, London W5 3PB
Founded: 1947
Focus: Hist 12710

Political and Economic Planning (PEP), 12 Upper Belgrave St, London SW1X 8BB
T: (071) 2355271
Founded: 1931; Members: 800
Focus: Poli Sci; Econ 12711

Political Studies Association of the United Kingdom (PSAUK), c/o Dept of Politics, Queen's University, Belfast BT7 1NN
T: (0232) 245133 ext 3288
Founded: 1950; Members: 800
Pres: T. Smith; Gen Secr: David Farrell
Focus: Poli Sci
Periodicals
Newsletter (weekly)
Political Studies (5 times annually)
Politics (3 times annually) 12712

Politics Association, 16 Gower St, London WC1E 6DP
T: (071) 3231131
Founded: 1969; Members: 760
Pres: Prof. D. Kauanagh; Gen Secr: S. Marshall
Focus: Poli Sci
Periodicals
Politics Association Resources Bank (weekly)
Talking Politics incorporating Teaching Politics (3 times annually) 12713

The Powys Society, Hamilton's, Kilmersdon BA3 5TE
T: (0761) 435134
Founded: 1967; Members: 300
Focus: Lit
Periodical
The Powys Journal (weekly) 12714

Prehistoric Society, c/o National Library of Air Photographs, Alexander House, 19 Fleming Way, Swindon SN1 2NG
T: (0793) 414100
Founded: 1935; Members: 2000
Pres: Prof. D. Harris; Gen Secr: Dr. B. Bewley
Focus: Hist
Periodical
Proceedings (weekly) 12715

Presbyterian Church of Wales Historical Society, The Manse, Caradog Rd, Aberystwyth SY23 3BU
T: 3391
Founded: 1914; Members: 650
Focus: Hist; Rel & Theol 12716

Primate Society of Great Britain, c/o Dr. Hilary Box, Psychology Dept, University of Reading, Reading
Focus: Bio
Periodical
Primate Eye (3 times annually) 12717

Printing Historical Society (PHS), c/o Saint Bride Institute, Bride Lane, Fleet St, London EC4Y 8EE
Founded: 1964; Members: 1000
Focus: Cultur Hist; Eng
Periodicals
Bulletin of the Printing Historial Society (weekly)
Journal of the Printing Historical Society (weekly) 12718

The Private Libraries Association (P.L.A.), Ravelston, South View Rd, Pinner HA5 3YD
Founded: 1956; Members: 1250
Pres: Peter Eton; Gen Secr: Frank Broomhead
Focus: Libraries & Bk Sci
Periodicals
The Private Library (weekly)
Private Press Books: Annual Bibliography 12719

Processors and Growers Research Organisation (PGRO), c/o Research Station, Great North Rd, Thornhaugh, Peterborough PE8 6HT
T: (0780) 782585
Founded: 1944
Focus: Hort
Periodical
Vegetable Grower (weekly) 12720

Procurators Fiscal Society, Sheriff Court, Dumbarton
T: 63266
Founded: 1930; Members: 80
Focus: Law 12721

Production Engineering Research Association (PERA), Melton Mowbray LE13 0PD
T: (0664) 501501
Founded: 1946; Members: 2000
Pres: R.A. Armstrong; Gen Secr: M.J. Kemm
Focus: Eng 12722

Production Mangagement Action Group (PROMAG), c/o Administrative Staff College, Greenlands, Henley-on-Thames, Oxon
T: Hambleden 454
Founded: 1975; Members: 60
Focus: Business Admin 12723

Professional Association of Teachers (PAT), 2 Saint James's Court, Friar Gate, Derby DE1 1BT
T: (0332) 372337; Fax: (0332) 290310
Founded: 1970; Members: 38000
Gen Secr: John R. Andrews
Focus: Educ
Periodical
Professional Teacher (5 times annually) . 12724

Psoriasis Association, 7 Milton St, Northampton NN2 7JG
T: (0604) 711129
Founded: 1968; Members: 10500
Focus: Derm
Periodical
Psoriasis (3 times annually) 12725

Psychiatric Rehabilitation Association (PRA), 21a Kingsland High St, London E8 2JS
T: (071) 2449753
Founded: 1961; Members: 6000
Focus: Psychiatry; Rehabil 12726

Psywar Society, 8 Ridgway Rd, Barton Seagrave, Kettering, Northants
T: (0536) 2921
Founded: 1958; Members: 123
Focus: Hist 12727

Pugin Society, 4 Boscombe Av, London E10 6HY
T: (081) 5393876
Founded: 1974
Focus: Arts 12728

Pugwash Conferences on Science and World Affairs, 9 Great Russell Mansions, 60 Great Russell St, London WC1B 3BE
T: (071) 4056661
Founded: 1957; Members: 30
Focus: Sci; Poli Sci 12729

Pure Water Preservation Society (PWPS), Lane End, Highlands Lane, Westfield, Woking GU22 9PU
T: (04862) 60385
Founded: 1967
Focus: Hydrology 12730

Pushkin Club, 46 Ladbroke Grove, London W11 2PA
T: (071) 7275311
Founded: 1954
Focus: Lit 12731

Quaker Esperanto Society, 69 Twemlow Parade, Morecambe LA3 2AL
Founded: 1921
Focus: Ling 12732

Queen's English Society (QES), Flat 8, Freshcliffe House East, Bury St, Guildford GU2 5AN
T: (0483) 450036
Founded: 1972; Members: 700
Pres: Geoffrey Talbot; Gen Secr: Michael Gorman
Focus: Ling; Lit
Periodical
Quest (weekly) 12733

Quekett Microscopical Club (QMC), c/o British Museum of Natural History, Cromwell Rd, London SW7 5BD
T: (081) 7788473
Founded: 1865; Members: 600
Focus: Nat Sci
Periodical
Microscopy QMC Journal (weekly) . . . 12734

Radical Philosophy Society (RPS), 9 Arley Park, Cotham, Bristol
Founded: 1973; Members: 14
Focus: Philos 12735

Radionic Association, Field House, Peaslake, Guildford GU5 9SS
T: Dorking 730080
Founded: 1960; Members: 475
Focus: Radiology 12736

Radio Society of Great Britain (R.S.G.B.), Lambda House, Cranborne Rd, Potters Bar EN6 3JE
Founded: 1913; Members: 32000
Gen Secr: Peter A. Kirby
Focus: Electric Eng
Periodical
Radio Communication (weekly) 12737

Radius / The Religious Drama Society of Great Britain, Christ Church and Upton Chapel, Kennington Rd, London SE1 7QP
T: (071) 4012422
Founded: 1929
Focus: Perf Arts
Periodical
Radius-Magazine (weekly) 12738

Radnorshire Society, c/o The Radnor College of Further Education, Spa Rd East, Llandrindod Wells LD1 5ES
T: (0597) 2618
Founded: 1930; Members: 550
Focus: Hist
Periodical
Transactions (weekly) 12739

Railway Club, 112 High Holborn, London WC1
Founded: 1899
Focus: Transport 12740

Railway Correspondence and Travel Society (R.C.T.S.), 158a North View Rd, Hornsey, London
T: (081) 3489081
Founded: 1928; Members: 5800
Focus: Transport 12741

Railway Development Association (RDA), 3 Hall Way, Purley, Surrey
T: 6606139
Founded: 1951; Members: 260
Focus: Transport 12742

RAPRA Technology, Shawbury, Shrewsbury SY4 4NR
T: (0939) 250383; Fax: (0939) 251118
Founded: 1919; Members: 600
Pres: Harry Jackson; Gen Secr: Dr. M.C. Copley
Focus: Materials Sci
Periodical
RAPRA News (weekly) 12743

Rare Poultry Society (R.P.S.), 8 Saint Thomas's Rd, Great Glenn LE8 0EG
Founded: 1969; Members: 445
Focus: Animal Husb
Periodicals
Breeders Lists (weekly)
Newsletter (weekly) 12744

The Ray Society, c/o The Natural History Museum, Cromwell Rd, London SW7 5BD
T: (071) 9389263; Fax: (071) 9389158
Founded: 1844; Members: 350
Pres: R.J. Lincoln; Gen Secr: N.J. Evans
Focus: Hist; Nat Sci 12745

Regency Society of Brighton and Hove, 33 Roedean Crescent, Brighton BN2 5RG
T: (0273) 66881
Founded: 1945; Members: 660
Focus: Archit 12746

Regional Studies Association (RSA), 15 Micawber St, London N1 7TB
T: (071) 4901128
Founded: 1965; Members: 800
Pres: Joe Davis; Gen Secr: Sally Hard
Focus: Urban Plan
Periodicals
Debates & Reviews (weekly)
Newsletter (weekly)
Regional Studies
Special Issue (weekly) 12747

Religious Society of Friends, 173-177 Euston Rd, London NW1 2BJ
T: (071) 3873601; Fax: (071) 3881977
Focus: Rel & Theol 12748

Remote Sensing Society (RSS), c/o Dept of Geography, University of Nottingham, Nottingham NG7 2RD
T: (0602) 515435; Fax: (0602) 515249
Founded: 1974; Members: 800
Focus: Geography; Cart; Computer & Info Sci
Periodicals
International Journal of Remote Sensing (weekly)
Proceedings of Annual Technical Meetings (weekly)
RSS Newsletter (weekly) 12749

Renal Association, c/o Renal Unit, Churchill Hospital, Headington, Oxford OX3
Founded: 1950; Members: 340
Gen Secr: Dr. C.G. Winearls
Focus: Med 12750

Research and Development Society, 47 Belgrave Sq, London SW1X 8QX
T: (071) 2356111; Fax: (071) 2596002
Founded: 1971; Members: 500
Gen Secr: Clive Jones
Focus: Sci; Eng 12751

Research Defence Society (RDS), 58 Great Marlborough St, London W1V 1DD
T: (071) 2872818
Founded: 1908; Members: 2565
Pres: Lord Adrian; Gen Secr: Dr. Mark Matfield
Focus: Sci
Periodicals
Conquest (weekly)
Newsletter (weekly) 12752

Research into Lost Knowledge Organisation (RILKO), 10 Kedleston Dr, Orpington BR5 2DR
T: (0689) 32265
Founded: 1969; Members: 420
Pres: John C. Irwin; Gen Secr: Brian Hargreaves
Focus: Rel & Theol; Philos; Archit
Periodical
Newsletter (weekly) 12753

Research Society for Natural Therapeutics, 8 Stokewood Rd, Bournemouth BH3 7NA
T: (0202) 25997
Focus: Therapeutics 12754

Richard III Society, 4 Oakley St, Chelsea, London SW3 5NN
T: (071) 3513391
Founded: 1924; Members: 4000
Focus: Hist
Periodical
The Ricardian (weekly) 12755

Richard Jefferies Society, 6 Chickerell Rd, Swindon SN3 2RQ
T: (0793) 21512
Founded: 1950; Members: 350
Focus: Lit
Periodicals
Annual Report and Bulletin (weekly)
Spring Newsletter (weekly) 12756

Right to a Comprehensive Education (RtCE), 4 Hammersmith Terrace, London W6 9TS
T: (081) 7489790
Founded: 1965; Members: 1500
Focus: Educ
Periodical
Comprehensive Education (3 times annually)
. . . . 12757

River Thames Society (RTS), 2 Ruskin Av, Kew, Richmond TW9 4DJ
T: (081) 8761520
Founded: 1962; Members: 2143
Focus: Hydrology; Ecology
Periodical
Lower Tideway Topics (weekly) 12758

Romantic Novelists' Association (RNA), 30 Langham House Close, Ham Common, Richmond TW10 7JE
T: (0748) 9409219
Founded: 1960
Focus: Lit 12759

Royal Academy of Arts in London (RA), Burlington House, Piccadilly, London W1V 0DS
T: (071) 4397438; Tx: 21812; Fax: 4340837
Founded: 1768
Pres: Roger de Grey; Gen Secr: Piers Rodgers
Focus: Arts
Periodicals
R A Illustrated (weekly)
R A Magazine (weekly) 12760

Royal Academy of Dancing (RAD), 36 Battersea Sq, London SW11 3RA
T: (071) 2230091; Fax: (071) 9243129
Founded: 1920; Members: 20000
Focus: Perf Arts
Periodical
Dancing Gazette (3 times annually) . . 12761

Royal Academy of Dramatic Art (RADA), 62-64 Gower St, London WC1E 6ED
T: (071) 6367076
Founded: 1904; Members: 200
Focus: Perf Arts 12762

Royal Academy of Engineering, 2 Little Smith St, London SW1P 3DL
T: (071) 2222688; Tx: 918358; Fax: (071) 2330054
Founded: 1976
Pres: Sir William Barlow; Gen Secr: G.A. Atkinson
Focus: Eng
Periodicals
Annual Report (weekly)
Newsletter (weekly) 12763

Royal Academy of Music (RAM), Marylebone Rd, London NW1 5HT
T: (071) 9355461
Founded: 1822
Focus: Music
Periodicals
Prospectus (weekly)
RAM Magazine (3 times annually) . . . 12764

Royal Aeronautical Society, 4 Hamilton Pl, London W1V 0BQ
T: (071) 4993515; Fax: (071) 4996230
Founded: 1866; Members: 20000
Pres: Dr. Geoffrey Pope; Gen Secr: R.J. Kennett
Focus: Aero
Periodicals
Aeronautical Journal (10 times annually)
Aerospace (10 times annually) 12765

The Royal African Society (RAS), SOAS, Thornhaugh St, Russell Sq, London WC1H 0XG
T: (071) 3236253
Founded: 1901; Members: 800
Pres: Prof. Kenneth Robinson; Gen Secr: Lindsay Allan
Focus: Ethnology
Periodicals
African Affairs
Journal (weekly) 12766

Royal Agricultural Society of England (RASE), c/o National Agricultural Centre, Stoneleigh Park CV8 2LZ
T: (0203) 696969; Fax: (0203) 696900
Founded: 1838; Members: 18000
Gen Secr: C.D. Runge
Focus: Agri
Periodicals
Diary of Agricultural Show Dates (weekly)
RASE Journal (weekly)
RASE News (weekly) 12767

Royal Anthropological Institute of Great Britain and Ireland (R.A.I.), 50 Fitzroy St, London W1P 5HS
T: (071) 3870455; Fax: (071) 3834235
Founded: 1843; Members: 2250
Pres: Prof. Ernest Gellner; Gen Secr: Jonathan Benthall
Focus: Anthro
Periodicals
Anthropological Index (weekly)
Anthropology Today (weekly)
MAN (weekly) 12768

Royal Archaeological Institute (R.A.I.), c/o Society of Antiquaries, Burlington House, Piccadilly, London W1V 0HS
Founded: 1844; Members: 1500
Pres: A.D. Sauders; Gen Secr: J.G. Coad
Focus: Archeol
Periodical
The Archaeological Journal (weekly) . . 12769

The Royal Asiatic Society (RAS), 60 Queen's Gardens, London W2 3AF
T: (071) 7244742
Founded: 1823; Members: 800
Pres: A.D.H. Bivar; Gen Secr: L. Collins
Focus: Ethnology
Periodical
Journal of the Royal Asiatic Society (3 times annually) 12770

Royal Association for Disability and Rehabilitation (RADAR), 25 Mortimer St, London W1N 8AB
T: (071) 6375400; Fax: (071) 6371827
Founded: 1977
Pres: Duke of Buccleuch; Gen Secr: Bert Massie
Focus: Rehabil
Periodicals
Bulletin (weekly)
Contact (weekly) 12771

Royal Astronomical Society (RAS), Burlington House, Piccadilly, London W1V 0NL
T: (071) 7344582; Fax: (071) 4940166
Founded: 1820; Members: 2800
Pres: Prof. Sir Martio Rees; Gen Secr: Prof. A.M. Cruise; Dr. K.A. Whaler; Prof. D.A. Williams
Focus: Astronomy
Periodicals
Geophysical Journal (weekly)
Monthly Notices (weekly)
Quarterly Journal (weekly) 12772

Royal Bath and West of England Society, The Showground, Shepton Mallet BA4 6QN
T: (0749) 823211; Fax: (0749) 823169
Founded: 1777; Members: 3500
Focus: Agri 12773

Royal Caledonian Horticultural Society (RCHS), 7 Melville Crescent, Edinburgh EH3 7NA
T: (031) 2252041
Founded: 1809; Members: 1330
Focus: Hort 12774

Royal Cambrian Academy of Art (R.C.A.), Crown Lane, High St, Conwy LL32 8BH
T: (0492) 593413
Founded: 1882; Members: 100
Pres: K. Williams; Gen Secr: Leonard H.S. Mercer
Focus: Arts 12775

Royal Celtic Society, 23 Rutland St, Edinburgh EH1 2RN
T: (031) 2286449; Fax: (031) 2296987
Founded: 1820; Members: 142
Focus: Ethnology 12776

Royal Choral Society (RCS), Royal Albert Hall, Kensington, London SW7 2AP
T: (071) 5845216
Founded: 1871; Members: 250
Focus: Music
Periodicals
Prospectus (weekly)
Voice 12777

The Royal College of Anaesthetists, 48-49 Russell Sq, London WC1B 4JY
T: (071) 8131900; Fax: (071) 8131876
Founded: 1988; Members: 5500
Pres: Prof. A.A. Spence; Gen Secr: Sir Geoffrey de Deney
Focus: Anesthetics
Periodical
Newsletter (weekly) 12778

Royal College of General Practitioners (R.C.G.P.), 14 Princes Gate, London SW7 1PU
T: (071) 5813232; Fax: (071) 2253047
Founded: 1952; Members: 16500
Pres: Dr. Colin Waine; Gen Secr: Dr. M. McBride
Focus: Med
Periodical
Journal of the Royal College of General Practitioners (weekly) 12779

Royal College of Midwives (RCM), 15 Mansfield St, London W1M 0BE
T: (071) 5806523
Founded: 1881; Members: 33000
Focus: Gynecology
Periodicals
Current Awareness Service (weekly)
Midwives Chronicle (weekly) 12780

Royal College of Music (R.C.M.), Prince Consort Rd, London SW7 2BS
T: (071) 5893643; Fax: 5897740
Founded: 1883; Members: 700
Focus: Music
Periodical
Royal College of Music Magazine (3 times annually) 12781

Royal College of Nursing of the United Kingdom (RCN), 20 Cavendish Sq, London W1M 0AB
T: (071) 4093333; Fax: (071) 4080190
Founded: 1916
Gen Secr: Christine Hancock
Focus: Med
Periodicals
Nursing Bibliography (weekly)
Nursing Standard (weekly)
Nursing the Elderly (10 times annually)
Paediatric Nursing (10 times annually)
Primary Health Care (10 times annually)
Senior Nurse (10 times annually) . . . 12782

Royal College of Obstetricians and Gynaecologists (R.C.O.G.), 27 Sussex Pl, London NW1 4RG
T: (071) 2625425; Fax: (071) 7247725
Founded: 1929; Members: 8000
Pres: G. Chamberlain; Gen Secr: P.A. Barnett
Focus: Gynecology
Periodical
Journal (weekly) 12783

The Royal College of Organists (RCO), 7 Saint Andrew St, London EC4A 3LQ
T: (071) 9363606
Founded: 1864; Members: 3200
Focus: Music
Periodicals
The Organist (weekly)
RCO Journal (weekly) 12784

Royal College of Pathologists, 2 Carlton House Terrace, London SW1Y 5AF
T: (071) 9305861; Fax: (071) 3210523
Founded: 1962; Members: 6350
Pres: Prof. A.J. Bellingham; Gen Secr: K. Lockyer
Focus: Pathology 12785

Royal College of Physicians and Surgeons of Glasgow, 234-242 Saint Vincent St, Glasgow G2 5RJ
T: (041) 2216072; Fax: (041) 2211804
Founded: 1599; Members: 6500
Pres: Dr. Robert Hume; Gen Secr: Dr. B.O. Williams
Focus: Surgery; Med; Dent 12786

Royal College of Physicians of Edinburgh (R.C.P.E.), 9 Queen St, Edinburgh EH2 1JQ
T: (031) 2257324; Fax: (031) 2203939
Founded: 1681
Gen Secr: Dr. J.L. Anderton
Focus: Physiology
Periodical
Proceedings of the Royal College of Physicians of Edinburgh (weekly) 12787

Royal College of Physicians of London (RCP), 11 Saint Andrew's Pl, London NW1 4LE
T: (071) 9351174; Fax: (071) 4875218
Founded: 1518; Members: 15000
Pres: Leslie A. Turnberg; Gen Secr: D.B. Lloyd
Focus: Educ; Med
Periodical
Journal of the Royal College of Physicians
. . . . 12788

The Royal College of Psychiatrists, 17 Belgrave Sq, London SW1X 8PG
T: (071) 2352351; Fax: (071) 2451231
Founded: 1971; Members: 8300
Pres: Dr. F. Caldicott; Gen Secr: V. Cameron
Focus: Psychiatry
Periodical
British Journal of Psychiatry (weekly) . . 12789

Royal College of Radiologists (RCR), 38 Portland Pl, London W1N 3DG
T: (071) 6364432
Founded: 1975; Members: 4100
Pres: C.H. Paine; Gen Secr: A.J. Cowles
Focus: Radiology
Periodicals
Clinical Oncology (weekly)
Clinical Radiology (weekly) 12790

Royal College of Surgeons of Edinburgh, Nicolson St, Edinburgh EH8 9DW
T: (031) 5566206
Founded: 1505
Pres: Prof. P.S. Boulter; Gen Secr: I.B. Maclead
Focus: Surgery
Periodical
Journal (weekly) 12791

The Royal College of Surgeons of England (R.C.S.E.), 35-43 Lincoln's Inn Fields, London WC2A 3PN
T: (071) 4053474
Founded: 1800; Members: 46000
Pres: Prof. N.L. Browse; Gen Secr: R.H.E. Duffett
Focus: Surgery
Periodicals
Almanack (weekly)
Annals and Bulletin (weekly)
Handbook (weekly) 12792

Royal College of Veterinary Surgeons (R.C.V.S.), 32 Belgrave Sq, London SW1X 8QP
T: (071) 2354971; Fax: (071) 2456100
Founded: 1844; Members: 16000
Pres: Barry Johnson; Gen Secr: Peter Woolley
Focus: Vet Med
Periodicals
Directory of Practices (weekly)
RCVS Newsletter (weekly)
Register of Members (weekly) 12793

Royal Commission for the Exhibition of 1851, c/o Imperial College of Science and Technology, Sherfield Bldg, London SW7
Pres: Philip Duke of Edinburgh; Gen Secr: M.C. Neale
Focus: Hist 12794

Royal Commission on Historical Manuscripts, Quality House, Quality Court, Chancery Lane, London WC2A 1HP
T: (071) 2421198; Fax: (071) 2421198
Founded: 1869; Members: 17
Gen Secr: C.J. Kitching
Focus: Hist
Periodicals
Annual Review (weekly)
Reports to the Crown (weekly) 12795

Royal Commonwealth Society, 18 Northumberland Av, London WC2N 5BJ
T: (071) 9306733; Fax: (071) 9309705
Founded: 1868; Members: 20000
Focus: Ethnology
Periodicals
Conference Reports
Library Notes
Newsletter (3 times annually) 12796

Royal Cornwall Agricultural Association (RCAA), The Showground, Wadebridge, Cornwall
T: (0208) 812183; Fax: (0208) 812713
Founded: 1793; Members: 6000
Gen Secr: C.P. Riddle
Focus: Agri 12797

Royal Drawing Society (RDS), 17 Carlton House Terrace, London SW1
Founded: 1888; Members: 175
Focus: Arts 12798

Royal Economic Society (R.E.S.), c/o Dept of Economics, Burbank College, University of London, 7-15 Gresse St, London W1P 1PA
Fax: (071) 7348760
Founded: 1890; Members: 3400
Pres: Prof. David Hendry; Gen Secr: Prof. Richard Portes
Focus: Econ
Periodicals
Economic Journal (weekly)
Newsletter (weekly) 12799

Royal Entomological Society of London (RES), 41 Queen's Gate, London SW7 5HU
T: (071) 5848361
Founded: 1883
Pres: Dr. M. Rothschild; Gen Secr: G.G. Bentley
Focus: Entomology
Periodicals
Antenna (weekly)
Ecological Entomology (weekly)
The Entomologist (weekly)
Handbooks for the Identification of British Insects (weekly)
Insect Molecular Biology (weekly)
Medical & Veterinary Entomology (weekly)
Physiological Entomology (weekly)
Symposium Volumes (weekly)
Systematic Entomology (weekly) 12800

Royal Environmental Health Institute of Scotland (R.E.H.I.S.), Virginia House, 62 Virginia St, Glasgow G1 1TX
T: (041) 5521533
Founded: 1891; Members: 427
Gen Secr: M.H. Cameron
Focus: Public Health
Periodicals
Annual Report (weekly)
Career Brochure
Environmental Health Scotland (weekly) . 12801

The Royal Faculty of Procurators in Glasgow, 12 Nelson Mandela Pl, Glasgow G2 1BT
T: (041) 3339104
Founded: 1668; Members: 592
Pres: Alexander Weatherhead; Gen Secr: John H. Sinclair
Focus: Law 12802

Royal Fine Art Commission, 7 Saint James's Sq, London SW1Y 4JU
T: (071) 8396537; Fax: (071) 8398475
Founded: 1924
Pres: Lord Saint John of Fawsley; Gen Secr: Sherban Cantacuzino
Focus: Arts 12803

Royal Fine Art Commission for Scotland, 9 Atholl Crescent, Edinburgh EH3 8HA
T: (031) 2291109; Fax: (031) 2296031
Founded: 1927; Members: 12
Gen Secr: Lord Charles Prosser
Focus: Fine Arts
Periodical
Report to Parliament (weekly) 12804

Royal Forestry Society of England, Wales and Northern Ireland (R.F.S.), 102 High St, Tring HP23 4AF
T: (0442) 822028; Fax: (0442) 890395
Founded: 1882; Members: 4400
Pres: Duke of Somerset; Gen Secr: Dr. J.E. Jackson
Focus: Forestry
Periodical
Journal (weekly) 12805

Royal Geographical Society (R.G.S.), 1 Kensington Gore, London SW7 2AR
T: (071) 5895466; Tx: 933669; Fax: (071) 5844447
Founded: 1830; Members: 11500
Pres: Earl Jellicoe; Gen Secr: Dr. John Hemming
Focus: Geography
Periodicals
Geographical Journal (3 times annually)
Geographical Magazine (weekly) 12806

Royal Glasgow Institute of the Fine Arts (RGI), 12 Sandyford Pl, Glasgow G3 7NE
T: (041) 2487411
Founded: 1861; Members: 1265
Focus: Fine Arts 12807

Royal Highland and Agricultural Society of Scotland (R.H.A.S.S.), Ingliston, Edinburgh EH28 8NF
T: (031) 3332444; Fax: (031) 3335236
Founded: 1784; Members: 15000
Gen Secr: H. Davies
Focus: Agri
Periodical
The Royal Highland News 12808

Royal Historical Society, c/o University College, Gower St, London WC1E 6BT
T: (071) 3877532
Founded: 1868
Pres: Prof. R.R. Davies; Gen Secr: J.N. McCarthy
Focus: Hist
Periodicals
Camden (weekly)
Transactions of the Royal Historical Society (weekly) 12809

Royal Horticultural Society (RHS), 80 Vincent Sq, London SW1P 2PE
T: (071) 8344333
Founded: 1804; Members: 160000
Pres: R.A.E. Herbert; Gen Secr: D.P. Hearn
Focus: Hort
Periodicals
Daffodils (weekly)
The Garden (weekly)
The Orchid Review (weekly)
The Plantsman (weekly)
Rhododendrons (weekly) 12810

Royal Incorporation of Architects in Scotland, 15 Rutland Sq, Edinburgh EH1 2BE
T: (031) 2297205, 2297545; Fax: (031) 2282188
Founded: 1916; Members: 3000
Pres: J.R. Laidlaw; Gen Secr: C. McKean
Focus: Archit
Periodicals
Practice Information (weekly)
Prospect (weekly) 12811

Royal Institute of British Architects (RIBA), 66 Portland Pl, London W1N 4AD
T: (071) 5805533; Fax: (071) 2551541
Founded: 1834; Members: 28000
Pres: Richard C. MacCormack; Gen Secr: Lord Rodgers of Quarrybank
Focus: Archit
Periodical
RIBA Journal (weekly) 12812

Royal Institute of International Affairs (RIIA), Chatham House, 10 Saint James's Sq, London SW1Y 4LE
T: (071) 9575700; Fax: (071) 9575710
Founded: 1920; Members: 3000
Pres: Lord Carrington; Lord Callaghan; Lord Jenkins of Hillhead; Gen Secr: Prof. Laurence Martin
Focus: Poli Sci
Periodicals
Chatham House Papers (weekly)
International Affairs (weekly)
RIIA Discussion Papers (weekly)
The World Today (weekly) 12813

Royal Institute of Navigation (RIN), c/o Royal Geographical Society, 1 Kensington Gore, London SW7 2AT
T: (071) 5895021; Fax: (071) 8238671
Founded: 1947; Members: 3400
Focus: Navig
Periodicals
Journal of Navigation (3 times annually)
Navigation News (weekly) 12814

Royal Institute of Oil Painters, 17 Carlton House Terrace, London SW1Y 5BD
T: (071) 9306844; Fax: (071) 8397830
Founded: 1882
Pres: Brian Bennett
Focus: Fine Arts 12815

Royal Institute of Painters in Water Colours, 17 Carlton House Terrace, London SW1Y 5BD
T: (071) 9306844; Fax: (071) 8397830
Founded: 1831; Members: 64
Pres: Ronald Maddox
Focus: Arts; Fine Arts
Periodical
Catalogue of Exhibitions (weekly) . . . 12816

The Royal Institute of Philosophy, 14 Gordon Sq, London WC1H 0AG
T: (071) 3874130
Founded: 1925; Members: 850
Pres: Lord Quinton; Gen Secr: Ingrid Purkiss
Focus: Philos
Periodicals
Lecture Series (weekly)
Philosophy (weekly) 12817

Royal Institute of Public Administration (RIPA), 3 Birdcage Walk, London SW1H 9JJ
T: (071) 2222248
Founded: 1922
Pres: Lord Barnett; Gen Secr: David Falcon
Focus: Poli Sci
Periodicals
Public Administration (weekly)
Public Administration and Development (weekly)
RIPA Report (weekly) 12818

The Royal Institute of Public Health and Hygiene, 28 Portland Pl, London W1N 4DE
T: (071) 5802731
Founded: 1886; Members: 2500
Focus: Public Health; Hygiene
Periodical
Health & Hygiene (weekly) 12819

The Royal Institution of Chartered Surveyors, 12 Great George St, London SW1P 3AD
T: (071) 2227000; Tx: 915443; Fax: (071) 2229430
Founded: 1868; Members: 67000
Pres: Clive Lewis; Gen Secr: Michael Pattison
Focus: Surveying
Periodicals
Chartered Quantity Surveyor (weekly)
Chartered Surveyor (weekly)
Surveying World (weekly) 12820

Royal Institution of Cornwall (RIC), c/o Royal Cornwall Museum, River St, Truro, Cornwall
T: (0872) 72205
Founded: 1818; Members: 1200
Focus: Sci; Archeol; Hist; Genealogy
Periodical
Journal of the Royal Institution of Cornwall (weekly) 12821

Royal Institution of Great Britain (R.I.G.B.), 21 Albemarle St, London W1X 4BS
T: (071) 4092992; Fax: (071) 6293569
Founded: 1799; Members: 2500
Pres: Duke of Kent; Gen Secr: Prof. K.J. Packer
Focus: Sci
Periodicals
Proceedings (weekly)
Record (weekly) 12822

The Royal Institution of Naval Architects, 10 Upper Belgrave St, London SW1X 8BQ
T: (071) 2354622; Fax: (071) 2456959
Founded: 1860; Members: 5600
Pres: P.J. Usher; Gen Secr: J. Rosewarn
Focus: Eng
Periodicals
The Naval Architect incorporating Warship Technology (weekly)
Ship and Boat International (weekly)
Ship Repair and Conversion Technology (weekly)
Transactions (weekly) 12823

Royal Institution of South Wales (RISW), c/o Swansea Museum, Victoria Rd, Swansea SA1 1SN
T: (0792) 653763
Founded: 1830; Members: 362
Focus: Sci
Periodical
Minerva (weekly) 12824

Royal Isle of Wight Agricultural Society, 66 Carisbrooke Rd, Newport PO30 1BW
T: 2925
Founded: 1882
Focus: Agri 12825

Royal Jersey Agricultural and Horticultural Society, c/o Agricultural Dept, Springfield, Saint Helier, Jersey
T: (0534) 37227; Fax: (0534) 24692
Founded: 1833; Members: 513
Focus: Agri; Hort
Periodical
Jersey at Home (weekly) 12826

Royal Lancashire Agricultural Society (RLAS), Ribby Hall, Wrea Green, Kirkham, Preston PR4 2RD
T: (0772) 3911
Founded: 1767; Members: 3423
Focus: Agri 12827

Royal Liverpool Philharmonic Society, Hope St, Liverpool L1 9BP
T: (051) 7092895; Tx: 6294669
Founded: 1840; Members: 2000
Focus: Music 12828

Royal London Aid Society (RLAS), 56-58 East India Dock Rd, London E14
T: (071) 9873864
Founded: 1939; Members: 1000
Focus: Sociology 12829

Royal Medical Society, Students' Centre, Bristol Sq, Edinburgh EH8 9AL
T: (031) 677969
Founded: 1737; Members: 900
Pres: Paul MacKenney; Gen Secr: Alexander Marshland
Focus: Med 12830

Royal Medico-Chirurgical Society of Glasgow, 5 Saint Vincent Pl, Glasgow
Founded: 1814
Focus: Surgery; Med 12831

Royal Meteorological Society, 104 Oxford Rd, Reading RG1 7LJ
T: (0734) 568500; Fax: (0734) 568571
Founded: 1850; Members: 3000
Pres: Dr. P.J. Mason; Gen Secr: D. Griggs
Focus: Geophys
Periodicals
International Journal of Climatology (weekly)
Quarterly Journal (weekly)
Weather (weekly) 12832

Royal Microscopical Society (RMS), 37-38 Saint Clements, Oxford OX4 1AJ
T: (0865) 248768; Fax: (0865) 791237
Founded: 1839; Members: 2500
Pres: Dr. H.Y. Elder; Gen Secr: P.B. Hirst
Focus: Nat Sci; Sci
Periodicals
Journal of Microscopy (weekly)
Proceedings (weekly) 12833

The Royal Musical Association (R.M.A.), c/o Faculty of Music, Saint Aldates, Oxford OX1 1DB
T: (0865) 276125; Fax: (0865) 276128
Founded: 1874; Members: 741
Pres: Dr. Stanley Sadie; Gen Secr: E. West
Focus: Music
Periodicals
Journal (weekly)
Research Chronicle (weekly) 12834

Royal Naval Bird Watching Society (RNBWS), 23 Saint Davids Rd, Southsea PO5 1QH
T: (0705) 822981
Founded: 1946; Members: 440
Focus: Ornithology; Navig
Periodicals
News Bulletins (weekly)
Sea Swallow (weekly) 12835

Royal Norfolk Agricultural Association (RNAA), Showground, Dereham Rd, New Costessey, Norwich NR5 0TT
T: (0603) 748931; Fax: (0603) 748229
Founded: 1847; Members: 4000
Focus: Agri
Periodical
Norfolk Mardler (weekly) 12836

Royal Numismatic Society, c/o Dept of Coins and Medals, British Museum, London WC1B 3DG
T: (071) 3238404
Founded: 1836; Members: 900
Pres: T.V. Buttrey; Gen Secr: J.E. Cribb; R.F. Bland
Focus: Numismatics
Periodicals
Coin Hoards
Numismatic Chronicle (weekly) 12837

Royal Pharmaceutical Society of Great Britain, 1 Lambeth High St, London SE1 7JN
T: (071) 7359141; Fax: (071) 7357629
Founded: 1841; Members: 38000
Pres: N. Wood; Gen Secr: J. Ferguson
Focus: Pharmacol
Periodicals
Annual Register of Pharmaceutical Chemists (weekly)
Journal of Pharmacy and Pharmacology (weekly)
The Pharmaceutical Journal (weekly) . . 12838

The Royal Philharmonic Society, 10 Stratford Pl, London W1N 9AE
T: (071) 4918110; Fax: (071) 4937463
Founded: 1813; Members: 550
Focus: Music
Periodical
Fanfare from the Royal Philharmonic Society (weekly) 12839

Royal Philosophical Society of Glasgow, 6 Hughenden Terrace, Glasgow W2
Founded: 1802; Members: 200
Pres: Jenny Johnston; Gen Secr: I. MacInnes
Focus: Philos 12840

The Royal Photographic Society (RPS), c/o RPS National Centre of Photography, The Octagon, Milsom St, Bath BA1 1DN
T: (0225) 462841; Fax: (0225) 448688
Founded: 1853; Members: 10500
Pres: Edwin Appleton; Gen Secr: Amanda Nevill
Focus: Photo
Periodicals
The Journal of Photographic Science
Photographic Abstracts (weekly)
The Photographic Journal (weekly) . . . 12841

Royal Physical Society of Edinburgh, c/o Dept of Genetics, University of Edinburgh, Edinburgh
Founded: 1771; Members: 160
Pres: Dr. H. Kacser
Focus: Physics
Periodical
Proceedings (weekly) 12842

Royal Sanitary Association of Scotland (R.S.A.S.), 62 Virginia St, Glasgow G1 1TX
T: (041) 5521533
Founded: 1875
Focus: Public Health 12843

Royal Scottish Academy (R.S.A.), The Mound, Edinburgh EH2 2EL
T: (031) 2256671; Fax: (031) 2252349
Founded: 1826; Members: 70
Pres: W.J.L. Baillie; Gen Secr: Ian McKenzie Smith
Focus: Sci
Periodical
Annual Report (weekly) 12844

Royal Scottish Forestry Society, Camsie House, Charlestown, Dunfermline KY11 3EE
T: (0383) 873014; Fax: (0383) 872863
Founded: 1854; Members: 1400
Pres: D.W. Goss; Gen Secr: M. Osborne
Focus: Forestry
Periodical
Scottish Forestry (weekly) 12845

Royal Scottish Geographical Society (R.S.G.S.), 50 George St, Glasgow G1 1QE
Founded: 1884; Members: 2500
Pres: John C. Bartholomew; Gen Secr: Alistair B. Cruickshank
Focus: Geography
Periodical
The Scottish Geographical Magazine (3 times annually) 12846

Royal Scottish Society of Arts (RSSA), 4 Alva St, Edinburgh EH3 6RE
T: (031) 5563818
Founded: 1821; Members: 205
Focus: Arts 12847

The Royal Society, 6 Carlton House Terrace, London SW1Y 5AG
T: (071) 8395561; Tx: 917876; Fax: (071) 9302170
Founded: 1660; Members: 1134
Pres: Sir Michael Atiyah; Gen Secr: P.T. Warren
Focus: Nat Sci
Periodicals
Annual Report (weekly)
Year Book (weekly) 12848

Royal Society for Asian Affairs, 2 Belgrave Sq, London SW1X 8PJ
T: (071) 2355122
Founded: 1901; Members: 1100
Pres: Lord Denman; Gen Secr: M. Fitzsimons
Focus: Ethnology
Periodical
Journal (3 times annually) 12849

Royal Society for Nature Conservation (RSNC), The Green, Witham Park, Lincoln LN5 4JR
T: (0522) 544400; Fax: (0522) 595325
Founded: 1912; Members: 204000
Pres: David Attenborough; Gen Secr: T.S. Cordy
Focus: Ecology
Periodical
Natural World: Annual Review (weekly) . 12850

Royal Society for the Encouragement of Arts, Manufactures and Commerce, 8 John Adam St, London WC2N 6EZ
T: (071) 9305115
Founded: 1754; Members: 11000
Pres: Philip Duke of Edinburgh; Gen Secr: Christopher Lucas
Focus: Arts; Econ
Periodical
Journal (weekly) 12851

Royal Society for the Prevention of Accidents (ROSPA), Cannon House, Priory Queensway, Birmingham B4 6BS
T: (021) 2002461; Tx: 336546; Fax: (021) 2001254
Founded: 1916
Focus: Safety
Periodicals
Care in the Home (weekly)
Care on the Road (weekly)
Occupational Safety & Health (weekly)
Safety Education (3 times annually) . . 12852

Royal Society for the Protection of Birds (R.S.P.B.), The Lodge, Sandy SG19 2DL
T: (0767) 80551; Fax: (0767) 692365
Founded: 1889; Members: 500000
Pres: Magnus Magnusson; Gen Secr: Ian Prestt
Focus: Ecology; Ornithology
Periodical
Birds (weekly) 12853

Royal Society of Arts (RSA), 6-8 John Adam St, London WC2N 6EZ
T: (071) 8392366
Founded: 1754; Members: 12000
Focus: Arts 12854

Royal Society of British Artists, 17 Carlton House Terrace, London SW1Y 5BD
T: (071) 9306844; Fax: (071) 8397830
Members: 119
Pres: Tom Coates
Focus: Arts; Fine Arts
Periodical
Annual Catalogue (weekly) 12855

Royal Society of British Sculptors, 108 Old Brompton Rd, London SW7 3RA
T: (071) 3738202; Fax: (071) 3739202
Founded: 1904; Members: 200
Pres: Philomena Davidson Davis; Gen Secr: Dr. Simon Hincks
Focus: Fine Arts 12856

Royal Society of Chemistry (RSC), Burlington House, Piccadilly, London W1 OBN
T: (071) 4378656; Tx: 268001; Fax: (071) 2879798
Founded: 1980; Members: 43500
Pres: Prof. C.W. Rees; Gen Secr: Dr. T.D. Inch
Focus: Chem
Periodicals
Analytical Abstracts
Chemical Business Bulletin
Chemical Business Newsbase
Chemical Business Update
Chemical Engineering Abstracts
Chemistry in Britain (weekly)
Current Biotechnology Abstracts
Education in Chemistry
Journal of the Chemical Society
Laboratory Hazards Bulletin
Mass Spectrometry Bulletin
Natural Product Updates
Professional Bulletin (9 times annually)
Specialist Periodical Reports 12857

The Royal Society of Edinburgh (R.S.E.), 22-24 George St, Edinburgh EH2 2PQ
T: (031) 2256057; Fax: (031) 2206889
Founded: 1783; Members: 1100
Pres: Sir Alastair Currie; Gen Secr: Prof. V.B. Proudfoot
Focus: Sci; Lit
Periodicals
Annual Report (weekly)
Proceedings A (Mathematics) (weekly)
Proceedings B (Biological Sciences) (weekly)
Transactions: Earth Sciences (weekly)
Year Book (weekly) 12858

The Royal Society of Health (R.S.H.), R.S.H. House, 38 Saint George Dr, London SW1V 4BH
T: (071) 6300121; Fax: (071) 9766847
Founded: 1876; Members: 11500
Pres: B.R. Saunders; Gen Secr: D. Goad
Focus: Public Health
Periodical
Journal (weekly) 12859

Royal Society of Literature of the United Kingdom, 1 Hyde Park Gardens, London W2 2LT
T: (071) 7235104
Founded: 1823; Members: 800
Pres: Lord Jenkins; Gen Secr: Maggie Parkham
Focus: Lit
Periodicals
Report
Transactions 12860

Royal Society of Marine Artists, 17 Carlton House Terrace, London SW1Y 5BD
T: (071) 9306844; Fax: (071) 8397830
Members: 49
Pres: Terence Storey
Focus: Fine Arts 12861

The Royal Society of Medicine (RSM), 1 Wimpole St, London W1M 8AE
T: (071) 4082119
Founded: 1805; Members: 17000
Pres: Sir John Pinker; Gen Secr: Dr. R.N. Thomson
Focus: Med
Periodicals
Annual Report (weekly)
Calendar (weekly) 12862

Royal Society of Portrait Painters, 17 Carlton House Terrace, London SW1Y 5BD
T: (071) 9306844; Fax: (071) 8397830
Founded: 1891; Members: 41
Pres: George J.D. Bruce
Focus: Arts; Fine Arts
Periodical
Catalogue of Annual Exhibition (weekly) . 12863

Royal Society of Tropical Medicine and Hygiene, 26 Portland Pl, London W1N 4EY
T: (071) 5802127; Fax: (071) 4361389
Founded: 1907; Members: 2617
Pres: Dr. Gordon C. Cook
Focus: Hygiene; Trop Med
Periodical
Transactions of the Royal Society of Tropical Medicine and Hygiene (weekly) 12864

Royal Society of Ulster Architects (RSUA), 51 Malone Rd, Belfast BT9 6RY
T: (0232) 668846
Founded: 1901
Focus: Archit 12865

Royal Statistical Society, 25 Enford St, London W1H 2BH
T: (071) 7235882; Fax: (071) 7061710
Founded: 1834; Members: 6000
Pres: Prof. D.J. Bartholomew
Focus: Stats
Periodicals
Journal of the Royal Statistical Society, Series A: Statistics in Society (3 times annually)
Journal of the Royal Statistical Society, Series B: Methodological (3 times annually)
Journal of the Royal Statistical Society, Series C: Applied Statistics (3 times annually)
Journal of the Royal Statistical Society, Series D: The Statistician 12866

Royal Stuart Society, 26 Ovse Walk, Huntingdon PE18 6OL
T: (0480) 459427
Founded: 1926
Focus: Hist
Periodicals
Royalist Focus (weekly)
Royal Stuart Papers (weekly)
Royal Stuart Review (weekly) 12867

The Royal Television Society (RTS), Holborn Hall, 100 Gray's Inn Rd, London WC1X 8AL
T: (071) 4301000; Fax: (071) 4300924
Founded: 1927; Members: 3500
Pres: Bill Cotton; Gen Secr: John Gau
Focus: Electric Eng
Periodical
Television (8 times annually) 12868

Royal Town Planning Institute (RTPI), 26 Portland Pl, London W1N 4BE
T: (071) 6369107; Fax: (071) 3231582
Founded: 1914; Members: 17000
Pres: Hazel McKay; Gen Secr: David R. Fryer
Focus: Urban Plan 12869

Royal Ulster Academy (RUA), 28 Woodeane, Lurgan, Armagh
Founded: 1879
Focus: Sci 12870

Royal Ulster Academy Association (R.U.A.A.), 3 Derryvolgie Park, Lambeg BT27 4DA
Founded: 1962; Members: 150
Pres: Jean Harrison; Gen Secr: Sally Wheeler
Focus: Sci 12871

Royal Ulster Agricultural Society (RUAS), The King's Hall, Balmoral, Belfast BT9 6GW
T: (0232) 665225
Founded: 1826; Members: 3500
Focus: Agri 12872

Royal Watercolour Society, c/o Bankside Gallery, 48 Hopton St, Blackfriars, London SE1 9JH
T: (071) 9287521; Fax: (071) 9282820
Founded: 1804; Members: 89
Pres: Leslie Worth; Gen Secr: Michael Spender
Focus: Fine Arts 12873

Royal Welsh Agricultural Society, Llanelwedd, Builth-Wells, Powys LD2 3SY
T: (0982) 553683; Fax: (0982) 553563
Founded: 1904; Members: 11000
Gen Secr: D. Walters
Focus: Agri
Periodical
Annual Journal (weekly) 12874

Royal West of England Academy, Queen's Rd, Clifton, Bristol BS8 1PX
Founded: 1844; Members: 200
Pres: Leonard Manasseh; Gen Secr: Jean McKinney
Focus: Arts 12875

The Royal Zoological Society of Scotland, c/o Scottish National Zoological Park, Murrayfield, Edinburgh EH12 6TS
T: (031) 3349171; Fax: (031) 3164050
Founded: 1909; Members: 10000
Pres: Viscount of Arbuthnott; Gen Secr: Prof. Roger J. Wheater
Focus: Zoology
Periodical
Ark File (weekly) 12876

RSNC Wildlife Trusts Partnership, The Green, Witham Park, Waterside South, Lincoln LN5 7JR
T: (0522) 544400; Fax: (0522) 511616
Founded: 1912; Members: 250000
Pres: Sir David Attenborough
Focus: Ecology 12877

Rugby Football Schools Union (RFSU), c/o Rugby Football Union, Twickenham TW1 1DZ
T: (081) 8928161; Fax: (081) 8929816
Founded: 1970; Members: 1500
Focus: Sports 12878

Rural Preservation Association (RPA), The Old Police Station, Lark Lane, Liverpool L17
T: (051) 7287011
Founded: 1975
Focus: Botany; Ecology
Periodicals
Natterjack (weekly)
RPA News (weekly) 12879

Rutland Agricultural Society, 8 High St, Oakham, Rutland LE15 6AL
T: (0572) 2018
Founded: 1830; Members: 500
Focus: Agri
Periodical
Schedule & Show Catalogue (weekly) . 12880

Safety and Reliability Society (SaRS), Cambridge Arcade, Lord St, POB 25, Southport PR8 1AS
T: (0925) 56106
Founded: 1980; Members: 150
Focus: Eng; Safety 12881

Saint Albans and Hertfordshire Architectural and Archaeological Society (SAHAAS), 24 Rose Walk, Saint Albans AL4 9AF
T: (0727) 853204
Founded: 1845; Members: 550
Pres: J.M. Kilvington; Gen Secr: B.E. Moody
Focus: Archeol; Archit; Hist
Periodical
Hertfordshire Archaeology 12882

Saint John's Hospital Dermatological Society, c/o Saint John's Dermatology Centre, Saint Thomas' Hospital, London SE1 7EH
T: (071) 9289292
Founded: 1911; Members: 500
Pres: Prof. R. Hay; Gen Secr: Dr. N. Walker
Focus: Derm
Periodical
Clinical and Experimental Dermatology (weekly) 12883

Saintpaulia and Houseplant Society, 33 Church Rd, Newbury Park, Ilford IG2 7ET
Founded: 1956; Members: 720
Focus: Hort
Periodical
Bulletin (weekly) 12884

Saltire Society, 9 Fountain Close, 22 High St, Edinburgh EH1 1TF
Founded: 1936; Members: 1600
Pres: Sir I. Noble; Gen Secr: Kathleen Munro
Focus: Arts 12885

SATRA Footwear Technology Centre, SATRA House, Rockingham Rd, Kettering NN16 9JH
T: (0536) 410000; Tx: 34323; Fax: (0536) 410626
Founded: 1919; Members: 1100
Pres: R.E. Whittaker; Gen Secr: John Butcher
Focus: Eng
Periodical
Footwear Business International (weekly) . 12886

Schizophrenia Association of Great Britain (SAGB), Bryn Hyfryd, The Crescent, Bangor LL57 2AG
T: (0248) 354048
Founded: 1970; Members: 2100
Pres: Dr. David Horrobin; Gen Secr: Gwynneth Hemmings
Focus: Psychiatry
Periodical
Newsletter (2-3 times annually) 12887

School Natural Science Society (SNSS), 22 Chada Av, Gillingham KT5 8QU
T: (0634) 51973
Founded: 1903; Members: 1400
Focus: Educ; Nat Sci
Periodical
Science Teaching (3 times annually) . . 12888

Science and Engineering Research Council (SERC), Polaris House, North Star Av, Swindon SN2 1ET
T: (0793) 411000; Tx: 449466; Fax: (0793) 411400
Founded: 1965
Pres: Prof. E.W.J. Mitchell; Gen Secr: J. Merchant
Focus: Astronomy; Bio; Chem; Eng; Math; Nucl Res; Physics
Periodicals
Annual Report (weekly)
SERC Bulletin (3 times annually) . . . 12889

Scientific Committee on Antarctic Research (SCAR), c/o Scott Polar Research Institute, Lensfield Rd, Cambridge CB2 1ER
T: (0223) 62061
Founded: 1958
Focus: Geophys; Geography 12890

Scientific Exploration Society (SES), Home Farm, Mildenhall, Marlborough SN8 2LF
T: (0672) 2994
Founded: 1969; Members: 154
Focus: Geography 12891

Scots Ancestry Research Society, 29b Albany St, Edinburgh EH1 3QN
T: (031) 5564220
Founded: 1945
Focus: Genealogy 12892

Scottish Arts Council (SAC), 12 Manor Pl, Edinburgh EH3 7DD
T: (031) 2266051; Fax: (031) 2259833
Founded: 1946; Members: 22
Pres: Dr. William Brown; Gen Secr: Seona Reid
Focus: Arts 12893

Scottish Association for Building Education and Training (SABET), c/o Reid Kerr College, Renfrew Rd, Paisley PA3 4DR
Founded: 1950
Focus: Adult Educ; Civil Eng 12894

The Scottish Association for Marine Sciemce (SAMS), c/o Dunstaffnage Marine Laboratory, Oban PA34 4AD
T: (0631) 62244; Fax: (0631) 65518
Founded: 1897; Members: 350
Pres: Dr. David Smith; Gen Secr: J.B.L. Matthews
Focus: Bio; Oceanography
Periodicals
Annual Report (weekly)
Newsletter 12895

Scottish Association for Mental Health (SAMH), 57 Melville St, Edinburgh EH3 7HL
T: (031) 2257682
Founded: 1923
Focus: Psychiatry 12896

Scottish Association of Advisers in Physical Education, c/o Educational Dept, Strathclyde Regional Council, Regional Offices, Dumbarton G82 3PU
T: 65151
Founded: 1939
Focus: Sports 12897

Scottish Association of Geography Teachers (SAGT), 19 Auchingramont Rd, Hamilton ML3 6JD
T: (0698) 29531
Founded: 1970; Members: 580
Focus: Geography; Educ 12898

Scottish Association of Local Government to Educational Psychologists (SALGEP), 13 Kelvinside Gardens East, Glasgow G20 6BE
T: (041) 9461075
Founded: 1974; Members: 160
Focus: Educ; Psych 12899

Scottish Association of Writers (SAW), 221 East Clyde St, Helensburgh Dunbartonshire
T: (041) 3343164
Founded: 1970
Focus: Lit 12900

Scottish Business Education Council, 22 Great King St, Edinburgh EH3 6QH
T: (031) 5774555
Focus: Adult Educ
Periodicals
Annual Report (weekly)
Business Education Guide (weekly)
Careers Leaflets (weekly)
Handbook (weekly) 12901

Scottish Catholic Historical Association (SCHA), c/o John S. Burns & Sons, 25 Finlas St, Glasgow G22 5DS
T: (041) 3368678
Founded: 1950
Focus: Hist; Rel & Theol 12902

Scottish Church History Society, 1 Denham Green Terrace, Edinburgh EH5 3PG
T: (031) 5524059
Founded: 1922; Members: 277
Pres: John McGaffrey; Gen Secr: Peter H. Donald
Focus: Hist; Rel & Theol
Periodical
Records (weekly) 12903

Scottish Economic Society (SES), c/o Dept of Economy, Adam Smith Bldg, Glasgow University, Glasgow G12 8RT
T: (041) 3398855; Tx: 777070; Fax: (041) 3304940
Founded: 1954; Members: 200
Pres: Prof. L. Hunter; Gen Secr: Dr. F. Hay
Focus: Econ
Periodical
Scottish Journal of Political Economy (weekly) 12904

Scottish Electrophysiological Society, c/o Physiology Dept, Bute Medical Buildings, University, Saint Andrews KY16 9TS
T: (0334) 76161; Tx: 76213
Focus: Electrochem 12905

Scottish Epilepsy Association (SEA), 48 Govan Rd, Glasgow G51 1JL
T: (041) 4274911
Founded: 1954
Focus: Pathology 12906

Scottish Field Studies Association (SFSA), Kindrogan House Field Centre, Enochdhu, Blairgowrie PH10 7PG
T: (0250881) 286; Fax: (0250881) 433
Founded: 1945; Members: 400
Pres: Dr. A. Pike
Focus: Nat Sci
Periodical
Annual Report (weekly) 12907

Scottish Further Education Association (SFEA), 111 Union St, Glasgow Gl 3SS
T: (041) 2210118
Founded: 1966
Focus: Adult Educ
Periodical
Sector 3 (10 times annually) 12908

Scottish Gaelic Texts Society (SGTS), c/o Dept of Celtic, The University, Glasgow G12 8QQ
T: (041) 3398855 ext 5425
Founded: 1934; Members: 150
Focus: Lit
Periodical
SGTS Publications (weekly) 12909

Scottish Genealogy Society, 15 Victoria Terrace, Edinburgh EH1 2JL
T: (031) 2203677
Founded: 1953
Focus: Genealogy
Periodicals
Pre-1855 Monumental Inscription Lists (weekly)
Register of Members' Interests (weekly)
The Scottish Genealogist (weekly) . . . 12910

Scottish History Society, c/o Dept of Scottish History, University of Edinburgh, Edinburgh E1
Founded: 1886; Members: 800
Pres: Prof. D.E.R. Watt; Gen Secr: Dr. E.P. Torrie
Focus: Hist 12911

Scottish Industrial Heritage Society (SSIA), 129 Fotheringay Rd, Glasgow G41 4LG
T: (041) 4231782
Founded: 1984; Members: 140
Focus: Archeol
Periodical
Newsletter (weekly) 12912

Scottish Inland Waterways Association (SIWA), 139 Old Dalkeith Rd, Edinburgh EH16 4SZ
T: (031) 6641070
Founded: 1971; Members: 240
Gen Secr: G.A. Hunter
Focus: Transport
Periodical
SIWA News (3 times annually) 12913

Scottish Language Society, 31 Braeside Park, Balloch, Inverness
T: Culloden Moor 689
Founded: 1969; Members: 485
Focus: Ling 12914

Scottish Library Association, Motherwell Business Centre, Coursington Rd, Motherwell ML1 1PW
T: (0698) 252526; Fax: (0698) 252057
Founded: 1908; Members: 2500
Gen Secr: Robert Craig
Focus: Libraries & Bk Sci
Periodical
Scottish Libraries (weekly) 12915

Scottish National Dictionary Association, 27 George Sq, Edinburgh EH8 9LD
T: (031) 6671011
Founded: 1929
Focus: Ling 12916

Scottish National Housing and Town Planning Council, 19 Bentinck Dr, Troon KA10 6HX
T: 314552
Founded: 1912; Members: 110
Focus: Urban Plan 12917

Scottish National Orchestra Society (SNOS), 150 Hope St, Glasgow C2
T: (041) 3327244
Focus: Music 12918

Scottish Ornithologists' Club (SOC), 21 Regent Terrace, Edinburgh EH7 5BT
T: (031) 5566042
Founded: 1936; Members: 2600
Focus: Ornithology
Periodicals
Scottish Bird News (weekly)
Scottish Bird Report (weekly)
Scottish Birds (weekly) 12919

Scottish Pharmaceutical Federation (SPF), 135 Buchanan St, Glasgow G1 2JQ
T: (041) 2211235
Founded: 1924
Focus: Pharmacol 12920

Scottish Physical Education Association (SPEA), c/o Jordanhill College of Education, Glasgow G13 1PP
T: (041) 9591232
Founded: 1973; Members: 1000
Focus: Sports 12921

Scottish Record Society, c/o Scottish History Dept, University of Glasgow, Glasgow G12 8QQ
T: (041) 3398855 ext 576
Founded: 1897; Members: 383
Focus: Hist; Genealogy 12922

Scottish Recreational Land Association (SRLA), 18 Abercromby Pl, Edingburgh EH3 6TV
T: (031) 5564466
Founded: 1972; Members: 99
Focus: Travel
Periodical
Newsletter (weekly) 12923

Scottish Reformation Society (SRS), 17 George IV Bridge, Edinburgh EH1 1EE
T: (031) 2251836
Founded: 1851; Members: 300
Focus: Rel & Theol
Periodical
The Bulwark (weekly) 12924

Scottish Rights of Way Society, 28 Rutland Sq, Edinburgh EH1 2B
Founded: 1845; Members: 1300
Focus: Law
Periodical
Annual Report (weekly) 12925

Scottish River Purification Boards Association (SRPBA), City Chambers, Glasgow G2 1DU
T: (041) 2219600
Founded: 1957; Members: 7
Focus: Ecology 12926

Scottish Secondary Teachers' Association (SSTA), 15 Dundas St, Edinburgh EH3 6QG
T: (031) 5565919, 5560605
Founded: 1946; Members: 7500
Focus: Educ
Periodicals
Bulletin (weekly)
Journal (weekly) 12927

Scottish Society for Crop Research, c/o Scottish Crop Research Institute, Invergowrie, Dundee DD2 5DA
T: (0382) 562731; Fax: (0382) 562426
Founded: 1981; Members: 300
Focus: Agri; Crop Husb
Periodical
Bulletin (weekly) 12928

Scottish Society for Northern Studies (SSNS), c/o School of Scottish Studies, University of Edinburgh, 27 George Sq, Edinburgh EH8 9LD
T: (031) 6504162
Founded: 1967; Members: 275
Focus: Ethnology; Archeol; Hist; Lit
Periodical
Northern Studies (weekly) 12929

Scottish Society for the Preservation of Historical Machinery, c/o Glasgow University Mechanical Engineering Research Annexe, 49 Spencer St, Glasgow G13
Focus: Eng; Preserv Hist Monuments; Cultur Hist 12930

Scottish Society of the History of Medicine (SSHM), c/o Grampian Health Board Offices, Woolmanhill, Aberdeen AB9 1EF
T: (0224) 663456; Fax: (0224) 840791
Founded: 1948; Members: 180
Focus: Med; Cultur Hist
Periodicals
Newsletter (weekly)
Report of Proceedings (weekly) 12931

Scottish Tartans Society, FONAB House, Pitlochry PH16 5ND
Founded: 1963; Members: 1086
Pres: Duke of Atholl
Focus: Hist
Periodicals
Proceedings (weekly)
Tartans (weekly) 12932

Scottish Teachers' Nursing Association, 45 West Nile St, Glasgow G1 2PT
T: (041) 2217788
Founded: 1936
Focus: Adult Educ; Med 12933

Scottish Text Society (STS), 27 George Sq, Edinburgh EH8 9LD
T: (031) 6504147
Founded: 1882; Members: 226
Focus: Lit
Periodical
Publications (weekly) 12934

Scottish Tramway Museum Society (STMS), POB 78, Glasgow G3 6ER
Founded: 1951; Members: 142
Focus: Transport
Periodical
Scottish Transport (weekly) 12935

Scottish Wildlife Trust (SWT), Cramond House, Kirk Cramond, Cramond Glebe Rd, Edinburgh EH4 6NS
T: (031) 3127765; Fax: (031) 3128705
Founded: 1964; Members: 19000
Pres: David Hughes Hallett
Focus: Ecology; Forestry
Periodical
Scottish Wildlife 12936

Seabird Group, c/o RSPB, The Lodge, Sandy SG19 2DL
Founded: 1966; Members: 300
Focus: Ornithology
Periodicals
Seabird: Annual Journal (weekly)
Seabird Group Newsletter (3 times annually) 12937

Selborne Society, 10 Sumbeam Cottages, Pollards Wood Rd, Limpsfield, Oxted RH8 0HY
Founded: 1885; Members: 700
Gen Secr: A.H. Austin
Focus: Hist; Nat Sci
Periodical
The Selborne Magazine (weekly) 12938

Selden Society, c/o Faculty of Law, Queen Mary College, Mile End Rd, London E1 4NS
T: (071) 9755136; Tx: 893750; Fax: (081) 9818733
Founded: 1887; Members: 1700
Pres: Peter Gerrard; Gen Secr: Victor Tunkel
Focus: Hist; Law
Periodicals
Early Law Reports
Legal Records (weekly) 12939

The Sempervivum Society, 11 Wingle Tye Rd, Burgess Hill RH15 9HR
T: (0444) 236848
Founded: 1970; Members: 300
Focus: Ecology
Periodical
Newsletter (3 times annually) 12940

Seventeenhundredandfourtyfive and National Military History Society, 45 Ann St, Edinburgh EH4 1PL
T: (031) 3325568
Founded: 1947; Members: 310
Focus: Hist; Military Sci 12941

Sexual Law Reform Society (SLRS), 31 Clapham Rd, London SW9 0JD
T: (071) 5820972
Founded: 1968; Members: 100
Focus: Law 12942

Shakespeare Authorship Society, 25 Montagu Sq, London W1M 1RE
T: (071) 4863764
Founded: 1922; Members: 180
Focus: Lit 12943

Shakespeare Birthplace Trust, c/o The Shakespeare Centre, Stratford-upon-Avon CV37 6QW
T: (0789) 204016; Fax: (0789) 296083
Founded: 1847
Gen Secr: Roger Pringle
Focus: Lit; Preserv Hist Monuments . . 12944

Shakespeare Institue, c/o University of Birmingham, Mason Croft, Church St, Stratford-upon-Avon CV37 6HP
T: (0789) 293138; Fax: (0789) 414992
Founded: 1952
Pres: Prof. Stanley Wells
Focus: Lit
Periodicals
Conference Report
Report of the International Shakespeare Conference 12945

Shakespeare Reading Society, 17 Lake Close, Lake Rd, London SW19 7EE
T: (081) 9473036
Founded: 1875; Members: 50
Focus: Lit 12946

Shaviana, High Orchard, 125 Markyate Rd, Dagenham, Essex
Founded: 1941
Gen Secr: Eric Ford
Focus: Lit
Periodicals
The Shavian
Shavian Tracts 12947

Sheffield Metallurgical and Engineering Association (SMEA), c/o British Steel Technical, Moorgate, Rotherham S60 3AR
T: (0709) 820166; Fax: (0709) 825337
Founded: 1963; Members: 600
Pres: Dr. R.A.E. Hooper; Gen Secr: M.J. Heesom
Focus: Metallurgy; Eng 12948

Sherlock Holmes Society of London, 3 Outram Rd, Southsea PO5 1QP
T: (0705) 812104
Founded: 195; Members: 1300
Gen Secr: G.S. Stavert
Focus: Lit
Periodical
Sherlock Holmes Journal (weekly) . . . 12949

Shetland Council of Social Service, 4b Market St, Lerwick ZE1 0JN
T: (0595) 3816
Founded: 1959; Members: 130
Focus: Sociology 12950

Shropshire Archaeological and Parish Register Society (SAS), c/o Much Wenlock Museum, High St, Much Wenlock TF13 6HR
T: (0952) 727773; Tx: 81604; Fax: (0223) 461522
Founded: 1877; Members: 313
Focus: Archeol; Hist
Periodical
Transactions (weekly) 12951

Simplified Speeling Society (SSS), 5 Gwyder Rd, Beckenham, Kent
T: 5841848
Founded: 1908
Focus: Ling 12952

Sira, South Hill, Chislehurst BR7 5EH
T: (081) 4672636; Fax: 4676515
Founded: 1918; Members: 85
Pres: Dr. J. Alvey; Gen Secr: R.A. Brook
Focus: Eng
Periodical
Spotlight (weekly) 12953

The Sleep-Learning Association (SLA), 64 Grange Rd, London W5 5BX
T: (081) 5797966
Founded: 1963; Members: 2770
Pres: Geoffrey G.L. Stocker; Gen Secr: John Leo
Focus: Bio 12954

Social History Society of the United Kingdom (SHS), c/o Centre for Social History, Lancaster University, Lancaster LA1 4YG
T: (0524) 592605; Fax: (0524) 846102
Founded: 1976; Members: 712
Gen Secr: Dr. Cynthia Hay
Focus: Hist
Periodical
Bulletin (weekly) 12955

Socialist Educational Association (SEA), 62 Thornhill Rd, Heaton Mersey, Stockport SK4 3HL
T: 4321409
Founded: 1960; Members: 650
Focus: Educ 12956

Socialist Health Association (SHA), 16 Charles Sq, London N1 6HP
T: (071) 4900057; Fax: (071) 4900057
Founded: 1930; Members: 1330
Pres: Dr. Douglas Naysmith; Gen Secr: Christine May
Focus: Public Health
Periodical
Socialism & Health (weekly) 12957

Socialist International Research Council (SIRC), 88a Saint John's Wood High St, London NW8 7SJ
T: (071) 5861101
Founded: 1971
Focus: Sci 12958

Social Policy Association (SAA), c/o School of Social and Professional Studies, University of Humberside, Inglemire Av, Hull HU6 7LU
Founded: 1968; Members: 320
Gen Secr: Dr. Gary Craig
Focus: Public Admin
Periodical
Social Policy Review (weekly) 12959

Social Research Association (SRA), c/o Dept of Applied Social Studies, Polytechnic of North London, Ladbroke House, Highbury Grove, London N5 2AD
T: (071) 6072789 ext 5082/5000
Founded: 1978; Members: 491
Focus: Sociology 12960

La Société Guernesiaise, Candie Gardens, Saint Peter Port GY1 1UG
T: (0481) 725093; Fax: (0481) 66217
Founded: 1882; Members: 1800
Gen Secr: F.G. Caldwell
Focus: Nat Sci; Hist; Archeol; Preserv Hist Monuments
Periodical
Transactions (weekly) 12961

Société Jersiaise, 7 Pier Rd, Saint Helier, Jersey
T: (0534) 58314; Fax: (0534) 888262
Founded: 1873
Focus: Nat Sci
Periodical
Bulletin (weekly) 12962

Society for African Church History, c/o Dept of Religious Studies, King's College, University of Aberdeen, Aberdeen AB9 2UB
T: (0224) 40241
Founded: 1963
Focus: Hist; Rel & Theol 12963

Society for Applied Bacteriology (SAB), c/o Faculty of Agriculture, Food and Land Use, University of Ayrmouth, Newton Abbot TQ12 6NQ
Founded: 1931; Members: 1650
Focus: Microbio
Periodicals
Journal of Applied Bacteriology (weekly)
Symposium Series (weekly)
Technical Series (weekly) 12964

Society for Army Historical Research, c/o The National Army Museum, London SW3 4HT
Founded: 1921; Members: 1000
Pres: Sir John Chapple; Gen Secr: Dr. Peter Boyden
Focus: Hist; Military Sci
Periodical
Journal (weekly) 12965

Society for Back Pain Research (SBPR), c/o Bone and Joint Research Unit, The London Hospital, London E1 1BB
T: (071) 2475454
Founded: 1971; Members: 130
Focus: Med 12966

Society for Computers and Law (SCL), 10 Hurle Crescent, Clifton, Bristol BS8 2TA
T: (0272) 237393; Fax: (0272) 239305
Founded: 1973; Members: 1550
Pres: Justice Brooke; Gen Secr: Ruth Baker
Focus: Law; Computer & Info Sci
Periodical
Computers & Law (weekly) 12967

Society for Cooperation in Russian and Soviet Studies, 320 Brixton Rd, London SW9 6AB
T: (071) 2742282; Tx: 888941; Fax: (071) 4890391
Founded: 1924; Members: 1500
Pres: William Bowring; Gen Secr: Jean Turner
Focus: Ethnology
Periodical
Newsletter (weekly) 12968

Society for Earthquake and Civil Engineering Dynamics (SECED), c/o Institution of Civil Engineers, 1-7 Great George St, London SW1
T: (071) 8399827; Fax: (071) 2227500
Members: 150
Focus: Geophys; Civil Eng 12969

Society for Education in Film and Television (SEFT), 63 Old Compton St, London W1V 5PM
T: (071) 7345455
Founded: 1950
Focus: Educ 12970

Society for Endocrinology, 17-18 The Courtyard, Woodlands, Almondsbury, Bristol BS12 4NQ
T: (0454) 616046; Fax: (0454) 616071
Founded: 1946; Members: 1434
Pres: I.W. Henderson; Gen Secr: B. Furr
Focus: Endocrinology
Periodicals
Endocrine-Related Cancer (weekly)
Journal of Endocrinology (weekly)
Journal of Molecular Endocrinology (weekly)
. . . . 12971

Society for Environmental Therapy (SET), c/o Dept of Bacteriology and Virology, University of Manchester, Manchester M13 9PT
T: (061) 2738241 ext 68
Founded: 1980; Members: 300
Focus: Toxicology; Immunology; Pharmacol
Periodical
Newsletter (weekly) 12972

The Society for Experimental Biology (SEB), Burlington House, Piccadilly, London WIV OLQ
T: (071) 4398732; Fax: (071) 2874786
Founded: 1923; Members: 2300
Focus: Bio
Periodicals
Journal of Experimental Botany (weekly)
The Plant Journal (weekly)
Seminar Series (2-3 times annually)
Symposia (weekly) 12973

Society for Folk Life Studies, c/o Ulster Folk and Transport Museum, Cultra Manor, Hollywood BT18 0EU
T: 5411
Founded: 1961; Members: 600
Focus: Ethnology 12974

Society for General Microbiology, Marlborough House, Basingstoke Rd, Spencers Wood, Reading RG7 1AE
T: (0734) 885577; Fax: (0734) 885656
Founded: 1945; Members: 5000
Pres: Prof. J.A. Jomes; Gen Secr: H.J. Bower
Focus: Microbio
Periodicals
Journal of General Virology (weekly)
Microbiology (weekly)
Society for General Microbiology Quarterly (weekly) 12975

Society for Italic Handwriting, 59a Arlington Rd, London NW1
T: (071) 3887117
Founded: 1952; Members: 1100
Focus: Ling 12976

Society for Latin American Studies (SLAS), c/o Dept of Politics, University of York, Heslington, York YO1 5DD
T: (0904) 59861
Founded: 1965; Members: 442
Focus: Ethnology
Periodicals
Bulletin of Latin American Research (weekly)
Newsletter (weekly) 12977

Society for Libyan Studies, c/o Institute of Archaeology, 31-34 Gordon Sq, London WC1H 0PY
T: (071) 3877050 ext 7495
Founded: 1969; Members: 325
Pres: Peter Tripp
Focus: Ethnology; Geology; Archeol; Geography; Hist
Periodical
Libyan Studies (weekly) 12978

Society for Lincolnshire History and Archaeology (S.L.H.A.), Jews' Court, Steep Hill, Lincoln LN2 1LS
T: (0522) 521337
Founded: 1974; Members: 850
Pres: Dr. Dorothy Owen; Gen Secr: Susan Smith
Focus: Archeol; Hist
Periodicals
Lincolnshire History and Archaeology (weekly)
Lincolnshire Past and Present (weekly) . 12979

Society for Low Temperature Biology (SLTB), c/o Institute of Biology, 20-22 Queensberry Pl, London SW7 2DZ
T: (071) 5818333
Founded: 1964; Members: 130
Gen Secr: Dr. W.J. Armitage
Focus: Bio 12980

Society for Medicine Research, c/o Institute of Biology, 20 Queensberry Pl, London SW7 2DZ
T: (071) 5818333
Founded: 1966
Focus: Pharmacol 12981

Society for Medieval Archaeology (SMA), c/o City of Lincoln Archaeology Unit, The Lawn, Union Rd, Lincoln LN1 3BG
Fax: (0522) 548089
Founded: 1957; Members: 1600
Pres: Dr. Michael Thompson; Gen Secr: Dr. Alan Vince
Focus: Archeol
Periodical
Medieval Archaeology (weekly) 12982

Society for Multivariate Experimental Psychology, c/o National Foundation for Educational Research, The Mere, Upton Park, Slough
Focus: Psych 12983

Society for Nautical Research (SNR), c/o National Maritime Museum, Greenwich, London SE10 9NF
T: (081) 8584422
Founded: 1910; Members: 2100
Pres: Lord Lewin of Greenwich; Gen Secr: Derek Law
Focus: Navig; Archeol; Hist
Periodical
The Mariners' Mirror (weekly) 12984

Society for New Testament Study, c/o Dept 07 Religious Studies, University of Lancaster, Lancaster LA1 4YG
T: (0524) 65201 ext 4420
Founded: 1938; Members: 800
Focus: Rel & Theol
Periodical
New Testament Studies (weekly) 12985

Society for Post-Medieval Archaeology (SPMA), c/o British Museum, Great Russell St, London WC1B 3DG
Founded: 1967; Members: 800
Pres: M. Ponsford; Gen Secr: Dr. David Gaimster
Focus: Archeol
Periodicals
Journal (weekly)
Newsheet (06) 12986

Society for Promoting Christian Knowledge (SPCK), Holy Trinity Church, Marylebone Rd, London NW1 4DU
T: (071) 3875282; Fax: (071) 3882352
Founded: 1698
Focus: Rel & Theol
Periodical
Theology (weekly) 12987

Society for Promotion of Educational Reform through Teacher Training (SPERTTT), c/o Sidney Webb College, 9-12 Barrett St, London W1M 6DE
T: (071) 4864771
Founded: 1969
Focus: Educ 12988

Society for Psychical Research, 49 Marloes Rd, London W8 6LA
T: (071) 9378984
Founded: 1882; Members: 1180
Pres: Prof. A.E. Roy
Focus: Parapsych
Periodicals
Journal of the Society for Psychical Research (weekly)
Proceedings (weekly)
The PSI Researcher (weekly) 12989

Society for Psychosomatic Research, c/o Dept of Psychiatry, Westminster Hospital, Horseferry Rd, London SW1P 2AP
Founded: 1955; Members: 100
Focus: Neurology 12990

Society for Radiological Protection (SRP), c/o National Radiological Protection Board, Chilton, Didcot OX11 0RQ
T: (0235) 831600; Fax: (0235) 833891
Founded: 1963
Focus: Radiology
Periodical
Journal of Radiological Protection (weekly) 12991

The Society for Renaissance Studies, Dept of Italian, Royal Holloway and Bedford New College, London TW20 0EX
Founded: 1967; Members: 640
Pres: Prof. Robert Knecht; Gen Secr: Dr. Letizia Panizza
Focus: Hist
Periodical
Bulletin of the Society for Renaissance Studies (weekly) 12992

Society for Research in the Psychology of Music and Music Education, c/o Faculty of Arts, North East London Polytechnic, Longsbridge Rd, Dagenham RM8 2AS
T: 5901640
Founded: 1972; Members: 330
Focus: Psych; Music; Educ 12993

Society for Research into Higher Education (SRHE), 344-354 Gray's Inn Rd, London WC1X 8BP
T: (071) 8377880; Fax: (071) 7130690
Founded: 1964; Members: 644
Pres: Sir Herman Bondi
Focus: Educ
Periodicals
Abstracts (3 times annually)
Higher Education (weekly)
International Newsletter (weekly)
Proceedings (weekly)
SRHE News (weekly)
Studies in Higher Education (3 times annually)
. . . . 12994

Society for Research into Hydrocephalus and Spina Bifida, c/o Booth Hall Children's Hospital, University of Manchester School of Medicine, Charlestown Rd, Blackley, Manchester M9 2AA
T: (061) 7957000; Fax: (061) 7415387
Founded: 1957; Members: 200
Focus: Pathology
Periodical
European Journal of Paediatric Surgery . 12995

Society for Social Medicine, 1 Wimpole St, London W1M 8AE
T: (071) 5802070
Founded: 1955; Members: 600
Focus: Med 12996

The Society for South Asian Studies / Incorporating Society for Afghan Studies, c/o The British Academy, 20-21 Cornwall Terrace, London NW1 4QP
Founded: 1973; Members: 200
Focus: Ethnology; Archeol; Hist; Arts
Periodical
South Asian Studies (weekly) 12997

Society for the Advancement of Anaesthesia in Dentistry (SAAD), 59 Summerlands Av, London W3 6EW
T: (081) 9936844
Focus: Anesthetics
Periodical
SAAD Digest (weekly) 12998

Society for the Advancement of Games and Simulations in Education and Training (SAGSET), c/o Centre for Extension Studies, University of Technology, Loughborough LE11 3TU
T: 6317
Founded: 1970; Members: 210
Focus: Adult Educ; Educ
Periodical
Simulation: Games for Learning (weekly) 12999

The Society for Theatre Research (STR), c/o Theatre Museum, 1E Tavistock St, London WC2E 7PA
Founded: 1948; Members: 800
Pres: Prof. Glynne Wickham; Gen Secr: Eileen Cottis
Focus: Perf Arts
Periodical
Theatre Notebook (3 times annually) . . 13000

Society for the Development of Techniques in Industrial Marketing (SDTIM), 9 Aston Rd, Nuneaton CV11 5EL
T: 67161
Founded: 1967
Focus: Advert 13001

Society for the Health Education (SHE), 69 Cornwall Gardens, London SW7 4BA
T: (071) 9372520
Founded: 1962; Members: 1077
Focus: Public Health; Educ 13002

Society for the History of Alchemy and Chemistry, c/o The Science Museum, South Kensington, London SW7 2DD
T: (071) 9388045; Tx: 21200; Fax: 9388118
Founded: 1937; Members: 250
Focus: Chem; Hist
Periodical
Ambix (3 times annually) 13003

Society for the History of Natural History, c/o British Museum (Natural History), London SW7 5BD
Founded: 1936; Members: 600
Focus: Hist; Nat Sci
Periodicals
Archives of Natural History (3 times annually)
Special Publications (weekly) 13004

Society for the Promotion of Hellenic Studies, 31-34 Gordon Sq, London WC1H 0PP
T: (071) 3877495
Founded: 1879; Members: 2800
Pres: Prof. J.P. Barron; Gen Secr: Dr. L. Rodley
Focus: Lit; Hist; Archeol
Periodical
The Journal of Hellenic Studies and Supplement Archaeological Reports (weekly) . . . 13005

Society for the Promotion of New Music, Francis House, Francis St, London SW1P 1DE
T: (071) 8289696; Fax: (071) 9319928
Founded: 1943; Members: 1000
Pres: Dame Elizabeth Maconchy
Focus: Music
Periodical
New: Notes (weekly) 13006

Society for the Promotion of Principles, 47 Long Mynd Rd, Northfield, Birmingham B31 1HJ
Members: 8
Focus: Philos 13007

Society for the Promotion of Roman Studies, 31-34 Gordon Sq, London WC1H 0PP
T: (071) 3878157
Founded: 1910; Members: 3500
Pres: Prof. T.P. Woseman
Focus: Hist; Archeol; Lit
Periodicals
Britannia (weekly)
Britannia Monographs (weekly)
Journal of Roman Studies (weekly)
Journal of Roman Studies Monographs (weekly)
. . . . 13008

Society for the Promotion of Vocational Training and Education, c/o South Bristol Technical College, Marksbury Rd, Bedminster, Bristol BS3 5JL
T: (0272) 661105
Founded: 1973
Focus: Adult Educ 13009

Society for the Protection of Ancient Buildings (S.P.A.B), 37 Spital Sq, London E1 6DY
T: (071) 3771644; Fax: (071) 2475296
Founded: 1877; Members: 6000
Pres: Duke of Grafton; Gen Secr: Philip Venning
Focus: Preserv Hist Monuments
Periodical
S.P.A.B. News (weekly) 13010

Society for the Protection of Science and Learning (SPSL), 3 Buckland Crescent, London NW3 5DH
T: (071) 7222095
Founded: 1933
Focus: Sci; Educ 13011

Society for the Social History of Medicine (SSHM), c/o Wellcome Unit for the History of Medicine, Maths Tower, The University, Manchester M13 9PL
T: (061) 2755910
Founded: 1969; Members: 600
Gen Secr: Mary Fissell
Focus: Hist; Med
Periodical
Social History of Medicine (3 times annually)
. . . . 13012

Society for the Study of Addiction (SSA), c/o Saint Christopher's Day Hospital, 52 Hurst Rd, Horsham RH12 2EP
T: (0403) 4367
Founded: 1884; Members: 400
Focus: Med 13013

Society for the Study of Fertility (SSF), 141 Newmarket Rd, Cambridge CB5 8HA
T: (0223) 351870
Founded: 1950; Members: 1250
Pres: Dr. J.R. Clarke; Gen Secr: Dr. S.R. Milligan
Focus: Bio; Med 13014

Society for the Study of Human Biology (SSHB), c/o London School of Hygiene and Tropical Medicine, Keppel St, London WC1E 7HT
T: (071) 3800599; Fax: (071) 3835859
Founded: 1958
Focus: Bio
Periodicals
Annals of Human Biology (weekly)
Symposia volumes (weekly) *13015*

Society for the Study of Inborn Errors of Metabolism (SSIEM), c/o Willink Laboratory, Royal Manchester Childrens Hospital, Manchester M27 1HA
T: (061) 7944696 ext 2138
Founded: 1963; Members: 700
Focus: Med
Periodical
Journal of Inherited Metabolic Disease (weekly) *13016*

Society for the Study of Information Transfer, 76c The Avenue, Beckenham BR3 2ES
Founded: 1973
Focus: Computer & Info Sci *13017*

Society for the Study of Labour History, c/o Polytechnic of Central London, Regent St, London W1A 8AL
T: (071) 5802021
Founded: 1960; Members: 1050
Focus: Hist; Sociology *13018*

Society for the Study of Medieval Languages and Literature (SSMLL), c/o Hertford College, Oxford OX1 3BW
Founded: 1932
Pres: Prof. Anne Hudson; Gen Secr: Dr. Charlotte Brewer
Focus: Ling; Lit
Periodical
Medium Aevum (weekly) *13019*

Society for the Study of Normal Psychology, 151 Talgarth Rd, London W14
Focus: Psych
Periodicals
The Bridge (weekly)
Fresh Morning of Life *13020*

Society for Underwater Technology (SUT), 76 Mark Ln, London EC3R 7JN
T: (071) 4810750; Fax: (071) 4814001
Founded: 1965; Members: 1000
Pres: M.G. Vallis; Gen Secr: D.R. Wardle
Focus: Eng
Periodical
Underwater Technology (weekly) *13021*

Society of Antiquaries of London (S.A.L.), Burlington House, Piccadilly, London W1V 0HS
T: (071) 7940193; Fax: (071) 2876964
Founded: 1707; Members: 2100
Pres: B.W. Cunliffe; Gen Secr: David M. Evans
Focus: Archeol
Periodicals
Antiquaries Journal (weekly)
Archaeologia (every 3-4 years) *13022*

Society of Antiquaries of Newcastle-upon-Tyne, Black Gate, Castle Garth, Newcastle-upon-Tyne NE1 1RQ
T: (091) 2615390
Founded: 1813; Members: 730
Focus: Archeol
Periodical
Archaeologia Aeliana (weekly) *13023*

Society of Antiquaries of Scotland, c/o National Museums of Scotland, York Bldgs, Queen St, Edinburgh EH2 1JD
Founded: 1780; Members: 2800
Pres: Dr. Anna Ritchie; Gen Secr: Fionna Ashmore
Focus: Archeol; Hist
Periodicals
Monographs (weekly)
Proceedings (weekly) *13024*

Society of Archer Antiquaries, 62 Lambert Rd, Bridlington YO16 5RD
Founded: 1956; Members: 400
Focus: Archeol *13025*

Society of Architectural and Industrial Illustrators, POB 22, Stroud GL5 3DH
T: (0453) 882563
Founded: 1975; Members: 359
Pres: Philip Crowe; Gen Secr: Eric Monk
Focus: Graphic & Dec Arts, Design
Periodicals
Newsletter
Yearbook *13026*

Society of Architectural Historians of Great Britain (SAHGB), 4 Woodlands Av, London N3 2NR
Founded: 1956; Members: 1100
Pres: John Newman; Gen Secr: Ruth Harman
Focus: Archit; Hist
Periodicals
Architectural History (weekly)
Newsletter *13027*

Society of Archivists (SA), c/o Information House, 20-24 Old St, London EC1V 9AP
T: (071) 2535087
Founded: 1947; Members: 866
Pres: H. Forde; Gen Secr: M. Turner
Focus: Archives
Periodical
Journal (weekly) *13028*

Society of Assistants Teaching in Preparatory Schools (SATIPS), 222 Hale Rd, Hale, Altrincham WA15 8EB
T: (061) 9802609
Founded: 1953; Members: 1200
Focus: Educ *13029*

Society of Authors, 84 Drayton Gardens, London SW10 9SB
T: (071) 3736642
Founded: 1884; Members: 5600
Focus: Lit; Music
Periodicals
The Author (weekly)
The Electronic Author (weekly) *13030*

Society of British Neurological Surgeons (SBNS), c/o Institute of Neurological Science, Southern General Hospital, 1345 Govan Rd, Glasgow G51 4TF
Fax: (041) 4452466
Founded: 1926; Members: 580
Pres: J. Brice; Gen Secr: T. Hide
Focus: Surgery; Neurology
Periodical
Journal of Neurology, Neurosurgery and Psychiatry (weekly) *13031*

Society of Business Economists (SBE), 11 Bay Tree Walk, Watford WD1 3RX
T: (0923) 237287
Founded: 1953; Members: 700
Pres: Sir D. Henderson; Gen Secr: Marian Marshall
Focus: Business Admin
Periodical
The Business Economist (3 times annually) *13032*

Society of Cartographers (SUC), c/o Dept of Geography and Topographic Science, University of Glasgow, Glasgow G12 8QQ
T: (041) 3398855; Fax: (041) 3304894
Founded: 1964; Members: 250
Pres: Michael Wood; Gen Secr: Michael C. Shand
Focus: Cart
Periodical
Bulletin of the Soiety of Cartographers (weekly) *13033*

Society of Chemical Industry (SCI), 14-15 Belgrave Sq, London SW1X 8PS
T: (071) 2353681; Fax: (071) 8231698
Founded: 1881; Members: 5500
Pres: J.M. Bruel; Gen Secr: J.J. Wren
Focus: Chem *13034*

Society of Chemical Industry (SCI), 15 Belgrave Sq, London SW1X 8PS
T: (071) 2353691; Fax: (071) 8231698
Founded: 1881; Members: 5400
Pres: Jean-Marc Bruel; Gen Secr: John Wren
Focus: Chem
Periodicals
Chemistry and Industry (weekly)
Chemistry and Industry (weekly)
Journal of Chemical Technology and Biotechnology (weekly)
Journal of Chemical Technology and Biotechnology (weekly)
Journal of the Science of Food and Agriculture (weekly)
Journal of the Science of Food and Agriculture (weekly)
Pesticide Science (weekly)
Pesticide Science (weekly)
Polymer International (weekly)
Polymer International (weekly) *13035*

Society of Chief Architects of Local Authorities (SCALA), POB 73, Wirral L63 0QG
T: (0517) 3344068
Founded: 1974; Members: 260
Focus: Archit
Periodicals
SCALA Conference Papers (weekly)
SCALA Maintenance Expenature Revue (weekly)
SCALA Newsletter (weekly)
SCALA Study Day Papers (weekly) . . *13036*

Society of Cirplanologists, 26 Roe Cross Green, Mottram, Hyde SK14 6LP
Founded: 1955; Members: 105
Focus: Hist; Rel & Theol *13037*

Society of Consulting Marine Engineers and Ship Surveyors, 6 Lloyds Av, London EC3N 3AX
T: (071) 4883010; Fax: (071) 4883010
Founded: 1920
Pres: Ian Burrows; Gen Secr: Peter Hicks
Focus: Eng *13038*

Society of Cosmetic Scientists, 57 Guildford St, Luton LU1 2NL
T: (0582) 26661; Fax: (0582) 405217
Members: 950
Gen Secr: Lorna Weston
Focus: Chem
Periodical
International Journal of Cosmetic Science (weekly) *13039*

Society of County Librarians (SCL), c/o County Library Headquarters, County Hall, Worcester WR5 2NP
T: (0905) 766230; Fax: (0905) 763000
Founded: 1966; Members: 47
Focus: Libraries & Bk Sci
Periodical
Information (weekly) *13040*

Society of County Museum Directors, c/o Kent County Museum Service, West Malling Air Station, West Malling ME19 6QE
T: (0732) 845845 ext 2114
Focus: Arts
Periodical
Guide to County Museum Services in England and Wales (weekly) *13041*

Society of Dairy Technology (S.D.T.), 72 Ermine St, Huntingdon PE18 6EZ
T: (0480) 450741; Fax: (0480) 431800
Founded: 1943; Members: 2000
Pres: Peter Fleming; Gen Secr: R.A. Gale
Focus: Dairy Sci
Periodical
Journal of Dairy Technology (weekly) . . *13042*

Society of Designer Craftsmen, 24 Rivington St, London EC2A 3DU
T: (071) 7393663
Founded: 1888; Members: 750
Pres: Lord Dainton; Gen Secr: Richard O'Donoghue
Focus: Graphic & Dec Arts, Design
Periodicals
The Designer-Craftsman (weekly)
News-Sheet (weekly) *13043*

The Society of Dyers and Colourists, Perkin House, 82 Grattan Rd, POB 244, Bradford BD1 2JB
T: (0274) 725138; Tx: 55293; Fax: (0274) 392888
Founded: 1884; Members: 3600
Pres: Prof. D.M. Lewis; Gen Secr: James D. Watson
Focus: Chem
Periodicals
Colour Index
Journal (10 times annually)
Review of Progress in Coloration and Related Topics (weekly) *13044*

The Society of Engineers, Guinea Wiggs, Nayland, Colchester CO6 4NF
T: (0206) 263332
Founded: 1854; Members: 3000
Pres: A.D. McLaren; Gen Secr: L. Wright
Focus: Eng
Periodical
Journal and Transactions (weekly) . . . *13045*

Society of Feed Technologists, 156 Oxford Rd, Reading, Berks
T: (0734) 595458
Members: 260
Focus: Agri *13046*

Society of Genealogists (SG), 14 Charterhouse Buildings, Goswell Rd, London, EC1M 7BA
T: (071) 2518799
Founded: 1911; Members: 13000
Pres: Michael Prince of Kent; Gen Secr: Anthony J. Camp
Focus: Genealogy
Periodicals
Computers in Genealogy (weekly)
Genealogists Magazine (weekly) *13047*

Society of Glass Technology (S.G.T.), 20 Hallam Gate Rd, Sheffield S10 5BT
T: (0742) 663168
Founded: 1916; Members: 1258
Pres: D. Rotherham; Gen Secr: W. Simpson
Focus: Materials Sci
Periodicals
Glass Technology (weekly)
Physics and Chemistry of Glasses (weekly) *13048*

Society of Health and Beauty Therapists, 77 New Bond St, London W1Y 9DB
T: (071) 4933321
Founded: 1962; Members: 5000
Focus: Public Health; Therapeutics . . . *13049*

Society of Hearing Aid Audiologists (S.H.A.A.), 54 Croham Manor Rd, South Croydon CR2 7BE
T: 6882503
Founded: 1954; Members: 600
Focus: Rehabil
Periodical
S.H.A.A. Journal (weekly) *13050*

Society of Homoeopaths, 59 Norfolk House Rd, London SW16
T: (081) 6773260
Founded: 1978; Members: 110
Focus: Homeopathy
Periodical
The Homoeopath (weekly) *13051*

Society of Indexers, 38 Rochester Rd, London NW1 9JJ
T: (071) 9167809
Founded: 1957; Members: 900
Pres: D. Crystal; Gen Secr: C. Troughton
Focus: Standards; Educ
Periodical
The Indexer: Micro Indexer (weekly) . . *13052*

Society of Industrial Tutors (SIT), c/o Teesside Polytechnic, Middlesbrough, Cleveland
T: (0642) 44176
Founded: 1968; Members: 330
Focus: Adult Educ *13053*

Society of King Charles the Martyr, c/o Saint Mary le Strand Parish Office, 171 Strand, London WC2R 2LS
Founded: 1894
Focus: Rel & Theol
Periodical
Church and King (weekly) *13054*

Society of Licensed Aircraft Engineers and Technologists, Grey Tiles, Kingston Hill, Kingston-upon-Thames KT2 7LW
Founded: 1944; Members: 6500
Pres: G.M. McCoombe; Gen Secr: J.B. Groves
Focus: Eng
Periodical
Tech Air (weekly) *13055*

The Society of Metaphysicians, Archers' Court, Stonestile Lane, The Ridge, Hastings TN35 4PG
T: (0424) 751577; Fax: (0424) 722387
Founded: 1944; Members: 62000
Pres: John Jacob Williamson; Gen Secr: M. Ross
Focus: Philos
Periodicals
Borderline Science Series: Esoteric Series (weekly)
Neo Metaphysical Digest (weekly)
Neometaphysical Series (weekly)
Neometaphysics & Current Affairs Series (weekly) *13056*

Society of Miniaturists, c/o Leslie Simpson, Ralston House, 41 Lister St, Ilkley LS29 9ET
T: (0943) 609075
Founded: 1895
Focus: Fine Arts *13057*

Society of Occupational Medicine (SOM), 6 Saint Andrew's Pl, London NW1 4LB
T: (071) 4862641; Fax: (071) 4860028
Founded: 1935; Members: 2000
Pres: Dr. I.S. Symington; Gen Secr: Dr. G. Smith
Focus: Med; Hygiene
Periodical
Occupational Medicine (weekly) *13058*

Society of Professional Engineers, c/o Parsifal College, 527 Finchley Rd, London NW3 7BG
Founded: 1969
Pres: Preecem M.J.; Gen Secr: K.A. Statham
Focus: Eng *13059*

Society of Protozoologists, British Section, c/o Dept of Anatomy, Guy's Hospital Medical School, London SE1 9RT
T: (071) 4077600
Founded: 1962; Members: 150
Focus: Zoology *13060*

The Society of Public Health, 31 Battye Av, Huddersfield HD4 5PW
Founded: 1856; Members: 1300
Pres: Dr. Irene Clarke; Gen Secr: Dr. P.A. Gardner
Focus: Public Health
Periodical
Public Health (weekly) *13061*

Society of Public Teachers of Law (SPTL), c/o School of Law, University of Buckingham, Buckingham MK18 1EG
T: (0280) 814080
Founded: 1908; Members: 1450
Focus: Educ; Law
Periodical
Legal Studies (3 times annually) *13062*

Society of Remedial Gymnasts and Recreational Therapies (SRG), c/o Combingo Training Institute, Cardiff University Hospital of Wales, Cardiff, Wales
Founded: 1946; Members: 700
Focus: Therapeutics
Periodical
Journal (weekly) *13063*

Society of Scribes and Illuminators (SSI), 54 Boileau Rd, London SW13 9BL
T: (081) 7489951
Founded: 1921; Members: 1868
Gen Secr: S. Cavendish
Focus: Libraries & Bk Sci
Periodical
Journal (3 times annually) *13064*

Society of Teachers in Business Education, 88 Springfield Rd, Sheffield S7 2GF
T: (0742) 363659
Founded: 1907; Members: 1600
Focus: Educ
Periodical
Focus on Business Education (3 times annually) *13065*

Society of Teachers of Speech and Drama (S.T.S.D.), Abbot's Lodging, Marshside, Canterbury CT3 4EF
T: (0227) 86346
Founded: 1948; Members: 950
Focus: Ling; Educ; Perf Arts *13066*

Society of Technical Analysts, 28 Panton St, Cambridge CB2 1DH
T: (0223) 356251; Fax: (0223) 329806
Founded: 1969; Members: 700
Focus: Eng 13067

Society of Thoracic and Cardiovascular Surgeons of Great Britain and Ireland, c/o Cardiothoracic Unit, Walsgrave Hospital, Coventry CV2 2DX
T: (0203) 613232
Founded: 1933; Members: 500
Focus: Cardiol; Surgery 13068

Society of Town Planning Technicians (STPT), c/o Royal Town Planning Institute, 26 Portland Pl, London W1N 4BE
T: (071) 6369107
Founded: 1972; Members: 350
Focus: Urban Plan
Periodical
Technical Planner (weekly) 13069

Society of Wildlife Artists, 17 Carlton House Terrace, London SW1Y 5BD
T: (071) 9306844; Fax: (071) 8397830
Founded: 1963; Members: 66
Pres: Robert Gillmor
Focus: Arts; Fine Arts
Periodical
Annual Catalogue (weekly) 13070

Society of Women Artists, 17 Carlton House Terrace, London SW1Y 5BD
T: (071) 9306844
Focus: Arts; Fine Arts
Periodical
Annual Catalogue (weekly) 13071

Society of Writers to Her Majesty's Signet, c/o Signet Library, Parliament Sq, Edinburgh EH1 1RF
T: (031) 2254923; Fax: (031) 2204016
Founded: 1594; Members: 900
Focus: Law; Hist 13072

Soil Association, 86 Colston St, Bristol BS1 5BB
T: (0272) 290661; Fax: (0272) 252504
Founded: 1946; Members: 5500
Focus: Agri
Periodical
The Living Earth (weekly) 13073

Somerset Archaeological and Natural History Society (S.A.N.H.S.), Taunton Castle, Taunton TA1 4AD
T: (0823) 72429
Founded: 1849; Members: 970
Focus: Archeol; Hist; Nat Sci 13074

Somerset Record Society, c/o The Local History Library, The Castle, Castle Green, Taunton TA1 4AD
T: (0823) 288871
Founded: 1889; Members: 250
Focus: Hist 13075

Sound Learning Society (SLS), 4 Plaitford Close, Rickmansworth WD3 1NJ
T: 75776
Founded: 1971; Members: 60
Focus: Educ 13076

South Bedfordshire Archaeological Society (SBAS), 27 Lords Lane, Bradwell, Great Yarmouth NR31 8NY
T: (0493) 68605
Founded: 1950; Members: 15
Focus: Archeol 13077

South-Eastern Union of Scientific Societies (SEUSS), 53 The Drive, Shoreham-by-Sea BN4 5GD
T: 2478
Founded: 1896; Members: 70
Focus: Sci 13078

Southern Arts Board (SAB), 13 Saint Clement St, Winchester SO23 9DQ
T: (0962) 855099
Founded: 1991; Members: 206
Focus: Arts
Periodicals
Artlook (weekly)
Spring Floor (3 times annually) 13079

Southern Skirmish Association (SoSkAN), 24 Adioham House, Pembury Rd, London E5
T: (081) 9864709
Founded: 1967; Members: 542
Focus: Hist 13080

South of England Agricultural Society, The Showground, Ardingly, Haywards Heath RH17 6TL
T: (0444) 892700; Fax: (0444) 892888
Founded: 1967; Members: 5000
Gen Secr: P.J.S. Salter
Focus: Agri
Periodical
Show Times (weekly) 13081

South Place Ethical Society (SPES) / International Humanist Centre, Conway Hall, 25 Red Lion Sq, London WC1R 4RL
T: (071) 2428032, 8317723
Founded: 1793; Members: 400
Gen Secr: Nina Khare
Focus: Philos; Educ
Periodical
The Ethical Record (10 times annually) . 13082

South Staffordshire Archaeological and Historical Society, 307 Erdington Rd, Aldridge, Walsall WS9 0SB
T: (0922) 52097
Founded: 1959; Members: 260
Gen Secr: J. Gould
Focus: Archeol; Hist
Periodical
Transactions of South Staffordshire Archaeological and Historical Society (weekly) 13083

South Wales Institute of Engineers (SWIE), Empire House, Mount Stuart Sq, Cardiff CF1 6DN
T: (0222) 481726; Fax: (0222) 451953
Founded: 1857; Members: 604
Pres: Prof. K.M. Brown; Gen Secr: R.E. Lindsay
Focus: Eng
Periodical
Proceedings (biennial) 13084

South Wiltshire Industrial Archaeology Society (SWIAS), Sandlewood, Portland Av, Salisbury SP2 8BS
T: (0722) 28831
Founded: 1966; Members: 65
Focus: Archeol 13085

The Spastics Society, 12 Park Crescent, London W1N 4EQ
T: (071) 6365020; Fax: 4362601
Founded: 1952; Members: 190
Pres: Duke of Westminster; Gen Secr: Martyn Dunleavy
Focus: Rehabil
Periodicals
Clinics in Developmental Medicine (weekly)
Developmental Medicine + Child Neurology (weekly)
Disability Now (weekly) 13086

Spohr Society of Great Britain, 123 Mount View Rd, Sheffield S8 8PJ
T: (0742) 585420
Founded: 1969; Members: 62
Gen Secr: C.H. Tutt
Focus: Music
Periodical
Spohr Journal (weekly) 13087

The Sports Council, 16 Upper Woburn Pl, London WC1H 0QP
T: (071) 3881277; Tx: 27830; Fax: (071) 3835740
Founded: 1972
Focus: Sports 13088

Spring Research and Manufacturers Association (S.R.A.M.A.), Henry St, Sheffield S3 7EQ
T: (0742) 760771; Tx: 547871; Fax: (0742) 726344
Founded: 1945
Gen Secr: A. Hooper
Focus: Eng; Metallurgy 13089

Staffordshire Agricultural Society, County Showground, Weston Rd, Stafford ST18 0BD
T: (0785) 58060; Fax: (0785) 46443
Founded: 1803; Members: 4200
Focus: Agri 13090

Staffordshire Parish Registers Society, 91 Brenton Rd, Penn WV4 5NS
T: (0902) 341885
Founded: 1901; Members: 180
Gen Secr: A.T.C. Lavender
Focus: Hist 13091

Staffordshire Record Society, c/o William Salt Library, Eastgate St, Stafford ST16 2LZ
Founded: 1879
Gen Secr: D.A. Johnson
Focus: Hist
Periodical
Collections for a History of Staffordshire (biennial) 13092

Stair Society, 16 Charlotte Sq, Edinburgh EH2 4YS
T: (031) 2258585; Tx: 727251
Founded: 1934; Members: 600
Pres: Lord Hope; Gen Secr: Ivor R. Guild
Focus: Hist; Law 13093

Standing Committee on Commonwealth Forestry, c/o Forestry Commission, 231 Corstorphine Rd, Edinburgh EH12 7AT
T: (031) 3340303; Tx: 727879; Fax: (031) 3340442
Focus: Forestry
Periodical
Newsletter (weekly) 13094

Standing Conference of Arts and Social Sciences, c/o Dr. S. Delamont, SOCAS, University of Wales College of Cardiff, 62 Park Pl, Cardiff CF1 3AS
Founded: 1984
Pres: Prof. J.H. Westergaard
Focus: Arts; Soc Sci 13095

The Standing Conference of National and University Libraries (SCONUL), 102 Euston St, London NW1 2HA
T: (071) 3870317
Founded: 1950
Gen Secr: Gillian M. Pentelow
Focus: Libraries & Bk Sci
Periodicals
Annual Review (weekly)
ISG News (weekly)
Newsletter of the SCONUL Advisory Committee on Manuscripts (weekly)
SCONULOG (weekly)
Solanus: Bulletin of ACOSEEM (weekly) . 13096

Standing Conference on Library Materials on Africa (SCOLMA), c/o Records Branch, Foreign and Commonwealth Space, Hanslope Park, Milton Keynes MK19 7BH
T: (0908) 511389; Fax: (0908) 511419
Founded: 1962; Members: 100
Gen Secr: M.A. Cousins
Focus: Libraries & Bk Sci
Periodical
African Research and Documentation (3 times annually) 13097

Standing International Committee for Mycenaean Studies, c/o Downing College, Cambridge
Focus: Hist 13098

Statute Law Society, 186 City Rd, London EC1V 2NU
T: (071) 2511644; Fax: 2500801
Founded: 1968; Members: 150
Gen Secr: Nigel Frudd
Focus: Law
Periodical
Statute Law Review (weekly) 13099

The Steel Construction Institute, Silwood Park, Ascot SL5 7QN
T: (0344) 23345; Tx: 846843; Fax: (0344) 22944
Founded: 1985
Focus: Materials Sci; Metallurgy
Periodical
New Steel Construction (weekly) 13100

Stephenson Locomotive Society (SLS), 2 Gainsborough Rd, London W4
Founded: 1909; Members: 1800
Focus: Transport 13101

The Stewart Society, 314 Leith Walk, Edinburgh EH6 5BU
T: (031) 5554640
Founded: 1899; Members: 610
Gen Secr: Muriel Walker
Focus: Genealogy
Periodical
The Stewarts (weekly) 13102

Stroke Association, CHSH House, Whitecross St, London EC1V 8JJ
T: (071) 4907999
Founded: 1899
Pres: Lord Smith of Marlow; Gen Secr: Sir David Atkinson
Focus: Cardiol 13103

Structural Fire Protection Association (SFPA), 37 Soho Sq, London W1
T: (071) 4377107
Founded: 1975; Members: 25
Focus: Civil Eng 13104

Suffolk Agricultural Association (SAA), Suffolk Showground, Bucklesham Rd, Ipswich IP3 8UH
T: (0473) 726847; Fax: (0473) 721973
Founded: 1831; Members: 3200
Focus: Agri 13105

Suffolk Institute of Archaeology and History, Oak Tree Farm, Hitcham, Ipswich IP7 7LS
T: (0284) 722022
Founded: 1848; Members: 850
Pres: J.M. Blatchly; Gen Secr: E.A. Martin
Focus: Hist; Archeol
Periodicals
Proceedings (weekly)
Suffolk Archaeology & History Newsletter (bi-annually) 13106

Suffolk Records Society (SRS), County Hall, Ipswich IP4 2JS
T: (0473) 230000
Founded: 1958; Members: 390
Pres: Sir Anthony Wagner; Gen Secr: C.J. Evans
Focus: Hist 13107

Sunday Shakespeare Society (SSS), Heathfield, Chestnut Walk, Felcourt, East Grinstead RH19 2LB
T: (0342) 870436
Founded: 1874; Members: 47
Pres: Dorothy Tutin; Gen Secr: Peter J. Cox
Focus: Lit 13108

Surgical Research Society (SRS), c/o Dept of Surgery, The General Hospital, Gwendolen Rd, Glasgow G4 0SF
T: (041) 5523535
Founded: 1953; Members: 400
Focus: Surgery 13109

Surrey Archaeological Society (S.A.S.), Castle Arch, Guildford GU1 3SX
T: (0483) 32454
Founded: 1854; Members: 1065
Focus: Archeol
Periodical
S.A.S. Collections (weekly) 13110

Surrey County Agricultural Society, Brook House, Mint St, Godalming GU7 1HE
T: (04868) 4651; Fax: (04868) 25697
Members: 1500
Focus: Agri 13111

Surrey Record Society (SRS), Castle Arch, Guildford GU1 3SX
Founded: 1913; Members: 210
Focus: Hist 13112

Sussex Archaeological Society (S.A.S.), Barbican House, 169 High St, Lewes BN7 1YE
T: (0273) 486290; Fax: (0273) 486990
Founded: 1846; Members: 2000
Focus: Archeol; Hist
Periodicals
Newsletter (3 times annually)
Sussex Archaeological Collections (weekly) 13113

Sussex Industrial Archaeology Society (SIAS), 42 Falmer Av, Saltdean, Brighton BN2 8FG
T: (0273) 303805
Founded: 1967; Members: 300
Focus: Archeol
Periodical
Sussex Industrial History (weekly) . . . 13114

The Swedenborg Society, 20 Bloomsbury Way, London WC1A 2TH
T: (071) 4057986
Founded: 1810; Members: 1000
Focus: Philos; Rel & Theol
Periodical
The Swedenborg Society Magazine (weekly) 13115

The Systematics Association, c/o Dept of Geological Sciences, Durham University, South Rd, Durham DH1 3LE
T: (091) 64971
Founded: 1937
Pres: Prof. A.J. Cain; Gen Secr: G.P. Larwood; F.A. Bisby
Focus: Bio 13116

The Tavistock Institute of Medical Psychology (T.I.M.P.), Tavistock Centre, Belsize Ln, London NW3 5BA
T: (071) 4357111; Fax: (071) 4351080
Founded: 1929; Members: 32
Gen Secr: G.C. Hume
Focus: Psych 13117

The Tennyson Society, c/o Tennyson Research Centre, Central Library, Free School Lane, Lincoln LN2 1EZ
T: (0522) 552851; Tx: 56306; Fax: (0522) 552858
Founded: 1960; Members: 550
Pres: Lord Tennyson; Gen Secr: J.K. Jefferson
Focus: Lit
Periodicals
Annual Report
Bulletin (3 times annually) 13118

Tensor Society of Great Britain (TSGB), 66 South Terrace, Surbiton KT6 6HU
T: (081) 3992724
Founded: 1950
Focus: Sci; Eng
Periodical
The Matrix and Tensor (weekly) 13119

The Textile Institute (TI), 10 Blackfriars St, Manchester M3 5DR
T: (061) 8348457; Tx: 668297; Fax: (061) 8353087
Founded: 1910; Members: 10258
Gen Secr: R.G. Denyer
Focus: Textiles
Periodicals
International Textile Calendar (weekly)
Journal of the Textile Institute (weekly)
Manual of Textile Technology (weekly)
Textile Horizons (weekly)
Textile Progress (weekly)
Textiles (weekly) 13120

Textile Research Council (TRC), 2 First Av, Sherwood Rise, Nottingham NG7 6JL
T: (0602) 63311
Founded: 1947; Members: 16
Focus: Textiles 13121

Theatre Arts Society (TAS), Wyndhams Theatre, Charing Cross Rd, London WC2 0DA
T: (071) 8362671
Founded: 1969; Members: 3000
Focus: Perf Arts 13122

Theatre Organ Preservation Society (TOPS), 8 Dale Court, Seymour Rd, Slough SL1 2NU
T: 36786
Founded: 1960; Members: 2000
Focus: Music 13123

Theosophical Society in England (TS), 50 Gloucester Pl, London W1H 3HJ
T: (071) 9359261
Founded: 1875; Members: 1750
Gen Secr: I. Hoskins
Focus: Philos; Rel & Theol; Sci; Psych; Soc Sci
Periodical
The Theosophical Journal (weekly) . . . 13124

Theosophical Society in Europe, 2 Tekels Park, Camberley, Surrey
Founded: 1903
Focus: Rel & Theol 13125

Theosophical Society in Scotland, 28 Great King St, Edinburgh EH3 6QH
T: (031) 5565385
Founded: 1895; Members: 100
Focus: Rel & Theol 13126

Thomas Paine Society, 43 Wellington Gardens, Selsey PO20 0RF
T: (0243) 605730
Founded: 1963; Members: 175
Pres: Michael Foot; Gen Secr: Eric Paine
Focus: Hist
Periodicals
Bulletin (weekly)
Newsletter (weekly) 13127

Thomas Tallis Society (TTS), 13 Albury St, London SE8 3PT
T: (081) 6918337; Fax: (081) 6918337
Founded: 1965; Members: 60
Pres: Philip Simms; Gen Secr: Deborah Sandringham
Focus: Music 13128

Thoracic Society, c/o Dept of Medicine, D Level, Center Block, Southampton
T: (0703) 777222
Founded: 1945; Members: 540
Focus: Pulmon Dis 13129

Thoresby Society, 23 Clarendon Rd, Leeds LS2 9NZ
Founded: 1889; Members: 420
Focus: Hist
Periodical
Publications (weekly) 13130

Thoroton Society of Nottinghamshire, Bromley House, Angel Row, Nottingham
T: (0602) 609384
Founded: 1897; Members: 650
Focus: Archeol; Hist 13131

Three Counties Agricultural Society (TCAS), The Showyard, Malvern WR13 6NW
T: (0684) 892751; Fax: (0684) 568236
Founded: 1797; Members: 8000
Focus: Agri 13132

Timber Research and Development Association (TRADA), TRADA House, Stocking Lane, Hughenden Valley, High Wycombe HP14 4ND
T: (0494) 563091; Fax: (0494) 565487
Founded: 1934
Focus: Forestry 13133

Time Series Analysis and Forecasting Society, 9 Ingham Grove, Lenton Gardens, Nottingham NG7 2LQ
T: (0602) 708085
Founded: 1979; Members: 2800
Pres: Dr. Oliver D. Anderson
Focus: Math
Periodicals
TSA&F Flyer (weekly)
TSA&F News (weekly) 13134

Tobacco Research Council, Glen House, Stag Pl, London SW1E 5AG
T: (071) 8282041; Fax: (071) 6309638
Focus: Agri 13135

Tolkien Society, 307 Dyke Rd, Hove BN3 6PD
Founded: 1969; Members: 1000
Focus: Lit
Periodicals
Amon Hen (weekly)
Mallorn (weekly) 13136

Town and Country Planning Association (TCPA), 17 Carlton House Terrace, London SW1Y 5AS
T: (071) 9308903; Fax: (071) 9303280
Founded: 1899; Members: 1300
Pres: Ralph Rockwood; Gen Secr: David Hall
Focus: Urban Plan
Periodicals
Community Network (weekly)
Planning Bulletin (weekly)
Town and Country Planning (weekly) . . 13137

Tramway Museum Society (TMS), c/o The National Tramway Museum, Crich, Matlock DE4 5DP
T: (0773) 852565; Fax: (0773) 852326
Founded: 1955
Focus: Transport
Periodical
TMS Journal (weekly) 13138

Trans-Antarctic Association (TAA), c/o British Antarctic Survey, High Cross, Madingley Rd, Cambridge CB3 0ET
T: (0223) 61188; Tx: 817725
Founded: 1960; Members: 3
Focus: Geography 13139

Transplantation Society, c/o Dept of Immunology, Saint Mary's Hospital Medical School, London W2 1PC
Founded: 1966
Focus: Surgery 13140

Transport Ticket Society, 18 Villa Rd, Luton LU2 7NT
T: (0582) 21774
Founded: 1946; Members: 500
Focus: Transport 13141

Trevithick Society, 23 Merrick Av, Truro, Cornwall
Founded: 1934; Members: 660
Focus: Hist 13142

Trinitarian Bible Society (TBS), 217 Kingston Rd, London SW19 3NN
T: (081) 5437857; Fax: (081) 5436370
Founded: 1831
Focus: Rel & Theol
Periodical
Record (weekly) 13143

Tuberous Sclerosis Association of Great Britain (TSA), Little Barnsley Farm, Catshill, Bromsgrove
Founded: 1977; Members: 1200
Focus: Pulmon Dis
Periodicals
TSA Fact Sheets (3 times annually)
TSA Scan (3 times annually) 13144

Turner Society, BCM Box Turner, London WC1N 3XX
Founded: 1975; Members: 580
Gen Secr: Alan Scrivener
Focus: Fine Arts
Periodical
Turner Society News (3 times annually) . 13145

Twentieth Century Society, 58 Crescent Lane, London SW4 9PU
Founded: 1979; Members: 450
Pres: John Harris; Gen Secr: Alan Powers
Focus: Preserv Hist Monuments
Periodical
Journal (weekly) 13146

TWI, Abington Hall, Cambridge CB1 6AL
T: (0223) 891162; Tx: 81183
Founded: 1968; Members: 7000
Gen Secr: A.B.M. Braithwaite
Focus: Eng
Periodicals
Connect (weekly)
Research Bulletin (weekly)
Weldasearch Industry News (weekly)
Welding Abstracts (weekly)
The Welding Institute News Video (weekly) 13147

Ulster Archaeological Society (UAS), c/o Dept of Archaeology, Queen's University, Belfast BT7 1NN
T: (0232) 245133
Founded: 1938
Focus: Archeol
Periodical
Ulster Journal of Archaeology (weekly) . 13148

Ulster Archery Association (UAA), 35 Norglen Dr, Belfast BT11 8DG
Founded: 1956
Focus: Archeol 13149

Ulster Architectural Heritage Society (UAHS), 185 Stranmills Rd, Belfast BT9 5DU
T: (0232) 660809
Founded: 1967; Members: 1100
Pres: C.E.B. Brett; Gen Secr: Joan Kinch
Focus: Archit; Preserv Hist Monuments
Periodical
Newsletter (weekly) 13150

Ulster Chemists' Association (UCA), 73 University St, Belfast BT7 1HL
T: (0232) 320787
Founded: 1901; Members: 654
Focus: Chem 13151

Ulster Folklife Society, c/o Ulster Folk and Transport Museum, Cultra Manor, Holywood BT18 0EU
T: (02317) 428428
Founded: 1962; Members: 175
Gen Secr: J.R.R. Adams
Focus: Ethnology
Periodical
Ulster Folklife 13152

The Ulster Genealogical and Historical Guild / The Ulster Historical Foundation, 12 College Sq East, Belfast BT1 6DO
T: (0232) 332288; Fax: (0232) 239885
Founded: 1956; Members: 5000
Pres: Sir Robert Kidd; Gen Secr: Dr. Brian Trainor
Focus: Genealogy; Hist
Periodicals
Directory of Irish Family History Research (weekly)
Familia: The Ulster Genealogical Review (weekly) 13153

Ulster Historical Foundation, 12 College Sq East, Belfast BT1 6DD
T: (0232) 332288; Fax: (0232) 239885
Founded: 1956; Members: 900
Focus: Hist
Periodicals
Biographical Series
Educational Series
Gravestone Inscription Series (weekly)
Historical Series 13154

Ulster Society for Irish Historical Studies (USIHS), 36 North Parade, Belfast BT7 2GG
T: (0232) 643229
Founded: 1936; Members: 85
Focus: Hist 13155

Ulster Society for the Preservation of the Countryside (USPC), West Winds, Carney Hill, Holywood BT18 0JR
T: 2300
Founded: 1937; Members: 500
Focus: Ecology
Periodical
The Countryside Recorder (weekly) . . . 13156

Ulster Society in London, 11 Berkeley St, London W1 6BU
Focus: Hist
Periodical
Newsletter (weekly) 13157

Ulster Teachers' Union (UTU), 94 Malone Rd, Belfast BT9 5HP
T: (0232) 662216; Fax: (0232) 663055
Founded: 1919; Members: 5500
Gen Secr: David Allen
Focus: Educ
Periodical
UTU News (3 times annually) 13158

UMIST Association, POB 88, Manchester M60 1QD
T: (061) 2003066; Fax: (061) 2368066
Members: 17500
Pres: J.C. Dwek
Focus: Eng
Periodical
Mainstream (weekly) 13159

Union of Educational Institutions (UEI), Norfolk House, Smallbrook Queensway, Birmingham B5 4NB
T: (021) 6438924
Founded: 1895
Focus: Educ 13160

Union of Lancashire and Cheshire Institutes (ULCI), Town Hall, Walkden Rd Worsley, Manchester M28 4QE
Founded: 1839
Focus: Educ
Periodicals
General Regulations & Examinations Timetable (weekly)
Regional Guide to Further Education Courses in the North West (weekly)
Regional List of Colleges in the North West (weekly) 13161

Unitarian Historical Society, 6 Ventnor Terrace, Edinburgh EH9 2BL
T: (031) 6674360
Founded: 1915; Members: 250
Focus: Hist
Periodical
Transactions (weekly) 13162

United Earth Sciences Exploration Group (UESEG), 8 Utton's Av, Leigh-on-Sea SS9 2EL
T: (0702) 710521
Founded: 1972; Members: 83
Focus: Geology
Periodical
Expedition Reports 13163

United Kingdom Association for European Law (U.K.A.E.L.), c/o King's College, Strand, London WC2R 2LS
T: (071) 2400206
Founded: 1975; Members: 300
Pres: Lord Slynn of Hadley; Gen Secr: Eva Evans
Focus: Law 13164

United Kingdom Home Economics Federation (UKHEF), 171 Whitechapel Rd, Cleckheaton BD19 6HW
T: (0274) 872488
Founded: 1954
Gen Secr: A.P. Cooper
Focus: Home Econ
Periodical
Bulletin: Conference Report (3 times annually) 13165

United Kingdom Institute for Conservation of Historic and Artistic Works, c/o Conservation Dept, Tate Gallery, 6 Whitehorse News, Westminster Bridge Rd, London SE1 7QD
Focus: Preserv Hist Monuments; Arts
Periodicals
Conservation News (3 times annually)
The Conservator (weekly) 13166

United Kingdom Reading Association (UKRA), c/o Warrington Road C.P. School, Naylor Rd, Widnes WA8 0BP
T: (051) 4202552
Founded: 1964; Members: 1500
Gen Secr: M. Cooper
Focus: Educ
Periodicals
Journal of Research in Reading (weekly)
Newsletter (3 times annually)
Reading (3 times annually) 13167

United Kingdom Science Park Association, Aston Science Park, Love Lane, Aston Triangle, Birmingham B7 4BJ
T: (021) 3590981; Fax: (021) 3335852
Founded: 1984; Members: 45
Focus: Sci
Periodicals
UKSPA News (weekly)
UKSPA Science Park Directory 13168

United Reformed Church History Society, 86 Tavistock Pl, London WC1H 9RT
T: (071) 9162020
Founded: 1972; Members: 600
Gen Secr: S.C. Orchard
Focus: Hist; Rel & Theol
Periodical
Journal (weekly) 13169

United Society for Christian Literature (USCL), Luke House, Farnham Rd, Guildford GU1 4XD
T: (0483) 557536
Founded: 1799
Focus: Lit 13170

United Society for the Propagation of the Gospel (USPG), 157 Waterloo Rd, London SE1 8XA
T: (081) 2988681
Founded: 1701; Members: 7000
Focus: Rel & Theol
Periodicals
Quarterly Intercession Paper (weekly)
Thinking Mission (weekly)
USPG Yearbook (weekly) 13171

United Society of Artists, York Hill Cottage, Alderney, Channel Islands
T: (048182) 2226
Gen Secr: Michael Palmer
Focus: Fine Arts
Periodical
Annual Catalogue (weekly) 13172

United World Colleges, London House, Mecklenburgh Sq, London WC1N 2AB
T: (071) 8332626; Tx: 296459; Fax: (071) 8373102
Founded: 1962; Members: 41
Focus: Educ
Periodical
UWC Journal (weekly) 13173

University Association for Contemporary European Studies (UACES), c/o King's College, London WC2R 2LS
Founded: 1968; Members: 360
Pres: Dr. William Paterson; Gen Secr: Eva Evans
Focus: Educ; Hist
Periodicals
Register of Courses in European Studies in British Universities and Polytechnics
Register of Current Research into European Integration (bi-annually) 13174

University of Bristol Spelaeological Society (U.B.S.S.), c/o Geography Dept, University Rd, Bristol BS8 1SP
Founded: 1919; Members: 350
Focus: Speleology; Archeol
Periodical
Proceedings U.B.S.S. (weekly) 13175

UV Spectrometry Group (UVG), c/o Pye Unicam Ltd, York St, Cambridge CB1 2PX
Founded: 1948; Members: 170
Focus: Physics 13176

Vasectomy Advancement Society of Great Britain (VASofGB), 1 Ravenscroft Court, 56 Ravenscroft Av, London NW11 8BA
T: (081) 4556541
Founded: 1972; Members: 150
Focus: Med 13177

Vegan Society, 7 Battle Rd, Saint Leonards-on-Sea TN37 7AA
T: (0424) 427393
Founded: 1944; Members: 4000
Gen Secr: R. Farhall
Focus: Ecology
Periodical
The Vegon (weekly) 13178

Vernacular Architecture Group (V.A.G.), c/o Bob Meeson, 16 Falna Crescent, Coton Green, Tamworth B79 8JS
T: (0827) 69434
Founded: 1954; Members: 581
Gen Secr: Bob Meeson
Focus: Archit
Periodical
Vernacular Architecture (weekly) 13179

Verulam Institute, Shopwyke Park, Chichester PO20 6BQ
T: (0243) 786863
Founded: 1971
Pres: Alexander Thynn; Gen Secr: Dr. A.W. Harrison-Barbet
Focus: Philos 13180

Veterinary History Society, 32 Belgrave Sq, London SW1X 8QP
T: (071) 2356568
Founded: 1962; Members: 130
Focus: Vet Med
Periodical
Veterinary History 13181

The Victoria Institute or Philosophical Society of Great Britain, Latchett Hall, Latchett Rd, London E18 1DL
T: (081) 5055224
Founded: 1865
Pres: Dr. David J.E. Ingram
Focus: Philos
Periodical
Faith and Thought Bulletin 13182

Victorian Military Society (VMS), Arm Farm Cottage, Blisworth Arm, Northampton NN7 3EF
T: (0604) 858647
Founded: 1975; Members: 850
Gen Secr: G. Dibley
Focus: Hist; Military Sci

Periodicals
Soldiers of the Queen (weekly)
Soldiers Small Book (weekly) 13183

Victorian Society, 1 Priory Gardens, Bedford Park, London W4 1TT
T: (081) 9941019; Fax: (081) 9954895
Founded: 1958; Members: 3250
Gen Secr: Teresa Sladen
Focus: Archit
Periodicals
Annual (weekly)
Newsletter (3 times annually) 13184

Viking Society for Northern Research, c/o Dept of Scandinavian Studies, University College, Gower St, London WC1E 6BT
T: (071) 3877050
Founded: 1892; Members: 650
Focus: Hist; Lit
Periodical
Saga-Book (weekly) 13185

Vintage Light Music Society (VLMS), 4 Harvest Bank Rd, West Wickham BR4 9DJ
T: 4625641
Founded: 1968; Members: 250
Focus: Music
Periodical
Vintage Light Music (weekly) 13186

Viola da Gamba Society, 93a Sutton Rd, London N10 1HH
T: (081) 8834677
Founded: 1948; Members: 250
Focus: Music
Periodicals
Chelys (weekly)
Newsletter (weekly)
Thematic Index of Music for Viols (weekly)
. . . . 13187

Viola d'Amore Society, 9 Hubbards Rd, Chorleywood WD3 5JL
T: (0923) 283526
Founded: 1965; Members: 125
Gen Secr: Ian Whire
Focus: Music 13188

Virgil Society, c/o Dept of Classics, King's College, Strand, London WC2R 2LS
T: (071) 8365454
Founded: 1944; Members: 700
Focus: Lit
Periodical
Proceedings of the Virgil Society . . . 13189

Visible Record and Minicomputer Society, 37 West St, Croydon CR0 1DJ
T: 6810417
Founded: 1971
Focus: Electric Eng; Electronic Eng . . 13190

Vivaldi Society, 67 Twyford Av, London N2
T: (081) 8835266
Focus: Music 13191

Wagner Society, 293 Regent St, London W1R 7PD
T: (071) 6314048; Tx: 267427
Founded: 1952; Members: 651
Focus: Music 13192

The Walpole Society, c/o c/o Dept of Prints and Drawings, British Museum, London WC1B 3DG
T: (071) 8238408, (081) 3630735
Founded: 1911; Members: 480
Focus: Hist; Arts 13193

Water Research Centre (WRC), Henley Rd, Medmenham, POB 16, Marlow, Bucks SL7 2HD
T: (049166) 531; Tx: 848632
Founded: 1974; Members: 500
Focus: Water Res; Ecology
Periodicals
Technical Reports (weekly)
WRC Information (weekly) 13194

Webb Society, 7 Elvendon Rd, London N13 4SJ
T: (081) 8887238
Founded: 1967; Members: 90
Focus: Astronomy 13195

Wedgwood Society, Roman Villa, Rockbourne, Fordingbridge SP6 3PG
T: Rockbourne 445
Founded: 1954; Members: 190
Focus: Arts 13196

Wellesbourne Vegetable Research Association, c/o Horticulture Research International, Warwick, Warwicks.
T: (0789) 470382
Founded: 1958; Members: 505
Focus: Hort 13197

Welsh Arts Council, Holst House, Museum Pl, Cardiff CF1 3NX
Pres: Mathew Prichard; Gen Secr: E. Jenkins
Focus: Arts
Periodical
CREFFT (weekly) 13198

Welsh Baptist Historical Society, 13 Ty'r-fran Av, Llanelli, Dyfed
T: 3244
Founded: 1901; Members: 400
Focus: Hist; Rel & Theol 13199

Welsh Federation of Head Teachers' Associations, 970 Llangyfelach Rd, Tirdeunaw, Swansea
T: (0792) 71104
Focus: Educ 13200

Welsh Folk Dance Society, Bryn Mair, Llanfair Caereinion PE7 B70
Focus: Music 13201

Welsh Folk Song Society (WFSS), Ynys Ceti, Penrhyn-Coch, Aberystwyth SY23 3EH
T: 36088
Focus: Music
Periodical
Canu Gwerin: Folk Song (weekly) . . . 13202

Welsh Library Association, c/o Dept of Information and Library Studies, Llanbadarn Fawr, Aberystwyth SY23 3AS
T: (0970) 622156; Tx: 35165
Founded: 1933; Members: 903
Gen Secr: G. Collins
Focus: Lit
Periodical
Y Ddolen (weekly) 13203

Welwyn Hall Research Association (W.H.R.A.), 11 White Lion House, Town Centre, Hatfield AL10 0JL
T: (07072) 71580
Founded: 1964; Members: 24
Focus: Archit 13204

Wesley Historical Society (WHS), 34 Spiceland Rd, Birmingham B31 1NJ
T: (021) 4754914
Founded: 1893; Members: 1000
Pres: A. Raymond George; Gen Secr: Dr. E.D. Graham
Focus: Hist; Lit
Periodical
Proceedings (3 times annually) 13205

Wildlife Sound Recording Society (WSRS), Chadswell, Sandy Lane, Rushmoor, Tilford, Farnham GU10 2ET
T: Frensham 2673
Founded: 1968; Members: 252
Focus: Zoology 13206

Wilhelm Furtwängler Society, 37 Chester Way, London SE11 4UR
Founded: 1967; Members: 210
Focus: Music
Periodical
Newsletter (weekly) 13207

William Morris Society, 26 Upper Mall, London W6 9TA
T: (081) 7413735
Founded: 1955; Members: 1900
Gen Secr: John Purkis
Focus: Arts; Hist; Lit; Poli Sci
Periodical
Journal (weekly) 13208

Wiltshire Archaeological and Natural History Society (W.A.N.H.S.), c/o The Museum, 41 Long St, Devizes SN10 1NS
T: (0380) 727369
Founded: 1853; Members: 1300
Focus: Archeol; Hist; Nat Sci
Periodical
Magazine (weekly) 13209

Wiltshire Record Society (W.R.S.), c/o County Record Office, County Hall, Trowbridge BA14 8JG
Founded: 1938; Members: 339
Pres: C. Elrington; Gen Secr: J.N. D'Arcy
Focus: Hist
Periodical
Wiltford Record Society: A Series of Historical Records of Local and National Interest (weekly)
. . . . 13210

Wireless Preservation Society (WPS), 32 Luccombe Rd, Shanklin
T: 2586
Founded: 1972
Focus: Electric Eng 13211

Wolverton and District Archaeological Society, 82 Clarence Rd, Stony Stratford, Milton Keynes MK11 1JD
T: (0908) 565481
Founded: 1955; Members: 120
Pres: Lady Markham; Gen Secr: A. Lambert
Focus: Archeol
Periodical
Newsletter (weekly) 13212

Woolhope Naturalists' Field Club, c/o Hereford Library, Broad St, Hereford HR4 9AU
T: (0432) 272456; Fax: (0432) 359668
Founded: 1851; Members: 800
Gen Secr: J.W. Tonkin
Focus: Archeol; Hist; Nat Sci
Periodical
Woolhope Club Transactions (weekly) . . 13213

Worcestershire Archaeological Society, 4 Orchard Rd, Malvern, Worcs
T: 4215
Founded: 1854; Members: 370
Focus: Archeol 13214

Workers' Educational Association (WEA), 17 Victoria Park Sq, London E2 9PB
T: (081) 9831515; Fax: (081) 9834840
Founded: 1903
Gen Secr: R. Lochrie
Focus: Educ
Periodicals
Studies for Trade Unionists (weekly)
Trade Union Studies Journal (weekly)
Women's Studies Newsletter (weekly) . . 13215

Workers' Music Association (WMA), 236 Westbourne Park Rd, London W11 1EL
T: (071) 7277005
Founded: 1936; Members: 321
Focus: Music 13216

World Bureau of Metal Statistics, 27a High St, London SG12 9BA
Founded: 1946
Gen Secr: J.L.T. Davies
Focus: Stats; Metallurgy
Periodicals
World Flow of Unwrought Aluminium (weekly)
World Flow of Unwrought Copper (weekly)
World Flow of Unwrought Lead (weekly)
World Flow of Unwrought Nickel (weekly)
World Flow of Unwrought Tin (weekly)
World Flow of Unwrought Zinc (weekly)
World Metal Statistics (weekly)
World Metal Statistics Yearbook (weekly)
World Stainless Steel Statistics (weekly)
World Tin Statistics (weekly)
World Wrought Copper Statistics (weekly) 13217

World Education Fellowship (WEF), 33 Kinnaird Av, London W4 3SH
T: (081) 9947258
Founded: 1921
Focus: Educ
Periodical
New Era in Education (3 times annually) 13218

World Endodontics Confederation, 13 Southwood Lawn Rd, London N6
T: (081) 3404393
Focus: Dent 13219

World Federation for Medical Education, c/o Dept of Psychiatry, University of Edinburgh, Morningside Park, Edinburgh EH10 5HF
T: (031) 4472011
Focus: Educ; Med
Periodical
Proceedings (weekly) 13220

World Federation of Societies of Anaesthesiologists (WFSA), c/o Dept of Anaesthetics, University of Wales College of Medicine, Heath Park, Cardiff, CF4 4XN
T: (0222) 743110; Fax: (0222) 747203
Founded: 1955; Members: 45000
Gen Secr: Prof. M.D. Vickers
Focus: Anesthetics
Periodical
Newsletter (weekly) 13221

World Medical Tennis Association, c/o Poole General Hospital, Poole, Dorset
Focus: Therapeutics 13222

World Rabbit Science Association (W.R.S.A.), Tyning House, Shurdington, Cheltenham GL51 5XF
Founded: 1976
Focus: Zoology 13223

World's Poultry Science Association, United Kingdom Branch, BOCM Silcock, Risborough Rd, Stoke Mandeville, Aylesbury HP22 5XW
Focus: Ornithology; Agri
Periodical
World's Poultry Sciene Journal (3 times annually)
. . . . 13224

World Union of Pythagorean Organizations (WUPO), 17 Longfield, Lutton, Devon
Founded: 1964
Focus: Philos 13225

World Veterinary Poultry Association (W.V.P.A.), c/o Houghton Poultry Research Station, Houghton, Huntingdon, Cambs
Founded: 1960; Members: 700
Focus: Vet Med
Periodical
Avian Pathology (weekly) 13226

World-wide Education Service of the Bell Educational Trust, 10 Barley Mow Passage, London W4 4PH
T: (081) 7478376
Founded: 1887
Focus: Educ 13227

Writers' Guild of Great Britain (WGGB), 430 Edgware Rd, London W2 1EH
T: (071) 7238074
Founded: 1959
Gen Secr: Alison V. Gray
Focus: Lit 13228

Yorkshire Agricultural Society, Great Yorkshire Showground, Hookstone Oval, Harrogate HG2 8PW
T: (0423) 561536
Founded: (0423) 531112
Focus: Agri 13229

Yorkshire Archaeological Society (Y.A.S.), Claremont, 23 Clarendon Rd, Leeds LS2 9NZ
T: (0532) 457910
Founded: 1863; Members: 1500
Gen Secr: N.A. Cookson
Focus: Archeol
Periodicals
Y.A.S. Parish Register Series (weekly)
Y.A.S. Record Series (weekly)
Y.A.S. Wakefield Court Rolls Series (weekly)
Yorkshire Archaeological Journal (weekly) 13230

Yorkshire Arts Association, Glyde House, Glydegate, Bradford BD5 0BQ
T: (0274) 723051/52
Founded: 1969; Members: 2650
Focus: Arts
Periodicals
Regional Arts and What's on Magazine
Yorkshire Artscene (10 times annually) . 13231

Yorkshire Dialect Society (Y.D.S.), c/o School of English, University, Leeds LS2 9JT
Founded: 1897; Members: 1230
Focus: Ling 13232

Yorkshire Geological Society (Y.G.S.), c/o S. Rogers, 4 Middledyke Ln, Cottingham HU16 4NH
Founded: 1837; Members: 1200
Pres: B. Young; Gen Secr: A.C. Benfield
Focus: Geology
Periodical
Proceedings (weekly) 13233

Yorkshire Philosophical Society (Y.P.S.), The Lodge, Museum Gardens, York YO1 2DR
T: (0904) 656713
Founded: 1822; Members: 420
Focus: Philos
Periodical
Annual Report (weekly) 13234

Zinc Development Association, 42 Weymouth St, London W1N 3LQ
T: (071) 4996636; Tx: 261286; Fax: (071) 4931555
Focus: Metallurgy 13235

Zoological Society of Glasgow and West of Scotland, c/o Glasgow Zoological Gardens, Calderpark Uddingston, Glasgow G71 7RZ
T: (041) 7711185
Founded: 1937; Members: 609
Focus: Zoology
Periodical
Annual Report (weekly) 13236

The Zoological Society of London (Z.S.L.), Regent's Park, London NW1 4RY
T: (071) 7223333; Tx: 265247; CA: Zoological London; Fax: (071) 4834436
Founded: 1826; Members: 5310
Pres: Sir John Chapple; Gen Secr: Prof. R. McNeil Alexander
Focus: Zoology
Periodicals
International Zoo Yearbook (weekly)
Journal of Zoology (weekly)
Nomenclator Zoologicus (weekly)
Symposia (weekly)
Zoological Record (weekly) 13237

Zoological Society of Northern Ireland, 33 Saratoga Av, Newtownards BT23 4BO
Founded: 1963; Members: 73
Focus: Zoology 13238

Uruguay

Academia Nacional de Ingeniería, Av Libertador Brigadier General Lavalleja 1464, Montevideo
Founded: 1965; Members: 32
Pres: Hector Fernández Guido; Gen Secr: Alberto Ponce Delgado
Focus: Eng 13239

Academia Nacional de Letras (ANL), 25 de Mayo 376, Montevideo
T: 952374
Founded: 1946; Members: 19
Pres: Arturo Sergio Visca; Gen Secr: Luis Bausero
Focus: Ling
Periodical
Boletín de la Academia Nacional de Letras (weekly) 13240

Academia Nacional de Medicina del Uruguay, 18 de Julio 2175, Montevideo
T: 401444
Founded: 1976; Members: 27
Focus: Med
Periodical
Boletín (weekly) 13241

Asociacíon de Bibliotecarios del Uruguay (ABU), Dante 2255, Casilla 1315, Montevideo
Founded: 1978; Members: 384
Pres: Martha Ottino
Focus: Libraries & Bk Sci
Periodical
Actualidades Bibliotecológicas (weekly) . . 13242

Asociación de Bibliotecólogos del Uruguay, Dante 2255, Casilla 1315, Montevideo
Founded: 1978; Members: 443
Pres: Susana Gil; Gen Secr: Sonia Tavarez
Focus: Libraries & Bk Sci

Periodicals
Actualidades Bibliotecológicas (weekly)
Panel de Noticias (weekly) 13243

Asociación de Ingenieros del Uruguay, Av Libertador Brigadier General Lavalleja 1464, Montevideo
Founded: 1950; Members: 800
Pres: Eduardo Crispo Ayala; Gen Secr: Ponciano J. Torrado
Focus: Eng
Periodical
Ingeniería 13244

Asociación de Química y Farmacia del Uruguay, Av Libertador Brigadier General Lavalleja 1464, Montevideo
Founded: 1888; Members: 600
Focus: Chem; Pharmacol 13245

Asociación Ondontológica Uruguaya, Durazno 937-939, Montevideo
T: 901572; CA: Odontur
Founded: 1946; Members: 3000
Pres: Dr. Luis P. Lauko; Gen Secr: Dr. José Llaguno
Focus: Dent
Periodical
Odontología Uruguaya: Revista Científica Gremial e Informativa (weekly) 13246

Asociación Rural del Uruguay, Uruguay 864, Montevideo
Founded: 1871
Focus: Agri 13247

Asociación Uruguaya de Escritores, Bartolomé Mitre 1260, Montevideo
Focus: Lit 13248

Association for the Study of Man Environment Relations (ASMER), Casilla 13125, Montevideo
Founded: 1985
Pres: E. Gudynas
Focus: Ecology 13249

Ateneo de Clínica Quirúrgica, Montevideo
Founded: 1934
Gen Secr: Prof. Dr. Eduardo Blanco Acevedo
Focus: Surgery
Periodical
Anales 13250

Centro de Investigaciones Agrícolas Alberto Boerger, La Estanzuela, Colonia
Founded: 1914
Focus: Agri 13251

Centro de Investigaciones y Estudios Familiares (CIEF), José Enrique Rodo 2115, Montevideo
T: (02) 400681
Founded: 1965; Members: 50
Focus: Family Plan
Periodical
Digesto Familiar (weekly) 13252

Centro de Nacionales y Comercio Internacional del Uruguay (CENCI), Misiones 1361, Casilla 1510, Montevideo
Founded: 1956; Members: 1200
Focus: Stats; Econ
Periodicals
Manual Práctico Aduanero (weekly)
Manual Práctico del Contribuyente (weekly)
Manual Práctico del Exportador (weekly)
Manual Práctico del Importador (weekly)
Manual Práctico Sudamericano del Transporte Internacional por Carretera de Carga y Pasajeros (weekly) 13253

Centro Interamericano de Investigación y Documentación sobre Formación Profesional (CINTERFOR), Colonia 993, Montevideo
Focus: Adult Educ 13254

Comisión Nacional de Energía Atómica (CNEA), Soriano 1014, Montevideo
T: 986783, 901944, 915916
Founded: 1955
Pres: Walter R. Cibils
Focus: Nucl Res 13255

Consejo Nacional de Educación, Centro Nacional de Información y Documentación (CONAE-CENID), Av del Libertador Brigadier General Lavalleja 2025, Montevideo
T: 905293, 983557
Founded: 1974
Focus: Educ 13256

Consejo Nacional de Higiene, c/o Ministerio de Salud Pública, Av 18 de Julio 1892, Montevideo
Focus: Hygiene 13257

Consejo Nacional de Investigaciones Científicas y Técnicas, Sarandí 450, Montevideo
Founded: 1961; Members: 7
Focus: Eng; Sci 13258

Gremial Uruguaya de Médicos Radiólogos, Av Libertador Brigadier General Lavalleja 1464, Montevideo
Founded: 1972; Members: 70
Focus: Radiology 13259

Instituto de Investigaciones Biológicas Clemente Estable, Av Italia 3318, Montevideo
T: 811012, 811662
Founded: 1927; Members: 81
Focus: Bio 13260

Instituto Histórico y Geográfico del Uruguay, Convención 1366, Casilla 10999, Montevideo
T: 914441
Founded: 1843; Members: 40
Pres: Prof. Edmundo Narancio; Gen Secr: Prof. Fernando O. Assuncao; Prof. Jorge A. Anselmi
Focus: Geography; Hist
Periodical
Revista del Instituto Histórico y Geográfico del Uruguay (weekly) 13261

Instituto Uruguayo de Normas Técnicas, Av Libertador Brigadier General Lavalleja, Montevideo
Founded: 1939; Members: 400
Focus: Standards; Eng 13262

Library Association of Uruguay, Ibicuy 1276, Montevideo
Focus: Libraries & Bk Sci 13263

Liga Uruguaya contra la Tuberculosis (LUCT), Magallanes 1320, Montevideo
T: 43570, 497869
Founded: 1902; Members: 7
Focus: Pulmon Dis 13264

Sociedad de Amigos de Arqueología, Buenos Aires 652, Casilla 399, Montevideo
Founded: 1926; Members: 86
Focus: Archeol
Periodical
Revista 13265

Sociedad de Biología de Montevideo, Casilla 567, Montevideo
Founded: 1927
Focus: Bio 13266

Sociedad de Cirugía del Uruguay, Av Libertador Brigadier General Lavalleja 1464, Montevideo
Founded: 1920; Members: 426
Pres: Dr. Milton Mazza; Gen Secr: Dr. Guillermo Piacenza
Focus: Surgery
Periodical
Cirúgia del Uruguay (weekly) 13267

Sociedad de Radiología del Uruguay, Av Libertador Brigadier General Lavalleja 1464, Montevideo
Founded: 1923; Members: 60
Focus: Radiology 13268

Sociedad Malacológica del Uruguay, Casilla 1401, Montevideo
Founded: 1957; Members: 210
Focus: Zoology 13269

Sociedad Uruguaya de Patología Clínica, Casilla 6147, Montevideo
T: 801515
Founded: 1954; Members: 140
Pres: Dr. Walter Alallon; Gen Secr: Dr. Carlos Krul
Focus: Pathology
Periodical
Revista 13270

Sociedad Uruguaya de Pediatría, Casilla 10906, Montevideo
Founded: 1927
Pres: Prof. Dr. I. Gentile Ramos; Gen Secr: Dr. Edmundo Batthyány
Focus: Pediatrics
Periodical
Archivos de Pediatría del Uruguay (weekly) 13271

Sociedad Zoológica del Uruguay, Casilla 399, Montevideo
Founded: 1961
Pres: Prof. Miguel A. Klappenbach
Focus: Zoology
Periodical
Boletín 13272

U.S.A.

AACE International, c/o Kenneth K. Humphreys, POB 1557, Morgantown, WV 26507-1557
T: (304) 296-8444; Fax: (304) 291-5728
Founded: 1956; Members: 6000
Gen Secr: Kenneth K. Humphreys
Focus: Eng
Periodicals
Cost Engineering (weekly)
Cost Engineers Notebook
Directory of Members (weekly)
Transactions (weekly) 13273

Aaron Burr Association (ABA), 4520 King Edward Ct, Annandale, VA 22003
T: (703) 256-7226
Founded: 1946; Members: 250
Gen Secr: Dr. David H. Burr
Focus: Hist
Periodicals
The Chronicle (weekly)
Newsletter 13274

Aboriginal Research Club (ARC), c/o Jerry Atkinson, Dearborn Historical Museum, 915 S Brady Rd, Dearborn, MI 48124
T: (313) 565-3000
Founded: 1940; Members: 30
Pres: Jerry Atkinson
Focus: Ethnology 13275

Abrasive Engineering Society (AES), 108 Elliott Dr, Butler, PA 16001-1118
T: (412) 282-6210; Fax: (412) 282-6210
Founded: 1957; Members: 400
Gen Secr: Ted Giese
Focus: Eng
Periodical
AES-Magazine (weekly) 13276

Academy for Educational Development (AED), 1255 23 St NW, Washington, DC 20037
T: (202) 862-1900; Tx: 197601; Fax: (202) 862-1947
Founded: 1961
Pres: Stephen F. Moseley
Focus: Educ
Periodicals
Academy News (3 times annually)
Newsletter 13277

Academy for Implants and Transplants (AIT), POB 223, Springfield, VA 22150
T: (703) 451-0001
Founded: 1972; Members: 268
Gen Secr: Anthony J. Viscido
Focus: Dent
Periodical
Implant Update (weekly) 13278

Academy for Interscience Methodology (AIM), 907 N Elm St, Ste 203, Hinsdale, IL 60521
T: (708) 654-0240
Founded: 1961; Members: 15
Gen Secr: Jeffrey E. Painter
Focus: Sci 13279

Academy of Ambulatory Foot Surgery (AAFS), POB 2730, Tuscaloosa, AL 35403
T: (205) 758-3678
Founded: 1972; Members: 1500
Gen Secr: Dr. Stanford Rosen
Focus: Surgery
Periodicals
Directory
Journal of the Academy of Ambulatory Foot Surgery
Newsletter (weekly) 13280

Academy of American Franciscan History (AAFH), 1712 Euclid Av, Berkeley, CA 94709
T: (510) 548-1755
Founded: 1944; Members: 301
Gen Secr: William Short
Focus: Hist; Rel & Theol
Periodical
The Americas (weekly) 13281

Academy of American Poets (AAP), 584 Broadway, Ste 1208, New York, NY 10012
T: (212) 274-0343; Fax: (212) 274-9427
Founded: 1934; Members: 2400
Gen Secr: Beth McCabe
Focus: Lit
Periodical
Booklist (weekly) 13282

Academy of Aphasia (AA), c/o Victoria A. Fromkin, Dept of Linguistics, UCLA, Los Angeles, CA 90024
T: (310) 206-3206
Founded: 1962; Members: 189
Pres: Victoria A. Fromkin
Focus: Logopedy 13283

Academy of Applied Science (AAS), 1 Maple St, Concord, NH 03301
T: (603) 225-2072
Founded: 1962; Members: 300
Gen Secr: Howard S. Curtis
Focus: Sci 13284

Academy of Arts and Sciences of the Americas (AASA), 3551 Main Hwy, Miami, FL 33133
T: (305) 444-3003
Founded: 1965; Members: 25
Pres: Julia Allen Field
Focus: Arts; Sci 13285

Academy of Criminal Justice Sciences (ACJS), 402 Nunn Hall, Northern Kentucky University, Highland Heights, KY 41099-5998
T: (606) 572-5634; Fax: (606) 572-6665
Founded: 1963; Members: 2400
Gen Secr: Patricia DeLancey
Focus: Educ; Law
Periodicals
ACJS Employment Bulletin (8 times annually)
ACJS Membership Directory (weekly)
ACJS Today (weekly)
Journal of Criminal Justice Education (weekly)
Justice Quarterly (weekly) 13286

Academy of Dental Materials (ADM), POB 21418, Charleston, SC 29413-1418
T: (803) 792-9802
Founded: 1940; Members: 300
Gen Secr: Betty Davis
Focus: Dent
Periodicals
ADM Newsletter (weekly)
Dental Materials (weekly) 13287

Academy of Dentistry for the Handicapped (ADH), 211 E Chicago Av, Ste 1616, Chicago, IL 60611
T: (312) 440-2661; Fax: (312) 440-7494
Founded: 1950; Members: 500
Gen Secr: John S. Rufkauskas
Focus: Dent
Periodicals
Interface (weekly)
Membership Referral Roster (weekly)
Special Care in Dentistry (weekly) . . . 13288

Academy of Dentistry International (ADI), 5125 Mac Arthur Blvd NW, Washington, DC 20016-3315
T: (202) 364-8349; Fax: (202) 244-6244
Founded: 1974; Members: 1500
Gen Secr: Henry J. Sazima
Focus: Dent
Periodicals
International Communicator (weekly)
Roster of ADI (weekly) 13289

Academy of Denture Prosthetics (ADP), 1515 116 Av NE, Ste 303, Bellevue, WA 98004
T: (206) 455-5661
Founded: 1918; Members: 92
Focus: Dent
Periodical
Journal of Prosthetic Dentistry (weekly) . 13290

Academy of Dispensing Audiologists (ADA), 3008 Millwood Av, Columbia, SC 29205
T: (803) 252-5646; Fax: (803) 765-0860
Founded: 1977; Members: 525
Gen Secr: Carol H. Davis
Focus: Otorhinolaryngology
Periodical
ADA Feedback (weekly) 13291

Academy of General Dentistry (AGD), 211 E Chicago Av, Ste 1200, Chicago, IL 60611
T: (312) 440-4300; Fax: (312) 440-0559
Founded: 1952; Members: 33000
Gen Secr: Harold E. Donnell
Focus: Dent
Periodical
AGD Impact (10 times annually) 13292

Academy of International Business (AIB), c/o School of Business Administration, Wayne State University, Detroit, MI 48202
T: (313) 577-4493; Fax: (313) 577-4641
Founded: 1959; Members: 2400
Gen Secr: Attila Yaprak
Focus: Econ
Periodicals
Journal of International Business Studies (weekly)
Membership Directory (weekly)
Newsletter (weekly) 13293

Academy of International Dental Studies (AIDS), c/o Vartan Ghugasian, D.M.D., 100 Boylston Av, Ste 806, Boston, MA 02116
T: (617) 542-0777
Founded: 1981; Members: 500
Pres: Vartan Ghugasian
Focus: Dent
Periodical
The Intercom (weekly) 13294

Academy of Legal Studies in Business (ALSB), c/o Dept of Finance, Miami University, 120 Upham Hall, Oxford, OH 45056
T: (513) 529-2945; Fax: (513) 529-6992
Founded: 1923; Members: 1300
Gen Secr: Dr. Daniel J. Herron
Focus: Law
Periodicals
American Business Law Journal (weekly)
Journal of Legal Studies Education (weekly)
Newsletter (weekly) 13295

Academy of Management (AM), 300 S Union, POB 209, Ada, OH 45810-0209
T: (419) 772-1953
Founded: 1936; Members: 8700
Gen Secr: Ken Cooper
Focus: Business Admin
Periodicals
Academy of Management Executive (weekly)
Academy of Management Journal (weekly)
Academy of Management Newsletter
Academy of Management Proceedings (weekly)
Academy of Management Review (weekly) 13296

Academy of Marketing Science (AMS), c/o School of Business Administration, University of Miami, POB 248012, Coral Gables, FL 33124
T: (305) 284-6673; Fax: (305) 284-3762
Founded: 1971; Members: 1200
Gen Secr: Harold W. Berkman
Focus: Business Admin
Periodicals
Academy of Marketing Science (weekly)
Academy of Marketing Science News (weekly)
Index of JAMS (weekly)
Journal of the Academy of Marketing Science 13297

Academy of Motion Picture Arts and Sciences (AMPAS), 8949 Wilshire Blvd, Beverly Hills, CA 90211
T: (310) 247-3000; Fax: (310) 247-9619
Founded: 1927; Members: 5210
Gen Secr: Bruce Davis
Focus: Cinema
Periodicals
Academy Players Directory (3 times annually)
Annual Index of Motion Picture Credits (weekly) 13298

Academy of Natural Sciences (ANSP), 1900 Benjamin Franklin Pkwy, Philadelphia, PA 19103-1195
T: (215) 299-1000; Fax: (215) 299-1028
Founded: 1812; Members: 9500

Pres: Dr. Keith S. Thomson
Focus: Nat Sci
Periodicals
Academy News
Proceedings of the Academy of Natural Sciences of Philadelphia (weekly) 13299

Academy of Operative Dentistry (AOD), 643 Broadway, Menomonie, WI 54751
T: (715) 235-7566
Founded: 1972; Members: 900
Gen Secr: R.J. Werner
Focus: Dent
Periodicals
Membership Roster
Operative Dentistry (weekly) 13300

Academy of Oral Dynamics (AOD), 5950 Elmer Derr Rd, Frederick, MD 21701
T: (301) 473-9719
Founded: 1950; Members: 75
Gen Secr: Joseph P. Skellchock
Focus: Dent 13301

Academy of Pharmacy Practice and Management (APPM), c/o American Pharmaceutical Association, 2215 Constitution Av NW, Washington, DC 20037
T: (202) 628-4410; Fax: (202) 783-2351
Founded: 1965; Members: 18500
Gen Secr: John M. Hammond
Focus: Pharmacol
Periodical
APPM Update (weekly) 13302

Academy of Political Science (APS), 475 Riverside Dr, Ste 1274, New York, NY 10115-1274
T: (212) 870-2500; Fax: (212) 870-2202
Founded: 1880; Members: 9000
Pres: Demetrios Caraley
Focus: Poli Sci
Periodicals
Political Science Quarterly (weekly)
Proceedings 13303

Academy of Psychosomatic Medicine (APM), 5824 N Magnolia, Chicago, IL 60660
T: (312) 784-2025
Founded: 1952; Members: 1000
Gen Secr: Evelyne A. Hallberg
Focus: Psych
Periodical
Psychosomatics (weekly) 13304

Academy of Rehabilitative Audiology (ARA), c/o Dept of Communication Disorders, University of Akron, Akron, OH 44325-3001
T: (216) 972-7883
Founded: 1966; Members: 300
Gen Secr: Dr. Sharon A. Lesner
Focus: Otorhinolaryngology
Periodicals
Journal of the Academy of Rehabilitative Audiology (weekly)
Membership Directory (weekly) 13305

Academy of Religion and Psychical Research (ARPR), POB 614, Bloomfield, CT 06002
T: (203) 242-4593
Founded: 1972; Members: 300
Gen Secr: Boyce Batey
Focus: Psych; Rel & Theol
Periodicals
ARPR Bulletin (weekly)
The Journal of Religion and Psychical Research (weekly) 13306

Academy of Science Fiction, Fantasy and Horror Films (ASFFHF), 334 W 54 St, Los Angeles, CA 90037
T: (213) 752-5811
Founded: 1972; Members: 3000
Pres: Dr. Donald A. Reed
Focus: Cinema
Periodicals
Newsletter (weekly)
Saturn (weekly) 13307

Academy of Scientific Hypnotherapy (ASH), POB 12041, San Diego, CA 92112
T: (619) 427-6225
Founded: 1977
Pres: William E. Kemery
Focus: Psych; Therapeutics
Periodicals
Bulletin
Hypnotherapy in Review 13308

Academy of Security Educators and Trainees (ASET), c/o Dr. Richard W. Kobetz, Rt 2, Box 3644, Berryville, VA 22511
T: (703) 955-1129
Members: 510
Gen Secr: Dr. Richard W. Kobetz
Focus: Educ; Eng
Periodicals
ASET Membership (weekly)
The Educator (weekly) 13309

Academy of Television Arts and Sciences (ATAS), 5220 Lankershim Blvd, North Hollywood, CA 91601
T: (818) 754-2800
Founded: 1948; Members: 6500
Gen Secr: James L. Loper
Focus: Journalism
Periodicals
Debut (weekly)
Emmy Directory (weekly)
Emmy Magazine (weekly) 13310

Academy of Veterinary Cardiology (AVC), 51 Atlantic Av, POB 208, Floral Park, NY 11002
T: (516) 358-4500
Founded: 1967; Members: 600
Gen Secr: Larry P. Tilley
Focus: Cardiol; Vet Med
Periodicals
Membership Directory (biennial)
Newsletter 13311

Accordion Teacher's Guild (ATG), c/o Joan Sommers, POB 22342, Kansas City, MO 64113
T: (913) 722-5625
Founded: 1941; Members: 100
Pres: Joan Sommers
Focus: Music; Educ
Periodical
Accordion Teacher's Guild Newsletter (10 times annually) 13312

Accountants Computer Users Technical Exchange (ACUTE), 8060 Knue Rd, Ste 127, Indianapolis, IN 46250
T: (317) 845-8702
Founded: 1965; Members: 385
Focus: Computer & Info Sci
Periodicals
The Account (weekly)
Membership Directory (weekly) 13313

Accreditation Association for Ambulatory Health Care (AAAHC), 9933 Lawler Av, Skokie, IL 60077-3708
T: (312) 676-9610; Fax: (708) 676-9628
Founded: 1979; Members: 12
Gen Secr: Christopher Damon
Focus: Med 13314

Accreditation Board for Engineering and Technology (ABET), 345 E 47 St, New York, NY 10017
T: (212) 705-7685; Tx: 5101012914; Fax: (212) 838-8062
Founded: 1932; Members: 26
Gen Secr: David R. Reyes-Guerra
Focus: Eng
Periodical
Accreditation Yearbook (weekly) 13315

Accrediting Bureau of Health Education Schools (ABHES), Oak Manor Office, 29089 U.S. 20 W, Elkhart, IN 46514
T: (219) 293-0124; Fax: (219) 295-8564
Founded: 1964; Members: 154
Gen Secr: Mary Lou Reed
Focus: Adult Educ; Med
Periodical
ABHES News 13316

Accrediting Commission on Education for Health Services Administration (ACEHSA), 1911 N Fort Myer Dr, Ste 503, Arlington, VA 22209
T: (703) 524-0511; Fax: (703) 525-4791
Founded: 1968; Members: 62
Gen Secr: Dr. Sherril B. Gelmon
Focus: Public Health
Periodical
Official List of Accredited Programs (weekly) 13317

Accrediting Council for Continuing Education and Training (ACCET), 600 E Main St, Ste 1425, Richmond, VA 23219
T: (804) 648-6742; Fax: (804) 780-0821
Founded: 1974; Members: 400
Pres: Roger J. Williams
Focus: Adult Educ
Periodicals
Directory of Accredited Noncollegiate Continuing Education Programs
Growing Edge (weekly) 13318

Accrediting Council on Education in Journalism and Mass Communications (ACEJMC), c/o School of Journalism, University of Kansas, Lawrence, KS 66045
T: (913) 864-3973
Founded: 1929; Members: 23
Gen Secr: Susanne Shaw
Focus: Adult Educ; Journalism
Periodical
Accredited Journalism and Mass Communications Education (weekly) 13319

ACEC Research and Management Foundation (ACEC/RMF), 1015 15 St NW, Ste 802, Washington, DC 20005
T: (202) 347-7474; Fax: (202) 898-0068
Founded: 1974
Gen Secr: Thomas E. Kern
Focus: Eng
Periodical
RMF Newsletter (weekly) 13320

Ackerman Institute for Family Therapy (AIFT), 149 E 78 St, New York, NY 10021
T: (212) 879-4900; Fax: (212) 744-0206
Founded: 1960
Gen Secr: Dr. Peter Steinglass
Focus: Therapeutics 13321

Acoustical Society of America (ASA), 500 Sunnyside Blvd, Woodbury, NY 11797
T: (516) 349-7800; Fax: (516) 349-7669
Founded: 1929; Members: 6700
Gen Secr: Charles E. Schmid
Focus: Acoustics
Periodical
Journal of the Acoustical Society of America (weekly) 13322

Actuarial Studies in Non-Life Insurance (ASTIN), c/o Wharton School, Dept of Insurance, University of Pennsylvania, 3641 Locust Walk, Philadelphia, PA 19104
Founded: 1957
Gen Secr: Prof. Jean Lemaire
Focus: Insurance 13323

Acupuncture International Association (AIA), 2330 S Brentwood Blvd, Saint Louis, MO 63144-2096
T: (314) 961-2300; Fax: (314) 961-9828
Founded: 1949; Members: 4827
Gen Secr: Dr. Carol Ann Lee
Focus: Med; Physical Therapy
Periodicals
Bulletin
Needlenews (weekly) 13324

Acupuncture Research Institute (ARI), 313 W Andrix St, Monterey Park, CA 91754
T: (213) 722-7353
Founded: 1972; Members: 750
Gen Secr: Louis Gasper
Focus: Med; Physical Therapy
Periodical
The Meridian (weekly) 13325

Addiction Research and Treatment Corporation (ARTC), 22 Chapel St, Brooklyn, NY 11201
T: (718) 260-2900
Founded: 1969
Gen Secr: Dr. Beny J. Primm
Focus: Med 13326

Adhesion Society (AS), c/o Dept of Materials Science, University of Cincinnati, Cincinnati, OH 45221-0012
T: (513) 556-3111; Fax: (801) 864-8312
Members: 300
Pres: F. James Boerio
Focus: Chem
Periodical
Newsletter (weekly) 13327

Adirondack Historical Association (AHA), c/o Adirondack Museum, Blue Mountain Lake, NY 12812-0099
T: (518) 352-7311; Fax: (518) 352-7603
Founded: 1948; Members: 121
Gen Secr: Jacqueline Bay
Focus: Hist
Periodical
Guide Line (weekly) 13328

Administrative Management Society (AMS), 1101 14 St NW, Washington, DC 20005-5601
T: (202) 371-8299; Fax: (202) 371-1090
Founded: 1919; Members: 4000
Gen Secr: M. Sutherland
Focus: Business Admin
Periodicals
AMS Insights (weekly)
Management Salaries Report (weekly)
Office, Professional and Data Processing Report (weekly) 13329

Adrenal Metabolic Research Society of the Hypoglycemia Foundation (AMRSHF), 153 Pawling Av, Troy, NY 12180
T: (518) 272-7154
Founded: 1956
Gen Secr: Marilyn H. Light
Focus: Med
Periodical
Homeostasis Quarterly (weekly) 13330

Adult Christian Education Foundation (ACEF), POB 8398, Madison, WI 53708
T: (608) 849-5933
Founded: 1959
Pres: Harley A. Swiggum
Focus: Adult Educ
Periodical
Annual Report (weekly) 13331

Advanced Technology Alert System (ATAS), c/o UNCSTD, 1 United Nations Plaza, New York, NY 10017
T: (212) 963-8435; Tx: 232422; Fax: (212) 963-4116
Founded: 1984
Focus: Eng
Periodical
ATAS Bulletin (weekly) 13332

Advanced Transit Association (ATRA), 9019 Hamilton Dr, Fairfax, VA 22031
T: (703) 591-8328
Founded: 1976; Members: 100
Gen Secr: Jarold A. Kieffer
Focus: Transport
Periodicals
ATRA Updates
Journal of Advanced Transportation (weekly) 13333

Advertising Research Foundation (ARF), 3 E 54 St, New York, NY 10022
T: (212) 751-5656; Fax: (212) 319-5265
Founded: 1936; Members: 355
Pres: Michael J. Naples
Focus: Advert
Periodicals
ARF Transcript Proceedings
Journal of Advertising Research (weekly) 13334

Advisory Committee on Science and Technology for Development (ACSTD), c/o United Nations, Room DCI-1068, 1 United Nations Plaza, New York, NY 10017
T: (212) 963-8813; Tx: 422311; Fax: (212) 963-1267
Founded: 1981; Members: 28
Gen Secr: Hiroko Morita-Lou
Focus: Develop Areas 13335

Aerospace Department Chairmen's Association (ADCA), c/o Walter Haisler, Dept of Aerospace Engineering, Texas A & M University, 701 H.R. Bright Bldg, College Station, TX 77843-3141
T: (409) 845-1600; Fax: (409) 845-9267
Founded: 1968; Members: 79
Pres: Walter Haisler
Focus: Aero 13336

Aerospace Education Foundation (AEF), 1501 Lee Hwy, Arlington, VA 22209-1198
T: (703) 247-5839; Fax: (703) 247-5853
Founded: 1956; Members: 2000
Gen Secr: Steven S. Lee
Focus: Educ; Aero
Periodicals
Foundation Forums (weekly)
Newsletter (weekly) 13337

Aerospace Electrical Society (AES), 700 E South St, Anaheim, CA 92805
T: (714) 778-1840; Fax: (714) 774-8659
Founded: 1941; Members: 250
Pres: Lloyd P. Appelmann
Focus: Electric Eng
Periodical
News & Views (weekly) 13338

Aerospace Medical Association (AsMA), 320 S Henry St, Alexandria, VA 22314
T: (703) 739-2240; Fax: (703) 739-9652
Founded: 1929; Members: 4200
Gen Secr: Russell B. Rayman
Focus: Med
Periodicals
Aviation, Space and Envivonmental Medicine (weekly)
Membership Directory (weekly)
Scientific Papers (weekly) 13339

Afghanistan Studies Association (ASA), c/o Center for Afghanistan Studies, University of Nebraska, ASH 238, Omaha, NE 68182-0006
T: (402) 554-2901; Fax: (402) 554-3242
Founded: 1971; Members: 125
Gen Secr: Thomas E. Gouttierre
Focus: Ethnology
Periodicals
Afghanistan Studies Journal (weekly)
ASA Newsletter (weekly) 13340

The Africa Fund (AF), 198 Broadway, New York, NY 10038
T: (212) 962-1210
Founded: 1966; Members: 5000
Gen Secr: Jennifer Davis
Focus: Poli Sci
Periodicals
Annual Report (weekly)
Southern Africa Literature List (weekly)
Southern Africa Perspectives Series
Unified List of United States Companies doing Business in South Africa (weekly) . . . 13341

African-American Institute (AAI), 833 United Nations Plaza, New York, NY 10017
T: (212) 949-5666; Tx: 666565; Fax: (212) 682-6174
Founded: 1953
Pres: Vivian Lowery Derryck
Focus: Poli Sci
Periodicals
Africa Report (weekly)
Annual Report (weekly)
Bulletin 13342

African American Museum (AAM), 1765 Crawford Rd, Cleveland, OH 44106
T: (216) 791-1700
Founded: 1953; Members: 37
Gen Secr: Dr. Eleanor Engram
Focus: Hist
Periodical
African American Museum Newsletter (weekly) 13343

African Heritage Center for African Dance and Music (AHCADM), 4018 Minnesota Av NE, Washington, DC 20019
T: (202) 399-5252
Founded: 1973; Members: 20
Gen Secr: Melvin Deal
Focus: Ethnology 13344

African Heritage Studies Association (AHSA), c/o Africana Studies and Research Institute, Queens College, Flushing, NY 11367
T: (718) 997-5478
Founded: 1969

Pres: Dr. W. Ofuatey-Kodjoe
Focus: Ethnology
Periodicals
International Journal of Africana Studies (3 times annually)
Newsletter (3 times annually) 13345

African Studies Association (ASA), c/o Dr. Edna Bay, Emory University, Credit Union Bldg, Atlanta, GA 30322
T: (404) 329-6410; Fax: (404) 329-6433
Founded: 1957; Members: 2700
Gen Secr: Dr. Edna Bay
Focus: Ethnology
Periodicals
African Studies Review (3 times annually)
ASA News (weekly)
History in Africa (weekly)
Issue (weekly) 13346

Afro-American Historical and Genealogical Society (AAHGS), POB 73086, Washington, DC 20056
T: (202) 234-5350
Founded: 1977; Members: 1000
Pres: Sylvia Cooke Martin
Focus: Genealogy; Hist
Periodicals
Journal (weekly)
Newsletter (weekly) 13347

Afro-Asian Center (AAC), POB 337, Saugerties, NY 12477
T: (914) 246-7828
Founded: 1972
Gen Secr: Robert Carroll
Focus: Ethnology 13348

Agency for Instructional Technology (AIT), Box A, Bloomington, IN 47402-0120
T: (812) 339-2203; Fax: (812) 333-4218
Founded: 1970
Gen Secr: Michael F. Sullivan
Focus: Educ
Periodical
TECHNOS (weekly) 13349

Agricultural Communicators in Education (ACE), c/o Ashley Wood, University of Florida, Bldg 116, 601 IFAS, Gainesville, FL 32611
T: (904) 392-9588; Fax: (904) 392-8583
Founded: 1913; Members: 700
Gen Secr: Ashley Wood
Focus: Agri
Periodicals
Journal of Applied Communications (weekly)
Newsletter (weekly) 13350

Agricultural History Society (AHS), 1301 New York Av NW, Rm 932, Washington, DC 20005-4788
T: (202) 219-0787
Founded: 1919; Members: 1400
Gen Secr: Wayne D. Rasmussen
Focus: Hist; Agri
Periodicals
Agricultural History (weekly)
Symposium Proceedings (weekly) . . . 13351

Agriservices Foundation, 648 W Sierra Av, POB 429, Clovis, CA 93612
T: (209) 299-2263; Fax: (209) 299-2098
Founded: 1964
Pres: Dr. Marion E. Ensminger
Focus: Agri 13352

Aid for International Medicine (AIM), POB 119, Rocklamd, DE 19732
T: (302) 655-8290
Founded: 1965
Focus: Med 13353

AIESEC – United States, 135 W 50 St, New York, NY 10020
T: (212) 757-3774; Tx: 220923; Fax: (212) 757-4062
Founded: 1948; Members: 5000
Pres: Chelle Izzi
Focus: Econ
Periodicals
AIESEC-U.S. Annual Report (weekly)
Linkletter (weekly) 13354

Air and Waste Management Association (AWMA), POB 2861, Pittsburgh, PA 15230
T: (412) 232-3444; Fax: (412) 232-3450
Founded: 1907; Members: 13000
Gen Secr: Martin E. Rivers
Focus: Ecology
Periodical
Journal of the Air and Waste Management Association (weekly) 13355

Aircraft Electronics Association (AEA), POB 1963, Independence, MO 64055-0963
T: (816) 373-6565; Fax: (816) 478-3100
Founded: 1958; Members: 862
Gen Secr: Monte Mitchell
Focus: Electronic Eng
Periodical
Avionics News (weekly) 13356

Airline Medical Directors Association (AMDA), c/o Nestor B. Kowalsky, American Airlines Medical Dept, POB 66033, Chicago, IL 60666
T: (312) 686-4192; Fax: (312) 686-6393
Founded: 1946; Members: 130
Gen Secr: Nestor B. Kowalsky
Focus: Med 13357

Air National Guard Optometric Society (ANGOS), c/o Lyman Nordan, 5517 Afton Dr, Birmingham, AL 35242
T: (205) 991-8663
Founded: 1975; Members: 50
Pres: Lyman Nordan
Focus: Optics
Periodical
Newsletter (3 times annually) 13358

Airship Association – U.S. (AA), 8512 Cedar St, Silver Spring, MD 20910
T: (301) 588-7916; Fax: (301) 588-2085
Founded: 1953
Gen Secr: Frank G. McGuire
Focus: Aero
Periodical
Advanced Lighter-Than-Air Review (weekly) 13359

Alcor Life Extension Foundation (ALEF), 12327 Doherty, Riverside, CA 92503
T: (714) 736-1703; Fax: (714) 736-5817
Founded: 1972; Members: 447
Pres: Carlos Mondragon
Focus: Med
Periodical
Cryonics (weekly) 13360

Alexander Graham Bell Association for the Deaf (AGBAD), 3417 Volta Pl NW, Washington, DC 20007
T: (202) 337-5220
Founded: 1890; Members: 5000
Gen Secr: Dr. Donna McCord Dickman
Focus: Educ Handic
Periodicals
Newsounds (10 times annually)
Our Kids Magazine (2-3 times annually) 13361

Alfred Adler Institute (AAI), 1780 Broadway, New York, NY 10019
T: (212) 974-0431
Founded: 1950
Gen Secr: Leo Rattner
Focus: Psych
Periodicals
Bulletin (weekly)
Journal of Individual Psychology (weekly)
Newsletter (weekly) 13362

ALI-ABA Committee on Continuing Professional Education, 4025 Chestnut St, Philadelphia, PA 19104-3099
T: (215) 243-1600; Fax: (215) 243-1664
Founded: 1947
Gen Secr: Paul A. Wolkin
Focus: Educ
Periodicals
ALI-ABA CLR Review (weekly)
ALI-ABA Course Materials Journal (weekly)
ALI-ABA Reporter (weekly) 13363

Alliance for Alternatives in Healthcare (AAH), POB 9178, Calabasas, CA 91372
T: (818) 702-0888; Fax: (818) 702-9954
Founded: 1983; Members: 1000
Pres: Steve Gorman
Focus: Marketing 13364

Alliance for Environmental Education (AEE), 51 Main St, POB 368, The Plains, VA 22171
T: (703) 253-5812; Fax: (703) 253-5811
Founded: 1972; Members: 300
Pres: Steven C. Kussmann
Focus: Ecology
Periodicals
Annual Report (weekly)
Network Exchange (weekly) 13365

Alliance to Save Energy (ASE), 1725 K St NW, Ste 509, Washington, DC 20006-1401
T: (202) 857-0666; Fax: (202) 331-9588
Founded: 1977; Members: 1500
Pres: William A. Nitze
Focus: Energy
Periodicals
Alliance Update (weekly)
Annual Report (weekly) 13366

ALSAC – Saint Jude Children's Research Hospital, 501 Saint Saint Jude Pl, POB 3704, Memphis, TN 38173-0704
T: (901) 522-9733
Founded: 1957; Members: 2000
Gen Secr: Richard C. Shadyac
Focus: Pediatrics
Periodicals
ALSAC News (weekly)
Partners in Hope (weekly) 13367

Alternative Energy Resources Organization (AERO), 44 N Last Chance Gulch, Helena, MT 59601
T: (406) 443-7272; Fax: (406) 442-9120
Founded: 1974; Members: 600
Gen Secr: Al Kurki
Focus: Energy; Econ; Agri
Periodicals
AERO Sun-Times (weekly)
Sustainable Farming (weekly) 13368

Alternative Sources of Energy (ASE), 620 Central Av N, Milaca, MN 56353
T: (612) 983-6892; Fax: (612) 983-6893
Founded: 1971
Gen Secr: Donald Marier
Focus: Energy
Periodicals
Independent Energy: The Business Magazine of the Independent Energy Industry (10 times annually)
Independent Energy: The Business Magazine of the Independent Power Industry (10 times annually) 13369

Alzheimer's Association, 919 N Michigan Av, Ste 1000, Chicago, IL 60611
T: (312) 335-8700; Fax: (312) 335-1110
Founded: 1980; Members: 30000
Pres: Edward Truschke
Focus: Med
Periodical
Alzheimer's Association Newsletter (weekly) 13370

Alzheimer's Disease International (ADI), 70 E Lake St, Chicago, IL 60601
T: (312) 853-3060
Founded: 1984; Members: 56
Gen Secr: Edward Truschke
Focus: Med
Periodical
ADI Global Perspective (weekly) 13371

AMA International, 135 W 50 St, New York, NY 10020
T: (212) 586-8100; Fax: (212) 713-1652
Founded: 1956; Members: 10000
Pres: Domenico A. Fanelli
Focus: Business Admin
Periodicals
Comp Flash (weekly)
Management Review (weekly)
Organizational Dynamics (weekly)
Personnel Magazine (weekly)
Supervision Management (weekly)
Trainer's Workshop (weekly) 13372

Ambulatory Pediatric Association (APA), 6728 Old McLean Village Dr, McLean, VA 22101
T: (703) 556-9222
Founded: 1960; Members: 1200
Gen Secr: M. Degnon
Focus: Pediatrics
Periodicals
Membership Directory (biennial)
Newsletter (3 times annually) 13373

Amdahl Users Group (AUG), c/o G.E. Frickle, 10900 NE Fourth St, Ste 1600, Bellevue, WA 98004
T: (206) 453-0271
Founded: 1976; Members: 350
Gen Secr: G.E. Frickle
Focus: Computer & Info Sci
Periodical
Directory (weekly) 13374

American Academy and Institute of Arts and Letters (AAIAL), 633 W 155 St, New York, NY 10032
T: (212) 368-5900; Fax: (212) 491-4615
Founded: 1898; Members: 250
Gen Secr: Virginia Dajani
Focus: Arts; Lit
Periodicals
Proceedings (weekly)
Yearbook (weekly) 13375

American Academy for Cerebral Palsy and Developmental Medicine (AACPDM), 1910 Byrd Av, Ste 100, POB 11086, Richmond, VA 23230-1086
T: (804) 228-0036; Fax: (804) 282-0090
Founded: 1947; Members: 1800
Gen Secr: John A. Hinckley
Focus: Med; Neurology
Periodicals
Journal of Developmental Medicine and Child Neurology (weekly)
Newsletter (weekly) 13376

American Academy for Jewish Research (AAJR), 3080 Broadway, New York, NY 10027
T: (212) 678-8864; Fax: (212) 678-8947
Founded: 1919; Members: 460
Gen Secr: C. Frost
Focus: Hist; Ethnology; Rel & Theol
Periodicals
Monograph Series
Monograph Studies
Proceedings of the American Academy for Jewish Research (weekly) 13377

American Academy for Plastics Research in Dentistry (AAPRD), 211 South St, Davison, MJ 48423
Founded: 1940; Members: 100
Focus: Dent 13378

American Academy of Actuaries (AAA), 1720 Eye St NW, Washington, DC 20006
T: (202) 223-8196; Fax: (202) 872-1948
Founded: 1965; Members: 10000
Gen Secr: J.J. Murphy
Focus: Insurance
Periodicals
Actuarial Update (weekly)
Contingencies (weekly)
Enrolled Actuaries Report (5 times annually)
. . . . 13379

American Academy of Advertising (AAA), c/o Dr. Robert L. King, School of Business, University of Richmond, Richmond, VA 23173
T: (804) 289-8902; Fax: (804) 289-8878
Founded: 1958; Members: 675
Gen Secr: Dr. Robert L. King
Focus: Advert
Periodicals
AAA Newsletter (weekly)
Journal of Advertising (weekly)
Proceedings of the Conference (weekly) . 13380

American Academy of Allergy and Immunology (AAAI), 611 E Wells St, Milwaukee, WI 53202
T: (414) 272-6071; Fax: (414) 276-3349
Founded: 1943; Members: 4500
Gen Secr: Donald L. McNeil
Focus: Immunology
Periodicals
American Academy of Allergy and Immunology Abstract Book (weekly)
News & Notes (weekly) 13381

American Academy of Arts and Sciences (AAAS), Northon's Woods, 136 Irving St, Cambridge, MA 02138
T: (617) 492-8800; Fax: (617) 576-5050
Founded: 1780; Members: 3100
Gen Secr: Joel Orlen
Focus: Arts; Sci
Periodicals
Bulletin (8 times annually)
Daedalus (weekly) 13382

American Academy of Child and Adolescent Psychiatry (AACAP), 3615 Wisconsin Av NW, Washington, DC 20016
T: (202) 966-7300; Fax: (202) 966-2891
Founded: 1953; Members: 5000
Gen Secr: Virginia Q. Anthony
Focus: Psychiatry
Periodicals
Journal of the AACAP (weekly)
Membership Directory
Newsletter (weekly) 13383

American Academy of Clinical Psychiatrists (AACP), POB 3212, San Diego, CA 92163
T: (619) 298-0538
Founded: 1975; Members: 600
Gen Secr: Alicia A. Munoz
Focus: Psychiatry
Periodicals
Annals of Clinical Psychiatry (weekly)
Clinical Psychiatry Quarterly (weekly) . . 13384

American Academy of Cosmetic Surgery (AACS), 159 E Live Oak Av, Ste 204, Arcadia, CA 91006-5249
T: (312) 527-6713; Fax: (818)447-7880
Founded: 1985; Members: 1250
Gen Secr: Cathey J. McMann
Focus: Surgery
Periodicals
American Journal of Cosmetic Surgery (weekly)
Membership Roster (weekly)
Newsletter (weekly) 13385

American Academy of Crisis Interveners (AACI), c/o Edward S. Rosenbluth, 215 Breckinridge Ln, Ste 102, Louisville, KY 40207
T: (502) 896-0200
Founded: 1977; Members: 200
Pres: Edward S. Rosenbluth
Focus: Psych 13386

American Academy of Dental Electrosurgery (AADE), POB 374, Planetarium Station, New York, NY 10024
T: (212) 595-1925
Founded: 1963; Members: 200
Gen Secr: Maurice J. Oringer
Focus: Surgery; Dent
Periodical
Current Events (weekly) 13387

American Academy of Dental Group Practice (AADGP), 5110 N 40 St, Ste 250, Phoenix, AZ 85018
T: (602) 381-1185; Fax: (602) 381-1093
Founded: 1973; Members: 1575
Focus: Dent
Periodicals
Directory (biennial)
Newsletter (weekly) 13388

American Academy of Dental Practice Administration (AADPA), c/o Kathleen Uebel, 1063 Whippoorwill Ln, Palatine, IL 60067
T: (708) 934-4404
Founded: 1958; Members: 250
Gen Secr: Kathleen Uebel
Focus: Dent
Periodicals
Communicator (weekly)
Essay Tapes (weekly)
Journal of Dental Practice Administration (weekly)
Proceedings (weekly) 13389

American Academy of Dermatology (AAD), 930 N Meacham Rd, Schaumburg, IL 60173-6016
T: (708) 330-9830; Fax: (708) 330-0050
Founded: 1938; Members: 8800
Gen Secr: Bradford W. Claxton
Focus: Derm
Periodicals
Bulletin (weekly)
Journal of the American Academy of Dermatology (weekly) 13390

American Academy of Environmental Engineers (AAEE), 130 Holiday Ct, Annapolis, MD 21401
T: (301) 266-3311; Fax: (301) 266-7653
Founded: 1955; Members: 2600

Gen Secr: William C. Anderson
Focus: Eng
Periodicals
Environmental Engineer (weekly)
Environmental Engineering Selection Guide (weekly) 13391

American Academy of Esthetic Dentistry (AAED), 500 N Michigan Av, Ste 1400, Chicago, IL 60611
T: (312) 661-1700
Founded: 1975; Members: 100
Gen Secr: Sharon Bennett
Focus: Dent
Periodical
Esthetics (3 times annually) 13392

American Academy of Facial Plastic and Reconstructive Surgery (AAFPRS), 1110 Vermont Av NW, Ste 220, Washington, DC 22005
T: (202) 842-4500; Fax: (202) 371-1514
Founded: 1964; Members: 3200
Gen Secr: Stephen C. Duffy
Focus: Surgery
Periodicals
Face Facts (weekly)
Facial Plastic Surgery Today (weekly)
Facial Plastic Times (weekly) 13393

American Academy of Family Physicians (AAFP), 8880 Ward Pkwy, Kansas City, MO 64114
T: (816) 333-9700
Founded: 1947; Members: 66000
Gen Secr: Robert Graham
Focus: Med; Physiology
Periodicals
AAFP Reporter (weekly)
American Family Physician (weekly) . . 13394

American Academy of Fixed Prosthodontics, c/o Dr. Robert S. Staffanou, 3302 Gaston Av, Rm 330, Dallas, TX 75246
T: (214) 828-8370
Founded: 1952; Members: 485
Gen Secr: Dr. Robert S. Staffanou
Focus: Dent
Periodicals
Journal of Prosthetic Dentisty (weekly)
Newsletter (weekly) 13395

American Academy of Forensic Sciences (AAFS), 410 N 21 St, Ste 203, Colorado Springs, CO 80901-0669
T: (719) 636-1100; Fax: (719) 636-1993
Founded: 1948; Members: 3500
Gen Secr: Anne H. Warren
Focus: Law
Periodicals
Journal of Forensic Sciences (weekly)
Membership Directory (weekly)
Newsletter (weekly) 13396

American Academy of Gnathologic Orthopedics (AAGO), POB 548, Richmond, TX 77406-0548
T: (713) 341-5250
Founded: 1970; Members: 400
Gen Secr: L.M. Alderson
Focus: Orthopedics
Periodicals
Journal (weekly)
Membership Roster (weekly) 13397

American Academy of Gold Foil Operators (AAGFO), 17922 Tall Grass Ct, Noblesville, IN 46060
T: (317) 274-8686; Fax: (317) 963-9143
Founded: 1952; Members: 400
Gen Secr: Dr. Ronald K. Harris
Focus: Dent
Periodicals
Journal of Operative Dentistry (weekly)
Roster (weekly) 13398

American Academy of Implant Dentistry (AAID), 211 E Chicago Av, Chicago, IL 60611
T: (312) 335-1550
Founded: 1952; Members: 2300
Gen Secr: Joyce Sigmon
Focus: Dent
Periodicals
Journal of Oral Implantology (weekly)
Newsletter (weekly) 13399

American Academy of Implant Prosthodontics (AAIP), 5555 Peachtree-Dunwoody Rd NE, Ste 140, Atlanta, GA 30342
T: (404) 847-9200; Fax: (404) 257-1201
Founded: 1980; Members: 300
Gen Secr: Donna P. Vaughn
Focus: Dent
Periodicals
Membership Directory
Newsletter (3 times annually) 13400

American Academy of Insurance Medicine (AAIM), c/o William J. Baker, 1 Monarch Pl, Springfield, MA 01133
T: (413) 784-6637
Founded: 1889; Members: 800
Gen Secr: William J. Baker
Focus: Insurance; Med
Periodicals
Journal of Life Insurance Medicine (weekly)
Transactions (weekly) 13401

American Academy of Legal and Industrial Medicine (AALIM), 810 Park Av, New York, NY 10021
T: (212) 570-2137
Founded: 1946; Members: 200
Focus: Med
Periodical
Compensation Medicine Newsletter . . . 13402

American Academy of Matrimonial Lawyers / AAML, 150 N Michigan Av, Ste 2040, Chicago, IL 60601
T: (312) 263-6477
Founded: 1962; Members: 1300
Gen Secr: Lorraine J. West
Focus: Law
Periodicals
Journal of the American Academy of Matrimonial Lawyers (weekly)
Newsletter 13403

American Academy of Maxillofacial Prosthetics (AAMP), c/o Dr. Carl Andres, 135 Bexhill Dr, Carmel, IN 46032
T: (317) 274-5628
Founded: 1953; Members: 148
Gen Secr: Dr. Carl Andres
Focus: Dent
Periodical
Journal of Prosthetic Dentistry (weekly) . 13404

American Academy of Mechanics (AAM), c/o University of California at San Diego, 9500 Gilman Dr, La Jolla, CA 92093-0411
T: (619) 534-2036; Fax: (619) 534-7078
Founded: 1969; Members: 1200
Gen Secr: S. Nemat-Nasser
Focus: Eng
Periodicals
Directory (weekly)
Mechanics (weekly) 13405

American Academy of Medical Administrators (AAMA), 30555 Southfield Rd, Ste 150, Southfield, MI 48076
T: (313) 540-4310
Founded: 1957; Members: 2000
Pres: Thomas R. O'Donovan
Focus: Med
Periodical
Executive (weekly) 13406

American Academy of Microbiology (AAM), 1325 Massachusetts Av NW, Washington, DC 20005
T: (202) 737-3600
Founded: 1955; Members: 1000
Gen Secr: Carol A. Colgan
Focus: Microbio
Periodical
Directory of Fellows of the American Academy of Microbiology (triennial) 13407

American Academy of Neurological Surgery (AANS), c/o Dept of Neurosurgery, Temple University Hospital, 3401 N Broad St, Philadelphia, PA 19140
T: (215) 221-4068
Founded: 1938; Members: 168
Gen Secr: Dr. William Buchheit
Focus: Surgery; Neurology 13408

American Academy of Nursing (AAN), 2420 Pershing Rd, Kansas City, MO 64108
T: (816) 474-5720; Fax: (816) 471-4903
Founded: 1973; Members: 800
Pres: Nancy Fugate Woods
Focus: Adult Educ; Med
Periodical
Nursing Outlook: Journal of the American Academy of Nursing (weekly) 13409

American Academy of Ophthalmology (AAO), 655 Beach St, San Francisco, CA 94109
T: (415) 561-8500; Fax: (415) 561-8533
Founded: 1896; Members: 17000
Gen Secr: Bruce E. Spivey
Focus: Ophthal
Periodicals
Argus (weekly)
Directory (bi-annually)
Ophthalmology (weekly) 13410

American Academy of Optometry (AAO), 4330 East West Hwy, Ste 1117, Bethesda, MD 20817
T: (301) 718-6500; Fax: (301) 656-0989
Founded: 1921; Members: 3850
Gen Secr: David Lewis
Focus: Ophthal
Periodical
Optometry and Vision Science (weekly) . 13411

American Academy of Oral and Maxillofacial Radiology (AAOMR), POB 31162, Aurora, CO 80041
T: (303) 699-0518
Founded: 1949; Members: 500
Gen Secr: Dr. Tom McDavid
Focus: Dent; Radiology
Periodical
Roster of Membership (weekly) 13412

American Academy of Oral Medicine (AAOM), 4143 Mischive, Houston, TX 77025
T: (713) 665-6029
Founded: 1946; Members: 800
Gen Secr: Granvil L. Hays
Focus: Otorhinolaryngology
Periodicals
Journal of Oral Medicine (weekly)
Newsletter (weekly) 13413

American Academy of Orofacial Pain (AAOP), 10 Joplin Ct, Lafayette, CA 94549
T: (510) 945-9298; Fax: (510) 945-9299
Founded: 1975; Members: 220
Gen Secr: Martha Boam
Focus: Dent
Periodical
Journal of Craniomandibular Disorders Head and Neck Pain 13414

American Academy of Orthodontics for the General Practitioner (AAOGP), 3953 N 76 St, Milwaukee, WI 53222
T: (414) 464-7870
Founded: 1959; Members: 250
Gen Secr: Maxine J. Hirmer
Focus: Dent
Periodicals
Academic Calendar (weekly)
International Journal of Orthodontics (weekly) 13415

American Academy of Orthopaedic Surgeons (AAOS), 6300 N River Rd, Rosemont, IL 60018-4226
T: (708) 823-7186; Fax: (708) 823-8125
Founded: 1933; Members: 15100
Gen Secr: Thomas C. Nelson
Focus: Surgery; Orthopedics
Periodical
Bulletin (weekly) 13416

American Academy of Orthotists and Prosthetists (AAOP), 1650 King St, Ste 500, Alexandria, VA 22314
T: (703) 836-7118; Fax: (703) 836-0838
Founded: 1970; Members: 1700
Gen Secr: Dr. Ian R. Horen
Focus: Rehabil
Periodical
Journal of Prosthetics and Orthotics: C.P.O. (weekly) 13417

American Academy of Osteopathy (AAO), 1127 Mount Vernon Rd, POB 750, Newark, OH 43058-0750
T: (614) 366-7911; Fax: (614) 366-0194
Founded: 1937; Members: 3500
Gen Secr: Stephen J. Noone
Focus: Med; Pathology
Periodical
AAO Journal (weekly) 13418

American Academy of Otolaryngic Allergy (AAOA), 8455 Colesville Rd, Ste 745, Silver Spring, MD 20910-9998
T: (301) 588-1800; Fax: (301) 588-2454
Founded: 1941; Members: 1900
Gen Secr: Donald J. Clark
Focus: Immunology; Otorhinolaryngology; Ophthal
Periodicals
Directory (weekly)
Newsletter (weekly)
Transactions (weekly) 13419

American Academy of Pediatric Dentistry (AAPD), 211 E Chicago Av, Ste 1036, Chicago, IL 60611
T: (312) 337-2169
Founded: 1947; Members: 3300
Gen Secr: Dr. John A. Bogert
Focus: Dent
Periodicals
Newsletter (weekly)
Pediatric Dentistry (weekly) 13420

American Academy of Pediatrics (AAP), 141 NW Point Blvd, POB 927, Elk Grove Village, IL 60009-0927
T: (312) 228-5005; Fax: (312) 228-5097
Founded: 1930; Members: 43000
Gen Secr: James E. Strain
Focus: Pediatrics
Periodicals
AAP News (weekly)
Pediatrics (weekly) 13421

American Academy of Periodontology (AAP), 737 N Michigan Av, Ste 800, Chicago, IL 60611-2615
T: (312) 787-5518; Fax: (312) 787-3670
Founded: 1914; Members: 6007
Gen Secr: Alice DeForest
Focus: Dent
Periodicals
Journal of Periodontology (weekly)
Newsletter (weekly)
Roster of Members (weekly) 13422

American Academy of Physical Education (AAPE), c/o Emily Haymes, 7720 Bass Ridge Trail, Tallahassee, FL 32312
T: (904) 893-2788
Founded: 1926; Members: 125
Gen Secr: Emily Haymes
Focus: Sports
Periodical
The Academy Papers (weekly) 13423

American Academy of Physical Medicine and Rehabilitation (AAPMR), 122 S Michigan Av, Ste 1300, Chicago, IL 60603
T: (312) 922-9366; Fax: (312) 922-6754
Founded: 1938; Members: 3200
Gen Secr: Ronald A. Heinrichs
Focus: Med; Rehabil
Periodicals
Academy PM&R News (weekly)
Archives of Physical Medicine and Rehabilitation (weekly) 13424

American Academy of Physician Assistants (AAPA), 950 N Washington St, Alexandria, VA 22314
T: (703) 836-2272; Fax: (703) 684-1924
Founded: 1968; Members: 15000
Gen Secr: Harry A. Bradley
Focus: Physiology
Periodicals
AAPA Bulletin (weekly)
Newsletter (weekly)
PA Job Find (weekly) 13425

American Academy of Physiologic Dentistry (AAPD), c/o William Koperud, 567 S Washington St, Naperville, IL 60540
T: (708) 355-2625
Founded: 1958; Members: 75
Gen Secr: William Kopperud
Focus: Dent 13426

American Academy of Podiatric Sports Medicine (AAPSM), 1729 Glastonberry Rd, Potomac, MD 20854
T: (301) 424-7440
Founded: 1970; Members: 800
Gen Secr: Larry I. Shane
Focus: Orthopedics
Periodicals
Membership Directory (weekly)
Newsletter (weekly) 13427

American Academy of Podiatry Administration (AAPO), 836 Farmington Av, Ste 105, West Hartford, CT 06119
T: (203) 236-2564; Fax: (203) 233-0251
Founded: 1961; Members: 200
Pres: Harvey Lederman
Focus: Public Admin
Periodical
AAPO Newsletter (weekly) 13428

American Academy of Political and Social Science (AAPSS), 3937 Chestnut St, Philadelphia, PA 19104
T: (215) 386-4594; Fax: (215) 386-4630
Founded: 1889; Members: 5000
Pres: Marvin E. Wolfgang
Focus: Poli Sci; Sociology
Periodical
The Annals (weekly) 13429

American Academy of Psychiatrists in Alcoholism and Addictions (AAPAA), POB 376, Greenbelt, MD 20770
T: (301) 220-0951; Fax: (301) 220-0941
Founded: 1985; Members: 1000
Gen Secr: Alice Conde
Focus: Psychiatry
Periodicals
American Journal of Addictions (weekly)
Newsletter 13430

American Academy of Psychiatry and the Law (AAPL), 891 Park Av, Baltimore, MD 21201
T: (301) 539-0379; Fax: (301) 331-1389
Founded: 1969; Members: 1430
Gen Secr: Jonas Rappeport
Focus: Psychiatry; Law
Periodicals
Bulletin of the American Academy of Psychiatry and the Law (weekly)
Newsletter (3 times annually) 13431

American Academy of Psychoanalysis (AAP), 47 E 19 St, New York, NY 10003
T: (212) 475-7980; Fax: (212) 475-8101
Founded: 1956; Members: 775
Gen Secr: Vivian Mendelsohn
Focus: Psychoan
Periodicals
The Academy Forum (weekly)
Journal of the American Academy of Psychoanalysis (weekly) 13432

American Academy of Psychotherapists (AAP), POB 607, Decatur, GA 30031
T: (404) 299-6336
Founded: 1955; Members: 725
Gen Secr: Nancy Hunt
Focus: Therapeutics
Periodicals
AAP Newsletter (weekly)
Directory (biennial)
Voices (weekly) 13433

American Academy of Religion (AAR), c/o James B. Wiggins, Dept of Religion, Syracuse University, 501 Hall of Languages, Syracuse, NY 13244-1170
T: (315) 443-4019; Fax: (315) 443-5390
Founded: 1909; Members: 5500
Gen Secr: James B. Wiggins
Focus: Rel & Theol
Periodicals
Academy Series (4-6 times annually)
Classical Resources Series (weekly)
Critical Review of Books (weekly)
Directory of Position-Holders
Journal (weekly)
Monograph Series (4-6 times annually)
Openings, Job Opportunities for Scholars of Religion (5 times annually)
Religious Studies (5 times annually) . . 13434

American Academy of Restorative Dentistry (AARD), c/o Donald H. Downs, 1235 Lake Plaza Dr, Colorado Springs, CO 80906
T: (719) 576-8840
Founded: 1928; Members: 285
Gen Secr: Donald H. Downs
Focus: Dent
Periodicals
Journal of Prosthetic Dentistry
Roster (weekly) 13435

American Academy of Safety Education (AASE), c/o Robert L. Baldwin, Central Missouri State University, Safety Center, Humphreys Bldg, Warrensburg, MO 64093
T: (816) 543-4281
Founded: 1962; Members: 107
Gen Secr: Robert L. Baldwin
Focus: Safety
Periodical
Ideas: Issues and Readings in Safety (weekly) 13436

American Academy of Sanitarians (AAS), c/o James W. Pees, 829 Brookside Dr, Miami, OK 74354
T: (918) 540-2025
Founded: 1966; Members: 400
Gen Secr: James W. Pees
Focus: Adult Educ; Med
Periodicals
Newsletter (weekly)
Register of Professional Sanitarians
Roster of Diplomates (weekly) 13437

American Academy of Sports Physicians (AASP), 17113 Gledhill St, Northridge, CA 91325
T: (818) 886-7891
Founded: 1979; Members: 150
Gen Secr: Janie Zimmer
Focus: Med; Sports
Periodical
Newsletter (weekly) 13438

American Academy of Teachers of Singing (AATS), c/o William Gephart, 75 Bank St, New York, NY 10014
T: (212) 242-1836
Founded: 1922; Members: 35
Gen Secr: William Gephart
Focus: Music; Educ 13439

American Academy of the History of Dentistry (AAHD), c/o Aletha A. Kowitz, 2400 Lakeview Av, Chicago, IL 60614
T: (312) 440-2642
Founded: 1951; Members: 600
Gen Secr: Aletha A. Kowitz
Focus: Hist; Dent
Periodicals
Bulletin of the History of Dentistry (weekly)
Newsletter of the American Academy of History of Dentistry (weekly) 13440

American Academy of Veterinary Dermatology (AAVD), c/o Dr. Nita Gulbas, Desert Sage Veterinary Clinic, 2249 W Bethany Home Rd, Phoenix, AZ 85015
T: (602) 433-0198
Founded: 1964; Members: 348
Pres: Dr. Nita Gulbas
Focus: Vet Med; Derm
Periodicals
Derm-Dialogue (weekly)
Membership Directory (weekly)
Newsletter (weekly) 13441

American Academy of Veterinary Nutrition (AAVN), c/o David J. Mayberry, 270 Spalding Cir, Athens, GA 30605
T: (404) 972-6941
Founded: 1956; Members: 320
Gen Secr: David J. Mayberry
Focus: Vet Med
Periodical
Membership Directory (weekly) 13442

American Academy on Mental Retardation (AAMR), c/o Jack A. Stark, Dept of Psychiatry, Creighton-Nebraska University, 2205 S Tenth St, Omaha, NE 68108
T: (402) 449-4783
Founded: 1960; Members: 250
Gen Secr: J.A. Stark
Focus: Psych
Periodical
Newsletter (weekly) 13443

American Accounting Association (AAA), 5717 Bessie Dr, Sarasota, FL 34233-2399
T: (813) 921-7747; Fax: (813) 923-4093
Founded: 1916; Members: 10000
Gen Secr: Paul L. Gerhardt
Focus: Business Admin
Periodicals
Accounting Review (weekly)
Newsletter 13444

American Aging Association (AGE), c/o Denham Harman, 600 S 42 St, Omaha, NE 68198-4635
T: (402) 559-4416; Fax: (402) 559-7330
Founded: 1970; Members: 500
Gen Secr: Denham Harman
Focus: Geriatrics
Periodical
AGE (weekly) 13445

American Agricultural Economics Association (AAEA), c/o Iowa State University, 80 Heady Hall, Ames, IA 50011-1070
T: (515) 294-8700; Fax: (515) 294-1234
Founded: 1910; Members: 4500
Gen Secr: Raymond R. Beneke
Focus: Agri; Econ
Periodicals
American Journal of Agricultural Economics (5 times annually)
Choices (weekly)
Newsletter (weekly) 13446

American Agricultural Law Association (AALA), c/o University of Arkansas, Leflar Law Center, Fayetteville, AR 72701
T: (501) 575-7389; Fax: (501) 575-5830
Founded: 1980; Members: 900
Gen Secr: William Babione
Focus: Agri; Law
Periodicals
Agricultural Law Update (weekly)
American Agricultural Law Association Membership Directory (biennial) 13447

American Allergy Association (AAA), POB 7273, Menlo Park, CA 94026
T: (415) 322-1663
Founded: 1978
Pres: Carol Rudoff
Focus: Immunology
Periodical
Living with Allergies (weekly) 13448

American Alliance for Health, Physical Education, Recreation and Dance (AAHPERD), 1900 Association Dr, Reston, VA 22091
T: (703) 476-3400; Fax: (703) 476-9527
Founded: 1885; Members: 42000
Gen Secr: Millie Puccio
Focus: Sports
Periodicals
Health Education (weekly)
Journal of Physical Education, Recreation & Dance (9 times annually)
Research Quarterly for Exercise & Sport (weekly)
Strategies (weekly)
Update (9 times annually) 13449

American Alliance for Theatre and Education (AATE), c/o Theatre Dept, Arizona State University, Tempe, AZ 85287-3411
T: (602) 965-6064; Fax: (602) 965-5351
Founded: 1987; Members: 1500
Gen Secr: Barbara Salisbury Wills
Focus: Perf Arts
Periodical
The O'Neill (weekly) 13450

American Animal Hospital Association (AAHA), POB 150899, Denver, CO 80215-0899
T: (303) 986-2800; Fax: (303) 986-1700
Founded: 1933; Members: 14000
Gen Secr: John W. Albers
Focus: Vet Med
Periodicals
AAHA Trends (weekly)
Journal of the AAHA (weekly)
Proceedings (weekly) 13451

American Anorexia/Bulimia Association (AA/BA), 418 E 76 St, New York, NY 10021
T: (212) 734-1114
Founded: 1978; Members: 2000
Gen Secr: Randy Wirth
Focus: Med
Periodical
Newsletter (weekly) 13452

American Anthropological Association (AAA), 1703 New Hampshire Av NW, Washington, DC 20009
T: (202) 232-8800; Fax: (202) 667-5345
Founded: 1902; Members: 10000
Focus: Anthro
Periodicals
American Anthropologist (weekly)
American Ethnologist (weekly)
Anthropology Newsletter (9 times annually) 13453

American Antiquarian Society (AAS), 185 Salisbury St, Worcester, MA 01609-1634
T: (508) 755-5221
Founded: 1812; Members: 564
Pres: Ellen S. Dunlap
Focus: Hist
Periodical
Proceedings of the American Antiquarian Society (weekly) 13454

American Apitherapy Society (AAS), c/o Dr. Christopher Kim, 252 Broad, Red Bank, NJ 07701
T: (908) 842-5700; Fax: (908) 530-7220
Founded: 1978; Members: 300
Pres: Dr. Christopher Kim
Focus: Therapeutics 13455

American Arbitration Association (AAA), 140 W 51 St, New York, NY 10020
T: (212) 484-4000; Fax: (212) 765-4874
Founded: 1926; Members: 7000
Pres: Robert Coulson
Focus: Law
Periodicals
Arbitration Journal (weekly)
Arbitration Times (weekly)
Newsletter (weekly) 13456

American Architectural Foundation (AAF), 1735 New York Av NW, Washington, DC 20006
T: (202) 626-7500; Fax: (202) 626-7420
Founded: 1942
Pres: Norman L. Koonce
Focus: Archit
Periodical
Forum (weekly) 13457

American Archives Association (ARA), 7979 Old Georgetown Rd, Ste 1000, Bethesda, MD 48823
T: (301) 654-4796
Focus: Archives 13458

American Art Therapy Association (AATA), 1202 Allanson Rd, Mundelein, IL 60060
T: (708) 949-6064; Fax: (708) 566-4580
Founded: 1969; Members: 3600
Gen Secr: Edward J. Stygar
Focus: Therapeutics
Periodicals
Journal (3 times annually)
Newsletter (weekly)
Proceedings of Annual Conference . . . 13459

American Association of Cardiovascular and Pulmonary Rehabilitation (AACVPR), 7611 Elmwood Av, Ste 201, Middleton, WI 53562
T: (608) 831-6989; Fax: (608) 831-5122
Founded: 1985; Members: 2500
Gen Secr: Jane C. Shepard
Focus: Cardiol; Rehabil
Periodical
Journal of Cardiopulmonary Rehabilitation (weekly) 13460

American Assembly (AA), c/o Barnard College, Columbia University, 412 Altschul Hall, New York, NY 10027
T: (212) 854-3456; Fax: (212) 662-3655
Founded: 1950
Pres: Daniel A. Sharp
Focus: Poli Sci
Periodical
Annual Report (weekly) 13461

American Assembly of Collegiate Schools of Business (AACSB), 605 Old Ballas Rd, Ste 220, Saint Louis, MO 63141-7077
T: (314) 872-8481; Fax: (314) 872-8495
Founded: 1916; Members: 900
Gen Secr: William K. Laidlaw
Focus: Adult Educ; Econ
Periodicals
Business and Management Education Funding Alert (5 times annually)
Newsline (weekly) 13462

American Association for Accreditation of Laboratory Animal Care (AAALAC), 11300 Rockville Pike, Ste 1211, Rockville, MD 20852-3035
T: (301) 231-5353; Fax: (301) 231-8282
Founded: 1965; Members: 36
Gen Secr: Albert E. New
Focus: Vet Med
Periodical
Communique 13463

American Association for Adult and Continuing Education (AAACE), 2101 Wilson Blvd, Ste 925, Arlington, VA 22201
T: (703) 522-2234; Fax: (703) 522-2250
Founded: 1982; Members: 4000
Gen Secr: Carolyn Berkowitz
Focus: Adult Educ
Periodicals
AAACE Newsletter (10 times annually)
Adult Education (weekly)
Lifelong Learning (8 times annually)
Membership Directory (weekly) 13464

American Association for Aerosol Research (AAAR), 6500 Glenway Av, Bldg D-7, Cincinnati, OH 452114438
T: (513) 661-3385; Fax: (513) 661-7195
Founded: 1981; Members: 800
Pres: Dr. John Seinfeld
Focus: Chem
Periodicals
Aerosol Science and Technology (8 times annually)
Newsletter (weekly) 13465

American Association for Agricultural Education (AAAE), c/o H. Dean Sutphin, Agricultural Education, Cornell University, Kennedy Hall, Ithaca, NY 14853-5901
T: (607) 255-2197
Founded: 1959; Members: 300
Focus: Agri; Educ
Periodicals
Directory
Journal (weekly)
Summaries of Studies (weekly) 13466

American Association for Applied Linguistics (AAAL), POB 24083, Oklahoma City, OK 73124
T: (405) 843-5113; Fax: (405) 843-4863
Founded: 1977; Members: 700
Gen Secr: Paul Angelis
Focus: Ling
Periodicals
Applied Linguistics (3 times annually)
Newsletter (3 times annually) 13467

American Association for Artificial Intelligence (AAAI), 445 Burgess Dr, Menlo Park, CA 94025
T: (415) 328-3123; Fax: (415) 321-4457
Founded: 1979; Members: 13000
Gen Secr: Carol M. Hamilton
Focus: Computer & Info Sci
Periodicals
AI Magazine (weekly)
Conference Proceedings (weekly)
Tutorial Syllabus (weekly) 13468

American Association for Budget and Program Analysis (AABPA), POB 1157, Falls Church, VA 22041
T: (703) 941-4300; Fax: (703) 941-1535
Founded: 1976; Members: 550
Gen Secr: Christine LaChance
Focus: Finance
Periodicals
Journal of Public Budgeting and Finance (weekly)
Newsletter (weekly) 13469

American Association for Cancer Education (AACE), c/o Dr. Robert M. Chamberlain, Dept of Epidemiology, 1515 Holcombe Blvd, Houston, TX 77030
T: (713) 792-3020; Fax: (713) 792-0807
Founded: 1966; Members: 550
Gen Secr: Dr. Robert M. Chamberlain
Focus: Cell Biol & Cancer Res
Periodical
Directory of Members (weekly) 13470

American Association for Cancer Research (AACR), 620 Chestnut St, Ste 816, Philadelphia, PA 19106-3483
T: (215) 440-9300; Fax: (215) 440-9313
Founded: 1907; Members: 8000
Gen Secr: Margaret Foti
Focus: Cell Biol & Cancer Res
Periodicals
Cancer Research (weekly)
Directory (weekly)
Proceedings (weekly) 13471

American Association for Career Education (AACE), 2900 Amby Pl, Hermosa Beach, CA 90254-2216
T: (310) 376-7378; Fax: (310) 374-1360
Founded: 1980; Members: 900
Pres: Dr. Pat Nellor Wickwire
Focus: Educ
Periodical
Newsletter 13472

American Association for Chinese Studies (AACS), 300 Bricker Hall, Ohio State University, Columbus, OH 43210
T: (614) 292-6681
Founded: 1958; Members: 357
Gen Secr: Prof. Wen-Lang Li
Focus: Ling; Adult Educ
Periodicals
Directory of Members
Journal of Chinese Studies (weekly)
Newsletter (weekly) 13473

American Association for Clinical Chemistry (AACC), 2029 K St NW, Washington, DC 20006
T: (202) 857-0717; Fax: (202) 887-5083
Founded: 1948; Members: 10000
Gen Secr: Richard Flaherty
Focus: Chem; Med
Periodicals
Annual Report (weekly)
Clinical Chemistry Journal (weekly)
Clinical Chemistry News (weekly)
Clinical Chemistry Reference Edition (weekly) 13474

American Association for Correctional Psychology (AACP), c/o Robert Smith, West Virginia Graduate College, Institute, WV 25112
T: (304) 766-1929; Fax: (304) 766-1942
Founded: 1953; Members: 400
Gen Secr: Robert R. Smith
Focus: Psych
Periodicals
Clinical Justice and Behavior (weekly)
Newsletter (weekly) 13475

American Association for Crystal Growth (AACG), c/o Dr. Anthony L. Gentile, POB 3233, Thousand Oaks, CA 91359-0233
T: (805) 492-7047; Fax: (805) 492-4062
Founded: 1969; Members: 700
Gen Secr: Dr. Anthony L. Gentile
Focus: Chem; Geology; Physics
Periodicals
Membership Directory (weekly)
Newsletter (weekly) 13476

American Association for Dental Research (AADR), 1111 14 St NW, Ste 1000, Washington, DC 20005
T: (202) 898-1050; Fax: (202) 789-1033
Founded: 1972; Members: 5000
Gen Secr: John J. Clarkson
Focus: Dent
Periodicals
Around (weekly)
Journal of Dental Research (weekly)
Membership Directory 13477

American Association for Functional Orthodontics (AAFO), 106 S Kent St, Winchester, VA 22601
T: (703) 662-2200

Founded: 1981; Members: 1800
Pres: Dr. Craig C. Stoner
Focus: Dent
Periodical
Functional Orthodontist (weekly) 13478

American Association for Geodetic Surveying (AAGS), 5410 Grosvenor Ln, Bethesda, MD 20814
T: (301) 493-0200; Fax: (301) 493-8245
Members: 1500
Gen Secr: John Lisack
Focus: Surveying 13479

American Association for Geriatric Psychiatry (AAGP), POB 376-A, Greenbelt, MD 20768
T: (301) 220-0952; Fax: (301) 220-0941
Founded: 1978; Members: 1468
Gen Secr: Alice Conde
Focus: Geriatrics; Psychiatry
Periodicals
Membership Directory
Newsletter (weekly) 13480

American Association for Hand Surgery (AAHS), 435 N Michigan Av, Ste 1717, Chicago, IL 60611-4067
T: (312) 644-0828; Fax: (312) 644-8557
Founded: 1970; Members: 1000
Gen Secr: Robert E. Becker
Focus: Surgery
Periodical
Hand Surgery Newsletter (3 times annually) 13481

American Association for Higher Education (AAHE), 1 Dupont Circle, Ste 360, Washington, DC 20036
T: (202) 293-6440; Fax: (202) 293-0073
Founded: 1870; Members: 7800
Pres: Russell Edgerton
Focus: Educ
Periodicals
AAHE Bulletin (weekly)
Change (weekly) 13482

American Association for Laboratory Accreditation (AALA), 656 Quince Orchard Rd, Gaithersburg, MD 20878-1409
T: (301) 670-1377; Fax: (301) 869-1495
Founded: 1978; Members: 260
Pres: John W. Locke
Focus: Chem
Periodicals
Directory of Accredited Laboratories (weekly)
Newsletter (weekly) 13483

American Association for Laboratory Animal Science (AALAS), 70 Timber Creek, Ste 5, Cordova, TN 38018
T: (901) 754-8620; Fax: (901) 753-0046
Founded: 1949; Members: 6000
Gen Secr: Donald W. Keene
Focus: Vet Med
Periodicals
Bulletin (weekly)
Laboratory Animal Science (weekly) . . 13484

American Association for Marriage and Family Therapy (AAMFT), 1100 17 St NW, Washington, DC 20036
T: (202) 452-0109; Fax: (202) 223-2329
Founded: 1942; Members: 19000
Gen Secr: Mark R. Ginsberg
Focus: Therapeutics
Periodicals
Journal (weekly)
Membership Directory (bi-annually)
Newspaper (weekly) 13485

American Association for Medical Transcription (AAMT), POB 576187, Modesto, CA 95357
T: (209) 551-0883; Fax: (209) 551-9317
Founded: 1978; Members: 9000
Gen Secr: Claudia Tessier
Focus: Med
Periodical
Journal (weekly) 13486

American Association for Music Therapy (AAMT), POB 50012, Valley Forge, PA 19484
T: (215) 265-4006; Fax: (215) 265-4006
Founded: 1971; Members: 650
Gen Secr: Katie H. Opher
Focus: Therapeutics
Periodicals
International Newsletter (weekly)
Music Therapy (weekly)
Tuning In (weekly) 13487

American Association for Paralegal Education (AAFPE), POB 40244, Overland, KS 66204
T: (913) 381-4458; Fax: (913) 381-9308
Founded: 1981; Members: 330
Gen Secr: Sandra L. Sabanske
Focus: Educ
Periodicals
Journal of Paralegal Education and Practice (weekly)
The Paralegal Educator (weekly) 13488

American Association for Pediatric Ophthalmology and Strabismus (AAPO&S), POB 193832, San Francisco, CA 94119
T: (415) 561-8505; Fax: (415) 561-8575
Founded: 1974; Members: 570
Gen Secr: Sue A. Brown
Focus: Ophthal
Periodicals
Journal (weekly)
Membership Directory (weekly) 13489

American Association for Rehabilitation Therapy (AART), POB 93, North Little Rock, AR 72115
Founded: 1950; Members: 400
Gen Secr: Clarence English
Focus: Rehabil
Periodicals
Directory (weekly)
Journal (3 times annually)
Newsletter (weekly)
Rehabilitation Bulletin (weekly) 13490

American Association for Social Psychiatry (AASP), 2020 K St NW, Ste 810, Washington, DC 20006
T: (202) 466-6739
Founded: 1971; Members: 500
Gen Secr: Eliot Sorel
Focus: Psychiatry
Periodicals
The American Journal of Social Psychiatry (weekly)
Newsletter (weekly) 13491

American Association for State and Local History (AASLH), 530 Church St, Ste 600, Nashville, TN 37219
T: (615) 255-2971; Fax: (615) 255-2979
Founded: 1940; Members: 6200
Gen Secr: Patricia G. Michael
Focus: Hist
Periodicals
History News (weekly)
History News Dispatch (weekly) 13492

American Association for Study of Neoplastic Diseases (AASND), 1768 S Belvoir Rd, South Euclid, OH 44121
T: (216) 382-0031
Founded: 1929; Members: 361
Gen Secr: Mrs. Robert H. Jackson
Focus: Cell Biol & Cancer Res 13493

American Association for Study of the United States in World Affairs (USWA), 3813 Annandale Rd, Annandale, VA 22003
T: (703) 256-8761
Founded: 1948; Members: 1500
Pres: Prof. Gilbert P. Richardson
Focus: Int'l Relat
Periodical
U.S. in World Affairs Journal (weekly) . 13494

American Association for Textile Technology (AATT), POB 99, Gastonia, NC 28053
T: (704) 824-3522; Fax: (704) 824-5358
Founded: 1933; Members: 500
Gen Secr: Jim H. Conner
Focus: Textiles
Periodicals
Annual Conference Proceedings
Membership Directory
Newsletter (weekly) 13495

American Association for the Advancement of Automotive Medicine, 2340 Des Plaines Av, Ste 106, Des Plaines, IL 60018
T: (708) 390-8927; Fax: (708) 390-9962
Founded: 1957; Members: 650
Gen Secr: Elaine Petrucelli
Focus: Med
Periodical
Accident Analysis and Prevention (weekly) 13496

American Association for the Advancement of Science (AAAS), 1333 H St NW, Washington, DC 20005
T: (202) 326-6400; Fax: (202) 289-4021
Founded: 1848; Members: 135296
Gen Secr: Richard S. Nicholson
Focus: Sci
Periodicals
Handbook (weekly)
Research and Development in Federal Budget (weekly)
Science (weekly)
Science Books and Films (5 times annually)
Science Education News (weekly) . . . 13497

American Association for the Advancement of Slavic Studies (AAASS), c/o Stanford University, 125 Panama St, Stanford, CA 94305
T: (415) 723-9668; Fax: (415) 725-7737
Founded: 1948; Members: 4000
Gen Secr: Dorothy Atkinson
Focus: Sci
Periodicals
American Bibliography of Slavic & East European Studies (weekly)
Directory of Members (triennially)
Directory of Programs in Societ & East European Studies (every 3 years)
Newsletter (weekly)
Slavic Review (weekly) 13498

American Association for the Comparative Study of Law (AACSL), c/o Edward D. Re, U.S. Court of International Trade, 1 Federal Plaza, New York, NY 10007
T: (212) 264-2800
Founded: 1951; Members: 55
Gen Secr: Joseph DeLombardi
Focus: Law
Periodical
The American Journal of Comparative Law (weekly) 13499

American Association for the History of Medicine (AAHM), c/o School of Medicine, Boston University, 80 E Concord St, Boston, MA 02118-2394
T: (617) 638-4328; Fax: (617) 638-4329
Founded: 1925; Members: 1300
Gen Secr: J. Worth Estes
Focus: Med; Cultur Hist
Periodicals
Bulletin of the History of Medicine (weekly)
Membership Directory (biennial)
Newsletter (3 times annually) 13500

American Association for the Study of Headache (AASH), 875 Kings Hwy, Ste 200, West Deptford, NJ 08096
T: (609) 845-0322; Fax: (609) 384-5811
Founded: 1959; Members: 950
Gen Secr: Robert K. Talley
Focus: Med; Neurology
Periodical
Headache (10 times annually) 13501

American Association for the Study of Liver Diseases (AASLD), 6900 Grove Rd, Thorofare, NJ 08086
T: (609) 848-1000; Fax: (609) 848-5274
Founded: 1950; Members: 1450
Gen Secr: Susan Nelson
Focus: Intern Med
Periodicals
Hepatology (weekly)
Journal (weekly)
Newsletter (2-3 times annually) 13502

American Association for Thoracic Surgery (AATS), 13 Elm St, POB 1565, Manchester, MA 01944
T: (508) 526-8330; Fax: (508) 526-4018
Founded: 1917; Members: 900
Gen Secr: William T. Maloney
Focus: Surgery
Periodical
Journal of Thoracic and Cardiovascular Surgery (weekly) 13503

American Association for Vocational Instructional Materials (AAVIM), 745 Gaines School Rd, Athens, GA 30605
T: (706) 543-7557; Fax: (706) 613-6779
Founded: 1949; Members: 8
Gen Secr: George W. Smith
Focus: Adult Educ 13504

American Association for World Health (AAWH), 1129 20 St NW, Ste 400, Washington, DC 20036
T: (202) 466-5883
Founded: 1953; Members: 1200
Gen Secr: Richard L. Wittenberg
Focus: Public Health
Periodicals
Newsletter (weekly)
World Health Magazine (weekly) 13505

American Association of Anatomists (AAA), c/o Dr. Robert D. Yates, Tulane Medical Center, 1430 Tulane Av, New Orleans, LA 70112
T: (504) 584-2727; Fax: (504) 584-1687
Founded: 1888; Members: 2661
Gen Secr: Dr. Robert D. Yates
Focus: Anat
Periodicals
American Journal of Anatomy (weekly)
Anatomical News (weekly)
Anatomical Record (weekly) 13506

American Association of Avian Pathologists (AAAP), c/o New Bolton Center, University of Pennsylvania, Kennett Sq, PA 19348
T: (215) 444-4282; Fax: (215) 444-5387
Founded: 1957; Members: 1016
Gen Secr: Dr. Robert J. Eckroade
Focus: Vet Med; Pathology
Periodical
Avian Diseases (weekly) 13507

American Association of Behavioral Therapists, POB 767156, Roswell, GA 30076-7156
T: (404) 642-4063
Founded: 1987; Members: 955
Gen Secr: Dan J. Allen
Focus: Therapeutics 13508

American Association of Bible Colleges (AABC), POB 1523, Fayetteville, AR 72702
T: (501) 521-8164; Fax: (501) 521-9202
Founded: 1947; Members: 107
Gen Secr: Randall Bell
Focus: Rel & Theol
Periodical
AABC Newsletter (3 times annually) . . 13509

American Association of Blacks in Energy (AABE), 801 Pennsylvania Av NW, Ste 250, Washington, DC 20003
T: (202) 547-9378
Founded: 1977; Members: 800
Pres: Erskine E. Cade
Focus: Energy
Periodicals
Bulletin (weekly)
Energy News (weekly) 13510

American Association of Blood Banks (AABB), 8101 Glenbrook Rd, Bethesda, MD 20814-2749
T: (301) 907-6977; Fax: (301) 907-6895
Founded: 1947; Members: 11500
Gen Secr: Joel M. Solomom
Focus: Public Health
Periodicals
News Briefs (weekly)
Transfusion (weekly) 13511

American Association of Botanical Gardens and Arboreta (AABGA), 786 Church Rd, Wayne, PA 19087
T: (215) 688-1120; Fax: (215) 293-0149
Founded: 1940; Members: 1800
Gen Secr: Susan H. Lathrop
Focus: Botany
Periodicals
Arboretum and Botanical Garden Bulletin (weekly)
Newsletter (weekly)
Proceedings (weekly) 13512

American Association of Bovine Practitioners (AABP), c/o Dr. Harold E. Amstutz, Box 2319, West Lafayette, IN 47906
T: (317) 494-8560; Fax: (317) 494-9353
Founded: 1965; Members: 5000
Gen Secr: Dr. Harold E. Amstutz
Focus: Vet Med
Periodicals
Bovine Practitioner (weekly)
Directory (weekly)
Newsletter (weekly)
Proceeding of Annual Meeting (weekly) . 13513

American Association of Cereal Chemists (AACC), 3340 Pilot Knob Rd, Saint Paul, MN 55121-2097
T: (612) 454-7250; Fax: (612) 454-0766
Founded: 1915; Members: 4000
Gen Secr: Steven C. Nelson
Focus: Chem
Periodicals
Cereal Chemistry (weekly)
Cereal Foods World (weekly) 13514

American Association of Certified Allergists (AACA), 800 E Northwest Hwy, Ste 1080, Palatine, IL 60067
T: (708) 359-3919
Founded: 1968; Members: 510
Gen Secr: Joseph J. Lotharius
Focus: Med 13515

American Association of Certified Orthoptists (AACO), c/o Hermann Eye Center, 6411 Fannin, Houston, TX 77030-1697
T: (713) 797-1777
Founded: 1940; Members: 400
Pres: Patricia Jenkins
Focus: Orthopedics
Periodicals
American Orthoptic Journal (weekly)
Directory (weekly)
The Prism (weekly) 13516

American Association of Chairmen of Departments of Psychiatry (AACDP), c/o Frederick G. Guggenheim, University of Arkansas Medical Sciences Center, 4301 W Markham, Little Rock, AR 72205
T: (501) 686-5483; Fax: (501) 686-8154
Founded: 1967; Members: 136
Gen Secr: Frederick G. Guggenheim
Focus: Psychiatry
Periodical
Membership List 13517

American Association of Christian Schools (AACS), POB 1088, Fairfax, VA 22030
T: (703) 818-7150; Fax: (703) 818-7281
Founded: 1972; Members: 1250
Gen Secr: Dr. Carl Herbster
Focus: Educ
Periodicals
The Administrator (weekly)
Directory (weekly)
Newsletter (weekly) 13518

American Association of Colleges for Teacher Education (AACTE), 1 Dupont Circle NW, Ste 610, Washington, DC 20036
T: (202) 293-2450; Fax: (202) 457-8095
Founded: 1948; Members: 721
Gen Secr: David G. Imig
Focus: Adult Educ; Educ
Periodicals
AACTE Policy Papers
The Journal of Teacher Education . . . 13519

American Association of Colleges of Nursing (AACN), 1 Dupont Circle NW, Ste 530, Washington, DC 20036
T: (202) 463-6930; Fax: (202) 785-8320
Founded: 1969; Members: 414
Gen Secr: Dr. Geraldine Bednash
Focus: Adult Educ; Med
Periodicals
Annual Report (weekly)
Journal of Professional Nursing (weekly)
Newsletter (weekly) 13520

American Association of Colleges of Osteopathic Medicine (AACOM), 6110 Executive Blvd, Rockville, MD 20852
T: (301) 468-0990
Founded: 1898; Members: 15
Gen Secr: Sherry R. Arnstein
Focus: Pathology

Periodicals
Annual Organizational Guide (weekly)
Annual Statistical Report (weekly) . . . 13521

American Association of Colleges of Pharmacy (AACP), 1426 Prince St, Alexandria, VA 22314
T: (703) 739-2330
Founded: 1900; Members: 2100
Gen Secr: Carl E. Trinca
Focus: Adult Educ; Pharmacol
Periodicals
AACP News (weekly)
American Journal of Pharmaceutical Education (weekly)
Pharmacy School Admission Requirements (weekly) 13522

American Association of Colleges of Podiatric Medicine (AACPM), 1350 Piccard Dr, Ste 322, Rockville, MD 2085
T: (301) 990-7400; Fax: (301) 990-2807
Founded: 1932
Pres: Anthony J. McNevin
Focus: Adult Educ; Orthopedics
Periodical
Newsletter (3 times annually) 13523

American Association of Community and Junior Colleges (AACJC), c/o National Center for Higher Education, 1 Dupont Circle NW, Ste 410, Washington, DC 20036-1176
T: (202) 728-0200; Fax: (202) 833-2467
Founded: 1920; Members: 1113
Pres: David Pierce
Focus: Educ
Periodicals
Community, Technical and Junior College Journal (weekly)
Community, Technical and Junior College Times (weekly)
International Update (weekly) 13524

American Association of Community Theatre (AACT), c/o L. Ross Rowland, 8209 N Costa Mesa Dr, Muncie, IN 47303
T: (317) 288-0144
Founded: 1986; Members: 700
Gen Secr: L. Ross Rowland
Focus: Perf Arts
Periodicals
AACT Directory of Community Theatres in the United States
Spotlight (weekly)
Theatre Crafts 13525

American Association of Critical-Care Nurses (AACN), 101 Columbia, Aliso Viejo, CA 92656
T: (714) 362-2000; Fax: (714) 362-2020
Founded: 1969; Members: 77000
Gen Secr: Sarah Sanford
Focus: Med
Periodicals
AACN Clinical Issues in Critical Care Nursing (weekly)
American Journal of Critical Care (weekly)
Critical Care Nurse (weekly) 13526

American Association of Dental Examiners (AADE), 211 E Chigago Av, Ste 844, Chigago, IL 60611
T: (312) 699-7900
Founded: 1883; Members: 850
Gen Secr: Molly S. Nadler
Focus: Dent
Periodicals
Bulletin (3-4 times annually)
Proceedings (weekly) 13527

American Association of Dental Schools (AADS), 1625 Massachusetts Av NW, Washington, DC 20036
T: (202) 667-9433; Fax: (202) 667-0642
Founded: 1923; Members: 3600
Gen Secr: Preston Littleton
Focus: Adult Educ; Dent
Periodicals
Admissions Requirements at U.S. and Canadian Dental Schools (weekly)
Directory of Institutional Members (weekly)
Journal of Dental Education (weekly)
Journal of Public Health Dentistry (weekly) 13528

American Association of Diabetes Educators (AADE), 444 N Michigan Av, Ste 1240, Chicago, IL 60611-3901
T: (312) 644-2233; Fax: (312) 644-4411
Founded: 1974; Members: 7000
Gen Secr: Kate Doyle
Focus: Educ; Diabetes
Periodicals
The Diabetes Educator (weekly)
Newsletter (weekly) 13529

American Association of Electrodiagnostic Medicine (AAEM), 21 Second St SW, Ste 103, Rochester, MN 55902
T: (507) 288-0100
Founded: 1953; Members: 2700
Gen Secr: Ella M. Van Laningham
Focus: Neurology; Rehabil
Periodicals
AAEE Annual Courses
AAEE Annual Meetings
AAEE Case Reports
AAEE Minimonographs 13530

American Association of Endodontists (AAE), 211 E Chicago Av, Ste 1501, Chicago, IL 60611
T: (312) 266-7255; Fax: (312) 266-9867
Founded: 1943; Members: 3950
Gen Secr: Irma S. Kudo
Focus: Dent
Periodicals
Communique (3 times annually)
Journal (weekly)
Membership Roster (weekly) 13531

American Association of Engineering Societies (AAES), 1111 19 St NW, Ste 608, Washington, DC 20036
T: (202) 296-2237; Fax: (202) 296-1151
Founded: 1979; Members: 22
Gen Secr: Mitchell Bradley
Focus: Eng
Periodicals
Engineering and Technology Degrees (weekly)
Engineering and Technology Enrollments (weekly) 13532

American Association of Entrepreneurial Dentists (AAED), 420 Magazine St, Tupelo, MS 38801
T: (601) 842-1036
Founded: 1983
Gen Secr: Dr. Charles E. Moore
Focus: Dent
Periodical
Entrepreneurial News 13533

American Association of Equine Practitioners (AAEP), 4075 Iron Works Pike, Lexington, KY 40511
T: (606) 233-0147; Fax: (606) 233-1968
Founded: 1955; Members: 5000
Gen Secr: Gary L. Carpenter
Focus: Vet Med
Periodicals
Directory (weekly)
Newsletter (weekly)
Proceedings (weekly) 13534

American Association of Feed Microscopists (AAFM), c/o Patricia Ramsey, 3292 Meadowview Rd, Sacramento, CA 95832
T: (916) 427-4997
Founded: 1953; Members: 210
Gen Secr: Patricia Ramsey
Focus: Food
Periodical
Newsletter (weekly) 13535

American Association of Genito-Urinary Surgeons (AAGUS), c/o Baylor College of Medicine, 6560 Fannin, Ste 1004, Houston, TX 77030
T: (713) 798-4001
Founded: 1886; Members: 180
Gen Secr: C. Eugene Carlton
Focus: Surgery; Urology 13536

American Association of Gynecological Laparoscopists (AAGL), 13021 E Florence Av, Santa Fe Springs, CA 90670
T: (310) 946-8774
Founded: 1972; Members: 5200
Pres: Jordan M. Phillips
Focus: Gynecology
Periodical
News-Scope (3 times annually) 13537

American Association of Handwriting Analysts (AAHA), 820 W Maple, Hinsdale, IL 60521
T: (708) 325-2266
Founded: 1962; Members: 400
Pres: Liz Mills
Focus: Graphology
Periodicals
Annals
Directory
Newsletter (weekly) 13538

American Association of Homeopathic Pharmacists (AAHP), POB 24, Norwood, PA 19074
Founded: 1922; Members: 20
Pres: Jack Craig
Focus: Pharmacol 13539

American Association of Hospital Dentists (AAHD), 211 E Chicago Av, Chicago, IL 60611
T: (312) 440-2661; Fax: (312) 440-7494
Founded: 1960; Members: 900
Gen Secr: John S. Rutkauskas
Focus: Dent
Periodicals
Bulletin of Hospital Dental Practice (weekly)
Special Care in Dentistry (weekly) . . . 13540

American Association of Hospital Podiatrists (AAHP), 420 74 St, Brooklyn, NY 11209
T: (718) 836-1017
Founded: 1950; Members: 800
Gen Secr: Dr. Louis J. Arancia
Focus: Orthopedics
Periodicals
Hospital Podiatrist (weekly)
Newsletter (weekly) 13541

American Association of Housing Educators (AAHE), c/o College of Architecture, Texas A & M University, College Station, TX 77843-3137
T: (409) 845-0986; Fax: (409) 845-4491
Founded: 1965; Members: 350
Gen Secr: Paul Woods
Focus: Educ
Periodicals
Housing and Society (3 times annually)
Newsletter (2-3 times annually)
Proceedings (weekly) 13542

American Association of Immunologists (AAI), 9650 Rockville Pike, Bethesda, MD 20814
T: (301) 530-7178; Fax: (301) 571-1816
Founded: 1913; Members: 5500
Gen Secr: Raymond A. Palmer
Focus: Immunology
Periodical
Journal of Immunology (weekly) 13543

American Association of Industrial Veterinarians (AAIV), c/o Caroline & Co, 1015 E Broadway, Columbia, MO 65201
T: (314) 449-3109
Founded: 1954; Members: 450
Gen Secr: Curt Schafer
Focus: Vet Med
Periodicals
AAIV Highlights (3 times annually)
Directory (weekly) 13544

American Association of Laban Movement Analysts (AALMA), c/o Laban/Bartenieff Institute for Movement Studies, 11 E Fourth St, New York, NY 10003
T: (212) 477-4299
Founded: 1976; Members: 400
Focus: Sports; Psych
Periodicals
Directory (weekly)
Newsletter (3 times annually) 13545

American Association of Language Specialists (TAALS), 1000 Connecticut Av NW, Ste 9, Washington, DC 20036
T: (202) 986-1542
Founded: 1957; Members: 200
Pres: Harris Coulter
Focus: Ling
Periodical
Yearbook (weekly) 13546

American Association of Law Libraries (AALL), 53 W Jackson Blvd, Ste 940, Chicago, IL 60604
T: (312) 939-4764; Fax: (312) 431-1097
Founded: 1906; Members: 4800
Gen Secr: Judith Genesen
Focus: Libraries & Bk Sci
Periodicals
Directory of Law Libraries (weekly)
Index to Foreign Legal Periodicals (weekly)
Law Library Journal (weekly) 13547

American Association of Medical Assistants (AAMA), 20 N Wacker Dr, Ste 1575, Chicago, IL 60606-2903
T: (312) 899-1500; Fax: (312) 899-1259
Founded: 1956; Members: 12000
Gen Secr: Donald A. Balasa
Focus: Med
Periodicals
Byline: AAMA (weekly)
The Professional Medical Assistant (weekly) 13548

American Association of Medical Milk Commissions (AAMMC), c/o Paul Fleiss, 1824 N Hillhurst Av, Los Angeles, CA 90027
T: (213) 664-1977
Founded: 1907; Members: 35
Pres: Paul Fleiss
Focus: Med
Periodical
Methods and Standards for the Production of Certified Milk (weekly) 13549

American Association of Medical Society Executives (AAMSE), 515 N State St, Chicago, IL 60610
T: (312) 464-2555; Fax: (312) 464-2467
Founded: 1947; Members: 1000
Gen Secr: Robin Kriegel
Focus: Med
Periodicals
Hotline (weekly)
The Medical Executive (weekly) 13550

American Association of Medico-Legal Consultants (AAMC), The Barclay, Rittenhouse Sq, Philadelphia, PA 19103
T: (215) 545-6363
Founded: 1972; Members: 1200
Pres: Evelyn M. Goldstein
Focus: Law 13551

American Association of Mental Health Professionals in Corrections (AAMHPC), c/o John S. Zil, POB 163359, Sacramento, CA 95816-9359
T: (707) 864-0910; Fax: (707) 864-0910
Founded: 1940; Members: 2000
Pres: John S. Zil
Focus: Psych; Psychiatry
Periodical
Corrective and Social Psychiatry and Journal of Behavior Technology Methods and Therapy (weekly) 13552

American Association of Meta-Science, POB 1182, Huntsville, AL 35807
T: (205) 881-7165
Founded: 1977; Members: 200
Pres: William F. Sowder
Focus: Sci 13553

American Association of Museums (AAM), 1225 Eye St NW, Ste 200, Washington, DC 20005
T: (202) 289-1818; Fax: (202) 289-6578
Founded: 1906; Members: 11205
Gen Secr: Edward H. Able
Focus: Arts
Periodicals
Aviso (weekly)
Museum News (weekly) 13554

American Association of Neurological Surgeons (AANS), 22 S Washington St, Ste 100, Park Ridge, IL 60068
T: (708) 692-9500; Fax: (708) 692-2589
Founded: 1931; Members: 3470
Gen Secr: Carl H. Hauber
Focus: Neurology; Surgery
Periodicals
Bulletin (weekly)
Journal of Neurosurgery (weekly)
Neurosurgical Topics (weekly) 13555

American Association of Neuropathologists (AANP), c/o Jeannette J. Townsend, Dept of Pathology, University of Utah School of Medicine, 50 N Medical Dr, Salt Lake City, UT 84132
T: (901) 581-2507; Fax: (901) 585-3831
Founded: 1924; Members: 660
Gen Secr: Jeannette J. Townsend
Focus: Pathology; Neurology
Periodicals
Journal of Neuropathology and Experimental Neurology (weekly)
Roster of Members (weekly) 13556

American Association of Oral and Maxillofacial Surgeons (AAOMS), 9700 W Bryn Mawr Av, Rosemont, IL 60018-5701
T: (708) 678-6200; Fax: (708) 678-6286
Founded: 1918; Members: 5891
Gen Secr: Bernard J. Degen
Focus: Dent
Periodicals
Forum (weekly)
Journal of Oral and Maxillofacial Surgery (weekly) 13557

American Association of Orthodontists (AAO), 401 N Lindbergh Blvd, Saint Louis, MO 63141-7816
T: (314) 993-1700; Fax: (314) 997-1745
Founded: 1901; Members: 11000
Gen Secr: Ronald S. Moen
Focus: Dent
Periodicals
American Journal of Orthodontics and Dentofacial Orthopedics (weekly)
Membership Directory (bi-annually)
Orthodontic Bulletin (weekly) 13558

American Association of Orthomolecular Medicine (AAOM), 900 N Federal Hwy, Ste 330, Boca Raton, FL 33432
T: (305) 393-6167
Founded: 1971; Members: 310
Gen Secr: Mary Roddy Haggerty
Focus: Psychiatry
Periodical
Journal of Orthomolecular Psychiatry and Medicine (weekly) 13559

American Association of Orthopedic Medicine (AAOrthMed), 5147 Lewiston Rd, Lewiston, NY 14092
T: (716) 284-5777; Fax: (716) 688-9633
Founded: 1982; Members: 300
Gen Secr: Don Fraser
Focus: Orthopedics
Periodicals
AAOM Membership Directory (weekly)
AAOM News (weekly) 13560

American Association of Pathologists (AAP), 9650 Rockville Pike, Bethesda, MD 20814-3993
T: (301) 530-7130; Fax: (301) 571-1879
Founded: 1976; Members: 2000
Gen Secr: Frances A. Pitlick
Focus: Pathology
Periodicals
AAP Newsletter (weekly)
American Journal of Pathology (weekly) . 13561

American Association of Petroleum Geologists (AAPG), Box 979, Tulsa, OK 74101
T: (918) 584-2555; Fax: (918) 584-0469
Founded: 1917; Members: 33000
Gen Secr: Fred A. Dix
Focus: Geology
Periodicals
Bulletin (weekly)
Explorer (weekly)
Geobyte (weekly) 13562

American Association of Philosophy Teachers (AAPT), c/o University of Oklahoma, POB 26901, Oklahoma City, OK 73190
T: (405) 271-2111
Founded: 1979; Members: 300
Gen Secr: Richard A. Wright
Focus: Educ; Philos
Periodical
AAPT News (3 times annually) 13563

American Association of Phonetic Sciences (AAPS), POB 14095, University Station, Gainesville, FL 32604
T: (904) 392-2046; Fax: (904) 382-6170
Founded: 1973; Members: 200

Gen Secr: Dr. W.S. Brown
Focus: Ling
Periodical
Newsletter (weekly) 13564

American Association of Physical Anthropologists (AAPA), c/o Dr. Joyce E. Sirianni, Dept of Anthropology, 380 MFAC, SUNY Buffalo, Buffalo, NY 14261
T: (716) 645-2942; Fax: (716) 645-3808
Founded: 1930; Members: 1500
Gen Secr: Dr. Joyce E. Sirianni
Focus: Anthro
Periodicals
American Journal of Physical Anthropology (weekly)
Career Information Bulletin 13565

American Association of Physicists in Medicine (AAPM), 355 E 45 St, New York, NY 10017
T: (212) 661-9404; Fax: (212) 661-7026
Founded: 1958; Members: 3400
Pres: Faiz Khan; Gen Secr: Elaine P. Osterman
Focus: Physiology
Periodicals
Medical Physics (weekly)
Physics in Medicine & Biology (weekly) . 13566

American Association of Physics Teachers (AAPT), 5112 Berwyn Rd, College Park, MD 20740
T: (301) 345-4200; Fax: (301) 345-1857
Founded: 1930; Members: 11000
Gen Secr: Dr. Bernard V. Khoury
Focus: Educ; Physics
Periodicals
American Journal of Physics (weekly)
The Physics Teacher (9 times annually) 13567

American Association of Plastic Surgeons (AAPS), c/o Stephen H. Miller, 10666 N Torrey Pines Rd, La Jolla, CA 92037
T: (619) 554-9940; Fax: (619) 791-1723
Founded: 1921; Members: 425
Gen Secr: Stephen H. Miller
Focus: Surgery 13568

American Association of Poison Control Centers (AAPCC), c/o Dr. Ted Tong, Arizona Poison and Drug Information Center, Health Sciences Center, Room 3204K, 1501 N Campbell, Tucson, AZ 85725
T: (602) 626-1587; Fax: (602) 626-4063
Founded: 1958; Members: 1000
Gen Secr: Dr. Ted Talibuong
Focus: Toxicology
Periodicals
Annual Report (weekly)
Membership Directory (weekly) 13569

American Association of Presidents of Independent Colleges and Universities (AAPICU), c/o Pepperdine University, 24255 Pacific Coast Hwy, Malibu, CA 90263
T: (310) 456-4451
Founded: 1968; Members: 180
Gen Secr: K. Thomas
Focus: Sci
Periodical
Private Higher Education (weekly) . . . 13570

American Association of Professional Hypnotherapists (AAPH), POB 29, Boones Mill, VA 24065
T: (703) 334-3035
Founded: 1980; Members: 1550
Gen Secr: William S. Brink
Focus: Therapeutics
Periodical
Hypnotherapy Today (weekly) 13571

American Association of Professors of Yiddish (AAPY), NSF 350, Queens College, Flushing, NY 11367
T: (718) 997-3622
Founded: 1974; Members: 300
Gen Secr: Joseph C. Landis
Focus: Ling; Educ
Periodicals
Yiddish (weekly)
Yiddish Studies Newsletter (weekly) . . 13572

American Association of Pro Life Obstetricians and Gynecologists (AAPLOG), 850 Elm Grove Rd, Elm Grove, WI 53122
T: (414) 789-7984; Fax: (414) 782-8788
Founded: 1973; Members: 800
Pres: David V. Foley
Focus: Gynecology
Periodical
Newsletter (weekly) 13573

American Association of Psychiatric Administrators (AAPA), c/o Dr. Dave Davis, 1938 Peachtree Rd NW, Ste 505, Atlanta, GA 30304-1253
T: (404) 355-2914; Fax: (404) 355-2917
Founded: 1960; Members: 300
Pres: Dr. Dave Davis
Focus: Psychiatry
Periodicals
Journal
List of Members (weekly)
Newsletter (weekly) 13574

American Association of Psychiatric Services for Children (AAPSC), 1200C Scottsville Rd, Ste 225, Rochester, NY 14624
T: (716) 235-6910; Fax: (716) 235-0654
Founded: 1948; Members: 165
Gen Secr: Dr. Sydney Koret
Focus: Psychiatry
Periodicals
Journal of Child Psychiatry and Human Development (weekly)
Membership Directory (bi-annually)
Newsletter (weekly) 13575

American Association of Public Health Dentistry (AAPHD), 10619 Jousting Ln, Richmond, VA 23235
T: (804) 272-8344
Founded: 1937; Members: 650
Pres: Jack Dillenberg
Focus: Dent
Periodicals
Communique (weekly)
Journal of Public Health Dentistry (weekly) 13576

American Association of Public Health Physicians (AAPHP), c/o Armand Start, Dept of Family Practice, 777 South Mills St, Madison, WI 53715
T: (608) 263-1326; Fax: (608) 263-5813
Founded: 1954; Members: 200
Gen Secr: Armand Start
Focus: Public Health; Physiology
Periodical
Bulletin 13577

American Association of Railway Surgeons (AARS), POB 503, Daleville, VA 24083
T: (703) 992-2513
Founded: 1888; Members: 500
Gen Secr: Lynne Harris
Focus: Surgery
Periodical
Newsletter (weekly) 13578

American Association of School Librarians (AASL), 50 E Huron St, Chicago, IL 60611
T: (312) 944-6780; Fax: (312) 664-7459
Founded: 1951; Members: 7611
Gen Secr: Ann C. Weeks
Focus: Libraries & Bk Sci
Periodicals
AASL Presidential Hotline (weekly)
School Library Media (weekly) 13579

American Association of School Personnel Administrators (AASPA), 2330 Alhambra Blvd, Sacramento, CA 95817
T: (916) 736-2000; Fax: (916) 736-3322
Founded: 1940; Members: 1700
Gen Secr: H. Salinger
Focus: Educ
Periodicals
Bulletin (weekly)
Membership Roster (weekly)
Report (weekly) 13580

American Association of Sex Educators, Counselors and Therapists (AASECT), 435 N Michigan Av, Ste 1717, Chicago, IL 60611
T: (312) 644-0828; Fax: (312) 644-8557
Founded: 1967; Members: 2600
Gen Secr: Cynthia A. Larson
Focus: Educ
Periodicals
Contemporary Sexuality Newsletter (weekly)
Journal of Sex Education & Therapy (weekly) 13581

American Association of Small Ruminant Practitioners (AASRP), 1675 Ellis Hollow Rd, Ithaca, NY 14850
T: (607) 539-6181; Fax: (607) 539-6181
Founded: 1969; Members: 1200
Gen Secr: Dr. Phyllis Larsen
Focus: Vet Med
Periodical
Wool and Wattles (weekly) 13582

American Association of State Colleges and Universities (AASCU), 1 Dupont Circle NW, Ste 700, Washington, DC 20036-1192
T: (202) 293-7070; Fax: (202) 296-5819
Founded: 1961; Members: 409
Pres: James Appleberry
Focus: Sci
Periodicals
Membership List (weekly)
Memo: To The President (weekly) . . . 13583

American Association of Suicidology (AAS), 2459 S Ash, Denver, CO 80222
T: (303) 692-0985; Fax: (303) 756-3299
Founded: 1967; Members: 1200
Gen Secr: Julie Perlman
Focus: Psychiatry
Periodicals
Newslink (weekly)
Suicide and Life (weekly) 13584

American Association of Teachers of Arabic (AATA), c/o Brigham Young University, 280 HRCB, Provo, UT 84602
T: (801) 378-4684
Founded: 1963; Members: 180
Gen Secr: Dilworth B. Parkinson
Focus: Ling; Educ
Periodicals
Al-Arabiyya (weekly)
Bulletin of the Pan American Health Organization 13585

American Association of Teachers of Esperanto (AATE), c/o Dorothy Holland-Kaupp, 5140 San Lorenzo Dr, Santa Barbara, CA 93111-2521
T: (805) 967-5241
Founded: 1961; Members: 90
Gen Secr: Dorothy Holland-Kaupp
Focus: Ling; Educ
Periodical
Bulteno (weekly) 13586

American Association of Teachers of French (AATF), 57 E Armory Av, Champaign, IL 61820
T: (217) 333-2842; Fax: (217) 333-2842
Founded: 1927; Members: 11000
Gen Secr: Fred M. Jenkins
Focus: Ling; Educ
Periodicals
French Review (weekly)
National Bulletin (weekly) 13587

American Association of Teachers of German (AATG), 112 Haddontowne Ct, Cherry Hill, NJ 08034
T: (609) 795-5553; Fax: (609) 795-9398
Founded: 1926; Members: 7300
Gen Secr: Helene Zimmer-Loew
Focus: Ling; Educ
Periodicals
The German Quarterly (weekly)
Newsletter (5 times annually)
Rundbrief (weekly)
Die Unterrichtspraxis (weekly) 13588

American Association of Teachers of Italian (AATI), c/o Albert N. Mancini, Dept of French and Italian, Ohio State University, Columbus, OH 43210
T: (614) 292-2273
Founded: 1924; Members: 1500
Pres: Albert N. Mancini
Focus: Ling; Educ
Periodicals
Italica (weekly)
Newsletter (weekly) 13589

American Association of Teachers of Spanish and Portuguese (AATSP), POB 6349, Mississippi State, MS 39762
T: (601) 325-2041; Fax: (601) 325-3299
Founded: 1917; Members: 13000
Gen Secr: James R. Chatham
Focus: Ling; Educ
Periodicals
Directory (weekly)
Hispania (weekly) 13590

American Association of Teachers of Turkish (AATT), c/o Dept of Near Eastern Studies, Princeton University, 110 Jones Hall, Princeton, NJ 08544-1008
T: (609) 424-2686
Founded: 1985; Members: 85
Gen Secr: Prof. Erika Gilson
Focus: Educ; Ling
Periodical
AATT Newsletter (weekly) 13591

American Association of Textile Chemists and Colorists (AATCC), POB 12215, Research Triangle Park, NC 27709-2215
T: (919) 549-8141; Fax: (919) 549-8933
Founded: 1921; Members: 7188
Gen Secr: William R. Martin
Focus: Chem; Textiles
Periodicals
AATCC Technical Manual (weekly)
Textile Chemist and Colorist (weekly) . . 13592

American Association of Tissue Banks (AATB), 1350 Beverly Rd, Ste 220-A, McLean, VA 22101
T: (703) 827-9582
Founded: 1976; Members: 800
Gen Secr: Jeanne Mowe
Focus: Surgery
Periodicals
Membership Directory
Newsletter (weekly) 13593

American Association of University Professors (AAUP), 1012 14 St, Ste 500, Washington, DC 20005
T: (202) 737-5900
Founded: 1915; Members: 42000
Gen Secr: Ernst Benjamin
Focus: Adult Educ
Periodicals
Academe (weekly)
Collective Bargaining Newsletter (weekly) 13594

American Association of University Women (AAUW), 1111 16 St NW, Washington, DC 20036
T: (202) 785-7700; Fax: (202) 872-1425
Founded: 1881; Members: 135000
Gen Secr: Anne L. Bryant
Focus: Adult Educ
Periodicals
Action Alert (weekly)
Graduate Woman (9 times annually)
Leader in Action (weekly)
On Campus with Women (weekly) . . . 13595

American Association of University Women Educational Foundation (AAUWEF), 1111 16 St NW, Washington, DC 20036
T: (202) 785-7700; Fax: (202) 872-1425
Founded: 1958
Gen Secr: Anne L. Bryant
Focus: Educ 13596

American Association of Variable Star Observers (AAVSO), 25 Birch St, Cambridge, MA 02138
T: (617) 354-0484; Fax: (617) 354-0665
Founded: 1911; Members: 1300
Gen Secr: Dr. Janet A. Mattei
Focus: Astronomy
Periodicals
AAVSO Circular (weekly)
AAVSO Report (weekly)
Bulletin (weekly)
Journal of the AAVSO (weekly)
Solar Bulletin (weekly) 13597

American Association of Veterinary Anatomists (AAVA), c/o Dr. Paul F. Rumph, Dept of Anatomy and Histology, College of Veterinary Medicine, Auburn University, Auburn, AL 36849
T: (205) 844-6743
Founded: 1949; Members: 250
Gen Secr: Dr. Paul F. Rumph
Focus: Vet Med; Anat
Periodicals
Directory (biennial)
Newsletter (weekly) 13598

American Association of Veterinary Laboratory Diagnosticians (AAVLD), c/o Dr. Harvey Gosser, Veterinary Medical Diagnostic Laboratory, POB 6023, Columbia, MO 65205
T: (314) 882-6811; Fax: (314) 882-1411
Founded: 1957; Members: 700
Gen Secr: Dr. Harvey Gosser
Focus: Vet Med
Periodicals
Journal of Veterinary Diagnostic Investigation (weekly)
Newsletter (weekly) 13599

American Association of Veterinary Parasitologists (AAVP), c/o Dr. S.D. Folz, The Upjohn Company, Dept 9690-190-40, Kalamazoo, MI 49001
T: (616) 385-6523; Fax: (616) 385-6707
Founded: 1956; Members: 450
Gen Secr: Dr. S.D. Folz
Focus: Vet Med
Periodicals
Directory (Biennial)
Newsletter (weekly)
Proceedings (weekly) 13600

American Association of Veterinary State Boards (AAVSB), POB 633, Jefferson City, MD 65102
T: (314) 751-0031; Fax: (314) 751-4176
Founded: 1957; Members: 275
Gen Secr: G.T. Barrows
Focus: Vet Med 13601

American Association of Wildlife Veterinarians (AAWV), c/o Dr. Mike Miller, Colorado Div of Wildlife, 317 W Prospect, Fort Collins, CO 80526-2097
T: (303) 484-2836
Founded: 1979; Members: 350
Gen Secr: Dr. Mike Miller
Focus: Vet Med
Periodicals
Membership Directory
Newsletter (weekly) 13602

American Association of Women Dentists (AAWD), 401 N Michigan Av, Chicago, IL 60611-4267
T: (312) 644-6610; Fax: (312) 245-1084
Founded: 1921; Members: 2000
Gen Secr: Christine Norris
Focus: Dent
Periodicals
Chronicle (weekly)
Newsletter (weekly) 13603

American Association of Women in Community and Junior Colleges (AAWCJC), c/o Middlesex Community College, 100 Training Hill Rd, Middletown, CT 06457
T: (203) 344-3001; Fax: (203) 344-7488
Founded: 1973; Members: 2300
Pres: Leila Gonzalez Sullivan
Focus: Educ
Periodicals
AAWCJC Journal (weekly)
AAWCJC Quarterly (weekly) 13604

American Association of Zoo Keepers (AAZK), c/o Topeka Zoological Park, 635 SW Gage Blvd, Topeka, KS 66606
T: (913) 272-5821; Fax: (913) 272-2539
Founded: 1967; Members: 2700
Pres: Ed Hansen
Focus: Zoology
Periodical
Animal Keeper's Forum (weekly) 13605

American Association of Zoological Parks and Aquariums (AAZPA), c/o Oglebay Park, Rte 88, Wheeling, WV 26003
T: (304) 242-2160; Fax: (304) 242-2283
Founded: 1924; Members: 6200
Gen Secr: Robert O. Wagner
Focus: Zoology

Periodicals
Annual Conference Proceedings (weekly)
Regional Conference Proceedings (weekly)
Zoological Parks & Aquariums in the Americas (bi-annually) 13606

American Association of Zoo Veterinarians (AAZV), c/o Dr. Wilbur Amand, Philadelphia Zoological Garden, 3400 Girard Av, Philadelphia, PA 19104-1196
T: (215) 387-8733; Fax: (215) 387-8733
Founded: 1947; Members: 950
Gen Secr: Dr. Wilbur Amand
Focus: Vet Med
Periodicals
AAZV Newsletter (weekly)
Conference Proceedings (weekly)
Journal of Zoo and Wildlife Medicine (weekly)
Membership Directory (weekly) 13607

American Association on Mental Retardation (AAMR), 1719 Kalorama Rd NW, Washington, DC 200009
T: (202) 387-1968
Founded: 1876; Members: 9500
Gen Secr: M. Doreen Croser
Focus: Psych; Educ Handic; Soc Sci; Med
Periodical
American Journal of Mental Deficiency (weekly) 13608

American Astronautical Society (AAS), 6352 Rolling Mill Pl, Ste 102, Springfield, VA 22152
T: (703) 866-0020; Fax: (703) 866-3526
Founded: 1952; Members: 1600
Gen Secr: Carolyn F. Brown
Focus: Aero
Periodicals
Journal of the Astronautical Sciences (weekly)
Newsletter (weekly) 13609

American Astronomical Society (AAS), c/o Dr. Peter B. Boyce, 1630 Connecticut Av NW, Ste 200, Washington, DC 20009
T: (202) 328-2010; Fax: (202) 234-2560
Founded: 1899; Members: 5000
Gen Secr: Dr. Peter B. Boyce
Focus: Astronomy
Periodicals
Astronomical Journal (weekly)
Astrophysical Journal (3 times annually)
Bulletin (weekly) 13610

American Auditory Society (AAS), 1966 Inwood Rd, Dallas, TX 75235
T: (214) 330-4203; Fax: (214) 905-3022
Founded: 1976; Members: 2688
Gen Secr: Ross J. Roeser
Focus: Otorhinolaryngology
Periodicals
The Bulletin of the AAS (3 times annually)
Ear and Hearing (weekly) 13611

American Automatic Control Council (AACC), c/o EECS, Northwestern University, 2145 Sheridan Rd, Evanston, IL 60208-3118
T: (708) 491-3641; Fax: (708) 491-4455
Founded: 1957; Members: 7
Gen Secr: A.H. Haddad
Focus: Eng
Periodicals
Newsletter (weekly)
Proceedings of the American Control Conference (weekly) 13612

American Aviation Historical Society (AAHS), 2333 Otis St, Santa Ana, CA 92704
T: (714) 549-4818
Founded: 1956; Members: 4332
Pres: Harry Gann
Focus: Aero
Periodicals
Catalog (weekly)
Index
Journal (weekly)
Newsletter (weekly) 13613

American Bach Foundation (ABF), 1211 Potomac St NW, Washington, DC 20007
T: (202) 338-1111
Founded: 1958
Pres: Raissa Tselentis Chadwell
Focus: Music 13614

American Bamboo Society (ABS), 666 Wagnon Rd, Sebastopol, CA 95472
T: (707) 823-5866; Fax: (707) 829-8106
Founded: 1979; Members: 750
Gen Secr: Gerald Bol
Focus: Botany
Periodicals
Journal of the American Bamboo Society (weekly)
Newsletter (weekly) 13615

American Baptist Historical Society (ABHS), 1106 S Goodman St, Rochester, NY 14620
T: (716) 473-1740
Founded: 1853; Members: 270
Gen Secr: James Lynch
Focus: Hist; Rel & Theol
Periodicals
American Baptist Quarterly (weekly)
The Associate 13616

American Bar Association (ABA), 750 N Lake Shore Dr, Chicago, IL 60611
T: (312) 988-5000; Fax: (312) 988-6281
Founded: 1878; Members: 375000
Gen Secr: David J.A. Hayes
Focus: Law
Periodicals
ABA Journal (weekly)
Administrative Law Review (weekly)
Antitrust (3 times annually)
Antitrust Law Journal (weekly)
Barrister (weekly)
The Brief (weekly)
Business Lawyer (weekly)
China Law Reporter (weekly)
Communications Law (weekly)
The Compleat Lawyer (weekly)
Criminal Justice (weekly)
The Entertainment and Sports Lawyer (weekly)
Environmental Law (weekly)
Family Advocate (weekly)
Family Law Quarterly (weekly)
Fidelity and Security News (weekly)
Franchise Law Journal (weekly)
Human Rights (3 times annually)
Intelligence Report (weekly)
International Lawyer (weekly)
The Judge's Journal (weekly)
Jurimetrics: Journal of Law, Science and Technology (weekly)
Juvenile and Child Welfare Law Reporter (weekly)
Labor Lawyer (weekly)
Law Practice Management (8 times annually)
Letter (weekly)
Litigation (weekly)
Probate and Property (weekly)
Public Contract Law Journal (weekly)
Student Lawyer (weekly)
The Tax Lawyer (weekly)
Tort and Insurance Law Journal (weekly)
The Urban Lawyer (weekly) 13617

American Beethoven Society (ABS), c/o Center for Beethoven Studies, San Jose State University, 1 Washington Sq, San Jose, CA 95192-0171
T: (408) 924-4590; Fax: (408) 924-4365
Founded: 1985; Members: 300
Gen Secr: Dr. William Meredith
Focus: Music
Periodical
The Beethoven Newsletter (3 times annually) 13618

American Behavioral Science Association, c/o Prof. Harold Takooshian, 113 W 60 St, Rm 916, New York, NY 10023
T: (212) 636-6393; Fax: (212) 581-1284
Founded: 1987
Gen Secr: Prof. Harold Takooshian
Focus: Behav Sci 13619

American Behcet's Association (ABA), POB 54063, Minneapolis, MN 55454-0063
Founded: 1986
Pres: Susan Sternfels
Focus: Med
Periodical
Newsletter (weekly) 13620

American Biblical Encyclopedia Society (ABES), 24 W Maple Av, Monsey, NY 10952
T: (914) 352-4609
Founded: 1930
Gen Secr: Irving Fredman
Focus: Rel & Theol 13621

American Birding Association (ABA), POB 6599, Colorado Springs, CO 80934
T: (719) 634-7736; Fax: (719) 471-4722
Founded: 1969; Members: 10000
Pres: Allan R. Keith
Focus: Ornithology
Periodicals
Birding (weekly)
Directory 13622

American Black Chiropractors Association (ABCA), 1918 E Grand Blvd, Saint Louis, MO 63107
T: (314) 531-0615
Founded: 1980
Gen Secr: Dr. Bobby Westbrooks
Focus: Med 13623

American Blood Commission (ABC), c/o Paul Mehl, American Hospital Association, 840 N Lakeshore Dr, Chicago, IL 60611
T: (312) 280-6000
Founded: 1975; Members: 32
Gen Secr: Paul Mehl
Focus: Public Health; Hematology
Periodical
Report (weekly) 13624

American Board for Certification in Orthotics and Prosthetics (ABC), 1650 King St, Ste 500, Alexandria, VA 22314
T: (703) 836-7114; Fax: (703) 836-0838
Founded: 1948; Members: 3200
Gen Secr: Dr. Ian R. Horen
Focus: Rehabil
Periodical
Registry of Accredited Facilities and Certified Practitioners (weekly) 13625

American Board of Abdominal Surgery (ABAS), 675 Main St, Melrose, MA 02176
T: (617) 665-6101
Founded: 1957; Members: 1865
Gen Secr: Blaise F. Alfano
Focus: Surgery 13626

American Board of Allergy and Immunology (ABAI), c/o University City Science Center, 3624 Market St, Philadelphia, PA 19104
T: (215) 349-9466
Founded: 1972
Gen Secr: Herbert C. Mansmann
Focus: Immunology 13627

American Board of Anesthesiology (ABA), 100 Constitution Plaza, Hartford, CT 06103
T: (203) 522-9857
Founded: 1938; Members: 12
Gen Secr: Francis P. Hughes
Focus: Anesthetics 13628

American Board of Bioanalysis (ABB), 818 Olive, Ste 918, Saint Louis, MO 63101-1598
T: (314) 241-1445; Fax: (314) 241-1449
Founded: 1968
Gen Secr: David Birenbaum
Focus: Bio; Med
Periodical
Test of the Month (weekly) 13629

American Board of Colon and Rectal Surgery (ABCRS), 20600 Eureka Rd, Ste 713, Taylor, MI 48180
T: (313) 282-9400; Fax: (313) 282-9402
Founded: 1934
Gen Secr: Herand Abcarian
Focus: Surgery
Periodical
Newsletter to Diplomates (weekly) . . . 13630

American Board of Dental Public Health (ABDPH), c/o Stanley Lotzkar, 1321 NW 47 Terrace, Gainesville, FL 32605
T: (904) 378-6301
Founded: 1950; Members: 5
Gen Secr: Stanley Lotzkar
Focus: Dent
Periodical
Newsletter (weekly) 13631

American Board of Dermatology (ABD), c/o Henry Ford Hospital, Detroit, MI 48202
T: (313) 871-8739; Fax: (313) 872-3221
Founded: 1932; Members: 14
Gen Secr: Clarence S. Livingood
Focus: Derm
Periodical
Booklet of Information (weekly) 13632

American Board of Endodontics (ABE), 211 E Chicago Av, Ste 1501, Chicago, IL 60611
T: (312) 266-7310
Pres: Dr. Leif K. Bakland
Focus: Dent
Periodical
Membership Roster (weekly) 13633

American Board of Funeral Service Education (ABFSE), 14 Crestwood Dr, Cumberland Center, ME 04021
T: (207) 879-5715; Fax: (207) 829-4443
Founded: 1946; Members: 48
Gen Secr: Dr. Gordon S. Bigelow
Focus: Educ 13634

American Board of Health Physics (ABHP), 8000 Westpark Dr, Ste 130, McLean, VA 22102
T: (703) 790-1745; Fax: (703) 790-9063
Founded: 1960; Members: 8
Gen Secr: Richard J. Burk
Focus: Public Health 13635

American Board of Internal Medicine (ABIM), 3624 Market St, Philadelphia, PA 19104
T: (215) 243-1500
Founded: 1936; Members: 40
Pres: Harry R. Kimball
Focus: Intern Med
Periodical
Policies and Procedures (weekly) . . . 13636

American Board of Master Educators (ABME), 300 25 Av N, Ste 10, Nashville, TN 37203
T: (615) 327-2984
Founded: 1893; Members: 450
Gen Secr: Dr. Jack Miller
Focus: Educ
Periodical
Master Educator (weekly) 13637

American Board of Medical Toxicology (ABMT), c/o Primary Children's Medical Center, 100 N Medical Dr, Salt Lake City, UT 84113
T: (801) 588-3286; Fax: (801) 588-2435
Founded: 1968; Members: 100
Gen Secr: William Banner
Focus: Toxicology
Periodicals
American Board of Medical Toxicology Journal (weekly)
Veterinary and Human Toxicology (weekly) 13638

American Board of Neurological Surgery (ABNS), 6550 Fannin St, Houston, TX 77030-2722
T: (713) 790-6015
Founded: 1940; Members: 14
Gen Secr: Dr. Edward L. Seljeskog
Focus: Neurology; Surgery
Periodical
Newsletter to Diplomates (weekly) . . . 13639

American Board of Nuclear Medicine (ABNM), 900 Veteran Av, Los Angeles, CA 90024-1786
T: (310) 825-6787; Fax: (310) 825-9433
Founded: 1971; Members: 12
Pres: Joseph F. Ross
Focus: Nucl Med 13640

American Board of Nutrition (ABN), 9650 Rockville Pike, Bethesda, MD 20814
T: (301) 530-7110; Fax: (301) 571-1892
Founded: 1948; Members: 400
Gen Secr: Suzanne Beauchesne
Focus: Food 13641

American Board of Obstetrics and Gynecology (ABOG), 936 N 34 St, Ste 200, Seattle, WA 98103
T: (206) 547-4884
Founded: 1927; Members: 15
Gen Secr: James A. Merrill
Focus: Gynecology
Periodical
Bulletin (weekly) 13642

American Board of Ophthalmology (ABO), 111 Presidential Blvd, Ste 241, Bala Cynwyd, PA 19004
T: (215) 664-1175
Founded: 1916; Members: 17
Gen Secr: William H. Spencer
Focus: Ophthal 13643

American Board of Opticianry (ABO), 10341 Democracy Ln, Fairfax, VA 22030
T: (703) 691-8356; Fax: (703) 691-3929
Founded: 1947
Gen Secr: Nancy Roylance
Focus: Optics
Periodical
Bulletin of Information for Candidates . . 13644

American Board of Oral Pathology (AMBOP), 5401 W Kennedy Blvd, POB 25915, Tampa, FL 33622-5915
T: (813) 286-2444
Founded: 1948; Members: 260
Gen Secr: Clarita Wendrich
Focus: Pathology
Periodical
Oral Surgery, Oral Medicine, Oral Pathology (weekly) 13645

American Board of Orthodontics (ABO), 401 N Lindbergh Blvd, Ste 308, Saint Louis, MO 63141
T: (314) 432-6130
Founded: 1929; Members: 1695
Gen Secr: Dr. George D. Selfridge
Focus: Dent
Periodical
American Board of Orthodontics Directory (weekly) 13646

American Board of Orthopedic Surgery (ABOS), 400 Silver Cedar Ct, Chapel Hill, NC 27514
T: (919) 929-7103
Founded: 1934; Members: 12
Gen Secr: Donald B. Kettelkamp
Focus: Surgery; Orthopedics 13647

American Board of Otolaryngology (ABO), 5615 Kirby Dr, Ste 936, Houston, TX 77005
T: (713) 528-6200
Founded: 1924
Gen Secr: Dr. Robert W. Cantrell
Focus: Otorhinolaryngology
Periodical
Newsletter (weekly) 13648

American Board of Pathology (ABP), Lincoln Center, 5401 W Kennedy Blvd, POB 25915, Tampa, FL 33622
T: (813) 286-2444
Founded: 1936; Members: 12
Gen Secr: William H. Hartmann
Focus: Pathology
Periodicals
The American Board of Pathology (2-3 times annually)
Information Booklet (weekly) 13649

American Board of Pediatric Dentistry (ABPD), 1193 Woodgate Dr, Carmel, IN 46033
T: (317) 573-0877; Fax: (317) 846-7235
Founded: 1940; Members: 831
Gen Secr: James R. Roche
Focus: Dent 13650

American Board of Pediatrics (ABP), 111 Silver Cedar Ct, Chapel Hill, NC 27514
T: (919) 929-0461; Fax: (919) 929-9255
Founded: 1933; Members: 18
Pres: James A. Stockman
Focus: Pediatrics 13651

American Board of Periodontology (ABP), c/o Baltimore College of Dental Surgery, University of Maryland, 666 W Baltimore St, Baltimore, MD 21201
T: (410) 328-2432; Fax: (410) 328-0074
Founded: 1939
Gen Secr: Gerald M. Bowers
Focus: Dent 13652

American Board of Physical Medicine and Rehabilitation (ABPMR), 21 First St SW, Ste 674, Rochester, MN 55902
T: (507) 282-1776
Founded: 1947; Members: 4024

Gen Secr: Dr. Joachim L. Opitz
Focus: Physiology; Rehabil 13653

American Board of Plastic Surgery (ABPS), 1635 Market St, Ste 400, Philadelphia, PA 19103
T: (215) 587-9322
Founded: 1937; Members: 19
Gen Secr: Gilbert P. Gradinger
Focus: Surgery 13654

American Board of Podiatric Orthopedics (ABPO), 108 Orange St, Ste 6, Redlands, CA 92373
T: (714) 798-8910; Fax: (714) 793-7786
Founded: 1975; Members: 617
Gen Secr: Judith A. Baerg
Focus: Orthopedics
Periodicals
Directory of Diplomates (weekly)
Newsletter (weekly) 13655

American Board of Podiatric Surgery (ABPS), 1601 Dolores St, San Francisco, CA 94110-4906
T: (415) 826-3200; Fax: (415) 826-4640
Founded: 1975; Members: 3000
Gen Secr: John L. Bennett
Focus: Surgery; Orthopedics 13656

American Board of Preventive Medicine (ABPM), c/o Stanley R. Mohler, Dept of Community Health, School of Medicine, Wright State University, POB 927, Dayton, OH 45401
T: (513) 278-6915
Founded: 1948; Members: 6000
Gen Secr: Stanley R. Mohler
Focus: Med; Hygiene
Periodical
Bulletin 13657

American Board of Professional Psychology (ABPP), 2100 E Broadway, Ste 313, Columbia, MO 65201
T: (314) 875-1267; Fax: (314) 443-1199
Founded: 1947; Members: 3750
Gen Secr: Nicholas Palo
Focus: Psych
Periodical
Newsletter (weekly) 13658

American Board of Prosthodontics (ABP), c/o Dr. William D. Culpepper, POB 8437, Atlanta, GA 30306
T: (706) 876-2625; Fax: (404) 872-8804
Gen Secr: Dr. William D. Culpepper
Focus: Dent 13659

American Board of Psychiatry and Neurology (ABPN), 500 Lake Cook Rd, Ste 335, Deerfield, IL 60015
T: (708) 945-7900; Fax: (708) 945-1146
Founded: 1934; Members: 36960
Gen Secr: Stephen C. Scheiber
Focus: Neurology; Psychiatry 13660

American Board of Psychological Hypnosis (ABPH), c/o Billie S. Strauss, Humana Hospital, 2929 S Ellis, Chicago, IL 60616
T: (312) 791-3855
Founded: 1959; Members: 350
Gen Secr: Billie S. Strauss
Focus: Psych 13661

American Board of Radiology (ABR), 2301 W Big Beaver Rd, Ste 625, Troy, MI 48084
T: (313) 643-0300; Fax: (313) 643-0353
Founded: 1934; Members: 23
Gen Secr: Kenneth L. Krabbenhoft
Focus: Radiology 13662

American Board of Surgery (ABS), 1617 John F. Kennedy Blvd, Philadelphia, PA 19103
T: (215) 568-4000
Founded: 1937; Members: 28
Gen Secr: W.O. Griffen
Focus: Surgery
Periodical
Booklet of Information (weekly) 13663

American Board of Thoracic Surgery (ABTS), 1 Rotary Center, Ste 803, Evanston, IL 60201
T: (708) 475-1520; Fax: (708) 475-6240
Founded: 1948
Gen Secr: Richard J. Cleveland
Focus: Surgery 13664

American Board of Urology (ABU), 31700 Telegraph Rd, Ste 150, Bingham Farms, MI 48025
T: (313) 646-9720
Founded: 1935
Gen Secr: Alan D. Perlmutter
Focus: Urology 13665

American Board of Vocational Experts (ABVE), 3500 SW Sixth Av, Ste 100, Topeka, KS 66606-28706
T: (913) 232-9937; Fax: (913) 232-8815
Founded: 1979; Members: 380
Gen Secr: Kenneth E. Ogren
Focus: Educ
Periodicals
National Directory of Vocational Experts (weekly)
The Vocational Expert (weekly) 13666

American Boccaccio Association (ABA), c/o Michael Sherberg, Dept of Romance Languages and Literatures, Washington University, 1 Brookings Dr, Campus Box 1077, Saint Louis, MO 63130-4899
T: (314) 935-6628; Fax: (314) 726-3494
Members: 100
Gen Secr: Michael Sherberg
Focus: Lit
Periodical
Boccaccio Newsletter (weekly) 13667

American Boxwood Society (ABS), Box 85, Boyce, VA 22620
T: (703) 939-4646
Founded: 1961; Members: 800
Gen Secr: Katherine D. Ward
Focus: Forestry
Periodical
Boxwood Bulletin (weekly) 13668

American Brain Tumor Association (ABTA), 3725 N Talman Av, Chicago, IL 60618
T: (312) 286-5571; Fax: (312) 549-5561
Founded: 1973; Members: 20000
Gen Secr: Naomi Berkowitz
Focus: Neurology
Periodical
Message Line (weekly) 13669

American Broncho-Esophagological Association (ABEA), c/o Lauren D. Holinger, Dept of Otolaryngology, 2300 Children's Memorial Hospital Plaza, Chicago, IL 60614
T: (312) 880-4457
Founded: 1917; Members: 300
Gen Secr: Lauren D. Holinger
Focus: Otorhinolaryngology
Periodical
Transactions (weekly) 13670

American Bryological and Lichenological Society (ABLS), c/o Robert S. Egan, Dept of Biology, University of Nebraska at Omaha, Omaha, NE 68182-0072
T: (402) 554-2491
Founded: 1898; Members: 506
Gen Secr: Robert S. Egan
Focus: Botany
Periodical
The Bryologist (weekly) 13671

American Bureau for Medical Advancement in China (ABMAC), 2 E 103 St, New York, NY 10029
T: (212) 860-1990; Fax: (212) 860-1994
Founded: 1937; Members: 48
Gen Secr: Hope N.F. Philips
Focus: Med
Periodical
Annual Report (weekly) 13672

American Burn Association (ABA), c/o Andrew M. Munster, Baltimore Regional Burn Center, Francis Scott Key Hospital, 4940 Eastern Av, Baltimore, MD 21224
T: (800) 548-2876; Fax: (410) 550-1165
Members: 3500
Gen Secr: Andrew M. Munster
Focus: Traumatology
Periodicals
Burn Care Services in North America (weekly)
Journal of Burn Care and Rehabilitation (weekly) 13673

American-Canadian Genealogical Society (ACGS), POB 668, Manchester, NH 03105
T: (603) 622-1554
Founded: 1973; Members: 1900
Pres: Pauline Cusson
Focus: Genealogy
Periodical
The Genealogist (weekly) 13674

American Canal Society (ACS), c/o William H. Shank, 809 Rathton Rd, York, PA 17403
T: (717) 843-4035
Founded: 1972; Members: 850
Pres: William H. Shank
Focus: Eng; Hist
Periodical
American Canal Guides (weekly) 13675

American Cancer Society (ACS), 1599 Clifton Rd NE, Atlanta, GA 30329
T: (404) 320-3333; Fax: (404) 325-0230
Founded: 1913
Gen Secr: John R. Seffrin
Focus: Cell Biol & Cancer Res
Periodicals
Annual Report (weekly)
CA-A Cancer Journal for Clinicians (weekly)
Cancer Facts and Figures (weekly)
Cancer News (3 times annually)
Cancer Nursing News (weekly)
World Smoking and Health (3 times annually) 13676

American Carbon Society (ACS), c/o Carbone of America, 215 Stackpole St, Saint Marys, PA 15857
T: (814) 781-8410
Founded: 1957; Members: 500
Gen Secr: Dr. William A. Nystrom
Focus: Physics
Periodical
Extended Abstracts and Proceedings from Biennial Carbon Conferences (bi-annually) 13677

American Cartographic Association (ACA), c/o American Congress on Surveying and Mapping, 5410 Grosvenor Ln, Bethesda, MD 20814
T: (301) 493-0200
Members: 1600
Gen Secr: John Lisack
Focus: Cart 13678

American Catholic Esperanto Society (ACES), POB 626, La Jolla, CA 92038
Founded: 1967; Members: 25
Gen Secr: Alberta S. Casey
Focus: Ling
Periodical
Annual Report of the Representative (weekly) 13679

American Catholic Historical Association (ACHA), c/o The Catholic University of America, Mullen Library Rm 305, Washington, DC 20064
T: (202) 319-5079; Fax: (202) 319-4967
Founded: 1919; Members: 1100
Focus: Hist
Periodical
Catholic Historical Review (weekly) . . . 13680

American Catholic Philosophical Association (ACPA), c/o Catholic University of America, Administration Bldg, Rm 403, 620 Michigan Av NE, Washington, DC 20064
T: (202) 319-5518; Fax: (202) 319-5047
Founded: 1926; Members: 1100
Gen Secr: Dr. Michael Baur
Focus: Philos
Periodical
The New Scholasticism (weekly) 13681

American Celiac Society (ACS), 58 Mussano Ct, West Orange, NJ 07052
T: (201) 325-8837
Founded: 1970; Members: 4000
Gen Secr: Annette Bentley
Focus: Hematology; Gastroenter; Nutrition
Periodical
Newsletter (3 times annually) 13682

American Center for Design (ACD), 233 E Ontario, Ste 500, Chicago, IL 60611
T: (312) 787-2018; Fax: (312) 649-9518
Founded: 1927; Members: 3000
Gen Secr: Jane Dunne
Focus: Graphic & Dec Arts, Design
Periodical
Creative Communicator Magazine (weekly) 13683

American Center for the Alexander Technique (ACAT), 129 W 67 St, New York, NY 10023
T: (212) 799-0468
Founded: 1964; Members: 80
Gen Secr: Kathryn M. Miranda
Focus: Sports
Periodical
ACAT News 13684

American Ceramic Society (ACerS), 735 Ceramic Pl, Westerville, OH 43081
T: (614) 890-4700; Fax: (614) 899-6109
Founded: 1899; Members: 16800
Gen Secr: W. Paul Holbrook
Focus: Materials Sci
Periodicals
Advanced Ceramic Materials (weekly)
Cements Research Progress (weekly)
Ceramic Abstracts (weekly)
Ceramic Bulletin (weekly)
Ceramic Engineering and Science Proceedings (weekly)
Journal of the American Ceramic Society (weekly) 13685

American Cetacean Society (ACS), POB 2639, San Pedro, CA 90731
T: (310) 548-6279
Founded: 1967; Members: 2600
Gen Secr: Paul Gold
Focus: Bio; Water Res
Periodicals
Whale News (weekly)
The Whalewatcher (weekly) 13686

American Chamber of Commerce Researchers Association (ACCRA), c/o American Chamber of Commerce Executives, 4232 King St, Alexandria, VA 22302
T: (703) 998-4172; Fax: (703) 931-5624
Founded: 1960; Members: 450
Gen Secr: Deborah Mason
Focus: Econ
Periodicals
ACCRA Newsletter (weekly)
Inter City Cost of Living Index (weekly)
Membership Directory (weekly) 13687

American Chemical Society (ACS), 1155 16 St NW, Washington, DC 20036
T: (202) 872-4600; Fax: (202) 872-6067
Founded: 1876; Members: 144467
Gen Secr: John K. Crum
Focus: Chem
Periodicals
Accounts of Chemical Research
Analytical Chemistry
Biochemistry (weekly)
Bioconjugate Chemistry
Biotechnology Progress
Chemical & Engineering News
Chemical Research in Toxicology
Chemical Reviews
Chemistry of Materials
CHEMTECH
Energy & Fuels
Environmental Science & Technology
Industrial & Engineering Chemistry Research
Inorganic Chemistry
Journal of Agricultural and Food Chemistry (weekly)
Journal of Chemical and Engineering Data
Journal of Chemical Information and Computer Sciences
Journal of Medicinal Chemistry
The Journal of Organic Chemistry
Journal of Physical and Chemical Reference Data
The Journal of Physical Chemistry
Journal of the American Chemical Society
Langmuir (weekly)
Macromolecules (weekly)
Organometallics 13688

American Chestnut Foundation (ACF), c/o West Virginia University, 401 Brooks Hall, POB 6057, Morgantown, WV 26506-6057
T: (304) 293-3911; Fax: (304) 293-2872
Founded: 1983; Members: 3000
Gen Secr: William MacDonald
Focus: Botany
Periodical
Journal of the American Chestnut Foundation (weekly) 13689

American Chiropractic Association (ACA), 1701 Clarendon Blvd, Arlington, VA 22209
T: (703) 276-8800; Fax: (703) 243-2593
Founded: 1930; Members: 20000
Gen Secr: J. Ray Morgan
Focus: Orthopedics
Periodicals
Journal of Chiropractic (weekly)
Membership Directory (weekly) 13690

American Classical League (ACL), c/o Miami University, Oxford, OH 45056
T: (513) 529-7741; Fax: (513) 529-7742
Founded: 1919; Members: 3500
Pres: Harry Rutledge
Focus: Ling; Educ
Periodical
The Classical Outlook (weekly) 13691

American Cleft Palate-Craniofacial Association (ACPA), 1218 Grandview Av, Pittsburgh, PA 15211
T: (412) 481-1376; Fax: (412) 481-0847
Founded: 1943; Members: 2500
Pres: Alphonse R. Burdi; Gen Secr: Nancy C. Smythe
Focus: Otorhinolaryngology
Periodicals
Cleft Palate Journal (weekly)
Craniofacial: Cleft Palate Bibliography (weekly) 13692

American Clinical and Climatological Association (ACCA), c/o Dept of Medicine, Medical University of South Carolina, 171 Ashley Av, Charleston, SC 29425
T: (803) 792-2914; Fax: (803) 792-5265
Founded: 1884; Members: 375
Gen Secr: Dr. James Allen
Focus: Physical Therapy; Surgery
Periodical
Transactions (weekly) 13693

American Cocoa Research Institute (ACRI), 7900 Westpark Dr, Ste A-320, McLean, VA 22102
T: (703) 790-5011
Founded: 1945
Pres: Richard T. O'Connell
Focus: Coffee, Tea, Cocoa 13694

American College for Continuing Education (ACCE), POB 54, Union, NJ 07083
T: (908) 688-1616
Founded: 1968; Members: 250
Gen Secr: Dr. Jerome J. Erman
Focus: Orthopedics 13695

American College Health Association (ACHA), POB 28937, Baltimore, MD 21240-8937
T: (410) 859-1500; Fax: (410) 859-1510
Founded: 1920; Members: 3900
Gen Secr: Charles H. Hartman
Focus: Public Health
Periodicals
Action Newsletter (weekly)
Journal of American College Health (weekly) 13696

American College of Allergy and Immunology (ACAI), 800 E Northwest Hwy, Ste 1080, Palatine, IL 60067
T: (708) 359-2800; Fax: (708) 359-7367
Founded: 1987; Members: 3400
Gen Secr: James R. Slawny
Focus: Immunology
Periodicals
Immunology and Allergy Practice (weekly)
Membership Directory (every 3 years)
Newsletter (weekly) 13697

American College of Apothecaries (ACA), 205 Daingerfield Rd, Alexandria, VA 22314
T: (703) 684-8603; Fax: (703) 683-3619
Founded: 1940; Members: 1000
Gen Secr: Dr. D.C. Huffman
Focus: Pharmacol
Periodicals
Newsletter (weekly)
The Voice of the Pharmacist (weekly) . 13698

American College of Cardiology (ACC), 9111 Old Georgetown Rd, Bethesda, MD 20814
T: (301) 897-5400; Fax: (301) 897-9745
Founded: 1949; Members: 19617
Gen Secr: William D. Nelligan
Focus: Cardiol
Periodical
Journal of the American College of Cardiology (weekly) 13699

American College of Chest Physicians (ACCP), 3300 Dundee Rd, Northbrook, IL 60062
T: (708) 498-1400
Founded: 1935; Members: 15000
Gen Secr: Alvin Lever
Focus: Cardiol; Pulmon Dis
Periodicals
Bulletin (weekly)
Chest (weekly)
Membership Directory 13700

American College of Chiropractic Orthopedists (ACCO), c/o P.D. Rake, 1030 Broadway, Ste 101, El Centro, CA 92243
T: (619) 352-1452
Founded: 1964; Members: 755
Pres: P.D. Rake
Focus: Orthopedics
Periodical
Ortho-Briefs (weekly) 13701

American College of Clinical Pharmacology (ACCP), 175 Strafford Av, Ste 1, Wayne, PA 19087-3396
T: (215) 687-7711; Fax: (215) 687-7886
Founded: 1969; Members: 1165
Gen Secr: Joelle Bertolet
Focus: Pharmacol
Periodical
Journal of Clinical Pharmacology (weekly) 13702

American College of Cryosurgery (ACC), POB 4014, Schaumburg, IL 60168-4014
T: (708) 330-0230; Fax: (708) 869-4382
Founded: 1977; Members: 336
Pres: Jack D. Waller
Focus: Surgery 13703

American College of Dentists (ACD), 839-J Quince Orchard Blvd, Gaithersburg, MD 20878
T: (301) 986-0555; Fax: (301) 654-3275
Founded: 1920; Members: 5700
Gen Secr: Dr. Gordon H. Rovelstad
Focus: Dent
Periodicals
Journal (weekly)
News and Views (weekly) 13704

American College of Emergency Physicians (ACEP), POB 619911, Dallas, TX 75261-9911
T: (214) 550-0911; Fax: (214) 580-2816
Founded: 1968; Members: 15700
Gen Secr: Dr. Colin C. Rorrie
Focus: Med
Periodicals
ACEP News (weekly)
Annals of Emergency Medicine (weekly) . 13705

American College of Foot Orthopedists (ACFO), 108 Orange St, Ste 6, Redlands, CA 92373
T: (714) 798-8910
Founded: 1949; Members: 675
Gen Secr: Richard H. Baerg
Focus: Orthopedics
Periodical
Newsletter (weekly) 13706

American College of Foot Surgeons (ACFS), 444 N Northwest Hwy, Ste 155, Park Ridge, IL 60068
T: (708) 292-2237; Fax: (708) 292-2022
Founded: 1940; Members: 3000
Gen Secr: Cheryl Beversdorf
Focus: Surgery; Orthopedics
Periodical
Journal of Foot Surgery (weekly) . . . 13707

American College of Gastroenterology (ACG), 4900B S 31 St, Arlington, VA 22206-1656
T: (703) 820-7400; Fax: (703) 931-4520
Founded: 1932; Members: 4000
Gen Secr: Thomas F. Fise
Focus: Gastroenter
Periodical
American Journal of Gastroenterology (weekly) 13708

American College of General Practitioners in Osteopathic Medicine and Surgery (ACGPOMS), 330 E Algonquin, Arlington Heights, IL 60005
T: (312) 228-6090
Founded: 1950; Members: 9000
Gen Secr: George Nyhart
Focus: Med; Surgery; Pathology
Periodicals
Membership Directory (weekly)
Newsletter (weekly)
Official Journal (weekly) 13709

American College of Healthcare Executives (ACHE), 840 N Lake Shore Dr, Ste 1103W, Chicago, IL 60611
T: (312) 943-0544; Fax: (312) 943-3791
Founded: 1933; Members: 23000
Pres: Thomas C. Dolan
Focus: Adult Educ; Med
Periodicals
Healthcare Executive Magazine (weekly)
Hospital & Health Services Administration (weekly) 13710

American College of Heraldry (ACH), Drawer CG, Tuscaloosa, AL 35486-2870
Founded: 1972; Members: 1054
Pres: Dr. David Pittman
Focus: Genealogy
Periodicals
The Armiger's News (weekly)
The Heraldic Register of America (weekly) 13711

American College of International Physicians (ACIP), 711 Second St NE, Ste 200, Washington, DC 20002
T: (202) 544-7498; Fax: (202) 546-7105
Founded: 1975; Members: 1300
Gen Secr: Dale Dirks
Focus: Physics
Periodical
Bulletin (weekly) 13712

American College of Laboratory Animal Medicine (ACLAM), c/o Dept of Comparative Medicine, Milton S. Hershey Medical Center, Pennsylvania State University, POB 850, Hershey, PA 17033
T: (717) 531-8462
Founded: 1957; Members: 360
Focus: Vet Med
Periodicals
Membership Directory (weekly)
Newsletter (weekly) 13713

American College of Medical Group Administrators (ACMGA), 104 Inverness Terrace E, Englewood, CO 80112-5306
T: (303) 397-7869; Fax: (303) 799-1683
Founded: 1956; Members: 1700
Gen Secr: Fred E. Graham
Focus: Med; Dent
Periodical
College Review (weekly) 13714

American College of Mohs' Micrographic Surgery and Cutaneous Oncology (ACMMSCO), POB 4014, Schaumburg, IL 60168-4014
T: (708) 330-0230; Fax: (708) 869-4382
Founded: 1967; Members: 260
Gen Secr: Michael G. Thompson
Focus: Cell Biol & Cancer Res; Surgery
Periodical
Bulletin (3 times annually) 13715

American College of Musicians (ACM), 808 Rio Grande, POB 1807, Austin, TX 78767
T: (512) 478-5775
Founded: 1929; Members: 14867
Pres: Richard Allison
Focus: Music 13716

American College of Neuropsychiatrists (ACN), 28595 Orchard Lake Rd, Ste 200, Farmington Hills, MI 48334
T: (313) 553-9877; Fax: (313) 553-5957
Founded: 1937; Members: 420
Gen Secr: Louis E. Rentz
Focus: Psychiatry; Neurology
Periodicals
Directory (weekly)
Journal (weekly) 13717

American College of Neuropsychopharmacology (ACNP), c/o Vanderbilt University, POB 1823, Station B, Nashville, TN 37235
T: (615) 327-7200
Founded: 1961; Members: 606
Gen Secr: O. Ray
Focus: Pharmacol
Periodicals
Mailings (weekly)
Roster (weekly) 13718

American College of Nuclear Physicians (ACNP), 1101 Connecticut Av NW, Ste 700, Washington, DC 20036
T: (202) 857-1135; Fax: (202) 223-4579
Founded: 1974; Members: 1200
Gen Secr: Carol A. Lively
Focus: Nucl Med; Physiology
Periodicals
Directory (weekly)
Scanner (10 times annually) 13719

American College of Nurse-Midwives (ACNM), 1522 K St NW, Ste 1000, Washington, DC 20005
T: (202) 289-0171; Fax: (202) 289-4395
Founded: 1955; Members: 3000
Gen Secr: Ronald E. Nitzsche
Focus: Adult Educ; Med
Periodicals
ACNM Membership Directory (weekly)
Journal of Nurse-Midwifery (weekly)
Quickening (weekly) 13720

American College of Nutrition (ACN), 722 Robert E. Lee Dr, Wilmington, NC 28412-0927
T: (919) 452-1222; Fax: (919) 452-9130
Founded: 1959; Members: 1010
Gen Secr: Mildred S. Seelig
Focus: Hematology
Periodicals
Journal (weekly)
Newsletter (weekly) 13721

American College of Obstetricians and Gynecologists (ACOG), 409 12 St SW, Washington, DC 20024
T: (202) 638-5577
Founded: 1951; Members: 31000
Gen Secr: Warren H. Pearse
Focus: Gynecology
Periodicals
Newsletter (weekly)
Obstetrics and Gynecology (weekly) . . 13722

American College of Oral and Maxillofacial Surgeons (ACOMS), c/o James E. Bauerle, 1100 NW Loop 410, Ste 500, San Antonio, TX 78213-2266
T: (210) 344-5674; Fax: (210) 344-9754
Founded: 1975; Members: 1500
Gen Secr: James E. Bauerle
Focus: Doc
Periodical
Newsletter (weekly) 13723

American College of Orgonomy (ACO), POB 490, Princeton, NJ 08542
T: (908) 821-1144; Fax: (908) 821-0174
Founded: 1968; Members: 12
Pres: Peter A. Crist
Focus: Physiology
Periodical
Journal of Orgonomy (weekly) 13724

American College of Osteopathic Internists (ACOI), 300 Fifth St NE, Washington, DC 20002
T: (202) 546-0095; Fax: (202) 543-5584
Founded: 1943; Members: 1300
Gen Secr: Brian J. Donadio
Focus: Pathology
Periodicals
Directory (weekly)
Newsletter (weekly) 13725

American College of Osteopathic Obstetricians and Gynecologists (ACOOG), 900 Auburn Rd, Pontiac, MI 48342-3365
T: (313) 332-6360; Fax: (313) 332-4607
Founded: 1934; Members: 565
Gen Secr: J. Polsinelli
Focus: Gynecology
Periodicals
ACOOG Newsletter (weekly)
Directory of Members (weekly) 13726

American College of Osteopathic Pediatricians (ACOP), 172 W State St, Trenton, NJ 08608
T: (609) 393-3350
Members: 350
Gen Secr: Theresa E. Goeke
Focus: Pathology; Pediatrics
Periodicals
Membership Directory (weekly)
Newsletter (weekly) 13727

American College of Osteopathic Surgeons (ACOS), 123 N Henry St, Alexandria, VA 22314-2903
T: (703) 684-0416
Founded: 1927; Members: 1350
Gen Secr: Guy D. Beaumont
Focus: Surgery; Pathology
Periodicals
Membership Directory and By-laws (weekly)
News (weekly) 13728

American College of Physicians (ACP), Independence Mall W, Sixth St at Race, Philadelphia, PA 19106
T: (215) 351-2400; Fax: (215) 351-2448
Founded: 1915; Members: 70000
Gen Secr: John R. Ball
Focus: Physiology
Periodicals
Annals of Internal Medicine (weekly)
Directory
Medical Knowledge Self-Assessment
Observer (weekly) 13729

American College of Podiatric Radiologists (ACPR), 169 Lincoln Rd, Miami Beach, Fl 33139
T: (305) 531-9866
Founded: 1944; Members: 80
Gen Secr: Irving H. Block
Focus: Radiology
Periodicals
Newsletter (2-4 times annually)
Post Convention Reports (1-2 times annually) 13730

American College of Podopediatrics (ACP), 10515 Carnegie Av, Cleveland, OH 44106
T: (216) 231-3300; Fax: (216) 231-6537
Founded: 1977; Members: 200
Gen Secr: Donna Perzeski
Focus: Orthopedics
Periodical
Newsletter (weekly) 13731

American College of Preventive Medicine (ACPM), 1015 15 St NW, Ste 403, Washington, DC 20005
T: (202) 789-0003; Fax: (202) 289-8274
Founded: 1954; Members: 2000
Gen Secr: Hazel K. Keimowitz
Focus: Med; Hygiene
Periodicals
American Journal of Preventive Medicine
Membership Roster
Newsletter (weekly) 13732

American College of Prosthodontists (ACP), 1777 NE Loop 410, Ste 904, San Antonio, TX 78217
T: (512) 829-7236
Founded: 1970; Members: 2057
Gen Secr: Linda Wallenborn
Focus: Dent
Periodical
Newsletter (weekly) 13733

American College of Psychiatrists (ACP), POB 365, Greenbelt, MD 20768
T: (301) 345-3534; Fax: (301) 474-0219
Founded: 1963; Members: 800
Gen Secr: Alice Conde
Focus: Psychiatry
Periodical
ACP Newsletter (weekly) 13734

American College of Psychoanalysts (ACPA), 7081 Colton Blvd, Oakland, CA 94611
T: (510) 339-3723
Founded: 1969; Members: 200
Gen Secr: Angela Clark
Focus: Psychoan
Periodical
Bulletin (3 times annually) 13735

American College of Rheumatology (ACR), 60 Executive Park Dr S, Ste 150, Atlanta, GA 30329
T: (404) 633-3777; Fax: (404) 633-1870
Founded: 1934; Members: 4700
Gen Secr: Mark Andrejeski
Focus: Rheuma
Periodicals
ARA Membership Directory (weekly)
ARA Scientific Program (weekly)
Arthritis and Rheumatism (weekly)
Rheumatism Review (bi-annually) 13736

American College of Sports Medicine (ACSM), POB 1440, Indianapolis, IN 46206-1440
T: (317) 637-9200; Fax: (317) 634-7817
Founded: 1954; Members: 12800
Gen Secr: James R. Whitehead
Focus: Med; Physical Therapy
Periodicals
ACSM Membership Directory (weekly)
Medicine and Science in Sports and Exercise (weekly)
Sports Medicine Bulletin (weekly) . . . 13737

American College of Surgeons (ACS), 55 E Erie St, Chicago, IL 60611
T: (312) 664-4050; Fax: (312) 440-7014
Founded: 1913; Members: 52000
Gen Secr: Paul A. Ebert
Focus: Surgery
Periodicals
Bulletin (weekly)
Surgery, Gynecology and Obstetrics (weekly)
Surgical Forum (weekly)
Yearbook (weekly) 13738

American College of Veterinary Internal Medicine (ACVIM), 620 N Main St, Ste 301, Blacksburg, VA 24060
T: (703) 951-8543; Fax: (703) 951-4268
Founded: 1972; Members: 543
Gen Secr: Dr. Inge E. Pyle
Focus: Vet Med
Periodical
Proceedings of the Annual Veterinary Medical Forums (weekly) 13739

American College of Veterinary Pathologists (ACVP), 875 King Hwy, West Deptford, NJ 08096
T: (609) 848-7784; Fax: (609) 853-0411
Founded: 1948; Members: 921
Gen Secr: Dr. Gary Cockerell
Focus: Vet Med
Periodicals
Membership List (weekly)
Proceedings (weekly)
Veterinary Pathology (weekly) 13740

American College of Veterinary Radiology (ACVR), c/o Dr. M. Bernstein, POB 87, Glencoe, IL 60022
T: (708) 251-5517; Fax: (708) 446-8618
Founded: 1964; Members: 135
Gen Secr: Dr. M. Bernstein
Focus: Radiology; Vet Med
Periodical
Veterinary Radiology (weekly) 13741

American College Testing (ACT), Box 168, Iowa City, IA 52243
T: (319) 337-1000; Fax: (319) 337-1551
Founded: 1959
Pres: Richard L. Ferguson
Focus: Educ
Periodical
ACTivity (weekly) 13742

American Collegiate Retailing Association (ACRA), c/o Dr. Susan S. Fiorito, Florida State University, 318C Sandels, Rm 86B, Tallhassee, FL 32306-2033
T: (904) 644-9883
Founded: 1948; Members: 300
Pres: Dr. Susan S. Fiorito
Focus: Business Admin
Periodicals
Directory of Members (weekly)
Newsletter (weekly) 13743

American Committee for International Conservation (ACIC), c/o Roger E. McManus, Center for Marine Conservation, 1725 DeSales St NW, Ste 500, Washington, DC 20036
T: (202) 429-5609; Fax: (202) 872-0619
Founded: 1930; Members: 25
Gen Secr: Roger E. McManus
Focus: Bio; Ecology 13744

American Committee for the Weizmann Institute of Science (ACWIS), 51 Madison Av, Ste 117, New York, NY 10010
T: (212) 779-2500; Fax: (212) 779-3209
Founded: 1944
Gen Secr: Bernard N. Samers
Focus: Sci
Periodical
Weizmann Now (weekly) 13745

American Committee to Advance the Study of Petroglyphs and Pictographs (ACASPP), c/o Joseph J. Snyder, POB 158, Shepherdstown, WV 25443
T: (304) 876-9431; Fax: (304) 876-9431
Founded: 1979; Members: 237
Gen Secr: Joseph J. Snyder
Focus: Archeol
Periodicals
Membership Directory
Newsletter (weekly)
Rock Art (weekly) 13746

American Comparative Literature Association (ACLA), c/o Larry H. Peer, Comparative Literature Dept, Brigham Young University, Provo, UT 84602
T: (801) 378-5529; Fax: (801) 378-4649
Founded: 1960; Members: 800
Gen Secr: Larry H. Peer
Focus: Lit
Periodicals
Newsletter (weekly)
Yearbook of Comparative and General Literature (weekly) 13747

American Concrete Institute (ACI), POB 19150, Detroit, MI 48219
T: (313) 532-2600; Fax: (313) 538-0655
Founded: 1905; Members: 19500
Gen Secr: George F. Leyh
Focus: Civil Eng
Periodicals
ACI Materials Journal (weekly)
ACI Structural Journal (weekly)
Concrete Abstracts (weekly)
Concrete International: Design & Construction (weekly) 13748

American Conference for Irish Studies (ACIS), c/o English Dept, Indiana University, Fort Wayne, IN 46805
T: (219) 481-6765
Founded: 1959; Members: 1500
Gen Secr: Mary Helen Thuente
Focus: Ethnology
Periodical
Newsletter (3 times annually) 13749

American Conference of Academic Deans (ACAD), 1818 R St NW, Washington, DC 20009
T: (202) 387-3760
Founded: 1944; Members: 600
Gen Secr: Maria-Helena Price
Focus: Adult Educ
Periodical
Proceedings (weekly) 13750

American Conference of Governmental Industrial Hygienists (ACGIH), Bldg D-7, 6500 Glenway Av, Cincinnati, OH 45211-4438
Fax: (513) 661-7881
Founded: 1938; Members: 4600
Focus: Hygiene
Periodical
Applied Industrial Hygiene (weekly) . . . 13751

American Congress of Rehabilitation Medicine (ACRM), 130 S Michigan Av, Ste 1310, Chicago, IL 60603
T: (312) 922-9368
Founded: 1921; Members: 3300
Gen Secr: Ronald A. Henrichs
Focus: Rehabil
Periodicals
Archives of Physical Medicine and Rehabilitation (weekly)
Membership Directory (weekly) 13752

American Congress on Surveying and Mapping (ACSM), 5410 Grosvenor Ln, Bethesda, MD 20814-2122
T: (301) 493-0200; Fax: (301) 493-8245
Founded: 1941; Members: 10500
Gen Secr: John Lisack
Focus: Surveying; Cart
Periodicals
American Cartographer (weekly)
Bulletin (weekly)
News (weekly)
Surveying and Mapping (weekly)
Technical Papers (weekly) 13753

American Conservation Association (ACA), 30 Rockefeller Plaza, Rm 5402, New York, NY 10112
T: (212) 649-5600; Fax: (212) 649-5921
Founded: 1958
Gen Secr: George Lamb
Focus: Ecology 13754

American Consulting Engineers Council (ACEC), 1015 15 St NW, Washington, DC 20005
T: (202) 347-7474; Fax: (202) 347-7474
Founded: 1973; Members: 5200
Gen Secr: Howard M. Messner
Focus: Eng
Periodicals
Interpro (weekly)
The Last Word (weekly)
Membership Directory (weekly) 13755

American Correctional Association (ACA), 8025 Laurel Lakes Ct, Laurel, MD 20707-5075
T: (301) 206-5100; Fax: (301) 206-5061
Founded: 1870; Members: 25000
Gen Secr: James A. Gondles
Focus: Criminology; Educ
Periodicals
Corrections Today (weekly)
On the Line (weekly) 13756

American Correctional Health Services Association (ACHSA), POB 2307, Dayton, OH 45401-2307
T: (513) 223-9630; Fax: (513) 223-6307
Founded: 1975; Members: 1600
Gen Secr: Francine W. Rickenbach
Focus: Public Health
Periodical
Corhealth (weekly) 13757

American Coucil for an Energy Efficient Economy (ACEEE), 1001 Connecticut Av NW, Ste 801, Washington, DC 20036
T: (202) 429-8873; Fax: (202) 429-2248
Founded: 1980
Gen Secr: Howard Geller
Focus: Energy
Periodical
Conference Proceedings (bi-annually) . . 13758

American Council for Construction Education (ACCE), 901 Hudson Ln, Monroe, LA 71201
T: (318) 323-2413; Fax: (318) 323-2413
Founded: 1974; Members: 150
Gen Secr: Daniel E. Dupree
Focus: Educ
Periodical
Annual Report (weekly) 13759

American Council for the Arts (ACA), 1285 Av of the Americas, New York, NY 10019
T: (212) 245-4510; Fax: (212) 245-4513
Founded: 1960; Members: 2500
Pres: Milton Rhodes
Focus: Arts
Periodicals
ACA Update (weekly)
Vantage Point 13760

American Council of Applied Clinical Nutrition (ACACN), POB 509, Florissant, MO 63032
T: (314) 921-3997
Founded: 1974; Members: 500
Pres: Clarence T. Smith
Focus: Hematology 13761

American Council of Independent Laboratories (ACIL), 1629 K St NW, Washington, DC 20006
T: (202) 887-5872; Fax: (202) 887-0021
Founded: 1937; Members: 450
Gen Secr: Joseph F. O'Neil
Focus: Finance; Insurance
Periodicals
Directory (bi-annually)
Newsletter (weekly) 13762

American Council of Learned Societies (ACLS), 228 E 45 St, New York, NY 10017
T: (212) 697-1505
Founded: 1919; Members: 51
Pres: Stanley N. Katz
Focus: Soc Sci
Periodicals
Annual Report (weekly)
Newsletter (weekly) 13763

American Council of the International Institute of Welding (ACIIW), 550 NW LeJeune Rd, POB 351040, Miami, FL 33135
T: (305) 443-9353; Fax: (305) 443-7559
Founded: 1948
Gen Secr: H. Glenn Ziegenfuss
Focus: Metallurgy; Eng 13764

American Council on Consumer Interests (ACCI), 240 Stanley Hall, University of Missouri, Columbia, MO 65211
T: (314) 882-3817; Fax: (314) 884-4807
Founded: 1953; Members: 1600
Gen Secr: Anita Metzen
Focus: Econ
Periodicals
ACCI Conference Proceedings (weekly)
ACCI Newsletter (9 times annually)
Journal of Consumer Interest (weekly) . 13765

American Council on Education (ACE), 1 Dupont Circle NW, Ste 800, Washington, DC 20036
T: (202) 939-9300; Fax: (202) 833-4760
Founded: 1918; Members: 1853
Pres: Robert H. Atwell
Focus: Educ
Periodicals
Educational Record (weekly)
Higher Education – National Affairs (weekly) 13766

American Council on Pharmaceutical Education (ACPE), 311 W Superior St, Ste 512, Chicago, IL 60610
T: (312) 664-3575; Fax: (312) 664-4652
Founded: 1932; Members: 10
Gen Secr: Daniel A. Nona
Focus: Adult Educ; Pharmacol
Periodicals
Colleges and Schools of Pharmacy, Accredited Professional Degree Programs (weekly)
Continuing Pharmaceutical Education, Approved Providers (weekly) 13767

American Council on Rural Special Education (ACRES), c/o National Rural Development Institute, Western Washington University, 359 Miller Hall, Bellingham, WA 98225
T: (801) 650-5659; Fax: (209) 647-4845
Members: 1000
Focus: Educ
Periodicals
ACRES Membership Newsletter (weekly)
Rural Special Education Quarterly (weekly) 13768

American Council on Schools and Colleges, c/o Dr. Fredrick R. O'Keefe, 13014 N Dale Mabry, Ste 270, Tampa, FL 33618-2814
T: (813) 264-6700
Founded: 1927; Members: 18
Gen Secr: Dr. Fredrick R. O'Keefe
Focus: Educ 13769

American Council on the Teaching of Foreign Languages (ACTFL), 6 Executive Plaza, Yonkers, NY 10701-6801
T: (914) 963-8830; Fax: (914) 963-1275
Founded: 1967; Members: 7000
Gen Secr: C. Edward Scebold
Focus: Ling; Educ
Periodicals
ACTFL Foreign Language Education Series (weekly)
Foreign Language Annals (weekly) . . . 13770

American Counseling Association (ACA), 5999 Stevenson Av, Alexandria, VA 22304-3300
T: (703) 823-9800; Fax: (703) 823-0252
Founded: 1952; Members: 60000
Gen Secr: Dr. Theodore P. Remley
Focus: Educ
Periodicals
The Career Development (weekly)
Counseling and Values (3 times annually)
Counselor Education and Supervision (weekly)
Elementary School Guidance and Counseling (weekly)
Guidepost (18 times annually)
Journal for Specialists in Group Work (weekly)
The Journal of College Student Personnel (weekly)
The Journal of Counseling & Development (weekly)
Journal of Employment Counseling (weekly)
The Journal of Humanistic Education and Development (weekly)
Journal of Multicultural Counseling & Development Guidance (weekly)
Journal of Offender Counseling (weekly)
Measurement and Evaluation in Counseling & Development (weekly)
Rehabilitation Counseling Bulletin (weekly)
The School Counselor (5 times annually) 13771

American Criminal Justice Association – Lambda Alpha Epsilon (ACJA-LAE), POB 61047, Sacramento, CA 95860
T: (916) 484-6553; Fax: (916) 488-4757
Founded: 1937; Members: 4000
Gen Secr: Karen K. Campbell
Focus: Criminology; Law
Periodical
Journal of the American Criminal Justice Association (weekly) 13772

American Cryonics Society (ACS), POB 1509, Cupertino, CA 95015
T: (408) 734-4200; Fax: (408) 734-4441
Founded: 1969; Members: 204
Gen Secr: Jim Yount
Focus: Med
Periodicals
American Cryonics (weekly)
American Cryonics News (weekly) . . . 13773

American Crystallographic Association (ACA), c/o William L. Duax, POB 96, Ellicott Station, Buffalo, NY 14205-0096
T: (716) 856-9600; Fax: (716) 852-4846
Founded: 1949; Members: 2000
Gen Secr: William L. Duax
Focus: Geology; Chem; Physics; Biochem
Periodicals
ACA Monographs (weekly)
ACA Newsletter (weekly)
ACA Program and Abstracts (weekly)
ACA Transactions (weekly) 13774

American Dairy Science Association (ADSA), c/o Carl D. Johnson, 309 W Clark St, Champaign, IL 61820
T: (217) 356-3182
Founded: 1906; Members: 3200
Gen Secr: Carl D. Johnson
Focus: Food
Periodical
Journal of Dairy Science (weekly) . . . 13775

American Dance Therapy Association (ADTA), 2000 Century Plaza, Ste 108, Columbia, MD 21044
T: (301) 997-4040
Founded: 1966; Members: 1200
Pres: Susan Kleinman
Focus: Therapeutics
Periodicals
American Journal of Dance Therapy (weekly)
Newsletter (weekly) 13776

American Deafness and Rehabilitation Association (ADARA), POB 251554, Little Rock, AR 72225
T: (501) 663-7074; Fax: (501) 663-0336
Founded: 1966; Members: 1000
Pres: Greg Long
Focus: Rehabil
Periodicals
Journal of Rehabilitation of the Deaf (weekly)
Newsletter (weekly) 13777

American Dental Assistants Association (ADAA), 919 N Michigan Av, Ste 3400, Chicago, IL 60611
T: (312) 664-3327; Fax: (312) 664-5288
Founded: 1923; Members: 15000
Gen Secr: Lawrence H. Sepin
Focus: Dent
Periodical
The Dental Assistant (weekly) 13778

American Dental Association (ADA), 211 E Chicago Av, Chicago, IL 60611
T: (312) 440-2500
Founded: 1859; Members: 140000
Gen Secr: Thomas J. Ginley
Focus: Dent
Periodicals
Dental Abstracts (weekly)
Directory (weekly)
Index to Dental Literature (weekly)
Journal (weekly)
News (weekly)
Special Care in Dentistry (weekly) . . . 13779

American Dental Hygienists' Association (ADHA), 444 N Michigan Av, Ste 3400, Chicago, IL 60611
T: (312) 440-8929; Fax: (312) 440-8929
Founded: 1923; Members: 30000
Gen Secr: Kathleen Bell
Focus: Hygiene; Dent
Periodicals
Access (weekly)
Dental Hygiene (weekly) 13780

American Dental Society of Anesthesiology (ADSA), 211 E Chicago Av, Ste 948, Chicago, IL 60611
T: (312) 664-8270; Fax: (312) 642-9713
Founded: 1953; Members: 3200
Gen Secr: Peter C. Goulding
Focus: Anesthetics; Dent
Periodicals
ADSA Directory (weekly)
Anesthesia Progress (weekly)
Pulse (weekly) 13781

American Dermatological Association (ADA), c/o Dept of Dermatology, University Hospitals BT 2045-1, Iowa City, IA 52242
T: (319) 356-2274
Founded: 1876; Members: 350
Gen Secr: John S. Strauss
Focus: Derm 13782

American Dermatologic Society of Allergy and Immunology (ADSAI), c/o Dr. R.S. Rogers, Dept of Dermatology, Mayo Clinic, Rochester, MN 55905
T: (507) 284-2555; Fax: (507) 284-2072
Founded: 1974; Members: 150
Gen Secr: Dr. R.S. Rogers
Focus: Immunology 13783

American Design Drafting Association (ADDA), POB 799, Rockville, MD 20848-0799
T: (301) 460-6875; Fax: (301) 460-8591
Founded: 1959; Members: 2000
Gen Secr: Rachel H. Howard
Focus: Graphic & Dec Arts, Design
Periodical
Design and Drafting News (weekly) . . 13784

American Diabetes Association (ADA), 1660 Duke St, POB 25757, Alexandria, VA 22314
T: (703) 549-1500; Fax: (703) 836-7439
Founded: 1940; Members: 280000
Gen Secr: John H. Graham
Focus: Diabetes
Periodicals
Clinical Diabetes (weekly)
Diabetes (weekly)
Diabetes Care (weekly)
Forecast (weekly) 13785

American Dialect Society (ADS), c/o Allan A. Metcalf, Dept of English, MacMurray College, 1611 N Kent St, Jacksonville, IL 62650
T: (217) 479-7049
Founded: 1889; Members: 750
Gen Secr: Allan A. Metcalf
Focus: Ling

Periodicals
American Speech (weekly)
Newsletter (3 times annually)
Publication of the ADS 13786

American Dietetic Association (ADA), 216 W Jackson Blvd, Ste 800, Chicago, IL 60606
T: (312) 899-0040; Fax: (312) 899-1979
Founded: 1917; Members: 60000
Gen Secr: Beverly Bajus
Focus: Nutrition
Periodical
Journal (weekly) 13787

American Diopter and Decibel Society (ADDS), 3518 Fifth Av, Pittsburgh, PA 15213
T: (412) 647-2227; Fax: (412) 647-7964
Founded: 1960; Members: 200
Gen Secr: Albert W. Biglan
Focus: Ophthal
Periodicals
Directory (weekly)
Seminar Transactions of Meetings (bi-annually) 13788

American Economic Association (AEA), 2014 Broadway, Ste 305, Nashville, TN 37203-2418
T: (615) 322-2595; Fax: (615) 343-7590
Founded: 1885; Members: 26000
Gen Secr: Elton Hinshaw
Focus: Econ
Periodicals
American Economic Review (weekly)
Journal of Economic Literature (weekly)
Papers and Proceeding (weekly) 13789

American Educational Research Association (AERA), 1230 17 St NW, Washington, DC 20036
T: (202) 223-9485; Fax: (202) 775-1824
Founded: 1915; Members: 19000
Gen Secr: William J. Russell
Focus: Educ
Periodicals
American Educational Research Journal (weekly)
Educational Evaluation and Policy Analysis (weekly)
Educational Researcher (9 times annually)
Journal of Educational Statistics (weekly)
Review of Educational Research (weekly)
Review of Educational Research (weekly) 13790

American Educational Studies Association (AESA), c/o Dr. Kathryn Borman, Graduate Studies and Research, University of Cincinnati, Cincinnati, OH 45221
T: (513) 556-2256
Founded: 1968; Members: 650
Gen Secr: Dr. Kathryn Borman
Focus: Depth Psych
Periodicals
Educational Studies (weekly)
Newsletter (weekly) 13791

American Education Association (AEA), POB 463, Center Moriches, NY 11934
Founded: 1938
Pres: Rudolph P. Blaum
Focus: Educ
Periodical
Newsletter 13792

American Electroencephalographic Society (AEEGS), 1 Regency Dr, POB 30, Bloomfield, CT 06002
T: (203) 243-3977
Founded: 1946; Members: 1320
Gen Secr: J.T. Coleman
Focus: Neurology
Periodical
Journal of Clinical Neurophysiology (weekly) 13793

American Electrology Association (AEA), 106 Oak Ridge Rd, Trumbull, CT 06611
T: (203) 374-6667
Founded: 1958; Members: 2000
Pres: Teresa E. Petricca
Focus: Med
Periodicals
Electrolysis World (weekly)
Journal of the American Electroly Association (weekly)
Roster (weekly) 13794

American Electronics Association (AEA), 5201 Great America Pkwy, Ste 520, Santa Clara, CA 95054
T: (408) 987-4200; Fax: (408) 970-8565
Founded: 1943; Members: 3500
Pres: J. Richard Iverson
Focus: Electronic Eng
Periodicals
Directory (weekly)
Update (weekly) 13795

American Electroplaters and Surface Finishers Society (AESFS), 12644 Research Pkwy, Orlando, FL 32826
T: (407) 281-6441; Fax: (407) 281-6446
Founded: 1909; Members: 10000
Gen Secr: J. Howard Schumacher
Focus: Metallurgy
Periodical
Plating and Surface Finishing (weekly) . 13796

American Endodontic Society (AES), 1440 N Harbor Blvd, Ste 719, Fullerton, CA 92635
T: (714) 870-5590
Founded: 1969; Members: 10000
Gen Secr: Dr. Ramon Werts
Focus: Dent
Periodicals
Hotline
Newsletter (weekly) 13797

American Engineering Model Society (AEMS), c/o Engineering Center, 1 Walnut St, Boston, MA 02108
T: (617) 227-5551
Founded: 1969; Members: 400
Gen Secr: Paula Golden
Focus: Eng
Periodical
Annual Seminar Papers 13798

American Entomological Society (AES), c/o Academy of Natural Sciences of Philadelphia, 1900 Race St, Philadelphia, PA 19103
T: (215) 561-3978; Fax: (215) 299-1028
Founded: 1859; Members: 350
Gen Secr: Mildred G. Morgan
Focus: Entomology
Periodicals
Entomological News (5 times annually)
Memoirs
Transactions (weekly) 13799

American Epidemiological Society (AES), c/o Dr. Philip S. Brachman, School of Public Health, Emory University, 1599 Clifton Rd NE, Atlanta, GA 30329
T: (404) 727-0199
Founded: 1927; Members: 300
Gen Secr: Dr. Philip S. Brachman
Focus: Pathology 13800

American Epilepsy Society (AES), 638 Prospect Av, Hartford, CT 06105-4298
T: (203) 232-4825; Fax: (203) 232-0819
Founded: 1946; Members: 1550
Gen Secr: Susan C. Berry
Focus: Pathology
Periodical
Epilepsia (weekly) 13801

American Equilibration Society (AES), 8726 N Ferris Av, Morton Grove, IL 60053
T: (708) 965-2888; Fax: (708) 965-4888
Founded: 1955; Members: 1100
Gen Secr: Shel Marcus
Focus: Dent
Periodical
Newsletter (weekly) 13802

American Ethnological Society (AES), c/o Nancy McDowell, Franklin and Marshall College, POB 3003, Lancaster, PA 17604
T: (717) 291-3985
Founded: 1842; Members: 2628
Gen Secr: Nancy McDowell
Focus: Ethnology
Periodicals
American Ethnologist (weekly)
Monograph Series
Unit News in Anthropology Newsletter (weekly) 13803

American Family Therapy Association (AFTA), 2020 Pennsylvania Av NW, Ste 273, Washington, DC 20006
T: (202) 994-2776; Fax: (202) 994-4812
Founded: 1977; Members: 900
Gen Secr: B. Miles
Focus: Therapeutics
Periodicals
Membership Directory (weekly)
Newsletter (weekly) 13804

American Farm Bureau Research Foundation (AFBRF), 225 Touhy Av, Park Ridge, IL 60068
T: (312) 399-5700
Founded: 1967
Pres: Dean R. Kleckner
Focus: Agri 13805

American Federation for Clinical Research (AFCR), 6900 Grove Rd, Thorofare, NJ 08086
T: (609) 848-7072; Fax: (609) 384-6504
Founded: 1940; Members: 12500
Gen Secr: Jean M. Haddock
Focus: Med
Periodical
Clinical Research Journal (weekly) . . . 13806

American Federation of Arts (AFA), 2510 Channing Way, Ste 4, Berkeley, CA 94704
Founded: 1909; Members: 600
Focus: Arts
Periodical
AFA Newsletter (weekly) 13807

American Federation of Astrologers (AFA), POB 22040, Tempe, AZ 85285-2040
T: (602) 838-1751
Founded: 1938
Gen Secr: Robert W. Cooper
Focus: Parapsych
Periodical
Today's Astrologer (weekly) 13808

American Federation of Mineralogical Societies (AFMS), POB 26523, Oklahoma City, OK 73126-0523
T: (405) 794-7150
Founded: 1947; Members: 52000
Gen Secr: Dan McLennan
Focus: Mineralogy
Periodical
Newsletter (9 times annually) 13809

American Fern Society (AFS), 456 McGill Pl, Atlanta, GA 30312
T: (404) 525-3147
Founded: 1893; Members: 1000
Gen Secr: Dr. Richard L. Hauke
Focus: Botany
Periodicals
American Fern Journal (weekly)
Fiddlehead Forum (weekly) 13810

American Fertility Society (AFS), 2140 11 Av S, Ste 200, Birmingham, AL 35205-2800
T: (205) 933-8494; Fax: (205) 930-9904
Founded: 1944; Members: 10000
Gen Secr: Nancy C. Hayley
Focus: Gynecology; Urology; Vet Med
Periodicals
Fertility and Sterility (weekly)
Fertility News (weekly) 13811

American Film and Video Association (AFVA), 8050 Millawake, POB 48659, Niles, IL 60714
T: (708) 698-6440; Fax: (708) 352-7528
Founded: 1943; Members: 1400
Focus: Libraries & Bk Sci
Periodicals
Bulletin (weekly)
Evaluations (3 times annually)
Sightlines (weekly) 13812

American Film Institute (AFI), c/o John F. Kennedy Center for the Performing Arts, Washington, DC 20566
T: (202) 828-4000
Founded: 1967; Members: 135000
Gen Secr: Jean Firstenberg
Focus: Cinema
Periodicals
American Film (weekly)
Guide to College Film Courses (bi-annually) 13813

American Finance Association (AFA), c/o Graduate School of Business Administration, New York University, 100 Trinity Pl, New York, NY 10006
T: (212) 285-8915
Founded: 1940; Members: 7000
Gen Secr: Dr. Michael Keenan
Focus: Finance
Periodical
Journal of Finance (5 times annually) . 13814

American Fine Arts Society (AFAS), 215 W 57 St, New York, NY 10019
T: (212) 247-4510
Founded: 1889
Pres: Rosina A. Florio
Focus: Arts 13815

American Fisheries Society (AFS), 5410 Grosvenor Ln, Ste 110, Bethesda, MD 20814
T: (301) 897-8616; Fax: (301) 897-8096
Founded: 1870; Members: 8400
Gen Secr: Paul Brouha
Focus: Fisheries
Periodicals
Journal of Aquatic Animal Health (weekly)
North American Journal of Fisheries Management (weekly)
The Progressive Fish-Culturist (weekly)
Transactions of the American Fisheries Society (weekly) 13816

American Folklore Society (AFS), c/o Shalom Staub, Pennsylvania Heritage Affairs Commission, 309 Forum Bldg, Harrisburg, PA 17120
T: (717) 783-8625
Founded: 1888; Members: 1600
Gen Secr: Shalom Staub
Focus: Ethnology
Periodicals
American Folklore Society Newsletter (weekly)
Journal of American Folklore (weekly) . 13817

American Foreign Law Association (AFLA), c/o Richard E. Lutringer, Whitman & Ransom, 200 Park Av, New York, NY 10166
T: (212) 351-3277; Fax: (212) 351-3131
Founded: 1925; Members: 650
Pres: Richard E. Lutringer
Focus: Law
Periodicals
American Journal of Comparative Law (weekly)
Newsletter 13818

American Forensic Association (AFA), Box 256, River Falls, WI 54022
T: (715) 425-3198; Fax: (715) 425-9533
Founded: 1949; Members: 900
Gen Secr: James W. Pratt
Focus: Law
Periodicals
Journal (weekly)
Newsletter 13819

American Forestry Association (AFA), 1516 P St NW, Washington, DC 20005
T: (202) 667-3300; Fax: (202) 667-7751
Founded: 1875; Members: 34818
Gen Secr: R. Neil Sampson
Focus: Forestry
Periodicals
American Forests Magazine (weekly)
Resource Hotline (weekly) 13820

American Forum for Global Education (AFGE), 45 John St, Ste 908, New York, NY 10038
T: (212) 791-4132; Fax: (212) 791-4132
Founded: 1987
Pres: Andrew F. Smith
Focus: Educ
Periodical
Access (weekly) 13821

American Foundation for Aids Research (AmFAR), 5900 Wilshire Blvd, Los Angeles, CA 90036
T: (213) 857-5900
Founded: 1985
Pres: Mervyn F. Silverman
Focus: Immunology
Periodical
Aids/HIV Treatment Directory (weekly) . . 13822

American Foundation for Vision Awareness (AFVA), 243 N Lindbergh Blvd, Saint Louis, MO 63141
T: (314) 991-4100; Fax: (314) 991-4101
Founded: 1927; Members: 5900
Gen Secr: Huck Roberts
Focus: Ophthal
Periodical
In Focus (weekly) 13823

American Foundrymen's Society (AFS), 505 State St, Des Plaines, IL 60016-8399
T: (708) 824-0181; Fax: (708) 824-7848
Founded: 1896; Members: 13500
Gen Secr: Charles H. Jones
Focus: Metallurgy
Periodical
Modern Casting (weekly) 13824

American Fracture Association (AFA), POB 668, Bloomington, IL 61702-0668
T: (309) 663-6272
Founded: 1938; Members: 500
Gen Secr: Barbara J. Dehority
Focus: Surgery; Orthopedics
Periodicals
Directory
Orthopedic Transactions (weekly) 13825

American-French Genealogical Society (AFGS), POB 2113, Pawtucket, RI 02861
T: (401) 722-1100
Founded: 1978; Members: 1000
Pres: Janice Burkhart
Focus: Genealogy
Periodical
Je Me Souviens (weekly) 13826

American Friends of Cambridge University (AFCU), POB 7070, Arlington, VA 22207
T: (703) 777-8850
Founded: 1967; Members: 1000
Pres: Stephen C. Price
Focus: Educ
Periodical
Annual Report (weekly) 13827

American Friends of Lafayette (AFL), c/o Daniel A. Evans, Skillman Library, Lafayette College, Easton, PA 18042
T: (215) 250-5153
Founded: 1932; Members: 125
Gen Secr: Daniel A. Evans
Focus: Hist
Periodical
Gazette (weekly) 13828

American Gastroenterological Association (AGA), 7910 Woodmont Av, Ste 914, Bethesda, MD 20814
T: (301) 654-2055; Fax: (301) 654-5920
Founded: 1897; Members: 6400
Gen Secr: Robert B. Greenberg
Focus: Gastroenter
Periodicals
AGA News (weekly)
Gastroenterology (weekly)
Viewpoints on Digestive Diseases (5 times annually) 13829

American Genetic Association (AGA), POB 39, Buckeystown, MD 21717-0039
T: (301) 695-9292
Founded: 1903; Members: 1500
Gen Secr: Stephen O'Brien
Focus: Genetics
Periodical
Journal of Heredity (weekly) 13830

American Geographical Society (AGS), 156 Fifth Av, Ste 600, New York, NY 10010-7002
T: (212) 242-0214; Fax: (212) 989-1583
Founded: 1851; Members: 1800
Gen Secr: Mary Lynne Bird
Focus: Geography
Periodicals
Focus (weekly)
Geographical Review (weekly) 13831

American Geological Institute (AGI), 4220 King St, Alexandria, VA 22302
T: (703) 379-2480; Fax: (703) 379-7563
Founded: 1948; Members: 21
Gen Secr: Dr. Marcus E. Milling
Focus: Geology
Periodicals
Bibliography and Index of Geology (weekly)
Earth Science (weekly)
Geotimes (weekly) 13832

American Geophysical Union (AGU), 2000 Florida Av NW, Washington, DC 20009
T: (202) 462-6900; Fax: (202) 328-0566
Founded: 1919; Members: 28000
Gen Secr: A.F. Spilhaus
Focus: Geophys
Periodicals
Earth in Space (7 times annually)
Geophysical Research Letters (weekly)
Global Biogeochemical Cycyles (weekly)
Journal of Geophysical Research (weekly)
Paleoceanography (weekly)
Planetology Papers (weekly)
Radio Science (weekly)
Reviews of Geophysics and Space Physics (weekly)
Tectonics (weekly)
Tectonics (weekly)
Water Resources Research (weekly) . . 13833

American Geriatrics Society (AGS), 770 Lexington Av, Ste 300, New York, NY 10021
T: (212) 308-1414; Fax: (212) 832-8646
Founded: 1942; Members: 6022
Gen Secr: Linda Hiddemen
Focus: Geriatrics
Periodicals
Clinical Report on Aging (weekly)
Journal (weekly)
Newsletter (weekly) 13834

American Group Practice Association (AGPA), 1422 Duke St, Alexandria, VA 22314
T: (703) 838-0033; Fax: (703) 548-1890
Founded: 1949; Members: 300
Gen Secr: Dr. Donald W. Fisher
Focus: Med
Periodicals
Directory (weekly)
Executive News Service (weekly)
Group Practice Journal (weekly) 13835

American Group Psychotherapy Association (AGPA), 25 E 21 St, New York, NY 10010
T: (212) 477-2677; Fax: (212) 979-6627
Founded: 1942; Members: 3500
Gen Secr: Marsha Block
Focus: Therapeutics
Periodicals
International Journal of Group Psychotherapy (weekly)
Membership Directory (bi-annually)
Newsletter (weekly) 13836

American Guild of Hypnotherapists (AGH), 7117 Farnam St, Omaha, 68132
T: (402) 397-1500
Founded: 1975; Members: 1106
Pres: Reg Sheldrick
Focus: Psych
Periodical
American Journal of Hypnotherapy (weekly)
. . . . 13837

American Gynecological and Obstetrical Society (AGOS), c/o James R. Scott, University of Utah, 50 N Medical Dr, Salt Lake City, UT 84132
T: (801) 581-5501
Founded: 1981; Members: 300
Gen Secr: James R. Scott
Focus: Gynecology
Periodical
Transactions (weekly) 13838

American Harp Society (AHS), 6331 Quebec Dr, Hollywood, CA 90068
T: (213) 463-0716; Fax: (213) 464-2950
Founded: 1962; Members: 3300
Gen Secr: Dorothy Remsen
Focus: Music
Periodical
American Harp Journal (weekly) 13839

American Healing Association (AHA), c/o Brian Zink, 811 Ridge Dr, Glendale, CA 91206
Founded: 1975
Gen Secr: Brian Zink
Focus: Med
Periodicals
Directory of Healers and Counselors
Newsletter (weekly) 13840

American Health Information Management Association, 919 N Michigan Av, Ste 1400, Chicago, IL 60690
T: (312) 787-2672
Founded: 1928; Members: 31000
Gen Secr: Pam Wear
Focus: Med
Periodicals
The Gavel (weekly)
Journal (weekly) 13841

American Health Planning Association (AHPA), POB 770097, Oklahoma City, OK 73120
T: (405) 271-6868
Founded: 1970
Pres: Suzanne W. Nichols
Focus: Public Health
Periodicals
Directory of State and Local Health Planning Agencies
Today in Health Planning (weekly) . . . 13842

American Hearing Research Foundation (AHRF), 55 E Washington St, Ste 2022, Chicago, IL 60602
T: (312) 726-9670
Founded: 1956; Members: 1585
Gen Secr: William L. Lederer
Focus: Pathology
Periodicals
Newsletter (weekly)
Progress Report (weekly) 13843

American Heart Association (AHA), 7272 Greenville Av, Dallas, TX 75231-4596
T: (214) 373-6300; Fax: (214) 706-1341
Founded: 1924; Members: 200000
Gen Secr: Dudley H. Hafner
Focus: Cardiol
Periodicals
Arteriosclerosis
Cardiovascular Nursing
Circulation
Circulation Research
Current Concepts
Heart Disease and Stroke (weekly)
Hypertension
Stroke 13844

American Herb Association (AHA), POB 1673, Nevada City, CA 95959
Founded: 1981; Members: 1000
Gen Secr: K. Keville
Focus: Botany
Periodicals
Herb Gardens of the U.S. (weekly)
Newsletter (weekly) 13845

American Historical Association (AHA), 400 A St SE, Washington, DC 20003
T: (202) 544-2422; Fax: (202) 544-8307
Founded: 1884; Members: 15500
Gen Secr: Samuel R. Gammon
Focus: Hist
Periodicals
AHA Perspectives: Newsletter (9 times annually)
American Historical Review (8 times annually)
Recently Published Articles (3 times annually)
. . . . 13846

American Historical Society of Germans from Russia (AHSGR), 631 D St, Lincoln, NE 68502-1199
T: (402) 474-3363
Founded: 1968; Members: 6000
Gen Secr: Elizabeth K. Wilson
Focus: Folklore; Genealogy
Periodicals
Clues (weekly)
Journal (weekly)
Newsletter (weekly) 13847

American Hobbit Association (AHA), 730F Northland Rd, Cincinnati, OH 45240
T: (513) 742-4384
Founded: 1977; Members: 175
Gen Secr: Renee A. Alper
Focus: Lit
Periodical
The Rivendell Review (weekly) 13848

American Holistic Medical Association (AHMA), 4101 Lake Boone Tr, Ste 201, Raleigh, NC 27607
T: (919) 787-5146
Founded: 1978; Members: 400
Gen Secr: Sally Nicholson
Focus: Lit
Periodical
Holistic Medicine (weekly) 13849

American Holistic Medical Foundation (AHMF), 4101 Lake Boone Tr, Ste 201, Raleigh, NC 27607
T: (919) 787-5146; Fax: (919) 787-4816
Founded: 1974; Members: 450
Gen Secr: Sally Nicholson
Focus: Med
Periodical
Holistic Medicine (weekly) 13850

American Home Economics Association (AHEA), 1555 King St, Alexandria, VA 22314
T: (703) 706-4600
Founded: 1909; Members: 24000
Gen Secr: Dr. Mary Jane Kolar
Focus: Home Econ
Periodicals
AHEA Action (weekly)
Home Economics Research Journal (weekly)
Journal of Home Economics (weekly) . . 13851

American Horticultural Society (AHS), 7931 E Boulevard Dr, Alexandria, VA 22308
T: (703) 768-5700; Fax: (703) 765-6032
Founded: 1922; Members: 18800
Gen Secr: Helen L. Walutes
Focus: Hort
Periodical
American Horticulturist (weekly) 13852

American Horticultural Therapy Association (AHTA), 362A Christopher Av, Gaithersburg, MD 20879
T: (301) 948-3010
Founded: 1973; Members: 691
Gen Secr: Steven Davis
Focus: Therapeutics; Rehabil; Hort
Periodicals
Journal of Therapeutic Horticulture (weekly)
People Plant Connection (11 times annually)
. . . . 13853

American Hospital Association (AHA), 840 N Lake Shore Dr, Chicago, IL 60611
T: (312) 280-6000; Fax: (312) 280-5979
Founded: 1898; Members: 54500
Pres: Richard J. Davidson
Focus: Public Health
Periodicals
Cross Reference (The Hospital Manager)
Discharge Planning Update
Health Law Vigil
Hospital Literature Index
Hospital Technology Alerts
Hospital Technology Series
Infection Control Digest
Medical Staff Forum
Outreach
Promoting Health 13854

American Hungarian Educators' Association (AHEA), c/o Eniko Molnar Basa, 707 Snider Ln, Silver Spring, MD 20905
T: (301) 384-4657
Founded: 1975; Members: 250
Gen Secr: Eniko Molnar Basa
Focus: Educ
Periodical
The Educator (3 times annually) 13855

American Hungarian Library and Historical Society (AHLHS), 213 E 82 St, New York, NY 10021
T: (212) 744-5298
Founded: 1955; Members: 450
Pres: Otto Hamos
Focus: Hist 13856

American Hypnotists' Association (AHA), 1159 Green St, Ste 6, San Francisco, CA 94109
T: (415) 775-6130
Founded: 1959; Members: 401
Pres: Dr. Rafael M. Bertuccelli
Focus: Psych; Med 13857

American Indian Council of Architects and Engineers (AICAE), POB 230685, Tigard, OR 97223
T: (503) 684-5680; Fax: (503) 228-2058
Founded: 1976; Members: 30
Pres: Fred C. Cooper
Focus: Archit; Eng 13858

American Indian Culture Research Center (AICRC), Box 98, Blue Cloud Abbey, Marvin, SD 57251
T: (605) 432-5528; Fax: (605) 432-4754
Founded: 1967
Gen Secr: Stanislaus Maudlin
Focus: Hist 13859

American Indian Lore Association (AILA), 960 Walhonding Av, Logan, OH 43138
T: (614) 385-7136
Founded: 1957
Pres: Lelamd L. Conner
Focus: Ethnology 13860

American Indian Science and Engineering Society (AISES), 1630 30 St, Ste 301, Boulder, CO 80301
T: (303) 492-8658; Fax: (303) 492-3400
Founded: 1977; Members: 2000
Gen Secr: Norbert S. Hill
Focus: Eng; Sci
Periodicals
Annual Report (weekly)
Science Education Newsletter (weekly)
Winds of Change (weekly) 13861

American Industrial Health Council (AIHC), 1330 Connecticut Av NW, Ste 300, Washington, DC 20036
T: (202) 659-0060
Founded: 1977; Members: 60
Gen Secr: Ronald A. Lang
Focus: Cell Biol & Cancer Res; Public Health
Periodical
Newsletter (weekly) 13862

American Industrial Hygiene Association (AIHA), 345 White Pond Dr, POB 8390, Akron, OH 44320
T: (216) 873-2442; Fax: (216) 873-1642
Founded: 1939; Members: 9500
Gen Secr: O. Gordon Banks
Focus: Hygiene
Periodicals
Directory (weekly)
Journal (weekly) 13863

American Institute for Archaeological Research (AIAR), 24 Cross Rd, Mount Vernon, NH 03057
T: (603) 673-3005
Founded: 1982; Members: 250
Gen Secr: Dorothy L. Hayden
Focus: Archeol
Periodicals
Newsletter (weekly)
On Site (weekly) 13864

American Institute for Chartered Property Casuality Underwriters (AICPCU), 720 Providence Rd, POB 3016, Malvern, PA 19355-0716
T: (215) 644-2100; Fax: (215) 251-9995
Founded: 1972
Pres: Dr. N.A. Baglini
Focus: Insurance
Periodical
Institute Insights (weekly) 13865

American Institute for Conservation of Historic and Artistic Works (AIC), 1400 16 St NW, Ste 340, Washington, DC 20036
T: (202) 232-6636; Fax: (202) 232-6630
Founded: 1972; Members: 2700
Gen Secr: Sarah Rosenberg
Focus: Hist; Arts
Periodicals
Journal (weekly)
Newsletter (weekly) 13866

American Institute for Contemporary German Studies (AICGS), 11 Dupont Circle NW, Ste 350, Washington, DC 20036
T: (202) 332-9312; Fax: (202) 265-8531
Founded: 1983
Gen Secr: Robert G. Livingston
Focus: Ling 13867

American Institute for Economic Research (AIER), Great Barrington, MA 01230
T: (413) 528-1216
Founded: 1933; Members: 8000
Pres: Dr. Robert A. Gilmour
Focus: Econ
Periodicals
Economic Education Bulletin (weekly)
Research Report (weekly) 13868

American Institute for Foreign Study (AIFS), 102 Greeenwich Av, Greenwich, CT 06830
T: (203) 869-9090; Fax: (203) 869-1173
Founded: 1964; Members: 300000
Pres: Robert G. Brennan
Focus: Educ 13869

American Institute for Patristic and Byzantine Studies (AIPBS), R.R. 1, Box 353A, Minuet Ln, Kingston, NY 12401
T: (915) 336-8797
Founded: 1967; Members: 390
Pres: Dr. Constantine N. Tsirpanlis
Focus: Cultur Hist 13870

American Institute for Verdi Studies (AIVS), c/o Dept of Music, Faculty of Arts and Science, New York University, 24 Waverly Pl, Rm 268, New York, NY 10003
T: (212) 998-2587; Fax: (212) 995-4147
Founded: 1976; Members: 300
Gen Secr: Dr. Martin Chusid
Focus: Music
Periodical
Verdi Newsletter (weekly) 13871

American Institute of Aeronautics and Astronautics (AIAA), 370 L'Enfant Promenade SW, Washington, DC 20024
T: (202) 646-7400; Fax: (202) 646-7508
Founded: 1963; Members: 45000
Gen Secr: C. Durocher
Focus: Aero
Periodicals
Aerospace America (weekly)
AIAA Journal (weekly)
International Aerospace Abstracts (weekly)
Journal of Aircraft (weekly)
Journal of Guidance, Control and Dynamics (weekly)
Journal of Propulsion and Power (weekly)
Journal of Spacecraft and Rockets (weekly)
Journal of Thermophysics and Heat Transfer (weekly) 13872

American Institute of Aeronautics and Astronautics – Technical Information Division (AIAATID), 555 W 57 St, Ste 1200, New York, NY 10019
T: (212) 247-6500; Fax: (212) 582-4861
Founded: 1963; Members: 45
Gen Secr: Barbara Lawrence
Focus: Eng 13873

American Institute of Architects (AIA), 1735 New York Av NW, Washington, DC 20006
T: (202) 626-7300; Fax: (202) 783-8247
Founded: 1857; Members: 57000
Pres: James P. Cramer
Focus: Archit
Periodicals
Architecture (weekly)
Memo (weekly) 13874

American Institute of Biological Sciences (AIBS), 730 11 St NW, Washington, DC 20001-4521
T: (202) 628-1500; Fax: (202) 628-1509
Founded: 1947; Members: 7200
Gen Secr: Charles M. Chambers
Focus: Bio
Periodicals
BioScience (11 times annually)
Forum (weekly)
Membership Directory (weekly) 13875

American Institute of Biomedical Climatology (AIBC), 1023 Welsh Rd, Philadelphia, PA 19115
T: (215) 673-8368; Fax: (215) 579-1494
Founded: 1958; Members: 60
Gen Secr: Richmond G. Kent
Focus: Bio
Periodical
ABC Bulletin (weekly) 13876

American Institute of Building Design (AIBD), POB 2249, Shelton, CT 06484
T: (203) 227-3640
Founded: 1958; Members: 1000
Gen Secr: Tammy J. Thomas
Focus: Civil Eng; Archit
Periodicals
Bulletin
Professional Designers Newsletter (weekly)
Roster of Members (weekly) 13877

American Institute of Chemical Engineers (AIChE), 345 E 47 St, New York, NY 10017
T: (212) 705-7338; Fax: (212) 752-3294
Founded: 1908; Members: 54000
Gen Secr: Richard E. Emmert
Focus: Chem
Periodicals
Chemical Engineering Progress (weekly)
Environmental Progress (weekly)
International Chemical Engineering (weekly)
Plant/Operations Progress (weekly) . . . 13878

American Institute of Chemists (AIC), 7315 Wisconsin Av NW, Bethesda, MD 20814
T: (301) 652-2447
Founded: 1923; Members: 4000
Focus: Chem
Periodicals
The Chemist (11 times annually)
Professional Directory (weekly) 13879

American Institute of Constructors (AIC), 9887 N Gandy, Ste 104, Saint Petersburg, FL 33702
T: (813) 578-0317; Fax: (813) 578-9982
Founded: 1971; Members: 1600
Gen Secr: Cheryl P. Harris
Focus: Civil Eng
Periodicals
American Professional Constructor (weekly)
Newsletter (weekly)
Register (bi-annually) 13880

American Institute of Graphic Arts (AIGA), 1059 Third Av, New York, NY 10021
T: (212) 752-0813; Fax: (212) 755-6749
Founded: 1914; Members: 6700
Gen Secr: Caroline Hightower
Focus: Graphic & Dec Arts, Design
Periodical
AIGA Journal of Graphic Design (weekly) 13881

American Institute of Homeopathy (AIH), 1585 Glencoe St, Ste 44, Denver, CO 80220-1338
T: (303) 898-5477
Founded: 1844; Members: 160
Gen Secr: Karen Kaiser Nossaman
Focus: Homeopathy
Periodicals
Journal of the American Institute of Homeopathy (weekly)
Newsletter (weekly) 13882

American Institute of Indian Studies (AILS), 1130 E 59 St, Rm 412, Chicago, IL 60637
T: (312) 702-8638
Founded: 1961; Members: 47
Pres: Joseph W. Elder
Focus: Ethnology
Periodicals
Annual Report (weekly)
Ethnomusicology Newsletter 13883

American Institute of Iranian Studies (AIIS), c/o Dr. Marilyn R. Waldman, 106 Dulles Hall, Ohio State University, 230 W 17 Av, Columbus, OH 43210-1311
T: (614) 292-1265; Fax: (614) 292-2282
Founded: 1967; Members: 20
Pres: Dr. Marilyn R. Waldman
Focus: Ethnology
Periodical
Newsletter 13884

American Institute of Islamic Studies (AIIS), POB 100398, Denver, CO 80250
T: (303) 936-0108
Founded: 1965
Gen Secr: Charles L. Geddes
Focus: Ethnology; Hist; Rel & Theol
Periodical
Bibliographic Series 13885

American Institute of Life Threatening Illness and Loss (FT), 630 W 168 St, New York, NY 10032
T: (212) 928-2066
Founded: 1967; Members: 400
Gen Secr: Maxine Lazarus
Focus: Psychiatry
Periodicals
Advances in Thanatology (weekly)
Archives of the Foundation of Thanatology (weekly)
Thanatology Abstracts (weekly) 13886

American Institute of Management (AIM), POB 7039, Quincy, MA 02269
T: (617) 472-0277
Founded: 1948; Members: 1500
Pres: Barbara C. Doll
Focus: Business Admin
Periodicals
The Associates Digest (weekly)
The Executive Counselor (weekly)
National Biographies
The Presidents Journal (weekly)
The Institute Bulletin (weekly) 13887

American Institute of Mining, Metallurgical and Petroleum Engineers (AIME), 345 E 47 St, New York, NY 10017
T: (212) 705-7695; Fax: (212) 371-9622
Founded: 1871; Members: 73000
Gen Secr: Alfred Weiss
Focus: Mining 13888

American Institute of Musical Studies (AIMS), 3500 Maple Av, Ste 120, Dallas, TX 75219-3901
T: (214) 528-9234
Founded: 1969
Gen Secr: Nora Sands
Focus: Music
Periodical
AIMS Bulletin (biennial) 13889

American Institute of Nutrition (AIN), 9650 Rockville Pike, Bethesda, MD 20814-3990
T: (301) 530-7050; Fax: (301) 571-1892
Founded: 1928; Members: 2770
Gen Secr: Richard G. Allison
Focus: Nutrition
Periodical
Journal of Nutrition (weekly) 13890

American Institute of Oral Biology (AIOB), POB 9481, South Laguna, CA 92677
T: (714) 499-1286
Founded: 1943
Gen Secr: Barbara P. Ward
Focus: Dent 13891

American Institute of Parliamentarians (AIP), 203 W Wayne, Ste 312, Fort Wayne, IN 46802
T: (219) 422-3680
Founded: 1958; Members: 1350
Gen Secr: Karen Magdich
Focus: Poli Sci
Periodical
Parliamentary Journal (weekly) 13892

American Institute of Physics (AIP), 335 E 45 St, New York, NY 10017
T: (212) 661-9404; Fax: (212) 949-0473
Founded: 1931; Members: 10
Gen Secr: Kenneth W. Ford
Focus: Physics
Periodicals
The Astronomical Journal (weekly)
Current Physics Index (weekly)
Journal of Applied Physics: Microfiche Edition (weekly)
Journal of Chemical Physics: Microfiche Edition (weekly)
Journal of Lightware Technology: Microfiche Edition (weekly)
Journal of Physical and Reference Data (weekly)
Journal of Vacuum Science and Technology A&B (8 times annually)
Mathematical Physics: Microfiche Edition (weekly)
Medical Physics (weekly)
Optics Letters: Microfiche Edition (weekly)
Optics News (weekly)
Physical Review A, B, C, D: Microfiche Edition(s) (72 times annually)
Physical Review Abstracts (weekly)
Physical Review Letters: Microfiche Edition (52 times annually)
Physics Briefs (weekly)
Physics Briefs of Fluids: Microfiche Edition (weekly)
Physics Today: Microfiche Edition (weekly)
Reviews of Modern Physics: Microfiche Edition (weekly)
Reviews of Scientific Instruments: Microfiche Edition (weekly) 13893

American Institute of Plant Engineers (AIPE), 8180 Corporate Paril Dr, Ste 306, Cincinnati, OH 45242
T: (513) 489-2473; Fax: (513) 247-7422
Founded: 1954; Members: 10000
Gen Secr: Michael J. Tillar
Focus: Eng
Periodicals
AIPE Facilities Management, Operations and Engineering (weekly)
AIPE Newsline (5 times annually) . . . 13894

American Institute of Steel Construction (AISC), 1 E Wacker Dr, Ste 3100, Chicago, IL 60601-2001
T: (312) 670-2400; Fax: (312) 670-5403
Founded: 1921; Members: 2512
Pres: Neil W. Zundel
Focus: Civil Eng
Periodicals
Engineering Journal (weekly)
Modern Steel Construction (weekly) . . . 13895

American Institute of the History of Pharmacy (AIHP), Pharmacy Bldg, 425 N Chater St, Madison, WI 53706
T: (608) 262-5378
Founded: 1941; Members: 1200
Gen Secr: Gregory J. Higby
Focus: Pharmacol; Cultur Hist
Periodical
Pharmacy in History (weekly) 13896

American Institute of Timber Construction (AITC), 11818 SE Mill Plain Blvd, Ste 407, Vancouver, WA 98684-5092
T: (206) 254-9132; Fax: (206) 254-9456
Founded: 1952; Members: 19
Gen Secr: Matthew Mathias
Focus: Eng
Periodical
Lam Lines (weekly) 13897

American Institute of Ultrasound in Medicine (AIUM), 11200 Rockville Pike, Ste 205, Rockville, MD 29852-3139
T: (301) 881-2486; Fax: (301) 881-7303
Founded: 1951; Members: 10000
Gen Secr: James S. Packer
Focus: Med
Periodicals
AIUM Newsletter (weekly)
Journal of Ultrasound in Medicine (weekly)
Membership Directory (bi-annually)
Scientific Proceedings (weekly) 13898

American Institutes for Research in the Behavioral Sciences (AIR), 3333 K St NW, Ste 300, Washington, DC 20007
T: (202) 342-5000; Fax: (202) 342-5033
Founded: 1946
Pres: Dr. David A. Goslin
Focus: Behav Sci 13899

American Intellectual Property Law Association (AIPLA), 2001 Jefferson Davis Hwy, Ste 203, Arlington, VA 22202
T: (703) 415-0780; Fax: (703) 415-0786
Founded: 1897; Members: 7200
Gen Secr: Martha R. Morales
Focus: Law
Periodicals
AIPLA Bulletin (5 times annually)
AIPLA Journal (weekly) 13900

American Irish Bicentennial Committee (AIBC), 3917 Moss Dr, Annandale, VA 22003
T: (703) 354-4721
Founded: 1973; Members: 350
Gen Secr: Joseph F. O'Connor
Focus: Hist
Periodical
Stars and Harp (weekly) 13901

American Irish Historical Society (AIHS), 991 Fifth Av, New York, NY 10028
T: (212) 288-2263; Fax: (212) 628-7927
Founded: 1897; Members: 500
Gen Secr: Thomas M. Horan
Focus: Hist
Periodical
The Recorder (weekly) 13902

American Iron and Steel Institute (AISI), 1101 17 St NW, Washington, DC 20036-4700
T: (202) 452-7100; Fax: (202) 463-6573
Founded: 1908; Members: 1280
Pres: Milton Deaner
Focus: Metallurgy 13903

American Italian Historical Association (AIHA00), 209 Flagg Pl, Staten Island, NY 10304
T: (718) 667-6628
Founded: 1966; Members: 500
Pres: Rudolph Juliani
Focus: Hist
Periodical
Newsletter (weekly) 13904

American Jewish Historical Society (AJHS), 2 Thornton Rd, Waltham, MA 02154
T: (617) 891-8110; Fax: (617) 899-9208
Founded: 1892; Members: 3300
Pres: Ronald C. Curhan
Focus: Hist
Periodicals
American Jewish History (weekly)
Heritage (bi-annually) 13905

American Joint Committee on Cancer (AJCC), 55 E Erie St, Chicago, IL 60611
T: (312) 664-4050; Fax: (312) 440-7144
Founded: 1959; Members: 40
Gen Secr: Oliver H. Beahrs
Focus: Cell Biol & Cancer Res 13906

American Judges Association (AJA), 300 Newport Av, Williamsburg, VA 23187-8798
T: (804) 253-2000
Founded: 1959; Members: 2800
Gen Secr: Shelley R. Rockwell
Focus: Law
Periodical
Court Review (weekly) 13907

American Judicature Society (AJS), 25 E Washington St, Ste 1600, Chicago, IL 60602
T: (312) 558-6900; Fax: (312) 558-9175
Founded: 1913; Members: 22000
Gen Secr: Frances K. Zemans
Focus: Law
Periodical
Judicature (weekly) 13908

American Kinesiotherapy Association, c/o Ed Reiling, POB 611, Wright Brothers Station, Dayton, OH 45409
Founded: 1946; Members: 1000
Gen Secr: David Ser
Focus: Rehabil
Periodical
Journal (weekly) 13909

American Laryngological Association (ALA), c/o H. Bryan Neel, 200 SW First St, Rochester, MN 55905
T: (507) 284-2369; Fax: (507) 284-2072
Founded: 1879; Members: 190
Gen Secr: H. Bryan Neel
Focus: Otorhinolaryngology; Surgery
Periodical
Transactions (weekly) 13910

American Laryngological, Rhinological and Otological Society (ALROS), c/o Lilibet Coe, POB 155, East Greenville, PA 18041
T: (215) 679-7180; Fax: (215) 679-8160
Founded: 1895; Members: 750
Gen Secr: Lilibet Coe
Focus: Otorhinolaryngology 13911

American Law Institute (ALI), 4025 Chestnut St, Philadelphia, PA 19104-3099
T: (215) 243-1600; Fax: (215) 243-1664
Founded: 1923; Members: 3342
Gen Secr: Paul A. Wolkin
Focus: Law
Periodical
ALI Reporter 13912

American Leather Chemists Association (ALCA), c/o University of Cincinnati, Location 14, Cincinnati, OH 45221
T: (513) 556-1197
Founded: 1903; Members: 700
Gen Secr: Velma Becker
Focus: Chem
Periodical
Journal of the American Leather Chemists Association (weekly) 13913

American Legal Studies Association (ALSA), c/o Program in Law, Policy and Society, Northeastern University, 341 Cushing Hall, Boston, MA 02115
T: (617) 437-5211; Fax: (617) 437-4691
Founded: 1975; Members: 427
Gen Secr: Leonard G. Buckle
Focus: Law
Periodical
Legal Studies Forum: A Journal of Interdisciplinary Legal Studies (weekly) 13914

American Leprosy Missions (ALM), 1 ALM Way, Greenville, SC 29601
T: (803) 271-7040; Fax: (803) 271-7062
Founded: 1906
Pres: Thomas F. Frist
Focus: Derm
Periodical
Word and Deed (weekly) 13915

American Library Association (ALA), 50 E Huron St, Chicago, IL 60611
T: (312) 944-6780; Fax: (312) 280-3255
Founded: 1876; Members: 55000
Gen Secr: Peggy Sullivan
Focus: Libraries & Bk Sci; Computer & Info Sci
Periodicals
ALA Handbook of Organization and Membership Directory (weekly)
American Libraries (weekly)
Booklist (weekly)
Choice (11 times annually)
Library Video Magazine (weekly)
Washington Newsletter (weekly) 13916

American Library Trustee Association (ALTA), 50 E Huron St, Chicago, IL 60611
T: (312) 280-2161; Fax: (312) 280-3257
Founded: 1890; Members: 1880
Gen Secr: Susan Roman
Focus: Libraries & Bk Sci
Periodicals
Newsletter (weekly)
Trustee Digest (weekly) 13917

American Liszt Society (ALS), c/o Fernando Laires, 210 Devonshire Dr, Rochester, NY 14625-1905
T: (716) 586-9922; Fax: (716) 381-1144
Founded: 1964; Members: 550
Pres: Fernando Laires
Focus: Music
Periodicals
Journal of the American Liszt Society (weekly)
Newsletter of the American Liszt Society (weekly) 13918

American Literary Translators Association (ALTA), c/o University of Texas at Dallas, POB 830688, Richardson, TX 75083-0688
T: (214) 690-2093; Fax: (214) 690-2989
Founded: 1978; Members: 1000
Gen Secr: Nancy Sabin
Focus: Lit
Periodicals
Newsletter (weekly)
Translation Review (3 times annually) . . 13919

American Littoral Society (ALS), c/o Sandy Hook, Highlands, NJ 07732
T: (908) 291-0055
Founded: 1961; Members: 9000
Gen Secr: D.W. Bennett
Focus: Zoology
Periodicals
Coastal Reporter (weekly)
Underwater Naturalist (weekly) 13920

American Longevity Association (ALA), 1000 W Carson St, Torrance, CA 90509
T: (310) 544-7057
Founded: 1980
Pres: Robert J. Morin
Focus: Geriatrics

Periodical
Longevity Letter (weekly) 13921

American Lunar Society (ALS), POB 209, East Pittsburgh, PA 15112
T: (608) 837-6054
Founded: 1982; Members: 120
Pres: David O. Darling
Focus: Astronomy
Periodical
Selenology (weekly) 13922

American Lung Association (ALA), 1740 Broadway, New York, NY 10019-4374
T: (212) 315-8700; Fax: (212) 265-5642
Founded: 1904; Members: 9500
Gen Secr: John R. Garrison
Focus: Pulmon Dis
Periodical
American Review of Respiratory Diseases (weekly) 13923

American Malacological Union (AMU), POB 30, North Myrtle Beach, SC 29582
T: (803) 249-1651
Founded: 1931; Members: 750
Gen Secr: Richard E. Petit
Focus: Zoology
Periodicals
American Malacological Bulletin (weekly)
Membership List (weekly)
Newsletter (weekly) 13924

American Management Association (AMA), 135 W 50 St, New York, NY 10020-1201
T: (212) 586-8100; Fax: (212) 903-8168
Founded: 1923; Members: 70000
Pres: David Fagiano
Focus: Business Admin
Periodicals
Compensation and Benefits Review (weekly)
Comp Flash (weekly)
Management Review (weekly)
The President (weekly) 13925

American Marketing Association (AMA), 250 S Wacker Dr, Ste 200, Chicago, IL 60606
T: (312) 648-0536; Fax: (312) 993-7542
Founded: 1915; Members: 53000
Gen Secr: Jeffrey Heilbrunn
Focus: Marketing
Periodicals
AMA Annual Membership Roster & International Buyers' Guide (weekly)
Journal of Marketing (weekly)
Journal of Marketing Research (weekly)
Marketing News (weekly) 13926

American Massage Therapy Association (AMTA), 1130 W North Shore Av, Chicago, IL 60626-4670
Fax: (312) 761-0009
Founded: 1943; Members: 14000
Pres: Elliot Greene
Focus: Therapeutics
Periodicals
Massage Journal (weekly)
Yearbook and Registry 13927

American Mathematical Association of Two Year Colleges (AMATYC), c/o Mott Community College, 1401 E Court St, Flint, MI 48503
T: (313) 232-3980
Founded: 1974; Members: 2800
Pres: Karen Sharp
Focus: Math; Educ
Periodicals
AMATYC News (3 times annually)
AMATYC Review (weekly)
Annual Developmental Mathematics Committee Report (weekly)
Newsletter (weekly) 13928

American Mathematical Society (AMS), POB 6248, Providence, RI 02940
T: (401) 455-4000; Fax: (401) 455-4004
Founded: 1888; Members: 26000
Gen Secr: Dr. William H. Jaco
Focus: Math
Periodicals
Abstracts of Papers Presented to the American Mathematical Society (7 times annually)
Bulletin of the American Mathematical Society (weekly)
Current Mathematical Publications (weekly)
Journal of the American Mathematical Society (weekly)
Mathematical Reviews (weekly)
Mathematics of Computation (weekly)
Memoirs (weekly)
Notices (8 times annually)
Proceedings of the American Mathematical Society (weekly)
Transactions (weekly) 13929

American Matthay Association (AMA), c/o Thomas C. Davis, 46 Poplar St, Greentree, Pittsburgh, GA 15205
T: (412) 921-3095
Founded: 1925; Members: 170
Pres: Thomas C. Davis
Focus: Music
Periodical
Matthay News (weekly) 13930

American Meat Science Association (AMSA), c/o National Live Stock and Meat Board, 444 N Michigan Av, Chicago, IL 60611
T: (312) 467-5520
Founded: 1964; Members: 900
Gen Secr: H. Kenneth Johnson
Focus: Food
Periodical
Proceedings of Reciprocal Meat Conference (weekly) 13931

American Medical Associaction Auxiliary (AMAA), 515 N State St, Chicago, IL 60610
T: (312) 464-4470; Fax: (312) 464-4184
Founded: 1922; Members: 70000
Gen Secr: Hazel J. Lewis
Focus: Public Health
Periodical
Facets (weekly) 13932

American Medical Association (AMA), 515 N State St, Chicago, IL 60610
T: (312) 464-4818; Fax: (312) 464-4184
Founded: 1847; Members: 271000
Gen Secr: James S. Todd
Focus: Med
Periodicals
AJDC: American Journal of Diseases of Children (weekly)
American Medical News (weekly)
Archives of Dermatology (weekly)
Archives of General Psychiatry (weekly)
Archives of Internal Medicine (weekly)
Archives of Neurology (weekly)
Archives of Ophthalmology (weekly)
Archives of Otolaryngology, Head & Neck Surgery (weekly)
Archives of Pathology & Laboratory Medicine (weekly)
Archives of Surgery (weekly)
JAMA: The Journal of the American Medical Association (weekly) 13933

American Medical Association Education and Research Foundation (AMA-ERF), 515 N State St, Chicago, IL 60610
T: (312) 464-5000; Fax: (312) 464-4184
Founded: 1962
Gen Secr: James S. Todd
Focus: Educ; Med 13934

American Medical Electroencephalographic Association (AMEEGA), 850 Elm Grove Rd, Elm Grove, WI 53122
T: (414) 797-7800; Fax: (414) 782-8788
Founded: 1964; Members: 850
Gen Secr: Robert H. Herzog
Focus: Neurology
Periodical
Clinical EEG (weekly) 13935

American Medical Fly Fishing Association (AMFFA), c/o Veryl Frye, POB 768, Lock Haven, PA 17745
T: (717) 769-7375
Founded: 1969; Members: 220
Gen Secr: Veryl Frye
Focus: Ecology; Med
Periodical
Newsletter 13936

American Medical Informatics Association (AMIA), 4915 Saint Elmo Av, Ste 302, Bethesda, MD 20814
T: (301) 657-1291
Founded: 1981; Members: 850
Gen Secr: Gail Mutnik
Focus: Med; Computer & Info Sci
Periodical
Proceedings (weekly) 13937

American Medical Technologists (AMT), 710 Higgins Rd, Park Ridge, IL 60068
T: (708) 823-5169; Fax: (708) 823-0458
Founded: 1939; Members: 22150
Gen Secr: Gerard P. Boe
Focus: Med
Periodicals
AMT Events (weekly)
Journal (weekly) 13938

American Medical Women's Association (AMWA), 801 N Fairfax St, Ste 400, Alexandria, VA 22314
T: (703) 838-0500; Fax: (703) 549-3864
Founded: 1915; Members: 11000
Gen Secr: Eileen McGrath
Focus: Med
Periodicals
Journal (weekly)
Quarterly Newsletter (weekly) 13939

American Medical Writers' Association (AMWA), 9650 Rockville Pike, Bethesda, MD 20814-3998
T: (301) 493-0003; Fax: (301) 493-0005
Founded: 1940; Members: 3700
Gen Secr: Lillian Sablack
Focus: Lit; Med
Periodicals
AMWA Freelance Directory of Medical Communication Services (weekly)
AMWA Journal (weekly)
AMWA Membership Directory (weekly) . 13940

American Mensa, 2626 E 14 St, Brooklyn, NY 11235
T: (718) 934-3700
Founded: 1960; Members: 55000
Gen Secr: Sheila Skolnik
Focus: Philos
Periodicals
Interloc (10 times annually)
Mensa Bulletin (10 times annually)
Mensa Research Journal (3 times annually) 13941

American Merchant Marine Library Association (AMMLA), 1 World Trade Center, Ste 2161, New York, NY 10048
T: (212) 775-1038; Fax: (212) 432-5492
Founded: 1921
Gen Secr: Roger T. Korner
Focus: Libraries & Bk Sci
Periodical
AMMLA Annual Report (weekly) 13942

American Meteorological Society (AMS), 45 Beacon St, Boston, MA 02108
T: (617) 227-2425; Fax: (617) 742-8718
Founded: 1919; Members: 11000
Gen Secr: Richard E. Hallgren
Focus: Geophys
Periodicals
AMS Newsletter
Bulletin (weekly)
Journal of Applied Meteorology (weekly)
Journal of Atmospheric and Oceanic Technology (weekly)
Journal of Climate (weekly)
Journal of Physical Oceanography (weekly)
Journal of the Atmospheric Sciences (weekly)
Meteorological and Geoastrophysical Abstracts (weekly)
Monthly Weather Review (weekly)
Weather and Forecasting (weekly) . . . 13943

American Meteor Society (AMS), c/o Dept of Physics and Astronomy, State University College, 1 College Cir, Geneseo, NY 14454
T: (716) 245-5282; Fax: (716) 243-1901
Founded: 1911; Members: 100
Gen Secr: David D. Meisel
Focus: Geophys
Periodicals
Annual Report (weekly)
Meteor News (weekly) 13944

American Microchemical Society (AMS), c/o Leonard Klein, FMC Corp, POB 8, Princeton, NJ 08543
T: (609) 951-3422; Fax: (609) 951-3809
Founded: 1935; Members: 150
Gen Secr: Leonard Klein
Focus: Chem
Periodical
Microchemical Journal (weekly) 13945

American Microscopical Society (AMS), c/o Donald A. Munson, Dept of Biology, Chestertown, MD 21620
T: (410) 778-2800; Fax: (410) 778-0151
Founded: 1878; Members: 697
Gen Secr: Melvin W. Denner
Focus: Electronic Eng
Periodical
Transactions of the American Microscopical Society (weekly) 13946

American Montessori Society (AMS), 150 Fifth Av, Ste 203, New York, NY 10011
T: (212) 924-3209
Founded: 1960; Members: 12000
Gen Secr: Michael Eanes
Focus: Educ
Periodicals
Annual Report (weekly)
Constructive Triangle (weekly)
School Directory (weekly) 13947

American Mosquito Control Association (AMCA), POB 5416, Lake Charles, LA 70606
T: (318) 474-2723; Fax: (318) 478-9434
Founded: 2000; Members: 2000
Gen Secr: Mark Vinsand
Focus: Entomology
Periodicals
Mosquito News Journal of the American Mosquito Control Association (weekly)
Mosquito Systematics (3 times annually) . 13948

American Musical Instrument Society (AMIS), c/o The Shrine to Music Museum, 414 E Clark St, Vermillion, SD 57069
T: (605) 677-5306; Fax: (605) 677-5073
Founded: 1971; Members: 850
Gen Secr: Margaret Downie Banks
Focus: Music
Periodicals
Journal (weekly)
Membership Directory (weekly)
Newsletter (3 times annually) 13949

American Music Conference (AMC), 5140 Av Encinas, Carlsbad, CA 92008-4391
T: (619) 431-9124; Fax: (619) 438-7327
Founded: 1947; Members: 175
Gen Secr: Karl Bruhn
Focus: Music
Periodical
Music USA-Industry Statistics (weekly) . 13950

The American Musicological Society (AMS), c/o University of Pennsylvania, 201 S 34 St, Philadelphia, PA 19104-6313
T: (215) 898-8698
Founded: 1934; Members: 3500
Gen Secr: Alvin H. Johnson
Focus: Music
Periodicals
Directory (weekly)
Newsletter (weekly) 13951

American Name Society (ANS), c/o Prof. Wayne H. Finke, Dept of Modern Languages and Comparative Literature, Baruch College, POB 340, New York, NY 10010
T: (212) 387-1570; Fax: (212) 387-1591
Founded: 1951; Members: 900
Gen Secr: Prof. Wayne H. Finke
Focus: Ling
Periodical
Names (weekly) 13952

American Naprapathic Association (ANA), c/o Roy P. Krueger, 5913 W Montrose Av, Chicago, IL 60634
T: (312) 685-6020
Founded: 1909; Members: 300
Gen Secr: Roy P. Krueger
Focus: Med
Periodicals
Directory
The Voice of Naprapathy (weekly) . . . 13953

American National Standards Institute (ANSI), 11 W 42 St, New York, NY 10036
T: (212) 642-4900; Fax: (212) 398-0023
Founded: 1918; Members: 1250
Pres: Manuel Peralta
Focus: Standards
Periodicals
Catalog of Standards (weekly)
Progress Report
Standards Action (weekly) 13954

American Natural Hygiene Society (ANHS), POB 30630, Tampa, FL 33630
T: (813) 855-6607
Founded: 1948; Members: 6500
Gen Secr: James Michael Lennon
Focus: Hygiene
Periodical
Health Science Magazine (weekly) . . . 13955

American Nature Study Society (ANSS), c/o John A. Gustafson, 5881 Cold Brook Rd, Homer, NY 13077
T: (607) 749-3655
Founded: 1908; Members: 850
Gen Secr: Dr. John A. Gustafson
Focus: Nat Sci
Periodicals
Nature Study (weekly)
News (weekly) 13956

American Nepal Education Foundation (ANEF), 2790 Cape Meares Loop, Tillamook, OR 9714
T: (503) 842-4024; Fax: (503) 842-4654
Founded: 1955; Members: 350
Gen Secr: Hugh B. Wood
Focus: Educ 13957

American Neurological Association (ANA), 2221 University Av SE, Ste 350, Minneapolis, MN 55414
T: (612) 378-3290
Founded: 1875; Members: 860
Gen Secr: Linda J. Wilkerson
Focus: Neurology
Periodicals
Abstract Program (weekly)
Neurology (weekly) 13958

American Nuclear Energy Council (ANEC), 410 First St SE, Washington, DC 20003
T: (202) 484-2670
Founded: 1975; Members: 125
Pres: Edward M. Davis
Focus: Nucl Res 13959

American Nuclear Society (ANS), 555 N Kensington Av, La Grange Park, IL 60525
T: (312) 352-6611; Fax: (312) 352-0499
Founded: 1954; Members: 16000
Gen Secr: James G. Toscas
Focus: Nucl Res
Periodicals
Fusion Technology (weekly)
Nuclear News (weekly)
Nuclear Science & Engineering (weekly)
Nuclear Technology (weekly) 13960

American Numismatic Association (ANA), 818 N Cascade Av, Colorado Springs, CO 80903-3279
T: (719) 632-2646; Fax: (719) 634-4085
Founded: 1891; Members: 30000
Gen Secr: Robert J. Leuver
Focus: Numismatics
Periodical
The Numismatist (weekly) 13961

American Numismatic Society (ANS), Broadway between 155 & 156 Sts, New York, NY 10032
T: (212) 234-3130; Fax: (212) 234-3381
Founded: 1858; Members: 2424
Gen Secr: Leslie A. Elam
Focus: Numismatics
Periodicals
Ancient Coins in North American Collections (weekly)
ANS Museum Notes (weekly)
ANS Numismatic Studies (weekly)
Numismatic Literature (weekly)
Numismatic Notes and Monographs (weekly)

Sylloge Nummorum Graecorum/The Collection of the American Numismatic Society (weekly) *13962*

American Occupational Therapy Association (AOTA), 1383 Piccard Dr, Ste 301, Rockville, MD 20850-1725
T: (301) 948-9626; Fax: (301) 948-5529
Founded: 1917; Members: 45000
Gen Secr: Jeannette Bair
Focus: Adult Educ; Therapeutics
Periodical
American Journal of Occupational Therapy (weekly) *13963*

American Oil Chemists' Society (AOCS), POB 3489, Champaign, IL 61826-3489
T: (217) 359-2344; Fax: (217) 351-8091
Founded: 1909; Members: 4200
Gen Secr: James Lyon
Focus: Chem
Periodicals
Journal: JAOCS (weekly)
Lipids (weekly) *13964*

American Ophthalmological Society (AOS), c/o W. Banks Anderson, Eye Center, Duke University, Durham, NC 27710
T: (919) 684-5365
Founded: 1864; Members: 225
Gen Secr: W. Banks Anderson
Focus: Ophthal
Periodical
Transactions of the American Ophthalmological Society (weekly) *13965*

American Optometric Association (AOA), 243 N Lindbergh Blvd, Saint Louis, MO 63141
T: (314) 991-4100; Fax: (314) 991-4101
Founded: 1898; Members: 29000
Gen Secr: Earle L. Hunter
Focus: Ophthal
Periodicals
Journal (weekly)
News (weekly) *13966*

American Orchid Society (AOS), 6000 S Olive Av, West Palm Beach, FL 33405
T: (305) 585-8666; Fax: (305) 585-0654
Founded: 1921; Members: 27000
Gen Secr: Lee S. Cooke
Focus: Botany
Periodical
Bulletin (weekly) *13967*

American Oriental Society (AOS), c/o Harlan Hatcher Library 111E, University of Michigan, Ann Arbor, MI 48109-1205
T: (313) 747-4760
Founded: 1842; Members: 1400
Gen Secr: Jonathan Rodgers
Focus: Ethnology; Hist; Lit; Ling; Arts; Archeol
Periodicals
American Oriental Series
American Oriental Series Essays
Journal (weekly) *13968*

American Ornithologists' Union (AOU), c/o National Museum of Natural History, Smithsonian Institution, Washington, DC 20560
T: (202) 357-2051
Founded: 1883; Members: 5000
Gen Secr: Mary V. McDonald
Focus: Ornithology
Periodicals
The Autz (weekly)
Ornithological Monographs (weekly)
Ornithological Newsletter (weekly) . . . *13969*

American Orthodontic Society (AOS), 9550 Forest Ln, Ste 215, Dallas, TX 75243
T: (214) 343-0805
Founded: 1974; Members: 1900
Gen Secr: Murray Forsvall
Focus: Dent
Periodicals
Newsletter (weekly)
Technique Directory *13970*

American Orthopaedic Association (AOA), 6300 N River Rd, Ste 300, Rosemont, IL 60018-4263
T: (703) 318-7330
Founded: 1887; Members: 300
Gen Secr: Hildegard Klemm
Focus: Orthopedics
Periodical
The Manual of Orthopaedic Surgery . . *13971*

American Orthopaedic Foot and Ankle Society (AOFAS), 6300 N River Rd, Rosemont, IL 60018
T: (708) 698-1626; Fax: (708) 823-0536
Founded: 1969; Members: 550
Gen Secr: S. Ojenus
Focus: Orthopedics
Periodicals
In-Stride (weekly)
Journal of the Foot and Ankle (weekly) . *13972*

American Orthopaedic Society for Sports Medicine (AOSSM), 2250 E Devon Av, Ste 115, Des Plaines, IL 60018
T: (708) 803-8700; Fax: (708) 803-8653
Founded: 1972; Members: 1000
Gen Secr: Carol A. Rosegay
Focus: Orthopedics
Periodical
American Journal of Sports Medicine (weekly) *13973*

American Orthopsychiatric Association (AOA), 19 W 44 St, New York, NY 10036
T: (212) 354-5770; Fax: (212) 302-9463
Founded: 1923; Members: 10000
Gen Secr: Ernest Herman
Focus: Psychiatry; Psych; Pediatrics; Educ; Soc Sci; Law
Periodical
American Journal of Orthopsychiatry (weekly) *13974*

American Orthoptic Council (AOC), c/o Leslie France, 3914 Nakoma Rd, Madison, WI 53711
T: (608) 233-5383; Fax: (608) 263-7694
Members: 19
Gen Secr: Leslie France
Focus: Ophthal
Periodical
American Orthoptic Journal (weekly) . . *13975*

American Orthotic and Prosthetic Association (AOPA), 1650 King St, Ste 500, Alexandria, VA 22314
T: (703) 836-7116; Fax: (703) 836-0838
Founded: 1917; Members: 1250
Gen Secr: Dr. Ian R. Horen
Focus: Rehabil
Periodicals
Almanac (weekly)
Journal of Orthotics and Prosthetics (weekly) *13976*

American Osteopathic Academy of Addictionology, 270 Tavistock, Cherry Hill, NJ 08034-4017
T: (609) 795-0026; Fax: (609) 354-8029
Founded: 1986; Members: 225
Pres: William Vilensky
Focus: Orthopedics *13977*

American Osteopathic Academy of Sclerotherapy (AOAS), 107 Maple Av, Wilmington, DE 19809
T: (302) 792-0641
Founded: 1954; Members: 125
Gen Secr: Judy Wilbank
Focus: Therapeutics; Pathology
Periodical
Get the Point! (weekly) *13978*

American Osteopathic Academy of Sports Medicine (AOASM), 7611 Elmwood Av, Ste 201, Middleton, WI 53562
T: (608) 831-4400; Fax: (608) 831-5122
Founded: 1975; Members: 750
Gen Secr: Alice Halbrow
Focus: Sports; Med
Periodical
Journal of Osteopathic Sports Medicine (weekly) *13979*

American Osteopathic Association (AOA), 142 E Ontario St, Chicago, IL 60611
T: (312) 280-5800; Fax: (312) 280-5893
Founded: 1897; Members: 21735
Focus: Pathology
Periodicals
The D.O. (weekly)
Journal of AOA (weekly)
Yearbook & Directory (weekly) *13980*

American Osteopathic College of Allergy and Immunology (AOCAI), 3030 N Hayden, Scottsdale, AZ 85251
T: (602) 949-8898
Founded: 1974; Members: 75
Gen Secr: William Higgins
Focus: Immunology
Periodical
Newsletter (weekly) *13981*

American Osteopathic College of Anesthesiologists (AOCA), 17201 E US Hwy 40, Independence, MO 64055
T: (816) 373-4700; Fax: (816) 373-1529
Founded: 1952; Members: 500
Gen Secr: A.A. Mannarelli
Focus: Anesthetics; Pathology
Periodicals
Membership Directory (weekly)
Newsletter (3 times annually) *13982*

American Osteopathic College of Dermatology (AOCD), 1900 The Exchange, Ste 160, Atlanta, GA 30339-2022
T: (404) 953-0802; Fax: (404) 955-5538
Founded: 1955; Members: 130
Gen Secr: Cathy M. Garris
Focus: Derm
Periodicals
Directory (weekly)
Newsletter (weekly) *13983*

American Osteopathic College of Pathologists (AOCP), c/o Joan Gross, 12368 NW 13 Ct, Pembroke Pines, FL 33026
T: (305) 432-9640; Fax: (305) 432-9640
Founded: 1954; Members: 180
Gen Secr: Joan Gross
Focus: Pathology
Periodicals
American Osteopathic College of Pathologists Directory (weekly)
Nova (weekly) *13984*

American Osteopathic College of Proctology (AOCPr), 1020 Galloping Hill Rd, Union, NJ 07083
T: (908) 687-2062
Members: 155
Gen Secr: Dr. Z. Brody
Focus: Pathology; Gastroenter *13985*

American Osteopathic College of Radiology (AOCR), 119 E Second St, Milan, MO 63556
T: (816) 265-4011; Fax: (816) 265-3494
Founded: 1940; Members: 600
Gen Secr: Pamela A. Smith
Focus: Educ
Periodical
Viewbox (weekly) *13986*

American Osteopathic College of Rehabilitation Medicine (AOCRM), 9058 W Church, Des Plaines, IL 60016
T: (708) 699-0048; Fax: (708) 296-1366
Founded: 1954; Members: 125
Gen Secr: Julie Pickett
Focus: Rehabil; Pathology
Periodicals
Annual Directory (weekly)
Newsletter *13987*

American Otological Society (AOS), c/o Loyola University Medical School, 2160 S First Av, Maywood, IL 60153
T: (708) 216-8526; Fax: (919) 748-4204
Founded: 1868; Members: 220
Gen Secr: Dr. Gregory Matz
Focus: Otorhinolaryngology
Periodical
Transactions (weekly) *13988*

American Pancreatic Association (APA), c/o Surgical Service, VA Hospital, 16111 Plummer St, Sepulveda, CA 91343
T: (818) 895-9461; Fax: (818) 895-9535
Founded: 1970; Members: 325
Gen Secr: Howard A. Reber
Focus: Gastroenter *13989*

American Paper Institute (API), 260 Madison Av, New York, NY 10016
T: (212) 340-0600
Founded: 1964; Members: 166
Pres: Red Cavaney
Focus: Eng
Periodicals
Capacity Report (weekly)
Report (weekly)
Statistical Summary (weekly)
Statistics of Paper and Paperboard (weekly) *13990*

American Parkinson Disease Association (APDA), 60 Bay St, Ste 401, Staten Island, NY 10301
T: (718) 981-8001; Fax: (718) 981-4399
Founded: 1961; Members: 800
Pres: Mario Esposito
Focus: Neurology; Psychiatry
Periodicals
Annual Report (weekly)
Newsletter (weekly) *13991*

American Peace Society (APS), 1319 18 St NW, Washington, DC 20036-1802
T: (202) 296-6267
Founded: 1828; Members: 110
Pres: Dr. Evron M. Kirkpatrick
Focus: Poli Sci
Periodical
World Affairs (weekly) *13992*

American Peanut Research and Education Society (APRES), 376 AG Hall, Oklahoma State University, Stillwater, OK 74078
T: (405) 744-6421; Fax: (405) 744-5269
Founded: 1969; Members: 550
Gen Secr: J.R. Sholar
Focus: Food
Periodicals
Peanut Research (weekly)
Peanut Science (weekly)
Proceedings (weekly) *13993*

American Pediatric Society (APS), POB 675, Elk Grove Village, IL 60009-0675
T: (708) 427-0206; Fax: (708) 427-1305
Founded: 1888; Members: 1400
Gen Secr: Catherine De Angelis
Focus: Pediatrics
Periodicals
Pediatric Research-Program Issue (weekly)
Program and Abstracts of Annual Meeting *13994*

American Petroleum Institute (API), 1220 L St NW, Washington, DC 20005
T: (202) 682-8000; Fax: (202) 682-8030
Founded: 1919; Members: 250
Pres: Charles J. DiBona
Focus: Energy
Periodicals
API Report (weekly)
Directory (weekly)
Imported Crude Oil and Petroleum Products (weekly)
Inventories of Natural Gas Liquids and Liquefied Refinery Gases (weekly) *13995*

American Pharmaceutical Association (APhA), 2215 NW Constitution Av, Washington, DC 20037
T: (202) 628-4410; Fax: (202) 783-2351
Founded: 1852; Members: 40000
Gen Secr: Dr. John A. Gans
Focus: Pharmacol
Periodicals
Academy Reporter (weekly)
American Pharmacy (weekly)
Journal of Pharmaceutical Sciences (weekly)
The Pharmacy Student (weekly)
Pharmacy Today (weekly) *13996*

American Philological Association (APA), c/o Dept of Classics, College of the Holy Cross, Worcester, MA 01610-2395
T: (508) 793-2203; Fax: (508) 793-3428
Founded: 1869; Members: 3000
Gen Secr: Prof. William Ziobro
Focus: Lit
Periodicals
American Classical Studies (weekly)
Newsletter (weekly)
Philological Monographs (weekly)
Textbook Series (weekly)
Transactions (weekly) *13997*

American Philosophical Association (APA), c/o David A. Hoekema, University of Delaware, Newark, DE 19716
T: (302) 831-1112; Fax: (302) 831-8690
Founded: 1900; Members: 9000
Gen Secr: David A. Hoekema
Focus: Philos
Periodicals
APA Newsletter (weekly)
Jobs for Philosophers (5 times annually) *13998*

American Philosophical Society (APS), 104 S Fifth St, Philadelphia, PA 19106-3387
T: (215) 440-3400; Fax: (215) 440-3436
Founded: 1743; Members: 690
Gen Secr: Herman H. Goldstine
Focus: Hist; Sci; Archeol
Periodicals
Proceedings (weekly)
Transactions (8 times annually) *13999*

American Physical Society (APS), 335 E 45 St, New York, NY 10017-3483
T: (212) 682-7341; Fax: (212) 687-2532
Founded: 1899; Members: 43000
Gen Secr: N.N. Richard Werthamer
Focus: Physics
Periodicals
Bulletin (weekly)
Membership Directory (bi-annually)
Physical Review (weekly)
Physical Review Letters (weekly)
Reviews of Modern Physics (weekly) . . *14000*

American Physical Therapy Association (APTA), 1111 N Fairfax St, Alexandria, VA 22314
T: (703) 684-2782
Founded: 1921; Members: 53456
Gen Secr: William D. Coughlan
Focus: Therapeutics
Periodicals
Clinical Management in Physical Therapy (weekly)
Physical Therapy (weekly)
Physical Therapy Resource and Buyer's Guide (weekly)
Progress Report (weekly) *14001*

American Physicians Association of Computer Medicine (APACM), 10 N Main St, Pittsford, NY 14534
T: (716) 586-8159
Founded: 1984; Members: 350
Pres: Lawrence B. Tillis
Focus: Med; Computer & Info Sci . . . *14002*

American Physicians Fellowship for Medicine in Israel (APF), 2001 Beacon ST, Brookline, MA 02146
T: (617) 232-5382
Founded: 1950; Members: 8500
Gen Secr: Daniel Goldfarb
Focus: Physiology
Periodicals
Koroth (weekly)
News (weekly) *14003*

American Physicians Poetry Association (APPA), 230 Toll Dr, Southampton, PA 18966
T: (215) 364-2990
Founded: 1976; Members: 150
Gen Secr: Richard A. Lippin
Focus: Toxicology
Periodicals
Membership List (weekly)
Newsletter (weekly) *14004*

American Physiological Society (APS), 9650 Rockville Pike, Bethesda, MD 20814
T: (301) 530-7164; Fax: (301) 571-1814
Founded: 1887; Members: 6600
Gen Secr: Martin Frank
Focus: Physiology
Periodicals
Advances in Physiology Education (weekly)
American Journal of Physiology (weekly)
American Journal of Physiology: Cell Physiology (weekly)
American Journal of Physiology: Endocrinology and Metabolism (weekly)
American Journal of Physiology: Gastrointestinal and Liver Physiology (weekly)
American Journal of Physiology: Heart and Circulatory Physiology (weekly)
American Journal of Physiology: Lung Cellular and Molecular Physiology (weekly)
American Journal of Physiology: Regulatory, Integrative and Comparative Physiology (weekly)

American Journal of Physiology: Renal, Fluid and Electrolyte Physiology (weekly)
Journal of Applied Physiology (weekly)
Journal of Neurophysiology (weekly)
News in Physiological Sciences (weekly)
Physiological Reviews (weekly)
The Physiologist (weekly) 14005

American Phytopathological Society (APS), 3340 Pilot Knob Rd, Saint Paul, MN 55121-2097
T: (612) 454-7250; Fax: (612) 454-0766
Founded: 1908; Members: 4474
Gen Secr: Steven C. Nelson
Focus: Botany
Periodicals
Biological and Cultural Tests for Control of Plant Diseases (weekly)
Molecular Plant-Microbe Interactions (weekly)
Phytopathology (weekly)
Plant Disease (weekly) 14006

American Planning Association (API), 1776 Massachusetts Av NW, Washington, DC 20036
T: (202) 872-0611; Fax: (202) 872-0643
Founded: 1909; Members: 27000
Gen Secr: Israel Stollman
Focus: Econ
Periodicals
JobMart (22 times annually)
Land Use Law and Zoning Digest (weekly)
PAS Memo (weekly)
Planning Advisory Service Report (8 times annually)
Planning Magazine (weekly)
Zoning News (weekly) 14007

American Podiatric Medical Association (APMA), 9312 Old Georgetown Rd, Bethesda, MD 20814
T: (301) 571-9200; Fax: (301) 530-2752
Founded: 1912; Members: 9200
Gen Secr: Frank J. Malouff
Focus: Orthopedics
Periodicals
Journal (weekly)
News (weekly) 14008

American Polar Society (APS), c/o Raz Mereier, 108 Scott Hall, Columbus, OH 43210
T: (614) 292-6531; Fax: (614) 292-4697
Founded: 1934; Members: 2060
Gen Secr: Raz Mereier
Focus: Geography
Periodical
The Polar Times (weekly) 14009

American Political Science Association (APSA), 1527 New Hampshire Av NW, Washington, DC 20036
T: (202) 483-2512; Fax: (202) 483-2657
Founded: 1903; Members: 12000
Gen Secr: Catherine E. Rudder
Focus: Poli Sci
Periodicals
The American Political Science Review (weekly)
News for Teachers of Political Science (weekly)
PS (weekly) 14010

American Pomological Society (APS), 102 Tyson Bldg, University Park, PA 16802
T: (519) 738-2251
Founded: 1848; Members: 1000
Gen Secr: Dr. Robert M. Grassweller
Focus: Hort; Food
Periodical
Fruit Varieties Journal (weekly) 14011

American Poultry Historical Society (APHS), c/o G. Carpenter, Div of Animal and Veterinary Science, University of West Virginia, Agricultural Science Bldg, Morgantown, WV 26506-6108
T: (304) 293-5229; Fax: (304) 293-6954
Founded: 1952; Members: 240
Gen Secr: G. Carpenter
Focus: Agri; Hist
Periodical
Newsletter (weekly) 14012

American Press Institute (API), 11690 Sunrise Valley Dr, Reston, VA 22091
T: (703) 620-3611; Fax: (703) 620-5814
Gen Secr: William L. Winter
Focus: Journalism
Periodical
Bulletin (weekly) 14013

American Printing History Association (APHA), POB 4922, Grand Central Station, New York, NY 10163
Founded: 1974; Members: 1000
Gen Secr: Stephen Crook
Focus: Libraries & Bk Sci; Hist
Periodicals
APHA Newsletter (weekly)
Printing History Journal (weekly) 14014

American Production and Inventory Control Society (APICS), 500 W Annandale Rd, Falls Church, VA 22046
T: (703) 237-8344; Fax: (703) 237-1071
Founded: 1957; Members: 70000
Gen Secr: Paul Sheehan
Focus: Econ
Periodicals
Journal (weekly)
News Production and Inventary Management (weekly) 14015

American Prosthodontic Society (APS), 919 N Michigan Av, Ste 460, Chicago, IL 60611
T: (312) 944-7618
Founded: 1928; Members: 1300
Gen Secr: Howard J. Harvey
Focus: Dent
Periodical
Journal of Prosthetic Dentistry (weekly) . 14016

American Psychiatric Association (APA), 1400 K St NW, Washington, DC 20005
T: (202) 682-6000; Fax: (202) 682-6114
Founded: 1844; Members: 39000
Gen Secr: Melvin Sabshin
Focus: Psychiatry
Periodicals
American Journal of Psychiatry (weekly)
American Psychiatric Press Review of Psychiatry (weekly)
Hospital & Community Psychiatry (weekly) 14017

American Psychoanalytic Association (APsaA), 309 E 49 St, New York, NY 10017
T: (212) 752-0450; Fax: (212) 593-1571
Founded: 1911; Members: 3025
Gen Secr: Helen Fischer
Focus: Psychiatry
Periodicals
Journal (weekly)
Newsletter (weekly) 14018

American Psychological Association (APA), 750 First St NE, Washington, DC 20002-4242
T: (202) 336-5500
Founded: 1892; Members: 70000
Gen Secr: R.D. Fowler
Focus: Psych
Periodicals
American Psychologist (weekly)
APA Monitor (weekly)
Behavioral Neuroscience (weekly)
Comparative Psychology and Behavior (weekly)
Contemporary Psychology (weekly)
Developmental Psychology (weekly)
Journal of Abnormal Psychology (weekly)
Journal of Applied Psychology (weekly)
Journal of Consulting and Clinical Psychology (weekly)
Journal of Counseling Psychology (weekly)
Journal of Educational Psychology (weekly)
Journal of Experimental Psychology: Animal Behavior Processes (weekly)
Journal of Experimental Psychology: Human Perception and Performance (weekly)
Journal of Experimental Psychology: Learning, Memory and Cognition (weekly)
Journal of Experiment Psychology: General (weekly)
Journal of Personality and Social Psychology (weekly)
Professional Psychology (weekly)
Psychological Abstracts (weekly)
Psychological Bulletin (weekly)
Psychological Documents (weekly)
Psychological Review (weekly)
PsycSCAN: Applied Psychology (weekly)
PsycSCAN: Clinical Psychology (weekly)
PsycSCAN: Development Psychology (weekly)
PsycSCAN: LD/MR (weekly)
PsycSCAN Series from PsycINFO (weekly) 14019

American Psychology-Law Society (AP-LS), c/o Dr. Tom Grisso, Dept of Psychology, University of Massachusetts Medical Center, 55 Lake Av N, Worcester, MA 01655
T: (508) 856-3625
Founded: 1968; Members: 1700
Pres: Dr. Tom Grisso
Focus: Psych; Law
Periodicals
Journal of Law and Human Behavior (weekly)
Newsletter (weekly) 14020

American Psychopathological Association (APPA), c/o Dr. Ellen Frank, Western Psychiatric Institute and Clinic, 3811 O'Hara, Pittsburgh, PA 15213
T: (412) 624-2383; Fax: (412) 624-9155
Founded: 1912; Members: 500
Gen Secr: Dr. Ellen Frank
Focus: Psychiatry
Periodicals
Comprehensive Psychiatry (weekly)
Proceedings of Annual Meeting 14021

American Public Health Association (APHA), 1015 15 St NW, Washington, DC 20005
T: (202) 789-5600; Fax: (202) 789-5681
Founded: 1872; Members: 31500
Gen Secr: William H. McBeath
Focus: Public Health
Periodicals
American Journal of Public Health (weekly)
The National Health (weekly) 14022

American Quaternary Association (AMQUA), c/o Julie Brigham-Grette, Dept of Geology and Geography, University of Massachusetts, Amherst, MA 01003
T: (413) 545-4840; Fax: (413) 545-1200
Founded: 1969; Members: 1200
Gen Secr: Julie Brigham-Grette
Focus: Nat Sci
Periodical
Newsletter (weekly) 14023

American Quilt Study Group (AQSG), 660 Mission St, Ste 400, San Francisco, CA 94105-4007
Founded: 1980; Members: 700
Gen Secr: Sarah K. Howard
Focus: Hist
Periodical
Blanket Statements (weekly) 14024

American Radium Society (ARS), 1101 Market St, Philadelphia, PA 19107
T: (215) 574-3179; Fax: (215) 928-0153
Founded: 1916; Members: 850
Gen Secr: Suzanne Bohn
Focus: Radiology
Periodical
Membership Directory (weekly) 14025

American Real Estate and Urban Economics Association (AREUEA), c/o School of Business, Indiana University, Bloomington, IN 47405
T: (812) 855-7794; Fax: (812) 855-8679
Founded: 1965; Members: 1260
Gen Secr: George H. Lentz
Focus: Econ
Periodicals
Journal (weekly)
Newsletter (weekly) 14026

American Registry of Radiologic Technologists (ARRT), 1255 Northland Dr, Mendota Heights, MN 55120
T: (612) 687-0048
Founded: 1922; Members: 186000
Gen Secr: Jerry B. Reid
Focus: Radiology
Periodical
Annual Directory of Registered Technologists (weekly) 14027

American Rehabilitation Counseling Association (ARCA), c/o American Counseling Association, 5999 Stevenson Av, Alexandria, VA 22304
T: (703) 823-9800; Fax: (703) 823-0252
Founded: 1958; Members: 3000
Gen Secr: Dr. Theodore P. Remley
Focus: Rehabil
Periodicals
Newsletter (weekly)
Rehabilitation Counseling Bulletin (weekly) 14028

American Research Institute in Turkey (ARIT), c/o University Museum, 33 and Spruce Sts, Philadelphia, PA 19104
T: (215) 898-3473
Founded: 1964; Members: 32
Pres: G. Kenneth Sams
Focus: Ethnology
Periodical
Newsletter 14029

American Revolution Round Table (ARRT), 1 Ascan Av, Forest Hills, NY 11375
T: (718) 793-1505
Founded: 1958; Members: 120
Pres: Jane Lawton
Focus: Hist
Periodical
Newsletter (5 times annually) 14030

American Rhinologic Society (ARS), c/o Frank Lucente, Long Island College Hospital, Brooklyn, NY 11201
T: (718) 780-1281; Fax: (718) 802-1036
Founded: 1954; Members: 604
Gen Secr: Frank Lucente
Focus: Otorhinolaryngology
Periodicals
Journal of International Rhinology (1-4 times annually)
Membership Directory (bi-annually)
Newsletter (4-5 times annually) 14031

American Risk and Insurance Association (ARIA), c/o Dr. Patricia Cheshier, Dept of Management, School of Business, 6000 J St, Sacramento, CA 95819
T: (916) 278-6609; Fax: (916) 278-5437
Founded: 1932; Members: 2000
Gen Secr: Dr. Patricia Cheshier
Focus: Insurance
Periodical
The Journal of Risk and Insurance (weekly) 14032

American Rock Art Research Association (ARARA), POB 65, San Miguel, CA 93451
T: (805) 467-3704; Fax: (805) 467-2532
Gen Secr: A.J. Bock
Focus: Archeol 14033

American Roentgen Ray Society (ARRS), c/o Paul R. Fullagar, 1891 Preston White Dr, Reston, VA 22091
T: (703) 648-8992; Fax: (703) 264-8863
Founded: 1900; Members: 6150
Gen Secr: Paul R. Fullagar
Focus: X-Ray Tech
Periodicals
American Journal of Roentgenology (weekly)
Membership Directory (weekly) 14034

American Schizophrenia Association (ASA), 900 N Federal Hwy, Ste 330, Boca Raton, FL 33432
T: (305) 393-6167
Founded: 1964; Members: 3000
Gen Secr: Mary Roddy Haggerty
Focus: Psychiatry 14035

American School Health Association (ASHA), 7263 State Rte 43, POB 708, Kent, OH 44240
T: (216) 678-1601; Fax: (216) 678-4526
Founded: 1927; Members: 4000
Gen Secr: Dana A. Davis
Focus: Public Health
Periodical
Journal of School Health (10 times annually) 14036

American Schools Association (ASA), 3069 Amwiler Rd, Ste 4, Atlanta, GA 30360
T: (404) 449-7141
Founded: 1914
Pres: Carl M. Dye
Focus: Educ
Periodical
Directory of College Transfer Information (bi-annually) 14037

American Security Council (ASC), c/o Washington Communications Center, Boston, VA 22713
T: (703) 547-1776
Founded: 1955; Members: 325700
Pres: John M. Fisher
Focus: Safety
Periodical
National Security Report (weekly) . . . 14038

American Security Council Foundation (ASCF), c/o Washington Communications Center, Boston, VA 22713
T: (703) 547-1776
Founded: 1958
Pres: John M. Fisher
Focus: Safety 14039

Americans for Energy Independence (AFEI), 1629 K St NW, Ste 602, Washington, DC 20006
T: (202) 466-2105
Founded: 1975
Gen Secr: Elihu Bergman
Focus: Energy
Periodical
Newsletter (weekly) 14040

American Siam Society, 633 24 St, Santa Monica, CA 90402-3135
T: (213) 393-1176
Founded: 1956; Members: 32
Pres: H. Carroll Parish
Focus: Ethnology 14041

American Social Health Association (ASHA), POB 13827, Research Triangle Park, NC 27709
T: (919) 361-8400; Fax: (919) 361-8425
Founded: 1914; Members: 14000
Gen Secr: Peggy Clarke
Focus: Public Health
Periodical
Helper (weekly) 14042

American Society for Adolescent Psychiatry (ASAP), 4330 East West Hwy, Ste 1117, Bethesda, MD 20814
T: (301) 718-6502; Fax: (301) 656-0989
Founded: 1967; Members: 1500
Gen Secr: David Lewis
Focus: Psychiatry
Periodicals
Annals of Adolescent Psychiatry (weekly)
Newsletter (3 times annually) 14043

American Society for Advancement of Anesthesia in Dentistry (ASAAD), c/o Dr. Louis L. Zall, 11245 W Atlantic Blvd, Apt 301, Coral Springs, FL 33071-5112
Founded: 1929; Members: 400
Gen Secr: Dr. Louis L. Zall
Focus: Anesthetics
Periodicals
Modern Pain Control (weekly)
Pain Control in Dentistry (weekly)
Proceedings (bi-annually)
Transcripts (weekly) 14044

American Society for Aesthetic Plastic Surgery (ASAPS), 3922 Atlantic Av, Long Beach, CA 90807
T: (310) 595-4275; Fax: (310) 427-2234
Founded: 1967; Members: 1052
Gen Secr: Robert G. Stanton
Focus: Surgery
Periodical
Aesthetic Surgery (3 times annually) . . 14045

American Society for Artificial Internal Organs (ASAIO), POB C, Boca Raton, FL 33429-0468
T: (407) 391-8589
Founded: 1955; Members: 1650
Gen Secr: Dr. Brendan P. Teehan
Focus: Eng; Surgery
Periodicals
Abstracts (weekly)
Author Index
Primers (weekly)
Transactions (weekly) 14046

American Society for Biochemistry and Molecular Biology (ASBMC), 9650 Rockville Pike, Bethesda, MD 20814-3996
T: (301) 530-7145; Fax: (301) 571-1824
Founded: 1906; Members: 8900
Gen Secr: Charles C. Hancock
Focus: Biochem
Periodical
Journal of Biological Chemistry (3 times monthly) 14047

American Society for Bone and Mineral Research (ASBMR), 1101 Connecticut Av NW, Ste 700, Washington, DC 20036
T: (202) 857-1161; Fax: (202) 223-4579
Founded: 1977; Members: 1850
Pres: Mark Haussler
Focus: Biochem
Periodicals
Membership Roster (weekly)
Newsletter (3 times annually) *14048*

American Society for Cell Biology (ASCB), 9650 Rockville Pike, Bethesda, MD 20814
T: (301) 530-7153; Fax: (301) 530-7139
Founded: 1960; Members: 7100
Gen Secr: Dorothea C. Wilson
Focus: Cell Biol & Cancer Res
Periodicals
Journal of Cell Biology (weekly)
Newsletter (weekly) *14049*

American Society for Clinical Investigation (ASCI), 6900 Grove Rd, Thorofare, NJ 08086
T: (609) 848-1000; Fax: (609) 848-5247
Founded: 1909; Members: 2400
Gen Secr: Andrew Schafer
Focus: Med
Periodical
Journal of Clinical Investigation (weekly) . *14050*

American Society for Clinical Nutrition (ASCN), 9650 Rockville Pike, Bethesda, MD 20814-3998
T: (301) 530-7110; Fax: (301) 571-1892
Founded: 1959; Members: 1200
Gen Secr: S. Stephen Schiaffino
Focus: Food
Periodical
The American Journal of Clinical Nutrition (weekly) *14051*

American Society for Colposcopy and Cervical Pathology (ASCCP), c/o American College of Obstetricians amd Gynecologists, 409 12 St SW, Washington, DC 20024
Fax: (202) 484-5107
Founded: 1964; Members: 2000
Gen Secr: Kathleen Poole
Focus: Gynecology
Periodical
The Colposcopist (weekly) *14052*

American Society for Conservation Archaeology (ASCA), c/o Anthony Lutonski, 435 Montano Rd NE, Albuquerque, NM 87107
T: (505) 761-8792; Fax: (505) 761-8911
Founded: 1974; Members: 300
Gen Secr: Anthony Lutonski
Focus: Archeol
Periodical
ASCA Report (weekly) *14053*

American Society for Cybernetics (ASC), c/o Dept of Decision Sciences, George Mason University, Fairfax, VA 22030
T: (800) 422-2319; Fax: (800) 422-2319
Founded: 1964; Members: 250
Gen Secr: Stephen Ruth
Focus: Cybernetics
Periodicals
Conference Proceedings (weekly)
Cybernetic (weekly)
Newsletter (weekly) *14054*

American Society for Cytotechnology (ASCT), 920 Paverstone Dr, Raleigh, NC 27614
T: (919) 848-9911; Fax: (919) 848-9853
Founded: 1979; Members: 1500
Gen Secr: Margaret A. Bundy
Focus: Cell Biol & Cancer Res
Periodical
Newsletter (weekly) *14055*

American Society for Dental Aesthetics (ASDA), 635 Madison Av, New York, NY 10022
T: (212) 751-3263; Fax: (212) 308-5182
Founded: 1978; Members: 109
Gen Secr: Diana Okula
Focus: Dent
Periodical
Newsletter (weekly) *14056*

American Society for Dermatologic Surgery (ASDS), POB 4014, Schaumburg, IL 60204
T: (708) 330-0230; Fax: (708) 869-4382
Founded: 1970; Members: 2157
Gen Secr: Lawrence F. Rosenthal
Focus: Derm; Surgery
Periodicals
Journal of Dermatologic Surgery and Oncology (weekly)
Roster (weekly) *14057*

American Society for Eighteenth-Century Studies (ASECS), c/o Utah State University, USU CC108, Logan, UT 84322-3730
T: (801) 750-4065; Fax: (801) 750-4065
Founded: 1969; Members: 2000
Gen Secr: Jeffrey Smitten
Focus: Hist; Lit; Sci; Poli Sci; Philos
Periodicals
Eighteenth-Century Studies (weekly)
Studies in Eighteenth-Century Culture (weekly) *14058*

American Society for Engineering Management (ASEM), POB 867, Annapolis, MD 21401
T: (410) 263-7065
Founded: 1979; Members: 800
Gen Secr: Richard Jedlicka
Focus: Eng
Periodicals
Engineering Management International (weekly)
Newsletter (weekly)
Proceedings of Annual Meeting *14059*

American Society for Ethnohistory (ASE), c/o Newberry Library, 60 W Walton St, Chicago, IL 60610
T: (219) 875-7237
Founded: 1953; Members: 1300
Gen Secr: William O. Autry
Focus: Hist; Ethnology; Cultur Hist
Periodical
Ethnohistory (weekly) *14060*

American Society for Gastrointestinal Endoscopy (ASGE), 13 Elm St, POB 1565, Manchester, MA 01944
T: (508) 526-8330; Fax: (508) 526-4018
Founded: 1941; Members: 5000
Gen Secr: William T. Maloney
Focus: Gastroenter
Periodical
Gastrointestinal Endoscopy (weekly) . . . *14061*

American Society for Geriatric Dentistry (ASGD), 211 E Chicago Av, Chicago, IL 60611
T: (312) 440-2661
Founded: 1965; Members: 450
Gen Secr: Paul Van Ostenberg
Focus: Dent; Geriatrics
Periodical
Newsletter (weekly) *14062*

American Society for Head and Neck Surgery (ASHNS), c/o John Hopkins Hospital, Carnegie Bldg, Rm 466, POB 41402, Baltimore, MD 21203
T: (410) 955-3669; Fax: (410) 955-0035
Founded: 1959; Members: 600
Gen Secr: Dr. Charles Cummings
Focus: Surgery; Otorhinolaryngology
Periodical
Archives and Otolaryngology (weekly) . . *14063*

American Society for Healthcare Education and Training of the American Hospital Association (ASHET), 840 N Lake Shore Dr, Chicago, IL 60611
T: (312) 280-3556
Founded: 1970; Members: 1500
Gen Secr: Beverly J. Rogers
Focus: Educ; Public Health
Periodicals
Healthcare Education Dataline (3 times annually)
Journal of Healthcare Education and Training *14064*

American Society for Healthcare Human Resources Administration (ASHHRA), c/o American Hospital Association, 840 N Lake Shore Dr, Chicago, IL 60611
T: (312) 280-6722; Fax: (312) 280-4152
Founded: 1964; Members: 2550
Gen Secr: Beverly J. Rogers
Focus: Med
Periodicals
Hospital (weekly)
Human Resources Administration (weekly) *14065*

American Society for Horticultural Science (ASHS), 113 S West St, Ste 400, Alexandria, VA 22314-2824
T: (703) 836-4606; Fax: (703) 836-2024
Founded: 1903; Members: 5000
Gen Secr: Skip McAfee
Focus: Hort
Periodicals
ASHS Newsletter (weekly)
HortScience (weekly)
HortTechnology (weekly)
Journal of the American Society for Horticultural Science (weekly) *14066*

American Society for Hospital Engineering (ASHE), c/o American Hospital Association, 840 N Lake Shore Dr, Chicago, IL 60611
T: (312) 280-5223; Fax: (312) 280-6786
Founded: 1962; Members: 5500
Gen Secr: Nancy Montenegro
Focus: Med; Eng
Periodicals
Hospital Codes and Standards Update Letters (4-6 times annually)
Hospital Engineering Bulletin (weekly)
Technical Documents (weekly) *14067*

American Society for Information Science (ASIS), 8720 Georgia Av, Ste 501, Silver Spring, MD 20910-3602
T: (301) 495-0900; Fax: (301) 495-0810
Founded: 1937; Members: 4000
Gen Secr: Richard B. Hill
Focus: Computer & Info Sci
Periodicals
Bulletin (weekly)
Jobline (weekly)
Journal (weekly) *14068*

American Society for Laser Medicine and Surgery (ASLMS), 2404 Stewart Sq, Wausau, WI 54401
T: (715) 845-9283; Fax: (715) 848-2493
Founded: 1980; Members: 1728
Gen Secr: Ellet H. Drake
Focus: Radiology; Surgery
Periodicals
Lasers in Surgery and Medicine (weekly)
Official ASLMS Newsletter (weekly) . . . *14069*

American Society for Legal History (ASLH), c/o Prof. M. de Landon, Dept of History, University of Mississippi, University, MS 38677
T: (601) 232-7148; Fax: (601) 232-5918
Founded: 1956; Members: 1100
Gen Secr: Prof. M. de Landon
Focus: Law
Periodicals
Law and History Review (weekly)
Newsletter (weekly) *14070*

American Society for Mass Spectrometry (ASMS), 815 Don Gaspar, Santa Fe, NM 82501
T: (505) 989-4517
Founded: 1969; Members: 2000
Gen Secr: Judith A. Sjoberg
Focus: Astronomy *14071*

American Society for Medical Technology (ASMT), 7910 Woodmont Av, Ste 1301, Bethesda, MD 20814
T: (301) 657-2768; Fax: (301) 657-2909
Founded: 1932; Members: 20000
Gen Secr: Lynn P. Robinson
Focus: Med; Eng
Periodical
Clinical Laboratory Science (weekly) . . *14072*

American Society for Microbiology (ASM), 1325 Massachusetts Av NW, Washington, DC 20005
T: (202) 737-3600; Fax: (202) 887-0327
Founded: 1899; Members: 39000
Gen Secr: Michael I. Goldberg
Focus: Microbio
Periodicals
Antinicrobial Agents and Chemotherapy (weekly)
Applied and Environmental Microbiology (weekly)
ASM News (weekly)
Clinical Microbiology Review (weekly)
International Journal of Systematic Bacteriology (weekly)
Journal of Bacteriology (weekly)
Journal of Clinical Microbiology (weekly)
Journal of Virology (weekly)
Microbiological Review (weekly) *14073*

American Society for Neurochemistry (ASN), c/o Dept of Pathology Research, Veterans Medical Center, 3801 Miranda Av, Palo Alto, CA 94304
T: (415) 493-5000; Fax: (415) 725-7023
Founded: 1969; Members: 1018
Gen Secr: Dr. Lawrence F. Eng
Focus: Chem; Neurology
Periodicals
Membership Directory (weekly)
Newsletter (weekly)
Transactions (weekly) *14074*

American Society for Nondestructive Testing (ASNT), 1711 Arlingate Ln, POB 28518, Columbus, OH 43228-0518
T: (614) 274-6003; Fax: (614) 274-6899
Founded: 1941; Members: 11000
Gen Secr: Timothy B. Strawn
Focus: Materials Sci
Periodicals
Materials Evaluation (weekly)
Research in Nondestructive Evaluation (weekly) *14075*

American Society for Parenteral and Enteral Nutrition (ASPEN), 8630 Fenton St, Ste 412, Silver Spring, MD 20910-3805
T: (301) 587-6315; Fax: (301) 587-2365
Founded: 1975; Members: 7200
Gen Secr: Barney Sellers
Focus: Hematology
Periodicals
Journal of Parenteral and Enteral Nutrition (weekly)
Nutrition in Clinical Practice (weekly) . . *14076*

American Society for Pharmacology and Experimental Therapeutics (ASPET), 9650 Rockville Pike, Bethesda, MD 20814-3995
T: (301) 530-7060
Founded: 1908; Members: 4200
Gen Secr: Kay A. Croker
Focus: Pharmacol
Periodicals
Drug Metabolism and Disposition (weekly)
Journal of Pharmacology and Experimental Therapeutics (weekly)
Molecular Pharmacology (weekly)
Pharmacological Reviews (weekly)
The Pharmacologist (weekly) *14077*

American Society for Pharmacy Law (ASPL), c/o Donald A. Dee, POB 2184, Vienna, VA 22183
T: (701) 281-0107
Founded: 1974; Members: 800
Gen Secr: Donald A. Dee
Focus: Law; Pharmacol
Periodicals
ASPL Membership Directory (weekly)
RX Ipsa Loquitur (weekly) *14078*

American Society for Photobiology (ASP), Biotech Park, 1021 15 St, Ste 9, Augusta, GA 30901
T: (706) 722-7511; Fax: (706) 721-3048
Founded: 1972; Members: 1606
Gen Secr: Sherwood M. Reichard
Focus: Bio
Periodicals
Directory and Constitution (bi-annually)
Newsletter (weekly)
Photochemistry and Photobiology (weekly) *14079*

American Society for Photogrammetry and Remote Sensing, 5410 Grosvenor Ln, Ste 210, Bethesda, MD 20814-2160
T: (301) 493-0290; Fax: (301) 493-0208
Founded: 1934; Members: 8200
Gen Secr: William D. French
Focus: Surveying
Periodical
Photogrametric Engineering and Remote Sensing (weekly) *14080*

American Society for Political and Legal Philosophy (ASPLP), c/o Prof. Kenneth Winston, Wheaton College, Knapton Hall, Norton, MA 02766
T: (508) 285-7722
Founded: 1955; Members: 500
Gen Secr: Prof. Kenneth Winston
Focus: Philos; Poli Sci *14081*

American Society for Psychical Research (ASPR), 5 W 73 St, New York, NY 10023
T: (212) 799-5050; Fax: (212) 496-2497
Founded: 1885; Members: 2000
Gen Secr: Patrice Keane
Focus: Psych
Periodicals
ASPR Newsletter (weekly)
Journal of the American Society for Psychical Research (weekly) *14082*

American Society for Psychoprophylaxis in Obstetrics (ASPO/Lamaze), 1101 Connecticut Av NW, Ste 700, Washington, DC 20036
T: (202) 857-1128; Fax: (202) 223-4579
Founded: 1960; Members: 5000
Gen Secr: Linda Harmon
Focus: Gynecology
Periodicals
Annual Directory (weekly)
Genesis (weekly) *14083*

American Society for Public Administration (ASPA), 1120 G St NW, Ste 700, Washington, DC 20005
T: (202) 393-7878; Fax: (202) 393-7878
Founded: 1939; Members: 13000
Gen Secr: John P. Thomas
Focus: Public Admin
Periodicals
Public Administration Review (weekly)
Public Administration Times (weekly) . . *14084*

American Society for Quality Control (ASQC), 611 E Wisconsin Av, Milwaukee, WI 53202
T: (414) 272-8575; Fax: (414) 272-1734
Founded: 1946; Members: 68500
Gen Secr: Paul E. Borawski
Focus: Materials Sci
Periodicals
Journal of Quality Technology (weekly)
Quality Engineering (weekly)
Quality Progress (weekly)
The Quality Review (weekly)
Technometrics (weekly) *14085*

American Society for Stereotactic and Functional Neurosurgery (ASSFN), c/o Philip L. Gildenberg, 6560 Fannin, Ste 1530, Houston, TX 77030
T: (713) 790-0795
Founded: 1968; Members: 200
Gen Secr: Philip L. Gildenberg
Focus: Neurology
Periodical
Applied Neurophysiology (weekly) . . . *14086*

American Society for Surgery of the Hand (ASSH), 3025 S Parker Rd, Ste 65, Aurora, CO 80014
T: (303) 755-4588; Fax: (303) 755-8702
Founded: 1946; Members: 1180
Gen Secr: Gail M. Gorman
Focus: Surgery
Periodical
Journal of Hand Surgery (weekly) . . . *14087*

American Society for Technion-Israel Institute of Technology (ASTITT), 810 Seventh Av, New York, NY 10019
T: (212) 262-6200; Fax: (212) 262-6155
Founded: 1940; Members: 24000
Gen Secr: Melvyn H. Bloom
Focus: Eng
Periodicals
Technion Magazine (weekly)
Technion USA Magazine (weekly)
Update Newsletter (weekly) *14088*

American Society for Testing and Materials (ASTM), 1916 Race St, Philadelphia, PA 19103-1187
T: (215) 299-5400; Fax: (215) 977-8679
Founded: 1898; Members: 33000
Pres: James A. Thomas
Focus: Standards
Periodical
Annual Book of ASTM Standards (weekly) *14089*

American Society for Theatre Research (ASTR), c/o Theatre Dept, University of Rhode Island, Kingston, RI 02881-0824
T: (401) 792-2706; Fax: (401) 792-7198
Founded: 1956; Members: 700
Gen Secr: Gordon Armstrong
Focus: Perf Arts
Periodicals
Newsletter (weekly)
Theatre Survey (weekly) *14090*

American Society for Therapeutic Radiology and Oncology (ASTRO), 1101 Market St, Ste 1400, Philadelphia, PA 19107
T: (215) 574-3180; Fax: (215) 928-0153
Founded: 1955; Members: 4200
Gen Secr: Frances Glica
Focus: Radiology; Cell Biol & Cancer Res
Periodicals
ASTRO Newsletter
The International Journal of Radiation Oncology, Biology and Physics *14091*

American Society for the Study of Orthodontics (ASSO), 50-12 204 St, Oakland Gardens, NY 11364
T: (212) 224-8898
Founded: 1945; Members: 300
Gen Secr: Daisy N. Buchalter
Focus: Dent
Periodicals
ASSO Newsletter (weekly)
International Journal of Orthodontics (3 times annually) *14092*

American Society for Training and Development (ASTD), 1640 King St, Box 1443, Alexandria, VA 22313
T: (703) 683-8100; Fax: (703) 683-8103
Founded: 1944; Members: 55000
Gen Secr: Curtis E. Plott
Focus: Educ; Econ
Periodicals
Journal (weekly)
National Report on Human Resources (weekly) *14093*

American Society for Value Inquiry (ASVI), c/o Prof. Tom Regan, Dept of Philosophy, North Carolina State University, Raleigh, NC 14201
T: (716) 881-3200
Founded: 1970; Members: 200
Pres: Prof. Tom Regan
Focus: Philos
Periodicals
Journal of Value Inquiry (weekly)
Newsletter (weekly) *14094*

American Society of Abdominal Surgery (ASAS), 675 Main St, Melrose, MA 02176
T: (617) 665-6102
Founded: 1959; Members: 9300
Gen Secr: Blaise F. Alfano
Focus: Surgery
Periodicals
Abdominal Surgery (weekly)
The Surgeon *14095*

American Society of Addiction Medicine (ASAM), 5225 Wisconsin Av, Ste 409, Washington, DC 20015
T: (202) 244-8948; Fax: (202) 537-7252
Founded: 1954; Members: 3500
Pres: Anthony B. Radcliffe
Focus: Med
Periodicals
Journal of Addictive Diseases (weekly)
Newsletter (weekly) *14096*

American Society of Agricultural Engineers (ASAE), 2950 Niles Rd, Saint Joseph, MI 49085-9659
T: (616) 429-0300; Fax: (616) 429-3852
Founded: 1907; Members: 9000
Gen Secr: Roger Castensen
Focus: Agri; Eng
Periodicals
Agricultural Engineering Magazine (weekly)
Newsletter (weekly)
Transactions of the ASAE (weekly) . . . *14097*

American Society of Agronomy (ASA), 677 S Segoe Rd, Madison, WI 53711
T: (608) 273-8080; Fax: (608) 273-2021
Founded: 1907; Members: 12500
Gen Secr: Robert F. Barnes
Focus: Agri
Periodicals
Agronomy Abstracts (weekly)
Agronomy Journal (weekly)
Agronomy News (weekly)
Journal of Agronomic Education (weekly)
Journal of Environmental Quality (weekly)
Journal of Production Agriculture (weekly) *14098*

American Society of Anesthesiologists (ASA), 520 N Northwest Hwy, Park Ridge, IL 60068-2573
T: (708) 825-5586; Fax: (708) 825-1682
Founded: 1905; Members: 28000
Gen Secr: Glenn W. Johnson
Focus: Anesthetics
Periodicals
Anesthesiology (weekly)
ASA Newsletter (weekly) *14099*

American Society of Animal Science (ASAS), c/o Carl D. Johnson, 309 W Clark St, Campaign, IL 61820
T: (217) 356-3182
Founded: 1908; Members: 4588
Gen Secr: Carl D. Johnson
Focus: Zoology
Periodicals
ASAS Handbook and Membership Directory (biannually)
Combined Abstracts (weekly)
Journal of Animal Science (weekly) . . *14100*

American Society of Bakery Engineers (ASBE), 2 N Riverside Plaza, Rm 1733, Chicago, IL 60606
T: (312) 332-2246
Founded: 1924; Members: 2700
Pres: Robert A. Fischer
Focus: Food
Periodicals
Letter (weekly)
Proceedings (weekly) *14101*

American Society of Body and Design Engineers (ASBDE), Wilshire Office Center, Ste 3031, 24634 Five Mile, Redford, MI 48239
T: (313) 532-6100
Founded: 1946; Members: 1500
Gen Secr: H. Eby
Focus: Eng
Periodicals
Body Engineering Journal (weekly)
Directory (weekly)
Newsletter (7 times annually)
Proceedings (weekly) *14102*

American Society of Brewing Chemists (ASBC), 3340 Pilot Knob Rd, Saint Paul, MN 55121-2097
T: (612) 454-7250; Fax: (612) 454-0766
Founded: 1934; Members: 600
Gen Secr: Steven C. Nelson
Focus: Chem
Periodical
Journal of the ASBC (weekly) *14103*

American Society of Cataract and Refractive Surgery (ASCRS), 4000 Lugato Rd, Ste 850, Fairfax, VA 22033
T: (703) 591-2220; Fax: (703) 591-0614
Founded: 1974; Members: 4500
Gen Secr: David A. Karcher
Focus: Surgery
Periodical
Journal of Cataract and Refractive Surgery (weekly) *14104*

American Society of Certified Engineering Technicians (ASCET), POB 371474, El Paso, TX 79937
T: (915) 591-5115
Founded: 1964; Members: 3000
Gen Secr: Kurt H. Schuler
Focus: Eng
Periodicals
Annual Report (weekly)
Certified Engineering Technician Magazine (weekly)
President's Message (weekly) *14105*

American Society of Church History (ASCH), 328 Deland Av, Indialantic, FL 32903
T: (407) 768-8306
Founded: 1888; Members: 1550
Gen Secr: William B. Miller
Focus: Hist
Periodical
Church History (weekly) *14106*

American Society of Civil Engineers (ASCE), c/o Kelly Cunningham, 1015 15 St NW, Ste 600, Washington, DC 20005
T: (202) 705-7406
Founded: 1852; Members: 110000
Gen Secr: Dr. Edward O. Pfrang
Focus: Civil Eng
Periodicals
Aerospace Journal (weekly)
ASCE News (weekly)
Civil Engineering (weekly)
Cold Regions Journal (weekly)
Computing in Civil Engineering Journal (weekly)
Construction Journal (weekly)
Energy Journal (weekly)
Energy Mechnics Journal (weekly)
Environment Journal (weekly)
Geotechnical Journal (weekly)
Hydraulic Journal (weekly)
Irrigation/Drainage Journal (weekly)
Management in Engineering Journal (weekly)
Materials in Civil Engineering Journal (weekly)
Performance of Constructed Facilities Journal (weekly)
Professional Journal (weekly)
Structural Journal (weekly)
Surveying Journal (weekly)
Transportation/Pipeline Journal (weekly)
Urban Planning Journal (weekly)
Water Resources Journal (weekly)
Waterway/Port/Coastal Journal (weekly) . *14107*

American Society of Clinical Hypnosis (ASCH), 2200 E Devon Av, Ste 291, Des Plaines, IL 60018-4534
T: (708) 297-3317; Fax: (708) 297-7309
Founded: 1957; Members: 4000
Gen Secr: William F. Hoffman
Focus: Med; Psych
Periodicals
American Journal of Clinical Hypnosis (weekly)
Directory (weekly)
News Letter (8 times annually) *14108*

American Society of Clinical Oncology (ASCO), 435 N Michigan Av, Ste 1717, Chicago, IL 60611-4067
T: (312) 644-0828; Fax: (312) 644-8557
Founded: 1964; Members: 8744
Gen Secr: James B. Gantenberg
Focus: Cell Biol & Cancer Res
Periodicals
Directory (biennial)
Journal of Clinical Oncology (weekly) . . *14109*

American Society of Clinical Pathologists (ASCP), 2100 W Harrison St, Chicago, IL 60612
T: (312) 738-1336; Fax: (312) 738-1619
Founded: 1922; Members: 50000
Gen Secr: Robert C. Rock
Focus: Pathology
Periodicals
American Journal of Clinical Pathology (weekly)
Laboratory Medicine (weekly) *14110*

American Society of Colon and Rectal Surgeons (ASCRS), 800 E Northwest Hwy, Ste 1080, Palatine, IL 60067
T: (708) 359-9184; Fax: (708) 359-7367
Founded: 1899; Members: 1500
Gen Secr: James R. Slawny
Focus: Surgery
Periodical
Diseases of the Colon and Rectum (weekly) *14111*

American Society of Consultant Pharmacists (ASCP), 1321 Duke St, Alexandria, VA 22314-3563
T: (703) 739-1300; Fax: (703) 739-1321
Founded: 1969; Members: 5000
Gen Secr: R. Timothy Webster
Focus: Pharmacol
Periodicals
Clinical Consult Newsletter (weekly)
The Consultant Pharmacist Journal (weekly)
Update Newsletter (weekly) *14112*

American Society of Contemporary Medicine and Surgery (ASCMS), 233 E Erie St, Ste 710, Chicago, IL 60611
T: (312) 951-1400
Founded: 1968; Members: 8000
Gen Secr: John G. Bellows
Focus: Surgery; Med
Periodical
Comprehensive Therapy (weekly) *14113*

American Society of Contemporary Ophthalmology (ASCO), 233 E Erie St, Ste 710, Chicago, IL 60611
T: (312) 951-1400; Fax: (800) 621-4002
Founded: 1966; Members: 6000
Gen Secr: John G. Bellows
Focus: Ophthal
Periodicals
Annals of Ophthalmology (weekly)
Glancoma (weekly) *14114*

American Society of Criminology (ASC), 1314 Kinnear Rd, Ste 212, Columbus, OH 43212
T: (614) 292-9207
Founded: 1941; Members: 2150
Gen Secr: Joseph E. Scott
Focus: Criminology
Periodicals
The Criminologist Newsletter (weekly)
Criminology (weekly) *14115*

American Society of Cytology (ASC), 1015 Chestnut St, Ste 1518, Philadelphia, PA 19107
T: (215) 922-3880
Founded: 1951; Members: 3600
Gen Secr: Y.S. Erozan
Focus: Cell Biol & Cancer Res
Periodicals
Acta Cytologica (weekly)
The Cytotechnologist's Bulletin (weekly) . *14116*

American Society of Danish Engineers (ASDE), c/o T.H. Storm, POB 606, Larchmont, NY 10538
T: (914) 632-7632; Fax: (914) 632-7631
Founded: 1930; Members: 200
Gen Secr: T.H. Storm
Focus: Eng
Periodical
Newsletter (5 times annually) *14117*

American Society of Dentistry for Children (ASDC), 211 E Chicago Av, Ste 1430, Chicago, IL 60611
T: (312) 943-1244; Fax: (312) 943-3541
Founded: 1927; Members: 10000
Gen Secr: Carol Teuscher
Focus: Dent
Periodicals
ASDC Newsletter
Journal of Dentistry for Children *14118*

American Society of Dermatopathology (ASD), 550 N Broadway, Ste 408, Baltimore, MD 21205
T: (410) 955-2332; Fax: (410) 955-2445
Founded: 1962; Members: 880
Gen Secr: David G. Uffer
Focus: Derm; Pathology *14119*

American Society of Echocardiography (ASE), 4101 Lake Boone Trail, Ste 201, Raleigh, NC 27607
T: (919) 787-5181; Fax: (919) 787-4916
Founded: 1976; Members: 5000
Gen Secr: Sharon Perry
Focus: Cardiol
Periodical
Journal of the American Society of Echocardiography *14120*

American Society of Electroneurodiagnostic Technologists (ASET), 204 W Seventh St, Carroll, IA 51401-2317
T: (712) 792-2978; Fax: (712) 792-6962
Founded: 1959; Members: 2650
Gen Secr: M. Fran Pedelty
Focus: Neurology
Periodicals
American Journal of EEG Technology (weekly)
ASET Newsletter (weekly) *14121*

American Society of Extra-Corporeal Technology (AmSECT), 11480 Sunset Hills Rd, Ste 100E, Reston, VA 22090
T: (703) 435-8556; Fax: (703) 435-0056
Founded: 1964; Members: 3000
Gen Secr: George M. Cate
Focus: Cardiol
Periodicals
Journal of Extra-Corporal Technology (weekly)
Perfusion Life (11 times annually) . . . *14122*

American Society of Forensic Odontology (ASFO), c/o Dr. E. Steven Smith, Northwestern University Dental School, 240 E Huron, Chicago, IL 60611
T: (312) 908-5966
Founded: 1966; Members: 450
Gen Secr: Dr. E. Steven Smith
Focus: Dent; Forensic Med
Periodicals
Membership Directory (weekly)
Newsletter (weekly) *14123*

American Society of Gas Engineers (ASGE), POB 31756, Independence, OH 44131
T: (216) 328-0322
Founded: 1954; Members: 550
Gen Secr: John H. Fitzgerald
Focus: Energy
Periodical
Digest (weekly) *14124*

American Society of Genealogists (ASG), 1328 Stimmel Rd, Columbus, OH 43223-2917
T: (614) 272-0234
Founded: 1940; Members: 50
Gen Secr: Cameron Allen
Focus: Genealogy *14125*

American Society of Geolinguistics (ASG), c/o Dr. Jesse Levitt, 485 Brooklawn Av, Fairfield, CT 06432
T: (203) 333-8920
Founded: 1964; Members: 120
Gen Secr: Dr. Jesse Levitt
Focus: Ling
Periodical
Geolinguistics (weekly) *14126*

American Society of Golf Course Architects (ASGCA), 221 N LaSalle St, Chicago, IL 60601
T: (312) 372-7090; Fax: (312) 372-6160
Members: 115
Gen Secr: Paul Fullmer
Focus: Archit
Periodical
ASGCA Newsletter (weekly) *14127*

American Society of Group Psychotherapy and Psychodrama (ASGP&P), 6728 Old McLean Village Dr, McLean, VA 22101
T: (703) 556-9222
Founded: 1942; Members: 800
Gen Secr: George K. Degnon
Focus: Therapeutics
Periodical
Group Psychodrama (weekly) *14128*

American Society of Hand Therapists (ASHT), 1002 Vandora Springs Rd, Ste 101, Garner, NC 27529
Fax: (919) 779-5642
Founded: 1977; Members: 1529
Gen Secr: Charlene Barbour
Focus: Physical Therapy
Periodical
Newsletter (weekly) *14129*

American Society of Heating, Refrigerating and Air-Conditioning Engineers (ASHRAE), 1791 Tullie Cir NE, Atlanta, GA 30329
T: (404) 636-8400; Fax: (404) 321-5478
Founded: 1894; Members: 55000
Gen Secr: Frank M. Coda
Focus: Air Cond
Periodicals
Handbook (weekly)
Journal (weekly)
Transactions (weekly) *14130*

American Society of Hematology (ASH), 1101 Connecticut Av NW, Washington, DC 20036-4303
T: (202) 857-1118; Fax: (202) 857-1164
Founded: 1957; Members: 4600
Gen Secr: Michael L. Payne
Focus: Hematology

Periodicals
Blood (weekly)
Directory of Members
Meeting Program (weekly) 14131

American Society of Hospital Pharmacists (ASHP), 4630 Montgomery Av, Bethesda, MD 20814
T: (301) 657-3000; Fax: (301) 652-8278
Founded: 1942; Members: 25000
Gen Secr: Joseph A. Oddis
Focus: Pharmacol
Periodicals
American Hospital Formulary Service (weekly)
American Journal of Hospital Pharmacy (weekly)
Clinical Pharmacy (weekly)
International Pharmaceutical Abstracts (weekly) 14132

American Society of Human Genetics (ASHG), 9650 Rockville Pike, Bethesda, MD 20814-3998
T: (301) 571-1825; Fax: (301) 530-7079
Founded: 1948; Members: 4500
Gen Secr: Elaine Strass
Focus: Genetics
Periodical
American Journal of Human Genetics (weekly) 14133

American Society of Ichthyologists and Herpetologists (ASIH), c/o Dept of Zoology, Southern Illinois University, Carbondale, IL 62901-6501
T: (618) 453-4221
Founded: 1913; Members: 3600
Gen Secr: B.M. Burr
Focus: Nat Sci; Zoology
Periodical
Copeia (weekly) 14134

American Society of Indexers (ASI), POB 386, Port Aransas, TX 78373
T: (512) 749-4052; Fax: (512) 749-6334
Founded: 1968; Members: 950
Pres: Linda K. Fetters
Focus: Libraries & Bk Sci
Periodicals
ASI Newsletter (5 times annually)
The Indexer (weekly)
Register of Indexers (weekly) 14135

American Society of Internal Medicine (ASIM), 2011 Pennsylvania Av NW, Ste 800, Washington, DC 20006-1808
T: (202) 835-2746; Fax: (202) 835-0443
Founded: 1956; Members: 26000
Gen Secr: Alan R. Nelson
Focus: Intern Med
Periodicals
The Internist: Health Policy in Practice (weekly)
Internist's Intercom (weekly) 14136

American Society of International Law (ASIL), 2223 Massachusetts Av NW, Washington, DC 20008
T: (202) 265-4313; Fax: (202) 797-7133
Founded: 1906; Members: 4300
Gen Secr: Barry Carter
Focus: Law
Periodicals
American Journal of International Law (weekly)
Newsletter (weekly) 14137

American Society of Irrigation Consultants (ASIC), 425A Oak St, Brentwood, CA 94513
T: (510) 516-1124; Fax: (510) 516-1301
Founded: 1970; Members: 185
Gen Secr: Wanda M. Sarsfield
Focus: Agri
Periodicals
ASIC Newsletter (weekly)
Bulletin (weekly)
Membership Roster (weekly) 14138

American Society of Landscape Architects (ASLA), 4401 Connecticut Av NW, Washington, DC 20008-2302
T: (202) 686-2752; Fax: (202) 686-1001
Founded: 1899; Members: 10500
Gen Secr: David Bohardt
Focus: Archit
Periodicals
Land: Landscape Architectural News Digest (weekly)
Landscape Architecture (weekly)
LATIS: Landscape Architecture Technical Information Series (weekly) 14139

American Society of Law and Medicine (ASLM), 765 Commonwealth Av, Boston, MA 02215
T: (617) 262-4990; Fax: (617) 437-7596
Founded: 1972; Members: 4500
Gen Secr: Larry Gostin
Focus: Med; Law; Forensic Med
Periodicals
American Journal of Law and Medicine (weekly)
Law, Medicine and Health Care (weekly) 14140

American Society of Limnology and Oceanography (ASLO), c/o Virginia Institute of Marine Science, College of William and Mary, Rte 1208, Gloucester Point, VA 23062
T: (804) 642-7242; Fax: (804) 357-0422
Founded: 1948; Members: 3500
Gen Secr: Polly A. Penhale
Focus: Hydrology; Oceanography
Periodical
Limnology and Oceanography (weekly) . 14141

American Society of Mammalogists (ASM), c/o Dept of Zoology, Brigham Young University, 501 Widtsoe Bldg, Provo, UT 84602
T: (801) 378-2492
Founded: 1919; Members: 3600
Gen Secr: Dr. H. Duane Smith
Focus: Zoology
Periodicals
Journal of Mammalogy (weekly)
Mammalian Species (weekly) 14142

American Society of Master Dental Technologists (ASMDT), POB 248, Oakland Gardens, NY 11364
T: (718) 428-0075
Founded: 1976; Members: 125
Gen Secr: Sue Heppenheimer
Focus: Dent 14143

American Society of Maxillofacial Surgeons (ASMS), 444 E Algonquin Rd, Arlington Heights, IL 6005
T: (708) 228-9900
Founded: 1947; Members: 385
Focus: Surgery 14144

American Society of Mechanical Engineers (ASME), 345 E 47 St, New York, NY 10017
T: (212) 705-7722; Fax: (212) 705-7674
Founded: 1880; Members: 118934
Gen Secr: Dr. David Belden
Focus: Eng
Periodicals
Applied Mechanics (weekly)
Applied Mechanics Review (weekly)
Biomechanical Engineering (weekly)
Dynamic Systems, Measurement and Control (weekly)
Energy Resources Technology (weekly)
Engineering for Gas Turbines and Power (weekly)
Engineering for Industry (weekly)
Engineering Materials and Technology (weekly)
Fluids Engineering (weekly)
Heat Transfer (weekly)
The Journal of Electronic Packaging (weekly)
Manufacturing Review (weekly)
Mechanical Design (weekly)
Mechanical Engineering (weekly)
Offshore Mechanics and Arctic Engineering (weekly)
Pressure Vessel Technology (weekly)
Solar Energy Engineering (weekly)
Tribology (weekly)
Turbomachinery (weekly)
Vibration and Acoustics (weekly) 14145

American Society of Naturalists (ASN), c/o Dr. Peter C. Chabora, Dept of Biology, Queens College of CUNY, Flushing, NY 11367
T: (718) 997-3400
Founded: 1883; Members: 1746
Gen Secr: Dr. Peter C. Chabora
Focus: Bio
Periodical
Records of the American Society of Naturalists (tri-annually) 14146

American Society of Naval Engineers (ASNE), 1452 Duke St, Alexandria, VA 22314
T: (703) 836-6727; Fax: (703) 836-7491
Founded: 1888; Members: 8000
Gen Secr: Charles L. Smith
Focus: Navig; Eng
Periodical
Naval Engineers Journal (weekly) . . . 14147

American Society of Nephrology (ASN), 1101 Connecticut Av NW, Ste 700, Washington, DC 20036
T: (202) 857-1190; Fax: (202) 223-4579
Founded: 1966; Members: 4600
Gen Secr: Judith Walker
Focus: Med; Geriatrics
Periodicals
Abstracts and Program (weekly)
Kidney International (weekly)
Membership Directory 14148

American Society of Neuroradiology (ASNR), 2210 Midwest Rd, Ste 207, Oak Brook, IL 60521
T: (708) 574-0220; Fax: (708) 574-0661
Founded: 1962; Members: 2028
Gen Secr: Patricia Grosshauser
Focus: Radiology
Periodicals
American Journal of Neuroradiology (weekly)
Membership Roster (weekly) 14149

American Society of Notaries (ASN), 918 16 St NW, Washington, DC 20006
T: (202) 955-6162; Fax: (202) 655-6163
Founded: 1965; Members: 24000
Gen Secr: Eugene E. Hines
Focus: Law
Periodical
American Notary (weekly) 14150

American Society of Papyrologists (ASP), c/o Dept of Greek and Latin, Wayne State University, Detroit, MI 48202
T: (313) 577-3032
Founded: 1961; Members: 175
Pres: James G. Keenan
Focus: Hist
Periodical
Bulletin (weekly) 14151

American Society of Parasitologists (ASP), c/o Dr. Lillian F. Mayberry, Dept of Biological Sciences, University of Texas at El Paso, El Paso, TX 79968-0519
T: (915) 747-5844; Fax: (915) 747-5111
Founded: 1924; Members: 1400
Gen Secr: Dr. Lillian F. Mayberry
Focus: Microbio
Periodicals
The Journal of Parasitology (weekly)
Newsletter (weekly) 14152

American Society of Pharmacognosy (ASP), c/o Dr. William J. Keller, School of Pharmacy, Northeast Louisiana University, Monroe, LA 71209
T: (318) 342-1690
Founded: 1959; Members: 900
Gen Secr: Dr. William J. Keller
Focus: Pharmacol
Periodicals
Journal of Natural Products (weekly)
Newsletter (weekly) 14153

American Society of Physoanalytic Physicians (ASPA), 4808 Jasmine Dr, Rockville, MD 20853
T: (301) 929-1623
Founded: 1962; Members: 300
Gen Secr: Janice S. Wright
Focus: Psych
Periodicals
Newsletter (weekly)
Proceedings (weekly) 14154

American Society of Plant Physiologists (ASPP), 15501 Monona Dr, Rockville, MD 20855-2768
T: (301) 251-0560; Fax: (301) 279-2996
Founded: 1924; Members: 5200
Gen Secr: Dr. Melvin J. Josephs
Focus: Botany
Periodicals
Newsletter (weekly)
Plant Physiology (weekly) 14155

American Society of Plant Taxonomists (ASPT), c/o Dept of Botany, University of Georgia, Athens, GA 30602
T: (706) 542-1802
Founded: 1937; Members: 1200
Gen Secr: Dr. Samuel Jones
Focus: Botany
Periodicals
Systematic Botany (weekly)
Systematic Botany Monographs 14156

American Society of Plastic and Reconstructive Surgeons (ASPRS), 444 E Algonquin Rd, Arlington Heights, IL 60006
T: (312) 228-9900; Fax: (312) 228-9131
Founded: 1931; Members: 5000
Gen Secr: D. Fellers
Focus: Surgery
Periodicals
Journal of Plastic and Reconstructive Surgery (weekly)
Plastic and Reconstructive Surgeons combined Roster (weekly)
Plastic Surgery News (weekly) 14157

American Society of Plumbing Engineers (ASPE), 3617 Thousand Oaks Blvd, Westlake, CA 91362
T: (805) 495-7120; Fax: (805) 495-4861
Founded: 1964; Members: 6400
Gen Secr: John S. Shaw
Focus: Eng
Periodicals
Data Book (bi-annually)
Plumbing Engineer (weekly) 14158

American Society of Podiatric Dermatology (ASPD), c/o Dr. Steven Berlin, Podiatry Association, 1 N Main, Bel Air, MD 21014
T: (301) 879-1212
Founded: 1914; Members: 267
Pres: Dr. Steven Berlin
Focus: Derm
Periodical
Newsletter (weekly) 14159

American Society of Primatologists (ASP), c/o Dept OB/GYN, Rush-Presbyterian Saint Lukes Medical Center, 1653 W Congress Pkwy, Chicago, IL 60612
T: (312) 942-2152
Founded: 1976; Members: 600
Pres: Dr. Richard G. Rawlins
Focus: Anthro
Periodical
Bulletin (weekly) 14160

American Society of Psychopathology of Expression (ASPE), 74 Lawton St, Brookline, MA 02146
T: (617) 738-9821
Founded: 1964; Members: 137
Focus: Pathology; Psych
Periodicals
Annual Proceedings (weekly)
Psychiatry and Art 14161

American Society of Questioned Document Examiners (ASQDE), 402 Great Falls Rd, Rockville, MD 20850
T: (301) 294-0417
Founded: 1942; Members: 110
Pres: Gideon Epstein
Focus: Law 14162

American Society of Radiologic Technologists (ASRT), 15000 Central Av SE, Albuquerque, NM 87123-3909
T: (505) 298-4500; Fax: (505) 298-5063
Founded: 1920; Members: 18000
Gen Secr: Ward M. Keller
Focus: Radiology
Periodicals
Journal
Scanner 14163

American Society of Regional Anesthesia (ASRA), 1910 Byrd Av, POB 11086, Richmond, VA 23230-1086
T: (804) 282-0010; Fax: (804) 282-0090
Founded: 1974; Members: 6000
Gen Secr: John A. Hinckley
Focus: Anesthetics
Periodical
Regional Anesthesia (weekly) 14164

American Society of Safety Engineers (ASSE), 1800 E Oakton St, Des Plaines, IL 60018
T: (708) 692-4121; Fax: (708) 296-3769
Founded: 1911; Members: 27000
Gen Secr: Judy T. Neel
Focus: Safety
Periodicals
Conference Proceedings (weekly)
Professional Safety (weekly) 14165

American Society of Sanitary Engineering (ASSE), POB 40362, Bay Village, OH 44140
T: (216) 835-3040
Founded: 1906; Members: 2700
Gen Secr: Gael H. Dunn
Focus: Eng; Hygiene
Periodicals
A.S.S.E. Year Book (weekly)
Standards Report (weekly) 14166

American Society of Sephardic Studies (ASOSS), 500 W 185 St, New York, NY 10033
T: (212) 960-5236; Fax: (212) 960-0288
Founded: 1963; Members: 117
Gen Secr: Dr. M. Mitchell Serels
Focus: Hist
Periodical
The Sephardic Scholar 14167

American Society of Sugar Beet Technologists (ASSBT), 90 Madison, Ste 208, Denver, CO 80206
T: (303) 321-1520
Founded: 1935; Members: 600
Gen Secr: Thomas K. Schwartz
Focus: Food
Periodical
Journal of Sugar Beet Research (weekly) 14168

American Society of Swedish Engineers (ASSE), c/o Erik Mortensen, 106 Morning Side Dr, Apt 88, New York, NY 10027
T: (212) 666-1626
Founded: 1888; Members: 150
Pres: Erik Mortensen
Focus: Eng
Periodical
Membership Directory (weekly) 14169

American Society of Trial Consultants, c/o Dept of Speech and Mass Communication, Towson State University, Towson, MD 21204
T: (410) 830-2448
Founded: 1982; Members: 225
Gen Secr: Ronald J. Matlon
Focus: Behav Sci; Law
Periodicals
Court Call (weekly)
Directory (weekly) 14170

American Society of Tropical Medicine and Hygiene (ASTMH), 8000 Westpark Dr, Ste 130, McLean, VA 22102
T: (703) 790-1745; Fax: (703) 790-9063
Founded: 1952; Members: 2300
Gen Secr: Richard J. Burk
Focus: Hygiene; Trop Med
Periodicals
American Journal of Tropical Medicine and Hygiene (weekly)
Tropical Medicine and Hygiene News (weekly) 14171

American Society of Veterinary Ophthalmology (ASVO), 1528 Shalamar, Stillwater, OK 74074
T: (405) 377-2134
Founded: 1957; Members: 250
Gen Secr: A.J. Quinn
Focus: Vet Med; Ophthal
Periodical
Newsletter 14172

American Society of Veterinary Physiologists and Pharmacologists (ASVPP), c/o Dr. J.V. Kitzman, College of Veterinary Medicine, Mississippi State University, Drawer V, Mississippi State, MS 39762
T: (601) 325-1130
Founded: 1946; Members: 200
Gen Secr: Dr. J.V. Kitzman
Focus: Vet Med 14173

American Society of Zoologists (ASZ), 401 N Michigan Av, Chicago, IL 60651-4267
T: (312) 527-6697
Founded: 1890; Members: 3700
Gen Secr: Laura Jungen

Focus: Zoology
Periodical
American Zoologist (weekly) *14174*

American Sociological Association (ASA), 1722 N St NW, Washington, DC 20036
T: (202) 833-3410
Founded: 1905; Members: 12300
Gen Secr: William V. D'Antonio
Focus: Sociology
Periodicals
American Sociological Review (weekly)
Contemporary Sociology: A Journal of Reviews (weekly)
Directory of Departments (weekly)
Directory of Members (weekly)
Footnotes (9 times annually)
Guide to Graduate Departments of Sociology (weekly)
Journal of Health and Social Behavior (weekly)
Social Psychology (weekly)
Sociological Theory (weekly)
Sociology of Education (weekly)
Teaching Sociology (weekly) *14175*

American Software Users Group (ASUG), 401 N Michigan, Chicago, IL 60611
T: (312) 644-6610; Fax: (312) 245-1084
Founded: 1980; Members: 300
Gen Secr: Nicholas Leever
Focus: Computer & Info Sci *14176*

American Solar Energy Society (ASES), 2400 Central Av, Boulder, CO 80301
T: (303) 443-3130; Fax: (303) 443-3212
Founded: 1970; Members: 2000
Pres: Larry Sherwood
Focus: Energy
Periodical
Solar Today (weekly) *14177*

American Speech-Language-Hearing Association (ASHA), 10801 Rockville Pike, Rockville, MD 20852
T: (301) 897-5700; Fax: (301) 571-0457
Founded: 1925; Members: 67000
Gen Secr: Frederick T. Spahr
Focus: Logopedy
Periodicals
ASHA (weekly)
ASHA Monograph (weekly)
ASHA Report (weekly)
Journal of Speech and Hearing Disorders (weekly)
Journal of Speech and Hearing Research (weekly)
Language, Speech and Hearing Services in Schools (weekly) *14178*

American Spelean History Association (ASHA), 711 E Atlantic Av, Altoona, PA 16602
T: (814) 946-3155
Founded: 1968; Members: 150
Gen Secr: Jack H. Speece
Focus: Geology; Hist
Periodical
Journal of Spelean History (weekly) . . *14179*

American Spinal Injury Association (ASIA), 250 E Superior, Rm 619, Chicago, IL 60611
T: (312) 908-3425; Fax: (312) 908-1819
Founded: 1973; Members: 450
Gen Secr: Marianne G. Kaplen
Focus: Pathology; Physiology
Periodical
ASIA Bulletin (weekly) *14180*

American Sports Education Institute (ASEI), 200 Castlewood Dr, North Palm Beach, FL 33408
T: (407) 842-3600; Fax: (407) 863-8984
Founded: 1977; Members: 20000
Gen Secr: Michael May
Focus: Sports
Periodical
Booster (weekly) *14181*

American Statistical Association (ASA), 1429 Duke St, Alexandria, VA 22314-3402
T: (703) 684-1221; Fax: (703) 684-2037
Founded: 1839; Members: 17000
Gen Secr: Barbara A. Bailar
Focus: Stats
Periodicals
The American Statistician (weekly)
AMSTAT News (10 times annually)
Current Index to Statistics: Applications, Methods and Theory (weekly)
Journal of ASA (weekly)
Journal of Business & Economic Statistics (weekly)
Journal of Educational Statistics (weekly)
Proceedings (weekly)
Technometrics (weekly) *14182*

American String Teachers Association (ASTA), 4020 McEwen, Dallas, TX 75244
T: (214) 233-3116; Fax: (214) 490-4219
Founded: 1946; Members: 8300
Gen Secr: Madeleine Crouch
Focus: Music; Educ
Periodical
American String Teacher (3 times annually) *14183*

American Student Dental Association (ASDA), 211 E Chicago Av, Ste 840, Chicago, IL 60611
T: (312) 440-2795; Fax: (312) 440-2820
Founded: 1971; Members: 13500
Gen Secr: Karen S. Cervenka
Focus: Dent
Periodicals
Dental Student Handbook (weekly)
Dentistry (weekly)
News (weekly) *14184*

American Studies Association (ASA), c/o University of Maryland, 2101 S Campus, Surge Bldg, College Park, MD 20742-7711
T: (301) 405-1364
Founded: 1951; Members: 3500
Gen Secr: John F. Stephens
Focus: Hist; Lit; Arts
Periodicals
American Quarterly (weekly)
American Studies Association Newsletter (weekly) *14185*

American Surgical Association (ASA), c/o George F. Sheldon, University of North Carolina at Chapel Hill, 136 Burnett-Womack, Box CB 7245, Chapel Hill, NC 27599-7245
T: (919) 966-6320; Fax: (919) 966-6009
Founded: 1880; Members: 949
Gen Secr: George F. Sheldon
Focus: Surgery
Periodicals
Annals of Surgery (weekly)
Transactions (weekly) *14186*

American Swedish Institute (ASI), 2600 Park Av, Minneapolis, MN 55407
T: (612) 871-4907; Fax: (612) 871-8682
Founded: 1929; Members: 6700
Gen Secr: Bruce Karstadt
Focus: Ethnology *14187*

American Technical Education Association (ATEA), c/o North Dakota State College of Science, Wahpeton, ND 58076
T: (701) 671-2240; Fax: (701) 671-2145
Founded: 1928; Members: 2500
Gen Secr: Betty Krump
Focus: Adult Educ; Eng
Periodical
ATEA Journal (weekly) *14188*

American Teilhard Association (ATA), 40 Hillside Ln, Syosset, NY 11791
T: (516) 921-8655
Founded: 1964
Gen Secr: Jim McPartlin
Focus: Rel & Theol; Futurology
Periodicals
Teilhard Perspective (weekly)
Teilhard Studies (weekly) *14189*

American Theatre Critics Association (ATCA), c/o Clara Hieronymus, The Tennessean, 2200 Hemingway Dr, Nashville, TN 37215
T: (615) 665-0595; Fax: (615) 259-8057
Founded: 1974; Members: 275
Gen Secr: Clara Hieronymus
Focus: Perf Arts
Periodical
Newsletter (weekly) *14190*

American Theatre Organ Society (ATOS), POB 417490, Sacramento, CA 95841
T: (916) 962-1019
Founded: 1955; Members: 6000
Gen Secr: Douglas C. Fish
Focus: Music
Periodical
Theatre Organ (weekly) *14191*

American Theological Library Association (ATLA), 820 Church St, Ste 300, Evanston, IL 60201-3707
T: (708) 869-7788
Founded: 1947; Members: 654
Gen Secr: Joanne Juhnke
Focus: Libraries & Bk Sci; Rel & Theol
Periodicals
Index to Book Reviews in Religion: (IBRR) (weekly)
Newsletter (weekly)
Religion Index One: Periodicals (weekly)
Religion Index Two: Multi-Author Works: (RIT) (weekly)
Research in Ministry: (RIM) (weekly)
Summary of Proceedings of the Annual Conference (weekly) *14192*

American Theological Society - Midwest Division, c/o North Park Theological Seminary, 3225 W Foster Av, Chicago, IL 60625
T: (312) 583-2700
Founded: 1927; Members: 121
Gen Secr: Dr. John Weborg
Focus: Rel & Theol
Periodical
Membership Directory *14193*

American Thoracic Society (ATS), 1740 Broadway, New York, NY 10019-4374
T: (212) 315-8700; Fax: (212) 265-5642
Founded: 1905; Members: 10500
Gen Secr: Marilyn Hansen
Focus: Intern Med
Periodicals
American Journal of Respiratory Cell and Molecular Biology
American Review of Respiratory Disease (weekly) *14194*

American Thyroid Association (ATA), c/o Walter Reed Army Medical Center, Washington, DC 20307-5001
T: (202) 882-7717; Fax: (202) 882-7813
Founded: 1923; Members: 650
Gen Secr: Leonard Wartofsky
Focus: Endocrinology *14195*

American Tinnitus Association (ATA), POB 5, Portland, OR 97207
T: (503) 248-9985
Founded: 1971; Members: 150000
Gen Secr: Gloria E. Reich
Focus: Otorhinolaryngology
Periodical
Tinnitus Today (weekly) *14196*

American Tolkien Society (ATS), c/o Phil Helms, POB 373, Highland, MI 48357-0373
T: (813) 585-0985
Founded: 1975; Members: 475
Pres: Phil Helms
Focus: Lit
Periodical
Minas Tirith Evening Star (weekly) . . . *14197*

American Trauma Society (ATS), 8903 Presidential Pkwy, Ste 512, Upper Marlboro, MD 20772
T: (301) 420-4189; Fax: (301) 420-0617
Founded: 1968; Members: 2500
Gen Secr: Harry Teter
Focus: Traumatology
Periodicals
ATS Hotline (weekly)
Directory of Specialty Referral Centers
EMS News (weekly) *14198*

American Type Culture Collection (ATCC), 12301 Parklawn Dr, Rockville, MD 20852
T: (301) 881-2600; Fax: (301) 231-5826
Founded: 1925
Gen Secr: Robert E. Stevenson
Focus: Bio
Periodical
Newsletter (weekly) *14199*

American Underground-Space Association (AUA), 511 11 Av S, Box 320, Minneapolis, MN 55415
T: (612) 339-5403; Fax: (612) 339-5403
Founded: 1975; Members: 400
Gen Secr: Susan R. Nelson
Focus: Archit; Eng
Periodical
AVA News (weekly) *14200*

American Urological Association (AUA), 1120 N Charles St, Baltimore, MD 21201
T: (410) 727-1100
Founded: 1902; Members: 9000
Gen Secr: G. James Gallagher
Focus: Urology
Periodicals
AUA Today (3 times annually)
Journal of Urology (weekly)
Membership Roster (biennial) *14201*

American Vacuum Society (AVS), 335 E 45 St, New York, NY 10017
T: (212) 661-9404
Founded: 1953; Members: 5200
Gen Secr: Marion Churchill
Focus: Physics
Periodical
Journal of Vacuum Science and Technology (JVST) (weekly) *14202*

American Venereal Disease Association (AVDA), POB 1753, Baltimore, MD 21203-1753
T: (205) 934-5191; Fax: (205) 955-7889
Founded: 1934; Members: 1150
Gen Secr: Edward W. Hook
Focus: Venereology
Periodical
Sexually Transmitted Diseases (weekly) . *14203*

American Veterinary Medical Association (AVMA), 930 N Meacham Rd, Schaumburg, IL 60196
T: (708) 605-8070; Fax: (708) 330-2862
Founded: 1863; Members: 52000
Gen Secr: A. Roland Dommert
Focus: Vet Med
Periodicals
American Journal of Veterinary Research (weekly)
Journal (weekly) *14204*

American Veterinary Society of Animal Behavior (AVSAB), c/o Dr. Ilang Reisner, Animal Behavior Clinic, Cornell University, Ithaca, NY 14853-6401
Founded: 1975; Members: 150
Gen Secr: Dr. Ilang Reisner
Focus: Vet Med *14205*

American Vocational Association (AVA), 1410 King St, Alexandria, VA 22314
T: (703) 683-3111; Fax: (703) 683-7424
Founded: 1925; Members: 46000
Gen Secr: Charles H. Buzzell
Focus: Adult Educ
Periodicals
Job Market Update (bi-annually)
Update (weekly)
Vocational Education Journal (weekly) . . *14206*

American Vocational Education Personnel Development Association (AVEPDA), c/o Dr. Juanita Bice, Dept of Vocational-Technical Education, 1500 W Seventh Av, Stillwater, OK 74074
T: (405) 743-5535; Fax: (405) 743-5541
Founded: 1972; Members: 250
Gen Secr: Dr. Juanita Bice
Focus: Educ *14207*

American Vocational Education Research Association (AVERA), c/o Curt Finch, Div of Vocational and Technical Education, Virginia Tech, Blacksburg, VA 24061-0254
T: (703) 231-8175; Fax: (703) 231-8175
Founded: 1966; Members: 500
Pres: Curt Finch
Focus: Adult Educ
Periodicals
The Beacon (weekly)
Journal of Vocational Education Research (weekly) *14208*

American Water Resources Association (AWRA), 5410 Grosvenor Ln, Ste 220, Bethesda, MD 20814-2192
T: (301) 493-8600; Fax: (301) 493-5844
Founded: 1964; Members: 3800
Gen Secr: Kenneth D. Reid
Focus: Water Res
Periodicals
Hydata News and Views (weekly)
Proceedings (weekly)
Water Resources Bulletin (weekly) . . . *14209*

American Welding Society (AWS), 550 LeJeune Rd NW, POB 351040, Miami, FL 33135
T: (305) 443-9353; Fax: (305) 443-7559
Founded: 1919; Members: 37000
Gen Secr: Dr. Frank G. De Laurier
Focus: Metallurgy
Periodical
Welding Journal (weekly) *14210*

American Wind Energy Association (AWEA), 777 N Capitol St NE, Ste 805, Washington, DC 20002
T: (202) 408-8988; Fax: (202) 408-8536
Founded: 1974
Gen Secr: Randall Swisher
Focus: Energy
Periodicals
AWEA Wind Energy (weekly)
Windletter (8 times annually) *14211*

American Wood-Preservers' Association (AWPA), POB 286, Woodstock, MD 21163-0286
T: (410) 465-3169; Fax: (410) 465-3195
Founded: 1904; Members: 1700
Gen Secr: John F. Hall
Focus: Forestry *14212*

American Wood Preservers Institute (AWPI), 1945 Old Gallows Rd, Ste 550, Vienna, VA 22182
T: (703) 893-4005; Fax: (703) 893-8942
Founded: 1921; Members: 150
Pres: Victor Lindenheim
Focus: Ecology
Periodicals
Environmental Report (weekly)
Wood Preserving Statistics (weekly) . . *14213*

Amerind Foundation (AF), POB 248, Dragoon, AZ 85609
T: (602) 586-3666; Fax: (602) 586-3667
Founded: 1937
Gen Secr: Anne I. Woosley
Focus: Ethnology; Archeol *14214*

ANAD - National Association of Anorexia Nervosa and Associated Disorders, Box 7, Highland Park, IL 60035
T: (708) 831-3438; Fax: (708) 433-4632
Founded: 1976
Gen Secr: Vivian Meehan
Focus: Med *14215*

Ancient Astronaut Society (AAS), 1921 Saint Johns Av, Highland Park, IL 60035
T: (708) 295-8899
Founded: 1973; Members: 6000
Pres: Gene M. Philips
Focus: Aero
Periodical
Ancient Skies (weekly) *14216*

An Claidheamh Soluis - The Irish Arts Center, 553 W 51 St, New York, NY 10019
T: (212) 757-3318; Fax: (212) 247-0930
Founded: 1972; Members: 3500
Gen Secr: Nye Heron
Focus: Arts
Periodicals
An Gael
Litir Nuachta (weekly) *14217*

Animal Behavior Society (ABS), c/o Dr. Janis W. Driscoll, Dept of Psychology, University of Colorado at Denver, POB 173364, Denver, CO 80217-3364
T: (303) 556-2557; Fax: (303) 556-4861
Founded: 1964; Members: 2600
Gen Secr: Dr. Janis W. Driscoll
Focus: Zoology; Behav Sci
Periodicals
Animal Behavior (weekly)
Newsletter (weekly) *14218*

Animal Medical Center (AMC), 510 E 62 St, New York, NY 10021
T: (212) 838-8100; Fax: (212) 832-9630
Founded: 1910
Gen Secr: William J. Kay
Focus: Vet Med
Periodical
The Center Scope (3 times annually) . . *14219*

Animal Nutrition Research Council (ANRD), c/o Roger L. Garrett, Diversified Laboratories Inc, 3810 Concorde Pkwy, Chantilly, VA 22030
T: (703) 222-8700
Founded: 1939; Members: 421
Gen Secr: Roger L. Garrett
Focus: Vet Med
Periodical
Annual Proceedings (weekly) 14220

Animal Protection Institute of America (API), POB 22505, Sacramento, CA 95822
T: (916) 731-5521; Fax: (916) 731-4467
Founded: 1968; Members: 150000
Gen Secr: D. Fisher
Focus: Ecology
Periodicals
Apevine (weekly)
Emergency Update
Mainstream Magazine (weekly)
New Paths (weekly) 14221

Annual Reviews (AR), 4139 El Camino Way, POB 10139, Palo Alto, CA 94303-0897
T: (415) 493-4400; Fax: (415) 855-9815
Pres: Dr. Robert H. Haynes; Gen Secr: John S. McNeil
Focus: Nat Sci
Periodicals
Annual Review of Anthropology (weekly)
Annual Review of Astronomy & Astrophysics (weekly)
Annual Review of Biochemistry (weekly)
Annual Review of Biophysics & Biophysical Chemistry (weekly)
Annual Review of Cell Biology (weekly)
Annual Review of Computer Science (weekly)
Annual Review of Earth & Planetary Sciences (weekly)
Annual Review of Ecology & Systematics (weekly)
Annual Review of Energy (weekly)
Annual Review of Entomology (weekly)
Annual Review of Fluid Mechanics (weekly)
Annual Review of Genetics (weekly)
Annual Review of Immunology (weekly)
Annual Review of Materials Science (weekly)
Annual Review of Medicine (weekly)
Annual Review of Microbiology (weekly)
Annual Review of Neuroscience (weekly)
Annual Review of Nuclear & Particle Science (weekly)
Annual Review of Nutrition (weekly)
Annual Review of Pharmacology & Toxicology (weekly)
Annual Review of Physical Chemistry (weekly)
Annual Review of Physiology (weekly)
Annual Review of Phytopathology (weekly)
Annual Review of Plant Physiology & Plant Molecular Biology (weekly)
Annual Review of Psychology (weekly)
Annual Review of Public Health (weekly)
Annual Review of Sociology (weekly) . . 14222

Annular Bearing Engineers Committee (ABEC), 1101 Connecticut Av NW, Ste 700, Washington, DC 20036-4303
T: (202) 429-5155; Fax: (202) 223-4579
Gen Secr: Gary T. Satterfield
Focus: Eng 14223

Anonymous Arts Recovery Society (AARS), 380 W Broadway, New York, NY 10012
T: (212) 431-3600
Founded: 1960
Pres: Ivan C. Karp
Focus: Arts 14224

Anonymous Families History Project (AFHP), c/o Prof. Tamara K. Hareven, Dept of Individual and Family Studies, University of Delaware, 101 Alison Hall, Newark, DE 19716
T: (302) 831-6522
Founded: 1971
Gen Secr: Prof. Tamara K. Hareven
Focus: Hist 14225

Antarctica and Southern Ocean Coalition (ASOC), 707 D St SE, Washington, DC 20003
T: (202) 544-0236
Founded: 1978; Members: 220
Gen Secr: Jim Barnes
Focus: Geography 14226

Antenna Measurement Techniques Association (AMTA), 6065 Roswell Rd, Ste 2252, Atlanta, GA 30328
T: (214) 952-2896
Founded: 1979; Members: 500
Gen Secr: Tom Watson
Focus: Electric Eng
Periodicals
Call for Papers (weekly)
Newsletter (weekly) 14227

Anthology Film Archives (AFA), 32-34 Second Av, New York, NY 10003
T: (212) 505-5181; Fax: (212) 477-2714
Founded: 1970; Members: 600
Gen Secr: Jonas Mekas
Focus: Cinema
Periodical
Film Culture 14228

Anthropology Film Center Foundation (AFCF), POB 493, Santa Fe, NM 87501
T: (505) 983-4127
Founded: 1965
Gen Secr: Joan S. Williams
Focus: Anthro 14229

AOAC International (AOAC), 2200 Wilson Blvd, Ste 400, Arlington, VA 22201-3301
T: (703) 522-3032; Fax: (703) 522-5468
Founded: 1884; Members: 3700
Gen Secr: Ronald R. Christensen
Focus: Chem
Periodicals
Journal of the AOAC (weekly)
The Referee (weekly) 14230

AOIP, 231 W 29 St, Ste 1205, New York, NY 10001
T: (212) 967-4008; Fax: (212) 971-4682
Founded: 1980; Members: 94
Gen Secr: Benjamin Wright
Focus: Lit 14231

APEC, Miami Valley Tower, Ste 2100, 40 W Fourth St, Dayton, OH 45402
T: (513) 228-2602
Founded: 1966; Members: 200
Gen Secr: Doris J. Wallace
Focus: Eng
Periodicals
Journal (weekly)
Membership Directory (weekly) 14232

Applied Technology Council (ATC), 555 Twin Dolphin Dr, Ste 550, Redwood City, CA 94065
T: (415) 595-1542; Fax: (415) 593-2320
Founded: 1971; Members: 12
Gen Secr: Christopher Rojahn
Focus: Nat Sci 14233

Aquatic Plant Management Society (APMS), POB 2695, Washington, DC 20013
T: (301) 330-8831; Fax: (301) 926-2355
Founded: 1961; Members: 600
Gen Secr: William N. Rushing
Focus: Botany
Periodicals
Directory of Membership APMS
Journal of Aquatic Plant Management (weekly)
Newsletter 14234

Aquatic Research Institute (ARI), 2242 Davis Ct, Hayward, CA 94545
T: (510) 785-2216; Fax: (510) 784-0945
Founded: 1962
Gen Secr: Dr. Robert R. Rofen
Focus: Oceanography
Periodicals
Aquariculture (weekly)
Aquatica 14235

Archaeological Conservancy (AC), 415 Orchard Dr, Santa Fe, NM 87501
T: (505) 982-3278
Founded: 1979; Members: 12000
Pres: Mark Michel
Focus: Archeol
Periodical
Newsletter (weekly) 14236

Archaeological Institute of America (AIA), 675 Commonwealth Av, Boston, MA 02215
T: (617) 353-9361; Fax: (617) 353-6550
Founded: 1879; Members: 11200
Gen Secr: Mark Meister
Focus: Archeol
Periodicals
American Journal of Archaeology (weekly)
Archaeology (weekly) 14237

Archeology Section (AS), c/o American Anthropological Association, 1703 New Hampshire Av NW, Washington, DC 20009
T: (202) 232-8800; Fax: (202) 667-5345
Founded: 1984; Members: 1989
Pres: Richard Ford
Focus: Archeol 14238

Architectural League of New York (ALNY), 457 Madison Av, New York, NY 10022
T: (212) 753-1722
Founded: 1881; Members: 1000
Gen Secr: Rosalie Genevro
Focus: Archit 14239

Archives of American Art (AAA), c/o Smithsonian Institution, Rm 331, Eighth and G Sts NW, Washington, DC 20560
T: (202) 357-2781
Founded: 1954; Members: 2000
Focus: Arts
Periodical
Archives of American Art Journal (weekly) 14240

Arica Institute (AI), 150 Fifth Av, Ste 912, New York, NY 10011
T: (212) 807-9600
Founded: 1971; Members: 800
Focus: Psych
Periodical
Arica Newsletter (weekly) 14241

Ark-La-Tex Genealogical Association (ALTGA), POB 4462, Shreveport, LA 71134-0462
T: (318) 798-7108
Founded: 1955; Members: 500
Pres: Carroll H. Goyne
Focus: Genealogy
Periodical
The Genie (weekly) 14242

Armed Forces Civilian Instructors Association (AFCIA), 98 Sanger Pl, Sa Ysidro, CA 92073
T: (619) 428-1274
Founded: 1985; Members: 270
Pres: Michael Fleming
Focus: Military Sci 14243

Armed Forces Institute of Pathology (AFIP), 6825 16 St NW, Washington, DC 20306-6000
T: (202) 576-2904; Fax: (202) 576-2164
Founded: 1862
Focus: Pathology
Periodical
Letter (weekly) 14244

Armed Forces Optometric Society (AFOS), 205 Walnut Dr, Harrison, AR 72601
T: (501) 741-4312
Founded: 1970; Members: 600
Gen Secr: James F. Hudson
Focus: Ophthal
Periodicals
Newsletter (weekly)
Roster (weekly) 14245

Armenian Educational Foundation (AEF), 600 W Broadway, Ste 130, Glendale, CA 91204
T: (818) 242-4154
Founded: 1950; Members: 100
Focus: Educ 14246

Armenian Literary Society (ALS), 77 Everett Rd, Demarest, NJ 07627
T: (201) 767-1494
Founded: 1956; Members: 300
Gen Secr: Arthur Hamparian
Focus: Lit
Periodical
Kir ou Kirk (weekly) 14247

ARMS Firms Users Association (AFUA), c/o David Stiteler, 10 E Gratiot Av, Mount Clemens, MI 48043
T: (313) 469-6065
Founded: 1979; Members: 186
Gen Secr: David Stiteler
Focus: Computer & Info Sci
Periodical
Newsletter (weekly) 14248

Art Institute of Light (AIL), POB 429, Gap, PA 17527-0429
T: (717) 768-8255
Founded: 1930
Pres: Thomas C. Wilfred
Focus: Arts 14249

Art Libraries Society/North America (ARLIS/NA), 3900 E Timrod St, Tucson, AZ 85711
T: (602) 881-8479; Fax: (602) 322-6778
Founded: 1972; Members: 1340
Gen Secr: Pamela J. Parry
Focus: Libraries & Bk Sci
Periodicals
ARLIS/NA Update (weekly)
Art Documentation (weekly) 14250

Arts International Program of the Institute of International Education, c/o Institute of International Education, 809 United Nations Plaza, New York, NY 10017
T: (212) 984-5370; Fax: (212) 984-5574
Founded: 1981; Members: 250
Gen Secr: Jane M. Gullong
Focus: Arts; Educ 14251

Asian Literature Division (of MLA) (ALD), c/o Dept of East Asian Languages and Culture, Indiana University, Bloomington, IN 47405
T: (812) 855-5339; Fax: (812) 855-6402
Founded: 1953
Focus: Lit 14252

The Asia Society (AS), 725 Park Av, New York, NY 10021
T: (212) 288-6400; Fax: (212) 517-8315
Founded: 1956; Members: 5600
Pres: Nicolas Platt
Focus: Ethnology; Poli Sci
Periodicals
Annual Report (weekly)
Focus on Asian Studies (3 times annually)
Newsletter (weekly) 14253

ASME International Gas Turbine Institute (IGTI), 6085 Barfield Rd, Ste 207, Atlanta, GA 30328
T: (404) 847-0072; Tx: 707340; Fax: (404) 847-0151
Founded: 1986; Members: 27
Focus: Eng
Periodicals
Global Gas Turbine News (weekly)
International Gas Turbine and Aeroengine Technology Report (weekly) 14254

ASM International, 9639 Kinsman, Materials Park, OH 44073-0002
T: (216) 338-5151; Fax: (216) 338-4634
Founded: 1920; Members: 52000
Gen Secr: Edward L. Langer
Focus: Metallurgy
Periodicals
Advanced Materials and Processes (weekly)
ASM News (weekly)
Bulletin of Alloy Phase Diagrams (weekly)
International Materials Reviews (weekly)
Journal of Heat Treating (weekly)
Journal of Materials Engineering (weekly)
Journal of Materials Shaping Technology (weekly)
Metallurgical Transactions A (weekly)
Metallurgical Transactions B (weekly) . . 14255

The Aspen Institute (TAI), Carmichael Rd, Wye Center, POB 222, Queenstown, MD 21658
T: (410) 820-5461; Fax: (410) 820-5460
Founded: 1949
Pres: David T. McLaughlin
Focus: Sociology
Periodical
Aspen Institute Monograph 14256

Associated Business Writers of America (ABWA), 1450 S Havana, Ste 620, Aurora, CO 80012
T: (303) 751-7844; Fax: (303) 751-8593
Founded: 1945; Members: 150
Gen Secr: Sandy Whelchel
Focus: Lit
Periodicals
Authorship (weekly)
Flash Market News (weekly) 14257

Associated Colleges of the Midwest (ACM), 18 S Michigan Av, Ste 1010, Chicago, IL 60603
T: (312) 263-5000; Fax: (312) 263-5879
Founded: 1958; Members: 14
Pres: Elizabeth R. Hayford
Focus: Educ 14258

Associated Laboratories (AL), 500 S Vermont St, Palatine, IL 60067
T: (708) 358-7400; Fax: (708) 358-7082
Founded: 1958; Members: 16
Pres: Robert F. Hutchison
Focus: Water Res 14259

Associated Schools of Construction (ASC), POB 834, Peoria, IL 61652
T: (309) 677-2943
Founded: 1965; Members: 89
Gen Secr: F. Eugene Rebholz
Focus: Civil Eng; Adult Educ
Periodicals
Annual Report (weekly)
Proceedings of the Annual Meeting (weekly) 14260

Associated Universities (AUI), 1400 16 St NW, Ste 730, Washington, DC 20036
T: (202) 462-1676
Founded: 1946
Pres: Dr. Robert E. Hughes
Focus: Sci 14261

Association for Academic Surgery (AAS), 420 Delaware St SE, Box 242 UMHC, Minneapolis, MN 55455
T: (612) 626-1999; Fax: (612) 625-8496
Founded: 1966; Members: 2400
Gen Secr: Dr. David Dunn
Focus: Surgery
Periodical
Journal of Surgical Research (weekly) . 14262

Association for Advancement of Behavior Therapy (AABT), 15 W 36 St, New York, NY 10018
T: (212) 279-7970; Fax: (212) 239-8038
Founded: 1966; Members: 4300
Gen Secr: Mary Jane Eimer
Focus: Psychiatry; Behav Sci
Periodicals
The Behavior Therapist (10 times annually)
Behavior Therapy (weekly)
Membership Directory (bi-annually) . . . 14263

Association for Advancement of Psychoanalysis (AAP), 329 E 62 St, New York, NY 10021
T: (212) 838-8044
Founded: 1941; Members: 80
Pres: Barbara E. Bell
Focus: Psychoan
Periodical
The American Journal of Psychoanalysis (weekly) 14264

Association for Advancement of Psychology (AAP), POB 38129, Colorado Springs, 80937
T: (719) 520-0688; Fax: (719) 520-0375
Founded: 1974; Members: 6000
Gen Secr: R.H. Wright
Focus: Psych
Periodical
Advance (weekly) 14265

Association for Applied Psychophysiology and Biofeedback (AAPB), 10200 W 44 Av, Ste 304, Wheat Ridge, CO 80033
T: (303) 422-8436; Fax: (303) 422-8894
Founded: 1969; Members: 2100
Gen Secr: Francine Butler
Focus: Physiology
Periodical
Proceedings of the Annual Meeting (weekly) 14266

Association for Arid Lands Studies (AALS), c/o International Center Arid and Semi-Arid Land Studies, Texas Technical University, Box 41036, Lubbock, TX 79409-1036
T: (806) 742-2218; Fax: (806) 742-1954
Founded: 1977; Members: 250
Gen Secr: Dr. Idris, R. Traylor
Focus: Ecology
Periodicals
AALS Newsletter (weekly)
Abstracts (weekly)
Forum (weekly) 14267

Association for Asian Studies (AAS), c/o University of Michigan, 1 Lane Hall, Ann Arbor, MI 48109
T: (313) 665-2490; Fax: (313) 665-3801
Founded: 1941; Members: 7000
Gen Secr: Peter Gosling
Focus: Ethnology
Periodicals
Asian Studies Newsletter (5 times annually)
Bibliography of Asian Studies (weekly)
Doctoral Dissertations on Asia (weekly)
Journal of Asian Studies (weekly)
Monograph Series (weekly) 14268

Association for Assessment in Counseling (AAC), c/o American Counseling Association, 5999 Stevenson Av, Alexandria, VA 22304
T: (703) 823-9800; Fax: (703) 823-0252
Founded: 1965; Members: 1684
Gen Secr: Dr. Theodore P. Remley
Focus: Law
Periodicals
Measurement and Evaluation in Counseling and Development (weekly)
Newsnotes (weekly) 14269

Association for Astrological Psychology (AAP), 360 Quietwood Dr, San Rafael, CA 94903
T: (415) 382-0304
Founded: 1987; Members: 300
Pres: Glenn Perry
Focus: Psych 14270

Association for Behavior Analysis (ABA), c/o Western Michigan University, 258 Wood Hall, Kalamazoo, MI 49008
T: (616) 387-4494; Fax: (616) 387-4457
Founded: 1974; Members: 2000
Gen Secr: William K. Redmon
Focus: Psychoan
Periodicals
Membership Directory (bi-annually)
Newsletter (3 times annually)
Program Book (weekly) 14271

Association for Biology Laboratory Education (ABLE), c/o Cornell University, 1130 Comstock Hall, Ithaca, NY 14853
T: (607) 255-3007; Fax: (607) 255-8088
Founded: 1979; Members: 500
Pres: Dr. Jon Glase
Focus: Bio
Periodical
Tested Studies for Laboratory Teaching (weekly) 14272

Association for Birth Psychology (ABP), 444 E 82 St, New York, NY 10028
T: (212) 988-6617
Founded: 1978; Members: 352
Gen Secr: Leslie Feher
Focus: Psych
Periodical
Birth Psychology Bulletin (weekly) . . . 14273

Association for Bridge Construction and Design (ABCD), POB 23264, Pittsburgh, PA 15222
T: (412) 281-9900; Fax: (412) 281-2056
Founded: 1976; Members: 300
Pres: Gary A. Runco
Focus: Archit 14274

Association for Business Communication, c/o Dept of Management, College of Business, University of North Texas, Denton, TX 76203
T: (817) 565-4423; Fax: (817) 565-4930
Founded: 1935; Members: 2500
Gen Secr: John D. Pettit
Focus: Econ
Periodicals
The Bulletin of the Association for Business Communication (weekly)
The Journal of Business Communication (weekly) 14275

Association for Canadian Studies in the United States (ACSUS), 1 Dupont Circle, Ste 620, Washington, DC 20036
T: (202) 887-6375; Fax: (202) 296-8379
Founded: 1971; Members: 1300
Gen Secr: David N. Biette
Focus: Educ
Periodicals
American Review of Canadian Studies (3 times annually)
Canadian Studies Update 14276

Association for Chemoreception Sciences (AChemS), c/o Panacea Associates, 229 Westridge Dr, Tallahassee, FL 32304
T: (904) 576-5530; Fax: (215) 898-2084
Founded: 1979; Members: 649
Pres: Dr. John Caprio
Focus: Chem 14277

Association for Childbirth at Home, International (ACHI), POB 430, Glendale, CA 91205
T: (213) 667-0839
Founded: 1972; Members: 30000
Pres: Tonya Brooks
Focus: Gynecology
Periodicals
Birth Notes (weekly)
Founders Letter (weekly) 14278

Association for Childhood Education International (ACEI), 11501 Georgia Av, Ste 312, Wheaton, MD 20902
T: (301) 942-2443
Founded: 1931; Members: 17000
Gen Secr: Gerald Odland
Focus: Educ
Periodicals
ACEI Exchange (weekly)
Childhood Education (5 times annually)
Journal of Research in Childhood Education (bi-annually) 14279

Association for Child Psychoanalysis (ACP), c/o Rachel May, POB 366, Great Falls, VA 22066
T: (703) 759-6698; Fax: (703) 759-6783
Founded: 1965; Members: 500
Gen Secr: Rachel May
Focus: Psychoan
Periodicals
Membership Roster (biennial)
Newsletter (weekly) 14280

Association for Communication Administration (ACA), 5105 Backlick Rd, Annadale, VA 22003
T: (502) 762-3411
Founded: 1971; Members: 700
Gen Secr: Jim Gaudino
Focus: Comm Sci
Periodicals
Bulletin (weekly)
Communication Careers (biennial)
Directory of Radio-TV-Film Programs
Directory of Theatre Programs 14281

Association for Comparative Economic Studies (ACES), c/o Michael R. Dohan, Dept of Economics, Queens College, CUNY, Flushing, NY 11367
T: (718) 997-5461; Fax: (718) 997-5535
Founded: 1972; Members: 600
Gen Secr: Michael R. Dohan
Focus: Econ
Periodical
ACES Bulletin (3-4 times annually) . . . 14282

Association for Computational Linguistics (ACL), c/o Dr. Donald E. Walker, Bellcore, MRE 2A379, 445 South St, Box 1910, Morristown, NJ 07960
T: (201) 829-4312; Fax: (201) 292-0067
Founded: 1962; Members: 2000
Gen Secr: Dr. Donald E. Walker
Focus: Ling
Periodicals
Computational Linguistics (weekly)
The Finite String (weekly) 14283

Association for Computers and the Humanities (ACH), c/o Randall Jones, Brigham Young University, 2054 JKHB, Provo, UT 84602
T: (801) 378-3513; Fax: (801) 378-4649
Founded: 1978; Members: 350
Gen Secr: Randall Jones
Focus: Computer & Info Sci
Periodicals
Computers and the Humanities (weekly)
Newsletter (weekly) 14284

Association for Computing Machinery (ACM), 1515 Broadway, New York, NY 10036
T: (212) 869-7440; Fax: (212) 944-1318
Founded: 1947; Members: 75000
Gen Secr: Joseph De Blasi
Focus: Computer & Info Sci
Periodicals
ACM Guide to Computing Literature (Guide) (weekly)
ACM Transactions on Computer Systems (TOCS) (weekly)
ACM Transactions on Database Systems (TODS) (weekly)
ACM Transactions on Graphics (TOG) (weekly)
ACM Transactions on Mathematical Software (TOMS) (weekly)
ACM Transactions on Office Information Systems (TOOIS) (weekly)
ACM Transactions on Programming Languages & Systems (TOPLAS) (weekly)
Communications of the ACM (weekly)
Computing Reviews (weekly)
Computing Surveys (weekly)
Journal of the ACM (weekly) 14285

Association for Continuing Higher Education (ACHE), c/o Dr. Scott E. Evenbeck, Indiana Purdue University at Indianapolis, 620 Union Dr, Rm 143N, Indianapolis, IN 46202
T: (317) 274-2637; Fax: (317) 274-4016
Founded: 1939; Members: 1238
Gen Secr: Dr. Scott E. Evenbeck
Focus: Adult Educ
Periodicals
5 Minutes with ACHE (9 times annually)
Journal of Continuing Higher Education (weekly) 14286

Association for Counselor Education and Supervision (ACES), c/o American Counseling Association, 5999 Stevenson Av, Alexandria, VA 22304
T: (703) 823-9800; Fax: (703)823-0252
Founded: 1964; Members: 2940
Gen Secr: Dr. Theodore P. Remley
Focus: Educ
Periodicals
Counselor Education and Supervision (weekly)
Newsletter (weekly) 14287

Association for Creative Change (ACC), 6600 Grand Av S, Minneapolis, MN 55423-2333
T: (812) 861-7589
Founded: 1968; Members: 400
Gen Secr: Arleta Bredehoft
Focus: Psych
Periodicals
Journal of Religion & the Applied Behavioral Sciences (3 times annually)
Newsletter (weekly) 14288

Association for Documentary Editing (ADE), c/o Harriet F. Simon, Center for Dewey Studies, Southern Illinois University, Carbondale, IL 62901
T: (618) 453-2629
Founded: 1978; Members: 450
Gen Secr: Harriet F. Simon
Focus: Libraries & Bk Sci
Periodicals
Directory (weekly)
Documentary Editing (weekly) 14289

Association for Educational Communications and Technology (AECT), 1025 Vermont Av NW, Ste 820, Washington, DC 20005
T: (202) 347-7834; Fax: (202) 347-7834
Founded: 1923; Members: 5000
Gen Secr: Stan Zenor
Focus: Educ
Periodicals
Educational Communication & Technology Journal (weekly)
Instructional Innovator Magazine (8 times annually)
Journal of Instructional Development (weekly) 14290

Association for Education and Rehabilitation of the Blind and Visually Impaired (AER), 206 N Washington St, Ste 320, Alexandria, VA 22314
T: (703) 548-1884
Founded: 1984; Members: 5500
Gen Secr: Kathleen Megivern
Focus: Educ Handic
Periodicals
AER Report (weekly)
Education of the Visually Handicapped (weekly)
Job Exchange (weekly) 14291

Association for Education in Journalism and Mass Communication (AEJMC), c/o College of Journalism, University of South Carolina, 1621 College St, Columbia, SC 29208
T: (803) 777-2005; Fax: (803) 777-4728
Founded: 1912; Members: 2800
Gen Secr: Jennifer H. McGill
Focus: Adult Educ; Journalism
Periodicals
Journalism Educator (weekly)
Journalism Monographs (weekly)
Newsletter (weekly) 14292

Association for Education of Teachers in Science (AETS), c/o Auburn University, 5040 Haley Center, Auburn, AL 36849
T: (205) 844-6799
Founded: 1930; Members: 677
Gen Secr: Bill Baird
Focus: Educ; Sci 14293

Association for Equine Sports Medicine (AESM), c/o Veterinary Practice Publishing Company, POB 4457, Santa Barbara, CA 93160-4457
T: (805) 965-1028; Fax: (805) 965-0722
Founded: 1982; Members: 400
Focus: Vet Med 14294

Association for Evolutionary Economics (AFEE), c/o Dept of Economics, College of Business Administration, University of Nebraska, Lincoln, NE 68588
T: (402) 472-3967
Founded: 1963; Members: 2000
Gen Secr: Ann M. May
Focus: Econ; Sociology
Periodical
Journal of Economic Issues (weekly) . . 14295

Association for Faculty in the Medical Humanities (AFMH), 6728 Old McLean Village Dr, McLean, VA 22101
T: (703) 556-9222
Founded: 1983; Members: 350
Gen Secr: George K. Degnon
Focus: Med 14296

Association for Finishing Processes of the Society of Manufacturing Engineers (AFP/SME), 1 SME Dr, POB 930, Dearborn, MI 48121-0930
T: (313) 271-1500; Fax: (313) 271-2861
Founded: 1975; Members: 3000
Gen Secr: C. Willetts
Focus: Eng
Periodical
The Finishing Line (weekly) 14297

Association for General and Liberal Studies (AGLS), c/o Dr. Bruce Busby, Ohio Dominican College, 1216 Sunbury Rd, Columbus, OH 43219
T: (614) 251-4634
Founded: 1961; Members: 539
Gen Secr: Dr. Bruce Busby
Focus: Educ
Periodicals
Directory of Members (weekly)
Newsletter (weekly)
Perspectives (3 times annually) 14298

Association for Gerontology in Higher Education (AGHE), 1001 Connecticut Av NW, Ste 410, Washington, DC 20036
T: (202) 429-9277
Founded: 1974; Members: 320
Focus: Geriatrics
Periodicals
AGHE Exchange Newsletter (weekly)
Brief Bibliography Series (3-4 times annually)
National Directory of Educational Programs in Gerontology (4-5 times annually) 14299

Association for Gnotobiotics (AG), 65 Brooklea Dr, East Aurora, NY 14062-1917
T: (716) 655-1680
Founded: 1961; Members: 415
Gen Secr: Dr. Patricia M. Bealmear
Focus: Vet Med
Periodicals
Membership Directory (weekly)
Newsletter (weekly) 14300

Association for Group Psychoanalysis and Process (AGPP), c/o Dr. Milton M. Berger, 501 E 79 St, New York, NY 10021
T: (212) 288-2297
Founded: 1957; Members: 25
Pres: Dr. Milton M. Berger
Focus: Psychoan; Psych 14301

Association for Hospital Medical Education (AHME), 1101 Connecticut Av NW, Ste 700, Washington, DC 20036
T: (202) 857-1196; Fax: (202) 223-4579
Founded: 1954; Members: 700
Gen Secr: Michael S. Hamm
Focus: Adult Educ; Med
Periodicals
Membership Directory (bi-annually)
Newsletter (weekly) 14302

Association for Humanistic Education (AHE), c/o University of Wyoming, Box 3374, Laramie, WY 82071-3374
T: (307) 766-5329
Founded: 1976; Members: 400
Gen Secr: Tracy Cross
Focus: Humanities; Educ
Periodicals
Celebrations (weekly)
Journal of Humanistic Education (weekly)
Membership Roster (weekly) 14303

Association for Humanistic Education and Development (AHEAD), c/o American Counseling Association, 5999 Stevenson Av, Alexandria, VA 22304
T: (703) 823-9800
Founded: 1951; Members: 2272
Gen Secr: Dr. Theodore P. Remley
Focus: Educ
Periodicals
Infochange (6-8 times annually)
Journal of Humanistic Education and Development (weekly) 14304

Association for Humanistic Psychology (AHP), 1772 Vallejo St, Ste 3, San Francisco, CA 94123
T: (415) 346-7929; Fax: (415) 346-7993
Founded: 1962; Members: 3500
Gen Secr: Jody Timms
Focus: Psych
Periodicals
AHP Perspective (weekly)
Journal of Humanistic Psychology (weekly) 14305

Association for Institutional Research (AIR), c/o Florida State University, 314 Stone Bldg, Tallahassee, FL 32306-3038
T: (904) 644-4470; Fax: (904) 644-8824
Founded: 1965; Members: 2500
Gen Secr: Terrence R. Russell
Focus: Educ
Periodicals
Directory (weekly)
Journal (8 times annually)
Newsletter (weekly)
Professional File (weekly)
Research Monograph (weekly) 14306

Association for Intelligent Systems Technology (AIST), 2-212 Center for Science and Technology, Syracuse, NY 13244-4100
T: (315) 443-3771
Founded: 1986; Members: 110
Pres: Charles Saylor
Focus: Electronic Eng 14307

Association for International Practical Training (AIPT), 10400 Little Patuxent Pkwy, Ste 250, Columbia, MD 21044-3510
T: (410) 997-2200; Fax: (410) 992-3924
Founded: 1950
Gen Secr: Robert M. Sprinkle
Focus: Educ
Periodical
International IAESTE Annual Report (weekly) 14308

Association for Investment Management and Research (AIMR), 5 Boar's Head Ln, POB 3668, Charlottesville, VA 22903
T: (804) 977-6600; Fax: (804) 977-1103
Founded: 1947; Members: 22500

Pres: Darwin M. Bayston
Focus: Finance 14309

Association for Jewish Demography and Statistics – American Branch, 4320 Cedarhurst Circle, Los Angeles, CA 90027
T: (213) 469-4976
Founded: 1957; Members: 50
Gen Secr: Dr. Fred Massarik
Focus: Ethnology; Stats 14310

Association for Jewish Studies (AJS), c/o Widener Library M, Harvard University, Cambridge, MA 02138
Founded: 1969; Members: 1300
Gen Secr: Charles Berlin
Focus: Rel & Theol
Periodicals
AJS Newsletter (weekly)
AJS Review (weekly) 14311

Association for Korean Studies (AKS), 30104 Av Tranquila, Rancho Palos Verdes, CA 91754
T: (213) 544-1267
Founded: 1974; Members: 35
Pres: John Song
Focus: Ethnology 14312

Association for Library and Information Science Education (ALISE), 4101 Lake Boone Tr, Ste 201, Raleigh, NC 27607
T: (919) 787-5181; Fax: (919) 787-4916
Founded: 1915; Members: 722
Gen Secr: Sally Nicholson
Focus: Adult Educ; Libraries & Bk Sci
Periodicals
Journal of Education for Library and Information Science (5 times annually)
Library and Information Science Education Statistical Report (weekly) 14313

Association for Library Collections and Technical Services (ALCTS), c/o American Library Association, 50 E Huron St, Chicago, IL 60611
T: (312) 280-5305; Fax: (312) 280-3257
Founded: 1957; Members: 6000
Gen Secr: Karen Muller
Focus: Libraries & Bk Sci
Periodicals
ALCTS Newsletter (8 times annually)
Library Resources and Technical Services (weekly) 14314

Association for Library Service to Children (ALSC), 50 E Huron St, Chicago, IL 60611
T: (312) 280-2163
Founded: 1901; Members: 3665
Gen Secr: Susan Roman
Focus: Libraries & Bk Sci
Periodicals
Journal of Youth Services in Libraries (weekly)
Newsletter (weekly) 14315

Association for Living Historical Farms and Agricultural Museums (ALHFAM), Rte 14, Box 214, Santa Fe, NM 87505
T: (505) 471-2261
Founded: 1970; Members: 1000
Gen Secr: George B. Paloheimo
Focus: Agri
Periodicals
Convention Proceedings (weekly)
Living Historical Farms Bulletin (weekly)
Membership List (weekly) 14316

Association for Macular Diseases (AMD), 210 E 64 St, New York, NY 10021
T: (212) 605-3719
Founded: 1977; Members: 4000
Pres: Nikolai Stevenson
Focus: Ophthal 14317

Association for Mexican Cave Studies (AMCS), POB 7037, Austin, TX 78712
T: (512) 452-5709
Founded: 1963
Gen Secr: Terry Raines
Focus: Geology
Periodicals
Bulletin
Newsletter
Postcard 14318

Association for Population/Family Planning Libraries and Information Centers International (APLIC), c/o William Record, Association for Vol. Surgical Contraception, 79 Madison Av, New York, NY 10016
T: (212) 561-8040; Fax: (212) 779-9439
Founded: 1968; Members: 115
Gen Secr: William Record
Focus: Libraries & Bk Sci
Periodicals
APLIC Communicator (weekly)
Proceedings of Annual Conference (weekly) 14319

Association for Practical Theology (APT), c/o George S. Worcester, Eden Theological Seminary, 475 E Lodgewood, Saint Louis, MO 63119
T: (314) 961-3627; Fax: (314) 961-5738
Founded: 1950; Members: 75
Gen Secr: George S. Worcester
Focus: Rel & Theol 14320

Association for Practitioners in Infection Control (APIC), 505 E Hawley St, Mundelein, IL 60060
T: (708) 949-6052; Fax: (708) 566-7282
Founded: 1972; Members: 9000
Gen Secr: Robert B. Willis
Focus: Med
Periodical
American Journal of Infection Control (weekly) 14321

Association for Progressive Communications (APC), c/o IGC, 18 De Boom St, San Francisco, CA 94107
T: (415) 442-0220; Tx: 15405417; Fax: (415) 546-1794
Members: 8000
Focus: Comm Sci 14322

Association for Psychoanalytic Medicine (APM), 4560 Delafield Av, New York, NY 10471
T: (212) 548-6088; Fax: (212) 548-8302
Founded: 1945; Members: 241
Pres: Dr. Otto Kernberg
Focus: Psychoan
Periodicals
Bulletin (weekly)
Roster (bi-annually) 14323

Association for Quality and Participation (AQP), 801B W Eighth St, Ste 501, Cincinnati, OH 45203-1601
T: (513) 381-0070; Fax: (513) 381-0070
Founded: 1977; Members: 8000
Gen Secr: Cathy Kramer
Focus: Materials Sci
Periodicals
Circle Report (weekly)
Conference Proceedings (weekly)
Quality Circle (weekly)
Quality Circles Journal (weekly) 14324

Association for Recorded Sound Collections (ARSC), POB 10162, Silver Spring, MD 20914
T: (301) 593-6552
Founded: 1966; Members: 1000
Gen Secr: Phillip Rochlin
Focus: Libraries & Bk Sci; Archives
Periodicals
Journal (weekly)
Membership Directory (bi-annually)
Newsletter (weekly)
Recorded Sound 14325

Association for Religious and Value Issues in Counseling (ARVIC), c/o American Counseling Association, 5999 Stevenson Av, Alexandria, VA 22304
T: (703) 823-9800; Fax: (703) 823-0252
Founded: 1955; Members: 4126
Gen Secr: Dr. Theodore P. Remley
Focus: Educ
Periodicals
Counseling and Values (weekly)
Interaction (weekly) 14326

Association for Research, Administration, Professional Councils and Societies (ARAPCS), c/o AAHPERD, 1900 Association Dr, Reston, VA 22091
T: (703) 476-3430; Fax: (703) 476-9527
Founded: 1949; Members: 7800
Gen Secr: Dr. Raymond A. Ciszek
Focus: Sports
Periodical
Research Quarterly for Exercise and Sport (weekly) 14327

Association for Research and Enlightenment (ARE), Atlantic Av, POB 595, Virginia Beach, VA 23451
T: (804) 428-3588
Founded: 1931; Members: 70000
Pres: Charles Thomas Cayce
Focus: Parapsych
Periodicals
Magazine Venture Inward (weekly)
Perspective on Consciousness and Psi Research (weekly) 14328

Association for Research in Nervous and Mental Disease (ARNMD), 1 Gustave Levy Pl, Box 1052, New York, NY 10029
T: (212) 348-8133; Fax: (212) 831-1816
Founded: 1920; Members: 950
Gen Secr: Ivan Bodis-Wollner
Focus: Psychiatry; Neurology
Periodical
Proceedings (weekly) 14329

Association for Research in Vision and Ophthalmology (ARVO), 9650 Rockville Pike, Bethesda, MD 20814-3998
T: (301) 571-1844; Fax: (301) 571-8311
Founded: 1928; Members: 7600
Gen Secr: Joanne G. Angle
Focus: Ophthal
Periodical
Investigative Ophthalmology and Vision Science (weekly) 14330

Association for Research of Childhood Cancer (AROCC), POB 251, Buffalo, NY 14225-0251
T: (716) 681-4433
Founded: 1970; Members: 1000
Gen Secr: Charles Moll
Focus: Cell Biol & Cancer Res; Pediatrics
Periodical
Newsletter (weekly) 14331

Association for Science, Technology and Innovation (ASTI), POB 1242, Arlington, VA 22210
T: (301) 985-7989
Founded: 1978; Members: 100
Pres: John Aje
Focus: Nat Sci; Eng
Periodical
Newsletter (weekly) 14332

Association for Social Anthropology in Oceania (ASAO), c/o Dr. Juliana Flinn, Dept of Sociology and Anthropology, University of Arkansas at Little Rock, 2801 S University, Little Rock, AR 72204
T: (501) 569-3173
Founded: 1967; Members: 300
Gen Secr: Dr. Juliana Flinn
Focus: Anthro
Periodical
Newsletter (weekly) 14333

Association for Social Economics (ASE), c/o Louisiana Technical University, Box 10318, Ruston, LA 71272
T: (318) 257-3701; Fax: (318) 257-4253
Founded: 1941; Members: 500
Gen Secr: Edward L. O'Boyle
Focus: Econ
Periodicals
Forum (weekly)
Review of Social Economy (weekly) . . 14334

Association for Specialists in Group Work (ASGW), c/o American Counseling Assoiation, 5999 Stevenson Av, Alexandria, VA 22304-3300
T: (703) 823-9800; Fax: (703) 823-0252
Founded: 1974; Members: 5393
Gen Secr: Dr. Theordore P. Remley
Focus: Educ
Periodicals
Journal of Specialists in Group Work (weekly)
Newsletter (3 times annually) 14335

Association for Supervision and Curriculum Development (ASCD), 1250 N Pitt St, Alexandria, VA 22314-1403
T: (703) 549-9110; Fax: (703) 549-3891
Founded: 1943; Members: 150000
Gen Secr: Gene R. Carter
Focus: Educ
Periodicals
ASCD Update (8 times annually)
ASCD Yearbook (weekly)
Educational Leadership (8 times annually) 14336

Association for Surgical Education (ASE), c/o Merrill T. Dayton, University Medical Center, 50 N Medical Dr, Rm 3B 312, Salt Lake City, UT 84132
T: (801) 581-7123
Founded: 1980; Members: 500
Gen Secr: Merrill T. Dayton
Focus: Educ; Surgery 14337

Association for Symbolic Logic (ASL), c/o Dept of Mathematics, University of Illinois, 1409 W Green St, Urbana, IL 61801
T: (217) 333-3410
Founded: 1936; Members: 1450
Gen Secr: C. Ward Henson
Focus: Logic
Periodical
Journal of Symbolic Logic (weekly) . . . 14338

Association for Systems Management (ASM), 1433 W Bagley Rd, POB 38370, Cleveland, OH 44138-0370
T: (216) 243-6900; Fax: (216) 234-2930
Founded: 1947; Members: 7000
Focus: Business Admin 14339

Association for Technology in Music Instruction (ATMI), c/o Gary Karpinski, School of Music, University of Oregon, Eugene, OR 97403-1225
T: (503) 346-5651
Founded: 1975; Members: 300
Gen Secr: Gary Karpinski
Focus: Music 14340

Association for the Advancement of Baltic Studies (AABS), 111 Knob Hill Rd, Hackettstown, NJ 07840
T: (908) 852-5258; Fax: (908) 852-3233
Founded: 1968; Members: 1000
Gen Secr: Janis Gaigulis
Focus: Ethnology
Periodicals
AABS Newsletter (weekly)
Journal of Baltic Studies (weekly) . . . 14341

Association for the Advancement of Health Education (AAHE), 1900 Association Dr, Reston, VA 22091
T: (703) 476-3437
Founded: 1937; Members: 11000
Gen Secr: B.J. Smith
Focus: Public Health; Educ
Periodicals
Health Education (weekly)
HE-XTRA (3-4 times annually) 14342

Association for the Advancement of International Education (AAIE), c/o Lewis A. Grell, Westminster College, Thompson House, New Wilmington, PA 16172
T: (412) 946-7192; Fax: (412) 946-7194
Founded: 1966; Members: 500
Gen Secr: Lewis A. Grell
Focus: Educ 14343

Association for the Advancement of Policy, Research and Development in the Third World, POB 70257, Washington, DC 20024-0257
T: (202) 723-7010; Fax: (202) 723-7010
Founded: 1981
Gen Secr: Dr. Mekki Mtewa
Focus: Poli Sci; Develop Areas
Periodical
Journal for Inter-Regional Cooperation and Development (weekly) 14344

Association for the Advancement of Psychotherapy (AAP), c/o Paul V. Trad, 53 E Tenth St, New York, NY 10003
T: (212) 529-1087
Founded: 1939; Members: 400
Gen Secr: Paul V. Trad
Focus: Therapeutics
Periodical
American Journal of Psychotherapy (weekly) 14345

Association for Theatre in Higher Education (ATHE), c/o Theatre Service, POB 15282, Evansville, IN 47716-0282
T: (812) 474-0549; Fax: (812) 476-4168
Founded: 1986; Members: 1950
Pres: Carole Brendt
Focus: Educ 14346

Association for the Behavioral Sciences and Medical Education (ABSAME), 6728 Old McLean Village Dr, McLean, VA 22101
T: (703) 556-9222
Founded: 1970; Members: 117
Gen Secr: Carol Ann Kiner
Focus: Behav Sci; Educ 14347

Association for the Bibliography of History (ABH), c/o Charles A. d'Aniello, Lockwood Memorial Library, State University of New York at Buffalo, Amherst, NY 14260
T: (716) 636-2817; Fax: (716) 636-3859
Founded: 1978; Members: 400
Gen Secr: Charles A. d'Aniello
Focus: Hist
Periodicals
ABH Bulletin (weekly)
Membership Directory (weekly) 14348

Association for the Care of Children's Health (ACCH), 7910 Woodmont Av, Ste 300, Bethesda, MD 20814
T: (301) 654-6549; Fax: (301) 986-4553
Founded: 1965; Members: 4300
Gen Secr: William Sciarillo
Focus: Pediatrics
Periodicals
ACCH Network (weekly)
ACCH News (weekly)
Children's Health Care (weekly)
Directory of Psychosocial Policy and Programs 14349

Association for the Development of Computer-Based Instructional Systems (ADCIS), 1601 W Fifth Av, Ste 111, Columbus, OH 43212
T: (614) 487-1528; Fax: (614) 487-1528
Founded: 1968; Members: 1800
Gen Secr: Carol Norris
Focus: Computer & Info Sci
Periodical
Journal of Computer-Based Instructions (weekly) 14350

Association for the Development of Human Potential (ADHP), 2328 W Pacific Av, Spokane, WA 99204
T: (509) 838-6652; Fax: (509) 838-6652
Founded: 1970; Members: 25
Pres: Margaret White
Focus: Educ; Psych 14351

Association for the Development of Religious Information Systems, c/o Dept of Social and Cultural Sciences, Marquette University, Milwaukee, WI 53233
T: (414) 288-6838
Founded: 1971
Gen Secr: Dr. David O. Moberg
Focus: Comm Sci 14352

Association for the Preservation of Virginia Antiquities (APVA), 2300 E Grace St, Richmond, VA 23223
T: (804) 648-1889; Fax: (804) 775-0802
Founded: 1889; Members: 6000
Gen Secr: Peter D. Grover
Focus: Preserv Hist Monuments
Periodicals
Discovery (weekly)
Newsletter – APVA News (3 times annually) 14353

Association for the Sociological Study of Jewry (ASSJ), c/o J. Winter, Dept of Sociology, Box 5302, New London, CT06320
T: (203) 439-2241; Fax: (203) 439-2700
Founded: 1971; Members: 300
Gen Secr: J. Winter
Focus: Sociology
Periodicals
Journal (weekly)
Newsletter (3 times annually) 14354

Association for the Sociology of Religion (ASR), c/o Barbara J. Denison, Lebanon Valley College, 931 Harrisburg, Ste 10, Lancaster, PA 17603
T: (717) 399-4419; Fax: (717) 867-6124
Founded: 1939; Members: 633
Gen Secr: Barbara J. Denison
Focus: Sociology; Rel & Theol; Anthro
Periodical
Sociological Analysis (weekly) 14355

Association for the Study of Afro-American Life and History (ASALH), 1407 14 St NW, Washington, DC 20005
T: (202) 667-2822
Founded: 1915; Members: 10000
Gen Secr: Karen A. McRae
Focus: Ethnology; Hist
Periodicals
Black History Kit (weekly)
Journal of Negro History (weekly)
Negro History Bulletin (weekly) 14356

Association for the Study of Dada and Surrealism (ASDS), c/o George H. Bauer, Dept of French and Italian, University of Southern California, Taper 126, University Park, Los Angeles, CA 90089
T: (310) 546-1438; Fax: (213) 746-7297
Founded: 1964; Members: 150
Pres: George H. Bauer
Focus: Arts; Lit
Periodical
Dada-Surrealism (weekly) 14357

Association for the Study of Higher Education (ASHE), c/o D. Stanley Carpenter, Dept of Educational Administration, Texas A & M University, College Station, TX 77843
T: (409) 845-0393; Fax: (409) 845-6129
Founded: 1972; Members: 800
Gen Secr: D. Stanley Carpenter
Focus: Educ
Periodical
Review of Higher Education (weekly) . . 14358

Association for the Study of Jewish Languages, 67-07 215 St, Oakland Garden, NY 11364-2523
Founded: 1979
Focus: Ling 14359

Association for the Treatment of Sexual Abusers (ABTSA), POB 66029, Portland, OR 97266
T: (503) 494-6144
Founded: 1985; Members: 250
Gen Secr: Sharon Siebert
Focus: Behav Sci 14360

Association for the World University (AWU), 847 N 28 St, Allentown, PA 18164
T: (215) 432-3124
Founded: 1959; Members: 80
Pres: A. Calamidas
Focus: Adult Educ
Periodical
Newsletter (weekly) 14361

Association for Transpersonal Psychology (ATP), POB 3049, Stanford, CA 94309
T: (415) 327-2066
Founded: 1971; Members: 2600
Gen Secr: Miles A. Vich
Focus: Psych
Periodicals
Journal on Transpersonal Psychology (weekly)
Listing of Professional Members (weekly)
Newsletter (weekly) 14362

Association for Tropical Biology (ATB), c/o J.S. Denslow, Dept of EEO Biology, Tulane University, New Orleans, LA 70118
T: (504) 865-5191; Fax: (504) 862-9706
Founded: 1963; Members: 1350
Gen Secr: J.S. Denslow
Focus: Bio 14363

Association for University Business and Economic Research (AUBER), c/o Terry Creeth, IBRC, 801 W Michigan, Indianapolis, IN 46202-5151
T: (317) 274-2204
Founded: 1947; Members: 170
Gen Secr: Terry Creeth
Focus: Business Admin; Econ
Periodicals
Bibliography of Publications (weekly)
Membership Directory (weekly)
Newsletter (weekly) 14364

Association for Women Geoscientists (AWG), c/o Dept of Geology, Macalester College, 1600 Grand Av, Saint Paul, MN 55015-1899
T: (612) 696-6448; Fax: (612) 686-6122
Founded: 1977; Members: 900
Pres: Janet L. Wright
Focus: Geology
Periodical
Gaea (weekly) 14365

Association for Women in Computing (AWC), 41 Sutter St, Ste 1006, San Francisco, CA 94104
T: (415) 905-4663
Founded: 1978; Members: 650
Pres: Cheryl Deichter
Focus: Computer & Info Sci
Periodicals
Conference Proceedings (weekly)
Directory (weekly)
Newsletter (weekly) 14366

Association for Women in Mathematics (AWM), c/o University of Maryland, 4114 Computer and Space Science Bldg, College Park, MD 20742-2461
T: (301) 405-7892
Founded: 1971; Members: 4000
Gen Secr: Ginny Reinhart
Focus: Math
Periodical
Newsletter (weekly) 14367

Association for Women in Psychology (AWP), c/o Angela R. Gillem, 526 W Sedgwick St, Philadelphia, PA 19119
Founded: 1969; Members: 1500
Gen Secr: Angela R. Gillem
Focus: Psych
Periodical
Newsletter (weekly) 14368

Association for Women in Science (AWIS), 1522 K St NW, Ste 820, Washington, DC 20005
T: (202) 408-0742; Fax: (202) 408-8321
Founded: 1971; Members: 3700
Gen Secr: Catherine Didion
Focus: Sci
Periodicals
AWIS Legislative Update (weekly)
Newsletter (weekly)
Resources for Women in Science Series (weekly)
. . . . 14369

Association for Women Veterinarians (AWV), c/o Chris Stone Payne, 32205 Allison Dr, Union City, CA 94587
T: (510) 471-8379
Founded: 1947; Members: 625
Gen Secr: Chris Stone Payne
Focus: Vet Med
Periodical
AWV Bulletin (weekly) 14370

Association Henri Capitant (AHC), c/o Lousiana State University Law Center, Baton Rouge, LA 70903
T: (504) 388-1126; Fax: (504) 388-3677
Founded: 1935; Members: 200
Pres: Alain Levasseur
Focus: Law 14371

Association of Academic Health Centers (AAHC), 1400 16 St NW, Ste 410, Washington, DC 20036
T: (202) 265-9600
Founded: 1969; Members: 102
Pres: R.J. Bulger
Focus: Public Health 14372

Association of Academic Health Sciences Library Directors (AAHSLD), c/o Sandra Wilson, HAM-TMC Library, 1133 M.D. Anderson Blvd, Houston, TX 77030
T: (713) 790-7060; Fax: (713) 790-7052
Founded: 1978; Members: 126
Pres: Judith Messerle
Focus: Libraries & Bk Sci
Periodicals
AAHSLD News (weekly)
Annual Report (weekly) 14373

Association of Academic Physiatrists (AAP), 5987 E 71 St, Ste 112, Indianapolis, IN 46220
T: (317) 845-4200; Fax: (317) 845-4299
Founded: 1967; Members: 1000
Gen Secr: Carolyn L. Braddom
Focus: Rehabil 14374

Association of Accredited Cosmetology Schools (AACS), 5201 Leesburg Pike, Ste 205, Falls Church, VA 22041
T: (703) 845-1333; Fax: (703) 845-1336
Founded: 1956; Members: 1000
Pres: Ronald E. Smith
Focus: Adult Educ
Periodicals
NAACS News (weekly)
Washington Report (weekly) 14375

Association of Advanced Rabbinical and Talmudic Schools (AARTS), 175 Fifth Av, New York, NY 10010
T: (212) 477-0950; Fax: (212) 533-5335
Founded: 1971; Members: 54
Gen Secr: Dr. Bernard Fryshman
Focus: Educ; Rel & Theol
Periodical
Annual Handbook 14376

Association of Air Medical Services (AAMS), 35 S Raymond Av, Ste 205, Pasadena, CA 91105
T: (818) 793-1232; Fax: (818) 793-1039
Founded: 1980; Members: 300
Gen Secr: Nina Merrill
Focus: Med
Periodicals
Aeromedical Journal (weekly)
Membership Directory (weekly) 14377

Association of American Colleges (AAC), 1818 R St NW, Washington, DC 20009
T: (202) 387-3760; Fax: (202) 265-9532
Founded: 1915; Members: 635
Pres: Paula P. Brownlee
Focus: Educ
Periodicals
Liberal Education (5 times annually)
On Campus with Women (weekly) . . . 14378

Association of American Geographers (AAG), 1710 16 St NW, Washington, DC 20009-3198
T: (202) 234-1450; Fax: (202) 234-2744
Founded: 1904; Members: 6400
Gen Secr: Ronald F. Abler
Focus: Geography
Periodicals
Annals (weekly)
Newsletter (10 times annually)
The Professional Geographer (weekly) . 14379

Association of American Indian Physicians (AAIP), 10015 S Pennsylvania, Oklahoma City, OK 73159
T: (405) 692-1202; Fax: (405) 692-1577
Founded: 1971; Members: 152
Gen Secr: Terry Hunter
Focus: Physiology
Periodical
Newsletter (weekly) 14380

Association of American Jurists (AAJ), POB 673, Berkeley, CA 94701
T: (510) 848-0599; Fax: (510) 848-6008
Founded: 1974
Gen Secr: Camilo Perez Bustillo
Focus: Law; Prom Peace
Periodical
Association of American Jurists Newsletter (weekly) 14381

Association of American Law Schools (AALS), 1201 Connecticut Av NW, Ste 800, Washington, DC 20036-2605
T: (202) 296-8851; Fax: (202) 296-8869
Founded: 1900; Members: 158
Gen Secr: Betsy Levin
Focus: Adult Educ; Law
Periodicals
Journal of Legal Education (weekly)
Newsletter (weekly)
Placement Bulletin (weekly) 14382

Association of American Medical Colleges (AAMC), 2450 N St NW, Washington, DC 20037
T: (202) 828-0400; Fax: (202) 785-5027
Founded: 1876; Members: 2200
Pres: Robert G. Petersdorf
Focus: Adult Educ; Med
Periodicals
Curriculum Directory (weekly)
Journal of Medical Education (weekly)
Weekly Report (weekly) 14383

Association of American Physicians (AAP), c/o Krannert Institute of Cardiology, Indiana University School of Medicine, 1111 W Tenth St, Indianapolis, IN 46202-4800
T: (317) 630-6468
Founded: 1886; Members: 1200
Focus: Physiology
Periodical
Transactions on the Association of American Physicians (weekly) 14384

Association of American Physicians and Surgeons (AAPS), 1601 N Tucson Blvd, Ste 9, Tucson, AZ 85716
T: (602) 327-4885; Fax: (602) 290-9674
Founded: 1943
Gen Secr: Jane M. Orient
Focus: Surgery; Physiology
Periodical
AAPS News (weekly) 14385

Association of American Schools in South America (AASSA), 6972 NW 50 St, Miami, FL 33166
T: (305) 594-3936; Fax: (305) 940-5772
Founded: 1961; Members: 28
Gen Secr: James W. Morris
Focus: Educ 14386

Association of American Seed Control Officials (AASCO), c/o David Turner, Oregon Dept of Agriculture, 635 Capitol St NE, Salem, OR 97310-0110
T: (503) 378-3774; Fax: (503) 373-1479
Founded: 1956
Gen Secr: David Turner
Focus: Agri 14387

Association of American State Geologists (AASG), c/o Morris W. Leighton, Illinois Geological Survey, 615 E Peabody Dr, Champaign, IL 61820
T: (217) 333-5111; Fax: (217) 244-7004
Founded: 1908; Members: 51
Pres: Morris W. Leighon
Focus: Geology
Periodical
Journal (weekly) 14388

Association of American Universities (AAU), 1 Dupont Circle NW, Ste 730, Washington, DC 20036
T: (202) 466-5030
Founded: 1900; Members: 59
Pres: Cornelius J. Pings
Focus: Sci 14389

Association of American Veterinary Medical Colleges (AAVMC), 1101 Vermont Av, Ste 710, Washington, DC 20005
T: (202) 371-9195; Fax: (202) 842-4360
Founded: 1966; Members: 41
Gen Secr: Dr. B.E. Hooper
Focus: Adult Educ; Vet Med
Periodical
Journal of Veterinary Medical Education (weekly)
. . . . 14390

Association of Analytical Chemists (ANACHEM), 2017 Hyde Park Rd, Detroit, MI 48207
T: (313) 393-3685
Founded: 1941; Members: 250
Gen Secr: Dr. Edward J. Havlena
Focus: Chem 14391

Association of Ancient Historians (AAH), c/o Prof. Stanley M. Burstein, Dept of History, California State University at Los Angeles, Los Angeles, CA 90032
T: (213) 343-2020
Founded: 1974; Members: 650
Pres: Prof. Stanley M. Burstein
Focus: Hist
Periodical
Newsletter (3 times annually) 14392

Association of Applied Insect Ecologists (AAIE), 1008 Tenth St, Ste 549, Sacramento, CA 95814
T: (916) 444-5224
Founded: 1967; Members: 300
Gen Secr: John F. Plain
Focus: Entomology; Ecology
Periodical
AAIE Bulletin (weekly) 14393

Association of Arab-American University Graduates (AAUG), 556 Trapelo Rd, Belmont, MA 02178
T: (617) 484-5483; Fax: (617) 484-5489
Founded: 1967
Focus: Adult Educ; Poli Sci; Ethnology
Periodicals
Arab Studies (weekly)
Newsletter (weekly) 14394

Association of Asphalt Paving Technologists (AAPT), 1983 Sloan Pl, Maplewood, MN 55117-2004
T: (612) 776-7703
Founded: 1924; Members: 860
Gen Secr: Eugene L. Skok
Focus: Civil Eng
Periodical
Asphalt Paving Technology (weekly) . . 14395

Association of Avian Veterinarians (AAV), POB 811720, Boca Raton, FL 33481
T: (407) 393-8901; Fax: (407) 393-8902
Founded: 1980; Members: 3000
Gen Secr: Adina Rae Freedman
Focus: Vet Med 14396

Association of Aviation Psychologists (AAP), c/o Dr. Tom McCloy, U.S. Air Force Academy, 2354 Fairchild Dr, Ste 6L47, Colorado Springs, CO 80864-6228
T: (719) 472-3860
Founded: 1964; Members: 350
Gen Secr: Dr. Tom McCloy
Focus: Psych 14397

Association of Balloon and Airship Constructors (ABAC), POB 90864, San Diego, CA 92169
T: (619) 270-4049; Fax: (619) 270-4049
Founded: 1974; Members: 250
Gen Secr: Donald E. Woodward
Focus: Aero
Periodical
Aerostation (weekly) 14398

Association of Biological Collections Appraisers (ABCA), 3493 Greenfield Pl, Carmel, CA 93923
T: (408) 624-5677
Founded: 1980; Members: 34
Pres: Frank P. Sala
Focus: Bio 14399

Association of Biomedical Communication Directors (ABCD), c/o Maurice G. Sherrard, Educational Resoures, PSU, POB 850, Hershey, PA 17033
T: (717) 531-8345; Fax: (717) 531-8576
Founded: 1972; Members: 87
Gen Secr: Maurice G. Sherrard
Focus: Comm Sci 14400

Association of Black Anthropologists (ABA), c/o American Anthropological Association, 1703 New Hampshire Av NW, Washington, DC 20009
T: (202) 232-8800; Fax: (202) 667-5345
Founded: 1970; Members: 130
Pres: Faye V. Harrison
Focus: Anthro 14401

Association of Black Psychologists (ABP), POB 55999, Washington, DC 20040-5999
T: (202) 722-0808; Fax: (202) 722-5941
Founded: 1968; Members: 650
Focus: Psych
Periodicals
Journal of Black Psycholoy (weekly)
NIA (weekly)
Psych Discourse (weekly) 14402

Association of Black Sociologists (ABS), c/o Howard University, POB 302, Washington, DC 20069
T: (202) 806-6853; Fax: (202) 806-4893

Founded: 1968; Members: 400
Pres: Walter Allen
Focus: Sociology
Periodicals
ABS Newsletter (weekly)
Roster of Membership (weekly) 14403

Association of Black Women in Higher Education (ABWHE), c/o Lenore R. Gall, 234 Hudson Av, Albany, NY 12210
T: (518) 472-1791
Founded: 1979; Members: 350
Pres: Lenore R. Gall
Focus: Educ 14404

Association of Bone and Joint Surgeons (ABJS), 6300 N River Rd, Ste 727, Rosemont, IL 60018-4226
T: (708) 698-1628; Fax: (708) 823-0536
Founded: 1947; Members: 185
Gen Secr: Carole Murphy
Focus: Surgery; Orthopedics
Periodical
Clinical Orthopedics and Related Research (8 times annually) 14405

Association of Business Officers of Preparatory Schools (ABOPS), c/o Aubrey K. Loomis, Loomis Chaffee School, Windsor, CT 06095
T: (203) 688-4937
Founded: 1924; Members: 36
Pres: Aubrey K. Loomis
Focus: Econ 14406

Association of Caribbean Studies (ACS), POB 22202, Lexington, KY 40522
T: (606) 257-6966; Fax: (606) 257-1074
Founded: 1978; Members: 1200
Gen Secr: Dr. O.R. Dathorne
Focus: Ethnology 14407

Association of Catholic Colleges and Universities (ACCU), 1 Dupont Circle, Ste 650, Washington, DC 20036
T: (202) 457-0650; Fax: (202) 728-0977
Founded: 1899; Members: 206
Gen Secr: Benito Lopez
Focus: Educ 14408

Association of Catholic Diocesan Archivists (ACDA), 955 Alton Rd, East Lansing, MI 48823
T: (517) 351-7216
Founded: 1981; Members: 180
Pres: George C. Michalek
Focus: Archives 14409

Association of Chairmen of Departments of Mechanics (ACDM), c/o Dr. Martin A. Eisenberg, Dept of Aereo, Mechanical and Engineering Science, University of Florida, 231 Aerospace Bldg, Gainesville, FL 32611-2031
T: (904) 392-0961
Founded: 1969; Members: 103
Pres: Dr. Martin A. Eisenberg
Focus: Eng
Periodical
Newsletter (weekly) 14410

Association of Chiropractic Colleges (ACC), 2005 Via Barrett, POB 367, San Lorenzo, CA 94580
T: (510) 276-9013; Fax: (510) 276-4893
Founded: 1977; Members: 18
Focus: Educ; Med 14411

Association of Christian Librarians (ACL), POB 4, Cedarville, OH 45314
T: (513) 766-7842; Fax: (513) 766-2337
Founded: 1957; Members: 300
Gen Secr: W. Bueschlen
Focus: Libraries & Bk Sci
Periodical
Christian Librarian (weekly) 14412

Association of Clinical Scientists (ACS), POB 1292, Farmington, CT 06034
T: (203) 679-2154; Fax: (203) 679-2154
Founded: 1949; Members: 658
Gen Secr: Dr. F. William Sunderman
Focus: Med
Periodical
Annals of Clinical Laboratory Science (weekly) 14413

Association of College and Research Libraries (ACRL), 50 E Huron St, Chicago, IL 60611
T: (312) 280-2516; Fax: (312) 440-9374
Founded: 1889; Members: 10600
Gen Secr: Altea Jenkins
Focus: Libraries & Bk Sci
Periodicals
Choice (11 times annually)
College and Research Libraries (weekly)
College and Research Libraries News (11 times annually)
Fast Job Listing Service (weekly)
Publications in Librarianship
Rare Books and Manuscripts Librarianship (weekly) 14414

Association of College and University Telecommunications Administrators (ACUTA), Lexington Financial Center, Ste 2420, Lexington, KY 40507
T: (606) 252-2882; Fax: (606) 252-5673
Founded: 1971; Members: 1200
Gen Secr: D. Combs
Focus: Comm
Periodicals
ACUTA News (weekly)
Membership Directory (weekly) 14415

Association of Collegiate Schools of Architecture (ACSA), 1735 New York Av NW, Washington, DC 20006
T: (202) 785-2324; Fax: (202) 626-7421
Founded: 1912
Gen Secr: Richard McCommons
Focus: Archit; Adult Educ
Periodicals
ACSA News (weekly)
Journal of Architectural Education (weekly) 14416

Association of Collegiate Schools of Planning (ACSP), c/o Dept of Urban and Regional Planning, University of Wisconsin at Madison, 925 Bascom Mall, Madison, WI 53706
T: (608) 262-1004; Fax: (608) 262-8307
Founded: 1959; Members: 1517
Pres: Jerome Kaufman
Focus: Urban Plan
Periodicals
Directory (bi-annually)
Journal of Planning Education and Research (weekly) 14417

Association of Community Cancer Centers (ACCC), 11600 Nebel St, Ste 201, Rockville, MD 20852
T: (301) 984-9496; Fax: (301) 770-1949
Founded: 1974; Members: 715
Gen Secr: Lee E. Mortenson
Focus: Cell Biol & Cancer Res 14418

Association of Community College Trustees (ACCT), 1740 N St NW, Washington, DC 20036
T: (202) 775-4667; Fax: (202) 223-1297
Founded: 1969; Members: 780
Gen Secr: Ray Taylor
Focus: Educ
Periodicals
ACCT-O-Line (weekly)
Advisor (weekly)
Trustee Quarterly (weekly) 14419

Association of Community Tribal Schools (ACTS), c/o Dr. Roger Bordeaux, 616 Fourth Av W, Sisseton, SD 57262-1349
T: (605) 698-3112; Fax: (605) 658-7686
Founded: 1982; Members: 30
Gen Secr: Dr. Roger Bordeaux
Focus: Educ 14420

Association of Computer Users (ACU), 21154A Berkeley, POB 2189, Berkeley, CA 94702-0189
T: (510) 549-4300; Fax: (510) 549-4331
Founded: 1979; Members: 1000
Gen Secr: Steve Gordon
Focus: Computer & Info Sci
Periodicals
Business Tools (weekly)
Executive Computing Newsletter (weekly)
PC Review (weekly) 14421

Association of Consulting Chemists and Chemical Engineers (ACC&CE), 295 Madison Av, New York, NY 10017
T: (212) 983-3160; Fax: (212) 983-3161
Founded: 1928; Members: 125
Gen Secr: Elizabeth K. Jones
Focus: Chem
Periodical
Your Consultant (weekly) 14422

Association of Continuing Legal Education Administrators (ACLEA), c/o American Bar Association, 541 N Fairbanks Ct, Chicago, IL 60611
T: (312) 998-6196; Fax: (312) 988-5368
Founded: 1965; Members: 264
Gen Secr: James J. Parente
Focus: Educ; Law 14423

Association of Cytogenetic Technologists (ACT), 1021 15 St, Ste 9, Augusta, GA 30901
T: (706) 722-7511; Fax: (706) 721-3048
Founded: 1975; Members: 1500
Gen Secr: Ann Cork
Focus: Bio
Periodical
Karyogram (weekly) 14424

Association of Data Communications Users (ADCU), POB 385728, Bloomington, MN 55438
T: (612) 881-6803; Fax: (612) 881-6709
Founded: 1975; Members: 275
Gen Secr: August H. Blegen
Focus: Computer & Info Sci
Periodicals
Membership Mailing (weekly)
Newsletter (weekly) 14425

Association of Defense Trial Attorneys (ADTA), 600 Bank One Bldg, Peoria, IL 61602
T: (309) 676-0400
Founded: 1937; Members: 700
Gen Secr: Gary M. Peplow
Focus: Insurance
Periodicals
Membership Roster (weekly)
Newsletter 14426

Association of Departments of English (ADE), 10 Astor Pl, New York, NY 10003
T: (212) 614-6317
Founded: 1962; Members: 840
Gen Secr: David Laurence
Focus: Ling
Periodicals
ADE Bulletin (3 times annually)
MLA/ADE Job Information List (5 times annually) 14427

Association of Departments of Foreign Languages (ADFL), 10 Astor Pl, New York, NY 10003
T: (212) 614-6319
Founded: 1969; Members: 1050
Gen Secr: John Cross
Focus: Ling; Adult Educ
Periodicals
ADFL Bulletin (3 times annually)
MLA/ADFL Job Information List (5 times annually) 14428

Association of Disciples for Theological Discussion (ADTD), c/o Div of Higher Education, 11780 Borman Dr, Ste 100, Saint Louis, MO 63146
T: (314) 991-3000
Founded: 1954; Members: 25
Gen Secr: John M. Imbler
Focus: Rel & Theol
Periodical
Papers (weekly) 14429

Association of Earth Science Editors (AESE), c/o M. Adkins-Heljeson, Kansas Geological Survey, 1930 Constant Av, Lawrence, KS 66047
T: (913) 864-3965
Founded: 1967; Members: 400
Gen Secr: M. Adkins-Heljeson
Focus: Geology
Periodicals
Blueline (weekly)
Membership Directory (weekly) 14430

Association of Energy Engineers (AEE), 4025 Pleasantdale Rd, Ste 420, Atlanta, GA 30340
T: (404) 447-5083; Fax: (404) 446-3969
Founded: 1977; Members: 8000
Gen Secr: Albert Thumann
Focus: Energy
Periodicals
Directory
Newsletter 14431

The Association of Engineering Firms Practicing in the Geosciences (ASFE), 8811 Colesville Rd, Ste G106, Silver Spring, MD 20910
T: (301) 565-2733; Fax: (301) 589-2017
Founded: 1969; Members: 350
Gen Secr: John P. Bachner
Focus: Agri
Periodical
Newslook (8 times annually) 14432

Association of Engineering Geologists (AEG), 323 Boston Post Rd, Ste 2D, Sudbury, MA 01776
T: (508) 443-4639
Founded: 1957; Members: 3200
Gen Secr: Edwin A. Blackey
Focus: Geology
Periodical
Bulletin of the Association of Engineering Geologists (weekly) 14433

Association of Environmental Engineering Professors (AEEP), c/o Prof. C. Robert Baillod, Michigan Technological University, 1400 Townsend Dr, Houghton, MI 49931
T: (906) 487-2530
Founded: 1963; Members: 500
Pres: Prof. C. Robert Baillod
Focus: Eng
Periodicals
Membership List (weekly)
Newsletter (3 times annually) 14434

Association of Episcopal Colleges (AEC), 815 Second Av, New York, NY 10017-4594
T: (212) 986-0989; Fax: (212) 986-5039
Founded: 1962; Members: 12
Pres: Dr. Linda A. Chisholm
Focus: Educ; Rel & Theol
Periodicals
President's Report (weekly)
Views and Views (weekly) 14435

Association of Federal Communications Consulting Engineers (AFCCE), POB 19333, 20 St Station, Washington, DC 20036
T: (202) 223-6700
Founded: 1948; Members: 200
Focus: Electric Eng 14436

Association of Field Ornithologists (AFO), c/o Elissa Landre, Broadmoor Wildlife Sanctuary, Massachusetts Audubon Society, 280 Eliot St, South Natick, MA 01760
T: (508) 655-2296
Founded: 1924; Members: 1400
Gen Secr: Elissa Landre
Focus: Ornithology
Periodical
Journal of Field Ornithology (weekly) . . 14437

Association of Firearm and Tool Mark Examiners (AFTE), 7857 Esterel Dr, La Jolla, CA 92037
T: (619) 453-0847
Founded: 1969; Members: 500
Gen Secr: Eugenia Bell
Focus: Eng
Periodical
Journal (weekly) 14438

Association of Food and Drug Officials (AFDO), POB 3425, York, PA 17402
T: (717) 757-2888; Fax: (717) 755-8089
Founded: 1897; Members: 600
Gen Secr: Whitney Almquist
Focus: Food
Periodicals
Bulletin (5 times annually)
Journal of the Association of Food and Drug Officials (5 times annually)
News and Views (weekly) 14439

Association of Forensic Document Examiners (AFDE), 3813 Sheridan Av S, Minneapolis, MN 55410
T: (612) 922-3060
Founded: 1985; Members: 20
Gen Secr: Anne B. Hooten
Focus: Graphology; Forensic Med; Criminology 14440

Association of Graduate Liberal Studies Programs (AGLSP), c/o Dr. Charles R. Strain, DePaul University, 2323 N Seminary Av, Chicago, IL 60614
T: (202) 362-8246; Fax: (202) 362-5481
Founded: 1975; Members: 89
Gen Secr: Dr. Charles R. Strain
Focus: Educ; Adult Educ
Periodical
Newsletter 14441

Association of Graduate Schools in Association of American Universities (AGS), 1 Dupont Circle NW, Ste 730, Washington, DC 20036
T: (202) 466-5030
Founded: 1948; Members: 56
Gen Secr: John Vaughn
Focus: Educ
Periodical
Newsletter 14442

The Association of Higher Education Facilities Officers (APPA), 1446 Duke St, Alexandria, VA 22314-3492
T: (703) 684-1446; Fax: (703) 549-2772
Founded: 1914; Members: 4300
Gen Secr: Walter A. Schaw
Focus: Sports
Periodicals
APPA Newsletter (weekly)
Facilities Management Magazine (weekly) 14443

Association of Independent Colleges and Schools (AICS), 1 Dupont Circle NW, Ste 350, Washington, DC 20036
T: (202) 659-2460
Founded: 1962; Members: 1100
Pres: John G. Pucciano
Focus: Educ
Periodicals
Capital Comments (weekly)
Compass (weekly) 14444

Association of Information and Dissemination Centers (ASIDIC), POB 8105, Athens, GA 30603
T: (706) 542-6820
Founded: 1968; Members: 85
Gen Secr: J. Webb
Focus: Computer & Info Sci
Periodical
ASIDIC Newsletter (weekly) 14445

Association of Internal Management Consultants (AIMC), c/o Margaret M. Custer, POB 304, East Bloomfield, NY 14443
T: (716) 657-7878
Founded: 1971; Members: 280
Gen Secr: Margaret M. Custer
Focus: Business Admin 14446

Association of International Colleges and Universities (AICU), 1301 S Noland Rd, Independence, MO 64055
T: (816) 461-3633; Fax: (816) 461-3634
Founded: 1973; Members: 4011
Pres: Dr. John Wayne Johnston
Focus: Sci
Periodical
Directory (weekly) 14447

Association of International Health Researchers (AIHR), 2665 Pleasant Valley Rd, Mobile, AL 36606
T: (205) 473-3946
Founded: 1982; Members: 123
Pres: Dr. Roy E. Kadel
Focus: Public Health 14448

Association of Iron and Steel Engineers (AISE), 3 Gateway Center, Ste 2350, Pittsburgh, PA 15222
T: (412) 281-6323; Fax: (412) 281-4657
Founded: 1907; Members: 10000
Gen Secr: Lawrence G. Maloney
Focus: Eng; Metallurgy
Periodicals
Directory Iron and Steel Plants (weekly)
Iron and Steel Engineer (weekly) . . . 14449

Association of Jesuit Colleges and Universities (AJCU), 1424 16 St NW, Ste 504, Washington, DC 20036
T: (202) 667-3889; Fax: (202) 328-8643
Founded: 1970; Members: 28

Pres: Paul S. Tipton
Focus: Educ; Rel & Theol; Sci
Periodical
Higher Education Report (10 times annually) . . . 14450

Association of Jewish Libraries (AJL), c/o National Foundation for Jewish Culture, 330 Seventh Av, New York, NY 10001
T: (212) 381-6440
Founded: 1966; Members: 1000
Pres: Dr. Ralph Simon
Focus: Libraries & Bk Sci
Periodicals
AJL Newsletter (weekly)
Judaica Librarianship (bi-annually) . . . 14451

Association of Legal Administrators (ALA), 175 E Hawthorn Pkwy, Ste 325, Vernon Hills, IL 60061-1428
T: (708) 816-1212
Founded: 1971; Members: 8800
Gen Secr: Peter R. Bonavich
Focus: Law
Periodicals
ALA News (weekly)
Human Resources (weekly)
Management and Administration (weekly) 14452

Association of Life Insurance Counsel (ALIC), c/o Frank D. Casciano, 520 Broad St, Newark, NJ 01702-3184
T: (201) 481-4526; Fax: (201) 268-4335
Founded: 1913; Members: 1000
Gen Secr: Frank D. Casciano
Focus: Insurance
Periodicals
Membership Directory (weekly)
Papers (weekly)
Proceedings . . . 14453

Association of Lutheran College Faculties (ALCF), c/o B. Stratton, Augsburg College, 731 21 Av S, Minneapolis, MN 55454
T: (612) 330-1063
Founded: 1935; Members: 150
Pres: B. Stratton
Focus: Educ . . . 14454

Association of Lutheran Secondary Schools (ALSS), c/o Edward Reitz, Lutheran High School, 5401 Lucas, Saint Lucas, MO 63121
T: (314) 389-3100
Members: 60
Pres: Edward Reitz
Focus: Educ . . . 14455

Association of Management (AM), Rte 17, POB 1301, Grafton, VA 23692-1301
T: (804) 479-5363; Fax: (804) 479-0656
Founded: 1975; Members: 2800
Gen Secr: W.A. Hamel
Focus: Psych
Periodicals
HRMOB News (weekly)
Journal of Information Resources Management Systems (weekly)
Journal of Management in Practice (weekly) . . . 14456

Association of Management Consultants (AMC), 521 Fifth Av, New York, NY 10175
T: (212) 697-8262
Founded: 1959; Members: 445
Pres: Ed Hendricks
Focus: Business Admin . . . 14457

Association of Maternal and Child Health Programs (AMCHP), 1350 Connecticut Av NW, Ste 803, Washington, DC 20036
T: (202) 775-0436; Fax: (202) 775-0061
Founded: 1944; Members: 300
Gen Secr: Catherine Hess
Focus: Public Health . . . 14458

Association of Medical Education and Research in Substance Abuse (AMERSA), c/o Center for Alcohol and Addiction Studies, Brown University, Box G-BH, Providence, RI 02912
T: (401) 863-7791; Fax: (401) 863-3510
Founded: 1976; Members: 500
Pres: David C. Lewis
Focus: Educ; Med . . . 14459

Association of Medical Rehabilitation Administrators (AMRA), c/o Children's Hospital and Health Center, 8001 Frost Av, San Diego, CA 92123
T: (619) 576-5843; Fax: (619) 422-3106
Founded: 1953; Members: 250
Gen Secr: Carl Schneiderman
Focus: Rehabil
Periodicals
Directory (weekly)
Quarterly Bulletin (weekly) . . . 14460

Association of Medical School Pediatric Department Chairmen (AMSPDC), c/o Michael A. Simmons, Primary Medical Children's Center, 100 N Medical Dr, Salt Lake City, UT 84113
T: (801) 588-2365
Founded: 1961; Members: 147
Gen Secr: Michael A. Simmons
Focus: Pediatrics
Periodical
AMSPDC Membership List (weekly) . . 14461

Association of Mental Health Administrators (AMHA), 60 Revere Dr, Ste 500, Northbrook, IL 60062
T: (708) 480-9626; Fax: (708) 480-9282
Founded: 1959; Members: 1450
Gen Secr: Ellen Locurto
Focus: Psychiatry
Periodicals
Journal (weekly)
Newsletter (weekly) . . . 14462

Association of Mercy Colleges (AMC), c/o Dr. William P. Garvey, Mercyhurst College, 501 E 38 Blvd, Erie, PA 16501
T: (814) 825-0269; Fax: (814) 641-5596
Founded: 1982; Members: 18
Pres: Dr. William P. Garvey
Focus: Educ . . . 14463

Association of Military Colleges and Schools of the United States (AMCS), 9115 Mc Nair Dr, Alexandria, VA 22309
T: (703) 360-1678
Founded: 1914; Members: 45
Gen Secr: Willard W. Scott
Focus: Adult Educ; Military Sci
Periodicals
Membership List (weekly)
Newsletter (8 times annually) . . . 14464

Association of Military Surgeons of the U.S. (AMSUS), 9320 Old Georgetown Rd, Bethesda, MD 20814
T: (301) 897-8800; Fax: (301) 530-5446
Founded: 1891; Members: 15000
Gen Secr: Max Bralliar
Focus: Surgery
Periodical
Military Medicine (weekly) . . . 14465

Association of Minicomputers Users (AMU), 363 E Central St, Franklin, MA 02038
T: (617) 520-1555; Fax: (617) 520-1558
Founded: 1978; Members: 1000
Gen Secr: Raymond P. Wenig
Focus: Computer & Info Sci
Periodicals
Mini-Beacon (weekly)
Minicomputer Software Quarterly (weekly) 14466

Association of Minority Health Professions Schools (AMHPS), 711 Second St NE, Ste 200, Washington, DC 20002
T: (202) 544-7499; Fax: (202) 546-7105
Founded: 1978; Members: 8
Gen Secr: Dale P. Dirks
Focus: Educ; Public Health . . . 14467

Association of Muslim Scientists and Engineers (AMSE), POB 38, Plainfield, IN 46168
T: (317) 839-8157
Founded: 1969; Members: 640
Gen Secr: I.J. Unus
Focus: Eng; Sci
Periodicals
ASME Membership Directory (weekly)
International Journal of Science and Technology (weekly)
Newsletter (weekly)
Proceedings of Conference (weekly) . . 14468

Association of Muslim Social Scientists (AMSS), POB 669, Herndon, VA 22070
T: (703) 471-1133; Fax: (703) 471-3922
Founded: 1972; Members: 580
Gen Secr: S.M. Syeed
Focus: Sociology
Periodicals
American Journal of Islamic Social Sciences (weekly)
Bulletin (weekly)
Newsletter (weekly)
Proceedings of Conferences (weekly) . . 14469

Association of Naval R.O.T.C. Colleges and Universities, 2000 Perrine St, Lafayette, IN 47904
T: (317) 742-1641
Founded: 1946; Members: 63
Gen Secr: Dorothy A. Bolder
Focus: Educ; Navig . . . 14470

Association of Official Racing Chemists (AORC), POB 19232, Portland, OR 97280
T: (503) 644-9224; Fax: (503) 626-7039
Founded: 1947; Members: 140
Gen Secr: Margaret A. Sullivan
Focus: Chem . . . 14471

Association of Official Seed Analysts (AOSA), c/o Larry J. Prentice, 268 Plant Science, IANR-UNL, Lincoln, NE 68583-0911
T: (402) 472-8649; Fax: (402) 472-7204
Founded: 1908; Members: 250
Gen Secr: Larry J. Prentice
Focus: Agri
Periodicals
Journal of Seed Technology (weekly)
Newsletter (3 times annually) . . . 14472

Association of Old Crows (AOC), 1000 N Payne St, Alexandria, VA 22314
T: (703) 549-1600
Founded: 1964; Members: 23000
Gen Secr: Gus Slayton
Focus: Electronic Eng
Periodical
Journal of Electronic Defense (weekly) . 14473

Association of Optometric Educators (AOE), c/o Dr. D. Bezan, NSU College of Optometry, Tahlequah, OK 74464
T: (918) 456-5111
Founded: 1972; Members: 100
Pres: Dr. D. Bezan
Focus: Adult Educ; Ophthal . . . 14474

Association of Orthodox Jewish Scientists (AOJS), 3 W 16 St, New York, NY 10011
T: (212) 229-2340
Founded: 1948; Members: 1500
Gen Secr: Joel Schwartz
Focus: Rel & Theol; Sci
Periodicals
Intercom (weekly)
Membership Directory
Newsletter (5 times annually)
Proceedings (weekly) . . . 14475

Association of Orthodox Jewish Teachers (AOJT), 1577 Coney Island Av, Brooklyn, NY 11230
T: (718) 258-3585; Fax: (718) 258-3586
Founded: 1964; Members: 5000
Gen Secr: Dr. Philip J. Kipust
Focus: Educ
Periodicals
Journal (weekly)
Newspaper . . . 14476

Association of Osteopathic State Executive Directors (AOSED), 455 Capitol Mall, Ste 225, Sacramento, CA 95814
T: (916) 447-2004
Founded: 1918; Members: 112
Gen Secr: Matt Weyuker
Focus: Pathology
Periodicals
Directory of Osteopathic Publications (bi-annually)
Newsletter (weekly) . . . 14477

Association of Overseas Educators (AOE), c/o Dr. Robert L. Morris, Indiana University of Pennsylvania, Indiana, PA 15705
T: (412) 357-2295
Founded: 1955
Gen Secr: Dr. Robert L. Morris
Focus: Educ
Periodicals
Directory (weekly)
News (weekly) . . . 14478

Association of Pacific Systematists (APS), c/o Dept of Botany, Bishop Museum, POB 19000-A, Honolulu, HI 96817-0916
Founded: 1982
Gen Secr: S.H. Sohmer
Focus: Botany
Periodical
Association of Pacific Systematists Newsletter . . . 14479

Association of Pathology Chairmen (APC), c/o Dr. Frances A. Pitlick, 9650 Rockville Pike, Bethesda, MD 20814-3993
T: (301) 571-1880; Fax: (301) 571-1879
Founded: 1967; Members: 138
Gen Secr: Dr. Frances A. Pitlick
Focus: Pathology
Periodical
Newsletter (weekly) . . . 14480

Association of Philippine Physicians in America (APPA), 2717 W Olive Av, Ste 200, Burbank, CA 91505
T: (818) 843-8616
Founded: 1972; Members: 3000
Gen Secr: Dr. Fred Quevedo
Focus: Physiology
Periodicals
APPA Quarterly (weekly)
Leadership Roster of Officers (weekly)
Newsletter (weekly) . . . 14481

Association of Physician Assistant Programs (APAP), 950 N Washington St, Alexandria, VA 22314
T: (703) 836-2272
Founded: 1972; Members: 53
Gen Secr: Harry Bradley
Focus: Physiology
Periodicals
Conference Proceedings (weekly)
National Directory of Physician Assistant Programs (weekly)
Newsletter (weekly) . . . 14482

Association of Physician's Assistants in Cardiovascular Surgery (APACVS), 2000 Tate Springs Rd, POB 2242, Lynchburg, VA 24501-2242
T: (800) 528-1506
Founded: 1981; Members: 675
Pres: J. Richard Milam
Focus: Cardiol; Surgery . . . 14483

Association of Presbyterian Colleges and Universities (APCU), 100 Witherspoon St, Rm 1050, Louisville, KY 40202-1396
T: (502) 569-5606
Founded: 1983; Members: 68
Gen Secr: Duncan S. Ferguson
Focus: Educ . . . 14484

Association of Professional Baseball Physicians (APBP), c/o Dr. Harvey O'Phelan, 2600 E Franklin Av, Minneapolis, MN 55406-1192
T: (612) 339-8976
Founded: 1970; Members: 36
Pres: Dr. Harvey O'Phelan
Focus: Physical Therapy; Physiology . . 14485

Association of Professional Energy Managers (APEM), 3104 O St, Ste 301, Sacramento, CA 95816
T: (916) 454-4338; Fax: (916) 736-3581
Founded: 1983; Members: 600
Pres: Tim Speller
Focus: Energy
Periodicals
Membership Directory (weekly)
The Professional Energy Manager (weekly) 14486

Association of Professional Genealogists (APG), 3421 M St NW, Ste 236, Washington, DC 20007-3552
T: (703) 920-2385
Founded: 1979; Members: 600
Pres: Shirley L. Wilcox
Focus: Genealogy
Periodical
APG Quarterly (weekly) . . . 14487

Association of Professional Material Handling Consultants (APMHC), 8720 Red Oak Blvd, Ste 224, Charlotte, NC 28217
T: (704) 529-1725; Fax: (704) 525-2880
Founded: 1960; Members: 40
Gen Secr: Peter Youngs
Focus: Materials Sci . . . 14488

Association of Professors of Gynecology and Obstetrics (APGO), 409 12 St SW, Washington, DC 20024
T: (202) 863-2507; Fax: (202) 863-2514
Founded: 1962; Members: 156
Gen Secr: Donna Wachter
Focus: Gynecology
Periodicals
Academic Position Report (weekly)
Membership Directory (weekly)
Newsletter (weekly) . . . 14489

Association of Professors of Medicine (APM), 1101 Connecticut Av NW, Ste 700, Washington, DC 20036
T: (202) 857-1158
Founded: 1954; Members: 125
Gen Secr: James Terwilliger
Focus: Med . . . 14490

Association of Research Directors (ARD), c/o Leon G. Chavous, South Carolina State University, 300 College St NE, POB 7265, Orangeburg, SC 29117
T: (803) 536-8928
Founded: 1977; Members: 17
Gen Secr: Leon G. Chavous
Focus: Business Admin
Periodical
Membership List (weekly) . . . 14491

Association of Research Libraries (ARL), 1527 New Hampshire Av NW, Washington, DC 20036
T: (202) 232-2466; Fax: (202) 462-7849
Founded: 1932; Members: 119
Gen Secr: Duane Webster
Focus: Libraries & Bk Sci
Periodicals
ARL Annual Solary Sunvey (weekly)
ARL Newsletter (5 times annually)
ARL Statistics (weekly)
Minutes of the Art Meetings (weekly) . . 14492

Association of Safety Council Executives (ASCE), c/o Larry Schenck, 3710 NW 51 St. Ste A, Gainesville, FL 32606
T: (904) 377-2566; Fax: (904) 377-7544
Founded: 1936; Members: 250
Pres: Larry Schenck
Focus: Safety . . . 14493

Association of School Business Officials International (ASBO), 11401 N Shore Dr, Reston, VA 22090
T: (703) 478-0405; Fax: (703) 478-0205
Founded: 1910; Members: 6900
Gen Secr: Don I. Tharpe
Focus: Educ
Periodicals
ASBO Accents (weekly)
School Business Affairs (weekly) . . . 14494

Association of Schools and Colleges of Optometry (ASCO), 6110 Executive Blvd, Ste 690, Rockville, MD 20852
T: (301) 231-5944; Fax: (301) 770-1828
Founded: 1941; Members: 19
Gen Secr: Martin A. Wall
Focus: Adult Educ; Ophthal
Periodicals
Information for Applicants to Schools and Colleges of Optometry (weekly)
Journal of Optometric Education (weekly)
Proceedings (weekly) . . . 14495

Association of Schools of Allied Health Professions (ASAHP), 1101 Connecticut Av NW, Ste 700, Washington, DC 20036
T: (202) 857-1150
Founded: 1967; Members: 1220
Gen Secr: Carolyn Delpolito
Focus: Public Health
Periodicals
Allied Health Trends (weekly)
Directory (weekly)
Journal of Allied Health (weekly) . . . 14496

Association of Schools of Journalism and Mass Communication (ASJMC), c/o Jennifer H. McGill, University of South Carolina, 1621 College St, Columbia, SC 29208-0251
T: (803) 777-2005; Fax: (803) 777-4728
Founded: 1917; Members: 195
Gen Secr: Jennifer H. McGill
Focus: Journalism
Periodicals
JMC Administrator (weekly)
Journalism Educator (weekly) *14497*

Association of Schools of Public Health (ASPH), 1015 15 St NW, Ste 404, Washington, DC 20005
T: (202) 842-4668
Founded: 1941; Members: 24
Gen Secr: Michael K. Gemmell
Focus: Adult Educ; Public Health
Periodical
Graduate Education for Public Health (bi-annually) *14498*

Association of Science Museum Directors (ASMD), c/o Chicago Academy of Sciences, 2001 N Clark St, Chicago, IL 60614
T: (312) 549-0606
Founded: 1960; Members: 92
Focus: Sci
Periodical
Science Museum News *14499*

Association of Science-Technology Centers (ASTC), 1025 Vermont Av NW, Ste 500, Washington, DC 20005
T: (202) 783-7200
Founded: 1973; Members: 417
Gen Secr: Bonnie Van Dorn
Focus: Eng; Sci
Periodicals
ASTC Newsletter (weekly)
ASTC Staff Directory (weekly)
Science Center Exhibitions Calendar (weekly) *14500*

Association of Seventh-Day Adventist Educators (ASDAE), c/o General Conf., Dept of Education, 12501 Old Columbia Pike, Silver Spring, MD 20904
T: (301) 680-6443; Fax: (301) 680-6090
Founded: 1970; Members: 600
Gen Secr: Gordon Madgwick
Focus: Educ *14501*

Association of Social and Behavioral Scientists (ASBS), c/o Dr. Jacqueline Rovse, Dept of History, American University, 4400 Massachusetts Av, Washington, DC 20016-8038
T: (202) 885-2461
Founded: 1935; Members: 200
Gen Secr: Dr. Jacqueline Rovse
Focus: Sociology; Behav Sci
Periodical
Journal of Social and Behavioral Sciences (weekly) *14502*

Association of Southern Baptist Colleges and Schools (ASBCS), 901 Commerce, Ste 600, Nashville, TN 37203
T: (615) 244-2362; Fax: (615) 242-2153
Founded: 1915; Members: 140
Gen Secr: Arthur L. Walker
Focus: Educ
Periodicals
Directory of Southern Baptist Colleges and Schools (biennial)
The Southern Baptist Educator (weekly) . *14503*

Association of Space Explorers U.S.A- (ASE-USA), 35 White St, San Francisco, CA 94109
T: (415) 931-0585
Founded: 1985; Members: 210
Pres: John Fabian
Focus: Astrophys *14504*

Association of Specialized and Cooperative Library Agencies (ASCLA), c/o American Library Association, 50 E Huron St, Chicago, IL 60611
T: (312) 280-4399; Fax: (312) 280-3257
Founded: 1978; Members: 1432
Gen Secr: Andrew M. Hansen
Focus: Libraries & Bk Sci
Periodical
Interface (weekly) *14505*

Association of State and Interstate Water Pollution Control Administrators (ASIWPCA), 750 First St NE, Ste 910, Washington, DC 20002
T: (202) 898-0905
Founded: 1960; Members: 57
Gen Secr: Roberta J. Savage
Focus: Ecology
Periodicals
Annual Report (weekly)
Membership Directory (weekly)
Positions Statement (weekly) *14506*

Association of State and Territorial Dental Directors (ASTDD), c/o Dr. Robert Isman, POB 942732, Sacramento, CA 94234-7320
T: (916) 322-4933
Members: 53
Gen Secr: Dr. Robert Isman
Focus: Dent *14507*

Association of State and Territorial Directors of Public Health Education (ASTDPHE), c/o Nancy Miller, Dept of Health, 201 Business Loop 70 W, Columbia, MO 65203
T: (314) 876-3250; Fax: (314) 443-3592
Founded: 1946; Members: 76
Pres: Nancy Miller
Focus: Public Health
Periodicals
Conference Call (weekly)
Proceedings (biennial) *14508*

Association of State and Territorial Health Officials (ASTHO), 415 Second St NE, Ste 200, Washington, DC 20002
T: (202) 546-5400
Founded: 1942; Members: 55
Gen Secr: George K. Degnon
Focus: Public Health
Periodicals
Conference Proceedings (weekly)
Membership Directory (bi-annually)
Newsletter (weekly) *14509*

Association of State Supervisors of Mathematics (ASSM), c/o Charles Watson, Dept of Education, 4 Capitol Mall, Little Rock, AR 72201
T: (501) 682-4474; Fax: (501) 682-4618
Founded: 1959; Members: 200
Gen Secr: Charles Watson
Focus: Math *14510*

Association of Surgical Technologists (AST), 7108C S Alton Way, Englewood, CO 80112
T: (303) 694-9130; Fax: (303) 694-9169
Founded: 1969; Members: 14600
Gen Secr: William J. Teutsch
Focus: Surgery; Eng
Periodical
The Surgical Technologist (weekly) . . . *14511*

Association of Systematics Collections (ASC), 730 11 St NW, Washington, DC 20001
T: (202) 347-2850; Fax: (202) 347-0072
Founded: 1972; Members: 100
Gen Secr: Dr. K. Elaine Hoagland
Focus: Bio
Periodical
ASC Newsletter (weekly) *14512*

Association of Teacher Educators (ATE), 1900 Association Dr, Ste ATE, Reston, VA 22091
T: (703) 620-3110; Fax: (703) 620-3110
Founded: 1920; Members: 4000
Gen Secr: Dr. Gloria Cherney
Focus: Adult Educ; Educ *14513*

Association of Teachers of Japanese (ATJ), c/o Japanese Program, Middlebury College, Middlebury, VT 05753
T: (802) 388-3711
Founded: 1962; Members: 550
Pres: Hiroshi Miyaji
Focus: Ling; Educ
Periodicals
ATJ Newsletter (3 times annually)
Journal of the Association of Teachers of Japanese (weekly) *14514*

Association of Teachers of Maternal and Child Health (ATMCH), 6505 Alvorado Rd, Ste 205, San Diego, CA 92120-5011
T: (619) 594-4493; Fax: (619) 594-4570
Founded: 1968; Members: 200
Pres: Albert Chang
Focus: Public Health *14515*

Association of Teachers of Preventive Medicine (ATPM), c/o Kay B. Doggett, 1015 15 St NW, Ste 405, Washington, DC 20005
T: (202) 682-1698
Founded: 1942; Members: 700
Gen Secr: Kay B. Doggett
Focus: Adult Educ; Hygiene
Periodicals
American Journal of Preventive Medicine (weekly)
Newsletter (weekly)
Perspectives on Prevention (weekly) . . *14516*

Association of Teachers of Technical Writing (ATTW), c/o Dr. Dan Jones, Dept of English, University of Central Florida, POB 25000, Orlando, FL 79409
T: (407)823-2212
Founded: 1973; Members: 1200
Gen Secr: Dr. Dan Jones
Focus: Educ *14517*

Association of the German Nobility in North America (DAGNA) / Deutsche Adels-Gesellschaft in Nord-Amerika, 1101 W Second St, Benicia, CA 94510
T: (707) 745-1605
Founded: 1980; Members: 48
Pres: Gilbert von Studnitz
Focus: Genealogy
Periodicals
Der Adelsbote
Review (weekly) *14518*

Association of the Health Occupations Teacher Educators (AHOTE), c/o North Carolina State University, 502 Poe Hall, Raleigh, NC 27695-7801
T: (919) 737-2234; Fax: (814) 863-7532
Founded: 1978; Members: 28
Pres: Dr. Beverly Richards
Focus: Educ; Public Health *14519*

Association of the Institute for Certification of Computer Professionals (AICCP), 2200 E Devon Av, Ste 268, Des Plaines, IL 60018
T: (312) 299-4227; Fax: (312) 299-4280
Founded: 1982; Members: 45000
Gen Secr: George R. Eggert
Focus: Computer & Info Sci
Periodicals
AICCP Newsletter (weekly)
Bulletin (weekly) *14520*

Association of Theological Schools (ATS), 10 Summit Park Dr, Pittsburgh, PA 15275-1103
T: (412) 788-6505
Founded: 1918; Members: 211
Gen Secr: James L. Waits
Focus: Educ; Rel & Theol
Periodical
Theological Education (weekly) *14521*

Association of Third World Affairs (ATWA), 1629 K St NW, Ste 802, Washington, DC 20006
T: (202) 331-8455; Fax: (202) 785-3607
Founded: 1967; Members: 1000
Gen Secr: Dr. Lorna Hahn
Focus: Develop Areas
Periodicals
Monographs
Third World Forum (weekly) *14522*

Association of Trial Lawyers of America (ATLA), 1050 31 St NW, Washington, DC 20007
T: (202) 965-3500; Fax: (202) 337-0977
Founded: 1946; Members: 56000
Gen Secr: Thomas Henderson
Focus: Law
Periodicals
Advocate (10 times annually)
Directory (weekly)
Law Reporter (10 times annually)
Product Liability Law Reporter (10 times annually)
Professional Negligence Law Reporter (10 times annually)
Trial (weekly) *14523*

Association of United States Members of the International Institute of Space Law (AUSMIISL), c/o Kenneth Schwetje, 15397 Autumn Ln, Dumfries, VA 22026
T: (703) 680-0863
Members: 110
Pres: Kenneth Schwetje
Focus: Law *14524*

Association of Universities for Research in Astronomy (AURA), 1625 Massachusetts Av NW, Ste 701, Washington, DC 20036
T: (202) 483-2101; Fax: (202) 483-2106
Founded: 1957; Members: 22
Pres: Dr. Goetz K. Oertel
Focus: Astronomy *14525*

Association of University Anesthesiologists (AUA), c/o Michael J. Bishop, 2033 Sixth Av, Seattle, WA 98121
T: (206) 441-6020
Founded: 1953; Members: 600
Gen Secr: Michael J. Bishop
Focus: Anesthetics
Periodical
Directory *14526*

Association of University Architects (AUA), c/o Yale University School of Medicine, 333 Cedar St, New Haven, CT 06510
T: (203) 785-4667
Founded: 1955; Members: 100
Gen Secr: A.G. Kellog
Focus: Archit
Periodical
Newsletter (weekly) *14527*

Association of University Professors of Ophthalmology (AUPO), POB 420369, San Francisco, CA 94142-0369
T: (415) 561-8548
Founded: 1966; Members: 205
Gen Secr: Robert E. Kalina
Focus: Ophthal
Periodicals
Bulletin (weekly)
Membership Directory *14528*

Association of University Programs in Health Administration (AUPHA), 1911 N Fort Myer Dr, Ste 503, Arlington, VA 22209
T: (703) 524-5500
Founded: 1948; Members: 1200
Pres: Gary L. Filerman
Focus: Public Health
Periodicals
Health Services Administration Education Directory (bi-annually)
Journal of Health Administration Education (weekly)
Staff Report (weekly) *14529*

Association of University Radiologists (AUR), 1891 Preston White Dr, Reston, VA 22091
T: (703) 648-8900; Fax: (703) 648-9176
Founded: 1953; Members: 1600
Gen Secr: Sheila A. Aubin
Focus: Radiology
Periodical
Investigative Radiology (weekly) *14530*

Association of University Related Research Parks (AURRP), 4500 S Lakeshore Dr, Ste 475, Tempe, AZ 85282
T: (602) 752-2002; Fax: (602) 752-2003
Founded: 1986; Members: 285
Gen Secr: Chris Boettcher
Focus: Sci *14531*

Association of University Summer Sessions (AUSS), c/o Dr. Leslie J. Coyne, Summer Sessions/Special Programs, Indiana University, 254 Maxwell Hall, Bloomington, IN 47405
T: (812) 335-5048; Fax: (812)855-3815
Founded: 1925; Members: 50
Gen Secr: Dr. Leslie J. Coyne
Focus: Adult Educ; Sci *14532*

Association of Visual Science Librarians (AVSL), c/o Madelyn Hall, Good Samaritan Hospital and Medical Center, 1040 NW 22 Av, Portland, OR 97210
T: (503) 229-7678; Fax: (503) 790-1201
Founded: 1968; Members: 70
Gen Secr: Madelyn Hall
Focus: Libraries & Bk Sci
Periodical
Membership List (weekly) *14533*

Association of Vitamin Chemists (AVC), c/o Deborah Becker, 3 Lake Dr, Northville, IL 60093
T: (708) 998-7261
Founded: 1943; Members: 125
Pres: Deborah Becker
Focus: Chem; Food *14534*

Association of Youth Museums (AYM), c/o Jeanne Finan, The Children's Museum of Memphis, 2525 Central Av, Memphis, TN 38104
T: (901) 458-2678; Fax: (901) 458-4033
Founded: 1962; Members: 310
Gen Secr: Jeanne Finan
Focus: Arts
Periodical
Hand to Hand (weekly) *14535*

Association on Boarding Schools (ABS), c/o NAIS, 75 Federal St, Boston, MA 02110
T: (617) 451-2444
Founded: 1976; Members: 241
Pres: Richard H. Cowan
Focus: Educ
Periodicals
Boarding School Life (weekly)
Boarding Schools Directory (weekly) . . *14536*

Association to Advance Ethical Hypnosis (AAEH), 2675 Oakwood Dr, Cuyahoga Falls, OH 44221
T: (216) 923-8880; Fax: (216) 923-8880
Founded: 1955; Members: 1500
Gen Secr: N.R. Orndorff
Focus: Psychoan
Periodical
Suggestion (weekly) *14537*

Asthma and Allergy Foundation of America (AAFA), 1125 15 St NW, Ste 502, Washington, DC 20005
T: (202) 466-7643; Fax: (202) 466-8940
Founded: 1953
Gen Secr: Mary Worstell
Focus: Med *14538*

Astrologers' Guild of America (AGA), 5 Fair Meadow Dr, Brewster, NY 10509
T: (914) 279-4935
Founded: 1927
Pres: Joelle K.D. Mahoney
Focus: Astronomy
Periodical
Astrological Review (weekly) *14539*

Astronomical League (AL), c/o Berton Stevens, 2112 E Kingfisher Ln, Rolling Meadows, IL 60008
T: (708) 398-0562
Founded: 1946; Members: 11100
Gen Secr: Berton Stevens
Focus: Astronomy
Periodicals
Convention Proceedings (weekly)
The Reflector (weekly) *14540*

Astronomical Society of the Pacific (ASP), 390 Ashton Av, San Francisco, CA 94112
T: (415) 337-1100; Fax: (415) 337-5205
Founded: 1889; Members: 7000
Gen Secr: Andrew Fraknoi
Focus: Astronomy
Periodicals
Information Pockets on Astronomy (weekly)
Mercury (weekly)
Universe in the Classroom (weekly) . . *14541*

Astro-Psychology Institute (API), 2640 Greenwich, Ste 403, San Francisco, CA 94123
T: (415) 921-1192
Founded: 1979
Gen Secr: Milo Kovar
Focus: Psych *14542*

ATAS Technology Assessment Network, c/o UNCSTD, 1 United Nations Plaza, New York, NY 10017
T: (212) 9638435; Tx: 232422
Founded: 1987
Focus: Eng *14543*

Atex Newspaper Users Group (ANUG), c/o Audrey Novak, City News Bureau of Chicago, 35 E Walker Dr, Ste 792, Chicago, IL 60601
T: (215) 854-4607; Fax: (215) 854-4737
Founded: 1975; Members: 223
Pres: Audrey Novak
Focus: Computer & Info Sci
Periodicals
Membership Directory (weekly)
Weekly Wire (weekly) 14544

Athenaeum of Philadelphia (PAT), 219 S Sixth St, E Washington Sq, Philadelphia, PA 19106
T: (215) 925-2688; Fax: (215) 925-3755
Founded: 1814; Members: 1200
Gen Secr: Dr. Roger W. Moss
Focus: Arts; Lit; Archit
Periodicals
Athenaeum Annotations
Athenaeum Architectural Archives
Book List (weekly) 14545

Atlantic Council of the United States (ACUS), 1616 H St NW, Washington, DC 20006
T: (202) 347-9353
Founded: 1961
Gen Secr: Rozanne Ridgway
Focus: Poli Sci
Periodicals
Atlantic Community News (weekly)
The Atlantic Community Quarterly (weekly)
Issues and Options (weekly) 14546

Atlantic Economic Society (AES), c/o Southern Illinois University at Edwardsville, Box 1101, Edwardsville, IL 62026-1101
T: (618) 692-2291; Fax: (618) 692-3400
Founded: 1973; Members: 1000
Gen Secr: Dr. John M. Virgo
Focus: Econ
Periodical
Atlantic Economic Journal (weekly) . . . 14547

Audio Engineering Society (AES), 60 E 42 St, Rm 2520, New York, NY 10065
T: (212) 661-8528; Fax: (212) 682-0477
Founded: 1948; Members: 10100
Gen Secr: Donald J. Plunkett
Focus: Eng
Periodical
Journal of the AES (10 times annually) . 14548

Audubon Naturalist Society of the Central Atlantic States (ANS), 8940 Jones Mill Rd, Chevy Chase, MD 20815
T: (301) 652-9188
Founded: 1897; Members: 14000
Gen Secr: Ken Nicholls
Focus: Nat Sci; Educ
Periodical
Audubon Naturalist News (10 times annually)
. . . . 14549

Augustana Historical Society (AHS), c/o Augustana College Library, 3435 Ninth 1/2 Av, Rock Island, IL 61201
T: (309) 794-7317
Founded: 1930; Members: 225
Gen Secr: Judy Belan
Focus: Hist
Periodical
Newsletter (weekly) 14550

Augustan Reprint Society (ARS), 56 E 13 St, New York, NY 10003
T: (212) 777-4700
Founded: 1946; Members: 700
Gen Secr: Gabriele Hornstein
Focus: Lit 14551

Augustan Society (AS), POB P, Torrance, CA 90508-0210
T: (213) 320-7766; Fax: (213) 530-7530
Founded: 1957; Members: 4195
Pres: Dr. Robert Cleve
Focus: Genealogy
Periodicals
The Augustan (weekly)
Roll of Arms 14552

August Derleth Society (ADS), 61 Teecomwas Dr, Uncasville, CT 06382
T: (203) 848-0636
Founded: 1977; Members: 225
Pres: Richard H. Fawcett
Focus: Lit
Periodical
Newsletter (weekly) 14553

Authors Guild (AG), 330 W 42 St, New York, NY 10036
T: (212) 563-5904; Fax: (212) 564-8363
Founded: 1912; Members: 6500
Gen Secr: Helen A. Stephenson
Focus: Lit
Periodical
The Authors Guild Bulletin (weekly) . . 14554

Authors League of America (ALA), 330 W 42 St, New York, NY 10036
T: (212) 564-8350
Founded: 1912; Members: 14700
Gen Secr: Helen A. Stephenson
Focus: Lit
Periodicals
The Authors Guild Bulletin (weekly)
The Dramatist Guild Quarterly (weekly) . 14555

Autism Services Center (ASC), 605 Ninth St, POB 507, Huntington, WV 25710-0507
T: (304) 525-8014; Fax: (304) 525-8026
Founded: 1979
Gen Secr: Ruth C. Sullivan
Focus: Psych 14556

Autism Society of America (ASA), 8601 Georgia Av, Ste 503, Silver Spring, MD 20910
T: (301) 565-0433; Fax: (301) 565-0834
Founded: 1965; Members: 9000
Gen Secr: Veronica Zysk
Focus: Psych 14557

Automotive Market Research Council (AMRC), POB 13966, Research Triangle Park, NC 27709
T: (919) 549-4800
Founded: 1966; Members: 100
Pres: George Steward
Focus: Marketing
Periodicals
Automotive Data Bibliography (bi-annually)
Newsletter (weekly)
Semiannual Vehicle Forecasts (weekly) . 14558

Autoprep 5000 Users Group (APUG), 83 Gerber Rd W, South Windsor, CT 06074
T: (203) 644-1551
Founded: 1983; Members: 240
Focus: Computer & Info Sci
Periodical
Newsletter 14559

Auxiliary of the American Osteopathic Association (AAOA), 142 E Ontario St, Chicago, IL 60611
T: (312) 280-5819
Founded: 1940; Members: 5000
Gen Secr: Bridget Price
Focus: Pathology
Periodicals
Annual Report (weekly)
Newsletter (weekly)
Record (weekly)
Roster of Affiliates (weekly) 14560

Auxiliary to the American Pharmaceutical Association, c/o Dottie Kay, 171 Boxfield Rd, Pittsburgh, PA 15241
Founded: 1936; Members: 400
Pres: Dottie Kay
Focus: Pharmacol
Periodicals
APHA Auxiliary Newsletter (weekly)
Journal (2-4 times annually) 14561

Aviation Maintenance Foundation International (AMFI), POB 2826, Redmond, WA 98073
T: (206) 828-3917; Fax: (206) 827-6895
Founded: 1971; Members: 6000
Pres: Richard S. Kost
Focus: Aero
Periodicals
AMFI Industry News (weekly)
AMFI Industry Survey Report (weekly)
Professionalism (weekly)
World of Aviation Maintenance (weekly) . 14562

Aviation Safety Institute (ASI), Box 304, Worthington, OH 43085
T: (614) 885-4242; Fax: (614) 885-5891
Founded: 1973; Members: 500
Pres: John B. Galipault
Focus: Safety; Aero
Periodical
Monitor (weekly) 14563

Aviation/Space Writers Association (AWA), 17 S High St, Ste 1200, Columbus, OH 43215
T: (614) 221-1900; Fax: (614) 221-1989
Founded: 1938; Members: 700
Gen Secr: Madeline Mohanco Field
Focus: Lit
Periodicals
Newsletter (weekly)
Yearbook and Directory (weekly) 14564

Aviation Technician Education Council (ATEC), 2090 Wexford Ct, Harrisburg, PA 17112-1579
T: (717) 540-7121
Founded: 1961; Members: 167
Gen Secr: Dr. R. Dumaresq
Focus: Aero; Adult Educ
Periodical
Newsletter (5 times annually) 14565

Avicultural Society of America (ASA), c/o Joe Krader, 24692 Paseo de Toronto, Yorba Linda, CA 92687-5115
T: (714) 996-5538
Founded: 1927; Members: 1000
Pres: Joe Krader
Focus: Ornithology
Periodical
Avicultural Bulletin (weekly) 14566

AVKO Educational Research Foundation (AVKOERF), 3084 W Willard Rd, Birch Run, MI 48415
T: (313) 686-9283; Fax: (313) 686-1101
Founded: 1974; Members: 500
Gen Secr: Don McCabe
Focus: Educ 14567

Baker Street Irregulars (BSI), 34 Pierson Av, Norwood, NJ 07648
T: (201) 768-2241
Founded: 1934; Members: 275
Pres: Thomas L. Stix
Focus: Lit
Periodical
The Baker Street Journal (weekly) . . . 14568

Ball Manufacturers Engineers Committee (BMEC), 1101 Connecticut Av NW, Ste 700, Washington, DC 20036-4303
T: (202) 429-5155; Fax: (202) 223-4579
Gen Secr: Gary T. Satterfield
Focus: Eng 14569

Basque Educational Organization (BEO), POB 640037, San Francisco, CA 94164-0037
T: (415) 583-4035; Fax: (415) 753-0298
Founded: 1983; Members: 7
Gen Secr: Martin Minaberry
Focus: Ling; Educ 14570

Bay Area Physicians for Human Rights (BAPHR), 4111 18 St, San Francisco, CA 94114
T: (415) 558-9353; Fax: (415) 558-0466
Founded: 1977; Members: 350
Pres: Leonard Simpson
Focus: Law; Physiology
Periodical
BAPHRON (weekly) 14571

Beef Improvement Federation (BIF), c/o Extension Animal Science, University of Georgia, Athens, GA 30602
T: (706) 542-2584; Fax: (706) 542-9316
Founded: 1967; Members: 79
Gen Secr: Dr. Charles A. McPeake
Focus: Vet Med
Periodical
Update (weekly) 14572

Behavioral Pharmacology Society (BPS), c/o Larry D. Byrd, Div of Behavioral Biology, Yerkes Regional Primate Research Center, Emory University, Atlanta, GA 30322
T: (404) 727-7730; Fax: (404) 727-7845
Founded: 1957; Members: 200
Pres: Larry D. Byrd
Focus: Pharmacol 14573

Behavioral Research Council (BRC), Division St, Great Barrington, MA 01230
T: (413) 528-1217; Fax: (413) 528-0103
Founded: 1960
Pres: Dr. Robert A. Gilmour
Focus: Psych; Behav Sci 14574

Behavior Genetics Association (BGA), c/o George Vogler, Div of Biostatistics, Washington University School of Medicine, 660 S Euclid, Saint Louis, MO 63110
T: (316) 362-3642; Fax: (316) 362-2994
Founded: 1971; Members: 330
Gen Secr: George Vogler
Focus: Genetics
Periodical
Behavior Genetics (weekly) 14575

Belgian American Educational Foundation (BAEF), 195 Church St, New Haven, CT 06510
T: (203) 777-5765
Founded: 1920; Members: 250
Pres: Emile L. Boulpaep
Focus: Educ 14576

Bernard Herrmann Society (BHS), c/o Kevin Fahey, 5223 Denny Av, North Hollywood, CA 91601
T: (818) 769-4568
Founded: 1980; Members: 100
Pres: Kevin Fahey
Focus: Music 14577

The Bernard Shaw Society (BSS), POB 1159, Madison Square Station, New York, NY 10159-1159
T: (212) 982-9885
Founded: 1962; Members: 300
Gen Secr: Douglas Laurie
Focus: Lit
Periodical
The Independent Shavian (3 times annually)
. . . . 14578

Bertrand Russell Society (BRS), c/o Michael Rockler, 4036 Emerson, Skokie, IL 60076
T: (708) 679-5305
Founded: 1974; Members: 300
Pres: Michael Rockler
Focus: Lit; Philos
Periodical
Russell Society News (weekly) 14579

Bet Nahrain (BN), POB 4116, Modesto, CA 95352
T: (209) 522-3229; Fax: (209) 538-2795
Founded: 1974; Members: 5000
Gen Secr: Dr. Sargon Dadesho
Focus: Ling; Ethnology
Periodical
Bet-Nahrain Journal (weekly) 14580

Better Education thru Simplified Spelling (BETSS), 300 Riverfront, Ste 26-H, Detroit, MI 48226-4516
T: (313) 645-5131
Founded: 1978
Pres: Charles F. Kleber
Focus: Ling; Educ 14581

Better Hearing Institute (BHI), POB 1840, Washington, DC 20013
T: (703) 642-0580; Fax: (703) 750-9302
Founded: 1973
Gen Secr: Joseph J. Rizzo
Focus: Otorhinolaryngology 14582

Better Vision Institute (BVI), 1800 N Kent St, Ste 904, Rosslyn, VA 22209
T: (703) 243-1508; Fax: (703) 243-1537
Founded: 1929; Members: 3500
Gen Secr: Susan Burton
Focus: Ophthal
Periodical
Newsletter (weekly) 14583

Bibliographical Society of America (BSA), POB 397, Grand Central Station, New York, NY 10163
T: (212) 995-9151
Founded: 1904; Members: 1300
Gen Secr: Marjory Zaik
Focus: Libraries & Bk Sci
Periodical
Papers (weekly) 14584

Bibliographical Society of the University of Virginia (BSUV), c/o Ray W. Frantz, Alderman Library, University of Virginia, Charlottesville, VA 22903
T: (804) 924-7013
Founded: 1947; Members: 1000
Gen Secr: Ray W. Frantz
Focus: Libraries & Bk Sci
Periodical
Studies in Bibliography (weekly) 14585

Bioelectromagnetics Society (BEMS), 120 W Church St, Ste 4, Frederick, MD 21701
T: (301) 663-4252; Fax: (301) 663-0043
Founded: 1978; Members: 700
Gen Secr: Dr. William D. Wisecup
Focus: Electronic Eng
Periodical
Bioelectromagnetics (weekly) 14586

Bio-Integral Resource Center (BIRC), POB 7414, Berkeley, CA 94707
T: (415) 524-2567; Fax: (415) 524-1758
Founded: 1978; Members: 4000
Gen Secr: Sheila Daar
Focus: Agri; Nat Sci
Periodicals
Common Sense Pest Control (weekly)
IPM Practitioner (10 times annually) . . 14587

Biological Anthropological Section (BIO), c/o Dr. David Frayer, Dept of Anthropology, University of Kansas, 622 Fraser, Lawrence, KS 66045-2110
T: (913) 864-4103; Fax: (913) 864-5224
Founded: 1984; Members: 512
Pres: Dr. David Frayer
Focus: Anthro 14588

Biological Institute of Tropical America (BIOTA), POB 2585, Menlo Park, CA 94026-2585
T: (415) 593-9024
Founded: 1981
Pres: Herman G. Real
Focus: Bio 14589

Biological Photographic Association (BPA), 115 Stoneridge Dr, Chapel Hill, NC 27514
T: (919) 967-8247; Fax: (919) 967-8246
Founded: 1931; Members: 1200
Gen Secr: Thomas P. Hurtgen
Focus: Bio; Photo
Periodical
Journal (weekly) 14590

Biological Stain Commission (BSC), POB 626, Rochester, NY 14642
T: (716) 275-3197; Fax: (716) 271-1346
Founded: 1922; Members: k7580
Gen Secr: Dr. Eric A. Schenk
Focus: Bio
Periodical
Stain Technology (weekly) 14591

Biomass Energy Research Association (BERA), 1825 K St NW, Ste 503, Washington, DC 20006
T: (202) 785-2856; Fax: (202) 223-4625
Founded: 1982; Members: 200
Pres: Dr. Donald L. Klass
Focus: Energy
Periodicals
BioEnergy Update (weekly)
Biologue (weekly) 14592

Biomedical Engineering Society (BMES), POB 2399, Culver City, CA 90231
T: (310) 618-9322
Founded: 1968; Members: 1700
Gen Secr: Fred J. Weibell
Focus: Med; Eng; Bio
Periodicals
Annals of Biomedical Engineering (weekly)
Bulletin (weekly) 14593

Biometric Society, 1429 Duke St, Ste 401, Alexandria, VA 22314
T: (703) 836-8311
Founded: 1947; Members: 6800
Gen Secr: R. Mead
Focus: Bio
Periodical
Biometrics (weekly) 14594

Biometric Society, Western North American Region, c/o Robert Cochran, Dept of Statistics, University of Wyoming, Laramie, WY 82071
T: (307) 766-3341; Fax: (307) 766-3927
Founded: 1948; Members: 650
Gen Secr: Robert Cochran
Focus: Bio
Periodicals
Biometric Bulletin (10 times annually)
Biometrics (weekly) 14595

Biophysical Society (BPS), c/o Emily M. Gray, 9650 Rockville Pike, Rm 0512, Bethesda, MD 20814
T: (301) 530-7114; Fax: (301) 530-7133
Founded: 1957; Members: 4000
Gen Secr: Emily M. Gray
Focus: Biophys 14596

BITNET, c/o Dept of Computing and Information Technology, Princeton University, 220 Nassau Hall, Princeton, NJ 08544
T: (609) 452-5601; Fax: (609) 452-3602
Founded: 1981; Members: 1350
Pres: Ira Fuchs
Focus: Computer & Info Sci 14597

Black Filmmakers Hall of Fame (BFHFI), 405 14 St, Ste 515, Oakland, CA 94612
T: (510) 465-0804; Fax: (510) 839-9858
Founded: 1973; Members: 500
Pres: Mary P. Smith
Focus: Cinema
Periodical
Souvenir Catalogue (weekly) 14598

Black Lung Association (BLA), c/o Bill Bailey, Box 872, Crab Orchard, WV 25827
T: (304) 252-9654
Founded: 1968; Members: 73000
Gen Secr: Bill Bailey
Focus: Pulmon Dis 14599

Black Psychiatrists of America (BPA), c/o Dr. Isaac Slaughter, 2730 Adeline St, Oakland, CA 94607
T: (510) 465-1800
Founded: 1968; Members: 550
Gen Secr: Dr. Isaac Slaughter
Focus: Psychiatry
Periodical
BPA Quarterly (weekly) 14600

Black Resources Information Coordinating Services (BRICS), 614 Howard Av, Tallahassee, FL 32304
T: (904) 576-7522
Founded: 1972; Members: 670
Pres: Emily A. Copeland
Focus: Genealogy
Periodicals
Brics Bracs (weekly)
Media Showcase (weekly)
Newsletter (weekly) 14601

Black Women's Educational Alliance (BWEA), 6625 Greene St, Philadelphia, PA 19119
Founded: 1976; Members: 300
Pres: Deidre Farmbey
Focus: Educ 14602

Bloomsday Club (BC), c/o Old York Books, 122 French St, New Brunswick, NJ 08901
T: (201) 249-0430
Founded: 1975; Members: 150
Gen Secr: E.T. Cecile Hopkins
Focus: Lit 14603

B'Nai B'Rith International Commission on Continuing Jewish Education (BBICCJE), 1640 Rhode Island Av NW, Washington, DC 20036
T: (202) 857-6580; Fax: (202) 857-0980
Founded: 1959
Gen Secr: Dr. Michael Neiditch
Focus: Educ 14604

Bockus International Society of Gastroenterology, 2946 Woodford Dr, La Jolla, CA 92037
T: (619) 453-7166
Founded: 1958
Gen Secr: Dr. William S. Haubrich
Focus: Gastroenter 14605

Bolivarian Society of the United States, 7 E 51 St, New York, NY 10022
T: (212) 826-1660
Founded: 1941; Members: 60
Gen Secr: Pedro M. Rincones
Focus: Hist 14606

Book Industry Study Group (BISG), 160 Fifth Av, New York, NY 10010
T: (212) 929-1393; Fax: (212) 989-7542
Founded: 1976; Members: 220
Gen Secr: Sandra K. Paul
Focus: Lit 14607

Bostonian Society (BS), 206 Washington St, Old State House, Boston, MA 02109
T: (617) 720-1713; Fax: (617) 720-3289
Founded: 1881; Members: 1200
Gen Secr: Joan C. Hull
Focus: Hist
Periodical
Proceedings 14608

Boston Theological Institute (BTI), 210 Herrick Rd, Newton Centre, MA 02159
T: (617) 527-4880; Fax: (617) 527-1073
Founded: 1967
Gen Secr: Dr. Rodney Petersen
Focus: Rel & Theol
Periodicals
Catalog (weekly)
Faculty Directory (weekly)
Journal (weekly)
Newsletter (weekly) 14609

Botanical Society of America (BSA), c/o Christopher Haufle, Dept of Botany, University of Kansas, Haworth Hall, Lawrence, KS 66045-2106
T: (913) 864-4301; Fax: (913) 864-5321
Founded: 1906; Members: 2600
Gen Secr: Christopher Haufle
Focus: Botany
Periodicals
American Journal of Botany (10 times annually)
Plant Science Bulletin (weekly) 14610

Brain Information Service (BIS), c/o School of Medicine, UCLA, Los Angeles, CA 90024-1746
T: (213) 825-3417; Fax: (213) 206-3499
Founded: 1964
Gen Secr: Dr. Michael H. Chase
Focus: Neurology 14611

Bram Stoker Memorial Association (BSMA), 29 Washington Sq W, New York, NY 10011
T: (212) 982-6754
Founded: 1985; Members: 336
Gen Secr: Jenny O'Casey
Focus: Lit 14612

Brandeis-Bardin Institute (BBI), 1101 Peppertree Ln, Brandeis, CA 93064
T: (805) 582-4450; Fax: (805) 526-1398
Founded: 1941; Members: 1000
Gen Secr: Dr. Alvin Mars
Focus: Educ
Periodical
News (weekly) 14613

Breast Cancer Advisory Center (BCAC), POB 224, Kensington, MD 20895
Founded: 1975
Gen Secr: Rose Kushner
Focus: Cell Biol & Cancer Res 14614

The Bridge, 248 W 108 St, New York, NY 10025
T: (212) 663-3000; Fax: (212) 663-3181
Founded: 1954; Members: 350
Gen Secr: Murray Itzkowitz
Focus: Therapeutics 14615

British Schools and Universities Foundation (BSUF), 575 Madison Av, Ste 1006, New York, NY 10022-2511
T: (212) 662-5576
Founded: 1961; Members: 350
Pres: Alan C. Root
Focus: Educ 14616

Broadcast Education Association (BEA), 1771 N St NW, Washington, DC 20036-2891
T: (202) 429-5355
Founded: 1955; Members: 950
Gen Secr: Louisa A. Nielsen
Focus: Adult Educ; Journalism
Periodicals
Feedback (weekly)
Journal of Broadcasting & Electronic Media (weekly) 14617

Browning Institute (BI), POB 2983, Grand Central Station, New York, NY 10163-2983
T: (316) 221-2779
Founded: 1971; Members: 350
Pres: Philip Kelley
Focus: Lit 14618

Brown Lung Association (BLA), POB 7583, Greenville, SC 29610
T: (803) 269-8048
Founded: 1975; Members: 8000
Pres: Lemar Case
Focus: Pulmon Dis 14619

Bruckner Society of America (BSA), 2150 Dubuque Rd, Iowa City, IA 52240
T: (319) 351-5758
Founded: 1932
Pres: Charles L. Eble
Focus: Music
Periodical
Chord and Discord 14620

Buffalo Bill Memorial Association (BBHC), POB 1000, Cody, WY 82414
T: (307) 587-4771
Founded: 1927
Gen Secr: Peter H. Hassrick
Focus: Cultur Hist
Periodical
Newsletter (weekly) 14621

Building Officials and Code Administrators International (BOCA), 4051 W Flossmoor Rd, Country Club Hills, IL 60478-5795
T: (708) 799-2300; Fax: (708) 799-4981
Founded: 1915; Members: 12000
Focus: Civil Eng
Periodical
BOCA Magazine 14622

Bureau of Professional Education of the American Osteopathic Association (BPEAOA), c/o Douglas Ward, American Osteopathic Association, 142 E Ontario St, Chicago, IL 60611
T: (312) 280-5800; Fax: (312) 280-5803
Gen Secr: Douglas Ward
Focus: Pathology 14623

Burlesque Historical Society (BHS), c/o Exotic World, 29053 Wild Rd, Helendale, CA 92342
T: (619) 243-5261
Founded: 1963; Members: 500
Gen Secr: Dixie Evans
Focus: Perf Arts
Periodical
The Legend of Jennie Lee 14624

Burns Society of the City of New York (BSCNY), c/o W. Smith, Gold Impressions, 1050 W Jericho Turnpike, Smithtown, NY 11787
T: (516) 864-1136; Fax: (516) 864-1162
Founded: 1871; Members: 100
Gen Secr: W. Smith
Focus: Lit
Periodical
Robert Burns Chronicle (weekly) 14625

Business Council for Effective Literacy (BCEL), 1221 Av of the Americas, New York, NY 10020
T: (212) 512-2466
Founded: 1983
Gen Secr: Gail Spangenberg
Focus: Lit 14626

Business-Higher Education Forum (B-HEF), 1 Dupont Circle NW, Ste 800, Washington, DC 20036
T: (202) 939-9345; Fax: (202) 833-4723
Founded: 1978; Members: 90
Gen Secr: Don M. Blandin
Focus: Educ; Econ 14627

Business History Conference (BHC), c/o William J. Hausman, Dept of Economics, College of William and Mary, Williamsburg, VA 23187-8795
T: (804) 221-2381; Fax: (804) 221-2390
Founded: 1954; Members: 450
Gen Secr: William J. Hausman
Focus: Econ
Periodicals
Conference Newsletter (weekly)
Economic and Business History (weekly)
List of Members (biennial) 14628

Business Professionals of America (BPA), 5454 Cleveland Av, Columbus, OH 43231
T: (614) 895-7277
Founded: 1966; Members: 54000
Pres: Gary L. Hannah
Focus: Educ; Business Admin 14629

Byron Society (BS), 259 New Jersey Av, Collingswood, NJ 08108
T: (212) 496-1274
Founded: 1971; Members: 2000
Gen Secr: Marsha M. Manns
Focus: Lit 14630

Cajal Club (CC), c/o Dr. David Whitlock, Dept of Cellular and Structural Biology, University of Colorado Health Sciences Center, 4200 E Ninth Av, Denver, CO 80262
T: (303) 270-8201; Fax: (303) 270-5969
Founded: 1947; Members: 450
Gen Secr: Dr. David Whitlock
Focus: Neurology
Periodicals
History of Cajal Club
Proceedings (weekly) 14631

California Rug Study Society (CRSS), 6091 Claremont Av, Oakland, CA 94618
T: (510) 654-4207
Founded: 1981; Members: 15
Gen Secr: Jan David Winitz
Focus: Arts 14632

Calorimetry Conference (CC), c/o Dept of Chemistry, Brigham Young University, Provo, UT 84602
T: (801) 378-2302; Fax: (801) 378-5474
Founded: 1947; Members: 250
Gen Secr: Dr. J. Boerio-Goates
Focus: Energy 14633

Cambodian Buddhist Society (CBS), 13800 New Hampshire Av, Silver Spring, MD 20904
T: (301) 622-6544
Founded: 1978; Members: 1000
Gen Secr: Dr. M. Tarun Khemradhipati
Focus: Rel & Theol 14634

Canadian-American Committee (CAC), c/o National Planning Association, 1424 16 St NW, Ste 700, Washington, DC 20036
T: (202) 265-7685; Fax: (202) 797-5516
Founded: 1957; Members: 130
Gen Secr: Jonathan Lemco
Focus: Poli Sci 14635

Canal Society of New York State (CSNYS), 311 Montgomery St, Syracuse, NY 13202
T: (315) 428-1862
Founded: 1956; Members: 235
Pres: Thomas X. Grasso
Focus: Hist 14636

Cancer Care (CC), 1180 Av of the Americas, New York, NY 10036
T: (212) 221-3300; Fax: (212) 719-0263
Founded: 1944; Members: 20000
Gen Secr: Diane Blum
Focus: Cell Biol & Cancer Res 14637

Cancer Control Society (CCS), 2043 N Berendo St, Los Angeles, CA 90027
T: (213) 663-7801
Founded: 1973; Members: 5500
Pres: Norman Fritz
Focus: Cell Biol & Cancer Res
Periodicals
Cancer Book House List (bi-annually)
Doctor and Clinic Directory (weekly)
Patient Directory (weekly) 14638

Cancer Federation, 21250 Box Springs Rd, Moreno Valley, CA 92387
T: (800) 982-3270; Fax: (714) 682-0169
Founded: 1977; Members: 1500
Gen Secr: John Steinbacher
Focus: Cell Biol & Cancer Res
Periodicals
Challenge (weekly)
Newsletter (weekly)
Proceedings of Annual Riverside Conference on Stress, Emotions and Cancer (weekly) . 14639

Cancer Guidance Institute (CGI), 1323 Forbes Av, Ste 200, Pittsburgh, PA 15219
T: (412) 261-2211
Founded: 1981; Members: 1000
Gen Secr: Estelle Weissburg
Focus: Cell Biol & Cancer Res 14640

Candlelighters Childhood Cancer Foundation (CCCF), 7910 Woodmont Av, Ste 460, Bethesda, MD 20814
T: (301) 657-8401; Fax: (301) 657-8319
Founded: 1976; Members: 33000
Gen Secr: James R. Kitterman
Focus: Cell Biol & Cancer Res 14641

Canon Law Society of America (CLDA), c/o Catholic University, 431 Caldwell Hall, Washington, DC 20064
T: (202) 269-3491; Fax: (202) 319-5719
Founded: 1939; Members: 1900
Gen Secr: Patrick Cogan
Focus: Rel & Theol; Law
Periodicals
Convention Proceedings (weekly)
Membership Directory
Newsletter (weekly) 14642

Cardiovascular Credentialing International (CCI), POB 611, Wright Bros Station, Dayton, OH 45409-0611
T: (513) 268-0268; Fax: (513) 293-0958
Founded: 1988
Gen Secr: Julia Dow
Focus: Cardiol 14643

Cardiovascular System Dynamics Society, c/o Likoff Cardiovascular Institute, Hahnemann University, Broad and Vine Sts, Philadelphia, PA 19102-1192
T: (215) 448-1703, 448-7314
Founded: 1976
Gen Secr: Prof. J. Yasha Kresh
Focus: Cardiol 14644

Career College Association (CCA), 750 First St NE, Ste 900, Washington, DC 20002
T: (202) 336-6700; Fax: (202) 336-6828
Founded: 1991; Members: 1900
Pres: Stephen J. Blair
Focus: Adult Educ; Business Admin; Eng
Periodicals
Career Training (weekly)
CCA News (weekly)
Washington Update (weekly) 14645

Carnegie Foundation for the Advancement of Teaching (CFAT), 5 Ivy Ln, Princeton, NJ 08540
T: (609) 452-1780
Founded: 1905
Pres: Ernest L. Boyer
Focus: Educ 14646

Carnegie Institution of Washington (CIW), 1530 P St NW, Washington, DC 20005
T: (202) 387-6400; Fax: (202) 387-8092
Founded: 1902
Pres: Maxine F. Singer
Focus: Physics; Astronomy; Botany; Bio; Geology
Periodical
Year Book (weekly) 14647

Catalysis Society (North America), c/o Dr. William J. Linn, 1311 Circle Dr, West Chester, PA 19382
T: (215) 793-2831; Fax: (215) 793-2831
Founded: 1956; Members: 2200
Gen Secr: Dr. William J. Linn
Focus: Chem
Periodical
Newsletter (weekly) 14648

Catecholamine Club (CC), c/o Walter R. Dixon, Dept of Pharmacology and Toxicology, School of Pharmacy, University of Kansas, Lawrence, KS 66045
T: (913) 864-3951
Founded: 1969; Members: 350
Gen Secr: Walter R. Dixon
Focus: Chem 14649

Catgut Acoustical Society (CAS), c/o Carleen Maley Hutchins, 112 Essex Av, Montclair, NJ 07042
Founded: 1963; Members: 800
Gen Secr: Carleen Maley Hutchins
Focus: Music
Periodical
Journal of the Catgut Acoustical Society (weekly) . . . 14650

Catholic Audio-Visual Educators Association (CAVE), c/o John Manear, POB 9257, Pittsburgh, PA 15224
T: (412) 561-3583
Founded: 1948; Members: 2000
Gen Secr: John Manear
Focus: Educ
Periodicals
CAVE Evaluation
Newsletter (weekly) . . . 14651

Catholic Biblical Association of America (CBA), c/o Catholic University, 620 Michigan Av NE, Washington, DC 20064
T: (202) 319-5519; Fax: (202) 319-4479
Founded: 1936; Members: 1200
Gen Secr: Joseph Jensen
Focus: Educ; Rel & Theol
Periodicals
The Catholic Biblical Quarterly (weekly)
Old Testament Abstracts (3 times annually) . . . 14652

Catholic Commission on Intellectual and Cultural Affairs (CCICA), c/o La Salle University, 1900 W Olney Av, POB 673, Philadelphia, PA 19141-1199
Founded: 1946; Members: 300
Gen Secr: Daniel Burke
Focus: Humanities
Periodical
CCICA Annual (weekly) . . . 14653

Catholic Fine Arts Society (CFAS), c/o Jean Dominici de Maria, Molloy College, 1000 Hampstead Av, Rockville Centre, NY 11570
T: (516) 678-5460; Fax: (516) 678-7295
Founded: 1955; Members: 200
Pres: Jean Dominici de Maria
Focus: Fine Arts
Periodicals
Membership List (weekly)
Newsletter (weekly) . . . 14654

Catholic Health Association of the United States (CHA), 4455 Woodson Rd, Saint Louis, MO 63134-0889
T: (314) 427-2500; Fax: (314) 427-0029
Founded: 1915; Members: 1200
Pres: John E. Curley
Focus: Public Health
Periodicals
Catholic Health World: Official Assoc. Newspaper (weekly)
Guidebook of Catholic Healthcare Facilities (weekly)
Health Progress: Official Journal (10 times annually) . . . 14655

Catholic Library Association (CLA), 461 W Lancaster Av, Haverford, PA 19041
T: (215) 649-5250; Fax: (215) 896-1991
Founded: 1921; Members: 3100
Gen Secr: Anthony Prete
Focus: Libraries & Bk Sci
Periodicals
Catholic Library World (weekly)
Catholic Periodical and Literature Index (weekly)
Handbook and Membership Directory (weekly) . . . 14656

Catholic Medical Mission Board (CMMB), 10 W 17 St, New York, NY 10011-5765
T: (212) 242-7757; Fax: (212) 807-9161
Founded: 1928
Pres: James J. Yannarell
Focus: Med . . . 14657

Catholic Theological Society of America (CTSA), c/o La Salle University, POB 119, Philadelphia, PA 19141
T: (215) 951-1335; Fax: (215) 951-1488
Founded: 1946; Members: 1400
Gen Secr: Michael McGinniss
Focus: Rel & Theol
Periodical
Proceedings (weekly) . . . 14658

Caucus for a New Political Science (CNPS), c/o Government Dept, Suffolk University, Boston, MA 02108-2770
T: (617) 573-8126
Founded: 1967; Members: 350
Pres: John Ehrenberg
Focus: Soc Sci
Periodical
New Political Science (weekly) . . . 14659

CDS International (CDSI), 330 Seventh Av, New York, NY 10001
T: (212) 760-1400; Fax: (212) 268-1288
Founded: 1968
Gen Secr: Wolfgang Linz
Focus: Econ . . . 14660

Cecchetti Council of America (CCA), 770 Greenhills Dr, Ann Arbor, MI 48105
T: (313) 769-2719
Founded: 1939; Members: 500
Pres: Jane C. Miller
Focus: Perf Arts
Periodical
CCA Newsletter (weekly) . . . 14661

CEDAM International, 1 Fox Rd, Croton on Hudson, NY 10520
T: (914) 271-5365; Fax: (914) 271-4723
Founded: 1967; Members: 1000
Pres: Richard M. Sammon
Focus: Archeol
Periodicals
Bulletin (weekly)
Reef Report (weekly) . . . 14662

Center for Advanced Study in the Behavioral Sciences (CASBS), 202 Junipero Serra Blvd, Stanford, CA 94305
T: (415) 321-2052; Fax: (415) 321-1192
Founded: 1954; Members: 50
Gen Secr: Philip E. Converse
Focus: Sci; Humanities; Stats; Med; Behav Sci
Periodical
Annual Report (weekly) . . . 14663

Center for American Archaeology (CAA), POB 366, Kampsville, IL 62053
T: (618) 653-4316
Founded: 1953; Members: 545
Pres: Jane E. Buikstra
Focus: Archeol
Periodicals
Center for American Archaeology Annual Report (weekly)
Center for American Archaeology Newsletter (weekly)
Kampsville Archaeological Center Research Series
Kampsville Archaeological Center Technical Report . . . 14664

Center for Applied Linguistics (CAL), 1118 22 St NW, Washington, DC 20037
T: (202) 429-9292; Fax: (202) 429-9766
Founded: 1959; Members: 80
Pres: Sara E. Melendez
Focus: Ling . . . 14665

Center for Applied Research in the Apostolate (CARA), c/o Georgetown University, POB 1601, Washington, DC 20057
T: (202) 687-8080; Fax: (202) 687-8083
Founded: 1945; Members: 8
Gen Secr: Edward C. Foster
Focus: Rel & Theol
Periodicals
Cara Seminary Directory (weekly)
Cara Seminary Forum (weekly) . . . 14666

Center for Arab-Islamic Studies (CAIS), POB 678, Brattleboro, VT 05301
T: (802) 257-0872
Founded: 1980
Pres: Samir Abed-Rabbo
Focus: Ling . . . 14667

Center for Attitudinal Healing (CAH), 19 Main St, Tiburon, CA 94920
T: (415) 435-5022; Fax: (415) 435-5085
Founded: 1975; Members: 158
Gen Secr: Phoebe Lauren
Focus: Public Health . . . 14668

Center for Austrian Studies (CAS), c/o University of Minnesota, 314 Social Sciences, 267 19 Av S, Minneapolis, MN 55455
T: (612) 624-9811; Fax: (612) 626-2242
Founded: 1977
Gen Secr: Prof. David Good
Focus: Ethnology . . . 14669

Center for Auto Safety (CAS), 2001 S St NW, Ste 410, Washington, DC 20009-1160
T: (202) 328-7700
Founded: 1970; Members: 12000
Pres: Clarence M. Ditlow
Focus: Safety . . . 14670

Center for Chinese Research Materials (CCRM), POB 3090, Oakton, VA 22124
T: (703) 281-7731
Founded: 1967
Gen Secr: Pingfeng Chi
Focus: Ethnology; Libraries & Bk Sci
Periodical
Newsletter . . . 14671

Center for Community Change (CCC), 1000 Wisconsin Av NW, Washington, DC 20007
T: (202) 342-0519
Founded: 1968
Pres: Pablo Eisenberg
Focus: Public Health; Sociology
Periodicals
Friday Report (weekly)
Monitor (4-5 times annually) . . . 14672

Center for Computer-Assisted Legal Instruction (CALI), 229 19 Av S, Minneapolis, MN 55455
T: (612) 625-3419
Founded: 1982
Gen Secr: Ron W. Staudt
Focus: Educ; Law; Computer & Info Sci 14673

Center for Cuban Studies (CCS), 124 W 23 St, New York, NY 10011
T: (212) 242-0559
Founded: 1972; Members: 2000
Gen Secr: Sandra Levinson
Focus: Adult Educ
Periodicals
Bilingual Books and Document Series (bi-annually)
Cuba Update (weekly)
Newsletter (weekly) . . . 14674

Center for Death Education and Research (CDER), c/o University of Minnesota, 1167 Social Science Bldg, 267 19 Av S, Minneapolis, MN 55455
T: (612) 624-1895
Founded: 1969
Gen Secr: Robert L. Fulton
Focus: Educ . . . 14675

Center for Design Planning (CDP), 2300 E Mallory St, Pensacola, FL 32503
T: (904) 432-8478; Fax: (904) 438-1662
Founded: 1973
Gen Secr: Prof. Harold Lewis Malt
Focus: Urban Plan . . . 14676

Center for Early Adolescence (CEA), c/o University of North Carolina at Chapel Hill, D-2 Carr Mill Town Center, Carrboro, NC 27510
T: (919) 966-1148; Fax: (919) 966-7657
Founded: 1978; Members: 6000
Gen Secr: Frank A. Loda
Focus: Educ
Periodical
Common Focus (weekly) . . . 14677

Center for Economic Conversion (CEC), 222 View St, Ste C, Mountain View, CA 94041
T: (415) 968-8798; Fax: (415) 968-1126
Founded: 1975; Members: 2400
Gen Secr: Michael Closson
Focus: Econ
Periodicals
Plowshare Press (weekly)
Positive Alternatives (weekly) . . . 14678

Center for Energy Policy and Research (CEPR), c/o New York Institute of Technology, Old Westbury, NY 11568
T: (516) 686-7578
Founded: 1975
Gen Secr: Gale T. Spak
Focus: Energy . . . 14679

Center for Field Research (CFR), 680 Mount Auburn St, POB 403, Watertown, MA 02272
T: (617) 926-8200; Fax: (617) 926-8532
Founded: 1971
Pres: Brian A. Rosborough
Focus: Sci . . . 14680

Center for Global Education (CGE), c/o Augsburg College, 731 21 Av S, Minneapolis, MN 55454
T: (612) 330-1159; Fax: (612) 330-1695
Founded: 1982
Gen Secr: Joel Mugge
Focus: Educ . . . 14681

Center for Hospitality Research and Service, c/o Dept of Hotel, Restaurant and Institutional Management, Virginia Polytechnic Institute and State University, Blacksburg, VA 24061
T: (703) 231-5515
Founded: 1982
Gen Secr: Michael D. Olsen
Focus: Food; Home Econ . . . 14682

Center for International Development and Environment, 1709 New York AV NW, Washington DC 20006
T: (202) 662-2532; Tx: 64414; Fax: (202) 638-0036
Founded: 1988
Focus: Develop Areas; Ecology . . . 14683

Center for Management Development (CMD), 135 W 50 St, New York, NY 10020
T: (212) 586-8100; Fax: (212) 903-8168
Founded: 1923; Members: 77000
Gen Secr: Holly D. Lee
Focus: Business Admin . . . 14684

Center for Marine Conservation (CMC), 1725 DeSales NW, Ste 500, Washington, 20036
T: (202) 429-5609; Fax: (202) 872-0619
Founded: 1972
Gen Secr: Roger E. McManus
Focus: Ecology
Periodicals
Directory of Environmental Education Resources
Directory of Marine Education Resources
Report (weekly) . . . 14685

Center for Medical Consumers and Health Care Information (CMC), 237 Thompson St, New York, NY 10012
T: (212) 674-7105; Fax: (212) 674-7100
Founded: 1976; Members: 1500
Gen Secr: Arthur A. Levin
Focus: Public Health
Periodical
Health Facts (weekly) . . . 14686

Center for Medieval and Early Renaissance Studies (CEMERS), c/o State University of New York at Binghamton, POB 6000, Binghamton, NY 13902-6000
T: (607) 777-2730
Founded: 1966; Members: 75
Gen Secr: Robin S. Oggins
Focus: Hist
Periodicals
Acta (weekly)
Mediaevalia (weekly)
Old English Newsletter (weekly)
Proceedings of Conference (weekly) . . . 14687

Center for Medieval and Renaissance Studies (CMRS), c/o Ohio State University, 306 Dulles Hall, 230 W 17 Av, Columbus, OH 43210
T: (614) 292-7495
Founded: 1965; Members: 100
Gen Secr: Eve Levin
Focus: Hist
Periodical
Nouvelles (weekly) . . . 14688

Center for Migration Studies of New York (CMS), 209 Flagg Pl, Staten Island, NY 10304
T: (718) 351-8800; Fax: (718) 667-4598
Founded: 1964
Gen Secr: Lydio F. Tomasi
Focus: Ethnology
Periodicals
International Migration Review (weekly)
Migration World (weekly)
Newsletter (weekly)
Proceedings of Legal Conference (weekly) 14689

Center for Neo-Hellenic Studies (CNHS), 1010 W 22 St, Austin, TX 78705
T: (512) 477-5526
Founded: 1965; Members: 150
Gen Secr: E.G. Arnakis
Focus: Hist
Periodicals
Bulletin (weekly)
Neo-Hellenika (weekly) . . . 14690

Center for Oceans Law and Policy, c/o School of Law, University of Virginia, 580 Nancy Rd, Charlottesville, VA 22901
T: (804) 924-7441
Founded: 1976
Pres: Prof. John Norton Moore
Focus: Oceanography; Law . . . 14691

Center for Philosophy, Law, Citizenship (CPLC), 15 Knapp Hall, SUNY, Farmingdale, NY 11735
T: (516) 420-2047; Fax: (516) 420-2698
Founded: 1972; Members: 800
Pres: Prof. James P. Friel
Focus: Philos; Law; Poli Sci
Periodicals
AITIA Magazine (3 times annually)
Perspective in Philosophy . . . 14692

Center for Process Studies (CPS), 1325 College Av, Claremont, CA 91711
T: (714) 626-3521
Founded: 1974; Members: 1000
Gen Secr: Prof. David Griffin
Focus: Philos
Periodicals
Center for Process Studies Newsletter (weekly)
Process Studies (weekly) . . . 14693

Center for Public Justice (CPJ), 321 Eighth St NE, Washington, DC 20002
T: (202) 546-0489
Founded: 1973
Gen Secr: James W. Skillen
Focus: Poli Sci
Periodical
Public Justice Report (10 times annually) 14694

Center for Research in Ambulatory Health Care Administration (CRAHCA), 104 Inverness Terrace E, Englewood, CO 80112-5306
T: (303) 397-7879; Fax: (303) 799-1683
Founded: 1973
Gen Secr: Steven S. Lazarus
Focus: Public Health; Med . . . 14695

Center for Research Libraries (CRL), 6050 S Kenwood Av, Chicago, IL 60637
T: (312) 955-4545; Fax: (312) 955-4339
Founded: 1949; Members: 135
Pres: Donald B. Simpson
Focus: Libraries & Bk Sci
Periodicals
Annual Report (weekly)
Center for Reseach Libraries Directory (weekly)
Focus (weekly)
Handbook (biennial) . . . 14696

Center for Responsive Psychology (CRP), c/o Brooklyn College, CUNY, Brooklyn, NY 11210
T: (718) 951-5000
Founded: 1972; Members: 37
Gen Secr: Robert Buckhout
Focus: Psych
Periodical
Social Action and the Law (weekly) . . 14697

Center for Safety in the Arts (CSA), 5 Beekman St, New York, NY 10038
T: (212) 227-6220
Founded: 1977; Members: 45000
Gen Secr: Michael McCann
Focus: Safety
Periodical
Newsletter (10 times annually) . . . 14698

Center for Science in the Public Interest (CSPI), 1875 Connecticut Av NW, Washington, DC 20009
T: (202) 332-9110; Fax: (202) 265-4954
Founded: 1971; Members: 250000
Gen Secr: Michael Jacobson
Focus: Materials Sci; Sci
Periodical
Nutrition Action Healthletter (10 times annually) . . . 14699

Center for Short Lived Phenomena (CSLP), POB 199, Harvard Square Station, Cambridge, MA 02238
T: (617) 492-3310
Founded: 1968
Gen Secr: Richard Golob
Focus: Physics
Periodical
Oil Spill Intelligence Report (weekly) . . 14700

Center for Socialist History (CSH), 2633 Etna St, Berkeley, CA 94704
T: (510) 843-4658
Founded: 1981
Gen Secr: Ernest Haberkern
Focus: Socialism
Periodical
Journal of Socialist History 14701

Center for Social Research and Education (CSRE), 2940 16 St, Ste 102, San Francisco, CA 94103
T: (415) 255-2296; Fax: (415) 255-2298
Members: 30
Gen Secr: L.A. Kauffman
Focus: Sociology
Periodicals
Public Sphere (weekly)
Socialist Review (weekly) 14702

Center for Studies in Criminal Justice (CSCJ), c/o University of Chicago Law School, 1111 E 60 St, Chicago, IL 60637
T: (312) 702-9493; Fax: (312) 702-0730
Founded: 1965
Pres: Stephen J. Schulhofer
Focus: Law 14703

Center for Sustainable Agriculture (CSA), 2318 Bree Ln, Davis, CA 95616
T: (916) 756-7177; Fax: (916) 756-7188
Founded: 1977; Members: 275000
Gen Secr: David Katz
Focus: Agri
Periodical
Washington Newsletter (weekly) 14704

Center for Sutton Movement Writing, POB 7344, Newport Beach, CA 92660
T: (714) 644-1686
Founded: 1973; Members: 300
Gen Secr: Valerie J. Sutton
Focus: Educ
Periodical
The Sign Writer (2-4 times annually) . . 14705

Center for the Study of Human Rights (CSHR), c/o Columbia University, 1108 International Affairs Bldg, New York, NY 10027
T: (212) 854-2479; Fax: (212) 864-4847
Founded: 1977; Members: 34
Gen Secr: Dr. J. Paul Martin
Focus: Poli Sci; Prom Peace
Periodicals
Annual Report (weekly)
Newsletter (weekly) 14706

Center for the Study of Parent Involvement (CSPI), c/o JFK University, 370 Camino Pablo, Orinda, CA 94563
T: (510) 254-0110; Fax: (510) 254-4870
Founded: 1973
Pres: Daniel Safran
Focus: Educ 14707

Center for the Study of the Presidency (CSP), 208 E 75 St, New York, NY 10021
T: (212) 249-1200; Fax: (212) 628-9503
Founded: 1965; Members: 5000
Pres: R. Gordon Hoxie
Focus: Poli Sci
Periodicals
Center House Bulletin (weekly)
Presidential Studies Quarterly
Proceedings (weekly) 14708

Center for Urban and Regional Studies (CURS), c/o University of North Carolina at Chapel Hill, Hickerson House, Campus Box 3410, Chapel Hill, NC 27599-3410
T: (919) 962-3074; Fax: (919) 962-2518
Founded: 1957; Members: 6
Gen Secr: Michael A. Stegman
Focus: Urban Plan 14709

Center for Urban Black Studies (CUBS), c/o Graduate Theological Union, 2465 LeConte Av, Berkeley, CA 94709
T: (415) 841-8401
Founded: 1969
Gen Secr: Dorsay O. Blake
Focus: Sociology
Periodical
Newsletter 14710

Center for War, Peace and the News Media (CWPNM), c/o New York University, 10 Washington Pl, New York, NY 10003
T: (212) 998-7960; Fax: (212) 995-4143
Founded: 1985
Gen Secr: Robert Karl Manoff
Focus: Poli Sci 14711

Center for War/Peace Studies (CW/PS), 218 E 18 St, New York, NY 10003
T: (212) 475-1077; Fax: (212) 260-6384
Founded: 1966; Members: 2000
Gen Secr: Richard Hudson
Focus: Military Sci; Prom Peace
Periodicals
Global Report: Newsletter (weekly)
Special Studies (weekly) 14712

Center on Education and Training for Employment (CETE), c/o Ohio State University, 1900 Kenny Rd, Columbus, OH 43210
T: (614) 292-4353; Fax: (614) 292-1260
Founded: 1965
Gen Secr: Dr. Ray B. Ryan
Focus: Adult Educ
Periodicals
Centergram (weekly)
Vocational Educator (weekly) 14713

Centers and Regional Associations (CARA), c/o Mediaeval Academy of America, 1430 Massachusetts Av, Cambridge, MA 02138
T: (617) 491-1622
Founded: 1968; Members: 87
Pres: Prof. Paul Szarmach
Focus: Hist 14714

Central Organization for Jewish Education (COJE), 770 Eastern Pkwy, Brooklyn, NY 11213
T: (718) 774-4000
Founded: 1943; Members: 600000
Gen Secr: Dr. Nissan Mindel
Focus: Educ 14715

Central Society for Clinical Research (CSCR), c/o Dr. John P. Phair, Dept of Medicine, Northwestern University Medical School, 303 E Chicago Av, Chicago, IL 60611
T: (312) 951-5610
Founded: 1928; Members: 1294
Gen Secr: Dr. John P. Phair
Focus: Med
Periodical
The Journal of Laboratory and Clinical Medicine (weekly) 14716

Central States Anthropological Society (CSAS), c/o Robert Kegerleis, Dayton Museum of Natural History, 2600 Dewese Pkwy, Dayton, OH 45414
T: (513) 275-7431
Founded: 1921; Members: 500
Gen Secr: Dr. Robert Kegerleis
Focus: Anthro
Periodicals
Central States Anthropology Society Bulletin
Current Issues in Anthropology Newsletter (weekly) 14717

Ceramic Educational Council (CEC), 735 Ceramic Pl, Westerville, OH 43081
T: (614) 890-4700; Fax: (614) 899-6109
Founded: 1938; Members: 360
Gen Secr: W. Paul Holbrook
Focus: Educ 14718

C.G. Jung Foundation for Analytical Psychology, 28 E 39 St, New York, NY 10016
T: (212) 697-6430; Fax: (212) 953-3989
Founded: 1963; Members: 2500
Gen Secr: Dr. A. Maidenbaum
Focus: Psychoan; Psych; Anthro
Periodicals
Annual Report (weekly)
Quadrant (weekly) 14719

Character Education Institute (CEI), 8918 Tesoro Dr, Ste 220, San Antonio, TX 78217-6253
T: (210) 829-1727; Fax: (210) 829-1729
Founded: 1942
Pres: Young Jay Mulkey
Focus: Educ 14720

Charles Ives Society, c/o School of Music, Indiana University, Bloomington, IN 47405
T: (718) 780-5655
Founded: 1973
Pres: J. Peter Burkholder
Focus: Music 14721

Charles S. Peirce Society (CPS), c/o Christian J.W. Kloesel, Peirce Edition Project, 425 University Blvd, Indianapolis, IN 46202-5140
T: (317) 274-2173; Fax: (317) 274-2347
Founded: 1946; Members: 590
Gen Secr: Christian J.W. Kloesel
Focus: Philos
Periodical
Transactions (weekly) 14722

Chautauqua Literary and Scientific Circle (CLSC), c/o Chautauqua Institution, 1 Ames Av, Chautauqua, NY 14722
T: (716) 357-6200
Founded: 1878; Members: 2000
Gen Secr: Norman Peberson
Focus: Lit 14723

Chemical Industry Institute of Toxicology (CIIT), POB 12137, Research Triangle Park, NC 27709
T: (919) 541-2070
Founded: 1974; Members: 50
Pres: Dr. Roger O. McClellan
Focus: Toxicology
Periodical
CIIT Activities (weekly) 14724

Chemical Management and Resources Association (CMRA), 60 Gay St, Ste 702, Staten Island, NY 10301
T: (718) 876-8800
Founded: 1940; Members: 900
Gen Secr: Mary J. Carrick
Focus: Marketing
Periodical
Chemical Marketing and Management (weekly) 14725

Chemists' Club, 295 Madison Av, New York, NY 10017
T: (212) 532-7649; Fax: (212) 532-7649
Founded: 1898; Members: 2500
Gen Secr: L. John Polite
Focus: Chem
Periodical
Newsletter (weekly) 14726

Cherokee National Historical Society (CNHS), POB 515, Tahlequah, OK 74465
T: (918) 456-6007
Founded: 1963; Members: 900
Gen Secr: Marilyn G. Moss
Focus: Hist
Periodical
The Columns (weekly) 14727

Chesapeake and Ohio Historical Society (COHS), POB 79, Clifton Forge, VA 24422
T: (703) 862-2210
Founded: 1969; Members: 2500
Pres: Thomas W. Dixon
Focus: Hist
Periodical
Chesapeake and Ohio Historical Magazine (weekly) 14728

Chief Officers of State Library Agencies (COSLA), c/o Barratt Wilkins, Council of State Govts., POB 11910, Lexington, KY 11910
T: (606) 231-1925; Fax: (606) 231-1928
Founded: 1973; Members: 51
Gen Secr: Nancy Zussy
Focus: Libraries & Bk Sci 14729

Childbirth without Pain Education Association (CWPEA), 20134 Snowden, Detroit, MI 48235-1170
T: (313) 341-3816
Founded: 1958; Members: 3000
Gen Secr: Flora Hommel
Focus: Gynecology
Periodical
Memo (weekly) 14730

Child Care Employee Project (CCEP), 6536 Telegraph Av, Ste 201A, Oakland, CA 94609
T: (510) 653-9889; Fax: (510) 653-8385
Founded: 1977; Members: 1200
Gen Secr: M. Whitebook
Focus: Educ 14731

Child Neurology Society (CNS), 475 Cleveland Av, Ste 220, Saint Paul, MN 55104-5051
T: (612) 641-1584; Fax: (612) 641-1634
Founded: 1971; Members: 1000
Gen Secr: Mary Currey
Focus: Neurology; Pediatrics 14732

Children's Literature Association (ChLA), POB 138, Battle Creek, MI 49016
T: (616) 965-8180
Founded: 1972; Members: 700
Pres: Sarah Smedman
Focus: Lit
Periodicals
Children's Literature: An International Journal (weekly)
Journal (weekly)
Proceedings (weekly) 14733

China Institute in America (CIA), 125 E 65 St, New York, NY 10021
T: (212) 744-8181; Fax: (212) 628-4159
Founded: 1926; Members: 1100
Pres: Charles P. Wang
Focus: Ethnology
Periodicals
Annual Report (weekly)
Art Exhibition Catalogs (weekly)
Bulletin (weekly)
Journal of Inner Asian Studies (weekly)
School Catalog (weekly) 14734

China Medical Board of New York (CMBNY), 750 Third Av, New York, NY 10017
T: (212) 682-8000; Fax: (212) 949-8726
Founded: 1928; Members: 11
Pres: William D. Sawyer
Focus: Med
Periodical
Annual Report (weekly) 14735

Chinese-American Librarians Association (CALA), c/o Sheila Lai, CSU at Sacramento, 2000 Jed Smith Dr, Sacramento, CA 95819-6039
T: (916) 278-6201; Fax: (916) 363-0868
Founded: 1973; Members: 500
Gen Secr: Sheila Lai
Focus: Libraries & Bk Sci
Periodicals
Journal of Library and Information Science (weekly)
Newsletter (3 times annually) 14736

Chinese Culture Association (CCA), POB 1272, Palo Alto, CA 94302-1272
T: (415) 948-2251
Founded: 1966; Members: 2000
Gen Secr: Prof. P.F. Tao
Focus: Humanities
Periodicals
Journal (2-3 times annually)
Newsletter (4-5 times annually) 14737

Chinese-English Translation Assistance Group (CETA), POB 400, Kensington, MD 20895
T: (301) 946-7006
Founded: 1971
Gen Secr: Jim Mathias
Focus: Ling
Periodical
Bulletin 14738

Chinese Historical Society of America (CHSA), 650 Commercial St, San Francisco, CA 94111
T: (415) 391-1188
Founded: 1963; Members: 500
Pres: Enid N. Lim
Focus: Ethnology; Hist
Periodicals
Bulletin (weekly)
Chinese America: History and Perspectives 14739

Chinese Language Teachers Association (CLTA), c/o Dept of Foreign Languages, Kalamazoo College, 1200 Academy St, Kalamazoo, MI 49006
T: (616) 337-7001; Fax: (616) 337-7251
Founded: 1963; Members: 700
Gen Secr: Prof. Madeline Chu
Focus: Ling; Educ
Periodical
Journal (3 times annually) 14740

Chinese Musical and Theatrical Association (CMTA), 24 Pell St, New York, NY 10013
T: (212) 385-3531
Founded: 1934; Members: 150
Gen Secr: Stanley S. Chiu
Focus: Music; Perf Arts 14741

Chinese Music Society of North America (CMSNA), 1 Heritage Plaza, POB 5275, Woodridge, IL 60517
T: (708) 910-1551; Fax: (708) 910-1561
Founded: 1976; Members: 500
Pres: Sin-Yan Shen
Focus: Music; Acoustics
Periodical
Chinese Music (weekly) 14742

Christian Association for Psychological Studies (CAPS), c/o Dr. Robert R. King, POB 890279, Temecula, CA 92589-0279
T: (714) 695-2277; Fax: (714) 695-3431
Founded: 1956; Members: 2400
Gen Secr: Dr. Robert R. King
Focus: Psych
Periodical
Journal of Psychology and Christianity (weekly) 14743

Christian College Coalition (CCC), 329 Eighth St NE, Washington, DC 20002
T: (202) 546-8713; Fax: (202) 546-8913
Founded: 1976; Members: 84
Pres: Dr. Myron Augsburger
Focus: Educ
Periodicals
Christian College News (weekly)
Guide to Christian Colleges (weekly) . . 14744

Christian Dental Society (CDS), c/o Richard Haw, POB 177, Sumner, IA 50674
T: (319) 578-5232
Founded: 1962; Members: 300
Gen Secr: Richard Haw
Focus: Dent 14745

Christian Literacy Associates (CLA), 541 Perry Hwy, Pittsburgh, PA 15229
T: (412) 364-3777
Founded: 1977; Members: 3200
Gen Secr: Dr. William E. Kofmehl
Focus: Lit 14746

Christian Research Institute International (CRI), POB 500, San Juan Capistrano, CA 92693-0500
Founded: 1960
Gen Secr: Hendrik Hanegraaff
Focus: Rel & Theol 14747

Christian Schools International (CSI), 3350 E Paris Av SE, POB 8709, Grand Rapids, MI 49512
T: (616) 957-1070; Fax: (616) 957-5022
Founded: 1920; Members: 450
Gen Secr: S.D. Haan
Focus: Educ
Periodicals
Christian Home and School (8 times annually)
Directory (weekly)
Intercom (5 times annually) 14748

Christopher Morley Knothole Association (CMKA), c/o Bryant Library, Paper Mill Rd, Roslyn, NY 11576
T: (516) 621-2240
Founded: 1961; Members: 180
Pres: Peter Cohn
Focus: Lit 14749

Chronic Fatigue Syndrome Society (CFSS), POB 230108, Portland, OR 97223
T: (503) 684-5261
Founded: 1985; Members: 15000
Gen Secr: D.E. Wilcox
Focus: Med 14750

Church and Synagogue Library Association (CSLA), POB 19357, Portland, OR 97280
T: (503) 244-6919
Founded: 1967; Members: 1900
Gen Secr: Lorraine E. Burson
Focus: Libraries & Bk Sci
Periodical
Church and Synagogue Libraries (weekly) 14751

Cinemists 63, 1810 N Cherokee, Ste 214, Los Angeles, CA 90028
T: (213) 469-9696
Founded: 1963; Members: 243
Gen Secr: Donald Deschner
Focus: Cinema 14752

Circulo de Radioterapeutas Ibero-Latinoamericanos (CRILA), c/o Dept of Radioterapy, M.D. Anderson Cancer Centre, 1515 Holcombe Blvd, Houston, TX 77030
Founded: 1931
Gen Secr: Luuis Delclos
Focus: Therapeutics 14753

Circus Historical Society (CHS), 3477 Vienna Court, Westerville, OH 43081
Founded: 1939; Members: 1400
Gen Secr: Dale C. Haynes
Focus: Hist 14754

Cities in Schools (CIS), 401 Wythe St, Ste 200, Alexandria, VA 22314
T: (703) 519-8999; Fax: (703) 519-7213
Founded: 1974
Pres: William E. Milliken
Focus: Educ 14755

Citizens for a Better Environment (CBE), 407 S Dearborn, Ste 1775, Chicago, IL 60605
T: (312) 939-1530; Fax: (312) 939-2536
Founded: 1971; Members: 30000
Pres: William Davis
Focus: Ecology
Periodical
CBE Environmental Review (weekly) . . 14756

City of Hope (CH), 1500 E Duarte Rd, Duarte, CA 91010
T: (818) 359-8111; Fax: (818) 301-8115
Founded: 1913
Gen Secr: Dr. Sanford M. Shapero
Focus: Med 14757

Civil War Press Corps (CWPC), 7674 Heriot Dr, Fayetteville, NC 28311
T: (919) 488-0598; Fax: (919) 396-1900
Founded: 1958; Members: 500
Gen Secr: Joseph A. Malcom
Focus: Hist
Periodical
Civil War Byline (8 times annually) . . 14758

Civil War Round Table Associates (CWRTA), POB 7388, Little Rock, AR 72217
T: (501) 225-3996; Fax: (501) 225-5167
Founded: 1968; Members: 1500
Pres: Jerry L. Russell
Focus: Hist
Periodical
Civil War Round Table Digest (weekly) . 14759

Classical America (CA), 227 E 50 St, New York, NY 10022
T: (212) 753-4376
Founded: 1968; Members: 700
Pres: Henry H. Reed
Focus: Arts 14760

Clay Minerals Society (CMS), POB 4417, Boulder, CO 80306
T: (303) 444-6505
Founded: 1963; Members: 950
Gen Secr: Jo Eberl
Focus: Mineralogy
Periodicals
Clays and Clay Minerals (weekly)
Newsletter (weekly) 14761

Clinical Chemistry Data Communication Group (CCDCG), c/o University of Minnesota Medical School, 420 Delaware St, POB 198, Minneapolis, MN 55455
T: (612) 625-0932; Tx: 9105762955; Fax: (612) 625-0617
Founded: 1975
Gen Secr: Ellis S. Benson
Focus: Computer & Info Sci
Periodical
CCDCG Newsletter (3 times annually) . 14762

Clinical Orthopedic Society (COS), POB 727, Bloomington, IL 61704
T: (309) 662-0454
Founded: 1912; Members: 700
Gen Secr: Barbara Dehority
Focus: Orthopedics
Periodical
Directory (weekly) 14763

Coalition for International Cooperation and Peace (CICP), 301 E 45 St, New York, NY 10017
T: (212) 983-3353
Founded: 1975; Members: 500
Gen Secr: Dr. Harry H. Lerner
Focus: Prom Peace
Periodicals
Conference Reports (weekly)
Lifelines (weekly)
Newsletter (weekly) 14764

Coalition for the Advancement in Jewish Education (CAJE), 261 W 35 St, Flat 12A, New York, NY 10001
T: (212) 268-4210; Fax: (212) 268-4214
Founded: 1976; Members: 3800
Gen Secr: Eliot G. Spack
Focus: Educ 14765

Coastal Engineering Research Council (CERC), c/o Edge & Associates, 79 Anson St, Charleston, SC 29401
T: (803) 723-4864
Founded: 1950
Gen Secr: Billy L. Edge
Focus: Oceanography; Eng
Periodical
Proceedings (biennial) 14766

The Coastal Society (TCS), POB 2081, Gloucester, MA 01930-2081
T: (508) 281-9209
Founded: 1975; Members: 400
Gen Secr: Thomas E. Bigford
Focus: Ecology
Periodicals
Bulletin (weekly)
Directory 14767

Coblentz Society (CS), c/o Robert W. Hannah, Perkin-Elmer Corp, Main Av, Norwalk, CT 06859-0284
T: (203) 834-4628
Founded: 1954; Members: 600
Gen Secr: Robert W. Hannah
Focus: Physics
Periodical
Coblentz Society Mailings (weekly) . . . 14768

Cognitive Science Society (CSS), c/o Alan M. Lesgold, Learning Research and Development Center, University of Pittsburgh, Pittsburgh, PA 15260
T: (412) 624-7046; Fax: (412) 624-9149
Members: 800
Gen Secr: Alan M. Lesgold
Focus: Psych 14769

College Art Association (CAA), 275 Seventh Av, New York, NY 10001
T: (212) 691-1051; Fax: (212) 627-2381
Founded: 1911; Members: 13500
Gen Secr: Susan Ball
Focus: Arts
Periodicals
The Art Bulletin (weekly)
Art Journal (weekly)
Newsletter (weekly)
Positions Listing (5 times annually) . . . 14770

The College Board (TCB), 45 Columbus Av, New York, NY 10023
T: (212) 713-8000; Fax: (212) 713-8255
Founded: 1900; Members: 2800
Pres: Donald M. Stewart
Focus: Educ
Periodicals
College Board News (weekly)
College Board Review (weekly)
College Times (weekly)
Going Right on (weekly)
Membership Directory (weekly)
Proceedings (weekly) 14771

College English Association (CEA), c/o John J. Joyce, Nazareth College of Rochester, 4245 East Av, Rochester, NY 14618
T: (716) 586-2525
Founded: 1939; Members: 1500
Gen Secr: John J. Joyce
Focus: Ling
Periodicals
The CEA Critic (weekly)
The CEA Forum (weekly) 14772

College Language Association (CLA), c/o Lucy C. Grigsby, Clark Atlanta University, James P. Brawley Dr at Fair St SW, Atlanta, GA 30314
T: (404) 808-8524; Fax: (404) 880-8222
Founded: 1937; Members: 350
Gen Secr: Lucy C. Grigsby
Focus: Ling 14773

College Music Society (CMS), 202 W Spruce, Missoula, MT 59802
T: (406) 721-9616
Founded: 1947; Members: 4500
Gen Secr: Robby D. Gunstream
Focus: Music
Periodicals
Bibliographies in American Music (weekly)
Directory of Music Faculties in Colleges and Universities, U.S. and Canada
Symposium 14774

College of American Pathologists (CAP), 325 Waukegan Rd, Northfield, IL 60093-2750
T: (708) 446-8800; Fax: (708) 446-8807
Founded: 1947; Members: 13500
Gen Secr: Lee Van Bremen
Focus: Pathology
Periodicals
CAP Today (weekly)
Directory (weekly)
Job Placement Bulletin
Newsletter 14775

College of Optometrists in Vision Development (COVD), POB 285, Chula Vista, CA 92012
T: (714) 425-6191
Founded: 1970; Members: 1400
Gen Secr: Robert M. Wold
Focus: Ophthal
Periodicals
Fellow Directory (weekly)
Journal of Optometric Vision Development (weekly)
Newsletter (weekly) 14776

College Press Service (CPS), c/o Diana Smith, 64 E Concord St, Orlando, FL 32801
T: (407) 839-5754; Fax: (407) 839-5794
Founded: 1971; Members: 620
Gen Secr: Diana Smith
Focus: Journalism
Periodical
College Press Service (weekly) 14777

College Reading and Learning Association (CRLA), c/o Becky Johnen, Chemeketa Community College, POB 14007, Salem, OR 97309
T: (503) 399-2556; Fax: (503) 399-5214
Founded: 1967; Members: 1000
Gen Secr: Becky Johnen
Focus: Educ
Periodicals
Journal of College Reading and Learning (weekly)
WCRLA Newsletter (weekly) 14778

College Theology Society (CTS), c/o Dept of Theology, University of San Diego, Alcala Park, San Diego, CA 92110
T: (619) 260-4600
Founded: 1954; Members: 960
Pres: Dr. Joan Leonard
Focus: Rel & Theol
Periodical
Horizons (weekly) 14779

College-University Resource Institute (CURI), 1001 Connecticut Av, Ste 901, Washington, DC 20036
T: (202) 659-2104; Fax: (202) 835-1159
Founded: 1982
Pres: Mark Gelber
Focus: Educ 14780

Collegium Internationale Chirurgiae Digestivae (CICD), c/o Dept of Surgery, Medical College of Wisconsin, 9200 W Wisconsin Av, Milwaukee, WI 53226
T: (414) 454-5705; Fax: (414) 259-9225
Founded: 1969; Members: 2600
Gen Secr: Robert E. Condon
Focus: Surgery
Periodical
World Journal of Surgery (weekly) . . . 14781

Colonial Society of Massachusetts (CSM), 87 Mount Vernon St, Boston, MA 02108
T: (617) 227-2782
Founded: 1892; Members: 240
Gen Secr: John W. Tyler
Focus: Hist
Periodical
New England (weekly) 14782

Colorado River Association (CRA), 417 S Hill St, Rm 1024, Los Angeles, CA 90013
T: (213) 626-4621; Fax: (213) 624-2987
Members: 25
Gen Secr: Lowell O. Weeks
Focus: Water Res
Periodical
Newsletter (weekly) 14783

Commission for the Scientific and Technological Development of Central America and Panama, c/o Dept of Scientific and Technological Affairs, OAS General Secretariat, 1889 F St NW, Washington, DC 20006
Founded: 1976; Members: 6
Gen Secr: Orlando Mason
Focus: Develop Areas; Eng 14784

Commission on Accreditation of Rehabilitation Facilities (CARF), 101 N Wilmot Rd, Ste 500, Tucson, AZ 85711
T: (602) 748-1212
Founded: 1966
Gen Secr: Alan H. Toppel
Focus: Rehabil
Periodical
Report (weekly) 14785

Commission on Gay/Lesbian Issues in Social Work Education, c/o Council on Social Work Education, 1600 Duke St, Alexandria, VA 22314
T: (703) 683-8080; Fax: (703) 683-8099
Founded: 1980; Members: 9
Gen Secr: Dean Pierce
Focus: Educ; Soc Sci 14786

Commission on Mental and Physical Disability Law (CMPDL), c/o American Bar Association, 1800 M St NW, Washington, DC 20036
T: (202) 331-2240; Fax: (202) 331-2220
Founded: 1976
Gen Secr: John Parry
Focus: Law
Periodicals
Disability Law Reporter (weekly)
Mental and Physical Disability Law Reporter (weekly) 14787

Commission on Pastoral Research (COMMISS), 7135 Minstrel Way, Ste 101, Columbia, MD 21045
T: (410) 290-5995; Fax: (410) 290-7844
Founded: 1972; Members: 8
Pres: Dr. Jack Compton
Focus: Law
Periodical
Abstracts of Research in Pastoral Care and Counseling (weekly) 14788

Commission on Professional and Hospital Activities (CPHA), 2929 Plymouth Rd, Ste 208, POB 304, Ann Arbor, MI 48106-0304
T: (313) 995-9800; Fax: (313) 973-2791
Founded: 1953
Pres: Dr. William Jessee
Focus: Med
Periodical
Length of Stay, by Diagnosis, by Operation (weekly) 14789

Commission on Professionals in Science and Technology (CPST), 1500 Massachusetts Av NW, Ste 831, Washington, DC 20005
T: (202) 223-6995
Founded: 1953; Members: 525
Gen Secr: Betty M. Vetter
Focus: Eng
Periodicals
Professional Women and Minorities: A Manpower Data Resource Service (weekly)
Salaries of Scientists, Engineers and Technicians (biennial)
Scientific, Engineering, Technical Manpower Comments (weekly)
Supply and Demand for Scientists and Engineers (weekly) 14790

Commission to Study the Organization of Peace (CSOP), c/o Prof. Louis B. Sohn, National Law Center, George Washington University, 720 20 St, Washington, DC 20052
T: (202) 994-7390
Founded: 1939; Members: 90
Pres: Prof. Louis B. Sohn
Focus: Prom Peace 14791

Committee for Better Transit (CBT), POB 3106, Long Island City, NY 11103
T: (718) 278-0650
Founded: 1962; Members: 350
Pres: Dr. Stephen B. Dobrow
Focus: Transport
Periodical
Notes from Unterground (weekly) . . . 14792

Committee for Economic Development (CED), 2000 L St NW, Ste 700, Washington, DC 20036
T: (202) 296-5860; Fax: (202) 223-0776
Founded: 1942; Members: 225
Pres: S. Hurwitz
Focus: Econ 14793

Committee for Freedom of Choice in Medicine (CFCM), 1180 Walnut Av, Chula Vista, CA 92011
T: (619) 429-8200; Fax: (619) 429-8004
Founded: 1972; Members: 30000
Pres: Mike Culbert
Focus: Cell Biol & Cancer Res 14794

Committee for Hispanic Arts and Research (CHAR), POB 12865, Austin, TX 78711
T: (512) 479-6397
Founded: 1980; Members: 250
Gen Secr: Romeo Rodriguez
Focus: Arts
Periodical
Arriba (weekly) 14795

Committee for the Implementation of the Standardized Yiddish Orthography (CISYO), 200 W 72 St, Ste 40, New York, NY 10023
T: (212) 787-6675
Founded: 1958
Pres: Dr. M. Schaechter
Focus: Ling 14796

Committee for the Promotion of Medical Research, 191 Hayward St, Yonkers, NY 10704
T: (914) 968-0262
Founded: 1944
Gen Secr: Ellen M. Cosgrove
Focus: Med 14797

Committee for the Scientific Investigation of Claims of the Paranormal (CSICOP), POB 703, Buffalo, NY 14226
T: (716) 636-1425; Fax: (716) 636-1733
Founded: 1976; Members: 150
Gen Secr: Paul Kurtz
Focus: Psych
Periodicals
Skeptical Briefs (weekly)
The Skeptical Inquirer (weekly) 14798

Committee on Allied Health Education and Accreditation (CAHEA), 515 N State St, Chicago, IL 60610
T: (312) 464-4660; Fax: (312) 464-4184
Founded: 1976; Members: 14
Gen Secr: John J. Fauser
Focus: Educ; Public Health 14799

Committee on Changing International Realities (CIR), c/o National Planning Association, 1424 16 St NW, Ste 700, Washington, DC 20036
T: (202) 265-7685
Founded: 1975; Members: 60
Gen Secr: Dr. Richard S. Belous
Focus: Int'l Relat 14800

Committee on Continuing Education for School Personnel (CCESP), c/o Dr. George Sisko, Academic Services, Kean College of New Jersey, Union, NJ 07083
T: (201) 527-2161
Founded: 1922; Members: 42
Gen Secr: Dr. George Sisko
Focus: Educ 14801

Committee on Earth Observation Satellites (CEOS), c/o Office of Space Science and Applications, National Aeronautics and Space Administration, Washington, DC 20546
Members: 13#
Gen Secr: Dr. Shelby Tilford
Focus: Aero 14802

Committee on Institutional Cooperation (CIC), 302 E John St, Ste 1705, Champaign, IL 61820
T: (217) 333-8475; Fax: (217) 244-7127
Founded: 1958; Members: 12
Gen Secr: Roger G. Clark
Focus: Educ
Periodicals
Biennial Report (biennial)
CIC Directory of Minority Ph.D. Candidates and Recipients (weekly)
President's Newsletter 14803

Committee on Political Education, AFL-CIO (COPE), 815 16 St NW, Washington, DC 20006
T: (202) 637-5101
Founded: 1955; Members: 50
Gen Secr: John Perkins
Focus: Poli Sci
Periodical
Political Memo form COPE 14804

Committee on Research Materials on Southeast Asia (CORMOSEA), c/o Association for Asian Studies, University of Michigan, 1 Lane Hall, Ann Arbor, MI 48109
T: (313) 665-2490
Founded: 1969; Members: 12
Gen Secr: Fe Susan T. Go
Focus: Libraries & Bk Sci
Periodical
CORMOSEA Bulletin (weekly) 14805

Committee on the Role and Status of Women in Educational Research and Development, c/o Emily Lowe Brizendine, Dept of Educational Leadership, California State University, Hayward, CA 94542
T: (510) 658-7013; Fax: (510) 727-2283
Founded: 1972; Members: 7
Pres: Emily Lowe Brizendine
Focus: Educ 14806

Committee on the Teaching of Science (CTS), c/o Dept of Geography, Western Michigan University, Kalamazoo, MI 49008-5053
T: (616) 387-3429
Founded: 1961
Gen Secr: Prof. J.P. Stoltman
Focus: Educ 14807

Community Associations Institute (CAI), 1630 Duke St, Alexandria, VA 22314
T: (703) 548-8600; Fax: (703) 684-1581
Founded: 1973; Members: 13000
Gen Secr: Barbara Byrd-Lawler
Focus: Urban Plan
Periodicals
Common Ground (weekly)
Law Reporter (weekly)
News (weekly)
Report (weekly) 14808

Community College Association for Instruction and Technology (CCAIT), c/o New Mexico Military Institute, 101 W College Blvd, Roswell, NM 88201-5173
T: (505) 624-8381; Fax: (505) 624-8390
Founded: 1971; Members: 350
Gen Secr: Bernard Fradkin
Focus: Educ 14809

Community Development Society (CDS), c/o Dr. Del Yoder, 1123 N Water St, Milwaukee, WI 53202
T: (414) 276-8788; Fax: (414) 276-7704
Founded: 1969; Members: 1100
Gen Secr: Dr. Del Yoder
Focus: Sociology
Periodicals
Journal of the Community Development Society (biennial)
Vanguard (weekly) 14810

Community Environmental Council (CEC), 930 Miramonte Dr, Santa Barbara, CA 93109
T: (805) 963-0583; Fax: (805) 962-9080
Founded: 1970; Members: 750
Gen Secr: J. Clark
Focus: Ecology
Periodical
Newsletter (weekly) 14811

Community for Religious Research and Education (CRRE), c/o Michael McKale, Religious Studies Dept, Saint Francis College, Loretto, PA 15940
T: (814) 472-3396
Founded: 1973; Members: 1000
Pres: Michael McKale
Focus: Rel & Theol; Poli Sci; Sociology
Periodical
Radical Religion (weekly) 14812

Community Guidance Service (CGS), 133 E 73 St, New York, NY 10021
T: (212) 988-4800
Founded: 1953
Focus: Therapeutics 14813

Community Nutrition Institute (CNI), 2001 S St NW, Ste 530, Washington, DC 20009
T: (202) 462-4700
Founded: 1970
Gen Secr: Rodney E. Leonard
Focus: Hematology
Periodical
Nutrition Week (weekly) 14814

Company of Military Historians (CMH), N Main St, Westbrook, CT 06498
T: (203) 399-9460; Fax: (203) 399-9320
Founded: 1951; Members: 2500
Gen Secr: William R. Reid
Focus: Military Sci
Periodicals
Military Collector and Historian (weekly)
Military Uniforms in America (weekly) . . 14815

Comparative and International Education Society (CIES), c/o SUNY at Buffalo, 428 Baldy Hall, Buffalo, NY 14260
T: (716) 645-2487
Founded: 1956; Members: 2800
Gen Secr: P. Altbach
Focus: Educ
Periodicals
Comparative Education Review (weekly)
Directory (weekly)
Newsletter (weekly) 14816

Computer Aided Manufacturing International (CAM-I), 1250 E Copeland Rd, Ste 500, Arlington, TX 76011
T: (817) 860-1654; Fax: (817) 275-6450
Founded: 1972; Members: 100
Pres: Dale Hartman
Focus: Computer & Info Sci 14817

Computer Law Association (CLA), 3028 Javier Rd, Ste 500E, Fairfax, VA 22031
T: (703) 560-7747; Fax: (703) 207-7028
Founded: 1971; Members: 1350
Gen Secr: Barbara G. Fieser
Focus: Law
Periodicals
Computer Law Association Newsletter (weekly)
International Update Newsletter (weekly)
Membership Directory (weekly) 14818

Concern, 1794 Columbia Rd NW, Washington, DC 20009
T: (202) 328-8160; Fax: (202) 387-3378
Founded: 1970; Members: 2100
Gen Secr: Susan Boyd
Focus: Ecology 14819

Conduit, c/o University of Iowa, Oakdale Campus, Iowa City, IA 52242
T: (319) 335-4100; Fax: (319) 335-4077
Founded: 1971; Members: 15
Gen Secr: P. Trotter
Focus: Educ 14820

Confederate Memorial Literary Society (CMLS), c/o The Museum of the Confederacy, 1201 E Clay St, Richmond, VA 23219
T: (804) 649-1861; Fax: (804) 644-7150
Founded: 1890; Members: 3000
Gen Secr: Robin Reed
Focus: Lit
Periodicals
Interpretative Catalogue (weekly)
Journal (weekly) 14821

Conference Board of Associated Research Councils (CBARC), c/o American Council of Learned Societies, 228 E 45 St, New York, NY 10017
T: (202) 697-1505; Fax: (202) 949-5058
Focus: Sci 14822

Conference Board of the Mathematical Sciences (CBMS), 1529 18 St NW, Washington, DC 20036
T: (202) 293-1170; Fax: (202) 265-2384
Founded: 1960; Members: 15
Gen Secr: Ronald C. Rosier
Focus: Math 14823

Conference for Chinese Oral and Performing Literature (CHINOPERL), c/o Susan Blader, Asian Studies Program, Dartmouth College, 6191 Bartlett Hall, Hanover, NH 03755
T: (603) 646-3478
Founded: 1969; Members: 200
Pres: Susan Blader
Focus: Lit
Periodical
CHINOPERL Papers (weekly) 14824

Conference Group on French Politics and Society (CGFPS), c/o Center for European Studies, Harvard University, 27 Kirkland St, Cambridge, MA 02138
T: (617) 495-4303; Fax: (617) 495-4303
Founded: 1974; Members: 300
Gen Secr: Prof. George Ross
Focus: Poli Sci
Periodicals
French Politics and Society (weekly)
Research and Teaching Register 14825

Conference Group on German Politics (CGGP), POB 345, Durham, NH 03824
T: (603) 862-1778
Founded: 1968; Members: 250
Gen Secr: Prof. G.K. Romoser
Focus: Poli Sci
Periodicals
Directory of Current Research
Newsletter 14826

Conference Group on Italian Politics and Society (CONGRIPS), c/o Dept of Political Science, DePaul University, 2323 N Seminary Av, Rm 524, Chicago, IL 60614
T: (312) 362-6824
Founded: 1975; Members: 210
Gen Secr: Robert Leonardi
Focus: Poli Sci
Periodicals
Italian Politics (weekly)
Membership List (weekly)
Newsletter (weekly) 14827

Conference of Chief Justices (CCJ), c/o National Center for State Courts, 300 Newport Av, Williamsburg, VA 23187
T: (804) 253-2000
Founded: 1949; Members: 58
Pres: Robert N.C. Rex
Focus: Law 14828

Conference of Consulting Actuaries (CCA), 475 N Martingale Rd, Ste 800, Schaumburg, IL 60173
T: (312) 706-3535
Founded: 1950; Members: 1021
Pres: Mary S. Riebold
Focus: Insurance
Periodical
Proceedings (weekly) 14829

Conference of Educational Administrators Serving the Deaf (CEASD), c/o Ken Rislov, Arizona School for the Deaf and Blind, POB 5545, Tucson, AZ 85703-0545
T: (602) 770-3700
Founded: 1868; Members: 355
Pres: Ken Rislov
Focus: Educ Handic
Periodicals
Advocate for Education of the Deaf (weekly)
American Annals of the Deaf (5 times annually)
Newsletter (weekly) 14830

Conference of Podiatry Executives (COPE), 5310 McKitrick Blvd, Columbus, OH 43235
T: (614) 457-6269; Fax: (614) 457-3375
Founded: 1960; Members: 22
Gen Secr: Gary L. Fetgatter
Focus: Orthopedics 14831

Conference of Public Health Laboratorians (COPHL), POB 9083, Austin, TX 78766
T: (512) 458-7318
Founded: 1920; Members: 286
Gen Secr: Charles E. Sweet
Focus: Public Health
Periodical
Newsletter (weekly) 14832

Conference of Public Health Veterinarians (CPHV), c/o Dr. Tom Eng, 1919 N Wayne St, Arlington, VA 22201
T: (517) 332-7671
Founded: 1946; Members: 220
Gen Secr: Dr. Tom Eng
Focus: Vet Med
Periodical
Newsletter (weekly) 14833

Conference of Research Workers in Animal Diseases (CRWAD), c/o Dr. Robert Ellis, Dept of Microbiology, Colorado State University, Fort Collins, CO 80523
T: (303) 491-5740; Fax: (303) 491-1815
Founded: 1919; Members: 700
Gen Secr: Dr. Robert Ellis
Focus: Vet Med 14834

Conference on Asian History (CAH), c/o Prof. George M. Wilson, East Asian Studies Center, Indiana University, Bloomington, IN 47405
T: (812) 855-3765; Fax: (812) 855-7765
Founded: 1953; Members: 100
Pres: Prof. George M. Wilson
Focus: Hist 14835

Conference on Christianity and Literature (CCL), c/o Prof. Jewel Spears Brooker, Eckerd College, Saint Petersburg, FL 33711
T: (813) 864-8281; Fax: (813) 866-2304
Founded: 1956; Members: 1400
Gen Secr: Prof. Jewel Spears Brooker
Focus: Lit; Educ; Rel & Theol
Periodical
Christianity and Literature (weekly) . . . 14836

Conference on College Composition and Communication (CCCC), 1111 W Kenyon Rd, Urbana, IL 61801
T: (217) 328-3870; Fax: (217) 328-9645
Founded: 1949; Members: 7000
Gen Secr: Miles Myers
Focus: Educ
Periodical
College Composition and Communication (weekly) 14837

Conference on Consumer Finance Law (CCFL), c/o Lawrence X. Pusateri, Peterson & Ross, 200 E Randolph Dr, Ste 7300, Chicago, IL 60601
T: (312) 861-1400; Fax: (312) 565-0832
Founded: 1927; Members: 584
Gen Secr: Lawrence X. Pusateri
Focus: Law
Periodical
Quarterly Report (weekly) 14838

Conference on English Education (CEE), 1111 Kenyon Rd, Urbana, IL 61801
T: (217) 328-3870; Fax: (217) 328-9645
Founded: 1963; Members: 2000
Gen Secr: Miles Myers
Focus: Ling; Educ
Periodical
English Education (weekly) 14839

Conference on Latin American History (CLAH), c/o Donna J. Guy, Latin American Area Center, University of Arizona, 1522 E Drachman, Tucson, AZ 85721
T: (602) 622-4002; Fax: (602) 622-0177
Founded: 1926; Members: 850
Gen Secr: Donna J. Guy
Focus: Hist
Periodical
CLAH Newsletter (weekly) 14840

Congress of Lung Association Staff (CLAS), 1726 M St NW, Ste 902, Washington, DC 20036
T: (202) 785-3355; Fax: (202) 452-1805
Founded: 1912; Members: 800
Gen Secr: Janet Widmer
Focus: Pulmon Dis
Periodicals
Membership Directory (weekly)
Newsletter (weekly) 14841

Congress of Neurological Surgeons (CNS), c/o Thomas G. Saul, 506 Oak St, Cincinnati, OH 45219
T: (513) 872-2657; Fax: (513) 872-2597
Founded: 1951; Members: 2700
Gen Secr: Thomas G. Saul
Focus: Neurology; Surgery
Periodicals
Clinical Neurosurgery (weekly)
Neurosurgery (weekly)
Newsletter (weekly) 14842

Congress on Research in Dance (CORD), c/o Dept of Dance, SUNY at Brockport, Brockport, NY 14420
T: (716) 395-2590; Fax: (716) 395-5397
Founded: 1965; Members: 900
Gen Secr: Penelope Hanstein
Focus: Perf Arts
Periodical
Dance Research Journal (biennial) . . . 14843

Conseil International d'Etudes Francophones (CIEF), c/o Dept of French, Montclair State College, Upper Montclair, NJ 07043
T: (201) 893-4283; Fax: (201) 666-3715
Members: 300
Gen Secr: Maurice Cagnon
Focus: Ling 14844

Conservation International (CI), 1015 18 St NW, Washington, DC 20036
T: (202) 429-5660; Tx: 9102409104; Fax: (202) 887-5188
Founded: 1987
Pres: Prof. Russell A. Mittermeier
Focus: Ecology 14845

Consortium for Graduate Study in Management (CGSM), 12855 N Outer Forty Rd, Saint Louis, MO 63141
T: (314) 935-6324; Fax: (314) 935-5014
Founded: 1967; Members: 9
Gen Secr: Wallace L. Jones
Focus: Business Admin 14846

Consortium for International Studies Education (CISE), 199 W Tenth Av, Columbus, OH 43201
T: (614) 422-1681
Founded: 1972; Members: 20
Gen Secr: Dr. James E. Harf
Focus: Educ 14847

Consortium of College and University Media Centers, c/o Media Resources Center, Iowa State University, 121 Pearson Hall, Ames, IA 50011
T: (515) 294-8022; Fax: (515) 294-8089
Founded: 1971; Members: 300
Gen Secr: Don Rieck
Focus: Cinema
Periodicals
CUFC Leader (weekly)
Educational Film Locator 14848

Consortium on Peace Research, Education and Development (COPRED), c/o George Mason University, 4400 University Dr, Fairfax, VA 22030
T: (703) 993-3639
Founded: 1970; Members: 800
Gen Secr: Marie A. Dugan
Focus: Poli Sci
Periodicals
Directory of Peace Studies Programs
Peace Chronicle (weekly) 14849

Construction Specifications Institute (CSI), 601 Madison St, Alexandria, VA 22314-1791
T: (703) 684-0300; Fax: (703) 684-0465
Founded: 1948; Members: 17000
Gen Secr: Joseph A. Gascoigne
Focus: Civil Eng
Periodical
The Construction Specifier (weekly) . . . 14850

Construction Writers Association (CWA), c/o M. McIntyre, Box 30, Aldie, VA 22001
T: (703) 771-4133
Founded: 1957; Members: 150
Gen Secr: M. McIntyre
Focus: Lit
Periodical
Newsletter (weekly) 14851

Consultative Group on International Agricultural Research (CGIAR), 1818 H St NW, Washington, DC DC 20433
T: (202) 473-8951; Tx: 82987; Fax: (202) 334-8750
Founded: 1971; Members: 26
Gen Secr: Alexander von der Osten
Focus: Agri 14852

Consumer Education Research Center (CERC), 350 Scotland Rd, Orange, NJ 07050
T: (201) 676-6663; Fax: (201) 676-3241
Founded: 1969; Members: 18169
Gen Secr: Robert L Berko
Focus: Educ; Econ
Periodical
Caveat Emptor (weekly) 14853

Consumer Energy Council of America Research Foundation (CECA/RF), 2000 L St NW, Ste 802, Washington, DC 20036
T: (202) 659-0404; Fax: (202) 659-0407
Founded: 1973; Members: 70
Gen Secr: Ellen Berman
Focus: Energy 14854

Contact Literacy Center (CLC), POB 81826, Linoln, NE 86501
T: (402) 464-0602; Fax: (402) 464-5931
Founded: 1978
Gen Secr: Emily Herrick
Focus: Lit 14855

Continuing Library Education Network and Exchange Round Table (CLENERT), c/o American Library Association, 50 E Huron St, Chicago, IL 60611
T: (312) 280-4278; Fax: (312) 280-3256
Founded: 1984; Members: 378
Gen Secr: Margaret Myers
Focus: Libraries & Bk Sci 14856

Convention of American Instructors of the Deaf (CAID), c/o Dr. Stephanie Polowe, POB 9887, LBJ 2264, Rochester, NY 14623-0887
T: (716) 475-6201
Founded: 1850; Members: 1200
Pres: Dr. Stephanie Polowe
Focus: Educ Handic
Periodicals
American Annals of the Deaf (5 times annually)
Newsletter (weekly)
Proceedings of the Convention of American Instructors of the Deaf 14857

Cooling Tower Institute (CTI), POB 73383, Houston, TX 77273
T: (713) 583-4087; Fax: (713) 537-1721
Founded: 1950; Members: 400
Gen Secr: Dorothy Garrison
Focus: Eng
Periodicals
CTI News (weekly)
Journal of the CTI (weekly) 14858

Cooper Ornithological Society (COS), c/o Martin L. Morton, Biology Dept, Occidental College, Los Angeles, CA 90041
T: (310) 259-2674
Founded: 1893; Members: 2200
Gen Secr: Martin L. Morton
Focus: Ornithology
Periodicals
Condor (weekly)
The Flock (every 2-3 years) 14859

Coordinating Committee on Women in the Historical Profession (CCWHP/CGWH) / Conference Group on Women's History, c/o Barbara Winslow, 124 Park Pl, Brooklyn, NY 11217
T: (718) 638-3227; Fax: (718) 499-7595
Founded: 1969; Members: 800
Gen Secr: Barbara Winslow
Focus: Hist
Periodical
CCWHP/CGWH Newsletter (weekly) . . . 14860

Coordinating Research Council (CRC), 219 Perimeter Center Pkwy, Ste 400, Atlanta, GA 30346
T: (404) 396-3400
Founded: 1942; Members: 1000
Gen Secr: A.E. Zengel
Focus: Eng 14861

Copyright Society of the U.S.A. (CSUSA), c/o School of Law, Columbia University, 435 W 116 St, New York, NY 10017
T: (212) 854-7696; Fax: (212) 854-7946
Founded: 1953; Members: 1000
Pres: Walter J. Josiah
Focus: Law
Periodical
Journal (weekly) 14862

Coronary Club (CC), 9500 Euclid Av, Cleveland, OH 44106
T: (216) 444-3690
Founded: 1969; Members: 9000
Gen Secr: Kathryn E. Ryan-Muldoon
Focus: Cardiol 14863

Corporate Data Exchange (CDE), 255 Broadway, Ste 2625, New York, NY 10007
T: (212) 962-2980
Founded: 1975
Pres: Michael Locker
Focus: Econ 14864

Correctional Education Association (CEA), 8025 Laurel Lakes Center, Laurel, MD 20707
T: (301) 490-1440; Fax: (301) 206-5061
Founded: 1945; Members: 3200
Gen Secr: Dr. Stephen J. Steurer
Focus: Educ
Periodicals
Journal of Correctional Education (weekly)
Newsletter (weekly)
Yearbook 14865

Costume Society of America (CSA), 55 Edgewater Dr, POB 73, Earleville, MD 21919-0073
T: (410) 275-2329; Fax: (410) 275-8936
Founded: 1973; Members: 1605
Pres: Inez Brooks-Myers
Focus: Textiles 14866

Council for Advancement and Support of Education (CASE), 11 Dupont Circle NW, Ste 400, Washington, DC 20036
T: (202) 328-5900; Fax: (202) 387-4973
Founded: 1974; Members: 2800
Pres: Peter Buchanan
Focus: Educ 14867

Council for Agricultural Science and Technology (CAST), 137 Lynn Av, Ames, IA 50010-7197
T: (515) 292-2125
Founded: 1972; Members: 4000
Gen Secr: Dr. Stanley P. Wilson
Focus: Agri
Periodicals
Comments
News from CAST (weekly)
Papers
Reports
Science of Food and Agriculture (weekly)
Special Publications 14868

Council for American Private Education (CAPE), 1726 M St NW, Ste 1102, Washington, DC 20036
T: (202) 659-0016; Fax: (202) 659-0018
Founded: 1971; Members: 14
Gen Secr: Joyce G. McCray
Focus: Educ
Periodicals
Directory of Private Schools
Outlook (weekly) 14869

Council for Basic Education (CBE), 725 15 St NW, Washington, DC 20005
T: (202) 347-4171; Fax: (202) 347-5047
Founded: 1956; Members: 4000
Gen Secr: A. Graham Down
Focus: Educ
Periodicals
Basic Education (10 times annually)
Basic Education: Issues, Answers and Facts (weekly)
Independent Study in the Humanities: Directory of Fellows (weekly) 14870

Council for Biomedical Communications Associations, c/o Stewart White, Biomedical Communications, University of Michigan, 1327 Jones Dr, Ste 104, Ann Arbor, MI 48105
T: (313) 998-6140; Fax: (313) 998-6150
Founded: 1970; Members: 5
Gen Secr: Stewart White
Focus: Comm Sci 14871

Council for Children with Behavioral Disorders (CCBD), c/o Council for Exceptional Children, 1920 Association Dr, Reston, VA 22091-1589
T: (703) 620-3660; Fax: (703) 264-9494
Founded: 1962; Members: 8500
Pres: Dr. Douglas Cullinan
Focus: Educ Handic; Behav Sci
Periodicals
Behavioral Disorders (weekly)
CCBD Newsletter (weekly) 14872

Council for Educational Development and Research (CEDaR), 200 L St NW, Ste 601, Washington, DC 20036
T: (202) 223-1593
Founded: 1971; Members: 14
Gen Secr: Dena Stoner
Focus: Educ
Periodicals
Directory (weekly)
Newsletter (weekly)
R & D Preview (weekly) 14873

Council for Elementary Science International (CESI), 11 Marion Rd, Westport, CT 06880
T: (203) 226-4938
Founded: 1920; Members: 1600
Pres: Eileen Bingtson
Focus: Sci
Periodicals
CESI Directory (weekly)
CESI News (weekly)
CESI Source Books (1-2 times annually)
Monograph (weekly) 14874

Council for European Studies (CES), c/o Columbia University, Schermerhorn Hall, Box 44, New York, NY 10027
T: (212) 854-4172; Fax: (212) 749-0397
Founded: 1970; Members: 830
Gen Secr: Dr. Ioannis Sinanoglou
Focus: Soc Sci; Hist
Periodical
European Studies Newsletter (weekly) . . 14875

Council for Indian Education (CIE), 517 Rimrock Rd, Billings, MT 59102
T: (406) 252-7451
Founded: 1970; Members: 100
Pres: Dr. H. Gilliland
Focus: Educ 14876

Council for Interinstitutional Leadership (CIL), c/o Fred Baus, 8016 State Line Rd, Ste 205, Leawood, KS 66208
T: (913) 341-4141; Fax: (913) 341-4141
Founded: 1975; Members: 50
Gen Secr: Fred Baus
Focus: Educ
Periodicals
Consortium Directory (biennial)
Newsletter (3 times annually) 14877

Council for Jewish Education (CJE), 426 W 58 St, New York, NY 10019
T: (212) 713-0290; Fax: (212) 586-9579
Founded: 1926; Members: 350
Gen Secr: Philip Gorodetzer
Focus: Educ
Periodicals
Jewish Education (weekly)
Membership Directory (biennial)
Sheviley Hahinuch (weekly) 14878

Council for Medical Affairs (CFMA), POB 10944, Chicago, IL 60610
T: (312) 464-4655; Fax: (312) 464-5830
Founded: 1980; Members: 15
Gen Secr: Ira Singer
Focus: Med 14879

Council for Museum Anthropology (CMA), c/o Thomas H. Wilson, Center for African Art, 54 E 68 St, New York, NY 10021
T: (212) 861-1200
Members: 350
Pres: Thomas H. Wilson
Focus: Anthro
Periodical
Museum Anthropology (weekly) 14880

Council for Philosophical Studies (CPS), c/o Jules Coleman, Yale Law School, 127 Wall, New Haven, CT 06520
T: (203) 432-4842; Fax: (203) 432-8260
Founded: 1965; Members: 15
Gen Secr: Jules Coleman
Focus: Philos 14881

Council for Religion in Independent Schools (CRIS), POB 40613, Washington, DC 20016
T: (202) 342-1661
Founded: 1898; Members: 450
Gen Secr: Daniel R. Heischman
Focus: Educ; Rel & Theol
Periodical
Newsletter (weekly) 14882

Council for Research in Music Education (CRME), c/o School of Music, University of Illinois, 1114 W Nevada, Urbana, IL 61801
T: (217) 333-1027
Founded: 1963; Members: 150
Gen Secr: Marilyn P. Zimmerman
Focus: Music; Educ
Periodical
Bulletin (weekly) 14883

Council for Research in Values and Philosophy (RVP), POB 261, Cardinal Station, Washington, DC 20064
T: (202) 319-5636; Fax: (202) 319-6089
Founded: 1983
Gen Secr: Prof. George F. McLean
Focus: Philos 14884

Council for Sex Information and Education (CSIE), 2272 Colorado Blvd, Los Angeles, CA 90041
Founded: 1977
Gen Secr: Joel Adams
Focus: Public Health 14885

Council for the Advancement of Science Writing (CASW), POB 404, Greenlawn, NY 11740
T: (516) 757-5664; Fax: (516) 757-0069
Founded: 1959; Members: 26
Gen Secr: Ben Patrusky
Focus: Sci 14886

Council for Tobacco Research – U.S.A. (CTR-USA), 900 Third Av, New York, NY 10022
T: (212) 421-8885
Founded: 1954
Pres: R.F. Gertenbach
Focus: Agri
Periodical
Report (weekly) 14887

Council of 1890 College Presidents (CCP), c/o Langston University, Langston, OK 73050
T: (405) 466-3202; Fax: (405) 466-3461
Founded: 1913; Members: 18
Gen Secr: Dr. William B. DeLauder
Focus: Educ 14888

Council of Administrators of Special Education (CASE), 615 16 St NW, Albuquerque, NM 87104
T: (505) 243-7622; Fax: (505) 247-4822
Founded: 1951; Members: 4700
Gen Secr: Dr. Jo Thomason
Focus: Educ
Periodicals
CASE in Point (3 times annually)
Newsletter (5 times annually) 14889

Council of Biology Editors (CBE), Ste 200, 111 E Wacker Dr, Chicago, IL 60601-4298
T: (312) 616-0800
Founded: 1957; Members: 1000
Gen Secr: Cindy Clark
Focus: Bio
Periodical
CBE Views (weekly) 14890

Council of Chief State School Officers (CCSSO), 1 Massachusetts Av NW, Ste 700, Washington, DC 20001
T: (202) 408-5505; Fax: (202) 393-1228
Founded: 1927; Members: 57
Gen Secr: Gordon M. Ambach
Focus: Educ
Periodicals
Directory of State Education Agencies (weekly)
Membership List (weekly)
State Education Indicators Report (weekly) 14891

Council of Colleges of Arts and Sciences (CCAS), c/o Ohio State University, 186 University Hall, 230 N Oval Mall, Columbus, OH 43210
T: (614) 292-1882; Fax: (614) 292-8666
Founded: 1965; Members: 375
Gen Secr: Richard J. Hopkins
Focus: Arts; Sci
Periodicals
CCAS Newsletter (weekly)
Membership Directory (weekly) 14892

Council of Community Blood Centers (CCBC), 725 15 St NW, Ste 700, Washington, DC 20005
T: (202) 393-5725; Fax: (202) 393-1282
Founded: 1962; Members: 51
Gen Secr: James L. MacPherson
Focus: Med
Periodical
CCBC Newsletter (weekly) 14893

Council of Consulting Organizations (IMC), 521 Fifth Av, New York, NY 10175
T: (212) 697-9693
Founded: 1968; Members: 2200
Gen Secr: Bruce MacGregor
Focus: Business Admin
Periodicals
ACME Survey of Key Management Information (weekly)
Directory of Members (weekly) 14894

Council of Engineering and Scientific Society Executives (CESSE), 2000 Florida Av NW, Washington, DC 20009
T: (212) 549-8141
Founded: 1949; Members: 600
Pres: William R. Martin
Focus: Eng; Sci
Periodicals
The Quill (3 times annually)
Yearbook (weekly) 14895

Council of Graduate Schools (CGS), 1 Dupont Circle NW, Ste 430, Washington, DC 20036
T: (202) 223-3791
Founded: 1960; Members: 400
Pres: Jules B. Lapidus
Focus: Educ
Periodicals
Brochure (weekly)
Newsletter (weekly) 14896

Council of International Programs (CIP), 500 23 St NW, Ste B902, Washington, DC 20037
T: (202) 429-4351; Fax: (202) 429-4354
Founded: 1956; Members: 7000
Gen Secr: Sheldon Siegel
Focus: Sociology
Periodicals
Directory of Former Participants
Newsletter 14897

Council of Landscape Architectural Registration Boards (CLARB), 12700 Fair Lakes Cir, Ste 110, Fairfax, VA 22033
T: (703) 818-1300; Fax: (703) 818-1309
Founded: 1961; Members: 40
Gen Secr: Clarence L. Chaffee
Focus: Archit 14898

Council of Mennonite Colleges (CMC), c/o L. Balzer, Tabor College, Hillsboro, KS 67063
T: (316) 947-3121; Fax: (316) 947-2607
Founded: 1942; Members: 11
Pres: L. Balzer
Focus: Educ 14899

Council of National Library and Information Associations (CNLIA), 1700 18 St NW, Ste B-1, Washington, DC 20009
Founded: 1942; Members: 20
Pres: Kathleen Haefligger
Focus: Libraries & Bk Sci
Periodicals
Roster (weekly)
Update (weekly) 14900

Council of Planning Librarians (CPL), 1313 E 60 St, Chicago, IL 60637
T: (312) 947-2163
Founded: 1957; Members: 120
Pres: M. Kay Mowery
Focus: Libraries & Bk Sci; Public Admin; Urban Plan
Periodicals
Bibliographies (weekly)
CPL Newsletter (weekly)
Directory of Planning Libraries
Membership Directory (weekly) 14901

Council of Scientific Society Presidents (CSSP), 1155 16 St NW, Ste 100, Washington, DC 20036
T: (202) 872-4452; Fax: (202) 872-4615
Founded: 1973; Members: 58
Gen Secr: Dr. John D. Holmfeld
Focus: Sci; Educ 14902

Council of Societies for the Study of Religion (CSSR), c/o Valparaiso University, Valparaiso, IN 46383
T: (219) 464-5515; Fax: (219) 464-6714
Founded: 1969
Gen Secr: David G. Truemper
Focus: Rel & Theol
Periodicals
CSSR Bulletin (weekly)
Religious Studies Review (weekly) . . . 14903

Council of Teaching Hospitals (COTH), 2450 N St NW, Washington, DC 20037
T: (202) 828-1125
Founded: 1965; Members: 400
Gen Secr: Robert M. Dickler
Focus: Med; Adult Educ
Periodicals
COTH Report (weekly)
Directory of Educational Services and Programs (weekly) 14904

Council of the Great City Schools (CGCS), 1413 K St NW, Ste 400, Washington, DC 20005
T: (202) 371-0163
Founded: 1961; Members: 45
Gen Secr: Michael D. Casserly
Focus: Educ 14905

Council of Undergraduate Research (CUR), c/o John Stevens, University of North Carolina, Asheville, NC 28804
T: (704) 251-6006
Founded: 1978; Members: 1800
Gen Secr: John Stevens
Focus: Educ 14906

Council on Accreditation of Nurse Anesthesia Educational Programs/Schools, 216 Higgins Rd, Park Ridge, IL 60068-5790
T: (708) 692-7050; Fax: (708) 692-6968
Founded: 1975; Members: 100
Gen Secr: Betty Horton
Focus: Educ; Anesthetics 14907

Council on Anthropology and Education (CAE), 1703 New Hampshire Av NW, Washington, DC 20009
T: (202) 232-8800
Founded: 1968; Members: 921
Pres: David Fetterman
Focus: Anthro; Educ
Periodical
Anthropology and Education (weekly) . . 14908

Council on Arteriosclerosis of the American Heart Association (CAAHA), 7320 Greenville Av, Dallas, TX 75231
T: (214) 706-1293; Fax: (214) 706-1341
Founded: 1946; Members: 1014
Gen Secr: Richard W. Saint Clair
Focus: Cardiol
Periodicals
Arteriosclerosis (weekly)
Newsletter (weekly) 14909

Council on Chiropractic Education (CCE), 4401 Westown Pkwy, Ste 120, West Des Moines, IA 50266
T: (515) 226-9001; Fax: (515) 226-9031
Founded: 1971; Members: 23
Gen Secr: Dr. Ralph G. Miller
Focus: Educ; Med 14910

Council on Chiropractic Orthopedics (CCO), 190 E 100, Provo, UT 84604
T: (801) 373-2240
Founded: 1967; Members: 891
Gen Secr: Phil L. Alken
Focus: Orthopedics
Periodicals
Directory (weekly)
Orthopedic Briefs (weekly) 14911

Council on Chiropractic Physiological Therapeutics (CCPT), c/o Charles Brandstetter, 203 N Holmes Av, Idaho Falls, ID 83401
T: (208) 522-2591
Founded: 1920; Members: 200
Pres: Charles Brandstetter
Focus: Therapeutics 14912

Council on Clinical Optometric Care (CCOC), 243 N Lindbergh Blvd, Saint Louis, MO 63141
T: (314) 991-4100
Founded: 1967; Members: 10
Gen Secr: Barbara Nicholson
Focus: Ophthal 14913

Council on Diagnostic Imaging, POB 1655, Ashtabula, OH 44004
T: (216) 466-5881
Founded: 1936; Members: 2000
Gen Secr: Clark L. McClain
Focus: X-Ray Tech
Periodical
Roentgenological Briefs (weekly) 14914

Council on Education for Public Health (CEPH), 1015 15 St NW, Washington, DC 20005
T: (202) 789-1050
Founded: 1974; Members: 42
Gen Secr: Patricia P. Evans
Focus: Public Health 14915

Council on Education of the Deaf (CED), 800 Florida Av NE, Washington, DC 20002
T: (202) 651-5508; Fax: (202) 651-5508
Founded: 1960; Members: 12
Gen Secr: Don Hicks
Focus: Educ Handic
Periodicals
CED Approved Programs (weekly)
Newsletter (weekly)
Standards for the Certification of Teachers of the Hearing Impaired
Standards for the Evaluation of Programs for the Preparation of Teachers for the Hearing Impaired 14916

Council on Electrolysis Education (CEE), 46 S Holmes St, Memphis, TN 38111
T: (901) 458-1431
Founded: 1972
Pres: Dorothy Graves
Focus: Educ 14917

Council on Governmental Relations (COGR), 1 Dupont Circle NW, Ste 425, Washington, DC 20036
T: (202) 331-1803; Fax: (202) 331-8483
Founded: 1948; Members: 135
Gen Secr: Milton Goldberg
Focus: Poli Sci 14918

Council on Health Information and Education (CHIE), 2272 Colorado Blvd, Los Angeles, CA 90041
Founded: 1978
Gen Secr: D. Andre
Focus: Public Health 14919

Council on Hotel, Restaurant and Institutional Education (CHRIE), 1200 17 St NW, Washington, DC 20036-3097
T: (202) 331-5990; Fax: (202) 785-2511
Founded: 1946; Members: 2200
Gen Secr: Douglas E. Adair
Focus: Adult Educ
Periodicals
CHRIE Communique (weekly)
Hospitality Educator (weekly)
Journal of Hospitality Education (3 times annually) 14920

Council on Interracial Books for Children (CIBC), 1841 Broadway, Rm 608, New York, NY 10023
T: (212) 757-5339
Founded: 1965
Focus: Libraries & Bk Sci
Periodical
Bulletin (8 times annually) 14921

Council on Legal Education Opportunity (CLEO), 1800 M St NW, Ste 160, Washington, DC 20036
T: (202) 785-4840
Founded: 1968
Gen Secr: Denise W. Purdie
Focus: Adult Educ; Law 14922

Council on Library-Media Technical Assistants (COLT), c/o Margaret Barron, Library-Media Technology Dept, Cyahoga Community College, 2900 Community College Av, Cleveland, OH 44115
T: (216) 987-4296; Fax: (216) 987-4404
Founded: 1965; Members: 700
Gen Secr: Margaret Barron
Focus: Libraries & Bk Sci
Periodicals
Membership Directory and Data Book (biennial)
Newsletter (weekly) 14923

Council on Library Resources (CLR), 1785 Massachusetts Av NW, Ste 313, Washington, DC 20036
T: (202) 483-7474; Fax: (202) 483-6410
Founded: 1956; Members: 11
Pres: W. David Penniman
Focus: Libraries & Bk Sci
Periodicals
Annual Report (weekly)
Recent Developments (weekly) 14924

Council on Medical Education of the American Medical Association (CME-AMA), 515 N State St, Chicago, IL 60610
T: (312) 464-4804; Fax: (312) 464-5830
Founded: 1847; Members: 12
Gen Secr: Carlos J.M. Martin
Focus: Adult Educ; Med
Periodicals
Allied Health Education Directory (weekly)
Directory of Residency Training Programs (weekly) 14925

Council on National Literatures (CNL), POB 81, Whitestone, NY 11357
T: (212) 767-8380; Fax: (212) 767-8380
Founded: 1974; Members: 1200
Pres: Dr. Anne Paolucci
Focus: Lit
Periodicals
CNL/Review of Books (weekly)
CNL/World Report (weekly) 14926

Council on Nutritional Anthropology (CNA), 1703 New Hampshire Av NW, Washington, DC 20009
T: (202) 232-8800
Founded: 1973; Members: 328
Pres: Dr. Roberta Baer
Focus: Anthro 14927

Council on Peace Research in History (CPRH), c/o Melvin Small, Dept of History, Wayne State University, Detroit, MI 48202
T: (313) 577-6138
Founded: 1963; Members: 250
Pres: Melvin Small
Focus: Hist; Prom Peace 14928

Council on Podiatric Medical Education (CPME), 9312 Old Georgetown Rd, Bethesda, MD 20814-1621
T: (301) 571-9200; Fax: (301) 530-2752
Founded: 1918; Members: 11
Gen Secr: Jay Levrio
Focus: Adult Educ 14929

Council on Postsecondary Accreditation (COPA), 1 Dupont Circle NW, Ste 305, Washington, DC 20036
T: (202) 452-1433; Fax: (202) 331-9571
Founded: 1975
Pres: Kenneth L. Perrin
Focus: Educ
Periodicals
Accreditation (weekly)
Balance Wheel for Accreditation (weekly)
President's Bulletin (8 times annually) . . 14930

Council on Resident Education in Obstetrics and Gynecology (CREOG), 409 12 St SW, Washington, DC 20024
T: (202) 863-2554; Fax: (202) 484-5107
Founded: 1967; Members: 450
Gen Secr: D. Nehra
Focus: Gynecology 14931

Council on Social Work Education (CSWE), 1600 Duke St, Ste 300, Alexandria, VA 22314
T: (703) 683-8080; Fax: (703) 683-8099
Founded: 1952; Members: 5000
Gen Secr: Donald W. Beless
Focus: Adult Educ; Sociology
Periodicals
Journal of Education for Social Work (3 times annually)
Social Work Education Reporter (3 times annually)
Statistics on Social Work Education in the United States (weekly)
Summary Information on Master of Social Work Programs (weekly) 14932

Council on Tall Buildings and Urban Habitat (CTBUH), c/o Fritz Engineering Laboratory, Lehigh University, 13 Packer Av, Bethlehem, PA 18015
T: (215) 758-3515; Fax: (215) 758-4522
Founded: 1969; Members: 1200
Gen Secr: Dr. Lynn S. Beedle
Focus: Archit
Periodicals
Brochure (weekly)
The Times (3-4 times annually) 14933

Council on Technology Teacher Education (CTTE), c/o Gerald Jennings, College of Technology, Eastern Michigan University, Ypsilanti, MI 48197
T: (313) 487-4330
Founded: 1950; Members: 900
Gen Secr: Gerald Jennings
Focus: Adult Educ; Arts
Periodicals
Membership Directory (weekly)
Newsletter (weekly) 14934

Country Day School Headmasters Association of the U.S. (CDSHA), c/o Charlotte Country Day School, 1440 Carmel Rd, Charlotte, NC 28226
T: (704) 366-1241; Fax: (704) 364-4110
Founded: 1912; Members: 100
Pres: M. Gragg
Focus: Educ 14935

The Cousteau Society (TCS), 870 Greenbriar Cir, Ste 402, Chesapeake, VA 23320
T: (804) 523-9335; Fax: (804) 523-2747
Founded: 1973; Members: 300000
Gen Secr: Jacques-Yves Cousteau
Focus: Ecology
Periodicals
Calypso Log (weekly)
Dolphin Log (weekly) 14936

Cranial Academy (CA), 3500 Depauw Blvd, Indianapolis, IN 46268-1136
T: (317) 879-0713; Fax: (317) 879-0718
Founded: 1946; Members: 800
Gen Secr: Patricia Crampton
Focus: Med
Periodicals
Directory (weekly)
News Letter (weekly) 14937

Creation Research Society (CRS), POB 28473, Kansas City, MO 64118
T: (816) 746-5300
Founded: 1963; Members: 1800
Gen Secr: Glen Wolfrom
Focus: Sci
Periodical
Creation Research Society Quarterly (weekly) 14938

Creative Time (CT), 131 W 24 St, New York, NY 10011-1942
T: (212) 206-6674; Fax: (212) 571-2206
Founded: 1973
Gen Secr: C. Scott Brown
Focus: Arts 14939

Creek Indian Memorial Association (CIMA), c/o Creek Council House Museum, Town Sq, Okmulgee, OK 74447
T: (918) 756-2324
Founded: 1923; Members: 113
Gen Secr: Tommy A. Steinsiek
Focus: Hist 14940

Critical Mass Energy Project of Public Citizen (CMEPPC), 215 Pennsylvania Av SE, Washington, DC 20003
T: (202) 546-4996; Fax: (202) 547-7392
Founded: 1974
Gen Secr: Jonatham Becker
Focus: Energy
Periodical
Critical Mass Energy Bulletin (weekly) . 14941

Crop Science Society of America (CSSA), 677 S Segoe Rd, Madison, WI 53711
T: (608) 273-8080; Fax: (608) 273-2021
Founded: 1955; Members: 5500
Gen Secr: Robert F. Barnes
Focus: Agri; Ecology
Periodicals
Agronomy Journal (weekly)
Crop Science (weekly)
Journal of Agronomic Education (weekly)
Journal of Environmental Quality (weekly)
Journal of Production Agriculture (weekly)
Soil Science Society of America Journal (weekly) 14942

Cross-Examination Debate Association (CEDA), c/o Dept of Speech, California State University at Northridge, Northridge, CA 91330
T: (818) 885-2633
Founded: 1971; Members: 320
Gen Secr: Mike Bartanen
Focus: Ling 14943

Cryogenic Engineering Conference (CEC), c/o Dr. P. Kittel, MS 244-10, NASA/AMES, Moffet Field, CA 94035
T: (415) 604-4297
Founded: 1954; Members: 4000
Gen Secr: Dr. P. Kittel
Focus: Eng
Periodical
Advances in Cryogenic Engineering (biennial) 14944

Cryogenic Society of America (CSA), c/o Laurie Huget, Huget Advertising, 1033 South Blvd, Oak Park, IL 60302
T: (708) 383-6220; Fax: (708) 383-9337
Founded: 1964; Members: 400
Gen Secr: Laurie Huget
Focus: Physics
Periodical
The Cold Facts (weekly) 14945

CUBE, 100 Maple Park Blvd, Ste 152, Saint Clair Shores, MI 48081
T: (313) 771-9300; Fax: (313) 771-9598
Founded: 1976; Members: 3000
Gen Secr: Mary Lou Regan
Focus: Computer & Info Sci
Periodical
Newsletter (weekly) 14946

Czechoslovak Society of Arts and Sciences (CSAS), c/o F. Marlow, 4064 Woodcliff Rd, Sherman Oaks, CA 91403
T: (818) 784-0970; Fax: (818) 788-7063
Founded: 1958; Members: 1500
Gen Secr: F. Marlow
Focus: Arts; Sci
Periodicals
Bulletin (3 times annually)
Kosmas (weekly)
Promeny (weekly)
Zpravy (weekly) 14947

Dairy Research Foundation (DRINC), 95 King St, Oak Grove Village, IL 60007
T: (708) 228-7742
Founded: 1969; Members: 350000
Focus: Dairy Sci
Periodical
Dairy Research Digest (weekly) 14948

Dairy Society International (DSI), c/o George W. Weigold, 7185 Ruritan Dr, Chambersburg, PA 17201
T: (717) 375-4392
Founded: 1946; Members: 500
Gen Secr: George W. Weigold
Focus: Dairy Sci
Periodicals
Bulletin
Market Frontier News (weekly)
Report to Members 14949

Dalcroze Society of America (DSA), c/o Dr. Julia Schnebly-Black, University of Washington, 3871 45 Av NE, Seattle, WA 98105
T: (206) 527-7034
Founded: 1967
Pres: Dr. Julia Schnebly-Black
Focus: Music 14950

Damien Dutton Society for Leprosy Aid (DDSLA), 616 Bedford Av, Bellmore, NY 11710
T: (516) 221-5829
Founded: 1944; Members: 30000
Pres: Howard E. Crouch
Focus: Derm
Periodical
Damien-Dutton Call (weekly) 14951

Dana Center for Preventive Ophthalmology (DCPO), c/o Wilmer Eye Institute, Room 120, Johns Hopkins University, 600 N Wolfe St, Baltimore, MD 21205
T: (301) 955-2777; Tx: 5106003055; Fax: (301) 955-2542
Founded: 1979
Gen Secr: Dr. Harry Quigley
Focus: Ophthal 14952

Dance Films Association (DFA), 1133 Broadway, Rm 507, New York, NY 10010
T: (212) 727-0764
Founded: 1956
Pres: Susan Braun
Focus: Cinema
Periodical
Dance on Camera News (weekly) . . . 14953

Dannemiller Memorial Educational Foundation (DMEF), 12500 Network Blvd, Ste 101, San Antonio, TX 78249-3302
T: (210) 641-8311
Founded: 1971; Members: 30
Gen Secr: Bill Vaughn
Focus: Anesthetics
Periodical
Anesthesia File (weekly) 14954

Dante Society of America (DSA), 61 Kirkland St, Cambridge, MA 02138
T: (617) 495-0738; Fax: (617) 495-0730
Founded: 1881; Members: 450
Gen Secr: Richard H. Lansing
Focus: Lit
Periodical
Dante Studies (weekly) 14955

Deafness Research Foundation (DRF), 9 E 38 St, New York, NY 10016
T: (212) 684-6556; Fax: (212) 779-2125
Founded: 1958; Members: 2400
Gen Secr: M.H. Jacoby
Focus: Otorhinolaryngology 14956

Decision Sciences Institute (DSI), University Plaza, Atlanta, GA 30303
T: (404) 651-4000
Founded: 1969; Members: 5500
Gen Secr: Carol J. Latta
Focus: Econ
Periodicals
Decision Line (5 times annually)
Decision Sciences (weekly) 14957

Delta Dental Plans Association (DDPA), 211 E Chicago Av, Ste 800, Chicago, IL 60611
T: (312) 337-4707
Founded: 1965; Members: 47
Gen Secr: James E. Bonk
Focus: Dent
Periodicals
Delta Dictum (weekly)
Dozzle (weekly)
Newsletter (weekly) 14958

Deltiologists of America (D. of A.), POB 8, Norwood, PA 19074
T: (215) 485-8572
Founded: 1966; Members: 800
Pres: James L. Lowe
Focus: Hist; Arts
Periodical
Postcard Classics (weekly) 14959

Dental Assisting National Board (DANB), 216 E Ontario St, Chicago, IL 60611
T: (312) 642-3368
Founded: 1948
Gen Secr: Richard D. Hengl
Focus: Dent 14960

Dental Health International (DHI), 847 S Milledge Av, Athens, GA 30605
T: (404) 546-1715
Founded: 1973
Pres: Barry Simmons
Focus: Dent 14961

Dermatology Foundation (DF), 1566 Sherman Av, Ste 302, Evanston, IL 60201-4802
T: (312) 328-2256
Founded: 1964; Members: 3300
Gen Secr: Sandra Rahn Goldman
Focus: Derm 14962

DES Action USA, 1615 Broadway, Ste 510, Oakland, CA 94617
T: (510) 465-4011; Fax: (510) 465-4815
Founded: 1977
Pres: Andrea Goldstein
Focus: Cell Biol & Cancer Res
Periodical
DES Action Voice (weekly) 14963

Desert Botanical Garden (DBG), 1201 N Galvin Pkwy, Phoenix, AZ 85008
T: (602) 941-1225; Fax: (602) 991-8933
Founded: 1937; Members: 5500
Gen Secr: Robert G. Breunig
Focus: Botany; Ecology
Periodicals
Agave (weekly)
Saguaroland Bulletin (weekly) 14964

Desert Fishes Council, POB 337, Bishop, CA 93515
T: (619) 872-8751; Fax: (619) 872-8751
Founded: 1970; Members: 500
Gen Secr: E.P. Pister
Focus: Ecology
Periodical
Proceedings of the Desert Fishes Council (weekly) 14965

Desert Protective Council (DPC), POB 2312, Valley Center, CA 92082-2312
T: (619) 397-4264
Founded: 1954; Members: 400
Focus: Agri; Ecology
Periodical
El Paisano (weekly) 14966

Desert Tortoise Council (DTC), POB 1738, Palm Desert, CA 92261-1738
T: (619) 341-8449
Founded: 1975; Members: 175
Gen Secr: T. Correll
Focus: Zoology
Periodical
Proceedings of Symposium (weekly) . . 14967

Design/Build Institute, 5186 Lynn Rd, Greendale, WI 53129
T: (414) 421-4447
Founded: 1983; Members: 200
Gen Secr: Carl W. Bauman
Focus: Archit
Periodical
Design/Build Digest 14968

Designs for Change (DFC), 220 S State St, Ste 1900, Chicago, IL 60604
T: (312) 922-0317; Fax: (312) 922-0317
Founded: 1977
Gen Secr: Dr. Donald Moore
Focus: Educ 14969

D.H. Lawrence Society of North America, c/o Lydia Blanchard, Dept of English, Southwest Texas State University, San Marcos, TX 7866
T: (512) 245-2111
Founded: 1975; Members: 200
Pres: Lydia Blanchard
Focus: Lit 14970

Dickens Society (DS), c/o Dept of Humanities, Worcester Polytechnic Institute, Worcester, MA 01609-2280
T: (508) 831-5572; Fax: (508) 831-5483
Founded: 1970; Members: 500
Gen Secr: Joel J. Brattin
Focus: Lit
Periodical
Dickens Quarterly (weekly) 14971

Dictionary Society of North America (DSNA), c/o Prof. Louis T. Milic, 1983 E 24 St, Cleveland, OH 44115-2403
T: (216) 687-4830
Founded: 1975; Members: 550
Gen Secr: Prof. Louis T. Milic
Focus: Libraries & Bk Sci 14972

Dietary Managers Association (DMA), 400 E 22 St, Lombard, IL 60148
T: (708) 932-1444
Founded: 1960; Members: 12000
Gen Secr: William S. Saint John
Focus: Nutrition; Educ
Periodicals
Flyer (weekly)
Issues (weekly) 14973

Direct Marketing Educational Foundation (DMEF), 6 E 43 St, New York, NY 10017-4646
T: (212) 768-7277; Fax: (212) 599-1268
Founded: 1965; Members: 500
Pres: Dr. Richard L. Montesi
Focus: Educ; Business Admin 14974

Disability Insurance Training Council (DITC), 1000 Connecticut Av NW, Ste 1111, Washington, DC 20036
T: (202) 223-5533; Fax: (202) 785-2274
Founded: 1951; Members: 11000
Gen Secr: William F. Flood
Focus: Adult Educ 14975

Disability Rights Center (DRC), 2500 Q St NW, Ste 121, Washington, DC 20007
T: (202) 337-4119
Founded: 1976
Gen Secr: Janine Bertram
Focus: Rehabil 14976

Distillers Feed Research Council (DFRC), POB 7805, Des Moines, IA 50322
T: (515) 243-0757
Founded: 1947; Members: 19
Gen Secr: Robert H. Hatch
Focus: Food
Periodical
Proceedings (weekly) 14977

Distributive Education Clubs of America (DECA), 1908 Association Dr, Reston, VA 22091
T: (703) 860-5000
Founded: 1946; Members: 180000
Gen Secr: Edward Davis
Focus: Educ
Periodicals
Guide (weekly)
New Dimensions (weekly)
Newsletter (weekly) 14978

Division for Early Childhood, c/o Council for Exceptional Children, 1920 Association Dr, Reston, VA 22091
T: (703) 620-3660; Fax: (703) 264-9494
Founded: 1973; Members: 7000
Pres: Dr. Christine Salisbury
Focus: Educ Handic
Periodicals
The DEC Communicator (weekly)
Journal of Early Intervention (weekly) . . 14979

Division of Applied Experimental and Engineering Psychologists (DAEEP), c/o Jefferson Koonce, Institute of Aviation, University of Illinois at Willard Airport, Savoy, IL 61874
T: (217) 244-8601; Fax: (217) 244-8761
Founded: 1957; Members: 600
Gen Secr: Jefferson Koonce
Focus: Psych
Periodicals
Applied Experimental and Engineering Psychology (weekly)
Newsletter (weekly) 14980

Division on Mental Retardation of the Council for Exceptional Children (CEC-MR), c/o Dr. Dana M. Anderson, 245 Cedar Springs Dr, Athens, GA 30605
T: (703) 546-6132; Fax: (703) 542-2321
Founded: 1963; Members: 8120
Gen Secr: Dr. Dana M. Anderson
Focus: Educ Handic
Periodicals
CEC-M Report (biennial)
Education and Training in Mental Retardation (weekly) 14981

Django Reinhardt Society (DRS), 10 W Jackson Av, Middletown, NY 10940
T: (914) 342-1696
Founded: 1985; Members: 200
Pres: Mike Peters
Focus: Music 14982

Dozenal Society of America (DSA), c/o Fred Newhall, Dept of Mathematics, Nassau Community College, Garden City, NY 11530
T: (516) 669-0273
Founded: 1935; Members: 144
Pres: Fred Newhall
Focus: Math; Standards
Periodical
Duodecimal Bulletin (3 times annually) . 14983

Drama Tree (DT), 158 W 15 St, New York, NY 10011
T: (212) 620-0855
Founded: 1968; Members: 70
Gen Secr: Anthony Mannino
Focus: Perf Arts 14984

Drawing Society, 15 Penn Plaza, 415 Seventh Av, New York, NY 10001
T: (212) 563-4822; Fax: (212) 563-4829
Founded: 1959; Members: 1226
Pres: Paul Cummings
Focus: Fine Arts
Periodical
Drawing (weekly) 14985

Drug and Alcohol Nursing Association (DANA), 720 Light St, Baltimore, MD 21230
T: (410) 752-3318; Fax: (410) 752-8295
Founded: 1979; Members: 340
Gen Secr: J.A. Mellendick
Focus: Med 14986

Drug Information Association (DIA), POB 3113, Maple Glen, PA 19002
T: (215) 628-2288
Founded: 1965; Members: 6000
Gen Secr: Erich F. Lukas
Focus: Pharmacol
Periodicals
Journal (weekly)
Membership Directory (weekly) 14987

Ductile Iron Pipe Research Association (DIPRA), 245 Riverchase Pkwy E, Ste O, Birmingham, AL 35244
T: (205) 988-9870; Fax: (205) 988-9822
Founded: 1915; Members: 7
Pres: Troy F. Stroud
Focus: Eng 14988

Ductile Iron Society (DIS), 28938 Lorain Rd, Ste 202, North Olmsted, OH 44070
T: (216) 734-8040
Founded: 1958; Members: 80
Gen Secr: John V. Hall
Focus: Metallurgy
Periodicals
Ductile Iron News (3 times annually)
Ductile Technical Notes (3 times annually)
Membership Directory (weekly) 14989

The Duke Ellington Society (TDES), POB 31, Church St Station, New York, NY 10008-0031
T: (718) 463-0085
Founded: 1958; Members: 400
Pres: Morris Hodara
Focus: Music
Periodical
TDES Newsletter (10 times annually) . . 14990

Dyleague, 5900 Canoga Av, Ste 100, Woodland Hills, CA 91367
T: (818) 716-1616
Founded: 1992; Members: 600
Gen Secr: Phyllis Genszler
Focus: Computer & Info Sci
Periodical
Proceedings (weekly) 14991

Dynamics International Gardening Association (DIGA), Drawer 1165, Asheboro, NC 27204-1165
Founded: 1950; Members: 18230
Gen Secr: William A. Barnes
Focus: Hort
Periodicals
Garden Today News (weekly)
Tips 14992

Dystonia Medical Research Foundation (DMRF), 8383 Wilshire Blvd, Beverly Hills, CA 90211
T: (312) 755-0198
Founded: 1977; Members: 10000
Gen Secr: Nancy Harris
Focus: Neurology
Periodical
Newsletter (weekly) 14993

Dystrophic Epidermolysis Bullosa Research Association of America (DEBRA), 141 Fifth Av, Ste 7-S, New York, NY 10010
T: (212) 995-2220
Founded: 1979
Gen Secr: Miram Fedar
Focus: Derm 14994

EAA Aviation Foundation (EAAAF), POB 3065, Oshkosh, WI 54903-3065
T: (414) 426-4800
Founded: 1962
Pres: Tom Proberezny
Focus: Aero 14995

Earl Warren Legal Training Program (EWLTP), 99 Hudson St, Ste 1600, New York, NY 10013
T: (212) 219-1900
Founded: 1972
Gen Secr: G. Michael Bagley
Focus: Law 14996

Early Settlers Association of the Western Reserve (ESAWR), 24740 Antler Dr, North Olmsted, OH 44070
T: (216) 777-7088
Founded: 1879; Members: 500
Pres: Raymond J. Koch
Focus: Hist
Periodicals
Annals
The Pioneer (weekly)
Roster 14997

Early Sites Research Society (ESRS), c/o James Whittall, Long Hill, Rowley, MA 01969
T: (508) 948-2410; Fax: (508) 948-7270
Founded: 1973; Members: 225
Gen Secr: James Whittall
Focus: Archeol
Periodicals
Bulletin (weekly)
Newsletter (weekly) 14998

Earthmind, POB 743, Mariposa, CA 95338
T: (408) 336-5026
Founded: 1972
Pres: Michael A. Hackleman
Focus: Ecology

Periodical
Newsletter 14999

Earthquake Engineering Research Institute (EERI), 499 14 St, Ste 320, Oakland, CA 94612-1902
T: (510) 451-0905; Fax: (510) 451-5411
Founded: 1949; Members: 2250
Gen Secr: Susan K. Tubbesing
Focus: Eng; Geophys
Periodicals
Earthquake Spectra (weekly)
EERI Newsletter (weekly) 15000

Earthrise, c/o John Danner, 2120 E 59 Pl, Tulsa, OK 74106
T: (918) 742-5866
Founded: 1972; Members: 300
Gen Secr: John Danner
Focus: Futurology 15001

Eastern Bird Banding Association (EBBA), c/o Hannah B. Suthers, 4 View Point Dr, Hopewell, NJ 08525
T: (609) 466-1871
Founded: 1923; Members: 800
Gen Secr: Hannah B. Suthers
Focus: Ornithology
Periodical
North American Bird Bander (weekly) . . 15002

Eastern Finance Association (EFA), c/o Dr. Lon M. Carnes, Dept of Finance, Georgia Southern University, POB 8151, Statesboro, GA 30460
T: (912) 681-5437
Founded: 1965; Members: 1900
Gen Secr: Dr. Lon M. Carnes
Focus: Finance 15003

East-West Center (EWC), 1777 East-West Rd, Honolulu, HI 96848
T: (808) 944-7111; Fax: (808) 944-7970
Founded: 1960; Members: 2003
Focus: Poli Sci; Econ 15004

East-West Center Institute of Economic Development and Policy (IEDP), 1777 East-West Rd, Honolulu, HI 96848
T: (808) 944-7111; Tx: 989171; Fax: (808) 944-7970
Gen Secr: Dr. Seiji Naya
Focus: Poli Sci; Develop Areas
Periodicals
Asia-Pacific Report
Resource Systems Institute Bulletin . . . 15005

Ecological Society of America (ESA), c/o Center for Environmental Studies, Arizona State University, Tempe, AZ 85287-3211
T: (602) 965-3000
Founded: 1915; Members: 6100
Gen Secr: Duncan T. Patten
Focus: Ecology
Periodicals
Bulletin (weekly)
Ecological Monographs (weekly)
Ecology (weekly) 15006

Ecology Center (EC), 2530 San Pablo Av, Berkeley, CA 94702
T: (415) 548-2220
Founded: 1969; Members: 1500
Gen Secr: Robert Huang
Focus: Ecology
Periodical
Newsletter (weekly) 15007

EcoNet, c/o IGC, 18 De Boom St, San Francisco, CA 94107
T: (415) 442-0220; Tx: 154205417
Focus: Ecology 15008

Econometric Society (ES), c/o Dept of Economics, Northwestern University, Evanston, IL 60208-2600
T: (708) 491-3615
Founded: 1930; Members: 7000
Gen Secr: Julie P. Gordon
Focus: Stats; Math
Periodical
Econometrica (weekly) 15009

Economic Development Institute (EDI), 1818 H St NW, Washington, DC 20433
T: (202) 477-1234; Tx: 4400098; Fax: (202) 676-0858
Founded: 1955; Members: 149
Gen Secr: A. Golan
Focus: Develop Areas 15010

Economic History Association (EHA), c/o Dept of History, George Washington University, Washington, DC 20052
T: (202) 994-6052
Founded: 1941; Members: 3300
Gen Secr: William H. Becker
Focus: Hist; Econ 15011

ECRI, 5200 Butler Pike, Plymouth Meeting, PA 19462
T: (215) 825-6000; Fax: (215) 834-1275
Founded: 1955; Members: 2500
Pres: Joel J. Nobel
Focus: Med
Periodicals
Health Device Alerts (weekly)
Health Devices (weekly)
Health Devices Sourcebook (weekly)
Issues in Health Care Technology (weekly)
. . . . 15012

Edgar Allan Poe Society of Baltimore (EAPSB), c/o Carol Peirce, University of Baltimore, 1420 N Charles St, Baltimore, MD 21201
T: (410) 661-1180
Founded: 1923; Members: 350
Pres: Carol Peirce
Focus: Lit 15013

Edison Birthplace Association (EBA), c/o Edison Birthplace Museum, 9 Edison Dr, POB 451, Milan, OH 44846
T: (419) 499-2135
Founded: 1951; Members: 250
Pres: Robert K.L. Wheeler
Focus: Preserv Hist Monuments 15014

Edison Electric Institute (EEI), 701 Pennsylvania Av NW, Washington, DC 20004-2696
T: (202) 508-5000
Founded: 1933; Members: 190
Pres: Thomas R. Kuhn
Focus: Electric Eng
Periodicals
Electrical Report (weekly)
Rate Book (weekly)
Statistical Report (weekly)
Statistical Yearbook (weekly) 15015

Editorial Projects in Education (EPE), 4301 Connecticut Av NW, Ste 250, Washington, DC 20008
T: (202) 364-4114; Fax: (202) 364-1039
Founded: 1959
Gen Secr: Virginia Edwards
Focus: Journalism
Periodical
Education Week (40 times annually) . . 15016

Educational Center for Applied Ekistics (ECAE), 1900 DeKalb Av NE, Atlanta, GA 30307
T: (404) 378-2219
Founded: 1977
Gen Secr: Dr. Lorraine Wilson
Focus: Educ
Periodical
Ekistical Education (weekly) 15017

Educational Commission for Foreign Medical Graduates (ECFMG), 3624 Market St, Philadelphia, PA 19104
T: (215) 386-5900
Founded: 1956; Members: 20
Pres: Marjorie P. Wilson
Focus: Med; Adult Educ
Periodicals
ECFMG Annual Report (weekly)
ECFMG Information Booklet (weekly) . . 15018

Educational Foundation for the Fashion Industries (EFFI), 227 W 27 St, New York, NY 10001
T: (212) 760-7641
Founded: 1944
Gen Secr: Richard Streiter
Focus: Educ; Textiles 15019

Educational Leadership Institute (ELI), POB 11411, Shorewood, WI 53211
T: (414) 289-0706
Founded: 1980
Gen Secr: Dr. Jeremy J. Lietz
Focus: Educ 15020

Educational Planning (EP), c/o Robert H. Beach, Memphis State University, South Campus Bldg 48, Rm 102, Memphis, TN 38152
T: (901) 678-4204
Founded: 1970; Members: 350
Gen Secr: Robert H. Beach
Focus: Educ
Periodicals
Directory (weekly)
Educational Planning (weekly) 15021

Educational Planning Institute (EPI), 161 W 12 St, New York, NY 10011
T: (212) 807-7877; Fax: (212) 807-7884
Founded: 1971
Gen Secr: Anita Moses
Focus: Educ 15022

Educational Records Bureau (ERB), 140 W 65 St, New York, NY 10023
T: (212) 873-9108
Founded: 1927; Members: 1100
Pres: Dr. P. Norwood
Focus: Educ
Periodicals
Catalog of Programs and Services (weekly)
Newsletter (weekly)
Prospectus (weekly) 15023

Educational Research Analysts (EdReAn), POB 7518, Longview, TX 75607-7518
T: (903) 753-5993
Founded: 1961
Pres: Mel Gabler
Focus: Educ
Periodical
Newsletter (weekly) 15024

Educational Research Service (ERS), 2000 Clarendon Blvd, Arlington, VA 22201
T: (703) 243-2100; Fax: (703) 243-1985
Founded: 1973; Members: 2600
Pres: Glen E. Robinson
Focus: Educ
Periodical
Bulletin (10 times annually) 15025

Educational Testing Service (ETS), Rosedale Rd, Princeton, NJ 08541
T: (609) 921-9000; Fax: (609) 734-5410
Founded: 1947
Gen Secr: Eleanor V. Horne
Focus: Educ 15026

Education Commission of the States (ECS), 707 17 St, Ste 2700, Denver, CO 80202-3427
T: (303) 299-3600; Fax: (303) 296-8332
Founded: 1966; Members: 60
Pres: Frank Newman
Focus: Educ
Periodical
State Education Leader (weekly) 15027

Education Development Center (EDC), 55 Chapel St, Newton, MA 02160
T: (617) 969-7100; Fax: (617) 244-3436
Founded: 1958
Pres: Janet Whitla
Focus: Educ
Periodical
Annual Report (weekly) 15028

Education Turnkey Systems, 256 N Washington St, Falls Church, VA 22046
T: (703) 536-2313; Fax: (703) 536-3225
Founded: 1979
Gen Secr: Charles Blaschke
Focus: Computer & Info Sci 15029

Education Writers Association (EWA), 1001 Connecticut Av NW, Ste 310, Washington, DC 20036
T: (202) 429-9680; Fax: (202) 872-4016
Founded: 1947; Members: 550
Gen Secr: Lisa J. Walker
Focus: Lit; Educ
Periodicals
The Education Reporter (weekly)
Membership Directory (weekly) 15030

Educators' ad hoc Committee on Copyright Law, c/o August W. Steinhilber, National School Boards Association, 1680 Duke St, Alexandria, VA 22314
T: (703) 838-6710
Founded: 1963
Pres: August W. Steinhilber
Focus: Law 15031

Electric Auto Association (EAA), 1249 Lane St, Belmont, CA 94002-3756
T: (415) 591-6698
Founded: 1967; Members: 1500
Pres: John Newell
Focus: Transport; Auto Eng
Periodical
News (weekly) 15032

Electricity Consumers Resource Council (ELCON), 1333 H St NW, Washington, DC 20005
T: (202) 682-1390
Founded: 1976; Members: 21
Gen Secr: John A. Anderson
Focus: Electric Eng
Periodical
Report (weekly) 15033

Electric Vehicle Association (EVA), 20823 Stevens Creek Blvd, Ste 440, Cupertino, CA 95014
T: (408) 253-5262; Fax: (408) 253-9704
Founded: 1990; Members: 50
Gen Secr: Roland J. Risser
Focus: Electric Eng 15034

Electrochemical Society (ECS), 10 S Main St, Pennington, NJ 08534
T: (609) 737-1902
Founded: 1902; Members: 6000
Gen Secr: R.J. Calvo
Focus: Electrochem
Periodicals
Directory (weekly)
Journal (weekly) 15035

Electronic Music Consortium (EMC), c/o Dr. Thomas Wells, School of Music, Ohio State University, Columbus, OH 43210
T: (614) 292-7837
Founded: 1977; Members: 56
Pres: Dr. Thomas Wells
Focus: Music 15036

Elizabeth Linington Society (ELS), 1223 Glen Terrace, Glassboro, NJ 08028-1315
T: (609) 589-1571
Founded: 1983; Members: 150
Pres: Rinehart S. Potts
Focus: Lit 15037

Elm Research Institute (ERI), Main St, Harrisville, NH 03450
T: (603) 827-3048; Fax: (603) 827-3794
Founded: 1967; Members: 2000
Gen Secr: John P. Hansel
Focus: Ecology
Periodical
Press Release (weekly) 15038

Employee Benefit Research Institute (EBRI), 2121 K St NW, Ste 600, Washington, DC 20037
T: (202) 659-0670; Fax: (202) 775-6312
Founded: 1978
Pres: D.L. Salisbury
Focus: Sociology
Periodicals
Employee Benefit Notes (weekly)
Issue Briefs (weekly)
Proceedings (biennial) 15039

Emulsion Polymers Institute (EPI), c/o Lehigh University, 111 Research Dr, Bethlehem, PA 18015
T: (215) 758-3590; Fax: (215) 758-5880
Founded: 1975
Gen Secr: Mohamed A. El-Aasser
Focus: Chem; Eng; Materials Sci
Periodical
Graduate Research Progress Reports (weekly)
. . . . 15040

Endocrine Society (ES), 9650 Rockville Pike, Bethesda, MD 20814-3998
T: (301) 571-1800; Fax: (301) 571-1869
Founded: 1918; Members: 7000
Gen Secr: Scott Hunt
Focus: Endocrinology
Periodicals
Endocrine Review (weekly)
Endocrinology (weekly)
Journal of Clinical Endocrinology and Metabolism (weekly)
Molecular Endocrinology (weekly)
Newsletter 15041

Endometriosis Association (EA), 8585 N 76 Pl, Milwaukee, WI 53223
T: (414) 355-2200
Founded: 1980
Gen Secr: Mary Lou Ballweg
Focus: Gynecology
Periodical
Newsletter (weekly) 15042

Energy Research Institute (ERI), 6850 Rattlesnake Hammock Rd, Hwy 951, Naples, FL 33962
T: (813) 793-1922; Fax: (813) 793-1260
Founded: 1980; Members: 4500
Pres: J.C. Caruthers
Focus: Energy
Periodicals
Directory
Report 15043

Engineering Manpower Commission (EMC), 1111 19 St NW, Ste 608, Washington, DC 20036-3690
T: (202) 296-2237; Fax: (202) 296-1151
Founded: 1950; Members: 35
Gen Secr: R.A. Ellis
Focus: Eng
Periodical
Engineering and Technology Degrees (weekly)
. . . . 15044

English Institute (EI), c/o Center for Literary and Cultural Studies, Harvard University, 61 Kirkland St, Cambridge, MA 02138
T: (617) 496-1006
Founded: 1938; Members: 5000
Gen Secr: Lydia Alix Fillingham
Focus: Lit
Periodical
Selected Essays (weekly) 15045

Entomological Society of America (ESA), 9301 Annapolis Rd, Ste 300, Lanham, MD 20706-3115
T: (301) 731-4535
Founded: 1953; Members: 9200
Gen Secr: W. Darryl Hansen
Focus: Entomology
Periodicals
Annals (weekly)
Annual Review of Entomology (weekly)
Bulletin (weekly)
Environmental Entomology (weekly)
ESA Newsletter (weekly)
Journal of Economic Entomology (weekly)
Journal of Medical Entomology (weekly) . 15046

Environic Foundation International, 916 Saint Vincent St, South Bend, IN 46617
T: (219) 233-3357
Founded: 1970
Pres: Patrick Horsbrugh
Focus: Eng 15047

Environmental Defense Fund (EDF), 257 Park Av S, New York, NY 10010
T: (212) 505-2100; Fax: (212) 505-2375
Founded: 1967; Members: 200000
Gen Secr: Frederic D. Krupp
Focus: Ecology
Periodicals
Annual Report (weekly)
EDF Letter (weekly) 15048

Environmental Design Research Association (EDRA), POB 24083, Oklahoma City, OK 73124
T: (405) 848-4863
Founded: 1968; Members: 1000
Gen Secr: J. Thomas May
Focus: Anthro; Archit; Econ; Eng; Geography; Urban Plan
Periodical
Design Research News (weekly) 15049

Environmental Law Institute (ELI), 1616 P St NW, Ste 200, Washington, DC 20036
T: (202) 328-5150
Founded: 1969
Pres: J. William Futrell
Focus: Ecology; Law

Periodicals
ELI Associates Newsletter (weekly)
Environmental Law Reporter (weekly)
National Wetlands Newsletter (weekly) . 15050

Environmental Mutagen Society (EMS), 1730 N Lynn St, Arlington, VA 22209
T: (703) 525-1191; Fax: (703) 276-8196
Founded: 1969; Members: 1100
Gen Secr: Richard A. Guggolz
Focus: Bio
Periodical
Environmental and Molecular Mutagenesis (8 times annually) 15051

Environmental Technology Seminar (ETS), POB 391, Bethpage, NY 11714
T: (516) 931-3200
Founded: 1969
Pres: Jean Wood
Focus: Ecology
Periodical
The Environmentalist 15052

Epigraphic Society (ES), 6625 Bamburgh Dr, San Diego, CA 92117
T: (619) 571-1344; Fax: (619) 571-1124
Founded: 1974; Members: 800
Pres: Dr. Barry Fell
Focus: Archeol
Periodical
ESOP Annual (weekly) 15053

Ernest Bloch Society (EBS), 34844 Old Stage Rd, Gualala, CA 95445
T: (707) 884-3473
Founded: 1968; Members: 120
Gen Secr: Lucienne Bloch Dimitroff
Focus: Music
Periodical
Ernest Bloch Society Bulletin (weekly) . 15054

Esalen Institute (EI), Big Sur, CA 93920
T: (408) 667-3000; Fax: (408) 667-2724
Founded: 1962
Pres: Steve Donovan
Focus: Psych
Periodical
Catalog (3 times annually) 15055

Esperantic Studies Foundation (ESF), 3900 Northampton St NW, Washington, DC 20015-2851
T: (202) 362-3963; Fax: (202) 363-6899
Founded: 1968
Gen Secr: Dr. E. James Lieberman
Focus: Ling 15056

Esperanto League for North America (ELNA), POB 1129, El Cerrito, CA 94530
T: (415) 653-0998
Founded: 1952; Members: 1080
Gen Secr: Mike Donohoo
Focus: Ling
Periodicals
Catalog (weekly)
ELNA-Adresaro (weekly)
ELNA Newsletter (weekly) 15057

Esperanto Librarians International, Apt 101, 1517 Raymond Dr, Naperville, IL 60563
Founded: 1984; Members: 150
Gen Secr: Douglas Portman
Focus: Libraries & Bk Sci 15058

Essentia, 3821 Paseo del Prado, Boulder, CO 80301
T: (303) 443-3484
Founded: 1969; Members: 4
Gen Secr: Olena Gilbert
Focus: Sci 15059

Estonian Learned Society of America (ELSA), c/o Estonian Educational Society, Estonian House, 243 E 34 St, New York, NY 10016
T: (416) 485-5559
Founded: 1950; Members: 200
Gen Secr: Dr. Tonu Parming
Focus: Ethnology 15060

Estuarine Research Federation (ERF), POB 544, Crownsville, MD 21032-0544
T: (410) 266-5489
Founded: 1969; Members: 900
Gen Secr: Jerome Williams
Focus: Oceanography
Periodicals
Estuaries (weekly)
Newsletter (weekly) 15061

Ethnic Materials and Information Exchange Round Table (EMIERT), c/o American Library Association, 50 E Huron St, Chicago, IL 60611
T: (312) 280-4295; Fax: (312) 280-3256
Founded: 1971; Members: 700
Pres: Patricia F. Beilke
Focus: Libraries & Bk Sci
Periodical
EMIERT Bulletin (weekly) 15062

Etruscan Foundation (EF), 161 Country Club Dr, Grosse Pointe Farms, MI 48236
T: (313) 886-6654
Founded: 1959; Members: 300
Pres: Ferdinand Cinelli
Focus: Archeol
Periodicals
Etruscans (weekly)
Friends of Spannocchia (weekly)
Newsletter 15063

Eugene O'Neill Society (EOS), c/o Dept of English, University of Rhode Island, Kingston, RI 02881
T: (401) 792-5931
Founded: 1978; Members: 225
Gen Secr: Jordan Y. Miller
Focus: Lit 15064

European Community Studies Association (ECSA), c/o Dept of Political Science, Cleveland State University, Cleveland, OH 44115
T: (216) 687-3660
Founded: 1988
Gen Secr: Prof. Leon Hurwitz
Focus: Poli Sci 15065

Evangelical Lutheran Education Association (ELEA), 6020 Radford Av, North Hollywood, CA 91606
T: (818) 752-1019
Founded: 1961; Members: 351
Gen Secr: Julie Sieger
Focus: Educ
Periodical
Views and Vision (9 times annually) . . 15066

Evelyn Waugh Society (EWS), c/o English Dept, Nassau Community College, State University of New York, Garden City, NY 11530
T: (516) 222-7187
Founded: 1967; Members: 201
Pres: Dr. Paul A. Doyle
Focus: Lit
Periodical
Evelyn Waugh Newsletter (3 times annually) 15067

Exodus Trust (ET), 1523 Franklin St, San Francisco, CA 94109
T: (415) 928-1133
Founded: 1968
Pres: Ted McIlvenna
Focus: Educ 15068

Expanded Shale Clay and Slate Institute (ESCSI), 2225 E Murray Holladay Rd, Salt Lake City, UT 84117
T: (801) 272-7070; Fax: (801) 272-3377
Founded: 1952; Members: 27
Gen Secr: John P. Ries
Focus: Civil Eng
Periodicals
Information Sheet
Membership Roster (weekly)
Special Bulletin 15069

Experimental Aircraft Association (EAA), EAA Aircraft Center, POB 3086, Oshkosh, WI 54903-3086
T: (414) 426-4800; Fax: (414) 426-4828
Members: 128000
Pres: Tom Poberezny
Focus: Aero
Periodicals
Chapter Bulletin (weekly)
The EAA Experimenter (weekly)
Sport Aerobatics (weekly)
Sport Aviation (weekly)
The Vintage Airplane (weekly) 15070

Experiments in Art and Technology (EAT), c/o Billie Kluver, 69 Apple Tree Rd, Berkeley Heights, NJ 07922
T: (212) 285-1690
Founded: 1966
Pres: Billie Kluver
Focus: Arts; Eng 15071

The Explorers Club (EC), 46 E 70 St, New York, NY 10021
T: (212) 628-8383; Fax: (212) 288-4449
Founded: 1904; Members: 3000
Gen Secr: Eileen Harsch
Focus: Geography
Periodicals
Explorers Journal (weekly)
Newsletter (weekly) 15072

Extractive Metallurgy Institute (EMI), 1926 Cliff Rd, Point Roberts, WA 98281
T: (206) 945-0557; Fax: (206) 945-0556
Founded: 1981
Pres: Dr. Douglas J. Robinson
Focus: Metallurgy 15073

Ezra Pound Society (EPS), c/o University of Maine, 302 Neville Hall, Orono, ME 04469
T: (207) 581-3814
Founded: 1978
Gen Secr: Burton Hatlen
Focus: Lit
Periodicals
Paideuma (3 times annually)
Sagetrieb (3 times annually) 15074

Family Health International (FHI), POB 13950, Durham, NC 27709
T: (919) 544-7040; Fax: (919) 544-7261
Founded: 1971; Members: 230
Pres: Dr. Malcolm Potts
Focus: Family Plan; Pediatrics; Public Health
Periodicals
network (weekly)
network en espanol (weekly)
network en français (weekly) 15075

Farm Foundation (FF), 1211 W 22 St, Ste 216, Oak Brook, IL 60521
T: (708) 571-9393; Fax: (708) 571-9580
Founded: 1933
Gen Secr: Walter J. Armbruster
Focus: Agri 15076

Federal Bar Association (FBA), 1815 H St NW, Ste 408, Washington, DC 20006
T: (202) 638-0252; Fax: (202) 775-0295
Founded: 1920; Members: 15000
Gen Secr: John Blanche
Focus: Law
Periodicals
Federal Bar News and Journal (10 times annually)
Membership Directory 15077

Federated Council of Beth Jacob Schools (FCBJS), 142 Broome St, New York, NY 10002
T: (212) 473-4500
Founded: 1940; Members: 100
Gen Secr: Israel Garber
Focus: Educ 15078

Federation for Unified Science Education (FUSE), c/o Capital University, 231 Battele Hall of Science, Columbus, OH 43209
T: (614) 236-6816
Founded: 1966; Members: 450
Gen Secr: Victor M. Showalter
Focus: Educ
Periodicals
Prism II (weekly)
Proceedings of Annual Conference (bi-annually) 15079

Federation of American Scientists (FAS), 307 Massachusetts Av NE, Washington, DC 20002
T: (202) 546-3300
Founded: 1945; Members: 4000
Pres: Jeremy J. Stone
Focus: Sci
Periodical
FAS Public Incerest Report (10 times annually) 15080

Federation of American Societies for Experimental Biology (FASEB), 9650 Rockville Pike, Bethesda, MD 20814-3998
T: (301) 530-7000; Fax: (301) 530-7001
Founded: 1913; Members: 8
Gen Secr: Michael J. Jackson
Focus: Bio
Periodical
The FASEB Journal (weekly) 15081

Federation of Analytical Chemistry and Spectroscopy Societies (FACSS), c/o Jo Ann Brown, POB 1405, Frederick, MD 21702
T: (301) 846-4797
Founded: 1972; Members: 7
Gen Secr: Jo Ann Brown
Focus: Chem; Physics 15082

Federation of Behavioral, Psychological and Cognitive Sciences (FBPCS), c/o David H. Johnson, 750 First NW, Ste 5004, Washington, DC 20002-5004
T: (202) 336-5920; Fax: (202) 336-5920
Founded: 1980; Members: 16
Gen Secr: David H. Johnson
Focus: Psych
Periodicals
Annual Report (weekly)
Federation News (weekly) 15083

Federation of Government Information Processing Councils (FGIPC), c/o Virginia McCormick, U.S. General Services Administration, 4KTH-A, 401 W Peachtre St, Atlanta, GA 30365-2550
T: (404) 331-5106
Founded: 1978; Members: 12000
Gen Secr: Virginia McCormick
Focus: Public Admin; Computer & Info Sci
Periodicals
Directory (weekly)
FedFacts (weekly) 15084

Federation of Insurance and Corporate Counsel (FICC), c/o Joseph R. Olshan, POB 111, Walpole, MA 02081
T: (508) 668-6859; Fax: (508) 668-6892
Founded: 1935; Members: 1300
Gen Secr: Joseph R. Olshan
Focus: Insurance 15085

Federation of Materials Societies (FMS), 1707 L St NW, Ste 333, Washington, DC 20036
T: (202) 296-9282; Fax: (202) 833-3014
Founded: 1972; Members: 13
Gen Secr: Betsy Houston
Focus: Materials Sci
Periodicals
Conference Proceedings/Materials (bi-annually)
News (weekly) 15086

Federation of Orthodontic Associations (FOA), c/o Dr. David H. Watson, 3953 N 76 St, Milwaukee, WI 53222
T: (414) 464-7870
Founded: 1969
Gen Secr: Dr. David H. Watson
Focus: Dent 15087

Federation of Prosthodontic Organizations (FPO), 211 E Chicago Av, Ste 948, Chicago, IL 60611
T: (312) 642-7538
Founded: 1965; Members: 16
Gen Secr: Peter C. Goulding
Focus: Dent
Periodicals
Directory (weekly)
Newsletter (weekly) 15088

Federation of Societies for Coatings Technology (FSCT), 492 Norristown Rd, Blue Bell, PA 18422
T: (215) 940-0777; Fax: (215) 940-0292
Founded: 1922; Members: 7300
Gen Secr: Robert F. Ziegler
Focus: Eng; Civil Eng
Periodical
Journal of Coatings Technology (weekly) 15089

Federation of State Medical Boards of the United States (FSMB), 6000 Western Pl, Ste 707, Fort Worth, TX 76107
T: (817) 735-8445
Founded: 1912; Members: 67
Gen Secr: Dr. James Winn
Focus: Med
Periodicals
Federation Bulletin (weekly)
FSMB Newsletter (weekly)
Handbook (weekly) 15090

Federation of Trainers and Training Programs in Psychodrama (FTTPP), 2829 W Northwest Hwy, Ste 940, Dallas, TX 75220
T: (214) 352-9213
Founded: 1977; Members: 30
Gen Secr: Shirley A. Barclay
Focus: Therapeutics; Psych
Periodicals
Membership Directory (biennial)
Newsletter (weekly) 15091

Fiber Society (FS), POB 625, Princeton, NJ 08542
T: (609) 924-3150; Fax: (609) 683-7836
Founded: 1941; Members: 400
Gen Secr: Dr. George Lamb
Focus: Chem; Physics; Eng
Periodical
Membership Directory (weekly) 15092

Financial Management Association (FMA), c/o School of Business, University of South Florida, Tampa, FL 33620-5500
T: (813) 974-2084; Fax: (813) 974-3318
Founded: 1970; Members: 12600
Gen Secr: Jack S. Rader
Focus: Econ
Periodicals
Careers in Finance
Financial Management (weekly)
Financial Management Collection (tri-annually)
Membership/Professional Directory . . . 15093

Financial Marketing Association (FMA), POB 14167, Madison, WI 53714
T: (608) 271-2664; Fax: (608) 271-2303
Founded: 1983; Members: 600
Gen Secr: Fran Zaugg
Focus: Marketing 15094

Finnish-American Historical Society of the West (FAHSW), POB 5522, Portland, OR 97208
T: (503) 654-0448
Founded: 1962; Members: 400
Gen Secr: Roy Schulbach
Focus: Hist
Periodical
FINNAM Newsletter (weekly) 15095

Firearms Research and Identification Association (FRIA), 17524 Colima Rd, Ste 360, Rowland Heights, CA 91748
T: (714) 598-8919; Fax: (714) 598-5666
Founded: 1978; Members: 16
Pres: John Armand Caudron
Focus: Military Sci 15096

First – Foundation for Ichthyosis and Related Skin Types, POB 20921, Raleigh, NC 27619-0921
T: (919) 782-5728
Founded: 1981; Members: 3000
Pres: Ellen Rowe
Focus: Med
Periodical
Ichthyosis Focus (weekly) 15097

Flag Research Center (FRC), POB 580, Winchester, MA 01890
T: (617) 729-9410; Fax: (617) 721-4817
Founded: 1962; Members: 1160
Gen Secr: Dr. Whitney Smith
Focus: Hist
Periodicals
The Flag Bulletin (weekly)
National Flags (weekly) 15098

Flat Earth Research Society International (FERSI), POB 2533, Lancaster, CA 93539
T: (805) 727-1635
Founded: 1800; Members: 2800
Pres: Charles K. Johnson
Focus: Geology
Periodical
Flat Earth News (weekly) 15099

Flight Safety Foundation, 2200 Wilson Blvd, Ste 500, Arlington, VA 22201
T: (703) 522-8300; Fax: (703) 525-6047
Founded: 1945; Members: 540
Pres: John H. Enders
Focus: Safety; Aero 15100

Flower Essence Society (FES), POB 459, Nevada City, CA 95959
T: (916) 265-9163; Fax: (916) 265-6467
Founded: 1979; Members: 25000
Gen Secr: Patricia A. Kaminski
Focus: Med
Periodicals
Directory
The Flower Essence Society Newsletter (weekly) . . . *15101*

Fluid Power Society (FPS), 2433 N Mayfair Rd, Ste 111, Milwaukee, WI 53226
T: (414) 257-0910; Fax: (414) 257-4092
Founded: 1957; Members: 2440
Pres: Frederick A. Brown
Focus: Energy
Periodical
Newsletter (10 times annually) . . . *15102*

Flying Chiropractors Association (FCA), 7301 Hasbrook Av, Philadelphia, PA 19111
T: (215) 722-7200
Founded: 1968; Members: 300
Gen Secr: Dr. W.J. Quinlan
Focus: Med . . . *15103*

Flying Dentists Association (FDA), 4700 Chamblee-Dunwoody Rd, Dunwoody, GA 30338
T: (404) 457-1351; Fax: (404) 458-0890
Founded: 1960; Members: 500
Gen Secr: Dr. Max J. Cohen
Focus: Dent
Periodical
Flight Watch (weekly) . . . *15104*

Flying Physicians Association (FPA), POB 17841, Kansas City, MO 64134
T: (816) 763-9336
Founded: 1954; Members: 1200
Gen Secr: Carol Laurie
Focus: Physiology . . . *15105*

Flying Veterinarians Association (FVA), POB 1081, Columbia, MO 65205
T: (314) 882-7228; Fax: (314) 882-2950
Founded: 1966; Members: 328
Gen Secr: Robert C. McClure
Focus: Vet Med
Periodicals
Directory (weekly)
Newsletter . . . *15106*

F. Marion Crawford Memorial Sociey (FMCMS), c/o Jesse Knight, 2148 Av de los Flores, Santa Clara, CA 95054
Founded: 1975; Members: 35
Gen Secr: Jesse Knight
Focus: Lit . . . *15107*

Food and Nutrition Board (FNB), c/o Institute of Medicine, 2101 Constitution Av NW, Washington, DC 20418
T: (202) 334-1732
Founded: 1940; Members: 17
Gen Secr: Catherine E. Woteki
Focus: Food; Nutrition; Biochem
Periodicals
Activities Report
Directory (weekly) . . . *15108*

Food Distribution Research Society (FDRS), c/o Dept of Food and Resource Economics, University of Delaware, 213 Townsend Hall, Newark, DE 19717-1303
T: (302) 831-1320; Fax: (302) 831-3651
Founded: 1967; Members: 400
Gen Secr: J. Richard Bacon
Focus: Food
Periodical
Journal of Food Distribution Research (weekly) . . . *15109*

Foreign Policy Association (FPA), 729 Seventh Av, New York, NY 10019
T: (212) 764-4050; Fax: (212) 302-6123
Founded: 1918
Pres: R.T. Curran
Focus: Poli Sci
Periodical
Headline Series (5 times annually) . . . *15110*

Foreign Services Research Institute (FSRI), POB 6317, Washington, DC 20015-0317
Founded: 1974
Pres: John E. Whiteford
Focus: Hist; Philos; Sociology; Poli Sci . *15111*

Forest History Society (FHS), 701 Vickers Av, Durham, NC 27701
T: (919) 682-9319
Founded: 1946; Members: 1750
Gen Secr: Harold K. Steen
Focus: Forestry; Cultur Hist
Periodicals
Forest & Conservation History (weekly)
Forest History Cruiser (weekly)
Guides to Forest and Conservation History of North America (weekly) . . . *15112*

Forest Products Research Society (FPRS), 2801 Marshall Court, Madison, WI 53705
T: (608) 231-1361; Fax: (608) 231-2152
Founded: 1947; Members: 3500
Pres: Arthur B. Brauner
Focus: Forestry
Periodicals
Forest Products Journal (weekly)
FPRS Technical Newsletter (weekly) . . *15113*

The Forum, POB 5915, Santa Fe, NM 87502
T: (505) 983-3962
Founded: 1970; Members: 70
Gen Secr: Carol B. Knight
Focus: Psych
Periodical
The Torch (weekly) . . . *15114*

The Forum for Health Care Planning, 1101 Connecticut Av NW, Ste 700, Washington, DC 20036
T: (202) 857-1162
Founded: 1950; Members: 500
Gen Secr: Cornelia Hinz
Focus: Med
Periodicals
Membership Directory (weekly)
Newsletter (weekly) . . . *15115*

Forum for Medical Affairs (FMA), c/o Medical Society of the State of New York, 420 Lakeville Rd, Lake Success, NY 11042
T: (516) 488-6100
Founded: 1944; Members: 800
Gen Secr: Donald F. Foy
Focus: Med . . . *15116*

Forum International: International Ecosystems University (IEU), 91 Gregory Ln, Pleasant Hill, CA 94523
T: (510) 671-2900; Fax: (510) 946-1500
Founded: 1965; Members: 32000
Gen Secr: Dr. Nicholas D. Hetzer
Focus: Ecology
Periodical
Ecosphere (weekly) . . . *15117*

Fostoria Glass Society of America (FGSA), POB 826, Moundsville, WV 26041
T: (513) 335-6643
Founded: 1979; Members: 800
Pres: Ron Hufford
Focus: Graphic & Dec Arts, Design
Periodical
Facets of Fostoria (weekly) . . . *15118*

Foundation for Accounting Education (FAE), 200 Park Av, New York, NY 10166
T: (212) 973-8300; Fax: (212) 972-5714
Founded: 1972
Gen Secr: Robert L. Gray
Focus: Business Admin; Educ . . . *15119*

Foundation for Advancement in Cancer Therapy (FACT), POB 1242, Old Chelsea Station, New York, NY 10113
T: (212) 741-2790
Founded: 1971
Pres: Ruth Sackman
Focus: Cell Biol & Cancer Res; Therapeutics . . . *15120*

Foundation for Biomedical Research (FBR), 818 Connecticut Av NW, Ste 303, Washington, DC 20006
T: (202) 457-0654
Founded: 1981
Pres: F.L. Trull
Focus: Vet Med . . . *15121*

Foundation for Chiropractic Education and Research (FCER), 1701 Clarendon Blvd, Arlington, VA 22209
T: (703) 276-7445; Fax: (703) 276-8178
Founded: 1944; Members: 6000
Gen Secr: Steve Seater
Focus: Orthopedics
Periodicals
Advance (weekly)
Spinal Manipulation (weekly)
Staying Well (weekly) . . . *15122*

Foundation for Educational Futures (FEF), POB 2463, Charlotte, NC 28226
T: (704) 541-6256
Founded: 1983
Gen Secr: Thomas G. Voss
Focus: Educ . . . *15123*

Foundation for Exceptional Children (FEC), 1920 Association Dr, Reston, VA 22091
T: (703) 620-1054
Founded: 1971; Members: 1000
Gen Secr: Ken Collins
Focus: Educ Handic . . . *15124*

Foundation for Field Research (FFR), POB 2010, Alpine, CA 91003
T: (619) 445-9264; Fax: (619) 445-1893
Founded: 1982; Members: 4000
Gen Secr: Thomas J Banks
Focus: Nat Sci
Periodical
Explorer News (weekly) . . . *15125*

Foundation for Interior Design Education Research (FIDER), 60 Monroe Center NW, Grand Rapids, MI 49503
T: (616) 458-0400; Fax: (616) 458-0460
Founded: 1971
Gen Secr: K. Dunn
Focus: Educ; Graphic & Dec Arts, Design *15126*

Foundation for International Cooperation (FIC), 1237 S Western Av, Park Ridge, IL 60068
T: (708) 518-0934
Founded: 1960
Gen Secr: Irene B. Horst
Focus: Educ . . . *15127*

Foundation for Latin American Anthropological Research (FLAAR), c/o Dr. Nicholas M. Hellmuth, Brevard Community College, 1519 Clearlake Rd, Cocoa, FL 32922
T: (407) 632-1111; Fax: (407) 633-4565
Founded: 1969
Pres: Dr. Nicholas M. Hellmuth
Focus: Anthro . . . *15128*

Foundation for Microbiology (FFM), c/o Byron H. Waksman, 300 E 54 St, Ste 5K, New York, NY 10022
T: (212) 759-8729
Founded: 1951; Members: 9
Pres: Byron H. Waksman
Focus: Microbio
Periodical
Report (weekly) . . . *15129*

Foundation for Mideast Communication (FMC), POB 5264, Beverly Hills, CA 90210
T: (310) 820-8749; Fax: (310) 820-8759
Founded: 1983
Gen Secr: Michael Lame
Focus: Ethnology . . . *15130*

Foundation for Research in the Afro-American Creative Arts (FRAACA), PO Drawer 1, Cambria Heights, NY 11411
Founded: 1971; Members: 1000
Pres: Prof. Joseph Southern
Focus: Arts . . . *15131*

Foundation for Science and the Handicapped (FSH), 236 Grand St, Morgantown, WV 26505-7509
T: (304) 293-5201
Founded: 1978; Members: 265
Gen Secr: E.C. Keller
Focus: Educ Handic
Periodical
Newsletter (weekly) . . . *15132*

Foundation for the Advancement of Chiropractic Tenets and Science (FACTS), c/o International Chiropractors Association, 1110 N Glebe Rd, Ste 1000, Arlington, VA 22201
T: (703) 528-5000
Founded: 1972
Focus: Med . . . *15133*

Foundation for the Support of International Medical Training (FSIMT), 417 Center St, Lewiston, NY 14092
T: (716) 754-4883; Fax: (716) 754-4883
Founded: 1960; Members: 6000000
Gen Secr: M.A. Uffer
Focus: Med; Nucl Med
Periodicals
Brochures (weekly)
Directory (weekly)
Immunization Chart (weekly)
Malaria Risk Chart (weekly)
Traveller Clinical Chart (weekly) . . . *15134*

Francis Bacon Foundation (FBF), 655 N Dartmouth Av, Claremont, CA 91711
T: (714) 624-6305
Founded: 1938
Pres: Elizabeth S. Wrigley
Focus: Lit . . . *15135*

Franklin and Eleanor Roosevelt Institute, 511 Albany Post Rd, Hyde Park, NY 12538
T: (914) 229-5321; Fax: (914) 229-9046
Founded: 1987
Gen Secr: John F. Sears
Focus: Hist
Periodical
The View From Hyde Park (3 times annually) . . . *15136*

Frank Lloyd Wright Foundation (FLWF), Taliesin W, Scottsdale, AZ 85261
T: (602) 860-2700; Fax: (602) 391-4009
Founded: 1932
Gen Secr: Richard Carney
Focus: Archit
Periodical
Friends of Taliesin (3 times annually) . *15137*

Frederick Douglass Memorial and Historical Association (FDMHA), c/o Mary E.C. Gregory, 10594 Twin Rivers Rd, Apt E-1, Columbia, MD 21044
T: (410) 854-2938
Founded: 1900
Pres: Mary E.C. Gregory
Focus: Hist . . . *15138*

Friars Club (FC), 57 E 55 St, New York, NY 10022
T: (212) 751-7272; Fax: (212) 355-0217
Founded: 1904; Members: 1400
Gen Secr: Jean Pierre Trebot
Focus: Perf Arts . . . *15139*

Friends Historical Association (FHA), c/o Quaker Collection, Haverford College Library, Haverford, PA 19041
T: (215) 896-1161
Founded: 1873; Members: 800
Gen Secr: E. Potts Brown
Focus: Hist; Rel & Theol
Periodical
Quaker History (weekly) . . . *15140*

Friends of American Art in Religion (FAAR), c/o Eugene V. Clark, 143 E 43 St, New York, NY 10017
T: (212) 682-5722
Founded: 1970; Members: 100
Gen Secr: E.V. Clark
Focus: Arts . . . *15141*

Friends of Cast Iron Architecture (FCIA), 235 E 87 St, Rm 6C, New York, NY 10128
T: (212) 369-6004
Founded: 1970; Members: 1000
Pres: Margot Gayle
Focus: Preserv Hist Monuments . . . *15142*

Friends of George Sand (FGS), c/o Hofstra Cultural Center, Hofstra University, Hempstead, NY 11550
T: (516) 463-5669; Fax: (516) 564-4297
Founded: 1976; Members: 319
Gen Secr: Natalie Datlof
Focus: Lit . . . *15143*

Friends of the Lincoln Museum (FLC), c/o Abraham Lincoln Museum, Lincoln Memorial University, Harrogate, TN 37752
T: (615) 869-6235; Fax: (615) 869-6370
Founded: 1898
Gen Secr: Stephen G. Hague
Focus: Hist
Periodical
Lincoln Herald (weekly) . . . *15144*

Friends of the Sea Lion Marine Mammal Center (FSLMMC), 20612 Laguna Canyon Rd, Laguna Beach, CA 92651
T: (714) 494-3050
Founded: 1971; Members: 2500
Gen Secr: W.H. Ford
Focus: Zoology . . . *15145*

The Galactic Society (TGS), Box 326, Rock Hill, SC 29731
T: (803) 328-0705
Founded: 1989
Pres: J.G. Bowman
Focus: Aero . . . *15146*

Gas Research Institute (GRI), 8600 W Bryn Mawr Av, Chicago, IL 60631
T: (312) 399-8100; Fax: (312) 399-8170
Founded: 1976; Members: 319
Pres: Stephen D. Ban
Focus: Energy
Periodical
Gas Research Institute Digest (weekly) . *15147*

General Anthropology Division (GAD), c/o American Anthropological Association, 1703 New Hampshire Av NW, Washington, DC 20009
T: (202) 232-8800
Founded: 1984; Members: 4877
Pres: Conrad Kottak
Focus: Anthro
Periodical
American Anthropologist . . . *15148*

General Educational Development Institute (GEDI), 16211 Sixth Av NE, Seattle, WA 98155
T: (206) 362-2055
Founded: 1975
Gen Secr: Chuck Herring
Focus: Educ
Periodical
Adult Learner (weekly) . . . *15149*

General Society of Mechanics and Tradesmen (GSMT), 20 W 44 St, New York, NY 10036
T: (212) 840-1840
Founded: 1785
Gen Secr: Marshall D. Jacks
Focus: Eng . . . *15150*

Genetics Society of America (GSA), 9650 Rockville Pike, Bethesda, MD 20814-3998
T: (301) 571-1825
Founded: 1932; Members: 3700
Gen Secr: Elaine Strass
Focus: Genetics
Periodicals
Genetics (weekly)
Membership Directory (biennial) . . . *15151*

Geochemical Society (GS), c/o Dr. S.B. Shirey, Dept of Terrestrial Magnetism, 5241 Broad Branch Rd NW, Washington, DC 20015
T: (202) 686-4370
Founded: 1955; Members: 1800
Gen Secr: Dr. S.B. Shirey
Focus: Geology; Chem; Geophys; Mineralogy
Periodicals
The Geochemical News (weekly)
Geochimica et Cosmochimica Acta (weekly) . . . *15152*

Geological Society of America (GSA), 3300 Penrose Pl, POB 9140, Boulder, CO 80301-9140
T: (303) 447-2020; Fax: (303) 447-1133
Founded: 1888; Members: 16000
Gen Secr: F. Michael Wahl
Focus: Geology
Periodicals
The Geological Society of America Bulletin (weekly)
Geology (weekly)
GSA News & Information (weekly) . . . *15153*

George C. Marshall Foundation (GCMF), Drawer 1600, Lexington, VA 24450
T: (703) 463-7103; Fax: (703) 464-5229
Founded: 1953
Pres: Ronald F. Marryott
Focus: Hist; Military Sci
Periodical
The Papers of George Catlett Marshall . *15154*

Geothermal Resources Council (GRC), POB 1350, Davis, CA 95617
T: (916) 758-2360; Fax: (916) 758-2839
Founded: 1972; Members: 1048
Gen Secr: David N. Anderson
Focus: Geology
Periodicals
Bulletin (11 times annually)
Membership Roster (weekly) *15155*

German-American Information and Education Association (GIEA), POB 10888, Burke, VA 22015
T: (703) 425-0707
Founded: 1986
Pres: Stanley Rittenhouse
Focus: Educ; Ling *15156*

German American World Society (GAWS), 529A Central Av, Jersey City, NJ 07307
T: (201) 420-0159; Fax: (201) 420-0469
Founded: 1980
Gen Secr: John R. Lawrora
Focus: Hist *15157*

Gerontological Society of America, 1275 K St NW, Ste 350, Washington, DC 20005
T: (202) 842-1275; Fax: (202) 842-1150
Founded: 1945; Members: 7000
Gen Secr: Dr. Paul A. Kerschner
Focus: Geriatrics
Periodicals
The Gerontologist (weekly)
The Journal of Gerontology (weekly) . . *15158*

Gerson Institute (GI), POB 430, Bonita, CA 91908
T: (619) 267-1150
Founded: 1977; Members: 2500
Pres: Charlotte Gerson
Focus: Med *15159*

Gilbert and Sullivan Society (GSS), c/o Frances Yasprica, 1351 65 St, Brooklym, NY 11219
T: (718) 259-6431
Founded: 1936; Members: 400
Gen Secr: Frances Yasperica
Focus: Music
Periodical
The Palace Peeper (10 times annually) . *15160*

Glass Art Society (GAS), 1305 Fourth Av, Ste 711, Seattle, WA 98101-2401
T: (206) 382-1305; Fax: (206) 382-2630
Founded: 1971; Members: 1300
Gen Secr: Alice Rooney
Focus: Graphic & Dec Arts, Design
Periodicals
Journal (weekly)
Resource Directory (weekly) *15161*

Glazounov Society (GS), 17320 Park Av, Sonoma, CA 95476
T: (707) 996-1653
Founded: 1985; Members: 48
Pres: Donald J. Venturini
Focus: Music *15162*

Global Education Associates (GEA), 475 Riverside Dr, Ste 456, New York, NY 10115
T: (212) 870-3290; Fax: (212) 870-2055
Founded: 1973; Members: 8000
Pres: Gerald F. Mische
Focus: Educ
Periodical
Breakthrough (weekly) *15163*

Global Learning (GL), 1018 Stuyvesant Av, Union, NJ 07083
T: (908) 964-1114; Fax: (908) 964-6335
Founded: 1974
Gen Secr: Jeffrey L. Brown
Focus: Educ
Periodical
Global Learning Teacher Education Manual *15164*

Global Options (GO), POB 40601, San Francisco, CA 94140
T: (415) 550-1703
Founded: 1977; Members: 53
Pres: Cecilia O'Leary
Focus: Socialism; Sociology; Poli Sci; Econ
Periodicals
Central America Education Project (weekly)
Contra Watch (weekly)
Crime and Social Justice (weekly) . . . *15165*

Goethe Society of North America (GSNA), c/o German Dept, University of California, Irvine, CA 92717
T: (714) 856-6406
Founded: 1979; Members: 200
Gen Secr: Frederick Amrine
Focus: Lit *15166*

Goudy Society (GS), c/o Tany, 408 Eighth Av, New York, NY 10001
T: (212) 629-3232; Fax: (212) 465-2012
Founded: 1965; Members: 100
Pres: Steven J. Kennedy
Focus: Eng *15167*

Governmental Research Association (GRA), c/o Samford University, 315 Samford Hall, Birmingham, AL 35229
T: (205) 870-2482
Founded: 1914; Members: 165
Pres: James W. Williams
Focus: Poli Sci; Public Admin
Periodicals
Bibliography of Governmental Research Directory (biennial)
Reporter (weekly) *15168*

Graduate Management Admission Council (GMAC), 11601 Wilshire Blvd, Ste 760, Los Angeles, CA 90025
T: (310) 478-1433; Fax: (310) 473-2388
Founded: 1970; Members: 110
Pres: William Broesamle
Focus: Business Admin *15169*

Graduate Record Examinations Board (GRE BOARD), c/o Educational Testing Service 33-V, Princeton, NJ 08541
T: (609) 951-6506; Fax: (609) 951-1090
Founded: 1966
Gen Secr: Charlotte V. Kuh
Focus: Adult Educ
Periodicals
Board Newsletter (weekly)
General Test Practice Book (weekly)
GRE Information Bulletin (weekly)
Guide to the Graduate Record Examinations Program (weekly) *15170*

Graduates of Italian Medical Schools (GIMS), c/o Mario E. Milite, 360 E 194 St, Bronx, NY 10458
T: (718) 364-3164
Founded: 1966; Members: 500
Gen Secr: Mario E. Milite
Focus: Adult Educ; Med
Periodicals
Membership Directory (biennial)
Newsletter (5 times annually) *15171*

Graphic Communications Association (GCA), 100 Daingerfield Rd, Alexandria, VA 22314
T: (703) 519-8160; Fax: (703) 548-2867
Founded: 1966; Members: 350
Pres: Norman W. Scharpf
Focus: Computer & Info Sci; Eng
Periodicals
Newsletter (weekly)
Spectrum Proceedings (weekly) *15172*

Gravure Association of America (GAA), 1200A Scottsville Rd, Rochester, NY 14624
T: (716) 436-2159; Fax: (716) 463-7689
Founded: 1987; Members: 250
Gen Secr: Cheryl Kasunich
Focus: Eng
Periodicals
Gravure Environmental (1-2 times annually)
Membership Roster (weekly)
Newsletter (1-2 times annually) *15173*

Great Lakes Colleges Association (GLCA), 2929 Plymouth Rd, Ste 207, Ann Arbor, MI 48105-3206
T: (313) 761-4833; Fax: (313) 761-3939
Founded: 1961; Members: 12
Pres: Carol J. Guardo
Focus: Educ
Periodical
Faculty Newsletter (5 times annually) . . *15174*

Great Lakes Commission (GLC), 400 Fourth St, Ann Arbor, MI 48103-4816
T: (313) 665-9135; Fax: (313) 665-4370
Founded: 1955; Members: 36
Gen Secr: Dr. Michael J. Donahue
Focus: Water Res; Navig
Periodicals
Great Lakes Research Checklist (weekly)
Membership List
Minutes of Regular Meeting (weekly) . . *15175*

Great Lakes Historical Society (GLHS), c/o Great Lakes Historical Society Museum, 480 Main St, Vermilion, OH 44089
T: (216) 967-3467; Fax: (216) 967-1519
Founded: 1944; Members: 3000
Pres: Dr. Timothy Runyan
Focus: Hist
Periodicals
Chadburn (weekly)
Inland Seas (weekly) *15176*

Great Lakes Maritime Institute (GLMI), 100 Strand Dr, 1 Belle Isle, Detroit, MI 48207
T: (313) 267-6440
Founded: 1948; Members: 1700
Gen Secr: John Polacsek
Focus: Navig
Periodical
Telescope (weekly) *15177*

Great Plains Agricultural Council (GPAC), c/o Dept of Agricultural and Resource Economics, Colorado State University, Fort Collins, CO 80523-0002
T: (303) 491-7370
Founded: 1946; Members: 27
Gen Secr: Dr. Melvin D. Skold
Focus: Agri
Periodical
Proceedings (weekly) *15178*

Group for the Advancement of Psychiatry (GAP), POB 28218, Dallas, TX 75228
T: (214) 388-1310
Founded: 1946; Members: 300
Gen Secr: Alice Conde Martinez
Focus: Psychiatry *15179*

Group for the Use of Psychology in History (GUPH), c/o History Program, Sangamon State University, 383 Brookens, Springfield, IL 62794-9243
T: (217) 786-6778
Founded: 1972; Members: 600
Gen Secr: Larry Shiner
Focus: Hist; Psych
Periodical
The Psychohistory Review (3 times annually) *15180*

Group Health Association of America (GHAA), 1129 20 St NW, Ste 600, Washington, DC 20036
T: (202) 778-3200
Founded: 1959; Members: 1000
Pres: James F. Doherty
Focus: Public Health
Periodicals
Group Health News (weekly)
Journal (weekly)
Medical Directors Conference Proceedings (weekly) *15181*

Gulf and Caribbean Fisheries Institute (GCFI), 38 Wentworth St, Charleston, SC 29401
T: (803) 577-5697; Fax: (803) 577-5697
Founded: 1948; Members: 750
Gen Secr: Melvin Goodwin
Focus: Fisheries
Periodical
Proceedings (weekly) *15182*

Gypsy Lore Society (GLS), 5607 Greenleaf Rd, Cheverly, MD 20785
T: (301) 341-1261
Founded: 1977; Members: 250
Gen Secr: Sheila Salo
Focus: Ethnology
Periodicals
Membership Directory (weekly)
Newsletter (weekly) *15183*

Haiku Society of America (HSA), c/o Japan Society, Japan House, 333 E 47 St, New York, NY 10017
Founded: 1968; Members: 500
Pres: Francine Porad
Focus: Lit
Periodical
Frogpond (weekly) *15184*

Haitian Coalition on Aids (HCA), 50 Court St, Ste 605, Brooklyn, NY 11201
T: (718) 855-0972; Fax: (718) 855-0972
Founded: 1983; Members: 70
Gen Secr: Y. Rosemond
Focus: Immunology *15185*

Handwriting Analysts – International (HAI), 1504 W 29 ST, Davenport, IA 52804
T: (319) 391-7350
Founded: 1964; Members: 80
Pres: Robert B. Martin
Focus: Graphology
Periodical
Graphological Forum (3 times annually) . *15186*

Hardwood Research Council (HRC), POB 34518, Memphis, TN 38184-0518
T: (901) 377-1824; Fax: (901) 382-6419
Founded: 1953; Members: 1200
Gen Secr: John A. Picher
Focus: Forestry
Periodical
Proceedings: Annual Hardwood Symposium *15187*

Harry S. Truman Library Institute for National and International Affairs, US 24 Hwy and Delaware St, Independence, MO 64050-1798
T: (816) 833-1400; Fax: (816) 833-4368
Founded: 1957; Members: 2000
Gen Secr: Dr. Benedict K. Zobrist
Focus: Int'l Relat; Poli Sci
Periodical
Newsletter (weekly) *15188*

Harvard Environmental Law Society (ELS), c/o Harvard Law School, 201 Austin, Cambridge, MA 02138
T: (617) 495-3125
Founded: 1945; Members: 250
Pres: Mike Wall
Focus: Ecology *15189*

Harvey Society (HS), c/o Peter Palese, Div of Microbiology, Mount Sinai School of Medicine, 1 Gusgaze Levy Pl, New York, NY 10029
T: (212) 241-7318
Founded: 1905; Members: 1600
Gen Secr: Peter Palese
Focus: Lit; Med
Periodical
Harvey Lectures (weekly) *15190*

Haunt Hunters (HH), 2188 Sycamore Hill Ct, Chesterfield, MO 63017
T: (314) 831-1379
Founded: 1965; Members: 300
Gen Secr: Gordon J. Hoener
Focus: Psych *15191*

Headmasters Association (HA), c/o Agnes Underwood, National Cathedral School, Mount Saint Alban NW, Washington, DC 20016
T: (202) 537-6334; Fax: (202) 658-1378
Founded: 1893; Members: 263
Gen Secr: Agnes Underwood
Focus: Educ
Periodical
Membership List (weekly) *15192*

Health Industry Business Communications Council (HIBCC), 5110 N 40 St, Ste 250, Phoenix, AZ 85018
T: (602) 381-1091; Fax: (602) 381-1093
Founded: 1984; Members: 1000
Focus: Computer & Info Sci *15193*

Health Optimizing Institute (HOI), POB 1233, Del Mar, CA 92014
T: (619) 481-7751
Founded: 1978
Gen Secr: Gail Slavin
Focus: Public Health *15194*

Health Physics Society (HPS), 8000 Westpark Dr, Ste 130, McLean, VA 22102
T: (703) 790-1745
Founded: 1956; Members: 6890
Gen Secr: Richard J. Burk
Focus: Public Health; Physics
Periodicals
Health Physics Journal (weekly)
Membership Handbook (weekly)
Newsletter (weekly) *15195*

Health Sciences Communications Association (HESCA), 6728 Old McLean Village Dr, McLean, VA 22101
T: (703) 556-9324
Founded: 1956; Members: 400
Gen Secr: Carol Ann Kiner
Focus: Public Health
Periodicals
Feedback (weekly)
Journal of Biocommunications (weekly) . *15196*

Health Sciences Consortium (HSC), 201 Silver Cedar Ct, Chapel Hill, NC 27514
T: (919) 942-8731; Fax: (919) 942-3689
Founded: 1971; Members: 1000
Gen Secr: Frank B. Penta
Focus: Public Health *15197*

Hear Center, 301 E Del Mar Blvd, Pasadena, CA 91101
T: (213) 681-4641
Founded: 1954
Gen Secr: Josephine F. Wilson
Focus: Otorhinolaryngology
Periodical
The Listener (weekly) *15198*

Heart Disease Research Foundation (HDRF), 50 Court St, Rm 306, Brooklyn, NY 11201
T: (718) 649-6210
Founded: 1962
Gen Secr: Dr. Y. Omura
Focus: Cardiol *15199*

Hebrew Culture Foundation (HCF), 110 E 59 St, New York, NY 10022
T: (212) 752-0600
Founded: 1955
Gen Secr: Dr. Herman L. Sainer
Focus: Cultur Hist *15200*

Hegel Society of America (HSA), c/o Dept of Philosophy, Villanova University, Villanova, PA 19085
T: (215) 645-4747
Founded: 1969; Members: 500
Gen Secr: Lawrence S. Stepelevich
Focus: Philos
Periodical
The Owl of Minerva (weekly) *15201*

Hemingway Society (HS), c/o Prof. Robert W. Lewis, Dept of English, University of North Dakota, Grand Forks, ND 58202-8237
T: (701) 777-3321; Fax: (701) 777-3650
Founded: 1980; Members: 500
Pres: Prof. Robert W. Lewis
Focus: Lit *15202*

Herbert Hoover Presidential Library Association (HHPLA), POB 696, West Branch, IA 52358
T: (319) 643-5327
Founded: 1954
Gen Secr: F. Forbes Olberg
Focus: Libraries & Bk Sci
Periodical
Newsletter (weekly) *15203*

Herpes Resource Center – American Social Health Association, POB 13827, Research Triangle Park, NC 27709
T: (919) 361-8400; Fax: (919) 361-8425
Founded: 1979; Members: 28000
Gen Secr: Peggy Clarke
Focus: Med *15204*

Herpetologists' League, c/o Texas Natural Heritage Program, Texas Park and Wildlife Dept, 3000 Interstate Hwy S, Ste 100, Austin, TX 78704
T: (512) 449-4311; Fax: (512) 389-4394
Founded: 1936; Members: 2000
Gen Secr: Andrew H. Price
Focus: Zoology
Periodicals
Herpetologica (weekly)
Herpetological Monographs *15205*

Higher Education Consortium for Urban Affairs (HECUA), c/o Hamline University, 1536 Hewitt Av, Saint Paul, MN 55104
T: (612) 646-8832; Fax: (612) 659-9421
Founded: 1971; Members: 17

Gen Secr: Dr. G. Hesser
Focus: Educ 15206

Higher Education Panel (HEP), c/o Elaine El-Khawas, American Council on Education, 1 Dupont Circle, Washington, DC 20036
T: (202) 939-9445; Fax: (202) 833-4760
Founded: 1971; Members: 666
Gen Secr: Elaine El-Khawas
Focus: Educ 15207

Higher Education Resource Services (HERS), c/o Wellesley College, Cheever House, Wellesley, MA 02181-8259
T: (617) 283-2529; Fax: (617) 283-3645
Founded: 1972
Gen Secr: Cynthia Secor
Focus: Educ 15208

High Frontier (HF), 2800 Shirlington Rd, Ste 405A, Arlington, VA 22206
T: (703) 671-4111; Fax: (703) 931-6432
Founded: 1982
Pres: Daniel O. Graham
Focus: Aero 15209

Highway Users Federation for Safety and Mobility (HUF), 1776 Massachussets Av NW, Washington, DC 20036
T: (202) 857-1200; Fax: (202) 857-1220
Founded: 1970; Members: 400
Pres: Lester P. Lamm
Focus: Safety 15210

Himalayan International Institute of Yoga Science and Philosophy of the U.S.A., RR 1, Box 400, Honesdale, PA 18431
T: (717) 253-5551; Fax: (717) 253-9078
Founded: 1971; Members: 2500
Gen Secr: Rudolph M. Ballentine
Focus: Philos; Psych
Periodicals
Dawn (weekly)
Himalayan Institute Quarterly (weekly)
Research Bulletin (weekly) 15211

Hispanic Institute (HI), 612 W 116 St, New York, NY 10027
T: (212) 854-4187
Founded: 1920; Members: 189
Gen Secr: Susana Redondo de Feldman
Focus: Ethnology
Periodical
Revista Hispánica Moderna (weekly) . . 15212

Hispanic Society of America (HSA), 613 W 155 St, New York, NY 10032
T: (212) 926-2234
Founded: 1904; Members: 400
Gen Secr: Theodore S. Beardsley
Focus: Hist; Arts; Lit; Ling
Periodical
Hispanic Review (weekly) 15213

Histadruth Ivrith of America (HIA), 47 W 34 St, New York, NY 10001
T: (212) 629-9443
Founded: 1916; Members: 10000
Gen Secr: Dr. Aviva S. Barzel
Focus: Ling
Periodicals
Hadoar (weekly)
Lamishpacha (weekly) 15214

Histamine Research Society of North America (HRSNA), c/o Timothy Sullivan, Dept of International Medicine and Allergy, University of Texas Southwestern Medical Center, 5223 Harry Hines Blvd, Dallas, TX 75235-8859
T: (214) 688-3004
Founded: 1946; Members: 60
Pres: Timothy Sullivan
Focus: Chem; Immunology 15215

Histochemical Society (HCS), POB 294, Woods Hole, MA 02543
T: (508) 457-7680
Founded: 1950; Members: 550
Gen Secr: Morton D. Maser
Focus: Chem
Periodicals
The Journal of Histochemistry and Cytochemistry (weekly)
Membership Roster (weekly) 15216

Historical Committee of the Mennonite Church (MHC), 1700 S Main St, Goshen, IN 46526
T: (219) 535-7477; Fax: (219) 535-7660
Members: 8
Gen Secr: L. Miller
Focus: Hist; Rel & Theol
Periodical
Mennonite Historical Bulletin (weekly) . . 15217

Historical Society of Washington, DC (HSWDC), 1307 New Hampshire Av NW, Washington, DC 20036
T: (202) 785-2068
Founded: 1894; Members: 1800
Gen Secr: John V. Alviti
Focus: Hist
Periodicals
Calendar of Events (weekly)
Newsletter (3 times annually)
Records of Columbia Historical Society (weekly)
. . . . 15218

History of Dermatology Society (HDS), 1819 J.F. Kennedy Blvd, Ste 465, Philadelphia, PA 19103
T: (215) 563-8333
Founded: 1973; Members: 120
Pres: Lawrence C. Parish
Focus: Derm 15219

History of Economics Society (HES), c/o J. Patrick Raines, Economics Dept, University of Richmond, Richmond, VA 23173
T: (804) 289-8566
Founded: 1972; Members: 600
Gen Secr: J. Patrick Raines
Focus: Econ
Periodical
Bulletin (weekly) 15220

History of Education Society (HES), c/o C.H.Edson, College of Education, University of Oregon, Eugene, OR 97403
T: (503) 346-1367; Fax: (503) 346-5174
Founded: 1960; Members: 450
Gen Secr: C.H. Edson
Focus: Hist; Educ
Periodical
History of Education (weekly) 15221

History of Science Society (HSS), 35 Dean St, Worcester, MA 01609
T: (508) 831-5712
Founded: 1924; Members: 3000
Gen Secr: Michael M. Sokal
Focus: Hist; Sci
Periodicals
Isis (weekly)
Isis Critical Bibliography (weekly)
Newsletter (weekly)
Osiris (weekly)
Resource Letters 15222

Hohenzollern Society (HS), 82 Atlantic St, Keyport, NJ 07735
T: (908) 739-1799
Founded: 1983
Gen Secr: J.J. Daub
Focus: Hist 15223

Holistic Dental Association (HDA), c/o Dr. Paul Plowman, 4801 Richmond Sq, Oklahoma City, OK 73118
T: (405) 840-5600
Founded: 1980; Members: 200
Pres: Dr. Paul Plowman
Focus: Dent 15224

Holistic Health Havens (HHH), 3419 Thom Blvd, Las Vegas, NV 89130
T: (702) 873-4542
Founded: 1980
Pres: Dr. Joseph M. Kadans
Focus: Public Health 15225

Home Economics Education Association (HEEA), POB 603, Gainesville, VA 22065
T: (703) 349-4676
Founded: 1927; Members: 3700
Gen Secr: Stephanie H. Price
Focus: Educ; Home Econ
Periodical
Home Economics Educator (weekly) . . 15226

Homeopathic Council for Research and Education (HCRE), c/o William Bergman, 50 Park Av, New York, NY 10016
T: (212) 684-2290
Founded: 1965; Members: 5
Gen Secr: William Bergman
Focus: Homeopathy 15227

Horace Mann League of the U.S.A. (HML), POB 252, Summit, NJ 07902
T: (908) 273-8743
Founded: 1922; Members: 500
Gen Secr: Dr. Paul W. Rossey
Focus: Educ
Periodical
Newsletter 15228

Horatio Alger Society (HAS), 4907 Allison Dr, Lansing, MI 48910
T: (517) 882-3203
Founded: 1961; Members: 273
Gen Secr: Carl T. Hartmann
Focus: Lit
Periodical
The Newsboy (weekly) 15229

Horticultural Research Institute (HRI), 1250 I St NW, Ste 500, Washington, DC 20005
T: (202) 789-2900; Fax: (202) 789-1893
Founded: 1962; Members: 300
Gen Secr: Robert Dolibois
Focus: Hort
Periodicals
Directory for the Nursery Industry and Related Associations (weekly)
Journal of Environmental Horticulture (weekly)
Research Letter (weekly) 15230

Hospital Research and Educational Trust (HRET), 840 N Lake Shore Dr, Chicago, IL 60611
T: (312) 280-6620
Founded: 1944
Gen Secr: Deborah Bohr
Focus: Med
Periodicals
Economic Trends (5 times annually)
Health Services Research (weekly) . . . 15231

House Ear Institute (HEI), 2100 W Third St, Los Angeles, CA 90057
T: (213) 483-4431; Fax: (213) 413-6739
Founded: 1946
Gen Secr: James Boswell
Focus: Otorhinolaryngology
Periodicals
Oto Review (weekly)
Research Bulletin (weekly) 15232

Human Biology Council (HBC), c/o Dr. Gary D. James, Cardiovascular Center, New York Hospital, Cornell Medical Center, 520 E 70 St, New York, NY 10021
T: (212) 746-2191; Fax: (212) 746-8451
Founded: 1974
Gen Secr: Dr. Gary James
Focus: Bio 15233

Human/Dolphin Foundation (H/DF), 33307 Decker School Rd, Malibu, CA 90265
T: (213) 457-9600
Founded: 1976; Members: 2036
Gen Secr: John C. Lilly
Focus: Zoology 15234

Human Factors Society (HFS), POB 1369, Santa Monica, CA 90406-1369
T: (310) 394-1811; Fax: (310) 394-2410
Founded: 1957; Members: 4800
Gen Secr: Lynn Strother
Focus: Physiology; Eng; Psych
Periodicals
Bulletin (weekly)
Directory & Yearbook (weekly)
Human Factors (weekly)
Proceedings of The Human Factors Society (weekly) 15235

Human Lactation Center (HLC), 666 Sturges Hwy, Westport, CT 06880
T: (203) 259-5995; Fax: (203) 259-7667
Founded: 1975; Members: 1000
Gen Secr: Dana Raphael
Focus: Public Health
Periodical
The Lactation Review 15236

Human Relations Area Files (HRAF), POB 2054, Yale Station, New Haven, CT 06520
T: (203) 777-2334; Fax: (203) 777-2337
Founded: 1949; Members: 289
Pres: Melvin Ember
Focus: Anthro; Geography; Sociology; Psych
Periodical
Behavior Science Research: HRAF Journal ot Comparative Studies (weekly) 15237

Human Resource Certification Institute, 606 N Washington St, Alexandria, VA 22314
T: (703) 548-3440; Fax: (703) 836-0367
Founded: 1975; Members: 7500
Pres: Michael R. Losey
Focus: Business Admin 15238

Human Resource Planning Society (HRPS), 41 E 42 St, Ste 1509, New York, NY 10017
T: (212) 490-6387; Fax: (212) 682-6851
Founded: 1977; Members: 2150
Gen Secr: Dr. Steven J. Noble
Focus: Business Admin 15239

Human Resources Research Organization (HumRRO), 66 Canal Center Pl, Ste 400, Alexandria, VA 22314
T: (703) 549-3611; Fax: (703) 549-9025
Founded: 1951
Pres: William C. Osborn
Focus: Sociology
Periodical
Bibliography 15240

Huntington's Disease Society of America (HDSA), 140 W 22 St, New York, NY 10011-2420
T: (212) 242-1968; Fax: (212) 243-2443
Founded: 1986; Members: 37000
Gen Secr: Steve Bajardi
Focus: Neurology
Periodical
The Matter (weekly) 15241

Huxley Institute for Biosocial Research (HIBR), 900 N Federal Hwy, Ste 330, Boca Raton, FL 33432
T: (407) 393-6167
Founded: 1971; Members: 3000
Gen Secr: Mary R. Haggerty
Focus: Biochem; Psych
Periodicals
Newsletter (weekly)
Orthomolecular Psychiatry (weekly) . . . 15242

Hydroponic Society of America (HSA), POB 6067, Concord, CA 94524
T: (510) 682-4193; Fax: (510) 827-2504
Founded: 1979; Members: 600
Gen Secr: Gene Brisbon
Focus: Hort
Periodicals
Directory of Suppliers (weekly)
The Hydroponic/Soilless Grower (weekly)
Proceedings (weekly) 15243

Ibsen Society of America (ISA), 3 DeKalb Hall, Pratt Institute, Brooklyn, NY 11205
T: (718) 636-3794
Founded: 1978; Members: 200
Pres: Rolf Fjelde
Focus: Lit 15244

IEEE Control Systems Society, c/o Institute of Electrical and Electronics Engineers, 345 E 47 St, New York, NY 10017
T: (212) 705-7900; Fax: (212) 705-4929
Members: 9820
Pres: William S. Levine
Focus: Electric Eng
Periodicals
Control Systems Magazine (weekly)
Transactions on Automatic Control (weekly)
. . . . 15245

IEEE Engineering in Medicine and Biology Society (EMBS), c/o Institute of Electrical and Electronics Engineers, 345 E 47 St, New York, NY 10017
T: (212) 705-7900; Fax: (212) 705-4929
Members: 7026
Focus: Eng; Med; Bio
Periodicals
Engineering in Medicine and Biology Magazine (weekly)
Transaction on Biomedical Engineering (weekly)
. . . . 15246

IEEE Ultrasonics, Ferroelectrical and Frequency Control Society (UFFCS), c/o Donald C. Malocha, ECE Dept, University of Central Florida, Orlando, FL 32816-2450
T: (407) 823-2414; Fax: (407) 823-5835
Members: 2352
Pres: James Greenleaf
Focus: Electric Eng
Periodical
Transactions on Ultrasonics, Ferroelecrics and Frequency Control (weekly) 15247

Illuminating Engineering Society of North America (IESNA), 345 E 47 St, New York, NY 10017
T: (212) 705-7913; Fax: (212) 705-7641
Founded: 1906; Members: 9600
Gen Secr: William Hanley
Focus: Eng; Electric Eng
Periodicals
Journal of the I.E.S. (weekly)
Lighting Design + Application Magazine (weekly)
. . . . 15248

Immigration History Society (IHS), c/o Prof. John Bodnar, Dept of History, Indiana University, Bloomington, IN 47405
T: (812) 855-4491
Founded: 1965; Members: 830
Pres: Prof. John Bodnar
Focus: Hist
Periodicals
Immigration History Newsletter (weekly)
Journal of American Ethnic History (weekly)
. . . . 15249

Independent Association of Questioned Document Examiners (IAQDE), 403 W Washington, Red Oak, IA 51566
T: (712) 623-9130
Founded: 1969; Members: 175
Pres: Robert P. Larson
Focus: Law
Periodical
Insight (weekly) 15250

Independent Citizens Research Foundation for the Study of Degenerative Diseases (ICRFSDD), POB 97, Ardsley, NY 10502
T: (914) 478-1862
Founded: 1957
Gen Secr: Dorothea P. Seeber
Focus: Med 15251

Independent Computer Consultants Association (ICCA), 933 Gardenview Office Pkwy, Saint Louis, MO 63141
T: (314) 997-4633; Fax: (314) 567-5133
Founded: 1976; Members: 1700
Gen Secr: Carolyn Karelitz
Focus: Computer & Info Sci
Periodicals
Conference Proceedings (weekly)
The Independent (weekly) 15252

Independent Educational Services (IES), 353 Nassau St, Princeton, NJ 08542
T: (609) 921-6195
Founded: 1968
Gen Secr: Richard Belding
Focus: Educ
Periodicals
Ideas Newsletter (weekly)
TREE (5 times annually) 15253

Independent Research Libraries Association (IRLA), c/o Werner Gundersheimer, The Folger Shakespeare Library, 201 E Capitol St SE, Washington, DC 20003
T: (202) 544-4600; Fax: (202) 544-4623
Founded: 1972; Members: 15
Pres: Werner Gundersheimer
Focus: Libraries & Bk Sci
Periodical
Directory 15254

Independent Scholars of Asia (ISA), 2321 Russell, Berkeley, CA 94705
T: (510) 849-3791
Founded: 1977; Members: 186
Gen Secr: Dr. Ruth-Inge Heinze
Focus: Ethnology

Periodicals
Newsletter (3 times annually)
Proceedings (weekly)
Trance and Healing in Southeast Asia Today . . . 15255

Indian Dental Association U.S.A. (IDA), 146-02 89 Av, Jamaica, NY 11435
T: (718) 523-8438
Founded: 1983; Members: 345
Pres: Dr. Bhasker Patel
Focus: Dent . . . 15256

Indians Into Medicine (INMED), c/o School of Medicine, University of North Dakota, 501 N Columbia Rd, Grand Forks, ND 58203
T: (701) 777-3037; Fax: (701) 777-3277
Founded: 1973
Gen Secr: Gary D. Farris
Focus: Med . . . 15257

Industrial Development Research Council (IDRC), 40 Technology Park, Norcross, GA 30092
T: (404) 446-8955; Fax: (404) 662-8950
Founded: 1961; Members: 1600
Gen Secr: Prentice Knight
Focus: Business Admin; Econ
Periodicals
IDRC Communicator (weekly)
Industrial Development (weekly) . . . 15258

Industrial Mathematics Society (IMS), POB 159, Roseville, MI 48066
T: (313) 771-0403
Founded: 1949; Members: 100
Gen Secr: Dr. Robert Schmidt
Focus: Math
Periodical
Industrial Mathematics (weekly) . . . 15259

Industrial Relations Research Association (IRRA), c/o University of Wisconsin, 7226 Social Science Bldg, Madison, WI 53706
T: (608) 262-2762; Fax: (608) 262-4747
Founded: 1947; Members: 5000
Gen Secr: Kay B. Hutchinson
Focus: Business Admin; Poli Sci; Sociology; Econ
Periodicals
Annual Research Volume (weekly)
Newsletter (weekly)
Proceedings of IRRA Annual Meeting (weekly)
Proceedings of IRRA Spring Meeting (weekly) . . . 15260

Industrial Research Institute (IRI), 1550 M St NW, Washington, DC 20005
T: (202) 872-6350; Fax: (202) 872-6356
Founded: 1938; Members: 260
Gen Secr: Charles F. Larson
Focus: Eng
Periodicals
Annual Report (weekly)
Research Management (weekly) . . . 15261

Infectious Diseases Society of America (IDSA), c/o Vincent T. Andriole, Yale University School of Medicine, 333 Cedar St, New Haven, CT 06510-8056
T: (203) 785-6782; Fax: (203) 785-6179
Founded: 1963; Members: 3900
Gen Secr: Vincent T. Andriole
Focus: Immunology
Periodicals
Journal of Infectious Diseases (weekly)
Membership Roster (weekly)
Reviews of Infectious Diseases (weekly) . 15262

Inland Bird Banding Association (IBBA), Rte 2, Box 26, Wisner, NE 68791
T: (402) 529-6679
Founded: 1922; Members: 1000
Focus: Ornithology
Periodicals
Inland Bird Banding Newsletter (weekly)
North American Bird Bander (weekly) . . 15263

Inroads, 1221 Locust St, Ste 800, Saint Louis, MO 63103
T: (314) 241-7488; Fax: (314) 241-9325
Founded: 1970
Pres: Reginald D. Dickson
Focus: Econ . . . 15264

Institute for Advanced Research in Asian Science and Medicine (IARASM), POB 555, Garden City, NY 11530
T: (516) 248-0930; Fax: (516) 248-0930
Founded: 1972
Gen Secr: Dr. John J. Kao
Focus: Med; Sci
Periodical
American Journal of Chinese Medicine (weekly) . . . 15265

Institute for Advanced Studies in the Theatre Arts (IASTA), 310 W 56 St, New York, NY 10019
T: (212) 581-3133; Fax: (212) 581-3133
Founded: 1958; Members: 200
Pres: John D. Mitchell
Focus: Perf Arts . . . 15266

Institute for Advanced Studies of World Religions (IASWR), RD 2, Rte 301, Carmel, NY 10512
T: (914) 225-1445; Fax: (914) 225-1485
Founded: 1970
Pres: C.T. Shen
Focus: Rel & Theol
Periodicals
Asian Religious Studies Information
Buddhist Text Information
IASWR Conference Proceedings . . . 15267

Institute for Alternative Agriculture (IAA), 9200 Edmonston Rd, Ste 117, Greenbelt, MD 20770
T: (301) 441-8777
Founded: 1982; Members: 2000
Gen Secr: I. Garth Youngberg
Focus: Agri . . . 15268

Institute for American Indian Studies (IAIS), 38 Curtis St, POB 1260, Washington, CT 06793-0260
T: (203) 868-0518; Fax: (203) 868-1649
Founded: 1971; Members: 1350
Gen Secr: Alberto C. Meloni
Focus: Archeol
Periodicals
Artifacts (5 times annually)
Research Report (weekly) . . . 15269

Institute for Biblical Research (IBR), c/o Dr. Gerald Hawthorne, Wheaton College, Wheaton, IL 60187
T: (708) 752-5913; Fax: (708) 752-5788
Founded: 1973; Members: 300
Gen Secr: Dr. Gerald Hawthorne
Focus: Rel & Theol
Periodical
Newsletter (weekly) . . . 15270

Institute for Certification of Computer Professionals (ICCP), 2200 E Devon Av, Ste 268, Des Plaines, IL 60018
T: (312) 299-4227; Fax: (312) 299-4280
Founded: 1973; Members: 14
Gen Secr: George R. Eggert
Focus: Computer & Info Sci
Periodical
Newsletter (weekly) . . . 15271

Institute for Childhood Resources (INICR), c/o Dr. Stevanne Auerbach, 220 Montgomery St, San Francisco, CA 94104
T: (415) 864-1169
Founded: 1975
Gen Secr: Dr. Stevanne Auerbach
Focus: Educ . . . 15272

Institute for Community Design Analysis (ICDA), 66 Clover Dr, Great Neck, NY 11021
T: (516) 773-4727; Fax: (516) 482-5254
Founded: 1974
Pres: Oscar Newman
Focus: Archit . . . 15273

Institute for Defense Analyses (IDA), 1801 N Beauregard, Alexandria, VA 22311
T: (703) 845-2300; Fax: (703) 845-2588
Founded: 1956
Gen Secr: Larry D. Welch
Focus: Military Sci . . . 15274

Institute for Development of Educational Activities (IDEA), 259 Regency Ridge, Dayton, OH 45459
T: (513) 434-6969; Fax: (513) 434-5203
Founded: 1965
Pres: John M. Bahner
Focus: Educ . . . 15275

Institute for Econometric Research (IER), 3471 N Federal Hwy, Fort Lauderdale, FL 33306
T: (305) 563-9000; Fax: (305) 563-9003
Founded: 1971; Members: 1500
Pres: Norman G. Fosback
Focus: Stats; Econ
Periodicals
Income and Safety (weekly)
The Insiders (weekly)
Market Logic (weekly)
Mutual Fund Forecaster (weekly)
New Issues (weekly) . . . 15276

Institute for Economic Analysis (IEA), 508 Thayer Av, Silver Springs, MD 20910
T: (301) 588-4569
Founded: 1974
Pres: John S. Atlee
Focus: Econ
Periodical
Pocket Charts (weekly) . . . 15277

Institute for Educational Leadership (IEL), 1001 Connecticut Av NW, Ste 310, Washington, DC 20036
T: (202) 822-8405
Founded: 1971
Pres: Michael Usdan
Focus: Educ
Periodical
Newsletter (tri-annually) . . . 15278

Institute for Expressive Analysis (IEA), c/o Dr. Arthur Robbins, 325 West End Av, New York, NY 10023
T: (212) 362-5085
Founded: 1976; Members: 64
Gen Secr: Dr. Arthur Robbins
Focus: Therapeutics
Periodical
Journal (weekly) . . . 15279

Institute for Fluitronics Education, POB 106, Elm Grove, WI 53122-0106
T: (414) 782-0410; Fax: (414) 786-0419
Founded: 1971
Gen Secr: Russell W. Henke
Focus: Educ . . . 15280

Institute for Food and Development Policy (IFDP), 145 Ninth St, San Francisco, CA 94103
T: (415) 864-8555; Fax: (415) 864-3909
Founded: 1975; Members: 17000
Gen Secr: W. Bello
Focus: Sociology; Econ; Nutrition
Periodicals
Development Reports (weekly)
Food First News (weekly) . . . 15281

Institute for Gravitational Strain Pathology (IGSP), POB 526, Rangeley, ME 04970
T: (207) 864-5511
Founded: 1957
Gen Secr: Gertrude Jungmann
Focus: Pathology . . . 15282

Institute for Hospital Clinical Nursing Education (IHCNE), c/o American Hospital Association Center for Nursing, 840 N Lake Shore Dr, Chicago, IL 60611
T: (312) 280-6432; Fax: (312) 280-5995
Founded: 1967; Members: 165
Gen Secr: Marcia R. Hagopian
Focus: Adult Educ; Med
Periodical
The Assembly Communique (weekly) . . 15283

Institute for Intercultural Studies (IIS), 165 E 72 St, Ste 1B, New York, NY 10021
T: (212) 737-1011
Founded: 1943; Members: 6
Gen Secr: Ann Brownell Sloane
Focus: Hist . . . 15284

Institute for Labor and Mental Health (ILMH), 3137 Telegraph Av, Oakland, CA 94609
T: (510) 653-6166
Founded: 1977; Members: 50
Gen Secr: Dr. Richard Epstein
Focus: Psychiatry
Periodicals
Directory
Occupational Stress (weekly) . . . 15285

Institute for Mediterranean Affairs (IMA), 428 E 83 St, New York, NY 10028
T: (212) 988-1725
Founded: 1957; Members: 250
Pres: Seymour Maxwell Finger
Focus: Poli Sci
Periodical
Mediterranean Survey (weekly) . . . 15286

Institute for Palestine Studies (IPS), 3501 M St NW, Washington, DC 20007
T: (202) 342-3990; Fax: (202) 342-3927
Founded: 1963; Members: 40
Gen Secr: Dr. Philip Mattar
Focus: Hist; Poli Sci
Periodicals
Journal of Palestine Studies (weekly)
Revue d'Etudes Palestiniennes (weekly) . 15287

Institute for Philosophy and Public Policy (IPPP), c/o University of Maryland, College Park, MD 20742
T: (301) 405-4759; Fax: (301) 314-9346
Founded: 1976
Gen Secr: Mark Sagoff
Focus: Poli Sci; Philos
Periodical
Report from the Institute for Philosophy and Public Policy (weekly) . . . 15288

Institute for Policy Studies (IPS), 1601 Connecticut Av NW, Washington, DC 20009
T: (202) 234-9382
Founded: 1963
Gen Secr: Michael Shuman
Focus: Sociology; Philos; Sci; Hist; Soc Sci . . . 15289

Institute for Psychohistory (IP), POB 401, New York, NY 10024
T: (212) 799-2294; Fax: (212) 799-2294
Founded: 1976; Members: 50
Gen Secr: Lloyd de Mause
Focus: Hist; Psych
Periodical
The Journal of Psychohistory (weekly) . 15290

Institute for Public Relations Research and Education (IPRRE), 3800 S Tamiami Trail, Ste N, Sarasota, FL 34239
T: (813) 955-5577; Fax: (813) 954-5898
Founded: 1956
Pres: James L. Tolley
Focus: Educ . . . 15291

Institute for Rational-Emotive Therapy (IRET), 45 E 65 St, New York, NY 10021
T: (212) 535-0822
Founded: 1968
Gen Secr: Janet L. Wolfe
Focus: Therapeutics; Psych
Periodical
Journal of Rational-Emotive Therapy (weekly) . . . 15292

Institute for Reality Therapy (IRT), 7301 Medical Center Dr, Ste 104, Canoga Park, CA 91307
T: (818) 888-0688
Founded: 1967
Pres: William Glasser
Focus: Psych
Periodicals
Journal of Reality Therapy (weekly)
Newsletter (weekly) . . . 15293

Institute for Research in Hypnosis and Psychotherapy (IRHP), 1991 Broadway, Apt 18B, New York, NY 10023
T: (212) 874-5290; Fax: (212) 238-1422
Founded: 1954
Gen Secr: Dr. Milton V. Kline
Focus: Psych
Periodical
Morton Prince Digest of Hypnotherapy and Hypnoanalysis (weekly) . . . 15294

Institute for Responsive Education (IRE), 605 Commonwealth Av, Boston, MA 02215
T: (617) 353-3309; Fax: (617) 353-8444
Founded: 1973
Pres: Don Davies
Focus: Educ
Periodical
Equity and Choice (3 times annually) . . 15295

Institute for Social Research (ISR), c/o Marlene Smith, Institute for Social Research, University of Michigan, Ann Arbor, MI 48106
T: (313) 764-8363; Fax: (313) 747-4575
Founded: 1948
Gen Secr: Marlene Smith
Focus: Poli Sci; Sociology; Econ; Psych
Periodicals
Economic Outlook USA (weekly)
Newsletter (weekly)
Survey Research Center Monographs (8-15 times annually)
The Research News (weekly) . . . 15296

Institute for Southern Studies (ISS), POB 531, Durham, NC 27702
T: (919) 419-8311; Fax: (919) 419-8315
Founded: 1970; Members: 8000
Gen Secr: I. Madison
Focus: Hist . . . 15297

Institute for Studies in American Music (ISAM), c/o Conservatory of Music, Brooklyn College, Brooklyn, NY 11210
T: (718) 780-5655; Fax: (718) 951-6140
Founded: 1971; Members: 3900
Gen Secr: H. Wiley Hitchcock
Focus: Music
Periodical
ISAM Newsletter (weekly) . . . 15298

Institute for the Advancement of Engineering (IAE), 24300 Calvert St, Woodland Hills, CA 91367-1113
T: (818) 992-8292; Fax: (818) 992-8292
Founded: 1967; Members: 2400
Gen Secr: Lloyd W. Higginbotham
Focus: Eng . . . 15299

Institute for the Advancement of Human Behavior (IAHB), POB 7226, Stanford, CA 94309
T: (415) 851-8411; Fax: (415) 851-0406
Founded: 1977
Pres: G.W. Piaget
Focus: Med . . . 15300

Institute for the Development of Emotional and Life Skills (IDEALS), POB 391, State College, PA 16804
T: (814) 237-4805
Founded: 1972; Members: 12
Gen Secr: Patricia A. Yoder
Focus: Public Health . . . 15301

Institute for the Development of Indian Law (IDIL), c/o K.K. Kickingbird, School of Law, Oklahoma City University, 2501 N Blackwelder, Oklahoma City, OK 73106
T: (405) 521-5188
Founded: 1971
Gen Secr: K.K. Kickingbird
Focus: Law
Periodical
American Indian Journal (weekly) . . . 15302

Institute for the Future (IF), 2744 Sand Hill Rd, Menlo Park, CA 94025-7020
T: (415) 854-6322; Fax: (415) 854-7850
Founded: 1968
Pres: J. Ian Morrison
Focus: Futurology
Periodical
Perspectives (weekly) . . . 15303

Institute for Theological Encounter with Science and Technology (ITEST), 221 N Grand Blvd, Saint Louis, MO 63103
T: (314) 658-2703
Founded: 1968; Members: 575
Gen Secr: Robert A. Brungs
Focus: Rel & Theol
Periodical
Bulletin (weekly) . . . 15304

Institute for the Study of Human Knowledge (ISHK), POB 176, Los Altos, CA 94023
T: (415) 948-9428
Founded: 1969
Pres: Robert E. Ornstein
Focus: Sci . . . 15305

Institute for the Study of Man (ISM), 6861 Elm St, Ste 4-H, McLean, VA 22101
T: (703) 442-8010
Founded: 1975
Gen Secr: Roger Pearson
Focus: Anthro
Periodical
Journal of Indo-European Studies (weekly) 15306

Institute for Twenty-First Century Studies (ITFCS), 1611 N Kent St, Ste 204, Arlington, VA 22209-2111
T: (703) 841-0048; Fax: (703) 841-0050
Founded: 1983
Gen Secr: Gerald O. Barney
Focus: Lit
Periodical
Proceedings 15307

Institute for Urban Design (IUD), 4253 Karensue Av, San Diego, CA 92122
T: (619) 455-1251
Founded: 1979; Members: 500
Gen Secr: Ann Ferebee
Focus: Archit
Periodicals
Project Monograph: Urban Design Case Studies (weekly)
Urban Design International (weekly)
Urban Design Update (weekly) 15308

Institute of American Indian Arts (IAIA), POB 20007, Santa Fe, NM 87504
T: (505) 988-6463; Fax: (505) 988-6446
Founded: 1962
Pres: Tijerina, Kathryn Harris
Focus: Arts
Periodical
Faculty Handbook (weekly) 15309

Institute of Andean Research (IAR), c/o Craig Morris, Dept of Anthropology, American Museum of Natural History, Central Park W at 79 St, New York, NY 10024
T: (212) 769-5883; Fax: (212)769-5334
Founded: 1937; Members: 25
Gen Secr: Craig Morris
Focus: Anthro 15310

Institute of Andean Studies (IAS), POB 9307, Berkeley, CA 94709
T: (510) 525-7816
Founded: 1960; Members: 181
Gen Secr: John H. Rowe
Focus: Archeol
Periodical
Nawpa Pacha (weekly) 15311

Institute of Certified Financial Planners (ICFP), 7600 E Eastman Av, Ste 301, Denver, CO 80231
T: (303) 751-7600; Fax: (303) 751-1037
Founded: 1973; Members: 7300
Gen Secr: Brent A. Neiser
Focus: Finance 15312

Institute of Chinese Culture (ICC), 86 Riverside Dr, New York, NY 10024
T: (212) 787-6969
Founded: 1944; Members: 100
Gen Secr: Dr. Liang-Chien Cha
Focus: Ethnology 15313

Institute of Civil War Studies (ICWS), c/o Dept of History, Meramec College, 11333 Big Bend, Saint Louis, MO 63122
T: (601) 966-7687
Founded: 1974
Gen Secr: Prof. Alexander C. Niven
Focus: Hist 15314

Institute of Cultural Affairs (ICA), 4750 N Sheridan Rd, Chicago, IL 60640
T: (312) 769-6363; Fax: (312) 769-1144
Founded: 1973; Members: 1500
Pres: Raymond Caruso
Focus: Sociology; Psych
Periodical
Highlights: Edges (weekly) 15315

Institute of Early American History and Culture (IEAHC), POB 220, Williamsburg, VA 23185
T: (804) 221-1110
Founded: 1943
Gen Secr: Ronald Hoffman
Focus: Cultur Hist; Hist
Periodicals
Newsletter
William and Mary Quarterly (weekly) . . 15316

The Institute of Electrical and Electronics Engineers (IEEE), 345 E 47 St, New York, NY 10017
T: (212) 705-7900
Founded: 1963; Members: 274000
Gen Secr: Eric Herz
Focus: Electronic Eng; Electric Eng
Periodicals
Aerospace and Electronic Systems Magazine (weekly)
ASSP Magazine (weekly)
Circuits and Devices Magazine (weekly)
Communications Magazine (weekly)
Computer Applications in Power Magazine (weekly)
Computer Graphics and Applications Magazine (weekly)
Computer Magazine (weekly)
Control Systems Magazine (7 times annually)
Design and Test of Computers Magazine (weekly)
Electrical Insulation Magazine (weekly)
Electron Device Letters (weekly)
Engineering in Medicine and Biology Magazine (weekly)
Engineering Management Review (weekly)
Expert Magazine: Intelligent Systems and their Applications (weekly)
Journal of Electronic Materials (weekly)
Journal of Lightwave Technology (weekly)
Journal of Oceanic Engineering (weekly)
Journal of Quantum Electronics (weekly)
Journal of Solid-State Circuits (weekly)
Journal on Selected Areas in Communications (9 times annually)
Micro Magazine (weekly)
Network Magazine: The Magazine of Computer Communications (weekly)
Photonics Technology Letters (weekly)
Power Engineering Review (weekly)
Software Magazine (weekly)
Technology and Society Magazine (weekly)
Transactions on Acoustics, Speech and Signal Processing (weekly)
Transactions on Aerospace and Electronic Systems (weekly)
Transactions on Antennas and Propagation (weekly)
Transactions on Automatic Control (weekly)
Transactions on Biomedical Engineering (weekly)
Transactions on Broadcasting (weekly)
Transactions on Circuits and Systems (weekly)
Transactions on Communications (weekly)
Transactions on Components, Hybrids and Manufacturing Technology (weekly)
Transactions on Computer-Aided Design of Integrated Circuits (weekly)
Transactions on Computers (weekly)
Transactions on Consumer Electronics (weekly)
Transactions on Education (weekly)
Transactions on Electrical Insulation (weekly)
Transactions on Electromagnetic Compatibility (weekly)
Transactions on Electron Devices (weekly)
Transactions on Energy Conversion (weekly)
Transactions on Engineering Management (weekly)
Transactions on Geoscience and Remote Sensing (weekly)
Transactions on Industrial Electronics (weekly)
Transactions on Industry Applications (weekly)
Transactions on Information Theory (weekly)
Transactions on Instrumentation and Measurement (weekly)
Transactions on Kowledge and Data Engineering (weekly)
Transactions on Magnetics (weekly)
Transactions on Medical Imaging (weekly)
Transactions on Microwave Theory and Techniques (weekly)
Transactions on Nuclear Science (weekly)
Transactions on Pattern Analysis and Machine Intelligence (weekly)
Transactions on Plasma Science (weekly)
Transactions on Power Delivery (weekly)
Transactions on Power Electronics (weekly)
Transactions on Power Systems (weekly)
Transactions on Professional Communication (weekly)
Transactions on Reliability (5 times annually)
Transactions on Robotics and Automation (weekly)
Transactions on Semiconductor Manufacturing (weekly)
Transactions on Software Engineering (weekly)
Transactions on Systems, Man and Cybernetics (weekly)
Transactions on Ultrasonics, Ferroelectrics and Frequency Control (weekly)
Transactions on Vehicular Technology (weekly)
Translation Journal on Magnetics in Japan (weekly) 15317

Institute of Electrology Educators (IEE), c/o Wallace A. Roberts, POB 211, Bellingham, MA 02019
Founded: 1979; Members: 55
Pres: Wallace A. Roberts
Focus: Educ; Med 15318

Institute of Environmental Sciences (IES), 940 E Northwest Hwy, Mount Prospect, IL 60056
T: (312) 255-1561
Founded: 1959; Members: 4200
Gen Secr: Janet A. Ehmann
Focus: Nat Sci
Periodicals
Journal of Environmental Sciences (weekly)
Newsletter (weekly) 15319

Institute of Financial Education (IFE), 111 E Wacker Dr, Chicago, IL 60601-4680
T: (312) 946-8800; Fax: (312) 946-8802
Founded: 1922; Members: 50000
Pres: Gale Menele
Focus: Adult Educ; Finance
Periodical
Membership Newsletter (weekly) 15320

Institute of Food Technologists (IFT), 221 N LaSalle St, Ste 300, Chicago, IL 60601
T: (312) 782-8424; Fax: (312) 782-8348
Founded: 1939; Members: 23000
Gen Secr: Daniel Weber
Focus: Food
Periodicals
Food Technology (weekly)
Journal of Food Science (weekly)
Membership Directory (weekly) 15321

Institute of Gas Technology (IGT), 3424 S State St, Chicago, IL 60616
T: (312) 567-3650; Fax: (312) 567-5209
Founded: 1941; Members: 170
Pres: Bernard S. Lee
Focus: Eng; Energy
Periodicals
Energy Statistics (weekly)
Gas Abstracts (weekly)
International Gas Technology Highlights (weekly) 15322

Institute of General Semantics (IGS), 163 Engle St, Englewood, NJ 07631
T: (201) 568-0551; Fax: (201) 569-1793
Founded: 1938; Members: 500
Gen Secr: Marjorie S. Zelner
Focus: Ling
Periodicals
General Semantics Bulletin (weekly)
Newsletter (weekly) 15323

Institute of Human Origins (IHO), 2453 Ridge Rd, Berkeley, CA 94709
T: (510) 845-0333; Fax: (510) 845-9453
Members: 800
Pres: Donald C. Johanson
Focus: Anthro 15324

Institute of Industrial Engineers (IIE), 25 Technology Park, Norcross, GA 30092
T: (404) 449-0460; Fax: (404) 263-8532
Founded: 1948; Members: 35000
Gen Secr: Gregory Balestrero
Focus: Eng
Periodicals
The Engineering Economist (weekly)
Industrial Engineering (weekly)
Industrial Management (weekly)
Transactions (weekly) 15325

Institute of International Education (IIE), 809 United Nations Plaza, New York, NY 10017
T: (212) 883-8200; Fax: (212) 984-5452
Founded: 1919; Members: 350
Pres: Richard Krasho
Focus: Educ 15326

Institute of Judicial Administration (IJA), c/o New York School of Law, 1 Washington Sq S, New York, NY 10012
T: (212) 998-6288
Founded: 1952; Members: 1000
Gen Secr: Prof. Samuel Estreicher
Focus: Law
Periodical
Report (weekly) 15327

Institute of Laboratory Animal Resources (ILAR), c/o National Research Council, 2101 Constitution Av NW, Washington, DC 20418
T: (202) 334-2590; Fax: (202) 334-1639
Founded: 1952
Gen Secr: Thomas L. Wolfle
Focus: Vet Med
Periodical
ILAR News (weekly) 15328

Institute of Lithuanian Studies (ILS), c/o Dr. Violeta Kelertas, Dept of Slavic and Baltic Studies, University of Illinois at Chicago, POB 4348, Chicago, IL 60680
T: (312) 996-4412
Founded: 1951; Members: 135
Pres: Dr. Violeta Kelertas
Focus: Ethnology
Periodicals
Lithuanian Studies
Proceedings of Conferences of the Institute of Lithuanian Studies: Lituanistikos Instituto Suvaziavimo Darbai (biennial) 15329

The Institute of Management Sciences (TIMS), 290 Westminster St, Providence, RI 02903
T: (401) 274-2525; Fax: (401) 274-3189
Founded: 1953; Members: 8066
Focus: Business Admin
Periodicals
Information Systems Research (weekly)
Interfaces (weekly)
Management Science (weekly)
Marketing Science (weekly)
Mathematics of Operations Research (weekly)
Organization Science (weekly) 15330

Institute of Mathematical Statistics (IMS), 3401 Investment Blvd, Ste 7, Hayward, CA 94545
T: (510) 783-8141
Founded: 1935; Members: 1000000
Gen Secr: J.L. Gonzalez
Focus: Stats
Periodicals
The Annals of Probability (weekly)
The Annals of Statistics (weekly)
The IMS Bulletin (weekly)
Statistical Science (weekly) 15331

Institute of Medicine (IOM), 2101 Constitution Av NW, Washington, DC 20418
T: (202) 334-2138
Founded: 1970; Members: 500
Gen Secr: Queta Bond
Focus: Med 15332

Institute of Nautical Archaeology (INA), PO Drawer HG, College Station, TX 77841
T: (409) 845-6694; Fax: (409) 845-6399
Founded: 1973; Members: 900
Pres: Robert K. Vincent
Focus: Archeol 15333

Institute of Navigation (ION), 1800 Diagonal Rd, Ste 480, Alexandria, VA 22314
T: (703) 683-7101; Fax: (703) 683-7105
Founded: 1945; Members: 4000
Pres: Philip W. Ward
Focus: Navig
Periodical
Navigation (weekly) 15334

Institute of Noise Control Engineering (INCE), POB 3206, Arlington Branch, Poughkeepsie, NY 12603
T: (914) 462-4006
Founded: 1971; Members: 250
Gen Secr: George C. Maling
Focus: Public Health
Periodicals
Noise Control Engineering Journal (weekly)
Noise/News (weekly) 15335

Institute of Nuclear Materials Management (INMM), 60 Revere Dr, Ste 500, Northbrook, IL 60062
T: (312) 480-9080
Founded: 1958; Members: 800
Gen Secr: Barbara Scott
Focus: Nucl Res
Periodicals
Journal of Nuclear Materials Management (weekly)
Proceedings of Annual Meeting 15336

Institute of Outdoor Drama (IOD), c/o University of North Carolina, CB 3240 Nations Bank Plaza, Chapel Hill, NC 27599-3240
T: (919) 962-1328
Founded: 1963
Gen Secr: Scott J. Parker
Focus: Perf Arts
Periodical
Newsletter (weekly) 15337

Institute of Public Administration (IPA), 55 W 44 St, New York, NY 10036
T: (212) 730-5486; Fax: (212) 398-9305
Founded: 1906
Pres: Dwight Ink
Focus: Poli Sci; Public Admin 15338

Institute of Textile Technology (ITT), POB 391, Charlottesville, VA 22902
T: (804) 296-5511; Fax: (804) 296-2957
Founded: 1944
Pres: Charles G. Tewksbury
Focus: Textiles
Periodical
Textile Technology Digest (weekly) . . . 15339

Institute of the American Musical (IAM), 121 N Detroit St, Los Angeles, CA 90036
T: (213) 934-1221
Founded: 1972
Pres: Miles Kreuger
Focus: Music 15340

Institute of the Great Plains (IGP), c/o Museum of the Great Plains, Steve Wilson Dr, POB 68, Lawton, OK 73502
T: (405) 581-3460
Founded: 1960; Members: 814
Gen Secr: Steve Wilson
Focus: Hist
Periodicals
Great Plains Journal (weekly)
Museum of the Great Plains Newsletter . 15341

Institute of Transportation Engineers (ITE), 525 School St SW, Ste 410, Washington, DC 20024-2797
T: (202) 554-8050; Fax: (202) 863-5486
Founded: 1930; Members: 11000
Gen Secr: Thomas W. Brahms
Focus: Transport
Periodicals
Directory (weekly)
Journal (weekly) 15342

Institute of World Affairs (IWA), Twin Lakes, Salisbury, CT 06068
T: (203) 824-5135; Fax: (203) 824-7884
Founded: 1924
Gen Secr: Bradford P. Johnson
Focus: Sci 15343

Institute on Hospital and Community Psychiatry (IHCP), 1400 K St NW, Ste 505, Washington, 20005
T: (202) 682-6174
Founded: 1949; Members: 1000
Gen Secr: Sandra M. Hass-Yamhure
Focus: Psychiatry 15344

Insulated Cable Engineers Association (ICEA), POB 440, South Yarmouth, MA 02664
T: (617) 394-4424; Fax: (617) 394-1194
Founded: 1925; Members: 100
Gen Secr: Edward E. McIlveen
Focus: Electronic Eng; Eng 15345

Insurance Information Institute (III), 110 William St, New York, NY 10038
T: (212) 669-9200; Fax: (212) 732-1916
Founded: 1959; Members: 250
Pres: Gordon C. Stewart
Focus: Insurance
Periodicals
Data Base Reports (weekly)
Insurance Facts (weekly)
Insurance Review (weekly) 15346

Intelligent Buildings Institute (IBI), 2101 L St NW, Ste 300, Washington, DC 20037
T: (202) 457-1988; Fax: (202) 457-1989
Founded: 1986; Members: 300
Gen Secr: Sherry Hunt
Focus: Archit
Periodicals
Directory (weekly)
IBI News (weekly) 15347

Inter-American Association of Sanitary Engineering and Environmental Sciences (IAASEES), 18729 Considine Dr, Brookeville, MD 20833
T: (301) 492-7686; Fax: (301) 492-7285
Founded: 1946; Members: 5000
Gen Secr: Dr. Richard F. Cole
Focus: Eng; Hygiene
Periodicals
Ingeniera Sanitaria (weekly)
Newsletter (weekly) 15348

Inter-American Bar Association (IABA), 815 15 St NW, Ste 921, Washington, DC 20005
T: (202) 393-1217; Fax: (202) 393-1241
Founded: 1940; Members: 3090
Gen Secr: J.A. Toro
Focus: Law
Periodicals
Membership Directory (weekly)
Quarterly Newsletter (weekly) 15349

Inter-American College Association (IACA), 1832 Stratford Pl, Pomona, CA 91768
T: (714) 629-1460
Founded: 1962; Members: 25
Gen Secr: E. Soteris Wallace
Focus: Educ
Periodicals
Listing of Inter-American Graduate Programs (biennial)
Progress Bulletin (weekly) 15350

Inter-American Commercial Arbitration Commission (I-ACAC), OAS Administrative Bldg, 19 and Constitution Av NW, Rm 211, Washington, DC 20006
T: (202) 458-3249; Fax: (202) 828-0157
Founded: 1934
Gen Secr: Charles R. Norberg
Focus: Law
Periodical
Inter-American Arbitration (weekly) . . . 15351

Inter-American Council for Education, Science and Culture (CIECC), c/o Organization of American States, 1889 F St NW, Washington, DC 20006
T: (202) 458-3783
Founded: 1948; Members: 32
Gen Secr: Juan Carlos Torchia Estrada
Focus: Educ 15352

Inter-American Music Council (IAM), c/o Prof. E. Paesky, 1889 F St NW, Ste 230-C, Washington, DC 20006
T: (202) 458-3158; Fax: (202) 458-3967
Founded: 1956; Members: 32
Gen Secr: Prof. E. Paesky
Focus: Music 15353

Inter-American Safety Council (IASC), 33 Park Pl, Englewood, NJ 07631
T: (201) 871-0004; Fax: (201) 871-2074
Founded: 1938; Members: 3200
Gen Secr: Glen E. Mickey
Focus: Safety
Periodicals
Noticias de Seguridad (weekly)
El Supervisor (weekly)
Usted y su Familia (weekly) 15354

Inter-American Tropical Tuna Comission, c/o Scripps Institution of Oceanography, 8604 La Jolla Shores Dr, La Jolla, CA 92037
T: (619) 546-7100; Fax: (619) 546-7133
Founded: 1950
Gen Secr: Shigeto Hase
Focus: Fisheries; Zoology
Periodicals
Annual Report (weekly)
Bulletin (weekly)
Special Report (weekly) 15355

Interchurch Medical Assistance (IMA), College Av at Blue Ridge, POB 429, New Windsor, MD 21776
T: (410) 635-8720; Fax: (410) 635-8726
Founded: 1961; Members: 14
Gen Secr: Paul Derstine
Focus: Med 15356

Intercollegiate Broadcasting System (IBS), Box 592, Vails Gate, NY 12584-0592
T: (914) 565-6710; Fax: (914) 561-6932
Founded: 1940; Members: 650
Pres: Jeffrey N. Tellis
Focus: Educ
Periodical
Journal of College Radio (weekly) . . . 15357

Intercultural Development Research Association (IDRA), 5835 Callaghan Rd, Ste 350, San Antonio, TX 78228
T: (210) 684-8180; Fax: (210) 684-5389
Founded: 1974
Gen Secr: Dr. Maria Robledo Montecal
Focus: Ling 15358

Interfaith Forum on Religion, Art and Architecture (IFRAA), 1777 Church St NW, Washington, DC 20036
T: (215) 646-6799
Founded: 1978; Members: 500
Gen Secr: Henry Jung
Focus: Archit
Periodicals
Faith and Forum (weekly)
Newsletter (weekly) 15359

Interior Design Educators Council (IDEC), 14252 Culver Dr, Ste A-311, Irvine, CA 92714
T: (714) 551-1622
Founded: 1962; Members: 350
Pres: Joy Dohr
Focus: Adult Educ; Graphic & Dec Arts, Design
Periodicals
Journal of Interior Design Education and Research (weekly)
Record (weekly) 15360

Intermedia, 475 Riverside Dr, Rm 670, New York, NY 10115
T: (212) 870-2376; Fax: (212) 870-2055
Founded: 1970; Members: 30
Gen Secr: David W. Briddell
Focus: Lit; Comm Sci 15361

International Academy at Santa Barbara (IASB), 800 Garden, Ste D, Santa Barbara, CA 93101-1552
T: (805) 965-5010; Fax: (805) 965-6071
Founded: 1960
Gen Secr: S.J. Shaffer
Focus: Sci
Periodicals
Current World Leaders (8 times annually)
Energy Review (weekly)
Environmental Periodicals Bibliography (weekly)
. . . . 15362

International Academy of Behavioral Medicine (IABMCP), 6750 Hillcrest Plaza, Ste 304, Dallas, TX 75230
T: (214) 458-8333; Fax: (214) 490-5228
Founded: 1988; Members: 1100
Gen Secr: George Mount
Focus: Med
Periodical
Newsletter (weekly) 15363

International Academy of Chest Physicians and Surgeons (IACPS), c/o American College of Chest Physicians, 3300 Dundee Rd, Northbrook, IL 60062
T: (708) 498-1400; Fax: (708) 698-1791
Founded: 1941; Members: 15000
Gen Secr: Dr. Alfred Soffer
Focus: Surgery
Periodicals
Journal (weekly)
Member Bulletin
Membership Directory (weekly) 15364

International Academy of Management – U.S. Branch (IAM), c/o Dr. David Fagiano, American Management Association, 135 W 50 St, New York, NY 10020
T: (212) 586-8100
Founded: 1958; Members: 228
Pres: Dr. David Fagiano
Focus: Business Admin 15365

International Academy of Myodontics (IAM), c/o Harry N. Cooperman, 800 Airport Blvd, Doylestown, PA 18901
T: (215) 345-1149; Fax: (215) 609-2588
Founded: 1970; Members: 1100
Pres: Harry N. Cooperman
Focus: Dent 15366

International Academy of Nutrition and Preventive Medicine, POB 18433, Asheville, NC 28814-0433
T: (704) 258-3243
Founded: 1971; Members: 400
Gen Secr: Carroll Thompson
Focus: Med; Hygiene; Public Health
Periodicals
Conference Proceedings (weekly)
Directory (weekly)
Journal (weekly)
Lay Newsletter (10 times annually)
Professional Newsletter (weekly) 15367

International Academy of Pathology (IAP), c/o Jack M. Layton, 2509 N Campbell Av, Ste 300, Tucson, AZ 85719
T: (602) 626-6843; Fax: (602) 795-5895
Founded: 1906; Members: 8300
Gen Secr: Jack M. Layton
Focus: Pathology
Periodical
International Pathology (weekly) 15368

International Academy of Trial Lawyers (IATL), 4 N Second St, Ste 175, San Jose, CA 95113
T: (408) 275-6767; Fax: (408) 275-6767
Founded: 1954; Members: 554
Gen Secr: Barbara V. Laskin
Focus: Law 15369

International Aids Prospective Epidemiology Network (INAPEN), c/o David G. Ostrow, 155 N Harbor Dr, Chicago, IL 60601
T: (312) 565-2103; Fax: (312) 565-2109
Founded: 1984; Members: 100
Pres: David G. Ostrow
Focus: Immunology 15370

International Alban Berg Society (IABS), c/o John Frisch, 33 W 42 St, New York, NY 10036
T: (212) 642-2389; Fax: (212) 642-2642
Founded: 1966; Members: 300
Gen Secr: John Frisch
Focus: Music 15371

International Aloe Science Council (IASC), 1033 La Posada Dr, Ste 220, Austin, TX 78752-3824
T: (512) 454-8626; Fax: (512) 454-3036
Founded: 1981; Members: 30
Gen Secr: Don McCullough
Focus: Pharmacol 15372

International Anesthesia Research Society (IARS), 2 Summit Park Dr, Ste 140, Cleveland, OH 44131-2553
T: (216) 642-1124; Fax: (216) 642-1127
Founded: 1922; Members: 15000
Gen Secr: Anne F. Maggiore
Focus: Anesthetics
Periodical
Anesthesia and Analgesia (weekly) . . . 15373

International Arthur Schnitzler Research Association (IASRA), c/o Dept of Foreign Languages, California State University, San Bernardino, CA 92407
T: (714) 880-5851
Founded: 1961; Members: 700
Gen Secr: J.B. Johns
Focus: Lit; Cultur Hist
Periodical
Modern Austrian Literature (weekly) . . . 15374

International Association for Aquatic Animal Medicine (IAAAM), c/o Beverly Dixon, POB 3157, San Leandro, CA 94578
T: (510) 881-3422; Fax: (510) 727-2035
Founded: 1969; Members: 450
Gen Secr: Beverly Dixon
Focus: Vet Med
Periodicals
IAAAM Conference Abstract (weekly)
Newsletter (weekly) 15375

International Association for Comparative Research on Leukemia and Related Diseases (IACRLRD), 300 W Tenth Av, Ste 1132, Columbus, OH 43210
T: (614) 293-3067; Fax: (614) 293-3132
Founded: 1963; Members: 500
Gen Secr: Dr. David S. Yohn
Focus: Hematology
Periodical
Symposium Proceedings (biennial) . . . 15376

International Association for Computer Information Systems (IACIS), c/o Dr. Susan Haugen, Dept of Accountancy, University of Wisconsin at Eau Claire, Eau Claire, WI 54702
T: (715) 836-2952
Founded: 1960; Members: 700
Gen Secr: Dr. Susan Haugen
Focus: Computer & Info Sci; Educ
Periodical
The Journal of Computer Information Systems (weekly) 15377

International Association for Computing in Education (ISCE), c/o University of Oregon, 1787 Agate St, Eugene, OR 97403
T: (503) 346-4414; Fax: (503) 346-5890
Founded: 1989; Members: 2000
Gen Secr: Maia S. Howes
Focus: Computer & Info Sci; Educ
Periodicals
Handbook and Directory
Journal (weekly)
Newsletter 15378

International Association for Dental Research (IADR), 1111 14 St NW, Ste 1000, Washington, DC 20005
T: (202) 898-1050; Fax: (202) 789-1033
Founded: 1920; Members: 9500
Gen Secr: John J. Clarkson
Focus: Dent
Periodicals
IADR Newsletter (weekly)
Journal of Dental Research (weekly) . . 15379

International Association for Ecology (INTECOL), Drawer E, Aiken, SC 29802
T: (803) 725-2472
Founded: 1967; Members: 1300
Gen Secr: Dr. Rebecca R. Sharitz
Focus: Ecology
Periodicals
Ecology International (1-2 times annually)
INTECOL Newsletter (weekly) 15380

International Association for Energy Economics (IAEE), 28790 Chagrin Blvd, Ste 300, Cleveland, OH 44122
T: (216) 464-5365; Fax: (216) 464-5363
Founded: 1977; Members: 3000
Gen Secr: David Williams
Focus: Energy
Periodicals
The Energy Journal (weekly)
The LAEE Membership Directory (weekly) 15381

International Association for Great Lakes Research (IAGLR), 2200 Bonisteel Blvd, Ann Arbor, MI 48109-2099
T: (313) 747-1673; Fax: (313) 747-2748
Founded: 1967; Members: 1000
Gen Secr: Steve Schneider
Focus: Geography
Periodicals
Journal of Great Lakes Research (weekly)
Lakes Letter (weekly) 15382

International Association for Healthcare Security and Safety (IAHS2), POB 637, Lombard, IL 60148
T: (708) 953-0990; Fax: (708) 957-1786
Founded: 1968; Members: 1500
Pres: Linda Glasson
Focus: Med
Periodicals
Healthcare Protection Management (weekly)
Membership Directory (weekly)
Newsletter (weekly)
Training Bulletin 15383

International Association for Housing Science (IAHS), POB 340254, Coral Gables, FL 33134
T: (305) 348-3171; Fax: (305) 348-3797
Founded: 1972; Members: 600
Pres: Dr. Oktay Ural
Focus: Archit
Periodical
International Journal for Housing Science (weekly)
. . . . 15384

International Association for Hydrogen Energy (IAHE), POB 248266, Coral Gables, FL 33124
T: (305) 284-4666; Fax: (305) 284-4792
Founded: 1974; Members: 2500
Pres: T. Nejat Veziroglu
Focus: Energy
Periodical
International Journal of Hydrogen Energy (weekly)
. . . . 15385

International Association for Learning Laboratories (IALL), c/o Dr. Robin Lawrason, Media Learning Center, Temple University, Philadelphia, PA 19122
T: (215) 787-4758; Fax: (215) 787-3731
Founded: 1965; Members: 470
Gen Secr: Dr. Robin Lawrason
Focus: Ling; Educ; Adult Educ; Eng
Periodical
The Journal of Educational Techniques and Technologies (weekly) 15386

International Association for Mathematical Geology (IAMG), c/o Dr. Richard B. McCammon, U.S. Geological Survey, National Center 920, Reston, VA 22092
T: (703) 648-6150; Fax: (703) 648-6057
Founded: 1968; Members: 600
Pres: Dr. Richard B. McCammon
Focus: Geology
Periodicals
Computer & Geosciences (weekly)
Mathematical Geology (weekly) 15387

International Association for Near-Death Studies (IANDS), POB 7767, Philadelphia, PA 19101-7767
Founded: 1981; Members: 1200
Pres: Elizabeth W. Fenske
Focus: Parapsych
Periodicals
Journal of Near-Death Studies (weekly)
Revitalized Signs Newsletter (weekly) . . 15388

International Association for Personnel Women (IAPW), POB 969, Andover, MA 01810-0017
T: (508) 474-0750
Founded: 1950; Members: 1500
Pres: Brenda Jackson
Focus: Business Admin
Periodicals
Journal (weekly)
Membership Roster (weekly)
Newsletter (weekly) 15389

International Association for Philosophy of Law and Social Philosophy – American Section (AMINTAPHIL), c/o Robert Moffat, Coillege of Law, University of Florida, Gainesville, FL 32611-2038
T: (904) 392-2211; Fax: (904) 392-8727
Founded: 1963; Members: 350
Gen Secr: Robert Moffat
Focus: Philos 15390

International Association for Research in Income and Wealth (IARIW), c/o Dept of Economics, New York Univerity, 269 Mercer St, Rm 700, New York, NY 10003
T: (212) 924-4386; Fax: (212) 366-5067
Founded: 1947; Members: 400
Gen Secr: Jane Forman
Focus: Econ
Periodicals
Membership Directory
The Review of Income and Wealth (weekly)
. . . . 15391

International Association for the Advancement of Earth and Environmental Sciences (IAAEES), c/o Dr. Musa Qutub, Geography and Environmental Studies Dept, Northeastern Illinois University, 5500 N Saint Louis Av, Chicago, IL 60625
T: (312) 794-2628; Fax: (708) 824-8436
Founded: 1972; Members: 10000
Pres: Dr. Musa Qutub
Focus: Ecology

Periodical
Environmental Resource Magazine (weekly) 15392

International Association for the Physical Sciences of the Ocean (IAPSO), c/o Dr. Robert E. Stevenson, POB 1161, Del Mar, CA 92014-1161
T: (619) 481-0850; Fax: (619) 481-6938
Founded: 1919; Members: 6000
Gen Secr: Dr. Robert E. Stevenson
Focus: Oceanography
Periodicals
Procès-Verbaux (every two years)
Publication Scientifique (weekly) 15393

International Association for the Properties of Water and Steam (IAPWS), c/o Dr. Barry Dooley, EPRI, POB 10412, Palo Alto, CA 94303
T: (415) 855-2458
Founded: 1971; Members: 9
Gen Secr: Dr. Barry Dooley
Focus: Standards; Physics
Periodical
Releases and Guidelines 15394

International Association for the Study of Cooperation in Education (IASCE), POB 1582, Santa Cruz, CA 95061-1582
T: (408) 426-7926; Fax: (408) 426-3360
Founded: 1979; Members: 3500
Gen Secr: Nancy Graves
Focus: Educ
Periodicals
Cooperative Learning (weekly)
IASCE Newsletter (weekly) 15395

International Association of Agricultural Economists (IAAE), 1211 W 22 St, Ste 216, Oak Brook, IL 60521
T: (708) 571-9393; Fax: (708) 571-9580
Founded: 1929; Members: 1850
Gen Secr: Dr. Walter J. Armbruster
Focus: Agri
Periodical
Proceedings 15396

International Association of Allergology and Clinical Immunology (IAACI), 611 E Wells St, Milwaukee, WI 53202
T: (414) 276-6445; Fax: (414) 276-3349
Founded: 1945; Members: 17000
Gen Secr: Rick Iber
Focus: Immunology
Periodical
Allergy & Clinical Immunology News (weekly) 15397

International Association of Boards of Examiners in Optometry (IAB), 4330 East West Hwy, Ste 1117, Bethesda, MD 20814
T: (301) 718-6506; Fax: (301) 656-0989
Founded: 1919; Members: 54
Gen Secr: Ronald W. Jones
Focus: Adult Educ; Ophthal
Periodical
State Board Forum (3 times annually) . 15398

International Association of Buddhist Studies (IABS), c/o Institute of Buddhist Studies, 1900 Addison St, Berkeley, CA 94704
T: (415) 642-3547
Founded: 1976; Members: 500
Gen Secr: Lou Lancaster
Focus: Rel & Theol
Periodical
Journal of the International Association of Buddhist Studies (weekly) 15399

International Association of Campus Law Enforcement Administrators (IACLEA), c/o Peter J. Berry, 638 Prospect Av, Hartford, CT 06105-4298
T: (203) 233-4531; Fax: (203) 232-0819
Founded: 1958; Members: 1200
Gen Secr: Peter J. Berry
Focus: Law
Periodicals
Campus Law Enforcement Journal (weekly)
Conference Proceedings (weekly)
Membership Directory (weekly) 15400

International Association of Career Consulting Firms (IACCF), 1730 N Lynn St, Ste 502, Arlington, VA 22209
T: (703) 525-1191; Fax: (202) 833-3014
Founded: 1987; Members: 35
Gen Secr: William Drohan
Focus: Business Admin 15401

International Association of Coroners and Medical Examiners (IACME), 6913 W Plank Rd, Peoria, IL 61604
T: (309) 697-8100
Founded: 1938; Members: 335
Gen Secr: Herbert H. Buzbee
Focus: Med
Periodical
Newsletter (biennial) 15402

International Association of Counseling Services (IACS), 101 S Whiting St, Ste 211, Alexandria, VA 22304
T: (703) 823-9840; Fax: (703) 823-9843
Members: 250
Gen Secr: Nancy E. Roncketti
Focus: Educ
Periodicals
Counseling Services
Directory of Counseling Services 15403

International Association of Defense Counsel (IADC), 20 N Wacker Dr, Ste 3100, Chicago, IL 60606
T: (212) 368-1494
Founded: 1920; Members: 2500
Gen Secr: Richard J. Hayes
Focus: Law; Insurance
Periodicals
Committee Newsletter (weekly)
IADC News (weekly)
Insurance Counsel Journal (weekly)
Membership Directory (weekly) 15404

International Association of Educational Peace Officers (IAEPO), c/o Alan Bragg, Spring I.S.D. Police Dept, 15330A Kuykendahl, Houston, TX 77090
T: (713) 893-7473; Fax: (713) 444-6173
Founded: 1977; Members: 127
Gen Secr: Alan Bragg
Focus: Educ 15405

International Association of Educators for World Peace (IAEWP), POB 3282, Mastin Lake Station, Huntsville, AL 35810-0282
T: (205) 534-5501; Fax: (205) 536-1018
Founded: 1969; Members: 18500
Gen Secr: Charles Mercieca
Focus: Educ; Poli Sci; Prom Peace
Periodicals
Circulation Newsletter (weekly)
Directory (weekly)
Peace Progress (weekly) 15406

International Association of Forensic Toxicologists, c/o Dr. V. Spiehler, 422 Tustin, Newport Beach, CA 92663
Founded: 1963; Members: 1000
Pres: Prof. R.A. de Zeeuw; Gen Secr: Dr. V. Spiehler
Focus: Toxicology
Periodical
Bulletin (weekly) 15407

International Association of Human-Animal Interaction Organizations (IAHAIO), 321 Burnett Av S, Ste 303, POB 1080, Renton, WA 98057
T: (206) 226-7357; Fax: (206) 235-1076
Founded: 1974
Gen Secr: Linda M. Hines
Focus: Behav Sci 15408

International Association of Jazz Educators (IAJE), POB 724, Manhattan, KS 66502
T: (913) 776-8744
Founded: 1968; Members: 6800
Gen Secr: Bill McFarlin
Focus: Music; Educ
Periodical
Jazz Educators Journal (weekly) 15409

International Association of Knowledge Engineers (IAKE), 973D Russell Av, Gaithersburg, MD 20879
T: (301) 926-2577; Fax: (301) 926-4243
Founded: 1987; Members: 2500
Gen Secr: Julie Walker
Focus: Electronic Eng 15410

International Association of Laryngectomees (IAL), c/o American Cancer Society, 1599 Clifton Rd NE, Atlanta, GA 30329
T: (404) 320-3333
Founded: 1952; Members: 30000
Focus: Otorhinolaryngology
Periodicals
Directory (weekly)
News (weekly) 15411

International Association of Music Libraries – United States Branch (IAML-US), c/o Music Library, City College, Convent Av at 138 St, New York, NY 10031
Founded: 1955; Members: 415
Pres: Neil N. Ratliff
Focus: Libraries & Bk Sci 15412

International Association of Ocular Surgeons (IAOS), 233 E Erie St, Ste 710, Chicago, IL 60611
T: (312) 951-1400
Founded: 1981; Members: 900
Gen Secr: John G. Bellows
Focus: Surgery; Ophthal
Periodical
Journal of Ocular Therapy and Surgery (weekly) 15413

International Association of Optometric Executives (IAOE), 120 N Third, Bismarck, ND 58501-3860
T: (701) 258-6766; Fax: (701) 258-9005
Members: 75
Gen Secr: Mark Landreth
Focus: Ophthal 15414

International Association of Orthodontics (IAO), 1100 Lake St, Oak Park, IL 60301
T: (708) 445-0320; Fax: (708) 445-0321
Founded: 1961; Members: 1800
Gen Secr: Joanna Carey
Focus: Dent 15415

International Association of Physical Education and Sports for Girls and Women (IAPESGW), c/o Ruth Schellberg, 50 Skyline Dr, Mankato, MN 56001
T: (507) 345-3665
Founded: 1949; Members: 400
Gen Secr: Ruth Schellberg
Focus: Sports
Periodical
Report Following Congresses 15416

International Association of School Librarianship (IASL), POB 1486, Kalamazoo, MI 49005
T: (616) 343-5728
Founded: 1971; Members: 875
Gen Secr: Dr. Jean E. Lowrie
Focus: Libraries & Bk Sci
Periodicals
Annual Conference Proceedings (weekly)
Newsletter (weekly) 15417

International Association of Theoretical and Applied Limnology (IATAL), c/o Dr. Robert G. Wetzel, Dept of Biology, University of Alabama, Tuscaloosa, AL 35487-0344
T: (205) 438-1793
Founded: 1922; Members: 3000
Gen Secr: Dr. Robert G. Wetzel
Focus: Water Res 15418

International Atherosclerosis Society (IAS), c/o Barbara Gordin, 6565 Fannin St, Houston, TX 77030
T: (713) 790-4226; Fax: (713) 793-1080
Founded: 1979; Members: 7000
Gen Secr: Barbara Gordin
Focus: Cardiol 15419

International Bio-Environmental Foundation (IBEF), c/o Steven A. Ross, 15300 Ventura Blvd, Ste 405, Sherman Oaks, CA 91403
T: (818) 907-5483
Founded: 1978; Members: 400
Gen Secr: Steven A. Ross
Focus: Ecology 15420

International Bird Rescue Research Center (IBRRC), 699 Potter St, Berkeley, CA 94710
T: (510) 841-9086; Fax: (510) 841-9089
Founded: 1971; Members: 800
Gen Secr: Jay Holcomb
Focus: Ecology; Ornithology
Periodical
International Bird Rescue Newsletter (weekly) 15421

International Brecht Society (IBS), c/o Prof. Ward B. Lewis, Dept of Germanic and Slavic Languages, University of Georgia, Athens, GA 30602
T: (706) 542-3663
Founded: 1968; Members: 300
Gen Secr: Prof. Ward B. Lewis
Focus: Lit
Periodicals
Brecht Yearbook (weekly)
Communications (weekly) 15422

International Bridge, Tunnel and Turnpike Association (IBTTA), 2120 L St NW, Ste 305, Washington, DC 20037
T: (202) 659-4620; Fax: (202) 659-0500
Founded: 1932; Members: 260
Gen Secr: Neil D. Schuster
Focus: Civil Eng
Periodical
Tollways (weekly) 15423

International Bronchoesophagological Society (IBES), c/o David R. Sanderson, Mayo Clinic Scottsdale, 13400 E Shea Blvd, Scottsdale, AZ 85259
T: (602) 391-8000; Fax: (602) 391-7006
Members: 400
Gen Secr: David R. Sanderson
Focus: Otorhinolaryngology 15424

International Bulb Society (IBS), POB 4928, Culver City, CA 90230-4928
T: (310) 827-3229; Fax: (714) 856-8511
Founded: 1933; Members: 500
Gen Secr: M. Howe
Focus: Botany
Periodicals
Herbertia (weekly)
Newsletter of APLS (weekly) 15425

International Bundle Branch Block Association (IBBBA), 6631 W 83 St, Los Angeles, CA 90045-2899
T: (310) 670-9132
Founded: 1979
Gen Secr: Rita Kurtz Lewis
Focus: Med
Periodical
Heartbeat (weekly) 15426

International Cardiology Foundation (ICF), c/o American Heart Association, 7272 Greenville Av, Dallas, TX 75231-4599
T: (214) 373-6300
Founded: 1957; Members: 2500
Gen Secr: R. Starke
Focus: Cardiol 15427

International Catholic Esperanto Association, c/o Mubarak Anwar Amar, Saint Matthew's Roman Catholic Church, 1303 Lincolnshire Dr, Champaign, IL 61821
T: (217) 359-4224
Founded: 1910; Members: 1300
Gen Secr: Mubarak Anwar Amar
Focus: Ling
Periodicals
Espero Katolika (weekly)
Jarliboro (weekly) 15428

International Center for Law in Development (ICLD), 777 United Nations Plaza, New York, NY 10017
T: (212) 687-0036; Fax: (212) 972-9878
Founded: 1978
Pres: Clarence J. Dias
Focus: Law 15429

International Center for Research on Women (ICRW), 1717 Massachusetts Av NW, Washington, DC 20036
T: (202) 797-0007; Fax: (202) 797-0007
Founded: 1976
Gen Secr: Mayra Buvinic
Focus: Sociology; Econ
Periodical
Series of Occasional Papers (weekly) . . 15430

International Center for the Solution of Environmental Problems (ICSEP), 535 Lovett Blvd, Houston, TX 77006
T: (713) 527-8711; Fax: (713) 527-8025
Founded: 1976
Gen Secr: Joseph L. Goldman
Focus: Ecology 15431

International Center of Medieval Art (ICMA), The Cloisters, Fort Tryon Park, New York, NY 10040
T: (212) 928-1146
Founded: 1956; Members: 1200
Pres: Walter Cahn
Focus: Arts
Periodicals
Gesta (weekly)
Newsletter (3 times annually) 15432

International Center of Photography (ICP), 1130 Fifth Av, New York, NY 10128
T: (212) 860-1777; Fax: (212) 360-6490
Founded: 1974; Members: 6800
Gen Secr: C. Capa
Focus: Photo 15433

International Childbirth Education Association (ICEA), POB 20048, Minneapolis, MN 55420
T: (612) 854-8660
Founded: 1960; Members: 12000
Gen Secr: Doris Olson
Focus: Gynecology
Periodicals
ICEA Bookmarks (3 times annually)
International Journal of Childbirth Education (weekly)
Membership Directory (weekly) 15434

International Chiropractors Association (ICA), 1110 N Glebe Rd, Ste 1000, Arlington, VA 22201
T: (703) 528-5000
Founded: 1926; Members: 6000
Gen Secr: Ronald Hendrickson
Focus: Orthopedics
Periodicals
ICA Today: Newsletter (weekly)
International Review of Chiropractic (weekly)
Membership Directory (weekly) 15435

International Christian Esperanto Association (ICEA), c/o Edwin C. Harler, 47 Hardy Rd, Levittown, PA 19056-1311
Founded: 1911; Members: 1300
Gen Secr: Edwin C. Harler
Focus: Ling
Periodical
Dia Regno: Divine Kingdom (10-11 times annually) 15436

International Christian Leprosy Mission (ICLM), POB 23353, Portland, OR 97281-3353
T: (503) 244-5935
Founded: 1943; Members: 12
Pres: Lauritz P. Pillers
Focus: Hygiene
Periodical
Global Missions (weekly) 15437

International College of Dentists (ICD), 51 Monroe St, Ste 1501, Rockville, MD 20850-2421
T: (301) 251-8861; Fax: (301) 738-9143
Founded: 1928; Members: 7200
Gen Secr: Richard G. Shaffer
Focus: Dent
Periodicals
Annual News Letter (weekly)
Roster 15438

International College of Surgeons (ICS), 1516 N Lake Shore Dr, Chicago, IL 60610
T: (312) 642-3555; Fax: (312) 787-1624
Founded: 1935; Members: 14000
Gen Secr: Elisabeth Braam
Focus: Surgery
Periodical
International Surgery (weekly) 15439

International Commission for the Prevention of Alcoholism and Drug Dependency (ICPADD), 12501 Old Columbia Pike, Silver Spring, MD 20904
T: (301) 680-6719; Fax: (301) 680-6090
Founded: 1952; Members: 250
Gen Secr: Thomas R. Neslund
Focus: Public Health
Periodical
Quarterly Bulletin (weekly) 15440

International Commission on Physics Education (ICPE), c/o Prof. E.L. Jossem, Dept of Physics, Ohio State University, 174 W 18 Av, Columbus, OH 43210-1106
T: (614) 292-6959; Fax: (614) 292-7557
Founded: 1960; Members: 13
Pres: Prof. E.L. Jossem
Focus: Educ; Physics
Periodical
ICPE Newletter (weekly) 15441

International Commission on Radiation Units and Measurements (ICRU), 7910 Woodmont Av, Ste 800, Bethesda, MD 20814
T: (301) 657-2652; Fax: (301) 907-8768
Founded: 1925; Members: 101
Gen Secr: W. Roger Ney
Focus: Radiology
Periodical
ICRU Report 15442

International Commission on the History of the Geological Sciences, c/o Harvard-Smithsonian Center for Astrophysics, 60 Garden St, Cambridge, MA 02138
T: (617) 495-7270; Fax: (617) 495-7001
Founded: 1967; Members: 119
Pres: Prof. Dr. David F. Branagan; Gen Secr: Dr. Ursula B. Marvin
Focus: Hist; Geology
Periodical
INHIGEO Newsletter (weekly) 15443

International Communications Association (ICA), 12750 Merit Dr, Ste 710 LB-89, Dallas, TX 75251
T: (214) 233-3889; Fax: (214) 233-2813
Founded: 1948; Members: 740
Gen Secr: Robert Eilers
Focus: Journalism; Comm Sci
Periodicals
Communication Theory (weekly)
Communication Yearbook
Human Communication Research (weekly)
Membership Directory (bi-annually)
Newsletter (weekly) 15444

International Conference of Building Officials (ICBO), 5360 S Workman Mill Rd, Whittier, CA 90601
T: (310) 699-0541; Fax: (310) 692-3853
Founded: 1922; Members: 15000
Gen Secr: James E. Bihr
Focus: Civil Eng
Periodicals
Building Standards (weekly)
Membership Roster (weekly)
Newsletter (weekly) 15445

International Conference on Mechanics in Medicine and Biology (ICMMB), c/o University of Michigan, 2150 G.G. Brown Bldg, Ann Arbor, MI 48109
T: (313) 764-9910; Fax: (313) 747-3170
Founded: 1977
Gen Secr: Wen-Jei Yang
Focus: Eng; Med; Bio
Periodicals
Digest (bi-annually)
Directory of Conference Participants (biennial) 15446

International Congress of Oral Implantologists (ICOI), 248 Lorraine Av, Upper Montclair, NJ 07043
T: (201) 783-6300; Fax: (201) 783-1175
Founded: 1975; Members: 3000
Focus: Dent
Periodicals
Bulletin
Implantologist: The International Journal of Oral Implantology (weekly)
Membership Directory (weekly) 15447

International Congress on High-Speed Photography and Photonics (ICHSPP), 136 S Garfield Av, Janesville, WI 53545
T: (608) 752-5581
Founded: 1952; Members: 31
Gen Secr: William G. Hyzer
Focus: Optics; Photo
Periodical
Proceedings (biennial) 15448

International Copper Association (ICA), 260 Madison Av, New York, NY 10016
T: (212) 251-7240; Fax: (212) 251-7245
Founded: 1960; Members: 42
Pres: Lennart A. Gustafsson
Focus: Metallurgy
Periodical
Report (weekly) 15449

International Correspondence Society of Allergists (ICSA), 5811 Outlook Dr, Shawnee Mission, KS 66202
T: (913) 432-0625; Fax: (913) 432-5833
Founded: 1936; Members: 415
Focus: Immunology
Periodical
The Allergy Letters (weekly) 15450

International Cost Engineering Council (ICEC), c/o American Association of Cost Engineers, POB 1557, Morgantown, WV 26507-1557
T: (304) 296-8444
Founded: 1956; Members: 15
Gen Secr: Kenneth K. Humphreys
Focus: Public Admin 15451

International Council for Health, Physical Education and Recreation (ICHPER), 1900 Association Dr, Reston, VA 22091
T: (703) 476-3486; Fax: (703) 476-9527
Founded: 1958; Members: 600
Gen Secr: Dr. Dong Ja Yang
Focus: Public Health; Physical Therapy
Periodicals
Congress Proceedings (bi-annually)
International Journal of Physical Education (weekly) 15452

International Council for Pressure Vessel Technology, c/o Prof. G.E.O. Widera, Marquette University, 1515 W Wisconsin Av, Milwaukee, WI 53233
T: (414) 288-7259
Founded: 1969; Members: 2500
Pres: Prof. G.E.O. Widera
Focus: Eng 15453

The International Council for Traditional Music (ICTM), c/o Center for Ethnomusicology, Columbia University, New York, NY 10027
Founded: 1947; Members: 1300
Gen Secr: Dieter Christensen
Focus: Music
Periodicals
Bulletin ICTM (weekly)
Yearbook for Traditional Music (weekly) . 15454

International Council – National Academy of Television Arts and Sciences (IC/NATAS), 142 W 57 St, New York, NY 10019
T: (212) 489-6969; Fax: (212) 489-6557
Founded: 1968; Members: 75
Gen Secr: Richard Carlton
Focus: Cinema 15455

International Council of Library Association Executives (ICLAE), c/o Mary Sue Ferrell, California Library Association, 717 K St, Ste 300, Sacramento, CA 95814
T: (916) 447-8541
Founded: 1970; Members: 29
Pres: Mary Sue Ferrell
Focus: Libraries & Bk Sci
Periodical
Newsletter 15456

International Council of Museums – Committee of the American Association of Museums (AAM/ICOM), 1225 Eye St NW, Washington, DC 20005
T: (202) 289-1818; Fax: (202) 289-6578
Founded: 1946; Members: 800
Gen Secr: Dr. Mary Louise Wood
Focus: Arts 15457

International Council of Psychologists (ICP), c/o Patricia J. Fontes, POB 62, Hopkinton, RI 02833-0062
T: (401) 377-3092; Fax: (401) 377-6013
Founded: 1942; Members: 1700
Gen Secr: Patricia J. Fontes
Focus: Psych
Periodicals
Directory (biennial)
International Psychologist (weekly) . . . 15458

International Council of the Museum of Modern Art, 11 W 53 St, New York, NY 10019
T: (212) 708-9470
Founded: 1953; Members: 155
Gen Secr: Waldo Rasmussen
Focus: Arts 15459

International Council on Education for Teaching (ICET), 2009 N 14 St, Ste 609, Arlington, VA 22201
T: (703) 525-5253; Fax: (703) 351-9381
Founded: 1953
Gen Secr: Sandra Klassen
Focus: Adult Educ; Educ
Periodicals
International Yearbook on Teacher Education
Newsletter (weekly)
Proceedings (weekly) 15460

International Courtly Literature Society (ICLS), c/o Michigan State University, East Lansing, MI 48824
T: (517) 355-0769
Founded: 1973; Members: 700
Gen Secr: Joseph T. Snow
Focus: Lit
Periodical
Encomia: Bibliographical Bulletin of the International Courtly Literature Society (weekly) 15461

International Crane Foundation (ICF), E-11376 Shady Lane Rd, Baraboo, WI 53913-9778
T: (608) 356-9462; Fax: (608) 356-9465
Founded: 1973; Members: 5000
Gen Secr: George W. Archibald
Focus: Agri
Periodical
ICF Bugle (weekly) 15462

International Criminal Law Commission (ICLC), 733 El Rancho Rd, Santa Barbara, CA 93108
T: (805) 565-1949; Fax: (805) 565-1536
Founded: 1965; Members: 35
Gen Secr: A. Ballester
Focus: Law
Periodical
The Establishment of an International Criminal Court (weekly) 15463

International Dark-Sky Association (IDSA), 3545 N Stewart, Tucson, AZ 85716
T: (602) 795-1381
Founded: 1988; Members: 1100
Gen Secr: Dr. David L. Crawford
Focus: Astronomy 15464

International Dental Health Foundation (IDHF), 11484 Washington Plaza, Ste 307, Reston, VA 22090
T: (703) 471-8349
Founded: 1981; Members: 450
Gen Secr: Patricia L. Cartwright
Focus: Dent 15465

International Design Conference in Aspen (IDCA), POB 664, Aspen, CO 81612
T: (303) 925-2257; Fax: (303) 925-2257
Founded: 1950
Pres: Gianfranco Zaccai
Focus: Graphic & Dec Arts, Design . . 15466

International Dostoevsky Society (IDS), c/o Prof. Nadine Natov, 3707 Emily St, Kensington, MD 20895
T: (301) 933-2945
Founded: 1971; Members: 250
Gen Secr: Prof. Nadine Natov
Focus: Lit
Periodicals
Dostoevsky Studies (weekly)
Newsletter (weekly)
Proceedings of the International Symposia (every 3-4 years) 15467

International Education Research Foundation (IERF), POB 66940, Los Angeles, CA 90066
T: (310) 397-3655; Fax: (310) 397-7686
Founded: 1969
Gen Secr: Jasmin Saidi
Focus: Educ 15468

International Electronics Packaging Society (IEPS), 114 N Hale St, Wheaton, IL 60187-5113
T: (708) 260-1044; Fax: (708) 260-0867
Founded: 1977; Members: 1500
Gen Secr: William D. Ashman
Focus: Electronic Eng
Periodicals
News (weekly)
Newsletter (weekly) 15469

International Embryo Transfer Society (IETS), 309 W Clark St, Champaign, IL 61820
T: (217) 356-3182; Fax: (217) 398-4119
Founded: 1974; Members: 850
Focus: Vet Med 15470

International Epidemiological Association (IEA), c/o Roger Detels, Dept of Epidemiology, School of Public Health, University of California at Los Angeles, Los Angeles, CA 90024
T: (310) 206-2837; Fax: (310) 206-6039
Founded: 1950; Members: 1900
Gen Secr: Roger Detels
Focus: Med
Periodicals
International Journal of Epidemiology (weekly)
Membership Directory 15471

International Erosion Control Association (IECA), Box 4904, Steamboat Springs, CO 80477
T: (303) 879-3010; Fax: (303) 879-8563
Founded: 1972; Members: 400
Gen Secr: Ben Northcutt
Focus: Geomorph; Agri
Periodicals
Membership Directory (weekly)
Newsletter (weekly) 15472

International Federation for Family Life Promotion (IFFLP), 1511 K St NW, Ste 326, Washington, DC 20005
T: (202) 783-0137; Fax: (202) 783-7351
Founded: 1974; Members: 130
Gen Secr: Claude A. Lanctot
Focus: Sociology; Family Plan
Periodicals
Message to Members (weekly)
Newsletter/Bulletin (weekly) 15473

International Federation of Societies for Electron Microscopy (IFSEM), c/o Gareth Thomas, Dept of Materials Science and Engineering, University of California, 280 Hearst Mining Bldg, Berkeley, CA 94720
T: (510) 642-3813; Fax: (510) 486-5933
Founded: 1962; Members: 42
Pres: Gareth Thomas
Focus: Optics 15474

International Federation of Vexillological Associations (FIAV), POB 580, Winchester, MA 01890
T: (617) 729-9410
Founded: 1967; Members: 25
Gen Secr: Dr. Whitney Smith
Focus: Hist
Periodicals
Info FIAV (biennial)
Report of the International Congresses of Vexillology (biennial) 15475

International Fertilizer Development Center (IFDC), POB 2040, Muscle Shoals, AL 35662
T: (205) 381-6600; Fax: (205) 381-7408
Founded: 1974
Pres: Dr. Amit H. Roy
Focus: Agri 15476

International Film Seminars (IFS), 305 W 21 St, New York, NY 10011
T: (212) 727-7262; Fax: (212) 691-9565
Founded: 1960
Gen Secr: Sally Berger
Focus: Cinema 15477

International Foundation for Art Research (IFAR), 46 E 70 St, New York, NY 10021
T: (212) 879-1780; Fax: (212) 734-4174
Founded: 1968
Gen Secr: Dr. Constance Lowenthal
Focus: Arts
Periodical
IFAR Reports (weekly) 15478

International Foundation for Homeopathy (IFH), 2366 Eastlake Av E, Seattle, WA 98102
T: (206) 324-8230
Founded: 1978; Members: 2000
Gen Secr: Fred Bishop
Focus: Homeopathy
Periodical
Resonance (weekly) 15479

International Galdos Association (IGA), c/o G. Gullon, Dept of Spanish, University of California at Davis, Davis, CA 95616
T: (916) 752-0837
Founded: 1979; Members: 205
Pres: G. Gullon
Focus: Lit 15480

International Graphic Arts Education Association (IGAEA), 4615 Forbes Av, Pittsburgh, PA 15213
T: (412) 682-5170
Founded: 1923; Members: 700
Pres: Michael Stinnett
Focus: Adult Educ; Graphic & Dec Arts, Design
Periodicals
The Communicator (7 times annually)
Membership Directory (weekly)
Visual Communications Journal (weekly) . 15481

International Graphoanalysis Society (IGAS), 111 N Canal St, Chicago, IL 60606
T: (312) 930-9446
Founded: 1929; Members: 50000
Pres: Kathleen Kusta
Focus: Graphology
Periodical
The Journal of Graphoanalysis (weekly) . 15482

International Health Evaluation Association (IHEA), 90 W Montgomery Av, Ste 340, Rockville, MD 20850
T: (301) 762-6050; Fax: (301) 762-7127
Founded: 1971; Members: 300
Gen Secr: Harold A. Timken
Focus: Public Health
Periodical
Newsletter (weekly) 15483

International Health Society (IHS), 1001 E Oxford Lane, Cherry Hills Village, Englewood, CO 80110
T: (303) 789-3003
Founded: 1944; Members: 500
Gen Secr: Franklin L. Bowling
Focus: Adult Educ; Public Health
Periodicals
Bulletin (weekly)
Directory 15484

International Hospital Federation (IHF), c/o American Hospital Association, 840 N Lake Shore Dr, Chicago, IL 60611
T: (312) 280-6000
Founded: 1947; Members: 1788
Gen Secr: Jose Gonzalez
Focus: Med
Periodicals
Membership List (weekly)
Official Yearbook (weekly)
World Hospitals (weekly) 15485

International Human Powered Vehicle Association (IHPVA), POB 51255, Indianapolis, IN 46251
T: (317) 876-9478; Fax: (317) 876-9478
Founded: 1975; Members: 2000
Pres: M. Daily
Focus: Physiology
Periodicals
Human Power (weekly)
Newsletter (weekly) 15486

International Human Resources, Business and Legal Research Association (IHRBLR), POB 9478, Washington, DC 20016
T: (301) 948-5876
Founded: 1981; Members: 12
Pres: Dr. M. de Lafayette
Focus: Econ; Law 15487

International Imagery Association (IIA), POB 1046, Bronx, NY 10471
Founded: 1979
Pres: Akhter Ahsen
Focus: Psych; Lit
Periodicals
Imagery Today: Newsletter (weekly)
International Imagery Bulletin (weekly)
Journal of Mental Imagery (weekly) . . 15488

International Institute for Bioenergetic Analysis (IIBA), 144 E 36 St, New York, NY 10016
T: (212) 532-7742; Fax: (212) 532-5331

Founded: 1956; Members: 1000
Gen Secr: Alexander Lowen
Focus: Psychiatry
Periodicals
Bioenergetic Analysis
Membership Directory 15489

International Institute of Rural Reconstruction – U.S. Chapter (IIRR), 475 Riverside Dr, Rm 1270, New York, NY 10015
T: (212) 870-2992; Fax: (212) 870-2981
Founded: 1960
Focus: Sociology; Poli Sci; Public Health; Develop Areas; Educ
Periodicals
IIRR Report (weekly)
International Sharing (3 times annually)
Rural Reconstruction Review (weekly) . . 15490

International Insurance Society (IIS), POB 870223, Tuscaloosa, AL 35487
T: (205) 348-8974; Fax: (205) 348-8973
Founded: 1965; Members: 1200
Gen Secr: Mary Bickley Silberberg
Focus: Insurance
Periodicals
Governor's Journal (weekly)
International Insurance Society Yearbook (weekly) 15491

International Isotope Society (IIS), c/o Sandoz Research Institute, 403/116 Rte 10, East Hanover, NJ 07936
T: (201) 503-8853; Fax: (609) 235-1360
Founded: 1986; Members: 600
Focus: Biochem 15492

International John Steinbeck Society (IJSS), c/o English Dept, Ball State University, Muncie, IN 47306
T: (317) 285-5688
Founded: 1966; Members: 650
Pres: Dr. Tetsumaro Hayashi
Focus: Lit
Periodicals
Steinbeck Monograph Series (weekly)
Steinbeck Quarterly (weekly) 15493

International Joseph Diseases Foundation (IJDF), POB 2550, Livermore, CA 94551-2550
Founded: 1977
Gen Secr: Rose Marie Silva
Focus: Med
Periodical
Newsletter (weekly) 15494

International Kirlian Research Association (IKRA), 2202 Quentin Rd, Brooklyn, NY 11229
T: (718) 339-3888
Founded: 1975; Members: 200
Pres: B.G. Shafiroff
Focus: Electric Eng
Periodical
Communications and Acta Electrografica (weekly) 15495

International Law Institute (ILI), 1615 New Hampshire Av NW, Washington, DC 20009
T: (202) 483-3036; Fax: (202) 483-3029
Founded: 1955
Gen Secr: Stuart Kerr
Focus: Law
Periodicals
Journal of Law and Technology (3 times annually)
Wallenberg Papers on International Finance (weekly) 15496

International Lead Zinc Research Organization (ILZRO), 2525 Meridian Pkwy, Ste 100, POB 12036, Research Triangle Park, NC 27709
T: (919) 361-4647; Fax: (919) 361-1957
Founded: 1958; Members: 77
Pres: Jerome F. Cole
Focus: Metallurgy
Periodicals
Environmental Bulletin (weekly)
R & D Focus (weekly) 15497

International Magnesium Association (IMA), 1303 Vincent Pl, Ste 1, McLean, VA 22101
T: (703) 442-8888; Fax: (703) 821-1824
Founded: 1943; Members: 91
Gen Secr: Byron B. Clow
Focus: Metallurgy
Periodicals
Magnesium (9 times annually)
Proceedings (weekly) 15498

International Maledicta Society (IMS), POB 14123, Santa Rosa, CA 95402-6123
T: (707) 523-4761
Founded: 1975; Members: 3000
Pres: Reinhold A. Aman
Focus: Ling
Periodical
Maledicta: The International Journal of Verbal Aggression (weekly) 15499

International Management Council (IMC), 430 S 20 St, Ste 3, Omaha, NE 68102
T: (402) 345-1904; Fax: (402) 345-4480
Founded: 1934; Members: 6000
Gen Secr: John K. Shepherd
Focus: Business Admin 15500

International Metallographic Society (IMS), POB 2489, Columbus, OH 43216-2489
T: (614) 424-4278
Founded: 1967; Members: 600
Pres: Dr. Ed McGrath
Focus: Metallurgy
Periodicals
Directory (weekly)
Metallography (weekly)
Sliplines (weekly) 15501

International Microwave Power Institute (IMPI), 13542 Union Village Cir, Clifton, VA 22024
T: (703) 830-5588; Fax: (703) 830-0281
Founded: 1966; Members: 8000
Gen Secr: Robert C. LaGasse
Focus: Eng; Energy
Periodicals
Journal of Microwave Power
Microwave World 15502

International Myopia Prevention Association (IMPA), RD 5, Box 171, Ligonier, PA 15658
T: (0412) 238-2101
Founded: 1974; Members: 100
Pres: Donald S. Rehm
Focus: Ophthal 15503

International New Thought Alliance (INTA), 5003 E Broadway Rd, Mesa, AZ 85206
T: (602) 830-2461; Fax: (602) 830-2461
Founded: 1914
Gen Secr: Mimi Ronnie
Focus: Philos
Periodical
New Thought Quarterly (weekly) 15504

International Numismatic Society Authentication Bureau (INSAB), POB 33134, Philadelphia, PA 19142
T: (215) 365-0752; Fax: (215) 365-0752
Founded: 1976; Members: 500
Gen Secr: Charles R. Hoskins
Focus: Numismatics
Periodicals
Insight (weekly)
Newsletter 15505

International Omega Association (IOA), 1720 S Eads St, POB 2324, Arlington, VA 22202-0324
T: (604) 276-4626
Founded: 1975; Members: 400
Pres: Ian S. Anderson
Focus: Navig
Periodical
Proceedings (weekly) 15506

International Organization for Mycoplasmology (IOM), c/o Dr. George E. Kenny, Dept of Pathobiology, School of Public Health, University of Washington, Seattle, WA 98195
T: (206) 543-1036; Fax: (206) 543-3873
Founded: 1974; Members: 450
Gen Secr: Dr. George E. Kenny
Focus: Microbio
Periodicals
Congress Abstracts (biennial)
Congress Proceedings (biennial)
Newsletter (weekly) 15507

International Organization for the Education of the Hearing Impaired (IOEHI), c/o Alexander Graham Bell Association for the Deaf, 3417 Volta Pl NW, Washington, DC 20007
T: (202) 337-5220
Founded: 1967; Members: 450
Pres: Elizabeth Cole
Focus: Educ 15508

International Organization for the Study of Group Tensions (IOSGT), 240 E 76 St, Apt 1-B, New York, NY 10021
T: (212) 628-1797
Founded: 1970; Members: 150
Pres: Dr. Benjamin B. Wolman
Focus: Psych
Periodical
International Journal of Group Tensions (weekly) 15509

International Percy Grainger Society (IPGS), 7 Cromwell Pl, White Plains, NY 10601-5005
T: (212) 877-8953
Founded: 1964; Members: 400
Pres: Rolf K. Stang
Focus: Music 15510

International Personnel Management Association (IPMA), 1617 Duke St, Alexandria, VA 22314
T: (703) 549-7100; Fax: (703) 684-0948
Founded: 1973; Members: 6500
Gen Secr: Donald K. Tichenor
Focus: Business Admin
Periodicals
Agency Issues (weekly)
IPMA News (weekly)
Membership Directory (weekly)
Public Personnel Management (weekly) . 15511

International Phenomenological Society (IPS), c/o Philosophy and Phenomenological Research, Brown University, POB 1947, Providence, RI 02912
T: (401) 863-3215; Fax: (401) 863-2719
Founded: 1939
Pres: Prof. Ernest Sosa
Focus: Philos
Periodical
Philosophy and Phenomenological Research (weekly) 15512

International Planetarium Society (IPS), c/o Hansen Planetarium, 15 S State St, Salt Lake City, UT 84111
T: (801) 538-2104
Founded: 1970; Members: 400
Pres: John Pogue
Focus: Astronomy; Aero
Periodicals
Directory of Planetaria and Planetarians (weekly)
Planetarian (weekly) 15513

International Plant Propagators Society (IPPS), c/o Center for Urban Horticulture, University of Washington, Seattle, WA 98195
T: (206) 543-8602; Fax: (206) 685-2692
Founded: 1950; Members: 3200
Gen Secr: John A. Wott
Focus: Botany
Periodical
North American Plant Propagator (weekly) 15514

International Precious Metals Institute (IPMI), 4905 Tilghman St, Ste 160, Allentown, PA 18104
T: (215) 395-9700; Fax: (215) 395-5855
Founded: 1976; Members: 1000
Gen Secr: David E. Lundy
Focus: Metallurgy
Periodicals
Membership Directory (weekly)
Precious Metals News amd Reviews (weekly) 15515

International Professional Surrogates Association (IPSA), POB 74156, Los Angeles, CA 90004
T: (213) 469-4720
Founded: 1973; Members: 50
Pres: Valri Jean Swift
Focus: Med
Periodical
Newsletter (weekly) 15516

International Psychohistorical Association (IPA), POB 314, New York, NY 10024
T: (201) 891-4980
Founded: 1976; Members: 200
Gen Secr: Henry Lawton
Focus: Hist; Psych
Periodical
Psychohistory News (weekly) 15517

International Reading Association (IRA), 800 Barksdale Rd, Newark, DE 19714-8139
T: (302) 731-1600; Fax: (302) 731-1057
Founded: 1956; Members: 93000
Gen Secr: Alan E. Farstrup
Focus: Educ
Periodicals
Journal of Reading (8 times annually)
Lectura Vida (weekly)
Reading Research Quarterly (weekly)
Reading Teacher (9 times annually) . . 15518

International Reference Organization in Forensic Medicine and Sciences (INFORM), c/o Dr. William G. Eckert, POB 8282, Wichita, KS 67208
T: (316) 685-7612
Founded: 1966; Members: 1200
Gen Secr: Dr. William G. Eckert
Focus: Forensic Med
Periodicals
Conference Proceeding (weekly)
Criminalist's Sourcebook (weekly)
Letter (weekly) 15519

International Rescue and Emergency Care Association (IRECA), 627 Thompson Way, Covington, VA 24426-2179
T: (612) 941-2926
Founded: 1948; Members: 3000
Gen Secr: Carol Moss
Focus: Med; Safety
Periodical
International Rescuer (weekly) 15520

International Schools Services (ISS), 15 Roszel Rd, POB 5910, Princeton, NJ 08543
T: (609) 452-0990; Fax: (609) 452-2690
Founded: 1955
Pres: William P. Davison
Focus: Educ
Periodicals
ISS Directory of Overseas Schools (weekly)
NewsLinks (weekly) 15521

International Society for Artificial Organs (ISAO), 8937 Euclid Av, Cleveland, OH 44106
T: (216) 421-0757; Fax: (216) 421-1652
Founded: 1977; Members: 1000
Gen Secr: Dr. Yikihiko Nose
Focus: Med 15522

International Society for Astrological Research (ISAR), POB 38613, Los Angeles, CA 90038-0613
T: (818) 333-8702; Fax: (818) 461-3417
Members: 600
Gen Secr: Carol Tebbs
Focus: Parapsych
Periodicals
Kosmos (weekly)
Newsletter (weekly) 15523

International Society for British Genealogy and Family History (ISBGFH), POB 3115, Salt Lake City, UT 84110-3115
T: (216) 234-7508
Founded: 1979; Members: 1000
Gen Secr: Gracelouise Sims Moore
Focus: Genealogy
Periodical
Newsletter (weekly) 15524

International Society for Burn Injuries (ISBI), c/o John A. Boswick, 2005 Franklin St, Ste 660, Denver, CO 80205
T: (303) 839-1694
Founded: 1965; Members: 1300
Gen Secr: John A. Boswick
Focus: Hygiene
Periodicals
Bulletin and Clinical Review of Burn Injuries (weekly)
Journal Burns (weekly)
Membership Directory of ISBI (weekly) . 15525

International Society for Business Education – United States Chapter (ISBE), 1914 Association Dr, Reston, VA 22091
T: (703) 860-8300; Fax: (703) 620-8300
Members: 750
Gen Secr: Dr. Janet M. Treichel
Focus: Adult Educ; Econ 15526

International Society for Cardiovascular Surgery (ICVS), 13 Elm St, POB 1565, Manchester, MA 01944
T: (508) 526-8330; Fax: (508) 526-4018
Founded: 1951; Members: 2500
Gen Secr: William T. Maloney
Focus: Cardiol; Surgery 15527

International Society for Chronobiology (ISC), c/o Dr. Dora K. Hayes, 9105 Shasta Ct, Fairfax, VA 22031
T: (301) 504-9170; Fax: (301) 504-9062
Founded: 1937; Members: 680
Gen Secr: Dr. Dora K. Hayes
Focus: Bio
Periodicals
Chronobiologia (weekly)
International Journal of Chronobiology (weekly) 15528

International Society for Clinical Laboratory Technology (ISCLT), 818 Olive St, Ste 918, Saint Louis, MO 63101-1598
T: (314) 241-1445; Fax: (314) 241-1449
Founded: 1962; Members: 7500
Gen Secr: David Birenbaum
Focus: Med; Eng
Periodical
ISCLT Newsletter (weekly) 15529

International Society for General Semantics (ISGS), POB 728, Concord, CA 94522
T: (510) 798-0311
Founded: 1943; Members: 2400
Gen Secr: Paul Dennithorne-Johnston
Focus: Ling 15530

International Society for Human Ethology (ISHE), c/o Glenn E. Weisfeld, Dept of Psychology, Wayne State University, Detroit, MI 48202
T: (313) 577-2835; Fax: (313) 577-7636
Founded: 1973; Members: 400
Gen Secr: Glenn E. Weisfeld
Focus: Behav Sci 15531

International Society for Humor Studies (ISHS), c/o Don L.F. Nilsen, English Dept, Arizona State University, Tempe, AZ 85287-0302
T: (602) 965-7952; Fax: (602) 894-6457
Founded: 1980; Members: 400
Gen Secr: Don L.F. Nilsen
Focus: Ling; Lit
Periodical
Humor: International Journal of Humor Research (weekly) 15532

International Society for Hybrid Microelectronics (ISHM), 1861 Wiehle Av, Ste 260, POB 2698, Reston, VA 22090
T: (703) 471-0066; Fax: (703) 471-1937
Founded: 1967; Members: 7000
Gen Secr: Walter H. Biddle
Focus: Electronic Eng
Periodicals
Inside ISHM (weekly)
International Journal for Hybrid Microelectronics (weekly)
International Symposium Technical Proceedings (weekly) 15533

International Society for Intercultural Education, Training and Research (SIETAR/INTL), c/o Adriana Arzac, 733 15 St NW, Ste 900, Washington, DC 20005
T: (202) 737-5000; Fax: (202) 737-5553
Founded: 1974; Members: 2200
Gen Secr: Adriana Arzac
Focus: Educ 15534

International Society for Labor Law and Social Security – U.S. National Branch (ISLLSS), 29 Broadway, New York, NY 10006
T: (212) 943-2470
Founded: 1961; Members: 245
Gen Secr: Alfred Giardino
Focus: Law

Periodicals
Comparative Labor Law Journal (weekly)
Proceedings of the Congress 15535

International Society for Neoplatonic Studies (ISNS), c/o Dept of Philosophy, Old Dominion University, Norfolk, VA 23529-0083
T: (804) 683-3861; Fax: (804) 683-3241
Founded: 1973; Members: 600
Gen Secr: R. Baine Harris
Focus: Philos 15536

International Society for Pediatric Neurosurgery (ISPN), c/o M.L. Walker, 100 N Medical Center, Salt Lake City, UT 84113
T: (801) 588-3400
Founded: 1972; Members: 160
Gen Secr: M.L. Walker
Focus: Surgery; Pediatrics
Periodical
Child's Brain Nervous System (weekly) . 15537

International Society for Philosophical Enquiry (ISPE), c/o Betty Hansen, 277 Washington Blvd, Hudson, NY 12534
T: (518) 828-1996
Founded: 1974; Members: 530
Pres: Betty Hansen
Focus: Philos
Periodicals
Membership Roster (weekly)
Telicom (10 times annually) 15538

International Society for Plant Molecular Biology (ISPMB), c/o Biochemistry Dept, University of Georgia, Athens, GA 30602-7229
T: (706) 542-3239; Fax: (706) 542-2090
Founded: 1982; Members: 1850
Gen Secr: Diane Tyner
Focus: Bio
Periodicals
Directory of Members (biennial)
Plant Molecular Biology (weekly)
Plant Molecular Biology Reporter (weekly) 15539

International Society for Research on Aggression (ISRA), c/o L. Rowell Huesmann, Dept of Psychology, University of Illinois at Chicago, POB 4348, Chicago, IL 60680
T: (312) 996-7000; Fax: (312) 413-4122
Founded: 1970; Members: 450
Gen Secr: L. Rowell Huesmann
Focus: Psych; Behav Sci
Periodicals
Agressive Behavior (weekly)
Bulletin (3 times annually) 15540

International Society for Terrain-Vehicle Systems (ISTVS), 72 Lyme Rd, Hanover, NH 03755-1290
T: (603) 646-4362
Founded: 1962; Members: 350
Gen Secr: Ronald A. Liston
Focus: Eng
Periodicals
Journal of Terramechanics (weekly)
Newsletter (weekly) 15541

International Society for the History of Ideas (ISHI), c/o Sidney Axinn, Dept of Philosophy, Temple University, Anderson Hall 739, Philadelphia, PA 19122
T: (215) 787-8293
Founded: 1959; Members: 45
Gen Secr: Sidney Axinn
Focus: Cultur Hist; Philos
Periodical
Journal of the History of Ideas (weekly) 15542

International Society for the Study of Expressionism (ISSE-ETMS) / Ernst Toller Memorial Society, POB 20183, Cincinnati, OH 45220-0183
Founded: 1979
Gen Secr: Eva Lachman-Kalitzki
Focus: Lit 15543

International Society for the Systems Sciences (ISSS), c/o College of Business, Idaho State University, Box 8793, Pocatello, ID 83209
T: (208) 233-6521; Fax: (208) 236-4367
Founded: 1954; Members: 900
Pres: Dr. Ian Mitroff
Focus: Sci
Periodicals
General Systems Bulletin (weekly)
General Systems Yearbook (weekly)
Proceedings of Annual Meetings (weekly) 15544

International Society of Air Safety Investigators (ISASI), Technology Trading Park, 5 Export Dr, Sterling, VA 20174-4421
T: (703) 430-9668; Fax: (703) 450-1745
Founded: 1964; Members: 1190
Gen Secr: Robert A. Patterson
Focus: Safety
Periodicals
The Forum (weekly)
Membership Roster (biennial) 15545

International Society of Arboriculture (ISA), POB GG, Savoy, IL 61874
T: (217) 328-2032; Fax: (217) 328-7483
Founded: 1924; Members: 6500
Gen Secr: William P. Kruidener
Focus: Bio
Periodicals
Journal of Arboriculture (weekly)
Membership Roster
Yearbook (weekly) 15546

International Society of Chemical Ecology (ISCE), c/o Dr. John Romeo, Dept of Biology, University of South Florida, Tampa, FL 33620
T: (813) 974-2336; Fax: (813) 974-5273
Founded: 1983; Members: 750
Gen Secr: Dr. John Romeo
Focus: Ecology
Periodicals
ISCE Newsletter
Journal of Chemical Ecology (weekly)
Proceedings of the Annual Meeting . . 15547

International Society of Cryptozoology (ISC), POB 43070, Tucson, AZ 85733
T: (602) 884-8369
Founded: 1982; Members: 800
Gen Secr: J. Richard Greenwell
Focus: Zoology
Periodicals
Cryptozoology (weekly)
The ISC Newsletter (weekly) 15548

International Society of Dermatologic Surgery (ISDS), POB 4014, Schaumburg, IL 60204
T: (312) 330-0230; Fax: (708) 869-4382
Founded: 1976; Members: 1000
Gen Secr: Lawrence E. Rosenthal
Focus: Derm; Surgery 15549

International Society of Dermatology: Tropical, Geographic and Ecologic (ISD), 200 First St SW, Rochester, MN 55905
T: (507) 284-3736
Founded: 1957; Members: 3000
Gen Secr: S.A. Muller
Focus: Derm 15550

International Society of Differentiation (ISD), c/o Dept of Genetics and Cell Biology, University of Minnesota, 1445 Gortner Av, Saint Paul, MN 55108-1095
T: (612) 625-5754; Fax: (612) 625-5754
Founded: 1970; Members: 300
Gen Secr: Robert G. McKinnell
Focus: Bio
Periodicals
Conference Proceedings
Differentation (weekly)
Newsletter 15551

International Society of Hematology, c/o Dr. Robert A. Kyle, Mayo Clinic, 920 Hilton, Rochester, NY 55905
Fax: (507) 284-0043
Founded: 1948; Members: 2600
Gen Secr: Dr. Robert A. Kyle
Focus: Hematology
Periodical
Newsletter (weekly) 15552

International Society of Parametric Analysts (ISPA), POB 1056, Germantown, MD 20875-1056
T: (301) 353-1840; Fax: (301) 353-0514
Founded: 1979; Members: 800
Pres: Mel Eisman
Focus: Finance
Periodicals
Journal of Parametrics (weekly)
Parametric World (weekly) 15553

International Society of Performing Arts Administrators (ISPAA), 4920 Plainfield NE, Ste 3, Grand Rapids, MI 49505
T: (616) 364-3000; Fax: (616) 364-9010
Founded: 1948; Members: 600
Gen Secr: Michael C. Hardy
Focus: Arts
Periodical
Forum Newsletter (weekly) 15554

International Society of Reproductive Medicine (ISRM), c/o Donald C. McEwen, 11 Furman Ct, Rancho Mirage, CA 92270
T: (619) 340-5080; Fax: (619) 340-6920
Members: 300
Gen Secr: Donald C. McEwen
Focus: Med 15555

International Society on Metabolic Eye Disease (ISMED), 1125 Park Av, New York, NY 10128
T: (212) 427-1246
Founded: 1971; Members: 600
Gen Secr: Heskel M. Haddad
Focus: Ophthal
Periodical
Metabolic, Pediatric and Systemic Ophthalmology (weekly) 15556

International Society on Thrombosis and Hemostasis (ISTH), c/o Medical School, University of North Carolina, Chapel Hill, NC 27599-7035
T: (919) 929-3807; Fax: (919) 929-3935
Founded: 1969; Members: 1400
Gen Secr: Harold R. Roberts
Focus: Hematology
Periodical
Thrombosis and Hemostasis Journal . . 15557

International Studies Association (ISA), c/o Brigham Young University, David M. Kennedy Center 216 HRCB, Provo, UT 84602
T: (801) 378-5459; Fax: (801) 378-7075
Founded: 1959; Members: 2900
Gen Secr: W.L. Hollist
Focus: Sci
Periodicals
International Studies Newsletter (10 times annually)
International Studies Notes (3 times annually)
International Studies Quarterly (weekly)
New Dimensions in International Studies (weekly) 15558

International Survey Library Association (ISLA), c/o Roper Center, University of Connecticut, POB 440, Storrs, CT 06268
T: (203) 486-4440
Founded: 1964; Members: 57
Gen Secr: Everett C. Ladd
Focus: Libraries & Bk Sci
Periodical
Data Acquisitions Catalog (weekly) . . . 15559

International Technology Education Association (ITEA), 1914 Association Dr, Reston, VA 22091-1502
T: (703) 860-2100
Founded: 1939; Members: 6700
Gen Secr: Kendall N. Starkweather
Focus: Arts
Periodicals
Journal of Technology Education (weekly)
Teacher Education Newsletter
Technology Teacher (8 times annually) . 15560

International Technology Education Association – Council for Supervisors (ITEA-CS), c/o George R. Willcox, Virginia Dept of Education, POB 2120, Richmond, VA 23216-2120
T: (804) 225-2020; Fax: (804) 371-0249
Founded: 1951; Members: 300
Gen Secr: George R. Willcox
Focus: Arts
Periodical
Super Link (3 times annually) 15561

International Tele-Education (INTEL-ED), 4619 Larchwood Av, Philadelphia, PA 19143-2107
T: (215) 898-8918
Founded: 1977
Gen Secr: Tom Naff
Focus: Educ 15562

International Test and Evaluation Association (ITEA), 4400 Fair Lakes Ct, Fairfax, VA 22033-3899
T: (703) 631-6220; Fax: (703) 631-6221
Founded: 1980; Members: 2000
Gen Secr: Alan Plishker
Focus: Eng
Periodicals
Journal of Test and Evaluation (weekly)
Symposia Proceedings (weekly) 15563

International Textile and Apparel Association (ITAA), POB 1360, Monument, CO 80132
T: (719) 488-3716
Founded: 1944; Members: 950
Gen Secr: Sandra S. Hutton
Focus: Educ; Textiles 15564

International Thespian Society (ITS), 3368 Central Pkwy, Cincinnati, OH 45225-2392
T: (513) 559-1996
Founded: 1929; Members: 28000
Gen Secr: Ronald L. Longstreth
Focus: Perf Arts
Periodicals
Dramatics Magazine (9 times annually)
ITS Today (3-4 times annually)
State Directors' Newsletter (weekly)
Super Trouper (3-4 times annually) . . 15565

International Transactional Analysis Association (ITAA), 1772 Vallejo St, San Francisco, CA 94123
T: (415) 885-5992; Fax: (415) 885-5998
Founded: 1958; Members: 7000
Gen Secr: Susan Sevilla
Focus: Psychiatry
Periodicals
Script Newsletter (9 times annually)
Transactional Analysis Journal (weekly) . 15566

International Tree Crops Institute U.S.A. (ITCI), POB 4460, Davis, CA 95617
T: (916) 753-4535; Fax: (916) 753-4535
Founded: 1977
Gen Secr: Miles L. Nerwin
Focus: Forestry
Periodical
Technical Papers 15567

International Tsunami Information Center (ITIC), POB 50027, Honolulu, HI 96850-4993
T: (808) 541-1658; Fax: (808) 541-1678
Founded: 1965; Members: 750
Gen Secr: Dr. George Pararas-Carayannis
Focus: Oceanography
Periodicals
Newsletter (3 times annually)
Report (weekly) 15568

International Turfgrass Society (ITS), c/o International Turfgrass, Crop and Soil Environmental Science, Virginia Polytechnic, Blacksburg, VA 24061-0403
T: (703) 961-6482; Fax: (703) 231-3431
Founded: 1969; Members: 350
Gen Secr: John R. Hall
Focus: Agri
Periodical
Proceedings of Conference 15569

International Veterinary Acupuncture Society (IVAS), c/o Dr. Meredith L. Snader, 2140 Conestoga Rd, Chester Springs, PA 19425
T: (215) 827-7245; Fax: (215) 687-3605
Founded: 1974; Members: 400
Gen Secr: Dr. Meredith L. Snader
Focus: Vet Med
Periodical
Newsletter (weekly) 15570

International Visual Literacy Association (IVLA), c/o Alice D. Walker, Instructional Development Learning Resource Center, Virginia Polytechnic Institute and State University, Old Security Bldg, Blacksburg, VA 24061-0232
T: (703) 231-8992
Founded: 1968; Members: 200
Gen Secr: Alice D. Walker
Focus: Lit 15571

International Water Resources Association (IWRA), c/o University of Illinois, 205 N Mathews Av, Urbana, IL 61801
T: (217) 333-0536; Fax: (217) 244-6633
Founded: 1972; Members: 1400
Gen Secr: Glenn E. Stout
Focus: Water Res
Periodical
Water International (weekly) 15572

International Weed Science Society (IWSS), c/o Oregon State University, 100 Gilmore Annex, Corvallis, OR 97331-3904
T: (503) 737-3541; Fax: (503) 737-3080
Founded: 1976; Members: 630
Gen Secr: Susan Larson
Focus: Agri 15573

International Wizard of Oz Club (IWOC), POB 95, Kinderhook, IL 62345
Founded: 1957; Members: 2500
Gen Secr: Fred M. Meyer
Focus: Lit
Periodicals
The Baum Bugle (3 times annually)
Membership Directory (weekly)
Oz Trading Post (weekly) 15574

International Women's Anthropology Conference (IWAC), c/o Anthropology Dept, New York University, 25 Waverly Pl, New York, NY 10003
T: (212) 998-8550
Founded: 1978; Members: 450
Gen Secr: Dr. Linda Basch
Focus: Anthro
Periodicals
Bulletin
IWAC Newsletter (weekly) 15575

International Women's Health Coalition (IWHC), 24 E 21 St, New York, NY 10010
T: (212) 979-8500; Fax: (212) 979-9009
Founded: 1980
Pres: Joan B. Dunlop
Focus: Public Health 15576

Interprofessional Council on Environmental Design (ICED), c/o Curtis Deane, ASCE, 1015 15 St NW, Ste 600, Washington, DC 20005
T: (202) 789-2200; Fax: (202) 289-6797
Founded: 1963; Members: 8
Gen Secr: Curtis Deane
Focus: Ecology 15577

Inter-Society Color Council (ISCC), c/o Dr. Danny C. Rich, Data Color International, 5 Princess Rd, Lawrenceville, NJ 08648
T: (609) 895-7427; Fax: (609) 895-7461
Founded: 1931; Members: 900
Gen Secr: Dr. Danny C. Rich
Focus: Chem
Periodical
Newsletter (weekly) 15578

Intersociety Committee on Pathology Information (ICPI), 4733 Bethesda Av, Ste 700, Bethesda, MD 20814
T: (301) 656-2944; Fax: (301) 656-3179
Founded: 1957; Members: 5
Gen Secr: Eileen M. Lavine
Focus: Pathology
Periodical
Directory of Pathology Training Programs (weekly) 15579

Interstate Conference on Water Policy (ICWP), 955 L'Enfant Plaza SW, Washington, DC 20024
T: (202) 466-7287; Fax: (202) 646-6210
Founded: 1959; Members: 60
Pres: J. Randy Young
Focus: Water Res
Periodicals
Annual Report (weekly)
Membership Directory (weekly)
News-In-Brief (weekly)
Policy Statement (weekly)
Proceedings
Washington Report (weekly) 15580

Interstate Postgraduate Medical Association of North America (IPMANA), POB 5474, Madison, WI 53705
T: (608) 257-1401; Fax: (608) 257-1401
Founded: 1916
Gen Secr: H.B. Maroney
Focus: Adult Educ; Med 15581

Intertel, POB 1083, Tulsa, OK 74101-1083
T: (918) 583-2928
Founded: 1966; Members: 1700
Gen Secr: Lynn Chambers
Focus: Philos
Periodicals
Integra (weekly)
Membership List (weekly) 15582

Inter-University Consortium for Political and Social Research (ICPSR), POB 1248, Ann Arbor, MI 48106
T: (313) 764-2570
Founded: 1962; Members: 340
Gen Secr: Richard Rockwell
Focus: Poli Sci; Sociology
Periodicals
Annual Report (weekly)
Bulletin (weekly)
Guide to Resources and Services (weekly) 15583

IRI Research Institute, 169 Greenwich Av, POB 1276, Stamford, CT 06904-1276
T: (203) 327-5985; Fax: (203) 359-1595
Founded: 1950
Pres: Jerome F. Harrington
Focus: Agri
Periodical
Bulletin 15584

Irish American Cultural Institute (IACI), c/o University of Saint Thomas, 2115 Summit Av, Saint Paul, MN 55105
T: (612) 647-5678; Fax: (612) 647-5678
Founded: 1962
Pres: James Rogers
Focus: Hist; Lit
Periodical
Eire-Ireland (weekly) 15585

Iron Overload Diseases Association (IOD), 433 Westwind Dr, North Palm Beach, FL 33408
T: (407) 840-8512; Fax: (407) 842-9881
Founded: 1981; Members: 2000
Pres: Roberta Crawford
Focus: Hematology
Periodical
Ironic Blood: Newsletter (weekly) 15586

Italian American Librarians Caucus (IALC), 6 Peter Cooper Rd, Apt 11G, New York, NY 10010
T: (212) 228-8438
Founded: 1974; Members: 200
Pres: C. Michael Diodati
Focus: Libraries & Bk Sci
Periodicals
Bulletin (weekly)
Writings on Italian Americans (biennial) . 15587

Italian Historical Society of America (IHS), 111 Columbia Heights, Brooklyn, NY 11201
T: (718) 852-2929; Fax: (718) 855-3925
Founded: 1949; Members: 1900
Gen Secr: Dr. John J. LaCorte
Focus: Hist
Periodicals
Italian-American Newsletter
Italian-American Review (weekly) 15588

Jack London Research Center (JLRC), 14300 Arnold Dr, POB 337, Glen Ellen, CA 95442
T: (707) 996-2888
Founded: 1971
Gen Secr: Russ Kingman
Focus: Lit 15589

Jack Point Preservation Society (JPPS), POB 179, New Ellenton, SC 29809
T: 803) 652-3492
Founded: 1985; Members: 26
Gen Secr: Cheryl W. Duval
Focus: Music 15590

James Branch Cabell Society (JBCS), HC 63, Box 70A, Alstead, NH 03602
T: (603) 835-6436
Founded: 1965; Members: 200
Gen Secr: Paul Spencer
Focus: Lit
Periodical
Kalki: Studies in James Branch Cabell . 15591

James Joyce Society (JJS), 41 W 47 St, New York, NY 10036
T: (212) 719-4448
Founded: 1947; Members: 250
Gen Secr: Philip Lyman
Focus: Lit
Periodical
James Joyce Journal (weekly) 15592

James Willard Schultz Society (JWSS), 135 Wildwood Dr, New Bern, NC 28562
Founded: 1976; Members: 300
Gen Secr: David C. Andrews
Focus: Lit 15593

Japanese American Society for Legal Studies (JASLS), c/o Prof. Daniel H. Foote, University of Washington Law School, 110 NE Campus Pkwy, Seattle, WA 98105
T: (206) 685-1897; Fax: (206) 543-5671
Founded: 1964; Members: 1250
Gen Secr: Prof. Daniel H. Foote
Focus: Law
Periodical
Law in Japan (weekly) 15594

Jargon Society (JS), 411 N Cherry St, Winston-Salem, NC 27101
T: (919) 724-7619; Fax: (919) 725-3568
Founded: 1951; Members: 100
Pres: Jonathan Williams
Focus: Ling 15595

Jean Piaget Society (JPSSSKD) / Society for the Study of Knowledge and Development, c/o Dept of Psychology, State University of New York at Buffalo, Buffalo, NY 14260
T: (716) 636-3674
Founded: 1970; Members: 500
Pres: Jack Meacham
Focus: Psych
Periodicals
The Genetic Epistemologist (weekly)
Symposium Proceedings (weekly) . . . 15596

Jefferson Center for Character Education (JCCE), 202 S Lake Av, Pasadena, CA 91101
T: (818) 792-8130; Fax: (818) 792-8364
Founded: 1963; Members: 500
Gen Secr: B. David Brooks
Focus: Sociology; Psych
Periodical
Research Letter (weekly) 15597

Jefferson Davis Association (JDA), c/o Rice University, POB 1892, Houston, TX 77251
T: (713) 527-4990
Founded: 1963
Pres: Frank E. Vandiver
Focus: Hist
Periodical
The Papers of Jefferson Davis (biennial) 15598

Jesse Stuart Foundation (JSF), POB 391, Ashland, KY 41114
T: (606) 329-5233
Founded: 1979; Members: 5000
Gen Secr: Dr. James M. Gifford
Focus: Lit 15599

Jesuit Philosophical Association of the United States and Canada (JPA), c/o Arthur Madigan, Dept of Philosophy, Boston College, Chestnut Hill, MA 02167
T: (617) 552-3847
Founded: 1935; Members: 120
Gen Secr: Arthur Madigan
Focus: Philos
Periodical
Proceedings of the Jesuit Philosophical Association (weekly) 15600

Jesuit Secondary Education Association (JSEA), 1425 16 St NW, Ste 300, Washington, DC 20036
T: (202) 667-3888; Fax: (202) 328-9212
Founded: 1970; Members: 46
Pres: Carl E. Meirose
Focus: Educ
Periodicals
Annual Directory of Jesuit High Schools and Universities (weekly)
News Bulletin (weekly) 15601

Jesuit Seismological Association (JSA), c/o Weston Observatory, Boston College, 381 Concord Rd, Weston, MA 02193-1340
T: (617) 899-0950
Founded: 1925; Members: 12
Gen Secr: James McCaffrey
Focus: Geology 15602

Jewish Academy of Arts and Sciences (JAAS), 888 Seventh Av, New York, NY 10017
Founded: 1927; Members: 375
Focus: Arts; Sci
Periodical
Bulletin 15603

Jewish Braille Institute of America (JBI), 110 E 30 St, New York, NY 10016
T: (212) 889-2525; Fax: (212) 689-3692
Founded: 1931
Gen Secr: Gerald M. Kass
Focus: Rehabil 15604

Jewish Education Service of North America (JESNA), 730 Broadway, New York, NY 10003-9540
T: (212) 529-2000
Founded: 1939; Members: 1000
Gen Secr: Dr. Jonathan Woocher
Focus: Educ
Periodicals
Jewish Education Directory (tri-annually)
Pedagogic Reporter (weekly)
Trends (weekly) 15605

Jewish Educators Assembly (JEA), 15 E 26 St, Ste 1350A, New York, NY 10010
T: (212) 532-4949
Founded: 1951; Members: 400
Gen Secr: Bernard D. Troy
Focus: Educ
Periodicals
Observer (weekly)
Yearbook (weekly) 15606

Jewish Lawyers Guild (JLG), 299 Broadway, New York, NY 10007
T: (212) 227-8075; Fax: (212) 227-8075
Founded: 1962; Members: 750
Pres: Barbara R. Kapnick
Focus: Law
Periodical
Confrontation 15607

Jewish Librarians Task Force (JLTF), c/o Sylvia Eisen, 690 Anderson Av, Franklin Square, NY 11010
T: (516) 486-3748
Founded: 1975; Members: 100
Gen Secr: Sylvia Eisen
Focus: Libraries & Bk Sci
Periodical
Newsletter (weekly) 15608

Jewish Pharmaceutical Society of America (JPSA), 525 Ocean Pkwy, Brooklyn, NY 11218
T: (718) 436-8320
Founded: 1950; Members: 100
Pres: Max Grossman
Focus: Pharmacol
Periodical
The Jewish Pharmacist (weekly) 15609

Jewish Teachers Association – Morim (JTA-M), 45 E 33 St, Ste 604, New York, NY 10016
T: (212) 684-0556
Founded: 1924; Members: 8000
Pres: Phyllis L. Pullman
Focus: Educ
Periodical
Morim Bulletin (weekly) 15610

John Burroughs Association (JBA), c/o American Museum of Natural History, Central Park W at 79 St, New York, NY 10024
T: (212) 769-5169; Fax: (212) 769-5233
Founded: 1921; Members: 429
Gen Secr: Lisa Breslof
Focus: Nat Sci
Periodicals
Bulletin
Wakerobin (weekly) 15611

John Dewey Society (JDS), c/o Dr. Robert C. Morris, School of Education, University of Indianapolis, 1400 E Hanna Av, Indianapolis, IN 46227
T: (317) 788-3286
Founded: 1935; Members: 400
Gen Secr: Dr. Robert C. Morris
Focus: Educ
Periodicals
Current Issues (weekly)
Insights (biennial)
John Dewey Lectures (weekly) 15612

John Ericsson Society (JES), c/o Church of Sweden, 5 E 48 St, New York, NY 10017
T: (212) 308-9580; Fax: (212) 644-9024
Founded: 1907; Members: 100
Gen Secr: Sten-Sture Nordin
Focus: Hist 15613

The Johnsonians, c/o S.R. Parks, 1914 Yale Station, New Haven, CT 06520
T: (203) 432-2967
Founded: 1946; Members: 50
Gen Secr: S.R. Parks
Focus: Lit 15614

Joint Center for Political and Economic Studies (JCPES), 1090 Vermont Av NW, Ste 1100, Washington, DC 20005
T: (202) 789-3500; Fax: (202) 789-6390
Founded: 1970
Pres: Eddie N. Williams
Focus: Poli Sci; Econ
Periodicals
Focus (weekly)
National Roster of Black Elected Officials (weekly) 15615

Joint Commission on Accreditation of Healthcare Organizations (JCAHO), 1 Renaissance Blvd, Oak Brook Terrace, IL 60181
T: (708) 916-5600; Fax: (708) 916-5644
Founded: 1951
Pres: Dennis O'Leary
Focus: Med
Periodicals
Clinical Standards Digest
Joint Commission Perspectives
PTSM Series
Quality Review Bulletin 15616

Joint Committee of the States to Study Alcoholic Beverage Laws, c/o National Alcoholic Beverage Control Association, 4216 King St W, Alexandria, VA 22302
T: (703) 578-4200; Fax: (703) 820-3551
Gen Secr: Paul C. Dufek
Focus: Law 15617

Joint Council of Allergy and Immunology (JCAI), POB 4620, Arlington Heights, IL 60006
T: (708) 359-3090
Founded: 1975; Members: 2850
Gen Secr: Joseph J. Lotharius
Focus: Immunology 15618

Joint Industry Council (JIC), c/o IACUB, 702 Bloomington Rd, Champaign, IL 61820
T: (217) 359-2469
Founded: 1978; Members: 10
Gen Secr: Richard Newman
Focus: Standards 15619

Joint Institute for Laboratory Astrophysics (JILA), c/o University of Colorado, POB 440, Boulder, CO 80309-0440
T: (303) 492-7789; Fax: (303) 492-5235
Founded: 1962; Members: 225
Gen Secr: W. Patrick McInerny
Focus: Astrophys 15620

Joint Review Committee for Respiratory Therapy Education (JRCRTE), 1701 W Euless Blvd, Ste 200, Euless, TX 76040
T: (817) 283-2835
Founded: 1963; Members: 13
Gen Secr: Philip A. Von der Heydt
Focus: Therapeutics 15621

Joint Review Committee for the Ophthalmic Medical Personnel (JRCOMP), 2025 Woodlane Dr, Saint Paul, MN 55125-2995
T: (612) 731-2944
Members: 5
Pres: Lisa P. Rovick
Focus: Ophthal 15622

Joint Review Committee on Educational Programs for the EMT-Paramedic (JRCEMT-P), 1701 W Euless Blvd, Ste 200, Euless, TX 76040
T: (817) 283-2836
Founded: 1979
Gen Secr: Philip A. Von der Heydt
Focus: Educ; Med 15623

Joint Review Committee on Education in Diagnostic Medical Sonography (JRCDMS), 20 N Wacker Dr, Ste 900, Chicago, IL 60606
T: (312) 704-5151; Fax: (312) 704-5304
Founded: 1979
Gen Secr: Marilyn Fay
Focus: Educ; Med 15624

Joint Review Committee on Education in Radiologic Technology (JRCERT), 20 E Wacker Dr, Ste 900, Chicago, IL 60606
T: (312) 704-5300; Fax: (312) 704-5304
Founded: 1969
Gen Secr: Marilyn Fay
Focus: Educ; Radiology 15625

Joslin Diabetes Center (JDC), 1 Joslin Pl, Boston, MA 02215
T: (617) 732-2400; Fax: (617) 732-2562
Founded: 1968
Pres: Dr. Kenneth E. Quickel
Focus: Diabetes 15626

Journalism Association of Community Colleges (JACC), c/o Paul DeBolt, Contra Costa College, 2600 Mission Bell Dr, San Pablo, CA 94806
T: (510) 235-7800; Fax: (510) 236-6768
Founded: 1957; Members: 70
Gen Secr: Paul DeBolt
Focus: Adult Educ; Journalism
Periodicals
Directory (weekly)
Idea Exchange (weekly)
Newsletter (weekly) 15627

Journalism Education Association (JEA), c/o Kansas State University, Kedzie Hall 103, Manhattan, KS 66506-1505
T: (913) 532-5532; Fax: (913) 532-7309
Founded: 1924; Members: 1700
Gen Secr: Linda S. Puntney
Focus: Adult Educ; Journalism
Periodical
Communication: Journalism Education Today (weekly) 15628

Jozef Pilsudski Institute of America for Research in the Modern History of Poland, 381 Park Av S, New York, NY 10016
T: (212) 683-4342
Founded: 1943; Members: 1000
Pres: Stanislaw Jordanowski
Focus: Hist
Periodical
Wiadomosci Instytutu J. Pilsudskiego (weekly) 15629

Jugoslavia Study Group (JSG), 1514 N Third Av, Wausau, WI 54401
T: (715) 675-2833
Founded: 1974; Members: 40
Pres: Michael Lenard
Focus: Hist; Ethnology
Periodical
The Trumpeter (weekly) 15630

Junior Achievement (JA), 1 Education Way, Colorado Springs, CO 80906
T: (719) 540-8000; Fax: (719) 540-9150
Founded: 1919; Members: 1200000
Pres: Karl Flemke
Focus: Adult Educ; Econ
Periodicals
Junior Achievement Annual Report (weekly)
Partners (weekly) 15631

Junior Classical League (JCL), c/o Miami University, Oxford, OH 45056
T: (513) 529-7741; Fax: (513) 529-7741
Founded: 1936; Members: 50000
Gen Secr: Geri Dutra
Focus: Ling
Periodical
Torch: U.S. (weekly) 15632

Juvenile Diabetes Foundation International (JDFI), 432 Park Av S, New York, NY 10016-8013
T: (212) 889-7575; Fax: (212) 725-7259
Founded: 1970; Members: 115
Gen Secr: Ken Farber
Focus: Diabetes 15633

Kafka Society of America (KSA), c/o Prof. Maria Luise Caputo-Mayr, Dept of Germanic and Slavic Languages, Temple University, AB Bldg 335, Philadelphia, PA 19122
T: (215) 787-8282
Founded: 1975; Members: 400
Gen Secr: Prof. Maria Luise Caputo-Mayr
Focus: Lit
Periodical
Journal of the Kafka Society *15634*

Kate Greenaway Society (KGS), POB 8, Norwood, PA 19074
T: (215) 485-8572
Founded: 1971; Members: 450
Gen Secr: James L. Lowe
Focus: Lit
Periodical
Under the Window (weekly) *15635*

Keats-Shelley Association of America (KSAA), c/o New York Public Library, Rm 226, Fifth Av and 42 St, New York, NY 10018
T: (212) 764-0655
Founded: 1948; Members: 1000
Gen Secr: Dr. Robert A. Hartley
Focus: Lit
Periodical
Keats-Shelley Journal (weekly) *15636*

Keyboard Teachers Association International (KTAI), c/o Dr. Albert DeVito, 361 Pin Oak Ln, Westbury, NY 11590-1941
T: (516) 333-3236
Founded: 1963
Pres: Dr. Albert DeVito
Focus: Music; Educ
Periodical
Keyboard Teacher (weekly) *15637*

Kipling Society of North America (KS), c/o Dr. E. Karim, Dept of English, Rockford College, Rockford, IL 61101
T: (815) 226-4183; Fax: (815) 226-4119
Founded: 1986; Members: 200
Gen Secr: Dr. E. Karim
Focus: Lit
Periodical
Kipling Journal (weekly) *15638*

Kosciuszko Foundation (KF), 15 E 65 ST, New York, NY 10021
T: (212) 734-2130; Fax: (212) 628-4552
Founded: 1925; Members: 4000
Pres: Joseph E. Gore
Focus: Hist
Periodical
Newsletter (weekly) *15639*

Kroeber Anthropological Society (KAS), c/o Dept of Anthropology, University of California, Kroeber Hall 232, Berkeley, CA 94720
T: (415) 642-6932
Founded: 1949; Members: 500
Focus: Anthro
Periodical
Papers (weekly) *15640*

Kurt Weill Foundation for Music (KWFM), 7 E 20 St, New York, NY 10003-1106
T: (212) 505-5240
Founded: 1962
Gen Secr: David Farneth
Focus: Music *15641*

Laban/Bartenieff Institute of Movement Studies (LBIMS), 11 E Fourth St, New York, NY 10003
T: (212) 477-4299; Fax: (212) 477-3702
Founded: 1978; Members: 500
Gen Secr: Suzanne Youngerman
Focus: Sports; Psych
Periodicals
Membership Directory (weekly)
Newsletter (weekly) *15642*

Laboratory Animal Management Association (LAMA), c/o Fred Douglas, POB 1744, Silver Spring, MD 20915
T: (317) 494-7592; Fax: (317) 494-0781
Founded: 1984; Members: 435
Pres: Fred Douglas
Focus: Vet Med *15643*

Labor Policy Association (LPA), 1015 15 St NW, Washington, DC 20005
T: (202) 789-8670; Fax: (202) 789-0064
Members: 180
Gen Secr: Jeffrey C. McGuiness
Focus: Poli Sci *15644*

Labor Research Association (LRA), 145 E 28 St, New York, NY 10001-6191
T: (212) 714-1677; Fax: (212) 714-1674
Founded: 1927; Members: 3000
Gen Secr: Gregory Tarpinian
Focus: Econ
Periodicals
Economic Notes (weekly)
Trade Union Adviser (weekly) *15645*

Landscape Architecture Foundation (LAF), 4401 Connecticut Av NW, Washington, DC 20008
T: (202) 686-0068; Fax: (202) 686-1001
Founded: 1966
Gen Secr: Mary Hanson
Focus: Archit
Periodical
Agora *15646*

Laser Institute of America (LIA), 12424 Research Pkwy, Ste 130, Orlando, FL 32826
T: (407) 380-1553; Fax: (407) 380-5588
Founded: 1968; Members: 1700
Gen Secr: Peter M. Baker
Focus: Eng
Periodical
Journal of Laser Applications (weekly) . *15647*

Latin American Studies Association (LASA), c/o William Pitt, University of Pittsburgh, Pittsburgh, PA 15260
T: (412) 648-7929; Fax: (412) 624-7145
Founded: 1966; Members: 3000
Gen Secr: Reid Reading
Focus: Anthro; Poli Sci; Hist; Econ; Humanities
Periodicals
Lasa Forum/Newsletter (weekly)
Latin American Research Review (3 times annually)
Professional Journal (3 times annually) . *15648*

Laubach Literacy Action (LLA), 1320 Jamesville Av, Box 131, Syracuse, NY 13210
T: (315) 422-9121
Founded: 1968; Members: 55000
Gen Secr: Dr. Peter Waite
Focus: Lit
Periodicals
Directory (weekly)
Literacy Advance (weekly) *15649*

Laubach Literacy International (LLI), 1320 Jamesville Av, Box 131, Syracuse, NY 13210
T: (315) 422-9121; Fax: (315) 422-6369
Founded: 1955; Members: 80000
Pres: Robert F. Caswell
Focus: Lit *15650*

Laughter Therapy (LT), 2359 Nichols Canyon Rd, Los Angeles, CA 90046
T: (213) 851-3394
Founded: 1981
Pres: Allen A. Funt
Focus: Psych; Psychiatry *15651*

Laura Ingalls Wilder Memorial Society (LIWMS), POB 269, Peplin, WI 54759
T: (715) 442-3161
Founded: 1974; Members: 425
Gen Secr: Mary Fayerweather
Focus: Lit *15652*

Law and Society Association (LSA), c/o University of Massachusetts, Hampshire House, Amherst, MA 01003
T: (413) 545-4617; Fax: (413) 545-1640
Founded: 1964; Members: 1200
Gen Secr: Ronald Pipkin
Focus: Law; Sociology
Periodical
Law & Society Review (weekly) *15653*

Law of the Sea Institute (LSI), c/o Richardson School of Law, University of Hawaii, 2515 Dole St, Rm 208, Honolulu, HI 96822
T: (808) 956-6750; Fax: (808) 956-6402
Founded: 1965; Members: 300
Gen Secr: Dr. John Craven
Focus: Law
Periodicals
Occasional Papers
Proceedings *15654*

Law School Admission Council – Law School Admission Services (LSAC/LSAS), POB 40, Newtown, PA 18940
T: (215) 968-1101
Founded: 1948; Members: 190
Pres: Elizabeth Moody
Focus: Law
Periodicals
Annual Report (weekly)
Law School Admission Bulletin (weekly)
Prelaw Handbook (weekly) *15655*

Lawyers' Committee for Civil Rights under Law (LCCRUL), 1400 Eye St NW, Ste 400, Washington, DC 20005
T: (202) 371-1212
Founded: 1963; Members: 165
Gen Secr: Barbara R. Arnwine
Focus: Law
Periodical
Committee Report (weekly) *15656*

League for Industrial Democracy (LID), 815 15 St NW, Ste 511, Washington, DC 20005
T: (202) 638-1515; Fax: (202) 347-5585
Founded: 1905; Members: 1500
Gen Secr: Rita Freedman
Focus: Poli Sci; Econ
Periodical
Solidarnosc Bulletin (weekly) *15657*

Lentz Peace Research Laboratory (LPRL), 8242 Buchanan, Saint Louis, 63114
T: (314) 429-6433
Founded: 1945; Members: 100
Gen Secr: Miranda Duncan
Focus: Poli Sci; Psych; Soc Sci; Prom Peace
Periodical
Peace Research (3 times annually) . . *15658*

Leo Baeck Institute (LBI), 129 E 73 St, New York, NY 10021
T: (212) 744-6400; Fax: (212) 988-1305
Founded: 1955
Gen Secr: Robert A. Jacobs
Focus: Hist
Periodicals
Bulletin (weekly)
LBI News (weekly)
Library & Archives News (weekly)
Yearbook (weekly) *15659*

Leonardo – The International Society for the Arts, Sciences and Technology, 672 S Van Ness, San Francisco, CA 94110
T: (415) 431-7414; Fax: (415) 431-5737
Founded: 1968; Members: 700
Pres: Roger Malina
Focus: Arts; Sci; Eng *15660*

Leonard Wood Memorial – American Leprosy Foundation (LWM), 11600 Nebel St, Ste 210, Rockville, MD 20852
T: (301) 984-1336
Founded: 1928
Gen Secr: Gerald P. Walsh
Focus: Derm *15661*

Leschetizky Association (LA), c/o Arminda Canteros, 333 W 86 St, New York, NY 10024
T: (212) 496-7912
Founded: 1942; Members: 300
Pres: Arminda Canteros
Focus: Music
Periodical
News Bulletin (weekly) *15662*

Lessing Society (LS), c/o German Dept, University of Cincinnati, Cincinnati, OH 45221-0372
T: (513) 556-2744
Founded: 1966; Members: 350
Gen Secr: Edward P. Harris
Focus: Lit
Periodical
Lessing Yearbook (weekly) *15663*

Leukemia Society of America (LSA), 600 Third Av, New York, NY 10016
T: (212) 573-8484; Fax: (212) 972-5776
Founded: 1949
Pres: Rudolph F. Badum
Focus: Hematology
Periodical
Society News (weekly) *15664*

Lewis Carroll Society of North America (LCNA), 617 Rockford Rd, Silver Spring, MD 20902
T: (301) 593-7077
Founded: 1974; Members: 350
Pres: Charles Lovett
Focus: Lit
Periodical
Knight Letter (weekly) *15665*

Lexington Group in Transportation History (LGTH), c/o Dept of History, Saint Cloud State University, Saint Cloud, MN 56301
T: (612) 255-4906; Fax: (612) 654-5198
Founded: 1942; Members: 460
Gen Secr: Don L. Hofsommer
Focus: Cultur Hist; Transport
Periodical
Lexington Quarterly (weekly) *15666*

Liaison Committee on Medical Education (LCME), c/o American Medical Association, 515 N State St, Chicago, IL 60610
T: (312) 464-4657; Fax: (312) 464-4184
Founded: 1942; Members: 18
Gen Secr: Harry S. Jonas
Focus: Adult Educ; Med *15667*

Library Administration and Management Association (LAMA), 50 E Huron St, Chicago, IL 60611
T: (312) 944-6780; Fax: (312) 280-3257
Founded: 1957; Members: 5223
Gen Secr: Karen Muller
Focus: Libraries & Bk Sci
Periodicals
Library Administration and Management (weekly)
Library Buildings Consultants List (biennial) *15668*

Library and Information Technology Association (LITA), 50 E Huron St, Chicago, IL 60611-2795
T: (312) 280-4270; Fax: (312) 280-3257
Founded: 1966; Members: 5000
Gen Secr: Linda J. Knutson
Focus: Libraries & Bk Sci
Periodicals
Information Technology and Libraries (weekly)
Newsletter (weekly) *15669*

Library Public Relations Council (LPRC), 2 Jean Walling Civic Center, East Brunswick NJ 08816
T: (908) 390-6761
Founded: 1939; Members: 300
Pres: Sharon Karmazin
Focus: Libraries & Bk Sci *15670*

Life Insurance Marketing and Research Association (LIMRA), Box 208, Hartford, CT 06141
T: (203) 677-0033; Fax: (203) 678-0187
Founded: 1945; Members: 579
Pres: Ernest E. Cragg
Focus: Insurance
Periodical
Proceedings *15671*

Lincoln Center for the Performing Arts (LCPA), 70 Lincoln Center Plaza, New York, NY 10023-6583
T: (212) 875-5000
Founded: 1956; Members: 2500
Pres: Nathan Leventhal
Focus: Perf Arts
Periodical
Lincoln Center Calendar of Events (weekly) *15672*

Lincoln Institute for Research and Education (LIRE), 1001 Connecticut Av NW, Ste 1135, Washington, DC 20036
T: (202) 223-5112
Founded: 1978
Pres: J.A. Parker
Focus: Educ; Poli Sci
Periodical
Lincoln Review (weekly) *15673*

Linguistic Society of America (LSA), 1325 18 St NW, Ste 211, Washington, DC 20036
T: (202) 835-1714
Founded: 1924; Members: 7000
Gen Secr: Frederick J. Newmeyer
Focus: Ling
Periodicals
Annual Meeting Handbook
Bulletin (weekly)
Language (weekly) *15674*

Literacy and Evangelism International (LEI), 1800 S Jackson, Tulsa, OK 74107
T: (918) 585-3826; Fax: (918) 585-3224
Founded: 1967
Gen Secr: Robert Rice
Focus: Lit; Rel & Theol *15675*

Literacy Volunteers of America (LVA), 5795 S Widewaters Pkwy, Syracuse, NY 13214
T: (315) 445-8000; Fax: (315) 445-8006
Founded: 1962; Members: 100000
Pres: J. Crouch
Focus: Lit *15676*

Longfellow Society of Sudbury and Wayside Inn (LSSWI), c/o Petra Barraford, POB 138, Sherborn, MA 01770
Founded: 1972; Members: 25
Pres: Nora Barraford
Focus: Lit
Periodical
Longfellow Journal of Poetry (weekly) . . *15677*

Louisa May Alcott Memorial Association (LMAMA), POB 343, Concord, MA 01742
T: (508) 369-4118
Founded: 1911; Members: 57
Gen Secr: Stephanie M. Upton
Focus: Lit *15678*

Luso-American Education Foundation (LAEF), POB 1768, Oakland, CA 94604
T: (510) 452-4465; Fax: (510) 452-1617
Members: 275
Pres: Eduardo Eusebio
Focus: Educ *15679*

Lutheran Church Library Association (LCLA), 122 W Franklin Av, Minneapolis, MN 55404
T: (612) 870-3623
Founded: 1958; Members: 1750
Gen Secr: Leanna Kloempken
Focus: Libraries & Bk Sci
Periodical
Lutheran Libraries (weekly) *15680*

Lutheran Educational Conference of North America (LECNA), 122 C St NW, Ste 300, Washington, DC 20001
T: (202) 783-7505; Fax: (202) 783-7502
Founded: 1910; Members: 50
Gen Secr: Don Stoike
Focus: Educ
Periodicals
Lutheran Higher Education Directory (weekly)
Papers and Proceedings (weekly) . . . *15681*

Lutheran Education Association (LEA), 7400 Augusta Blvd, River Forest, IL 60305
T: (708) 209-3343
Founded: 1942; Members: 3850
Gen Secr: Barbara Goodwin
Focus: Educ *15682*

Lutheran Historical Conference (LHC), c/o August R. Suelflow, 801 DeMun Av, Saint Louis, MO 63105
T: (314) 721-5934
Founded: 1962; Members: 150
Pres: August R. Suelflow
Focus: Hist; Rel & Theol
Periodicals
Minutes of Biennial Conference
Newsletter (3-4 times annually) *15683*

Lyman A. Brewer International Surgical Society, c/o S. Jeanne Davies, White Memorial Medical Center, 1720 Brooklyn Av, Los Angeles, CA 90033
T: (213) 268-5000
Founded: 1981; Members: 150
Gen Secr: S. Jeanne Davies
Focus: Surgery
Periodical
American Journal of Surgery (weekly) . . *15684*

The Madison Project (TMP), c/o Robert B. Davis, Graduate School of Education, Rutgers University, 10 Seminary Pl, New Brunswick, NJ 08903
T: (908) 932-8847; Fax: (908) 932-8206
Founded: 1957

Gen Secr: Robert B. Davis
Focus: Math 15685

The Magnolia Society (TMS), c/o Phelan A. Bright, 907 S Chestnut St, Hammond, LA 70403
T: (504) 542-9477
Founded: 1961; Members: 425
Gen Secr: Phelan A. Bright
Focus: Botany
Periodicals
Magnolia (weekly)
Membership Roster (biennial) 15686

Management Association for Private Photogrammetric Surveyors (MAPPS), c/o John M. Palatiello Associates, 12020 Sunrise Valley Dr, Ste 100, Reston, VA 22091
T: (703) 391-2739; Fax: (703) 476-2217
Founded: 1982; Members: 120
Gen Secr: John M. Palatiello
Focus: Surveying
Periodicals
Flightline (weekly)
MAPPS Capability Survey (weekly) . . . 15687

Mandala Society (MS), POB 1233, Del Mar, CA 92014
T: (619) 481-7751
Founded: 1972
Pres: David J. Harris
Focus: Philos
Periodical
Holistic Education Series (weekly) . . . 15688

Manufacturers Standardization Society of the Valve and Fittings Industry (MSS), 127 Park St NE, Vienna, VA 22180
T: (703) 281-6613
Founded: 1924; Members: 75
Gen Secr: O. Thornton
Focus: Standards 15689

Maria Mitchell Association (MMA), 2 Vestal St, Nantucket, MA 02554
T: (508) 228-9198
Founded: 1903; Members: 1700
Gen Secr: Kathryn K. Pochman
Focus: Astronomy 15690

Marine Technology Society (MTS), 1828 L St NW, Ste 900, Washington, DC 20036-5104
T: (202) 775-5966; Fax: (202) 429-9417
Founded: 1963; Members: 3000
Gen Secr: Martin J. Finerty
Focus: Oceanography; Eng
Periodicals
Marine Technology Society Journal (weekly)
Membership Directory
Newsletter (weekly) 15691

Marketing Research Association (MRA), 2189 Silas Deane Hwy, Ste 5, Rocky Hill, CT 06067
T: (203) 257-4008; Fax: (203) 257-3990
Founded: 1954; Members: 2300
Gen Secr: Betsy J. Peterson
Focus: Marketing
Periodicals
Alert (weekly)
Applied Market Research: A Journal for Practitioners (weekly)
Research Service Directory (weekly) . . 15692

Marketing Science Institute (MSI), 1000 Massachusetts Av, Cambridge, MA 02138
T: (617) 491-2060; Fax: (617) 491-2065
Founded: 1961; Members: 64
Pres: H. Paul Root
Focus: Marketing
Periodicals
Newsletter (weekly)
Research Briefs (weekly) 15693

Mark Twain Association (MTA), 245 W 25 St, Apt 6C, New York, NY 10001
T: (212) 255-9640
Founded: 1926; Members: 200
Pres: Alice Chapman Dauer
Focus: Lit 15694

Mark Twain Research Foundation (MTRF), POB 13, Perry, MO 63462-0013
Founded: 1939; Members: 400
Gen Secr: Chester L. Davis
Focus: Lit
Periodical
The Twainian (weekly) 15695

Marlowe Society of America (MSA), c/o Dr. Constance Brown Kuriyama, Dept of English, Texas Technical University, Lubbock, TX 79409
T: (806) 742-2501
Founded: 1976; Members: 150
Gen Secr: Dr. Constance Brown Kuriyama
Focus: Lit 15696

Marquandia Society (MS), 209 Indian Springs Rd, Williamsburg, VA 23185
T: (804) 229-7049
Founded: 1971; Members: 15
Gen Secr: John R. Thelin
Focus: Hist; Lit
Periodical
Newsletter 15697

The Masonry Society (TMS), 2619 Spruce St, Ste B, Boulder, CO 80302-3808
T: (303) 939-9700; Fax: (303) 444-3239
Founded: 1977; Members: 1500
Gen Secr: Dr. James L. Noland
Focus: Civil Eng
Periodical
Journal (weekly) 15698

Massenet Society (MS), c/o The Treasurer, 9 Drury Ln, Fort Lee, NJ 07024
T: (201) 224-4526
Founded: 1977; Members: 75
Pres: Robert A. Frone
Focus: Music
Periodical
Massenet Newsletter (weekly) 15699

Material Handling Management Society (MHMS), 8720 Red Oak Blvd, Ste 224, Charlotte, NC 28217
T: (704) 525-4667; Fax: (704) 525-2880
Founded: 1949; Members: 1500
Gen Secr: Peter Youngs
Focus: Materials Sci
Periodicals
IMMS Membership Directory (weekly)
Materials Handling Outlook (weekly) . . 15700

Materials Properties Council (MPC), 345 E 47 St, New York, NY 10017
T: (212) 705-7693; Fax: (212) 752-4929
Founded: 1966; Members: 800
Gen Secr: Dr. Martin Prager
Focus: Metallurgy
Periodical
Annual Report (weekly) 15701

Maternity Center Association (MCA), 48 E 92 St, New York, NY 10128
T: (212) 369-7300; Fax: (212) 369-8747
Founded: 1918; Members: 600
Gen Secr: Ruth W. Lubic
Focus: Gynecology
Periodical
Special Delivery (weekly) 15702

Mathematical Association of America (MAA), 1529 18 St NW, Washington, DC 20036
T: (202) 387-5200; Fax: (202) 265-2384
Founded: 1915; Members: 32000
Gen Secr: Marcia P. Sward
Focus: Math
Periodicals
American Mathematical (weekly)
College Mathematics Journal (3 times annually)
Mathematics Magazine (5 times annually) 15703

Math/Science Network, 678 13 St, Ste 100, Oakland, CA 94612
T: (510) 893-6284
Founded: 1975; Members: 275
Focus: Math
Periodical
Broadcast (weekly) 15704

Maximilian Numismatic and Historical Society (Max Society), c/o Don Bailey, PO Drawer L, Tekonsha, MI 49092
T: (517) 767-4760
Founded: 1967; Members: 60
Gen Secr: Don Bailey
Focus: Numismatics; Hist 15705

Max Steiner Memorial Society (MSMS), POB 45713, Los Angeles, CA 90045-0713
Founded: 1965
Pres: Albert K. Bender
Focus: Music 15706

The Media Institute (TMI), 1000 Potomac St NW, Ste 204, Washington, DC 20007
T: (202) 298-7512; Fax: (202) 337-7092
Founded: 1976; Members: 350
Pres: Patrick D. Maines
Focus: Comm Sci; Mass Media 15707

Media Research Directors Association (MRDA), c/o Elliot Gluskin, Ladies Home Journal, 100 Park Av, New York, NY 10016
T: (212) 351-3611
Founded: 1947; Members: 135
Pres: Elliot Gluskin
Focus: Mass Media
Periodicals
MRDA Membership Directory (weekly)
MRDA News and Views (5 times annually)
. . . . 15708

Medical Letter (ML), 1000 Main St, New Rochelle, NY 10801
T: (914) 235-0500; Fax: (914) 576-3377
Founded: 1959
Gen Secr: Mark Abramowicz
Focus: Pharmacol
Periodical
Medical Letter on Drugs and Therapeutics (weekly) 15709

Medical Library Association (MLA), 6 N Michigan Av, Ste 300, Chicago, IL 60602
T: (312) 419-9094; Fax: (312) 419-8950
Founded: 1898; Members: 5000
Gen Secr: Carla J. Funk
Focus: Libraries & Bk Sci
Periodicals
Bulletin (weekly)
Current Catalog Proof Sheets (weekly)
MLA News (weekly) 15710

Medical Mycological Society of the Americas (MMSA), c/o Dr. Michael G. Rinaldi, Dept of Pathology, University of Texas Health Science Center, 7703 Floyd Curl Dr, San Antonio, TX 78284-7750
T: (210) 567-4132; Fax: (210) 567-6729
Founded: 1966; Members: 408
Gen Secr: Dr. Michael G. Rinaldi
Focus: Botany, Specific
Periodicals
Bulletin (weekly)
Directory 15711

Medical Seminars International (MSI), 9800D Topanga Cyn, Ste 232, Chatsworth, CA 91311
Fax: (818) 774-0244
Gen Secr: Mary Ullman
Focus: Med 15712

Medical Society of the United States and Mexico (MSUSM), c/o Arizona Medical Association, 810 W Bethany Home Rd, Phoenix, AZ 85013
T: (602) 246-8901
Founded: 1954; Members: 400
Focus: Med
Periodical
Directory (weekly) 15713

Medieval Academy of America (MAA), 1430 Massachusetts Av, Cambridge, MA 02138
T: (617) 491-1622
Founded: 1925; Members: 3900
Gen Secr: Luke Wenger
Focus: Hist; Lit; Arts; Philos
Periodical
Speculum: A Journal of Medieval Studies (weekly)
. . . . 15714

Medtner Society, United States of America, 555 W Madison Tower 1, Rm 706, Chicago, IL 60611
T: (312) 902-4899
Founded: 1983; Members: 30
Gen Secr: Dmitry Feofanov
Focus: Music 15715

Melville Society (MS), c/o Stanton Garner, 1016 Live Oak Ln, Arlington, TX 76012
T: (817) 265-1305
Founded: 1945; Members: 750
Gen Secr: Stanton Garner
Focus: Lit
Periodical
Melville Society Extracts (weekly) . . . 15716

Mencken Society (MS), POB 16218, Baltimore, MD 21210
T: (301) 377-2333
Founded: 1974; Members: 375
Pres: Arthur Gutman
Focus: Lit 15717

Mended Hearts (MH), 7272 Greenville Av, Dallas, TX 75231
T: (214) 706-1442
Founded: 1951; Members: 24000
Gen Secr: D. Bonham
Focus: Cardiol
Periodical
Heartbeat (weekly) 15718

Mental Health Materials Center (MHMC), POB 304, Bronxville, NY 10708
T: (914) 337-6596
Founded: 1953
Pres: Alex Sareyan
Focus: Psychiatry 15719

Mental Research Institute (MRI), 555 Middlefield Rd, Palo Alto, CA 94301
T: (415) 321-3055; Fax: (415) 321-3785
Founded: 1959
Gen Secr: Judith E. Foddrill
Focus: Med; Physiology
Periodical
Newsletter (weekly) 15720

Metal Treating Institute (MTI), 302 Third St, Ste 1, Neptune Beach, FL 32266
T: (904) 249-0448; Fax: (904) 249-0459
Founded: 1933; Members: 360
Gen Secr: M. Lance Miller
Focus: Metallurgy 15721

Metaphysical Society of America (MSA), c/o Brian J. Martine, Dept of Philosophy, University of Alabama at Huntsville, Huntsville, AL 35899
T: (205) 895-6555
Founded: 1950; Members: 600
Gen Secr: Brian J. Martine
Focus: Philos
Periodicals
Announcements (weekly)
Membership List
Presidential Address (weekly) 15722

Meteoritical Society (MS), c/o Roger Hewins, Dept of Geology, Rutgers University, Busch Campus, Wright Laboratories, POB 1179, Piscataway, NJ 08855-1179
T: (908) 932-3232; Fax: (908) 932-3374
Founded: 1933; Members: 900
Gen Secr: Roger Hewins
Focus: Geophys
Periodicals
Geochimica et Cosmochimica Acta (weekly)
Meteoritics (weekly) 15723

Metropolitan Association of Urban Designers and Environmental Planners (MAUDEP), POB 722, Church Street Station, New York, NY 10008
T: (212) 747-1755
Founded: 1968; Members: 30000
Gen Secr: Robert Schumacher
Focus: Urban Plan
Periodical
Newsletter 15724

Metropolitan College Mental Health Association (MCMHA), c/o Mary Jane Goodloe, Parsons School of Design, 66 Fifth Av, New York, NY 10011
T: (212) 741-8656
Founded: 1969; Members: 150
Gen Secr: Mary Jane Goodloe
Focus: Psychiatry
Periodicals
Directory of College Mental Health Services in Greater New York City
MCMHA Newsletter (3 times annually) . 15725

Michael E. Debakey International Surgical Society (MEDISS), c/o Kenneth L. Mattox, Dept of Surgery, 1 Baylor Plaza, Houston, TX 77030
T: (713) 798-4557
Founded: 1976; Members: 540
Gen Secr: Kenneth L. Mattox
Focus: Cardiol 15726

Microbeam Analysis Society (MAS), c/o VCH Publishers, 303 NW 12 Av, Deerfield Beach, FL 33442-1788
T: (800) 367-8249; Fax: (800) 428-8201
Founded: 1967; Members: 1000
Gen Secr: John Small
Focus: Physics; Med
Periodical
Micronews (weekly) 15727

Microneurography Society (MNS), c/o Cardiovascular Physiology, V.A. Medical Center, Richmond, VA 23249
T: (804) 230-0001
Founded: 1981; Members: 60
Gen Secr: Dwain L. Eckberg
Focus: Neurology
Periodical
Directory (weekly) 15728

Microscopy Society of America (MSA), POB MSA, Woods Hole, MA 02543
T: (508) 540-7639; Fax: (508) 548-9053
Founded: 1942; Members: 4800
Gen Secr: Dr. Patricia G. Calarco
Focus: Electronic Eng; Optics
Periodicals
Bulletin (weekly)
Directory (biennial)
Proceedings of Annual Meeting (weekly) . 15729

Mid-Continent Railway Historical Society (MCRHS), POB 55, North Freedom, WI 53951
T: (608) 522-4261
Founded: 1959; Members: 625
Gen Secr: Charles Kratz
Focus: Cultur Hist
Periodicals
Membership Roster
Railway Gazette (weekly) 15730

Middle Atlantic Planetarium Society (MAPS), c/o Steven R. Mitch, Benedum Science Center, Glebay Pk, Wheeling, WV 26003
T: (304) 242-3000
Founded: 1965; Members: 200
Gen Secr: Steven R. Mitch
Focus: Astronomy
Periodical
Constellation (weekly) 15731

Middle East Institute (MEI), 1761 N St NW, Washington, DC 20036-2882
T: (202) 785-1141; Fax: (202) 331-8861
Founded: 1946; Members: 1500
Pres: Robert V. Keeley
Focus: Educ
Periodical
The Middle East Journal (weekly) . . . 15732

Middle East Librarians' Association (MELA), c/o Michael Hopper, Main Library, University of California, Santa Barbara, CA 93106
T: (805) 893-3454; Fax: (805) 893-4676
Founded: 1972; Members: 200
Gen Secr: Michael Hopper
Focus: Libraries & Bk Sci
Periodical
Notes (3 times annually) 15733

Middle East Research and Information Project (MERIP), 1500 Massachusetts Av NW, Ste 119, Washington, DC 20005
T: (202) 223-3677
Founded: 1971; Members: 15
Gen Secr: Peggy Hutchison
Focus: Ethnology; Poli Sci
Periodical
Middle East Report (9 times annually) . 15734

Middle East Studies Association of North America (MESA), 1232 N Cherry Av, Tucson, AZ 85721
T: (602) 621-5850; Fax: (602) 321-7752
Founded: 1966; Members: 2300
Gen Secr: Anne H. Betteridge
Focus: Educ; Int'l Relat
Periodicals
Abstracts of Papers Delivered at Annual Meeting (weekly)
Bulletin (weekly)
Directory of Graduate and Undergraduate Programs and Courses in Middle East Studies in the U.S., Canada and Abroad
International Journal of Middle East Studies (weekly)

Newsletter
Roster of MESA Fellows (weekly) . . . 15735

Middle States Association of Colleges and Schools (MSA), 3624 Market St, Philadelphia, PA 19104
T: (215) 662-5600; Fax: (215) 662-5950
Founded: 1887; Members: 3500
Gen Secr: Kathleen A. Torpy
Focus: Educ
Periodicals
Accredited Membership List (weekly)
Higher Commission Newsletter (weekly)
Middle States Association Newsletter (weekly) . . . 15736

Midwest Archives Conference (MAC), c/o Kraft General Foods Archives, 6350 Kirk St, Morton Grove, IL 60053
T: (708) 998-2981
Founded: 1972; Members: 1100
Gen Secr: B. Haglund Tousey
Focus: Archives . . . 15737

Midwest Railway Historical Society (MRHS), 533 W Glencoe, Palatine, IL 60067
Founded: 1968
Gen Secr: Louis Fuchs
Focus: Hist . . . 15738

Military Operations Research Society (MORS), Landmark Towers, 101 S Whiting St, Ste 202, Alexandria, VA 22304
T: (703) 751-7290; Fax: (703) 751-8171
Founded: 1957; Members: 2700
Gen Secr: Richard I. Wiles
Focus: Military Sci
Periodical
Phalanx (weekly) . . . 15739

Milton Helpern Institute of Forensic Medicine (MHIFM), 520 First Av, New York, NY 10016
T: (212) 447-2318; Fax: (212) 447-2094
Founded: 1968; Members: 405
Gen Secr: Dr. Charles S. Hirsch
Focus: Forensic Med
Periodical
The International Microform Journal of Legal Medicine and Forensic Sciences (weekly) 15740

Milton H. Erickson Foundation (MHEF), 3606 N 24 St, Phoenix, AZ 85016
T: (602) 956-6196; Fax: (602) 956-0519
Founded: 1979
Gen Secr: Jeffrey K. Zeig
Focus: Psych; Psychiatry
Periodicals
The Ericksonian Monographs (3 times annually)
Newsletter (3 times annually) . . . 15741

Milton Society of America (MSA), c/o Duquesne University, Pittsburgh, PA 15282
T: (412) 434-6420
Founded: 1948; Members: 500
Gen Secr: Albert C. Labriola
Focus: Lit
Periodical
Bulletin (weekly) . . . 15742

Mineralogical Society of America (MSA), 1130 17 St NW, Ste 330, Washington, DC 20036
T: (202) 775-4344; Fax: (202) 775-0018
Founded: 1919; Members: 3000
Gen Secr: Susan L. Myers
Focus: Mineralogy
Periodicals
American Mineralogist (weekly)
Lattice (weekly)
Mineralogical Abstracts (weekly) . . . 15743

Minerals, Metals and Materials Society, 420 Commonwealth Dr, Warrendale, PA 15086
T: (412) 776-9000; Fax: (412) 776-3770
Founded: 1957; Members: 12000
Gen Secr: Alexander R. Scott
Focus: Metallurgy; Mineralogy; Materials Sci
Periodicals
Journal of Electronic Materials (weekly)
Journal of Metals (weekly)
Membership Directory (weekly)
Transactions A (weekly)
Transactions B (weekly) . . . 15744

Mining and Metallurgical Society of America (MMSA), 9 Escalle Ln, Larkspur, CA 94939
T: (415) 924-7441; Fax: (415) 924-7463
Founded: 1908; Members: 315
Gen Secr: Robert M. Crum
Focus: Mining; Metallurgy
Periodicals
Bulletin
News-Letter (weekly) . . . 15745

Missile, Space and Range Pioneers (MSRP), POB 5227, Patrick Air Force Base, FL 32925
T: (305) 494-4001
Founded: 1966; Members: 1400
Gen Secr: John Butler
Focus: Hist; Eng . . . 15746

Mission Doctors Association (MDA), 1531 W Ninth St, Los Angeles, CA 90015
T: (818) 285-8868; Fax: (818) 309-1716
Founded: 1957
Pres: Richard Mason
Focus: Med . . . 15747

Moby Dick Academy (MDA), POB 236, Ocean Park, WA 98640
T: (206) 665-4577
Founded: 1981
Gen Secr: Mary Tufts
Focus: Educ
Periodicals
Corn-line (weekly)
FCLA Teachers and their Areas of Expertise (weekly)
Touchstone (weekly) . . . 15748

Model Secondary School for the Deaf (MSSD), c/o Gallaudet University, 800 Florida Av NE, Washington, DC 20002
T: (202) 651-5346
Founded: 1969
Gen Secr: Dr. Michael Deninger
Focus: Educ Handic . . . 15749

Modern Language Association of America (MLAA), 10 Astor Pl, New York, NY 10003
T: (212) 475-9500; Fax: (212) 477-9863
Founded: 1883; Members: 30000
Gen Secr: Phyllis Franklin
Focus: Ling
Periodicals
MLA Directory of Periodicals
MLA International Bibliography (weekly)
PMLA (weekly) . . . 15750

Modern Poetry Association (MPA), 60 W Walton St, Chicago, IL 60610
T: (312) 280-4870
Founded: 1941
Gen Secr: Joseph Parisi
Focus: Lit
Periodical
Poetry Magazine (weekly) . . . 15751

Mongolia Society, c/o Indiana University, 321-322 Goodbody Hall, Bloomington, IN 47405
T: (812) 855-4078; Fax: (812) 855-7500
Founded: 1961; Members: 400
Gen Secr: Susie Drost
Focus: Ethnology
Periodicals
Mongolian Studies: Journal of the Mongolia Society
The Mongolia Society Bulletin
The Mongolia Society Newsletter . . . 15752

The Monroe Institute (TMI), Rte 1, Box 175, Faber, VA 22938
T: (804) 361-1252
Founded: 1971; Members: 1500
Gen Secr: Nancy McMoneagle
Focus: Educ
Periodical
Breakthrough (weekly) . . . 15753

Moody Institute of Science (MIS), 820 N LaSalle Dr, Chicago, IL 60610-3284
T: (312) 329-2190; Fax: (312) 329-4496
Founded: 1945
Gen Secr: Barbara J. Goodwin
Focus: Sci . . . 15754

Mormon History Association (MHA), POB 7010, University Station, Provo, UT 84602
T: (801) 378-4048
Founded: 1965; Members: 750
Gen Secr: Jessie Embry
Focus: Hist; Rel & Theol
Periodical
Journal of Mormon History (weekly) . . 15755

Motorcycle Safety Foundation (MSF), 2 Jenner St, Ste 150, Irvine, CA 92718-3899
T: (714) 727-3227; Fax: (714) 727-4217
Founded: 1973; Members: 5
Gen Secr: Alan R. Isley
Focus: Safety
Periodicals
Annual Report (weekly)
Safe Cycling (weekly) . . . 15756

MTM Association for Standards and Research (MTMASR), 1411 Peterson Av, Park Ridge, IL 60068
T: (708) 823-7120; Fax: (708) 823-2319
Founded: 1951; Members: 1000
Gen Secr: Dirk J. Rauglas
Focus: Physiology
Periodical
MTM Journal (weekly) . . . 15757

Museum Association of the American Frontier (MAAF), HC 74, Box 18, Chadron, NE 69337
T: (308) 432-3843
Founded: 1949; Members: 2100
Gen Secr: Charles E. Hanson
Focus: Hist
Periodical
Museum of the Fur Trade Quarterly (weekly) . . . 15758

Museum Computer Network (MCN), 5001 Baum Blvd, Pittsburgh, PA 15213-1851
T: (412) 681-1818; Fax: (412) 681-5758
Founded: 1967; Members: 750
Gen Secr: Lynn Cox
Focus: Computer & Info Sci
Periodical
Spectra (weekly) . . . 15759

Museum Education Roundtable (MER), POB 23664, Washington, DC 20026
T: (202) 296-2294; Fax: (202) 223-9533
Founded: 1969; Members: 600
Gen Secr: A.T. Stephens
Focus: Educ
Periodical
Journal of Museum Education Anthology 15760

Music and Arts Society of America (MASA), POB 9751, Washington, DC 20016
T: (301) 990-1426
Founded: 1985
Gen Secr: Dr. M. de Lafayette
Focus: Arts; Music . . . 15761

Music Critics Association (MCA), 7 Pine Ct, Westfield, NJ 07090
T: (908) 233-8468; Fax: (908) 233-8468
Founded: 1957; Members: 254
Pres: Nancy Malitz
Focus: Music; Journalism
Periodicals
Membership List (weekly)
Newsletter (3 times annually) . . . 15762

Music Educators National Conference (MENC), 1902 Association Dr, Reston, VA 22091
T: (703) 860-4000; Fax: (703) 860-1531
Founded: 1907; Members: 62000
Gen Secr: John J. Mahlmann
Focus: Music; Educ
Periodicals
General Music Today (3 times annually)
Journal of Research in Music Education (weekly)
Music Educators Journal (9 times annually)
Newsletter (weekly) . . . 15763

Music Library Association (MLA), POB 487, Canton, MA 02021
T: (617) 828-8450; Fax: (617) 828-8915
Founded: 1931; Members: 1800
Gen Secr: James Henderson
Focus: Libraries & Bk Sci
Periodicals
Index Series
Music Cataloging Bulletin (weekly)
Newsletter (weekly)
Notes (weekly)
Technical Reports . . . 15764

Music Teachers National Association (MTNA), 617 Vine St, Ste 1432, Cincinnati, OH 45202
T: (513) 421-1420; Fax: (513) 421-2503
Founded: 1876; Members: 25000
Gen Secr: Ronald Molen
Focus: Music; Educ
Periodicals
American Music Teacher Magazine (weekly)
Directory of Nationally Certified Teachers (weekly) . . . 15765

Mycological Society of America (MSA), c/o Mary E. Palm, BARC-West, Rm 329, Beltsville, MD 20705
T: (301) 504-5327; Fax: (301) 504-5435
Founded: 1931; Members: 1400
Gen Secr: Mary E. Palm
Focus: Botany, Specific
Periodicals
Mycologia (weekly)
Mycologia Memoirs (weekly)
Newsletter (weekly) . . . 15766

Mystic Seaport Museum (MSM), 50 Greenmanville Av, Mystic, CT 06355-0990
T: (203) 572-0711; Fax: (203) 572-5328
Members: 22000
Gen Secr: J. Revell Carr
Focus: Hist; Navig
Periodical
The LOG (weekly) . . . 15767

Mythopoeic Society (MS), POB 6707, Altadena, CA 91003
T: (818) 571-7727
Founded: 1967; Members: 650
Pres: Glen H. Goodknight
Focus: Lit
Periodicals
Mythic Circle (weekly)
Mythlore (weekly)
Mythprint (weekly) . . . 15768

Nathaniel Hawthorne Society (NHS), c/o Hawthorne-Longfellow Library, Bowdoin College, Brunswick, ME 04011
T: (207) 725-3281
Founded: 1974; Members: 400
Gen Secr: Arthur Monke
Focus: Lit
Periodical
Nathaniel Hawthorne Review (weekly) . . 15769

National Academic Advising Association (NACADA), c/o Kansas State University, 2323 Anderson Av, Ste 226, Manhattan, KS 66502
T: (913) 532-5717; Fax: (913) 532-7732
Founded: 1979; Members: 3100
Gen Secr: Roberta Flaherty
Focus: Sci
Periodicals
NACADA Journal (weekly)
NACADA Newsletter (weekly)
Proceedings of Annual Conference . . . 15770

National Academy of Code Administration (NACA), 138 E Court St, Ste 803, Cincinnati, OH 45202
T: (513) 632-8643
Founded: 1970; Members: 900
Gen Secr: Ralph W. Liebing
Focus: Civil Eng
Periodical
Newsletter: Blueprint (weekly) . . . 15771

National Academy of Design (NAD), 1083 Fifth Av, New York, NY 10128
T: (212) 369-4880; Fax: (212) 360-6795
Founded: 1825; Members: 450
Gen Secr: Edward Gallagher
Focus: Graphic & Dec Arts, Design
Periodicals
Academy Bulletin (weekly)
Academy Calendar (weekly) . . . 15772

National Academy of Education (NAE), c/o School of Education, Stanford University, 507G CERAS, Stanford, CA 94305-3084
T: (415) 725-1003; Fax: (415) 723-7235
Founded: 1965; Members: 125
Gen Secr: Debbie Leong
Focus: Educ . . . 15773

National Academy of Engineering (NAE), 2101 Constitution Av NW, Washington, DC 20418
T: (202) 334-3200; Fax: (202) 334-1684
Founded: 1964; Members: 1525
Gen Secr: William C. Salmon
Focus: Eng . . . 15774

National Academy of Opticianry (NAO), 10111 Martin Luther King jr. Hwy, Ste 112, Bowie, MD 20720
T: (301) 577-4828; Fax: (301) 577-3880
Founded: 1973; Members: 6000
Gen Secr: Floyd H. Holmgrain
Focus: Optics
Periodical
Academy Newsletter (weekly) . . . 15775

National Academy of Public Administration (NAPA), 1120 G St NW, Ste 850, Washington, DC 20005
T: (202) 347-3190; Fax: (202) 393-0993
Founded: 1967; Members: 400
Pres: R. Scott Fosler
Focus: Public Admin
Periodical
Annual Report (weekly) . . . 15776

National Academy of Recording Arts and Sciences (NARAS), 303 N Glenoaks Blvd, Ste 140, Burbank, CA 91502-1178
T: (213) 849-1313
Founded: 1957; Members: 7000
Gen Secr: Christine M. Farnon
Focus: Music; Eng
Periodicals
Grammy Pulse (weekly)
Program Book (weekly) . . . 15777

National Academy of Western Art (NAWA), 1700 NE 63 St, Oklahoma City, OK 73111
T: (405) 478-2250; Fax: (405) 478-4714
Founded: 1973; Members: 57
Pres: Ed Muno
Focus: Arts . . . 15778

National Accreditation Council for Environmental Health Science and Protection (NACEHSP), c/o National Environmental Health Association, 720 S Colorado Blvd, Ste 970, Denver, CO 80222
T: (303) 756-9090
Founded: 1969; Members: 6000
Pres: Dr. Gary Silverman
Focus: Public Health . . . 15779

National Accrediting Agency for Clinical Laboratory Sciences (NAACLS), 8410 W Bryn Mawr Av, Ste 670, Chicago, IL 60631
T: (312) 714-8880
Founded: 1973; Members: 734
Focus: Med; Eng
Periodical
Newsletter (2-4 times annually) . . . 15780

National Accrediting Commission of Cosmetology Arts and Sciences (NACCAS), 901 N State St, Arlington, VA 22203
T: (703) 527-7600; Fax: (703) 527-8811
Founded: 1969; Members: 32
Gen Secr: Mark Gross
Focus: Adult Educ
Periodicals
Directory of Accredited Cosmetology Schools (weekly)
NACCAS Review (weekly) . . . 15781

National Adult Vocational Education Association (NAVEA), c/o Carolyn S. Gasiorek, EHOVE Career Center, 316 W Mason Rd, Milan, OH 44846
T: (419) 499-4663; Fax: (419) 499-4076
Founded: 1978; Members: 500
Gen Secr: Carolyn S. Gasiorek
Focus: Adult Educ . . . 15782

National Alliance for Safe Schools (NASS), POB 30177, Bethesda, MD 20824
T: (301) 907-7888
Founded: 1977; Members: 25
Gen Secr: Robert J. Rubel
Focus: Educ . . . 15783

National Alliance of Black School Educators (NABSE), 2816 Georgia Av NW, Washington, DC 20001
T: (202) 483-1549; Fax: (202) 483-8323
Founded: 1970; Members: 5000
Gen Secr: William J. Saunders
Focus: Educ
Periodicals
Membership Roster (weekly)
News Briefs 15784

National Alliance of Media Arts Centers (NAMAC), 1212 Broadway, Oakland, CA 94612
T: (510) 451-2717; Fax: (510) 834-3741
Founded: 1978; Members: 300
Gen Secr: Julian Low
Focus: Arts 15785

National Alopecia Areata Foundation (NAAF), POB 150760, San Rafael, CA 94915-0760
T: (415) 456-4644; Fax: (415) 456-4274
Founded: 1981; Members: 5000
Gen Secr: Vicki Kalabokes
Focus: Med
Periodical
National Alopecia Areata Foundation Newsletter (weekly) 15786

National Anorexic Aid Society (NAAS), 1925 E Dublin-Granville Rd, Columbus, OH 43229
T: (614) 436-1112
Founded: 1977; Members: 500
Gen Secr: Arline Iannicello
Focus: Med 15787

National Architectural Accrediting Board (NAAB), 1735 New York Av NW, Washington, DC 20006
T: (202) 783-2007; Fax: (202) 626-7421
Founded: 1940; Members: 11
Gen Secr: John M. Maudin-Jeronimo
Focus: Archit
Periodicals
Criteria and Procedures (biennial)
List of Accredited Programs in Architecture (weekly) 15788

National Archives and Records Administration Volunteer Association (NARAVA), c/o N.E. Office of Public Programs, National Archives, Eighth at Pennsylvania Av NW, Washington, DC 20408
T: (202) 501-5205
Founded: 1976; Members: 170
Pres: J. Weimer
Focus: Archives 15789

National Art Education Association (NAEA), 1916 Association Dr, Reston, VA 22091-1590
T: (703) 860-8000; Fax: (703) 860-2960
Founded: 1947; Members: 13000
Gen Secr: Thomas A. Hatfield
Focus: Arts; Educ
Periodicals
Art Education (weekly)
Newsletter (weekly)
Studies in Art Education (weekly) . . . 15790

National Association for Ambulatory Care (NAFAC), 21 Michigan St, Grand Rapids, MI 49503
T: (616) 949-2138
Founded: 1981; Members: 600
Pres: John A. Rupke
Focus: Med 15791

National Association for Applied Arts, Science and Education (NAAASE), 412 E Burleigh St, Milwaukee, WI 53212
T: (414) 264-2455
Founded: 1956; Members: 250
Gen Secr: R.A. Kurth
Focus: Arts; Lit
Periodical
Journal of Applied Arts, Science and Education (weekly) 15792

National Association for Armenian Studies and Research (NAASR), 395 Concord Av, Belmont, MA 02178-3049
T: (617) 489-1610
Founded: 1955; Members: 1200
Gen Secr: M.S. Young
Focus: Ethnology
Periodicals
AICS Bulletin (weekly)
NAASR Newsletter (weekly) 15793

National Association for Bilingual Education (NABE), 810 First St NE, Washington, DC 20002
T: (202) 898-1829
Founded: 1975; Members: 3000
Gen Secr: James J. Lyons
Focus: Ling; Educ
Periodicals
Journal (3 times annually)
Newsletter (10 times annually) 15794

National Association for Biomedical Research (NABR), 818 Connecticut Av NW, Ste 303, Washington, DC 20006
T: (202) 857-0540; Fax: (202) 659-1902
Founded: 1985; Members: 400
Gen Secr: Frankie L. Trull
Focus: Med
Periodicals
Regulatory Alert (6-10 times annually)
Report (weekly)
Update (15-18 times annually) 15795

National Association for Business Teacher Education (NABTE), 1914 Association Dr, Reston, VA 22091
T: (703) 860-8300; Fax: (703) 620-4483
Founded: 1927; Members: 210
Pres: Dr. Janet M. Treichel
Focus: Adult Educ
Periodical
NABTE Review (weekly) 15796

National Association for Core Curriculum (NACC), c/o Kent State University, 404 White Hall, Kent, OH 44242-0001
T: (216) 672-2792
Founded: 1953; Members: 200
Gen Secr: Gordon F. Vars
Focus: Educ
Periodical
The Core Teacher (weekly) 15797

National Association for Equal Opportunity in Higher Education (NAFEO), 420 12 St NE, Washington, DC 20002
T: (202) 543-9111; Fax: (202) 543-9113
Founded: 1969; Members: 117
Pres: Dr. Samuel L. Myers
Focus: Educ
Periodicals
Inroads (weekly)
NAFEO/AID Update (weekly) 15798

National Association for Ethnic Studies (NAES), c/o Dept of English, Arizona State University, Tempe, AZ 85287
T: (602) 965-2197; Fax: (602) 965-2012
Founded: 1975; Members: 300
Gen Secr: Gretchen Bataille
Focus: Ethnology
Periodicals
The Ethnic Reporter (weekly)
Explorations in Ethnic Studies (weekly)
Explorations in Sights and Sounds (weekly) 15799

National Association for Family and Community Education, 5100 S Atlanta, Tulsa, OK 74105-6600
T: (918) 749-8383; Fax: (918) 749-8703
Founded: 1936; Members: 400000
Pres: Judy Weinkauf
Focus: Home Econ
Periodicals
Handbook (weekly)
The Homemaker Update (weekly) . . . 15800

National Association for Hearing and Speech Action (NAHSA), 10801 Rockville Pike, Rockville, MD 20852
T: (301) 897-8682
Founded: 1919; Members: 2000
Gen Secr: Russell L. Malone
Focus: Logopedy 15801

National Association for Human Development (NAHD), 1424 16 St NW, Ste 102, Washington, DC 20036
T: (202) 328-2191
Founded: 1974
Gen Secr: Anne Radd
Focus: Geriatrics
Periodical
Digest (weekly) 15802

National Association for Humane and Environmental Education (NAHEE), POB 362, East Haddam, CT 06423-0362
T: (203) 434-8666; Fax: (203) 434-9579
Founded: 1974; Members: 2000
Gen Secr: Patty A. Finch
Focus: Educ
Periodicals
Children and Animals (weekly)
Kind News (5 times annually) 15803

National Association for Humanities Education (NAHE), c/o Dr. James Mehl, Missouri Western State College, 4525 Downs Dr, Saint Joseph, MO 64507
T: (816) 271-4333
Founded: 1967; Members: 400
Focus: Humanities; Educ
Periodical
Humanities Education (weekly) 15804

National Association for Industry-Education Cooperation (NAIEC), 235 Hendricks Blvd, Buffalo, NY 14226-3304
T: (716) 834-7047; Fax: (716) 834-7047
Founded: 1948; Members: 1180
Pres: Dr. Donald M. Clark
Focus: Adult Educ; Econ
Periodicals
Journal
Newsletter (weekly) 15805

National Association for Interpretation (NAI), POB 1892, Fort Collins, CO 80522
T: (303) 491-6434; Fax: (303) 491-2255
Founded: 1988; Members: 1900
Gen Secr: Judith K. Giles
Focus: Nat Sci
Periodicals
Journal of Interpretation (weekly)
National Workshop Proceedings (weekly) . 15806

National Association for Legal Support of Alternative Schools (NALSAS), POB 2823, Santa Fe, NM 87501
T: (505) 471-6928
Founded: 1973; Members: 6000
Gen Secr: Ed Nagel
Focus: Educ
Periodical
Tidbits (weekly) 15807

National Association for Outlaw and Lawman History (NOLA), c/o Richard J. Miller, 615C N Eighth St, Killeen, TX 76541
T: (817) 634-8300
Founded: 1974; Members: 365
Gen Secr: Richard J. Miller
Focus: Cultur Hist; Hist
Periodicals
Index to NOLA Publications (biennial)
Newsletter (weekly)
Quarterly (weekly) 15808

National Association for Physical Education in Higher Education (NAPEHE), c/o Dept of HUP, San Jose State University, San Jose, CA 95192
T: (408) 924-3029; Fax: (408) 924-3053
Founded: 1978; Members: 500
Gen Secr: Gail G. Becker
Focus: Sports
Periodicals
Newsletter (weekly)
Quest (weekly) 15809

National Association for Poetry Therapy (NAPT), c/o Peggy Osna Heller, 7715 White Rim Terrace, Potomac, MD 20854
T: (301) 299-8330; Fax: (301) 299-8330
Founded: 1981; Members: 175
Pres: Peggy Osna Heller
Focus: Therapeutics
Periodicals
The Journal of Poetry Therapy
NAPT News Letter 15810

National Association for Practical Nurse Education and Service (NAPNES), 1400 Spring St, Ste 310, Silver Spring, MD 20910
T: (301) 588-2491; Fax: (301) 588-2839
Founded: 1941; Members: 30000
Gen Secr: John H. Word
Focus: Adult Educ; Med
Periodicals
Career Directory
Journal of Practical Nursing (weekly) . . 15811

National Association for Research in Science Teaching (NARST), c/o Dr. John R. Staverm Kansas State University, 219 Bluemont Hall, Manhattan, KS 66506
T: (913) 532-6294; Fax: (913) 532-7304
Founded: 1928; Members: 1100
Gen Secr: Dr. John R. Staver
Focus: Educ
Periodicals
Abstracts of Papers Presented to Annual Meeting (weekly)
Journal of Research in Science Teaching (weekly)
Membership Directory (weekly)
Newsletter (weekly) 15812

National Association for Rural Mental Health (ARMH), 301 E Armour Bldg, Ste 420, Kansas City, MO 64111
T: (816) 756-3140
Founded: 1977; Members: 300
Gen Secr: Teresa Cavender
Focus: Psychiatry
Periodical
Rural Community Health Newsletter (weekly) 15813

National Association for Search and Rescue (NASAR), POB 3709, Fairfax, VA 22038
T: (703) 352-1349; Fax: (703) 352-0309
Founded: 1970; Members: 3100
Gen Secr: Peggy McDonald
Focus: Med
Periodicals
Response (weekly)
Update (weekly) 15814

National Association for Sport and Physical Education (NASPE), 1900 Association Dr, Reston, VA 22091
T: (703) 476-3410; Fax: (703) 476-9527
Founded: 1974; Members: 30000
Gen Secr: Dr. Judith C. Young
Focus: Sports 15815

National Association for the Advancement of Black Americans in Vocational Education (NAABAVE), c/o Dr. Ethel O. Washington, 5057 Woodward, Rm 976, Detroit, MI 48202
T: (313) 494-1660; Fax: (313) 494-1535
Founded: 1977; Members: 900
Pres: Dr. Ethel O. Washington
Focus: Educ 15816

National Association for the Advancement of Psychoanalysis and The American Boards for Accreditation and Certification (NAAPABAC), 80 Eighth Av, Ste 1501, New York, NY 10011-1501
T: (212) 741-0515; Fax: (212) 741-0515
Founded: 1972; Members: 1550
Gen Secr: Margery Quackenbush
Focus: Psychoan 15817

National Association for the Education of Young Children (NAEYC), 1834 Connecticut Av NW, Washington, DC 20009
T: (202) 232-8777; Fax: (202) 328-1846
Founded: 1926; Members: 75000
Gen Secr: Dr. Marilyn M. Smith
Focus: Educ
Periodical
Young Children (weekly) 15818

National Association for the Exchange of Industrial Resources (NAEIR), 560 McClure St, POB 8076, Galesburg, IL 61402
T: (309) 343-0704; Fax: (309) 343-0862
Founded: 1977; Members: 7000
Pres: Gary C. Smith
Focus: Educ; Econ
Periodicals
Gift Catalog (weekly)
Membership Directory (weekly)
NAEIR News (weekly) 15819

National Association for the Practice of Anthropology (NAPA), c/o American Anthropological Association, 1703 New Hampshire Av NW, Washington, DC 20009
T: (202) 232-8800
Founded: 1983; Members: 887
Pres: Shirley Fiske
Focus: Anthro
Periodicals
Bulletin Series
Cultural Experts Directory 15820

National Association for Trade and Industrial Education (NATIE), POB 1665, Leesburg, VA 22075
T: (703) 777-1740
Founded: 1974; Members: 1400
Gen Secr: Dr. Ethel M. Smith
Focus: Adult Educ
Periodicals
NewsNotes (weekly)
State Supervisors/Consultants of Trade and Industrial Education (weekly) 15821

National Association for Veterinary Acupuncture (NAVA), 951 W Bastan Chury Rd, Fullerton, CA 92635
T: (714) 871-3000
Founded: 1973; Members: 60
Pres: Richard S. Glassberg
Focus: Vet Med 15822

National Association for Women in Education (NAWE), 1325 18 St NW, Ste 210, Washington, DC 20036-6511
T: (202) 659-9330; Fax: (202) 457-0946
Founded: 1916; Members: 1900
Gen Secr: Dr. Patricia A. Rueckel
Focus: Educ
Periodicals
Directory (weekly)
Journal (weekly)
Newsletter (weekly) 15823

National Association for Year-Round Education (NCYRE), c/o Dr. Charles Ballinger, POB 711386, San Diego, CA 92171-1386
T: (619) 276-5296; Fax: (619) 571-5754
Founded: 1972; Members: 1250
Gen Secr: Dr. Charles Ballinger
Focus: Educ
Periodicals
Directory of Year-Round Schools (weekly)
The Year-Rounder (weekly) 15824

National Association of Academic Advisors for Athletics, c/o University of Virginia, POB 3785, Charlottesville, VA 22903
T: (804) 982-5300
Founded: 1975; Members: 400
Focus: Sports; Educ
Periodical
Academic Athletic Journal (weekly) . . . 15825

National Association of Academies of Science (NAAS), c/o Dept of Mathematical Sciences, University of South Carolina, 133 Willard Bldg, Columbia, SC 29208
T: (803) 777-7007; Fax: (402) 399-2686
Founded: 1926; Members: 45
Pres: Dr. Don Jordan
Focus: Sci
Periodicals
Directory and Proceedings (weekly)
Newsletter (weekly) 15826

National Association of Advisers for the Health Professions (NAAHP), POB 5017, Station A, Champaign, IL 61825-5017
T: (217) 333-0090; Fax: (217) 333-0122
Founded: 1974; Members: 1275
Gen Secr: Julian M. Frankenberg
Focus: Educ; Med 15827

National Association of Alcoholism and Drug Abuse Counselors (NAADAC), 3717 Columbia Pike, Ste 300, Arlington, VA 22204
T: (703) 920-4644; Fax: (703) 920-4672
Founded: 1972; Members: 15000
Gen Secr: Linda Kaplan
Focus: Soc Sci 15828

National Association of Baptist Professors of Religion (NABPR), c/o Richard F. Wilson, Mercer University, Macon, GA 31207
T: (912) 752-2755
Founded: 1927; Members: 650
Gen Secr: Richard F. Wilson
Focus: Rel & Theol
Periodicals
Bibliographic Series (weekly)
Dissertation Series (weekly)
Monograph/Special Studies Series (weekly)
Perspectives in Religious Studies (weekly) 15829

National Association of Bar Executives (NABE), c/o Div for Bar Services, 541 N Fairbanks, Chicago, IL 60611-3314
T: (312) 988-5346; Fax: (312) 988-5492
Founded: 1941; Members: 550
Gen Secr: H. Maria Enright
Focus: Law
Periodicals
Bar Leader (weekly)
Roster and Operations Manual (weekly) . *15830*

National Association of Biology Teachers (NABT), 11250 Roger Bacon Dr, Reston, VA 22090
T: (703) 471-1134
Founded: 1938; Members: 7000
Gen Secr: Patricia J. McWethy
Focus: Bio; Educ
Periodicals
The American Biology Teacher (8 times annually)
News + Views (5 times annually) . . . *15831*

National Association of Black Professors (NABP), POB 526, Chrisfield, MD 21817
T: (410) 968-2393
Founded: 1974; Members: 135
Pres: Dr. Sarah Miles Woods
Focus: Adult Educ *15832*

National Association of Blind Teachers (NABT), c/o American Council of the Blind, 1155 15 St NW, Ste 720, Washington, DC 20005
T: (202) 467-5081; Fax: (202) 476-5085
Founded: 1971; Members: 190
Pres: Harvey Miller
Focus: Educ
Periodical
The Blind Teacher (weekly) *15833*

National Association of Boards of Pharmacy (NABP), 700 Busse Hwy, Park Ridge, IL 60068
T: (708) 698-6227
Founded: 1904; Members: 56
Gen Secr: Carmen A. Catizone
Focus: Pharmacol
Periodicals
Newsletter (weekly)
State Board Newsletter (weekly) *15834*

National Association of Bond Lawyers (NABL), POB 397, Hinsdale, IL 60522
T: (708) 920-0160
Founded: 1979; Members: 2750
Gen Secr: Rita J. Carlson
Focus: Law
Periodicals
Directory (weekly)
Newsletter (weekly) *15835*

National Association of Business Education State Supervisors (NABESS), c/o Janet Gandy, Arizona Dept of Education, 1535 W Jefferson, Phoenix, AZ 85007
T: (602) 542-5350; Fax: (602) 542-1849
Founded: 1965; Members: 125
Gen Secr: Janet Gandy
Focus: Educ; Econ *15836*

National Association of Children's Hospitals and Related Institutions (NACHRI), 401 Wythe St, Alexandria, VA 22314
T: (703) 684-1355; Fax: (703) 684-1589
Founded: 1968; Members: 110
Pres: Lawrence C. McAndrews
Focus: Pediatrics
Periodicals
Guide to Children's Hospitals (weekly)
Newsletter (weekly) *15837*

National Association of Classroom Educators in Business Education (NACEBE), c/o Janet H. Auten, Watauga High School, Hwy 105 S, Boone, NC 28607
T: (704) 264-2407; Fax: (704) 264-9030
Founded: 1968; Members: 700
Pres: Janet H. Auten
Focus: Educ; Econ *15838*

National Association of College and University Attorneys (NACUA), 1 Dupont Circle, Ste 620, Washington, DC 20036
T: (202) 833-8390; Fax: (202) 296-8379
Founded: 1961; Members: 2500
Gen Secr: Phillip M. Grier
Focus: Law
Periodical
Journal of College and University Law . *15839*

National Association of College and University Business Officers (NACUBO), 1 Dupont Circle NW, Ste 500, Washington, DC 20036
T: (202) 861-2500; Fax: (202) 861-2583
Founded: 1950; Members: 2100
Pres: C.L. Harris
Focus: Econ
Periodicals
Annual Report (weekly)
Business Officer *15840*

National Association of College Deans, Registrars and Admissions Officers (NACDRAO), 917 Dorsett St, Albany, GA 31701
T: (912) 435-4945
Founded: 1925; Members: 325
Gen Secr: Helen Mayes
Focus: Educ
Periodicals
NACDRAO Directory (weekly)
Newsletter (weekly)
Proceedings (weekly) *15841*

National Association of Colleges and Teachers of Agriculture (NACTA), c/o Dr. Jack C. Everly, 608 W Vermont, Urbana, IL 61801
T: (217) 344-5738
Founded: 1955; Members: 1500
Gen Secr: Dr. Jack C. Everly
Focus: Educ; Agri
Periodical
NACTA Journal (weekly) *15842*

National Association of College Wind and Percussion Instructors (NACWPI), c/o Div of Fine Arts, Northeast Missouri State University, Kirksville, MO 63501
T: (816) 785-4442; Fax: (816) 785-4181
Founded: 1952; Members: 1200
Gen Secr: Dr. Richard Weerts
Focus: Music
Periodical
Journal (weekly) *15843*

National Association of Community Health Centers (NACHC), 1330 New Hampshire Av NW, Ste 122, Washington, DC 20036
T: (202) 659-8008; Fax: (202) 659-8519
Founded: 1970; Members: 950
Gen Secr: Thomas Van Coverden
Focus: Public Health
Periodicals
Community Health Guides (weekly)
Community Health Listing (weekly)
Washington Update (weekly) *15844*

National Association of Corrosion Engineers (NACE), POB 218340, Houston, TX 77218
T: (713) 492-0535; Fax: (713) 492-8254
Founded: 1943; Members: 15000
Gen Secr: G.M. Shankel
Focus: Metallurgy
Periodicals
Corrosion (weekly)
Corrosion Abstracts (weekly)
Materials Performance (weekly) *15845*

National Association of County Engineers (NACE), c/o Charles E. Wiles, 440 First St NW, Washington, DC 20001
T: (202) 393-5041; Fax: (202) 393-2630
Founded: 1956; Members: 1700
Gen Secr: Charles E. Wiles
Focus: Eng
Periodical
Newsletter (weekly) *15846*

National Association of County Planners (NACP), c/o National Association of Counties, 440 First St NW, Washington, DC 20001
T: (202) 393-6226; Fax: (202) 393-2630
Founded: 1965; Members: 500
Gen Secr: Haron Battle
Focus: Urban Plan
Periodical
Roster (biennial) *15847*

National Association of Dental Assistants (NADA), 900 S Washington St, Falls Church, VA 22046
T: (703) 237-8616
Founded: 1974; Members: 4000
Pres: Joseph Salta
Focus: Dent *15848*

National Association of Dental Laboratories (NADL), 3801 Mount Vernon Av, Alexandria, VA 22305-2491
T: (703) 683-5263; Fax: (703) 549-4788
Founded: 1951; Members: 3100
Gen Secr: Robert W. Stanley
Focus: Dent
Periodicals
Executive Information Series
Trends and Techniques (10 times annually)
Who's Who in the Dental Laboratory Industry (weekly) *15849*

National Association of Disability Examiners (NADE), POB 4188, Frankfort, KY 40603
T: (502) 875-8388
Founded: 1963; Members: 2373
Pres: M. Marshall
Focus: Med *15850*

National Association of Dramatic and Speech Arts (NADSA), 208 Cherokee Dr, Blacksburg, VA 24060
T: (703) 231-5805
Founded: 1936; Members: 500
Gen Secr: Dr. H.D. Flowers
Focus: Perf Arts
Periodicals
Encore (biennial)
NADSA Conference Directory (weekly)
NADSA Update (weekly)
Newsletter (weekly) *15851*

National Association of Educational Buyers (NAEB), 450 Wireless Blvd, Hauppauge, NY 11788
T: (516) 273-2600; Fax: (516) 273-2305
Founded: 1920; Members: 2200
Gen Secr: Neil D. Markee
Focus: Educ *15852*

National Association of Elementary School Principals (NAESP), 1615 Duke St, Alexandria, VA 22314
T: (703) 684-3345; Fax: (703) 548-6021
Founded: 1921; Members: 26000
Gen Secr: Dr. Samuel G. Sava
Focus: Educ
Periodicals
Communicator (10 times annually)
Principal (5 times annually) *15853*

National Association of Emergency Medical Technicians (NAEMT), 9140 Ward Park, Kansas City, MO 64114
T: (816) 444-3500; Fax: (816) 444-0330
Founded: 1975; Members: 3000
Gen Secr: L. Thomas
Focus: Med
Periodicals
Newsletter (weekly)
Perspective (weekly) *15854*

National Association of Employers on Health Care Action (NAEHCA), 240 Crandon Blvd, Ste 110, POB 220, Key Biscayne, FL 33149
T: (305) 361-2810; Fax: (305) 361-2842
Founded: 1976; Members: 100
Pres: Ruth H. Stack
Focus: Public Health
Periodicals
Health Directions Letter (weekly)
Managed Care-HMOs (weekly) *15855*

National Association of Environmental Professionals (NAEP), 5165 MacArthur Blvd NW, POB 9400, Washington, DC 20016-3315
T: (202) 966-1500; Fax: (202) 966-1977
Founded: 1975; Members: 3000
Gen Secr: Susan Eisenberg
Focus: Ecology
Periodicals
The Environmental Professional
Newsletter (weekly) *15856*

National Association of Episcopal Schools (NAES), 815 Second Av, New York, NY 10017-4594
T: (212) 922-5173; Fax: (212) 949-6781
Founded: 1954; Members: 360
Gen Secr: Ann M. Gordon
Focus: Educ
Periodicals
Directory of Episcopal Church Schools
NAES Journal (weekly)
Newsletter (weekly) *15857*

National Association of Extension 4-H Agents (NAE4-HA), c/o Peggy M. Adkins, University of Georgia, Hoke Smith Annex, Athens, GA 30602
T: (706) 978-8171; Fax: (404) 978-8171
Founded: 1946; Members: 3452
Gen Secr: Peggy M. Adkins
Focus: Educ; Agri
Periodicals
Journal of Extension (weekly)
Membership Report (weekly)
NAE4-HA Membership Directory (weekly)
News and Views (weekly) *15858*

National Association of Federal Veterinarians (NAFV), 1101 Vermont Av NW, Ste 710, Washington, DC 20005-3521
T: (202) 289-6334
Founded: 1918; Members: 1600
Gen Secr: Edward L. Menning
Focus: Vet Med
Periodical
The Federal Veterinarian (weekly) . . . *15859*

National Association of Flight Instructors (NAFI), Ohio State University Airport, POB 793, Columbus, OH 43017
T: (614) 889-6148; Fax: (614) 889-2610
Founded: 1966; Members: 2000
Pres: Jack J. Eggspuehler
Focus: Aero
Periodical
Newsletter (weekly) *15860*

National Association of Geology Teachers (NAGT), c/o Dr. Robert Christman, Dept of Geology, Western Washington University, Bellingham, WA 98225
T: (206) 676-3587; Fax: (206) 647-7295
Founded: 1938; Members: 1800
Gen Secr: Dr. Robert Christman
Focus: Geology; Educ
Periodicals
Journal of Geological Education (5 times annually)
Membership Directory *15861*

National Association of Health Career Schools (NAHCS), 10963 Saint Charles Rock Rd, Saint Ann, MO 63074
T: (314) 739-4450
Founded: 1980
Gen Secr: L. Hicks
Focus: Adult Educ; Public Health
Periodical
Bulletin (weekly) *15862*

National Association of Health Services Executives (NAHSE), 10320 Little Patuxent Pkwy, Ste 1106, Columbia, MD 21044
T: (202) 628-3953; Fax: (202) 628-3958
Founded: 1968; Members: 500
Gen Secr: O. Jenkins
Focus: Public Health
Periodical
Notes (weekly) *15863*

National Association of Hebrew Day School Administrators (NAHDSA), 1114 J Av, Brooklyn, NY 11230
T: (718) 258-7767
Members: 400
Gen Secr: Dov Milians
Focus: Educ
Periodical
Directory *15864*

National Association of Hebrew Day School PTA'S, 160 Broadway, New York, NY 10038
T: (212) 227-1000; Fax: (212) 406-6934
Founded: 1947; Members: 300
Gen Secr: Bernice Brand
Focus: Educ
Periodical
National PTA Bulletin (3 times annually) *15865*

National Association of Home and Workshop Writers (NAH & WW), c/o Alfred Lees, 140 Nassau St, Rm 9B, New York, NY 10038
T: (212) 267-1153
Founded: 1973; Members: 60
Gen Secr: Alfred Lees
Focus: Lit
Periodical
NAH & WW Newsletter (weekly) *15866*

National Association of Independent Colleges and Universities (NAICU), 122 C St NW, Ste 750, Washington, DC 20001
T: (202) 347-7512
Founded: 1976; Members: 817
Pres: Richard F. Rosser
Focus: Sci *15867*

National Association of Independent Schools (NAIS), 75 Federal, Boston, MA 02110
T: (617) 451-2444; Fax: (617) 482-3913
Founded: 1962; Members: 1050
Pres: Peter Relic
Focus: Educ
Periodical
Independent School (3 times annually) . *15868*

National Association of Industrial and Technical Teacher Educators (NAITTE), c/o Dept of Technological and Adult Education, University of Tennessee, 402 Claxton Addition, Knoxville, TN 37996-3400
T: (615) 974-2574; Fax: (615) 974-2048
Founded: 1937; Members: 850
Gen Secr: Gregory C. Petty
Focus: Adult Educ; Eng
Periodicals
Directory (weekly)
Journal of Industrial Teacher Education (weekly)
News and Views (3 times annually) . . *15869*

National Association of Legal Assistants (NALA), 1601 S Main St, Ste 300, Tulsa, OK 74119
T: (918) 587-6828; Fax: (918) 582-6772
Founded: 1975; Members: 3500
Gen Secr: M. Dover
Focus: Law
Periodical
Facts & Findings (weekly) *15870*

National Association of Management and Technical Assistance Centers (NAMTAC), 733 15 St NW, Ste 917, Washington, DC 20005
T: (202) 347-6740; Fax: (202) 347-6740
Founded: 1980; Members: 160
Gen Secr: Harold W. Williams
Focus: Business Admin *15871*

National Association of Marine Surveyors (NAMS), POB 9306, Chesapeake, VA 23321-9306
T: (800) 822-6267
Founded: 1960; Members: 400
Gen Secr: Kim I. MacCartney
Focus: Surveying; Navig
Periodicals
Annual Membership List (weekly)
NAMS Newsletter (weekly) *15872*

National Association of Medical Examiners (NAME), 1402 S Grand Blvd, Saint Louis, MO 63104
T: (314) 577-8298; Fax: (314) 772-1307
Founded: 1966
Gen Secr: Judy Thomas
Focus: Med
Periodical
American Journal of Forensic Medicine and Pathology (weekly) *15873*

National Association of Music Executives in State Universities (NAMESU), c/o Dr. Dorothy Payne, School of Music, University of Arizona, Tucson, AZ 85721
T: (602) 621-7023; Fax: (602) 621-8118
Founded: 1935; Members: 50
Gen Secr: Dr. Dorothy Payne
Focus: Music *15874*

National Association of Nutrition and Aging Services Programs (NANASP), 2675 44 St NW, Ste 305, Grand Rapids, MI 49509
T: (616) 530-3250; Fax: (616) 531-3103
Founded: 1977; Members: 1000
Gen Secr: Connie Benton Wolfe
Focus: Nutrition; Geriatrics

Periodicals
Annual Report (weekly)
Monthly Membership Updates (weekly)
NANASP News (weekly)
Special Bulletin (weekly) 15875

National Association of Optometrists and Opticians (NAOO), 18903 S Miles Rd, Cleveland, OH 44128
T: (216) 475-8925
Founded: 1960; Members: 13225
Gen Secr: Franklin D. Rozak
Focus: Ophthal; Optics 15876

National Association of Pastoral Musicians (NPM), 225 Sheridan St NW, Washington, DC 20011-1492
T: (202) 723-5800; Fax: (202) 723-2262
Founded: 1976; Members: 8500
Gen Secr: Virgil C. Funk
Focus: Music
Periodicals
Pastoral Musician's Notebook (weekly)
Pastoral Music Magazine (weekly) . . . 15877

National Association of Power Engineers (NAPE), 5-7 Springfield St, Chicopee, MA 01013
T: (413) 592-6273; Fax: (413) 592-1998
Founded: 1882; Members: 6500
Gen Secr: William Judd
Focus: Energy
Periodical
National Engineer (weekly) 15878

National Association of Principals of Schools for Girls (NAPSG), 4050 Little River Rd, Hendersonville, NC 28739
T: (704) 693-1490; Fax: (704) 693-1490
Founded: 1920; Members: 600
Gen Secr: Nancy E. Kussrow
Focus: Educ
Periodical
Proceedings (weekly) 15879

National Association of Private Schools for Exceptional Children (NAPSEC), 1525 K St NW, Ste 1032, Washington, DC 20005
T: (202) 408-3338; Fax: (202) 408-3340
Founded: 1971; Members: 200
Gen Secr: Sherry L. Kolbe
Focus: Educ
Periodicals
Directory (biennial)
Newsbriefs (3 times annually)
Newsletter (3 times annually) 15880

National Association of Professional Educators (NAPE), 412 First St SE, Washington, DC 20003
T: (202) 484-8969
Founded: 1972
Gen Secr: Philip Strittmatter
Focus: Educ 15881

National Association of Professors of Hebrew (NAPH), c/o University of Wisconsin, 1346 Van Hise Hall, Madison, WI 53705
T: (608) 262-2968
Founded: 1950; Members: 450
Gen Secr: G. Moragh
Focus: Ling; Lit
Periodicals
Hebrew Studies Journal (weekly)
Iggeret (weekly) 15882

National Association of Rehabilitation Facilities (NARF), POB 17675, Washington, DC 20041
T: (703) 648-9300; Fax: (703) 648-0346
Founded: 1969; Members: 812
Gen Secr: James Studzinski
Focus: Rehabil
Periodical
Rehabilitation Review (weekly) 15883

National Association of School Psychologists (NASP), 8455 Colesville Rd, Ste 1000, Silver Spring, MD 20910
T: (301) 608-0500; Fax: (301) 608-2514
Founded: 1969; Members: 15520
Gen Secr: Margaret Gibelman
Focus: Psych
Periodicals
Directory (biennial)
Journal Review (weekly)
Newsletter (8 times annually) 15884

National Association of School Safety and Law Enforcement Officers, c/o Peter Blauvelt, 507 Largo Rd, Upper Marlboro, MD 20772
T: (301) 336-5400; Fax: (301) 808-1213
Founded: 1970; Members: 100
Pres: Peter Blauvelt
Focus: Educ; Law
Periodicals
Membership Directory (weekly)
President's Newsletter (weekly)
School Security Journal (weekly) 15885

National Association of Schools and Colleges of the United Methodist Church (NASCUMC), POB 871, Nashville, TN 37202-0871
T: (615) 340-7399; Fax: (615) 340-7048
Founded: 1940; Members: 128
Gen Secr: Ken Yamada
Focus: Educ 15886

National Association of Schools of Art and Design (NASAD), 11250 Roger Bacon Dr, Reston, VA 22090
T: (703) 437-0700
Founded: 1944; Members: 172
Gen Secr: Samuel Hope
Focus: Graphic & Dec Arts, Design; Educ
Periodicals
Directory (weekly)
Handbook 15887

National Association of Schools of Music (NASM), 11250 Roger Bacon Dr, Reston, VA 22090
T: (703) 437-0700
Founded: 1924; Members: 550
Gen Secr: Samuel Hope
Focus: Music; Educ
Periodicals
Directory (weekly)
Handbook (biennial)
Proceedings (weekly) 15888

National Association of Schools of Public Affairs and Administration (NASPAA), 1120 G St NW, Ste 730, Washington, DC 20005
T: (202) 628-8965; Fax: (202) 626-4978
Founded: 1970; Members: 223
Gen Secr: Alfred M. Zuck
Focus: Public Admin 15889

National Association of Schools of Theatre (NAST), 11250 Roger Bacon Dr, Reston, VA 22090
T: (703) 437-0700
Founded: 1969; Members: 83
Gen Secr: Samuel Hope
Focus: Educ 15890

National Association of Science Writers (NASW), POB 294, Greenlawn, NY 11740
T: (516) 757-5664; Fax: (516) 757-0069
Founded: 1934; Members: 1715
Gen Secr: Diane McGurgan
Focus: Sci; Journalism
Periodical
NASW Newsletter (weekly) 15891

National Association of Secondary School Principals (NASSP), 1904 Association Dr, Reston, VA 22091
T: (703) 860-0200; Fax: (703) 476-5432
Founded: 1916; Members: 43000
Gen Secr: Dr. Timothy J. Dyer
Focus: Educ
Periodicals
Bulletin (9 times annually)
Curriculum Report (5 times annually)
Legal Memorandum (5 times annually)
Newsleader (9 times annually)
Practitioner (weekly)
Schools in the Middle (weekly)
School Technology News (weekly)
Student Acitivities Magazine (weekly)
Tips for Principals 15892

National Association of State Archeologists (NASA), c/o Robert L. Brooks, Oklahoma Archeological Survey, 1808 Newton Dr, Rm 116, Norman, OK 73019
T: (405) 325-7211; Fax: (405) 325-7604
Founded: 1979; Members: 53
Gen Secr: Robert L. Brooks
Focus: Archeol
Periodicals
Directory (weekly)
Newsletter (weekly) 15893

National Association of State Boards of Education (NASBE), 1012 Cameron St, Alexandria, VA 22314
T: (703) 684-4000; Fax: (703) 836-2313
Founded: 1958; Members: 590
Gen Secr: Gene Wilhoit
Focus: Educ 15894

National Association of State Development Agencies (NASDA), 750 First St NE, Ste 710, Washington, DC 20002
T: (202) 898-1302
Founded: 1946; Members: 250
Gen Secr: Miles Friedman
Focus: Econ
Periodicals
Directory of Development Agencies and Officials (weekly)
Legislative Watch
NASDA Letter (every 6-8 weeks)
Trade Monitor (weekly) 15895

National Association of State Directors of Special Education (NASDSE), 1800 Diagonal Rd, Ste 320, Alexandria, VA 22314
T: (703) 519-3800; Fax: (703) 519-3808
Founded: 1938; Members: 2100
Gen Secr: William Schipper
Focus: Educ
Periodicals
Counterpoint (weekly)
Liaison Bulletin (weekly) 15896

National Association of State Directors of Teacher Education and Certification (NASDTEC), c/o Dr. Donald Hair, 3600 Whitman Av N, Ste 105, Seattle, WA 98103
T: (206) 547-0437; Fax: (206) 548-0116
Founded: 1922; Members: 100
Gen Secr: Dr. Donald Hair
Focus: Adult Educ; Educ
Periodical
Roster (weekly) 15897

National Association of State Directors of Vocational Technical Education (NASDVTE), 1616 P St NW, Ste 340, Washington, DC 20036
T: (202) 328-0216; Fax: (202) 797-3756
Founded: 1920; Members: 185
Gen Secr: Madeleine B. Hemmings
Focus: Adult Educ 15898

National Association of State Educational Media Professionals (NASEMP), c/o Betty Latture, Tennessee Dept of Education, Cordell Hall Bldg, Nashville, TN 37219
T: (615) 741-0874; Fax: (615) 741-6236
Founded: 1976; Members: 110
Focus: Educ 15899

National Association of State Mental Health Program Directors (NASMHPD), 66 Canal Center Plaza, Ste 302, Alexandria, VA 22314
T: (703) 739-9333; Fax: (703) 548-9517
Founded: 1963; Members: 55
Gen Secr: Harry C. Schnibbe
Focus: Psychiatry 15900

National Association of State Mental Retardation Program Directors, 113 Oronoco St, Alexandria, VA 22314
T: (703) 683-4202
Founded: 1963; Members: 53
Gen Secr: Robert M. Gettings
Focus: Psychiatry
Periodicals
Community Management Initiative Reports
Federal Funding Inquiry Reports
New Directions (weekly) 15901

National Association of State Park Directors (NASPD), 126 Mill Branch Rd, Tallahassee, FL 32312
T: (904) 893-4959
Founded: 1982; Members: 50
Gen Secr: Ney C. Landrum
Focus: Astrophys 15902

National Association of State Public Health Veterinarians (NASPHV), c/o Ohio Dept of Health, POB 118, Columbus, OH 43266-0118
T: (614) 466-0283; Fax: (614) 644-7740
Founded: 1953; Members: 50
Gen Secr: Kathy Smith
Focus: Vet Med
Periodical
Annual National Compendium of Animal Rabies Control (weekly) 15903

National Association of State Supervisors and Directors of Secondary Education (NASSDSE), c/o Wyland Borth, School Accreditation Dept, 700 Governor's Dr, Pierre, SD 57501
T: (605) 773-4709
Members: 75
Pres: Wyland Borth
Focus: Educ 15904

National Association of State Supervisors of Vocational Home Economics (NASSVHE), c/o Judith Heatherly, Texas Education Agency, 1701 N Congress Av, Austin, TX 78701
T: (512) 463-9454
Members: 175
Pres: Judith Heatherly
Focus: Home Econ; Educ
Periodicals
Directory (weekly)
Newsletter (2-3 times annually) 15905

National Association of State Units on Aging (NASUA), 2033 K St NW, Ste 304, Washington, DC 20006
T: (202) 785-0707
Founded: 1964; Members: 57
Gen Secr: Daniel A. Quirk
Focus: Geriatrics 15906

National Association of State Universities and Land-Grant Colleges (NASULGC), 1 Dupont Circle NW, Ste 710, Washington, DC 20036-1191
T: (202) 778-0818; Fax: (202) 296-6456
Founded: 1962; Members: 160
Pres: C. Peter Magrath
Focus: Educ; Sci
Periodicals
Annual Report (weekly)
The Green Sheet (10 times annually)
Washington Gleanings (weekly) 15907

National Association of Supervisors of Agricultural Education (NASAE), c/o State Dept of Education, 65 S Front St, Rm 911, Columbus, OH 43266-0308
T: (614) 466-3076
Founded: 1962; Members: 156
Pres: Robert Sommers
Focus: Adult Educ; Agri 15908

National Association of Supervisors of Business Education (NASBE), c/o Jean Lane, Fort Worth Independent School District, 3210 W Lancaster, Fort Worth, TX 76107
T: (817) 878-3741
Founded: 1955; Members: 300
Pres: Jean Lane
Focus: Business Admin; Adult Educ
Periodical
Newsletter (weekly) 15909

National Association of Teacher Educators for Business Education (NATEBE), c/o Dr. Lonnie Echpernacht, Business Education Dept, University of Missouri at Columbia, 303 Hill Hall, Columbia, MO 65211
T: (314) 882-2377
Founded: 1970; Members: 350
Pres: Dr. Lonnie Echpernacht
Focus: Educ; Econ 15910

National Association of Teachers of Singing (NATS), 2800 University Blvd N, Jacksonville, FL 32211
T: (904) 744-9022; Fax: (904) 744-9022
Founded: 1944; Members: 5000
Gen Secr: Dr. William A. Vessels
Focus: Music; Educ
Periodicals
Inter NOS (3 times annually)
Journal (5 times annually)
Membership Directory (biennial) 15911

National Association of Test Directors (NATD), c/o Kevin M. Matter, 4700 S Yosemite St, Englewood, CO 80111
T: (303) 773-1184; Fax: (303) 773-9370
Founded: 1985; Members: 250
Gen Secr: Kevin M. Matter
Focus: Educ 15912

National Association of University Women (NAUW), 1553 Pine Forest Dr, Tallahassee, FL 32301
T: (904) 878-4660
Founded: 1923; Members: 4000
Pres: Ruth R. Corbin
Focus: Educ
Periodicals
Bulletin (biennial)
Directory of Branch Presidents and Members (weekly)
Journal of the National Association of University Women (biennial) 15913

National Association of Vocational Education Special Needs Personnel (NAVESNP), c/o Pennsylvania State University, 101 Ostermayer, McKeesport, PA 15132
T: (412) 675-9065; Fax: (412) 675-9067
Founded: 1973; Members: 2200
Gen Secr: Eleanor Bicanich
Focus: Educ 15914

National Association of Women Lawyers (NAWL), 750 N Lake Shore Dr, Chicago, IL 60611
T: (312) 988-6186; Fax: (312) 988-6281
Founded: 1911; Members: 1200
Gen Secr: Patricia O'Mahoney
Focus: Law
Periodicals
Membership Directory (biennial)
Presidents Newsletter (weekly)
Women Lawyers Journal (weekly) . . . 15915

National Association on Drug Abuse Problems (PACT/NADAP), 335 Lexington Av, New York, NY 10017
T: (212) 986-1170; Fax: (212) 697-2939
Founded: 1971; Members: 27
Pres: Warren F. Pelton
Focus: Public Health; Rehabil
Periodical
NADAP News/Report (weekly) 15916

National Assoiation of Self-Instructional Language Programs (NASILP), c/o Critical Languages Center, Temple University, 022-38 Anderson Hall, Philadelphia, PA 19122
T: (215) 787-1715; Fax: (215) 787-3731
Founded: 1971; Members: 145
Gen Secr: Dr. John B. Means
Focus: Ling 15917

National Avionics Society (NAS), 400 W Chester, Lafayette, CO 80026
T: (303) 673-0409
Founded: 1973; Members: 204
Gen Secr: John Gera
Focus: Astrophys
Periodical
NAS Avionics Newsletter (weekly) . . . 15918

National Black MBA Association (NBMBAA), 180 N Michigan Av, Ste 1820, Chicago, IL 60601
T: (312) 236-2622; Fax: (312) 236-4131
Founded: 1871; Members: 2000
Pres: Derryl L. Reed
Focus: Educ
Periodical
National Black MBA Association Newsletter (weekly) 15919

National Black Music Caucus of the Music Educators National Conference (NBMC), c/o Dr. Willis Patterson, University of Michigan, Ann Arbor, MI 48109
T: (313) 764-0586
Founded: 1972
Gen Secr: Dr. Willis Patterson
Focus: Music 15920

National Black Women's Health Project (NBWHP), 1237 Ralph David Abernathy Blvd SW, Atlanta, GA 30310
T: (404) 758-9590; Fax: (404) 752-6756
Founded: 1981; Members: 2000
Gen Secr: Cynthia Newbille-Marsh
Focus: Public Health 15921

National Board for Certification in Dental Technology, 3801 Mount Vernon Av, Alexandria, VA 22305
T: (703) 683-5310
Founded: 1958
Gen Secr: Sandra Stewart
Focus: Dent 15922

National Board for Certification of Dental Laboratories (NBCDL), 3801 Mount Vernon Av, Alexandria, VA 22305
T: (703) 683-5263; Fax: (703) 549-4788
Founded: 1979; Members: 600
Gen Secr: Robert W. Stanley
Focus: Dent 15923

National Board for Respiratory Care (NBRC), 8310 Nieman Rd, Lenexa, KS 66214
T: (913) 599-4200
Founded: 1960; Members: 90000
Gen Secr: Steven K. Bryant
Focus: Therapeutics
Periodicals
Annual Directory (weekly)
Newsletter (weekly) 15924

National Board of Examiners in Optometry (NBEO), 5530 Wisconsin Av, Ste 805, Washington, DC 20815
T: (301) 652-5192; Fax: (301) 907-0013
Founded: 1951; Members: 8
Gen Secr: Dr. Norman E. Wallis
Focus: Optics
Periodicals
Cardichte's Guide (weekly)
Newsletter (weekly)
Report & Boara Activities (biennial)
Report & Examinations (weekly) 15925

National Board of Medical Examiners (NBME), 3930 Chestnut St, Philadelphia, PA 19104
T: (215) 590-9500; Fax: (215) 590-9555
Founded: 1915; Members: 75
Pres: L. Thompson Bowles
Focus: Med
Periodicals
Annual Report (weekly)
The National Board Examiner (weekly) . 15926

National Bureau of Economic Research (NBER), 1050 Massachusetts Av, Cambridge, MA 02138
T: (617) 868-3900; Fax: (617) 868-2742
Founded: 1920
Pres: Martin Feldstein
Focus: Econ 15927

National Burn Information Exchange (NBIE), c/o University of Michigan Burn Center, 1500 E Medical Center Dr, Rm 1B401, Ann Arbor, MI 48109-0033
T: (313) 936-9666
Founded: 1964; Members: 137
Gen Secr: Jorge Rodriguez
Focus: Derm 15928

National Burn Victim Foundation (NBVF), 32-34 Scotland Rd, Orange, NJ 07050
T: (201) 676-7700; Fax: (201) 673-6353
Founded: 1974; Members: 26
Pres: Harry J. Gaynor
Focus: Derm 15929

National Business Education Association (NBEA), 1914 Association Dr, Reston, VA 22091
T: (703) 860-8300
Founded: 1892; Members: 18000
Gen Secr: Dr. Janet M. Treichel
Focus: Adult Educ; Business Admin
Periodical
Business Education Forum (8 times annually)
. . . . 15930

National Cancer Center (NCC), 88 Sunnyside Blvd, Plainview, NY 11803
T: (516) 349-0610
Founded: 1953
Gen Secr: Regina English
Focus: Cell Biol & Cancer Res 15931

National Captioning Institute (NCI), 5203 Leesburg Pike, Ste 1500, Falls Church, VA 22041
T: (703) 998-2400; Fax: (703) 998-2458
Founded: 1979
Pres: John Ball
Focus: Educ Handic 15932

National Catholic Business Education Association (NCBEA), c/o Richard F. Reicherter, POB 1847, Emporia, KS 66801
T: (316) 343-8463
Founded: 1945; Members: 1000
Gen Secr: Richard F. Reicherter
Focus: Adult Educ
Periodical
Review (weekly) 15933

National Catholic Conference for Interracial Justice (NCCIJ), 3033 Fourth St NE, Washington, DC 20017-1102
T: (202) 529-6480
Founded: 1969; Members: 1151
Gen Secr: Jerome B. Ernst
Focus: Law
Periodical
Commitment (weekly) 15934

National Catholic Educational Association (NCEA), 1077 30 St NW, Ste 100, Washington, 20007
T: (202) 337-6232; Fax: (202) 333-6706
Founded: 1904; Members: 20000
Pres: Catherine T. McNamee
Focus: Educ
Periodicals
Notes (5 times annually)
Private School Law Digest (5 times annually)
Update (weekly) 15935

National Catholic Forensic League (NCFL), c/o Richard Gaudette, 21 Nancy Rd, Milford, MA 01757
T: (617) 473-0438
Founded: 1952; Members: 775
Gen Secr: Richard Gaudette
Focus: Law
Periodical
NCFL Newsletter (3 times annually) . . 15936

National Catholic Pharmacists Guild of the United States (NCPG), 1012 Surrey Hills Dr, Saint Louis, MO 63117-1438
T: (314) 645-0085
Founded: 1962; Members: 400
Gen Secr: John P. Winkelmann
Focus: Pharmacol
Periodical
The Catholic Pharmacist (weekly) . . . 15937

National Center for Appropriate Technology (NCAT), POB 3838, Butte, MT 59702
T: (406) 494-4572; Fax: (406) 494-2905
Founded: 1976
Pres: George Turman
Focus: Energy 15938

National Center for Automated Information Research (NCAIR), 165 E 72 St, Ste 1B, New York, NY 10021
T: (212) 249-0760
Founded: 1966; Members: 25
Gen Secr: Ann Brownell Sloane
Focus: Law; Computer & Info Sci . . . 15939

National Center for Business and Economic Communication (NCBEC), c/o The American University, 4400 Massachusetts Av, Washington, DC 20016
T: (202) 885-6167
Founded: 1979
Gen Secr: Louis M. Kohlmeier
Focus: Econ 15940

National Center for Computer Crime Data (NCCCD), 1222 17 Av, Ste B, Santa Cruz, CA 95062
T: (408) 475-4457; Fax: (408) 475-5336
Founded: 1978
Gen Secr: Jay J. Bloom Becker
Focus: Criminology; Computer & Info Sci
Periodicals
Annual Statistical Report (weekly)
Computer Crime Chronicles
Computer Crime Law Reporter (weekly) . 15941

National Center for Disability Services (NCDS), 201 I.U. Willets Rd W, Albertson, NY 11507
T: (516) 747-5400; Fax: (516) 747-5378
Founded: 1952
Pres: Dr. Edwin W. Martin
Focus: Educ; Rehabil 15942

National Center for Fair and Open Testing, 342 Broadway, Cambridge, MA 02139
T: (617) 864-4810; Fax: (617) 497-2224
Founded: 1985
Gen Secr: C. Schuman
Focus: Educ 15943

National Center for Homeopathy (NCH), 801 N Fairfax St, Ste 306, Alexandria, VA 22314
T: (703) 548-7790
Founded: 1974; Members: 3500
Focus: Homeopathy
Periodicals
Directory of Homeopathic Physicians in the U.S. (biennial)
Homeopathy Today (weekly) 15944

National Center for Housing Management (NCHM), 1275 K St NW, Ste 700, Washington, DC 20005-4006
T: (202) 872-1717
Founded: 1972
Pres: Roger G. Stevens
Focus: Business Admin
Periodicals
Certified Occupancy Specialists (weekly)
Occupancy Update (weekly) 15945

National Center for Law and Deafness (NCLD), 8000 Florida Av NE, Washington, DC 20002
T: (202) 651-5373; Fax: (202) 651-5381
Founded: 1975
Focus: Law; Otorhinolaryngology 15946

National Center for State Courts (NCSC), 300 Newport Av, Williamsburg, VA 23187
T: (804) 253-2000; Fax: (804) 220-0449
Founded: 1971
Pres: Larry L. Sipes
Focus: Law
Periodicals
Justice System Journal (3 times annually)
Report and Master Calendar (weekly)
State Court Journal (weekly)
Survey of Judicial Salaries (weekly) . . 15947

National Center for Youth Law (NCYL), 114 Sansome St, Ste 900, San Francisco, CA 94104
T: (415) 543-3307; Fax: (415) 956-9024
Founded: 1978
Gen Secr: John O'Toole
Focus: Law
Periodical
Youth Law News (weekly) 15948

National Center on Arts and the Aging (NCAA), c/o National Council on the Aging, 409 Third St SW, Washington, DC 20024
T: (202) 479-1200; Fax: (202) 424-9046
Founded: 1973
Gen Secr: Sylvia Riggs Liroff
Focus: Geriatrics
Periodical
Collage: Cultural Enrichment and Older Adults (weekly) 15949

National Center on Institutions and Alternatives (NCIA), 635 Slaters Ln, Ste G-100, Alexandria, VA 22314
T: (703) 684-0373; Fax: (703) 684-6037
Founded: 1979
Pres: Jerome G. Miller
Focus: Sociology; Law 15950

National Certification Agency for Medical Lab Personnel (NCA), 7910 Woodmont Av, Ste 7301, Bethesda, MD 20814
T: (301) 654-1622
Founded: 1977; Members: 65000
Gen Secr: Andrea Guevara
Focus: Med; Eng 15951

National Character Laboratory (NCL), c/o A.J. Stuart, 4635 Leeds Av, El Paso, TX 79903
T: (915) 562-5046; Fax: (915) 562-5046
Founded: 1971; Members: 75
Pres: A.J. Stuart
Focus: Anthro; Psych
Periodical
Newsletter (weekly) 15952

National Coalition for Literacy (NCL), 1800 M St NW, Washington, DC 20036
T: (202) 331-2287
Founded: 1981; Members: 27
Gen Secr: Richard P. Lynch
Focus: Lit 15953

National Coalition for Public Education and Religious Liberty (NCPERL), 100 Maryland Av NE, Washington, DC 20002
T: (202) 547-5050; Fax: (202) 544-7213
Founded: 1974; Members: 31
Gen Secr: A. Must
Focus: Educ; Rel & Theol 15954

National Coalition for Women and Girls in Education (NCWGE), c/o Displaced Homemakers Network, 1625 K St NW, Ste 300, Washington, DC 20006
T: (202) 467-6346
Founded: 1975; Members: 60
Gen Secr: Jill Miller
Focus: Educ 15955

National Coalition of Alternative Community Schools (NCACS), 58 Schoolhouse Rd, Summertown, TN 38483
T: (615) 964-3670
Founded: 1976; Members: 1500
Gen Secr: Michael Traugot
Focus: Educ
Periodicals
National Coalition News (weekly)
Newsletter (weekly)
There Ought to be Free Choice 15956

National College of District Attorneys (NCDA), c/o Law Center, University of Houston, Houston, TX 77204-6380
T: (713) 747-6232
Founded: 1969
Gen Secr: John Jay Douglass
Focus: Law 15957

National College of Foot Surgeons (NCFS), c/o Dr. Albert Apkarian, POB 264, Woodland Hills, CA 71765-0264
Founded: 1960
Gen Secr: Dr. Albert Apkarian
Focus: Surgery; Orthopedics
Periodical
Journal (weekly) 15958

National Collegiate Conference Association (NCCA), 84 N Allen St, Albany, NY 12203
T: (212) 397-1954
Founded: 1947; Members: 2140
Gen Secr: Raymond J. Freda
Focus: Educ 15959

National Commission for Cooperative Education (NCCE), 501 Stearns Center, 360 Huntington Av, Boston, MA 02115
T: (617) 437-3778; Fax: (617) 437-3444
Founded: 1962; Members: 214
Pres: Dr. Jane Scarborough
Focus: Educ
Periodical
Co-op Education Undergraduate Program Directory (weekly) 15960

National Committee for Clinical Laboratory Standards (NCCLS), 771 E Lancaster Av, Villanova, PA 19085
T: (215) 525-2435; Fax: (215) 527-8399
Founded: 1968; Members: 1300
Gen Secr: John V. Bergen
Focus: Med 15961

National Committee for Latin and Greek (NCLG), c/o Virginia Berrett, 6669 Vinahaven, Cypress, CA 90630
T: (714) 894-0938
Founded: 1978; Members: 15
Pres: Virginia Barrett
Focus: Ling 15962

National Committee for Quality Health Care (NCQHC), 1500 K St NW, Ste 360, Washington, DC 20005
T: (202) 347-5731; Fax: (202) 347-5836
Founded: 1978; Members: 151
Pres: Pamela G. Bailey
Focus: Public Health
Periodicals
Capital Outlock (4-8 times annually)
Quality Outlook (weekly) 15963

National Committee for the Furtherance of Jewish Education (NCFJE), 824 Eastern Pkwy, Brooklyn, NY 11213
T: (718) 735-0200; Fax: (718) 735-4455
Founded: 1940; Members: 400
Pres: Sholem B. Hecht
Focus: Educ 15964

National Committee on the Treatment of Intractable Pain (NCTIP), c/o Wayne Coy, Cohn & Marks, 1333 New Hampshire Av NW, Washington, DC 20036
T: (202) 452-4836; Fax: (202) 293-4827
Founded: 1977; Members: 3500
Pres: Wayne Coy
Focus: Med; Anesthetics
Periodical
Newsletter (2-3 times annually) 15965

National Community Education Association (NCEA), 801 N Fairfax St, Ste 209, Alexandria, VA 22314
T: (703) 683-6232; Fax: (703) 683-0161
Founded: 1966; Members: 1500
Gen Secr: S. Jewell-Kelly
Focus: Educ
Periodicals
Community Education Journal (weekly)
Community Education Today (weekly)
Membership Directory (weekly) 15966

National Conference of Bankruptcy Judges (NCBJ), c/o Paul Mannes, U.S. Bancrupty Court, 451 Hungerford Dr, Rockville, MD 20850
T: (301) 443-7010
Founded: 1926; Members: 275
Pres: Paul Mannes
Focus: Law 15967

National Conference of Black Lawyers (NCBL), 2 W 125 St, New York, NY 10027
T: (212) 864-4000
Founded: 1968; Members: 1000
Gen Secr: A. Ayetero
Focus: Law
Periodical
NCBL Notes (weekly) 15968

National Conference of Commissioners on Uniform State Laws (NCCUSL), 676 N Saint Clair, Ste 1700, Chicago, IL 60611
T: (312) 915-0195; Fax: (312) 915-0187
Founded: 1892; Members: 307
Gen Secr: Edith O. Davies
Focus: Law
Periodical
Handbook and Proceedings (weekly) . . 15969

National Conference of Regulatory Utility Commission Engineers (NCRUCE), c/o Glynn Blanton, Tennessee Public Service Commission, 460 James Robertson Pkwy, Nashville, TN 37219
T: (615) 741-2844
Founded: 1923; Members: 150
Gen Secr: Glynn Blanton
Focus: Standards 15970

National Conference of Standards Laboratories (NCSL), 1800 30 St, Ste 305B, Boulder, CO 80301
T: (303) 440-3339; Fax: (303) 440-3384
Founded: 1961; Members: 1200
Gen Secr: Wilbur J. Anson
Focus: Standards
Periodicals
Conference Proceedings (weekly)
Newsletter (weekly)
Training Information Directory (weekly) . 15971

National Conference of States on Building Codes and Standards (NCSBCS), 505 Huntmar Park Dr, Ste 210, Herndon, VA 22070
T: (703) 437-0100; Fax: (703) 481-3596
Founded: 1967; Members: 250
Gen Secr: Robert C. Wible
Focus: Civil Eng; Standards 15972

National Conference of Yeshiva Principals (NCYP), 160 Broadway, New York, NY 10038
T: (212) 227-1000; Fax: (212) 406-6934
Founded: 1947; Members: 1000
Gen Secr: A. Moshe Possick
Focus: Educ

Periodicals
Machberes Hamenahel (weekly)
Newsletter (weekly) 15973

National Conference on Fluid Power (NCFP), 3333 N Mayfair Rd, Milwaukee, WI 53222
T: (414) 778-3368; Fax: (414) 778-3361
Founded: 1945
Gen Secr: William H. Prueser
Focus: Energy
Periodical
National Conference on Fluid Power (weekly) 15974

National Conference on Research in English (NCRE), c/o University of Massachusetts, 223 Furcolo Hall, Amherst, MA 01003
T: (413) 545-4247
Founded: 1937; Members: 475
Pres: David Bloom
Focus: Ling; Educ
Periodicals
Directory (biennial)
Newsletter (weekly) 15975

National Conference on the Advancement of Research (NCAR), c/o A.W. Betts, Southwest Research Institute, PO Drawer 28510, San Antonio, TX 78228
T: (210) 522-2202; Fax: (210) 520-5505
Founded: 1947; Members: 260
Gen Secr: A.W. Betts
Focus: Sci
Periodical
Proceedings (weekly) 15976

National Conference on Weights and Measures (NCWM), POB 4025, Gaithersburg, MD 20885
T: (301) 975-4009; Fax: (301) 926-0647
Founded: 1905
Gen Secr: A.D. Tholen
Focus: Standards
Periodicals
NBS Handbook (weekly)
Report (weekly) 15977

National Congress for Community Economic Development (NCCED), 1875 Connecticut Av NW, Ste 524, Washington, DC 20009
T: (202) 234-5009
Founded: 1970; Members: 320
Pres: Robert Zdenek
Focus: Sociology; Econ
Periodical
Resources for Community-Based Economic Development (weekly) 15978

National Congress of Inventors Organizations (NCIO), POB 6158, Rheem Valley, CA 94570
T: (510) 376-7541; Fax: (510) 376-7762
Members: 63
Pres: Norman C. Parrish
Focus: Sci 15979

National Consortium for Black Professional Development (NCBPD), POB 18308, Louisville, KY 40218-0308
T: (502) 896-2838
Founded: 1974; Members: 57
Gen Secr: Hanford D. Stafford
Focus: Adult Educ
Periodical
Science and Engineering Newsletter (weekly) 15980

National Consortium for Child Mental Health Services (NCCMHS), 3615 Wisconsin Av NW, Washington, DC 20016
T: (202) 966-7300
Founded: 1971; Members: 20
Gen Secr: Virginia Q. Anthony
Focus: Psychiatry 15981

National Consortium of Arts and Letters for Historically Black Colleges and Universities (NCALHBCU), c/o Dr. Walter Anderson, 2555 Pennsylvania Av NW, Ste 818, Washington, DC 20037
T: (202) 833-1327
Founded: 1984; Members: 38
Gen Secr: Dr. Walter F. Anderson
Focus: Educ 15982

National Consumer Law Center (NCLC), 11 Beacon St, Boston, MA 02108
T: (617) 523-8010; Fax: (617) 523-7398
Founded: 1969
Gen Secr: Willard Ogburn
Focus: Law
Periodicals
NCLC Energy Update (weekly)
NCLC Reports (weekly) 15983

National Council for Accreditation of Teacher Education (NCATE), 2010 Massachusetts Av NW, Ste 200, Washington, DC 20036-1023
T: (202) 466-7496; Fax: (202) 296-6620
Founded: 1954; Members: 556
Pres: Arthur E. Wise
Focus: Adult Educ; Educ
Periodical
Annual List of Accredited Institutions (weekly) 15984

National Council for Black Studies (NCBS), c/o Ohio State University, 1030 Lincoln Tower, 1800 Cannon Dr, Columbus, OH 43210
T: (614) 292-1035; Fax: (614) 292-2713
Founded: 1975; Members: 500
Gen Secr: Jacqueline E. Wade
Focus: Educ 15985

National Council for Culture and Art (NCCA), 1600 Broadway, Ste 611C, New York, NY 10019
T: (212) 757-7933
Founded: 1980; Members: 1500
Pres: Robert H. LaPrince
Focus: Arts 15986

National Council for Environmental Balance (NCEB), 4169 Westport Rd, POB 7732, Louisville, KY 40257-0732
T: (502) 896-8731
Founded: 1972; Members: 1320
Pres: I.W. Tucker
Focus: Ecology
Periodical
Newsletter Energy & Environment Alert: Timely Environmental Pamphlets & Booklets (weekly) 15987

National Council for Geographic Education (NCGA), c/o Indiana University of Pennsylvania, 16A Leonard Hall, Indiana, PA 15705
T: (412) 357-6290; Fax: (412) 357-7708
Founded: 1915; Members: 3700
Gen Secr: Ruth I. Shirey
Focus: Geography; Educ
Periodicals
Journal of Geography (weekly)
Perspective (5 times annually) 15988

National Council for Labor Reform (NCLR), 406 S Plymouth Ct, Chicago, IL 60605
T: (312) 427-0206
Founded: 1969; Members: 3000
Gen Secr: Thomas Hugh Latimer
Focus: Poli Sci; Law 15989

National Council for Research and Planning (NCRP), c/o Dr. Edith Carter, Radford University, POB 6924, Radford, VA 24142
T: (703) 639-1263
Founded: 1977; Members: 256
Pres: Dr. Edith Carter
Focus: Educ
Periodicals
Community College Journal for Research and Planning (weekly)
New Directors for Two-Year Colleges (biennial)
Newsletter (weekly) 15990

National Council for Textile Education (NCTE), POB 391, Charlottesville, VA 22902
T: (804) 296-5511; Fax: (804) 296-2957
Founded: 1933; Members: 30
Pres: Charles G. Tewksbury
Focus: Adult Educ; Textiles 15991

National Council for the Social Studies (NCSS), 3501 Newark St NW, Washington, DC 20016
T: (202) 966-7840; Fax: (202) 966-2061
Founded: 1921; Members: 26000
Gen Secr: Frances Haley
Focus: Sociology
Periodicals
Bulletin (3 times annually)
Social Education (7 times annually)
Social Studies Professionals (5 times annually) 15992

National Council for the Traditional Arts (NCTA), 1320 Fenwick Ln, Ste 200, Silver Spring, MD 20910
T: (301) 565-0654; Fax: (301) 565-0472
Founded: 1934
Gen Secr: Joseph T. Wilson
Focus: Arts 15993

National Council for Torah Education (NCTE), c/o Religious Zionists of America, 25 W 26 St, New York, NY 10010
T: (212) 689-1414
Founded: 1939
Gen Secr: M. Golombek
Focus: Educ; Rel & Theol 15994

National Council for Urban Economic Development (CUED), 1730 K St NW, Ste 915, Washington, DC 20006
T: (202) 223-4735; Fax: (202) 223-4745
Founded: 1967; Members: 1200
Gen Secr: Jeff Finkle
Focus: Urban Plan; Econ
Periodicals
Commentary (weekly)
Legislative Report
Urban Economic Developments: Newsletter (weekly) 15995

National Council of Athletic Training, c/o National Association for Sport and Physical Education, 1900 Association Dr, Reston, VA 22091
T: (703) 476-3410; Fax: (703) 476-9527
Founded: 1976; Members: 2000
Gen Secr: Dr. Judith Young
Focus: Sports
Periodicals
Directory (weekly)
Newsletter 15996

National Council of BIA Educators (NCBIAE), 6001 Marble NE, Ste 10, Albuquerque, NM 87110
T: (505) 266-6638; Fax: (505) 266-1967
Founded: 1967; Members: 150
Gen Secr: Fannie Bahe
Focus: Educ 15997

National Council of Examiners for Engineering and Surveying (NCEE), POB 1686, Clemson, SC 29633
T: (803) 654-6824; Fax: (803) 654-6033
Founded: 1920; Members: 68
Gen Secr: Roger B. Stricklin
Focus: Eng 15998

National Council of Forestry Association Executives (NCFAE), c/o MFPC, 146 State St, Augusta, ME 04330
T: (207) 622-9288
Founded: 1949; Members: 100
Gen Secr: Ted Johnston
Focus: Forestry 15999

National Council of Guilds for Infant Survival (NCIS), POB 3586, Davenport, IA 52808
T: (319) 322-4870
Founded: 1964; Members: 200
Gen Secr: Melanie Pangburn
Focus: Pediatrics
Periodical
Newsletter (weekly) 16000

National Council of Intellectual Property Law Associations (NCIPLA), 2001 Jefferson Davis Hwy, Ste 203, Arlington, VA 22202
T: (703) 415-0780
Founded: 1934; Members: 53
Pres: Allen R. Jensen
Focus: Law
Periodicals
Chairman's Letter (weekly)
Newsletter (weekly) 16001

National Council of Juvenile and Family Court Judges (NCJFCJ), c/o University of Nevada, POB 8970, Reno, NV 89507
T: (702) 784-6012; Fax: (702) 784-6628
Founded: 1937; Members: 2500
Gen Secr: Louis W. McHardy
Focus: Law
Periodicals
Juvenile and Family Court Journal (weekly)
Juvenile and Family Court Newsletter (8 times annually)
Juvenile and Family Law Digest (weekly) 16002

National Council of Local Administrators of Vocational Education and Practical Arts (NCLA), c/o Trade and Technical Education, Board of Education, City of New York, 66 Rugby Rd, Brooklyn, NY 11226
T: (212) 282-3269
Founded: 1942; Members: 1000
Gen Secr: Dr. Harry Lewis
Focus: Adult Educ 16003

National Council of State Directors of Community and Junior Colleges, c/o James H. Folkening, Michigan Dept of Education, POB 30008, Lansing, MI 48909
T: (517) 373-3360; Fax: (517) 373-2759
Members: 38
Pres: Dale Campbell
Focus: Educ 16004

National Council of State Education Associations (NCSEA), 1201 NW 16 St, Washington, DC 20036
T: (202) 822-7745
Founded: 1966; Members: 130
Gen Secr: Larry Diebold
Focus: Educ 16005

National Council of State Pharmacy Executives (NCSPE), c/o Al Melbane, POB 151, Chapel Hill, NC 27514-0151
Fax: (919) 968-9430
Founded: 1927; Members: 51
Gen Secr: Al Melbane
Focus: Pharmacol 16006

National Council of State Supervisors of Foreign Languages (NCSSFL), c/o Dept of Education, 4 Capitol Mall, Little Rock, AR 72201-1071
T: (501) 682-4398; Fax: (501) 682-4618
Founded: 1960; Members: 67
Gen Secr: Walter H. Bartz
Focus: Ling; Educ 16007

National Council of State Supervisors of Music (NCSSM), c/o Linda Mercer, Ohio Dept of Education, 65 S Front St, Columbus, OH 43266-0308
T: (614) 466-2211; Fax: (614) 752-8148
Founded: 1938; Members: 31
Gen Secr: Bob Gross
Focus: Music; Educ 16008

National Council of Supervisors of Mathematics (NCSM), POB 10667, Golden, CO 80401
T: (414) 229-4844; Fax: (414) 229-4666
Founded: 1968; Members: 2000
Pres: Henry S. Kepner
Focus: Math 16009

National Council of Teachers of English (NCTE), 1111 Kenyon Rd, Urbana, IL 61801
T: (217) 328-3870; Fax: (217) 328-9645
Founded: 1911; Members: 120000
Gen Secr: Miles Myers
Focus: Ling; Educ
Periodicals
College English (8 times annually)
English Journal (8 times annually)
Language Arts (8 times annually) . . . 16010

National Council of Teachers of Mathematics (NCTM), 1906 Association Dr, Reston, VA 22091-1593
T: (703) 620-9840; Fax: (703) 476-2970
Founded: 1920; Members: 98000
Gen Secr: Dr. James D. Gates
Focus: Math; Educ
Periodicals
Arithmetic Teacher (9 times annually)
Mathematics Teacher (9 times annually)
News Bulletin (9 times annually) 16011

National Council of the Paper Industry for Air and Stream Improvement (NCASI), 260 Madison Av, New York, NY 10016
T: (212) 532-9000; Fax: (212) 779-2849
Founded: 1943; Members: 100
Pres: Dr. I. Gellman
Focus: Eng
Periodical
Technical Bulletin (20 times annually) . . 16012

National Council of University Research Administrators (NCURA), 1 Dupont Circle NW, Ste 220, Washington, DC 20036
T: (202) 466-3894
Founded: 1960; Members: 1900
Gen Secr: Natalie Kirkman
Focus: Sci
Periodicals
Newsletter (weekly)
Research Management Review (weekly) . 16013

National Council of Urban Education Associations (NCUEA), c/o National Education Association, 1201 16 St NW, Washington, DC 20036
T: (202) 822-7137
Founded: 1964; Members: 201
Pres: Bruce Colwell
Focus: Educ 16014

National Council on Agricultural Life and Labor Research Fund (NCALL), 20 E Division St, POB 1092, Dover, DE 19903
T: (302) 678-9400; Fax: (302) 678-9058
Founded: 1976; Members: 1200
Gen Secr: Joe L. Myer
Focus: Archit
Periodical
NCALL News (weekly) 16015

National Council on Alcoholism and Drug Dependence (NCADD), 12 W 21 St, New York, NY 10010
T: (212) 206-6770; Fax: (212) 645-1690
Founded: 1944
Pres: Paul Wood
Focus: Public Health
Periodical
Annual Report (weekly) 16016

National Council on Crime and Delinquency (NCCD), 685 Market St, San Francisco, CA 94105
T: (415) 896-6223; Fax: (415) 956-1559
Founded: 1907; Members: 11000
Pres: Barry A. Krisberg
Focus: Criminology
Periodicals
Crime and Delinquency (weekly)
Criminal Justice Newsletter (weekly)
Journal of Research in Crime and Delinquency (weekly) 16017

National Council on Economic Education (NCEE), 432 Park Av S, New York, NY 10016
T: (212) 685-5499; Fax: (212) 213-2872
Founded: 1949
Pres: Stephen Buckles
Focus: Educ; Econ
Periodicals
Annual Report (weekly)
Directory of Affiliated Councils and Centers (weekly)
Economic Education Update (weekly) . . 16018

National Council on Employment Policy (NCEP), 1717 K St NW, Ste 1200, Washington, DC 20006
T: (202) 833-2530
Founded: 1964; Members: 7
Gen Secr: S.A. Levitan
Focus: Poli Sci; Econ 16019

National Council on Family Relations (NCFR), 3989 Central Av NE, Ste 550, Minneapolis, MN 55421
T: (612) 781-9331; Fax: (612) 781-9348
Founded: 1938; Members: 3900
Gen Secr: Mary Jo Czaplewski
Focus: Sociology; Family Plan
Periodicals
Family Relations (weekly)
Journal of Marriage and the Family (weekly)
Newsletter (weekly) 16020

National Council on Gene Resources (NCGR), 1738 Thousand Oaks Blvd, Berkeley, CA 94707
T: (510) 524-8973
Founded: 1980
Gen Secr: Dr. David Kafton
Focus: Genetics 16021

National Council on Health Laboratory Services (NCHLS), c/o Patricia Ashton, 708 Wendy Way, Durham, NC 27712
T: (919) 490-3713
Founded: 1952; Members: 22
Gen Secr: Patricia Ashton
Focus: Public Health
Periodical
Membership Organization List (weekly) . 16022

National Council on Measurement in Education (NCME), 1230 17 St NW, Washington, DC 20036
T: (202) 223-9318; Fax: (202) 775-1824
Founded: 1938; Members: 2100
Gen Secr: William J. Russell
Focus: Educ
Periodicals
Educational Measurement: Issues and Practice (weekly)
Journal of Educational Measurement (weekly) 16023

National Council on Public History (NCPH), c/o Elizabeth B. Monroe, Indiana University, 425 University Blvd, Indianapolis, IN 46202
T: (317) 274-2716; Fax: (317) 274-2347
Founded: 1979; Members: 1100
Gen Secr: Elizabeth B. Monroe
Focus: Hist; Sociology 16024

National Council on Radiation Protection and Measurements (NCRPM), 7910 Woodmont Av, Ste 800, Bethesda, MD 20814
T: (301) 657-2652; Fax: (301) 907-8768
Founded: 1929; Members: 75
Gen Secr: W. Roger Ney
Focus: Radiology
Periodicals
Lauriston S. Taylor Lectures (weekly)
NCRP Commentaries
NCRP Reports
Proceedings of the Annual Meeting (weekly) 16025

National Council on Religion and Public Education (NCRPE), c/o Iowa State University, N155 Lagomarcino Hall, Ames, IA 50011
T: (515) 294-2881; Fax: (515) 294-6206
Founded: 1971; Members: 550
Gen Secr: Charles R. Kniker
Focus: Rel & Theol; Educ 16026

National Crime Prevention Institute (NCPI), c/o Dept of Justice Administration, University of Louisville, Shelby Campus, Louisville, KY 40292-0001
T: (502) 588-6987; Fax: (502) 588-6990
Founded: 1971
Gen Secr: Wilbur Rykert
Focus: Criminology
Periodical
NCPI Hotline (weekly) 16027

National Criminal Defense College (NCDC), c/o Mercer Law School, Macon, GA 31207
T: (912) 746-4151; Fax: (912) 743-0160
Founded: 1985
Gen Secr: Deryl D. Dantzler
Focus: Law; Criminology 16028

National Dairy Council (NDC), 10255 Higgins, Rosemont, IL 60018
T: (708) 803-2000
Founded: 1915; Members: 24
Gen Secr: Thomas Gallagher
Focus: Dairy Sci
Periodicals
Dairy Council Digest (weekly)
Nutrition Education Materials Catalog (weekly)
Nutrition News (weekly) 16029

National Dental Assistants Association (NDAA), c/o Elizabeth Brezill, 5506 Connecticut Av NW, Ste 24, Washington, DC 20015
T: (202) 244-7555; Fax: (202) 244-5992
Founded: 1964; Members: 500
Pres: Elizabeth Brezill
Focus: Dent 16030

National Dental Association (NDA), 5506 Connecticut Av NW, Ste 24, Washington, DC 20015
T: (202) 244-7555; Fax: (202) 244-5992
Founded: 1913; Members: 2500
Gen Secr: Robert S. Johns
Focus: Dent
Periodicals
Journal (weekly)
Newsletter (weekly) 16031

National Dental Hygienists' Association (NDHA), 5506 Conncecticut Av NW, Stes 24-25, Washington, DC 20015
T: (202) 244-7555; Fax: (202) 244-5992
Founded: 1932; Members: 50
Pres: Dr. Carolyn Roundtree
Focus: Dent; Hygiene 16032

National Denturist Association (NDA), POB 40307, Portland, OR 97240-0307
T: (310) 867-0658
Founded: 1975
Gen Secr: Dr. James Davis
Focus: Dent 16033

National Digestive Diseases Information Clearinghouse (NDDIC), 9000 Rockville Pike, POB NDDIC, Bethesda, MD 20892
T: (301) 468-6344
Founded: 1980
Gen Secr: Elizabeth H. Singer
Focus: Gastroenter
Periodical
Directory of Digestive Diseases Organizations (weekly) 16034

National District Attorneys Association (NDAA), 1033 N Fairfax St, Ste 200, Alexandria, VA 22314
T: (703) 549-9222; Fax: (703) 836-3195
Founded: 1950; Members: 7000
Gen Secr: N. Flanagan
Focus: Law
Periodical
The Prosecutor (weekly) 16035

National Dry Bean Council (NDBC), 1101 Connecticut Av NW, Ste 700, Washington, DC 20036
T: (202) 857-1169
Founded: 1950; Members: 12
Pres: Philip Kimball
Focus: Food 16036

National Earth Science Teachers Association (NESTA), c/o Michael Burton, 340 Prairiewood Cir, Fargo, ND 58103
T: (701) 241-9818; Fax: (701) 241-9818
Founded: 1983; Members: 1500
Pres: Michael Burton
Focus: Geology; Educ 16037

National Economic Association (NEA), c/o Alfred L. Edwards, School of Business, University of Michigan, Ann Arbor, MI 48109-1234
T: (313) 763-0121
Founded: 1969
Gen Secr: Alfred L. Edwards
Focus: Finance 16038

National Education Association (NEA), 1201 16 St NW, Washington, DC 20036
T: (202) 833-4000
Founded: 1857; Members: 2000000
Gen Secr: Don Cameron
Focus: Educ
Periodicals
NEA Today (weekly)
Today's Eduction (weekly) 16039

National Energy Management Institute, 601 N Fairfax St, Ste 160, Alexandria, VA 22314
T: (703) 739-7100; Fax: (703) 683-7615
Founded: 1981
Gen Secr: James T. Golden
Focus: Energy
Periodical
News from EMI (weekly) 16040

National Energy Resources Organization (NERO), 919 18 St NW, Ste 450, Washington, DC 20006
T: (202) 466-6535; Fax: (703) 739-9248
Founded: 1975; Members: 450
Pres: Anson Franklin
Focus: Energy 16041

National Engineering Consortium (NEC), 303 E Wacker Dr, Ste 740, Chicago, IL 60601
T: (312) 938-3500; Fax: (312) 938-8787
Founded: 1944
Gen Secr: Robert M. Janowiak
Focus: Electronic Eng
Periodical
Conference Proceedings (weekly) 16042

National Environmental Health Association (NEHA), 720 S Colorado Blvd, Ste 970, Denver, CO 80222
T: (303) 756-9090; Fax: (303) 691-9490
Founded: 1930; Members: 5700
Gen Secr: Nelson E. Fabian
Focus: Public Health
Periodical
Journal of Environmental Health (weekly) 16043

National Federation Interscholastic Music Association (NFIMA), 11724 NW Plaza Cir, POB 20626, Kansas City, MO 64195-0626
T: (816) 464-5400; Fax: (816) 464-5571
Founded: 1983; Members: 605
Gen Secr: Richard Fawcett
Focus: Music 16044

National Federation of Abstracting and Information Services (NFAIS), 1429 Walnut St, Philadelphia, PA 19102
T: (215) 563-2406; Fax: (215) 563-2848
Founded: 1958; Members: 74
Gen Secr: Ann Marie Cunningham
Focus: Educ
Periodicals
Membership Directory (weekly)
NFAIS Newsletter (weekly)
Report Series (weekly) 16045

National Federation of Catholic Physicians Guilds (NFCPG), 850 Elm Grove Rd, Elm Grove, WI 53122
T: (414) 784-3435; Fax: (414) 782-8788
Founded: 1932; Members: 3500
Gen Secr: Robert H. Herzog
Focus: Physiology
Periodical
The Linacre (weekly) 16046

National Federation of Modern Language Teachers Associations (NFMLTA), c/o Elizabeth H. Hoffman, 659 N 57 Av, Omaha, NE 68132
T: (402) 551-6290
Founded: 1916; Members: 15
Gen Secr: Elizabeth H. Hoffman
Focus: Ling; Educ
Periodical
The Modern Language Journal (weekly) . 16047

National Federation of Societies for Clinical Social Work (NFSCSW), POB 3740, Arlington, VA 22203
T: (703) 522-3866; Fax: (703) 522-3866
Founded: 1971
Gen Secr: Linda B. O'Leary
Focus: Sociology
Periodicals
Clinical Social Work Journal (weekly)
National News (weekly) 16048

National Federation of State Poetry Societies (NFSPS), c/o Wanda Blaisdell, 2664 Shemrock, Ogden, VT 84403
T: (801) 393-7562
Founded: 1959; Members: 7500
Pres: Wanda Blaisdell
Focus: Lit
Periodicals
Prize Poems (weekly)
Strophes (weekly) 16049

National FFA Organization (NFFAO), 5632 Mount Vernon Memorial Hwy, POB 15160, Alexandria, VA 22309-0160
T: (703) 360-3600; Fax: (703) 360-5524
Founded: 1928; Members: 385376
Gen Secr: C. Coleman Harris
Focus: Adult Educ; Agri
Periodicals
Between Issues (weekly)
National Future Farmer Magazine (weekly)
Update (weekly) 16050

National Fire Protection Association (NFPA), 1 Batterymarch Park, Quincy, MA 02269-9101
T: (617) 770-3000; Fax: (617) 770-0700
Founded: 1896; Members: 57000
Pres: George D. Miller
Focus: Safety
Periodicals
Catalogs of Publications and Visual Aids (weekly)
Fire Command (weekly)
Fire Journal (weekly)
Fire News (8 times annually)
Fire Protection Reference Directory (weekly)
Fire Technology (weekly)
National Fire Codes (weekly)
Technical Committe Reports (weekly)
Yearbook (weekly) 16051

National Fisheries Contaminant Research Center (NFCRC), c/o Fish and Wildlife Service, U.S. Dept of the Interior, 4200 New Haven Rd, Columbia, MO 65201
T: (314) 875-5399; Fax: (314) 876-1896
Gen Secr: Dr. Richard Schoettger
Focus: Ecology; Fisheries 16052

National Fisheries Education and Research Foundation (NFERF), 1525 Wilson Blvd, Ste 500, Arlington, VA 22209
T: (703) 524-9216; Fax: (703) 524-4619
Founded: 1979
Gen Secr: Adrian Paradin
Focus: Fisheries 16053

National Forensic Association (NFA), c/o Dr. Christina Reynolds, Dept of Speech Communication, Otterbein College, Westerville, OH 43081
T: (614) 898-1753; Fax: (614) 898-1200
Founded: 1974; Members: 300
Pres: Dr. Christina Reynolds
Focus: Ling 16054

National Forensic League (NFL), POB 38, Ripon, WI 54971
T: (414) 748-6206
Founded: 1925; Members: 798000
Gen Secr: James M. Copeland
Focus: Law
Periodical
Rostrum (weekly) 16055

National Foundation for Advancement in the Arts (NFAA), 3915 Biscayne Blvd, Miami, FL 33137
T: (305) 573-0490; Fax: (305) 573-4870
Founded: 1981; Members: 750
Pres: Dr. William H. Banchs
Focus: Arts 16056

National Foundation for Brain Research (NFBR), 1250 24 St NW, St 300, Washington, DC 20037
T: (202) 293-5453; Fax: (202) 466-2888
Founded: 1953; Members: 230
Gen Secr: Lawrence S. Hoffheimer
Focus: Pathology
Periodicals
Annual Report (weekly)
Newsletter 16057

National Foundation for Children's Hearing Education and Research (CHEAR), 928 McLean Av, Yonkers, NY 10704
T: (914) 237-2676
Founded: 1969; Members: 4000
Pres: Philip B. Miller
Focus: Educ; Otorhinolaryngology . . . 16058

National Foundation for Non-Invasive Diagnostics (NFNID), 103 Carnegie Center, Ste 311, Princeton, NJ 08540
T: (609) 520-1300; Fax: (609) 452-8544
Founded: 1977
Gen Secr: Philip B. Papier
Focus: Med 16059

National Foundation of Dentistry for the Handicapped (NFDH), 1600 Stout St, Ste 1420, Denver, CO 80202
T: (303) 573-0264; Fax: (303) 573-0267
Founded: 1974
Gen Secr: Larry Coffee
Focus: Dent 16060

National Fund for Medical Education (NFME), 35 Kneeland St, Boston, MA 02111
T: (617) 956-8404
Founded: 1949
Pres: Norman S. Stearns
Focus: Educ; Med 16061

National Genealogical Society (NGS), 4527 17 St N, Arlington, VA 22207-2399
T: (703) 525-0050; Fax: (703) 525-0052
Founded: 1903; Members: 13000
Gen Secr: J.K. Findeis
Focus: Genealogy
Periodicals
National Genealogical Society Quarterly (weekly)
Newsletter 16062

National Geographic Society (NGS), 17 and M Sts NW, Washington, DC 20036
T: (202) 857-7000
Founded: 1888; Members: 10000000
Pres: Gilbert M. Grosvenor
Focus: Geography
Periodicals
National Geographic (weekly)
National Geographic World (weekly) . . 16063

National Geriatrics Society (NGS), 1200 W Crooked Lake Pl, Eustis, FL 32736-6433
Founded: 1952
Gen Secr: Dr. John Eckhardt
Focus: Geriatrics
Periodical
Modern Geriatric Topics (weekly) . . . 16064

National Grants Management Association (NGMA), 501 Capitol Ct NE, Ste 100, Washington, DC 20002
T: (202) 546-6993; Fax: (202) 546-2121
Founded: 1978; Members: 400
Gen Secr: Jeffrey C. Smith
Focus: Finance
Periodical
Grants and Assistance News (weekly) . 16065

National Guild of Catholic Psychiatrists (NGCP), c/o Taylor Manor Hospital, 4100 College Av, Ellicott City, MD 21041-0396
T: (410) 465-3322; Fax: (410) 461-7075
Founded: 1949; Members: 75
Pres: Robert J. McAllister
Focus: Psychiatry
Periodical
Bulletin (weekly) 16066

National Guild of Community Schools of the Arts (NGCSA), POB 8018, Englewood, NJ 07631
T: (201) 871-3337; Fax: (201) 871-7639
Founded: 1937; Members: 300
Gen Secr: Lolita Mayadas
Focus: Arts; Music; Educ
Periodical
Guildletter (weekly) 16067

National Guild of Piano Teachers (NGPT), 808 Rio Grande, POB 1807, Austin, TX 78767
T: (512) 478-5775
Founded: 1929; Members: 11868
Pres: Richard Allison
Focus: Music; Educ 16068

National Health Council (NHC), 1730 M St NW, Ste 500, Washington, DC 20036
T: (202) 785-3913
Founded: 1920; Members: 113
Pres: Joseph C. Isaacs
Focus: Public Health 16069

National Health Federation (NHF), POB 688, Monrovia, CA 91016
T: (818) 357-2181; Fax: (818) 303-0642
Founded: 1955; Members: 20000
Pres: Dr. H. Couger
Focus: Public Health
Periodical
Health Freedom News (11 times annually) 16070

National Health Law Program (NHeLP), 2639 S La Cienega Blvd, Los Angeles, CA 90034
T: (310) 204-6010; Fax: (310) 204-0891
Members: 8
Gen Secr: Laurence M. Lavin

Focus: Public Health
Periodical
Health Advocate (weekly) 16071

National Health Lawyers Association (NHLA), 1120 Connecticut Av NW, Ste 950, Washington, DC 20036
T: (202) 833-1100; Fax: (202) 833-1105
Founded: 1971; Members: 7000
Gen Secr: Mary Lou King
Focus: Law; Public Health
Periodicals
Digest Index (weekly)
Health Law Digest (weekly)
Health Lawyers News Report (weekly)
Register (weekly) 16072

National Hearing Aid Society (NHAS), 20361 Middlebelt Rd, Livonia, MI 48152
T: (313) 478-2610; Fax: (313) 478-4520
Founded: 1951; Members: 4000
Pres: Robin Holm
Focus: Educ Handic
Periodicals
Audecibel (weekly)
Confidential Report to Members (weekly)
Directory of Certified Hearing Aid Audiologists (weekly) 16073

National Hearing Conservation Association (NHCA), 431 E Locust, Des Moines, IA 50309
T: (515) 243-1558; Fax: (515) 243-2049
Founded: 1977; Members: 550
Gen Secr: Michele Johnson
Focus: Otorhinolaryngology 16074

National Heart Research Project (NHRP), 306 W Joppa Rd, Baltimore, MD 21204
T: (410) 494-0300
Founded: 1982
Pres: Frederick C. Ruof
Focus: Cardiol 16075

National Home Study Council (NHSC), 1601 18 St NW, Ste 2, Washington, DC 20009
T: (202) 234-5100
Founded: 1926; Members: 92
Gen Secr: Michael P. Lambert
Focus: Educ 16076

National Hormone and Pituitary Program (NHPP), 685 Lofstrand Dr, Rockville, MD 20850
T: (301) 340-9245; Fax: (301) 837-9566
Founded: 1963
Gen Secr: Dr. Terry Taylor
Focus: Med; Microbio 16077

National Housing Conference (NHC), 1126 16 St NW, Ste 211, Washington, DC 20036
T: (202) 223-4844; Fax: (202) 331-7731
Founded: 1931; Members: 700
Pres: Kathleen A. Boland
Focus: Archit
Periodicals
Newsletter (weekly)
NHC Policy and Resolutions (weekly) . . 16078

National Humane Education Society (NHES), 15B Catoctin Cir SE, Ste 207, Leesburg, VA 22075
T: (703) 777-8319
Founded: 1948; Members: 200000
Pres: Anna Briggs
Focus: Educ
Periodical
NHES Quarterly Journal (weekly) . . . 16079

National Humanities Center (NHC), 7 Alexander Dr, POB 12256, Research Triangle Park, NC 27709
T: (919) 549-0661
Founded: 1976; Members: 40
Gen Secr: W. Robert Connor
Focus: Humanities
Periodical
Newsletter (3 times annually) 16080

National Hydropower Association (NHA), 555 13 St NW, Ste 900, Washington, DC 20004
T: (202) 637-8115
Founded: 1983; Members: 75
Gen Secr: Linda Church Ciocci
Focus: Energy; Water Res
Periodicals
Hydro Regulatory Report (weekly)
NHA News from Washington (weekly) . 16081

National Identification Program for the Advancement of Women in Higher Education Administration (NIP), c/o Office of Women in Higher Education, American Council on Education, 1 Dupont Circle NW, Washington, DC 20036
T: (202) 939-9390; Fax: (202) 833-4760
Founded: 1977
Gen Secr: Donna Shavlik
Focus: Educ 16082

National Immigration Law Center (NILC), 1636 W Eighth St, Ste 215, Los Angeles, CA 90017
T: (213) 487-2531; Fax: (213) 384-4899
Founded: 1977
Gen Secr: Charles Wheeler
Focus: Law
Periodical
Legalization Update (weekly) 16083

National Indian Education Association (NIEA), 1819 H St NW, Ste 800, Washington, DC 20006
T: (202) 835-3001
Founded: 1970; Members: 2000
Pres: Donna Rhodes
Focus: Educ
Periodical
Indian Education Newsletter (weekly) . . 16084

National Industrial Zoning Committee (NIZC), 1858 Chatfield Rd, Columbus, OH 43221
T: (614) 488-9001
Founded: 1948; Members: 8
Gen Secr: James M. Jennings
Focus: Econ 16085

National Information Center for Educational Media (NICEM), POB 40130, Albuquerque, NM 87196
T: (505) 265-3591; Fax: (505) 256-1080
Founded: 1967
Pres: Marjorie M.K. Hlava
Focus: Educ 16086

National Information Service for Earthquake Engineering (NISEE), c/o Earthquake Engineering Research Center, University of California, 1301 S 46 St, Richmond, CA 94804
T: (510) 231-9554; Fax: (510) 231-9471
Founded: 1971
Gen Secr: Jack P. Moehle
Focus: Eng; Geophys
Periodicals
Abstract Journal in Earthquake Engineering (weekly)
Computer Applications Program Catalog (weekly)
Current Abstract Update Service (weekly)
Customized Current Information Service (weekly)
Earthquake Engineering Research Center Reports (weekly)
Library Acquisitions Alert (weekly) . . . 16087

National Institute for Burn Medicine (NIBM), 909 E Ann St, Ann Arbor, MI 48104
T: (313) 769-9000; Fax: (313) 769-9009
Founded: 1968
Gen Secr: Claudella A. Jones
Focus: Med
Periodical
National Burn Information Exchange Newsletter (2-3 times annually) 16088

National Institute for Certification in Engineering Technologies (NICET), 1420 King St, Alexandria, VA 22314-2794
T: (703) 684-2835
Founded: 1981
Gen Secr: John D. Antrim
Focus: Eng
Periodical
Newsletter 16089

National Institute for Farm Safety (NIFS), 205 Agriculture Engineering, Columbia, MO 65211
T: (314) 882-2731; Fax: (314) 882-1115
Founded: 1962; Members: 204
Gen Secr: David E. Baker
Focus: Safety 16090

National Institute for Public Policy (NIPP), 3031 Javier Rd, Fairfax, VA 22031
T: (703) 698-0563
Founded: 1981
Pres: Keith B. Payne
Focus: Poli Sci 16091

National Institute for Rehabilitation Engineering (NIRE), POB T, Hewitt, NJ 07421
T: (201) 853-6585
Founded: 1967
Gen Secr: Donald Selwyn
Focus: Rehabil; Eng 16092

National Institute of Ceramic Engineers (NICE), 735 Ceramic Pl, Westerville, OH 43081
T: (614) 890-4700; Fax: (614) 899-6109
Founded: 1938; Members: 2086
Gen Secr: W. Paul Holbrook
Focus: Eng
Periodicals
Ceramic Abstracts (weekly)
Ceramic Bulletin (weekly)
Ceramic Engineering and Science Proceedings (weekly)
Ceramic Source (weekly)
Journal of the American Ceramic Society incorporating Advanced Ceramic Materials (weekly) 16093

National Institute of Management Counsellors (NIMC), POB 193, Great Neck, NY 11022
T: (516) 482-5683
Founded: 1954; Members: 250
Gen Secr: Willard Warren
Focus: Business Admin 16094

National Institute of Science (NIS), c/o Dr. Arthur C. Washington, Tennessee State University, 3500 John Merritt Blvd, Nashville, TN 37209-1561
T: (615) 320-3410
Founded: 1942; Members: 843
Gen Secr: Dr. Arthur C. Washington
Focus: Sci
Periodicals
Newsletter (weekly)
Transactions (weekly) 16095

National Institute of Social Sciences (NISS), c/o Reed Foundation, 444 Madison Av, Ste 2901, New York, NY 10022
T: (212) 223-1330
Founded: 1912; Members: 1000
Pres: J. Sinclair Armstrong
Focus: Sociology 16096

National Institute of Steel Detailing (NISD), c/o Gunther Baresel, 300 S Harbor Blvd, Ste 500, Anaheim, CA 92805
T: (714) 776-3200; Fax: (714) 776-1255
Founded: 1969; Members: 141
Pres: Gunther Baresel
Focus: Metallurgy 16097

National Jewish Center for Immunology and Respiratory Medicine (NJCIRM), 1400 Jackson St, Denver, CO 80206
T: (303) 388-4461
Founded: 1978; Members: 1200
Pres: Leonard Perlmutter
Focus: Med
Periodicals
Annual Report (weekly)
Lung Line Letter (weekly)
Medical Scientific Update (weekly)
New Directions (weekly) 16098

National Judicial College (NJC), c/o Judicial College, University of Nevada, Reno, NV 89557
T: (702) 784-6747; Fax: (702) 784-4234
Founded: 1963
Gen Secr: V. Robert Payant
Focus: Law 16099

National League of American PEN Women (NLAPW), 1300 17 St NW, Washington, DC 20036
T: (202) 785-1997
Founded: 1897; Members: 5000
Pres: Muriel C. Freeman
Focus: Arts; Music; Lit
Periodicals
The PEN Woman (9 times annually)
Roster (biennial) 16100

National Leukemia Association (NLA), 585 Stewart Av, Ste 536, Garden City, NY 11530
T: (516) 222-1944; Fax: (516) 222-0457
Founded: 1965
Gen Secr: Allan D. Weinberg
Focus: Cell Biol & Cancer Res 16101

National Librarians Association (NLA), POB 486, Alma, MI 48801
T: (517) 463-7227; Fax: (517) 463-8694
Founded: 1975; Members: 150
Gen Secr: Peter Dollard
Focus: Libraries & Bk Sci
Periodical
The National Librarian (weekly) 16102

National Marine Educators Association (NMEA), POB 512515, Pacific Grove, CA 93950
T: (408) 648-4841; Fax: (408) 372-8471
Founded: 1976; Members: 1200
Gen Secr: Michael Rigsby
Focus: Adult Educ
Periodicals
Current: The Journal of Marine Education (weekly)
NMEA News (weekly) 16103

National Mastitis Council (NMC), c/o Anne Saeman, 1840 Wilson Blvd, Ste 400, Arlington, VA 22201
T: (703) 243-8268; Fax: (703) 931-4520
Founded: 1961; Members: 1900
Gen Secr: Anne Saeman
Focus: Vet Med
Periodical
Annual Meeting Proceedings 16104

National Materials Advisory Board (NMAB), 2101 Constitution Av NW, Washington, DC 20418
T: (202) 334-3505
Founded: 1951; Members: 16
Gen Secr: Klaus M. Zwilsky
Focus: Materials Sci
Periodical
Newsletter (weekly) 16105

National Medical and Dental Association (NMDA), 9412 Academy Rd, Philadelphia, PA 19114
T: (215) 676-2242
Founded: 1910; Members: 600
Gen Secr: Dorothy Czarnecki
Focus: Med; Dent
Periodical
Bulletin (weekly) 16106

National Medical Association (NMA), 1012 Tenth St NW, Washington, DC 20001
T: (202) 347-1895; Fax: (202) 842-3293
Founded: 1895; Members: 14500
Gen Secr: William C. Garrett
Focus: Med
Periodicals
Journal of the National Medical Association (weekly)
National Medical Association Newsletter (weekly) 16107

National Medical Fellowships (NMF), 254 W 31 St, New York, NY 10001
T: (212) 714-0933; Fax: (212) 239-9718
Founded: 1946
Pres: Leon Johnson
Focus: Educ; Med 16108

National Mental Health Association (NMHA), 1021 Prince St, Alexandria, VA 22314-2971
T: (703) 684-7722; Fax: (703) 684-5968
Founded: 1909
Pres: John H. Horner
Focus: Psychiatry
Periodical
Focus (weekly) 16109

National Middle School Association (NMSA), 4807 Evanswood Dr, Columbus, OH 43229
T: (614) 848-8211; Fax: (614) 848-4301
Founded: 1973; Members: 10000
Gen Secr: Ron Williamson
Focus: Educ
Periodicals
Middle Ground (weekly)
Middle School Journal (weekly)
Target (weekly) 16110

National Multiple Sclerosis Society (NMSS), 733 Third Av, New York, NY 10017
T: (212) 986-3240; Fax: (212) 986-7981
Founded: 1946; Members: 470000
Pres: Michael Dugan
Focus: Med; Pathology
Periodical
Inside MS (weekly) 16111

National Music Council (NMC), POB 5551, Englewood, NJ 07631-5551
T: (201) 877-9088
Founded: 1940; Members: 50
Gen Secr: Keith J. King
Focus: Music
Periodical
NMC News 16112

National Old Timers' Association of the Energy Industry (NOTAEI), POB 168, Mineola, NY 11501
T: (516) 431-4668; Fax: (516) 431-9850
Founded: 1926; Members: 1000
Gen Secr: John M. Sibarium
Focus: Energy
Periodicals
Hour Glass (weekly)
Technica Manuel (5 times annually) . . 16113

National Opera Association (NOA), c/o Robert Murray, 212 Texas St, Ste 101, Shreveport, LA 71101-3249
Founded: 1955; Members: 900
Pres: Robert Murray
Focus: Music
Periodicals
NOA Newsletter (weekly)
Opera Journal (weekly) 16114

National Optometric Association (NOA), 1489 E Livingston Av, Columbus, OH 43205
T: (614) 253-5593
Founded: 1969; Members: 350
Gen Secr: Dr. Clayton Hicks
Focus: Ophthal
Periodical
Newsletter (weekly) 16115

National Orchestral Association (NOA), 475 Riverside Dr, Rm 249, New York, NY 10115
T: (212) 870-2009; Fax: (212) 870-2009
Founded: 1930; Members: 75
Pres: Frances J. Kennedy
Focus: Music
Periodicals
Fact Sheet (weekly)
NOA News (weekly) 16116

National Organization for Continuing Education of Roman Catholic Clergy (NOCERCC), 1337 W Ohio St, Chicago, IL 60622
T: (312) 226-1890
Founded: 1973; Members: 425
Gen Secr: Peter A. Fitzpatrick
Focus: Educ
Periodicals
News Notes (weekly)
Resources (weekly) 16117

National Organization for Rare Disorders (NORD), POB 8923, New Fairfield, CT 06812-1783
T: (203) 746-6518; Fax: (203) 746-6481
Founded: 1983; Members: 40000
Gen Secr: A.S. Meyers
Focus: Med 16118

National Organization of Minority Architects (NOMA), c/o School of Architecture and Planning, Howard University, 2366 Sixth St, Washington, DC 20059
T: (804) 788-0338; Fax: (804) 649-8502
Founded: 1971; Members: 250
Pres: Robert L. Easter
Focus: Archit
Periodicals
Newsletter (weekly)
Roster of Minority Firms 16119

National Organization on Legal Problems of Education (NOLPE), 3601 W 29 St, Ste 223, Topeka, KS 66614
T: (913) 273-3550
Founded: 1954; Members: 1800
Gen Secr: Floyd Delon
Focus: Law; Educ

Periodicals
Notes (weekly)
School Law Reporter (weekly)
School Law Update (weekly) 16120

National Organization to Insure a Sound-Controlled Environment (NOISE), 1620 Eye St NW, Ste 300, Washington, DC 20006
T: (202) 682-3901; Fax: (202) 293-3109
Founded: 1969
Gen Secr: Charles Price
Focus: Public Health
Periodicals
Newsletter (weekly)
Officers and Directors (weekly) 16121

National Orthotic and Prosthetic Research Institute (NOPRI), POB 491, Lenox Hill, NY 10021
T: (212) 755-3366
Founded: 1969; Members: 10
Pres: Ralph Florio
Focus: Rehabil 16122

National Osteopathic Guild Association (NOGA), c/o Auxiliary to the American Osteopathic Association, 142 E Ontario St, Chicago, IL 60611
T: (312) 280-5819
Founded: 1955; Members: 1800
Pres: N. Teague
Focus: Pathology
Periodical
Newsletter (weekly) 16123

National Perinatal Association (NPA), 3500 E Fletcher Av, Ste 525, Tampa, FL 33613
T: (813) 971-1008; Fax: (813) 971-9306
Founded: 1976; Members: 5000
Gen Secr: Julie A. Leachman
Focus: Gynecology
Periodicals
Newsletter (weekly)
Proceedings (weekly) 16124

National Pest Control Association (NPCA), 8100 Oak St, Dunn Loring, VA 22027
T: (703) 573-8330; Fax: (703) 573-4116
Founded: 1933; Members: 2300
Gen Secr: Harvey S. Gold
Focus: Microbio
Periodicals
Pest Management (weekly)
Roster of Members (weekly)
Technical Release (weekly) 16125

National Pharmaceutical Association (NPhA), c/o College of Pharmacy and Pharmacological Sciences, Howard University, 2726 12 St NE, Washington, DC 20018
T: (202) 529-7747
Founded: 1947; Members: 325
Gen Secr: Dr. Terry Smith Moore
Focus: Pharmacol
Periodical
Journal (weekly) 16126

National Photography Instructors Association (NPIA), 1255 Hill Dr, Eagle Rock, CA 90041
T: (213) 254-1549; Fax: (213) 254-1549
Gen Secr: Pierre Odier
Focus: Photo 16127

National Piano Foundation (NPF), c/o Donald W. Dillon Associates, 4020 McEwen St, Ste 105, Dallas, TX 75244
T: (214) 233-9107; Fax: (214) 490-4219
Founded: 1962
Gen Secr: Donald W. Dillon
Focus: Music
Periodical
NPF Piano News (weekly) 16128

National Plant Board (NPB), c/o Alabama Dept of Agriculture, POB 3336, Montgomery, AL 36109-0336
T: (205) 242-2656; Fax: (205) 240-3103
Founded: 1925; Members: 51
Pres: Guy Karr
Focus: Agri
Periodicals
Minutes (weekly)
Proceedings of the Annual Meeting . . 16129

National Podriatic Medical Association (NPMA), c/o Raymond E. Lee, 1638 E 87 St, Chicago, IL 60617
T: (312) 374-1616
Founded: 1971; Members: 200
Gen Secr: Raymond E. Lee
Focus: Orthopedics
Periodicals
Annual Seminar Ad Book
Newsletter (weekly) 16130

National Psoriasis Foundation (NPF), 6443 SW Beaverton Hwy, Ste 210, Portland, OR 97221
T: (503) 297-1545; Fax: (503) 292-9341
Founded: 1968; Members: 20000
Gen Secr: Gail M. Zimmerman
Focus: Derm 16131

National Psychological Association for Psychoanalysis (NPAP), 150 W 13 St, New York, NY 10011
T: (212) 924-7440; Fax: (212) 989-7543
Founded: 1946; Members: 355
Gen Secr: S. Coppersmith
Focus: Psychoan
Periodicals
Bulletin (biennial)
News & Reviews (weekly)
The Psychoanalytic Review (weekly) . . 16132

National Ramah Commission (NRC), 3080 Broadway, New York, NY 10027
T: (212) 678-8881; Fax: (212) 749-8251
Founded: 1953; Members: 45
Gen Secr: Sheldon Dorph
Focus: Rel & Theol 16133

National Reading Conference (NRC), 11 E Hubbard St, Ste 200, Chicago, IL 60611
T: (312) 329-2512; Fax: (312) 329-9131
Founded: 1950; Members: 1000
Gen Secr: Judith C. Burnison
Focus: Educ
Periodicals
Journal of Reading Behavior (weekly)
Yearbook (weekly) 16134

National Records Management Council (NAREMCO), 60 E 42 St, New York, NY 10165
T: (212) 697-0290
Founded: 1948
Pres: Alan A. Andolsen
Focus: Public Admin; Business Admin
Periodicals
Information Alert (weekly)
NAREMCO Report (weekly) 16135

National Registry in Clinical Chemistry (NRCC), 1155 16 St NW, Washington, DC 20036
T: (202) 745-1698; Fax: (202) 452-2116
Founded: 1967; Members: 1200
Gen Secr: Gilbert E. Smith
Focus: Chem
Periodical
Directory (weekly) 16136

National Rehabilitation Association (NRA), 633 S Washington St, Alexandria, VA 22314
T: (703) 836-0850; Fax: (703) 836-0848
Founded: 1925; Members: 17000
Gen Secr: Dr. Ann Tourigny
Focus: Rehabil
Periodicals
Journal of Rehabilitation (weekly)
Newsletter (8 times annually) 16137

National Research Council (NRC), 2101 Constitution Av NW, Washington, DC 20418
T: (202) 334-2000; Fax: (202) 334-2158
Founded: 1916; Members: 9500
Gen Secr: Philip M. Smith
Focus: Sci; Eng
Periodical
News Report (10 times annually) . . . 16138

National Research Council on Peace Strategy (NRCPS), 241 W 12 St, New York, NY 10014
T: (212) 675-3839
Founded: 1961; Members: 60
Pres: Harold Taylor
Focus: Prom Peace 16139

National Resident Matching Program (NRMP), 2450 N St NW, Ste 201, Washington, DC 20037-1141
T: (202) 828-0676
Founded: 1951; Members: 30000
Gen Secr: Phyllis Weiland
Focus: Educ; Med 16140

National Resource Center for Consumers of Legal Services (NRCCLS), POB 340, Gloucester, VA 23061
T: (804) 693-9330; Fax: (804) 693-7363
Founded: 1977
Gen Secr: William A. Bolger
Focus: Law
Periodical
Legal Plan Letter (weekly) 16141

National Rural and Small Schools Consortium (NRSSC), c/o National Rural Development Institute, Western Washington University, 359 Miller Hall, Bellingham, WA 98225
T: (801) 585-5659; Fax: (801) 647-4845
Founded: 1985; Members: 3000
Pres: Doris Helge
Focus: Educ 16142

National Rural Development Institute (NRDI), c/o Western Washington University, 359 Miller Hall, Bellingham, WA 98225
T: (206) 676-3576; Fax: (801) 650-5659
Founded: 1974
Gen Secr: Doris Helge
Focus: Agri 16143

National Rural Education Association (NREA), c/o Joseph T. Newlin, Colorado State University, 230 Education Bldg, Fort Collins, CO 80523
T: (303) 491-7022; Fax: (303) 491-1317
Founded: 1907; Members: 1000
Gen Secr: Joseph T. Newlin
Focus: Educ 16144

National Safety Council (NSC), 1121 Spring Lake Dr, Itasca, IL 60143-3201
T: (708) 285-1121; Fax: (708) 285-1315
Founded: 1913; Members: 12000
Pres: T.C. Gilchest
Focus: Safety
Periodicals
Family Safety and Health (weekly)
Fleet Safety Newsletter (weekly)
Forest Industries Newsletter (weekly)
Glass and Ceramics Newsletter (weekly)
Health Care Newsletter (weekly)
Journal of Safety Research (weekly)
Metals Newsletter (weekly)
Mining Newsletter (weekly)
Petroleum Newsletter (weekly)
Power Press and Forging Newsletter (weekly)
Printing and Publishing Newsletter (weekly)
Public Employee Newsletter (weekly)
Public Utilities Newsletter (weekly)
Railroad Newsletter (weekly)
Safe Driver (weekly)
Safety and Health (weekly)
Safe Worker (weekly)
Traffic Safety (weekly) 16145

National Safety Management Society (NSMS), 12 Pickens Ln, Weaverville, NC 28787
T: (704) 645-5229; Fax: (704) 645-5229
Founded: 1968; Members: 750
Focus: Safety
Periodicals
Focus (weekly)
Insights (weekly)
Safety Management Monograph (weekly)
Update Newsletter (10 times annually) . 16146

National School Boards Association (NSBA), 1680 Duke St, Alexandria, VA 22314
T: (703) 838-6722; Fax: (703) 683-7590
Founded: 1940; Members: 53
Gen Secr: Dr. Thomas A. Shannon
Focus: Educ
Periodicals
The American School Board Journal (weekly)
Beliefs and Policies (weekly)
The Executive Educator (weekly)
Inquiry and Analysis (weekly)
Leadership Reports (weekly)
School Board News (weekly)
Updating School Board Policies (weekly) 16147

National School Public Relations Association (NSPRA), 1501 Lee Hwy, Ste 201, Arlington, VA 22209
T: (703) 528-5840; Fax: (703) 528-7017
Founded: 1935; Members: 2000
Gen Secr: Richard D. Bagin
Focus: Educ
Periodicals
Education USA (weekly)
Paragraphs (weekly)
Starts in the Classroom (weekly) . . . 16148

National Schools Committee for Economic Education (NSCEE), 86 Valley Rd, POB 295, Cos Cob, CT 06807
T: (203) 869-1706
Founded: 1953; Members: 465
Pres: John G. Murphy
Focus: Educ; Econ 16149

National Science Supervisors Association (NSSA), c/o Dr. Kenneth R. Roy, 330 Hubbard St, Glastonbury, CT 06033
T: (203) 633-5231; Fax: (203) 659-3366
Founded: 1958; Members: 1200
Gen Secr: Dr. Kenneth R. Roy
Focus: Educ
Periodicals
Conference Model Guidelines (weekly)
Directory (weekly)
Newsletter (weekly)
Proceedings of Summer Conference (weekly)
. . . . 16150

National Science Teachers Association (NSTA), 1742 Connecticut Av NW, Washington, DC 20009-1171
T: (202) 328-5800; Fax: (202) 328-0974
Founded: 1895; Members: 63000
Gen Secr: Bill G. Aldridge
Focus: Educ
Periodicals
Journal of College Science Teaching (weekly)
Science and Children (8 times annually)
Science Scope (8 times annually)
The Science Teacher (9 times annually) 16151

National Sculpture Society (NSS), 15 E 26 St, New York, NY 10010
T: (212) 889-6960; Fax: (212) 545-0779
Founded: 1893; Members: 4350
Gen Secr: Gwen Pier
Focus: Fine Arts
Periodicals
National Sculpture Review (weekly)
Newsletter (weekly) 16152

National Shellfisheries Association (NSA), c/o Dr. Stephen Tettelbach, National Science Div, Southampton College, Long Island University, Southampton, NY 11968
T: (516) 283-4000
Founded: 1909; Members: 900
Gen Secr: Dr. Stephen Tettelbach
Focus: Fisheries
Periodical
Journal of Shellfish Research (weekly) . 16153

National Society for Cardiovascular Technology (NSCT), 10500 Wakeman Dr, Fredericksburg, VA 22407
T: (703) 891-0079; Fax: (703) 898-8869
Founded: 1966; Members: 2500
Gen Secr: Peggy McElgunn
Focus: Cardiol 16154

National Society for Experiential Education (NSEE), 3509 Haworth Dr, Ste 207, Raleigh, NC 27609-7229
T: (919) 787-3263
Founded: 1978; Members: 1300
Gen Secr: Allen Wutzdorff
Focus: Educ
Periodicals
Experiential Education (weekly)
The National Directory of Internships (weekly)
. . . . 16155

National Society for Performance and Instruction (NSPI), 1300 L St NW, Ste 1250, Washington, DC 20005
T: (202) 408-7969; Fax: (202) 408-7972
Founded: 1962; Members: 5000
Gen Secr: Paul Tremper
Focus: Educ
Periodicals
Official International Membership Directory (weekly)
Performance Instruction (10 times annually)
. . . . 16156

National Society for the Preservation of Covered Bridges (NSPCB), 44 Cleveland Av, Worcester, MA 01603
T: (617) 756-4516
Founded: 1948; Members: 1004
Gen Secr: A.L. Ellsworth
Focus: Preserv Hist Monuments
Periodicals
Bulletin (weekly)
Covered Bridge Topics (weekly)
Notices (weekly) 16157

National Society for the Study of Education (NSSE), 5835 Kimbark Av, Chicago, IL 60637
T: (312) 702-1582; Fax: (312) 702-0248
Founded: 1901; Members: 2300
Gen Secr: Kenneth J. Rehage
Focus: Educ
Periodical
Yearbook (weekly) 16158

National Society of Professional Engineers (NSPE), 1420 King St, Alexandria, VA 22314
T: (703) 684-2800; Fax: (703) 836-4875
Founded: 1934; Members: 75000
Gen Secr: Donald G. Weinert
Focus: Eng
Periodicals
Directory of Professional Engineers in Private Practice (biennial)
Engineering Times (weekly) 16159

National Society of Professional Sanitarians (NSPS), 1224 Hoffman Dr, Jefferson City, MO 65101
T: (314) 751-6095
Founded: 1956; Members: 500
Gen Secr: John G. Norris
Focus: Public Health
Periodical
The Professional Sanitarian (weekly) . . 16160

National Space Club (NSC), 655 15 St NW, Washington, DC 20005
Founded: 1957; Members: 1700
Gen Secr: Rory Heydon
Focus: Astrophys
Periodical
Newsletter (weekly) 16161

National Space Society (NSS), 922 Pennsylvania Av SE, Washington, DC 20003
T: (202) 543-1900; Fax: (202) 543-1995
Founded: 1987; Members: 30000
Gen Secr: Lori B. Garver
Focus: Astrophys
Periodical
Space World (weekly) 16162

National Speleological Society (NSS), Cave Av, Huntsville, AL 35810-4431
T: (205) 852-1300; Fax: (205) 851-9241
Founded: 1941; Members: 10600
Pres: Jeanne Gurnee
Focus: Speleology
Periodicals
Membership List (weekly)
Monthly News (weekly)
NSS Bulletin (weekly) 16163

National Standard Plumbing Code Committee (NSPCC), 180 S Washington St, POB 6808, Falls Church, VA 22040
T: (703) 237-8100; Fax: (703) 237-7442
Founded: 1970; Members: 12
Gen Secr: Agnes Hoyt
Focus: Standards 16164

National Student Nurses' Association (NSNA), 555 W 57 St, Ste 1327, New York, NY 10019
T: (212) 581-2211; Fax: (212) 581-2368
Founded: 1953; Members: 35000
Gen Secr: Robert V. Piemonte
Focus: Adult Educ; Med
Periodical
Imprint (5 times annually) 16165

National Student Speech Language Hearing Association (NSSLHA), 10801 Rockville Pike, Rockville, MD 20852
T: (301) 897-5700; Fax: (301) 571-0457
Founded: 1972; Members: 12500

Gen Secr: Amy K. Harbison
Focus: Logopedy
Periodicals
Clinical Series (biennial)
Journal (weekly) 16166

National Study of School Evaluation (NSSE), c/o Prof. Vernon D. Pace, School of Education, Indiana University, Bloomington, IN 47405
T: (812) 855-7185; Fax: (812) 855-3044
Founded: 1933; Members: 6
Gen Secr: Prof. Vernon D. Pace
Focus: Educ 16167

National Tax Association – Tax Institute of America (NTA-TIA), 5310 E Main St, Ste 104, Columbus, OH 43213
T: (614) 864-1221
Founded: 1973; Members: 1800
Gen Secr: Frederick D. Stocker
Focus: Finance; Law; Econ
Periodicals
National Tax Journal (weekly)
Proceedings of the Annual Conference on Taxation (weekly) 16168

National Technical Association (NTA), POB 7045, Washington, DC 20032-0145
T: (202) 829-6100; Fax: (202) 684-3952
Founded: 1926; Members: 1500
Pres: C. Stewart
Focus: Eng
Periodicals
Journal (weekly)
Newsletter (weekly) 16169

National Technical Services Association (NTSA), 325 S Patrick St, Ste 104, Alexandria, VA 22314-3501
T: (703) 684-4722; Fax: (703) 684-7627
Founded: 1961; Members: 150
Gen Secr: Laura McGuire Mackail
Focus: Eng
Periodicals
Membership Roster (weekly)
Reporter (weekly) 16170

National Telemedia Council (NTC), 120 E Wilson St, Madison, WI 53703
T: (608) 257-7712; Fax: (608) 257-7714
Founded: 1953
Gen Secr: M. Rowe
Focus: Mass Media
Periodicals
Look-Listen Project Report (weekly)
Notable Programs (weekly)
Telemedium (4-6 times annually) 16171

National Theatre Institute (NTI), 305 Great Neck Rd, Waterford, CT 06385
T: (203) 443-7139; Fax: (203) 443-9653
Founded: 1970
Gen Secr: Richard Digby Day
Focus: Perf Arts
Periodical
Alumnae Newsletter (weekly) 16172

National Therapeutic Recreation Society (NTRS), 2775 S Quincy St, Ste 300, Arlington, VA 22206
T: (703) 820-4940; Fax: (703) 671-6772
Founded: 1966; Members: 3200
Gen Secr: Rikki S. Epstein
Focus: Therapeutics
Periodicals
Journal of Leisure Research (weekly)
NTRS Newsletter (weekly)
Parks and Recreation (weekly)
Therapeutic Recreation Journal (weekly) . 16173

National Tumor Registrars Association (NTRA), 505 E Hawley St, Mundelein, IL 60060
T: (708) 566-0833
Founded: 1974; Members: 1700
Gen Secr: Robert B. Willis
Focus: Cell Biol & Cancer Res
Periodicals
Abstract (weekly)
Membership Roster (weekly)
Proceedings/Annual Report 16174

National University Continuing Education Association (NUCEA), 1 Dupont Circle, Ste 615, Washington, DC 20036
T: (202) 659-3130; Fax: (202) 785-0374
Founded: 1915; Members: 2170
Gen Secr: Kay J. Kohl
Focus: Adult Educ
Periodicals
Guide to Independent Study through Correspondence Instruction (biennial)
Newsletter (weekly) 16175

National Urban League (NUL), 500 E 62 St, New York, NY 10021
T: (212) 310-9000; Fax: (212) 593-8250
Founded: 1910; Members: 50000
Pres: John E. Jacob
Focus: Law
Periodicals
State of Black America (weekly)
The Urban League News (weekly)
The Urban League Review (weekly) . . 16176

National Vocational Agricultural Teachers' Association (NVATA), POB 15450, Alexandria, VA 22309
T: (703) 780-1862
Founded: 1948; Members: 9000
Gen Secr: Bob Graham
Focus: Adult Educ; Agri
Periodical
News and Views of NVATA (weekly) . . 16177

National Vocational Technical Educational Foundation (NVTEF), 1420 16 St NW, Ste 301, Washington, DC 20036
T: (202) 328-0216; Fax: (202) 797-3756
Founded: 1979; Members: 43
Gen Secr: Madeleine B. Hemmings
Focus: Educ 16178

National Water Resources Association (NWRA), 3800 N Fairfax Dr, Ste 4, Arlington, VA 22203
T: (703) 524-1544; Fax: (703) 524-1548
Founded: 1932; Members: 4800
Gen Secr: Thomas F. Donnelly
Focus: Water Res
Periodicals
National Waterline (weekly)
National Water Resources Association Directory (biennial) 16179

National Wildlife Federation (NWF), 1400 16 St NW, Washington, DC 20036-2266
T: (202) 797-6800
Founded: 1936; Members: 6200000
Pres: Lynn Bowersox
Focus: Ecology
Periodicals
International Wildlife (weekly)
National Wildlife (weekly) 16180

National Women's Health Network (NWHN), 1325 G St NW, Washington, DC 20005
T: (202) 347-1140; Fax: (202) 347-1140
Founded: 1976; Members: 15000
Gen Secr: Beverly Baker
Focus: Public Health
Periodicals
Network News (weekly)
Newsalerts 16181

National Women's Studies Association (NWSA), c/o Deborah Louis, University of Maryland, College Park, MD 20742-1325
T: (301) 405-5573
Founded: 1977; Members: 4000
Gen Secr: Deborah Louis
Focus: Adult Educ
Periodicals
NWSA Action (weekly)
NWSA Journal (weekly)
Women's Studies Program Directory (biennial) 16182

National Writers Club (NWC), 1450 S Havana St, Ste 620, Aurora, CO 80012
T: (303) 751-7844; Fax: (303) 751-8593
Founded: 1937; Members: 4000
Gen Secr: Sandy Whelchel
Focus: Lit
Periodicals
Authorship (weekly)
Flash Market News (weekly)
Market Update (weekly)
NWC Newsletter (10 times annually)
Professional Freelance Directory (weekly) 16183

National Writing Project (NWP), c/o School of Education, University of California, 615 University Hall, Berkeley, CA 94720
T: (510) 642-0963; Fax: (510) 643-6239
Members: 166
Gen Secr: James R. Gray
Focus: Educ
Periodical
The Quarterly (weekly) 16184

Nation's Report Card – National Assembly of Educational Progress (NRC), POB 6710, Princeton, NJ 08541
T: (800) 223-0267; Fax: (609) 734-1878
Founded: 1964
Gen Secr: Ina Mullis
Focus: Educ 16185

Native Seeds/Search (NS/S), 2509 N Campbell Av, Tucson, AZ 85719
T: (602) 327-9123
Founded: 1982; Members: 3200
Pres: Gary Nabhan
Focus: Agri
Periodicals
Fall Harvest Catalog (weekly)
Seedhead News (weekly)
Seedlisting (weekly) 16186

Natural Resources Defense Council (NRDC), 40 W 20 St, New York, NY 10011
T: (212) 727-2700; Fax: (212) 727-1773
Founded: 1970; Members: 170000
Gen Secr: John H. Adams
Focus: Ecology
Periodicals
Amicus Journal (weekly)
NRDC Newsline (5 times annually) . . . 16187

The Nature Conservancy (TNC), 1815 N Lynn St, Arlington, VA 22209
T: (703) 841-5300; Fax: (703) 841-1283
Founded: 1951; Members: 588000
Gen Secr: John C. Sawhill
Focus: Ecology
Periodical
The Nature Conservancy Magazine (weekly) 16188

Near East Archaeological Society (NEAS), c/o Dr. W. Harold Mare, Covenant Theological Seminary, 12330 Conway Rd, Saint Louis, MO 63141
T: (314) 434-4044
Founded: 1960; Members: 200
Pres: Dr. W. Harold Mare
Focus: Archeol
Periodicals
NEAS Bulletin (weekly)
NEAS Newsletter (weekly) 16189

Near East College Association (NECA), 850 Third Av, New York, NY 10022
T: (212) 319-2453; Fax: (212) 752-6971
Founded: 1919; Members: 6
Gen Secr: Peter Rupprecht
Focus: Educ 16190

Nepal Studies Association (NSA), c/o Asian Survey Office, University of California, 6701 San Pablo Av, Rm 408, Oakland, CA 94608
T: (510) 642-0978
Founded: 1972; Members: 400
Gen Secr: Leo Rose
Focus: Ethnology
Periodical
Himalayan Research Bulletin (3 times annually) 16191

The Network, 5420 Mayfield Rd, Ste 205, Cleveland, OH 44124
T: (216) 442-5600; Fax: (216) 449-3227
Founded: 1975; Members: 400
Pres: Irwin Friedman
Focus: Energy
Periodical
Newsletter 16192

Neurosurgical Society of America (NSA), c/o Dr. Russell W. Hardy, University Neurosurgeons of Cleveland, 2074 Abington Rd, Cleveland, OH 44106
T: (216) 844-5949; Fax: (216) 844-3014
Founded: 1948; Members: 162
Gen Secr: Dr. Russell W. Hardy
Focus: Surgery; Neurology 16193

Newcomen Society of the United States (NSUS), 412 Newcomen Rd, Exton, PA 19341
T: (215) 363-6600; Fax: (215) 363-0612
Founded: 1923; Members: 15000
Pres: James E. Fritz
Focus: Hist
Periodical
Monographs of Corporate and Institutional Histories (weekly) 16194

New England Antiquities Research Association (NEARA), c/o Ann Humphrey, 10 Loring Av, Kingston, MA 02364
T: (617) 585-4666
Founded: 1964; Members: 350
Gen Secr: Ann Humphrey
Focus: Antique
Periodical
Journal (weekly) 16195

New England Association of Schools and Colleges (NEASC), 15 High St, Winchester, MA 01890
T: (617) 729-6762
Founded: 1885; Members: 1600
Gen Secr: Richard J. Bradley
Focus: Educ
Periodicals
Membership Directory (weekly)
Newsletter (3 times annually) 16196

New England Historic Genealogical Society (NEHGS), 99-101 Newbury St, Boston, MA 02116
T: (617) 536-5740
Founded: 1845; Members: 13000
Gen Secr: Ralph J. Crandall
Focus: Genealogy
Periodical
New England Historical and Genealogical Register (weekly) 16197

New World Foundation (NWF), 100 E 85 St, New York, NY 10028
T: (212) 249-1023
Founded: 1954
Pres: Colin Greer
Focus: Soc Sci 16198

New York Academy of Sciences (NYAS), 2 E 63 St, New York, NY 10021
T: (212) 838-0230; Fax: (212) 888-2894
Founded: 1817; Members: 40000
Gen Secr: Rodney W. Nichols
Focus: Sci 16199

New York Browning Society (NYBS), POB 2911, New York, NY 10185
T: (215) 667-5941
Founded: 1907; Members: 65
Pres: Richard S. Kennedy
Focus: Lit
Periodicals
Bulletin (weekly)
Directory (biennial) 16200

New York C.S. Lewis Society (NYCSLS), c/o James Como, York College, Jamaica, NY 11451
T: (718) 262-2400
Founded: 1969; Members: 600
Gen Secr: James Como
Focus: Lit 16201

New York Drama Critics Circle (NYDCC), c/o Michael Kuchwara, Associated Press, 50 Rockefeller Plaza, New York, NY 10163
T: (212) 621-1841
Founded: 1935; Members: 23
Gen Secr: Michael Kuchwara
Focus: Perf Arts; Journalism 16202

New York Financial Writers' Association (NYFWA), POB 21, Syosset, NY 11791
T: (516) 921-7766; Fax: (516) 921-5762
Founded: 1938; Members: 390
Gen Secr: Joyce Spartonos
Focus: Journalism; Finance
Periodical
Directory (weekly) 16203

New York Genealogical and Biographical Society (NYGBS), 122 E 58 St, New York, NY 10022-1939
T: (212) 755-8532
Founded: 1869; Members: 1500
Gen Secr: William P. Johns
Focus: Genealogy
Periodical
New York Genealogical and Biographical Record (weekly) 16204

New York Pigment Club (NYPC), c/o Henry A. Souverein, BASF Corporation, 1255 Broad St, Clifton, NJ 07015
T: (201) 365-3400
Founded: 1953; Members: 100
Pres: Henry A. Souverein
Focus: Chem
Periodical
Membership Roster (biennial) 16205

Nieman Foundation (NF), c/o Walter Lippmann House, Harvard University, 1 Francis Av, Cambridge, MA 02138
T: (617) 495-2237
Founded: 1938; Members: 910
Focus: Ling 16206

Noah Worcester Dermatological Society (NWDS), 9500 Kenwood Rd, Cincinnati, OH 45242
T: (513) 891-8045
Founded: 1958; Members: 170
Gen Secr: K. William Kitzmiller
Focus: Derm
Periodical
Membership Roster (weekly) 16207

Nockian Society (NS), 42 Leathers St, Fort Mitchell, KY 41017
T: (606) 341-4841
Founded: 1963; Members: 750
Pres: Robert M. Thornton
Focus: Lit 16208

North American Academy of Ecumenists (NAAE), c/o Ernest R. Falardeau, 1818 Coal Pl SE, Albuquerque, NM 87106-4095
T: (505) 242-3462; Fax: (505) 256-0071
Founded: 1967; Members: 375
Pres: Ernest R. Falardeau
Focus: Rel & Theol 16209

North American Academy of Muscoloskeletal Medicine (NAAMM), 7611 Elmwood Av, Ste 202, Middleton, WI 53562
T: (800) 992-2063; Fax: (608) 839-5122
Founded: 1965; Members: 350
Gen Secr: Barbara Johns
Focus: Med
Periodicals
NAAMM Membership Roster (weekly)
NAAMM Newsletter (weekly) 16210

North American Association for Environmental Education (NAEE), 1255 23 St NW, Ste 400, Washington, DC 20037
T: (202) 467-8754; Fax: (202) 862-1947
Founded: 1971; Members: 1600
Gen Secr: Edward McCrea
Focus: Educ
Periodicals
The Environmental Communicator (weekly)
Monograph Series (1-2 times annually) . 16211

North American Association of Professors of Christian Education (NAPCE), 850 N Grove Av, Ste C, Elgin, IL 60120
T: (708) 741-2400; Fax: (708) 741-0595
Founded: 1947; Members: 250
Gen Secr: Dr. Dennis Williams
Focus: Educ
Periodicals
Christian Education Journal (weekly)
Newsletter (weekly) 16212

North American Association of Summer Sessions (NAASS), c/o Michael U. Nelson, 11728 Summerhaven Dr, Saint Louis, MO 63146
T: (314) 872-8406
Founded: 1964; Members: 425
Gen Secr: Michael U. Nelson
Focus: Educ 16213

North American Clinical Dermatologic Society (NACDS), c/o John W. White, Mayo Clinic, 4500 San Pablo Rd, Jacksonville, FL 32082
T: (908) 223-2000
Founded: 1959; Members: 180
Gen Secr: John W. White
Focus: Derm

Periodicals
Cutis (weekly)
Program (weekly) 16214

North American Conference on British Studies, c/o Prof. George Behlmer, Dept of History, University of Washington, Seattle, WA 98195
T: (206) 543-5790; Fax: (206) 543-9451
Founded: 1951; Members: 850
Gen Secr: Prof. George Behlmer
Focus: Ethnology; Hist
Periodicals
Albion (weekly)
Journal of British Studies (weekly) . . . 16215

North American Dostoevsky Society (NADS), c/o Roger Andersen, Dept of Russian and Eastern Studies, University of Kentucky, 1055 Patterson Office Tower, Lexington, KY 40506-0027
T: (606) 257-3761
Founded: 1970; Members: 120
Pres: Roger Andersen
Focus: Lit 16216

North American Electric Reliability Council (NERC), 101 College Rd E, Princeton, NJ 08540-6601
T: (609) 452-8060; Fax: (609) 452-9550
Founded: 1968; Members: 9
Pres: Michael R. Gent
Focus: Electric Eng
Periodicals
Annual Report (weekly)
Electricity Supply and Demand (weekly)
Reliability Assessment (weekly) 16217

North American Medical/Dental Association (NAMDA), POB 1982, Newport Beach, CA 92663
T: (714) 642-7689
Founded: 1968; Members: 10000
Pres: Dorene M. Christensen
Focus: Med; Dent 16218

North American Mycological Association (NAMA), 3556 Oakwood St, Ann Arbor, MI 48104-5213
T: (313) 971-2552
Founded: 1959; Members: 1700
Gen Secr: Dr. Kenneth W. Cochran
Focus: Botany, Specific
Periodicals
McIlvainea (weekly)
The Mycophile (weekly) 16219

North American Primary Care Research Group (NAPCRG), c/o Medical College of Virginia, POB 251, Richmond, VA 23298
T: (804) 786-9625; Fax: (804) 786-5856
Founded: 1972; Members: 500
Gen Secr: Robert B. Williams
Focus: Med 16220

North American Simulation and Gaming Association (NASAGA), c/o Dr. John del Regato, Pentathlon Institute, POB 20590, Indianapolis, IN 20590
T: (317) 782-1553
Founded: 1974; Members: 200
Gen Secr: Dr. John del Regato
Focus: Educ
Periodicals
Conference Program (weekly)
Newsletter 16221

North American Society for Pediatric Gastroenterology and Nutrition (NASPGN), c/o Judith M. Sondheimer, Children's Hospital, 1056 E 19 Av, Denver, CO 80218
T: (303) 861-6669
Founded: 1970; Members: 350
Gen Secr: Judith M. Sondheimer
Focus: Pediatrics; Gastroenter
Periodicals
NASPG Membership Directory (weekly)
Newsletter of the NASPG (4-6 times annually)
. . . . 16222

North American Society for Sport History (NASSH), c/o Pennsylvania State University, 101 White Bldg, University Park, PA 16802
T: (814) 865-2416
Founded: 1972; Members: 950
Gen Secr: Ronald A. Smith
Focus: Sports; Cultur Hist
Periodicals
Journal of Sport History (3 times annually)
Newsletter (weekly)
Proceedings (weekly) 16223

North American Society of Adlerian Psychology (NASAP), 202 S State St, Ste 1212, Chicago, IL 60604
T: (312) 939-0834
Founded: 1951; Members: 1200
Gen Secr: Neva L. Hefner
Focus: Psych
Periodicals
Calendar-Newsletter (weekly)
Individual Psychology: The Journal of Adlerian Theory, Research & Practice (weekly)
Membership List (biennial) 16224

North American Society of Pacing and Electrophysiology (NASPE), 377 Elliot St, Newton Upper Falls, MA 02164
T: (617) 244-7300; Fax: (617) 244-3920
Founded: 1979; Members: 1600
Gen Secr: Carol J. McGlinchey
Focus: Physiology; Electric Eng
Periodicals
NASPE News (weekly)
PACE (weekly) 16225

North American Thermal Analysis Society (NATAS), c/o Jeanne Taborsky, Nova Pharmaceutical Co, 6200 Freeport Center, Buffalo, NY 21224
T: (410) 558-9522; Fax: (410) 522-7000
Founded: 1968; Members: 750
Gen Secr: Jeanne Taborsky
Focus: Energy
Periodicals
Conference Proceedings (weekly)
Membership Directory (weekly)
Notes (weekly) 16226

North American Vexillological Association (NAVA), 1977 N Olden Av, Ste 225, Trenton, NJ 08618
T: (408) 295-1425
Founded: 1967; Members: 400
Pres: S. Guenter
Focus: Hist
Periodical
NAVA News (weekly) 16227

North Central Association of Colleges and Schools (NCACS), c/o Commission on Schools, Arizona State University, Tempe, AZ 85287-3011
T: (602) 965-8700; Fax: (602) 965-9423
Founded: 1895; Members: 8200
Gen Secr: Dr. Kenneth F. Gose
Focus: Educ
Periodical
NCA Quarterly (weekly) 16228

North Central Conference on Summer Schools (NCCSS), c/o Dr. Roger Swanson, Summer Sessions, University of Wisconsin at River Falls, River Falls, WI 54022
T: (715) 425-3256
Founded: 1949; Members: 150
Gen Secr: Dr. Roger Swanson
Focus: Educ 16229

Northeast Conference on the Teaching of Foreign Languages (NEC), 200 Twin Oaks Terrace, South Burlington, VT 05403
T: (802) 863-9939; Fax: (802) 863-0475
Founded: 1954; Members: 39700
Gen Secr: Elizabeth Holekamp
Focus: Ling; Educ
Periodicals
Conference Program (weekly)
Newsletter (weekly)
Reports (weekly) 16230

Northern Libraries Colloquy (NLC), c/o Nancy Lesh, University of Alaska at Anchorage Library, 3211 Providence Dr, Anchorage, AK 99509
T: (907) 786-1877
Founded: 1970
Gen Secr: Nancy Lesh
Focus: Libraries & Bk Sci
Periodicals
Directory of Polar and Regions Library Resources
Northern Libraries Bulletin 16231

Northwest Association of Schools and Colleges (NASC), c/o Boise University, 528 Education Bldg, 1910 University Dr, Boise, ID 83725
T: (208) 385-1596
Founded: 1917; Members: 1385
Gen Secr: David Steadman
Focus: Educ
Periodicals
Directory of Accredited and Affiliated Institutions (weekly)
Newsletter (weekly) 16232

Norwegian-American Historical Association (NAHA), c/o Saint Olaf College, 1510 Saint Olaf Av, Northfield, MN 55057-1097
T: (507) 646-3221
Founded: 1925; Members: 1700
Gen Secr: Lloyd Hustvedt
Focus: Hist
Periodicals
Norwegian-American Studies (weekly)
Travel and Description Series (weekly) . 16233

NTL Institute for Applied Behavioral Science, 1240 N Pitt St, Ste 100, Alexandria, VA 22314-1403
T: (703) 548-1500; Fax: (703) 684-1256
Founded: 1947; Members: 450
Pres: L. Joseph
Focus: Behav Sci
Periodical
The Journal of Applied Behavioral Science
. . . . 16234

Numerical Control Society/AIM Tech, 67 Alexander Dr, POB 12277, Research Triangle Park, NC 27709
T: (815) 399-8700; Fax: (815) 399-7279
Founded: 1962; Members: 650
Gen Secr: Mike Mastroianni
Focus: Eng
Periodical
Integrated Manufacturing 16235

Numismatic Literary Guild (NLG), 12 Abbington Terrace, Glen Rock, NJ 07452
T: (201) 612-0482; Fax: (201) 612-9581
Founded: 1968; Members: 360
Gen Secr: Ed Reiter
Focus: Numismatics
Periodical
Newsletter (weekly) 16236

Nutrition Institute of America (NIA), 200 W 86 St, Ste 17A, New York, NY 10024
T: (212) 799-2234
Founded: 1974; Members: 6
Pres: Leonard L. Steinman
Focus: Nutrition 16237

Nuttall Ornithological Club (NOC), c/o Museum of Comparative Zoology, Harvard University, Cambridge, MA 02138
T: (617) 495-2471
Founded: 1873; Members: 130
Gen Secr: R.A. Paynter
Focus: Ornithology 16238

Oak Ridge Associated Universities (ORAU), POB 117, Oak Ridge, TN 37831-0117
T: (615) 576-3146
Founded: 1946; Members: 59
Pres: Dr. J.M. Veigel
Focus: Energy
Periodical
Annual Report (weekly) 16239

Ocean Education Project (OEP), c/o Sam Levering, Rte 2, POB 308, Ararat, VA 24053
T: (703) 755-3592
Founded: 1973
Gen Secr: Sam Levering
Focus: Educ 16240

Oceanic Educational Foundation (OEF), 3710 Whispering Ln, Falls Church, VA 22041
T: (703) 256-0279
Founded: 1970; Members: 6000
Pres: G.M. Slonim
Focus: Educ 16241

Oceanic Society (OS), 218 D St SE, Washington, DC 20003
T: (202) 544-2600
Founded: 1969; Members: 40000
Pres: Clifton Curtis
Focus: Oceanography
Periodical
Newsletter 16242

Oceanic Society Expeditions (OSE), Fort Mason Center, Bldg E, San Francisco, CA 94123
T: (415) 441-1106; Fax: (415) 474-3395
Founded: 1972; Members: 70000
Pres: Birgit Winning
Focus: Oceanography
Periodicals
Oceanic Expeditions Brochure (weekly)
Oceans Magazine (weekly) 16243

Odyssey Institute Corporation (OCI), 5 Hedley Farms Rd, Westport, CT 06880
T: (203) 255-4198; Fax: (203) 255-3006
Founded: 1975
Pres: J. Densen-Gerber
Focus: Rehabil
Periodical
Odyssey Journal (weekly) 16244

Office for Advancement of Public Black Colleges of the National Association of State Universities and Land Grant Colleges (OAPBC), 1 Dupont Circle NW, Ste 710, Washington, DC 20036-1191
T: (202) 778-0818
Founded: 1968; Members: 35
Gen Secr: Dr. N. Joyce Payne
Focus: Educ 16245

Office of International Education of the American Council on Education, 1 Dupont Circle NW, Ste 800, Washington, DC 20036
T: (202) 939-9313; Fax: (202) 833-4760
Founded: 1979
Gen Secr: Barbara Turlington
Focus: Educ 16246

Office of Management Services (OMS), 1527 New Hampshire Av NW, Washington, DC 20036
T: (202) 232-8656; Fax: (202) 462-7846
Founded: 1970; Members: 119
Gen Secr: Susan Jurow
Focus: Libraries & Bk Sci
Periodical
SPEC Kit (10 times annually) 16247

Office Systems Research Association (OSRA), c/o Administration Office System, Southwest Missouri State University, 901 S National Av, Springfield, MO 65804-0089
T: (417) 836-5616; Fax: (417) 836-6337
Founded: 1980; Members: 350
Gen Secr: Heidi Perreault
Focus: Business Admin 16248

Oncology Nursing Society (ONS), 501 Holiday Dr, Pittsburgh, PA 15220
T: (412) 921-7373; Fax: (412) 921-6565
Founded: 1975; Members: 22000
Pres: Carol Curtis
Focus: Cell Biol & Cancer Res
Periodicals
Oncology Nursing Forum (weekly)
Proceedings of Annual Congress 16249

Open Space Institute (OSI), 145 Main St, New York, NY 10562
T: (914) 762-4630
Founded: 1974
Pres: Christopher J. Elliman
Focus: Ecology
Periodical
Farmland Preservation Directory 16250

Opera America, 777 14 St NW, Ste 520, Washington, DC 20005
T: (202) 347-9262; Fax: (202) 393-0735
Founded: 1970; Members: 800
Pres: Marc A. Scorca
Focus: Music
Periodicals
Bulletin (weekly)
Intercompany Announcements (10 times annually)
Membership Directory (weekly)
Repertoire Survey (weekly) 16251

Operation Enterprise (OE), c/o American Management Association, POB 88, Hamilton, NY 13346
T: (315) 824-2000; Fax: (315) 824-2000
Founded: 1963
Gen Secr: Andrew J. Mason
Focus: Business Admin
Periodical
Operation Enterprise Newsletter (weekly) . 16252

Operations Research Society of America (ORSA), 1314 Guilford Av, Baltimore, MD 21202
T: (410) 528-4146; Fax: (410) 528-8556
Founded: 1952; Members: 8300
Gen Secr: Patricia H. Morris
Focus: Business Admin
Periodicals
Interfaces (weekly)
Mathematics of Operations Research (weekly)
Operations Research (weekly)
ORSA Journal on Computing (weekly) . 16253

Ophthalmic Research Institute (ORI), 433 East West Hwy, Ste 1117, Bethesda, MD 20814
T: (301) 718-6524; Fax: (301) 656-0989
Founded: 1972; Members: 84
Gen Secr: Dr. Norman E. Wallis
Focus: Ophthal 16254

Optical Society of America (OSA), 2010 Massachusetts Av NW, Washington, DC 20036
T: (202) 223-8130; Fax: (202) 223-1096
Founded: 1916; Members: 11000
Gen Secr: Jarus W. Quinn
Focus: Optics
Periodicals
Applied Optics (36 times annually)
Atmospheric Optics (weekly)
Journal of Lightwave Technology (weekly)
Journal of the Optical Society of America A (weekly)
Journal of the Optical Society of America B (weekly)
Optics and Spectroscpy (weekly)
Optics Letters (weekly)
Optics News (weekly) 16255

Opticians Association of America (OAA), 10341 Democracy Ln, Fairfax, VA 22030
T: (703) 691-8355; Fax: (703) 691-3929
Founded: 1926; Members: 700
Gen Secr: Paul Houghland
Focus: Optics
Periodicals
Guild Quarterly (weekly)
OAA News (8 times annually)
State Leadership Bulletin (weekly) . . . 16256

Optometric Historical Society (OHS), 243 N Lindbergh Blvd, Saint Louis, MO 63141
T: (314) 991-4100; Fax: (314) 991-4101
Founded: 1969
Gen Secr: T. David Williams
Focus: Ophthal
Periodical
Newsletter (weekly) 16257

Oral History Association (OHA), 1093 Broxton Av, Los Angeles, CA 90024
T: (310) 825-0597; Fax: (310) 206-1864
Founded: 1966; Members: 1400
Gen Secr: Richard Candida Smith
Focus: Hist
Periodicals
Annual Report and Membership Directory (weekly)
Newsletter (weekly)
Oral History Review (weekly) 16258

Order of the Indian Wars (OIW), POB 7401, Little Rock, AR 72217
T: (501) 225-3996; Fax: (501) 225-5167
Founded: 1979; Members: 750
Pres: Jerry L. Russell
Focus: Hist
Periodicals
Communique (weekly)
Journal of the Order of Indian Wars (weekly)
. . . . 16259

Organ Historical Society (OHS), POB 26811, Richmond, VA 23261
T: (804) 353-9226; Fax: (804) 353-9266
Founded: 1956; Members: 2900
Gen Secr: William T. Van Pelt
Focus: Music
Periodicals
Organ Handbook (weekly)
The Tracker (weekly) 16260

Organizational Behavior Teaching Society (OBTS), c/o College of Business Administration, University of Oklahoma, 4 Adams Hall, Norman, OK 73019
T: (405) 325-2931; Fax: (405) 325-7688
Founded: 1973

Gen Secr: Peter Frost
Focus: Psych; Educ 16261

Organization Development Institute (ODI), 11234 Walnut Ridge Rd, Chesterland, OH 44026
T: (216) 461-4333; Fax: (216) 729-9319
Founded: 1968; Members: 488
Gen Secr: Dr. Donald W. Cole
Focus: Business Admin
Periodicals
The Organization Development Journal (weekly)
Organizations and Change (weekly) . . 16262

Organization for Flora Neotropica (OFN), c/o New York Botanical Garden, Bronx, NY 10458-5126
T: (718) 220-8742; Fax: (718) 220-6504
Founded: 1964; Members: 150
Gen Secr: Dr. Scott A. Mori
Focus: Botany
Periodical
Flora Neotropica (5 times annually) . . 16263

Organization for Tropical Studies (OTS), c/o North American Office, POB DM, Duke Station, Durham, NC 27706
T: (919) 684-5774; Fax: (919) 684-5661
Founded: 1963; Members: 50
Gen Secr: Donald E. Stone
Focus: Bio 16264

Organization of American Historians (OAH), 112 N Bryan St, Bloomington, IN 47408
T: (812) 855-7311
Founded: 1907; Members: 12000
Gen Secr: A.A. Jones
Focus: Hist 16265

Organization of American Kodaly Educators (OAKE), c/o Music Dept, Nicholls State University, POB 2017, Thibodaux, LA 70310
T: (504) 448-4602; Fax: (504) 448-4927
Members: 1800
Gen Secr: James Fields
Focus: Music; Educ
Periodical
The Kodaly Envoy (weekly) 16266

Organization of Professional Acting Coaches and Teachers (OPACT), 3968 Eureka Dr, Studio City, CA 91604
T: (213) 877-4988; Fax: (213) 877-4988
Founded: 1980; Members: 12
Pres: Lilyan Chauvin
Focus: Educ 16267

Orthodontic Education and Research Foundation (OERF), 3556 Caroline St, Saint Louis, MO 63104
T: (314) 577-8189
Founded: 1957; Members: 500
Gen Secr: Peter G. Sotiropoulos
Focus: Dent; Educ 16268

Orthodox Theological Society in America (OTSA), c/o Thomas Fitzgerald, Holy Cross School of Theology, 50 Goddard Av, Brookline, MA 02146
T: (617) 731-3500
Founded: 1968; Members: 95
Pres: Thomas Fitzgerald
Focus: Rel & Theol
Periodicals
Bulletin (weekly)
Membership Directory
Presidents Report (weekly) 16269

Orthopedic Foundation for Animals (OFA), 2300 Nifong Blvd, Columbia, MO 65201
T: (314) 442-0418
Founded: 1967
Gen Secr: Dr. E.A. Corley
Focus: Vet Med 16270

Orthopedic Research Society (ORS), 6300 N River Rd, Ste 727, Rosemont, IL 60018-4238
T: (708) 698-1625; Fax: (708) 823-0536
Founded: 1954; Members: 1500
Gen Secr: Karen Jared
Focus: Orthopedics
Periodicals
Journal of Orthopedic Research (weekly)
Proceedings of Annual Meeting 16271

Orton Dyslexia Society (ODS), 8600 LaSalle Rd, Ste 382, Baltimore, MD 21286-2044
T: (410) 296-0232
Founded: 1949; Members: 9000
Gen Secr: Rosemary F. Bowler
Focus: Med 16272

Osteopathic College of Ophthalmology and Otorhinolaryngology (OCOO), c/o Dayton District Academy of Osteopathic Medicine, 405 Grand Av, Dayton, OH 45405
T: (513) 226-3438
Founded: 1916; Members: 385
Gen Secr: George Saul
Focus: Otorhinolaryngology; Ophthal
Periodicals
Newsletter (weekly)
Photo Roster 16273

Otosclerosis Study Group (OSG), 6465 S Yale, Ste 202, Tulsa, OK 74136
T: (918) 481-2753
Founded: 1947; Members: 128
Gen Secr: Roger Wehrs
Focus: Otorhinolaryngology 16274

Outdoor Education Association (OEA), c/o Dr. Edward J. Ambry, 143 Fox Hill Rd, Denville, NJ 07834
T: (201) 627-7214
Founded: 1940; Members: 200
Pres: Dr. Edward J. Ambry
Focus: Educ
Periodical
Extending Education 16275

Outdoor Ethics Guild (OEG), c/o Bruce Bandurski, General Delivery, Bucks Harbor, ME 04618
Founded: 1967
Pres: Bruce Bandurski
Focus: Ecology 16276

Outer Critics Circle (OCC), c/o Marjorie Gunner, 101 W 57 St, New York, NY 10019
T: (212) 765-8557
Founded: 1950; Members: 70
Pres: Marjorie Gunner
Focus: Lit; Perf Arts; Journalism 16277

Overseas Education Association (OEA), 1201 16 St NW, Washington, DC 20036
T: (202) 822-7850
Founded: 1956; Members: 6700
Gen Secr: Ronald R. Austin
Focus: Educ
Periodicals
Journal (weekly)
OEA Leader
OEA News 16278

Pacific Arts Association (PAA), 900 S Beretania St, Honolulu, HI 96814
T: (808) 538-3693; Fax: (808) 521-6591
Founded: 1974; Members: 300
Gen Secr: Roger G. Rose
Focus: Arts 16279

Pacific Dermatologic Association (PDA), POB 4014, Schaumburg, IL 60204
T: (312) 330-0230; Fax: (312) 869-4382
Founded: 1948; Members: 1200
Pres: Faye D. Arundell
Focus: Derm
Periodical
Transactions (weekly) 16280

Pacific Rocket Society (PRS), 1825 N Oxnard Blvd, Ste 24, Oxnard, CA 93030
T: (805) 983-1947
Founded: 1946; Members: 100
Pres: George Morgan
Focus: Eng 16281

Pacific Science Association (PSA), POB 17801, Honolulu, HI 96817
T: (808) 848-4139; Fax: (808) 841-8968
Founded: 1920; Members: 1500
Gen Secr: L.G. Eldredge
Focus: Sci
Periodical
Information Bulletin (weekly) 16282

Pacific Seabird Group (PSG), c/o Savannah River Ecology Laboratory, PO Drawer E, Aiken, SC 29801
T: (803) 725-2475
Founded: 1972; Members: 450
Pres: Dr. D. Michael Hay
Focus: Ornithology
Periodicals
Bulletin (weekly)
Membership Directory 16283

Pacific Studies Center (PSC), 222B View St, Mountain View, CA 94041
T: (415) 969-1545; Fax: (415) 968-1126
Founded: 1969
Gen Secr: Leonard M. Siegel
Focus: Poli Sci
Periodicals
California Military Monitor (weekly)
Global Electronics (weekly) 16284

Packaging Education Foundation (PEF), 481 Carlisle Dr, Herndon, VA 22070
T: (703) 318-8975; Fax: (703) 318-0310
Founded: 1957
Gen Secr: William C. Pflaum
Focus: Materials Sci
Periodicals
Leader Newsletter (weekly)
PEF Leader (weekly) 16285

Paleontological Research Institution (PRI), 1259 Trumansburg Rd, Ithaca, NY 14850
T: (607) 273-6623; Fax: (607) 273-6620
Founded: 1932; Members: 700
Gen Secr: Dr. Warren D. Allmon
Focus: Paleontology
Periodicals
Bulletin of American Paleontology (weekly)
Paleontographica Americana 16286

Paleontological Society (PS), c/o Donald L. Wolberg, New Mexico Bureau of Mines and Mineral Resources, Socorro, NM 87801
T: (505) 835-5140
Founded: 1908; Members: 1725
Gen Secr: Donald L. Wolberg
Focus: Paleontology
Periodicals
Journal of Paleontology (6 times annually)
Memoirs in Paleontology (weekly)
Paleobiology (weekly) 16287

Pan American Association of Biochemical Societies (PAABS), c/o Dr. Marino Martinez-Carrion, School of Basic Life Sciences, University of Missouri at Kansas City, 109 Biological Sciences Bldg, Kansas City, MO 64110
T: (816) 235-2249; Fax: (816) 235-5158
Founded: 1969; Members: 14
Gen Secr: Dr. Marino Martinez-Carrion
Focus: Biochem 16288

Pan-American Association of Ophthalmology (PAAO), 1301 S Bowen Rd, Ste 365, Arlington, TX 76013
T: (817) 265-2831; Fax: (817) 275-3961
Founded: 1939; Members: 8000
Gen Secr: Teresa J. Bradshaw
Focus: Ophthal
Periodical
Insights (weekly) 16289

Pan-American Biodeterioration Society (PABS), c/o Charles O'Rear, Dept of Forensic Sciences, George Washington University, Washington, DC 20052
T: (202) 994-7319; Fax: (202) 994-0458
Founded: 1985; Members: 100
Gen Secr: Charles O'Rear
Focus: Bio
Periodicals
Conference Proceedings
Newsletter (weekly) 16290

Panamerican Cultural Circle (PCC), 16 Malvern Pl, Verona, NJ 07044
T: (201) 239-3125
Founded: 1963; Members: 800
Gen Secr: Elio Alba-Buffill
Focus: Hist; Lit
Periodicals
Circulo Poético (weekly)
Circulo: Revista de Cultura (weekly) . . 16291

Pan American Health Organization (PAHO), 525 23 St NW, Washington, DC 20037
T: (202) 861-3200
Founded: 1902; Members: 38
Gen Secr: Dr. Carlyle Guerra de Macedo
Focus: Public Health
Periodicals
Bulletin of the Pan American Health Organization (weekly)
Disaster Preparedness in the Americas (weekly)
Epidemiological Bulletin (weekly)
EPI Newsletter (weekly) 16292

Pan-American Medical Association (PAMA), c/o F.C. Fenig, 745 Fifth Av, Ste 403, New York, NY 10151
T: (212) 753-6033
Founded: 1925; Members: 6000
Gen Secr: F.C. Fenig
Focus: Med 16293

Pan-Pacific Surgical Association (PPSA), 733 Bishop St, Honolulu, HI 96813
T: (808) 523-8978; Fax: (808) 599-3991
Founded: 1929; Members: 2716
Gen Secr: Gayle Yosheda
Focus: Surgery
Periodical
Newsletter (weekly) 16294

Parapsychological Association (PA), POB 12236, Research Triangle Park, NC 27709
T: (919) 688-8241; Fax: (919) 683-4338
Founded: 1957; Members: 280
Gen Secr: Dr. Richard Broughton
Focus: Parapsych
Periodical
Proceedings (weekly) 16295

Parker Chiropractic Resource Foundation (PCRF), POB 40444, Fort Worth, TX 76140
T: (817) 293-6444; Fax: (817) 293-0776
Founded: 1951; Members: 30000
Pres: Dr. W. Karl Parker
Focus: Med 16296

Parkinson's Educational Program – U.S.A. (PEP/USA), 3900 Birch St, Newport Beach, CA 92660
T: (714) 250-2975; Fax: (714) 250-8530
Founded: 1982
Gen Secr: Charlotte Jayne
Focus: Educ
Periodicals
PEP Exchange (weekly)
Physicians Referral List (weekly)
Speakers Forum 16297

Passaic River Coalition (PRC), 246 Madisonville Rd, Basking Ridge, NJ 07920
T: (201) 766-7550; Fax: (201) 766-7550
Founded: 1971; Members: 2500
Gen Secr: Ella F. Filippone
Focus: Ecology
Periodicals
Goals & Strategies (weekly)
Groundwater News (weekly)
Vibes from the Libe (weekly) 16298

Pattern Recognition Society (PRS), c/o National Biomedical Research Foundation, Georgetown University Medical Center, 3900 Reservoir Rd NW, Washington, DC 20007
T: (202) 687-2121
Founded: 1966; Members: 550
Gen Secr: Robert S. Ledley
Focus: Sci
Periodical
Pattern Recognition (weekly) 16299

Peace Science Society (PSS), c/o Dept of Political Science, SUNY at Binghamton, POB 6000, Binghamton, NY 13902-6000
T: (607) 777-4398; Fax: (607) 777-4000
Founded: 1963; Members: 500
Gen Secr: Prof. Stuart A. Bremer
Focus: Prom Peace
Periodical
Conflict Management and Peace Science (weekly) 16300

PEN American Center (PENAC), 568 Broadway, New York, NY 10012
T: (212) 334-1660; Fax: (212) 334-2181
Founded: 1921; Members: 2500
Gen Secr: Karen Kennerly
Focus: Lit
Periodicals
Grants and Awards Available to American Writers (biennial)
Newsletter (weekly) 16301

People United for Rural Education (PURE), RR, Box 41, Alden, IA 50006
T: (515) 855-4206
Founded: 1977; Members: 3500
Pres: Fred Erickson
Focus: Educ 16302

Percussive Arts Society (PAS), POB 25, Lawton, OK 73502
T: (405) 353-1455; Fax: (405) 353-1456
Founded: 1960; Members: 5500
Gen Secr: S. Beck
Focus: Music
Periodicals
Percussion News (weekly)
Percussive Notes (5 times annually)
Research Edit (weekly) 16303

Performing and Visual Arts Society (PAVAS), POB 102, Kinnelon, NJ 07405
T: (201) 838-5360
Founded: 1967; Members: 400
Pres: Paul De Francis
Focus: Arts 16304

Perlite Institute (PI), 88 New Dorp Plaza, Staten Island, NY 10306
T: (718) 351-5723; Fax: (718) 351-5725
Founded: 1949; Members: 65
Gen Secr: William C. Hall
Focus: Mining 16305

Permanent International Association of Navigation Congresses – United States Section (PIANC), c/o CECW-RK, 20 Massachusetts Av NW, Washington, DC 20314-1000
T: (202) 504-4312; Fax: (202) 272-8839
Founded: 1902; Members: 600
Gen Secr: A.K. Du Wayne
Focus: Navig
Periodical
Newsletter (weekly) 16306

Peruvian Heart Association (PHA), 38760 Northwood Dr, Wadsworth, IL 60083
T: (708) 249-1900; Fax: (708) 249-2772
Founded: 1967; Members: 400
Gen Secr: Luis Vasquez
Focus: Cardiol 16307

PFB Project, c/o Robert B. Fitzpatrick, 4801 Massachusetts Av NW, Ste 400, Washington, DC 20016-2087
T: (202) 364-8710
Founded: 1978
Focus: Derm 16308

P.G. Wodehouse Society (PGWS), 530 Georgina Av, Santa Monica, CA 90402
T: (310) 828-4492
Founded: 1979; Members: 48
Pres: Jeremy H. Thompson
Focus: Lit 16309

Philip C. Jessup International Law Moot Court Competition (PCJILMCC), c/o International Law Students Association, 2223 Massachusetts Av NW, Washington, DC 20008
T: (202) 265-4375; Fax: (202) 265-0386
Founded: 1959; Members: 1200
Gen Secr: Brett J. Lorenzen
Focus: Law
Periodicals
ILSA Journal of International Law (weekly)
Jessup Competition Compendium (weekly) 16310

Philip Jose Farmer Society (PJFS), 1901 Bittersweet Dr, Champaign, IL 61821-6371
T: (217) 356-5235
Founded: 1978; Members: 150
Gen Secr: George H. Scheetz
Focus: Lit 16311

Philosophical Research Society (PRS), 3910 Los Feliz Blvd, Los Angeles, CA 90027
T: (213) 663-2167; Fax: (213) 663-9443
Founded: 1934
Pres: A. Kunkin
Focus: Philos
Periodical
Journal (weekly) 16312

Philosophic Society for the Study of Sport (PSSS), c/o Prof. Janet M. Oussaty, Kean College of New Jersey, Union, NJ 07083
T: (908) 527-2101; Fax: (908) 355-5143
Founded: 1972; Members: 150
Gen Secr: Prof. Janet M. Oussaty
Focus: Philos; Sports
Periodical
Journal of the Philosophy of Sport (weekly) . . . 16313

Philosophy of Science Association (PSA), c/o Prof. Peter D. Asquith, Dept of Philosophy, Michigan State University, 503 S Kedzie Hall, East Lansing, MI 48824
T: (517) 353-9392
Founded: 1934; Members: 1000
Gen Secr: Prof. Peter D. Asquith
Focus: Philos
Periodical
Philosophy of Science (weekly) . . . 16314

Phoenix Society for Burn Survivors (PSBS), 11 Rust Hill Rd, Levittown, PA 19056
T: (215) 946-4788; Fax: (215) 946-4788
Founded: 1977; Members: 6500
Gen Secr: Alan J. Breslau
Focus: Rehabil . . . 16315

Photographic Art and Science Foundation (PASF), 111 Stratford, Des Plaines, IL 60016-2105
T: (708) 824-6855
Founded: 1965; Members: 375
Pres: Frederick Quellmalz
Focus: Photo . . . 16316

Physicians Forum (PF), 1507 53 St, Ste 155, Chicago, IL 60615
T: (312) 922-1968; Fax: (312) 633-6442
Founded: 1939; Members: 1000
Pres: Raymond Demers
Focus: Physiology
Periodicals
Newsletter (weekly)
Physicians Forum Bulletin (weekly) . . . 16317

Phytochemical Society of North America (PSNA), c/o Dr. Helen M. Haberman, Dept of Biological Sciences, Goucher College, Towson, MD 21204
T: (410) 337-6303; Fax: (410) 337-6123
Founded: 1960; Members: 405
Gen Secr: Dr. Helen M. Haberman
Focus: Botany
Periodicals
PSNA Newsletter (weekly)
Recent Advances in Phytochemistry (weekly) . . . 16318

Pierre Fauchard Academy (PFA), c/o Dr. Richard Kozal, 8021 W 79 St, Justice, IL 60458-1607
T: (708) 594-5884; Fax: (708) 496-1066
Founded: 1936; Members: 5000
Gen Secr: Dr. Richard Kozal
Focus: Dent
Periodicals
Dental World (weekly)
Membership Roster . . . 16319

Pilgrim Society (PS), c/o Pilgrim Hall Museum, 75 Court St, Plymouth, MA 02360
T: (508) 746-1620
Founded: 1820; Members: 750
Gen Secr: U. Ernest Buchner
Focus: Hist
Periodical
Pilgrim Society Notes (weekly) . . . 16320

Pioneer America Society (PAS), c/o Charles F. Calkins, Dept of Geography, Carroll College, Waukesha, WI 53181
T: (415) 524-7212
Founded: 1967; Members: 500
Gen Secr: Charles F. Calkins
Focus: Hist; Folklore
Periodicals
Material Culture (3 times annually)
Newsletter (weekly)
P.A.S.T.: Pioneer America Society Transactions (weekly) . . . 16321

Pirandello Society of America (PSA), c/o Dr. Anne Paolucci, Dept of English, Saint John's University, Jamaica, NY 11439
T: (718) 767-8380; Fax: (718) 767-8380
Founded: 1958; Members: 500
Pres: Dr. Anne Paolucci
Focus: Lit
Periodical
PSA (3-4 times annually) . . . 16322

Planetary Society (PS), 65 N Catalina Av, Pasadena, CA 91106
T: (818) 793-5100; Fax: (818) 793-5528
Founded: 1980; Members: 100000
Gen Secr: Dr. Louis Friedman
Focus: Astronomy
Periodical
Planetary Report (weekly) . . . 16323

Planned Parenthood Federation of America (PPFA), 810 Seventh Av, New York, NY 10019
T: (212) 541-7800; Fax: (212) 245-1845
Founded: 1916; Members: 169
Pres: David J. Andrews
Focus: Med; Sociology; Family Plan
Periodicals
Affiliates Directory (weekly)
Annual Report (weekly) . . . 16324

Plastics Institute of America (PIA), 277 Fairfield Rd, Ste 100, Fairfield, NJ 07004-1932
T: (201) 808-5950; Fax: (201) 808-5953
Founded: 1961; Members: 50
Gen Secr: Dr. William Sacks
Focus: Eng
Periodicals
Institute Report (weekly)
Pipeline (weekly) . . . 16325

Plastic Surgery Research Council (PSRC), c/o W. Thomas Lawrence, University of North Carolina, Wing D-Medical School, CB 7195, Chapel Hill, NC 27599
T: (919) 966-3080
Founded: 1955; Members: 232
Gen Secr: W. Thomas Lawrence
Focus: Surgery
Periodical
Meeting Abstracts (weekly) . . . 16326

Play Schools Association (PSA), 9 E 38 St, New York, NY 10016
T: (212) 725-6540; Fax: (212) 532-9674
Founded: 1917
Gen Secr: Joseph Corrado
Focus: Educ . . . 16327

Plywood Research Foundation (PRF), POB 11700, Tacoma, WA 98411
T: (206) 565-6600; Fax: (206) 565-1524
Founded: 1944; Members: 178
Gen Secr: Marc J. Mullins
Focus: Materials Sci . . . 16328

Poe Foundation (PF), 1914-1916 Main St, Richmond, VA 23223
T: (804) 648-5523
Founded: 1921
Gen Secr: James R. Furqueron
Focus: Lit
Periodical
The Poe Messenger (weekly) . . . 16329

Poe Studies Association (PSA), c/o Prof. Dennis W. Eddings, English Dept, Western Oregon State College, Monmouth, OR 97361
Founded: 1971; Members: 206
Gen Secr: Prof. Dennis W. Eddings
Focus: Lit . . . 16330

Poetry Society of America (PSA), 15 Gramercy Park, New York, NY 10003
T: (212) 254-9628
Founded: 1910; Members: 2500
Gen Secr: Elise Paschen
Focus: Lit
Periodical
Newsletter (3 times annually) . . . 16331

Policy Studies Organization (PSO), c/o University of Illinois, 361 Lincoln Hall, 702 S Wright St, Urbana, IL 61801
T: (217) 359-8541; Fax: (217) 244-5712
Founded: 1971; Members: 2200
Gen Secr: Stuart Nagel
Focus: Poli Sci
Periodicals
Policy Studies Journal (weekly)
Policy Studies Review (weekly) . . . 16332

Polish American Historical Association (PAHA), 984 N Milwaukee Av, Chicago, IL 60622
T: (313) 384-3352
Founded: 1941; Members: 600
Focus: Hist
Periodicals
Newsletter (weekly)
Polish American Studies (weekly) . . . 16333

Polish Genealogical Society (PGS), 984 N Milwaukee Av, Chicago, IL 60622
T: (312) 384-3352
Founded: 1978; Members: 760
Pres: Stanley R. Schmidt
Focus: Genealogy
Periodical
PGS Newsletter (weekly) . . . 16334

Polish Institute of Arts and Sciences in America (PIASA), 208 E 30 St, New York, NY 10016
T: (212) 686-4164; Fax: (212) 545-1130
Founded: 1942; Members: 1500
Gen Secr: Dr. Thaddeus V. Gromada
Focus: Arts; Sci
Periodical
Polish Review (weekly) . . . 16335

Popular Culture Association (PCA), Popular Culture Center, Bowling Green State University, Bowling Green, OH 43403
T: (419) 372-7861; Fax: (419) 372-8095
Founded: 1969; Members: 2500
Gen Secr: Ray B. Browne
Focus: Hist
Periodical
Journal of Popular Culture (weekly) . . . 16336

Population Association of America (PAA), 1722 N St NW, Washington, DC 20036
T: (202) 429-0891; Fax: (202) 785-0146
Founded: 1931; Members: 2700
Gen Secr: Jen L. Suter
Focus: Sociology
Periodicals
Demography (weekly)
PAA Affairs (weekly)
Population Index (weekly) . . . 16337

Population Council (PC), 1 Dag Hammarskjold Plaza, New York, NY 10017
T: (212) 339-0500; Fax: (212) 755-6052
Founded: 1952
Pres: Margaret Catley-Carlson
Focus: Sociology
Periodicals
Annual Report (weekly)
Population and Development Review (weekly)
Studies in Family Planning (weekly) . . . 16338

Population Crisis Committee (PCC), 1120 NW 19 St, Ste 550, Washington, DC 20036
T: (202) 659-1833; Fax: (202) 293-1795
Founded: 1965; Members: 1500
Pres: J. Joseph Speidel
Focus: Sociology . . . 16339

Population Reference Bureau (PRB), 1875 Connecticut Av NW, Ste 520, Washington, DC 20009
T: (202) 483-1100; Fax: (202) 328-3937
Founded: 1929; Members: 4200
Pres: Dr. Thomas W. Merrick
Focus: Sociology
Periodicals
Interchange (weekly)
Intercom (weekly)
Population Bulletin (weekly)
US Population Data Sheet (weekly) . . . 16340

Postal History Society (PHS), 8207 Daren Ct, Pikesville, MD 21208
T: (410) 653-0665
Founded: 1951; Members: 600
Gen Secr: Kalman V. Illyefawi
Focus: Cultur Hist
Periodical
Postal History Journal (3 times annually) 16341

Postgraduate Center for Mental Health (PCMH), 124 E 28 St, New York, NY 10016
T: (212) 689-7700; Fax: (212) 576-4194
Founded: 1945
Pres: David A. Glazer
Focus: Psychiatry
Periodicals
Dynamic Psychotherapy Journal (weekly)
New Outlook (weekly)
Pathways (weekly) . . . 16342

Potash and Phosphate Institute (PPI), 2801 Buford Hwy NE, Atlanta, GA 30329
T: (404) 634-4274; Fax: (404) 636-8733
Founded: 1935; Members: 18
Pres: D.W. Dibb
Focus: Agri
Periodicals
Better Crops International (weekly)
Better Crops with Plant Food (weekly) . 16343

Potato Association of America (PAA), c/o David Curnen, Rte 1, Box 115, Hancock, kWI 54943
T: (715) 249-5712; Fax: (715) 249-5850
Founded: 1913; Members: 1100
Gen Secr: David Curwen
Focus: Food; Agri . . . 16344

Poultry Science Association (PSA), 309 W Clark St, Champaign, IL 61820
T: (217) 356-3182; Fax: (217) 398-4119
Founded: 1908; Members: 1600
Gen Secr: C.D. Johnson
Focus: Zoology
Periodical
Poultry Science (weekly) . . . 16345

Powys Society of North America (PSNA), c/o English Dept, Valparaiso University, Valparaiso, IN 46383
T: (219) 464-5069
Founded: 1983; Members: 150
Gen Secr: Richard Maxwell
Focus: Lit . . . 16346

Practical Allergy Research Foundation (PARF), POB 60, Buffalo, NY 14223
T: (716) 875-5578; Fax: (716) 877-8475
Pres: Doris J. Rapp
Focus: Med . . . 16347

Practising Law Institute (PLI), 810 Seventh Av, New York, NY 10019
T: (212) 765-5700; Fax: (212) 265-4742
Founded: 1933; Members: 50000
Gen Secr: Victor J. Rubino
Focus: Law
Periodical
Business Accounting for Lawyers Newsletter . . . 16348

Precision Chiropractic Research Society (PCRS), 1412 Alta Mesa Way, Brea, CA 92621
T: (213) 694-4181
Founded: 1976; Members: 200
Pres: Dr. A.C. Fulkerson
Focus: Orthopedics . . . 16349

Precision Measurements Association (PMA), 3685 Motor Av, Ste 240, Los Angeles, CA 90034
T: (310) 287-0941; Fax: (310) 287-1851
Founded: 1959; Members: 600
Gen Secr: Robert Myers
Focus: Eng
Periodical
Newsnotes (weekly) . . . 16350

Presbyterian Historical Society (PHS), 425 Lombard St, Philadelphia, PA 19147
T: (215) 627-1852; Fax: (215) 627-0509
Founded: 1852; Members: 1100
Gen Secr: Frederick J. Heuser
Focus: Hist
Periodical
Journal of Presbyterian History (weekly) . 16351

Price-Pottenger Nutrition Foundation (PPNF), POB 2614, La Mesa, CA 91943-2614
T: (619) 582-4168
Founded: 1952; Members: 750
Gen Secr: Marion Patricia Connolly
Focus: Food
Periodical
Membership Journal (weekly) . . . 16352

Primitive Art Society of Chicago (PAS), POB 1840, Chicago, IL 60690
T: (312) 280-0208
Founded: 1977; Members: 100
Pres: Judith Prichard
Focus: Arts . . . 16353

Print Council of America (PCA), c/o The Baltimore Museum of Art, Art Museum Dr, Baltimore, MD 21218
T: (410) 396-6345; Fax: (410) 396-6562
Founded: 1956; Members: 150
Pres: Jay McKean Fisher
Focus: Eng . . . 16354

Probe Ministries International (PMI), 1900 Firman Dr, Ste 100, Richardson, TX 75081-6796
T: (214) 480-0240; Fax: (214) 644-9664
Founded: 1973
Pres: Jimmy Williams
Focus: Educ
Periodical
Spiritual Fitness in Business (weekly) . . 16355

Professional Football Researches Association (PFRA), 12870 Rte 30, North Huntington, PA 15642
T: (412) 863-6345
Founded: 1979; Members: 220
Gen Secr: Bob Carroll
Focus: Sports
Periodicals
Coffin Corner (weekly)
End of Year Journal (weekly)
Membership Directory (weekly) . . . 16356

Professional Numismatists Guild (PNG), c/o Paul L. Koppenhaver, POB 430, Van Nuys, CA 91408
T: (818) 781-1764
Founded: 1954; Members: 300
Gen Secr: Paul L. Koppenhaver
Focus: Numismatics . . . 16357

Professional Psychics United (PPU), c/o Phyllis Allen, 7115 W North Av, Oak Park, IL 60302
T: (312) 877-2662; Fax: (312) 693-6737
Founded: 1977; Members: 350
Pres: Phyllis Allen
Focus: Parapsych . . . 16358

Professors of Curriculum (PC), c/o Marcella Kysilka, College of Education, University of Central Florida, POB 2500, Orlando, FL 32816-0250
T: (407) 823-2011; Fax: (407) 823-5135
Members: 100
Gen Secr: Marcella Kysilka
Focus: Educ
Periodical
Directory . . . 16359

Program for Appropriate Technology in Health (PATH), 4 Nickerson St, Seattle, WA 98109
T: (206) 285-3500; Fax: (206) 285-6619
Founded: 1981
Pres: Gordon W. Perkin
Focus: Public Health; Eng . . . 16360

Project Magic (PM), c/o Kansas Rehabilitation Hospital, 1504 SW Eighth St, Topeka, KS 66606
T: (913) 235-6600
Founded: 1982
Gen Secr: Julie DeJean
Focus: Rehabil . . . 16361

Project Management Institute (PMI), POB 43, Drexel Hill, PA 19026-3190
T: (215) 622-1796; Fax: (215) 622-5640
Founded: 1969; Members: 8000
Gen Secr: Richard Colosi
Focus: Business Admin
Periodical
Project Management Journal (5 times annually) . . . 16362

Project on Equal Education Rights (PEER), c/o NOW LDEF, 99 Hudsom St, New York, NY 10013
T: (212) 925-6635; Fax: (212) 226-1066
Founded: 1974
Gen Secr: W. Grady Truely
Focus: Educ
Periodical
Equal Education Alert (weekly) . . . 16363

Protein Society, 9650 Rockville Pike, Bethesda, MD 20814
T: (301) 530-7003; Fax: (206) 685-9359
Founded: 1986; Members: 2100
Focus: Biochem . . . 16364

Proust Research Association (PRA), c/o J. Theodore Johnson, Dept of French and Italian, University of Kansas, Lawrence, KS 66045
T: (913) 864-3388
Founded: 1967; Members: 250
Gen Secr: J. Theodore Johnson
Focus: Lit
Periodical
Newsletter (weekly) 16365

Psi Chi – The National Honor Society in Psychology, 201 Frazier Av, Ste F, Chattanooga, TN 37405
T: (615) 756-2044
Founded: 1929; Members: 235000
Gen Secr: Kay Wilson
Focus: Psych
Periodicals
Handbook (biennial)
Newsletter (weekly) 16366

Psoriasis Research Association (PRA), 107 Vista del Grande, San Carlos, CA 94070
T: (415) 593-1394
Founded: 1952
Gen Secr: Diane B. Mullins
Focus: Derm
Periodicals
Abstracts (weekly)
Bulletin 16367

Psoriasis Research Institute (PRI), 600 Town and Country Village, Palo Alto, CA 94301
T: (415) 326-1848; Fax: (415) 326-1262
Founded: 1979
Pres: Eugene M. Farber
Focus: Derm 16368

Psychic Science International Special Interest Group (PSISIG), 7514 Belleplain Dr, Huber Heights, OH 45424-3229
T: (513) 236-0361; Fax: (513) 236-0361
Founded: 1976; Members: 130
Pres: Richard Allen Strong
Focus: Psych 16369

Psychologists Interested in Religious Issues (PIRI), c/o Dr. Margaret Gorman, Theology Dept, Boston College, Chestnut Hill, MA 02167
T: (617) 552-3880; Fax: (617) 552-8604
Founded: 1948; Members: 1175
Gen Secr: Dr. Margaret Gorman
Focus: Rel & Theol; Psych
Periodical
Newsletter (weekly) 16370

Psychology Society (PS), 100 Beekman St, New York, NY 10038-1810
T: (212) 285-1872
Founded: 1960; Members: 3200
Gen Secr: Dr. Pierre C. Haber
Focus: Psych
Periodicals
Membership List (biennial)
PS Quarterly (weekly) 16371

Psychometric Society (PS), c/o Dr. Terry Ackerman, 260C Education Bldg, 1310 S Sixth St, Champaign, IL 61820-6990
T: (217) 244-3361; Fax: (217) 244-7620
Founded: 1935; Members: 2200
Gen Secr: Dr. Terry Ackerman
Focus: Psych
Periodical
Psychometrika (weekly) 16372

Psychonomic Society (PS), c/o Dr. Cynthia H. Null, POB 7104, San Jose, CA 95150-7104
T: (415) 604-1260; Fax: (415) 604-3323
Founded: 1959; Members: 2400
Gen Secr: Dr. Cynthia H. Null
Focus: Psych
Periodicals
Animal Learning and Behavior (weekly)
Behavior Research Methods (weekly)
Bulletin (weekly)
Memory & Cognition (weekly)
Perception and Psychophysics (weekly)
Psychobiology (weekly) 16373

Public Citizen Health Research Group (PCHRG), 2000 P St NW, Washington, DC 20036
T: (202) 833-3000
Founded: 1971
Gen Secr: Sidney M. Wolfe
Focus: Public Health
Periodical
Health Letter (weekly) 16374

Public Law Education Institute (PLEI), 1601 Connecticut Av NW, Ste 450, Washington, DC 20009
T: (202) 232-1400
Founded: 1968
Pres: Thomas P. Alder
Focus: Law 16375

Public Leadership Education Network (PLEN), 1001 Connecticut Av NW, Ste 925, Washington, DC 20036
T: (202) 872-1585
Founded: 1978; Members: 17
Gen Secr: Marianne Alexander
Focus: Educ 16376

Public Library Association (PLA), c/o American Library Association, 50 E Huron St, Chicago, IL 60611
T: (312) 280-5752; Fax: (312) 280-5029
Founded: 1944; Members: 7300
Gen Secr: George Needham
Focus: Libraries & Bk Sci
Periodicals
Public Libraries (weekly)
Public Library Reporter (weekly) 16377

Public Relations Society of America (PRSA), 33 Irving Pl, New York, NY 10003-2376
T: (212) 995-2230; Fax: (212) 995-0757
Founded: 1947; Members: 15462
Gen Secr: Elizabeth Ann Kovacs
Focus: Public Health
Periodical
Health Academy News (weekly) 16378

Public Responsibility in Medicine and Research (PRIM&R), 132 Boylston St, Boston, MA 02116
T: (617) 423-4112; Fax: (617) 423-1185
Founded: 1974; Members: 65
Gen Secr: Joan Rachlin
Focus: Med
Periodical
Conference Report (weekly) 16379

Public Service Research Council (PSRC), 1761 Business Center Dr, Ste 230, Reston, VA 22090
T: (703) 438-3966; Fax: (703) 438-3935
Founded: 1973; Members: 50000
Pres: David Y. Denholm
Focus: Public Admin
Periodicals
The Government Union Critique (weekly)
Government Union Review (weekly) . . 16380

Public Works Historical Society (PWHS), 1801 Maple, Evanton, IL 60637
T: (708) 491-5829; Fax: (312) 667-2304
Founded: 1975; Members: 1900
Gen Secr: Howard Rosen
Focus: Hist
Periodicals
Essay Series (weekly)
Newsletter (weekly)
Oral History (weekly) 16381

Rabbinic Center for Research and Counseling (RCRC), 128 E Dudley Av, Westfield, NJ 07090
T: (908) 233-0419
Founded: 1970
Gen Secr: Irwin H. Fishbein
Focus: Therapeutics 16382

R.A. Bloch Cancer Foundation, 4410 Main, Kansas City, MO 64111
T: (816) 932-8453; Fax: (816) 753-5346
Founded: 1980
Gen Secr: A.J. Yarmot
Focus: Cell Biol & Cancer Res 16383

The Radiance Technique Association International (TRTAI), POB 40570, Saint Petersburg, FL 33743-0570
T: (813) 347-3421
Founded: 1980
Gen Secr: Dr. Barbara Ray
Focus: Med
Periodical
The Radiance Technique Journal . . . 16384

Radiation Research Society (RRS), 1819 Preston White Dr, Reston, VA 22091
T: (703) 648-3780; Fax: (703) 648-9176
Founded: 1952; Members: 2025
Gen Secr: Laura Fleming Jones
Focus: Radiology
Periodicals
Radiation Research (weekly)
RRS News (weekly) 16385

Radiation Therapy Oncology Group (RTOG), c/o American College of Radiology, 1101 Market St, Philadelphia, PA 19107
T: (215) 574-3150
Founded: 1971; Members: 120
Gen Secr: Nancy Smith
Focus: Cell Biol & Cancer Res; Radiology; Therapeutics 16386

Radio and Television Research Council (RTRC), 245 Fifth Av, Ste 2103, New York, NY 10016
T: (212) 481-3038; Fax: (212) 481-3071
Founded: 1941; Members: 200
Pres: Terry Drucker
Focus: Mass Media 16387

Radiological Society of North America (RSNA), 2021 Spring Rd, Ste 600, Oak Brook, IL 60521
T: (708) 571-2670; Fax: (708) 571-7837
Founded: 1915; Members: 26000
Gen Secr: D.J. Stauffer
Focus: Radiology
Periodicals
Radio Graphics (weekly)
Radiology (weekly)
RSNA Today (weekly) 16388

Radio Technical Commission for Aeronautics (RTCA), 1140 Connecticut Av NW, Ste 1020, Washington, DC 20036
T: (202) 833-9339; Fax: (202) 833-9434
Founded: 1935; Members: 103
Gen Secr: David S. Watrous
Focus: Electric Eng; Aero
Periodical
RTCA Digest (weekly) 16389

Radix Institute (RI), Sery Pl, Ste 102, Granbury, TX 76049
T: (817) 326-5670
Founded: 1960; Members: 96
Gen Secr: William H. Thrash
Focus: Psych
Periodicals
Calendar of Events (biennial)
List of Radix Teachers Worldwide (weekly)
Radix Journal (weekly) 16390

Radix Teachers Association (RTA), 5310 Harvest Hill Rd, Dallas, TX 75230-5805
T: (214) 661-1746
Founded: 1981; Members: 77
Pres: Stephen Atkinson
Focus: Educ 16391

Railroad Station Historical Society (RSHS), 430 Ivy Av, Crete, NE 68333
T: (402) 826-3356
Founded: 1967; Members: 400
Gen Secr: Janet L.C. Rapp
Focus: Hist
Periodicals
Monograph (weekly)
The Bulletin (weekly) 16392

Ralph Waldo Emerson Memorial Association (RWEMA), c/o J.M. Forbes & Co, 79 Milk St, Boston, MA 02109
T: (617) 423-5705
Founded: 1930; Members: 15
Gen Secr: Roger L. Gregg
Focus: Lit 16393

Rare Earth Research Conference (RERC), c/o Prof. Lance DeLong, Dept of Physics and Astronomy, University of Kentucky, Lexington, KY 40506
T: (606) 257-4775; Fax: (606) 258-2846
Founded: 1960; Members: 325
Pres: Prof. Lance DeLong
Focus: Sci
Periodical
Proceedings of Conference (biennial) . . 16394

R. Austin Freeman Society (RAFS), 121 Follen Rd, Lexington, MA 02173
T: (617) 861-8705
Founded: 1982; Members: 200
Pres: John J. McAleer
Focus: Lit 16395

Reading is Fundamental (RIF), 600 Maryland Av SW, Ste 500, Washington, DC 20024
T: (202) 287-3220; Fax: (202) 287-3196
Founded: 1966; Members: 3559
Pres: Ruth P. Graves
Focus: Ling 16396

Reading Reform Foundation (RRF), POB 98785, Tacoma, WA 98498-0785
T: (206) 588-3436; Fax: (206) 582-7877
Founded: 1961; Members: 258
Gen Secr: M. Hinds
Focus: Ling
Periodical
The Reading Informator (weekly) 16397

Read Natural Childbirth Foundation (RNCF), POB 150956, San Rafael, CA 94915
T: (415) 456-8462
Founded: 1978; Members: 30
Pres: Margaret B. Farley
Focus: Gynecology 16398

Reference and Adult Services Division (of ALA) (RASD), c/o American Library Association, 50 E Huron St, Chicago, IL 60611
T: (312) 280-4398; Fax: (312) 545-2433
Founded: 1972; Members: 5500
Gen Secr: Andrew M. Hansen
Focus: Libraries & Bk Sci
Periodicals
RASD Update (weekly)
RQ (weekly) 16399

Reforma: National Association to Promote Library Services to the Spanish-Speaking, c/o Mario M. Gonzalez, Office of Special Services, New York Public Library, 455 Fifth Av, New York, NY 10016
T: (212) 340-0988
Founded: 1971; Members: 700
Pres: Mario M. Gonzalez
Focus: Libraries & Bk Sci
Periodicals
Membership Directory (biennial)
Reforma (weekly) 16400

Regional Education Board of the Christian Brothers, c/o Christian Brothers Conference, 100 De La Salle Dr, Romeoville, IL 60441
T: (815) 838-3336; Fax: (815) 838-7092
Members: 16
Gen Secr: Robert McCann
Focus: Educ 16401

Regional Institute of Social Welfare Research (RISWR), POB 152, Athens, GA 30603
T: (404) 546-0798
Founded: 1970
Pres: Dr. George Thomas
Focus: Sociology
Periodicals
Annual Report (weekly)
Checkpoints (weekly) 16402

Regional Science Association (RSA), c/o Observatory, University of Illinois, 901 S Mathews, Urbana, IL 61801
T: (217) 333-8904
Founded: 1954; Members: 2500
Gen Secr: Dr. G. Hewings
Focus: Urban Plan 16403

Registered Medical Assistants of American Medical Technologists (RMAAMT), 710 Higgins Rd, Park Ridge, IL 60068-5765
T: (708) 823-5169; Fax: (708) 823-0458
Founded: 1976; Members: 9000
Gen Secr: Gerard G. Boe
Focus: Med
Periodicals
AMT Directory (weekly)
Vital Signs (weekly) 16404

Rehabilitation Information Round Table (RIRT), c/o Phyllis Quinn, American Physical Therapy Association, 1111 N Fairfax St, Alexandria, VA 22314
T: (703) 706-3210
Founded: 1979; Members: 60
Gen Secr: Phyllis Quinn
Focus: Rehabil
Periodical
Newsletter (weekly) 16405

Rehabilitation International (RI), 25 E 21 St, New York, NY 10010
T: (212) 420-1500; Fax: (212) 505-0871
Founded: 1992; Members: 145
Gen Secr: Susan R. Hammerman
Focus: Rehabil
Periodicals
International Journal of Rehabilitation Research (weekly)
International Rehabilitation Review
One-in-Ten (weekly)
Rehabilitation (weekly) 16406

Reinforced Concrete Research Council (RCRC), 205 N Mathews Av, Urbana, IL 61801
T: (217) 333-7384
Founded: 1948; Members: 40
Gen Secr: Sharon L. Wood
Focus: Civil Eng 16407

Religious Research Association (RRA), c/o Catholic University of America, 108 Marist Hall, Washington, DC 20064
T: (202) 319-5447
Founded: 1959; Members: 400
Pres: James Davidson
Focus: Sociology; Rel & Theol
Periodical
Review of Religious Research (weekly) . 16408

Religious Speech Communication Association (RSCA), c/o Jessica L. Rousselow, Taylor University, Upland, IN 46989
T: (317) 998-5280; Fax: (317) 998-5569
Founded: 1973; Members: 265
Gen Secr: Jessica L. Rousselow
Focus: Ling 16409

Renaissance English Text Society (RETS), c/o Arthur F. Kinney, Dept of English, University of Massachusetts, Amherst, MA 01003
T: (413) 256-8648
Founded: 1959; Members: 200
Pres: Arthur F. Kinney
Focus: Lit 16410

Renaissance Society of America (RSA), 24 W 12 St, New York, NY 10011
T: (212) 998-3797; Fax: (212) 995-4205
Founded: 1954; Members: 3000
Gen Secr: Margaret L. King
Focus: Hist
Periodical
Renaissance Quarterly (weekly) 16411

Research and Engineering Council of the Graphic Arts Industry (RECGAI), Marshallton Bldg, POB 639, Chadds Ford, PA 19317
T: (215) 388-7394; Fax: (215) 388-2708
Founded: 1950; Members: 325
Gen Secr: Fred M. Rogers
Focus: Eng
Periodical
The Review: Recent Patents of Interest to the Graphic Arts Industry (weekly) 16412

Research Association of Minority Professors (RAMP), PO Drawer 67, Prairie View, TX 77446
T: (409) 857-4710
Founded: 1982; Members: 111
Focus: Educ 16413

Research Council on Structural Connections (RCSC), c/o Sagent & Lyndy, 55 E Monroe St, Chicago, IL 60603
T: (312) 269-2424
Founded: 1946; Members: 56
Gen Secr: Prof. James Doyle
Focus: Eng 16414

Research Discussion Group (RSD), c/o Jeffrey S. Fedan, Dept of Pharmacology and Toxicology, West Virginia University Medical Center, Morgantown, WV 26506
T: (304) 293-4449
Founded: 1957; Members: 50
Gen Secr: Jeffrey S. Fedan
Focus: Med 16415

Research Libraries Group (RLG), 1200 Villa St, Mountain View, CA 94041-1100
T: (415) 962-9951; Fax: (415) 964-0943
Founded: 1974; Members: 120
Pres: James P. Michalko
Focus: Libraries & Bk Sci
Periodicals
Research Libraries Group Annual Report (weekly)
Research Libraries Group Directory (weekly)
Research Libraries Group News (3 times annually) 16416

Research Society for Victorian Periodicals (RSVP), c/o Prof. B.Q. Schmidt, Dept of English, Southern Illinois University, Edwardsville, IL 62026-1436
T: (618) 692-2326
Founded: 1969; Members: 600
Gen Secr: Prof. B.Q. Schmidt
Focus: Lit; Journalism
Periodical
Victorian Periodicals Review (weekly) . . 16417

Research Society on Alcoholism (RSA), 4314 Medical Pkwy, Austin, TX 78756
T: (512) 454-0022; Fax: (512) 454-0022
Founded: 1976; Members: 870
Gen Secr: D. Sharp
Focus: Soc Sci 16418

Research to Prevent Blindness (RPB), 598 Madison Av, New York, NY 10022
T: (212) 752-4333
Founded: 1960
Pres: David F. Weeks
Focus: Ophthal
Periodicals
Annual Report (weekly)
Progress Report (weekly) 16419

Rhetoric Society of America (RSA), c/o Kathleen Welsh, Dept of English, University of Oklahoma, Norman, OK 73019
T: (405) 325-2004
Founded: 1968; Members: 450
Gen Secr: Michael Halloran
Focus: Humanities
Periodical
Rhetoric Society Quarterly (weekly) . . . 16420

Richard III Society, POB 13786, New Orleans, LA 70185
T: (504) 827-0161; Fax: (504) 822-7599
Founded: 1924; Members: 750
Gen Secr: Carole M. Rike
Focus: Hist
Periodicals
The Ricardian (weekly)
The Ricardian Register (weekly) 16421

Robinson Jeffers Committee (RJC), c/o Robert J. Brophy, Dept of English, California State University, 1250 Bellflower Blvd, Long Beach, CA 90840-2403
T: (310) 985-4235; Fax: (310) 985-2269
Founded: 1962; Members: 3
Gen Secr: Robert J. Brophy
Focus: Lit 16422

Robotics International of the Society of Manufacturing Engineers (RI/SME), 1 SME Dr, POB 930, Dearborn, MI 48121
T: (313) 271-1500; Fax: (313) 271-2861
Founded: 1980; Members: 7800
Gen Secr: Robert A. Ankrapp
Focus: Electronic Eng; Computer & Info Sci
Periodical
Robotics Today Magazine (weekly) . . . 16423

Rodale Institute (RI), 222 Main St, Emmaus, PA 18049
T: (215) 967-8405; Fax: (215) 967-8959
Founded: 1982; Members: 75000
Pres: John Haberern
Focus: Agri 16424

Rodeo Historical Society (RHS), 1700 NE 63 St, Oklahoma City, OK 73111
T: (405) 478-2250; Fax: (405) 478-4714
Founded: 1968; Members: 1200
Gen Secr: B. Byron Price
Focus: Hist 16425

Rogers Group (RG), 4932 Prince George Av, Beltsville, MD 20705
T: (301) 937-7899
Founded: 1970; Members: 110
Gen Secr: George C. Humphrey
Focus: Arts
Periodical
Newsletter (weekly) 16426

Rolf Institute (RI), POB 1868, Boulder, CO 80306
T: (303) 449-5903; Fax: (303) 449-5978
Founded: 1971; Members: 745
Gen Secr: Daniel J. Kuchars
Focus: Physical Therapy
Periodicals
Bulletin of Structural Integration (weekly)
Directory of Certified Rolfers and Movement Teachers (weekly)
National Listing of Rolfers (weekly) . . . 16427

Roller Bearing Engineers Committee (RBEC), 1101 Connecticut Av NW, Ste 700, Washington, DC 20036-4303
T: (202) 429-5155; Fax: (202) 223-4579
Gen Secr: Gary T. Satterfield
Focus: Eng 16428

Rural Community Assistance Program (RCAP), 602 S King St, Ste 402, Leesburg, VA 22075
T: (703) 771-8636
Founded: 1973; Members: 13
Gen Secr: Kathleen Stanley
Focus: Water Res
Periodicals
Directory of Network Organizations (weekly)
Rural Water News (weekly) 16429

Rural Sociological Society (RSS), c/o Dept of Sociology, Montana State University, Wilson Hall, Bozeman, MT 59717
T: (406) 994-5248
Founded: 1937; Members: 1050
Gen Secr: Patrick C. Jobes
Focus: Sociology
Periodicals
The Rural Sociologist (weekly)
Rural Sociology (weekly) 16430

Rushlight Club (RC), 1657 The Fairway, Ste 196, Jenkintown, PA 19046
Founded: 1932
Pres: Dr. R.M. Winborne
Focus: Hist
Periodical
The Rushlight (weekly) 16431

Sacro Occipital Research Society International (SORSI), POB 8245, Prairie Village, KS 66208
T: (913) 384-2748
Founded: 1958; Members: 1500
Pres: Dr. C. DeCamp
Focus: Orthopedics
Periodical
The Source (weekly) 16432

SAE International, 400 Commonwealth Dr, Warrendale, PA 15096-0001
T: (412) 776-4841; Fax: (412) 776-5760
Founded: 1905; Members: 60000
Gen Secr: Max E. Rumbaugh
Focus: Eng
Periodicals
Aerospace Engineering (weekly)
Automotive Engineering (weekly)
Handbook (weekly)
SAE Update (weekly) 16433

Safe Association, 4995 Scotts Valley Rd, POB 490, Yoncalla, OR 97499-0490
T: (503) 849-2977; Fax: (503) 849-2997
Founded: 1960; Members: 900
Gen Secr: Jean Benton
Focus: Eng; Transport
Periodicals
Safe Journal (weekly)
Symposium Proceedings (weekly) . . . 16434

Safety Equipment Institute (SEI), 1901 N Moore St, Ste 808, Arlington, VA 22209
T: (703) 525-3354; Fax: (703) 528-2148
Pres: Thomas G. Augherton
Focus: Safety
Periodical
SEI Certified Product List (weekly) . . . 16435

Salt Institute (SI), 700 N Fairfax, Ste 600, Alexandria, VA 22314-2040
T: (703) 549-4648; Fax: (703) 548-2194
Founded: 1914; Members: 25
Pres: Richard L. Hanneman
Focus: Mineralogy
Periodicals
Agriculture Digest (weekly)
Highway Digest (weekly) 16436

Salzburg Seminar (SS), c/o The Marble Works, Box 886, Middlebury, VT 05753
T: (802) 388-0007; Fax: (802) 388-1030
Founded: 1947
Focus: Hist 16437

San Francisco Aids Foundation (SFAF), POB 426182, San Francisco, CA 94142-6182
T: (415) 864-5855; Fax: (415) 552-8619
Founded: 1982
Gen Secr: Pat Christen
Focus: Immunology 16438

San Martin Society of Washington, DC, POB 33, McLean, VA 22101
T: (703) 883-0950; Fax: (703) 243-1020
Founded: 1977; Members: 305
Pres: Christian Garcia-Godoy
Focus: Hist
Periodical
San Martin News 16439

Saving and Preserving Arts and Cultural Environments (SPACES), 1804 N Van Ness, Los Angeles, CA 90028
T: (213) 463-1629
Founded: 1978; Members: 250
Pres: Seymour Rosen
Focus: Arts 16440

Scandinavian Seminar (SS), 24 Dickinson St, Amherst, MA 01002
T: (413) 253-9736; Fax: (413) 253-5282
Founded: 1949
Gen Secr: Mary S. Cattani
Focus: Ethnology 16441

School Management Study Group (SMSG), 860 18 Av, Salt Lake City, UT 84103
T: (801) 532-5340; Fax: (801) 484-2089
Founded: 1969; Members: 400
Gen Secr: Donald Thomas
Focus: Educ
Periodical
Newsletter 16442

School Science and Mathematics Association (SSMA), c/o Bowling Green State University, 126 Life Science Bldg, Bowling Green, OH 43403
T: (419) 372-7393; Fax: (419) 372-2327
Founded: 1903; Members: 1100
Gen Secr: Darrell W. Fyffe
Focus: Math; Educ
Periodicals
Convention Program (weekly)
Newsletter (weekly)
School Science and Mathematics (8 times annually) 16443

Science Fiction Research Association (SFRA), c/o Peter Lowentrout, 5225 Saint George Rd, Westminster, CA 92683
T: (714) 897-9060
Founded: 1970; Members: 450
Pres: Peter Lowentrout
Focus: Lit
Periodical
Book Review Journal (10 times annually) 16444

Science Service (SS), 1719 N St NW, Washington, DC 20036
T: (202) 785-2255; Fax: (202) 785-1243
Founded: 1921
Pres: Dr. Alfred S. McLaren
Focus: Educ 16445

Scientific Committee on Oceanic Research, c/o Dept of Earth and Planetary Sciences, Johns Hopkins University, Baltimore, MD 21218
T: (410) 516-4070; Tx: 7401472; Fax: (410) 516-7933
Founded: 1957
Gen Secr: Elizabeth Gross
Focus: Oceanography 16446

Scientists and Engineers for Secure Energy (SE2), 570 Seventh Av, Ste 1007, New York, NY 10018
T: (212) 840-6595; Fax: (212) 840-6597
Founded: 1976; Members: 1200
Gen Secr: Dr. Miro M. Todorovich
Focus: Energy 16447

Scientists Center for Aminal Welfare (SCAW), 4805 Saint Elmo Av, Bethesda, MD 20814
T: (301) 654-6390; Fax: (301) 907-3993
Founded: 1978; Members: 2000
Gen Secr: Lee Kruusch
Focus: Vet Med
Periodical
Newsletter (weekly) 16448

Scoliosis Research Society (SRSO), 6300 River Rd, Ste 727, Rosemont, IL 60018-4226
T: (708) 698-1628; Fax: (708) 823-0536
Founded: 1966; Members: 445
Gen Secr: Carole Murphy
Focus: Orthopedics 16449

Scribes, c/o Wake Forest University School of Law, POB 7206, Reynolds Station, Winston-Salem, NC 27109
T: (919) 759-5440; Fax: (919) 759-6077
Founded: 1952; Members: 800
Gen Secr: Thomas M. Steele
Focus: Lit; Law
Periodical
The Scrivener (weekly) 16450

Sea Education Association (SEA), POB 6, Woods Hole, MA 02543
T: (508) 540-3954; Fax: (508) 457-4673
Founded: 1971
Gen Secr: Rafe E.A. Parker
Focus: Adult Educ
Periodicals
Annual Report (weekly)
Following Sea (weekly) 16451

Sea Grant Association (SGA), c/o Dr. James J. Sullivan, Sea Grant College Program, University of California at San Diego, La Jolla, CA 92093
T: (619) 534-4440; Fax: (619) 534-2231
Founded: 1970; Members: 39
Pres: Dr. James J. Sullivan
Focus: Oceanography 16452

Search Foundation, POB 43388, Birmingham, AL 35243
T: (205) 991-9516; Fax: (205) 991-2807
Pres: John M. Bradley
Focus: Archeol 16453

Search Group (SGI), 7311 Greenhaven Dr, Ste 145, Sacramento, CA 95831
T: (916) 392-2550; Fax: (916) 392-8440
Founded: 1969; Members: 60
Gen Secr: Gary R. Cooper
Focus: Law
Periodicals
Interface (weekly)
Report (weekly) 16454

Sea Shepherd Conservation Society (SSCS), 1314 Second St, Santa Monica, CA 90401
T: (310) 394-3198; Fax: (310) 394-0360
Founded: 1977; Members: 18000
Pres: Paul Watson
Focus: Ecology
Periodical
Newsletter (weekly) 16455

Secondary School Admission Test Board (SSATB), 12 Stockton St, Princeton, NJ 08540
T: (609) 683-4440
Founded: 1957; Members: 650
Gen Secr: Regan Kenyon
Focus: Educ
Periodicals
Bulletin of Information (weekly)
Newsletter (weekly)
SSATB Network Directory (weekly) . . . 16456

Section for Rehabilitation Hospitals and Programs (SRHP), c/o American Hospital Association, 840 N Lake Shore Dr, Chicago, IL 60611
T: (312) 280-6671; Fax: (312) 280-6252
Founded: 1984; Members: 890
Gen Secr: Susanne Sonik
Focus: Rehabil 16457

Section on Women in Legal Education of the AALS, c/o Association of American Law Schools, 1201 Connecticut Av NW, Ste 800, Washington, DC 20036
T: (202) 296-8851
Pres: Mary Becker
Focus: Law
Periodical
Newsletter (weekly) 16458

Seingalt Society (SS), 555 13 Av, Salt Lake City, UT 84103
T: (801) 532-2204
Founded: 1985; Members: 45
Pres: Tom Vitelli
Focus: Lit 16459

Seismological Society of America (SSA), El Cerrito Professional Bldg, Ste 201, El Cerrito, CA 94530
T: (415) 525-5474; Fax: (415) 525-7204
Founded: 1906; Members: 1800
Gen Secr: Susan B. Newman
Focus: Geology; Geophys
Periodical
Bulletin (weekly) 16460

Seminar on the Acquisition of Latin American Library Materials (SALALM), c/o General Library, University of New Mexico, Albuquerque, NM 87131
T: (505) 277-5102
Founded: 1956; Members: 500
Gen Secr: Sharon Moynahan
Focus: Libraries & Bk Sci
Periodicals
Bibliography of Latin American and Caribbean Bibliographies (weekly)
Microfilming Projects Newsletter (weekly)
Papers of the Seminar on the Acquisition of Latin American Library Materials (weekly)
SALALM Bibliography and Reference Series (3 times annually)
SALALM Newsletter (weekly) 16461

Semiotic Society of America (SSA), c/o Applied Behavioral Sciences Dept, University of California, Davis, CA 95616
T: (916) 752-6437
Founded: 1975; Members: 780
Gen Secr: Dean MacConnell
Focus: Ling
Periodical
American Journal of Semiotics (weekly) . 16462

Senior Scholars (SS), c/o Kathy Manos, Office of Continuing Education, Case Western Reserve University, 10900 Euclid Av, Cleveland, OH 44106-7116
T: (216) 368-2090; Fax: (216) 368-2091
Founded: 1972; Members: 130
Gen Secr: Kathy Manos
Focus: Adult Educ 16463

Sex Information and Education Council of the U.S. (SIECUS), 130 W 42 St, Ste 2500, New York, NY 10036
T: (212) 819-9770; Fax: (212) 819-9776
Founded: 1964; Members: 3600
Gen Secr: Debra Haffner
Focus: Educ
Periodicals
SIECUS Newsletter (weekly)
SIECUS Report (weekly) 16464

SFI Foundation, 15708 Pomerado Rd, Poway, CA 92064
T: (619) 451-8868; Fax: (619) 451-9268
Founded: 1978
Pres: Arnold S. Kuhns
Focus: Eng 16465

Shakespeare Association of America (SAA), c/o Dept of English, Southern Methodist University, Dallas, TX 75275
Founded: 1972; Members: 800
Gen Secr: Dr. Nancy Elizabeth Hodge
Focus: Lit
Periodicals
Bulletin (weekly)
Directory of Members (weekly) 16466

Shakespeare Data Bank (SDB), 1217 Ashland Av, Evanston, IL 60202
T: (708) 475-7550; Fax: (708) 475-2415
Founded: 1984; Members: 160
Gen Secr: Dr. Louis Marder
Focus: Lit 16467

Shakespeare Society of America (SSA), 1107 N Kings Rd, West Hollywood, CA 90069
T: (213) 654-623
Founded: 1967; Members: 1500
Pres: R. Thad Taylor
Focus: Lit 16468

Sherwood Anderson Society (SAS), c/o Dept of English, Virginia Tech, Blacksburg, VA 24061-0112
T: (703) 231-6501
Founded: 1975; Members: 150
Gen Secr: Charles E. Modlin
Focus: Lit
Periodicals
Bibliography (weekly)
Winesburg Eagle (weekly) 16469

Shevchenko Scientific Society (SSS), 63 Fourth Av, New York, NY 10003
T: (212) 254-5130; Fax: (212) 254-5139
Founded: 1873; Members: 600
Pres: Leonid Rudnytzky
Focus: Sci
Periodicals
Memoirs
Nationalities Papers (weekly)
Newsletter (weekly)
Proceedings
Ukrainian Bibliographical Quarterly
Ukrainska Knyha: Ukrainian Book (weekly) 16470

Sierra Club (SC), 730 Polk St, San Francisco, CA 94109
T: (415) 776-2211; Fax: (415) 776-0350
Founded: 1892; Members: 650000
Gen Secr: Michael Fischer
Focus: Ecology
Periodicals
National News Report
Sierra 16471

SIGMA Xi – The Scientific Research Society, 99 Alexander Dr, POB 13975, Research Triangle Park, NC 27708
T: (919) 549-4691
Founded: 1886; Members: 103000
Gen Secr: Dr. John F. Aheame
Focus: Humanities; Nat Sci
Periodical
American Scientist (weekly) 16472

Sigmund Freud Archives (SFA), c/o Harold P. Blum, 23 The Hemlocks, Roslyn, NY 11576
T: (516) 621-6850; Fax: (516) 621-3014
Founded: 1951; Members: 18
Gen Secr: Harold P. Blum
Focus: Archives; Psych 16473

Signal Processing Society, c/o Institute of Electrical and Electronics Engineers, 345 E 47 St, New York, NY 10017
T: (212) 705-7900; Fax: (212) 705-4929
Members: 13383
Focus: Electric Eng 16474

Simian Society of America (SSA), 3625 Watson Rd, Saint Louis, MO 63109
T: (314) 647-6218
Founded: 1957; Members: 400
Gen Secr: J. Paulette
Focus: Zoology
Periodicals
Membership Roster (weekly)
The Simian (weekly) 16475

Sir Thomas Beecham Society, c/o Charles Niss, 664 S Irena Av, Redondo Beach, CA 90277
T: (213) 540-7265
Founded: 1964; Members: 720
Gen Secr: Charles Niss
Focus: Music 16476

Sister Kenny Institute (SKI), 800 E 28 St, Minneapolis, MN 55407
T: (612) 863-4457
Founded: 1942
Gen Secr: Nancy Rehkamp
Focus: Rehabil
Periodical
The Independent (weekly) 16477

Skin Cancer Foundation (SCF), 245 Fifth Av, Ste 2402, New York, NY 10016
T: (212) 725-5176; Fax: (212) 725-5751
Founded: 1977
Pres: Perry Robins
Focus: Cell Biol & Cancer Res 16478

Sleep Research Society (SRS), c/o Gary Richardson, Endocrine Div, Brigham and Women's Hospital, 221 Longwood Av, Boston, MA 02115
T: (617) 732-7996; Fax: (617) 732-4015
Founded: 1961; Members: 428
Gen Secr: Gary Richardson
Focus: Psychiatry
Periodicals
Sleep (weekly)
Sleep Research (weekly) 16479

Small Towns Institute (STI), Third amd Poplar St, POB 517, Ellensburg, WA 98926
T: (509) 925-1830
Founded: 1969; Members: 1850
Gen Secr: Kenneth Munsell
Focus: Urban Plan
Periodical
Small Town (weekly) 16480

Smithsonian Institution (S.I.), 1000 Jefferson Dr SW, Washington, DC 20560
T: (202) 357-2700
Founded: 1846
Gen Secr: Robert McCormick Adams
Focus: Humanities; Nat Sci; Educ
Periodicals
Smithsonian Contributions to Anthropology (weekly)
Smithsonian Contributions to Botany (weekly)
Smithsonian Contributions to Paleobiology (weekly)
Smithsonian Contributions to the Earth Sciences (weekly)
Smithsonian Contributions to the Marine Sciences (weekly)
Smithsonian Contributions to Zoology (weekly)
Smithsonian Folklife Studies (weekly)
Smithsonian Studies in History and Technology (weekly)
Smithsonian Studies to Air and Space (weekly) 16481

Social Psychiatry Research Institute (SPRI), 150 E 69 St, Ste 2H, New York, NY 10021
T: (212) 628-4800; Fax: (212) 249-8546
Founded: 1970
Pres: Ari Kiev
Focus: Psychiatry 16482

Social Responsibilities Round Table (SRRT), c/o American Library Association, 50 E Huron St, Chicago, IL 60611
T: (312) 280-4294
Founded: 1969; Members: 1500
Gen Secr: Stephen J. Stillwell
Focus: Libraries & Bk Sci
Periodical
SRRT Newsletter (weekly) 16483

Social Science Education Consortium (SSEC), 3300 Mitchell Ln, Ste 240, Boulder, CO 80301-2272
T: (303) 492-8154; Fax: (303) 449-3925
Founded: 1963; Members: 130
Gen Secr: James R. Giese
Focus: Educ; Sociology; Soc Sci . . . 16484

Social Science Research Council (SSRC), 605 Third Av, New York, NY 10158
T: (212) 661-0280; Fax: (212) 370-7896
Founded: 1923
Pres: David Featherman
Focus: Soc Sci
Periodicals
Annual Report (weekly)
Items (weekly) 16485

Social Sciences Services and Resources (SSSR), POB 153, Wasco, IL 60183-0153
T: (708) 897-5345
Founded: 1973
Gen Secr: Jack F. Kinton
Focus: Econ; Sociology; Public Admin; Business Admin; Develop Areas
Periodicals
American Communities Tomorrow (weekly)
Best Books in the Social Sciences (weekly) 16486

Social Welfare History Group (SWHG), c/o John Herrick, School of Social Work, Michigan State University, 212 Baker Hall, East Lansing, MI 48824
T: (517) 353-8620
Founded: 1956; Members: 250
Pres: John Herrick
Focus: Sociology; Hist
Periodical
Newsletter 16487

Société des Professeurs Français et Francophones en Amérique (SPFFA), 22 E 60 St, New York, NY 10022
T: (212) 410-7405
Founded: 1904; Members: 1800
Pres: Jean Macary
Focus: Ling; Educ
Periodical
Bulletin (weekly) 16488

Society for Academic Emergency Medicine (SAEM), 900 W Ottawa, Lansing, MI 48915
T: (517) 485-5484; Fax: (517) 485-0801
Founded: 1975; Members: 2400
Gen Secr: Mary Ann Schroop
Focus: Adult Educ; Med
Periodicals
Annals of Emergency Medicine (weekly)
Directory (weekly)
Status Report (weekly) 16489

Society for Adolescent Medicine (SAM), 19401 E 40 Hwy, Ste 120, Independence, MO 64055
Founded: 1968; Members: 1100
Gen Secr: Dr. M. Susan Jay
Focus: Pediatrics
Periodical
Journal of Adolescent Health Care (weekly) 16490

Society for Advancement of Management (SAM), 126 Lee Av, Ste 11, Box 889, Vinton, VA 24179
T: (703) 342-5563; Fax: (703) 342-6413
Founded: 1912; Members: 11000
Gen Secr: Joseph L. Bush
Focus: Business Admin
Periodicals
SAM Advanced Management Journal (weekly)
SAM Focus on Management (weekly) . . 16491

Society for American Archaeology (SAA), 808 17 St NW, Ste 200, Washington, DC 20006
T: (202) 223-9774
Founded: 1934; Members: 5200
Gen Secr: Jerome A. Miller
Focus: Archeol
Periodicals
American Antiquity (weekly)
Bulletin (weekly) 16492

Society for Ancient Greek Philosophy (SAGP), c/o Anthony Preus, Dept of Philosophy, Binghamton University, Binghamton, NY 13902-6000
T: (607) 777-2886; Fax: (607) 777-4000
Founded: 1953; Members: 500
Gen Secr: Anthony Preus
Focus: Philos
Periodical
Essays in Ancient Greek Philosophy (weekly) 16493

Society for Applied Anthropology (SAA), POB 24083, Oklahoma City, OK 73124
T: (405) 843-5113
Founded: 1941; Members: 1850
Pres: Carole E. Hill
Focus: Anthro
Periodicals
Human Organization (weekly)
Practicing Anthropology (weekly) 16494

Society for Applied Learning Technology (SALT), 50 Culpeper St, Warrenton, VA 22186
T: (703) 347-0055; Fax: (703) 349-3169
Founded: 1972; Members: 875
Pres: Raymond G. Fox
Focus: Educ
Periodicals
Conference Proceedings (weekly)
Journal of Educational Technology Systems (weekly) 16495

Society for Applied Spectroscopy (SAS), 198 Thomas Johnson Dr, Ste 2, Frederick, MD 21702
T: (301) 694-8122; Fax: (301) 694-6860
Founded: 1958; Members: 6500
Gen Secr: Jo Ann Brown
Focus: Physics
Periodicals
Applied Spectroscopy (8 times annually)
Newsletter (weekly) 16496

Society for Armenian Studies (SAS), c/o Armenian Research Center, University of Michigan, 4901 Evergreen Rd, Dearborn, MI 48128-1491
T: (313) 593-5181; Fax: (313) 593-5452
Founded: 1974; Members: 172
Pres: Dr. Dennis Papazian
Focus: Anthro 16497

Society for Asian and Comparative Philosophy (SACP), c/o Ashok Malhotra, Dept of Philosophy, State University of New York at Oneonta, Oneonta, NY 13820
T: (607) 431-3220
Founded: 1968; Members: 315
Gen Secr: Ashok Malhotra
Focus: Philos 16498

Society for Asian Art (SAA), c/o Asian Art Museum, Golden Gate Park, San Francisco, CA 94118
T: (415) 387-5675
Founded: 1958; Members: 1200
Gen Secr: Delia Rodriguez
Focus: Arts
Periodical
The Society for Asian Art Newsletter (weekly) 16499

Society for Asian Music (SAM), c/o Dept of Asian Studies, Cornell University, 388 Rockefeller Hall, Ithaca, NY 14853
T: (607) 255-4097; Fax: (607) 255-1454
Founded: 1959; Members: 500
Gen Secr: Marty Hatch
Focus: Music
Periodical
Asian Music (weekly) 16500

Society for Austrian and Habsburg History (SAHH), c/o John Spielman, History Dept, Haverford College, Haverford, PA 19041-1392
T: (215) 896-1075; Fax: (215) 896-1495
Founded: 1957; Members: 450
Gen Secr: John Spielman
Focus: Hist
Periodical
Austrian History Yearbook (weekly) . . . 16501

Society for Biomaterials (SB), 6524 Walker St, Ste 215, Minneapolis, MN 55426
T: (612) 927-8108; Fax: (612) 658-2921
Founded: 1974; Members: 1200
Gen Secr: Rosalee M. Lee
Focus: Med; Eng
Periodicals
Journal of Applied Biomaterials (weekly)
Journal of Biomedical Materials Research (weekly)
The Torch (2-3 times annually) 16502

Society for Business Ethics (SBE), c/o Dept of Philosophy, Loyola University of Chicago, 6525 N Sheridan Rd, Chicago, IL 60626
T: (312) 508-2725; Fax: (312) 508-8509
Founded: 1979; Members: 800
Gen Secr: Patricia Werhane
Focus: Econ 16503

Society for Calligraphy (SC), POB 64174, Los Angeles, CA 90064
T: (310) 306-2326
Founded: 1974; Members: 1194
Gen Secr: Sue Perez
Focus: Graphic & Dec Arts, Design
Periodical
Newsletter (4-6 times annually) 16504

Society for Cardiac Angiography and Interventions (SCA&I), POB 7849, Breckenridge, CO 80424
T: (303) 453-1773; Fax: (303) 453-2636
Founded: 1978; Members: 800
Gen Secr: Justine J. Parker
Focus: Cardiol 16505

Society for Ch'ing Studies (SCS), c/o Div of Humanities, California Institute of Technology, Pasadena, CA 91125
T: (818) 356-3830; Fax: (818) 405-9841
Founded: 1965; Members: 500
Gen Secr: James Lee
Focus: Hist
Periodical
Late Imperial China (weekly) 16506

Society for Cinema Studies (SCS), c/o Gorham Kindem, Dept of Radio, TV and Motion Picture, University of North Carolina, Swain Hall, Chapel Hill, NC 27599-6235
T: (919) 962-2311
Founded: 1959; Members: 900
Gen Secr: Gorham Kindem
Focus: Cinema
Periodical
Cinema Journal (weekly) 16507

Society for Clinical and Experimental Hypnosis (SCEH), 128A Kings Park Dr, Liverpool, NY 13090
T: (315) 652-7299
Founded: 1949; Members: 1100
Gen Secr: Marion Kenn
Focus: Med; Psych; Therapeutics
Periodicals
International Journal of Clinical and Experimental Hypnosis (weekly)
SCEH Newsletter (weekly) 16508

Society for Clinical Trials (SCT), 600 Wyndhurst Av, Baltimore, MD 21210
T: (410) 433-4722; Fax: (410) 435-8631
Founded: 1978; Members: 1450
Pres: Curt D. Furberg
Focus: Med 16509

Society for College and University Planning (SCUP), c/o University of Michigan, 2026M School of Education Bldg, 610 E University Av, Ann Arbor, MI 48109
T: (313) 763-4776; Fax: (313) 764-2510
Founded: 1965; Members: 2700
Gen Secr: Mary Ann Armour
Focus: Adult Educ
Periodicals
Membership Roster (weekly)
News (weekly)
Planning for Higher Education (weekly) . 16510

Society for Commercial Archeology (SCA), c/o National Museum of American History, Rm 5010, Washington, DC 20560
T: (202) 882-5424
Founded: 1977; Members: 800
Gen Secr: Rebecca A. Shiffer
Focus: Archeol 16511

Society for Computer Simulation International (SCSI), POB 17900, San Diego, CA 92117-7900
T: (619) 277-3888; Fax: (619) 277-3930
Founded: 1952; Members: 2000
Gen Secr: C.G. Stockton
Focus: Computer & Info Sci
Periodicals
Simulation (weekly)
Transactions (weekly) 16512

Society for Creative Anachronism (SCA), c/o Hilary Powers, 385 Palm Av, Oakland, CA 94610
T: (510) 834-1066
Founded: 1966; Members: 22000
Pres: Hilary Powers
Focus: Cultur Hist
Periodicals
Compleat Anachronist (weekly)
Tournaments Illuminated (weekly) 16513

Society for Cryobiology (SC), c/o Federation of American Societies for Experimental Biology, 9650 Rockville Pike, Bethesda, MD 20814
T: (301) 530-7120; Fax: (301) 530-7001
Founded: 1964; Members: 450
Gen Secr: Austin H. Henry
Focus: Bio

Periodical
Cryobiology: International Journal of Low Temperature Biology and Medicine (weekly) . . . 16514

Society for Cultural Anthropology (SCA), 1703 New Hampshire Av NW, Washington, DC 20009
T: (202) 232-8800; Fax: (202) 667-5345
Founded: 1983; Members: 764
Pres: James Peacock
Focus: Anthro
Periodical
Cultural Anthropology (weekly) . . . 16515

Society for Developmental Biology (SDB), POB 40741, Washington, DC 20016
Founded: 1939; Members: 1600
Gen Secr: Holly Schauer
Focus: Bio
Periodicals
Developmental Biology (weekly)
Symposium Volume (weekly) . . . 16516

Society for Economic Botany (SEB), c/o American Archaeology Div, University of Missouri, 103 Swallow Hall, Columbia, MO 65211
T: (314) 882-3038; Fax: (314) 882-9410
Founded: 1959; Members: 800
Gen Secr: Dr. Deborah Pearsall
Focus: Botany
Periodicals
Economic Botany (weekly)
SEB Newsletter (weekly) . . . 16517

Society for Educational Reconstruction (SER), c/o Dr. Angela Raffel, 701 Abbey Ln, New York, NY 10989
T: (203) 333-1243
Founded: 1969; Members: 100
Gen Secr: Dr. Angela Raffel
Focus: Educ
Periodical
SER in Action Newsletter (weekly) . . . 16518

Society for Epidemiologic Research (SER), c/o L. Stallones, Dept of Environmental Health, Colorado State University, Fort Collins, CO 80523
T: (303) 491-0735; Fax: (303) 491-2940
Founded: 1967; Members: 3500
Gen Secr: L. Stallones
Focus: Immunology
Periodicals
Abstracts (weekly)
Newsletter (weekly) . . . 16519

Society for Ethnomusicology (SEM), c/o Indiana University, 005 Morrison Hall, Bloomington, IN 47405
T: (812) 855-6672
Founded: 1955; Members: 2200
Gen Secr: Shelly Kennedy
Focus: Music; Ethnology
Periodicals
Ethnomusicology (3 times annually)
Newsletter (3 times annually) . . . 16520

Society for Experimental and Descriptive Malacology (SEDM), POB 3037, Ann Arbor, MI 48106
T: (313) 764-0470
Founded: 1967; Members: 600
Pres: John B. Burch
Focus: Zoology
Periodicals
Malacological Review (weekly)
Walkerana-Proceedings . . . 16521

Society for Experimental Biology and Medicine (SEBM), 1300 New York Av, New York, NY 10021
Founded: 1903; Members: 1800
Gen Secr: Felice O'Grady
Focus: Bio; Med
Periodical
Proceedings of the Society for Experimental Biology and Medicine (weekly) . . . 16522

Society for Experimental Mechanics (SEM), 7 School St, Bethel, CT 06801
T: (203) 790-6373; Fax: (203) 790-4472
Founded: 1943; Members: 2607
Gen Secr: K.A. Galione
Focus: Eng
Periodicals
Experimental Mechanics (weekly)
Experimental Techniques (weekly)
International Journal of Experimental and Analytical Modal Analysis (weekly) . . . 16523

Society for French American Cultural Services and Educational Aid (FACSEA), 972 Fifth Av, New York, NY 10021
T: (212) 439-1439; Fax: (212) 439-1455
Founded: 1955
Gen Secr: Pascal Chrobocinski
Focus: Educ
Periodicals
Documentation (weekly)
Newsletter (weekly) . . . 16524

Society for French Historical Studies (SFHS), c/o Dept of History, University of Iowa, Iowa City, IA 52242
T: (319) 335-2330
Founded: 1955; Members: 1450
Gen Secr: Sarah Hanley
Focus: Hist . . . 16525

Society for General Music (SGM), c/o Music Educators National Conference, 1902 Association Dr, Reston, VA 22091
T: (703) 860-4000
Founded: 1982; Members: 3100
Gen Secr: John J. Mahlmann
Focus: Music
Periodical
Soundings (weekly) . . . 16526

Society for German-American Studies (SGAS), c/o Dr. Don Heinrich Tolzmann, Central Library, University of Cincinnati, Cincinnati, OH 45221
T: (513) 556-1859
Founded: 1968; Members: 12500
Pres: Dr. Don Heinrich Tolzmann
Focus: Hist
Periodicals
Newsletter (weekly)
Yearbook (weekly) . . . 16527

Society for Gynecologic Investigation (SGI), 409 12 St SW, Washington, DC 20024
T: (202) 863-2544; Fax: (202) 544-0453
Founded: 1953; Members: 600
Gen Secr: Ava Tayman
Focus: Gynecology
Periodical
Gynecologic Investigation (weekly) . . . 16528

Society for Historians of American Foreign Relations (SHAFR), c/o Allan Spetter, Dept of History, Wright State University, Dayton, OH 45435
T: (513) 873-3110
Founded: 1967; Members: 1200
Gen Secr: Allan Spetter
Focus: Hist
Periodicals
Diplomatic History (weekly)
Newsletter (weekly)
Roster and Research List (biennial) . . . 16529

Society for Historical Archaeology (SHA), POB 30446, Tucson, AZ 85751-0446
T: (602) 886-8006; Fax: (602) 886-0182
Founded: 1967; Members: 2000
Gen Secr: Stephanie H. Rodeffer
Focus: Archeol
Periodicals
Historical Archaeology (weekly)
Newsletter (weekly) . . . 16530

Society for History Education (SHE), c/o California State University at Long Beach, Long Beach, CA 90840
T: (310) 985-1653; Fax: (310) 985-5431
Founded: 1972; Members: 2000
Pres: Simeon Crowther
Focus: Hist; Educ
Periodicals
The History Teacher (weekly)
Network News Exchange (weekly) . . . 16531

Society for Hospital Social Work Directors (SHSWD), c/o American Hospital Association, 840 N Lake Shore Dr, Chicago, IL 60611
T: (312) 280-6414
Founded: 1966; Members: 2475
Gen Secr: Harry Bryan
Focus: Public Health; Sociology
Periodical
Discharge Planning Update (weekly) . . . 16532

Society for Humanistic Anthropology (SHA), c/o Paul Stoller, Dept of Anthropology and Sociology, West Chester University, West Chester, PA 19383
T: (215) 436-2884
Founded: 1971; Members: 300
Gen Secr: Paul Stoller
Focus: Anthro
Periodical
Quarterly of Humanistic Anthropology (weekly) . . . 16533

Society for Humanistic Judaism (SHJ), 28611 W 12 Mile Rd, Farmington Hills, MI 48334
T: (313) 478-7610; Fax: (313) 477-9014
Founded: 1970; Members: 2000
Gen Secr: Miriam Jerris
Focus: Rel & Theol
Periodicals
Humanistic Judaism (weekly)
Newsletter (3-4 times annually) . . . 16534

Society for Human Resource Management (SHRM), 606 N Washington St, Alexandria, VA 22314
T: (703) 548-3440; Fax: (703) 836-0367
Founded: 1948; Members: 50000
Gen Secr: Michael R. Losey
Focus: Business Admin
Periodicals
HR Magazine (weekly)
Personnel Administrator (weekly)
Resource (weekly) . . . 16535

Society for Industrial and Applied Mathematics (SIAM), 3600 University City Science Center, Philadelphia, PA 19104-2688
T: (215) 382-9800; Fax: (215) 386-7999
Founded: 1952; Members: 7500
Gen Secr: I. Edward Block
Focus: Math
Periodicals
Journal of Scientific and Statistical Computing (weekly)
Journal on Applied Mathematics (weekly)
Journal on Computing (weekly)
Journal on Control (weekly)
Journal on Mathematical Analysis (weekly)
Journal on Numerical Analysis (weekly)
News (weekly)
SIAM Journal on Discrete Mathematics (weekly)
SIAM Journal on Matrix Analysis (weekly) 16536

Society for Industrial Archaeology (SIA), c/o National Museum of American History, Rm 5014, Washington, DC 20560
Founded: 1971; Members: 1650
Pres: Amy Federman
Focus: Archeol
Periodicals
Bibliography
Journal (weekly)
Newsletter (weekly) . . . 16537

Society for Industrial Microbiology (SIM), POB 12534, Arlington, VA 22209-8534
T: (703) 941-5373; Fax: (703) 941-8790
Founded: 1949; Members: 2122
Gen Secr: Ann Kulback
Focus: Microbio
Periodicals
Developments in Industrial Microbiology (weekly)
Journal of Industrial Microbiology (weekly) 16538

Society for Information Display (SID), 8055 W Manchester Av, Ste 615, Playa Del Rey, CA 90293
T: (310) 305-1502; Fax: (310) 305-1433
Founded: 1962; Members: 3100
Gen Secr: Deborah Lally
Focus: Computer & Info Sci
Periodicals
Proceedings (weekly)
Quarterly Proceedings (weekly) . . . 16539

Society for Information Management (SIM), 401 N Michigan Av, Chicago, IL 60611-4267
T: (312) 644-6610; Fax: (312) 321-6869
Founded: 1968; Members: 2500
Gen Secr: Henry Givray
Focus: Business Admin
Periodicals
Member Forum (weekly)
MIS Quarterly (weekly)
Proceedings (weekly) . . . 16540

Society for International Numismatics (SIN), POB 943, Santa Monica, CA 90406
T: (213) 399-1085
Founded: 1955; Members: 450
Gen Secr: Daniel Murdock
Focus: Numismatics
Periodical
SINformation (weekly) . . . 16541

Society for Invertebrate Pathology (SIP), c/o FASEB, 9650 Rockville Pike, Bethesda, MD 20814
T: (301) 530-7120; Fax: (301) 530-7120
Founded: 1967; Members: 723
Pres: Dr. C.C. Payne
Focus: Vet Med
Periodicals
Abstracts of Symposia (weekly)
Directory for Invertebrate Pathology
Newsletter (weekly) . . . 16542

Society for Investigative Dermatology (SID), c/o David R. Bickers, Dept of Dermatology, University Hospital, 2074 Abington Rd, Cleveland, OH 44106
T: (216) 844-3682; Fax: (216) 844-8993
Founded: 1937; Members: 2300
Gen Secr: Angela Welsh
Focus: Derm
Periodical
The Journal of Investigative Dermatology 16543

Society for Iranian Studies (SIS), c/o Middle East Institute, Columbia University, 420 W 118 St, Rm 1113, New York, NY 10027
T: (212) 854-5284
Founded: 1967; Members: 450
Gen Secr: Hamid Dabashi
Focus: Ethnology
Periodicals
Iranian Studies (weekly)
SIS Newsletter . . . 16544

Society for Italian Historical Studies (SIHS), c/o Boston College, Chestnut Hill, MA 02167
T: (617) 552-3814
Founded: 1955; Members: 350
Gen Secr: Alan J. Reinerman
Focus: Hist
Periodicals
Membership List (weekly)
Newsletter (weekly) . . . 16545

Society for Latin American Anthropology (SLAA), 1703 New Hampshire Av NW, Washington, DC 20009
T: (202) 232-8800
Founded: 1969; Members: 610
Gen Secr: Paul Doughty
Focus: Anthro
Periodical
Proceedings (weekly) . . . 16546

Society for Leukocyte Biology, c/o Dr. Sherwood M. Reichard, Medical College of Georgia, 1120 15 St, POB 3044, Augusta, GA 30912
T: (706) 721-2601; Fax: (706) 721-3048
Founded: 1954; Members: 1025
Gen Secr: Dr. Sherwood M. Reichard
Focus: Pathology; Physiology
Periodicals
Directory and Constitution (weekly)
RES (weekly) . . . 16547

Society for Life History Research (SLHR), c/o Dept of Psychiatry, Washington University, 4940 Children's Pl, Saint Louis, MO 63110
T: (314) 362-2469
Founded: 1970; Members: 700
Gen Secr: Dr. Lee Robins
Focus: Psychiatry . . . 16548

Society for Linguistic Anthropology (SLA), 1703 New Hampshire Av NW, Washington, DC 20009
T: (202) 232-8800; Fax: (202) 667-5345
Founded: 1983; Members: 512
Pres: Judith Irvine
Focus: Anthro . . . 16549

Society for Magnetic Resonance Imaging (SMRI), 213 W Institute Pl, Ste 501, Chicago, IL 60610
T: (312) 751-2590; Fax: (312) 951-6474
Founded: 1982; Members: 1700
Gen Secr: K. Coe
Focus: Physics
Periodical
Magnetic Resonance Imaging (weekly) . 16550

Society for Medical Anthropology (SMA), 1703 New Hampshire Av NW, Washington, DC 20009
T: (202) 232-8800
Founded: 1971; Members: 1647
Pres: Shirley Lindenbaum
Focus: Anthro
Periodical
Medical Anthropology Quarterly (weekly) . 16551

Society for Mining, Metallurgy and Exploration, POB 625002, Littleton, CO 80162-5002
T: (303) 973-9550; Fax: (303) 973-3845
Founded: 1871; Members: 20000
Gen Secr: Tom Hendricks
Focus: Mining
Periodicals
Minerals and Metallurgical Processing
Mining Engineering (weekly) . . . 16552

Society for Music Teacher Education (SMTE), c/o Music Educators National Conference, 1902 Association Dr, Reston, VA 22091
T: (703) 860-4000
Founded: 1983
Gen Secr: John J. Mahlmann
Focus: Educ; Music . . . 16553

Society for Natural Philosophy (SNP), c/o Prof. Chi-Sing Man, University of Kentucky, Lexington, KY 40506
Founded: 1963; Members: 375
Gen Secr: Prof. Chi-Sing Man
Focus: Philos
Periodical
List of Members and Officers (weekly) . 16554

Society for Neuroscience (SN), 11 Dupont Circle, Ste 500, Washington, DC 20036
T: (202) 462-6688
Founded: 1969; Members: 18000
Gen Secr: Helen Tus
Focus: Neurology
Periodicals
Directory (weekly)
The Journal of Neuroscience (weekly)
Neuroscience Newsletter (weekly)
Neuroscience Training Programs in North America (biennial) . . . 16555

Society for New Language Study (SNLS), POB 10596, Denver, CO 80210
T: (303) 777-6115
Founded: 1972; Members: 12
Gen Secr: Raymond P. Tripp
Focus: Ling
Periodical
Proceedings (weekly) . . . 16556

Society for Nutrition Education (SNE), 2001 Killebrew Dr, Ste 340, Minneapolis, MN 55425-1882
T: (612) 854-0035; Fax: (612) 854-7869
Founded: 1967; Members: 2500
Gen Secr: Darlene Lansing
Focus: Nutrition
Periodicals
Journal of Nutrition Education (weekly)
SNE Exchange (weekly) . . . 16557

Society for Obstetric Anesthesia and Perinatology (SOAP), c/o Dept of Anesthesiology, Baylor Medical College, 6550 Fannin, Ste 1003, Houston, TX 77030
T: (713) 798-5519; Fax: (713) 793-0117
Founded: 1969; Members: 366
Gen Secr: Dr. J. Longmire
Focus: Anesthetics; Gynecology
Periodical
Newsletter (weekly) . . . 16558

Society for Occlusal Studies (SOS), c/o Dr. Bernard Williams, 1010 Carondelet Dr, Kansas City, MO 64114
T: (816) 941-0509
Founded: 1964; Members: 800
Pres: Dr. Bernard Williams
Focus: Dent
Periodicals
Newsletter (weekly)
Roster (weekly) 16559

Society for Occupational and Environmental Health (SOEH), 6728 Old McLean Village Dr, McLean, VA 22101
T: (703) 556-9222
Founded: 1972; Members: 350
Focus: Pathology
Periodical
Letter (weekly) 16560

Society for Pediatric Dermatology (SPD), c/o James E. Rasmussen, University of Michigan Hospitals, 1910 Taubman Health Care Center, Ann Arbor, MI 48109-0314
T: (313) 936-4086; Fax: (313) 936-6395
Founded: 1975; Members: 450
Gen Secr: James E. Rasmussen
Focus: Derm; Pediatrics
Periodical
Newsletter (weekly) 16561

Society for Pediatric Psychology (SPP), c/o Suzanne Bennett-Johnson, Health Sciences Center, University of Florida, Box J234, Gainesville, FL 32610
T: (904) 392-3611
Founded: 1968; Members: 980
Pres: Suzanne Bennett-Johnson
Focus: Psych
Periodical
Journal of Pediatric Psychology (weekly) 16562

Society for Pediatric Radiology (SPR), 2021 Spring Rd, Ste 600, Oak Brook, IL 60521
T: (708) 571-2197; Fax: (708) 571-7837
Founded: 1958; Members: 800
Gen Secr: Jennifer Boylan
Focus: Pediatrics; Radiology
Periodical
Membership Directory (weekly) 16563

Society for Pediatric Research (SPR), 141 Northwest Point Blvd, POB 675, Elk Grove Village, IL 60009-0675
T: (708) 427-0205
Founded: 1929; Members: 2000
Gen Secr: Debbie Anagnostelis
Focus: Pediatrics
Periodical
Pediatric Research-Program Issue (weekly) 16564

Society for Pediatric Urology (SPU), c/o Children's Hospital and Medical Center, POB 5371, Seattle, WA 98105
T: (206) 527-3950; Fax: (206) 527-3966
Founded: 1941; Members: 300
Gen Secr: Dr. Michael Mitchell
Focus: Pediatrics; Urology
Periodicals
Breakthroughs (weekly)
For a Change (weekly) 16565

Society for Personality Assessment (SPA), 7901 Fourth St N, Ste 210, Saint Petersburg, FL 33702
T: (813) 577-9583; Fax: (813) 577-9583
Founded: 1938; Members: 3000
Gen Secr: Carl E. Mullis
Focus: Psych
Periodical
Journal of Personality Assessment (weekly)
. . . . 16566

Society for Phenomenology and Existential Philosophy (SPEP), c/o Leonore Langsdorf, Dept of Speech Communication, Southern Illinois University, Carbondale, IL 62901
T: (618) 453-2291; Fax: (219) 239-8209
Founded: 1962; Members: 1500
Gen Secr: Leonore Langsdorf
Focus: Philos
Periodicals
Newsletter (weekly)
Volume of Essays (weekly) 16567

Society for Philosophy and Public Affairs (SPPA), c/o Joan Callahan, Dept of Philosophy, University of Kentucky, Lexington, KY 40506
T: (606) 257-1861
Founded: 1969; Members: 300
Gen Secr: Joan Callahan
Focus: Philos; Poli Sci
Periodicals
Having Children
Philosophy and Political Action
Philosophy, Morality and International Affairs
The Sporadical (weekly) 16568

Society for Philosophy of Religion (SPR), c/o Frank R. Harrison, Dept of Philosophy, University of Georgia, Peabody Hall, Athens, GA 30602
T: (706) 542-2823
Founded: 1940; Members: 125
Gen Secr: Frank R. Harrison
Focus: Philos; Rel & Theol 16569

Society for Photographic Education (SPE), POB 222116, Dallas, TX 75222-2116
T: (214) 943-8442
Founded: 1963; Members: 1600
Gen Secr: Judith Thorpe
Focus: Photo; Adult Educ
Periodicals
Exposure (weekly)
Membership Directory (weekly)
Newsletter (5 times annually) 16570

Society for Psychological Anthropology (SPA), 1703 New Hampshire Av NW, Washington, DC 20009
T: (202) 232-8800
Founded: 1976; Members: 847
Pres: Richard A. Shweder
Focus: Anthro
Periodical
Ethos (weekly) 16571

Society for Psychophysiological Research (SPR), c/o Robert J. Gatchel, Psychology Dept, University of Texas Southwestern Medical Center, 5323 Harry Hines Blvd, Dallas, TX 75235-9044
T: (214) 688-7100
Founded: 1960; Members: 906
Gen Secr: Robert J. Gatchel
Focus: Psych; Physiology; Psychiatry; Med
Periodical
Psychophysiology (weekly) 16572

Society for Public Health Education (SOPHE), 2001 Addison St, Ste 220, Berkeley, CA 94704
T: (510) 644-9242; Fax: (510) 644-9319
Founded: 1950; Members: 1200
Gen Secr: James P. Lovegren
Focus: Public Health
Periodicals
Health Education Quarterly (weekly)
SOPHE News and Views (weekly) . . . 16573

Society for Range Management (SRM), 1839 York St, Denver, CO 80206
T: (303) 355-7070
Founded: 1948; Members: 5700
Gen Secr: Peter V. Jackson
Focus: Business Admin
Periodicals
Journal of Range Management (weekly)
Mini-Directory (weekly)
Rangelands (weekly)
SRM Notes 16574

Society for Reformation Research (SRR), c/o Center for Reformation Research, 6477 San Benito Av, Saint Louis, MO 63105
T: (314) 727-6655
Founded: 1947; Members: 400
Pres: H.C.E. Midelfort
Focus: Hist 16575

Society for Research in Child Development (SRCD), c/o University of Chicago Press, 5720 S Woodlawn Av, Chicago, IL 60637
T: (312) 702-7470; Fax: (312) 702-0694
Founded: 1933; Members: 4500
Gen Secr: Barbara Kahn
Focus: Educ
Periodicals
Child Development (weekly)
Child Development Abstracts & Bibliography (3 times annually)
Monographs (weekly) 16576

Society for Risk Analysis (SRA), 8000 Westpark Dr, Ste 130, McLean, VA 22102
T: (703) 790-1745; Fax: (703) 790-9063
Founded: 1981; Members: 2000
Gen Secr: Richard J. Burk
Focus: Math
Periodicals
Newsletter (weekly)
Risk Analysis Journal (weekly) 16577

Society for Sedimentary Geology, POB 4756, Tulsa, OK 74159-0756
T: (918) 743-9765; Fax: (918) 743-2498
Founded: 1926; Members: 5522
Gen Secr: Robin J. Dixon
Focus: Mineralogy; Paleontology; Econ
Periodicals
Journal of Sedimentary Paleontology (weekly)
Palaios (weekly)
SEPM Newsletter (3 times annually) . . 16578

Society for Slovene Studies (SSS), c/o Dept of Slavic Languages and Literatures, Indiana University, Bloomington, IN 47405
T: (812) 855-2608; Fax: (812) 855-6884
Founded: 1973; Members: 300
Pres: Dr. Henry R. Cooper
Focus: Lit; Ling; Hist; Anthro; Ethnology
Periodicals
Journal of Slovene Studies (weekly)
Letter (biennial) 16579

Society for Social Studies of Science (SSSS), c/o Wesley Shrum, Dept of Sociology, Louisiana State University, Baton Rouge, LA 70803
T: (504) 388-1645
Founded: 1975; Members: 500
Gen Secr: Wesley Shrum
Focus: Soc Sci
Periodical
Science and Technology Studies (weekly) 16580

Society for South India Studies (SSIS), c/o Dr. George L. Hart, Dept of South and Southeast Asian Studies, University of California, 1203 Dwinelle Hall, Berkeley, CA 94720
T: (510) 642-4564
Founded: 1968; Members: 200
Gen Secr: Dr. George L. Hart
Focus: Educ 16581

Society for Spanish and Portuguese Historical Studies (SSPHS), c/o Paul Freedman, Dept of History, Vanderbilt University, Nashville, TN 37235
T: (615) 322-2575; Fax: (615) 343-8028
Founded: 1969; Members: 450
Gen Secr: Paul Freedman
Focus: Hist
Periodicals
Membership Directory (biennial)
Society for Spanish and Portuguese Historical Studies Bulletin (3 times annually) . . . 16582

Society for Surgery of the Alimentary Tract (SSAT), c/o John H.C. Ranson, Dept of Surgery, New York University Medical Center, 530 First Av, Ste 6B, New York, NY 10016
T: (212) 263-6387; Fax: (212) 263-7606
Founded: 1960; Members: 1117
Gen Secr: John H.C. Ranson
Focus: Surgery 16583

Society for Technical Communication (STC), 901 N Stuart St, Ste 304, Arlington, VA 22203
T: (703) 522-4114; Fax: (703) 522-2075
Founded: 1960; Members: 17000
Gen Secr: William C. Stolgitis
Focus: Eng
Periodicals
Intercom Newsletter (weekly)
Proceedings of Annual Conference
Technical Communication (weekly) . . . 16584

Society for the Advancement of Ambulatory Care (SAAC), 1330 New Hampshire Av NW, Ste 121, Washington, DC 20036
T: (202) 659-8008; Fax: (202) 659-8519
Founded: 1981; Members: 508
Pres: Robert Russell
Focus: Med 16585

Society for the Advancement of American Philosophy (SAAP), c/o Dept of Philosophy, Seattle University, Seattle, WA 98122
T: (206) 296-5467; Fax: (206) 296-5997
Founded: 1972; Members: 900
Gen Secr: Kenneth W. Stikkers
Focus: Philos
Periodicals
Membership Directory
SAAP Newsletter (weekly) 16586

Society for the Advancement of Behavior Analysis (SABA), c/o Western Michigan University, 260 Wood Hall, Kalamazoo, MI 49008
T: (616) 387-4495
Founded: 1980; Members: 150
Gen Secr: William K. Redmon
Focus: Psychoan
Periodical
The Behavior Analyst (weekly) 16587

Society for the Advancement of Education (SAE), 99 W Hawthorne Av, Valley Stream, NY 11580
T: (516) 568-9191
Founded: 1939; Members: 3000
Pres: Stanley Lehrer
Focus: Educ
Periodical
USA Today (weekly) 16588

Society for the Advancement of Material and Process Engineering (SAMPE), POB 2459, Covina, CA 91722
T: (818) 331-0616; Fax: (818) 332-8929
Founded: 1944; Members: 11000
Gen Secr: Dr. C.L. Hamermesh
Focus: Materials Sci
Periodicals
National SAMPE Technical Conference Series (weekly)
SAMPE Journal (weekly)
SAMPE Quarterly (weekly)
The Science of Advaneed Materials + Process Engineering (weekly) 16589

Society for the Advancement of Scandinavian Study (SASS), c/o Steven P. Sondrap, Brigham Young University, 3003 JKHB, Provo, UT 84602
T: (801) 378-5598; Fax: (801) 378-4649
Founded: 1911; Members: 800
Gen Secr: Steven P. Sondrap
Focus: Ethnology
Periodical
Scandinavian Studies (weekly) 16590

Society for the Anthropology of Europe (SAE), 1703 New Hampshire Av NW, Washington, DC 20009
T: (202) 232-8800
Founded: 1986; Members: 530
Pres: Michael Herzfeld
Focus: Anthro 16591

Society for the Arts, Religion and Contemporary Culture (ARC), c/o Mark Harvey, POB 8721, JFK Station, Boston, MA 02114
T: (617) 253-8778
Founded: 1961; Members: 160
Gen Secr: Mark Harvey
Focus: Arts; Rel & Theol
Periodical
SEEDBED (biennial) 16592

Society for the Comparative Study of Society and History (CSSH), c/o University of Michigan, 102 Rackham Bldg, Ann Arbor, MI 48109-1070
T: (313) 764-6362; Fax: (313) 763-2447
Founded: 1958
Pres: Prof. Raymond Grew
Focus: Hist; Sociology; Anthro
Periodical
Comparative Studies in Society and History (weekly) 16593

Society for the Conservation of Bighorn Sheep (SCBS), 2106 E Mesita Av, West Covina, CA 91791
T: (818) 331-6241; Fax: (818) 393-1173
Members: 350
Pres: John Carnakis
Focus: Animal Husb
Periodical
Sheepherder (weekly) 16594

Society for the Furtherance and Study of Fantasy and Science Fiction, POB 1624, Madison, WI 53701-1624
T: (608) 233-5640
Founded: 1977; Members: 75
Gen Secr: Pat Hario
Focus: Cinema; Lit
Periodicals
Cube (weekly)
New Moon
SF3 Locater List (weekly) 16595

Society for the History of Czechoslovak Jews (SHCJ), 8708 Santiago St, Holliswood, NY 11423
T: (718) 468-6844
Founded: 1961; Members: 200
Pres: Lewis Weiner
Focus: Hist
Periodicals
The Jews of Czechoslovakia
Review I
Review II 16596

Society for the History of Discoveries (SHD), 6300 Waterway Dr, Falls Church, VA 22044-1316
T: (703) 256-9217
Founded: 1960; Members: 375
Gen Secr: Eric W. Wolf
Focus: Hist
Periodical
Terrae Incognitae (weekly) 16597

Society for the History of Technology (SHOT), c/o Dept of Social Sciences, Michigan Technological University, 1400 Townsend Dr, Houghton, MI 49931-1295
T: (906) 487-2459; Fax: (906) 487-2824
Founded: 1958; Members: 2800
Gen Secr: Bruce E. Seeley
Focus: Cultur Hist; Eng
Periodicals
SHOT Newsletter (weekly)
Technology and Culture (weekly) 16598

Society for the History of the Germans in Maryland (SHGM), POB 22585, Baltimore, MD 21203
Founded: 1886; Members: 250
Pres: Gerard W. Wittstadt
Focus: Hist
Periodical
The Report: A Journal of German American History (biennial) 16599

Society for the Humanities (SH), c/o Andrew D. White House, Cornell University, 27 East Av, Ithaca, NY 14853
T: (607) 255-4086
Founded: 1966
Gen Secr: Jonathan Culler
Focus: Humanities 16600

Society for the Investigation of Recurring Events (SIRE), c/o Peter G. Constable, POB 274, Iselin, NJ 08830
T: (800) 992-9982
Founded: 1957; Members: 300
Pres: Peter G. Constable
Focus: Hist
Periodicals
Abstract (weekly)
Bulletin (weekly) 16601

Society for the Philosophical Study of Marxism (SPSM), c/o Prof. Howard L. Parsons, Dept of Philosophy, University of Bridgeport, Bridgeport, CT 06601-2449
T: (203) 576-4212
Founded: 1962; Members: 50
Pres: Prof. Howard L. Parsons
Focus: Philos
Periodical
SPSM Newsletter (biennial) 16602

Society for the Preservation of American Business History (SPABH), Drawer JH, Williamsburg, VA 23187-3632
T: (804) 220-3838
Founded: 1981; Members: 100

Pres: George H.H. Garrison
Focus: Econ; Business Admin 16603

Society for the Preservation of English Language and Literature (SPELL), POB 118, Waleska, GA 30183
T: (706) 479-8685
Founded: 1984; Members: 2000
Pres: Richard Dowis
Focus: Ling; Lit
Periodical
Spell/Binder (weekly) 16604

Society for the Preservation of New England Antiquities (SPNEA), 141 Cambridge St, Boston, MA 02114
T: (617) 227-3956
Founded: 1910; Members: 3000
Gen Secr: Jane Nylander
Focus: Preserv Hist Monuments
Periodicals
House Guide (weekly)
Newsletter 16605

Society for the Promotion of Science and Scholarship (SPOSS), 4139 El Camino Way, POB 10139, Palo Alto, CA 94303-0897
T: (415) 493-4400
Founded: 1975
Gen Secr: Dr. Janet Gardiner
Focus: Nat Sci 16606

Society for the Protection of Old Fishes (SPOOF), c/o School of Fisheries, University of Washington, Seattle, WA 98195
T: (206) 778-7397; Fax: (206) 685-7471
Founded: 1967; Members: 200
Pres: Dr. George W. Brown
Focus: Fisheries
Periodicals
Membership List (weekly)
Newsletter 16607

Society for the Psychological Study of Social Issues (SPSSI), POB 1248, Ann Arbor, MI 48106-1248
T: (313) 662-9130; Fax: (313) 662-5607
Founded: 1936; Members: 3000
Gen Secr: Sandy West
Focus: Psych; Sociology
Periodicals
Journal of Social Issues (weekly)
Newsletter (3 times annually) 16608

Society for Theriogenology (ST), POB 2118, Hastings, NE 68902
T: (402) 463-0392; Fax: (402) 461-4103
Founded: 1954; Members: 2508
Gen Secr: Don Ellerbee
Focus: Vet Med
Periodicals
Newsletter (weekly)
Proceedings of Annual Meeting 16609

Society for the Scientific Study of Religion (SSSR), c/o Purdue University, 193 Pierce Hall, West Lafayette, IN 47907-1305
T: (317) 494-6286
Founded: 1949; Members: 1600
Gen Secr: Edward Lehman
Focus: Rel & Theol; Philos; Behav Sci
Periodical
Journal for the Scientific Study of Religion (weekly) 16610

Society for the Scientific Study of Sex (SSSS), c/o Howard J. Ruppel, POB 208, Mount Vernon, IA 52314
T: (319) 895-8407; Fax: (319) 895-6203
Founded: 1957; Members: 1100
Gen Secr: Howard J. Ruppel
Focus: Bio
Periodical
The Journal of Sex Research (weekly) . 16611

Society for the Study of Amphibians and Reptiles (SSAR), c/o Dr. Douglas Taylor, Dept of Zoology, Miami University, Oxford, OH 45056
T: (513) 529-4901; Fax: (513) 529-6900
Founded: 1958; Members: 2700
Gen Secr: Dr. Douglas Taylor
Focus: Zoology
Periodicals
Herpetological Circulars
Herpetological Review (weekly)
Journal of Herpetology (weekly) 16612

Society for the Study of Blood (SSB), c/o Dr. Ian Yudelman, Brooklyn VA Medical Center, 800 Poly Pl, Brooklyn, NY 11209
T: (718) 836-6600
Founded: 1945; Members: 200
Gen Secr: Dr. Ian Yudelman
Focus: Hematology 16613

Society for the Study of Breast Disease (SSBD), 3409 Worth, Dallas, TX 75246
T: (214) 821-2962; Fax: (214) 827-7032
Founded: 1976; Members: 250
Gen Secr: G.N. Peters
Focus: Intern Med 16614

Society for the Study of Early China (SSEC), c/o Institute of East Asian Studies, University of California, Berkeley, CA 94720
T: (510) 643-6325; Fax: (510) 643-7062
Founded: 1975; Members: 350
Gen Secr: David N. Keightley
Focus: Hist
Periodical
Early China (weekly) 16615

Society for the Study of Evolution (SSE), c/o Allen Press, POB 368, Lawrence, KS 66044
Fax: (608) 262-7509
Founded: 1946; Members: 2500
Gen Secr: Donald M. Waller
Focus: Bio
Periodicals
Evolution (weekly)
Membership Directory 16616

Society for the Study of Male Psychology and Physiology (SSMPP), c/o Jerry Bergman, 321 Iuka, Montpellier, OH 43543
T: (419) 485-3602
Founded: 1975; Members: 180
Gen Secr: Jerry Bergman
Focus: Psych 16617

Society for the Study of Process Philosophies (SSPP), c/o Dr. Joseph Grange, Dept of Philosophy, University of Southern Maine, 96 Falmouth St, Portland, ME 04103
T: (207) 780-4259
Founded: 1966; Members: 350
Gen Secr: Dr. Joseph Grange
Focus: Philos 16618

Society for the Study of Reproduction (SSR), 309 W Clark St, Champaign, IL 61820
T: (217) 356-3182; Fax: (217) 398-4119
Founded: 1967; Members: 1450
Focus: Med
Periodical
Biology of Reproduction (10 times annually) 16619

Society for the Study of Social Biology (SSSB), c/o Robin U. Loomis, East-West Population Institute, 1777 East-West Rd, Honolulu, HI 96848
T: (808) 944-7444; Fax: (808) 944-7490
Founded: 1926; Members: 400
Gen Secr: Robert D. Retherford
Focus: Bio
Periodical
Social Biology (weekly) 16620

Society for the Study of Social Problems (SSSP), c/o Dept of Sociology, University of Tennessee, 906 McClung Tower, Knoxville, TN 37996-0490
T: (615) 974-3620; Fax: (615) 974-8546
Founded: 1951; Members: 1546
Gen Secr: Thomas C. Hood
Focus: Sociology
Periodicals
Social Problems (5 times annually)
The Newsletter 16621

Society for the Study of Southern Literature (SSSL), c/o David C. Estes, Dept of English, Loyola University, New Orleans, LA 70118
T: (504) 865-2476
Founded: 1968; Members: 425
Gen Secr: David C. Estes
Focus: Lit
Periodical
News-Letter (weekly) 16622

Society for the Study of Symbolic Interaction (SSSI), 4975 Dover St NE, Saint Petersburg, FL 33703
T: (813) 525-2896
Founded: 1975; Members: 600
Gen Secr: Donna Kelleher Darden
Focus: Sociology
Periodicals
Journal (weekly)
Newsletter (weekly) 16623

Society for Urban Anthropology (SUA), 1703 New Hampshire Av NW, Washington, DC 20009
T: (202) 232-8800; Fax: (202) 667-5345
Founded: 1979; Members: 1000
Pres: Kathleen Logan
Focus: Anthro 16624

Society for Values on Higher Education (SVHE), c/o Georgetown University, POB B2814, Washington, DC 20057
T: (202) 687-3653; Fax: (202) 687-7084
Founded: 1923; Members: 1375
Gen Secr: Kathleen McGrory
Focus: Educ
Periodical
Soundings: An Interdisciplinary Journal (weekly) 16625

Society for Vascular Surgery (SVS), 13 Elm St, POB 1565, Manchester, MA 01944
T: (508) 526-8330; Fax: (508) 526-4018
Founded: 1945; Members: 500
Gen Secr: Joseph M. Stone
Focus: Surgery
Periodical
Journal of Vascular Surgery (weekly) . . 16626

Society for Vector Ecology (SVE), POB 87, Santa Ana, CA 92702
T: (714) 971-2421; Fax: (714) 971-3940
Founded: 1968; Members: 820
Gen Secr: Gilbert L. Challet
Focus: Ecology
Periodical
Bulletin of The Society for Vector Ecology (weekly) 16627

Society for Visual Anthropology (SVA), c/o Joanna C. Scherer, Smithsonian Institution, Natural Science Bldg, Rm 85, Washington, DC 20560
T: (202) 357-1861; Fax: (202) 357-2801
Founded: 1968; Members: 525
Pres: Joanna C. Scherer
Focus: Anthro
Periodicals
Directory of Visual Anthropologists
Newsletter (weekly) 16628

Society for Women in Philosophy – Eastern Division (SWIP), c/o Dr. Patrice Di Quinzio, Philosophy Dept, University of Scranton, Scranton, PA 18510
T: (717) 941-7757
Founded: 1971; Members: 850
Gen Secr: Dr. Patrice Di Quinzio
Focus: Philos
Periodicals
Hypatia (weekly)
Newsletter (weekly) 16629

Society for Women in Philosophy – Pacific Division (SWIP), c/o Rita Manning, Dept of Philosophy, San Jose State University, San Jose, CA 95192
T: (408) 924-4501
Founded: 1975; Members: 140
Gen Secr: Rita Manning
Focus: Philos
Periodical
Newsletter (weekly) 16630

Society of Actuaries (SOA), 475 N Martingale Rd, Ste 800, Schaumburg, IL 60173-2226
T: (708) 706-3500; Fax: (708) 706-3599
Founded: 1949; Members: 13839
Gen Secr: John E. O'Connor
Focus: Insurance
Periodicals
The Actuary (10 times annually)
Record (weekly)
Transactions (weekly) 16631

Society of Allied Weight Engineers (SAWE), 344 E J St, Chula Vista, CA 92010
T: (619) 427-8262
Founded: 1939; Members: 1150
Gen Secr: Fred H. Wetmore
Focus: Eng
Periodicals
Newsletter (weekly)
Weight Engineering Journal (3 times annually) 16632

Society of American Archivists (SAA), 600 S Federal St, Ste 504, Chicago, IL 60605
T: (312) 922-0140; Fax: (312) 347-1452
Founded: 1936; Members: 4500
Gen Secr: Anne P. Diffendal
Focus: Libraries & Bk Sci; Archives
Periodicals
American Archivist (weekly)
Membership Directory (weekly)
Newsletter (weekly) 16633

Society of American Foresters (SAF), 5400 Grosvenor Ln, Washington, DC 20814
T: (301) 897-8720; Fax: (301) 897-3690
Founded: 1900; Members: 190000
Gen Secr: William H. Banzhaf
Focus: Forestry
Periodicals
Forest Science (weekly)
Journal of Forestry (weekly)
Northern Journal of Applied Forestry (weekly)
Southern Journal of Applied Forestry (weekly)
Western Journal of Applied Forestry (weekly) 16634

Society of American Historians (SAH), c/o Butler Library, Columbia University, New York, NY 10027
T: (212) 854-2221; Fax: (212) 932-0602
Founded: 1939; Members: 350
Gen Secr: Mark Carnes
Focus: Hist 16635

Society of American Law Teachers (SALT), c/o Law School, New York University, 40 Washington Sq S, New York, NY 10012
T: (212) 998-6265; Fax: (212) 995-3156
Founded: 1974; Members: 500
Gen Secr: Sylvia A. Law
Focus: Educ; Law 16636

Society of American Registered Architects (SARA), 1245 S Highland Av, Lombard, IL 60148
T: (708) 932-4622
Founded: 1956; Members: 800
Gen Secr: Stanley D. Banash
Focus: Archit
Periodicals
Practicing Architect (weekly)
Practicing Architect Magazine (weekly)
Sarascope Newsletter (weekly) 16637

Society of American Value Engineers (SAVE), 60 Revere Dr, Ste 500, Northbrook, IL 60062
T: (708) 480-9080
Founded: 1959; Members: 1600
Gen Secr: Barbara Scott
Focus: Standards
Periodicals
Interactions (weekly)
Value World (weekly) 16638

Society of Architectural Historians (SAH), 1232 Pine St, Philadelphia, PA 19107
T: (215) 735-0224; Fax: (215) 735-2590
Founded: 1940; Members: 3600
Gen Secr: David Bahlman
Focus: Archit; Hist
Periodicals
Journal of the SAH (weekly)
Newsletter (weekly)
Preservation Forum (weekly) 16639

Society of Basque Studies in America (SBSA), c/o Ignacio R.M. Galbis, 19 Colonial Gardens, Brooklyn, NY 11209
T: (718) 745-1141
Founded: 1978; Members: 1000
Gen Secr: Ignacio R. Galbis
Focus: Ling 16640

Society of Behavioral Medicine (SBM), 103 S Adams St, Rockville, MD 20850
T: (301) 251-2790; Fax: (301) 279-6749
Founded: 1978; Members: 2500
Gen Secr: Judith C. Woodward
Focus: Behav Sci
Periodicals
Annals of Behavioral Medicine (weekly)
Behavioral Medicine Abstracts (weekly) . 16641

Society of Biblical Literature (SBL), 1549 Clairmont Rd, Ste 204, Decatur, GA 30033-4635
T: (404) 636-4744
Founded: 1880; Members: 5200
Gen Secr: David J. Lull
Focus: Lit; Rel & Theol
Periodicals
Journal of Biblical Literature (weekly)
Semeia (weekly) 16642

Society of Biological Psychiatry (SBP), c/o Mental Health Clinical Research Center, Utah Southwestern Mental Health Center, 5959 Harrry Hines Blvd, Ste 600, Dallas, TX 75235
T: (214) 688-8222; Fax: (214) 688-4278
Founded: 1945; Members: 900
Gen Secr: A. John Rush
Focus: Psychiatry
Periodical
Biological Psychiatry (weekly) 16643

Society of Broadcast Engineers (SBE), POB 20450, Indianapolis, IN 46220
T: (317) 253-1640; Fax: (317) 253-0418
Founded: 1963; Members: 5600
Gen Secr: John L. Poray
Focus: Electric Eng 16644

Society of Cable Television Engineers (SCTE), 669 Exton Commons, Exton, PA 19341
T: (215) 363-6888; Fax: (215) 363-5898
Founded: 1969; Members: 10000
Pres: William M. Riker
Focus: Eng; Mass Media
Periodical
The Interval (weekly) 16645

Society of Carbide and Tool Engineers (SCTE), c/o American Society for Metals, Metals Park, OH 44073
T: (216) 388-5151
Founded: 1947; Members: 1500
Pres: Carl Wegner
Focus: Eng
Periodicals
Carbide and Tool Journal (weekly)
Proceedings (weekly) 16646

Society of Cardiovascular and Interventional Radiology (SCVIR), 10201 Lee Hwy, Ste 160, Fairfax, VA 22030
T: (703) 691-1805; Fax: (703) 691-1855
Founded: 1973; Members: 2000
Gen Secr: Todd M. Palmquist
Focus: Radiology
Periodical
SCVIR Newsletter (weekly) 16647

Society of Cardiovascular Anesthesiologists (SCA), 1910 Byrd Av, POB 11086, Richmond, VA 23230-1086
T: (804) 282-0084; Fax: (804) 282-0090
Members: 4000
Gen Secr: John A. Hinckley
Focus: Anesthetics
Periodical
SCA Newsletter (weekly) 16648

Society of Christian Ethics (SCE), c/o School of Theology, Boston University, 745 Commonwealth Av, Boston, MA 02215
T: (617) 353-7322
Founded: 1959; Members: 950
Gen Secr: John Cartwright
Focus: Philos 16649

Society of Composers, POB 296, Old Chelsea Station, New York, NY 10013-0296
T: (718) 899-2605
Founded: 1966; Members: 900
Gen Secr: Martin Gonzalez
Focus: Music
Periodicals
Journal of Music Scores (weekly)
Newsletter (weekly)
Record Series (weekly) 16650

Society of Cosmetic Chemists (SCC), 1995 Broadway, Ste 1701, New York, NY 10023
T: (212) 874-0600
Founded: 1945; Members: 3250

Gen Secr: Theresa Cesario
Focus: Chem
Periodicals
Journal (weekly)
Newsletter (weekly) 16651

Society of Cost Estimating and Analysis (SCEA), 101 S Whiting St, Ste 201, Alexandria, VA 22304
T: (703) 751-8069; Fax: (703) 461-7328
Founded: 1990; Members: 4000
Gen Secr: L.T. Baseman
Focus: Finance
Periodicals
Journal of Cost Analysis (weekly)
Newsletter (weekly) 16652

Society of Critical Care Medicine (SCCM), 8101 E Kaiser Bldv, Anaheim, CA 92808
T: (714) 282-6000; Fax: (714) 870-5243
Founded: 1970; Members: 4000
Gen Secr: Norma Shoemaker
Focus: Med
Periodicals
Critical Care Medicine (weekly)
Critical Care: State of the Art (weekly)
Newsmagazine Concern (weekly) 16653

Society of Diagnostics Medical Sonographers (SDMS), 12770 Coit Rd, Ste 508, Dallas, TX 75251
T: (214) 239-7367; Fax: (214) 239-7378
Founded: 1970; Members: 10000
Gen Secr: Gwen Grim
Focus: Med
Periodicals
Journal of Diagnostic Medical Sonography (weekly)
Newsletter (weekly) 16654

Society of Economic Geologists (SEG), 5805 S Rapp St, Ste 209, Littleton, CO 80120
T: (303) 797-0332; Fax: (303) 797-0417
Founded: 1920; Members: 2700
Gen Secr: John A. Thoms
Focus: Geology; Econ
Periodicals
Economic Geology (8 times annually)
Membership List (biennial) 16655

Society of Engineering Illustrators (SEI), c/o Robert A. Clarke, 1818 Englewood, Madison Heights, MI 48071
T: (313) 588-2776
Founded: 1954; Members: 80
Gen Secr: Robert A. Clarke
Focus: Eng
Periodical
The Fine Line (weekly) 16656

Society of Engineering Science (SES), c/o J. Mark Duva, Dept of Applied Mathematics, University of Virginia, 111G Olsson Hall, Charlottesville, VA 22903
T: (804) 924-1029; Fax: (804) 924-6270
Founded: 1963; Members: 400
Gen Secr: J. Mark Duva
Focus: Eng 16657

Society of Ethnic and Special Studies (SESS), c/o Southern Illinois University at Edwardsville, POB 1652, Edwardsville, IL 62026
T: (618) 692-2042
Founded: 1973; Members: 400
Pres: Dr. Emil F. Jason
Focus: Ethnology; Educ
Periodical
Journal (weekly) 16658

Society of Ethnobiology (SE), c/o Catherine Fowler, Dept of Anthropology, University of Nevada, Reno, NV 89557
T: (702) 784-4686; Fax: (702) 784-1300
Founded: 1978; Members: 450
Gen Secr: Catherine Fowler
Focus: Bio
Periodical
Journal of Ethnobiology (weekly) 16659

Society of Experimental Psychologists (SEP), c/o Dr. Herschel W. Leibowitz, Dept of Psychology, Pennsylvania State University, University Park, PA 16802-3106
T: (814) 863-1735; Fax: (814) 863-7002
Members: 180
Gen Secr: Dr. Herschel W. Leibowitz
Focus: Psych
Periodical
Annual Report (weekly) 16660

Society of Experimental Test Pilots, POB 986, Lancaster, CA 93534
T: (805) 942-9574; Fax: (805) 940-0398
Members: 1900
Gen Secr: Thomas H. Smith
Focus: Aero 16661

Society of Exploration Geophysicists (SEG), POB 702740, Tulsa, OK 74170
T: (918) 493-3516; Fax: (918) 493-2074
Founded: 1930; Members: 15000
Gen Secr: John Hyden
Focus: Geophys
Periodicals
Geophysics (weekly)
Roster (weekly) 16662

Society of Federal Linguists (SFL), POB 7765, Washington, DC 20044
T: (202) 324-2989
Founded: 1930; Members: 100
Pres: Bonnie L. Jenkins
Focus: Ling
Periodicals
Membership Directory (weekly)
Newsletter (weekly) 16663

Society of Fire Protection Engineers (SFPE), 1 Liberty Sq, Boston, MA 02109-9825
T: (617) 482-0686; Fax: (617) 482-8184
Founded: 1950; Members: 3250
Gen Secr: D. Peter Lund
Focus: Safety
Periodical
SFPE Journal of Fire Protection Engineering (weekly) 16664

Society of Flavor Chemists (SFC), c/o Denise McCafferty, McCormick & Co., 204 Wight Av, Hunt Valley, MD 21031
T: (410) 771-7491
Founded: 1959; Members: 300
Gen Secr: Denise McCafferty
Focus: Chem
Periodical
Newsletter (weekly) 16665

Society of Flight Test Engineers (SFTE), POB 4047, Lancaster, CA 93539
T: (805) 538-9715; Fax: (805) 538-9715
Founded: 1968; Members: 1100
Gen Secr: Dianne Van Norman
Focus: Eng
Periodicals
Flight Test News (weekly)
Membership Directory (biennial) 16666

Society of Freight Car Historians (SFCH), POB 2480, Monrovia, CA 91017
Founded: 1980; Members: 1000
Pres: David G. Casdorph
Focus: Transport
Periodicals
Freight Cars Journal (weekly)
Freight Cars Journal Monograph (6-10 times annually) 16667

Society of Gastroenterology Nurses and Associates (SGNA), 1070 Sibley Tower, Rochester, NY 14604
T: (716) 546-7241; Fax: (716) 546-5141
Founded: 1974; Members: 6000
Gen Secr: Margaret M. Crevey
Focus: Gastroenter
Periodical
Journal (weekly) 16668

Society of General Physiologists (SGP), POB 257, Woods Hole, MA 02543
T: (508) 540-6719; Fax: (508) 540-0155
Founded: 1946; Members: 1000
Pres: Dr. Douglas D. Eaton
Focus: Physiology
Periodical
Journal of General Physiology (weekly) . 16669

Society of Glass Science and Practices (SGSP), POB 166, Clarksburg, WV 26301
T: (412) 225-4400
Founded: 1969; Members: 150
Gen Secr: Dr. Paul Pakula
Focus: Materials Sci 16670

Society of Head and Neck Surgeons (SHNS), 4900B S 31 St, Arlington, VA 22206
T: (703) 820-7400; Fax: (703) 931-4520
Founded: 1954; Members: 700
Gen Secr: Ernst T. Bomar
Focus: Surgery 16671

Society of Imaging Science and Technology (SPSE), 7003 Kilworth Ln, Springfield, VA 22151
T: (703) 642-9090; Fax: (703) 642-9094
Founded: 1947; Members: 3000
Gen Secr: Calva A. Lotridge
Focus: Photo
Periodicals
Journal of Imaging Technology (weekly)
Photographic Imaging Science (weekly) . 16672

Society of Independent Professional Earth Scientists (SIPES), 4925 Greenville Av, Ste 170, Dallas, TX 75206
T: (214) 363-1780; Fax: (214) 363-8195
Founded: 1963; Members: 1400
Gen Secr: Diane Finstrom
Focus: Geology
Periodical
Newsletter (weekly) 16673

Society of Insurance Research (SIR), POB 72221, Marietta, GA 30007
T: (404) 518-0453; Fax: (404) 518-1998
Founded: 1970; Members: 600
Gen Secr: Stanley M. Hopp
Focus: Insurance
Periodical
Research Review (weekly) 16674

Society of Jewish Science (SJS), POB 11803, Plainview, NY 11803
T: (516) 349-0022
Founded: 1922
Gen Secr: David Goldstein
Focus: Sci
Periodical
Jewish Science Interpreter (8 times annually) 16675

Society of Logistics Engineers (SOLE), 8100 Professional Pl, Ste 211, New Carrollton, MD 20785
T: (301) 459-8446; Fax: (301) 459-1522
Members: 9000
Gen Secr: Norman Michaud
Focus: Business Admin
Periodicals
Annals (weekly)
Member's Handbook and Membership Directory (weekly)
Soletter (weekly)
Spectrum (weekly) 16676

Society of Manufacturing Engineers (SME), 1 SME Dr, POB 930, Dearborn, MI 48121
T: (313) 271-1500; Fax: (313) 271-2861
Founded: 1932; Members: 75000
Gen Secr: Philip Trimble
Focus: Eng 16677

Society of Medical Consultants to the Armed Forces (SMCAF), 4301 James Bridge Rd, Bethesda, MD 20814-4799
T: (301) 295-3903
Founded: 1945; Members: 1000
Gen Secr: Anne Hufman
Focus: Med
Periodicals
Newsletter (3 times annually)
Roster 16678

Society of Medical Jurisprudence (SMJ), c/o Lea S. Singer, POB 1304, New York, NY 10008
T: (212) 473-0523
Founded: 1883; Members: 300
Pres: Lea S. Singer
Focus: Law; Med
Periodical
Bulletin (weekly) 16679

Society of Motion Picture and Television Art Directors (SMPTAD), 11365 Ventura Blvd, Ste 315, Studio City, CA 91604
T: (818) 762-9995
Founded: 1960; Members: 400
Gen Secr: Gene Allen
Focus: Cinema 16680

Society of Motion Picture and Television Engineers (SMPTE), 595 W Hartsdale Av, White Plains, NY 10607
T: (914) 761-1100; Fax: (914) 761-3115
Founded: 1916; Members: 9000
Gen Secr: Lynette H. Robinson
Focus: Cinema
Periodicals
Directory (weekly)
Journal (weekly)
News and Notes (weekly) 16681

Society of Multivariate Experimental Psychology (SMEP), c/o Dr. Jack McArdle, Dept of Psychology, University of Virginia, 102 Gilmer Hall, Charlottesville, VA 22903
Founded: 1960; Members: 65
Gen Secr: Dr. Jack McArdle
Focus: Psych
Periodical
Multivariate Behavioral Research (weekly) 16682

Society of Naval Architects and Marine Engineers (SNAME), 601 Pavonia Av, Ste 400, Jersey City, NJ 07306
T: (201) 798-4800; Fax: (201) 798-4975
Founded: 1893; Members: 10598
Gen Secr: Francis M. Cagliari
Focus: Eng; Archit
Periodicals
Journal of Ship Production (weekly)
Journal of Ship Research (weekly)
Marine Technology (weekly)
Transactions (weekly) 16683

Society of Nematologists (SON), c/o R.N. Huehel, Nematology Laboratory, USDA, Bldg 011A BARC-W, Beetsville, MD 20705
T: (301) 344-3081; Fax: (301) 344-2016
Founded: 1961; Members: 860
Gen Secr: R.N. Huehel
Focus: Zoology
Periodicals
Annual Meeting Presentations Abstracts (weekly)
Journal of Nematology (weekly)
Nematology Newsletter (weekly) 16684

Society of Neurological Surgeons (SNS), 750 Washington St, POB 178, Boston, MA 02111
T: (617) 956-5858
Founded: 1920; Members: 215
Gen Secr: William Shucart
Focus: Surgery; Neurology 16685

Society of Neurosurgical Anesthesia and Critical Care (SNACC), 11512 Allecingie Pkwy, Richmond, VA 23235
T: (804) 379-5513
Founded: 1973; Members: 613
Gen Secr: Jerry Wilhoit
Focus: Neurology
Periodicals
Comprehensive Bibliography in Neuroanesthesia (weekly)
Newsletter (3-4 times annually) 16686

Society of Nuclear Medicine (SNM), 136 Madison Av, New York, NY 10016-67760
T: (212) 889-0717; Fax: (212) 545-0221
Founded: 1954; Members: 12000
Gen Secr: Torry S. Sansone
Focus: Nucl Med 16687

Society of Pelvic Surgeons (SPS), c/o Dr. Carmel Cohen, Dept of OB-GYN, Mount Sinai School of Medicine, 1176 Fifth Av, New York, NY 10029
T: (212) 241-6554; Fax: (212) 360-6917
Founded: 1952; Members: 125
Gen Secr: Dr. Carmel Cohen
Focus: Surgery 16688

Society of Petroleum Engineers (SPE), POB 833836, Richardson, TX 75083-3836
T: (214) 952-9393; Fax: (214) 952-9435
Founded: 1922; Members: 53000
Gen Secr: Dan K. Adamson
Focus: Petrochem
Periodicals
Enhanced Oil-Recovery Field Reports (weekly)
Journal of Petroleum Technology (weekly)
SPE Drilling Engineering (weekly)
SPE Formation Evaluation (weekly)
SPE Production Engineering (weekly)
SPE Reservoir Engineering (weekly) . . 16689

Society of Philatelists and Numismatists (SPAN), 1929 Millis St, Montebello, CA 90640
T: (213) 724-1595
Members: 300
Gen Secr: Joe R. Ramos
Focus: Numismatics
Periodical
Expansion (weekly) 16690

Society of Philippine Surgeons of America (SPSA), c/o Antonio T. Donato, 1125 S Jefferson St, Roanoke, VA 24016
T: (703) 982-1141
Founded: 1972; Members: 471
Pres: Antonio T. Donato
Focus: Surgery
Periodicals
Directory (weekly)
Philippine Surgeon (weekly) 16691

Society of Plastics Engineers (SPE), 14 Fairfield Dr, Brookfield, CT 06804-0403
T: (203) 775-0471; Fax: (203) 775-8490
Founded: 1942; Members: 37000
Gen Secr: Robert D. Forger
Focus: Eng
Periodicals
Journal of Vinyl Technology (weekly)
Plastics Engineering (weekly)
Polymer Composites (weekly)
Polymer Engineering (18 times annually) 16692

Society of Professional Archaeologists (SOPA), c/o Kansas State History Society, 120 W Tenth, Topeka, KS 66612
T: (913) 296-2625; Fax: (913) 296-1005
Founded: 1976; Members: 650
Pres: Larry Banks
Focus: Archeol
Periodicals
Directory of Professional Archaeoloists (weekly)
Newsletter (weekly) 16693

Society of Professional Well Log Analysts (SPWLA), 6001 Gulf Freeway, Ste C129, Houston, TX 77023
T: (713) 928-8925; Fax: (713) 928-9061
Founded: 1959; Members: 4000
Gen Secr: Vicki J. King
Focus: Geology; Eng; Energy
Periodicals
Annual Transactions (weekly)
The Log Analyst (weekly) 16694

Society of Professors of Child and Adolescent Psychiatry (SPCAP), 3615 Wisconsin Av NW, Washington, DC 20016-3007
T: (202) 966-7300
Founded: 1969; Members: 160
Focus: Psychiatry 16695

Society of Professors of Education (SPE), c/o Richard Wisniewski, College of Education, University of Tennessee, Knoxville, TN 37996
T: (615) 974-2201; Fax: (615) 974-8718
Founded: 1902; Members: 274
Gen Secr: Richard Wisniewski
Focus: Educ
Periodicals
DeGarmo Lectures (weekly)
Membership Directory
Monograph Series (weekly) 16696

Society of Prospective Medicine (SPM), POB 55110, Indianapolis, IN 46205-0110
T: (317) 549-3600; Fax: (317) 549-3670
Founded: 1972; Members: 250
Gen Secr: Pamela S. Hall
Focus: Med
Periodicals
An Ounce of Prevention (weekly)
Membership Directory (weekly) 16697

Society of Protozoologists (SP), c/o Baruch College, 17 Lexington Av, POB 199, New York, NY 10010
T: (212) 387-1230; Fax: (212) 387-1235
Founded: 1947; Members: 1125
Pres: Kwang W. Jeon
Focus: Zoology 16698

Society of Research Administrators (SRA), 500 N Michigan Av, Ste 1400, Chicago, IL 60611-3796
T: (312) 661-1700; Fax: (312) 661-0769
Founded: 1967; Members: 2650
Gen Secr: Joan Carter
Focus: Sci
Periodicals
Journal (weekly)
Membership Directory (weekly)
Newsletter (weekly) *16699*

Society of Rheology (SOR), c/o American Institute of Physics, 335 E 45 St, New York, NY 10017
T: (212) 661-9404; Fax: (212) 949-0473
Founded: 1929; Members: 1200
Gen Secr: Dr. Kenneth Ford
Focus: Bio; Chem; Physics
Periodicals
Journal of Rheology (8 times annually)
Rheology Bulletin (weekly) *16700*

Society of Soft Drink Technologists (SSDT), 113 Heron Point Ln, Hartfield, VA 23071
T: (804) 776-9230; Fax: (804) 776-9230
Founded: 1953; Members: 950
Gen Secr: Anthony Meushaw
Focus: Food
Periodical
Annual Meeting Proceedings (weekly) . . *16701*

Society of State Directors of Health, Physical Education and Recreation (SSDHPER), 9805 Hillridge Dr, Kensington, MD 20895
T: (301) 949-0709
Founded: 1926; Members: 180
Gen Secr: Simon A. McNeely
Focus: Sports
Periodicals
Directory (weekly)
Newsletter *16702*

Society of Surgical Oncology (SSO), c/o James R. Slawny, 800 E Northwest Hwy, Ste 1080, Palatine, IL 60067
T: (708) 359-4605
Founded: 1940; Members: 980
Gen Secr: James R. Slawny
Focus: Cell Biol & Cancer Res *16703*

Society of Systematic Biologists (SSB), c/o National Museum of Natural History, Washington, DC 20560
T: (202) 357-2964
Founded: 1948; Members: 1550
Gen Secr: Elizabeth A. Kellogg
Focus: Bio
Periodical
Systematic Zoology (weekly) *16704*

Society of Teachers in Education of Professional Photography (STEPP), 371 Greenport Dr, West Carrollton, OH 45449
T: (513) 859-0180
Founded: 1978; Members: 75
Pres: Jay E. Vada
Focus: Educ; Photo
Periodicals
Journal (weekly)
Membership Directory (weekly) *16705*

Society of Teachers of Family Medicine (STFM), 8880 Ward Pkwy, POB 8729, Kansas City, MO 64114
T: (816) 333-9700; Fax: (816) 333-3884
Founded: 1968; Members: 3200
Gen Secr: Roger A. Sherwood
Focus: Family Plan; Med
Periodical
Family Medicine: Journal (weekly) . . . *16706*

Society of Thoracic Surgeons (STS), 401 N Michigan Av, Chicago, IL 60611-4267
T: (312) 644-6610; Fax: (312) 527-6635
Founded: 1964; Members: 3228
Gen Secr: Walter G. Purcell
Focus: Surgery
Periodical
Annals of Thoracic Surgery (weekly) . . *16707*

Society of Toxicology (SOT), 1101 14 St NW, Ste 1100, Washington, DC 20005-5601
T: (202) 371-1393; Fax: (202) 371-1090
Founded: 1961; Members: 3400
Pres: Joan W. Cassedy
Focus: Toxicology
Periodicals
Fundamental and Applied Toxicology (weekly)
Newsletter (weekly)
Toxicology and Applied Pharmacology (15 times annually) *16708*

Society of Turkish Architects, Engineers and Scientists in America (STAESA), 821 U.N. Plaza, New York, NY 10017
T: (212) 682-7688; Fax: (609) 275-5357
Founded: 1970; Members: 500
Pres: Erhan Atay
Focus: Eng; Archit
Periodicals
MIM Membership Directory
MIM News Bulletin (weekly) *16709*

Society of United States Air Force Flight Surgeons, POB 35387, Brooks Air Force Base, TX 78235
T: (210) 536-3646
Founded: 1960; Members: 15000
Gen Secr: B. Zwart
Focus: Surgery
Periodical
Society of USAF Flight Surgeons Newsletter *16710*

Society of University Surgeons (SUS), c/o Linda M. Graham, POB 7069, New Haven, CT 06519
T: (203) 932-0541
Founded: 1938; Members: 1215
Gen Secr: Linda M. Graham
Focus: Surgery *16711*

Society of Vertebrate Paleontology (SVP), c/o University of Nebraska, W 436 Nebraska Hall, Lincoln, NE 68588-0542
T: (402) 472-4604; Fax: (402) 472-8949
Founded: 1940; Members: 1300
Gen Secr: Dr. Robert M. Hunt
Focus: Paleontology
Periodicals
Bibliography of Fossil Vertebrates (weekly)
Journal of Vertebrate Paleontology (weekly)
News Bulletin (3 times annually) *16712*

Society of Women Engineers (SWE), 345 E 47 St, Rm 305, New York, NY 10017
T: (212) 705-7855; Fax: (212) 319-0947
Founded: 1950; Members: 15000
Gen Secr: B.J. Harrod
Focus: Eng
Periodical
U.S. Woman Engineer (weekly) *16713*

Society of Wood Science and Technology (SWST), 1 Gifford Pinchot Dr, Madison, WI 53705
T: (608) 231-9347; Fax: (608) 231-9592
Founded: 1958; Members: 400
Gen Secr: Vicki L. Claas
Focus: Forestry
Periodicals
Newsletter (weekly)
Wood and Fiber Science (weekly) . . . *16714*

Soiety for Northwestern Vertebrate Biology (SNVB), c/o John Pierce, Wildlife Management Div, Washington State Department of Wildlife, 600 Capitol Way N, Olympia, WA 98501-1091
T: (206) 753-2868
Founded: 1920; Members: 400
Gen Secr: John Pierce
Focus: Ornithology; Zoology
Periodical
Northwestern Naturalist (3 times annually) *16715*

Soil and Water Conservation Society (SWCS), 7515 NE Ankeny Rd, Ankeny, IA 50021
T: (515) 289-2331; Fax: (515) 289-1227
Founded: 1945; Members: 13000
Gen Secr: Douglas M. Kleine
Focus: Agri; Ecology
Periodical
Journal of Soil and Water Conservation (weekly) *16716*

Soil Science Society of America (SSSA), 677 S Segoe Rd, Madison, WI 53711
T: (608) 273-8080; Fax: (608) 273-2021
Founded: 1936; Members: 6200
Gen Secr: Robert F. Barnes
Focus: Agri
Periodicals
Journal of Environmental Quality (weekly)
Journal of Production Agriculture (weekly) *16717*

Solartherm, 1315 Apple Av, Silver Spring, MD 20910
T: (301) 587-8686
Founded: 1977; Members: 600
Pres: Dr. Carl Schleicher
Focus: Energy *16718*

Solid Waste Association of North America (SWANA), 8750 Georgia Av, Ste 140, Silver Spring, MD 20910
T: (301) 585-2898; Fax: (301) 589-7068
Founded: 1961; Members: 7000
Gen Secr: H.L. Hickman
Focus: Ecology
Periodical
Municipal Solid Waste News (weekly) . . *16719*

Solomon Schecher Day School Association (SSDSA), 155 Fifth Av, New York, NY 10010
T: (212) 260-8450
Founded: 1964; Members: 63
Pres: Joel Roseman
Focus: Educ *16720*

Sonneck Society (SS), POB 476, Canton, MA 02021
T: (617) 828-8450; Fax: (617) 828-8915
Founded: 1974; Members: 960
Gen Secr: Kate Keller
Focus: Music
Periodicals
American Music (weekly)
Membership Directory (weekly)
Newsletter (3 times annually) *16721*

South American Explorers Club (SAEC), 126 Indian Creek Rd, Ithaca, NY 14850
T: (607) 277-0488
Founded: 1977; Members: 6500
Pres: Don Montague
Focus: Bio; Anthro; Archeol; Geography
Periodical
South American Explorer (weekly) . . . *16722*

Southern Association of Colleges and Schools (SACS), 1866 Southern Ln, Decatur, GA 30033
T: (404) 679-4500; Fax: (404) 329-6598
Founded: 1895; Members: 11811
Focus: Educ
Periodicals
Membership List
Newsletter (7 times annually)
Proceedings *16723*

Southern Building Code Congress – International (SBCCI), 900 Montclair Rd, Birmingham, AL 35213
T: (205) 591-1853; Fax: (205) 592-7001
Founded: 1940; Members: 7300
Gen Secr: William J. Tangye
Focus: Civil Eng *16724*

Southern Historical Association (SHA), c/o Dept of History, University of Georgia, Athens, GA 30602
T: (706) 542-8848; Fax: (706) 542-2455
Founded: 1934; Members: 4500
Gen Secr: William F. Holmes
Focus: Hist
Periodical
The Journal of Southern History (weekly) *16725*

Southern Humanities Conference (SHC), c/o Dr. John Phillips, University of Tennessee at Chattanooga, Chattanooga, TN 37403
T: (615) 755-4153
Founded: 1947; Members: 200
Gen Secr: Dr. John Phillips
Focus: Humanities
Periodicals
Humanities in the South (weekly)
Southern Humanities Review (weekly) . . *16726*

Southern Regional Council (SRC), 1900 Rhodes Haverty Bldg, 134 Peachtree St NW, Atlanta, GA 30303-1925
T: (404) 522-8764; Fax: (404) 522-8791
Founded: 1944; Members: 120
Gen Secr: Stephen T. Suitts
Focus: Sociology
Periodicals
Legislative Report (weekly)
Southern Changes (weekly)
SRC Home Record (weekly) *16727*

Southern Society of Genealogists (SSG), c/o F. Stewart, RFD 5, Box 109, Piedmont, AL 36272
T: (205) 447-2939
Founded: 1962; Members: 50
Pres: F. Stewart
Focus: Genealogy
Periodicals
Bulletin
Directory (weekly) *16728*

Southern States Communication Association (SSCA), c/o Dr. Susan A. Siltanen, University of Southern Mississippi, POB 5131, Southern Station, Hattiesburg, MS 39406-5131
T: (601) 266-4271
Founded: 1930; Members: 1300
Gen Secr: Dr. Susan A. Siltanen
Focus: Ling
Periodical
Southern Speech Communication Journal (weekly) *16729*

Southwestern Legal Foundation (SWLF), POB 830707, Richardson, TX 75083-0707
T: (214) 690-2370; Fax: (214) 690-2458
Founded: 1947; Members: 700
Pres: J. David Ellwanger
Focus: Law
Periodicals
Labor Law Developments (weekly)
Oil and Gas Reporter (weekly)
Patent Law Annual (weekly)
Private Investors Abroad: Problems and Solutions in International Business (weekly) . . . *16730*

Southwest Parks and Monuments Association (SPMA), 221 N Court Av, Tucson, AZ 85701
T: (602) 622-1999; Fax: (602) 623-9519
Founded: 1937; Members: 259
Gen Secr: T.J. Priehs
Focus: Travel *16731*

Space Settlement Studies Program (SSSP), c/o Niagara University, Niagara University, NY 14109
T: (716) 285-1212
Founded: 1977; Members: 1000
Gen Secr: Stewart B. Whitney
Focus: Aero *16732*

Space Studies Institute (SSI), POB 82, Princeton, NJ 08542
T: (609) 921-0377; Fax: (609) 921-0389
Founded: 1977; Members: 5000
Gen Secr: Bettie Greber
Focus: Astrophys
Periodical
Update (weekly) *16733*

Special Interest Group for Computer Personnel Research (SIGCPR), c/o Raymond McLeod, College of Business, Texas A & M University, College Station, TX 77841
T: (409) 845-3139; Fax: (409) 845-5653
Founded: 1962; Members: 460
Gen Secr: Raymond McLeod
Focus: Computer & Info Sci
Periodical
Newsletter (weekly) *16734*

Special Interest Group for Computer Uses in Education (SIGCUE), c/o Lloyd P. Rieber, Educational Technology Program, Texas A & M University, College Station, TX 77843-4232
T: (409) 845-7806
Members: 1699
Pres: Lloyd P. Rieber
Focus: Computer & Info Sci *16735*

Special Interest Group for Symbolic and Algebraic Manipulation (SIGSAM), c/o Prof. Paul S. Wang, Dept of Mathematics and Computer Science, Kent State University, Kent, OH 44242
T: (216) 672-2249; Fax: (216) 672-7824
Founded: 1967; Members: 1285
Gen Secr: Prof. Paul S. Wang
Focus: Math
Periodicals
Bulletin (weekly)
Proceedings of Symposia (biennial) . . . *16736*

Special Interest Group on Artificial Intelligence (SIGART), c/o Association for Computing Machinery, 1515 Broadway, New York, NY 10036
T: (212) 869-7440; Fax: (212) 302-5826
Members: 8255
Pres: Stuart C. Shapiro
Focus: Computer & Info Sci
Periodical
SIGART Newsletter *16737*

Special Interest Group on Biomedical Computing (SIGBIO), c/o William E. Hammond, Duke University Medical Center, Box 2914, Durham, NC 27710
T: (919) 684-6421; Fax: (919) 684-8675
Founded: 1967; Members: 1006
Pres: William E. Hammond
Focus: Computer & Info Sci
Periodical
Newsletter (weekly) *16738*

Special Interest Group on Information Retrieval (SIGIR), c/o Dept of Computer Science, Virginia Polytechnic Institute, Blackburg, VA 24061-0106
T: (703) 231-5113; Fax: (703) 545-1249
Founded: 1966; Members: 1845
Pres: Ed Fox
Focus: Computer & Info Sci
Periodical
Forum (weekly) *16739*

Special Libraries Association (SLA), 1700 18 St NW, Washington, DC 20009-2508
T: (202) 234-4700; Fax: (202) 265-9317
Founded: 1909; Members: 13500
Gen Secr: David R. Bender
Focus: Libraries & Bk Sci
Periodicals
SpeciaList (weekly)
Special Libraries (weekly)
Who's Who in Special Libraries (weekly) *16740*

Speech Communication Association (SCA), 5105 E Backlick Rd, Bldg E, Annandale, VA 22003
T: (703) 750-0533; Fax: (703) 914-9471
Founded: 1914; Members: 6000
Gen Secr: James L. Gaudino
Focus: Ling
Periodicals
Communication Education (weekly)
Communication Monographs (weekly)
Critical Studies In Mass Communication (weekly)
Journal of Speech (weekly)
Spectra (11 times annually) *16741*

Spenser Society (SS), c/o John Webster, Dept of English, University of Washington, Seattle, WA 98195
T: (206) 543-2690
Members: 200
Gen Secr: John Webster
Focus: Lit *16742*

Spirit and Breath Association (SBA), 8210 Elmwood Av, Ste 209, Skokie, IL 60077
T: (312) 673-1384
Founded: 1979; Members: 250
Pres: Morton Liebling
Focus: Cell Biol & Cancer Res *16743*

Standards Engineering Society (SES), 11 W Monument Av, Ste 510, POB 2307, Dayton, OH 45401-2307
T: (513) 223-2410; Fax: (513) 223-6307
Founded: 1947; Members: 750
Gen Secr: Francine W. Rickenbach
Focus: Standards
Periodicals
Membership Directory (weekly)
Proceedings (weekly)
Standards Engineering (weekly) *16744*

Steel Founders' Society of America (SFSA), Cast Metals Federation Bldg, 455 State St, Des Plaines, IL 60016
T: (708) 299-9160; Fax: (708) 299-3105
Founded: 1902; Members: 69
Gen Secr: Raymond W. Monroe
Focus: Metallurgy

Periodicals
Casteel Magazine (weekly)
Directory of Steel Foundries (biennial) . 16745

Steel Structures Painting Council (SSPC), 4400 Fifth Av, Pittsburgh, PA 15213-2683
T: (412) 268-3327; Fax: (412) 268-7048
Founded: 1950; Members: 6500
Gen Secr: Dr. Bernard R. Appleman
Focus: Chem; Metallurgy
Periodicals
Journal of Protective Coatings and Linings (weekly)
SSPC Members Directory (weekly)
Steel Structures Painting Bulletin 16746

Stonehenge Study Group (SSG), 2261 Las Positas Rd, Santa Barbara, CA 93105-4116
T: (805) 687-9350
Founded: 1970
Gen Secr: Donald L. Cyr
Focus: Astronomy; Geology; Archeol
Periodical
Stonehenge Viewpoint (weekly) 16747

Structural Stability Research Council (SSRC), c/o Fritz Engineering Laboratory, Lehigh University, 13 E Packer Av, Bethlehem, PA 18015
T: (215) 758-3522; Fax: (215) 758-4522
Founded: 1944; Members: 400
Gen Secr: Dr. Lynn S. Beedle
Focus: Eng
Periodical
SSRC Annual Technical Session Proceedings (weekly) 16748

Student National Medical Association (SNMA), 1012 Tenth St NW, Washington, DC 20001
T: (202) 371-1616
Founded: 1964; Members: 2600
Pres: Colin Ottey
Focus: Educ; Med 16749

Study Group for Mathematical Learning (SGML), c/o Robert B. Davis, 501 S First Av, Highland Park, NJ 08904
T: (908) 545-4960; Fax: (908) 932-8206
Founded: 1956
Pres: Robert B. Davis
Focus: Math 16750

Subterranean Sociological Association (SSA), c/o Dept of Sociology, Eastern Michigan University, Ypsilanti, MI 48197
T: (517) 522-3551
Founded: 1975; Members: 200
Pres: Marcello Truzzi
Focus: Sociology
Periodical
The Subterranean Sociology Newsletter . 16751

Sulphur Institute (SI), 1140 Connecticut Av NW, Ste 612, Washington, DC 20036
T: (202) 331-9660; Fax: (202) 293-2940
Founded: 1960; Members: 35
Pres: Robert J. Morris
Focus: Agri; Chem; Eng
Periodical
Sulphur in Agriculture (1-2 times annually) 16752

Sumi-E Society of America (SSA), 135 Fort Williams Pkwy, Alexandria, VA 22304
T: (703) 823-6574
Founded: 1962; Members: 450
Gen Secr: Wayne C. Shaffer
Focus: Arts
Periodicals
Catalogue (weekly)
Membership List
Sumi-E (weekly) 16753

Superstition Mountain Historical Society (SMHS), POB 3845, Apache Junction, AZ 85217-3845
T: (602) 983-4888
Founded: 1980; Members: 270
Gen Secr: Larry Hedrick
Focus: Hist 16754

Supreme Court Historical Society (SCHS), 111 Second St NE, Washington, DC 20002
T: (202) 543-0400
Founded: 1974; Members: 4200
Gen Secr: David Pride
Focus: Hist
Periodicals
Annual Report (weekly)
Journal of Supreme Court History (weekly)
Newsletter (weekly) 16755

Surveyors Historical Society (SHS), POB 11154, Lansing, MI 48901
T: (206) 378-2300
Founded: 1977; Members: 500
Gen Secr: John Thalacker
Focus: Surveying
Periodicals
Backsight (weekly)
Membership Directory (biennial) 16756

Suzuki Association of the Americas (SAA), POB 17310, Boulder, CO 80308
T: (303) 444-0948
Founded: 1972; Members: 5000
Gen Secr: Robert Klein Reinsager
Focus: Music 16757

Swedish-American Historical Society (SAHS), 5125 N Spaulding Av, Chicago, IL 60625
T: (312) 583-5722
Founded: 1948; Members: 1200
Gen Secr: Timothy J. Johnson
Focus: Hist
Periodical
The Swedish-American Historical Quarterly (weekly) 16758

Swedish Colonial Society (SCS), c/o Wallace F. Richter, 336 S Devon Av, Philadelphia, PA 19087
T: (215) 688-1766
Founded: 1908; Members: 400
Pres: Wallace F. Richter
Focus: Hist
Periodical
Directory 16759

Swedish Women's Educational Association International (SWEA), 7414 Herschel Av, POB 2585, La Jolla, CA 92038-2585
T: (619) 459-8435
Founded: 1979; Members: 4500
Gen Secr: Boel Alkdal
Focus: Educ
Periodical
SWEA Forum (3 times annually) . . . 16760

Swiss-American Historical Society (SAHS), c/o Prof. E. Schmocker, 6440 N Bosworth Av, Chicago, IL 60626
T: (312) 262-8336
Founded: 1927; Members: 375
Pres: Prof. E. Schmocker
Focus: Hist
Periodical
Review (3 times annually) 16761

Tamarind Institute (TI), 108 Cornell Dr SE, Albuquerque, NM 87106
T: (505) 277-3901; Fax: (505) 277-3920
Founded: 1970
Gen Secr: Marjorie Devon
Focus: Arts
Periodical
The Tamarind Papers (weekly) 16762

Tarleton Foundation (TF), 50 Francisco, Ste 103, San Francisco, CA 94123
T: (415) 989-2810
Founded: 1978; Members: 3000
Gen Secr: Julie Begley
Focus: Zoology
Periodicals
Whale Center Newsletter (weekly)
Whales Tales (weekly) 16763

Tax Analysts (TA), 6830 N Fairfax Dr, Arlington, VA 22213
T: (703) 533-4400; Fax: (703) 533-4444
Founded: 1970
Gen Secr: Thomas F. Field
Focus: Law; Stats
Periodicals
Highlights and Documents
Microfiche Tax Data Base (weekly)
The Tax Directory (weekly)
Tax Notes 16764

Teachers Educational Council – Association of Accredited Cosmetology Schools (TEC/AACS), 5201 Leesburg Pike, Ste 205, Falls Church, VA 22041
T: (703) 845-1333; Fax: (703) 845-1336
Founded: 1956; Members: 600
Pres: Ronald E. Smith
Focus: Adult Educ
Periodicals
Date (weekly)
NAACS News (weekly)
Washington Update (weekly) 16765

Teachers of English to Speakers of Other Languages (TESOL), 1600 Cameron St, Ste 300, Alexandria, VA 22314-2751
T: (703) 836-0774; Fax: (703) 836-7864
Founded: 1966; Members: 23000
Gen Secr: Susan C. Bayley
Focus: Ling; Educ
Periodicals
TESOL Newsletter (weekly)
TESOL Quarterly (weekly) 16766

Technical Association of the Graphic Arts (TAGA), POB 9887, Rochester, NY 14623-0887
T: (716) 272-0557; Fax: (716) 475-2250
Founded: 1948; Members: 1300
Gen Secr: Karen Lawrence
Focus: Eng; Graphic & Dec Arts, Design
Periodicals
Newsletter (weekly)
Proceedings (weekly) 16767

Technical Marketing Society of America (TMSA), 4383 Via Majorca, Cypress, CA 90680
T: (714) 821-8672
Founded: 1975; Members: 5000
Gen Secr: James A. Pearson
Focus: Marketing
Periodicals
Aeorospace Market Outlook (weekly)
Newsletter (weekly) 16768

Technology Transfer Society (TTS), 611 N Capitol Av, Indianapolis, IN 46204
T: (317) 262-5022; Fax: (317) 262-5044
Founded: 1975; Members: 650
Focus: Eng
Periodicals
Journal of Technology Transfer (weekly)
TSQUARED (weekly) 16769

Tennessee Folklore Society (TFS), c/o Middle Tennessee State University, POB 529, Murfreesboro, TN 37132
T: (615) 898-2576
Founded: 1934; Members: 500
Gen Secr: Dr. Charles Wolfe
Focus: Cultur Hist 16770

Teratology Society (TS), 9650 Rockville Pike, Bethesda, MD 20814
T: (301) 571-1841; Fax: (301) 571-1852
Founded: 1960; Members: 750
Gen Secr: Barbara Ventura
Focus: Toxicology
Periodical
Teratology: The International Journal of Abnormal Development (weekly) 16771

Theatre Guild (TG), 226 W 47 St, New York, NY 10036
T: (212) 869-5470; Fax: (212) 869-5463
Founded: 1919; Members: 105000
Pres: Philip Langner
Focus: Perf Arts 16772

Theatre Historical Society (THS), 152 N York Rd, Ste 200, Elmhurst, IL 60126
T: (708) 782-1800; Fax: (708) 782-1802
Founded: 1969; Members: 1000
Gen Secr: William T. Benedict
Focus: Hist; Perf Arts
Periodicals
Directory
Marquee (weekly) 16773

Theatre in Education (TE), POB 357, Old Lyme, CT 06371
T: (203) 434-2480
Founded: 1956
Gen Secr: Lyn Ely
Focus: Educ; Perf Arts 16774

Theatre Library Association (TLA), 111 Amsterdam Av, Rm 513, New York, NY 10023
T: (212) 870-1670
Founded: 1937; Members: 500
Gen Secr: Richard M. Buck
Focus: Libraries & Bk Sci; Perf Arts
Periodicals
Broadside: Newsletter of the TLA (weekly)
Performing Arts Resources (weekly) . . 16775

Theodore Roosevelt Association (TRA), POB 719, Oyster Bay, NY 11771
T: (516) 922-1221; Fax: (516) 922-0364
Founded: 1919; Members: 2000
Gen Secr: John A. Gable
Focus: Hist
Periodical
Journal (weekly) 16776

Theodor Herzl Institute (THI), 110 E 59 St, New York, NY 10022
T: (212) 339-6038; Fax: (212) 318-6175
Founded: 1955; Members: 1000
Gen Secr: Jacques Torczyner
Focus: Educ
Periodicals
Annual Season Preview (weekly)
Herzl Institute Bulletin (weekly) 16777

Thomas Wolfe Society (TWS), c/o Dr. William P. Brown, 4302 Fordham Ct, Richmond, VA 23236
T: (804) 276-1335
Founded: 1979; Members: 600
Gen Secr: Dr. William P. Brown
Focus: Lit
Periodical
The Thomas Wolfe Review (weekly) . . 16778

Thoreau Lyceum, 156 Belknap St, Concord, MA 01742
T: (617) 369-5912
Founded: 1966; Members: 1500
Gen Secr: Anne McGrath
Focus: Lit
Periodical
The Concord Saunterer (3 times annually) 16779

Thoreau Society (TS), 156 Belknap St, Concord, MA 01742
T: (508) 369-5912
Founded: 1941; Members: 1500
Gen Secr: D. Bradley
Focus: Lit
Periodicals
Bulletin (weekly)
The Concord Saunterer (weekly) 16780

Thorne Ecological Institute (TEI), 5398 Manhattan Circle, Boulder, CO 80303
T: (303) 499-3647
Founded: 1954
Gen Secr: Susan Q. Foster
Focus: Ecology
Periodical
Update (weekly) 16781

Tibet Society (TS), POB 1968, Bloomington, IN 47402
T: (812) 335-8222
Founded: 1966; Members: 450
Pres: T.J. Norbu
Focus: Ethnology
Periodicals
Journal of the Tibet Society (weekly)
Tibet Society Bulletin (weekly) 16782

Tin Research Institute (TRI), 1353 Perry St, Columbus, OH 43201
T: (614) 424-6200; Fax: (614) 424-6924
Founded: 1949
Gen Secr: William B. Hampshire
Focus: Metallurgy
Periodicals
Annual Report (weekly)
Tin and its Use (weekly) 16783

Tissue Culture Association (TCA), 8815 Centre Park Dr, Ste 210, Columbia, MD 21045
T: (410) 992-0946; Fax: (410) 992-0949
Founded: 1946; Members: 2500
Gen Secr: Marietta W. Ellis
Focus: Med 16784

Torrey Botanical Club (TBC), c/o Dr. H.D. Hammond, New York Botanical Garden, Bronx, NY 10458
T: (718) 220-8987; Fax: (718) 220-6504
Founded: 1860
Pres: Dr. H.D. Hammond
Focus: Botany
Periodical
Bulletin of the TBC (weekly) 16785

Touch for Health Foundation (THF), 1200 N Lake Av, Ste A, Pasadena, CA 91104
T: (818) 794-1181; Fax: (818) 798-7895
Founded: 1974; Members: 1500
Pres: John Thie
Focus: Med
Periodicals
In Touch for Health (weekly)
Touch for Health Directory (weekly)
Touch for Health Journal (weekly)
Touch for Health Times (weekly) . . . 16786

Tourette Syndrome Association (TSA), 40-42 Bell Blvd, Bayside, NY 11361
T: (718) 224-2999; Fax: (718) 279-9596
Founded: 1972; Members: 30000
Gen Secr: Steven M. Friedlander
Focus: Neurology; Psychiatry
Periodical
TSA Newsletter (weekly) 16787

Transplantation Society (TS), c/o Ronald M. Ferguson, Dept of Surgery, 1654 Upham Dr, Rm 259, Columbus, OH 43210
T: (614) 293-8545; Fax: (614) 293-4670
Founded: 1966; Members: 1400
Gen Secr: Ronald M. Ferguson
Focus: Med
Periodicals
Transplantation (weekly)
Transplantation Proceedings (weekly) . . 16788

Transportation Alternatives (TA), 92 Saint Marks Pl, New York, NY 10009
T: (212) 475-4600
Founded: 1973; Members: 1500
Gen Secr: Jonathan Orcutt
Focus: Transport 16789

Transportation Research Board (TRB), 2101 Constitution Av NW, Washington, DC 20418
T: (202) 334-2934; Fax: (202) 334-2003
Founded: 1920; Members: 5600
Gen Secr: Thomas B. Deen
Focus: Transport
Periodicals
National Cooperative Highway Research Program Report
National Cooperative Transit Research and Development Program Reports (weekly)
NCHRP Synthesis of Highway Practice
NCTRP Research Result Digest (weekly)
NCTRP Synthesis of Transit Practice (weekly)
Special Report
Transportation Research Circular
Transportation Research Record 16790

Tree-Ring Society (TRS), c/o Tree-Ring Research Laboratory, University of Arizona, Tucson, AZ 85721
T: (602) 621-2191; Fax: (602) 621-8229
Founded: 1934; Members: 320
Gen Secr: Jeffrey S. Dean
Focus: Forestry
Periodical
Tree-Ring Bulletin (weekly) 16791

Tripoli Rocketry Association (TRA), POB 339, Kenner, LA 70065-0339
T: (504) 467-1967
Founded: 1985; Members: 1892
Pres: Charles Rogers
Focus: Educ; Eng
Periodical
Tripolitan (weekly) 16792

Trout Unlimited (TU), 800 Follin Ln, Ste 250, Vienna, VA 22180
T: (703) 281-1100
Founded: 1959; Members: 65000
Gen Secr: Charles F. Gauvin
Focus: Fisheries; Water Res
Periodicals
Chapter and Council Handbook (weekly)
Lines to Leaders (weekly)
Trout Magazine (weekly) 16793

The Trumpeter Swan Society (TTSS), 3800 County Rd 24, Maple Plain, MN 55359
T: (612) 476-4663; Fax: (612) 476-1514
Founded: 1968; Members: 500
Gen Secr: Donna Compton
Focus: Zoology
Periodicals
Conference Proceedings and Papers (biennial)
Newsletter (weekly) 16794

Turkish American Physicians Association (TAPA), c/o Dr. Cemil Bikmen, 222 Middle Country Rd, Smithtown, NY 11787
T: (516) 724-0777
Founded: 1969; Members: 1260
Gen Secr: Dr. Cemil Bikmen
Focus: Physiology
Periodical
Membership Roster (biennial) 16795

Ukrainian Academy of Arts and Sciences in the U.S. (UAAS), 206 W 100 St, New York, NY 10025
T: (212) 222-1866
Founded: 1950; Members: 223
Gen Secr: Prof. William Omelchenko
Focus: Arts; Sci 16796

Ukrainian Engineers' Society of America (UESA), c/o George Honczarenko, 300 Winston Dr, Apt 116, Cliffside Park, NJ 07010
T: (201) 224-9862
Founded: 1948; Members: 900
Pres: George Honczarenko
Focus: Eng
Periodicals
Bulletin (weekly)
Ukrainian Engineers News (weekly) . . . 16797

Ukrainian Institute of America (UIA), 2 E 79 St, New York, NY 10021
T: (212) 288-8660
Founded: 1948; Members: 450
Pres: Walter Baranetsky
Focus: Ethnology
Periodical
Newsletter 16798

Ukrainian Medical Association of North America (UMANA), 2247 W Chicago Av, Chicago, IL 60622
T: (312) 278-6262
Founded: 1950; Members: 1000
Gen Secr: Andrew Lincky
Focus: Med
Periodicals
Medical Journal (weekly)
Newsletter to Membership (weekly) . . . 16799

Ukrainian Political Science Association in the United States (UPSA), POB 12963, Philadelphia, PA 19108
Founded: 1970; Members: 104
Gen Secr: Petro Diachenko
Focus: Poli Sci
Periodical
Newsletter (weekly) 16800

ULI – The Urban Land Institute, 625 Indiana Av NW, Washington, DC 20004
T: (202) 624-7000; Fax: (202) 624-7140
Founded: 1936; Members: 15000
Gen Secr: Rick Rosan
Focus: Urban Plan
Periodicals
Development Trends (weekly)
Dollars and Cents of Shopping Centers (tri-annually)
Land Use Digest (weekly)
Market Profiles (weekly)
Project Reference File (weekly)
Urban Land Magazine (weekly) 16801

Underground Technology Research Council Education Committee (CTUC), c/o Vincent Tirolo, Civil Engineering Dept, Pratt Institute, 200 Willoughby Av, Brooklyn, NY 11205
T: (718) 636-3461
Founded: 1969; Members: 10
Pres: Vincent Tirolo
Focus: Civil Eng 16802

Undersea and Hyperbaric Medical Society (UHMS), 9650 Rockville Pike, Bethesda, MD 20814
T: (301) 571-1818; Fax: (301) 571-1815
Founded: 1967; Members: 2500
Gen Secr: Leon J. Greenbaum
Focus: Med
Periodicals
Journal of Hyperbaric Medicine (weekly)
Pressure: Newsletter (weekly)
Undersea Biomedical Research: Journal (weekly)
Undersea & Hyperbaric Medicine: Abstracts from the Literature Abstract Journal (weekly) . 16803

Union Institute (UI), 440 E McMillan St, Cincinnati, OH 45206-1947
T: (513) 861-6400; Fax: (513) 861-0779
Founded: 1964
Pres: Robert T. Conley
Focus: Educ; Sci
Periodical
The Network (weekly) 16804

Unitarian Universalist Association of Congregations, c/o Washington Office for Social Justice, 100 Maryland Av NE, Rm 106, Washington, DC 20002
T: (202) 547-0254; Fax: (202) 544-2854
Founded: 1975
Gen Secr: Robert Z. Alpern
Focus: Sociology
Periodical
Ethics and Action (weekly) 16805

Unitarian Universalist Historical Society (UUHS), 535 Canton Av, Milton, MA 02186
T: (617) 698-6329
Founded: 1978; Members: 600
Pres: Mark Harris
Focus: Hist
Periodicals
Journal
Newsletter
Proceedings (biennial) 16806

United Board for Christian Higher Education in Asia (UBCMEA), 475 Riverside Dr, Rm 1221, New York, NY 10115
T: (212) 870-2608; Fax: (212) 870-2322
Founded: 1932; Members: 45
Pres: Dr. David W. Vikner
Focus: Educ
Periodicals
Annual Report (weekly)
New Horizons (3 times annually) . . . 16807

United Cancer Council (UCC), 8009 Fishback Rd, Indianapolis, IN 46298-1047
Founded: 1963; Members: 39
Pres: Randall B. Grove
Focus: Cell Biol & Cancer Res
Periodical
Coordinator (weekly) 16808

United Cerebral Palsy Associations (UPCA), 1522 K St NW, Ste 1112, Washington, DC 20005
T: (202) 842-1266; Fax: (202) 842-3519
Founded: 1948; Members: 155
Gen Secr: John D. Kemp
Focus: Med 16809

United Cerebral Palsy Research and Educational Foundation (UCPREF), 7 Penn Plaza, Ste 804, New York, NY 10001
T: (212) 268-5962; Fax: (212) 268-5960
Founded: 1955
Gen Secr: Leon Sternfeld
Focus: Med; Educ 16810

United Engineering Trustees (UET), 345 E 47 St, New York, NY 10017
T: (212) 705-7828; Fax: (212) 705-7441
Founded: 1904
Gen Secr: Jerome Sishel
Focus: Eng 16811

United Nations Committee on the Peaceful Uses of Outer Space (COPUOS), c/o Office for Outer Space Affairs, United Nations, New York, NY 10017
T: (212) 963-6051; Fax: (212) 963-7998
Founded: 1959; Members: 53
Pres: Peter Hohenfellner
Focus: Poli Sci
Periodicals
Coordination of Outer Space Research (weekly)
Highlights in Space Technology (weekly)
Progress of Space Research (weekly)
Report of the COPUOS (weekly) . . . 16812

United New Conservationists (UNC), POB 362, Campbell, CA 95009
T: (408) 241-5769
Founded: 1969; Members: 58
Pres: Lilyann Brannon
Focus: Ecology
Periodicals
Membership Roster (weekly)
UN-Common Group (10 times annually) . 16813

United Ostomy Association (UOA), 36 Executive Park, Ste 120, Irvine, CA 92714
T: (714) 660-8624; Fax: (714) 660-9262
Founded: 1962; Members: 50000
Gen Secr: Darlene Smith
Focus: Gastroenter
Periodicals
Ostomy Quarterly (weekly)
The Phoenix (weekly) 16814

United States Antarctic Research Program (USARP), c/o Guy G. Guthridge, Polar Information Program, National Science Foundation, Washington, DC 20550
T: (202) 357-7817; Fax: (202) 357-9422
Founded: 1959
Gen Secr: Guy G. Guthridge
Focus: Geography; Bio; Oceanography; Astronomy
Periodicals
Antarctic Journal of the United States (weekly)
Current Antarctic Literature (weekly) . . 16815

United States Branch of the International Committee for the Defense of the Breton Language (US ICDBL), c/o L. Kuter, 169 Greenwood Av, Jenkintown, PA 19046
T: (215) 886-6361
Founded: 1981; Members: 150
Gen Secr: L. Kuter
Focus: Ling 16816

United States Capitol Historical Society (USCHS), 200 Maryland Av NE, Washington, DC 20002
T: (202) 543-8919
Founded: 1962; Members: 12000
Gen Secr: Cornelius W. Heine
Focus: Hist
Periodicals
The Capitol Dome (weekly)
Congress and the Presidency (weekly) . 16817

United States Committee of the International Association of Art (USCIAA), POB 28068, Central Station, Washington, DC 20038-8068
T: (202) 628-9633
Founded: 1952; Members: 12
Gen Secr: Catherine Auth
Focus: Arts
Periodical
Information Bulletin (weekly) 16818

United States Committee on Irrigation and Drainage (USCID), 1616 17 St, Denver, CO 80202
T: (303) 628-5430; Fax: (303) 628-5451
Founded: 1952; Members: 700
Gen Secr: Larry D. Stephens
Focus: Water Res
Periodical
ICID Bulletin (weekly) 16819

United States Committee on Large Dams (USCOLD), 1616 17 St, Denver, CO 80202
T: (303) 628-5430; Fax: (303) 628-5431
Founded: 1928; Members: 1150
Gen Secr: Larry D. Stephens
Focus: Civil Eng
Periodicals
Membership Directory (weekly)
Newsletter (3 times annually) 16820

United States Council for Energy Awareness (USCEA), 1776 Eye St NW, Ste 400, Washington, DC 20006-3708
T: (202) 293-0770; Fax: (202) 785-4019
Founded: 1987; Members: 405
Pres: J. Philip Bayne
Focus: Nucl Res
Periodicals
INFO (weekly)
Nuclear Industry (weekly) 16821

United States Energy Association (USEA), 1620 Eye St, Ste 210, Washington, DC 20006
T: (202) 331-0415; Fax: (202) 331-0418
Founded: 1924; Members: 135
Gen Secr: Barry K. Worthington
Focus: Energy 16822

United States Federation for Culture Collections (USFCC), c/o Marianna Jackson, Dept 47P, AP9A, Abbott Laboratories, Abbott Park, IL 60064
T: (708) 937-8764; Fax: (708) 938-6603
Founded: 1970; Members: 260
Gen Secr: Marianna Jackson
Focus: Bio
Periodicals
Directory (weekly)
Newsletter (weekly) 16823

United States Federation of Scholars and Scientists (USFSS), c/o Prof. Roger Dittmann, Dept of Physics, California State University at Fullerton, Fullerton, CA 92634
T: (714) 773-3421; Fax: (714) 449-5810
Founded: 1938
Gen Secr: Prof. Roger Dittmann
Focus: Sci
Periodical
Scientific World (weekly) 16824

United States Institute for Theatre Technology (USITT), 10 W 19 St, Ste 5A, New York, NY 10011
T: (212) 924-9088; Fax: (212) 924-9343
Founded: 1960; Members: 3500
Gen Secr: Sarah Nash Gates
Focus: Eng; Perf Arts; Educ
Periodicals
Theatre Design and Technology (weekly)
USITT Newsletter (weekly) 16825

United States Metric Association (USMA), 10245 Andasol Av, Los Angeles, CA 91325-1504
T: (818) 363-5606; Fax: (818) 368-7443
Founded: 1915; Members: 1500
Gen Secr: Valerie Antoine
Focus: Sci; Math
Periodical
USMA Newsletter (weekly) 16826

United States-Mexico Border Health Association (USMBHA), 6006 N Mesa, Ste 600, El Paso, TX 79912
T: (915) 581-6645; Fax: (915) 833-4768
Founded: 1943; Members: 2300
Gen Secr: Herbert H. Ortega
Focus: Public Health
Periodicals
Border Health Journal (weekly)
Epidemiological Bulletin (weekly) 16827

United States National Committee for Byzantine Studies (USNCBS), c/o John Barker, Dept of History, University of Wisconsin, Madison, WI 53706
T: (608) 263-1800
Founded: 1962; Members: 140
Gen Secr: John Barker
Focus: Hist 16828

United States National Committee of the International Union of Radio Science (USNC-URSI), c/o National Research Council, 2101 Constitution Av NW, Washington, DC 20418
T: (202) 334-3520; Fax: (202) 334-2791
Founded: 1919; Members: 30
Gen Secr: Dr. Robert L. Riemer
Focus: Electric Eng
Periodical
Program and Abstract of Meetings (weekly)
. . . . 16829

United States National Committee on Theoretical and Applied Mechanics (USNC/TAM), c/o National Research Council, 2101 Constitution Av NW, Washington, DC 20418
T: (202) 334-3142; Fax: (202) 334-2571
Founded: 1947; Members: 29
Gen Secr: Dana Caines
Focus: Eng; Physics
Periodical
Proceedings 16830

United States National Society for the International Society of Soil Mechanics and Foundation Engineering (USNSISSMMFE), c/o Prof. Harvey E. Wahls, Civil Engineering Dept, North Carolina State University, Box 7908, Raleigh, NC 27695
T: (919) 515-7344; Fax: (919) 515-7908
Founded: 1947; Members: 4500
Gen Secr: Prof. Harvey E. Wahls
Focus: Eng
Periodical
Geotechnical News (weekly) 16831

United States Pharmacopeial Convention (USP), 12601 Twinbrook Pkwy, Rockville, MD 20852
T: (301) 881-0666; Fax: (301) 816-8247
Founded: 1820; Members: 391
Gen Secr: Jerome A. Halperin
Focus: Pharmacol
Periodicals
About your Medicines (biennial)
Advice for the Patient (weekly)
Drug Information for the Health Care Professional (weekly)
Pharmacopeial Forum (weekly)
USP DI Review (weekly)
U.S.P. Dispensing Information (weekly) . 16832

United States Physical Therapy Association (USPTA), 1803 Avon Ln, Arlington Heights, IL 60004
Founded: 1970; Members: 12700
Pres: James J. McCoy
Focus: Physical Therapy
Periodical
Journal 16833

United States Space Education Association (USSEA), 746 Turnpike Rd, Elizabethtown, PA 17022-1161
T: (717) 367-3265
Founded: 1973; Members: 1000
Pres: Stephen M. Cobaugh
Focus: Aero
Periodicals
Space Age Times: The Intemntional Publication of Space News, Benefits and Education (weekly)
Update (weekly) 16834

United States Sports Academy (USSA), 1 Academy Dr, Daphne, AL 36526
T: (205) 626-3303; Fax: (205) 626-3874
Founded: 1972
Pres: Dr. Thomas P. Rosandich
Focus: Sports 16835

United Synagogue Commission on Jewish Education (USACOJE), 155 Fifth Av, New York, NY 10010
T: (212) 260-8450; Fax: (212) 353-9439
Members: 39
Gen Secr: Dr. Robert Abramson
Focus: Educ
Periodicals
In Your Hands (weekly)
Your Child (weekly) 16836

Universal Serials and Book Exchange (USBE), 2969 W 25 St, Cleveland, OH 44113
T: (216) 241-6960; Fax: (216) 241-6966
Founded: 1948; Members: 18500
Pres: John T. Zubal
Focus: Libraries & Bk Sci
Periodicals
Annual Report (weekly)
Newsletter (weekly) 16837

Universities Council on Water Resources (UCOWR), c/o Southern Illinois University, 4543 Faner Hall, Carbondale, IL 62901
T: (618) 536-7571; Fax: (618) 453-2671
Founded: 1962; Members: 97
Gen Secr: Dr. Duane D. Baumann
Focus: Water Res
Periodicals
Newsletter (weekly)
Proceedings of Annual Meetings 16838

Universities Research Association (URA), 1111 19 St NW, Ste 400, Washington, DC 20036
T: (202) 293-1382; Fax: (202) 293-5012
Founded: 1965; Members: 78
Pres: Dr. John S. Toll
Focus: Physics 16839

Universities Space Research Association (USRA), American City Bldg, Ste 212, Columbia, MD 21044
T: (410) 730-2656; Fax: (410) 730-3496
Founded: 1969; Members: 75

Gen Secr: W.D. Cummings
Focus: Astronomy
Periodicals
Lunar and Planetary Information Bulletin (weekly)
News and Notes (weekly) 16840

University and College Designers Association (UCDA), 615 1/2 Roosevelt Rd, Ste 2, Walkerton, IN 46574
T: (219) 586-2988; Fax: (219) 586-3599
Founded: 1967; Members: 1130
Gen Secr: Tracy Erdelyi
Focus: Adult Educ; Graphic & Dec Arts, Design
Periodical
Designer (weekly) 16841

University and College Labor Education Association (UCLEA), c/o Labor Education Dept, Rutgers State University, Ryders Ln and Clifton Av, New Brunswick, NJ 08903
T: (812) 855-9082; Fax: (812) 855-9779
Founded: 1959; Members: 373
Pres: Sue Schurman
Focus: Educ 16842

University Aviation Association (UAA), 3410 Skyway Dr, Opelika, AL 36801
T: (205) 844-2434
Founded: 1948; Members: 540
Gen Secr: Gary W. Kiteley
Focus: Aero
Periodicals
Newsletter (weekly)
Proceedings (weekly) 16843

University Consortium for Instructional Development and Technology (UCIDT), c/o Dr. Kent L. Gustafson, Dept of Instructional Technology, University of Georgia, 607 Aderhold Hall, Athens, GA 30602
T: (706) 542-3810; Fax: (706) 542-2321
Founded: 1967; Members: 12
Gen Secr: Dr. Kent L. Gustafson
Focus: Educ; Adult Educ 16844

University Corporation for Atmospheric Research (UCAR), POB 3000, Boulder, CO 80307-3000
T: (303) 497-1673
Founded: 1960; Members: 59
Pres: Dr. Richard A. Anthes
Focus: Geophys
Periodicals
Annual Report (weekly)
Newsletter (weekly) 16845

University Council for Educational Administration (UCEA), c/o Pennsylvania State University, 212 Rackley Bldg, University Park, PA 16802-3200
T: (814) 863-7916; Fax: (814) 863-7918
Founded: 1959; Members: 51
Gen Secr: Patrick B. Forsyth
Focus: Educ
Periodicals
Educational Administration Abstracts (weekly)
Journal of Educational Equity and Leadership (weekly)
Review (4-5 times annually) 16846

University Film and Video Association (UFVA), c/o Donald J. Zirpola, Communication Arts Dept, Loyola Marymount University, Los Angeles, CA 90045
T: (310) 338-3033; Fax: (310) 641-3964
Founded: 1947; Members: 800
Pres: Donald J. Zirpola
Focus: Cinema
Periodicals
Digest (weekly)
Journal of Film and Video (weekly)
Membership Directory (biennial) 16847

University Photographers Association of America (UPAA), c/o Brigham Young University, Provo, UT 84602
T: (801) 378-7322; Fax: (801) 378-7377
Founded: 1961; Members: 300
Pres: Mark Philbrick
Focus: Photo
Periodical
Newsletter (weekly) 16848

University Professors for Academic Order (UPAO), 7938 Bayberry Dr, Alexandria, VA 22306-3215
T: (703) 768-6198
Founded: 1970; Members: 300
Gen Secr: Donald Senese
Focus: Adult Educ
Periodical
Universitas (weekly) 16849

Urban Affairs Association (UAA), c/o Mary Helen Callahan, University of Delaware, Newark, DE 19716
T: (302) 831-1681
Founded: 1969; Members: 355
Gen Secr: Mary Helen Callahan
Focus: Urban Plan
Periodicals
Communication (weekly)
Directory of University Urban Programs
Journal of Urban Affairs (weekly) . . . 16850

Urban and Regional Information Systems Association (URISA), 900 Second St NE, Ste 304, Washington, DC 20002
T: (202) 289-1685; Fax: (202) 842-1850
Founded: 1963; Members: 3800
Gen Secr: Thomas M. Palmerlee
Focus: Computer & Info Sci
Periodicals
Annual Conference Proceedings
Directory (weekly)
URISA News (5 times annually) 16851

Urban Libraries Council (ULC), 500 E Marilyn Av, State College, PA 18601
T: (814) 237-0194; Fax: (814) 237-2687
Founded: 1971; Members: 185
Gen Secr: Keith Doms
Focus: Libraries & Bk Sci
Periodical
Urban Libraries Exchange (weekly) . . . 16852

Use, 4611 Assembly Dr, Ste J, Lanham, MD 20706-4371
T: (301) 577-1881; Fax: (301) 577-8312
Founded: 1975; Members: 600
Gen Secr: Cathy M. Ewbank
Focus: Doc
Periodicals
Administrative Minutes (weekly)
Conference Software Notes (weekly)
Newsletter (weekly)
Technical Papers (weekly) 16853

Vachel Lindsay Association (VLA), 603 S Fifth St, POB 9356, Springfield, IL 62791-9356
T: (217) 632-7284
Founded: 1946; Members: 500
Gen Secr: Sheila Z. Beebe
Focus: Lit 16854

Valley Forge Historical Society (VFHS), POB 122, Valley Forge, PA 19481
T: (215) 783-0535
Founded: 1918; Members: 900
Gen Secr: Margaret Conner
Focus: Hist 16855

Vergilian Society (VS), c/o Emory University, Box 23085, Atlanta, GA 30322
T: (404) 373-7436; Fax: (404) 727-0223
Founded: 1937; Members: 1000
Gen Secr: Herbert W. Benario
Focus: Hist
Periodicals
The Augustan Age (weekly)
Newsletter (weekly)
Vergilius (weekly) 16856

Vernacular Architecture Forum (VAF), 109 Brandon Rd, Baltimore, MD 21212
T: (410) 296-7538
Founded: 1980; Members: 700
Gen Secr: Peter Kurtze
Focus: Archit 16857

Vesterheim Genealogical Center (VGC), 415 W Main St, Madison, WI 53703
T: (608) 255-2224; Fax: (608) 255-6842
Founded: 1975; Members: 1900
Gen Secr: Gerhard B. Naeseth
Focus: Genealogy
Periodical
Norwegian Tracks: Newsletter (weekly) . 16858

Veterinary Cancer Society (VCS), c/o Dr. Robert Rosenthal, 2816 Monroe Av, Rochester, NY 14618
T: (716) 271-7700; Fax: (716) 271-7815
Founded: 1974; Members: 500
Gen Secr: Dr. Robert Rosenthal
Focus: Vet Med 16859

Veterinary Orthopaedic Society (VOS), POB 9491, Salt Lake City, UT 84109-0491
T: (801) 484-8912
Founded: 1972; Members: 500
Gen Secr: Nancy Bunker
Focus: Vet Med 16860

Victorian Society in America (VSA), 219 S Sixth St, Philadelphia, PA 19106
T: (215) 627-4252
Founded: 1966; Members: 2800
Gen Secr: Judith Snyder
Focus: Hist
Periodicals
Classic America (weekly)
The Victorian (weekly) 16861

VIM, POB 11903, Saint Paul, MN 55111-0903
T: (612) 482-4868; Fax: (612) 482-4876
Founded: 1978; Members: 165
Gen Secr: Sandra R. Rowe
Focus: Computer & Info Sci
Periodicals
Conference Proceedings (weekly)
Cyberline (weekly) 16862

Virchow-Pirquet Medical Society (VPMS), c/o Otto Kestler, 158 E 84 St, New York, NY 10028-2005
T: (212) 734-0682
Founded: 1975; Members: 250
Gen Secr: Otto Kestler
Focus: Med
Periodical
Proceedings 16863

Virginia Woolf Society (VWS), 1545 18 St NW, Washington, DC 20036
T: (202) 462-8005
Founded: 1975; Members: 300
Pres: Prof. Karen L. Levenback
Focus: Lit
Periodical
Virginia Woolf Miscellany (weekly) . . . 16864

Vladimir Nabokov Society (VNS), c/o Slavic Languages and Literatures, University of Kansas, Lawrence, KS 66045
T: (913) 864-3313
Founded: 1978; Members: 280
Gen Secr: Prof. Stephen Parker
Focus: Lit 16865

Vocational Industrial Clubs of America (VICA), POB 3000, Leesburg, VA 22075
T: (703) 777-8810; Fax: (703) 777-8999
Founded: 1965; Members: 272000
Gen Secr: Stephen Denby
Focus: Adult Educ
Periodicals
Journal (weekly)
Professional News (5 times annually) . . 16866

Vocational Instructional Materials Section (VIM), c/o Harley Schlichting, University of Missouri, 10 London Hall, Columbia, MO 65211
T: (314) 882-2884
Founded: 1969; Members: 85
Pres: Harley Schlichting
Focus: Educ 16867

Volunteer Committees of Art Museums of Canada and the United States (VCAMCUS), c/o Roberta Walton, RR3, Box 198, Sheridan, IN 46069
Members: 96
Gen Secr: Roberta Walton
Focus: Arts
Periodicals
Conference Report
Directory
News (weekly) 16868

Von Braun Astronomical Society (VBAS), POB 1142, Huntsville, AL 35807
T: (205) 539-0316
Founded: 1954; Members: 150
Pres: Sandra Sherman
Focus: Astronomy
Periodical
Via Stellaris (weekly) 16869

Wagner Society of New York (WSNY), POB 949, Ansonia Station, New York, NY 10023
T: (212) 749-4561
Founded: 1977; Members: 1000
Pres: Nathalie D. Wagner
Focus: Music 16870

Walter Bagehot Research Council on National Sovereignty (WBRC), POB 81, Whitestone, NY 11357
T: (718) 767-8380; Fax: (718) 767-8380
Founded: 1972; Members: 800
Pres: Henry Paolucci
Focus: Poli Sci
Periodical
State of the Nation 16871

Washington Journalism Center (WJC), 2600 Virginia Av NW, Ste 502, Washington, DC 20037
T: (202) 337-3603; Fax: (202) 298-7312
Founded: 1965
Gen Secr: Don Campbell
Focus: Journalism 16872

Water Environment Federation (WEF), 601 Wythe St, Alexandria, VA 22314-1994
T: (703) 684-2400; Fax: (703) 684-2492
Founded: 1928; Members: 38000
Gen Secr: Dr. Q. Brown
Focus: Eng; Ecology
Periodicals
Highlights (weekly)
Journal of the Water Pollution Federation (weekly)
Operations Forum (weekly) 16873

Water Quality Research Council (WQRC), 4151 Naperville Rd, Lisle, IL 60532
T: (708) 505-0160; Fax: (708) 505-9637
Founded: 1950; Members: 2600
Gen Secr: Peter J. Censky
Focus: Water Res; Hygiene 16874

Water Resources Congress (WRC), 2300 Clarendon Blvd, Ste 404, Arlington, VA 22201-3367
T: (703) 525-4881; Fax: (703) 527-1693
Founded: 1971; Members: 300
Gen Secr: Kathleen A. Phelps
Focus: Water Res
Periodicals
Hotline
Platform (weekly)
Washington Report (weekly) 16875

Weather Modification Association (WMA), POB 8116, Fresno, CA 93747
T: (209) 434-3486; Fax: (209) 291-5579
Founded: 1951
Gen Secr: Hilda Duckering
Focus: Geophys
Periodicals
Journal of Weather Modification (weekly)
Newsletter (weekly) 16876

Weed Science Society of America (WSSA), 309 W Clark St, Champaign, IL 61820
T: (217) 356-3182; Fax: (217) 398-4119
Founded: 1950; Members: 2300
Gen Secr: Robert A. Schmidt
Focus: Agri
Periodicals
Abstracts (weekly)
Newsletter (weekly)
Weed Science (weekly)
Weeds Technology (weekly) 16877

Welding Research Council (WRC), 345 E 47 St, Rm 1301, New York, NY 10017
T: (212) 705-7956; Fax: (212) 371-9622
Founded: 1935; Members: 400
Pres: Dr. Martin Prager
Focus: Eng
Periodicals
Bulletin (weekly)
Progress Reports (weekly)
Welding Research Abroad (weekly)
Welding Research News (weekly)
Yearbook (weekly) 16878

Wenner-Gren Foundation for Anthropological Research (WGFAR), 220 Fifth Av, New York, NY 10001-7708
T: (212) 683-5000
Founded: 1941
Pres: Dr. S. Silverman
Focus: Anthro
Periodicals
Annual Report (weekly)
Current Anthropology (weekly) 16879

Western Association of Map Libraries (WAML), c/o H. Fox, 1883 Ashcroft, Clovis, CA 93611
T: (209) 294-0177
Founded: 1967; Members: 226
Gen Secr: H. Fox
Focus: Libraries & Bk Sci; Cart
Periodical
Information Bulletin (3 times annually) . 16880

Western Bird Banding Association (WBBA), c/o Dr. Carl D. Barrentine, Dept of Biology, California State University, Bakersfield, CA 93311-1099
T: (805) 664-3179; Fax: (805) 664-3194
Founded: 1925; Members: 400
Pres: Dr. Carl D. Barrentine
Focus: Ornithology
Periodical
North American Bird Bander (weekly) . . 16881

Western College Association (WCA), c/o Mills College, POB 9990, Oakland, CA 94613
T: (510) 632-5000; Fax: (510) 632-8361
Founded: 1924; Members: 181
Gen Secr: Stephen S. Weiner
Focus: Educ
Periodical
Addresses and Proceedings (weekly) . . 16882

Western History Association (WHA), c/o University of New Mexico, 1080 Mesa Vista Hall, Albuquerque, NM 87131-1181
T: (505) 277-5234; Fax: (505) 277-6023
Founded: 1962; Members: 2000
Gen Secr: Paul Hutton
Focus: Hist
Periodicals
Montana (weekly)
Western Historical Quarterly (weekly) . . 16883

Western Literature Association (WLA), c/o English Dept, Utah State University, Logan, UT 84322-3200
T: (801) 750-1603
Founded: 1966; Members: 500
Gen Secr: Dana Brunvand
Focus: Lit
Periodical
Western American Literature (weekly) . . 16884

Western Society of Malacologists (WSM), c/o Dr. Henry W. Chaney, 1633 Posilipo Ln, Santa Barbara, CA 93108
T: (805) 969-1434; Fax: (805) 963-9679
Founded: 1968; Members: 300
Gen Secr: Dr. Henry W. Chaney
Focus: Zoology
Periodical
Annual Report (weekly) 16885

Western Society of Naturalists (WSN), c/o Dr. Michael Foster, Moss Landing Marine Laboratories, POB 450, Moss Landing, CA 95039
T: (408) 755-8650; Fax: (408) 753-2826
Founded: 1911; Members: 2100
Gen Secr: Dr. Michael Foster
Focus: Nat Sci
Periodicals
Abstracts of Contributed Papers (weekly)
Newsletter (weekly) 16886

Western Surgical Association (WSA), c/o Mayo Clinic, 200 First St SW, Rochester, MN 55905
T: (507) 284-3364
Founded: 1891; Members: 600
Gen Secr: J.A. Van Heerden
Focus: Surgery
Periodicals
Newsletter (weekly)
Program (weekly)
Transactions (weekly) 16887

Western Veterinary Conference (WVC), 2425 E Oquendo Rd, Las Vegas, NV 89120
T: (702) 739-6698
Founded: 1928; Members: 3800
Gen Secr: Kenneth Weide
Focus: Vet Med

Periodical
Directory (weekly) 16888

W.E. Upjohn Institute for Employment Research, 300 S Westnedge Av, Kalamazoo, MI 49007-4686
T: (616) 343-5541; Fax: (616) 343-3308
Founded: 1945
Gen Secr: Robert G. Spiegelman
Focus: Rehabil 16889

Whaling Museum Society (WMS), Main St, POB 25, Cold Spring Harbor, NY 11724
T: (516) 367-3418
Founded: 1936; Members: 1000
Gen Secr: Ann M. Gill
Focus: Hist; Navig
Periodicals
Annual Report (weekly)
Newsletter (weekly) 16890

White House Historical Association (WHHA), 740 Jackson Pl NW, Washington, DC 20503
T: (202) 737-8292; Fax: (202) 789-0440
Founded: 1961
Gen Secr: Bernard R. Meyer
Focus: Hist 16891

Whooping Crane Conservation Association (WCCA), 1007 Carmel Av, Lafayette, LA 70501
T: (318) 234-6339
Founded: 1961; Members: 700
Gen Secr: Mary L. Courville
Focus: Ornithology
Periodicals
Grus Americana (weekly)
Membership Directory (weekly) 16892

Wilbur Hot Springs Health Sanctuary (WHSHS), Star Rte, Wilbur Springs, CA 95987
T: (916) 473-2306
Founded: 1972; Members: 16
Gen Secr: Dr. Richard Louis Miller
Focus: Psych 16893

Wilderness Education Association (WEA), 20 Winona Av, POB 89, Saranac Lake, NY 12983
T: (518) 891-2915
Founded: 1977; Members: 1500
Gen Secr: Mark Wagstaff
Focus: Educ
Periodical
Newsletter (weekly) 16894

Wilderness Medical Society (WMS), POB 2463, Indiamapolis, IN 46206
T: (317) 631-1745; Fax: (317) 634-7817
Founded: 1983; Members: 2800
Gen Secr: D.M. Simpkins
Focus: Med
Periodical
Wilderness Medicine (weekly) 16895

The Wilderness Society (TWS), 900 17 St NW, Washington, DC 20006-2596
T: (202) 833-2300; Fax: (202) 429-3958
Founded: 1935; Members: 310000
Pres: George T. Frampton
Focus: Ecology
Periodicals
Annual Report (weekly)
Wilderness (weekly) 16896

Wilderness Watch (WW), POB 782, Sturgeon Bay, WI 54235
T: (414) 743-1238
Founded: 1969
Pres: Jerome O. Gandt
Focus: Ecology
Periodical
Watch It (weekly) 16897

Wild Goose Association (WGA), 150 S Plains Rd, The Plains, OH 45780
Founded: 1972; Members: 700
Pres: Robert W. Lilley
Focus: Navig
Periodicals
Newsletter (weekly)
Radionavigation Journal (weekly) 16898

Wildlife Management Institute (WMI), 1101 14 St NW, Ste 725, Washington, DC 20005
T: (202) 371-1808; Fax: (202) 408-5059
Founded: 1911
Pres: Rollin D. Sparrowe
Focus: Ecology
Periodical
Outdoor News Bulletin (weekly) 16899

Wildlife Preservation Trust International (WPTI), 3400 W Girard Av, Philadelphia, PA 19104
T: (215) 222-3636; Fax: (215) 222-2191
Founded: 1971; Members: 3000
Gen Secr: Dr. Jerry Eberhart
Focus: Ecology
Periodical
On the Edge (3 times annually) 16900

The Wildlife Society (TWS), 5410 Grosvenor Ln, Bethesda, MD 20814-2197
T: (301) 897-9770; Fax: (301) 530-2471
Founded: 1937; Members: 8600
Gen Secr: Harry E. Hodgdon
Focus: Ecology
Periodicals
Wildlife Monographs (weekly)
Wildlifer (weekly)
Wildlife Society Bulletin (weekly) 16901

Willem Mengelberg Society (WMS), 1408A Marshall St, Manitowoc, WI 54220-5140
Founded: 1970
Gen Secr: Ronald Klett
Focus: Music
Periodical
Newsletter (weekly) 16902

William Hunter Society (WHS), 24 Kay St, Newport, RI 02840
T: (401) 846-7711
Founded: 1974
Gen Secr: Howard Browne
Focus: Hist 16903

William Morris Society in the United States (WMS/US), c/o Mark Samuels Lasner, 1870 Wyoming Av NW, Apt 101, Washington, DC 20009
T: (202) 745-1927
Founded: 1956; Members: 400
Gen Secr: Mark Samuels Lasner
Focus: Lit 16904

Wilson Ornithological Society (WOS), c/o Museum of Zoology, University of Michigan, Ann Arbor, MI 48109-1079
T: (313) 764-0457
Founded: 1888; Members: 1592
Pres: Dr. Richard Banks
Focus: Ornithology
Periodical
The Wilson Bulletin (weekly) 16905

Wilson's Disease Association (WDA), POB 75324, Washington, DC 20013
T: (703) 636-3014
Founded: 1979; Members: 150
Pres: Carol A. Terry
Focus: Intern Med
Periodical
Newsletter (weekly) 16906

Winrock International Institute for Agricultural Development (WIIAD), Petit Jean Mountain, Rte 3, Box 376, Morrilton, AR 72110-9543
T: (501) 727-5435; Fax: (501) 727-5242
Founded: 1985
Pres: Robert D. Havener
Focus: Agri
Periodical
Annual Report (weekly) 16907

Women and Mathematics Education (WME), c/o Charlene Morrow, Mount Holyoke College, 302 Shattuck Hall, South Hadley, MA 01075
T: (413) 538-2608
Founded: 1978; Members: 500
Gen Secr: Charlene Morrow
Focus: Math; Educ
Periodicals
Newsletter (3 times annually)
WME Directory (weekly)
Women and Mathematics Education of Girls and Women 16908

Women Educators (WE), c/o Renee Martin, College of Education, University of Toledo, Toledo, OH 43636-3390
T: (419) 537-4337; Fax: (419) 267-1052
Founded: 1973; Members: 300
Pres: Renee Martin
Focus: Educ
Periodicals
Annual Awards Report (weekly)
Newsletter (weekly) 16909

Women in Aerospace (WIA), 6352 Rolling Mill Pl, Springfield, VA 22152
T: (703) 644-7875; Fax: (703) 866-3526
Founded: 1985; Members: 300
Pres: Susan Brand
Focus: Aero 16910

Women in Broadcast Technology (WBT), 2435 Spaulding St, Berkeley, CA 94703
T: (510) 540-8640
Founded: 1983; Members: 50
Focus: Electric Eng 16911

Women's Auxiliary of the ICA (WAICA), 1925 E Apple Av, Muskegon, MI 49442
T: (616) 777-2622
Founded: 1951; Members: 500
Pres: Alice Peterson
Focus: Med 16912

Women's Caucus for Art (WCA), c/o Moore College of Art, 20 and The Parkway, Philadelphia, PA 19103
T: (215) 854-0922
Founded: 1972; Members: 3500
Gen Secr: Essie Karp
Focus: Arts
Periodicals
Honors Catalog (weekly)
National Network Directory (biennial)
National Update (weekly) 16913

Women's Caucus for Political Science (WCPS), c/o Karen O'Connor, Dept of Political Science, Emory University, Atlanta, GA 30322
T: (404) 727-6572; Fax: (404) 874-6925
Founded: 1969; Members: 900
Pres: Karen O'Connor
Focus: Poli Sci
Periodicals
Membership Directory (weekly)
WCPS Quarterly (weekly) 16914

Women's Caucus for the Modern Languages (WCML), c/o Emily Toth, Dept of English-Women's Studies, Louisiana State University, Baton Rouge, LA 70803
Founded: 1970; Members: 700
Gen Secr: Emily Toth
Focus: Ling
Periodical
Concerns (3 times annually) 16915

Women's Caucus of the Endocrine Society (WCES), c/o Dept of Physiology, School of Medicine, University of Marylamd, 655 W Baltimore St, Baltimore, MD 21201
T: (410) 328-3851
Founded: 1975; Members: 850
Gen Secr: Phyllis Wise
Focus: Endocrinology 16916

Women's Classical Caucus (WCC), c/o Prof. Barbara McManus, 5 Chester Dr, Rye, NY 10580
T: (914) 698-8798
Founded: 1972; Members: 500
Gen Secr: Prof. Barbara McManus
Focus: Humanities
Periodical
Newsletter (weekly) 16917

Women's College Coalition (WCC), 1090 Vermont Av NW, Washington, DC 20005
T: (202) 789-2556; Fax: (202) 842-4032
Founded: 1972; Members: 63
Gen Secr: Jadwiga S. Sebrechts
Focus: Educ 16918

Women's Computer Literacy Center (WCLC), 870 Market St, San Francisco, CA 94102
T: (415) 641-7007
Founded: 1982
Gen Secr: Heidi Steele
Focus: Computer & Info Sci 16919

Wordsworth-Coleridge Association, c/o Marlon Ross, Dept of English, University of Michigan, Ann Arbor, MI 48109
Founded: 1970; Members: 300
Pres: Marlon Ross
Focus: Lit
Periodical
The Wordsworth Circle (weekly) 16920

World Aquaculture Society (WAS), c/o Louisiana State University, 143 J.M. Parker Coliseum, Baton Rouge, LA 70803
T: (504) 388-3137; Fax: (504) 388-3493
Founded: 1970; Members: 2300
Gen Secr: Juliette Massey
Focus: Bio; Oceanography
Periodicals
Abstracts (weekly)
Directory (weekly)
Journal of the World Aquaculture Society (weekly)
World Aquaculture (weekly) 16921

World Archaeological Society (WAS), HCR 1, Box 445, Hollister, MO 65672
T: (417) 334-2377
Founded: 1971
Gen Secr: Ron Miller
Focus: Archeol
Periodicals
Special Publications (weekly)
WAS Newsletter (weekly) 16922

World Association of Veterinary Anatomists (WAVA), c/o Dept of Veterinary Anatomy, Purdue University, West Lafayette, IN 47907
T: (317) 494-7882
Founded: 1957; Members: 350
Gen Secr: Melvin W. Stromberg
Focus: Vet Med
Periodicals
Anatomia, Histologia, Embryologia (weekly)
WAVA-News (weekly) 16923

World Communication Association (WCA), c/o Ronald L. Applbaum, Westfield State College, Westfield, MA 01086
T: (413) 568-3311; Fax: (413) 562-3613
Founded: 1968
Gen Secr: Ronald L. Applbaum
Focus: Comm Sci 16924

World Education Fellowship, United States Section (WEF), c/o Dr. Mildred Haipt, College of New Rochelle, 29 Castle Pl, New Rochelle, NY 10805
T: (914) 654-5578
Founded: 1921; Members: 75
Pres: Dr. Mildred Haipt
Focus: Educ
Periodicals
The New Era (weekly)
U.S. Section News 16925

World Federation of Public Health Associations (WFPHA), c/o American Public Health Association, 1015 15 St NW, Washington, DC 20005
T: (202) 789-5696; Fax: (202) 789-5681
Founded: 1967; Members: 45
Gen Secr: Diane Kuntz
Focus: Public Health
Periodicals
Annual Report (weekly)
Salubritas (weekly) 16926

World Future Society (WFS), 7910 Woodmont Av, Ste 450, Bethesda, MD 20814
T: (301) 656-8274; Fax: (301) 951-0394
Founded: 1966; Members: 30000
Pres: Edward S. Cornish
Focus: Futurology
Periodicals
Futures Research Quarterly (weekly)
Future Survey: A Monthly Abstract of Books, Articles and Reports concerning Forecasts, Trends and Ideas about the Future (weekly)
The Futurist: A Journal of Forecasts, Trends and Ideas about the Future (weekly) 16927

World Hemophilia Aids Center (WHAC), 10 Congress St, Ste 340, Pasadena, CA 91105
T: (818) 577-4366; Fax: (818) 796-2875
Founded: 1983
Gen Secr: Terry Hall
Focus: Hematology
Periodical
Hemophilia World (weekly) 16928

World Organization for Human Potential (WOHP), 8801 Stenton Av, Philadelphia, PA 19118
T: (215) 233-2050; Fax: (215) 233-3940
Founded: 1968
Pres: Neil Harvey
Focus: Rehabil
Periodical
The IN-Report (weekly) 16929

World Population Society (WPS), 1333 H St NW, Ste 106, Washington, DC 20005
T: (202) 898-1303; Fax: (202) 861-0621
Founded: 1973; Members: 1100
Gen Secr: Frank H. Oram
Focus: Sociology
Periodical
Proceedings of Annual Conferences . . 16930

World Rehabilitation Fund (WRF), 386 Park Av, Ste 500, New York, NY 10016
T: (212) 725-7875; Fax: (212) 725-8402
Founded: 1970; Members: 27
Pres: Howard A. Rusk
Focus: Rehabil 16931

World Society for Stereotactic and Functional Neurosurgery (WSSFN), c/o Dr. Philip L. Gildenberg, 6560 Fannin, Ste 1530, Houston, TX 77030
T: (713) 790-0795
Founded: 1963; Members: 600
Gen Secr: Dr. Philip L. Gildenberg
Focus: Surgery; Neurology
Periodicals
Applied Neurophysiology (weekly)
Studies in Stereoencephalotomy 16932

World Space Foundation, POB Y, South Pasadena, CA 91031-1000
T: (818) 357-2878
Founded: 1979
Pres: Robert L. Staehle
Focus: Aero 16933

World's Poultry Science Association – U.S.A. Branch (WPSA), c/o Dr. Irvin L. Peterson, U.S. Dept of Agriculture, Federal Center Bldg, Rm 769, Hyattsville, MD 20782
T: (301) 436-7768; Fax: (301) 436-6465
Founded: 1965; Members: 640
Gen Secr: Dr. Irvin L. Peterson
Focus: Ornithology
Periodical
Journal (3 times annually) 16934

World University Service (WUS), 6530 Kissena Blvd, Flushing, NY 11367
T: (718) 520-7334; Fax: (718) 520-7241
Founded: 1920; Members: 60
Pres: Dr. George Priestly
Focus: Adult Educ 16935

World War Two Studies Association, c/o Prof. D. Clayton James, Dept of History and Politics, Virginia Military Institute, Lexington, VA 24450
T: (703) 464-7243
Founded: 1967; Members: 368
Gen Secr: Prof. D. Clayton James
Focus: Hist
Periodical
Newsletter of the American Committee on the History of the Second World War (weekly)
. . . . 16936

Wound, Ostomy and Continence Nurses – An Association of E.T. Nurses, 2755 Bristol St, Ste 110, Costa Mesa, CA 92626
T: (714) 476-0268
Founded: 1969; Members: 2400
Gen Secr: Debi Dahlman
Focus: Med 16937

W.T. Bandy Center for Baudelaire Studies (CBS), c/o Vanderbilt University, POB 6325, Station B, Nashville, TN 37235
T: (615) 322-2800
Founded: 1968; Members: 450
Gen Secr: Claude Pichois
Focus: Lit
Periodical
Bulletin Baudelairien (weekly) 16938

XPLOR International, 2550 Via Tejon, Ste 3L, POB 1501, Palos Verdes Estates, CA 90274
T: (310) 373-3633; Fax: (310) 375-4240
Members: 1400
Gen Secr: Keith T. Davidson
Focus: Nat Sci
Periodicals
Proceedings of European Conference (weekly)
Proceedings of Worldwide Conference (weekly)
. . . . *16939*

Yellowstone-Bighorn Research Association (YBRA), 7314 Charolais St, Billings, MT 59106
T: (406) 652-1760
Founded: 1931; Members: 350
Gen Secr: Betsy Campen
Focus: Geology *16940*

YIVO Institute for Jewish Research, 1048 Fifth Av, New York, NY 10028
T: (212) 535-6700; Fax: (212) 734-1062
Founded: 1925
Focus: Ethnology
Periodicals
News of the YIVO (weekly)
Yidishe Shprakh (weekly) *16941*

Yosemite Association (YA), POB 545, Yosemite National Park, CA 95389
T: (209) 379-2646; Fax: (209) 379-2486
Founded: 1920; Members: 6000
Pres: Steve Medley
Focus: Nat Sci
Periodical
Newsletter *16942*

Yuki Teikei Haiku Society of the United States and Canada, 1020 S Eighth St, San Jose, CA 95112
T: (408) 297-0692
Founded: 1979; Members: 70
Gen Secr: Kiyoshi Tokutomi
Focus: Lit
Periodicals
Geppo Haiku Journal (weekly)
Haiku Journal (weekly)
Members Anthology (weekly) *16943*

Zane Grey's West Society (ZGWS), c/o Carolyn Timmerman, 708 Warwick Av, Fort Wayne, IN 46825
T: (219) 484-2904
Founded: 1983; Members: 300
Gen Secr: Carolym Timmerman
Focus: Lit *16944*

Zionist Archives and Library of World Zionist Organization – American Section, 110 E 59 St, New York, NY 10022
T: (212) 753-2167
Founded: 1939
Gen Secr: Esther Togman
Focus: Archives; Libraries & Bk Sci . . *16945*

Uzbekistan

Scientific Industrial Association „Biology", Ul Khodzhaeva 28, 700125 Tashkent
T: 625821
Pres: O.D. Zhalilov
Focus: Bio *16946*

Scientific Industrial Association „Solar Physics", Ul Timiryazeva 2b, 700084 Tashkent
T: 331271
Pres: K.G. Gulamov
Focus: Physics *16947*

Uzbek Academy of Sciences, Ul Gogolya 70, 700000 Tashkent
T: 386847
Members: 144
Pres: M.S. Salokhitdinov; Gen Secr: K.G. Gulamov
Focus: Sci
Periodicals
Doklady: Bulletin
Fan va Turmush: Life and Science
Geliozekhnika: Helio Engineering
Izvestiya AN Uzbekistana-Tekhnicheskie Nauki: Journal of the Uzbek Academy of Sciences: Engineering Sciences
Khimiya Prirodnykh Soedinenii: Chemistry of Natural Compounds
Nauki v Uzbekistane: Social Sciences in Uzbekistan
Uzbekskii Geologicheskii Zhurnal: Uzbek Geological Journal
Uzbekskii Khimicheskii Zhurnal: Uzbek Chemical Journal
Uzbek Tili va Adabieti: Uzbek Language and Literature
Zapiski Uzbekskogo Otdelenia Vsesoyuznogo Mineralohicheskogo Obshchestva: Notes of the Uzbek Division of the All-Union Mineralogy Society *16948*

Vatican City

Accademia Romana di S. Tommaso d'Aquino e di Religione Cattolica, Piazza della Cancelleria 1, 00186 Roma
Founded: 1879; Members: 70
Pres: Cardinal Mario Luigi Ciappi; Gen Secr: Luigi Bogliolo
Focus: Rel & Theol *16949*

Collegium Cultorum Martyrum, Via Napoleone III 1, 00185 Roma
Founded: 1879; Members: 750
Gen Secr: Pietro Pozzi
Focus: Hist; Rel & Theol *16950*

Commission Théologique Internationale / International Theological Commission, Palazzo del S. Uffizio, 00120 Roma
T: Palazzo del (06) 6984753
Founded: 1969
Gen Secr: Georges Cottier
Focus: Rel & Theol *16951*

Pontificia Academia Scientiarum / Pontifical Academy of Sciences, Casina Pio IV, 00120 Roma
T: (06) 69883451, 69883195; Tx: 2024; Fax: (06) 69885218
Founded: 1936; Members: 80
Pres: Prof. Nicola Cabibbo; Gen Secr: Renato Dardozzi
Focus: Sci *16952*

Pontificia Accademia dell'Immacolata (PAJ), Via del Serafico 1, 00142 Roma
Founded: 1835; Members: 8
Pres: Cardinal Andrea M. Deskur; Gen Secr: Prof. Dr. Lorenzo Di Fonzo
Focus: Rel & Theol *16953*

Pontificia Accademia Mariana Internationalis (PAMI), Via Merulana 124, 00185 Roma
T: (06) 70373267; Fax: (06) 70373400
Founded: 1946; Members: 378
Pres: P. Melada
Focus: Rel & Theol *16954*

Pontificia Accademia Romana di Archeologia, Palazzo della Cancelleria Apostolica, 00186 Roma
Founded: 1810; Members: 111
Pres: Carlo Pietrangeli; Gen Secr: Silvio Panciera
Focus: Archeol
Periodicals
Memorie
Rendiconti *16955*

Pontificia Accademia Teologica Romana, Piazza S. Giovanni in Laterano 4, 00184 Roma
Founded: 1718; Members: 74
Pres: Cardinal William W. Baum; Gen Secr: Antonio Piolanti
Focus: Rel & Theol *16956*

Pontificia Insigne Accademia Artistica dei Virtuosi al Pantheon, Palazzo della Canceleria Apostolica, 00186 Roma
Founded: 1543
Pres: Armando Schiavo; Gen Secr: Dr. Vitaliano Tiberia
Focus: Arts *16957*

Venezuela

Academia de Ciencias Físicas, Matemáticas y Naturales, Av Universidad-Bolsa a San Francisco, Apdo 1421, Caracas 1010
T: (02) 4834133; Tx: 416611
Founded: 1917; Members: 80
Pres: Paul Lustgarten; Gen Secr: José M. Carrillo
Focus: Math; Nat Sci; Physics
Periodical
Boletín de la Academia de Ciencias Fı1sicas, Matemáticas y Naturales (3 times annually)
. . . . *16958*

Academia de Ciencias Politícas y Sociales, Palacio de las Academias, Bolsa a San Francisco, Caracas 1010
Founded: 1917; Members: 15
Pres: Dr. Pascual Venegas Filardo; Gen Secr: Dr. Victor M. Alvarez
Focus: Poli Sci; Sociology
Periodical
Boletín *16959*

Academia de Historia del Zulia, c/o Academia de Bellas Artes, Maracaibo
Founded: 1940; Members: 12
Pres: Abrahán Belloso; Gen Secr: Aniceto Ramirez y Astier
Focus: Hist
Periodical
Boletín *16960*

Academia Nacional de la Historia, Palacio de las Academias, Bolsa a San Franciso, Caracas 1010
Founded: 1888
Pres: Dr. Blas Bruni Celli; Gen Secr: Dr. Carlos Felice Cardot
Focus: Hist
Periodicals
Anuario
Boletín
Memorias *16961*

Academia Nacional de Medicina, Bolsa a San Franciso, Caracas, 1010
Founded: 1904
Pres: Dr. Augusto León; Gen Secr: Dr. J. Morales Rocha
Focus: Med
Periodical
Gaceta Medica de Caracas (weekly) . . *16962*

Academia Venezolana de la Lengua, Bolsa a San Francisco, Caracas 1010
T: 411795
Founded: 1883
Pres: Pedro Díaz Seijas; Gen Secr: Luis Beltrán Guerrero
Focus: Ling
Periodicals
Boletín (weekly)
Clásicos Venezolanos *16963*

Action Committee for the Establishment of a Latin American Network of Technological Information (RITLA), c/o SELA, Torre Europa, Av Francisco de Miranda, Caracas 1010A
T: (02) 9514233; Tx: 232944, 24615
Founded: 1979; Members: 4
Focus: Computer & Info Sci *16964*

Action for National Information Systems, c/o Ministerio de Información y Turismo, Altamira 68644, Caracas
Focus: Computer & Info Sci *16965*

Asociación Cultural Humboldt, Av Leonardo da Vinci, Colinas de Bello Monte, Apdo 60501, Chacao, Caracas
Founded: 1949; Members: 450
Focus: Hist; Lit; Poli Sci; Ethnology; Geography
Periodical
Boletín (weekly) *16966*

Asociación de Agrimensores de Venezuela, c/o Colegio de Ingenieros de Venezuela, Caracas
Focus: Agri *16967*

Asociación de Linguistica y Filologia de América Latina (ALFAL), Av Rio Orinoco 12-41, Cumbres de Curumo, Caracas 108
Founded: 1962; Members: 500
Focus: Ling; Lit *16968*

Asociación Interamericana de Bibliotecarios y Documentalistas Agrícolas, Filial Venezuela, c/o Universidad de Los Andes, Apdo 106, Mérida
Focus: Libraries & Bk Sci; Doc *16969*

Asociación Nacional de Escritores Venezolanos, Velázquez a Miseria 22, Apdo 429, Caracas
Founded: 1935
Focus: Lit *16970*

Asociación Psiquiátrica de la América Latina (APAL), Apdo 3380, Caracas
Founded: 1951; Members: 500
Focus: Psychiatry *16971*

Asociación Venezolana Amigos del Arte Colonial Caracas, Quinta de Anauco, Av Panteon San Bernardino, Caracas
T: (02) 518650
Founded: 1942; Members: 242
Pres: Gloria S. de Egui; Gen Secr: Carlos F. Duarte
Focus: Arts *16972*

Asociación Venezolana de Facultades de Medicina (AVEFAM), Av Las Ciencias, Los Chaguaramos, Apdo 50681, Caracas
T: 6614585
Focus: Med *16973*

Asociación Venezolana de Geología, Minería y Petróleo, Apdo Este 4000, Caracas
Founded: 1948
Focus: Mining; Geology *16974*

Asociacíon Venezolana de Ingeniería Eléctrica y Mecánica (AVIEM), c/o Colegio de Ingenieros de Venezuela, Apdo 6255, Caracas 1050
T: 5713657
Founded: 1956; Members: 3500
Focus: Electric Eng; Eng *16975*

Asociación Venezolana de Ingeniería Sanitaria y Ambiental Caracas, c/o Colegio de Ingenieros de Venezuela, Apdo 6255, Caracas
Focus: Eng *16976*

Asociación Venezolana para el Avance de la Ciencia (ASOVAC), Apdo del Este 61843, Caracas
Founded: 1950; Members: 3000
Focus: Sci
Periodical
Acta Científica Venezolana: Multidisciplinary Scientific Journal (weekly) *16977*

Ateneo Venezolano de Morfología, El Rosal 1060, Caracas
Founded: 1963
Focus: Bio *16978*

Biosciences Information Network for Latin America and the Caribbean (BINLAC), Apdo 21827, Caracas 1020
T: (02) 749543; Tx: 21338; Fax: (02) 691957
Founded: 1987
Focus: Bio *16979*

Caribbean Institute for Social Formation (CARISFORM), c/o UTAL, CLAT'S Headquarter, POB 6681, Caracas 1010
Members: 30
Gen Secr: A.C. Scoop
Focus: Soc Sci; Educ *16980*

Center for OPEC Studies, Torre Oeste, Piso 16, Parque Central, Caracas 1010
T: (02) 5076657; Tx: 21692; Fax: (02) 5754386
Founded: 1981
Gen Secr: Rebecca Sánchez
Focus: Energy *16981*

Centro de Estudios Venezolanos Indigenas, Apdo 261, Caracas
Founded: 1943; Members: 50
Focus: Ethnology *16982*

Centro de Historia del Táchira, Carrera 4, No 13-68, San Cristóbal, Táchira
T: 433079
Founded: 1942; Members: 25
Pres: Raúl Méndez-Moncada; Gen Secr: Horacio Moreno
Focus: Hist
Periodical
Boletín del Centro de Historia del Táchira (weekly) *16983*

Centro de Historia Larense, Calle 22, Diagonal a Plaza Lara, Apdo 406, Barquisimeto, Distr. Iribarren
Founded: 1941; Members: 12
Focus: Hist *16984*

Centro Histórico del Zulia, c/o Academia de Bellas Artes, Maracaibo
Founded: 1940
Focus: Hist *16985*

Centro Histórico Sucrense, Cumana
Founded: 1945
Focus: Hist *16986*

Centro Interamericano de Desarrollo Integras de Aguas y Tierra (CIDIAT), Parque La Isla, Apdo 219, Mérida
T: 23220
Focus: Ecology *16987*

Centro Interamericano para el Desarrollo Regional (CINDER), Calle 69, No 15D-32, Apdo 1304, Maracaibo 4001
T: (061) 516953, 517336; Tx: 61101; CA: CINDER; Fax: (061) 523554
Founded: 1976; Members: 14
Pres: Carmelo Contreras Barroza; Gen Secr: Rafael Pina Perez
Focus: Econ; Public Admin
Periodical
Boletín Informativo (weekly) *16988*

Centro Latinoamericano de Adminstración para el Desarrollo (CLAD) / Latinamerican Centre for Development Administration, Calle Herrera Toro, Qta Clad, Sector Los Naranjos, Las Mercedes, Apdo 4181, Caracas 1010
T: (02) 924064, 923297, 925953; Tx: 29076; Fax: (02) 918427
Founded: 1974; Members: 24
Gen Secr: Carlos Blanco
Focus: Public Admin
Periodicals
Boletín de Resumenes (weekly)
Boletín Informativo del CLAD (weekly)
Reforma y Democracia: Revista del CLAD (weekly)
Servicio de Información al Dia en Gestión Pública (weekly) *16989*

Centro Latinoamericano de Creación e Investigación Teatral (CELCIT), Av Juan Germán Rocio 9, Quinta Marisela, San Bernadino, Caracas 101
T: (02) 511675
Founded: 1977
Gen Secr: Luis Molina López *16990*

Centro Nacional de Investigaciones Agropecuarias (CENIAP) / National Centre of Agricultural Research, Zona Universitaria, Via El Limon, Apdo 4653, Maracay 2101
T: (043) 452491; Fax: (043) 454320
Founded: 1937; Members: 2239
Pres: Dr. Claudio Chicco
Focus: Agri
Periodicals
Agronomica Tropical
Veterinaria Tropical (weekly) *16991*

Colegio de Abogados del Distrito Federal, Apdo 347, Caracas
Founded: 1788; Members: 2000
Focus: Law *16992*

Colegio de Economistas del Distrito Federal y Estado Miranda, Calle Vicuña, Urb. Valle Arriba, Caracas
T: 918787
Founded: 1958
Focus: Econ *16993*

Colegio de Farmacéuticos del Distrito Federal y Estado Miranda, Urbanización Las Mercedes, Caracas 1060
Founded: 1949; Members: 1200
Pres: Dr. Pedro Rodriguez Murillo; Gen Secr: Dr. Esther V. de Pérez B.
Focus: Pharmacol
Periodical
Colfar *16994*

Colegio de Médicos del Distrito Federal, Pl de Bellas Artes, Caracas
Founded: 1942; Members: 2800
Focus: Med *16995*

Colegio de Médicos del Estado Anzoátegui, Apdo 84, Barcelona
Focus: Med 16996

Colegio de Médicos del Estado Mérida, Mérida
Focus: Med 16997

Colegio de Médicos del Estado Miranda, Av El Golf, Qta. La Setentiseis, El Bosque, Caracas
Founded: 1944; Members: 3100
Focus: Med 16998

Comité para la Defensa de las Lenguas Indígenas de América Latina y el Caribe, Urbanización California Norte, Av La Haya, Quinta Lilibeth, Caracas
Focus: Ling 16999

Consejo de Desarrollo Científico y Humanístico, c/o Universidad Central de Venezuela, Caracas
Fax: (02) 2851104
Founded: 1958
Focus: Sci; Humanities 17000

Consejo Nacional de Investigaciones Científicas y Tecnológicas (CONICIT), Los Ruices, Apdo 70617, Caracas
T: 349621
Founded: 1967
Focus: Sci; Eng 17001

Consejo Nacional de la Cultura (CONAC), Av Principal de Chuao, Edificio Los Roques, Apdo 50 995, Caracas
Founded: 1975
Focus: Hist 17002

Consejo Venezolano del Niño, Av San Martín, Apdo 1209, Caracas
Founded: 1951
Focus: Educ 17003

Consejo Zuliano de Planificación y Coordinación (CONZUPLAN), Av 5 de Julio, Edificio Las Laras, Apdo 1053, Maracaibo
T: 80754/55; Tx: 62364
Founded: 1964; Members: 100
Focus: Business Admin; Urban Plan . . 17004

Federación Médica Venezolana, Av El Golf, El Bosque
Founded: 1954
Focus: Med 17005

Federación Panamericana de Asociaciones de Facultades y Escuelas de Medicina (FEPAFEM) / Panamerican Federation of Associations of Medical Schools, Apdo 60411, Caracas 1060
T: 936271, 930875; Tx: 24627; Fax: 936346
Founded: 1962; Members: 353
Pres: Gaspar Garcia de Paredes; Gen Secr: Pablo A. Pulido
Focus: Med
Periodicals
Boletín FEPAFEM: PAFAMS Bulletin (weekly)
FEPAFEM Informa: PAFAMS Newsletter (weekly)
. . . . 17006

Federación Venezolana de Camaras y Asociaciones de Comercio y Producción, Av El Empalme, El Bosque, Apdo 2568, Caracas
T: 719742
Founded: 1973
Focus: Econ 17007

Sociedad Amigos del Museo de Bellas Artes, c/o Museo de Bellas Artes, Parque los Caobos, Caracas
Founded: 1957; Members: 250
Pres: Mimi de Herrera Uslar; Gen Secr: Ana de Besson
Focus: Fine Arts 17008

Sociedad Bolivariana de Venezuela, Apdo 874, Caracas
Founded: 1938; Members: 180
Focus: Hist 17009

Sociedad de Ciencias Naturales La Salle, Av Boyacá 3, Apdo 1930, Caracas 101
Founded: 1940
Focus: Nat Sci 17010

Sociedad de Obstetricia y Ginecología de Venezuela, Maternidad Concepción Palacios, Av San Martín, Apdo 20081, Caracas 1020
Pres: Dr. Saúl Kizer
Focus: Gynecology
Periodical
Revista de Obstetricia y Ginecología de Venezuela 17011

Sociedad de Tisiología y Neumonología de Venezuela, c/o Instituto Nacional de Tuberculosis, El Algodonal, Antimano
Founded: 1937
Focus: Pulmon Dis 17012

Sociedad Latinoamericana de Farmacología, Apdo 2455, Caracas
Founded: 1964; Members: 250
Focus: Pharmacol 17013

Sociedad Médica, Ciudad Bolivar, Bolivar
Focus: Med 17014

Sociedad Odontológica Zuliana de Prótesis Maracaibo, Edificio Cruz Roja, Local 1, Av 11, Maracaibo
Founded: 1964
Focus: Dent 17015

Sociedad Venezolana de Angiologia, c/o Colegio de Médicos del DF, Pl de Bellas Artes, Los Chaguaramos, Caracas
Focus: Hematology 17016

Sociedad Venezolana de Cardiología (SVC), Urb. Santa Fé, Av José María Vargas, Torre del Colegio, Apdo 80917, Caracas 1080
Members: 196
Pres: Dr. Simon Muñoz Armas; Gen Secr: Dr. Domingo Navarro Dona
Focus: Cardiol 17017

Sociedad Venezolana de Ciencias Naturales, Calle Cumaco con Arichuna El Marqués, Apdo 1521, Caracas 1070
T: 217653, 217780, 217579
Founded: 1931; Members: 980
Focus: Nat Sci
Periodicals
Boletín (weekly)
Newsletter (weekly) 17018

Sociedad Venezolana de Cirugía, Torre del Colegio, Av José María Vargas, Urb. Santa Fé, Caracas 1080
Founded: 1945
Pres: Dr. A. Diez; Gen Secr: Dr. I.J. Salas M.
Focus: Surgery 17019

Sociedad Venezolana de Cirugía Plástica y Reconstrucción, Edificio Bucaral, Av La Salle, Caracas
Focus: Surgery 17020

Sociedad Venezolana de Cirurgía Ortopédica y Traumatología, c/o Colegio de Médicos del DF, Pl Las Tres Gracías, Los Ghaguaramós, Caracas
Founded: 1949; Members: 197
Focus: Surgery; Orthopedics; Traumatology 17021

Sociedad Venezolana de Entomología, Apdo 4579, Maracay
Fax: (043) 453242
Founded: 1964
Focus: Entomology
Periodical
Boletin de Entomologia Venezolana (weekly)
. . . . 17022

Sociedad Venezolana de Hematología, c/o Hospital Vargas, San José, Caracas
Focus: Hematology 17023

Sociedad Venezolana de Historia de la Medicina, Palacio de las Academias, Bolsa a San Francisco, Caracas 101
Focus: Med; Cultur Hist 17024

Sociedad Venezolana de Ingenieros de Minas y Metalúrgicos, Apdo 18223, Caracas 1010
Founded: 1958; Members: 400
Pres: Luis Francisco Rivera Infante
Focus: Mining
Periodical
Fusión (weekly) 17025

Sociedad Venezolana de Medicina del Trabajo y Deportes, c/o Clínica Luis Razetti, Los Caobos, Caracas
Focus: Med; Hygiene 17026

Sociedad Venezolana de Oftalmología, Apdo Este 50150A, Caracas 1050
Founded: 1953; Members: 600
Pres: Dr. Henry Lugo Romero; Gen Secr: Dr. Rafael Cortéz
Focus: Ophthal
Periodical
Revista Oftalmológica Venezolana (weekly) 17027

Sociedad Venezolana de Oncología, c/o Centro Médico de Caracas, San Bernardino, Caracas
Pres: Dr. Víctor Brito
Focus: Cell Biol & Cancer Res 17028

Sociedad Venezolana de Otorinolaringología, Av Cajigal, San Bernardino, Apdo 40174, Caracas 1011
Pres: Dr. François Conde Jahn; Gen Secr: Dr. Germán Tovar Bustamante
Focus: Otorhinolaryngology
Periodical
Acta Venezolana de ORL (weekly) . . . 17029

Sociedad Venezolana de Psiquiatría y Neurología, Apdo 3380, Caracas
Focus: Neurology; Psychiatry 17030

Sociedad Venezolana de Puericultura y Pediatría, Av Libertador, Edificio La Linea, Caracas 105
Focus: Pediatrics 17031

Sociedad Venezolana de Radiología, c/o Policlínica Méndez Gimón, Av Andrés Bello, Caracas
Focus: Radiology 17032

Sociedad Venezolana de Tisiología y Neumonología, c/o Sanatorio Simón Bolívar, Carretera de Antímano, Caracas
Founded: 1937
Focus: Pulmon Dis 17033

Sociedad Venezolana de Urología, Apdo 75988, Caracas 1071
Founded: 1940; Members: 200
Pres: Dr. Darío Pisani Méndez; Gen Secr: Dr. Felipe Pulido Bueno
Focus: Urology
Periodical
Revista Venezolana de Urología 17034

Vietnam

Vietnamese Asscoiation of Morphology, c/o General Association of Medicine, 68 Ba Trieu St, Hanoi
T: 52323
Founded: 1967
Focus: Pathology 17035

Vietnamese Association of Acupuncture, c/o General Association of Medicine, 68 Ba Trieu St, Hanoi
T: 52323
Founded: 1968
Focus: Pathology 17036

Vietnamese Association of Agriculture, c/o Vietnam Union of Scientific and Technical Associations, 53 Nguyen Du, Hanoi
T: 57785
Focus: Agri 17037

Vietnamese Association of Anaesthesiology, c/o General Association of Medicine, 68 Ba Trieu St, Hanoi
T: 52323
Founded: 1978
Focus: Anesthetics 17038

Vietnamese Association of Anti-Contagious Diseases, c/o General Association of Medicine, 68 Ba Trieu St, Hanoi
T: 52323
Founded: 1978
Focus: Immunology 17039

Vietnamese Association of Anti-Tuberculosis and Lung Diseases, c/o General Association of Medicine, 68 Ba Trieu St, Hanoi
T: 52323
Founded: 1961
Focus: Pulmon Dis 17040

Vietnamese Association of Architects, c/o Vietnam Union of Scientific and Technical Associations, 53 Nguyen Du, Hanoi
T: 57785
Focus: Archit 17041

Vietnamese Association of Biology, c/o Vietnam Union of Scientific and Technical Associations, 53 Nguyen Du, Hanoi
T: 57785
Focus: Bio 17042

Vietnamese Association of Chemistry, c/o Vietnam Union of Scientific and Technical Associations, 53 Nguyen Du, Hanoi
T: 57785
Focus: Chem 17043

Vietnamese Association of Dermatology, c/o General Association of Medicine, 68 Ba Trieu St, Hanoi
T: 52323
Founded: 1961
Focus: Derm 17044

Vietnamese Association of Engineering, c/o Vietnam Union of Scientific and Technical Associations, 53 Nguyen Du, Hanoi
T: 57785
Focus: Eng 17045

Vietnamese Association of Forensic Scientists, c/o General Association of Medicine, 68 Ba Trieu St, Hanoi
T: 52323
Founded: 1978
Focus: Forensic Med; Law 17046

Vietnamese Association of Forestry, c/o Vietnam Union of Scientific and Technical Associations, 53 Nguyen Du, Hanoi
T: 57785
Focus: Forestry 17047

Vietnamese Association of Geography, c/o Vietnam Union of Scientific and Technical Associations, 53 Nguyen Du, Hanoi
T: 57785
Focus: Geography 17048

Vietnamese Association of Geology, c/o Vietnam Union of Scientific and Technical Associations, 53 Nguyen Du, Hanoi
T: 57785
Focus: Geology 17049

Vietnamese Association of Internal Medicine, c/o General Association of Medicine, 68 Ba Trieu St, Hanoi
T: 52323
Founded: 1960
Focus: Intern Med 17050

Vietnamese Association of Mathematics, c/o Vietnam Union of Scientific and Technical Associations, 53 Nguyen Du, Hanoi
T: 57785
Focus: Math 17051

Vietnamese Association of Medical Biochemistry, c/o General Association of Medicine, 68 Ba Trieu St, Hanoi
T: 52323
Founded: 1963
Focus: Biochem 17052

Vietnamese Association of Neurology, Psychiatry and Neurosurgery, c/o General Association of Medicine, 68 Ba Trieu St, Hanoi
T: 52323
Founded: 1962
Focus: Neurology; Psychiatry; Surgery . . 17053

Vietnamese Association of Obstetrics, Gynaecology and Family Planning, c/o General Association of Medicine, 68 Ba Trieu St, Hanoi
T: 52323
Founded: 1961
Focus: Gynecology; Family Plan 17054

Vietnamese Association of Odonto-Maxillo-Facial Medicine, c/o General Association of Medicine, 68 Ba Trieu St, Hanoi
T: 52323
Founded: 1960
Focus: Dent 17055

Vietnamese Association of Ophthalmology, c/o General Association of Medicine, 68 Ba Trieu St, Hanoi
T: 52323
Founded: 1960
Focus: Ophthal 17056

Vietnamese Association of Oto-Rhino-Laryngology, c/o National ENT Institute of Vietnam, Bach Mai, Hanoi
T: 52323
Focus: Otorhinolaryngology 17057

Vietnamese Association of Paediatrics, c/o General Association of Medicine, 68 Ba Trieu St, Hanoi
T: 52323
Founded: 1961
Focus: Pediatrics 17058

Vietnamese Association of Pharmacy, c/o General Association of Medicine, 68 Ba Trieu St., Hanoi
T: 52323
Founded: 1960
Focus: Pharmacol 17059

Vietnamese Association of Physics, c/o Vietnam Union of Scientific and Technical Associations, 53 Nguyen Du, Hanoi
T: 57785
Focus: Physics 17060

Vietnamese Association of Physiology, c/o General Association of Medicine, 68 Ba Trieu St, Hanoi
T: 52323
Founded: 1969; Members: 231
Pres: Prof. N.T.G. Trong; Gen Secr: Prof. T. Uyen
Focus: Physiology
Periodical
Sinh-Ly-Hoc (weekly) 17061

Vietnamese Association of Prophylactic Hygiene, c/o General Association of Medicine, 68 Ba Trieu St, Hanoi
T: 52323
Founded: 1961
Focus: Hygiene 17062

Vietnamese Association of Radio-Electronics, c/o Vietnam Union of Scientific and Technical Associations, 53 Nguyen Du, Hanoi
T: 57785
Focus: Electronic Eng 17063

Vietnamese Association of Radiology, c/o General Association of Medicine, 68 Ba Trieu St, Hanoi
T: 52323
Founded: 1961
Focus: Radiology 17064

Vietnamese Association of Serum and Blood Transfusion, c/o General Association of Medicine, 68 Ba Trieu St, Hanoi
T: 52323
Founded: 1978
Focus: Intern Med 17065

Vietnamese Association of Surgery, c/o General Association of Medicine, 68 Ba Trieu St, Hanoi
T: 52323
Founded: 1961
Focus: Surgery 17066

Vietnamese Associaton of Cast and Metallurgy, c/o Vietnam Union of Scientific and Technical Associations, 53 Nguyen Du, Hanoi
T: 57785
Focus: Metallurgy 17067

Vietnamese Fine Arts Association, c/o Writers and Artists Union, 51 Tran Hung Dao St, Hanoi
T: 52694
Focus: Fine Arts 17068

Vietnamese Photographers Association, c/o Writers and Artists Union, 51 Tran Hung Dao St, Hanoi
T: 52694
Focus: Photo 17069

Vietnamese Writers' Association, c/o Writers and Artists Union, 51 Tran Hung Dao St, Hanoi
T: 52694
Focus: Lit 17070

Vietnam Union of Sientific and Technical Associations, 53 Nguyen Du, Hanoi
T: 57785
Founded: 1983
Focus: Sci; Eng 17071

Writers and Artists Union, 51 Tran Hung Dao St, Hanoi
T: 52694
Founded: 1957
Pres: Ch Huy Can
Focus: Lit; Arts 17072

Virgin Islands (U.S.)

Caribbean Food Crops Society (CFCS), c/o University of the Virgin Islands, POB 10000, Saint Croix, VI 00850
T: (809) 778-0246; Fax: (809) 778-6570
Founded: 1962
Gen Secr: Kofi Boateng
Focus: Crop Husb
Periodical
CFCS Newsletter (3 times annually) . . 17073

Caribbean Natural Resources Institute (CANARI), 1104 Strand St, Christiansted, Saint Croix, VI 00820
T: (809) 773-9854; Fax: (809) 773-5770
Founded: 1977; Members: 18
Gen Secr: Allen D. Putney
Focus: Geology 17074

Consortium of Caribbean Universities for Resource Management, c/o University of the Virgin Islands, VI 00802, Saint Thomas
T: (809) 776-0200 ext 1343
Founded: 1988; Members: 15
Gen Secr: Dr. L.E. Ragster
Focus: Educ; Business Admin 17075

Yugoslavia

Akademia e Shkencave dhe e Arteve e Kosovës / Akademija Nauka i Umetnosti Kosova / Academy of Sciences and Arts of Kosovo, Milladin Popoviqi 10, 38000 Priština
Founded: 1976
Pres: Dr. Vukasin Filipović; Gen Secr: Osman Imami
Focus: Ling; Lit; Arts; Soc Sci; Nat Sci
Periodicals
Acta Biologiae et Medicinae Experimentalis
Bibliografia/Bibliografija
Botime të Veçanta/Posebna Izdanja: Monographs
Kërkime/Istraživanja: Research
Libërshënuesi/Spomenica: Diary
Studime/Studije: Studies
Vjetari/Godisnjak: Annual Report 17076

Association of Jurists of Serbia, Proleterskih Brigada 74, 11000 Beograd
Founded: 1946
Focus: Law
Periodical
Pravni Život 17077

Association of Pharmacists of Serbia, Terazije 12, 11000 Beograd
Founded: 1946
Focus: Pharmacol
Periodicals
Arhiv za Farmaciju
Bilten 17078

Association of the Mathematicians', Physicists' and Astronomers' Societies of Yugoslavia, c/o Akademija Nauka SAP Kosovo, POB 194, 38000 Priština
Founded: 1950
Focus: Math; Physics; Astronomy
Periodicals
Bilten: Astronomy
Fizika, Matematičko-Fizički List za Učenike Srednjih Škola
Matematica Balkanica
Matematički List za Učenike Osnovnih Škola
Topologija 17079

Crnogorska Akademija Nauka i Umjetnosti (CANU) / Montenegrin Academy of Sciences and Arts, Rista Stijovića 5, 81000 Podgorica
T: (081) 31095
Pres: D. Vukotić; Gen Secr: B. Gluščević
Focus: Arts; Sci
Periodicals
Glasnik: Review
Godišnjak CANU (weekly) 17080

Društvo Arhitekata Srbije, Kneza Miloša 7, 11000 Beograd
T: (011) 330059
Focus: Archit 17081

Društvo Arhivskih Radnika Srbije, Karnegijeva 2, 11000 Beograd
T: (011) 323132
Focus: Archives 17082

Društvo Bibliotekara Vojvodine, c/o Bibliothèque Centrale de la Faculté des Lettres et de Sciences, Université, 21000 Novi Sad
Focus: Libraries & Bk Sci 17083

Društvo Istoričara Srbije / Historical Society of Serbia, Cika Ljubina 18-20, 11000 Beograd
Founded: 1948; Members: 1500
Focus: Hist
Periodical
Istoriski glasnik (weekly) 17084

Društvo Lekara Vojvodine, Vase Stajica 9, 21000 Novi Sad
Focus: Med
Periodical
Medicinski pregled: Journal for General Medicine (weekly) 17085

Društvo Ljekara Crne Gore, Krusevac bb, 81000 Podgorica
T: 41620
Focus: Med 17086

Društvo Matematičara Srbije / Society of Mathematicians of Serbia, Knez Mihailova 35, 11000 Beograd
T: (011) 638263
Founded: 1948
Focus: Math
Periodical
Matematički Vesnik (weekly) 17087

Društvo Psihologa Srbije, Drusvina 7, 11000 Beograd
T: (011) 339685
Focus: Psych 17088

Društvo Veterinara Srbije, Bulevar JNA 18, 11000 Beograd
T: (011) 684597
Focus: Vet Med 17089

Društvo za Srpski Jezik i Književnost / Society of Serbian Language and Literature, c/o Belgrade University, 11000 Beograd
Founded: 1910
Focus: Ling; Lit
Periodical
Pritozi za Knjizevnost, Jezik, Istorija i Folklor 17090

Društvo za Srpskohrvatski Jezik i Književnost, Knez Mihailova 35, 11000 Beograd
T: (011) 630089
Focus: Ling; Lit 17091

Economists' Society of Serbia, Nusićeva 6, 11000 Beograd
Founded: 1944
Focus: Econ
Periodical
Ekonomika preduzeća (weekly) 17092

Filozofsko Društvo Srbije, Studentski Trg 1, 11000 Beograd
Focus: Philos 17093

Jugoslovenski Centar za Tehničku i Naučnu Dokumentaciju / Yugoslav Centre for Technical and Scientific Documentation, S. Penezića-Krcuna 29-31, POB 724, 11000 Beograd
T: (011) 644250; Tx: 12497
Founded: 1952
Focus: Doc; Eng
Periodicals
Bulletin of Documentation
Informatika
Scientific and Professional Meetings in Yugoslavia and Foreign Countries 17094

Jugoslovenski Savez za Zavarivanje (JSZ) / Yugoslav Welding Association, Svetozara Markovića 56, 11000 Beograd
T: (011) 687221, 682963; Tx: 12247
Founded: 1953; Members: 6
Focus: Eng
Periodicals
Varilna tehnika (weekly)
Zavarivač (weekly)
Zavarivanje (weekly) 17095

Jugoslovenski Zavod za Produktivnost Rada, Uzun Mirkova 1, 11000 Beograd
T: (011) 637978
Focus: Business Admin 17096

Jugoslovenski Zavod za Standardizaciju, Slobodana Penezića-Krcuna 35, POB 933, 11000 Beograd
T: (011) 643557
Focus: Standards 17097

Jugoslovensko Društvo za Fiziologiju, c/o Zavod za Biohemiju, Medicinski Fakultet, 21000 Novi Sad
Focus: Physiology 17098

Jugoslovensko Društvo za Mehaniku, Kneza Milosa 9-13, 11000 Beograd
T: (011) 342273
Focus: Eng 17099

Jugoslovensko Društvo za Proučavanje Zemljišta, Nemanjina 6, 11080 Zemun
Founded: 1953
Focus: Agri
Periodicals
Agrohemija (weekly)
Zemljišste i biljka (3 times annually) . . 17100

Jugoslovensko Naučno Voársko Društvo, Vojvode Stepe 5, 32000 Čačak
T: (032) 47411 ext 13
Founded: 1954
Focus: Hort
Periodical
Jugoslovensko Voćarstvo: Journal of Yugoslav Pomology (weekly) 17101

Jugoslovensko Udruženje za Filozofiju, Studebtski Trg 1, 11000 Beograd
T: (011) 638104
Focus: Philos 17102

Jugoslovensko Udruženje za Sociologiju, Studentski Trg 1, 11000 Beograd
Founded: 1954
Focus: Sociology
Periodical
Sociology (weekly) 17103

Matica Srpska / Serbian Society, Ul Matice Sprske 1, 21000 Novi Sad
Founded: 1826
Focus: Ling; Lit
Periodicals
Letopis Matice Srpske
Proceedings 17104

Nikola Tesla Association of Societies for Promotion of Technical Sciences in Yugoslavia, POB 359, 11000 Beograd
Focus: Eng 17105

Pedagogical Society of Yugoslavia, Moše Pijade 12, 11000 Beograd
Founded: 1952
Focus: Educ
Periodicals
Pedagogícja
Predškolsko Dete 17106

Savez Arheološki Društava Jugoslavije, Cara Urosa 20, 11000 Beograd
Members: 300
Focus: Archeol 17107

Savez Astronautičkih i Raketnih Organizacija Jugoslavije (SAROJ), Bulevar Revolucije 44, 11000 Beograd
T: (011) 33404
Founded: 1953; Members: 475
Focus: Aero 17108

Savez Bibliotečkih Radnika Srbije / Union of Serbian Library Workers, Skerlićeva 1, 11000 Beograd
Tx: 12208
Founded: 1947
Focus: Libraries & Bk Sci
Periodical
Bibliotekar (weekly) 17109

Savez Društava Arhiviskih Radnika Jugoslavije (SDAR), Karnegijeva 2, 11000 Beograd
Founded: 1953; Members: 950
Focus: Archives 17110

Savez Društava Istoričara Jugoslavije, Karnegijeva 2, 11000 Beograd
T: (011) 338431
Founded: 1951
Focus: Hist 17111

Savez Društava Matematičara, Fizičara i Astronoma Jugoslavije (SDMFAJ), c/o Institut za Matematiku, Dr. Ilije Djuricica 4, 21000 Novi Sad
Founded: 1950; Members: 2007
Focus: Math; Physics; Astronomy . . . 17112

Savez Društava Veterinara Jugoslavije, Bulevar JNA 18, 11000 Beograd
T: (011) 684597
Focus: Vet Med 17113

Savez Društava za Strane Jezike i Knjizevnosti Jugoslavije, Knez Mihajlova 35, 11000 Beograd
Focus: Ling; Lit; Educ 17114

Savez Društava Zubarskih Radnika Jugoslavije, Mose Pijade 12, 11000 Beograd
T: (011) 339928
Focus: Dent 17115

Savez Ekonomista Jugoslavije / Yugoslav Economists' Association, Nusiceva 6, 11000 Beograd
T: (011) 334417
Founded: 1945
Focus: Econ
Periodical
Ekonomist (weekly) 17116

Savez Lekarskih Društava Jugoslavije (SLD), Zeleni Venac 1, 11000 Beograd
T: (011) 320992
Founded: 1946
Focus: Med 17117

Savez Muzejskih Društava Jugoslavije / Federation of Museums Associations, Vuka Karaclžića 18, 11000 Beograd
Focus: Archives
Periodical
Muzeji 17118

Savez Pedagoških Društava Jugoslavije, Mose Pijade 12, 11000 Beograd
T: (011) 339381
Founded: 1950
Focus: Educ 17119

Savez Pedagoških Društava Srbije / Federation of Pedagogical Societies of Serbia, Terazije 26, 11000 Beograd
T: (011) 325569
Founded: 1923; Members: 800
Focus: Educ
Periodicals
Nastava i va spitanje (5 times annually)
Pedagoška Biblioteka 17120

Savez Udruženja Pravnika Jugoslavije, Proleterskih Brigada 74, 11000 Beograd
T: (011) 441910
Founded: 1947; Members: 10000
Focus: Law 17121

Savez Udruženja za Krivično Pravo i Kriminolgiju Jugoslavije (SUKKJ), Gracanicka 18, 11000 Beograd
T: (011) 626322
Founded: 1958; Members: 2500
Focus: Law; Criminology 17122

Serbian Association against Cancer, Knez Mihailova 2, 11000 Beograd
Founded: 1967
Focus: Cell Biol & Cancer Res 17123

Sindikat Radnicka Drustvenih Delatnosti Jugoslavije, Trg Marksa i Engelsa 5, 11000 Beograd
T: (011) 332953
Members: 30000
Focus: Educ 17124

Srpska Akademija Nauka i Umetnosti (SANU) / Serbian Academy of Sciences and Arts, Knez Mihailova ul 35, POB 366, 11000 Beograd
T: (011) 187144
Founded: 1886; Members: 70
Pres: D. Kanazir; Gen Secr: Dejan Medaković
Focus: Sci; Arts
Periodicals
Glas: Review
Godišnjak: Yearbook (weekly)
Posebna izdanja: Monographs
Spomenik: Monument
Srpski etnografski zbornik: Serbian Ethnographic Collection 17125

Srpsko Biološko Društvo / Serbian Biological Society, Kneza Miloša 101, 11000 Beograd
T: (011) 682966 ext 303
Founded: 1947; Members: 700
Focus: Bio
Periodicals
Archive of Biological Science
Contemporary Biology (weekly) 17126

Srpsko Geografsko Društvo / Serbian Geographical Society, Studentski trg 3, 11000 Beograd
Founded: 1910; Members: 1350
Focus: Geography
Periodicals
Bulletin (weekly)
Globus (weekly)
Terre et Hommes (weekly) 17127

Srpsko Geološko Društvo / Serbian Geological Society, Kamenicka 6, POB 227, 11000 Beograd
Founded: 1891; Members: 602
Focus: Geology
Periodical
Zapisnici Srpskog Geološkog Društva (weekly) 17128

Srpsko Hemijsko Društvo / Serbian Chemical Society, Karnegijeva 4, POB 462, 11000 Beograd
T: (011) 328583
Founded: 1897; Members: 2200
Focus: Chem
Periodicals
Chemical Review (weekly)
Journal of the Serbian Chemical Society (weekly) 17129

Srpsko Lekarsko Društvo, Zeleni Venac 1, 11000 Beograd
T: (011) 327181
Founded: 1872
Focus: Med
Periodicals
Gastroenterohepatološki Arhiv (weekly)
Kardiologija (weekly)
List Lekar (weekly)
Srpski arhiv za celokupno lekarstvo (weekly)
Stomatološki glasnik Srbije (weekly) . . 17130

Udruženje Univerzitetskih Nastavnika i Van-Univerzitetskih Naučnik Radnika, Safarikova 7, 11000 Beograd
T: (011) 324939
Focus: Educ; Adult Educ 17131

Unija Bioloških Naučnih Društava Jugoslavije, Nemanjina 6, POB 127, 11080 Zemun
Focus: Bio 17132

Union of Engineers and Technicians of Serbia, Kneza Miloša 7a, 11000 Beograd
T: (011) 330067
Founded: 1945; Members: 8014
Focus: Eng
Periodical
Tehnika (weekly) 17133

Union of Engineers and Technicians of Yugoslavia, Kneza Miloša 9-11, POB 187, 11000 Beograd
T: (011) 335816; Tx: 72905
Founded: 1841
Focus: Eng
Periodicals
It Novine (weekly)
Tehnika (weekly) 17134

Union of Jurists' Associations of Yugoslavia, Proleterskih Brigade 74, 11000 Beograd
Founded: 1947; Members: 30000
Focus: Law
Periodical
The New Yugoslav Law 17135

Urbanisticki Savez Jugoslavije, Bul Avnoj 3, 21000 Novi Sad
Focus: Urban Plan 17136

Yugoslav Economists' Association, Nušićva 6, 11000 Beograd
Founded: 1945
Focus: Econ
Periodical
Ekonomist (weekly) 17137

Zajednica Universitéta Jugoslavije / Association of Yugoslav Universities, Palmotićeva 22, 11000 Beograd
T: (011) 334524
Founded: 1957
Focus: Educ; Adult Educ
Periodical
University Today (weekly) 17138

Zaire

African Bureau of Educational Sciences, BP 1764, Kinshasa 1
T: (012) 22006
Founded: 1973; Members: 39
Focus: Educ
Periodicals
Annuaire Africain des Sciences de l'Education (3 times annually)
Bulletin d'Information (weekly)
Répertoire Africain des Institutions de Recherche (weekly)
Revue Africaine des Sciences de l'Education (weekly) 17139

Association des Institutions d'Enseignement Théologique en Afrique Occidentale (ASTHEOL-WEST), c/o Faculté de Théologie, BP 4745, Kinshasa
Focus: Rel & Theol; Educ 17140

Association Zaïroise des Archivistes, Bibliothécaires et Documentalistes, BP 805, Kinshasa
Founded: 1973
Focus: Archives; Libraries & Bk Sci; Doc 17141

Central African Regional Branch of the International Council on Archives (CENARBICA), c/o Archives Nationales, BP 3428, Kinshasa
T: (012) 31083
Founded: 1982
Gen Secr: Lumenga-Neso Kiobe
Focus: Archives 17142

Centrale des Enseignants Zairois (CEZ), BP 8814, Kinshasa
Founded: 1957
Focus: Educ 17143

Centre for the Coordination of Research and Documentation in Social Science for Sub-Saharan Africa / Centre de Coordination des Recherches et de Documentation en Sciences Sociales desservant l'Afrique Sub-Saharienne, BP 836, Kinshasa
T: (012) 27003; Tx: 2151
Founded: 1974
Gen Secr: Prof. Lapika Dimomfu
Focus: Soc Sci
Periodical
CERDAS Liaison (weekly) 17144

Conférence des Recteurs des Universités Francophones d'Afrique (CRUFA), c/o Présidence des Universités du Zaïre, BP 13399, Kinshasa
T: (012) 78681; Tx: 21215
Founded: 1978
Gen Secr: Thsishiku Tshibangu
Focus: Educ 17145

Société des Historiens Zairois, BP 7246, Lubumbashi
Founded: 1974
Pres: Prof. N.E. Nziem; Gen Secr: Prof. Dr. T.-B.M. Kabet
Focus: Hist
Periodical
Likundoli (weekly) 17146

Zambia

Adult Education Association of Zambia (AEAZ), POB 2379, Lusaka
Founded: 1968; Members: 200
Focus: Adult Educ 17147

Africa Association for Liturgy, Music and Arts (AFALMA), c/o United Church of Zambia, POB 50122, Lusaka
Gen Secr: Bwalya S. Chuba
Focus: Rel & Theol; Arts; Music . . . 17148

Africa Literature Centre (ALC), POB 21319, Kitwe
Founded: 1959
Focus: Lit 17149

Agricultural Research Council of Zambia, POB 2218, Lusaka
Founded: 1967
Focus: Agri 17150

Association of Medical Schools in Africa (AMSA), c/o Faculty of Medicine, University of Zambia, POB 50110, Lusaka
T: (01) 252641; Tx: 44370
Founded: 1964
Pres: Prof. Kopauo Mukelabai
Focus: Educ; Med 17151

The Engineering Institution of Zambia, POB 34730, Lusaka
Founded: 1955; Members: 1500
Pres: S.K. Tamele; Gen Secr: E.A. Kashita
Focus: Eng
Periodical
Journal (weekly) 17152

International Red Locust Control Organisation for Central and Southern Africa (IRLCO-CSA), POB 240252, Ndola
T: 612433; Tx: 30072
Founded: 1970; Members: 10
Gen Secr: S. Moobola
Focus: Entomology
Periodicals
Annual Report (weekly)
Quarterly Report (weekly)
Scientific Papers 17153

National Council for Scientific Research, Chelston, POB CH 158, Lusaka
T: (01) 281081
Founded: 1967; Members: 20
Gen Secr: Dr. S.M. Silangwa
Focus: Sci
Periodicals
Zambia Journal of Science and Technology (weekly)
Zambia Science Abstracts (weekly) . . . 17154

National Food and Nutrition Commission (N.F.N.C.), POB 32669, Lusaka
T: (01) 211724; CA: Fonutcom, Lusaka
Founded: 1967; Members: 5
Pres: A. Ndalama; Gen Secr: A.B. Vermoer
Focus: Nutrition
Periodical
Nutrition News (3 times annually) . . . 17155

National Monuments Commission, POB 60124, Livingstone
T: (03) 320481
Founded: 1948
Gen Secr: N.M. Katanekwa
Focus: Ecology; Preserv Hist Monuments
Periodicals
Annual Report (weekly)
Archaeologia Zambiana Newsletter (weekly)
Research Publications 17156

Wildlife Conservation Society of Zambia (WCSZ), POB 30255, Lusaka
T: (01) 254226; Tx: 44810
Founded: 1953; Members: 1500
Pres: Prof. A.A. Siwela; Gen Secr: A.J. Scott
Focus: Ecology
Periodical
Black Lechwe (weekly) 17157

Zambia Library Association, POB 32839, Lusaka
Pres: C. Zulu; Gen Secr: W.C. Mulalami
Focus: Libraries & Bk Sci
Periodicals
Journal (weekly)
Newsletter (weekly) 17158

Zimbabwe

African Organization for Research and Training in Cancer (AORTIC), c/o Dept of Medicine, University of Zimbabwe, POB A178, Avondale, Harare
Founded: 1982
Pres: C.F. Kire
Focus: Intern Med
Periodical
AORTIC Bulletin (weekly) 17159

African Regional Computer Confederation (ARCC), POB 8385, Causeway, Harare
T: (04) 752657
Focus: Computer & Info Sci 17160

African Rehabilitation Institute (ARI), POB 4056, Harare
T: (04) 731083
Founded: 1985; Members: 12
Gen Secr: J.D. Bukutu
Focus: Rehabil
Periodical
African Rehabilitation Journal (weekly) . . 17161

Agricultural Research Council of Zimbabwe, POB 8108, Causeway, Harare
Founded: 1970; Members: 12
Pres: K.D. Kirkman
Focus: Agri
Periodicals
Annual Report (weekly)
Technical Report 17162

Arts Association Harare, POB 4011, Harare
Founded: 1968
Pres: H. Marsh; Gen Secr: E. Strannix
Focus: Arts 17163

Botanical Society of Zimbabwe, POB 461, Harare
Founded: 1934
Gen Secr: J.R. James
Focus: Botany 17164

Commonwealth Association for the Education and Training of Adults (CAETA), c/o Dept of Adult Education, University of Zimbabwe, POB MP 167, Mount Pleasant, Harare
T: (04) 303211 ext 1528; Tx: 26580; Fax: (04) 732828
Founded: 1985; Members: 660
Gen Secr: Dr. Meshack Matshazi
Focus: Adult Educ
Periodical
CAETA Newsletter (weekly) 17165

Conference of African Theological Institutions (CATI), POB MP 36, Mount Pleasant, Harare
T: (04) 303211
Founded: 1980
Gen Secr: Dr. A. Moyo
Focus: Rel & Theol 17166

Crop Science Society of Zimbabwe, Union Av, POB UA 409, Harare
Focus: Agri 17167

Dental Association of Zimbabwe, POB 3303, Harare
Focus: Dent 17168

Ecumenical Documentation and Information Centre for Eastern and Southern Africa (EDICESA), POB H 94, Harare
T: (04) 50311; Tx: 26206
Founded: 1987
Gen Secr: Hartwig Liebich
Focus: Rel & Theol 17169

Geographical Association of Zimbabwe, c/o Dept of Geography, University of Zimbabwe, POB MP 167, Mount Pleasant, Harare
T: (04) 303211
Members: 200
Pres: S.C. Jackson; Gen Secr: E. Manaka
Focus: Geography
Periodicals
Geographical Education Magazine
Proceedings 17170

Geological Society of Zimbabwe, POB 8427, Causeway, Harare
Focus: Geology 17171

Kirk Biological Society, c/o Div of Biological Sciences, Univeristy of Zimbabwe, POB MP 167, Mount Pleasant, Harare
Focus: Bio 17172

The Literature Bureau, Zimbabwe, POB 8137, Causeway, Harare
Founded: 1954
Gen Secr: B.C. Chitsike
Focus: Lit
Periodical
Bureau Bulletin (weekly) 17173

Lowveld Natural History Branch, Wildlife Society of Zimbabwe, POB 81, Chiredzi
Founded: 1968
Pres: C. Stockil; Gen Secr: M. Davy
Focus: Hist; Nat Sci
Periodicals
The Hartebeest (weekly)
Newsletter (weekly) 17174

Mennel Society, c/o Dept of Geology, University of Zimbabwe, POB MP 167, Mount Pleasant, Harare
Founded: 1964; Members: 30
Pres: Prof. J.F. Wilson
Focus: Geology
Periodical
Detritus 17175

Ornithological Association of Zimbabwe, POB 8382, Causeway, Harare
T: (04) 48347; Tx: 4614
Pres: R. Butler
Focus: Ornithology
Periodical
Honeyguide (weekly) 17176

PEN Centre of Zimbabwe, POB 1900, Harare
Gen Secr: Nora S. Kane
Focus: Lit 17177

Pharmaceutical Society of Zimbabwe, POB 8520, Causeway, Harare
Pres: M. Torongo; Gen Secr: A. Palmer
Focus: Pharmacol 17178

Pre-History Society of Zimbabwe, POB 876, Harare
Founded: 1958
Focus: Hist; Anthro
Periodical
Zimbabwe Prehistory 17179

Scientific Council of Zimbabwe, POB 8510, Causeway, Harare
Founded: 1964
Focus: Sci 17180

Standards Association of Central Africa, 17 Coventry Rd, Workington, POB 2259, Harare
T: (04) 760258; CA: SACA
Founded: 1957; Members: 322
Focus: Standards
Periodicals
Annual Report (weekly)
Central African Specifications and Codes of Practice 17181

Survey Institute of Zimbabwe, POB 3869, Harare
Founded: 1967; Members: 150
Pres: R.J. Good; Gen Secr: R.F. Goodwin
Focus: Surveying 17182

Wildlife Society of Zimbabwe, POB 3497, Harare
Founded: 1927; Members: 1500
Pres: I.G. Cormack; Gen Secr: S.M. Brookes-Ball
Focus: Ecology
Periodical
Zimbabwe Wildlife (weekly) 17183

Zimbabwe Agricultural Society, POB 442, Harare
Tx: 22221
Focus: Agri 17184

Zimbabwe Association for Science Education, 16 Dirriemuur Dr, Marlborough, Harare
T: (04) 32867
Members: 210
Focus: Sci; Educ 17185

Zimbabwe Library Association, POB 3133, Harare
Founded: 1961
Pres: G.C. Mots; Gen Secr: D.E. Barron
Focus: Libraries & Bk Sci
Periodical
Zimbabwe Librarian 17186

Zimbabwe Medical Association, POB 3671, Harare
T: (04) 720731
Pres: Dr. D.M. Sadza; Gen Secr: R.D. Martin
Focus: Med 17187

Zimbabwe Scientific Association, POB 978, Harare
Founded: 1899; Members: 350
Pres: Dr. P.R. Morgan
Focus: Sci
Periodicals
Transactions of the Zimbabwe Sientific Association (5 times annually)
Zimbabwe Science News (weekly) . . . 17188

Zimbabwe Veterinary Association, POB 8387, Causeway, Harare
Pres: Dr. L.V. Orsmond; Gen Secr: Dr. A.J.M. Isdale
Focus: Vet Med
Periodical
Zimbabwe Veterinary Journal (weekly) . . 17189

Zimbabwe Writers Union, POB MP 167, Mount Pleasant, Harare
Members: 250
Pres: C. Hove; Gen Secr: M. Zimunya
Focus: Lit 17190

Alphabetical Index to Association Names

Alphabetisches Verzeichnis der Gesellschaftsnamen

AA, London 11205, 11245
AA, Los Angeles 13283
AA, New York 13461
AA, Silver Spring 13359
AAA, London 11200
AAA, Menlo Park 13448
AAA, New Orleans 13506
AAA, New York 13456
AAA, Richmond 13380
AAA, Sarasota 13444
AAA, Washington 13379, 13453, 14240
A.A.A.A., Buenos Aires 00054
AAACE, Arlington 13464
AAACU, Laguna 09495
AAAE, Ithaca 13466
AAAF, Paris 04144
AAAHC, Skokie 13314
AAAI, Menlo Park 13468
AAAI, Milwaukee 13381
AAAL, Oklahoma City 13467
AAALAC, Rockville 13463
A.A.A.N.Z., Clayton 00212
AAAP, Kennett Sq 13507
AAAP, Miaoli 03236
AAAR, Cincinnati 13465
AAAS, Cambridge 13382
AAAS, Washington 13497
AAASA, Addis Ababa 03942
AAASS, Stanford 13498
A.A.B., Bologna 07261
AAB, Littlehampton 11301
AA/BA, New York 13452
AABB, Bethesda 13511
AABC, Fayetteville 13509
AABE, Manila 09494
AABE, Washington 13510
AABGA, Wayne 13512
AABNF, Nairobi 08459
AABP, West Lafayette 13513
AABPA, Falls Church 13469
AABS, Hackettstown 14341
AABT, New York 14263
AAC, Alexandria 14269
AAC, Saugerties 13348
AAC, Washington 14378
AACA, Palatine 13515
AACAP, Washington 13383
AACB, Maylands 00263
AACC, Evanston 13612
AACC, Saint Paul 13514
AACC, Washington 13474
AACDP, Little Rock 13517
AACE, Hermosa Beach 13472
AACE, Houston 13470
AACE, London 11257
AACE, Nairobi 08460
AACE International, Morgantown 13273
AACG, Thousand Oaks 13476
Aachener Geschichtsverein, Aachen 05028
AACI, Louisville 13386
AACJC, Washington 13524
AACN, Aliso Viejo 13526
AACN, Washington 13520
AACO, Houston 13516
AACOM, Rockville 13521
AACP, Alexandria 13522
AACP, Institute 13475
AACP, San Diego 13384
AACPDM, Richmond 13376
AACPM, Rockville 13523
AACR, Philadelphia 13471
AACS, Aberdeen 11303
AACS, Arcadia 13385
AACS, Columbus 13473
AACS, Fairfax 13518
AACS, Falls Church 14375
AACSB, Saint Louis 13462
AACSL, New York 13499
AACT, Muncie 13525
AACTE, Washington 13519
AACVPR, Middleton 13460
AAD, Schaumburg 13390
AADE, Chicago 13529
AADE, Chigago 13527
AADE, New York 13387
AADGP, Phoenix 13388
AADPA, Palatine 13389
AADR, Washington 13477
AADS, Washington 13528
AADTS, Camberley 11298
AAE, Barcelona 10245
AAE, Chicago 13531
AAEA, Ames 13446
AAEA, Tunis 11127
AAED, Chicago 13392
AAED, Tupelo 13533
AAEE, Annapolis 13391
AÄGP, Düsseldorf 05071
AAEH, Cuyahoga Falls 14537
AAEM, Rochester 13530
AAEP, Lexington 13534
A.A.E.S., Armidale 00255
AAES, Mogerhanger 11299
AAES, Washington 13532
AAF, Washington 13457
AAFA, Washington 14538
AAFH, Berkeley 13281
AAFM, Sacramento 13535
AAFO, Winchester 13478
AAFP, Kansas City 13394
AAFPE, Overland 13488
AAFPRS, Washington 13393
AAFS, Colorado Springs 13396
AAFS, Tuscaloosa 13280
A.A.G., Parkville 00264
AAG, Washington 14379
AAGFO, Noblesville 13398
AAGL, Santa Fe Springs 13537
AAGO, Richmond 13397
AAGP, Greenbelt 13480
AAGS, Bethesda 13479
AAGUS, Houston 13536
AAH, Calabasas 13364
A.A.H., Canberra 00252
AAH, Grimsby 11259
AAH, Los Angeles 14392
AAHA, Denver 13451
AAHA, Hinsdale 13538
AAHC, Washington 14372
AAHD, Chicago 13440, 13540
AAHE, College Station 13542
AAHE, Reston 14342
AAHE, Washington 13482
AAHGS, Washington 13347
AAHM, Boston 13500
AAHP, Brooklyn 13541
AAHP, Norwood 13539
AAHPERD, Reston 13449
AAHPSSS, Sydney 00237
AAHS, Chicago 13481
AAHS, Santa Ana 13613
AAHSLD, Houston 14373
AAI, Bethesda 13543
AAI, Bruxelles 01105
AAI, Dublin 06924
AAI, Hereford 11302
AAI, New York 13342, 13362
AAIAL, New York 13375
AAID, Chicago 13399
AAIE, New Wilmington 14343
AAIE, Sacramento 14393
AAIM, Springfield 13401
AAIP, Atlanta 13400
AAIP, Oklahoma City 14380
A.A.I.P.D., Napoli 07130
AAIS, Nairobi 08463
AAIV, Columbia 13544
AAJ, Berkeley 14381
AAJ, Buenos Aires 00052
AAJR, New York 13377
AALA, Fayetteville 13447
AALA, Gaithersburg 13483
AALAE, Nairobi 08461
AALAS, Cordova 13484
AALCC, New Dehli 06719
AALIM, New York 13402
AALL, Chicago 13547
AALMA, New York 13545
AALS, Lubbock 14267
AALS, Washington 14382
AAM, Cleveland 13343
AAM, La Jolla 13405
AAM, Washington 13407, 13554
AAMA, Chicago 13548
AAMA, Southfield 13406
AAMC, Philadelphia 13551
AAMC, Washington 14383
AAMFT, Washington 13485
AAMHA, Sousse 11134
AAMHPC, Sacramento 13552
AAM/ICOM, Washington 15457
AAML, Chicago 13403
AAMMC, Los Angeles 13549
AAMO, Hong Kong 06588
AAMP, Carmel 13404
AAMR, Omaha 13443
AAMR, Washington 13608
AAMS, Pasadena 14377
AAMSE, Chicago 13550
AAMT, Modesto 13486
AAMT, Valley Forge 13487
AAMTI, Alexandria 03880
A.A.M.U.S., Milano 07127
AAN, Kansas City 13409
A.A.N., Melbourne 00265
A.A.N.H.S., Ayr 11419
AANP, Salt Lake City 13556
AANS, Park Ridge 13555
AANS, Philadelphia 13408
AAO, Bethesda 13411
AAO, Newark 13418
AAO, Saint Louis 13558
AAO, San Francisco 13410
AAOA, Chicago 14560
AAOA, Silver Spring 13419
AAOGP, Milwaukee 13415
AAOH, Bangkok 11078
AAOM, Boca Raton 13559
AAOM, Houston 13413
AAOMR, Aurora 13412
AAOMS, Rosemont 13557
AAOP, Alexandria 13417
AAOP, Lafayette 13414
AAOrthMed, Lewiston 13560
AAOS, Rosemont 13416
A.A.O.T., Alphington 00266
AAOU, Seoul 08518
AAP, Bethesda 13561
AAP, Chicago 13422
AAP, Colorado Springs 14265, 14397
AAP, Decatur 13433
AAP, Elk Grove Village 13421
AAP, Indianapolis 14374, 14384
AAP, Lisboa 09653
AAP, Melbourne 00238
AAP, New York 13282, 13432, 14264, 14345
AAP, San Rafael 14270
AAPA, Alexandria 13425
AAPA, Atlanta 13574
AAPA, Buffalo 13565
AAPA, Cairo 03863
AAPAA, Greenbelt 13430
AAPAM, Addis Ababa 03934
AAPB, Wheat Ridge 14266
AAPCC, Tucson 13569
AAPD, Chicago 13420
AAPD, Naperville 13426
AAPE, Tallahassee 13423
AAPG, Tulsa 13562
AAPH, Boones Mill 13571
AAPH, Manila 09490
AAPHD, Richmond 13576
AAPHP, Madison 13577
AAPICU, Malibu 13570
AAPL, Baltimore 13431
AAPLOG, Elm Grove 13573
AAPM, New York 13566
AAPMR, Chicago 13424
AAPO, West Hartford 13428
AAPO&S, San Francisco 13489
AAPRD, Davison 13378
AAPS, Gainesville 13564
AAPS, Lagos 09212
AAPS, La Jolla 13568
AAPS, Tucson 14385
AAPSC, Rochester 13575
AAPSM, Potomac 13427
AAPSS, Philadelphia 13429
AAPT, College Park 13567
AAPT, Maplewood 14395
AAPT, Oklahoma City 13563
AAPY, Flushing 13572
AAR, Syracuse 13434
AARD, Colorado Springs 13435
AARINENA, Roma 07252
AAROI, Napoli 07255
Aaron Burr Association, Annandale 13274
AARS, Daleville 13578
AARS, New York 14224
AARS, Tokyo 08097
AART, North Little Rock 13490
AARTS, New York 14376
AARU, Amman 08451
AAS, Ann Arbor 14268
AAS, Canberra 00250
AAS, Concord 13284
AAS, Dallas 13611
A.A.S., Darlinghurst 00253
AAS, Denver 13584
AAS, Highland Park 14216
A.A.S., Menai Bridge 11230
AAS, Miami 13437
AAS, Minneapolis 14262
AAS, Nairobi 08458
AAS, Red Bank 13455
AAS, Springfield 13609
AAS, Washington 13610
AAS, Worcester 13454
A.A.S.A., Chieti 07136
AASA, Miami 13285
AASCO, Salem 14387
AASCU, Washington 13583
AASE, Addis Ababa 03935
AASE, Warrensburg 13436
AASECT, Chicago 13581
AASG, Champaign 14388
AASH, West Deptford 13501
AASL, Chicago 13579
AASLD, Thorofare 13502
AASLH, Nashville 13492
AASND, South Euclid 13493
AASP, Northridge 13438
AASP, Washington 13491
AASPA, Sacramento 13580
AASR, Saint Lucia 00261
AASRP, Ithaca 13582
AASSA, Miami 14386
AASSREC, New Delhi 06732
AATA, Mundelein 13459
AATA, Provo 13585
AATB, McLean 13593
AATCC, Research Triangle Park 13592
AATE, Santa Barbara 13586
AATE, Tempe 13450
AATF, Champaign 13587
AATG, Cherry Hill 13588
AATI, Columbus 13589
AATO, Accra 06475
AATS, Manchester 13503
AATS, New York 13439
AATSP, Mississippi State 13590
AATT, Gastonia 13495
AATT, Princeton 13591
AAU, Accra 06477
A.A.U., Cairo 03878
AAU, Washington 14389
AAUG, Belmont 14394
AAUP, Washington 13594
AAUW, Washington 13595
AAUWEF, Washington 13596
AAV, Boca Raton 14396
AAVA, Auburn 13598
AAVD, Phoenix 13441
AAVIM, Athens 13504
AAVLD, Columbia 13599
AAVMC, Washington 14390
AAVN, Athens 13442
AAVP, Kalamazoo 13600
AAVSB, Jefferson City 13601
AAVSO, Cambridge 13597
AAWCJC, Middletown 13604
AAWD, Chicago 13603
AAWH, Washington 13505
AAWORD, Dakar 10021
AAWV, Fort Collins 13602
AAZK, Topeka 13605
AAZPA, Wheeling 13606
AAZV, Philadelphia 13607
AB, Stuttgart 05066
ABA, Annandale 13274
ABA, Baltimore 13673
ABA, Chicago 13617
ABA, Colorado Springs 13622
ABA, Hartford 13628
ABA, Heidelberg 05102
ABA, Kalamazoo 14271
ABA, Minneapolis 13620
ABA, Saint Louis 13667
ABA, Washington 14401
ABAC, San Diego 14398
ABAH, Tegucigalpa 06582
ABAI, Philadelphia 13627
ABAPSTAS, London 11306
ABAS, Melrose 13626
ABB, Saint Louis 13629
ABC, Alexandria 13625
ABC, Chicago 13624
ABC, London 11307
ABCA, Carmel 14399
ABCA, Saint Louis 13623
A.B.C.A., Vancouver 01915
A.B.C.C., London 11308
ABCD, Hershey 14400
ABCD, Pittsburgh 14274
ABCRS, Taylor 13630
ABD, Bruxelles 01106
ABD, Detroit 13632
ABDPH, Gainesville 13631
A.B.D.R.-B.V.A.R., Bruxelles 01107
ABE, Bruxelles 01108
ABE, Chicago 13633
ABEA, Chicago 13670
ABEC, Washington 14223
ABEF, Paris 04171
ABEGS, Riyadh 09997
Aberdeen-Angus Cattle Society, Perth 11194
Abergavenny and Border Counties Agricultural Society, Abergavenny 11195
ABES, Monsey 13621
ABES, Rio de Janeiro 01588
ABESAO, Dakar 10017
ABESC, Brasília 01589
ABET, New York 13315
A.B.F., Padova 07268
ABF, Paris 04169
ABF, Washington 13614
ABFSE, Cumberland Center 13634
ABG, Dagenham 11312
ABGRA, Buenos Aires 00077
ABH, Amherst 14348
ABH, London 11355
ABHES, Elkhart 13316
ABHP, McLean 13635
ABHS, Rochester 13616
ABIISE, Lima 09442
ABIM, Philadelphia 13636
ABJS, Rosemont 14405
ABLE, Ithaca 14272
ABLS, Omaha 13671
ABM, Boston 11261
ABM, Thaining 05139
ABMAC, New York 13672
ABME, Nashville 13637
ABMS, São Paulo 01592
ABMT, Salt Lake City 13638
ABN, Bethesda 13641
ABN, Bonn 05081
ABN, Dakar 10008
ABNM, Los Angeles 13640
ABNS, Houston 13639
ABO, Bala Cynwyd 13643
ABO, Fairfax 13644
ABO, Houston 13648
ABO, Saint Louis 13646
ABOG, Seattle 13642
ABOP, Amsterdam 08820
ABOPS, Windsor 14406
Aboriginal Research Club, Dearborn 13275
Abortion Law Reform Association, London 11196
ABOS, Chapel Hill 13647
ABP, Atlanta 13659
ABP, Baltimore 13652
ABP, Chapel Hill 13651
ABP, New York 14273
ABP, Tampa 13649
ABP, Washington 14402
ABPC, Birmingham 11315
ABPC, Bruxelles 01109
ABPD, Carmel 13650
ABPH, Chicago 13661
ABPM, Dayton 13657
ABPMR, Rochester 13653
ABPN, Deerfield 13660
ABPN, London 11314
ABPO, Redlands 13655
ABPP, Columbia 13658
ABPS, Philadelphia 13654
ABPS, San Francisco 13656
ABR, Troy 13662
Abrasive Engineering Society, Butler 13276
A.B.R.E., Brescia 07262
ABRO, Edinburgh 11234
ABS, Boston 14536
ABS, Boyce 13668
ABS, Denver 14218
ABS, Philadelphia 13663
ABS, San Jose 13618
ABS, Sebastopol 13615
ABS, Washington 14403
ABSAME, McLean 14347
ABSW, London 11316
ABT, Southampton 11305
ABTA, Chicago 13669
ABTAPL, Cambridge 11318
ABTS, Evanston 13664
ABTSA, Portland 14360
ABU, Bingham Farms 13665
ABU, Montevideo 13242
ABUEN, León 09203
ABVE, Topeka 13666
ABWA, Aurora 14257
Abwassertechnische Vereinigung e.V., Sankt Augustin 05029
ABWHE, Albany 14404
ABYDAP, Lima 09448
AC, København 03558
AC, Santa Fe 14236
ACA, Alexandria 13698, 13771
ACA, Annadale 14281
ACA, Arlington 13690
ACA, Bethesda 13678
ACA, Buffalo 13774
A.C.A., Clayton 00243
ACA, Laurel 13756
ACA, London 11332
ACA, New York 13754, 13760
ACA, Ottawa 01917
ACACN, Florissant 13761
ACAD, Washington 13750
Academia Alagoana de Letras, Maceió 01560
Academia Amazonense de Letras, Manaus 01561
Academia Antioqueña de Historia, Medellín 03334
Academia Argentina de Cirugía, Buenos Aires 00037
Academia Argentina de Letras, Buenos Aires 00038
Academia Belgica, Roma 07121
Academia Boliviana, La Paz 01528
Academia Boyacense de Historia, Tunja 03335
Academia Brasileira de Ciência da Administração, Rio de Janeiro 01562
Academia Brasileira de Ciências, Rio de Janeiro 01563
Academia Brasileira de Letras, Rio de Janeiro 01564
Academia Cachoeirense de Letras, Cachoeira de Itapemerim 01565
Academia Campinense de Letras, Campinas 01566
Academia Cardinalis Bessarionis, Roma 07122
Academia Catarinense de Letras, Florianópolis 01567
Academia Cearense de Letras, Fortaleza 01568
Academia Chilena de Bellas Artes, Santiago 03007
Academia Chilena de Ciencias, Santiago 03008
Academia Chilena de Ciencias Sociales, Politicas y Morales, Santiago 03009
Academia Chilena de la Historia, Santiago 03010
Academia Chilena de la Lengua, Santiago 03011
Academia Chilena de Medicina, Santiago 03012
Academia Colombiana de Ciencias Exactas, Físicas y Naturales, Bogotá 03336
Academia Colombiana de Historia, Bogotá 03337
Academia Colombiana de Jurisprudencia, Bogotá 03338
Academia Colombiana de la Lengua, Bogotá 03339
Academia Costarricense de la Lengua, San José 03427
Academia Costarricense de Periodoncia, San José 03428
Academia Cubana de la Lengua, La Habana 03468
Academia das Ciências de Lisboa, Lisboa 09633
Academia de Arte Dramática, Hermosillo 08677
Academia de Artes Plásticas, Guanajuato 08678
Academia de Artes Plásticas, Hermosillo 08679
Academia de Buenas Letras de Barcelona, Barcelona 10238
Academia de Ciencias de Cuba, La Habana 03469
Academia de Ciencias Exactas, Físicas, Químicas y Naturales de Zaragoza, Zaragoza 10239
Academia de Ciencias Físicas, Matemáticas y Naturales, Caracas 16958
Academia de Ciencias Históricas de Monterrey, Monterrey 08680
Academia de Ciencias Médicas, Córdoba 00039
Academia de Ciencias Médicas de Bilbao, Bilbao 10240
Academia de Ciencias Médicas, Físicas y Naturales de Guatemala, Guatemala City 06553
Academia de Ciencias Politícas y Sociales, Caracas 16959
Acadèmia de Ciències Mèdiques de Catalunya i de Baleares, Barcelona 10241
Academia de Dramática, Guanajuato 08681
Academia de Estomatología del Perú, Lima 09437

Academia de Geografía e Historia de Costa Rica, San José 03429
Academia de Geografía e Historia de Guatemala, Guatemala City 06554
Academia de Historia de Cartagena de Indias, Cartagena 03340
Academia de Historia del Norte de Santander, Cúcuta 03341
Academia de Historia del Zulia, Maracaibo 16960
Academia de la Investigación Científica, México 08682
Academia de la Lengua Maya Quiché, Quezaltenango 06555
Academia de la Lengua y Cultura Guaraní, Asunción 09424
Academia de Letras, João Pessoa 01569
Academia de Letras da Bahia, Salvador 01570
Academia de Letras de Piauí, Teresina 01571
Academia de Letras e Artes do Planalto, Luziânia 01572
Academia de Medicina de São Paulo, São Paulo 01573
Academia de Música, Guanajuato 08683
Academia de Música, Hermosillo 08684
Academia de Stiinte Agricole si Silvice, Bucuresti 09734
Academia de Stiinte Medicale, Bucuresti 09735
Academia de Stiinte Sociale si Politice, Bucuresti 09736
Academia Dominicana de la Historia, Santo Domingo 03829
Academia Dominicana de la Lengua, Santo Domingo 03830
Academia Ecuatoriana de la Lengua, Quito 03839
Academia Ecuatoriana de Medicina, Quito 03840
Academia Española de Dermatología y Sifilografía, Madrid 10242
Academia Europaea, London 11197
Academia Filipina, Manila 09488
Academia Gentium Pro Pace, Roma 07123
Academia Guatemalteca de la Lengua, Guatemala City 06556
Academia Historica, Taipei 03226, 03227
Academia Hondureña, Tegucigalpa 06580
Academia Hondureña de Geografía e Historia, Tegucigalpa 06581
Academia Iberoamericana y Filipina de Historia Postal, Madrid 10243
Academia Matogrossense de Letras, Cuiabá 01574
Academia Médico-Quirúrgica Española, Madrid 10244
Academia Mexicana de Jurisprudencia y Legislación, México 08685
Academia Mexicana de la Historia, México 08686
Academia Mexicana de la Lengua, México 08687
Academia Miniera de Letras, Belo Horizonte 01575
Academia Nacional de Agronomía y Veterinaria, Buenos Aires 00040
Academia Nacional de Belas-Artes, Lisboa 09634
Academia Nacional de Bellas Artes, Buenos Aires 00041
Academia Nacional de Ciencias, México 08688
Academia Nacional de Ciencias de Bolivia, La Paz 01529
Academia Nacional de Ciencias de Buenos Aires, Buenos Aires 00042
Academia Nacional de Ciencias de Córdoba, Córdoba 00043
Academia Nacional de Ciencias de Panamá, Panamá City 09405
Academia Nacional de Ciencias Económicas, Buenos Aires 00044
Academia Nacional de Ciencias Exactas, Físicas y Naturales, Buenos Aires 00045
Academia Nacional de Ciencias Exactas, Físicas y Naturales de Lima, Lima 09438
Academia Nacional de Ciencias Morales y Políticas, Buenos Aires 00046
Academia Nacional de Derecho y Ciencias Sociales, Buenos Aires 00047
Academia Nacional de Derecho y Ciencias Sociales (Córdoba), Córdoba 00048
Academia Nacional de Farmacia, Rio de Janeiro 01576
Academia Nacional de Filosofía, Managua 09201
Academia Nacional de Geografía, Buenos Aires 00049
Academia Nacional de Historia y Geografía, México 08689
Academia Nacional de Ingeniería, Montevideo 13239
Academia Nacional de la Historia, Buenos Aires 00050
Academia Nacional de la Historia, Caracas 16961
Academia Nacional de la Historia, La Paz 01530
Academia Nacional de Letras, Montevideo 13240
Academia Nacional de Medicina, Bogotá 03342
Academia Nacional de Medicina, Buenos Aires 00051
Academia Nacional de Medicina, Caracas 16962
Academia Nacional de Medicina, Lima 09439
Academia Nacional de Medicina, México 08690
Academia Nacional de Medicina, Rio de Janeiro 01577
Academia Nacional de Medicina del Uruguay, Montevideo 13241
Academia Nicaragüense de la Lengua, Managua 09202
Academia Ophthalmologica Internationalis, Albi 04083
Academia Panameña de la Historia, Panamá City 09406
Academia Panameña de la Lengua, Panamá City 09407
Academia Paraibana de Letras, João Pessoa 01578
Academia Paulista de Letras, São Paulo 01579
Academia Pernambucana de Letras, Recife 01580
Academia Peruana de Cirurgía, Lima 09440
Academia Peruana de la Lengua, Lima 09441
Academia Petrarca di Lettere, Arti e Scienze, Arezzo 07124
Academia Portuguesa da História, Lisboa 09635
Academia Puertorriqueña de la Historia, Santurce 09718
Academia Puertorriqueña de la Lengua Española, San Juan 09719
Academia Riograndense de Letras, Pôrto Alegre 01581
Academia Romana, Bucuresti 09737
Academia Salvadoreña, San Salvador 03916
Academia Salvadoreña de la Historia, San Salvador 03917
Academia Sinica, Taipei 03228
Academia Venezolana de la Lengua, Caracas 16963
Academic Circle of Tel Aviv, Tel Aviv 07039
L'Académie Arabe de Damas, Damascus 11036
Académie Canadienne de Medécine Sportive, Vanier City 02008
Académie Canadienne d'Histoire de la Pharmacie, Toronto 02009
Académie Commerciale Internationale, Paris 04084
Académie d'Agriculture de France, Paris 04085
Académie d'Architecture, Paris 04086
Académie d'Arles, Arles 04087
Académie de Chirurgie, Paris 04088
Académie de la Réunion, Saint-Denis 04089
Académie de Marine, Paris 04090
Académie de Nîmes, Nîmes 04091
Académie de Pharmacie de Paris, Paris 04092
Académie des Beaux-Arts, Paris 04093
Académie des Belles-Lettres, Sciences et Arts de La Rochelle, La Rochelle 04094
Académie des Inscriptions et Belles-Lettres, Paris 04095
Académie des Jeux Floraux, Toulouse 04096
Académie des Lettres du Québec, Montréal 01814
Académie des Lettres, Sciences et Arts d'Amiens, Amiens 04097
Académie d'Espéranto, Paris 04471
Académie des Sciences, Paris 04098
Académie des Sciences, Agriculture, Arts et Belles-Lettres d'Aix, Aix-en-Provence 04099
Académie des Sciences, Arts et Belles-Lettres de Dijon, Dijon 04100
Académie des Sciences, Belles-Lettres et Arts de Clermont, Clermont-Ferrand 04101
Académie des Sciences, Belles-Lettres et Arts de Lyon, Lyon 04102
Académie des Sciences, Belles-Lettres et Arts de Rouen, Rouen 04103
Académie des Sciences, Belles-Lettres et Arts de Savoie, Chambéry 04104
Académie des Sciences d'Outre-Mer, Paris 04105
Académie des Sciences et Lettres de Montpellier, Montpellier 04106
Académie des Sciences, Lettres et Arts d'Arras, Arras 04107
Académie des Sciences, Lettres et Arts de Marseille, Marseille 04108
Académie des Sciences Morales et Politiques, Paris 04109
Académie Diplomatique Internationale, Paris 04110
Académie du Cinéma Canadien, Toronto 01815
Académie Européenne d'Allergologie et d'Immunologie Clinique, Bruxelles 01090
Académie Européenne d'Anesthésiologie, Strasbourg 04477
Académie Européenne des Ecrivains Publics, Bruxelles 01091
Académie Européenne du Cinéma et de la Télévision, Bruxelles 01267
Académie Française, Paris 04111
Académie Goncourt, Paris 04112
Académie Internationale de Céramique, Genève 10606
Académie Internationale de Cytologie, Freiburg 05984
Académie Internationale de Science Politique et d'Histoire Constitutionnelle, Paris 04113
Académie Internationale d'Héraldique, Genève 10607
Académie Malgache, Antananarivo 08620
Académie Mallarmé, Paris 04114
Académie Montaigne, Sillé-le-Guillaume 04115
Académie Nationale de Chirurgie Dentaire, Paris 04116
Académie Nationale de Danse, Paris 04117
Académie Nationale de Médecine, Paris 04118
Académie Nationale de Metz, Metz 04119
Académie Nationale des Sciences, Belles-Lettres et Arts de Bordeaux, Bordeaux 04120
Académie Polonaise des Sciences, Paris 04121
Académie Régionale des Sciences et Techniques de la Mer d'Abidjan, Abidjan 08056
Académie Royale d'Archéologie de Belgique, Bruxelles 01092
Académie Royale de Langue et de Littérature Françaises, Bruxelles 01093
Académie Royale de Médecine de Belgique, Bruxelles 01094
Académie Royale des Arts du Canada, Toronto 02864
Académie Royale des Sciences, des Lettres et des Beaux-Arts de Belgique, Bruxelles 01095
Académie Royale des Sciences d'Outre-Mer, Bruxelles 01096
Académie Vétérinaire de France, Paris 04122
Academi Gymreig, Cardiff 11198
Academy for Educational Development, Washington 13277
Academy for Implants and Transplants, Springfield 13278
Academy for Interscience Methodology, Hinsdale 13279
Academy of Agricultural and Forest Sciences, Bucuresti 09734
Academy of Agricultural Science, Pyongyang 08508
Academy of Agricultural Sciences, Sofia 01741
Academy of Ambulatory Foot Surgery, Tuscaloosa 13280
Academy of American Franciscan History, Berkeley 13281
Academy of American Poets, New York 13282
Academy of Aphasia, Los Angeles 13283
Academy of Applied Science, Concord 13284
Academy of Architectural Engineering, Beijing 03105
Academy of Arts, Berlin 05048
Academy of Arts and Sciences of the Americas, Miami 13285
Academy of Building Materials, Beijing 03106
Academy of Canadian Cinema, Toronto 01815
Academy of Cement Research, Beijing 03107
Academy of Chemical Engineering, Beijing 03108
Academy of Criminal Justice Sciences, Highland Heights 13286
Academy of Dental Materials, Charleston 13287
Academy of Dentistry for the Handicapped, Chicago 13288
Academy of Dentistry International, Washington 13289
Academy of Denture Prosthetics, Bellevue 13290
Academy of Dispensing Audiologists, Columbia 13291
Academy of Fisheries, Sinpo City 08509
Academy of Forestry Science, Pyongyang 08510
Academy of General Dentistry, Chicago 13292
Academy of Highway Sciences, Beijing 03109
Academy of Hydrotechnology, Beijing 03110
Academy of International Business, Detroit 13293
Academy of International Dental Studies, Boston 13294
Academy of Legal Studies in Business, Oxford 13295
Academy of Letters, Islamabad 09356
Academy of Light Industry Science, Pyongyang 08511
Academy of Management, Ada 13296
Academy of Marketing Science, Coral Gables 13297
Academy of Medical Sciences, Bucuresti 09735
Academy of Medical Sciences, Pyongyang 08512
Academy of Medicine, Toronto 01816
Academy of Medicine, Singapore, Singapore 10033
Academy of Motion Picture Arts and Sciences, Beverly Hills 13298
Academy of Natural Sciences, Philadelphia 13299
Academy of Non-Ferrous Metallurgical Design, Beijing 03111
Academy of Operative Dentistry, Menomonie 13300
Academy of Oral Dynamics, Frederick 13301
Academy of Petroleum Research, Beijing 03112
Academy of Pharmacy Practice and Management, Washington 13302
Academy of Political Science, New York 13303
Academy of Psychosomatic Medicine, Chicago 13304
Academy of Railway Sciences, Pyongyang 08513
Academy of Rehabilitative Audiology, Akron 13305
Academy of Religion and Psychical Research, Bloomfield 13306
Academy of Science Fiction, Fantasy and Horror Films, Los Angeles 13307
Academy of Sciences, Pyongyang 08514
Academy of Sciences, Tallinn 03933
Academy of Sciences, Ulan Bator 08798
Academy of Sciences and Arts of Kosovo, Priština 17076
Academy of Sciences and Technology in Berlin, Berlin 05051
Academy of Scientific Hypnotherapy, San Diego 13308
Academy of Scientific Research and Technology, Cairo 03859
Academy of Security Educators and Trainees, Berryville 13309
Academy of Social and Political Sciences, Bucuresti 09736
Academy of Social Sciences, Pyongyang 08515
Academy of Television Arts and Sciences, North Hollywood 13310
Academy of the Arabic Language, Cairo 03860
Academy of the Hebrew Language, Jerusalem 07040
Academy of the Social Sciences in Australia, Canberra 00211
Academy of Veterinary Cardiology, Floral Park 13311
The Academy of Zoology, Agra 06708
ACADI, Khartoum 10424
ACAFAM, Tegucigalpa 06583
ACAHN, Guatemala City 06558
ACAI, Palatine 13697
ACARTSOD, Tripoli 08566
ACAS, Bruxelles 01113
ACAS, Panamá City 09408
ACASPP, Shepherdstown 13746
ACAT, New York 13684
ACB, London 11324
ACB, Paris 04174
ACC, Bethesda 13699
ACC, Minneapolis 14288
ACC, San Lorenzo 14411
ACC, Schaumburg 13703
ACCA, Charleston 13693
Accademia Agraria, Pesaro 07125
Accademia Albertina di Belle Arti e Liceo Artistico, Torino 07126
Accademia Ambrosiana Medici Umanisti e Scrittori, Milano 07127
Accademia Americana, Roma 07128
Accademia Anatomico-Chirurgica, Perugia 07129
Accademia Artistica Internazionale Pinocchio d'Oro, Napoli 07130
Accademia Biella Cultura, Biella 07131
Accademia Corale Stefano Tempia, Torino 07132
Accademia Cosentina, Cosenza 07133
Accademia Culturale d'Europa, Bassano Romano 07134
Accademia Culturale di Rapallo, Rapallo 07135
Accademia degli Abruzzi per le Scienze e le Arti, Chieti 07136
Accademia degli Euteleti, San Miniato 07137
Accademia degli Incamminati, Modigliana 07138
Accademia degli Incolti, Roma 07139
Accademia degli Ottimi, Roma 07140
Accademia degli Sbalzati, Sansepolcro 07141
Accademia dei Filedoni, Perugia 07142
Accademia dei Filodrammatici, Milano 07143
Accademia dei Filopatridi (Rubiconia), Savignano sul Rubicone 07144
Accademia dei Gelati, Scanno 07145
Accademia dei Sepolti, Volterra 07146
Accademia della Crusca, Firenze 07147
Accademia delle Scienze dell'Istituto di Bologna, Bologna 07148
Accademia delle Scienze di Ferrara, Ferrara 07149
Accademia delle Scienze di Torino, Torino 07150
Accademia delle Scienze e delle Arti degli Ardenti di Viterbo, Viterbo 07151
Accademia delle Scienze Mediche di Palermo, Palermo 07152
Accademia di Agricoltura di Torino, Torino 07153
Accademia di Agricoltura Scienze e Lettere, Verona 07154
Accademia di Belle Arti, Firenze 07155
Accademia di Belle Arti, Milano 07156
Accademia di Belle Arti, Perugia 07157
Accademia di Belle Arti, Ravenna 07158
Accademia di Belle Arti e Liceo Artistico, Bologna 07159
Accademia di Belle Arti e Liceo Artistico, Carrara 07160
Accademia di Belle Arti e Liceo Artistico, Lecce 07161
Accademia di Belle Arti e Liceo Artistico, Napoli 07162
Accademia di Belle Arti e Liceo Artistico, Palermo 07163
Accademia di Belle Arti e Liceo Artistico, Roma 07164
Accademia di Belle Arti e Liceo Artistico, Venezia 07165
Accademia di Costume e di Moda, Libero Istituto di Studi Superiori di Belle Arti, Roma 07166
Accademia di Danimarca, Roma 07167
Accademia di Francia, Roma 07168
Accademia di Medicina di Torino, Torino 07169
Accademia di Paestum Eremo Italico, Mercato San Severino 07170
Accademia di Relazioni Pubbliche, Roma 07171
Accademia di Romania, Roma 07172
Accademia di Scienze, Lettere e Arti, Lucca 07173
Accademia di Scienze, Lettere e Arti, Udine 07174
Accademia di Scienze, Lettere e Belle Arti degli Zelanti e dei Dafnici, Acireale 07175
Accademia di Scienze, Lettere ed Arti, Palermo 07176
Accademia Economico Agraria dei Georgofili, Firenze 07177
Accademia Etrusca, Cortona-Arezzo 07178
Accademia Euro-Afro-Asiatica del Turismo, Catania 07179
Accademia Europea Dentisti Implantologi, Milano 07180
Accademia Filarmonica di Bologna (Reale), Bologna 07181
Accademia Filarmonica di Verona, Verona 07182
Accademia Filarmonica Romana, Roma 07183
Accademia Fulginia di Arti, Lettere, Scienze, Foligno 07184
Accademia Georgica, Treia 07185
Accademia Gioenia di Scienze Naturali, Catania 07186
Accademia Gli Amici Dei Sacri Lari, Bergamo 07187
Accademia Il Tetradramma, Roma 07188
Accademia Internazionale d'Arte Moderna, Roma 07189
Accademia Internazionale della Tavola Rotonda, Milano 07190
Accademia Internazionale di Medicina Legale e di Medicina Sociale, Roma 07191
Accademia Internazionale per le Scienze Economiche, Sociali e Sanitarie, Roma 07192
Accademia Italiana di Medicina Omeopatica Hahnemanniana, Roma 07193
Accademia Italiana di Scienze Forestali, Firenze 07194
Accademia Italiana di Stenografia e di Dattilografia Giuseppe Aliprandi, Firenze 07195
Accademia Italiana di Storia della Farmacia, Pisa 07196
Accademia Italiana di Studi Filatelici e Numismatici, Reggio Emilia 07197
Accademia Lancisiana di Roma, Roma 07198
Accademia Letteraria Italiana, Arcadia, Roma 07199
Accademia Ligure di Scienze e Lettere, Genova 07200
Accademia Ligustica de Belle Arti, Genova 07201
Accademia Lunigianese di Scienze Giovanni Capellini, La Spezia 07202
Accademia Medica, Genova 07203
Accademia Medica di Roma, Roma 07204
Accademia Medica Lombarda, Milano 07205
Accademia Medica Pistoiese Filippo Pacini, Pistoia 07206

Accademia Medico-Chirurgica del Piceno, Ancona *07207*
Accademia Medico-Fisica Fiorentina, Firenze *07208*
Accademia Musicale Chigiana, Siena *07209*
Accademia Musicale Ottorino Respighi, Roma *07210*
Accademia Nazionale dei Lincei, Roma *07211*
Accademia Nazionale dei Sartori, Roma *07212*
Accademia Nazionale delle Scienze, detta dei XL, Roma *07213*
Accademia Nazionale di Agricoltura, Bologna *07214*
Accademia Nazionale di Arte Drammatica Silvio d'Amico, Roma *07215*
Accademia Nazionale di Belle Arti di Parma, Parma *07216*
Accademia Nazionale di Danza, Roma *07217*
Accademia Nazionale di Entomologia, Firenze *07218*
Accademia Nazionale di Marina Mercantile, Genova *07219*
Accademia Nazionale di San Luca, Roma *07220*
Accademia Nazionale di Santa Cecilia, Roma *07221*
Accademia Nazionale di Scienze, Lettere e Arti, Modena *07222*
Accademia Nazionale Italiana di Entomologia, Firenze *07223*
Accademia Nazionale Virgiliana di Scienze, Lettere ed Arti di Mantova, Mantova *07224*
Accademia Olimpica, Vicenza *07225*
Accademia Polacca delle Scienze, Roma *07226*
Accademia Pomposiana, Codigoro *07227*
Accademia Pontaniana, Napoli *07228*
Accademia Pratese di Medicina e Scienze, Prato *07229*
Accademia Prenestina del Cimento di Musica, Lettere, Scienze, Arti Visive e Figurative, Palestrina *07230*
Accademia Raffaello, Urbino *07231*
Accademia Romana di Cultura, Roma *07232*
Accademia Romana di Scienze Mediche e Biologiche, Roma *07233*
Accademia Romana di S. Tommaso d'Aquino e di Religione Cattolica, Roma *16949*
Accademia Roveretana degli Agiati, Rovereto *07234*
Accademia Salentina di Lettere ed Arti, Lecce *07235*
Accademia Scientifica, Letteraria, Artistica del Frignano Lo Scoltenna, Pievepelago *07236*
Accademia Senese degli Intronati, Siena *07237*
Accademia Simba, Roma *07238*
Accademia Spagnola di Belle Arti, Roma *07239*
Accademia Spoletina, Spoleto *07240*
Accademia Tedesca Villa Massimo, Roma *07241*
Accademia Tiberina, Roma *07242*
Accademia Toscana di Scienze e Lettere La Colombaria, Firenze *07243*
Accademia Universale Citta' Eterna, Roma *07244*
Accademia Universale Guglielmo Marconi, Roma *07245*
Accademia Valdarnese del Poggio, Montevarchi *07246*
ACCC, Rockville *14418*
ACC&CE, New York *14422*
ACCD, Cairo *03870*
ACCE, Monroe *13759*
ACCE, Nairobi *08466*
ACCE, Union *13695*
ACCE/ICDR, Nairobi *08457*
ACCE Institute for Communication Development and Research, Nairobi *08457*
ACCET, Richmond *13318*
ACCH, Bethesda *14349*
ACCI, Columbia *13765*
Accident Prevention Association of Manitoba, Winnipeg *01817*
ACCIS, Genève *10608*
ACCO, El Centro *13701*
Accordion Teacher's Guild, Kansas City *13312*
Accountants Computer Users Technical Exchange, Indianapolis *13313*
Accountants Study Group of the EEC, London *11199*
Accounting Association of Australia and New Zealand, Clayton *00212*
ACCP, Northbrook *13700*
ACCP, Wayne *13702*
ACCRA, Alexandria *13687*
Accra Regional Maritime Academy, Accra *06469*
Accreditation Association for Ambulatory Health Care, Skokie *13314*
Accreditation Board for Engineering and Technology, New York *13315*
Accrediting Bureau of Health Education Schools, Elkhart *13316*
Accrediting Commission on Education for Health Services Administration, Arlington *13317*
Accrediting Council for Continuing Education and Training, Richmond *13318*
Accrediting Council for Theological Education in Africa, Kaduna *09209*
Accrediting Council on Education in Journalism and Mass Communications, Lawrence *13319*
ACCT, Washington *14419*
ACCU, Washington *14408*
ACD, Chicago *13683*
ACD, Gaithersburg *13704*
ACDA, East Lansing *14409*
ACDM, Gainesville *14410*
ACDP, Braddon *00277*
ACDS, London *11231*
ACE, Bruxelles *01101*
ACE, Gainesville *13350*
ACE, Kingston *08068*
ACE, London *11206*
ACE, Washington *13766*
ACEC, Washington *13755*
ACEC Research and Management Foundation, Washington *13320*
ACEC/RMF, Washington *13320*
ACEEE, Washington *13758*
ACEF, Madison *13331*
ACEHSA, Arlington *13317*
ACEI, Wheaton *14279*
ACEID, Bangkok *11079*
ACEJMC, Lawrence *13319*
ACELF, Sillery *01873*
ACEP, Dallas *13705*
ACER, Hawthorn *00279*
ACerS, Westerville *13685*
ACES, Alexandria *14287*
ACES, Cairo *03868*
ACES, Flushing *14282*
ACES, La Jolla *13679*
ACF, Morgantown *13689*
ACFD, Lagos *09213*
ACFO, Redlands *13706*
ACFS, Park Ridge *13707*
ACG, Arlington *13708*
ACG, Inverness *11229*
ACGC, Nedlands *00222*
ACGIH, Cincinnati *13751*
ACGPOMS, Arlington Heights *13709*
ACGS, Manchester *13674*
ACH, Provo *14284*
ACH, Tuscaloosa *13711*
ACHA, Baltimore *13696*
ACHA, Washington *13680*
ACHE, Chicago *13710*
ACHE, Indianapolis *14286*
AChemS, Tallahassee *14277*
ACHI, Glendale *14278*
ACHICE, Santiago *03018*
A.Ch.M., Santiago *03014*
ACHPER, Hindmarsh *00280*
ACHSA, Dayton *13757*
ACI, Detroit *13748*
ACIC, Washington *13744*
ACID, Toronto *01872*
ACIIW, Miami *13764*
ACIL, Washington *13762*
ACIP, Washington *13712*
ACIS, Fort Wayne *13749*
ACISN, Napoli *07263*
ACJA-LAE, Sacramento *13772*
ACJS, Highland Heights *13286*
Ackerman Institute for Family Therapy, New York *13321*
ACL, Cedarville *14412*
ACL, Morristown *14283*
ACL, Oxford *13691*
ACLA, Provo *13747*
ACLALS, Kingston *08067*
ACLAM, Hershey *13713*
ACLEA, Chicago *14423*
ACLS, New York *13763*
ACM, Austin *13716*
ACM, Chicago *14258*
ACM, New York *14285*
ACMGA, Englewood *13714*
ACML, Kuwait *08544*
ACML, Ottawa *01924*
ACMMSCO, Schaumburg *13715*
ACMS, Canberra *00273*
ACMS, Dakar *10009*
A.C.M.S., Paris *04295*
ACN, Farmington Hills *13717*
ACN, Wilmington *13721*
ACNM, Washington *13720*
ACNP, Nashville *13718*
ACNP, Washington *13719*
ACO, Ibadan *09215*
ACO, Princeton *13724*
ACODA, Winksele *01143*
A.C.O.G., Melbourne *00275*
ACOG, Washington *13722*
ACOI, Washington *13725*
ACOM, Bogotá *03346*
ACOMS, San Antonio *13723*
ACOOG, Pontiac *13726*
ACOP, Trenton *13727*
ACOPS, London *11207*
ACORD, New Delhi *06721*
ACOS, Alexandria *13728*
ACOS, Zaghouan *11128*
Acoustical Society of America, Woodbury *13322*
Acoustical Society of China, Beijing *03113*
Acoustical Society of Japan, Tokyo *08353*
Acoustical Society of Scandinavia, Lyngby *03780*
Acoustical Society of Sweden, Göteborg *10533*
A.C.P., Brighton *11325*
ACP, Cleveland *13731*
ACP, Great Falls *14280*
ACP, Greenbelt *13734*
A.C.P., Parkville *00276*
ACP, Philadelphia *13729*
ACP, San Antonio *13733*
ACPA, Oakland *13735*
ACPA, Pittsburgh *13692*
ACPA, Washington *13681*
ACPE, Chicago *13767*
ACPF, Tokyo *08094*
ACPM, Washington *13732*
ACPP, London *11262*
ACPR, Miami Beach *13730*
ACPT, Jakarta *06880*
ACR, Atlanta *13736*
ACRA, Tallhassee *13743*
ACRES, Bellingham *13768*
A.Cr.I., Genova *07265*
ACRI, McLean *13694*
ACRL, Chicago *14414*
ACRM, Chicago *13752*
ACS, Atlanta *13676*
ACS, Chicago *13738*
ACS, Cupertino *13773*
ACS, Farmington *14413*
ACS, Lexington *14407*
ACS, Montréal *01899*
ACS, Saint Marys *13677*
ACS, San Pedro *13686*
ACS, Washington *13688*
ACS, West Orange *13682*
ACS, York *13675*
ACSA, Washington *14416*
ACSAD, Damascus *11038*
ACSM, Bethesda *13753*
ACSM, Indianapolis *13737*
ACSP, Madison *14417*
ACSTD, New York *13335*
ACSUS, Washington *14276*
ACT, Augusta *14424*
ACT, Iowa City *13742*
ACT, Jakarta *06873*
ACT, London *11323*
ACT, Loughborough *11320*
ACT, Manila *09496*
ACTEA, Kaduna *09209*
ACTFL, Yonkers *13770*
Action Against Allergy, London *11200*
Action Committee for the Establishment of a Latin American Network of Technological Information, Caracas *16964*
Action for Dysphasic Adults, London *11201*
Action for National Information Systems, Caracas *16965*
Action for Rational Drugs in Asia, Penang *08626*
Action in International Medicine, London *11202*
Action pour un Droit Foncier Libéral, Zürich *10611*
Acton Society Trust, London *11203*
ACTS, Nairobi *08464*
ACTS, Sisseton *14420*
Actuarial Institute of the Republic of China, Taipei *03229*
The Actuarial Society of Finland, Helsinki *04005*
Actuarial Studies in Non-Life Insurance, Philadelphia *13323*
Actuarieel Genootschap, Amsterdam *08818*
ACU, Berkeley *14421*
A.C.U., London *11329*
ACUM, Tel Aviv *07112*
Acupuncture Foundation of Canada, Markham *01818*
Acupuncture International Association, Saint Louis *13324*
Acupuncture Research Institute, Monterey Park *13325*
Acupuncture Scientific and Clinical Advisory Group, Bristol *11204*
ACURIL, San Juan *09723*
ACUS, Washington *14546*
ACUTA, Lexington *14415*
ACUTE, Indianapolis *13313*
A.C.V., Paris *04147*
ACV, Regensburg *05073*
ACVIM, Blacksburg *13739*
ACVP, West Deptford *13740*
ACVR, Glencoe *13741*
ACW, Newcastle-upon-Tyne *11331*
ACWIS, New York *13745*
ACWRTUK, Richmond *11224*
ADA, Alexandria *13785*
ADA, Chicago *13779, 13787*
ADA, Columbia *13291*
ADA, Iowa City *13782*
ADA, London *11201*
A.D.A., Saint Leonards *00282*
ADAA, Chicago *13778*
Adalbert Stifter-Gesellschaft, Wien *00531*
ADARA, Little Rock *13777*
ADBU, Paris *04130*
ADCA, College Station *13336*
ADCIS, Columbus *14350*
ADCU, Bloomington *14425*
ADDA, Rockville *13784*
Addiction Research and Treatment Corporation, Brooklyn *13326*
Addiction Research Foundation, Toronto *01819*
ADDS, Pittsburgh *13788*
ADE, Carbondale *14289*
ADE, New York *14427*
ADE, Paris *04209*
ADEE, Madrid *10282*
ADELF, Le Kremlin-Bicêtre *04181*
A.D.E.L.F., Paris *04180*
ADENA, Madrid *10280*
A.D.E.R.P., Paris *04337*
A.D.E.S., Dundee *11336*
ADETIEF TECHNIQUE, Paris *04201*
ADF, Paris *04155*
ADFL, New York *14428*
ADH, Chicago *13288*
ADHA, Chicago *13780*
Adhesion Society, Cincinnati *13327*
ADHILAC, Paris *04139*
ADHP, Spokane *14351*
ADI, Chicago *13371*
adi, Kiel *05077*
A.D.I., Paris *04110*
ADI, Roma *07273, 07305*
ADI, Washington *13289*
A.D.I.G., Roma *07387*
ADIPA, Kuala Lumpur *08639*
A.D.I.R.I, Bucuresti *09740*
Adirondack Historical Association, Blue Mountain Lake *13328*
ADLAF, Eichstätt *05096*
ADM, Charleston *13287*
A.D.M., Palermo *07369*
Administration Universitaire Francophone et Européenne en Médecine et Odontologie, Paris *04123*
Administrative Management Society, Washington *13329*
Administrative Sciences Association of Canada, Montréal *01820*
ADP, Bellevue *13290*
ADP, Horbury *11339*
ADPC, Bangkok *11080*
Adrenal Metabolic Research Society of the Hypoglycemia Foundation, Troy *13330*
Adriatic Society of Sciences, Trieste *07769*
ADS, Jacksonville *13786*
ADS, Ljubljana *10096*
ADS, London *01860*
ADS, Uncasville *14553*
ADSA, Champaign *13775*
ADSA, Chicago *13781*
ADSAI, Rochester *13783*
ADSGM, Manila *09509*
ADSS, Stockport *11337*
ADSW, Melrose *11338*
ADTA, Columbia *13776*
ADTA, Peoria *14426*
ADTD, Saint Louis *14429*
Adult Christian Education Foundation, Madison *13331*
Adult Education Association of Guyana, Georgetown *06575*
Adult Education Association of Zambia, Lusaka *17147*
Advanced Informatics in Medicine in Europe, Bruxelles *01097*
Advanced Technology Alert System, New York *13332*
Advanced Transit Association, Fairfax *13333*
Advertising Association, London *11205*
Advertising Research Foundation, New York *13334*
Advisory Centre for Education, London *11206*
Advisory Committee for the Coordination of Information Systems, Genève *10608*
Advisory Committee on Protection of the Sea, London *11207*
Advisory Committee on Science and Technology for Development, New York *13335*
Advisory Group for Aerospace Research and Development, Neuilly-sur-Seine *04124*
AE, Bruxelles *01099*
AEA, Center Moriches *13792*
AEA, Independence *13356*
AEA, Nashville *13789*
AEA, Noordwijk *08823*
AEA, Santa Clara *13795*
AEA, Trumbull *13794*
AEA, York *11215*
AEAC, Madrid *10256*
AEAZ, Lusaka *17147*
A.E.B., Bruxelles *01117*
AEB, Quito *03843*
AEC, New York *14435*
A.E.C., Paris *04203*
A.E.C.A.P., Bruxelles *01126*
AECS, Arnhem *08826*
AECT, Washington *14290*
AECTS, Zürich *10672*
AED, Washington *13277*
AEDA, Montesson *04129*
AEDBF, Bruxelles *01131*
AEDE, Paris *04206*
A.E.D.I., Milano *07180*
AEE, Atlanta *14431*
AEE, The Plains *13365*
AEEA, Kassel *05154*
AEEF, Bruxelles *01123*
AEEGS, Bloomfield *13793*
AEEMTRC, Jakarta *06874*
AEEP, Houghton *14434*
AEESEAP, Christchurch *09118*
AEEVTPLF, Dakar *10018*
AEF, Arlington *13337*
AEF, Berlin *06161*
AEF, Glendale *14246*
AEF, London *11218*
AEFA, Bad Honnef *05156*
AEFMA, Bruxelles *01129*
AEG, Sudbury *14433*
AEGS, London *11342*
AEI, Milano *07411*
AEI, Roma *07275*
Äidinkielen opettajain liitto, Helsinki *03948*
AEJMC, Columbia *14292*
ÄKNo, Düsseldorf *05039*
AELFA, Barcelona *10263*
AEM, Madrid *10264*
AEMS, Boston *13798*
AEO, Madrid *10265*
ÄOL, Helsinki *03948*
AEOM, Chichester *11344*
AEP, Bruxelles *01091*
AEP, Durham *11341*
A.E.P.C., Madrid *10272*
AEQCT, Barcelona *10268*
AER, Alexandria *14291*
AERA, Washington *13790*
AERC, Nairobi *08467*
Aerial Survey, Photogrammetric and Remote Sensing Research Group, Durban *10118*
AERO, Helena *13368*
Aerodynamics Research Society, Beijing *03114*
Aeronautical Society of India, New Delhi *06709*
Aeronautical Society of South Africa, Bryanston *10119*
The Aeroplane Collection, Altrinchem *11208*
Aerospace Department Chairmen's Association, College Station *13336*
Aerospace Education Foundation, Arlington *13337*
Aerospace Electrical Society, Anaheim *13338*
Aerospace Medical Association, Alexandria *13339*
AERSG, Nairobi *08468*
A.E.R.Z.A.P., Louvain-la-Neuve *01151*
Ärztegemeinschaft im Katholischen Akademieverband der Erzdiözese Wien, Wien *00532*
Ärztegesellschaft Innsbruck, Innsbruck *00533*
Ärztekammer Berlin, Berlin *05030*
Ärztekammer Bremen, Bremen *05031*
Ärztekammer des Saarlandes, Saarbrücken *05032*
Ärztekammer des Saarlandes, Abteilung Zahnärzte, Saarbrücken *05033*
Ärztekammer Frankfurt, Frankfurt *05034*
Ärztekammer für Kärnten, Klagenfurt *00534*
Ärztekammer für Niederösterreich, Wien *00535*
Ärztekammer für Wien, Wien *00536*
Ärztekammer Hamburg, Hamburg *05035*
Ärztekammer Land Brandenburg, Cottbus *05036*
Ärztekammer Mecklenburg-Vorpommern, Rostock *05037*
Ärztekammer Niedersachsen, Hannover *05038*
Ärztekammer Nordrhein, Düsseldorf *05039*
Ärztekammer Sachsen-Anhalt, Magdeburg *05040*
Ärztekammer Schleswig-Holstein, Bad Segeberg *05041*
Ärztekammer Westfalen-Lippe, Münster *05042*
Ärztekommission für Rettungswesen SRK, Bern *10609*
Ärztliche Gesellschaft für Physiotherapie, Kneippärztebund e.V., Bad Münstereifel *05043*
Ärztliche Gesellschaft für Physiotherapie, Österreichischer Kneippärztebund, Neunkirchen *00537*
AES, Anaheim *13338*
AES, Atlanta *13800*
AES, Belfast *11214*
AES, Butler *13276*
AES, Edwardsville *14547*
A.E.S., Feltham *11223*
AES, Fullerton *13797*
AES, Hartford *13801*
AES, Lancaster *13803*
AES, Morton Grove *13802*
AES, New Delhi *06722*
AES, New York *14548*
AES, Philadelphia *13799*
AESA, Cincinnati *13791*
AESA, Parkville *00213*
AESE, Lawrence *14430*
AESFS, Orlando *13796*
AESM, Santa Barbara *14294*
AESOP, Nijmegen *08827*
AESSEA, Manila *09489*
AES, UK, Sevenoaks *11416*
AET, Strasbourg *04207*
AETFAT, Rosières *01153*
The Aetherius Society, London *11209*

AETR, Madrid *10270*
AETS, Auburn *14293*
A.E.T.T., Axminster *11263*
A.E.W.V.R., Droitwich Spa *11287*
AF, Dragoon *14214*
AF, New York *13341*
AFA, Berkeley *13807*
AFA, Bloomington *13825*
AFA, New York *13814, 14228*
AFA, Paris *04212*
AFA, River Falls *13819*
AFA, Tempe *13808*
AFA, Washington *13820*
AFALMA, Lusaka *17148*
AFAS, New York *13815*
AFASIC, London *11258*
AFBRF, Park Ridge *13805*
AFC, Paris *04213*
AFCCE, Washington *14436*
AFCF, Santa Fe *14229*
AFCIA, Sa Ysidro *14243*
AFCMA, Seoul *08519*
AFCR, Swindon *11213*
AFCR, Thorofare *13806*
AFCU, Arlington *13827*
AFDE, Minneapolis *14440*
AFDO, York *14439*
AFE, Paris *04218*
AFEA, Paris *04237*
AfeB, Heidelberg *05133*
A.F.E.C., Paris *04245*
AFEC, Sèvres *04249*
AFEE, Limoges *04244*
AFEE, Lincoln *14295*
AFEI, Washington *14040*
AFELSH, Dakar *10019*
AFEO, Bangkok *11067*
A.F.E.Q., Caen *04246*
AFES, Jakarta *06875*
A.F.E.S., Plaisir *04247*
AFF, Paris *04239*
AFG, Paris *04216*
AFGE, New York *13821*
Afghanistan Academy of Sciences, Kabul *00001*
Afghanistan Studies Association, Omaha *13340*
AFGR, Paris *04217*
AFGS, Pawtucket *13826*
AFHB, Kuala Lumpur *08628*
AFHP, Newark *14225*
AFI, Washington *13813*
AFICCA, Cannes *04321*
AFICTIC, Lyon *04232*
AFIDES, Montréal *01908*
AFIGAP, Brie-Comte-Robert *04251*
AFIP, Washington *14244*
AFIRD, Paris *04233*
AFL, Easton *13828*
AFLA, New York *13818*
AFM, Kuala Lumpur *08630*
AFMC, Tokyo *08099*
AFME, Asmara *03936*
AFMH, McLean *14296*
AFMR, Abidjan *08058*
AFMS, Oklahoma City *13809*
AFNETA, Ibadan *09223*
AFNOR, Paris La Défence *04221*
AFO, South Natick *14437*
AFOD, Damascus *11041*
AföG, Düsseldorf *05056*
AFOPDA, Abidjan *08059*
AFOS, Harrison *14245*
AFPE, Paris *04241*
A.F.P.I.A., Paris *04325*
A.F.P.I.C., Paris *04312*
A.F.P.I.C, Paris *04324*
AFPMH, Kuala Lumpur *08627*
AFP/SME, Dearborn *14297*
AFREM, Paris *04224*
Africa AIDS Research Network, Dakar *10007*
Africa Association for Liturgy, Music and Arts, Lusaka *17148*
The Africa Fund, New York *13341*
Africa Genetics Association, Benin City *09210*
Africa Institute of South Africa, Pretoria *10120*
Africa Kyokai, Tokyo *08090*
Africa Leadership Forum, Abeokuta *09211*
Africa Literature Centre, Kitwe *17149*
African Academy of Sciences, Nairobi *08458*
African-American Institute, New York *13342*
African American Museum, Cleveland *13343*
African and Malagasy Council on Higher Education, Ouagadougou *01804*
African and Mauritian Institute of Statistics and Applied Economics, Kigali *09993*
African Association for Biological Nitrogen Fixation, Nairobi *08459*
African Association for Correspondence Education, Nairobi *08460*
African Association for Literacy and Adult Education, Nairobi *08461*
African Association for Public Administration and Management, Addis Ababa *03934*
African Association for the Study of Liver Diseases, Cape Town *10121*
African Association of Dermatology, Nairobi *08462*
African Association of Education for Development, Lomé *11103*
African Association of Insect Scientists, Nairobi *08463*
African Association of Microbiology and Food Hygiene, Sousse *11134*
African Association of Political Science, Lagos *09212*
African Association of Science Editors, Addis Ababa *03935*
African Biosciences Network, Dakar *10008*
African Bureau of Educational Sciences, Kinshasa *17139*
African Centre for Applied Research and Training in Social Development, Tripoli *08566*
African Centre for Democracy and Human Rights Studies, Banjul *05005*
African Centre for Fertilizer Development, Lagos *09213*
African Centre for Monetary Studies, Dakar *10009*
African Centre for Technology Studies, Nairobi *08464*
African Commission on Agricultural Statistics, Accra *06470*
African Commission on Mathematics Education, Nairobi *08465*
African Council for Communication Education, Nairobi *08466*
African Council for Social and Human Sciences, Dakar *10010*
African Council of Food and Nutrition Sciences, Harare *09214*
African Curriculum Organization, Ibadan *09215*
African Economic Research Consortium, Nairobi *08467*
African Elephant and Rhino Specialist Group, Nairobi *08468*
African Feed Resources Research Network, Harare *09216*
African Forestry and Wildlife Commission, Accra *06471*
African Forum for Mathematical Ecology, Asmara *03936*
African Geographers' Association, Cotonou *01514*
African Gerontological Society, Legon *06472*
African Heritage Center for African Dance and Music, Washington *13344*
African Heritage Studies Association, Flushing *13345*
African Institute for Economic and Social Development, Abidjan *08057*
African Institute for Economic Development and Planning, Dakar *10011*
African Institute for Higher Technical Training and Research, Nairobi *08469*
African Library Association of South Africa, Pietersburg *10122*
African Literature Association, Edmonton *01821*
African Mathematical Union, Ibadan *09217*
African Medical and Research Foundation, Nairobi *08470*
African Mountain Association, Asmara *03937*
African Music Rostrum, Abidjan *08058*
African Network for Integrated Development, Dakar *10012*
African Network for the Development of Ecological Agriculture, Accra *06473*
African Network of Administrative Information, Tangiers *08799*
African Network of Scientific and Technological Institutions, Nairobi *08471*
African Oil Palm Development Association, Abidjan *08059*
African Organization for Research and Training in Cancer, Harare *17159*
African Paediatric Club, Kenema *10029*
African Participatory Research Network, Dar es Salaam *11046*
African Peace Research Institute, Lagos *09218*
African Regional Centre for Engineering Design and Manufacturing, Ibadan *09219*
African Regional Centre for Technology, Dakar *10013*
African Regional Computer Confederation, Harare *17160*
African Regional Network for Microbiology, Okigwi *09220*
African Regional Organization for Standardization, Nairobi *08472*
African Rehabilitation Institute, Harare *17161*
African Small Ruminant Research Network, Addis Ababa *03938*
The African Society, Cairo *03861*
African Society for Environmental Studies Programme, Nairobi *08473*
African Society of International and Comparative Law, London *11210*
African Soil Science Association, Cairo *03862*
African Standing Conference on Bibliographic Control, Dakar *10014*
African Statistical Association, Lagos *09221*
African Studies Association, Atlanta *13346*
African Studies Association of the United Kingdom, London *11211*
African Technical Association, Paris *04125*
African Training and Research Centre for Women, Addis Ababa *03939*
African Training and Research Centre in Administration for Development, Tangiers *08800*
African Training Centre for Literacy and Adult Education, Nairobi *08474*
African Trypanotolerant Livestock Network, Nairobi *08475*
African Union for Scientific Development, Accra *06474*
African Union for the Management of Development Banks, Cotonou *01515*
African Union of Sports Medicine, Algiers *08060*
African Water Network, Nairobi *08476*
African Women Jurists Federation, Libreville *05004*
Africa Regional Centre for Information Science, Ibadan *09222*
Africa Society of Japan, Tokyo *08090*
Afro-American Historical and Genealogical Society, Washington *13347*
Afro-Asian Center, Saugerties *13348*
Afro-Asian Philosophy Association, Cairo *03863*
Afro-Asian Writers' Permanent Bureau, Cairo *03864*
AFRONUS, Harare *09214*
AFS, Atlanta *13810*
AFS, Bethesda *13816*
AFS, Birmingham *13811*
AFS, Des Plaines *13824*
AFS, Harrisburg *13817*
AFS, Manila *09497*
AFSA, Lagos *09221*
AFSAU, Nairobi *08479*
A.F.S.E., Toulouse *04226*
AFSP, Paris *04227*
AFTA, Washington *13804*
AFTE, Baghdad *06908*
AFTE, La Jolla *14438*
AFUA, Mount Clemens *14248*
AFVA, Niles *13812*
AFVA, Saint Louis *13823*
AFW, Bad Harzburg *05055*
AfW, Stuttgart *05144*
AFWC, Accra *06471*
AG, Amsterdam *08818*
AG, East Aurora *14300*
A.G., Heidelberg *05157*
AG, New York *14554*
AGA, Bethesda *13829*
AGA, Brewster *14539*
AGA, Buckeystown *13830*
AGAB, Luzern *10792*
AGARD, Neuilly-sur-Seine *04124*
AgaS, München *05101*
AGAVA, Wien *00548*
AGBAD, Washington *13361*
AGCCPFI, Paris *04252*
AGD, Chicago *13292*
AGE, Bangkok *11081*
AGE, Omaha *13445*
A.GE.I., Roma *07270*
AGEMUS Arbeitsgemeinschaft Evolution, Menschheitszukunft und Sinnfragen, Wien *00538*
Agence Internationale de l'Energie, Paris *04126*
Agence Latinoaméricaine d'Information, Quito *03841*
Agence Spatiale Européenne, Paris *04127*
Agency for Cooperation and Research in Development, London *11212*
Agency for Instructional Technology, Bloomington *13349*
Agency for Tele-Education in Canada, Toronto *01822*
Agency for the Safety of Aerial Navigation in Africa, Dakar *10015*
A.GEN.P.P., Roma *07123*
A.G.E.R., Bucuresti *09741*
AGF, Bonn *05086*
AGF, Detmold *05115*
AGF, Paris *04185*
AGFW, Frankfurt *05098*
AGGS, Bern *10614*
AGH, Omaha *13837*
AGHE, Washington *14299*
AGHG, Guatemala City *06554*
AGHTM, Paris *04253*
AGI, Alexandria *13832*
A.G.I., Ancona *07283*
AGI, Köln *05119*
AGI, Padova *07278*
AGI, Roma *07280*
AGI, Zürich *10615*
AGICOA, Genève *10626*
AGID, Bangkok *11088*
AGL, Bonn *05108*
AGLINET, Roma *07247*
AGLS, Columbus *14298*
AGLSP, Chicago *14441*
AG.MA, Frankfurt *05124*
AGMANZ, Wellington *09116*
A.G.M.F., Paris *04254*
AGMÖ, Wien *00549*
AGO, Vallendar *05088*
AGOS, Salt Lake City *13838*
AGPA, Alexandria *13835*
AGPA, New York *13836*
AGPP, New York *14301*
AGRA, Horsham *11345*
Agrarian Society, Valletta *08657*
Agrarsoziale Gesellschaft e.V., Göttingen *05044*
AGRE, La Plaine-Saint-Denis *04360*
Agricultural and Food Research Council, Swindon *11213*
Agricultural Chemistry Society of China, Taipei *03230*
Agricultural Communicators in Education, Gainesville *13350*
Agricultural Economics Society, Belfast *11214*
Agricultural Economics Society of South East Asia, Manila *09489*
Agricultural Education Association, York *11215*
Agricultural Engineering Society, Australia, Parkville *00213*
Agricultural Extension Association of China, Taipei *03231*
Agricultural Guidance Research Institute, Dokki *03865*
Agricultural History Society, Washington *13351*
Agricultural Information Development Scheme, Bangkok *11064*
Agricultural Institute of Canada, Ottawa *01823*
Agricultural Law Association, Wageningen *09094*
Agricultural Libraries Network, Roma *07247*
Agricultural Research Corporation, Wad Medani *10423*
The Agricultural Research Council of Norway, Oslo *09286*
Agricultural Research Council of Zambia, Lusaka *17150*
Agricultural Research Council of Zimbabwe, Harare *17162*
Agricultural Research Institute of Northern Ireland, Hillsborough *11216*
Agricultural Research Organization, Bet Dagan *07041*
Agricultural Science Association, Dublin *06921*
Agricultural Science Society of Thailand, Bangkok *11065*
Agricultural Society of Iceland, Reykjavik *06679*
Agricultural Society of Kenya, Nairobi *08477*
Agricultural Society of Trinidad and Tobago, Port of Spain *11105*
Agriculture Association of China, Taipei *03232*
Agri-Horticultural Society of India, Calcutta *06710*
Agri-Horticultural Society of Madras, Madras *06711*
Agriservices Foundation, Clovis *13352*
Agro-Industrial Society, Moskva *09799*
Agronomes sans Frontières, Paris *04128*
Agronomiliitto, Helsinki *03949*
Agronomy Society of New Zealand, Canterbury *09115*
Agrupación Astronáutica Española, Barcelona *10245*
Agrupación de Bibliotecas para la Información Socio-Económica, Lima *09442*
AGS, New York *13831, 13834*
AGS, Washington *14442*
AGSIDC, Baghdad *06909*
AGST, Graz *00568*
AGU, Bonn *05112*
AGU, Washington *13833*
AGW, Wien *00544*
AGZ, Zürich *10616*
AHA, Baghdad *06910*
AHA, Blue Mountain Lake *13328*
AHA, Chicago *13854*
AHA, Cincinnati *13848*
AHA, Dallas *13844*
AHA, Glendale *13840*
AHA, Nevada City *13845*
AHA, San Francisco *13857*
AHA, Washington *13846*
AHC, Baton Rouge *14371*
AHCADM, Washington *13344*
AHE, Laramie *14303*
AHEA, Alexandria *13851*
AHEA, Silver Spring *13855*
AHEAD, Alexandria *14304*
AHI, Aichi *08100*
AHILA, Brazzaville *03422*
AHIS, Windor *11352*
AHLHS, New York *13856*
AHMA, Raleigh *13849*
AHME, Washington *14302*
Ahmedabad Textile Industry's Research Association, Ahmadabad *06712*
AHMF, Raleigh *13850*
AHMPS, London *11351*
AHOTE, Raleigh *14519*
AHP, San Francisco *14305*
AHPA, Oklahoma City *13842*
AHRF, Chicago *13843*
AHRTAG, London *11239*
AHS, Alexandria *13852*
AHS, Hollywood *13839*
AHS, Rock Island *14550*
A.H.S., Wadhurst *11238*
AHS, Washington *13351*
AHSA, Flushing *13345*
AHSGR, Lincoln *13847*
AHST, Warwick *11354*
AHTA, Gaithersburg *13853*
AI, New York *14241*
AIA, Boston *14237*
AIA, Paris *04137*
A.I.A., Pisa *07299*
AIA, Saint Louis *13324*
AIA, Telford *11266*
AIA, Washington *13874*
AIAA, Washington *13872*
AIAATID, New York *13873*
AIAC, Roma *07286*
AIAF, Tunis *11135*
A.I.A.M., Roma *07189*
AIAP, Paris *04265*
AIAR, Mount Vernon *13864*
A.I.A.S., Trieste *07335*
AIAS, Wien *00552*
AIB, Detroit *13293*
AIB, Roma *07291*
AIBC, Annandale *13901*
AIBC, Philadelphia *13876*
AIBD, Shelton *13877*
AIBM, Berlin *06003*
AIBS, Washington *13875*
AIC, Bethesda *13879*
AIC, Namur *01135*
AIC, Napoli *07301*
A.I.C., Pisa *07300*
AIC, Saint Petersburg *13880*
AIC, Washington *13866*
AICA, Bonn *05996*
AICAE, Tigard *13858*
AICARDES, Tunis *11137*
AICCP, Des Plaines *14520*
AICGS, Washington *13867*
AIChE, New York *13878*
AICL, Paris *04267*
AICPCU, Malvern *13865*
AICQ, Milano *07338*
AICRC, Marvin *13859*
AICS, Washington *14444*
AICT, Paris *04266*
AICTC, Milano *07302*
AICU, Independence *14447*
AIDAA, Roma *07297*
AID Auswertungs- und Informationsdienst für Ernährung, Landwirtschaft und Forsten e.V., Bonn *05045*
AIDCOM, Kuala Lumpur *08634*
AIDE, Louvain-la-Neuve *01136*
AIDELF, Paris *04268*
Aid for International Medicine, Rocklamd *13353*
AIDI, Roma *07307*
AIDMO, Baghdad *06911*
AIDP, Pau *04256*
AIDS, Bath *11265*
AIDS, Boston *13294*
AIDUM, Aulnay-sous-Bois *04290*
AIE, Paris *04126*
AIEA, Jerusalem *07045*
AIEAS, Paris *04272*
AIECE, Louvain-la-Neuve *01119*
AIEE, Genève *10621*
AIEF, Paris *04273*
A.I.E.G.L., Paris *04262*
AIEJI, Pau *04271*
AIEO, Oegstgeest *08824*
AIEPAD, Córdoba *00083*
AIER, Great Barrington *13868*
A.I.E.S., Perugia *07341*
AIESEC – United States, New York *13354*
AIESEP, Liège *01137*
AIESI, Paris *04270*
AIF, Paris *04187, 04579*
AIFRO, Rennes *04287*
AIFS, Greenwich *13869*
AIFSPR, Torino *07308*
AIFT, New York *13321*
AIGA, México *08693*
AIGA, New York *13881*
A.I.G.D., Roma *07332*
AIGR, Padova *07309*
AIGYPFB, La Paz *01533*
AIH, Denver *13882*
AIHA, Akron *13863*
AIHA00, Staten Island *13904*
AIHC, Washington *13862*
AIHD, Nakhon Pathom *11069*
AIHP, Madison *13896*
AIHR, Mobile *14448*
AIHS, New York *13902*
AIHTTR, Nairobi *08469*
AII, Roma *07284*
A.I.I.A., Canberra *00304*
AIIG, Trieste *07293*
AIII, Charbonnières-les-Bains *04284*
A.I.I.M.B., Napoli *07312*
AIIS, Columbus *13884*
AIIS, Denver *13885*
AIIS, Toronto *01861*
AIISUP, Paris *04283*
AIIT, Paris *04258*
AIJD, Bruxelles *01139*
AIL, Gap *14249*
AILA, Logan *13860*
AILS, Chicago *13883*
AIM, Bruxelles *01097*
AIM, Hinsdale *13279*
AIM, London *11202*
AIM, Manila *09498*
A.I.M., Milano *07315*
AIM, Quincy *13887*
AIM, Rocklamd *13353*

AIMAS, Roma 07313
AIMBE, Paris 04260
AIMC, East Bloomfield 14446
AIME, New York 13888
AIMEA, Bruxelles 01142
AIMLC, Mayaguez 09724
AIMM, Parkville 00307
AIMOH, Roma 07193
AIMR, Charlottesville 14309
AIMS, Dallas 13889
AIN, Alexandria 03872
AIN, Bethesda 13890
A.I.N., Sydney 00308
AINA, Calgary 01871
AINDT, Parkville 00294
AINF, Seclin 04294
AINSE, Menai 00309
AIOB, South Laguna 13891
A.I.O.M., Milano 07317
AIORMS, Lausanne 10631
AIOSP, Strassen 08578
AIP, Fort Wayne 13892
AIP, New York 13893
AIP, Parkville 00310
A.I.P.A., Roma 07345
AIPBS, Kingston 13870
AIPE, Cincinnati 13894
AIPLA, Arlington 13900
AIPPHi, Minden 05155
AIPS, Bruxelles 01144
A.I.P.S., Ferrara 07319
A.I.P.S., Sydney 00311
AIPT, Columbia 14308
A.I.R., East Melbourne 00313
AIR, Tallahassee 14306
AIR, Washington 13899
AIRAH, Parkville 00314
Air and Waste Management Association, Pittsburgh 13355
AIRBM, Catania 07320
Aircraft Building Society, Moskva 09800
Aircraft Electronics Association, Independence 13356
Aircraft Research Association, Bedford 11217
AIRE, Milano 07336
Aire Méditerranéenne et Latinoaméricaine, Bruxelles 01098
Airfields Environment Federation, London 11218
A.I.R.I., Roma 07339
AIRIEL, Milano 07340
AIRIT, Paris 04263
Airline Medical Directors Association, Chicago 13357
A.I.R.M.N., Roma 07321
Air National Guard Optometric Society, Birmingham 13358
AIRO, Genova 07322
A.I.R.P., Bologna 07318
Airship Association, London 11219
Airship Association – U.S., Silver Spring 13359
A.I.S., Pisa 07294
AIS, Roma 07324
A.I.S.A., Genova 07351
AISA, Nairobi 08480
AISC, Chicago 13895
AISC, Roma 07347
A.I.S.D., Firenze 07344
AISE, Paris 04278
AISE, Pittsburgh 14449
AISES, Boulder 13861
A.I.S.E.S.S., Roma 07192
A.I.S.F., Pisa 07196
A.I.S.I., Milano 07350
AISI, Roma 07346
AISI, Washington 13903
AISJ, Paris 04276
AISLF, Toulouse 04277
AISM, Milano 07333
A.I.S.M., Milano 07348
AISM, Roma 07349
AISPS, Roma 07323
A.I.S.S., Torino 07327
AIST, Syracuse 14307
AIT, Bloomington 13349
AIT, Manila 09499
AIT, Springfield 13278
A.I.T.A., Milano 07329
AITC, Vancouver 13897
AITEC, Roma 07352
AITES, Bron 04280
A.I.T.O., Roma 07330
AITP, Milano 07331
AIU, Paris 04281
AIUM, Rockville 13898
AIUS, Canberra 00315
AIVS, New York 13871
AJA, Williamsburg 13907
AJA5, Parkville 00296
AJBD, Augsburg 05105
AJCC, Chicago 13906
AJCU, Washington 14450
AJHS, Waltham 13905
Ajia Chosakai, Tokyo 08091
Ajiakurabu, Tokyo 08092
Ajia Seikei Gakkai, Tokyo 08093
AJL, New York 14451
AJS, Cambridge 14311
AJS, Chicago 13908
AJSEJT, Durham 11358
AJVPA, Madrid 10274
Akademia e Shkencave dhe e Arteve e Kosovës, Priština 17076
Akademie der Arbeit in der Universität, Frankfurt 05046
Akademie der bildenden Künste in Wien, Wien 00539
Akademie der Diözese Rottenburg-Stuttgart, Stuttgart 05047
Akademie der Künste, Berlin 05048
Akademie der Wissenschaften in Göttingen, Göttingen 05049
Akademie der Wissenschaften und der Literatur zu Mainz, Mainz 05050
Akademie der Wissenschaften zu Berlin, Berlin 05051
Akademie für Allgemeinmedizin, Graz 00540
Akademie für Arbeit und Sozialwesen, Saarbrücken 05052
Akademie für das Grafische Gewerbe, München 05053
Akademie für Fernstudium und Weiterbildung Bad Harzburg, Bad Harzburg 05054
Akademie für Fort- und Weiterbildung von Fach- und Führungskräften, Wuppertal 06310
Akademie für Führungskräfte der Wirtschaft e.V., Bad Harzburg 05055
Akademie für öffentliches Gesundheitswesen in Düsseldorf, Düsseldorf 05056
Akademie für Organisation, Bonn 05057
Akademie für Politische Bildung, Tutzing 05058
Akademie für Publizistik in Hamburg e.V., Hamburg 05059
Akademie für Raumforschung und Landesplanung, Hannover 05060
Akademie für Sozialarbeit der Stadt Wien, Wien 00541
Akademie gemeinnütziger Wissenschaften zu Erfurt e.V., Erfurt 05061
Akademie Klausenhof, Hamminkeln 05062
Akademie Kontakte der Kontinente, Bonn 05063
Akademie Remscheid für musische Bildung und Medienerziehung e.V., Remscheid 05064
Akademiet for de skønne Kunster, København 03556
Akademiet for de tekniske Videnskaber, Lyngby 03557
Akademija Nauka i Umetnosti Kosova, Priština 17076
Akademikernes Centralorganisation, København 03558
Akademische Arbeitsgemeinschaft für Volkskunde, Wien 00542
Akademischer Verein Hütte, Berlin 05065
Akademisch-soziale Arbeitsgemeinschaft Österreichs, Wien 00543
Akademisk Arkitektforening, København 03602
Akademi Teknologi Kulit, Jogjakarta 06872
Akadimia Athinon, Athinai 06498
AKOR SRK, Bern 10609
AKS, Rancho Palos Verdes 14312
AKThB, Paderborn 05122
Aktion 100, Verein zur Förderung der Sicherheit im Strassenverkehr, Wabern 10610
Aktion Bildungsinformation e.V., Stuttgart 05066
Aktion Freiheitliche Bodenordnung, Zürich 10611
Aktion Psychisch Kranke, Bonn 05067
Aktionsgemeinschaft Natur- und Umweltschutz Baden-Württemberg, Stuttgart 05068
Aktionsgemeinschaft Soziale Marktwirtschaft e.V., Tübingen 05069
AL, Palatine 14259
AL, Rolling Meadows 14540
ALA, Chicago 13916
ALA, Edmonton 01821
ALA, Jakarta 06876
ALA, New York 13923, 14555
ALA, Rochester 13910
ALA, Torrance 13921
ALA, Vernon Hills 14452
ALADA, Madrid 10275
ALADEFE, Quito 03846
ALAETS, Saõ Luís 01583
ALAFO, Medellín 03353
ALAI, Paris 04299
ALAI, Quito 03841
ALAS, Córdoba 00088
ALAS, México 08695
ALASA, Pietersburg 10122
Ala – Schweizerische Gesellschaft für Vogelkunde und Vogelschutz, Sempach 10612
Albanian Academy of Sciences, Tirana 00003
Alberta Association of College Librarians, Calgary 01824
Alberta Association of Library Technicians, Edmonton 01825
Alberta Association of Rehabilitation Centres, Calgary 01826
Alberta Association of Social Workers, Edmonton 01827
Alberta Association on Gerontology, Edmonton 01828
Alberta Chiropractic Association, Edmonton 01829
Alberta Construction Association, Edmonton 01830
Alberta Dental Association, Edmonton 01831
Alberta Educational Communications Corporation, Edmonton 01832
Alberta Forest Development Research Trust Fund, Spruce Grove 01833
Alberta Genealogical Society, Edmonton 01834
Alberta Health Records Association, Edmonton 01835
Alberta Heritage Foundation for Medical Research, Edmonton 01836
Alberta Historical Resources Foundation, Calgary 01837
Alberta Law Foundation, Calgary 01838
Alberta Lung Association, Edmonton 01839
Alberta Medical Council, Calgary 01840
Alberta Museums Association, Calgary 01841
Alberta Psychiatric Association, Edmonton 01842
Alberta Public Health Association, Olds 01843
Alberta Registered Dietitians Association, Calgary 01844
Alberta Research Council, Edmonton 01845
Alberta Safety Council, Edmonton 01846
Alberta School Trustees' Association, Edmonton 01847
Alberta Society of Professional Biologists, Edmonton 01848
Alberta Sulphur Research, Calgary 01849
Alberta Teachers' Association, Edmonton 01850
Alberta Thoracic Society, Edmonton 01851
Alberta Veterinary Medical Association, Edmonton 01852
Albertus-Magnus-Institut, Bonn 05070
ALC, Buenos Aires 00089
ALC, Kitwe 17149
ALCA, Cincinnati 13913
ALCECOOP, Rosario 00086
ALCES, Beech Grove 11362
ALCF, Minneapolis 14454
ALCL, Barking 11363
Alcor Life Extension Foundation, Riverside 13360
ALCTS, Chicago 14314
Alcuin Club, Runcorn 11220
ALD, Bloomington 14252
ALEF, Riverside 13360
ALEFPA-Europe, Bruxelles 01146
ALER, Quito 03845
ALERT, Addis Ababa 03940
Alexander Graham Bell Association for the Deaf, Washington 13361
Alexandria Medical Association, Alexandria 03866
ALF, Kokkedal 03565
ALFAL, Caracas 16968
ALFAL, Santiago 03019
A.L.F.E., Paris 04298
Alfred Adler Institute, New York 13362
Algemene Bond ter Bevordering van Beroepsonderwijs, 's-Gravenhage 08819
Algemene Bond van Onderwijzend Personeel, Amsterdam 08820
Algemene Nederlandse Vereniging voor Wijsbegeerte, Uithoorn 08821
Algoma Lung Association, Sault Sainte Marie 01853
ALHFAM, Santa Fe 14316
ALI, Philadelphia 13912
ALI-ABA Committee on Continuing Professional Education, Philadelphia 13363
ALIC, Newark 14453
ALISE, Raleigh 14313
ALJA, Edinburgh 11361
Alkan Society, Salisbury 11221
A.L.K.D., Luxembourg 08579
ALL, Chard 11270
All Africa Leprosy and Rehabilitation Training Center, Addis Ababa 03940
All Africa Teachers' Organization, Accra 06475
Allahabad Mathematical Society, Allahabad 06713
ALLC, Bangor 11271
Allergy Association of Calgary, Calgary 01854
Allergy Foundation of Canada, Saskatoon 01855
Allergy Information Association, Etobicoke 01856
Alley Farming Network for Tropical Africa, Ibadan 09223
Allgemeine Ärztliche Gesellschaft für Psychotherapie, Düsseldorf 05071
Allgemeine Anthroposophische Gesellschaft, Dornach 10613
Allgemeine Geschichtforschende Gesellschaft der Schweiz, Bern 10614
Allgemeine Gesellschaft für Philosophie in Deutschland e.V., Giessen 05072
Allgemeiner Cäcilien-Verband für Deutschland, Regensburg 05073
Alliance for Alternatives in Healthcare, Calabasas 13364
Alliance for Environmental Education, The Plains 13365
Alliance Graphique Internationale, Zürich 10615
Alliance to Save Energy, Washington 13366
All India Fine Arts and Crafts Society, New Delhi 06714
All-Pakistan Educational Conference, Karachi 09357
ALM, Greenville 13915
ALNY, New York 14239
ALPA, Porto Alegre 01585
ALPE, Madrid 10262
ALPP, Porto Alegre 01584
ALRA, London 11196
ALROS, East Greenville 13911
ALS, Demarest 14247
ALS, East Pittsburgh 13922
ALS, Highlands 13920
ALS, Rochester 13918
ALSA, Boston 13914
ALSAC – Saint Jude Children's Research Hospital, Memphis 13367
ALSB, Oxford 13295
ALSC, Chicago 14315
ALSS, Saint Lucas 14455
ALT, Leicester 11360
ALTA, Chicago 13917
ALTA, Richardson 13919
Alternative Energy Resources Organization, Helena 13368
Alternative Society, Kidlington 11222
Alternative Sources of Energy, Milaca 13369
ALTGA, Shreveport 14242
Alzheimer Europe, Bruxelles 01099
Alzheimer's Association, Chicago 13370
Alzheimer's Disease International, Chicago 13371
Alzheimer Society of Manitoba, Winnipeg 01857
AM, Ada 13296
AM, Grafton 14456
AMA, Asmara 03937
AMA, Barton 00322
AMA, Chicago 13926, 13933
AMA, New York 13925
AMA, Pittsburgh 13930
AMAA, Chicago 13932
AMA-ERF, Chicago 13934
AMAI, Bruxelles 01100
AMA International, New York 13372
Amalgamated Bermuda Union of Teachers, Hamilton 01517
Amateur Entomologists' Society, Feltham 11223
Amateur Society of Basque Language, Tbilisi 05006
AMATYC, Flint 13928
AMBAC, México 08699
AMBOP, Tampa 13645
Ambulatory Pediatric Association, McLean 13373
AMC, Belconnen 00321
AMC, Carlsbad 13950
AMC, Erie 14463
AMC, New York 14219, 14457
AMCA, Lake Charles 13948
AMCHP, Washington 14458
AMCS, Alexandria 14464
AMCS, Austin 14318
AMD, New York 14317
AMDA, Chicago 13357
AMDA, Okayana 08113
Amdahl Users Group, Bellevue 13374
AMDE, Belfast 11272
AMDI, Roma 07357
AMEE, Dundee 11273
AMEEGA, Elm Grove 13935
AMELA, Bruxelles 01098
A.M.E.R., Roma 07358
American Academy and Institute of Arts and Letters, New York 13375
American Academy for Cerebral Palsy and Developmental Medicine, Richmond 13376
American Academy for Jewish Research, New York 13377
American Academy for Plastics Research in Dentistry, Davison 13378
American Academy of Actuaries, Washington 13379
American Academy of Advertising, Richmond 13380
American Academy of Allergy and Immunology, Milwaukee 13381
American Academy of Arts and Sciences, Cambridge 13382
American Academy of Child and Adolescent Psychiatry, Washington 13383
American Academy of Clinical Psychiatrists, San Diego 13384
American Academy of Cosmetic Surgery, Arcadia 13385
American Academy of Crisis Interveners, Louisville 13386
American Academy of Dental Electrosurgery, New York 13387
American Academy of Dental Group Practice, Phoenix 13388
American Academy of Dental Practice Administration, Palatine 13389
American Academy of Dermatology, Schaumburg 13390
American Academy of Environmental Engineers, Annapolis 13391
American Academy of Esthetic Dentistry, Chicago 13392
American Academy of Facial Plastic and Reconstructive Surgery, Washington 13393
American Academy of Family Physicians, Kansas City 13394
American Academy of Fixed Prosthodontics, Dallas 13395
American Academy of Forensic Sciences, Colorado Springs 13396
American Academy of Gnathologic Orthopedics, Richmond 13397
American Academy of Gold Foil Operators, Noblesville 13398
American Academy of Implant Dentistry, Chicago 13399
American Academy of Implant Prosthodontics, Atlanta 13400
American Academy of Insurance Medicine, Springfield 13401
American Academy of Legal and Industrial Medicine, New York 13402
American Academy of Matrimonial Lawyers, Chicago 13403
American Academy of Maxillofacial Prosthetics, Carmel 13404
American Academy of Mechanics, La Jolla 13405
American Academy of Medical Administrators, Southfield 13406
American Academy of Microbiology, Washington 13407
American Academy of Neurological Surgery, Philadelphia 13408
American Academy of Nursing, Kansas City 13409
American Academy of Ophthalmology, San Francisco 13410
American Academy of Optometry, Bethesda 13411
American Academy of Oral and Maxillofacial Radiology, Aurora 13412
American Academy of Oral Medicine, Houston 13413
American Academy of Orofacial Pain, Lafayette 13414
American Academy of Orthodontics for the General Practitioner, Milwaukee 13415
American Academy of Orthopaedic Surgeons, Rosemont 13416
American Academy of Orthotists and Prosthetists, Alexandria 13417
American Academy of Osteopathy, Newark 13418
American Academy of Otolaryngic Allergy, Silver Spring 13419
American Academy of Pediatric Dentistry, Chicago 13420
American Academy of Pediatrics, Elk Grove Village 13421
American Academy of Periodontology, Chicago 13422
American Academy of Physical Education, Tallahassee 13423
American Academy of Physical Medicine and Rehabilitation, Chicago 13424
American Academy of Physician Assistants, Alexandria 13425
American Academy of Physiologic Dentistry, Naperville 13426
American Academy of Podiatric Sports Medicine, Potomac 13427
American Academy of Podiatry Administration, West Hartford 13428
American Academy of Political and Social Science, Philadelphia 13429
American Academy of Psychiatrists in Alcoholism and Addictions, Greenbelt 13430
American Academy of Psychiatry and the Law, Baltimore 13431
American Academy of Psychoanalysis, New York 13432
American Academy of Psychotherapists, Decatur 13433
American Academy of Religion, Syracuse 13434
American Academy of Restorative Dentistry, Colorado Springs 13435
American Academy of Safety Education, Warrensburg 13436
American Academy of Sanitarians, Miami 13437
American Academy of Sports Physicians, Northridge 13438
American Academy of Teachers of Singing, New York 13439
American Academy of the History of Dentistry, Chicago 13440
American Academy of Veterinary Dermatology, Phoenix 13441
American Academy of Veterinary Nutrition, Athens 13442
American Academy on Mental Retardation, Omaha 13443
American Accounting Association, Sarasota 13444
American Aging Association, Omaha 13445
American Agricultural Economics Association, Ames 13446
American Agricultural Law Association, Fayetteville 13447
American Allergy Association, Menlo Park 13448
American Alliance for Health, Physical Education, Recreation and Dance, Reston 13449

American Alliance for Theatre and Education, Tempe 13450
American and Canadian Underwater Certifications, Burlington 01858
American Animal Hospital Association, Denver 13451
American Anorexia/Bulimia Association, New York 13452
American Anthropological Association, Washington 13453
American Antiquarian Society, Worcester 13454
American Apitherapy Society, Red Bank 13455
American Arbitration Association, New York 13456
American Architectural Foundation, Washington 13457
American Archives Association, Bethesda 13458
American Art Therapy Association, Mundelein 13459
American Association of Cardiovascular and Pulmonary Rehabilitation, Middleton 13460
American Assembly, New York 13461
American Assembly of Collegiate Schools of Business, Saint Louis 13462
American Association for Accreditation of Laboratory Animal Care, Rockville 13463
American Association for Adult and Continuing Education, Arlington 13464
American Association for Aerosol Research, Cincinnati 13465
American Association for Agricultural Education, Ithaca 13466
American Association for Applied Linguistics, Oklahoma City 13467
American Association for Artificial Intelligence, Menlo Park 13468
American Association for Budget and Program Analysis, Falls Church 13469
American Association for Cancer Education, Houston 13470
American Association for Cancer Research, Philadelphia 13471
American Association for Career Education, Hermosa Beach 13472
American Association for Chinese Studies, Columbus 13473
American Association for Clinical Chemistry, Washington 13474
American Association for Correctional Psychology, Institute 13475
American Association for Crystal Growth, Thousand Oaks 13476
American Association for Dental Research, Washington 13477
American Association for Functional Orthodontics, Winchester 13478
American Association for Geodetic Surveying, Bethesda 13479
American Association for Geriatric Psychiatry, Greenbelt 13480
American Association for Hand Surgery, Chicago 13481
American Association for Higher Education, Washington 13482
American Association for Laboratory Accreditation, Gaithersburg 13483
American Association for Laboratory Animal Science, Cordova 13484
American Association for Marriage and Family Therapy, Washington 13485
American Association for Medical Transcription, Modesto 13486
American Association for Music Therapy, Valley Forge 13487
American Association for Paralegal Education, Overland 13488
American Association for Pediatric Ophthalmology and Strabismus, San Francisco 13489
American Association for Rehabilitation Therapy, North Little Rock 13490
American Association for Social Psychiatry, Washington 13491
American Association for State and Local History, Nashville 13492
American Association for Study of Neoplastic Diseases, South Euclid 13493
American Association for Study of the United States in World Affairs, Annandale 13494
American Association for Textile Technology, Gastonia 13495
American Association for the Advancement of Automotive Medicine, Des Plaines 13496
American Association for the Advancement of Science, Washington 13497
American Association for the Advancement of Slavic Studies, Stanford 13498
American Association for the Comparative Study of Law, New York 13499
American Association for the History of Medicine, Boston 13500
American Association for the Study of Headache, West Deptford 13501
American Association for the Study of Liver Diseases, Thorofare 13502
American Association for Thoracic Surgery, Manchester 13503
American Association for Vocational Instructional Materials, Athens 13504
American Association for World Health, Washington 13505
American Association of Anatomists, New Orleans 13506
American Association of Avian Pathologists, Kennett Sq 13507
American Association of Behavioral Therapists, Roswell 13508
American Association of Bible Colleges, Fayetteville 13509
American Association of Blacks in Energy, Washington 13510
American Association of Blood Banks, Bethesda 13511
American Association of Botanical Gardens and Arboreta, Wayne 13512
American Association of Bovine Practitioners, West Lafayette 13513
American Association of Cereal Chemists, Saint Paul 13514
American Association of Certified Allergists, Palatine 13515
American Association of Certified Orthoptists, Houston 13516
American Association of Chairmen of Departments of Psychiatry, Little Rock 13517
American Association of Christian Schools, Fairfax 13518
American Association of Colleges for Teacher Education, Washington 13519
American Association of Colleges of Nursing, Washington 13520
American Association of Colleges of Osteopathic Medicine, Rockville 13521
American Association of Colleges of Pharmacy, Alexandria 13522
American Association of Colleges of Podiatric Medicine, Rockville 13523
American Association of Community and Junior Colleges, Washington 13524
American Association of Community Theatre, Muncie 13525
American Association of Critical-Care Nurses, Aliso Viejo 13526
American Association of Dental Examiners, Chigago 13527
American Association of Dental Schools, Washington 13528
American Association of Diabetes Educators, Chicago 13529
American Association of Electrodiagnostic Medicine, Rochester 13530
American Association of Endodontists, Chicago 13531
American Association of Engineering Societies, Washington 13532
American Association of Entrepreneurial Dentists, Tupelo 13533
American Association of Equine Practitioners, Lexington 13534
American Association of Feed Microscopists, Sacramento 13535
American Association of Genito-Urinary Surgeons, Houston 13536
American Association of Gynecological Laparoscopists, Santa Fe Springs 13537
American Association of Handwriting Analysts, Hinsdale 13538
American Association of Homeopathic Pharmacists, Norwood 13539
American Association of Hospital Dentists, Chicago 13540
American Association of Hospital Podiatrists, Brooklyn 13541
American Association of Housing Educators, College Station 13542
American Association of Immunologists, Bethesda 13543
American Association of Industrial Veterinarians, Columbia 13544
American Association of Jurists, Buenos Aires 00052
American Association of Laban Movement Analysts, New York 13545
American Association of Language Specialists, Washington 13546
American Association of Law Libraries, Chicago 13547
American Association of Medical Assistants, Chicago 13548
American Association of Medical Milk Commissions, Los Angeles 13549
American Association of Medical Society Executives, Chicago 13550
American Association of Medico-Legal Consultants, Philadelphia 13551
American Association of Mental Health Professionals in Corrections, Sacramento 13552
American Association of Meta-Science, Huntsville 13553
American Association of Museums, Washington 13554
American Association of Neurological Surgeons, Park Ridge 13555
American Association of Neuropathologists, Salt Lake City 13556
American Association of Oral and Maxillofacial Surgeons, Rosemont 13557
American Association of Orthodontists, Saint Louis 13558
American Association of Orthomolecular Medicine, Boca Raton 13559
American Association of Orthopedic Medicine, Lewiston 13560
American Association of Pathologists, Bethesda 13561
American Association of Petroleum Geologists, Tulsa 13562
American Association of Philosophy Teachers, Oklahoma City 13563
American Association of Phonetic Sciences, Gainesville 13564
American Association of Physical Anthropologists, Buffalo 13565
American Association of Physicists in Medicine, New York 13566
American Association of Physics Teachers, College Park 13567
American Association of Physics Teachers (Ontario), Kingston 01859
American Association of Plastic Surgeons, La Jolla 13568
American Association of Poison Control Centers, Tucson 13569
American Association of Presidents of Independent Colleges and Universities, Malibu 13570
American Association of Professional Hypnotherapists, Boones Mill 13571
American Association of Professors of Yiddish, Flushing 13572
American Association of Pro Life Obstetricians and Gynecologists, Elm Grove 13573
American Association of Psychiatric Administrators, Atlanta 13574
American Association of Psychiatric Services for Children, Rochester 13575
American Association of Public Health Dentistry, Richmond 13576
American Association of Public Health Physicians, Madison 13577
American Association of Railway Surgeons, Daleville 13578
American Association of School Librarians, Chicago 13579
American Association of School Personnel Administrators, Sacramento 13580
American Association of Sex Educators, Counselors and Therapists, Chicago 13581
American Association of Small Ruminant Practitioners, Ithaca 13582
American Association of State Colleges and Universities, Washington 13583
American Association of Suicidology, Denver 13584
American Association of Teachers of Arabic, Provo 13585
American Association of Teachers of Esperanto, Santa Barbara 13586
American Association of Teachers of French, Champaign 13587
American Association of Teachers of German, Cherry Hill 13588
American Association of Teachers of Italian, Columbus 13589
American Association of Teachers of Spanish and Portuguese, Mississippi State 13590
American Association of Teachers of Turkish, Princeton 13591
American Association of Textile Chemists and Colorists, Research Triangle Park 13592
American Association of Tissue Banks, McLean 13593
American Association of University Professors, Washington 13594
American Association of University Women, Washington 13595
American Association of University Women Educational Foundation, Washington 13596
American Association of Variable Star Observers, Cambridge 13597
American Association of Veterinary Anatomists, Auburn 13598
American Association of Veterinary Laboratory Diagnosticians, Columbia 13599
American Association of Veterinary Parasitologists, Kalamazoo 13600
American Association of Veterinary State Boards, Jefferson City 13601
American Association of Wildlife Veterinarians, Fort Collins 13602
American Association of Women Dentists, Chicago 13603
American Association of Women in Community and Junior Colleges, Middletown 13604
American Association of Zoo Keepers, Topeka 13605
American Association of Zoological Parks and Aquariums, Wheeling 13606
American Association of Zoo Veterinarians, Philadelphia 13607
American Association on Mental Retardation, Washington 13608
American Astronautical Society, Springfield 13609
American Astronomical Society, Washington 13610
American Auditory Society, Dallas 13611
American Automatic Control Council, Evanston 13612
American Aviation Historical Society, Santa Ana 13613
American Bach Foundation, Washington 13614
American Bamboo Society, Sebastopol 13615
American Baptist Historical Society, Rochester 13616
American Bar Association, Chicago 13617
American Beethoven Society, San Jose 13618
American Behavioral Science Association, New York 13619
American Behcet's Association, Minneapolis 13620
American Biblical Encyclopedia Society, Monsey 13621
American Birding Association, Colorado Springs 13622
American Black Chiropractors Association, Saint Louis 13623
American Blood Commission, Chicago 13624
American Board for Certification in Orthotics and Prosthetics, Alexandria 13625
American Board of Abdominal Surgery, Melrose 13626
American Board of Allergy and Immunology, Philadelphia 13627
American Board of Anesthesiology, Hartford 13628
American Board of Bioanalysis, Saint Louis 13629
American Board of Colon and Rectal Surgery, Taylor 13630
American Board of Dental Public Health, Gainesville 13631
American Board of Dermatology, Detroit 13632
American Board of Endodontics, Chicago 13633
American Board of Funeral Service Education, Cumberland Center 13634
American Board of Health Physics, McLean 13635
American Board of Internal Medicine, Philadelphia 13636
American Board of Master Educators, Nashville 13637
American Board of Medical Toxicology, Salt Lake City 13638
American Board of Neurological Surgery, Houston 13639
American Board of Nuclear Medicine, Los Angeles 13640
American Board of Nutrition, Bethesda 13641
American Board of Obstetrics and Gynecology, Seattle 13642
American Board of Ophthalmology, Bala Cynwyd 13643
American Board of Opticianry, Fairfax 13644
American Board of Oral Pathology, Tampa 13645
American Board of Orthodontics, Saint Louis 13646
American Board of Orthopedic Surgery, Chapel Hill 13647
American Board of Otolaryngology, Houston 13648
American Board of Pathology, Tampa 13649
American Board of Pediatric Dentistry, Carmel 13650
American Board of Pediatrics, Chapel Hill 13651
American Board of Periodontology, Baltimore 13652
American Board of Physical Medicine and Rehabilitation, Rochester 13653
American Board of Plastic Surgery, Philadelphia 13654
American Board of Podiatric Orthopedics, Redlands 13655
American Board of Podiatric Surgery, San Francisco 13656
American Board of Preventive Medicine, Dayton 13657
American Board of Professional Psychology, Columbia 13658
American Board of Prosthodontics, Atlanta 13659
American Board of Psychiatry and Neurology, Deerfield 13660
American Board of Psychological Hypnosis, Chicago 13661
American Board of Radiology, Troy 13662
American Board of Surgery, Philadelphia 13663
American Board of Thoracic Surgery, Evanston 13664
American Board of Urology, Bingham Farms 13665
American Board of Vocational Experts, Topeka 13666
American Boccaccio Association, Saint Louis 13667
American Boxwood Society, Boyce 13668
American Brain Tumor Association, Chicago 13669
American Broncho-Esophagological Association, Chicago 13670
American Bryological and Lichenological Society, Omaha 13671
American Bureau for Medical Advancement in China, New York 13672
American Burn Association, Baltimore 13673
American-Canadian Genealogical Society, Manchester 13674
American Canal Society, York 13675
American Cancer Society, Atlanta 13676
American Carbon Society, Saint Marys 13677
American Cartographic Association, Bethesda 13678
American Catholic Esperanto Society, La Jolla 13679
American Catholic Historical Association, Washington 13680
American Catholic Philosophical Association, Washington 13681
American Celiac Society, West Orange 13682
American Center for Design, Chicago 13683
American Center for the Alexander Technique, New York 13684
American Ceramic Society, Westerville 13685
American Cetacean Society, San Pedro 13686
American Chamber of Commerce Researchers Association, Alexandria 13687
American Chemical Society, Washington 13688
American Chestnut Foundation, Morgantown 13689
American Chiropractic Association, Arlington 13690
American Civil War Round Table, United Kingdom, Richmond 11224
American Classical League, Oxford 13691
American Cleft Palate-Craniofacial Association, Pittsburgh 13692
American Clinical and Climatological Association, Charleston 13693
American Cocoa Research Institute, McLean 13694
American College for Continuing Education, Union 13695
American College Health Association, Baltimore 13696
American College of Allergy and Immunology, Palatine 13697
American College of Apothecaries, Alexandria 13698
American College of Cardiology, Bethesda 13699
American College of Chest Physicians, Northbrook 13700
American College of Chiropractic Orthopedists, El Centro 13701
American College of Clinical Pharmacology, Wayne 13702
American College of Cryosurgery, Schaumburg 13703
American College of Dentists, Gaithersburg 13704
American College of Emergency Physicians, Dallas 13705
American College of Foot Orthopedists, Redlands 13706
American College of Foot Surgeons, Park Ridge 13707
American College of Gastroenterology, Arlington 13708
American College of General Practitioners in Osteopathic Medicine and Surgery, Arlington Heights 13709
American College of Healthcare Executives, Chicago 13710
American College of Heraldry, Tuscaloosa 13711
American College of International Physicians, Washington 13712
American College of Laboratory Animal Medicine, Hershey 13713
American College of Medical Group Administrators, Englewood 13714
American College of Mohs' Micrographic Surgery and Cutaneous Oncology, Schaumburg 13715
American College of Musicians, Austin 13716
American College of Neuropsychiatrists, Farmington Hills 13717
American College of Neuropsychopharmacology, Nashville 13718
American College of Nuclear Physicians, Washington 13719
American College of Nurse-Midwives, Washington 13720
American College of Nutrition, Wilmington 13721
American College of Obstetricians and Gynecologists, Washington 13722
American College of Oral and Maxillofacial Surgeons, San Antonio 13723
American College of Orgonomy, Princeton 13724
American College of Osteopathic Internists, Washington 13725

American College of Osteopathic Obstetricians and Gynecologists, Pontiac 13726
American College of Osteopathic Pediatricians, Trenton 13727
American College of Osteopathic Surgeons, Alexandria 13728
American College of Physicians, Philadelphia 13729
American College of Podiatric Radiologists, Miami Beach 13730
American College of Podopediatrics, Cleveland 13731
American College of Preventive Medicine, Washington 13732
American College of Prosthodontists, San Antonio 13733
American College of Psychiatrists, Greenbelt 13734
American College of Psychoanalysts, Oakland 13735
American College of Rheumatology, Atlanta 13736
American College of Sports Medicine, Indianapolis 13737
American College of Surgeons, Chicago 13738
American College of Veterinary Internal Medicine, Blacksburg 13739
American College of Veterinary Pathologists, West Deptford 13740
American College of Veterinary Radiology, Glencoe 13741
American College Testing, Iowa City 13742
American Collegiate Retailing Association, Tallhassee 13743
American Committee for International Conservation, Washington 13744
American Committee for the Weizmann Institute of Science, New York 13745
American Committee to Advance the Study of Petroglyphs and Pictographs, Shepherdstown 13746
American Comparative Literature Association, Provo 13747
American Concrete Institute, Detroit 13748
American Conference for Irish Studies, Fort Wayne 13749
American Conference of Academic Deans, Washington 13750
American Conference of Governmental Industrial Hygienists, Cincinnati 13751
American Congress of Rehabilitation Medicine, Chicago 13752
American Congress on Surveying and Mapping, Bethesda 13753
American Conservation Association, New York 13754
American Consulting Engineers Council, Washington 13755
American Correctional Association, Laurel 13756
American Correctional Health Services Association, Dayton 13757
American Coucil for an Energy Efficient Economy, Washington 13758
American Council for Construction Education, Monroe 13759
American Council for the Arts, New York 13760
American Council of Applied Clinical Nutrition, Florissant 13761
American Council of Independent Laboratories, Washington 13762
American Council of Learned Societies, New York 13763
American Council of the International Institute of Welding, Miami 13764
American Council on Consumer Interests, Columbia 13765
American Council on Education, Washington 13766
American Council on Pharmaceutical Education, Chicago 13767
American Council on Rural Special Education, Bellingham 13768
American Council on Schools and Colleges, Tampa 13769
American Council on the Teaching of Foreign Languages, Yonkers 13770
American Counseling Association, Alexandria 13771
American Criminal Justice Association – Lambda Alpha Epsilon, Sacramento 13772
American Cryonics Society, Cupertino 13773
American Crystallographic Association, Buffalo 13774
American Dairy Science Association, Champaign 13775
American Dance Therapy Association, Columbia 13776
American Deafness and Rehabilitation Association, Little Rock 13777
American Dental Assistants Association, Chicago 13778
American Dental Association, Chicago 13779
American Dental Hygienists' Association, Chicago 13780
American Dental Society of Anesthesiology, Chicago 13781
American Dental Society of Europe, London 11225
American Dermatological Association, Iowa City 13782
American Dermatologic Society of Allergy and Immunology, Rochester 13783
American Design Drafting Association, Rockville 13784
American Diabetes Association, Alexandria 13785
American Dialect Society, Jacksonville 13786
American Dialect Society, London 01860
American Dietetic Association, Chicago 13787
American Diopter and Decibel Society, Pittsburgh 13788
American Economic Association, Nashville 13789
American Educational Research Association, Washington 13790
American Educational Studies Association, Cincinnati 13791
American Education Association, Center Moriches 13792
American Electroencephalographic Society, Bloomfield 13793
American Electrology Association, Trumbull 13794
American Electronics Association, Santa Clara 13795
American Electroplaters and Surface Finishers Society, Orlando 13796
American Endodontic Society, Fullerton 13797
American Engineering Model Society, Boston 13798
American Entomological Society, Philadelphia 13799
American Epidemiological Society, Atlanta 13800
American Epilepsy Society, Hartford 13801
American Equilibration Society, Morton Grove 13802
American Ethnological Society, Lancaster 13803
American European Dietetic Association, Montesson 04129
American Family Therapy Association, Washington 13804
American Farm Bureau Research Foundation, Park Ridge 13805
American Federation for Clinical Research, Thorofare 13806
American Federation of Arts, Berkeley 13807
American Federation of Astrologers, Tempe 13808
American Federation of Mineralogical Societies, Oklahoma City 13809
American Fern Society, Atlanta 13810
American Fertility Society, Birmingham 13811
American Film and Video Association, Niles 13812
American Film Institute, Washington 13813
American Finance Association, New York 13814
American Fine Arts Society, New York 13815
American Fisheries Society, Bethesda 13816
American Folklore Society, Harrisburg 13817
American Foreign Law Association, New York 13818
American Forensic Association, River Falls 13819
American Forestry Association, Washington 13820
American Forum for Global Education, New York 13821
American Foundation for Aids Research, Los Angeles 13822
American Foundation for Vision Awareness, Saint Louis 13823
American Foundrymen's Society, Des Plaines 13824
American Fracture Association, Bloomington 13825
American-French Genealogical Society, Pawtucket 13826
American Friends of Cambridge University, Arlington 13827
American Friends of Lafayette, Easton 13828
American Gastroenterological Association, Bethesda 13829
American Genetic Association, Buckeystown 13830
American Geographical Society, New York 13831
American Geological Institute, Alexandria 13832
American Geophysical Union, Washington 13833
American Geriatrics Society, New York 13834
American Group Practice Association, Alexandria 13835
American Group Psychotherapy Association, New York 13836
American Guild of Hypnotherapists, Omaha 13837
American Gynecological and Obstetrical Society, Salt Lake City 13838
American Harp Society, Hollywood 13839
American Healing Association, Glendale 13840
American Health Information Management Association, Chicago 13841
American Health Planning Association, Oklahoma City 13842
American Hearing Research Foundation, Chicago 13843
American Heart Association, Dallas 13844
American Herb Association, Nevada City 13845
American Historical Association, Washington 13846
American Historical Society of Germans from Russia, Lincoln 13847
American Hobbit Association, Cincinnati 13848
American Holistic Medical Association, Raleigh 13849
American Holistic Medical Foundation, Raleigh 13850
American Home Economics Association, Alexandria 13851
American Horticultural Society, Alexandria 13852
American Horticultural Therapy Association, Gaithersburg 13853
American Hospital Association, Chicago 13854
American Hungarian Educators' Association, Silver Spring 13855
American Hungarian Library and Historical Society, New York 13856
American Hypnotists' Association, San Francisco 13857
American Indian Council of Architects and Engineers, Tigard 13858
American Indian Culture Research Center, Marvin 13859
American Indian Lore Association, Logan 13860
American Indian Science and Engineering Society, Boulder 13861
American Industrial Health Council, Washington 13862
American Industrial Hygiene Association, Akron 13863
American Institute for Archaeological Research, Mount Vernon 13864
American Institute for Chartered Property Casuality Underwriters, Malvern 13865
American Institute for Conservation of Historic and Artistic Works, Washington 13866
American Institute for Contemporary German Studies, Washington 13867
American Institute for Economic Research, Great Barrington 13868
American Institute for Foreign Study, Greenwich 13869
American Institute for Patristic and Byzantine Studies, Kingston 13870
American Institute for Verdi Studies, New York 13871
American Institute of Aeronautics and Astronautics, Washington 13872
American Institute of Aeronautics and Astronautics – Technical Information Division, New York 13873
American Institute of Architects, Washington 13874
American Institute of Biological Sciences, Washington 13875
American Institute of Biomedical Climatology, Philadelphia 13876
American Institute of Building Design, Shelton 13877
American Institute of Chemical Engineers, New York 13878
American Institute of Chemists, Bethesda 13879
American Institute of Constructors, Saint Petersburg 13880
American Institute of Graphic Arts, New York 13881
American Institute of Homeopathy, Denver 13882
American Institute of Indian Studies, Chicago 13883
American Institute of Iranian Studies, Columbus 13884
American Institute of Iranian Studies, Toronto 01861
American Institute of Islamic Studies, Denver 13885
American Institute of Life Threatening Illness and Loss, New York 13886
American Institute of Management, Quincy 13887
American Institute of Mining, Metallurgical and Petroleum Engineers, New York 13888
American Institute of Musical Studies, Dallas 13889
American Institute of Nutrition, Bethesda 13890
American Institute of Oral Biology, South Laguna 13891
American Institute of Parliamentarians, Fort Wayne 13892
American Institute of Physics, New York 13893
American Institute of Plant Engineers, Cincinnati 13894
American Institute of Steel Construction, Chicago 13895
American Institute of the History of Pharmacy, Madison 13896
American Institute of Timber Construction, Vancouver 13897
American Institute of Ultrasound in Medicine, Rockville 13898
American Institutes for Research in the Behavioral Sciences, Washington 13899
American Intellectual Property Law Association, Arlington 13900
American Irish Bicentennial Committee, Annandale 13901
American Irish Historical Society, New York 13902
American Iron and Steel Institute, Washington 13903
American Italian Historical Association, Staten Island 13904
American Jewish Historical Society, Waltham 13905
American Joint Committee on Cancer, Chicago 13906
American Judges Association, Williamsburg 13907
American Judicature Society, Chicago 13908
American Kinesiotherapy Association, Dayton 13909
American Laryngological Association, Rochester 13910
American Laryngological, Rhinological and Otological Society, East Greenville 13911
American Law Institute, Philadelphia 13912
American Leather Chemists Association, Cincinnati 13913
American Legal Studies Association, Boston 13914
American Leprosy Missions, Greenville 13915
American Library Association, Chicago 13916
American Library Trustee Association, Chicago 13917
American Liszt Society, Rochester 13918
American Literary Translators Association, Richardson 13919
American Littoral Society, Highlands 13920
American Longevity Association, Torrance 13921
American Lunar Society, East Pittsburgh 13922
American Lung Association, New York 13923
American Malacological Union, North Myrtle Beach 13924
American Management Association, New York 13925
American Management Association / International, Bruxelles 01100
American Marketing Association, Chicago 13926
American Massage Therapy Association, Chicago 13927
American Mathematical Association of Two Year Colleges, Flint 13928
American Mathematical Society, Providence 13929
American Matthay Association, Pittsburgh 13930
American Meat Science Association, Chicago 13931
American Medical Associaction Auxiliary, Chicago 13932
American Medical Association, Chicago 13933
American Medical Association Education and Research Foundation, Chicago 13934
American Medical Electroencephalographic Association, Elm Grove 13935
American Medical Fly Fishing Association, Lock Haven 13936
American Medical Informatics Association, Bethesda 13937
American Medical Technologists, Park Ridge 13938
American Medical Women's Association, Alexandria 13939
American Medical Writers' Association, Bethesda 13940
American Mensa, Brooklyn 13941
American Merchant Marine Library Association, New York 13942
American Meteorological Society, Boston 13943
American Meteor Society, Geneseo 13944
American Microchemical Society, Princeton 13945
American Microscopical Society, Chestertown 13946
American Montessori Society, New York 13947
American Mosquito Control Association, Lake Charles 13948
American Musical Instrument Society, Vermillion 13949
American Music Conference, Carlsbad 13950
The American Musicological Society, Philadelphia 13951
American Name Society, New York 13952
American Naprapathic Association, Chicago 13953
American National Standards Institute, New York 13954
American Natural Hygiene Society, Tampa 13955
American Nature Study Society, Homer 13956
American Nepal Education Foundation, Tillamook 13957
American Neurological Association, Minneapolis 13958
American Nuclear Energy Council, Washington 13959
American Nuclear Society, La Grange Park 13960
American Numismatic Association, Colorado Springs 13961
American Numismatic Society, New York 13962
American Occupational Therapy Association, Rockville 13963
American Oil Chemists' Society, Champaign 13964
American Ophthalmological Society, Durham 13965
American Optometric Association, Saint Louis 13966
American Orchid Society, West Palm Beach 13967
American Oriental Society, Ann Arbor 13968
American Ornithologists' Union, Washington 13969
American Orthodontic Society, Dallas 13970
American Orthopaedic Association, Rosemont 13971
American Orthopaedic Foot and Ankle Society, Rosemont 13972
American Orthopaedic Society for Sports Medicine, Des Plaines 13973
American Orthopsychiatric Association, New York 13974
American Orthoptic Council, Madison 13975
American Orthotic and Prosthetic Association, Alexandria 13976
American Osler Society, Hamilton 01862
American Osteopathic Academy of Addictionology, Cherry Hill 13977
American Osteopathic Academy of Sclerotherapy, Wilmington 13978
American Osteopathic Academy of Sports Medicine, Middleton 13979
American Osteopathic Association, Chicago 13980
American Osteopathic College of Allergy and Immunology, Scottsdale 13981
American Osteopathic College of Anesthesiologists, Independence 13982
American Osteopathic College of Dermatology, Atlanta 13983
American Osteopathic College of Pathologists, Pembroke Pines 13984
American Osteopathic College of Proctology, Union 13985
American Osteopathic College of Radiology, Milan 13986
American Osteopathic College of Rehabilitation Medicine, Des Plaines 13987
American Otological Society, Maywood 13988
American Pancreatic Association, Sepulveda 13989
American Paper Institute, New York 13990
American Parkinson Disease Association, Staten Island 13991
American Peace Society, Washington 13992
American Peanut Research and Education Society, Stillwater 13993
American Pediatric Society, Elk Grove Village 13994
American Petroleum Institute, Washington 13995
American Pharmaceutical Association, Washington 13996
American Philological Association, Worcester 13997
American Philosophical Association, Newark 13998
American Philosophical Society, Philadelphia 13999
American Physical Society, New York 14000
American Physical Therapy Association, Alexandria 14001
American Physicians Association of Computer Medicine, Pittsford 14002
American Physicians Fellowship for Medicine in Israel, Brookline 14003
American Physicians Poetry Association, Southampton 14004
American Physiological Society, Bethesda 14005
American Phytopathological Society, Saint Paul 14006
American Planning Association, Washington 14007
American Podiatric Medical Association, Bethesda 14008
American Polar Society, Columbus 14009
American Political Science Association, Washington 14010

American Pomological Society, University Park 14011
American Poultry Historical Society, Morgantown 14012
American Press Institute, Reston 14013
American Printing History Association, New York 14014
American Production and Inventory Control Society, Falls Church 14015
American Prosthodontic Society, Chicago 14016
American Psychiatric Association, Washington 14017
American Psychoanalytic Association, New York 14018
American Psychological Association, Washington 14019
American Psychology-Law Society, Worcester 14020
American Psychopathological Association, Pittsburgh 14021
American Public Health Association, Washington 14022
American Quaternary Association, Amherst 14023
American Quilt Study Group, San Francisco 14024
American Radium Society, Philadelphia 14025
American Real Estate and Urban Economics Association, Bloomington 14026
American Registry of Radiologic Technologists, Mendota Heights 14027
American Rehabilitation Counseling Association, Alexandria 14028
American Research Institute in Turkey, Philadelphia 14029
American Revolution Round Table, Forest Hills 14030
American Rhinologic Society, Brooklyn 14031
American Risk and Insurance Association, Sacramento 14032
American Rock Art Research Association, San Miguel 14033
American Roentgen Ray Society, Reston 14034
American Schizophrenia Association, Boca Raton 14035
American School Health Association, Kent 14036
American Schools Association, Atlanta 14037
American Security Council, Boston 14038
American Security Council Foundation, Boston 14039
Americans for Energy Independence, Washington 14040
American Siam Society, Santa Monica 14041
American Social Health Association, Research Triangle Park 14042
American Society for Adolescent Psychiatry, Bethesda 14043
American Society for Advancement of Anesthesia in Dentistry, Coral Springs 14044
American Society for Aesthetic Plastic Surgery, Long Beach 14045
American Society for Aesthetics, Edmonton 01863
American Society for Artificial Internal Organs, Boca Raton 14046
American Society for Biochemistry and Molecular Biology, Bethesda 14047
American Society for Bone and Mineral Research, Washington 14048
American Society for Cell Biology, Bethesda 14049
American Society for Clinical Investigation, Thorofare 14050
American Society for Clinical Nutrition, Bethesda 14051
American Society for Colposcopy and Cervical Pathology, Washington 14052
American Society for Conservation Archaeology, Albuquerque 14053
American Society for Cybernetics, Fairfax 14054
American Society for Cytotechnology, Raleigh 14055
American Society for Dental Aesthetics, New York 14056
American Society for Dermatologic Surgery, Schaumburg 14057
American Society for Eighteenth-Century Studies, Logan 14058
American Society for Engineering Management, Annapolis 14059
American Society for Ethnohistory, Chicago 14060
American Society for Gastrointestinal Endoscopy, Manchester 14061
American Society for Geriatric Dentistry, Chicago 14062
American Society for Head and Neck Surgery, Baltimore 14063
American Society for Healthcare Education and Training of the American Hospital Association, Chicago 14064
American Society for Healthcare Human Resources Administration, Chicago 14065
American Society for Horticultural Science, Alexandria 14066
American Society for Hospital Engineering, Chicago 14067
American Society for Information Science, Silver Spring 14068
American Society for Information Science (Western Canada Chapter), Edmonton 01864
American Society for Laser Medicine and Surgery, Wausau 14069
American Society for Legal History, University 14070
American Society for Mass Spectrometry, Santa Fe 14071
American Society for Medical Technology, Bethesda 14072
American Society for Metals, Winnipeg 01865
American Society for Microbiology, Washington 14073
American Society for Neurochemistry, Palo Alto 14074
American Society for Nondestructive Testing, Columbus 14075
American Society for Parenteral and Enteral Nutrition, Silver Spring 14076
American Society for Pharmacology and Experimental Therapeutics, Bethesda 14077
American Society for Pharmacy Law, Vienna 14078
American Society for Photobiology, Augusta 14079
American Society for Photogrammetry and Remote Sensing, Bethesda 14080
American Society for Political and Legal Philosophy, Norton 14081
American Society for Psychical Research, New York 14082
American Society for Psychoprophylaxis in Obstetrics, Washington 14083
American Society for Public Administration, Washington 14084
American Society for Quality Control, Milwaukee 14085
American Society for Stereotactic and Functional Neurosurgery, Houston 14086
American Society for Surgery of the Hand, Aurora 14087
American Society for Technion-Israel Institute of Technology, New York 14088
American Society for Testing and Materials, Philadelphia 14089
American Society for Theatre Research, Kingston 14090
American Society for Therapeutic Radiology and Oncology, Philadelphia 14091
American Society for the Study of Orthodontics, Oakland Gardens 14092
American Society for Training and Development, Alexandria 14093
American Society for Value Inquiry, Raleigh 14094
American Society of Abdominal Surgery, Melrose 14095
American Society of Addiction Medicine, Washington 14096
American Society of Agricultural Engineers, Saint Joseph 14097
American Society of Agronomy, Madison 14098
American Society of Anesthesiologists, Park Ridge 14099
American Society of Animal Science, Campaign 14100
American Society of Bakery Engineers, Chicago 14101
American Society of Body and Design Engineers, Redford 14102
American Society of Brewing Chemists, Saint Paul 14103
American Society of Cataract and Refractive Surgery, Fairfax 14104
American Society of Certified Engineering Technicians, El Paso 14105
American Society of Church History, Indialantic 14106
American Society of Civil Engineers, Washington 14107
American Society of Clinical Hypnosis, Des Plaines 14108
American Society of Clinical Oncology, Chicago 14109
American Society of Clinical Pathologists, Chicago 14110
American Society of Colon and Rectal Surgeons, Palatine 14111
American Society of Consultant Pharmacists, Alexandria 14112
American Society of Contemporary Medicine and Surgery, Chicago 14113
American Society of Contemporary Ophthalmology, Chicago 14114
American Society of Criminology, Columbus 14115
American Society of Cytology, Philadelphia 14116
American Society of Danish Engineers, Larchmont 14117
American Society of Dentistry for Children, Chicago 14118
American Society of Dermatopathology, Baltimore 14119
American Society of Echocardiography, Raleigh 14120
American Society of Electroneurodiagnostic Technologists, Carroll 14121
American Society of Extra-Corporeal Technology, Reston 14122
American Society of Forensic Odontology, Chicago 14123
American Society of Gas Engineers, Independence 14124
American Society of Genealogists, Columbus 14125
American Society of Geolinguistics, Fairfield 14126
American Society of Golf Course Architects, Chicago 14127
American Society of Group Psychotherapy and Psychodrama, McLean 14128
American Society of Hand Therapists, Garner 14129
American Society of Heating, Refrigerating and Air-Conditioning Engineers, Atlanta 14130
American Society of Hematology, Washington 14131
American Society of Hospital Pharmacists, Bethesda 14132
American Society of Human Genetics, Bethesda 14133
American Society of Ichthyologists and Herpetologists, Carbondale 14134
American Society of Indexers, Port Aransas 14135
American Society of Internal Medicine, Washington 14136
American Society of International Law, Washington 14137
American Society of Irrigation Consultants, Brentwood 14138
American Society of Landscape Architects, Washington 14139
American Society of Law and Medicine, Boston 14140
American Society of Limnology and Oceanography, Gloucester Point 14141
American Society of Mammalogists, Provo 14142
American Society of Master Dental Technologists, Oakland Gardens 14143
American Society of Maxillofacial Surgeons, Arlington Heights 14144
American Society of Mechanical Engineers, New York 14145
American Society of Naturalists, Flushing 14146
American Society of Naval Engineers, Alexandria 14147
American Society of Nephrology, Washington 14148
American Society of Neuroradiology, Oak Brook 14149
American Society of Notaries, Washington 14150
American Society of Papyrologists, Detroit 14151
American Society of Parasitologists, El Paso 14152
American Society of Pharmacognosy, Monroe 14153
American Society of Physoanalytic Physicians, Rockville 14154
American Society of Plant Physiologists, Rockville 14155
American Society of Plant Taxonomists, Athens 14156
American Society of Plastic and Reconstructive Surgeons, Arlington Heights 14157
American Society of Plumbing Engineers, Westlake 14158
American Society of Podiatric Dermatology, Bel Air 14159
American Society of Primatologists, Chicago 14160
American Society of Psychopathology of Expression, Brookline 14161
American Society of Questioned Document Examiners, Rockville 14162
American Society of Radiologic Technologists, Albuquerque 14163
American Society of Regional Anesthesia, Richmond 14164
American Society of Safety Engineers, Des Plaines 14165
American Society of Sanitary Engineering, Bay Village 14166
American Society of Sephardic Studies, New York 14167
American Society of Sugar Beet Technologists, Denver 14168
American Society of Swedish Engineers, New York 14169
American Society of Trial Consultants, Towson 14170
American Society of Tropical Medicine and Hygiene, McLean 14171
American Society of Veterinary Ophthalmology, Stillwater 14172
American Society of Veterinary Physiologists and Pharmacologists, Mississippi State 14173
American Society of Zoologists, Chicago 14174
American Sociological Association, Washington 14175
American Software Users Group, Chicago 14176
American Solar Energy Society, Boulder 14177
American Speech-Language-Hearing Association, Rockville 14178
American Spelean History Association, Altoona 14179
American Spinal Injury Association, Chicago 14180
American Sports Education Institute, North Palm Beach 14181
American Statistical Association, Alexandria 14182
American String Teachers Association, Dallas 14183
American Student Dental Association, Chicago 14184
American Studies Association, College Park 14185
American Surgical Association, Chapel Hill 14186
American Swedish Institute, Minneapolis 14187
American Technical Education Association, Wahpeton 14188
American Teilhard Association, Syosset 14189
American Theatre Critics Association, Nashville 14190
American Theatre Organ Society, Sacramento 14191
American Theological Library Association, Evanston 14192
American Theological Society – Midwest Division, Chicago 14193
American Thoracic Society, New York 14194
American Thyroid Association, Washington 14195
American Tinnitus Association, Portland 14196
American Tolkien Society, Highland 14197
American Trauma Society, Upper Marlboro 14198
American Type Culture Collection, Rockville 14199
American Underground-Space Association, Minneapolis 14200
American Urological Association, Baltimore 14201
American Vacuum Society, New York 14202
American Venereal Disease Association, Baltimore 14203
American Veterinary Medical Association, Schaumburg 14204
American Veterinary Society of Animal Behavior, Ithaca 14205
American Vocational Association, Alexandria 14206
American Vocational Education Personnel Development Association, Stillwater 14207
American Vocational Education Research Association, Blacksburg 14208
American Water Resources Association, Bethesda 14209
American Welding Society, Miami 14210
American Wind Energy Association, Washington 14211
American Wood-Preservers' Association, Woodstock 14212
American Wood Preservers Institute, Vienna 14213
Amerind Foundation, Dragoon 14214
AMERSA, Providence 14459
AMES, Liverpool 11365
AMEWPR, Kuala Lumpur 08638
AmFAR, Los Angeles 13822
AMFFA, Lock Haven 13936
AMFI, Redmond 14562
AMHA, Northbrook 14462
AMHPS, Washington 14467
AMIA, Bethesda 13937
AMIC, Singapore 10035
Amicale des Directeurs de Bibliothèques Universitaires, Paris 04130
Amicale Internationale de Phytosociologie, Bailleul 04131
Amici Thomae Mori, Angers 04132
Amics dels Museus de Catalunya, Barcelona 10246
Amigos de la Ciudad, La Paz 01531
AMIN, Kuwait 08545
AMINTAPHIL, Gainesville 15390
AMIS, Vermillion 13949
Amis de Guy de Maupassant, Paris 04133
Amis de Rimbaud, Paris 04134
A.M.I.S.I., Milano 07356
AMJ, Barnet 11366
AMM, Winnipeg 01935
AMMF, Ottawa 01913
AMMLA, New York 13942
A.M.O.R., Roma 07210
AMP, Cambrai 04326
AMPAS, Beverly Hills 13298
AMPT, Ickenham 11274
AMQUA, Amherst 14023
AMR, Paris 04152
AMR, Wien 00566
AMRA, Dakar 10022
AMRA, San Diego 14460
AMRC, Research Triangle Park 14558
AMREF, Nairobi 08470
AMRO, Hopton 11353
AMRSHF, Troy 13330
AMS, Boston 13943
AMS, Cairo 03873
AMS, Chestertown 13946
AMS, Coral Gables 13297
AMS, Geneseo 13944
A.M.S., Hobart 00320
AMS, London 11228, 11413
AMS, New York 13947
AMS, Philadelphia 13951
AMS, Princeton 13945
AMS, Providence 13929
AMS, Washington 13329
AMSA, Chicago 13931
AMSA, Lusaka 17151
A.M.S.E., Gent 01148
AMSE, Plainfield 14468
AmSECT, Reston 14122
AMSME, Alexandria 03881
AMSPDC, Salt Lake City 14461
AMSS, Herndon 14469
AMSUS, Bethesda 14465
AMT, Park Ridge 13938
AMTA, Atlanta 14227
AMTA, Chicago 13927
AMTIESA, Arusha 11047
AMU, Franklin 14466
AMU, Ibadan 09217
AMU, North Myrtle Beach 13924
AMU, Paris 04153
A.M.V.M.I., Maisons Alfort 04304
AMWA, Alexandria 13939
AMWA, Bethesda 13940
Amyotrophic Lateral Sclerosis Society of British Columbia, Vancouver 01866
Amyotrophic Lateral Sclerosis Society of Canada, Toronto 01867
ANA, Bologna 07214
ANA, Chicago 13953
ANA, Colorado Springs 13961
ANA, Minneapolis 13958
ANABAD, Madrid 10257
ANACHEM, Detroit 14391
ANAD – National Association of Anorexia Nervosa and Associated Disorders, Highland Park 14215
Anaesthetic Research Society, Liverpool 11226
ANAI, Roma 07359
ANAI, Tangiers 08799
ANAIC, Perth 00223
ANAS, Paris 04306
Anatomical Society of Great Britain and Ireland, Saint Andrews 11227
Anatomische Gesellschaft, Lübeck 05074
ANBS, Singapore 10036
ANC, Guilden Morden 11372
ANCEFN, Buenos Aires 00045
Ancient Astronaut Society, Highland Park 14216
The Ancient Iran Cultural Society, Teheran 06896
Ancient Monuments Society, London 11228
An Claidheamh Soluis – The Irish Arts Center, New York 14217
ANCOLD, Brisbane 00323
An Comunn Gaidhealach, Inverness 11229
ANDEA, Accra 06473
Andean Commission of Jurists, Lima 09443
Andean Institute for Population Studies and Development, Lima 09444
Andean Institute of Social Studies, Lima 09445
Andean Institute of Popular Arts, Quito 03842
Andean Technological Information System, Lima 09446
Andhra Historical Research Society, Rajahmundry 06715
ANDIS, Roma 07367
ANE, Barcelona 10278
ANEA, Lima 09453
ANEC, Washington 13959
ANEF, Tillamook 13957
ANEJ, Paris 04310
ANENA, Grenoble 04316
ANET, Roma 07370
ANFFAS, Roma 07371
Anglesey Antiquarian Society and Field Club, Menai Bridge 11230
Anglo-Continental Dental Society, London 11231
Anglo-European College of Chiropractic, Bournemouth 11232
Anglo-Norman Text Society, London 11233
ANGOC, Manila 09500
ANGOS, Birmingham 13358
ANHS, Tampa 13955
ANHSO, Oxford 11255
ANHSO, Wareham 11370
ANIAI, Roma 07374
ANID, Dakar 10012
ANID, Napoli 07376
ANIEST, Roma 07377
A.N.I.M., Roma 07375
A.N.I.M.A., Milano 07373
Animal Behavior Society, Denver 14218
Animal Breeding Research Organization, Edinburgh 11234
Animal Diseases Research Association, Edinburgh 11235
Animal Health Trust, Newmarket 11236
Animal Medical Center, New York 14219
Animal Nutrition Research Council, Chantilly 14220

Animal Protection Institute of America, Sacramento 14221
ANIPLA, Milano 07378
ANL, Montevideo 13240
Annual Reviews, Palo Alto 14222
Annular Bearing Engineers Committee, Washington 14223
Anonymous Arts Recovery Society, New York 14224
Anonymous Families History Project, Newark 14225
ANP, Berlin 05158
ANPGP, Milano 07364
ANPI, Ottignies 01149
ANPO, Kendal 11371
ANPUR, Roma 07383
ANRD, Chantilly 14220
A.N.R.T, Paris 04305
ANS, Chevy Chase 14549
ANS, La Grange Park 13960
ANS, New York 13952, 13962
ANS, Sydney 00324
ANSI, New York 13954
ANSI, Roma 07381
ANSP, Philadelphia 13299
ANSS, Homer 13956
ANSTI, Nairobi 08471
Antarctica and Southern Ocean Coalition, Washington 14226
Antenna Measurement Techniques Association, Atlanta 14227
Anthology Film Archives, New York 14228
Anthropological Society of Bombay, Bombay 06716
Anthropological Society of New South Wales, Sydney 00214
Anthropological Society of Nippon, Tokyo 08315
Anthropological Society of Queensland, Saint Lucia 00215
Anthropological Society of South Australia, Adelaide 00216
Anthropological Society of Western Australia, Nedlands 00217
Anthropological Survey of India, Calcutta 06717
Anthropologische Gesellschaft in Wien, Wien 00544
Anthropology Film Center Foundation, Santa Fe 14229
Anthropos Institut, Sankt Augustin 05075
Anthroposophical Society in Canada, Toronto 01868
Anthroposophical Society in Great Britain, London 11237
Anthroposophische Gesellschaft in Deutschland, Stuttgart 05076
Anthroposophische Gesellschaft in Österreich, Wien 00545
Anthroposophische Gesellschaft in Wien, Wien 00546
Antiquarian and Numismatic Society of Montréal, Montréal 02918
Antiquarian Horological Society, Wadhurst 11238
Antiquarische Gesellschaft in Zürich, Zürich 10616
Anton-Bruckner-Institut Linz, Linz 00547
Antonín Dvořák Society, Praha 03554
ANTS, London 11233
Anuarios de Filosofía y Letras, México 08691
ANUG, Chicago 14544
A.N.V.W., Uithoorn 08821
Anwenderverband Deutscher Informationsverarbeiter e.V., Kiel 05077
ANZAAS, Sydney 00258
AOA, Chicago 13980
AOA, London 11374
AOA, New York 13974
AOA, Rosemont 13971
AOA, Saint Louis 13966
AOAC, Arlington 14230
AOAC International, Arlington 14230
AOAS, Wilmington 13978
AOASM, Middleton 13979
AOC, Alexandria 14473
AOC, Madison 13975
AOCA, Independence 13982
AOCAI, Scottsdale 13981
AOCD, Atlanta 13983
AOCP, Pembroke Pines 13984
AOCPr, Union 13985
AOCR, Milan 13986
AOCRM, Des Plaines 13987
AOCS, Champaign 13964
AOD, Frederick 13301
AOD, Menomonie 13300
AOE, Indiana 14478
AOE, Tahlequah 14474
AÖG, Wien 00560
AOFAS, Rosemont 13972
AOFOG, Singapore 10034
AOI, Firenze 07385
AOIP, New York 14231
AOJS, New York 14475
AOJT, Brooklyn 14476
AOOI, Roma 07384
AOPA, Alexandria 13976
AORC, Portland 14471
AORTIC, Harare 17159
AOS, Ann Arbor 13968
AOS, Bucuresti 09743
AOS, Dallas 13970
AOS, Durham 13965
AOS, Hamilton 01862
AOS, Maywood 13988
AOS, West Palm Beach 13967
AOSA, Lincoln 14472
AOSCE, Hong Kong 06587
Aosdana, Dublin 06922
AOSED, Sacramento 14477
AOSSM, Des Plaines 13973
A.O.T., Cambridge 11400
AOTA, Nagasaki 08095
AOTA, Rockville 13963
AOTA, Yaoundé 01812
AOU, Washington 13969
APA, Buenos Aires 00094
A.P.A., Lima 09455
A.P.A., London 11382
APA, McLean 13373
APA, Newark 13998
APA, Sepulveda 13989
APA, Washington 14017, 14019
APA, Worcester 13997
APAA, Tokyo 08105
APAC, Dakar 10020
APACM, Pittsford 14002
APACVS, Lynchburg 14483
APADI, Jakarta 06885
APAL, Caracas 16971
APALMS, Singapore 10038
APANF, Cotonou 01516
APAO, Tokyo 08111
APAP, Alexandria 14482
APASWE, Bundoora 00220
APB, São Paulo 01605
APBB, München 05089
APBD, Bruxelles 01156
APBP, Minneapolis 14485
APC, Bethesda 14480
APC, Broxborn 11378
APC, Lisburn 11379
APC, San Francisco 14322
APCAS, Bangkok 11072
APCC, Sydney 00225
APCChE, Barton 00224
APCO, Tokyo 08104
APCP, México 08706
A.P.C.T., Saint Leonards-on-Sea 11375
APCTT, Bangalore 06720
APCU, Jakarta 06877
APCU, Louisville 14484
APDA, Staten Island 13991
APDA, Tokyo 08106
APDF, Singapore 10039
A.P.D.P., Lisboa 09650
APE, Lisboa 09642
APEA, Huntington 11275
APEA, Sydney 00327
APEC, Dayton 14232
APED, Bangkok 11074
APEE, Lisboa 09652
APEM, Sacramento 14486
APESS, Luxembourg 08576
APF, Beijing 03116
APF, Brookline 14003
APF, Porto 09646
APFA, Khartoum 10425
APFAN, Canberra 00228
APFC, Bangkok 11085
APFCB, Adelaide 00226
APFHRM, Hong Kong 06591
APFOCC, Tokyo 08096
APFTC, Bangkok 11083
APG, Washington 14487
APGO, Washington 14489
APHA, New York 14014
APhA, Washington 13996
APHA, Washington 14022
A.P.H.G., Evry 04197
APHS, Morgantown 14012
API, Kuwait 08546
API, Madrid 10252
API, New York 13990
API, Reston 14013
API, Sacramento 14221
API, San Francisco 14542
API, Washington 13995, 14007
APIC, Colombo 10406
APIC, Mundelein 14321
APICS, Falls Church 14015
APIMO, Roma 07398
Apimondia, Roma 07248
APINESS, Bangkok 11086
APINMAP, Bangkok 11076
APJT, Edinburgh 11381
APL, São Paulo 01579
APLA, Halifax 01962
APLAR, Semarang 06883
APLF, Strasbourg 04192
APLIC, New York 14319
APLIC, Ottawa 01938
AP-LS, Worcester 14020
APLV, Paris 04195
APM, Chicago 13304
APM, Lisboa 09647
APM, New York 14323
APM, Washington 14490
APMA, Bethesda 14008
APMEP, Paris 04196
APMHC, Charlotte 14488
APMS, Washington 14234
APO, Tokyo 08107
APOAC, Taipei 03241
APOCB, Okayama 08101
APÖ, Wien 00556
APORS, Tokyo 08112
Apothecaries' Hall, Dublin 06923
APPA, Alexandria 14443
APPA, Burbank 14481
APPA, Paris 04327
APPA, Pittsburgh 14021
APPA, Southampton 14004
APPEAL, Bangkok 11087
APPEN, Penang 08637
APPITA, Melbourne 00508
Applied Scientific Research Corporation of Thailand, Bangkok 11066
Applied Technology Council, Redwood City 14233
APPM, Washington 13302
APPPC, Bangkok 11073
Appropriate Health Resources and Technologies Action Group, London 11239
Appropriate Technology Study Centre for Latin America, Valparaiso 03013
APPROTECH ASIA, Manila 09493
APPS, Adelaide 00246
A.P.P.S., Bruxelles 01150
APPSA, Singapore 10040
APPSGAN, East Perth 00227
APPTEA, Manila 09507
APQCO, Diliman 09508
A.P.R., Bucuresti 09744
APRA, Christchurch 09117
APRES, Stillwater 13993
APRI, Lagos 09218
A.P.R.S., Edinburgh 11288
APS, Bethesda 14005
APS, Chicago 14016
APS, Columbus 14009
APS, Elk Grove Village 13994
APS, Honolulu 14479
APS, New York 13303, 14000
APS, Philadelphia 13999
APS, Saint Paul 14006
APS, University Park 14011
APS, Washington 13992
APSA, Canberra 00247
APSA, Edinburgh 11276
APSA, Saint Lucia 00245
APSA, Washington 14010
APsaA, New York 14018
APSC, Seoul 08522
APSGB, Northampton 11376
A.P.S.M., Monaco 08794
APSSEAR, Manila 09510
APT, Saint Louis 14320
APT, Southsea 11377
APTA, Alexandria 14001
APTI, Baghdad 06914
APTIRC, Singapore 10042
APUG, South Windsor 14559
APVA, Richmond 14353
APWLD, Kuala Lumpur 08636
APWSS, Laguna 09501
AQP, Cincinnati 14324
AQSG, San Francisco 14024
Aquatic Plant Management Society, Washington 14234
Aquatic Research Institute, Hayward 14235
Aquatic Sciences and Fisheries Information System, Roma 07249
Aqua Viva, Schweizerische Aktionsgemeinschaft zur Erhaltung der Flüsse und Seen, Bern 10617
AR, Palo Alto 14222
ARA, Akron 13305
ARA, Bedford 11217
ARA, Bethesda 13458
ARA, Camarines Sur 09502
ARAB, Bruxelles 01157
The Arab Academy of Damascus, Damascus 11036
Arab Academy of Music, Baghdad 06905
Arab Administrative Development Organization, Amman 08449
Arab Aerospace Educational Organization, Cairo 03867
Arab Agronomists Union, Damascus 11037
Arab Atomic Energy Agency, Tunis 11127
Arab Biosciences Network, Jubaiha 08450
Arab Bureau for Prevention of Crime, Baghdad 06906
Arab Bureau of Education for the Gulf States, Riyadh 09997
Arab Center for Energy Studies, Cairo 03868
Arab Center for Medical Literature, Kuwait 08544
Arab Center for the Studies of Arid Zones and Dry Lands, Damascus 11038
Arab Centre for Agricultural Documentation and Information, Khartoum 10424
Arab Centre for Information Studies on Population, Development and Construction, Damascus 11039
Arab Commission for International Law, Cairo 03869
Arab Committee for Ottoman Studies, Zaghouan 11128
Arab Council for Childhood and Development, Cairo 03870
Arab Council for Medical Specialization, Damascus 11040
Arab Dental Federation, Baghdad 06907
Arab Federation for Technical Education, Baghdad 06908
Arab Federation for the Organs of the Deaf, Damascus 11041
Arab Federation of Construction, Wood and Building Materials, Tripoli 08567
Arab Federation of Sports Medicine, Tunis 11129
Arab Gulf States Information Documentation Centre, Baghdad 06909
Arab Higher Committee for Pharmacological Affairs, Cairo 03871
Arab Historians Association, Baghdad 06910
Arab Industrial Development and Mining Organization, Baghdad 06911
Arab Information Network for Terminology, Tunis 11130
Arab Institute for Forestry and Ranges, Lattakia 11042
Arab Institute for Training and Research in Statistics, Baghdad 06912
Arab Institute of Navigation, Alexandria 03872
Arab League Educational, Cultural and Scientific Organization, Tunis 11131
Arab Literacy and Adult Education Organization, Baghdad 06913
Arab Management Society, Cairo 03873
Arab Maritime Transport Academy, Alexandria 03874
Arab Medical Information Network, Kuwait 08545
Arab Medical Union, Tunis 11132
Arab Music Rostrum, Paris 04135
ARABN, Jubalha 08450
Arab Organization for Standardization and Metrology, Cairo 03875
Arab Petroleum Training Institute, Baghdad 06914
Arab Planning Institute, Kuwait 08546
Arab Regional Branch of the International Council on Archives, Tunis 11133
Arab Regional Office of the United Schools International, Gudaibiya 01053
Arab Research Centre, London 11240
ARABSAT, Riyadh 09998
Arab Satellite Communications Organization, Riyadh 09998
Arab Scientific Advisory Committee for Blood Transfer, Cairo 03876
Arab Scientific and Technical Information Network, Algiers 00008
Arab Security Studies and Training Center, Riyadh 09999
Arab States Regional Broadcasting Training Centre, Damascus 11043
Arab States Regional Centre for Functional Literacy in Rural Areas, Menoufia 03877
ARABTERM, Tunis 11130
Arab Union for Cement and Building Materials, Damascus 11044
Arab Union of Veterinary Surgeons, Baghdad 06915
Arab Urban Development Institute, Riyadh 10000
Arab World Institute, Paris 04136
ARADO, Amman 08449
ARAHE, Seoul 08523
ARAN, London 11289
ARAPCS, Reston 14327
ARARA, San Miguel 14033
ARB, Pôrto Alegre 01609
Arbeitsausschuß Wälzlager im DIN Deutsches Institut für Normung e.V., Köln 05078
Arbeitsgemeinschaft Allensbach e.V., Allensbach 05079
Arbeitsgemeinschaft Allergiekrankes Kind, Herborn 05080
Arbeitsgemeinschaft audiovisueller Archive Österreichs, Wien 00548
Arbeitsgemeinschaft beruflicher und ehrenamtlicher Naturschutz, Bonn 05081
Arbeitsgemeinschaft Bremer Schule e.V., Bremen 05082
Arbeitsgemeinschaft der Archive und Bibliotheken in der evangelischen Kirche, Nürnberg 05083
Arbeitsgemeinschaft der Deutschen Werkkunstschulen, Bremen 05084
Arbeitsgemeinschaft der Direktoren der Institute für Leibesübungen an Universitäten und Hochschulen der Bundesrepublik Deutschland, Karlsruhe 05085
Arbeitsgemeinschaft der Grossforschungs-Einrichtungen, Bonn 05086
Arbeitsgemeinschaft der kirchlichen Büchereiverbände, Bonn 05087
Arbeitsgemeinschaft der Kunstbibliotheken, Roma 07250
Arbeitsgemeinschaft der Musikerzieher Österreichs, Wien 00549
Arbeitsgemeinschaft der Ordenshochschulen, Vallendar 05088
Arbeitsgemeinschaft der Parlaments- und Behördenbibliotheken, München 05089
Arbeitsgemeinschaft der Regionalbibliotheken, Hannover 05090
Arbeitsgemeinschaft der Seminarlehrer im Saarländischen Philologenverband, Ottweiler 05091
Arbeitsgemeinschaft der Sozialdemokraten im Gesundheitswesen, Bonn 05092
Arbeitsgemeinschaft der Spezialbibliotheken e.V., Leverkusen 05093
Arbeitsgemeinschaft der Verbände Gemeinnütziger Privatschulen in der Bundesrepublik, Köln 05094
Arbeitsgemeinschaft der Wirtschaft für Produktdesign und Produktplanung e.V., Essen 05159
Arbeitsgemeinschaft der wissenschaftlichen Institute des Handwerks der EG-Länder, München 05095
Arbeitsgemeinschaft Deutsche Lateinamerika-Forschung, Eichstätt 05096
Arbeitsgemeinschaft deutscher wirtschaftswissenschaftlicher Forschungsinstitute e.V., München 05097
Arbeitsgemeinschaft Fernwärme e.V., Frankfurt 05098
Arbeitsgemeinschaft Freier Schulen, Vereinigungen und Verbände gemeinnütziger Schulen in freier Trägerschaft, Berlin 05099
Arbeitsgemeinschaft für Abfallwirtschaft, Köln 05100
Arbeitsgemeinschaft für angewandte Sozialforschung, München 05101
Arbeitsgemeinschaft für betriebliche Altersversorgung e.V., Heidelberg 05102
Arbeitsgemeinschaft für Deutschdidaktik, Klagenfurt 00550
Arbeitsgemeinschaft für Elektronenoptik e.V., Karlsruhe 05103
Arbeitsgemeinschaft für Historische Sozialkunde, Wien 00551
Arbeitsgemeinschaft für interdisziplinäre angewandte Sozialforschung, Wien 00552
Arbeitsgemeinschaft für Jugendhilfe, Bonn 05104
Arbeitsgemeinschaft für juristisches Bibliotheks- und Dokumentationswesen, Augsburg 05105
Arbeitsgemeinschaft für Kieferchirurgie innerhalb der Deutschen Gesellschaft für Zahn-, Mund- und Kieferheilkunde, Kiel 05106
Arbeitsgemeinschaft für klinische Ernährung, Wien 00553
Arbeitsgemeinschaft für Krebsbekämpfung des Landes Niedersachsen e.V., Hannover 05107
Arbeitsgemeinschaft für Landschaftsentwicklung, Bonn 05108
Arbeitsgemeinschaft für medizinisches Bibliothekswesen, Mannheim 05109
Arbeitsgemeinschaft für Neurobiologie, Wien 00554
Arbeitsgemeinschaft für orale Implantologie, Zürich 10618
Arbeitsgemeinschaft für Osteuropaforschung, Tübingen 05110
Arbeitsgemeinschaft für Präventivpsychologie, Wien 00555
Arbeitsgemeinschaft für Psychotechnik in Österreich, Wien 00556
Arbeitsgemeinschaft für sparsamen und umweltfreundlichen Energieverbrauch e.V., Hamburg 05111
Arbeitsgemeinschaft für theoretische und klinische Leistungsmedizin der Hochschullehrer Österreichs, Graz 00557
Arbeitsgemeinschaft für Umweltfragen e.V., Bonn 05112
Arbeitsgemeinschaft für Verhaltensmodifikation, Salzburg 00558
Arbeitsgemeinschaft für wirtschaftliche Verwaltung e.V., Eschborn 05113
Arbeitsgemeinschaft für Wissenschaft und Politik, Innsbruck 00559
Arbeitsgemeinschaft für zeitgemässes Bauen e.V., Kiel 05114
Arbeitsgemeinschaft Getreideforschung e.V., Detmold 05115
Arbeitsgemeinschaft Grünland und Futterbau in der Gesellschaft für Pflanzenbauwissenschaften, Giessen 05116
Arbeitsgemeinschaft Hauswirtschaft e.V., Bonn 05117
Arbeitsgemeinschaft Historischer Kommissionen und Landesgeschichtlicher Institute e.V., Marburg 05118
Arbeitsgemeinschaft Industriebau, Köln 05119
Arbeitsgemeinschaft industrieller Forschungsvereinigungen „Otto von Guericke" e.V., Köln 05120
Arbeitsgemeinschaft Kartoffelforschung e.V., Detmold 05121
Arbeitsgemeinschaft katholisch-theologischer Bibliotheken, Paderborn 05122
Arbeitsgemeinschaft Korrosion e.V., Frankfurt 05123
Arbeitsgemeinschaft Media-Analyse e.V., Frankfurt 05124
Arbeitsgemeinschaft Österreichischer Entomologen, Wien 00560
Arbeitsgemeinschaft Personalwesen im Österreichischen Produktivitäts- und Wirtschaftlichkeits-Zentrum, Wien 00561
Arbeitsgemeinschaft Personenzentrierte Psychotherapie und Gesprächsführung, Wien 00562
Arbeitsgemeinschaft Sozialwissenschaftlicher Institute e.V., Bonn 05125

Arbeitsgemeinschaft Spina bifida und Hydrocephalus e.V., Dortmund *05126*
Arbeitsgemeinschaft Verstärkte Kunststoffe e. V., Frankfurt *05127*
Arbeitsgemeinschaft Versuchsreaktor, Düsseldorf *05128*
Arbeitsgemeinschaft Währungsethik, Köln *05129*
Arbeitsgemeinschaft wildbiologischer und jagdkundlicher Forschungsstätten, Bonn *05130*
Arbeitsgemeinschaft wissenschaftliche Literatur e.V., Frankfurt *05131*
Arbeitsgemeinschaft zur Erforschung der Ärztlichen Allgemeinpraxis, Brunn *00563*
Arbeitsgemeinschaft zur Verbesserung der Agrarstruktur in Hessen, Wiesbaden *05132*
Arbeitsgruppe für empirische Bildungsforschung e.V., Heidelberg *05133*
Arbeitsgruppe für strukturelle Molekularbiologie in der Max-Planck-Gesellschaft, Hamburg *05134*
Arbeitskreis Bildung und Politik Rheinland, Remagen *05135*
Arbeitskreis Chemische Industrie, Köln *05136*
Arbeitskreis der Wiener Altgermanisten, Wien *00564*
Arbeitskreis der Wiener Skandinavisten, Wien *00565*
Arbeitskreis deutscher Bildungsstätten e.V., Bonn *05137*
Arbeitskreis Ethnomedizin, Hamburg *05138*
Arbeitskreis für Betriebsführung München, Thaining *05139*
Arbeitskreis für die Bundesrepublik Deutschland e.V., Düsseldorf *05149*
Arbeitskreis für Hochschuldidaktik, Hamburg *05140*
Arbeitskreis für Medizinische Geographie im Zentralverband der deutschen Geographen, Heidelberg *05141*
Arbeitskreis für neue Methoden in der Regionalforschung, Wien *00566*
Arbeitskreis für Ost-West-Fragen, Vlotho *05142*
Arbeitskreis für Schulmusik und allgemeine Musikpädagogik e.V., Würzburg *05143*
Arbeitskreis für Tibetische und Buddhistische Studien, Wien *00567*
Arbeitskreis für Wehrforschung, Stuttgart *05144*
Arbeitskreis Gesundheitskunde e.V., Sankt Georgen *05145*
Arbeitskreis katholischer Schulen in freier Trägerschaft in der Bundesrepublik Deutschland, Bonn *05146*
Arbeitskreis Rhetorik in Wirtschaft, Politik und Verwaltung, Bonn *05147*
Arbeitswissenschaft im Landbau e.V., Stuttgart *05148*
Arbeit und Leben, Düsseldorf *05149*
Arbeo-Gesellschaft e.V., Bachenhausen *05150*
ARBICA, Tunis *11133*
Arbitrators' Institute of Canada, Toronto *01869*
Arboricultural Association, Romsey *11241*
ARC, Boston *16592*
ARC, Brighton *11415*
ARC, Chesterfield *11250*
ARC, Dearborn *13275*
ARCA, Alexandria *14028*
ARCASIA, Karachi *09358*
ARCC, Harare *17160*
ARCCOH, Roorkee *06725*
ARCEDEM, Ibadan *09219*
Archaeological and Anthropological Society of Victoria, Melbourne *00218*
Archaeological Conservancy, Santa Fe *14236*
Archaeological Foundation, Bruxelles *01361*
Archaeological Institute of America, Boston *14237*
Archaeological Society of Macedonia, Skopje *08618*
Archaeologiki Hetairia, Athinai *06499*
Archäologische Gesellschaft zu Berlin, Berlin *05151*
Archaeoloische Gesellschaft Steiermark, Graz *00568*
Archeology Section, Washington *14238*
Archief- en Biblioteekwezen in Belgie, Bruxelles *01102*
Architects and Surveyors Institute, Chippenham *11242*
Architects Council of Europe, Bruxelles *01101*
Architects Regional Council Asia, Karachi *09358*
Architectural and Archaeological Society for the County of Buckingham, Aylesbury *11243*
Architectural and Archaeological Society of Durham and Northumberland, Durham City *11244*
Architectural Association, London *11245*
Architectural Association of Ireland, Dublin *06924*
Architectural Association of Israel, Tel Aviv *07042*
Architectural Intitute of British Columbia, Vancouver *01870*
Architectural League of New York, New York *14239*
The Architectural Society of China, Beijing *03115*
Architektenraad, Amsterdam *08822*
Archival Association, Helskini *03950*
Archives et Bibliothèques de Belgique, Bruxelles *01102*
Archives of American Art, Washington *14240*
ARCIS, Ibadan *09222*
ARCSS, Cairo *03885*
ARCT, Dakar *10013*
Arctic Institute, København *03560*
Arctic Institute of North America, Calgary *01871*
ARD, Orangeburg *14491*
ARDA, Penang *08626*
ARDE, Roma *07400*
ARDI, Bruxelles *01159*
ARE, Cumnor Hill *11279*
ARE, Sheffield *11278*
ARE, Virginia Beach *14328*
AREA, Camberwell *00332*
ARELS, London *11383*
AREUEA, Bloomington *14026*
ARF, New York *13334*
Argentine Association of Geophysicists and Geodesists, Buenos Aires *00065*
Argentine Biophysical Society, Buenos Aires *00157*
Argentine Centre of Engineering, Buenos Aires *00100*
Argentine Electrotechnical Association, Buenos Aires *00080*
Argentine Entomological Society, La Plata *00197*
Argentine Society of Endocrinology and Metabolism, Buenos Aires *00165*
Argentine Society of Geographical Studies, Buenos Aires *00167*
Argentine Society of Physiological Sciences, Buenos Aires *00161*
Argentinian Geological Association, Buenos Aires *00082*
Arhivsko Društvo Slovenije, Ljubljana *10096*
ARI, Harare *17161*
ARI, Hayward *14235*
ARI, Monterey Park *13325*
ARIA, Sacramento *14032*
ARIC, Saint-Denis *04330*
Arica Institute, New York *14241*
ARIPS, Molinetto di Mazzano *07274*
Aristotelian Society, London *11246*
ARIT, Philadelphia *14029*
Arkistoyhdistys, Helskini *03950*
Arkitektafélag Islands, Reykjavik *06677*
Arkivarforeningen, Oslo *09247*
Arkivforeningen, København *03559*
Ark-La-Tex Genealogical Association, Shreveport *14242*
Arktisk Institut, København *03560*
ARL, Hannover *05060*
ARL, Washington *14492*
ARLIS/NA, Tucson *14250*
ARLIS UK & Ireland, Bromsgrove *11251*
ARLO, Baghdad *06913*
A.R.L.T., Woking *11290*
Armagh Field Naturalists Society, Armagh *11247*
ARMB, Bruxelles *01094*
Armed Forces Civilian Instructors Association, Sa Ysidro *14243*
Armed Forces Institute of Pathology, Washington *14244*
Armed Forces Optometric Society, Harrison *14245*
Armenian Academy of Sciences, Yerewan *00203*
Armenian Artistic Union, Cairo *03878*
Armenian Educational Foundation, Glendale *14246*
Armenian Literary Society, Demarest *14247*
ARMH, Kansas City *15813*
ARMS, London *11385*
Arms and Armour Society of Great Britain, Staines *11248*
ARMS Firms Users Association, Mount Clemens *14248*
ARNM, Okigwi *09220*
ARNMD, New York *14329*
Arnold Bennett Literary Society, Stoke-on-Trent *11249*
Arnold-Bergstraesser-Institut für kulturwissenschaftliche Forschung e.V., Freiburg *05152*
AROCC, Buffalo *14331*
A.R.O.E.V.E.N., Paris *04346*
A.R.P., Roma *07171*
ARPLA, Bangkok *11077*
A.R.P.L.O.E.V., Paris *04331*
ARPR, Bloomfield *13306*
ARR, Northwood *11277*
ARRG, Witham *11281*
ARRS, Reston *14034*
ARRT, Forest Hills *14030*
ARRT, Mendota Heights *14027*
ARS, Brooklyn *14031*
ARS, New York *14551*
ARS, Philadelphia *14025*
ARS, Remscheid *05064*
ARSC, Silver Spring *14325*
ARSO, Nairobi *08472*
ARTC, Brooklyn *13326*
ARTDO, Manila *09503*
Art Galleries and Museums Association of New Zealand, Wellington *09116*
The Arthritis and Rheumatism Council, Chesterfield *11250*
Arthur Rubinstein International Music Society, Tel Aviv *07043*
Arthur Schnitzler-Institut, Wien *00569*
Art Institute of Light, Gap *14249*
Artists' Association of the 18th of November, København *03762*
Art Libraries Society/North America, Tucson *14250*
Art Libraries Society of the United Kingdom and Ireland, Bromsgrove *11251*
Arts Association, London *11252*
Arts Association Harare, Harare *17163*
Arts Association of Namibia, Windhoek *08814*
Arts Council, Dublin *06925*
Arts Council, Reykjavik *06695*
Arts Council of Australia, Sydney *00219*
Arts Council of Ghana, Accra *06476*
Arts Council of Great Britain, London *11253*
Arts Council of Northern Ireland, Belfast *11254*
Arts Council of Pakistan, Karachi *09359*
Arts International Program of the Institute of International Education, New York *14251*
Art Society of India, Bombay *06718*
ARVAC, Wivenhoe *11280*
ARVIC, Alexandria *14326*
ARVO, Bethesda *14330*
A.S., Ascot *11417*
AS, Cincinnati *13327*
AS, Kidlington *11222*
AS, New York *14253*
AS, Torrance *14552*
AS, Washington *14238*
ASA, Alexandria *14182*
ASA, Atlanta *13346, 14037*
ASA, Boca Raton *14035*
ASA, Chapel Hill *14186*
ASA, College Park *14185*
ASA, Edmonton *01863*
ASA, London *11392*
ASA, Macquarie *00231*
ASA, Madison *14098*
ASA, Omaha *13340*
ASA, Paddington *00347*
ASA, Park Ridge *14099*
ASA, Silver Spring *14557*
A.S.A., Strawberry Hills *00349*
ASA, Washington *14175*
ASA, Woodbury *13322*
ASA, Yorba Linda *14566*
ASAAD, Coral Springs *14044*
ASAB, Egham *11291*
ASAB-VEBI, Bruxelles *01103*
ASAC, Tokyo *08098*
ASAE, Saint Joseph *14097*
ASAFED, Lomé *11103*
ASAHP, Washington *14496*
ASAIHL, Bangkok *11089*
ASAIO, Boca Raton *14046*
ASALH, Washington *14356*
ASAM, Washington *14096*
ASANAL, Kuala Lumpur *08633*
ASAO, Little Rock *14333*
ASAO, Newark *11390*
ASAP, Bethesda *14043*
ASAP, Mount Barker *00348*
ASAPE, Tsukuba *08109*
ASAPS, Long Beach *14045*
ASAS, Bucuresti *09734*
ASAS, Campaign *14100*
ASAS, Melrose *14095*
ASAUK, London *11211*
ASBAH, Peterborough *11284*
ASBC, Saint Paul *14103*
ASBCS, Nashville *14503*
ASBDE, Redford *14102*
ASBE, Chicago *14101*
Asbestos International Association, Paris *04137*
ASbH, Dortmund *05126*
ASB Management-Zentrum-Heidelberg e.V., Heidelberg *05153*
ASBMB, Edwardstown *00338*
ASBMC, Bethesda *14047*
ASBMR, Washington *14048*
ASBO, Reston *14494*
ASBS, Washington *14502*
ASC, Boston *14038*
ASC, Columbus *14115*
ASC, Fairfax *14054*
ASC, Huntington *14556*
ASC, Peoria *14260*
ASC, Philadelphia *14116*
ASC, Washington *14512*
ASCA, Albuquerque *14053*
ASCA, Taipei *03237*
ASCB, Bethesda *14049*
ASCCP, Washington *14052*
ASCD, Alexandria *14336*
ASCE, Gainesville *14493*
ASCE, Washington *14107*
ASCET, El Paso *14105*
ASCF, Boston *14039*
ASCH, Des Plaines *14108*
ASCH, Indialantic *14106*
ASCHB, London *11285*
ASCI, Thorofare *14050*
Asciación Indeginista del Paraguay, Asunción *09425*
ASCLA, Chicago *14505*
ASCMS, Chicago *14113*
ASCN, Bethesda *14051*
ASCO, Chicago *14109, 14114*
ASCO, Rockville *14495*
ASCOBIC, Dakar *10014*
ASCOLBI, Bogotá *03343*
ASCP, Alexandria *14112*
ASCP, Chicago *14110*
ASCRS, Fairfax *14104*
ASCRS, Palatine *14111*
ASCT, Raleigh *14055*
ASCUN, Bogotá *03348*
ASD, Baltimore *14119*
ASDA, Chicago *14184*
ASDA, Genève *10632*
ASDA, New York *14056*
ASDAE, Silver Spring *14501*
ASDC, Chicago *14118*
ASDE, Larchmont *14117*
ASDS, Los Angeles *14357*
ASDS, Schaumburg *14057*
ASE, Chicago *14060*
A.S.E., Edinburgh *11414*
A.S.E., Hatfield *11282*
ASE, Milaca *13369*
ASE, Paris *04127*
ASE, Raleigh *14120*
ASE, Ruston *14334*
ASE, Salt Lake City *14337*
A.S.E., Subiaco *00352*
ASE, Washington *13366*
ASEAMS, Manila *09511*
ASEAN Association for Planning and Housing, Manila *09490*
ASEAN Council of Teachers, Jakarta *06873*
ASEANCUPS, Sungai Ehsan *08640*
ASEAN Energy Management and Research Training Centre, Jakarta *06874*
ASEAN Federation for Psychiatric and Mental Health, Kuala Lumpur *08627*
ASEAN Federation of Endocrine Societies, Jakarta *06875*
ASEAN Federation of Engineering Organizations, Bangkok *11067*
ASEAN Federation of Plastic Surgeons, Bangkok *11068*
ASEAN Food Handling Bureau, Kuala Lumpur *08628*
ASEAN Institute for Health Development, Nakhon Pathom *11069*
ASEAN Institute for Physics, Kuala Lumpur *08629*
ASEAN Institute of Forest Management, Kuala Lumpur *08630*
ASEANIP, Kuala Lumpur *08629*
ASEAN Law Association, Jakarta *06876*
ASEAN Neurological Society, Manila *09491*
ASEAN Otorhinolaryngological Head and Neck Federation, Bangkok *11070*
ASEAN Plant Quarantine Centre and Training Institute, Serdang *08631*
ASEANPOPIN, Jakarta *06878*
ASEAN Population Coordination Unit, Jakarta *06877*
ASEAN Population Information Network, Jakarta *06878*
ASEAN Sub Committee on Non-Conventional Energy Research, Bangkok *11071*
ASEAN Training Centre for Preventive Drug Education, Manila *09492*
ASECIC, Madrid *10259*
ASECS, Logan *14058*
ASEI, North Palm Beach *14181*
ASEIBI, Medellín *03349*
ASEM, Annapolis *14059*
ASEM, Quito *03844*
ASEP, Bangkok *11084*
ASEREP, Toulouse *04204*
ASES, Boulder *14177*
ASESP, Nairobi *08473*
ASET, Berryville *13309*
ASET, Carroll *14121*
ASE-USA, San Francisco *14504*
ASFE, Silver Spring *14432*
ASFEC, Menoufia *03877*
ASFFHF, Los Angeles *13307*
ASFIS, Roma *07249*
ASFO, Chicago *14123*
ASG, Bonn *05092*
ASG, Columbus *14125*
ASG, Fairfield *14126*
ASG, Göttingen *05044*
ASGCA, Chicago *14127*
ASGD, Chicago *14062*
ASGE, Independence *14124*
ASGE, Manchester *14061*
ASGP, Sheffield *11292*
ASGP&P, McLean *14128*
ASGW, Alexandria *14335*
ASH, San Diego *13308*
ASH, Washington *14131*
ASHA, Altoona *14179*
ASHA, Kent *14036*
ASHA, Research Triangle Park *14042*
ASHA, Rockville *14178*
ASHE, Chicago *14067*
ASHE, College Station *14358*
ASHET, Chicago *14064*
ASHG, Bethesda *14133*
ASHHRA, Chicago *14065*
Ashmolean Natural History Society of Oxfordshire, Oxford *11255*
ASHNS, Baltimore *14063*
ASHP, Bethesda *14132*
ASHRAE, Atlanta *14130*
ASHS, Alexandria *14066*
ASHT, Garner *14129*
ASI, Bonn *05125*
ASI, Chippenham *11242*
A.S.I., Milano *07405*
ASI, Minneapolis *14187*
ASI, Paris *04349*
ASI, Port Aransas *14135*
ASI, Worthington *14563*
ASIA, Chicago *14180*
Asia and Oceania Federation of Nuclear Medicine and Biology, Taipei *03233*
Asia and Oceania Federation of Obstetrics and Gynecology, Singapore *10034*
Asia and Oceania Society for Comparative Endocrinology, Hong Kong *06587*
Asia and Pacific Association for the Control of Tobacco, Taipei *03234*
Asia and Pacific Commission on Agricultural Statistics, Bangkok *11072*
Asia and Pacific Plant Protection Commission, Bangkok *11073*
Asia and Pacific Programme of Educational Innovation for Development, Bangkok *11074*
Asia Crime Prevention Foundation, Tokyo *08094*
Asia Foundation, Bangkok *11075*
Asia Foundation, Islamabad *09360*
The Asia Foundation, Jakarta *06879*
Asia Foundation, Kabul *00002*
Asia Foundation, Kuala Lumpur *08632*
Asia Foundation, Seoul *08517*
The Asia Foundation, Taipei *03235*
Asian Affairs Research Council, Tokyo *08091*
Asian-African Legal Consultative Committee, New Dehli *06719*
Asian Alliance of Appropriate Technology Practitioners, Manila *09493*
Asian and Oceanian Thyroid Association, Nagasaki *08095*
Asian and Pacific Association for Social Work Education, Bundoora *00220*
Asian and Pacific Centre for Transfer of Technology, Bangalore *06720*
Asian and Pacific Federation of Organizations for Cancer Research and Control, Tokyo *08096*
Asian and Pacific Information Network on Medicinal and Aromatic Plants, Bangkok *11076*
Asian and Pacific Project for Labour Administration, Bangkok *11077*
Asian Association for Biology Education, Manila *09494*
Asian Association of Agricultural Colleges and Universities, Laguna *09495*
Asian Association of Management Organizations, Hong Kong *06588*
Asian Association of National Languages, Kuala Lumpur *08633*
Asian Association of Occupational Health, Bangkok *11078*
Asian Association of Open Universities, Seoul *08518*
Asian Association on Remote Sensing, Tokyo *08097*
Asian-Australasian Society of Neurological Surgeons, Brisbane *00221*
Asian-Australian Association of Animal Production Societies, Miaoli *03236*
Asian Center for Organization, Research and Development, New Delhi *06721*
Asian Centre of Educational Innovation for Development, Bangkok *11079*
Asian Club, Tokyo *08092*
Asian Confederation of Physical Therapy, Jakarta *06880*
Asian Confederation of Teachers, Manila *09496*
Asian Coordinating Group for Chemistry, Nedlands *00222*
Asian Council of Securities Analysts, Tokyo *08098*
Asian Crystallographic Association, Taipei *03237*
Asian Dermatological Association, Hong Kong *06589*
Asian Disaster Preparedness Center, Bangkok *11080*
Asian Ecological Society, Taichung *03238*
Asian Environmental Society, New Delhi *06722*
Asian Federation of Catholic Medical Associations, Seoul *08519*
Asian Federation of Societies for Ultrasound in Medicine and Biology, Seoul *08520*
Asian Fisheries Society, Manila *09497*
Asian Fluid Mechanics Committee, Tokyo *08099*
Asian Food Council, Taipei *03239*
Asian Foundation for the Prevention of Blindness, Hong Kong *06590*
Asian Geotechnical Engineering Information Center, Bangkok *11081*
Asian Grain Legumes Network, Patancheru *06723*
Asian Health Institute, Aichi *08100*

Asian Institute for Development Communication, Kuala Lumpur 08634
Asian Institute of Management, Manila 09498
Asian Institute of Technology Library and Regional Documentation Centre, Bangkok 11082
Asian Institute of Tourism, Manila 09499
Asian Literature Division (of MLA), Bloomington 14252
Asian Mass Communication Research and Information Centre, Singapore 10035
Asian Music Rostrum, Paris 04138
Asian Network for Analytical and Inorganic Chemistry, Perth 00223
Asian Network for Biological Sciences, Singapore 10036
Asian Network for Industrial Technology Information and Extension, Singapore 10037
Asian Network of Human Resource Development Planning Institutes, New Delhi 06724
Asian NGO Coalition for Agrarian Reform and Rural Development, Manila 09500
Asian-Pacific Association for Laser Medicine and Surgery, Singapore 10038
Asian Pacific Association for the Study of the Liver, Seoul 08521
Asian Pacific Confederation of Chemical Engineering, Barton 00224
Asian-Pacific Corrosion Control Organization, Sydney 00225
Asian Pacific Dental Federation, Singapore 10039
Asian Pacific Federation of Clinical Biochemistry, Adelaide 00226
Asian Pacific Federation of Human Resource Management, Hong Kong 06591
Asian-Pacific Federation of Therapeutic Communities, Bangkok 11083
Asian-Pacific Organization for Cell Biology, Okayama 08101
Asian-Pacific Political Science Association, Singapore 10040
Asian Pacific Section of the International Confederation for Plastic and Reconstructive Surgery, Singapore 10041
Asian-Pacific Society of Cardiology, Seoul 08522
Asian Pacific Society of Respirology, Tokyo 08102
Asian-Pacific Tax and Investment Research Centre, Singapore 10042
Asian Pacific University Presidents Conference, Tokyo 08103
Asian-Pacific Weed Science Society, Laguna 09501
Asian Packaging Federation, Beijing 03116
Asian Packaging Information Centre, Colombo 10406
Asian Pan-Pacific Society for Paediatric Gastroenterology and Nutrition, East Perth 00227
Asian Parasite Control Organization, Tokyo 08104
Asian Patent Attorneys Association, Tokyo 08105
Asian Peace Research Association, Christchurch 09117
Asian Physics Education Network, Jakarta 06881
Asian Population and Development Association, Tokyo 08106
Asian Productivity Organization, Tokyo 08107
Asian Recycling Association, Camarines Sur 09502
Asian Regional Association for Home Economics, Seoul 08523
Asian Regional Cooperative Project on Food Irradiation, Wien 00570
Asian Regional Coordinating Committee on Hydrology, Roorkee 06725
Asian Regional Training and Development Organization, Manila 09503
Asian Rice Farming Network, Manila 09504
Asian Rural Institute – Rural Leaders Training Center, Tochigi 08108
Asian Society for Adapted Physical Education and Exercise, Tsukuba 08109
Asian Society for Environmental Protection, Bangkok 11084
Asian Society for Sport Psychology, Tokyo 08110
Asian Society of Agricultural Economists, Seoul 08524
Asian Society of Oto-Rhino-Laryngology, Jakarta 06882
Asian-South Pacific Bureau of Adult Education, Colombo 10407
Asian Surgical Association, Hong Kong 06592
Asian Vegetable Research and Development Center, Tainan 03240
Asian Wetland Bureau, Kuala Lumpur 08635
Asian Women's Research and Action Network, Manila 09505
Asia-Oceania Association of Otolaryngological Societies, Bombay 06726
Asia Pacific Academy of Ophthalmology, Tokyo 08111
Asia Pacific Food Analysis Network, Canberra 00228
Asia-Pacific Forestry Commission, Bangkok 11085
Asia Pacific Forum on Women, Law and Development, Kuala Lumpur 08636
Asia Pacific Grouping of Consulting Engineers, Manila 09506
Asia-Pacific Information Network in Social Sciences, Bangkok 11086
Asia-Pacific Lawyers Association, Seoul 08525
Asia Pacific League against Rheumatism, Semarang 06883
Asia-Pacific Office Automation Council, Taipei 03241
Asia-Pacific People's Environment Network, Penang 08637
Asia Pacific Physics Teachers and Educators Association, Manila 09507
Asia-Pacific Programme of Education for All, Bangkok 11087
Asia Pacific Quality Control Organization, Diliman 09508
The Asia Society, New York 14253
Asia Soil Conservation Network for the Humid Tropics, Jakarta 06884
Asia Theological Association, Bangalore 06727
Asiatic Society, Calcutta 06728
Asiatic Society of Bombay, Bombay 06729
ASIC, Brentwood 14138
ASIDIC, Athens 14445
ASIH, Carbondale 14134
ASIL, Washington 14137
ASIM, Washington 14136
A.S.I.P., Roma 07395
ASIS, Silver Spring 14068
ASISTI, Bourges 04141
ASIWPCA, Washington 14506
ASJMC, Columbia 14497
ASL, Urbana 14338
ASLA, Alderley 00336
A.S.L.A., Palermo 07402
ASLA, Washington 14139
ASLH, University 14070
ASLHC, Edinburgh 11389
ASLIB, London 11267
Aslib, The Association for Information Management, London 11256
ASLM, Boston 14140
ASLMS, Wausau 14069
ASLO, Gloucester Point 14141
ASLP, Manila 09512
ASLS, Aberdeen 11283
A.S.M., Bucuresti 09735
ASM, Cleveland 14339
A.S.M., Parkville 00342
ASM, Provo 14142
ASM, Tübingen 05069
ASM, Washington 14073
AsMA, Alexandria 13339
A.S.M.A.F., Paris 04348
ASMB, Nicosia 03487
ASMD, Chicago 14499
ASMDT, Oakland Gardens 14143
ASME, Dundee 11293
ASME, New York 14145
ASMECCANICA, Milano 07368
ASME International Gas Turbine Institute, Atlanta 14254
ASMER, Montevideo 13249
ASM International, Materials Park 14255
ASM International-European Council, Bruxelles 01104
ASMR, Paris 04138
ASMS, Arlington Heights 14144
ASMS, Santa Fe 14071
ASMT, Bethesda 14072
ASN, Flushing 14146
ASN, Palo Alto 14074
ASN, Washington 14148, 14150
ASNE, Alexandria 14147
ASNEMGE, London 11369
ASNHS, Lancing 11386
ASNIBI, Managua 09204
ASNR, Oak Brook 14149
ASNT, Columbus 14075
ASO, London 11294
ASOC, Washington 14226
Asociación Archivística Argentina, Buenos Aires 00053
Asociacion Argentina Amigos de la Astronomía, Buenos Aires 00054
Asociación Argentina de Alergía e Inmunología, Buenos Aires 00055
Asociación Argentina de Astronomía, La Plata 00056
Asociación Argentina de Bibliotecas y Centros de Información Cientificis y Tecnical, Buenos Aires 00057
Asociación Argentina de Bibliotecas y Centros de Información Científicos y Técnicos, Buenos Aires 00058
Asociación Argentina de Biología y Medicina Nuclear, Buenos Aires 00059
Asociación Argentina de Ciencias Naturales, Buenos Aires 00060
Asociación Argentina de Cirugía, Buenos Aires 00061
Asociación Argentina de Ecología, Córdoba 00062
Asociación Argentina de Estudios Americanos, Buenos Aires 00063
Asociación Argentina de Farmacia y Bioquímica Industrial, Buenos Aires 00064
Asociación Argentina de Geofísicos y Geodestas, Buenos Aires 00065
Asociación Argentina de la Ciencia del Suelo, Buenos Aires 00066
Asociación Argentina de la Cultura Inglesa, Buenos Aires 00067
Asociación Argentina del Frío, Buenos Aires 00068
Asociación Argentina de Micología, Buenos Aires 00069
Asociación Argentina de Ortopedía y Traumatología, Buenos Aires 00070
Asociación Argentina para el Estudio Científico de la Deficiencia Mental, Buenos Aires 00071
Asociación Argentina para el Progreso de las Ciencias, Buenos Aires 00072
Asociación Bernardino Rivadavia, Bahía Blanca 00073
Asociación Bibliotecológica Guatemalteca, Guatemala City 06557
Asociación Bioquímica Argentina, Buenos Aires 00074
Asociación Centroamericana de Anatomía, San Salvador 03918
Asociación Centroamericana de Historia Natural, Guatemala City 06558
Asociación Centroamericana de Sociología, Panamá City 09408
Asociación Chilena de Microbiología, Santiago 03014
Asociación Chilena de Sismología e Ingeniería Antisísmica, Santiago 03015
Asociación Chilena para la Investigación y Desarrollo del Hormigón Estructural, Santiago 03016
Asociación Científica Argentino-Alemana, Buenos Aires 00075
Asociación Colombiana de Bibliotecarios, Bogotá 03343
Asociación Colombiana de Facultades de Medicina, Bogotá 03344
Asociación Colombiana de Fisioterapía, Bogotá 03345
Asociación Colombiana de Museos, Institutos y Casas de Cultura, Bogotá 03346
Asociación Colombiana de Sociedades Científicas, Bogotá 03347
Asociación Colombiana de Universidades, Bogotá 03348
Asociación Costarricense de Bibliotecarios, San José 03430
Asociación Costarricense de Cirurgíca, San José 03431
Asociación Costarricense de Pediatría, San José 03432
Asociación Cubana de Bibliotecarios, La Habana 03470
Asociación Cultural Humboldt, Caracas 16966
Asociación Dante Alighieri, Buenos Aires 00076
Asociación de Agrimensores de Venezuela, Caracas 16967
Asociación de Arquitectos de Bolivia, La Paz 01532
Asociación de Artistas Aficionados, Lima 09447
Asociación de Bibliotecarios de Chile, Santiago 03017
Asociación de Bibliotecarios de El Salvador, San Salvador 03919
Asociación de Bibliotecarios del Paraguay, Asunción 09426
Asociacíon de Bibliotecarios del Uruguay, Montevideo 13242
Asociación de Bibliotecarios Graduados de la República Argentina, Buenos Aires 00077
Asociación de Bibliotecarios Graduados del Istmo de Panamá, Panamá City 09409
Asociación de Bibliotecarios Universitarios del Paraguay, Asunción 09427
Asociación de Bibliotecarios y Archiveros de Honduras, Tegucigalpa 06582
Asociación de Bibliotecarios y Documentalistas Agrícolas del Perú, Lima 09448
Asociación de Bibliotecas Universitarias y Especializadas de Nicaragua, León 09203
Asociación de Bibliotecólogos del Uruguay, Montevideo 13243
Asociación de Cardiología, San José 03433
Asociación de Ciencias Naturales del Litoral, Santo Tomé 00078
Asociación de Egresados de la Escuela Interamericana de Bibliotecología de la Universidad de Antioquia, Medellín 03349
Asociación de Escritores y Artistas Españoles, Madrid 10247
Asociación de Facultades de Medicina, Tegucigalpa 06583
Asociación de Historadores Latinoamericanos y del Caribe, Paris 04139
Asociación de Informatica y Computación en Educación, Santiago 03018
Asociación de Ingenieros Civiles del Perú, Lima 09449
Asociación de Ingenieros del Uruguay, Montevideo 13244
Asociación de Ingenieros y Arquitectos de México, México 08692
Asociación de Ingenieros y Geólogos de Yacimientos Petrolíferos Fiscales Bolivianos, La Paz 01533
Asociación de Investigación Técnica de la Industria Papelera Española, Madrid 10248
Asociación del Cuerpo Nacional Veterinario, Madrid 10249
Asociación de Lingüística y Filología de América Latina, Santiago 03019
Asociación de Linguistica y Filologia de América Latina, Caracas 16968
Asociación de Maestros de Puerto Rico, Hato Rey 09720
Asociación de Medicina Aeronáutica y Espacial, Barcelona 10250
Asociación de Medicina Interna, San José 03434
Asociación de Obstetricia y Ginecología, San José 03435
Asociación de Ortodoncistas de Guatemala, Guatemala City 06559
Asociación de Peritos Agrícolas del Estado, Madrid 10251
Asociación de Personal Investigador del CSIC, Madrid 10252
Asociación de Psicólogos de Puerto Rico, Rio Piedras 09721
Asociación de Química y Farmacia del Uruguay, Montevideo 13245
Asociación de Radiólogos de América Central y Panamá, San Salvador 03920
Asociación de Universidades Confiadas a la Compañía de Jesús en América Latina, Bogotá 03350
Asociación Dominicana de Bibliotecarios, Santo Domingo 03831
Asociación Dominicana pro Bienestar de la Familia, Santo Domingo 03832
Asociación Ecuatoriana de Bibliotecarios, Quito 03843
Asociación Ecuatoriana de Museos, Quito 03844
Asociación Electrotécnica Argentina, Buenos Aires 00079, 00080
Asociación Electrotécnica Española, Barcelona 10253
Asociación Electrotécnica Peruana, Lima 09450
Asociación Española Astronáutica, Madrid 10254
Asociación Española contra el Cancer, Barcelona 10255
Asociación Española de Amigos de los Castillos, Madrid 10256
Asociación Española de Bibliotecarios, Archiveros, Museologos y Documentalistas, Madrid 10257
Asociación Española de Biopatología Clínica, Madrid 10258
Asociación Española de Cine Científico, Madrid 10259
Asociación Española de Cirujanos, Madrid 10260
Asociación Española de Gemología, Barcelona 10261
Asociación Española de la Lucha contra la Poliomielitis, Madrid 10262
Asociación Española de Logopedia, Foniatria y Audiología, Barcelona 10263
Asociación Española de Marketing, Madrid 10264
Asociación Española de Orientalistas, Madrid 10265
Asociación Española de Pediatría, Madrid 10266
Asociación Española de Psicoterapia Analítica, Madrid 10267
Asociación Española de Químicos y Coloristas Textiles, Barcelona 10268
Asociación Española de Técnicos de Cerveza y Malta, Madrid 10269
Asociación Española de Técnicos de Radiología, Madrid 10270
Asociación Española para el Estudio Científico del Retraso Mental, Madrid 10271
Asociación Española para el Progreso de las Ciencias, Madrid 10272
Asociación Física Argentina, Villa Elisa 00081
Asociación Geológica Argentina, Buenos Aires 00082
Asociación Iberoamericana de Educación Superior a Distancia, Madrid 10273
Asociación Iberoamericana de Estudio de los Problemas del Alcohol y la Droga, Córdoba 00083
Asociación Interamericana de Bibliotecarios y Documentalistas Agrícolas, Filial Venezuela, Mérida 16969
Asociación Interamericana de Escritores, Buenos Aires 00084
Asociación Interamericana de Gastroentereología, México 08693
Asociación Interamericana de Ingenieria Sanitaria y Ambiental, São Paulo 01582
Asociación Internacional de Hidatidología, Buenos Aires 00085
Asociación Internacional de Pediatría, Paris 04261
Asociación Internacional Veterinaria de Producción Animal, Madrid 10274
Asociación Latinoamericana Científico de Plantas, Lima 09451
Asociación Latinoamericana de Biomatemática, Santiago 03020
Asociación Latinoamericana de Centros de Educación, Rosario 00086
Asociación Latinoamericana de Derecho Aeronáutico y Espacial, Madrid 10275
Asociación Latinoamericana de Ecodesarrollo, Mexicali 08694
Asociación Latinoamericana de Educación Radiofónica, Quito 03845
Asociacion Latinoamericana de Escuelas de Cirurgía Dental, Guatemala City 06560
Asociación Latinoamericana de Escuelas de Trabajo Social, São Luís 01583
Asociación Latinoamericana de Escuelas y Facultades de Enfermería, Quito 03846
Asociación Latinoamericana de Física Médica, Cali 03351
Asociación Latinoamericana de Micología, Habana 03471
Asociación Latinoamericana de Paleobotánica y Palinología, Porto Alegre 01584
Asociación Latinoamericana de Sociedades de Biología y Medicina Nuclear, Buenos Aires 00087
Asociación Latinoamericana de Sociología, Córdoba 00088
Asociación Latinoamericana de Sociología, México 08695
Asociación Latinoamericana para la Calidad, Buenos Aires 00089
Asociación Latinoamericana para la Producción Animal, Porto Alegre 01585
Asociación Latinoamericana para Microbiologa, Bogotá 03352
Asociación Lationoamericana de Facultades de Odontologia, Medellín 03353
Asociación Médica Argentina, Buenos Aires 00090
Asociación Médica del Hospital Beistegui, México 08696
Asociación Médica de Santiago, Santiago de los Caballeros 03833
Asociación Médica Dominicana, Santo Domingo 03834
Asociación Médica Franco-Mexicana, México 08697
Asociación Médica Peruana Daniel A. Carrión, Lima 09452
Asociación Mexicana de Administración Científica, México 08698
Asociación Mexicana de Bibliotecarios, AC, México 08699
Asociación Mexicana de Facultades y Escuelas de Medicina, San Luis de Potosí 08700
Asociación Mexicana de Géologos Petroleros, México 08701
Asociación Mexicana de Ginecologia y Obstetricia, México 08702
Asociación Mexicana de Orquideologia, México 08703
Asociación Mexicana de Profesores de Microbiología y Parasitología en Escuelas de Medicina, Guadalajara 08704
Asociación Mundial de Veterinaria, Madrid 10405
Asociación Nacional de Escritores Venezolanos, Caracas 16970
Asociación Nacional de Escritores y Artistas, Lima 09453
Asociación Nacional de Químicos de España, Madrid 10276
Asociación Nacional de Universidades e Institutos de Enseñanza Superior, México 08705
Asociación Nacional de Veterinarios Titulares, Madrid 10277
Asociación Nicaragüense de Bibliotecarios, Managua 09204
Asociación Nórdica de Intercambio en Educación Popular con América Latina, Kungälv 10436
Asociación Numismática Española, Barcelona 10278
Asociación Odontológica Argentina, Buenos Aires 00091
Asociación Ondontológica Uruguaya, Montevideo 13246
Asociación Ornitológica del Plata, Buenos Aires 00092
Asociación pala la Lucha la Parálisis Infantil, Buenos Aires 00093
Asociación Paleontológica Argentina, Buenos Aires 00094
Asociación Panameña de Bibliotecarios, Panamá City 09410
Asociación Panamericana de Cirurgía Pediátrica, México 08706

Asociación Panamericana de Oftalmología, Panamá City *09411*
Asociación para el Progreso de la Dirección, Madrid *10279*
Asociación para la Defensa de la Naturaleza, Madrid *10280*
Asociación Pediátrica de Guatemala, Guatemala City *06561*
Asociación Peruana de Archiveros, Lima *09454*
Asociación Peruana de Astronomía, Lima *09455*
Asociación Peruana de Bibliotecarios, Lima *09456*
Asociación Psicoanalítica Argentina, Buenos Aires *00095*
Asociación Psiquiátrica de la América Latina, Caracas *16971*
Asociación Química Argentina, Buenos Aires *00096*
Asociación Química Española de la Industria del Cuero, Barcelona *10281*
Asociación Rural del Uruguay, Montevideo *13247*
Asociación Uruguaya de Escritores, Montevideo *13248*
Asociación Venezolana Amigos del Arte Colonial Caracas, Caracas *16972*
Asociación Venezolana de Facultades de Medicina, Caracas *16973*
Asociación Venezolana de Geología, Minería y Petróleo, Caracas *16974*
Asociacion Venezolana de Ingeniería Eléctrica y Mecánica, Caracas *16975*
Asociación Venezolana de Ingeniería Sanitaria y Ambiental Caracas, Caracas *16976*
Asociación Venezolana para el Avance de la Ciencia, Caracas *16977*
Asociata Internationalà de Studi Romåne, Paris *04140*
Asociatia Bibliotecarilor din Romania, Bucuresti *09738*
Asociaţia Cineaştilor din Romania, Bucuresti *09739*
Asociatia de Drept International si Relatii Internationale din Romania, Bucuresti *09740*
Asociaţia Generala a Economiştilor din Romania, Bucuresti *09741*
Asociatia Juristilor din Romania, Bucuresti *09742*
Asociatia Oamenilor de Stiinta din Romania, Bucuresti *09743*
Asociaţia Psihiatricǎ Românǎ, Bucuresti *09744*
Asociatia Psihologilor din Romania, Bucuresti *09745*
Asociatia Romana de Ştiinte Politice, Bucuresti *09746*
Asocio de Studato Internacia pri Spiritaj kaj Teologiaj Instruoj, Bourges *04141*
ASOCON, Jakarta *06884*
ASODOBI, Santo Domingo *03831*
ASOM, Paris *04105*
ASOR, Melbourne *00343*
Asosiasi Perpustakaan, Arsip dan Dokumentasi Indonesia, Jakarta *06885*
ASOSS, New York *14167*
ASOVAC, Caracas *16977*
ASP, Augusta *14079*
ASP, Chicago *14160*
ASP, Detroit *14151*
ASP, El Paso *14152*
A.S.P., Melbourne *00358*
ASP, Monroe *14153*
ASP, Paris *04143*
ASP, San Francisco *14541*
ASPA, Napoli *07401*
ASPA, Rockville *14154*
ASPA, Washington *14084*
ASPAC, Manila *09506*
ASpB, Leverkusen *05093*
ASPBAE, Colombo *10407*
ASPD, Bel Air *14159*
ASPE, Brookline *14161*
ASPE, Westlake *14158*
ASPEI, Mangilao *06552*
ASPEN, Jakarta *06881*
ASPEN, Silver Spring *14076*
The Aspen Institute, Queenstown *14256*
ASPET, Bethesda *14077*
ASPH, Washington *14498*
ASPHER, Bristol *11387*
ASPHER, Maastricht *08829*
ASPL, Vienna *14078*
ASPLP, Norton *14081*
ASPO/Lamaze, Washington *14083*
A.S.P.P., Canberra *00357*
ASPP, Rockville *14155*
ASPR, New York *14082*
ASPRS, Arlington Heights *14157*
ASPT, Athens *14156*
ASQC, Milwaukee *14085*
ASQDE, Rockville *14162*
ASR, Lancaster *14355*
ASRA, Richmond *14164*
ASRO, London *11393*
ASRT, Albuquerque *14163*
ASSA, Canberra *00211*
Assam Research Society, Gauhati *06831*
ASSBB, Milano *07393*
ASSBT, Denver *14168*
AssCPHO's, Malmesbury *11333*
ASSE, Bay Village *14166*
ASSE, Des Plaines *14165*
ASSE, New York *14169*
ASSFN, Houston *14086*
ASSGB, Grantown-on-Spey *11391*
ASSH, Aurora *14087*
ASSITEB, Le Perreux *04279*
ASSITEJ, København *03562*
ASSJ, New London *14354*
ASSM, Little Rock *14510*
AssNAS, Roma *07360*
ASSO, Oakland Gardens *14092*
Assocation of Canadian Industrial Designers, Toronto *01872*
Assocation of Japanese Geographers, Tokyo *08289*
Associação Bahiana de Medicina, Salvador *01586*
Associação Brasileira de Educadores Lassalistas, São Carlos *01587*
Associação Brasileira de Engenharia Sanitária e Ambiental, Rio de Janeiro *01588*
Associação Brasileira de Escolas Superiores Católicas, Brasília *01589*
Associação Brasileira de Farmacêuticos, Rio de Janeiro *01590*
Associação Brasileira de Imprensa, Rio de Janairo *01591*
Associação Brasileira de Mecânica dos Solos, São Paulo *01592*
Associação Brasileira de Química, Rio de Janeiro *01593*
Associação Católica Interamericana de Filosofia, Rio de Janeiro *01594*
Associação Central de Agricultura Portuguesa, Lisboa *09636*
Associação das Universidades de Lingua Portuguesa, Lisboa *09637*
Associação de Educação Católica do Brasil, Brasília *01595*
Associação de Ensino, Marília *01596*
Associação de Ensino, Ribeirão Preto *01597*
Associação de Ensino e Cultura Urubupungá, Pereira Barreto *01598*
Associação dos Advogados, São Paulo *01599*
Associação dos Arqueólogos Portugueses, Lisboa *09638*
Associação dos Técnicos e Auxiliares de Radiologia de Portugal, Lisboa *09639*
Associação Educacional Presidente Kennedy, Guarulhos *01600*
Associação Internacional de Críticos de Arte, Rio de Janeiro *01601*
Associação Itaquerense de Ensino, Itaquera *01602*
Associação Médica Brasileira, São Paulo *01603*
Associação Pananericana de Medicina Social, Rio de Janeiro *01604*
Associação Paulista de Bibliotecários, São Paulo *01605*
Associação Paulista de Medicina, São Paulo *01606*
Associação Pernambucana de Bibliotecarios, Recife *01607*
Associação Portuguesa de Bibliotecários, Arquivistas e Documentalistas, Lisboa *09640*
Associação Portuguesa de Economistas, Lisboa *09641*
Associação Portuguesa de Escritores, Lisboa *09642*
Associação Portuguesa de Estudos Clássicos, Coimbra *09643*
Associação Portuguesa de Fisioterapeutas, Lisboa *09644*
Associação Portuguesa de Fotogrametria, Lisboa *09645*
Associação Portuguesa de Fundição, Porto *09646*
Associação Portuguesa de Management, Lisboa *09647*
Associação Portuguesa de Odontologia, Lisboa *09648*
Associação Portuguesa para a Qualidade Industrial, Lisboa *09649*
Associação Protectora dos Diabéticos de Portugal, Lisboa *09650*
Associação Prudentina de Educação e Cultura, Presidente Prudente *01608*
Associação Rio-Grandense de Bibliotecarios, Pôrto Alegre *01609*
Associação Técnica da Indústria do Cimento, Lisboa *09651*
Associação Tibiriça de Educação, São Paulo *01610*
Associação Universitaria Santa Ursula, Rio de Janeiro *01611*
Associaçção de Geografia Teorética, Rio Claro *01612*
Associació Cultural i Artística Els Esquirols, La Massana *00027*
Associació per la Defensa de la Natura, Andorra la Vella *00028*
Associated Business Writers of America, Aurora *14257*
Associated Colleges of the Midwest, Chicago *14258*
Associated European Human Biological Centres, Paris *04142*
Associated Laboratories, Palatine *14259*
Associated Schools of Construction, Peoria *14260*
Associated Schools Project in Education for International Cooperation and Peace, Paris *04143*
The Associated Scientific and Technical Societies of South Africa, Marshalltown *10123*
Associated Universities, Washington *14261*
Association Actuarielle Internationale, Bruxelles *01105*
Association Aéronautique et Astronautique de France, Paris *04144*
Association Africaine de Microbiologie et d'Hygiène Alimentaire, Sousse *11134*
Association Anthroposophique en France, Paris *04145*
Association Belge de Documentation, Bruxelles *01106*
Association Belge de Droit Rural, Bruxelles *01107*
Association Belge de l'Eclairage, Bruxelles *01108*
Association Belge de Linguistique Appliquée; ABLA, Bruxelles *01162*
Association Belge de Photographie et de Cinématographie, Bruxelles *01109*
Association Belge de Radioprotection, Bruxelles *01110*
Association Belge des Analystes Financiers, Bruxelles *01111*
Association Belge d'Hygiène et de Médecine Sociale, Bruxelles *01112*
Association Botanique du Canada, Ottawa *02083*
Association Canadienne d'Acoustique, Ottawa *02012*
Association Canadienne de Cartographie, Calgary *02088*
Association Canadienne de Dermatologie, Vancouver *02116*
Association Canadienne de Droit Maritime, Montréal *02194*
Association Canadienne d'Education, Toronto *02122*
Association Canadienne d'Education de Langue Française, Sillery *01873*
Association Canadienne de Gastroentérologie, Montréal *02051*
Association Canadienne de Justice Pénale, Ottawa *02113*
Association Canadienne de la Construction, Ottawa *02104*
Association Canadienne de la Gestion de Recherche, Mississauga *02243*
Association Canadienne de l'Alarme et de la Securité, Willowdale *02018*
Association Canadienne de la Narcolepsie, Toronto *02037*
Association Canadienne de l'Electricité, Montréal *02123*
Association Canadienne de Linguistique, Toronto *02191*
Association Canadienne de Médecine Physique et de Réadaptation, London *02064*
Association Canadienne de Microbiologie Clinique et des Maladies, Willowdale *02027*
Association Canadienne de Normalisation, Rexdale *02301*
Association Canadienne de Philosophie, Ottawa *02224*
Association Canadienne de Prévention des Incendies, Ottawa *02463*
Association Canadienne de Protection Médicale, Ottawa *02198*
Association Canadienne de Psychothérapie de Groupe, Montréal *02150*
Association Canadienne de Recherches Dentaires, London *02029*
Association Canadienne de Recherhe et d'Education pour la Paix, Brandon *02221*
Association Canadienne d'Ergonomie, Mississauga *02508*
Association Canadienne des Anatomistes, London *02046*
Association Canadienne des Bibliothèques Musicales, Ottawa *02059*
Association Canadienne des Chercheurs en Education, Ottawa *02121*
Association Canadienne de Science Politique, Ottawa *02229*
Association Canadienne des Collèges Carrière, Brantford *01919*
Association Canadienne des Diététistes, Toronto *02118*
Association Canadienne des Doyens d'Education, Vancouver *02049*
Association Canadienne des Doyens des Facultés des Lettres et des Sciences, Charlottetown *02048*
Association Canadienne des Ecoles Universitaires de Nursing, Ottawa *02072*
Association Canadienne des Ecoles Universitaires de Réadaptation, Montréal *02073*
Association Canadienne de Sécurité Incendie, Willowdale *02135*
Association Canadienne des Educateurs en Radiodiffusion, Burnaby *01994*
Association Canadienne de Sémiotique, Kingston *02248*
Association Canadienne des Etudes Africaines, Toronto *02045*
Association Canadienne des Etudes Latino-Américaines, Vancouver *02055*
Association Canadienne des Etudes Latino-Américaines et Caraibes, Ottawa *02036*
Association Canadienne des Etudes Prospectives, Montréal *02032*
Association Canadienne des Géographes, Montréal *02052*
Association Canadienne des Ludothèques et des Centres de Resources pour la Famille, Ottawa *02070*
Association Canadienne des Optométristes, Ottawa *02060*
Association Canadienne des Paraplégiques, Toronto *02220*
Association Canadienne des Pathologistes, Kingston *02063*
Association Canadienne d'Espéranto, Montréal *02125*
Association Canadienne des Physiciens, Ottawa *02065*
Association Canadienne des Professeurs d'Immersion, Ottawa *02053*
Association Canadienne des Professeurs d'Université, Ottawa *02074*
Association Canadienne des Radiologistes, Montréal *02066*
Association Canadienne des Sciences de l'Information, Perth *02034*
Association Canadienne des Sciences du Sport, Gloucester *02069*
Association Canadienne des Thérapeutes du Sport, Alorchester *02078*
Association Canadienne des Travailleurs Sociaux, Ottawa *02068*
Association Canadienne des Vétérinaires, Ottawa *02315*
Association Canadienne d'Histoire Ferroviaire, Saint Constant *02241*
Association Canadienne du Gaz, Don Mills *02143*
Association Canadienne du Génie Eolien, London *02318*
Association Canadienne d'Urbanisme, Regina *02370*
Association Canadienne du Transport Urbain, Toronto *02313*
Association Canadienne en Psychopédagogie, London *02031*
Association Canadienne-Française pour l'Avancement des Sciences, Montréal *01874, 01875*
Association Canadienne pour la Recherche dans l'Industrie Sidérurgique, Toronto *02303*
Association Canadienne pour la Recherche sur la Pollution de l'Eau et sa Maîtrise, Montréal *02076*
Association Canadienne pour la Santé, l'Education Physique et la Récréation, Vanier *02033*
Association Canadienne pour la Technologie des Animaux de Laboratoire, Edmonton *02035*
Association Canadienne pour l'Avancement des Etudes Néerlandaises, Windsor *02041*
Association Canadienne pour l'Education Pastorale, Toronto *02038*
Association Canadienne pour le Qualité de l'Eau, Waterloo *02316*
Association Canadienne pour l'Etude de l'Administration Scolaire, Halifax *02043*
Association Canadienne pour l'Etude des Fondemants de l'Education, Kingston *02050*
Association Canadienne pour l'Etude du Quaternaire, Waterloo *02239*
Association Catéchétique Nationale pour l'Audio Visuel, Paris *04146*
Association Centrale des Assistants Sociaux, Bruxelles *01113*
Association Centrale des Vétérinaires, Paris *04147*
Association Chiropratique Canadienne, Toronto *02095*
Association d'Art des Universités du Canada, Halifax *02973*
Association de Directeurs d'Education en Nouvelle-Ecosse, Sydney *01936*
Association de Documentation pour l'Industrie Nationale, Paris *04148*
Association de E1coles Forestières Universitaires du Canada, Thunder Bay *01946*
Association de Gérontologie du 13e, Paris *04149*
Association de la Chirurgie Infantile Canadienne, London *02061*
Association de l'Economie des Institutions, Paris *04150*
Association de l'Education Morale de la Jeunesse, Paris *04151*
Association de l'Enseignement du Nouveau-Québec, Sainte-Foy *01876*
Association de l'Europe Occidentale pour la Psychologie Aéronautique, Bruxelles *01114*
Association de Médecine Rurale, Paris *04152*
Association de Médecine Urbaine, Paris *04153*
Association d'Enseignement Féminin Professionnel et Ménager, Paris *04154*
Association Dentaire Canadienne, Ottawa *02114*
Association Dentaire Française, Paris *04155*
Association de Paralysie Cérébrale du Québec, Québec *01877*
Association de Prévention des Accidents dans l'Industrie Forestière, North Bay *02471*
Association de Psychologie du Travail de Langue Française, Versailles *04156*
Association de Psychologie Scientifique de Langue Française, Paris *04157*
Association de Radiologie d'Afrique Francophone, Dakar *10016*
Association de Réadaption Psychopédagogique et Scolaire, Paris *04158*
Association de Recherche et d'Etudes Catéchétiques, Paris *04159*
Association de Recherche et d'Expression dans l'Art, Paris *04160*
Association de Recherches Universitaires Géographiques et Cartographiques, Paris *04161*
Association des Actuaires Diplomés de l'Institut de Science Financière et d'Assurances, Paris *04162*
Association des Amateurs de la Musique Andalouse, Casablanca *08801*
Association des Amis d'Alfred de Vigny, Paris *04163*
Association des Amis de Miguel Angel Asturias, Paris *04164*
Association des Anatomistes, Vandoeuvre-les-Nancy *04165*
Association des Architectes Paysagistes du Canada, Ottawa *02285*
Association des Archivistes Français, Paris *04166*
Association des Artistes Sculpteurs, Architectes, Graveurs et Dessinateurs, Paris *04167*
Association des Auteurs de Films, Paris *04168*
Association des Bibliothécaires et du Personnel des Bibliothèques des Ministères de Belgique, Bruxelles *01115*
Association des Bibliothécaires Français, Paris *04169*
Association des Bibliothécaires Parlementaires du Canada, Ottawa *01938*
Association des Bibliothécaires Suisses, Bern *11017*
Association des Bibliothèques de Judaïca et Hébraïca en Europe, Paris *04170*
Association des Bibliothèques de la Santé du Canada, Ottawa *02153*
Association des Bibliothèques de l'Enseignement Supérieure de l'Afrique de l'Ouest Francophone, Dakar *10017*
Association des Bibliothèques de Recherche du Canada, Toronto *02067*
Association des Bibliothèques d'Ottawa-Hull, Ottawa *02590*
Association des Bibliothèques Ecclésiastiques de France, Paris *04171*
Association des Bibliothèques Internationales, Genève *10619*
Association des Bureaux d'Information des Collèges et Universités du Canada, Halifax *01920*
Association des Cartothèques Canadiennes, Ottawa *01924*
Association des Centres Médicaux de la Seine, Paris *04172*
Association des Chimistes, Ingénieurs et Cadres des Industries Agricoles et Alimentaires, Paris *04173*
Association des Chirurgiens Généraux de la Province de Québec, Montréal *01878*
Association des Collèges Communautaires du Canada, Toronto *01921*
Association des Collèges du Québec, Montréal *01879*
Association des Commissaires d'Ecoles du Manitoba, Winnipeg *02608*
Association des Commissions des Accidents du Travail du Canada, Scarborough *01949*
Association des Compositeurs, Auteurs et Editeurs du Canada, Toronto *02373*
Association des Conseillers en Orientation de l'Ontario, Don Mills *02794*
Association des Conseils Asiatiques pour la Recherche en Sciences Sociales, New Delhi *06732*
Association des Conseils Scolaires de la Nouvelle-Ecosse, Halifax *02744*
Association des Conservateurs de Bibliothèques, Paris *04174*
Association des Critiques de Théâtre du Canada, Toronto *02307*
Association des Dermatologistes et Syphiligraphes de Langue Française, Créteil *04175*
Association des Designers Industriels du Canada, Rexdale *01923*
Association des Désigners Industriels du Canada, Toronto *01872*
Association des Diplômés de l'Ecole de Bibliothécaires-Documentalistes, Paris *04176*

Association des Diplômés de Polytechnique, Montréal 01880
Association des Diplomés en Histoire de l'Art et Archéologie de l'Université Catholique de Louvain, Louvain-la-Neuve 01116
Association des Directeurs des Centres Universitaires d'Administration des Entreprises, Paris 04177
Association des Directeurs et Coordonnateurs des Programmes de Journalisme des Universités Canadienne, London 01930
Association des Directeurs Généraux d'Ecoles du Nouveau-Brunswick, Rexton 02693
Association des Doyens de Pharmacie du Canada, Toronto 01929
Association des Ecoles Alternatives et Indépendantes de l'Ontario, Toronto 02756
Association des Ecoles de Service Social de la Région Parisienne, Paris 04178
Association des Ecoles d'Optométrie du Canada, Waterloo 01943
Association des Ecoles Internationales, Genève 10620
Association des Ecoles Privées du Québec, Beaconsfield 02854
Association des Ecrivains Belges de Langue Française, Bruxelles 01117
Association des Ecrivains Combattants, Paris 04179
Association des Ecrivains de Langue Française, Paris 04180
Association de Securité des Exploitations Forestières du Québec, Québec 01881
Association de Securité des Industriels Forestiers du Québec, Québec 01882
Association de Securité des Pâtes et Papiers du Québec, Québec 01883
Association des Enseignants du Lakeshore, Dollard des Ormeaux 02574
Association des Enseignants Franco-Ontariens, Ottawa 01884
Association des Enseignants Francophones du Nouveau-Brunswick, Fredericton 01885
Association des Epidémiologists de Langue Française, Le Kremlin-Bicêtre 04181
Association des Etablissements d'Enseignement Vétérinaire Totalement ou Partiellement de Langue Française, Dakar 10018
Association des Etudes Tsiganes, Paris 04182
Association des Experts-Comptables Internationaux, Gateshead 11359
Association des Facultés d'Agriculture d'Afrique, Rabat 08803
Association des Facultés de Médecine du Canada, Ottawa 01925
Association des Facultés Dentaires du Canada, Edmonton 01922
Association des Facultés de Pharmacie du Canada, Vancouver 01933
Association des Facultés ou Etablissements de Lettres et Sciences Humaines des Universités d'Expression Française, Dakar 10019
Association des Femmes Africaines pour la Recherche sur le Développement, Dakar 10021
Association des Françaises Diplômées des Universités, Paris 04183
Association des Géographes de l'Est, Nancy 04184
Association des Géographes Français, Paris 04185
Association des Géologues Arabes, Baghdad 06916
Association des Grands Conseils Scolaires de l'Ontario, Toronto 01934
Association des Hautes Etudes Hospitalières, Paris 04186
Association des Hypnologues du Canada, Montréal 02158
Association des Industriels de France contre les Accidents du Travail, Paris 04187
Association des Informaticiens de Langue Française, Paris 04188
Association des Ingénieurs Hongrois Canadiens, Don Mills 02510
Association des Instituteurs Polonais au Canada, Toronto 02822
Association des Institutions de Niveaux Préscolaire et Elementaire du Québec, Montréal 01886
Association des Institutions d'Enseignement Secondaire, Montréal 01887
Association des Institutions d'Enseignement Théologique en Afrique Centrale, Yaoundé 01811
Association des Institutions d'Enseignement Théologique en Afrique Occidentale, Kinshasa 17140
Association des Instituts Africains de Formation Maritime, Alexandria 03880
Association des Instituts d'Etudes Européennes, Genève 10621
Association des Instituts et des Centres Arabes pour les Recherches Economiques et Sociales, Tunis 11137
Association des Internes et Anciens Internes en Pharmacie des Hôpitaux de Nancy, Nancy 04189
Association des Langues et Civilisations, Paris 04190
Association des Léprologues de Langue Française, Paris 04191
Association des Médecins de Langue Française du Canada, Montréal 01888
Association des Médecins de Travail du Québec, Montréal 01889
Association des Médecins et Médecins-Dentistes du Grand-Duché de Luxembourg, Luxembourg 08575
Association des Media et de la Technologie en Education au Canada, Toronto 01901
Association des Medias Régionaux Anglophones du Québec, Anne de Bellevue 01940
Association des Médicins du Travail du Canada, Toronto 02752
Association des Musées Canadiens, Ottawa 02201
Association des Palynologues de Langue Française, Strasbourg 04192
Association des Pédiatres d'Afrique Noire Francophone, Cotonou 01516
Association des Pédiatres de Langue Française, Clamart 04193
Association des Physiologistes, Aubière 04194
Association des Professeurs d'Allemand des Universités Canadiennes, Edmonton 02075
Association des Professeurs de Français de la Saskatchewan, Saskatoon 01890
Association des Professeurs de Français des Universités et Collèges Canadiens, Victoria 01891
Association des Professeurs de Français en Afrique, Khartoum 10425
Association des Professeurs de Langues Vivantes de l'Enseignement Public, Paris 04195
Association des Professeurs de l'Enseignement Secondaire et Supérieur du Grand-Duché de Luxembourg, Luxembourg 08576
Association des Professeurs de Mathématiques de l'Enseignement Publique, Paris 04196
Association des Professeurs des Bibliotechniciens d'Ontario, Welland 02759
Association des Professeurs d'Histoire et de Géographie de l'Enseignement Public, Evry 04197
Association des Professeurs en Alimentation du Québec, Montréal 01892
Association des Professionnelles Africaines de la Communication, Dakar 10020
Association des Professionnels de l'Information et de la Documentation, Paris 04198
Association des Psychiatres du Canada, Ottawa 02231
Association des Psychiatres du Nouveau-Brunswick, Saint John 02690
Association des Psychiatres du Québec, Montréal 01893
Association des Régistraires des Universités et Collèges du Canada, Ottawa 01941
Association des Sciences Administratives du Canada, Montréal 01820
Association des Services de Réhabilitation Sociale du Québec, Montréal 01894
Association des Services Géologiques Africains, Orléans 04317
Association des Sociétés Scientifiques Médicales Belges, Bruxelles 01118
Association des Sociétés Suisses des Professeurs de Langues Vivantes, Luzern 10622
Association des Techniciens et Ingénieurs Sanitaires Africains, Ouagadougou 01805
Association des Universités Africaines, Accra 06477
Association des Universités Arabes, Amman 08451
Association des Universités de l'Afrique de l'Est et du Sud, Kampala 11182
Association des Universités de l'Amazonie, Belém 01613
Association des Universités Partiellement ou Entièrement de Langue Française, Montréal 01895
Association des Universités Populaires Suisses, Bern 10999
Association de Transfusion Sanguine, Paris 04199
Association d'Etudes Canadiennes, Montréal 01899
Association d'Etudes pour l'Expansion de la Recherche Scientifique, Paris 04200
Association d'Etudes Techniques des Industries de l'Estampage et de la Forge, Paris 04201
Association d'Histoire et de Science Politique, Wabern 10623
Association d'Instituts Européens de Conjoncture Economique, Louvain-la-Neuve 01119
Association d'Orthophonie et d'Audiologie du Nouveau-Brunswick, Fredricton 02695
Association du Barreau Canadien, Ottawa 02081
Association du Counselling des Collèges et Universités du Canada, Victoria 02311
Association du Planning Familial des Sept-Iles, Sept-Iles 01896
Association du Sport Scolaire et Universitaire, Paris 04202
Association du Syndrome de Turner, Downsview 02971
Association du Syndrôme de Turner du Québec, Boucherville 01897
Association Ecologie et Population, Zollikofen 11022
Association Episcopale Catéchistique, Paris 04203
Association Euro-Arabe de Juristes, Bruxelles 01120
Association Européenne Camac, Louvain-la-Neuve 01121
Association Européenne contre les Maladies à Virus, Bruxelles 01122
Association Européenne d'Analyse Transactionnelle, Genève 10676
Association Européenne d'Anthropologie, Bruxelles 01270
Association Européenne de Catalyse, Namur 01299
Association Européenne de Comptabilité, Bruxelles 01268
Association Européenne d'Economistes Agricoles, Bruxelles 01286
Association Européenne de la Recherche en Economie Industrielle, Bruxelles 01279
Association Européenne de l'Ethnie Française, Bruxelles 01123
Association Européenne de Lutte contre la Rétinite Pigmentaire et ses Syndromes, Bruxelles 01272
Association Européenne de Méthodes Médicales Nouvelles, Luxembourg 08577
Association Européenne de Psychiatrie, Luxembourg 08580
Association Européenne de Psychologie Sociale Expérimentale, Louvain-la-Neuve 01291
Association Européenne de Recherche en Avtivité Adaptée, Bruxelles 01280
Association Européenne de Recherches et d'Echanges Pédagogiques, Toulouse 04204
Association Européenne des Affaires Internationales, Bruxelles 01124
Association Européenne des Audioprothésistes, Fleurus 01292
Association Européenne des Barreaux des Courts Suprêmes, Grimbergen 01125
Association Européenne des Centres d'Audiologie, Saint-Etienne 04492
Association Européenne des Centres de Dissémination des Informations Scientifiques, Luxembourg 08583
Association Européenne des Centres de Lutte contre les Poisons, Bruxelles 01126
Association Européenne des Centres d'Ethique Médicale, Bruxelles 01288
Association Européenne des Centres Nationaux de Productivité, Bruxelles 01127
Association Européenne des Chirurgiens Plastiques, Paris 04489
Association Européenne des Conservatoires, Académies de Musique et Musikhochschulen, Angers 04205
Association Européenne des Ecoles et Collèges d'Optométrie, Bures-sur-Yvette 04318
Association Européenne des Enseignants, Section Française, Paris 04206
Association Européenne des Etablissements d'Enseignement Vétérinaire, Maisons-Alfort 04485
Association Européenne des Festivals de Musique, Genève 10679
Association Européenne des Institutions d'Aménagement Rural, Bruxelles 01128
Association Européenne des Institutions d'Aménagement Rural, Bruxelles 01275
Association Européenne des Institutions d'Enseignement Supérieur, Bruxelles 01278
Association Européenne des Instituts de Recherche et de Formation en Matière de Développement, Genève 10677
Association Européenne des Métaux, Bruxelles 01294
Association Européenne des Musées de l'Histoire des Sciences Médicales, Denicé 04488
Association Européenne des Organisations de Recherche sous Contrat, Plaisir 04484
Association Européenne des Podologues, Bruxelles 01295
Association Européenne des Polyoléfines Textiles, Bruxelles 01281
Association Européenne des Sciences et Techniques de la Mer, Talence 04487
Association Européenne de Thermographie, Strasbourg 04207
Association Européenne d'Etude des Traumatisés Crâniens et de leur Réinsertion, Bruxelles 01301
Association Européenne d'Etudes Bourguignonnes, Bruxelles 01274
Association Européenne d'Etudes Chinoises, Paris 04481
Association Européenne d'Urologie, Paris 04490
Association Européenne F. Matthias Alexander, Bruxelles 01129
Association Européenne Francophone pour les Etudes Bahá'íes, Bern 10624
Association Européenne pour l'Administration de la Recherche Industrielle, Paris 04208
Association Européenne pour l'Analyse Transculturelle de Groupe, Bruxelles 01130
Association Européenne pour la Promotion de l'Hygiène des Mains, Bruxelles 01283
Association Européenne pour la Réduction de la Pollution due aux Fibres, Bruxelles 01271
Association Européenne pour le Droit Bancaire et Financier, Bruxelles 01131
Association Européenne pour l'Education aux Médias Audiovisuels, Bruxelles 01273
Association Européenne pour l'Enseignement de l'Architecture, Kassel 05154
Association Européenne pour l'Enseignement de l'Architecture, Louvain-la-Neuve 01287
Association Européenne pour l'Enseignment de la Théorie du Droit, Bruxelles 01285
Association Européenne pour les Etudes d'Opinion et de Marketing, Amsterdam 08892
Association Européenne pour les Graphiques sur l'Ordinateur, Aire-la-Ville 10673
Association Européenne pour le Transfert des Technologies, de l'Innovation et de l'Information Industrielle, Luxembourg 08582
Association Européenne pour l'Etude de l'Alimentation et du Développement de l'Enfant, Paris 04209
Association Européenne pour l'Etude des Problèmes de Sécurité dans la Fabrication et l'Emploi des Poudres Propulsives, Bruxelles 01284
Association Européenne pour l'Etude du Rêve, Ile-Saint-Denis 04483
Association Européenne pour l'Information et les Bibliothèques de Santé, Bruxelles 01276
Association Européenne pour l'Information sur le Développement Local, Bruxelles 01277
Association for 18th Century Studies, Hamilton 01898
Association for Academic Surgery, Minneapolis 14262
Association for Adult and Continuing Education, London 11257
Association for Advancement of Behavior Therapy, New York 14263
Association for Advancement of Psychoanalysis, New York 14264
Association for Advancement of Psychology, Colorado Springs 14265
Association for All Speech Impaired Children, London 11258
Association for Applied Hypnosis, Grimsby 11259
Association for Applied Psychophysiology and Biofeedback, Wheat Ridge 14266
Association for Arid Lands Studies, Lubbock 14267
Association for Asian Studies, Ann Arbor 14268
Association for Assessment in Counseling, Alexandria 14269
Association for Astrological Psychology, San Rafael 14270
Association for Behavior Analysis, Kalamazoo 14271
Association for Biology Laboratory Education, Ithaca 14272
Association for Birth Psychology, New York 14273
Association for Brain Damaged Children, Coventry 11260
Association for Bridge Construction and Design, Pittsburgh 14274
Association for British Music, Boston 11261
Association for Business Communication, Denton 14275
Association for Canadian Studies, Montréal 01899
Association for Canadian Studies in the United States, Washington 14276
Association for Chemoreception Sciences, Tallahassee 14277
Association for Childbirth at Home, International, Glendale 14278
Association for Childhood Education International, Wheaton 14279
Association for Child Psychoanalysis, Great Falls 14280
Association for Child Psychology and Psychiatry, London 11262
Association for Commonwealth Language and Literature Studies, Kingston 08067
Association for Communication Administration, Annadale 14281
Association for Comparative Economic Studies, Flushing 14282
Association for Computational Linguistics, Morristown 14283
Association for Computational Linguistics-Europe, Lugano 10625
Association for Computers and the Humanities, Provo 14284
Association for Computing Machinery, New York 14285
Association for Computing Machinery, Verdun 01900
Association for Continuing Higher Education, Indianapolis 14286
Association for Cooperation in Banana Research in the Caribbean and Tropical America, Panamá City 09412
Association for Counselor Education and Supervision, Alexandria 14287
Association for Creative Change, Minneapolis 14288
Association for Dental Education in Europe, Madrid 10282
Association for Documentary Editing, Carbondale 14289
Association for Eastern and Southeastern European Studies, Linz 00628
Association for Educational and Training Technology, Axminster 11263
Association for Educational Communications and Technology, Washington 14290
Association for Education and Rehabilitation of the Blind and Visually Impaired, Alexandria 14291
Association for Education in Journalism and Mass Communication, Columbia 14292
Association for Education of Teachers in Science, Auburn 14293
Association for Education through Art, Republic of China, Taichung Hsien 03242
Association for Engineering Education in South and Central Asia, Mysore 06730
Association for Engineering Education in Southeast Asia and the Pacific, Christchurch 09118
Association for Equine Sports Medicine, Santa Barbara 14294
Association for Europeam Astronauts, Noordwijk 08823
Association for European Training for Employees in Technology, Bruxelles 01132
Association for Evolutionary Economics, Lincoln 14295
Association for Faculty in the Medical Humanities, McLean 14296
Association for Finishing Processes of the Society of Manufacturing Engineers, Dearborn 14297
Association for General and Liberal Studies, Columbus 14298
Association for Gerontology in Higher Education, Washington 14299
Association for Gnotobiotics, East Aurora 14300
Association for Group and Individual Psychotherapy, London 11264
Association for Group Psychoanalysis and Process, New York 14301
Association for Gypsy Studies, Paris 04182
Association for Health Information and Libraries in Africa, Brazzaville 03422
Association for Hospital Medical Education, Washington 14302
Association for Humanistic Education, Laramie 14303
Association for Humanistic Education and Development, Alexandria 14304
Association for Humanistic Psychology, San Francisco 14305
Association for Independent Disabled Self-Sufficiency, Bath 11265
Association for Industrial Archaeology, Telford 11266
Association for Information Management, London 11267
Association for Institutional Research, Tallahassee 14306
Association for Intelligent Systems Technology, Syracuse 14307
Association for International Cancer Research, Cameron-Markby 11268
Association for International Practical Training, Columbia 14308
Association for Investment Management and Research, Charlottesville 14309
Association for Jewish Demography and Statistics – American Branch, Los Angeles 14310
Association for Jewish Studies, Cambridge 14311
Association for Korean Studies, Rancho Palos Verdes 14312
Association for Language Learning, Rugby 11269
Association for Latin Liturgy, Chard 11270
Association for Library and Information Science Education, Raleigh 14313

Association for Library Collections and Technical Services, Chicago 14314
Association for Library Service to Children, Chicago 14315
Association for Literary and Linguistic Computing, Bangor 11271
Association for Living Historical Farms and Agricultural Museums, Santa Fe 14316
Association for Macular Diseases, New York 14317
Association for Media and Technology in Education, Toronto 01901
Association for Medical Deans in Europe, Belfast 11272
Association for Medical Education in Europe, Dundee 11273
Association for Medical Education in South-East Asia, New Delhi 06731
Association for Medical Education in the Middle East, Alexandria 03879
Association for Medical Education in the Western Pacific Region, Kuala Lumpur 08638
Association for Medical Physics Technology, Ickenham 11274
Association for Mexican Cave Studies, Austin 14318
Association for Nordic Dialysis and Transplant Personnel, Trondheim 09248
Association for Petroleum and Explosives Administration, Huntington 11275
Association for Physics and Chemistry Education of the Republic of China, Taipei 03243
Association for Population/Family Planning Libraries and Information Centers International, New York 14319
Association for Practical Theology, Saint Louis 14320
Association for Practitioners in Infection Control, Mundelein 14321
Association for Professionals in Services for Adolescents, Edinburgh 11276
Association for Progressive Communications, San Francisco 14322
Association for Psychoanalytic Medicine, New York 14323
Association for Quality and Participation, Cincinnati 14324
Association for Radiation Research, Northwood 11277
Association for Recorded Sound Collections, Silver Spring 14325
Association for Recurrent Education, Sheffield 11278
Association for Religious and Value Issues in Counseling, Alexandria 14326
Association for Religious Education, Cumnor Hill 11279
Association for Research, Administration, Professional Councils and Societies, Reston 14327
Association for Research and Enlightenment, Virginia Beach 14328
Association for Research in Nervous and Mental Disease, New York 14329
Association for Research in the Voluntary and Community Sector, Wivenhoe 11280
Association for Research into Restricted Growth, Witham 11281
Association for Research in Vision and Ophthalmology, Bethesda 14330
Association for Research of Childhood Cancer, Buffalo 14331
Association for Science Documents Information, Tokyo 08128
The Association for Science Education, Hatfield 11282
Association for Science, Technology and Innovation, Arlington 14332
Association for Scottish Literary Studies, Aberdeen 11283
Association for Social Anthropology in Oceania, Little Rock 14333
Association for Social Economics, Ruston 14334
Association for Social Psychology, Toronto 01902
Association for Social Work Education in Africa, Addis Ababa 03941
Association for Socio-Economic Development in China, Hsinchu 03244
Association for Specialists in Group Work, Alexandria 14335
Association for Spina Bifida and Hydrocephalus, Peterborough 11284
Association for Studies in the Conservation of Historic Buildings, London 11285
Association for Supervision and Curriculum Development, Alexandria 14336
Association for Surgical Education, Salt Lake City 14337
Association for Symbolic Logic, Urbana 14338
Association for Systems Management, Cleveland 14339
Association for Teacher Education in Africa, Kampala 11181
Association for Teacher Education in Africa, Nairobi 08478
Association for Teacher Education in Europe, Bruxelles 01133
Association for Teaching Psychology, Leicester 11286
Association for Technology in Music Instruction, Eugene 14340
Association for the Advancement of Agricultural Sciences in Africa, Addis Ababa 03942
Association for the Advancement of Baltic Studies, Hackettstown 14341
Association for the Advancement of Christian Scholarship, Toronto 01903
Association for the Advancement of Health Education, Reston 14342
Association for the Advancement of International Education, New Wilmington 14343
Association for the Advancement of Policy, Research and Development in the Third World, Washington 14344
Association for the Advancement of Psychotherapy, New York 14345
Association for the Advancement of Scandinavian Studies in Canada, Guelph 01904
Association for the Advancement of Science in Canada, Ottawa 01905
Association for the Advancement of Science in Israel, Ramat-Gan 07044
Association for Theatre in Higher Education, Evansville 14346
Association for the Behavioral Sciences and Medical Education, McLean 14347
Association for the Bibliography of History, Amherst 14348
Association for the Care of Children's Health, Bethesda 14349
Association for the Development of Computer-Based Instructional Systems, Columbus 14350
Association for the Development of Human Potential, Spokane 14351
Association for the Development of Religious Information Systems, Milwaukee 14352
Association for the Education and Welfare of the Visually Handicapped, Droitwich Spa 11287
Association for the Epidemiological Study and Assessment of Disasters in Developing Countries, Bruxelles 01134
Association for the Expansion of International Roles of the Languages of Continental Europe, Paris 04210
The Association for the Geological Collaboration in Japan, Tokyo 08116
Association for the International Collective Management of Audiovisual Works, Genève 10626
Association for the Introduction of New Biological Nomenclature, Kalmthout 01154
Association for Theological Education in South East Asia, Singapore 10043
Association for the Preservation of Virginia Antiquities, Richmond 14353
The Association for the Protection of Rural Scotland, Edinburgh 11288
Association for the Reduction of Aircraft Noise, London 11289
Association for the Reform of the Latin Teaching, Woking 11290
Association for the Sociological Study of Jewry, New London 14354
Association for the Sociology of Religion, Lancaster 14355
Association for the Study of Afro-American Life and History, Washington 14356
Association for the Study of Animal Behaviour, Egham 11291
Association for the Study of Canadian Radio and Television, Montréal 01906
Association for the Study of Dada and Surrealism, Los Angeles 14357
Association for the Study of German Politics, Sheffield 11292
Association for the Study of Higher Education, College Station 14358
Association for the Study of Jewish Languages, Oakland Garden 14359
Association for the Study of Man Environment Relations, Montevideo 13249
Association for the Study of Medical Education, Dundee 11293
Association for the Study of Obesity, London 11294
Association for the Study of the World Refugee Problem, A.W.R., Roma 07392
Association for the Taxonomic Study of the Flora of Tropical Africa, Zomba 08621
Association for the Teaching of the Social Sciences, Saint Albans 11295
Association for the Treatment of Sexual Abusers, Portland 14360
Association for the World University, Allentown 14361
Association for Transpersonal Psychology, Stanford 14362
Association for Tropical Biology, New Orleans 14363
Association for University Business and Economic Research, Indianapolis 14364
Association for Veterinary Clinical Pharmacology and Therapeutics, Dorking 11296
Association for Women Geoscientists, Saint Paul 14365
Association for Women in Computing, San Francisco 14366
Association for Women in Mathematics, College Park 14367
Association for Women in Psychology, Philadelphia 14368
Association for Women in Science, Washington 14369
Association for Women Veterinarians, Union City 14370
Association for World Education, Snedsted 03561
Association Française d'Acupuncture, Paris 04211
Association Française d'Astronomie, Paris 04212
Association Française de Chirurgie, Paris 04213
Association Française de Formation, Paris 04214
Association Française de Formation, Coopération, Promotion et Animation d'Entreprises, Paris 04215
Association Française de Gemmologie, Paris 04216
Association Française de Génie Rural, Paris 04217
Association Française de l'Eclairage, Paris 04218
Association Française de l'Ecole Paysanne, Paris 04219
Association Française de Management, Paris 04220
Association Française de Normalisation, Paris La Défence 04221
Association Française de Prévention des Accidents de Travail et Incendie, Paris 04222
Association Française de Psychiatrie, Ville d'Avray 04223
Association Française de Recherches d'Essais sur les Matériaux et les Constructions, Paris 04224
Association Française des Arabisants, Paris 04225
Association Française de Science Economique, Toulouse 04226
Association Française de Science Politique, Paris 04227
Association Française des Conseils Scolaires de l'Ontario, Ottawa 01907
Association Française des Enseignants de Français, Paris 04228
Association Française des Femmes Médecins, Paris 04229
Association Française des Hémophiles, Paris 04230
Association Française des Historiens Economistes, Paris 04231
Association Française des Ingénieurs, Chimistes et Techniciens des Industries du Cuir, Lyon 04232
Association Française des Instituts de Recherche sur le Développement, Paris 04233
Association Française des Professeurs de Langues Vivantes, Paris 04234
Association Française des Sociétés d'Etudes et de Conseils Exportatrices, Paris 04235
Association Française d'Etude des Relations Professionnelles, Fresnes 04236
Association Française d'Etudes Américaines, Paris 04237
Association Française d'Observateurs d'Etoiles Variables, Strasbourg 04238
Association Française du Froid, Paris 04239
Association Française Inter Médicale, Paris 04240
Association Française pour la Protection des Eaux, Paris 04241
Association Française pour la Recherche et la Création Musicales, Paris 04242
Association Française pour le Développement de la Stomatologie, Paris 04243
Association Française pour l'Etude des Eaux, Limoges 04244
Association Française pour l'Etude du Cancer, Paris 04245
Association Française pour l'Etude du Quaternaire, Caen 04246
Association Française pour l'Etude du Sol, Plaisir 04247
Association Française pour l'Information en Economie Ménagère, Paris 04248
Association Francophone d'Education Comparée, Sèvres 04249
Association Francophone de Spectrométrie des Masses Solides, Lannion 04250
Association Francophone Internationale des Directeurs d'Etablissements Scolaires, Montréal 01908
Association Francophone Internationale des Groupes d'Animation de la Paraplégie, Brie-Comte-Robert 04251
Association Générale des Conservateurs des Collections Publiques de France, Paris 04252
Association Générale des Hygiénistes et Techniciens Municipaux, Paris 04253
Association Générale des Médecins de France, Paris 04254
Association Géologique du Canada, Saint John's 02482
Association Guillaume Budé, Paris 04255
Association Henri Capitant, Baton Rouge 14371
Association Historique Internationale de l'Océan Indien, Sainte-Clotilde 09733
Association in Scotland to Research into Astronautics, Wishaw 11297
Association Internationale de Cybernétique, Namur 01135
Association Internationale de Droit Economique, Louvain-la-Neuve 01136
Association Internationale de Droit Pénal, Pau 04256
Association Internationale d'Education Périnatale, Calgary 02545
Association Internationale de Géodésie, Paris 04257
Association Internationale de la Securité Sociale, Genève 10627
Association Internationale de l'Inspection du Travail, Paris 04258
Association Internationale de Littérature Comparée, Paris 04259
Association Internationale de Médecine et de Biologie de l'Environnement, Paris 04260
Association Internationale de Pédagogie Universitaire, Montréal 01909
Association Internationale de Pédiatrie, Paris 04261
Association Internationale d'Epigraphie Grecque et Latine, Paris 04262
Association Internationale de Psychothérapie de Groupe, Montréal 02541
Association Internationale de Recherche en Informatique Toxicologique, Paris 04263
Association Internationale des Amis de Vasile Stanciu, Paris 04264
Association Internationale des Archives Francophones, Tunis 11135
Association Internationale des Arts Plastiques, Paris 04265
Association Internationale des Critiques de Théâtre, Paris 04266
Association Internationale des Critiques Littéraires, Paris 04267
Association Internationale des Démographes de Langue Française, Paris 04268
Association Internationale des Directeurs d'Ecoles Hôtelières, Lausanne 10628
Association Internationale des Docteurs en Economie du Tourisme, Aix-en-Provence 04269
Association Internationale des Ecoles des Sciences de l'Information, Paris 04270
Association Internationale des Ecoles Supérieures d'Education Physique, Liège 01137
Association Internationale des Educateurs de Jeunes Inadaptés, Pau 04271
Association Internationale des Etudes Arméniennes, Jerusalem 07045
Association Internationale des Etudes Coptes, Louvain-la-Neuve 01138
Association Internationale des Etudes de l'Asie du Sud-Est, Paris 04272
Association Internationale des Etudes Françaises, Paris 04273
Association Internationale des Juristes Démocrates, Bruxelles 01139
Association Internationale des Laboratoires Textiles Lainiers, Bruxelles 01140
Association Internationale des Mathématiques et Calculateurs en Simulation, Liège 01141
Association Internationale des Métiers et Enseignements d'Art, Bruxelles 01142
Association Internationale de Sociologie, Madrid 10330
Association Internationale des Ponts et Charpentes (AIPC), Zürich 10742
Association Internationale des Professeurs de Langue et Littérature Russes, Paris 04274
Association Internationale des Professeurs de Philosophie, Minden 05155
Association Internationale des Professeurs et Maîtres de Conférences des Universités, Nancy 04275
Association Internationale d'Essais de Semences, Zürich 10745
Association Internationale des Sciences Juridiques, Paris 04276
Association Internationale des Sociologues de Langue Française, Toulouse 04277
Association Internationale des Statisticiens d'Enquêtes, Paris 04278
Association Internationale des Techniciens Biologistes de Langue Française, Le Perreux 04279
Association Internationale des Travaux en Souterrain, Bron 04280
Association Internationale des Universités, Paris 04281
Association Internationale d'Etudes Occitanes, Oegstgeest 08824
Association Internationale d'Histoire Economique, Paris 04282
Association Internationale d'Information Scolaire, Universitaire et Professionnelle, Paris 04283
Association Internationale d'Irradiation Industrielle, Charbonnières-les-Bains 04284
Association Internationale d'Océanographie Médicale, Nice 04285
Association Internationale Données pour le Développement, Marseille 04286
Association Internationale d'Orientation Scolaire et Professionelle, Strassen 08578
Association Internationale du Théâtre pour l'Enfance et la Jeunesse, København 03562
Association Internationale du Théâtre pour l'Enfance et la Jeunesse, Montréal 01910
Association Internationale Francophone de Recherche Odontologique, Rennes 04287
Association Internationale pour la Coopération et le Développement en Afrique Australe, Winksele 01143
Association Internationale pour la Protection de la Propriété Industrielle, Zürich 10743
Association Internationale pour le Développement de l'Odonto-Stomatologie Tropicale, Bordeaux 04288
Association Internationale pour le Développement des Gommes Naturelles, Neuilly-sur-Seine 04289
Association Internationale pour le Développement des Universités Internationales et Mondiales, Aulnay-sous-Bois 04290
Association Internationale pour le Management du Sport, Saint-Michel 04291
Association Internationale pour le Progrès Social, Bruxelles 01144
Association Internationale pour l'Etude de la Paléontologie Humaine, Paris 04292
Association Internationale pour l'Etude du Comportement des Conuducteurs, Paris 04293
Association Internationale pour l'Histoire Contemporaine de l'Europe, Genève 10629
Association Internationale pour l'Utilisation des Langues Régionales à l'Ecole, Liège 01145
Association Interprofessionnelle de France, Seclin 04294
Association Interprofessionnelle des Centres Médicaux et Sociaux de la Région Parisienne, Paris 04295
Association Interprofessionnelle pour la Formation Permanente dans le Commerce Textile, Paris 04296
Association Laïque pour l'Education et la Formation Professionnelle des Adolescents en Europe, Bruxelles 01146
Association Les Amis de Gustave Courbet, Ornans 04297
Association Libanaise des Sciences Juridiques, Beirut 08560
Association Linguistique Franco-Européenne, Paris 04298
Association Littéraire et Artistique Internationale, Paris 04299
Association Luxembourgeoise des Kinésithérapeutes Diplomés, Luxembourg 08579
Association Lyonnaise de Criminologie et Anthropologie Sociale, Lyon 04300
Association Maghrébine des Etudes de la Population, Tunis 11136
Association Marc Bloch, Paris 04301
Association Maria Montessori, Paris 04302
Association Mathématique du Québec, L'Assomption 01911
Association Médicale Canadienne, Ottawa 02197
Association Médicale de la Défense du Canada, Ottawa 02413
Association Médicale du Québec, Montréal 01912
Association Médicale Européenne, Bruxelles 01147
Association Médico-Sociale Protestante de Langue Française, Alfortville 04303
Association Métrique Canadienne, Fonthill 02200
Association Mondiale des Médecins Francophones, Ottawa 01913
Association Mondiale des Sciences de l'Education, Gent 01148
Association Mondiale des Vétérinaires Microbiologistes, Immunologistes et Spécialistes des Maladies Infectieuses, Maisons Alfort 04304
Association Monégasque de Préhistoire, Monaco 08794
Association Montessori Internationale, Amsterdam 08825
Association Musées du Nouveau-Brunswick, Saint John 01914

Association Museums of New Brunswick, Saint John 01914
Association Nationale d'Art Photographique, Scarborough 02668
Association Nationale de la Recherche Technique, Paris 04305
Association Nationale des Assistants de Service Social, Paris 04306
Association Nationale des Bibliothécaires, Paris 04307
Association Nationale des Cours Professionnels pour les Préparateurs en Pharmacie, Paris 04308
Association Nationale des Docteurs en Droit, Paris 04309
Association Nationale des Educateurs de Jeunes Inadaptés, Paris 04310
Association Nationale des Géographes Marocains, Rabat 08802
Association Nationale des Professeurs en Economie Sociale et Familiale, Paris 04311
Association Nationale pour la Formation et la Promotion Professionnelle dans l'Industrie et le Commerce de la Chaussure et des Cuirs et Peaux, Paris 04312
Association Nationale pour la Protection contre l'Incendie, Ottignies 01149
Association Nationale pour la Protection des Eaux, Paris 04313
Association Nationale pour la Protection des Villes d'Art, Paris 04314
Association Nationale pour la Réhabilitation Professionelle par le Travail Protégé, Paris 04315
Association Nationale pour l'Etude de la Neige et des Avalanches, Grenoble 04316
Association Nucléaire Canadienne, Toronto 02211
Association Oecuménique des Théologiens Africains, Yaoundé 01812
Association of Academic Health Centers, Washington 14372
Association of Academic Health Sciences Library Directors, Houston 14373
Association of Academic Physiatrists, Indianapolis 14374
Association of Accredited Cosmetology Schools, Falls Church 14375
Association of Advanced Rabbinical and Talmudic Schools, New York 14376
Association of Advisers in Design and Technical Studies, Camberley 11298
Association of Advisers on Education in International Religious Congregations, Roma 07251
Association of African Faculties of Agriculture, Rabat 08803
Association of African Geological Surveys, Orléans 04317
Association of African Maritime Training Institutes, Alexandria 03880
Association of African Universities, Accra 06477
Association of African Women for Research and Development, Dakar 10021
Association of Agricultural Education Staffs, Mogerhanger 11299
Association of Agricultural Research Institutions in the Near East and North Africa, Roma 07252
Association of Air Medical Services, Pasadena 14377
Association of all Medical Faculties in the Federal Republic of Germany, Münster 06117
Association of Amazonian Universities, Belém 01613
Association of American Colleges, Washington 14378
Association of American Geographers, Washington 14379
Association of American Indian Physicians, Oklahoma City 14380
Association of American Jurists, Berkeley 14381
Association of American Law Schools, Washington 14382
Association of American Medical Colleges, Washington 14383
Association of American Physicians, Indianapolis 14384
Association of American Physicians and Surgeons, Tucson 14385
Association of American Schools in South America, Miami 14386
Association of American Seed Control Officials, Salem 14387
Association of American State Geologists, Champaign 14388
Association of American Universities, Washington 14389
Association of American Veterinary Medical Colleges, Washington 14390
Association of Anaesthetists of Great Britain and Ireland, London 11300
Association of Analytical Chemists, Detroit 14391
Association of Ancient Historians, Los Angeles 14392
Association of Animal Husbandry and Veterinary Medicine of Taiwan, Nantou Hsien 03245
The Association of Applied Biologists, Littlehampton 11301
Association of Applied Insect Ecologists, Sacramento 14393
Association of Arab-American University Graduates, Belmont 14394
Association of Arab Geologists, Baghdad 06916
Association of Arab Institutes and Centres for Economic and Social Research, Tunis 11137
Association of Arab Universities, Amman 08451
Association of Art Historians in Poland, Warszawa 09616
Association of Art Institutions, Hereford 11302
Association of Arts and Letters, Athinai 06500
Association of Arts Centres in Scotland, Aberdeen 11303
Association of Asian-Pacific Operational Research Societies, Tokyo 08112
Association of Asian Social Science Research Councils, New Delhi 06732
Association of Asphalt Paving Technologists, Maplewood 14395
Association of Australasian and Pacific Area Police Medical Officers, Melbourne 00229
Association of Avian Veterinarias, Boca Raton 14396
Association of Aviation Psychologists, Colorado Springs 14397
Association of Balloon and Airship Constructors, San Diego 14398
Association of Basic Science Teachers in Dentistry, Glasgow 11304
Association of Beauty Teachers, Southampton 11305
Association of Biological Collections Appraisers, Carmel 14399
Association of Biomedical Communication Directors, Hershey 14400
Association of Black Anthropologists, Washington 14401
Association of Black Psychologists, Washington 14402
Association of Black Sociologists, Washington 14403
Association of Black Women in Higher Education, Albany 14404
Association of Blind and Partially-Sights Teachers and Students, London 11306
Association of Bone and Joint Surgeons, Rosemont 14405
Association of British Climatologists, London 11307
Association of British Columbia Archivists, Vancouver 01915
Association of British Columbia School Superintendents, Osoyoos 01916
Association of British Correspondence Colleges, London 11308
Association of British Dental Surgery Assistants, Fleetwood 11309
Association of British Dispensing Opticians, London 11310
Association of British Forensic Specialists, Moseley 11311
Association of British Geodesists, Dagenham 11312
Association of British Neurologists, London 11313
Association of British Paediatric Nurses, London 11314
Association of British Pewter Craftsmen, Birmingham 11315
Association of British Science Writers, London 11316
Association of British Spectroscopists, Kent 11317
Association of British Theological and Philosophical Libraries, Cambridge 11318
Association of Building Engineers, Northampton 11319
Association of Business Officers of Preparatory Schools, Windsor 14406
Association of Canadian Archivists, Ottawa 01917
Association of Canadian Bible Colleges, Kitchener 01918
Association of Canadian Career Colleges, Brantford 01919
Association of Canadian College and University Information Bureaus, Halifax 01920
Association of Canadian Community Colleges, Toronto 01921
Association of Canadian Faculties of Dentistry, Edmonton 01922
Association of Canadian Industrial Designers, Rexdale 01923
Association of Canadian Map Libraries, Ottawa 01924
Association of Canadian Medical Colleges, Ottawa 01925
Association of Canadian Universities for Northern Studies, Ottawa 01926
Association of Canadian University and College Teachers of French, Victoria 01891
Association of Canadian University Planning Programs, Halifax 01927
Association of Career Teachers, Loughborough 11320
Association of Caribbean Economists, Kingston 08068
Association of Caribbean Historians, Nassau 01050
Association of Caribbean Studies, Lexington 14407
Association of Caribbean Universities and Research Institutes, San Juan 09722
Association of Caribbean University, Research and Institutional Libraries, San Juan 09723
Association of Catholic Colleges and Universities, Washington 14408
Association of Catholic Diocesan Archivists, East Lansing 14409
Association of Cereal Research, Detmold 05115
Association of Chairmen of Departments of Mechanics, Gainesville 14410
Association of Chartered Engineers in Iceland, Reykjavik 06706
Association of Chief Architects of Scottish Local Authorities, Lanark 11321
Association of Child Education of the Republic of China, Taipei 03246
Association of Child Psychotherapists, London 11322
Association of Chiropractic Colleges, San Lorenzo 14411
Association of Christian Librarians, Cedarville 14412
Association of Christian Teachers, London 11323
Association of Clinical Biochemists, London 11324
Association of Clinical Pathologists, Brighton 11325
Association of Clinical Scientists, Farmington 14413
Association of College and Research Libraries, Chicago 14414
Association of College and University Telecommunications Administrators, Lexington 14415
Association of Colleges for Further and Higher Education, Swindon 11326
Association of Collegiate Schools of Architecture, Washington 14416
Association of Collegiate Schools of Planning, Madison 14417
Association of Commonwealth Archivists and Records Managers, London 11327
Association of Commonwealth Teachers, London 11328
Association of Commonwealth Universities, London 11329
Association of Community Cancer Centers, Rockville 14418
Association of Community College Trustees, Washington 14419
Association of Community Health Councils for England and Wales, London 11330
Association of Community Tribal Schools, Sisseton 14420
Association of Community Workers, Newcastle-upon-Tyne 11331
Association of Computer Users, Berkeley 14421
Association of Concern for Ultimate Reality and Meaning, Toronto 01928
Association of Consulting Actuaries, London 11332
Association of Consulting Chemists and Chemical Engineers, New York 14422
Association of Consulting Engineers of Ireland, Dublin 06926
Association of Continuing Legal Education Administrators, Chicago 14423
Association of County Public Health Officers, Malmesbury 11333
Association of Crematorium Medical Referees, Westcliff-on-Sea 11334
Association of Cytogenetic Technologists, Augusta 14424
The Association of Danish Biologists, Birkerød 03727
Association of Danish Graduates in Forestry, Klampenborg 03605
Association of Danish Graduates in Horticulture, Klampenborg 03631
Association of Danish Music Libraries, København 03651
Association of Danish Physiotherapists, København 03606
Association of Data Communications Users, Bloomington 14425
Association of Deans of Pharmacy of Canada, Toronto 01929
Association of Deans of Southeast Asian Graduate Schools of Management, Manila 09509
Association of Defense Trial Attorneys, Peoria 14426
Association of Dental Hospitals of Great Britain and Northern Ireland, Birmingham 11335
Association of Departments of English, New York 14427
Association of Departments of Foreign Languages, New York 14428
Association of Development Research and Training Institutes of Asia and the Pacific, Kuala Lumpur 08639
Association of Directors of Education in Scotland, Dundee 11336
Association of Directors of Journalism Programs in Canadian Universities, London 01930
Association of Directors of Social Services, Stockport 11337
Association of Directors of Social Work, Melrose 11338
Association of Disabled Professionals, Horbury 11339
Association of Disciples for Theological Discussion, Saint Louis 14429
Association of District Medical Officers, Bath 11340
Association of Earth Science Editors, Lawrence 14430
Association of Eastern and Southern African Universities, Kampala 11182
Association of Economic Jurisprudence, Tokyo 08160
Association of Economic Scientific Institutions, Moskva 09801
Association of Educational Psychologists, Durham 11341
Association of Educators of Gifted, Talented and Creative Children in B.C., Coquitlam 01931
Association of Energy Engineers, Atlanta 14431
The Association of Engineering Firms Practicing in the Geosciences, Silver Spring 14432
Association of Engineering Geologists, Sudbury 14433
Association of Engineers, Lisboa 09664
Association of Engineers and Architects in Israel, Tel Aviv 07046
Association of Engineers and Technicians of Slovenia, Ljubljana 10097
Association of Environmental Engineering Professors, Houghton 14434
Association of Episcopal Colleges, New York 14435
Association of European Cancer Leagues, København 03563
Association of European Conjuncture Institutes, Louvain-la-Neuve 01119
Association of European Correspondence Schools, Arnhem 08826
Association of European Federations of Agro-Engineers, Bad Honnef 05156
Association of European Geological Societies, London 11342
Association of European Latin Americanist Historians, Liverpool 11343
Association of European Open Air Museums, Chichester 11344
Association of European Operational Research Societies, Bologna 07253
Association of European Paediatric Cardiologists, Helsinki 03951
Association of European Psychiatrists, Luxembourg 08580
Association of European Schools and Colleges of Optometry, Bures-sur-Yvette 04318
Association of European Schools of Planning, Nijmegen 08827
Association of Exploration Geochemists, Rexdale 01932
Association of Faculties of Pharmacy of Canada, Vancouver 01933
Association of Faculties of Science in African Universities, Nairobi 08479
Association of Federal Communications Consulting Engineers, Washington 14436
Association of Field Ornithologists, South Natick 14437
Association of Finnish Authors, Helsinki 04038
Association of Finnish Chemical Societies, Helsinki 04036
Association of Firearm and Tool Mark Examiners, La Jolla 14438
Association of Folklorists of Macedonia, Skopje 08619
Association of Food and Drug Officials, York 14439
Association of Forensic Document Examiners, Minneapolis 14440
Association of French-Language Leprologists, Paris 04191
Association of French-Speaking and European University Administrations in Medicine and Odontology, Paris 04123
Association of French-Speaking Computer Professionals and Users, Paris 04188
Association of French-Speaking Dermatologists and Syphilligraphers, Créteil 04175
Association of French-Speaking Planetariums, Strasbourg 04319
Association of French-Speaking Psychologists of the Working Environment, Versailles 04156
Association of French Teachers in Africa, Khartoum 10425
Association of Genealogists and Record Agents, Horsham 11345
Association of General Practitioner Hospitals, Lichfield 11346
Association of General Surgeons of the Province of Quebec, Montréal 01878
Association of Geoscientists for International Development, Bangkok 11088
Association of German Engineers, Düsseldorf 06377
Association of Governing Bodies of Girls Public Schools, Petersfield 11347
Association of Governing Bodies of Public Schools (Boys), Petersfield 11348
Association of Graduate Liberal Studies Programs, Chicago 14441
Association of Graduate Schools in Association of American Universities, Washington 14442
Association of Greek Chemists, Athinai 06519
Association of Hairdressing Teachers in Colleges of Further Education, Balby 11349
Association of Head and Neck Oncologists of Great Britain, London 11350
Association of Headmistresses of Preparatory Schools, London 11351
Association of Heads of Independent Schools, Windor 11352
Association of Health Care Information and Medical Records Officers, Hopton 11353
Association of Health Service Treasurers, Warwick 11354
The Association of Higher Education Facilities Officers, Alexandria 14443
Association of Higher Education Libraries in French-Speaking West Africa, Dakar 10017
Association of Hispanists of Great Britain and Ireland, London 11355
Association of Historians of Latin America and the Caribbean, Paris 04139
Association of Historic Councils and Regional History Institutes, Marburg 05118
Association of Hungarian Librarians, Budapest 06644
Association of Hungarian Writers, Budapest 06639
Association of Independent Colleges and Schools, Washington 14444
Association of Independent Libraries, Manchester 11356
Association of Independent Research and Technology Organisations, Cambridge 11357
Association of Indian Universities, New Delhi 06733
Association of Information and Dissemination Centers, Athens 14445
Association of Institute and School of Education In-Service Tutors, Durham 11358
Association of Institutes of European Studies, Genève 10621
Association of Internal Management Consultants, East Bloomfield 14446
Association of International Accountants, Gateshead 11359
Association of International Colleges and Universities, Independence 14447
Association of International Consultants on Human Rights, Genève 10630
Association of International Development of Natural Gums, Neuilly-sur-Seine 04289
Association of International Education, Tokyo 08232
Association of International Health Researchers, Mobile 14448
Association of International Industrial Irradiation, Charbonnières-les-Bains 04284
Association of International Law, Moskva 09802
Association of International Law, Tokyo 08172
Association of International Libraries, Genève 10619
Association of International Schools in Africa, Nairobi 08480
Association of Irish Headmistresses, Dun Laoghaire 06927
Association of Irish Jurists, Dublin 06928
Association of Irish Musical Societies, Dublin 06929
Association of Iron and Steel Engineers, Pittsburgh 14449
Association of Jesuit Colleges and Universities, Washington 14450
Association of Jewish Libraries, New York 14451
Association of Jurists of Serbia, Beograd 17077
Association of Large School Boards in Ontario, Toronto 01934
Association of Latin American Lawyers for the Defense of Human Rights, São Paulo 01614
The Association of Law Teachers, Leicester 11360
Association of Lecturers in Accountancy, Edinburgh 11361
Association of Lecturers in Colleges of Education in Scotland, Beech Grove 11362
Association of Legal Administrators, Vernon Hills 14452
Association of Libraries of Judaica and Hebraica in Europe, Paris 04170

Association of Life Insurance Counsel, Newark 14453
Association of Life Insurance Medicine of Japan, Tokyo 08207
Association of London Chief Librarians, Barking 11363
Association of Lutheran College Faculties, Minneapolis 14454
Association of Lutheran Secondary Schools, Saint Lucas 14455
Association of Maize Researchers in Africa, Dakar 10022
Association of Management, Grafton 14456
Association of Management Consultants, New York 14457
Association of Management Development Institutions in Asia, Hyderabad 06734
Association of Management Training Institutions of Eastern and Southern Africa, Arusha 11047
Association of Manitoba Museums, Winnipeg 01935
Association of Manufacturing Chemists, London 11364
Association of Marine Engineering Schools, Liverpool 11365
Association of Marine Laboratories of the Caribbean, Mayaguez 09724
Association of Maternal and Child Health Programs, Washington 14458
Association of Mathematicians, Physicists and Astronomers of Slovenia, Ljubljana 10098
Association of Meat Inspectors, Barnet 11366
Association of Medical Advisers to the Pharmaceutical Industry, London 11367
Association of Medical Directors of the Australian Pharmaceutical Industry, West Ryde 00230
Association of Medical Doctors for Asia, Okayana 08113
Association of Medical Education and Research in Substance Abuse, Providence 14459
Association of Medical Rehabilitation Administrators, San Diego 14460
Association of Medical School Pediatric Department Chairmen, Salt Lake City 14461
Association of Medical Schools in Africa, Lusaka 17151
Association of Medical Schools in the Middle East, Alexandria 03881
Association of Mental Health Administrators, Northbrook 14462
Association of Mercy Colleges, Erie 14463
Association of Military Colleges and Schools of the United States, Alexandria 14464
Association of Military Surgeons of the U.S., Bethesda 14465
Association of Minicomputers Users, Franklin 14466
Association of Minority Health Professions Schools, Washington 14467
Association of Municipal Engineers, London 11368
Association of Museums Societies of Croatia, Zagreb 03466
Association of Muslim Scientists and Engineers, Plainfield 14468
Association of Muslim Social Scientists, Herndon 14469
Association of National European and Mediterranean Societies of Gastroenterology, London 11369
Association of National Health Service Officers, Wareham 11370
Association of National Park Officers, Kendal 11371
Association of Naval R.O.T.C. Colleges and Universities, Lafayette 14470
Association of Noise Consultants, Guilden Morden 11372
Association of Nordic Paper Historians, Stockholm 10437
Association of Northumberland Local History Societies, Newcastle-upon-Tyne 11373
The Association of Norwegian Visual Artists, Oslo 09297
Association of Nova Scotia Education Administrators, Sydney 01936
The Association of Obstetrics and Gynecology of the Republic of China, Taipei 03247
Association of Occupational Therapists of Manitoba, Winnipeg 01937
Association of Official Architects, London 11374
Association of Official Racing Chemists, Portland 14471
Association of Official Seed Analysts, Lincoln 14472
Association of Old Crows, Alexandria 14473
Association of Ophthalmic Opticians of Ireland, Dublin 06930
Association of Ophthalmists, Teheran 06897
Association of Optometric Educators, Tahlequah 14474
Association of Orientalists, Moskva 09803
Association of Orthodox Jewish Scientists, New York 14475
Association of Orthodox Jewish Teachers, Brooklyn 14476
Association of Osteopathic State Executive Directors, Sacramento 14477
Association of Overseas Educators, Indiana 14478
Association of Pacific Systematists, Honolulu 14479
Association of Paediatric Education in Europe, Lisboa 09652
Association of Painting Craft Teachers, Saint Leonards-on-Sea 11375
Association of Parliamentary Librarians in Canada, Ottawa 01938
Association of Pathology Chairmen, Bethesda 14480
Association of Pediatric Societies of the Southeast Asian Region, Manila 09510
Association of Pharmacists of Serbia, Beograd 17078
Association of Philippine Physicians in America, Burbank 14481
Association of Physician Assistant Programs, Alexandria 14482
Association of Physician's Assistants in Cardiovascular Surgery, Lynchburg 14483
Association of Police Surgeons, Northampton 11376
Association of Political Sciences, Moskva 09804
Association of Polytechnic Teachers, Southsea 11377
Association of Presbyterian Colleges and Universities, Louisville 14484
Association of Principals of Colleges, Broxborn 11378
Association of Principals of Colleges (Northern Ireland Branch), Lisburn 11379
Association of Professional Baseball Physicians, Minneapolis 14485
Association of Professional Energy Managers, Sacramento 14486
Association of Professional Genealogists, Washington 14487
Association of Professional Geologists and Geophysicists of Quebec, Montréal 01952
Association of Professional Material Handling Consultants, Charlotte 14488
Association of Professional Scientists and Technologists, London 11380
Association of Professors of Gynecology and Obstetrics, Washington 14489
Association of Professors of Medicine, Washington 14490
Association of Psychiatrists in Training, Edinburgh 11381
Association of Psychologists of Nova Scotia, Millstream 01939
Association of Public Analysts, London 11382
Association of Quebec Regional English Media, Anne de Bellevue 01940
Association of Recognised English Language Services, London 11383
Association of Registrars of Universities and Colleges of Canada, Ottawa 01941
Association of Religious in Education, London 11384
Association of Religious Writers, Jerusalem 07047
Association of Research Directors, Orangeburg 14491
Association of Researchers in Medicine and Science, London 11385
Association of Research Libraries, Washington 14492
Association of Research Library Employees, Nacka 10604
Association of Roman Ceramic Archaeologists, Nijmegen 08828
Association of Safety Council Executives, Gainesville 14493
Association of School Business Officials International, Reston 14494
Association of School Business Officials of Saskatchewan, Yorkton 01942
Association of School Natural History Societies, Lancing 11386
Association of Schools and Colleges of Optometry, Rockville 14495
Association of Schools of Allied Health Professions, Washington 14496
Association of Schools of Journalism and Mass Communication, Columbia 14497
Association of Schools of Optometry of Canada, Waterloo 01943
Association of Schools of Public Health, Washington 14498
Association of Schools of Public Health in the European Region, Bristol 11387
Association of Schools of Public Health in the European Region, Maastricht 08829
Association of Science Museum Directors, Chicago 14499
Association of Sciences and Art, Bitola 08600
Association of Science-Technology Centers, Washington 14500
Association of Scientific, Technical and Managerial Staffs, London 11388
Association of Scottish Local Health Councils, Edinburgh 11389
Association of Secondary Teachers, Ireland, Dublin 06931
Association of Seventh-Day Adventist Educators, Silver Spring 14501
Association of Show and Agricultural Organisations, Newark 11390
Association of Sinologists, Moskva 09805
Association of Ski Schools in Great Britain, Grantown-on-Spey 11391
Association of Slovak Mathematicians and Physicists, Bratislava 10057
Association of Small Public Libraries of Ontario, Tillsonburg 01944
Association of Social and Behavioral Scientists, Washington 14502
Association of Social Anthropologists of the Commonwealth, London 11392
Association of Social Research Organisation, London 11393
Association of Southeast Asian Institutions of Higher Learning, Bangkok 11089
Association of Southeast Asian Marine Scientists, Manila 09511
Association of South-East Asian Nation Countries' Union of Polymer Science, Sungai Ehsan 08640
Association of Southern Baptist Colleges and Schools, Nashville 14503
Association of South Pacific Environmental Institutions, Mangilao 06552
Association of Space Explorers U.S.A-, San Francisco 14504
Association of Specialized and Cooperative Library Agencies, Chicago 14505
Association of Special Libraries of the Philippines, Manila 09512
Association of Sports Medicine of the Balkan, Nicosia 03487
Association of State and Interstate Water Pollution Control Administrators, Washington 14506
Association of State and Territorial Dental Directors, Sacramento 14507
Association of State and Territorial Directors of Public Health Education, Columbia 14508
Association of State and Territorial Health Officials, Washington 14509
Association of State Supervisors of Mathematics, Little Rock 14510
Association of Surgeons of East Africa, Nairobi 08481
Association of Surgeons of Great Britain and Ireland, London 11394
Association of Surgeons of India, Madras 06735
Association of Surgeons of South Africa, Johannesburg 10124
Association of Surgical Technologists, Englewood 14511
Association of Swimming Therapy, Shrewsbury 11395
Association of Systematics Collections, Washington 14512
Association of Teacher Educators, Reston 14513
Association of Teachers of Japanese, Middlebury 14514
Association of Teachers of Management, London 11396
Association of Teachers of Maternal and Child Health, San Diego 14515
Association of Teachers of Mathematics, Derby 11397
Association of Teachers of Preventive Medicine, Washington 14516
Association of Teachers of Printing and Allied Subjects, Liverpool 11398
Association of Teachers of Technical Writing, Orlando 14517
Association of the Friends of Miguel Angel Asturias, Paris 04164
Association of the Geographical Societies of Slovenia, Ljubljana 10115
Association of the German Nobility in North America, Benicia 14518
Association of the Health Occupations Teacher Educators, Raleigh 14519
Association of the Institute for Certification of Computer Professionals, Des Plaines 14520
Association of the Mathematicians', Physicists' and Astronomers' Societies of Yugoslavia, Priština 17079
Association of Theological Institutes in the Middle East, Beirut 08561
Association of Theological Institutions of Eastern Africa, Nairobi 08482
Association of Theological Schools, Pittsburgh 14521
Association of the Slovak Librarians and Information Workers, Bratislava 10095
Association of Third World Affairs, Washington 14522
Association of Track and Field Statisticians, Salisbury 11399
Association of Trial Lawyers of America, Washington 14523
Association of Tutors, Cambridge 11400
Association of Tutors in Adult Education, Nottingham 11401
Association of United States Members of the International Institute of Space Law, Dumfries 14524
Association of Universities and Colleges of Canada, Ottawa 01945
Association of Universities for Research in Astronomy, Washington 14525
Association of Universities of Bangladesh, Dhaka 01064
Association of University Anesthesiologists, Seattle 14526
Association of University Architects, New Haven 14527
Association of University Forestry Schools of Canada, Thunder Bay 01946
Association of University of New Brunswick Teachers, Fredericton 01947
Association of University Professors of Ophthalmology, San Francisco 14528
Association of University Programs in Health Administration, Arlington 14529
Association of University Radiation Protection Officers, Cardiff 11402
Association of University Radiologists, Reston 14530
Association of University Related Research Parks, Tempe 14531
Association of University Summer Sessions, Bloomington 14532
Association of University Teachers, London 11403
Association of University Teachers in Accounting, Sheffield 11404
Association of University Teachers (Scotland), Glasgow 11405
Association of Veterinary Anaesthetists of Great Britain and Ireland, Bury Saint Edmunds 11406
Association of Veterinary Teachers and Research Workers, Edinburgh 11407
Association of Vice-Principals in Colleges, Maidenhead 11408
Association of Visual Science Librarians, Portland 14533
Association of Vitamin Chemists, Northville 14534
Association of Voluntary Aided Secondary Schools, Beckenham 11409
Association of Wholly or Partially French Language Universities, Montréal 01948
Association of Workers' Compensation Boards of Canada, Scarborough 01949
Association of Workers for Maladjusted Children, Kettering 11410
Association of Young Irish Archaeologists, Belfast 11411
Association of Youth Museums, Memphis 14535
Association of Yugoslav Universities, Beograd 17138
Association Olympique Internationale pour la Recherche Médico-Sportive, Lausanne 10631
Association on Boarding Schools, Boston 14536
Association on Marginal Literature and Art, Leuven 01373
Association Ontarienne de Gérontologie, Waterloo 02779
Association Ontarienne des Agents de l'Administration Scolaire, Toronto 02758
Association Paritaire de Prévention pour la Santé et la Securité du Travail, Montréal 01950
Association Pharmaceutique Canadienne, Ottawa 02223
Association Philotechnique, Paris 04320
Association pour la Fondation Internationale du Cinéma et de la Communication Audiovisuelle, Cannes 04321
Association pour la Formation aux Professions Immobilières, Paris 04322
Association pour la Formation des Cadres de l'Industrie et du Commerce, Paris 04323
Association pour la Formation Européenne des Travailleurs aux Technologies, Bruxelles 01132
Association pour la Formation Professionnelle dans les Industries Céréalières, Paris 04324
Association pour la Formation Professionnelle dans les Industries de l'Ameublement, Paris 04325
Association pour la Médiathèque Public, Cambrai 04326
Association pour la Prévention de la Pollution Atmosphérique, Paris 04327
Association pour la Promotion de la Pédagogie Nouvelle, Paris 04328
Association pour la Promotion des Publications Scientifiques, Bruxelles 01150
Association pour la Recherche et le Développement en Informatique Chimique, Paris 04329
Association pour la Recherche Interculturelle, Saint-Denis 04330
Association pour la Rééducation de la Parole et du Langage Oral et Ecrit et de la Voix, Paris 04331
Association pour l'Avancement des Sciences au Canada, Ottawa 01905
Association pour l'Avancement des Sciences et des Techniques de la Documentation, Montréal 01951
Association pour le Développement de la Formation Professionnelle Continue dans les Industries Lourdes du Bois, Paris 04332
Association pour le Développement de la Formation Professionnelle dans les Transports, Paris 04333
Association pour le Développement de la Recherche en Toxicologie Expérimentale, Paris 04334
Association pour le Développement de la Stomatologie, Paris 04335
Association pour le Développement de la Traduction Automatique et de Linguistique Appliqué, Paris 04336
Association pour le Développement de l'Enseignement et des Recherches Scientifiques auprès des Universités de la Région Parisienne, Paris 04337
Association pour le Développement des Etudes Biologiques en Psychiatrie, Paris 04338
Association pour le Développement des Relations Médicales entre la France et les Pays Etrangers, Paris 04339
Association pour le Développement des Techniques de Transport, d'Environnement et de Circulation, Paris 04340
Association pour le Développement du Droit Mondial, Paris 04341
Association pour l'Education Permanente dans des Universités du Canada, Ottawa 02044
Association pour l'Enseignement de l'Assurance, Paris 04342
Association pour l'Enseignement Social en Afrique, Addis Ababa 03941
Association pour les Etudes et Recherches de Zoologie Appliquée et de Phytopathologie, Louvain-la-Neuve 01151
Association pour les Etudes sur la Radiotélévision Canadienne, Montréal 01906
Association pour l'Etude et l'Evaluation Epidémiologiques des Désastres dans les Pays en Voie de Développement, Bruxelles 01152
Association pour l'Etude Taxonomique de la Flore d'Afrique Tropicale, Rosières 01153
Association pour l'Expansion du Rôle International des Langues d'Europe Continentale, Paris 04210
Association pour l'Innovation Scientifique, Paris 04343
Association pour l' Introduction de la Nomenclature Biologique Nouvelle, Kalmthout 01154
Association Professionnelle Belge des Pédiatres, Bruxelles 01155
Association Professionnelle des Bibliothécaires et Documentalistes, Bruxelles 01156
Association Professionnelle des Géologues et des Géophysiciens du Québec, Montréal 01952
Association Provinciale des Enseignants Protestants du Québec, Dollard des Ormeaux 02849
Association Psychoanalytique de France, Paris 04344
Association Pulmonaire, Ottawa 02595
Association Pulmonaire du Nouveau-Brunswick, Fredericton 02687
Association Pulmonaire du Québec, Québec 01953
Association Québécoise des Archivistes Médicales, Rock-Forest 01954
Association Québécoise des Techniques de l'Eau, Montréal 01955
Association Régionale d'Education Permanente, Paris 04345
Association Régionale des Oeuvres Educatives et des Vacances de l'Education Nationale, Paris 04346
Association Régionale d'Informations Sociales, Paris 04347
Association Royale des Actuaires Belges, Bruxelles 01157
Association Royale des Demeures Historiques de Belgique, Bruxelles 01158
Association Scientifique des Médecins Acupuncteurs de France, Paris 04348
Association Stomatologique Internationale, Paris 04349
Association Suisse de Droit Aérien et Spatial, Genève 10632
Association Suisse de Géographie, Zürich 10998
Association Suisse de l'Arbitrage, ZZürich 10950
Association Suisse de Politique Sociale, Zürich 10951
Association Suisse de Science Politique, Genève 10633
Association Suisse des Ecoles Hôtelières, Glion-sur-Montreux 10634
Association Suisse des Ingénieurs Agronomes et des Ingénieurs en

Technologie Alimentaire, Zollikofen 10907
Association Suisse des Médecins Assistants et Chefs de Clinique, Bern 11005
Association Suisse des Romanistes, Luzern 10903
Association Suisse pour l'Etude de l'Antiquité, Basel 10939
Association Technique de la Fonderie, Paris 04350
Association Technique de la Réfrigération et de l'Equipement Ménager, Paris 04351
Association Technique de la Sidérurgie Française, Paris la Défense 04352
Association Technique de l'Industrie du Gaz en France, Paris 04353
Association Technique de l'Industrie Papetière, Paris 04354
Association Technique Maritime et Aéronautique, Paris 04355
Association Technique pour l'Etude de la Gestion des Institutions Publiques et des Entreprises Privées, Paris 04356
Association to Advance Ethical Hypnosis, Cuyahoga Falls 14537
Association to Combat Huntington's Chorea, Hinckley 11412
Association Togolaise pour le Développement de la Documentation, des Bibliothèques, Archives et Musées, Lomé 11104
Association Tunisienne des Bibliothécaires, Documentalistes et Archivistes, Tunis 11138
Association Universitaire Canadienne d'Etudes Nordiques, Ottawa 01926
Association Universitaire pour la Diffusion Internationale de la Recherche, Paris 04357
Association Universitaire pour le Développement de l'Enseignement et de la Culture en Afrique et à Madagascar, Paris 04358
Association Zaïroise des Archivistes, Bibliothécaires et Documentalistes, Kinshasa 17141
Associazione Alessandro Scarlatti, Napoli 07254
Associazione Anestesisti Rianimatori Ospedalieri Italiani, Napoli 07255
Associazione Archaeologica Romana, Roma 07256
Associazione Archeologica Allumiere Adolfo Klitsche de la Grange, Allumiere 07257
Associazione Archeologica Centumcellae, Civitavecchia 07258
Associazione Archeologica Romana, Roma 07259
Associazione Archivistica Ecclesiastica, Roma 07260
Associazione Astrofili Bolognesi, Bologna 07261
Associazione Bresciana di Ricerche Economiche, Brescia 07262
Associazione Campana degli Insegnanti di Scienze Naturali, Napoli 07263
Associazione Centri di Orientamento Scolastico Professionale e Sociale, Roma 07264
Associazione Criogenica Italiana, Genova 07265
Associazione degli Africanisti Italiani, Pavia 07266
Associazione degli Statistici, Udine 07267
Associazione dei Biologi delle Facoltá di Farmacia, Padova 07268
Associazione dei Critici Letterari Italiani, Roma 07269
Associazione dei Geografi Italiani, Roma 07270
Associazione di Cultura Lao Silesu, Iglesias 07271
Associazione di Cultura Romana Te Roma Sequor, Roma 07272
Associazione Dietetica Italiana, Roma 07273
Associazione di Ricerca e Interventi Psicosociali e Psicoterapeutici, Molinetto di Mazzano 07274
Associazione Educatrice Italiana, Roma 07275
Associazione Forense Italiana, Roma 07276
Associazione Forestale Italiana, Roma 07277
Associazione Genetica Italiana, Padova 07278
Associazione Geo-Archeologica Italiana, Roma 07279
Associazione Geofisica Italiana, Roma 07280
Associazione Geotecnica Italiana, Roma 07281
Associazione Giacomo Boni per la Difesa dei Monumenti di Roma Antica, Roma 07282
Associazione Grafologica Italiana, Ancona 07283
Associazione Idrotecnica Italiana, Roma 07284
Associazione Internazionale Centro Studi di Storia e Documentazione delle Regioni, Reggio Emilia 07285
Associazione Internazionale di Archeologia Classica, Roma 07286
Associazione Internazionale di Diritto Nucleare, Roma 07287
Associazione Internazionale di Poesia, Roma 07288
Associazione Internazionale Giuristi Italia-USA, Roma 07289
Associazione Internazionale per lo Studio del Diritto Canonico, Roma 07290
Associazione Italiana Biblioteche, Roma 07291
Associazione Italiana Condizionamento dell'Aria, Riscaldamento e Refrigerazione, Milano 07292
Associazione Italiana degli Insegnanti di Geografia, Trieste 07293
Associazione Italiana degli Slavisti, Pisa 07294
Associazione Italiana dei Chimici del Cuoio, Torino 07295
Associazione Italiana dei Giuristi Europei, Roma 07296
Associazione Italiana di Aeronautica e Astronautica, Roma 07297
Associazione Italiana di Anestesia Odontostomatologica, Bologna 07298
Associazione Italiana di Anglistica, Pisa 07299
Associazione Italiana di Cardiostimolazione, Pisa 07300
Associazione Italiana di Cartografia, Napoli 07301
Associazione Italiana di Chimica Tessile e Coloristica, Milano 07302
Associazione Italiana di Cinematografia Scientifica, Roma 07303
Associazione Italiana di Cultura Classica, Firenze 07304
Associazione Italiana di Dietetica e Nutrizione Clinica, Roma 07305
Associazione Italiana di Diritto Marittimo, Roma 07306
Associazione Italiana di Documentazione e di Informazione, Roma 07307
Associazione Italiana di Fisica Sanitaria e di Protezione contro le Radiazioni, Torino 07308
Associazione Italiana di Genio Rurale, Padova 07309
Associazione Italiana di Immuno-Oncologia Clinico-Pratica, Mantova 07310
Associazione Italiana di Ingegneria Chimica, Milano 07311
Associazione Italiana di Ingegneria Medica e Biologica, Napoli 07312
Associazione Italiana di Medicina Aeronautica e Spaziale, Roma 07313
Associazione Italiana di Medicina dell'Assicurazione Vita, Roma 07314
Associazione Italiana di Metallurgia, Milano 07315
Associazione Italiana di Microbiologia Applicata, Milano 07316
Associazione Italiana di Oncologia Medica, Milano 07317
Associazione Italiana di Protezione contro le Radiazione, Bologna 07318
Associazione Italiana di Psicologia dello Sport, Ferrara 07319
Associazione Italiana di Radiobiologia Medica, Catania 07320
Associazione Italiana di Radiologia e Medicina Nucleare, Roma 07321
Associazione Italiana di Ricerca Operativa, Genova 07322
Associazione Italiana di Scienze Politiche e Sociali, Roma 07323
Associazione Italiana di Sociologia, Roma 07324
Associazione Italiana di Strumentisti, Milano 07325
Associazione Italiana di Studio delle Relazioni Industriali, Roma 07326
Associazione Italiana di Studi Semiotici, Torino 07327
Associazione Italiana di Tecnica Navale, Genova 07328
Associazione Italiana di Tecnologia Alimentare, Milano 07329
Associazione Italiana di Terapia Occupazionale, Roma 07330
Associazione Italiana di Terapie Psicologiche, Milano 07331
Associazione Italiana Giuristi Democratici, Roma 07332
Associazione Italiana per gli Studi di Marketing, Milano 07333
Associazione Italiana per la Difesa degli Interessi di Diabetici, Roma 07334
Associazione Italiana per l'Analisi delle Sollecitazioni, Trieste 07335
Associazione Italiana per la Promozione degli Studi e delle Ricerche per l'Edifizia, Milano 07336
Associazione Italiana per la Psicologia Umanistica e Transpersonale, Roma 07337
Associazione Italiana per la Qualità, Milano 07338
Associazione Italiana per la Ricerca Industriale, Roma 07339
Associazione Italiana per la Ricerca nell'Impiego degli Elastomeri, Milano 07340
Associazione Italiana per L'Educazione Sanitaria, Perugia 07341
Associazione Italiana per le Ricerche di Storia del Cinema, Roma 07342
Associazione Italiana per l'Informatica et il Calcolo Automatico, Milano 07343
Associazione Italiana per lo Studio del Dolore, Firenze 07344
Associazione Italiana per lo Studio della Psicologia Analitica, Roma 07345
Associazione Italiana per lo Sviluppo Internazionale, Roma 07346
Associazione Italiana Santa Cecilia per la Musica Sacra, Roma 07347
Associazione Italiana Scientifica di Metapsichica, Milano 07348
Associazione Italiana Sclerosi Multipla, Roma 07349
Associazione Italiana Socioanalisi Individuale, Milano 07350
Associazione Italiana Studi Americanisti, Genova 07351
Associazione Italiana Tecnico-Economica del Cemento, Roma 07352
Associazione Italiana tra Foniatri e Logopedisti, Padova 07353
Associazione La Nostra Famiglia, Ponte Lambro 07354
Associazione Ligure per lo Studio e la Divulgazione dell'Astronomia e dell'Astronautica, Genova 07410
Associazione Medica Italiana di Idroclimatologia, Talassologia e Terapia Fisica, Roma 07355
Associazione Medica Italiana per lo Studio della Ipnosi, Milano 07356
Associazione Medici Dentisti Italiani, Roma 07357
Associazione Micologica ed Ecologica Romana, Roma 07358
Associazione Nazionale Archivistica Italiana, Roma 07359
Associazione Nazionale Assistenti Sociali, Roma 07360
Associazione Nazionale degli Urbanisti, Treviso 07361
Associazione Nazionale dei Musei di Enti Locali e Istituzionali, Padova 07362
Associazione Nazionale dei Musei Italiani, Roma 07363
Associazione Nazionale dei Periti Grafici a Base Psicologica, Milano 07364
Associazione Nazionale del Libero Pensiero Giordano Bruno, Roma-Prati 07365
Associazione Nazionale di Ingegneria Nucleare, Roma 07366
Associazione Nazionale di Ingegneria Sanitaria, Roma 07367
Associazione Nazionale di Meccanica, Milano 07368
Associazione Nazionale Disegno di Macchine, Palermo 07369
Associazione Nazionale Esercenti Teatri, Roma 07370
Associazione Nazionale Famiglie di Fanciulli e Adulti Subnormali, Roma 07371
Associazione Nazionale Filosofia Arti Scienze, Bologna 07372
Associazione Nazionale Industria Meccanica Varia ed Affine, Milano 07373
Associazione Nazionale Ingegneri ed Architetti Italiani, Roma 07374
Associazione Nazionale Ingegneri Minerari, Roma 07375
Associazione Nazionale Insegnanti di Disegno, Napoli 07376
Associazione Nazionale Italiana Esperti Scientifici del Turismo, Roma 07377
Associazione Nazionale Italiana per l'Automazione, Milano 07378
Associazione Nazionale per i Centri Storico-Artistici, Gubbio 07379
Associazione Nazionale per il Progresso della Scuola Italiana, Roma 07380
Associazione Nazionale per la Scuola Italiana, Roma 07381
Associazione Nazionale per lo Studio dei Problemi del Credito, Roma 07382
Associazione Nazionale Professori Universitari di Ruolo, Roma 07383
Associazione Otologica Ospedaliera Italiano, Roma 07384
Associazione Ottica Italiana, Firenze 07385
Associazione Pedagogica Italiana, Chiusi Città 07386
Associazione per il Diabete Infantile e Giovanile, Roma 07387
Associazione per il Superamento delle Barriere Linguistiche in Europa, Verona 07661
Associazione per Imola Storico-Artistica, Imola 07388
Associazione per la Conservazione delle Tradizioni Popolari, Palermo 07389
Associazione per l'Agricoltura Biodinamica, Milano 07390
Associazione per le Previsioni Econometriche, Bologna 07391
Associazione per lo Studio del Problema Mondiale dei Rifugiati, Sezione Italiana, Roma 07392
Associazione per lo Sviluppo degli Studi di Banca e Borsa, Milano 07393
Associazione per lo Sviluppo delle Scienze Religiose in Italia, Bologna 07394
Associazione per lo Sviluppo dell'Istruzione e della Formazione Professionale, Roma 07395
Associazione per lo Sviluppo di Studi e Ricerche nell'Industria Tessile Laniera Oreste Rivetti, Biella 07396
Associazione Piemontese di Studi Filosofici, Biella 07397
Associazione Professionale Italiana Medici Oculisti, Roma 07398
Associazione Psicanalitica Italiana, Milano 07399
Associazione Romana di Entomologia, Roma 07400
Associazione Scientifica di Produzione Animale, Napoli 07401
Associazione Siciliana per le Lettere e le Arti, Palermo 07402
Associazione Sociologi Lucani, Lauria Inferiore 07403
Associazione Sole Italico, Sulmona 07404
Associazione Studi sull'Informazione, Milano 07405
Associazione Tecnica Italiana per la Cellulosa e la Carta, Milano 07406
Associazione Tecnica Italiana per la Cinematografia, Roma 07407
Associazione Teologica Italiana per lo Studio della Morale, Novara 07408
Associazione Termotecnica Italiana, Torino 07409
Associazione Urania, Genova 07410
Associação dos Arquitectos Portugueses, Lisboa 09653
Associtation de la Paralysie Cérébrale en Alberta, Calgary 02333
Assoiation of Portuguese-Speaking Universities, Lisboa 09637
A.S.S.P., Bucuresti 09736
ASSS, Cairo 03862
ASSSI, Saint Lucia 00359
ASSTC, Riyadh 09999
Assurance Medical Society, London 11413
ASSURBANISTI, Treviso 07361
AST, Englewood 14511
AST, Shrewsbury 11395
ASTA, Dallas 14183
A.S.T.A., Warradale 00337
ASTC, Washington 14500
ASTD, Alexandria 14093
ASTDD, Sacramento 14507
ASTDPHE, Columbia 14508
ASTED, Montréal 01951
A.St.G., Wien 00531
ASTHEOL-CENTRAL, Yaoundé 01811
ASTHEOL-WEST, Kinshasa 17140
Asthma and Allergy Foundation of America, Washington 14538
Asthma Association of Canada, Toronto 01956
ASTHO, Washington 14509
ASTI, Arlington 14332
ASTIN, Philadelphia 13323
ASTINET, Algiers 00008
ASTITT, New York 14088
ASTM, Philadelphia 14089
ASTMH, McLean 14171
ASTR, Kingston 14090
ASTRA, Wishaw 11297
ASTRO, Philadelphia 14091
Astrologers' Guild of America, Brewster 14539
Astronautical Society of the Republic of China, Taipei 03248
Astronomical and Geodesical Society, Moskva 09806
Astronomical Association of Indonesia, Jakarta 06886
Astronomical Data Centre, Strasbourg 04359
Astronomical League, Rolling Meadows 14540
Astronomical Society of Australia, Macquarie 00231
Astronomical Society of Bermuda, Hamilton 01518
Astronomical Society of Edinburgh, Edinburgh 11414
Astronomical Society of India, Hyderabad 06736
Astronomical Society of Japan, Tokyo 08268
Astronomical Society of Queensland, Boondall 00232
Astronomical Society of South Australia, Adelaide 00233
Astronomical Society of Southern Africa, Observatory 10125
Astronomical Society of Tasmania, Launceston 00234
Astronomical Society of the Pacific, San Francisco 14541
Astronomical Society of the Republic of China, Taipei 03249
Astronomical Society of Victoria, Melbourne 00235
Astronomical Society of Western Australia, Perth 00236
Astronomische Gesellschaft e.V., Heidelberg 05157
Astronomisk Selskab, København 03564
Astro-Psychology Institute, San Francisco 14542
ASTS, Marshalltown 10123
ASUE, Hamburg 05111
ASUG, Chicago 14176
ASV, Melbourne 00235
ASVI, Raleigh 14094
ASVO, Stillwater 14172
ASVPP, Mississippi State 14173
A.S.W.A., Nedlands 00217
ASWEA, Addis Ababa 03941
ASZ, Chicago 14174
ATA, Bangalore 06727
ATA, Paris 04125
ATA, Portland 14196
ATA, Syosset 14189
ATA, Washington 14195
ATAE, Nottingham 11401
ATAS, New York 13332
ATAS, North Hollywood 13310
ATAS Technology Assessment Network, New York 14543
ATB, New Orleans 14363
ATC, Redwood City 14233
ATCA, Nashville 14190
ATCC, Rockville 14199
ATE, Reston 14513
ATEA, Wahpeton 14188
ATEC, Harrisburg 14565
ATEC, Paris 04340
ATEE, Bruxelles 01133
ATEGIPE, Paris 04356
Atelier, Alexandria 03882
ATENA, Genova 07328
Ateneo Barcelonés, Barcelona 10283
Ateneo Científico, Literario y Artístico, Madrid 10284
Ateneo Científico, Literario y Artístico, Mahón 10285
Ateneo de Ciencias y Artes de Chiapas, Tuxtla Gutiérrez 08707
Ateneo de Clínica Quirúrgica, Montevideo 13250
Ateneo de El Salvador, San Salvador 03921
Ateneo de Macoris, San Pedro de Macoris 03835
Ateneo de Medicina de Sucre, Sucre 01534
Ateneo de Ponce, Ponce 09725
Ateneo Puertorriqueño, San Juan 09726
Ateneo Venezolano de Morfología, Caracas 16978
ATESEA, Singapore 10043
Atex Newspaper Users Group, Chicago 14544
ATF, Paris 04350
ATFS, Salisbury 11399
ATG, Kansas City 13312
ATG, Paris 04353
ATHE, Evansville 14346
Athenaeum of Philadelphia, Philadelphia 14545
Athens Centre of Ekistics, Athinai 06501
ATI, Torino 07409
ATIC, Lisboa 09651
A.T.I.C., Roma 07407
ATICLCA, Milano 07406
ATIEA, Nairobi 08482
ATIME, Beirut 08561
ATIP, Paris 04354
ATIRA, Ahmadabad 06712
ATISA, Ouagadougou 01805
A.T.I.S.M., Novara 07408
ATJ, Middlebury 14514
ATKL, Graz 00557
ATLA, Evanston 14192
ATLA, Washington 14523
Atlantic Cerebral Palsy Association, Fredericton 01957
Atlantic Conference on Learning Disabilities, Dartmouth 01958
Atlantic Council of the United States, Washington 14546
Atlantic Economic Society, Edwardsville 14547
Atlantic Gas Research Rxchange, La Plaine-Saint-Denis 04360
Atlantic Petroleum Association, Halifax 01959
Atlantic Planners Institute, Halifax 01960
Atlantic Provinces Council on the Sciences, St. John's 01961
Atlantic Provinces Library Association, Halifax 01962
Atlantis Research Centre, Brighton 11415
ATM, Angers 04132
A.T.M., Derby 11397
ATM, London 11396
ATMA, Paris 04355
ATMCH, San Diego 14515
ATMI, Eugene 14340
Atomic Energy Council, Taipei 03250
Atomic Energy Society of Japan, Tokyo 08203
ATOS, Sacramento 14191
ATP, Stanford 14362
ATPAS, Liverpool 11398
ATPM, Washington 14516
ATPsych, Leicester 11286
ATRA, Fairfax 13333
ATRCW, Addis Ababa 03939
A.T.R.E.M., Paris 04351
ATS, Helsinki 04007
ATS, Highland 14197
ATS, New York 14194
ATS, Paris la Défense 04352
ATS, Pittsburgh 14521

ATS, Upper Marlboro *14198*
ATSS, Saint Albans *11295*
ATTW, Orlando *14517*
ATV, Lyngby *03557*
ATV, Sankt Augustin *05029*
ATWA, Washington *14522*
AUA, Baltimore *14201*
AUA, Minneapolis *14200*
AUA, New Haven *14527*
AUA, Seattle *14526*
AUBER, Indianapolis *14364*
AUCBM, Damascus *11044*
Auckland Medical Research Foundation, Auckland *09119*
AUDECAM, Paris *04358*
AUDI, Riyadh *10000*
Audio Engineering Society, New York *14548*
Audio Engineering Society, British Section, Sevenoaks *11416*
Audiologopaedisk Forening, Kokkedal *03565*
Audubon Naturalist Society of the Central Atlantic States, Chevy Chase *14549*
AUFEMO, Paris *04123*
AUG, Bellevue *13374*
Augustana Historical Society, Rock Island *14550*
Augustan Reprint Society, New York *14551*
Augustan Society, Torrance *14552*
August Derleth Society, Uncasville *14553*
AUI, Washington *14261*
AULLA, Sydney *00248*
AULP, Lisboa *09637*
AUPELF, Montréal *01895, 01948*
AUPHA, Arlington *14529*
AUPO, San Francisco *14528*
AUPS, Bern *10999*
AUR, Reston *14530*
AURA, Washington *14525*
AUREG, Paris *04161*
AURPO, Cardiff *11402*
AURRP, Tempe *14531*
AUSD, Accra *06474*
AUSJAL, Bogotá *03350*
AUSMIISL, Dumfries *14524*
AUSS, Bloomington *14532*
Ausschuss Normenpraxis im DIN Deutsches Institut für Normung e.V., Berlin *05158*
AusSI, Melbourne *00354*
Australasian Association for the History, Philosophy and Social Studies of Science, Sydney *00237*
Australasian Association of Philosophy, Melbourne *00238*
The Australasian Ceramic Society, Sydney *00239*
Australasian College of Dermatologists, Gladesville *00240*
Australasian College of Physical Scientists and Engineers in Medicine, Melbourne *00241*
Australasian College of Venereologists, NSW 2000 *00242*
Australasian Corrosion Association, Clayton *00243*
The Australasian Institute of Mining and Metallurgy, Parkville *00244*
Australasian Pharmaceutical Science Association, Saint Lucia *00245*
Australasian Plant Pathology Society, Adelaide *00246*
Australasian Political Studies Association, Canberra *00247*
Australasian Universities Language and Literature Association, Sydney *00248*
Australia Council, Redfern *00249*
Australian Academy of Science, Canberra *00250*
Australian Academy of Technological Sciences and Engineering, Parkville *00251*
Australian Academy of the Humanities, Canberra *00252*
Australian Acoustical Society, Darlinghurst *00253*
The Australian Agricultural Council, Canberra *00254*
Australian Agricultural Economics Society, Armidale *00255*
Australian and New Zealand Association for Canadian Studies, North Ryde *00256*
Australian and New Zealand Association for Medieval and Renaissance Studies, Sydney *00257*
Australian and New Zealand Association for the Advancement of Science, Sydney *00258*
Australian and New Zealand Society of Nuclear Medicine, Darlinghurst *00259*
Australian and New Zealand Solar Energy Society, Caulfield East *00260*
Australian Association for the Study of Religions, Saint Lucia *00261*
Australian Association of Adult and Community Education, Canberra *00262*
Australian Association of Clinical Biochemists, Maylands *00263*
Australian Association of Clinical Biochemists – New Zealand Branch, Hamilton *09120*
Australian Association of Gerontology, Parkville *00264*
Australian Association of Neurologists, Melbourne *00265*
Australian Association of Occupational Therapists, Alphington *00266*
Australian Association of Social Workers, North Richmond *00267*
Australian Atomic Energy Commission Research Establishment, Sutherland *00268*
Australian Bar Association, Melbourne *00269*
Australian Biochemical Society, Canberra *00270*
Australian Bird Study Association, Sydney South *00271*
Australian Cancer Society, Sydney *00272*
Australian Clay Minerals Society, Canberra *00273*
Australian College of Education, Deakin *00274*
The Australian College of Obstetricians and Gynaecologists, Melbourne *00275*
Australian College of Paediatrics, Parkville *00276*
Australian Committee of Directors of Principals, Braddon *00277*
Australian Conservation Foundation, Fitzroy *00278*
The Australian Council for Educational Research, Hawthorn *00279*
The Australian Council for Health, Physical Education and Recreation, Hindmarsh *00280*
Australian Council of National Trusts, Civic Square *00281*
Australian Dental Association, Saint Leonards *00282*
Australian Dental Association, Northern Territory Branch, Darwin *00283*
Australian Dental Association, Queensland Branch, Bowen Hills *00284*
Australian Dental Association, South Australian Branch, Unley *00285*
Australian Dental Association, Tasmania Branch, New Town *00286*
Australian Dental Association, Victorian Branch, Toorak *00287*
Australian Dental Association, Western Australia Branch, West Perth *00288*
The Australian Entomological Society, Burnley *00289*
Australian Fabian Society, Melbourne *00290*
Australian Federation for Medical and Biological Engineering, Parkville *00291*
Australian Geography Teachers' Association, Adelaide *00292*
Australian Geomechanics Society, Barton *00293*
Australian Institute for Non-Destructive Testing, Parkville *00294*
Australian Institute of Aboriginal Studies, Canberra City *00295*
Australian Institute of Agricultural Science, Parkville *00296*
Australian Institute of Archaeology, Melbourne *00297*
Australian Institute of Cartographers, Brisbane *00298*
Australian Institute of Credit Management, Artarmon *00299*
Australian Institute of Energy, Wahroonga *00300*
Australian Institute of Food Science and Technology, Pymble *00301*
Australian Institute of Homoeopathy, Roseville *00302*
Australian Institute of Industrial Psychology, Sydney *00303*
Australian Institute of International Affairs, Canberra *00304*
Australian Institute of Management, Saint Kilda *00305*
Australian Institute of Marine Science, Townsville *00306*
Australian Institute of Mining and Metallurgy, Parkville *00307*
Australian Institute of Navigation, Sydney *00308*
Australian Institute of Nuclear Science and Engineering, Menai *00309*
Australian Institute of Physics, Parkville *00310*
Australian Institute of Political Science, Sydney *00311*
Australian Institute of Quantity Surveyors, Deakin *00312*
Australian Institute of Radiography, East Melbourne *00313*
The Australian Institute of Refrigeration, Air Conditioning and Heating, Parkville *00314*
Australian Institute of Urban Studies, Canberra *00315*
Australian Institute of Valuers and Land Economists, Deakin *00316*
Australian Law Librarians' Group, Melbourne *00317*
Australian Library and Information Association, Canberra *00318*
Australian Mammal Society, Lyneham *00319*
The Australian Mathematical Society, Hobart *00320*
Australian Mathematics Competition Committee, Belconnen *00321*
Australian Medical Association, Barton *00322*
Australian National Committee on Large Dams, Brisbane *00323*
Australian Numismatic Society, Sydney *00324*
Australian Optometrical Association, Carlton South *00325*
Australian Orthopaedic Association, Sydney *00326*
Australian Petroleum Exploration Association, Sydney *00327*
Australian Physiological and Pharmacological Society, Melbourne *00328*
Australian Physiotherapy Association, Concord *00329*
Australian Postgraduate Federation in Medicine, Camperdown *00330*
Australian Psychological Society, Parkville *00331*
Australian Remedial Education Association, Camberwell *00332*
Australian Research Council, Canberra *00333*
Australian Research Grants Committee, Woden *00334*
Australian Road Research Board, Melbourne *00335*
Australian School Library Association, Alderley *00336*
Australian Science Teachers Association, Warradale *00337*
Australian Society for Biochemistry and Molecular Biology, Edwardstown *00338*
Australian Society for Fish Biology, Queenscliff *00339*
Australian Society for Limnology, Abbotsford *00340*
Australian Society for Medical Research, Sydney *00341*
Australian Society for Microbiology, Parkville *00342*
Australian Society for Operations Research, Melbourne *00343*
Australian Society for Parasitology, Canberra *00344, 00345*
Australian Society for Reproductive Biology, Newcastle *00346*
Australian Society of Anaesthesists, Paddington *00347*
Australian Society of Animal Production, Mount Barker *00348*
Australian Society of Authors, Strawberry Hills *00349*
Australian Society of Clinical Hypnotherapy, Eastwood *00350*
Australian Society of Cosmetic Chemists, Kingsgrove *00351*
Australian Society of Endodontology, Subiaco *00352*
Australian Society of Herpetologists, Lyneham *00353*
Australian Society of Indexers, Melbourne *00354*
Australian Society of Orthodontists, Perth *00355*
Australian Society of Periodontology, Melbourne *00356*
Australian Society of Plant Physiologists, Canberra *00357*
Australian Society of Prosthodontists, Melbourne *00358*
Australian Society of Soil Science, Saint Lucia *00359*
Australian Sociological Association, Bundoora *00360*
Australian Veterinary Association, Artarmon *00361*
Australian Vice-Chancellors' Committee, Canberra *00362*
Australian Wool Corporation, Parkville *00363*
Australian Zinc Development Association, Melbourne *00364*
Australia-Pacific Society for Management Studies, Gold Coast *00365*
Austria Esperanto Federacio, Wien *00571*
Austrian Association for American Studies, Salzburg *00753*
Austrian Committee for Sociology of Sport, Wien *00883*
Austrian Conference on Regional Planning, Wien *00887*
Austrian Economic Association, Wien *00711*
Austrian Electrotechnical Association, Wien *00904*
Austrian Esperanto Society, Wien *00571*
Austrian Hemophilia Society, Wien *00858*
Austrian Institute of East and South-East European Studies, Wien *00915*
Austrian Society for Communication, Wien *00803*
Austrian Society of Tribology, Wien *00923*
Austrotransplant – Österreichische Gesellschaft für Transplantation, Transfusion und Genetik, Wien *00572*
AUT, London *11403*
AUTA, Sheffield *11404*
Auteursunie, Amsterdam *08830*
Authors Guild, New York *14554*
Authors' Guild of Ireland, Dublin *06932*
Authors League of America, New York *14555*
Autism Services Center, Huntington *14556*
Autism Society of America, Silver Spring *14557*
Automazione Energia Informazione, Milano *07411*
Automobile and Road Building Society, Moskva *09807*
Automobile, General Engineering and Mechanical Operations Union, Dublin *06933*
Automotive Market Research Council, Research Triangle Park *14558*
Autoprep 5000 Users Group, South Windsor *14559*
Autorenkreis Linz, Linz *00573*
A.U.T.(S.), Glasgow *11405*
Auxiliary of the American Osteopathic Association, Chicago *14560*
Auxiliary to the American Pharmaceutical Association, Pittsburgh *14561*
AV, Wien *00886*
AVA, Alexandria *14206*
AVA, Artarmon *00361*
AVA, Bury Saint Edmunds *11406*
AVA, Wiesbaden *05132*
AVASS, Beckenham *11409*
AVC, Floral Park *13311*
AVC, Northville *14534*
AVDA, Baltimore *14203*
AVEFAM, Caracas *16973*
AVEPDA, Stillwater *14207*
AVERA, Blacksburg *14208*
Aviation Maintenance Foundation International, Redmond *14562*
Aviation Research and Development Institute, Bruxelles *01159*
Aviation Safety Institute, Worthington *14563*
Aviation/Space Writers Association, Columbus *14564*
Aviation Technician Education Council, Harrisburg *14565*
The Avicultural Society, Ascot *11417*
Avicultural Society of America, Yorba Linda *14566*
AVIEM, Caracas *16975*
AVKO Educational Research Foundation, Birch Run *14567*
AVKOERF, Birch Run *14567*
AVM, Salzburg *00558*
AVMA, Schaumburg *14204*
AVPC, Maidenhead *11408*
AVR, Düsseldorf *05128*
AVRDC, Tainan *03240*
AVS, New York *14202*
AVSAB, Ithaca *14205*
AVSL, Portland *14533*
A.V.T.R.W., Edinburgh *11407*
AWA, Columbus *14564*
AWB, Kuala Lumpur *08635*
AWC, San Francisco *14366*
AWE, Snedsted *03561*
AWEA, Washington *14211*
AWG, Saint Paul *14365*
AWIS, Washington *14369*
AWM, College Park *14367*
AWMA, Pittsburgh *13355*
AWMC, Kettering *11410*
A Woman's Right to Choose Campaign, London *11196*
AWP, Philadelphia *14368*
AWPA, Woodstock *14212*
AWPI, Vienna *14213*
AW produktplanung, Essen *05159*
AWRA, Bethesda *14209*
AWRAN, Manila *09505*
AWS, Miami *14210*
AWU, Allentown *14361*
AWV, Eschborn *05113*
AWV, Union City *14370*
AWV-Fachausschuss Mikrofilm/Optische Informationssysteme, Eschborn *05160*
AYJA, Belfast *11411*
AYM, Memphis *14535*
Ayrshire Agricultural Association, Ayr *11418*
Ayrshire Archaeological and Natural History Society, Ayr *11419*
Azerbaijan Academy of Sciences, Baku *01036*
Azerbaijan Mathematics Society, Baku *01037*
Azerbaijan Petroleum Academy, Baku *01038*
Azerbaijan Physical Society, Baku *01039*

BAA, Lomdon *11481*
BAA, London *11530*
BAAL, Clevedon *11486*
B.A.A.L.P.E., Exmouth *11500*
BAAR, London *11473*
B.A.A.S., London *11495*
B.A.A.S., Newcastle-upon-Tyne *11485*
BAAT, Brighton *11501*
BABS, Cambridge *11487*
BAC, Rugby *11491*
BACA, Manchester *11503*
BACG, Towcester *11505*
Bach-Verein Köln e.V., Köln *05161*
BACIE, London *11490*
BACR, London *11489*
BACS, London *11488, 11504*
Baden-Württembergischer Sportärzteverband, Bezirksgruppe Süd-Baden, Freiburg *05162*
Badischer Landesverein für Naturkunde und Naturschutz e.V., Freiburg *05163*
BAEE, Bruxelles *01166*
BAEF, New Haven *14576*
BÄK, Köln *05229*
BÄK NW, Stuttgart *05211*
BAEM, London *11484*
B.A.F.M., Leeds *11498*
BAFM, Manchester *11506*
B.A.F.S., London *11472*
B.A.F.T.A., London *11470*
BAFTA, London *11471*
BAG, Brampton *11482*
Bahamas Historical Society, Nassau *01051*
Bahamas National Trust, Nassau *01052*
BAHOH, London *12155*
BAHP, London *11508*
Bahrain Arts Society, Manama *01054*
Bahrain Bar Society, Manama *01055*
Bahrain Computer Society, Manama *01056*
Bahrain Contemporary Art Association, Manama *01057*
Bahrain Medical Society, Manama *01058*
Bahrain Society of Engineers, Manama *01059*
Bahrain Society of Sociologists, Manama *01060*
Bahrain Writers and Literators Association, Manama *01061*
Bahrein Historical and Archaeological Society, Manama *01062*
BAHS, Reading *11474*
Baker Street Irregulars, Norwood *14568*
Bal Bhavan Society, Calcutta *06737*
Bal Bhavan Society, New Delhi *06738*
BALI, Keighley *11510*
Balint Society, London *11420*
Balkan Medical Union, Bucuresti *09747*
Balkan Physical Union, Thessaloniki *06502*
Ball Manufacturers Engineers Committee, Washington *14569*
Baltic Marine Biologists, Lysekil *10438*
Baltic Marine Environment Protection Commission, Helsinki *03952*
Baltische Gesellschaft in Deutschland e. V., München *05164*
Baluchi Academy, Quetta *09361*
BAMRG, Henley-on-Thames *11475*
Bangla Academy, Dhaka *01065*
Bangladesh Academy of Sciences, Dhaka *01066*
Bangladesh Council of Scientific and Industrial Research, Dhaka *01067*
Bangladesh Economic Association, Dhaka *01068*
Bankakademie e.V., Frankfurt *05165*
Baños Biological Club, Laguna *09521*
BANS, London *11511*
Banting Research Foundation, Toronto *01963*
The Bantock Society, London *11421*
BAO, London *11515*
BAO, Maidenhead *11514*
BAOS, London *11513*
BAOT, London *11512*
BAP, London *11494*
BAPC, Telford *11531*
BAPHR, San Francisco *14571*
B.A.P.S., Edinburgh *11517*
BAPS, London *11519*
The Baptist Historical Society, Oxford *11422*
Bar Association for Commerce, Finance and Industry, London *11423*
Bar Association of India, New Dehli *06739*
B.A.R.B., Norfolk *11529*
Barbados Association of Medical Practitioners, Saint Michael *01072*
Barbados Astronomical Society, Bridgetown *01073*
Barbados Museum and Historical Society, Bridgetown *01074*
Barbados Pharmaceutical Society, Bridgetown *01075*
Barbirolli Society, Retford *11424*
Barcelona Athenaeum, Barcelona *10283*
Barcelona College of Lawyers, Barcelona *10288*
Barge and Canal Development Association, Wakefield *11425*
Barra Mexicana-Colegio de Abogados, México *08708*
BAS, Bombay *06744*
BAS, Bridgetown *01073*
BASA, London *11521*
Basic Education Resource Centre, Nairobi *08483*
Basic English Foundation, London *11426*
BASICS, Ipswich *11492*
BASM, New Malden *11524*
BASO, London *11525*
BASP, Kew *11522*
Basque Educational Organization, San Francisco *14570*
BASRA, Watford *11476*
BASW, Birmingham *11523*
Bataafsch Genootschap der Proefondervindelijke Wijsbegeerte, Rotterdam *08831*
Bath and Camerton Archaeological Society, Bath *11427*
BATOD, Cheadle Hulme *11526*

Battelle-Institut e.V., Frankfurt 05166
Battery Vehicle Society, Peacehaven 11428
BAUS, London 11528
BAW, Bremen 05222
Bay Area Physicians for Human Rights, San Francisco 14571
Bayerische Akademie der Schönen Künste, München 05167
Bayerische Akademie der Werbung, München 05168
Bayerische Akademie der Wissenschaften, München 05169
Bayerische Akademie für Arbeits- und Sozialmedizin, München 05170
Bayerische Botanische Gesellschaft, München 05171
Bayerische Kommission für die Internationale Erdmessung, München 05172
Bayerische Krebsgesellschaft e.V., München 05173
Bayerische Landesärztekammer, München 05174
Bayerische Landeszahnärztekammer, München 05175
Bayerischer Holzwirtschaftsrat, München 05176
Bayerischer Landesverein für Familienkunde e.V., München 05177
Bayerischer Lehrer- und Lehrerinnenverband, München 05178
Bayerische Röntgengesellschaft, Fürth 05179
Bayerischer Sportärzteverband e.V., München 05180
Bayerischer Volkshochschulverband e.V., München 05181
BBAC, Leicester 11533
BBCS, Colchester 11536
BBG, München 05171
BBHC, Cody 14621
BBI, Brandeis 14613
BBICCJE, Washington 14604
BBK, Bonn 05241
BBMRA, London 11534
B.B.O., London 11532
BBRS, Leicester 11453
BBS, Cardiff 11535
BBS, Dundee 11711
BBW, Eschborn 05239
bbw, München 05218
BC, New Brunswick 14603
BCA, Cambridge 11454
BCA, Crowthorne 11541
BCAC, Kensington 14614
B.C.A.P., Vancouver 01971
BCAS, London 11546
BCDA, Wakefield 11425
BCEL, New York 14626
B.C.R.A., Bristol 11540
BCS, Kingswinford 11452
BCS, London 11547
BCS, Lymington 11538
B.C.S., Stoke-on-Trent 11542
B.C.S.H., London 11545
BC Society, Hassocks 11429
B.C.S.O.T., Burnaby 01989
BCT, Droitwich 11539
BDA, Birmingham 11552
BDA, Bonn 05226
BDA, London 11549
B.D.A., London 11551
BDA, Reading 11553
BDAA, Stourbridge 11442
BDB, Bonn 05227
BDC, Hamburg 05195
BDD, Sofia 01747
B.D.H.A., London 11550
BDK, Hannover 05228
BDP, Bonn 05204
BDS, Bielefeld 05205
BDZ, Köln 05261
BEA, Leeds 11558
BEA, Washington 14617
Bedfordshire Archaeological Council, Bedford 11430
Bedfordshire Historical Record Society, Bedford 11431
Beef Improvement Federation, Athens 14572
BEGS, Higher Poynton 11477
Behavioral Pharmacology Society, Atlanta 14573
Behavioral Research Council, Great Barrington 14574
Behavior Genetics Association, Saint Louis 14575
Beijing Academy of Agricultural Sciences, Beijing 03117
Beijing Academy of Coal Mine Design, Beijing 03118
Beijing Academy of Hydroelectrical Engineering Design, Beijing 03119
Beilstein-Institut für Literatur der Organischen Chemie, Frankfurt 05182
Belarussian Academy of Sciences, Minsk 01088
Belfast Natural History and Philosophical Society, Belfast 11432
Belgian American Educational Foundation, New Haven 14576
Belgian Center for Corrosion Study, Bruxelles 01184
Belgian Physical Society, Bruxelles 01442
Belgian Society of Anaesthesia and Resuscitation, Bierges 01424
Belgische Natuurkundige Vereniging, Bruxelles 01442
Belgische Neus-, Keel- en Oorheelkundige Vereniging, Brugge 01160
Belgische Unie van Landmeters en Meetkundigen-Schatters van Onroerende Goederen, Bruxelles 01493
Belgische Vereniging van Toeristische Schrijvers en Journalisten, Vilvoorde 01494
Belgische Vereniging voor Aardrijkskundige Studies, Heverlee 01161
Belgische Vereniging voor Agrarisch Recht, Bruxelles 01107
Belgische Vereniging voor Anesthesie en Reanimatie, Bierges 01424
Belgische Vereniging voor Psychoanalyse, Loverval 01444
Belgische Vereniging voor Stralingsbescherming, Bruxelles 01110
Belgische Vereniging voor Toegepaste Linguistiek, Bruxelles 01162
Belgische Vereniging voor Tropische Geneeskunde, Antwerpen 01163
Belgische Wetenschappelijke Vereniging voor Neurochirurgie, Brugge 01164
Belgisch Instituut voor Arbeidsverhoudingen, Leuven 01165
Belorussian Agricultural Academy, Gorkii 01089
BEMS, Frederick 14586
B.E.N.A., Higham Ferrers 11617
Benelux Association of Energy Economists, Bruxelles 01166
Benelux Phlebology Society, Bruxelles 01167
Benelux Society for Microcirculation, Maastricht 08832
Benesh Institute of Choreology, London 11433
Bengal Natural History Society, Darjeeling 06740
Benslow Music Trust, Hitchin 11434
BEO, San Francisco 14570
BERA, Hato Rey 09727
BERA, Washington 14592
BERC, Nairobi 08483
Bergbau-Forschung, Essen 05183
Bergianska Stiftelsen, Stockholm 10439
Bergischer Geschichtsverein e. V., Wuppertal 05184
Bergius Foundation, Stockholm 10439
Berlin-Brandenburgische Akademie der Wissenschaften, Berlin 05185
Berliner Arbeitskreis Information, Berlin 05186
Berliner Gesellschaft für Anthropologie, Ethnologie und Urgeschichte, Berlin 05187
Berliner Mathematische Gesellschaft e.V., Berlin 05188
Berliner Medizinische Gesellschaft, Berlin 05189
Berliner Orthopädische Gesellschaft e.V., Berlin 05190
Berliner Sportärztebund e.V., Berlin 05191
Berlin Society for Anthropology, Ethnology and Prehistory, Berlin 05187
Bermuda Audubon Society, Hamilton 01519
Bermuda Historical Society, Hamilton 01520
Bermuda Medical Society, Paget 01521
The Bermuda National Trust, Hamilton 01522
Bermuda Society of Arts, Hamilton 01523
Bermuda Technical Society, Hamilton 01524
Bermuda Tuberculosis, Cancer and Health Association, Hamilton 01525
Bernard Herrmann Society, North Hollywood 14577
The Bernard Shaw Society, New York 14578
Bernische Botanische Gesellschaft, Bern 10635
Bernoulli Society for Mathematical Statistics and Probability, Voorburg 08833
Bertrand Russell Society, Manchester 11435
Bertrand Russell Society, Skokie 14579
Berufsverband der Ärzte für Orthopädie e.V., Karlsruhe 05192
Berufsverband der Augenärzte Deutschlands e.V., Düsseldorf 05193
Berufsverband der Berliner Hals-, Nasen-, Ohren-Ärzte, Berlin 05194
Berufsverband der Deutschen Chirurgen, Hamburg 05195
Berufsverband der Deutschen Dermatologen e.V., Pforzheim 05196
Berufsverband der Deutschen Fachärzte für Urologie, Hamburg 05197
Berufsverband der Deutschen Radiologen und Nuklearmediziner e.V., München 05198
Berufsverband der Deutschen Urologen e.V., Dorfen 05199
Berufsverband der Kinderärzte Deutschlands e.V., Köln 05200
Berufsverband der Lehrer an Gehörlosen- und Schwerhörigenschulen, Hamburg 05202
Berufsverband der Praktischen Ärzte und Ärzte für Allgemeinmedizin Deutschlands e.V., Köln 05201
Berufsverband deutscher Arbeitsmediziner, Karlsruhe 06333
Berufsverband Deutscher Graphologen, Schwerte 05450
Berufsverband Deutscher Hörgeschädigtenpädagogen, Hamburg 05202
Berufsverband Deutscher Nervenärzte e.V., Frankfurt 05203
Berufsverband Deutscher Psychologen e.V., Bonn 05204
Berufsverband Deutscher Soziologen e.V., Bielefeld 05205
Berufsverband freiberuflich tätiger Tierärzte Österreichs, Irdning 00574
Berufsverband Geprüfter Graphologen/Psychologen e.V., München 05206
Berufsverband Österreichischer Psychologen, Wien 00575
Berwickshire Naturalists' Club, Eyemouth 11436
BES, Bologna 07412
B.E.S., London 11443, 11554, 11557
Beta Beta Delta, Stuttgart 05207
Bet Nahrain, Modesto 14580
Beton-Verein Berlin e.V., Berlin 05208
Betriebswirtschafts-Akademie e.V., Wiesbaden 05209
BETSS, Detroit 14581
Better Education thru Simplified Spelling, Detroit 14581
Better Hearing Institute, Washington 14582
Better Vision Institute, Rosslyn 14583
Bezirksärztekammer Koblenz, Koblenz 05210
Bezirksärztekammer Nordwürttemberg, Stuttgart 05211
Bezirksärztekammer Pfalz, Neustadt 05212
Bezirksärztekammer Trier, Trier 05213
Bezirkszahnärztekammer Koblenz, Koblenz 05214
Bezirkszahnärztekammer Pfalz, Ludwigshafen 05215
Bezirkszahnärztekammer Rheinhessen, Mainz 05216
Bezirkszahnärztekammer Trier, Trier 05217
BFBS, Swindon 11478
BFFS, London 11560
BFHFI, Oakland 14598
BFI, London 11562
BFMIRA, Leatherhead 11564
BFÖ, Irdning 00574
BFR, Stockholm 10524
BFWG, London 11561
BGA, Saint Louis 14575
BGG/P, München 05206
BGRG, London 11566
BGS, Bovey Tracey 11569
BGS, London 11567
B.G.S., London 11568
BGS, Maidenhead 11570
BGS, Sofia 01749
B.H.A., London 11572
BHA, London 11574, 11577
Bharata Ganita Parisad, Lucknow 06741
Bharata Itihasa Samshodhaka Mandala, Poona 06742
BHAVS, Stanford-in-the-Vale 11509
BHC, Williamsburg 14628
B-HEF, Washington 14627
BHI, Upton 11573
BHI, Washington 14582
BHRA, Bedford 11575
BHRA, Somersham 11576
BHRA Fluid Engineering, Bedford 11437
BHRS, Bedford 11431
BHS, Helendale 14624
B.H.S., London 11571
BHS, North Hollywood 14577
B.H.S., Oxford 11422
BI, New York 14618
BIALL, Birmingham 11479
Białostockie Towarzystwo Naukowe, Bialystok 09539
BIAS, Bristol 11469
BIBIC, Bridgwater 11579
The Bibliographical Society, London 11438
Bibliographical Society of America, New York 14584
Bibliographical Society of Australia and New Zealand, Canberra 00366
Bibliographical Society of Canada, Toronto 01964
Bibliographical Society of the University of Virginia, Charlottesville 14585
Bibliotecarios Agriolas Colombiano, Bogotá 03354
BIBRA Toxicology International, Carshalton 11439
BICS, Northampton 11580
BIE, Genève 10639
BIF, Athens 14572
Bigaku-Kai, Tokyo 08114
Bihar Research Society, Patna 06743
BIHR, London 11581
B.I.I.C.L., London 11583
BIIT, Southampton 11582
Bijutsu-shi Gakkai, Tokyo 08115
B.I.L.A., London 11591
BILD, Paris 04361, 04362
Bildungswerk der Bayerischen Wirtschaft e.V., München 05218
Bildungswerk der Konrad-Adenauer-Stiftung, Politische Akademie der Konrad-Adenauer-Stiftung, Wesseling 05219
Bildungswerk der Nordrhein-Westfälischen Wirtschaft e.V., Schwelm 05220
BINLAC, Caracas 16979
BIO, Lawrence 14588
The Biochemical Society, London 11440
Biochemical Society, Moskva 09808
Biochemical Society of Israel, Rehovot 07048
Biodeterioration Society, Kew 11441
Bio-dynamic Agricultural Association, Stourbridge 11442
Bioelectrochemical Society, Bologna 07412
Bioelectromagnetics Society, Frederick 14586
Bio-Integral Resource Center, Berkeley 14587
Biokemisk Forening, København 03566
Biological Anthropological Section, Lawrence 14588
Biological Council of Canada, Ottawa 01965
Biological Engineering Society, London 11443
Biological Engineering Society, Moskva 09809
Biological Institute of Tropical America, Menlo Park 14589
Biological Photographic Association, Chapel Hill 14590
Biological Society of China, Taipei 03251
Biological Stain Commission, Rochester 14591
Biologisk Selskab, København 03567
Biomass Energy Institute, Winnipeg 01966
Biomass Energy Research Association, Washington 14592
Biomedical Engineering Society, Culver City 14593
Biometric Society, Alexandria 14594
Biometric Society, Western North American Region, Laramie 14595
Biophysical Society, Bethesda 14596
Biophysical Society of China, Beijing 03120
Biophysical Society of Japan, Osaka 08255
Biorhythmic Research Association, Loughborough 11444
Bio-Rhythm Research and Information Centre, Brisbane 00367
BIOS, Roma 07491
Biosciences Information Network for Latin America and the Caribbean, Caracas 16979
BIOTA, Menlo Park 14589
Biotechnology and Biological Sciences Research Council, Reading 12238
Biotechnology Research for Innovation, Development and Growth in Europe, Bruxelles 01168
BIPM, Sèvres 04363
B.I.P.P., London 11587
BIR, Bruxelles 01174
B.I.R., London 11589
BIRC, Berkeley 14587
Bird Life International, Cambridge 11445
Bird Observers Club of Australia, Nunawading 00368
Bird Strike Committee Europe, København 03568
Birks Family Foundation, Montréal 01967
Birmingham and Midland Institute, Birmingham 11446
Birmingham and Warwickshire Archaeological Society, Birmingham 11447
Birmingham Bibliographical Society, Birmingham 11448
The Birmingham Metallurgical Association, Birmingham 11449
Birmingham Natural History Society, Birmingham 11450
Birmingham Transport Historical Group, Folkestone 11451
BIS, Los Angeles 14611
BIS, Welwyn Garden City 11578
BIS, Worthing 11596
BISG, New York 14607
Bisnuth Institute, Grimbergen 01169
B.I.S.T., Sevenoaks 11590
BIT, Bruxelles 01175
BITNET, Princeton 14597
BJCERJ, Slough 11593
BJF, Frankfurt 05254
BKED, Essen 05268
BKPA, Bordon 11598
BKSTS, London 11599
BLA, Crab Orchard 14599
BLA, Greenville 14619
Blackcountry Society, Kingswinford 11452
Black Filmmakers Hall of Fame, Oakland 14598
Black Lung Association, Crab Orchard 14599
Black Psychiatrists of America, Oakland 14600
Black Resources Information Coordinating Services, Tallahassee 14601
Black Women's Educational Alliance, Philadelphia 14602
Blair Bell Research Society, Leicester 11453
Bliss Classification Association, Cambridge 11454
BLL, Bonn 05265
BLLV, München 05178
BLNN, Freiburg 05163
Bloomsday Club, New Brunswick 14603
BLS, Huddersfield 11465
BLZK, München 05175
BMA, London 11607
BMAA, Hull 11605
BMB, Lysekil 10438
BMEC, Washington 14569
BMES, Culver City 14593
BMG, Teddington 11606
BMHS, Bridgetown 01074
BMHS, London 11611
BMI, Birmingham 11446
B.M.S., Portsmouth 11614
BMS, Upminster 11613
BMSS, London 11610
BMT Cortec, Wallsend 11455
BN, Modesto 14580
B.N.A., Amsterdam 08947
B'Nai B'Rith International Commission on Continuing Jewish Education, Washington 14604
BNC, Eyemouth 11436
BNES, London 11620
BNF, London 11622
BNF, Wantage 11456
BNF Metals Technology Centre, Wantage 11456
BNFMRA, Wantage 11619
BNHPS, Belfast 11432
BNHS, Birmingham 11450
BNHS, Frinton-on-Sea 11616
BNOA, London 11618
BNS, Amsterdam 08835
B.N.S., Pontyclun 11621
BOA, London 11627
Boarding Schools Association, Reigate 11457
Board of Governors of the European Schools, Bruxelles 01170
Bobechko Foundation, Toronto 01968
B.O.C., Nunawading 00368
BOC, Oakham 11625
BOCA, Country Club Hills 14622
Bockus International Society of Gastroenterology, La Jolla 14605
B.Ö.P., Wien 00575
Bőr-, Cipő-, és Bőrfeldolgozóipari Tudományos Egyesület, Budapest 06596
B.O.H.S., London 11623
Bókavardafélag Islands, Reykjavik 06678
Bolivarian Society of the United States, New York 14606
Bolyai János Matematikai Társulat, Budapest 06597
Bombay Art Society, Bombay 06744
Bombay Historical Society, Bombay 06745
Bombay Medical Union, Bombay 06746
Bombay Natural History Society, Bombay 06747
Bombay Textile Research Association, Bombay 06748
Bond Heemschut, Amsterdam 08834
Bond van Nederlandse Stedebouwkundigen, Amsterdam 08835
Bone and Tooth Society, Oxford 11458
Bonn Institute for Economic and Social Research, Bonn 05977
Book Industry Study Group, New York 14607
Bookplate Society, London 11459
Book Trust, London 11460
Border Union Agricultural Society, Kelso 11461
Born-Bunge Research Foundation, Wilrijk 01362
Borthwick Institute of Historical Research, York 11462
BOS, Birmingham 11624
B.O.S., London 11628
Bostonian Society, Boston 14608
Boston Theological Institute, Newton Centre 14609
Botanical Institute, São Paulo 01666
Botanical Society of America, Lawrence 14610
Botanical Society of China, Beijing 03121
Botanical Society of Israel, Beer Sheva 07049
Botanical Society of Japan, Tokyo 08383
Botanical Society of Lund, Lund 10498
The Botanical Society of Scotland, Edinburgh 11463
The Botanical Society of South Africa, Claremont 10126
Botanical Society of the British Isles, London 11464
Botanical Society of Zimbabwe, Harare 17164
Botanical Survey of India, Calcutta 06749
Botany 2000-Asia, Kensington 00369

The Botswana Society, Gaborone 01559
B.O.U., Tring 11626
BPA, Chapel Hill 14590
BPA, Columbus 14629
BPA, Köln 05201
BPA, London 11629
BPA, Oakland 14600
BPA, Teddington 12576
BPEAOA, Chicago 14623
BPMF, London 11633
BPMTG, Yeading 11637
BPS, Atlanta 14573
BPS, Bethesda 14596
BPS, Chorley 11634
BPS, Leicester 11636
B.P.S., London 11630, 11631
BPS, London 11632
BRA, Birmingham 11641
BRA, London 11638
Brain Information Service, Los Angeles 14611
Bram Stoker Memorial Association, New York 14612
The Bram Stoker Society, Dublin 06934
Branche Belge de la Société de Chimie Industrielle, Bruxelles 01171
Branch Line Society, Huddersfield 11465
Brandeis-Bardin Institute, Brandeis 14613
Brandförsvarsföreningen, Stockholm 10440
Brandteknisk Selskab, København 03569
Braunschweigische Wissenschaftliche Gesellschaft, Braunschweig 05221
Brazilian Academy of Administration, Rio de Janeiro 01562
Brazilian Academy of Sciences, Rio de Janeiro 01563
Brazilian Agricultural Research Corporation, Campo Grande 01648
Brazilian Genetics Society, Ribeirão Preto 01711
Brazilian Institute of Economics, Rio de Janeiro 01664
Brazilian Society of Dermatology, Rio de Janeiro 01708
BRC, Great Barrington 14574
Breast Cancer Advisory Center, Kensington 14614
Bremer Ausschuss für Wirtschaftsforschung, Bremen 05222
Bremer Gesellschaft für Wirtschaftsforschung e.V., Bremen 05223
Bremer Sportärztebund, Bremen 05224
Brewery History Society, Sutton 11466
Brewing and Malting Barley Research Institute, Winnipeg 01969
BRF, Nutfield 11467
BRF International, Nutfield 11467
BRICS, Tallahassee 14601
BRIDGE, Bruxelles 01168
The Bridge, New York 14615
Bristol and Gloucestershire Archaeological Society, Gloucester 11468
Bristol Industrial Archaeological Society, Bristol 11469
Brite-EuRam II, Bruxelles 01172
British Academy, London 11470
British Academy of Film and Television Arts, London 11471
British Academy of Forensic Sciences, London 11472
British Acupuncture Association and Register, London 11473
British Agricultural History Society, Reading 11474
British Agricultural Marketing Research Group, Henley-on-Thames 11475
British American Scientific Research Association, Watford 11476
British and European Geranium Society, Higher Poynton 11477
British and Foreign Bible Society, Swindon 11478
British and Irish Association of Law Librarians, Birmingham 11479
British Arachnological Society, Banff 11480
British Archaeological Association, Lomdon 11481
British Artists in Glass, Brampton 11482
British Arts, London 11483
British Association for Accident and Emergency Medicine, London 11484
British Association for American Studies, Newcastle-upon-Tyne 11485
British Association for Applied Linguistics, Clevedon 11486
British Association for Brazing and Soldering, Cambridge 11487
British Association for Canadian Studies, London 11488
British Association for Cancer Research, London 11489
British Association for Commercial and Industrial Education, London 11490
British Association for Counselling, Rugby 11491
British Association for Immediate Care, Ipswich 11492
British Association for Information and Libraries Education and Research, Sheffield 11493
British Association for Psychopharmacology, London 11494
British Association for the Advancement of Science, London 11495
British Association for the History of Religions, Cardiff 11496
British Association for the Study of Religions, Cardiff 11497
British Association in Forensic Medicine, Leeds 11498
British Association of Academic Phoneticians, Glasgow 11499
British Association of Advisers and Lecturers in Pysical Education, Exmouth 11500
British Association of Art Therapists, Brighton 11501
British Association of Blind Esperantists, Middlesborough 11502
British Association of Clinical Anatomists, Manchester 11503
British Association of Cosmetic Surgeons, London 11504
British Association of Crystal Growth, Towcester 11505
British Association of Friends of Museums, Manchester 11506
British Association of Hair Transplant Surgeons, Birmingham 11507
British Association of Homoeopathic Pharmacists, London 11508
British Association of Homoeopathic Veterinary Surgeons, Stanford-in-the-Vale 11509
British Association of Landscape Industries, Keighley 11510
British Association of Numismatic Societies, London 11511
British Association of Occupational Therapists, London 11512
British Association of Oral and Maxillo-Facial Surgeons, London 11513
British Association of Orthodontists, Maidenhead 11514
British Association of Otolaryngologists, London 11515
British Association of Paediatric Nephrology, Birmingham 11516
British Association of Paediatric Surgeons, Edinburgh 11517
British Association of Picture Libraries, London 11518
British Association of Plastic Surgeons, London 11519
British Association of Psychotherapists, London 11520
British Association of Seed Analysts, London 11521
British Association of Social Psychiatry, Kew 11522
British Association of Social Workers, Birmingham 11523
British Association of Sport Medicine, New Malden 11524
British Association of Surgical Oncology, London 11525
British Association of Teachers of the Deaf, Cheadle Hulme 11526
British Association of the Experiment in International Living, Malvern 11527
British Association of Urological Surgeons, London 11528
British Association Representing Breeders, Norfolk 11529
British Astronomical Association, London 11530
British Aviation Preservation Council, Telford 11531
British Ballet Organization, London 11532
British Balloon and Airship Club, Leicester 11533
British Brush Manufacturers Research Association, London 11534
British Bryological Society, Cardiff 11535
British Butterfly Conservation Society, Colchester 11536
British Cardiac Society, London 11537
British Caribbean Veterinary Association, Kingston 08069
British Cartographic Society, Lymington 11538
British Caspian Trust, Droitwich 11539
British Cave Research Association, Bristol 11540
British Cement Association, Crowthorne 11541
British Ceramic Society, Stoke-on-Trent 11542
British Ceramic Tile Council, Stoke-on-Trent 11543
British College of Optometrists, London 11544
British Columbia and Yukon Heart Foundation, Vancouver 01970
British Columbia Association of Podiatry, Vancouver 01971
British Columbia Association of Speech, Vancouver 01972
British Columbia Cancer Foundation, Vancouver 01973
British Columbia Corrections Association, Vancouver 01974
British Columbia Council for Leadership in Educational Administration, Richmond 01975
British Columbia Drama Association, Victoria 01976
British Columbia Epilepsy Society, Vancouver 01977
British Columbia Genealogical Society, Richmond 01978
British Columbia Historical Association, Nanaimo 01979
British Columbia Library Trustees' Association, Victoria 01980
British Columbia Lung Association, Vancouver 01981
British Columbia Medical Association, Vancouver 01982
British Columbia Museums Association, Victoria 01983
British Columbia Parents in Crisis Society, Burnaby 01984
British Columbia Parkinsons's Disease Association, Vancouver 01985
British Columbia Pharmacists' Society, Vancouver 01986
British Columbia Pollution Control Association, West Vancouver 01987
British Columbia Research Council, Vancouver 01988
British Columbia Society of Occupational Therapists, Burnaby 01989
British Columbia Speleological Federation, Gold River 01990
British Columbia Teacher-Librarians' Association, Vancouver 01991
British Columbia Teachers' Federation, Vancouver 01992
British Columbia Thoracic Society, Vancouver 01993
British Committee for Standards in Haematology, London 11545
British Compressed Air Society, London 11546
The British Computer Society, London 11547
The British Cryogenics Council, London 11548
British Dental Association, London 11549
British Dental Hygienists Association, London 11550
British Diabetic Association, London 11551
The British Dietetic Association, Birmingham 11552
British Dyslexia Association, Reading 11553
The British Ecological Society, London 11554
British Educational Management and Administration Society, Oxford 11555
British Electrophoresis Society, Aberdeen 11556
British Endodontic Society, London 11557
British Epilepsy Association, Leeds 11558
British Federation of Festivals, Macclesfield 11559
British Federation of Film Societies, London 11560
British Federation of Women Graduates, London 11561
British Film Institute, London 11562
British Fluoridation Society, Alderley Edge 11563
British Food Manufacturing Industries Research Association, Leatherhead 11564
British Geological Survey, Keyworth 11565
British Geomorphological Research Group, London 11566
British Geotechnical Society, London 11567
British Geriatrics Society, London 11568
British Goat Society, Bovey Tracey 11569
British Grassland Society, Maidenhead 11570
British Herpetological Society, London 11571
The British Homoeopathic Association, London 11572
British Horological Institute, Upton 11573
British Humanist Association, London 11574
British Hydromechanics Research Association, Bedford 11575
British Hypnosis Research Association, Somersham 11576
British Hypnotherapy Association, London 11577
British Ichthyological Society, Welwyn Garden City 11578
British Institute for Brain Injured Children, Bridgwater 11579
The British Institute of Cleaning Science, Northampton 11580
British Institute of Human Rights, London 11581
British Institute of Industrial Therapy, Southampton 11582
British Institute of International and Comparative Law, London 11583
British Institute of Learning Disabilities, Kidderminster 11584
British Institute of Musculoskeletal Medicine, Northwood 11585
British Institute of Persian Studies, London 11586
British Institute of Persian Studies, Teheran 06898
British Institute of Practical Psychology, London 11587
British Institute of Professional Photography, Ware 11588
British Institute of Radiology, London 11589
British Institute of Surgical Technologists, Sevenoaks 11590
British Insurance Law Association, London 11591
British Interlingua Society, Sheffield 11592
British Internal Combustion Engine Research Institute, Slough 11593
British International Studies Association, Bailrigg 11594
British Interplanetary Society, London 11595
British Iris Society, Worthing 11596
British Jazz Society, Twickenham 11597
British Kidney Patient Association, Bordon 11598
British Kinematograph Sound and Television Society, London 11599
British Laboratory Animals Veterinary Association, Salisbury 11600
British Leather Confederation, Northampton 11601
British Leprosy Relief Association, Colchester 11602
British Library National Sound Archive, London 11603
British Lichen Society, London 11604
British Marine Aquarist Association, Hull 11605
British Measures Group, Teddington 11606
British Medical Association, London 11607
British Microcirculation Society, London 11608
British Migraine Association, West Byfeet 11609
British Model Soldier Society, London 11610
British Morgan Horse Society, London 11611
British Museum Society, London 11612
British Music Society, Upminster 11613
British Mycological Society, Portsmouth 11614
British National Committee on Space Research, London 11615
British Natural Hygiene Society, Frinton-on-Sea 11616
British Naturalists' Association, Higham Ferrers 11617
British Naturopathic and Osteopathic Association, London 11618
British Non-Ferrous Metals Research Association, Wantage 11619
British Nuclear Energy Society, London 11620
British Numismatic Society, Pontyclun 11621
British Nutrition Foundation, London 11622
British Occupational Hygiene Society, London 11623
British Origami Society, Birmingham 11624
British Ornithologists' Club, Oakham 11625
British Ornithologists' Union, Tring 11626
British Orthopaedic Association, London 11627
British Orthoptic Society, London 11628
British Paediatric Association, London 11629
British Pharmacological Society, London 11630
British Photobiology Society, London 11631
British Phycological Society, London 11632
British Postgraduate Medical Federation, London 11633
British Postmark Society, Chorley 11634
British Psychoanalytical Society, London 11635
British Psychological Society, Leicester 11636
The British Puppet and Model Theatre Guild, Yeading 11637
British Records Association, London 11638
British Record Society, London 11639
British Retinitis Pigmentosa Society, Buckingham 11640
British Robot Association, Birmingham 11641
British Safety Council, London 11642
British Schools and Universities Foundation, New York 14616
British Schools Exploring Society, London 11643
British Science Fiction Association, Wantage 11644
British Shell Collectors Club, New Maldon 11645
British Small Animal Veterinary Association, Cheltenham 11646
British Social Biology Council, London 11647
British Society for Agricultural Labour Science, Reading 11648
British Society for Allergy and Clinical Immunology, Manchester 11649
British Society for Antimicrobial Chemotherapy, Birmingham 11650
British Society for Cell Biology, Cambridge 11651
British Society for Clinical Cytology, Manchester 11652
British Society for Developmental Biology, Aberdeen 11653
British Society for Eighteenth Century Studies, Aberdeen 11654
British Society for Electronic Music, London 11655
British Society for Haematology, London 11656
The British Society for Immunology, East Grinstead 11657
British Society for Middle Eastern Studies, Cambridge 11658
British Society for Music Therapy, East Barnet 11659
British Society for Oral Medicine, London 11660
British Society for Parasitology, Reading 11661
The British Society for Phenomenology, Cardiff 11662
British Society for Plant Pathology, Leeds 11663
British Society for Research on Ageing, Manchester 11664
British Society for Restorative Dentistry, London 11665
British Society for Rheumatology, London 11666
British Society for Social Responsibility in Science, London 11667
British Society for Strain Measurement, Newcastle-upon-Tyne 11668
British Society for Surgery of the Hand, London 11669
British Society for the History of Mathematics, Milton Keynes 11670
British Society for the History of Medicine, London 11671
British Society for the History of Pharmacy, Edinburgh 11672
The British Society for the History of Science, Faringdon 11673
British Society for the Philosophy of Science, London 11674
British Society for the Promotion of Vegetable Research, Wellesbourne 11675
British Society for the Study of Infection, London 11676
British Society for the Study of Mental Subnormality, Cranage 11677
British Society for the Study of Orthodontics, London 11678
British Society of Aesthetics, Nottingham 11679
British Society of Animal Production, Penicuik 11680
British Society of Audiology, Reading 11681
The British Society of Dowsers, Ashford 11682
British Society of Flavourists, Haverhill 11683
British Society of Gastroenterology, London 11684
British Society of Hypnotherapists, London 11685
British Society of Medical and Dental Hypnosis, London 11686
British Society of Painters in Oil, Pastels and Acrylic, Ilkley 11687
British Society of Rheology, Hitchin 11688
British Society of Scientific Glassblowers, Saint Helens 11689
British Society of Soil Science, Reading 11690
The British Sociological Association, Durham 11691
British Standards Institution, London 11692
British Theatre Association, London 11693
British Theatre Institute, London 11694
British Tinnitus Association, London 11695
British Trolleybus Society, Reading 11696
British Trust for Ornithology, Thetford 11697
British Tunnelling Society, London 11698
British Universities Association of Slavists, Reading 11699
British Universities Film and Video Council, London 11700
British Universities Industrial Relations Association, Manchester 11701
British Urban and Regional Information System Association, Winchester 11702
British Veterinary Association, London 11703
British Veterinary Radiology Association, Hatfield 11704
British Veterinary Zoological Society, Studham 11705
British Vexillological Society, Beckenham 11706
British Watercolour Society, Ilkley 11707
British Wind Energy Association, Milton Keynes 11708
British Wood Preserving and Damp-proofing Association, London 11709

British Zeolite Association, London 11710
Brittle Bone Society, Dundee 11711
Broadcast Education Association, Washington 14617
Broadcast Education Association of Canada, Burnaby 01994
Broadcast Research Council of Canada, Toronto 01995
Brodie Club, Toronto 01996
The Brontë Society, Keighley 11712
Browning Institute, New York 14618
Browning Society of London, London 11713
Brown Lung Association, Greenville 14619
BRPS, Buckingham 11640
BRS, London 11639
BRS, Skokie 14579
Bruce, Dufferin, Grey Lung Association, Owen Sound 01997
Bruckner Society of America, Iowa City 14620
Brunel Society, Bristol 11714
BS, Boston 14608
BS, Collingswood 14630
BSA, Durham 11691
BSA, Iowa City 14620
BSA, Lawrence 14610
BSA, New York 14584
BSA, Nottingham 11679
B.S.A., Reading 11681
BSA, Reigate 11457
BSA, Zürich 10638
BSAC, Birmingham 11650
BSACI, Manchester 11649
BSALS, Reading 11648
BSAP, Penicuik 11680
BSAVA, Cheltenham 11646
B.S.B.C., London 11647
B.S.B.I, London 11464
BSC, London 11642
BSC, Rochester 14591
B.S.C.B., Cambridge 11651
BSCC, Manchester 11652
BSCC, New Maldon 11645
BSCE, København 03568
BSCNY, Smithtown 14625
B.S.D., Ashford 11682
BSDB, Aberdeen 11653
BSECS, Aberdeen 11654
BSEM, London 11655
BSES, London 11643
BSF, Haverhill 11683
BSFA, Wantage 11644
B.S.G., London 11684
B.S.H., London 11656
BSH, London 11685
BSHP, Edinburgh 11672
BSHS, Faringdon 11673
BSI, East Grinstead 11657
BSI, London 11692
BSI, Norwood 14568
BSMA, New York 14612
BSMT, East Barnet 11659
BSN, Sofia 01760
B.S.P., Cardiff 11662
BSPP, Leeds 11663
BSPS, London 11674
BSRA, Manchester 11664
B.S.R.D., London 11665
BSRIA, Bracknell 11716
BSS, Dublin 06934
B.S.S., Edinburgh 11463
BSS, New York 14578
BSSG, Saint Helens 11689
B.S.S.M., Newcastle-upon-Tyne 11668
BSSMS, Cranage 11677
BSSO, London 11678
BSSRS, London 11667
BSSS, Reading 11690
BSUF, New York 14616
BSUV, Charlottesville 14585
B.T.A., London 11693
BTA, London 11695
BTHG, Folkestone 11451
BTI, London 11694
BTI, Newton Centre 14609
BTN, Bydgoszcz 09540
BTO, Thetford 11697
BTRA, Bombay 06748
BTS, Reading 11696
BUAS, Kelso 11461
BUAS, Reading 11699
Budapest Union for the International Recognition of the Deposit of Microorganisms for the Purposes of Patent Procedure, Genève 10636
Buddhist Academy of Ceylon, Colombo 10408
The Buddhist Society, London 11715
Buffalo Bill Memorial Association, Cody 14621
BUFVC, London 11700
Building Centre, Hørsholm 03571
Building Materials Federation, Dublin 06935
Building Officials and Code Administrators International, Country Club Hills 14622
Building Services Research and Information Association, Bracknell 11716
BUIRA, Manchester 11701
Bulgarian Academy of Sciences, Sofia 01742
Bulgarian Association of Penal Law, Sofia 01743
Bulgarian Astronautical Society, Sofia 01744
Bulgarian Biochemical and Biophysical Society, Sofia 01745
Bulgarian Botanical Society, Sofia 01746
Bulgarian Dermatological Society, Sofia 01747
Bulgarian Geographical Society, Sofia 01748
Bulgarian Geological Society, Sofia 01749
The Bulgarian Gynecological and Obstetrical Society, Sofia 01750
Bulgarian Historical Society, Sofia 01751
Bulgarian Nutrition Society, Sofia 01752
Bulgarian Philosophical Society, Sofia 01753
Bulgarian Scientific Pharmaceutical Association, Sofia 01754
Bulgarian Society for Microbiology, Sofia 01755
Bulgarian Society of Anaesthesiology and Resuscitation, Sofia 01756
Bulgarian Society of Cardiology, Sofia 01757
Bulgarian Society of Electroencephalography, Electromyography and Clinical Neurophysiology, Sofia 01758
Bulgarian Society of Natural History, Sofia 01759
Bulgarian Society of Neurosurgery, Sofia 01760
Bulgarian Society of Physiological Sciences, Sofia 01761
Bulgarian Sociological Society, Sofia 01762
Bulgarian Soil Society, Sofia 01763
Bulgarian Union of Public Libraries, Sofia 01764
Búnadarfélag Islands, Reykjavik 06679
BUND, Bonn 05266
Bund der Freien Waldorfschulen e.V., Stuttgart 05225
Bund Deutscher Architekten, Bonn 05226
Bund Deutscher Baumeister, Architekten und Ingenieure, Bonn 05227
Bund Deutscher Kunsterzieher e.V., Hannover 05228
Bundesärztekammer, Köln 05229
Bundesakademie für musikalische Jugendbildung, Trossingen 05230
Bundesakademie für öffentliche Verwaltung, Bonn 05231
Bundes-Arbeitsgemeinschaft Akademischer Räte in der Bundesrepublik, München 05232
Bundesarbeitsgemeinschaft der katholisch-kirchlichen Büchereiarbeit, Bonn 05233
Bundesarbeitsgemeinschaft für Rehabilitation, Frankfurt 05234
Bundesarbeitsgemeinschaft katholischer Familienbildungsstätten, Düsseldorf 05235
Bundesarbeitsgemeinschaft Schule-Wirtschaft, Köln 05236
Bundesarbeitsgemeinschaft zur Förderung haltungsgefährdeter Kinder und Jugendlicher e.V., Mainz 05237
Bundesarchitektenkammer, Bonn 05238
Bundesausschuß Betriebswirtschaft, Eschborn 05239
Bundesausschuss für Wissenschaft und Bildung des Deutschen Sportbundes, Frankfurt 05240
Bundeskammer der Tierärzte Österreichs, Wien 00576
Bundesverband Bildender Künstler Bundesrepublik Deutschland e.V., Bonn 05241
Bundesverband der beamteten Tierärzte, Heinsberg 05242
Bundesverband der freiberuflichen und unabhängigen Sachverständigen für das Kraftfahrzeugwesen e.V., Königswinter 05243
Bundesverband der Friedrich-Bödecker-Kreise e.V., Mainz 05244
Bundesverband der öffentlichen angestellten und vereidigten Chemiker e.V., Hamburg 05245
Bundesverband der Pneumologen, Mülheim 05246
Bundesverband der Vertrauens- und Rentenversicherungsärzte, Aurich 05247
Bundesverband Deutscher Ärzte für Mund-Kiefer-Gesichtschirurgie e.V., Hamburg 05248
Bundesverband Deutscher Ärzte für Naturheilverfahren e.V., München 05249
Bundesverband Deutscher Leibeserzieher, Stuttgart 05250
Bundesverband Deutscher Privatschulen, Frankfurt 05251
Bundesverband evangelischer Einrichtungen und Dienste, Hannover 05689
Bundesverband für den Selbstschutz, Bonn 05252
Bundesverband für Gesundheitsförderung, Bad Wörishofen 06073
Bundesverband Hilfe für das autistische Kind e.V., Hamburg 05253
Bundesverband Jugend und Film e.V., Frankfurt 05254
Bundesverband Katholischer Ingenieure und Wirtschaftler Deutschlands, Bonn 05255
Bundesverband Legasthenie e.V., Hannover 05256
Bundesvereinigung für Gesundheit e.V., Bonn 05257
Bundesvereinigung Kulturelle Jugendbildung e.V., Remscheid 05258
Bundesvereinigung Lebenshilfe für geistig Behinderte e.V., Marburg 05259
Bundesvereinigung Logistik e.V., Bremen 05260
Bundeszahnärztekammer, Köln 05261
Bund Freiheit der Wissenschaft e.V., Bonn 05262
Bund für Deutsche Schrift und Sprache, Ahlhorn 05263
Bund für freie und angewandte Kunst e.V., Darmstadt 05264
Bund für Lebensmittelrecht und Lebensmittelkunde e.V., Bonn 05265
Bund für Umwelt und Naturschutz Deutschland e.V., Bonn 05266
Bund für Umwelt und Naturschutz Deutschland, Landesverband Hessen e.V., Mörfelden-Walldorf 05267
Bund für vereinfachte rechtschreibung, Zürich 10637
Bund katholischer Erzieher Deutschlands, Essen 05268
Bund Naturschutz in Bayern e.V., München 05269
Bund Schweizer Architekten, Zürich 10638
Bund Technischer Experten e.V., Bremen 05270
Bureau Canadien de l'Education Internationale, Ottawa 02084
Bureau Canadien de Soudage, Mississauga 02317
Bureau Canadien pour l'Avancement de Musique, Toronto 02085
Bureau des Archives et Bibliothèques de Brazzaville, Brazzaville 03423
Bureau Européen des Langues moins Répandues, Bruxelles 01303
Bureau International d'Audiophonologie, Bruxelles 01173
Bureau International de Documentation, Paris 04361
Bureau International d'Education, Genève 10639
Bureau International de la Paix, Genève 10640
Bureau International de la Récupération, Bruxelles 01174
Bureau International de Liaison et de Documentation, Paris 04362
Bureau International des Poids et Mesures, Sèvres 04363
Bureau International Technique de l'ABS, Bruxelles 01175
Bureau International Technique des Polyesters Insaturés, Bruxelles 01176
Bureau International Technique du Spathfluor, Bruxelles 01177
Bureau National d'Examen Dentaire du Canada, Ottawa 02671
Bureau of Professional Education of the American Osteopathic Association, Chicago 14623
Bureau Permanent des Congrès Internationaux des Sciences Généalogique et Héraldique, København 03570
BURISA, Winchester 11702
Burlesque Historical Society, Helendale 14624
Burma Medical Research Council, Rangoon 08811
Burma Research Society, Rangoon 08812
The Burns Federation, Kilmarnock 11717
Burns Society of the City of New York, Smithtown 14625
Business Council for Effective Literacy, New York 14626
Business Education Research of America, Hato Rey 09727
Business-Higher Education Forum, Washington 14627
Business History Conference, Williamsburg 14628
Business Professionals of America, Columbus 14629
Buteshire Natural History Society, Bute 11718
BVA, Düsseldorf 05193
BVA, London 11703
BVDN, Frankfurt 05203
BVFI, Reykjavik 06678
BVI, Rosslyn 14583
BVL, Bremen 05260
BVÖ, Leoben 00952
BVRA, Hatfield 11704
BVS, Beckenham 11706
BVS, Peacehaven 11428
BVSK, Königswinter 05243
BVTL, Bruxelles 01162
BWA, Wiesbaden 05209
B.W.A.S., Birmingham 11447
BWEA, Philadelphia 14602
BWG, Braunschweig 05221
BWG, Wien 00729
BWPA, London 11709
Bydgoskie Towarzystwo Naukowe, Bydgoszcz 09540
Byggecentrum, Hørsholm 03571
Byron Society, Collingswood 14630
Byron Society, London 11719
BZA, London 11710

CA, Indianapolis 14937
CA, New York 14760
CAA, Kampsville 14664
CAA, London 02046
CAA, New York 14770
CAA, Thames Ditton 11780
CAAHA, Dallas 14909
CAARC, London 11807
CAAS, Burnaby 02026
CAB, Aberdeen 11814
CAB, Bruxelles 01211
C.A.B. International, Wallingford 11720
CAB International Bureau of Crop Protection, Wallingford 11721
CAB International Bureau of Horticulture and Plantation Crops, Wallingford 11722
C.A.B. International Bureau of Horticulture and Plantation Crops, Wallingford 11723
CAB International Bureau of Nutrition, Wallingford 11724
CAB International Bureau of Plant Breeding and Genetics, Wallingford 11725, 11726
C.A.B. International Department of Dairy Science and Technology, Wallingford 11727
C.A.B. International Department of Soils, Wallingford 11728
CAB International Division of Animal Health and Medical Parasitology, Wallingford 11729
C.A.B. International Division of Animal Production, Wallingford 11730
CAB International Forestry Bureau, Wallingford 11731
C.A.B. International Institute of Entomology, London 11732
CAC, Washington 14635
CACC, London 11853
CACT, Saint George's 06551
CACYT, Lima 09465
CAD, London 11808
CADAS, Coventry 11865
CADEICA, Bruxelles 01243
Cadmium Pigments Association, Bruxelles 01178
CAE, Washington 14908
CAEC, Glostrup 03574
CAEJC, London 02364
CAEMC, Bruxelles 01219
Caernarvonshire Historical Society, Caernarfon 11733
CAETA, Harare 17165
CAFRA, Tunapuna 11108
CAFS, Nairobi 08484
CAG/ACG, Montréal 02052
CAGRIS, Saint Augustine 11115
CAGS, Beijing 03139
CAH, Bloomington 14835
CAH, Tiburon 14668
CAHEA, Chicago 14799
C.A.H.S., Willowdale 02080
CAI, Alexandria 14808
CAI, Buenos Aires 00100
CAI, Monte Carlo 08795
CAICYT, Buenos Aires 00099
CAID, Rochester 14857
Cairo Demographic Centre, Cairo 03883
Cairo Odontological Society, Cairo 03884
CAIS, Brattleboro 14667
Cajal Club, Denver 14631
CAJE, New York 14765
CAL, Washington 14665
CALA, Sacramento 14736
C.A.L.A.C.S., Ottawa 02036
CALC, Edmonton 02365
Calcutta Mathematical Society, Calcutta 06750
Calcutta Statistical Association, Calcutta 06751
Calgary Society for the Treatment of Autism, Calgary 01998
Calgary Zoological Society, Calgary 01999
CALI, Minneapolis 14673
California Rug Study Society, Oakland 14632
CALL, North York 02056
Calorimetry Conference, Provo 14633
CAM, Calgary 02366
CAM, Mons 01206
CAMA, Ottawa 02350
CAMAS, Dar es Salaam 11048
Cambodian Buddhist Society, Silver Spring 14634
Cambrian Archaeological Association, Newport 11734
Cambridge Antiquarian Society, Cambridge 11735
Cambridge Bibliographical Society, Cambridge 11736
Cambridge Philosophical Society, Cambridge 11737
Cambridge Refrigeration Technology, Cambridge 11738
Cambridge Society for Industrial Archaeology, Cambridge 11739
CAMHDD, Sheffield 11809
CAM-I, Arlington 14817
CAML, Ottawa 02059
Campagne d'Education Civique Européenne, Genève 10641
Campaign for the Advancement of State Education, London 11740
Campaign for the Protection of Rural Wales, Welshpool 11741
Campden Food and Drink Research Association, Chipping Campden 11742
CAMRODD, Jong Bloed 09112
Canada Council, Ottawa 02000
Canada Safety Council, Ottawa 02001
Canadian Academic Accounting Association, Toronto 02002
Canadian Academy of Engineering, Ottawa 02003
Canadian Academy of Medical Illustrators, Toronto 02004
Canadian Academy of Periondontology, Toronto 02005
Canadian Academy of Recording Arts and Sciences, Toronto 02006
Canadian Academy of Restorative Dentistry, London 02007
Canadian Academy of Sport Medicine, Vanier City 02008
Canadian Academy of the History of Pharmacy, Toronto 02009
Canadian Academy of Urological Surgeons, Vancouver 02010
Canadian Acid Precipitation Foundation, Toronto 02011
Canadian Acoustical Association, Ottawa 02012
Canadian Active Health Foundation, Vancouver 02013
Canadian Advanced Technology Association, Ottawa 02014
Canadian Advertising Research Foundation, Toronto 02015
Canadian Aeronautics and Space Institute, Ottawa 02016
Canadian Agricultural Economics and Farm Management Society, Ottawa 02017
Canadian Alarm and Security Association, Willowdale 02018
Canadian Aldeburgh Foundation, Toronto 02019
Canadian-American Committee, Washington 14635
Canadian Amphibian and Reptile Conservation Society, Mississauga 02020
Canadian Anaesthetists' Society, Toronto 02021
Canadian Apparel Manufacturers Institute, Ottawa 02022
Canadian Arctic Resources Committee, Ottawa 02023
Canadian Art Museums Directors' Organization, Victoria 02024
Canadian Association for Adult Education, Toronto 02025
Canadian Association for American Studies, Burnaby 02026
Canadian Association for Clinical Micro and Infectious Diseases, Willowdale 02027
Canadian Association for Corporate Growth, Toronto 02028
Canadian Association for Dental Research, London 02029
Canadian Association for Distance Education, Waterloo 02030
Canadian Association for Educational Psychology, London 02031
Canadian Association for Future Studies, Montréal 02032
Canadian Association for Health Physical Education and Recreation, Vanier 02033
Canadian Association for Information Science, Perth 02034
Canadian Association for Laboratory Animal Science, Edmonton 02035
Canadian Association for Latin American and Caribbean Studies, Ottawa 02036
Canadian Association for Narcolepsy, Toronto 02037
Canadian Association for Pastoral Education, Toronto 02038
Canadian Association for Production and Inventory Control, Islington 02039
Canadian Association for Scottish Studies, Guelph 02040
Canadian Association for the Advancement of Netherlandic Studies, Windsor 02041
Canadian Association for the Social Studies, Fredericton 02042
Canadian Association for the Study of Educational Administration, Halifax 02043
Canadian Association for University Continuing Education, Ottawa 02044
Canadian Association of African Studies, Toronto 02045
Canadian Association of Anatomists, London 02046
Canadian Association of Business Education Teachers, Medicine Hat 02047
Canadian Association of Deans of Arts and Sciences, Charlottetown 02048

Canadian Association of Deans of Education, Vancouver *02049*
Canadian Association of Foundations of Education, Kingston *02050*
Canadian Association of Gastroenterology, Montréal *02051*
Canadian Association of Geographers, Montréal *02052*
Canadian Association of Immersion Teachers, Ottawa *02053*
Canadian Association of Independent Schools, Toronto *02054*
Canadian Association of Latin American Studies, Vancouver *02055*
Canadian Association of Law Libraries, North York *02056*
Canadian Association of Library Schools, Halifax *02057*
Canadian Association of Marketing Research Organizations, Toronto *02058*
Canadian Association of Music Libraries, Ottawa *02059*
Canadian Association of Optometrists, Ottawa *02060*
Canadian Association of Paediatric Surgeons, London *02061*
Canadian Association of Parapsychologists, Ottawa *02062*
Canadian Association of Pathologists, Kingston *02063*
Canadian Association of Physical Medicine and Rehabilitation, London *02064*
Canadian Association of Physicists, Ottawa *02065*
Canadian Association of Radiologists, Montréal *02066*
Canadian Association of Research Libraries, Toronto *02067*
Canadian Association of Social Workers, Ottawa *02068*
Canadian Association of Sport Sciences, Gloucester *02069*
Canadian Association of Toy Libraries and Parent Resources Centres, Ottawa *02070*
Canadian Association of University Development Officers, Ottawa *02071*
Canadian Association of University Schools of Nursing, Ottawa *02072*
Canadian Association of University Schools of Rehabilitation, Montréal *02073*
Canadian Association of University Teachers, Ottawa *02074*
Canadian Association of University Teachers of German, Edmonton *02075*
Canadian Association on Water Pollution Research and Control, Montréal *02076*
Canadian Astronomical Society, Ottawa *02077*
Canadian Athletic Therapists Association, Alorchester *02078*
Canadian Authors Association, Ottawa *02079*
Canadian Aviation Historical Society, Willowdale *02080*
Canadian Bar Association, Ottawa *02081*
Canadian Biochemical Society, London *02082*
Canadian Botanical Association, Ottawa *02083*
Canadian Bureau for International Education, Ottawa *02084*
Canadian Bureau for the Advancement of Music, Toronto *02085*
Canadian Canon Law Society, Ottawa *02086*
Canadian Cardiovascular Society, Westmount *02087*
Canadian Cartographic Association, Calgary *02088*
Canadian Catholic Historical Association (English Section), London *02089*
Canadian Celtic Arts Association, Toronto *02090*
Canadian Centre for Toxicology, Guelph *02091*
Canadian Ceramic Society, Willowdale *02092*
Canadian Certified General Accountants Research Foundation, Vancouver *02093*
Canadian Children's Book Centre, Toronto *02094*
Canadian Chiropractic Association, Toronto *02095*
Canadian Classification Research Group, London *02096*
Canadian Coalition for Nuclear Responsibility, Montréal *02097*
Canadian Coalition on Acid Rain, Toronto *02098*
Canadian College of Medical Geneticists, Calgary *02099*
Canadian College of Teachers, Edmonton *02100*
Canadian Committee on Early Childhood, Edmonton *02101*
Canadian Conference of the Arts, Ottawa *02102*
Canadian Conference on Continuing Education in Pharmacy, Saint John *02103*
Canadian Construction Association, Ottawa *02104*
Canadian Council for European Affairs, Saskatoon *02105*
Canadian Council of Library Schools, Vancouver *02106*
Canadian Council of Professional Engineers, Ottawa *02107*
Canadian Council of Teachers of English, Oakville *02108*
Canadian Council of University Biology Chairmen, Saskatoon *02109*
Canadian Council of University Physical Education Administrators, Edmonton *02110*
Canadian Council on Social Development, Ottawa *02111*
Canadian Credit Institute Educational Foundation, Mississauga *02112*
Canadian Criminal Justice Association, Ottawa *02113*
Canadian Dental Association, Ottawa *02114*
Canadian Dental Research Foundation, Ottawa *02115*
Canadian Dermatological Association, Vancouver *02116*
Canadian Diabetes Association, Toronto *02117*
Canadian Dietetic Association, Toronto *02118*
Canadian Drilling Research Association, Calgary *02119*
Canadian Economics Association, Montréal *02120*
Canadian Educational Researchers' Association, Ottawa *02121*
Canadian Education Association, Toronto *02122*
Canadian Electrical Association, Montréal *02123*
Canadian Energy Research Institute, Calgary *02124*
Canadian Esperanto Association, Montréal *02125*
Canadian Esperanto Youth, Montréal *02126*
Canadian Ethnic Studies Association, Toronto *02127*
Canadian Federation for the Humanities, Ottawa *02128*
Canadian Federation of Biological Societies, Ottawa *02129*
Canadian Federation of Deans of Management and Administrative Studies, Ottawa *02130*
Canadian Federation of University Women, Nanaimo *02131*
Canadian Fertility and Andrology Society, Montréal *02132*
Canadian Fertilizer Institute, Ottawa *02133*
Canadian Film Institute, Ottawa *02134*
Canadian Fire Safety Association, Willowdale *02135*
Canadian Flexible Packaging Institute, Toronto *02136*
Canadian Folk Arts Council, Toronto *02137*
Canadian Folk Music Society, Calgary *02138*
Canadian Forestry Association, Ottawa *02139*
Canadian Foundation for Economic Education, Toronto *02140*
Canadian Foundation for the Advancement of Pharmacy, Toronto *02141*
Canadian Foundation for the Study of Infant Deaths, Toronto *02142*
Canadian Gas Association, Don Mills *02143*
Canadian Gemmological Association, Toronto *02144*
Canadian General Standards Board, Hull *02145*
Canadian Geoscience Council, Waterloo *02146*
Canadian Geotechnical Society, Rexdale *02147*
Canadian Geriatrics Research Society, Toronto *02148*
Canadian Graphic Arts Institute, Toronto *02149*
Canadian Group Psychotherapy Association, Montréal *02150*
Canadian Health Care Material Management Association, High River *02151*
Canadian Health Education Society, Toronto *02152*
Canadian Health Libraries Association, Ottawa *02153*
Canadian Hearing Society Foundation, Toronto *02154*
Canadian Heart Foundation, Ottawa *02155*
Canadian Hematology Society, Edmonton *02156*
Canadian Historical Association, Ottawa *02157*
Canadian Hypnotherapy Association, Montréal *02158*
Canadian Image Processing and Pattern Recognition Society, Toronto *02159*
Canadian Industrial Arts Association, Burnaby *02160*
Canadian Industrial Computer Society, Ottawa *02161*
Canadian Institute for Environmental Law and Policy, Toronto *02162*
Canadian Institute for Historical Microreproductions, Ottawa *02163*
Canadian Institute for Middle East Research, Calgary *02164*
Canadian Institute for Organization Management, Ottawa *02165*
Canadian Institute for Studies in Telecommunications, Pierrefonds *02166*
Canadian Institute for the Administration of Justice, Montréal *02167*
Canadian Institute of Chartered Accountants, Toronto *02168*
Canadian Institute of Child Health, Ottawa *02169*
Canadian Institute of Cost Reduction, Ottawa *02170*
Canadian Institute of Credit and Financial Management, Mississauga *02171*
Canadian Institute of Facial Plastic Surgery, Toronto *02172*
Canadian Institute of Financial Planning, Toronto *02173*
Canadian Institute of Food Science and Technology, Ottawa *02174*
Canadian Institute of Forestry, Ottawa *02175*
Canadian Institute of Hypnotism, Sainte Anne de Bellevue *02176*
Canadian Institute of International Affairs, Toronto *02177*
Canadian Institute of Management, Willowdale *02178*
Canadian Institute of Metalworking, Hamilton *02179*
Canadian Institute of Mining and Metallurgy, Montréal *02180*
Canadian Institute of Planners, Ottawa *02181*
Canadian Institute of Religion and Gerontology, Toronto *02182*
Canadian Institute of Stress, Toronto *02183*
Canadian Institute of Surveying, Ottawa *02184*
Canadian Institute of Treated Wood, Ottawa *02185*
Canadian Institute of Ukrainian Studies, Edmonton *02186*
Canadian Law and Society Association, Downsview *02187*
Canadian Law Information Council, Ottawa *02188*
Canadian Library Association, Ottawa *02189*
Canadian Lime Institute, Ingerson *02190*
Canadian Linguistic Association, Toronto *02191*
Canadian Lung Association, Ottawa *02192*
Canadian Man-Computer Communications Society, Toronto *02193*
Canadian Maritime Law Association, Montréal *02194*
Canadian Mathematical Society, Ottawa *02195*
Canadian Medical and Biological Engieneering Society, Ottawa *02196*
Canadian Medical Association, Ottawa *02197*
Canadian Medical Protective Association, Ottawa *02198*
Canadian Meteorological and Oceanographic Society, Newmarket *02199*
Canadian Metric Association, Fonthill *02200*
Canadian Museums Association, Ottawa *02201*
Canadian Music Centre, Toronto *02202*
Canadian Music Council, Ottawa *02203*
Canadian National Committee for The International Geographical Union, Vanier *02204*
Canadian National Committee of World Energy Conference, Ottawa *02205*
Canadian National Institute for the Blind, Toronto *02206*
Canadian Natural Hygiene Society, Toronto *02207*
Canadian Nautical Research Society, Ottawa *02208*
The Canadian Neurological Society, Montréal *02209*
Canadian Neurosurgical Society, Toronto *02210*
Canadian Nuclear Association, Toronto *02211*
Canadian Numismatic Association, Barrie *02212*
Canadian Operational Research Society, Ottawa *02213*
Canadian Oral History Association, Ottawa *02214*
Canadian Orthoptic Society, Montréal *02215*
Canadian Osteogenesis Imperfecta Society, Toronto *02216*
Canadian Osteopathic Aid Society, London *02217*
Canadian Osteopathic Association, London *02218*
Canadian Paediatric Society, Ottawa *02219*
Canadian Paraplegic Association, Toronto *02220*
Canadian Peace Research and Education Association, Brandon *02221*
Canadian Petroleum Law Foundation, Calgary *02222*
Canadian Pharmaceutical Association, Ottawa *02223*
Canadian Philosophical Association, Ottawa *02224*
Canadian Physiological Society, Kingston *02225*
Canadian Phytopathological Society, Lethbridge *02226*
Canadian Podiatry Association, Ottawa *02227*
Canadian Polish Research Institute, Toronto *02228*
Canadian Political Science Association, Ottawa *02229*
Canadian Psoriasis Foundation, Ottawa *02230*
Canadian Psychiatric Association, Ottawa *02231*
Canadian Psychiatric Research Foundation, Toronto *02232*
Canadian Psychoanalytic Society, Montréal *02233*
Canadian Psychological Association, Old Chelsea *02234*
Canadian Public Health Association, Ottawa *02235*
Canadian Public Relations Society, Ottawa *02236*
Canadian Public Relations Society (Nova Scotia), Halifax *02237*
Canadian Public Relations Society (Ottawa), Ottawa *02238*
Canadian Public Relations Society (Québec), Montréal *02916*
Canadian Quaternary Association, Waterloo *02239*
Canadian Radiation Protection Association, Ottawa *02240*
Canadian Railroad Historical Association, Saint Constant *02241*
Canadian Religious Conference, Ottawa *02380*
Canadian Research Institute for the Advancement of Women, Ottawa *02242*
Canadian Research Management Association, Mississauga *02243*
Canadian Research Society for Children's Literature, London *02244*
Canadian Rodeo Historical Association, Cochrane *02245*
Canadian Schizophrenia Foundation, Burnaby *02246*
Canadian Science and Technology Historical Association, North York *02247*
Canadian Semiotic Association, Kingston *02248*
Canadian Sheet Steel Building Institute, Willowdale *02249*
Canadian Sickle Cell Society, Toronto *02250*
Canadian Society for Cellular and Molecular Biology, Québec *02251*
Canadian Society for Civil Engineering, Ottawa *02252*
Canadian Society for Clinical Investigation, Montréal *02253*
Canadian Society for Color, Ottawa *02254*
Canadian Society for Education through Art, Malton *02255*
Canadian Society for Electrical and Computer Engineering, Ottawa *02256*
Canadian Society for Endocrinology and Metabolism, Montréal *02257*
Canadian Society for Horticultural Science, Kentville *02258*
Canadian Society for Immunology, Toronto *02259*
Canadian Society for International Health, Ottawa *02260*
Canadian Society for Italian Studies, Ottawa *02261*
Canadian Society for Mechanical Engineering, Ottawa *02262*
Canadian Society for Nondestructive Testing, Hamilton *02263*
Canadian Society for Nutritional Sciences, Halifax *02264*
Canadian Society for Psychomotor Learning and Sport Psychology, Kingston *02265*
Canadian Society for the Computational Study of Intelligence, Toronto *02266*
Canadian Society for the History and Philosophy of Science, Montréal *02267*
Canadian Society for the Study of Education, Ottawa *02268*
Canadian Society for the Study of Higher Education, Ottawa *02269*
Canadian Society for the Weizmann Institute of Science, Downsview *02270*
Canadian Society for Training and Development, Ottawa *02271*
Canadian Society of Agricultural Engineering, Ottawa *02272*
Canadian Society of Agronomy, Ottawa *02273*
Canadian Society of Animal Science, Ottawa *02274*
Canadian Society of Biblical Studies, Calgary *02275*
Canadian Society of Cardiology Technologists, Winnipeg *02276*
Canadian Society of Children's Authors, Illustrators and Performers, Toronto *02277*
Canadian Society of Church History, Halifax *02278*
Canadian Society of Cinematographers, Toronto *02279*
Canadian Society of Clinical Neurophysiologists, Ottawa *02280*
Canadian Society of Cytology, Hamilton *02281*
Canadian Society of Electroencephalographers, Electromyographers and Clinical Neurophysiologists, Sherbrooke *02282*
Canadian Society of Engineering Management, Ottawa *02283*
Canadian Society of Environmental Biologists, Toronto *02284*
Canadian Society of Landscape Architects, Ottawa *02285*
Canadian Society of Microbiologists, Ottawa *02286*
Canadian Society of Military Medals and Insignia, Thorold *02287*
Canadian Society of Orthopaedic Technologists, Agincourt *02288*
Canadian Society of Otolaryngology, Head and Neck Surgery, Islington *02289*
Canadian Society of Painters in Watercolour, Toronto *02290*
Canadian Society of Petroleum Geologists, Calgary *02291*
Canadian Society of Plant Physiologists, Guelph *02292*
Canadian Society of Plastic Surgeons, Willowdale *02293*
Canadian Society of Pulmonary and Cardiovascular Technology, Toronto *02294*
Canadian Society of Safety Engineering, Mississauga *02295*
Canadian Society of Soil Science, Ottawa *02296*
Canadian Society of Zoologists, Toronto *02297*
Canadian Sociology and Anthropology Association, Montréal *02298*
Canadian Speech and Hearing Society, Edmonton *02299*
Canadian Spice Association, Montréal *02300*
Canadian Standards Association, Rexdale *02301*
Canadian Steel Environmental Association, Toronto *02302*
Canadian Steel Industry Research Association, Toronto *02303*
Canadian Stroke Recovery Association, Don Mills *02304*
Canadian Teachers' Federation, Ottawa *02305*
Canadian Technion Society, Montréal *02306*
Canadian Theatre Critics Association, Toronto *02307*
Canadian Thoracic Society, Ottawa *02308*
Canadian Tinplate Recycling Council, Hamilton *02309*
Canadian Universities and Colleges Conference Officers Association, London *02310*
Canadian University and College Counselling Association, Victoria *02311*
Canadian University Music Society, Windsor *02312*
Canadian Urban Transit Association, Toronto *02313*
Canadian Urological Association, Halifax *02314*
Canadian Veterinary Medical Association, Ottawa *02315*
Canadian Water Quality Association, Waterloo *02316*
Canadian Welding Bureau, Mississauga *02317*
Canadian Wind Engineering Association, London *02318*
Canadian Writers' Foundation, Ottawa *02319*
Canal Society of New York State, Syracuse *14636*
CANARI, Saint Croix *17074*
CANASA, Willowdale *02018*
Cancer Association of South Africa, Braamfontein *10127*
Cancer Care, New York *14637*
Cancer Control Society, Los Angeles *14638*
Cancer Federation, Moreno Valley *14639*
Cancerfonden – Riksföreningen mot Cancer, Stockholm *10441*
Cancer Guidance Institute, Pittsburgh *14640*
Cancer Research Campaign, London *11743*
Cancer Research Society, Montréal *02320*
Cancer Society of Finland, Helsinki *04061*
CANDI, West Moorings *11109*

Candlelighters Childhood Cancer Foundation, Bethesda 14641
Canon Law Society of America, Washington 14642
CANSCAIP, Toronto 02277
Canterbury and York Society, Twickenham 11744
Canterbury Medical Research Foundation, Christchurch 09121
CANU, Podgorica 17080
CAP, Kingston 08076
CAP, Northfield 14775
CAP, Ottawa 02367
CAPA, Nairobi 08488
CAPAC, Toronto 02373
CAPAM, London 11810
CAPE, Washington 14869
CAPPI, Manila 09515
CAPS, Temecula 14743
CARA, Cambridge 14714
CARA, Washington 14666
CARAT AFRICA, Nairobi 08485
CARBICA, Port of Spain 11118
Carbohydrate and Lipid Metabolism Research Group, Johannesburg 10128
CARCAE, Saint John's 00035
CARDA, Castries 09994
CARDA, Saint Michael 01078
CARDATS, Saint George's 06550
CARDI, Saint Augustine 11106
The Cardiac Society of Australia and New Zealand, Sydney 00370
Cardiology Technicians Associations of British Columbia, Vancouver 02321
Cardiology Technologists Association of Ontario, Downsview 02322
Cardiovascular Credentialing International, Dayton 14643
Cardiovascular System Dynamics Society, Philadelphia 14644
Career College Association, Washington 14645
Careers Research and Advisory Centre, Cambridge 11745
CARF, Toronto 02015
CARF, Tucson 14785
Caribbean Agricultural and Rural Development, Advisory and Training Service, Saint George's 06550
Caribbean Agricultural Research and Development Institute, Saint Augustine 11106
Caribbean Agro-Economic Society, Saint Augustine 11107
Caribbean Appropriate Technology Centre, Bridgetown 01076
Caribbean Association for Feminist Research and Action, Tunapuna 11108
Caribbean Association for Rehabilitation Therapists, Castries 09994
Caribbean Association of Catholic Teachers, Saint George's 06551
Caribbean Association of Law Librarians, Cave Hill 01077
Caribbean Association of Nutritionists and Dieticians, West Moorings 11109
Caribbean Association on Mental Retardation and other Developmental Disabilities, Jong Bloed 09112
Caribbean Atlantic Regional Dental Association, Saint Michael 01078
Caribbean Centre for Development Administration, Saint Michael 01079
Caribbean Conservation Association, Bridgetown 01080
Caribbean Council for Science and Technology, Port of Spain 11110
Caribbean Council of Engineering Organizations, Kingston 08070
Caribbean Council of Legal Education, Kingston 08071
Caribbean Documentation Centre, Port of Spain 11111
Caribbean Energy Information System, Kingston 08072
Caribbean Environmental Health Institute, Castries 09995
Caribbean Examinations Council, Saint Michael 01081
Caribbean Federation for Mental Health, Port of Spain 11112
Caribbean Food and Nutrition Institute, Kingston 08073
Caribbean Food Crops Society, Saint Croix 17073
Caribbean Hospitality Training Centre, Santurce 09728
Caribbean Industrial Research Institute, Tunapuna 11113
Caribbean Information System for Economic and Social Planning, Port of Spain 11114
Caribbean Information System for the Agricultural Sciences, Saint Augustine 11115
Caribbean Institute for Meteorology and Hydrology, Bridgetown 01082
Caribbean Institute for Social Formation, Caracas 16980
Caribbean Institute of Mass Communications, Mona 08074
Caribbean Institute of Perinatology, Groningen 08836
Caribbean Law Institute, Bridgetown 01083
Caribbean Management Development Association, Saint Michael 01084
Caribbean Meteorological Organization, Port of Spain 11116
Caribbean Natural Resources Institute, Saint Croix 17074
Caribbean Network of Educational Innovation for Development, Saint Michael 01085
Caribbean Plant Protection Commission, Arima 11117
Caribbean Policy Development Centre, Bridgetown 01086
Caribbean Regional Branch of the International Council on Archives, Port of Spain 11118
Caribbean Regional Council for Adult Education, Saint John's 00035
Caribbean Regional Drug Testing Laboratory, Kingston 08075
Caribbean Sports Medicine Association, Port of Spain 11119
Caribbean Studies Association, Rio Piedras 09729
Caribbean Technical Cooperation Network on Upper Watershed Management, Santiago 03021
CARICAD, Saint Michael 01079
CARIMAC, Mona 08074
CARIRI, Tunapuna 11113
CARIS, Roma 07631
CARISFORM, Caracas 16980
CARISPLAN, Port of Spain 11114
Carl-Cranz-Gesellschaft e.V., Weßling 05271
Carl Duisberg Gesellschaft e.V., Köln 05272
Carl Johans Förbundet, Uppsala 10442
Carlyle Society of Edinburgh, Edinburgh 11746
Carnegie Foundation for the Advancement of Teaching, Princeton 14646
Carnegie Institution of Washington, Washington 14647
CARNEID, Saint Michael 01085
CAROLL, Cave Hill 01077
Carpathian Balkan Geological Association, Sofia 01765
CAS, Bodmin 11844
C.A.S., Cambridge 11735
C.A.S., Chester 11773
CAS, London 11834
C.A.S., London 11840
CAS, Minneapolis 14669
CAS, Montclair 14650
C.A.S., Nanjing 03165
CAS, Ottawa 02363
CAS, Praha 03502
CAS, Washington 14670
Casa de la Cultura de Occidente, Quezaltenango 06562
Casa de la Cultura Ecuatoriana, Quito 03847
CASAS, Kuala Lumpur 08641
CASBS, Stanford 14663
CASE, Albuquerque 14889
C.A.S.E., London 11740
CASE, Washington 14867
CASMA, Port of Spain 11119
CASME, London 11812
CAST, Ames 14868
Castings Technology International, Sheffield 11747
CASW, Greenlawn 14886
C.A.T., Rieti 07415
Catalan Agricultural Institute, Barcelona 10313
Catalysis Society (North America), West Chester 14648
Catalysis Society of Japan, Tokyo 08422
CATC, Bridgetown 01076
CATCM, Beijing 03129
Catecholamine Club, Lawrence 14649
CATER, Loja 03848
Catgut Acoustical Society, Montclair 14650
Catholic Audio-Visual Educators Association, Pittsburgh 14651
Catholic Biblical Association of America, Washington 14652
Catholic Commission on Intellectual and Cultural Affairs, Philadelphia 14653
Catholic Fine Arts Society, Rockville Centre 14654
Catholic Health Association of the United States, Saint Louis 14655
Catholic International Education Office, Bruxelles 01179
Catholic Library Association, Haverford 14656
Catholic Media Council, Aachen 05273
Catholic Medical Mission Board, New York 14657
Catholic Office for Information on European Problems, Bruxelles 01180
Catholic Public Schools Inter-Society Committee, Prince George 02323
Catholic Record Society, London 11748
Catholic Theological Society of America, Philadelphia 14658
CATI, Harare 17166
CATRA, Sheffield 11872
Caucus for a New Political Science, Boston 14659
CAUSN, Ottawa 02072
CAVE, Pittsburgh 14651
CBA, Ottawa 02083
CBA, Washington 14652
CBA, York 11847
CBA/ABC, Ottawa 02081
CBARC, New York 14822
CBDA, Brasília 01639
CBE, Chicago 14756, 14890
CBE, Washington 14870
CBGA, Sofia 01765
CBMS, Washington 14823
CBNM, Geel 01183
CBO, Kiel 05278
CBPM, Bruxelles 01209
CBS, London 02082
CBS, Nashville 16938
CBS, Silver Spring 14634
CBT, Long Island City 14792
CBTIF, Bruxelles 01210
CC, Cleveland 14863
CC, Denver 14631
CC, Lawrence 14649
CC, New York 14637
CC, Provo 14633
CCA, Ann Arbor 14661
CCA, Bridgetown 01080
CCA, Nottingham 11832
CCA, Palo Alto 14737
CCA, Schaumburg 14829
CCA, Washington 14645
CCAIT, Roswell 14809
CCAMLR, Hobart 00373
CCAS, Columbus 14892
CCB, Stockholm 10445
CCBC, Washington 14893
CCBD, Reston 14872
CCC, London 11854
CCC, Washington 14672, 14744
CCCC, Urbana 14837
CCCF, Bethesda 14641
CCDCG, Minneapolis 14762
CCDLNE, Jeddah 10002
CCE, West Des Moines 14910
CCEA, Armidale 00375
CCEO, Kingston 08070
CCEP, Oakland 14731
CCES, Beijing 03130
CCES, Bruxelles 01250
CCESP, Union 14801
CCFL, Chicago 14838
CCGTM, London 11816
CCI, Dayton 14643
CCICA, Philadelphia 14653
CCIRN, Amsterdam 08846
CCJ, Williamsburg 14828
CCL, Saint Petersburg 14836
CCMRC, Kingston 08077
CCMS, Bruxelles 01239
CCO, Provo 14911
CCOC, Saint Louis 14913
CCP, Langston 14888
CCPS, Bern 10657
CCPT, Idaho Falls 14912
CCRM, Oakton 14671
CCS, Los Angeles 14638
CCS, New York 14674
CCSP, Capo di Ponte 07417
CCSSO, Washington 14891
CCST, Port of Spain 11110
CCTE, Oakville 02108
CCU, London 11815
CCWHP/CGWH, Brooklyn 14860
C.D.A., Milano 07523
CDC, Cairo 03883
CDC, Port of Spain 11111
CDE, New York 14864
CDER, Minneapolis 14675
CDET, London 11849
CDP, Pemsacola 14676
CDS, Milwaukee 14810
CDS, Sumner 14745
C.D.S.-A.A., Bolzano 07426
CDSHA, Charlotte 14935
CDSI, New York 14660
CDS International, New York 14660
CDSO, London 11817
CEA, Carrboro 14677
CEA, Laurel 14865
CEA, Rochester 14772
CEAA, Bruxelles 01187
CEA/ACE, Toronto 02122
CEADS, Bogotá 03355
CEASD, Tucson 14830
CEBELCOR, Bruxelles 01184
CEC, London 11841
CEC, Moffet Field 14944
CEC, Mountain View 14678
CEC, Santa Barbara 14811
CEC, Westerville 14718
CECA/RF, Washington 14854
CECC, Frankfurt 05274
Cecchetti Council of America, Ann Arbor 14661
C.E.C.I.L., Milano 07460
CECIOS, Eschborn 05676
CEC-MR, Athens 14981
CECOM, Krakow 09541
CECUA, Bussum 08845
CED, Washington 14793, 14916
CEDA, Melbourne 00374
CEDA, Northridge 14943
CEDAM International, Croton on Hudson 14662
CEDaR, Washington 14873
C.E.D.C., Roma 07463
CEDIAS, Paris 04369
C.E.D.I.E.S., Casablanca 08804
CEE, Memphis 14917
CEE, Reading 11852
CEE, Urbana 14839
CEEC, Paris 08592
CEEP, Torino 07559
CEHI, Castries 09995
CEHMN, Enghien 01216
CEI, Genève 10649
C.E.I., Milano 07613
CEI, San Antonio 14720
CEIS, Kingston 08072
Ce.I.S., Roma 07496
C.E.L.A.C., Paris 04394
CELADE, Santiago 03026
CELAM, Bogotá 03365
CELC, London 11819, 11820
CELCIT, Caracas 16990
CELEX, Bruxelles 01181
C.E.L.I.D.A., Saint-Quentin 04396
C.E.L.I.H.M., Saint-Dizier 04395
CELIM, Montpellier 04451
CEM, Derby 11783
CEME, Roma 07520
CEMERS, Binghamton 14687
CEMI, Bruxelles 01213
CEMLA, México 08716
CEMS, Jouy-en-Josas 04432
CEN, Bruxelles 01208
CENARBICA, Kinshasa 17142
CEN/CENELEC, Bruxelles 01182
CENCI, Montevideo 13253
CENELEC, Bruxelles 01226
CENELEC Electronic Components Committee, Frankfurt 05274
CENIAP, Maracay 16991
CENICAFE, Chinchina 03361
CENID, Santiago 03027
CENTA, Santa Tecla 03926
Center for Advanced Study in the Behavioral Sciences, Stanford 14663
Center for Agricultural Research for the Brazilian Savannahs, Planaltina 01625
Center for American Archaeology, Kampsville 14664
Center for Andean Regional Studies Bartolomé de las Casas, Cusco 09457
Center for Applied Linguistics, Washington 14665
Center for Applied Research in the Apostolate, Washington 14666
Center for Arab-Islamic Studies, Brattleboro 14667
Center for Attitudinal Healing, Tiburon 14668
Center for Austrian Studies, Minneapolis 14669
Center for Auto Safety, Washington 14670
Center for Bioethics, Montréal 02327
Center for Chinese Research Materials, Oakton 14671
Center for Community Change, Washington 14672
Center for Computer-Assisted Legal Instruction, Minneapolis 14673
Center for Cuban Studies, New York 14674
Center for Death Education and Research, Minneapolis 14675
Center for Design Planning, Pemsacola 14676
Center for Early Adolescence, Carrboro 14677
Center for Economic and Social Studies in the Third World, México 08709
Center for Economic Conversion, Mountain View 14678
Center for Energy Policy and Research, Old Westbury 14679
Center for Field Research, Watertown 14680
Center for Global Education, Minneapolis 14681
Center for Hospitality Research and Service, Blacksburg 14682
Center for Integrated Social Development, Rosario 00097
Center for International Development and Environment, Washington 14683
Center for International Research on Economic Tendency Surveys, München 05275
Center for Management Development, New York 14684
Center for Marine Conservation, Washington 14685
Center for Medical Consumers and Health Care Information, New York 14686
Center for Medieval and Early Renaissance Studies, Binghamton 14687
Center for Medieval and Renaissance Studies, Columbus 14688
Center for Meso American Studies on Appropriate Technology, Guatemala City 06563
Center for Migration Studies of New York, Staten Island 14689
Center for Neo-Hellenic Studies, Austin 14690
Center for Oceans Law and Policy, Charlottesville 14691
Center for OPEC Studies, Caracas 16981
Center for Philosophy, Law, Citizenship, Farmingdale 14692
Center for Process Studies, Claremont 14693
Center for Public Justice, Washington 14694
Center for Research and Documentation on the World Language Problem, Rotterdam 08837
Center for Research in Ambulatory Health Care Administration, Englewood 14695
Center for Research in Islamic Education, Mecca 10001
Center for Research Libraries, Chicago 14696
Center for Responsive Psychology, Brooklyn 14697
Center for Safety in the Arts, New York 14698
Center for Science in the Public Interest, Washington 14699
Center for Short Lived Phenomena, Cambridge 14700
Center for Socialist History, Berkeley 14701
Center for Social Research and Education, San Francisco 14702
Center for Studies in Criminal Justice, Chicago 14703
Center for Studies, Research and Training in International Understanding and Cooperation, Paris 04364
Center for Sustainable Agriculture, Davis 14704
Center for Sutton Movement Writing, Newport Beach 14705
Center for the Study of Human Rights, New York 14706
Center for the Study of Parent Involvement, Orinda 14707
Center for the Study of the Presidency, New York 14708
Center for Urban and Regional Studies, Chapel Hill 14709
Center for Urban Black Studies, Berkeley 14710
Center for War, Peace and the News Media, New York 14711
Center for War/Peace Studies, New York 14712
Center of Coordination and Diffusion of Latin American Studies, México 08710
Center of Information and Research SIIS, San Sebastián 10286
Center of Regional Cooperation in Adult Education for Latin America, Patzcuaro 08711
Center on Education and Training for Employment, Columbus 14713
Centers and Regional Associations, Cambridge 14714
Centraal Bureau voor Genealogie, 's-Gravenhage 08838
Central African Regional Branch of the International Council on Archives, Kinshasa 17142
Central American Association of Sociology, Panamá City 09408
Central American Dermatological Society, San Salvador 03922
Central American Federation of Sports Medicine, Heredía 03443
Central American Institute for Business Administration, Alajuela 03436
Central American Institute of Public Administration, San José 03437
Central American Institute of Social Studies, San José 03438
Central American, Mexican and Caribbean Network for Bean Research, San José 03439
Central American Paediatric Society, San Salvador 03923
Central American Public Health Council, San Salvador 03924
Central American Research Institute for Industry, Guatemala City 06564
Central American Society of Pharmacology, Panamá City 09413
Central American University Confederation, San José 03440
Central Bureau for Educational Visits and Exchanges, London 11749
Central Bureau for Genealogy, 's-Gravenhage 08838
Central Bureau for Nuclear Measurements, Geel 01183
Central Committee for Norwegian Research, Oslo 09268
Central Council of Physical Recreation, London 11750
Central Council of Scientific and Technical Unions, Sofia 01766
Central Council of the Academies for Architecture, Amsterdam 09070
Centrale de l'Enseignement du Québec, Montréal 02324
Centrale des Enseignants Zairois, Kinshasa 17143
Central European Mass Communication Research Documentation Centre, Krakow 09541
Central Ontario Industrial Relations Institute, Toronto 02325
Central Organization for Jewish Education, Brooklyn 14715
Central Research Organization, Rangoon 08813
Central Scotland Aviation Group, Edinburgh 11751
Central Society for Clinical Research, Chicago 14716

Central South Academy of Industrial Architecture, Hupei *03122*
Central States Anthropological Society, Dayton *14717*
Central Sterilising Club, Kingston-upon-Thames *11752*
Central Veterinary Institute, Budapest *06668*
Centre Africain d'Etudes Supérieures en Gestion, Dakar *10023*
Centre Belge d'Etude de la Corrosion, Bruxelles *01184*
Centre Belge pour la Gestion de la Qualité, Bruxelles *01185*
Centre Canadien de Toxicologie, Guelph *02091*
Centre d'Animation, de Développement et de Recherche en Education, Montréal *02326*
Centre de Bioéthique, Montréal *02327*
Centre de Coopération Internationale en Recherche Agronomique pour le Développement, Paris *04365*
Centre de Coopération pour les Recherches Scientifiques Relatives au Tabac, Paris *04366*
Centre de Coordination des Etudes et des Recherches sur les Infrastructures et les Equipements, Alger *00009*
Centre de Coordination des Recherches et de Documentation en Sciences Sociales desservant l'Afrique Sub-Saharienne, Kinshasa *17144*
Centre de Musique Canadienne, Toronto *02202*
Centre de Perfectionnement des Industries Textiles Rhône-Alpes, Lyon *04367*
Centre d'Epidémiologie, Statistique et Information Sanitaire, Bobo-Dioulasso *01806*
Centre de Promotion de l'Enseignement Catholique en Europe, Bruxelles *01204*
Centre de Recherches Anthropologiques, Préhistoriques et Ethnographiques, Alger *00010*
Centre de Recherches en Economie Appliquée pour le Développement, Alger *00011*
Centre de Recherches Océanographiques, Abidjan *08061*
Centre de Recherches Océanographiques et des Pêches, Alger *00012*
Centre de Recherches pour le Développement International, Ottawa *02552*
Centre de Recherches sur les Resources Biologiques Terrestres, Alger *00013*
Centre de Recherches Zootechniques, Bouaké *08062*
Centre de Rencontres et d'Etudes des Dirigeants des Administrations Fiscales, Paris *04368*
Centre des Sciences et de la Technologie Nucléaires, Alger *00014*
Centre des Sciences Humaines, Abidjan *08063*
Centre d'Etudes, de Documentation, d'Information et d'Action Sociales, Paris *04369*
Centre d'Etudes, de Documentations et d'Informations Economiques et Sociales, Casablanca *08804*
Centre d'Histoire Militaire et d'Etudes de Défense Nationale, Montpellier *04370*
Centre Européen de Diffusion de la Culture, Roma *07463*
Centre Européen de Formation des Statisticiens Economistes des Pays en Voie de Développement, Malakoff *04371*
Centre Européen de Recherches sur les Congrégations et Ordres Religieux, Saint-Etienne *04372*
Centre Européen de Réflexion et d'Etude en Thermodynamique, Lausanne *10642*
Centre Européen des Silicones, Bruxelles *01186*
Centre Européen pour l'Etude de l'Argumentation, Bruxelles *01187*
Centre for African Family Studies, Nairobi *08484*
Centre for Agricultural Research in Suriname, Paramaribo *10430*
Centre for Alternative Technology, Machynlleth *11753*
Centre for Analysis of Economic Affairs, Rio de Janeiro *01619*
Centre for Constitutional Studies, Madrid *10287*
Centre for Coordination of Research of the International Federation of Catholic Universities, Roma *07413*
Centre for Documentation, Research and Training on the Islands of the South West Indian Ocean, Rose Hill *08673*
Centre for Human Evolution Studies, Roma *07414*
Centre for Iberian Studies, Keele *11754*
Centre for Latin American Monetary Studies, México *08712*
Centre for Linguistic and Historical Studies by Oral Tradition, Niamey *09207*
Centre for Mathematics and Computer Science, Amsterdam *08841*
Centre for Monarchial Studies, Toronto *02328*
Centre for Plant Breeding and Reproduction Research, Wageningen *08840*
Centre for Research on European Women, Bruxelles *01188*
Centre for Research on Latin America and the Caribbean, Toronto *02329*
Centre for Research on the Epidemiology of Disasters, Bruxelles *01189*
Centre for Social Research and Documentation for the Arab Region, Cairo *03885*
Centre for the Advancement and Study of the European Currency, Lyon *04373*
Centre for the Coordination of Research and Documentation in Social Science for Sub-Saharan Africa, Kinshasa *17144*
Centre for the Independence of Judges and Lawyers, Genève *10643*
Centre for the Study of Education in Developing Countries, 's-Gravenhage *08839*
Centre for the Study of International Relations, Stockholm *10443*
Centre Français de Droit Comparé, Paris *04374*
Centre Franco-Ontarien de Resources Pédagogiques, Ottawa *02330*
Centre International de Criminologie Comparée, Montréal *02543*
Centre International de Cyto-Cybernétique, Paris *04375*
Centre International de Documentation Marguerite Yourcenar, Bruxelles *01190*
Centre International de Formation et de Recherche en Population et Développement en Association avec les Nations Unies, Louvain-la-Neuve *01191*
Centre International de l'Enfance, Paris *04376*
Centre International de Liaison des Ecoles de Cinéma et de Télévision, Bruxelles *01192*
Centre International de Perfectionnement Professionnel et Technique, Torino *07709*
Centre International de Recherches et d'Information sur l'Economie Publique, Sociale et Coopérative, Liège *01193*
Centre International de Recherches Glyptographiques, Braine-le-Château *01194*
Centre International de Recherche sur le Bilinguisme, Sainte Foy *02544*
Centre International des Langues, Littératures et Traditions d'Afrique au Service du Développement, Louvain-la-Neuve *01195*
Centre International d'Etude de la Peinture Médiévale des Bassins de l'Exant et de la Meuse, Bruxelles *01196*
Centre International d'Etude des Textiles Anciens, Lyon *04377*
Centre International d'Etude de Tantale et de Niobium, Bruxelles *01197*
Centre International d'Etudes de la Formation Religieuse, Bruxelles *01198*
Centre International d'Etudes, de Recherche et d'Action pour le Développemnt, Bruxelles *01199*
Centre International d'Etudes du Lindane, Bruxelles *01200*
Centre International d'Etudes Latines, Paris *04378*
Centre International d'Etudes Romanes, Paris *04379*
Centre International du Film pour l'Enfance et la Jeunesse, Paris *04380*
Centre International Scolaire de Correspondance Sonore, Sainte-Savine *04381*
Centre Muraz, Bobo-Dioulasso *01807*
Centre National d'Astronomie, d'Astrophysique et de Géophysique, Alger *00015*
Centre National de Documentation, Rabat *08805*
Centre National de Documentation et de Recherche en Pédagogie, Alger *00016*
Centre National de Documentation Scientifique et Technique, Bruxelles *01201, 01202*
Centre National de la Recherche Scientifique, Paris *04382*
Centre National de Recherches de Logique, Bruxelles *01203*
Centre National de Recherches et d'Application des Géosciences, Alger *00017*
Centre National de Recherches et Expérimentations Forestières, Alger *00018*
Centre National de Recherche sur les Zones Arides, Alger *00019*
Centre National de Traduction et de Terminologie Arabe, Alger *00020*
Centre National d'Etudes et de Recherches pour l'Aménagement du Territoire, Alger *00021*
Centre National d'Etudes Spatiales, Paris *04383*
Centre Naturopa, Strasbourg *04384*
Centre of Economic and Social Studies and Experiments in Western Africa, Bobo-Dioulasso *01808*
Centre of Promotion of Catholic Education in Europe, Bruxelles *01204*
Centre pour l'Etude des Problèmes du Monde Musulman Contemporain, Bruxelles *01205*
Centre Régional de Formation et d'Application en Agrométéorologie et Hydrologie Opérationnelle, Niamey *09208*
Centre Suisse de Documentation en Matière d'Enseignement et d'Education, Grand-Saconnex *10644*
Centre Technique et de Promotion des Laitiers Sidérurgiques, Paris La Défense *04385*
Centre Technique Forestier Tropical, Abidjan *08064*
Centre Technique Forestier Tropical, Pointe-Noire *03424*
Centro Académico Hugo Simas, Curitiba *01615*
Centro Andino de Tecnología Rural, Loja *03848*
Centro Appenninico del Terminillo Carlo Jucci, Rieti *07415*
Centro Argentino de Espeleología, Buenos Aires *00098*
Centro Argentino de Información Científica y Tecnológica, Buenos Aires *00099*
Centro Argentino de Ingenieros, Buenos Aires *00100*
Centro Asturiano, La Habana *03472*
Centro Bibliografica Francescano, Benevento *07416*
Centro Brasileiro de Estudos, Campinas *01616*
Centro Brasileiro de Pesquisas Fisicas, Rio de Janeiro *01617*
Centro Camuno di Studi Preistorici ed Etnologici, Capo di Ponte *07417*
Centro Científico y Técnico Francés en México, México *08713*
Centro Coordinador y Difusor de los Estudios Latinoamericanos, México *08710*
Centro Cultural de Botucatu, Botucatu *01618*
Centro d'Arte e di Cultura, Bologna *07418*
Centro de Análise Conjuntura Econômica, Rio de Janeiro *01619*
Centro de Aperfeiçoamento e Especialização Médica, Rio de Janeiro *01620*
Centro de Biomédica de Campina Grande, Campina Grande *01621*
Centro de Ciências, Letras e Artes, Campinas *01622*
Centro de Documentação Cinetífica e Técnica, Lisboa *09663*
Centro de Documentación Bibliotecológica, Bahía Blanca *00101*
Centro de Educación en Administración de Salud, Bogotá *03355*
Centro de Estudios Constitucionales, Madrid *10287*
Centro de Estudios de Demografía Histórica de América Latina, São Paulo *01623*
Centro de Estudios de Historia y Organización de la Ciencia Carlos J. Finlay, La Habana *03473*
Centro de Estudios Económicos Sociales, Buenos Aires *00102*
Centro de Estudios Económicos y Sociales del Tercer Mundo, México *08714*
Centro de Estudios Educativos, México *08715*
Centro de Estudios e Investigaciones Geotécnicas, San Salvador *03925*
Centro de Estudios Histórico-Militares del Perú, Lima *09458*
Centro de Estudios Médicos Ricardo Moreno Cañas, San José *03441*
Centro de Estudios Monetarios Latinoamericanos, México *08716*
Centro de Estudios sobre Desarrollo Económico, Bogotá *03356*
Centro de Estudios Urbanos y Regionales, Buenos Aires *00103*
Centro de Estudios Venezolanos Indigenas, Caracas *16982*
Centro de Estudos de História e Cartografia Antiga, Lisboa *09654*
Centro de Estudos e Pesquisas em Administração, Porto Alegre *01624*
Centro de Historia del Táchira, San Cristóbal *16983*
Centro de Historia Larense, Barquisimeto *16984*
Centro de Información Científica y Humanística, México *08717*
Centro de Información e Divulgación Agropecuario, La Habana *03474*
Centro de Investigación de Biologia Marina, San Martín *00104*
Centro de Investigación Documentaria, Buenos Aires *00105*
Centro de Investigaciones Agrícolas Alberto Boerger, La Estanzuela *13251*
Centro de Investigaciones Bella Vista, Bella Vista *00106*
Centro de Investigaciones de Recursos Naturales, Castelar *00107*
Centro de Investigaciones Económicas, Buenos Aires *00108*
Centro de Investigaciones Históricas, Guayaquil *03849*
Centro de Investigaciones Pesqueras, La Habana *03475*
Centro de Investigaciones y Estudios Familiares, Montevideo *13252*
Centro de Investigación Forestal, La Habana *03476*
Centro de Investigación Minera y Metalúrgica, Santiago *03022*
Centro de Investigación y Acción Social, Bogotá *03357*
Centro de Investigación y de Estudios Avanzados del Instituto Politécnico Nacional, México *08718*
Centro de Investigación y Restauración de Bienes Monumentales del Instituto Nacional de Cultura, Lima *09459*
Centro del PEN Internacional, Lima *09460*
Centro de Medicina Nuclear, Buenos Aires *00109*
Centro de Nacionales y Comercio Internacional del Uruguay, Montevideo *13253*
Centro de Navegación Transatlántica, Buenos Aires *00110*
Centro de Perfeccionamiento, Experimentación e Investigaciones Pedagógicas, Santiago *03023*
Centro de Pesquisa Agropecuária dos Cerrados, Planaltina *01625*
Centro de Pesquisa Agropecuária do Trópico-Árido, Petrolina *01626*
Centro de Pesquisa Agropecuária do Trópico Umido, Belém *01627*
Centro de Pesquisas de Geografia do Brasil, Rio de Janeiro *01628*
Centro de Pesquisas Folclóricas, Rio de Janeiro *01629*
Centro di Demodossalogia, Roma *07419*
Centro di Documentazione, Pistoia *07420*
Centro di Documentazione Economica per Giornalisti, Roma *07421*
Centro di Documentazione e di Iniziativa Politica, Roma *07422*
Centro di Documentazione e Promozione Archeologica, Roma *07423*
Centro di Documentazione Giornalistica, Roma *07424*
Centro di Documentazione Statistica Internazionale, Bologna *07425*
Centro di Documentazione Storica per l'Alto Adige, Bolzano *07426*
Centro di Documentazione, Studi e Ricerche Jacques Maritain, Rimini *07427*
Centro di Formazione e Studi per il Mezzogiorno, Roma *07428*
Centro di Informazioni Sterilizzazione e Aborto, Milano *07429*
Centro di Musicologia Walter Stauffer, Cremona *07430*
Centro di Ricerca Applicata e Documentazione, Udine *07431*
Centro di Ricerca e di Studio sul Movimento dei Disciplinati, Perugia *07432*
Centro di Ricerca Pergamene Medievali e Protocolli Notarili, Roma *07433*
Centro di Ricerca per il Teatro, Milano *07434*
Centro di Ricerche, Prato *08037*
Centro di Ricerche Biopsichiche di Padova, Padova *07435*
Centro di Studi Aziendali e Amministrativi, Cremona *07436*
Centro di Studi Bonaventuriani, Bagnoregio *07437*
Centro di Studi Chimico-Fisici di Macromolecole Sintetiche e Naturali, Genova *07438*
Centro di Studi e Applicazioni di Organizzazione Aziendale della Produzione e dei Trasporti, Torino *07439*
Centro di Studi e di Ricerche per la Psicoterapia della Coppia e della Famiglia, Roma *07440*
Centro di Studi e Documentazione delle Ricerche sulla Didattica dell'Educazione Fisica e dello Sport, Napoli *07441*
Centro di Studi e Ricerche di Museologia Agraria, Milano *07442*
Centro di Studi e Ricerche sulla Nutrizione e Sugli Alimenti, Parma *07443*
Centro di Studi Etruschi, Orvieto *07444*
Centro di Studi Filologici e Linguistici Siciliani, Palermo *07445*
Centro di Studi Filosofici di Gallarate, Padova *07446*
Centro di Studi Grafici, Milano *07447*
Centro di Studi Metodologici, Torino *07448*
Centro di Studio e Documentazione sul Vietnam e il Terzo Mondo, Milano *07449*
Centro di Studi per la Storia dell'Architettura, Roma *07450*
Centro di Studi per l'Educazione Fisica, Bologna *07451*
Centro di Studi Pratici di Agricoltura, Fondazione Fratelli Gustavo e Severino Navarra, Ferrara *07452*
Centro di Studi Preistorici e Archeologici, Varese *07453*
Centro di Studi Sociali e Politici Lorenzo Milani, Arezzo *07454*
Centro di Studi Storici ed Etnografici del Piceno, Ascoli Piceno *07455*
Centro di Studi Storici Maceratesi, Macerata *07456*
Centro di Studi sul Teatro Medioevale e Rinascimentale, Viterbo *07457*
Centro di Vita Europea, Roma *07458*
Centro Emilia-Romagna per la Storia del Giornalismo, Bologna *07459*
Centro Esperantista Contro l'Imperialismo Linguistico, Milano *07460*
Centro Europeo per il Progresso della Scuola, Roma *07461*
Centro Europeo per il Progresso Economico e Sociale, Roma *07462*
Centro Europeo per la Diffusione della Cultura, Roma *07463*
Centro Histórico del Zulia, Maracaibo *16985*
Centro Histórico Sucrense, Cumana *16986*
Centro I.C.T.A., Bologna *07530*
Centro Informazione Farine e Pane, Roma *07464*
Centro Informazioni e Studi sulla Comunita' Europea, Milano *07465*
Centro Intelectual Galindo, La Paz *01535*
Centro Interamericano de Desarrollo Integras de Aguas y Tierra, Mérida *16987*
Centro Interamericano de Enseñanza de Estadistica, Santiago *03024*
Centro Interamericano de Fotointerpretación, Bogotá *03358*
Centro Interamericano de Investigación y Documentación sobre Formación Profesional, Montevideo *13254*
Centro Interamericano de Vivienda y Planeamiento, Bogotá *03359*
Centro Interamericano para el Desarrollo Regional, Maracaibo *16988*
Centro Internacional de Agricultura Tropical, Cali *03360*
Centro Internacional de Mejoramiento de Maíz y Trigo, México *08719*
Centro Internazionale della Pace, Torino *07466*
Centro Internazionale di Documentazione e Comunicazione, Roma *07467*
Centro Internazionale di Ipnosi Medica e Psicologica, Milano *07468*
Centro Internazionale di Ricerche sulle Strutture Ambiente Pio Manzù, Verucchio *07469*
Centro Internazionale di Scienze Meccaniche, Udine *07470*
Centro Internazionale di Studi Archeologici Maiuri, Ercolano *07471*
Centro Internazionale di Studi di Architettura Andrea Palladio, Vicenza *07472*
Centro Internazionale di Studi e Documentazione sulle Comunità Europee, Milano *07473*
Centro Internazionale di Studi Rosminiani, Stresa *07474*
Centro Internazionale di Studi Sardi, Cagliari *07475*
Centro Internazionale di Studi Umanistici, Roma *07476*
Centro Internazionale Magistrati Luigi Severini, Perugia *07477*
Centro Internazionale per gli Studi sulla Irrigazione, Verona *07478*
Centro Internazionale per l'Avanzamento della Ricerca e dell'Educazione, Vigliano Biellese *07479*
Centro Internazionale per le Communicazione Sociali, Roma *07480*
Centro Internazionale per l'Educazione Artistica della Fondazione Giorgio Cini, Venezia *07481*
Centro Internazionale per l'Iniziativa Giuridica, Roma *07482*
Centro Internazionale per lo Studio dei Papiri Ercolanesi, Napoli *07483*
Centro Internazionale Ricerche sulle Strutture Ambientali Pio Manzu, Verucchio *07484*
Centro Internazionale Sonnenberg per l'Italia, Torino *07485*
Centro Internazionale Studi Famiglia, Milano *07486*
Centro Internazionale Studi Musicali, Roma *07487*
Centro Internazionale Studi Umanistici Scientifici Psicologici, Roma *07488*
Centro Interuniversitario de Desarollo, Santiago *03025*
Centro Isec, Iniziative per Studi e Convegni, Roma *07489*
Centro Italiano di Antropologia Culturale, Roma *07490*
Centro Italiano di Biostatistica, Roma *07491*
Centro Italiano di Musica Antica, Roma *07492*

Centro Italiano di Parapsicologia, Napoli 07493
Centro Italiano di Ricerche e d'Informazione sull'Economia delle Imprese Pubbliche e di Pubblico Interesse, Milano 07494
Centro Italiano di Sessuologia, Roma 07495
Centro Italiano di Solidarieta', Roma 07496
Centro Italiano di Studi Aziendali, Milano 07497
Centro Italiano di Studi di Diritto dell'Energia, Roma 07498
Centro Italiano di Studi e Programmazioni per la Pesca, Roma 07499
Centro Italiano di Studi Europei, Roma 07500
Centro Italiano di Studi Finanziari, Roma 07501
Centro Italiano di Studi Sociali Economici e Giuridici, Roma 07502
Centro Italiano per lo Studio e lo Sviluppo dell'Agopuntura Moderna e dell'Altra Medicina, Vicenza 07503
Centro Italiano per lo Studio e lo Sviluppo della Psicoterapia e dell'Autogenes Training, Padova 07504
Centro Italiano Ricerca e Informazione Economica, Roma 07505
Centro Italiano Ricerche e Studi Assicurativi, Roma 07506
Centro Italiano Ricerche e Studi Trasporto Aereo, Roma 07507
Centro Italiano Studi Containers, Genova 07508
Centro Italiano Studi Politici Economici Sociali, Roma 07509
Centro Italiano Studi sull'Arte dello Spettacolo, Roma 07510
Centro Italiano Sviluppo Impieghi Acciaio, Milano 07511
Centro Italo-Nipponico di Studi Economici, Milano 07512
Centro Latinoamericana de Estudios y Difusión de la Construcción en Tierra, Lima 09461
Centro Latinoamericano de Adminstración para el Desarrollo, Caracas 16989
Centro Latinoamericano de Creación e Investigación Teatral, Caracas 16990
Centro Latinoamericano de Demografía, Santiago 03026
Centro Latino Americano de Física, Rio de Janeiro 01630
Centro Lattiero Caseario di Assistenza e Sperimentazione Antonio Bizzozero, Parma 07513
Centro Ligure di Storia Sociale, Genova 07514
Centro Lincoln, Buenos Aires 00111
Centro Luigi Lavazza per gli Studi e Ricerche sul Caffe', Torino 07515
Centro Médico Federal del Azuay, Cuenca 03850
Centro Nacional de Cálcula, México 08720
Centro Nacional de Ciencias y Tecnologías Marinas, Veracruz 08721
Centro Nacional de Documentación Científica y Tecnológica, La Paz 01536
Centro Nacional de Documentación e Información Educativa, Buenos Aires 00112
Centro Nacional de Documentación e Información Educativa, La Paz 01537
Centro Nacional de Informação Científica em Microbiologia, Rio de Janeiro 01631
Centro Nacional de Información de Ciencias Médicas, La Habana 03477
Centro Nacional de Información y Documentación, Santiago 03027
Centro Nacional de Investigaciones Agropecuarias, Maracay 16991
Centro Nacional de Investigaciones Científicas, La Habana 03478
Centro Nacional de Investigaciones de Café, Chinchina 03361
Centro Nacional de Patologia Animal, Lima 09462
Centro Nacional de Pesquisa de Arroz e Feijão, Goiânia 01632
Centro Nacional de Pesquisa de Gado de Corte, Campo Grande 01648
Centro Nacional de Pesquisa de Mandioca e Fruticultura, Cruz das Almas 01633
Centro Nacional de Pesquisa de Milho e Sorgo, Sete Lagoas 01634
Centro Nacional de Pesquisa de Seringueira, Manaus 01635
Centro Nacional de Pesquisa de Soja, Londrina 01636
Centro Nacional de Pesquisa do Algodão, Campina Grande 01637
Centro Nacional de Tecnologia Agropecuario, Santa Tecla 03926
Centro Nazionale di Dietobiologia ed Igiene della Alimentazione, Milano 07516
Centro Nazionale di Studi Urbanistici, Roma 07517
Centro para el Desarrollo de la Capacidad Nacional de Investigación, Panamá City 09414
Centro Paraguayo de Estudios de Desarrollo Económico y Social, Asunción 09428
Centro Pedagógico y Cultural de Portales, Cochabamba 01538
Centro per gli Studi di Tecnica Navale, Genova 07518
Centro per gli Studi e le Applicazioni delle Risorse Energetiche, Rovigo 07519
Centro per gli Studi sui Mercati Esteri, Roma 07520
Centro per gli Studi sui Sistemi Distributivi e il Turismo, Milano 07521
Centro per la Diffusione del Libro Lucano, Potenza 07522
Centro per la Documentazione Automatica, Milano 07523
Centro per la Riforma del Diritto di Famiglia, Milano 07524
Centro per la Statistica Aziendale, Firenze 07525
Centro per la Storia della Tradizione Aristotelica Nel Veneto, Padova 07526
Centro per lo Studio dei Dialetti Veneti dell'Istria, Trieste 07527
Centro Piombinese di Studi Storici, Piombino 07528
Centro Polesano di Studi Storici, Archeologici ed Etnografici, Rovigo 07529
Centro Provinciale Impiego Combinato Tecniche Agricole, Bologna 07530
Centro Psico-Pedagogico Didattico, Bologna 07531
Centro Regional de Educación de Adultos y Alfabetización Funcional para América Latina, Pátzcuaro 08722
Centro Regional de Pesquisas Educacionais do Sul, Porto Alegre 01638
Centro Regionale di Studi Sociali V.G. Galati, Lamezia Terme 07532
Centro Regional para el Fomento del Libro en América Latina, Bogotá 03362
Centro Ricerche Applicazione Bioritmo, Roma 07533
Centro Ricerche Archeologiche e Scavi di Torino per il Medio Oriente e l'Asia, Torino 07534
Centro Ricerche Cosmetologiche, Roma 07535
Centro Ricerche Didattiche Ugo Morin, Paderno del Grappa 07536
Centro Ricerche di Storia e Arte Bitontina, Bitonto 07537
Centro Ricerche Economiche ed Operative della Cooperazione, Napoli 07538
Centro Ricerche Economiche Sociologiche e di Mercato nell'Edilizia, Roma 07539
Centro Ricerche Metapsichiche e Psicofoniche, Fermo 07540
Centro Ricerche Socio-Religiose, Napoli 07541
Centro Ricerche Urbanistiche e di Progettazione, Milano 07542
Centro Romano per lo Studio dei Problemi di Attualita' Sociale, Roma 07543
Centro Rossiniano di Studi, Pesaro 07544
Centro Sperimentale di Cinematografia, Roma 07545
Centro Sperimentale Italiano di Giornalismo, Milano 07546
Centro Sperimentale Metallurgico, Roma 07547
Centro Studi Archaeologici di Boscoreale, Boscotrecase e Trecase, Boscoreale 07548
Centro Studi Assicurativi Piero Sacerdoti, Milano 07549
Centro Studi Cinematografici, Roma 07550
Centro Studi della Cooperazione nel Veneto, Vicenza 07551
Centro Studi di Diritto del Lavoro, Milano 07552
Centro Studi di Diritto Fluviale e della Navigazione Interna, Venezia 07553
Centro Studi di Diritto Sportivo, Vicenza 07554
Centro Studi di Economia Applicata all'Ingegneria, Bari 07555
Centro Studi di Economia Applicata all'Ingegneria, Napoli 07556
Centro Studi di Estimo e di Economia Territoriale, Firenze 07557
Centro Studi di Poesia e di Storia delle Poetiche, Roma 07558
Centro Studi di Politica Economica, Torino 07559
Centro Studi di Psicologia e Sociologia Applicate ad Indirizzo Adleriano, Roma 07560
Centro Studi di Psicoterapia e Psicologia Clinica, Genova 07561
Centro Studi Diritto Comunitario, Roma 07562
Centro Studi di Storia Locale, Massa 07563
Centro Studi e Applicazioni in Tecnologie Avanzate, Bari 07564
Centro Studi e Archivio della Comunicazione, Università di Parma, Parma 07565
Centro Studi Economici, Roma 07566
Centro Studi Economici e Sociali Giuseppe Toniolo, Pisa 07567
Centro Studi Economici per l'Alta Italia, Milano 07568
Centro Studi ed Esperienze Scout Baden Powell, Roma 07569
Centro Studi e di Educazione Civica Enrico Mattei, Torino 07570
Centro Studi e Documentazione della Cultura Armena, Milano 07571
Centro Studi e Indagini sull'Opinione Pubblica, Roma 07419
Centro Studi e Iniziative Pier Santi Mattarella, Trapani 07572
Centro Studi e Ricerche per la Conoscenza della Liguria Attraverso le Testimonianze dei Viaggiatori Stranieri, Genova 07573
Centro Studi e Ricerche sui Rapporti Umani, Roma 07574
Centro Studi Filippo e Marta Larizza per la Formazione Permanente degli Educatori e per la Prevenzione del Disadattamento Giovanile, Bassano del Grappa 07575
Centro Studi Mario Mazza, Genova 07576
Centro Studi Mutualistici Emancipazione e Partecipazione, Roma 07577
Centro Studi Nord e Sud, Napoli 07578
Centro Studi Parlamentari, Roma 07579
Centro Studi per il Mezzogiorno A. Ajon, Acireale 07580
Centro Studi per il Progresso della Educazione Sanitaria e del Diritto Sanitario, Roma 07581
Centro Studi per la Programmazione Economica e Sociale, Roma 07582
Centro Studi per la Valorizzazione delle Risorse del Mezzogiorno, Napoli 07583
Centro Studi Piero Gobetti, Torino 07584
Centro Studi Pietro Mancini, Cosenza 07585
Centro Studi Politici e Sociali Alcide de Gasperi, Cosenza 07586
Centro Studi Politico-Sociali Achille Grandi, Milano 07587
Centro Studi Problemi Medici, Milano 07588
Centro Studi Ricerche e Documentazioni per l'Agricoltura Siciliana, Roma 07589
Centro Studi Ricerche Ligabue, Venezia 07590
Centro Studi Russia Cristiana, Milano 07591
Centro Studi Santa Veronica Giuliani, Città di Castello 07592
Centro Studi Storici di Mestre, Mestre 07593
Centro Studi Storici Sociali, Bologna 07594
Centro Studi sulla Resistenza, Urbino 07595
Centro Studi Wilhelm Reich, Napoli 07596
Centro Studi Zingari, Roma 07597
Centro Superiore di Logica e Scienze Comparate, Bologna 07598
Centro Sviluppo Impiego Diesel, Trieste 07599
Centro Thomas Mann, Roma 07600
Centro Vasco, México 08723
Centrum voor de Studie van het Onderwijs in Ontwikkelingslanden, 's-Gravenhage 08839
Centrum voor Plantenveredelings- en Reproduktieonderzoek, Wageningen 08840
Centrum voor Wiskunde en Informatica, Amsterdam 08841
CEOCOR, Liège 01222
CEOS, Washington 14802
CEOS, Winchester 11860
CEP, Paris 04387
CEPC, Strasbourg 04402
CEPCEO, Bruxelles 01223
C.E.P.E.S., Roma 07462
CEPH, Washington 14915
CEPITRA, Lyon 04367
CEPO, Worlester 11861
CEPR, Old Westbury 14679
C.E.P.S.A., Roma 07560
Ce.Psi.Pe.Di., Bologna 07531
CEQ, Montréal 02324
Ceramic Educational Council, Westerville 14718
Ceramics and Stone Accident Prevention Association, Toronto 02331
The Ceramic Society of Japan, Tokyo 08369
CERAM Research, Stoke-on-Trent 11755
CERC, Charleston 14766
CERC, Orange 14853
Cercle Archéologique de Mons, Mons 01206
Cercle Benelux d'Histoire de la Pharmacie, Kortrijk 01207
Cercle de les Arts i de les Lletres, Escaldes-Engordany 00029
Cercle d'Etudes Architecturales, Paris 04386
Cercle d'Etudes Numismatiques, Bruxelles 01208
Cercle d'Etudes Pédiatriques, Paris 04387
Cercles des Jeunes Naturalistes, Montréal 02332
CERCOR, Saint-Etienne 04372
CERD, Bruxelles 01227
Cerebral Palsy Association in Alberta, Calgary 02333
Cerebral Palsy Association of Manitoba, Winnipeg 02334
Cerebral Palsy Association of Newfoundland, St. John's 02335
Cerebral Palsy Association of Nova Scotia, Halifax 02336
Cerebral Palsy Association of Prince Edward Island, Charlottetown 02337
Cerebral Palsy Association of Prince George and District, Prince George 02338
Cerebral Palsy Association of Quebec, Québec 01877
CERET, Lausanne 10642
CERFCI, Paris 04364
Ce.Ri.Me.Ps., Fermo 07540
CERLAC, Toronto 02329
CERLAL, Bogotá 03362
CERP-EDUCATION, Gent 01290
C.E.R.S.G., Bologna 07459
CERVED, Roma 07991
CES, Bruxelles 01186, 01225
CES, New York 14875
CESA, Toronto 02127
CESAG, Dakar 10023
CESD, Malakoff 04371
CE.S.DI.S., Vicenza 07554
CESDIT, Milano 07521
CESDOC, Grand-Saconnex 10644
CESE, Bruxelles 01242
CE.S.E., Roma 07566
Ce.S.E.T., Firenze 07557
CESI, Westport 14874
CE.S.I.D., Trieste 07599
CESIS, Bobo-Dioulasso 01806
Česká Akademie Věd, Praha 03500
Česká Akademie Zemědělských Věd, Praha 03501
Česká Archeologická Společnost, Praha 03502
Česká Biologicka Společnost, Brno 03503
Česká Botanická Společnost, Praha 03504
Česká Farmaceutická Společnost, Praha 03505
Česká Geografická Společnost, Praha 03506
Česká Geologická Společnost, Praha 03507
Česká Historická Společnost, Praha 03508
Česká Lékařská Společnost J.E. Purkyně, Praha 03509
Česká Meteorologická Společnost, Praha 03510
Česká Národopisná Spolecnost, Praha 03511
Česká Oftalmologická Společnost, Praha 03512
Česká Orientalistická Společnost, Praha 03513
Česká Pediatrická Společnost, Praha 03514
Česká Psychologická Společnost, Praha 03515
Česká Radiologická Společnost, Praha 03516
Česká Společnost Anesteziologie a Resuscitace, Praha 03517
Česká Společnost Biochemická, Praha 03518
Česká Společnost Bioklimatologická, Praha 03519
Česká Společnost Chemická, Praha 03520
Česká Společnost Ekonomická, Praha 03521
Česká Společnost Entomologická, Praha 03522
Česká Společnost Fyziologie a Patologie Dýchání, Praha 03523
Česká Společnost Histo- a Cytochemnická, Praha 03524
Česká Společnost Kybernetiku a Informatiku, Praha 03525
Česká Spolecnost Mikrobiologická, Praha 03526
Česká Společnost Parasitologická, Praha 03527
Česká Společnost pro Mechaniku, Praha 03528
Česká Společnost pro Politické Vědy, Praha 03529
Česká Společnost pro Vědy Zemědělské, Lesnické, Veterinární a Potravinářské, Praha 03530
Česká Společnost Zoologická, Praha 03531
Česká Stomatologická Společnost, Praha 03532
Česká Védecká Společnost pro Mykologii, Praha 03533
Česká Vědecká Společnost pro Mykologii, Praha 03534
Česká Vědecko-Technická Společnost pro Geodezii a Kartografii, Praha 03535
Česká Vědeckotechnická Společnost, Praha 03536
CESP, Bruxelles 01245
CESSE, Washington 14895
CESSTEM, México 08709
CE.S.VIET., Milano 07449
CETA, Kensington 14738
CETE, Columbus 14713
CE.TE.NA., Genova 07518
CETP, Paris 04436
CEUR, Buenos Aires 00103
Ceylon Gemmologists Association, Colombo 10409
Ceylon Geographical Society, Colombo 10410
Ceylon Humanist Society, Jaffna 10411
Ceylon Institute of World Affairs, Colombo 10412
Ceylon Palaeological Society, Colombo 10413
Ceylon Society of Arts, Colombo 10414
CEZ, Kinshasa 17143
CF, Stockholm 10588
C.F.A., Oxford 11821
C.F.A., Taipei 03268
CFAS, Rockville Centre 14654
CFAT, Princeton 14646
CFB, Brasília 01643
CFBS, Ottawa 02129
CFCM, Chula Vista 14794
CFCS, Saint Croix 17073
CFDRA, Chipping Campden 11742
CFDT, Douala 01813
CFH, Ottawa 02128
CFMA, Chicago 14879
CFNI, Kingston 08073
CFR, Uppsala 10446
CFR, Watertown 14680
CFSS, Portland 14750
CGA, Nicosia 03491
CGB, Montréal 02368
CGCS, Washington 14905
CGE, Minneapolis 14681
CGFPS, Cambridge 14825
CGGP, Durham 14826
CGI, Pittsburgh 14640
CGI, Torino 07614
CGIAR, Washington 14852
C.G. Jung Foundation for Analytical Psychology, New York 14719
CGS, Beijing 03171
CGS, New York 14813
CGS, Washington 14896
CGSM, Saint Louis 14846
CH, Duarte 14757
CHA, Saint Louis 14655
Chakoten Dansk Militaershistorisk Selskab, København 03572
Challenger Society for Marine Science, Godalming 11756
Chamber of Architects and Civil Engineers, Paceville 08658
Chamber of Geological Engineers of Turkay, Ankara 11147
Chambre Belge des Pédicures Médicaux, Bruxelles 01209
Chambre Belge des Traducteurs, Interprètes et Philologues, Bruxelles 01210
Chambre des Architectes de Belgique, Bruxelles 01211
Chambre Syndicale des Sociétés d'Etudes et de Conseils, Paris 04388
Champlain Society, Toronto 02339
Changchun Academy of Electrical Power Engineering Design, Chilin 03123
CHAR, Austin 14795
Character Education Institute, San Antonio 14720
Charles Darwin Foundation for the Galapagos Isles, Bruxelles 01212
Charles H. Ivey Foundation, Willowdale 02340
Charles Ives Society, Bloomington 14721
The Charles Lamb Society, Richmond 11757
Charles Rennie Mackintosh Society, Glasgow 11758
Charles S. Peirce Society, Indianapolis 14722
Chartered Association of Certified Accountants, London 11759
The Chartered Institute of Bankers, London 11760
The Chartered Institute of Building, Ascot 11761
Chartered Institute of Management Accountants, London 11762
Chartered Institute of Marketing, Maidenhead 11763
Chartered Institute of Patent Agents, London 11764
Chartered Institute of Public Finance and Accountancy, London 11765
The Chartered Institute of Transport, London 11766
Chartered Institute of Transport in Australia, Sydney 00371
Chartered Institution of Building Services Engineers, London 11767
Chartered Insurance Institute, London 11768
Chartered Society of Designers, London 11769
The Chartered Society of Physiotherapy, London 11770
CHAS, Cork 06938

Chautauqua Literary and Scientific Circle, Chautauqua *14723*
CHEAR, Yonkers *16058*
CHEEC, Bruxelles *01238*
Chekiang Academy of Agricultural Sciences, Chechiang *03124*
Chemical Engineering Research Group, Durban *10129*
Chemical Industries Accident Prevention Association, Toronto *02341*
Chemical Industry and Engineering Society of China, Beijing *03125*
Chemical Industry Institute of Toxicology, Research Triangle Park *14724*
Chemical Institute of Canada, Ottawa *02342*
Chemical Management and Resources Association, Staten Island *14725*
Chemicals Notation Association, Dartford *11771*
Chemical Societies of the Nordic Countries, Stockholm *10444*
Chemical Society of Finland, Helsinki *03960*
Chemical Society of Japan, Tokyo *08318*
Chemisch-Physikalische Gesellschaft in Wien, Wien *00577*
Chemists' Club, New York *14726*
Chemometrics Society, Sandwich *11772*
Cherokee National Historical Society, Tahlequah *14727*
CHES, Toronto *02152*
Chesapeake and Ohio Historical Society, Clifton Forge *14728*
The Chester Archaeological Society, Chester *11773*
Chester Society of Natural Science, Literatur and Art, Chester *11774*
The Chest, Heart and Stroke Association, London *11775*
CHIE, Los Angeles *14919*
Chief Officers of State Library Agencies, Lexington *14729*
Chigaku Dantai Kenkyu-Kai, Tokyo *08116*
Childbirth without Pain Education Association, Detroit *14730*
Child Care Employee Project, Oakland *14731*
Child Development Centre Society of Fort Saint John and District, Fort Saint John *02343*
Child Neurology Society, Saint Paul *14732*
Children's Broadcast Institute, Toronto *02344*
Children's Literature Association, Battle Creek *14733*
Children's Oncology Care of Ontario, Toronto *02345*
Children's Rehabilitation and Cerebral Palsy Association, Vancouver *02346*
Chilean Association of Seismology and Anti-Seismic Engineering, Santiago *03015*
Chilean Librarianship Association, Santiago *03030*
Chilean Society of Cardiology and Cardiovascular Surgery, Santiago *03068*
Chilean Society of Obstetrics and Gynecology, Santiago *03078*
Chiltern Society, Great Missenden *11776*
China Academy, Hwa Kang *03252*
China Academy of Railway Sciences, Beijing *03126*
China Academy of Traditional Chinese Medicine, Beijing *03127*
China Association for Science and Technology, Beijing *03128*
China Association of the Five Principles of Administrative Authority, Taipei *03253*
China Association of Traditional Chinese Medicine, Beijing *03129*
China Civil Engineering Society, Beijing *03130*
China Coal Society, Beijing *03131*
China Computer Federation, Beijing *03132*
China Education Society, Taipei *03254*
China-Europe Management Institute, Bruxelles *01213*
China Institute in America, New York *14734*
China International Education Research Association, Taipei *03255*
China Law Society, Beijing *03133*
China Medical Board of New York, New York *14735*
China National Association of Literature and the Arts, Taipei *03256*
China Policy Study Group, London *11777*
China Social Education Society, Taipei *03257*
China Society, London *11778*
The China Society, Singapore *10044*
The China Society, Taipei *03258*
China Society of Fisheries, Beijing *03134*
China Spiritual Therapy Study Association, Taipei *03259*
Chinese Abacus Association, Beijing *03135*
Chinese Academy of Agricultural Sciences, Beijing *03136*
Chinese Academy of Coal Mining Sciences, Beijing *03137*
Chinese Academy of Forestry, Beijing *03138*
Chinese Academy of Geological Sciences, Beijing *03139*
Chinese Academy of Medical Sciences, Beijing *03140*
Chinese Academy of Meteorological Sciences, Beijing *03141*
Chinese Academy of Sciences, Beijing *03142*
Chinese Academy of Sciences – Anhwei Branch, Anhui *03143*
Chinese Academy of Sciences – Central South Branch, Kuangtung *03144*
Chinese Academy of Sciences – Chekiang Branch, Chechiang *03145*
Chinese Academy of Sciences – East China Branch, Shanghai *03146*
Chinese Academy of Sciences – Fukien Branch, Fuchien *03147*
Chinese Academy of Sciences – Hopeh Branch, Hopei *03148*
Chinese Academy of Sciences – Kiangsu Branch, Chiangsu *03149*
Chinese Academy of Sciences – Kirin Branch, Chilin *03150*
Chinese Academy of Sciences – Northwestern Branch, Shanhsi *03151*
Chinese Academy of Sciences – Shansi Branch, Shanhsi *03152*
Chinese Academy of Sciences – Shantung Branch, Shantung *03153*
Chinese Academy of Sciences – Sinkiang Branch, Hsinchiang *03154*
Chinese Academy of Social Sciences, Beijing *03155*
Chinese Academy of Space Technology, Beijing *03156*
Chinese-American Librarians Association, Sacramento *14736*
Chinese Anti-Cancer Association, Tian Jing *03157*
Chinese Anti-Tuberculosis Society, Beijing *03158*
Chinese Archives Society, Beijing *03159*
Chinese Association for Folklore, Taipei *03260*
Chinese Association for the Advancement of Science, Taipei *03261*
Chinese Association of Agricultural Science Societies, Beijing *03160*
Chinese Association of Animal Science and Veterinary Medicine, Beijing *03161*
Chinese Association of Automation, Beijing *03162*
The Chinese Association of Fire Protection, Beijing *03163*
Chinese Association of Integrated Traditional and Western Medicine, Beijing *03164*
Chinese Association of Psychological Testing, Taipei *03262*
Chinese Astronomical Society, Nanjing *03165*
Chinese Biochemical Society, Shanghai *03166*
Chinese Buddhist Association, Taipei *03263*
Chinese Center, International PEN, Taipei *03264*
Chinese Chemical Society, Beijing *03167*
Chinese Chemical Society, Taipei *03265*
Chinese Classical Music Association, Taipei *03266*
Chinese Culture Association, Palo Alto *14737*
Chinese Electrotechnical Society, Beijing *03168*
Chinese Engineering Graphics Society, Wuhan *03169*
Chinese-English Translation Assistance Group, Kensington *14738*
Chinese Film Critic's Association of China, Taipei *03267*
Chinese Forestry Association, Taipei *03268*
Chinese Foundrymen Association, Kaohsiung *03269*
Chinese Geological Society, Beijng *03170*
Chinese Geophysical Socitey, Beijing *03171*
Chinese Historical Association, Taipei *03270*
Chinese Historical Society of America, San Francisco *14739*
Chinese Home Education Promotion Association, Taipei *03271*
Chinese Hydraulic Engineering Society, Beijing *03172*
Chinese Information Processing Society, Beijing *03173*
Chinese Institute of Civil and Hydraulic Engineering, Taipei *03272*
The Chinese Institute of Engineers, Taipei *03273*
Chinese Institute of Mining and Metallurgical Engineers, Taipei *03274*
Chinese Language and Research Centre, Singapore *10045*
Chinese Language Society, Taipei *03275*
Chinese Language Teachers Association, Kalamazoo *14740*
Chinese Light Industry Society, Beijing *03174*
Chinese Mathematical Society, Beijing *03175*
Chinese Mathematical Society, Taipei *03276*
Chinese Mechanical Engineering Society, Beijing *03176*
Chinese Mechanics Society, Beijing *03177*
Chinese Medical Association, Taipei *03277*
Chinese Medical History Association, Taipei *03278*
Chinese Medical Society, Beijing *03178*
Chinese Medical Woman's Association, Taipei *03279*
Chinese Musical and Theatrical Association, New York *14741*
Chinese Music Society of North America, Woodridge *14742*
Chinese National Association for Mental Hygiene, Taipei *03280*
Chinese National Foreign Relations Association, Taipei *03281*
Chinese Physiological Society, Taipei *03282*
Chinese Psychological Association, Taipei *03283*
Chinese Society for EC Studies, Shanghai *03179*
Chinese Society for Electronic Data Processing, Taipei *03284*
Chinese Society for Future Studies, Beijing *03180*
Chinese Society for Materials Science, Hsinchu *03285*
Chinese Society for Oceanology and Limnology, Quingdao *03181*
Chinese Society of Aeronautics and Astronautics, Beijing *03182*
Chinese Society of Agricultural Machinery, Beijing *03183*
Chinese Society of Anatomy, Beijing *03184*
Chinese Society of Astronautics, Beijing *03185*
Chinese Society of Budgetary Management, Taipei *03286*
Chinese Society of Chemical Engineering, Beijing *03186*
Chinese Society of Electronics, Beijing *03187*
Chinese Society of Engineering Thermophysics, Beijing *03188*
Chinese Society of Environmental Sciences, Beijing *03189*
Chinese Society of Forestry, Beijing *03190*
Chinese Society of Geodesy, Photogrammetry and Cartography, Beijing *03191*
Chinese Society of High Energy Physics, Beijing *03192*
Chinese Society of History of Science, Bejing *03193*
Chinese Society of International Law, Taipei *03287*
Chinese Society of Library Science, Beijing *03194*
The Chinese Society of Metals, Beijing *03195*
Chinese Statistical Association, Taipei *03288*
The Chinese Taipei Pediatric Association, Taipei *03289*
Chinese Women Writer's Association, Taipei *03290*
Chinese Youth Academic Research Association, Taipei *03291*
CHINOPERL, Hanover *14824*
Chippendale Society, Leeds *11779*
Chiropractic Advancement Association, Thames Ditton *11780*
Chirurgiese Vereniging van Suid-Africa, Johannesburg *10124*
ChLA, Battle Creek *14733*
CHMM, Montpellier *04370*
Choir Schools Association, Wells *11781*
The Chopin Society, Wembley Park *11782*
CHRIE, Washington *14920*
Christian Association for Psychological Studies, Temecula *14743*
Christian College Coalition, Washington *14744*
Christian Dental Society, Sumner *14745*
Christian Education Movement, Derby *11783*
Christian Literacy Associates, Pittsburgh *14746*
Christian Medical Commission, Genève *10651*
Christian Medical Society, Islington *02347*
Christian Research Institute International, San Juan Capistrano *14747*
Christian Schools International, Grand Rapids *14748*
Christliche Gesundheitskommission, Genève *10651*
Christlicher Lehrer- und Erzieherverein der Schweiz, Horw *10645*
Christopher Morley Knothole Association, Roslyn *14749*
Chronic Fatigue Syndrome Society, Portland *14750*
CHS, Caernarfon *11733*
CHS, Leigh-on-Sea *11836*
CHS, Westerville *14754*
CHSA, San Francisco *14739*
CHTI, Santurce *09728*
Church and Synagogue Library Association, Portland *14751*
Church Education Society for Ireland, Dublin *06936*
Church History Society, Sandvika *09273*
Church Library Association, Aylmer *02348*
Church Music Association, London *11784*
Church Organization Research and Advisory Trust of Africa, Nairobi *08485*
Chusei Tetsugakkai, Tokyo *08117*
CI, Washington *14845*
CIA, Hamilton *02362*
CIA, New York *14734*
CIAF, Bogotá *03358*
CIAMAN, Pierrelatte *04437*
CIAS, Bogotá *03357*
CIAS, Cleveland *11793*
CIAT, Cali *03360*
CIAT Andean Zone Network for Bean Research, Lima *09463*
CIBC, New York *14921*
CIC, Champaign *14803*
CICA, Cambridge *11838*
C.I.C. CYB., Paris *04375*
CICD, Milwaukee *14781*
CICP, New York *14764*
CICRED, Paris *04430*
CIC/SLS, Gunnislake *11866*
CIDD, Paris *04452*
CIDEAFA, Rabat *08807*
CIDEP, Louvain-la-Neuve *01191*
CIDIAT, Mérida *16987*
CIDIE, Nairobi *08487*
C.I.D.I.P., Roma *07422*
CIDMEF, Tours *04441*
CIDMY, Bruxelles *01190*
CIDSS, Paris *04409*
CIE, Billings *14876*
CIE, Wien *00578*
CIEC, Bogotá *03367*
CIECC, Washington *15352*
C.I.ED.ART., Venezia *07481*
CIEF, Montevideo *13252*
CIEF, Upper Montclair *14844*
CIEL, Bruxelles *01200*
C.I.E.L., Paris *04378*
CIEM, Québec *02384*
CIENES, Santiago *03024*
CIER, Paris *04379*
CIES, Buffalo *14816*
C.I.E.S., Perugia *07615*
CIESM, Monaco *08796*
CIETA, Lyon *04377*
CIFA, Bruxelles *01229*
CIFDUF, Aix-en-Provence *04442*
CIFEJ, Paris *04380*
C.I.F.H., Paris *04399*
C.I.F.I., Roma *07607*
CIFPEF, Paris *04443*
CIFST, Ottawa *02174*
CIGB, Paris *04425*
CIGRE, Paris *04445*
CIHS, Innsbruck *00579*
C.I.I., London *11768*
C.I.I.G., Roma *07482*
CIIT, Research Triangle Park *14724*
CIJL, Genève *10643*
CIL, Leawood *14877*
C.I.L.A.D., Lisboa *09655*
CILECT, Bruxelles *01192*
CILTADE, Louvain-la-Neuve *01195*
CIM, Montréal *02180*
CIM, Paris *04454*
CIMA, Okmulgee *14940*
C.I.M.A., Roma *07492*
CIME, Paris *04457*
CIMH, Bridgetown *01082*
CIMME, Taipei *03274*
CIMMYT, México *08719*
CIMS, Roma *07630*
CINDA, Santiago *03025*
CINDER, Maracaibo *16988*
Cinema Organ Society, Cambridge *11785*
Cinema Theatre Association, Teddington *11786*
Cinémathèque Québécoise, Montréal *02349*
Cinema Workers' Association of Romania, Bucuresti *09739*
Cinemists 63, Los Angeles *14752*
Cineteca Italiana Archivio Storico del Film, Milano *07601*
CINP, Milano *07610*
C.I.N.S.E., Milano *07512*
CINTERAD, Bruxelles *01199*
CINTERFOR, Montevideo *13254*
CIOHL, Paris *04426*
CIP, Groningen *08836*
C.I.P., Napoli *07493*
C.I.P., Torino *07466*
CIP, Washington *14897*
CIPA, London *11764*
CIPAC, London *11795*
CIPAC, Wädenswil *10647*
CIPAT, Lausanne *10656*
CIPEL, Lausanne *10650*
CIPFA, London *11765*
CIP International Study and Research Center in Psychosynteresis, Lausanne *10646*
CIPL, Paris *04407*
CIPSH, Paris *04455*
CIR, Pierrefitte *04410*
CIR, Washington *14800*
CIRAD, Paris *04365*
Circle of Modern Philologists, Praha *03548*
Circle of Musical Culture, Macao *08591*
Circle of State Librarians, London *11787*
Circolo Filosofico di Studi Tomistici, Roma *07602*
Circolo Giuridico Italiano, Roma *07603*
Circolo Speleologico Romano, Roma *07604*
Círculo de Bellas Artes, La Paz *01539*
Circulo de Cultura Musical, Macao *08591*
Circulo de Radioterapeutas Ibero-Latinoamericanos, Houston *14753*
Círculo Médico de Córdoba, Córdoba *00113*
Círculo Médico de Rosario, Rosario *00114*
Circus Historical Society, Westerville *14754*
CIRET, München *05275*
CIRG, Braine-le-Château *01194*
CIRIA, London *11839*
CIRIEC, Liège *01193*
C.I.R.I.E.C., Milano *07494*
CIRIEC, Roma *07505*
CIRP, Paris *04391*
C.I.R.S.A., Roma *07506*
C.I.R.S.A., Verucchio *07484*
CIRSTA, Roma *07507*
CIRUISEF, Talence *04446*
CIS, Alexandria *14755*
C.I.S., Roma *07495*
C.I.S.A., Milano *07429, 07497*
CISAC, Paris *04438*
C.I.S.A.S., Roma *07510*
CISBH, Bruxelles *01230*
CISCo, Genova *07508*
CISCOM, Roma *07480*
CISCS, Sainte-Savine *04381*
C.I.S.D.C.E., Milano *07473*
CISDEN, Roma *07498*
C.I.S.D.O., Roma *07617*
CISE, Columbus *14847*
C.I.S.E., Roma *07500*
C.I.S.F., Milano *07486*
CISH, Paris *04408*
C.I.S.I.A., Milano *07511*
CISM, Ottawa *02184*
C.I.S.M., Roma *07487*
CISM, Udine *07470*
CISMEC, Milano *07465*
CISP, Roma *07616*
C.I.S.P.E., Napoli *07483*
C.I.S.P.E.S., Roma *07509*
C.I.S.P.P., Roma *07499*
CISPR, London *12395*
C.I.S.S., Cagliari *07475*
CISS, Paris *04459*
C.I.S.S.A.M., Vicenza *07503*
C.I.S.S.E.G., Roma *07502*
C.I.S.S.P.A.T., Padova *07504*
CISYO, New York *14796*
CIT, London *11766*
CITA, México *08728*
CITEF, Villeurbanne *04444*
CITEP, London *11823*
Cities in Schools, Alexandria *14755*
Citizens for a Better Environment, Chicago *14756*
Citizens Protection Society, London *11788*
City and Guilds of London Institute, London *11789*
City of Hope, Duarte *14757*
City of Stoke on Trent Museum Archaeological Society, Stoke-on-Trent *11790*
CIUS, Paris *04460*
CIUTI, Antwerpen *01248*
The Civic Trust, London *11791*
Civil Aviation Medical Association, Ottawa *02350*
Civil Engineering Society, Moskva *09810*
Civil War Press Corps, Fayetteville *14758*
Civil War Round Table Associates, Little Rock *14759*
CIW, Washington *14647*
CJE, New York *14878*
CJN, Montréal *02332*
CLA, Atlanta *14773*
CLA, Fairfax *14818*
CLA, Haverford *14656*
CLA, London *11825*
CLA, Ottawa *02189, 02192*
CLA, Pittsburgh *14746*
CLA, Toronto *02191*
CLAD, Caracas *16989*
CLAF, Rio de Janeiro *01630*
CLAFIC, Santiago *03046*
CLAH, Tucson *14840*
C.L.A.H.S., Dundalk *06939*
CLAIS, México *08725*
CLARB, Fairfax *14898*
CLAS, Washington *14841*
CLASA, Bogotá *03369*

Classical America, New York *14760*
The Classical Association, Cambridge *11792*
The Classical Association of Ceylon, Colombo *10415*
Classical Association of Finland, Helsingin Yliopisto *03978*
Classical Association of Ghana, Accra *06478*
Classical Association of South Africa, Stellenbosch *10130*
The Classical Society of Japan, Kyoto *08366*
Clay Minerals Society, Boulder *14761*
CLC, Linoln *14855*
CLDA, Washington *14642*
CLE, Kingston *08071*
C.L.E., London *11857*
CLEA, London *11826*
Clean Air Society of Australia and New Zealand, Eastwood *00372*
Clean Air Society of Australia and New Zealand (N.Z. Branch), Wellington *09122*
CLEDTIERRA, Lima *09461*
CLENERT, Chicago *14856*
CLEO, Washington *14922*
Cleveland Industrial Archaeology Society, Cleveland *11793*
Climate Network Africa, Nairobi *08486*
Climatology Research Group, Johannesburg *10131*
Clinical Chemistry Data Communication Group, Minneapolis *14762*
Clinical Orthopedic Society, Bloomington *14763*
Clinical Paediatric Club of Israel, Zerifin *07050*
Clinical Research Society of Toronto, Toronto *02351*
Clinical Theology Association, Oxford *11794*
CLR, Washington *14924*
C.L.S.A., Sydney *00379*
CLSC, Chautauqua *14723*
CLTA, Kalamazoo *14740*
Club Européen d'Histoire de la Neurologie, Lyon *04389*
Club of Rome, Paris *04390*
Club Turati, Milano *07605*
CMA, London *11784*
CMA, New York *14880*
CMA, Ottawa *02201*
CMA, Taipei *03277*
CMAAO, Manila *09514*
CMBNY, New York *14735*
CMC, Genève *10651*
CMC, Hillsboro *14899*
CMC, New York *14686*
CMC, Ottawa *02203*
CMC, Washington *14685*
CMD, New York *14684*
CMDA, Camborne *11843*
CMDA, Saint Michael *01084*
CMEA Coordination Centre for Study of World Oceans Development of Techniques for Exploration and Utilization of Resources, Moskva *09811*
CME-AMA, Chicago *14925*
CMEPPC, Washington *14941*
CMES, Beijing *03176*
CMH, Westbrook *14815*
CMHA, Camon Downs *11842*
CMKA, Roslyn *14749*
C.M.L., Latina *07611*
CMLS, Richmond *14821*
CMMB, New York *14657*
CMO, Port of Spain *11116*
CMPDL, Washington *14787*
CMRA, Staten Island *14725*
CMRS, Columbus *14688*
CMS, Boulder *14761*
C.M.S., Calcutta *06750*
CMS, Missoula *14774*
CMS, Staten Island *14689*
CMSNA, Woodridge *14742*
CMTA, New York *14741*
CMU, Bari *07625*
CNA, Bruxelles *01233*
CNA, Dartford *11771*
CNA, Washington *14927*
CNAAG, Alger *00015*
C.N.A.M., Fresnes *04236*
CNBOS, Bruxelles *01232*
CNDRP, Alger *00016*
CNDST, Bruxelles *01201, 01202*
CNEA, Montevideo *13255*
CNERAT, Alger *00021*
CNFRA, Paris *04419*
C.N.G.R.S.R., Bucuresti *09748*
CNHS, Austin *14690*
CNHS, Tahlequah *14727*
CNI, Washington *14814*
CNIC, La Habana *03478*
CNIE, Buenos Aires *00119*
CNL, Whitestone *14926*
CNLIA, Washington *14900*
CNOF, Paris *04220*
CNPS, Boston *14659*
CNR, Roma *07628*
CNREF, Alger *00018*
CNRS, Paris *04382*
CNRS, Vandoeuvre-les-Nancy *04165*
CNRZA, Alger *00019*
CNS, Cincinnati *14842*
CNS, Nicosia *03493*
CNS, Saint Paul *14732*
CNSD, Paris *04439*
CNTMR, Paris *04412*
CNTTA, Alger *00020*
Coalition Clean Baltic, Stockholm *10445*
Coalition for International Cooperation and Peace, New York *14764*
Coalition for the Advancement in Jewish Education, New York *14765*
Coastal Engineering Research Council, Charleston *14766*
The Coastal Society, Gloucester *14767*
Coblentz Society, Norwalk *14768*
COBRIG, London *11856*
COCESNA, Guatemala City *06566*
COCTA, Oslo *09249*
CODATA, Paris *04420, 04431*
CODECI, Santiago *03049*
CODEFF, Santiago *03042*
CODESRIA, Dakar *10026*
Codex Coordinating Committee for Africa, Roma *07606*
Cognitive Science Society, Pittsburgh *14769*
COGR, Washington *14918*
COHEHRE, Gent *01255*
COHS, Clifton Forge *14728*
COI, Paris *04428*
Coimbra Group, Bruxelles *01214*
COJE, Brooklyn *14715*
COL, Vancouver *02369*
COLCIENCIAS, Bogotá *03374*
Colegio Colombiano de Cirujanos, Bogotá *03363*
Colegio de Abogados, Concepción *03028*
Colegio de Abogados de Barcelona, Barcelona *10288*
Colegio de Abogados de la Ciudad de Buenos Aires, Buenos Aires *00115*
Colegio de Abogados del Distrito Federal, Caracas *16992*
Colegio de Architectos de Chile, Santiago *03029*
Colegio de Arquitectos de Cataluña, Barcelona *10289*
Colegio de Arquitectos del Perú, Lima *09464*
Colegio de Bibliotecarios Colombianos, Bogotá *03364*
Colegio de Bibliotecarios de Chile, Santiago *03030*
Colegio de Dentistas de Chile, Concepción *03031*
Colegio de Economistas del Distrito Federal y Estado Miranda, Caracas *16993*
Colegio de Farmacéuticos del Distrito Federal y Estado Miranda, Caracas *16994*
Colegio de Graduados en Ciencias Económicas, Buenos Aires *00116*
Colegio de Ingenieros de Chile, Concepción *03032*
Colegio de Ingenieros de Guatemala, Guatemala City *06565*
Colegio de Médicos del Distrito Federal, Caracas *16995*
Colegio de Médicos del Estado Anzoátegui, Barcelona *16996*
Colegio de Médicos del Estado Mérida, Mérida *16997*
Colegio de Médicos del Estado Miranda, Caracas *16998*
Colegio de Quimico-Farmacéuticos de Chile, Santiago *03033*
Colegio de Químicos Farmacéuticos, Concepción *03034*
Colegio de Técnicos, Concepción *03035*
Colegio Farmacéutico de Chile, Concepción *03036*
Colégio Ibero-Latino-Americano de Dermatologia, Lisboa *09655*
Colegio Médico de El Salvador, San Salvador *03927*
El Colegio Nacional, México *08724*
Colegio Notarial, Barcelona *10290*
Colegio Oficial de Ingenieros Industriales de Cataluña, Barcelona *10291*
Collaborative International Pesticides Analytical Council, Wädenswil *10647*
Collaborative International Pesticides Analytic Council, London *11795*
College Art Association, New York *14770*
The College Board, New York *14771*
Collège Canadien de Généticians Médicaux, Calgary *02099*
Collège Canadien des Enseignants, Edmonton *02100*
Collège des Médecins de Famille du Canada, Willowdale *02354*
Collège des Psychologues du Nouveau-Brunswick, Fredericton *02361*
College English Association, Rochester *14772*
Collège Européen des Technologies, Arlom *01215*
Collège Européen d'Hygiène et de Médecine Naturelles, Enghien *01216*
Collège International pour l'Etude Scientifique des Techniques de Production Mécanique, Paris *04391*
College Language Association, Atlanta *14773*
College Music Society, Missoula *14774*
College of American Pathologists, Northfield *14775*
College of Architects of Catalonia, Barcelona *10289*
College of Dental Surgeons of British Columbia, Vancouver *02352*
College of Dental Surgeons of Saskatchewan, Saskatoon *02353*
College of Europe, Brugge *01217*
College of Family Physicians of Canada, Willowdale *02354*
The College of Medicine of South Africa, Rondebosch *10132*
College of Notaries, Barcelona *10290*
College of Ophthalmologists, London *11796*
College of Optometrists in Vision Development, Chula Vista *14776*
College of Physicians and Surgeons of Alberta, Edmonton *02355*
College of Physicians and Surgeons of British Columbia, Vancouver *02356*
College of Physicians and Surgeons of Manitoba, Winnipeg *02357*
College of Physicians and Surgeons of New Brunswick, Saint John *02358*
College of Physicians and Surgeons of Ontario, Toronto *02359*
College of Physicians and Surgeons of Saskatchewan, Saskatoon *02360*
The College of Preceptors, Epping *11797*
College of Psychologists of New Brunswick, Fredericton *02361*
The College of Radiographers, London *11798*
College Press Service, Orlando *14777*
College Reading and Learning Association, Salem *14778*
Collège Royal des Chirurgiens Dentistes du Canada, Toronto *02868*
Collège Royal des Médecins et Chirurgiens du Canada, Ottawa *02869*
College Theology Society, San Diego *14779*
College-University Resource Institute, Washington *14780*
Collegio degli Ingegneri Ferroviari Italiani, Roma *07607*
Collegio dei Tecnici dell' Acciaio, Milano *07608*
Collegium Biologicum Europa, Roma *07609*
Collegium Cultorum Martyrum, Roma *16950*
Collegium Internationale Allergologicum, Hamilton *02362*
Collegium Internationale Chirurgiae Digestivae, Milwaukee *14781*
Collegium Internationale Neuro-Psychopharmacologicum, Milano *07610*
Collegium Musicum di Latina, Latina *07611*
Collegium Palynologicum Scandinavicum, Aarhus *03573*
Collegium Romanicum, Lausanne *10648*
Colombian Institute for the Development of Higher Education, Bogotá *03389*
Colombian Institute of Pediatric Oncology, Bogotá *03387*
Colombian Society of Obstetrics and Gynecology, Bogotá *03410*
Colombo Plan Staff College for Technician Education, Manila *09513*
Colonial Society of Massachusetts, Boston *14782*
Colorado River Association, Los Angeles *14783*
Colour Group (Great Britain), Harrow *11799*
COLT, Cleveland *14923*
Columbian Society of Chemists and Chemical Engineers, Bogotá *03414*
Combat, Hinckley *11412*
Combustion Society of Japan, Kyoto *08346*
Combustion Society of Japan, Tokyo *08345*
COMEDA, Asnières *04462*
COMETEC-GAZ, Bruxelles *01224*
Comisión Bolivana de Energía Nuclear, La Paz *01540*
Comisión Chilena de Energía Nuclear, Santiago *03037*
Comisión de Energía Atómica de Costa Rica, San José *03442*
Comisión de Investigaciones Científicas de la Provincia de Buenos Aires, La Plata *00117*
Comisión de Planeamiento y Coordinación de las Universidades Bolvianas, Cochabamba *01541*
Comisión Ecuatoriana de Energía Atómica, Quito *03851*
Comisión Episcopal Latino Americano, Bogotá *03365*
Comisión Latinoamericana de Investigadores en Sorgo, México *08725*
Comisión Nacional de Arqueología y Monumentos Históricos, Panamá City *09415*
Comisión Nacional de Energía Atómica, Buenos Aires *00118*
Comisión Nacional de Energía Atómica, Montevideo *13255*
Comisión Nacional de Investigaciones Espaciales, Buenos Aires *00119*
Comisión Nacional de Museos y de Monumentos y Lugares Históricos, Buenos Aires *00120*
Comisión Nacional Protectora de Bibliotecas Populares, Buenos Aires *00121*
Comisión Panamericana de Normas Técnicas, Buenos Aires *00122*
Comisión Paraguaya de Documentación e Información, Asunción *09429*
Comisíon Económica para América Latina y el Caribe, Santiago *03038*
Comisíon Nacional de Investigación Científica y Tecnológica, Santiago *03039*
Comissão Brasileira de Documentâcão Agrícola, Brasília *01639*
Comissão Nacional de Folclore, Rio de Janeiro *01640*
Comitato dei Geografi Italiani, Roma *07612*
Comitato Elettrotecnico Italiano, Milano *07613*
Comitato Glaciologico Italiano, Torino *07614*
Comitato Italiano per l'Educazione Sanitaria, Perugia *07615*
Comitato Italiano per lo Studio dei Problemi della Popolazione, Roma *07616*
Comitato Italiano per lo Studio del Dolore in Oncologia, Roma *07617*
Comitato Nazionale Italiana per l'Organizzazione Scientifica, Roma *07618*
Comitato Nazionale Italiano, Milano *07713*
Comitato per Bologna Storica-Artistica, Bologna *07619*
Comitato Tecnico Permanente per lo Studio della Legislazione e dei Rimedi, Roma *07685*
Comitato Termotecnico Italiano, Torino *07620*
Comité Andorrà de Ciències Històriques, Andorra la Vella *00030*
Comité Arctique International, Monte Carlo *08795*
Comité Belge d'Histoire des Sciences, Bruxelles *01218*
Comité Chileno Veterinario de Zootecnia, Santiago *03040*
Comité d'Associations Européennes de Médecins Catholiques, Bruxelles *01219*
Comité d'Education Sanitaire et Sociale de la Pharmacie Française, Paris *04392*
Comité de Liaison des Architectes de l'Europe Unie, Bruxelles *01220*
Comité de Liaison des Géomètres-Experts Européens, London *11800*
Comité Dentaire de Liaison de la Communauté Européenne, London *11801*
Comité d'Entente des Ecoles de Formation en Economie Sociale Familiale, Paris *04393*
Comité des Cancérologues de la Communauté Européenne, Bruxelles *01221*
Comité d'Etude de la Corrosion et de la Protection des Canalisations, Liège *01222*
Comité d'Etude des Producteurs de Charbon d'Europe Occidentale, Bruxelles *01223*
Comité d'Etudes Economiques de l'Industrie du Gaz, Bruxelles *01224*
Comité d'Etudes et de Liaison des Amendements Calcaires, Paris *04394*
Comité d'Etudes et de Liaison Interprofessionnel de la Haute-Marne, Saint-Dizier *04395*
Comité d'Etudes et de Liaison Interprofessionnel du Département de l'Aisne, Saint-Quentin *04396*
Comité d'Etudes Fiscales et Contentieuses, Paris *04397*
Comité d'Etudes pour un Nouveau Contrat Social, Paris *04398*
Comité du Développement de l'Éducation Internationale de l'Ontario, Toronto *02550*
Comité du Film Ethnographique, Paris *04399*
Comité Economique et Social des Communautés Européennes, Bruxelles *01225*
Comité Euro-International du Béton, Lausanne *10667*
Comité Européen de Normalisation Electrotechnique, Bruxelles *01226*
Comité Européen de Recherche et de Développement, Bruxelles *01227*
Comité Européen Permanent de Recherches pour la Protection des Populations contre les Risques d'Intoxication à Long Terme, Paris *04400*
Comité Européen pour l'Education des Enfants et Adolescents Précoces, Doués, Talentueux, Nîmes *04401*
Comité Européen pour les Problèmes Criminels, Strasbourg *04402*
Comité Français d'Education et d'Assistance de l'Enfance Déficiente, Paris *04403*
Comité Français de l'Electricité, Paris-La Défense *04404*
Comité Historique du Centre-Est, Lyon *04405*
Comité International de l'AISS pour la Prévention des Risques Professionnels du Bâtiment et des Travaux Publics, Boulogne-Billancourt *04406*
Comité International de Médecine Militaire, Liège *01228*
Comité International de Paléographie Latine, Paris *04407*
Comité International de Recherche et d'Etude de Facteurs de l'Ambiance, Bruxelles *01229*
Comité International des Sciences Historiques, Paris *04408*
Comité International de Standardisation en Biologie Humaine, Bruxelles *01230*
Comité International d'Etude des Géants Processionnels, Bruxelles *01231*
Comité International d'Etudes Morisques, Zaghouan *11139*
Comité International d'Histoire de l'Art, Utrecht *08842*
Comité International Permanent des Etudes Mycéniennes, Vitoria *10292*
Comité International pour l'Information et la Documentation en Sciences Sociales, Paris *04409*
Comité International Rom, Pierrefitte *04410*
Comité Mondial pour la Recherche Spatiale, Paris *04411*
Comité Nacional de Cristalografía, Buenos Aires *00123*
Comité Nacional de Geografía, Geodesía y Geofísica, Santiago *03041*
Comité Nacional del Consejo Internacional de Museos, Bogotá *03366*
Comité Nacional de Lucha contra el Cáncer, Bogotá *03367*
Comité Nacional Español del Consejo Internacional de la Música, Madrid *10293*
Comité Nacional Pro Defensa de la Fauna y Flora, Santiago *03042*
Comité National Belge de l'Organisation Scientifique, Bruxelles *01232*
Comité National contre les Maladies Respiratoires et la Tuberculose, Paris *04412*
Comité National de Géographie du Maroc, Rabat *08806*
Comité National de l'Enseignement Libre, Paris *04413*
Comité National des Conseillers de l'Enseignement Technique, Paris *04414*
Comité National des Ecoles Françaises de Service Social, Paris *04415*
Comité National des Musées, Tunis *11140*
Comité National Français de Géodesie et Géophysique, Paris *04416*
Comité National Français de Géographie, Paris *04417*
Comité National Français de Mathématiciens, Paris *04418*
Comité National Français des Recherches Antarctiques, Paris *04419*
Comité National pour l'Etude et la Prévention de l'Alcoolisme et des Autres Toxicomanies, Bruxelles *01233*
Comité Oceanográfico Nacional, Valparaíso *03043*
Comité para la Defensa de las Lenguas Indígenas de América Latina y el Caribe, Caracas *16999*
Comité pour les Données Scientifiques et Technologiques, Paris *04420*
Comité Scientifique Inter-Africain Post-Récolte, Dakar *10024*
Comitetul National al Geologilor din Romania, Bucuresti *09748*
Comité Universitaire d'Information Pédagogique, Paris *04421*
COMLA, Mandeville *08078*
Commemorative Collectors Society, Nottingham *11802*
Commision on Mining, Baku *01040*
COMMISS, Columbia *14788*
Commission Belge de Bibliographie, Bruxelles *01234*
Commission de la Carte Géologique du Monde, Paris *04422*
Commission de l'Enseignement Supérieur des Provinces Maritimes, Fredericton *02640*
Commission de l'Enseignement Supérieur en Biologie, Paris *04423*
Commissione di Studio dei Fenomeni di Corrosione Elettrolitica, Bologna *07621*
Commission Electrotechnique Internationale, Genève *10649*
Commission for a Linguistic Atlas of Europe, Moskva *09812*
Commission for Archaeology, Khartoum *10426*
Commission for Atomic Energy, Moskva *09813*
Commission for Computer Technology, Moskva *09814*
Commission for Computer Technology, Novosibirsk *09815*

Commission for Controlling the Desert Locust in the Eastern Region of its Distribution Area in South West Asia, Roma 07622
Commission for Controlling the Desert Locust in the Near East, Jeddah 10002
Commission for Co-ordination of Research in State Nature Reserves, Moskva 09816
Commission for Ecology, Moskva 09817
Commission for International Scientific Links, Moskva 09818
Commission for International Tectonic Maps, Moskva 09819
Commission for Links in Social Science with the American Council of Learned Societies, Moskva 09820
Commission for Links with US Research Establishments in the Use of Technology and New Communications Technologies in Education, Moskva 09821
Commission for Management of the Development of Cities, Moskva 09822
Commission for Mechanics and Physics of Polymers, Moskva 09823
Commission for Meteorites and Space Dust, Novosibirsk 09824
Commission for Nuclear Physics, Moskva 09825
Commission for Philology and Phonetics, Moskva 09826
Commission for Scientific and Technical Co-operation of the Academy of Sciences and Organizations of Moscow Oblast, Moskva 09827
Commission for Socio-ecological Research, Moskva 09828
Commission for Space Toponomy, Moskva 09829
Commission for Synchroton Irradiation, Moskva 09830
Commission for the Conservation of Antarctic Marine Living Resources, Hobart 00373
Commission for the Co-ordination of Co-operation of Humanities Institutions of the Russian Academy of Sciences with UNESCO, Moskva 09831
Commission for the Development of Scientific Co-operation with Great Britain, Moskva 09832
Commission for the Effective Use of Shales in the Russian Economy, Moskva 09833
Commission for the European (Vienna) Centre for Co-ordination of Research and Documentation in Social Sciences, Moskva 09834
Commission for the Geological Map of the World, Paris 04422
Commission for the History of Geological Knowledge and Geological Study of the Russian Federation, Moskva 09835
Commission for the History of Philology, Moskva 09836
Commission for the Processing of the Scientific Legacy of Academician V.I. Vernadsky, Moskva 09837
Commission for the Prospects of Development of Science in the Russian Federation, Moskva 09838
Commission for the Protection of Natural Waters, Moskva 09839
Commission for the Scientific and Technological Development of Central America and Panama, Washington 14784
Commission for the Study of Production Forces and Natural Resources, Tbilisi 05007
Commission for the Study of Productive Forces and Natural Resources, Baku 01041
Commission for the Study of Productive Forces and Natural Resources, Moskva 09840
Commission for the Study of the Arctic, Moskva 09841
Commission for the Use of Computers and for Raising the Qualifications of Computer Users, Moskva 09842
Commission for the World's Oceans, Moskva 09843
Commission for Work with Young People, Moskva 09844
Commission Grand-Ducale d'Instruction, Luxembourg 08581
Commission Interaméricaine d'Arbitrage Commercial (Section Canadienne), Ottawa 02535
Commission Internationale de Bibliographie, Paris 04424
Commission Internationale de l'Eclairage, Wien 00578
Commission Internationale de Marketing, Brixham 11803
Commission Internationale des Grands Barrages, Paris 04425
Commission Internationale d'Etudes Historiques Latinoaméricaines et des Caraïbes, Paris 04426
Commission Internationale d'Histoire des Mouvements Sociaux et des Structures Sociales, Paris 04427
Commission Internationale d'Histoire du Sel, Innsbruck 00579
Commission Internationale pour la Protection des Eaux du Léman contre la Pollution, Lausanne 10650
Commission Internationale pour la Protection du Lac de Constance, München 05276
Commission Internationale pour l'Exploration Scientifique de la Mer Méditerranée, Monaco 08796
Commission Médicale Chrétienne, Genève 10651
Commission Mixte sur les Aspects Internationaux de l'Arriération Mentale, Bruxelles 01235
Commission Océanographique Intergouvernementale, Paris 04428
Commission of Socialist Teachers of the European Community, Zutphen 08843
Commission of Studies for Latin American Church History, México 08726
Commission of the European Communities Liaison Committee of Historians, Louvain-la-Neuve 01236
Commission of the History of Historiography, Budapest 06598
Commission on Accreditation of Rehabilitation Facilities, Tucson 14785
Commission on Biosphere and Ecology Research, Tbilisi 05008
Commission on Computer Technology, Yerevan 00204
Commission on Gay/Lesbian Issues in Social Work Education, Alexandria 14786
Commission on International Scientific Contacts, Baku 01042
Commission on Livestock Development in Latin America and the Caribbean, Santiago 03044
Commission on Mental and Physical Disability Law, Washington 14787
Commission on Mountain Mud Flows, Baku 01043
Commission on Nature Conservation, Baku 01044
Commission on Pastoral Research, Columbia 14788
Commission on Professional and Hospital Activities, Ann Arbor 14789
Commission on Professionals in Science and Technology, Washington 14790
Commission on Terminology, Bishkek 09845
Commission on the Caspian Sea, Baku 01045
Commission Permanente des Groupements Professionnels d'Assistants Sociaux, Bruxelles 01237
Commission pour l'Encouragement de la Recherche Scientifique, Bern 10652
Commission Théologique Internationale, Roma 16951
Commission to Study the Organization of Peace, Washington 14791
Committee for Better Transit, Long Island City 14792
Committee for Co-ordination of Investigations of the Lower Mekong Basin, Bangkok 11090
Committee for Economic Development, Washington 14793
Committee for Economic Development of Australia, Melbourne 00374
Committee for European Marine Biology Symposia, Brest 04429
Committee for Freedom of Choice in Medicine, Chula Vista 14794
Committee for Hispanic Arts and Research, Austin 14795
Committee for International Cooperation in National Research in Demography, Paris 04430
Committee for Mapping the Flora of Europe, Helsinki 03953
Committee for Russian Scientists in Defence of Peace and against Nuclear War, Moskva 09846
Committee for Systems Analysis, Moskva 09847
Committee for the Coordination of Construction of Scientific Apparatus and Antomation of Research Work, Moskva 09848
Committee for the European Development of Science and Technology, Roma 07623
Committee for the Implementation of the Standardized Yiddish Orthography, New York 14796
Committee for the Promotion of Medical Research, Yonkers 14797
Committee for the Scientific Investigation of Claims of the Paranormal, Buffalo 14798
Committee for UNESCO Programme „Man and Biosphere", Moskva 09849
Committee of Directors of Polytechnics, London 11804
Committee of International Development Institutions on the Environment, Nairobi 08487
Committee of the Acta-Endocrinologica Countries, Glostrup 03574
Committee of the International Programme „Litosphere", Moskva 09850
Committee of the International Programme of Geological Correlation, Moskva 09851
Committee of Vice-Chancellors and Principals of the Universities of the United Kingdom, London 11805
Committee on Allied Health Education and Accreditation, Chicago 14799
Committee on Atlantic Studies, Ottawa 02363
Committee on Changing International Realities, Washington 14800
Committee on Conceptual and Terminological Analysis, Oslo 09249
Committee on Continuing Education for School Personnel, Union 14801
Committee on Data for Science and Technology, Paris 04420, 04431
Committee on Earth Observation Satellites, Washington 14802
Committee on Family Research, Uppsala 10446
Committee on Forest Development in the Tropics, Roma 07624
Committee on Higher Education in the European Community, Bruxelles 01238
Committee on Institutional Cooperation, Champaign 14803
Committee on Meteorites, Moskva 09852
Committee on Petrography, Moskva 09853
Committee on Political Education, AFL-CIO, Washington 14804
Committee on Research Materials on Southeast Asia, Ann Arbor 14805
Committee on Science and Technology in Developing Countries, Madras 06752
Committee on Space Research, Paris 04411
Committee on the Challenges of Modern Society, Bruxelles 01239
Committee on the Role and Status of Women in Educational Research and Development, Hayward 14806
Committee on the Teaching of Science, Kalamazoo 14807
Committee on UNESCO Programme „Man and Biosphere", Yerevan 00205
Committee on Water Research, Delft 08844
Common Office for European Training, Mons 01240
Commons, Open Spaces and Footpaths Preservation Society, Henley-on-Thames 11806
Commonwealth Advisory Aeronautical Research Council, London 11807
Commonwealth Association for Development, London 11808
Commonwealth Association for Education in Journalism and Communication, London 02364
Commonwealth Association for Mental Handicap and Developmental Disabilities, Sheffield 11809
Commonwealth Association for Public Administration and Management, London 11810
Commonwealth Association for the Education and Training of Adults, Harare 17165
Commonwealth Association of Architects – Education Committee, London 11811
Commonwealth Association of Legislative Counsel, Edmonton 02365
Commonwealth Association of Museums, Calgary 02366
Commonwealth Association of Planners, Kingston 08076
Commonwealth Association of Planners, Ottawa 02367
Commonwealth Association of Polytechnics in Africa, Nairobi 08488
Commonwealth Association of Science and Mathematics Educators, London 11812
Commonwealth Association of Scientific Agricultural Societies, Kuala Lumpur 08641
Commonwealth Association on Surveying and Land Economy, London 11813
Commonwealth Bureau of Nutrition, Aberdeen 11814
Commonwealth Caribbean Medical Research Council, Kingston 08077
Commonwealth Commercial Crime Unit, London 11815
Commonwealth Consultative Group on Technology Management, London 11816
Commonwealth Council for Educational Administration, Armidale 00375
Commonwealth Defence Science Organisation, London 11817
Commonwealth Dental Association, London 11818
Commonwealth Education Liaison Committee, London 11819
Commonwealth Education Programme, London 11820
Commonwealth Forestry Association, Oxford 11821
Commonwealth Geographical Bureau, Montréal 02368
Commonwealth Heraldry Board, Auckland 09123
Commonwealth Independent Centre for the Study of the South African Economy and International Finance, London 11822
Commonwealth Industrial Training and Experience Programme, London 11823
Commonwealth Institute, London 11824
Commonwealth Institute of Valuers, Sydney 00376
Commonwealth Lawyers' Association, London 11825
Commonwealth Legal Education Association, London 11826
Commonwealth Library Association, Mandeville 08078
Commonwealth Medical Association, London 11827
Commonwealth Music Association, London 11828
Commonwealth of Learning, Vancouver 02369
Commonwealth Pharmaceutical Association, London 11829
Commonwealth Scientific and Industrial Research Organisation, Dickson 00377
Commonwealth Secretariat, Education Programme, Human Resource, Development Group, London 11830
Commonwealth Trust, London 11831
Commonwealth Veterinary Association, Scotsburn 00378
Community Associations Institute, Alexandria 14808
Community College Association for Instruction and Technology, Roswell 14809
Community Development Society, Milwaukee 14810
Community Environmental Council, Santa Barbara 14811
Community for Religious Research and Education, Loretto 14812
Community Guidance Service, New York 14813
Community Network for European Education and Training, Liège 01241
Community Nutrition Institute, Washington 14814
Community of European Management Schools, Jouy-en-Josas 04432
Community of Mediterranean Universities, Bari 07625
Community Planning Association of Canada, Regina 02370
Community Planning Association of Nova Scotia, Dartmouth 02371
COMNET, Liège 01241
Compagnie des Experts Architectes près la Cour d'Appel de Paris, Paris 04433
Compagnie Française pour le Développement des Fibres Textiles, Douala 01813
The Company Chemists' Association, Nottingham 11832
Company of Military Historians, Westbrook 14815
Comparative and International Education Society, Buffalo 14816
Comparative Education Society in Europe, Bruxelles 01242
Comparative Federation and Federalism Research Committee, Windsor 02372
Comparative Literature Society of Japan, Tokyo 08304
Composers, Authors and Publishers Association of Canada, Toronto 02373
The Composers' Guild of Great Britain, London 11833
Computer Aided Manufacturing International, Arlington 14817
Computer Arts Society, London 11834
Computer Communications Institute, Willowdale 02374
Computer Law Association, Fairfax 14818
Computer Law Association, Toronto 02375
CONAC, Caracas 17002
CONACYT, México 08729
CONAE-CENID, Montevideo 13256
CONAMOH, Panamá City 09415
Concern, Washington 14819
Conchological Society of Great Britain and Ireland, Ilford 11835
Conduit, Iowa City 14820
Confederación Boliviana de Odontólogos, La Paz 01542
Confederación Centroamericana de Medicina del Deporte, Heredía 03443
Confederación de Educadores Americanos, México 08727
Confederación de Educadores de América Latina, Santiago 03045
Confederación Interamericana de Educación Católica, Bogotá 03368
Confederacion Latinoamericana de Sociedades de Anestesiología, Bogotá 03369
Confederación Latinoamericano de Fisioterapía y Kinesiología, Santiago 03046
Confederación Panamericana de Medicina Deportiva, Porto Alegre 01641
Confederación Sudamericana de Medicina del Deporte, Buenos Aires 00124
Confederación Universitaria Boliviana, La Paz 01543
Confederate Historical Society, Leigh-on-Sea 11836
Confederate Memorial Literary Society, Richmond 14821
Confédération d'Associations d'Ecoles Indépendantes de la Communauté Européenne, Bruxelles 01243
Confédération des Sociétés Scientifiques Françaises, Paris 04434
Confédération des Syndicats Médicaux Français, Paris 04435
Confédération Européenne des Syndicats Nationaux et Associations Professionnelles de Pédiatres, Bruxelles 01244
Confédération Européenne pour la Thérapie Physique, Paris 04436
Confédération Internationale des Associations de Médecines Alternatives Naturelles, Pierrelatte 04437
Confédération Internationale des Sociétés d'Auteurs et Compositeurs, Paris 04438
Confédération Nationale des Syndicats Dentaires, Paris 04439
Confederation of African Medical Associations and Societies, Dar es Salaam 11048
Confederation of Alberta Faculty Associations, Edmonton 02376
Confederation of British Industry, London 11837
Confederation of European Laryngectomees, Bebra 05277
Confederation of European Specialists in Paediatrics, Bruxelles 01245
Confederation of Medical Associations in Asia and Oceania, Manila 09514
Conference Board of Associated Research Councils, New York 14822
Conference Board of Canada, Ottawa 02377
Conference Board of the Mathematical Sciences, Washington 14823
Conférence Canadienne des Arts, Ottawa 02102
Conférence des Directeurs des Ecoles Techniques Supérieures de Suisse, Winterthur 10653
Conférence des Facultés et Ecoles de Médecine d'Afrique d'Expression Française, Dakar 10025
Conférence des Présidents d'Universités, Paris 04440
Conférence des Recteurs des Universités Belges, Bruxelles 01246
Conférence des Recteurs des Universités Francophones d'Afrique, Kinshasa 17145
Conférence des Regions de l'Europe du Nord-Ouest, Brugge 01249
Conférence des Statisticiens Européens, Genève 10654
Conférence Diplomatique de Droit Maritime International, Bruxelles 01247
Conference for Chinese Oral and Performing Literature, Hanover 14824
Conference Group on French Politics and Society, Cambridge 14825
Conference Group on German Politics, Durham 14826
Conference Group on Italian Politics and Society, Chicago 14827
Conference Group on Women's History, Brooklyn 14860
Conférence Internationale des Directeurs et Doyens des Etablissements d'Enseignement Supérieur et Facultés d'Expression Française des Sciences de l'Agriculture et de l'Alimentation, Rabat 08807
Conférence Internationale des Doyens des Facultés de Médecine d'Expression Française, Tours 04441
Conférence Internationale des Facultés de Droit ayant en Commun l'Usage du Français, Aix-en-Provence 04442
Conférence Internationale des Facultés, Instituts et Ecoles de Pharmacie d'Expression Française, Paris 04443
Conférence Internationale des Formations d'Ingénieurs et Techniciens d'Expression Française, Villeurbanne 04444
Conférence Internationale des Grands Réseaux Electriques à Haute Tension, Paris 04445
Conférence Internationale des Responsables des Universités et Instituts à Dominante Scientifique et Technique d'Expression Française, Talence 04446
Conférence Internationale Permanente de Directeurs d'Instituts Universitaires pour la Formation de Traducteurs et d'Interprètes, Antwerpen 01248
Conference of African Theological Institutions, Harare 17166
Conference of Alberta School Superintendents, Calgary 02378

Conference of Asian-Pacific Pastoral Institutes, Manila *09515*
Conference of Baltic Oceanographers, Kiel *05278*
Conference of Chief Justices, Williamsburg *14828*
Conference of Consulting Actuaries, Schaumburg *14829*
Conference of Danish Rectors, København *03796*
Conference of Defence Associations Institute, Ottawa *02379*
Conference of Educational Administrators Serving the Deaf, Tucson *14830*
Conference of European Computer User Associations, Bussum *08845*
Conference of European Statisticians, Genève *10654*
Conference of Latin American Data-Processing Authorities, Buenos Aires *00125*
Conference of Podiatry Executives, Columbus *14831*
Conference of Public Health Laboratorians, Austin *14832*
Conference of Public Health Veterinarians, Arlington *14833*
Conference of Regions of North-West-Europe, Brugge *01249*
Conference of Research Workers in Animal Diseases, Fort Collins *14834*
Conference of the Baltic Oceanographers, Norrköping *10447*
Conference on Asian History, Bloomington *14835*
Conference on Christianity and Literature, Saint Petersburg *14836*
Conference on College Composition and Communication, Urbana *14837*
Conference on Consumer Finance Law, Chicago *14838*
Conference on English Education, Urbana *14839*
Conference on Latin American History, Tucson *14840*
Conference on the Development and the Planning of Urban Transport in Developing Countries, La Défense *04447*
Conférence Religieuse Canadienne, Ottawa *02380*
Conferencia de Rectores de las Universidades del Estado, Córdoba *10294*
Conferentie voor Regionale Ontwikkeling in Noord-West-Europa, Brugge *01249*
Confucius-Mencius Society of the Republic of China, Taipei *03292*
Congregational Libraries Association of British Columbia, Victoria *02381*
Congrès de Psychiatrie et de Neurologie de Langue Française, Limoges *04448*
Congrès des Psychoanalystes de Langues Romanes, Paris *04449*
Congrès International de Médecine Légale et de Médecine Sociale de Langue Française, Paris *04450*
Congreso de Poesía de Puerto Rico, Mayaguez *09730*
Congreso Internacional de Americanistas, La Plata *00126*
Congresso da América do Sul da Zoologia, São Paulo *01642*
Congress of Lung Association Staff, Washington *14841*
Congress of Neurological Surgeons, Cincinnati *14842*
Congress on Research in Dance, Brockport *14843*
CONGRIPS, Chicago *14827*
CONICET, Buenos Aires *00130*
CONICIT, Caracas *17001*
Conité Européen Lex Informatica Mercatoriaque, Montpellier *04451*
Conradh na Gaeilge, Dublin *06937*
Conseil Canadian de Biologie, Ottawa *01965*
Conseil Canadien de Développement Social, Ottawa *02111*
Conseil Canadien de la Documentation Juridique, Ottawa *02188*
Conseil Canadien de la Médecine Sportive, Vanier *02957*
Conseil Canadien de la Musique, Ottawa *02203*
Conseil Canadien des Administrateurs Universitaires en Education Physique, Edmonton *02110*
Conseil Canadien des Affaires Européennes, Saskatoon *02105*
Conseil Canadien des Arts Populaires, Toronto *02137*
Conseil Canadien des Ecoles de Bibliothécaires, Vancouver *02106*
Conseil Canadien des Normes, Ottawa *02958*
Conseil Canadienne de la Sécurité, Ottawa *02001*
Conseil Consultatif Economique et Social de l'Union Economique Benelux, Bruxelles *01250*
Conseil de Directeurs de Département de Chimie d'Universités Canadiennes, Burnaby *02402*
Conseil de la Conservation du Nouveau-Brunswick, Fredericton *02388*
Conseil de l'Homosexualité et la Religion, Winnipeg *02409*
Conseil de Recherche en Réassurance, Toronto *02858*
Conseil de Recherche et de Productivité du Nouveau-Brunswick, Fredericton *02691*
Conseil de Recherches en Sciences Naturelles et en Génie du Canada, Ottawa *02675*
Conseil de Recherches Médicales du Canada, Ottawa *02646*
Conseil des Arts du Canada, Ottawa *02000*
Conseil des Doyens des Facultes de Droit du Canada, Ottawa *02401*
Conseil des Ecoles Privées, Montréal *02406*
Conseil de Securité d'Ottawa-Carleton, Ottawa *02809*
Conseil de Securité du Nouveau-Brunswick, Fredericton *02692*
Conseil des Organisations Internationales des Sciences Médicales, Genève *10655*
Conseil des Provinces Atlantiques pour les Sciences, St. John's *01961*
Conseil des Recherches en Pêche et en Agro-Alimentaire du Québec, Québec *02382*
Conseil des Recteurs des Institutions Universitaires Francophones de Belgique, Bruxelles *01251*
Conseil des Recteurs des Universités de l'Ouest Canadien, Winnipeg *02408*
Conseil des Sciences du Canada, Ottawa *02907*
Conseil des Universités de l'Ontario, Toronto *02403*
Conseil des Universités du Québec, Sainte Foy *02383*
Conseil du Canada d'Informations et Education Sexuelles, Toronto *02912*
Conseil Géoscientifique Canadien, Waterloo *02146*
Conseil International d'Education des Adultes, Toronto *02548*
Conseil International d'Education Mésologique, Québec *02384*
Conseil International de la Danse, Paris *04452*
Conseil International de la Langue Française, Paris *04453*
Conseil International de la Musique, Paris *04454*
Conseil International de la Philosophie et des Sciences Humaines, Paris *04455*
Conseil International des Associations de Bibliothèques de Théologie, Köln *05279*
Conseil International des Monuments et des Sites, Paris *04456*
Conseil International des Moyens du Film d'Enseignement, Paris *04457*
Conseil International des Musées, Paris *04458*
Conseil International des Sciences Sociales, Paris *04459*
Conseil International des Unions Scientifiques, Paris *04460*
Conseil International d'Etudes Francophones, Upper Montclair *14844*
Conseil International pour l'Information Scientifique et Technique, Paris *04461*
Conseil International sur les Mathématiques dans les Pays en Voie de Développement, Tunis *11141*
Conseil International sur les Problèmes de l'Alcoolisme et des Toxicomanies, Lausanne *10656*
Conseil Interuniversitaire de la Communauté Française, Bruxelles *01252*
Conseil Interuniversitaire sur les Echanges Académiques avec l'URSS et l'Europe de l'Est, Waterloo *02564*
Conseil Médical du Canada, Ottawa *02643*
Conseil Mondial d'Ethique des Droits de l'Animal, Asnières *04462*
Conseil National de Design, Ottawa *02672*
Conseil National de la Recherche Scientifique et Technique, Brazzaville *03425*
Conseil National de l'Ordre des Médecins, Bruxelles *01253*
Conseil National de Recherche, Ottawa *02673*
Conseil National des Recherches Scientifiques, Port-au-Prince *06579*
Conseil Ontarien de Recherche Pédagogiques, Ajax *02774*
Conseil Ontarien des Affaires Universitaires, Toronto *02771*
Conseil Panafricain pour la Protection de l'Environnement et le Développement, Nouakchott *08672*
Conseil Québecoise pour l'Enfance et la Jeunesse, Montréal *02385*
Conseil Supérieur de l'Education, Sainte Foy *02386*
Conseil Supérieur de Statistique, Bruxelles *01254*
Conseil Universitaire des Directeurs de Biologie du Canada, Saskatoon *02109*
Consejo Andino de Cienccia y Tecnología, Lima *09465*
Consejo de Desarrollo Científico y Humanístico, Caracas *17000*
Consejo de Economía Nacional, Panamá City *09416*
Consejo de Rectores de Universidades Chilenas, Santiago *03047*
Consejo de Universidades, Madrid *10295*
Consejo Federal de Inversiones, Buenos Aires *00127*
Consejo General de Colegios Oficiales de Diplomados en Trabajo Social, Madrid *10296*
Consejo General de Colegios Oficiales de Doctores y Licenciados en Filosofía y Letras y en Ciencias, Madrid *10297*
Consejo General de Colegios Oficiales de Farmacéuticos, Madrid *10298*
Consejo General de Colegios Oficiales de Odontólogos y Estomatólogos de España, Madrid *10299*
Consejo General de Colegios Veterinarios de España, Madrid *10300*
Consejo General de Oficiales de Ingenieros Técnicos Agrícolas de España, Madrid *10301*
Consejo Interamericano de Archiveros, México *08728*
Consejo Internacional de Administración Científica, Buenos Aires *00128*
Consejo Inter-Universitario Nacional, Buenos Aires *00129*
Consejo Nacional de Archivos, Bogotá *03370*
Consejo Nacional de Ciencia, Panamá City *09417*
Consejo Nacional de Ciencia y Tecnologia, México *08729*
Consejo Nacional de Educación, La Paz *01544*
Consejo Nacional de Educación, Centro Nacional de Información y Documentación, Montevideo *13256*
Consejo Nacional de Higiene, Montevideo *13257*
Consejo Nacional de Investigaciones Científicas y Técnicas, Buenos Aires *00130*
Consejo Nacional de Investigaciones Científicas y Técnicas, Montevideo *13258*
Consejo Nacional de Investigaciones Científicas y Tecnológicas, Caracas *17001*
Consejo Nacional de la Cultura, Caracas *17002*
Consejo Nacional de la Universidad Peruana, Lima *09466*
Consejo Nacional de Política Económica Planeación, Bogotá *03371*
Consejo Nacional de Tuberculosis Paulina Aldina, La Habana *03479*
Consejo Superior de los Colegios de Arquitectos de España, Madrid *10302*
Consejo Superior Universitario Centroemericano, San José *03444*
Consejo Venezolano del Niño, Caracas *17003*
Consejo Zuliano de Planificación y Coordinación, Maracaibo *17004*
Conselho de Reitores das Universidades Portuguesas, Lisboa *09656*
Conselho Federal de Biblioteconomia, Brasília *01643*
Conselho Federal de Educação, Rio de Janeiro *01644*
Conselho Nacional de Desenvolvimento Científico e Tecnológico, Brasília *01645*
Conselho Nacional Serviço Social, Rio de Janeiro *01646*
Conservation and Development Association of Saskatchewan, Canora *02387*
Conservation Council of New Brunswick, Fredericton *02388*
Conservation Council of Ontario, Toronto *02389*
Conservation International, Washington *14845*
Consiglio Nazionale degli Architetti, Roma *07626*
Consiglio Nazionale dei Chimici, Roma *07627*
Consiglio Nazionale delle Ricerche, Roma *07628*
Consiglio Nazionale Forense, Roma *07629*
Consociatio Internationalis Musicae Sacrae, Roma *07630*
Consortium-Distance Education Network, Sainte-Foy *02390*
Consortium for Graduate Study in Management, Saint Louis *14846*
Consortium for International Studies Education, Columbus *14847*
Consortium of Caribbean Universities for Resource Management, Saint Thomas *17075*
Consortium of College and University Media Centers, Ames *14848*
Consortium of Institutions of Higher Education in Health and Rehabilitation in Europe, Gent *01255*
Consortium of Ontario Public Alternative Schools, Toronto *02391*
Consortium on Peace Research, Education and Development, Fairfax *14849*
Construction Industry Computing Association, Cambridge *11838*
Construction Industry Research and Information Association, London *11839*
Construction Management Institute, Willowdale *02392*
Construction Safety Association of Ontaria, Toronto *02393*
Construction Specifications Institute, Alexandria *14850*
Construction Writers Association, Aldie *14851*
Consultative Council for Postal Studies, Bern *10657*
Consultative Group on International Agricultural Research, Washington *14852*
Consultative Working Group for the Preparation of New Questions of Longterm Prospects of Development of Energy, Moskva *09854*
CONSUMED, Buenos Aires *00124*
Consumer Education Research Center, Orange *14853*
Consumer Energy Council of America Research Foundation, Washington *14854*
The Contact Lens Society of Australia, Sydney *00379*
Contact Literacy Center, Linoln *14855*
Contemporary Art Society, London *11840*
Contemporary Art Society of Australia, Sydney *00380*
CONTICYT, Santiago *03039*
Continuing Library Education Network and Exchange Round Table, Chicago *14856*
Convention of American Instructors of the Deaf, Rochester *14857*
CONZUPLAN, Maracaibo *17004*
Cooling Tower Institute, Houston *14858*
Cooperation Centre for Scientific Research Relative to Tobacco, Paris *04366*
Cooperation for Open Systems Interconnection Networking in Europe, Bruxelles *01256*
Cooperative Institute for Innovation, Roma *07725*
Cooperative League of the Republic of China, Taipei *03293*
Cooperative Program for Monitoring and Evaluation of the Long-range Transmission of Air Pollutants in Europe, Genève *10658*
Cooperative Program in Technological Research and Higher Education in Southeast Asia and the Pacific, Jakarta *06887*
COOPERE, Bruxelles *01257*
Cooper Ornithological Society, Los Angeles *14859*
Coordenação de Folclore e Cultura Popular, Rio de Janeiro *01647*
Coordinating Committee for Intercontinental Research Networking, Amsterdam *08846*
Coordinating Committee on Women in the Historical Profession, Brooklyn *14860*
Coordinating European Council for the Development of Performance Tests for Lubricants and Engine Fuels, London *11841*
Coordinating Research Council, Atlanta *14861*
Coordination and Cooperation Organization for the Control of the Major Endemic Diseases, Bobo-Dioulasso *01809*
Co-ordination Committee for Computer Technology, Moskva *09855*
Co-ordination Council for Information on Achievements, Moskva *09856*
Co-ordination Council for Scientific Problems Linked with Ecological Consequences of the Use of New Technological Systems, Moskva *09857*
Coordination et Promotion de l'Enseignement de la Réligion en Europe, Bruxelles *01257*
Coordination Office of Paediatric Endocrine Societies, Petah Tikva *07051*
COPA, Washington *14930*
COPAMADE, Porto Alegre *01641*
COPANT, Buenos Aires *00122*
COPE, Columbus *14831*
COPE, Washington *14804*
Copernicus Astronomical Centre, Warszawa *09542*
COPES, Petah Tikva *07051*
COPHL, Austin *14832*
COPRED, Fairfax *14849*
COPUOS, New York *16812*
Copyright Society of the U.S.A., New York *14862*
COR, Paris *04390*
CORD, Brockport *14843*
CORESTA, Paris *04366*
CORFO, Santiago *03048*
Cork Historical and Archaeological Society, Cork *06938*
CORMOSEA, Ann Arbor *14805*
Cornish Methodist Historical Association, Carnon Downs *11842*
Cornish Mining Development Association, Camborne *11843*
Cornwall Archaeological Society, Bodmin *11844*
Coronary Club, Cleveland *14863*
Corporación Centroamericana de Servicios de Navegación Aéreal, Guatemala City *06566*
Corporación de Fomento de la Producción, Santiago *03048*
Corporación Latinoamericana de Investigación para el Desarrollo del Sector Rural y Zona Costeros, Bogotá *03372*
Corporación para el Desarrollo de la Ciencia, Santiago *03049*
Corporación Toesca para el Desarrollo de la Arquitectura, Santiago *03050*
Corporate Data Exchange, New York *14864*
Corporation of Professional Social Workers of the Province of Quebec, Montréal *02399*
Corporation Professionnelle des Ergothèrapeutes du Québec, Montréal *02394*
Corporation Professionnelle des Médecins du Québec, Montréal *02395*
Corporation Professionnelle des Médecins Vétérinaires du Québec, Saint Hyacinthe *02396*
Corpòration Professionnelle des Orthophonistes et Audiologistes du Québec, Montréal *02397*
Corporation Professionnelle des Psychologues du Québec, Montréal *02398*
Corporation Professionnelle des Travailleurs Sociaux du Québec, Montréal *02399*
Correctional Education Association, Laurel *14865*
Corrosion Society of Finland, Helsinki *04043*
CORS, Ottawa *02213*
COS, Bloomington *14763*
COS, Cambridge *11785*
COS, Los Angeles *14859*
ČOS, Praha *03512*
COSINE, Bruxelles *01256*
COSLA, Lexington *14729*
COSPAS-SARSAT, London *11845*
COSPES, Roma *07264*
Costarrican's Pediatric Association, San José *03432*
COSTED, Madras *06752*
The Costume Society, London *11846*
Costume Society of America, Earleville *14866*
Cot Death Research and Support, London *12089*
COTH, Washington *14904*
Cotton Research Corporation, Kampala *11183*
Council for Advancement and Support of Education, Washington *14867*
Council for Agricultural Science and Technology, Ames *14868*
Council for American Private Education, Washington *14869*
Council for Basic Education, Washington *14870*
Council for Biomedical Communications Associations, Ann Arbor *14871*
Council for British Archaeology, York *11847*
Council for British Geography, London *11848*
Council for Children with Behavioral Disorders, Reston *14872*
Council for Dance Education and Training, London *11849*
Council for Educational Development and Research, Washington *14873*
Council for Education in World Citizenship, London *11850*
Council for Elementary Science International, Westport *14874*
Council for Environmental Conservation, London *11851*
Council for Environmental Education, Reading *11852*
Council for European Studies, New York *14875*
Council for Forestry and Agricultural Research, Stockholm *10520*
Council for Indian Education, Billings *14876*
Council for Interinstitutional Leadership, Leawood *14877*
Council for International Congresses of Dipterology, Edmonton *02400*
Council for International Congresses of Entomology, Canberra *00381*
Council for International Co-operation in Social Sciences, Moskva *09858*
Council for International Organizations of Medical Sciences, Genève *10655*
Council for Jewish Education, New York *14878*
Council for Links between the Academy of Sciences and Higher Education, Moskva *09859*
Council for Medical Affairs, Chicago *14879*

Council for Metrological Provision and Standardization, St. Petersburg *09860*
Council for Museum Anthropology, New York *14880*
Council for Philosophical Studies, New Haven *14881*
Council for Religion in Independent Schools, Washington *14882*
Council for Research in Music Education, Urbana *14883*
Council for Research in Values and Philosophy, Washington *14884*
Council for Scientific and Industrial Research, Accra *06479*
Council for Sex Information and Education, Los Angeles *14885*
Council for Study of Productive Forces, Alma-Ata *08455*
Council for the Accreditation of Correspondence Colleges, London *11853*
Council for the Advancement of Science Writing, Greenlawn *14886*
Council for the Care of Churches, London *11854*
Council for the Development of Economic and Social Research in Africa, Dakar *10026*
Council for the Protection of Rural England, London *11855*
Council for the Study of Productive Forces, Moskva *09861*
Council for Tobacco Research – U.S.A., New York *14887*
Council of 1890 College Presidents, Langston *14888*
Council of Administrators of Special Education, Albuquerque *14889*
Council of Adult Education, Melbourne *00382*
Council of Agriculture, Taipei *03294*
Council of Australian Food Technology Associations, North Sydney *00383*
Council of Biology Editors, Chicago *14890*
Council of British Geography, London *11856*
Council of Canadian Law Deans, Ottawa *02401*
Council of Canadian University Chemistry Chairmen, Burnaby *02402*
Council of Chief State School Officers, Washington *14891*
Council of Colleges of Arts and Sciences, Columbus *14892*
Council of Community Blood Centers, Washington *14893*
Council of Consulting Organizations, New York *14894*
Council of Directors of Institutes of Tropical Medicine in Europe, Budapest *06599*
Council of Engineering and Scientific Society Executives, Washington *14895*
Council of Graduate Schools, Washington *14896*
Council of International Programs, Washington *14897*
Council of Landscape Architectural Registration Boards, Fairfax *14898*
Council of Legal Education, London *11857*
Council of Libraries, Tirana *00004*
Council of Mennonite Colleges, Hillsboro *14899*
Council of National Library and Information Associations, Washington *14900*
Council of Nordic Teachers' Associations, Stockholm *10448*
Council of Ontario Universities, Toronto *02403*
Council of Organisations of Musicians in the Netherlands, Amsterdam *09002*
Council of Pacific Arts, Nouméa *09113*
Council of Parent Participation Pre-Schools in British Columbia, Burnaby *02404*
Council of Planning Librarians, Chicago *14901*
Council of Prairie University Libraries, Calgary *02405*
Council of Private Technical Schools, Montréal *02406*
Council of Professors of Building, Birmingham *11858*
Council of Regents for Colleges of Applied Arts and Technology, Toronto *02407*
Council of Scientific and Industrial Research, New Delhi *06753*
Council of Scientific Medical Societies, Moskva *09862*
Council of Scientific Society Presidents, Washington *14902*
Council of Societies for the Study of Religion, Valparaiso *14903*
Council of Teaching Hospitals, Washington *14904*
Council of the Great City Schools, Washington *14905*
Council of Undergraduate Research, Asheville *14906*
Council of Universities of Quebec, Sainte Foy *02383*
Council of Western Canadian University Presidents, Winnipeg *02408*
Council on Accreditation of Nurse Anesthesia Educational Programs/Schools, Park Ridge *14907*
Council on Anthropology and Education, Washington *14908*
Council on Arteriosclerosis of the American Heart Association, Dallas *14909*
Council on Chiropractic Education, West Des Moines *14910*
Council on Chiropractic Orthopedics, Provo *14911*
Council on Chiropractic Physiological Therapeutics, Idaho Falls *14912*
Council on Clinical Optometric Care, Saint Louis *14913*
Council on Co-ordinating Scientific Studies of the Georgian Language, Tbilisi *05009*
Council on Diagnostic Imaging, Ashtabula *14914*
Council on Education for Public Health, Washington *14915*
Council on Education of the Deaf, Washington *14916*
Council on Electrolysis Education, Memphis *14917*
Council on Exploitation of Sciemtific Equipment, Baku *01046*
Council on Governmental Relations, Washington *14918*
Council on Health Information and Education, Los Angeles *14919*
Council on Homosexuality and Religion, Winnipeg *02409*
Council on Hotel, Restaurant and Institutional Education, Washington *14920*
Council on Informatics, Yerevan *00206*
Council on International Cooperation in Research and Uses of Outer Space, Moskva *09863*
Council on Interracial Books for Children, New York *14921*
Council on Legal Education Opportunity, Washington *14922*
Council on Libraries, Riga *08556*
Council on Library-Media Technical Assistants, Cleveland *14923*
Council on Library Resources, Washington *14924*
Council on Medical Education of the American Medical Association, Chicago *14925*
Council on National Literatures, Whitestone *14926*
Council on Nutritional Anthropology, Washington *14927*
Council on Peace Research in History, Detroit *14928*
Council on Podiatric Medical Education, Bethesda *14929*
Council on Postsecondary Accreditation, Washington *14930*
Council on Resident Education in Obstetrics and Gynecology, Washington *14931*
Council on Social Work Education, Alexandria *14932*
Council on Tall Buildings and Urban Habitat, Bethlehem *14933*
Council on Technology Teacher Education, Ypsilanti *14934*
Council on the History of Natural Sciences and Technology, Tbilisi *05010*
Council on Use of Atomic Energy and Technology, Yerevan *00207*
Country Day School Headmasters Association of the U.S., Charlotte *14935*
County, City and Borough Architects' Association, Harrow *11859*
County Education Officers' Society, Winchester *11860*
County Emergency Planning Officers Society, Worlester *11861*
County Louth Archaeological and Historical Society, Dundalk *06939*
County Museum Society, Kidderminster *11862*
County of York Law Association, Toronto *02410*
County Planning Officers Society, Leicester *11863*
County Surveyors Society, Northampton *11864*
The Cousteau Society, Chesapeake *14936*
COVD, Chula Vista *14776*
Coventry and District Archaeological Society, Coventry *11865*
COWAR, Delft *08844*
Cowethas Lacha Sten Cernow, Gunnislake *11866*
CPA, London *11829*
CPA, Ottawa *02224*
CPA, Taipei *03283*
CPAC, Planaltina *01625*
CPATU, Belém *01627*
CPDC, Bridgetown *01086*
CPE, Salvador *01651*
CPEIP, Santiago *03023*
CPHA, Ann Arbor *14789*
CPhA, Ottawa *02223*
CPHV, Arlington *14833*
CPJ, Washington *14694*
CPL, Chicago *14901*
CPLC, Farmingdale *14692*
CPME, Bethesda *14929*
CPPC, Arima *11117*
CPPED, Nouakchott *08672*
CPRH, Detroit *14928*
CPRO, Wageningen *08840*
CPRW, Welshpool *11741*
CPS, Aarhus *03573*
CPS, Claremont *14693*
CPS, Indianapolis *14722*
CPS, London *11788*
CPS, New Haven *14881*
CPS, Orlando *14777*
CPSC, Manila *09513*
CPSG, London *11777*
C.P.S.S.A.E., Rovigo *07529*
CPST, Washington *14790*
CR, Lausanne *10648*
CRA, Los Angeles *14783*
C.R.A.B., Roma *07533*
CRAC, Cambridge *11745*
C.R.A.D., Udine *07431*
Crafts Council of Western India, Bombay *06754*
CRAG, Alger *00017*
CRAHCA, Englewood *14695*
Cranial Academy, Indianapolis *14937*
CRAPE, Alger *00010*
C.R.B., Padova *07435*
CRBT, Alger *00013*
CRC, Atlanta *14861*
CRC, London *11743*
C.R.D.M., Paderno del Grappa *07536*
CRDTL, Kingston *08075*
C.R.E.A.D., Alger *00011*
Creation Research Society, Kansas City *14938*
Creative Time, New York *14939*
CRED, Bruxelles *01189*
CREDAF, Paris *04368*
Creek Indian Memorial Association, Okmulgee *14940*
CREFAL, Pátzcuaro *08722*
C.R.E.O.C., Napoli *07538*
CREOG, Washington *14931*
C.R.E.S.M.E., Roma *07539*
CREW, Bruxelles *01188*
CRI, San Juan Capistrano *14747*
C.R.I.F., Zwijnaarde *01258*
CRILA, Houston *14753*
Criminal Law Society of Japan, Tokyo *08327*
CRIS, Washington *14882*
Critical Mass Energy Project of Public Citizen, Washington *14941*
Critics' Circle, London *11867*
Critics' Guild, London *11868*
CRL, Chicago *14696*
CRLA, Salem *14778*
CRME, Urbana *14883*
CRM Society, Glasgow *11758*
Crnogorska Akademija Nauka i Umjetnosti, Podgorica *17080*
CRO, Abidjan *08061*
Croatiam Geographic Society, Zagreb *03453*
Croatian Academy of Sciences and Arts, Zagreb *03459*
Croatian Library Association, Zagreb *03460*
Croatian Medical Association, Zagreb *03454*
Croatian Numismatic Society, Zagreb *03461*
Croatian Pharmaceutical Society, Zagreb *03455*
Croatian Society of Natural Sciences, Zagreb *03462*
CROP, Alger *00012*
Crop Science Society of America, Madison *14942*
Crop Science Society of China, Beijing *03196*
Crop Science Society of Japan, Tokyo *08252*
Crop Science Society of the Philippines, Laguna *09516*
Crop Science Society of Zimbabwe, Harare *17167*
Cross and Cockade, Great Britain Society of World Ware One Aero Historians, Farnborough *11869*
Cross-Examination Debate Association, Northridge *14943*
Crown Imperial Society, Southall *11870*
CRP, Brooklyn *14697*
CRRE, Loretto *14812*
CRS, Kansas City *14938*
C.R.S., London *11748*
C.R.S.A.B., Bitonto *07537*
CRSS, Oakland *14632*
CRT, Cambridge *11738*
C.R.T., Milano *07434*
CRUCH, Santiago *03047*
CRUFA, Kinshasa *17145*
CRWAD, Fort Collins *14834*
Cryogenic Engineering Conference, Moffet Field *14944*
Cryogenic Society of America, Oak Park *14945*
Crystallographic Society of Japan, Tokyo *08330*
CS, Norwalk *14768*
CS, Toronto *02339*
CSA, Davis *14704*
CSA, Earleville *14866*
CSA, New York *14698*
CSA, Oak Park *14945*
CSA, Rio Piedras *09729*
CSAA, Beijing *03182*
CSAG, Edinburgh *11751*
C.S.A.O., Torino *07439*
C.S.A.R.E., Rovigo *07519*
CSAS, Dayton *14717*
CSAS, Ottawa *02274*
CSAS, Sherman Oaks *14947*
C.S.A.T.A., Bari *07564*
ČSAV, Praha *03500*
ČSBKS, Praha *03519*
C.S.C., Roma *07545, 07550*
CSCJ, Chicago *14703*
CSCR, Chicago *14716*
CSDCA, Milano *07571*
C.S.E.F., Bologna *07451*
C.S.E.I., Bari *07555*
C.S.E.I., Napoli *07556*
CSH, Berkeley *14701*
CSHR, New York *14706*
CSI, Alexandria *14850*
CSI, Grand Rapids *14748*
CSI, Toronto *02259*
CSIA, Cambridge *11739*
CSICOP, Buffalo *14798*
CSIE, Los Angeles *14885*
C.S.I.P.M., Trapani *07572*
CSIR, Accra *06479*
CSIR, Pretoria *10133*
C.S.I.R.O., Dickson *00377*
CSL, London *11787*
CSLA, Ottawa *02285*
CSLA, Portland *14751*
CSLP, Cambridge *14700*
C.S.L.S.C., Bologna *07598*
CSM, Boston *14782*
CSM, Ottawa *02286*
CSM, Praha *03526*
CSME, Ottawa *02262*
CSMF, Paris *04435*
CSNYS, Syracuse *14636*
C.S.O.L., Quingdao *03181*
CSOP, Washington *14791*
CSP, London *11770*
CSP, New York *14708*
C.S.P.E.S., Roma *07582*
CSPI, Orinda *14707*
CSPI, Washington *14699*
C.S.P.M., Milano *07588*
CSPWC, Toronto *02290*
CSR, Roma *07604*
CSRE, San Francisco *14702*
C.S.S., Bruxelles *01254*
C.S.S., Mestre *07593*
CSS, Northampton *11864*
CSS, Pittsburgh *14769*
CSSA, Madison *14942*
CSSE, Ottawa *02268*
CSSF, Paris *04255, 04434, 04703*
CSSH, Ann Arbor *16593*
CSSP, Washington *14902*
CSSR, Valparaiso *14903*
CSTEC, Zutphen *08843*
CSTN, Alger *00014*
CSUCA, San José *03444*
CSUSA, New York *14862*
CSWE, Alexandria *14932*
CT, New York *14939*
CTA, Milano *07608*
CTA, Oxford *11794*
CTA, Teddington *11786*
CTBUH, Bethlehem *14933*
CTI, Houston *14858*
C.T.I., Torino *07620*
C.T.P.L., Paris La Défense *04385*
CTR-USA, New York *14887*
CTS, Kalamazoo *14807*
CTS, Ottawa *02308*
CTS, San Diego *14779*
CTSA, Philadelphia *14658*
CTTE, Ypsilanti *14934*
CTUC, Brooklyn *16802*
CUB, La Paz *01543*
CUBE, Saint Clair Shores *14946*
CUBS, Berkeley *14710*
CUED, Washington *15995*
Cultural Heritage Council, Rijswijk *09064*
Cultus et Lectura Patrum, Roma *07122*
Cumberland and Westmorland Antiquarian and Archaeological Society, Kendal *11871*
Cumberland County Family Planning Association, Amherst *02411*
CUR, Asheville *14906*
CURI, Washington *14780*
Current Agricultural Research Information System, Roma *07631*
CURS, Chapel Hill *14709*
Cusanus-Institut, Trier *05280*
Cutlery and Allied Trades Research Association, Sheffield *11872*
CVA, Scotsburn *00378*
CVCP, London *11805*
CWA, Aldie *14851*
C.W.A.A.S., Kendal *11871*
CWPC, Fayetteville *14758*
CWPEA, Detroit *14730*
CWPNM, New York *14711*
CW/PS, New York *14712*
CWRTA, Little Rock *14759*
CXC, Saint Michael *01081*
Cymdeithas Cysyllyiadau Cyhoeddus Cymru, Cardiff *11873*
Cymdeithas Ddrama Cymru, Cardiff *11905*
Cymdeithas Hanes Sir Ddinbych, Denbigh *11874*
Cymdeithas yr Iaith Gymraeg, Aberystwyth *11875*
Cyprus Astronautical Society, Limassol *03488*
Cyprus Civil Engineers and Architects Association, Nicosia *03489*
Cyprus Economic Association, Nicosia *03490*
Cyprus Geographical Association, Nicosia *03491*
Cyprus Joint Technical Council, Nicosia *03492*
Cyprus Numismatic Society, Nicosia *03493*
Cyprus Ophthalmological Society, Larnaca *03494*
Cyprus Photogrammetric Society, Nicosia *03495*
Cyprus Research Centre, Nicosia *03496*
Cystic Fibrosis Association of Ireland, Dublin *06940*
Cystic Fibrosis Foundation of Alberta, Edmonton *02412*
Cystic Fibrosis Research Trust, Bromley *11876*
Czech Academy of Sciences, Praha *03500*
Czech Anatomical Society, Praha *03537*
Czech Archaeological Society, Praha *03502*
Czech Biochemical Society, Praha *03518*
Czech Chemical Society, Praha *03520*
Czech Dermatological Society, Praha *03538*
Czech Economic Association, Praha *03521*
Czech Entomological Society, Praha *03522*
Czech Ethnographical Society, Praha *03511*
Czech Geological Society, Praha *03507*
Czech Historical Society, Praha *03508*
Czech Medical Association J.E. Purkyně, Praha *03509*
Czech Meterological Society, Praha *03510*
Czech Oncologic Society, Praha *03539*
Czech Ophthalmological Society, Praha *03512*
Czechoslovak Society of Arts and Sciences, Sherman Oaks *14947*
Czech Otolaryngological Society, Praha *03540*
Czech Pharmaceutical Association, Praha *03505*
Czech Psychological Association, Praha *03515*
Czech Radiological Society, Praha *03516*
Czech Scientific Society for Mycology, Praha *03533*
Czech Society for Agriculture, Veterinary Sciences and Food Technology, Praha *03530*
Czech Society for Cybernetics and Information Sciences, Praha *03525*
Czech Society for Eastern Studies, Praha *03513*
Czech Society for Histochemistry and Cytochemistry, Praha *03524*
Czech Society for Mechanics, Praha *03528*
Czech Society for Microbiology, Praha *03526*
Czech Society for Mycology, Praha *03534*
Czech Society for Political Sciences, Praha *03529*
Czech Society for Sciences and Technology, Praha *03536*
Czech Society of Anaesthesiology and Resuscitation, Praha *03517*
Czech Society of Bioclimatology, Praha *03519*
Czech Society of Cardiology, Praha *03541*
Czech Society of Parasitology, Praha *03527*
Czech Society of Respiratory Physiology and Pathology, Praha *03523*
Czech Society of Stomatology, Praha *03532*
Czech Surgical Society, Praha *03542*
Czech Zoological Society, Praha *03531*

D.A., Exeter *11890*
DA, Klampenborg *03588*
D.A.A., Canberra *00385*
DAAD, Bonn *05496*
DAB, Nürnberg *05495*
Dachorganisation der österreichischen Behindertenverbände, Wien *00726*
Dachverband Psychosozialer Hilfsvereinigungen e.V., Bonn *05281*
Dachverband Schweizerischer Lehrerinnen und Lehrer, Zürich *10659*
Dachverband wissenschaftlicher Gesellschaften der Agrar-, Forst-, Ernährungs-, Veterinär- und Umweltforschung e.V., Frankfurt *05282*
DADA, London *11886*
DAEEP, Savoy *14980*
Dag Hammarskjöld Foundation, Uppsala *10449*
DAGNA, Benicia *14518*
DAGV, Jülich *05294*

DAH, Düren 05467
Daini-Tokyo Bar Association, Tokyo 08118
Daini-Tokyo Bengoshikai Toshokan, Tokyo 08118
Dairy Engineerings Association, Odense 03649
Dairy Industry Association of Australia, Highett 00384
Dairy Research Foundation, Oak Grove Village 14948
Dairy Society International, Chambersburg 14949
DAL, Düsseldorf 05499
DAL, København 03602
Dalcroze Society of America, Seattle 14950
Damien Dutton Society for Leprosy Aid, Bellmore 14951
DANA, Baltimore 14986
Dana Center for Preventive Ophthalmology, Baltimore 14952
Danatom, Haslev 03575
DANB, Chicago 14960
Dance Films Association, New York 14953
DANFIP, København 03618
Danish Academy of Technical Sciences, Lyngby 03557
Danish Acoustical Society, Lyngby 03589
Danish Arts and Crafts Association, København 03609
Danish Association for International Cooperation, København 03772
Danish Association of Logopedics and Phoniatrics, København 03677
Danish Association of Medical Imaging, Sønderborg 03613
The Danish Association of Medical Specialists, København 03734
The Danish Association of School Librarians, Viby 03586
Danish Association of Social Workers, København 03691
Danish Biological Society, København 03567
Danish Committee for International Historical Cooperation, København 03608
Danish Composers' Society, København 03639
The Danish Confederation of Professional Associations, København 03558
Danish Epilepsy Society, Hvidovre 03612
Danish Forestry Society, Frederiksberg 03690
Danish Gerontological Society, Naestved 03627
Danish Land Development Service, Viborg 03705
Danish Library Association, Ballerup 03576
Danish Medical Society, København 03647
Danish Military Technical Society, København 03773
Danish Mycological Society, Søborg 03737
Danish Ophthalmological Society, København 03658
Danish Pharmacists Association, København 03616
Danish Research Library Association, Lyngby 03578
Danish Society, Reykjavik 06680
Danish Society for Photogrammetry and Surveying, Ålborg 03675
Danish Society for the Conservation of Nature, København 03583
Danish Society of Cardiology, Herlev 03597
Danish Society of Domestic Crafts, Kerteminde 03632
Danish Society of EEG and Clinical Neurophysiology, Gentofte 03808
Danish Society of Food Science and Technology, København 03768
Danish Society of Gastroenterology, Hvidovre 03623
Danish Society of Heating, Ventilating and Airconditioning Engineers, Lyngby 03699
Danish Society of Pathology, København 03686
Danish Society of Soil Science, Frederiksberg 03620
Danish Speeck and Hearing Association, Kokkedal 03565
Danish Town Planning Institution, København 03596
Danish Veterinary History Society, Egtved 03698
Danish Water Supply Association, Viby 03697
Danish Writers' Association, København 03622
Danmarks Biblioteksforening, Ballerup 03576
Danmarks Farmaceutiske Selskab, København 03577
Danmarks Forskningsbiblioteksforening, Lyngby 03578
Danmarks Handelsskoleforening, København 03579
Danmarks Jurist- og Økonomforbund, København 03580
Danmarks Laererforening, København 03581
Danmarks Mikrobiologiske Selskab, København 03582
Danmarks Naturfredningsforening, København 03583
Danmarks Naturvidenskabelige Samfund, Roskilde 03584
Danmarks Realskoleforening, Kolding 03585
Danmarks Skolebibliotekarforening, Viby 03586
Danmarks Skolebiblioteksforening, København 03587
Dannemiller Memorial Educational Foundation, San Antonio 14954
Dansk Agronomforening, Klampenborg 03588
Dansk Akustisk Selskab, Lyngby 03589
Dansk Anaestesiologisk Selskab, Arhus 03590
Dansk Astronautisk Forening, København 03591
Dansk Automationsselskab, Københan 03592
Dansk Betonforening, København 03593
Dansk Biologisk Selskab, København 03594
Dansk Botanisk Forening, København 03595
Dansk Byplanlaboratorium, København 03596
Dansk Cardiologisk Selskab, Herlev 03597
Dansk Cerealforening, Lyngby 03598
Dansk Dataforening, København 03599
Dansk Dendrologisk Forening, Hørsholm 03600
Dansk Dermatologisk Selskab, København 03601
Danske Arkitekters Landsforbund, København 03602
Danske Dermato-Venerologers Organisation, Alborg 03603
Danske Fødsels- og Kvindelaegers Organisation, Søborg 03604
Danske Forstkandidaters Forening, Klampenborg 03605
Danske Fysioterapeuter, København 03606
Danske Interne Medicineres Organisation, København 03607
Danske Komité for Historikernes Internationale Samarbejde, København 03608
Danske Kunsthåndvårkeres Landssammenslutning, København 03609
Danske Lunglaegers Organisation, Holbaek 03610
Danske Nervelaegers Organisation, Odense 03611
Dansk Epilepsiforening, Hvidovre 03612
Danske Radiologers Organisation, Sønderborg 03613
Danske Selskab i Reykjavik, Reykjavik 06680
Danske Veterinårhygiejnikeres Organisation, Slagense 03614
Dansk Exlibris Selskab, Brœnshœj 03615
Dansk Farmaceutforening, København 03616
Dansk Farmacihistorisk Selskab, København 03617
Dansk Føderation for Informationsbehandling og Virksomhedsstyring, København 03618
Dansk Forening for Europaret, København 03619
Dansk Forening for Jordbundsvidenskab, Frederiksberg 03620
Dansk Forening for Retssociologi, København 03621
Dansk Forfatterforening, København 03622
Dansk Gastroenterologisk Selskab, Hvidovre 03623
Dansk Geofysisk Forening, København 03624
Dansk Geologisk Forening, København 03625
Dansk Geoteknisk Forening, Lyngby 03626
Dansk Gerontologisk Selskab, Naestved 03627
Dansk Grafologisk Selskab, Holte 03628
Dansk Haematologisk Selskab, Hellerup 03629
Dansk Historielaererforening, Vordingborg 03630
Dansk Hortonomforening, Klampenborg 03631
Dansk Hufliddsselskab, Kerteminde 03632
Dansk Idraetslaererforening, Nyborg 03633
Dansk Industrimedicinsk Selskab, Charlottenlund 03634
Dansk Ingeniørforening, København 03635
Dansk Kerneteknisk Selskab, København 03636
Dansk Kirurgisk Selskab, Hørsholm 03637
Dansk Køleforening, Lyngby 03638
Dansk Komponistforening, København 03639
Dansk Kriminalistforening, København 03640
Dansk Kulturhistorisk Museumsforening, Fur 03641
Dansk Kunstmuseumsforening, København 03642
Dansk Laererforeningen, København 03643
Dansk Lokalhistorisk Forening, Bagsvaerd 03644
Dansk Mathematisk Forening, København 03645
Dansk Medicinsk-Historisk Selskab, København 03646
Dansk Medicinsk Selskab, København 03647
Dansk Medikoteknisk Selskab, Holte 03648
Dansk Mejeringeniør Forening, Odense 03649
Dansk Metallurgisk Selskab, Lyngby 03650
Dansk Musikbiblioteksforening, København 03651
Dansk Musikpaedagogisk Forening, København 03652
Dansk Naturhistorisk Forening, København 03653
Dansk Nefrologisk Selskab, Arhus 03654
Dansk Neurologisk Selskab, Hellerup 03655
Dansk Numismatisk Forening, Taastrup 03656
Dansk Odontologisk Selskab, Klampenborg 03657
Dansk Oftalmologisk Selskab, København 03658
Dansk Økologisk Forening Oikos, København 03659
Dansk Ornithologisk Forening, København 03660
Dansk Ortopaedisk Selskab, København 03661
Dansk Oto-laryngologisk Selskab, Hellerup 03662
Dansk Pediatrisk Selskab, Virum 03663
Dansk Pneumologisk Selskab, Allrød 03664
Dansk Pressehistorisk Selskab, København 03665
Dansk Psykoanalytisk Selskab, København 03666
Dansk Psykolog Forening, Valby 03667
Dansk Radiologisk Selskab, Herlev 03668
Dansk Reumatologisk Selskab, København 03669
Dansk Selskab for Akupunktur, København 03670
Dansk Selskab for Allergologi og Immunologi, København 03671
Dansk Selskab for Almen Medicin, Greve 03672
Dansk Selskab for Bygningsstatik, Lyngby 03673
Dansk Selskab for Cancerforskning, København 03674
Dansk Selskab for Fotogrammetri og Landmåling, Ålborg 03675
Dansk Selskab for Intern Medicin, Gentofte 03676
Dansk Selskab for Logopaedi og Foniatri, København 03677
Dansk Selskab for Materialprøvning og -forskning, København 03678
Dansk Selskab for Musikforskning, København 03679
Dansk Selskab for Obstetrik og Gynaekologi, København 03680
Dansk Selskab for Oldtids- og Middelalderforskning, København 03681
Dansk Selskab for Oligofreniforskning, København 03682
Dansk Selskab for Operationsanalyse, Hornbaek 03683
Dansk Selskab for Optometri, Vedbaek 03684
Dansk Selskab for Opvarmnings- og Ventilationsteknik, København 03685
Dansk Selskab for Patologi, København 03686
Dansk Selskab for Social Medicin, Odense 03687
Dansk Selskab for Teoretisk Statistik, København 03688
Dansk Skattevidenskabelig Forening, København 03689
Dansk Skovforening, Frederiksberg 03690
Dansk Socialradgiverforening, København 03691
Dansk Sociologisk Selskab, København 03692
Dansk Sprogvaern, Vaerløse 03693
Dansk Svejseteknisk Landsforening, Broendby 03694
Dansk Tandlaegeforening, København 03695
Dansk Teknisk Laererforening, København 03696
Dansk Vandteknisk Forening, Viby 03697
Dansk Veterinårhistorisk Samfund, Egtved 03698
Dante Society of America, Cambridge 14955
Danubian League against Thrombosis and Hemorrhagic Disorders, Istanbul 11144
danvak VVS Teknisk Forening, Lyngby 03699
Dartmoor Pony Society, Newton Abbot 11877
Dartmoor Preservation Association, Plymouth 11878
D.A.S., Uttoxeter 11882
DASA, Houghton 10134
DAS, Arhus 03590
DAS, Exeter 11889
DAS, Lyngby 03589
DASt, Köln 05500
Data for Development International Association, Marseille 04286
DAtF, Bonn 05569
DAu, Københan 03592
DAV, Hannover 05501
DAW, Cardiff 11905
DB, Ballerup 03576
DBG, Frankfurt 05296
DBG, Phoenix 14964
dbi, Berlin 05570
DBS, Essen 05504
DBS, København 03704
DBS, Ljubljana 10111
DBV, Berlin 05506
DBV, Wiesbaden 05505
DCF, Lyngby 03598
DCPO, Baltimore 14952
DCS, Herlev 03597
DCWAB, London 11892
DDA, Essen 05507
DdD, Vanlöse 03701
DDH, Viborg 03705
DDPA, Chicago 14958
DDS, København 03601
DDSLA, Bellmore 14951
Deafness Research Foundation, New York 14956
DEBRA, New York 14994
DECA, Reston 14978
DECHEMA, Frankfurt 05334
Decision Sciences Institute, Atlanta 14957
Decus Europe, Genève 10663
Dedicated Road Infrastructure for Vehicle Safety in Europe, Bruxelles 01259
Defence Housing Society, Karachi 09362
Defence Medical Association of Canada, Ottawa 02413
DeGePo, Bremerhaven 05409
Delhi Library Association, Dehli 06755
Delius Society, Chatham 11879
Delphinium Society, Farnham 11880
Delta Dental Plans Association, Chicago 14958
Deltiologists of America, Norwood 14959
Demeure Historique, Paris 04463
Denbighshire Historical Society, Denbigh 11874
Den Danske Aktuarforening, Lyngby 03700
Den Danske Dyrlaegeforening, Vanlöse 03701
Den Danske Historiske Forening, København 03702
Den Geofysiske Kommisjon, Oslo 09250
Denki Kagaku Kyokai, Tokyo 08119
Denki Tsushin Kyokai, Tokyo 08120
Den Norske Aktuarforening, Oslo 09251
Den Norske Historiske Forening, Oslo 09252
Den Norske Laegeforening, Lysaker 09253
Den Norske Mikrobionomforening, Oslo 09254
Den Norske Tannlegeforening, Oslo 09255
Den Norske Veterinaerforening, Oslo 09256
Den Polytekniske Forening, Oslo 09257
Denshi Joho Tsushin Gakkai, Tokyo 08121
Dental Assisting National Board, Chicago 14960
Dental Association of Malta, Saint Andrews 08659
Dental Association of Prince Edward Island, Charlottetown 02414
Dental Association of South Africa, Houghton 10134
Dental Association of Zimbabwe, Harare 17168
The Dental Council, Dublin 06941
Dental Health International, Athens 14961
Dentists' Liaison Committee for the EEC, London 11881
Denturist Society of Alberta, Sherwood Park 02415
Denturist Society of Nova Scotia, Halifax 02416
Departamento Administrativo Nacional de Estadística, Bogotá 03373
Departamento de Antropología e Historia de Nayarit, Tepic 08730
Departamento de Educación Audiovisual, México 08731
Departamento de Estudios Etnográficos y Coloniales, Santa Fé 00131
Departamento de Estudios Históricos Navales, Buenos Aires 00132
Departamento de Historia del Arte Diego Velázquez, Madrid 10303
Departamento de Microbiología, Castelar 00133
Department of Naval History Studies, Buenos Aires 00132
Departments of Education Correspondence Schools Association (Canada), Barrhead 02417
Dept of Ethnographical and Colonial Studies, Santa Fé 00131
Deputazione di Storia Patria per gli Abruzzi, L'Aquila 07632
Deputazione di Storia Patria per il Friuli, Udine 07633
Deputazione di Storia Patria per la Calabria, Reggio Calabria 07634
Deputazione di Storia Patria per la Lucania, Potenza 07635
Deputazione di Storia Patria per la Sardegna, Cagliari 07636
Deputazione di Storia Patria per la Toscana, Firenze 07637
Deputazione di Storia Patria per le Antiche Province Modenesi, Modena 07638
Deputazione di Storia Patria per le Marche, Ancona 07639
Deputazione di Storia Patria per le Province di Romagna, Bologna 07640
Deputazione di Storia Patria per le Province Parmensi, Parma 07641
Deputazione di Storia Patria per le Venezie, Venezia 07642
Deputazione di Storia Patria per l'Umbria, Perugia 07643
Deputazione Provinciale Ferrarese di Storia Patria, Ferrara 07644
Deputazione Reggiana di Storia Patria, Reggio Emilia 07645
Deputazione Subalpina di Storia Patria, Torino 07646
Derbyshire Archaeological Society, Uttoxeter 11882
Dermatology Foundation, Evanston 14962
DES Action USA, Oakland 14963
Desert Botanical Garden, Phoenix 14964
Desert Fishes Council, Bishop 14965
Desert Locust Control Organization for Eastern Africa, Nairobi 08489, 08490
Desert Protective Council, Valley Center 14966
Desert Tortoise Council, Palm Desert 14967
Design and Industries Association, London 11883
Design/Build Institute, Greendale 14968
Design Council, London 11884
The Design Council – Scotland, Glasgow 11885
Designers and Art Directors Association of London, London 11886
Design Research Centre for the Gold, Silver and Jewellery Industries, London 11887
Designs for Change, Chicago 14969
Design Zentrum Nordrhein-Westfalen e.V., Essen 05283
DESY, Hamburg 05575
Det Danske Afrika Selskab, Hellerup 03703
Det Danske Bibelselskab, København 03704
Det Danske Hedeselskab, Viborg 03705
Det Danske Orgelselskab, Vanløse 03706
Det Danske Shakespeare Selskab, København 03707
Det Danske Sprog- og Litteraturselskab, København 03708
Det Grønlandske Selskab, Charlottenlund 03709
Det Kongelige Danske Videnskabernes Selskab, København 03710
Det Kongelige Nordiske Oldskriftselskab, København 03711
Det Kongelige Norske Videnskabers Selskab, Trondheim 09258
Det Krigsvidenskabelige Selskab, Frederiksberg 03712
Det Laerde Selskab i Arhus, Arhus 03713
Det Medicinske Selskab i København, København 03714
Det Norske Geografiske Selskap, Oslo 09259
Det Norske Hageselskap, Oslo 09260
Det Norske Medicinske Selskab, Oslo 09261
Det Norske Samlaget, Oslo 09262
Det Norske Skogselskap, Oslo 09263
Det Norske Videnskaps-Akademi, Oslo 09264
Det Udenrigspolitiske Selskab, København 03715
Deuqua, Hannover 05493
Deutsche Adels-Gesellschaft in Nord-Amerika, Benicia 14518
Deutsche Akademie der Darstellenden Künste e.V., Frankfurt 05284
Deutsche Akademie der Naturforscher Leopoldina, Halle 05285
Deutsche Akademie für Kinder- und Jugendliteratur e.V., Würzburg 05286
Deutsche Akademie für medizinische Fortbildung, Kassel 05287

Deutsche Akademie für Nuklearmedizin, Hannover *05288*
Deutsche Akademie für Sprache und Dichtung e.V., Darmstadt *05289*
Deutsche Akademie für Städtebau und Landesplanung e.V., München *05290*
Deutsche Akademie für Verkehrswissenschaft e.V., Hamburg *05291*
Deutsche Akademie Villa Massimo, Roma *07241*
Deutsche Arbeitsgemeinschaft für Paradontologie, Goslar *05292*
Deutsche Arbeitsgemeinschaft genealogischer Verbände e.V., Brühl *05293*
Deutsche Arbeitsgemeinschaft Vakuum, Jülich *05294*
Deutsche Botanische Gesellschaft e.V., Göttingen *05295*
Deutsche Bunsen-Gesellschaft für Physikalische Chemie e.V., Frankfurt *05296*
Deutsche Dendrologische Gesellschaft, Trier *05297*
Deutsche Dermatologische Gesellschaft, Kiel *05298*
Deutsche Diabetes-Gesellschaft, Bad Oeynhausen *05299*
Deutsche EEG-Gesellschaft, Berlin *05300*
Deutsche Elektrotechnische Kommission im DIN und VDE, Frankfurt *05301*
Deutsche Exlibris Gesellschaft e.V., Konstanz *05302*
Deutsche farbwissenschaftliche Gesellschaft e.V., Berlin *05303*
Deutsche Film- und Fernsehakademie Berlin, Berlin *05304*
Deutsche Forschungsgemeinschaft, Bonn *05305*
Deutsche Forschungsgesellschaft für Oberflächenbehandlung e.V., Düsseldorf *05306*
Deutsche Forschungs- und Versuchsanstalt für Luft- und Raumfahrt e.V., Köln *05307*
Deutsche Gartenbauwissenschaftliche Gesellschaft e.V., Hannover *05308*
Deutsche Gemmologische Gesellschaft e.V., Idar-Oberstein *05309*
Deutsche Geodätische Kommission, München *05310*
Deutsche Geologische Gesellschaft, Hannover *05311*
Deutsche Geophysikalische Gesellschaft e.V., Münster *05312*
Deutsche Gesellschaft für Aesthetische Medizin, Berlin *05313*
Deutsche Gesellschaft für Agrarrecht, Bonn *05314*
Deutsche Gesellschaft für Allergieforschung, Tübingen *05315*
Deutsche Gesellschaft für Allergie- und Immunitätsforschung, Bochum *05316*
Deutsche Gesellschaft für allgemeine und angewandte Entomologie e.V., Dossenheim *05317*
Deutsche Gesellschaft für Amerikastudien e.V., Nürnberg *05318*
Deutsche Gesellschaft für Anästhesiologie und Intensivmedizin, Nürnberg *05319*
Deutsche Gesellschaft für Analytische Psychologie e.V., Berlin *05320*
Deutsche Gesellschaft für angewandte Optik e.V., Jena *05321*
Deutsche Gesellschaft für Angiologie, Esslingen *05322*
Deutsche Gesellschaft für Anthropologie, Freiburg *05323*
Deutsche Gesellschaft für Arbeitsmedizin e.V., München *05324*
Deutsche Gesellschaft für Asienkunde e.V., Hamburg *05325*
Deutsche Gesellschaft für Auswärtige Politik e.V., Bonn *05326*
Deutsche Gesellschaft für Bauingenieurwesen e.V., Karlsruhe *05327*
Deutsche Gesellschaft für Baukybernetik e.V, Holzminden *05328*
Deutsche Gesellschaft für Baurecht e.V., Frankfurt *05329*
Deutsche Gesellschaft für Betriebswirtschaft e.V., Berlin *06277*
Deutsche Gesellschaft für Bevölkerungswissenschaft e.V., Wiesbaden *05330*
Deutsche Gesellschaft für Biomedizinische Technik e.V., Berlin *05331*
Deutsche Gesellschaft für Biophysik, Garching *05332*
Deutsche Gesellschaft für Bluttransfusion und Immunhämatologie e.V., Frankfurt *05333*
Deutsche Gesellschaft für Chemisches Apparatewesen, Chemische Technik und Biotechnologie e.V., Frankfurt *05334*
Deutsche Gesellschaft für Chirurgie, Bonn *05335*
Deutsche Gesellschaft für Christliche Kunst e.V., München *05336*
Deutsche Gesellschaft für Chronometrie e.V., Stuttgart *05337*
Deutsche Gesellschaft für das Badewesen e.V., Essen *05338*
Deutsche Gesellschaft für die Bekämpfung der Muskelkrankheiten e.V., Freiburg *05339*
Deutsche Gesellschaft für die Vereinten Nationen e.V., Bonn *05340*
Deutsche Gesellschaft für Dokumentation e.V., Frankfurt *05341*
Deutsche Gesellschaft für Dynamische Psychiatrie, München *05342*
Deutsche Gesellschaft für Edelsteinkunde, Idar-Oberstein *05309*
Deutsche Gesellschaft für Elektronenmikroskopie e.V., Berlin *05343*
Deutsche Gesellschaft für Endokrinologie, Berlin *05344*
Deutsche Gesellschaft für Erd- und Grundbau e.V., Essen *05345*
Deutsche Gesellschaft für Ernährung e.V., Frankfurt *05346*
Deutsche Gesellschaft für Erziehungswissenschaft, Berlin *05347*
Deutsche Gesellschaft für Fettwissenschaft e.V., Münster *05348*
Deutsche Gesellschaft für Filmdokumentation, Wiesbaden *05349*
Deutsche Gesellschaft für Film- und Fernsehforschung, München *05350*
Deutsche Gesellschaft für Forschung im Graphischen Gewerbe, München *05351*
Deutsche Gesellschaft für Galvano- und Oberflächentechnik e.V., Düsseldorf *05352*
Deutsche Gesellschaft für Gartenkunst und Landschaftspflege e.V., Karlsruhe *05353*
Deutsche Gesellschaft für Gerontologie, Lübeck *05354*
Deutsche Gesellschaft für Geschichte der Medizin, Naturwissenschaft und Technik e.V., München *05355*
Deutsche Gesellschaft für Gesundheitsvorsorge e.V., Leverkusen *05356*
Deutsche Gesellschaft für Gynäkologie und Geburtshilfe, Amberg *05357*
Deutsche Gesellschaft für Hämatologie und Onkologie e.V., München *05358*
Deutsche Gesellschaft für Hals-Nasen-Ohren-Heilkunde, Kopf- und Hals-Chirurgie, Bonn *05359*
Deutsche Gesellschaft für Heereskunde e.V., Beckum *05360*
Deutsche Gesellschaft für Herpetologie und Terrarienkunde e.V., Rheinbach *05361*
Deutsche Gesellschaft für Herz- und Kreislaufforschung, Düsseldorf *05362*
Deutsche Gesellschaft für Holzforschung e.V., München *05363*
Deutsche Gesellschaft für Hopfenforschung e.V., Wolnzach *05364*
Deutsche Gesellschaft für Hydrokultur e.V., Herten *05365*
Deutsche Gesellschaft für Hygiene und Mikrobiologie e.V., Heidelberg *05366*
Deutsche Gesellschaft für Innere Medizin, Wiesbaden *05367*
Deutsche Gesellschaft für Internistische Intensivmedizin, Hamburg *05368*
Deutsche Gesellschaft für Kartographie e.V., Berlin *05369*
Deutsche Gesellschaft für Kieferorthopädie e.V., Würzburg *05370*
Deutsche Gesellschaft für Kinderheilkunde, Hannover *05371*
Deutsche Gesellschaft für Kinder- und Jugendpsychiatrie e.V., Marburg *05372*
Deutsche Gesellschaft für Kommunikationsforschung, München *05373*
Deutsche Gesellschaft für Laboratoriumsmedizin e.V., Düsseldorf *05374*
Deutsche Gesellschaft für Lichtforschung, Hanau *05375*
Deutsche Gesellschaft für Logistik e.V., Dortmund *05376*
Deutsche Gesellschaft für Luft- und Raumfahrt e.V., Bonn *05377*
Deutsche Gesellschaft für Luft- und Raumfahrtmedizin e.V., Ulm *05378*
Deutsche Gesellschaft für Lungenkrankheiten und Tuberkulose, Freiburg *05379*
Deutsche Gesellschaft für Manuelle Medizin e.V., Boppard *05380*
Deutsche Gesellschaft für Materialkunde e.V., Oberursel *05381*
Deutsche Gesellschaft für Medizinische Informatik, Biometrie und Epidemiologie, Köln *05382*
Deutsche Gesellschaft für Medizinische Soziologie, Ulm *05383*
Deutsche Gesellschaft für Missionswissenschaft, Heidelberg *05384*
Deutsche Gesellschaft für Moor- und Torfkunde, Hannover *05385*
Deutsche Gesellschaft für Mund-, Kiefer- und Gesichtschirurgie, Bonn *05386*
Deutsche Gesellschaft für Neurochirurgie, Essen *05387*
Deutsche Gesellschaft für Neurologie, Würzburg *05388*
Deutsche Gesellschaft für Neuropathologie und Neuroanatomie e.V., München *05389*
Deutsche Gesellschaft für Neuroradiologie, Würzburg *05390*
Deutsche Gesellschaft für Nuklearmedizin, Bonn *05391*
Deutsche Gesellschaft für Orthopädie und Traumatologie e.V., Frankfurt *05392*
Deutsche Gesellschaft für Ortung und Navigation e.V., Düsseldorf *05393*
Deutsche Gesellschaft für Osteuropakunde e.V., Berlin *05394*
Deutsche Gesellschaft für Parasitologie e.V., Marburg *05395*
Deutsche Gesellschaft für Parodontologie, Hamburg *05396*
Deutsche Gesellschaft für Pathologie, Frankfurt *05397*
Deutsche Gesellschaft für Perinatale Medizin, Berlin *05398*
Deutsche Gesellschaft für Personalführung e.V., Düsseldorf *05399*
Deutsche Gesellschaft für Pharmakologie und Toxikologie, Darmstadt *05400*
Deutsche Gesellschaft für Phlebologie, Norderney *05401*
Deutsche Gesellschaft für Photogrammetrie und Fernerkundung, Neubiberg *05402*
Deutsche Gesellschaft für Photographie e.V., Köln *05403*
Deutsche Gesellschaft für Physikalische Medizin und Rehabilitation, Hannover *05404*
Deutsche Gesellschaft für Pilzkunde, Karlsruhe *05405*
Deutsche Gesellschaft für Plastische und Wiederherstellende Chirurgie e.V., Rotenburg *05406*
Deutsche Gesellschaft für Pneumologie, Greifenstein *05407*
Deutsche Gesellschaft für Poesie- und Bibliotherapie e.V., Köln *05408*
Deutsche Gesellschaft für Polarforschung, Bremerhaven *05409*
Deutsche Gesellschaft für Psychiatrie und Nervenheilkunde e.V., Köln *05410*
Deutsche Gesellschaft für Psychologie e.V., Münster *05411*
Deutsche Gesellschaft für Psychosomatische Medizin e.V., München *05412*
Deutsche Gesellschaft für Publizistik- und Kommunikationswissenschaft e.V., Eichstätt *05413*
Deutsche Gesellschaft für Qualität e.V., Frankfurt *05414*
Deutsche Gesellschaft für Qualitätsforschung (Pflanzliche Nahrungsmittel) e.V., Freising *05415*
Deutsche Gesellschaft für Rechtsmedizin, Köln *05416*
Deutsche Gesellschaft für Rheumatologie e.V., Bad Bramstedt *05417*
Deutsche Gesellschaft für Säugetierkunde e.V., Tübingen *05418*
Deutsche Gesellschaft für Sexualforschung e.V., Hamburg *05419*
Deutsche Gesellschaft für Sexualpädagogik und Sexualberatung e.V., Frankfurt *06240*
Deutsche Gesellschaft für Sonnenenergie e.V., München *05420*
Deutsche Gesellschaft für Soziale Psychiatrie e.V., Köln *05421*
Deutsche Gesellschaft für Sozialmedizin und Prävention e.V., Bochum *05422*
Deutsche Gesellschaft für Sozialpädiatrie e.V., München *05423*
Deutsche Gesellschaft für Soziologie, Mannheim *05424*
Deutsche Gesellschaft für Sportmedizin e.V., Heidelberg *05539*
Deutsche Gesellschaft für Sprachheilpädagogik e.V., Berlin *05425*
Deutsche Gesellschaft für Sprach- und Stimmheilkunde, Münster *05426*
Deutsche Gesellschaft für Sprachwissenschaft, Passau *05427*
Deutsche Gesellschaft für Sprechwissenschaft und Sprecherziehung e.V., Münster *05428*
Deutsche Gesellschaft für Suchtforschung und Suchttherapie e.V., Hamm *05429*
Deutsche Gesellschaft für Technische Zusammenarbeit, Eschborn *05430*
Deutsche Gesellschaft für Thorax-, Herz- und Gefässchirurgie, Bad Nauheim *05431*
Deutsche Gesellschaft für Unfallheilkunde e.V., Frankfurt *05432*
Deutsche Gesellschaft für Urologie, Hannover *05433*
Deutsche Gesellschaft für Verdauungs- und Stoffwechselkrankheiten, Frankfurt *05434*
Deutsche Gesellschaft für Verhaltenstherapie e.V., Tübingen *05435*
Deutsche Gesellschaft für Versicherungsmathematik e.V., Bonn *05436*
Deutsche Gesellschaft für Völkerkunde e.V., Freiburg *05437*
Deutsche Gesellschaft für Völkerrecht, Bonn *05438*
Deutsche Gesellschaft für Volkskunde e.V., Göttingen *05439*
Deutsche Gesellschaft für Wehrtechnik e.V., Bonn *05440*
Deutsche Gesellschaft für Wirtschaftliche Fertigung und Sicherheitstechnik e.V., Kaarst *05441*
Deutsche Gesellschaft für Wohnmedizin und Bauhygiene e.V., Spöck-Stutensee *05442*
Deutsche Gesellschaft für Zahnärztliche Prothetik und Werkstoffkunde e.V., Hannover *05443*
Deutsche Gesellschaft für Zahnerhaltung, Berlin *05444*
Deutsche Gesellschaft für Zahn-, Mund- und Kieferheilkunde, Düsseldorf *05445*
Deutsche Gesellschaft für Zerstörungsfreie Prüfung e.V., Berlin *05446*
Deutsche Gesellschaft für Züchtungskunde e.V., Bonn *05447*
Deutsche Gesellschaft zur Förderung der Gehörlosen und Schwerhörigen e.V., München *05448*
Deutsche Glastechnische Gesellschaft e.V., Frankfurt *05449*
Deutsche Graphologische Vereinigung e.V., Schwerte *05450*
Deutsche Gruppenpsychotherapeutische Gesellschaft e.V., Berlin *05451*
Deutsche Hämophilieberatung, Marl-Hüls *05452*
Deutsche Hämophiliegesellschaft zur Bekämpfung von Blutungskrankheiten e.V., Hamburg *05453*
Deutsche Hauptstelle gegen die Suchtgefahren e.V., Hamm *05454*
Deutsche Ileostomie-Kolostomie-Urostomie-Vereinigung e.V., Freising *05455*
Deutsche Jazz-Föderation e.V., Frankfurt *05456*
Deutsche Kakteen-Gesellschaft e.V., Ovelgönne *05457*
Deutsche Kautschuk-Gesellschaft e.V., Frankfurt *05458*
Deutsche Keramische Gesellschaft e.V., Köln *05459*
Deutsche Kommission für Ingenieurausbildung, Düsseldorf *05460*
Deutsche Krebsgesellschaft e.V., Frankfurt *05461*
Deutsche Landjugend-Akademie Fredeburg, Schmallenberg *05462*
Deutsche Landwirtschafts-Gesellschaft e.V., Frankfurt *05463*
Deutsche Lichttechnische Gesellschaft e.V., Berlin *05464*
Deutsche Malakozoologische Gesellschaft, Frankfurt *05465*
Deutsche Mathematiker-Vereinigung e.V., Freiburg *05466*
Deutsche Medizinische Arbeitsgemeinschaft für Herd- und Regulationsforschung e.V., Düren *05467*
Deutsche MERU Gesellschaft, Bissendorf *05468*
Deutsche Meteorologische Gesellschaft e.V., Traben-Trarbach *05469*
Deutsche Meteorologische Gesellschaft e.V., Zweigverein Hamburg, Hamburg *05470*
Deutsche Mineralogische Gesellschaft e.V., Münster *05471*
Deutsche Montessori Gesellschaft, Würzburg *05472*
Deutsche Morgenländische Gesellschaft e.V., Heidelberg *05473*
Deutsche Mozart-Gesellschaft e.V., Augsburg *05474*
Deutsche Multiple Sklerose Gesellschaft Bundesverband e.V., Hannover *05475*
Deutsche Neurovegetative Gesellschaft, Düsseldorf *05476*
Deutsche Numismatische Gesellschaft, Speyer *05477*
Deutsche Ophthalmologische Gesellschaft Heidelberg, Heidelberg *05478*
Deutsche Orchideen-Gesellschaft, Sottrum *05479*
Deutsche Orient-Gesellschaft e.V., Berlin *05480*
Deutsche Ornithologen-Gesellschaft e.V., Radolfzell *05481*
Deutsche Parlamentarische Gesellschaft e.V., Bonn *05482*
Deutsche Paul-Tillich-Gesellschaft e.V., Göttingen *05483*
Deutsche Pharmakologische Gesellschaft e.V., Wuppertal *05484*
Deutsche Pharmazeutische Gesellschaft e.V., Eschborn *05485*
Deutsche Phono-Akademie e.V., Hamburg *05486*
Deutsche Physikalische Gesellschaft e.V., Bad Honnef *05487*
Deutsche Physiologische Gesellschaft e.V., Heidelberg *05488*
Deutsche Phytomedizinische Gesellschaft e.V., Mainz *05489*
Deutsche Planungsgesellschaft EG Bonn, Bonn *05490*
Deutsche Psychoanalytische Gesellschaft e.V., München *05491*
Deutsche Psychoanalytische Vereinigung e.V., Berlin *05492*
Deutsche Quartärvereinigung, Hannover *05493*
Deutscher Ärztinnenbund e.V., Köln *05494*
Deutscher Akademikerinnenbund e.V., Nürnberg *05495*
Deutscher Akademischer Austauschdienst, Bonn *05496*
Deutscher Altphilologen-Verband, Puchheim *05497*
Deutscher Arbeitsgerichtsverband e.V., Köln *05498*
Deutscher Arbeitsring für Lärmbekämpfung e.V., Düsseldorf *05499*
Deutscher Ausschuss für Stahlbau, Köln *05500*
Deutscher Autoren-Verband e.V., Hannover *05501*
Deutscher Bäderverband e.V., Bonn *05502*
Deutscher Berufsverband der Hals-Nasen-Ohrenärzte e.V., Neumünster *05503*
Deutscher Berufsverband der Sozialarbeiter und Sozialpädagogen e.V., Essen *05504*
Deutscher Beton-Verein e.V., Wiesbaden *05505*
Deutscher Bibliotheksverband e.V., Berlin *05506*
Deutscher Dampfkesselausschuss, Essen *05507*
Deutscher Diabetiker-Verband e.V., Kaiserslautern *05508*
Deutsche Religionsgeschichtliche Studiengesellschaft, Saarbrücken *05509*
Deutscher Erfinderring e.V., Nürnberg *05510*
Deutscher Esperanto-Bund e.V., Bonn *05785*
Deutscher Forstverein e.V., Frankfurt *05511*
Deutscher Forstwirtschaftsrat e.V., Rheinbach *05512*
Deutscher Germanistenverband, Aachen *05513*
Deutsche Rheologische Gesellschaft e.V., Berlin *05514*
Deutsche Rheuma-Liga Bundesverband e.V., Bonn *05515*
Deutscher Hochschulverband, Bonn *05516*
Deutscher Holzwirtschaftsrat, Wiesbaden *05517*
Deutsche Richterakademie, Trier *05518*
Deutscher Juristen-Fakultätentag, Würzburg *05519*
Deutscher Juristentag e.V., Bonn *05520*
Deutscher Kälte- und Klimatechnischer Verein e.V., Stuttgart *05521*
Deutscher Kassenarztverband e.V., Gross-Gerau *05522*
Deutscher Kommunikationsverband BDW e.V., Bonn *05523*
Deutscher Komponisten-Verband e.V., Berlin *05524*
Deutscher Künstlerbund e.V., Berlin *05525*
Deutscher Lehrerverband Niedersachsen, Hannover *05526*
Deutscher Markscheider-Verein e.V., Herne *05527*
Deutscher Medizinischer Informationsdienst e.V., Frankfurt *05528*
Deutscher Museumsbund e.V., Karlsruhe *05529*
Deutscher Musikrat e.V., Nationalkomitee der Bundesrepublik Deutschland im Internationalen Musikrat, Bonn *05530*
Deutscher Naturheilbund e.V., Crailsheim *05531*
Deutscher Naturkundeverein e.V., Stuttgart *05532*
Deutscher Naturschutzring e.V., Bundesverband für Umweltschutz, Bonn *05533*
Deutscher Nautischer Verein von 1868 e.V., Hamburg *05534*
Deutsche Röntgengesellschaft, Neu-Isenburg *05535*
Deutscher Philologen-Verband e.V., Unterhaching *05536*
Deutscher Politologen-Verband e.V., Bonn *05537*
Deutscher Rat für Landespflege, Bonn *05538*
Deutscher Rechtshistorikertag, Graz *00580*
Deutscher Sportärztebund, Heidelberg *05539*
Deutscher Sportlehrerverband e.V., Wetzlar *05540*
Deutscher Stahlbau-Verband, Köln *05541*
Deutscher Stenografielehrerverband e.V., Hamburg *05542*
Deutscher Tonkünstlerverband e.V., München *05543*
Deutscher Verband Evangelischer Büchereien e.V., Göttingen *05544*
Deutscher Verband Farbe, Berlin *05545*
Deutscher Verband Forstlicher Forschungsanstalten, Trippstadt *05546*
Deutscher Verband für Angewandte Geographie e.V., Hamburg *05547*
Deutscher Verband für das Skilehrwesen e.V., Oberstdorf *05548*
Deutscher Verband für Materialforschung und -prüfung e.V., Berlin *05549*

Deutscher Verband für Physiotherapie, Köln 05550
Deutscher Verband für Schweisstechnik e.V., Düsseldorf 05551
Deutscher Verband für Wasserwirtschaft und Kulturbau e.V., Bonn 05552
Deutscher Verband für Wohnungswesen, Städtebau und Raumplanung e.V., Bonn 05553
Deutscher Verband Technischer Assistenten in der Medizin e.V., Hamburg 05554
Deutscher Verband technisch-wissenschaftlicher Vereine, Düsseldorf 05555
Deutscher Verein des Gas- und Wasserfaches e.V., Eschborn 05556
Deutscher Verein für Internationales Seerecht e.V., Hamburg 05557
Deutscher Verein für Kunstwissenschaft e.V., Berlin 05558
Deutscher Verein für Vermessungswesen e.V., Heidelberg 05559
Deutscher Verein für Versicherungswissenschaft e.V., Berlin 05560
Deutscher Verkehrssicherheitsrat e.V., Bonn 05561
Deutscher Volkshochschul-Verband e.V., Bonn 05562
Deutscher Werkbund e.V., Frankfurt 05563
Deutscher Wissenschaftler Verband, Goslar 05564
Deutscher Zentralausschuss für Chemie, Frankfurt 05565
Deutscher Zentralverein Homöopathischer Ärzte e.V., Bonn 05566
Deutsches Anwaltsinstitut e.V., Bochum 05567
Deutsches Archäologisches Institut, Berlin 05568
Deutsches Atomforum e.V., Bonn 05569
Deutsches Bibliotheksinstitut, Berlin 05570
Deutsche Schillergesellschaft e.V., Marbach 05571
Deutsche Sekretärinnen-Akademie, Düsseldorf 05572
Deutsche Sektion der International Association of Consulting Actuaries, Grünwald 06401
Deutsche Sektion der Internationalen Liga gegen Epilepsie, Bielefeld 05573
Deutsche Sektion des Internationalen Instituts für Verwaltungswissenschaften, Bonn 05574
Deutsches Elektronen-Synchrotron, Hamburg 05575
Deutsches Forum für Entwicklungspolitik, Bonn 05576
Deutsche Shakespeare-Gesellschaft, Weimar 05577
Deutsche Shakespeare-Gesellschaft West e.V., Bochum 05578
Deutsches Handwerksinstitut e.V., München 05579
Deutsches High-Fidelity Institut e.V., Frankfurt 05580
Deutsches Institut für Ärztliche Mission, Tübingen 05581
Deutsches Institut für angewandte Kommunikation und Projektförderung e.V., Bonn 05582
Deutsches Institut für Betriebswirtschaft e.V., Frankfurt 05583
Deutsches Institut für Filmkunde e.V., Frankfurt 05584
Deutsches Institut für Internationale Pädagogische Forschung, Frankfurt 05585
Deutsches Institut für medizinische Dokumentation und Information, Köln 05586
Deutsches Institut für Urbanistik, Berlin 05587
Deutsches Institut für Vormundschaftswesen, Heidelberg 05588
Deutsches Institut für Wirtschaftsforschung, Berlin 05589
Deutsches Jugendinstitut e.V., München 05590
Deutsches Komitee Instandhaltung e.V., Düsseldorf 05591
Deutsches Krebsforschungszentrum, Heidelberg 05592
Deutsches Kunststoff-Institut, Darmstadt 05593
Deutsches Kupfer-Institut e.V., Berlin 05594
Deutsches Nationales Komitee des Weltenergierats, Düsseldorf 05595
Deutsches Nationalkomitee für Denkmalschutz, Bonn 05596
Deutsches Optisches Komitee, München 05597
Deutsches Orient-Institut, Hamburg 05598
Deutsche Statistische Gesellschaft, Konstanz 05599
Deutsches Textilforschungszentrum Nord-West e.V., Krefeld 05600
Deutsche Stiftung für internationale Entwicklung, Bonn 05601
Deutsche Straßenliga e.V., Bonn 05602
Deutsche Studiengesellschaft für Publizistik, Stuttgart 05603
Deutsches Übersee-Institut, Hamburg 05604
Deutsches wissenschaftliches Steuerinstitut der Steuerberater und Steuerbevollmächtigten e.V., Bonn 05605
Deutsches Wollforschungsinstitut, Aachen 05606
Deutsches Zentralinstitut für soziale Fragen, Berlin 05607
Deutsches Zentralkomitee zur Bekämpfung der Tuberkulose, Mainz 05608
Deutsches Zentrum für Altersfragen e.V., Berlin 05609
Deutsche Tierärzteschaft e.V., Bonn 05610
Deutsche Tropenmedizinische Gesellschaft e.V., Frankfurt 05611
Deutsche Vereinigung für die Rehabilitation Behinderter e.V., Heidelberg 05612
Deutsche Vereinigung für Finanzanalyse und Anlageberatung e.V., Dreieich 05613
Deutsche Vereinigung für gewerblichen Rechtsschutz und Urheberrecht e.V., Köln 05614
Deutsche Vereinigung für Internationales Steuerrecht, Köln 05615
Deutsche Vereinigung für internationales Steuerrecht im Verband der Fiscal Association, Bayerische Sektion e.V., München 05616
Deutsche Vereinigung für Parlamentsfragen e.V., Bonn 05617
Deutsche Vereinigung für Politische Wissenschaft, Darmstadt 05618
Deutsche Vereinigung für Religionsgeschichte, Hannover 05619
Deutsche Vereinigung für Sportwissenschaft, Hamburg 05620
Deutsche Vereinigung für Verbrennungsforschung e.V., Essen 05621
Deutsche Vereinigung zur Bekämpfung der Viruskrankheiten e.V., München 05622
Deutsche Vereinigung zur Förderung der Weiterbildung von Führungskräften, Köln 06455
Deutsche Verkehrswissenschaftliche Gesellschaft e.V., Bergisch Gladbach 05623
Deutsche Veterinärmedizinische Gesellschaft e.V., Giessen 05624
Deutsche Werbewissenschaftliche Gesellschaft e.V., Bonn 05625
Deutsche Wissenschaftliche Kommission für Meeresforschung, Hamburg 05626
Deutsche Zeitungswissenschaftliche Vereinigung e.V., München 05627
Deutsche Zentrale für Volksgesundheitspflege e.V., Frankfurt 05628
Deutsche Zoologische Gesellschaft e.V., Bonn 05629
Deutsch-Pazifische Gesellschaft e.V., München 05630
Deutschschweizerischer Sprachverein, Bremgarten 10660
Deutschschweizerisches PEN-Zentrum, Bern 10661
Deutschsprachige Arbeitsgemeinschaft für Handchirurgie, Hamburg 05631
Development Innovations and Networks, Genève 10662
Development Study Center, Rehovot 07052
Devon and Cornwall Record Society, Exeter 11888
Devon Archaeological Society, Exeter 11889
The Devonshire Association for the Advancement of Science, Literature and Art, Exeter 11890
DF, Evanston 14962
D.F., København 03606
DFA, New York 14953
DFC, Chicago 14969
DFE, København 03619
DFF, Klampenborg 03605
DFG, Bonn 05305
DFRA, Sheffield 11907
DFRC, Des Moines 14977
DFV, Frankfurt 05511
DfwG, Berlin 05303
DFWR, Rheinbach 05512
DGaaE, Dossenheim 05317
DGAI, Nürnberg 05319
DGaO, Jena 05321
DGAP, Berlin 05320
DGAP, Bonn 05326
DGBK, Holzminden 05328
DGC, Stuttgart 05337
DGD, Frankfurt 05341
DGDP, München 05342
DGE, Berlin 05343
DGE, Frankfurt 05346
DGEG, Essen 05345
DGF, København 03624, 03625
DGF, Lyngby 03626
DGfA, Nürnberg 05318
DGfdB, Essen 05338
DGfE, Berlin 05347
DGfH, München 05363
DGfH, Wolnzach 05364
DGfHK, Beckum 05360
D.G.f.K., Berlin 05369
DGFL, Dortmund 05376
DGfPs, Münster 05411
DGfS, Hamburg 05419
DGfZ, Bonn 05447
DGG, Berlin 05451
DGG, Frankfurt 05449
D.G.G., Hannover 05311
DGG, Lübeck 05354
DGG, Münster 05312
DGGL, Karlsruhe 05353
DGGV, Leverkusen 05356
DGHM, Heidelberg 05366
DGHT, Rheinbach 05361
DGK, München 05310
DGLR, Bonn 05377
DGLRM, Ulm 05378
DGM, Oberursel 05381
DGMKG, Bonn 05386
DGMW, Heidelberg 05384
DGO, Berlin 05394
DGO, Düsseldorf 05352
DGON, Düsseldorf 05393
DGOT, Frankfurt 05392
DGP, Greifenstein 05407
DGP, Hamburg 05396
DGP, Marburg 05395
DGPB, Köln 05408
DGPh, Köln 05403
DGPM, München 05412
DGPN, Köln 05410
DGQ, Frankfurt 05414
DGQ, Freising 05415
DGS, Berlin 05425
DGS, Charlottenlund 03709
DGS, Holte 03628
DGS, Hvidovre 03623
DGS, München 05420
DGSP, Köln 05421
DGSP, München 05423
DGSS, Münster 05428
DGV, Freiburg 05437
DGV, Göttingen 05439
DGV, Schwerte 05450
DGVT, Tübingen 05435
DGW, Kaarst 05441
DGZfP, Berlin 05446
DGZPW, Hannover 05443
DH, Klampenborg 03631
DHDS, Southampton 11899
DHFI, Frankfurt 05580
DHI, Athens 14961
DHI, München 05579
D.H. Lawrence Society of North America, San Marcos 14970
DHS, Hamm 05454
DHS, Hellerup 03629
DIA, London 11883
DIA, Maple Glen 14987
Dialogue et Coopération, Paris 04464
DIB, Frankfurt 05583
Dickens Fellowship, Boulogne-sur-Mer 04465
Dickens Fellowship, London 11891
Dickens Society, Worcester 14971
Dictionary Society of North America, Cleveland 14972
Dietary Managers Association, Lombard 14973
Dietitians Association of Australia, Canberra 00385
DIF, Frankfurt 05584
Diffuse Obstructive Pulmonary Syndrome Research Group, Tygerberg 10135
DIGA, Asheboro 14992
DIGAP, Bury 11893
Digital Equipment Computer Users Society, Genève 10663
DIMDI, Köln 05586
D.I. Mendeleev Chemical Society, Moskva 09864
DIMO, København 03607
DIN Deutsches Institut für Normung e.V., Berlin 05632
DIPF, Frankfurt 05585
Diplomatic Academy of Vienna, Wien 00581
Diplomatic and Commonwealth Writers Association of Britain, London 11892
Diplomatische Akademie Wien, Wien 00581
DIPRA, Birmingham 14988
Direcção Provincial dos Servicos de Geologia e Minas de Angola, Luanda 00033
Dirección General de Estadística y Censos, San Salvador 03928
Dirección General de Geología y Minas, Quito 03852
Dirección General de Investigaciones Agronómicas, Santa Tecla 03929
Direct Investigation Group on Aerial Phenomena, Bury 11893
Direct Marketing Educational Foundation, New York 14974
DIS, North Olmsted 14989
Disability Insurance Training Council, Washington 14975
Disability Rights Center, Washington 14976
Disabled Living Foundation, London 11894
Disinfected Mail Study Circle, London 11895
Distillers Feed Research Council, Des Moines 14977
Distributive Education Clubs of America, Reston 14978
Distributive Trades Education and Training Council, London 11896
District Planning Officers Society, Crowborough 11897
DITC, Washington 14975
Division de Chimie Physique de la Société Française de Chimie, Paris 04466
Division for Early Childhood, Reston 14979
Division of Applied Experimental and Engineering Psychologists, Savoy 14980
Division of Energy Technology, Pretoria 10136
Division on Mental Retardation of the Council for Exceptional Children, Athens 14981
DIW, Berlin 05589
Django Reinhardt Society, Middletown 14982
DJI, München 05590
DKE, Frankfurt 05301
DKFZ, Heidelberg 05592
DKG, Frankfurt 05458
DKG, Köln 05459
DKIN, Düsseldorf 05591
D.K.L., København 03609
DKM, Fur 03641
D.K.N.V.S., Trondheim 09258
DKV, Stuttgart 05521
DLA, Schmallenberg 05462
DLCOEA, Nairobi 08489
DLF, København 03581
DLF, London 11894
DLG, Frankfurt 05463
DL/SWA, Roma 07622
DMA, Lombard 14973
DMB, Karlsruhe 05529
DMEF, New York 14974
DMEF, San Antonio 14954
DMF, København 03645
DMG, Augsburg 05474
DMG, Frankfurt 05465
DMG, Hamburg 05470
DMG, Münster 05471
DMG, Traben-Trarbach 05469
DMI, Frankfurt 05528
DMpF, København 03652
DMR, Bonn 05530
DMRF, Beverly Hills 14993
DMS, København 03647
DMS, Lyngby 03650
DMSC, London 11895
DMSG, Hannover 05475
DMV, Freiburg 05466
DMV, Herne 05527
DN, København 03583
DNK, Düsseldorf 05595
DNLF, Lysaker 09253
DNM, Oslo 09254
DNO, Odense 03611
DNR, Bonn 05533
DNS, Hellerup 03655
DNV, Oslo 09256
Doboku-Gakkai, Tokyo 08122
Documentation and Research Centre, Roma 07647
DOF, København 03660
D. of A., Norwood 14959
DOG, Berlin 05480
DOG, Heidelberg 05478
Dokumentationsarchiv des österreichischen Widerstandes, Wien 00582
Dokumentationsring Elektrotechnik, Erlangen 05633
Dokumentationsring Pädagogik, Frankfurt 05634
Dokumentationsstelle für neuere österreichische Literatur, Wien 00583
Dolmetsch Foundation, Chichester 11898
Dolmetsch Historical Dance Society, Southampton 11899
The Donizetti Society, London 11900
Donner Canadian Foundation, Toronto 02418
DOPAED, Frankfurt 05634
DOPAED Coordination office, Frankfurt 05634
Dorothy L. Sayers Society, Hurstpierpoint 11901
DORS, Hornbaek 03683
Dorset Natural History and Archaeological Society, Dorchester 11902
Dorset Record Society, Dorchester 11903
DOS, Hellerup 03662
DOS, København 03661
Doshitsu Kogakkai, Tokyo 08123
Dozenal Society of America, Garden City 14983
The Dozenal Society of Great Britain, Moulsford 11904
dp, Bonn 05490, 05537
DP, Valby 03667
DPA, Plymouth 11878
DPC, Valley Center 14966
DPD, Marseille 04286
DPG, Bad Honnef 05487
DPG, Mainz 05489
DPG, München 05491, 05630
DPhG, Eschborn 05485
DPhV, Unterhaching 05536
DPOS, Crowborough 11897
DPS, Virum 03663
DPV, Berlin 05492
Drama Association of Wales, Cardiff 11905
Drama Board Association, Chilehurst 11906
Dramatiker-Union e.V., Berlin 05635
Dramatists' League of Finland, Helsinki 04054
Drama Tree, New York 14984
Drawing Society, New York 14985
DRC, Washington 14976
DRF, New York 14956
DRG, Berlin 05514
DRG, Neu-Isenburg 05535
DRINC, Oak Grove Village 14948
DRIVE, Bruxelles 01259
DRL, Bonn 05538
DRO, Sønderborg 03613
Drop Forging Research Association, Sheffield 11907
DRS, Dorchester 11903
DRS, Herlev 03668
DRS, Middletown 14982
Drug and Alcohol Nursing Association, Baltimore 14986
Drug Information Association, Maple Glen 14987
Društvo Arhitekata Srbije, Beograd 17081
Društvo Arhivskih Radnika Srbije, Beograd 17082
Društvo Bibliotekara Vojvodine, Novi Sad 17083
Društvo Ekonomista Hrvatske, Zagreb 03456
Društvo Istoričara Srbije, Beograd 17084
Društvo Lekara Vojvodine, Novi Sad 17085
Društvo Ljekara Crne Gore, Podgorica 17086
Društvo Matematičara Srbije, Beograd 17087
Društvo Matematikov, Fizikov i Astronomov Slovenije, Ljubljana 10098
Društvo na Istoričarite na Umetnosta od Makedonija, Skopje 08593
Društvo na Kompozitorite na Makedonija, Skopje 08594
Društvo na Likovnite Umetnici na Makedonija, Skopje 08595
Društvo na Literaturnite Preveduvači na Makedonija, Skopje 08596
Društvo na Muzejskite Rabotnici na Makedonija, Skopje 08597
Društvo na Pisatelite na Makedonija, Skopje 08598
Društvo Psihologa Srbije, Beograd 17088
Društvo Slovenskih Skladateljev, Ljubljana 10099
Društvo Veterinara Srbije, Beograd 17089
Društvo za Filozofija, Sociologija i Politikologija na Makedonija, Skopje 08599
Društvo za Medicinsku i Biološku Tehniku, Ljubljana 10100
Društvo za Nauka i Umetnost, Bitola 08600
Drustvo za Proučavanje i Unapredenje Pomorstva, Rijeka 03457
Društvo za Srpski Jezik i Književnost, Beograd 17090
Društvo za Srpskohrvatski Jezik i Književnost, Beograd 17091
Dryland Professional Network, London 11908
Dry Stone Walling Association, Oswestry 11909
DS, Worcester 14971
DSA, Cambridge 14955
DSA, Garden City 14983
DSA, Seattle 14950
DSÄB, Heidelberg 05539
DSAM, Greve 03672
DSBy, Lyngby 03673
DSC, Rehovot 07052
DSF, København 03587
D.S.F., København 03669
DSF, København 03689, 03691
DSFL, Ålborg 03675
DSG, Marbach 05571
DSGB, Moulsford 11904
DSI, Atlanta 14957
DSI, Chambersburg 14949
DSIM, Gentofte 03676
DSL, København 03708
DSLV, Wetzlar 05540
DSNA, Cleveland 14972
DSO, Vedbaek 03684
DSOM, København 03681
DSOV, København 03685
D.S.P.C., Reggio Calabria 07634
DSS, Ljubljana 10099
DSSV, Bremgarten 10660
DStG, Konstanz 05599
DSTS, København 03688
DSTV, Köln 05541
DT, Bonn 05610
DT, New York 14984
DTC, Palm Desert 14967
DTF, København 03695
DTG, Frankfurt 05611
DTKV, München 05543
DTL, København 03696
DTL, Lyngby 03716
DTL – Dansk Forening for Information og Dokumentation, Lyngby 03716
DU, Berlin 05635

Dublin University Biological Association, Dublin *06942*
Ductile Iron Pipe Research Association, Birmingham *14988*
Ductile Iron Society, North Olmsted *14989*
The Dugdale Society, Stratford-upon-Avon *11910*
The Duke Ellington Society, New York *14990*
Dumfriesshire and Galloway Natural History and Antiquarian Society, Dumfries *11911*
Dunedin Society, Glasgow *11912*
Durham County Local History Society, Durham *11913*
Dutch Association for Animal Production, Wassenaar *09046*
Dutch Association for Medical Education, Utrecht *09026*
Dutch Centre for Public Libraries and Literature, 's-Gravenhage *08980*
Dutch Dental Organization, Nieuwegein *08983*
Dutch Society of Educational Psychologists, Utrecht *09009*
Dutch Society of Sciences, Haarlem *08910*
DVAG, Hamburg *05547*
DVF, Berlin *05545*
DVFA, Dreieich *05613*
DVFFA, Trippstadt *05546*
DVfVW, Berlin *05560*
DVG, Giessen *05624*
DVGW, Eschborn *05556*
DVIS, Hamburg *05557*
DVM, Berlin *05549*
DVO, Slagense *03614*
DVPW, Darmstadt *05618*
DVR, Bonn *05561*
DVRG, Hannover *05619*
DVS, Düsseldorf *05551*
DVS, Oberstdorf *05548*
DVT, Düsseldorf *05555*
DVV, Bonn *05562*
DVV, München *05622*
DVW, Heidelberg *05559*
DVWG, Bergisch Gladbach *05623*
DVWK, Bonn *05552*
DWG, Bonn *05625*
DWI, Aachen *05606*
DWT, Bonn *05440*
Dyleague, Woodland Hills *14991*
Dynamics International Gardening Association, Asheboro *14992*
Dyslexia Institute, Staines *11914*
Dystonia Medical Research Foundation, Beverly Hills *14993*
Dystonia Medical Research Foundation, Vancouver *02419*
Dystrophic Epidermolysis Bullosa Research Association of America, New York *14994*
DZfCh, Frankfurt *05565*
DZG, Bonn *05629*
DZK, Mainz *05608*
DZV, Frankfurt *05628*

EA, Milwaukee *15042*
EAA, Belmont *15032*
EAA, Bruxelles *01268*
EAA, Düsseldorf *05659*
EAA, Edinburgh *11931*
EAA, Oshkosh *15070*
EAAAF, Oshkosh *14995*
EAA Aviation Foundation, Oshkosh *14995*
EAACI, Roma *07662*
EAAE, Bruxelles *01286*
EAAFRO, Arusha *11051*
EAAM, Dunstable *11986*
EAAP-FAO Global Data Bank for Animal Genetic Resources, Hannover *05636*
EAAS, Lyon *04480*
EAATEE, Njombe *11049*
EAB, Berlin *05647*
EAB, London *11973*
E.A. Baker Foundation for Prevention of Blindness, Toronto *02420*
EABS, Bruxelles *01274*
EABS, Milano *07663*
EABT, Harrow *11987*
EAC, Köln *05670*
EACE, Amsterdam *08852*
EACE, Groningen *08851*
EACHR, København *03721*
EACL, Aachen *05661*
EACR, Nottingham *11988*
EACRO, Plaisir *04484*
EACROTANAL, Zanzibar *11054*
EACS, Paris *04481*
EACTA, Papworth Everard *11994*
EADA, Brisbane *00386*
EADE, Leiden *08864*
EADTU, Heerlen *08865*
EAEC, London *11985*
EAEC, Nairobi *08494*
EAEE, Zagreb *03458*
E.A.E.G., Zeist *08867*
EAEN, Nairobi *08500*
EAES, Paris *04491*
EAESP, Amsterdam *08866*
EAEVE, Maisons-Alfort *04485*
EAFA, Frankfurt *05660*
EAFE, Southsea *11995*
EAFP, Bruxelles *01271*
EAGE, Amsterdam *08853*
EAGLE, 's-Gravenhage *08854*
EAHC, Stratford Saint Mary *11976*
EAHIL, Bruxelles *01276*
EAHP, Aarhus *03722*
EAIE, Amsterdam *08856*
EAJS, Leiden *08857*
EALA, London *11984*
EALA, Nairobi *08496*
EALE, Stockholm *10451*
EALJS, Cairo *03889*
EAMDA, Dublin *06946*
EAMF, Genève *10679*
EAMRC, Arusha *11050*
EANHS, Nairobi *08493*
E.A.N.H.S., Nairobi *08497*
EANM, London *11998*
EANS, Berlin *05669*
EAO, Nonnweiler *05648*
EAPA, Breukelen *08850*
EAPCA, Saint-Etienne *04492*
EAPE, Newcastle-upon-Tyne *11990*
EAPG, Zeist *08869*
EAPI, Manila *09517*
EAPM, London *11991*
EAPR, Wageningen *08860*
EAPS, 's-Gravenhage *08859*
EAPSB, Baltimore *15013*
EAR, Leuven *01296*
EARDHE, Berlin *05662*
EARIE, Bruxelles *01279*
EARLI, Turku *03956*
Earl Warren Legal Training Program, New York *14996*
Early Childhood Education Society of the Republic of China, Taipei *03295*
Early English Text Society, Oxford *11915*
Early Settlers Association of the Western Reserve, North Olmsted *14997*
Early Sites Research Society, Rowley *14998*
EARN, Orsay *04476*
EAROPH, New Delhi *06756*
EARSeL, Napoli *07665*
Earthmind, Mariposa *14999*
Earthnet Programme Office, Frascati *07648*
Earthquake Engineering Research Institute, Oakland *15000*
Earthrise, Tulsa *15001*
Earthwatch, Nairobi *08491*
EAS, Colchester *11977*
EAS, London *12002*
EASD, Düsseldorf *05666, 05667*
EASD, Ile-Saint-Denis *04483*
EASE, London *12000*
EASE, Stuttgart *05663*
EASL, Amsterdam *08863*
EASL, Kampala *11185*
EASSP, Bruxelles *01284*
East Africa Association for Theological Education by Extension, Njombe *11049*
East Africa Medical Research Council, Arusha *11050*
East African Academy, Nairobi *08492*
East African Agricultural Economics Society, Kampala *11184*
East African Agriculture and Forestry Research Organization, Arusha *11051*
East Africa Natural History Society, Nairobi *08493*
East African Engineering Consultants, Nairobi *08494*
East African Industrial Research Organization, Nairobi *08495*
East African Library Association, Nairobi *08496*
East African Literature Bureau, Dar es Salaam *11052*
East African Natural History Society Nairobi, Nairobi *08497*
East African School of Librarianship, Kampala *11185*
East African Wild Life Society, Nairobi *08498*
East and Southeast Asia Federation of Soil Science Societies, Tokyo *08124*
East Asian Bird Protection Society, Tokyo *08125*
East Asian Pastoral Institute, Manila *09517*
East Asian Seas Action Plan, Nairobi *08499*
EASTC, Dar es Salaam *11055*
East, Central and Southern African College of Nursing, Arusha *11053*
Eastern Africa Environment Network, Nairobi *08500*
Eastern African Centre for Research on Oral Traditions and African National Languages, Zanzibar *11054*
Eastern Africa Statistical Training Centre, Dar es Salaam *11055*
Eastern and Southern African Management Institute, Arusha *11056*
Eastern and Southern African Mineral Resources Development Centre, Dodoma *11057*
Eastern and Southern African Regional Branch of the International Council on Archives, Nairobi *08501*
Eastern and Southern African Universities Research Programme, Dar es Salaam *11058*
Eastern Arts Association, Cambridge *11916*
Eastern Bird Banding Association, Hopewell *15002*
Eastern Caribbean Institute of Agriculture and Forestry, Arima *11120*
Eastern Counties Lung Association, Cornwall *02421*
Eastern Dredging Association, Brisbane *00386*
Eastern Finance Association, Statesboro *15003*
Eastern Mediterranean Hand Society, Cairo *03886*
Eastern Newfoundland Engineering Society, St. John's *02422*
Eastern Québec Teachers Association, Sainte Foy *02423*
Eastern Regional Organization for Planning and Housing, New Delhi *06756*
Eastern Regional Organization for Public Administration, Manila *09518*
Eastern Townships Association of Teachers, Cookshire *02424*
East Hertfordshire Archaeological Society, Ware *11917*
East London History Society, London *11918*
East Lothian Antiquarian and Field Naturalists' Society, Dunbar *11919*
East Midlands Arts Board, Loughborough *11920*
East of England Agricultural Society, Peterborough *11921*
East Riding Archaeological Society, Hull *11922*
East-West Center, Honolulu *15004*
East-West Center Institute of Economic Development and Policy, Honolulu *15005*
East-West Sign Language Association, Tokyo *08126*
EASVO, Rotterdam *08870*
EAT, Berkeley Heights *15071*
EATCS, Paderborn *05665*
EATLT, Bruxelles *01285*
EATP, Bruxelles *01281*
EATWOT, Port Harcourt *09225*
EAVPT, Utrecht *08872*
EAVSoM, Firenze *07664*
E.A.W.L.S., Nairobi *08498*
EBA, Leuven *01304*
EBA, Luxembourg *08584*
EBA, Milan *15014*
EBAD, Dakar *10027*
EBAE, Amersfoort *08873*
EBAG, Bonn *05649*
EBBA, Hopewell *15002*
EBBS, Oslo *09265*
EBCD, Bruxelles *01302*
EBEN, Bruxelles *01305*
EBIS, Bruxelles *01301*
EBM, Dortmund *05668*
EBMT, Wien *00587*
EBRI, Washington *15039*
E.B.S., Edinburgh *11932*
EBS, Gualala *15054*
EBSA, Cambridge *11979*
EBZ, Idstein *05650*
EC, Berkeley *15007*
EC, New York *15072*
ECA, London *11941*
ECACC, Salisbury *12009*
ECAE, Atlanta *15017*
ECAS, Strasbourg *04499*
ECAST, Paris *04494*
ECBA, Namur *01326*
ECBA, Roma *07674*
ECBO, Milano *07668*
ECC, Genève *10683*
ECCA, Bruxelles *01317*
ECCAI, Povo *07676*
ECCE, London *12012*
Ecclesiastical History Society, Glasgow *11923*
Ecclesiological Society, London *11924*
ECCLS, Coventry *12020*
ECCO, Graz *00589*
ECCOMAS, Bruxelles *01325*
ECCS, Bruxelles *01332*
ECDPM, Maastricht *08876*
ECDS, Belfast *12018*
ECES, Torino *07672*
ECETOC, Bruxelles *01315*
ECFA, Aachen *05675*
ECFMG, Philadelphia *15018*
ECG, Nijmegen *08878*
ECGS, Bruxelles *01310*
ECIAF, Arima *11120*
ECIM, Bruxelles *01335*
ECIS, Bruxelles *01321*
ECIS, Petersfield *12022*
ECISS, Bruxelles *01322*
ECL, København *03563*
ECLA, 's-Gravenhage *08882*
ECLE, Praha *03543*
ECLG, Louvain-la-Neuve *01330*
ECLSO, London *12019*
ECMI, Lahti *03957*
ECMMR, Brixham *12010*
ECMRA, London *12008*
ECNAIS, Slagelse *03724*
ECNP, Hillerød *03723*
ECOD, Strasbourg *04498*
ECODU, Diepenbeek *01331*
ECOG, Rotterdam *08879*
Ecole de Bibliothécaires, Archivistes et Documentalistes, Dakar *10027*
Ecole Internationale de Bordeaux, Talence *04467*
Ecole Internationale d'Informatique de l'AFCET, Paris *04468*
Ecological and Toxicological Association of the Dyestuffs Manufacturing Industry, Basel *10664*
Ecological Institute, Xalapa *08735*
Ecological Physics Research Group, Bedford *11925*
Ecological Society of America, Tempe *15006*
Ecological Society of Australia, Canberra *00387*
Ecological Society of China, Beijing *03197*
The Ecological Society of Japan, Sendai *08365*
Ecological Society of Nigeria, Lagos *09224*
Ecology Center, Berkeley *15007*
EcoNet, San Francisco *15008*
Econometric Society, Evanston *15009*
Economic and Social Committee of the European Communities, Bruxelles *01225*
Economic and Social Science Research Association, London *11926*
Economic Commission for Latin America and the Caribbean, Santiago *03038*
Economic Council of Canada, Ottawa *02425*
Economic Development Institute, Washington *15010*
Economic History Association, Washington *15011*
Economic History Society, Heslington *11927*
Economic Institute for the Building Industry, Amsterdam *09071*
Economic Research Assocation, Tokyo *08158*
Economic Research Centre, Singapore *10046*
The Economics and Business Education Association, Hassocks *11928*
Economic Society of Australia, Sydney *00388*
Economic Society of Finland, Helsinki *03954*
Economic Society of Ghana, Accra *06480*
The Economic Society of Malta, Paceville *08660*
Economic Society of South Africa, Pretoria *10137*
Economics Society, Moskva *09865*
Economics Society of Alberta, Calgary *02426*
Economic Study Association, London *11929*
Economists Advisory Group, London *11930*
Economists' Society of Croatia, Zagreb *03456*
Economists' Society of Serbia, Beograd *17092*
ECOO, London *12023*
ECOPOP, Zollikofen *11022*
ECOPS, Strasbourg *04497*
ECOR, London *11957*
ECP, Bruxelles *01306*
ECPA, Bruxelles *01336*
ECPR, Colchester *12016, 12017*
ECPRD, Luxembourg *08585*
ECQAC, Frankfurt *05638*
ECRAM, Bruxelles *01263*
ECRCBC, Bruxelles *01264*
ECRI, Plymouth Meeting *15012*
ECRO, Zürich *10681*
ECS, Denver *15027*
ECS, Oxford *12007*
ECS, Pennington *15035*
ECSA, Cleveland *15065*
ECSACON, Arusha *11053*
ECSS, Paris *04496*
ECSSID, Wien *00590*
ECSWPR, Wien *00588*
ECTARC, Llangollen *12006*
ECTP, London *12024*
Ecuadorian Association of Museums, Quito *03844*
Ecumenical Association of Third World Theologians, Port Harcourt *09225*
Ecumenical Documentation and Information Centre for Eastern and Southern Africa, Harare *17169*
Ecumenical Institute Bossey, Céligny *10665*
Ecumenical Institute for Theological Research, Jerusalem *07053*
ECVAM, Ispra *07670*
ECWS, Maastricht *08877*
E.D.A., Birmingham *11943*
EDA, Le Mesnil-Saint-Denis *04503*
EDA, Wisbech *11942*
EDC, Newton *15028*
EDECN, Alkmaar *08883*
EDEN, Budapest *06603*
EDEN, Glasgow *12026*
EDF, New York *15048*
Edgar Allan Poe Society of Baltimore, Baltimore *15013*
EDI, Washington *15010*
EDICESA, Harare *17169*
Edinburgh Architectural Association, Edinburgh *11931*
Edinburgh Bibliographical Society, Edinburgh *11932*
Edinburgh Festival Society, Edinburgh *11933*
The Edinburgh Highland Reel and Strathspey Society, Edinburgh *11934*
Edinburgh Mathematical Society, Edinburgh *11935*
Edinburgh Medical Missionary Society, Edinburgh *11936*
Edinburgh Obstetrical Society, Edinburgh *11937*
Edinburgh Royal Choral Union, Edinburgh *11938*
Edinburgh Sir Walter Scott Club, Edinburgh *11939*
Edison Birthplace Association, Milan *15014*
Edison Electric Institute, Washington *15015*
Editorial Projects in Education, Washington *15016*
Edmonton Hundred Historical Society, London *11940*
EDRA, Oklahoma City *15049*
EdReAn, Longview *15024*
EDSA, Verviers *01337*
EDTA/ERA, Parma *07678*
Educational and Development Foundation of the Latin American Confederation of Credit Unions, Panamá City *09418*
Educational Center for Applied Ekistics, Atlanta *15017*
Educational Centres Association, London *11941*
Educational Commission for Foreign Medical Graduates, Philadelphia *15018*
Educational Development Association, Wisbech *11942*
Educational Drama Association, Birmingham *11943*
Educational Foundation for the Fashion Industries, New York *15019*
Educational Innovation Programme for Development in the Arab States, Kuwait *08547*
The Educational Institute of Design, Craft and Technology, Melton Mowbray *11944*
Educational Institute of Scotland, Edinburgh *11945*
Educational Leadership Institute, Shorewood *15020*
Educational Planning, Memphis *15021*
Educational Planning Institute, New York *15022*
Educational Records Bureau, New York *15023*
Educational Research Analysts, Longview *15024*
Educational Research Service, Arlington *15025*
Educational Television Association, York *11946*
Educational Testing Service, Princeton *15026*
Education Commission of the States, Denver *15027*
Education Development Center, Newton *15028*
Education for All Network, Paris *04469*
Education Information Network in the European Community, Bruxelles *01260*
Education Relations Commission of Ontario, Toronto *02427*
Education Turnkey Systems, Falls Church *15029*
Education Writers Association, Washington *15030*
Educators' ad hoc Committee on Copyright Law, Alexandria *15031*
EDUC International, Roma *07251*
Edwardian Studies Association, Dagenham *11947*
EEA, Bruxelles *01341*
EEA, Leuven *01338*
EEB, Bruxelles *01342*
EEB, Schwalbach *05655*
EEC, Brugge *01266*
EEG Society, Birmingham *11953*
EEI, Washington *15015*
EEMS, Edinburgh *12028*
E.E.M.U.A., London *11959*
EEP, Amsterdam *08884*
EERI, Oakland *15000*
EERO, Wageningen *08885*
E.E.S., London *11948*
EESDA, Crowthorne *12027*
EETS, Oxford *11915*
EF, Grosse Pointe Farms *15063*
EFA, Statesboro *15003*
EFAD, Oss *08888*
EFAMRO, Amsterdam *08886*
EFAR, Paris *04504*
EFB, Frankfurt *05651*
EFC, Frankfurt *05678*
EFCIW, Frankfurt *05652*
EFCNS, Bruxelles *01348*
EFCS, Düsseldorf *05679*
EFDSS, London *11961*
EFECOT, Bruxelles *01346*
EFEM, Antwerpen *01349*
EFFI, New York *15019*
EFK, Frankfurt *05653*
EFPS, Stockholm *10455*
EFTA, Bruxelles *01345*
EGCS, Chesham *11963*
EGOS, Plymouth *12030*

EGS, London 11962
The Egypt Exploration Society, London 11948
Egyptian Agricultural Organization, Cairo 03887
Egyptian Association for Psychological Studies, Cairo 03888
Egyptian Association of Archives, Librarianship and Information Science, Cairo 03889
Egyptian Botanical Society, Cairo 03890
Egyptian Dental Federation, Cairo 03891
Egyptian Horticultural Society, Cairo 03892
Egyptian Medical Association, Cairo 03893
Egyptian Organization for Biological Products and Vaccines, Giza 03894
Egyptian School Library Association, Cairo 03895
The Egyptian Society for the Dissemination of Universal Culture and Knowledge, Cairo 03896
Egyptian Society of Dairy Science, Cairo 03897
Egyptian Society of Engineers, Cairo 03898
Egyptian Society of International Law, Cairo 03899
Egyptian Society of Medicine and Tropical Hygiene, Alexandria 03900
Egyptian Society of Political Economy, Statistics and Legislation, Cairo 03901
EHA, Washington 15011
EHAS, Ware 11917
EHMA, Dublin 06947
E.H.S., Glasgow 11923
EI, Big Sur 15055
EI, Cambridge 15045
EIB, Talence 04467
Eichendorff-Gesellschaft e.V., Ratingen 05637
EIDCT, Melton Mowbray 11944
Eidgenössischer Musikverband, Dagmersellen 10666
Eighteen Nineties Society, London 11949
EIL, Malvern 11527
E.I.P.A., Milano 07654
EIPDAS, Kuwait 08547
EIRMA, Paris 04208
EIS, Edinburgh 11945
EISCAT, Kiruna 10456
EKIS, Verona 07661
Ekonomiska Samfundet i Finland, Helsinki 03954
ELA, Addis Ababa 03943
ELCON, Washington 15033
El-Djazairia El-Mossilia, Alger 00022
ELEA, North Hollywood 15066
Electoral Reform Society of Great Britain and Ireland, London 11950
Electric Auto Association, Belmont 15032
Electricity Consumers Resource Council, Washington 15033
Electric Railway Society, Sutton Coldfield 11951
Electric Vehicle Association, Cupertino 15034
Electric Vehicle Association of Canada, Ottawa 02428
Electric Vehicle Development Group, London 11952
Electrochemical Society, Pennington 15035
The Electrochemical Society of India, Bangalore 06757
Electroencephalographic Society, Birmingham 11953
Electrolysis Association of Canada, Toronto 02429
Electronic Components Quality Assurance Committee, Frankfurt 05638
Electronic Music Consortium, Columbus 15036
The Electron Microscopy Society of Southern Africa, Durban 10138
Electro Physiological Technologists' Association, London 11954
Electro-Technical Council of Ireland, Dublin 06943
Elektrotechnischer Verein Berlin e.V., Berlin 05639
Elektroteknisk Forening, Hellerup 03717
Elgar Society, Woldingham 11955
Elgar Society (London), Northwood 11956
ELI, Shorewood 15020
ELI, Washington 15050
Elizabeth Fry Society of Alberta (Edmonton), Edmonton 02430
Elizabeth Fry Society of Manitoba, Winnipeg 02431
Elizabeth Fry Society of New Brunswick, Moncton 02432
Elizabeth Fry Society of Saskatchewan, Saskatoon 02433
Elizabeth Linington Society, Glassboro 15037
Elliniki Anaisthisiologiki Etaireia, Athinai 06503
Elliniki Astronautiki Etaireia, Athinai 06504
Elliniki Cheirourgiki Etaireia, Athinai 06505
Elliniki Geografiki Etaireia, Athinai 06506
Elliniki Kardiologiki Etairia, Athinai 06507
Elliniki Ktiniatriki Eteria, Athinai 06508
Elliniki Laografiki Etaireia, Athinai 06509
Elliniki Mathimatiki Eteria, Athinai 06510
Elliniki Microbiologiki Etaireia, Athinai 06511
Elliniki Nomismatiki Etaireia, Athinai 06512
Elliniki Paidiatriki Etairia, Athinai 06513
Elliniki Pharmakeutiki Etaireia, Athinai 06514
Elliniki Spilaiologiki Etaireia, Athinai 06515
Ellinikon Kentron Paragogikotitos, Athinai 06516
Ellinikos Organismos Tupopoiesseos, Athinai 06517
Elm Research Institute, Harrisville 15038
ELNA, El Cerrito 15057
ELOT, Athinai 06517
Elpeka, Athinai 06516
ELS, Cambridge 15189
ELS, Glassboro 15037
ELSA, New York 15060
EMBRAPA, Barreiras 01731
EMBRAPA, Campina Grande 01637
EMBRAPA, Campo Grande 01648
EMBRAPA, Campos 01732
EMBRAPA, Corumba 01733
EMBRAPA, Cruz das Almas 01633
EMBRAPA, Dourados 01734
EMBRAPA, Goiânia 01632, 01736
EMBRAPA, Londrina 01636
EMBRAPA, Manaus 01635, 01737
EMBRAPA, Nova Iguaçu 01735
EMBRAPA, Petrolina 01626
EMBRAPA, Ponta Grossa 01738
EMBRAPA, Porto Velho 01739
EMBRAPA, Teresina 01740
EMBS, Brest 04429
EMBS, New York 15246
EMC, Columbus 15036
EMC, Washington 15044
EMEP, Genève 10658
EMG, Rheine 05654
EMI, Point Roberts 15073
EMIERT, Chicago 15062
EMM, Bucuresti 09749
EMMS, Edinburgh 11936
Employee Benefit Research Institute, Washington 15039
Empresa Brasileira de Pesquisa Agropecuária, Campo Grande 01648
EMS, Arlington 15051
E.M.S., Edinburgh 11935
EMS, Thessaloniki 06525
Emulsion Polymers Institute, Bethlehem 15040
Endocrine Society, Bethesda 15041
Endocrine Society of Australia, Camperdown 00389
Endometriosis Association, Milwaukee 15042
Energiagazdálkodási Tudományos Egyesület, Budapest 06600
Energietechnische Gesellschaft im VDE, Frankfurt 05640
Energie- und Umweltzentrum am Deister e.V., Springe-Eldagsen 05641
Energy Probe Research Foundation, Toronto 02434
Energy Research Institute, Naples 15043
Enertek, Pretoria 10136
Engei Gakkai, Kyoto 08127
Engineering and Scientific Association of Ireland, Mount Merrion 06944
Engineering Committee on Oceanic Resources, London 11957
The Engineering Council, London 11958
Engineering Equipment and Materials Users' Association, London 11959
Engineering Institute of Canada, Ottawa 02435
The Engineering Institution of Zambia, Lusaka 17152
Engineering Manpower Commission, Washington 15044
English Academy of Southern Africa, Wits 10139
English Association, London 11960
English Association – Sydney Branch, Rozelle 00390
The English Folk Dance and Song Society, London 11961
English Goethe Society, London 11962
English Guernsey Cattle Society, Chesham 11963
English Institute, Cambridge 15045
English Literary Society of Japan, Tokyo 08196
English Place-Name Society, Nottingham 11964
English Speaking Board (International), Southport 11965
English-Speaking Board (International), Southport 11966
English Spelling Association, London 11967
English Westerners Society, London 11968
ENIOS, Roma 07618
Enosis Ellinon Bibliothekarion, Athinai 06518
Enosis Ellinon Chimikon, Athinai 06519
Enosis Ellinon Physikon, Athinai 06520
Enosis Hellinon Mousourgon, Athinai 06521
E.N.S., Bern 10685
ENSIC, Bangkok 11091
ENT, Colchester 11978
Ente Autonomo La Biennale de Venezia, Venezia 07649
Ente di Unificazione Navale, Genova 07650
Ente Eugenio e Claudio Faina per l'Istruzione Professionale Agraria, Roma 07651
Ente Fauna Siciliana, Noto 07652
Ente Friulano di Economia Montana, Udine 07653
Ente Istruzione Professionale Artigiana, Milano 07654
Ente Nazionale Francesco Petrarca, Padova 07655
Ente Nazionale Italiano di Unificazione, Milano 07656
Entente Médicale Méditerranéenne, Bucuresti 09749
Ente per la Valorizzazione dei Vini Astigiani, Asti 07657
Ente per le Nuove Tecnologie, l'Energia e l'Ambiente, Roma 07658
Ente Studi Economici per la Calabria, Cosenza 07659
Entomological Society, St. Petersburg 09866
Entomological Society of America, Lanham 15046
Entomological Society of British Columbia, Victoria 02436
Entomological Society of Canada, Ottawa 02437
Entomological Society of China, Beijing 03198
The Entomological Society of Finland, Helsinki 04029
Entomological Society of Japan, Tokyo 08341
Entomological Society of Manitoba, Winnipeg 02438
Entomological Society of Moldova, Chişinău 08778
Entomological Society of New South Wales, Sydney 00391
Entomological Society of New Zealand, Upper Hutt 09124
Entomological Society of Nigeria, Zaria 09226
Entomological Society of Ontario, Guelph 02439
Entomological Society of Quebec, Sainte-Foy 02921
Entomological Society of Queensland, Brisbane 00392
Entomological Society of Saskatchewan, Indian Head 02440
Entomological Society of Southern Africa, Pretoria 10140
Entomologisk Forening, København 03718
Environic Foundation International, South Bend 15047
Environmental Council, London 11969
Environmental Defense Fund, New York 15048
Environmental Design Research Association, Oklahoma City 15049
Environmental Development in the Third World, Dakar 10028
Environmental Law Institute, Washington 15050
Environmental Mutagen Society, Arlington 15051
Environmental Sanitation Information Center, Bangkok 11091
Environmental Technology Seminar, Bethpage 15052
Environmental Training Network for Latin America and the Caribbean, México 08732
Eötvös Loránd Fizikai Társulat, Budapest 06601
EOS, Kingston 15064
E.O.S., London 12036
EP, Memphis 15021
EPA, Amsterdam 08891
EPE, Washington 15016
EPF, Gouda 08890
The Ephemera Society, London 11970
Ephsoc, London 11970
EPI, Bethlehem 15040
EPI, New York 15022
Epigraphic Society, San Diego 15053
Epilepsy Prince Edward Island, Richmond 02441
Epimelitirion Technikon tis Ellados, Athinai 06522
Epitéstudományi Egyesület, Budapest 06602
EPNS, Nottingham 11964
EPO, Frascati 07648
EPPO, Paris 04479
EPS, Egham 12044
EPS, Orono 15074
E.P.T.A., London 11954
Equipe 7 des Arts Visuels, México 08733
Equipe de Recherche Associée au C.N.R.S.-Laboratoire de Génétique, Talence 04470
ERA, Leatherhead 11971
Eranos Vindobonensis, Wien 00584
ERAS, Hull 11922
ERA Technology, Leatherhead 11971
ERB, New York 15023
ERCU, Edinburgh 11938
ERECO, Bruxelles 01339
EREV, Hannover 05689
ERF, Crownsville 15061
Ergonomics Society, Loughborough 11972
Ergonomics Society of Australia, Fortitude Valley 00393
ERI, Harrisville 15038
ERI, Naples 15043
ERIS, Guatemala City 06567
Ernest Bloch Society, Gualala 15054
Ernst Barlach Gesellschaft e.V., Hamburg 05642
Ernst-Mach-Institut, Freiburg 05643
Ernst Toller Memorial Society, Cincinnati 15543
EROPA, Manila 09518
ERS, Arlington 15025
ERS, Sutton Coldfield 11951
ES, Bethesda 15041
ES, Evanston 15009
ES, Loughborough 11972
ES, San Diego 15053
ESA, Camperdown 00389
E.S.A., Canberra 00387
ESA, Lanham 15046
ESA, Stevenston 12041
ESA, Tempe 15006
ESAFS, Tokyo 08124
Esalen Institute, Big Sur 15055
ESAMI, Arusha 11056
ESAMRDC, Dodoma 11057
ESARBICA, Nairobi 08501
ESARDA, Ispra 07681
ESAURP, Dar es Salaam 11058
ESAWR, North Olmsted 14997
E.S.B., Southport 11966
E.S.C., Rotterdam 08893
ESCAP/FAO/UNIDO Fertilizer, Development and Information Network for Asia and the Pacific, Bangkok 11092
ESCAP/WMO Typhoon Committee, Manila 09519
ESCSI, Salt Lake City 15069
Escuela Regional de Ingeniería Sanitaria, Guatemala City 06567
ESDUCK, Cairo 03896
E.S.E.C., Cosenza 07659
ESF, Washington 15056
ESLA, Cairo 03895
ESMU, Bruxelles 01312
ESOMAR, Amsterdam 08892
Espace Vidéo Européen, Dublin 06945
Esperantic Studies Foundation, Washington 15056
Esperantist Ornithologists' Association, Nynäshamn 10450
Esperanto Academy, Paris 04471
Esperanto-Asocio de Britio, London 11973
Esperanto League for North America, El Cerrito 15057
Esperanto Librarians International, Naperville 15058
Esperanto Teachers Association, Leicester 11974
Esperanto Writers' Association, Amsterdam 08847
ESPN, Zürich 10686
ESRIN, Frascati 07660
ESRS, Rowley 14998
Essentia, Boulder 15059
Essex Agricultural Society, Chelmsford 11975
Essex Archaeological and Historical Congress, Stratford Saint Mary 11976
Essex Society for Archaeology and History, Colchester 11977
Essex Wildlife Trust, Colchester 11978
ESSRA, London 11926
Estonian Jurists' Association, Mississauga 02442
Estonian Learned Society of America, New York 15060
Estonian Teachers' League, Don Mills 02443
Estuarine and Brackish-Water Sciences Association, Cambridge 11979
Estuarine Research Federation, Crownsville 15061
ET, San Francisco 15068
E.T.A., Leicester 11974
E.T.A., York 11946
ETAD, Basel 10664
E.T.A. Hoffmann-Gesellschaft e.V., Bamberg 05644
Etaireia Byzantinon kai Metabyzantinon Meleton, Athinai 06523
Etaireia Byzantinon Spoudon, Athinai 06524
Etaireia Makedonikon Spoudon, Thessaloniki 06525
Etaireia Odontostomatologikis Ereunis, Athinai 06526
ETAPC, Antwerpen 01352
ETE, Budapest 06600
Ethiopian Library Association, Addis Ababa 03943
Ethiopian Medical Association, Addis Ababa 03944
Ethnic Materials and Information Exchange Round Table, Chicago 15062
Ethnographic and Folk Culture Society, Lucknow 06758
Ethnological Society of China, Taipei 03296
Etruscan Foundation, Grosse Pointe Farms 15063
ETS, Bethpage 15052
ETS, Princeton 15026
Etudes Préhistoriques, Sainte-Foy-lès-Lyon 04472
ETUI, Bruxelles 01353
EUBS, Aberdeen 12042
EUCARPIA, Dronten 08861
EUCARPIA, Wageningen 08862
EUCLID, London 11980
EUDISED, Strasbourg 04502
EUFMD, Roma 07673
Eugene O'Neill Society, Kingston 15064
EUPO, Gent 01328
EurACS, Antwerpen 01289
EURAPS, Paris 04489
EURASAP, London 11993
EURASHE, Bruxelles 01278
EURASIP, Lausanne 10674
Euratom Scientific and Technical Committee, Bruxelles 01261
EUREG, Paris 04482
Eureka Organization, Bruxelles 01262
EURO, Bologna 07253
EuroACE, London 11992
Euro-African Association for the Anthropology of Social Change and Development, Montpellier 04473
EUROBITUME, Bruxelles 01300
EuroCentre-PHC, Perugia 07669
Euro-China Research Association in Management, Bruxelles 01263
Euro-China Research Centre for Business Cooperation, Bruxelles 01264
Euro Chlor, Bruxelles 01265
Eurocoast, Marseille 04474
Euroenviron, København 03719
Euro-Handelsinstitut e.V., Köln 05645
Euro-International Committee for Concrete, Lausanne 10667
EUROLAT – European Network for Studies on Laterites and Tropical Environment, Strasbourg 04475
Euromech, Cambridge 12033
Euro-Mediterranean Centre on Marine Contamination Hazards, Valletta 08661
EUROMET, Braunschweig 05674
EUROMICRO, Apeldoorn 08858
EUROMINERALS – Confederation of Learned/Engineering Societies in the Mineral Industry, London 11981
Europa Club, Sezione Italiana, Verona 07661
Europäische Akademie Bayern e.V., München 05646
Europäische Akademie Berlin e.V., Berlin 05647
Europäische Akademie Otzenhausen e.V., Nonnweiler 05648
Europäische Autorenvereinigung Die Kogge, Minden 05671
Europäische Bildungs- und Aktionsgemeinschaft, Bonn 05649
Europäische Bildungs- und Begegnungszentren, Idstein 05650
Europäische Föderation Biotechnologie, Frankfurt 05651
Europäische Föderation für Chemie-Ingenieur-Wesen, Frankfurt 05652
Europäische Föderation Korrosion, Frankfurt 05653
Europäische Gesellschaft für Diabetologie, Düsseldorf 05666
Europäische Gesellschaft für Schriftpsychologie und Schriftexpertise, Zürich 10668
Europäische Märchengesellschaft e.V., Rheine 05654
Europäische Organisation für Qualität, Bern 10669
Europäische Rektorenkonferenz, Genève 10670
Europäischer Erzieherbund e.V., Sektion Deutschland, Schwalbach 05655
Europäischer Verband für Produktivitätsförderung, Hamburg 05656
Europäisches Astronautenzentrum, Köln 05670
Europäische Staatsbürger-Akademie e.V., Bocholt 05657
Europäisches Zentrum für die Bildung im Versicherungswesen, Sankt Gallen 10671
Europäisches Zentrum für die Förderung der Berufsbildung, Berlin 05673
Europäisches Zentrum für Wohlfahrtspolitik und Sozialforschung, Wien 00585
Europäische Vereinigung der Datenbanken in der Aus- und Weiterbildung, Berlin 05664
Europäische Vereinigung für Eigentumsbildung, Hermann-Lindrath-Gesellschaft e.V., Hannover 05658
Europa Esperanto-Centro, Brugge 01266
Europa Nostra/IBI, 's-Gravenhage 08848
Europe 2000, Son en Breugel 08849
Europeam Federation of Endocrine Societies, London 11982
European Academic and Research Network, Orsay 04476
European Academy for Film and Television, Bruxelles 01267
European Academy of Allergology and Clinical Immunology, København 03720

European Academy of Allergology and Clinical Immunology, Roma 07662
European Academy of Anaesthesiology, Strasbourg 04477
European Academy of Facial Surgery, London 11983
European Academy of Public Writers, Bruxelles 01091
European Accounting Association, Bruxelles 01268
European Advisory Committee on Health Research, København 03721
European Aggregate Association, Paris 04478
European Air Law Assciation, London 11984
European Airlines Electronic Committee, London 11985
European Alliance of Muscular Dystrophy Associations, Dublin 06946
European Alliance of Safe Meat, Bruxelles 01269
European Aluminium Association, Düsseldorf 05659
European Aluminium Foil Association, Frankfurt 05660
European and Mediterranean Plant Protection Organization, Paris 04479
European Anthropological Association, Bruxelles 01270
European Asphalt Pavement Association, Breukelen 08850
European Association against Fibre Pollution, Bruxelles 01271
European Association against Pigmentary Dystrophy and its Syndromes, Bruxelles 01272
European Association for American Studies, Lyon 04480
European Association for Animal Production, Roma 07688
European Association for Aquatic Mammals, Dunstable 11986
European Association for Audiovisual Media Education, Bruxelles 01273
European Association for Behaviour Therapy, Harrow 11987
European Association for Bioeconomic Studies, Milano 07663
European Association for Burgundy Studies, Bruxelles 01274
European Association for Cancer Education, Groningen 08851
European Association for Cancer Research, Nottingham 11988
European Association for Cardio-Thoracic Surgery, Zürich 10672
European Association for Catholic Adult Education, Linz 00586
European Association for Chinese Law, Aachen 05661
European Association for Chinese Studies, Paris 04481
European Association for Cognitive Ergonomics, Amsterdam 08852
European Association for Computer Graphics, Aire-la-Ville 10673
European Association for Country Planning Institutions, Bruxelles 01275
European Association for Cranio-Maxillo-Facial Surgery, London 11989
European Association for Earthquake Engineering, Zagreb 03458
European Association for Environmental History, Helsinki 03955
European Association for Evolutionary Political Economy, Newcastle-upon-Tyne 11990
European Association for Gastroenterology and Endoscopy, Amsterdam 08853
European Association for Grey Literature Exploitation, 's-Gravenhage 08854
European Association for Gynaecologists and Obstetricians, Nijmegen 08855
European Association for Haematopathology, Aarhus 03722
European Association for Health Information and Libraries, Bruxelles 01276
European Association for Information on Local Development, Bruxelles 01277
European Association for Institutions of Higher Education, Bruxelles 01278
European Association for International Education, Amsterdam 08856
European Association for Japanese Studies, Leiden 08857
European Association for Microprocessing and Microprogramming, Apeldoorn 08858
European Association for Personnel Management, London 11991
European Association for Population Studies, 's-Gravenhage 08859
European Association for Potato Research, Wageningen 08860
European Association for Research and Development in Higher Education, Berlin 05662
European Association for Research in Industrial Economics, Bruxelles 01279
European Association for Research into Adapted Physical Activity, Bruxelles 01280
European Association for Research on Learning and Instruction, Turku 03956
European Association for Research on Plant Breeding, Dronten 08861
European Association for Research on Plant Breeding, Wageningen 08862
European Association for Signal Processing, Lausanne 10674
European Association for Special Education, Stuttgart 05663
European Association for Textile Polyolefins, Bruxelles 01281
European Association for the Advancement of Radiation Curing by UV, EB and Laser Beams, Fribourg 10675
European Association for the Conservation of Energy, London 11992
European Association for the Development of Databases in Education and Training, Berlin 05664
European Association for the International Space Year, Paris 04482
European Association for Theoretical Computer Science, Paderborn 05665
European Association for the Promotion of Poetry, Leuven 01282
European Association for the Promotion of the Hand Hygiene, Bruxelles 01283
European Association for the Science of Air Pollution, London 11993
European Association for the Study of Diabetes, Düsseldorf 05666, 05667
European Association for the Study of Dreans, Ile-Saint-Denis 04483
European Association for the Study of Safety Problems in the Production and Use of Propellant Powders, Bruxelles 01284
European Association for the Study of the Liver, Amsterdam 08863
European Association for the Teaching of Architecture, Kassel 05154
European Association for the Teaching of Legal Theory, Bruxelles 01285
European Association for the Transfer of Technologies, Innovations and Industrial Information, Luxembourg 08582
European Association for the Visual Studies of Man, Firenze 07664
European Association for Transactional Analysis, Genève 10676
European Association of Agricultural Economists, Bruxelles 01286
European Association of Architectural Education, Louvain-la-Neuve 01287
European Association of Business and Management Teachers, Dortmund 05668
European Association of Cardiothoracic Anaesthesiologists, Papworth Everard 11994
European Association of Centers of Medical Ethics, Bruxelles 01288
European Association of Classification Societies, Antwerpen 01289
European Association of Coleopterology, Barcelona 10304
European Association of Contract Research Organizations, Plaisir 04484
European Association of Development Research and Training, Genève 10677
European Association of Diabetes Educators, Leiden 08864
European Association of Distance Teaching Universities, Heerlen 08865
European Association of Education and Research in Public Relations, Gent 01290
European Association of Establishments for Veterinary Education, Maisons-Alfort 04485
European Association of Experimental Social Psychology, Amsterdam 08866
European Association of Experimental Social Psychology, Louvain-la-Neuve 01291
European Association of Exploration Geophysicists, Zeist 08867
European Association of Fisheries Economists, Southsea 11995
European Association of Football Teams Physicians, Barcelona 10305
European Association of Geochemistry, Paris 04486
European Association of Hearing Aid Audiologists, Fleurus 01292
European Association of Historical Associations, Louvain-la-Neuve 01293
European Association of Information Services, Luxembourg 08583
European Association of Internal Medicine, Brighton 11996, 11997
European Association of Labour Economists, Stockholm 10451
European Association of Law and Economics, Maastricht 08868
European Association of Marine Sciences and Techniques, Talence 04487
European Association of Metals, Bruxelles 01294
European Association of Museums of the History of Medical Sciences, Denicé 04488
European Association of Music Conservatories, Academies and High Schools, Bern 10678
European Association of Music Festivals, Genève 10679
European Association of National Productivitn Centres, Bruxelles 01127
European Association of Neurosurgical Societies, Berlin 05669
European Association of Nuclear Medicine, London 11998
European Association of Organic Geochemists, Newcastle-upon-Tyne 11999
European Association of Perinatal Medicine, Uppsala 10452
European Association of Petroleum Geoscientists and Engineers, Zeist 08869
European Association of Plastic Surgeons, Paris 04489
European Association of Podologists, Bruxelles 01295
European Association of Poisons Control Centres, Bruxelles 01126
European Association of Radiology, Leuven 01296
European Association of Remote Sensing Laboratories, Napoli 07665
European Association of Science Editors, London 12000
European Association of Senior Hospital Physicians, Udine 07666
European Association of Social Medicine, Torino 07667
European Association of South Asian Archaeologists, Cambridge 12001
European Association of State Veterinary Officers, Rotterdam 08870
European Association of Teachers, Beek-Ubbergen 08871
European Association of Urology, Paris 04490
European Association of Users of Satellites in Training and Education Programmes, Bruxelles 01297
European Association of Veterinary Anatomists, Gent 01298
European Association of Veterinary Pharmacology and Toxicology, Utrecht 08872
European Association on Catalysis, Namur 01299
European Astronaut Centre, Köln 05670
European Atherosclerosis Society, London 12002
The European-Atlantic Movement, Exeter 12003
European Atomic Energy Society, Paris 04491
European Audio Phonological Centers Association, Saint-Etienne 04492
European Authors' Association Die Kogge, Minden 05671
European Baptist Theological Teachers' Conference, Hamburg 05672
European Bioelectromagnetics Association, Madrid 10306
European Bitumen Association, Bruxelles 01300
European Bone Marrow Transplant Group, Wien 00587
European Brachytherapy Group, Villejuif 04493
European Brain and Behaviour Society, Oslo 09265
European Brain Injury Society, Bruxelles 01301
European Bureau for Conservation and Development, Bruxelles 01302
European Bureau of Adult Education, Amersfoort 08873
European Bureau of Lesser-Used Languages, Bruxelles 01303
European Burns Association, Leuven 01304
European Business Associates On-line, Luxembourg 08584
European Business Ethics Network, Bruxelles 01305
European Calcified Tissue Society, Amsterdam 08874
European Camac Association, Louvain-la-Neuve 01121
European Cancer Prevention Organizazion, Bruxelles 01306
European Capitals Universities Network, Bruxelles 01307
European Carbon Black Centre, Bruxelles 01308
European Cartographic Institute, 's-Gravenhage 08875
European Cell Biology Organization, Milano 07668
European Centre for Advanced Studies in Thermodynamics, Paris 04494
European Centre for Development Policy Management, Maastricht 08876
European Centre for Disaster Medicine, San Marino 09996
European Centre for Ethnolinguistic Cartography, Bruxelles 01309
European Centre for Geodynamics and Seismology, Bruxelles 01310
European Centre for Higher Education, Bucuresti 09750
European Centre for Insurance Education and Training, Sankt Gallen 10680
European Centre for Leisure and Education, Praha 03543
European Centre for Medium-Range Weather Forecasts, Reading 12004
European Centre for Parliamentary Research and Documentation, Luxembourg 08585
European Centre for Plastics in the Environment, Bruxelles 01311
European Centre for Pollution Research, London 12005
European Centre for Professional Training in Environment and Tourism, Madrid 10307
European Centre for Regional Development, Strasbourg 04495
European Centre for Research and Development in Primary Health Care, Perugia 07669
European Centre for Social Welfare Policy and Research, Wien 00588
European Centre for Strategic Management of Universities, Bruxelles 01312
European Centre for the Development of Vocational Training, Berlin 05673
European Centre for the Validation of Alternative Testing Methods, Ispra 07670
European Centre for Traditional and Regional Cultures, Llangollen 12006
European Centre for Training Craftsmen in the Conservation of the Architectural Heritage, Venezia 07671
European Centre for Work and Society, Maastricht 08877
European Centre of Environmental Studies, Torino 07672
European Centre of Ophthalmology, Bruxelles 01313
European Centre of Studies on Linear Alkylbenzene, Bruxelles 01314
European Cetacean Society, Oxford 12007
European Chapter of Combinatorial Optimization, Graz 00589
European Chemical Industry Ecology and Toxicology Centre, Bruxelles 01315
European Chemical Marketing Association, London 12008
European Chemoreception Research Organization, Zürich 10681
European Chlorinated Solvents Association, Bruxelles 01316
European Coil Coating Association, Bruxelles 01317
European Collaboration on Measurement Standards, Braunschweig 05674
European Collection of Animal Cell Cultures, Salisbury 12009
European College of Gerodontology, Nijmegen 08878
European College of Marketing and Marketing Research, Brixham 12010
European College of Neuropsychopharmacology, Hillerød 03723
European College of Obstetrics and Gynaecology, Rotterdam 08879
European Commission for the Control of Foot-and-Mouth Disease, Roma 07673
European Committee for Biological Effects of Carbon Black, Brighton 12011
European Committee for Catholic Education, Bruxelles 01318
European Committee for Civil Engineers, London 12012
European Committee for EC Agricultural Engineers, Bruxelles 01319
European Committee for Electrotechnical Standardization, Bruxelles 01320
European Committee for Future Accelerators, Aachen 05675
European Committee for Interoperable Systems, Bruxelles 01321
European Committee for Iron and Steel Standardization, Bruxelles 01322
European Committee for Scientific and Cultural Relations with Romania, Bucuresti 09751
European Committee for the Advancement of Thermal Sciences and Heat Transfer, Delft 08880
European Committee for the Study of Salt, Paris 04496
European Committee for Treatment and Research in Multiple Sclerosis, Melsbroek 01323
European Committee of Construction Economists, Paris 08592
European Committee of Organic Surfactants and their Intermediates, Bruxelles 01324
European Committee on Computational Methods in Applied Sciences, Bruxelles 01325
European Committee on Crime Problems, Strasbourg 04402
European Committee on Ocean and Polar Sciences, Strasbourg 04497
European Committee on Radiopharmaceuticals, Rotterdam 08881
European Communities Biologists Association, Roma 07674
European Communities Chemistry Council, London 12013
European Communities Clinical Chemistry Committee, London 12014
European Community Biologists Association, Namur 01326
European Community Network of the National Academic Recognition Information Centres, Bruxelles 01327
European Community Studies Association, Cleveland 15065
European Community University Professors in Ophthalmology, Gent 01328
European Company Lawyers Association, 's-Gravenhage 08882
European Confederation of Huntington Associations, Hinckley 12015
European Consortium for Church and State Research, Firenze 07675
European Consortium for Mathematics in Industry, Lahti 03957
European Consortium for Ocean Drilling, Strasbourg 04498
European Consortium for Political Research, Colchester 12016, 12017
European Consulting Engineering Network, Bruxelles 01329
European Consumer Law Group, Louvain-la-Neuve 01330
European Contact Dermatitis Society, Belfast 12018
European Contact Lens Society of Ophthalmologists, London 12019
European Control Data Users Group, Diepenbeek 01331
European Convention for Constructional Steelwork, Bruxelles 01332
European Cooperation in Social Science Information and Documentation, Wien 00590
European Cooperation in the Field of Scientific and Technical Research, Bruxelles 01333
European Cooperative Research Network on Olives, Cordoba 10308
European Coordinating Committee for Artificial Intelligence, Povo 07676
European Coordination Centre for Research and Documentation in Social Sciences, Wien 00591
European Council for Clinical and Laboratory Standardization, Coventry 12020
European Council for Industrial Marketing, Brixham 12021
European Council for Nondestructive Testing, Zenica 01558
European Council for Rural Law, Gent 01334
European Council for Social Research on Latin America, Roma 07677
European Council of Coloproctology, Genève 10682
European Council of Integrated Medicine, Bruxelles 01335
European Council of International Schools, Petersfield 12022
European Council of Management, Eschborn 05676
European Council of National Associations of Independent Schools, Slagelse 03724
European Council of Optics and Optometry, London 12023
European Council of Town Planners, London 12024
European Council on African Studies, Strasbourg 04499
European Council on Chiropractic Education, Poitiers 04500
European Council on Environmental Law, Strasbourg 04501
European Crop Protection Association, Bruxelles 01336
European Crystallographic Committee, Genève 10683
European Cytoskeletal Club, Genève 10684
European Dental Society, Pinner 12025
European Design Education Network, Glasgow 12026
European Development Education Curriculum Network, Alkmaar 08883
European Dialysis and Transplant Association – European Renal Association, Parma 07678
European Distance Education Network, Budapest 06603
European Documentation and Information System for Education, Strasbourg 04502
European Down's Syndrome Association, Verviers 01337
European Drosophila Centre, Umea 10453
European Dyslexia Association – International Organization for Specific Learning Disabilities, Le Mesnil-Saint-Denis 04503
European Ecological Federation, Lund 10454
European Economic Association, Leuven 01338
European Economic Research and Advisory Consortium, Bruxelles 01339
European Educational Association for News Distribution, Gerpinnes 01340
European Electrostatic Discharge Association, Crowthorne 12027
European Endangered Species Programme, Amsterdam 08884

European Environment Agency, Bruxelles 01341
European Environmental Bureau, Bruxelles 01342
European Environmental Mutagen Society, Edinburgh 12028
European Environmental Research Organization, Wageningen 08885
European Environment Information and Observation Network, Bruxelles 01343
European Extruded Polystyrene Insulation Board Association, Bruxelles 01344
European Family Therapy Association, Bruxelles 01345
European Fast Reactor Association, Eggenstein 05677
European Federation for Physical Medicine and Rehabilitation, Gent 01357
European Federation for the Advancement of Anaesthesia in Dentistry, Bologna 07679
European Federation for the Education of Children of Occupational Travellers, Bruxelles 01346
European Federation of AIDS Research, Paris 04504
European Federation of Animal Health, Bruxelles 01347
European Federation of Associations of Industrial Safety and Medical Officers, Leicester 12029
European Federation of Associations of Market Research Organizations, Amsterdam 08886
European Federation of Biotechnology, Frankfurt 05651
European Federation of Branches of the World's Poultry Science Association, Beekbergen 08887
European Federation of Chemical Engineering, Frankfurt 05652
European Federation of Child Neurology Societies, Bruxelles 01348
European Federation of Corrosion, Frankfurt 05653, 05678
European Federation of Cytology Societies, Düsseldorf 05679
European Federation of Energy Management Associations, Antwerpen 01349
European Federation of Productivity Services, Stockholm 10455
European Federation of the Associations of Dieticians, Oss 08888
European Finance Association, Bruxelles 01350
European Financial Society, Paris 04780
European Grassland Federation, 's-Gravenhage 08889
European Group for Organization Studies, Plymouth 12030
European Healthcare Management Association, Dublin 06947
European Incoherent Scatter Scientific Association, Kiruna 10456
European Industrial Marketing Research Society, Nuneaton 12031
European Industrial Research Management Association, Paris 04208
European Institute of Environmental Cybernetics, Athinai 06527
European International Business Association, Bruxelles 01124
European Marketing Research Board International, London 12032
European Mechanics Council, Cambridge 12033
The European Movement, London 12034
European Network of Scientific Information Referral Centres, Frascati 07680
European Nuclear Society, Bern 10685
European Organization for Caries Research, Leeds 12035
European Organization for Quality, Bern 10669
European Orthodontic Society, London 12036
European Packaging Federation, Gouda 08890
European Photochemistry Association, Amsterdam 08891
European Power Electronics and Drives Association, Bruxelles 01351
European Psycho-Analytical Federation, London 12037
European Safeguards Research and Development Association, Ispra 07681
European Society for Clinical Investigation, London 12038
European Society for Clinical Respiratory Physiology, Milano 07682
European Society for Comparative Endocrinology, Reading 12039
European Society for Engineering Education, Bruxelles 01460
European Society for Opinion and Marketing Research, Amsterdam 08892
European Society for Pediatric Nephrology, Zürich 10686
European Society for the Study of Ultrasonics, Roma 07683
European Society of Cardiology, Rotterdam 08893
European Society of Cardio-Vascular Radiology and Interventional Radiology, Lyon 04778
European Society of Handwriting Psychology, Zürich 10668
European Society of Hypnosis in Psychotherapy and Psychosomatic Medicine-Italian Constituent Society, Verona 07684
European Society of Nematologists, Invergowrie 12040
European Society of Neuroradiology, Homburg 05680
European Society of Pathology, Milano 07803
European Space Agency, Paris 04127
European Space Association, Stevenston 12041
European Technical Association for Protective Coatings, Antwerpen 01352
European Thermographic Association, Strasbourg 04207
European Trade Union Institute, Bruxelles 01353
European Underseas Bio-Medical Society, Aberdeen 12042
European Weed Research Society, Leverkusen 05681
EUROSTEP, Bruxelles 01297
EUROTALENT, Nîmes 04401
EUROTHERM, Delft 08880
EURYDICE, Bruxelles 01260
EUSIREF, Frascati 07680
E.u.U.Z., Springe-Eldagsen 05641
EVA, Cupertino 15034
EVAC, Ottawa 02428
Evangelical Lutheran Education Association, North Hollywood 15066
Evangelische Akademie Bad Boll, Bad Boll 05682
Evangelische Akademie Baden, Karlsruhe 05683
Evangelische Akademie Hofgeismar, Hofgeismar 05684
Evangelische Akademie in Wien, Wien 00592
Evangelische Akademie Loccum, Rehburg-Loccum 05685
Evangelische Akademien in Deutschland e.V., Bad Boll 05686
Evangelische Akademie Tutzing, Tutzing 05687
Evangelische Akademikerschaft in Deutschland, Stuttgart 05688
Evangelischer Erziehungs-Verband e.V., Hannover 05689
Evangelische Sozialakademie Friedewald, Friedewald 05690
Evangelische Studiengemeinschaft e.V., Heidelberg 05691
EVDG, London 11952
EVE, Dublin 06945
Evelyn Waugh Society, Garden City 15067
E.V.V.A., Asti 07657
EWA, Washington 15030
E.W. Bickle Foundation, Toronto 02444
EWC, Honolulu 15004
EWLTP, New York 14996
EWRS, Leverkusen 05681
EWS, Garden City 15067
EWS, London 11968
EWSLA, Tokyo 08126
Exeter Industrial Archaeology Group, Exeter 12043
EXIBA, Bruxelles 01344
Exodus Trust, San Francisco 15068
Expanded Polystyrene Association of Canada, Don Mills 02445
Expanded Shale Clay and Slate Institute, Salt Lake City 15069
Expéditions Polaires Françaises, Paris 04505
Experimental Aircraft Association, Oshkosh 15070
Experimental Aircraft Association of Canada, Mount Albert 02446
Experimental Psychology Society, Egham 12044
Experiments in Art and Technology, Berkeley Heights 15071
The Explorers Club, New York 15072
Extractive Metallurgy Institute, Point Roberts 15073
Ezra Pound Society, Orono 15074

FA, Leeuwarden 08899
FAA, Zürich 10688
FAAR, New York 15141
FAB, Bruxelles 01360
Fabian Society, London 12045
Fabric Care Research Association, Harrogate 12046
Fachgruppe Carosserie- und Fahrzeugtechnik des STV, Sissach 10687
Fachgruppe für Arbeiten im Ausland des SIA, Zürich 10688
Fachgruppe für Architektur des SIA, Zürich 10689
Fachgruppe für Betriebstechnik des STV, Winterthur 10690
Fachgruppe für Brückenbau und Hochbau des SIA, Zürich 10691
Fachgruppe für das Management im Bauwesen des SIA, Zürich 10692
Fachgruppe für Elektronik des STV, Lausanne 10693
Fachgruppe für industrielles Bauen des SIA, Zürich 10694
Fachgruppe für Raumplanung und Umwelt des SIA, Zürich 10695
Fachgruppe für Untertagbau des SIA, Zürich 10696
Fachgruppe für Verfahrenstechnik des SIA, Basel 10697
Fachgruppe für Vermessung und Kulturtechnik (Deutsch-Schweiz) des STV, Weiningen 10698
Fachinformationszentrum Technik e.V., Frankfurt 05692
Fachschaft Berliner Chirurgen e.V., Berlin 05693
Fachverband Deutscher Heilpraktiker e.V., Bonn 05694
Fachverband Moderne Fremdsprachen, Augsburg 05695
Fachverband Pulvermetallurgie, Hagen 05696
Fachverband Textilunterricht e.V., Münster 05697
FACSEA, New York 16524
FACSS, Frederick 15082
FACT, New York 15120
FACTS, Arlington 15133
Faculty of Actuaries in Scotland, Edinburgh 12047
Faculty of Advocates, Edinburgh 12048
Faculty of Astrological Studies, London 12049
Faculty of Building, Elstree 12050
Faculty of Dental Surgery, London 12051
Faculty of Homoeopathy, London 12052
Faculty of Royal Designers for Industry, London 12053
Faculty of Teachers in Commerce, Sutton Coldfield 12054
FAE, New York 15119
FAFS, London 12057
FAGEC, Bastia 04509
FAHSW, Portland 15095
FAI, Paris 04506
Faipari Tudományos Egyesület, Budapest 06604
Fair Organ Preservation Society, Norwich 12055
F.A.K., Auckland Park 10141
Famiglia e Libertà, Roma 07685
Family Health International, Durham 15075
The Family Planning Association, London 12056
Family Planning Resource Team, Sydney 02447
FAP, São Paulo 01650
Farmaceutsko Društvo na Makedonija, Skopje 08601
Farm and Food Society, London 12057
Farm Buildings Association, Oxford 12058
Farm Foundation, Oak Brook 15076
FAS, Bologna 07372
FAS, København 03734
FAS, Washington 15080
FASEB, Bethesda 15081
FAST, Milano 07686
FAT, Frankfurt 05742
Fauna and Flora Preservation Society, London 12059
Faust-Gesellschaft, Mühlacker 05698
FB, Elstree 12050
FBA, Oxford 12058
FBA, Washington 15077
FBCE, Hornchurch 12071
FBCSM, Braine l'Allend 01355
FBEP, Bruxelles 01354
FBF, Claremont 15135
FBH, Zürich 10691
FBPCS, Washington 15083
FBR, London 12088
FBR, Washington 15121
FBW, Wiesbaden 05700
FC, New York 15139
FCA, Philadelphia 15103
FCBJS, New York 15078
F.C.D.E.C., Milano 07699
FCER, Arlington 15122
FCIA, New York 15142
FC & PMS, Portsmouth 12086
FCRA, Harrogate 12046
FDA, Dunwoody 15104
FDA, München 05760
FDC, København 03728
FDMHA, Columbia 15138
FDRS, Newark 15109
FEBAB, São Paulo 01649
FEBS, Sofia 01767
FEC, Reston 15124
Federação Brasileira de Associaeões de Bibliotecarios, São Paulo 01649
Federación Argentina de Asociaciones de Anestesiología, Buenos Aires 00134
Federación del Patronato del Enfermo de Lepra de la República Argentina, Buenos Aires 00135
Federación de Sociedades Latinoamericanas del Cáncer, Lima 09467
Federación de Universidades Privadas de América Central, Guatemala City 06568
Federación de Urbanismo y de la Vivienda, Madrid 10309
Federación Española de Religiosos de Enseñanza, Madrid 10310
Federación Gremial, Santiago 03101
Federación Interamericana del Instituto de Enseñanza Publicitaria, Buenos Aires 00136
Federación Lanera Argentina, Buenos Aires 00137
Federación Latinoamericana de Parasitología, México 08734
Federación Latinoamericana de Parasitología, Santiago 03051
Federación Médica Peruana, Lima 09468
Federación Médica Venezolana, El Bosque 17005
Federación Nacional de Médicos del Ecuador, Quito 03853
Federación Odontológica de Centro América y Panamá, Panamá City 09419
Federación Panamericana de Asociaciones de Facultades y Escuelas de Medicina, Caracas 17006
Federación Universitaria Argentina, Buenos Aires 00138
Federación Universitaria del Paraguay, Asunción 09430
Federación Venezolana de Camaras y Asociaciones de Comercio y Producción, Caracas 17007
FEDERAGRONOMI, Roma 07692
Federal Bar Association, Washington 15077
Federal Centre of Forest Operations and Techniques, Groß-Umstadt 06089
Federal Chamber of Technical and Scientific Societies, Budapest 06665
Federal Library Association, Rawalpindi 09363
Federal Trust for Education and Research, London 12060
Federasie van Afrikaanse Kultuurvereniginge, Auckland Park 10141
Federasie van Rapportryerskorpse, Auckland Park 10142
Federated Council of Beth Jacob Schools, New York 15078
Federatie van Organisaties van Bibliotheek-, Informatie-, Dokumentatiewezen, 's-Gravenhage 08894
Fédération Aéronautique Internationale, Paris 04506
Fédération Autonome de l'Education Nationale, Paris 04507
Fédération Belge d'Education Physique, Bruxelles 01354
Fédération Belge des Chambres Syndicales de Médecins, Braine l'Allend 01355
Fédération Belge des Sociétés Scientifiques, Bruxelles 01356
Fédération Canadienne de Doyens de Gestion et d'Administration, Ottawa 02130
Fédération Canadienne des Enseignants, Ottawa 02305
Fédération Canadienne des Etudes Humaines, Ottawa 02128
Fédération Canadienne des Femmes Diplômées des Universités, Nanaimo 02131
Fédération Canadienne des Sciences Sociales, Ottawa 02914
Fédération Canadienne des Sociétés de Biologie, Ottawa 02129
Fédération Caribe de Santé Mentale, Fort de France 04508
Fédération d'Associations et Groupements pour les Etudes Corses, Bastia 04509
Fédération de l'Education Nationale, Paris 04510
Fédération des Affaires Sociales, Montréal 02448
Fédération des Amicales des Documentalistes et Bibliothécaires de l'Education Nationale, Paris 04511
Fédération des Associations de Professeurs des Universités du Québec, Montréal 02449
Fédération des Associations de Professeurs d'Université du Nouveau-Brunswick, Fredericton 02459
Fédération des Chambres Syndicales des Chirurgiens Dentistes de la Région de Paris, Paris 04512
Fédération des Collèges d'Enseignement Général et Professionel, Montréal 02450
Fédération des Commissions Scolaires Catholiques du Québec, Sainte Foy 02451
Fédération des Ecoles Indépendantes du Canada, Edmonton 02457
Fédération des Ecoles Privées de la Suisse Romande, Rolle 10699
Fédération des Enseignants de l'Ontario, Toronto 02801
Fédération des Enseignants du Nouveau-Brunswick, Fredericton 02696
Fédération des Femmes Médecins du Canada, Ottawa 02458
Fédération des Gynécologues et Obstétriciens de Langue Française, Paris 04513
Fédération des Médecins de France, Paris 04514
Fédération des Naturalistes du Nouveau-Brunswick, Saint John 02680
Fédération des Professionnelles et Professionnels de Cégeps et de Collèges, Montréal 02452
Fédération des Sociétés d'Agriculture de la Suisse Romande, Lausanne 10700
Fédération des Sociétés Françaises de Généalogie, d'Héraldique et de Sigillographie, Versailles 04515
Fédération des Sociétés Historiques et Archéologiques de Paris et de l'Ile de France, Paris 04516
Fédération des Sociétés Savantes de la Charente-Maritime, La Rochelle 04517
Fédération des Syndicats Dentaires Libéraux, Paris 04518
Fédération des Syndicats Pharmaceutiques de France, Paris 04519
Fédération du Québec pour le Planning des Naissances, Montréal 02453
Fédération Européenne de Médecine Physique et Réadaption, Gent 01357
Fédération Européenne de Zootechnie, Roma 07688
Federation for Ulster Local Studies, Belfast 12061
Federation for Unified Science Education, Columbus 15079
Fédération Française d'Education Physique et de Gymnastique Volontaire, Paris 04520
Fédération Française de Sociétés des Sciences Naturelles, Paris 04521
Fédération Française de Spéléologie, Paris 04522
Fédération Française des Sociétés d'Amis de Musées, Paris 04523
Fédération Française des Sociétés de Protection de la Nature, Paris 04524
Fédération Française des Sociétés de Sciences Naturelles, Paris 04525
Fédération Française d'Etudes et de Sports Sous Marins, Marseille 04526
Fédération Générale des Syndicats de Biologistes, Paris 04527
Fédération Historique de Provence, Marseille 04528
Fédération Hospitalière de France, Paris 04529
Fédération Internationale Catholique d'Education Physique et Sportive, Paris 04530
Fédération Internationale d'Education Physique, Cheltenham 12062
Fédération Internationale des Associations d'Etudes Classiques, Bellevue 10701
Fédération Internationale des Associations d'Instituteurs, Paris 04531
Fédération Internationale des Centres d'Entraînement aux Méthodes d'Education Active, Paris 04532
Fédération Internationale des Femmes Diplômées des Universités, Genève 10702
Fédération Internationale des Instituts de Recherches Socio-Religieuses, Louvain-la-Neuve 01358
Fédération Internationale des Mouvements d'Ecole Moderne, Cannes-La Bocca 04533
Fédération Internationale des Professeurs de Français, Sèvres 04534
Fédération Internationale des Professeurs de l'Enseignement Secondaire Officiel, Paris 04535
Fédération Internationale des Sociétés de Philosophie, Fribourg 10703
Fédération Internationale des Universités Catholiques, Paris 04536
Fédération Internationale d'Information et de Documentation, 's-Gravenhage 08895
Fédération Internationale pour l'Economie Familiale, Paris 04537
Fédération Internationale pour l'Education des Parents, Sèvres 04538
Fédération Métallurgique Française, Paris 04539
Fédération Mondiale pour la Santé Mentale, Vancouver 02998
Fédération Nationale Aéronautique, Paris 04540
Fédération Nationale de l'Enseignement Moyen Catholique, Bruxelles 01359
Fédération Nationale des Associations de Chimie de France, Paris 04541
Fédération Nationale des Costumes Suisses, Burgdorf 10935
Fédération Nationale des Enseignants et des Enseignantes du Québec, Montréal 02454
Fédération Nationale des Syndicats Départementaux de Médecins Electro-Radiologistes Qualifiés, Paris 04542
Federation of American Scientists, Washington 15080
Federation of American Societies for Experimental Biology, Bethesda 15081
Federation of Anaesthesiologists and Reanimatologists, Moskva 09867

Federation of Analytical Chemistry and Spectroscopy Societies, Frederick 15082
Federation of Australian Scientific and Technological Societies, Canberra 00394
Federation of Behavioral, Psychological and Cognitive Sciences, Washington 15083
Federation of British Artists, London 12063
The Federation of Danish Architects, København 03602
Federation of English Speaking Catholic Teachers, Montréal 02455
Federation of European Biochemical Societies, Sofia 01767
Federation of European Veterinaries in Industry and Research, Hounslow 12064
Federation of Family History Societies, Birmingham 12065
Federation of French-Language Gynecologists and Obstetricians, Paris 04513
Federation of Government Information Processing Councils, Atlanta 15084
Federation of Hungarian Medical Societies, Budapest 06650
Federation of Independent School Associations in British Columbia, Vancouver 02456
Federation of Independent Schools in Canada, Edmonton 02457
Federation of Indian Library Associations, Patiala 06759
Federation of Insurance and Corporate Counsel, Walpole 15085
Federation of Irish Film Societies, Dublin 06948
Federation of Library Information and Documentation Organisations, 's-Gravenhage 08894
Federation of London Area Dental Committees, London 12066
Federation of Materials Societies, Washington 15086
Federation of Medical Women of Canada, Ottawa 02458
Federation of Museums Associations, Beograd 17118
Federation of New Brunswick Faculty Associations, Fredericton 02459
Federation of Old Cornwall Societies, Launceston 12067
Federation of Orthodontic Associations, Milwaukee 15087
Federation of Pedagogical Societies of Serbia, Beograd 17120
Federation of Polish Medical Societies, Warszawa 09631
Federation of Prosthodontic Organizations, Chicago 15088
Federation of Societies for Coatings Technology, Blue Bell 15089
Federation of State Medical Boards of the United States, Fort Worth 15090
Federation of Trainers and Training Programs in Psychodrama, Dallas 15091
Federation of Veterinarians of the EEC, London 12068
Federation of Women Teachers' Associations of Ontario, Toronto 02460
The Federation of Zoological Gardens of Great Britain and Ireland, London 12069
Fédération Québécoise des Directeurs et Directrices d'Ecole, Anjou 02461
Fédération Royale des Sociétés d'Architectes de Belgique, Bruxelles 01360
Fédération Suisse des Ecoles Privées, Genève 10704
Federazione delle Associazioni Scientifiche e Tecniche, Milano 07686
Federazione delle Istituzioni Antropologiche Italiane, Genova 07687
Federazione Europea di Zootecnia, Roma 07688
Federazione Italiana contro la Tubercolosi e le Malattie Polmonari Sociali, Roma 07689
Federazione Italiana dei Cineclub, Roma 07690
Federazione Italiana delle Scienze e delle Attività Motorie, Roma 07691
Federazione Italiana Dottori in Agraria e Forestali, Roma 07692
Federazione Medico-Sportiva Italiana, Roma 07693
Federazione Nazionale Collegi Tecnici di Radiologia Medica, Mestre 07694
Federazione Nazionale degli Ordini dei Medici Chirurghi e Odontoiatri, Roma 07695
Federazione Nazionale degli Ordini dei Veterinari Italiani, Roma 07696
Federazione Nazionale Insegnanti Educazione Fisica, Roma 07697
Federazione Nazionale pro Natura, Bologna 07698
FEDIC, Roma 07690
FEF, Charlotte 15123
FEKI, Reykjavik 06682
Félag Bókavarda í Rannsókarbókasöfnum, Reykjavik 06681
Félag Enskukennara á Islandi, Reykjavik 06682
Félag Háls-, Nef- og Eyrnalaekna, Reykjavik 06683
Félag Islenskra Röntgenlaekna, Reykjavik 06684
Félag Islenskra Tryggingastaerdfraedinga, Reykjavik 06685
Félag Menntaskólakennara, Reykjavik 06686
The Fellowship for Freedom in Medicine, Storrington 12070
Fellowship of Australian Writers, Sydney 00395
Fellowship of British Christian Esperantists, Hornchurch 12071
Fellowship of Christian Writers, London 12072
Fellowship of Postgraduate Medicine, London 12073
FEN, Paris 04510
FEO, Liverpool 12080
FEPAFEM, Caracas 17006
Ferdinand Tönnies-Gesellschaft e.V., Kiel 05699
FERE, Madrid 10310
FERES, Louvain-la-Neuve 01358
Ferrous Metallurgy Society, Moskva 09868
FERSI, Lancaster 15099
FES, London 12083
FES, Nevada City 15101
FEVIR, Hounslow 12064
F.E.Z., Roma 07688
FF, Oak Brook 15076
FFEPGV, Paris 04520
FfH, Berlin 05738
FFHS, Birmingham 12065
FFM, New York 15129
FFM, Storrington 12070
FFPS, London 12059
FFR, Alpine 15125
FFS, Paris 04522
FFSAM, Paris 04523
FFSPN, Paris 04524
FFTU, Amsterdam 08933
FGD, Frankfurt 05718
FGG, Frankfurt 05758
FGH, Mannheim 05709
FGIPC, Atlanta 15084
FGOLF, Paris 04513
FGS, Hempstead 15143
FGSA, Moundsville 15118
FGU, Zürich 10696
FGV, Rio de Janeiro 01654
FGW, Remscheid 05716
FGW, Wien 00597
FHA, Haverford 15140
FHI, Durham 15075
FHS, Clwyd 12079
FHS, Durham 15112
FHS, London 12095
FIAB, Recklinghausen 05728
F.I.A.I., Genova 07687
FIAV, Winchester 15475
Fiber Society, Princeton 15092
FIC, Park Ridge 15127
FICC, Walpole 15085
FICEMEA, Paris 04532
FICEP, Paris 04530
FID, 's-Gravenhage 08895
FIDER, Grand Rapids 15126
FIEC, Bellevue 10701
FIEF, Paris 04537
Field Naturalists' Club of Victoria, South Yarra 00396
The Field Naturalists' Society of South Australia, Adelaide 00397
Field Studies Council, Shrewsburry 12074
FIEN, Roma 07700
FIEP, Cheltenham 12062
F.I.E.P., Sèvres 04538
FIFDU, Genève 10702
FIGO, London 12356
The Fiji Law Society, Suva 03946
Fiji Society, Suva 03947
Filmbewertungsstelle Wiesbaden, Wiesbaden 05700
Filmkritiker Kooperative, München 05701
Filologisk-Historiske Samfund, København 03725
Filozofsko Društvo Srbije, Beograd 17093
Filtration Society, Leicester 12075
FIMEN, Cannes-La Bocca 04533
Finance and Taxation Institute, Bonn 05945
Finance Association of China, Taipei 03297
Financial Management Association, Tampa 15093
Financial Marketing Association, Madison 15094
Finlands Brandvärnsförbund, Helsinki 04055
Finlands Fysioterapeutförbund, Helsinki 04018
Finlands Svenska Författareförening, Helsinki 03958
Finlands Svenska Lärarförbund, Helsinki 03959
Finnish Academy of Science and Letters, Helsinki 04000
The Finnish Academy of Technology, Helsinki 04074
Finnish-American Historical Society of the West, Portland 15095
Finnish Association of Academic Agronomists, Helsinki 03949
Finnish Association of Allergology and Immunology, Helsinki 04006
The Finnish Association of Graduate Engineers, Helsinki 04072
The Finnish Association of Physiotherapists, Helsinki 04018
Finnish Centre for Radiation and Nuclear Safety, Helsinki 03989
Finnish Council Council for Information Provision, Helsinki 04076
Finnish Demographic Society, Helsinki 04066
Finnish Dental Society, Helsinki 04025
Finnish Economic Association, Helsinki 03973
Finnish Egyptological Society, Helsinki 04011
Finnish Fire Protection Association, Helsinki 04055
The Finnish Folk High School Association, Helsinki 04032
Finnish Geotechnical Society, Espoo 04023
Finnish Language Society, Turku 04037
The Finnish Library Association, Helsinki 04039
Finnish Literature Society, Helsinki 04001
Finnish Mathematical Society, Helsingin Yliopisto 04049
The Finnish Medical Society Duodecim, Helsinki 03997
Finnish Museums Association, Helsinki 04052
Finnish Oriental Society, Helsingin Yliopisto 04031
Finnish Pharmacists' Association, Helsinki 04015
Finnish Physical Society, Helsingin Yliopisto 04019
The Finnish Physiological Society, Oulu 04017
Finnish Political Science Association, Helsingin Yliopisto 04082
Finnish Research Library Association, Helsinki 04064
Finnish Society for Church History, Helsingin Yliopisto 04040
The Finnish Society for Economic Research, Helsinki 04071
The Finnish Society for Educational Research, Oulu 04035
Finnish Society for Information Services, Espoo 04077
Finnish Society for Medical Physics and Medical Engineering, Tampere 03980
Finnish Society of Gastroenterology, Helsinki 04020
The Finnish Society of Sciences and Letters, Helsinki 03996
Finnish Standards Association, Helsinki 04059
The Finnish Statistical Society, Helsinki 04065
Finska Betongföreningen, Helsinki 04010
Finska Kemistsamfundet, Helsinki 03960
Finska Konstföreningen, Helsinki 04062
Finska Läkaresällskapet, Helsinki 03961
F.I.P.E.S.O., Paris 04535
FIPF, Sèvres 04534
FIR, Aachen 05731
FIRA, Stevenage 12096
Firearms Research and Identification Association, Rowland Heights 15096
Fire Fighters Burn Treatment Society, Edmonton 02462
Fire Prevention Canada Association, Ottawa 02463
Fire Protection Association, London 12076
First – Foundation for Ichthyosis and Related Skin Types, Raleigh 15097
F.I.S.A.M., Roma 07691
Fish and Seafood Association of Ontario, Toronto 02464
Fisheries Association of British Columbia, Vancouver 02465
Fisheries Society of Nigeria, Lagos 09227
Fisheries Society of the British Isles, Cambridge 12077
FISP, Fribourg 10703
FIT, Reykjavik 06685
FIUC, Paris 04536
FIW, Gräfelfing 05732
Fizik Tedavi ve Rehabilitasyon Derneği, Istanbul 11145
FL, København 03732
FLA, Buenos Aires 00137
FLAAR, Cocoa 15128
Flag Institute, Chester 12078
Flag Research Center, Winchester 15098
FLAP, México 08734
Flat Earth Research Society International, Lancaster 15099
FLC, Harrogate 15144
Flemish Interuniversity Council, Bruxelles 01508
Flemish Museums Association, Antwerpen 01509
Flight Safety Foundation, Arlington 15100
Flintshire Historical Society, Clwyd 12079
F. Liszt Society, Budapest 06611
Flora Europaea Organization, Liverpool 12080
Flour, Fodder and Grain Storage Society, Moskva 09869
Flour Milling and Baking Research Association, Rickmansworth 12081
Flower Essence Society, Nevada City 15101
FLT, Frankfurt 05749
Fluid Power Society, Milwaukee 15102
FLWF, Scottsdale 15137
Flygtekniska Föreningen, Stockholm 10457
Flying Chiropractors Association, Philadelphia 15103
Flying Dentists Association, Dunwoody 15104
Flying Physicians Association, Kansas City 15105
Flying Veterinarians Association, Columbia 15106
FM, Reykjavik 06686
FMA, Lake Success 15116
FMA, Madison 15094
FMA, Tampa 15093
F. Marion Crawford Memorial Sociey, Santa Clara 15107
FMB, Zürich 10692
FMBRA, Rickmansworth 12081
FMC, Beverly Hills 15130
FMCMS, Santa Clara 15107
FMD, Dublin 06950
FMF, Augsburg 05695
FMF, Paris 04514
FMS, Washington 15086
FMSI, Roma 07693
FNB, Washington 15108
FNCA, Köln 06156
F.N.C.V., South Yarra 00396
FNF, Berlin 06167
FNFW, Berlin 06169
FNIEF, Roma 07697
FNKä, Köln 06181
FNL, Berlin 06189
FNL, London 12094
FNOVI, Roma 07696
FNSSA, Adelaide 00397
FNTh, Berlin 06154
FOA, Milwaukee 15087
FOBID, 's-Gravenhage 08894
FOCS, Launceston 12067
Föderation der Internationalen Donausymposia über Diabetes mellitus, Wien 00593
Förbundet Sveriges Arbetsterapeuter, Nacka 10458
Fördergemeinschaft für Absatz- und Werbeforschung, Frankfurt 05702
Fördergemeinschaft für das Süddeutsche Kunststoff-Zentrum e.V., Würzburg 05703
Förderungsgemeinschaft der Kartoffelwirtschaft e.V., Munster 05704
Föreningen för Vattenhygien, Vällingby 10459
Föreningen Svenska Tonsättare, Stockholm 10460
Företagsekonomiska Föreningen, Stockholm 10461
FOGRA Forschungsgesellschaft Druck e.V., München 05705
Folklore of Ireland Society, Dublin 06949
Folklore Society, London 12082
Folklore Society, Sliema 08663
FOM, Utrecht 09078
FOM Institute of Atomic and Molecular Physics, Amsterdam 08896
FOM-Instituut voor Atoom- en Molecuulfysica, Amsterdam 08896
FOM-Instituut voor Plasmafysica Rijnhuizen, Nieuwegein 08897
Fondation Archéologique, Bruxelles 01361
Fondation Born-Bunge pour la Recherche, Wilrijk 01362
Fondation Canadienne d'Education Economique, Toronto 02140
Fondation Canadienne de Recherche en Publicité, Toronto 02015
Fondation Canadienne des Maladies du Coeur, Ottawa 02155
Fondation Canadienne Donner, Toronto 02418
Fondation Canadienne du Parkinson, Toronto 02811
Fondation de la Famille Molson, Montréal 02660
Fondation de Recherche de l'Association des Comptables Généraux Licenciés du Canada, Vancouver 02093
Fondation de Recherches sur les Blessures de la Route au Canada, Ottawa 02968
Fondation du Canada pour les Maladies Thyroïdiennes, Kingston 02963
Fondation du Nouveau-Brunswick des Maladies du Coeur, Saint John 02684
Fondation du Québec des Maladies du Coeur, Montréal 02466
Fondation E.A. Baker pour la Prévention de la Cécité, Toronto 02420
Fondation Egyptologique Reine Elisabeth, Bruxelles 01363
Fondation Fernand Lazard, Bruxelles 01364
Fondation Francqui, Bruxelles 01365
Fondation Hindemith, Vevey 10705
Fondation J. Armand Bombardier, Valcourt 02467
Fondation Justine Lacoste-Beaubien, Montréal 02468
Fondation J.W. McConnell, Montréal 02572
Fondation Lionel Groulx, Outremont 02469
Fondation Médicale Reine Elisabeth, Bruxelles 01366
Fondation Nationale des Sciences Politiques, Paris 04543
Fondation pour la Recherche Sociale, Paris 04544
Fondation pour la Science – Centre International de Synthèse, Paris 04545
Fondation RP pour la Recherche sur les Yeux, Toronto 02872
Fondation Saint-John Perse, Aix-en-Provence 04546
Fondation Schiller Suisse, Zumikon 10925
Fondation Sciences Jeunesse, Ottawa 03002
Fondation Scolaire de l'Institut Canadien du Crédit, Mississauga 02112
Fondation Suisse pour la Psychologie Appliquée, Fribourg 10932
Fondation Universitaire, Bruxelles 01367
Fondazione Centro di Documentazione Ebraica Contemporanea, Milano 07699
Fondo Colombiano de Investigaciones Cientificas y Proyectos Especiales Francisco José de Caldas, Bogotá 03374
Fondo Nacional de las Artes, Buenos Aires 00139
FONOLA-SOL, Kungälv 10436
Food and Nutrition Board, Washington 15108
Food Distribution Research Society, Newark 15109
Food Education Society, London 12083
Food Products Accident Prevention Association, Toronto 02470
Food Technology Association of New South Wales, Sydney 00398
Food Technology Association of Queensland, Brisbane 00399
Food Technology Association of Tasmania, Hobart 00400
Food Technology Association of Western Australia, Perth 00401
FOPS, Norwich 12055
FORCE Institutes, Glostrup 03726
FORCE Institutterne, Glostrup 03726
Foreign Affairs Circle, Richmond 12084
Foreign Policy Association, New York 15110
Foreign Services Research Institute, Washington 15111
Foreningen af Danske Biologer., Birkerød 03727
Foreningen af Danske Civiløkonomer, København 03728
Foreningen af Danske Kunstmuseer, Fur 03729
Foreningen af Danske Museumsmaend, København 03730
Foreningen af Geografilaerere ved de Gymnasiale Uddannelser, Ribe 03731
Foreningen af Licentiater, København 03732
Foreningen af Medarbejdere ved Danmarks Forskningsbiblioteker, Lyngby 03733
Foreningen af Speciallåger, København 03734
Foreningen for Danske Lanbrugsskoler, Greve 03735
Foreningen for National Kunst, København 03736
Foreningen til Norske Fortidsminnesmerkers Bevaring, Oslo 09266
Foreningen til Svampekundskabens Fremme, Søborg 03737
The Forensic Science Society, Harrogate 12085
The Forest Association of Iceland, Reykjavik 06698
Forest History Society, Durham 15112
Forest Products Accident Prevention Association, North Bay 02471
Forest Products Research Society, Madison 15113
Forest Research Center, La Habana 03476
Forestry Association of Nigeria, Ibadan 09228
Forestry Canada – Ontario Region, Sault Sainte Marie 02472
FORMEZ, Roma 07428
FORRAD, Milano 08000
Forschungsgemeinschaft Angewandte Geophysik e.V., Hannover 05706
Forschungsgemeinschaft Eisenhüttenschlacken, Duisburg 05707
Forschungsgemeinschaft Feuerfest e.V., Bonn 05708
Forschungsgemeinschaft für Erkrankungen des Bewegungsapparates, Wien 00594

Forschungsgemeinschaft für Hochspannungs- und Hochstromtechnik e.V., Mannheim 05709
Forschungsgemeinschaft für Nationalökonomie, Sankt Gallen 10706
Forschungsgemeinschaft für technisches Glas e.V., Wertheim 05710
Forschungsgemeinschaft für Verpackungs- und Lebensmitteltechnik e.V., München 05711
Forschungsgemeinschaft Industrieofenbau e.V., Frankfurt 05712
Forschungsgemeinschaft Kalk und Mörtel e.V., Köln 05713
Forschungsgemeinschaft Kraftpapiere und Papiersäcke, Wiesbaden 05714
Forschungsgemeinschaft Naturstein-Industrie e.V., Bonn 05715
Forschungsgemeinschaft Werkzeuge und Werkstoffe e.V., Remscheid 05716
Forschungsgemeinschaft Zink e.V., Düsseldorf 05717
Forschungsgesellschaft Druckmaschinen e.V., Frankfurt 05718
Forschungsgesellschaft für Agrarpolitik und Agrarsoziologie e.V., Bonn 05719
Forschungsgesellschaft für das Verkehrs- und Straßenwesen im ÖIAV, Wien 00595
Forschungsgesellschaft für Psycho-Elektronik und Kybernetik, Wien 00596
Forschungsgesellschaft für Strassen- und Verkehrswesen e.V., Köln 05720
Forschungsgesellschaft für Wohnen, Bauen und Planen, Wien 00597
Forschungsgesellschaft Kunststoffe e.V., Darmstadt 05721
Forschungsgesellschaft Landschaftsentwicklung Landschaftsbau e.V., Bonn 05722
Forschungsgesellschaft Stahlverformung e.V., Hagen 05723
Forschungsgesellschaft Steinzeugindustrie e.V., Köln 05724
Forschungs-Gesellschaft Verfahrenstechnik e.V., Düsseldorf 05725
Forschungsgruppe für Anthropologie und Religionsgeschichte e.V., Saarbrücken 05726
Forschungsgruppe Köln, Köln 05727
Forschungsinstitut für Arbeiterbildung e.V., Recklinghausen 05728
Forschungsinstitut für Gesellschaftspolitik und beratende Sozialwissenschaft e.V., Göttingen 05729
Forschungsinstitut für Pigmente und Lacke e.V., Stuttgart 05730
Forschungsinstitut für Rationalisierung e.V., Aachen 05731
Forschungsinstitut für Wärmeschutz e.V., Gräfelfing 05732
Forschungsinstitut zu Politik und Gesellschaft überseeischer Länder, Freiburg 05152
Forschungskuratorium Gesamttextil, Eschborn 05733
Forschungskuratorium Maschinenbau e.V., Frankfurt 05734
Forschungsrat Kältetechnik e.V., Frankfurt 05735
Forschungsstelle des Bundesverbandes der Deutschen Ziegelindustrie e.V., Bonn 05736
Forschungsstelle für Acetylen, Dortmund 05737
Forschungsstelle für den Handel Berlin e.V., Berlin 05738
Forschungsstelle für internationale Agrar- und Wirtschaftsentwicklung e.V., Heidelberg 05739
Forschungsstiftung der VGB Technische Vereinigung der Großkraftwerksbetreiber e.V., Essen 06432
Forschungsverband für den Handelsvertreter- und Handelsmaklerberuf, Köln 05740
Forschungsvereinigung Antriebstechnik e.V., Frankfurt 05741
Forschungsvereinigung Automobiltechnik e.V., Frankfurt 05742
Forschungsvereinigung der Deutschen Asphaltindustrie e.V., Offenbach 05743
Forschungsvereinigung der Gipsindustrie e.V., Darmstadt 05744
Forschungsvereinigung der Rheinischen Bimsindustrie e.V., Neuwied 05745
Forschungsvereinigung Elektrotechnik beim ZVEI e.V., Frankfurt 05746
Forschungsvereinigung Feinmechanik und Optik e.V., Köln 05747
Forschungsvereinigung für angewandte Schloß-, Beschlag- und präventive Sicherheitstechnik e.V., Velbert 05748
Forschungsvereinigung für Luft- und Trocknungstechnik e.V., Frankfurt 05749
Forschungsvereinigung Kalk-Sand e.V., Hannover 05750
Forschungsvereinigung Porenbeton e.V., Wiesbaden 05751
Forschungsvereinigung Programmiersprachen für Fertigungseinrichtungen e.V., Aachen 05752
Forschungsvereinigung Schweißen und Schneiden e.V., Düsseldorf 05753
Forschungsvereinigung Verbrennungskraftmaschinen e.V., Frankfurt 05754
Forschungsvereinigung Ziegelindustrie, Bonn 05755
Forschungszentrum des Deutschen Schiffbaues e.V., Hamburg 05756
Forskningsrådsnämnden, Stockholm 10462
Fort Cumberland and Portsmouth Militaria Society, Portsmouth 12086
Fortress Study Group, Newport 12087
The Forum, Santa Fe 15114
The Forum for Health Care Planning, Washington 15115
Forum for Medical Affairs, Lake Success 15116
Forum International: International Ecosystems University, Pleasant Hill 15117
Forum Italiano dell'Energia Nucleare, Roma 07700
Fostoria Glass Society of America, Moundsville 15118
Foundation Centre of Research and Study, Salvador 01651
Foundation for Accounting Education, New York 15119
Foundation for Advancement in Cancer Therapy, New York 15120
Foundation for Biomedical Research, Washington 15121
Foundation for Business Responsibilities, London 12088
Foundation for Chiropractic Education and Research, Arlington 15122
Foundation for Educational Futures, Charlotte 15123
Foundation for Exceptional Children, Reston 15124
Foundation for Field Research, Alpine 15125
Foundation for Fundamental Research on Matter, Utrecht 09078
Foundation for Independent Research on Technology and Health, Goodwood 02473
Foundation for Indonesian Standardisation, Bandung 06895
Foundation for Interior Design Education Research, Grand Rapids 15126
Foundation for International Cooperation, Park Ridge 15127
Foundation for Latin American Anthropological Research, Cocoa 15128
Foundation for Microbiology, New York 15129
Foundation for Mideast Communication, Beverly Hills 15130
Foundation for Research in the Afro-American Creative Arts, Cambria Heights 15131
Foundation for Research, Science and Technology, Wellington 09125
Foundation for Science and the Handicapped, Morgantown 15132
The Foundation for Scientific and Industrial Research at the Norwegian Institute of Technology, Trondheim 09353
Foundation for the Advancement of Chiropractic Tenets and Science, Arlington 15133
Foundation for the Advancement of Tropical Research, 's-Gravenhage 09079
Foundation for the Study of Infant Deaths, London 12089
Foundation for the Support of International Medical Training, Lewiston 15134
Foundation of the Canadian College of Health Service Executives, Ottawa 02474
FPA, Kansas City 15105
FPA, London 12056, 12076
FPA, New York 15110
Fpm, Hagen 05696
FPO, Chicago 15088
FPRS, Madison 15113
FPS, Milwaukee 15102
FRAACA, Cambria Heights 15131
Fränkische Geographische Gesellschaft e.V., Erlangen 05757
France Intec, Paris 04547
Francis Bacon Foundation, Claremont 15135
The Francis Bacon Society, London 12090
Franco-British Society, London 12091
Frankfurter Geographische Gesellschaft e.V., Frankfurt 05758
Franklin and Eleanor Roosevelt Institute, Hyde Park 15136
Frank Lloyd Wright Foundation, Scottsdale 15137
Frankslaererforeningen, København 03738
Franz Schmidt-Gesellschaft, Wien 00598
Fraser Institute, Vancouver 02475
Fraunhofer-Gesellschaft zur Förderung der angewandten Forschung e.V., München 05759
FRC, Winchester 15098
Frederick Chopin Society, Warszawa 09620
Frederick Douglass Memorial and Historical Association, Columbia 15138
Freier Deutscher Autorenverband e.V., München 05760
Freier Verband Deutscher Zahnärzte e.V., Bonn 05761
Freie Vereinigung von Fachleuten öffentlicher Verkehrsbetriebe, Gelsenkirchen 05762
French Association for Standardization, Paris La Défence 04221
French-Language Society for Reanimation, Caen 04685
French Society for Clinical Sexology, Paris 04842
French Society of Genetics, Orléans 04801
French Youth Esperanto Organization, Paris 04612
Freshwater Biological Association, Ambleside 12092
Freundeskreis Deutscher Auslandsschulen e.V., Bonn 05763
FRIA, Rowland Heights 15096
Friars Club, New York 15139
The Friedrich Ebert Foundation, Bonn 05764
Friedrich-Ebert-Stiftung e.V., Bonn 05764
Friedrich Hebbel-Gesellschaft, Wien 00599
Friends Historical Association, Haverford 15140
Friends Historical Society, London 12093
Friends of American Art in Religion, New York 15141
Friends of Cast Iron Architecture, New York 15142
Friends of George Sand, Hempstead 15143
Friends of Medieval Dublin, Dublin 06950
Friends of the Lincoln Museum, Harrogate 15144
Friends of the Museums Association, Barcelona 10246
The Friends of the National Collections of Ireland, Dublin 06951
Friends of The National Libraries, London 12094
Friends of the Sea Lion Marine Mammal Center, Laguna Beach 15145
Fries Genootschap van Geschied-, Oudheid- en Taalkunde, Leeuwarden 08898
Frisian Academy, Leeuwarden 08899
FRN, Stockholm 10462
Frobenius-Gesellschaft e.V., Frankfurt 05765
Frontinus-Gesellschaft e.V., Bergisch Gladbach 05766
Fryske Akademy, Leeuwarden 08899
FS, Leicester 12075
FS, Princeton 15092
FSA, Nacka 10458
FSAI, Lachen 11001
F.S.A.S.R., Lausanne 10700
FSBI, Cambridge 12077
FSC, Shrewsburry 12074
FSCT, Blue Bell 15089
F.S.D.L., Paris 04518
FSF, Helsinki 03958
FSG, Newport 12087
FSH, Morgantown 15132
FSIMT, Lewiston 15134
FSKZ, Würzburg 05703
FSL, Helsinki 03959
FSLMMC, Laguna Beach 15145
FSMB, Fort Worth 15090
FSRI, Washington 15111
F.S.T., Stockholm 10460
FT, New York 13886
FTF, Stockholm 10457
FtG, Wertheim 05710
FTTPP, Dallas 15091
FUA, Buenos Aires 00138
Fuldaer Geschichtsverein e.V., Fulda 05767
FUNAI, Brasília 01660
Fundação Antonio Prudente, São Paulo 01650
Fundação Calouste Gulbenkian, Lisboa 09657
Fundação Centro de Pesquisas e Estudos, Salvador 01651
Fundação de Estudos Sociais do Paraná, Curitiba 01652
Fundação Educacional de Fortaleza, Fortaleza 01653
Fundação Getulio Vargas, Rio de Janeiro 01654
Fundação Instituto Brasileiro de Geografia e Estatística, Rio de Janeiro 01655
Fundação Instituto Tecnológico do Estado de Pernambuco, Recife 01656
Fundação João Pineiro, Belo Horizonte 01657
Fundação Joaquim Nabuco, Recife 01658
Fundação Moinho Santista, São Paulo 01659
Fundação Nacional do Indio, Brasília 01660
Fundación Gildemeister, Santiago 03052
Fundación Juan March, Madrid 10311
Fundación Miguel Lillo, San Miguel de Tucumán 00140
Fundación OFA para el Avance de las Ciencias Biomédicas, Bogotá 03375
Fundación Universitaria Simon I. Patiño, Cochabamba 01545
FUNDAJ, Recife 01658
FUPAC, Guatemala City 06568
Furniture History Society, London 12095
Furniture Industry Research Association, Stevenage 12096
FUSE, Columbus 15079
FV, Gelsenkirchen 05762
FVA, Columbia 15106
FVH, Vällingby 10459
FVS, Wien 00595
FVV, Frankfurt 05754
Fylkingen, Stockholm 10463
Fysikkforeningen, Oslo 09267
Fysisk Forening, København 03739

G.A., London 12112
G.A., Sheffield 12110
GAA, Rochester 15173
GAD, Washington 15148
GADEF, Paris 04553
Gaelic League, Dublin 06937
Gaelic League of Scotland, Glasgow 12097
G.A. Frecker Association on Gerontology, St. John's 02476
GAI, Bad Homburg 05797
G.A.I., Roma 07701
Gairdner Foundation, Willowdale 02477
Gakujutsu Bunken Fukyu-Kai, Tokyo 08128
The Galactic Society, Rock Hill 15146
The Galpin Society for the Study of Musical Instruments, Leicester 12098
Galton Institute, London 12099
The Game Conservancy Trust, Fordingbridge 12100
GAMM, Hamburg 05798
GAP, Dallas 15179
Garrod Association of Canada, Halifax 02478
G.A.S., Glasgow 12118
GAS, Seattle 15161
Gas Research Institute, Chicago 15147
Gastroenterologická Společnost, Praha 03544
Gaswärme-Institut e.V., Essen 05768
GAWS, Jersey City 15157
GBA, Petersfield 11348
GBCh, Tutzing 05805
GBDL, Karlsruhe 05804
GBGSA, Petersfield 11347
GBO, Bruxelles 01369
GBS, Bruxelles 01370
GCA, Alexandria 15172
GCFI, Charleston 15182
GCMF, Lexington 15154
G.C.R.O., Reading 12102
Gdánskie Towarzystwo Naukowe, Gdánsk 09543
GdB, Eschborn 05792
G.D.C., London 12104
GDCh, Frankfurt 05793
GDD, Bonn 05806
GdM, Wien 00612
GDMB, Clausthal-Zellerfeld 05794
GDNÄ, Leverkusen 05795
GDPA, Manchester 12105
GEA, New York 15163
Gebrüder Schrammel-Gesellschaft, Wien 00600
GEDI, Seattle 15149
GEDOK, Wuppertal 06338
Gedverndarfélag Islands, Reykjavik 06687
GE.FE.BI., Graz 00635
GEFIU, Frankfurt 05822
GEG, Basel 10709
GEGZ, Zürich 10708
Gelre, Arnhem 08900
Gemeinnütziger Verein zur Durchführung von Lehr- und Forschungsaufgaben an der Wirtschaftsuniversität Wien, Wien 00601
Gemeinnütziger Verein zur Förderung von Philosophie und Theologie e.V., Bornheim 05769
Gemeinschaft katholischer Studierender und Akademiker, Viernheim 05770
Gemeinschaftsausschuss Kaltformgebung e.V., Düsseldorf 05771
Gemeinschaftswerk der Evangelischen Publizistik e.V., Frankfurt 05772
Gemeinschaft zur Förderung der privaten deutschen Pflanzenzüchtung e.V., Bonn 05773
Gemmological Association and Gem Testing Laboratory of Great Britain, London 12101
Gemmological Association of Germany, Idar-Oberstein 05309
Gemmological Society of Finland, Helsinki 04021
Gemologisches Institut, Idar-Oberstein 05774
Genealogical Association of Nova Scotia, Halifax 02479
Genealogical Institute of the Maritimes, Halifax 02480
Genealogical Office, Dublin 06952
Genealogical Society, Tokyo 08162
Genealogical Society of Finland, Helsinki 04060
Genealogical Society of South Africa, Kelvin 10143
Genealogische Gesellschaft Hamburg e.V., Hamburg 05775
Genealogiska Samfundet i Finland, Helsinki 04060
General Anthropology Division, Washington 15148
General Council and Register of Osteopaths, Reading 12102
General Council of Official Colleges of Doctors and Licenciates in Philosophy, Letters and Science, Madrid 10297
General Council of the Bar, London 12103
General Dental Council, London 12104
General Dental Practitioner's Association, Manchester 12105
General Educational Development Institute, Seattle 15149
General Medical Council, London 12106
General Organization for Housing, Building and Planning Research, Cairo 03902
General Society of Mechanics and Tradesmen, New York 15150
General Society of Spanish Authors, Madrid 10398
General Studies Association, York 12107
The General Teaching Council for Scotland, Edinburgh 12108
The Genetical Society of Great Britain, Edinburgh 12109
Genetics Society of America, Bethesda 15151
Genetics Society of Canada, Ottawa 02481
Genetics Society of China, Beijing 03199
The Genetics Society of Israel, Neve Ya'ar 07054
Genetics Society of Japan, Shizuoka 08209
Genetics Society of Nigeria, Ibadan 09229
Genootschap Amstelodamum, Amsterdam 08901
Genootschap Architectura et Amicitia, Amsterdam 08902
Genootschap ter Bevordering van Melkkunde, Rijswijk 08903
Genootschap ter Bevordering von Natuur-, Genees- en Heelkunde, Amsterdam 08904
Genootschap voor Wetenschappelijke Filosofie, Paterswolde 08905
Geochemical Society, Washington 15152
Geodetic Society of Japan, Ibaraki 08385
Geodéziai és Kartográfiai Egyesület, Budapest 06605
Geofysiikan Seura, Espoo 03962
Geografiläramas Riksförening, Lund 10464
Geografiska Förbundet, Stockholm 10465
Geografsko Društvo na Makedonija, Skopje 08602
The Geographical Association, Sheffield 12110
Geographical Association of Zimbabwe, Harare 17170
Geographical Society of China, Beijing 03200
The Geographical Society of China, Taipei 03298
Geographical Society of Finland, Helsingin Yliopisto 04047
Geographical Society of India, Calcutta 06760
The Geographical Society of Ireland, Galway 06953
Geographical Society of Macedonia, Skopje 08602
Geographical Society of New South Wales, Gladesville 00402
Geographic Society of Moldova, Chişinău 08779
Geographische Gesellschaft, München 05776
Geographische Gesellschaft Bern, Bern 10707
Geographische Gesellschaft Bremen, Bremen 05777
Geographische Gesellschaft in Hamburg e.V., Hamburg 05778
Geographische Gesellschaft zu Hannover, Hannover 05779
Geographisch-Ethnographische Gesellschaft Zürich, Zürich 10708
Geographisch-Ethnologische Gesellschaft Basel, Basel 10709
Geography Teachers' Association of New South Wales, Rozelle 00403
GEO-KART, Oberkochen 05780
Geologian Tutkimuskeskus, Espoo 03963
Geological Association of Canada, Saint John's 02482
Geological, Mining and Metallurgical Society of India, Calcutta 06761
Geological, Mining and Metallurgical Society of Liberia, Monrovia 08563
The Geological Society, London 12111
Geological Society, Moskva 09870
Geological Society, Uppsala 10466
Geological Society of America, Boulder 15153

Geological Society of Australia, Sydney *00404*
Geological Society of China, Taipei *03299*
Geological Society of Jamaica, Kinston *08079*
Geological Society of Japan, Tokyo *08290*
Geological Society of New Zealand, Lower Hutt *09126*
Geological Society of Poland, Kraków *09572*
The Geological Society of South Africa, Linden *10144*
Geological Society of Zimbabwe, Harare *17171*
Geological Survey of Finland, Espoo *03963*
Geological Survey of Ghana, Accra *06481*
Geological Survey of India, Calcutta *06762*
Geological Survey of Malawi, Zomba *08622*
Geological Survey of Malaysia, Ipoh *08642*
Geological Survey of Malaysia, Kuching *08643*
Geological Survey of New South Wales, Sydney *00405*
Geological Survey of Nigeria, Kaduna South *09230*
Geological Survey of Norway, Trondheim *09283*
Geological Survey of Western Australia, East Perth *00406*
Die Geologiese Vereniging van Suid-Afrika, Linden *10144*
Geologiliitto, Helsinki *03964*
Geologische Vereinigung e.V., Mendig *05781*
Geologisch Mijnbouwkundige Dienst, Paramaribo *10431*
Geologiska Föreningen, Uppsala *10466*
Geologists' Association, London *12112*
Geologists' Association of Albania, Tirana *00007*
Geomorphological Commission, Moskva *09871*
Geophysical Commission, Oslo *09250*
Geophysical Society of Finland, Espoo *03962*
Georg-Agricola-Gesellschaft zur Förderung der Geschichte der Naturwissenschaften und der Technik e.V., Düsseldorf *05782*
George C. Marshall Foundation, Lexington *15154*
George Eliot Fellowship, Coventry *12113*
Georg-Friedrich-Händel-Gesellschaft e.V., Halle *05783*
Georgian Academy of Sciences, Tbilisi *05011*
Georgian Bio-Mecico-Technical Society, Tbilisi *05012*
Georgian Botanical Society, Tbilisi *05013*
Georgian Commission on Archaeology, Tbilisi *05014*
Georgian Geographical Society, Tbilisi *05015*
Georgian Geological Society, Tbilisi *05016*
Georgian History Society, Tbilisi *05017*
Georgian National Committee on UNESCO Long-Term Programme „Man and the Biosphere", Tbilisi *05018*
Georgian National Speleological Society, Tbilisi *05019*
Georgian Philosophy Centre, Tbilisi *05020*
Georgian Society of Biochemistry, Tbilisi *05021*
Georgian Society of Genetics and Selectionists, Tbilisi *05022*
Georgian Society of Helminthologists, Tbilisi *05023*
Georgian Society of Patho-Anatomists, Tbilisi *05024*
Georgian Society of Physiologists, Tbilisi *05025*
Georgian Society of Psychologists, Tbilisi *05026*
Georg-von-Vollmar-Akademie e.V., München *05784*
Geothermal Resources Council, Davis *15155*
Gépipari Tudományos Egyesület, Budapest *06606*
German Aerospace Research Establishment, Köln *05307*
Germana Esperanto Asocio r.a., Bonn *05785*
German-American Information and Education Association, Burke *15156*
German American World Society, Jersey City *15157*
German Association for Asian Studies, Hamburg *05325*
German Association for Water Resources and Land Improvement, Bonn *05552*
German Association of Biologists, Hamburg *06350*
German-Canadian Historical Association, Montréal *02483*
Germania Judaica, Köln *05786*
German Institute for Medical Documentation and Information, Köln *05586*
German Institute for Urban Affairs, Berlin *05587*
German Library Institute, Berlin *05570*
German Society for Bog and Peat Research, Hannover *05385*
German Society for Linguistics, Passau *05427*
German Society of Anaesthesiology and Intensive Medicine, Nürnberg *05319*
Gerontological Society of America, Washington *15158*
Gerontology and Geriatrics Society, Kiev *11189*
Gerontology Association of Nova Scotia, Wolfville *02484*
Gerson Institute, Bonita *15159*
Gesamtverein der deutschen Geschichts- und Altertumsvereine, Köln *05787*
Geschichtsverein für Kärnten, Klagenfurt *00602*
Geschiedkundige Vereniging Die Haghe, 's-Gravenhage *08906*
Gesellschaft anthroposophischer Ärzte, Stuttgart *05788*
Gesellschaft der Ärzte für Erfahrungsheilkunde e.V., Heidelberg *05789*
Gesellschaft der Ärzte in Vorarlberg, Dornbirn *00603*
Gesellschaft der Ärzte in Wien, Wien *00604*
Gesellschaft der Bibliophilen e.V., Höchberg *05790*
Gesellschaft der Chirurgen in Wien, Wien *00605*
Gesellschaft der Freunde alter Musikinstrumente, Binningen *10710*
Gesellschaft der Freunde der Biologischen Station Wilhelminenberg, Wien *00606*
Gesellschaft der Freunde der Neuen Galerie, Graz *00607*
Gesellschaft der Freunde der Österreichischen Nationalbibliothek, Wien *00608*
Gesellschaft der Freunde des Kunsthistorischen Institutes der Karl-Franzens-Universität in Graz, Graz *00609*
Gesellschaft der Geologie- und Bergbaustudenten in Österreich, Wien *00610*
Gesellschaft der Kunstfreunde, Wien *00611*
Gesellschaft der Musikfreunde in Wien, Wien *00612*
Gesellschaft der Musik- und Kunstfreunde Heidelberg e.V., Heidelberg *05791*
Gesellschaft des Bauwesens e.V., Eschborn *05792*
Gesellschaft Deutscher Chemiker e.V., Frankfurt *05793*
Gesellschaft Deutscher Metallhütten- und Bergleute e.V., Clausthal-Zellerfeld *05794*
Gesellschaft Deutscher Naturforscher und Ärzte e.V., Leverkusen *05795*
Gesellschaft für Agrargeschichte, Stuttgart *05796*
Gesellschaft für angewandte Informatik, Bad Homburg *05797*
Gesellschaft für Angewandte Mathematik und Mechanik, Hamburg *05798*
Gesellschaft für Anlagen- und Reaktorsicherheit, Köln *05799*
Gesellschaft für Anthropologie und Humangenetik, Bremen *05800*
Gesellschaft für Arbeitsrecht und Sozialrecht, Linz *00613*
Gesellschaft für Arbeitswissenschaft e.V., Dortmund *05801*
Gesellschaft für Arzneipflanzenforschung e.V., Kleinrinderfeld *05802*
Gesellschaft für Arzneipflanzenforschung e.V., Zürich *10711*
Gesellschaft für bedrohte Völker e.V., Göttingen *05803*
Gesellschaft für Bibliothekswesen und Dokumentation des Landbaues, Karlsruhe *05804*
Gesellschaft für Biologische Chemie e.V., Tutzing *05805*
Gesellschaft für biologische und psychosomatische Medizin, Wien *00614*
Gesellschaft für Chemiewirtschaft, Wien *00615*
Gesellschaft für das Recht der Ostkirchen, Wien *00616*
Gesellschaft für das Schweizerische Landesmuseum, Zürich *10712*
Gesellschaft für Datenschutz und Datensicherung e.V., Bonn *05806*
Gesellschaft für Deutsche Postgeschichte, Frankfurt *05807*
Gesellschaft für deutsche Sprache e.V., Wiesbaden *05808*
Gesellschaft für deutsche Sprache und Literatur in Zürich, Zürich *10713*
Gesellschaft für Deutschlandforschung e.V., Berlin *05809*
Gesellschaft für Dezentralisierte Energiewirtschaft e.V., Ludwigsburg *05810*
Gesellschaft für die EDV-Ausbildung, Darmstadt *05941*
Gesellschaft für die Geschichte des Protestantismus in Österreich, Wien *00617*
Gesellschaft für die Geschichte und Bibliographie des Brauwesens e.V., Berlin *05811*
Gesellschaft für Elektrische Hochleistungsprüfungen, Frankfurt *05812*
Gesellschaft für empirische soziologische Forschung e.V., Nürnberg *05813*
Gesellschaft für Epilepsieforschung e.V., Bielefeld *05814*
Gesellschaft für Erdkunde zu Berlin, Berlin *05815*
Gesellschaft für Erdkunde zu Köln e.V., Köln *05816*
Gesellschaft für Erd- und Völkerkunde, Bonn *05817*
Gesellschaft für Erd- und Völkerkunde zu Stuttgart e.V., Stuttgart *05818*
Gesellschaft für Ernährungsbiologie e.V., München *05819*
Gesellschaft für Ernährungsphysiologie, Frankfurt *05820*
Gesellschaft für Familienkunde in Franken e.V., Nürnberg *05821*
Gesellschaft für Finanzwirtschaft in der Unternehmensführung e.V., Frankfurt *05822*
Gesellschaft für Forschung auf biophysikalischen Grenzgebieten, Binningen *10714*
Gesellschaft für Forschungen zur Aufführungspraxis, Graz *00618*
Gesellschaft für Führungspraxis und Personalentwicklung, Zürich *10715*
Gesellschaft für Ganzheitsforschung, Wien *00619*
Gesellschaft für Geistesgeschichte e.V., Erlangen *05823*
Gesellschaft für Geographie und Geologie Bochum, Bochum *05824*
Gesellschaft für Geschichte der Neuzeit, Salzburg *00620*
Gesellschaft für Geschichte und Kultur e.V., Bonn *05825*
Gesellschaft für Goldschmiedekunst, Hanau *05826*
Gesellschaft für Historische Waffen- und Kostümkunde, Hannover *05827*
Gesellschaft für Immunologie e.V., Marburg *05828*
Gesellschaft für Informatik e.V., Bonn *05829*
Gesellschaft für Informationsverarbeitung in der Landwirtschaft, Stuttgart *05830*
Gesellschaft für Informationsvermittlung und Technologieberatung, München *05831*
Gesellschaft für Innere Medizin in Wien, Wien *00621*
Gesellschaft für Input-Output-Analyse, Wien *00622*
Gesellschaft für Interkulturelle Germanistik e.V., Bayreuth *05832*
Gesellschaft für internationale Geldgeschichte, Frankfurt *05833*
Gesellschaft für internationale Sprache e.V., Reinbek *05834*
Gesellschaft für Klassifikation e.V., Aachen *05835*
Gesellschaft für Klassische Philologie in Innsbruck, Innsbruck *00623*
Gesellschaft für Konsum-, Markt- und Absatzforschung, Nürnberg *05836*
Gesellschaft für Kulturpsychologie, Salzburg *00624*
Gesellschaft für Landeskunde, Linz *00717*
Gesellschaft für Literatur in Nordrhein-Westfalen e.V., Münster *05837*
Gesellschaft für Logotherapie und Existenzanalyse, Wien *00625*
Gesellschaft für Lungen- und Atmungsforschung e.V., Bochum *05838*
Gesellschaft für Manuelle Lymphdrainage nach Dr. Vodder, Walchsee *00626*
Gesellschaft für Mathematik und Datenverarbeitung, Sankt Augustin *05839*
Gesellschaft für mathematische Forschung e.V., Oberwolfach *05840*
Gesellschaft für Mittelrheinische Kirchengeschichte, Koblenz *05841*
Gesellschaft für Musikforschung e.V., Kassel *05842*
Gesellschaft für Naturkunde in Württemberg e.V., Stuttgart *05843*
Gesellschaft für Natur- und Völkerkunde Ostasiens e.V., Hamburg *05844*
Gesellschaft für Neue Musik e.V., Frankfurt *05845*
Gesellschaft für öffentliche Wirtschaft e.V., Berlin *05846*
Gesellschaft für österreichische Kulturgeschichte, Eisenstadt *00627*
Gesellschaft für Organisation e.V., Giessen *05847*
Gesellschaft für Ost- und Südostkunde, Linz *00628*
Gesellschaft für Pädagogik und Information e.V., Paderborn *05848*
Gesellschaft für Pädiatrische Radiologie, Würzburg *05849*
Gesellschaft für Phänomenologie und Kritische Anthropologie, Wien *00629*
Gesellschaft für Photographie und Geschichte, Wien *00630*
Gesellschaft für politische Aufklärung, Wien *00631*
Gesellschaft für prä- und postoperative Tumortherapie e.V., Undenheim *05850*
Gesellschaft für praktische Energiekunde e.V., München *05851*
Gesellschaft für Programmierte Instruktion und Mediendidaktik e.V., Giessen *05852*
Gesellschaft für Psychotherapie, Psychosomatik und Medizinische Psychologie e.V., Lipzig *05853*
Gesellschaft für publizistische Bildungsarbeit e.V., Hagen *05854*
Gesellschaft für rationale Verkehrspolitik e.V., Düsseldorf *05855*
Gesellschaft für Rationelle Energieverwendung e.V., Berlin *05856*
Gesellschaft für Rechtsvergleichung e.V., Göttingen *05857*
Gesellschaft für Regionalforschung e.V., Bonn *05858*
Gesellschaft für Salzburger Landeskunde, Salzburg *00632*
Gesellschaft für Schweizerische Kunstgeschichte, Bern *10716*
Gesellschaft für Sicherheitswissenschaft e.V., Wuppertal *05859*
Gesellschaft für Sozial- und Wirtschaftsgeschichte, Heidelberg *05860*
Gesellschaft für sozialwissenschaftliche Forschung in der Medizin, Freiburg *05861*
Gesellschaft für Soziologie an der Universität Graz, Graz *00633*
Gesellschaft für Strahlen- und Umweltforschung, Ergersheim *05862*
Gesellschaft für Strategische Unternehmensführung, Innsbruck *00634*
Gesellschaft für Systementwicklung und Informationsverarbeitung, Köln *05940*
Gesellschaft für technisch-wissenschaftliche Weiterbildung, Weßling *05271*
Gesellschaft für Technologiefolgenforschung e.V., Berlin *05863*
Gesellschaft für Tribologie e.V., Moers *05864*
Gesellschaft für Übernationale Zusammenarbeit e.V., Bonn *05865*
Gesellschaft für Unternehmensgeschichte e.V., Köln *05866*
Gesellschaft für Ursachenforschung bei Verkehrsunfällen e.V., Köln *05867*
Gesellschaft für vergleichende Felsbildforschung, Graz *00635*
Gesellschaft für vergleichende Kunstforschung, Wien *00636*
Gesellschaft für Versicherungswissenschaft und -gestaltung e.V., Köln *05868*
Gesellschaft für Versuchstierkunde, Füllinsdorf *10717*
Gesellschaft für Wirbelsäulenforschung, Frankfurt *05869*
Gesellschaft für Wirtschaftskunde e.V., Siegen *05870*
Gesellschaft für Wirtschafts- und Sozialwissenschaften, Mannheim *05871*
Gesellschaft für wissenschaftliche Gesprächspsychotherapie e.V., Köln *05872*
Gesellschaft für Wissenschaft und Leben im Rheinisch-Westfälischen Industriegebiet e.V., Essen *05873*
Gesellschaft für Wohnungsrecht und Wohnungswirtschaft Köln e.V., Köln *05874*
Gesellschaft Information Bildung e.V., Frankfurt *05875*
Gesellschaft österreichischer Chemiker, Wien *00637*
Gesellschaft Österreichischer Nervenärzte und Psychiater, Wien *00638*
Gesellschaft Pro Vindonissa, Brugg *10718*
Gesellschaft Rheinischer Ornithologen e.V., Düsseldorf *05876*
Gesellschaft schweizerischer Amtsärzte, Bern *10719*
Gesellschaft Schweizerischer Maler, Bildhauer und Architekten, Muttenz *10720*
Gesellschaft Schweizerischer Tierärzte, Bern *10721*
Gesellschaft schweizerischer Zeichenlehrer, Einsiedeln *10722*
Gesellschaft Sozialwissenschaftlicher Infrastruktureinrichtungen e.V., Mannheim *05877*
Gesellschaft zum Studium strukturpolitischer Fragen e.V., Bonn *05878*
Gesellschaft zum Studium und zur Erneuerung der Struktur der Rechtsordnung, Wien *00639*
Gesellschaft zur Erforschung des Markenwesens e.V., Wiesbaden *05879*
Gesellschaft zur Erforschung slawischer Sprachen und Kulturen, Graz *00640*
Gesellschaft zur Erforschung und Förderung der österreichischen Bundesstaatsidee und des österreichischen Nationalbewusstseins, Wien *00936*
Gesellschaft zur Errichtung der Akademie für Allgemeinmedizin, Graz *00641*
Gesellschaft zur Förderung der Erforschung der Zuckerkrankheit e.V., Düsseldorf *05880*
Gesellschaft zur Förderung der finanzwissenschaftlichen Forschung e.V., Köln *05881*
Gesellschaft zur Förderung der industriellen Pflanzenbaus, Wien *00642*
Gesellschaft zur Förderung der Lufthygiene und Silikoseforschung e.V., Düsseldorf *05882*
Gesellschaft zur Förderung der Metallforschung, Düsseldorf *06296*
Gesellschaft zur Förderung der Segelflugforschung e.V., Freiburg *05883*
Gesellschaft zur Förderung der Spektrochemie und angewandten Spektroskopie e.V., Dortmund *05884*
Gesellschaft zur Förderung der Wissenschaftlichen Forschung über das Spar- und Girowesen e.V., Bonn *05885*
Gesellschaft zur Förderung der wissenschaftlichen Zusammenarbeit mit der Universität Tel-Aviv, Bonn *05886*
Gesellschaft zur Förderung des Unternehmernachwuchses e.V., Baden-Baden *05887*
Gesellschaft zur Förderung Frankfurter Malerei des 19. und 20. Jahnhunderts e.V., Frankfurt *05888*
Gesellschaft zur Förderung Pädagogischer Forschung e.V., Frankfurt *05889*
Gesellschaft zur Förderung Slawistischer Studien, Wien *00643*
Gesellschaft zur Förderung von Nordamerika-Studien an der Universität Wien, Wien *00644*
Gesellschaft zur Herausgabe des Corpus Catholicorum, Münster *05890*
Gesellschaft zur Herausgabe von Denkmälern der Tonkunst in Österreich, Wien *00645*
GES-Gesellschaft für elektronische Systemforschung, Allensbach *05891*
GESIS, Mannheim *05877*
Gesundheitspolitische Gesellschaft e.V., Kiel *05892*
Gesundheitstechnische Gesellschaft e.V., Berlin *05893*
GF, Stockholm *10465*
GfA, Dortmund *05801*
GfA, Graz *00618*
GfA, Kleinrinderfeld *05802*
GFBG, Binningen *10714*
GfdS, Wiesbaden *05808*
GFE, Berlin *05815*
G.f.G., Wien *00619*
GfK, Nürnberg *05836*
GfKl, Aachen *05835*
GfM, Kassel *05842*
GFM-GETAS Gesellschaft für Marketing-, Kommunikations- und Sozialforschung, Hamburg *05894*
GFP, Bonn *05773*
GfP, Zürich *10715*
GFPE, München *05851*
GFPF, Frankfurt *05889*
GfR, Bonn *05858*
GfS, Wuppertal *05859*
GfT, Moers *05864*
GFU, Baden-Baden *05887*
GfürO, Giessen *05847*
GG, Berlin *05893*
GG, Mainz *05902*
G.G.A., Accra *06485*
GGA, Churcham *12123*
GGH, Hannover *05779*
GGL, London *12134*
GHAA, Washington *15181*
Ghana Academy of Arts and Sciences, Accra *06482*
Ghana Association of Writers, Accra *06483*
Ghana Bar Association, Accra *06484*
Ghana Geographical Association, Accra *06485*
Ghana Library Association, Accra *06486*
Ghana Library Board, Accra *06487*
Ghana Meteorological Services Department, Accra *06488*
Ghana Science Association, Accra *06489*
Ghana Sociological Association, Accra *06490*
Ghaqda Bibljotekarji, Floriana *08662*
Ghaqda Tal Folklor, Sliema *08663*
Gh.B., Floriana *08662*
G.H.S., Cardiff *12116*
GI, Bonita *15159*
GI, Bonn *05829*
GIB, Frankfurt *05875*
Gibraltar Ornithological and Natural History Society, Gibraltar *06495*
The Gibraltar Society, Gibraltar *06496*
Gibraltar Teachers' Association, Gibraltar *06497*
GIE, Idar-Oberstein *05774*
GIEA, Burke *15156*

GIG, Bayreuth 05832
GIG, Frankfurt 05833
GIL, Stuttgart 05830
GILA, Pisa 07702
Gilbert and Sullivan Society, Brooklym 15160
Gilbert and Sullivan Society of Edinburgh, Edinburgh 12114
Gilde Internationaler Edelsteinexperten, Idar-Oberstein 05774
GIMS, Bronx 15171
Girl's Schools Association, Oxford 12115
GIRSO, Bruxelles 01371, 01372
G.K. Chesterton Society, Saskatoon 02485
GKE, Budapest 06605
GKSS, Geesthacht 05895
GKSS-Forschungszentrum Geesthacht, Geesthacht 05895
GL, Union 15164
GLA, Georgetown 06577
Glamorgan History Society, Cardiff 12116
GLAR, Buenos Aires 00141
Glasgow Agricultural Society, Ayr 12117
Glasgow Archaeological Society, Glasgow 12118
Glasgow Mathematical Association, Glasgow 12119
Glasgow Obstetrical and Gynaecological Society, Glasgow 12120
Glass Art Society, Seattle 15161
Glazounov Society, Sonoma 15162
GLC, Ann Arbor 15175
GLCA, Ann Arbor 15174
Glenbow Alberta Institute, Calgary 02486
Glenn Miller Society, London 12121
GLHS, Vermilion 15176
GLMI, Detroit 15177
Global Education Associates, New York 15163
Global Learning, Union 15164
Global Options, San Francisco 15165
Gloucestershire Society for Industrial Archaeology, Dursley 12122
GLS, Cheverly 15183
GLS, Sandon near Stafford 12139
GMA, Düsseldorf 06330
GMA, Glasgow 12119
GMAC, Los Angeles 15169
GMDS, Köln 05382
GMEA, Nantes 04557
GMF, Bonn 06410
GMPCA, Rennes 04550
GO, San Francisco 15165
Görres-Gesellschaft zur Pflege der Wissenschaft, Köln 05896
Göteborgs Läkarförening, Göteborg 10467
Goethe-Gesellschaft in Weimar e.V., Weimar 05897
Goethe Society of North America, Irvine 15166
Göttinger Arbeitskreis, Göttingen 05898
GöWG, Berlin 05846
Good Gardeners' Association, Churcham 12123
Goodwin's Foundation, Ottawa 02487
Gooseberry Society, Barnet 12124
Gosford District Historical Research and Heritage Association, Hardy's Bay 00407
Gottfried-Wilhelm-Leibniz-Gesellschaft e.V., Hannover 05899
Goudy Society, New York 15167
Governmental Research Association, Birmingham 15168
Gower Society, Swansea 12125
GPAC, Fort Collins 15178
GPrÖ, Wien 00617
GRA, Birmingham 15168
Graduate Management Admission Council, Los Angeles 15169
Graduate Record Examinations Board, Princeton 15170
Graduates of Italian Medical Schools, Bronx 15171
Graduate Teachers' Association, Paceville 08664
Grain, Feed and Fertilizer Accident Prevention Association, Toronto 02488
Grape and Wine Institute of British Columbia, Kelowna 02489
Graphic Arts Association of China, Taipei 03300
Graphic Communications Association, Alexandria 15172
Grassland Society of Victoria, Parkville 00408
Gravure Association of America, Rochester 15173
GRC, Davis 15155
Greater London Federation of Parent-Teacher Associations, London 12126
Great Lakes Colleges Association, Ann Arbor 15174
Great Lakes Commission, Ann Arbor 15175
Great Lakes Historical Society, Vermilion 15176
Great Lakes Maritime Institute, Detroit 15177
Great Plains Agricultural Council, Fort Collins 15178
Great Western Society, Didcot 12127
GRE BOARD, Princeton 15170
Greek Atomic Energy Commission, Athinai 06533
The Greek Institute, London 12128
Greek Library Association, Athinai 06518
Greek National Committee for Astronomy, Athinai 06528
Greek National Committee for Space Research, Athinai 06529
Greek National Committee for the Quiet Sun International Years, Athinai 06530
Greek Playwrights' Association, Athinai 06536
Greek Teachers Association, Toronto 02490
Gregor Mendel-Gesellschaft Wien, Wien 00646
Gremial Uruguaya de Médicos Radiólogos, Montevideo 13259
Grémio Literario Carlos Ferreira, Amparo 01661
Grémio Literario e Comercial Portugués, Belém 01662
G.R.E.P., Paris 04548
GRI, Chicago 15147
Grillparzer-Gesellschaft, Wien 00647
Group-Analytic Society, London 12129
Group and Association of County Medical Officers of Health of England and Wales, Stafford 12130
Groupe Belge d'Etude de l'Arriération Mentale, Bruxelles 01368
Groupe Canadien pour la Recherche en Classification, London 02096
Groupe de Liaison des Historiens près la Commission des Communautés Européennes, Louvain-la-Neuve 01236
Groupe de Recherche et pour l'Education et la Prospective, Paris 04548
Groupe de Recherche Génétique Epidémiologique, Paris 04549
Groupe des Méthodes Pluridisciplinaires Contribuant à l'Archéologie, Rennes 04550
Groupe d'Etude et de Synthèse des Microstructures, Paris 04551
Groupe Leibniz, Grenoble 04552
Groupement Belge des Omnipraticiens, Bruxelles 01369
Groupement d'Artistes et d'Ecrivains, Alexandria 03882
Groupement des Associations Dentaires Francophones, Paris 04553
Groupement des Bureaux Médicaux, Paris 04554
Groupement des Unions Professionnelles Belges de Médecins Spécialistes, Bruxelles 01370
Groupement d'Etudes et de Réalisations Médicales, Paris 04555
Groupement Industriel Européen d'Etudes Spatiales, Paris 04556
Groupement International pour la Recherche Scientifique en Odontologie et en Stomatologie, Bruxelles 01371
Groupement International pour la Recherche Scientifique en Stomatologie et Odontologie, Bruxelles 01372
Groupement Médical d'Etudes sur l'Alcoolisme, Nantes 04557
Groupement Médical Saint-Augustin, Paris 04558
Groupement National d'Etude des Médecins du Bâtiment et des Travaux Publics, Paris 04559
Groupement Professionnel National de l'Informatique, Paris 04560
Groupement Romand du Marketing, Lausanne 10723
Groupe Phonétique de Paris, Paris 04561
Groupe pour l'Avancement des Sciences Analytiques, Paris 04562
Groupe Rhône-Alpes de Recherche et d'Etudes en Gestion, Lyon 04563
Group for the Advancement of Psychiatry, Dallas 15179
Group for the Study of Irish Historic Settlement, Newtonabbey 12131
Group for the Use of Psychology in History, Springfield 15180
Group Health Association of America, Washington 15181
Group of the Far Eastern Division, Moskva 09872
GRS, Köln 05799
Grupo Bibliográfico Nacional de la República Dominicana, Santo Domingo 03836
Grupo de Biblioteca y Documentación, Bogotá 03376
Grupo Latinoamericano de R.I.L.E.M., Buenos Aires 00141
Grupo Nacional de Radiología, La Habana 03480
Gruppe Ökologie, Hannover 05900
Gruppe Olten, Schweizer Autorengruppe, Tägerwilen 10724
Gruppi Archeologici d'Italia, Roma 07701
Gruppo Italiano di Linguistica Applicata, Pisa 07702
GRV, Düsseldorf 05855
GS, New York 15167
GS, Sonoma 15162
GS, Washington 15152
GSA, Bethesda 15151
GSA, Boulder 15153
GSA, Oxford 12115
G.S.A., York 12107
G.S.C., Beijng 03170
GSF, Ergersheim 05862
G.S.I., Galway 06953
GSIHS, Newtonabbey 12131
GSK, Bern 10716
GSLK, Salzburg 00632
GSMBA, Muttenz 10720
GSMT, New York 15150
GSNA, Irvine 15166
GSNZ, Lower Hutt 09126
GSS, Brooklym 15160
GSSA, Linden 10144
GST, Bern 10721
G.S.V., Parkville 00408
GTA, Gibraltar 06497
GTE, Budapest 06606
GTF, Berlin 05863
GTN, Gdánsk 09543
GTW, Reading 12137
GTZ, Eschborn 05430
Guild of Catholic Doctors, Bristol 12132
Guild of Church Musicians, Blechingley 12133
Guild of Guide Lecturers, London 12134
Guild of Health, London 12135
Guild of Pastoral Psychology, London 12136
Guild of Travel Writers, Reading 12137
Gujarat Research Society, Bombay 06763
Gulf and Caribbean Fisheries Institute, Charleston 15182
GUPH, Springfield 15180
Gustav Freytag Gesellschaft e.V., Ratingen 05901
Gutenberg-Gesellschaft, Mainz 05902
GUVU, Köln 05867
Guyana Institute of International Affairs, Georgetown 06576
Guyana Library Association, Georgetown 06577
Guyana Society, Georgetown 06578
GV, Füllinsdorf 10717
GVC/VDI-Gesellschaft Verfahrenstechnik und Chemieingenieurwesen, Düsseldorf 05903
GVG, Köln 05868
Gwent Wildlife Trust, Monmouth 12138
GwG, Köln 05872
GWL, Essen 05873
GWS, Didcot 12127
Gymnasieskolernes Laererforening, København 03740
Gymnasieskolernes Tysklaererforening, Nykøbing 03741
Gypsy Lore Society, Cheverly 15183
Gypsy Lore Society, Sandon near Stafford 12139

HA, London 12172
HA, Washington 15192
H.A.A., Port Elizabeth 10146
HAA, Redhill 12171
HAB, Frankfurt 05912
HACSG, Chichester 12190
Haemophilia Society, London 12140
Hafenbautechnische Gesellschaft e.V., Hamburg 05904
H.A.F.S., Harbin 03201
HAG, Basel 10728
Hague Academy of International Law, 's-Gravenhage 08907
Hahn-Schickard-Gesellschaft für angewandte Forschung e.V., Stuttgart 05905
HAI, Davenport 15186
Haiku Society of America, New York 15184
HAIL, 's-Gravenhage 08907
Haitian Coalition on Aids, Brooklyn 15185
The Hakluyt Society, London 12141
Haldane Society, London 12142
Halle Concerts Society, Manchester 12143
Hamber Foundation, Vancouver 02491
Hamburger Autorenvereinigung e.V., Hamburg 05906
Hamburger Gesellschaft für Völkerrecht und Auswärtige Politik e.V., Hamburg 05907
Hamdard Foundation, Karachi 09364
Hamilton Academy of Medicine, Hamilton 02492
Hamilton Foundation, Hamilton 02493
Hamilton-Wentworth Lung Association, Hamilton 02494
Hampshire Field Club and Archaeological Society, Winchester 12144
Handotai Kenkyu Shinkokai Handotai Kenkyusho Toshoshitsu, Sendai 08129
Handwriting Analysts – International, Davenport 15186
Hannoversches Forschungsinstitut für Fertigungsfragen e.V., Hannover 05908
Hansard Society for Parliamentary Government, London 12145
Hanshin Doitsubungakukai, Kobe 08130
Hardwood Research Council, Memphis 15187
Hardy Plant Society, West Ewell 12146
The Harleian Society, London 12147
Harold Crabtree Foundation, Ottawa 02495
Harry S. Truman Library Institute for National and International Affairs, Independence 15188
Hartmannbund, Bonn 05909
Harvard Environmental Law Society, Cambridge 15189
The Harveian Society of London, London 12148
Harvey Society, New York 15190
Harzverein für Geschichte und Altertumskunde, Goslar 05910
H.A.S., Hawick 12151
HAS, Inverness 12153
HAS, Lansing 15229
HATRA, Nottingham 12149
Haunt Hunters, Chesterfield 15191
Haus des Meeres Vivarium Wien, Wien 00648
Havergal Brian Society, Watford 12150
Hawick Archaeological Society, Hawick 12151
Hawk and Owl Trust, London 12152
Hawke's Bay Medical Research Foundation, Napier 09127
HBC, New York 15233
H.B.S., Norwich 12157
HBS, Watford 12150
HCA, Brooklyn 15185
HCF, New York 15200
HCRE, New York 15227
HCS, Woods Hole 15216
H.C.S.A., Overton 12183
HDA, Oklahoma City 15224
H/DF, Malibu 15234
HDRA, Coventry 12158
HDRF, Brooklyn 15199
HDS, Philadelphia 15219
HDSA, New York 15241
Headmasters Association, Washington 15192
Head Teachers' Association of Scotland, Inverness 12153
Health Industry Business Communications Council, Phoenix 15193
Health Labour Relations Association of British Columbia, Vancouver 02496
Health Libraries Association of British Columbia, Vancouver 02497
Health Optimizing Institute, Del Mar 15194
Health Physics Society, McLean 15195
Health Research Board, Dublin 06954
Health Sciences Communications Association, McLean 15196
Health Sciences Consortium, Chapel Hill 15197
Health Service Social Worker Group, Barnet 12154
Hear Center, Pasadena 15198
Hearing Concern, London 12155
Heart and Stroke Foundation of Alberta, Calgary 02498
Heart Disease Research Foundation, Brooklyn 15199
Heather Society, Ipswich 12156
Hebrew Culture Foundation, New York 15200
Hebrew Writers Association in Israel, Tel Aviv 07055
HECUA, Saint Paul 15206
HEEA, Gainesville 15226
Hegel Society of America, Villanova 15201
HEI, Los Angeles 15232
Heidelberger Akademie der Wissenschaften, Heidelberg 05911
Heilongjiang Academy of Forestry Sciences, Harbin 03201
Heimverband Schweiz, Zürich 10725
HELCOM, Helsinki 03952
Hellenic Association of University Women, Athinai 06531
Hellenic Geographical Society, Athinai 06506
Hellenic Institute of International and Foreign Law, Athinai 06532
Hellenic Numismatic Society, Athinai 06512
Hellenic Organization for Standardization, Athinai 06517
Hellenic Society of Ptolemaic Egypt, Alexandria 03903
Hellenic Veterinary Medical Society, Athinai 06508
Helliniki Epitropi Atomikis Energhias, Athinai 06533
Helminthological Society of India, Mathura 06764
Hemingway Society, Grand Forks 15202
Henry Bradshaw Society, Norwich 12157
Henry Doubleday Research Association, Coventry 12158
Henry Williamson Society, Freshwater 12159
Henty Society, Cheltenham 12160
HEP, Washington 15207
Heraldisch-Genealogische Gesellschaft Adler, Wien 00649
Heraldisk Selskab, Bagsvaerd 03823
The Heraldry Society, London 12161
Heraldry Society of Canada, Ottawa 02499
Heraldry Society of Scotland, Edinburgh 12162
Heraldry Society of Southern Africa, Claremont 10145
Herbert Hoover Presidential Library Association, West Branch 15203
Herpes Resource Center – American Social Health Association, Research Triangle Park 15204
Herpetological Association of Africa, Port Elizabeth 10146
Herpetologists' League, Austin 15205
HERS, Wellesley 15208
Hertfordshire Local History Council, Hitchin 12163, 12164
Hertfordshire Natural History Society and Field Club, Potters Bar 12165
HES, Eugene 15221
HES, Richmond 15220
HESCA, McLean 15196
Hesketh Hubbard Art Society, London 12166
Hessische Akademie für Bürowirtschaft e.V., Frankfurt 05912
Hessische Krebsgesellschaft e.V., Marburg 05913
Hessischer Philologen-Verband, Wiesbaden 05914
Hetaireia Hellenon Philologon, Athinai 06534
Hetairia Hellinon Logotechnon, Athinai 06535
Hetairia Hellinon Thetricon Syngrapheon, Athinai 06536
HF, Arlington 15209
HFF, Hannover 05908
HFS, Santa Monica 15235
H.G. Bertram Foundation, Hamilton 02500
HGTA, Ayr 12181
H.G. Wells Society, Moulton Park 12167
HH, Chesterfield 15191
HHH, Las Vegas 15225
HHPLA, West Branch 15203
HI, New York 15212
HIA, New York 15214
HIBCC, Phoenix 15193
HIBR, Boca Raton 15242
Hid Islenska Fornleifafélag, Reykjavik 06688
Hid Islenska Náttúrufraedifélag, Reykjavik 06689
High Council of Arts and Literature, Cairo 03904
High Council of Culture, Cairo 03905
Higher Council for Promotion of Arts and Letters, Riyadh 10003
Higher Education Consortium for Urban Affairs, Saint Paul 15206
Higher Education Funding Council for England, Bristol 12168
Higher Education Panel, Washington 15207
Higher Education Resource Services, Wellesley 15208
High Frontier, Arlington 15209
The Highland Association, Inverness 11229
High Pressure Technology Association, Leeds 12169
Highway Users Federation for Safety and Mobility, Washington 15210
Hikaku-ho Gakkai, Tokyo 08131
Himalayan International Institute of Yoga Science and Philosophy of the U.S.A., Honesdale 15211
Himpunam Pustakawan Chusus Indonesia, Jakarta 06888
Hiradástechnikai Tudományos Egyesület, Budapest 06607
Hiroshima Philosphical Society, Hiroshima-ken 08133
Hiroshima Shikagu Kenkyukai, Hiroshima 08132
Hiroshima Tetsugakkai, Hiroshima-ken 08133
Hispanic and Luso-Brazilian Council, London 12170
Hispanic Institute, New York 15212
Hispanic Society of America, New York 15213
HISPO, Wabern 10623
Histadruth Ivrith of America, New York 15214
Histamine Research Society of North America, Dallas 15215
Histochemical Society, Woods Hole 15216
Historian Ystäväin Liitto, Tampere 03965
Historic Aircraft Association, Redhill 12171
Historical and Ethnological Society of Greece, Athinai 06537
Historical and Scientific Society of Manitoba, Winnipeg 02501
Historical Association, London 12172
Historical Association of Kenya, Nairobi 08502
Historical Association of Oman, Ruwi 09355
Historical Committee of the Mennonite Church, Goshen 15217
The Historical Metallurgy Society, Swansea 12173
Historical Newspaper Service, New Barnet 12174
Historical Research Commission of Taiwan, Taipei 03301
The Historical Society, Reykjavik 06700
Historical Society of Alberta, Calgary 02502

Historical Society of Ghana, Accra 06491
Historical Society of Israel, Jerusalem 07056
Historical Society of Japan, Tokyo 08417
Historical Society of Mecklenburg Upper Canada, Toronto 02503
Historical Society of Nigeria, Lagos 09231
Historical Society of Serbia, Beograd 17084
Historical Society of Sierra Leone, Freetown 10030
Historical Society of the Church in Wales, Carmarthen 12175
Historical Society of the Gatineau, Old Chelsea 02504
Historical Society of the Methodist Church in Wales, Pwllheli 12176
Historical Society of Trinidad and Tobago, Port of Spain 11121
Historical Society of Washington, DC, Washington 15218
Historic Breechloading Smallarms Association, London 12177
Historic Society of Lancashire and Cheshire, Liverpool 12178
Historisch-Antiquarischer Verein Heiden, Heiden 10726
Historische Kommision der Deutschen Gesellschaft für Erziehungswissenschaft, Hannover 05915
Historische Kommission des Börsenvereins des Deutschen Buchhandels, Frankfurt 05916
Historische Kommission zu Berlin, Berlin 05917
Historische Kommission zur Erforschung des Pietismus an der Universität Münster, Münster 05918
Historische Landeskommission für Steiermark, Graz 00650
Historischer Verein, Vaduz 08569
Historischer Verein Bamberg, Bamberg 05919
Historischer Verein der Pfalz e.V., Speyer 05920
Historischer Verein des Kantons Bern, Bern 10727
Historischer Verein Dillingen an der Donau, Dillingen 05921
Historischer Verein für die Saargegend e.V., Saarbrücken 05922
Historischer Verein für Hessen, Darmstadt 05923
Historischer Verein für Oberfranken e.V., Bayreuth 05924
Historischer Verein für Schwaben, Augsburg 05925
Historischer Verein für Steiermark, Graz 00651
Historischer Verein für Württembergisch Franken, Schwäbisch Hall 05926
Historischer Verein Rupertiwinkel e.V., Laufen 05927
Historische und Antiquarische Gesellschaft zu Basel, Basel 10728
Historisch Genootschap De Maze, Rotterdam 08908
Historisch Genootschap Roterodamum, (010) Rotterdam 08909
Historisk Samfund for Sønderjylland, Abenrå 03742
Historisk-Topografisk Selskab, Frederiksberg 03743
History of Dermatology Society, Philadelphia 15219
History of Economics Society, Richmond 15220
History of Education Society, Eugene 15221
History of Education Society, Leicester 12179
History of Science Society, Worcester 15222
HKLA, Hong Kong 06593
HLABC, Vancouver 02497
HLB, Bonn 05928
HLC, Westport 15236
HLHC, Hitchin 12163, 12164
H.M.E., Athinai 06510
HML, Summit 15228
HMS, Swansea 12173
H.M.W., Haarlem 08910
HND, Zagreb 03461
Hochschullehrerbund e.V., Bonn 05928
Hochschulrektorenkonferenz, Bonn 05929
Hölderlin-Gesellschaft, Tübingen 05930
Hogaku Kyokai, Tokyo 08134
Hohenzollerischer Geschichtsverein, Sigmaringen 05931
Hohenzollern Society, Keyport 15223
HOI, Del Mar 15194
Hokkaido Economic Federation, Sapporo 08135
Hokkaido Keizai Rengokai, Sapporo 08135
Holistic Dental Association, Oklahoma City 15224
Holistic Health Havens, Las Vegas 15225
Hollandsche Maatschappij der Wetenschappen, Haarlem 08910
Home Economics Education Association, Gainesville 15226
Homeopathic Council for Research and Education, New York 15227
Honan Academy of Agricultural Sciences, Honan 03202
Hong Kong Library Association, Hong Kong 06593
Honorable Society of King's Inns, Dublin 06955
Honourable Society of Cymmrodorion, London 12180
Honours Graduate Teachers' Association, Ayr 12181
Hopeh Academy of Agricultural Sciences, Hopei 03203
Horace Mann League of the U.S.A., Summit 15228
Horatian Society, London 12182
Horatio Alger Society, Lansing 15229
Horological Guild of Australasia, Sydney 00409
Horticultural Research Institute, Washington 15230
Hosei-shi Gakkai, Tokyo 08136
Hosokai, Tokyo 08137
Hospital for Sick Children Foundation, Toronto 02505
Hospital Medical Records Institute, Don Mills 02506
Hospital Research and Educational Trust, Chicago 15231
Hospitals Consultants and Specialists Association, Overton 12183
Hotel, Catering and Institutional Management Associaton, London 12184
House Ear Institute, Los Angeles 15232
Housing and Urban Development Association of Nova Scotia, Halifax 02507
Housman Society, Bromsgrove 12185
Hovedkomiteen for Norsk Forskning, Oslo 09268
Høyokoleutdannedes Forbund, Oslo 09269
HPCI, Jakarta 06888
HPS, McLean 15195
HPS, West Ewell 12146
H.P.T.A., Leeds 12169
HRAF, New Haven 15237
HRC, Memphis 15187
HRET, Chicago 15231
HRI, Washington 15230
HRPS, New York 15239
HRSNA, Dallas 15215
Hrvatska Akademija Znanosti i Umjetnosti, Zagreb 03459
Hrvatsko Bibliotekarsko Društvo, Zagreb 03460
Hrvatsko Numizmatičko Društvo, Zagreb 03461
Hrvatsko Prirodoslovno Društvo, Zagreb 03462
HS, Bromsgrove 12185
HS, Grand Forks 15202
HS, Keyport 15223
HS, New York 15190
HSA, Concord 15243
HSA, New York 15184, 15213
HSA, Villanova 15201
HSC, Chapel Hill 15197
HSFR, Stockholm 10468
H.S.L.C., Liverpool 12178
HSS, Worcester 15222
HSSL, Freetown 10030
HSWDC, Washington 15218
HTG, Hamburg 05904
Hüttentechnische Vereinigung der Deutschen Glasindustrie e.V., Frankfurt 05932
HUF, Washington 15210
Huguenot Society of Great Britain and Ireland, London 12186
Human Biology Council, New York 15233
Human/Dolphin Foundation, Malibu 15234
Human Factors Association of Canada, Mississauga 02508
Human Factors Society, Santa Monica 15235
The Human Geographical Society of Japan, Kyoto 08148
Humanistische Union e.V., München 05933
Humanistisk-Samhällsvetenskapliga Forskningsradet, Stockholm 10468
Human Lactation Center, Westport 15236
Human Nutrition Research Council of Ontario, Stittsville 02509
Human Relations Area Files, New Haven 15237
Human Resource Certification Institute, Alexandria 15238
Human Resource Planning Society, New York 15239
Human Resources Research Organization, Alexandria 15240
Human Sciences Research Council, Pretoria 10147
Humboldt-Gesellschaft für Wissenschaft, Kunst und Bildung e.V., Mannheim 05934
HumRRO, Alexandria 15240
Hunan Academy of Agricultural Sciences, Changsha 03204
Hungarian Academy of Sciences, Budapest 06661
Hungarian Association for the Protection of Industrial Property, Budapest 06637
Hungarian Association of Gerontology, Budapest 06632
Hungarian Astronautical Society, Budapest 06617
Hungarian Biochemical Society, Budapest 06620
Hungarian Biological Society, Budapest 06621
Hungarian Biophysical Society, Budapest 06619
Hungarian Canadian Engineers' Association, Don Mills 02510
Hungarian Chemical Society, Budapest 06643
Hungarian Diabetes Association, Budapest 06623
Hungarian Electrotechnical Association, Budapest 06624
Hungarian Ethnographical Society, Budapest 06647
Hungarian Forestry Association, Budapest 06669
Hungarian Geological Society, Budapest 06635
Hungarian Historical Society, Budapest 06659
Hungarian Hydrological Society, Budapest 06634
Hungarian Lawyers Association, Budapest 06640
Hungarian Linguistic Society, Budapest 06649
Hungarian Meteorological Society, Budapest 06645
Hungarian Mining and Metallurgical Society, Budapest 06670
Hungarian Music Council, Budapest 06663
Hungarian PEN Club, Budapest 06652
Hungarian Pharmaceutical Society, Budapest 06633
Hungarian Physiological Society, Budapest 06626
Hungarian Psychological Association, Budapest 06653
Hungarian Scientific Society for the Food Industry, Budapest 06625
Hungarian Society for Trauma Surgery, Budapest 06660
Hungarian Society of Agricultural Sciences, Budapest 06671
Hungarian Society of Anatomists, Histologists and Embryologists, Budapest 06615
Hungarian Society of Archaeology and History of Fine Arts, Budapest 06655
Hungarian Society of Cardiology, Budapest 06641
Hungarian Society of Microbiology, Budapest 06646
Hungarian Society of Textile Technology and Science, Budapest 06675
Hungarian Theatre Institute, Budapest 06658
Hunter Archaeological Society, Sheffield 12187
Hunterian Society, London 12188
Huntingdonshire Local History Society, Thurleigh 12189
Huntington's Disease Society of America, New York 15241
Huntington Society of Canada, Cambridge 02511
Huxley Institute for Biosocial Research, Boca Raton 15242
HVG, Frankfurt 05932
HWWA – Institut für Wirtschaftsforschung – Hamburg, Hamburg 05935
Hyderabad Educational Conference, Hyderabad 06765, 06766
Hydrobiological Society, Moskva 09873
Hydrobiological Society of Moldova, Chişinău 08780
Hydroponic Society of America, Concord 15243
Hyomen Gijutsu Kyokai, Tokyo 08138
Hyper Active Children's Support Group, Chichester 12190

IAA, Greenbelt 15268
IAA, Paris 04583
IAAAM, San Leandro 15375
IAACI, Milwaukee 15397
IAAE, Oak Brook 15396
IAAEES, Chicago 15392
IAAL, Edinburgh 12327
IAASEES, Brookeville 15348
IAB, Bethesda 15398
IABA, Washington 15349
IABG, Edinburgh 02540
IABMCP, Dallas 15363
IABS, Berkeley 15399
IABS, Hassocks 12328
IABS, New York 15371
IABSE, Zürich 10742
IAC, Freiburg 05984
I.A.C.A., Grünwald 06401
IACA, Pomona 15350
I-ACAC, Washington 15351
IACCF, Arlington 15401
IACCP, Kingston 02539
IACI, Saint Paul 15585
IACIS, Eau Claire 15377
IACLEA, Hartford 15400
IACME, Peoria 15402
IACPS, Northbrook 15364
IACRLRD, Columbus 15376
IACS, Alexandria 15403
I.A.C.S., Calcutta 06770
IADA, Marburg 05989
IADC, Chicago 15404
IADR, Washington 15379
IAE, Woodland Hills 15299
I.A.E.A., New Delhi 06768
IAEA, Wien 00663
IAEC, Chalfont Saint Gills 12329
IAEE, Cleveland 15381
I.A.E.G., Paris 04581
IAEPO, Houston 15405
IAEWP, Huntsville 15406
I.A.G., Canberra 00412
IAG, México 08741
IAGLR, Ann Arbor 15382
IAGOD, Jeseník 03546
IAGP, Montréal 02541
IAgrE, Bedford 12292
IAHA, Manila 09520
IAHAIO, Renton 15408
IAHB, Stanford 15300
IAHE, Coral Gables 15385
I.A.H.P., Roma 07337
IAHR, Delft 08918
IAHS, Coral Gables 15384
IAHS, Paris 04578
IAHS2, Lombard 15383
IAI, London 12321
IAIA, Santa Fe 15309
IAIS, Washington 15269
IAJE, Manhattan 15409
IAKE, Gaithersburg 15410
IAL, Arlon 01375
IAL, Atlanta 15411
IAL, Liège 01376
IALC, New York 15587
IALL, Philadelphia 15386
IAM, Doylestown 15366
IAM, Kassel 05994
IAM, Los Angeles 15340
IAM, New York 15365
IAM, Orpington 12212
IAM, 's-Gravenhage 08917
IAM, Washington 15353
IAMCR, Leicester 12324
IAMFE, Uppsala 10474
IAMG, Reston 15387
IAMLT, Stockholm 10471
IAML-US, New York 15412
IAMPTH, Ottawa 02542
IAMS, Utrecht 08919
IANDS, Philadelphia 15388
I.A.O., Firenze 07724
IAO, Oak Park 15415
IAO, Toronto 02534
IAOE, Bismarck 15414
IAOS, Chicago 15413
IAP, Lisboa 09660
IAP, Praha 03545
IAP, Tucson 15368
IAPA, Dublin 06975
IAPESGW, Mankato 15416
IAPESGW, Viborg 03744
IAPI, Dublin 06958
IAPS, London 12194
IAPSO, Del Mar 15393
IAPW, Andover 15389
IAPWS, Palo Alto 15394
IAQDE, Red Oak 15250
IAR, New York 15310
IARASM, Garden City 15265
IARIW, New York 15391
IARP, Tokyo 08141
IARS, Cleveland 15373
IAS, Berkeley 15311
IAS, Halifax 02538
IAS, Houston 15419
IASA, Stockholm 10473
IASB, Santa Barbara 15362
IASC, Austin 15372
IASC, Englewood 15354
IASC, México 08751
IASC, Toronto 02515
IASC, Voorburg 08921
IASCE, Santa Cruz 15395
IASI, Panamá City 09420
IASL, Kalamazoo 15417
IASLIC, Calcutta 06775
IASP, Wien 00685
IASRA, San Bernardino 15374
IASS, Madrid 10328
IASS, Norwich 12325
IASSW, Wien 00662
IASTA, New York 15266
IASWR, Carmel 15267
IAT, Staines 12213
IAT, Sydney 00424
IATAL, Tuscaloosa 15418
IATEFL, Whitstable 12333
IATL, San Jose 15369
IATLIS, Nagpur 06777
IATSS, Tokyo 08142
IATUL, Edinburgh 12334
IAUPE, Marazion 12335
IAW, Tübingen 05950
IAWA, Utrecht 08925
IAWL, Roma 07705
IAWQ, London 12337
IBA, London 12338
IBBA, Wisner 15263
IBBBA, Los Angeles 15426
IBBY, Basel 10734
IBECC, Rio de Janeiro 01665
IBEF, Sherman Oaks 15420
Ibero-American Association for the Study of Alcohol and Drug Problems, Córdoba 00083
IBES, Scottsdale 15424
IBFD, Amsterdam 08926
IBI, Washington 15347
IBN, Bruxelles 01379
IBP, Bruxelles 01391
IBP, La Paz 01546
IBPGR, Roma 07708
IBR, Wheaton 15270
IBRA, Bruxelles 01380
IBRA, Cardiff 12339
IBRRC, Berkeley 15421
IBS, Aachen 05990
IBS, Athens 15422
IBS, Bruxelles 01378
IBS, Culver City 15425
IBS, Vails Gate 15357
I.B.S.A., Bruxelles 01382
Ibsen Society of America, Brooklyn 15244
IBTTA, Washington 15423
ICA, Arlington 15435
ICA, Chicago 15315
ICA, Dallas 15444
ICA, Guadalajara 08742
ICA, México 08743
ICA, Monterrey 08744
ICA, Mosquera 03390
ICA, New York 15449
ICA, Paris 04585
ICA, São Paulo 01694
ICAITI, Guatemala City 06569
ICAN, Bogotá 03380
ICAP, San José 03445
ICASE, Hong Kong 06594
ICBF, Bogotá 03381
ICBO, Whittier 15445
ICC, Genève 10735
ICC, Madrid 10329
ICC, New York 15313
ICCA, Saint Louis 15252
ICCC, Montréal 02543
ICCP, Des Plaines 15271
ICCP, Groningen 08931
ICD, Rockville 15438
ICDA, Great Neck 15273
ICDE, Montréal 02549
ICE, London 12294
ICE, Loughborough 12206
ICEA, Levittown 15436
ICEA, Minneapolis 15434
ICEA, South Yarmouth 15345
ICEC, Morgantown 15451
ICECU, San José 03446
ICED, Washington 15577
I.C.E.I., Milano 07726
Icelandic Architects' Association, Reykjavik 06677
Icelandic Composers' Society, Reykjavik 06705
Icelandic Dental Associatio, Reykjavik 06703
Icelandic Medical Association, Reykjavik 06694
The Icelandic Natural History Society, Reykjavik 06692
Icelandic Society of Anesthesiologists, Reykjavik 06702
Icelandic Society of Radiology, Reykjavik 06684
The Icelandic Teachers Association, Reykjavik 06693
ICEP, London 12354
I.C.E.P.S., Roma 07751
ICES, København 03746
ICET, Arlington 15460
ICETEX, Bogotá 03382
ICF, Baraboo 15462
ICF, Dallas 15427
ICF, Sendai 08143
ICFES, Bogotá 03389
ICFP, Denver 15312
ICHC, Leiden 08930
ICHPER, Reston 15452
ICHSPP, Janesville 15448
ICIE, Roma 07725
ICKL, London 12347
ICLAE, Sacramento 15456
ICLAS, Kuopio 03966
ICLC, Santa Barbara 15463
ICLD, New York 15429
ICLM, Portland 15437
ICLS, East Lansing 15461
ICM, London 12226
ICM, Stamford 12230
ICMA, New York 15432
ICMMB, Ann Arbor 15446
I.C.M.R., New Dehli 06787
IC/NATAS, New York 15455
ICNCP, Wageningen 08928
ICN-MHN, Bogotá 03391
ICOHTEC, Toronto 02547
ICOI, Upper Montclair 15447
ICOLPE, Bogotá 03388
ICOM, Bogotá 03366
I.C.O.M., Milano 07713
ICOM, Paris 04458
ICOMIDC, Tunis 11141
ICOM Museums/Musées Canada, Ottawa 02512
Icon, Leuven 01373
ICONTEC, Bogotá 03386
ICOS, Leuven 01394
ICP, Hopkinton 15458
ICP, New York 15433
ICPADD, Silver Spring 15440
ICPBR, Lusignan 04584
ICPE, Columbus 15441
ICPI, Bethesda 15579

ICPM, Cambridge 12344
ICPS, London 12340
ICPSR, Ann Arbor 15583
ICR, Tunbridge Wells 12207
ICRFSDD, Ardsley 15251
ICRPMA, Roma 07712
ICRU, Bethesda 15442
ICRW, Washington 15430
ICS, Chicago 15439
ICS, Haifa 07069
ICSA, Shawnee Mission 15450
ICSEP, Houston 15431
ICSK, Seoul 08526
ICSSPE, Jyväskylä 03967
ICSSR, New Dehli 06788
ICSW, Wien 00664
ICT, Beaconsfield 12225
ICT, Poznan 09544
ICT, Ripley 12223
ICTM, New York 15454
ICTP, Trieste 07711
I.C.U., Roma 07752
ICUAE, Fredericton 02546
ICUMSA, Braunschweig 05985
ICVS, Manchester 15527
I.C.W.A., New Delhi 06789
ICWP, Washington 15580
ICWS, Saint Louis 15314
ICZN, London 12345
IDA, Alexandria 15274
IDA, Dublin 06981
IDA, Jamaica 15256
IDA, New Dehli 06790
I.D.A., Tel Aviv 07076
Idarah-i-Yadgar-i-Ghalib, Karachi 09365
IDCA, Aspen 15466
IDEA, Dayton 15275
IDEALS, State College 15301
IDEC, Irvine 15360
I.D.E.F., Paris 04572
IDF, Bruxelles 01395
IDF, London 12348
IDHF, Reston 15465
IDIL, Oklahoma City 15302
IDOC, Roma 07467
IDPM, London 12231
IDRA, San Antonio 15358
IDRC, Norcross 15258
IDS, Kensington 15467
IDSA, New Haven 15262
IDSA, Tucson 15464
IE, London 12236
IEA, London 12233
IEA, Los Angeles 15471
IEA, New York 15279
I.E.A., Paris 04586
IEA, Silver Springs 15277
IEA, Stockholm 10470
IEAHC, Williamsburg 15316
IEAJ, Wembley 12350
IECA, Steamboat Springs 15472
IECI, Valkenswaard 08911
IEDP, Honolulu 15005
IEE, Bellingham 15318
IEE, London 12296
IEEE, New York 15317
IEEE Control Systems Society, New York 15245
IEEE Engineering in Medicine and Biology Society, New York 15246
IEEE Ultrasonics, Ferroelectrical and Frequency Control Society, Orlando 15247
IEEIE, London 12298
IEH, London 12301
IEI, Dublin 06968
IEIAS, Marcinelle 01385
IEL, Washington 15278
IEN, Torino 07728
I.E.P., Lahore 09369
IEPS, Wheaton 15469
IER, Fort Lauderdale 15276
IERF, Los Angeles 15468
I.E.S., Jerusalem 07071
IES, Mount Prospect 15319
IES, Princeton 15253
IESNA, New York 15248
IETS, Champaign 15470
IEU, Pleasant Hill 15117
IF, Menlo Park 15303
IFA, London 12362, 12363
IFA, Ruislip 12353
IfaA, Köln 05946
IFAC, Laxenburg 00686
IfAPF, Seelze 05948
IFAR, New York 15478
IFCB, London 12355
IFDC, Muscle Shoals 15476
IFDP, San Francisco 15281
IFE, Chicago 15320
IFE, Leicester 12303
I.F.E., Paris 04568
IFEU, Heidelberg 05958
IFFLP, Washington 15473
IFGB, Edinburgh 12221
IFH, Seattle 15479
IFHP, 's-Gravenhage 08932
IFIS, Reading 12364
IFLA, 's-Gravenhage 08934
IFM, Holmer 12237
I.F.M.S.S., London 12357
Ifo Institute for Economic Research, München 05936
Ifo Institut für Wirtschaftsforschung e.V., München 05936
IFOS, Nijmegen 08935
IFPA, Dublin 06983
IFPA, London 12199
IFPNT, London 12358
IFRA, Darmstadt 05938
IFRAA, Washington 15359
IfS, Kiel 05969
IFS, London 12208
IFS, New York 15477
IFSCC, Luton 12359
IFSEM, Berkeley 15474
IFSM, Slough 12360
IFST, London 12239
IFSW, Oslo 09270
IFT, Chicago 15321
IFTR, Lancaster 12352
IFUT, Dublin 06984
IFVTCC, Reinach 10737
IGA, Davis 15480
IGA, Frankfurt 05939
IGA, Innsbruck 00653
IGAEA, Pittsburgh 15481
IGAS, Chicago 15482
IGE, London 12304
IGEB, Graz 00672
IGG, Innsbruck 00654
I.G.H.B., Salvador 01672
IGJ, Graz 00671
IGMG, Wien 00673
IGMI, Firenze 07729
IGMW, Basel 10738
IGP, Lawton 15341
IGS, Cambridge 12366
IGS, Dublin 06986
IGS, Englewood 15323
IGS, Jerusalem 07072
IGS, Tel Aviv 07074
IGSP, Rangeley 15282
IGT, Chicago 15322
IGTI, Atlanta 14254
IGU, Bonn 06007
IGW, Innsbruck 00656
IHA, Harfsen 08937
I.H.B.R., Bruxelles 01386
IHCNE, Chicago 15283
IHCP, Washington 15344
IHE, Hale Barns 12241
IHEA, Rockville 15483
IHEU, Utrecht 08936
IHF, Chicago 15485
I.H.G./R.N., Natal 01683
IHGS, Canterbury 12242
IHGSC, Florianópolis 01677
IHO, Berkeley 15324
IHP, Paris 04577
IHPVA, Indianapolis 15486
IHRBLR, Washington 15487
IHS, Bloomington 15249
IHS, Brooklyn 15588
IHS, Englewood 15484
IHS, London 12368
IIA, Bronx 15488
IIASA, Laxenburg 00687
IIBA, New York 15489
I.I.C., Genova 07730
IIC, London 12369
IICA, San José 03447
IIDU, San Remo 07731
IIE, Dublin 06962
IIE, New York 15326
IIE, Norcross 15325
IIF, Paris 04575
III, New York 15346
IIM, Calcutta 06796
IIN, Roma 07737
I.I.P., Roma 07739
I.I.P., Tel Aviv 07078
IIPE, Paris 04573
I.I. Polzunov Science Production Association for Research and Design of Power Equipment, St. Petersburg 09874
I.I.P.U., Roma 07738
IIRA, Genève 10750
I.I.R.B., Bruxelles 01387
IIRR, New York 15490
IIS, East Hanover 15492
IIS, New York 15284
IIS, Tuscaloosa 15491
IISA, Bruxelles 01388
IISE, Hull 12370
IISG, Amsterdam 08916
IISL, Bordighera 07732
IISM, Roma 07744
IIT, Paris 04576
IJA, London 12249
IJA, New York 15327
IJDF, Livermore 15494
IJO, Roma 07714
IJSS, Muncie 15493
Ikatan Dokter Indonesia, Jakarta 06889
Ikatan Pustakawan Indonesia, Jakarta 06890
Ikomasan Astronomical Society, Nara Ken 08139
Ikomasan Tenmon Kyokai, Nara Ken 08139
IKRA, Brooklyn 15495
ILA, Delhi 06799
ILA, London 12373
I.L.A., Roma 07715
ILA, Tel Aviv 07080
ILAM, Reading 12250
ILAR, Washington 15328
ILCA, Addis Ababa 03945
ILDIS, Quito 03854
ILDM, Corby 12252
Ileostomy Association of Great Britain and Ireland, Mansfield 12191
ILI, Washington 15496
Illuminating Engineering Institute of Japan, Tokyo 08424
Illuminating Engineering Society of North America, New York 15248
ILMH, Oakland 15285
ILS, Chicago 15329
ILS, Woking 12200
ILSGB, Barnsley 12371
ILZRO, Research Triangle Park 15497
ILZSG, London 12374
IMA, Dublin 06992
IMA, Marburg 06009
IMA, McLean 15498
IMA, New Dehli 06801
IMA, New Windsor 15356
IMA, New York 15286
IMA, Southend-on-Sea 12258
I.M.A., Tel Aviv 07082
IMACS, Liège 01141
IMAG, Wageningen 08912
IMBM, Colwyn Bay 12263
I.M.C., Dublin 06990
IMC, New York 14894
IMC, Omaha 15500
IMCB, Buckingham 12375
IMCS, Flyford Flavell 12376
IMD, Lausanne 10751
IMechE, London 12307
IMEKO, Budapest 06608
IMI, Dublin 06989
IMI, Egham 12378
IMIE, London 12308
IMinE, Doncaster 12310
IMLS, London 12261
IMM, Corby 12257
IMM, London 12309
Immigration History Society, Bloomington 15249
Immuno, Heidelberg 05937
IMPA, Ligonier 15503
IMPI, Clifton 15502
IMRA, Lichfield 12201
IMS, Columbus 15501
IMS, Enfield 12254
IMS, Hayward 15331
IMS, Roseville 15259
IMS, Santa Rosa 15499
IMU, Rio de Janeiro 01695
IMZ, Wien 00682
INA, College Station 15333
I.N.A., Roma 07747
INADES, Abidjan 08065
INAPEN, Chicago 15370
INBEL, Institut Belge d'Information et de Documentation, Bruxelles 01374
INCA-FIEJ Research Association, Darmstadt 05938
INCAP, Guatemala City 06570
INCE, Poughkeepsie 15335
INCORA, Bogotá 03385
Incorporated Association of Architects, Weston Favell 12192
Incorporated Association of Organists, Redditch 12193
Incorporated Association of Preparatory Schools, London 12194
Incorporated Law Society of Ireland, Dublin 06956
Incorporated Society of Musicians, London 12195
Incorporated Society of Registered Naturopaths, Edinburgh 12196
Incorporating Society for Afghan Studies, London 12997
Independent Association of Questioned Document Examiners, Red Oak 15250
Independent Association of Telecommunications Users, London 12197
Independent Citizens Research Foundation for the Study of Degenerative Diseases, Ardsley 15251
Independent Computer Consultants Association, Saint Louis 15252
Independent Educational Services, Princeton 15253
Independent Petroleum Association of Canada, Calgary 02513
Independent Research Libraries Association, Washington 15254
Independent Scholars of Asia, Berkeley 15255
Independent Schools Association of British Columbia, Vancouver 02514
Independent Schools Joint Committee, Petersfield 12198
Indexing and Abstracting Society of Canada, Toronto 02515
Indian Academy of Sciences, Bangalore 06767
Indian Adult Education Association, New Delhi 06768
Indian Anthropological Association, Delhi 06769
Indian Association for the Cultivation of Science, Calcutta 06770
Indian Association of Academic Librarians, New Delhi 06771
Indian Association of Biological Sciences, Calcutta 06772
Indian Association of Geohydrologists, Calcutta 06773
Indian Association of Parasitologists, Calcutta 06774
Indian Association of Special Libraries and Information Centres, Calcutta 06775
Indian Association of Systematic Zoologists, Calcutta 06776
Indian Association of Teachers of Library Science, Nagpur 06777
Indian Biophysical Society, Calcutta 06778
Indian Botanical Society, Madras 06779
Indian Brain Research Association, Calcutta 06780
Indian Cancer Society, Bombay 06781
Indian Ceramic Society, Calcutta 06782
Indian Chemical Society, Calcutta 06783
Indian College Library Association, Hyderabad 06784
Indian Council of Agricultural Research, New Delhi 06785
Indian Council of Historical Research, New Delhi 06786
Indian Council of Medical Research, New Delhi 06787
Indian Council of Social Science Research, New Dehli 06788
Indian Council of World Affairs, New Delhi 06789
Indian Dairy Association, New Dehli 06790
Indian Dental Association U.S.A., Jamaica 15256
Indian Economic Association, Dehli 06791
Indian Fine Arts Society, Singapore 10047
Indian Folklore Society, Calcutta 06792
Indian Geographical Society, Madras 06793
Indian Geologists' Association, Chandigarh 06794
Indian Institute of Architects, Bombay 06795
Indian Institute of Metals, Calcutta 06796
Indian Jute Industries' Research Association, Calcutta 06797
Indian Law Institute, New Delhi 06798
Indian Library Association, Delhi 06799
Indian Mathematical Society, New Delhi 06800
The Indian Medical Association, New Dehli 06801
Indian Musicological Society, Baroda 06802
Indian National Academy of Engineering, New Delhi 06803
Indian National Foundation, Brasília 01660
Indian National Science Academy, New Delhi 06804
Indian Optometric Association, New Delhi 06805
Indian Pharmaceutical Association, Bombay 06806
Indian Phytopathological Society, New Dehli 06807
Indian Political Science Association, Madras 06808
Indian Psychoanalytical Society, Calcutta 06809
Indian Psychometric and Educational Research Association, Patna 06810
Indian Public Health Association, Calcutta 06811
Indian Rubber Manufacturers Research Association, Thane 06812
Indian Science Congress Association, Calcutta 06813
Indians Into Medicine, Grand Forks 15257
Indian Society for Nuclear Techniques in Agriculture and Biology, New Delhi 06814
The Indian Society of Agricultural Economics, Bombay 06815
Indian Society of Criminology, Madras 06816
Indian Society of Earthquake Technology, Roorkee 06817
Indian Society of Genetics and Plant Breeding, New Dehli 06818
Indian Society of Oriental Art, Calcutta 06819
Indian Society of Soil Science, New Delhi 06820
Indian Space Research Organization, Bangalore 06821
Indian Standards Institution, New Dehli 06822
Indo-British Historical Society, Madras 06823
Indonesian Institute of Engineers, Jakarta Pusat 06893
Indonesian Library Association, Jakarta 06890
Indo-Pacific Prehistory Association, Canberra 00410
Industrial Accident Prevention Association, Toronto 02516
Industrial Development Research Council, Norcross 15258
The Industrial Explosives Society of Japan, Tokyo 08168
Industrial Fire Protection Association of Great Britain, London 12199
Industrial Locomotive Society, Woking 12200
Industrial Marketing Research Association, Lichfield 12201
Industrial Mathematics Society, Roseville 15259
Industrial Medical Association, Haifa 07057
Industrial Relations Research Association, Madison 15260
Industrial Research Institute, Washington 15261
Industrial Society, London 12202
Industrial Unit of Tribology, Leeds 12203
Industrie-Gemeinschaft Aerosole e.V., Frankfurt 05939
INE, Madrid 10326
Infectious Diseases Society of America, New Haven 15262
INFODAS, Köln 05940
INFORM, Wichita 15519
Informatica, Darmstadt 05941
Information Processing Society of Japan, Tokyo 08140
Information Resource Management Association of Canada, Toronto 02517
Informationstechnische Gesellschaft im VDE, Frankfurt 05942
Ingenjörsvetenskapsakademien, Stockholm 10469
INGENOMINAS, Bogotá 03393
INICR, San Francisco 15272
INIDEP, Mar del Plata 00144
Inland Bird Banding Association, Wisner 15263
Inland Fisheries Society of India, Barrackpore 06824
INMED, Grand Forks 15257
INMM, Northbrook 15336
Inner Mongolian Academy of Agricultural Sciences and Animal Husbandry, Neimengku 03205
Inner Mongolian Academy of Social Sciences, Huhehot 03206
Innovation Management Institute of Canada, Ottawa 02518
Innsbrucker Arbeitskreis für Psychoanalyse, Psychoanalytisches Forschungs- und Ausbildungsinstitut, Innsbruck 00652
Innsbrucker Germanistische Arbeitsgemeinschaft, Innsbruck 00553
Innsbrucker Gesellschaft zur Pflege der Geisteswissenschaften, Innsbruck 00654
Innsbrucker Sprachwissenschaftliche Gesellschaft, Innsbruck 00655
INO, Firenze 07748
INPES, Bogotá 03395
Inroads, Saint Louis 15264
INSAB, Philadelphia 15505
INSIVUMEH, Guatemala 06571
Institiúid Ceimice Na hEireann, Dublin 06961
Institiuid Teangeolaiochta Eireann, Dublin 06957
Institución Fernando el Católico, Zaragoza 10312
Institut Aéronautique et Spatial du Canada, Ottawa 02016
Institut Africain, Mouyondzi 03426
Institut Africain pour le Développement Economique et Social, Abidjan 08065
Institut Agrícola Català de Sant Isidre, Barcelona 10313
Institut Agricole du Canada, Ottawa 01823
Institut Archéologique du Luxembourg, Arlon 01375
Institut Archéologique Liégeois, Liège 01376
Institut Belge de Droit Comparé, Bruxelles 01377
Institut Belge de la Soudure, Bruxelles 01378
Institut Belge de Normalisation, Bruxelles 01379
Institut Belge de Régulation et d'Automatisme, Bruxelles 01380
Institut Belge des Hautes Etudes Chinoises, Bruxelles 01381
Institut Belge des Sciences Administratives, Bruxelles 01382
Institut Canadien de Chirurgie Plastique Faciale, Toronto 02172
Institut Canadien d'Education des Adultes, Montréal 02519
Institut Canadien de Gestion, Willowdale 02178
Institut Canadien de Gestion de l'Innovation, Ottawa 02518
Institut Canadien de la Santé Infantile, Ottawa 02169
Institut Canadien de la Tole d'Acier pour le Bâtiment, Willowdale 02249
Institut Canadien de l'Emballage Souple, Toronto 02136
Institut Canadien de Microréproductions Historiques, Ottawa 02163
Institut Canadien de Québec, Québec 02520
Institut Canadien de Recherches pour l'Avancement de la Femme, Ottawa 02242
Institut Canadien des Bois Traités, Ottawa 02185
Institut Canadien de Science et Technologie Alimentaire, Ottawa 02174
L'Institut Canadien des Engrais, Ottawa 02133
Institut Canadien des Moulages d'Acier, Ottawa 02961

Institut Canadien des Urbanistes, Ottawa *02181*
Institut Canadien du Droit et de la Politique de l'Environnement, Toronto *02162*
Institut Canadien du Film, Ottawa *02134*
Institut Canadienne d'Hypnotisme, Sainte Anne de Bellevue *02176*
Institut Canadien pour les Etudes des Télécommunications, Pierrefonds *02166*
Institut d'Administration Publique du Canada, Toronto *02531*
Institut de France, Paris *04564*
Institut d'Egypte, Cairo *03906*
Institut de la Publicité Canadienne, Toronto *02527*
Institut del Teatre, Barcelona *10314*
Institut de Radiotélédiffusion pour Enfants, Toronto *02344*
Institut der deutschen Wirtschaft e.V., Köln *05943*
Institut de Recherches Politiques, Halifax *02525*
Institut de Recherches Psychologiques, Montréal *02521*
Institut der Hessischen Volkshochschulen, Frankfurt *05944*
Institut des Actuaires Français, Paris *04565*
Institut des Agronomes du Nouveau-Brunswick, Fredericton *02686*
Institut des Arbitres du Canada, Toronto *01869*
Institut des Comptables Agréés de l'Ontario, Toronto *02528*
Institut des Sciences Historiques, Paris *04566*
Institut d'Estudis Catalans, Barcelona *10315*
Institut d'Histoire de l'Amérique Française, Montréal *02522*
Institut d'Horlogerie du Canada, Montréal *02523*
Institut du Congrès des Associations de Défense, Ottawa *02379*
Institute for Advanced Research in Asian Science and Medicine, Garden City *15265*
Institute for Advanced Studies in the Theatre Arts, New York *15266*
Institute for Advanced Studies of World Religions, Carmel *15267*
Institute for Aerospace Studies, Downsview *02524*
Institute for Alternative Agriculture, Greenbelt *15268*
Institute for American Indian Studies, Washington *15269*
Institute for Animal Health, Newbury *12204*
Institute for Animal Health Pirbright, Pirbright Woking *12205*
Institute for Applied Social Sciences, Nijmegen *08915*
Institute for Biblical Research, Wheaton *15270*
Institute for Certification of Computer Professionals, Des Plaines *15271*
Institute for Childhood Resources, San Francisco *15272*
Institute for Community Design Analysis, Great Neck *15273*
Institute for Consumer Ergonomics, Loughborough *12206*
Institute for Cultural Research, Tunbridge Wells *12207*
Institute for Defense Analyses, Alexandria *15274*
Institute for Development of Educational Activities, Dayton *15275*
Institute for Econometric Research, Fort Lauderdale *15276*
Institute for Economic Analysis, Silver Springs *15277*
Institute for Educational Leadership, Washington *15278*
Institute for Esperanto in Commerce and Industry, Valkenswaard *08911*
Institute for Expressive Analysis, New York *15279*
Institute for Fiscal Studies, London *12208*
Institute for Fluitronics Education, Elm Grove *15280*
Institute for Food and Development Policy, San Francisco *15281*
Institute for Gravitational Strain Pathology, Rangeley *15282*
Institute for Hospital Clinical Nursing Education, Chicago *15283*
Institute for Intercultural Studies, New York *15284*
Institute for International Economic Cooperation and Development, Roma *07751*
Institute for Labor and Mental Health, Oakland *15285*
Institute for Mediterranean Affairs, New York *15286*
Institute for Palestine Studies, Washington *15287*
Institute for Perception Research, Eindhoven *08914*
Institute for Philosophy and Public Policy, College Park *15288*
Institute for Policy Studies, Washington *15289*
Institute for Psychohistory, New York *15290*
Institute for Public Relations Research and Education, Sarasota *15291*
Institute for Rational-Emotive Therapy, New York *15292*
Institute for Reality Therapy, Canoga Park *15293*
Institute for Research in Hypnosis and Psychotherapy, New York *15294*
Institute for Research on Public Policy, Halifax *02525*
Institute for Responsive Education, Boston *15295*
Institute for Scientific Information, Uxbridge *12209*
Institute for Sex Education and Research, Birmingham *12210*
Institute for Social Research, Ann Arbor *15296*
Institute for Southern Studies, Durham *15297*
Institute for Standardization, Moskva *09875*
Institute for Studies in American Music, Brooklyn *15298*
Institute for the Advancement of Engineering, Woodland Hills *15299*
Institute for the Advancement of Human Behavior, Stanford *15300*
Institute for the Development of Emotional and Life Skills, State College *15301*
Institute for the Development of Indian Law, Oklahoma City *15302*
Institute for the Encyclopedia of U.R.A.M., Toronto *02526*
Institute for the Future, Menlo Park *15303*
Institute for Theological Encounter with Science and Technology, Saint Louis *15304*
Institute for the Protection of Cultural Monuments of Macedonia, Skopje *08606*
Institute for the Study of Human Knowledge, Los Altos *15305*
Institute for the Study of Man, McLean *15306*
Institute for Twenty-First Century Studies, Arlington *15307*
Institute for Urban Design, San Diego *15308*
Institute of Actuaries, London *12211*
The Institute of Actuaries of Australia, Saint Ives *00411*
Institute of Administrative Management, Orpington *12212*
Institute of Advertising Practitioners in Ireland, Dublin *06958*
Institute of Agricultural Engineering, Wageningen *08912*
Institute of American Indian Arts, Santa Fe *15309*
Institute of Andean Research, New York *15310*
Institute of Andean Studies, Berkeley *15311*
Institute of Arab Music, Alexandria *03907*
Institute of Arab Music, Cairo *03908*
Institute of Architectural and Associated Technology, Dublin *06959*
Institute of Asphalt Technology, Staines *12213*
Institute of Australian Geographers, Canberra *00412*
Institute of Bankers in Scotland, Edinburgh *12214*
Institute of Bankers in South Africa, Johannesburg *10148*
Institute of Biology, London *12215*
Institute of British Geographers, London *12216*
Institute of Business Administration, Wigan *12217*
Institute of Canadian Advertising, Toronto *02527*
Institute of Ceramics, Stoke-on-Trent *12218*
Institute of Certified Financial Planners, Denver *15312*
Institute of Chartered Accountants in England and Wales, London *12219*
Institute of Chartered Accountants in Ireland, Dublin *06960*
Institute of Chartered Accountants of India, New Delhi *06825*
Institute of Chartered Accountants of Ontario, Toronto *02528*
Institute of Chartered Accountants of Scotland, Edinburgh *12220*
Institute of Chartered Foresters, Edinburgh *12221*
Institute of Chartered Secretaries and Administrators, London *12222*
The Institute of Chemistry of Ireland, Dublin *06961*
Institute of Chinese Culture, New York *15313*
Institute of Civil War Studies, Saint Louis *15314*
Institute of Clay Technology, Ripley *12223*
Institute of Community Studies, London *12224*
Institute of Concrete Technology, Beaconsfield *12225*
Institute of Construction Management, London *12226*
Institute of Contemporary Arts, London *12227*
Institute of Contemporary History and Wiener Library, London *12228*
Institute of Corrosion, Leighton Buzzard *12229*
Institute of Cost and Management Accountants of Pakistan, Karachi *09366*
Institute of Credit Management, Stamford *12230*
Institute of Cultural Affairs, Chicago *15315*
Institute of Data Processing Management, London *12231*
Institute of Directors, London *12232*
Institute of Early American History and Culture, Williamsburg *15316*
Institute of Eastern Culture, Tokyo *08434*
Institute of Economic Affairs, London *12233*
The Institute of Electrical and Electronics Engineers, New York *15317*
Institute of Electrical and Electronics Engineers, Thornhill *02529*
Institute of Electrical Engineers of Japan, Tokyo *08119*
Institute of Electrology Educators, Bellingham *15318*
Institute of Electrolysis, Manchester *12234*
Institute of Electronics, Information and Communication Engineers of Japan, Tokyo *08121*
Institute of Energy, London *12235*
Institute of Energy (New Zealand Section), Wellington *09128*
Institute of Engineers and Technicians, London *12236*
Institute of Environmental Sciences, Mount Prospect *15319*
Institute of Financial Education, Chicago *15320*
Institute of Fisheries Management, Holmer *12237*
Institute of Food Research, Reading *12238*
Institute of Food Science and Technology of the United Kingdom, London *12239*
Institute of Food Technologists, Chicago *15321*
Institute of Gas Technology, Chicago *15322*
Institute of General Semantics, Englewood *15323*
The Institute of Group-Analysis, Group-Analysis Society (London), London *12240*
Institute of Health Education, Hale Barns *12241*
Institute of Heraldic and Genealogical Studies, Canterbury *12242*
Institute of Hispanic Art, Barcelona *10316*
Institute of Hospital Engineering, Portsmouth *12243*
Institute of Housing, Coventry *12244*
Institute of Human Origins, Berkeley *15324*
Institute of Hydrology, Wallingford *12245*
Institute of Industrial Engineers, Dublin *06962*
Institute of Industrial Engineers, Norcross *15325*
Institute of Industrial Science, Tokyo *08408*
Institute of Information Scientists, London *12246*
Institute of International Education, New York *15326*
Institute of International Sociology, Gorizia, Gorizia *07727*
Institute of Inventors, London *12247*
Institute of Investment Management and Research, Bromley *12248*
Institute of Islamic Culture, Lahore *09367*
Institute of Jamaica, Kingston *08080*
Institute of Jewish Affairs, London *12249*
Institute of Judicial Administration, New York *15327*
Institute of Laboratory Animal Resources, Washington *15328*
Institute of Leisure and Amenity Management, Reading *12250*
Institute of Linguists, London *12251*
Institute of Lithuanian Studies, Chicago *15329*
Institute of Logistics, Corby *12252*
Institute of Management, Corby *12253*
Institute of Management Consultants in Ireland, Dublin *06963*
The Institute of Management Sciences, Providence *15330*
Institute of Management Services, Enfield *12254*
Institute of Marine Engineers, London *12255*
Institute of Materials, London *12256*
Institute of Materials Handling, South Melbourne *00413*
Institute of Materials Management, Corby *12257*
Institute of Mathematical Statistics, Hayward *15331*
Institute of Mathematics and its Applications, Southend-on-Sea *12258*
Institute of Measurement and Control, London *12259*
Institute of Media Executives, Willenhall *12260*
Institute of Medical Laboratory Sciences, London *12261*
Institute of Medicine, Washington *15332*
Institute of Mental Health Research and Postgraduate Training, Parkville *00414*
The Institute of Metals, London *12262*
Institute of Metals and Materials Australasia, Parkville *00415*
Institute of Moralogy, Chiba *08183*
Institute of Municipal Building Management, Colwyn Bay *12263*
Institute of Municipal Management, South Melbourne *00416*
Institute of Nautical Archaeology, College Station *15333*
Institute of Navigation, Alexandria *15334*
Institute of Noise Control Engineering, Poughkeepsie *15335*
Institute of Nuclear Materials Management, Northbrook *15336*
Institute of Nutrition of Central America and Panama, Guatemala City *06570*
Institute of Occupational Medicine, Edinburgh *12264*
Institute of Outdoor Drama, Chapel Hill *15337*
Institute of Packaging, Melton Mowbray *12265*
Institute of Personnel Management, London *12266*
Institute of Petroleum, London *12267*
Institute of Photographic Technology, Melbourne *00417*
Institute of Physics, London *12268*
Institute of Physics, Singapore, Singapore *10048*
Institute of Plasmaphysics Rijnhuizen, Nieuwegein *08897*
Institute of Population Registration, London *12269*
Institute of Printing, Tunbridge Wells *12270*
Institute of Professional Investigators, Blackburn *12271*
Institute of Professional Libraries of Ontario, Toronto *02530*
Institute of Psycho-Analysis, London *12272*
Institute of Psychological Research, Montréal *02521*
Institute of Psycho-Sexual Medicine, London *12273*
Institute of Public Administration, Dublin *06964*
Institute of Public Administration, New York *15338*
Institute of Public Administration of Canada, Toronto *02531*
Institute of Public Affairs, Jolimont *00418*
Institute of Purchasing and Supply, Ascot *12274*
Institute of Quarrying, Nottingham *12275*
Institute of Race Relations, London *12276*
Institute of Refrigeration, Carshalton *12277*
Institute of Sales and Marketing Management, Royal Leamington Spa *12278*
Institute of Science Technology, Lichfield *12279*
Institute of Scientific and Technical Communicators, London *12280*
Institute of Sheet Metal Engineering, Birmingham *12281*
Institute of Small Business, London *12282*
Institute of South African Architects, Houghton *10149*
Institute of Southeast Asian Studies, Singapore *10049*
Institute of South West African Architects, Windhoek *08815*
Institute of Sports Medicine, London *12283*
Institute of Statisticians, Preston *12284*
Institute of Supervisory Management, Lichfield *12285*
Institute of Taxation in Ireland, Dublin *06965*
Institute of Technicians in Venereology, Orsett *12286*
Institute of Textile Science, Montréal *02532*
Institute of Textile Technology, Charlottesville *15339*
Institute of the American Musical, Los Angeles *15340*
Institute of the Great Plains, Lawton *15341*
Institute of Training and Development, Marlow *12287*
Institute of Translation and Interpreting, London *12288*
Institute of Transport, Beverly Hills *00419*
Institute of Transportation Economics, Tokyo *08443*
Institute of Transportation Engineers, Washington *15342*
Institute of Trichologists, London *12289*
Institute of Value Management, Buxton *12290*
Institute of Water Resources Management and Air Pollution Control, Köln *05962*
Institute of Wood Science, High Wycombe *12291*
Institute of World Affairs, Salisbury *15343*
Institute on Hospital and Community Psychiatry, Washington *15344*
Institut Européen d'Ecologie et de Cancérologie, Bruxelles *01383*
Institut Européen des Armes de Chasse et de Sport, Liège *01384*
Institut Européen Interuniversitaire de l'Action Sociale, Marcinelle *01385*
Institut Finanzen und Steuern e.V., Bonn *05945*
Institut Forestier du Canada, Ottawa *02175*
Institut Français d'Analyse de Groupe et de Psychodrame, Paris *04567*
Institut Français de l'Energie, Paris *04568*
Institut Français d'Histoire Sociale, Paris *04569*
Institut für angewandte Arbeitswissenschaft e.V., Köln *05946*
Institut für Angewandte Geodäsie, Frankfurt *05947*
Institut für Angewandte Pädagogische Forschung e.V., Seelze *05948*
Institut für angewandte Verbraucherforschung e.V., Köln *05949*
Institut für Angewandte Wirtschaftsforschung, Tübingen *05950*
Institut für angewandte Wirtschaftsforschung im Mittelstand, Düsseldorf *05951*
Institut für Auslandsbeziehungen, Stuttgart *05952*
Institut für bankhistorische Forschung e.V., Frankfurt *05953*
Institut für Bauforschung e.V., Hannover *05954*
Institut für Chemiefasern, Denkendorf *05955*
Institut für den Wissenschaftlichen Film, Göttingen *05956*
Institut für Deutsches und Internationales Baurecht e.V., Frankfurt *05957*
Institut für Energie- und Umweltforschung Heidelberg e.V., Heidelberg *05958*
Institut für Europäische Umweltpolitik, Bonn *05959*
Institut für Film und Bild in Wissenschaft und Unterricht, Grünwald *05960*
Institut für Gesellschaftswissenschaften Walberberg e.V., Bonn *05961*
Institut für gewerbliche Wasserwirtschaft und Luftreinhaltung e.V., Köln *05962*
Institut für Grenzgebiete der Wissenschaft, Innsbruck *00656*
Institut für Handwerkswirtschaft München, München *05963*
Institut für Kunststoffverarbeitung, -anwendung und -prüfung, Würzburg *06308*
Institut für Länderkunde e.V., Leipzig *05964*
Institut für nationale und internationale Fleisch- und Ernährungswirtschaft, Heidelberg *05965*
Institut für Neue Musik und Musikerziehung e.V., Darmstadt *05966*
Institut für Neue Technische Form, Darmstadt *05967*
Institut für ökologische Forschung und Bildung e.V., Münster *05968*
Institut für Österreichische Musikdokumentation, Wien *00657*
Institut für Österreichkunde, Wien *00658*
Institut für Schadenverhütung und Schadenforschung der öffentlich-rechtlichen Versicherer e.V., Kiel *05969*
Institut für Sozialarbeit und Sozialpädagogik, Frankfurt *05970*
Institut für Sozialdienste, Bregenz *00659*
Institut für Soziales Design – Entwicklung und Forschung, Wien *00660*
Institut für Sozialforschung und Sozialwirtschaft e.V., Saarbrücken *05971*
Institut für Städtebau, Wohnungswirtschaft und Bausparwesen e.V., Bonn *05972*
Institut für Technik der Betriebsführung im Handwerk, Karlsruhe *05973*
Institut für technische Weiterbildung Berlin e.V., Berlin *05974*
Institut für Textil- und Faserforschung Stuttgart, Denkendorf *05975*
Institut für Urheber- und Medienrecht e.V., München *05976*
Institut für Wirtschaft und Gesellschaft Bonn e.V., Bonn *05977*
Institut für Wissenschaftliche Zusammenarbeit, Tübingen *05978*
Institut für Wissenschaft und Kunst, Wien *00661*
Institut für Zeitungsforschung, Dortmund *05979*

Institut für Ziegelforschung Essen e.V., Essen 05980
Institut Géographique National, Paris 04570
Institut Historique Belge de Rome, Bruxelles 01386
Institut International d'Administration Publique, Paris 04571
Institut International de Droit d'Expression et d'Inspiration Françaises, Paris 04572
Institut International de Planification de l'Education, Paris 04573
Institut International de Psychologie et de Psychothérapie Charles Baudouin, Coppet 10729
Institut International de Recherches Betteravières, Bruxelles 01387
Institut International des Droits de l'Homme, Strasbourg 04574
Institut International des Sciences Administratives, Bruxelles 01388
Institut International du Fer et de l'Acier, Bruxelles 01389
Institut International du Froid, Paris 04575
Institut International du Théâtre, Paris 04576
Institutiom of Surveyors – Australia, Canberra 00420
Institution of Agricultural Engineers, Bedford 12292
The Institution of Chemical Engineers, Rugby 12293
Institution of Civil Engineers, London 12294
Institution of Civil Engineers (Republic of Ireland), Killiney 06966
Institution of Corrosion Science and Technology, Leighton Buzzard 12295
Institution of Electrical Engineers, Lahore 09368
Institution of Electrical Engineers, London 12296
Institution of Electrical Engineers (Irish Branch), Dun Laoghaire 06967
Institution of Electronics, Rochdale 12297
Institution of Electronics and Electrical Incorporated Engineers, London 12298
Institution of Engineering Designers, Westbury 12299
Institution of Engineers amd Shipbuilders in Scotland, Glasgow 12300
Institution of Engineers – Australia, Barton 00421
Institution of Engineers of Ireland, Dublin 06968
Institution of Engineers, Pakistan, Lahore 09369
Institution of Environmental Health Officers, London 12301
Institution of Environmental Sciences, London 12302
Institution of Fire Engineers, Leicester 12303
Institution of Gas Engineers, London 12304
Institution of Highways and Transportation, London 12305
Institution of Lighting Engineers, Rugby 12306
Institution of Mechanical Engineers, London 12307
Institution of Mechanical Incorporated Engineers, London 12308
Institution of Mining and Metallurgy, London 12309
Institution of Mining Engineers, Doncaster 12310
Institution of Nuclear Engineers, London 12311
Institution of Occupational Safety and Health, Leicester 12312
Institution of Plant Engineers, London 12313
Institution of Polish Engineers in Great Britain, London 12314
Institution of Post Office Electrical Engineers, London 12315
Institution of Professional Engineers New Zealand, Wellington 09129
Institution of Radio and Electronics Engineers – Australia, Sydney 00422
Institution of Railway Signal Engineers, Dawlish 12316
Institution of Structural Engineers, London 12317
Institution of Water and Environment Management, London 12318
Institut Militaire de Québec, Québec 02533
Institut National Canadien pour les Aveugles, Toronto 02206
Institut National d'Archéologie et d'Art, Tunis 11142
Institut National de Cinématographie Scientifique de Belgique, Bruxelles 01390
Institut National Genevois, Genève 10730
Institut Neue Wirtschaft e.V., Hamburg 05981
Instituto Açoriano de Cultura, Angra do Heroismo 09658
Instituto Amatller de Arte Hispánico, Barcelona 10316
Instituto Archeológico, Histórico e Geográfico Pernambucano, Recife 01663
Instituto Aula de Mediterráneo, Valencia 10317
Instituto Boliviano del Petróleo, La Paz 01546
Instituto Bonaerense de Numismática y Antigüedades, Buenos Aires 00142
Instituto Brasileiro de Economia, Rio de Janeiro 01664
Instituto Brasileiro de Educâcão, Ciencia e Cultura, Rio de Janeiro 01665
Instituto Caro y Cuervo, Bogotá 03377
Instituto Central de Medicina Legal, Bogotá 03378
Instituto Centroamericano de Administración Pública, San José 03445
Instituto Centroamericano de Extensión de la Cultura, San José 03446
Instituto Centroamericano de Investigación y Tecnología, Guatemala City 06569
Instituto Colombiano Agropecuario, Bogotá 03379
Instituto Colombiano de Antropología, Bogotá 03380
Instituto Colombiano de Bienestar Familiar, Bogotá 03381
Instituto Colombiano de Crédito Educativo y Estudios Técnicos en el Exterior, Bogotá 03382
Instituto Colombiano de Cultura, Bogotá 03383
Instituto Colombiano de Cultura Hispánica, Bogotá 03384
Instituto Colombiano de la Reforma Agraria, Bogotá 03385
Instituto Colombiano de Normas Técnicas, Bogotá 03386
Instituto Colombiano de Oncologia Pediatrica, Bogotá 03387
Instituto Colombiano de Pedagogía, Bogotá 03388
Instituto Colombiano para el Fomento de la Educación Superior, Bogotá 03389
Instituto Colombinao Agropecuario, Mosquera 03390
Instituto de Botânica, São Paulo 01666
Instituto de Chile, Santiago 03053
Instituto de Ciencias Naturales, Bogotá 03391
Instituto de Coímbra, Coímbra 09659
Instituto de Ecología, Xalapa 08735
Instituto de Engenharia de São Paulo, São Paulo 01667
Instituto de España, Madrid 10318
Instituto de Estudios Africanos, Madrid 10319
Instituto de Estudios Asturianos, Oviedo 10320
Instituto de Estudios y Publicaciones Juan Molina, Santiago 03054
Instituto de Farmacología Española, Madrid 10321
Instituto de Filologia, Madrid 10322
Instituto de Historia y Cultura Naval, Madrid 10323
Instituto de Ingenieros de Chile, Santiago 03055
Instituto de Investigaciones Agropecuarias, Santiago 03056
Instituto de Investigaciones Biológicas Clemente Estable, Montevideo 13260
Instituto de la Engeniería de España, Madrid 10324
Instituto de Literatura, La Plata 00143
Instituto de Nutrición de Centro América y Panamá, Guatemala City 06570
Instituto de Planejamento de Pernambuco, Recife 01668
Instituto de Salud Publica de Chile, Santiago 03057
Instituto de Zoonosis e Investigación Pecuaria, Lima 09469
Instituto do Ceará, Fortaleza 01669
Instituto dos Actuarios Portugueses, Lisboa 09660
Instituto dos Advogados Brasileiros, Rio de Janeiro 01670
Instituto Genealógico Brasileiro, São Paulo 01671
Instituto Geográfico e Histórico da Bahia, Salvador 01672
Instituto Geográfico e Histórico do Amazonas, Manaus 01673
Instituto Geográfico Nacional, Lima 09470
Instituto Geográfico Nacional, Madrid 10325
Instituto Histórico de Alagoas, Maceío 01674
Instituto Histórico e Geográfico Brasileiro, Rio de Janeiro 01675
Instituto Histórico e Geográfico de Goiás, Goiânia 01676
Instituto Histórico e Geográfico de Santa Catarina, Florianópolis 01677
Instituto Histórico e Geográfico de Santos, Santos 01678
Instituto Histórico e Geográfico de São Paulo, São Paulo 01679
Instituto Histórico e Geográfico de Sergipe, Aracajú 01680
Instituto Histórico e Geográfico do Maranhão, São Luís 01681
Instituto Histórico e Geográfico do Pará, Belém 01682
Instituto Histórico e Geográfico do Rio Grande do Norte, Natal 01683
Instituto Histórico e Geográfico do Rio Grande do Sul, Porto Alegre 01684
Instituto Histórico e Geográfico Paraibano, João Pessoa 01685
Instituto Histórico, Geográfico e Etnográfico Paranaense, Curitiba 01686
Instituto Histórico y Geográfico del Uruguay, Montevideo 13261
Instituto Hondureño de Cultura Interamericana, Tegucigalpa 06584
Instituto Indigenista Interamericano, México 08736
Instituto Interamericano de Cooperación para la Agricultura, San José 03447
Instituto Latinoamericano de Investigaciones Sociales, Quito 03854
Instituto Latinoamericano de las Naciones Unidas para la Prevención del Delito y Tratamiento del Delincuente, San José 03448
Instituto Latinoamericano de Planificación Económica y Social, Santiago 03058
Instituto Nacional de Bellas Artes, México 08737
Instituto Nacional de Cancerología, Bogotá 03392
Instituto Nacional de Estadística, Madrid 10326
Instituto Nacional de Estudos e Pesquisas Educacionais, Brasília 01687
Instituto Nacional de Higiene, México 08738
Instituto Nacional de Investigaciones Agrícolas, México 08739
Instituto Nacional de Investigaciones Geológico-Mineras, Bogotá 03393
Instituto Nacional de Investigación y Desarrollo Pesquero, Mar del Plata 00144
Instituto Nacional de Medicina Legal, Bogotá 03394
Instituto Nacional de Meteorología, Madrid 10327
Instituto Nacional de Parasitología, Asunción 09431
Instituto Nacional de Pesquisas da Amazonia, Manaus 01688
Instituto Nacional de Pesquisas Hidroviarias, Rio de Janeiro 01689
Instituto Nacional de Salud, Bogotá 03395
Instituto Nacional de Sismología, Vulcanología, Meteorología e Hidrología, Guatemala 06571
Instituto Nacional de Tecnologia, Rio de Janeiro 01690
Instituto Nacional de Tecnologia Industrial, Buenos Aires 00145
Instituto Nacional de Vitivinicultura, Mendoza 00146
Instituto Nacional do Cancer, Rio de Janeiro 01691
Instituto Nacional do Livro, Brasília 01692
Instituto Nacional do Livro, Rio de Janeiro 01693
Instituto Oceanográfico de Valparaíso, Valparaíso 03059
Instituto Panamericano de Geografía e Historia, México 08740
Instituto para la Integración de América Latina, Buenos Aires 00147
Instituto Peruano de Ingenieros Mecánicos, Lima 09471
Instituto Português de Arqueologia, História e Etnografia, Lisboa 09661
Instituto Uruguayo de Normas Técnicas, Montevideo 13262
Institut Potasse et Phosphate du Canada, Etobicoke 02826
Institut Professionnel de la Fonction Publique du Canada, Ottawa 02846
Institut Romand de Recherches et de Documentation Pédagogiques, Neuchâtel 10731
Institut Royal Belge du Pétrole, Bruxelles 01391
Institut Royal d'Architecture du Canada, Ottawa 02862
Institut Royal des Relations Internationales, Bruxelles 01392
Institut Suisse de Recherches Expérimentales sur le Cancer, Epalinges 10732
Institut Suisse pour l'Etude de l'Art, Zürich 10927
Institut Supérieur de Perfectionnement des Cadres I.S.P.C., Ottawa 02845
Instituut voor Mechanisatie, Arbeid en Gebouwen, Wageningen 08912
Instituut voor Onderwijskundige Dienstverlening, Nijmegen 08913
Instituut voor Perceptie Onderzoek, Eindhoven 08914
Instituut voor Toegepaste Sociale Wetenschapen, Nijmegen 08915
Inst M E, Willenhall 12260
Inst SMM, Royal Leamington Spa 12278
Insulated Cable Engineers Association, South Yarmouth 15345
Insurance Information Institute, New York 15346
Insurance Institute of Ireland, Dublin 06969
Insurers' Advisory Organization, Toronto 02534
INT, Rio de Janeiro 01690
INTA, Castelar 00133
INTA, Mesa 15504
INTECOL, Aiken 15380
INTEL-ED, Philadelphia 15562
Intellectual Society of Libya, Tripoli 08568
Intelligent Buildings Institute, Washington 15347
Inter-African Scientific Committee Post-Harvest, Dakar 10024
Inter-American Association of Agricultural Librarians and Documentalists, Turrialba 03449
Inter-American Association of Sanitary and Environmental Engineering, São Paulo 01582
Inter-American Association of Sanitary Engineering and Environmental Sciences, Brookeville 15348
Inter-American Bar Association, Washington 15349
Inter-American College Association, Pomona 15350
Inter-American Commercial Arbitration Commission, Washington 15351
Inter-American Commercial Arbitration Commission (Canadian Section), Ottawa 02535
Inter-American Council for Education, Science and Culture, Washington 15352
Inter-American Indian Institute, México 08736
Inter-American Institute for Cooperation on Agriculture, San José 03447
Inter-American Music Council, Washington 15353
Inter-American Safety Council, Englewood 15354
Inter-American Statistical Institute, Panamá City 09420
Inter-American Tropical Tuna Comission, La Jolla 15355
Interchurch Medical Assistance, New Windsor 15356
Intercollegiate Broadcasting System, Vails Gate 15357
Intercontinental Church Society, London 12319
INTERCOSMOS, Moskva 09863
Intercultural Development Research Association, San Antonio 15358
Interdepartmental Council for Study of Ethnic Problems, Yerevan 00208
Interdepartmental Council on Physicochemical Biology and Biotechnology, Yerevan 00209
Interessengemeinschaft deutschsprachiger Autoren, Eutin 05982
Interessengemeinschaft für Lederforschung und Häuteschädenbekämpfung im Verband der deutschen Lederindustrie e.V., Frankfurt 05983
Interessenverband der Musiklehrer an allgemeinbildenden Schulen und der diese Berufsgruppe ausbildenden Hochschullehrer, Mainz 06361
Interfaith Forum on Religion, Art and Architecture, Washington 15359
Intergovernmental Council for the International Hydrological Programme, Paris 04577
Intergovernmental Oceanographic Commission, Paris 04428
Interior Design Educators Council, Irvine 15360
Interior Forest Labour Relations Association, Kelowna 02536
Intermedia, New York 15361
Intermediate Technology Development Group, Rugby 12320
Internationaal Instituut voor Sociale Geschiedenis (Stichting), Amsterdam 08916
International Academy at Santa Barbara, Santa Barbara 15362
International Academy of Behavioral Medicine, Dallas 15363
International Academy of Ceramics, Genève 10606
International Academy of Chest Physicians and Surgeons, Northbrook 15364
International Academy of Chest Physicians and Surgeons of the American College of Chest Physicians, Republic of China Chapter, Taipei 03302
International Academy of Cytology, Freiburg 05984
International Academy of Indian Culture, New Delhi 06826
International Academy of Management, 's-Gravenhage 08917
International Academy of Management – U.S. Branch, New York 15365
International Academy of Myodontics, Doylestown 15366
International Academy of Nutrition and Preventive Medicine, Asheville 15367
International Academy of Pathology, Tucson 15368
International Academy of the History of Science, Paris 04578
International Academy of Trial Lawyers, San Jose 15369
International African Institute, London 12321
International African Migratory Locust Organisation, Bamako 08656
International Agricultural Exchange (Saskatchewan), Maidstone 02537
International Aids Prospective Epidemiology Network, Chicago 15370
International Alban Berg Society, New York 15371
International Aloe Science Council, Austin 15372
International Anatomical Nomenclature Committee, London 12322
International Anesthesia Research Society, Cleveland 15373
International Arthurian Society – North American Branch, Halifax 02538
International Arthur Schnitzler Research Association, San Bernardino 15374
International Association for a Natural Economic Order, Aarau 10744
International Association for Aquatic Animal Medicine, San Leandro 15375
International Association for Bridge and Structural Engineering, Zürich 10742
International Association for Cereal Science and Technology, Schwechat 00669
International Association for Comparative Research on Leukemia and Related Diseases, Columbus 15376
International Association for Computer Information Systems, Eau Claire 15377
International Association for Computing in Education, Eugene 15378
International Association for Contemporary History of Europe, Genève 10629
International Association for Cross-Cultural Psychology, Kingston 02539
International Association for Cybernetics, Namur 01135
International Association for Dental Research, Washington 15379
International Association for Ecology, Aiken 15380
International Association for Energy Economics, Cleveland 15381
International Association for Esperanto in Libraries, London 12323
International Association for Great Lakes Research, Ann Arbor 15382
International Association for Healthcare Security and Safety, Lombard 15383
International Association for Housing Science, Coral Gables 15384
International Association for Hydraulic Research, Delft 08918
International Association for Hydrogen Energy, Coral Gables 15385
International Association for Learning Laboratories, Philadelphia 15386
International Association for Mass Communication Research, Leicester 12324
International Association for Mathematical Geology, Reston 15387
International Association for Mathematics and Computers in Simulation, Liège 01141
International Association for Media in Science, Utrecht 08919
International Association for Near-Death Studies, Philadelphia 15388
International Association for Official Statistics, Voorburg 08920
International Association for Personnel Women, Andover 15389
International Association for Philosophy of Law and Social Philosophy – American Section, Gainesville 15390
International Association for Religion and Parapsychology, Tokyo 08141
International Association for Research in Income and Wealth, New York 15391
International Association for Romanian Studies, Paris 04140
International Association for Scandinavian Studies, Norwich 12325
International Association for Shell and Spatial Structures, Madrid 10328
International Association for Statistical Computing, Voorburg 08921
International Association for Suicide Prevention, Wien 00685
International Association for the Advancement of Earth and Environmental Sciences, Chicago 15392
International Association for the Development of International and World Universities, Aulnay-sous-Bois 04290
International Association for the Development of Tropical Odonto-Stomatology, Bordeaux 04288
The International Association for the Evaluation of Educational Achievement, 's-Gravenhage 08922

International Association for the Evaluation of Educational Achievement, Stockholm *10470*
International Association for the History of Glass, Amsterdam *08923*
International Association for the Physical Sciences of the Ocean, Del Mar *15393*
International Association for the Properties of Water and Steam, Palo Alto *15394*
International Association for the Protection of Industrial Property, Zürich *10743*
International Association for the Study of Canon Law, Roma *07703*
International Association for the Study of Cooperation in Education, Santa Cruz *15395*
International Association for the Study of Insurance Economics, Genève *10733*
International Association for Veterinary Homeopathy, Milano *07704*
International Association for Water Law, Roma *07705*
International Association Futuribles, Paris *04579*
International Association of Agricultural Economists, Oak Brook *15396*
International Association of Agricultural Information Specialists, Montpellier *04580*
International Association of Agricultural Museums, Reading *12326*
International Association of Allergology and Clinical Immunology, Milwaukee *15397*
International Association of Applied Linguistics, Edinburgh *12327*
International Association of Applied Psychology, Nijmegen *08924*
International Association of Biblicists and Orientalists, Ravenna *07706*
International Association of Biological Standardization, Hassocks *12328*
International Association of Boards of Examiners in Optometry, Bethesda *15398*
International Association of Botanical Gardens, Edinburgh *02540*
International Association of Buddhist Studies, Berkeley *15399*
International Association of Campus Law Enforcement Administrators, Hartford *15400*
International Association of Career Consulting Firms, Arlington *15401*
International Association of Coroners and Medical Examiners, Peoria *15402*
International Association of Counseling Services, Alexandria *15403*
International Association of Defense Counsel, Chicago *15404*
International Association of Democratic Lawyers, Bruxelles *01139*
International Association of Educational Peace Officers, Houston *15405*
International Association of Educators for World Peace, Huntsville *15406*
International Association of Engineering Geology, Paris *04581*
International Association of Engineering Geology, Sezione Italiana, Bari *07707*
International Association of Environmental Coordinators, Chalfont Saint Gills *12329*
International Association of Environmental Mutagen Societies, Adelaide *00423*
International Association of Forensic Toxicologists, Newport Beach *15407*
International Association of French-Language Demographers, Paris *04268*
International Association of French-Speaking Directors of Educational Institutions, Montréal *01908*
International Association of Geodesy, Paris *04257*
International Association of Gerontology, México *08741*
International Association of Group Psychotherapy, Montréal *02541*
International Association of Historians of Asia, Manila *09520*
International Association of Human-Animal Interaction Organizations, Renton *15408*
International Association of Institutes of Navigation, London *12330*
International Association of Jazz Educators, Manhattan *15409*
International Association of Knowledge Engineers, Gaithersburg *15410*
International Association of Labour Inspections, Paris *04258*
International Association of Laryngectomees, Atlanta *15411*
International Association of Legal Science, Paris *04276*
International Association of Master Penmen and Teachers of Handwriting, Ottawa *02542*
International Association of Medical Laboratory Technologists, Stockholm *10471*
International Association of Medical Oceanography, Nice *04285*
International Association of Music Libraries, Archives and Documentation Centres, Stockholm *10472*
International Association of Music Libraries, Archives and Documentation Centres (United Kingdom Branch), Exeter *12331*
International Association of Music Libraries – United States Branch, New York *15412*
International Association of Ocular Surgeons, Chicago *15413*
International Association of Optometric Executives, Bismarck *15414*
International Association of Orthodontics, Oak Park *15415*
International Association of Paediatric Dentistry, London *12332*
International Association of Physical Education and Sports for Girls and Women, Mankato *15416*
International Association of Physical Education and Sports for Girls and Women, Viborg *03744*
International Association of Planetology, Praha *03545*
International Association of Sanskrit Studies, Paris *04582*
International Association of School Librarianship, Kalamazoo *15417*
International Association of Schools of Social Work, Wien *00662*
International Association of Sound Archives, Stockholm *10473*
International Association of South-East European Studies, Bucuresti *09752*
International Association of Survey Statisticians, Paris *04278*
International Association of Teachers of English as a Foreign Language, Whitstable *12333*
International Association of Teachers of Philosophy, Minden *05155*
International Association of Teachers of Russian Language and Literature, Paris *04274*
International Association of Technological University Libraries, Edinburgh *12334*
International Association of Theoretical and Applied Limnology, Tuscaloosa *15418*
International Association of Traffic and Safety Sciences, Tokyo *08142*
International Association of Trichologists, Sydney *00424*
International Association of Universities, Paris *04281*
International Association of University Professors of English, Marazion *12335*
International Association of Volcanology and Chemistry of the Earth's Interior JAVCEJ, Leeds *12336*
International Association of Wood Anatomists, Utrecht *08925*
International Association on Mechanization of Field Experiments, Uppsala *10474*
International Association on the Genesis of Ore Deposits, Jeseník *03546*
International Association on Water Quality, London *12337*
International Astronautical Association, Paris *04583*
International Astronomical Union, Paris *04971*
International Atherosclerosis Society, Houston *15419*
International Atomic Energy Agency, Wien *00663*
International Banking Research Institute, København *03745*
International Bar Association, London *12338*
International Bee Research Association, Cardiff *12339*
International Bibliographic Commission, Paris *04424*
International Bio-Environmental Foundation, Sherman Oaks *15420*
International Bird Rescue Research Center, Berkeley *15421*
International Board for Plant Genetic Resources, Roma *07708*
International Board on Books for Young People, Basel *10734*
International Brecht Society, Athens *15422*
International Bridge, Tunnel and Turnpike Association, Washington *15423*
International Bronchoesophagological Society, Scottsdale *15424*
International Bulb Society, Culver City *15425*
International Bundle Branch Block Association, Los Angeles *15426*
International Bureau of Education, Genève *10639*
International Bureau of Fiscal Documentation, Amsterdam *08926*
International Cardiology Foundation, Dallas *15427*
International Catholic Esperanto Association, Champaign *15428*
International Catholic Movement for Intellectual and Cultural Affairs/ICMICA-Pax Romana, Genève *10765*
International Center for Comparative Criminology, Montréal *02543*
International Center for Law in Development, New York *15429*
International Center for Research on Bilingualism, Sainte Foy *02544*
International Center for Research on Women, Washington *15430*
International Center for the Solution of Environmental Problems, Houston *15431*
International Center of Medical and Psychological Hypnosis, Milano *07468*
International Center of Medieval Art, New York *15432*
International Center of Photography, New York *15433*
International Centre for Advanced Technical and Vocational Training, Torino *07709*
International Centre for Mechanical Sciences, Udine *07470, 07710*
International Centre for Studies in Religious Education, Bruxelles *01198*
International Centre for Theoretical Physics, Trieste *07711*
International Centre for the Study of Medieval Painting in the Schelde and the Meuse Valleys, Bruxelles *01196*
International Centre of Films for Children and Young People, Paris *04380*
International Cerebral Palsy Society, London *12340*
International Childbirth Education Association, Calgary *02545*
International Childbirth Education Association, Minneapolis *15434*
International Chiropractors Association, Arlington *15435*
International Christian Esperanto Association, Alphen *08969*
International Christian Esperanto Association, Levittown *15436*
International Christian Leprosy Mission, Portland *15437*
International College of Dentists, Rockville *15438*
International College of Surgeons, Chicago *15439*
International Colour Association, Eindhoven *08927*
International Commission for Orders of Chivalry, Edinburgh *12341*
International Commission for Plant-Bee Relationships, Lusignan *04584*
International Commission for the History of Representative and Parliamentary Institutions, Brighton *12342*
International Commission for the Nomenclature of Cultivated Plants, Wageningen *08928, 08929*
International Commission for the Prevention of Alcoholism and Drug Dependency, Silver Spring *15440*
International Commission for the Protection of Alps, Vaduz *08570*
International Commission for the Protection of the Rhine against Pollution, Koblenz *05993*
International Commission for Uniform Methods of Sugar Analysis, Braunschweig *05985*
International Commission of Sugar Technology, Rain *05986*
International Commission on Illumination, Wien *00578*
International Commission on Large Dams, Paris *04425*
International Commission on Occupational Health, Singapore *10050*
International Commission on Physics Education, Columbus *15441*
International Commission on Physics Education, Malvern *12343*
International Commission on Polar Meteorology, Cambridge *12344*
International Commission on Radiation Units and Measurements, Bethesda *15442*
International Commission on the History of the Geological Sciences, Cambridge *15443*
International Commission on Trichinellosis, Poznan *09544*
International Commission on Zoological Nomenclature, London *12345*
International Commitee for Histochemistry and Cytochemistry, Leiden *08930*
International Committee for Recording the Productivity of Milk Animals, Roma *07712*
International Committee of Dialectologists, Leuven *01393*
International Committee of Military Medicine, Liège *01228*
International Committee on Veterinary Anatomical Nomenclature, Zürich *10748*
International Communication Agency, Guadalajara *08742*
International Communication Agency, México *08743*
International Communication Agency, Monterrey *08744*
International Communication Agency, São Paulo *01694*
International Communications Association, Dallas *15444*
International Comparative Literature Association, Paris *04259*
International Computing Centre, Genève *10735*
International Confederation of Theatre, Moskva *09876*
International Conference of Building Officials, Whittier *15445*
International Conference of Historians of the Labour Movement, Wien *00684*
International Conference on Large High Voltage Electric Systems, Paris *04445*
International Conference on Mechanics in Medicine and Biology, Ann Arbor *15446*
International Conference on Social Science and Medicine, Brighton *12346*
International Congress of Carboniferous Stratigraphy and Geology, Madrid *10329*
International Congress of Oral Implantologists, Upper Montclair *15447*
International Congress of University Adult Education, Fredericton *02546*
International Congress on Fracture, Sendai *08143*
International Congress on High-Speed Photography and Photonics, Janesville *15448*
International Cooperation in History of Technology Committee, Toronto *02547*
International Copper Association, New York *15449*
International Copyright Society, München *05987*
International Correspondence Society of Allergists, Shawnee Mission *15450*
International Cost Engineering Council, Morgantown *15451*
International Council for Adult Education, Toronto *02548*
International Council for Children's Play, Groningen *08931*
International Council for Distance Education, Montréal *02549*
International Council for Educational Media, Paris *04457*
International Council for Health, Physical Education and Recreation, Reston *15452*
International Council for Laboratory Animal Science, Kuopio *03966*
International Council for Pressure Vessel Technology, Milwaukee *15453*
International Council for Scientific and Technical Information, Paris *04461*
International Council for the Exploration of the Sea, København *03746*
The International Council for Traditional Music, New York *15454*
International Council – National Academy of Television Arts and Sciences, New York *15455*
International Council of Associations for Science Education, Hong Kong *06594*
International Council of Environmental Law, Bonn *05998*
International Council of Kinetography, London *12347*
International Council of Library Association Executives, Sacramento *15456*
International Council of Museums, Milano *07713*
International Council of Museums, Andorran National Committee, Andorra la Vella *00031*
International Council of Museums – Committee of the American Association of Museums, Washington *15457*
International Council of Onomastic Sciences, Leuven *01394*
International Council of Psychologists, Hopkinton *15458*
International Council of Scientific Unions, Paris *04460*
International Council of Sport Science and Physical Education, Jyväskylä *03967*
International Council of the Museum of Modern Art, New York *15459*
International Council on Alcohol and Addictions, Lausanne *10656*
International Council on Archives, Paris *04585*
International Council on Education for Teaching, Arlington *15460*
International Council on Monuments and Sites, Paris *04456*
International Council on Social Welfare, Wien *00664*
International Courtly Literature Society, East Lansing *15461*
International Crane Foundation, Baraboo *15462*
International Criminal Law Commission, Santa Barbara *15463*
International Cultural Society of Korea, Seoul *08526*
International Dance Council, Paris *04452*
International Dark-Sky Association, Tucson *15464*
International Dental Federation, London *12348*
International Dental Health Foundation, Reston *15465*
International Design Conference in Aspen, Aspen *15466*
International Development Education Committee of Ontario, Toronto *02550*
International Development Education Resources Association, Vancouver *02551*
International Development Program of Australian Universities and Colleges, Canberra *00425*
International Development Research Centre, Ottawa *02552*
International Diabetes Federation, Bruxelles *01395*
International Documentation and Communication Center, Roma *07467*
International Dostoevsky Society, Kensington *15467*
International Drivers Behaviour Research Association, Paris *04293*
Internationale Akademie für Keramik, Genève *10606*
Internationale Akademie für Pathologie, Deutsche Abteilung e.V., Bonn *05988*
Internationale Albrechtsberger-Gesellschaft, Klosteneuburg *00665*
Internationale Alpenschutzkommission, Vaduz *08570*
Internationale Arbeitsgemeinschaft der Archiv-, Bibliotheks- und Graphikrestauratoren, Marburg *05989*
Internationale Architekten-Union, Sektion Schweiz, Zürich *10736*
Internationale Biometrische Gesellschaft, Deutsche Region, Aachen *05990*
Internationale Bruckner-Gesellschaft, Wien *00666*
Internationale Chopin-Gesellschaft in Wien, Wien *00667*
International Economic Association, Paris *04586*
International Economic History Association, Paris *04282*
Internationale Coronelli-Gesellschaft für Globen- und Instrumentenkunde, Wien *00668*
International Educational and Cultural Association, Bruxelles *01396*
International Education Association of China, Taipei *03303*
International Education Research Foundation, Los Angeles *15468*
Internationale Föderation der Vereine der Textilchemiker und Coloristen, Reinach *10737*
Internationale Gesellschaft für Geschichte der Pharmazie, Bremen *05991*
Internationale Gesellschaft für Getreidewissenschaft und -technologie, Schwechat *00669*
Internationale Gesellschaft für Ingenieurpädagogik, Klagenfurt *00670*
Internationale Gesellschaft für Jazzforschung, Graz *00671*
Internationale Gesellschaft für Musikwissenschaft, Basel *10738*
Internationale Gesellschaft zur Erforschung und Förderung der Blasmusik, Graz *00672*
Internationale Gewässerschutzkommission für den Bodensee, München *05276*
Internationale Gustav Mahler Gesellschaft, Wien *00673*
Internationale Heinrich Schütz-Gesellschaft e.V., Kassel *05992*
Internationale Hugo Wolf-Gesellschaft, Wien *00674*
Internationale Kommission zum Schutze des Rheins gegen Verunreinigung, Koblenz *05993*
International Electronics Packaging Society, Wheaton *15469*
International Electrotechnical Commission, Genève *10649*
International Embryo Transfer Society, Champaign *15470*
International Enamellers Institute, Ripley near Derby *12349*
International Energy Agency, Paris *04126*
Internationale Nestroy-Gesellschaft, Wien *00675*
Internationale Paracelsus-Gesellschaft, Salzburg *00676*
International Epidemiological Association, Los Angeles *15471*
Internationaler Arbeitskreis für Musik e.V., Kassel *05994*
Internationaler Arbeitskreis Sonnenberg, Braunschweig *05995*
Internationaler Kunstkritikerverband, Sektion der Bundesrepublik Deutschland e.V., Bonn *05996*
Internationaler Museumsrat, Deutsches Nationalkomitee, München *05997*
International Erosion Control Association, Steamboat Springs *15472*
Internationaler Rat für Umweltrecht, Bonn *05998*
Internationaler Verband Forstlicher Forschungsanstalten, Wien *00677*
Internationale Schönberg-Gesellschaft, Mödling *00678*
Internationales Dokumentations- und Studienzentrum für Jugendkonflikte, Wuppertal *05999*
Internationales Forschungszentrum für Grundfragen der Wissenschaften, Salzburg *00679*

Internationales Institut für den Frieden, Wien 00680
Internationales Institut für Jugendliteratur und Leseforschung, Wien 00681
Internationales Institut für Öffentliche Finanzen, Saarbrücken 06000
Internationales Institut für Traditionelle Musik e.V., Berlin 06001
Internationales Kali-Institut, Basel 10739
Internationales Katholisches Büro für Unterricht und Erziehung, Bruxelles 01179
Internationales Komitee Giessereitechnischer Vereinigungen, Zürich 10740
Internationales Musikzentrum, Wien 00682
International Esperanto-Association of Jurists, Wembley 12350
Internationale Stiftung Mozarteum, Salzburg 00683
Internationales Zentralinstitut für das Jugend- und Bildungsfernsehen, München 06002
Internationale Tagung der Historiker der Arbeiterbewegung, Wien 00684
Internationale Union Demokratisch-Sozialistischer Erzieher, Wildegg 10741
Internationale Vereinigung der Musikbibliotheken, Musikarchive und Musikdokumentationszentren, Gruppe Bundesrepublik Deutschland, Berlin 06003
Internationale Vereinigung für Brückenbau und Hochbau (IVBH), Zürich 10742
Internationale Vereinigung für Geschichte und Gegenwart der Druckkunst, Mainz 05902
Internationale Vereinigung für gewerblichen Rechtsschutz, Zürich 10743
Internationale Vereinigung für Natürliche Wirtschaftsordnung, Aarau 10744
Internationale Vereinigung für Rechts- und Sozialphilosophie e.V., Göttingen 06004
Internationale Vereinigung für Saatgutprüfung, Zürich 10745
Internationale Vereinigung für Selbstmordprophylaxe, Wien 00685
Internationale Vereinigung für Vegetationskunde, Göttingen 06005
Internationale Vereinigung gegen den Lärm, Luzern 10746
Internationale Vereinigung von Versicherungsjuristen (A.I.D.A.), Deutsche Landesgruppe, Berlin 06006
Internationale Vereinigung wissenschaftlicher Fremdenverkehrsexperten, Sankt Gallen 10747
Internationale Veterinäranatomische Nomenklatur-Kommission, Zürich 10748
International Federation for Family Life Promotion, Washington 15473
International Federation for Home Economics, Paris 04537
International Federation for Housing and Planning, 's-Gravenhage 08932
International Federation for Information and Documentation, 's-Gravenhage 08895
International Federation for Medical and Biological Engineering, Ottawa 02553
International Federation for Modern Languages and Literatures, Belfast 12351
International Federation for Psychotherapy, Bern 10749
International Federation for Theatre Research, Lancaster 12352
International Federation of Airworthiness, Ruislip 12353
International Federation of Anti-Leprosy Associations, London 12354
International Federation of Associations of Textile Chemists and Colourists, Reinach 10737
International Federation of Automatic Control, Laxenburg 00686
International Federation of Catholic Universities, Paris 04536
International Federation of Cell Biology, London 12355
International Federation of Free Teachers' Unions, Amsterdam 08933
International Federation of Gynecology and Obstetrics, London 12356
International Federation of Library Associations and Institutions, 's-Gravenhage 08934
International Federation of Multiple Sclerosis Societies, London 12357
International Federation of Ophthalmological Societies, Nijmegen 08935
International Federation of Practitioners of Natural Therapeutics, London 12358
International Federation of Secondary Teachers, Paris 04535
International Federation of Social Workers, Oslo 09270
International Federation of Societies for Electron Microscopy, Berkeley 15474
International Federation of Societies of Cosmetic Chemists, Luton 12359
International Federation of Sportive Medicine, Slough 12360
International Federation of Surgical Colleges, Edinburgh 12361
International Federation of the Societies of Classical Studies, Bellevue 10701
International Federation of Training Centres for the Promotion of Progressive Education, Paris 04532
International Federation of University Women, Genève 10702
International Federation of Vexillological Associations, Winchester 15475
International Federation on Ageing, Montréal 02554
International Federation on Aging – European Office, London 12362
International Fertilizer Development Center, Muscle Shoals 15476
International Filariasis Association, London 12363
International Film Seminars, New York 15477
International Food Information Service, Reading 12364
International Foundation for Art Research, New York 15478
International Foundation for Homeopathy, Seattle 15479
International French-Language Congresses of Forensic and Social Medicine, Paris 04450
International Galdos Association, Davis 15480
International Genetics Federation, Cambridge 12365
International Geographical Union, Bonn 06007
International Glaciological Society, Cambridge 12366
International Graphic Arts Education Association, Pittsburgh 15481
International Graphoanalysis Society, Chicago 15482
International Gypsy Committee, Pierrefitte 04410
International Health Evaluation Association, Rockville 15483
International Health Society, Englewood 15484
International Hospital Federation, Chicago 15485
International House Association, Taipei Chapter, Taipei 03304
International Humanist and Ethical Union, Utrecht 08936
International Humanist Centre, London 13082
International Human Powered Vehicle Association, Indianapolis 15486
International Human Resources, Business and Legal Research Association, Washington 15487
International Huntington Association, Harfsen 08937
International Hydrofoil and Multihull Society, Great Yarmouth 12367
International Hydrofoil Society, London 12368
International Hydrographic Organization, Monte Carlo 08797
International Imagery Association, Bronx 15488
International Industrial Relations Association, Genève 10750
International Institute for Applied Systems Analysis, Laxenburg 00687
International Institute for Bioenergetic Analysis, New York 15489
International Institute for Conservation of Historic and Artistic Works, London 12369
International Institute for Educational Planning, Paris 04573
International Institute for Management Development, Lausanne 10751
International Institute for Organizational and Social Development, Leuven 01397
International Institute for Peace, Wien 00680
International Institute for Sugar Beeb Research, Bruxelles 01387
International Institute of Administrative Sciences, Bruxelles 01388
International Institute of Cellular and Molecular Pathology, Bruxelles 01398
International Institute of Human Rights, Strasbourg 04574
International Institute of Public Finance, Saarbrücken 06008
International Institute of Refrigeration, Paris 04575
International Institute of Rural Reconstruction – U.S. Chapter, New York 15490
International Institute of Social Economics, Hull 12370
International Institute of Social History, Amsterdam 08916
International Insurance Society, Tuscaloosa 15491
International Iron and Steel Institute, Bruxelles 01389
International Isotope Society, East Hanover 15492
International John Steinbeck Society, Muncie 15493
International Joseph Diseases Foundation, Livermore 15494
International Juridical Organization for Environment and Development, Roma 07714
International Kirlian Research Association, Brooklyn 15495
International Language Society of Great Britain, Barnsley 12371
International Language Union, Harrogate 12372
International Law Association, London 12373
International Law Association: Danish Branch, København 03747
International Law Association: Finnish Branch, Helsinki 03968
International Law Association, Österreichischer Zweigverein, Wien 00688
International Law Association, Sezione Italiana, Roma 07715
International Law Association, Swedish Branch, Stockholm 10475
International Law Institute, Utrecht 09108
International Law Institute, Washington 15496
International Lead and Zinc Study Group, London 12374
International Lead Zinc Research Organization, Research Triangle Park 15497
International League of Esperantist Teachers, Massa 07716
International League of Societies for Persons with Mental Handicap, Bruxelles 01399
International Livestock Centre for Africa, Addis Ababa 03945
International Magnesium Association, McLean 15498
International Maledicta Society, Santa Rosa 15499
International Management Centre from Buckingham, Buckingham 12375
International Management Council, Omaha 15500
International Map Collector's Society, Flyford Flavell 12376
International Mathematical Union, Rio de Janeiro 01695
International Measurement Confederation, Budapest 06608
International Medical Society of Paraplegia, Aylesbury 12377
International Metallographic Society, Columbus 15501
International Microwave Power Institute, Clifton 15502
International Mineralogical Association, Marburg 06009
International Montessori Association, Amsterdam 08825
International Musicological Society, Basel 10738
International Mycological Institute, Egham 12378
International Myopia Prevention Association, Ligonier 15503
International New Thought Alliance, Mesa 15504
International Numismatic Commission, London 12379
International Numismatic Society Authentication Bureau, Philadelphia 15505
International Office for Audiophonology, Bruxelles 01173
International Omega Association, Arlington 15506
International Organisation for the Study of the Endurance of Wire Ropes, Delft 08938
International Organization for Biological Control of Noxious Animals and Plants, Zürich 10752
International Organization for Mycoplasmology, Seattle 15507
International Organization for Septuagint and Cognate Studies, Toronto 02555
International Organization for Standardization, Genève 10753
International Organization for the Education of the Hearing Impaired, Washington 15508
International Organization for the Study of Group Tensions, New York 15509
International Organization for the Study of the Old Testament, Cambridge 12380
International Organization of Palaeobotany, London 12381
International Palaeontological Association, Sapporo 08144
International Peace Bureau, Genève 10640
International Peat Society, Jyskä 03969
International Peat Society (Canadian National Committee), Halifax 02556
International Pediatric Association, Paris 04261
International PEN Centre, Reykjavik 06690
International P.E.N. Club, Kingston 08081
International PEN Club, Flemish Centre, Dilbeek 01400
International PEN Club, French Speaking Branch, Kraainem 01401
International P.E.N. Writers Association, London 12382
International Percy Grainger Society, White Plains 15510
International Personnel Management Association, Alexandria 15511
International Petroleum Industry Environmental Conservation Association, London 12383
International Phenomenological Society, Providence 15512
International Phonetic Association, Leeds 12384
International Planetarium Society, Salt Lake City 15513
International Plant Propagators Society, Seattle 15514
International Political Science Association, Oslo 09271
International Poplar Commission, Roma 07717
International Potash Institute, Basel 10739
International Precious Metals Institute, Allentown 15515
International Professional Surrogates Association, Los Angeles 15516
International Project Management Association, Zürich 10754
International Psychoanalytical Association, London 12385
International Psychohistorical Association, New York 15517
International Public Relations Association, Genève 10755
International Radiation Protection Association, Eindhoven 08939
International Reading Association, Newark 15518
International Red Locust Control Organisation for Central and Southern Africa, Ndola 17153
International Reference Organization in Forensic Medicine and Sciences, Wichita 15519
International Rescue and Emergency Care Association, Covington 15520
International Research Centre on the Habitat Pio Manzu, Verucchio 07484
International Research Council on Biokinetics of Impacts, Bron 04587
International Research Group on Wood Preservation, Stockholm 10476
International Rhinologic Society, Bruxelles 01402
International Rubber Study Group, Wembley 12386
International Schools Association, Genève 10620
International Schools Services, Princeton 15521
International Seaweed Association, São Paulo 01696
International Secretariat for University Study of Education, Gent 01403
International Seed Testing Association, Zürich 10745
International Seismological Centre, Newbury 12387
International Society Against Breast Cancer, Paris 04884
International Society and Federation of Cardiology, ISFC, Genève 10973
International Society for Applied Ethology, Edinburgh 12388
International Society for Artificial Organs, Cleveland 15522
International Society for Astrological Research, Los Angeles 15523
International Society for British Genealogy and Family History, Salt Lake City 15524
International Society for Burn Injuries, Denver 15525
International Society for Business Education – United States Chapter, Reston 15526
International Society for Cardiovascular Surgery, Manchester 15527
International Society for Chronobiology, Fairfax 15528
International Society for Clinical Laboratory Technology, Saint Louis 15529
International Society for Fat Research, Karlshamn 10477
International Society for Folk-Narrative Research, Turku 03970
International Society for General Relativity and Gravitation, Bern 10756
International Society for General Semantics, Concord 15530
International Society for Group Activity in Education, Schriesheim 06010
International Society for Heart Research, Winnipeg 02557
International Society for Horticultural Science, Wageningen 08940
International Society for Human Ethology, Detroit 15531
International Society for Humor Studies, Tempe 15532
International Society for Hybrid Microelectronics, Reston 15533
International Society for Intercultural Education, Training and Research, Washington 15534
International Society for Labor Law and Social Security – U.S. National Branch, New York 15535
International Society for Medical and Psychological Hypnosis, Milano 07718
International Society for Microelectronics, Harbertonford 12389
International Society for Neoplatonic Studies, Norfolk 15536
International Society for Oriental Research, Istanbul 11148
International Society for Pathophysiology, Moskva 09877
International Society for Pediatric Neurosurgery, Salt Lake City 15537
International Society for Philosophical Enquiry, Hudson 15538
International Society for Plant Molecular Biology, Athens 15539
International Society for Prosthetics and Orthotics, København 03748
International Society for Research in Palmistry, Westmount 02558
International Society for Research on Aggression, Chicago 15540
International Society for Research on Civilization Diseases and on Environment, Bruxelles 01464
International Society for Rock Mechanics, Lisboa 09662
International Society for Soilless Culture, Wageningen 08941
International Society for Soil Mechanics and Foundation Engineering, Cambridge 12390
International Society for Terrain-Vehicle Systems, Hanover 15541
International Society for the History of Ideas, Philadelphia 15542
International Society for the Promotion and Investigation of Band Music, Graz 00672
International Society for the Psychology of Writing, Milano 07816
International Society for the Study of Expressionism, Cincinnati 15543
International Society for the Study of Infectious and Parasitic Diseases, Torino 07719
International Society for the Systems Sciences, Pocatello 15544
International Society for Twin Studies, Roma 07720, 07818
International Society of Air Safety Investigators, Sterling 15545
International Society of Arboriculture, Savoy 15546
International Society of Art and Psychopathology, Paris 04879
International Society of Audiology, London 12391
International Society of Biometeorology, Sainte Anne de Bellevue 02559
International Society of Blood Transfusion, Les Ulis 04882
International Society of Chemical Ecology, Tampa 15547
International Society of City and Regional Planners, 's-Gravenhage 08942
International Society of Classical Bibliography, Paris 04875
International Society of Computerized and Quantitative EMG, Jerusalem 07058
International Society of Criminology, Paris 04877
International Society of Cryptozoology, Tucson 15548
International Society of Dermatologic Surgery, Schaumburg 15549
International Society of Dermatology: Tropical, Geographic and Ecologic, Rochester 15550
International Society of Developmental Biologists, Göttingen 06011
International Society of Development Biologists, Southampton 12392
International Society of Differentiation, Saint Paul 15551
International Society of Electrophysiological Kinesiology, Montréal 02560
International Society of Hematology, Rochester 15552
International Society of Internal Medicine, Langenthal 10757, 10977
International Society of Libraries and Museums of the Performing Arts, København 03749
International Society of Neuropathology, London 12393
International Society of Paediatric Oncology, 's-Hertogenbosch 08943
International Society of Parametric Analysts, Germantown 15553
International Society of Performing Arts Administrators, Grand Rapids 15554
International Society of Plant Morphologists, Dehli 06827
International Society of Radiographers and Radiological Technicians, Cardiff 12394
International Society of Reproductive Medicine, Rancho Mirage 15555

International Society of Soil Science, Wien 00689
International Society of Surgery, Reinach 10976
International Society of Theoretical and Experimental Hypnosis, Milano 07721
International Society on Metabolic Eye Disease, New York 15556
International Society on Thrombosis and Hemostasis, Chapel Hill 15557
International Sociological Association, Madrid 10330
International Solar Energy Society, Sezione Italiana, Napoli 07722
International Special Committee on Radio Interference, London 12395
International Statistical Institute, Voorburg 08944
International Stomatological Association, Paris 04349
International Studies Association, Provo 15558
International Study Group for Mathematics Learning, London 12396
International Study Group for Steroid Hormones, Roma 07723
International Survey Library Association, Storrs 15559
International Tamil League, Madras 06828
International Tar Conference, London 12397
International Technical and Scientific Organization for Soaring Flight, Wessling 08945
International Technology Education Association, Reston 15560
International Technology Education Association – Council for Supervisors, Richmond 15561
International Telecommunication Union, Genève 10758
International Tele-Education, Philadelphia 15562
International Test and Evaluation Association, Fairfax 15563
International Textile and Apparel Association, Monument 15564
International Theatre Institute, Toronto 02561
International Theological Commission, Roma 16951
International Thespian Society, Cincinnati 15565
International Towing Tank Conference, Madrid 10331
International Transactional Analysis Association, San Francisco 15566
International Tree Crops Institute U.S.A., Davis 15567
International Tsunami Information Center, Honolulu 15568
International Tunnelling Association, Bron 04280
International Turfgrass Society, Blacksburg 15569
International Union against Cancer, Genève 10993
International Union Against the Venereal Diseases and the Treponematoses, London 12398
International Union for Conservation of Nature and Natural Resources, Gland 10759
International Union for Electroheat, Paris-La Défense 04983
International Union for Health Education, Paris 04588
International Union for Oriental and Asian Studies, Saint-Maur 04986
International Union for Quaternary Research, Egham 12399
International Union for the Liberty of Education, Paris 04990
International Union for the Protection of New Varieties of Plants, Genève 10994
International Union for the Scientific Study of Population, Liège 01404
International Union of Academies, Bruxelles 01492
International Union of Air Pollution Prevention and Environmental Protection Associations, Brighton 12400
International Union of Architects, Paris 04985
International Union of Biological Sciences, Paris 04988
International Union of Crystallography, Chester 12401
International Union of Food Science and Technology, Dublin 06970
International Union of Forestry Research Organisations, Wien 00677
International Union of Independent Laboratories, Elstree 12402
International Union of Microbiological Societies, Strasbourg 04589
International Union of Nutritional Sciences, London 12403
International Union of Pure and Applied Chemistry, Oxford 12404
International Union of Pure and Applied Chemistry (Canadian National Committee), Edmonton 02562
International Union of Pure and Applied Physics, Québec 02563
International Union of Women Architects, Paris 04987
International Veterinary Acupuncture Society, Chester Springs 15570
International Visual Literacy Association, Blacksburg 15571
International Waterfowl and Wetlands Research Bureau, Slimbridge 12405
International Water Resources Association, Urbana 15572
International Weed Science Society, Corvallis 15573
International Wizard of Oz Club, Kinderhook 15574
International Women's Anthropology Conference, New York 15575
International Women's Health Coalition, New York 15576
International Wool Secretariat, London 12406
International Wool Study Group, London 12407
International Work Group for Indigenous Affairs, København 03750
Inter Nationes e.V., Bonn 06012
INTERNET, Zürich 10754
Interparlamentarische Arbeitsgemeinschaft, Bonn 06013
Interplanetary Space Travel Research Association, London 12408
Interprofessional Council on Environmental Design, Washington 15577
Inter-Society Color Council, Lawrenceville 15578
Intersociety Committee on Pathology Information, Bethesda 15579
Interstate Conference on Water Policy, Washington 15580
Interstate Postgraduate Medical Association of North America, Madison 15581
Intertel, Tulsa 15582
Inter-Union Commission on Frequency Allocations for Radio Astronomy and Space Science, Epping 00426
Inter-University Consortium for Political and Social Research, Ann Arbor 15583
Inter-University Council for East Africa, Kampala 11186
Inter-University Council on Academic Exchanges with the USSR and Eastern Europe, Waterloo 02564
INTI, Buenos Aires 00145
Intitut für Baustoffprüfung und Fußbodenforschung, Troisdorf 06014
Intitut für Bildungsmedien e.V., Frankfurt 06015
I.N.T.O., Dublin 06994
Intractable Pain Society of Great Britain and Ireland, Derby 12409
INU, Roma 07749
INucE, London 12311
INWO, Aarau 10744
IOA, Arlington 15506
IOD, Chapel Hill 15337
I.O.D., Leuven 01397
IOD, London 12232
IOD, North Palm Beach 15586
IOEHI, Washington 15508
IÖM, Wien 00657
I of E, Manchester 12234
IOM, Seattle 15507
IOM, Washington 15332
ION, Alexandria 15334
I.O.P., London 12381
IOP, Melton Mowbray 12265
IOP, Tunbridge Wells 12270
I.O.S., Jerusalem 07084
IOSCS, Toronto 02555
IOSGT, New York 15509
IOSOT, Cambridge 12380
IP, London 12267
IP, New York 15290
I.P.A., Bombay 06806
IPA, Dublin 06964
IPA, Jolimont 00418
IPA, Leeds 12384
IPA, London 12385
IPA, New York 15338, 15517
IPA, Sapporo 08144
IPAC, Calgary 02513
IPAC, Toronto 02531
IPC, Dublin 06996
IPC, Roma 07717
IPE, London 12313
IPG, Salzburg 00676
IPGH, México 08740
IPGS, White Plains 15510
IPI, Blackburn 12271
IPI, Jakarta 06890
IPIECA, London 12383
IPM, London 12266
IPMA, Alexandria 15511
IPMANA, Madison 15581
IPMI, Allentown 15515
IPO, Eindhoven 08914
I.P.O., Roma 07754
IPOEE, London 12315
IPPA, Canberra 00410
I.P. Pavlov Physiological Society, Moskva 09878
IPPP, College Park 15288
IPPS, Seattle 15514
IPR, London 12269
IPRA, Eindhoven 08939
IPRA, Genève 10755
IPRRE, Sarasota 15291
IPS, Ascot 12274
IPS, Derby 12409
IPS, Haifa 07086
IPS, Jyskä 03969
I.P.S., New Dehli 06807
IPS, Providence 15512
IPS, Salt Lake City 15513
IPS, Washington 15287, 15289
IPSA, Los Angeles 15516
IPSA, Oslo 09271
IPSJ, Tokyo 08140
Iqbal Academy, Lahore 09370
IRA, Newark 15518
Iranian Society of Microbiology, Teheran 06899
Iraq Academy, Baghdad 06917
Iraqi Medical Society, Baghdad 06918
IRCOBI, Bron 04587
IRDP, Neuchâtel 10731
IRE, Boston 15295
IRECA, Covington 15520
IRET, New York 15292
IRG, Stockholm 10476
IRHP, New York 15294
IRI, Washington 15261
IRI Research Institute, Stamford 15584
Irish Academy of Letters, Dublin 06971
Irish American Cultural Institute, Saint Paul 15585
Irish Association for Cultural, Economic and Social Relations, Belfast 12410
Irish Association for Economic Geology, Dublin 06972
Irish Association of Civil Liberty, Dublin 06973
Irish Association of Curriculum Development, Dublin 06974
Irish Association of Professional Archaeologists, Dublin 06975
Irish Association of Social Workers, Dublin 06976
Irish Astronomical Society, Dublin 06977
Irish Cancer Society, Dublin 06978
Irish Commercial Horticultural Association, Dublin 06979
Irish Computer Society, Dublin 06980
Irish Dental Association, Dublin 06981
Irish Epilepsy Association, Dublin 06982
Irish Family Planning Association, Dublin 06983
Irish Federation of University Teachers, Dublin 06984
Irish Film Society, Dublin 06985
Irish Georgian Society, Dublin 06986
Irish Grassland and Animal Production Association, Belclare, Tuam 06987
Irish Heritage, London 12411
Irish Institute of Purchasing and Materials Management, Dublin 06988
Irish Management Institute, Dublin 06989
Irish Manuscripts Commission, Dublin 06990
Irish Maritime Law Association, Dublin 06991
Irish Medical Association, Dublin 06992
Irish Mining and Quarrying Society, Dublin 06993
Irish National Teachers' Organisation, Dublin 06994
Irish PEN, Bray 06995
Irish Productivity Centre, Dublin 06996
Irish Psychoanalytical Association, Dublin 06997
Irish Quality Control Association, Dublin 06998
Irish Science Teachers Association, Dublin 06999
The Irish Society for Design and Craftwork, Dublin 07000
Irish Society of Arts and Commerce, Dublin 07001
Irish Textiles Federation, Dublin 07002
Irish Timber Growers Association, Dublin 07003
Irish Veterinary Association, Dublin 07004
Irish Welding Association, Dublin 07005
Irish Wildbird Conservancy, Monkstown 07006
IRLA, Washington 15254
IRLCO-CSA, Ndola 17153
IRMAC, Toronto 02517
I.R.M.R.A., Thane 06812
The Iron and Steel Institute of Japan, Tokyo 08392
Iron Overload Diseases Association, North Palm Beach 15586
IRP, Montréal 02521
IRRA, Madison 15260
IRRI, Bruxelles 01392
IRS, Bruxelles 01402
IRSE, Dawlish 12316
IRSG, Wembley 12386
IRT, Canoga Park 15293
ISA, Berkeley 15255
ISA, Brooklyn 15244
I.S.A., Haifa 07097
I.S.A., Imola 07388
ISA, London 11967
ISA, Madrid 10330
ISA, Provo 15558
ISA, São Paulo 01696
ISA, Savoy 15546
I.S.A.A., Tel Aviv 07095
ISAE, Bombay 06815
ISAM, Brooklyn 15298
ISAO, Cleveland 15522
ISAR, Los Angeles 15523
Isarel Plastics and Rubber Center, Haifa 07059
ISASI, Sterling 15545
ISB, London 12282
ISBE, Reston 15526
ISBGFH, Salt Lake City 15524
ISBI, Denver 15525
ISC, Fairfax 15528
ISC, Newbury 12387
ISC, Tucson 15548
ISCA, Calcutta 06813
ISCC, Lawrenceville 15578
ISCE, Eugene 15378
ISCE, Tampa 15547
ISCLT, Saint Louis 15529
ISCN, Jerusalem 07098
ISD, Rochester 15550
ISD, Saint Paul 15551
ISD, Wien 00660
ISDB, Southampton 12392
ISDS, Göttingen 06011
ISDS, Schaumburg 15549
ISEK, Montréal 02560
ISER, Birmingham 12210
I.S.E.S., Napoli 07722
ISF, Karlshamn 10477
ISFNR, Turku 03970
ISG, Innsbruck 00655
ISG, Kassel 05992
ISGE, Schriesheim 06010
ISGML, London 12396
ISGS, Concord 15530
ISGSH, Roma 07723
ISHE, Detroit 15531
ISHI, Philadelphia 15542
ISHK, Los Altos 15305
ISHM, Reston 15533
ISHR, Winnipeg 02557
ISHS, Tempe 15532
I.S.H.S., Wageningen 08940
ISI, New Dehli 06822
ISI, Voorburg 08944
I.S.I.G., Gorizia 07727
ISIM, Langenthal 10757
ISJC, Petersfield 12198
ISLA, Storrs 15559
The Islamic Association, Manama 01063
Islamic Research Association, Bombay 06829
Island Association of Rehabilitation Workshops, Charlottetown 02565
The Islandic Literary Society, Reykjavik 06691
Islandic Research Librarians Association, Reykjavik 06681
Islandske Litteratursamfund i København, Hillerød 03751
Islenzka bókmennatafélag, Reykjavik 06691
Islenzka nátturúfrádifélag, Hid, Reykjavik 06692
Isle of Man Natural History and Antiquarian Society, Douglas 12412
Isle of Wight Natural History and Archaeological Society, Shanklin 12413
ISLIC, Tel Aviv 07104
ISLLSS, New York 15535
ISM, Jerusalem 07094
ISM, Lichfield 12285
ISM, London 12283
ISM, McLean 15306
ISME, Birmingham 12281
ISMED, New York 15556
ISMEO, Roma 07742
ISMPH, Milano 07718
ISNS, Norfolk 15536
I.S.O.A., Calcutta 06819
ISoCaRP, 's-Gravenhage 08942
ISOSC, Wageningen 08941
Isotype Institute, London 12414
ISPA, Germantown 15553
ISPAA, Grand Rapids 15554
ISPE, Hudson 15538
ISPI, Milano 07750
ISPMB, Athens 15539
ISPN, Salt Lake City 15537
ISPO, København 03748
ISR, Ann Arbor 15296
ISRA, Chicago 15540
The Israel Academy of Sciences and Humanities, Jerusalem 07060
Israel Association for Applied Animal Genetics, Haifa 07061
Israel Association for Asian Studies, Jerusalem 07062
Israel Association for Physical Medicine and Rheumatology, Tel Aviv 07063
Israel Association of Archaeologists, Jerusalem 07064
Israel Association of General Practitioners, Hof Ha-Karmel 07065
Israel Association of Plastic Surgeons, Ramat-Gan 07066
Israel Atomic Energy Commission, Tel Aviv 07067
Israel Bar Association, Tel Aviv 07068
Israel Crystallographic Society, Haifa 07069
Israel Dermatological Society, Petah Tikva 07070
Israel Exploration Society, Jerusalem 07071
Israel Geographical Society, Jerusalem 07072
Israel Geological Society, Jerusalem 07073
Israel Gerontological Society, Tel Aviv 07074
Israel Heart Society, Jerusalem 07075
Israeli Dental Association, Tel Aviv 07076
Israeli Neurological Association, Tel Hashomer 07077
The Israel Institute of Productivity, Tel Aviv 07078
Israeli Urological Association, Tel Aviv 07079
Israel Library Association, Tel Aviv 07080
Israel Mathematical Union, Beer Sheva 07081
Israel Medical Association, Tel Aviv 07082
Israel Music Institute, Tel Aviv 07083
Israel Oriental Society, Jerusalem 07084
Israel Pediatric Association, Petach Tikva 07085
Israel Physical Society, Haifa 07086
Israel Political Science Association, Haifa 07087
Israel Prehistoric Society, Jerusalem 07088
Israel Radiological Society, Haifa 07089
Israel Society for Biblical Research, Jerusalem 07090
Israel Society for Experimental Biology and Medicine, Rehovoth 07091
Israel Society for Gastroenterology, Tel Aviv 07092
Israel Society for Hematology and Blood Transfusion, Jerusalem 07093
Israel Society for Microbiology, Jerusalem 07094
Israel Society of Aeronautics and Astronautics, Tel Aviv 07095
Israel Society of Allergology, Tel Aviv 07096
Israel Society of Anesthesiologists, Haifa 07097
Israel Society of Clinical Neurophysiology, Jerusalem 07098
Israel Society of Criminology, Jerusalem 07099
Israel Society of Logic and Philosophy of Science, Jerusalem 07100
Israel Society of Pathologists, Petan Tiqva 07101
Israel Society of Soil Mechanics and Foundation Engineering, Tel Aviv 07102
Israel Society of Soil Science, Rehovoth 07103
Israel Society of Special Libraries and Information Centres, Tel Aviv 07104
Israel Veterinary Medical Association, Tel Aviv 07105
ISRM, Lisboa 09662
ISRM, Rancho Mirage 15555
ISRN, Edinburgh 12196
ISRO, Bangalore 06821
ISRRT, Cardiff 12394
ISS, Durham 15297
ISS, Princeton 15521
ISSA, Keele 11754
ISSE-ETMS, Cincinnati 15543
ISSMFE, Cambridge 12390
ISSS, Pocatello 15544
IST, Lichfield 12279
Istanbul Dişhekimleri Odasi, Istanbul 11146
ISTC, London 12280
ISTEH, Milano 07721
ISTH, Chapel Hill 15557
Istituto Agronomico per l'Oltremare, Firenze 07724
Istituto Cooperativo per l'Innovazione, Roma 07725
Istituto Cooperazione Economica Internazionale, Milano 07726
Istituto di Sociologia Internazionale di Gorizia, Gorizia 07727
Istituto Elettrotecnico Nazionale Galileo Ferraris, Torino 07728
Istituto Geografico Militare, Firenze 07729
Istituto Histórico e Geográfico do Espirito Santo, Vitória 01697
Istituto Internazionale delle Comunicazioni, Genova 07730
Istituto Internazionale di Diritto Umanitario, San Remo 07731
Istituto Internazionale di Studi Liguri, Bordighera 07732
Istituto Internazionale di Vulcanologia, Catania 07733
Istituto Internazionale per l'Unificazione del Diritto Privato, Roma 07734
Istituto Italiano degli Attuari, Roma 07735
Istituto Italiano della Saldatura, Genova 07736
Istituto Italiano di Numismatica, Roma 07737
Istituto Italiano di Paleontologia Umana, Roma 07738
Istituto Italiano di Pubblicismo, Roma 07739
Istituto Italiano di Storia della Chimica, Roma 07740
Istituto Italiano di Studi Germanici, Roma 07741
Istituto Italiano per il Medio ed Estremo Oriente, Roma 07742

Istituto Italiano per la Storia Antica, Roma 07743
Istituto Italiano per la Storia della Musica, Roma 07744
Istituto Italiano Studi di Ipnosi Clinica e Psicoterapia, Verona 07745
Istituto Italo-Africano, Roma 07746
Istituto Nazionale delle Assicurazioni, Roma 07747
Istituto Nazionale di Ottica, Firenze 07748
Istituto Nazionale di Urbanistica, Roma 07749
Istituto Nazionale per la Ricerca Applicata ed il Trasferimento Tecnologico della Lega Nazionale delle Cooperative e Mutue, Roma 07725
Istituto per gli Studi di Politica Internazionale, Milano 07750
Istituto per la Cooperazione Economica Internazionale e i Problemi di Sviluppo, Roma 07751
Istituto per la Cooperazione Universitaria, Roma 07752
Istituto per la Storia del Risorgimento Italiano, Roma 07753
Istituto per l'Oriente, Roma 07754
Istituto Storico Italiano per l'Età moderna e contemporanea, Roma 07755
ISTOR, Roma 07753
Istoriki kai Ethnologiki Etaireia tis Ellados, Athinai 06537
ISTRA, London 12408
ISTS, Roma 07720
ISTVS, Hanover 15541
IT, London 12289
ITAA, Monument 15564
ITAA, San Francisco 15566
ITALCONSULT, Roma 07756
Italian American Librarians Caucus, New York 15587
Italian Association for the Promotion of Building Research and Studies, Milano 07336
Italian Center of Culture Anthropology, Roma 07490
Italian Historical Society of America, Brooklyn 15588
Italia Nostra – Associazione Nazionale per la Tutela del Patrimonio Storico Artistico e Naturale della Nazione, Roma 07757
Italian Society of Agricultural Genetics, Foggia 07880
Italian Society of Agronomy, Bologna 07830
Italian Society of Swine Pathology and Breeding, Parma 07929
Italian Urological Association, Roma 07957
ITALSIEL, Roma 07758
Itaria Gakkai, Kyoto 08145
ITC, London 12397
ITCI, Davis 15567
ITD, Marlow 12287
ITDG, Rugby 12320
ITE, Washington 15342
ITEA, Fairfax 15563
ITEA, Reston 15560
ITEA-CS, Richmond 15561
ITECA, Bruxelles 01396
ITEP, Recife 01656
ITEST, Saint Louis 15304
ITFCS, Arlington 15307
ITG, Frankfurt 05942
ITH, Wien 00684
ITIC, Honolulu 15568
ITS, Blacksburg 15569
ITS, Cincinnati 15565
ITS, Nijmegen 08915
ITT, Charlottesville 15339
ITU, Genève 10758
ITW, Berlin 05974
IUCN, Gland 10759
IUCr, Chester 12401
IUD, San Diego 15308
IUDSE, Wildegg 10741
IUMS, Strasbourg 04589
IUNS, London 12403
IUPAC, Oxford 12404
IUPAP, Québec 02563
IUS Federación Universitaria Nacional, Bogotá 03396
IUT, Leeds 12203
IUVDT, London 12398
IVA, Stockholm 10469
IVANK, Zürich 10748
IVAS, Chester Springs 15570
IVfgR, Zürich 10743
IVLA, Blacksburg 15571
IVM, Buxton 12290
IVV, Göttingen 06005
IWA, Salisbury 15343
IWAC, New York 15575
IWG Bonn, Bonn 05977
IWGJA, København 03750
IWHC, New York 15576
IWK, Wien 00661
IWM, Düsseldorf 05951
IWOC, Kinderhook 15574
IWRA, Urbana 15572
IWRB, Slimbridge 12405
IWS, London 12406
IWSc, High Wycombe 12291
I.W.S.G., London 12407
IWSS, Corvallis 15573
IZIP, Lima 09469
JA, Colorado Springs 15631
JAAS, New York 15603
JACC, San Pablo 15627
Jack London Research Center, Glen Ellen 15589
Jack Miner Migratory Bird Foundation, Kingsville 02566
Jack Point Preservation Society, New Ellenton 15590
JACT, London 12422
Jajasan Dana Normalisasi Indonesia, Bandung 06891
Jamaica Historical Society, Kingston 08082
Jamaica Library Association, Kingston 08083
Jamaican Association of Sugar Technologists, Mandeville 08084
Jamaica National Trust Commission, Kingston 08085
Jamaican Geographical Society, Kingston 08086
Jamarska zveza Slovenije, Ljubljana 10101
James Branch Cabell Society, Alstead 15591
James Joyce Society, New York 15592
James Willard Schultz Society, New Bern 15593
Jamiyat-ul-Falah, Karachi 09371
Jammu and Kashmir Academy of Art, Culture and Languages, Srinagar 06830
Jana Marca Marci Spectroscopic Society, Praha 03553
Jane Austen Society, Alton 12415
Jane Austen Society of North America, North Vancouver 02567, 02568
János Bolyai Mathematical Society, Budapest 06597
Japan Adhesive Industry Association, Tokyo 08259
Japana Esperanto-Instituto, Tokyo 08198
Japan Agricultural Law Association, Tokyo 08399
Japan Anti-Tuberculosis Assocation, Tokyo 08163
Japan Association for Philosophy of Science, Tokyo 08150
Japan Association for Quaternary Research, Tokyo 08292
Japan Association of Applied Psychology, Tokyo 08246
Japan Association of Automatic Control Engineers, Kyoto 08314
Japan Association of Civil Procedure Law, Tokyo 08181
Japan Association of Economic Geographers, Tokyo 08157
Japan Association of Economics and Econometrics, Tokyo 08404
Japan Association of International Relations, Tokyo 08339
Japan Association of Private Law, Tokyo 08373
Japan Association of Sociology of Law, Tokyo 08307
Japan Bearing Industrial Assocation, Tokyo 08188
Japan Bronco-Esophagological Society, Tokyo 08332
Japan Copper Development Association, Tokyo 08195
Japan Diabetes Society, Tokyo 08393
Japan Documentation Society, Tokyo 08296
Japan Dry Battery Assocation, Tokyo 08219
Japan Economic Policy Association, Tokyo 08328
The Japan Electric Association, Tokyo 08293
Japan Electronic Industry Developent Assocation, Tokyo 08192
Japanese American Society for Legal Studies, Seattle 15594
Japanese American Society for Legal Studies, Tokyo 08185
Japanese Archaeologists Association, Tokyo 08228
Japanese Art History Society, Tokyo 08115
Japanese Association for Dental Science, Tokyo 08376
Japanese Association for Fire Science and Engineering, Tokyo 08220
Japanese Association for Legal Philosophy, Kyoto 08309
Japanese Association for Religious Studies, Tokyo 08266
Japanese Association of Anatomiats, Tokyo 08320
Japanese Association of Fiscal Studies, Tokyo 08277
Japanese Association of Indian and Buddhist Studies, Tokyo 08211
Japanese Association of Mineralogists, Petrologists and Economic Geologists, Sendai 08300
Japanese Association of Museums, Tokyo 08205
Japanese Association of Refrigeration, Tokyo 08248
Japanese Association of Transportation Medicine, Tokyo 08233
Japanese Cancer Association, Tokyo 08299
Japanese Circulation Society, Kyoto 08215
Japanese Correctional Assocation, Tokyo 08177
Japanese Forestry Society, Tokyo 08358
Japanese Institute of Light Metals, Tokyo 08156
Japanese Labour Law Association, Tokyo 08359
Japanese Leprosy Association, Tokyo 08356
Japanese Literature Association, Tokyo 08282
Japanese Nursing Assocation, Tokyo 08324
Japanese Orthopaedic Association, Tokyo 08363
Japanese Pharmacological Society, Tokyo 08273
Japanese Political Science Association, Tokyo 08368
Japanese Psychological Association, Tokyo 08262
The Japanese Section of the Combustion Institute, Kyoto 08346
The Japanese Society for Aesthetics, Tokyo 08114
The Japanese Society for Animal Psychology, Ibaraki 08193
Japanese Society for Bacteriology, Tokyo 08361
Japanese Society for Dental Health, Tokyo 08230
Japanese Society for Ethics, Tokyo 08249
Japanese Society for Horticultural Science, Kyoto 08127
Japanese Society for Hygiene, Tokyo 08197
Japanese Society for Public Administration, Tokyo 08302
Japanese Society for Stereotactic Functional Neurosurgery, Tokyo 08391
Japanese Society for Theatre Research, Tokyo 08298
Japanese Society for the Study of Education, Tokyo 08237
The Japanese Society for Tuberculosis, Tokyo 08329
Japanese Society of Allergology, Tokyo 08187
Japanese Society of Applied Entomology and Zoology, Tokyo 08355
Japanese Society of Breeding, Tokyo 08210
Japanese Society of Conservative Dentistry, Tokyo 08375
Japanese Society of Dental Radiology, Tokyo 08374
Japanese Society of Electron Microscopy, Tokyo 08294
Japanese Society of Environment Control in Biology, Tokyo 08256
Japanese Society of Ethnology, Tokyo 08182
Japanese Society of French Language and Literature, Tokyo 08199
The Japanese Society of Gastroenterology, Tokyo 08381
Japanese Society of German Literature, Tokyo 08194
Japanese Society of Internal Medicine, Tokyo 08243
Japanese Society of Irrigatin, Drainage and Reclamation Engineering, Tokyo 08398
Japanese Society of Limnology, Tokyo 08357
Japanese Society of Neurology, Tokyo 08378
Japanese Society of Nuclear Medicine, Tokyo 08218
Japanese Society of Parasitology, Tokyo 08334
Japanese Society of Pathology, Tokyo 08286
Japanese Society of Pediatric Surgeons, Tokyo 08264
Japanese Society of Pharmacognosy, Tokyo 08389
Japanese Society of Physical Education, Tokyo 08390
Japanese Society of Plant Physiologists, Kyoto 08384
Japanese Society of Psychiatry and Neurology, Tokyo 08257
Japanese Society of Scientific Fisheries, Tokyo 08388
Japanese Society of Sericultural Science, Ibaraki 08254
The Japanese Society of Social Psychology, Tokyo 08260
Japanese Society of Soil Mechanics and Foundation Engineering, Tokyo 08123
Japanese Society of Starch Science, Ibaraki 08191
Japanese Society of Tropical Medicine, Nagasaki 08244
Japanese Society of Veterinary Science, Tokyo 08214
The Japanese Society of Western History, Osaka 08367
Japanese Society of Zootechnical Science, Tokyo 08287
Japanese Sociological Society, Tokyo 08370
Japanese Standards Assocation, Tokyo 08223
Japanese Stomatological Society, Tokyo 08231
The Japanese Technical Association of the Pulp and Paper Industry, Tokyo 08153
The Japanese Urological Association, Tokyo 08305
Japan Esperanto Institute, Tokyo 08198
Japan Federation of Engineering Societies, Tokyo 08336
Japan Gas Association, Tokyo 08201
Japan Gastroenterological Endoscopy Society, Tokyo 08263
Japan Geriatrics Society, Tokyo 08251
Japan Haematological Society, Kyoto 08331
Japan Information Center of Science and Technology, Tokyo 08216
Japan Institute of Metals, Sendai 08333
Japan Institute of Navigation, Tokyo 08337
Der Japanische Verein für Germanistik, Kobe 08130
Japan Library Association, Tokyo 08271
Japan Management Association, Tokyo 08350
The Japan Medical Association, Tokyo 08212
Japan Neuroscience Society, Tokyo 08379
Japan Neurosurgical Society, Tokyo 08351
Japan Oil Chemists' Society, Tokyo 08396
Japan Orthodontic Society, Tokyo 08240
Japan Pediatric Society, Tokyo 08265
Japan PEN Club, Tokyo 08247
Japan Prosthodontic Society, Tokyo 08310
Japan Public Health Association, Tokyo 08343
Japan Public Law Association, Tokyo 08227
Japan Radiation Research Society, Chiba-shi 08308
Japan Radioisotope Association, Tokyo 08186
Japan Radiological Society, Tokyo 08311
Japan Society for Aeronautical and Space Sciences, Tokyo 08340
Japan Society for Analytical Chemistry, Tokyo 08189
Japan Society for Bioscience, Biotechnology and Agrochemistry, Tokyo 08348
The Japan Society for Heat Treatment, Tokyo 08347
Japan Society for Science Education, Tokyo 08319
Japan Society for Studies in Journalism and Mass Communication, Tokyo 08377
Japan Society for the History of Social and Economic Thought, Kobe 08159
Japan Society for the Promotion of Machine Industry, Economic Research Institute, Tokyo 08165
Japan Society for the Promotion of Science, Tokyo 08200
Japan Society for the Study of Business Administration, Tokyo 08326
Japan Society for the Study of Educational Sociology, Tokyo 08238
Japan Society of Aeronautical and Space Science, Tokyo 08338
Japan Society of Anesthesiology, Tokyo 08241
Japan Society of Applied Physics, Tokyo 08402
Japan Society of Blood Transfusion, Tokyo 08397
The Japan Society of Christian Culture, Tokyo 08225
Japan Society of Civil Engineers, Tokyo 08122
Japan Society of Clinical Pathology, Tokyo 08250
Japan Society of Commercial Sciences, Tokyo 08380
Japan Society of Comparative Law, Tokyo 08131
The Japan Society of Human Genetics, Tokyo 08316
Japan Society of Library Science, Tokyo 08394
Japan Society of Lubrication Engineers, Tokyo 08317
Japan Society of Mathematical Education, Tokyo 08387
Japan Society of Mechanical Engineers, Tokyo 08222
Japan Society of Obstetrics and Gynaecology, Tokyo 08362
Japan Society of Photogrammetry and Remote Sensing, Tokyo 08372
Japan Society of Taoistic Research, Tokyo 08297
Japan Statistical Society, Tokyo 08270
Japan Sugar Refiners' Association, Tokyo 08409
Japan Surgical Society, Tokyo 08301
Japan Techno-Economics Society, Tokyo 08146
Japan Transportation Association, Tokyo 08234
Japan Weather Association, Tokyo 08147
Japan Welding Society, Tokyo 08446
Japan Wood Research Society, Tokyo 08344
Jargon Society, Winston-Salem 15595
J. Armand Bombardier Foundation, Valcourt 02467
JAS, Hœjbjerg 03754
JAS, Nottingham 12424
JASLS, Seattle 15594
JASNA, North Vancouver 02567
JATES, Tokyo 08146
Jazz Centre Society, London 12416
JBA, New York 15611
JBCS, Alstead 15591
JBI, New York 15604
JCAHO, Oak Brook Terrace 15616
JCAI, Arlington Heights 15618
JCCE, Pasadena 15597
JCL, Oxford 15632
JCPES, Washington 15615
JCS, London 12416
JDA, Houston 15598
JDC, Boston 15626
JDFI, New York 15633
JDS, Indianapolis 15612
JEA, Manhattan 15628
JEA, New York 15606
Jean Piaget Society, Buffalo 15596
Jednota Českých Matematikû a Fysikû, Praha 03547
Jednota Slovenských Matematikov a Fyzikov, Bratislava 10057
Jefferson Center for Character Education, Pasadena 15597
Jefferson Davis Association, Houston 15598
Jeoloji Mühendisleri Odasi, Ankara 11147
Jersey Association of the National Association of Schoolmasters, Saint Saviour's 12417
Jerusalem Philosophical Society, Jerusalem 07106
JES, New York 15613
JESNA, New York 15605
Jesse Stuart Foundation, Ashland 15599
Jesuit Philosophical Association of the United States and Canada, Chestnut Hill 15600
Jesuit Secondary Education Association, Washington 15601
Jesuit Seismological Association, Weston 15602
Jeunes Biologistes du Québec, Saint-Fidèle 02569
Jeunesse Espérantiste Canadienne, Montréal 02126
Jeunesse Intellectuelle, Bruxelles 01405
Jeunesses Littéraires de France, Paris 04590
Jeunesses Musicales de France, Paris 04591
Jewish Academy of Arts and Sciences, New York 15603
Jewish Braille Institute of America, New York 15604
Jewish Education Service of North America, New York 15605
Jewish Educators Assembly, New York 15606
Jewish Historical Society of England, London 12418
Jewish Lawyers Guild, New York 15607
Jewish Librarians Task Force, Franklin Square 15608
Jewish Pharmaceutical Society of America, Brooklyn 15609
Jewish Teachers Association – Morim, New York 15610
JHSE, London 12418
JIC, Champaign 15619
JICST, Tokyo 08216
JILA, Boulder 15620
Jilin Academy of Agricultural Sciences, Gongzhuling 03207
Jinbun Chiri Gakkai, Kyoto 08148
Jishin Gakkai, Tokyo 08149
JJAP, Tokyo 08402
JJG, Hamburg 06016
JJS, New York 15592
JLA, Kingston 08083
JLG, New York 15607
JLRC, Glen Ellen 15589
JLTF, Franklin Square 15608
JNICT/CDCT, Lisboa 09663
Joachim Jungius-Gesellschaft der Wissenschaften e.V., Hamburg 06016
Joanneum-Verein, Graz 00690
Joaquim Nabuco Foundation, Recife 01658
Johannes-Althusius-Gesellschaft e.V., Münster 06017
Johann Gottfried Herder-Forschungsrat e.V., Marburg 06018
Johann-Joseph-Fux-Gesellschaft, Graz 00691
Johann Strauß-Gesellschaft, Wien 00692
Johann Strauss Society of Great Britain, Caterham 12419
John Burroughs Association, New York 15611
John Dewey Society, Indianapolis 15612
John Ericsson Society, New York 15613
The Johnsonians, New Haven 15614
Johnson Society, Lichfield 12420

Johnson Society of London, Hindhead 12421
John von Neumann Computer Society, Budapest 06666
Joint Association for Medical Librarianship, Mannheim 05109
The Joint Association of Classical Teachers, London 12422
Joint Association of Regional Libraries, Hannover 05090
Joint Center for Political and Economic Studies, Washington 15615
Joint Commission on Accreditation of Healthcare Organizations, Oak Brook Terrace 15616
Joint Committee of the Nordic Natural Science Research Councils, Oslo 09272
Joint Committee of the States to Study Alcoholic Beverage Laws, Alexandria 15617
Joint Committee on Climatic Changes and the Ocean, Paris 04592
Joint Council of Allergy and Immunology, Arlington Heights 15618
Joint Council of Scientific Societies, Yeoville 10150
Joint Industry Council, Champaign 15619
Joint Institute for Laboratory Astrophysics, Boulder 15620
Joint Review Committee for Respiratory Therapy Education, Euless 15621
Joint Review Committee for the Ophthalmic Medical Personnel, Saint Paul 15622
Joint Review Committee on Educational Programs for the EMT-Paramedic, Euless 15623
Joint Review Committee on Education in Diagnostic Medical Sonography, Chicago 15624
Joint Review Committee on Education in Radiologic Technology, Chicago 15625
Jordan Library Association, Amman 08452
Jordan Research Council, Amman 08453
The Josephine Butler Society, Hatfield 12423
Joslin Diabetes Center, Boston 15626
Journalism Association of Community Colleges, San Pablo 15627
Journalism Education Association, Manhattan 15628
Jozef Pilsudski Institute of America for Research in the Modern History of Poland, New York 15629
JPA, Chestnut Hill 15600
J.P. Bickell Foundation, Toronto 02570
JPPS, New Ellenton 15590
JPSA, Brooklyn 15609
JPSSSKD, Buffalo 15596
JRCDMS, Chicago 15624
JRCEMT-P, Euless 15623
JRCERT, Chicago 15625
JRCOMP, Saint Paul 15622
JRCRTE, Euless 15621
JS, Winston-Salem 15595
JSA, Weston 15602
JSEA, Washington 15601
JSF, Ashland 15599
JSFK, Arhus 03753
JSG, Wausau 15630
JSRS, Beckenham 12425
JSZ, Beograd 17095
JTA-M, New York 15610
Jugoslavia Study Group, Wausau 15630
Jugoslovenski Centar za Tehničku i Naučnu Dokumentaciju, Beograd 17094
Jugoslovenski Savez za Zavarivanje, Beograd 17095
Jugoslovenski Zavod za Produktivnost Rada, Beograd 17096
Jugoslovenski Zavod za Standardizaciju, Beograd 17097
Jugoslovensko Društvo za Fiziologiju, Novi Sad 17098
Jugoslovensko Društvo za Mehaniku, Beograd 17099
Jugoslovensko Društvo za Proučavanje Zemljišta, Zemun 17100
Jugoslovensko Naučno Voársko Društvo, Čačak 17101
Jugoslovensko Udruženje za Filozofiju, Beograd 17102
Jugoslovensko Udruženje za Sociologiju, Beograd 17103
Julio de Urquijo Seminary of Basque Philology, San Sebastián 10368
Jung-Stilling-Gesellschaft e.V., Siegen 06019
Junior Achievement, Colorado Springs 15631
Junior Astronomical Society, Nottingham 12424
Junior Classical League, Oxford 15632
Junta de Control de Energía Atómica, Lima 09472
Junta de Historia Eclesiástica Argentina, Buenos Aires 00148
Junta Nacional de Folclore, Bogotá 03397
Junta Nacional de Investigação Científica e Tecnológica, Lisboa 09663
Junta Nacional de Planeamiento Económico, San José 03450
Juridiska Föreningen i Finland, Helsinki 03971
Juridisk Forening, København 03752
Jurisprudence Association, Tokyo 08134
Jury System Reform Society, Beckenham 12425
Juvenile Diabetes Foundation, Willowdale 02571
Juvenile Diabetes Foundation International, New York 15633
J.W. MacConnell Foundation, Montréal 02572
JWSS, New Bern 15593
Jydsk Selskab for Fysik og Kemi, Arhus 03753
Jysk Arkaeologisk Selskab, Hœjbjerg 03754
Jysk Forening for Naturvidenskab, Arhus 03755
Jysk Selskab for Historie, Arhus 03756
JZS, Ljubljana 10101

Kärntner Juristische Gesellschaft, Klagenfurt 00693
Kafka Society of America, Philadelphia 15634
Kagaku Kisoron Gakkai, Tokyo 08150
Kaigai Nogyo Kaihatsu Kyokai, Tokyo 08151
Kajima Heiwa Kenkyujo, Tokyo 08152
Kajima Institute of International Peace, Tokyo 08152
Kalevalaseura, Helsinki 03972
Kallitechnikon Epimelitirion Ellados, Athinai 06538
Kamarupa Anusandhan Samiti, Gauhati 06831
Kami Parupu Gijutsu Kyokai, Tokyo 08153
Kansai Economic Federation, Osaka 08154
Kansai Keizai Rengokai, Osaka 08154
The Kansai Society of Naval Architects, Osaka 08155
Kansai Zosen Kyokai, Osaka 08155
Kansantaloudellinen Yhdistys, Helsinki 03973
Kansantaloustieten Professorien ja Dosenttien Yhidistys, Jyväskylä 03974
Kansanvalistusseura, Helsinki 03975
Kant-Gesellschaft, Bonn 06020
KANTL, Gent 01408
Karachi Theosophical Society, Karachi 09372
Karnatak Historical Research Society, Dharwar 06832
Karolinska Förbundet, Stockholm 10478
Kartografiska Sällskapet, Gävle 10479
KAS, Berkeley 15640
K.A.S., Gravesend 12428
KASKA-NHISKA, Antwerpen 01406
Kassenärztliche Bundesvereinigung, Köln 06021
Kassenärztliche Vereinigung Bayerns, München 06022
Kassenärztliche Vereinigung Berlin, Berlin 06023
Kassenärztliche Vereinigung Bremen, Bremen 06024
Kassenärztliche Vereinigung Hamburg, Hamburg 06025
Kassenärztliche Vereinigung Hessen, Frankfurt 06026
Kassenärztliche Vereinigung Koblenz, Koblenz 06027
Kassenärztliche Vereinigung Niedersachsen, Hannover 06028
Kassenärztliche Vereinigung Nordbaden, Karlsruhe 06029
Kassenärztliche Vereinigung Nordrhein, Düsseldorf 06030
Kassenärztliche Vereinigung Nord-Württemberg, Stuttgart 06031
Kassenärztliche Vereinigung Pfalz, Neustadt 06032
Kassenärztliche Vereinigung Rheinhessen, Mainz 06033
Kassenärztliche Vereinigung Saarland, Saarbrücken 06034
Kassenärztliche Vereinigung Schleswig-Holstein, Bad Segeberg 06035
Kassenärztliche Vereinigung Südbaden, Freiburg 06036
Kassenärztliche Vereinigung Südwürttemberg, Tübingen 06037
Kassenärztliche Vereinigung Trier, Trier 06038
Kassenärztliche Vereinigung Westfalen-Lippe, Dortmund 06039
Kassenzahnärztliche Bundesvereinigung, Köln 06040
Kassenzahnärztliche Vereinigung Berlin, Berlin 06041
Kassenzahnärztliche Vereinigung für den Regierungsbezirk Freiburg, Freiburg 06042
Kassenzahnärztliche Vereinigung für den Regierungsbezirk Karlsruhe, Mannheim 06043
Kassenzahnärztliche Vereinigung für den Regierungsbezirk Stuttgart, Stuttgart 06044
Kassenzahnärztliche Vereinigung für den Regierungsbezirk Tübingen, Tübingen 06045
Kassenzahnärztliche Vereinigung Hamburg, Hamburg 06046
Kassenzahnärztliche Vereinigung Hessen, Frankfurt 06047
Kassenzahnärztliche Vereinigung im Lande Bremen, Bremen 06048
Kassenzahnärztliche Vereinigung Koblenz-Trier, Koblenz 06049
Kassenzahnärztliche Vereinigung Niedersachsen, Hannover 06050
Kassenzahnärztliche Vereinigung Nordrhein, Düsseldorf 06051
Kassenzahnärztliche Vereinigung Pfalz, Ludwigshafen 06052
Kassenzahnärztliche Vereinigung Rheinhessen, Mainz 06053
Kassenzahnärztliche Vereinigung Saarland, Saarbrücken 06054
Kassenzahnärztliche Vereinigung Schleswig-Holstein, Kiel 06055
Kassenzahnärztliche Vereinigung Westfalen-Lippe, Münster 06056
Katalyse-Umweltgruppe Köln e.V., Köln 06057
Kate Greenaway Society, Norwood 15635
Katholiek Documentatie Centrum, Nijmegen 08946
Katholische Ärztearbeit Deutschlands, Bonn 06058
Katholische Akademie in Bayern, München 06059
Katholische Akademie Trier, Trier 06060
Katholische Akademikerarbeit Deutschlands, Bonn 06061
Katholische Bundesarbeitsgemeinschaft für Erwachsenenbildung, Bonn 06062
Katholische Erwachsenenbildung im Lande Niedersachsen e.V., Hannover 06063
Katholische Erziehergemeinschaft in Bayern, München 06064
Katholische Juristenarbeit Deutschlands, Bonn 06065
Katholische Pädagogenarbeit Deutschlands, Bonn 06066
Katholischer Akademikerverband, Bonn 06067
Katholischer Akademischer Ausländer-Dienst, Bonn 06068
Katholisches Institut für Medieninformation e.V., Köln 06069
KAV, Bonn 06067
Kazakh Academy of Sciences, Alma-Ata 08456
KBE, Bonn 06062
KBV, Köln 06021
KDA, Nairobi 08503
KDC, Nijmegen 08946
KDVS, København 03710
Keats-Shelley Association of America, New York 15636
Keats-Shelley Memorial Association, Leamington Spa 12426
Keats-Shelley Memorial Association, Roma 07759
K.E.E., Athinai 06538
KEG, München 06064
Keikinzoku Gakkai, Tokyo 08156
Keizai Chiri Gakkai, Tokyo 08157
Keizai Chosa Kai Shiryoshitsu, Tokyo 08158
Keizaigaku-shi Gakkai, Kobe 08159
Keizai-ho Gakkai, Tokyo 08160
Keizai Tokei Kenkyukai, Tokyo 08161
Keizu Kyokai Toshokan, Tokyo 08162
Kekkaku Yobokai Kekkaku Kenkyusho Toshoshitsu, Tokyo 08163
KELI, Alphen 08969
Kemisk Forening, København 03757
Ken-Ikai Foundation, Tokyo 08164
Ken-i Kai Ganka Toshokan, Tokyo 08164
Kennarasamband Islands, Reykjavik 06693
Kent and Sussex Poetry Society, Hadlow 12427
Kent Archaeological Society, Gravesend 12428
Kent County Agricultural Society, Maidstone 12429
Kentron Epistemonikon Erevnon, Nicosia 03496
Kenya Dental Association, Nairobi 08503
Kenya Library Association, Nairobi 08504
Kenya National Academy of Sciences, Nairobi 08505
Kestner-Gesellschaft, Hannover 06070
Keyboard Teachers Association International, Westbury 15637
KF, New York 15639
KGAA, Uppsala 10484
KGS, Norwood 15635
KGvL, Wageningen 08962
K.I., Reykjavik 06693
Kiangsu Academy of Agricultural Sciences, Chiangsu 03208
Kikai Shinko Kyokai Keizai Kenkyo-sho Shiryoshitsu, Tokyo 08165
Kilvert Society, Worcester 12430
King Abdul Aziz Research Centre, Riyadh 10004
Kinsmen Rehabilitation Foundation of British Columbia, Vancouver 02573
KiPHi, Graz 00799
The Kipling Society, Haslemere 12431
Kipling Society of North America, Rockford 15638
Kirjallisuudentutkijain Seura, Helsinki 03976
Kirjastonhoitajaliitto, Helsinki 03977
Kirk Biological Society, Harare 17172
Kirkehistorisk Samfunn, Sandvika 09273
Kirkeligt Centrum, Hellerup 03758
KITLV, Leiden 08964
Klvl, 's-Gravenhage 08963
KJG, Klagenfurt 00693
Klagenfurter Sprachwissenschaftliche Gesellschaft, Klagenfurt 00694
Klassillis-Filologinen Yhdistys, Helsingin Yliopisto 03978
Klassisk-filologiska Föreningen, Helsingin Yliopisto 03978
Klinische Forschungsgruppe für Multiple Sklerose, Würzburg 06071
Klinische Forschungsgruppe für Reproduktionsmedizin, Münster 06072
KMA, Seoul 08537
KMA, Stockholm 10487
K.M.T.P., 's-Gravenhage 08948
K.N.A.G., Amsterdam 08965
K.N.B.V., Nijmegen 08954
KNCV, 's-Gravenhage 08955
Kneipp-Bund e.V., Bad Wörishofen 06073
KNGMG, Haarlem 08967
KNMG, Utrecht 08951
K.N.M.v.D., Utrecht 08957
KNNV, Utrecht 08958
KNVL, 's-Gravenhage 08960
Kobunshi Gakkai, Tokyo 08166
Koeki Jigyo Gakkai, Tokyo 08167
Kölner Bibliothek zur Geschichte des deutschen Judentums e.V., Köln 05786
Közlekedéstudományi Egyesület, Budapest 06609
K.O.G., Amsterdam 08968
Kogyo Kayaku Kyokai, Tokyo 08168
Kokka Gakkai, Tokyo 08169
Koko Eisei Gakkai, Tokyo 08170
Kokugo Gakkai, Tokyo 08171
Kokusaiho Gakkai, Tokyo 08172
Kokusai Keizai Gakkai, Tokyo 08173
KOLIS, Seoul 08535
Kollegium der Medizinjournalisten, Oberaudorf 06074
Kommission für Alte Geschichte und Epigraphik des Deutschen Archäologischen Instituts, München 06075
Kommission für Erforschung der Agrar- und Wirtschaftsverhältnisse des Europäischen Ostens e.V., Giessen 06076
Kommission für Geschichte des Parlamentarismus und der politischen Parteien e.V., Bonn 06077
Kommission für geschichtliche Landeskunde in Baden-Württemberg, Stuttgart 06078
Kommission für Neuere Geschichte Österreichs, Salzburg 00695
Kommission Reinhaltung der Luft im VDI und DIN, Düsseldorf 06079
Kommission Sportmed des Schweizerischen Landesverbandes für Sport, Bern 10760
Kommittén för Petrokemisk Forskning och Utveckling, Stockholm 10480
Kommunale Bibliotekarbeiderers Forening, Kolbotn 09274
Kommunalpolitische Vereinigung der CDU und CSU Deutschland, Bonn 06080
Konferenz der deutschen Akademien der Wissenschaften, Mainz 06081
Konferenz der Landesfilmdienste in der Bundesrepublik Deutschland e.V., Bonn 06082
Konferenz Schweizerischer Gymnasialrektoren, Nyon 10761
Konferenz Schweizerischer Handelsschulrektoren, Chiasso 10762
Kongelige Danske Geografiske Selskab, København 03759
Kongelige Danske Landhusholdningsselskab, Frederiksberg 03760
Kongreßgesellschaft für ärztliche Fortbildung e.V., Berlin 06083
Koninklijke Academie en Nationaal Hoger Instituut voor Schone Kunsten, Antwerpen 01406
Koninklijke Academie voor Geneeskunde van België, Bruxelles 01407
Koninklijke Academie voor Nederlandse Taal- en Letterkunde, Gent 01408
Koninklijke Academie voor Wetenschappen, Letteren en Schone Kunsten van België, Bruxelles 01409
Koninklijke Belgische Vereniging voor Dierkunde, Bruxelles 01489
Koninklijke Maatschappij tot Bevordering der Bouwkunst, Bond van Nederlandse Architekten, Amsterdam 08947
Koninklijke Maatschappij Tuinbouw en Plantkunde, 's-Gravenhage 08948
Koninklijke Maatschappij voor Dierkunde van Antwerpen, Antwerpen 01410
Koninklijke Maatschappij voor Natuurkunde onder de Zinspreuk Diligentia, 's-Gravenhage 08949
Koninklijke Nederlandsche Heidemaatschappij, Arnhem 08950
Koninklijke Nederlandsche Maatschappij tot Bevordering der Geneeskunst, Utrecht 08951
Koninklijke Nederlandse Akademie van Wetenschappen, Amsterdam 08952
Koninklijke Nederlandse Bosbouw Vereniging, Arnhem 08953
Koninklijke Nederlandse Botanische Vereniging, Nijmegen 08954
Koninklijke Nederlandse Chemische Vereniging, 's-Gravenhage 08955
Koninklijke Nederlandse Maatschappij ter Bevordering der Pharmacie, 's-Gravenhage 08956
Koninklijke Nederlandse Maatschappij voor Diergeneeskunde, Utrecht 08957
Koninklijke Nederlandse Natuurhistorische Vereniging, Utrecht 08958
Koninklijke Nederlandse Toonkunstenaars-vereniging, Amsterdam 08959
Koninklijke Nederlandse Vereniging voor Luchtvaart, 's-Gravenhage 08960
Koninklijke Vereniging der Belgische Actuarissen, Bruxelles 01157
Koninklijke Vereniging van Leraren en Onderwijzers in de Lichamelijke Opvoeding, Zeist 08961
Koninklijke Vereniging voor Natuur- en Stedeschoon, Antwerpen 01411
Koninklijke Vereniging voor Vogel- en Natuurstudie de Wielewaal, Turnhout 01412
Koninklijk Genootschap voor Landbouwwetenschap, Wageningen 08962
Koninklijk Instituut van Ingenieurs, 's-Gravenhage 08963
Koninklijk Instituut voor Taal-, Land- en Volkenkunde, Leiden 08964
Koninklijk Nederlands Aardrijkskundige Genootschap, Amsterdam 08965
Koninklijk Nederlandsch Genootschap voor Geslacht- en Wapenkunde, 's-Gravenhage 08966
Koninklijk Nederlands Geologisch Mijnbouwkundig Genootschap, Haarlem 08967
Koninklijk Oudheidkundig Genootschap Amsterdam, Amsterdam 08968
Konsthistoriska Sällskapet, Stockholm 10481
Kontext – Institut für Kommunikations- und Textanalysen, Wien 00696
Korányi Sandor Társaság, Budapest 06610
Korea Branch of the Royal Asiatic Society, Seoul 08527
Korean Association for the Biological Sciences, Seoul 08528
Korean Association of Sinology, Seoul 08529
Korean Chemical Society, Seoul 08530
Korean Economic Association, Seoul 08531
Korean Forestry Society, Kyonggido 08532
Korean Geographical Society, Seoul 08533
Korean Historical Association, Seoul 08534
Korean Library and Information Science Society, Seoul 08535
Korean Library Association, Seoul 08536
Korean Medical Association, Seoul 08537
Korean Micro-Library Association, Seoul 08538
Korean Psychological Association, Seoul 08539
The Korean Society of Pharmacology, Seoul 08540
Kosciuszko Foundation, New York 15639
Kotikielen Seura, Helsingin Yliopisto 03979
Krahuletz-Gesellschaft, Eggenburg 00697
Kristaina Esperantista Ligo Internacia, Alphen 08969
Kroeber Anthropological Society, Berkeley 15640
Kruh Moderních Filologu Při, Praha 03548
Kružok Moderných Filológov, Bratislava 10058
KS, Rockford 15638
KSA, Philadelphia 15634
KSAA, New York 15636
KSLA, Stockholm 10489
K-SMA, Leamington Spa 12426
KSMA, Roma 07759
KTAI, Westbury 15637
KTBL, Darmstadt 06088
KTE, Budapest 06609
Kuki-Chowa Eisei Kogakkai, Tokyo 08174
Kulturgeschichtliche Gesellschaft am Landesmuseum Joanneum, Graz 00698
Kulturpolitische Gesellschaft e.V., Hagen 06084
Kungliga Akademien för de fria Konsterna, Stockholm 10482
Kungliga Fysiografiska Sällskapet i Lund, Lund 10483
Kungliga Gustav Adolfs Akademien, Uppsala 10484
Kungliga Humanistiska Vetenskaps-Samfundet i Uppsala, Uppsala 10485

Kungliga Krigsvetenskapsakademien, Stockholm 10486
Kungliga Musikaliska Akademien, Stockholm 10487
Kungliga Samfundet för Utgivande av Handskrifter rörande Skandinaviens Historia, Stockholm 10488
Kungliga Skogs- och Lantbruksakademien, Stockholm 10489
Kungliga Vetenskapsakademien, Stockholm 10490
Kungliga Vetenskaps- och Vitterhets-Samhället i Göteborg, Göteborg 10491
Kungliga Vetenskaps-Societeten i Uppsala, Uppsala 10492
Kungliga Vitterhets Historie och Antikvitets Akademien, Stockholm 10493
Kunstforeningen, København 03761
Kunsthistorische Gesellschaft, Wien 00699
Kunsthistorische Gesellschaft an der Universität Graz, Graz 00700
Kunstnerforeningen af 18. November, København 03762
Kunstverein Sankt Gallen, Sankt Gallen 10763
Kuratorium der Deutschen Wirtschaft für Berufsbildung, Bonn 06085
Kuratorium für Forschung und Technik der Zellstoff- und Papierindustrie, Bonn 06086
Kuratorium für Kulturbauwesen, Hannover 06087
Kuratorium für Technik und Bauwesen in der Landwirtschaft e.V., Darmstadt 06088
Kuratorium für Verkehrssicherheit, Wien 00701
Kuratorium für Waldarbeit und Forsttechnik e.V., Groß-Umstadt 06089
Kurt Weill Foundation for Music, New York 15641
KVA, Stockholm 10490
KVHAA, Stockholm 10493
Kwangtung Academy of Agricultural Sciences, Kuangtung 03209
KWF, Groß-Umstadt 06089
KWFM, New York 15641
Kyoikushi Gakkai, Tokyo 08175
Kyoiku Tetsugakkai, Tokyo 08176
Kyosei Kyokai Kyosei Toshokan, Tokyo 08177
Kyoto Daigaku Keizai Gakkai, Kyoto 08178
Kyoto Tetsugakkai, Kyoto 08179
Kyrgyz Academy of Sciences, Bishkek 08548
Kyrgyz Commission on Earthquake Forecasting, Bishkek 08549
Kyrgyz Genetics and Selection Society, Bishkek 08550
Kyrgyz Geographical Society, Bishkek 08551
KZBV, Köln 06040

L.A., London 12450
LA, New York 15662
L.A.A., Edmonton 02589
Laban/Bartenieff Institute of Movement Studies, New York 15642
Laboratoire Européen pour la Physique des Particules, Genève 10764
Laboratory Animal Management Association, Silver Spring 15643
Laboratory Animal Science Association, London 12432
Labor Policy Association, Washington 15644
Labor Research Association, New York 15645
LAC, Taipei 03305
Lace Research Association, Nottingham 12433
Lace Society, Wrexham 12434
Lääketieteellisen Fysiikan ja Tekniikan Yhdistys, Tampere 03980
LAEF, Oakland 15679
Ländliche Erwachsenenbildung in Niedersachsen e.V., Hannover 06090
Lärarförbundet, Stockholm 10494
Lärarnas Riksförbund, Stockholm 10495
LAF, Washington 15646
L.A.G.B., Glasgow 12453
LAHS, Leicester 12445
Lakeland Dialect Society, Carlisle 12435
Lakeshore Teachers Association, Dollard des Ormeaux 02574
LALAT, Buenos Aires 00149
L.A.M., Roma 07978
LAMA, Chicago 15668
LAMA, Silver Spring 15643
LAMAS, London 12460
Lambeg Industrial Research Association, Lisburn 12436
Lancashire and Cheshire Fauna Society, Manchester 12437
Lancashire Dialect Society, Stockport 12438
Lancashire Parish Register Society, Hemel Hempstead 12439
Landelijke Huisartsen Vereniging, Utrecht 08970
Landelijke Specialisten Vereniging, Utrecht 08971
Landesärztekammer Baden-Württemberg, Stuttgart 06091
Landesärztekammer Hessen, Frankfurt 06092
Landesärztekammer Rheinland-Pfalz, Mainz 06093
Landesärztekammer Thüringen, Jena 06094
Landesarbeitsgemeinschaft Jugend und Literatur NRW e.V., Pulheim-Brauweiler 06095
Landesverband der Volkshochschulen Niedersachsens e.V., Hannover 06096
Landesverband der Volkshochschulen Schleswig-Holsteins e.V., Kiel 06097
Landesverband der Volkshochschulen von Nordrhein-Westfalen e.V., Dortmund 06098
Landesverein für Höhlenkunde in Wien und Niederösterreich, Wien 00702
Landeszahnärztekammer Baden-Württemberg, Stuttgart 06099
Landeszahnärztekammer Hessen, Frankfurt 06100
Landeszahnärztekammer Rheinland-Pfalz, Mainz 06101
Landscape Architecture Foundation, Washington 15646
The Landscape Institute, London 12440
Landschaftsverband Westfalen-Lippe, Münster 06434
Landsforeningen af Foldterapeuter, København 03763
Landsforeningen for Polio-, Trafik- og Ulykkesskadede, Hellerup 03764
Landsforeningen for Sukkersyge, Odense 03765
Landsforeningen til Kraeftens Bekaempelse, København 03766
Landslaget for Lokalhistorie, Dragvoll 09275
Landslaget for Språklig Samling, Oslo 09276
Landslaget Musikk i Skolen, Oslo 09277
Language Pathologists and Audiologists, Vancouver 01972
Languages of Instruction Commission of Ontario, Toronto 02575
Lao Buddhist Fellowship, Vientiane 08552
Laos-China Association, Vientiane 08553
Laos-Soviet Association, Vientiane 08554
Laos-Viet Nam Association, Vientiane 08555
Lapidary, Rock and Mineral Society of British Columbia, Delta 02576
LAS, Singapore 10051
LASA, London 12432
LASA, Pittsburgh 15648
Laser Institute of America, Orlando 15647
LASRA, Palmerston North 09163
LAST, London 12461
Lastentarhanopettajaliitto, Helsinki 03981
Latin-American Association, Tokyo 08403
Latin American Association for Aeronautical and Space Law, Madrid 10275
Latin American Association for Animal Production, Porto Alegre 01585
Latin American Association for Ecological Development, Mexicali 08694
Latin American Association for Education by Radio, Quito 03845
Latin American Association for Medical Physics, Cali 03351
Latin American Association for Quality, Buenos Aires 00089
Latin American Association of Biomathematics, Santiago 03020
Latin American Association of Cooperative Centres of Education, Rosario 00086
Latin American Association of Nursing Science Schools and Departments, Quito 03846
Latin American Association of Paleobotany and Palynology, Porto Alegre 01584
Latin American Association of Physiological Sciences, Santiago 03060
Latin American Association of Schools of Social Work, Saõ Luís 01583
Latinamerican Centre for Development Administration, Caracas 16989
Latin American Centre for Physics, Rio de Janeiro 01630, 01698
Latin American Demographic Centre, Santiago 03026
Latin American Social Science Research Institute, Quito 03854
Latin American Studies Association, Pittsburgh 15648
Latin Language Mathematicians' Group, Bucuresti 09753
Latin-Mediterranean Society of Pharmacy, Bologna 07760
Latvian Academy of Medicine, Riga 08557
Latvian Academy of Sciences, Riga 08558
Laubach Literacy Action, Syracuse 15649
Laubach Literacy International, Syracuse 15650
Laubach Literacy of Canada (Prince Edward Island), Charlottetown 02577
Laughter Therapy, Los Angeles 15651
Laura Ingalls Wilder Memorial Society, Peplin 15652
Law and Society Association, Amherst 15653
Law Council of Australia, Braddon 00427
Law Foundation of Nova Scotia, Halifax 02578
Law of the Sea Institute, Honolulu 15654
Law School Admission Council – Law School Admission Services, Newtown 15655
Law Society, London 12441
Law Society of Finland, Helsinki 03971
Law Society of Manitoba, Winnipeg 02579
Law Society of Newfoundland, Saint John's 02580
Law Society of New South Wales, Sydney 00428
Law Society of Saskatchewan, Regina 02581
Law Society of Scotland, Edinburgh 12442
Law Society of the Northwest Territories, Yellowknife 02582
Law Society of Upper Canada, Toronto 02583
Lawyers' Association, Tokyo 08137
Lawyers' Committee for Civil Rights under Law, Washington 15656
LBF, Lund 10498
LBG, Wien 00703
LBI, New York 15659
LBIMS, New York 15642
LCCRUL, Washington 15656
LCLA, Minneapolis 15680
LCME, Chicago 15667
LCNA, Silver Spring 15665
LCPA, New York 15672
LDS, Carlisle 12435
L.D.S., Stockport 12438
LEA, River Forest 15682
League for Industrial Democracy, Washington 15657
League for the Exchange of Commonwealth Teachers, London 12443
League of Canadian Poets, Toronto 02584
League of Education Administrators, Directors and Superintendents, Warren 02585
League of Greek Composers, Athinai 06521
League of Jurists' Associations of Croatia, Zagreb 03463
Learning Disabilities Association of Canada, Ottawa 02586
Learning Resources Council, Edmonton 02587
LEB, Hannover 06090
Lebanese Library Association, Beirut 08562
LECNA, Washington 15681
LECT, London 12443
Leeds Philosophical and Literary Society, Leeds 12444
Lega Italiana per la Lotta contro i Tumori, Roma 07761
Legal History Association, Tokyo 08136
Legitimerade Sjukgymnasters Riksförbund, Stockholm 10496
LEI, Tulsa 15675
Leicestershire Archaeological and Historical Society, Leicester 12445
Leicestershire Local History Council, Leicester 12446
Leisure Studies Association, Edinburgh 12447
Leiterkreis der Katholischen Akademien, Schwerte 06102
Leitstelle des DOPAED, Frankfurt 05634
Lentz Peace Research Laboratory, Saint Louis 15658
Leo Baeck Institute, New York 15659
Leon and Thea Koerner Foundation, Vancouver 02588
Leonardo – The International Society for the Arts, Sciences and Technology, San Francisco 15660
Leonard Wood Memorial – American Leprosy Foundation, Rockville 15661
Lepidopterologisk Forening, Holte 03767
LEPRA, Colchester 11602
Lernen Fördern – Bundesverband zur Förderung Lernbehinderter e.V., Köln 06103
Leschetizky Association, New York 15662
Lessing Society, Cincinnati 15663
Leukemia Society of America, New York 15664
Levnedsmiddelselskabet, København 03768
LEVS, København 03768
Lewis Carroll Society, Chislehurst Common 12448
Lewis Carroll Society of North America, Silver Spring 15665
Lexington Group in Transportation History, Saint Cloud University 15666
L.F.H.M., Paris 04596
L.F.S.E.P., Paris 04593
LGOG, Maastricht 08972
LGTH, Saint Cloud University 15666
LGU, Vaduz 08571
LHC, Saint Louis 15683
LI, London 12440
L.I., Reykjavik 06694
LIA, Orlando 15647
Liaison Committee on Medical Education, Chicago 15667
Liaoning Academy of Agricultural Sciences, Liaoning 03210
Liberal Industrial Relations Association, Darwen 12449
Liberia Arts and Crafts Association, Monrovia 08564
Library Administration and Management Association, Chicago 15668
Library and Information Technology Association, Chicago 15669
Library Association, London 12450
Library Association of Alberta, Edmonton 02589
Library Association of Antigua and Barbuda, Saint John's 00036
Library Association of Bangladesh, Dhaka 01069
Library Association of Barbados, Bridgetown 01087
The Library Association of China, Taipei 03305
Library Association of Cyprus, Nicosia 03497
Library Association of Ireland, Dublin 07007
Library Association of Ottawa-Hull, Ottawa 02590
Library Association of Singapore, Singapore 10051
Library Association of Slovenia, Ljubljana 10111
Library Association of the Democratic People's Republic of Korea, Pyongyang 08516
Library Association of Trinidad and Tobago, Port of Spain 11122
Library Association of Uruguay, Montevideo 13263
The Library Association – Personnel, Training and Education Group, London 12451
Library Public Relations Council, East Brunswick 15670
Library Science Alumni Association, Toronto 02591
Library Technicians Association of British Columbia, Vancouver 02592
LID, Washington 15657
Lidhjy e Shkrimtareve dhe e Artisteve te Shqiperise, Tirana 00005
Liechtensteinische Gesellschaft für Umweltschutz, Vaduz 08571
Liechtensteinischer Ärzteverein, Vaduz 08572
LIEEP, Paris 04597
Lietuviu Tautodailes Institutas, Toronto 02594
Life Insurance Marketing and Research Association, Hartford 15671
Liga Argentina contra la Tuberculosis, Buenos Aires 00149
Liga Colombiana de Lucha Contra el Cancer, Bogotá 03398
Liga Marítima de Chile, Valparaíso 03061
Liga Nacional de Higiene y Profilaxia Social, Lima 09473
Liga Uruguaya contra la Tuberculosis, Montevideo 13264
Ligue de Sécurite du Québec, Montréal 02593
Ligue Française contre la Sclérose en Plaques, Paris 04593
Ligue Française de l'Enseignement et de l'Education Permanente, Paris 04594
Ligue Française de l'Enseignement Oroleis de Paris, Paris 04595
Ligue Française d'Hygiène Mentale, Paris 04596
Ligue Internationale de l'Enseignement, de l'Education et de la Culture Populaire, Paris 04597
Ligue Internationale pour l'Ordre Economique Naturel, Aarau 10744
Ligue Luxembourgeoise pour la Protection de la Nature et des Oiseaux, Luxembourg 08586
Ligue Nationale Belge contre l'Epilepsie, Bruxelles 01413
Ligue Nationale contre le Cancer, Paris 04598
Ligue pour la Protection des Oiseaux, Rochefort 04599
Limburgs Geschied- en Oudheidkundig Genootschap, Maastricht 08972
LIMRA, Hartford 15671
Lincoln Center for the Performing Arts, New York 15672
Lincoln Institute for Research and Education, Washington 15673
Lincoln Record Society, Lincoln 12452
Linguistics Association of Great Britain, Glasgow 12453
Linguistics Institute of Ireland, Dublin 06957
Linguistic Society of America, Washington 15674
Linguistic Society of India, Poona 06833
Linguistic Society of Japan, Tokyo 08202
Linnean Society of London, London 12454
Linnean Society of New South Wales, Milsons Point 00429
LIRA, Darwen 12449
LIRA, Lisburn 12436
LIRE, Washington 15673
List Gesellschaft e.V., Bochum 06104
Liszt Ferenc Társaság, Budapest 06611
The Liszt Society, London 12455
LITA, Chicago 15669
Literacy and Evangelism International, Tulsa 15675
Literacy Volunteers of America, Syracuse 15676
Literarischer Verein in Stuttgart e.V., Stuttgart 06105
Literarisches Colloquium, Berlin 06106
Literárněvědná Společnost, Praha 03549
Literary and Philosophical Society of Liverpool, Wallasey 12456
Literary and Philosophical Society of Newcastle upon Tyne, Newcastle-upon-Tyne 12457
The Literary Research Society, Helsinki 03976
Literary Society, Praha 03549
The Literature Bureau, Zimbabwe, Harare 17173
Lithuanian Academy of Sciences, Vilnius 08573
The Lithuanian Folk Art Institute, Toronto 02594
Litteraturfrämjandet, Stockholm 10497
LIWMS, Peplin 15652
LLA, Syracuse 15649
LLI, Syracuse 15650
LLPNO, Luxembourg 08586
LMAMA, Concord 15678
LMS, Hisings Kärra 10515
L.M.S., London 12462
LMS, London 12463
LNHS, Frinton-on-Sea 12465
LNPIB, Drumnadrochit 12459
Local Population Studies Society, London 12458
Loch Ness Phenomena Investigation Bureau, Drumnadrochit 12459
Lódzkie Towarzystwo Naukowe, Lódź 09545
Loeknafélag Islands, Reykjavik 06694
Lok Virsa, Islamabad 09373
London and Middlesex Archaeological Society, London 12460
London Association of Science Teachers, London 12461
London Mathematical Society, London 12462
London Medieval Society, London 12463
London Metallurgical Society, Wallington 12464
London Natural History Society, Frinton-on-Sea 12465
London Orchestral Association, London 12466
London Record Society, London 12467
London Society, London 12468
London Subterranean Survey Association, London 12469
London Topographical Society, Richmond 12470
London Underground Railway Society, Morden 12471
London Wargames Section, Barnet 12472
London Welsh Association, London 12473
Longfellow Society of Sudbury and Wayside Inn, Sherborn 15677
Los Baños Biological Club, Laguna 09521
Louisa May Alcott Memorial Association, Concord 15678
Lowveld Natural History Branch, Wildlife Society of Zimbabwe, Chiredzi 17174
LPA, Washington 15644
LPO, Rochefort 04599
LPRC, East Brunswick 15670
LPRL, Saint Louis 15658
LPRS, Hemel Hempstead 12439
LPSS, London 12458
LR, Stockholm 10495
LRA, New York 15645
LRA, Nottingham 12433
L.R.S., Lincoln 12452
L.R.S., London 12467
LS, Cincinnati 15663
LSA, Amherst 15653
LSA, Edinburgh 12447
LSA, New York 15664
LSA, Washington 15674
LSAC/LSAS, Newtown 15655
LSI, Honolulu 15654
LSS, Oslo 09276
LSSA, London 12469
LSSWI, Sherborn 15677
LSV, Utrecht 08971
LT, Los Angeles 15651
LTN, Lódź 09545
LTS, Richmond 12470
LTS, Stenløse 03769
Lubelskie Towarzystwo Naukowe, Lublin 09546
Lubuskie Towarzystwo Naukowe, Zielona Góra 09547
LUCT, Montevideo 13264

Ludwig Boltzmann-Gesellschaft, Österreichische Vereinigung zur Förderung der wissenschaftlichen Forschung, Wien 00703
Lugvaartkundige Vereniging van Suid-Afrika, Bryanston 10119
Lunds Botaniska Förening, Lund 10498
Lunds Matematiska Sällskap, Lund 10499
Lung Association, Ottawa 02595
Lung Association, Kawartha Pine Ridge Region, Peterborough 02596
Luso-American Education Foundation, Oakland 15679
Lute Society, Richmond 12474
Lutheran Church Library Association, Minneapolis 15680
Lutheran Educational Conference of North America, Washington 15681
Lutheran Education Association, River Forest 15682
Lutheran Historical Conference, Saint Louis 15683
LVA, Syracuse 15676
LWA, London 12473
LWM, Rockville 15661
LWS, Barnet 12472
Lyman A. Brewer International Surgical Society, Los Angeles 15684
Lysteknisk Selskab, Stenløse 03769

MA, Essen 06108
MAA, Cambridge 15714
MAA, Washington 15703
MAAF, Chadron 15758
MAARA, Derby 12525
Maataloustuottajain Keskusliitto, Helsinki 03982
Maatschappij Arti et Amicitiae, Amsterdam 08973
Maatschappij der Nederlandse Letterkunde, Leiden 08974
Maatschappij tot Bevordering der Toonkunst, Amsterdam 08975
MAB, Paris 04621
MAC, Morton Grove 15737
MacBride Museum Society, Whitehorse 02597
Macedonian Academy of Sciences and Art, Skopje 08603
Macedonian Geological Society, Skopje 08604
Macmillan Cancer Fund, London 12475
The Madison Project, New Brunswick 15685
Madras Literary Society and Auxiliary of the Royal Asiatic Society, Madras 06834
MAE, Budapest 06612
Maghreb Studies Association, London 12476
The Magnolia Society, Hammond 15686
Magyar Agrártudományi Egyesület, Budapest 06612
Magyar Allergologiai és Klinikai Immunológiai Társaság, Kékestetõ 06613
Magyar Altalános Orvosok Tudományos Egyesülete, Budapest 06614
Magyar Anatómusok, Histologusok és Embryologusok Társasága, Budapest 06615
Magyar Angiologiai Társaság, Budapest 06616
Magyar Asztonautikai Egyesület, Budapest 06617
Magyar Balneológiai Egyesület, Budapest 06618
Magyar Biofizikai Társaság, Budapest 06619
Magyar Biokemiai Társaság, Budapest 06620
Magyar Biológiai Társaság, Budapest 06621
Magyar Dermatologiai Társulat, Budapest 06622
Magyar Diabetes Társaság, Budapest 06623
Magyar Elektrotechnikai Egyesület, Budapest 06624
Magyar Elemezésipari Tudományos Egyesület, Budapest 06625
Magyar Elettani Társaság, Budapest 06626
Magyar Farmakológiai Társaság, Budapest 06627
Magyar Fógorvosok Egyesülete, Budapest 06628
Magyar Földrajzi Társaság, Budapest 06629
Magyar Gastroenterologiai Társaság, Budapest 06630
Magyar Geofizikusok Egyesülete, Budapest 06631
Magyar Gerontológiai Társaság, Budapest 06632
Magyar Gyógyszerészeti Társaság, Budapest 06633
Magyar Hidrológiai Társaság, Budapest 06634
Magyarhoni Földtani Társulat, Budapest 06635
Magyar Immunológiai Társaság, Budapest 06636
Magyar Iparjogvédelmi Egyesület, Budapest 06637
Magyar Irodalomtörténeti Társaság, Budapest 06638
Magyar Irók Szövetsége, Budapest 06639
Magyar Jogász Egylet, Budapest 06640
Magyar Kardiologusok Társasága, Budapest 06641
Magyar Karszt- és Barlangkutató Társulat, Budapest 06642
Magyar Kémikusok Egyesülete, Budapest 06643
Magyar Könyvtárosok Egyesülete, Budapest 06644
Magyar Meteorológiai Társaság, Budapest 06645
Magyar Mikrobiológiai Társaság, Budapest 06646
Magyar Néprajzi Társasag, Budapest 06647
Magyar Numizmatikai Társulat, Budapest 06648
Magyar Nyelvtudományi Társaság, Budapest 06649
Magyar Orvostudományi Társaságok és Egyesületek Szövetsége, Budapest 06650
Magyar Pathológusok Társasága, Budapest 06651
Magyar PEN Club, Budapest 06652
Magyar Pszichológiai Társaság, Budapest 06653
Magyar Radiológusok Társasága, Budapest 06654
Magyar Régészeti és Müvészettörténeti Társulat, Budapest 06655
Magyar Rovartani Társaság, Budapest 06656
Magyar Sebész Társaság, Budapest 06657
Magyar Szinházi Intézet, Budapest 06658
Magyar Történelmi Társulat, Budapest 06659
Magyar Traumatológus Társaság, Budapest 06660
Magyar Tudományos Akadémia, Budapest 06661
Magyar Urbanisztikai Társaság, Budapest 06662
Magyar Zenei Tanács, Budapest 06663
Maha Bodhi Society of Ceylon, Colombo 10416
Maha Bodhi Society of Sri Lanka (U.K.), London 12477
MAI, New Dehli 06837
Mainzer Altertumsverein, Mainz 06107
MAJ, Kingston 08087
Makedonska Akademija na Naukite i Umetnostite, Skopje 08603
Makedonsko Geološko Društvo, Skopje 08604
Makedonsko Lekarsko Društvo, Skopje 08605
MAKIT, Kékestetõ 06613
Malacological Society of Australia, Melbourne 00430
Malacological Society of China, Taipei 03306
Malacological Society of Japan, Tokyo 08217
The Malacological Society of London, Canterbury 12478
Malawi Library Association, Zomba 08623
Malayian Nature Society, Kuala Lumpur 08644
Malaysian Biochemical Society, Kuala Lumpur 08645
Malaysian Branch of the Royal Asiatic Society, Kuala Lumpur 08646
Malaysian Historical Society, Kuala Lumpur 08647
Malaysian Institute of Architects, Kuala Lumpur 08648
Malaysian Library Association, Kuala Lumpur 08649
Malaysian Medical Association, Kuala Lumpur 08650
Malaysian Rubber Producers' Research Association, Hertford 12479
Malaysian Rubber Research and Development Board, Kuala Lumpur 08651
Malaysian Scientific Association, Kuala Lumpur 08652
Malaysian Society of Anaesthesiologists, Kuala Lumpur 08653
Malaysian Zoological Society, Kuala Lumpur 08654
Malone Society, Oxford 12480
The Malraux Society, Edmonton 02598
Malta Geographical Society, Gzira 08665
Malta Historical Society, Qormi 08666
Malta Union of Teachers, Valletta 08667
MAM, Paceville 08668
MANA, Warwick 12547
Management Akademie, Essen 06108
Management Association for Private Photogrammetric Surveyors, Reston 15687
Management Research Groups, Corby 12481
Management Systems Training Council, Birmingham 12482
Manchester Geographical Society, Manchester 12483
Manchester Literary and Philosophical Society, Manchester 12484
Manchester Medical Society, Manchester 12485
Manchester Region Industrial Archaeology Society, Disley 12486
Manchester Statistical Society, Manchester 12487
Mandala Society, Del Mar 15688
Manitoba Archaeological Society, Winnipeg 02599
Manitoba Arts Council, Winnipeg 02600
Manitoba Association for Art Education, Winnipeg 02601
Manitoba Association for Multicultural Education, Winnipeg 02602
Manitoba Association for the Promotion of Ancestral Languages, Winnipeg 02603
Manitoba Association of Cardiology Technologists, Winnipeg 02604
Manitoba Association of Library Technicians, Winnipeg 02605
Manitoba Association of Mathematics Teachers, Winnipeg 02606
Manitoba Association of Medical Radiation Technologists, Winnipeg 02607
Manitoba Association of School Trustees, Winnipeg 02608
Manitoba Association of Social Workers, Winnipeg 02609
Manitoba Association on Gerontology, Winnipeg 02610
Manitoba Cancer Treatment and Research Foundation, Winnipeg 02611
Manitoba Cerebral Palsy Association, Winnipeg 02612
Manitoba Committee on Children and Youth, Winnipeg 02613
Manitoba Dental Association, Winnipeg 02614
Manitoba Environmental Council, Winnipeg 02615
Manitoba Epilepsy Association, Winnipeg 02616
Manitoba Genealogical Society, Winnipeg 02617
Manitoba Health Libraries Association, Winnipeg 02618
Manitoba Health Records Association, Winnipeg 02619
Manitoba Heart Foundation, Winnipeg 02620
Manitoba Historical Society, Winnipeg 02621
Manitoba Indian Education Association, Winnipeg 02622
Manitoba Institute of Agrologists, Winnipeg 02623
Manitoba Institute of Registered Social Workers, Winnipeg 02624
Manitoba Library Association, Winnipeg 02625
Manitoba Library Trustee Association, Neepawa 02626
Manitoba Lung Association, Winnipeg 02627
Manitoba Medical Association, Winnipeg 02628
Manitoba Naturalists Society, Winnipeg 02629
Manitoba Paramedical Association, Winnipeg 02630
Manitoba Safety Council, Winnipeg 02631
Manitoba School Library Audio-Visual Association, Winnipeg 02632
Manitoba Social Science Teachers' Association, Winnipeg 02633
Manitoba Society for Training and Development, Winnipeg 02634
Manitoba Society of Criminology, Winnipeg 02635
Manitoba Society of Occupational Therapists, Winnipeg 02636
Manitoba Speech and Hearing Association, Winnipeg 02637
Manitoba Teachers' Society, Winnipeg 02638
Manitoba Veterinary Medical Association, Winnipeg 02639
Manorial Society of Great Britain, London 12488
Manpower Society, London 12489
Mansfield Law Club, London 12490
MANU, Skopje 08603
Manufacturers Standardization Society of the Valve and Fittings Industry, Vienna 15689
The Manx Gaelic Society, Douglas 12491
Manyo Gakkai, Osaka 08180
MAOTE, Budapest 06614
Mapping and Prospecting Engineering Society, Moskva 09879
MAPPS, Reston 15687
Map Record and Issue Office, Dehra Dun 06863
MAPS, Wheeling 15731
Marburger Bund, Köln 06109
Margarine-Institut für gesunde Ernährung, Hamburg 06110
Maria Mitchell Association, Nantucket 15690
Marine Biological Association of the United Kingdom, Plymouth 12492
Marine Conservation Society, Kempley 12493
Marinehistorisk Selskab, København 03770
Marine Librarians Association, London 12494
Marine Standardization Office, Genova 07650
Marine Technology Society, Washington 15691
Maritime Institute of Ireland, Dun Laoghaire 07008
Maritime Law Association of Japan, Tokyo 08321
Maritime Provinces Higher Education Commission, Fredericton 02640
Maritime Trust, London 12495
Marketing Research Association, Rocky Hill 15692
Marketing Science Institute, Cambridge 15693
The Marketing Society, Dublin 07009
Mark Twain Association, New York 15694
Mark Twain Research Foundation, Perry 15695
Marlowe Society, London 12496
Marlowe Society of America, Lubbock 15696
Marquandia Society, Williamsburg 15697
Marquetry Society, Saint Albans 12497
Martins Sarmento Society, Guimarães 09687
MAS, Deerfield Beach 15727
MAS, Liverpool 12521, 12522
M.A.S.A., Lynnwood 10151
MASA, Washington 15761
Masarykova Česká Sociologická Společnost, Praha 03550
Masaryk's Czech Sociological Association, Praha 03550
The Masonry Society, Boulder 15698
Massenet Society, Fort Lee 15699
MATE, Budapest 06664
Matematiklaererforeningen, Birkerød 03771
Matematiska Föreningen, Uppsala 10500
Material Handling Management Society, Charlotte 15700
Materials Properties Council, New York 15701
Maternity Center Association, New York 15702
Mathematical Association, Leicester 12498
Mathematical Association of America, Washington 15703
Mathematical Instruction Subcommittee, London 12499
Mathematical Society of Japan, Tokyo 08386
Mathematische Gesellschaft in Hamburg, Hamburg 06111
Mathematisch-Naturwissenschaftlicher Fakultätentag, Hamburg 06112
Mathematisch-Physikalische Gesellschaft in Innsbruck, Innsbruck 00704
Math/Science Network, Oakland 15704
Matica Srpska, Novi Sad 17104
Matice Moravská, Brno 03551
National Scientific Medical Society of Forensic Medical Officers, Moskva 09880
MAUDEP, New York 15724
Max Bell Foundation, Toronto 02641
Max-Eyth-Gesellschaft für Agrartechnik e.V., Darmstadt 06113
Maximilian Numismatic and Historical Society, Tekonsha 15705
Max-Planck-Gesellschaft zur Förderung der Wissenschaften e.V., München 06114
Max Society, Tekonsha 15705
Max Steiner Memorial Society, Los Angeles 15706
MB, Köln 06109
MBA, Plymouth 12492
MBFT, Budapest 06619
MBSOGB, Cambridge 12548
MBT, Budapest 06621
MCA, New York 15702
MCA, Westfield 15762
MCLA, London 12502
MCMHA, New York 15725
MCN, Pittsburgh 15759
MCRHS, North Freedom 15730
MCS, Kempley 12493
MCU, London 12534
MDA, Los Angeles 15747
MDA, Ocean Park 15748
MDU, London 12503
Mechanical Engineering Society, Moskva 09881
Mechanics' Institute of Montréal, Montréal 02642
Medau Society of Great Britain and Northern Ireland, London 12500
Médecins sans Frontières, Paris 04600
Mediacult, Internationales Forschungsinstitut für Medien, Kommunikation und kulturelle Entwicklung, Wien 00705
The Media Institute, Washington 15707
Media Research Directors Association, New York 15708
Media Studies Association, Leeds 12501
MEDICA, Deutsche Gesellschaft zur Förderung der Medizinischen Diagnostik e.V., Stuttgart 06115
Medical Academy, Sofia 01768
The Medical and Dental Council of Nigeria, Lagos 09232
Medical Association, Lisboa 09665
Medical Association of Jamaica, Kingston 08087
Medical Association of Malta, Paceville 08668
The Medical Association of South Africa, Lynnwood 10151
Medical Association of Thailand, Bangkok 11093
Medical Contact Lens Association, London 12502
The Medical Council, Dublin 07010
Medical Council of Canada, Ottawa 02643
Medical Council of India, New Dehli 06835
Medical Council of New Brunswick, Saint John 02644
Medical Council of Prince Edward Island, Charlottetown 02645
The Medical Defence Union, London 12503
Medical Engineering Society, Moskva 09882
Medical Foundation, Sydney 00431
Medical Graduates' Association of the University of the Witwatersrand, Johannesburg 10152
Medical Letter, New Rochelle 15709
Medical Library Association, Chicago 15710
Medical Mycological Society of the Americas, San Antonio 15711
Medical Nomenclature Society of Iran, Isfahan 06900
Medical Officers of Schools Association, London 12504
Medical Practices Committee, London 12505
Medical Practitioners' Union, London 12506
Medical Protection Society, London 12507
Medical Research Council, Johannesburg 10153
Medical Research Council, London 12508
Medical Research Council, Stockholm 10501
Medical Research Council Laboratories, Kingston 08088
Medical Research Council of Canada, Ottawa 02646
Medical Research Council of Ireland, Dublin 07011
Medical Research Council of New Zealand, Auckland 09130
Medical Research Society, London 12509
Medical Sciences Historical Society, Liverpool 12510
Medical Seminars International, Chatsworth 15712
Medical Society for the Study of Radiesthesia, Bournemouth 12511
Medical Society for the Study of Venereal Diseases, London 12512
Medical Society of Bergen, Bergen 09278
The Medical Society of Copenhagen, København 03714
Medical Society of Finland, Helsinki 03961
The Medical Society of London, London 12513
Medical Society of Macedonia, Skopje 08605
Medical Society of Nova Scotia, Halifax 02647
Medical Society of Prince Edward Island, Charlottetown 02648
Medical Society of the United States and Mexico, Phoenix 15713
Medical Society of Victoria, Parkville 00432
Medical Staff Council, Winnipeg 02649
Medical Symposium Association of Canada, Edmonton 02650
Medical Union, Dublin 07012
Medical Women's Federation, London 12514
Medical Women's International Association, Köln 06116
Medicinska Forskningsrådet, Stockholm 10501
Medicinske Selskap i Bergen, Bergen 09278
Medico-Legal Society, London 12515
Medico-Legal Society of Japan, Tokyo 08306
Medico Legal Society of New South Wales, Saint Leonards 00433
Die Mediese Vereniging van Suid-Afrika, Lynnwood 10151
Medieval Academy of America, Cambridge 15714
Medieval Combat Society, London 12516
Medieval Settlement Research Group, Taunton 12517
MEDISS, Houston 15726

Medizinische Gesellschaft für Oberösterreich, Linz 00706
Medizinischer Fakultätentag der Bundesrepublik Deutschland, Münster 06117
Medizinisch Pharmazeutische Studiengesellschaft e.V., Bonn 06118
Medizinisch-wissenschaftiche Information, Heidelberg 05937
Medtner Society, United States of America, Chicago 15715
MEE, Budapest 06624
Mehran Library Association, Hyderabad 09374
MEI, Washington 15732
MELA, Santa Barbara 15733
Mellemfolkeligt Samvirke, København 03772
Melville Society, Arlington 15716
MENC, Reston 15763
Mencken Society, Baltimore 15717
Mended Hearts, Dallas 15718
Mendeleev Chemical Society, Chişinău 08781
Mennel Society, Harare 17175
Menntamálaráð, Reykjavik 06695
Men of the Stones, Stamford 12518
Mental Health Association of Ireland, Dublin 07013
Mental Health Materials Center, Bronxville 15719
Mental Research Institute, Palo Alto 15720
MER, Washington 15760
Mercia Cinema Society, Birmingham 12519
Méréstechnikai és Automatizálási Tudományos Egyesület, Budapest 06664
Merioneth Agricultural Society, Dolgellau 12520
MERIP, Washington 15734
Merseyside Archaeological Society, Liverpool 12521
Merseyside Aviation Society, Liverpool 12522
Mervyn Peake Society, London 12523
MES, Burton-on-Trent 12531
MESA, Tucson 15735
Metals Society, London 12524
Metal Treating Institute, Neptune Beach 15721
Metaphysical Society of America, Huntsville 15722
METE, Budapest 06625
Meteoritical Society, Piscataway 15723
Meteorological Society of Japan, Tokyo 08335
Meteorological Society of New Zealand, Wellington 09131
The Meteorological Society of the Republic of China, Taipei 03307
Metropolitan Association of Urban Designers and Environmental Planners, New York 15724
Metropolitan College Mental Health Association, New York 15725
Metropolitan Research Trust, Canberra City 00434
Metropolitan Toronto Zoological Society, West Hill 02651
Mexcian Academy of Scientific Research, México 08682
Mexican Librarians Association, México 08699
Mexican Nuclear Society, México 08774
Mexican Society for Plant Breeding and Genetics, Chapingo 08762
Mexican Society of Cardiology, México 08758
MFE, Budapest 06628
MFQ, Nanterre 04601
MFT, Budapest 06627, 06629
MFT, Münster 06117
MGE, Budapest 06631
MGH, Hamburg 06111
MH, Dallas 15718
MHA, Provo 15755
MHC, Goshen 15217
MHEF, Phoenix 15741
MHIFM, New York 15740
MHMC, Bronxville 15719
MHMS, Charlotte 15700
MHRA, London 12535
MHS, Bromley 12527
MHS, Qormi 08666
Michael E. Debakey International Surgical Society, Houston 15726
Micmac Association of Cultural Studies, Sydney 02652
Microbeam Analysis Society, Deerfield Beach 15727
Microbiological Society, Moskva 09883
Microbiological Society of Moldova, Chişinău 08782
Microneurography Society, Richmond 15728
Microscopical Society of Canada, Toronto 02653
Microscopy Society of America, Woods Hole 15729
Mid-Continent Railway Historical Society, North Freedom 15730
Middle Atlantic Planetarium Society, Wheeling 15731
Middle East Institute, Washington 15732
Middle East Librarians' Association, Santa Barbara 15733
Middle East Research and Information Project, Washington 15734
Middle East Studies Association of North America, Tucson 15735
Middle States Association of Colleges and Schools, Philadelphia 15736
Midlands Asthma and Allergy Research Association, Derby 12525
Midwest Archives Conference, Morton Grove 15737
Midwest Railway Historical Society, Palatine 15738
MIE, Budapest 06637
Migraine Foundation, Toronto 02654
Mijnbouwkundige Vereeniging, Delft 08976
Mikrographische Gesellschaft, Wien 00707
Militaerteknisk Forening, København 03773
Military Heraldry Society, Southall 12526
Military Historical Society, Bromley 12527
The Military History Society of Ireland, Dublin 07014
Military Operations Research Society, Alexandria 15739
Milletlerarasi Sark Tetkikleri Cemiyeti, Istanbul 11148
Milton Helpern Institute of Forensic Medicine, New York 15740
Milton H. Erickson Foundation, Phoenix 15741
Milton Society of America, Pittsburgh 15742
Mind Association, Saint Andrews 12528
MIND – National Association for Mental Health, London 12529
Mineralogical Association of Canada, Ottawa 02655
Mineralogical Society of America, Washington 15743
Mineralogical Society of Great Britain and Ireland, London 12530
Mineralogical Society of India, Mysore 06836
Minerals Engineering Society, Burton-on-Trent 12531
Minerals, Metals and Materials Society, Warrendale 15744
Mining and Metallurgical Society of America, Larkspur 15745
Mining Association of British Columbia, Vancouver 02656
Mining Association of Manitoba, Winnipeg 02657
Mining Engineering Society, Moskva 09884
Mining Institute of Scotland, Edinburgh 12532
Mining Society of Nova Scotia, Glace Bay 02658
Minji Soshoho Gakkai, Tokyo 08181
Minority Rights Group, London 12533
Mintek, Randburg 10154
Minzokugaku Shinkokai, Tokyo 08182
MIRA, Nuneaton 12540
Miramichi Historical Society, Newcastle 02659
MIS, Chicago 15754
MIS, London 12499
Missile, Space and Range Pioneers, Patrick Air Force Base 15746
Mission Doctors Association, Los Angeles 15747
MJSZ, Budapest 06640
MKE, Budapest 06643, 06644
ML, New Rochelle 15709
MLA, Canton 15764
MLA, Chicago 15710
MLA, London 12494
MLAA, New York 15750
MLC, London 12490
MMA, Guildford 12549
MMA, Kuala Lumpur 08650
MMA, Nantucket 15690
MMM, Edmonton 02598
MMS, Whitehorse 02597
MMSA, Larkspur 15745
MMSA, San Antonio 15711
MMT, Budapest 06646
MNS, Richmond 15728
MNS, Winnipeg 02629
MNyT, Budapest 06649
Moby Dick Academy, Ocean Park 15748
Model Secondary School for the Deaf, Washington 15749
Modern Churchpeople's Union, London 12534
Modern Fine Arts Association of Southern Taiwan, Tainan 03308
Modern Humanities Research Association, London 12535
Modern Language Association of America, New York 15750
Modern Language Society, Helsingin Yliopisto 04081
Modern Poetry Association, Chicago 15751
Moinho Santista Foundation, São Paulo 01659
Moldovan Academy of Sciences, Chişinău 08783
Moldovan Society of Animal Protection, Chişinău 08784
Moldovan Sociological Association, Beltsi 08785
Molson Family Foundation, Montréal 02660
Mommsen-Gesellschaft, Kiel 06119
Mongolia Society, Bloomington 15752
The Monroe Institute, Faber 15753
Montanhistorischer Verein für Österreich, Leoben-Donawitz 00708
Montenegrin Academy of Sciences and Arts, Podgorica 17080
The Montessori Society A.M.I. (UK), London 12536
Monumenta Germaniae Historica, München 06120
Monumental Brass Society, London 12537
Monumentenraad, Zeist 08977
Monuments and Historic Buildings Council, Zeist 08977
Moody Institute of Science, Chicago 15754
Moralogy Kenkyusho, Chiba 08183
Moray Society, Elgin 12538
Mormon History Association, Provo 15755
The Morris Ring, Ipswich 12539
MORS, Alexandria 15739
MOS, Valletta 08669
M.O.S.A., London 12504
Moscow Science-Production Association, Moskva 09885
Moscow Society of Naturalists, Moskva 09886
Moses Mendelssohn Zentrum für europäisch-jüdische Studien, Potsdam 06121
Motorcycle Safety Foundation, Irvine 15756
Motor Industry Research Association, Nuneaton 12540
Motor Vehicle Safety Association, Mississauga 02661
Mouvement Français pour la Qualité, Nanterre 04601
Mouvement International des Intellectuels Catholiques/MIIC-Pax Romana, Genève 10765
Mouvement Universel de la Responsabilité Scientifique, Paris 04602
MPA, Chicago 15751
MPC, New York 15701
MPG, München 06114
MPS, London 12507, 12523
MPU, London 12506
MRA, Rocky Hill 15692
MRC, London 12508
MRC, Ottawa 02646
M.R.C.I., Dublin 07011
MRC LABS, Kingston 08088
MRCNZ, Auckland 09130
MRDA, New York 15708
MRG, Corby 12481
MRG, London 12533
MRHS, Palatine 15738
MRI, Palo Alto 15720
M.R.I.A.S., Disley 12486
MRPRA, Hertford 12479
MRS, London 12509
MS, Altadena 15768
MS, Arlington 15716
MS, Baltimore 15717
MS, Del Mar 15688
MS, Fort Lee 15699
MS, London 12524
MS, Piscataway 15723
MS, Stamford 12518
MS, Williamsburg 15697
MSA, Beltsville 15766
MSA, Huntsville 15722
MSA, Leeds 12501
MSA, Lubbock 15696
MSA, Philadelphia 15736
MSA, Pittsburgh 15742
MSA, Washington 15743
MSA, Woods Hole 15729
MSF, Irvine 15756
MSGB, London 12488
MSI, Cambridge 15693
MSI, Chatsworth 15712
M.S.I. Foundation, Edmonton 02662
MSM, Mystic 15767
MSMS, Los Angeles 15706
MSOT, Winnipeg 02636
MSRP, Patrick Air Force Base 15746
MSS, Manchester 12487
MSS, Vienna 15689
MSSD, Washington 15749
MSSR, Bournemouth 12511
MSSVD, London 12512
MSTC, Birmingham 12482
MSUSM, Phoenix 15713
MTA, New York 15694
MtF, København 03773
MTG, München 06122
MTI, Neptune Beach 15721
MTK, Helsinki 03982
MTMASR, Park Ridge 15757
MTM Association for Standards and Research, Park Ridge 15757
MTNA, Cincinnati 15765
MTRF, Perry 15695
MTS, Washington 15691
Münchner Tierärztliche Gesellschaft, München 06122
Münchner Dermatologische Gesellschaft, München 06123
Münchner Entomologische Gesellschaft e.V., München 06124
Münchner Kreis, München 06125
Müszaki és Természettudományi Egyesületek Szövetségi Kamarája, Budapest 06665
Mukden Academy of Chemical Engineering, Liaoning 03211
Multicultural History Society of Ontario, Toronto 02663
Multiple Dwelling Standards Association, Willowdale 02664
The Multiple Sclerosis Society of Great Britain and Northern Ireland, London 12541
Municipal Economy Soiety, Moskva 09887
Municipal Librarians' Association, Kolbotn 09274
MURS, Paris 04602
Muscular Dystrophy Group of Great Britain & Northern Ireland, London 12542
Musealverein in Hallstatt, Hallstatt 00709
Museovirasto, Helsinki 03983
Museum Association of China, Taipei 03309
Museum Association of the American Frontier, Chadron 15758
Museum Attendants Association, Old Portsmouth 12543
Museum Computer Network, Pittsburgh 15759
Museum Council, Moskva 09888
Museum Education Roundtable, Washington 15760
Museums and Galleries Commission, London 12544
The Museums Association, London 12545
Museums Association of Australia, Melbourne 00435
Museums Association of India, New Dehli 06837
Museums Association of Israel, Tel Aviv 07107
Museums Association of Saskatchewan, Regina 02665
Museum Society of Macedonia, Skopje 08597
Museums-Verein Stillfried, Stillfried 00710
Museum Training Institute, Bradford 12546
Music Advisers' National Association, Warwick 12547
Musical Art Association, Stockholm 10502
Musical Box Society of Great Britain, Cambridge 12548
Music and Arts Society of America, Washington 15761
The Music Association of Ireland, Dublin 07015
Music Association of Korea, Seoul 08541
Music Critics Association, Westfield 15762
Music Educators National Conference, Reston 15763
Music Library Association, Canton 15764
Music Masters' and Mistresses' Association, Guildford 12549
Musicological Society of Australia, Canberra 00436
The Musicological Society of Japan, Tokyo 08352
Music Society, Reykjavik 06704
Music Teachers National Association, Cincinnati 15765
Musikaliska Konstföreningen, Stockholm 10502
MUT, Budapest 06662
MUT, Valletta 08667
MWF, London 12514
MWIA, Köln 06116
Myasthenia Gravis Foundation of British Columbia, Vancouver 02666
Mycological Society of America, Beltsville 15766
Mycological Society of Japan, Tokyo 08224
Mysore Horticultural Society, Bangalore 06838
Mystic Seaport Museum, Mystic 15767
Mythic Society, Bangalore 06839
Mythopoeic Society, Altadena 15768

NA, Knaresborough 12551
NAA, Köln 06149
NAAASE, Milwaukee 15792
NAAB, Washington 15788
NAABAVE, Detroit 15816
NAACLS, Chicago 15780
NAADAC, Arlington 15828
NAAE, Albuquerque 16209
NAAF, San Rafael 15786
NAAHP, Champaign 15827
NAAMM, Middleton 16210
NAAPABAC, New York 15817
NAAS, Columbia 15826
NAAS, Columbus 15787
NAASR, Belmont 15793
NAASS, Saint Louis 16213
NABD, Berlin 06152
NABE, Chicago 15830
NABE, Washington 15794
NABESS, Phoenix 15836
NABL, Hinsdale 15835
NABP, Chrisfield 15832
NABP, Park Ridge 15834
NABPR, Macon 15829
NABR, Washington 15795
NABSE, Washington 15784
NABT, Reston 15831
NABT, Washington 15833
NABTE, Reston 15796
NACA, Cincinnati 15771
NACA, Johannesburg 10155
NACADA, Manhattan 15770
NACC, Kent 15797
NACCAS, Arlington 15781
NACDRAO, Albany 15841
NACDS, Jacksonville 16214
NACE, Houston 15845
NACE, Washington 15846
NACEBE, Boone 15838
NACEHSP, Denver 15779
NACF, London 12556
NACGT, Townbridge 12565
NACHC, Washington 15844
NACHRI, Alexandria 15837
NACP, Washington 15847
NACTA, Urbana 15842
NACUA, Washington 15839
NACUBO, Washington 15840
NACWPI, Kirksville 15843
NAD, New York 15772
NADA, Falls Church 15848
NADA, Romford 12566
NADE, Frankfort 15850
NADL, Alexandria 15849
NADS, Lexington 16216
NADSA, Blacksburg 15851
NAE, Stanford 15773
NAE, Washington 15774
NAE4-HA, Athens 15858
NAEA, Reston 15790
NAEB, Hauppauge 15852
NA EBM, Düsseldorf 06162
NAEE, Walsall 12557
NAEE, Washington 16211
NAEHCA, Key Biscayne 15855
NAEIR, Galesburg 15819
NAEMT, Kansas City 15854
NAEP, Washington 15856
NAES, New York 15857
NAES, Tempe 15799
NAESP, Alexandria 15853
NAEYC, Washington 15818
NAF, Oslo 09280
NAFA, Köln 06166
NAFAC, Grand Rapids 15791
NAFE, Bexleyheath 12558
NAFEO, Washington 15798
NAFI, Columbus 15860
NAFuO, Pforzheim 06168
NAFV, Washington 15859
NAGT, Bellingham 15861
NAHA, Northfield 16233
NAHCS, Saint Ann 15862
NAHD, Washington 15802
NAHDSA, Brooklyn 15864
NAHE, Saint Joseph 15804
NAHEE, East Haddam 15803
NAHSA, Rockville 15801
NAHSE, Columbia 15863
NAHT, Haywards Heath 12567
Nah- und Mittelost-Verein e.V., Hamburg 06126
NAH & WW, New York 15866
NAI, Fort Collins 15806
NAICU, Washington 15867
NAIEC, Buffalo 15805
NAIS, Boston 15868
NAITTE, Knoxville 15869
Nakladni Zavod Matice Hrvatske, Zagreb 03464
NAL, Berlin 06188
NAL, Oslo 09296
NALA, Rugby 12568
NALA, Tulsa 15870
NALSAS, Santa Fe 15807
NAM, Frankfurt 06190
NAMA, Ann Arbor 16219
NAMAC, Oakland 15785
NAMDA, Newport Beach 16218
NAME, Saint Louis 15873
NAME, Slough 12559
NAMed, Berlin 06193
Names Society, Thames Ditton 12550
NAMESU, Tucson 15874
NAMS, Chesapeake 15872
NAMTAC, Washington 15871
Nanaimo Neurological and Cerebral Palsy Association, Nanaimo 02667
NANASP, Grand Rapids 15875
NAO, Bowie 15775
NAOE, Sheffield 12560
NAOO, Cleveland 15876
NAPA, Washington 15776, 15820
NAPAEO, Beverley 12569
NAPCE, Elgin 16212
NAPCRG, Richmond 16220
NAPE, Chicopee 15878
NAPE, Washington 15881
NAPEHE, San Jose 15809
NAPH, Madison 15882
NAPNES, Silver Spring 15811
Napoleonic Association, Knaresborough 12551
Napoleonic Society, London 12552
NAPSEC, Washington 15880
NAPSG, Hendersonville 15879
NAPT, Potomac 15810
NARAS, Burbank 15777
NARAVA, Washington 15789

NAREMCO, New York 16135
NARF, Washington 15883
NARIC, Bruxelles 01327
Narrow Gauge Railway Society, Huddersfield 12553
NARST, Manhattan 15812
NAS, Berlin 06204
NAS, Earls Barton 12631
NAS, Lafayette 15918
NAS, London 12575
NAS, Lyngby 03780
NASA, Norman 15893
NASAD, Reston 15887
NASAE, Columbus 15908
NASAGA, Indianapolis 16221
NASAP, Chicago 16224
NASAR, Fairfax 15814
NASBE, Alexandria 15894
NASBE, Fort Worth 15909
NASC, Boise 16232
NASCUMC, Nashville 15886
NASDA, Washington 15895
NASDSE, Alexandria 15896
NASDTEC, Seattle 15897
NASDVTE, Washington 15898
NASEES, Milton Keynes 12562
NASEMP, Nashville 15899
NASILP, Philadelphia 15917
NASK, Berlin 06205
NASM, Reston 15888
NASMHPD, Alexandria 15900
N.A.S.O., Birmingham 12554
NASP, Silver Spring 15884
NASPAA, Washington 15889
NASPD, Tallahassee 15902
NASPE, Newton Upper Falls 16225
NASPE, Reston 15815
NASPGN, Denver 16222
NASPHV, Columbus 15903
NASS, Bethesda 15783
NASSDSE, Pierre 15904
NASSH, University Park 16223
NASSP, Reston 15892
NASSVHE, Austin 15905
NAST, Reston 15890
NASUA, Washington 15906
NASULGC, Washington 15907
NASUWT, Birmingham 12570
NASW, Greenlawn 15891
NATA, Chatswood 00437
NATAS, Buffalo 16226
NATD, Englewood 15912
N.A.T.E., Sheffield 12564
NATEBE, Columbia 15910
NATFHE, London 12571
Nathaniel Hawthorne Society, Brunswick 15769
NATHE, London 12573
NATIE, Leesburg 15821
Nationaal Centrum voor Wetenschappelijke en Technische Documentatie; NCWTD, Bruxelles 01201
National Academic Advising Association, Manhattan 15770
National Academy of Art, New Dehli 06840
National Academy of Arts, Seoul 08542
National Academy of Code Administration, Cincinnati 15771
National Academy of Design, New York 15772
National Academy of Education, Stanford 15773
National Academy of Engineering, Washington 15774
The National Academy of Exact, Physical and Natural Sciences of Argentina, Buenos Aires 00045
National Academy of Letters, New Dehli 06854
National Academy of Opticianry, Bowie 15775
National Academy of Public Administration, Washington 15776
National Academy of Recording Arts and Sciences, Burbank 15777
National Academy of Sciences, Allahabad 06841
National Academy of Sciences, Seoul 08543
National Academy of Western Art, Oklahoma City 15778
National Accreditation Council for Environmental Health Science and Protection, Denver 15779
National Accrediting Agency for Clinical Laboratory Sciences, Chicago 15780
National Accrediting Commission of Cosmetology Arts and Sciences, Arlington 15781
National Adult School Organisation, Birmingham 12554
National Adult Vocational Education Association, Milan 15782
National Alliance for Safe Schools, Bethesda 15783
National Alliance of Black School Educators, Washington 15784
National Alliance of Media Arts Centers, Oakland 15785
National Alopecia Areata Foundation, San Rafael 15786
National Anorexic Aid Society, Columbus 15787
National Anti-Vivisection Society, London 12555
National Architectural Accrediting Board, Washington 15788
National Archives and Records Administration Volunteer Association, Washington 15789
National Art-Collections Fund, London 12556
National Art Education Association, Reston 15790
National Arts Foundation, Buenos Aires 00139
National Assocation of Agrarian Structure Improvement, Tokyo 08447
National Association for Ambulatory Care, Grand Rapids 15791
National Association for Applied Arts, Science and Education, Milwaukee 15792
National Association for Armenian Studies and Research, Belmont 15793
National Association for Bilingual Education, Washington 15794
National Association for Biomedical Research, Washington 15795
National Association for Business Teacher Education, Reston 15796
National Association for Clean Air, Johannesburg 10155
National Association for Core Curriculum, Kent 15797
National Association for Environmental Education, Walsall 12557
National Association for Equal Opportunity in Higher Education, Washington 15798
National Association for Ethnic Studies, Tempe 15799
National Association for Family and Community Education, Tulsa 15800
National Association for Film in Education, Bexleyheath 12558
National Association for Hearing and Speech Action, Rockville 15801
National Association for Human Development, Washington 15802
National Association for Humane and Environmental Education, East Haddam 15803
National Association for Humanities Education, Saint Joseph 15804
National Association for Industry-Education Cooperation, Buffalo 15805
National Association for Interpretation, Fort Collins 15806
National Association for Legal Support of Alternative Schools, Santa Fe 15807
National Association for Multiracial Education, Slough 12559
National Association for Outdoor Education, Sheffield 12560
National Association for Outlaw and Lawman History, Killeen 15808
National Association for Photographic Art, Scarborough 02668
National Association for Physical Education in Higher Education, San Jose 15809
National Association for Poetry Therapy, Potomac 15810
National Association for Practical Nurse Education and Service, Silver Spring 15811
National Association for Research in Science Teaching, Manhattan 15812
National Association for Road Safety Instruction in Schools, London 12561
National Association for Rural Mental Health, Kansas City 15813
National Association for Search and Rescue, Fairfax 15814
National Association for Soviet and East European Studies, Milton Keynes 12562
National Association for Sport and Physical Education, Reston 15815
National Association for the Advancement of Black Americans in Vocational Education, Detroit 15816
National Association for the Advancement of Psychoanalysis and The American Boards for Accreditation and Certification, New York 15817
National Association for the Education of Young Children, Washington 15818
National Association for the Exchange of Industrial Resources, Galesburg 15819
National Association for the Practice of Anthropology, Washington 15820
National Association for the Relief of Pagets Disease, Manchester 12563
National Association for the Teaching of English, Sheffield 12564
National Association for Trade and Industrial Education, Leesburg 15821
National Association for Veterinary Acupuncture, Fullerton 15822
National Association for Women in Education, Washington 15823
National Association for Year-Round Education, San Diego 15824
National Association of Academic Advisors for Athletics, Charlottesville 15825
National Association of Academies of Science, Columbia 15826
National Association of Advisers for the Health Professions, Champaign 15827
National Association of Alcoholism and Drug Abuse Counselors, Arlington 15828
National Association of Baptist Professors of Religion, Macon 15829
National Association of Bar Executives, Chicago 15830
National Association of Biology Teachers, Reston 15831
National Association of Black Professors, Chrisfield 15832
National Association of Blind Teachers, Washington 15833
National Association of Boards of Pharmacy, Park Ridge 15834
National Association of Bond Lawyers, Hinsdale 15835
National Association of Business Education State Supervisors, Phoenix 15836
National Association of Careers and Guidance Teachers, Townbridge 12565
National Association of Chemists, Madrid 10276
National Association of Children's Hospitals and Related Institutions, Alexandria 15837
National Association of Classroom Educators in Business Education, Boone 15838
National Association of College and University Attorneys, Washington 15839
National Association of College and University Business Officers, Washington 15840
National Association of College Deans, Registrars and Admissions Officers, Albany 15841
National Association of Colleges and Teachers of Agriculture, Urbana 15842
National Association of College Wind and Percussion Instructors, Kirksville 15843
National Association of Community Health Centers, Washington 15844
National Association of Corrosion Engineers, Houston 15845
National Association of County Engineers, Washington 15846
National Association of County Planners, Washington 15847
National Association of Dental Assistants, Falls Church 15848
National Association of Dental Laboratories, Alexandria 15849
National Association of Disability Examiners, Frankfort 15850
National Association of Drama Advisers, Romford 12566
National Association of Dramatic and Speech Arts, Blacksburg 15851
National Association of Educational Buyers, Hauppauge 15852
National Association of Elementary School Principals, Alexandria 15853
National Association of Emergency Medical Technicians, Kansas City 15854
National Association of Employers on Health Care Action, Key Biscayne 15855
National Association of Environmental Professionals, Washington 15856
National Association of Episcopal Schools, New York 15857
National Association of Extension 4-H Agents, Athens 15858
National Association of Federal Veterinarians, Washington 15859
National Association of Flight Instructors, Columbus 15860
National Association of Geology Teachers, Bellingham 15861
National Association of Head Teachers, Haywards Heath 12567
National Association of Health Career Schools, Saint Ann 15862
National Association of Health Services Executives, Columbia 15863
National Association of Hebrew Day School Administrators, Brooklyn 15864
National Association of Hebrew Day School PTA'S, New York 15865
National Association of Home and Workshop Writers, New York 15866
National Association of Independent Colleges and Universities, Washington 15867
National Association of Independent Schools, Boston 15868
National Association of Industrial and Technical Teacher Educators, Knoxville 15869
National Association of Language Advisers, Rugby 12568
National Association of Legal Assistants, Tulsa 15870
National Association of Management and Technical Assistance Centers, Washington 15871
National Association of Marine Surveyors, Chesapeake 15872
National Association of Medical Examiners, Saint Louis 15873
National Association of Music Executives in State Universities, Tucson 15874
National Association of Nutrition and Aging Services Programs, Grand Rapids 15875
National Association of Optometrists and Opticians, Cleveland 15876
National Association of Pastoral Musicians, Washington 15877
National Association of Power Engineers, Chicopee 15878
National Association of Principal Agricultural Education Officers, Beverley 12569
National Association of Principals of Schools for Girls, Hendersonville 15879
National Association of Private Schools for Exceptional Children, Washington 15880
National Association of Professional Educators, Washington 15881
National Association of Professors of Hebrew, Madison 15882
National Association of Rehabilitation Facilities, Washington 15883
National Association of Schoolmasters, Birmingham 12570
National Association of School Psychologists, Silver Spring 15884
National Association of School Safety and Law Enforcement Officers, Upper Marlboro 15885
National Association of Schools and Colleges of the United Methodist Church, Nashville 15886
National Association of Schools of Art and Design, Reston 15887
National Association of Schools of Music, Reston 15888
National Association of Schools of Public Affairs and Administration, Washington 15889
National Association of Schools of Theatre, Reston 15890
National Association of Science Writers, Greenlawn 15891
The National Association of Scientists, Pretoria 10156
National Association of Secondary School Principals, Reston 15892
National Association of State Archeologists, Norman 15893
National Association of State Boards of Education, Alexandria 15894
National Association of State Development Agencies, Washington 15895
National Association of State Directors of Special Education, Alexandria 15896
National Association of State Directors of Teacher Education and Certification, Seattle 15897
National Association of State Directors of Vocational Technical Education, Washington 15898
National Association of State Educational Media Professionals, Nashville 15899
National Association of State Mental Health Program Directors, Alexandria 15900
National Association of State Mental Retardation Program Directors, Alexandria 15901
National Association of State Park Directors, Tallahassee 15902
National Association of State Public Health Veterinarians, Columbus 15903
National Association of State Supervisors and Directors of Secondary Education, Pierre 15904
National Association of State Supervisors of Vocational Home Economics, Austin 15905
National Association of State Units on Aging, Washington 15906
National Association of State Universities and Land-Grant Colleges, Washington 15907
National Association of Supervisors of Agricultural Education, Columbus 15908
National Association of Supervisors of Business Education, Fort Worth 15909
National Association of Swedish Architects, Stockholm 10535
National Association of Teacher Educators for Business Education, Columbia 15910
National Association of Teachers in Further and Higher Education, London 12571
National Association of Teachers of Singing, Jacksonville 15911
National Association of Teachers of the Mentally Handicapped, Aylesbury 12572
National Association of Test Directors, Englewood 15912
National Association of Testing Authorities, Australia, Chatswood 00437
National Association of University Women, Tallahassee 15913
National Association of Vocational Education Special Needs Personnel, McKeesport 15914
National Association of Women in Construction, Edmonton 02669
National Association of Women Lawyers, Chicago 15915
National Association on Drug Abuse Problems, New York 15916
National Associations of Teachers of Home Economics and Technology, London 12573
National Assoiation of Self-Instructional Language Programs, Philadelphia 15917
National Asthma Campaign, London 12574
National Audio Visual Association of Canada, Toronto 02670
National Audio-Visual Education Association of China, Taipei 03310
National Autistic Society, London 12575
National Avionics Society, Lafayette 15918
National Back Pain Association, Teddington 12576
National Bar Association, Taipei 03311
National Bible Society of Scotland, Edinburgh 12577
National Black MBA Association, Chicago 15919
National Black Music Caucus of the Music Educators National Conference, Ann Arbor 15920
National Black Women's Health Project, Atlanta 15921
National Board for Certification in Dental Technology, Alexandria 15922
National Board for Certification of Dental Laboratories, Alexandria 15923
National Board for Respiratory Care, Lenexa 15924
National Board of Antiquities, Helsinki 03983
National Board of Examiners in Optometry, Washington 15925
National Board of Medical Examiners, Philadelphia 15926
National Book Council of Pakistan, Islamabad 09375
National Bureau of Economic Research, Cambridge 15927
National Burn Information Exchange, Ann Arbor 15928
National Burn Victim Foundation, Orange 15929
National Business Education Association, Reston 15930
National Campaign for Nursery Education, London 12578
National Cancer Center, Plainview 15931
National Cancer Center, Tokyo 08184
National Captioning Institute, Falls Church 15932
National Cartographic Centre, Teheran 06901
National Catholic Business Education Association, Emporia 15933
National Catholic Conference for Interracial Justice, Washington 15934
National Catholic Educational Association, Washington 15935
National Catholic Forensic League, Milford 15936
National Catholic Pharmacists Guild of the United States, Saint Louis 15937
National Caving Association, Derby 12579
National Center for Appropriate Technology, Butte 15938
National Center for Automated Information Research, New York 15939
National Center for Business and Economic Communication, Washington 15940
National Center for Computer Crime Data, Santa Cruz 15941
National Center for Disability Services, Albertson 15942
National Center for Fair and Open Testing, Cambridge 15943
National Center for Homeopathy, Alexandria 15944
National Center for Housing Management, Washington 15945
National Center for Law and Deafness, Washington 15946
National Center for State Courts, Williamsburg 15947
National Center for Youth Law, San Francisco 15948
National Center on Arts and the Aging, Washington 15949
National Center on Institutions and Alternatives, Alexandria 15950
National Centre for Educational Research, Cairo 03909
National Centre for Language Development, Jakarta 06894
National Centre of Agricultural Research, Maracay 16991
National Certification Agency for Medical Lab Personnel, Bethesda 15951
National Character Laboratory, El Paso 15952
National Christian Education Council, Redhill 12580
National Coalition for Literacy, Washington 15953

National Coalition for Public Education and Religious Liberty, Washington *15954*
National Coalition for Women and Girls in Education, Washington *15955*
National Coalition of Alternative Community Schools, Summertown *15956*
National College of District Attorneys, Houston *15957*
National College of Foot Surgeons, Woodland Hills *15958*
National Collegiate Conference Association, Albany *15959*
National Commission for Cooperative Education, Boston *15960*
National Commission for the Protection of Public Libraries, Buenos Aires *00121*
National Commission on Space Research, Buenos Aires *00119*
National Committee for Clinical Laboratory Standards, Villanova *15961*
National Committee for Latin and Greek, Cypress *15962*
National Committee for Quality Health Care, Washington *15963*
National Committee for the Collection and Assessment of Numerical Data in Science and Technology, Moskva *09889*
National Committee for the Defence of Fauna and Flora, Santiago *03042*
National Committee for the Furtherance of Jewish Education, Brooklyn *15964*
National Committee for Thermal Analysis, Moskva *09890*
National Committee of Biochemists, Moskva *09891*
National Committee of Biologists, Moskva *09892*
National Committee of Chemists, Moskva *09893*
National Committee of Danish Art Societies, Niva *03801*
National Committee of Finno-Ugric Philologists, Moskva *09894*
National Committee of Geologists, Moskva *09895*
National Committee of History and Philosophy of Natural Science and Technology, Moskva *09896*
National Committee of Mathematicians, Moskva *09897*
National Committee of Russian Historians, Moskva *09898*
National Committee of Slavonic Philologists, Moskva *09899*
National Committee of the International Council of Scientific Unions, Moskva *09900*
National Committee of the International Music Council, Madrid *10293*
National Committee of the International Scientific Radio Union, Moskva *09901*
National Committee of the Pacific Ocean Research Association, Moskva *09902*
National Committee of the Scientific Committee for Problems of the Environment, Moskva *09903*
National Committee of Turkish Philologists, Moskva *09904*
National Committee on Automatic Control, Moskva *09905*
National Committee on the International Biological Programme, Moskva *09906*
National Committee on Theoretical and Applied Mechanics, Moskva *09907*
National Committee on the Treatment of Intractable Pain, Washington *15965*
National Community Education Association, Alexandria *15966*
National Conference of Bankruptcy Judges, Rockville *15967*
National Conference of Black Lawyers, New York *15968*
National Conference of Commissioners on Uniform State Laws, Chicago *15969*
National Conference of Regulatory Utility Commission Engineers, Nashville *15970*
National Conference of Standards Laboratories, Boulder *15971*
National Conference of States on Building Codes and Standards, Herndon *15972*
National Conference of University Professors, Uxbridge *12581*
National Conference of Yeshiva Principals, New York *15973*
National Conference on Fluid Power, Milwaukee *15974*
National Conference on Research in English, Amherst *15975*
National Conference on the Advancement of Research, San Antonio *15976*
National Conference on Weights and Measures, Gaithersburg *15977*
National Congress for Community Economic Development, Washington *15978*
National Congress of Inventors Organizations, Rheem Valley *15979*
National Consortium for Black Professional Development, Louisville *15980*
National Consortium for Child Mental Health Services, Washington *15981*
National Consortium of Arts and Letters for Historically Black Colleges and Universities, Washington *15982*
National Consumer Law Center, Boston *15983*
National Council for Accreditation of Teacher Education, Washington *15984*
National Council for Black Studies, Columbus *15985*
National Council for Culture and Art, New York *15986*
National Council for Educational Technology, Coventry *12582*
National Council for Environmental Balance, Louisville *15987*
National Council for Geographic Education, Indiana *15988*
National Council for Labor Reform, Chicago *15989*
National Council for Research, Khartoum *10427*
National Council for Research and Development, Jerusalem *07108*
National Council for Research and Planning, Radford *15990*
National Council for Scientific Research, Lusaka *17154*
National Council for Textile Education, Charlottesville *15991*
National Council for the Social Studies, Washington *15992*
National Council for the Traditional Arts, Silver Spring *15993*
National Council for Torah Education, New York *15994*
National Council for Urban Economic Development, Washington *15995*
National Council of Applied Economic Research, New Delhi *06842*
National Council of Athletic Training, Reston *15996*
National Council of BIA Educators, Albuquerque *15997*
National Council of Corrosion Societies, London *12583*
National Council of Educational Research and Training, New Dehli *06843*
National Council of Examiners for Engineering and Surveying, Clemson *15998*
National Council of Forestry Association Executives, Augusta *15999*
National Council of Guilds for Infant Survival, Davenport *16000*
National Council of Intellectual Property Law Associations, Arlington *16001*
National Council of Jewish Religious Day Schools, London *12584*
National Council of Juvenile and Family Court Judges, Reno *16002*
National Council of Local Administrators of Vocational Education and Practical Arts, Brooklyn *16003*
National Council of Psycho-Therapists, London *12585*
National Council of Scientific and Technical Research, Buenos Aires *00130*
National Council of Scientific and Technological Development, Brasília *01645*
National Council of State Directors of Community and Junior Colleges, Lansing *16004*
National Council of State Education Associations, Washington *16005*
National Council of State Pharmacy Executives, Chapel Hill *16006*
National Council of State Supervisors of Foreign Languages, Little Rock *16007*
National Council of State Supervisors of Music, Columbus *16008*
National Council of Supervisors of Mathematics, Golden *16009*
National Council of Teachers of English, Urbana *16010*
National Council of Teachers of Mathematics, Reston *16011*
National Council of the Paper Industry for Air and Stream Improvement, New York *16012*
National Council of University Research Administrators, Washington *16013*
National Council of Urban Education Associations, Washington *16014*
National Council on Agricultural Life and Labor Research Fund, Dover *16015*
National Council on Alcoholism and Drug Dependence, New York *16016*
National Council on Crime and Delinquency, San Francisco *16017*
National Council on Economic Education, New York *16018*
National Council on Employment Policy, Washington *16019*
National Council on Family Relations, Minneapolis *16020*
National Council on Gene Resources, Berkeley *16021*
National Council on Health Laboratory Services, Durham *16022*
National Council on Measurement in Education, Washington *16023*
National Council on Public History, Indianapolis *16024*
National Council on Radiation Protection and Measurements, Bethesda *16025*
National Council on Religion and Public Education, Ames *16026*
National Crime Prevention Institute, Louisville *16027*
National Criminal Defense College, Macon *16028*
National Culture Commission, Bangkok *11094*
National Dairy Council, Rosemont *16029*
National Dental Assistants Association, Washington *16030*
National Dental Association, Washington *16031*
National Dental Examining Board of Canada, Ottawa *02671*
National Dental Hygienists' Association, Washington *16032*
National Denturist Association, Portland *16033*
National Design Council, Ottawa *02672*
National Development Association, Dublin *07016*
National Digestive Diseases Information Clearinghouse, Bethesda *16034*
National District Attorneys Association, Alexandria *16035*
National Dry Bean Council, Washington *16036*
National Earth Science Teachers Association, Fargo *16037*
National Economic Association, Ann Arbor *16038*
National Eczema Society, London *12586*
National Educational Information System, Santiago *03064*
National Education Association, Tewin Wood *12587*
National Education Association, Washington *16039*
National Education Society of Ceylon, Colombo *10417*
Nationalekonomiska Föreningen, Stockholm *10503*
National Energy Management Institute, Alexandria *16040*
National Energy Resources Organization, Washington *16041*
National Engineering Consortium, Chicago *16042*
National Environmental Health Association, Denver *16043*
Nationale Vereniging voor Economisch Onderwijs, Bergen *08978*
National Federation Interscholastic Music Association, Kansas City *16044*
National Federation of Abstracting and Information Services, Philadelphia *16045*
National Federation of Catholic Physicians Guilds, Elm Grove *16046*
National Federation of Continuative Teachers' Associations, Bexley *12588*
National Federation of Modern Language Teachers Associations, Omaha *16047*
National Federation of Sailing Schools, Lymington *12589*
National Federation of Societies for Clinical Social Work, Arlington *16048*
National Federation of State Poetry Societies, Ogden *16049*
National Federation of Voluntary Literacy Schemes, London *12590*
National FFA Organization, Alexandria *16050*
National Fire Protection Association, Quincy *16051*
National Fisheries Contaminant Research Center, Columbia *16052*
National Fisheries Education and Research Foundation, Arlington *16053*
National Food and Nutrition Commission, Lusaka *17155*
National Forensic Association, Westerville *16054*
National Forensic League, Ripon *16055*
National Foundation for Advancement in the Arts, Miami *16056*
National Foundation for Brain Research, Washington *16057*
National Foundation for Children's Hearing Education and Research, Yonkers *16058*
National Foundation for Educational Research in England and Wales, Slough *12591*
National Foundation for Non-Invasive Diagnostics, Princeton *16059*
National Foundation of Dentistry for the Handicapped, Denver *16060*
National Foundry and Engineering Training Association, Glasgow *12592*
National Fund for Medical Education, Boston *16061*
National Genealogical Society, Arlington *16062*
National Geographical Society of India, Varanasi *06844*
National Geographic Society, Washington *16063*
National Geriatrics Society, Eustis *16064*
National Grants Management Association, Washington *16065*
National Guild of Catholic Psychiatrists, Ellicott City *16066*
National Guild of Community Schools of the Arts, Englewood *16067*
National Guild of Piano Teachers, Austin *16068*
National Health and Medical Research Council, Camberra *00438*
National Health Council, Washington *16069*
National Health Federation, Monrovia *16070*
National Health Law Program, Los Angeles *16071*
National Health Lawyers Association, Washington *16072*
National Hearing Aid Society, Livonia *16073*
National Hearing Conservation Association, Des Moines *16074*
National Heart Research Project, Baltimore *16075*
National Home Study Council, Washington *16076*
National Hormone and Pituitary Program, Rockville *16077*
National Housing and Town Planning Council, London *12593*
National Housing Conference, Washington *16078*
National Humane Education Society, Leesburg *16079*
National Humanities Center, Research Triangle Park *16080*
National Hungarian Cecilia Society, Budapest *06671*
National Hydropower Association, Washington *16081*
National Identification Program for the Advancement of Women in Higher Education Administration, Washington *16082*
National Immigration Law Center, Los Angeles *16083*
National Immunological Society, Moskva *09908*
National Indian Education Association, Washington *16084*
National Industrial Zoning Committee, Columbus *16085*
National Information and Documentation Centre, Cairo *03910*
National Information Center for Educational Media, Albuquerque *16086*
National Information Service for Earthquake Engineering, Richmond *16087*
National Institute for Burn Medicine, Ann Arbor *16088*
National Institute for Certification in Engineering Technologies, Alexandria *16089*
National Institute for Compilation and Translation, Taipei *03312*
National Institute for Educational Research, Brasilia *01687*
National Institute for Farm Safety, Columbia *16090*
National Institute for Medical Research, London *12594*
National Institute for Public Policy, Fairfax *16091*
National Institute for Rehabilitation Engineering, Hewitt *16092*
National Institute for Social Work, London *12595*
National Institute of Adult Continuing Education (England and Wales), Leicester *12596*
National Institute of Agricultural Botany, Cambridge *12597*
National Institute of Ceramic Engineers, Westerville *16093*
National Institute of Design, Ahmedabad *06845*
National Institute of Economic and Social Research, London *12598*
National Institute of Folk and Traditional Heritage, Islamabad *09373*
National Institute of Industrial Psychology, Slough *12599*
National Institute of Industrial Technology, Buenos Aires *00145*
National Institute of Management Counsellors, Great Neck *16094*
National Institute of Medical Herbalists, Exeter *12600*
National Institute of Science, Nashville *16095*
National Institute of Social Sciences, New York *16096*
National Institute of Steel Detailing, Anaheim *16097*
National Institute of Technology, Rio de Janeiro *01690*
National Institute of Viti-Viniculture, Mendoza *00146*
National Jewish Center for Immunology and Respiratory Medicine, Denver *16098*
National Judicial College, Reno *16099*
National League of American PEN Women, Washington *16100*
National Leukemia Association, Garden City *16101*
National Librarians Association, Alma *16102*
National Library Service, Belize City *01513*
National Marine Educators Association, Pacific Grove *16103*
National Mastitis Council, Arlington *16104*
National Materials Advisory Board, Washington *16105*
National Materials Handling Centre, Bedford *12601*
National Medical and Dental Association, Philadelphia *16106*
National Medical and Technical Scientific Society, Moskva *09909*
National Medical Association, Washington *16107*
National Medical Fellowships, New York *16108*
National Mental Health Association, Alexandria *16109*
National Middle School Association, Columbus *16110*
National Monuments Commission, Livingstone *17156*
National Multiple Sclerosis Society, New York *16111*
National Music Council, Englewood *16112*
National Music Council of China, Taipei *03313*
Nationalökonomische Gesellschaft, Wien *00711*
National Office for Research and Special Libraries, Oslo *09344*
Nationaløkonomisk Forening, København *03774*
National Old Timers' Association of the Energy Industry, Mineola *16113*
National Opera Association, Shreveport *16114*
National Operatic and Dramatic Association, London *12602*
National Ophthalmological Society, Moskva *09910*
National Optometric Association, Columbus *16115*
National Orchestral Association, New York *16116*
National Organization for Continuing Education of Roman Catholic Clergy, Chicago *16117*
National Organization for Rare Disorders, New Fairfield *16118*
National Organization of Minority Architects, Washington *16119*
National Organization on Legal Problems of Education, Topeka *16120*
National Organization to Insure a Sound-Controlled Environment, Washington *16121*
National Orthotic and Prosthetic Research Institute, Lenox Hill *16122*
National Osteopathic Guild Association, Chicago *16123*
National Perinatal Association, Tampa *16124*
National Pest Control Association, Dunn Loring *16125*
National Pharmaceutical Association, Washington *16126*
National Pharmaceutical Society, Moskva *09911*
National Photography Instructors Association, Eagle Rock *16127*
National Piano Foundation, Dallas *16128*
National Plant Board, Montgomery *16129*
National Podriatic Medical Association, Chicago *16130*
National Productivity Council, New Delhi *06846*
National Psoriasis Foundation, Portland *16131*
National Psychological Association for Psychoanalysis, New York *16132*
National Pure Water Association, Ellesmere *12603*
National Ramah Commission, New York *16133*
National Reading Conference, Chicago *16134*
National Records Management Council, New York *16135*
National Registry in Clinical Chemistry, Washington *16136*
National Rehabilitation Association, Alexandria *16137*
National Research Centre, Cairo *03911*
The National Research Council, Reykjavik *06696*
National Research Council, Washington *16138*
National Research Council of Canada, Ottawa *02673*
National Research Council of the Philippines, Manila *09522*
National Research Council on Peace Strategy, New York *16139*
National Resident Matching Program, Washington *16140*
National Resource Center for Consumers of Legal Services, Gloucester *16141*
National Romanian Society for Soil Science, Bucuresti *09790*
National Rural and Small Schools Consortium, Bellingham *16142*
National Rural Development Institute, Bellingham *16143*
National Rural Education Association, Fort Collins *16144*
National Safety Council, Dublin *07017*

National Safety Council, Itasca 16145
National Safety Management Society, Weaverville 16146
National Schizophrenia Fellowship, Surbiton 12604
National School Boards Association, Alexandria 16147
National School Public Relations Association, Arlington 16148
National Schools Committee for Economic Education, Cos Cob 16149
National Science Council, Taipei 03314
National Science Council of Pakistan, Islamabad 09376
National Science Information Center, Bruxelles 01201
National Science Supervisors Association, Glastonbury 16150
National Science Teachers Association, Washington 16151
National Scientific Medical Society of Anatomists, Histologists and Embryologists, Moskva 09912
National Scientific Medical Society of Endocrinologists, Moskva 09913
National Scientific Medical Society of Gastroenterologists, Moskva 09914
National Scientific Medical Society of Haematologists and Transfusiologists, Moskva 09915
National Scientific Medical Society of History of Medicine, Moskva 09916
National Scientific Medical Society of Hygienists, Moskva 09917
National Scientific Medical Society of Infectionists, Moskva 09918
National Scientific Medical Society of Medical Geneticists, Moskva 09919
National Scientific Medical Society of Nephrologists, Moskva 09920
National Scientific Medical Society of Neuropathologists and Psychiatrists, Moskva 09921
National Scientific Medical Society of Neurosurgeons, Moskva 09922
National Scientific Medical Society of Obstetricians and Gynaecologists, Moskva 09923
National Scientific Medical Society of Oncologists, St. Petersburg 09924
National Scientific Medical Society of Oto-Rhino-Laryngologists, Moskva 09925
National Scientific Medical Society of Paediatricians, Moskva 09926
National Scientific Medical Society of Phtisiologists, Moskva 09927
National Scientific Medical Society of Physical Therapists and Health-Resort Physicians, Moskva 09928
National Scientific Medical Society of Physicians-Analysts, Moskva 09929
National Scientific Medical Society of Physicians in Curative Physical Culture and Sports Medicine, Moskva 09930
National Scientific Medical Society of Rheumatologists, Moskva 09931
National Scientific Medical Society of Roentgenologists and Radiologists, Moskva 09932
National Scientific Medical Society of Stomatologists, Moskva 09933
National Scientific Medical Society of Surgeons, Moskva 09934
National Scientific Medical Society of Toxicologists, St. Petersburg 09935
National Scientific Medical Society of Traumatic Surgeons and Orthopaedists, Moskva 09936
National Scientific Medical Society of Venereologists and Dermatologists, Moskva 09937
National Scientific Medical Soiety of Therapists, Moskva 09938
National Scientific Society of Biomedical Physics and Engineering, Sofia 01769
National Scientific Society of Urological Surgeons, Moskva 09939
National Sculpture Society, New York 16152
National Secular Society, London 12605
National Service of Geology and Mining, Santiago 03063
National Shellfisheries Association, Southampton 16153
National Society (Church of England), London 12606
National Society for Cardiovascular Technology, Fredericksburg 16154
National Society for Clean Air, Brighton 12607
National Society for Educational and Cultural Advancement, Montréal 02934
National Society for Education in Art and Design, Corsham 12608
National Society for Epilepsy, Chalfont Saint Peter 12609
National Society for Experiential Education, Raleigh 16155
National Society for Performance and Instruction, Washington 16156
National Society for Research into Allergy, Hinckley 12610
National Society for the Preservation of Covered Bridges, Worcester 16157
National Society for the Study of Education, Chicago 16158
National Society for Transplant Surgery, Cardiff 12611
National Society of Fine Arts, Lisboa 09688
National Society of General Practice of Romania, Bucuresti 09789
National Society of Polio and Accident Victims, Hellerup 03764
National Society of Professional Engineers, Alexandria 16159
National Society of Professional Sanitarians, Jefferson City 16160
National Space Club, Washington 16161
National Space Society, Washington 16162
National Speleological Society, Huntsville 16163
National Standard Plumbing Code Committee, Falls Church 16164
National Strict Baptist Sunday School Association, Bishops Stortford 12612
National Student Nurses' Association, New York 16165
National Student Speech Language Hearing Association, Rockville 16166
National Study of School Evaluation, Bloomington 16167
National Tax Association – Tax Institute of America, Columbus 16168
National Tax Research Association of China, Taipei 03315
National Technical Association, Washington 16169
National Technical Services Association, Alexandria 16170
National Telemedia Council, Madison 16171
National Theatre Institute, Waterford 16172
National Therapeutic Recreation Society, Arlington 16173
National Trolleybus Association, London 12613
National Trust for Places of Historic Interest or Natural Beauty, London 12614
National Trust for Scotland, Edinburgh 12615
National Tumor Registrars Association, Mundelein 16174
National Union of Teachers, London 12616
National University Continuing Education Association, Washington 16175
National Urban League, New York 16176
National Vocational Agricultural Teachers' Association, Alexandria 16177
National Vocational Technical Educational Foundation, Washington 16178
National Water Resources Association, Arlington 16179
National Waterways Transport Association, London 12617
National Wildlife Federation, Washington 16180
National Women's Health Network, Washington 16181
National Women's Studies Association, College Park 16182
National Writers Club, Aurora 16183
National Writing Project, Berkeley 16184
National Young Writers Association of China, Taipei 03316
National Zoological Association of Great Britain, Bury Saint Edmunds 12618
Nation's Report Card – National Assembly of Educational Progress, Princeton 16185
Native Plants Preservation Society of Victoria, Toorak 00439
Native Seeds/Search, Tucson 16186
NATMH, Aylesbury 12572
NATS, Jacksonville 15911
Natural Environment Research Council, Swindon 12619
Natural History Society of Northumbria, Newcastle-upon-Tyne 12620
Natural History Society of Prince Edward Island, Charlottetown 02674
Les Naturalistes Belges, Bruxelles 01414
Naturalistes Parisiens, Paris 04603
Natural Products Chemistry Research Group, Johannesburg 10157
Natural Resources Conservation League of Victoria, Springvale South 00440
Natural Resources Defense Council, New York 16187
The Natural Resources Research Organization, Jerusalem 07109
Natural Sciences and Engineering Research Council of Canada, Ottawa 02675
The Nature Conservancy, Arlington 16188
Nature Conservancy of Canada, Toronto 02676
Naturforschende Gesellschaft Bamberg e.V., Viereth-Trunstadt 06127
Naturforschende Gesellschaft des Kantons Glarus, Glarus 10766
Naturforschende Gesellschaft Freiburg, Freiburg 06128
Naturforschende Gesellschaft in Basel, Riehen 10767
Naturforschende Gesellschaft in Bern, Bern 10768
Naturforschende Gesellschaft in Zürich, Forch 10769
Naturforschende Gesellschaft Luzern, Luzern 10770
Naturforschende Gesellschaft Schaffhausen, Schaffhausen 10771
Naturforschende Gesellschaft Solothurn, Brugglen 10772
Naturhistorische Gesellschaft Hannover, Hannover 06129
Naturhistorische Gesellschaft Nürnberg e.V., Nürnberg 06130
Naturhistorischer Verein der Rheinlande und Westfalens, Bonn 06131
Naturhistorisk Forening for Jylland, Arhus 03775
Naturhistorisk Forening for Nordsjaelland, Hillerød 03776
Naturschutzbund Deutschland e.V., Bonn 06132
Naturvetenskapliga Forskningsrådet, Stockholm 10504
Naturwissenschaftliche Gesellschaft Winterthur, Wiesendangen 10773
Naturwissenschaftlicher und Historischer Verein für das Land Lippe e.V., Detmold 06133
Naturwissenschaftlicher Verein für das Fürstentum Lüneburg von 1851 e.V., Lüneburg 06134
Naturwissenschaftlicher Verein für Kärnten, Klagenfurt 00712
Naturwissenschaftlicher Verein für Steiermark, Graz 00713
Naturwissenschaftlicher Verein in Hamburg, Hamburg 06135
Naturwissenschaftlicher Verein zu Bremen, Bremen 06136
Naturwissenschaftlich-medizinischer Verein in Innsbruck, Innsbruck 00714
Naturwissenschaftlich-Medizinische Vereinigung in Salzburg, Salzburg 00715
Natuurhistorisch Genootschap in Limburg, Maastricht 08979
The Nautical Institute, London 12621
NAUW, Tallahassee 15913
NAV, Köln 06137, 06213
NAVA, Fullerton 15822
NAVA, Trenton 16227
NAVEA, Milan 15782
NAVESNP, McKeesport 15914
NAVF, Oslo 09281
NAVp, Berlin 06214
NAVS, London 12555
NAV-Virchowbund, Köln 06137
Navy Records Society, London 12622
NAW, Berlin 06217
NAWA, Oklahoma City 15778
NAWE, Washington 15823
NAWL, Chicago 15915
NBBI, 's-Gravenhage 08981
NBCDL, Alexandria 15923
NBD, København 03781
NBEA, Reston 15930
NBEO, Washington 15925
NBER, Cambridge 15927
NBF, Oslo 09293, 09294
NBIE, Ann Arbor 15928
NBMBAA, Chicago 15919
NBMC, Ann Arbor 15920
NBME, Philadelphia 15926
NBRC, Lenexa 15924
NBSS, Edinburgh 12577
NBVF, Orange 15929
NBWHP, Atlanta 15921
NCA, Bethesda 15951
NCAA, Washington 15949
NCACS, Summertown 15956
NCACS, Tempe 16228
NCADD, New York 16016
NCAER, New Delhi 06842
NCAIR, New York 15939
NCALHBCU, Washington 15982
NCALL, Dover 16015
NCAR, San Antonio 15976
NCASI, New York 16012
NCAT, Butte 15938
NCATE, Washington 15984
NCBEA, Emporia 15933
NCBEC, Washington 15940
NCBIAE, Albuquerque 15997
NCBJ, Rockville 15967
NCBL, New York 15968
NCBPD, Louisville 15980
NCBS, Columbus 15985
NCC, Plainview 15931
NCCA, Albany 15959
NCCA, New York 15986
NCCCD, Santa Cruz 15941
NCCD, San Francisco 16017
NCCE, Boston 15960
NCCED, Washington 15978
NCCIJ, Washington 15934
NCCLS, Villanova 15961
NCCMHS, Washington 15981
NCCSS, River Falls 16229
NCCUSL, Chicago 15969
NCDA, Houston 15957
NCDC, Macon 16028
NCDS, Albertson 15942
NCEA, Alexandria 15966
NCEA, Washington 15935
NCEB, Louisville 15987
NCEC, Redhill 12580
NCEE, Clemson 15998
NCEE, New York 16018
NCEP, Washington 16019
NCFAE, Augusta 15999
NCFJE, Brooklyn 15964
NCFL, Milford 15936
NCFP, Milwaukee 15974
NCFR, Minneapolis 16020
NCFS, Woodland Hills 15958
NCGA, Indiana 15988
NCGR, Berkeley 16021
NCH, Alexandria 15944
NCHLS, Durham 16022
NCHM, Washington 15945
NCI, Falls Church 15932
NCIA, Alexandria 15950
NCIO, Rheem Valley 15979
NCIPLA, Arlington 16001
NCIS, Davenport 16000
NCJFCJ, Reno 16002
NCL, El Paso 15952
NCL, Washington 15953
NCLA, Brooklyn 16003
NCLC, Boston 15983
NCLD, Washington 15946
NCLG, Cypress 15962
NCLR, Chicago 15989
NCME, Washington 16023
NCNE, London 12578
NCPERL, Washington 15954
NCPG, Saint Louis 15937
NCPH, Indianapolis 16024
NCPI, Louisville 16027
NCQHC, Washington 15963
NCRD, Jerusalem 07108
NCRE, Amherst 15975
NCRP, Radford 15990
NCRPE, Ames 16026
NCRPM, Bethesda 16025
NCRUCE, Nashville 15970
NCSBCS, Herndon 15972
NCSC, Williamsburg 15947
NCSEA, Washington 16005
NCSL, Boulder 15971
NCSM, Golden 16009
NCSPE, Chapel Hill 16006
NCSS, Washington 15992
NCSSFL, Little Rock 16007
NCSSM, Columbus 16008
NCTA, Silver Spring 15993
NCTE, Charlottesville 15991
NCTE, New York 15994
NCTE, Urbana 16010
NCTIP, Washington 15965
NCTM, Reston 16011
NCUEA, Washington 16014
NCURA, Washington 16013
NCWGE, Washington 15955
NCWM, Gaithersburg 15977
NCYL, San Francisco 15948
NCYP, New York 15973
NCYRE, San Diego 15824
NDA, Portland 16033
NDA, Washington 16031
NDAA, Alexandria 16035
NDAA, Washington 16030
NDBC, Washington 16036
NDC, Rosemont 16029
NDDIC, Bethesda 16034
NDHA, Washington 16032
NDR, Berlin 06160
NDV, Wageningen 08991
NEA, Ann Arbor 16038
NEA, Tewin Wood 12587
NEA, Washington 16039
NEARA, Kingston 16195
Near East Archaeological Society, Saint Louis 16189
Near East College Association, New York 16190
NEAS, Saint Louis 16189
NEASC, Winchester 16196
NEC, Chicago 16042
NEC, South Burlington 16230
NECA, New York 16190
Nederlands Bibliotheek en Lektuur Centrum, 's-Gravenhage 08980
Nederlands Bureau voor Bibliotheekwezen en Informatieverzorging, 's-Gravenhage 08981
Nederlandsche Internisten Vereeniging, Utrecht 08982
Nederlandsche Maatschappij tot Bevordering der Tandheelkunde, Nieuwegein 08983
Nederlandsche Vereniging voor Druk- en Boekkunst, Haarlem 08984
Nederlandsche Vereniging voor Levensverzekeringgeneeskunde, Rotterdam 08985
Nederlandse Anatomen Vereniging, Amsterdam 08986
Nederlandse Astronomenclub, Leiden 08987
Nederlandse Bond voor Natuurgeneeswijze, Amsterdam 08988
Nederlands Economisch Instituut, Rotterdam 08989
Nederlandse Dendrologische Vereniging, Boskoop 08990
Nederlandse Dierkundige Vereniging, Wageningen 08991
Nederlandse Entomologische Vereniging, Amsterdam 08992
Nederlandse Genealogische Vereniging, Amsterdam 08993
Nederlandse Genetische Vereniging, Leiden 08994
Nederlandse Museumvereniging, Enkhuizen 08995
Nederlandse Mycologische Vereniging, Wyster 08996
Nederlandse Organisatie voor Internationale Samenwerking in het Hoger Onderwijs, 's-Gravenhage 08997
Nederlandse Organisatie voor Wetenschappelijk Onderzoek, 's-Gravenhage 08998
Nederlandse Ornithologische Unie, Odijk 08999
Nederlandse Patholoog Anatomen Vereniging, Maastricht 09000
Nederlandse Sint-Gregorius Vereniging ter Bevordering van Liturgische Muziek, Utrecht 09001
Nederlandse Toonkunstenaarsraad, Amsterdam 09002
Nederlandse Tuinbouwraad, 's-Gravenhage 09003
Nederlandse Vereniging van Artsen voor Revalidatie en Physische Geneeskunde, Groningen 09004
Nederlandse Vereniging van Bibliothecarissen, Documentalisten en Literatuuronderzoekers, Schelluinen 09005
Nederlandse Vereniging van Gieterijtechnici, Zoetermeer 09006
Nederlandse Vereniging van Laboratoriumartsen, Leeuwarden 09007
Nederlandse Vereniging van Neurochirurgen, Rotterdam 09008
Nederlandse Vereniging van Pedagogen, Onderwijskundigen en Andragologen, Utrecht 09009
Nederlandse Vereniging van Radiologische Laboranten, Utrecht 09010
Nederlandse Vereniging van Specialisten in den Dento-Maxillaire Orthopaedie, Nijmegen 09011
Nederlandse Vereniging van Tandartsen, Maarssen 09012
Nederlandse Vereniging van Wiskundeleraren, 's-Gravenhage 09013
Nederlandse Vereniging voor Afvalwaterbehandeling en Waterkwaliteitsbeheer, Rijswijk 09014
Nederlandse Vereniging voor Algemene Gezondheidszorg, Rotterdam 09015
Nederlandse Vereniging voor Anesthesiologie, Utrecht 09016
Nederlandse Vereniging voor Cardiologie, Nieuwegein 09017
Nederlandse Vereniging voor Gastro-Enterologie, Haarlem 09018
Nederlandse Vereniging voor Geodesie, Apeldoorn 09019
Nederlandse Vereniging voor Heelkunde, Utrecht 09020
Nederlandse Vereniging voor Internationaal Recht, 's-Gravenhage 09021
Nederlandse Vereniging voor Kindergeneeskunde, Amsterdam 09022
Nederlandse Vereniging voor Logica en Wijsbegeerte der Exacte Wetenschappen, Utrecht 09023
Nederlandse Vereniging voor Luchtvaarttechniek, Amsterdam 09024
Nederlandse Vereniging voor Management, 's-Gravenhage 09025
Nederlandse Vereniging voor Medisch Onderwijs, Utrecht 09026
Nederlandse Vereniging voor Microbiologie, Bilthoven 09027
Nederlandse Vereniging voor Mondziekten en Kaakchirurgie, 's-Gravenhage 09028
Nederlandse Vereniging voor Neurologie, Utrecht 09029
Nederlandse Vereniging voor Obstetrie en Gynaecologie, Utrecht 09030
Nederlandse Vereniging voor Orthodontische Studie, Gorinchem 09031
Nederlandse Vereniging voor Parasitologie, Nijmegen 09032
Nederlandse Vereniging voor Pathologie, Rotterdam 09033
Nederlandse Vereniging voor Personeelbeleid, Utrecht 09034
Nederlandse Vereniging voor Produktieleiding, 's-Gravenhage 09035
Nederlandse Vereniging voor Psychiatrie, Utrecht 09036
Nederlandse Vereniging voor Psychotherapie, Utrecht 09037
Nederlandse Vereniging voor Thoraxchirurgie, Groningen 09038
Nederlandse Vereniging voor Toegepaste Taalwetenschap, Utrecht 09039
Nederlandse Vereniging voor Tropische Geneeskunde, 's-Gravenhage 09040
Nederlandse Vereniging voor Urologie, Utrecht 09041
Nederlandse Vereniging voor Veiligheidskunde, Amsterdam 09042
Nederlandse Vereniging voor Zeegeschiedenis, Leiden 09043
Nederlandse Vereniging voor Dermatologie en Venereologie, Breda 09044
Nederlandse Werkgroep van Praktizijns in de Natuurlijke Geneeskunst, Leusden 09045
Nederlandse Zootechnische Vereniging, Wassenaar 09046

Nederlands Filosofisch Genootschap, Zeist 09047
Nederlands Genootschap van Leraren, Dordrecht 09048
Nederlands Genootschap voor Anthropologie, Amsterdam 09049
Nederlands Genootschap voor Fysiotherapie, Amersfoort 09050
Nederlands Historisch Genootschap, 's-Gravenhage 09051
Nederlands Huisartsen Genootschap, Utrecht 09052
Nederlands Instituut voor Internationale Betrekkingen Clingendael, 's-Gravenhage 09053
Nederlands Instituut voor Marketing, Amsterdam 09054
Nederlands Oogheelkundig Gezelschap, Maastricht 09055
Nederlands Psychoanalytisch Genootschap, Utrecht 09056
Nederlands-Zuidafrikaanse Vereniging, Amsterdam 09057
Nederlanedse Orthopaedische Vereniging, Nijmegen 09058
NEHA, Denver 16043
NEHGS, Boston 16197
Neonatal Society, Oxford 12623
Nepal Studies Association, Oakland 16191
Nephrologischer Arbeitskreis Saar-Pfalz-Mosel e.V., Kaiserslautern 06138
NERC, Princeton 16217
NERC, Swindon 12619
NERO, Washington 16041
NESTA, Fargo 16037
Netherlands Agronomic-Historical Foundation, Groningen 09074, 09075
Netherlands Association for the Philosophy of Law, Amstelveen 09106
Netherlands Branch of the International Law Association, 's-Gravenhage 09021
Netherlands Cancer Institute, Amsterdam 09085
Netherlands Centre of the International PEN, Maastricht 09059
Netherlands Economic Institute, Rotterdam 08989
Netherlands Film and TV Academy, Hilversum 09076
Netherlands Foundation for the Advancement of Tropical Research (WOTRO), Paramaribo 10433
Netherlands Institute for Art History, 's-Gravenhage 09066
Netherlands Institute of International Relations Clingendael, 's-Gravenhage 09053
Netherlands Institute of Marketing, Amsterdam 09054
Netherlands Organisation for International Cooperation in Higher Education, 's-Gravenhage 08997
Netherlands Organization for Libraries and Information Services, 's-Gravenhage 08981
Netherlands Organization for Scientific Research, 's-Gravenhage 08998
Netherlands Society for Microbiology, Bilthoven 09027
The Netherlands Society for Nature and Environment, Utrecht 09073
Netherlands Society for the Art of Printing and Bookproduction, Haarlem 08984
Netherlands Society on Safety Engineering, Amsterdam 09042
Netherlands South African Society, Amsterdam 09057
The Network, Cleveland 16192
Neue Bachgesellschaft e.V., Leipzig 06139
Neue Kriminologische Gesellschaft, Tübingen 06140
Neue Schweizerische Chemische Gesellschaft, Basel 10774
Neumann János Számítógéptudományi Társaság, Budapest 06666
Neurosurgical Society of America, Cleveland 16193
NEV, Amsterdam 08992
Newark and Nottinghamshire Agricultural Society, Newark 12624
New Brunswick Association of Pathologists, Saint John 02677
New Brunswick Dental Society, Rothesay 02678
New Brunswick Dietetic Association, Saint John 02679
New Brunswick Federation of Naturalists, Saint John 02680
New Brunswick Genealogical Society, Fredericton 02681
New Brunswick Gerontology Association, Saint John 02682
New Brunswick Health Records Association, Saint John 02683
New Brunswick Heart Foundation, Saint John 02684
New Brunswick Historical Society, Saint John 02685
New Brunswick Institute of Agrologists, Fredericton 02686
New Brunswick Lung Association, Fredericton 02687
New Brunswick Medical Society, Fredericton 02688
New Brunswick Music Educators' Association, Quispamsis 02689
New Brunswick Psychiatric Association, Saint John 02690
New Brunswick Research and Productivity Council, Fredericton 02691
New Brunswick Safety Council, Fredericton 02692
New Brunswick School Superintendents Association, Rexton 02693
New Brunswick Society of Occupational Therapists, Fredericton 02694
New Brunswick Speech and Hearing Association, Fredricton 02695
New Brunswick Teachers' Federation, Fredericton 02696
New Brunswick Veterinary Medical Association, Fredericton 02697
The Newcomen Society, London 12625
Newcomen Society of the United States, Exton 16194
New England Antiquities Research Association, Kingston 16195
New England Association of Schools and Colleges, Winchester 16196
New England Historic Genealogical Society, Boston 16197
New English Art Club, London 12626
Newfoundland and Labrador Association of Occupational Therapists, Saint John's 02698
Newfoundland and Labrador Drama Society, Saint John's 02699
Newfoundland and Labrador Veterinary Medical Association, Mount Pearl 02700
Newfoundland Association of Medical Radiation Technologists, Saint John's 02701
Newfoundland Association of Social Workers, Saint John's 02702
Newfoundland Dietetic Association, Saint John's 02703
Newfoundland Forest Protection Association, Corner Brook 02704
Newfoundland Historical Society, Saint John's 02705
Newfoundland Lung Association, Saint John's 02706
Newfoundland Medical Association, Saint John's 02707
Newfoundland Medical Board, Saint John's 02708
Newfoundland Medical Council, Saint John's 02709
Newfoundland Music Educators' Association, Corner Brook 02710
Newfoundland Speech and Hearing Association, Saint John's 02711
Newfoundland Teachers' Association, Saint John's 02712
Newfoundland Thoracic Society, Saint John's 02713
The New Philosophical Society, Istanbul 11179
Newspaper Research Centre, Toronto 02714
New World Foundation, New York 16198
New York Academy of Sciences, New York 16199
New York Browning Society, New York 16200
New York C.S. Lewis Society, Jamaica 16201
New York Drama Critics Circle, New York 16202
New York Financial Writers' Association, Syosset 16203
New York Genealogical and Biographical Society, New York 16204
New York Pigment Club, Clifton 16205
New Zealand Academy of Fine Arts, Wellington 09132
New Zealand Archaeological Association, Dunedin North 09133
New Zealand Association of Clinical Biochemists, Wellington 09134
New Zealand Association of Scientists, Wellington 09135
New Zealand Association of Soil Conservators, Blenheim 09136
New Zealand Biochemical Society, Canterbury 09137
New Zealand Book Council, Wellington 09138
New Zealand Cartographic Society, Thorndon 09139
New Zealand Computer Society, Wellington 09140
New Zealand Council for Educational Research, Wellington 09141
New Zealand Dairy Technology Society, Waitoa 09142
New Zealand Dietetic Association, Wellington 09143
New Zealand Ecological Society, Christchurch 09144
New Zealand Electronics Institute, Auckland 09145
New Zealand Fertiliser Manufacturers Research Association, Auckland 09146
New Zealand Genetical Society, Christchurch 09147
New Zealand Geographical Society, Christchurch 09148
New Zealand Geophysical Society, Wellington 09149
New Zealand Historical Association, Christchurch 09150
New Zealand Historic Places Trust, Wellington 09151
New Zealand Hydrological Society, Wellington 09152
New Zealand Institute of Agricultural Science, Christchurch 09153
New Zealand Institute of Architects, Wellington 09154
New Zealand Institute of Chemistry, Wellington 09155
New Zealand Institute of Food Science and Technology, Christchurch 09156
New Zealand Institute of Forestry, Christchurch 09157
New Zealand Institute of International Affairs, Wellington 09158
The New Zealand Institute of Management, Wellington 09159
New Zealand Institute of Physics, Auckland 09160
New Zealand Institute of Surveyors, Wellington 09161
New Zealand Law Society, Wellington 09162
New Zealand Leather and Shoe Research Association, Palmerston North 09163
New Zealand Library Association, Wellington 09164
New Zealand Limnological Society, Rotorua 09165
New Zealand Maori Arts and Crafts Institute, Rotorua 09166
New Zealand Marine Sciences Society, Wellington 09167
New Zealand Mathematical Society, Dunedin 09168
New Zealand Medical Association, Wellington 09169
New Zealand Microbiological Society, Dunedin 09170
The New Zealand National Society for Earthquake Engineering, Wellington 09171
New Zealand Pottery and Ceramics Research Association, Lower Hutt 09172
New Zealand Psychological Society, Wellington 09173
New Zealand Society for Electron Microscopy, Auckland 09174
New Zealand Society for Horticultural Science, Havelock North 09175
New Zealand Society for Parasitology, Hamilton 09176
New Zealand Society of Animal Production, Hamilton 09177
New Zealand Society of Dairy Science and Technology, Edgecumbe 09178
New Zealand Society of Plant Physiologists, Palmerston North 09179
New Zealand Society of Soil Science, Canterbury 09180
New Zealand Statistical Association, Wellington 09181
New Zealand Veterinary Association, Wellington 09182
N.E.Z.S., Chester 12640
NF, Cambridge 16206
NF, Oslo 09308
NFA, Westerville 16054
NFAA, Miami 16056
NFAIS, Philadelphia 16045
NFBR, Washington 16057
NFCPG, Elm Grove 16046
NFCRC, Columbia 16052
NFCTA, Bexley 12588
NFDH, Denver 16060
NFER, Slough 12591
NFERF, Arlington 16053
NFETA, Glasgow 12592
NFF, Oslo 09298, 09300
NFFAO, Alexandria 16050
NFFR, Trondheim 09282
NFIMA, Kansas City 16044
NFIR, Oslo 09309
NFL, Ripon 16055
NFME, Boston 16061
NFMLTA, Omaha 16047
N.F.N.C., Lusaka 17155
NFNID, Princeton 16059
NFPA, Quincy 16051
NFS, Kjeller 09314
NFSCSW, Arlington 16048
NFSPS, Ogden 16049
NFSS, Lymington 12589
NFVLS, London 12590
NG, Berlin 06174
NGB, Bern 10768
NGCP, Ellicott City 16066
NGCSA, Englewood 16067
NGF, Oslo 09318
NGF, Trondheim 09316
NGG, Glarus 10766
NGL, Dordrecht 09048
NGL, Köln 06172
NGL, Luzern 10770
NGMA, Washington 16065
NGPT, Austin 16068
NGRS, Huddersfield 12553
NGS, Arlington 16062
NGS, Berlin 06173
NGS, Eustis 16064
NGS, Washington 16063
NGU, Trondheim 09283
N.G.v.F., Amersfoort 09050
NGZ, Forch 10769
NH, Frankfurt 06176
NHA, Washington 16081
NHAS, Livonia 16073
NHC, Research Triangle Park 16080
NHC, Washington 16069, 16078
NHCA, Des Moines 16074
NHeLP, Los Angeles 16071
NHES, Leesburg 16079
NHF, Monrovia 16070
NHF, Nordnes 09301
NHG, Hannover 06129
NHG, Maastricht 08979
NHG, Nürnberg 06130
N.H.G., 's-Gravenhage 09051
N.H.G., Utrecht 09052
NHLA, Washington 16072
NHPP, Rockville 16077
NHR, Berlin 06177
NHRP, Baltimore 16075
NHS, Brunswick 15769
NHSC, Washington 16076
NHTPC, London 12593
NHW, Berlin 06175
NI, Berlin 06179
NIA, New York 16237
NIAB, Cambridge 12597
NIACE, Leicester 12596
NIAS, København 03785
NIBM, Ann Arbor 16088
NIBR, Oslo 09321
NICE, Westerville 16093
NICEM, Albuquerque 16086
NICET, Alexandria 16089
Nichibei Hogakkai, Tokyo 08185
Nicholas Roerich Society, London 12627
NICOD, Belfast 12628
NICRA, Belfast 12637
NID, Ahmedabad 06845
NIDOC, Cairo 03910
NIEA, Washington 16084
Niederösterreichisches Bildungs- und Heimatwerk, Arbeitsgemeinschaft für Volkskunde, Wien 00716
Niedersächsische Krebsgesellschaft, Hannover 05107
Niedersächsischer Bund für freie Erwachsenenbildung e.V., Hannover 06141
Niedersächsischer Landesverband der Heimvolkshochschulen e.V., Hannover 06142
Nieman Foundation, Cambridge 16206
NIESR, London 12598
NIFS, Columbia 16090
Nigeria Educational Research Council, Lagos 09233
Nigerian Academy of Science, Akoka 09234
Nigerian Bar Association, Warri 09235
Nigerian Economic Society, Ibadan 09236
Nigerian Geographical Association, Ibadan 09237
Nigerian Institute of International Affairs, Lagos 09238
Nigerian Institute of Management, Lagos 09239
Nigerian Library Association, Lagos 09240
Nigerian Society for Microbiology, Ibadan 09241
Nigerian Veterinary Medical Association, Vom 09242
Nihon Aisotopu Kyokai, Tokyo 08186
Nihon Arerugi Gakkai, Tokyo 08187
Nihon Bearingu Kogyokai, Tokyo 08188
Nihon Bunseki Kagaku-Kai, Tokyo 08189
Nihon Chugoku Gakkai, Tokyo 08190
Nihon Denpun Gakkai, Ibaraki 08191
Nihon Denshi Kogyo Shinko Kyokai, Tokyo 08192
Nihon Dobutsu Shinri Gakkai, Ibaraki 08193
Nihon Dokubun Gakkai, Tokyo 08194
Nihon Do Senta, Tokyo 08195
Nihon Eibungakkai, Tokyo 08196
Nihon Eisei Gakkai, Tokyo 08197
Nihon Esperanto Gakkai, Tokyo 08198
Nihon Furansu-go Furansu-bun-gaku-kai, Tokyo 08199
Nihon Gakujutsu Shinko-kai, Tokyo 08200
Nihon Gas Kyokai Chosabu Chosaka, Tokyo 08201
Nihon Gengogakkai, Tokyo 08202
Nihon Genshiryoku Gakkai, Tokyo 08203
Nihon Gomu Kyokai, Tokyo 08204
Nihon Hakubutsukan Kyokai, Tokyo 08205
Nihon Hoken Gakkai, Tokyo 08206
Nihon Hoken Igakkai, Tokyo 08207
Nihon Hozon Shika Gakkai, Tokyo 08208
Nihon Iden Gakkai, Shizuoka 08209
Nihon Ikushu Gakkai, Tokyo 08210
Nihon Indogaku Bukkyogaku Kai, Tokyo 08211
Nihon Ishi-Kai, Tokyo 08212
Nihon Ishinkin Gakkai, Tokyo 08213
Nihon Jui Gakkai, Tokyo 08214
Nihon Junkanki Gakkai, Kyoto 08215
Nihon Kagaku Gijutsu Joho Sentah, Tokyo 08216
Nihon Kairui Gakkai, Tokyo 08217
Nihon Kakuigakukai, Tokyo 08218
Nihon Kandenchi Kogyokai, Tokyo 08219
Nihon Kasai Gakkai, Tokyo 08220
Nihon Kensetsu Kikaika Kyokai, Tokyo 08221
Nihon Kikai Gakkai, Tokyo 08222
Nihon Kikaku Kyokai Gakkoku Kikaku Raiburari, Tokyo 08223, 08223
Nihon Kin Gakkai, Tokyo 08224
Nihon Kirisutokyo Gakkai, Tokyo 08225
Nihon Kobutsu Gakkai, Tokyo 08226
Nihon Koho Gakkai, Tokyo 08227
Nihon Kokogaku Kyokai, Tokyo 08228
Nihon Koko Geka Gakkai, Tokyo 08229
Nihon Koku Eisei Gakkai, Tokyo 08230
Nihon Kokukai Gakkai, Tokyo 08231
Nihon Kokusai Kyoiku Kyokai, Tokyo 08232
Nihon Kotsu Igakkai, Tokyo 08233
Nihon Kotsu Kyokai Toshokan, Tokyo 08234
Nihon Kuho Gakkai, Tokyo 08235
Nihon Kyobu Geka Gakkai, Tokyo 08236
Nihon Kyoiku Gakkai, Tokyo 08237
Nihon Kyoiku-shakai Gakkai, Tokyo 08238
Nihon Kyoiku Shinri Gakkai, Tokyo 08239
Nihon Kyosei Shikagakkai, Tokyo 08240
Nihon Masui Gakkai, Tokyo 08241
Nihon Myakkan Gakkai, Tokyo 08242
Nihon Naika Gakkai, Tokyo 08243
Nihon Nettai Igakkai, Nagasaki 08244
Nihon Noshinkei Geka Gakkai, Tokyo 08245
Nihon Oyo Shinri-gakkai, Tokyo 08246
Nihon PEN Kurabu, Tokyo 08247
Nihon Reito Kyokai, Tokyo 08248
Nihon Rinrigakukai, Tokyo 08249
Nihon Rinsho Byori Gakkai, Tokyo 08250
Nihon Ronen Igakukai, Tokyo 08251
Nihon Sakumotsu Gakkai, Tokyo 08252
Nihon Sangyo Eisei Gakkai, Tokyo 08253
Nihon Sanshi Gakkai, Ibaraki 08254
Nihon Seibutsu Butsuri Gakkai, Osaka 08255
Nihon Seibutsu Kankyo Chosetsu Kenkyukai, Tokyo 08256
Nihon Seishin Shinkei Gakkai, Tokyo 08257
Nihon Sen'i Kikai Gakkai, Osaka 08258
Nihon Setchakuzai Kogyokai Toshoshitsu, Tokyo 08259
Nihon Shakai Shinri Gakkai, Tokyo 08260
Nihon Shinku Kyokai, Tokyo 08261
Nihon Shinrigakkai, Tokyo 08262
Nihon Shokaki Naishikyo Gakkai, Tokyo 08263
Nihon Shoni Geka Gakkai, Tokyo 08264
Nihon Shonika Gakkai, Tokyo 08265
Nihon Shukyo Gakkai, Tokyo 08266
Nihon Tairyoku Igakkai, Tokyo 08267
Nihon Tenmon Gakkai, Tokyo 08268
Nihon Tetsugakkai, Tokyo 08269
Nihon Tokei Gakkai, Tokyo 08270
Nihon Toshokan Kyokai, Tokyo 08271
Nihon Uirusu Gakkai, Tokyo 08272
Nihon Yakuri Gakkai, Tokyo 08273
Nihon Yosetsu Kyokai, Tokyo 08274
Nihon Zairyo Gakkai, Kyoto 08275
Nihon Zairyo Kyodo Gakkai, Sendai 08276
Nihon Zaisei Gakkai, Tokyo 08277
Nihon Zeiho Gakkai, Kyoto 08278
Nihon Zosen Gakkai, Tokyo 08279
NIIP, Slough 12599
Nikola Tesla Association of Societies for Promotion of Technical Sciences in Yugoslavia, Beograd 17105
NILC, Los Angeles 16083
NIM, Lagos 09239
NIMA, Amsterdam 09054
NIMC, Great Neck 16094
NIMH, Exeter 12600
NIN, Köln 06180
NIP, Washington 16082
NIPF, Belfast 12638
NIPP, Fairfax 16091
Nippon Afurika Gakkai, Tokyo 08280
Nippon Bitamin Gakkai, Kyoto 08281
Nippon Bungaku Kyokai, Tokyo 08282
Nippon Bunko Gakkai, Tokyo 08283
Nippon Butsuri Gakkai, Tokyo 08284
Nippon Butsuri-Kagaku Kenkyukai, Kyoto 08285
Nippon Byori Gakkai, Tokyo 08286
Nippon Chikusan Gakkai, Tokyo 08287
Nippon Chikyu Denki Ziki Gakkai, Tokyo 08288
Nippon Chiri Gakkai, Tokyo 08289
Nippon Chishitsu Gakkai, Tokyo 08290
Nippon Chô Gakkai, Tokyo 08291
Nippon Dai-Yonki Gakkai, Tokyo 08292
Nippon Denki Kyokai Chosabu Chosaka, Tokyo 08293
Nippon Denshi Kenbikyo Gakkai, Tokyo 08294
Nippon Dobutsu Gakkai, Tokyo 08295
Nippon Dokumenesyon Kyokai, Tokyo 08296
Nippon Dokyo Gakkai, Tokyo 08297
Nippon Engeki Gakkai, Tokyo 08298
Nippon Gan Gakkai, Tokyo 08299

Nippon Ganseki Kobutsu Kosho Gakkai, Sendai *08300*
Nippon Geka Gakkai, Tokyo *08301*
Nippon Gyosei Gakkai, Tokyo *08302*
Nippon Hifuka Gakkai, Tokyo *08303*
Nippon Hikaku Bungakukai, Tokyo *08304*
Nippon Hinyoki-ka Gakkai, Tokyo *08305*
Nippon Hoi Gakkai, Tokyo *08306*
Nippon Hoshakai Gakkai, Tokyo *08307*
Nippon Hoshasen Eikyo Gakkai, Chiba-shi *08308*
Nippon Hotetsu Gakkai, Kyoto *08309*
Nippon Hotetsu Shika Gakkai, Tokyo *08310*
Nippon Igaku Hoshasen Gakkai, Tokyo *08311*
Nippon Imono Kyokai, Tokyo *08312*
Nippon Jibi-Inkoka Gakkai, Tokyo *08313*
Nippon Jidoseigyo Kyokai, Kyoto *08314*
Nippon Jinruigaku Kai, Tokyo *08315*
Nippon Jinrui Iden Gakkai, Tokyo *08316*
Nippon Junkatsu Gakkai, Tokyo *08317*
Nippon Kagakukai, Tokyo *08318*
Nippon Kagaku Kyoiku Gakukai, Tokyo *08319*
Nippon Kaibo Gakkai, Tokyo *08320*
Nippon Kaiho Gakkai, Tokyo *08321*
Nippon Kaisui Gakkai, Tokyo *08322*
Nippon Kaiyo Gakkai, Tokyo *08323*
Nippon Kango Kyokai, Tokyo *08324*
Nippon Kazan Gakkai, Tokyo *08325*
Nippon Keiei Gakkai, Tokyo *08326*
Nippon Keiho Gakkai, Tokyo *08327*
Nippon Keizai Seisaku Gakkai, Tokyo *08328*
Nippon Kekkaku-byo Gakkai, Tokyo *08329*
Nippon Kessho Gakkai, Tokyo *08330*
Nippon Ketsueki Gakkai, Kyoto *08331*
Nippon Kikan-Shokudo-ka Gakkai, Tokyo *08332*
Nippon Kinzoku Gakkai, Sendai *08333*
Nippon Kisei-chu Gakkai, Tokyo *08334*
Nippon Kisho Gakkai, Tokyo *08335*
Nippon Kogakukai, Tokyo *08336*
Nippon Kokai Gakkai, Tokyo *08337*
Nippon Koku Gakkai, Tokyo *08338*
Nippon Kokusai Seiji Gakkai, Tokyo *08339*
Nippon Koku Uchu Gakkai, Tokyo *08340*
Nippon Kontyu Gakkai, Tokyo *08341*
Nippon Koseibutsu Gakkai, Tokyo *08342*
Nippon Koshu-Eisei Kyokai, Tokyo *08343*
Nippon Mokuzai Gakkai, Tokyo *08344*
Nippon Nensho Gakkai, Tokyo *08345*
Nippon Nensho Kenkyukai, Kyoto *08346*
Nippon Netsushori Gijutsu Kyokai, Tokyo *08347*
Nippon Nogei Kagaku Kai, Tokyo *08348*
Nippon Nogyo-Kisho Gakkai, Tokyo *08349*
Nippon Noritsu Kyokai Shiryoshitsu, Tokyo *08350*
Nippon No-Shinkei Gek Gakkai, Tokyo *08351*
Nippon Ongaku Gakkai, Tokyo *08352*
Nippon Onkyo Gakkai, Tokyo *08353*
Nippon Orient Gakkai, Tokyo *08354*
Nippon Oyo-Doboku-Konchu Gakkai, Tokyo *08355*
Nippon Rai Gakkai, Tokyo *08356*
Nippon Rikusui Gakkai, Tokyo *08357*
Nippon Ringakukai, Tokyo *08358*
Nippon Rodo-ho Gakkai, Tokyo *08359*
Nippon Rosiya Bungakkai, Tokyo *08360*
Nippon Saikingakkai, Tokyo *08361*
Nippon Sanka-Fujinka Gakkai, Tokyo *08362*
Nippon Seikei Geka Gakkai, Tokyo *08363*
Nippon Seiri Gakkai, Tokyo *08364*
Nippon Seitai Gakkai, Sendai *08365*
Nippon Seiyo Koten Gakkai, Kyoto *08366*
Nippon Seiyoshi Gakkai, Osaka *08367*
Nippon Seizi Gakkai, Tokyo *08368*
Nippon Seramikkusu Kyokai, Tokyo *08369*
Nippon Shakai Gakkai, Tokyo *08370*
Nippon Shashin Gakkai, Tokyo *08371*
Nippon Shashin Sokuryo Gakkai, Tokyo *08372*
Nippon Shiho Gakkai, Tokyo *08373*
Nippon Shika Hoshasen Gakkai, Tokyo *08374*
Nippon Shika Hozon Gakkai, Tokyo *08375*
Nippon Shika Igakkai, Tokyo *08376*
Nippon Shimbun Gakkai, Tokyo *08377*
Nippon Shinkei Gakkai, Tokyo *08378*
Nippon Shinkeikagaku Gakkai, Tokyo *08379*
Nippon Shogyo Gakkai, Tokyo *08380*
Nippon Shokaki-byo Gakkai, Tokyo *08381*
Nippon Shokubutsu-Byori Gakkai, Tokyo *08382*
Nippon Shokubutsu Gakkai, Tokyo *08383*
Nippon Shokubutsu Seiri Gakkai, Kyoto *08384*
Nippon Sokuchi Gakkai, Ibaraki *08385*
Nippon Sugaku Kai, Tokyo *08386*
Nippon Sugaku Kyoiku Gakkai, Tokyo *08387*
Nippon Suisan Gakkai, Tokyo *08388*
Nippon Syoyakugakkai, Tokyo *08389*
Nippon Taiiku Gakkai, Tokyo *08390*
Nippon Teiinoshujutsu Kenkyukai, Tokyo *08391*
Nippon Tekko Kyokai, Tokyo *08392*
Nippon Tonyo-byo Gakkai, Tokyo *08393*
Nippon Toshokan Gakkai, Tokyo *08394*
Nippon Yakugaku-Kai, Tokyo *08395*
Nippon Yukagaku Kyokai, Tokyo *08396*
Nippon Yuketsu Gakkai, Tokyo *08397*
NIRE, Hewitt *16092*
NIS, Nashville *16095*
NISD, Anaheim *16097*
NISEE, Richmond *16087*
NISS, New York *16096*
NISW, London *12595*
NIV, Utrecht *08982*
NIVE, 's-Gravenhage *09025*
NIWTA, Belfast *12639*
NIZC, Columbus *16085*
NJC, Reno *16099*
NJCIRM, Denver *16098*
NKe, Berlin *06183*
NKFO, Helsinki *03985*
NKS, Oslo *09323*
NKT, Berlin *06184*
NLA, Alma *16102*
NLA, Garden City *16101*
NLA, Lagos *09240*
NLAPW, Washington *16100*
NLC, Anchorage *16231*
NLFR, Taastrup *03788*
NLG, Glen Rock *16236*
NLI, Oslo *09327*
NLL, Trondheim *09326*
NMA, Washington *16107*
NMAB, Washington *16105*
NMC, Arlington *16104*
NMC, Englewood *16112*
NMDA, Philadelphia *16106*
NMEA, Pacific Grove *16103*
NMF, New York *16108*
NMF, Oslo *09330*
NMHA, Alexandria *16109*
NMHC, Bedford *12601*
N.M.L.L., Kristiansand *09304*
NMP, Berlin *06191*
NMSA, Columbus *16110*
NMSS, New York *16111*
NMV, Enkhuizen *08995*
N.M.V., Wyster *08996*
NNF, Århus *03789*
N.N.H.S., Northampton *12632*
NNML, Tromsø *09305*
NOA, Columbus *16115*
NOA, New York *16116*
NOA, Shreveport *16114*
Noah Worcester Dermatological Society, Cincinnati *16207*
Nobel Foundation, Stockholm *10505*
Nobelstiftelsen, Stockholm *10505*
NOC, Cambridge *16238*
NOCERCC, Chicago *16117*
Nockian Society, Fort Mitchell *16208*
NODA, London *12602*
NÖG, Köln *06164*
NÖG, Wien *00711*
NOG, Maastricht *09055*
NOGA, Chicago *16123*
Nogyo-Doboku Gakkai, Tokyo *08398*
Nogyo-Ho Gakkai, Tokyo *08399*
Nogyokikai Gakkai, Tokyo *08400*
NOISE, Washington *16121*
NOLA, Killeen *15808*
NOLPE, Topeka *16120*
NOMA, Washington *16119*
NOPRI, Lenox Hill *16122*
NORD, New Fairfield *16118*
Nordforsk, København *03777*
Nordic Association of Applied Geophysics, Luleå *10506*
Nordic Building Conference, København *03781*
Nordic Council for Scientific Information, Espoo *03984*
Nordic Federation for Medical Education, København *03778*
Nordic Institute for Studies in Urban and Regional Planning, Stockholm *10508*
Nordic Institute of Asian Studies, København *03785*
Nordic Road Safety Council, Oslo *09279*
Nordic Society for Cell Biology, København *03779*
Nordiska Afrikainstitutet, Uppsala *10507*
Nordiska Institutet för Samhällsplanering, Stockholm *10508*
Nordisk Akustik Selskab, Lyngby *03780*
Nordisk Anaesthesiologisk Forening, Oslo *09280*
Nordiska Samarbetskommittén för Internationell Politik, inklusive Konfliktoch Fredsforskning, Stockholm *10509*
Nordiska Samfundet för Latinamerikaforskning, Stockholm *10510*
Nordisk Byggedag, København *03781*
Nordisk Cerealist Foreningen, Lyngby *03782*
Nordisk Förening för Tillämpad Geofysik, Stockholm *10511*
Nordisk Forening for Celleforskning, København *03783*
Nordisk Forening for Rettssociologi, København *03784*
Nordisk Genetikerforening, Stockholm *10512*
Nordisk Institut for Asienstudier, København *03785*
Nordisk Institut for Teoretisk Atomfysik, København *03786*
Nordisk Kollegium for Fysisk Oceanografi, Helsinki *03985*
Nordisk Kollegium for Fysisk Oceanografi, København *03787*
Nordisk Laederforskningsråd, Taastrup *03788*
Nordisk Neurokirurgisk Forening, Århus *03789*
Nordisk Numismatisk Union, København *03790*
Nordisk Odontologisk Förening, Helsinki *03986*
Nordisk Statistisk Sekretariat, København *03791*
NORDITA, København *03786*
Nordrhein-Westfälische Gesellschaft für Urologie, Osnabrück *06143*
Nordwestdeutsche Gesellschaft für ärztliche Fortbildung e.V., Steinburg *06144*
Nordwestdeutsche Gesellschaft für innere Medizin, Hamburg *06145*
Nordwestdeutsche Vereinigung der Hals-Nasen-Ohrenärzte, Kiel *06146*
Norfolk and Norwich Archaeological Society, Norwich *12629*
Norfolk Record Society, Norwich *12630*
Norges Almenvitenskapelige Forskningsråd, Oslo *09281*
Norges Fiskeriforskningsråd, Trondheim *09282*
Norges Geologiske Undersökelse, Trondheim *09283*
Norges Geotekniske Institutt, Oslo *09284*
Norges Kunstnerråd, Oslo *09285*
Norges Landbruksvitenskapelige Forskningsråd, Oslo *09286*
Norges Standardiseringsforbund, Oslo *09287*
Norges Tekniske Vitenskapsakademi, Trondheim *09288*
Norges Teknisk-Naturvitenskapelige Forskningsråd, Oslo *09289*
Normenausschuss Akustik, Lärmminderung und Schwingungstechnik im DIN Deutsches Institut für Normung e.V., Berlin *06147*
Normenausschuss Anstrichstoffe und ähnliche Beschichtungsstoffe im DIN Deutsches Institut für Normung e.V., Berlin *06148*
Normenausschuss Armaturen im DIN Deutsches Institut für Normung e.V., Köln *06149*
Normenausschuss Bauwesen im DIN Deutsches Institut für Normung e.V., Berlin *06150*
Normenausschuss Bergbau im DIN Deutsches Institut für Normung e.V., Essen *06151*
Normenausschuss Bibliotheks- und Dokumentationswesen im DIN Deutsches Institut für Normung e.V., Berlin *06152*
Normenausschuss Bild und Film im DIN Deutsches Institut für Normung e.V., Berlin *06153*
Normenausschuss Bühnentechnik in Theatern und Mehrzweckhallen im DIN Deutsches Institut für Normung e.V., Berlin *06154*
Normenausschuss Bürowesen im DIN Deutsches Institut für Normung e.V., Berlin *06155*
Normenausschuss Chemischer Apparatebau im DIN Deutsches Institut für Normung e.V., Köln *06156*
Normenausschuss Dental im DIN Deutsches Institut für Normung e.V., Pforzheim *06157*
Normenausschuss Dichtungen im DIN Deutsches Institut für Normung e.V., Köln *06158*
Normenausschuss Druckgasanlagen im DIN Deutsches Institut für Normung e.V., Berlin *06159*
Normenausschuss Druck- und Reproduktionstechnik im DIN Deutsches Institut für Normung e.V., Berlin *06160*
Normenausschuss Einheiten und Formelgrössen im DIN Deutsches Institut für Normung e.V., Berlin *06161*
Normenausschuss Eisen-, Blech- und Metallwaren im DIN Deutsches Institut für Normung e.V., Düsseldorf *06162*
Normenausschuß Eisen und Stahl im DIN Deutsches Institut für Normung e.V., Düsseldorf *06163*
Normenausschuss Erdöl- und Erdgasgewinnung im DIN Deutsches Institut für Normung e.V., Köln *06164*
Normenausschuss Ergonomie im DIN Deutsches Institut für Normung e.V., Berlin *06165*
Normenausschuss Fahrräder im DIN Deutsches Institut für Normung e.V., Köln *06166*
Normenausschuss Farbe im DIN Deutsches Institut für Normung e.V., Berlin *06167*
Normenausschuss Feinmechanik und Optik im DIN Deutsches Institut für Normung e.V., Pforzheim *06168*
Normenausschuss Feuerwehrwesen im DIN Deutsches Institut für Normung e.V., Berlin *06169*
Normenausschuss Gastechnik im DIN Deutsches Institut für Normung e.V., Eschborn *06170*
Normenausschuss Giessereiwesen im DIN Deutsches Institut für Normung e.V., Köln *06171*
Normenausschuss Gleitlager im DIN Deutsches Insitut für Normung e.V., Köln *06172*
Normenausschuss Graphische Symbole im DIN Deutsches Institut für Normung e.V., Berlin *06173*
Normenausschuss Grundlagen der Normung im DIN Deutsches Institut für Normung e.V., Berlin *06174*
Normenausschuss Hauswirtschaft im DIN Deutsches Institut für Normung e.V., Berlin *06175*
Normenausschuss Heiz-, Koch- und Wärmegeräte im DIN Deutsches Institut für Normung e.V., Frankfurt *06176*
Normenausschuss Heiz- und Raumlufttechnik im DIN Deutsches Institut für Normung e.V., Berlin *06177*
Normenausschuss Holzwirtschaft und Möbel im DIN Deutsches Institut für Normung e.V., Köln *06178*
Normenausschuß Informationsverarbeitungssysteme im DIN Deutsches Institut für Normung e.V., Berlin *06179*
Normenausschuss Instandhaltung im DIN Deutsches Institut für Normung e.V., Köln *06180*
Normenausschuss Kältetechnik im DIN Deutsches Institut für Normung e.V., Köln *06181*
Normenausschuss Kautschuktechnik im DIN Deutsches Institut für Normung e.V., Frankfurt *06182*
Normenausschuss Kerntechnik im DIN Deutsches Institut für Normung e.V., Berlin *06183*
Normenausschuss Kommunale Technik im DIN Deutsches Institut für Normung e.V., Berlin *06184*
Normenausschuss Kraftfahrzeuge im DIN Deutsches Institut für Normung e.V., Frankfurt *06185*
Normenausschuss Kunststoffe im DIN Deutsches Institut für Normung e.V., Berlin *06186*
Normenausschuss Laborgeräte und Laboreinrichtungen im DIN Deutsches Institut für Normung e.V., Frankfurt *06187*
Normenausschuss Lebensmittel und Landwirtschaftliche Produkte im DIN Deutsches Institut für Normung e.V., Berlin *06188*
Normenausschuss Lichttechnik im DIN Deutsches Institut für Normung e.V., Berlin *06189*
Normenausschuss Maschinenbau im DIN Deutsches Institut für Normung e.V., Frankfurt *06190*
Normenausschuss Materialprüfung im DIN Deutsches Institut für Normung e.V., Berlin *06191*
Normenausschuss Mechanische Verbindungselemente im DIN Deutsches Institut für Normung e.V., Köln *06192*
Normenausschuss Medizin im DIN Deutsches Institut für Normung e.V., Berlin *06193*
Normenausschuss Nichteisenmetalle im DIN Deutsches Institut für Normung e.V., Köln *06194*
Normenausschuss Papier und Pappe im DIN Deutsches Institut für Normung e.V., Berlin *06195*
Normenausschuss Persönliche Schutzausrüstung und Sicherheitskennzeichnung im DIN Deutsches Insitut für Normung e.V., Berlin *06196*
Normenausschuss Pigmente und Füllstoffe im DIN Deutsches Institut für Normung e.V., Berlin *06197*
Normenausschuss Pulvermetallurgie im DIN Deutsches Institut für Normung e.V., Köln *06198*
Normenausschuss Rohre, Rohrverbindungen und Rohrleitungen im DIN Deutsches Institut für Normung e.V., Köln *06199*
Normenausschuss Rundstahlketten im DIN Deutsches Institut für Normung e.V., Köln *06200*
Normenausschuss Schienenfahrzeuge im DIN Deutsches Institut für Normung e.V., Kassel *06201*
Normenausschuss Schmiedetechnik im DIN Deutsches Institut für Normung e.V., Hagen *06202*
Normenausschuss Schmuck im DIN Deutsches Insitut für Normung e.V., Pforzheim *06203*
Normenausschuss Schweisstechnik im DIN Deutsches Institut für Normung e.V., Berlin *06204*
Normenausschuss Siebböden und Kornmessung im DIN Deutsches Institut für Normung e.V., Berlin *06205*
Normenausschuss Sport- und Freizeitgerät im DIN Deutsches Institut für Normung e.V., Köln *06206*
Normenausschuss Stahldraht und Stahldrahterzeugnisse im DIN Deutsches Institut für Normung e.V., Köln *06207*
Normenausschuss Terminologie im DIN Deutsches Institut für Normung e.V., Berlin *06208*
Normenausschuss Textil und Textilmaschinen im DIN Deutsches Institut für Normung e.V., Berlin *06209*
Normenausschuss Transportkette im DIN Deutsches Institut für Normung e.V., Berlin *06210*
Normenausschuss Überwachungsbedürftige Anlagen im DIN Deutsches Institut für Normung e.V., Köln *06211*
Normenausschuss Uhren und Schmuck im DIN Deutsches Institut für Normung e.V., Pforzheim *06212*
Normenausschuss Vakuumtechnik im DIN Deutsches Institut für Normung e.V., Köln *06213*
Normenausschuss Verpackungswesen im DIN Deutsches Institut für Normung e.V., Berlin *06214*
Normenausschuss Waagenbau im DIN Deutsches Institut für Normung e.V., Berlin *06215*
Normenausschuss Wärmebehandlungstechnik metallischer Werkstoffe im DIN Deutsches Institut für Normung e.V., Köln *06216*
Normenausschuss Wasserwesen im DIN Deutsches Institut für Normung e.V., Berlin *06217*
Normenausschuss Werkzeuge und Spannzeuge im DIN Deutsches Institut für Normung e.V., Köln *06218*
Normenausschuss Werkzeugmaschinen im DIN Deutsches Institut für Normung e.V., Frankfurt *06219*
Normenausschuss Zeichnungswesen im DIN Deutsches Institut für Normung e.V., Berlin *06220*
Norsk Anestesiologisk Forening, Gjövik *09290*
Norsk Arkeologisk Selskap, Oslo *09291*
Norsk Astronautisk Forening, Oslo *09292*
Norsk Avdeling av International Law Association, Oslo *09309*
Norsk Bibliotekforening, Oslo *09293*
Norsk Botanisk Forening, Oslo *09294*
Norske Akademi for Sprog og Litteratur, Oslo *09295*
Norske Arkitekters Landsforbund, Oslo *09296*
Norske Billedkunstnere, Oslo *09297*
Norske Fagbibliotek Forening, Oslo *09298*
Norske Forfatterforening, Oslo *09299*
Norske Fysioterapeuters Forbund, Oslo *09300*
Norske Havforskeres Forening, Nordnes *09301*
Norske Kunst- og Kulturhistoriske Museer, Oslo *09302*
Norske Meierifolks Landsforening, Oslo *09303*
Norske Musikklaereres Landsforbund, Kristiansand *09304*
Norske Naturhistoriske Museers Landsforbund, Tromsø *09305*
Norske Sivilingeniørers Forening, Oslo *09306*
Norske Siviløkonomers Forening, Oslo *09307*
Norsk Faglaererlag, Oslo *09308*
Norsk Forening for Internasjonal Rett, Oslo *09309*
Norsk Forening for Mikrobiologi, Aas *09310*
Norsk Forening mot Støy, Oslo *09311*
Norsk Forsikringsjuridisk Forening, Oslo *09312*
Norsk Fysiologisk Forening, Oslo *09313*
Norsk Fysisk Selskap, Kjeller *09314*
Norsk Gastroenterologisk Selskap, Oslo *09315*
Norsk Geofysisk Forening, Trondheim *09316*
Norsk Geologisk Forening, Trondheim *09317*
Norsk Geoteknisk Forening, Oslo *09318*
Norsk Heraldisk Forening, Oslo *09319*
Norsk Homøopatisk Pasientforening, Trondheim *09320*
Norsk Institutt for By- og Regionforskning, Oslo *09321*
Norsk Kirurgisk Forening, Lysaker *09322*
Norsk Kjemisk Selskap, Oslo *09323*
Norsk Korrosjonsteknisk Forening, Oslo *09324*
Norsk Laererlag, Oslo *09325*

Norsk Logopedlag, Trondheim 09326
Norsk Lokalhistorisk Institutt, Oslo 09327
Norsk Matematisk Forening, Oslo 09328
Norsk Metallurgisk Selskap, Oslo 09329
Norsk Meteorologforening, Oslo 09330
Norsk Musikkinformasjon, Oslo 09331
Norsk Naturforvalterforbund, Oslo 09332
Norsk Operasjonsanalyseforening, Oslo 09333
Norsk P.E.N., Oslo 09334
Norsk Radiologisk Forening, Oslo 09335
Norsk Regnesentral, Oslo 09336
Norsk Samfunnsgeografisk Forening, Dragvoll 09337
Norsk Senter for Samferdselsforskning, Oslo 09338
Norsk Slektshistorisk Forening, Oslo 09339
Norsk Tekstil Teknisk Forbund, Bergen 09340
North American Academy of Ecumenists, Albuquerque 16209
North American Academy of Muscoloskeletal Medicine, Middleton 16210
North American Association for Environmental Education, Washington 16211
North American Association of Professors of Christian Education, Elgin 16212
North American Association of Summer Sessions, Saint Louis 16213
North American Clinical Dermatologic Society, Jacksonville 16214
North American Conference on British Studies, Seattle 16215
North American Dostoevsky Society, Lexington 16216
North American Electric Reliability Council, Princeton 16217
North American Medical/Dental Association, Newport Beach 16218
North American Mycological Association, Ann Arbor 16219
North American Primary Care Research Group, Richmond 16220
North American Simulation and Gaming Association, Indianapolis 16221
North American Society for Pediatric Gastroenterology and Nutrition, Denver 16222
North American Society for Sport History, University Park 16223
North American Society of Adlerian Psychology, Chicago 16224
North American Society of Pacing and Electrophysiology, Newton Upper Falls 16225
North American Thermal Analysis Society, Buffalo 16226
North American Vexillological Association, Trenton 16227
Northamptonshire Archaeological Society, Earls Barton 12631
Northamptonshire Natural History Society and Field Club, Northampton 12632
Northamptonshire Record Society, Northampton 12633
North Central Association of Colleges and Schools, Tempe 16228
North Central Conference on Summer Schools, River Falls 16229
Northeast Conference on the Teaching of Foreign Languages, South Burlington 16230
Northern Horticultural Society, Harrogate 12634
Northern Ireland Association for Mental Health, Belfast 12635
Northern Ireland Chest, Heart and Stroke Association, Belfast 12636
Northern Ireland Civil Rights Association, Belfast 12637
Northern Ireland Polio Fellowship, Belfast 12638
Northern Ireland Women Teacher's Association, Belfast 12639
Northern Libraries Colloquy, Anchorage 16231
Northern Quebec Teaching Association, Sainte-Foy 01876
North Island/Laurentian Teachers Union, Chomedey-Laval 02715
The North of England Zoological Society, Chester 12640
North Okanagan Neurological Association, Vernon 02716
North Queensland Naturalists Club, Cairns 00441
North Somerset Agricultural Society, Bristol 12641
North Staffordshire Field Club, Stoke-on-Trent 12642
North Tyne and Redesdale Agricultural Society, Morpeth 12643
North Wales Agricultural Society, Portdinorwic 12644
North Wales Arts Association, Bangor 12645
Northwest Association of Schools and Colleges, Boise 16232
Northwestern Academy of Industrial Architecture, Shanhsi 03212
North Western Naturalists' Union, Stockport 12646
Northwestern Ontario Water and Waste Conference, Thunder Bay 02717
North Western Society for Industrial Archaeology and History, Liverpool 12647
North West Regional Association of Industrial Safety Groups, Preston 12648
Northwest Territories Law Foundation, Yellowknife 02718
Northwest Territories Library Association, Yellowknife 02719
Northwest Territories Music Educators' Association, Fort Smith 02720
Northwest Territories Teachers' Association, Yellowknife 02721
Northwest Territories Teachers Association Home Economics Council, Inuvik 02722
The Norwegian Academy of Science and Letters, Oslo 09264
Norwegian-American Historical Association, Northfield 16233
Norwegian Artists' Council, Oslo 09285
Norwegian Association of Agriculture Graduates, Oslo 09332
Norwegian Association of Logopedists, Trondheim 09326
The Norwegian Association of Meteorologists, Oslo 09330
Norwegian Astronautical Society, Oslo 09292
Norwegian Authors' Society, Oslo 09299
Norwegian Botanical Association, Oslo 09294
Norwegian Center for Transport Research, Oslo 09338
Norwegian Chemical Society, Oslo 09323
Norwegian Computing Centre, Oslo 09336
Norwegian Council of Cultural Affairs, Oslo 09341
The Norwegian Dental Association, Oslo 09255
The Norwegian Fisheries Research Council, Trondheim 09282
The Norwegian Forestry and Forest Industry Association, Aas 09348
Norwegian Geotechnical Institute, Oslo 09284
Norwegian Geotechnical Society, Oslo 09318
Norwegian Institute for Urban and Regional Research, Oslo 09321
Norwegian Institute of Local History, Oslo 09327
Norwegian Library Association, Oslo 09293
Norwegian Music Information Centre, Oslo 09331
Norwegian Physiotherapist Association, Oslo 09300
The Norwegian Pulp and Paper Research Institute, Oslo 09343
Norwegian Society of Chartered Engineers, Oslo 09306
Norwegian Standards Association, Oslo 09287
Norwegian Surgical Society, Lysaker 09322
Norwegian Union of Teachers, Oslo 09325
NOSALF, Stockholm 10510
NOTAEI, Mineola 16113
Nottingham and Nottinghamshire Field Club, Nottingham 12649
Nottinghamshire Local History Association, Nottingham 12650
N.O.U., Odijk 08999
N.O.V., Nijmegen 09058
Nova Scotia Association of Optometrists, Halifax 02723
Nova Scotia Association of Social Workers, Dartmouth 02724
Nova Scotia Barristers' Society, Halifax 02725
Nova Scotia Cardiology Technicians Association, Halifax 02726
Nova Scotia Chiropractic Association, Antigonish 02727
Nova Scotia College Conference, Kingston 02728
Nova Scotia Confederation of University Faculty Associations, Halifax 02729
Nova Scotia Criminology and Corrections Association, Halifax 02730
Nova Scotia Dental Association, Halifax 02731
Nova Scotia Dietetic Association, Halifax 02732
Nova Scotia Heart Foundation, Halifax 02733
Nova Scotia Institute of Agrologists, Truro 02734
Nova Scotia Library Association, Halifax 02735
Nova Scotia Lung Association, Halifax 02736
Nova Scotia Medical Council, Halifax 02737
Nova Scotia Mineral and Gem Society, Dartmouth 02738
Nova Scotia Music Educators' Association, Armdale 02739
Nova Scotian Institute of Science, Halifax 02740
Nova Scotia Pharmaceutical Society, Halifax 02741
Nova Scotia Research Foundation Corporation, Dartmouth 02742
Nova Scotia Safety Council, Halifax 02743
Nova Scotia School Boards Association, Halifax 02744
Nova Scotia Society of Medical Radiation Technologists, Armdale 02745
Nova Scotia Society of Occupational Therapists, Halifax 02746
Nova Scotia Teachers Union, Armdale 02747
Nova Scotia Thoracic Society, Halifax 02748
Nova Scotia Veterinary Medical Association, Kentville 02749
NPA, Tampa 16124
NPAP, New York 16132
NPAV, Maastricht 09000
NPB, Montgomery 16129
NPCA, Dunn Loring 16125
NPF, Berlin 06197
NPF, Dallas 16128
NPF, Portland 16131
N.P.G., Utrecht 09056
NPhA, Washington 16126
NPIA, Eagle Rock 16127
NPL, 's-Gravenhage 09035
NPM, Washington 15877
NPMA, Chicago 16130
NPS, Berlin 06196
NPu, Köln 06198
NPWA, Ellesmere 12603
N.Q.N.C., Cairns 00441
NRA, Alexandria 16137
NRC, Chicago 16134
NRC, New York 16133
NRC, Ottawa 02673
NRC, Princeton 16185
NRC, Washington 16138
NRCC, Washington 16136
NRCCLS, Gloucester 16141
N.R.C.L., Springvale South 00440
NRCP, Manila 09522
NRCPS, New York 16139
NRDC, New York 16187
NRDI, Bellingham 16143
NREA, Fort Collins 16144
NRF, Oslo 09335
NRK, Köln 06200
NRMP, Washington 16140
NRSSC, Bellingham 16142
NS, Fort Mitchell 16208
NSA, Cleveland 16193
NSA, Oakland 16191
NSA, Southampton 16153
NSBA, Alexandria 16147
NSBSSA, Bishops Stortford 12612
NSC, Itasca 16145
NSC, Washington 16161
NSCA, Brighton 12607
NSCEE, Cos Cob 16149
NSCT, Fredericksburg 16154
N.S.E., Chalfont Saint Peter 12609
NSEAD, Corsham 12608
NSEE, Raleigh 16155
NSERC, Ottawa 02675
NSF, Oslo 09287, 09307
NSF, Surbiton 12604
NSGF, Dragvoll 09337
NSIS, Halifax 02740
NSMS, Weaverville 16146
NSNA, New York 16165
NSPCB, Worcester 16157
NSPCC, Falls Church 16164
NSPE, Alexandria 16159
NSPI, Washington 16156
NSPRA, Arlington 16148
NSPS, Jefferson City 16160
NSRA, Hinckley 12610
NSS, Huntsville 16163
NSS, New York 16152
NS/S, Tucson 16186
NSS, Washington 16162
NSSA, Glastonbury 16150
NSSE, Bloomington 16167
NSSE, Chicago 16158
NSSLHA, Rockville 16166
NSTA, Washington 16151
NSUS, Exton 16194
NTA, London 12613
NTA, Washington 16169
NTA-TIA, Columbus 16168
NTC, Madison 16171
NTF, Oslo 09255
NTI, Waterford 16172
NTK, Berlin 06210
NTL Institute for Applied Behavioral Science, Alexandria 16234
NTNF, Oslo 09289
NTRA, Mundelein 16174
NTRS, Arlington 16173
NTSA, Alexandria 16170
NTTF, Bergen 09340
NUCEA, Washington 16175
NUFFIG, 's-Gravenhage 08997
NUL, New York 16176
Numerical Control Society/AIM Tech, Research Triangle Park 16235
Numismatic Educational Services Association, Barrie 02750
Numismatic Literary Guild, Glen Rock 16236
Numismatische Kommission der Länder in der Bundesrepublik Deutschland, Berlin 06221
N.U.T., London 12616
Nutrition Institute of America, New York 16237
Nutrition Society, London 12651
Nutrition Society of New Zealand, Dunedin 09183
Nutrition Society of Nigeria, Ile-Ife 09243
The Nutrition Society of Southern Africa, Medunsa 10158
Nuttall Ornithological Club, Cambridge 16238
NVA, Utrecht 09016
NVAG, Rotterdam 09015
NVATA, Alexandria 16177
NVG, Apeldoorn 09019
NVGE, Haarlem 09018
NVIR, 's-Gravenhage 09021
NVOS, Gorinchem 09031
NVP, Utrecht 09034
NVRL, Utrecht 09010
NVTEF, Washington 16178
N.V.T.G., 's-Gravenhage 09040
NVvGT, Zoetermeer 09006
NVVK, Amsterdam 09042
NVvL, Amsterdam 09024
NVvM, Bilthoven 09027
NVvN, Rotterdam 09008
NVvT, Maarssen 09012
NVvW, 's-Gravenhage 09013
NWB, Berlin 06215
NWC, Aurora 16183
NWDS, Cincinnati 16207
NWF, New York 16198
NWF, Washington 16180
NWHN, Washington 16181
NWM, Frankfurt 06219
NWNU, Stockport 12646
NWP, Berkeley 16184
NWP, Leusden 09045
NWRA, Arlington 16179
NWSA, College Park 16182
NWSIAH, Liverpool 12647
NWT, Köln 06216
NWTA, London 12617
NYAS, New York 16199
NYBS, New York 16200
NYCSLS, Jamaica 16201
NYDCC, New York 16202
NYFWA, Syosset 16203
NYGBS, New York 16204
NYPC, Clifton 16205
NZ, Berlin 06220
N.Z.A.A., Dunedin North 09133
NZEI, Auckland 09145
NZFMRA, Auckland 09146
NZHS, Wellington 09152
NZIAS, Christchurch 09153
N.Z.I.F.S.T., Christchurch 09156
N.Z.I.I.A., Wellington 09158
NZIM, Wellington 09159
NZIP, Auckland 09160
NZLA, Wellington 09164
NZSDST, Edgecumbe 09178
NZSSS, Canterbury 09180
N.Z.V.A., Wellington 09182

OAA, Fairfax 16256
OAGB, London 12664
OAH, Bloomington 16265
OAHS, Oxford 12669
OAJ, Helsinki 03987
OAKE, Thibodaux 16266
Oak Ridge Associated Universities, Oak Ridge 16239
Oaks Historical Society, The Oaks 00442
OAPBC, Washington 16245
OBC, Bruxelles 01415
Obec Architektů, Praha 03552
Oberösterreichischer Musealverein, Linz 00717
Oberrheinische Gesellschaft für Geburtshilfe und Gynäkologie, Tübingen 06222
OBTS, Norman 16261
OCA, Toronto 02765
OCC, New York 16277
OCCGE, Bobo-Dioulasso 01810
Occupational Health and Safety Council of Prince Edward Island, Charlottetown 02751
Occupational Medical Association of Canada, Toronto 02752
Ocean Education Project, Ararat 16240
Oceanic Educational Foundation, Falls Church 16241
Oceanic Society, Washington 16242
Oceanic Society Expeditions, San Francisco 16243
Oceanographical Society of Japan, Tokyo 08323
OCG, Wien 00735
OCI, Westport 16244
OCOO, Dayton 16273
OCS, London 12661
ODA, Manchester 12666
ODI, Chesterland 16262
Odinic Rite, London 12652
ODS, Baltimore 16272
ODUCAL, Buenos Aires 00150
Odyssey Institute Corporation, Westport 16244
OE, Hamilton 16252
OEA, Denville 16275
OEA, Washington 16278
ÖÄK, Wien 00721
ÖAGG, Wien 00882
ÖAKT, Salzburg 00728
ÖAL, Wien 00885
ÖANP, Wien 00725
ÖAV, Innsbruck 00881
ÖAV, Wien 00727
ÖAW, Wien 00723
ÖBG, Wien 00733
ÖDK, Wien 00736
ÖEG, Wien 00739
OEF, Falls Church 16241
ÖFVK, Graz 00891
OEG, Bucks Harbor 16276
ÖGA, Wien 00760
ÖGAM, Klagenfurt 00752
ÖGCF, Wien 00768
ÖGDI, Wien 00774
ÖGE, Wien 00778
ÖGEFUE, Wien 00856
ÖGEW, Wien 00777
ÖGFKM, Wien 00779
ÖGfM, Wien 00811
ÖGfRV, Wien 00827
ÖGfU, Wien 00837
ÖGFW, Wien 00755
ÖGG, Wien 00745, 00747
ÖGGH, Wien 00781
ÖGGL, Salzburg 00786
ÖGH, Wien 00791
ÖGHMP, Wien 00793
ÖGI, Leoben 00910
ÖGI, Linz 00795
ÖGIST, Wien 00794
ÖGKC, Wien 00802
ÖGMAC, Wien 00810
ÖGMS, Wien 00808
ÖGOR, Wien 00817
ÖGP, Wien 00819
ÖGRR, Wien 00826
ÖGS, Wien 00830, 00832, 00834, 00835
ÖGV, Wien 00843
ÖGW, Wien 00847
ÖGWT, Wien 00844
ÖHG, Wien 00858, 00859
OEI, Madrid 10333
ÖIAV, Wien 00893
ÖKG, Wien 00803
Öko-Institut, Institut für angewandte Ökologie e.V., Freiburg 06223
Ökosoziales Forum, Wien 00718
ÖLV, Wien 00896
ÖMG, Wien 00867, 00869, 00871
ÖNB, Salzburg 00898
ÖNG, Wien 00872
ÖNR, Graz 00868
ÖOG, Wien 00873
OEP, Ararat 16240
ÖPG, Graz 00879
ÖPG, Wien 00877
ÖPhG, Wien 00878
ÖPWZ, Wien 00917
OERF, Saint Louis 16268
ÖRK, Wien 00890
ÖROK, Wien 00887
ÖSGK, Wien 00921
ÖSLV, Wien 00902
Österreichische Ärztegesellschaft für Psychotherapie, Wien 00719
Österreichische Ärztegesellschaft zur Bekämpfung der cystischen Fibrose, Wien 00720
Österreichische Ärztekammer, Wien 00721
Österreichische ärztliche Gesellschaft für medizinisches und technisches Ozon, Wien 00722
Österreichische Akademie der Wissenschaften, Wien 00723
Österreichische Arbeitsgemeinschaft für morphologische und funktionelle Atheroskleroseforschung, Wien 00724
Österreichische Arbeitsgemeinschaft für Neuropsychiatrie und Psychologie des Kindes- und Jugendalters und verwandter Berufe, Wien 00725
Österreichische Arbeitsgemeinschaft für Rehabilitation, Wien 00726
Österreichische Arbeitsgemeinschaft für Volksgesundheit, Wien 00727
Österreichische Arbeitskreise für Tiefenpsychologie, Salzburg 00728
Österreichische Bankwissenschaftliche Gesellschaft an der Wirtschaftsuniversität Wien, Wien 00729
Österreichische Bibelgesellschaft, Wien 00730
Österreichische Biochemische Gesellschaft, Wien 00731
Österreichische Biophysikalische Gesellschaft, Wien 00732
Österreichische Bodenkundliche Gesellschaft, Wien 00733
Österreichische Byzantinische Gesellschaft, Wien 00734
Österreichische Computer-Gesellschaft, Wien 00735
Österreichische Dentistenkammer, Wien 00736
Österreichische Diabetikervereinigung, Wien 00737
Österreichische Entomologische Gesellschaft, Graz 00738
Österreichische Ethnologische Gesellschaft, Wien 00739
Österreichische Ethnomedizinische Gesellschaft, Wien 00740

Österreichische Exlibris-Gesellschaft, Wien 00741
Österreichische Forschungsgemeinschaft, Wien 00742
Österreichische Forschungsstiftung für Entwicklungshilfe, Wien 00743
Österreichische Gartenbau-Gesellschaft, Wien 00744
Österreichische Geographische Gesellschaft, Wien 00745
Österreichische Geographische Gesellschaft, Zweigverein Innsbruck, Innsbruck 00746
Österreichische Geologische Gesellschaft, Wien 00747
Österreichische Gesellschaft der Tierärzte, Wien 00748
Österreichische Gesellschaft für ärztliche Hypnose und autogenes Training, Wien 00749
Österreichische Gesellschaft für Agrar- und Umweltrecht, Wien 00750
Österreichische Gesellschaft für Akupunktur und Aurikulotherapie, Wien 00751
Österreichische Gesellschaft für Allgemeinmedizin, Klagenfurt 00752
Österreichische Gesellschaft für Amerikastudien, Salzburg 00753
Österreichische Gesellschaft für Anästhesiologie, Reanimation und Intensivtherapie, Wien 00754
Österreichische Gesellschaft für angewandte Fremdenverkehrswissenschaft, Wien 00755
Österreichische Gesellschaft für angewandte Zytologie, Graz 00756
Österreichische Gesellschaft für Angiologie, Wien 00757
Österreichische Gesellschaft für Arbeitsmedizin, Hall 00758
Österreichische Gesellschaft für Arbeitsrecht und Sozialrecht, Linz-Auhof 00759
Österreichische Gesellschaft für Archäologie, Wien 00760
Österreichische Gesellschaft für Architektur, Wien 00761
Österreichische Gesellschaft für Artificial Intelligence, Wien 00762
Österreichische Gesellschaft für Aussenpolitik und Internationale Beziehungen, Wien 00763
Österreichische Gesellschaft für Autogenes Training und allgemeine Psychotherapie, Wien 00764
Österreichische Gesellschaft für Balneologie und medizinische Klimatologie, Wien 00765
Österreichische Gesellschaft für Biomedizinische Technik, Wien 00766
Österreichische Gesellschaft für Bionome Psychotherapie, Wien 00767
Österreichische Gesellschaft für China-Forschung, Wien 00768
Österreichische Gesellschaft für Chirurgie, Wien 00769
Österreichische Gesellschaft für Chirurgische Forschung, Wien 00770
Österreichische Gesellschaft für Christliche Kunst, Wien 00771
Österreichische Gesellschaft für Denkmal- und Ortsbildpflege, Wien 00772
Österreichische Gesellschaft für Dermatologie und Venerologie, Graz 00773
Österreichische Gesellschaft für Dokumentation und Information, Wien 00774
Österreichische Gesellschaft für Elektroencephalographie und klinische Neurophysiologie, Innsbruck 00775
Österreichische Gesellschaft für Elektronenmikroskopie, Graz 00776
Österreichische Gesellschaft für Erdölwissenschaften, Wien 00777
Österreichische Gesellschaft für Ernährungsforschung, Wien 00778
Österreichische Gesellschaft für Filmwissenschaft, Kommunikations- und Medienforschung, Wien 00779
Österreichische Gesellschaft für Friedensforschung, Wien 00780
Österreichische Gesellschaft für Gastroenterologie und Hepatologie, Wien 00781
Österreichische Gesellschaft für Gefässchirurgie, Salzburg 00782
Österreichische Gesellschaft für Geriatrie und Gerontologie, Wien 00783
Österreichische Gesellschaft für gerichtliche Medizin, Wien 00784
Österreichische Gesellschaft für Geschichte der Pharmazie, Wien 00785
Österreichische Gesellschaft für Gesetzgebungslehre, Salzburg 00786
Österreichische Gesellschaft für Gruppendynamik und Organisationsberatung, Klagenfurt 00787
Österreichische Gesellschaft für Gynäkologie und Geburtshilfe, Wien 00788
Österreichische Gesellschaft für Hals-, Nasen- und Ohrenheilkunde, Kopf- und Halschirurgie, Wien 00789
Österreichische Gesellschaft für Hochschuldidaktik, Wien 00790
Österreichische Gesellschaft für Holzforschung, Wien 00791
Österreichische Gesellschaft für Humanökologie, Wien 00792
Österreichische Gesellschaft für Hygiene, Mikrobiologie und Präventivmedizin, Wien 00793
Österreichische Gesellschaft für Industrielle Strahltechnik, Wien 00794
Österreichische Gesellschaft für Informatik, Linz 00795
Österreichische Gesellschaft für Innere Medizin, Wien 00796
Österreichische Gesellschaft für internistische Intensivmedizin, Wien 00797
Österreichische Gesellschaft für Kinderchirurgie, Graz 00798
Österreichische Gesellschaft für Kinderphilosophie, Graz 00799
Österreichische Gesellschaft für Kinder- und Jugendheilkunde, Wien 00800
Österreichische Gesellschaft für Kirchenrecht, Wien 00801
Österreichische Gesellschaft für Klinische Chemie, Wien 00802
Österreichische Gesellschaft für Kommunikationsfragen, Wien 00803
Österreichische Gesellschaft für Laboratoriumsmedizin, Wien 00804
Österreichische Gesellschaft für Literatur, Wien 00805
Österreichische Gesellschaft für Logopädie, Phoniatrie und Pädaudiologie, Wien 00806
Österreichische Gesellschaft für Lungenerkrankungen und Tuberkulose, Wien 00807
Österreichische Gesellschaft für Medizinsoziologie, Wien 00808
Österreichische Gesellschaft für Meteorologie, Wien 00809
Österreichische Gesellschaft für Mikrochemie und analytische Chemie, Wien 00810
Österreichische Gesellschaft für Musik, Wien 00811
Österreichische Gesellschaft für Nephrologie, Linz 00812
Österreichische Gesellschaft für Neugriechische Studien, Wien 00813
Österreichische Gesellschaft für Neurochirurgie, Wien 00814
Österreichische Gesellschaft für Neuropathologie, Wien 00815
Österreichische Gesellschaft für Nuclearmedizin, Wien 00816
Österreichische Gesellschaft für Operations Research, Wien 00817
Österreichische Gesellschaft für Orthopädie und Orthopädische Chirurgie, Wien 00818
Österreichische Gesellschaft für Parapsychologie, Wien 00819
Österreichische Gesellschaft für Pathologie, Wien 00820
Österreichische Gesellschaft für Perinatale Medizin, Wien 00821
Österreichische Gesellschaft für Philosophie, Salzburg 00822
Österreichische Gesellschaft für Physikalische Medizin und Rehabilitation, Wien 00823
Österreichische Gesellschaft für Politikwissenschaft, Wien 00824
Österreichische Gesellschaft für Psychische Hygiene, Wien 00825
Österreichische Gesellschaft für Raumforschung und Raumplanung, Wien 00826
Österreichische Gesellschaft für Rechtsvergleichung, Wien 00827
Österreichische Gesellschaft für Religionswissenschaft, Salzburg 00828
Österreichische Gesellschaft für Schweisstechnik, Wien 00829
Österreichische Gesellschaft für Semiotik, Wien 00830
Österreichische Gesellschaft für Sexualforschung, Wien 00831
Österreichische Gesellschaft für Soziologie, Wien 00832
Österreichische Gesellschaft für Sprachheilpädagogik, Wien 00833
Österreichische Gesellschaft für Statistik, Wien 00834
Österreichische Gesellschaft für Strassenwesen, Wien 00835
Österreichische Gesellschaft für Tropenmedizin und Parasitologie, Wien 00836
Österreichische Gesellschaft für Unfallchirurgie, Wien 00837
Österreichische Gesellschaft für Unternehmensgeschichte, Wien 00838
Österreichische Gesellschaft für Urologie, Wien 00839
Österreichische Gesellschaft für Ur- und Frühgeschichte, Wien 00840
Österreichische Gesellschaft für Vakuumtechnik, Wien 00841
Österreichische Gesellschaft für Versicherungsfachwissen, Wien 00842
Österreichische Gesellschaft für Vogelkunde, Wien 00843
Österreichische Gesellschaft für Warenkunde und Technologie, Wien 00844
Österreichische Gesellschaft für Weltraumforschung, Innsbruck 00845
Österreichische Gesellschaft für Wirtschaftspolitik, Wien 00846
Österreichische Gesellschaft für Wirtschaftsraumforschung, Wien 00847
Österreichische Gesellschaft für Wirtschaftssoziologie, Wien 00848
Österreichische Gesellschaft für Wissenschaftsgeschichte, Wien 00849
Österreichische Gesellschaft für Zahn-, Mund- und Kieferheilkunde, Wien 00850
Österreichische Gesellschaft für Zeitgeschichte, Wien 00851
Österreichische Gesellschaft für und Institut für Umweltschutz, Umwelttechnologie und Umweltwissenschaften, Wien 00852
Österreichische Gesellschaft zum Studium der Sterilität und Fertilität, Wien 00853
Österreichische Gesellschaft zur Bekämpfung der Cystischen Fibrose, Wien 00854
Österreichische Gesellschaft zur Erforschung des 18. Jahrhunderts, Wien 00855
Österreichische Gesellschaft zur Föderung von Umweltschutz und Energieforschung, Wien 00856
Österreichische Gesellschaft zur Förderung medizin-meteorologischer Forschung in Österreich, Wien 00857
Österreichische Hämophilie-Gesellschaft, Wien 00858
Österreichische Himalaya-Gesellschaft, Wien 00859
Österreichische Humanistische Gesellschaft für die Steiermark, Graz 00860
Österreichische Kardiologische Gesellschaft, Wien 00861
Österreichische Kommission für Internationale Erdmessung, Wien 00862
Österreichische Krebshilfe – Österreichische Krebsgesellschaft, Wien 00863
Österreichische Krebshilfe Steiermark, Graz 00864
Österreichische Kulturgemeinschaft, Wien 00865
Österreichische Ludwig-Wittgenstein-Gesellschaft, Kirchberg 00866
Österreichische Mathematische Gesellschaft, Wien 00867
Österreichische medizinische Gesellschaft für Neuraltherapie nach Huneke-Regulationsforschung, Graz 00868
Österreichische Mineralogische Gesellschaft, Wien 00869
Österreichische Multiple Sklerosis Gesellschaft, Wien 00870
Österreichische Mykologische Gesellschaft, Wien 00871
Österreichische Numismatische Gesellschaft, Wien 00872
Österreichische Ophthalmologische Gesellschaft, Wien 00873
Österreichische Orchideen-Gesellschaft, Wien 00874
Österreichische Orient-Gesellschaft Hammer-Purgstall, Wien 00875
Österreichische Pädagogische Gesellschaft, Wien 00876
Österreichische Paläontologische Gesellschaft, Wien 00877
Österreichische Pharmazeutische Gesellschaft, Wien 00878
Österreichische Physikalische Gesellschaft, Graz 00879
Österreichische Physiologische Gesellschaft, Wien 00880
Österreichischer Alpenverein, Wissenschaftlicher Unterausschuss, Innsbruck 00881
Österreichischer Arbeitskreis für Gruppentherapie und Gruppendynamik, Wien 00882
Österreichischer Arbeitskreis für Soziologie des Sports und der Leibeserziehung, Wien 00883
Österreichischer Arbeitskreis für Stadtgeschichtsforschung, Linz 00884
Österreichischer Arbeitsring für Lärmbekämpfung, Wien 00885
Österreichischer Astronomischer Verein, Wien 00886
Österreichische Raumordnungskonferenz, Wien 00887
Österreichischer Burgenlandbund, Arbeitsgemeinschaft für Burgenländische Geschichte und Persönlichkeiten-Deutschtum in Ungarn, Eisenstadt 00888
Österreichischer Burgenverein, Grieskirchen 00889
Österreichische Rektorenkonferenz, Wien 00890
Österreichischer Fachverband für Volkskunde, Graz 00891
Österreichischer Fernschulverband, Wien 00892
Österreichischer Ingenieur- und Architektenverein, Wien 00893
Österreichischer Juristentag, Wien 00894
Österreichischer Kunsthistorikerverband, Wien 00895
Österreichischer Lehrerverband, Wien 00896
Österreichischer Museumsbund, Wien 00897
Österreichischer Naturschutzbund, Salzburg 00898
Österreichische Röntgengesellschaft – Gesellschaft für medizinische Radiologie und Nuklearmedizin, Wien 00899
Österreichischer PEN-Club, Wien 00900
Österreichischer Schriftstellerverband, Wien 00901
Österreichischer Sportlehrerverband, Wien 00902
Österreichischer Stahlbauverband, Wien 00903
Österreichischer Verband für Elektrotechnik, Wien 00904
Österreichischer Verband für Strahlenschutz, Seibersdorf 00905
Österreichischer Verein für Individualpsychologie, Wien 00906
Österreichischer Verein für Vermessungswesen und Photogrammetrie, Wien 00907
Österreichisches College: Collegegemeinschaft Wien, Wien 00908
Österreichisches Forschungsinstitut für Sparkassenwesen, Wien 00909
Österreichisches Giesserei-Institut, Verein für Praktische Giessereiforschung, Leoben 00910
Österreichisches Institut für Bibliographie, Wien 00911
Österreichisches Institut für Wirtschaftsforschung, Wien 00912
Österreichisches Lateinamerika-Institut, Wien 00913
Österreichisches Normungsinstitut, Wien 00914
Österreichisches Ost- und Südosteuropa-Institut, Wien 00915
Österreichische Sportwissenschaftliche Gesellschaft, Salzburg 00916
Österreichisches Produktivitäts- und Wirtschaftlichkeits-Zentrum, Wien 00917
Österreichisches Studienzentrum für Frieden und Konfliktforschung, Burg Schlaining 00918
Österreichische Statistische Gesellschaft, Wien 00919
Österreichische Studiengesellschaft für Kinderpsychoanalyse, Salzburg 00920
Österreichische Studiengesellschaft für Kybernetik, Wien 00921
Österreichisches Volksliedwerk, Wien 00922
Österreichische Tribologische Gesellschaft, Wien 00923
Österreichische Vereinigung der Zellstoff- und Papierchemiker und -techniker, Wien 00924
Österreichische Vereinigung für politische Wissenschaften, Wien 00925
Österreichische Verkehrswissenschaftliche Gesellschaft, Wien 00926
Österreichische Verwaltungswissenschaftliche Vereinigung, Wien 00927
Österreichische Werbewissenschaftliche Gesellschaft, Wien 00928
Österreichische wissenschaftliche Gesellschaft für Prophylaktische Medizin und Sozialhygiene, Wien 00929
ÖTG, Wien 00923
Oeuvre Belge du Cancer, Bruxelles 01415
ÖVE, Wien 00904
ÖVG, Wien 00926
ÖVS, Seibersdorf 00905
ÖZEPA, Wien 00924
OFA, Columbia 16270
Office Commun de Formation Européenne, Mons 01240
Office for Advancement of Public Black Colleges of the National Association of State Universities and Land Grant Colleges, Washington 16245
Office Généalogique et Héraldique de Belgique, Bruxelles 01416
Office Général du Bâtiment et des Travaux Publics, Paris 04604
Office International de l'Enseignement Catholique, Bruxelles 01417
Office National d'Etudes et des Recherches Aérospatiales, Chatillon 04605
Office National d'Information sur les Enseignements et les Professions, Paris 04606
Office of International Education of the American Council on Education, Washington 16246
Office of Management Services, Washington 16247
Office of the Atomic Energy Commission for Peace, Bangkok 11095
Office Systems Research Association, Springfield 16248
Oficina Internacional de Información y Observación del Español, Madrid 10332
Oficina Regional de Educación de la UNESCO para América Latina y el Caribe, Santiago 03062
OFN, Bronx 16263
OG, Wien 00930
O.G.B.T.P., Paris 04604
O.G.E., Paris 04609
OHA, Los Angeles 16258
OHI, Monte Carlo 08797
OHS, Richmond 16260
OHS, Saint Louis 16257
OIEC, Bruxelles 01417
Oikos, Lund 10513
Oil and Colour Chemists' Association, Wembley 12653
OIML, Paris 04615
OIPC, Petit-Lancy 10776
OIW, Little Rock 16259
Okinawa Kyokai Chosa Engoka, Tokyo 08401
Old Dublin Society, Dublin 07018
OMBKE, Budapest 06670
OMCE, Budapest 06671
OMM, Genève 10777
Omnibus Society, London 12654
Omospondia Didaskaliki Ellados, Athinai 06539
Omospondia Panellinios Syndesmon Dasoponon, Athinai 06540
OMRS, Croydon 12657
OMS, Washington 16247
ON, Wien 00914
Oncology Nursing Society, Pittsburgh 16249
O.N.I.S.E.P., Paris 04606
ONS, Pittsburgh 16249
ONS, Reading 12662
The Ontario Archaeological Society, Willowdale 02753
Ontario Association for Continuing Education, Toronto 02754
Ontario Association for Curriculum Development, London 02755
Ontario Association of Alternative and Independent Schools, Toronto 02756
Ontario Association of Archivists, Ottawa 02757
Ontario Association of Education Administrative Officials, Toronto 02758
Ontario Association of Library Technician Instructors, Welland 02759
Ontario Association of Library Technicians, Oakville 02760
Ontario Association of Property Standards Officers, Tillsonburg 02761
Ontario Athletic Therapists Association, Downsview 02762
Ontario Black History Society, Willowdale 02763
Ontario Cancer Treatment and Research Foundation, Toronto 02764
Ontario Chiropractic Association, Toronto 02765
Ontario College of Percussion, Toronto 02766
Ontario Community Development Society, Guelph 02767
Ontario Co-operative Education Association, Hamilton 02768
Ontario Council for Leadership in Educational Administration, Toronto 02769
Ontario Council of Health, Toronto 02770
Ontario Council on University Affairs, Toronto 02771
Ontario Deafness Research Foundation, Toronto 02772
Ontario Dental Association, Toronto 02773
Ontario Educational Research Council, Ajax 02774
Ontario Electric Railway Historical Association, Scarborough 02775
Ontario Family Studies Home Economics Educators Association, Thorold 02776
Ontario Farm Drainage Association, Chatham 02777
The Ontario Genealogical Society, Toronto 02778
Ontario Gerontology Association, Waterloo 02779
Ontario Group Psychotherapy Association, Etobicoke 02780
Ontario Health Record Association, Harrow 02781
Ontario Historical Society, Willowdale 02782
Ontario Institute of Agrologists, Guelph 02783
Ontario Library Association Literacy Guild, Toronto 02784
Ontario Lung Association, Toronto 02785
Ontario Music Educators' Association, London 02786
Ontario Osteopathic Association, Toronto 02787
Ontario Podiatry Association, Toronto 02788
Ontario Psychogeriatric Association, Kingston 02789
Ontario Public Interest Research Group, Ottawa 02790
Ontario Public School Boards' Association, North Bay 02791
Ontario Public School Teachers' Federation, Toronto 02792

Ontario Registered Music Teachers Association, Toronto 02793
Ontario School Counsellors' Association, Don Mills 02794
Ontario Secondary School Teachers' Federation, Toronto 02795
Ontario Society for Cable Television Engineering, Mississauga 02796
Ontario Society for Industrial Archaeology, Toronto 02797
Ontario Society of Clinical Hypnosis, Toronto 02798
Ontario Society of Occupational Therapists, Toronto 02799
The Ontario Speech and Hearing Association, Toronto 02800
Ontario Teachers' Federation, Toronto 02801
Ontario Teachers' Superannuation Commission, North York 02802
Ontario Thoracic Society, Toronto 02803
Ontario Veterinary Association, Guelph 02804
Ontario Vocational Educational Association, Caledonia 02805
OPACT, Studio City 16267
OPAKF, Budapest 06667
Open Space Institute, New York 16250
Opera America, Washington 16251
Opera Svizzera dei Monumenti d'Arte, Locarno 10775
Operational Research Society, Birmingham 12655
Operational Research Society of India, New Dehli 06847
Operational Research Society of New Zealand, Wellington 09184
Operation Enterprise, Hamilton 16252
Operation Eyesight Universal, Calgary 02806
Operations Research Society of America, Baltimore 16253
Opetusalan Ammattijarjestö, Helsinki 03987
Ophthalmic Research Institute, Bethesda 16254
Ophthalmological Society of East Africa, Nairobi 08506
Ophthalmological Society of Egypt, Cairo 03912
Ophthalmological Society of the Republic of China, Taipei 03317
Opolskie Towarzystwo Przyjaciót Nauk, Opole 09548
Optical, Acoustical and Filmtechnical Society, Budapest 06667
Optical Society of America, Washington 16255
Optical Society of India, Calcutta 06848
Opticians and Optometrists Association of New South Wales, Sydney 00443
Opticians Association of America, Fairfax 16256
Optikai, Akusztikai és Filmtechnikai Egyesület, Budapest 06667
Optometric Historical Society, Saint Louis 16257
OPUS, Esher 12659
O.R., London 12652
Oral and Dental Research Institute, Tygerberg 10159
Oral History Association, Los Angeles 16258
Oral History Society, Colchester 12656
ORAU, Oak Ridge 16239
ORCA, Leeds 12035
Orchester-Akademie des Berliner Philharmonischen Orchesters e.V., Berlin 06224
Ordem dos Engenheiros, Lisboa 09664
Ordem dos Médicos, Lisboa 09665
Order of the Indian Wars, Little Rock 16259
Orders and Medals Research Society, Croydon 12657
Orde van Architecten, Bruxelles 01418
Ordine Nazionale degli Attuari, Roma 07762
Ordre des Architectes, Bruxelles 01418
Ordre des Architectes, Paris 04607
Ordre des Chirurgiens Dentistes, Paris 04608
Ordre des Géomètres-Experts, Paris 04609
Ordre National des Chirurgiens Dentistes, Paris 04610
Ordre National des Médecins, Paris 04611
Organ Historical Society, Richmond 16260
Organic Federation, Malvern 12658
Organic Soil Association of South Africa, Parklands 10160
Organisation Canadienne pour l'Education Préscolaire, Edmonton 02101
Organisation de la Jeunesse Esperantiste Française, Paris 04612
Organisation des Directeurs des Musées d'Art Canadiens, Victoria 02024
Organisation des Musées Militaires du Canada, Ottawa 02807
Organisation Européenne pour l'Equipement de l'Aviation Civile, Paris 04613
Organisation Hydrographique Internationale, Monte Carlo 08797
Organisation Internationale contre le Trachome, Créteil 04614
Organisation Internationale de Métrologie Légale, Paris 04615
Organisation Internationale de Protection Civile, Petit-Lancy 10776
Organisation Internationale de Recherche sur la Cellule, Paris 04616
Organisation Météorologique Mondiale, Genève 10777
Organisation of Professional Users of Statistics, Esher 12659
Organisation of Teachers of Transport Studies, Southall 12660
Organisation Scientifique des Industries du Bâtiment, Paris 04617
Organização Guarão de Ensino, Guaratinguetá 01699
Organización de Estados Iberoamericanos para la Educación, la Ciencia y la Cultura, Madrid 10333
Organización de Estudios Tropicales, San José 03451
Organización de Universidades Católicas de América Latina, Buenos Aires 00150
Organización Médica Colegial-Consejo General de Colegios Médicos de España, Madrid 10334
Organizačné Stredisko Vedeckých Spoločností pri SAV, Bratislava 10059
Organizational Behavior Teaching Society, Norman 16261
Organization Centre of the Scientific Societies of the SAS, Bratislava 10059
Organization Development Institute, Chesterland 16262
Organization for Co-ordination and Co-operation in the Control of Major Endemic Diseases, Bobo-Dioulasso 01810
Organization for Flora Neotropica, Bronx 16263
Organization for Tropical Studies, Durham 16264
Organization of African Unity – Scientific, Technical and Research Commission, Lagos 09244
Organization of American Historians, Bloomington 16265
Organization of American Kodaly Educators, Thibodaux 16266
Organization of Latin American Jesuit Universities and Faculties, Bogotá 03350
Organization of Military Museums of Canada, Ottawa 02807
Organization of Nordic Teachers Associations, Oslo 09342
Organization of Professional Acting Coaches and Teachers, Studio City 16267
Organization of Research and Educational Planning, Teheran 06903
ORI, Bethesda 16254
Oriental Ceramic Society, London 12661
Orientalische Gesellschaft, Wien 00930
Oriental Numismatic Society, Reading 12662
Orientalsk Samfund, København 03792
Ornithological Association of Zimbabwe, Harare 17176
The Ornithological Society, Valletta 08669
Ornithological Society of Japan, Tokyo 08291
Ornithological Society of Moldova, Chişinău 08786
Ornithological Society of New Zealand, Wellington 09185
Ornithological Society of the Middle East, Sandy 12663
ORS, Birmingham 12655
O.R.S., Oxford 12670
ORS, Rosemont 16271
ORSA, Baltimore 16253
ORSI, New Dehli 06847
Országos Állategészségügyi Intézet, Budapest 06668
Országos Erdészeti Egyesület, Budapest 06669
Országos Magyar Bányászati és Kohászati Egyesület, Budapest 06670
Országos Magyar Cecília Társulat, Budapest 06671
Orthodontic Education and Research Foundation, Saint Louis 16268
Orthodox Theological Society in America, Brookline 16269
Orthopedic Foundation for Animals, Columbia 16270
Orthopedic Research Society, Rosemont 16271
Orton Dyslexia Society, Baltimore 16272
OS, London 12654
OS, Washington 16242
OSA, Washington 16255
OSE, San Francisco 16243
OSEA, Nairobi 08506
OSG, Tulsa 16274
OSI, Calcutta 06848
OSI, New York 16250
OSMA, Locarno 10775
OSME, Sandy 12663
OSNZ, Wellington 09185
OSRA, Springfield 16248
OSROC, Taipei 03318
Ost-Akademie e.V., Lüneburg 06225
Osteopathic Association of Great Britain, London 12664
Osteopathic College of Ophthalmology and Otorhinolaryngology, Dayton 16273
Osteopathic Medical Association, London 12665
Osteoporosis Society of Canada, Toronto 02808
Osteuropa-Institut München, München 06226
OSTIV, Wessling 08945
Oto-Laryngological Society of Israel, Tel Aviv 07110
The Otolaryngological Society of the Republic of China, Taipei 03318
Oto-Rhino-Laryngological Society of Japan, Tokyo 08313
Otosclerosis Study Group, Tulsa 16274
OTPN, Opole 09548
OTS, Durham 16264
OTSA, Brookline 16269
Ottawa-Carleton Safety Council, Ottawa 02809
Otto A. Friedrich-Kuratorium für Grundlagenforschung zur Eigentumspolitik, Köln 06227
OTTS, Southall 12660
OUAS, Oxford 12671
Oulun Luonnonystäväin Yhdistys, Oulu 03988
Outdoor Education Association, Denville 16275
Outdoor Ethics Guild, Bucks Harbor 16276
Outer Critics Circle, New York 16277
Outward Bound – Deutsche Gesellschaft für Europäische Erziehung e.V., München 06228
Overseas Agricultural Development Association, Tokyo 08151
Overseas Doctors Association in the UK, Manchester 12666
Overseas Education Association, Washington 16278
OWL, Monmouth 12138
Oxford Centre for Islamic Studies, Oxford 12667
Oxford Preservation Trust, Oxford 12668
Oxfordshire Architectural and Historical Society, Oxford 12669
Oxfordshire Record Society, Oxford 12670
Oxford University Archaeological Society, Oxford 12671
Oyo-buturi Gakkai, Tokyo 08402

PA, Research Triangle Park 16295
PAA, Hancock 16344
PAA, Honolulu 16279
PAA, Washington 16337
PAABS, Kansas City 16288
PAAO, Arlington 16289
PABC, Burnaby 02818
PABS, Washington 16290
Pacific Arts Association, Honolulu 16279
Pacific Coast Family Therapy Training Association, Vancouver 02810
Pacific Dermatologic Association, Schaumburg 16280
Pacific Rocket Society, Oxnard 16281
Pacific Science Association, Honolulu 16282
Pacific Seabird Group, Aiken 16283
Pacific Studies Center, Mountain View 16284
Packaging Education Foundation, Herndon 16285
PACRA, Lower Hutt 09172
PACT/NADAP, New York 15916
Pädagogische Arbeitsstelle des Deutschen Volkshochschul-Verbandes e.V., Frankfurt 06229
Pädagogisches Zentrum, Berlin 06230
Paedagogisk Forening, København 03793
Paediatric Society of Victoria, Parkville 00444
Paediatrics Society of Madrid and the Central Region, Madrid 10375
Paediatric Welfare and Research Association, Macclesfield 12672
PAHA, Chicago 16333
PAHO, Washington 16292
Paint Research Association, Teddington 12673
PAJ, Roma 16953
Pakistam Anti Tuberculosis Association, Karachi 09377
Pakistan Academy of Letters, Islamabad 09378
Pakistan Academy of Sciences, Islamabad 09379
Pakistan Atomic Energy Commission, Islamabad 09380
Pakistan Board for Advancement of Literature, Lahore 09381
Pakistan Concrete Institute, Karachi 09382
Pakistan Council of Architects and Town Planners, Karachi 09383
Pakistan Council of Scientific and Industrial Research, Karachi 09384
Pakistan Historical Society, Karachi 09385
Pakistan Institute of International Affairs, Karachi 09386
Pakistan Library Association, Karachi 09387
Pakistan Medical Association, Karachi 09388
Pakistan Medical Research Council, Karachi 09389
Pakistan Museum Association, Karachi 09390
Pakistan Philosophical Congress, Lahore 09391
Palaeontographical Society, Nottingham 12674
Palaeontological Association, Cardiff 12675
Palaeontological Society of China, Nanjing 03213
Palaeontological Society of Japan, Tokyo 08342
Paläontologische Gesellschaft, Frankfurt 06231
Paleontological Research Institution, Ithaca 16286
Paleontological Society, Socorro 16287
Palestine Exploration Fund, London 12676
Pali Text Society, London 12677
Palmerston North Medical Research Foundation, Palmerston North 09186
Palynological Society of India, Thiruvananltapuram 06849
PAMA, New York 16293
PAMI, Roma 16954
PAN, Warszawa 09549
Pan African Association of Neurological Sciences, Lagos 09245
Pan American Association of Biochemical Societies, Kansas City 16288
Pan-American Association of Ophthalmology, Arlington 16289
Pan American Association of Social Medicine, Rio de Janeiro 01604
Pan-American Biodeterioration Society, Washington 16290
Panamerican Cultural Circle, Verona 16291
Panamerican Federation of Associations of Medical Schools, Caracas 17006
Pan American Health Organization, Washington 16292
Pan American Institute of Geography and History, México 08740
Pan-American Medical Association, New York 16293
Pan American Standards Commission, Buenos Aires 00122
Pancyprian Medical Association, Nicosia 03498
Panellinios Enosis Technikon, Athinai 06541
Panellinios Omospondia Syndesmon Geoponon, Athinai 06542
Pan-Pacific Surgical Association, Honolulu 16294
Paper and Wood-Working Society, Moskva 09940
Papír- és Nyomdaipari Müszaki Egyesület, Budapest 06672
Papirindustriens Forskningsinstitutt, Oslo 09343
Papua New Guinea Library Association, Boroko 09421
Papua New Guinea Scientific Society, Boroko 09422
Papua New Guinea Teachers' Association, Boroko 09423
Parapsychological Association, Research Triangle Park 16295
PARF, Buffalo 16347
Paris et son Histoire, Paris 04618
Parker Chiropractic Resource Foundation, Fort Worth 16296
Parkinson Foundation of Canada, Toronto 02811
Parkinson's Disease Society of the United Kingdom, London 12678
Parkinson's Educational Program – U.S.A., Newport Beach 16297
Parliamentary and Scientific Committee, London 12679
Parrot Society, Bedford 12680
PAS, Chicago 16353
PAS, Lawton 16303
PAS, Waukesha 16321
PASF, Des Plaines 16316
Pashto Academy, Peshawar 09392
Passaic River Coalition, Basking Ridge 16298
Pastel Society, London 12681
PAT, Derby 12724
PAT, Philadelphia 14545
PATH, Seattle 16360
Pathological Society of Great Britain and Ireland, London 12682
Patronato de Biología Animal, Madrid 10335
Pattern Recognition Society, Washington 16299
PAVAS, Kinnelon 16304
Le Pays Bas-Normand, Flers 04619
PC, New York 16338
PC, Orlando 16359
PCA, Baltimore 16354
PCA, Bowling Green 16336
PCC, Verona 16291
PCC, Washington 16339
PCHRG, Washington 16374
PCJILMCC, Washington 16310
PCMH, New York 16342
PCO, 's-Gravenhage 09062
PCRF, Fort Worth 16296
PCRS, Brea 16349
PCSIR, Karachi 09384
PDA, Schaumburg 16280
PDMHS, Nottingham 12683
PDR, Wuppertal 06234
PDS, London 12678
Peace Science Society, Binghamton 16300
Peak District Mines Historical Society, Nottingham 12683
Peakland Archaeological Society, Stockport 12684
Pedagocical Society of Slovenia, Ljubljana 10116
Pedagogical Society of Yugoslavia, Beograd 17106
Pedagoško-Književni zbor, Savez Pedagoških Društava Hrvatske, Zagreb 03465
PEER, New York 16363
PEF, Herndon 16285
PEHLA, Frankfurt 05812
Pembrokeshire Agricultural Society, Haverfordwest 12685
Pembrokeshire Historical Society, Scloddau Fishguard 12686
PENAC, New York 16301
PEN All-India Centre, Bombay 06850
PEN American Center, New York 16301
PEN Centre, Athinai 06543
PEN Centre of Albania, Tirana 00006
PEN Centre of Lithuania, Vilnius 08574
PEN Centre of Moldova, Chişinău 08787
PEN Centre of Zimbabwe, Harare 17177
PEN Club, Madrid 10336
PEN Club Argentino, Buenos Aires 00151
PEN Club de Bolivia-Centro Internacional de Escritores, La Paz 01547
PEN Club de Colombia, Bogotá 03399
PEN Club de Suisse Romande, Carouge 10778
PEN Clube do Brasil-Associação Universal de Escritores, Rio de Janeiro 01700
Pendragon Society, Bristol 12687
P.E.N. English Centre, London 12688
P.E.N. International Centre, Roma 07763
PEN International, Centre Québécois, Montréal 02812
PEN International (Sydney Centre), Woollahra 00445
P.E.N. Maison Internationale, Paris 04620
P.E.N. New Zealand, Auckland 09187
Pennklubben, Stockholm 10514
P.E.N., Turkish Centre, Istanbul 11149
P.E.N. Yazarlar Derneği, Istanbul 11149
PEN Zentrum Bundesrepublik Deutschland, Darmstadt 06232
People United for Rural Education, Alden 16302
PEP, London 12711
PEP/USA, Newport Beach 16297
PERA, Melton Mowbray 12722
Percussive Arts Society, Lawton 16303
Performing and Visual Arts Society, Kinnelon 16304
Perkumpulan Penggemar Alam di Indonesia, Bogor 06892
Perlite Institute, Staten Island 16305
Permanent Commission of Gosplan, Moskva 09941
Permanent International Association of Navigation Congresses – United States Section, Washington 16306
Permanent International Committee of Linguists, Leiden 09060
Permanent Working Group of European Junior Hospital Doctors, Bern 10779
Persatuan Insinyur Indonesia, Jakarta Pusat 06893
Perthshire Society of Natural Science, Perth 12689
Peruvian Association of Agricultural Librarians and Documentalists, Lima 09448
Peruvian Heart Association, Wadsworth 16307
PESGB, London 12691
PET, Athinai 06541
PETA, Rozelle 00447
Peter-Schwingen-Gesellschaft e.V., Bonn 06233
Peter Warlock Society, London 12690
Petroleum and Gas Society, Moskva 09942
Petroleum Association of Japan, Tokyo 08412
Petroleum Exploration Society of Great Britain, London 12691
PF, Chicago 16317
PF, Richmond 16329
PFA, Justice 16319
PFB Project, Washington 16308
PFRA, North Huntington 16356
PGRO, Peterborough 12720
PGS, Chicago 16334
P.G. Wodehouse Society, Santa Monica 16309
PGWS, Santa Monica 16309
PGZ, Zürich 10780

PHA, Manila 09526
PHA, Wadsworth 16307
The Pharmaceutical Society of Denmark, København 03577
Pharmaceutical Society of Ghana, Accra 06492
Pharmaceutical Society of Ireland, Dublin 07019
The Pharmaceutical Society of Japan, Tokyo 08395
Pharmaceutical Society of Northern Ireland, Belfast 12692
Pharmaceutical Society of Zimbabwe, Harare 17178
Pharmacological Society of Canada, Vancouver 02813
Pharmacological Society of Macedonia, Skopje 08601
Pharmacy Association of Nova Scotia, Halifax 02814
Pharmacy Council of India, New Dehli 06851
Pharma-Dokumentationsring e.V., Oss 09061
Pharma-Dokumentationsring e.V., Wuppertal 06234
Philip C. Jessup International Law Moot Court Competition, Washington 16310
Philip Jose Farmer Society, Champaign 16311
Philippine Association of Nutrition, Manila 09523
Philippine Atomic Energy Commission, Quezon City 09524
Philippine Council of Chemists, Manila 09525
Philippine Historical Association, Manila 09526
Philippine Institute of Architects, Manila 09527
Philippine Library Association, Manila 09528
Philippine Medical Association, Quezon City 09529
Philippine Paediatric Society, Manila 09530
Philippine Pharmaceutical Association, Manila 09531
Philippine Society of Parasitology, Manila 09532
Philippine Veterinary Medical Association, Quezon City 09533
The Philological Society, London 12693
Philosophical Association of the Philippines, Manila 09534
Philosophical Research Society, Los Angeles 16312
Philosophical Society, Khartoum 10428
Philosophical Society, Tokyo 08429
The Philosophical Society of England, Brighton 12694
The Philosophical Society of Finland, Helsingin Yliopisto 04016
Philosophic Society for the Study of Sport, Union 16313
Philosophische Gesellschaft an der Universität Graz, Graz 00931
Philosophische Gesellschaft Innsbruck, Innsbruck 00932
Philosophische Gesellschaft in Salzburg, Salzburg 00933
Philosophische Gesellschaft Klagenfurt, Klagenfurt 00934
Philosophische Gesellschaft Wien, Wien 00935
Philosophischer Fakultätentag, Saarbrücken 06235
Philosophy and Humanities Society, Teheran 06902
Philosophy of Education Society of Great Britain, Coleraine 12695
Philosophy of Science Association, East Lansing 16314
Philosophy Society, Moskva 09943
Phoenix Society for Burn Survivors, Levittown 16315
Photo Electric Arts Foundation, Toronto 02815
The Photogrammetric Society, London 12696
Photographic Art and Science Foundation, Des Plaines 16316
The Photographic Historical Society of Canada, Toronto 02816
Photographic Society of Ireland, Dublin 07020
photokinonorm, Berlin 06153
PHS, London 12718
PHS, Philadelphia 16351
PHS, Pikesville 16341
Physical Education Association of Great Britain and Northern Ireland, Birmingham 12697
Physical Education Council of the New Brunswick Teachers Association, Moncton 02817
Physical Society of China, Taipei 03319
The Physical Society of Japan, Tokyo 08284
Physical Society of Moldova, Chişinău 08788
Physicians Forum, Chicago 16317
Physico-Chemical Society of Japan, Kyoto 08285
Physikalische Gesellschaft Zürich, Zürich 10780
Physikalisch-Medizinische Sozietät zu Erlangen, Erlangen 06236
Physiological and Biochemical Society of Southern Africa, Pretoria 10161
The Physiological Society, Oxford 12698
Physiological Society of Japan, Tokyo 08364
Physiological Society of New Zealand, Palmerston North 09188
Physiotherapy Association of British Columbia, Burnaby 02818
Physiotherapy Foundation of Canada, London 02819
Phytochemical Society of Europe, Swansea 12699
Phytochemical Society of North America, Towson 16318
Phytogeographical Society, Kyoto 08423
Phytopathological Society of Japan, Tokyo 08382
PI, Staten Island 16305
PIA, Fairfield 16325
PIANC, Washington 16306
PIASA, New York 16335
Pierre Fauchard Academy, Justice 16319
Pilgrim Society, Plymouth 16320
Pioneer America Society, Waukesha 16321
Pipe Roll Society, London 12700
Pira International, Leatherhead 12701
Pirandello Society of America, Jamaica 16322
PIRI, Chestnut Hill 16370
PJFS, Champaign 16311
PLA, Chicago 16377
P.L.A., Pinner 12719
Plainsong and Mediaeval Music Society, Cambridge 12702
Planalto Academy of Arts and Letters, Luziânia 01572
Planetary Society, Pasadena 16323
Planned Parenthood Federation of America, New York 16324
Planning Institute of British Columbia, Vancouver 02820
PLANTI, Serdang 08631
Plant Protection Society of Western Australia, Victoria Park 00446
The Plastics and Rubber Institute, London 12703
Plastics Institute of America, Fairfield 16325
Plastic Surgery Research Council, Chapel Hill 16326
Play Matters – The National Association of Toy and Leisure Libraries, London 12704
Play Schools Association, New York 16327
Playwriters Association of the Republic of China, Taipei 03320
PLEI, Washington 16375
PLEN, Washington 16376
PLI, New York 16348
PLMA, Paceville 08670
PLTR, Warszawa 09550
The Plymouth Athenaeum, Plymouth 12705
Plywood Research Foundation, Tacoma 16328
PM, Topeka 16361
PMA, Karachi 09388
PMA, Los Angeles 16350
P.M.A., Quezon City 09529
PMI, Drexel Hill 16362
PMI, Richardson 16355
PMRC, Karachi 09389
PNG, Van Nuys 16357
PNGLA, Boroko 09421
Poe Foundation, Richmond 16329
Poe Studies Association, Monmouth 16330
Poetry Society, London 12706
Poetry Society of America, New York 16331
Poets' Theatre Guild, South Croydon 12707
Policy Studies Organization, Urbana 16332
Polish American Historical Association, Chicago 16333
Polish Biochemical Society, Warszawa 09557
Polish Canadian Librarians Association, Toronto 02821
Polish Dental Association, Łódź 09606
Polish Dermatological Society, Poznań 09560
Polish Economic Society, Warszawa 09562
Polish Educational Society, London 12708
Polish Ethnological Society, Wroclaw 09584
Polish Genealogical Society, Chicago 16334
Polish Geographical Society, Warszawa 09571
Polish Institute of Arts and Sciences in America, New York 16335
Polish Librarians Association, Warszawa 09615
Polish Mathematical Society, Warszawa 09585
Polish Medical Association, London 12709
Polish Nautological Society, Gdynia 09592
Polish Neurological Society, Konstancin-Jeziorna 09594
Polish Oriental Society, Warszawa 09595
Polish Orthopedic and Traumatologic Society, Warszawa 09596
Polish Otolaryngological Society, Warszawa 09597
Polish Parasitological Society, Wroclaw 09598
Polish P.E.N. Centre, Warszawa 09610
Polish Physical Society, Warszawa 09569
The Polish Psychiatric Association, Warszawa 09602
Polish Psychological Association, Warszawa 09603
Polish Society for Immunology, Łódź 09579
Polish Society of Animal Production, Warszawa 09609
Polish Society of History of Medicine and Pharmacy, Warszawa 09577
Polish Society of Microbiology, Warszawa 09588
Polish Society of Pathologists, Łódź 09599
Polish Society of Theoretical and Applied Mechanics, Warszawa 09586
Polish Society of Veterinary Sciences, Warszawa 09591
Polish Sociological Association, Warszawa 09605
Polish Teachers' Association in Canada, Toronto 02822
Polish Underground Movement (1939-1945) Study Trust, London 12710
Political and Economic Planning, London 12711
Political Studies Association of the United Kingdom, Belfast 12712
Politics Association, London 12713
Politische Akademie Biggesee, Attendorn-Neulisternohl 06237
POLLICHIA e.V., Annweiler 06238
Pollution Control Association of Ontario, Aurora 02823
Pollution Probe Foundation, Toronto 02824
Pollution Research Group, Durban 10162
Polska Akademia Nauk, Roma 07226
Polska Akademia Nauk, Warszawa 09549
Polskie Lekarskie Towarzystwo Radiologiczne, Warszawa 09550
Polskie Towarzystwo Anatomiczne, Warszawa 09551
Polskie Towarzystwo Antropologiczne, Warszawa 09552
Polskie Towarzystwo Archeologiczne i Numizmatyczne, Warszawa 09553
Polskie Towarzystwo Astronautyczne, Warszawa 09554
Polskie Towarzystwo Astronomiczne, Warszawa 09555
Polskie Towarzystwo Balneologii, Bioklimatologii i Medycyny Fizykalnej, Poznan 09556
Polskie Towarzystwo Biochemiczne, Warszawa 09557
Polskie Towarzystwo Botaniczne, Warszawa 09558
Polskie Towarzystwo Chemiczne, Warszawa 09559
Polskie Towarzystwo Dermatologiczne, Poznań 09560
Polskie Towarzystwo Dermatologiczne, Warszawa 09561
Polskie Towarzystwo Ekonomiczne, Warszawa 09562
Polskie Towarzystwo Endokrynologiczne, Warszawa 09563
Polskie Towarzystwo Entomologiczne, Warszawa 09564
Polskie Towarzystwo Farmaceutyczne, Warszawa 09565
Polskie Towarzystwo Filologiczne, Warszawa 09566
Polskie Towarzystwo Filozoficzne, Warszawa 09567
Polskie Towarzystwo Fizjologiczne, Lublin 09568
Polskie Towarzystwo Fizyczne, Warszawa 09569
Polskie Towarzystwo Fizyki Medycznej, Warszawa 09570
Polskie Towarzystwo Geograficzne, Warszawa 09571
Polskie Towarzystwo Geologiczne, Kraków 09572
Polskie Towarzystwo Ginekologiczne, Warszawa 09573
Polskie Towarzystwo Gleboznawcze, Warszawa 09574
Polskie Towarzystwo Hematologiczne, Wroclaw 09575
Polskie Towarzystwo Higieny Psychicznej, Warszawa 09576
Polskie Towarzystwo Historii Medycyny i Farmacji, Warszawa 09577
Polskie Towarzystwo Historyczne, Warszawa 09578
Polskie Towarzystwo Immunologiczne, Łódź 09579
Polskie Towarzystwo Jezykoznawcze, Kraków 09580
Polskie Towarzystwo Kardiologiczne, Warszawa 09581
Polskie Towarzystwo Lekarski, Warszawa 09582
Polskie Towarzystwo Lesne, Warszawa 09583
Polskie Towarzystwo Ludoznawcze, Wroclaw 09584
Polskie Towarzystwo Matematyczne, Warszawa 09585
Polskie Towarzystwo Mechaniki Teoretycznej i Stosowanej, Warszawa 09586
Polskie Towarzystwo Medycyny Pracy, Sosnowiec 09587
Polskie Towarzystwo Mikrobiologów, Warszawa 09588
Polskie Towarzystwo Milosników Astronomii, Kraków 09589
Polskie Towarzystwo Mineralogiczne, Kraków 09590
Polskie Towarzystwo Nauk Weterynaryjnych, Warszawa 09591
Polskie Towarzystwo Nautologiczne, Gdynia 09592
Polskie Towarzystwo Neofilologiczne, Poznan 09593
Polskie Towarzystwo Neurologiczne, Konstancin-Jeziorna 09594
Polskie Towarzystwo Orientalistyczne, Warszawa 09595
Polskie Towarzystwo Ortopedyczne i Traumatologiczne, Warszawa 09596
Polskie Towarzystwo Otolaryngologiczne, Warszawa 09597
Polskie Towarzystwo Parazytologiczne, Wroclaw 09598
Polskie Towarzystwo Patologów, Łódź 09599
Polskie Towarzystwo Pediatryczne, Warszawa 09600
Polskie Towarzystwo Przyrodników im. Kopernika, Warszawa 09601
Polskie Towarzystwo Psychiatryczne, Warszawa 09602
Polskie Towarzystwo Psychologiczne, Warszawa 09603
Polskie Towarzystwo Semiotyczne, Warszawa 09604
Polskie Towarzystwo Socjologiczne, Warszawa 09605
Polskie Towarzystwo Stomatologiczne, Łódź 09606
Polskie Towarzystwo Urologiczne, Katowice 09607
Polskie Towarzystwo Zoologiczne, Wroclaw 09608
Polskie Towarzystwo Zootechniczne, Warszawa 09609
Polski Klub Literacki P.E.N., Warszawa 09610
Polski Komitet Pomiarów Automatyki, Warszawa 09611
Polymerteknisk Selskab, København 03794
Polynesian Society, Auckland 09189
Polyteknisk Forening, Lyngby 03795
Pontifical Academy of Sciences, Roma 16952
Pontificia Academia Scientiarum, Roma 16952
Pontificia Accademia dell'Immacolata, Roma 16953
Pontificia Accademia Mariana Internationalis, Roma 16954
Pontificia Accademia Romana di Archeologia, Roma 16955
Pontificia Accademia Teologica Romana, Roma 16956
Pontificia Insigne Accademia Artistica dei Virtuosi al Pantheon, Roma 16957
Popular Culture Association, Bowling Green 16336
Population Association of America, Washington 16337
Population Association of China, Taipei 03321
Population Association of New Zealand, Wellington 09190
Population Council, New York 16338
Population Crisis Committee, Washington 16339
Population Reference Bureau, Washington 16340
Portugese Society of Rheumatology, Lisboa 09714
Portuguese Archaeological, Historical and Ethnographical Institute, Lisboa 09661
Portuguese Architects' Association, Lisboa 09653
Portuguese Pharmaceutical Society, Lisboa 09684
Portuguese Road Safety Association, Lisboa 09666
Portuguese Society of Allergology and Clinical Immunology, Lisboa 09689
Portuguese Society of Physical Medicine and Rehabilitation, Lisboa 09704
Portuguese Society of Sociological Veterinary Studies, Lisboa 09716
Portuguese Society of Specialists in Small Animals, Lisboa 09699
Portuguese-Spanish-Latin American Anatomical Society, Lisboa 09673
Portuguese Veterinary Society of Comparative Anatomy, Lisboa 09715
Postakademie, Kleinheubach 06239
Postal History Society, Pikesville 16341
Postal History Society of Canada, Ottawa 02825
Postgraduate Center for Mental Health, New York 16342
Postgraduate Teaching and Research Organization, Rhode-Saint-Genese 01512
Potash and Phosphate Institute, Atlanta 16343
Potash and Phosphate Institute of Canada, Etobicoke 02826
Potato Association of America, Hancock 16344
Poultry Science Association, Champaign 16345
Power and Electrical Power Engineering Society, St. Petersburg 09944
The Powys Society, Kilmersdon 12714
Powys Society of North America, Valparaiso 16346
Poznánskie Towarzystwo Przyjaciól Nauk, Poznan 09612
PPFA, New York 16324
PPI, Atlanta 16343
PPNF, La Mesa 16352
PPSA, Honolulu 16294
PPU, Oak Park 16358
PRA, Lawrence 16365
PRA, London 12726
PRA, San Carlos 16367
PRA, Teddington 12673
Practical Allergy Research Foundation, Buffalo 16347
Practising Law Institute, New York 16348
PRB, Washington 16340
PRC, Basking Ridge 16298
Precision Chiropractic Research Society, Brea 16349
Precision Measurements Association, Los Angeles 16350
Preclinical Diagnostic Chemistry Research Group of the South African Medical Research Council, Congella 10163
Prehistoric Society, Swindon 12715
Pre-History Society of Zimbabwe, Harare 17179
Presbyterian Church of Wales Historical Society, Aberystwyth 12716
Presbyterian Historical Society, Philadelphia 16351
Press and Publishing Engineering Society, Moskva 09945
Pretoria Horticultural Society, Pretoria 10164
Prevenção Rodoviária Portuguesa, Lisboa 09666
PRF, Tacoma 16328
PRI, Ithaca 16286
PRI, London 12703
PRI, Palo Alto 16368
Price-Pottenger Nutrition Foundation, La Mesa 16352
Primary English Teaching Association, Rozelle 00447
Primate Behaviour Research Group, Johannesburg 10165
Primate Society of Great Britain, Reading 12717
Primitive Art Society of Chicago, Chicago 16353
PRIM&R, Boston 16379
Prince Edward Island Association of Community Schools, Belle River 02827
Prince Edward Island Association of Medical Radiation Technologists, Montague 02828
Prince Edward Island Association of Social Workers, Charlottetown 02829
Prince Edward Island Cerebral Palsy Association, Charlottetown 02830
Prince Edward Island Criminology and Corrections Association, Charlottetown 02831
Prince Edward Island Dietetic Association, Charlottetown 02832
Prince Edward Island Institute of Agrologists, Charlottetown 02833
Prince Edward Island Lung Association, Charlottetown 02834
Prince Edward Island Medical Society, Charlottetown 02835
Prince Edward Island Museum and Heritage Foundation, Charlottetown 02836
Prince Edward Island Physical Education Association, Charlottetown 02837
Prince Edward Island Psychiatric Association, Charlottetown 02838
Prince Edward Island Society of Occupational Therapists, Charlottetown 02839
Prince Edward Island Speech and Hearing Association, Charlottetown 02840
Prince Edward Island Teachers' Federation, Charlottetown 02841
Prince Edward Island Veterinary Medical Association, Charlottetown 02842
Print and Drawing Council of Canada, Toronto 02843
Print Council of America, Baltimore 16354
Printing Historical Society, London 12718
Printing Trades Accident Prevention Association, Toronto 02844

Prirodoslovno Društvo Slovenije v Ljubljani, Ljubljana 10102
The Private Libraries Association, Pinner 12719
Pro Austria, Wien 00936
Probe Ministries International, Richardson 16355
Processors and Growers Research Organisation, Peterborough 12720
Procurators Fiscal Society, Dumbarton 12721
Production Engineering Research Association, Melton Mowbray 12722
Production Mangagement Action Group, Henley-on-Thames 12723
Pro Familia, Frankfurt 06240
Professional Association of Teachers, Derby 12724
Professional Corporation of Physicians of Quebec, Montréal 02395
Professional Corporation of Speech Therapists and Audiologists of Quebec, Montréal 02397
Professional Development Institute, Ottawa 02845
Professional Football Researches Association, North Huntington 16356
Professional Institute of the Public Service of Canada, Ottawa 02846
Professional Librarians of Malta Association, Paceville 08670
Professional Marketing Research Society, Toronto 02847
Professional Numismatists Guild, Van Nuys 16357
Professional Psychics United, Oak Park 16358
Professors of Curriculum, Orlando 16359
Program for Appropriate Technology in Health, Seattle 16360
Programme Committee Surface Physics, Chemistry and Mechanics, Moskva 09946
Program on Man and the Biosphere, Paris 04621
Pro Helvetia, Zürich 10781
Project Magic, Topeka 16361
Project Management Institute, Drexel Hill 16362
Project on Equal Education Rights, New York 16363
PROMAG, Henley-on-Thames 12723
PROMETEIA, Bologna 07391
Prosthodontics Society of South Africa, Houghton 10166
Protein Society, Bethesda 16364
Protestant Institute for Interdisciplinary Studies, Heidelberg 05691
Protestants-Christelijke Onderwijsvakorganisatie, 's-Gravenhage 09062
Protozoological Society of Moldova, Chişinău 08789
Proust Research Association, Lawrence 16365
Provincial Association of Catholic Teachers (Québec), Montréal 02848
Provincial Association of Protestant Teachers of Québec, Dollard des Ormeaux 02849
Provincial Medical Board of Nova Scotia, Halifax 02850
Prozess- und Umwelttechnik, Düsseldorf 05903
PRP, Lisboa 09666
PRS, Los Angeles 16312
PRS, Oxnard 16281
PRS, Washington 16299
PRSA, New York 16378
Prüf- und Forschungsinstitut für die Schuhherstellung e.V., Pirmasens 06241
PS, Champaign 16372
PS, London 12706
PS, New York 16371
PS, Pasadena 16323
PS, Plymouth 16320
PS, San Jose 16373
PS, Socorro 16287
PSA, Champaign 16345
PSA, East Lansing 16314
PSA, Honolulu 16282
PSA, Jamaica 16322
PSA, Monmouth 16330
PSA, New York 16327, 16331
PSAUK, Belfast 12712
PSBS, Levittown 16315
PSC, Mountain View 16284
PSC, Nanjing 03213
PSE, Swansea 12699
PSG, Aiken 16283
PSI, Dublin 07020
Psi Chi – The National Honor Society in Psychology, Chattanooga 16366
PSISIG, Huber Heights 16369
PSNA, Towson 16318
PSNA, Valparaiso 16346
PSNI, Belfast 12692
PSNS, Perth 12689
PSO, Urbana 16332
Psoriasis Association, Northampton 12725
Psoriasis Research Association, San Carlos 16367
Psoriasis Research Institute, Palo Alto 16368
PSRC, Chapel Hill 16326
PSRC, Reston 16380
PSS, Binghamton 16300
PSSS, Union 16313
Psychiatric Rehabilitation Association, London 12726
Psychic Science International Special Interest Group, Huber Heights 16369
Psychobiologische Gesellschaft, Mülheim 06242
Psychological Association of South Africa, Pretoria 10167
Psychological Society of Saskatchewan, Regina 02851
Psychologists Interested in Religious Issues, Chestnut Hill 16370
Psychology Society, New York 16371
Psychometric Society, Champaign 16372
Psychonomic Society, San Jose 16373
Psywar Society, Kettering 12727
PTA, Warszawa 09554, 09555
PTAiN, Warszawa 09553
PTCh, Warszawa 09559
PTD, Warszawa 09561
PTE, Warszawa 09562, 09564
PTF, Warszawa 09566, 09567, 09569
PTG, Kraków 09572
PTG, Warszawa 09571
P.T.G., Warszawa 09574
PTH, Warszawa 09578
PTI, Lódź 09579
PTJ, Kraków 09580
P.T.K., Warszawa 09581
PTL, Warszawa 09582, 09583
PTL, Wroclaw 09584
PTM, Warszawa 09585
PTN, Gdynia 09592
P.T.N., Konstancin-Jeziorna 09594
PTN, Poznan 09593
PTNW, Warszawa 09591
PTO, Warszawa 09595
PTOL, Warszawa 09597
PTP, Warszawa 09600
P.T.P., Warszawa 09602
PTP, Warszawa 09603
PTPN, Poznan 09612
PTS, København 03794
PTS, Lódź 09606
PTS, London 12677
PTS, Warszawa 09604, 09605
PTU, Hellerup 03764
PTU, Katowice 09607
PTZ, Warszawa 09609
Public Administration Society of China, Taipei 03322
Public Affairs Council for Education, Sainte-Foy 02852
Public Citizen Health Research Group, Washington 16374
Public Law Education Institute, Washington 16375
Public Leadership Education Network, Washington 16376
Public Library Association, Chicago 16377
Public Relations Society of America, New York 16378
Public Responsibility in Medicine and Research, Boston 16379
Public Service Research Council, Reston 16380
Public Works Historical Society, Evanton 16381
Publizistische Medienplanung für Entwicklungsländer e.V., Aachen 05273
Pugin Society, London 12728
Pugwash Conferences on Science and World Affairs, London 12729
Pulp and Paper Research Institute of Canada, Pointe Claire 02853
PUMST, London 12710
Punjab Bureau of Education, Lahore 09393
Punjabi Adabi Academy, Lahore 09394
Punjab Text Board, Lahore 09395
Punjab University Historical Society, Lahore 09396
PURE, Alden 16302
Pure Water Preservation Society, Woking 12730
Pusat Pembinaan dan Pengembangan Bahasa, Jakarta 06894
Pushkin Club, London 12731
Pushkin Commission, St. Petersburg 09947
PVMA, Quezon City 09533
PWHS, Evanton 16381
PWPS, Woking 12730

QES, Guildford 12733
QMC, London 12734
Q.N.C., Brisbane 00449
Quaid-i-Azam Academy, Karachi 09397
Quaker Esperanto Society, Morecambe 12732
Quebec Association of Independent Schools, Beaconsfield 02854
Quebec Association of Social Rehabilitation Agencies, Montréal 01894
Quebec Family Planning Federation, Montréal 02453
Quebec Forest Industrial Safety Association, Québec 01882
Quebec Heart Foundation, Montréal 02466
Quebec Logging Safety Association, Québec 01881
Québec Lung Association, Québec 01953
Quebec Medical Association, Montréal 01912
Quebec Occupational Medical Association, Montréal 01889
Quebec Pulp and Paper Safety Association, Québec 01883
Quebec Safety League, Montréal 02593
Quebec Thoracic Society, Montréal 02927
Queen Elizabeth II Arts Council of New Zealand, Wellington 09191
Queen's English Society, Guildford 12733
Queensland Institute for Educational Research, Saint Lucia 00448
The Queensland Naturalists' Club, Brisbane 00449
Quekett Microscopical Club, London 12734

RA, London 12760
Raad vir Geesteswetenskaplike Navorsing, Pretoria 10147
Raad voor de Kunst, 's-Gravenhage 09063
Raad voor het Cultuurbeheer, Rijswijk 09064
Rabanus-Maurus-Akademie, Frankfurt 06243
Rabbinic Center for Research and Counseling, Westfield 16382
R.A. Bloch Cancer Foundation, Kansas City 16383
RACAB, Barcelona 10344
RACGP, Sydney 00462
R.A.C.J., Parkville 00461
R.A.C.M.Y.P., Madrid 10343
RACP, Sydney 00456
R.A.C.S., Melbourne 00458
RAD, London 12761
RADA, London 12762
RADAR, London 12771
The Radiance Technique Association International, Saint Petersburg 16384
Radiation Research Society, Reston 16385
Radiation Therapy Oncology Group, Philadelphia 16386
Radical Philosophy Society, Bristol 12735
Radio and Television Research Council, New York 16387
Radio Engineering, Electronics and Telecommunications Society, Moskva 09948
Radioisotope Society of the Philippines, Quezon City 09535
Radiological Protection Mexican Association, México 08772
Radiological Society of North America, Oak Brook 16388
The Radiological Society of the Republic of China, Taipei 03323
Radionic Association, Guildford 12736
Radio Society of Great Britain, Potters Bar 12737
Radio Technical Commission for Aeronautics, Washington 16389
Radius, London 12738
Radix Institute, Granbury 16390
Radix Teachers Association, Dallas 16391
Radnorshire Society, Llandrindod Wells 12739
RAFS, Lexington 16395
R.A.I., London 12768, 12769
Railroad Station Historical Society, Crete 16392
Railway Club, London 12740
Railway Correspondence and Travel Society, London 12741
Railway Development Association, Purley 12742
Rajasthan Academy of Science, Pilani 06852
Ralph Waldo Emerson Memorial Association, Boston 16393
RAM, London 12764
RAMP, Prairie View 16413
Rannsóknaráđ Rikisins, Reykjavik 06696
Rannsóknastofnun Fiskidnadarins, Reykjavik 06697
R.A.N.Z.C.P., Carlton 00460
R.A.O.U., Moonee Ponds 00459
RAPI, Hawthorn 00465
RAPRA Technology, Shrewsbury 12743
Rare Earth Research Conference, Lexington 16394
Rare Poultry Society, Great Glenn 12744
RAS, London 12766, 12770, 12772
RASD, Chicago 16399
RASE, Stoneleigh Park 12767
R.A.S.N.Z., Wellington 09195
Raten Amerika Kyokai Kenkyubu, Tokyo 08403
Rat für Formgebung, Frankfurt 06244
Rationalisierungs-Gemeinschaft Bauwesen im RKW, Eschborn 06245
Rationalisierungs-Gemeinschaft Verpackung im RKW, Eschborn 06246
Rationalisierungs-Kuratorium der Deutschen Wirtschaft e.V., Eschborn 06247
Rationalisierungs-Kuratorium der Deutschen Wirtschaft e.V., Landesgruppe Baden-Württemberg, Stuttgart 06248
Rationalisierungs-Kuratorium der Deutschen Wirtschaft e.V., Landesgruppe Berlin, Berlin 06249
Rationalisierungs-Kuratorium der Deutschen Wirtschaft e.V., Landesgruppe Bremen, Bremen 06250
Rationalisierungs-Kuratorium der Deutschen Wirtschaft e.V., Landesgruppe Hamburg, Hamburg 06251
Rationalisierungs-Kuratorium der Deutschen Wirtschaft e.V., Landesgruppe Hessen, Eschborn 06252
Rationalisierungs-Kuratorium der Deutschen Wirtschaft e.V., Landesgruppe Niedersachsen, Hannover 06253
Rationalisierungs-Kuratorium der Deutschen Wirtschaft e.V., Landesgruppe Nord-Ost, Kiel 06254
Rationalisierungs-Kuratorium der Deutschen Wirtschaft e.V., Landesgruppe Nordrhein-Westfalen, Düsseldorf 06255
Rationalisierungs-Kuratorium der Deutschen Wirtschaft e.V., Landesgruppe Rheinland-Pfalz, Mainz 06256
Rationalisierungs-Kuratorium für Landwirtschaft e.V., Osterrönfeld 06257
Rat von Sachverständigen für Umweltfragen, Wiesbaden 06258
R. Austin Freeman Society, Lexington 16395
The Ray Society, London 12745
Raziskovalna Skupnost Slovenije, Ljubljana 10103
RBEC, Washington 16428
RC, Jenkintown 16431
R.C., Milano 07591
R.C.A., Conwy 12775
RCAA, Wadebridge 12797
RCAP, Leesburg 16429
R.C.G.P., London 12779
RCHS, Edinburgh 12774
RCM, London 12780
R.C.M., London 12781
RCN, London 12782
RCO, London 12784
R.C.O.G., London 12783
RCP, London 12788
R.C.P.E., Edinburgh 12787
RCPSC, Ottawa 02869
RCR, London 12790
RCRC, Urbana 16407
RCRC, Westfield 16382
RCS, London 12777
RCSC, Chicago 16414
R.C.S.E., London 12792
RCSI, Dublin 07023
R.C.T.S., London 12741
R.C.V.S., London 12793
RDA, Purley 12742
R.D.S., Dublin 07024
RDS, London 12752, 12798
Reading is Fundamental, Washington 16396
Reading Reform Foundation, Tacoma 16397
Read Natural Childbirth Foundation, San Rafael 16398
Real Academia de Bellas Artes de la Purísima Concepción, Valladolid 10337
Real Academia de Bellas Artes de San Fernando, Madrid 10338
Real Academia de Bellas Artes de Santa Isabel de Hungría, Sevilla 10339
Real Academia de Bellas Artes de San Telmo, Málaga 10340
Real Academia de Bellas Artes y Ciencias Históricas de Toledo, Toledo 10341
Real Academia de Ciencias Exactas, Físicas y Naturales, Madrid 10342
Real Academia de Ciencias Morales y Políticas, Madrid 10343
Real Academia de Ciencias y Artes de Barcelona, Barcelona 10344
Real Academia de Córdoba de Ciencias, Bellas Letras y Nobles Artes, Córdoba 10345
Real Academia de Farmacia, Madrid 10346
Real Academia de Jurisprudencia y Legislación, Madrid 10347
Real Academia de la Historia, Madrid 10348
Real Academia de la Lengua Vasca, Bilbao 10349
Real Academia de Medicina de Sevilla, Sevilla 10350
Real Academia de Medicina y Cirugía de Palma de Mallorca, Palma de Mallorca 10351
Real Academia de Nobles y Bellas Artes de San Luis, Zaragoza 10352
Real Academia Española, Madrid 10353
Real Academia Gallega, La Coruña 10354
Real Academia Hispano-Americana, Cádiz 10355
Real Academia Nacional de Medicina, Madrid 10356
Real Academia Sevillana de Buenas Letras, Sevilla 10357
Real Instituto Arqueológico de Portugal, Lisboa 09667
Real Sociedad Arqueológica Tarraconense, Tarragona 10358
Real Sociedad Bascongada de los Amigos del Pais, San Sebastian 10359
Real Sociedade Arqueológica Lusitana, Santiago de Cacém 09668
Real Sociedad Económica de Amigos del País de Tenerife, Tenerife 10360
Real Sociedad Española de Física, Madrid 10361
Real Sociedad Española de Historia Natural, Madrid 10362
Real Sociedad Española de Química, Madrid 10363
Real Sociedad Fotográfica, Madrid 10364
Real Sociedad Geográfica, Madrid 10365
Real Sociedad Matemática Española, Madrid 10366
Real Sociedad Vascongada de los Amigos del País, San Sebastián 10367
RECGAI, Chadds Ford 16412
Rechts- und Staatswissenschaftliche Gesellschaft, Recklinghausen 06259
Rechts- und Staatswissenschaftliche Vereinigung Düsseldorf e.V., Düsseldorf 06260
Rectors' Conference, Bangkok 11096
Recycling Council of Ontario, Toronto 02855
REFA-Verband für Arbeitsstudien und Betriebsorganisation e.V., Darmstadt 06261
Reference and Adult Services Division (of ALA), Chicago 16399
Reforma: National Association to Promote Library Services to the Spanish-Speaking, New York 16400
Regency Society of Brighton and Hove, Brighton 12746
Regional Education Board of the Christian Brothers, Romeoville 16401
Regionale Organisation der FDI für Europa (ERO), Köln 06262
Regional Institute of Social Welfare Research, Athens 16402
Regional Maritime Academy, Abidjan, Abidjan 08056
Regional Science Association, Urbana 16403
Regional Studies Association, London 12747
Registered Medical Assistants of American Medical Technologists, Park Ridge 16404
Registered Music Teachers' Association, Winnipeg 02856
Regroupement pour la Surveillance du Nucléaire, Montréal 02097
Rehabilitation Information Round Table, Alexandria 16405
Rehabilitation International, New York 16406
R.E.H.I.S., Glasgow 12801
Reinforced Concrete Research Council, Urbana 16407
Reinforcing Steel Institute of Ontario, Willowdale 02857
Reinsurance Research Council, Toronto 02858
Rektorkollegiet, København 03796
The Religious Drama Society of Great Britain, London 12738
Religious Research Association, Washington 16408
Religious Society of Friends, London 12748
Religious Speech Communication Association, Upland 16409
Remote Sensing Society, Nottingham 12749
REMP, 's-Gravenhage 09065
Renaissance English Text Society, Amherst 16410
Renaissance, Schweizerischer Verband katholischer Akademiker-Gesellschaften, Luzern 10782
Renaissance Society of America, New York 16411
Renal Association, Oxford 12750
Republički Zavod za Zaštita na Spomenicite na Kulturata, Skopje 08606
Republic of China Society of Cardiology, Taipei 03324
RERC, Lexington 16394
Rersearch Council for Applied Mathematics, Moskva 09949
R.E.S., London 12799
RES, London 12800
Research and Development Society, London 12751
Research and Engineering Council of the Graphic Arts Industry, Chadds Ford 16412
Research and Productivity Council, Fredericton 02859
Research Association for the Fine Arts, Sofia 01770

Research Association of Minority Professors, Prairie View 16413
Research Association of Statistical Sciences, Fukuoka 08435
Research Association on Fundamental Problems of Technical Sciences, Sofia 01771
Research Community of Slovenia, Ljubljana 10103
The Research Council of Norway, Oslo 09272
Research Council on Structural Connections, Chicago 16414
Research Defence Society, London 12752
Research Designs and Standards Organization, Lucknow 06853
Research Discussion Group, Morgantown 16415
Research, Education and Assistance for Canadians with Herpes, Toronto 02860
Research Group for European Migration Problems, 's-Gravenhage 09065
Research Group for Lung Metabolism, Tygerberg 10168
Research Group for Neurochemistry, Tygerberg 10169
Research Institute for Foreign travellers in Italy, Genova 07573
Research into Lost Knowledge Organisation, Orpington 12753
Research Libraries Group, Mountain View 16416
Research Society for Natural Therapeutics, Bournemouth 12754
Research Society for Victorian Periodicals, Edwardsville 16417
Research Society of Pakistan, Lahore 09398
Research Society on Alcoholism, Austin 16418
Research to Prevent Blindness, New York 16419
RETS, Amherst 16410
Réunion Internationale des Laboratoires d'Essais et de Recherche sur les Matériaux et les Constructions, Cachan 04622
R.F.S., Tring 12805
RFSU, Twickenham 12878
RG, Beltsville 16426
RGI, Glasgow 12807
R.G.S., London 12806
R.G.S.A., Fortitude Valley 00468
R.H.A.S.S., Edinburgh 12808
Rheinaubund, Schweizerische Arbeitsgemeinschaft für Natur und Heimat, Schaffhausen 10783
Rheinische Naturforschende Gesellschaft, Mainz 06263
Rheinische Vereinigung für Volkskunde, Bonn 06264
Rheinisch-Westfälische Akademie der Wissenschaften, Düsseldorf 06265
Rheinisch-Westfälische Auslandsgesellschaft e.V., Dortmund 06266
Rheinisch-Westfälisches Institut für Wirtschaftsforschung, Essen 06267
Rheinisch-Westfälische Vereinigung für Lungen- und Bronchialheilkunde, Essen 06268
Rhetoric Society of America, Norman 16420
RHS, London 12810
RHS, Oklahoma City 16425
R.H.S.Q., Brisbane North Quay 00469
RHSV, Melbourne 00470
RI, Boulder 16427
RI, Emmaus 16424
RI, Granbury 16390
RI, New York 16406
R.I.A., Dublin 07028
RIAI, Dublin 07026
RIAM, Dublin 07029
RIBA, London 12812
RIC, Truro 12821
Richard III Society, London 12755
Richard III Society, New Orleans 16421
Richard Ivey Foundation, London 02861
Richard Jefferies Society, Swindon 12756
Richard-Wagner-Verband Bayreuth e.V., Bayreuth 06269
RIF, Washington 16396
R.I.G.B., London 12822
Right to a Comprehensive Education, London 12757
RIIA, London 12813
Rijksbureau voor Kunsthistorische Documentatie, 's-Gravenhage 09066
Riksbibliotektjenesten, Oslo 09344
Riksföreningen för Lärarna i Moderna Språk, Hisings Kärra 10515
RILEM, Cachan 04622
RILKO, Orpington 12753
RIN, London 12814
RIPA, London 12818
Riron Keiryo Keizai Gakkai, Tokyo 08404
RIRT, Alexandria 16405
RI/SME, Dearborn 16423
Rissho Koseikai, Tokyo 08405
Rissho Koseikai Fuzoku Kosei Toshokan, Tokyo 08405
RISW, Swansea 12824
RISWR, Athens 16402
RITLA, Caracas 16964
River Thames Society, Richmond 12758
RJC, Long Beach 16422
R.K.D., 's-Gravenhage 09066
RKW, Eschborn 06247
RLAS, London 12829
RLAS, Preston 12827
RLG, Mountain View 16416
R.M.A., Oxford 12834
RMAAMT, Park Ridge 16404
RMS, Oxford 12833
RNA, Richmond 12759
RNAA, Norwich 12836
RNBWS, Southsea 12835
RNCF, San Rafael 16398
RNSHS, Halifax 02870
Robinson Jeffers Committee, Long Beach 16422
Robotics International of the Society of Manufacturing Engineers, Dearborn 16423
ROCSOC, Taipei 03324
Rodale Institute, Emmaus 16424
Rodeo Historical Society, Oklahoma City 16425
Rogers Group, Beltsville 16426
Roland Eötvös Physical Society, Budapest 06601
Rolf Institute, Boulder 16427
Roller Bearing Engineers Committee, Washington 16428
Romanian Association of Political Sciences, Bucuresti 09746
Romanian General Association of Economists, Bucuresti 09741
Romanian Medical Association, Bucuresti 09754
Romanian Psychiatric Association, Bucuresti 09744
Romanian Society for Orthopaedics and Traumatology, Bucuresti 09794
Romanian Society of Neurosurgery, Bucuresti 09793
Romanian Support Society for People Suffering of Alzheimer Type Diseases, Bucuresti 09795
Romantic Novelists' Association, Richmond 12759
ROSPA, Birmingham 12852
Ross Dependency Research Committee, Wellington 09192
Royal Academy of Arts in London, London 12760
Royal Academy of Dancing, London 12761
Royal Academy of Dramatic Art, London 12762
Royal Academy of Engineering, London 12763
Royal Academy of Medicine, Dublin 07021
Royal Academy of Music, London 12764
Royal Aeronautical Society, London 12765
Royal Aeronautical Society (Australian Division), Mascot 00450
Royal Aeronautical Society, New Zealand Division, Wellington 09193
Royal Aeronautical Society, Southern Africa Division, Bryanston 10170
The Royal African Society, London 12766
Royal Agricultural and Horticultural Society of South Australia, Wayville 00451
Royal Agricultural Society of England, Stoneleigh Park 12767
Royal Agricultural Society of New Zealand, Wellington 09194
Royal Agricultural Society of Tasmania, Glenorchy 00452
The Royal Agricultural Society of Western Australia, Claremont 00453
Royal Anthropological Institute of Great Britain and Ireland, London 12768
Royal Archaeological Institute, London 12769
Royal Architectural Institute of Canada, Ottawa 02862
Royal Art Society of New South Wales, North Sydney 00454
The Royal Asiatic Society, London 12770
Royal Asiatic Society, Hong Kong Branch, Hong Kong 06595
Royal Asiatic Society of Sri Lanka, Colombo 10418
Royal Association for Disability and Rehabilitation, London 12771
Royal Astronomical Society, London 12772
Royal Astronomical Society of Canada, Toronto 02863
Royal Astronomical Society of New Zealand, Wellington 09195
Royal Australasian College of Dental Surgeons, Sydney 00455
Royal Australasian College of Physicians, Sydney 00456
Royal Australasian College of Radiologists, Millers Point 00457
Royal Australasian College of Surgeons, Melbourne 00458
Royal Australasian Ornithologists Union, Moonee Ponds 00459
The Royal Australian and New-Zealand College of Psychiatrists, Carlton 00460
Royal Australian Chemical Institute, Parkville 00461
Royal Australian College of General Practitioners, Sydney 00462
Royal Australian College of Ophthalmologists, Sydney 00463
Royal Australian Historical Society, Sydney 00464
Royal Australian Planning Institute, Hawthorn 00465
Royal Bath and West of England Society, Shepton Mallet 12773
Royal Belgian Society of Surgery, Bruxelles 01471
Royal Caledonian Horticultural Society, Edinburgh 12774
Royal Cambrian Academy of Art, Conwy 12775
Royal Canadian Academy of Arts, Toronto 02864
Royal Canadian Geographical Society, Vanier 02865
Royal Canadian Institute, Toronto 02866
Royal Canadian Military Institute, Toronto 02867
Royal Celtic Society, Edinburgh 12776
Royal Choral Society, London 12777
The Royal College of Anaesthetists, London 12778
Royal College of Dental Surgeons of Canada, Toronto 02868
Royal College of General Practitioners, London 12779
Royal College of Midwives, London 12780
Royal College of Music, London 12781
Royal College of Nursing – Australia, Melbourne 00466
Royal College of Nursing of the United Kingdom, London 12782
Royal College of Obstetricians and Gynaecologists, London 12783
The Royal College of Organists, London 12784
Royal College of Pathologists, London 12785
Royal College of Pathologists of Australasia, Surrey Hills 00467
Royal College of Physicians and Surgeons of Canada, Ottawa 02869
Royal College of Physicians and Surgeons of Glasgow, Glasgow 12786
Royal College of Physicians of Edinburgh, Edinburgh 12787
Royal College of Physicians of Ireland, Dublin 07022
Royal College of Physicians of London, London 12788
The Royal College of Psychiatrists, London 12789
Royal College of Radiologists, London 12790
Royal College of Surgeons in Ireland, Dublin 07023
Royal College of Surgeons of Edinburgh, Edinburgh 12791
The Royal College of Surgeons of England, London 12792
Royal College of Veterinary Surgeons, London 12793
Royal Commission for the Exhibition of 1851, London 12794
Royal Commission on Historical Manuscripts, London 12795
Royal Commonwealth Society, London 12796
Royal Commonwealth Society, Bermuda Branch, Devonshire 01526
Royal Cornwall Agricultural Association, Wadebridge 12797
The Royal Danish Academy of Sciences and Letters, København 03710
Royal Danish Agricultural Society, Frederiksberg 03760
Royal Drawing Society, London 12798
Royal Dublin Society, Dublin 07024
Royal Dutch Geographical Society, Amsterdam 08965
Royal Dutch Society for Natural History, Utrecht 08958
Royal Economic Society, London 12799
Royal Economic Society of Friends of Tenerife, Tenerife 10360
Royal Entomological Society of London, London 12800
Royal Environmental Health Institute of Scotland, Glasgow 12801
The Royal Faculty of Procurators in Glasgow, Glasgow 12802
Royal Fine Art Commission, London 12803
Royal Fine Art Commission for Scotland, Edinburgh 12804
Royal Forestry Society of England, Wales and Northern Ireland, Tring 12805
Royal Geographical Society, London 12806
Royal Geographical Society of Queensland, Fortitude Valley 00468
Royal Geological and Mining Society of the Netherlands, Haarlem 08967
Royal Glasgow Institute of the Fine Arts, Glasgow 12807
Royal Hibernian Academy of Arts, Dublin 07025
Royal Highland and Agricultural Society of Scotland, Edinburgh 12808
Royal Historical Society, London 12809
The Royal Historical Society of Queensland, Brisbane North Quay 00469
Royal Historical Society of Victoria, Melbourne 00470
Royal Horticultural Society, London 12810
Royal Horticultural Society of New South Wales, Sydney 00471
Royal Incorporation of Architects in Scotland, Edinburgh 12811
Royal Institute, Bangkok 11097
Royal Institute of Architects of Ireland, Dublin 07026
Royal Institute of British Architects, London 12812
Royal Institute of International Affairs, London 12813
Royal Institute of Linguistics and Anthropology, Leiden 08964
Royal Institute of Navigation, London 12814
Royal Institute of Oil Painters, London 12815
Royal Institute of Painters in Water Colours, London 12816
The Royal Institute of Philosophy, London 12817
Royal Institute of Public Administration, London 12818
The Royal Institute of Public Health and Hygiene, London 12819
Royal Institution of Chartered Surveyors, Dublin 07027
The Royal Institution of Chartered Surveyors, London 12820
Royal Institution of Cornwall, Truro 12821
Royal Institution of Engineers in the Netherlands, 's-Gravenhage 08963
Royal Institution of Great Britain, London 12822
The Royal Institution of Naval Architects, London 12823
Royal Institution of South Wales, Swansea 12824
Royal Irish Academy, Dublin 07028
The Royal Irish Academy of Music, Dublin 07029
Royal Isle of Wight Agricultural Society, Newport 12825
Royal Jersey Agricultural and Horticultural Society, Saint Helier 12826
Royal Lancashire Agricultural Society, Preston 12827
Royal Liverpool Philharmonic Society, Liverpool 12828
Royal London Aid Society, London 12829
Royal Medical Society, Edinburgh 12830
Royal Medico-Chirurgical Society of Glasgow, Glasgow 12831
Royal Melbourne Institute of Technology, Melbourne 00472
Royal Meteorological Society, Reading 12832
Royal Meteorological Society, Australian Branch, Parkville 00473
Royal Microscopical Society, Oxford 12833
The Royal Musical Association, Oxford 12834
Royal Naval Bird Watching Society, Southsea 12835
Royal Nepal Academy, Kathmandu 08817
Royal Netherlands Academy of Arts and Sciences, Amsterdam 08952
Royal Netherlands Association for Advancement of Pharmacy, 's-Gravenhage 08956
Royal Netherlands Association of Musicians, Amsterdam 08959
Royal Netherlands Society for Agricultural Science, Wageningen 08962
Royal Norfolk Agricultural Association, Norwich 12836
Royal Norwegian Council for Scientific and Industrial Research, Oslo 09289
The Royal Norwegian Society of Sciences and Letters, Trondheim 09258
Royal Nova Scotia Historical Society, Halifax 02870
Royal Numismatic Society, London 12837
Royal Pharmaceutical Society of Great Britain, London 12838
The Royal Philharmonic Society, London 12839
Royal Philosophical Society of Glasgow, Glasgow 12840
The Royal Photographic Society, Bath 12841
Royal Physical Society of Edinburgh, Edinburgh 12842
Royal Queensland Art Society, Brisbane 00474
Royal Sanitary Association of Scotland, Glasgow 12843
Royal Scientific Society, Amman 08454
Royal Scottish Academy, Edinburgh 12844
Royal Scottish Forestry Society, Dunfermline 12845
Royal Scottish Geographical Society, Glasgow 12846
Royal Scottish Society of Arts, Edinburgh 12847
The Royal Society, London 12848
Royal Society for Asian Affairs, London 12849
Royal Society for Nature Conservation, Lincoln 12850
Royal Society for the Encouragement of Arts, Manufactures and Commerce, London 12851
Royal Society for the Prevention of Accidents, Birmingham 12852
Royal Society for the Protection of Birds, Sandy 12853
The Royal Society of Antiquaries of Ireland, Dublin 07030
Royal Society of Arts, London 12854
Royal Society of Arts and Science of Göteborg, Göteborg 10491
Royal Society of Arts and Sciences of Mauritius, Réduit 08674
Royal Society of British Artists, London 12855
Royal Society of British Sculptors, London 12856
Royal Society of Canada, Ottawa 02871
Royal Society of Canberra, Canberra 00475
Royal Society of Chemistry, London 12857
The Royal Society of Edinburgh, Edinburgh 12858
The Royal Society of Health, London 12859
Royal Society of Literature of the United Kingdom, London 12860
Royal Society of Marine Artists, London 12861
The Royal Society of Medicine, London 12862
Royal Society of New South Wales, North Ryde 00476
Royal Society of New Zealand, Wellington 09196
Royal Society of Portrait Painters, London 12863
Royal Society of Queensland, Saint Lucia 00477
Royal Society of South Africa, Cape Town 10171
Royal Society of South Australia, Adelaide 00478
Royal Society of Tasmania, Hobart 00479
Royal Society of the Humanities at Uppsala, Uppsala 10485
Royal Society of Tropical Medicine and Hygiene, London 12864
Royal Society of Ulster Architects, Belfast 12865
Royal Society of Victoria, Melbourne 00480
Royal Society of Western Australia, Perth 00481
Royal South Australian Society of Arts, Adelaide 00482
Royal Statistical Society, London 12866
Royal Stuart Society, Huntingdon 12867
Royal Swedish Academy of Military Sciences, Stockholm 10486
The Royal Swedish Academy of Sciences, Stockholm 10490
The Royal Television Society, London 12868
Royal Town Planning Institute, London 12869
Royal Ulster Academy, Lurgan 12870
Royal Ulster Academy Association, Lambeg 12871
Royal Ulster Agricultural Society, Belfast 12872
Royal Watercolour Society, London 12873
Royal Welsh Agricultural Society, Powys 12874
Royal Western Australian Historical Society, Nedlands 00483
Royal West of England Academy, Bristol 12875
Royal Zoological Society, Amsterdam 09072
Royal Zoological Society of Antwerp, Antwerpen 01410
Royal Zoological Society of Ireland, Dublin 07031
Royal Zoological Society of New South Wales, Mosman 00484
The Royal Zoological Society of Scotland, Edinburgh 12876
Royal Zoological Society of South Australia, Adelaide 00485
RPA, Liverpool 12879
RPB, New York 16419
RP Eye Research Foundation, Toronto 02872
RPFI, Wien 00570
RPS, Bath 12841
RPS, Bristol 12735
R.P.S., Great Glenn 12744
RRA, Washington 16408
RRF, Tacoma 16397
RRS, Reston 16385
RSA, Austin 16418
R.S.A., Edinburgh 12844
RSA, London 12747, 12854

RSA, New York 16411
RSA, Norman 16420
RSA, Urbana 16403
R.S.A.I., Dublin 07030
R. Samuel McLaughlin Foundation, Toronto 02873
R.S.A.S., Glasgow 12843
R.S.A.T., Tarragona 10358
R.S.B.A.P., San Sebastian 10359
RSC, London 12857
RSCA, Upland 16409
RSD, Morgantown 16415
R.S.E., Edinburgh 12858
RSF, Madrid 10364
R.S.G.B., Potters Bar 12737
R.S.G.S., Glasgow 12846
R.S.H., London 12859
RSHS, Crete 16392
RSM, London 12862
R.S.M.E., Madrid 10366
RSNA, Oak Brook 16388
RSNC, Lincoln 12850
RSNC Wildlife Trusts Partnership, Lincoln 12877
RSNZ, Wellington 09196
RSP, Quezon City 09535
R.S.P.B., Sandy 12853
R.S.S., Amman 08454
RSS, Bozeman 16430
RSS, Nottingham 12749
RSSA, Edinburgh 12847
R.S.S.P.S.A.D., Bucuresti 09795
RSUA, Belfast 12865
RSVP, Edwardsville 16417
RTA, Dallas 16391
RTCA, Washington 16389
RtCE, London 12757
RTOG, Philadelphia 16386
RTPI, London 12869
RTRC, New York 16387
RTS, London 12868
RTS, Richmond 12758
RUA, Lurgan 12870
R.U.A.A., Lambeg 12871
RUAS, Belfast 12872
Rudolf Kassner-Gesellschaft, Wien 00937
Rugby Football Schools Union, Twickenham 12878
Rural Community Assistance Program, Leesburg 16429
Rural Development Council of Prince Edward Island, Charlottetown 02874
Rural Education and Development Association, Edmonton 02875
Rural Preservation Association, Liverpool 12879
Rural Sociological Society, Bozeman 16430
Rushlight Club, Jenkintown 16431
Russian Academy of Agricultural Sciences, Moskva 09950
Russian Academy of Arts, Moskva 09951
Russian Academy of Medical Sciences, Moskva 09952
Russian Academy of Pedagogical Sciences, Moskva 09953
Russian Academy of Sciences, Moskva 09954
Russian Association for Comparative Literature, Moskva 09955
Russian Botanical Society, St. Petersburg 09956
Russian Geographical Society, St. Petersburg 09957
Russian Linguistics Society, Moskva 09958
Russian Literary Society in Japan, Tokyo 08360
Russian Mineralogical Society, St. Petersburg 09959
Russian Palaeontological Society, St. Petersburg 09960
Russian Palestine Society, Moskva 09961
Russian PEN Centre, Moskva 09962
Russian Pharmacological Society, Moskva 09963
Russian Pugwash Committee, Moskva 09964
Russian Society of Genetics and Breeders, Moskva 09965
Russian Union of Composers, Moskva 09966
Rutland Agricultural Society, Rutland 12880
Ruusbroecgenootschap, Antwerpen 01419
RVP, Washington 14884
RWEMA, Boston 16393
R.Z.S.I., Dublin 07031

SA, Helsinki 04004
SA, London 13028
S.A., Paris 04631
SA, Zürich 10798
SAA, Boulder 16757
SAA, Chicago 16633
SAA, Dallas 16466
SAA, Hull 12959
SAA, Ipswich 13105
SAA, Oklahoma City 16494
SAA, San Francisco 16499
SAA, Washington 16492
SAAC, Washington 16585
SAAD, London 12998
SAAFoST, Edenvale 10179
SAAI, Buenos Aires 00153
SAAP, Seattle 16586
Saarländische Gesellschaft für zahnärztliche Fortbildung, Saarbrücken 06270
Saarländischer Gymnasiallehrerverband e.V., Losheim 06271
Saarländisch-Pfälzische Internistengesellschaft e.V., Ludwigshafen 06272
S.A.A.S., Vlaeberg 10178
S.A.A.T.V.E, Arcadia 10180
SAB, Newton Abbot 12964
SAB, Stockholm 10585
SAB, Winchester 13079
SABA, Kalamazoo 16587
SABAM, Bruxelles 01447
SABET, Paisley 12894
S.A.B.L., Bayeux 04714
SAC, Edinburgh 12893
SAC, Paris 04626
S.A.C.A.C., Pretoria 10187
S.A.C.D., Bruxelles 01455
SACF, Buenos Aires 00161
S.A.C.I., Yeoville 10186
SACO Sveriges Akademikers Centralorganisation, Stockholm 10516
SACP, Oneonta 16498
Sacro Occipital Research Society International, Prairie Village 16432
SACS, Decatur 16723
SAD, Praha 03554
SADAIC, Buenos Aires 00156
SADAS, Athinai 06544
SADE, Buenos Aires 00166
SADIO, Buenos Aires 00174
SADMN, Buenos Aires 00176
SAE, Parkville 00493
SAE, Valley Stream 16588
SAE, Washington 16591
SAEC, Ithaca 16722
Sächsische Akademie der Wissenschaften zu Leipzig, Leipzig 06273
Sächsische Landesärztekammer, Dresden 06274
SAE International, Warrendale 16433
SAEM, Lansing 16489
Säteilyturvakeskus, Helsinki 03989
S.A.F., Paris 04633
SAF, Paris 04692
SAF, Washington 16634
Safe Association, Yoncalla 16434
Safety and Reliability Society, Southport 12881
Safety Equipment Institute, Arlington 16435
Safe Water Association of New South Wales, Sydney 00486
SAFI, Linköping 10580
SAFJL, Hillcrest 10190
SAFR, Luzern 10795
SAG, Bern 10786
SAG, Meyrin 10967
SAGA, Genève 10791
SAGB, Bangor 12887
SAGP, Binghamton 16493
S.A.G.S., Wits 10192
SAGSET, Loughborough 12999
SAGT, Hamilton 12898
SAH, Buenos Aires 00172
SAH, New York 16635
SAH, Philadelphia 16639
SAHAAS, Saint Albans 12882
SAHGB, London 13027
SAHH, Haverford 16501
Sahitya Akademi, New Dehli 06854
SAHS, Chicago 16758, 16761
SAIAeE, Sunnyside 10195
SAIF, Pretoria 10225
Saint Albans and Hertfordshire Architectural and Archaeological Society, Saint Albans 12882
Saint George's Historical Society, Saint George 01527
Saint John's Hospital Dermatological Society, London 12883
Saintpaulia and Houseplant Society, Ilford 12884
S.A.I.P., Cape Town 10201
S.A.I.P., Faure 10200
SAJBJ, Pretoria 10193
SAJIB, Montréal 02917
SAJM, Neuheim 10794
S.A.K., Stockholm 10586
SAL, Cointe-Ougrée 01422
S.A.L., London 13022
S.A.L., Palma de Mallorca 10403
SALALM, Albuquerque 16461
Sales Research Club, Toronto 02876
SALGEP, Glasgow 12899
SALT, New York 16636
SALT, Warrenton 16495
Salt Institute, Alexandria 16436
Saltire Society, Edinburgh 12885
Salzburger Ärztegesellschaft, Salzburg 00938
Salzburger Arbeitskreis für Psychoanalyse, Salzburg 00939
Salzburger Institut für juristische Information und Fortbildung, Salzburg 00940
Salzburger Juristische Gesellschaft, Salzburg 00941
Salzburger Kulturvereinigung, Salzburg 00942
Salzburg Seminar, Middlebury 16437
SAM, Independence 16490
SAM, Ithaca 16500
SAM, Vinton 16491
SAMA, Sunnyside 10226
Samenwerkingsverband van de Universiteitsbibliotheken, de Koninklijke Bibliotheek en de Bibliotheek van de Koninklijke Nederlandse Akademie van Wetenschappen, Amsterdam 09067
Samfundet De Nio, Stockholm 10517
Samfundet for Dansk Genealogi og Personalhistorie, Holte 03797
Samfundet til Udgivelse af Dansk Musik, København 03798
Samfund til Udgivelse af Gammel Nordisk Litteratur, Valby 03799
SAMH, Edinburgh 12896
Sammenslutningen af Danmarks Forskningsbiblioteker, Lyngby 03800
Sammenslutningen af Danske Kunstforeninger, Niva 03801
Sammenslutningen af Lokalarkiver, Vejle 03802
Sammenslutningen af Medieforskere i Danmark, Alborg 03803
Sammenslutningen af Praktiserende Dyrlaeger, Vanløse 03804
Samostatná sekce Historického Klubu při ČSAV, Brno 03551
SAMPE, Covina 16589
SAMRA, Johannesburg 10204
SAMS, Oban 12895
SAMS, Pretoria 10205
SAMW, Basel 10787
SANCI, Pretoria 10207
Sandford Fleming Foundation, Waterloo 02877
San Francisco Aids Foundation, San Francisco 16438
Sangeet Natak Akademi, New Delhi 06855
S.A.N.G.O.R.M., Marshalltown 10208
S.A.N.H.S., Taunton 13074
San Martin Society of Washington, DC, McLean 16439
Sanskrit Academy, Madras 06856
SANTA, Johannesburg 10210
SANU, Beograd 17125
Sanyo Association for Advancement of Science and Technology, Okayama-ken 08406
San-yo Gijutsu Shinkokai, Okayama-ken 08406
S.A.O.A., Adelaide 00496
S.A.O.S., Johannesburg 10227
SAP, Coimbra 09674
S.A.P., Paris 04639
SAPD, Vanløse 03804
Sapporo Norin Gakkai, Sapporo 08407
Sapporo Society of Agriculture and Fisheries, Sapporo 08407
SAR, Stockholm 10535
SARA, Lombard 16637
SAROC, Taipei 03326
SAROJ, Beograd 17108
SaRS, Southport 12881
SAS, Blacksburg 16469
SAS, Dearborn 16497
SAS, Frederick 16496
SAS, Göteborg 10533
S.A.S., Guildford 13110
S.A.S., Lewes 13113
SAS, Much Wenlock 12951
SAS, Trieste 07769
SASA, Rosebank 10217
Saskatchewan Amyotrophic Lateral Sclerosis Society Foundation, Radville 02878
Saskatchewan Anti-Tuberculosis League, Fort San 02879
Saskatchewan Association of Library Technicians, Saskatoon 02880
Saskatchewan Association of Medical Radiation Technologists, Saskatoon 02881
Saskatchewan Association of Pathologists, Regina 02882
Saskatchewan Association of Social Workers, Regina 02883
Saskatchewan Association of Teachers of French, Saskatoon 01890
Saskatchewan Association of Teachers of German, Saskatoon 02884
Saskatchewan Association on Gerontology, Saskatoon 02885
Saskatchewan Cardiology Technicians Association, Saskatoon 02886
Saskatchewan Dietetic Association, Regina 02887
Saskatchewan Genealogical Society, Regina 02888
Saskatchewan Geological Society, Regina 02889
Saskatchewan High School Principals' Group, Saskatoon 02890
Saskatchewan Horticultural Association, Balcarres 02891
Saskatchewan Institute of Agrologists, Saskatoon 02892
Saskatchewan Library Association, Regina 02893
Saskatchewan Lung Association, Fort San 02894
Saskatchewan Medical Association, Saskatoon 02895
Saskatchewan Pharmaceutical Association, Regina 02896
Saskatchewan Psychiatric Association, Saskatoon 02897
Saskatchewan Registered Music Teachers' Association, Saskatoon 02898
Saskatchewan Research Council, Saskatoon 02899
Saskatchewan Safety Council, Regina 02900
Saskatchewan Society of Medical Laboratory Technologists, Saskatoon 02901
Saskatchewan Society of Occupational Therapists, Regina 02902
Saskatchewan Society of Osteopathic Physicians, Regina 02903
Saskatchewan Speech and Hearing Association, Regina 02904
Saskatchewan Teachers' Federation, Saskatoon 02905
Saskatchewan Veterinary Medical Association, Saskatoon 02906
SASMI, Roma 07765
SASMIRA, Bombay 06857
SASS, Provo 16590
SAT, Tours 04629
SAT-AMIKARO, Paris 04977
S.A.T.F, Paris 04699
SATIPS, Altrincham 13029
SATL, Helsinki 04008
SATRA Footwear Technology Centre, Kettering 12886
SATW, Zürich 10790
Saudi Biological Society, Riyadh 10005
SAV, Bratislava 10060
SAVE, Northbrook 16638
Savez Arheoloških Društava Jugoslavije, Beograd 17107
Savez Astronautičkih i Raketnih Organizacija Jugoslavije, Beograd 17108
Savez Bibliotečkih Radnika Srbije, Beograd 17109
Savez Društava Arhiviskih Radnika Jugoslavije, Beograd 17110
Savez Društava Istoričara Jugoslavije, Beograd 17111
Savez Društava Matematičara, Fizičara i Astronoma Jugoslavije, Novi Sad 17112
Savez Društava Veterinara Jugoslavije, Beograd 17113
Savez Društava za Strane Jezike i Knjizevnosti Jugoslavije, Beograd 17114
Savez Društava Zubarskih Radnika Jugoslavije, Beograd 17115
Savez Ekonomista Jugoslavije, Beograd 17116
Savez Lekarskih Društava Jugoslavije, Beograd 17117
Savez Muzejskih Društava Hrvatske, Zagreb 03466
Savez Muzejskih Društava Jugoslavije, Beograd 17118
Savez Pedagoških Društava Jugoslavije, Beograd 17119
Savez Pedagoških Društava Srbije, Beograd 17120
Savez Udruženja Pravnika Jugoslavije, Beograd 17121
Savez Udruženja za Krivično Pravo i Kriminolgiju Jugoslavije, Beograd 17122
SAVI, Laguna 09537
Saving and Preserving Arts and Cultural Environments, Los Angeles 16440
SAW, Helensburgh 12900
SAW, Leipzig 06273
SAWE, Chula Vista 16632
Sazemane Pachuhesh va Barnamerizi Amuzeshi, Teheran 06903
SAZU, Ljubljana 10105
SB, Minneapolis 16502
SBA, Skokie 16743
SBAP, Forch 10888
SBAS, Great Yarmouth 13077
SBCCI, Birmingham 16724
SBD, Rio de Janeiro 01708
SBE, Bruxelles 01446
SBE, Chicago 16503
SBE, Indianapolis 16644
SBE, Luzern 10889
SBE, São Paulo 01709
SBE, Watford 13032
S.B.F., Châtenay-Malabry 04635
SBF, Stockholm 10440, 10538
SBG, Bruxelles 01430
SBG, Zürich 10800
SBGE, Bruxelles 01429
S.B.I., Firenze 07773
SBL, Decatur 16642
SBM, Rockville 16641
SBME, Liège 01435
SBMI, Bruxelles 01433
SBNS, Glasgow 13031
S.B.P., Bruxelles 01442
SBP, Bruxelles 01445
SBP, Dallas 16643
SBPR, London 12966
SBS, Borås 10539
SBSA, Brooklyn 16640
SC, Bethesda 16514
SC, Los Angeles 16504
SC, San Francisco 16471
SCA, Annandale 16741
SCA, Oakland 16513
SCA, Richmond 16648
SCA, Washington 16511, 16515
S.C.A.B., Bruxelles 01453
SCA&I, Breckenridge 16505
SCALA, Wirral 13036
Scandinavian Association for Research on Latin America, Stockholm 10510
Scandinavian Dental Association, Turku 03990
Scandinavian Forum for Lipid Research and Technology, Göteborg 10518
Scandinavian Institute of African Studies, Uppsala 10507
Scandinavian Library Center, Ballerup 03805
Scandinavian Orthopaedic Association, Stockholm 10519
Scandinavian Seminar, Amherst 16441
Scandinavian Simulation Society, Espoo 03991
Scandinavian Society for Economic and Social History, Odense 03806
Scandinavian Society for Electron Microscopy, Oslo 09345
The Scandinavian Society of Anaesthesiologists, Oslo 09280
Scandinavian Sociological Association, København 03807
SCAR, Cambridge 12890
SCAUL, Nairobi 08507
SCAW, Bethesda 16448
SCBS, West Covina 16594
SCC, New York 16651
SCCM, Anaheim 16653
SCCS, Bogotá 03408
SCE, Boston 16649
SCEA, Alexandria 16652
SCEH, Liverpool 16508
SCF, New York 16478
SCHA, Glasgow 12902
Schiffbautechnische Gesellschaft e.V., Hamburg 06275
Schizophrenia Association of Great Britain, Bangor 12887
Schleswig-Holsteinische Gesellschaft für Zahn-, Mund- und Kieferheilkunde, Lübeck 06276
Schmalenbach-Gesellschaft, Berlin 06277
School Health Association of the Republic of China, Taipei 03325
School Library Association of the Northern Territory, Darwin 00487
School Management Study Group, Salt Lake City 16442
School Natural Science Society, Gillingham 12888
School Science and Mathematics Association, Bowling Green 16443
Schopenhauer-Gesellschaft e.V., Bonn 06278
SCHS, Washington 16755
Schutzgemeinschaft Alt Bamberg e.V., Bamberg 06279
Schweizer Heimatschutz, Zürich 10784
Schweizerische Ärztegesellschaft für Manuelle Medizin, Zürich 10785
Schweizerische Afrika-Gesellschaft, Bern 10786
Schweizerische Akademie der medizinischen Wissenschaften, Basel 10787
Schweizerische Akademie der Naturwissenschaften, Bern 10788
Schweizerische Akademie der Sozial- und Geisteswissenschaften, Bern 10789
Schweizerische Akademie der Technischen Wissenschaften, Zürich 10790
Schweizerische Akademische Gesellschaft der Anglisten, Genève 10791
Schweizerische Arbeitsgemeinschaft für akademische Berufs- und Studienberatung, Luzern 10792
Schweizerische Arbeitsgemeinschaft für hauswirtschaftliche Bildungs- und Berufsfragen, Zürich 10793
Schweizerische Arbeitsgemeinschaft für Jugendmusik und Musikerziehung, Neuheim 10794
Schweizerische Arbeitsgemeinschaft für Raumfahrt, Luzern 10795
Schweizerische Arbeitsgemeinschaft für Rehabilitation, Basel 10796
Schweizerische Arbeitsgemeinschaft für Schul- und Jugendzahnpflege, Langnau 10797
Schweizerische Asiengesellschaft, Zürich 10798
Schweizerische Astronomische Gesellschaft, Meyrin 10967
Schweizerische Bibliophilen-Gesellschaft, Zürich 10799
Schweizerische Botanische Gesellschaft, Zürich 10800
Schweizerische Diabetes-Gesellschaft, Zürich 10801
Schweizerische Direktoren-Konferenz gewerblicher Berufs- und Fachschulen, Luzern 10802
Schweizerische Dokumentationsstelle für Schul- und Bildungsfragen, Grand-Saconnex 10644
Schweizerische Energie-Stiftung, Zürich 10803
Schweizerische Entomologische Gesellschaft, Zürich 10804
Schweizerische Esperanto-Gesellschaft, Kriens 10805
Schweizerische Ethnologische Gesellschaft, Bern 10806

Schweizerische Franchising-Vereinigung, Zollikon *10807*
Schweizerische Geologische Gesellschaft, Fribourg *10975*
Schweizerische Gesellschaft der Kernfachleute, Bern *10808*
Schweizerische Gesellschaft für Agrarwirtschaft, Zürich *10809*
Schweizerische Gesellschaft für Astrophysik und Astronomie, Sauverny *10810*
Schweizerische Gesellschaft für Aussenpolitik, Lenzburg *10811*
Schweizerische Gesellschaft für Automatik, Bern *10812*
Schweizerische Gesellschaft für Balneologie und Bioklimatologie, Sankt Moritz *10813*
Schweizerische Gesellschaft für Bildungs- und Erziehungsfragen, Langnau *10814*
Schweizerische Gesellschaft für Chirurgie, Genève *10983*
Schweizerische Gesellschaft für cystische Fibrose, Spiegel *10815*
Schweizerische Gesellschaft für Dermatologie und Venereologie, Lausanne *10816*
Schweizerische Gesellschaft für die Rechte der Urheber musikalischer Werke, Zürich *10817*
Schweizerische Gesellschaft für Ernährungsforschung, Basel *10818*
Schweizerische Gesellschaft für Familienforschung, Basel *10819*
Schweizerische Gesellschaft für Feintechnik, Zürich *10820*
Schweizerische Gesellschaft für Gerontologie, Bern *10821*
Schweizerische Gesellschaft für Geschichte der Medizin und der Naturwissenschaften, Bern *10822*
Schweizerische Gesellschaft für Gynäkologie und Geburtshilfe, Lausanne *10823*
Schweizerische Gesellschaft für Innere Medizin, Langenthal *10824*
Schweizerische Gesellschaft für Kardiologie, Bern *10825*
Schweizerische Gesellschaft für Kartographie, Zürich *10826*
Schweizerische Gesellschaft für kaufmännisches Bildungswesen, Genève *10827*
Schweizerische Gesellschaft für Kommunikations- und Medienwissenschaft, Bern *10828*
Schweizerische Gesellschaft für Lebensmittel-Wissenschaft und -Technologie, Wädenswil *10829*
Schweizerische Gesellschaft für Lehrerinnen- und Lehrerbildung, Bern *10830*
Schweizerische Gesellschaft für Logik und Philosophie der Wissenschaften, Bern *10831*
Schweizerische Gesellschaft für Marketing, Zürich *10832*
Schweizerische Gesellschaft für Mikrobiologie, Lausanne *10833*
Schweizerische Gesellschaft für Mikrotechnik, Zürich *10834*
Schweizerische Gesellschaft für Muskelkranke, Zürich *10835*
Schweizerische Gesellschaft für Oberflächentechnik, Bettlach *10836*
Schweizerische Gesellschaft für Onkologie, Epalinges *10837*
Schweizerische Gesellschaft für orientalische Altertumswissenschaft, Genève *10838*
Schweizerische Gesellschaft für Orthopädie, Bern *10839*
Schweizerische Gesellschaft für Oto-Rhino-Laryngologie, Hals- und Gesichtschirurgie, Bern *10840*
Schweizerische Gesellschaft für Personalfragen, Zürich *10841*
Schweizerische Gesellschaft für Physikalische Medizin und Rehabilitation, Zürich *10842*
Schweizerische Gesellschaft für Pneumologie, Bern *10843*
Schweizerische Gesellschaft für Psychologie, Lausanne *10844*
Schweizerische Gesellschaft für Rheumatologie, Zürich *10845*
Schweizerische Gesellschaft für Skandinavische Studien, Zürich *10846*
Schweizerische Gesellschaft für Sozial- und Präventivmedizin, Bern *10847*
Schweizerische Gesellschaft für Soziologie, Zürich *10848*
Schweizerische Gesellschaft für Statistik und Volkswirtschaft, Bern *10849*
Schweizerische Gesellschaft für Theaterkultur, Bonstetten *10850*
Schweizerische Gesellschaft für Tropenmedizin und Parasitologie, Zürich *10851*
Schweizerische Gesellschaft für Umweltschutz, Zürich *10852*
Schweizerische Gesellschaft für Ur- und Frühgeschichte, Basel *10853*
Schweizerische Gesellschaft für Vakuum-Physik und -Technik, Zürich *10854*
Schweizerische Gesellschaft für Versicherungsrecht, Zürich *10855*
Schweizerische Gesellschaft für Volkskunde, Basel *10856*
Schweizerische Gesellschaft Pro Technorama, Winterthur *10857*
Schweizerische Graphologische Berufsvereinigung, Erlenbach *10858*
Schweizerische Hämophilie-Gesellschaft, Zürich *10859*
Schweizerische Heilpädagogische Gesellschaft, Bern *10860*
Schweizerische Hochschulkonferenz, Bern *10861*
Schweizerische Hochschulrektoren-Konferenz, Bern *10862*
Schweizerische Informationskonferenz, Basel *10863*
Schweizerische katholische Arbeitsgemeinschaft für Ausländerfragen, Luzern *10864*
Schweizerische Konferenz der Rektoren der Berufsschulen kaufmännischer Richtung, Zürich *10865*
Schweizerische Krebsliga, Bern *10866*
Schweizerische Kriminalistische Gesellschaft, Giubiasco *10867*
Schweizerische Lichttechnische Gesellschaft, Bern *10868*
Schweizerische Liga gegen den Lärm, Luzern *10869*
Schweizerische Liga gegen Epilepsie, Zürich *10870*
Schweizerische Management-Gesellschaft, Zürich *10871*
Schweizerische Mathematische Gesellschaft, Zürich *10872*
Schweizerische Multiple Sklerose Gesellschaft, Zürich *10873*
Schweizerische Musikforschende Gesellschaft, Basel *10874*
Schweizerische Neurologische Gesellschaft, Aarau *10875*
Schweizerische Normen-Vereinigung, Zürich *10876*
Schweizerische Numismatische Gesellschaft, Bern *10877*
Schweizerische Ophthalmologische Gesellschaft, Horgen *10878*
Schweizerische Paläontologische Gesellschaft, Genève *10879*
Schweizerische Philosophische Gesellschaft, Sankt Gallen *10880*
Schweizerische Physikalische Gesellschaft, Genève *10881*
Schweizerischer Altphilologen-Verband, Basel *10882*
Schweizerischer Anglistenverband, Lugano *10883*
Schweizerischer Anwaltsverband, Bern *10884*
Schweizerischer Apotheker-Verein, Bern-Liebefeld *10885*
Schweizerischer Ballettlehrer-Verband, Hombrechtikon *10886*
Schweizerischer Berufsverband diplomierter Sozialarbeiterinnen und Sozialpädagoginnen, Bern *10887*
Schweizerischer Berufsverband für Angewandte Psychologie, Forch *10888*
Schweizerischer Bund für Elternbildung, Luzern *10889*
Schweizerischer Bund für Naturschutz, Basel *10890*
Schweizerischer Chemiker-Verband, Binningen *10891*
Schweizerischer Elektrotechnischer Verein, Zürich *10892*
Schweizerischer Fachverband der diplomierten medizinischen Laborantinnen und Laboranten, Sankt Gallen *10893*
Schweizerischer Fachverband für Schweiss- und Schneidmaterial, Zürich *10894*
Schweizerischer Fachverband Sozialdienst in Spitälern, Bern *10895*
Schweizerischer Forstverein, Zürich *10896*
Schweizerischer Handelslehrerverein, Einsiedeln *10897*
Schweizerische Rheumaliga, Zürich *10898*
Schweizerischer Ingenieur- und Architekten-Verein, Zürich *10899*
Schweizerischer Musikpädagogischer Verband, Zürich *10900*
Schweizerischer Notaren-Verband, Bern *10901*
Schweizerischer Physiotherapeuten-Verband, Sempach-Stadt *10902*
Schweizerischer Romanistenverband, Luzern *10903*
Schweizerischer Schriftstellerinnen- und Schriftsteller-Verband, Zürich *10904*
Schweizerischer Technischer Verband, Zürich *10905*
Schweizerischer Verband der Akademikerinnen, Chur *10906*
Schweizerischer Verband der Ingenieur-Agronomen und der Lebensmittel-Ingenieure, Zollikofen *10907*
Schweizerischer Verband der Lehrer an kaufmännischen Berufsschulen, Buchrain *10908*
Schweizerischer Verband für Betriebsorganisation und Fertigungstechnik, Winterthur *10909*
Schweizerischer Verband für die Materialtechnik, Pully *10910*
Schweizerischer Verband für die Wärmebehandlung der Werkstoffe, Winterthur *10911*
Schweizerischer Verband für Fernunterricht, Kreuzlingen *10912*
Schweizerischer Verband für Frauenrechte, Lausanne *10913*
Schweizerischer Verband für Landtechnik, Riniken *10914*
Schweizerischer Verband für Sport in der Schule, Bern *10915*
Schweizerischer Verband kantonal approbierter Zahnärzte, Zürich *10916*
Schweizerischer Verband von Fachleuten für Alkoholgefährdeten- und Suchtkrankenhilfe, Zürich *10917*
Schweizerischer Verein der Chemiker-Coloristen, Oftringen *10918*
Schweizerischer Verein des Gas- und Wasserfaches, Zürich *10919*
Schweizerischer Verein für Schweisstechnik, Basel *10920*
Schweizerischer Verein für Vermessung und Kulturtechnik, Solothurn *10921*
Schweizerischer Verkehrssicherheitsrat, Bern *10922*
Schweizerischer Werkbund, Zürich *10923*
Schweizerischer Wissenschaftsrat, Bern *10924*
Schweizerische Schillerstiftung, Zumikon *10925*
Schweizerisches Institut für das Gesundheitswesen, Aarau *10926*
Schweizerisches Institut für Experimentelle Krebsforschung, Epalinges *10732*
Schweizerisches Institut für Kunstwissenschaft, Zürich *10927*
Schweizerisches Institut für Volkskunde, Basel *10928*
Schweizerisches Nationalkomitee für geistige Gesundheit, Zürich *10929*
Schweizerische Sprachwissenschaftliche Gesellschaft, Basel *10930*
Schweizerische Stiftung für Alpine Forschungen, Zürich *10931*
Schweizerische Stiftung für Angewandte Psychologie, Fribourg *10932*
Schweizerische Studiengruppe für Konsumentenfragen, Bern *10933*
Schweizerische Theologische Gesellschaft, Bern *10934*
Schweizerische Trachtenvereinigung, Burgdorf *10935*
Schweizerische Vereinigung der Lack- und Farben-Chemiker, Frauenfeld *10936*
Schweizerische Vereinigung der Versicherungsmathematiker, Zürich *10937*
Schweizerische Vereinigung diplomierter Chemiker HTL, Basel *10938*
Schweizerische Vereinigung für Altertumswissenschaft, Basel *10939*
Schweizerische Vereinigung für Atomenergie, Bern *10940*
Schweizerische Vereinigung für Datenverarbeitung, Zürich *10941*
Schweizerische Vereinigung für Dokumentation, Thun *10942*
Schweizerische Vereinigung für Ernährung, Zollikofen *10943*
Schweizerische Vereinigung für Erwachsenenbildung, Zürich *10944*
Schweizerische Vereinigung für Filmkultur, Bern *10945*
Schweizerische Vereinigung für Gesundheitstechnik, Zürich *10946*
Schweizerische Vereinigung für Internationales Recht, Zürich *10947*
Schweizerische Vereinigung für Kleintiermedizin, Bern *10948*
Schweizerische Vereinigung für Landesplanung, Bern *10949*
Schweizerische Vereinigung für Schiedsgerichtsbarkeit, ZZürich *10950*
Schweizerische Vereinigung für Sozialpolitik, Zürich *10951*
Schweizerische Vereinigung für Steuerrecht, Zürich *10952*
Schweizerische Vereinigung für Urheber- und Medienrecht, Zürich *10953*
Schweizerische Vereinigung für Weltraumtechnik, Bern *10954*
Schweizerische Vereinigung für Zukunftsforschung, Horgen *10955*
Schweizerische Vereinigung gegen Tuberkulose und Lungenkrankheiten, Bern *10956*
Schweizerische Vereinigung zum Schutz und zur Förderung des Berggebietes, Brig *10957*
Schweizerische Verkehrswirtschaftliche Gesellschaft, Sankt Gallen *10958*
Schweizerische Zahnärzte-Gesellschaft, Bern *10959*
Schweizerische Zentralstelle für Hochschulwesen, Bern *10960*
Schweizerische Zoologische Gesellschaft, Basel *10961*
Schweizer Musikrat, Aarau *10962*
SCI, London *13034, 13035*
SCI, Roma *07774*
Science and Engineering Research Council, Swindon *12889*
Science Council of Canada, Ottawa *02907*
Science et Paix, Toronto *02908*
Science Fiction Research Association, Westminster *16444*
Science for Peace, Toronto *02908*
Science Production Association Biotech, Baku *01047*
Science Service, Washington *16445*
Science Society of Thailand, Bangkok *11098*
Science Teachers Association of New South Wales, Rozelle *00488*
Science Teachers Association of Queensland, Spring Hill *00489*
Science Teachers' Association of Victoria, Parkville *00490*
The Science Writers' Association of South Africa, Johannesburg *10172*
Scientific Agricultural Society of Finland, Piikkio *04048*
Scientific and Technical Societies National Headquarters, Kiev *11190*
Scientific and Technical Union of Agricultural Specialists, Sofia *01772*
Scientific and Technical Union of Transport, Sofia *01773*
Scientific Association for Commercial and Consumer Law, Zürich *11030*
Scientific Association of the Bulgarian Neurologists, Sofia *01774*
Scientific Committee on Antarctic Research, Cambridge *12890*
Scientific Committee on Oceanic Research, Baltimore *16446*
Scientific Committee on Problems of the Environment, Paris *04623*
Scientific Council of Zimbabwe, Harare *17180*
Scientific Council on Complex Problems (Cybernetics), Baku *01048*
Scientific Exploration Society, Marlborough *12891*
Scientific Industrial Association „Biology", Tashkent *16946*
Scientific Industrial Association „Solar Physics", Tashkent *16947*
Scientific Information Centre at the Bulgarian Academy of Sciences, Sofia *01775*
Scientific, Literary and Artistic Athenaeum, Mahón *10285*
Scientific, Literary and Art Society, Las Palmas *10372*
Scientific Medical Society of Anatomists-Pathologists, Moskva *09967*
Scientific Research Council, Kingston *08089*
Scientific Society for Building, Budapest *06602*
Scientific Society for Telecommunication, Budapest *06607*
Scientific Society for Transport, Budapest *06609*
Scientific Society of Measurement and Automation, Budapest *06664*
Scientific Society of Mechanical Engineers, Budapest *06606*
Scientific Society of Organization and Management, Warszawa *09625*
Scientific Society of Pakistan, Karachi *09399*
The Scientific Society of Płock, Płock *09626*
Scientific Society of the Leather, Shoe and Allied Industries, Budapest *06596*
Scientific Society of the Timber Industry, Budapest *06604*
Scientific-Technical Association, St. Petersburg *09968*
Scientific-Technical Council on Computer Technology, Mathematical Modelling, Automation of Scientific Research and Instrument Making, Tbilisi *05027*
Scientists and Engineers for Secure Energy, New York *16447*
Scientists Center for Aminal Welfare, Bethesda *16448*
SCISP, Asunción *09432*
SCL, Bristol *12967*
SCL, Worcester *13040*
SCM, Bogotá *03409*
SCNCER, Bangkok *11071*
SCOG, Bogotá *03410*
SCOLARE, Liège *01145*
Scoliosis Research Society, Rosemont *16449*
SCOLMA, Milton Keynes *13097*
SCONUL, London *13096*
SCOPE, Paris *04623*
Scots Ancestry Research Society, Edinburgh *12892*
Scottish Arts Council, Edinburgh *12893*
Scottish Association for Building Education and Training, Paisley *12894*
The Scottish Association for Marine Science, Oban *12895*
Scottish Association for Mental Health, Edinburgh *12896*
Scottish Association of Advisers in Physical Education, Dumbarton *12897*
Scottish Association of Geography Teachers, Hamilton *12898*
Scottish Association of Local Government to Educational Psychologists, Glasgow *12899*
Scottish Association of Writers, Helensburgh *12900*
Scottish Business Education Council, Edinburgh *12901*
Scottish Catholic Historical Association, Glasgow *12902*
Scottish Church History Society, Edinburgh *12903*
Scottish Economic Society, Glasgow *12904*
Scottish Electrophysiological Society, Saint Andrews *12905*
Scottish Epilepsy Association, Glasgow *12906*
Scottish Field Studies Association, Blairgowrie *12907*
Scottish Further Education Association, Glasgow *12908*
Scottish Gaelic Texts Society, Glasgow *12909*
Scottish Genealogy Society, Edinburgh *12910*
Scottish History Society, Edinburgh *12911*
Scottish Industrial Heritage Society, Glasgow *12912*
Scottish Inland Waterways Association, Edinburgh *12913*
Scottish Language Society, Inverness *12914*
Scottish Library Association, Motherwell *12915*
Scottish National Dictionary Association, Edinburgh *12916*
Scottish National Housing and Town Planning Council, Troon *12917*
Scottish National Orchestra Society, Glasgow *12918*
Scottish Ornithologists' Club, Edinburgh *12919*
Scottish Pharmaceutical Federation, Glasgow *12920*
Scottish Physical Education Association, Glasgow *12921*
Scottish Record Society, Glasgow *12922*
Scottish Recreational Land Association, Edingburgh *12923*
Scottish Reformation Society, Edinburgh *12924*
Scottish Rights of Way Society, Edinburgh *12925*
Scottish River Purification Boards Association, Glasgow *12926*
Scottish Secondary Teachers' Association, Edinburgh *12927*
Scottish Society for Crop Research, Dundee *12928*
Scottish Society for Northern Studies, Edinburgh *12929*
Scottish Society for the Preservation of Historical Machinery, Glasgow *12930*
Scottish Society of the History of Medicine, Aberdeen *12931*
Scottish Tartans Society, Pitlochry *12932*
Scottish Teachers' Nursing Association, Glasgow *12933*
Scottish Text Society, Edinburgh *12934*
Scottish Tramway Museum Society, Glasgow *12935*
Scottish Wildlife Trust, Edinburgh *12936*
Scribes, Winston-Salem *16450*
SCS, Chapel Hill *16507*
SCS, Pasadena *16506*
SCS, Philadelphia *16759*
SCSI, San Diego *16512*
SCST, Brandon *02943*
SCT, Baltimore *16509*
SCT, Châtenay Malabry *04647*
SCTE, Exton *16645*
SCTE, Metals Park *16646*
Sculptors' Society of Canada, Toronto *02909*
SCUP, Ann Arbor *16510*
SCVIR, Fairfax *16647*
SDA, Regina *02887*
SDAR, Beograd *17110*
SDB, Evanston *16467*
SDB, Washington *16516*
SDF, Lyngby *03800*
SDG, Zürich *10801*
S.D.I, Firenze *07777*
SDMFAJ, Novi Sad *17112*
SDMS, Dallas *16654*
SDS, Stockholm *10541*
S.D.T., Huntingdon *13042*
SDTIM, Nuneaton *13001*
SE, Reno *16659*
SE2, New York *16447*
SEA, Charleville-Mézières *04728*
SEA, Glasgow *12906*
SEA, La Plata *00197*
SEA, Stockport *12956*
SEA, Woods Hole *16451*
Seabird Group, Sandy *12937*
SEACA, Quito *03855*
Sea Education Association, Woods Hole *16451*
Sea Grant Association, La Jolla *16452*
Search Foundation, Birmingham *16453*
Search Group, Sacramento *16454*
Sea Shepherd Conservation Society, Santa Monica *16455*
SEB, Columbia *16517*
SEB, London *12973*
SEB, Madrid *10378*
SEBM, New York *16522*
Secção de Farmacia Galénica, Curitiba *01701*
SECED, London *12969*
Secondary School Admission Test Board, Princeton *16456*

Secretaria do Patrimônio Histórico e Artístico Nacional, Rio de Janeiro 01702
Secretaria Permanente del Tratado General de Integración Económica Centroamericana, Guatemala City 06572
Sectie Operationele Research, Amsterdam 09068
Section for Rehabilitation Hospitals and Programs, Chicago 16457
Section on Women in Legal Education of the AALS, Washington 16458
SEDAR, Madrid 10377
SEDEIS, Neuilly-sur-Seine 04732
S.E.D.E.S, Paris 04750
SEDM, Ann Arbor 16521
SEDO, Madrid 10392
S.E.F., Campoformido 07798
SEF, Paris 04723
SEFCO, Saint-Jean d'Angely 04739
S.E.F.I.M., Paris 04738
SEFT, London 12970
SEG, Bern 10806
SEG, Littleton 16655
SEG, Tulsa 16662
SEG/SES, Zürich 10804
S.E.H., Madrid 10388
SEI, Arlington 16435
SEI, Madison Heights 16656
S.E.I.N., Paris 04673
Seingalt Society, Salt Lake City 16459
S.E.I.O., Madrid 10373
Seisan Gijutsu Kenkyusho, Tokyo 08408
Seismological Society of America, El Cerrito 16460
Seismological Society of Japan, Tokyo 08448
Seito Kogyokai Jimubu Chosaka, Tokyo 08409
S.E.J.E.F., Paris 04745
Sekai Keizai Chosakai Shiryoshitsu, Tokyo 08410
Sekiyu Gijutsu Kyokai, Tokyo 08411
Sekiyu Remmei Kohobu Shiryoka, Tokyo 08412
Sektionen for Klinisk Neurofysiologi, Gentofte 03808
SELAF, Paris 04746
Selborne Society, Oxted 12938
Selden Society, London 12939
SELF, Caen 04690
Selskabet for Danmarks Kirkehistorie, København 03809
Selskabet for Dansk Kulturhistorie, København 03810
Selskabet for Dansk Skolehistorie, Hellerup 03811
Selskabet for Dansk Teaterhistorie, København 03812
Selskabet for Filosofi og Psykologi, København 03813
Selskabet for Historie og Samfundsøkonomi, København 03814
Selskabet for Københavns Historie, København 03815
Selskabet for Naturlaerens Udbredelse, København 03816
Selskabet for Tekniske Uddanelsespørgsmal, København 03817
Selskabet til Udgivelse af Danske Mindesmaerker, Nivaa 03818
Selskab for Arbejdsmiljø, København 03819
Selskab for Nordisk Filologi, København 03820
Selskapet for Lyskultur, Sandvika 09346
Selskapet til Vitenskapenes Fremme, Bergen 09347
SEM, Bethel 16523
SEM, Bloomington 16520
SEM, Madrid 10391
Semiconductor Research Foundation, Sendai 08129
Seminario de Filología Vasca Julio de Urquijo, San Sebastián 10368
Seminar on the Acquisition of Latin American Library Materials, Albuquerque 16461
Semiotic Society of America, Davis 16462
The Sempervivum Society, Burgess Hill 12940
SENA, Bogotá 03401
Senckenbergische Naturforschende Gesellschaft, Frankfurt 06280
Sen-i Gakkai, Tokyo 08413
Senior Scholars, Cleveland 16463
Senshokutai Gakkai, Tokyo 08414
S.E.O., Papeete 05003
SEP, University Park 16660
SEPT, Paris 04726
SER, Fort Collins 16519
S.E.R., Madrid 10397
SER, New York 16518
S.E.R.B., Paris 04733
Serbian Academy of Sciences and Arts, Beograd 17125
Serbian Association against Cancer, Beograd 17123
Serbian Biological Society, Beograd 17126
Serbian Chemical Society, Beograd 17129
Serbian Geographical Society, Beograd 17127
Serbian Geological Society, Beograd 17128
Serbian Society, Novi Sad 17104
SERC, Swindon 12889
S.E.R.C.H., Paris 04735
Serena British Columbia, Surrey 02910
Serena Canada, Ottawa 02911
S.E.R.E.S., Paris 04734
SERGS, Pôrto Alegre 01721
S.E.R.M., Madrid 10396
SERNAGEOMIN, Santiago 03063
Servicio Cooperativo Interamericano de Salud Pública, Asunción 09432
Servicio de Documentación y Biblioteca, Santo Domingo 03837
Servicio de Endocrinología y Metabolismo, Buenos Aires 00152
Servicio de Investigación Prehistórica y Museo de Prehistoria, Valencia 10369
Servicio de Investigación y Promoción Agraria, Lima 09474
Servicio Interamericana de Geodesía, Bogotá 03400
Servicio Nacional de Aprendizaje, Bogotá 03401
Servicio Nacional de Geología y Minería, Santiago 03063
Servicio Técnico Interamericano de Cooperación Agrícola, Asunción 09433
SES, Charlottesville 16657
SES, Dayton 16744
SES, Glasgow 12904
SES, Helsinki 04011
SES, Marlborough 12891
SESEP, Paris 04736
SESS, Edwardsville 16658
SET, Manchester 12972
SEURI, Herblay 04754
SEUSS, Shoreham-by-Sea 13078
SEV, Zürich 10892
Seventeenhundredandfourtyfive and National Military History Society, Edinburgh 12941
Sex Information and Education Council of Canada, Toronto 02912
Sex Information and Education Council of the U.S., New York 16464
Sexual Law Reform Society, London 12942
SEZEB, Paris 04724
Sezione di Training Autogeno del Centro Internazionale de Ipnosi, Milano 07764
SF, Oslo 09349
S.F.A., Paris 04786
SFA, Roslyn 16473
SFAF, San Francisco 16438
SFAS, Zürich 10894
SFC, Hunt Valley 16665
S.F.C., Palmi 07808
SFC, Paris 04790
SFCH, Monrovia 16667
SFCM, Besançon 04795
SFCPR, Paris 04794
SFDAS, Paris 04800
SFDI, Strasbourg 04867
SFE, Paris 04805
SFEA, Glasgow 12908
S.F.E.N., Paris 04824
SFER, Paris 04798
SFF, Nacka 10547
S.F.F., Udine 07806
SFFF, Helsingin Yliopisto 03995
SFG, Orléans 04801
SFG, Paris 04804
SFGGE, Helsingborg 10581
S.F.H.M, Reims 04856
SFHOM, Paris 04857
SFHS, Iowa City 16525
S.F.I., Milano 07809
S.F.I., Paris 04860
SFI Foundation, Poway 16465
SFL, Helsinki 04015
SFL, Washington 16663
SFM, Paris 04806
S.F.M., Paris 04815
SFME, Courbevoie 04810
S.F.M.E., Ivry-sur-Seine 04818
SFMG, Paris 04811
SFMM, Paris 04816
SFMS, Paris 04808
SFMT, Paris 04809
SFN, Paris 04826
SFO, Paris 04861
SFP, Paris 04828, 04836
S.F.P., Paris 04838
SFPA, London 13104
SFPE, Boston 16664
S.F.P.H., Paris 04844
SFR, Göteborg 10542
SFRA, Westminster 16444
S.F.S., Paris 04843
SFS, Stockholm 10549
SFSA, Blairgowrie 12907
SFSA, Des Plaines 16745
S.F.S.P., Vandoeuvre lès Nancy 04840
SFT, Paris 04846
SFTE, Lancaster 16666
S.F.U., Paris 04847
S.F.V., Aosta 08027
S.F.V., Paris 04866
SFV, Zollikon 10807
SFV, Zürich 10896
SFY, Helsingin Yliopisto 04016
SFZ, Bratislava 10093
SG, London 13047
SGA, Bern 10812
SGA, La Jolla 16452
SGA, Lenzburg 10811
SGAE, Madrid 10398
SGAS, Cincinnati 16527
S.G.B., Liège 01461
SGB, São Paulo 01726
SGE, Basel 10818
SGF, Linköping 10552
SGF, Norrköping 10551
SGF, Paris 04868
SGFF, Basel 10819
SGFT, Zürich 10820
SGG, Bern 10821
SGGG, Lausanne 10823
S.G.I, Roma 07812
S.G.I., Roma 07813
SGI, Sacramento 16454
SGI, Washington 16528
SGK, Bern 10808
SGKM, Bern 10828
SGKV, Frankfurt 06299
SGL, Bern 10830
SGLWT, Wädenswil 10829
SGM, Reston 16526
SGMB, Rennes 04869
SGMK, Zürich 10835
SGML, Highland Park 16750
SGNA, Rochester 16668
SGO, Bern 10839
SGP, Bern 10843
SGP, Lima 09478
SGP, Woods Hole 16669
SGP, Zürich 10841
SGS, Vällingby 10553
SGSP, Clarksburg 16670
SGSS, Zürich 10846
S.G.T., Sheffield 13048
SGTK, Bonstetten 10850
SGTS, Glasgow 12909
SGU, Uppsala 10591
SGU, Zürich 10852
SH, Ithaca 16600
SHA, Athens 16725
SHA, London 12957
SHA, Moncton 02929
SHA, Tucson 16530
SHA, West Chester 16533
S.H.A.A., South Croydon 13050
SHAFR, Dayton 16529
Shah Waliullah Academy, Hyderabad 09400
Shakai Keizaishi Gakkai, Tokyo 08415
Shakai Seisaku Gakkai, Tokyo 08416
Shakespeare Association of America, Dallas 16466
Shakespeare Authorship Society, London 12943
Shakespeare Birthplace Trust, Stratford-upon-Avon 12944
Shakespeare Data Bank, Evanston 16467
Shakespeare Institue, Stratford-upon-Avon 12945
Shakespeare Reading Society, London 12946
Shakespeare Society of America, West Hollywood 16468
SHAL, Saint Julien les Metz 04763
Shanghai Academy of Agricultural Sciences, Shanghai 03214
Shanghai Academy of Hydroelectrical Engineering Design, Shanghai 03215
Shanghai Textile Engineering Society, Shanghai 03216
Shansi Academy of Agricultural Sciences, Shanhsi 03217
Shantung Academy of Agricultural Sciences, Shantung 03218
Shaviana, Dagenham 12947
SHC, Chattanooga 16726
SHCJ, Holliswood 16596
SHD, Falls Church 16597
S.H.D., Paris 04759
SHE, London 13002
SHE, Long Beach 16531
Sheffield Metallurgical and Engineering Association, Rotherham 12948
Shensi Academy of Agricultural Sciences, Shenhsi 03219
Sherlock Holmes Society of London, Southsea 12949
Sherwood Anderson Society, Blacksburg 16469
Shetland Council of Social Service, Lerwick 12950
Shevchenko Scientific Society, New York 16470
SHG, Zürich 10859
SHGM, Baltimore 16599
Shigaku-kai, Tokyo 08417
Shigaku Kenkyukai, Kyoto 08418
Shika Kiso Igakkai, Tokyo 08419
Shinto Gakkai, Tokyo 08420
Shinto Shukyo Gakkai, Tokyo 08421
Shipbuilding Engineering Society, St. Petersburg 09969
Shipowners Refrigerated Cargo Research Association, Cambridge 11738
SHJ, Farmington Hills 16534
SHK, Bern 10861
SHNS, Arlington 16671
Shokubai Gakkai, Tokyo 08422
Shokubutsu Bunrui Chiri Gakkai, Kyoto 08423
Shomei Gakkai, Tokyo 08424
Shoqata e Gjeologeve te Shqiperise, Tirana 00007
SHOT, Houghton 16598
S.H.P., Paris 04758
S.H.P.F., Paris 04663
SHRM, Alexandria 16535
Shropshire Archaeological and Parish Register Society, Much Wenlock 12951
SHS, Helsinki 04025
SHS, Lancaster 12955
SHS, Lansing 16756
SHS, Warszawa 09616
SHSWD, Chicago 16532
SI, Alexandria 16436
S.I., Washington 16481
SI, Washington 16752
SIA, Bologna 07830
S.I.A., Pisa 07834
SIA, Roma 07839
SIA, Singapore 10053
SIA, Washington 16537
SIA, Zürich 10899
SIAE, Firenze 07964
SIAM, Philadelphia 16536
The Siam Society, Bangkok 11099
Sian Academy of Electric Power Design, Shanhsi 03220
SIANS, Milano 07822
S.I.A.R., Bari 07836
SIAS, Brighton 13114
S.I.A.S.A., Napoli 07965
S.I.A.S.O., Viareggio 07962
S.I.A.S.P., Trieste 07820
S.I.A.T., Roma 07888
SIB, Firenze 07845
S.I.B., Siena 07842
SIBC, Paris 04875
S.I.B.M., Livorno 07843
S.I.B.S., Napoli 07844
SIC, Genève 10973
SIC, Paris 04877
SIC, Pavia 07849
SIC, Reinach 10976
SICOT, Bruxelles 01462
S.I.C.R.E.O., Roma 07823
SID, Cleveland 16543
SID, Playa Del Rey 16539
SID, Roma 08028
S.I.D.E.A., Bologna 07864
S.I.De.S., Milano 07975
SIDES, Roma 07861
SIDM, Bologna 07909
S.I.E., Milano 07869
SIE, Roma 07868
SIECCAN, Toronto 02912
SIECUS, New York 16464
SIEDS, Roma 07865
S.I.E.M., Milano 07967
S.I.E.P.M., Louvain-la-Neuve 01465
Sierra Club, San Francisco 16471
Sierra Leone Association of Archivists, Librarians and Information Scientists, Freetown 10031
Sierra Leone Science Association, Freetown 10032
S.I.E.S., Milano 07870
SIETAR/INTL, Washington 15534
SIFET, Milano 07782
S.I.F.E.T., Milano 07878
S.I.G., Pisa 07885
SIGA, Foggia 07880
SIGART, New York 16737
SIGBIO, Durham 16738
SIGCPR, College Station 16734
SIGCUE, College Station 16735
SIGE, Bologna 07879
SIGIR, Blackburg 16739
SIGM, Milano 07884
SIGMA – Salzburger Gesellschaft für Semiologie, Salzburg 00943
SIGMA Xi – The Scientific Research Society, Research Triangle Park 16472
Sigmund Freud Archives, Roslyn 16473
Sigmund Freud-Gesellschaft, Wien 00944
Signal Processing Society, New York 16474
SIGSAM, Kent 16736
SIHS, Chestnut Hill 16545
SIIA, Singapore 10054
S.I.I.I., Bari 07887
S.I.I.T.S. – A.I.C.T., Roma 07886
SIJIF, Salzburg 00940
SIL, Helsinki 04030
Silk and Art Silk Mills' Research Association, Bombay 06857
SIM, Arlington 16538
SIM, Chicago 16540
S.I.M., Milano 07892
SIMA, Milano 07316
S.I.M.A., Roma 07905
S.I.M.F.E.R., Moncalieri 07897
SIMI, Roma 07898
Simian Society of America, Saint Louis 16475
S.I.M.L.A., Roma 07899
SIMP, Milano 07908
SIMP, Roma 07901
Simplified Speeling Society, Beckenham 12952
S.I.M.P.S., Genova 07900
SIMS, Espoo 03991
SIMS, Torino 07902
S.I.M.S.I., Napoli 07903
SIMT, Roma 07894
S.I.M.T.A., Bologna 07958
SIN, Ancona 07910
SIN, Santa Monica 16541
Sindacato Autonomo Scuola Media Italiana, Roma 07765
Sindacato Nazionale Autori Drammatici, Roma 07766
Sindacato Nazionale Istruzione Artistica, Roma 07767
Sindacato Nazionale Scrittori, Roma 07768
Sindhi Adabi Board, Tamshoro 09401
Sindicato Nacional dos Odontologistas Portuguesas, Lisboa 09669
Sindicato Nacional dos Professionais de Serviço Social, Lisboa 09670
Sindicato Nacional dos Professores, Lisboa 09671
Sindicato Nacional dos Protésicos Dentários, Lisboa 09672
Sindikat Radnicka Drustvenih Delatnosti Jugoslavije, Beograd 17124
Sind Library Association, Hyderabad 09402
Singapore Association for the Advancement of Science, Singapore 10052
Singapore Institute of Architects, Singapore 10053
Singapore Institute of International Affairs, Singapore 10054
Singapore Medical Association, Singapore 10055
Singapore National Academy of Science, Singapore 10056
SINIE, Santiago 03064
The Sinological Society, Tokyo 08433
SINPI, Napoli 07913
S.I.N.S., Bologna 07915
SINTEF, Trondheim 09353
S.I.O.C.M.F., L'Aquila 07919
S.I.O.e Ch.C.-F., Roma 07924
S.I.O.G., Roma 07922
SIOI, Roma 07918
S.I.O.I., Roma 07969
S.I.O.P., Roma 07925
S.I.O.T., Roma 07921
SIP, Bethesda 16542
SIP, Milano 07934
S.I.P., Roma 07932
S.I.P., Valencia 10369
S.I.P.A., Perugia 07928
S.I.P.A.I., Cremona 07940
S.I.P.A.O.C., Catania 07930
SIPE, Paris 04879
SIPES, Dallas 16673
S.I.P.I., Milano 07937
S.I.P.S., Milano 07816
SIPS, Palermo 07938
S.I.Ps., Roma 07936
S.I.P.T., Firenze 07939
S.I.P.V., Roma 07931
SIR, Marietta 16674
Sira, Chislehurst 12953
SIRC, London 12958
SIRE, Iselin 16601
S.I.R.I., Milano 07966
Sir Joseph Flavelle Foundation, Toronto 02913
SIRMCE, Bruxelles 01464
Sir Thomas Beecham Society, Redondo Beach 16476
SIS, New York 16544
SIS, Roma 07951
SIS, Stockholm 10522
S.I.S.A., Bari 07972
S.I.S.A., Milano 07968
S.I.S.C.A., Genova 07970
SISF, Milano 07944
S.I.S.MED., Bologna 07952
S.I.S.S., Firenze 07826
Sistema Nacional de Información en Educación, Santiago 03064
Sister Kenny Institute, Minneapolis 16477
S.I.S.Vet, Brescia 07828
SIT, Middlesbrough 13053
SIT, Sainte-Thérèse de Blainville 02962
S.I.T.H., Castellamare di Stabia 07819
SITS, Les Ulis 04882
SIU, Paris 04883
S.I.U., Roma 07957
SIWA, Edinburgh 12913
SJFR, Stockholm 10520
SJS, Plainview 16675
SKAF, Luzern 10864
Skandinaviska Simuleringssällskapet, Espoo 03991
Skandinavisk Museumsforbund, Danish Section, Aalborg 03821
SKF, Uppsala 10583
SKHS, Helsingin Yliopisto 04040
SKI, Minneapolis 16477
SKIF, Stockholm 10600
Skin Cancer Foundation, New York 16478
SKJP, Solothurn 11019
Skogbrukets og Skogindustrienes Forskningsforening, Aas 09348
Skógraektarfélag Islands, Reykjavik 06698
Skogs- och Jordbrukets Forskningsråd, Stockholm 10520
Skolledarförbundet, Stockholm 10521
SKS, Helsinki 04001, 04002
Skurdlaeknafélag Islands, Reykjavik 06699
SKV, Salzburg 00942
SKZ, Würzburg 06308
SLA, London 12954
SLA, Washington 16549, 16740
SLAA, Washington 16546
SLANT, Darwin 00487

SLAS, York *12977*
Slavistično Društvo Slovenije, Ljubljana *10104*
SLC, Ballerup *03805*
SLD, Beograd *17117*
The Sleep-Learning Association, London *12954*
Sleep Research Society, Boston *16479*
SLF, Stockholm *10521*
SLG, Bern *10868*
SLgE, Zürich *10870*
S.L.H.A., Lincoln *12979*
SLHR, Saint Louis *16548*
S.L.I., Roma *07784*
Slovak Academy of Sciences, Bratislava *10060*
Slovak Anthropological Society, Bratislava *10061*
Slovak Archaeological Society, Nitra-Hrad *10062*
Slovak Architects Society, Bratislava *10094*
Slovak Association of Classical Philologists, Bratislava *10076*
Slovak Astronomical Society, Tatranská Lomnica *10063*
Slovak Biochemical Society, Bratislava *10064*
Slovak Bioclimatology Society, Bratislava *10065*
Slovak Biological Soiety, Bratislava *10066*
Slovak Botanical Society, Bratislava *10067*
Slovak Chemical Society, Bratislava *10068*
Slovak Demographic and Statistical Society, Bratislava *10069*
Slovak Economics Society, Bratislava *10070*
Slovak Education Society, Prešov *10083*
Slovak Entomology Society, Bratislava *10071*
Slovak Ethnography Society, Bratislava *10080*
Slovak Geographical Society, Bratislava *10072*
Slovak Geological Society, Bratislava *10073*
Slovak Historical Society, Bratislava *10074*
Slovak Linguistics Society, Bratislava *10075*
Slovak Literary Society, Bratislava *10078*
Slovak Medical Society, Bratislava *10077*
Slovak Meteorological Society, Bratislava *10079*
The Slovak Philosophical Association, Bratislava *10093*
Slovak Psychological Society, Bratislava *10085*
Slovak Society for Agriculture, Forestry and Food, Ivanka pri Dunaji *10091*
Slovak Society for Cybernetics and Informatics, Bratislava *10089*
Slovak Society for History of Science and Technology, Bratislava *10088*
Slovak Society for International Law, Bratislava *10087*
Slovak Society for Mechanics, Bratislava *10090*
Slovak Society for Oriental Studies, Bratislava *10081*
Slovak Society for Parasitology, Košice *10082*
Slovak Society of Pneumology and Phtisiology, Bratislava *10084*
Slovak Sociological Society, Bratislava *10086*
Slovak Zoological Society, Bratislava *10092*
Slovenian Academy of Sciences, Ljubljana *10105*
Slovenian Historical Association, Ljubljana *10117*
Slovenian Society, Ljubljana *10106*
Slovenian Society of Historians of Art, Ljubljana *10107*
Slovenská Akadémia Vied, Bratislava *10060*
Slovenska Akademija Znanosti in Umetnosti, Ljubljana *10105*
Slovenská Antropologická Spoločnost, Bratislava *10061*
Slovenská Archeologická Spoločnost, Nitra-Hrad *10062*
Slovenská Astronomická Spoločnost, Tatranská Lomnica *10063*
Slovenská Biochemiká Spoločnost, Bratislava *10064*
Slovenská Bioklimatologická Spoločnost, Bratislava *10065*
Slovenská Biologická Spoločnost, Bratislava *10066*
Slovenská Botanická Spoločnost, Bratislava *10067*
Slovenská Chemická Spoleĉnost, Bratislava *10068*
Slovenská Demografická a Štatistická Spoločnost, Bratislava *10069*
Slovenská Ekonomická Spoločnost, Bratislava *10070*
Slovenská Entonologická Spoločnost, Bratislava *10071*
Slovenská Geografická Spoločnost, Bratislava *10072*
Slovenská Geologická Spoločnost, Bratislava *10073*
Slovenská Historická Spoločnost, Bratislava *10074*
Slovenská Jazykovedná Spoločnost, Bratislava *10075*
Slovenská Jednota Klassických Filológov, Bratislava *10076*
Slovenská Lekárska Spoleĉnost, Bratislava *10077*
Slovenská Literárnovedná Spoločnost, Bratislava *10078*
Slovenska Matica, Ljubljana *10106*
Slovenská Meteorologická Spoločnost, Bratislava *10079*
Slovenská Národopisná Spoločnost, Bratislava *10080*
Slovenská Orientalistická Spoločnost, Bratislava *10081*
Slovenská Parazitologická Spoločnost, Košice *10082*
Slovenská Pedagogická Spoločnost, Prešov *10083*
Slovenská Pneumologická a Ftizeologická Spoleĉnost, Bratislava *10084*
Slovenská Psychologická Spoločnost, Bratislava *10085*
Slovenská Sociologická Spoločnost, Bratislava *10086*
Slovenská Spoločnost Medzinárodné Právo, Bratislava *10087*
Slovenská Spoločnost pre Dejiny Vied a Techniky pri SAV, Bratislava *10088*
Slovenská Spoločnost pre Kybernetiku a Informatiku, Bratislava *10089*
Slovenská Spoločnost pre Mechaniku, Bratislava *10090*
Slovenská Spoločnost pre Vedy Polnohospodarske, Lesnické a Potravinárské, Ivanka pri Dunaji *10091*
Slovenská Zoologická Spoločnost, Bratislava *10092*
Slovenské Filozofické Združenie, Bratislava *10093*
Slovensko Umetnostno Zgodovinsko Društvo, Ljubljana *10107*
SLRS, London *12942*
SLS, Bratislava *10077*
SLS, Helsinki *04067*
SLS, London *13101*
SLS, Rickmansworth *13076*
SLS, Stockholm *10560*
SLTB, London *12980*
SMA, Lincoln *12982*
S.M.A., México *08756*
SMA, Singapore *10055*
SMA, Washington *16551*
Small Towns Institute, Ellensburg *16480*
SMC, México *08758*
SMCAF, Bethesda *16678*
SME, Dearborn *16677*
SMEA, Rotherham *12948*
SMEP, Charlottesville *16682*
SMF, Boulogne-Billancourt *04896*
SMF, Montecillo *08763*
S.M.F., Paris *04890*
SMG, Zürich *10871, 10872*
SMHS, Apache Junction *16754*
SMIER, Paris *04892*
Smithsonian Institution, Washington *16481*
SMJ, New York *16679*
SML, Helsinki *04052*
SMM, México *08753*
SMOL, Helsinki *04053*
SMPTAD, Studio City *16680*
SMPTE, White Plains *16681*
SMRI, Chicago *16550*
SMS, Hergiswil *11007*
S.M.S., Roma *07577*
SMSG, Salt Lake City *16442*
SMSG, Zürich *10873*
SMSR, México *08772*
SMTE, Reston *16553*
SMY, Helsinki *04051*
SN, Washington *16555*
SNAC, Paris *04938*
SNACC, Richmond *16686*
SNAD, Roma *07766*
SNALC, Paris *04946*
SNAME, Jersey City *16683*
SNAS, Singapore *10056*
S.N.B.A., Paris *04901*
SNCF, Paris *04940*
S.N.E., Luxembourg *08590*
SNE, Minneapolis *16557*
S.N.E.S., Paris *04944*
SNF, Stockholm *10568*
SNFGE, Paris *04905*
SNG, Aarau *10875*
SNG, Bern *10788*
SNG, Frankfurt *06280*
SNHF, Paris *04904*
S.N.I., Firenze *07993*
S.N.I.A., Roma *07767*
SNL, Luxembourg *08587*
SNLS, Denver *16556*
SNM, New York *16687*
SNMA, Washington *16749*
SNMG, Bucuresti *09789*
SNMOF, Nancy *04956*
S.N.M.S., Paris *04952*
SNOS, Glasgow *12918*
SNP, Lexington *16554*
S.N.P.A.M., Paris *04965*
SNPDES, Paris *04964*
SNPEN, Paris *04966*
SNPF, Clermont-Ferrand *04963*
SNPH, Paris *04967*
SNPMT, Toulouse *04970*
SNPN, Paris *04899*
SNR, Bucuresti *09791*
SNR, London *12984*
S.N.R.S.S., Bucuresti *09790*
SNS, Boston *16685*
SNS, Roma *07768*
SNS, Stockholm *10528*
SNSS, Gillingham *12888*
SNTS, Paris *04903*
SNV, Zürich *10876*
SNVB, Olympia *16715*
SNV Studiengesellschaft Verkehr, Hamburg *06281*
SOA, Schaumburg *16631*
SOAP, Houston *16558*
SOBCOT, Hony-Esneux *01428*
SOC, Concepción *03103*
SOC, Edinburgh *12919*
SOCHIL, Santiago *03077*
Social Affairs Federation, Montréal *02448*
Social History Society of the United Kingdom, Lancaster *12955*
Socialist Educational Association, Stockport *12956*
Socialist Health Association, London *12957*
Socialist International Research Council, London *12958*
Socialpaedagogernes Landsforbund, København *03822*
Social Policy Association, Hull *12959*
Social Psychiatry Research Institute, New York *16482*
Social Research Association, London *12960*
Social Responsibilities Round Table, Chicago *16483*
Social Science Education Consortium, Boulder *16484*
Social Science Federation of Canada, Ottawa *02914*
Social Science Research Council, New York *16485*
Social Sciences and Humanities Research Council of Canada, Ottawa *02915*
Social Sciences Services and Resources, Wasco *16486*
Social Welfare History Group, East Lansing *16487*
Sociedad Agronómica de Chile, Santiago *03065*
Sociedad Agronómica Mexicana, México *08745*
Sociedad Amantes de la Luz, Santiago de los Caballeros *03838*
Sociedad Americana de Oftalmologia y Optometria, Bogotá *03402*
Sociedad Amigos del Museo de Bellas Artes, Caracas *17008*
Sociedad Anatómica Española, Granada *10370*
Sociedad Antioqueña de Ingenieros y Architectos, Medellín *03403*
Sociedad Argentina de Alergia e Inmunopatología, Buenos Aires *00153*
Sociedad Argentina de Anatomía Normal y Patológica, Buenos Aires *00154*
Sociedad Argentina de Antropología, Buenos Aires *00155*
Sociedad Argentina de Autores y Compositores de Música, Buenos Aires *00156*
Sociedad Argentina de Biofísica, Buenos Aires *00157*
Sociedad Argentina de Biología, Buenos Aires *00158*
Sociedad Argentina de Botánica, San Isidro *00159*
Sociedad Argentina de Cardiología, Buenos Aires *00160*
Sociedad Argentina de Ciencias Fisiológicas, Buenos Aires *00161*
Sociedad Argentina de Ciencias Neurológicas, Psiquiátricas y Neuroquirúrgicas, Buenos Aires *00162*
Sociedad Argentina de Criminología, Buenos Aires *00163*
Sociedad Argentina de Dermatología, Buenos Aires *00164*
Sociedad Argentina de Endocrinología y Metabolismo, Buenos Aires *00165*
Sociedad Argentina de Escritores, Buenos Aires *00166*
Sociedad Argentina de Estudios Geográficos, Buenos Aires *00167*
Sociedad Argentina de Farmacología y Terapéutica, Buenos Aires *00168*
Sociedad Argentina de Fisiología Vegetal, Castelar *00169*
Sociedad Argentina de Gastroenterología, Buenos Aires *00170*
Sociedad Argentina de Gerontología y Geriatría, Buenos Aires *00171*
Sociedad Argentina de Hematología, Buenos Aires *00172*
Sociedad Argentina de Investigación Clínica, Buenos Aires *00173*
Sociedad Argentina de Investigación Operativa, Buenos Aires *00174*
Sociedad Argentina de Leprología, Buenos Aires *00175*
Sociedad Argentina de Medicina Nuclear, Buenos Aires *00176*
Sociedad Argentina de Medicina Social, Buenos Aires *00177*
Sociedad Argentina de Neurología, Psiquiatría y Neurocirugía, Buenos Aires *00178*
Sociedad Argentina de Oftalmología, Buenos Aires *00179*
Sociedad Argentina de Patología, Buenos Aires *00180*
Sociedad Argentina de Pediatría, Buenos Aires *00181*
Sociedad Argentina de Psicología, Buenos Aires *00182*
Sociedad Argentina de Radiología, Buenos Aires *00183*
Sociedad Argentina de Sociología, Córdoba *00184*
Sociedad Argentina de Socorros, Esperanza *00185*
Sociedad Argentina para el Estudio de la Esterilidad, Buenos Aires *00186*
Sociedad Argentina para el Estudio del Cáncer, Buenos Aires *00187*
Sociedad Arqueológica de Bolivia, La Paz *01548*
Sociedad Arqueológica de la Serena, La Serena *03066*
Sociedad Astronómica de México, México *08746*
Sociedad Bolivariana de Venezuela, Caracas *17009*
Sociedad Boliviana de Cirugía, La Paz *01549*
Sociedad Boliviana de Salud Pública, La Paz *01550*
Sociedad Central de Arquitectos, Buenos Aires *00188*
Sociedad Centroamericana de Cardiología, Tegucigalpa *06585*
Sociedad Centroamericana de Dermatologia, Guatemala City *06573*
Sociedad Chilena de Cancerología, Santiago *03067*
Sociedad Chilena de Cardiología y Cirugía Cardiovascular, Santiago *03068*
Sociedad Chilena de Cirugía Plástica y Reparadora, Santiago *03069*
Sociedad Chilena de Entomología, Santiago *03070*
Sociedad Chilena de Física, Santiago *03071*
Sociedad Chilena de Gastroenterología, Santiago *03072*
Sociedad Chilena de Gerontología, Santiago *03073*
Sociedad Chilena de Hematología, Santiago *03074*
Sociedad Chilena de Historia Natural, Santiago *03075*
Sociedad Chilena de Historia y Geografía, Santiago *03076*
Sociedad Chilena de Lingüística, Santiago *03077*
Sociedad Chilena de Obstetricia y Ginecología, Santiago *03078*
Sociedad Chilena de Parasitología, Santiago *03079*
Sociedad Chilena de Patología de la Adaptación y del Mesenquima, Santiago *03080*
Sociedad Chilena de Pediatría, Santiago *03081*
Sociedad Chilena de Química, Concepción *03082*
Sociedad Chilena de Sanidad, Santiago *03083*
Sociedad Chilena de Tisiología y Enfermedades Broncopulmonares, Santiago *03084*
Sociedad Científica Chilena Claudio Gay, Santiago *03085*
Sociedad Científica de Chile, Santiago *03086*
Sociedad Científica Argentina, Buenos Aires *00189*
Sociedad Científica del Paraguay, Asunción *09434*
Sociedad Colombiana de Biologí, Bogotá *03404*
Sociedad Colombiana de Cancerología, Bogotá *03405*
Sociedad Colombiana de Cirugía, Bogotá *03406*
Sociedad Colombiana de Economistas, Bogotá *03407*
Sociedad Colombiana de la Ciencia del Suelo, Bogotá *03408*
Sociedad Colombiana de Matemáticas, Bogotá *03409*
Sociedad Colombiana de Obstetricia y Ginecologia, Bogotá *03410*
Sociedad Colombiana de Patología, Cali *03411*
Sociedad Colombiana de Pediatría, Bogotá *03412*
Sociedad Colombiana de Psiquiatría, Bogotá *03413*
Sociedad Colombiana de Químicos e Ingenieros Químicos, Bogotá *03414*
Sociedad Colombina de Cardiología, Bogotá *03415*
Sociedad Cubana de Historia de la Ciencia y de la Técnica, La Habana *03481*
Sociedad Cubana de Historia de la Medicina, La Habana *03482*
Sociedad Cubana de Ingenieros, La Habana *03483*
Sociedad de Amigos de Arqueología, Montevideo *13265*
Sociedad de Anatomía Normal y Patológica de Chile, Santiago *03087*
Sociedad de Anestesiología de Chile, Santiago *03088*
Sociedad de Anestiología de El Salvador, San Salvador *03930*
Sociedad de Antropología de Antioquía, Medellín *03416*
Sociedad de Biología de Chile, Santiago *03089*
Sociedad de Biología de Concepción, Concepción *03090*
Sociedad de Biología de Córdoba, Córdoba *00190*
Sociedad de Biología de Montevideo, Montevideo *13266*
Sociedad de Bioquímica de Concepción, Concepción *03091*
Sociedad de Ciencias Aranzadi, San Sebastián *10371*
Sociedad de Ciencias, Letras y Artes El Museo Canario, Las Palmas *10372*
Sociedad de Ciencias Naturales Caldas, Medellín *03417*
Sociedad de Ciencias Naturales La Salle, Caracas *17010*
Sociedad de Cirugía de Buenos Aires, Buenos Aires *00191*
Sociedad de Cirugía del Uruguay, Montevideo *13267*
Sociedad de Cirujanos de Chile, Santiago *03092*
Sociedad de Dermatología y Sifilografía, Buenos Aires *00192*
Sociedad de Educación, México *08747*
Sociedad de Estadística e Investigación Operativa, Madrid *10373*
Sociedad de Estudios Biológicos, México *08748*
Sociedad de Estudios Geográficos e Históricos, Santa Cruz de la Sierra *01551*
Sociedad de Genética de Chile, Santiago *03093*
Sociedad de Geografía e Historia de Honduras, Tegucigalpa *06586*
Sociedad de Ginecología y Obstetricia de El Salvador, San Salvador *03931*
Sociedad de Ingenieros del Perú, Lima *09475*
Sociedad de Medicina Interna de Buenos Aires, Buenos Aires *00193*
Sociedad de Medicina Interna de Córdoba, Córdoba *00194*
Sociedad de Medicina Veterinaria de Chile, Santiago *03094*
Sociedad de Obstetricia y Ginecología de Venezuela, Caracas *17011*
Sociedad de Oftalmología del Hospital de Oftalmológico de Nuestra Señora de la Luz, México *08749*
Sociedad de Oftalmología Nicaragüense, Managua *09205*
Sociedad de Otorinolaringología de Valparaíso, Valparaíso *03095*
Sociedad de Pediatría de Cochabamba, Cochabamba *01552*
Sociedad de Pediatría de Madrid y Castilla La Mancha, Madrid *10374*
Sociedad de Pediatría de Madrid y Región Centro, Madrid *10375*
Sociedad de Pediatría de Valparaíso, Valparaíso *03096*
Sociedad de Pediatría y Puericultura del Atlántico, Barranquilla *03418*
Sociedad de Pediatría y Puericultura del Paraguay, Asunción *09435*
Sociedad de Psicología Médica, Psicoanálisis y Medicina Psicosomática, Buenos Aires *00195*
Sociedad de Radiología de La Habana, La Habana *03484*
Sociedad de Radiología del Uruguay, Montevideo *13268*
Sociedad de Tisiología y Neumonología del Hospital Tornu y Dispensarios, Buenos Aires *00196*
Sociedad de Tisiología y Neumonología de Venezuela, Antimano *17012*
Sociedade Anatómica Luso-Hispano-Americana, Lisboa *09673*
Sociedade Anatómica Portuguesa, Coimbra *09674*
Sociedade Astronómica de Portugal, Lisboa *09675*
Sociedade Botânica do Brasil, Brasília *01703*
Sociedade Brasileira de Autores Teatrais, Rio de Janeiro *01704*
Sociedade Brasileira de Belas Artes, Rio de Janeiro *01705*
Sociedade Brasileira de Cartografia, Rio de Janeiro *01706*
Sociedade Brasileira de Cultura, São Paulo *01707*
Sociedade Brasileira de Dermatologia, Rio de Janeiro *01708*
Sociedade Brasileira de Entomologia, São Paulo *01709*
Sociedade Brasileira de Filosofia, Rio de Janeiro *01710*
Sociedade Brasileira de Genética, Ribeirão Preto *01711*

Sociedade Brasileira de Geografia, Rio de Janeiro 01712
Sociedade Brasileira de Geologia, São Paulo 01713
Sociedade Brasileira de Instrução, Botafogo 01714
Sociedade Brasileira de Microbiologia, Rio de Janeiro 01715
Sociedade Brasileira de Romanistas, Rio de Janeiro 01716
Sociedade Broteriana, Coimbra 09676
Sociedade Científica da Universidade Católica Portuguesa, Lisboa 09677
Sociedade Civil de Educação São Marcos, São Paulo 01717
Sociedad Económica de Amigos del País, La Habana 03485
Sociedad Ecuatoriana de Alergía y Ciencias Afinas, Quito 03855
Sociedad Ecuatoriana de Astronomía, Quito 03856
Sociedad Ecuatoriana de Pediatría, Guayaquil 03857
Sociedade das Ciências Médicas de Lisboa, Lisboa 09678
Sociedade de Biologia do Brasil, Rio de Janeiro 01718
Sociedade de Biologia do Rio Grande do Sul, Pôrto Alegre 01719
Sociedade de Ciências Agrárias de Portugal, Lisboa 09679
Sociedade de Cultura e Educação do Litoral Sul, Registro 01720
Sociedade de Engenharia do Rio Grande do Sul, Pôrto Alegre 01721
Sociedade de Ensino Piratininga, São Paulo 01722
Sociedade de Estudos Açoreanos Afonso Chaves, Ponta Delgada 09680
Sociedade de Estudos Técnicos SARL-SETEC, Lisboa 09681
Sociedade de Farmácia e Química de São Paulo, São Paulo 01723
Sociedade de Geografia de Lisboa, Lisboa 09682
Sociedade de Lingua Portuguesa, Lisboa 09683
Sociedade de Medicina de Alagoas, Maceió 01724
Sociedade de Pediatria da Bahia, Salvador 01725
Sociedade Farmacêutica Lusitana, Lisboa 09684
Sociedade Geográfica Brasileira, São Paulo 01726
Sociedade Geológica de Portugal, Lisboa 09685
Sociedade Histórica da Independência de Portugal, Lisboa 09686
Sociedade Martins Sarmento, Guimarães 09687
Sociedade Nacional de Agricultura, Rio de Janeiro 01727
Sociedade Nacional de Belas Artes, Lisboa 09688
Sociedad Entomológica Argentina, La Plata 00197
Sociedad Entomológica del Perú, Lima 09476
Sociedade Paranaense de Matemática, Curitiba 01728
Sociedade Portuguesa de Alergologia e Imunologia Clínica, Lisboa 09689
Sociedade Portuguesa de Anestesiologia, Lisboa 09690
Sociedade Portuguesa de Antropologia e Etnologia, Porto 09691
Sociedade Portuguesa de Autores, Lisboa 09692
Sociedade Portuguesa de Bioquímica, Lisboa 09693
Sociedade Portuguesa de Cardiologia, Lisboa 09694
Sociedade Portuguesa de Ciências Naturais, Lisboa 09695
Sociedade Portuguesa de Ciências Veterinárias, Lisboa 09696
Sociedade Portuguesa de Dermatologia e Venereologia, Lisboa 09697
Sociedade Portuguesa de Educação Médica, Lisboa 09698
Sociedade Portuguesa de Especialistas de Pequenos Animales, Lisboa 09699
Sociedade Portuguesa de Espeleologia, Lisboa 09700
Sociedade Portuguesa de Gastroenterologia, Lisboa 09701
Sociedade Portuguesa de Hemorreologia e Microcirculaçã0, Lisboa 09702
Sociedade Portuguesa de Higiene Alimentar, Lisboa 09703
Sociedade Portuguesa de Medicina Fisica e Reabilitação, Lisboa 09704
Sociedade Portuguesa de Neurologia e Psiquiatria, Lisboa 09705
Sociedade Portuguesa de Numismática, Porto 09706
Sociedade Portuguesa de Nutrição e Alimentação Animal, Lisboa 09707
Sociedade Portuguesa de Oftalmologia, Lisboa 09708
Sociedade Portuguesa de Ortopedia e Traumatologia, Lisboa 09709
Sociedade Portuguesa de Otorinolaringologia e Bronco-Esofagologia, Lisboa 09710
Sociedade Portuguesa de Pediatria, Lisboa 09711
Sociedade Portuguesa de Química, Lisboa 09712
Sociedade Portuguesa de Reprodução Animal, Lisboa 09713
Sociedade Portuguesa de Reumatologia, Lisboa 09714
Sociedade Portuguesa Veterinária de Anatomia Comparativa, Lisboa 09715
Sociedade Portuguesa Veterinária de Estudos Sociológicos, Lisboa 09716
Sociedade Portuguese de Estomatologia, Lisboa 09717
Sociedade Propagadora Esdeva, Juiz de Fora 01729
Sociedad Española de Acústica, Madrid 10376
Sociedad Española de Anestesiología y Reanimación, Madrid 10377
Sociedad Española de Bioquímica, Madrid 10378
Sociedad Española de Bromatología, Madrid 10379
Sociedad Española de Cerámica y Vidrio, Madrid 10380
Sociedad Española de Ciencias Fisiológicas, Madrid 10381
Sociedad Española de Cirugía Oral y Maxillofacial, Madrid 10382
Sociedad Española de Cirugía Ortopédica y Traumatología, Madrid 10383
Sociedad Española de Cirugía Plástica, Reparadora y Estética, Madrid 10384
Sociedad Española de Diabetes, Madrid 10385
Sociedad Española de Estudios Clásicos, Madrid 10386
Sociedad Española de Filosofía, Madrid 10387
Sociedad Española de Horticultura, Madrid 10388
Sociedad Española de Mecánica del Suelo y Cimentaciones, Madrid 10389
Sociedad Española de Medicina Psicosomática y Psicoterapía, Barcelona 10390
Sociedad Española de Microbiología, Madrid 10391
Sociedad Española de Optica, Madrid 10392
Sociedad Española de Otorinolaringología y Patología Cévico-Facial, Madrid 10393
Sociedad Española de Patología Digestiva y de la Nutrición, Madrid 10394
Sociedad Española de Psicoanàlisis, Barcelona 10395
Sociedad Española de Radiología Medica, Madrid 10396
Sociedad Española de Rehabilitación y Medicina Fisica, Madrid 10397
Sociedade Visconde de São Leopoldo, Santos 01730
Sociedad General de Autores de España, Madrid 10398
Sociedad General de Autores de la Argentina, Buenos Aires 00198
Sociedad Geográfica de Colombia, Bogotá 03419
Sociedad Geográfica de La Paz, La Paz 01553
Sociedad Geográfica de Lima, Lima 09477
Sociedad Geográfica Sucre, Sucre 01554
Sociedad Geográfica y de Historia Potosío, Potosí 01555
Sociedad Geológica Boliviana, La Paz 01556
Sociedad Geológica de Chile, Santiago 03097
Sociedad Geológica del Perú, Lima 09478
Sociedad Geológica Mexicana, México 08750
Sociedad Hebraica Argentina, Buenos Aires 00199
Sociedad Ibero-Americana de Estudios Numismáticos, Madrid 10399
Sociedad Interamericana de Cardiología, México 08751
Sociedad Jurídica de la Universidad Nacional, Bogotá 03420
Sociedad Latinoamericana de Alergología, México 08752
Sociedad Latinoamericana de Ciencias de los Suelos, Mendoza 00200
Sociedad Latinoamericana de Farmacología, Caracas 17013
Sociedad Latinoamericana de Farmacología, Quito 03858
Sociedad Latinoamericana de Investigación Pediátrica, Lima 09479
Sociedad Luso-Española de Neurocirugía, Madrid 10400
Sociedad Malacológica del Uruguay, Montevideo 13269
Sociedad Matemática Mexicana, México 08753
Sociedad Mayaguezana por Bellas Artes, Mayaguez 09731
Sociedad Médica, Bolivar 17014
Sociedad Médica de Concepción, Concepción 03098
Sociedad Médica del Hospital General, México 08754
Sociedad Médica del Hospital Oftalmológico de Nuestra Señora de la Luz, México 08755
Sociedad Médica de Salud Pública, San Salvador 03932
Sociedad Médica de Santiago, Santiago 03099
Sociedad Médica de Valparaíso, Valparaíso 03100
Sociedad Mexicana de Antropología, México 08756
Sociedad Mexicana de Bibliografía, México 08757
Sociedad Mexicana de Cardiología, México 08758
Sociedad Mexicana de Entomología, Chapingo 08759
Sociedad Mexicana de Estudios Psico-Pedagógicos, México 08760
Sociedad Mexicana de Eugenesia, México 08761
Sociedad Mexicana de Fitogenética, Chapingo 08762
Sociedad Mexicana de Fitopatología, Montecillo 08763
Sociedad Mexicana de Geografía y Estadística, México 08764
Sociedad Mexicana de Historia de la Ciencia y la Tecnología, México 08765
Sociedad Mexicana de Historia Natural, México 08766
Sociedad Mexicana de Historia y Filosofía de la Medicina, México 08767
Sociedad Mexicana de Neurologia y Psiquiatria, México 08768
Sociedad Mexicana de Nutrición y Endocrinología, México 08769
Sociedad Mexicana de Pediatría, México 08770
Sociedad Mexicana de Salud Pública, México 08771
Sociedad Mexicana de Seguridad Radiologica, México 08772
Sociedad Mexicana de Tisiología, México 08773
Sociedad Nacional Agraria, Lima 09480
Sociedad Nacional de Agricultura, Santiago 03101
Sociedad Nacional de Minería, Lima 09481
Sociedad Nacional de Minería, Santiago 03102
Sociedad Nicaragüense de Psiquiatría y Psicología, Managua 09206
Sociedad Nuclear Mexicana, México 08774
Sociedad Nuevoleonesa de Historia, Geografía y Estadística, Monterrey 08775
Sociedad Odontológica de Concepción, Concepción 03103
Sociedad Odontológica Zuliana de Prótesis Maracaibo, Maracaibo 17015
Sociedad Oftalmológico Hispano-Americano, Madrid 10401
Sociedad Peruana de Espeleología, Lima 09482
Sociedad Peruana de Eugenesia, Lima 09483
Sociedad Peruana de Historia de la Medicina, Lima 09484
Sociedad Peruana de Ortopedia y Traumatología, Lima 09485
Sociedad Peruana de Tisiología y Enfermedades Respiratorias, Lima 09486
Sociedad Pro-Arte Musical, Guatemala City 06574
Sociedad Puertorriqueña de Escritores, San Juan 09732
Sociedad Química de México, México 08776
Sociedad Química del Perú, Lima 09487
Sociedad Rural Argentina, Buenos Aires 00201
Sociedad Rural Boliviana, La Paz 01557
Sociedad Uruguaya de Patología Clínica, Montevideo 13270
Sociedad Uruguaya de Pediatría, Montevideo 13271
Sociedad Venezolana de Angiologia, Caracas 17016
Sociedad Venezolana de Cardiología, Caracas 17017
Sociedad Venezolana de Ciencias Naturales, Caracas 17018
Sociedad Venezolana de Cirugía, Caracas 17019
Sociedad Venezolana de Cirugía Plástica y Reconstrucción, Caracas 17020
Sociedad Venezolana de Cirugía Ortopédica y Traumatología, Caracas 17021
Sociedad Venezolana de Entomología, Maracay 17022
Sociedad Venezolana de Hematología, Caracas 17023
Sociedad Venezolana de Historia de la Medicina, Caracas 17024
Sociedad Venezolana de Ingenieros de Minas y Metalúrgicos, Caracas 17025
Sociedad Venezolana de Medicina del Trabajo y Deportes, Caracas 17026
Sociedad Venezolana de Oftalmología, Caracas 17027
Sociedad Venezolana de Oncología, Caracas 17028
Sociedad Venezolana de Otorinolaringología, Caracas 17029
Sociedad Venezolana de Psiquiatría y Neurología, Caracas 17030
Sociedad Venezolana de Puericultura y Pediatría, Caracas 17031
Sociedad Venezolana de Radiología, Caracas 17032
Sociedad Venezolana de Tisiología y Neumonología, Caracas 17033
Sociedad Venezolana de Urología, Caracas 17034
Sociedad Veterinaria de Zootecnica de España, Madrid 10402
Sociedad Zoológica del Uruguay, Montevideo 13272
Società Adriatica di Scienze, Trieste 07769
Società Archeologica Comense, Como 07770
Società Archeologica Viterbese Pro Ferento, Viterbo 07771
Società Astronomica Italiana, Milano 07772
Società Botanica Italiana, Firenze 07773
Società Chimica Italiana, Roma 07774
Società Consortile per Azioni per lo Sviluppo della Ricerca Farmaceutica, Pomezia 08035
Società Dalmata di Storia Patria, Roma 07775
Società Dante Alighieri, Roma 07776
Società Dantesca Italiana, Firenze 07777
Società Dauna di Cultura, Foggia 07778
Società degli Amici del Museo Civico di Storia Naturale Giacomo Doria, Genova 07779
Società degli Ingegneri e degli Architetti in Torino, Torino 07780
Società dei Naturalisti in Napoli, Napoli 07781
Società di Fotogrammetria e Topografia, Milano 07782
Società di Letture e Conversazioni Scientifiche, Genova 07783
Società di Linguistica Italiana, Roma 07784
Società di Medicina Legale e delle Assicurazioni, Roma 07785
Società di Minerva, Trieste 07786
Società di Ortopedia e Traumatologia dell'Istituto Meridionale ed Insulare, Napoli 07787
Società di Scienze Farmacologiche Applicate, Milano 07788
Società di Scienze Naturali del Trentino, Trento 07789
Società di Storia Patria di Terra di Lavoro, Caserta 07790
Società di Storia Patria per la Puglia, Bari 07791
Società di Storia Patria per la Sicilia Orientale, Catania 07792
Società di Studi Celestiniani, Isernia 07793
Società di Studi Geografici, Firenze 07794
Società di Studi Romagnoli, Cesena 07795
Società di Studi Trentini di Scienze Storiche, Trento 07796
Società di Studi Valdesi, Torre Pellice 07797
Società Ecologica Friulana, Campoformido 07798
Società Economica di Chiavari, Chiavari 07799
Società Emiliana Pro Montibus et Silvis, Bologna 07800
Società Entomologica Italiana, Genova 07801
Società Europea di Cultura, Venezia 07802
Società Europea di Patologia, Milano 07803
Società Farmaceutica del Mediterraneo Latino, Messina 07804
Società Filarmonica di Trento, Trento 07805
Società Filologica Friulana G.I. Ascoli, Udine 07806
Società Filologica Romana, Roma 07807
Società Filosofica Calabrese, Palmi 07808
Società Filosofica Italiana, Milano 07809
Società Filosofica Romana, Roma 07810
Società Gallaratese per gli Studi Patri, Gallarate 07811
Società Generale per Progettazioni, Consulenze e Partecipazioni, Roma 07756
Società Geografica Italiana, Roma 07812
Società Geologica Italiana, Roma 07813
Società Incoraggiamento Arti e Mestieri, Milano 07814
Società Internazionale di Diritto Penale Militare e di Diritto della Guerra, Gruppo Italiano, Roma 07815
Società Internazionale di Psicologia della Scrittura, Milano 07816
Società Internazionale di Studi Francescani, Assisi 07817
Società Internazionale di Studi Gemellari, Roma 07818
Società Internazionale di Tecnica Idrotermale, Castellamare di Stabia 07819
Società Istriana di Archeologia e Storia Patria, Trieste 07820
Società Italiana Amici dei Fiori, Firenze 07821
Società Italiana Attività Nervosa Superiore, Milano 07822
Società Italiana Calcolo Ricerca Economica Operativa, Roma 07823
Società Italiana degli Economisti, Genova 07824
Società Italiana della Continenza, Roma 07825
Società Italiana della Scienza del Suolo, Firenze 07826
Società Italiana della Trasfusione del Sangue, Torino 07827
Società Italiana delle Scienze Veterinarie, Brescia 07828
Società Italiana di Agopuntura, Torino 07829
Società Italiana di Agronomia, Bologna 07830
Società Italiana di Allergologia e Immunologia Clinica, Roma 07831
Società Italiana di Anatomia, Firenze 07832
Società Italiana di Anatomia Patologica, Messina 07833
Società Italiana di Andrologia, Pisa 07834
Società Italiana di Anestesia, Analgesia, Rianimazione e Terapia Intensiva, Firenze 07835
Società Italiana di Anestesiologia Rianimazione e Terapia del Dolore, Bari 07836
Società Italiana di Angiologia, Pisa 07837
Società Italiana di Antropologia ed Etnologia, Firenze 07838
Società Italiana di Audiologia e Foniatria, Roma 07839
Società Italiana di Aziendologia, Milano 07840
Società Italiana di Biochimica, Perugia 07841
Società Italiana di Biogeografia, Siena 07842
Società Italiana di Biologia Marina, Livorno 07843
Società Italiana di Biologia Sperimentale, Napoli 07844
Società Italiana di Biometria, Firenze 07845
Società Italiana di Buiatria, Torino 07846
Società Italiana di Cancerologia, Milano 07847
Società Italiana di Cardiologia, Roma 07848
Società Italiana di Chemioterapia, Pavia 07849
Società Italiana di Chirurgia, Roma 07850
Società Italiana di Chirurgia Cardiaca e Vascolare, Roma 07851
Società Italiana di Chirurgia Clinica, Roma 07852
Società Italiana di Chirurgia della Mano, Firenze 07853
Società Italiana di Chirurgia d'Urgenza, di Pronto Soccorso e di Terapia Intensiva Chirurgica, Milano 07854
Società Italiana di Chirurgia Estetica, Roma 07855
Società Italiana di Chirurgia Pediatrica, Roma 07856
Società Italiana di Chirurgia Plastica, Verona 07857
Società Italiana di Chirurgia Toracica, Roma 07858
Società Italiana di Citologia Clinica e Sociale, Roma 07859
Società Italiana di Criminologia, Roma 07860
Società Italiana di Dermatologia e Sifilografia, Roma 07861
Società Italiana di Diabetologia, Torino 07862
Società Italiana di Diabetologia e Endocrinologia Pediatrica, Milano 07863
Società Italiana di Economia Agraria, Bologna 07864
Società Italiana di Economia Demografia e Statistica, Roma 07865
Società Italiana di Elettroencefalografia e Neurofisiologia, Bologna 07866
Società Italiana di Ematologia, Milano 07867
Società Italiana di Endocrinologia, Roma 07868
Società Italiana di Ergonomia, Milano 07869
Società Italiana di Ergonomia Stomatologica, Milano 07870
Società Italiana di Farmacologia, Milano 07871

Società Italiana di Farmacologia Clinica, Pisa 07872
Società Italiana di Filosofia Giuridica e Politica, Roma 07873
Società Italiana di Fisica, Bologna 07874
Società Italiana di Fisiologia, Firenze 07875
Società Italiana di Foniatria, Catania 07876
Società Italiana di Fotobiologia, S. Maria di Galeria 07877
Società Italiana di Fotogrammetria e Topografia, Milano 07878
Società Italiana di Gastroenterologia, Bologna 07879
Società Italiana di Genetica Agraria, Foggia 07880
Società Italiana di Geofisica e Meteorologia, Genova 07881
Società Italiana di Gerontologia e Geriatria, Firenze 07882
Società Italiana di Ginecologia Pediatrica, Roma 07883
Società Italiana di Ginnastica Medica, Medicina Fisica e Riabilitazione, Milano 07884
Società Italiana di Glottologia, Pisa 07885
Società Italiana di Immunoematologia e Trasfusione del Sangue, Roma 07886
Società Italiana di Immunologia e di Immunopatologia, Bari 07887
Società Italiana di Ingegneria, Aerofotogrammetria e Topografia, Roma 07888
Società Italiana di Ippologia, Roma 07889
Società Italiana di Laringologia, Otologia, Rinologia e Patologia Cervico-Facciale, Roma 07890
Società Italiana di Liposcultura, Roma 07891
Società Italiana di Malacologia, Milano 07892
Società Italiana di Medicina del Lavoro e di Igiene Industriale, Pavia 07893
Società Italiana di Medicina del Traffico, Roma 07894
Società Italiana di Medicina e Igiene della Scuola, Milano 07895
Società Italiana di Medicina Estetica, Roma 07896
Società Italiana di Medicina Fisica e Riabilitazione, Moncalieri 07897
Società Italiana di Medicina Interna, Roma 07898
Società Italiana di Medicina Legale e delle Assicurazioni, Roma 07899
Società Italiana di Medicina Preventiva e Sociale, Genova 07900
Società Italiana di Medicina Psicosomatica, Roma 07901
Società Italiana di Medicina Sociale, Torino 07902
Società Italiana di Medicina Subacquea ed Iperbarica, Napoli 07903
Società Italiana di Mesoterapia, Roma 07904
Società Italiana di Meteorologia Applicata, Roma 07905
Società Italiana di Microangiologia e Microcorcolazione, Pisa 07906
Società Italiana di Microbiologia, Catania 07907
Società Italiana di Mineralogia e Petrologia, Milano 07908
Società Italiana di Musicologia, Bologna 07909
Società Italiana di Neurochirurgia, Ancona 07910
Società Italiana di Neurologia, Roma 07911
Società Italiana di Neuropediatria, Siena 07912
Società Italiana di Neuropsichiatria Infantile, Napoli 07913
Società Italiana di Neuroradiologia, Napoli 07914
Società Italiana di Neurosonologia, Bologna 07915
Società Italiana di Nipiologia, Roma 07916
Società Italiana di Nutrizione Umana, Roma 07917
Società Italiana di Odontoiatria Infantile, Roma 07918
Società Italiana di Odontostomatologia e Chirurgia Maxillo-Facciale, L'Aquila 07919
Società Italiana di Oncologia Ginecologica, Roma 07920
Società Italiana di Ortopedia e Traumatologia, Roma 07921
Società Italiana di Ostetricia e Ginecologia, Roma 07922
Società Italiana di Oto-Neuro-Oftalmologia, Bologna 07923
Società Italiana di Otorinolaringologia e Chirurgia Cervico Facciale, Roma 07924
Società Italiana di Otorinolaringologia Pediatrica, Roma 07925
Società Italiana di Parassitologia, Roma 07926
Società Italiana di Patologia, Milano 07927
Società Italiana di Patologia Aviare, Perugia 07928
Società Italiana di Patologia ed Allevamento dei Suini, Parma 07929
Società Italiana di Patologia e di Allevamento degli Ovini e dei Caprini, Catania 07930
Società Italiana di Patologia Vascolare, Roma 07931
Società Italiana di Pediatria, Roma 07932
Società Italiana di Pneumologia, Torino 07933
Società Italiana di Psichiatria, Milano 07934
Società Italiana di Psichiatria Biologica, Napoli 07935
Società Italiana di Psicologia, Roma 07936
Società Italiana di Psicologia Individuale, Milano 07937
Società Italiana di Psicologia Scientifica, Palermo 07938
Società Italiana di Psicosintesi Terapeutica, Firenze 07939
Società Italiana di Psicoterapia Analitica Immaginativa, Cremona 07940
Società Italiana di Radiologia Medica e di Medicina Nucleare, Milano 07941
Società Italiana di Reologia, Napoli 07942
Società Italiana di Reumatologia, Bari 07943
Società Italiana di Scienze Farmaceutiche, Milano 07944
Società Italiana di Scienze Fisiche e Matematiche Mathesis, Roma 07945
Società Italiana di Scienze Naturali, Milano 07946
Società Italiana di Senologia, Roma 07947
Società Italiana di Sessuologia Clinica, Roma 07948
Società Italiana di Sessuologia Medica, Roma 07949
Società Italiana di Sociologà, Roma 07950
Società Italiana di Statistica, Roma 07951
Società Italiana di Storia della Medicina, Bologna 07952
Società Italiana di Studi sul Secolo XVIII, Roma 07953
Società Italiana di Terapia Familiare, Roma 07954
Società Italiana di Tossicologia, Roma 07955
Società Italiana di Urodinamica, Bologna 07956
Società Italiana di Urologia, Roma 07957
Società Italiana Medica del Training Autogeno, Bologna 07958
Società Italiana Medici e Operatori Geriatrici, Firenze 07959
Società Italiana Medico-Chirurgica di Pronto Soccorso, Bologna 07960
Società Italiana Organi Artificiali, Roma 07961
Società Italiana per gli Archivi Sanitari Ospedalieri, Viareggio 07962
Società Italiana per il Progresso della Zootecnia, Milano 07963
Società Italiana per l'Antropologia e la Etnologia, Firenze 07964
Società Italiana per l'Archeologia e la Storia delle Arti, Napoli 07965
Società Italiana per la Robotica Industriale, Milano 07966
Società Italiana per l'Educazione Musicale, Milano 07967
Società Italiana per le Scienze Ambientali: Biometeorologia, Bioclimatologia ed Ecologia, Milano 07968
Società Italiana per l'Organizzazione Internazionale, Roma 07969
Società Italiana per lo Studio della Cancerogenesi Ambientale ed Epidemiologia dei Tumori, Genova 07970
Società Italiana per lo Studio della Fertilità e della Sterilità, Roma 07971
Società Italiana per lo Studio dell'Arteriosclerosi, Bari 07972
Società Italiana per lo Studio delle Sostanze Grasse, Milano 07973
Società Italiana per lo Studio e l'Applicazione del Pirodiserbo, Piacenza 07974
Società Italiana pro Deontologia Sanitaria, Milano 07975
Società Italiana Sistemi Informativi Elettronici, Roma 07758
Società Jonico-Salentina de Medicina e Chirurgia, Taranto 07976
Società Laziale Abruzzese di Medicina del Lavoro, Roma 07977
Società Laziale – Abruzzese Marchigiana Molisana di Ostetricia e Ginecologia, Roma 07978
Società Letteraria, Verona 07979
Società Ligure di Storia Patria, Genova 07980
Società Lombarda di Criminologia, Milano 07981
Società Mazziniana Pensiero e Azione, Roma 07982
Società Medica Chirurgica di Bologna, Bologna 07983
Società Medico-Chirugica, Bari 07984
Società Medico-Chirurgica di Ferrara, Ferrara 07985
Società Medico-Chirurgica di Modena, Modena 07986
Società Messinese di Storia Patria, Messina 07987
Società Napoletana di Chirurgia, Napoli 07988
Società Napoletana di Storia Patria, Napoli 07989
Società Naturalisti Veronesi F. Zorzi, Verona 07990
Società Nazionale di Informatica delle Camere di Commercio per la Gestione dei Centri Elettronici Reteconnessi Valutazione Elaborazione Dati, Roma 07991
Società Nazionale di Scienze, Lettere ed Arti, Ex Società Reale, Napoli 07992
Società Nucleare Italiana, Firenze 07993
Società Oftalmologica Italiana, Bologna 07994
Società Ornitologica Italiana, Ravenna 07995
Società Ornitologica Reggiana, Reggio Emilia 07996
Società Orticola Italiana, Firenze 07997
Società Pavese di Storia Patria, Pavia 07998
Società per gli Studi Storici, Archeologici ed Artistici della Provincia di Cuneo, Cuneo 07999
Società per la Formazione la Ricerca e l'Addestramento per le Aziende e le Organizzazioni, Milano 08000
Società per la Matematica e l'Economia Applicate, Roma 08001
Società per le Belle Arti ed Esposizione Permanente, Milano 08002
Società Piemontese, Ligure, Lombarda di Ortopedia e Traumatologia, Genova 08003
Società Pistoiese di Storia Patria, Pistoia 08004
Società Promotrice di Belle Arti, Napoli 08005
Società Reggiana d'Archeologia, Reggio Emilia 08006
Società Reggiana di Studi Storici, Reggio Emilia 08007
Società Retorumantscha, Chur 10963
Società Ricerche Impianti Nucleari, Vercelli 08008
Società Romana di Chirurgia, Roma 08009
Società Romana di Storia Patria, Roma 08010
Societas Amicorum Naturae Ouluensis, Oulu 03988
Società Sassarese per le Scienze Giuridiche, Sassari 08011
Società Savonese di Storia Patria, Savona 08012
Societas Biochemica, Biophysica et Microbiologica Fenniae, Helsingin Yliopisto 03992
Societas Biologica Fennica Vanamo, Helsingin Yliopisto 03993
Societas Chirurgica Hungarica, Budapest 06657
Societas Heraldica Scandinavica, Bagsvaerd 03823
Società Siciliana per la Storia Patria, Palermo 08013
Società Siracusana di Storia Patria, Siracusa 08014
Societas Linguistica Europaea, Wien 00945
Societas Logopedica Latina, Leuven 01420
Societas Medicinae Physicalis et Rehabilitationis Fenniae, Helsinki 03994
Societas Oto-Rhino-Laryngologia Latina, Montpellier 04624
Societas Paediatrica Japonica, Tokyo 08265
Società Speleologica Italiana, Milano 08015
Societas pro Fauna et Flora Fennica, Helsingin Yliopisto 03995
Societas Radiologorum Hungarorum, Budapest 06654
Societas Scientiarum Bialostocensis, Bialystok 09539
Societas Scientiarum Fennica, Helsinki 03996
Societas Scientiarum Plocensis, Płock 09626
Società Storica Catanese, Catania 08016
Società Storica di Terra d'Otranto, Lecce 08017
Società Storica Locarnese, Locarno 10964
Società Storica Lombarda, Milano 08018
Società Storica Novarese, Novara 08019
Società Storica Pisana, Pisa 08020
Societat Andorrana de Ciències, Andorra la Vella 00032
Societat Arqueològica Lul·liana, Palma de Mallorca 10403
Società Tarquiniense d'Arte e Storia, Tarquinia 08021
Societatea de Anestezie si Terapie Intensiva, Bucuresti 09755
Societatea de Balneologie, Bucuresti 09756
Societatea de Cardiologie, Bucuresti 09757
Societatea de Chirurgie, Bucuresti 09758
Societatea de Dermato-Venerologie, Bucuresti 09759
Societatea de Endocrinologie, Bucuresti 09760
Societatea de Farmacie, Bucuresti 09761
Societatea de Fiziologie, Bucuresti 09762
Societatea de Ftiziologie, Bucuresti 09763
Societatea de Gastroenterologie, Bucuresti 09764
Societatea de Gerontologie, Bucuresti 09765
Societatea de Histochimie si Citochimie, Bucuresti 09766
Societatea de Igiena si Sànàtate Publica, Bucuresti 09767
Societatea de Istorie Medicinei, Bucuresti 09768
Societatea de Medicinà Generalà, Bucuresti 09769
Societatea de Medicinà Internà, Bucuresti 09770
Societatea de Medicinà Sportiva, Bucuresti 09771
Societatea de Medici si Naturalisti, Iasi 09772
Societatea de Microbiologie, Bucuresti 09773
Societatea de Obstetricà si Ginecologie, Bucuresti 09774
Societatea de Oftalmologie, Bucuresti 09775
Societatea de Oto-Rino-Laringologie, Bucuresti 09776
Societatea de Patologie Infectioasà, Bucuresti 09777
Societatea de Pediatrie, Bucuresti 09778
Societatea de Radiologie, Bucuresti 09779
Societatea de Stiinte Biologice din Romania, Bucuresti 09780
Societatea de Stiinte Farmaceutice, Bucuresti 09781
Societatea de Stiinte Filologice din Romania, Bucuresti 09782
Societatea de Stiinte Fizice si Chimice din Romania, Bucuresti 09783
Societatea de Stiinte Geografice din Romania, Bucuresti 09784
Societatea de Stiinte Geologice din Romania, Bucuresti 09785
Societatea de Stiinte Matematice din Romania, Bucuresti 09786
Societatea de Stomatologie, Bucuresti 09787
Societatea de Studii Clasice din Romania, Bucuresti 09788
Societatea Nationala de Medicina Generala din Romania, Bucuresti 09789
Societatea Nationala Romana pentru Stiinta Solului, Bucuresti 09790
Societatea Numismatica Romana, Bucuresti 09791
Societatea Romana de Linguistica, Bucuresti 09792
Societatea Romana de Neurochirurgie, Bucuresti 09793
Societatea Română de Ortopedie şi Traumatologie, Bucuresti 09794
Societatea Romana de Sprijia a Virstnicilor Suferinzi de Afectiuni de tip Alzheimer, Bucuresti 09795
Società Teosofica in Italia, Trieste 08022
Società Tiburtina di Storia e d'Arte, Tivoli 08023
Società Ticinese di Scienze Naturali, Lugano 10965
Società Torricelliana di Scienze e Lettere, Faenza 08024
Società Toscana di Orticoltura, Firenze 08025
Società Toscana di Scienze Naturali, Pisa 08026
Société Académique, Neuchâtel 10966
Société Académique des Arts Libéraux de Paris, La Varenne Saint-Hilaire 04625
Société Africaine de Culture, Paris 04626
Société Anatole France, Paris 04627
Société Anatomique de Paris, Paris 04628
Société Archéologique d'Alexandrie, Alexandria 03913
Société Archéologique de France, Paris 04566
Société Archéologique de Namur, Namur 01421
Société Archéologique de Touraine, Tours 04629
Société Archéologique du Département de Constantine, Constantine 00023
Société Archéologique et Historique du Limousin, Limoges 04630
Société Asiatique de Paris, Paris 04631
Société Astronomique de Bordeaux, Bordeaux 04632
Société Astronomique de France, Paris 04633
Société Astronomique de Liège, Cointe-Ougrée 01422
Société Astronomique de Lyon, Saint-Genis-Laval 04634
Société Astronomique de Suisse, Meyrin 10967
Société Belge d'Allergologie et d'immunologie Clinique, Gent 01423
Société Belge d'Anesthésie et de Réanimation, Bierges 01424
Société Belge de Biochimie et de Biologie Moléculaire, Bruxelles 01425
Société Belge de Biologie Clinique, Charleroi 01426
Société Belge de Cardiologie, Bruxelles 01427
Société Belge de Chirurgie Orthopédique et de Traumatologie, Hony-Esneux 01428
Société Belge de Gastro-Entérologie, Bruxelles 01429
Société Belge de Géologie, Bruxelles 01430
Société Belge d'Electroencéphalographie et de Neurophysiologie Clinique, Ottignies 01431
Société Belge de Logique et de Philosophie des Sciences, Bruxelles 01432
Société Belge de Médecine Interne, Bruxelles 01433
Société Belge de Médecine Tropicale, Antwerpen 01163
Société Belge d'EMG, Bruxelles 01434
Société Belge de Microscopie Electronique, Liège 01435
Société Belge de Musicologie, Bruxelles 01436
Société Belge de Neurologie, Bruxelles 01437
Société Belge de Pédiatrie, Bruxelles 01438
Société Belge de Philosophie, Bruxelles 01439
Société Belge de Photogrammétrie, de Télédétection et de Cartographie, Bruxelles 01440
Société Belge de Physiologie et de Pharmacologie, Bruxelles 01441
Société Belge de Physique, Bruxelles 01442
Société Belge de Pneumologie, Bruxelles 01443
Société Belge de Psychoanalyse, Loverval 01444
Société Belge de Psychologie, Bruxelles 01445
Société Belge d'Ergologie, Bruxelles 01446
Société Belge des Auteurs, Compositeurs et Editeurs, Bruxelles 01447
Société Belge d'Etudes Byzantines, Bruxelles 01448
Société Belge d'Etudes Géographiques, Heverlee 01161
Société Belge de Vacuologie et de Vacuotechnique, Bruxelles 01449
Société Belge d'Histoire des Hôpitaux, Bruxelles 01450
Société Belge d'Ophthalmologie, Bruxelles 01451
Société Beneluxienne de Métallurgie, Bruxelles 01452
Société Bibliographique du Canada, Toronto 01964
Société Botanique de France, Châtenay-Malabry 04635
Société Bulgare de Gastroentérologie, Sofia 01776
Société Canadienne d'Agronomie, Ottawa 02273
Société Canadienne d'Assistance Ostéopathique, London 02217
Société Canadienne d'Astronomie, Ottawa 02077
Société Canadienne de Biochimie, London 02082
Société Canadienne de Cardiologie, Westmount 02087
Société Canadienne de Droit Canonique, Ottawa 02086
Société Canadienne de Fertilité et d'Andrologie, Montréal 02132
Société Canadienne de Génie Biomedical, Ottawa 02196
Société Canadienne de Génie Civil, Ottawa 02252
Société Canadienne de Génie Mécanique, Ottawa 02262
Société Canadienne de Génie Rural, Ottawa 02272
Société Canadienne de la Santé et de la Sécurité au Travail, Mississauga 02295
Société Canadienne de la Science du Sol, Ottawa 02296
Société Canadienne de la Sclérose Laterale Amyotrophique, Toronto 01867
Société Canadienne de l'Education par l'Art, Malton 02255
La Société Canadienne de Météorologie et d'Océanographie, Newmarket 02199

Société Canadienne de Musique Folklorique, Calgary *02138*
Société Canadienne d'Endocrinologie et Metabolisme, Montréal *02257*
Société Canadienne de Neurologie, Montréal *02209*
Société Canadienne de Pédiatrie, Ottawa *02219*
Société Canadienne de Physiologie, Kingston *02225*
Société Canadienne de Physiologie Végétale, Guelph *02292*
Société Canadienne de Phytopathologie, Lethbridge *02226*
Société Canadienne de Psychoanalyse, Montréal *02233*
Société Canadienne de Psychologie, Old Chelsea *02234*
Société Canadienne de Recherche Clinique, Montréal *02253*
Société Canadienne de Recherche en Geriatrie, Toronto *02148*
Société Canadienne des Anesthésistes, Toronto *02021*
Société Canadienne des Auteurs, Illustrateurs et Artistes pour Enfants, Toronto *02277*
Société Canadienne de Science Horticole, Kentville *02258*
Société Canadienne des Etudes Bibliques, Calgary *02275*
Société Canadienne des Microbiologistes, Ottawa *02286*
Société Canadienne de Sociologie et d'Anthropologie, Montréal *02298*
Société Canadienne des Recherches pour Littérature Enfantine, London *02244*
Société Canadienne des Relations Publiques,, Ottawa *02236*
Société Canadienne des Relations Publiques (Ottawa), Ottawa *02238*
Société Canadienne des Relations Publiques (Québec), Montréal *02916*
Société Canadienne des Technologistes en Orthopédie, Agincourt *02288*
Société Canadienne de Traitement de l'Image et de Reconnaissance des Structures, Toronto *02159*
Société Canadienne d'Etudes Ethniques, Toronto *02127*
Société Canadienne d'Hématologie, Edmonton *02156*
Société Canadienne d'Histoire Orale, Ottawa *02214*
Société Canadienne d'Immunologie, Toronto *02259*
Société Canadienne du Dialogue Hommemachine, Toronto *02193*
Société Canadienne en Education Sanitaire, Toronto *02152*
Société Canadienne Géotechnique, Rexdale *02147*
Société Canadienne l'Institut Weizmann des Sciences, Downsview *02270*
Société Canadienne Osteogenesis Imperfecta, Toronto *02216*
Société Canadienne Ostéopathique, London *02218*
Société Canadienne pour Etudes d'Intelligence par Ordinateur, Toronto *02266*
Société Canadienne pour la Couleur, Ottawa *02254*
Société Canadienne pour la Formation et le Développement, Ottawa *02271*
Société Canadienne pour l'Analyse de Documents, Toronto *02515*
Société Canadienne pour la Santé Internationale, Ottawa *02260*
Société Canadienne pour les Etudes Italiennes, Ottawa *02261*
Société Canadienne pour l'Etude de l'Education, Ottawa *02268*
Société Canadienne pour l'Etude de l'Enseignement Supérieur, Ottawa *02269*
Société Canadienne Technion, Montréal *02306*
Société Cartographique de France, Paris *04636*
Société Centrale d'Apiculture, Paris *04637*
Société Centrale d'Architecture de Belgique, Bruxelles *01453*
Société d'Agriculture, Sciences, Belles-Lettres et Arts d'Orléans, Orléans *04638*
Société d'Animation du Jardin et de l'Institut Botaniques, Montréal *02917*
Société d'Anthropologie de Paris, Paris *04639*
Société d'Archéologie et de Numismatique de Montréal, Montréal *02918*
Société d'Archéologie et d'Histoire de la Manche, Saint-Lo *04640*
Société d'Archéologie et d'Histoire de l'Aunis, La Rochelle *04641*
Société de Biogéographie, Paris *04642*
Société de Biologie, Paris *04643*
Société de Biométrie Humaine, Paris *04644*
Société de Chimie Biologique, Paris *04645*
Société de Chimie Industrielle, Paris *04646*
Société de Chimie Thérapeutique, Châtenay Malabry *04647*
Société de Chirurgie de Marseille, Marseille *04648*
Société de Chirurgie de Toulouse, Toulouse *04649*
Société de Chirurgie Thoracique et Cardio-Vasculaire de Langue Française, Le Plessis Robinson *04650*
Société d'Economie et de Science Sociale, Paris *04651*
Société d'Economie Politique, Paris *04652*
Société de Criminologie du Manitoba, Winnipeg *02635*
Société de Criminologie du Québec, Montréal *02919*
Société de Démographie Historique, Paris *04653*
Société de Génealogie de Québec, Québec *02920*
Société de Génétique du Canada, Ottawa *02481*
Société de Géographie, Paris *04654*
Société de Géographie Commerciale de Bordeaux, Bordeaux *04655*
Société de Géographie Commerciale de Paris, Paris *04656*
Société de Géographie de Genève, Genève *10968*
Société de Géographie de Lille, Lille *04657*
Société de Géographie de Lyon, Lyon *04658*
Société de Géographie de Toulouse, Toulouse *04659*
Société de la Flore Valdôtaine, Aosta *08027*
Société de l'Anémie Calciforme du Canada, Toronto *02250*
Société de Langue et de Littérature Wallonnes, Liège *01454*
Société de Législation Comparée, Paris *04660*
Société de l'Histoire de France, Paris *04661*
Société de l'Histoire de l'Art Français, Paris *04662*
Société de l'Histoire de l'Ile Maurice, Port Louis *08675*
Société de l'Histoire du Protestantisme Français, Paris *04663*
Société de Linguistique de Paris, Paris *04664*
Société de Linguistique Romane, Strasbourg *04665*
Société de l'Ostéoporose du Canada, Toronto *02808*
Société de Médecine, Chirurgie et Pharmacie de Toulouse, Toulouse *04666*
Société de Médecine de Strasbourg, Strasbourg *04667*
Société de Médecine et de Chirurgie de Bordeaux, Bordeaux *04668*
Société de Médecine et d'Hygiène du Travail, Paris *04669*
Société de Médecine Légale et de Criminologie de France, Paris *04670*
Société de Microscopie du Canada, Toronto *02653*
Société d'Emulation Historique et Littéraire d'Abbéville, Abbéville *04671*
Société de Musique des Universités Canadiennes, Windsor *02312*
Société de Mythologie Française, Paris *04672*
Société d'Encouragement pour l'Industrie Nationale, Paris *04673*
Société de Neurochirurgie de Langue Française, Lyon *04674*
Société de Neurophysiologie Clinique de Langue Française, Paris *04675*
Société de Neuropsychologie de Langue Française, Paris *04676*
Societe d'Entomologie du Quebec, Sainte-Foy *02921*
Société de Nutrition et de Diététique de Langue Française, Paris *04677*
Société de Pathologie Comparée, Paris *04678*
Société de Pathologie Exotique, Paris *04679*
Société de Pharmacie de Bordeaux, Bordeaux *04680*
Société de Pharmacie de Lyon, Lyon *04681*
Société de Pharmacie de Marseille, Marseille *04682*
Société de Philosophie de Toulouse, Toulouse *04683*
Société de Physique et d'Histoire Naturelle de Genève, Genève *10969*
Société de Protection des Plantes du Québec, Saint-Hyacinthe *02922*
Société de Psychologie Médicale de Langue Française, Paris *04684*
Société de Réanimation de Langue Française, Caen *04685*
Société de Recherches et d'Etudes Historiques Corses, Ajaccio *04686*
Société de Recherches Géophysiques, Paris *04687*
Société de Recherches Pharmaceutiques et Cientifiques, Paris *04688*
Société de Recherches Psychothérapiques de Langue Française, Versailles *04689*
Société de Recherche sur le Cancer, Montréal *02320*
Société d'Ergonomie de Langue Française, Caen *04690*
Société des Africanistes, Paris *04691*
Société des Agriculteurs de France, Paris *04692*
Société des Américanistes, Paris *04693*
Société des Amis de la Revue de Géographie de Lyon, Lyon *04694*
Société des Amis de Marcel Proust et des Amis d'Illiers-Combray, Paris *04695*
Société des Amis d'Eugène Delacroix, Paris *04696*
Société des Amis du Louvre, Paris *04697*
Société des Amis du Musée de l'Homme, Paris *04698*
Société des Anciens Textes Français, Paris *04699*
Société des Auteurs, Eecherchistes, Documentalistes et Compositeurs, Montréal *02923*
Société des Auteurs et Compositeurs Dramatiques, Bruxelles *01455*
Société des Auteurs et Compositeurs Dramatiques, Paris *04700*
Société des Bibliophiles de Guyenne, Bordeaux *04701*
Société des Bollandistes, Bruxelles *01456*
Société des Ecoles du Dimanche, Paris *04702*
Société des Ecrivains Canadiens, Québec *02924*
Société des Ergothérapeutes du Nouveau-Brunswick, Fredericton *02694*
Société des Etudes Ecossaises, Guelph *02040*
Société des Etudes Latines, Paris *04703*
Société des Etudes Mélanésiennes, Nouméa *09114*
Société des Etudes Océaniennes, Papeete *05003*
Société des Etudes Renaniennes, Paris *04704*
Société des Etudes Socialistes, Winnipeg *02925*
Société des Experts-Chimistes de France, Paris *04705*
Société des Explorateurs et des Voyageurs Français, Paris *04706*
Société des Gens de Lettres de France, Paris *04707*
Société des Historiens Zairois, Lubumbashi *17146*
Société des Lépidopteristes Français, Paris *04708*
Société des Lettres, Sciences et Arts de la Haute-Auvergne, Aurillac *04709*
Société des Médecins-Chefs des Compagnies Européennes d'Aviation, Paris *04710*
Société des Naturalistes Luxembourgeois, Luxembourg *08587*
Société des Obstétriciens et Gynécologues du Canada, Toronto *02946*
Société des Océanistes, Paris *04711*
Société des Peintres, Sculpteurs et Architectes Suisses, Muttenz *10720*
Société des Physiciens des Hôpitaux d'Expression Française, Liège *01457*
Société des Poètes Français, Paris *04712*
Société des Professeurs d'Allemand en Suisse Romande et Italienne, Genève *10970*
Société des Professeurs de Dessin et d'Arts Plastiques de l'Enseignement Secondaire, Paris *04713*
Société des Professeurs Français et Francophones en Amérique, New York *16488*
Société des Relations d'Affaires Hautes Etudes Commerciales, Montréal *02926*
Société des Sciences, Arts et Belles-Lettres de Bayeux, Bayeux *04714*
Société des Sciences et Arts, Saint-Denis *04715*
Société des Sciences et Lettres de Lódź, Lódź *09545*
Société des Sciences Historiques et Naturelles de la Corse, Bastia *04716*
Société des Sciences Médicales du Grand-Duché de Luxembourg, Luxembourg *08588*
Société des Sciences Naturelles de Bourgogne, Dijon *04717*
Société des Sciences Naturelles et Physiques du Maroc, Rabat *08808*
Société des Sciences Physiques et Naturelles de Bordeaux, Talence *04718*
Société des Sculpteurs du Canada, Toronto *02909*
Société de Statistique de Paris, Paris *04719*
Société de Statistique, d'Histoire et d'Archéologie de Marseille et de Provence, Marseille *04720*
Société de Stomatologie de France, Paris *04721*
Société de Technologie Agricole et Sucrière de l'Ile Maurice, Réduit *08676*
Société d'Ethnographie de Paris, Paris *04722*
Société d'Ethnologie Française, Paris *04723*
Société d'Ethnozoologie et d'Ethnobotanique, Paris *04724*
Société de Thoracologie du Quebec, Montréal *02927*
Société de Transplantation, Paris *04725*
Société d'Etude de Psychodrame Pratique et Théorique, Paris *04726*
Société d'Etude du Dix-Septième Siècle, Paris *04727*
Société d'Etudes Ardennaises, Charleville-Mézières *04728*
Société d'Etudes Dantesques, Nice *04729*
Société d'Etudes Economiques et Comptables, Paris *04730*
Société d'Etudes Economiques, Sociales et Statistiques du Maroc, Rabat *08809*
Société d'Etudes et de Contrôles Juridiques, Paris *04731*
Société d'Etudes et de Documentation Economiques, Industrielles et Sociales, Neuilly-sur-Seine *04732*
Société d'Etudes et de Recherches Biologiques, Paris *04733*
Société d'Etudes et de Recherches en Sciences Sociales, Paris *04734*
Société d'Etudes et de Recherches pour la Connaissance de l'Homme, Paris *04735*
Société d'Etudes et de Soins pour les Enfants Paralysés, Paris *04736*
Société d'Etudes Ferroviaires, Paris *04737*
Société d'Etudes Financières et Meunières, Paris *04738*
Société d'Etudes Folkloriques du Centre-Ouest, Saint-Jean d'Angely *04739*
Société d'Etudes Hispaniques et de Diffusion de la Culture Française à l'Etranger, Périgueux *04740*
Société d'Etudes Historiques, Paris *04741*
Société d'Etudes Italiennes, Paris *04742*
Société d'Etudes Jaurésiennes, Paris *04743*
Société d'Etudes Juives, Paris *04744*
Société d'Etudes Juridiques, Economiques et Fiscales, Paris *04745*
Société d'Etudes Latines de Bruxelles, Tournai *01458*
Société d'Etudes Linguistiques et Anthropologiques de France, Paris *04746*
Société d'Etudes Médiévales, Poitiers *04747*
Société d'Etudes Minières, Industrielles et Financières, Paris *04748*
Société d'Etudes Ornithologiques, Brunoy *04749*
Société d'Etudes pour le Développement Economique et Social, Paris *04750*
Société d'Etudes Psychiques, Nancy *04751*
Société d'Etudes Robespierristes, Paris *04752*
Société d'Etudes Romantiques, Clermont-Ferrand *04753*
Société d'Etudes Scientifiques et de Recherches, Herblay *04754*
Société d'Etudes Techniques, Paris *04755*
Société d'Histoire de Bordeaux, Bordeaux *04756*
Société d'Histoire de la Médecine Hébraïque, Paris *04757*
Société d'Histoire de la Pharmacie, Paris *04758*
Société d'Histoire de la Suisse Romande, Lausanne *10971*
Société d'Histoire du Droit, Paris *04759*
Société d'Histoire du Droit Normand, Caen *04760*
Société d'Histoire du Théâtre, Paris *04761*
Société d'Histoire et d'Archéologie, Genève *10972*
Société d'Histoire et d'Archéologie de Bretagne, Rennes *04762*
Société d'Histoire et d'Archéologie de la Lorraine, Saint Julien les Metz *04763*
Société d'Histoire et d'Archéologie Le Vieux Montmartre, Paris *04764*
Société d'Histoire Générale et d'Histoire Diplomatique, Paris *04765*
Société d'Histoire Moderne, Paris *04766*
Société d'Histoire Religieuse de la France, Paris *04767*
Société d'Horticulture et d'Acclimatation du Maroc, Casablanca *08810*
Société d'Hygiène International, Paris *04768*
Société d'Obstétrique et de Gynécologie de Marseille, Marseille *04769*
Société d'Obstétrique et de Gynécologie de Toulouse, Toulouse *04770*
Société d'Océanographie de France, Paris *04771*
Société d'Ophtalmologie de l'Est de la France, Nancy *04772*
Société d'Ophtalmologie de Lyon, Lyon *04773*
Société d'Ophtalmologie de Paris, Paris *04774*
Société d'Oto-Neuro-Ophthalmologie du Sud-Est de la France, Marseille *04775*
Société du Barreau du Haut Canada, Toronto *02583*
Société du Salon d'Automne, Paris *04776*
Société Entomologique de France, Paris *04777*
Société Entomologique d'Egypte, Cairo *03914*
Société Entomologique du Canada, Ottawa *02437*
Société Entomologique Suisse, Zürich *10804*
Société et Fédération Internationale de Cardiologie, Genève *10973*
Société Européenne de Cardiologie, Rotterdam *08893*
Société Européenne de Culture, Venezia *07802*
Société Européenne d'Energie Atomique, Paris *04491*
Société Européenne de Radiobiologie, Liège *01459*
Société Européenne de Radiologie Cardio-Vasculaire et de Radiologie d'Intervention, Lyon *04778*
Société Européenne de Radiologie Pédiatrique, Paris *04779*
Société Européenne pour la Formation des Ingénieurs, Bruxelles *01460*
Société Financière Européenne, Paris *04780*
Société Finno-Ougrienne, Helsinki *04003*
Société Française d'Acoustique, Paris *04781*
Société Française d'Allergologie, Paris *04782*
Société Française d'Anesthésie et de Réanimation, Paris *04783*
Société Française d'Angéiologie, Neuilly-sur-Seine *04784*
Société Française d'Archéocivilisation et de Folklore, Paris *04785*
Société Française d'Archéologie, Paris *04786*
Société Française d'Art Contemporain, Paris *04787*
Société Française de Biologie Clinique, Paris *04788*
Société Française de Cardiologie, Paris *04789*
Société Française de Céramique, Paris *04790*
Société Française de Chimie, Paris *04791*
Société Française de Chirurgie Orthopédique et Traumatologique, Paris *04792*
Société Française de Chirurgie Pédiatrique, Lyon *04793*
Société Française de Chirurgie Plastique et Reconstructive, Paris *04794*
Société Française de Chronométrie et de Microtechnique, Besançon *04795*
Société Française d'Ecologie, Brunoy *04796*
Société Française de Composition, Paris *04797*
Société Française d'Economie Rurale, Paris *04798*
Société Française de Dermatologie et de Syphiligraphie, Paris *04799*
Société Française de Droit Aérien et Spatial, Paris *04800*
Société Française de Génétique, Orléans *04801*
Société Française de Géographie Economique, Paris *04802*
Société Française de Graphologie, Paris *04803*
Société Française de Gynécologie, Paris *04804*
Société Française d'Egyptologie, Paris *04805*
Société Française de Malacologie, Paris *04806*
Société Française de Médecine Aérospatiale, Bretigny *04807*
Société Française de Médecine du Sport, Paris *04808*
Société Française de Médecine du Trafic, Paris *04809*
Société Française de Médecine Esthétique, Courbevoie *04810*
Société Française de Médecine Générale, Paris *04811*
Société Française de Médecine Orthopédique et Thérapeutique Manuelle, Paris *04812*
Société Française de Médecine Préventive et Sociale, Paris *04813*
Société Française de Médecine Psychosomatique, Paris *04814*
Société Française de Mesothérapie, Paris *04815*
Société Française de Métallurgie et de Matériaux, Paris *04816*
Société Française de Microbiologie, Paris *04817*
Société Française de Microscopie Electronique, Ivry-sur-Seine *04818*
Société Française de Minéralogie et de Cristallographie, Paris *04819*
Société Française de Musicologie, Paris *04820*

Société Française de Mycologie Médicale, Paris 04821
Société Française d'Endocrinologie, Paris 04822
Société Française de Néonatologie, Paris 04823
Société Française d'Energie Nucléaire, Paris 04824
Société Française de Neurologie, Paris 04825
Société Française de Numismatique, Paris 04826
Société Française de Pathologie Respiratoire, Paris 04827
Société Française de Pédagogie, Paris 04828
Société Française de Pédiatrie, Paris 04829
Société Française de Philosophie, Paris 04830
Société Française de Phlébologie, Paris 04831
Société Française de Phoniatrie, Paris 04832
Société Française de Photogrammétrie et de Télédétection, Saint-Mandé 04833
Société Française de Photographie et Cinématographie, Paris 04834
Société Française de Physiologie Végétale, Paris 04835
Société Française de Physique, Paris 04836
Société Française de Phytiatrie et de Phytopharmacie, Versailles 04837
Société Française de Psychologie, Paris 04838
Société Française de Radiologie Médicale, Paris 04839
Société Française de Santé Publique, Vandoeuvre lès Nancy 04840
Société Française de Sciences et Techniques Pharmaceutiques, Paris 04841
Société Française de Sexologie Clinique, Paris 04842
Société Française de Sociologie, Paris 04843
Société Française des Physiciens d'Hôpital, Paris 04844
Société Française des Professeurs de Russe, Paris 04845
Société Française des Thermiciens, Paris 04846
Société Française des Urbanistes, Paris 04847
Société Française de Thérapeutique et de Pharmacologe Clinique, Paris 04848
Société Française de Toxicologie, Paris 04849
Société Française d'Etude du Dix-Huitième Siècle, Pau 04850
Société Française d'Etudes des Phénomènes Psychiques, Paris 04851
Société Française d'Etudes et de Réalisations Cartographiques, Paris 04852
Société Française d'Etudes Juridiques, Paris 04853
Société Française d'Hématologie, Paris 04854
Société Française d'Héraldique et de Sigillographie, Paris 04855
Société Française d'Histoire de la Médecine, Reims 04856
Société Française d'Histoire d'Outre-Mer, Paris 04857
Société Française d'Hydrologie et de Climatologie Médicales, Paris 04858
Société Française d'Hygiène, de Médecine Sociale et de Génie Sanitaire, Vandoeuvre-lès-Nancy 04859
Société Française d'Ichtyologie, Paris 04860
Société Française d'Ophthalmologie, Paris 04861
Société Française d'Optique Physiologique, Paris 04862
Société Française d'Orthopédie, Paris 04863
Société Française d'Oto-Rhino-Laryngologie et de Pathologie Cervico-Faciale, Paris 04864
Société Française d'Urologie, Paris 04865
Société Française du Vide, Paris 04866
Société Française pour le Droit International, Strasbourg 04867
Société Fribourgeoise des Sciences Naturelles, Fribourg 10974
Société Généalogique Canadienne-Française, Montréal 02928
Société Généalogique du Nouveau-Brunswick, Fredericton 02681
Société Géographique Royale du Canada, Vanier 02865
Société Géologique de Belgique, Liège 01461
Société Géologique de France, Paris 04868
Société Géologique et Minéralogique de Bretagne, Rennes 04869
Société Géologique Suisse, Fribourg 10975
La Société Guernesiaise, Saint Peter Port 12961
Société Héraldique du Canada, Ottawa 02499
Société Historique Acadienne, Moncton 02929
Société Historique Algérienne, Alger 00024
Société Historique, Archéologique et Littéraire de Lyon, Lyon 04870
Société Historique de la Province de Maine, Le Mans 04871
Société Historique de la Saskatchewan, Regina 02930
Société Historique de Québec, Québec 02931
Société Historique du Bas-Limousin, Tulle 04872
Société Historique du Saguenay, Chicoutimi 02932
Société Historique et Archéologique du Périgord, Périgueux 04873
Société Huntington du Canada, Cambridge 02511
Société Hydrotechnique de France, Paris 04874
Société Internationale de Bibliographie Classique, Paris 04875
Société Internationale de Biologie Mathématique, Antony 04876
Société Internationale de Chirurgie, Reinach 10976
Société Internationale de Chirurgie Orthopédique et de Traumatologie, Bruxelles 01462
Société Internationale de Criminologie, Paris 04877
Société Internationale de Droit Pénal Militaire et de Droit de la Guerre, Bruxelles 01463
Société Internationale de Droit Penal Militaire et de Droit de la Guerre; Groupe Italiennne, Roma 07815
Société Internationale de la Tourbe (Le Comité National Canadien), Halifax 02556
Société Internationale de Médecine Interne, Langenthal 10977
Société Internationale de Musicologie, Basel 10738
Société Internationale de Podologie Médico-Chirurgicale, Cannes-la-Bocca 04878
Société Internationale de Psychopathologie de l'Expression, Paris 04879
Société Internationale de Psycho-Prophylaxie Obstétricale, Paris 04880
Société Internationale des Amis de Montaigne, Paris 04881
Société Internationale de Transfusion Sanguine, Les Ulis 04882
Société Internationale d'Urologie, Paris 04883
Société Internationale pour la Lutte contre le Cancer du Sein, Paris 04884
Société Internationale pour la Recherche sur les Maladies de Civilisation et sur l'Environnement, Bruxelles 01464
Société Internationale pour l'Enseignement Commercial, Odense 03824
Société Internationale pour l'Etude de la Philosophie Médiévale, Louvain-la-Neuve 01465
Société Italienne des Auteurs et Editeurs, Paris 04885
Société Jean-Jacques Rousseau, Genève 10978
Société Jersiaise, Saint Helier 12962
Société J. S. Bach, Paris 04886
Société Juridique et Fiscale de France, Levallois Perret 04887
Société Linnéenne de Provence, Marseille 04888
Société Linnéenne de Québec, Sainte Foy 02933
Société Longuédocienne de Géographie, Montpellier 04889
Société Luxembourgeoise de Radiologie, Luxembourg 08589
Société Mathématique de Belgique, Bruxelles 01466
Société Mathématique de France, Paris 04890
Société Mathématique du Canada, Ottawa 02195
Société Médicale de la Suisse Romande, Sion 10979
Société Médicale des Hôpitaux de Paris, Paris 04891
Société Médicale d'Imagerie, Enseignement et Recherche, Paris 04892
Société Médicale du Nouveau-Brunswick, Fredericton 02688
Société Médico-Chirurgicale des Hôpitaux et Formations Sanitaires des Armées, Paris 04893
Société Médico-Chirurgicale des Hôpitaux Libres, Paris 04894
Société Médico-Psychologique, Boulogne 04895
Société Météorologique de France, Boulogne-Billancourt 04896
Société Mycologique de France, Paris 04897
Société Nationale Académique de Cherbourg, Cherbourg 04898
Société Nationale de Diffusion Educative et Culturelle, Montréal 02934
Société Nationale de Laiterie, Bruxelles 01467
Société Nationale de Protection de la Nature, Paris 04899
Société Nationale des Architectes de France, Paris 04900
Société Nationale des Beaux-Arts, Paris 04901
Société Nationale des Sciences Naturelles et Mathématiques, Cherbourg 04902
Société Nationale de Transfusion Sanguine, Paris 04903
Société Nationale d'Horticulture de France, Paris 04904
Société Nationale Française de Gastro-Entérologie, Paris 04905
Société Nationale Française de Rééducation et Réadaption Fonctionnelles, Paris 04906
Société Neuchâteloise des Sciences Naturelles, Neuchâtel 10980
Société Odontologique de Paris, Paris 04907
Société Ornithologique de France, Paris 04908
Société Parisienne d'Etudes et de Recherches Foncières, Paris 04909
Société Parisienne d'Etudes Spéciales, Paris 04910
Société Pédagogique de la Suisse Romande, Carouge 10981
Société Philosophique de Louvain, Louvain-la-Neuve 01468
Société Phycologique de France, Paris 04911
Société pour la Conservation des Sites Naturels, Toronto 02676
Société pour la Protection des Paysages, Sites et Monuments, Paris 04912
Société pour le Développement Minier de la Côte d'Ivoire, Abidjan 08066
Société pour l'Etude de l'Architecture au Canada, Ottawa 02940
Société pour vaincre la Pollution, Montréal 02935
Société Professionnelle de Recherche en Marketing, Toronto 02847
Société Provençale de Pédiatrie, Marseille 04913
Société Psychoanalytique de Paris, Paris 04914
Société Québécoise d'Assainissement des Eaux, Montréal 02936
Société Racinienne, Neuilly-sur-Seine 04915
Société Romande d'Audiophonologie et de Pathologie du Langage, Delémont 10982
Société Royale Belge d'Anthropologie et de Préhistoire, Bruxelles 01469
Société Royale Belge d'Astronomie, de Météorologie et de Physique du Globe, Bruxelles 01470
Société Royale Belge de Chirurgie, Bruxelles 01471
Société Royale Belge de Dermatologie et de Vénérologie, Bruxelles 01472
Société Royale Belge de Géographie, Bruxelles 01473
Société Royale Belge de Gynécologie et d'Obstétrique, Bruxelles 01474
Société Royale Belge de Médecine Physique et de Réhabilitation, Bruxelles 01475
Société Royale Belge d'Entomologie, Bruxelles 01476
Société Royale Belge de Rheumatologie, Bruxelles 01477
Société Royale Belge des Electriciens, Bruxelles 01478
Société Royale Belge de Stomatologie et Chirurgie Maxillo-Faciale, Gent 01479
Société Royale d'Archéologie de Bruxelles, Bruxelles 01480
Société Royale d'Astronomie du Canada, Toronto 02863
Société Royale de Chimie, Bruxelles 01481
Société Royale d'Economie Politique de Belgique, Charleroi 01482
Société Royale de Médecine Mentale de Belgique, Bruxelles 01483
Société Royale des Amis du Musée Royal de l'Armée et d'Histoire Militaire, Bruxelles 01484
Société Royale des Beaux-Arts, Bruxelles 01485
Société Royale des Bibliophiles et Iconophiles de Belgique, Bruxelles 01486
Société Royale des Sciences de Liège, Liège 01487
Société Royale des Sciences Médicales et Naturelles de Bruxelles, Bruxelles 01488
Société Royale du Canada, Ottawa 02871
Société Royale Zoologique de Belgique, Bruxelles 01489
Société Saint-Simon, Sceaux 04916
Société Savoisienne d'Histoire et d'Archéologie, Chambéry 04917
Société Scientifique de Bretagne, Rennes 04918
Société Scientifique de Bruxelles, Namur 01490
Société Scientifique d'Hygiène Alimentaire, Paris 04919
Société Statistique du Canada, Hamilton 02960
Société Suisse de Cardiologie, Bern 10825
Société Suisse de Chirurgie, Genève 10983
Société Suisse de Droit Pénal, Giubiasco 10867
Société Suisse de Juristes, Lausanne 10984
Société Suisse de Linguistique, Basel 10930
Société Suisse de Médecine Sociale et Préventive, Bern 10847
Société Suisse de Numismatique, Bern 10877
Société Suisse de Pédagogie Musicale, Zürich 10900
Société Suisse de Pharmacie, Bern-Liebefeld 10885
Société Suisse de Philosophie, Sankt Gallen 10880
Société Suisse de Physique, Genève 10881
Société Suisse des Américanistes, Genève 10985
Société Suisse de Sociologie, Zürich 10848
Société Suisse des Professeurs de Français, Chancy 10986
Société Suisse des Professeurs de Musique de l'Enseignement Secondaire, Emmenbrücke 10987
Société Suisse des Professeurs de Sciences Commerciales, Einsiedeln 10897
Société Suisse de Statistique et d'Economie Politique, Bern 10849
Société Suisse d'Ethnologie, Bern 10806
Société Suisse d'Etudes Anglaises, Genève 10791
Société Suisse d'Héraldique, Genève 10988
Société Suisse d'Histoire de la Médicine et des Sciences Naturelles, Bern 10822
Société Suisse d'Orthopédie, Bern 10839
Société Suisse d'Oto-Rhino-Laryngologie et de Chirurgie Cervico-Faciale, Bern 10840
Société Suisse pour la Recherche en Education, Carouge 10989
Société Suisse pour l'Etude du Proche Orient Ancien, Genève 10838
Société Technique d'Etudes Mécaniques et d'Outillage, Paris 04920
Société Technique et Chimique de Sucrerie de Belgique, Bruxelles 01491
Société Théosophique de France, Paris 04921
Société Universitaire Européenne de Recherches Financières, Tilburg 09069
Société Vaudoise des Sciences Naturelles, Lausanne 10990
Société Vaudoise d'Histoire et d'Archéologie, Chavannes-près-Renens 10991
Société Vétérinaire Pratique de France, Paris 04922
Société Zoologique de France, Paris 04923
Société Zoologique de Québec, Charlesbourg 02937
Society for Academic Emergency Medicine, Lansing 16489
Society for Adolescent Medicine, Independence 16490
Society for Advancement of Management, Vinton 16491
Society for African Church History, Aberdeen 12963
Society for American Archaeology, Washington 16492
Society for Ancient Greek Philosophy, Binghamton 16493
Society for Applied Anthropology, Oklahoma City 16494
Society for Applied Bacteriology, Newton Abbot 12964
Society for Applied Learning Technology, Warrenton 16495
Society for Applied Spectroscopy, Frederick 16496
Society for Armenian Studies, Dearborn 16497
Society for Army Historical Research, London 12965
Society for Asian and Comparative Philosophy, Oneonta 16498
Society for Asian Art, San Francisco 16499
Society for Asian Music, Ithaca 16500
Society for Asian Political and Economic Studies, Tokyo 08093
Society for Austrian and Habsburg History, Haverford 16501
Society for Back Pain Research, London 12966
Society for Biomaterials, Minneapolis 16502
Society for Business Ethics, Chicago 16503
Society for Byzantine and Post-byzantine Studies, Athinai 06523
Society for Byzantine Studies, Athinai 06524
Society for Cable Television Engineering, Mississauga 02938
Society for Calligraphy, Los Angeles 16504
Society for Cardiac Angiography and Interventions, Breckenridge 16505
Society for Ch'ing Studies, Pasadena 16506
Society for Cinema Studies, Chapel Hill 16507
Society for Clinical and Experimental Hypnosis, Liverpool 16508
Society for Clinical Trials, Baltimore 16509
Society for College and University Planning, Ann Arbor 16510
Society for Commercial Archeology, Washington 16511
Society for Computers and Law, Bristol 12967
Society for Computer Simulation International, San Diego 16512
Society for Cooperation in Russian and Soviet Studies, London 12968
Society for Coptic Archaeology, Cairo 03915
Society for Creative Anachronism, Oakland 16513
Society for Cryobiology, Bethesda 16514
Society for Cultural Anthropology, Washington 16515
Society for Developmental Biology, Washington 16516
Society for Dissemination of Sciences, Budapest 06676
Society for Earthquake and Civil Engineering Dynamics, London 12969
Society for Economic Botany, Columbia 16517
Society for Educational Reconstruction, New York 16518
Society for Education in Film and Television, London 12970
Society for Endocrinology, Bristol 12971
Society for Endocrinology, Metabolism and Diabetes of Southern Africa, Sandton 10173
Society for Environmental Therapy, Manchester 12972
Society for Epidemiologic Research, Fort Collins 16519
Society for Ethnomusicology, Bloomington 16520
Society for Experimental and Descriptive Malacology, Ann Arbor 16521
Society for Experimental Biology, Johannesburg 10174
The Society for Experimental Biology, London 12973
Society for Experimental Biology and Medicine, New York 16522
Society for Experimental Mechanics, Bethel 16523
Society for Folk Life Studies, Hollywood 12974
Society for French American Cultural Services and Educational Aid, New York 16524
Society for French Historical Studies, Iowa City 16525
Society for General Microbiology, Reading 12975
Society for General Music, Reston 16526
Society for Geodesy and Cartography, Budapest 06605
Society for German-American Studies, Cincinnati 16527
Society for Gynecologic Investigation, Washington 16528
Society for Historians of American Foreign Relations, Dayton 16529
Society for Historical Archaeology, Tucson 16530
Society for History Education, Long Beach 16531
Society for Hospital Social Work Directors, Chicago 16532
Society for Humanistic Anthropology, West Chester 16533
Society for Humanistic Judaism, Farmington Hills 16534
Society for Human Resource Management, Alexandria 16535
Society for Human Settlements, Budapest 06662
Society for Indian and Northern Education, Saskatoon 02939
Society for Industrial and Applied Mathematics, Philadelphia 16536
Society for Industrial Archaeology, Washington 16537
Society for Industrial Microbiology, Arlington 16538
Society for Information Display, Playa Del Rey 16539
Society for Information Management, Chicago 16540
Society for International Development, Roma 08028
Society for International Numismatics, Santa Monica 16541
Society for Invertebrate Pathology, Bethesda 16542

Society for Investigative Dermatology, Cleveland 16543
Society for Iranian Studies, New York 16544
Society for Italian Historical Studies, Chestnut Hill 16545
Society for Italic Handwriting, London 12976
Society for Latin American Anthropology, Washington 16546
Society for Latin American Studies, York 12977
Society for Leukocyte Biology, Augusta 16547
Society for Libyan Studies, London 12978
Society for Life History Research, Saint Louis 16548
Society for Lincolnshire History and Archaeology, Lincoln 12979
Society for Linguistic Anthropology, Washington 16549
Society for Low Temperature Biology, London 12980
Society for Macedonian Studies, Thessaloniki 06525
Society for Magnetic Resonance Imaging, Chicago 16550
Society for Manyo Studies, Osaka 08180
Society for Medical Anthropology, Washington 16551
Society for Medicinal Plant Research, Kleinrinderfeld 05802
Society for Medicine Research, London 12981
Society for Medieval Archaeology, Lincoln 12982
Society for Metal Science, Brno 03555
Society for Mining, Metallurgy and Exploration, Littleton 16552
Society for Multivariate Experimental Psychology, Upton Park 12983
Society for Music Teacher Education, Reston 16553
Society for National Art, København 03736
Society for Natural Philosophy, Lexington 16554
Society for Natural Sciences of Slovenia, Ljubljana 10108
Society for Nautical Research, London 12984
Society for Near Eastern Studies in Japan, Tokyo 08354
Society for Neuroscience, Washington 16555
Society for New Language Study, Denver 16556
Society for New Testament Study, Lancaster 12985
Society for Nutrition Education, Minneapolis 16557
Society for Obstetric Anesthesia and Perinatology, Houston 16558
Society for Occlusal Studies, Kansas City 16559
Society for Occupational and Environmental Health, McLean 16560
Society for Organization and Management Science, Budapest 06673
Society for Pediatric Dermatology, Ann Arbor 16561
Society for Pediatric Psychology, Gainesville 16562
Society for Pediatric Radiology, Oak Brook 16563
Society for Pediatric Research, Elk Grove Village 16564
Society for Pediatric Urology, Seattle 16565
Society for Personality Assessment, Saint Petersburg 16566
Society for Phenomenology and Existential Philosophy, Carbondale 16567
Society for Philosophy and Public Affairs, Lexington 16568
Society for Philosophy of Religion, Athens 16569
Society for Philosophy, Sociology and Politics of Macedonia, Skopje 08599
Society for Photographic Education, Dallas 16570
The Society for Popular Culture, Helsinki 03975
Society for Post-Medieval Archaeology, London 12986
Society for Promoting Christian Knowledge, London 12987
Society for Promotion of Educational Reform through Teacher Training, London 12988
Society for Psychical Research, London 12989
Society for Psychological Anthropology, Washington 16571
Society for Psychophysiological Research, Dallas 16572
Society for Psychosomatic Research, London 12990
Society for Public Health Education, Berkeley 16573
Society for Radiological Protection, Didcot 12991
Society for Railway Transport, Moskva 09970
Society for Range Management, Denver 16574
Society for Reformation Research, Saint Louis 16575
The Society for Renaissance Studies, London 12992
Society for Research and Promotion of Maritime Sciences, Rijeka 03457
The Society for Research in Asiatic Music, Tokyo 08441
Society for Research in Child Development, Chicago 16576
Society for Research in the Psychology of Music and Music Education, Dagenham 12993
Society for Research into Higher Education, London 12994
Society for Research into Hydrocephalus and Spina Bifida, Manchester 12995
Society for Research on the Azores, Ponta Delgada 09680
Society for Risk Analysis, McLean 16577
Society for Sedimentary Geology, Tulsa 16578
Society for Slavonic Studies in Slovenia, Ljubljana 10104
Society for Slovene Studies, Bloomington 16579
Society for Social Medicine, London 12996
Society for Social Responsibility in Science (A.C.T.), Canberra 00491
Society for Social Studies of Science, Baton Rouge 16580
The Society for South Asian Studies, London 12997
Society for South India Studies, Berkeley 16581
Society for Spanish and Portuguese Historical Studies, Nashville 16582
Society for Surgery of the Alimentary Tract, New York 16583
Society for Technical Communication, Arlington 16584
Society for the Advancement of Ambulatory Care, Washington 16585
Society for the Advancement of American Philosophy, Seattle 16586
Society for the Advancement of Anaesthesia in Dentistry, London 12998
Society for the Advancement of Behavior Analysis, Kalamazoo 16587
Society for the Advancement of Education, Valley Stream 16588
Society for the Advancement of Games and Simulations in Education and Training, Loughborough 12999
Society for the Advancement of Material and Process Engineering, Covina 16589
Society for the Advancement of Research, Laguna 09536
Society for the Advancement of Scandinavian Study, Provo 16590
Society for the Advancement of Science, Bergen 09347
Society for the Advancement of the Vegetable Industry, Laguna 09537
Society for the Anthropology of Europe, Washington 16591
Society for the Arts, Religion and Contemporary Culture, Boston 16592
The Society for Theatre Research, London 13000
Society for the Comparative Study of Society and History, Ann Arbor 16593
Society for the Conservation of Bighorn Sheep, West Covina 16594
Society for the Development of Techniques in Industrial Marketing, Nuneaton 13001
Society for the Furtherance and Study of Fantasy and Science Fiction, Madison 16595
Society for the Health Education, London 13002
Society for the History of Alchemy and Chemistry, London 13003
Society for the History of Czechoslovak Jews, Holliswood 16596
Society for the History of Discoveries, Falls Church 16597
Society for the History of Natural History, London 13004
Society for the History of Technology, Houghton 16598
Society for the History of the Germans in Maryland, Baltimore 16599
Society for the Humanities, Ithaca 16600
Society for the Investigation of Recurring Events, Iselin 16601
Society for the Philosophical Study of Marxism, Bridgeport 16602
Society for the Preservation of American Business History, Williamsburg 16603
Society for the Preservation of Ancient Monuments in Norway, Oslo 09266
Society for the Preservation of English Language and Literature, Waleska 16604
Society for the Preservation of New England Antiquities, Boston 16605
Society for the Promotion and Improvement of Libraries, Karachi 09403
Society for the Promotion of Greek Education, Athinai 06546
Society for the Promotion of Hellenic Studies, London 13005
Society for the Promotion of New Music, London 13006
Society for the Promotion of Principles, Birmingham 13007
Society for the Promotion of Roman Studies, London 13008
Society for the Promotion of Science and Scholarship, Palo Alto 16606
Society for the Promotion of Vocational Training and Education, Bristol 13009
Society for the Protection of Ancient Buildings, London 13010
Society for the Protection of Nature in Israel, Tel Aviv 07111
Society for the Protection of Old Fishes, Seattle 16607
Society for the Protection of Science and Learning, London 13011
Society for the Psychological Study of Social Issues, Ann Arbor 16608
Society for Theriogenology, Hastings 16609
Society for the Scientific Study of Religion, West Lafayette 16610
Society for the Scientific Study of Sex, Mount Vernon 16611
Society for the Social History of Medicine, Manchester 13012
Society for the Study and Conservation of Nature, Valletta 08671
Society for the Study of Addiction, Horsham 13013
Society for the Study of Amphibians and Reptiles, Oxford 16612
Society for the Study of Architecture in Canada, Ottawa 02940
Society for the Study of Blood, Brooklyn 16613
Society for the Study of Breast Disease, Dallas 16614
Society for the Study of Early China, Berkeley 16615
Society for the Study of Evolution, Lawrence 16616
Society for the Study of Fertility, Cambridge 13014
Society for the Study of Human Biology, London 13015
Society for the Study of Inborn Errors of Metabolism, Manchester 13016
Society for the Study of Information Transfer, Beckenham 13017
Society for the Study of Japanese Language, Tokyo 08171
Society for the Study of Knowledge and Development, Buffalo 15596
Society for the Study of Labour History, London 13018
Society for the Study of Male Psychology and Physiology, Montpellier 16617
Society for the Study of Medieval Languages and Literature, Oxford 13019
Society for the Study of Normal Psychology, London 13020
Society for the Study of Process Philosophies, Portland 16618
Society for the Study of Reproduction, Champaign 16619
Society for the Study of Social Biology, Honolulu 16620
Society for the Study of Social Problems, Knoxville 16621
Society for the Study of Southern Literature, New Orleans 16622
Society for the Study of Symbolic Interaction, Saint Petersburg 16623
Society for Trade and Commerce, Moskva 09971
Society for Underwater Technology, London 13021
Society for Urban Anthropology, Washington 16624
Society for Values on Higher Education, Washington 16625
Society for Vascular Surgery, Manchester 16626
Society for Vector Ecology, Santa Ana 16627
Society for Visual Anthropology, Washington 16628
Society for Women in Philosophy – Eastern Division, Scranton 16629
Society for Women in Philosophy – Pacific Division, San Jose 16630
Society of Actuaries, Schaumburg 16631
Society of Agricultural Machinery, Tokyo 08400
Society of Agricultural Meteorology of Japan, Tokyo 08349
Society of Allied Weight Engineers, Chula Vista 16632
Society of American Archivists, Chicago 16633
Society of American Foresters, Washington 16634
Society of American Historians, New York 16635
Society of American Law Teachers, New York 16636
Society of American Registered Architects, Lombard 16637
Society of American Value Engineers, Northbrook 16638
The Society of Anaesthesiologist of the Republic of China, Taipei 03326
Society of Andrology, Pisa 07834
Society of Antiquaries of London, London 13022
Society of Antiquaries of Newcastle-upon-Tyne, Newcastle-upon-Tyne 13023
Society of Antiquaries of Scotland, Edinburgh 13024
Society of Archer Antiquaries, Bridlington 13025
Society of Architectural and Industrial Illustrators, Stroud 13026
Society of Architectural Historians, Philadelphia 16639
Society of Architectural Historians of Great Britain, London 13027
Society of Archivists, London 13028
Society of Art Historians of Macedonia, Skopje 08593
Society of Arts, Literature and Welfare, Chittagong 01070
Society of Assistants Teaching in Preparatory Schools, Altrincham 13029
Society of Australian Genealogists, Sydney 00492
Society of Authors, London 13030
Society of Authors, Composers and Music Publishers in Israel, Tel Aviv 07112
Society of Authors ZAIKS, Warszawa 09614
Society of Autmotive Engineering of China, Beijing 03221
Society of Automotive Engineers – Australasia, Parkville 00493
Society of Basque Studies in America, Brooklyn 16640
Society of Behavioral Medicine, Rockville 16641
Society of Biblical Literature, Decatur 16642
Society of Biological Chemists, India, Bangalore 06858
Society of Biological Psychiatry, Dallas 16643
Society of Botanists of Moldova, Chişinău 08790
Society of British Neurological Surgeons, Glasgow 13031
Society of Broadcast Engineers, Indianapolis 16644
Society of Bulgarian Anatomists, Histologists and Embryologists, Sofia 01777
Society of Bulgarian Chemists, Sofia 01778
Society of Bulgarian Physicists, Sofia 01779
Society of Bulgarian Psychologists, Sofia 01780
Society of Business Economists, Watford 13032
Society of Cable Television Engineers, Exton 16645
Society of Carbide and Tool Engineers, Metals Park 16646
Society of Cardiology, Moskva 09972
Society of Cardiovascular and Interventional Radiology, Fairfax 16647
Society of Cardiovascular Anesthesiologists, Richmond 16648
Society of Cartographers, Glasgow 13033
Society of Chemical Industry, London 13034, 13035
Society of Chemical Industry (Canadian Section), Mississauga 02941
Society of Chief Architects of Local Authorities, Wirral 13036
Society of Chinese Acupuncture and Cauterizing, Taipei 03327
Society of Chinese Constitutional Law, Taipei 03328
Society of Christian Ethics, Boston 16649
Society of Christian Schools in British Columbia, Surrey 02942
Society of Cirplanologists, Hyde 13037
Society of Commercial Seed Technologists, Brandon 02943
Society of Composers, New York 16650
Society of Composers, Authors and Music Publishers of Canada, Don Mills 02944
Society of Consulting Marine Engineers and Ship Surveyors, London 13038
Society of Contemporary Music and Intermedia Art, Stockholm 10463
Society of Cosmetic Chemists, New York 16651
Society of Cosmetic Scientists, Luton 13039
Society of Cost Estimating and Analysis, Alexandria 16652
Society of County Librarians, Worcester 13040
Society of County Museum Directors, West Malling 13041
Society of Critical Care Medicine, Anaheim 16653
Society of Cypriot Studies, Nicosia 03499
The Society of Czech Architects, Praha 03552
Society of Dairy Technology, Huntingdon 13042
Society of Designer Craftsmen, London 13043
Society of Designers in Ireland, Dublin 07032
Society of Diagnostics Medical Sonographers, Dallas 16654
The Society of Dyers and Colourists, Bradford 13044
Society of Economic Geologists, Littleton 16655
Society of Economics, Tokyo 08438
Society of Educational Philosophy, Tokyo 08176
Society of Engineering Illustrators, Madison Heights 16656
Society of Engineering Science, Charlottesville 16657
The Society of Engineers, Colchester 13045
Society of Engineers and Technicians of Macedonia, Skopje 08614
Society of Esaff Alkhairia, Mecca 10006
Society of Ethnic and Special Studies, Edwardsville 16658
Society of Ethnobiology, Reno 16659
Society of Experimental Psychologists, University Park 16660
Society of Experimental Test Pilots, Lancaster 16661
Society of Exploration Geophysicists, Tulsa 16662
Society of Federal Linguists, Washington 16663
Society of Feed Technologists, Reading 13046
The Society of Fermentation and Bioengineering, Japan, Osaka 08425
Society of Fibre Science and Technology, Tokyo 08413
Society of Finnish Composers, Helsinki 04058
Society of Fire Protection Engineers, Boston 16664
Society of Flavor Chemists, Hunt Valley 16665
Society of Flight Test Engineers, Lancaster 16666
Society of Foreign Language Teachers, Sofia 01781
The Society of Forestry in Finland, Helsinki 04050
Society of Freight Car Historians, Monrovia 16667
Society of Friends of Polish Language, Kraków 09624
Society of Gastroenterology Nurses and Associates, Rochester 16668
Society of Genealogists, London 13047
Society of General Physiologists, Woods Hole 16669
Society of Genetics of Moldova, Chişinău 08791
Society of Glass Science and Practices, Clarksburg 16670
Society of Glass Technology, Sheffield 13048
Society of Greek Men of Letters, Athinai 06535
Society of Greek Philologists, Athinai 06534
Society of Head and Neck Surgeons, Arlington 16671
Society of Health and Beauty Therapists, London 13049
Society of Hearing Aid Audiologists, South Croydon 13050
Society of Heating, Airconditioning and Sanitary Engineers of Japan, Tokyo 08174
Society of Herminthologists, Moskva 09973
Society of Homoeopaths, London 13051
Society of Hungarian Literary History, Budapest 06638
The Society of Icelandic Actuaries, Reykjavik 06685
Society of Imaging Science and Technology, Springfield 16672
Society of Independent Professional Earth Scientists, Dallas 16673
Society of Indexers, London 13052
Society of Industrial Tutors, Middlesbrough 13053
Society of Insurance Research, Marietta 16674
Society of Iranian Clinicians, Teheran 06904
Society of Iraqi Artists, Baghdad 06919
Society of Irish Foresters, Dublin 07033
Society of Japanese Historical Research, Tokyo 08426
Society of Japanese Virologists, Tokyo 08272
Society of Jewish Science, Plainview 16675
Society of Jurists of Slovenia, Ljubljana 10109
Society of King Charles the Martyr, London 13054
Society of Leather Technologists and Chemists, Botany 00494
Society of Liberian Authors, Monrovia 08565
Society of Licensed Aircraft Engineers and Technologists, Kingston-upon-Thames 13055

Society of Light Industry, Moskva 09974
Society of Literary Translators of Macedonia, Skopje 08596
Society of Logistics Engineers, New Carrollton 16676
Society of Macedonian Composers, Skopje 08594
Society of Malawi, Blantyre 08624
Society of Mammalogists, Moskva 09975
Society of Management Accountants of Canada, Hamilton 02945
Society of Manufacturing Engineers, Dearborn 16677
Society of Mathematicians and Computerists of Macedonia, Skopje 08610
Society of Mathematicians of Serbia, Beograd 17087
Society of Medical Consultants to the Armed Forces, Bethesda 16678
Society of Medical Jurisprudence, New York 16679
The Society of Metaphysicians, Hastings 13056
Society of Miniaturists, Ilkley 13057
Society of Motion Picture and Television Art Directors, Studio City 16680
Society of Motion Picture and Television Engineers, White Plains 16681
Society of Multivariate Experimental Psychology, Charlottesville 16682
Society of Natural History in Limburg, Maastricht 08979
Society of Naval Architects and Marine Engineers, Jersey City 16683
Society of Naval Architects of Japan, Tokyo 08279
Society of Nematologists, Beetsville 16684
Society of Neurological Surgeons, Boston 16685
Society of Neurosurgical Anesthesia and Critical Care, Richmond 16686
Society of Non-Ferrous Metallurgy, Moskva 09976
Society of Nuclear Medicine, New York 16687
Society of Obstetricians and Gynaecologists of Canada, Toronto 02946
Society of Occupational Medicine, London 13058
Society of Oriental Research, Kyoto 08442
Society of Ornithologists, Moskva 09977
Society of Orthopedic Surgeons of the Israel Medical Association, Tel Aviv 07113
Society of Pelvic Surgeons, New York 16688
Society of Petroleum Engineers, Richardson 16689
Society of Pharmaceutical Sciences, Bucuresti 09781
Society of Philatelists and Numismatists, Montebello 16690
Society of Philippine Surgeons of America, Roanoke 16691
Society of Philosophy and Social Science, Kingston 02947
Society of Photographic Science and Technology of Japan, Tokyo 08371
The Society of Physical Medicine and Rehabilitation, Istanbul 11145
The Society of Physicians and Natural Scientists, Iasi 09772
Society of Physiologists, Biochemists and Pharmacologists, Pretoria 10175
Society of Plant Physiologists of Moldova, Chişinău 08792
Society of Plastic Arts of Macedonia, Skopje 08595
Society of Plastics Engineers, Brookfield 16692
Society of Polymer Science, Tokyo 08166
Society of Professional Archaeologists, Topeka 16693
Society of Professional Engineers, London 13059
Society of Professional Well Log Analysts, Houston 16694
Society of Professors of Child and Adolescent Psychiatry, Washington 16695
Society of Professors of Education, Knoxville 16696
Society of Prospective Medicine, Indianapolis 16697
Society of Protozoologists, New York 16698
Society of Protozoologists, St. Petersburg 09978
Society of Protozoologists, British Section, London 13060
Society of Psychologists, Moskva 09979
The Society of Public Health, Huddersfield 13061
Society of Public Teachers of Law, Buckingham 13062
Society of Remedial Gymnasts and Recreational Therapies, Cardiff 13063
Society of Research Administrators, Chicago 16699
Society of Rheology, New York 16700
Society of Scribes and Illuminators, London 13064
Society of Sea Water Science, Japan, Tokyo 08322
Society of Serbian Language and Literature, Beograd 17090
Society of Slovene Composers, Ljubljana 10099
Society of Soft Drink Technologists, Hartfield 16701
Society of Sport Medicine, Sofia 01782
Society of State Directors of Health, Physical Education and Recreation, Kensington 16702
Society of Surgical Oncology, Palatine 16703
Society of Swedish Composers, Stockholm 10460
The Society of Swedish Literature in Finland, Helsinki 04067
Society of Systematic Biologists, Washington 16704
Society of Teachers in Business Education, Sheffield 13065
Society of Teachers in Education of Professional Photography, West Carrollton 16705
Society of Teachers of Family Medicine, Kansas City 16706
Society of Teachers of Speech and Drama, Canterbury 13066
Society of Technical Analysts, Cambridge 13067
Society of Terrestrial Magnetism and Electricity of Japan, Tokyo 08288
Society of the Chinese Borders History and Languages, Taipei 03329
Society of the Food Industry, Moskva 09980
Society of the Friends of History, Tampere 03965
Society of the Hungarian Radiologists, Budapest 06654
Society of the Instrument Building Industry, Moskva 09981
Society of the Irish Motor Industry, Dublin 07034
Society of the Timber and Forestry Industry, Moskva 09982
Society of Thoracic and Cardiovascular Surgeons of Great Britain and Ireland, Conventry 13068
Society of Thoracic Surgeons, Chicago 16707
Society of Town Planning Technicians, London 13069
Society of Toxicology, Washington 16708
Society of Turkish Architects, Engineers and Scientists in America, New York 16709
Society of Ukrainian Engineers and Associates in Canada, Toronto 02948
Society of United States Air Force Flight Surgeons, Brooks Air Force Base 16710
Society of University Surgeons, New Haven 16711
Society of Vertebrate Paleontology, Lincoln 16712
Society of Wildlife Artists, London 13070
Society of Women Artists, London 13071
Society of Women Engineers, New York 16713
Society of Wood Science and Technology, Madison 16714
Society of Writers of Macedonia, Skopje 08598
Society of Writers to Her Majesty's Signet, Edinburgh 13072
Society of Young Scientists, New Dehli 06859
Sociological Association, Moskva 09983
Sociological Association of Australia and New Zealand, Bathurst 00495
SODEMI, Abidjan 08066
Sögufélag, Reykjavík 06700
SOEH, McLean 16560
SOF, Paris 04908
SOFCOT, Paris 04792
S.O.G., Horgen 10878
SOG, München 06309
S.O.G.T., Toulouse 04770
S.O.I., Firenze 07997
S.O.I., Ravenna 07995
Soiety for Northwestern Vertebrate Biology, Olympia 16715
Soil and Water Conservation Society, Ankeny 16716
Soil Association, Bristol 13073
Soil Science Society, Moskva 09984
Soil Science Society of America, Madison 16717
Soil Science Society of Southern Africa, Pretoria 10176
Sojuz na Društvata na Arhivskite Rabotnici na Makedonija, Skopje 08607
Sojuz na Društvata na Bibliotekarite na Makedonija, Skopje 08608
Sojuz na Društvata na Istoričarite na Makedonija, Skopje 08609
Sojuz na Društvata na Matematičarite i Informatičarite na Makedonija, Skopje 08610
Sojuz na Društvata na Veterinarnite Lekari i Tehničari na Makedonija, Skopje 08611
Sojuz na Društvata za Makedonski Jazik i Literatura, Skopje 08612
Sojuz na Ekonomistite na Makedonija, Skopje 08613
Sojuz na Inženeri i Tehničari na Makedonija, Skopje 08614
Sojuz na Inženeri i Tehničari po Sumarstvo i Industrija za Prerabotka na Drvo na Makedonija, Skopje 08615
Sojuz na Združenijata na Pravnicite na Makedonija, Skopje 08616
Sojuz na Zemjodelskite Inženeri i Tehničari na SR Makedonija, Skopje 08617
Solartherm, Silver Spring 16718
SOLE, New Carrollton 16676
Solid Waste Association of North America, Silver Spring 16719
Solomon Schecher Day School Association, New York 16720
SOM, London 13058
SOMEA, Roma 08001
SOMEFI, Chapingo 08762
Somerset Archaeological and Natural History Society, Taunton 13074
Somerset Record Society, Taunton 13075
SON, Beetsville 16684
Sønderjyllands Amatørarkaeologer, Abenrå 03825
Sonnblick-Verein, Wien 00946
Sonneck Society, Canton 16721
SOP, Paris 04774
S.O.P., Paris 04907
SOPA, Topeka 16693
SOPEREF, Paris 04909
SOPHE, Berkeley 16573
SOR, Amsterdam 09068
SOR, New York 16700
S.O.R., Reggio Emilia 07996
SORA, Stockholm 10571
Sorbisches Institut e.V., Bautzen 06282
SORIN, Vercelli 08008
SOROT, Bucuresti 09794
SORSI, Prairie Village 16432
SOS, Kansas City 16559
Sosialøkonomenes Forening, Oslo 09349
SoSkAN, London 13080
SOT, Washington 16708
S.O.T.I.M.I., Napoli 07787
S.O.T.M.A.S., Stoke-on-Trent 11790
Sound Learning Society, Rickmansworth 13076
South African Academy of Science and Arts, Pretoria 10177
South African Archaeological Society, Vlaeberg 10178
The South African Association for Food Science and Technology, Edenvale 10179
South African Association for Technical and Vocational Education, Arcadia 10180
South African Association of Arts, Pretoria 10181
The South African Association of Botanists, Alice 10182
The South African Association of Physicists in Medicine and Biology, Pretoria 10183
South African Biological Society, Pretoria 10184
South African Ceramic Society, Northmead 10185
The South African Chemical Institute, Yeoville 10186
South African Council for Automation and Computation, Pretoria 10187
South African Crystallographic Society, Johannesburg 10188
South African Dietetics and Home Economics Association, Stellenbosch 10189
South African Filtration Society, Hillcrest 10190
South African Genetic Society, Stellenbosch 10191
South African Geographical Society, Wits 10192
South African Institute for Librarianship and Information Science, Pretoria 10193
The South African Institute for Medical Research, Johannesburg 10194
South African Institute of Aeronautical Engineers, Sunnyside 10195
South African Institute of Assayers and Analysts, Mashalltown 10196
The South African Institute of International Affairs, Braamfontein 10197
The South African Institute of Mining and Metallurgy, Marshalltown 10198
The South African Institute of Organization and Methods, Pretoria 10199
The South African Institute of Physics, Faure 10200
South African Institute of Printing, Cape Town 10201
South African Institute of Race Relations, Braamfontein 10202
The South African Institution of Civil Engineers, Yeoville 10203
South African Market Research Association, Johannesburg 10204
The South African Mathematical Society, Pretoria 10205
The South African Medical Research Council, Tygerberg 10206
The South African National Committee on Illumination, Pretoria 10207
South African National Group of the International Society for Rock Mechanics, Marshalltown 10208
South African National Multiple Sclerosis Society, Walmer 10209
South African National Tuberculosis Association, Johannesburg 10210
The South African Numismatic Society, Cape Town 10211
South African Optometric Association, Pretoria 10212
South African Orthopaedic Association, Durban 10213
South African PEN Centre, Claremont 10214
The South African Rheumatism and Arthritis Association, Johannesburg 10215
South African Society for Photogrammetry, Remote Sensing and Cartography, Newlands 10216
The South African Society of Anaesthesists, Rosebank 10217
South African Society of Otorhinolaryngology, Bellville 10218
The South African Society of Physiotherapy, Johannesburg 10219
South African Speech and Hearing Association, Braamfontein 10220
The South African Statistical Association, Sunnyside 10221
South African Veterinary Association, Monument Park 10222
South American Explorers Club, Ithaca 16722
South Australian Ornithological Association, Adelaide 00496
South Australian Science Teachers Association, Parkside 00497
South Bedfordshire Archaeological Society, Great Yarmouth 13077
Southeast Asian Society of Soil Engineering, Bangkok 11100
South-Eastern Union of Scientific Societies, Shoreham-by-Sea 13078
Southern Africa Cardiac Society, Durban 10223
Southern African Association for the Advancement of Science, Arcadia 10224
Southern African Institute of Forestry, Pretoria 10225
The Southern African Museums Association, Sunnyside 10226
Southern African Ornithological Society, Johannesburg 10227
Southern African Society of Aquatic Scientists, Rondebosch 10228
Southern Arts Board, Winchester 13079
Southern Association of Colleges and Schools, Decatur 16723
Southern Building Code Congress – International, Birmingham 16724
Southern Historical Association, Athens 16725
Southern Humanities Conference, Chattanooga 16726
Southern Regional Council, Atlanta 16727
Southern Skirmish Association, London 13080
Southern Society of Genealogists, Piedmont 16728
Southern States Communication Association, Hattiesburg 16729
South Indian Horticultural Association, Coimbatore 06860
South India Textile Research Association, Coimbatore 06861
South of England Agricultural Society, Haywards Heath 13081
South Place Ethical Society, London 13082
South Staffordshire Archaeological and Historical Society, Walsall 13083
South Wales Institute of Engineers, Cardiff 13084
South West Africa Scientific Society, Windhoek 08816
Southwestern Legal Foundation, Richardson 16730
Southwest Parks and Monuments Association, Tucson 16731
South Wiltshire Industrial Archaeology Society, Salisbury 13085
Sozialakademie Dortmund, Dortmund 06283
Sozialwissenschaftliche Arbeitsgemeinschaft, Wien 00947
Sozialwissenschaftliche Studiengesellschaft, Wien 00948
SP, New York 16698
SPA, Saint Petersburg 16566
SPA, Washington 16571
S.P.A.B, London 13010
SPABH, Williamsburg 16603
SPACES, Los Angeles 16440
Space Settlement Studies Program, Niagara University 16732
Space Studies Institute, Princeton 16733
SPAIC, Lisboa 09689
SPAN, Montebello 16690
Spanish Federation of Religious Teaching, Madrid 10310
Spanish Horticultural Society, Madrid 10388
Spanish Optical Society, Madrid 10392
Spanish Polio Association, Madrid 10262
Spanish Psychoanalytical Society, Barcelona 10395
Spanish Society for Microbiology, Madrid 10391
Spanish Society of Ceramic and Glass, Madrid 10380
Spanish Veterinary Society of Zootechnics, Madrid 10402
The Spastics Society, London 13086
SPCAP, Washington 16695
SPCK, London 12987
SPCP, Lyon 04793
SPD, Ann Arbor 16561
SPE, Brookfield 16692
SPE, Dallas 16570
SPE, Knoxville 16696
S.P.E., Lisboa 09717
SPE, Richardson 16689
SPEA, Glasgow 12921
Special Education Association of the Republic of China, Taipei 03330
Special Interest Group for Computer Personnel Research, College Station 16734
Special Interest Group for Computer Uses in Education, College Station 16735
Special Interest Group for Symbolic and Algebraic Manipulation, Kent 16736
Special Interest Group on Artificial Intelligence, New York 16737
Special Interest Group on Biomedical Computing, Durham 16738
Special Interest Group on Information Retrieval, Blackburg 16739
Special Libraries Association, Washington 16740
Spectroscopial Society of Japan, Tokyo 08283
Spectroscopic Society of South Africa, Pretoria 10229
Spectroscopy Society of Canada, Ottawa 02949
Speech and Hearing Association of Alberta, Edmonton 02950
Speech and Hearing Association of Nova Scotia, Halifax 02951, 02952
Speech Communication Association, Annandale 16741
Speech Foundation of Ontario, Toronto 02953
Spektroskopická Společnost Jana Marca Marci, Praha 03553
Speleological Association of Slovenia, Ljubljana 10101
SPELL, Waleska 16604
Spenser Society, Seattle 16742
SPEP, Carbondale 16567
SPERTTT, London 12988
SPES, London 13082
SPF, Glasgow 12920
SPF, Ville d'Avray 04927
SPFFA, New York 16488
SPHAN, Rio de Janeiro 01702
Spina Bifida and Hydrocephalus Association of Ontario, Toronto 02954
Spina Bifida Association of British Columbia, Surrey 02955
Spina Bifida Association of Canada, Winnipeg 02956
Spirit and Breath Association, Skokie 16743
SPM, Indianapolis 16697
SPMA, London 12986
SPMA, Tucson 16731
S.P.M.F.R., Lisboa 09704
SPN, Porto 09706
SPNEA, Boston 16605
SPNI, Tel Aviv 07111
Spohr Society of Great Britain, Sheffield 13087
Společnost Antonína Dvořáka, Praha 03554
Spolok Architektov Slovenska, Bratislava 10094
SPOOF, Seattle 16607
Sportärztebund Hamburg, Hamburg 06284
Sportärztebund Hessen, Frankfurt 06285
Sportärztebund Niedersachsen, Göttingen 06286
Sportärztebund Nordrhein, Leverkusen 06287
Sportärztebund Rheinland-Pfalz, Kaiserslautern 06288
Sportärzteverband Schleswig-Holstein, Kiel 06289
Sport Medicine Council of Canada, Vanier 02957
The Sports Council, London 13088
Sportwissenschaftliche Gesellschaft der Universität Graz, Graz 00949
SPOSS, Palo Alto 16606
SPP, Gainesville 16562
SPPA, Lexington 16568
SPQ, Lisboa 09712
SPR, Athens 16569
SPR, Dallas 16572
SPR, Elk Grove Village 16564
SPR, Lisboa 09714
SPR, Oak Brook 16563

S.P.R., Paris 04845
SPR, Voorburg 09080
Sprachverband Deutsch für ausländische Arbeitnehmer e.V., Mainz 06290
SPRI, New York 16482
Spring Research and Manufacturers Association, Sheffield 13089
SPS, New York 16688
SPSA, Roanoke 16691
SPSE, Springfield 16672
SPSL, London 13011
SPSM, Bridgeport 16602
SPSSI, Ann Arbor 16608
SPTL, Buckingham 13062
SPU, Seattle 16565
SPV, Sempach-Stadt 10902
SPWLA, Houston 16694
SRA, Chicago 16699
SRA, London 12960
SRA, McLean 16577
S.R.A.M.A., Sheffield 13089
SRAPL, Delémont 10982
SRBE, Bruxelles 01478
SRBG, Bruxelles 01473
SRC, Atlanta 16727
SRCD, Chicago 16576
SRF, Stockholm 10594
SRG, Cardiff 13063
SRHE, London 12994
SRHP, Chicago 16457
S.R.I., Napoli 07942
Sri Aurobindo Centre, New Delhi 06862
Sri Lanka Association for the Advancement of Science, Colombo 10419
Sri Lanka Library Association, Colombo 10420
Sri Lanka Medical Association, Colombo 10421
SRL, Bochum 06411
SRLA, Edingburgh 12923
S.R.L.F., Caen 04685
SRM, Denver 16574
SRP, Didcot 12991
SRPBA, Glasgow 12926
Srpska Akademija Nauka i Umetnosti, Beograd 17125
Srpsko Biološko Društvo, Beograd 17126
Srpsko Geografsko Društvo, Beograd 17127
Srpsko Geološko Društvo, Beograd 17128
Srpsko Hemijsko Društvo, Beograd 17129
Srpsko Lekarsko Društvo, Beograd 17130
SRR, Chur 10963
SRR, Saint Louis 16575
SRRT, Chicago 16483
SRS, Boston 16479
SRS, Edinburgh 12924
SRS, Glasgow 13109
SRS, Guildford 13112
SRS, Ipswich 13107
SRSO, Rosemont 16449
SS, Amherst 16441
SS, Canton 16721
SS, Cleveland 16463
SS, Middlebury 16437
SS, Salt Lake City 16459
SS, Seattle 16742
SS, Washington 16445
SSA, Alexandria 16753
S.S.A., Canberra 00499
SSA, Davis 16462
SSA, El Cerrito 16460
SSA, Horsham 13013
SSA, Saint Louis 16475
SSA, West Hollywood 16468
SSA, Ypsilanti 16751
SSAG, Stockholm 10573
SSAR, Oxford 16612
SSAT, New York 16583
SSATB, Princeton 16456
SSB, Brooklyn 16613
SSB, Washington 16704
SSBD, Dallas 16614
SSC, Hamilton 02960
S.S.C., Isernia 07793
SSCA, Hattiesburg 16729
SSCS, Santa Monica 16455
SSDHPER, Kensington 16702
SSDSA, New York 16720
SSDT, Hartfield 16701
SSE, Lawrence 16616
SSEC, Berkeley 16615
SSEC, Boulder 16484
S.S.F., Bucuresti 09782
SSF, Cambridge 13014
SSF, Helsinki 03996
S.S.F.A., Milano 07788
SSFF, Aas 09348
SSFODF, Paris 04928
S.S.G., Firenze 07794
SSG, Piedmont 16728
SSG, Santa Barbara 16747
SSHA, Paris 04919
SSHB, London 13015
SSHM, Aberdeen 12931
SSHM, Manchester 13012
SSHRC, Ottawa 02915
SSI, London 13064
SSI, Milano 08015
SSI, Princeton 16733
SSI, Stockholm 10574
SSIA, Glasgow 12912
SSIEM, Manchester 13016
SSIS, Berkeley 16581
SSISI, Dublin 07035
SSL, Locarno 10964
SSM, Bucuresti 09786
SSMA, Bowling Green 16443
SSMLL, Oxford 13019
SSMPP, Montpellier 16617
SSNS, Edinburgh 12929
SSO, Palatine 16703
SSPC, Pittsburgh 16746
SSPHS, Nashville 16582
SSPP, Portland 16618
S.S.R., Cesena 07795
SSR, Champaign 16619
SSR, Stockholm 10595
SSRC, Bethlehem 16748
SSRC, New York 16485
SSRS, Canberra 00491
SSS, Beckenham 12952
SSS, Bloomington 16579
SSS, East Grinstead 13108
SSS, Helsinki 04060
SSS, New York 16470
SSSA, Madison 16717
SSSAA, Cuneo 07999
SSSB, Honolulu 16620
SSSI, Saint Petersburg 16623
SSSL, New Orleans 16622
SSSP, Knoxville 16621
SSSP, Niagara University 16732
S.S.S.P., Savona 08012
SSSR, Wasco 16486
SSSR, West Lafayette 16610
SSSS, Baton Rouge 16580
SSSS, Mount Vernon 16611
S.S.S.S.A., Pretoria 10176
SSTA, Edinburgh 12927
SSV, Zürich 10904
ST, Hastings 16609
Staats- und Handelspolitische Gesellschaft, Recklinghausen 06291
Staats- und Wirtschaftspolitische Gesellschaft e.V., Hamburg 06292
S.T.A.B., Bologna 08029
Ständiger Arbeitsausschuss für die Tagungen der Nobelpreisträger in Lindau, Lindau 06293
Ständiger Ausschuss des Gesamtverbandes der Textilindustrie in der Bundesrepublik Deutschland - Gesamttextil- e.V., Eschborn 05733
Ständiger Ausschuss für Geographische Namen, Frankfurt 06294
STAESA, New York 16709
Staffordshire Agricultural Society, Stafford 13090
Staffordshire Parish Registers Society, Penn 13091
Staffordshire Record Society, Stafford 13092
StAGN, Frankfurt 06294
Stair Society, Edinburgh 13093
Standardiseringsforeningen, Oslo 09350
Standardiseringskommissionen i Sverige, Stockholm 10522
Standards Association of Australia, North Sydney 00498
Standards Association of Central Africa, Harare 17181
Standards Council of Canada, Ottawa 02958
Standards Engineering Society, Dayton 16744
Standards Engineering Society (Canadian Region), Nepean 02959
Standing Committee on Commonwealth Forestry, Edinburgh 13094
Standing Conference of African University Libraries, Nairobi 08507
Standing Conference of Arts and Social Sciences, Cardiff 13095
The Standing Conference of National and University Libraries, London 13096
Standing Conference on Library Materials on Africa, Milton Keynes 13097
Standing International Committee for Mycenaean Studies, Cambridge 13098
Stannary Law Society, Gunnislake 11866
S.T.A.Q., Spring Hill 00489
Statens Kulturråd, Stockholm 10523
Statens Råd för Byggnadsforskning, Stockholm 10524
Statistical and Social Inquiry Society of Ireland, Dublin 07035
Statistical Society, Stockholm 10525
Statistical Society of Australia, Canberra 00499
Statistical Society of Canada, Hamilton 02960
Statistics Norway, Oslo 09351
Statistiska Föreningen, Stockholm 10525
Statistiska Samfundet i Finland, Helsinki 04065
Statsøkonomisk Forening, Oslo 09352
Statsvetenskapliga Förbundet, Stockholm 10526
Statute Law Society, London 13099
STC, Arlington 16584
Steel Castings Institute of Canada, Ottawa 02961
The Steel Construction Institute, Ascot 13100
Steel Founders' Society of America, Des Plaines 16745
Steel Structures Painting Council, Pittsburgh 16746
Steirische Gesellschaft für Psychologie, Graz 00950
S.T.E.M.O., Paris 04920
Stephenson Locomotive Society, London 13101
STEPP, West Carrollton 16705
The Stewart Society, Edinburgh 13102
STF, Stockholm 10596
STFM, Kansas City 16706
STG, Hamburg 06275
SThG, Bern 10934
STI, Ellensburg 16480
S.T.I., Trieste 08022
Stichting Centrale Raad voor de Academies van Bouwkunst, Amsterdam 09070
Stichting Cultureel Centrum Suriname, Paramaribo 10432
Stichting Economisch Instituut voor de Bouwnijverheid, Amsterdam 09071
Stichting Koninklijk Zoölogisch Genootschap Natura Artis Magistra, Amsterdam 09072
Stichting Natuur en Milieu, Utrecht 09073
Stichting Nederlands Agronomisch-Historisch Instituut, Groningen 09074, 09075
Stichting Verenigd Nederlands Filminstituut, Hilversum 09076
Stichting voor de Technische Wetenschappen, Utrecht 09077
Stichting voor Fundamenteel Onderzoek der Materie, Utrecht 09078
Stichting voor Wetenschappelijk Ondersoek van de Tropen, Paramaribo 10433
Stichting voor Wetenschappelijk Onderzoek van de Tropen, 's-Gravenhage 09079
Stiftelsen for Industriell og Teknisk Forskning ved Norges Tekniske Høgskole, Trondheim 09353
Stifterverband für die Deutsche Wissenschaft, Essen 06295
Stifterverband Metalle, Düsseldorf 06296
Stiftung für Humanwissenschaftliche Grundlagenforschung, Zürich 10992
STKS, Helsinki 03999
STMS, Glasgow 12935
Stonehenge Study Group, Santa Barbara 16747
Stowarzyszenie Archiwistow Polskich, Warszawa 09613
Stowarzyszenie Autorów ZAIKS, Warszawa 09614
Stowarzyszenie Bibliotekarzy Polskich, Warszawa 09615
Stowarzyszenie Historyków Sztuki, Warszawa 09616
STP, London 12314
STPT, London 13069
STR, London 13000
The Strindberg Society, Kungsängen 10527
Strindbergssällskapet, Kungsängen 10527
Stroke Association, London 13103
Structural Fire Protection Association, London 13104
Structural Stability Research Council, Bethlehem 16748
STS, Chicago 16707
STS, Edinburgh 12934
S.T.S.D., Canterbury 13066
STSN, Pisa 08026
Student National Medical Association, Washington 16749
STUDES, Roma 07581
Studieförbundet Nämigsliv och Samhälle, Stockholm 10528
Studiengemeinschaft für Fertigbau e.V., Wiesbaden 06297
Studiengemeinschaft Holzleimbau e.V., Düsseldorf 06298
Studiengesellschaft für den kombinierten Verkehr e.V., Frankfurt 06299
Studiengesellschaft für Holzschwellenoberbau e.V., Wiesbaden 06300
Studiengesellschaft für Stahlleitplanken e.V., Siegen 06301
Studiengesellschaft für unterirdische Verkehrsanlagen e.V., Köln 06302
Studiengesellschaft Stahlanwendung e.V., Düsseldorf 06303
Studiengruppe Entwicklung Technischer Hilfsmittel für Behinderte, Heidelberg 06304
Studiengruppe für Sozialforschung e.V., Marquartstein 06305
Studiengruppe Unternehmer in der Gesellschaft, Königstein 06306
Studienkreis für Presserecht und Pressefreiheit, Stuttgart 06307
Studievereniging voor Psychical Research, Voorburg 09080
Studio Teologico Accademico Bolognese, Bologna 08029
Study Centre for Problems of the Contemporary Muslim World, Bruxelles 01205
Study Group for Mathematical Learning, Highland Park 16750
STUK, Helsinki 03989
Stupu, Stuttgart 05603
Stuttgart Literary Society, Stuttgart 06105
STUVA, Köln 06302
STV, Helsinki 04068
STV, Zürich 10905
STW, Utrecht 09077
Styrelsen för Teknisk Utveckling, Stockholm 10529
SUA, Washington 16624
Subterranean Sociological Association, Ypsilanti 16751
SUC, Glasgow 13033
Succulent Society of South Africa, Pretoria 10230
S.U.C.E.A., Sydney 00500
Sudan Library Association, Khartoum 10429
Süddeutsches Kunststoff-Zentrum, Würzburg 06308
Südosteuropa-Gesellschaft e.V., München 06309
SUERF, Tilburg 09069
Suffolk Agricultural Association, Ipswich 13105
Suffolk Institute of Archaeology and History, Ipswich 13106
Suffolk Records Society, Ipswich 13107
Sugar Industry Technologists, Sainte-Thérèse de Blainville 02962
Sugar Technologist's Association of Trinidad and Tobago, Port of Spain 11123
Suid-Afrikaanse Akademie vir Wetenskap en Kuns, Pretoria 10177
Suid-Afrikaanse Biologiese Vereniging, Pretoria 10184
Suid-Afrikaanse Vereniging vir Spraak- en Gehoorheelkunde, Braamfontein 10220
Suifumeitokukai Shokokan, Ibaraki-ken 08427
SUISA, Zürich 10817
SUKKJ, Beograd 17122
Sulphur Institute, Washington 16752
Sumi-E Society of America, Alexandria 16753
Sunday Shakespeare Society, East Grinstead 13108
Suomalainen Lääkäriseura Duodecim, Helsinki 03997
Suomalainen Lakimiesyhdistys, Helsinki 03998
Suomalainen Teologinen Kirjallisuusseura, Helsinki 03999
Suomalainen Tiedeakatemia, Helsinki 04000
Suomalaisen Kirjallisuuden Seura, Helsinki 04001
Suomalaisten Kemistien Seura, Helsinki 04002
Suomalais-Ugrilainen Seura, Helsinki 04003
Suomen Akatemia, Helsinki 04004
Suomen Aktuaariyhdistys, Helsinki 04005
Suomen Allergologi- ja Immunologiyhdistys, Helsinki 04006
Suomen Atomiteknillinen Seura, Helsinki 04007
Suomen Autoteknillinen Liitto, Helsinki 04008
Suomen Avaruustutkimusseura, Helsinki 04009
Suomen Betoniyhdistys, Helsinki 04010
Suomen Egyptologinen Seura, Helsinki 04011
Suomen Eksegeettinen Seura, Helsinki 04012
Suomen Eläinlääkäriliitto, Helsinki 04013
Suomen Farmaseuttinen Yhdistys, Helsingin Yliopisto 04014
Suomen Farmasialiitto, Helsinki 04015
Suomen Filosofinen Yhdistys, Helsingin Yliopisto 04016
Suomen Fysiologiyhdistys, Oulu 04017
Suomen Fysioterapeuttiliitto, Helsinki 04018
Suomen Fyysikkoseura, Helsingin Yliopisto 04019
Suomen Gastroenterologiayhdistys, Helsinki 04020
Suomen Gemmologinen Seura, Helsinki 04021
Suomen Geologinen Seura, Espoo 04022
Suomen Geoteknillinen Yhdistys, Espoo 04023
Suomen Hammaslääkäriliitto, Helsinki 04024
Suomen Hammaslääkäriseura, Helsinki 04025
Suomen Heraldinen Seura, Helsinki 04026
Suomen Historiallinen Seura, Helsinki 04027
Suomen Hitsausteknillinen Yhdistys, Helsinki 04028
Suomen Hyönteistieteellinen Seura, Helsinki 04029
Suomen Ilmailuliitto, Helsinki 04030
Suomen Itämainen Seura, Helsingin Yliopisto 04031
Suomen Kansanopistoyhdistys, Helsinki 04032
Suomen Kardiologinen Seura, Helsinki 04033
Suomen Kasvatusopillinen Yhdistys, Helsinki 04034
Suomen Kasvatustieteellinen Seura, Oulu 04035
Suomen Kemian Seura, Helsinki 04036
Suomen Kielen Seura, Turku 04037
Suomen Kirjailijaliitto, Helsinki 04038
Suomen Kirjastoseura, Helsinki 04039
Suomen Kirkkohistoriallinen Seura, Helsingin Yliopisto 04040
Suomen Kirurgiyhdistys, Helsinki 04041
Suomen Kliinisen Neurofysiologian Yhdistys, Helsinki 04042
Suomen Korroosioyhdists Sky, Helsinki 04043
Suomen Lainopillinen Yhdistys, Helsinki 04044
Suomen Lakimiesliitto, Helsinki 04045
Suomen Lintutieteellinen Yhdistys, Helsinki 04046
Suomen Maantieteellinen Seura, Helsingin Yliopisto 04047
Suomen Maataloustieteellinen Seura, Piikkio 04048
Suomen Matemaattinen Yhdistys, Helsingin Yliopisto 04049
Suomen Metsätieteellinen Seura, Helsinki 04050
Suomen Muinaismuistoyhdistys, Helsinki 04051
Suomen Museoliitto, Helsinki 04052
Suomen Musiikinopettajain Liitto, Helsinki 04053
Suomen Näytelmäkirjailijaliitto, Helsinki 04054
Suomen Palontorjuntaliitto, Helsinki 04055
Suomen Psykologiliitto, Helsinki 04056
Suomen Radiologiyhdistys, Helsinki 04057
Suomen Säveltäjät, Helsinki 04058
Suomen Standardisoimisliitto, Helsinki 04059
Suomen Sukututkimusseura, Helsinki 04060
Suomen Syöpäyhdistys, Helsinki 04061
Suomen Taideyhdistys, Helsinki 04062
Suomen Tekstiiliteknillinen Liitto, Tampere 04063
Suomen Tieteellinen Kirjastoseura, Helsinki 04064
Suomen Tilastoseura, Helsinki 04065
Suomen Väestötieteen Yhdistys, Helsinki 04066
Superstition Mountain Historical Society, Apache Junction 16754
Supreme Court Historical Society, Washington 16755
Surface Finishing Society of Japan, Tokyo 08138
Surgical Association of the Republic of China, Taipei 03331
Surgical Research Society, Glasgow 13109
Surrey Archaeological Society, Guildford 13110
Surrey County Agricultural Society, Godalming 13111
Surrey Record Society, Guildford 13112
Surtseyjarfélagid, Reykjavik 06701
Surtsey Research Society, Reykjavik 06701
Survey Institute of Zimbabwe, Harare 17182
Survey of India, Dehra Dun 06863
Surveyors Historical Society, Lansing 16756
SUS, Helsinki 04003
SUS, New Haven 16711
Sussex Archaeological Society, Lewes 13113
Sussex Industrial Archaeology Society, Brighton 13114
SUT, London 13021
Suzuki Association of the Americas, Boulder 16757
SVA, Bern 10940
SVA, Washington 16628
Svaefingalaeknafélag Islands, Reykjavik 06702
SVBF, Winterthur 10909
SVC, Caracas 17017
SVCC, Oftringen 10918
SVCT, Basel 10938
SVD, Thun 10942
SVE, Santa Ana 16627
SVE, Zollikofen 10943
SVEB, Zürich 10944
Svenska Akademien, Stockholm 10530
Svenska Akademiska Rektorskonferensen, Uppsala 10531
Svenska Aktuarieföreningen, Stockholm 10532
Svenska Akustiska Sällskapet, Göteborg 10533
Svenska Arkeologiska Samfundet, Stockholm 10534
Svenska Arkitekters Riksförbund, Stockholm 10535
Svenska Arkivsamfundet, Stockholm 10536
Svenska Astronomiska Sällskapet, Saltsjöbaden 10537
Svenska Bergmannaföreningen, Stockholm 10538
Svenska Bibliotekariesamfundet, Borås 10539
Svenska Botaniska Föreningen, Stockholm 10540
Svenska Dermatologiska Sällskapet, Stockholm 10541
Svenska Färgeritekniska Riksförbundet, Göteborg 10542

Svenska Föreningen för Ljuskultur, Stockholm 10543
Svenska Föreningen för Medicinsk Teknik och Fysik, Umea 10544
Svenska Föreningen för Mikrobiologi, Lund 10545
Svenska Försäkringsföreningen, Stockholm 10546
Svenska Folkbibliotekarie Förbundet, Nacka 10547
Svenska Fornminnesföreningen, Stockholm 10548
Svenska Fornskriftsällskapet, Stockholm 10549
Svenska Fysikersamfundet, Uppsala 10550
Svenska Geofysiska Föreningen, Norrköping 10551
Svenska Geotekniska Föreningen, Linköping 10552
Svenska Gymnastikläraresällskapet, Vällingby 10553
Svenska Historiska Föreningen, Stockholm 10554
Svenska Kemistsamfundet, Stockholm 10555
Svenska Klassikerförbundet, Stockholm 10556
Svenska Kriminalistföreningen, Stockholm 10557
Svenska Kyltekniska Föreningen, Järfälla 10558
Svenska Kyrkohistoriska Föreningen, Uppsala 10559
Svenska Läkaresällskapet, Stockholm 10560
Svenska Linné-Sällskapet, Uppsala 10561
Svenska Litteratursällskapet, Uppsala 10562
Svenska Litteratursällskapet i Finland, Helsinki 04067
Svenska Livsmedelstekniska Föreningen, Stockholm 10563
Svenska Matematikersamfundet, Umea 10564
Svenska Museiföreningen, Stockholm 10565
Svenska Nationalkomittén för Geologi, Uppsala 10566
Svenska Nationalkommittén för Kristallografi, Stockholm 10567
Svenska Naturskyddsföreningen, Stockholm 10568
Svenska Österbottens Litteraturförening, Krylbo 10569
Svenska Ogonläkareföreningen, Stockholm 10570
Svenska Operationsanalysföreningen, Stockholm 10571
Svenska Psykoanalytiska Föreningen, Stockholm 10572
Svenska Sällskapet för Antropologi och Geografi, Stockholm 10573
Svenska Samfundet för Informationsbehandling, Stockholm 10574
Svenska Samfundet för Musikforskning, Stockholm 10575
Svenska Tekniska Vetenskapsakademien i Finland, Helsinki 04068
Svenska Växtgeografiska Sällskapet, Uppsala 10576
Svenska Yrketsutbildningsföreningen, Ludvika 10577
Svensk Flyghistorisk Förening, Stockholm 10578
Svensk Förening för Anatomi, Umeå 10579
Svensk Förening för Anestesi och Intensivvård, Linköping 10580
Svensk Förening för Gastroenterologi och Gastrointestinal Endoskopi, Helsingborg 10581
Svensk Förening för Radiofysik, Danderyd 10582
Svensk Kirurgisk Förening, Uppsala 10583
Svensk Otolaryngologisk Förening, Stockholm 10584
Sveriges Allmänna Biblioteksförening, Stockholm 10585
Sveriges Allmänna Konstförening, Stockholm 10586
Sveriges Arkivtjänstemäns Förening, Nacka 10587
Sveriges Civilingenjörsförbund, Stockholm 10588
Sveriges Författarförbund, Stockholm 10589
Sveriges Gemmologiska Riksförening, Stockholm 10590
Sveriges Geologiska Undersökning, Uppsala 10591
Sveriges Museimannaförbund, Nacka 10592
Sveriges Praktiserande Läkares Förening, Stockholm 10593
Sveriges Rationaliseringsförbund, Stockholm 10594
Sveriges Socianomer Riksförbund, Stockholm 10595
Sveriges Tandläkarförbund, Stockholm 10596
Sveriges Vetenskapliga Specialbiblioteks Förening, Stockholm 10597
Sveriges Veterinärförbund, Stockholm 10598
Svetstekniska Föreningen, Stockholm 10599
SVF, Stockholm 10598, 10599
SVG, Zürich 10946
SVGW, Zürich 10919
SVHA, Chavannes-près-Renens 10991
SVHE, Washington 16625
SVK, Bern 10948
SVLFC, Frauenfeld 10936
SVLT, Riniken 10914
SVMT, Pully 10910
SVP, Lincoln 16712
S.V.P.F., Paris 04922
SVP Italia, Milano 08030
SVS, Manchester 16626
SVSF, Stockholm 10597
SVSS, Bern 10915
SVTL, Bern 10956
SVUM, Zürich 10953
SVVK, Solothurn 10921
SVW, Winterthur 10911
SWA, Wien 00947
SWANA, Silver Spring 16719
Swaziland Art Society, Mbabane 10434
Swaziland Library Association, Mbabane 10435
SWB, Zürich 10923
SWCS, Ankeny 16716
SWE, New York 16713
SWEA, La Jolla 16760
The Swedenborg Society, London 13115
The Swedish Academy of Engineering Sciences in Finland, Helsinki 04068
Swedish-American Historical Society, Chicago 16758
Swedish Archaeological Society, Stockholm 10534
The Swedish Archival Association, Stockholm 10536
Swedish Association of Consulting Engineers, Stockholm 10600
The Swedish Association of Graduate Engineers, Stockholm 10588
Swedish Association of Museum Curators, Nacka 10592
Swedish Association of Occupational Therapists, Nacka 10458
Swedish Association of Public-Library Employees, Nacka 10547
Swedish Association of University and Research Librarians, Borås 10539
Swedish Aviation Historical Society, Stockholm 10578
Swedish Cancer Society, Stockholm 10441
Swedish Colonial Society, Philadelphia 16759
Swedish Council for Building Research, Stockholm 10524
Swedish Council for Cultural Affairs, Stockholm 10523
Swedish Council for Planning and Coordination of Research, Stockholm 10462
Swedish Council for Research in the Humanities and Social Sciences, Stockholm 10468
The Swedish Dental Association, Stockholm 10596
Swedish Economic Association, Stockholm 10503
Swedish Geotechnical Society, Linköping 10552
The Swedish Institute of International Affairs, Stockholm 10603
Swedish Insurance Federation, Stockholm 10532
Swedish Insurance Society, Stockholm 10546
Swedish Linnaeus Society, Uppsala 10561
The Swedish Museums Association, Stockholm 10565
Swedish National Board for Technical Development, Stockholm 10529
Swedish Natural Science Research Council, Stockholm 10504
Swedish Österbottens Literary Assocoiation, Krylbo 10569
Swedish Physical Society, Uppsala 10550
Swedish Political Science Association, Stockholm 10526
Swedish Society for Anaesthesia and Intensive Care, Linköping 10580
Swedish Society for Medical Engineering and Medical Physics, Umea 10544
Swedish Society for Microbiology, Lund 10545
Swedish Society for Musicology, Stockholm 10575
The Swedish Society for Nature Conservation, Stockholm 10568
Swedish Society for Technical Documentation, Stockholm 10601
The Swedish Society of Aeronautics and Astronautics, Stockholm 10457
Swedish Society of Dermatology and Venerology, Stockholm 10541
The Swedish Society of Refrigeration, Järfälla 10558
Swedish Standards Institution, Stockholm 10522
Swedish Surgical Society, Uppsala 10583
Swedish Women's Educational Association International, La Jolla 16760
Swedish Writers' Union, Stockholm 10589
SWHG, East Lansing 16487
SWIAS, Salisbury 13085
SWIE, Cardiff 13084
SWIP, San Jose 16630
SWIP, Scranton 16629
Swiss-American Historical Society, Chicago 16761
Swiss Association for Education by Correspondence, Kreuzlingen 10912
Swiss Association for Nutrition, Zollikofen 10943
Swiss Association of Certified Chemists HTL, Basel 10938
Swiss Federation for Adult Education, Zürich 10944
Swiss Foundation for Alpine Research, Zürich 10931
Swiss Gas and Water Industry Association, SGWA, Zürich 10919
Swiss Institute for Experimental Cancer Research, Epalinges 10732
Swiss Management Association, Zürich 10871
Swiss Physical Society, Genève 10881
Swiss Political Science Association, Genève 10633
Swiss Society for Futures Research, Horgen 10955
Swiss Society for Rights of Authors of Musical Works, Zürich 10817
Swiss Society for the Study of the Ancient Near East, Genève 10838
SWLF, Richardson 16730
SWR, Bern 10924
SWS, Wien 00948
SWST, Madison 16714
SWT, Edinburgh 12936
Sydney University Chemical Engineering Association, Sydney 00500
Sydney University Chemical Society, Sydney 00501
Sydney University Medical Society, Sydney 00502
Sydney University Psychological Society, Sydney 00503
Sydney University Veterinary Society, Sydney 00504
Syllogos Architktonon Diplomatouchon Anotaton Scholon, Athinai 06544
Syllogos latrikos Panellinios, Athinai 06545
Syllogos pros Diadosin ton Hellenikon Grammaton, Athinai 06546
Syndicat des Chirurgiens Dentistes de Paris, Paris 04924
Syndicat des Ecrivains, Paris 04925
Syndicat des Enseignants, Paris 04926
Syndicat des Psychiatres Français, Ville d'Avray 04927
Syndicat des Spécialistes Français en Orthopédie Dento-Faciale, Paris 04928
Syndicat des Vétérinaires de la Région Parisienne, Paris 04929
Syndicat Général de l'Education Nationale, Paris 04930
Syndicat Général des Personnels de l'Education Nationale, Paris 04931
Syndicat National de l'Enseignement Supérieur, Paris 04932
Syndicat National de l'Enseignement Technique, Paris 04933
Syndicat National de l'Intendance de l'Education Nationale, Paris 04934
Syndicat National de l'Orthopédie Française, Paris 04935
Syndicat National des Allergologistes Français, Tours 04936
Syndicat National des Anesthésistes-Réanimateurs Français, Paris 04937
Syndicat National des Auteurs et Compositeurs de Musique, Paris 04938
Syndicat National des Chefs d'Etablissements d'Enseignement Libre, Paris 04939
Syndicat National des Chirurgiens Français, Paris 04940
Syndicat National des Chirurgiens Plasticiens, Paris 04941
Syndicat National des Collèges de la Région Parisienne, Paris 04942
Syndicat National des Collèges et des Lycées, Paris 04943
Syndicat National des Enseignants, Luxembourg 08590
Syndicat National des Enseignements de Second Degré, Paris 04944
Syndicat National des Instituteurs et Professeurs de Collège, Paris 04945
Syndicat National des Lycées et Collèges, Paris 04946
Syndicat National des Médecins Acupuncteurs de France, Paris 04947
Syndicat National des Médecins Anatomo-Cyto-Pathologistes Français, Paris 04948
Syndicat National des Médecins Biologistes, Paris 04949
Syndicat National des Médecins de Groupe, Paris 04950
Syndicat National des Médecins des Hôpitaux Publics, Paris 04951
Syndicat National des Médecins du Sport, Paris 04952
Syndicat National des Médecins Electro-Radiologistes Qualifiés, Paris 04953
Syndicat National des Médecins Français Spécialistes des Maladies du Coeur et des Vaisseaux, Paris 04954
Syndicat National des Médecins Homéopathes Français, Paris 04955
Syndicat National des Médecins Ostéothérapeutes Français, Nancy 04956
Syndicat National des Médecins Phlébologues Français, Paris 04957
Syndicat National des Médecins Rhumatologues, Antony 04958
Syndicat National des Médecins Spécialisés en Phoniatrie, Saint-Denis 04959
Syndicat National des Médecins Spécialistes de l'Endocrinologie et de la Nutrition, Paris 04960
Syndicat National des Médecins Spécialistes en Stomatologie et Chirurgie Maxillo-Faciale, Paris 04961
Syndicat National des Oto-Rhino-Laryngologistes Français, Paris 04962
Syndicat National des Pédiatres Français, Clermont-Ferrand 04963
Syndicat National des Personnels de Direction de l'Enseignement Secondaire, Paris 04964
Syndicat National des Professeurs d'Arts Martiaux, Paris 04965
Syndicat National des Professeurs des Ecoles Normales d'Instituteurs, Paris 04966
Syndicat National des Psychiatres des Hôpitaux, Paris 04967
Syndicat National des Vétérinaires Français, Paris 04968
Syndicat National Français des Dermatologistes et Vénéréologistes, Paris 04969
Syndicat National Professionnel des Médecins du Travail, Toulouse 04970
SYNEMA – Gesellschaft für Film und Medien, Wien 00951
S.Y.S., New Dehli 06859
The Systematics Association, Durham 13116
Systematics Association of New Zealand, Dunedin 09197
Szczecinskie Towarzystwo Naukowe, Szczecin 09617
Szervezési és Vezetési Tudományos Társaság, Budapest 06673
SZF, Horgen 10955
SZG, Basel 10961
Szilikátipari Tudományos Egyesület, Budapest 06674

TA, Arlington 16764
TA, New York 16789
TAA, Cambridge 13139
TAALS, Washington 13546
TAC, Altrincham 11208
Tähtitieteellinen Yhdistys Ursa, Helsinki 04069
TAGA, Rochester 16767
TAI, Queenstown 14256
Tajik Academy of Sciences, Dushanbe 11045
Taloushistoriallinen Yhdistys, Jyväskylä 04070
Taloustieteellinen Seura, Helsinki 04071
Tamarind Institute, Albuquerque 16762
Tamil Association, Thanjavur 06864
Tamil Language Society, Kuala Lumpur 08655
Die Tandheelkundige Vereniging van Suid-Afrika, Houghton 10134
Tannlaeknafélag Islands, Reykjavik 06703
Tanzania Commission for Science and Technology, Dar es Salaam 11059
Tanzania Library Association, Dar es Salaam 11060
Tanzania Medical Association, Dar es Salaam 11061
The Tanzania Society, Dar es Salaam 11062
Tanzania Veterinary Association, Morogoro 11063
TAP, Poznan 09618
TAPA, Smithtown 16795
TARD, Istanbul 11150
Tarleton Foundation, San Francisco 16763
TAS, London 13122
Tasmanian Geological Survey, Rosny 00505
Tasmanian Historical Research Association, Sandy Bay 00506
Tasmanian University Agricultural Science Society, Hobart 00507
The Tavistock Institute of Medical Psychology, London 13117
TAW, Wuppertal 06310
Tax Analysts, Arlington 16764
TBC, Bronx 16785
TBS, London 13143
TC, Manila 09519
TCA, Columbia 16784
TCAS, Malvern 13132
TCB, New York 14771
TChP, Warszawa 09619
TCPA, London 13137
TCS, Chesapeake 14936
TCS, Gloucester 14767
TDES, New York 14990
TDK, Ankara 11154
TE, Old Lyme 16774
Teachers Educational Council – Association of Accredited Cosmetology Schools, Falls Church 16765
Teachers of English to Speakers of Other Languages, Alexandria 16766
Teachers' Society, Baghdad 06920
The Teachers' Union, Oslo 09354
TEAM, Exeter 12003
Tea Research Foundation (Central Africa), Mulanje 08625
TEC/AACS, Falls Church 16765
Technical Association of the Australian and New Zealand Pulp and Paper Industry, Melbourne 00508
Technical Association of the Graphic Arts, Rochester 16767
Technical Association of the Paper and Printing Industry, Budapest 06672
Technical Marketing Society of America, Cypress 16768
Technion Research and Development Foundation Ltd, Haifa 07114
Technische Akademie Wuppertal e.V., Wuppertal 06310
Technische Fördergemeinschaft Holzsilo, Freiburg 06311
Technische Vereinigung der Großkraftwerksbetreiber e.V., Essen 06312
Technisch-Wissenschaftlicher Verein Bergmännischer Verband Österreichs, Leoben 00952
Technisch-wissenschaftlicher Verein Eisenhütte Österreich, Leoben 00953
Technology Foundation, Utrecht 09077
Technology Transfer Society, Indianapolis 16769
TECHNONET ASIA, Singapore 10037
Tecnagro, Roma 08031
Tecneco, Fano 08032
Tecnocasa, L'Aquila 08033
Tecnocentro Italiano, Milano 08034
Tecnofarmaci, Pomezia 08035
Tecnomare, Venezia 08036
Tecnotessile, Prato 08037
TEI, Boulder 16781
Tekniikan Akateemisten Liitto, Helsinki 04072
Teknikkonsulterna, Stockholm 10600
Teknillisten Oppilaitosten Opettajainliitto, Helsinki 04073
Teknillisten Tieteiden Akatemia, Helsinki 04074
Tekniska Föreningen i Finland, Helsinki 04075
Tekniska Litteratursällskapet, Stockholm 10601
Tekniska Samfundet i Göteborg, Göteborg 10602
Teknisk Skoleforening, Odense 03826
Tel Aviv Astronomical Association, Tel Aviv 07115
Telecommunication Society of Australia, Melbourne 00509
Television Academy of Arts and Sciences of the Republic of China, Taipei 03332
Tennessee Folklore Society, Murfreesboro 16770
The Tennyson Society, Lincoln 13118
Tensor Society, Chigasaki 08428
Tensor Society of Great Britain, Surbiton 13119
Teologisk Forening ved Københavns Universitet, København 03827
Teratology Society, Bethesda 16771
Teriological Society of Moldova, Chişinău 08793
Terminological Commission, Riga 08559
Terminological Committee, Baku 01049
TESOL, Alexandria 16766
Tetsugaku-kai, Tokyo 08429
Textile Academy, Beijing 03222
The Textile Institute, Manchester 13120
Textile Machinery Society of Japan, Osaka 08258
Textile Research Council, Nottingham 13121
Textile Society of Australia, Melbourne 00510
Textilipari Müszaki és Tudományos Egyesület, Budapest 06675
Textilnorm, Berlin 06209
TF, San Francisco 16763
T.F.I., Reykjavik 06703
TFS, Freiburg 06311
TFS, Murfreesboro 16770
TG, New York 16772
TGS, Rock Hill 15146
Thai-Bhara Cultural Lodge and Swami Satyananda Puri Foundation, Bangkok 11101
Thai Library Association, Bangkok 11102
Theater Instituut Nederland, Amsterdam 09081
Theatre Arts Society, London 13122
Theatre Guild, New York 16772
Theatre Historical Society, Elmhurst 16773
Theatre in Education, Old Lyme 16774
Theatre Library Association, New York 16775
Theatre Organ Preservation Society, Slough 13123

Theatre Union of the Russian Federation, Moskva 09985
Theatrical Institute, Barcelona 10314
Theodore Roosevelt Association, Oyster Bay 16776
Theodor Herzl Institute, New York 16777
The Theosophical Society, Madras 06865
Theosophical Society in England, London 13124
Theosophical Society in Europe, Camberley 13125
Theosophical Society in Ireland, Dublin 07036
Theosophical Society in Scotland, Edinburgh 13126
Theosophical Society of Ceylon, Colombo 10422
Theosophical Society of Trinidad, Guaico 11124
THF, Pasadena 16786
THI, New York 16777
Thomas-Morus-Akademie Bensberg, Bergisch Gladbach 06313
Thomas Paine Society, Selsey 13127
Thomas Tallis Society, London 13128
Thomas Wolfe Society, Richmond 16778
Thoracic Society, Southampton 13129
Thoreau Lyceum, Concord 16779
Thoreau Society, Concord 16780
Thoresby Society, Leeds 13130
Thorne Ecological Institute, Boulder 16781
Thoroton Society of Nottinghamshire, Nottingham 13131
Three Counties Agricultural Society, Malvern 13132
THS, Elmhurst 16773
THY, Turku 04078
Thyroid Foundation of Canada, Kingston 02963
TI, Albuquerque 16762
TI, Manchester 13120
T.I., Milano 08034
Tibet Society, Bloomington 16782
Tietohuollon Neuvottelukunta, Helsinki 04076
Tietopalveluseura, Espoo 04077
TiFC, Warszawa 09620
Tiger Hills Arts Association, Holland 02964
TII, Luxembourg 08582
Timber Research and Development Association, High Wycombe 13133
Time Series Analysis and Forecasting Society, Nottingham 13134
T.I.M.P., London 13117
TIMS, Providence 15330
Tin Research Institute, Columbus 16783
TIP, Wroclaw 09621
Tiroler Gesellschaft zur Förderung der Alterswisschaft und des Seniorenstudiums an der Universität Innsbruck, Innsbruck 00954
Tiroler Juristische Gesellschaft, Innsbruck 00955
Tissue Culture Association, Columbia 16784
TKD, Ankara 11165
TLA, Bangkok 11102
TLA, Dar es Salaam 11060
TLA, New York 16775
TLURS, Morden 12471
TMHiZK, Kraków 09623
TMI, Faber 15753
TMI, Washington 15707
TMJP, Kraków 09624
TMMB, Ankara 11167
TMP, New Brunswick 15685
TMS, Boulder 15698
TMS, Hammond 15686
TMS, Matlock 13138
TMSA, Cypress 16768
TMTE, Budapest 06675
TNC, Arlington 16188
TNT, Torun 09627
Toa Igaku Kyokai, Tokyo 08430
Toa Kumo Gakkai, Osaka 08431
Tobacco Research Council, London 13135
Tobago District Agricultural Society, Scarborough 11125
Tochi Seidoshi Gakkai, Tokyo 08432
Todai Chugoku Gakkai, Tokyo 08433
Toho Gakkai, Tokyo 08434
TOK, Istanbul 11171
Tokei Kagaku Kenkyukai, Fukuoka 08435
Tokyo Bar Association, Tokyo 08436
Tokyo Bengoshikai Toshokan, Tokyo 08436
Tokyo Chigaku Kyokai, Tokyo 08437
Tokyo Daigaku Keizai Gakkai, Tokyo 08438
Tokyo Geographical Society, Tokyo 08437
Tolkien Society, Hove 13136
Tónlistarfélagið, Reykjavik 06704
Tónskáldafélag Islands, Reykjavik 06705
TOOL, Helsinki 04073
TOPS, Slough 13123
Toronto Area Archivists Group, Toronto 02965
Toronto Psychoanalytic Society, Toronto 02966
Toronto Society of Financial Analysts, Toronto 02967
Torrey Botanical Club, Bronx 16785
Toshi Kaihatsu Kyokai Joho Sabisu Senta Shiryoshitsu, Tokyo 08439
Touch for Health Foundation, Pasadena 16786
Tourette Syndrome Association, Bayside 16787
Towarzystwo Anestezjologów Polskich, Poznan 09618
Towarzystwo Chirurgów Polskich, Warszawa 09619
Towarzystwo imienia Fryderyka Chopina, Warszawa 09620
Towarzystwo Internistów Polskich, Wroclaw 09621
Towarzystwo Literackie im. Adama Mickiewicza, Warszawa 09622
Towarzystwo Milosników Historii i Zabytków Krakowa, Kraków 09623
Towarzystwo Milosnikow Jezyka Polskiego, Kraków 09624
Towarzystwo Naukowe Organizacji i Kierownictwa, Warszawa 09625
Towarzystwo Naukowe Płockie, Płock 09626
Towarzystwo Naukowe w Toruniu, Torun 09627
Towarzystwo Przyjaciól Nauk w Przymyslu, Przemysl 09628
Towarzystwo Urbanistów Polskich, Warszawa 09629
Town and Country Planning Association, London 13137
Town and Country Planning Association of Victoria, Melbourne 00511
Toyo Gakujutsu Kyôkai, Tokyo 08440
Toyo Ongaku Gakkai, Tokyo 08441
Toyoshi Kenkyukai, Kyoto 08442
TPN, Przemysl 09628
TRA, Kenner 16792
TRA, Oyster Bay 16776
TRADA, High Wycombe 13133
Traffic Injury Research Foundation of Canada, Ottawa 02968
Tramway Museum Society, Matlock 13138
Trans-Antarctic Association, Cambridge 13139
Transplantation Society, Columbus 16788
Transplantation Society, London 13140
Transport 2000 Canada, Ottawa 02969
Transportation Alternatives, New York 16789
Transportation Research Board, Washington 16790
Transport Ticket Society, Luton 13141
Transvaal Underwater Research Group, Johannesburg 10231
TRB, Washington 16790
TRC, Nottingham 13121
Tree-Ring Society, Tucson 16791
The Tree Society of Southern Africa, Johannesburg 10232
Trevithick Society, Truro 13142
TRI, Columbus 16783
Trinidad Music Association, Port of Spain 11126
Trinitarian Bible Society, London 13143
Tripoli Rocketry Association, Kenner 16792
Tripura Library Association, Agartala 06866
Tropical Committee, Moskva 09986
The Tropical Grassland Society of Australia, Saint Lucia 00512
Tropical Science Center, San José 03452
TROPMEDEUROPE, Budapest 06599
Troubles d'Apprentissage – Association Canadienne, Ottawa 02586
Trout Unlimited, Vienna 16793
TRS, Tucson 16791
TRTAI, Saint Petersburg 16384
The Trumpeter Swan Society, Maple Plain 16794
Truss Plate Institute of Canada, Concord 02970
TS, Bethesda 16771
TS, Bloomington 16782
TS, Columbus 16788
TS, Concord 16780
TS, London 13124
TSA, Bayside 16787
TSA, Bromsgrove 13144
T.S.A., Melbourne 00509
TSE, Ankara 11174
TSGB, Surbiton 13119
TTA, Helsinki 04074
TTK, Ankara 11175
TTS, Indianapolis 16769
TTS, London 13128
TTSS, Maple Plain 16794
TU, Vienna 16793
TUA, London 12197
Tuberous Sclerosis Association of Great Britain, Bromsgrove 13144
Tudományos Ismeretterjesztö Társulat, Budapest 06676
Tübinger Förderkreis zur Erforschung der Troas – Freunde von Troia, Tübingen 06314
Türk Anesteziyoloji ve Reanimasyon Derneği, Istanbul 11150
Türk Biyoloji Derneği, Istanbul 11151
Türk Cerrahi Cemiyeti, Istanbul 11152
Türk Diabetler Cemiyeti, Istanbul 11153
Türk Dil Kurumu, Ankara 11154
Türk Eczacilari Birliği, Istanbul 11155
Türk Fizik Derneği, Istanbul 11156
Türk Gastroenterologii Derneği, Ankara 11157
Türk Halk Bilgisi Derneği, Istanbul 11158
Türk Hukuk Kurumu, Ankara 11159
Türkiye Aktüerler Cemiyeti, Istanbul 11160
Türkiye Fitopatolji Derneği, Izmir 11161
Türkiye Ziraat Odalari Birligi, Ankara 11162
Türk Kanser Arastirma ve Savas Kurumu, Ankara 11163
Türk Kardiyoloji Derneği, Istanbul 11164
Türk Kütüphaneciler Derneği, Ankara 11165
Türk Mikrobiyoloji Cemiyeti, Istanbul 11166
Türk Mühendis ve Mimar Odalari Birligi, Ankara 11167
Türk Nöro-Psikiyatri Derněi, Istanbul 11168
Türk Oftalmoloji Derneği, Istanbul 11169
Türk Ortopedi ve Travmatoloji Dernegi, Istanbul 11170
Türk Otomatik Kontrol Kurumu, Istanbul 11171
Türk Oto-Rino-Laringoloji Cemiyeti, Istanbul 11172
Türk Sirfi ve Tatbiki Matematik Derneği, Istanbul 11173
Türk Standartlari Enstitüsü, Ankara 11174
Türk Tarih Kurumu, Ankara 11175
Türk Tüberküloz Cemiyeti, Istanbul 11176
Türk Uroloji Dernegi, Istanbul 11177
Türk Veteriner Hekimleri Dernegi, Ankara 11178
T.U.P., Warszawa 09629
Turkish American Physicians Association, Smithtown 16795
Turkish Association of Orthopaedics and Traumatology, Istanbul 11170
Turkish Folklore Society, Istanbul 11158
Turkish Historial Society, Ankara 11175
Turkish Law Association, Ankara 11159
Turkish Librarians' Association, Ankara 11165
Turkish Linguistic Society, Ankara 11154
Turkish Microbiological Society, Istanbul 11166
Turkish Pharmaceutical Association, Istanbul 11155
Turkish Phytopathological Society, Izmir 11161
Turkish Society of Cardiology, Istanbul 11164
Turkish Society of Pure and Applied Mathematics, Istanbul 11173
Turkish Surgical Society, Istanbul 11152
Turkish Tuberculosis Society, Istanbul 11176
Turkish Urological Society, Istanbul 11177
Turkish Veterinary Medicine Association, Ankara 11178
Turkmen Academy of Sciences, Ashkhabad 11180
Turku Historical Society, Turku 04078
Turku Music Society, Turku 04079
Turner Society, London 13145
Turner's Syndrome Society, Downsview 02971
Turun Historiallinen Yhdistys, Turku 04078
Turun Soitannollinen Seura, Turku 04079
TVA, Morogoro 11063
Twentieth Century Society, London 13146
TWI, Cambridge 13147
TWS, Bethesda 16901
TWS, Richmond 16778
TWS, Washington 16896
Työtehoseura, Helsinki 04080

UAA, Belfast 13149
UAA, Newark 16850
UAA, Opelika 16843
UAAS, New York 16796
UACES, London 13174
UAHS, Belfast 13150
U.A.I., Bologna 08039
UAI, Bruxelles 01492
UAI, Paris 04971
UAMDB, Cotonou 01515
U.A.N., Roma 08038
UAS, Belfast 13148
UBCMEA, New York 16807
UBG, Bruxelles 01493
U.B.J.E.T., Vilvoorde 01494
U.B.N., Bologna 08041
U.B.S.S., Bristol 13175
UCA, Belfast 13151
U.C.A.D., Paris 04972
UCAR, Boulder 16845
UCC, Indianapolis 16808
UCDA, Walkerton 16841
UCEA, University Park 16846
UCIDT, Athens 16844
UCIIM, Roma 08042
U.C.I.P.E.M., Bologna 08043
UCLEA, New Brunswick 16842
UCOWR, Carbondale 16838
UCPREF, New York 16810
Udruženje Univerzitetskih Nastavnika i Van-Univerzitetskih Naučnik Radnika, Beograd 17131
UEA, Rotterdam 09083
Übernationale Vereinigung für Kommunikationsforschung e.V., München 06125
UEI, Birmingham 13160
UEMS, Bruxelles 01496
UEPG, Paris 04478
UESA, Cliffside Park 16797
UESEG, Leigh-on-Sea 13163
UET, New York 16811
U.F.A.S., Bruxelles 01495
UFFCS, Orlando 15247
U.F.O.D., Paris 04979
UFUCH, Santiago 03104
UFVA, Los Angeles 16847
UFY, Helsingin Yliopisto 04081
Uganda Library Association, Kampala 11187
The Uganda Society, Kampala 11188
U.G.C.I., Roma 08046
UGGI, Toulouse 04982
UHMS, Bethesda 16803
UI, Cincinnati 16804
U.I., Stockholm 10603
UIA, New York 16798
UIA, Paris 04985
UIBS, Paris 04988
UICC, Genève 10993
UIE, Paris-La Défense 04983
UIEOA, Saint-Maur 04986
UIFA, Paris 04987
U.I.L.D.M., Padova 08047
UILE, Paris 04990
UIMC, Bruxelles 01498
UISPP, Gent 01497
U.K.A.E.L., London 13164
UKB, Amsterdam 09067
UKHEF, Cleckheaton 13165
UKRA, Widnes 13167
Ukrainian Academy of Agrarian Sciences, Kiev 11191
Ukrainian Academy of Arts and Sciences in the U.S., New York 16796
Ukrainian Academy of Medical Sciences, Kiev 11192
Ukrainian Academy of Sciences, Kiev 11193
Ukrainian Engineers' Society of America, Cliffside Park 16797
Ukrainian Institute of America, New York 16798
Ukrainian Medical Association of North America, Chicago 16799
Ukrainian Political Science Association in the United States, Philadelphia 16800
ULA, Kampala 11187
ULC, State College 16852
ULCI, Manchester 13161
U.L.D.O., Roma 08044
ULI – The Urban Land Institute, Washington 16801
Ulster Archaeological Society, Belfast 13148
Ulster Archery Association, Belfast 13149
Ulster Architectural Heritage Society, Belfast 13150
Ulster Chemists' Association, Belfast 13151
Ulster Folklife Society, Holywood 13152
The Ulster Genealogical and Historical Guild, Belfast 13153
The Ulster Historical Foundation, Belfast 13153
Ulster Historical Foundation, Belfast 13154
Ulster Society for Irish Historical Studies, Belfast 13155
Ulster Society for the Preservation of the Countryside, Holywood 13156
Ulster Society in London, London 13157
Ulster Teachers' Union, Belfast 13158
Ultralight Aircraft Association of Canada, Ottawa 02972
ULVB, Pedraces 08052
UMANA, Chicago 16799
UMEC, Roma 08053
UMEMPS, Athinai 06547
UMI, Bologna 08048
U.M.I., Bologna 08049
UMIST Association, Manchester 13159
Unabhängiger Ärzteverband Deutschlands e.V., Köln 06315
UNAMAZ, Belém 01613
UN.A.R., Roma 08040
UNATEB, Le Perreux 04993
UNAV, Genova 07650
UNC, Campbell 16813
Underground Technology Research Council Education Committee, Brooklyn 16802
Undersea and Hyperbaric Medical Society, Bethesda 16803
UNI, Milano 07656
União dos Escritores Angolanos, Luanda 00034
UNICA, Bruxelles 01307
UNICA, San Juan 09722
Unidade de Pesquisa de Ambito Estadual em Barreiras, Barreiras 01731
Unidade de Pesquisa de Ambito Estadual em Campos, Campos 01732
Unidade de Pesquisa de Ambito Estadual em Corumba, Corumba 01733
Unidade de Pesquisa de Ambito Estadual em Dourados, Dourados 01734
Unidade de Pesquisa de Ambito Estadual em Itaguai, Nova Iguaçu 01735
Unidade de Pesquisa de Ambito Estadual em Itapirema, Goiânia 01736
Unidade de Pesquisa de Ambito Estadual em Manaus, Manaus 01737
Unidade de Pesquisa de Ambito Estadual em Ponta Grossa, Ponta Grossa 01738
Unidade de Pesquisa de Ambito Estadual em Porto Velho, Porto Velho 01739
Unidade de Pesquisa de Ambito Estadual em Teresina, Teresina 01740
Unija Bioloških Naučnih Društava Jugoslavije, Zemun 17132
Union Académique Internationale, Bruxelles 01492
Union Astronomique Internationale, Paris 04971
Union Belge des Géomètres-Experts Immobiliers, Bruxelles 01493
Union Belge des Journalistes et Ecrivains du Tourisme, Vilvoorde 01494
Union Centrale des Arts Décoratifs, Paris 04972
Unión de Escritores y Artistas de Cuba, La Habana 03486
Unión de Federaciones Universitárias Chilenas, Santiago 03104
Union des Associations d'Assistants Sociaux Francophones, Bruxelles 01495
Union des Biologistes de France, Paris 04973
Union des Ecrivains Algériens, Alger 00025
Union des Océanographes de France, Paris 04974
Union des Physiciens, Paris 04975
Union des Professeurs de Spéciales (Mathématiques et Sciences Physiques), Paris 04976
Union des Travailleurs Espérantistes des Pays de Langue Française, Paris 04977
Unión de Universidades de América Latina, México 08777
Unione Accademica Nazionale, Roma 08038
Unione Antropologica Italiana, Bologna 08039
Unione Associazioni Regionali, Roma 08040
Unione Bolognese Naturalisti, Bologna 08041
Unione Cattolica Italiana Insegnanti Medi, Roma 08042
Unione Consultori Italiani Prematrimoniali e Matrimoniali, Bologna 08043
Unione della Legion d'Oro, Roma 08044
Unione Erpetologica Italiana, Roma 08045
Unione Giuristi Cattolici Italiani, Roma 08046
Unione Italiana Lotta alla Distrofia Muscolare, Padova 08047
Unione Matematica Italiana, Bologna 08048
Unione Micologica Italiana, Bologna 08049
Unione Tecnica Italiana Farmacisti, Genova 08050
Union Européenne des Médecins Spécialistes, Bruxelles 01496
Union Fédérative des Sociétés d'Education Physique et de Préparation Militaire, Paris 04978
Union for Modern Philology, Bratislava 10058
Union Française des Organismes de Documentation, Paris 04979
Union Française pour l'Espéranto, Paris 04980
Union Générale des Auteurs et Musiciens Professionnels, Paris 04981
Union Generela di Ladins dla Dolomites, Ortisei-Urtijëi 08051
Union Géodésique et Géophysique Internationale, Toulouse 04982
Union Institute, Cincinnati 16804
Union Internationale contre le Cancer, Genève 10993
Union Internationale d'Electrothermie, Paris-La Défense 04983
Union Internationale de Phlébologie, Paris 04984
Union Internationale de Physique Pure et Appliquée, Québec 02563
Union Internationale des Architectes, Paris 04985
Union Internationale des Etudes Orientales et Asiatiques, Saint-Maur 04986
Union Internationale des Femmes Architectes, Paris 04987
Union Internationale des Instituts de Recherche Forestières, Wien 00677

Union Internationale des Laboratoires Independants, Elstree 12402
Union Internationale des Sciences Biologiques, Paris 04988
Union Internationale des Sciences Préhistoriques et Protohistoriques, Gent 01497
Union Internationale des Services Médicaux des Chemins de Fer, Bruxelles 01498
Union Internationale des Sociétés d'Aide à la Santé Mentale, Bordeaux 04989
Union Internationale pour la Liberté d'Enseignement, Paris 04990
Union Internationale pour la Protection des Obtentions Végétales, Genève 10994
Union Internationale Thérapeutique, Paris 04991
Union Ladins Val Badia, Pedraces 08052
Unión Matemática Argentina, Cordoba 00202
Union Médicale Algérienne, Alger 00026
Union Médicale Balkanique, Bucuresti 09747
Union Mondiale des Enseignants Catholiques, Roma 08053
Union Nationale des Arts Plastiques, Tunis 11143
Union Nationale des Médecins Spécialistes Confédérés, Paris 04992
Union Nationale des Techniciens Biologistes, Le Perreux 04993
Union Nationale Patronale des Prothésistes Dentaire, Paris 04994
Union of Agricultural Engineers and Technicians of Macedonia, Skopje 08617
Union of Architects in Bulgaria, Sofia 01783
Union of Arts of the Russian Federation, Moskva 10110
Union of Associations for Macedonian Language and Literature, Skopje 08612
Union of Associations of Jurists of Macedonia, Skopje 08616
Union of Associations of Veterinary Surgeons and Technicians of Macedonia, Skopje 08611
Union of Bulgarian Composers, Sofia 01784
Union of Bulgarian Film Makers, Sofia 01785
Union of Bulgarian Mathematicians, Sofia 01786
Union of Bulgarian Writers, Sofia 01787
Union of Chambers of Turkish Engineers and Architects, Ankara 11167
Union of Chemistry and the Chemical Industry, Sofia 01788
Union of Civil Engineering, Sofia 01789
Union of Czech Mathematicians and Physicists, Praha 03547
Union of Economics, Sofia 01790
Union of Economists of Macedonia, Skopje 08613
Union of Educational Institutions, Birmingham 13160
Union of Electronics, Electrical Engineering and Communications, Sofia 01791
Union of Energetics, Sofia 01792
Union of Engineers and Technicians of Serbia, Beograd 17133
Union of Engineers and Technicians of Yugoslavia, Beograd 17134
The Union of Finnish Lawyers, Helsinki 04045
Union of Forest Engineering, Sofia 01793
Union of Forestry Engineers and Technicians of Macedonia, Skopje 08615
Union of Geodesy and Cartography, Sofia 01794
Union of Jurists' Associations of Yugoslavia, Beograd 17135
Union of Lancashire and Cheshire Institutes, Manchester 13161
Union of Librarians' Associations of Macedonia, Skopje 08608
Union of Mechanical Engineering, Sofia 01795
Union of Middle East Mediterranean Paediatric Societies, Athinai 06547
Union of Mining Engineering, Geology and Metallurgy, Sofia 01796
Union of Rural Economics, Sofia 01797
Union of Russian Architects, Moskva 09987
Union of Russian Filmmakers, Moskva 09988
Union of Russian Writers, Moskva 09989
Union of Scientific and Learned Societies, Moskva 09990
Union of Scientific Medical Societies in Bulgaria, Sofia 01798
Union of Scientific Workers in Bulgaria, Sofia 01799
Union of Serbian Library Workers, Beograd 17109
Union of Societies of Archivists of Macedonia, Skopje 08607
Union of Societies of Historians of Macedonia, Skopje 08609
Union of Textiles, Clothing and Leather, Sofia 01800
Union of the Food Industry, Sofia 01801
Union of the Societies of Engineers and Technicians of Croatia, Zagreb 03467
Union of Translators' in Bulgaria, Sofia 01802
Union of Turkish Chambers of Agriculture, Ankara 11162
Union of Water Works, Sofia 01803
Union of Women Teachers, Birmingham 12570
Unión Parroquial del Sur, Bogotá 03421
Union Professionnelle Belge des Médécins Ophthalmologistes, Bruxelles 01499
Union Professionnelle des Professeurs, Cadres et Techniciens du Secrétariat et de la Comptabilité, Paris 04995
Union Radio-Scientifique Internationale, Bruxelles 01500
Union Royale Belge pour les Pays d'Outre-Mer et l'Europe Unie, Bruxelles 01501
Union Scientifique Continentale du Verre, Charleroi 01502
Unión Sudamericana de Asociaciones de Ingenieros, Asunción 09436
Union Suisse des Instituts de Formation en Pédagogie Curative (UIPC), Luzern 10995
Union Syndicale Nationale des Angiologues, Paris 04996
Union Syndicale Vétérinaire Belge, Bruxelles 01503
Union Technique de l'Automobile, du Motocycle et du Cycle, Paris 04997
Union Technique de l'Electricité, Paris-La Défense 04998
Unitarian Historical Society, Edinburgh 13162
Unitarian Universalist Association of Congregations, Washington 16805
Unitarian Universalist Historical Society, Milton 16806
Unitas Malacologica, Leiden 09082
United Board for Christian Higher Education in Asia, New York 16807
United Cancer Council, Indianapolis 16808
United Cerebral Palsy Associations, Washington 16809
United Cerebral Palsy Research and Educational Foundation, New York 16810
United Earth Sciences Exploration Group, Leigh-on-Sea 13163
United Engineering Trustees, New York 16811
United Kingdom Association for European Law, London 13164
United Kingdom Home Economics Federation, Cleckheaton 13165
United Kingdom Institute for Conservation of Historic and Artistic Works, London 13166
United Kingdom Reading Association, Widnes 13167
United Kingdom Science Park Association, Birmingham 13168
United Lodge of Theosophists, Bombay 06867
United Nations Committee on the Peaceful Uses of Outer Space, New York 16812
United New Conservationists, Campbell 16813
United Ostomy Association, Irvine 16814
United Reformed Church History Society, London 13169
United Schools International, New Dehli 06868
United Society for Christian Literature, Guildford 13170
United Society for the Propagation of the Gospel, London 13171
United Society of Artists, Alderney 13172
United States Antarctic Research Program, Washington 16815
United States Branch of the International Committee for the Defense of the Breton Language, Jenkintown 16816
United States Capitol Historical Society, Washington 16817
United States Committee of the International Association of Art, Washington 16818
United States Committee on Irrigation and Drainage, Denver 16819
United States Committee on Large Dams, Denver 16820
United States Council for Energy Awareness, Washington 16821
United States Energy Association, Washington 16822
United States Federation for Culture Collections, Abbott Park 16823
United States Federation of Scholars and Scientists, Fullerton 16824
United States Institute for Theatre Technology, New York 16825
United States Metric Association, Los Angeles 16826
United States-Mexico Border Health Association, El Paso 16827
United States National Committee for Byzantine Studies, Madison 16828
United States National Committee of the International Union of Radio Science, Washington 16829
United States National Committee on Theoretical and Applied Mechanics, Washington 16830
United States National Society for the International Society of Soil Mechanics and Foundation Engineering, Raleigh 16831
United States Pharmacopeial Convention, Rockville 16832
United States Physical Therapy Association, Arlington Heights 16833
United States Space Education Association, Elizabethtown 16834
United States Sports Academy, Daphne 16835
United Synagogue Commission on Jewish Education, New York 16836
United Technological Organizations of the Philippines, Manila 09538
United World Colleges, London 13173
Uniunea Arhitectilor din Romania, Bucuresti 09796
Uniunea Artistilor Plastici din Romania, Bucuresti 09797
Uniunea Scriitorilor din Romania, Bucuresti 09798
Universal Esperanto Association, Rotterdam 09083
Universal Movement for Scientific Responsibility, Paris 04602
Universal Serials and Book Exchange, Cleveland 16837
Universités Unies pour l'Environnement, Paris 04999
Universities Art Association of Canada, Halifax 02973
Universities Council on Water Resources, Carbondale 16838
Universities Research Association, Washington 16839
Universities Space Research Association, Columbia 16840
University and College Designers Association, Walkerton 16841
University and College Labor Education Association, New Brunswick 16842
University Association for Contemporary European Studies, London 13174
University Aviation Association, Opelika 16843
University Consortium for Instructional Development and Technology, Athens 16844
University Corporation for Atmospheric Research, Boulder 16845
University Council for Educational Administration, University Park 16846
University Film and Video Association, Los Angeles 16847
University Geographical Society, Nedlands 00513
University of Bristol Spelaeological Society, Bristol 13175
University of New South Wales Chemical Engineering Association, Kensington 00514
University Philosophical Society, Dublin 07037
University Photographers Association of America, Provo 16848
University Professors for Academic Order, Alexandria 16849
Un-yu Chosakyoku Johobu Toshoshitsu, Tokyo 08443
UOA, Irvine 16814
UPAA, Provo 16848
UPAO, Alexandria 16849
UPBMO, Bruxelles 01499
UPCA, Washington 16809
UPOV, Genève 10994
Upper Canada Society for the History and Philosophy of Science and Technology, Hamilton 02974
U.P.S., Dublin 07037
U.P.S., Paris 04976
UPSA, Philadelphia 16800
URA, Washington 16839
Urban Affairs Association, Newark 16850
Urban and Regional Information Systems Association, Washington 16851
Urban Developers' Association of Japan, Tokyo 08439
Urban Development Institute of Ontario, Toronto 02975
Urbanisticki Savez Jugoslavije, Novi Sad 17136
Urban Libraries Council, State College 16852
Urdu Academy, Bahawalpur 09404
URISA, Washington 16851
URSA, Helsinki 04069
URSA Astronomical Association, Helsinki 04069
URSI, Bruxelles 01500
USACOJE, New York 16836
USAI, Asunción 09436
USARP, Washington 16815
USBE, Cleveland 16837
USCEA, Washington 16821
USCHS, Washington 16817
USCIAA, Washington 16818
USCID, Denver 16819
USCL, Guildford 13170
USCOLD, Denver 16820
U.S.C.V., Charleroi 01502
Use, Lanham 16853
USEA, Washington 16822
USFCC, Abbott Park 16823
USFSS, Fullerton 16824
USI, New Dehli 06868
US ICDBL, Jenkintown 16816
USIHS, Belfast 13155
USITT, New York 16825
USMA, Los Angeles 16826
USMBHA, El Paso 16827
USNCBS, Madison 16828
USNC/TAM, Washington 16830
USNC-URSI, Washington 16829
USNSISSMMFE, Raleigh 16831
USP, Rockville 16832
USPC, Holywood 13156
USPG, London 13171
USPTA, Arlington Heights 16833
USRA, Columbia 16840
USSA, Daphne 16835
USSEA, Elizabethtown 16834
USVB, Bruxelles 01503
USWA, Annandale 13494
U.T.A.C., Paris 04997
UTE, Paris-La Défense 04998
UTIFar, Genova 08050
Utrikespolitiska Institutet, Stockholm 10603
Uttar Pradesh (India) Library Association, Lucknow 06869
UTU, Belfast 13158
UUHS, Milton 16806
Uusfilologinen Yhdistys, Helsingin Yliopisto 04081
UVG, Cambridge 13176
UV Spectrometry Group, Cambridge 13176
Uzbek Academy of Sciences, Tashkent 16948

Vachel Lindsay Association, Springfield 16854
VAF, Baltimore 16857
V.A.G., Tamworth 13179
Valley Forge Historical Society, Valley Forge 16855
Valtiotieteellinen Yhdistys, Helsingin Yliopisto 04082
VAN, Leeuwarden 09088
Vancouver Museum Association, Vancouver 02976
Vancouver Natural History Society, Vancouver 02977
Vancouver Safety Council, Vancouver 02978
Van Riebeeck Society, Cape Town 10233
VAÖ, Wien 00956
VAR, Wageningen 09094
Vasectomy Advancement Society of Great Britain, London 13177
VASofGB, London 13177
VBAS, Huntsville 16869
VBB, Reutlingen 06368
VBE, Bonn 06332
VBT, Nacka 10604
VBV, Bern 11013
VCAMCUS, Sheridan 16868
VCS, Rochester 16859
VCV, Gent 01507
VDA, München 06372
VDAJ, Bonn 06348
VDB, Mainz 06373
VDBiol, Hamburg 06350
VDBW, Karlsruhe 06333
VdDB, Regensburg 06369
VDD-Berufsverband Dokumentation, Information, Kommunikation, Bonn 06316
VDE, Frankfurt 06351
VDEh, Düsseldorf 06374
VDEh-Gesellschaft zur Förderung der Eisenforschung, Düsseldorf 06317
VDE/VDI-Gesellschaft Mikroelektronik, Frankfurt 06318
VDG, Bonn 06404
VDG, Düsseldorf 06376
VDI, Düsseldorf 06377
VDI-EKW, Düsseldorf 06322
VDI-Gesellschaft Agrartechnik, Düsseldorf 06319
VDI-Gesellschaft Bautechnik, Düsseldorf 06320
VDI-Gesellschaft Energietechnik, Düsseldorf 06321
VDI-Gesellschaft Entwicklung Konstruktion Vertrieb, Düsseldorf 06322
VDI-Gesellschaft Fahrzeugtechnik, Düsseldorf 06323
VDI-Gesellschaft Fördertechnik Materialfluss Logistik, Düsseldorf 06324
VDI-Gesellschaft Kunststofftechnik, Düsseldorf 06325
VDI-Gesellschaft Produktionstechnik, Düsseldorf 06326
VDI-Gesellschaft Technische Gebäudeausrüstung, Düsseldorf 06327
VDI-Gesellschaft Werkstofftechnik, Düsseldorf 06328
VDI-Kommission Lärmminderung, Düsseldorf 06329
VDI-Society for Civil Engineering, Düsseldorf 06320
VDI-Society for Technical Building Services, Düsseldorf 06327
VDI-TGA, Düsseldorf 06327
VDI/VDE-Gesellschaft Mess- und Automatisierungstechnik, Düsseldorf 06330
VDI/VDE-Gesellschaft Mikro- und Feinwerktechnik, Düsseldorf 06331
VDLUFA, Darmstadt 06354
VDR, Viersen 06358
V.D.S.S., Hamburg 06359
VdTÜV, Essen 06343
VDZ, Schladen 06378
Vedag, Tuggen 11000
Védecká Společnost pro Nauku o Kovech, Brno 03555
Vegan Society, Saint Leonards-on-Sea 13178
Vej- og Byplanforeningen, København 03828
Verband Bildung und Erziehung e.V., Bonn 06332
Verband der Ärzte Deutschlands e.V., Bonn 05909
Verband der Akademikerinnen Österreichs, Wien 00956
Verband der angestellten und beamteten Ärzte Deutschlands e.V., Köln 06109
Verband der Betriebs- und Werksärzte e.V., Karlsruhe 06333
Verband der Bibliotheken des Landes Nordrhein-Westfalen e.V., Witten 06334
Verband der deutschen Forscher auf dem Gebiete des griechisch-römischen Altertums, Kiel 06119
Verband der deutschen Höhlen- und Karstforscher e.V., Leinfelden-Echterdingen 06335
Verband der Deutschen Münzvereine e.V., Speyer 05477
Verband der diplomierten Physiotherapeuten Österreichs, Wien 00957
Verband der Dozenten an Deutschen Ingenieurschulen, Mainz 06336
Verband der Geistig Schaffenden Österreichs, Wien 00958
Verband der Gemeinde-Tierärzte Baden-Württembergs, Baden-Baden 06337
Verband der Gemeinschaften der Künstlerinnen und Kunstfreunde e.V., Wuppertal 06338
Verband der Geschichtslehrer Deutschlands, Bokholt-Hanredder 06339
Verband der Heilpädagogischen Ausbildungsinstitute der Schweiz, Luzern 10995
Verband der Historiker Deutschlands, Göttingen 06340
Verband der Kantonschemiker der Schweiz, Bern 10996
Verband der Lehrer für Bürowirtschaft, Kurzschrift und Maschinenschreiben, Hamburg 05542
Verband der leitenden Krankenhausärzte Deutschlands e.V., Düsseldorf 06341
Verband der Marktforscher Österreichs, Wien 00959
Verband der Materialprüfungsämter e.V., Braunschweig 06342
Verband der Museen der Schweiz, Solothurn 10997
Verband der niedergelassenen Ärzte Deutschlands e.V., Köln 06137
Verband der Österreicher zur Wahrung der Geschichte Österreichs, Wien 00960
Verband der österreichischen Neuphilologen, Wien 00961
Verband der Russischlehrer Österreichs, Wien 00962
Verband der Schweizer Geographen, Zürich 10998
Verband der Schweizerischen Volkshochschulen, Bern 10999
Verband der Technischen Überwachungs-Vereine e.V., Essen 06343
Verband der Tierheilpraktiker e.V., Meitingen 06344
Verband der Volkshochschulen des Landes Bremen, Bremen 06345
Verband der Volkshochschulen des Saarlandes e.V., Saarbrücken 06346
Verband der Volkshochschulen von Rheinland-Pfalz e.V., Mainz 06347
Verband der Volksliedwerke der Bundesländer, Wien 00922
Verband der wissenschaftlichen Gesellschaften Österreichs, Wien 00963
Verband Deutscher Agrarjournalisten e.V., Bonn 06348
Verband Deutscher Badeärzte e.V., Bad Oeynhausen 06349
Verband Deutscher Biologen e.V., Hamburg 06350
Verband Deutscher Elektrotechniker e.V., Frankfurt 06351
Verband Deutscher Hochschullehrer der Geographie, Heidelberg 06352
Verband Deutscher Kunsthistoriker e.V., Darmstadt 06353
Verband Deutscher Landwirtschaftlicher Untersuchungs- und Forschungsanstalten, Darmstadt 06354
Verband Deutscher Lehrer im Ausland e.V., Husum 06355

Verband deutscher Musikschulen e.V., Bonn 06356
Verband Deutscher Physikalischer Gesellschaften, Stuttgart 06357
Verband Deutscher Realschullehrer im Deutschen Beamtenbund, Viersen 06358
Verband Deutscher Schiffahrts-Sachverständiger e.V., Hamburg 06359
Verband Deutscher Schulgeographen e.V., Burgwedel 06360
Verband Deutscher Schulmusiker e.V., Mainz 06361
Verband deutscher Waldvogelpfleger und Vogelschützer, Mainz 06362
Verband deutscher Werkbibliotheken e.V., Leverkusen 06363
Verband Deutscher Zoodirektoren e.V., Heidelberg 06364
Verband deutschschweizerischer Ärzte-Gesellschaften, Tuggen 11000
Verband deutschsprachiger Schriftsteller in Israel, Tel Aviv 07116
Verband freierwerbender Schweizer Architekten, Lachen 11001
Verband für Bildungswesen, Wien 01007
Verband für medizinischen Strahlenschutz in Österreich, Wien 00964
Verband Hochschule und Wissenschaft im Deutschen Beamtenbund, Bonn 06365
Verband Jüdischer Lehrer und Kantoren der Schweiz, Zürich 11002
Verband Katholischer Landvolkshochschulen Deutschlands, Bad Honnef 06366
Verband Musikschulen Schweiz, Liestal 11003
Verband niederösterreichischer Volkshochschulen, Wien 00965
Verband österreichischer Archivare, Wien 00966
Verband Österreichischer Bildungswerke, Wien 00967
Verband Österreichischer Geschichtsvereine, Wien 00968
Verband Österreichischer Höhlenforscher, Wien 00969
Verband Österreichischer Kurärzte, Wien 00970
Verband Österreichischer Privat-Museen, Bad Winsbach-Neydharting 00971
Verband Österreichischer Sportärzte, Wien 00972
Verband Österreichischer Volksbüchereien und Volksbibliothekare, Wien 00973
Verband Österreichischer Volkshochschulen, Wien 00974
Verband Österreichischer Wirtschaftsakademiker, Wien 00975
Verband Physikalische Therapie, Hamburg 06367
Verband Schweizerischer Abwasserfachleute, Zürich 11004
Verband Schweizerischer Assistenz- und Oberärzte, Bern 11005
Verband Schweizerischer Ergotherapeuten, Zürich 11006
Verband Schweizerischer Marketing- und Sozialforscher, Hergiswil 11007
Verband Schweizerischer Vermessungs-Techniker, Burgdorf 11008
Verein der Bibliothekare an Öffentlichen Bibliotheken e.V., Reutlingen 06368
Verein der Diplom-Bibliothekare an wissenschaftlichen Bibliotheken e.V., Regensburg 06369
Verein der Freien Pädagogischen Akademie, Hedingen 11009
Verein der Freunde der im Mittelalter von Österreich aus besiedelten Sprachinseln, Wien 00976
Verein der Freunde des Radwerkes IV in Vordernberg, Vordernberg 00977
Verein der Mundartfreunde Österreichs, Wien 00978
Verein der Museumsfreunde in Wien, Wien 00979
Verein der Textilchemiker und Coloristen e.V., Heidelberg 06370
Verein der Zellstoff- und Papier-Chemiker und -Ingenieure e.V., Darmstadt 06371
Verein Deutscher Archivare, München 06372
Verein Deutscher Bibliothekare e.V., Mainz 06373
Verein Deutscher Eisenhüttenleute, Düsseldorf 06374
Verein Deutscher Emailfachleute e.V., Hagen 06375
Verein Deutscher Giessereifachleute e.V., Düsseldorf 06376
Verein Deutscher Ingenieure, Düsseldorf 06377
Verein Deutscher Zuckertechniker, Schladen 06378
Verein Forschung für das graphische Gewerbe, Wien 00980
Verein Freunde der Archäologie, Graz 00981
Verein Freunde der Völkerkunde, Wien 00982
Verein für bayerische Kirchengeschichte, Nürnberg 06379
Verein für Binnenschiffahrt und Wasserstraßen e.V., Duisburg 06380
Verein für das Forschungsinstitut für Edelmetalle und Metallchemie e.V., Schwäbisch Gmünd 06381
Verein für Familienforschung in Ost- und Westpreussen e.V., Hamburg 06382
Verein für Familien- und Wappenkunde in Württemberg und Baden e.V., Stuttgart 06383
Verein für Forstliche Standortskunde und Forstpflanzenzüchtung e.V., Freiburg 06384
Verein für Gerberei-Chemie und -Technik e.V., München 06385
Verein für Geschichte der Arbeiterbewegung, Wien 00983
Verein für Geschichte der Stadt Wien, Wien 00984
Verein für Geschichte des Hegaus e.V., Singen 06386
Verein für Geschichte und Landeskunde von Osnabrück, Osnabrück 06387
Verein für Kommunalwirtschaft und Kommunalpolitik e.V., Düsseldorf 06388
Verein für Kommunalwissenschaften e.V., Berlin 06389
Verein für Landeskunde von Niederösterreich, Wien 00985
Verein für Psychiatrie und Neurologie, Wien 00986
Verein für Sozial- und Wirtschaftspolitik, Wien 00987
Verein für technische Holzfragen e.V., Braunschweig 06390
Verein für Versicherungs-Wissenschaft und -Praxis Nordhessen e.V., Kassel 06391
Verein für Volkskunde in Wien, Wien 00988
Verein für Wasser-, Boden- und Lufthygiene e.V., Berlin 06392
Verein für Westfälische Kirchengeschichte, Bielefeld 06393
Vereinigung Bayerischer Augenärzte, München 06394
Vereinigung bildender Künstler, Wien 00989
Vereinigung Burgenländischer Geographen, Stegersbach 00990
Vereinigung der Ärzte der Medizinaluntersuchungsämter, Osnabrück 06395
Vereinigung der Bayerischen Chirurgen e.V., Altötting 06396
Vereinigung der Freunde antiker Kunst, Basel 11010
Vereinigung der Freunde der Mineralogie und Geologie e.V., Heidelberg 06397
Vereinigung der Gymnastiklehrer, Zürich 11011
Vereinigung der Hochschullehrer für Zahn-, Mund- und Kieferheilkunde, Frankfurt 06398
Vereinigung der Humanistischen Gesellschaften Österreichs, Graz 00991
Vereinigung der kooperativen Forschungsinstitute der Österreichischen Wirtschaft, Wien 00992
Vereinigung der Landesdenkmalpfleger in der Bundesrepublik Deutschland, Hannover 06399
Vereinigung der Lehrer für Maschinenschreiben und Bürotechnik im SKV, Luzern 11012
Vereinigung der Praktischen und Allgemeinärzte Bayerns e.V., München 06400
Vereinigung der unabhängigen freiberuflichen Versicherungs- und Wirtschaftsmathematiker in der Bundesrepublik Deutschland e.V., Grünwald 06401
Vereinigung der Versicherungs-Betriebswirte e.V., Köln 06402
Vereinigung deutscher Ärzte, Frankfurt 06403
Vereinigung Deutscher Gewässerschutz e.V., Bonn 06404
Vereinigung deutscher Landerziehungsheime, Berlin 06405
Vereinigung Deutscher Neuropathologen und Neuroanatomen, Giessen 06406
Vereinigung Freischaffender Architekten Deutschlands e.V., Bonn 06407
Vereinigung für Agrar- und Umweltrecht e.V., Bonn 05314
Vereinigung für angewandte Botanik, Göttingen 06408
Vereinigung für Angewandte Lagerstättenforschung, Leoben 00993
Vereinigung für Bankbetriebsorganisation e.V., Frankfurt 06409
Vereinigung für Berufsbildung der Schweizerischen Versicherungswirtschaft, Bern 11013
Vereinigung für die physiotherapeutischen Berufe e.V., Hamburg 06367
Vereinigung für Hydrogeologische Forschungen in Graz, Graz 00994
Vereinigung für Rechsstaat und Individualrechte, Solothurn 11014
Vereinigung für Walsertum, Brig 11015
Vereinigung für wissenschaftliche Grundlagenforschung, Graz 00995
Vereinigung Getreide-, Markt- und Ernährungsforschung e.V., Bonn 06410
Vereinigung Österreichischer Ärzte, Wien 00996
Vereinigung Österreichischer Bibliothekarinnen und Bibliothekare, Innsbruck 00997
Vereinigung Schweizerischer Archivare, Bern 11016
Vereinigung Schweizerischer Bibliothekare, Bern 11017
Vereinigung Schweizerischer Hochschuldozenten, Würenlos 11018
Vereinigung Schweizerischer Kinder- und Jugendpsychologen, Solothurn 11019
Vereinigung Schweizerischer Naturwissenschaftslehrer, Neerach 11020
Vereinigung Schweizerischer Petroleumgeologen und -Ingenieure, Gelterkinden 11021
Vereinigung Stadt-, Regional- und Landesplanung e.V., Bochum 06411
Vereinigung Süddeutscher Orthopäden e.V., Baden-Baden 06412
Vereinigung Südwestdeutscher Dermatologen, Freiburg 06413
Vereinigung Südwestdeutscher HNO-Ärzte, Köln 06414
Vereinigung Südwestdeutscher Radiologen und Nuklearmediziner, Sindelfingen 06415
Vereinigung Umwelt und Bevölkerung, Zollikofen 11022
Vereinigung Westdeutscher Hals-, Nasen- und Ohrenärzte, Köln 06416
Vereinigung zur Erforschung der Neueren Geschichte e.V., Bonn 06417
Vereinigung zur Förderung der technischen Optik e.V., Wetzlar 06418
Vereinigung zur Förderung des Deutschen Brandschutzes e.V., Altenberge 06419
Vereinigung zur Förderung des Instituts für Kunststoffverarbeitung in Industrie und Handwerk an der Rhein.-Westf. Technischen Hochschule Aachen e.V., Aachen 06420
Vereinigung zur Förderung des Straßen- und Verkehrswesen, Bonn 05602
Vereinigung zur Reform der Versorgung psychisch Kranker e.V., Bonn 05067
Verein katholischer deutscher Lehrerinnen e.V., Essen 06421
Verein katholischer Lehrerinnen der Schweiz, Hochdorf 11023
Verein Montandenkmal Altböckstein, Leoben 00998
Verein Muttersprache, Lang-Enzersdorf 00999
Verein Naturschutzpark e.V., Niederhaverbeck 06422
Verein Nordfriesisches Institut e.V., Bredstedt 06423
Verein Österreichischer Lebensmittel- und Biotechnologen, Wien 01000
Verein österreichischer Ledertechniker, Wien 01001
Verein Österreichischer Textilchemiker und Coloristen, Dornbirn 01002
Verein Schweizerdeutsch, Lachen 11024
Verein Schweizerischer Geschichtslehrer, Riehen 11025
Verein Tiroler Landesmuseum Ferdinandeum, Innsbruck 01003
Verein von Altertumsfreunden im Rheinlande, Bonn 06424
Verein Wiener Frauenverlag, Wien 01004
Verein zur Beratung bei Blutungskrankheiten e.V., Marl-Hüls 05452
Verein zur Erhaltung Historischer Bauten, Grieskirchen 00889
Verein zur Förderung der Augenoptik, Zürich 11026
Verein zur Förderung der deutschen Tanz- und Unterhaltungsmusik e.V., Bonn 06425
Verein zur Förderung der Gießerei-Industrie e.V., Düsseldorf 06426
Verein zur Förderung der Versicherungswissenschaft in Hamburg e.V., Hamburg 06427
Verein zur Förderung der Versicherungswissenschaft in München e.V., München 06428
Verein zur Förderung der Wasser- und Lufthygiene, Zürich 11027
Verein zur Förderung des physikalischen und chemischen Unterrichts, Wien 01005
Verein zur Verbreitung naturwissenschaftlicher Kenntnisse, Wien 01006
Vereniging der Antwerpsche Bibliophielen, Antwerpen 01504
Vereniging Het Nederlandsch Economisch-Historisch Archief, Amsterdam 09084
Vereniging Het Nederlands Kanker Instituut, Amsterdam 09085
Vereniging Leraars Aardrijkskunde, Heverlee 01505
Vereniging Natuurmonumenten, 's-Graveland 09086
Vereniging tot Beoefening van Gelderse Geschiedenis, Oudheidkunde en Recht, Arnhem 08900
Vereniging tot Bevordering der Homoeopathie in Nederland, 's-Gravenhage 09087
Vereniging van Archivarissen in Nederland, Leeuwarden 09088
Vereniging van Docenten in Geschiedenis en Staatsrichting in Nederland, Leidschendam 09089
Vereniging van Homoeopathische Artsen in Nederland, Utrecht 09090
Vereniging van Katholieke Leraren Sint-Bonaventura, Hengelo 09091
Vereniging van Medische Analisten, Utrecht 09092
Vereniging van Nederlandse Kunsthistorici, Groningen 09093
Vereniging van Religieus-Wetenschappelijke Bibliothecarissen, Sint Truiden 01506
Vereniging voor Agrarisch Recht, Wageningen 09094
Vereniging voor Arbeidsrecht, Woerden 09095
Vereniging voor Bouwrecht, 's-Gravenhage 09096
Vereniging voor Calvinistische Wijsbegeerte, Amsterdam 09097
Vereniging voor de Staathuishoudkunde, Delft 09098
Vereniging voor Filosofie-Onderwijs, Zoetermeer 09099
Vereniging voor het Invoeren van Nieuwe Biologische Nomenklatuur, Kalmthout 01154
Vereniging voor het Theologisch Bibliothecariaat, Nijmegen 09100
Vereniging voor Hoger Beroepsonderwijs, Amsterdam 09101
Vereniging voor Nederlandse Muziekgeschiedenis, Utrecht 09102
Vereniging voor Penningkunst, Alkmaar 09103
Vereniging voor Statistiek, Heiloo 09104
Vereniging voor Studie en Onderzoek over Fitopathologie en Toegepaste Zoologie, Louvain-la-Neuve 01151
Vereniging voor Veldbiologie, Utrecht 08958
Vereniging voor Wijsbegeerte te s'-Gravenhage, 's-Gravenhage 09105
Vereniging voor Wijsbegeerte van het Recht, Amstelveen 09106
Vereniging voor Zuivelindustrie en Melkhygiëne, 's-Gravenhage 09107
Vergilian Society, Atlanta 16856
Verkfraedingafélag Islands, Reykjavik 06706
Vernacular Architecture Forum, Baltimore 16857
Vernacular Architecture Group, Tamworth 13179
Versicherungswissenschaftlicher und Versicherungswirtschaftlicher Fördererverein e.V., Nürnberg 06429
Versicherungswissenschaftlicher Verein in Hamburg e.V., Hamburg 06430
Versuchsgrubengesellschaft, Dortmund 06431
Verulam Institute, Chichester 13180
Vesterheim Genealogical Center, Madison 16858
Vetenskapliga Bibliotekens Tjänstemannaförening, Nacka 10604
Veterinary Cancer Society, Rochester 16859
Veterinary Council, Dublin 07038
Veterinary History Society, London 13181
Veterinary Orthopaedic Society, Salt Lake City 16860
Veterinary Services Council, Wellington 09198
VFA, Bonn 06407
VFDB, Altenberge 06419
VFG, Wien 00980
VFHS, Valley Forge 16855
VFMG, Heidelberg 06397
VfW, Brig 11015
VFWL, Zürich 11027
VGB, Essen 06312
VGB-Forschungsstiftung, Essen 06432
VGC, Madison 16858
VGCT, München 06385
VGN, Leidschendam 09089
VGS, Zürich 11028
VGStW, Wien 00984
VHpA, Luzern 10995
VHW, Bonn 06365
VICA, Leesburg 16866
Victoria Foundation, Victoria 02979
Victoria-Haliburton Lung Association, Lindsay 02980
The Victoria Institute or Philosophical Society of Great Britain, London 13182
Victorian Artists' Society, East Melbourne 00515
Victorian Military Society, Northampton 13183
Victorian Post-Secondary Education Commission, Hawthorn 00516
Victorian Public Interest Research Group, Fitzroy 00517
Victorian Society, London 13184
Victorian Society in America, Philadelphia 16861
Victorian Society of Pathology and Experimental Medicine, Parkville 00518
Victorian Studies Association of Western Canada, Edmonton 02981
Vieilles Maisons Françaises, Paris 05000
Vienna Institute for Development and Cooperation, Wien 01019
Vietnamese Assocoiation of Morphology, Hanoi 17035
Vietnamese Association of Acupuncture, Hanoi 17036
Vietnamese Association of Agriculture, Hanoi 17037
Vietnamese Association of Anaesthesiology, Hanoi 17038
Vietnamese Association of Anti-Contagious Diseases, Hanoi 17039
Vietnamese Association of Anti-Tuberculosis and Lung Diseases, Hanoi 17040
Vietnamese Association of Architects, Hanoi 17041
Vietnamese Association of Biology, Hanoi 17042
Vietnamese Association of Chemistry, Hanoi 17043
Vietnamese Association of Dermatology, Hanoi 17044
Vietnamese Association of Engineering, Hanoi 17045
Vietnamese Association of Forensic Scientists, Hanoi 17046
Vietnamese Association of Forestry, Hanoi 17047
Vietnamese Association of Geography, Hanoi 17048
Vietnamese Association of Geology, Hanoi 17049
Vietnamese Association of Internal Medicine, Hanoi 17050
Vietnamese Association of Mathematics, Hanoi 17051
Vietnamese Association of Medical Biochemistry, Hanoi 17052
Vietnamese Association of Neurology, Psychiatry and Neurosurgery, Hanoi 17053
Vietnamese Association of Obstetrics, Gynaecology and Family Planning, Hanoi 17054
Vietnamese Association of Odonto-Maxillo-Facial Medicine, Hanoi 17055
Vietnamese Association of Ophthalmology, Hanoi 17056
Vietnamese Association of Oto-Rhino-Laryngology, Hanoi 17057
Vietnamese Association of Paediatrics, Hanoi 17058
Vietnamese Association of Pharmacy, Hanoi 17059
Vietnamese Association of Physics, Hanoi 17060
Vietnamese Association of Physiology, Hanoi 17061
Vietnamese Association of Prophylactic Hygiene, Hanoi 17062
Vietnamese Association of Radio-Electronics, Hanoi 17063
Vietnamese Association of Radiology, Hanoi 17064
Vietnamese Association of Serum and Blood Transfusion, Hanoi 17065
Vietnamese Association of Surgery, Hanoi 17066
Vietnamese Associaton of Cast and Metallurgy, Hanoi 17067
Vietnamese Fine Arts Association, Hanoi 17068
Vietnamese Photographers Association, Hanoi 17069
Vietnamese Writers' Association, Hanoi 17070
Vietnam Union of Sientific and Technical Associations, Hanoi 17071
Viking Society for Northern Research, London 13185
VIM, Columbia 16867
VIM, Saint Paul 16862
Vintage Light Music Society, West Wickham 13186
Viola da Gamba Society, London 13187
Viola d'Amore Society, Chorleywood 13188
Virchow-Pirquet Medical Society, New York 16863
Virgil Society, London 13189
Virginia Woolf Society, Washington 16864
Visible Record and Minicomputer Society, Croydon 13190
Vísindafélag Islands, Reykjavik 06707
Visual Arts Ontario, Toronto 02982
Vitamin Society of Japan, Kyoto 08281
Vivaldi Society, London 13191
VkdL, Essen 06421
V.K.L., Hengelo 09091
VKV, Poperinge 01511
VLA, Springfield 16854
Vlaamse Chemische Vereniging, Gent 01507
Vlaamse Interuniversitaire Raad, Bruxelles 01508
Vlaamse Museumvereniging, Antwerpen 01509
Vlaamse Vereniging voor Familiekunde, Merksen 01510
Vlaams Kinesitherapeuten Verbond, Poperinge 01511

Vladimir Nabokov Society, Lawrence 16865
VLMB, Luzern 11012
VLMS, West Wickham 13186
VLMV, Dornbirn 01008
VLP, Bern 10949
VMF, Paris 05000
VMÖ, Wien 00959
VMPA, Braunschweig 06342
VMS, Northampton 13183
VMSÖ, Wien 00964
VNK, Groningen 09093
VNM, Utrecht 09102
VNS, Lawrence 16865
Vocational Industrial Clubs of America, Leesburg 16866
Vocational Instructional Materials Section, Columbia 16867
VÖA, Wien 00966
VÖÄ, Wien 00996
VÖBB, Innsbruck 00997
VÖCH, Wien 00637
V.Ö.L.B., Wien 01000
VÖLT, Wien 01001
VÖN, Wien 00961
VÖTC, Dornbirn 01002
VÖWA, Wien 00975
Volcanological Society of Japan, Tokyo 08325
Volkenrechtelijk Instituut, Utrecht 09108
Volksgesundheit Schweiz, Zürich 11028
Volkshochschulverband Baden-Württemberg e.V., Leinfelden-Oberaichen 06433
Volkskundliche Kommission für Westfalen, Münster 06434
Volks- und Betriebswirtschaftliche Vereinigung im Rheinisch-Westfälischen Industriegebiet e.V., Duisburg 06435
Volkswirtschaftliche Gesellschaft Österreich, Wien 01007
Volunteer Committees of Art Museums of Canada and the United States, Sheridan 16868
Von Braun Astronomical Society, Huntsville 16869
Von Karman Institute for Fluid Dynamics, Rhode-Saint-Genese 01512
Vorarlberger Landesmuseumsverein, Dornbirn 01008
VOS, Salt Lake City 16860
VPMS, New York 16863
VRA, Groningen 09004
V.R.B., Sint Truiden 01506
VRÖ, Wien 00962
VRS, Cape Town 10233
VS, Atlanta 16856
VSA, Bern 11016
VSA, Philadelphia 16861
VSA, Zürich 11004
VSAO, Bern 11005
VSB, Bern 11017
VSB, Brig 10957
VSE, Zürich 11006
VSN, Neerach 11020
VSVM, Zürich 10937
VSVT, Burgdorf 11008
VTB, Nijmegen 09100
VTCC, Heidelberg 06370
VVB, Köln 06402
V.v.C.W., Amsterdam 09097
V.v.M.A., Utrecht 09092
V.V.S., Heiloo 09104
VVZM, 's-Gravenhage 09107
VWGÖ, Wien 00963
VWS, Washington 16864

WAEC, Accra 06493
Wagner Society, Amsterdam 09109
Wagner Society, London 13192
Wagner Society of New York, New York 16870
Wagnervereeniging, Amsterdam 09109
WAICA, Muskegon 16912
Waikato Geological and Lapidary Society, Hamilton 09199
Waka Bungakkai, Tokyo 08444
Waldorf School Association of Alberta, Edmonton 02983
Waldviertler Heimatbund, Wissenschaftliche Sektion, Horn 01009
Waldwirtschaftsverband Schweiz, Solothurn 11029
The Walpole Society, London 13193
Walter Bagehot Research Council on National Sovereignty, Whitestone 16871
Walter Buchebner Gesellschaft, Mürzzuschlag 01010
Walter Eucken Institut e.V., Freiburg 06436
WAML, Clovis 16880
W.A.N.H.S., Devizes 13209
WAS, Baton Rouge 16921
WAS, Hollister 16922
W.A.S.C., Perth 00525
Washington Journalism Center, Washington 16872
WASP, Tokyo 08445
Watchmaking Institute of Canada, Montréal 02523
Water Authorities Association of Victoria, Melbourne 00519
Water Environment Federation, Alexandria 16873
Waterloo Historical Society, Kitchener 02984
Water Management Society, Moskva 09991
Water Quality Research Council, Lisle 16874
Water Research Centre, Marlow 13194
Water Research Commission, Pretoria 10234
Water Research Foundation of Australia, Kingsford 00520
Water Resources Congress, Arlington 16875
Water Systems Research Group, Johannesburg 10235
WAVA, West Lafayette 16923
WBBA, Bakersfield 16881
WBG, Wien 01012, 01013
WBRC, Whitestone 16871
WBT, Berkeley 16911
WCA, Oakland 16882
WCA, Philadelphia 16913
WCA, Westfield 16924
WCC, Rye 16917
WCC, Washington 16918
WCCA, Lafayette 16892
WCES, Baltimore 16916
WCLC, San Francisco 16919
WCML, Baton Rouge 16915
WCPS, Atlanta 16914
WCSZ, Lusaka 17157
WDA, Washington 16906
WE, Toledo 16909
WEA, London 13215
WEA, Saranac Lake 16894
Weather Modification Association, Fresno 16876
Webb Society, London 13195
Wedgwood Society, Fordingbridge 13196
WEE, Heidelberg 06437
Weed Science Society of America, Champaign 16877
Weed Science Society of New South Wales, Haymarket 00521
Weed Science Society of South Australia, Glenside 00522
WEF, Alexandria 16873
WEF, London 13218
WEF, New Rochelle 16925
Welding Institute of Canada, Oakville 02985
Welding Research Council, New York 16878
Wellesbourne Vegetable Research Association, Warwick 13197
Wellington Medical Research Foundation, Wellington 09200
Welsh Academy, Cardiff 11198
Welsh Arts Council, Cardiff 13198
Welsh Baptist Historical Society, Llanelli 13199
Welsh Federation of Head Teachers' Associations, Swansea 13200
Welsh Folk Dance Society, Llanfair Caereinion 13201
Welsh Folk Song Society, Aberystwyth 13202
Welsh Library Association, Aberystwyth 13203
Welsh Public Relations Association, Cardiff 11873
Weltbund für Erneuerung der Erziehung, Deutschsprachige Sektion, Heidelberg 06437
Welwyn Hall Research Association, Hatfield 13204
Wenner-Gren Center Foundation for Scientific Research, Stockholm 10605
Wenner-Gren Foundation for Anthropological Research, New York 16879
WES, Taipei 03333
Wesley Historical Society, Birmingham 13205
West African Association of Agricultural Economists, Ibadan 09246
West African Examinations Council, Accra 06493
West African Science Association, Accra 06494
West Coast Environmental Law Association, Vancouver 02986
West Coast Environmental Law Research Foundation, Vancouver 02987
West Coast Library Association, Corner Brook 02988
Westdeutsche Gesellschaft für Familienkunde e.V., Köln 06438
Westdeutscher Medizinischer Fakultätentag, Erlangen 06439
Western Association of Map Libraries, Clovis 16880
Western Australian Mental Health Association, Subiaco 00523
Western Australian Naturalists' Club, Nedlands 00524
Western Australian Shell Club, Perth 00525
Western Bird Banding Association, Bakersfield 16881
Western Board of Music, Edmonton 02989
Western Canada Water and Wastewater Association, Calgary 02990
Western Canadian Society for Horticulture, Brooks 02991
Western Cleft Lip and Palate Association, Burnaby 02992
Western College Association, Oakland 16882
Western History Association, Albuquerque 16883
Western Literature Association, Logan 16884
Western Québec Teachers Association, Hull 02993
Western Society of Malacologists, Santa Barbara 16885
Western Society of Naturalists, Moss Landing 16886
Western Surgical Association, Rochester 16887
Western Veterinary Conference, Las Vegas 16888
Westfälische Gesellschaft für Zahn-, Mund- und Kieferheilkunde, Münster 06440
West- und Süddeutscher Verband für Altertumsforschung, Mainz 06441
W.E. Upjohn Institute for Employment Research, Kalamazoo 16889
WFNS, Nijmegen 09111
WFPHA, Washington 16926
WFS, Bethesda 16927
WFSA, Cardiff 13221
WFSF, Roma 08055
WFSS, Aberystwyth 13202
W.G., Amsterdam 09110
WGA, The Plains 16898
W. Garfield Weston Foundation, Toronto 02994
WGC, Stockholm 10605
WGFAR, New York 16879
WGfF, Köln 06438
WGGB, London 13228
WHA, Albuquerque 16883
WHAC, Pasadena 16928
Whaling Museum Society, Cold Spring Harbor 16890
WHHA, Washington 16891
White House Historical Association, Washington 16891
WHO, Genève 11033
Whooping Crane Conservation Association, Lafayette 16892
W.H.R.A., Hatfield 13204
WHS, Birmingham 13205
WHS, Newport 16903
WHSHS, Wilbur Springs 16893
WIA, Springfield 16910
Wiener Arbeitskreis für Psychoanalyse, Wien 01011
Wiener Beethoven-Gesellschaft, Wien 01012
Wiener Bibliophilen-Gesellschaft, Wien 01013
Wiener Gesellschaft der Hals-Nasen-Ohren-Ärzte, Wien 01014
Wiener Gesellschaft für Innere Medizin, Wien 01015
Wiener Gesellschaft für Theaterforschung, Wien 01016
Wiener Goethe-Verein, Wien 01017
Wiener Humanistische Gesellschaft, Wien 01018
Wiener Institut für Entwicklungsfragen und Zusammenarbeit, Wien 01019
Wiener Institut für Internationale Wirtschaftsvergleiche, Wien 01020
Wiener Juristische Gesellschaft, Wien 01021
Wiener Konzerthausgesellschaft, Wien 01022
Wiener Kulturkreis, Wien 01023
Wiener Medizinische Akademie für Ärztliche Fortbildung und Forschung, Wien 01024
Wiener Psychoanalytische Vereinigung, Wien 01025
Wiener Rechtsgeschichtliche Gesellschaft, Wien 01026
Wiener Schubertbund, Wien 01027
Wiener Secession, Wien 00989
Wiener Sprachgesellschaft, Wien 01028
Wiener Volkswirtschaftliche Gesellschaft, Wien 01029
WIIAD, Morrilton 16907
Wilbur Hot Springs Health Sanctuary, Wilbur Springs 16893
Wilderness Education Association, Saranac Lake 16894
Wilderness Medical Society, Indianapolis 16895
The Wilderness Society, Washington 16896
Wilderness Watch, Sturgeon Bay 16897
Wild Goose Association, The Plains 16898
Wildlife Conservation Society, Lavington 00526
Wildlife Conservation Society of Zambia, Lusaka 17157
Wildlife Management Institute, Washington 16899
Wildlife Preservation Society of Australia, Sydney 00527
Wildlife Preservation Society of Queensland, Brisbane 00528
Wildlife Preservation Trust International, Philadelphia 16900
The Wildlife Society, Bethesda 16901
Wildlife Society of Southern Africa, Linden 10236
Wildlife Society of Zimbabwe, Harare 17183
Wildlife Sound Recording Society, Farnham 13206
Wilfrid Israel House for Oriental Art and Studies, Hazorea 07117
Wilhelm-Busch-Gesellschaft e.V., Hannover 06442
Wilhelm Furtwängler Society, London 13207
Willem Mengelberg Society, Manitowoc 16902
William Hunter Society, Newport 16903
William Morris Society, London 13208
William Morris Society in the United States, Washington 16904
Wilson Ornithological Society, Ann Arbor 16905
Wilson's Disease Association, Washington 16906
Wiltshire Archaeological and Natural History Society, Devizes 13209
Wiltshire Record Society, Trowbridge 13210
Winnipeg Foundation, Winnipeg 02995
Winnipeg Society of Financial Analysts, Winnipeg 02996
Winrock International Institute for Agricultural Development, Morrilton 16907
WIPO, Genève 11034
Wireless Preservation Society, Shanklin 13211
Wirtschaftsakademie für Lehrer e.V., Bad Harzburg 06443
Wirtschaftspolitische Gesellschaft von 1947, Frankfurt 06444
Wirtschaftspolitischer Club Bonn e.V., Bonn 06445
Wirtschaftsverband für Geodäsie und Kartographie e.V., Oberkochen 05780
Wiskundig Genootschap, Amsterdam 09110
Wissenschaftliche Ärztegesellschaft Innsbruck, Innsbruck 01030
Wissenschaftliche Arbeitsgemeinschaft für Leibeserziehung und Sportmedizin in Innsbruck, Innsbruck 01031
Wissenschaftliche Gesellschaft an der Johann Wolfgang Goethe Universität Frankfurt am Main, Frankfurt 06446
Wissenschaftliche Gesellschaft der Ärzte in der Steiermark, Graz 01032
Wissenschaftliche Gesellschaft für Europarecht, Trier 06447
Wissenschaftliche Gesellschaft für Sport und Leibeserziehung am Institut für Sportwissenschaften der Universität Salzburg, Salzburg 01033
Wissenschaftliche Gesellschaft für Theologie e.V., Tübingen 06448
Wissenschaftlicher Verein – Arbeitsgemeinschaft für Reibungs- und Verschleissfragen, Wien 00923
Wissenschaftlicher Verein für Verkehrswesen e.V., Dortmund 06449
Wissenschaftliche Vereinigung deutscher, österreichischer und schweizerischer Kriminologen, Tübingen 06140
Wissenschaftliche Vereinigung für Augenoptik und Optometrie e.V., Mainz 06450
Wissenschaftliche Vereinigung zur Pflege des Wirtschafts- und Konsumentenschutzrechts, Zürich 11030
Wissenschaftsrat, Köln 06451
Wissenschaftszentrum Berlin, Berlin 06452
Wittheit zu Bremen e.V., Bremen 06453
WJC, Washington 16872
WJG, Wien 01021
WLA, Logan 16884
W.L.P.S. of A., Sydney 00527
WMA, Ferney-Voltaire 05002
WMA, Fresno 16876
WMA, London 13216
WME, South Hadley 16908
WMFT, Erlangen 06439
WMI, Washington 16899
WMS, Cold Spring Harbor 16890
WMS, Indianapolis 16895
WMS, Manitowoc 16902
WMS/US, Washington 16904
WOHP, Philadelphia 16929
Wolverton and District Archaeological Society, Milton Keynes 13212
Women and Mathematics Education, South Hadley 16908
Women Educators, Toledo 16909
Women in Aerospace, Springfield 16910
Women in Broadcast Technology, Berkeley 16911
Women's Auxiliary of the ICA, Muskegon 16912
Women's Caucus for Art, Philadelphia 16913
Women's Caucus for Political Science, Atlanta 16914
Women's Caucus for the Modern Languages, Baton Rouge 16915
Women's Caucus of the Endocrine Society, Baltimore 16916
Women's Classical Caucus, Rye 16917
Women's College Coalition, Washington 16918
Women's Computer Literacy Center, San Francisco 16919
Women's Literary Society, Athinai 06548
Woodworkers Accident Prevention Association, Toronto 02997
Woolhope Naturalists' Field Club, Hereford 13213
Worcestershire Archaeological Society, Malvern 13214
Wordsworth-Coleridge Association, Ann Arbor 16920
Work Efficiency Institute, Helsinki 04080
Workers' Educational Association, London 13215
Workers' Music Association, London 13216
Working Committee of the Scientific Institutes for Crafts in the EEC Countries, München 05095
World Aquaculture Society, Baton Rouge 16921
World Archaeological Society, Hollister 16922
World Association for Animal Production, Roma 08054
World Association for Educational Research, W.A.E.R., Gent 01148
World Association for the School as an Instrument of Peace, Genève 11031
World Association of Societies of Pathology (Anatomic and Clinical), Tokyo 08445
World Association of Veterinary Anatomists, West Lafayette 16923
World Bureau of Metal Statistics, London 13217
World Communication Association, Westfield 16924
World Confederation of Organizations of the Teaching Profession, Morges 11032
World Education Fellowship, London 13218
World Education Fellowship, United States Section, New Rochelle 16925
World Endodontics Confederation, London 13219
World Federation for Medical Education, Edinburgh 13220
World Federation for Mental Health, Vancouver 02998
World Federation of Associations of Pediatric Surgeons, Barcelona 10404
World Federation of Neurosurgical Societies, Nijmegen 09111
World Federation of Public Health Associations, Washington 16926
World Federation of Scientific Workers, Montreuil 05001
World Federation of Societies of Anaesthesiologists, Cardiff 13221
World Future Society, Bethesda 16927
World Future Studies Federation, Roma 08055
World Health Organization, Genève 11033
World Hemophilia Aids Center, Pasadena 16928
World Intellectual Property Organization, Genève 11034
World Medical Association, Ferney-Voltaire 05002
World Medical Tennis Association, Poole 13222
World Meteorological Organization, Genève 10777
World Organization for Human Potential, Philadelphia 16929
World Population Society, Washington 16930
World Psychiatric Association, Prilly 11035
World Rabbit Science Association, Cheltenham 13223
World Rehabilitation Fund, New York 16931
World Society for Ekistics, Athinai 06549
World Society for Stereotactic and Functional Neurosurgery, Houston 16932
World Space Foundation, South Pasadena 16933
World's Poultry Science Association, Australian Branch, Seven Hills 00529
World's Poultry Science Association, United Kingdom Branch, Aylesbury 13224
World's Poultry Science Association – U.S.A. Branch, Hyattsville 16934
World Union of Catholic Teachers, Roma 08053
World Union of Pythagorean Organizations, Lutton 13225
World University Service, Flushing 16935
World Veterinary Association, Madrid 10405
World Veterinary Poultry Association, Huntingdon 13226
World War Two Studies Association, Lexington 16936
World-wide Education Service of the Bell Educational Trust, London 13227
World Wide Ethical Society, Taipei 03333
World Wildlife Fund, Madrid 10280
WOS, Ann Arbor 16905
WOTRO, Paramaribo 10433
WOTRO, 's-Gravenhage 09079

Wound, Ostomy and Continence Nurses – An Association of E.T. Nurses, Costa Mesa *16937*
WPS, Shanklin *13211*
WPS, Washington *16930*
WPSA, Hyattsville *16934*
W.P.S.Q., Brisbane *00528*
WPTI, Philadelphia *16900*
WQRC, Lisle *16874*
WRC, Arlington *16875*
WRC, Marlow *13194*
WRC, New York *16878*
WRF, New York *16931*
Writers' and Artists' Association, Madrid *10247*
Writers and Artists Union, Hanoi *17072*
Writers Development Trust, Toronto *02999*
Writers' Guild of Great Britain, London *13228*
Writers' Union of the SRR, Bucuresti *09798*
Wroclawskie Towarzystwo Naukowe, Wroclaw *09630*
W.R.S., Trowbridge *13210*
W.R.S.A., Cheltenham *13223*
WSA, Rochester *16887*
W.S.E., Athinai *06549*
WSG, Wien *01028*
WSM, Santa Barbara *16885*
WSN, Moss Landing *16886*
WSNY, New York *16870*
WSRS, Farnham *13206*
WSSA, Champaign *16877*
WSSFN, Houston *16932*
W.T. Bandy Center for Baudelaire Studies, Nashville *16938*
WTN, Wroclaw *09630*
Württembergische Bibliotheksgesellschaft, Vereinigung der Freunde der Landesbibliothek, Stuttgart *06454*
Wuhsi Academy of Textile Engineering, Wuhsi *03223*
WUPO, Lutton *13225*
Wuppertaler Kreis e.V., Köln *06455*
WUS, Flushing *16935*
WVA, Madrid *10405*
WVAO, Mainz *06450*
WVC, Las Vegas *16888*
W.V.P.A., Huntingdon *13226*
WVV, Dortmund *06449*
WW, Sturgeon Bay *16897*

Xinjiang Academy of Agricultural Sciences, Urumqi *03224*
XPLOR International, Palos Verdes Estates *16939*

YA, Yosemite National Park *16942*
Yad Izhak Ben-Zvi, Jerusalem *07118*
Y.A.S., Leeds *13230*
Yayasan Dana Normalisasi Indonesia, Bandung *06895*
YBRA, Billings *16940*
Y.D.S., Leeds *13232*
Yellowstone-Bighorn Research Association, Billings *16940*
Yeni Felsefe Cemiyeti, Istanbul *11179*
Yerevan Academy of National Economy, Yerevan *00210*
Yeshivat Dvar Yerushalayim, Jerusalem *07119*
Y.G.S., Cottingham *13233*
YIVO Institute for Jewish Research, New York *16941*
Yorkshire Agricultural Society, Harrogate *13229*
Yorkshire Archaeological Society, Leeds *13230*
Yorkshire Arts Association, Bradford *13231*
Yorkshire Dialect Society, Leeds *13232*
Yorkshire Geological Society, Cottingham *13233*
Yorkshire Philosophical Society, York *13234*
York Technology Association, Markham *03000*
York-Toronto Lung Association, Toronto *03001*
Yosemite Association, Yosemite National Park *16942*
Yosetsu Gakkai, Tokyo *08446*
Youth Science Foundation, Ottawa *03002*
Y.P.S., York *13234*
Yünnan Academy of Agricultural Sciences, Yünnan *03225*
Yugoslav Centre for Technical and Scientific Documentation, Beograd *17094*
Yugoslav Economists' Association, Beograd *17116, 17137*
Yugoslav Welding Association, Beograd *17095*
Yuki Teikei Haiku Society of the United States and Canada, San Jose *16943*
Yukon Historical and Museums Association, Whitehorse *03003*
Yukon Teachers' Association, Whitehorse *03004*
Yukon Tuberculosis and Health Association, Whitehorse *03005*

ZADI, Bonn *06466*
Zahnärztekammer Berlin, Berlin *06456*
Zahnärztekammer Bremen, Bremen *06457*
Zahnärztekammer Hamburg, Hamburg *06458*
Zahnärztekammer Niedersachsen, Hannover *06459*
Zahnärztekammer Nordrhein, Düsseldorf *06460*
Zahnärztekammer Schleswig-Holstein, Kiel *06461*
Zahnärztekammer Westfalen-Lippe, Münster *06462*
Zajednica Universitéta Jugoslavije, Beograd *17138*
Zambia Library Association, Lusaka *17158*
Zane Grey's West Society, Fort Wayne *16944*
ZDAS, Ljubljana *10112*
Združenie na Arheolozite na Makedonija, Skopje *08618*
Združenie na Folkloristite na Makedonija, Skopje *08619*
Zenkoku Nogyokozo Kaizen Kyobai Chosabu, Tokyo *08447*
Zentralausschuss für Deutsche Landeskunde e.V., Frankfurt *06463*
Zentrale für Fallstudien e.V., Erftstadt *06464*
Zentralinstitut für Kunstgeschichte, München *06465*
Zentralstelle der Büchereiarbeit in der Evangelischen Kirche in Deutschland, Göttingen *05544*
Zentralstelle für Agrardokumentation und -information, Bonn *06466*
Zentralstelle für Pilzforschung und Pilzverwertung, Waghäusel *06467*
Zentralverband der Deutschen Geographen, Heidelberg *06468*
Zentralverband der Krankengymnasten/Physiotherapeuten e.V., Köln *05550*
Zentralvereinigung der Architekten Österreichs, Wien *01034*
ZGWS, Fort Wayne *16944*
ZIH, Warszawa *09632*
Zimbabwe Agricultural Society, Harare *17184*
Zimbabwe Association for Science Education, Harare *17185*
Zimbabwe Library Association, Harare *17186*
Zimbabwe Medical Association, Harare *17187*
Zimbabwe Scientific Association, Harare *17188*
Zimbabwe Veterinary Association, Harare *17189*
Zimbabwe Writers Union, Harare *17190*
Zinc Development Association, London *13235*
Zionist Archives and Library of World Zionist Organization – American Section, New York *16945*
Zisin Gakkai, Tokyo *08448*
Znanie, Moskva *09992*
Zoological Board of Victoria, Parkville *00530*
Zoological Society of Bangladesh, Dhaka *01071*
Zoological Society of Calcutta, Calcutta *06870*
Zoological Society of Glasgow and West of Scotland, Glasgow *13236*
Zoological Society of India, Calcutta *06871*
Zoological Society of Israel, Jerusalem *07120*
The Zoological Society of Japan, Tokyo *08295*
The Zoological Society of London, London *13237*
Zoological Society of Manitoba, Winnipeg *03006*
Zoological Society of Northern Ireland, Newtownards *13238*
Zoological Society of Southern Africa, Port Elizabeth *10237*
Zoologisch-Botanische Gesellschaft in Österreich, Wien *01035*
ZPTM, Warszawa *09631*
Zrzeszenie Polskich Towarzystw Medycznych, Warszawa *09631*
Z.S.L., London *13237*
ZSSA, Port Elizabeth *10237*
Zväz Slovenských Knihovníkov a Informatikov, Bratislava *10095*
Zveza Bibliotekarskih Društev Slovenije, Ljubljana *10111*
Zveza Društev Arhitektov Slovenije, Ljubljana *10112*
Zveza Društev za Varilno Tehniko Slovenije, Ljubljana *10113*
Zveza Ekonomistov Slovenije, Ljubljana *10114*
Zveza Geografskih Društev Slovenije, Ljubljana *10115*
Zveza Pedagoskih Društev Slovenije, Ljubljana *10116*
Zveza Zgodovinskih Društev Slovenije, Ljubljana *10117*
ZWV, München *05627*
Zydowski Instytut Historyczny w Polsce, Warszawa *09632*

Subject Index

Register nach Fachgebieten

Internationales Kunst-Adressbuch

International Directory of Arts

21. Edition 1993/94

21. Ausgabe 1993.
2 Bände. 1.816 Seiten.
Gebunden.
DM 348,– /öS 2.715,– /sFr 351,–.
ISBN 3-598-23070-2

Die vorliegende 21. Ausgabe 1993/94 des *Internationalen Kunst-Adressbuchs* enthält aktuelle und detaillierte Angaben zu mehr als **150.000 Adressen, Namen, Daten und Fakten** aus dem weiten Bereich der Kunst, des Kunsthandels und des Museumswesens.

Das weltweite Adressenmaterial ist nach Kapiteln unterteilt, innerhalb der Kapitel nach Ländern, innerhalb der Länder nach Orten und innerhalb der Orte namensalphabetisch.

Mehr als **16.000 Neueinträge** und ca. **25% Änderungen** garantieren die Aktualität dieses Nachschlagewerks. Das gesamte Anschriftenmaterial wurde im Rahmen einer weltweiten Fragebogenerhebung überprüft und aktualisiert. Darüber hinaus stellten zahlreiche nationale und internationale Verbände ihre neuesten Mitgliederlisten zur Verfügung, aus denen umfangreiches neues Anschriftenmaterial entnommen wurde.

Dieses Buch ist eine unentbehrliche Arbeitsunterlage für Museen, Bibliotheken, den Kunst- und Antiquitätenhandel, Künstler, sowie für jeden, der sich beruflich oder privat mit Kunst beschäftigt.

Band I:

Museen und öffentliche Galerien • Vereinigungen

Band II:

Antiquitätenhandel und Numismatik • Galerien • Auktionatoren • Restauratoren • Kunstverleger • Kunstzeitschriften • Antiquariate und Kunstbuchhandlungen

K · G · Saur Verlag
München · New Providence · London · Paris
Reed Reference Publishing
Postfach 70 16 20 · D-81316 München · Tel. (089) 7 69 02-0

German-English Concordance to Areas of Specialization
Deutsch-Englische Konkordanz der Fachgebiete

Akustik s. Acoustics
Anästhesiologie s. Anesthetics
Anatomie s. Anatomy
Anthropolgie s. Anthropology
Antiquitäten s. Antiquities
Archäologie s. Archeology
Architektur s. Architecture
Archivwesen s. Archives
Astronomie s. Astronomy
Astrophysik s. Astrophysics
Augenheilkunde s. Ophthalmology
Bautechnik s. Civil Engineering
Behindertenpädagogik s. Education of the Handicapped
Bergbau s. Mining
Bibliotheks- und Buchwesen s. Librarianship and Book Science
Biochemie s. Biochemistry
Biologie s. Biology
Biophysik s. Biophysics
Botanik s. Botany
Botanik, Systematische s. Botany, Specific
Chemie s. Chemistry
Chirurgie s. Surgery
Darstellende Künste, Theater s. Performing Arts, Theater
Denkmalschutz, Restaurierung s. Preservation of Historical Monuments, Restoration
Dermatologie s. Dermatology
Diabetes s. Diabetes
Dokumentation s. Documentation
Elektrochemie s. Electrochemistry
Elektronik s. Electronic Engineering
Elektrotechnik s. Electrical Engineering
Endokrinologie s. Endocrinology
Energiewesen s. Energy
Entwicklungshilfe s. Developing Areas
Ernährung s. Nutrition
Erwachsenenbildung s. Adult Education
Erziehung und Ausbildung s. Education
Familienplanung s. Family Planning
Filmkunst s. Cinematography
Finanzen s. Finance
Fischerei s. Fishery
Forstwirtschaft s. Forestry
Friedensforschung s. Promotion of Peace
Futurologie s. Futurology
Gartenbau s. Horticulture
Gastroenterologie s. Gastroenterology
Geisteswissenschaften, allgemeine s. Humanities, general
Genealogie, Heraldik s. Genealogy, Heraldry
Genetik s. Genetics
Geographie s. Geography
Geologie s. Geology
Geomorphologie s. Geomorphology
Geophysik s. Geophysics
Geriatrie s. Geriatrics
Gerichtsmedizin s. Forensic Medicine
Geschichte s. History
Gesundheitswesen s. Public Health
Graphische und Dekorative Künste, Design s. Graphic and Decorative Arts, Design
Graphologie s. Graphology
Gynäkologie s. Gynecology
Hals-Nasen-Ohrenheilkunde s. Otorhinolaryngology
Hämatologie s. Hematology
Handel s. Commerce
Hauswirtschaft s. Home Economics
Höhlenkunde s. Speleology
Homöopathie s. Homeopathy
Hydrologie s. Hydrology
Hygiene s. Hygiene
Immunologie s. Immunology
Informatik, Datenverarbeitung s. Computer and Information Science, Data Processing
Ingenieurwesen s. Engineering
Innere Medizin s. Internal Medicine
Insektenkunde s. Entomology
Internationale Beziehungen s. International Relations
Kaffee, Tee, Kakao s. Coffee, Tea, Cocoa
Kardiologie s. Cardiology
Kartographie s. Cartography
Kernforschung s. Nuclear Research
Klimatechnik s. Air Conditioning
Kommunikationswissenschaft s. Communication Science
Kraftfahrzeugbau s. Automotive Engineering
Kriminologie s. Criminology
Kulturgeschichte s. Cultural History, History of Civilization
Kunst s. Arts
Kybernetik s. Cybernetics
Landvermessung, Photogrammetrie s. Surveying, Photogrammetry
Landwirtschaft s. Agriculture
Linguistik s. Linguistics
Literatur s. Literature
Logik s. Logic
Logopädie s. Logopedics
Luftfahrt, Raumfahrttechnik s. Aeronautics, Aviation, Space Technology
Malerei, Bildhauerei s. Fine Arts
Marktforschung s. Marketing
Maschinenbau s. Machine Engineering
Massenmedien s. Mass Media
Mathematik s. Mathematics
Medizin s. Medicine
Meereskunde s. Oceanography, Marine Sciences
Metallurgie s. Metallurgy

Mikrobiologie s. Microbiology
Milchwirtschaft s. Dairy Sciences
Militärwissenschaft s. Military Science
Mineralogie s. Mineralogy
Musikwissenschaft s. Musicology
Nachrichtentechnik s. Communications
Nahrungsmittel s. Food
Naturwissenschaften, allgemeine s. Natural Sciences, general
Navigation s. Navigation
Neurologie s. Neurology
Normung s. Standardization
Nuklearmedizin s. Nuclear Medicine
Numismatik s. Numismatics
Nutzpflanzenzüchtung s. Crop Husbandry
Ökologie s. Ecology
Optik s. Optics
Ornithologie s. Ornithology
Orthopädie s. Orthopedics
Pädiatrie s. Pediatrics
Paläontologie s. Paleontology
Parapsychologie s. Parapsychology
Pathologie s. Pathology
Petrochemie s. Petrochemistry
Pharmakologie s. Pharmacology
Philosophie s. Philosophy
Photographie s. Photography
Physik s. Physics
Physiologie s. Physiology
Physiotherapie s. Physical Therapy
Politologie s. Political Science
Psychiatrie s. Psychiatry
Psychoanalyse s. Psychoanalysis
Psychologie s. Psychology
Publizistik s. Journalism
Pulmologie s. Pulmonary Disease
Radiologie s. Radiology
Recht s. Law
Rehabilitation s. Rehabilitation
Reisen und Tourismus s. Travel and Tourism
Religionsphilosophie, Theologie s. Religions and Theology
Rheumatologie s. Rheumatology
Rohstoffe s. Natural Resources
Röntgenologie s. X-Ray Technology
Sozialismus s. Socialism
Sozialwissenschaften s. Social Sciences
Soziologie s. Sociology
Sport s. Sports
Stadt- und Regionalplanung s. Urban and Regional Planning
Statistik s. Statistics
Stomatologie s. Stomatology
Tabak s. Tobacco
Textilien s. Textiles
Therapeutik s. Therapeutics
Tiefenpsychologie s. Depth Psychology
Tierzüchtung s. Animal Husbandry
Toxikologie s. Toxicology
Traumatologie s. Traumatology
Tropenmedizin s. Tropical Medicine
Unfallverhütung, Sicherheitstechnik s. Safety and Protection, Safety Engineering
Unternehmensführung, Betriebswirtschaft s. Business Administration, Management
Urologie s. Urology
Venerologie s. Venereology
Verhaltensforschung s. Behavioral Sciences
Verkehrswesen s. Transport and Traffic
Versicherung s. Insurance
Verwaltung s. Public Administration
Veterinärmedizin s. Veterinary Medicine
Völkerkunde s. Ethnology
Volkskunde s. Folklore
Wasserversorgung s. Water Resources
Wein und Weinbau s. Wines and Wine Making
Werbung s. Advertising
Werkstoffkunde s. Materials Science
Wirtschaft s. Economics
Wissenschaft, allgemeine s. Science, general
Zahnheilkunde s. Dentistry
Zellbiologie, Krebsforschung s. Cell Biology, Cancer Research
Zoologie s. Zoology

Accident Prevention s. Safety and Protection, Safety Engineering

Acoustics

Australia
Australian Acoustical Society, Darlinghurst *00253*

Canada
Canadian Acoustical Association, Ottawa *02012*

China, People's Republic
Acoustical Society of China, Beijing *03113*

Denmark
Dansk Akustisk Selskab, Lyngby *03589*
Nordisk Akustik Selskab, Lyngby *03780*

France
Société Française d'Acoustique, Paris *04781*

Germany
Deutsches High-Fidelity Institut e.V., Frankfurt *05580*

Japan
Nippon Onkyo Gakkai, Tokyo *08353*

Spain
Sociedad Española de Acústica, Madrid *10376*

Sweden
Svenska Akustiska Sällskapet, Göteborg *10533*

U.S.A.
Acoustical Society of America, Woodbury *13322*
Chinese Music Society of North America, Woodridge *14742*

Adult Education

Antigua and Barbuda
Caribbean Regional Council for Adult Education, Saint John's *00035*

Argentina
Consejo Inter-Universitario Nacional, Buenos Aires *00129*
Federación Universitaria Argentina, Buenos Aires *00138*

Australia
Australian Association of Adult and Community Education, Canberra *00262*
Australian Postgraduate Federation in Medicine, Camperdown *00330*
Council of Adult Education, Melbourne *00382*

Austria
European Association for Catholic Adult Education, Linz *00586*
Evangelische Akademie in Wien, Wien *00592*
International Association of Schools of Social Work, Wien *00662*
Verband niederösterreichischer Volkshochschulen, Wien *00965*
Verband Österreichischer Volkshochschulen, Wien *00974*
Wiener Medizinische Akademie für Ärztliche Fortbildung und Forschung, Wien *01024*

Belgium
Association Européenne des Affaires Internationales, Bruxelles *01124*
Association Laïque pour l'Education et la Formation Professionnelle des Adolescents en Europe, Bruxelles *01146*
Conseil Interuniversitaire de la Communauté Française, Bruxelles *01252*
Fondation Universitaire, Bruxelles *01367*
Société Européenne pour la Formation des Ingénieurs, Bruxelles *01460*
Vlaamse Interuniversitaire Raad, Bruxelles *01508*

Bolivia
Confederación Universitaria Boliviana, La Paz *01543*
Fundación Universitaria Simon I. Patiño, Cochabamba *01545*

Brazil
Fundação Centro de Pesquisas e Estudos, Salvador *01651*

Canada
Canadian Association for Adult Education, Toronto *02025*
Elizabeth Fry Society of Alberta (Edmonton), Edmonton *02430*
Elizabeth Fry Society of Manitoba, Winnipeg *02431*
Elizabeth Fry Society of New Brunswick, Moncton *02432*
Elizabeth Fry Society of Saskatchewan, Saskatoon *02433*
Institut Canadien d'Education des Adultes, Montréal *02519*
International Congress of University Adult Education, Fredericton *02546*
International Council for Adult Education, Toronto *02548*
Ontario Association for Continuing Education, Toronto *02754*

Colombia
Instituto Colombiano para el Fomento de la Educación Superior, Bogotá *03389*

Costa Rica
Consejo Superior Universitario Centroemericano, San José *03444*

Denmark
Danmarks Handelsskoleforening, København *03579*
Foreningen for Danske Lanbrugsskoler, Greve *03735*
Nordic Federation for Medical Education, København *03778*
Société Internationale pour l'Enseignement Commercial, Odense *03824*

Egypt
Arab States Regional Centre for Functional Literacy in Rural Areas, Menoufia *03877*

Ethiopia
Association for Social Work Education in Africa, Addis Ababa *03941*

Finland
Kansantaloustieten Professorien ja Dosenttien Yhidistys, Jyväskylä *03974*
Kansanvalistusseura, Helsinki *03975*
Suomen Kansanopistoyhdistys, Helsinki *04032*
Teknillisten Oppilaitosten Opettajainliitto, Helsinki *04073*

France
Association d'Enseignement Féminin Professionnel et Ménager, Paris *04154*
Association Française de Formation, Paris *04214*
Association Française de Formation, Coopération, Promotion et Animation d'Entreprises, Paris *04215*
Association Interprofessionnelle pour la Formation Permanente dans le Commerce Textile, Paris *04296*
Association Nationale des Cours Professionnels pour les Préparateurs en Pharmacie, Paris *04308*
Association Nationale pour la Formation et la Promotion Professionnelle dans l'Industrie et le Commerce de la Chaussure et des Cuirs et Peaux, Paris *04312*
Association pour la Formation aux Professions Immobilières, Paris *04322*
Association pour la Formation des Cadres de l'Industrie et du Commerce, Paris *04323*
Association pour la Formation Professionnelle dans les Industries Céréalières, Paris *04324*
Association pour la Formation Professionnelle dans les Industries de l'Ameublement, Paris *04325*
Association pour le Développement de la Formation Professionnelle Continue dans les Industries Lourdes du Bois, Paris *04332*
Association pour le Développement de la Formation Professionnelle dans les Transports, Paris *04333*
Association Universitaire pour le Développement de l'Enseignement et de la Culture en Afrique et à Madagascar, Paris *04358*
Comité d'Education Sanitaire et Sociale de la Pharmacie Française, Paris *04392*
Comité d'Etudes et de Liaison Interprofessionnel de la Haute-Marne, Saint-Dizier *04395*
Comité d'Etudes et de Liaison Interprofessionnel du Département de l'Aisne, Saint-Quentin *04396*
Comité Européen pour l'Education des Enfants et Adolescents Précoces, Doués, Talentueux, Nîmes *04401*
Comité National des Conseillers de l'Enseignement Technique, Paris *04414*
Comité Universitaire d'Information Pédagogique, Paris *04421*
Conférence des Présidents d'Universités, Paris *04440*
Fédération Internationale pour l'Education des Parents, Sèvres *04538*
Ligue Française de l'Enseignement et de l'Education Permanente, Paris *04594*
Office National d'Information sur les Enseignements et les Professions, Paris *04606*
Syndicat National de l'Enseignement Technique, Paris *04933*
Union Professionnelle des Professeurs, Cadres et Techniciens du Secrétariat et de la Comptabilité, Paris *04995*

Germany
Arbeitsgemeinschaft der Seminarlehrer im Saarländischen Philologenverband, Ottweiler *05091*
Arbeitskreis deutscher Bildungsstätten e.V., Bonn *05137*
Arbeitskreis für Hochschuldidaktik, Hamburg *05140*
Bayerischer Volkshochschulverband e.V., München *05181*
Bildungswerk der Bayerischen Wirtschaft e.V., München *05218*
Bildungswerk der Nordrhein-Westfälischen Wirtschaft e.V., Schwelm *05220*
Bundesakademie für öffentliche Verwaltung, Bonn *05231*
Bundesarbeitsgemeinschaft katholischer Familienbildungsstätten, Düsseldorf *05235*
Carl-Cranz-Gesellschaft e.V., Weßling *05271*
Deutsche Akademie für medizinische Fortbildung, Kassel *05287*
Deutsche Kommission für Ingenieurausbildung, Düsseldorf *05460*
Deutscher Berufsverband der Sozialarbeiter und Sozialpädagogen e.V., Essen *05504*
Deutscher Stenografielehrerverband e.V., Hamburg *05542*
Deutscher Volkshochschul-Verband e.V., Bonn *05562*
Deutsche Sekretärinnen-Akademie, Düsseldorf *05572*
Europäische Staatsbürger-Akademie e.V., Bocholt *05657*
Evangelische Akademie Bad Boll, Bad Boll *05682*
Evangelische Akademie Baden, Karlsruhe *05683*
Evangelische Akademie Hofgeismar, Hofgeismar *05684*
Evangelische Akademie Loccum, Rehburg-Loccum *05685*
Evangelische Akademien in Deutschland e.V., Bad Boll *05686*
Evangelische Akademie Tutzing, Tutzing *05687*
Fördergemeinschaft für das Süddeutsche Kunststoff-Zentrum e.V., Würzburg *05703*
Forschungsinstitut für Arbeiterbildung e.V., Recklinghausen *05728*
Friedrich-Ebert-Stiftung e.V., Bonn *05764*
Gesellschaft zur Förderung des Unternehmernachwuchses e.V., Baden-Baden *05887*
Hessische Akademie für Bürowirtschaft e.V., Frankfurt *05912*
Hochschulrektorenkonferenz, Bonn *05929*
Institut der Hessischen Volkshochschulen, Frankfurt *05944*
Institut für Sozialarbeit und Sozialpädagogik, Frankfurt *05970*
Institut für technische Weiterbildung Berlin e.V., Berlin *05974*
Katholische Akademie in Bayern, München *06059*
Katholische Akademie Trier, Trier *06060*
Katholische Bundesarbeitsgemeinschaft für Erwachsenenbildung, Bonn *06062*
Katholische Erwachsenenbildung im Lande Niedersachsen e.V., Hannover *06063*
Kongreßgesellschaft für ärztliche Fortbildung e.V., Berlin *06083*
Kuratorium der Deutschen Wirtschaft für Berufsbildung, Bonn *06085*
Ländliche Erwachsenenbildung in Niedersachsen e.V., Hannover *06090*
Landesverband der Volkshochschulen Niedersachsens e.V., Hannover *06096*
Landesverband der Volkshochschulen Schleswig-Holsteins e.V., Kiel *06097*
Landesverband der Volkshochschulen von Nordrhein-Westfalen e.V., Dortmund *06098*
Leiterkreis der Katholischen Akademien, Schwerte *06102*
Management Akademie, Essen *06108*
Mathematisch-Naturwissenschaftlicher Fakultätentag, Hamburg *06112*
Niedersächsischer Bund für freie Erwachsenenbildung e.V., Hannover *06141*
Niedersächsischer Landesverband der Heimvolkshochschulen e.V., Hannover *06142*
Nordwestdeutsche Gesellschaft für ärztliche Fortbildung e.V., Steinburg *06144*
Pädagogische Arbeitsstelle des Deutschen Volkshochschul-Verbandes e.V., Frankfurt *06229*
Postakademie, Kleinheubach *06239*
Süddeutsches Kunststoff-Zentrum, Würzburg *06308*
Thomas-Morus-Akademie Bensberg, Bergisch Gladbach *06313*
Verband Bildung und Erziehung e.V., Bonn *06332*
Verband der Dozenten an Deutschen Ingenieurschulen, Mainz *06336*
Verband der Volkshochschulen des Landes Bremen, Bremen *06345*
Verband der Volkshochschulen des Saarlandes e.V., Saarbrücken *06346*
Verband der Volkshochschulen von Rheinland-Pfalz e.V., Mainz *06347*
Verband Hochschule und Wissenschaft im Deutschen Beamtenbund, Bonn *06365*
Verband Katholischer Landvolkshochschulen Deutschlands, Bad Honnef *06366*
Vereinigung der Hochschullehrer für Zahn-, Mund- und Kieferheilkunde, Frankfurt *06398*
Volkshochschulverband Baden-Württemberg e.V., Leinfelden-Oberaichen *06433*
Westdeutscher Medizinischer Fakultätentag, Erlangen *06439*
Wuppertaler Kreis e.V., Köln *06455*

Greece
Hellenic Association of University Women, Athinai *06531*

Guyana
Adult Education Association of Guyana, Georgetown *06575*

India
Indian Adult Education Association, New Delhi *06768*
Indian Association of Teachers of Library Science, Nagpur *06777*

Iraq
Arab Literacy and Adult Education Organization, Baghdad *06913*

Italy
Accademia Culturale di Rapallo, Rapallo *07135*
Accademia degli Abruzzi per le Scienze e le Arti, Chieti *07136*
Associazione Nazionale Insegnanti di Disegno, Napoli *07376*
Associazione Nazionale Professori Universitari di Ruolo, Roma *07383*
Ente Istruzione Professionale Artigiana, Milano *07654*
International Centre for Advanced Technical and Vocational Training, Torino *07709*
Istituto per la Cooperazione Universitaria, Roma *07752*

Kenya
African Association for Literacy and Adult Education, Nairobi *08461*
African Training Centre for Literacy and Adult Education, Nairobi *08474*

Luxembourg
Association Internationale d'Orientation Scolaire et Professionelle, Strassen *08578*

Mexico
Center of Regional Cooperation in Adult Education for Latin America, Patzcuaro *08711*
Centro Regional de Educación de Adultos y Alfabetización Funcional para América Latina, Pátzcuaro *08722*

Netherlands
Algemene Bond ter Bevordering van Beroepsonderwijs, 's-Gravenhage *08819*
European Bureau of Adult Education, Amersfoort *08873*
Vereniging van Docenten in Geschiedenis en Staatsrichting in Nederland, Leidschendam *09089*
Vereniging voor Hoger Beroepsonderwijs, Amsterdam *09101*

Norway
Norsk Faglaererlag, Oslo *09308*

Paraguay
Federación Universitaria del Paraguay, Asunción *09430*

Puerto Rico
Business Education Research of America, Hato Rey *09727*

South Africa
South African Association for Technical and Vocational Education, Arcadia *10180*

Sri Lanka
Asian-South Pacific Bureau of Adult Education, Colombo *10407*

Switzerland
Association Internationale des Directeurs d'Ecoles Hôtelières, Lausanne *10628*
Association Suisse des Ecoles Hôtelières, Glion-sur-Montreux *10634*
Europäisches Zentrum für die Bildung im Versicherungswesen, Sankt Gallen *10671*
Konferenz Schweizerischer Handelsschulrektoren, Chiasso *10762*
Schweizerische Direktoren-Konferenz gewerblicher Berufs- und Fachschulen, Luzern *10802*
Schweizerische Gesellschaft für kaufmännisches Bildungswesen, Genève *10827*
Schweizerische Hochschulkonferenz, Bern *10861*
Schweizerische Konferenz der Rektoren der Berufsschulen kaufmännischer Richtung, Zürich *10865*
Schweizerischer Bund für Elternbildung, Luzern *10889*
Schweizerischer Handelslehrerverein, Einsiedeln *10897*
Schweizerischer Verband der Lehrer an kaufmännischen Berufsschulen, Buchrain *10908*
Schweizerische Vereinigung für Erwachsenenbildung, Zürich *10944*
Verband der Schweizerischen Volkshochschulen, Bern *10999*
Vereinigung der Gymnastiklehrer, Zürich *11011*
Vereinigung der Lehrer für Maschinenschreiben und Bürotechnik im SKV, Luzern *11012*
Vereinigung Schweizerischer Hochschuldozenten, Würenlos *11018*

Uganda
Inter-University Council for East Africa, Kampala *11186*

United Kingdom
Agricultural Education Association, York *11215*
Association for Adult and Continuing Education, London *11257*
Association for the Study of Medical Education, Dundee *11293*
Association of Agricultural Education Staffs, Mogerhanger *11299*
Association of Blind and Partially-Sights Teachers and Students, London *11306*
Association of Career Teachers, Loughborough *11320*
Association of Colleges for Further and Higher Education, Swindon *11326*
Association of Hairdressing Teachers in Colleges of Further Education, Balby *11349*
Association of Lecturers in Accountancy, Edinburgh *11361*
Association of Marine Engineering Schools, Liverpool *11365*
Association of Painting Craft Teachers, Saint Leonards-on-Sea *11375*
Association of Teachers of Printing and Allied Subjects, Liverpool *11398*
Association of Tutors, Cambridge *11400*
Association of Tutors in Adult Education, Nottingham *11401*
Association of University Teachers, London *11403*
British Association for Commercial and Industrial Education, London *11490*
Careers Research and Advisory Centre, Cambridge *11745*
Distributive Trades Education and Training Council, London *11896*
Faculty of Teachers in Commerce, Sutton Coldfield *12054*
Guild of Guide Lecturers, London *12134*
Institute of Bankers in Scotland, Edinburgh *12214*
Institute of Professional Investigators, Blackburn *12271*
International Association of University Professors of English, Marazion *12335*
International Federation on Aging – European Office, London *12362*
Management Systems Training Council, Birmingham *12482*
National Adult School Organisation, Birmingham *12554*
National Association of Careers and Guidance Teachers, Townbridge *12565*
National Association of Principal Agricultural Education Officers, Beverley *12569*

National Association of Teachers in Further and Higher Education, London *12571*
National Federation of Continuative Teachers' Associations, Bexley *12588*
National Federation of Voluntary Literacy Schemes, London *12590*
National Foundation for Educational Research in England and Wales, Slough *12591*
National Foundry and Engineering Training Association, Glasgow *12592*
National Institute of Adult Continuing Education (England and Wales), Leicester *12596*
Scottish Association for Building Education and Training, Paisley *12894*
Scottish Business Education Council, Edinburgh *12901*
Scottish Further Education Association, Glasgow *12908*
Scottish Teachers' Nursing Association, Glasgow *12933*
Society for the Advancement of Games and Simulations in Education and Training, Loughborough *12999*
Society for the Promotion of Vocational Training and Education, Bristol *13009*
Society of Industrial Tutors, Middlesbrough *13053*

Uruguay

Centro Interamericano de Investigación y Documentación sobre Formación Profesional, Montevideo *13254*

U.S.A.

Accrediting Bureau of Health Education Schools, Elkhart *13316*
Accrediting Council for Continuing Education and Training, Richmond *13318*
Accrediting Council on Education in Journalism and Mass Communications, Lawrence *13319*
Adult Christian Education Foundation, Madison *13331*
American Academy of Nursing, Kansas City *13409*
American Academy of Sanitarians, Miami *13437*
American Assembly of Collegiate Schools of Business, Saint Louis *13462*
American Association for Adult and Continuing Education, Arlington *13464*
American Association for Chinese Studies, Columbus *13473*
American Association for Vocational Instructional Materials, Athens *13504*
American Association of Colleges for Teacher Education, Washington *13519*
American Association of Colleges of Nursing, Washington *13520*
American Association of Colleges of Pharmacy, Alexandria *13522*
American Association of Colleges of Podiatric Medicine, Rockville *13523*
American Association of Dental Schools, Washington *13528*
American Association of University Professors, Washington *13594*
American Association of University Women, Washington *13595*
American College of Healthcare Executives, Chicago *13710*
American College of Nurse-Midwives, Washington *13720*
American Conference of Academic Deans, Washington *13750*
American Council on Pharmaceutical Education, Chicago *13767*
American Occupational Therapy Association, Rockville *13963*
American Technical Education Association, Wahpeton *14188*
American Vocational Association, Alexandria *14206*
American Vocational Education Research Association, Blacksburg *14208*
Associated Schools of Construction, Peoria *14260*
Association for Continuing Higher Education, Indianapolis *14286*
Association for Education in Journalism and Mass Communication, Columbia *14292*
Association for Hospital Medical Education, Washington *14302*
Association for Library and Information Science Education, Raleigh *14313*
Association for the World University, Allentown *14361*
Association of Accredited Cosmetology Schools, Falls Church *14375*
Association of American Law Schools, Washington *14382*
Association of American Medical Colleges, Washington *14383*
Association of American Veterinary Medical Colleges, Washington *14390*
Association of Arab-American University Graduates, Belmont *14394*
Association of Collegiate Schools of Architecture, Washington *14416*
Association of Departments of Foreign Languages, New York *14428*
Association of Graduate Liberal Studies Programs, Chicago *14441*
Association of Military Colleges and Schools of the United States, Alexandria *14464*
Association of Optometric Educators, Tahlequah *14474*
Association of Schools and Colleges of Optometry, Rockville *14495*
Association of Schools of Public Health, Washington *14498*
Association of Teacher Educators, Reston *14513*
Association of Teachers of Preventive Medicine, Washington *14516*
Association of University Summer Sessions, Bloomington *14532*
Aviation Technician Education Council, Harrisburg *14565*
Broadcast Education Association, Washington *14617*
Career College Association, Washington *14645*
Center for Cuban Studies, New York *14674*
Center on Education and Training for Employment, Columbus *14713*
Council of Teaching Hospitals, Washington *14904*
Council on Hotel, Restaurant and Institutional Education, Washington *14920*
Council on Legal Education Opportunity, Washington *14922*
Council on Medical Education of the American Medical Association, Chicago *14925*
Council on Podiatric Medical Education, Bethesda *14929*
Council on Social Work Education, Alexandria *14932*
Council on Technology Teacher Education, Ypsilanti *14934*
Disability Insurance Training Council, Washington *14975*
Educational Commission for Foreign Medical Graduates, Philadelphia *15018*
Graduate Record Examinations Board, Princeton *15170*
Graduates of Italian Medical Schools, Bronx *15171*
Institute for Hospital Clinical Nursing Education, Chicago *15283*
Institute of Financial Education, Chicago *15320*
Interior Design Educators Council, Irvine *15360*
International Association for Learning Laboratories, Philadelphia *15386*
International Association of Boards of Examiners in Optometry, Bethesda *15398*
International Council on Education for Teaching, Arlington *15460*
International Graphic Arts Education Association, Pittsburgh *15481*
International Health Society, Englewood *15484*
International Society for Business Education – United States Chapter, Reston *15526*
Interstate Postgraduate Medical Association of North America, Madison *15581*
Journalism Association of Community Colleges, San Pablo *15627*
Journalism Education Association, Manhattan *15628*
Junior Achievement, Colorado Springs *15631*
Liaison Committee on Medical Education, Chicago *15667*
National Accrediting Commission of Cosmetology Arts and Sciences, Arlington *15781*
National Adult Vocational Education Association, Milan *15782*
National Association for Business Teacher Education, Reston *15796*
National Association for Industry-Education Cooperation, Buffalo *15805*
National Association for Practical Nurse Education and Service, Silver Spring *15811*
National Association for Trade and Industrial Education, Leesburg *15821*
National Association of Black Professors, Chrisfield *15832*
National Association of Health Career Schools, Saint Ann *15862*
National Association of Industrial and Technical Teacher Educators, Knoxville *15869*
National Association of State Directors of Teacher Education and Certification, Seattle *15897*
National Association of State Directors of Vocational Technical Education, Washington *15898*
National Association of Supervisors of Agricultural Education, Columbus *15908*
National Association of Supervisors of Business Education, Fort Worth *15909*
National Business Education Association, Reston *15930*
National Catholic Business Education Association, Emporia *15933*
National Consortium for Black Professional Development, Louisville *15980*
National Council for Accreditation of Teacher Education, Washington *15984*
National Council for Textile Education, Charlottesville *15991*
National Council of Local Administrators of Vocational Education and Practical Arts, Brooklyn *16003*
National FFA Organization, Alexandria *16050*
National Marine Educators Association, Pacific Grove *16103*
National Student Nurses' Association, New York *16165*
National University Continuing Education Association, Washington *16175*
National Vocational Agricultural Teachers' Association, Alexandria *16177*
National Women's Studies Association, College Park *16182*
Sea Education Association, Woods Hole *16451*
Senior Scholars, Cleveland *16463*
Society for Academic Emergency Medicine, Lansing *16489*
Society for College and University Planning, Ann Arbor *16510*
Society for Photographic Education, Dallas *16570*
Teachers Educational Council – Association of Accredited Cosmetology Schools, Falls Church *16765*
University and College Designers Association, Walkerton *16841*
University Consortium for Instructional Development and Technology, Athens *16844*
University Professors for Academic Order, Alexandria *16849*
Vocational Industrial Clubs of America, Leesburg *16866*
World University Service, Flushing *16935*

Yugoslavia

Udruženje Univerzitetskih Nastavnika i Van-Univerzitetskih Naučnik Radnika, Beograd *17131*
Zajednica Universitéta Jugoslavije, Beograd *17138*

Zambia

Adult Education Association of Zambia, Lusaka *17147*

Zimbabwe

Commonwealth Association for the Education and Training of Adults, Harare *17165*

Advertising s. a. Marketing

Argentina

Federación Interamericana del Instituto de Enseñanza Publicitaria, Buenos Aires *00136*

Austria

Österreichische Werbewissenschaftliche Gesellschaft, Wien *00928*

Canada

Canadian Advertising Research Foundation, Toronto *02015*
Canadian Public Relations Society, Ottawa *02236*
Canadian Public Relations Society (Nova Scotia), Halifax *02237*
Canadian Public Relations Society (Ottawa), Ottawa *02238*
Institute of Canadian Advertising, Toronto *02527*
Société Canadienne des Relations Publiques (Québec), Montréal *02916*

Germany

Bayerische Akademie der Werbung, München *05168*
Deutscher Kommunikationsverband BDW e.V., Bonn *05523*
Deutsche Werbewissenschaftliche Gesellschaft e.V., Bonn *05625*
Fördergemeinschaft für Absatz- und Werbeforschung, Frankfurt *05702*
Forschungsverband für den Handelsvertreter- und Handelsmaklerberuf, Köln *05740*

Ireland

Institute of Advertising Practitioners in Ireland, Dublin *06958*
Institute of Public Administration, Dublin *06964*
The Marketing Society, Dublin *07009*

Spain

Asociación Española de Marketing, Madrid *10264*

Switzerland

International Public Relations Association, Genève *10755*

United Kingdom

Advertising Association, London *11205*
Cymdeithas Cysyllyiadau Cyhoeddus Cymru, Cardiff *11873*
Society for the Development of Techniques in Industrial Marketing, Nuneaton *13001*

U.S.A.

Advertising Research Foundation, New York *13334*
American Academy of Advertising, Richmond *13380*

Aeronautics, Aviation, Space Technology

Argentina

Comisión Nacional de Investigaciones Espaciales, Buenos Aires *00119*

Australia

Royal Aeronautical Society (Australian Division), Mascot *00450*

Belgium

Aviation Research and Development Institute, Bruxelles *01159*
Von Karman Institute for Fluid Dynamics, Rhode-Saint-Genese *01512*

Bulgaria

Bulgarian Astronautical Society, Sofia *01744*

Canada

Canadian Aeronautics and Space Institute, Ottawa *02016*
Canadian Aviation Historical Society, Willowdale *02080*
Experimental Aircraft Association of Canada, Mount Albert *02446*
Institute for Aerospace Studies, Downsview *02524*
Ultralight Aircraft Association of Canada, Ottawa *02972*

China, People's Republic

Chinese Academy of Space Technology, Beijing *03156*
Chinese Society of Aeronautics and Astronautics, Beijing *03182*
Chinese Society of Astronautics, Beijing *03185*

China, Republic

Astronautical Society of the Republic of China, Taipei *03248*

Cyprus

Cyprus Astronautical Society, Limassol *03488*

Egypt

Arab Aerospace Educational Organization, Cairo *03867*

France

Advisory Group for Aerospace Research and Development, Neuilly-sur-Seine *04124*
Agence Spatiale Européenne, Paris *04127*
Association Aéronautique et Astronautique de France, Paris *04144*
Association Technique Maritime et Aéronautique, Paris *04355*
Centre National d'Etudes Spatiales, Paris *04383*
European Association for the International Space Year, Paris *04482*
Fédération Aéronautique Internationale, Paris *04506*
Fédération Nationale Aéronautique, Paris *04540*
Groupement Industriel Européen d'Etudes Spatiales, Paris *04556*
International Astronautical Association, Paris *04583*
Office National d'Etudes et des Recherches Aérospatiales, Chatillon *04605*

Germany

Deutsche Forschungs- und Versuchsanstalt für Luft- und Raumfahrt e.V., Köln *05307*
Deutsche Gesellschaft für Luft- und Raumfahrt e.V., Bonn *05377*
European Astronaut Centre, Köln *05670*
Gesellschaft zur Förderung der Segelflugforschung e.V., Freiburg *05883*

Greece

Elliniki Astronautiki Etaireia, Athinai *06504*

India

Aeronautical Society of India, New Delhi *06709*

Israel

Israel Society of Aeronautics and Astronautics, Tel Aviv *07095*

Italy

Associazione Italiana di Aeronautica e Astronautica, Roma *07297*
Centro Italiano Ricerche e Studi Trasporto Aereo, Roma *07507*
ESRIN, Frascati *07660*

Netherlands

Association for European Astronauts, Noordwijk *08823*
Nederlandse Vereniging voor Luchtvaarttechniek, Amsterdam *09024*

New Zealand

Royal Aeronautical Society, New Zealand Division, Wellington *09193*

Norway

Norsk Astronautisk Forening, Oslo *09292*

Poland

Polskie Towarzystwo Astronautyczne, Warszawa *09554*

Russia

Aircraft Building Society, Moskva *09800*
Commission for Space Toponomy, Moskva *09829*
Council on International Cooperation in Research and Uses of Outer Space, Moskva *09863*

Senegal

Agency for the Safety of Aerial Navigation in Africa, Dakar *10015*

South Africa

Aeronautical Society of South Africa, Bryanston *10119*
Royal Aeronautical Society, Southern Africa Division, Bryanston *10170*
South African Institute of Aeronautical Engineers, Sunnyside *10195*

Spain

Agrupación Astronáutica Española, Barcelona *10245*
Asociación Española Astronáutica, Madrid *10254*

Sweden

Flygtekniska Föreningen, Stockholm *10457*
Svensk Flyghistorisk Förening, Stockholm *10578*

Switzerland

Schweizerische Arbeitsgemeinschaft für Raumfahrt, Luzern *10795*
Schweizerische Vereinigung für Weltraumtechnik, Bern *10954*

United Kingdom

The Aeroplane Collection, Altrinchem *11208*
Aircraft Research Association, Bedford *11217*
Airfields Environment Federation, London *11218*
Airship Association, London *11219*
British Aviation Preservation Council, Telford *11531*
British Balloon and Airship Club, Leicester *11533*
Central Scotland Aviation Group, Edinburgh *11751*
Commonwealth Advisory Aeronautical Research Council, London *11807*
COSPAS-SARSAT, London *11845*
Historic Aircraft Association, Redhill *12171*
Merseyside Aviation Society, Liverpool *12522*
Royal Aeronautical Society, London *12765*

U.S.A.

Aerospace Department Chairmen's Association, College Station *13336*
Aerospace Education Foundation, Arlington *13337*
Airship Association – U.S., Silver Spring *13359*
American Astronautical Society, Springfield *13609*
American Aviation Historical Society, Santa Ana *13613*

American Institute of Aeronautics and Astronautics, Washington *13872*
Ancient Astronaut Society, Highland Park *14216*
Association of Balloon and Airship Constructors, San Diego *14398*
Aviation Maintenance Foundation International, Redmond *14562*
Aviation Safety Institute, Worthington *14563*
Aviation Technician Education Council, Harrisburg *14565*
Committee on Earth Observation Satellites, Washington *14802*
EAA Aviation Foundation, Oshkosh *14995*
Experimental Aircraft Association, Oshkosh *15070*
Flight Safety Foundation, Arlington *15100*
The Galactic Society, Rock Hill *15146*
High Frontier, Arlington *15209*
International Planetarium Society, Salt Lake City *15513*
National Association of Flight Instructors, Columbus *15860*
Radio Technical Commission for Aeronautics, Washington *16389*
Society of Experimental Test Pilots, Lancaster *16661*
Space Settlement Studies Program, Niagara University *16732*
United States Space Education Association, Elizabethtown *16834*
University Aviation Association, Opelika *16843*
Women in Aerospace, Springfield *16910*
World Space Foundation, South Pasadena *16933*

Yugoslavia

Savez Astronautičkih i Raketnih Organizacija Jugoslavije, Beograd *17108*

Agriculture s. a. Dairy Science; Wines and Wine Making; Crop Husbandry; Animal Husbandry

Argentina

Academia Nacional de Agronomía y Veterinaria, Buenos Aires *00040*
Asociación Argentina de la Ciencia del Suelo, Buenos Aires *00066*
Instituto Nacional de Vitivinicultura, Mendoza *00146*
Sociedad Latinoamericana de Ciencias de los Suelos, Mendoza *00200*
Sociedad Rural Argentina, Buenos Aires *00201*

Australia

Agricultural Engineering Society, Australia, Parkville *00213*
The Australian Agricultural Council, Canberra *00254*
Australian Agricultural Economics Society, Armidale *00255*
Australian Institute of Agricultural Science, Parkville *00296*
Australian Society of Soil Science, Saint Lucia *00359*
Grassland Society of Victoria, Parkville *00408*
Royal Agricultural and Horticultural Society of South Australia, Wayville *00451*
Royal Agricultural Society of Tasmania, Glenorchy *00452*
The Royal Agricultural Society of Western Australia, Claremont *00453*
Tasmanian University Agricultural Science Society, Hobart *00507*
The Tropical Grassland Society of Australia, Saint Lucia *00512*

Austria

Gesellschaft zur Förderung der industriellen Pflanzenbaus, Wien *00642*
International Society of Soil Science, Wien *00689*

Belarus

Belorussian Agricultural Academy, Gorkii *01089*

Belgium

European Association of Agricultural Economists, Bruxelles *01286*
European Committee for EC Agricultural Engineers, Bruxelles *01319*
Institut International de Recherches Betteravières, Bruxelles *01387*

Bolivia

Sociedad Rural Boliviana, La Paz *01557*

Brazil

Centro de Pesquisa Agropecuária dos Cerrados, Planaltina *01625*
Centro de Pesquisa Agropecuária do Trópico-Árido, Petrolina *01626*
Centro de Pesquisa Agropecuária do Trópico Umido, Belém *01627*
Centro Nacional de Pesquisa de Arroz e Feijão, Goiânia *01632*
Centro Nacional de Pesquisa de Mandioca e Fruticultura, Cruz das Almas *01633*
Centro Nacional de Pesquisa de Milho e Sorgo, Sete Lagoas *01634*
Centro Nacional de Pesquisa de Soja, Londrina *01636*
Centro Nacional de Pesquisa do Algodão, Campina Grande *01637*
Empresa Brasileira de Pesquisa Agropecuária, Campo Grande *01648*
Instituto de Planejamento de Pernambuco, Recife *01668*
Sociedade Nacional de Agricultura, Rio de Janeiro *01727*

Bulgaria

Academy of Agricultural Sciences, Sofia *01741*
Bulgarian Soil Society, Sofia *01763*
Scientific and Technical Union of Agricultural Specialists, Sofia *01772*
Union of Rural Economics, Sofia *01797*

Canada

Agricultural Institute of Canada, Ottawa *01823*
Canadian Agricultural Economics and Farm Management Society, Ottawa *02017*
Canadian Fertilizer Institute, Ottawa *02133*
Canadian Society of Agricultural Engineering, Ottawa *02272*
Canadian Society of Agronomy, Ottawa *02273*
Canadian Society of Soil Science, Ottawa *02296*
Conseil des Recherches en Pêche et en Agro-Alimentaire du Québec, Québec *02382*
International Agricultural Exchange (Saskatchewan), Maidstone *02537*
International Development Research Centre, Ottawa *02552*
Manitoba Institute of Agrologists, Winnipeg *02623*
New Brunswick Institute of Agrologists, Fredericton *02686*
Nova Scotia Institute of Agrologists, Truro *02734*
Ontario Farm Drainage Association, Chatham *02777*
Ontario Institute of Agrologists, Guelph *02783*
Potash and Phosphate Institute of Canada, Etobicoke *02826*
Prince Edward Island Institute of Agrologists, Charlottetown *02833*
Saskatchewan Institute of Agrologists, Saskatoon *02892*
Society of Commercial Seed Technologists, Brandon *02943*

Chile

Instituto de Investigaciones Agropecuarias, Santiago *03056*
Sociedad Agronómica de Chile, Santiago *03065*
Sociedad Nacional de Agricultura, Santiago *03101*

China, People's Republic

Beijing Academy of Agricultural Sciences, Beijing *03117*
Chekiang Academy of Agricultural Sciences, Chechiang *03124*
Chinese Academy of Agricultural Sciences, Beijing *03136*
Chinese Association of Agricultural Science Societies, Beijing *03160*
Chinese Society of Agricultural Machinery, Beijing *03183*
Crop Science Society of China, Beijing *03196*
Honan Academy of Agricultural Sciences, Honan *03202*
Hopeh Academy of Agricultural Sciences, Hopei *03203*
Hunan Academy of Agricultural Sciences, Changsha *03204*
Inner Mongolian Academy of Agricultural Sciences and Animal Husbandry, Neimengku *03205*
Jilin Academy of Agricultural Sciences, Gongzhuling *03207*
Kiangsu Academy of Agricultural Sciences, Chiangsu *03208*
Kwangtung Academy of Agricultural Sciences, Kuangtung *03209*
Liaoning Academy of Agricultural Sciences, Liaoning *03210*
Shanghai Academy of Agricultural Sciences, Shanghai *03214*
Shansi Academy of Agricultural Sciences, Shanhsi *03217*
Shantung Academy of Agricultural Sciences, Shantung *03218*
Shensi Academy of Agricultural Sciences, Shenhsi *03219*
Xinjiang Academy of Agricultural Sciences, Urumqi *03224*
Yünnan Academy of Agricultural Sciences, Yünnan *03225*

China, Republic

Agricultural Extension Association of China, Taipei *03231*
Agriculture Association of China, Taipei *03232*
Asia and Pacific Association for the Control of Tobacco, Taipei *03234*
Council of Agriculture, Taipei *03294*

Colombia

Bibliotecarios Agríolas Colombiano, Bogotá *03354*
Centro Internacional de Agricultura Tropical, Cali *03360*
Centro Nacional de Investigaciones de Café, Chinchina *03361*
Instituto Colombiano Agropecuario, Bogotá *03379*
Instituto Colombiano de la Reforma Agraria, Bogotá *03385*
Instituto Colombinao Agropecuario, Mosquera *03390*
Sociedad Colombiana de la Ciencia del Suelo, Bogotá *03408*

Costa Rica

Instituto Interamericano de Cooperación para la Agricultura, San José *03447*

Cuba

Centro de Información e Divulgación Agropecuario, La Habana *03474*
Centro de Investigaciones Pesqueras, La Habana *03475*

Czech Republic

Česká Akademie Zemědělských Věd, Praha *03501*
Česká Společnost pro Vědy Zemědělské, Lesnické, Veterinámi a Potravinářské, Praha *03530*

Denmark

Dansk Agronomforening, Klampenborg *03588*
Dansk Cerealforening, Lyngby *03598*
Dansk Forening for Jordbundsvidenskab, Frederiksberg *03620*
Det Danske Hedeselskab, Viborg *03705*
Foreningen af Licentiater, København *03732*
Foreningen for Danske Lanbrugsskoler, Greve *03735*
Kongelige Danske Landhusholdningsselskab, Frederiksberg *03760*
Nordisk Cerealist Foreningen, Lyngby *03782*

Egypt

African Soil Science Association, Cairo *03862*
Agricultural Guidance Research Institute, Dokki *03865*
Egyptian Agricultural Organization, Cairo *03887*

El Salvador

Centro Nacional de Tecnología Agropecuario, Santa Tecla *03926*
Dirección General de Investigaciones Agronómicas, Santa Tecla *03929*

Ethiopia

African Small Ruminant Research Network, Addis Ababa *03938*
Association for the Advancement of Agricultural Sciences in Africa, Addis Ababa *03942*
International Livestock Centre for Africa, Addis Ababa *03945*

Finland

Agronomiliitto, Helsinki *03949*
Maataloustuottajain Keskusliitto, Helsinki *03982*
Suomen Maataloustieteellinen Seura, Piikkio *04048*

France

Académie d'Agriculture de France, Paris *04085*
Académie des Sciences, Agriculture, Arts et Belles-Lettres d'Aix, Aix-en-Provence *04099*
Agronomes sans Frontières, Paris *04128*
Association Française de Génie Rural, Paris *04217*
Association Française pour l'Etude du Sol, Plaisir *04247*
Centre de Coopération Internationale en Recherche Agronomique pour le Développement, Paris *04365*
Centre de Coopération pour les Recherches Scientifiques Relatives au Tabac, Paris *04366*
Comité d'Etudes et de Liaison des Amendements Calcaires, Paris *04394*
International Association of Agricultural Information Specialists, Montpellier *04580*
Société d'Agriculture, Sciences, Belles-Lettres et Arts d'Orléans, Orléans *04638*
Société des Agriculteurs de France, Paris *04692*
Société Française d'Economie Rurale, Paris *04798*

Germany

AID Auswertungs- und Informationsdienst für Ernährung, Landwirtschaft und Forsten e.V., Bonn *05045*
Arbeitsgemeinschaft Getreideforschung e.V., Detmold *05115*
Arbeitsgemeinschaft Grünland und Futterbau in der Gesellschaft für Pflanzenbauwissenschaften, Giessen *05116*
Arbeitsgemeinschaft Kartoffelforschung e.V., Detmold *05121*
Arbeitsgemeinschaft zur Verbesserung der Agrarstruktur in Hessen, Wiesbaden *05132*
Arbeitswissenschaft im Landbau e.V., Stuttgart *05148*
Association of European Federations of Agro-Engineers, Bad Honnef *05156*
Dachverband wissenschaftlicher Gesellschaften der Agrar-, Forst-, Ernährungs-, Veterinär- und Umweltforschung e.V., Frankfurt *05282*
Deutsche Gesellschaft für Agrarrecht, Bonn *05314*
Deutsche Gesellschaft für Hopfenforschung e.V., Wolnzach *05364*
Deutsche Landjugend-Akademie Fredeburg, Schmallenberg *05462*
Deutsche Landwirtschafts-Gesellschaft e.V., Frankfurt *05463*
Deutscher Verband für Wasserwirtschaft und Kulturbau e.V., Bonn *05552*
Förderungsgemeinschaft der Kartoffelwirtschaft e.V., Munster *05704*
Forschungsstelle für internationale Agrar- und Wirtschaftsentwicklung e.V., Heidelberg *05739*
Gesellschaft für Agrargeschichte, Stuttgart *05796*
Gesellschaft für Bibliothekswesen und Dokumentation des Landbaues, Karlsruhe *05804*
Gesellschaft für Informationsverarbeitung in der Landwirtschaft, Stuttgart *05830*
Kommission für Erforschung der Agrar- und Wirtschaftsverhältnisse des Europäischen Ostens e.V., Giessen *06076*
Kuratorium für Technik und Bauwesen in der Landwirtschaft e.V., Darmstadt *06088*
Max-Eyth-Gesellschaft für Agrartechnik e.V., Darmstadt *06113*
Rationalisierungs-Kuratorium für Landwirtschaft e.V., Osterrönfeld *06257*
VDI-Gesellschaft Agrartechnik, Düsseldorf *06319*
Verband Deutscher Agrarjournalisten e.V., Bonn *06348*
Verband Deutscher Landwirtschaftlicher Untersuchungs- und Forschungsanstalten, Darmstadt *06354*
Zentralstelle für Agrardokumentation und -information, Bonn *06466*

Ghana

African Commission on Agricultural Statistics, Accra *06470*
African Network for the Development of Ecological Agriculture, Accra *06473*

Greece

Panellinios Omospondia Syndesmon Geoponon, Athinai *06542*

Grenada

Caribbean Agricultural and Rural Development, Advisory and Training Service, Saint George's *06550*

Guyana

Guyana Society, Georgetown *06578*

Hungary

Magyar Agrártudományi Egyesület, Budapest *06612*

Iceland

Búnadarfélag Islands, Reykjavik *06679*

India

Agri-Horticultural Society of India, Calcutta *06710*
Agri-Horticultural Society of Madras, Madras *06711*
Indian Council of Agricultural Research, New Delhi *06785*
Indian Dairy Association, New Dehli *06790*
Indian Society for Nuclear Techniques in Agriculture and Biology, New Delhi *06814*
The Indian Society of Agricultural Economics, Bombay *06815*
Indian Society of Soil Science, New Delhi *06820*

Ireland

Agricultural Science Association, Dublin *06921*
Irish Grassland and Animal Production Association, Belclare, Tuam *06987*
National Development Association, Dublin *07016*

Israel

Agricultural Research Organization, Bet Dagan *07041*
Israel Society of Soil Mechanics and Foundation Engineering, Tel Aviv *07102*
Israel Society of Soil Science, Rehovoth *07103*

Italy

Accademia Agraria, Pesaro *07125*
Accademia degli Incamminati, Modigliana *07138*
Accademia di Agricoltura di Torino, Torino *07153*
Accademia di Agricoltura Scienze e Lettere, Verona *07154*
Accademia Economico Agraria dei Georgofili, Firenze *07177*
Accademia Georgica, Treia *07185*
Accademia Nazionale di Agricoltura, Bologna *07214*
Association of Agricultural Research Institutions in the Near East and North Africa, Roma *07252*
Associazione Italiana di Genio Rurale, Padova *07309*
Associazione per l'Agricoltura Biodinamica, Milano *07390*
Centro di Studi e Ricerche di Museologia Agraria, Milano *07442*
Centro di Studi Pratici di Agricoltura, Fondazione Fratelli Gustavo e Severino Navarra, Ferrara *07452*
Centro Provinciale Impiego Combinato Tecniche Agricole, Bologna *07530*
Centro Studi della Cooperazione nel Veneto, Vicenza *07551*
Centro Studi Ricerche e Documentazioni per l'Agricoltura Siciliana, Roma *07589*
Commission for Controlling the Desert Locust in the Eastern Region of its Distribution Area in South West Asia, Roma *07622*
Current Agricultural Research Information System, Roma *07631*
Ente Eugenio e Claudio Faina per l'Istruzione Professionale Agraria, Roma *07651*
Federazione Italiana Dottori in Agraria e Forestali, Roma *07692*
International Centre for Advanced Technical and Vocational Training, Torino *07709*
International Committee for Recording the Productivity of Milk Animals, Roma *07712*
Istituto Agronomico per l'Oltremare, Firenze *07724*
Società Italiana di Agronomia, Bologna *07830*
Società Italiana di Economia Agraria, Bologna *07864*
Tecnagro, Roma *08031*

Japan

Asian Rural Institute – Rural Leaders Training Center, Tochigi *08108*
Kaigai Nogyo Kaihatsu Kyokai, Tokyo *08151*
Nihon Sakumotsu Gakkai, Tokyo *08252*
Nihon Sanshi Gakkai, Ibaraki *08254*
Nippon Bitamin Gakkai, Kyoto *08281*
Nippon Nogei Kagaku Kai, Tokyo *08348*
Nippon Nogyo-Kisho Gakkai, Tokyo *08349*
Nogyo-Ho Gakkai, Tokyo *08399*
Sapporo Norin Gakkai, Sapporo *08407*
Seito Kogyokai Jimubu Chosaka, Tokyo *08409*
Tochi Seidoshi Gakkai, Tokyo *08432*
Zenkoku Nogyokozo Kaizen Kyobai Chosabu, Tokyo *08447*

Kenya

African Association for Biological Nitrogen Fixation, Nairobi *08459*
Agricultural Society of Kenya, Nairobi *08477*

Korea, Democratic People's Republic
Academy of Agricultural Science, Pyongyang *08508*

Korea, Republic
Asian Society of Agricultural Economists, Seoul *08524*

Macedonia
Sojuz na Zemjodelskite Inženeri i Tehničari na SR Makedonija, Skopje *08617*

Malaysia
Commonwealth Association of Scientific Agricultural Societies, Kuala Lumpur *08641*

Malta
Agrarian Society, Valletta *08657*

Mauritius
Société de Technologie Agricole et Sucrière de l'Ile Maurice, Réduit *08676*

Mexico
Centro Internacional de Mejoramiento de Maíz y Trigo, México *08719*
Instituto Nacional de Investigaciones Agrícolas, México *08739*
Sociedad Agronómica Mexicana, México *08745*

Morocco
Association of African Faculties of Agriculture, Rabat *08803*
Conférence Internationale des Directeurs et Doyens des Etablissements d'Enseignement Supérieur et Facultés d'Expression Française des Sciences de l'Agriculture et de l'Alimentation, Rabat *08807*

Netherlands
Centrum voor Plantenveredelings- en Reproduktieonderzoek, Wageningen *08840*
European Grassland Federation, 's-Gravenhage *08889*
Instituut voor Mechanisatie, Arbeid en Gebouwen, Wageningen *08912*
International Society for Soilless Culture, Wageningen *08941*
Koninklijk Genootschap voor Landbouwwetenschap, Wageningen *08962*
Stichting Nederlands Agronomisch-Historisch Instituut, Groningen *09074, 09075*

New Zealand
Agronomy Society of New Zealand, Canterbury *09115*
New Zealand Institute of Agricultural Science, Christchurch *09153*
New Zealand Society of Animal Production, Hamilton *09177*
New Zealand Society of Soil Science, Canterbury *09180*
Royal Agricultural Society of New Zealand, Wellington *09194*

Nigeria
African Centre for Fertilizer Development, Lagos *09213*
African Feed Resources Research Network, Harare *09216*
Alley Farming Network for Tropical Africa, Ibadan *09223*
West African Association of Agricultural Economists, Ibadan *09246*

Norway
Norges Landbruksvitenskapelige Forskningsråd, Oslo *09286*
Norsk Naturforvalterforbund, Oslo *09332*

Panama
Association for Cooperation in Banana Research in the Caribbean and Tropical America, Panamá City *09412*

Paraguay
Servicio Técnico Interamericano de Cooperación Agrícola, Asunción *09433*

Peru
Asociación de Bibliotecarios y Documentalistas Agrícolas del Perú, Lima *09448*
Servicio de Investigación y Promoción Agraria, Lima *09474*
Sociedad Nacional Agraria, Lima *09480*

Philippines
Agricultural Economics Society of South East Asia, Manila *09489*
Asian Association of Agricultural Colleges and Universities, Laguna *09495*
Asian NGO Coalition for Agrarian Reform and Rural Development, Manila *09500*
Asian-Pacific Weed Science Society, Laguna *09501*
Asian Rice Farming Network, Manila *09504*
Crop Science Society of the Philippines, Laguna *09516*
Society for the Advancement of Research, Laguna *09536*

Poland
Polskie Towarzystwo Gleboznawcze, Warszawa *09574*

Portugal
Associação Central de Agricultura Portuguesa, Lisboa *09636*
Sociedade de Ciências Agrárias de Portugal, Lisboa *09679*
Sociedade Portuguesa de Nutrição e Alimentação Animal, Lisboa *09707*
Sociedade Portuguesa de Reprodução Animal, Lisboa *09713*

Romania
Academia de Stiinte Agricole si Silvice, Bucuresti *09734*
Societatea Nationala Romana pentru Stiinta Solului, Bucuresti *09790*

Russia
Agro-Industrial Society, Moskva *09799*
Flour, Fodder and Grain Storage Society, Moskva *09869*
Russian Academy of Agricultural Sciences, Moskva *09950*

Saudi Arabia
Commission for Controlling the Desert Locust in the Near East, Jeddah *10002*

Senegal
Association of Maize Researchers in Africa, Dakar *10022*

Slovakia
Slovenská Spoločnost pre Vedy Polnohospodarske, Lesnické a Potravinárské, Ivanka pri Dunaji *10091*

South Africa
Organic Soil Association of South Africa, Parklands *10160*
Soil Science Society of Southern Africa, Pretoria *10176*

Spain
Asociación de Peritos Agrícolas del Estado, Madrid *10251*
Consejo General de Oficiales de Ingenieros Técnicos Agrícolas de España, Madrid *10301*
Institut Agrícola Català de Sant Isidre, Barcelona *10313*

Sudan
Agricultural Research Corporation, Wad Medani *10423*
Arab Centre for Agricultural Documentation and Information, Khartoum *10424*

Suriname
Centre for Agricultural Research in Suriname, Paramaribo *10430*

Sweden
International Association on Mechanization of Field Experiments, Uppsala *10474*
Kungliga Skogs- och Lantbruksakademien, Stockholm *10489*
Skogs- och Jordbrukets Forskningsråd, Stockholm *10520*

Switzerland
Collaborative International Pesticides Analytical Council, Wädenswil *10647*
Fédération des Sociétés d'Agriculture de la Suisse Romande, Lausanne *10700*
Internationales Kali-Institut, Basel *10739*
Internationale Vereinigung für Saatgutprüfung, Zürich *10745*
Schweizerische Gesellschaft für Agrarwirtschaft, Zürich *10809*
Schweizerischer Verband der Ingenieur-Agronomen und der Lebensmittel-Ingenieure, Zollikofen *10907*
Schweizerischer Verband für Landtechnik, Riniken *10914*
Union Internationale pour la Protection des Obtentions Végétales, Genève *10994*

Syria
Arab Agronomists Union, Damascus *11037*

Tanzania
East African Agriculture and Forestry Research Organization, Arusha *11051*

Thailand
Agricultural Science Society of Thailand, Bangkok *11065*
Asia and Pacific Commission on Agricultural Statistics, Bangkok *11072*
Asia and Pacific Plant Protection Commission, Bangkok *11073*
ESCAP/FAO/UNIDO Fertilizer, Development and Information Network for Asia and the Pacific, Bangkok *11092*

Trinidad and Tobago
Agricultural Society of Trinidad and Tobago, Port of Spain *11105*
Caribbean Agricultural Research and Development Institute, Saint Augustine *11106*
Caribbean Agro-Economic Society, Saint Augustine *11107*
Caribbean Information System for the Agricultural Sciences, Saint Augustine *11115*
Eastern Caribbean Institute of Agriculture and Forestry, Arima *11120*
Tobago District Agricultural Society, Scarborough *11125*

Turkey
Türkiye Ziraat Odalari Birligi, Ankara *11162*

Uganda
East African Agricultural Economics Society, Kampala *11184*

Ukraine
Ukrainian Academy of Agrarian Sciences, Kiev *11191*

United Kingdom
Abergavenny and Border Counties Agricultural Society, Abergavenny *11195*
Agricultural and Food Research Council, Swindon *11213*
Agricultural Economics Society, Belfast *11214*
Agricultural Education Association, York *11215*
Agricultural Research Institute of Northern Ireland, Hillsborough *11216*
Association of Agricultural Education Staffs, Mogerhanger *11299*
Association of Show and Agricultural Organisations, Newark *11390*
Ayrshire Agricultural Association, Ayr *11418*
Bio-dynamic Agricultural Association, Stourbridge *11442*
Border Union Agricultural Society, Kelso *11461*
British Agricultural History Society, Reading *11474*
British Association of Seed Analysts, London *11521*
British Grassland Society, Maidenhead *11570*
British Society for Agricultural Labour Science, Reading *11648*
British Society for Plant Pathology, Leeds *11663*
British Society of Animal Production, Penicuik *11680*
British Society of Soil Science, Reading *11690*
C.A.B. International, Wallingford *11720*
CAB International Bureau of Crop Protection, Wallingford *11721*
CAB International Bureau of Horticulture and Plantation Crops, Wallingford *11722*
CAB International Bureau of Plant Breeding and Genetics, Wallingford *11725, 11726*
C.A.B. International Department of Dairy Science and Technology, Wallingford *11727*
C.A.B. International Department of Soils, Wallingford *11728*
Collaborative International Pesticides Analytic Council, London *11795*
Commonwealth Association on Surveying and Land Economy, London *11813*
Commonwealth Bureau of Nutrition, Aberdeen *11814*
County Planning Officers Society, Leicester *11863*
East of England Agricultural Society, Peterborough *11921*
English Guernsey Cattle Society, Chesham *11963*
Essex Agricultural Society, Chelmsford *11975*
Farm and Food Society, London *12057*
Farm Buildings Association, Oxford *12058*
Glasgow Agricultural Society, Ayr *12117*
Henry Doubleday Research Association, Coventry *12158*
Institute for Animal Health, Newbury *12204*
Institute of Food Research, Reading *12238*
Institution of Agricultural Engineers, Bedford *12292*
International Association of Agricultural Museums, Reading *12326*
Kent County Agricultural Society, Maidstone *12429*
Merioneth Agricultural Society, Dolgellau *12520*
National Association of Principal Agricultural Education Officers, Beverley *12569*
Newark and Nottinghamshire Agricultural Society, Newark *12624*
North Somerset Agricultural Society, Bristol *12641*
North Tyne and Redesdale Agricultural Society, Morpeth *12643*
North Wales Agricultural Society, Portdinorwic *12644*
Pembrokeshire Agricultural Society, Haverfordwest *12685*
Royal Agricultural Society of England, Stoneleigh Park *12767*
Royal Bath and West of England Society, Shepton Mallet *12773*
Royal Cornwall Agricultural Association, Wadebridge *12797*
Royal Highland and Agricultural Society of Scotland, Edinburgh *12808*
Royal Isle of Wight Agricultural Society, Newport *12825*
Royal Jersey Agricultural and Horticultural Society, Saint Helier *12826*
Royal Lancashire Agricultural Society, Preston *12827*
Royal Norfolk Agricultural Association, Norwich *12836*
Royal Ulster Agricultural Society, Belfast *12872*
Royal Welsh Agricultural Society, Powys *12874*
Rutland Agricultural Society, Rutland *12880*
Scottish Society for Crop Research, Dundee *12928*
Society of Feed Technologists, Reading *13046*
Soil Association, Bristol *13073*
South of England Agricultural Society, Haywards Heath *13081*
Staffordshire Agricultural Society, Stafford *13090*
Suffolk Agricultural Association, Ipswich *13105*
Surrey County Agricultural Society, Godalming *13111*
Three Counties Agricultural Society, Malvern *13132*
Tobacco Research Council, London *13135*
World's Poultry Science Association, United Kingdom Branch, Aylesbury *13224*
Yorkshire Agricultural Society, Harrogate *13229*

Uruguay
Asociación Rural del Uruguay, Montevideo *13247*
Centro de Investigaciones Agrícolas Alberto Boerger, La Estanzuela *13251*

U.S.A.
Agricultural Communicators in Education, Gainesville *13350*
Agricultural History Society, Washington *13351*
Agriservices Foundation, Clovis *13352*
Alternative Energy Resources Organization, Helena *13368*
American Agricultural Economics Association, Ames *13446*
American Agricultural Law Association, Fayetteville *13447*
American Association for Agricultural Education, Ithaca *13466*
American Farm Bureau Research Foundation, Park Ridge *13805*
American Poultry Historical Society, Morgantown *14012*
American Society of Agricultural Engineers, Saint Joseph *14097*
American Society of Agronomy, Madison *14098*
American Society of Irrigation Consultants, Brentwood *14138*
Association for Living Historical Farms and Agricultural Museums, Santa Fe *14316*
Association of American Seed Control Officials, Salem *14387*
The Association of Engineering Firms Practicing in the Geosciences, Silver Spring *14432*
Association of Official Seed Analysts, Lincoln *14472*
Bio-Integral Resource Center, Berkeley *14587*
Center for Sustainable Agriculture, Davis *14704*
Consultative Group on International Agricultural Research, Washington *14852*
Council for Agricultural Science and Technology, Ames *14868*
Council for Tobacco Research – U.S.A., New York *14887*
Crop Science Society of America, Madison *14942*
Desert Protective Council, Valley Center *14966*
Farm Foundation, Oak Brook *15076*
Great Plains Agricultural Council, Fort Collins *15178*
Institute for Alternative Agriculture, Greenbelt *15268*
International Association of Agricultural Economists, Oak Brook *15396*
International Crane Foundation, Baraboo *15462*
International Erosion Control Association, Steamboat Springs *15472*
International Fertilizer Development Center, Muscle Shoals *15476*
International Turfgrass Society, Blacksburg *15569*
International Weed Science Society, Corvallis *15573*
IRI Research Institute, Stamford *15584*
National Association of Colleges and Teachers of Agriculture, Urbana *15842*
National Association of Extension 4-H Agents, Athens *15858*
National Association of Supervisors of Agricultural Education, Columbus *15908*
National FFA Organization, Alexandria *16050*
National Plant Board, Montgomery *16129*
National Rural Development Institute, Bellingham *16143*
National Vocational Agricultural Teachers' Association, Alexandria *16177*
Native Seeds/Search, Tucson *16186*
Potash and Phosphate Institute, Atlanta *16343*
Potato Association of America, Hancock *16344*
Rodale Institute, Emmaus *16424*
Soil and Water Conservation Society, Ankeny *16716*
Soil Science Society of America, Madison *16717*
Sulphur Institute, Washington *16752*
Weed Science Society of America, Champaign *16877*
Winrock International Institute for Agricultural Development, Morrilton *16907*

Venezuela
Asociación de Agrimensores de Venezuela, Caracas *16967*
Centro Nacional de Investigaciones Agropecuarias, Maracay *16991*

Vietnam
Vietnamese Association of Agriculture, Hanoi *17037*

Yugoslavia
Jugoslovensko Društvo za Proučavanje Zemljišta, Zemun *17100*

Zambia
Agricultural Research Council of Zambia, Lusaka *17150*

Zimbabwe
Agricultural Research Council of Zimbabwe, Harare *17162*
Crop Science Society of Zimbabwe, Harare *17167*
Zimbabwe Agricultural Society, Harare *17184*

Air Conditioning

Argentina
Asociación Argentina del Frío, Buenos Aires *00068*

Denmark
Dansk Selskab for Opvarmnings- og Ventilationsteknik, København *03685*

France
Association Française du Froid, Paris *04239*
Association Technique de la Réfrigération et de l'Equipement Ménager, Paris *04351*
Institut International du Froid, Paris *04575*

Germany

Arbeitsgemeinschaft Fernwärme e.V., Frankfurt *05098*
Deutscher Kälte- und Klimatechnischer Verein e.V., Stuttgart *05521*
Forschungsrat Kältetechnik e.V., Frankfurt *05735*
Forschungsvereinigung für Luft- und Trocknungstechnik e.V., Frankfurt *05749*
Gaswärme-Institut e.V., Essen *05768*

Italy

Associazione Criogenica Italiana, Genova *07265*
Associazione Italiana Condizionamento dell'Aria, Riscaldamento e Refrigerazione, Milano *07292*
Associazione Termotecnica Italiana, Torino *07409*
Comitato Termotecnico Italiano, Torino *07620*

Japan

Kuki-Chowa Eisei Kogakkai, Tokyo *08174*

U.S.A.

American Society of Heating, Refrigerating and Air-Conditioning Engineers, Atlanta *14130*

Allergology s. Immunology

Anatomy

Argentina

Sociedad Argentina de Anatomía Normal y Patológica, Buenos Aires *00154*

Bulgaria

Society of Bulgarian Anatomists, Histologists and Embryologists, Sofia *01777*

Canada

Canadian Association of Anatomists, London *02046*

Chile

Sociedad de Anatomía Normal y Patológica de Chile, Santiago *03087*

China, People's Republic

Chinese Society of Anatomy, Beijing *03184*

Czech Republic

Czech Anatomical Society, Praha *03537*

El Salvador

Asociación Centroamericana de Anatomía, San Salvador *03918*

France

Association des Anatomistes, Vandoeuvre-les-Nancy *04165*
Société Anatomique de Paris, Paris *04628*
Syndicat National des Médecins Anatomo-Cyto-Pathologistes Français, Paris *04948*

Georgia

Georgian Society of Patho-Anatomists, Tbilisi *05024*

Germany

Anatomische Gesellschaft, Lübeck *05074*
Gesellschaft für Wirbelsäulenforschung, Frankfurt *05869*
Vereinigung Deutscher Neuropathologen und Neuroanatomen, Giessen *06406*

Hungary

Magyar Anatómusok, Histologusok és Embryologusok Társasága, Budapest *06615*

Italy

Accademia Anatomico-Chirurgica, Perugia *07129*
Società Italiana di Anatomia, Firenze *07832*
Società Italiana di Anatomia Patologica, Messina *07833*

Japan

Nippon Kaibo Gakkai, Tokyo *08320*

Netherlands

Nederlandse Anatomen Vereniging, Amsterdam *08986*
Nederlandse Patholoog Anatomen Vereniging, Maastricht *09000*

Poland

Polskie Towarzystwo Anatomiczne, Warszawa *09551*

Portugal

Sociedade Anatómica Luso-Hispano-Americana, Lisboa *09673*
Sociedade Anatómica Portuguesa, Coimbra *09674*
Sociedade Portuguesa Veterinária de Anatomia Comparativa, Lisboa *09715*

Russia

National Scientific Medical Society of Anatomists, Histologists and Embryologists, Moskva *09912*
Scientific Medical Society of Anatomists-Pathologists, Moskva *09967*

Spain

Sociedad Anatómica Española, Granada *10370*

Sweden

Svensk Förening för Anatomi, Umeå *10579*

United Kingdom

Anatomical Society of Great Britain and Ireland, Saint Andrews *11227*
British Association of Clinical Anatomists, Manchester *11503*
Ergonomics Society, Loughborough *11972*
International Anatomical Nomenclature Committee, London *12322*

U.S.A.

American Association of Anatomists, New Orleans *13506*
American Association of Veterinary Anatomists, Auburn *13598*

Anesthetics

Argentina

Federación Argentina de Asociaciones de Anestesiología, Buenos Aires *00134*

Australia

Australian Society of Anaesthesists, Paddington *00347*

Austria

Österreichische Gesellschaft für Akupunktur und Aurikulotherapie, Wien *00751*
Österreichische Gesellschaft für Anästhesiologie, Reanimation und Intensivtherapie, Wien *00754*

Belgium

Société Belge d'Anesthésie et de Réanimation, Bierges *01424*

Bulgaria

Bulgarian Society of Anaesthesiology and Resuscitation, Sofia *01756*

Canada

Canadian Anaesthetists' Society, Toronto *02021*
Canadian Association for Narcolepsy, Toronto *02037*
Ontario Society of Clinical Hypnosis, Toronto *02798*

Chile

Sociedad de Anestesiología de Chile, Santiago *03088*

China, Republic

The Society of Anaesthesiologist of the Republic of China, Taipei *03326*
Society of Chinese Acupuncture and Cauterizing, Taipei *03327*

Colombia

Confederacion Latinoamericana de Sociedades de Anestesiología, Bogotá *03369*

Czech Republic

Česká Společnost Anesteziologie a Resuscitace, Praha *03517*

Denmark

Dansk Anaestesiologisk Selskab, Arhus *03590*

El Salvador

Sociedad de Anestiología de El Salvador, San Salvador *03930*

France

Association Française d'Acupuncture, Paris *04211*
Association Scientifique des Médecins Acupuncteurs de France, Paris *04348*
European Academy of Anaesthesiology, Strasbourg *04477*
Société Française d'Anesthésie et de Réanimation, Paris *04783*
Syndicat National des Anesthésistes-Réanimateurs Français, Paris *04937*
Syndicat National des Médecins Acupuncteurs de France, Paris *04947*

Germany

Deutsche Gesellschaft für Anästhesiologie und Intensivmedizin, Nürnberg *05319*

Greece

Elliniki Anaisthisiologiki Etaireia, Athinai *06503*

Iceland

Svaefingalaeknafélag Islands, Reykjavik *06702*

Israel

Israel Society of Anesthesiologists, Haifa *07097*

Italy

Associazione Anestesisti Rianimatori Ospedalieri Italiani, Napoli *07255*
Associazione Italiana di Anestesia Odontostomatologica, Bologna *07298*
European Federation for the Advancement of Anaesthesia in Dentistry, Bologna *07679*
Società Italiana di Agopuntura, Torino *07829*
Società Italiana di Anestesia, Analgesia, Rianimazione e Terapia Intensiva, Firenze *07835*
Società Italiana di Anestesiologia Rianimazione e Terapia del Dolore, Bari *07836*

Japan

Nihon Masui Gakkai, Tokyo *08241*

Malaysia

Malaysian Society of Anaesthesiologists, Kuala Lumpur *08653*

Netherlands

Nederlandse Vereniging voor Anesthesiologie, Utrecht *09016*

Norway

Nordisk Anaesthesiologisk Forening, Oslo *09280*
Norsk Anestesiologisk Forening, Gjövik *09290*

Poland

Towarzystwo Anestezjologów Polskich, Poznan *09618*

Portugal

Sociedade Portuguesa de Anestesiologia, Lisboa *09690*

Romania

Societatea de Anestezie si Terapie Intensiva, Bucuresti *09755*

Russia

Federation of Anaesthesiologists and Reanimatologists, Moskva *09867*

South Africa

The South African Society of Anaesthesists, Rosebank *10217*

Spain

Sociedad Española de Anestesiología y Reanimación, Madrid *10377*

Sweden

Svensk Förening för Anestesi och Intensivvård, Linköping *10580*

Turkey

Türk Anesteziyoloji ve Reanimasyon Derneği, Istanbul *11150*

United Kingdom

Acupuncture Scientific and Clinical Advisory Group, Bristol *11204*
Anaesthetic Research Society, Liverpool *11226*
Association of Anaesthetists of Great Britain and Ireland, London *11300*
European Association of Cardiothoracic Anaesthesiologists, Papworth Everard *11994*
Intractable Pain Society of Great Britain and Ireland, Derby *12409*
The Royal College of Anaesthetists, London *12778*
Society for the Advancement of Anaesthesia in Dentistry, London *12998*
World Federation of Societies of Anaesthesiologists, Cardiff *13221*

U.S.A.

American Board of Anesthesiology, Hartford *13628*
American Dental Society of Anesthesiology, Chicago *13781*
American Osteopathic College of Anesthesiologists, Independence *13982*
American Society for Advancement of Anesthesia in Dentistry, Coral Springs *14044*
American Society of Anesthesiologists, Park Ridge *14099*
American Society of Regional Anesthesia, Richmond *14164*
Association of University Anesthesiologists, Seattle *14526*
Council on Accreditation of Nurse Anesthesia Educational Programs/Schools, Park Ridge *14907*
Dannemiller Memorial Educational Foundation, San Antonio *14954*
International Anesthesia Research Society, Cleveland *15373*
National Committee on the Treatment of Intractable Pain, Washington *15965*
Society for Obstetric Anesthesia and Perinatology, Houston *16558*
Society of Cardiovascular Anesthesiologists, Richmond *16648*

Vietnam

Vietnamese Association of Anaesthesiology, Hanoi *17038*

Animal Husbandry

Australia

Australian Society of Animal Production, Mount Barker *00348*

Brazil

Asociación Latinoamericana para la Producción Animal, Porto Alegre *01585*

Chile

Commission on Livestock Development in Latin America and the Caribbean, Santiago *03044*

China, Republic

Asian-Australian Association of Animal Production Societies, Miaoli *03236*

Germany

EAAP-FAO Global Data Bank for Animal Genetic Resources, Hannover *05636*

Ireland

Irish Grassland and Animal Production Association, Belclare, Tuam *06987*

Italy

International Centre for Advanced Technical and Vocational Training, Torino *07709*
World Association for Animal Production, Roma *08054*

Russia

Russian Society of Genetics and Breeders, Moskva *09965*

Spain

Sociedad Veterinaria de Zootecnica de España, Madrid *10402*

United Kingdom

Aberdeen-Angus Cattle Society, Perth *11194*
Animal Breeding Research Organization, Edinburgh *11234*
C.A.B. International Division of Animal Production, Wallingford *11730*
Dartmoor Pony Society, Newton Abbot *11877*
Rare Poultry Society, Great Glenn *12744*

U.S.A.

Society for the Conservation of Bighorn Sheep, West Covina *16594*

Anthropology

Algeria

Centre de Recherches Anthropologiques, Préhistoriques et Ethnographiques, Alger *00010*

Argentina

Sociedad Argentina de Antropología, Buenos Aires *00155*

Australia

Anthropological Society of New South Wales, Sydney *00214*
Anthropological Society of Queensland, Saint Lucia *00215*
Anthropological Society of South Australia, Adelaide *00216*
Anthropological Society of Western Australia, Nedlands *00217*
Archaeological and Anthropological Society of Victoria, Melbourne *00218*
Indo-Pacific Prehistory Association, Canberra *00410*

Austria

Anthropologische Gesellschaft in Wien, Wien *00544*
Anthroposophische Gesellschaft in Österreich, Wien *00545*
Anthroposophische Gesellschaft in Wien, Wien *00546*
Gesellschaft für Phänomenologie und Kritische Anthropologie, Wien *00629*

Belgium

Centre pour l'Etude des Problèmes du Monde Musulman Contemporain, Bruxelles *01205*
European Anthropological Association, Bruxelles *01270*
Fondation Egyptologique Reine Elisabeth, Bruxelles *01363*
Société Royale Belge d'Anthropologie et de Préhistoire, Bruxelles *01469*

Brazil

Fundação Joaquim Nabuco, Recife *01658*
Fundação Nacional do Indio, Brasília *01660*
Instituto do Ceará, Fortaleza *01669*

Bulgaria

Society of Bulgarian Anatomists, Histologists and Embryologists, Sofia *01777*

Canada

Brodie Club, Toronto *01996*
Canadian Sociology and Anthropology Association, Montréal *02298*
The Ontario Archaeological Society, Willowdale *02753*
Vancouver Museum Association, Vancouver *02976*

Chile

Corporación para el Desarrollo de la Ciencia, Santiago *03049*

Colombia

Instituto Colombiano de Antropología, Bogotá *03380*
Sociedad de Antropología de Antioquía, Medellín *03416*

Denmark

International Work Group for Indigenous Affairs, København *03750*

Ecuador

Asociación Ecuatoriana de Museos, Quito *03844*

France

Association Lyonnaise de Criminologie et Anthropologie Sociale, Lyon *04300*
Conseil International des Sciences Sociales, Paris *04459*
Euro-African Asscociation for the Anthropology of Social Change and Development, Montpellier *04473*
Société d'Anthropologie de Paris, Paris *04639*
Société de Biométrie Humaine, Paris *04644*
Société des Amis du Musée de l'Homme, Paris *04698*
Société d'Etudes Linguistiques et Anthropologiques de France, Paris *04746*

Germany

Anthropos Institut, Sankt Augustin *05075*
Berliner Gesellschaft für Anthropologie, Ethnologie und Urgeschichte, Berlin *05187*
Deutsche Gesellschaft für Anthropologie, Freiburg *05323*
Forschungsgruppe für Anthropologie und Religionsgeschichte e.V., Saarbrücken *05726*
Gesellschaft für Anthropologie und Humangenetik, Bremen *05800*
Naturhistorische Gesellschaft Nürnberg e.V., Nürnberg *06130*

Greece
World Society for Ekistics, Athinai *06549*

India
Anthropological Society of Bombay, Bombay *06716*
Anthropological Survey of India, Calcutta *06717*
Indian Anthropological Association, Delhi *06769*

Israel
Israel Prehistoric Society, Jerusalem *07088*

Italy
Associazione Italiana Studi Americanisti, Genova *07351*
Centro Camuno di Studi Preistorici ed Etnologici, Capo di Ponte *07417*
Centro di Studi Preistorici e Archeologici, Varese *07453*
Centro Italiano di Antropologia Culturale, Roma *07490*
Centro Studi Ricerche Ligabue, Venezia *07590*
European Association for the Visual Studies of Man, Firenze *07664*
Federazione delle Istituzioni Antropologiche Italiane, Genova *07687*
Società Italiana di Antropologia ed Etnologia, Firenze *07838*
Società Italiana per l'Antropologia e la Etnologia, Firenze *07964*
Unione Antropologica Italiana, Bologna *08039*

Jamaica
Institute of Jamaica, Kingston *08080*

Japan
Nippon Jinruigaku Kai, Tokyo *08315*

Mexico
Departamento de Antropología e Historia de Nayarit, Tepic *08730*
Sociedad Mexicana de Antropología, México *08756*

Netherlands
Koninklijk Instituut voor Taal-, Land- en Volkenkunde, Leiden *08964*
Nederlands Genootschap voor Anthropologie, Amsterdam *09049*

New Zealand
Polynesian Society, Auckland *09189*

Poland
Polskie Towarzystwo Antropologiczne, Warszawa *09552*

Portugal
Sociedade Portuguesa de Antropologia e Etnologia, Porto *09691*

Slovakia
Slovenská Antropologická Spoločnost, Bratislava *10061*

Spain
Sociedad de Ciencias Aranzadi, San Sebastián *10371*

Sweden
Svenska Sällskapet för Antropologi och Geografi, Stockholm *10573*

United Kingdom
Association of Social Anthropologists of the Commonwealth, London *11392*
British Institute of Persian Studies, London *11586*
Royal Anthropological Institute of Great Britain and Ireland, London *12768*

U.S.A.
American Anthropological Association, Washington *13453*
American Association of Physical Anthropologists, Buffalo *13565*
American Society of Primatologists, Chicago *14160*
Anthropology Film Center Foundation, Santa Fe *14229*
Association for Social Anthropology in Oceania, Little Rock *14333*
Association for the Sociology of Religion, Lancaster *14355*
Association of Black Anthropologists, Washington *14401*
Biological Anthropological Section, Lawrence *14588*
Central States Anthropological Society, Dayton *14717*
C.G. Jung Foundation for Analytical Psychology, New York *14719*
Council for Museum Anthropology, New York *14880*
Council on Anthropology and Education, Washington *14908*
Council on Nutritional Anthropology, Washington *14927*
Environmental Design Research Association, Oklahoma City *15049*
Foundation for Latin American Anthropological Research, Cocoa *15128*
General Anthropology Division, Washington *15148*
Human Relations Area Files, New Haven *15237*
Institute for the Study of Man, McLean *15306*
Institute of Andean Research, New York *15310*
Institute of Human Origins, Berkeley *15324*
International Women's Anthropology Conference, New York *15575*
Kroeber Anthropological Society, Berkeley *15640*
Latin American Studies Association, Pittsburgh *15648*
National Association for the Practice of Anthropology, Washington *15820*
National Character Laboratory, El Paso *15952*
Society for Applied Anthropology, Oklahoma City *16494*
Society for Armenian Studies, Dearborn *16497*
Society for Cultural Anthropology, Washington *16515*
Society for Humanistic Anthropology, West Chester *16533*
Society for Latin American Anthropology, Washington *16546*
Society for Linguistic Anthropology, Washington *16549*
Society for Medical Anthropology, Washington *16551*
Society for Psychological Anthropology, Washington *16571*
Society for Slovene Studies, Bloomington *16579*
Society for the Anthropology of Europe, Washington *16591*
Society for the Comparative Study of Society and History, Ann Arbor *16593*
Society for Urban Anthropology, Washington *16624*
Society for Visual Anthropology, Washington *16628*
South American Explorers Club, Ithaca *16722*
Wenner-Gren Foundation for Anthropological Research, New York *16879*

Zimbabwe
Pre-History Society of Zimbabwe, Harare *17179*

Antiquities

Argentina
Instituto Bonaerense de Numismática y Antigüedades, Buenos Aires *00142*

New Zealand
Polynesian Society, Auckland *09189*

United Kingdom
Berwickshire Naturalists' Club, Eyemouth *11436*

U.S.A.
New England Antiquities Research Association, Kingston *16195*

Archeology

Algeria
Société Archéologique du Département de Constantine, Constantine *00023*

Australia
Anthropological Society of South Australia, Adelaide *00216*
Archaeological and Anthropological Society of Victoria, Melbourne *00218*
Australian Institute of Archaeology, Melbourne *00297*
Indo-Pacific Prehistory Association, Canberra *00410*

Austria
Archaeolische Gesellschaft Steiermark, Graz *00568*
Eranos Vindobonensis, Wien *00584*
Österreichische Gesellschaft für Archäologie, Wien *00760*
Verein Freunde der Archäologie, Graz *00981*
Verein Tiroler Landesmuseum Ferdinandeum, Innsbruck *01003*

Bahrain
Bahrein Historical and Archaeological Society, Manama *01062*

Barbados
Barbados Museum and Historical Society, Bridgetown *01074*

Belgium
Académie Royale d'Archéologie de Belgique, Bruxelles *01092*
Association des Diplomés en Histoire de l'Art et Archéologie de l'Université Catholique de Louvain, Louvain-la-Neuve *01116*
Cercle Archéologique de Mons, Mons *01206*
Fondation Archéologique, Bruxelles *01361*
Fondation Egyptologique Reine Elisabeth, Bruxelles *01363*
Institut Archéologique du Luxembourg, Arlon *01375*
Institut Archéologique Liégeois, Liège *01376*
Société Archéologique de Namur, Namur *01421*
Société d'Etudes Latines de Bruxelles, Tournai *01458*
Société Royale Belge d'Anthropologie et de Préhistoire, Bruxelles *01469*
Société Royale d'Archéologie de Bruxelles, Bruxelles *01480*

Bolivia
Sociedad Arqueológica de Bolivia, La Paz *01548*

Brazil
Instituto Archeológico, Histórico e Geográfico Pernambucano, Recife *01663*

Canada
Canadian Quaternary Association, Waterloo *02239*
Manitoba Archaeological Society, Winnipeg *02599*
The Ontario Archaeological Society, Willowdale *02753*
Ontario Society for Industrial Archaeology, Toronto *02797*
Société d'Archéologie et de Numismatique de Montréal, Montréal *02918*

Chile
Corporación para el Desarrollo de la Ciencia, Santiago *03049*
Sociedad Arqueológica de la Serena, La Serena *03066*

Cyprus
Society of Cypriot Studies, Nicosia *03499*

Czech Republic
Česká Archeologická Společnost, Praha *03502*

Denmark
Det Kongelige Nordiske Oldskriftselskab, København *03711*
Jysk Arkaeologisk Selskab, Hœjbjerg *03754*
Sønderjyllands Amatørarkaeologer, Abenrå *03825*

Egypt
Société Archéologique d'Alexandrie, Alexandria *03913*
Society for Coptic Archaeology, Cairo *03915*

Finland
Klassillis-Filologinen Yhdistys, Helsingin Yliopisto *03978*
Museovirasto, Helsinki *03983*
Suomen Egyptologinen Seura, Helsinki *04011*
Suomen Muinaismuistoyhdistys, Helsinki *04051*

France
Centre International d'Etudes Romanes, Paris *04379*
Fédération des Sociétes Historiques et Archéologiques de Paris et de l'Ile de France, Paris *04516*
Féderation des Sociétés Savantes de la Charente-Maritime, La Rochelle *04517*
International Association of Sanskrit Studies, Paris *04582*
Société Archéologique de Touraine, Tours *04629*
Société Archéologique et Historique du Limousin, Limoges *04630*
Société d'Archéologie et d'Histoire de la Manche, Saint-Lo *04640*
Société d'Archéologie et d'Histoire de l'Aunis, La Rochelle *04641*
Société des Américanistes, Paris *04693*
Société des Lettres, Sciences et Arts de la Haute-Auvergne, Aurillac *04709*
Société de Statistique, d'Histoire et d'Archéologie de Marseille et de Provence, Marseille *04720*
Société d'Histoire et d'Archéologie de Bretagne, Rennes *04762*
Société d'Histoire et d'Archéologie de la Lorraine, Saint Julien les Metz *04763*
Société Française d'Archéocivilisation et de Folklore, Paris *04785*
Société Française d'Archéologie, Paris *04786*
Société Française d'Egyptologie, Paris *04805*
Société Historique, Archéologique et Littéraire de Lyon, Lyon *04870*
Société Historique et Archéologique du Périgord, Périgueux *04873*
Société Savoisienne d'Histoire et d'Archéologie, Chambéry *04917*

French Polynesia
Société des Etudes Océaniennes, Papeete *05003*

Georgia
Georgian Commission on Archaeology, Tbilisi *05014*

Germany
Archäologische Gesellschaft zu Berlin, Berlin *05151*
Deutsche Orient-Gesellschaft e.V., Berlin *05480*
Deutsches Archäologisches Institut, Berlin *05568*
Harzverein für Geschichte und Altertumskunde, Goslar *05910*
Kommission für Alte Geschichte und Epigraphik des Deutschen Archäologischen Instituts, München *06075*
Mainzer Altertumsverein, Mainz *06107*
Mommsen-Gesellschaft, Kiel *06119*
Tübinger Förderkreis zur Erforschung der Troas – Freunde von Troia, Tübingen *06314*
West- und Süddeutscher Verband für Altertumsforschung, Mainz *06441*

Greece
Archaeologiki Hetairia, Athinai *06499*

Hungary
Magyar Régészeti és Müvészettörténeti Társulat, Budapest *06655*

Iceland
Hid Islenska Fornleifafélag, Reykjavik *06688*

India
Kamarupa Anusandhan Samiti, Gauhati *06831*

Iran
British Institute of Persian Studies, Teheran *06898*

Ireland
Cork Historical and Archaeological Society, Cork *06938*
County Louth Archaeological and Historical Society, Dundalk *06939*
Friends of Medieval Dublin, Dublin *06950*
Irish Association of Professional Archaeologists, Dublin *06975*
The Royal Society of Antiquaries of Ireland, Dublin *07030*

Israel
Israel Association of Archaeologists, Jerusalem *07064*
Israel Exploration Society, Jerusalem *07071*
Wilfrid Israel House for Oriental Art and Studies, Hazorea *07117*

Italy
Accademia di Paestum Eremo Italico, Mercato San Severino *07170*
Associazione Archaeologica Romana, Roma *07256*
Associazione Archeologica Allumiere Adolfo Klitsche de la Grange, Allumiere *07257*
Associazione Archeologica Centumcellae, Civitavecchia *07258*
Associazione Archeologica Romana, Roma *07259*
Associazione Geo-Archeologica Italiana, Roma *07279*
Associazione Internazionale di Archeologia Classica, Roma *07286*
Associazione Italiana Studi Americanisti, Genova *07351*
Centro Camuno di Studi Preistorici ed Etnologici, Capo di Ponte *07417*
Centro di Documentazione e Promozione Archeologica, Roma *07423*
Centro di Studi Preistorici e Archeologici, Varese *07453*
Centro Internazionale di Studi Archeologici Maiuri, Ercolano *07471*
Centro Polesano di Studi Storici, Archeologici ed Etnografici, Rovigo *07529*
Centro Ricerche Archeologiche e Scavi di Torino per il Medio Oriente e l'Asia, Torino *07534*
Centro Ricerche di Storia e Arte Bitontina, Bitonto *07537*
Centro Studi Archaeologici di Boscoreale, Boscotrecase e Trecase, Boscoreale *07548*
Centro Studi Ricerche Ligabue, Venezia *07590*
Gruppi Archeologici d'Italia, Roma *07701*
Società Archeologica Comense, Como *07770*
Società Archeologica Viterbese Pro Ferento, Viterbo *07771*
Società Istriana di Archeologia e Storia Patria, Trieste *07820*
Società Italiana per l'Archeologia e la Storia delle Arti, Napoli *07965*
Società per gli Studi Storici, Archeologici ed Artistici della Provincia di Cuneo, Cuneo *07999*
Società Reggiana d'Archeologia, Reggio Emilia *08006*
Società Savonese di Storia Patria, Savona *08012*
Unione Accademica Nazionale, Roma *08038*

Japan
Nihon Kokogaku Kyokai, Tokyo *08228*
Shigaku Kenkyukai, Kyoto *08418*

Macedonia
Združenie na Arheolozite na Makedonija, Skopje *08618*

Netherlands
Association of Roman Ceramic Archaeologists, Nijmegen *08828*
Fries Genootschap van Geschied-, Oudheid- en Taalkunde, Leeuwarden *08898*

New Zealand
New Zealand Archaeological Association, Dunedin North *09133*

Norway
Norsk Arkeologisk Selskap, Oslo *09291*

Panama
Comisión Nacional de Arqueología y Monumentos Históricos, Panamá City *09415*

Poland
Polskie Towarzystwo Archeologiczne i Numizmatyczne, Warszawa *09553*

Portugal
Associação dos Arqueólogos Portugueses, Lisboa *09638*
Real Instituto Arqueológico de Portugal, Lisboa *09667*
Real Sociedade Arqueológica Lusitana, Santiago de Cacém *09668*

Romania
Societatea de Studii Clasice din Romania, Bucuresti *09788*

Slovakia
Slovenská Archeologická Spoločnost, Nitra-Hrad *10062*

South Africa
South African Archaeological Society, Vlaeberg *10178*

Spain
Asociación Española de Bibliotecarios, Archiveros, Museologos y Documentalistas, Madrid *10257*
Real Sociedad Arqueológica Tarraconense, Tarragona *10358*
Sociedad de Ciencias Aranzadi, San Sebastián *10371*
Societat Arqueològica Lul·liana, Palma de Mallorca *10403*

Sudan
Commission for Archaeology, Khartoum *10426*

Sweden
Kungliga Vitterhets Historie och Antikvitets Akademien, Stockholm *10493*
Svenska Arkeologiska Samfundet, Stockholm *10534*

Svenska Fornminnesföreningen, Stockholm 10548

Switzerland

Fédération Internationale des Associations d'Etudes Classiques, Bellevue 10701
Gesellschaft Pro Vindonissa, Brugg 10718
Schweizerische Gesellschaft für Ur- und Frühgeschichte, Basel 10853
Schweizerische Vereinigung für Altertumswissenschaft, Basel 10939
Société d'Histoire et d'Archéologie, Genève 10972
Société Vaudoise d'Histoire et d'Archéologie, Chavannes-près-Renens 10991

Tunisia

Institut National d'Archéologie et d'Art, Tunis 11142

United Kingdom

Anglesey Antiquarian Society and Field Club, Menai Bridge 11230
Architectural and Archaeological Society for the County of Buckingham, Aylesbury 11243
Architectural and Archaeological Society of Durham and Northumberland, Durham City 11244
Association for Industrial Archaeology, Telford 11266
Association of Young Irish Archaeologists, Belfast 11411
Atlantis Research Centre, Brighton 11415
Ayrshire Archaeological and Natural History Society, Ayr 11419
Bath and Camerton Archaeological Society, Bath 11427
BC Society, Hassocks 11429
Bedfordshire Archaeological Council, Bedford 11430
Birmingham and Warwickshire Archaeological Society, Birmingham 11447
Blackcountry Society, Kingswinford 11452
Bristol and Gloucestershire Archaeological Society, Gloucester 11468
Bristol Industrial Archaeological Society, Bristol 11469
British Institute of Persian Studies, London 11586
Cambrian Archaeological Association, Newport 11734
Cambridge Antiquarian Society, Cambridge 11735
Cambridge Society for Industrial Archaeology, Cambridge 11739
The Chester Archaeological Society, Chester 11773
City of Stoke on Trent Museum Archaeological Society, Stoke-on-Trent 11790
Cleveland Industrial Archaeology Society, Cleveland 11793
Cornwall Archaeological Society, Bodmin 11844
Council for British Archaeology, York 11847
Coventry and District Archaeological Society, Coventry 11865
Cumberland and Westmorland Antiquarian and Archaeological Society, Kendal 11871
Derbyshire Archaeological Society, Uttoxeter 11882
Devon Archaeological Society, Exeter 11889
Dorset Natural History and Archaeological Society, Dorchester 11902
Dumfriesshire and Galloway Natural History and Antiquarian Society, Dumfries 11911
East Hertfordshire Archaeological Society, Ware 11917
East Lothian Antiquarian and Field Naturalists' Society, Dunbar 11919
East Riding Archaeological Society, Hull 11922
Essex Archaeological and Historical Congress, Stratford Saint Mary 11976
Essex Society for Archaeology and History, Colchester 11977
European Association of South Asian Archaeologists, Cambridge 12001
Exeter Industrial Archaeology Group, Exeter 12043
Flintshire Historical Society, Clwyd 12079
Glasgow Archaeological Society, Glasgow 12118
Gloucestershire Society for Industrial Archaeology, Dursley 12122
Hampshire Field Club and Archaeological Society, Winchester 12144
Hawick Archaeological Society, Hawick 12151
Hunter Archaeological Society, Sheffield 12187
Isle of Man Natural History and Antiquarian Society, Douglas 12412
Isle of Wight Natural History and Archaeological Society, Shanklin 12413
Kent Archaeological Society, Gravesend 12428
Leicestershire Archaeological and Historical Society, Leicester 12445
London and Middlesex Archaeological Society, London 12460
Manchester Region Industrial Archaeology Society, Disley 12486
Medieval Settlement Research Group, Taunton 12517
Merseyside Archaeological Society, Liverpool 12521
Moray Society, Elgin 12538
Norfolk and Norwich Archaeological Society, Norwich 12629
Northamptonshire Archaeological Society, Earls Barton 12631
North Staffordshire Field Club, Stoke-on-Trent 12642
North Western Society for Industrial Archaeology and History, Liverpool 12647
Oxfordshire Architectural and Historical Society, Oxford 12669
Oxford University Archaeological Society, Oxford 12671
Palestine Exploration Fund, London 12676
Peakland Archaeological Society, Stockport 12684
Royal Archaeological Institute, London 12769
Royal Institution of Cornwall, Truro 12821
Saint Albans and Hertfordshire Architectural and Archaeological Society, Saint Albans 12882
Scottish Industrial Heritage Society, Glasgow 12912
Scottish Society for Northern Studies, Edinburgh 12929
Shropshire Archaeological and Parish Register Society, Much Wenlock 12951
La Société Guernesiaise, Saint Peter Port 12961
Society for Libyan Studies, London 12978
Society for Lincolnshire History and Archaeology, Lincoln 12979
Society for Medieval Archaeology, Lincoln 12982
Society for Nautical Research, London 12984
Society for Post-Medieval Archaeology, London 12986
The Society for South Asian Studies, London 12997
Society for the Promotion of Hellenic Studies, London 13005
Society for the Promotion of Roman Studies, London 13008
Society of Antiquaries of London, London 13022
Society of Antiquaries of Newcastle-upon-Tyne, Newcastle-upon-Tyne 13023
Society of Antiquaries of Scotland, Edinburgh 13024
Society of Archer Antiquaries, Bridlington 13025
Somerset Archaeological and Natural History Society, Taunton 13074
South Bedfordshire Archaeological Society, Great Yarmouth 13077
South Staffordshire Archaeological and Historical Society, Walsall 13083
South Wiltshire Industrial Archaeology Society, Salisbury 13085
Suffolk Institute of Archaeology and History, Ipswich 13106
Surrey Archaeological Society, Guildford 13110
Sussex Archaeological Society, Lewes 13113
Sussex Industrial Archaeology Society, Brighton 13114
Thoroton Society of Nottinghamshire, Nottingham 13131
Ulster Archaeological Society, Belfast 13148
Ulster Archery Association, Belfast 13149
University of Bristol Spelaeological Society, Bristol 13175
Wiltshire Archaeological and Natural History Society, Devizes 13209
Wolverton and District Archaeological Society, Milton Keynes 13212
Woolhope Naturalists' Field Club, Hereford 13213
Worcestershire Archaeological Society, Malvern 13214
Yorkshire Archaeological Society, Leeds 13230

Uruguay

Sociedad de Amigos de Arqueología, Montevideo 13265

U.S.A.

American Committee to Advance the Study of Petroglyphs and Pictographs, Shepherdstown 13746
American Institute for Archaeological Research, Mount Vernon 13864
American Oriental Society, Ann Arbor 13968
American Philosophical Society, Philadelphia 13999
American Rock Art Research Association, San Miguel 14033
American Society for Conservation Archaeology, Albuquerque 14053
Amerind Foundation, Dragoon 14214
Archaeological Conservancy, Santa Fe 14236
Archaeological Institute of America, Boston 14237
Archeology Section, Washington 14238
CEDAM International, Croton on Hudson 14662
Center for American Archaeology, Kampsville 14664
Early Sites Research Society, Rowley 14998
Epigraphic Society, San Diego 15053
Etruscan Foundation, Grosse Pointe Farms 15063
Institute for American Indian Studies, Washington 15269
Institute of Andean Studies, Berkeley 15311
Institute of Nautical Archaeology, College Station 15333
National Association of State Archeologists, Norman 15893
Near East Archaeological Society, Saint Louis 16189
Search Foundation, Birmingham 16453
Society for American Archaeology, Washington 16492
Society for Commercial Archeology, Washington 16511
Society for Historical Archaeology, Tucson 16530
Society for Industrial Archaeology, Washington 16537
Society of Professional Archaeologists, Topeka 16693
South American Explorers Club, Ithaca 16722
Stonehenge Study Group, Santa Barbara 16747
World Archaeological Society, Hollister 16922

Vatican City

Pontificia Accademia Romana di Archeologia, Roma 16955

Yugoslavia

Savez Arheološki Društava Jugoslavije, Beograd 17107

Architecture s. a. Civil Engineering; Urban and Regional Planning

Argentina

Sociedad Central de Arquitectos, Buenos Aires 00188

Austria

Österreichische Gesellschaft für Architektur, Wien 00761
Österreichischer Ingenieur- und Architektenverein, Wien 00893
Zentralvereinigung der Architekten Österreichs, Wien 01034

Belgium

Architects Council of Europe, Bruxelles 01101
Chambre des Architectes de Belgique, Bruxelles 01211
Comité de Liaison des Architectes de l'Europe Unie, Bruxelles 01220
European Association of Architectural Education, Louvain-la-Neuve 01287
Fédération Royale des Sociétés d'Architectes de Belgique, Bruxelles 01360
Ordre des Architectes, Bruxelles 01418
Société Centrale d'Architecture de Belgique, Bruxelles 01453

Bolivia

Asociación de Arquitectos de Bolivia, La Paz 01532

Bulgaria

Union of Architects in Bulgaria, Sofia 01783

Canada

Architectural Intitute of British Columbia, Vancouver 01870
Canadian Society of Landscape Architects, Ottawa 02285
Royal Architectural Institute of Canada, Ottawa 02862
Society for the Study of Architecture in Canada, Ottawa 02940

Chile

Colegio de Architectos de Chile, Santiago 03029
Corporación Toesca para el Desarrollo de la Arquitectura, Santiago 03050

China, People's Republic

The Architectural Society of China, Beijing 03115
Central South Academy of Industrial Architecture, Hupei 03122
Northwestern Academy of Industrial Architecture, Shanhsi 03212

Colombia

Sociedad Antioqueña de Ingenieros y Architectos, Medellín 03403

Cyprus

Cyprus Civil Engineers and Architects Association, Nicosia 03489

Czech Republic

Obec Architektú, Praha 03552

Denmark

Byggecentrum, Hørsholm 03571
Danske Arkitekters Landsforbund, København 03602

France

Académie d'Architecture, Paris 04086
Association des Artistes Sculpteurs, Architectes, Graveurs et Dessinateurs, Paris 04167
Cercle d'Etudes Architecturales, Paris 04386
Compagnie des Experts Architectes près la Cour d'Appel de Paris, Paris 04433
Ordre des Architectes, Paris 04607
Société Nationale des Architectes de France, Paris 04900
Union Internationale des Architectes, Paris 04985
Union Internationale des Femmes Architectes, Paris 04987
Vieilles Maisons Françaises, Paris 05000

Germany

Association Européenne pour l'Enseignement de l'Architecture, Kassel 05154
Bund Deutscher Architekten, Bonn 05226
Bund Deutscher Baumeister, Architekten und Ingenieure, Bonn 05227
Bundesarchitektenkammer, Bonn 05238
Vereinigung Freischaffender Architekten Deutschlands e.V., Bonn 06407

Greece

Syllogos Architktonon Diplomatouchon Anotaton Scholon, Athinai 06544
World Society for Ekistics, Athinai 06549

India

Indian Institute of Architects, Bombay 06795

Ireland

Architectural Association of Ireland, Dublin 06924
Institute of Architectural and Associated Technology, Dublin 06959
Irish Georgian Society, Dublin 06986
Royal Hibernian Academy of Arts, Dublin 07025
Royal Institute of Architects of Ireland, Dublin 07026

Israel

Architectural Association of Israel, Tel Aviv 07042
Association of Engineers and Architects in Israel, Tel Aviv 07046

Italy

Associazione Nazionale Ingegneri ed Architetti Italiani, Roma 07374
Centro di Studi per la Storia dell'Architettura, Roma 07450
Centro Internazionale di Studi di Architettura Andrea Palladio, Vicenza 07472
Consiglio Nazionale degli Architetti, Roma 07626
Ente Autonomo La Biennale de Venezia, Venezia 07649
Istituto Cooperativo per l'Innovazione, Roma 07725
Società degli Ingegneri e degli Architetti in Torino, Torino 07780

Malaysia

Malaysian Institute of Architects, Kuala Lumpur 08648

Malta

Chamber of Architects and Civil Engineers, Paceville 08658

Mexico

Asociación de Ingenieros y Arquitectos de México, México 08692

Namibia

Institute of South West African Architects, Windhoek 08815

Netherlands

Architektenraad, Amsterdam 08822
Bond Heemschut, Amsterdam 08834
Genootschap Architectura et Amicitia, Amsterdam 08902
Koninklijke Maatschappij tot Bevordering der Bouwkunst, Bond van Nederlandse Architekten, Amsterdam 08947
Stichting Centrale Raad voor de Academies van Bouwkunst, Amsterdam 09070

New Zealand

New Zealand Institute of Architects, Wellington 09154

Norway

Norske Arkitekters Landsforbund, Oslo 09296

Pakistan

Architects Regional Council Asia, Karachi 09358
Pakistan Council of Architects and Town Planners, Karachi 09383

Peru

Colegio de Arquitectos del Perú, Lima 09464

Philippines

Philippine Institute of Architects, Manila 09527

Portugal

Associção dos Arquitectos Portugueses, Lisboa 09653

Romania

Uniunea Arhitectilor din Romania, Bucuresti 09796

Russia

Union of Russian Architects, Moskva 09987

Singapore

Singapore Institute of Architects, Singapore 10053

Slovakia

Spolok Architektov Slovenska, Bratislava 10094

Slovenia

Zveza Društev Arhitektov Slovenije, Ljubljana 10112

South Africa

Institute of South African Architects, Houghton 10149

Spain

Colegio de Arquitectos de Cataluña, Barcelona 10289
Consejo Superior de los Colegios de Arquitectos de España, Madrid 10302

Sweden

Statens Råd för Byggnadsforskning, Stockholm 10524
Svenska Arkitekters Riksförbund, Stockholm 10535
Svenska Fornminnesföreningen, Stockholm 10548

Switzerland

Bund Schweizer Architekten, Zürich 10638
Fachgruppe für Arbeiten im Ausland des SIA, Zürich 10688
Fachgruppe für Architektur des SIA, Zürich 10689
Gesellschaft Schweizerischer Maler, Bildhauer und Architekten, Muttenz 10720
Internationale Architekten-Union, Sektion Schweiz, Zürich 10736
Schweizerischer Ingenieur- und Architekten-Verein, Zürich 10899
Verband freierwerbender Schweizer Architekten, Lachen 11001

Turkey

Türk Mühendis ve Mimar Odalari Birliği, Ankara 11167

United Kingdom

Architects and Surveyors Institute, Chippenham *11242*
Architectural and Archaeological Society for the County of Buckingham, Aylesbury *11243*
Architectural and Archaeological Society of Durham and Northumberland, Durham City *11244*
Architectural Association, London *11245*
Association of Chief Architects of Scottish Local Authorities, Lanark *11321*
Association of Official Architects, London *11374*
British Institute of Persian Studies, London *11586*
Charles Rennie Mackintosh Society, Glasgow *11758*
Cinema Theatre Association, Teddington *11786*
Commonwealth Association of Architects – Education Committee, London *11811*
Council of Professors of Building, Birmingham *11858*
County, City and Borough Architects' Association, Harrow *11859*
Edinburgh Architectural Association, Edinburgh *11931*
Faculty of Building, Elstree *12050*
Farm Buildings Association, Oxford *12058*
Incorporated Association of Architects, Weston Favell *12192*
Men of the Stones, Stamford *12518*
Oxfordshire Architectural and Historical Society, Oxford *12669*
Regency Society of Brighton and Hove, Brighton *12746*
Research into Lost Knowledge Organisation, Orpington *12753*
Royal Incorporation of Architects in Scotland, Edinburgh *12811*
Royal Institute of British Architects, London *12812*
Royal Society of Ulster Architects, Belfast *12865*
Saint Albans and Hertfordshire Architectural and Archaeological Society, Saint Albans *12882*
Society of Architectural Historians of Great Britain, London *13027*
Society of Chief Architects of Local Authorities, Wirral *13036*
Ulster Architectural Heritage Society, Belfast *13150*
Vernacular Architecture Group, Tamworth *13179*
Victorian Society, London *13184*
Welwyn Hall Research Association, Hatfield *13204*

U.S.A.

American Architectural Foundation, Washington *13457*
American Indian Council of Architects and Engineers, Tigard *13858*
American Institute of Architects, Washington *13874*
American Institute of Building Design, Shelton *13877*
American Society of Golf Course Architects, Chicago *14127*
American Society of Landscape Architects, Washington *14139*
American Underground-Space Association, Minneapolis *14200*
Architectural League of New York, New York *14239*
Association for Bridge Construction and Design, Pittsburgh *14274*
Association of Collegiate Schools of Architecture, Washington *14416*
Association of University Architects, New Haven *14527*
Athenaeum of Philadelphia, Philadelphia *14545*
Council of Landscape Architectural Registration Boards, Fairfax *14898*
Council on Tall Buildings and Urban Habitat, Bethlehem *14933*
Design/Build Institute, Greendale *14968*
Environmental Design Research Association, Oklahoma City *15049*
Frank Lloyd Wright Foundation, Scottsdale *15137*
Institute for Community Design Analysis, Great Neck *15273*
Institute for Urban Design, San Diego *15308*
Intelligent Buildings Institute, Washington *15347*
Interfaith Forum on Religion, Art and Architecture, Washington *15359*
International Association for Housing Science, Coral Gables *15384*
Landscape Architecture Foundation, Washington *15646*
National Architectural Accrediting Board, Washington *15788*
National Council on Agricultural Life and Labor Research Fund, Dover *16015*
National Housing Conference, Washington *16078*
National Organization of Minority Architects, Washington *16119*
Society of American Registered Architects, Lombard *16637*
Society of Architectural Historians, Philadelphia *16639*
Society of Naval Architects and Marine Engineers, Jersey City *16683*
Society of Turkish Architects, Engineers and Scientists in America, New York *16709*
Vernacular Architecture Forum, Baltimore *16857*

Vietnam

Vietnamese Association of Architects, Hanoi *17041*

Yugoslavia

Društvo Arhitekata Srbije, Beograd *17081*

Archives

Argentina

Asociación Archivística Argentina, Buenos Aires *00053*

Austria

Arbeitsgemeinschaft audiovisueller Archive Österreichs, Wien *00548*
Verband österreichischer Archivare, Wien *00966*

Belgium

Archives et Bibliothèques de Belgique, Bruxelles *01102*
Vlaamse Museumvereniging, Antwerpen *01509*

Canada

Alberta Museums Association, Calgary *01841*
Association Museums of New Brunswick, Saint John *01914*
Association of British Columbia Archivists, Vancouver *01915*
Association of Canadian Archivists, Ottawa *01917*
Association Québécoise des Archivistes Médicales, Rock-Forest *01954*
British Columbia Museums Association, Victoria *01983*
Museums Association of Saskatchewan, Regina *02665*
Ontario Association of Archivists, Ottawa *02757*
Organization of Military Museums of Canada, Ottawa *02807*
Prince Edward Island Museum and Heritage Foundation, Charlottetown *02836*
Toronto Area Archivists Group, Toronto *02965*
Yukon Historical and Museums Association, Whitehorse *03003*

China, People's Republic

Chinese Archives Society, Beijing *03159*

Colombia

Consejo Nacional de Archivos, Bogotá *03370*

Congo

Bureau des Archives et Bibliothèques de Brazzaville, Brazzaville *03423*

Denmark

Arkivforeningen, København *03559*
Sammenslutningen af Lokalarkiver, Vejle *03802*

Ecuador

Asociación Ecuatoriana de Museos, Quito *03844*

Egypt

Egyptian Association of Archives, Librarianship and Information Science, Cairo *03889*

Finland

Arkistoyhdistys, Helskini *03950*

France

Association des Archivistes Français, Paris *04166*
International Council on Archives, Paris *04585*

Germany

Arbeitsgemeinschaft der Archive und Bibliotheken in der evangelischen Kirche, Nürnberg *05083*
Deutscher Museumsbund e.V., Karlsruhe *05529*
Internationale Arbeitsgemeinschaft der Archiv-, Bibliotheks- und Graphikrestauratoren, Marburg *05989*
Verein Deutscher Archivare, München *06372*

Honduras

Asociación de Bibliotecarios y Archiveros de Honduras, Tegucigalpa *06582*

Indonesia

Asosiasi Perpustakaan, Arsip dan Dokumentasi Indonesia, Jakarta *06885*

Italy

Associazione Archivistica Ecclesiastica, Roma *07260*
Associazione Nazionale Archivistica Italiana, Roma *07359*
Centro di Ricerca Pergamene Medievali e Protocolli Notarili, Roma *07433*
International Council of Museums, Milano *07713*

Japan

Nihon Hakubutsukan Kyokai, Tokyo *08205*

Kenya

Eastern and Southern African Regional Branch of the International Council on Archives, Nairobi *08501*

Macedonia

Društvo na Muzejskite Rabotnici na Makedonija, Skopje *08597*
Sojuz na Društvata na Arhivskite Rabotnici na Makedonija, Skopje *08607*

Netherlands

Vereniging van Archivarissen in Nederland, Leeuwarden *09088*

New Zealand

Art Galleries and Museums Association of New Zealand, Wellington *09116*

Norway

Arkivarforeningen, Oslo *09247*

Peru

Asociación Peruana de Archiveros, Lima *09454*

Poland

Stowarzyszenie Archiwistow Polskich, Warszawa *09613*

Portugal

Associação Portuguesa de Bibliotecários, Arquivistas e Documentalistas, Lisboa *09640*

Senegal

Ecole de Bibliothécaires, Archivistes et Documentalistes, Dakar *10027*

Slovenia

Arhivsko Društvo Slovenije, Ljubljana *10096*

Spain

Amics dels Museus de Catalunya, Barcelona *10246*
Asociación Española de Bibliotecarios, Archiveros, Museólogos y Documentalistas, Madrid *10257*

Sweden

International Association of Music Libraries, Archives and Documentation Centres, Stockholm *10472*
International Association of Sound Archives, Stockholm *10473*
Svenska Arkivsamfundet, Stockholm *10536*
Sveriges Arkivtjänstemäns Förening, Nacka *10587*

Switzerland

Vereinigung Schweizerischer Archivare, Bern *11016*

Togo

Association Togolaise pour le Développement de la Documentation, des Bibliothèques, Archives et Musées, Lomé *11104*

Trinidad and Tobago

Caribbean Regional Branch of the International Council on Archives, Port of Spain *11118*

Tunisia

Arab Regional Branch of the International Council on Archives, Tunis *11133*
Association Internationale des Archives Francophones, Tunis *11135*
Association Tunisienne des Bibliothécaires, Documentalistes et Archivistes, Tunis *11138*

United Kingdom

Association of Commonwealth Archivists and Records Managers, London *11327*
Society of Archivists, London *13028*

U.S.A.

American Archives Association, Bethesda *13458*
Association for Recorded Sound Collections, Silver Spring *14325*
Association of Catholic Diocesan Archivists, East Lansing *14409*
Midwest Archives Conference, Morton Grove *15737*
National Archives and Records Administration Volunteer Association, Washington *15789*
Sigmund Freud Archives, Roslyn *16473*
Society of American Archivists, Chicago *16633*
Zionist Archives and Library of World Zionist Organization – American Section, New York *16945*

Yugoslavia

Društvo Arhivskih Radnika Srbije, Beograd *17082*
Savez Društava Arhivskih Radnika Jugoslavije, Beograd *17110*
Savez Muzejskih Društava Jugoslavije, Beograd *17118*

Zaire

Association Zaïroise des Archivistes, Bibliothécaires et Documentalistes, Kinshasa *17141*
Central African Regional Branch of the International Council on Archives, Kinshasa *17142*

Arts
s. a. Fine Arts

Andorra

Associació Cultural i Artística Els Esquirols, La Massana *00027*
Cercle de les Arts i de les Lletres, Escaldes-Engordany *00029*
International Council of Museums, Andorran National Committee, Andorra la Vella *00031*

Argentina

Academia Nacional de Bellas Artes, Buenos Aires *00041*
Asociación Dante Alighieri, Buenos Aires *00076*
Comisión Nacional de Museos y de Monumentos y Lugares Históricos, Buenos Aires *00120*
Fondo Nacional de las Artes, Buenos Aires *00139*

Australia

Arts Council of Australia, Sydney *00219*
Australia Council, Redfern *00249*
Contemporary Art Society of Australia, Sydney *00380*
Museums Association of Australia, Melbourne *00435*
Royal Art Society of New South Wales, North Sydney *00454*
Royal Queensland Art Society, Brisbane *00474*
Royal South Australian Society of Arts, Adelaide *00482*
Victorian Artists' Society, East Melbourne *00515*

Austria

Akademie der bildenden Künste in Wien, Wien *00539*
Gesellschaft der Freunde der Neuen Galerie, Graz *00607*
Gesellschaft der Freunde des Kunsthistorischen Institutes der Karl-Franzens-Universität in Graz, Graz *00609*
Gesellschaft der Kunstfreunde, Wien *00611*
Gesellschaft für vergleichende Felsbildforschung, Graz *00635*
Gesellschaft für vergleichende Kunstforschung, Wien *00636*
Institut für Wissenschaft und Kunst, Wien *00661*
Joanneum-Verein, Graz *00690*
Krahuletz-Gesellschaft, Eggenburg *00697*
Kunsthistorische Gesellschaft, Wien *00699*
Kunsthistorische Gesellschaft an der Universität Graz, Graz *00700*
Musealverein in Hallstatt, Hallstatt *00709*
Museums-Verein Stillfried, Stillfried *00710*
Oberösterreichischer Musealverein, Linz *00717*
Österreichische Gesellschaft für Christliche Kunst, Wien *00771*
Österreichischer Kunsthistorikerverband, Wien *00895*
Österreichischer Museumsbund, Wien *00897*
Verband der Geistig Schaffenden Österreichs, Wien *00958*
Verband Österreichischer Privat-Museen, Bad Winsbach-Neydharting *00971*
Verein der Museumsfreunde in Wien, Wien *00979*
Vereinigung bildender Künstler, Wien *00989*
Verein Tiroler Landesmuseum Ferdinandeum, Innsbruck *01003*
Walter Buchebner Gesellschaft, Mürzzuschlag *01010*
Wiener Kulturkreis, Wien *01023*

Bahrain

Bahrain Arts Society, Manama *01054*
Bahrain Contemporary Art Association, Manama *01057*

Bangladesh

Society of Arts, Literature and Welfare, Chittagong *01070*

Barbados

Barbados Museum and Historical Society, Bridgetown *01074*

Belgium

Académie Royale des Sciences, des Lettres et des Beaux-Arts de Belgique, Bruxelles *01095*
Association des Diplomés en Histoire de l'Art et Archéologie de l'Université Catholique de Louvain, Louvain-la-Neuve *01116*
Association Internationale des Métiers et Enseignements d'Art, Bruxelles *01142*
Icon, Leuven *01373*
Institut Archéologique du Luxembourg, Arlon *01375*
Koninklijke Academie en Nationaal Hoger Instituut voor Schone Kunsten, Antwerpen *01406*

Bermuda

Bermuda Society of Arts, Hamilton *01523*

Bolivia

Círculo de Bellas Artes, La Paz *01539*

Brazil

Academia de Letras e Artes do Planalto, Luziânia *01572*
Associação Internacional de Críticos de Arte, Rio de Janeiro *01601*
Fundação Moinho Santista, São Paulo *01659*
Secretaria do Patrimônio Histórico e Artístico Nacional, Rio de Janeiro *01702*
Sociedade Brasileira de Belas Artes, Rio de Janeiro *01705*

Bulgaria

Research Association for the Fine Arts, Sofia *01770*

Canada

American Society for Aesthetics, Edmonton *01863*
Association of Manitoba Museums, Winnipeg *01935*
Canada Council, Ottawa *02000*
Canadian Association of Deans of Arts and Sciences, Charlottetown *02048*
Canadian Celtic Arts Association, Toronto *02090*
Canadian Conference of the Arts, Ottawa *02102*
Canadian Folk Arts Council, Toronto *02137*
Canadian Museums Association, Ottawa *02201*
Commonwealth Association of Museums, Calgary *02366*
Glenbow Alberta Institute, Calgary *02486*
ICOM Museums/Musées Canada, Ottawa *02512*
Manitoba Arts Council, Winnipeg *02600*
Manitoba Association for Art Education, Winnipeg *02601*
Micmac Association of Cultural Studies, Sydney *02652*
National Design Council, Ottawa *02672*
Photo Electric Arts Foundation, Toronto *02815*
Print and Drawing Council of Canada, Toronto *02843*
Royal Canadian Academy of Arts, Toronto *02864*
Sculptors' Society of Canada, Toronto *02909*
Tiger Hills Arts Association, Holland *02964*
Universities Art Association of Canada, Halifax *02973*
Visual Arts Ontario, Toronto *02982*

Chile
Academia Chilena de Bellas Artes, Santiago 03007

China, Republic
Association for Education through Art, Republic of China, Taichung Hsien 03242
China National Association of Literature and the Arts, Taipei 03256
Graphic Arts Association of China, Taipei 03300
Modern Fine Arts Association of Southern Taiwan, Tainan 03308
Museum Association of China, Taipei 03309
Television Academy of Arts and Sciences of the Republic of China, Taipei 03332

Colombia
Asociación Colombiana de Museos, Institutos y Casas de Cultura, Bogotá 03346
Comité Nacional del Consejo Internacional de Museos, Bogotá 03366
Instituto Colombiano de Cultura, Bogotá 03383

Croatia
Hrvatska Akademija Znanosti i Umjetnosti, Zagreb 03459
Savez Muzejskih Društava Hrvatske, Zagreb 03466

Cuba
Unión de Escritores y Artistas de Cuba, La Habana 03486

Denmark
Akademiet for de skønne Kunster, København 03556
Danske Kunsthåndvårkeres Landssammenslutning, København 03609
Dansk Kulturhistorisk Museumsforening, Fur 03641
Dansk Kunstmuseumsforening, København 03642
Foreningen af Danske Kunstmuseer, Fur 03729
Foreningen af Danske Museumsmaend, København 03730
Foreningen for National Kunst, København 03736
Kunstforeningen, København 03761
Kunstnerforeningen af 18. November, København 03762
Sammenslutningen af Danske Kunstforeninger, Niva 03801
Selskabet til Udgivelse af Danske Mindesmaerker, Nivaa 03818
Skandinavisk Museumsforbund, Danish Section, Aalborg 03821

Dominican Republic
Ateneo de Macoris, San Pedro de Macoris 03835
Sociedad Amantes de la Luz, Santiago de los Caballeros 03838

Ecuador
Andean Instiute of Popular Arts, Quito 03842

Egypt
Armenian Artistic Union, Cairo 03878
Atelier, Alexandria 03882
High Council of Arts and Literature, Cairo 03904
High Council of Culture, Cairo 03905

El Salvador
Ateneo de El Salvador, San Salvador 03921

Finland
Suomen Museoliitto, Helsinki 04052
Suomen Taideyhdistys, Helsinki 04062

France
Académie des Beaux-Arts, Paris 04093
Académie des Lettres, Sciences et Arts d'Amiens, Amiens 04097
Académie des Sciences, Agriculture, Arts et Belles-Lettres d'Aix, Aix-en-Provence 04099
Académie des Sciences, Arts et Belles-Lettres de Dijon, Dijon 04100
Académie des Sciences, Belles-Lettres et Arts de Clermont, Clermont-Ferrand 04101
Académie des Sciences, Belles-Lettres et Arts de Lyon, Lyon 04102
Académie des Sciences, Belles-Lettres et Arts de Rouen, Rouen 04103
Académie des Sciences, Belles-Lettres et Arts de Savoie, Chambéry 04104
Académie des Sciences, Lettres et Arts d'Arras, Arras 04107
Académie des Sciences, Lettres et Arts de Marseille, Marseille 04108
Académie Nationale des Sciences, Belles-Lettres et Arts de Bordeaux, Bordeaux 04120
Association de Recherche et d'Expression dans l'Art, Paris 04160
Association des Artistes Sculpteurs, Architectes, Graveurs et Dessinateurs, Paris 04167
Association Générale des Conservateurs des Collections Publiques de France, Paris 04252
Association Internationale des Arts Plastiques, Paris 04265
Association Les Amis de Gustave Courbet, Ornans 04297
Association Littéraire et Artistique Internationale, Paris 04299
Conseil International des Musées, Paris 04458
European Association for American Studies, Lyon 04480
Fédération Française des Sociétés d'Amis de Musées, Paris 04523
International Association of Sanskrit Studies, Paris 04582
Le Pays Bas-Normand, Flers 04619
Société Académique des Arts Libéraux de Paris, La Varenne Saint-Hilaire 04625
Société d'Agriculture, Sciences, Belles-Lettres et Arts d'Orléans, Orléans 04638
Société de l'Histoire de l'Art Français, Paris 04662
Société des Amis d'Eugène Delacroix, Paris 04696
Société des Professeurs de Dessin et d'Arts Plastiques de l'Enseignement Secondaire, Paris 04713
Société des Sciences, Arts et Belles-Lettres de Bayeux, Bayeux 04714
Société des Sciences et Arts, Saint-Denis 04715
Société d'Etude du Dix-Septième Siècle, Paris 04727
Société d'Etudes Juives, Paris 04744
Société d'Histoire et d'Archéologie Le Vieux Montmartre, Paris 04764
Société du Salon d'Automne, Paris 04776
Société Française d'Art Contemporain, Paris 04787
Société Française de Céramique, Paris 04790
Société Nationale des Beaux-Arts, Paris 04901
Union Centrale des Arts Décoratifs, Paris 04972

Germany
Akademie der Künste, Berlin 05048
Akademie Remscheid für musische Bildung und Medienerziehung e.V., Remscheid 05064
Arbeitsgemeinschaft der Deutschen Werkkunstschulen, Bremen 05084
Bayerische Akademie der Schönen Künste, München 05167
Bund Deutscher Kunsterzieher e.V., Hannover 05228
Bund für freie und angewandte Kunst e.V., Darmstadt 05264
Deutsche Akademie der Darstellenden Künste e.V., Frankfurt 05284
Deutsche Gesellschaft für Asienkunde e.V., Hamburg 05325
Deutsche Gesellschaft für Christliche Kunst e.V., München 05336
Deutscher Künstlerbund e.V., Berlin 05525
Deutscher Museumsbund e.V., Karlsruhe 05529
Deutscher Verein für Kunstwissenschaft e.V., Berlin 05558
Ernst Barlach Gesellschaft e.V., Hamburg 05642
Gesellschaft der Musik- und Kunstfreunde Heidelberg e.V., Heidelberg 05791
Humboldt-Gesellschaft für Wissenschaft, Kunst und Bildung e.V., Mannheim 05934
Internationale Arbeitsgemeinschaft der Archiv-, Bibliotheks- und Graphikrestauratoren, Marburg 05989
Internationaler Kunstkritikerverband, Sektion der Bundesrepublik Deutschland e.V., Bonn 05996
Internationaler Museumsrat, Deutsches Nationalkomitee, München 05997
Kestner-Gesellschaft, Hannover 06070
Mainzer Altertumsverein, Mainz 06107
Normenausschuss Bühnentechnik in Theatern und Mehrzweckhallen im DIN Deutsches Institut für Normung e.V., Berlin 06154
Südosteuropa-Gesellschaft e.V., München 06309
Verband Deutscher Kunsthistoriker e.V., Darmstadt 06353

Ghana
Arts Council of Ghana, Accra 06476
Ghana Academy of Arts and Sciences, Accra 06482

Greece
Association of Arts and Letters, Athinai 06500
Etaireia Byzantinon kai Metabyzantinon Meleton, Athinai 06523
Kallitechnikon Epimelitirion Ellados, Athinai 06538

Honduras
Instituto Hondureño de Cultura Interamericana, Tegucigalpa 06584

Iceland
Arkitektafélag Islands, Reykjavik 06677
Menntamálaráð, Reykjavik 06695

India
All India Fine Arts and Crafts Society, New Delhi 06714
Art Society of India, Bombay 06718
Bombay Art Society, Bombay 06744
Crafts Council of Western India, Bombay 06754
Indian Society of Oriental Art, Calcutta 06819
Jammu and Kashmir Academy of Art, Culture and Languages, Srinagar 06830
Museums Association of India, New Dehli 06837
National Academy of Art, New Dehli 06840

Iraq
Society of Iraqi Artists, Baghdad 06919

Ireland
Aosdana, Dublin 06922
Arts Council, Dublin 06925
The Friends of the National Collections of Ireland, Dublin 06951
The Irish Society for Design and Craftwork, Dublin 07000
Irish Society of Arts and Commerce, Dublin 07001
Photographic Society of Ireland, Dublin 07020
Royal Hibernian Academy of Arts, Dublin 07025

Israel
Museums Association of Israel, Tel Aviv 07107
Wilfrid Israel House for Oriental Art and Studies, Hazorea 07117

Italy
Academia Petrarca di Lettere, Arti e Scienze, Arezzo 07124
Accademia Albertina di Belle Arti e Liceo Artistico, Torino 07126
Accademia Biella Cultura, Biella 07131
Accademia degli Euteleti, San Miniato 07137
Accademia degli Incamminati, Modigliana 07138
Accademia degli Sbalzati, Sansepolcro 07141
Accademia dei Filodrammatici, Milano 07143
Accademia dei Filopatridi (Rubiconia), Savignano sul Rubicone 07144
Accademia dei Gelati, Scanno 07145
Accademia di Belle Arti, Firenze 07155
Accademia di Belle Arti, Milano 07156
Accademia di Belle Arti, Perugia 07157
Accademia di Belle Arti, Ravenna 07158
Accademia di Belle Arti e Liceo Artistico, Bologna 07159
Accademia di Belle Arti e Liceo Artistico, Carrara 07160
Accademia di Belle Arti e Liceo Artistico, Lecce 07161
Accademia di Belle Arti e Liceo Artistico, Napoli 07162
Accademia di Belle Arti e Liceo Artistico, Palermo 07163
Accademia di Belle Arti e Liceo Artistico, Roma 07164
Accademia di Belle Arti e Liceo Artistico, Venezia 07165
Accademia di Paestum Eremo Italico, Mercato San Severino 07170
Accademia di Scienze, Lettere e Arti, Udine 07174
Accademia di Scienze, Lettere e Belle Arti degli Zelanti e dei Dafnici, Acireale 07175
Accademia di Scienze, Lettere ed Arti, Palermo 07176
Accademia Fulginia di Arti, Lettere, Scienze, Foligno 07184
Accademia Il Tetradramma, Roma 07188
Accademia Internazionale d'Arte Moderna, Roma 07189
Accademia Internazionale della Tavola Rotonda, Milano 07190
Accademia Ligustica de Belle Arti, Genova 07201
Accademia Nazionale di Belle Arti di Parma, Parma 07216
Accademia Nazionale di San Luca, Roma 07220
Accademia Nazionale di Scienze, Lettere e Arti, Modena 07222
Accademia Nazionale Virgiliana di Scienze, Lettere ed Arti di Mantova, Mantova 07224
Accademia Olimpica, Vicenza 07225
Accademia Pomposiana, Codigoro 07227
Accademia Pontaniana, Napoli 07228
Accademia Prenestina del Cimento di Musica, Lettere, Scienze, Arti Visive e Figurative, Palestrina 07230
Accademia Romana di Cultura, Roma 07232
Accademia Salentina di Lettere ed Arti, Lecce 07235
Accademia Scientifica, Letteraria, Artistica del Frignano Lo Scoltenna, Pievepelago 07236
Accademia Senese degli Intronati, Siena 07237
Accademia Spagnola di Belle Arti, Roma 07239
Accademia Spoletina, Spoleto 07240
Accademia Tedesca Villa Massimo, Roma 07241
Accademia Universale Citta' Eterna, Roma 07244
Accademia Universale Guglielmo Marconi, Roma 07245
Associazione di Cultura Lao Silesu, Iglesias 07271
Associazione Nazionale dei Musei Italiani, Roma 07363
Associazione per Imola Storico-Artistica, Imola 07388
Associazione Siciliana per le Lettere e le Arti, Palermo 07402
Associazione Sole Italico, Sulmona 07404
Centro Camuno di Studi Preistorici ed Etnologici, Capo di Ponte 07417
Centro d'Arte e di Cultura, Bologna 07418
Centro Europeo per la Diffusione della Cultura, Roma 07463
Centro Internazionale per l'Educazione Artistica della Fondazione Giorgio Cini, Venezia 07481
Centro Studi e Documentazione della Cultura Armena, Milano 07571
Comitato per Bologna Storica-Artistica, Bologna 07619
Ente Autonomo La Biennale de Venezia, Venezia 07649
Sindacato Nazionale Istruzione Artistica, Roma 07767
Società Dauna di Cultura, Foggia 07778
Società Europea di Cultura, Venezia 07802
Società Incoraggiamento Arti e Mestieri, Milano 07814
Società Italiana per l'Archeologia e la Storia delle Arti, Napoli 07965
Società Nazionale di Scienze, Lettere ed Arti, Ex Società Reale, Napoli 07992
Società per gli Studi Storici, Archeologici ed Artistici della Provincia di Cuneo, Cuneo 07999
Società Tarquiniense d'Arte e Storia, Tarquinia 08021
Società Tiburtina di Storia e d'Arte, Tivoli 08023

Japan
Bijutsu-shi Gakkai, Tokyo 08115

Kenya
Kenya National Academy of Sciences, Nairobi 08505

Korea, Republic
National Academy of Arts, Seoul 08542

Laos
Lao Buddhist Fellowship, Vientiane 08552

Liberia
Liberia Arts and Crafts Association, Monrovia 08564

Macedonia
Društvo na Istoričarite na Umetnosta od Makedonija, Skopje 08593
Makedonska Akademija na Naukite i Umetnostite, Skopje 08603

Mauritius
Royal Society of Arts and Sciences of Mauritius, Réduit 08674

Mexico
Academia de Artes Plásticas, Guanajuato 08678
Academia de Artes Plásticas, Hermosillo 08679
Ateneo de Ciencias y Artes de Chiapas, Tuxtla Gutiérrez 08707
Equipe 7 des Arts Visuels, México 08733
Instituto Nacional de Bellas Artes, México 08737

Namibia
Arts Association of Namibia, Windhoek 08814

Netherlands
Comité International d'Histoire de l'Art, Utrecht 08842
Koninklijk Oudheidkundig Genootschap Amsterdam, Amsterdam 08968
Maatschappij Arti et Amicitiae, Amsterdam 08973
Nederlandse Museumvereniging, Enkhuizen 08995
Raad voor de Kunst, 's-Gravenhage 09063
Rijksbureau voor Kunsthistorische Documentatie, 's-Gravenhage 09066
Vereniging van Nederlandse Kunsthistorici, Groningen 09093

New Caledonia
Council of Pacific Arts, Nouméa 09113

New Zealand
Art Galleries and Museums Association of New Zealand, Wellington 09116
New Zealand Academy of Fine Arts, Wellington 09132
New Zealand Maori Arts and Crafts Institute, Rotorua 09166
Queen Elizabeth II Arts Council of New Zealand, Wellington 09191

Norway
Norges Kunstnerråd, Oslo 09285
Norske Kunst- og Kulturhistoriske Museer, Oslo 09302
Norwegian Council of Cultural Affairs, Oslo 09341

Pakistan
Arts Council of Pakistan, Karachi 09359
Pakistan Museum Association, Karachi 09390

Peru
Asociación de Artistas Aficionados, Lima 09447

Poland
Stowarzyszenie Historyków Sztuki, Warszawa 09616

Portugal
Academia Nacional de Belas-Artes, Lisboa 09634
Sociedade Nacional de Belas Artes, Lisboa 09688

Puerto Rico
Ateneo de Ponce, Ponce 09725
Ateneo Puertorriqueño, San Juan 09726
Sociedad Mayaguezana por Bellas Artes, Mayaguez 09731

Romania
Uniunea Artistilor Plastici din Romania, Bucuresti 09797

Russia
Museum Council, Moskva 09888
Russian Academy of Arts, Moskva 09951

Saudi Arabia
Higher Council for Promotion of Arts and Letters, Riyadh 10003
King Abdul Aziz Research Centre, Riyadh 10004

Singapore
Indian Fine Arts Society, Singapore 10047

Slovenia
Union of Arts of the Russian Federation, Moskva 10110

South Africa
Human Sciences Research Council, Pretoria 10147
South African Academy of Science and Arts, Pretoria 10177
South African Association of Arts, Pretoria 10181
The Southern African Museums Association, Sunnyside 10226

Spain
Ateneo Barcelonés, Barcelona 10283
Ateneo Científico, Literario y Artístico, Madrid 10284
Ateneo Científico, Literario y Artístico, Mahón 10285
Departamento de Historia del Arte Diego Velázquez, Madrid 10303

Fundación Juan March, Madrid 10311
Institución Fernando el Católico, Zaragoza 10312
Instituto Amatller de Arte Hispánico, Barcelona 10316
Real Academia de Bellas Artes de la Purísima Concepción, Valladolid 10337
Real Academia de Bellas Artes de San Fernando, Madrid 10338
Real Academia de Bellas Artes de Santa Isabel de Hungría, Sevilla 10339
Real Academia de Bellas Artes de San Telmo, Málaga 10340
Real Academia de Bellas Artes y Ciencias Históricas de Toledo, Toledo 10341
Real Academia de Ciencias y Artes de Barcelona, Barcelona 10344
Real Academia de Córdoba de Ciencias, Bellas Letras y Nobles Artes, Córdoba 10345
Real Academia de Nobles y Bellas Artes de San Luis, Zaragoza 10352
Real Sociedad Fotográfica, Madrid 10364
Sociedad de Ciencias, Letras y Artes El Museo Canario, Las Palmas 10372

Sri Lanka

Ceylon Society of Arts, Colombo 10414
Royal Asiatic Society of Sri Lanka, Colombo 10418

Suriname

Stichting Cultureel Centrum Suriname, Paramaribo 10432

Swaziland

Swaziland Art Society, Mbabane 10434

Sweden

Konsthistoriska Sällskapet, Stockholm 10481
Kungliga Akademien för de fria Konsterna, Stockholm 10482
Kungliga Vetenskaps- och Vitterhets-Samhället i Göteborg, Göteborg 10491
Svenska Museiföreningen, Stockholm 10565
Sveriges Allmänna Konstförening, Stockholm 10586
Sveriges Museimannaförbund, Nacka 10592

Switzerland

Académie Internationale de Céramique, Genève 10606
Gesellschaft für das Schweizerische Landesmuseum, Zürich 10712
Gesellschaft für Schweizerische Kunstgeschichte, Bern 10716
Gesellschaft schweizerischer Zeichenlehrer, Einsiedeln 10722
Kunstverein Sankt Gallen, Sankt Gallen 10763
Opera Svizzera dei Monumenti d'Arte, Locarno 10775
Pro Helvetia, Zürich 10781
Schweizerischer Werkbund, Zürich 10923
Schweizerisches Institut für Kunstwissenschaft, Zürich 10927
Verband der Museen der Schweiz, Solothurn 10997
Vereinigung der Freunde antiker Kunst, Basel 11010

Thailand

Royal Institute, Bangkok 11097

Togo

Association Togolaise pour le Développement de la Documentation, des Bibliothèques, Archives et Musées, Lomé 11104

Tunisia

Comité National des Musées, Tunis 11140

United Kingdom

Arts Association, London 11252
Arts Council of Great Britain, London 11253
Arts Council of Northern Ireland, Belfast 11254
Association of Art Institutions, Hereford 11302
Association of Arts Centres in Scotland, Aberdeen 11303
Association of Principals of Colleges (Northern Ireland Branch), Lisburn 11379
Birmingham and Midland Institute, Birmingham 11446
British Archaeological Association, Lomdon 11481
British Arts, London 11483
British Association of Friends of Museums, Manchester 11506
British Institute of Persian Studies, London 11586
British Museum Society, London 11612
British Watercolour Society, Ilkley 11707
Chester Society of Natural Science, Literatur and Art, Chester 11774
Chippendale Society, Leeds 11779
Contemporary Art Society, London 11840
County Museum Society, Kidderminster 11862
Designers and Art Directors Association of London, London 11886
Design Research Centre for the Gold, Silver and Jewellery Industries, London 11887
The Devonshire Association for the Advancement of Science, Literature and Art, Exeter 11890
Dunedin Society, Glasgow 11912
Eastern Arts Association, Cambridge 11916
East Midlands Arts Board, Loughborough 11920
Federation of British Artists, London 12063
Hesketh Hubbard Art Society, London 12166
Honourable Society of Cymmrodorion, London 12180
Institute of Contemporary Arts, London 12227
Museum Attendants Association, Old Portsmouth 12543
Museums and Galleries Commission, London 12544
The Museums Association, London 12545
Museum Training Institute, Bradford 12546
National Art-Collections Fund, London 12556
National Society for Education in Art and Design, Corsham 12608
New English Art Club, London 12626
North Wales Arts Association, Bangor 12645
Oriental Ceramic Society, London 12661
Pastel Society, London 12681
The Plymouth Athenaeum, Plymouth 12705
Pugin Society, London 12728
Royal Academy of Arts in London, London 12760
Royal Cambrian Academy of Art, Conwy 12775
Royal Drawing Society, London 12798
Royal Fine Art Commission, London 12803
Royal Institute of Painters in Water Colours, London 12816
Royal Scottish Society of Arts, Edinburgh 12847
Royal Society for the Encouragement of Arts, Manufactures and Commerce, London 12851
Royal Society of Arts, London 12854
Royal Society of British Artists, London 12855
Royal Society of Portrait Painters, London 12863
Royal West of England Academy, Bristol 12875
Saltire Society, Edinburgh 12885
Scottish Arts Council, Edinburgh 12893
The Society for South Asian Studies, London 12997
Society of County Museum Directors, West Malling 13041
Society of Wildlife Artists, London 13070
Society of Women Artists, London 13071
Southern Arts Board, Winchester 13079
Standing Conference of Arts and Social Sciences, Cardiff 13095
United Kingdom Institute for Conservation of Historic and Artistic Works, London 13166
The Walpole Society, London 13193
Wedgwood Society, Fordingbridge 13196
Welsh Arts Council, Cardiff 13198
William Morris Society, London 13208
Yorkshire Arts Association, Bradford 13231

U.S.A.

Academy of Arts and Sciences of the Americas, Miami 13285
American Academy and Institute of Arts and Letters, New York 13375
American Academy of Arts and Sciences, Cambridge 13382
American Association of Museums, Washington 13554
American Council for the Arts, New York 13760
American Federation of Arts, Berkeley 13807
American Fine Arts Society, New York 13815
American Institute for Conservation of Historic and Artistic Works, Washington 13866
American Oriental Society, Ann Arbor 13968
American Studies Association, College Park 14185
An Claidheamh Soluis – The Irish Arts Center, New York 14217
Anonymous Arts Recovery Society, New York 14224
Archives of American Art, Washington 14240
Art Institute of Light, Gap 14249
Arts International Program of the Institute of International Education, New York 14251
Association for the Study of Dada and Surrealism, Los Angeles 14357
Association of Youth Museums, Memphis 14535
Athenaeum of Philadelphia, Philadelphia 14545
California Rug Study Society, Oakland 14632
Classical America, New York 14760
College Art Association, New York 14770
Committee for Hispanic Arts and Research, Austin 14795
Council of Colleges of Arts and Sciences, Columbus 14892
Council on Technology Teacher Education, Ypsilanti 14934
Creative Time, New York 14939
Czechoslovak Society of Arts and Sciences, Sherman Oaks 14947
Deltiologists of America, Norwood 14959
Experiments in Art and Technology, Berkeley Heights 15071
Foundation for Research in the Afro-American Creative Arts, Cambria Heights 15131
Friends of American Art in Religion, New York 15141
Hispanic Society of America, New York 15213
Institute of American Indian Arts, Santa Fe 15309
International Center of Medieval Art, New York 15432
International Council of Museums – Committee of the American Association of Museums, Washington 15457
International Council of the Museum of Modern Art, New York 15459
International Foundation for Art Research, New York 15478
International Society of Performing Arts Administrators, Grand Rapids 15554
International Technology Education Association, Reston 15560
International Technology Education Association – Council for Supervisors, Richmond 15561
Jewish Academy of Arts and Sciences, New York 15603
Leonardo – The International Society for the Arts, Sciences and Technology, San Francisco 15660
Medieval Academy of America, Cambridge 15714
Music and Arts Society of America, Washington 15761
National Academy of Western Art, Oklahoma City 15778
National Alliance of Media Arts Centers, Oakland 15785
National Art Education Association, Reston 15790
National Association for Applied Arts, Science and Education, Milwaukee 15792
National Council for Culture and Art, New York 15986
National Council for the Traditional Arts, Silver Spring 15993
National Foundation for Advancement in the Arts, Miami 16056
National Guild of Community Schools of the Arts, Englewood 16067
National League of American PEN Women, Washington 16100
Pacific Arts Association, Honolulu 16279
Performing and Visual Arts Society, Kinnelon 16304
Polish Institute of Arts and Sciences in America, New York 16335
Primitive Art Society of Chicago, Chicago 16353
Rogers Book Club, Beltsville 16426
Saving and Preserving Arts and Cultural Environments, Los Angeles 16440
Society for Asian Art, San Francisco 16499
Society for the Arts, Religion and Contemporary Culture, Boston 16592
Sumi-E Society of America, Alexandria 16753
Tamarind Institute, Albuquerque 16762
Ukrainian Academy of Arts and Sciences in the U.S., New York 16796
United States Committee of the International Association of Art, Washington 16818
Volunteer Committees of Art Museums of Canada and the United States, Sheridan 16868
Women's Caucus for Art, Philadelphia 16913

Vatican City

Pontificia Insigne Accademia Artistica dei Virtuosi al Pantheon, Roma 16957

Venezuela

Asociación Venezolana Amigos del Arte Colonial Caracas, Caracas 16972

Vietnam

Writers and Artists Union, Hanoi 17072

Yugoslavia

Akademia e Shkencave dhe e Arteve e Kosovës, Priština 17076
Crnogorska Akademija Nauka i Umjetnosti, Podgorica 17080
Srpska Akademija Nauka i Umetnosti, Beograd 17125

Zambia

Africa Association for Liturgy, Music and Arts, Lusaka 17148

Zimbabwe

Arts Association Harare, Harare 17163

Astronomy

Algeria

Centre National d'Astronomie, d'Astrophysique et de Géophysique, Alger 00015

Argentina

Asociacíon Argentina Amigos de la Astronomía, Buenos Aires 00054
Asociación Argentina de Astronomía, La Plata 00056

Australia

Astronomical Society of Australia, Macquarie 00231
Astronomical Society of Queensland, Boondall 00232
Astronomical Society of South Australia, Adelaide 00233
Astronomical Society of Tasmania, Launceston 00234
Astronomical Society of Victoria, Melbourne 00235
Astronomical Society of Western Australia, Perth 00236
Inter-Union Commission on Frequency Allocations for Radio Astronomy and Space Science, Epping 00426

Austria

Österreichische Gesellschaft für Weltraumforschung, Innsbruck 00845
Österreichischer Astronomischer Verein, Wien 00886

Barbados

Barbados Astronomical Society, Bridgetown 01073

Belgium

Société Astronomique de Liège, Cointe-Ougrée 01422
Société Royale Belge d'Astronomie, de Météorologie et de Physique du Globe, Bruxelles 01470

Bermuda

Astronomical Society of Bermuda, Hamilton 01518

Canada

Canadian Astronomical Society, Ottawa 02077
Institut d'Horlogerie du Canada, Montréal 02523
Royal Astronomical Society of Canada, Toronto 02863

China, People's Republic

Chinese Astronomical Society, Nanjing 03165

China, Republic

Astronomical Society of the Republic of China, Taipei 03249

Denmark

Astronomisk Selskab, København 03564

Ecuador

Sociedad Ecuatoriana de Astronomía, Quito 03856

Finland

Tähtitieteellinen Yhdistys Ursa, Helsinki 04069

France

Agence Spatiale Européenne, Paris 04127
Association Française d'Astronomie, Paris 04212
Association Française d'Observateurs d'Etoiles Variables, Strasbourg 04238
Association of French-Speaking Planetariums, Strasbourg 04319
Astronomical Data Centre, Strasbourg 04359
Comité Mondial pour la Recherche Spatiale, Paris 04411
Société Astronomique de Bordeaux, Bordeaux 04632
Société Astronomique de France, Paris 04633
Société Astronomique de Lyon, Saint-Genis-Laval 04634
Union Astronomique Internationale, Paris 04971

Germany

Astronomische Gesellschaft e.V., Heidelberg 05157
Deutsche Gesellschaft für Chronometrie e.V., Stuttgart 05337

Greece

Greek National Committee for Astronomy, Athinai 06528
Greek National Committee for Space Research, Athinai 06529
Greek National Committee for the Quiet Sun International Years, Athinai 06530

India

Astronomical Society of India, Hyderabad 06736
Indian Space Research Organization, Bangalore 06821

Indonesia

Astronomical Association of Indonesia, Jakarta 06886

Ireland

Irish Astronomical Society, Dublin 06977

Israel

Tel Aviv Astronomical Association, Tel Aviv 07115

Italy

Associazione Astrofili Bolognesi, Bologna 07261
Associazione Urania, Genova 07410
Società Astronomica Italiana, Milano 07772

Japan

Ikomasan Tenmon Kyokai, Nara Ken 08139
Nihon Tenmon Gakkai, Tokyo 08268

Mexico

Sociedad Astronómica de México, México 08746

Netherlands

Nederlandse Astronomenclub, Leiden 08987

New Zealand

Royal Astronomical Society of New Zealand, Wellington 09195

Peru

Asociación Peruana de Astronomía, Lima 09455

Poland

Copernicus Astronomical Centre, Warszawa 09542
Polskie Towarzystwo Astronomiczne, Warszawa 09555
Polskie Towarzystwo Milosników Astronomii, Kraków 09589

Portugal

Sociedade Astronómica de Portugal, Lisboa 09675

Russia

Astronomical and Geodesical Society, Moskva 09806
Commission for Meteorites and Space Dust, Novosibirsk 09824

Slovakia

Slovenská Astronomická Spoločnost, Tatranská Lomnica 10063

Slovenia

Društvo Matematikov, Fizikov i Astronomov Slovenije, Ljubljana 10098

South Africa

Astronomical Society of Southern Africa, Observatory 10125

Sweden

Svenska Astronomiska Sällskapet, Saltsjöbaden 10537

Switzerland
Schweizerische Gesellschaft für Astrophysik und Astronomie, Sauverny 10810
Société Astronomique de Suisse, Meyrin 10967

United Kingdom
Association in Scotland to Research into Astronautics, Wishaw 11297
Astronomical Society of Edinburgh, Edinburgh 11414
British Astronomical Association, London 11530
British Interplanetary Society, London 11595
British National Committee on Space Research, London 11615
European Space Association, Stevenston 12041
Interplanetary Space Travel Research Association, London 12408
Junior Astronomical Society, Nottingham 12424
Royal Astronomical Society, London 12772
Science and Engineering Research Council, Swindon 12889
Webb Society, London 13195

U.S.A.
American Association of Variable Star Observers, Cambridge 13597
American Astronomical Society, Washington 13610
American Lunar Society, East Pittsburgh 13922
American Society for Mass Spectrometry, Santa Fe 14071
Association of Universities for Research in Astronomy, Washington 14525
Astrologers' Guild of America, Brewster 14539
Astronomical League, Rolling Meadows 14540
Astronomical Society of the Pacific, San Francisco 14541
Carnegie Institution of Washington, Washington 14647
International Dark-Sky Association, Tucson 15464
International Planetarium Society, Salt Lake City 15513
Maria Mitchell Association, Nantucket 15690
Middle Atlantic Planetarium Society, Wheeling 15731
Planetary Society, Pasadena 16323
Stonehenge Study Group, Santa Barbara 16747
United States Antarctic Research Program, Washington 16815
Universities Space Research Association, Columbia 16840
Von Braun Astronomical Society, Huntsville 16869

Yugoslavia
Association of the Mathematicians', Physicists' and Astronomers' Societies of Yugoslavia, Priština 17079
Savez Društava Matematičara, Fizičara i Astronoma Jugoslavije, Novi Sad 17112

Astrophysics

Australia
Astronomical Society of Australia, Macquarie 00231

Ghana
Ghana Meteorological Services Department, Accra 06488

Russia
Committee on Meteorites, Moskva 09852

Slovakia
Slovenská Meteorologická Spoločnost, Bratislava 10079

Switzerland
Schweizerische Gesellschaft für Astrophysik und Astronomie, Sauverny 10810

U.S.A.
Association of Space Explorers U.S.A-, San Francisco 14504
Joint Institute for Laboratory Astrophysics, Boulder 15620
National Association of State Park Directors, Tallahassee 15902
National Avionics Society, Lafayette 15918
National Space Club, Washington 16161
National Space Society, Washington 16162
Space Studies Institute, Princeton 16733

Automotive Engineering s. a. Machine Engineering

Australia
Society of Automotive Engineers – Australasia, Parkville 00493

Canada
Electric Vehicle Association of Canada, Ottawa 02428

Finland
Suomen Autoteknillinen Liitto, Helsinki 04008

France
Union Technique de l'Automobile, du Motocycle et du Cycle, Paris 04997

Germany
Bundesverband der freiberuflichen und unabhängigen Sachverständigen für das Kraftfahrzeugwesen e.V., Königswinter 05243
Forschungsvereinigung Automobiltechnik e.V., Frankfurt 05742
Forschungsvereinigung Verbrennungskraftmaschinen e.V., Frankfurt 05754
Normenausschuss Kraftfahrzeuge im DIN Deutsches Institut für Normung e.V., Frankfurt 06185
VDI-Gesellschaft Fahrzeugtechnik, Düsseldorf 06323

Ireland
Automobile, General Engineering and Mechanical Operations Union, Dublin 06933
Society of the Irish Motor Industry, Dublin 07034

Italy
Centro Sviluppo Impiego Diesel, Trieste 07599
Federazione Italiana delle Scienze e delle Attività Motorie, Roma 07691

Russia
Automobile and Road Building Society, Moskva 09807

Switzerland
Fachgruppe Carosserie- und Fahrzeugtechnik des STV, Sissach 10687

United Kingdom
Motor Industry Research Association, Nuneaton 12540

U.S.A.
Electric Auto Association, Belmont 15032

Aviation s. Aeronautics, Aviation, Space Technology

Bacteriology s. Microbiology

Banking s. Finance

Behavioral Sciences s. a. Psychology

Italy
International Society for Twin Studies, Roma 07720

South Africa
Primate Behaviour Research Group, Johannesburg 10165

United Kingdom
Association for the Study of Animal Behaviour, Egham 11291
Group-Analytic Society, London 12129
Institute for Sex Education and Research, Birmingham 12210
The Institute of Group-Analysis, Group-Analysis Society (London), London 12240

U.S.A.
American Behavioral Science Association, New York 13619
American Institutes for Research in the Behavioral Sciences, Washington 13899
American Society of Trial Consultants, Towson 14170
Animal Behavior Society, Denver 14218
Association for Advancement of Behavior Therapy, New York 14263
Association for the Behavioral Sciences and Medical Education, McLean 14347
Association for the Treatment of Sexual Abusers, Portland 14360
Association of Social and Behavioral Scientists, Washington 14502
Behavioral Research Council, Great Barrington 14574
Center for Advanced Study in the Behavioral Sciences, Stanford 14663
Council for Children with Behavioral Disorders, Reston 14872
International Association of Human-Animal Interaction Organizations, Renton 15408
International Society for Human Ethology, Detroit 15531
International Society for Research on Aggression, Chicago 15540
NTL Institute for Applied Behavioral Science, Alexandria 16234
Society for the Scientific Study of Religion, West Lafayette 16610
Society of Behavioral Medicine, Rockville 16641

Biochemistry

Argentina
Asociación Argentina de Farmacia y Bioquímica Industrial, Buenos Aires 00064
Asociación Bioquímica Argentina, Buenos Aires 00074

Australia
Australian Biochemical Society, Canberra 00270
Australian Society for Biochemistry and Molecular Biology, Edwardstown 00338

Austria
Internationale Gesellschaft für Getreidewissenschaft und -technologie, Schwechat 00669
Österreichische Biochemische Gesellschaft, Wien 00731

Belgium
Société Belge de Biochimie et de Biologie Moléculaire, Bruxelles 01425

Bulgaria
Bulgarian Biochemical and Biophysical Society, Sofia 01745
Federation of European Biochemical Societies, Sofia 01767

Canada
Canadian Biochemical Society, London 02082

Chile
Sociedad de Bioquímica de Concepción, Concepción 03091

China, People's Republic
Chinese Biochemical Society, Shanghai 03166

Czech Republic
Česká Společnost Biochemická, Praha 03518
Česká Společnost Parasitologická, Praha 03527

Denmark
Biokemisk Forening, København 03566

Finland
Societas Biochemica, Biophysica et Microbiologica Fenniae, Helsingin Yliopisto 03992

France
Société de Chimie Biologique, Paris 04645
Société Française de Biologie Clinique, Paris 04788

Georgia
Georgian Society of Biochemistry, Tbilisi 05021

Germany
Europäische Föderation Biotechnologie, Frankfurt 05651
Gesellschaft für Biologische Chemie e.V., Tutzing 05805

Hungary
Magyar Biokemiai Társaság, Budapest 06620

India
Society of Biological Chemists, India, Bangalore 06858

Israel
Biochemical Society of Israel, Rehovot 07048

Italy
Bioelectrochemical Society, Bologna 07412
Società Italiana di Biochimica, Perugia 07841
Società Italiana di Patologia Vascolare, Roma 07931

Malaysia
Malaysian Biochemical Society, Kuala Lumpur 08645

Mexico
Centro de Investigación y de Estudios Avanzados del Instituto Politécnico Nacional, México 08718

Netherlands
Vereniging Het Nederlands Kanker Instituut, Amsterdam 09085

New Zealand
Australian Association of Clinical Biochemists – New Zealand Branch, Hamilton 09120
New Zealand Association of Clinical Biochemists, Wellington 09134
New Zealand Biochemical Society, Canterbury 09137

Poland
Polskie Towarzystwo Biochemiczne, Warszawa 09557

Portugal
Sociedade Portuguesa de Bioquímica, Lisboa 09693

Russia
Biochemical Society, Moskva 09808
National Committee of Biochemists, Moskva 09891

Slovakia
Slovenská Biochemiká Spoločnost, Bratislava 10064

South Africa
Physiological and Biochemical Society of Southern Africa, Pretoria 10161
Society of Physiologists, Biochemists and Pharmacologists, Pretoria 10175

Spain
Sociedad Española de Bioquímica, Madrid 10378

United Kingdom
Association of Clinical Biochemists, London 11324
The Biochemical Society, London 11440
Institute for Animal Health Pirbright, Pirbright Woking 12205
Phytochemical Society of Europe, Swansea 12699

U.S.A.
American Crystallographic Association, Buffalo 13774
American Society for Biochemistry and Molecular Biology, Bethesda 14047
American Society for Bone and Mineral Research, Washington 14048
Food and Nutrition Board, Washington 15108
Huxley Institute for Biosocial Research, Boca Raton 15242
International Isotope Society, East Hanover 15492
Pan American Association of Biochemical Societies, Kansas City 16288
Protein Society, Bethesda 16364

Vietnam
Vietnamese Association of Medical Biochemistry, Hanoi 17052

Biology

Algeria
Centre de Recherches sur les Resources Biologiques Terrestres, Alger 00013

Argentina
Asociación Argentina de Biología y Medicina Nuclear, Buenos Aires 00059
Asociación Latinoamericana de Sociedades de Biología y Medicina Nuclear, Buenos Aires 00087
Centro de Investigación de Biologia Marina, San Martín 00104
Centro de Investigaciones Bella Vista, Bella Vista 00106
Sociedad Argentina de Biología, Buenos Aires 00158
Sociedad de Biología de Córdoba, Córdoba 00190

Armenia
Committee on UNESCO Programme „Man and Biosphere", Yerevan 00205
Interdepartmental Council on Physicochemical Biology and Biotechnology, Yerevan 00209

Australia
Australian Federation for Medical and Biological Engineering, Parkville 00291
Australian Society for Biochemistry and Molecular Biology, Edwardstown 00338
Australian Society for Reproductive Biology, Newcastle 00346
Bio-Rhythm Research and Information Centre, Brisbane 00367
International Association of Environmental Mutagen Societies, Adelaide 00423

Austria
Arbeitsgemeinschaft für Neurobiologie, Wien 00554
Gesellschaft der Freunde der Biologischen Station Wilhelminenberg, Wien 00606
Haus des Meeres Vivarium Wien, Wien 00648

Belgium
Association pour l' Introduction de la Nomenclature Biologique Nouvelle, Kalmthout 01154
Charles Darwin Foundation for the Galapagos Isles, Bruxelles 01212
Comité International de Standardisation en Biologie Humaine, Bruxelles 01230
European Community Biologists Association, Namur 01326
Société Belge de Biologie Clinique, Charleroi 01426
Société Européenne de Radiobiologie, Liège 01459

Brazil
Centro de Biomédica de Campina Grande, Campina Grande 01621
Instituto Nacional de Pesquisas da Amazonia, Manaus 01688
Sociedade de Biologia do Brasil, Rio de Janeiro 01718
Sociedade de Biologia do Rio Grande do Sul, Pôrto Alegre 01719
Sociedade de Farmácia e Química de São Paulo, São Paulo 01723

Canada
Alberta Society of Professional Biologists, Edmonton 01848
Biological Council of Canada, Ottawa 01965
Canadian Council of University Biology Chairmen, Saskatoon 02109
Canadian Federation of Biological Societies, Ottawa 02129
Canadian Medical and Biological Engineering Society, Ottawa 02196
Canadian Quaternary Association, Waterloo 02239
Canadian Society of Environmental Biologists, Toronto 02284
International Federation for Medical and Biological Engineering, Ottawa 02553
Jeunes Biologistes du Québec, Saint-Fidèle 02569
Pulp and Paper Research Institute of Canada, Pointe Claire 02853
Société Linnéenne de Québec, Sainte Foy 02933

Chile
Sociedad de Biología de Chile, Santiago 03089
Sociedad de Biología de Concepción, Concepción 03090

China, Republic
Asia and Oceania Federation of Nuclear Medicine and Biology, Taipei 03233
Biological Society of China, Taipei 03251

Colombia
Fundación OFA para el Avance de las Ciencias Biomédicas, Bogotá 03375
Sociedad Colombiana de Biologí, Bogotá 03404

Cuba
Centro Nacional de Investigaciones Científicas, La Habana *03478*

Czech Republic
Česká Biologicka Společnost, Brno *03503*
Česká Společnost Bioklimatologická, Praha *03519*

Denmark
Biologisk Selskab, København *03567*
Dansk Biologisk Selskab, København *03594*
Foreningen af Danske Biologer., Birkerød *03727*

Egypt
Egyptian Organization for Biological Products and Vaccines, Giza *03894*

Finland
Committee for Mapping the Flora of Europe, Helsinki *03953*
Oulun Luonnonystäväin Yhdistys, Oulu *03988*
Societas Biologica Fennica Vanamo, Helsingin Yliopisto *03993*

France
Associated European Human Biological Centres, Paris *04142*
Association Internationale de Médecine et de Biologie de l'Environnement, Paris *04260*
Association pour le Développement des Etudes Biologiques en Psychiatrie, Paris *04338*
Commission de l'Enseignement Supérieur en Biologie, Paris *04423*
Fédération Générale des Syndicats de Biologistes, Paris *04527*
Société Centrale d'Apiculture, Paris *04637*
Société de Biologie, Paris *04643*
Société de Biométrie Humaine, Paris *04644*
Société d'Etudes et de Recherches Biologiques, Paris *04733*
Société Internationale de Biologie Mathématique, Antony *04876*
Syndicat National des Médecins Biologistes, Paris *04949*
Union des Biologistes de France, Paris *04973*
Union Internationale des Sciences Biologiques, Paris *04988*
Union Nationale des Techniciens Biologistes, Le Perreux *04993*

Germany
Arbeitsgemeinschaft wildbiologischer und jagdkundlicher Forschungsstätten, Bonn *05130*
Arbeitsgruppe für strukturelle Molekularbiologie in der Max-Planck-Gesellschaft, Hamburg *05134*
Deutsche Gesellschaft für Biomedizinische Technik e.V., Berlin *05331*
Deutsche Gesellschaft für Polarforschung, Bremerhaven *05409*
Deutsche Röntgengesellschaft, Neu-Isenburg *05535*
Gesellschaft für Ernährungsbiologie e.V., München *05819*
Gesellschaft für Informationsvermittlung und Technologieberatung, München *05831*
Internationale Biometrische Gesellschaft, Deutsche Region, Aachen *05990*
International Society of Developmental Biologists, Göttingen *06011*
Psychobiologische Gesellschaft, Mülheim *06242*
Verband Deutscher Biologen e.V., Hamburg *06350*

Hungary
Magyar Anatómusok, Histologusok és Embryologusok Társasága, Budapest *06615*
Magyar Biológiai Társaság, Budapest *06621*
Magyar Gerontológiai Társaság, Budapest *06632*

Iceland
Hid Islenska Náttúrufraedifélag, Reykjavik *06689*

India
Indian Association of Biological Sciences, Calcutta *06772*
Indian Society for Nuclear Techniques in Agriculture and Biology, New Delhi *06814*
Society of Biological Chemists, India, Bangalore *06858*

Ireland
Dublin University Biological Association, Dublin *06942*

Israel
Israel Gerontological Society, Tel Aviv *07074*
Israel Society for Experimental Biology and Medicine, Rehovoth *07091*

Italy
Accademia Medica Lombarda, Milano *07205*
Accademia Romana di Scienze Mediche e Biologiche, Roma *07233*
Associazione Italiana di Ingegneria Medica e Biologica, Napoli *07312*
Associazione Italiana di Radiobiologia Medica, Catania *07320*
Centre for Human Evolution Studies, Roma *07414*
Centro Italiano di Biostatistica, Roma *07491*
Collegium Biologicum Europa, Roma *07609*
European Communities Biologists Association, Roma *07674*
Societá Internazionale di Studi Gemellari, Roma *07818*
Società Italiana di Biologia Marina, Livorno *07843*
Società Italiana di Biologia Sperimentale, Napoli *07844*
Società Italiana di Biometria, Firenze *07845*
Società Italiana di Fotobiologia, S. Maria di Galeria *07877*
Società Italiana di Microbiologia, Catania *07907*
Società Italiana di Psichiatria Biologica, Napoli *07935*

Japan
Nihon Ikushu Gakkai, Tokyo *08210*
Nihon Seibutsu Kankyo Chosetsu Kenkyukai, Tokyo *08256*
Nippon Nogei Kagaku Kai, Tokyo *08348*
Nippon Rikusui Gakkai, Tokyo *08357*
Senshokutai Gakkai, Tokyo *08414*

Jordan
Arab Biosciences Network, Jubaiha *08450*

Korea, Republic
Korean Association for the Biological Sciences, Seoul *08528*

Mexico
Sociedad de Estudios Biológicos, México *08748*

Moldova, Republic of
Hydrobiological Society of Moldova, Chişinău *08780*

Netherlands
Koninklijke Nederlandse Natuurhistorische Vereniging, Utrecht *08958*
Nederlandse Organisatie voor Wetenschappelijk Onderzoek, 's-Gravenhage *08998*

Philippines
Asian Association for Biology Education, Manila *09494*
Los Baños Biological Club, Laguna *09521*

Romania
Societatea de Stiinte Biologice din Romania, Bucuresti *09780*

Russia
Hydrobiological Society, Moskva *09873*
National Committee of Biologists, Moskva *09892*
National Committee on the International Biological Programme, Moskva *09906*
Society of Herminthologists, Moskva *09973*

Saudi Arabia
Saudi Biological Society, Riyadh *10005*

Senegal
African Biosciences Network, Dakar *10008*

Singapore
Asian Network for Biological Sciences, Singapore *10036*

Slovakia
Slovenská Biologická Spoločnost, Bratislava *10066*

South Africa
Society for Experimental Biology, Johannesburg *10174*
The South African Association of Physicists in Medicine and Biology, Pretoria *10183*
South African Biological Society, Pretoria *10184*

Sweden
Baltic Marine Biologists, Lysekil *10438*
Svenska Växtgeografiska Sällskapet, Uppsala *10576*

Switzerland
International Organization for Biological Control of Noxious Animals and Plants, Zürich *10752*
International Union for Conservation of Nature and Natural Resources, Gland *10759*

Turkey
Türk Biyoloji Derneği, Istanbul *11151*
Türkiye Fitopatolji Derneği, Izmir *11161*

United Kingdom
The Association of Applied Biologists, Littlehampton *11301*
Association of British Spectroscopists, Kent *11317*
BIBRA Toxicology International, Carshalton *11439*
Biorhythmic Research Association, Loughborough *11444*
British Photobiology Society, London *11631*
British Social Biology Council, London *11647*
British Society for Developmental Biology, Aberdeen *11653*
British Society for Research on Ageing, Manchester *11664*
European Committee for Biological Effects of Carbon Black, Brighton *12011*
Freshwater Biological Association, Ambleside *12092*
Institute of Biology, London *12215*
International Association of Biological Standardization, Hassocks *12328*
International Society of Development Biologists, Southampton *12392*
Marine Biological Association of the United Kingdom, Plymouth *12492*
Primate Society of Great Britain, Reading *12717*
Science and Engineering Research Council, Swindon *12889*
The Scottish Association for Marine Sciemce, Oban *12895*
The Sleep-Learning Association, London *12954*
The Society for Experimental Biology, London *12973*
Society for Low Temperature Biology, London *12980*
Society for the Study of Fertility, Cambridge *13014*
Society for the Study of Human Biology, London *13015*
The Systematics Association, Durham *13116*

Uruguay
Instituto de Investigaciones Biológicas Clemente Estable, Montevideo *13260*
Sociedad de Biología de Montevideo, Montevideo *13266*

U.S.A.
American Board of Bioanalysis, Saint Louis *13629*
American Cetacean Society, San Pedro *13686*
American Committee for International Conservation, Washington *13744*
American Institute of Biological Sciences, Washington *13875*
American Institute of Biomedical Climatology, Philadelphia *13876*
American Society for Photobiology, Augusta *14079*
American Society of Naturalists, Flushing *14146*
American Type Culture Collection, Rockville *14199*
Association for Biology Laboratory Education, Ithaca *14272*
Association for Tropical Biology, New Orleans *14363*
Association of Biological Collections Appraisers, Carmel *14399*
Association of Cytogenetic Technologists, Augusta *14424*
Association of Systematics Collections, Washington *14512*
Biological Institute of Tropical America, Menlo Park *14589*
Biological Photographic Association, Chapel Hill *14590*
Biological Stain Commission, Rochester *14591*
Biomedical Engineering Society, Culver City *14593*
Biometric Society, Alexandria *14594*
Biometric Society, Western North American Region, Laramie *14595*
Carnegie Institution of Washington, Washington *14647*
Council of Biology Editors, Chicago *14890*
Environmental Mutagen Society, Arlington *15051*
Federation of American Societies for Experimental Biology, Bethesda *15081*
Human Biology Council, New York *15233*
IEEE Engineering in Medicine and Biology Society, New York *15246*
International Conference on Mechanics in Medicine and Biology, Ann Arbor *15446*
International Society for Chronobiology, Fairfax *15528*
International Society for Plant Molecular Biology, Athens *15539*
International Society of Arboriculture, Savoy *15546*
International Society of Differentiation, Saint Paul *15551*
National Association of Biology Teachers, Reston *15831*
Organization for Tropical Studies, Durham *16264*
Pan-American Biodeterioration Society, Washington *16290*
Society for Cryobiology, Bethesda *16514*
Society for Developmental Biology, Washington *16516*
Society for Experimental Biology and Medicine, New York *16522*
Society for the Scientific Study of Sex, Mount Vernon *16611*
Society for the Study of Evolution, Lawrence *16616*
Society for the Study of Social Biology, Honolulu *16620*
Society of Ethnobiology, Reno *16659*
Society of Rheology, New York *16700*
Society of Systematic Biologists, Washington *16704*
South American Explorers Club, Ithaca *16722*
United States Antarctic Research Program, Washington *16815*
United States Federation for Culture Collections, Abbott Park *16823*
World Aquaculture Society, Baton Rouge *16921*

Uzbekistan
Scientific Industrial Association „Biology", Tashkent *16946*

Venezuela
Ateneo Venezolano de Morfología, Caracas *16978*
Biosciences Information Network for Latin America and the Caribbean, Caracas *16979*

Vietnam
Vietnamese Association of Biology, Hanoi *17042*

Yugoslavia
Srpsko Biološko Društvo, Beograd *17126*
Unija Bioloških Naučnih Društava Jugoslavije, Zemun *17132*

Zimbabwe
Kirk Biological Society, Harare *17172*

Biophysics

Argentina
Sociedad Argentina de Biofísica, Buenos Aires *00157*

Armenia
Interdepartmental Council on Physicochemical Biology and Biotechnology, Yerevan *00209*

Austria
Österreichische Biophysikalische Gesellschaft, Wien *00732*

Belgium
Société Belge de Biochimie et de Biologie Moléculaire, Bruxelles *01425*

Bulgaria
Bulgarian Biochemical and Biophysical Society, Sofia *01745*

China, People's Republic
Biophysical Society of China, Beijing *03120*

Finland
Societas Biochemica, Biophysica et Microbiologica Fenniae, Helsingin Yliopisto *03992*

France
Société Française de Biologie Clinique, Paris *04788*

Georgia
Commission on Biosphere and Ecology Research, Tbilisi *05008*
Georgian National Committee on UNESCO Long-Term Programme „Man and the Biosphere", Tbilisi *05018*

Germany
Deutsche Gesellschaft für Biophysik, Garching *05332*
Europäische Föderation Biotechnologie, Frankfurt *05651*

India
Indian Biophysical Society, Calcutta *06778*

Japan
Nihon Seibutsu Butsuri Gakkai, Osaka *08255*

Netherlands
Benelux Society for Microcirculation, Maastricht *08832*
Instituut voor Perceptie Onderzoek, Eindhoven *08914*

Slovakia
Slovenská Bioklimatologická Spoločnost, Bratislava *10065*

Switzerland
Gesellschaft für Forschung auf biophysikalischen Grenzgebieten, Binningen *10714*

U.S.A.
Biophysical Society, Bethesda *14596*

Book Science
s. Librarianship and Book Science

Botany

Argentina
Centro de Investigaciones Bella Vista, Bella Vista *00106*
Fundación Miguel Lillo, San Miguel de Tucumán *00140*
Sociedad Argentina de Botánica, San Isidro *00159*
Sociedad Argentina de Fisiología Vegetal, Castelar *00169*

Australia
Australasian Plant Pathology Society, Adelaide *00246*
Australian Society of Plant Physiologists, Canberra *00357*
Botany 2000-Asia, Kensington *00369*
Weed Science Society of New South Wales, Haymarket *00521*
Weed Science Society of South Australia, Glenside *00522*

Austria
Österreichische Mykologische Gesellschaft, Wien *00871*
Österreichische Orchideen-Gesellschaft, Wien *00874*
Zoologisch-Botanische Gesellschaft in Österreich, Wien *01035*

Belgium
Association pour l'Etude Taxonomique de la Flore d'Afrique Tropicale, Rosières *01153*

Brazil
Asociación Latinoamericana de Paleobotánica y Palinología, Porto Alegre *01584*
Instituto de Botânica, São Paulo *01666*
Instituto Nacional de Pesquisas da Amazonia, Manaus *01688*
International Seaweed Association, São Paulo *01696*
Sociedade Botânica do Brasil, Brasília *01703*

Bulgaria
Bulgarian Botanical Society, Sofia *01746*

Canada
Canadian Botanical Association, Ottawa *02083*
Canadian Phytopathological Society, Lethbridge *02226*
Canadian Society of Plant Physiologists, Guelph *02292*
International Association of Botanical Gardens, Edinburgh *02540*
Société d'Animation du Jardin et de l'Institut Botaniques, Montréal *02917*

China, People's Republic
Botanical Society of China, Beijing 03121

Colombia
Instituto de Ciencias Naturales, Bogotá 03391

Cuba
Asociación Latinoamericana de Micología, Habana 03471

Czech Republic
Česká Botanická Společnost, Praha 03504

Denmark
Dansk Botanisk Forening, København 03595

Egypt
Egyptian Botanical Society, Cairo 03890

Finland
Societas pro Fauna et Flora Fennica, Helsingin Yliopisto 03995

France
Amicale Internationale de Phytosociologie, Bailleul 04131
International Commission for Plant-Bee Relationships, Lusignan 04584
Société Botanique de France, Châtenay-Malabry 04635
Société des Lettres, Sciences et Arts de la Haute-Auvergne, Aurillac 04709
Société d'Ethnozoologie et d'Ethnobotanique, Paris 04724
Société Française de Physiologie Végétale, Paris 04835
Société Française de Phytiatrie et de Phytopharmacie, Versailles 04837
Société Phycologique de France, Paris 04911

Germany
Bayerische Botanische Gesellschaft, München 05171
Deutsche Botanische Gesellschaft e.V., Göttingen 05295
Deutsche Gesellschaft für Qualitätsforschung (Pflanzliche Nahrungsmittel) e.V., Freising 05415
Deutsche Orchideen-Gesellschaft, Sottrum 05479
European Weed Research Society, Leverkusen 05681
Internationale Vereinigung für Vegetationskunde, Göttingen 06005
Vereinigung für angewandte Botanik, Göttingen 06408

Gibraltar
Gibraltar Ornithological and Natural History Society, Gibraltar 06495

India
Botanical Survey of India, Calcutta 06749
Indian Botanical Society, Madras 06779
Indian Phytopathological Society, New Dehli 06807
Indian Society of Genetics and Plant Breeding, New Dehli 06818
International Society of Plant Morphologists, Dehli 06827
Palynological Society of India, Thiruvananltapuram 06849

Israel
Botanical Society of Israel, Beer Sheva 07049

Italy
Ente Fauna Siciliana, Noto 07652
Società Botanica Italiana, Firenze 07773
Società degli Amici del Museo Civico di Storia Naturale Giacomo Doria, Genova 07779
Società Italiana Amici dei Fiori, Firenze 07821
Società Italiana di Biogeografia, Siena 07842

Ivory Coast
African Oil Palm Development Association, Abidjan 08059

Japan
Nippon Shokubutsu Gakkai, Tokyo 08383
Nippon Shokubutsu Seiri Gakkai, Kyoto 08384
Shokubutsu Bunrui Chiri Gakkai, Kyoto 08423

Malta
Society for the Study and Conservation of Nature, Valletta 08671

Mexico
Asociación Mexicana de Orquideología, México 08703
Sociedad Mexicana de Fitogenética, Chapingo 08762
Sociedad Mexicana de Fitopatología, Montecillo 08763

Moldova, Republic of
Society of Botanists of Moldova, Chişinău 08790
Society of Plant Physiologists of Moldova, Chişinău 08792

Netherlands
European Association for Research on Plant Breeding, Wageningen 08862
International Commission for the Nomenclature of Cultivated Plants, Wageningen 08928, 08929
Koninklijke Maatschappij Tuinbouw en Plantkunde, 's-Gravenhage 08948
Koninklijke Nederlandse Botanische Vereniging, Nijmegen 08954

New Zealand
New Zealand Society of Plant Physiologists, Palmerston North 09179
Systematics Association of New Zealand, Dunedin 09197

Norway
Norsk Botanisk Forening, Oslo 09294

Peru
Asociación Latinoamericana Científico de Plantas, Lima 09451

Philippines
Asian-Pacific Weed Science Society, Laguna 09501

Poland
Polskie Towarzystwo Botaniczne, Warszawa 09558

Portugal
Sociedade Broteriana, Coimbra 09676

Russia
Russian Botanical Society, St. Petersburg 09956

Slovakia
Slovenská Botanická Spoločnosť, Bratislava 10067

South Africa
The Botanical Society of South Africa, Claremont 10126
The South African Association of Botanists, Alice 10182
Succulent Society of South Africa, Pretoria 10230

Sweden
Bergianska Stiftelsen, Stockholm 10439
Lunds Botaniska Förening, Lund 10498
Svenska Botaniska Föreningen, Stockholm 10540
Svenska Växtgeografiska Sällskapet, Uppsala 10576

Switzerland
Bernische Botanische Gesellschaft, Bern 10635
Schweizerische Botanische Gesellschaft, Zürich 10800

Thailand
Asian and Pacific Information Network on Medicinal and Aromatic Plants, Bangkok 11076

United Kingdom
Biodeterioration Society, Kew 11441
The Botanical Society of Scotland, Edinburgh 11463
Botanical Society of the British Isles, London 11464
British and European Geranium Society, Higher Poynton 11477
British Bryological Society, Cardiff 11535
British Lichen Society, London 11604
British Phycological Society, London 11632
British Society for Plant Pathology, Leeds 11663
British Society of Flavourists, Haverhill 11683
C.A.B. International Bureau of Horticulture and Plantation Crops, Wallingford 11723
CAB International Bureau of Plant Breeding and Genetics, Wallingford 11725, 11726
Flora Europaea Organization, Liverpool 12080
Institution of Plant Engineers, London 12313
International Organization of Palaeobotany, London 12381
National Institute of Agricultural Botany, Cambridge 12597
Phytochemical Society of Europe, Swansea 12699
Rural Preservation Association, Liverpool 12879

U.S.A.
American Association of Botanical Gardens and Arboreta, Wayne 13512
American Bamboo Society, Sebastopol 13615
American Bryological and Lichenological Society, Omaha 13671
American Chestnut Foundation, Morgantown 13689
American Fern Society, Atlanta 13810
American Herb Association, Nevada City 13845
American Orchid Society, West Palm Beach 13967
American Phytopathological Society, Saint Paul 14006
American Society of Plant Physiologists, Rockville 14155
American Society of Plant Taxonomists, Athens 14156
Aquatic Plant Management Society, Washington 14234
Association of Pacific Systematists, Honolulu 14479
Botanical Society of America, Lawrence 14610
Carnegie Institution of Washington, Washington 14647
Desert Botanical Garden, Phoenix 14964
International Bulb Society, Culver City 15425
International Plant Propagators Society, Seattle 15514
The Magnolia Society, Hammond 15686
Organization for Flora Neotropica, Bronx 16263
Phytochemical Society of North America, Towson 16318
Society for Economic Botany, Columbia 16517
Torrey Botanical Club, Bronx 16785

Zimbabwe
Botanical Society of Zimbabwe, Harare 17164

Botany, Specific

Argentina
Asociación Argentina de Micología, Buenos Aires 00069

Czech Republic
Česká Vědecká Společnost pro Mykologii, Praha 03533
Česká Vědecká Společnost pro Mykologii, Praha 03534

Denmark
Foreningen til Svampekundskabens Fremme, Søborg 03737

France
Société Française de Mycologie Médicale, Paris 04821
Société Mycologique de France, Paris 04897

Georgia
Georgian Botanical Society, Tbilisi 05013

Germany
Deutsche Gesellschaft für Pilzkunde, Karlsruhe 05405
Zentralstelle für Pilzforschung und Pilzverwertung, Waghäusel 06467

Italy
Associazione Micologica ed Ecologica Romana, Roma 07358
Unione Micologica Italiana, Bologna 08049

Japan
Nihon Ishinkin Gakkai, Tokyo 08213
Nihon Kin Gakkai, Tokyo 08224

Malaysia
ASEAN Plant Quarantine Centre and Training Institute, Serdang 08631

Netherlands
Nederlandse Mycologische Vereniging, Wyster 08996

United Kingdom
British Association Representing Breeders, Norfolk 11529
British Iris Society, Worthing 11596
British Mycological Society, Portsmouth 11614
Heather Society, Ipswich 12156
International Mycological Institute, Egham 12378

U.S.A.
Medical Mycological Society of the Americas, San Antonio 15711
Mycological Society of America, Beltsville 15766
North American Mycological Association, Ann Arbor 16219

Building and Construction s. Civil Engineering

Business Administration, Management

Argentina
Sociedad Argentina de Investigación Operativa, Buenos Aires 00174

Australia
Accounting Association of Australia and New Zealand, Clayton 00212
Australian Institute of Management, Saint Kilda 00305
Australia-Pacific Society for Management Studies, Gold Coast 00365
Institute of Public Affairs, Jolimont 00418

Austria
Arbeitsgemeinschaft Personalwesen im Österreichischen Produktivitäts- und Wirtschaftlichkeits-Zentrum, Wien 00561
Gesellschaft für Strategische Unternehmensführung, Innsbruck 00634
Österreichische Gesellschaft für Operations Research, Wien 00817
Österreichische Gesellschaft für Warenkunde und Technologie, Wien 00844
Österreichisches Produktivitäts- und Wirtschaftlichkeits-Zentrum, Wien 00917
Österreichische Verwaltungswissenschaftliche Vereinigung, Wien 00927

Barbados
Caribbean Centre for Development Administration, Saint Michael 01079
Caribbean Management Development Association, Saint Michael 01084

Belgium
China-Europe Management Institute, Bruxelles 01213
Euro-China Research Association in Management, Bruxelles 01263
Euro-China Research Centre for Business Cooperation, Bruxelles 01264
European Accounting Association, Bruxelles 01268
European Association for Research in Industrial Economics, Bruxelles 01279
European Federation of Energy Management Associations, Antwerpen 01349
Institut Belge des Sciences Administratives, Bruxelles 01382
International Institute for Organizational and Social Development, Leuven 01397

Brazil
Associação Prudentina de Educação e Cultura, Presidente Prudente 01608

Canada
Canadian Academic Accounting Association, Toronto 02002
Canadian Certified General Accountants Research Foundation, Vancouver 02093
Canadian Institute for Organization Management, Ottawa 02165
Canadian Institute of Chartered Accountants, Toronto 02168
Canadian Institute of Management, Willowdale 02178
Canadian Operational Research Society, Ottawa 02213
Canadian Research Management Association, Mississauga 02243
Canadian Society of Engineering Management, Ottawa 02283
Conference Board of Canada, Ottawa 02377
Construction Management Institute, Willowdale 02392
Innovation Management Institute of Canada, Ottawa 02518
Society of Management Accountants of Canada, Hamilton 02945

Chile
Corporación de Fomento de la Producción, Santiago 03048

China, Republic
Asia-Pacific Office Automation Council, Taipei 03241
China Association of the Five Principles of Administrative Authority, Taipei 03253
Cooperative League of the Republic of China, Taipei 03293
Public Administration Society of China, Taipei 03322

Costa Rica
Central American Institute for Business Administration, Alajuela 03436

Denmark
Dansk Selskab for Operationsanalyse, Hornbaek 03683
Foreningen af Danske Civiløkonomer, København 03728

Egypt
Arab Management Society, Cairo 03873

Finland
Työtehoseura, Helsinki 04080

France
Association des Directeurs des Centres Universitaires d'Administration des Entreprises, Paris 04177
Association Française de Management, Paris 04220
Association Internationale de l'Inspection du Travail, Paris 04258
Association Internationale des Docteurs en Economie du Tourisme, Aix-en-Provence 04269
Association Internationale pour le Management du Sport, Saint-Michel 04291
Association Technique pour l'Etude de la Gestion des Institutions Publiques et des Entreprises Privées, Paris 04356
Centre de Rencontres et d'Etudes des Dirigeants des Administrations Fiscales, Paris 04368
Community of European Management Schools, Jouy-en-Josas 04432

Germany
Akademie für Führungskräfte der Wirtschaft e.V., Bad Harzburg 05055
Akademie für Organisation, Bonn 05057
Arbeitsgemeinschaft für wirtschaftliche Verwaltung e.V., Eschborn 05113
Arbeitskreis für Betriebsführung München, Thaining 05139
ASB Management-Zentrum-Heidelberg e.V., Heidelberg 05153
Betriebswirtschafts-Akademie e.V., Wiesbaden 05209
Bundesausschuß Betriebswirtschaft, Eschborn 05239
Deutsche Gesellschaft für Personalführung e.V., Düsseldorf 05399
Deutsches Institut für Betriebswirtschaft e.V., Frankfurt 05583
Euro-Handelsinstitut e.V., Köln 05645
Europäischer Verband für Produktivitätsförderung, Hamburg 05656
European Association of Business and Management Teachers, Dortmund 05668
European Council of Management, Eschborn 05676
Forschungsinstitut für Rationalisierung e.V., Aachen 05731
Gesellschaft für Arbeitswissenschaft e.V., Dortmund 05801
Gesellschaft für Finanzwirtschaft in der Unternehmensführung e.V., Frankfurt 05822
Gesellschaft für Informationsvermittlung und Technologieberatung, München 05831
Gesellschaft für Organisation e.V., Giessen 05847
Institut für Technik der Betriebsführung im Handwerk, Karlsruhe 05973
Normenausschuss Bürowesen im DIN Deutsches Institut für Normung e.V., Berlin 06155
Rationalisierungs-Gemeinschaft Bauwesen im RKW, Eschborn 06245
Rationalisierungs-Gemeinschaft Verpackung im RKW, Eschborn 06246
Rationalisierungs-Kuratorium der Deutschen Wirtschaft e.V., Eschborn 06247
Rationalisierungs-Kuratorium der Deutschen Wirtschaft e.V., Landesgruppe Baden-Württemberg, Stuttgart 06248
Rationalisierungs-Kuratorium der Deutschen Wirtschaft e.V., Landesgruppe Berlin, Berlin 06249

Rationalisierungs-Kuratorium der Deutschen Wirtschaft e.V., Landesgruppe Bremen, Bremen 06250
Rationalisierungs-Kuratorium der Deutschen Wirtschaft e.V., Landesgruppe Hamburg, Hamburg 06251
Rationalisierungs-Kuratorium der Deutschen Wirtschaft e.V., Landesgruppe Hessen, Eschborn 06252
Rationalisierungs-Kuratorium der Deutschen Wirtschaft e.V., Landesgruppe Niedersachsen, Hannover 06253
Rationalisierungs-Kuratorium der Deutschen Wirtschaft e.V., Landesgruppe Nord-Ost, Kiel 06254
Rationalisierungs-Kuratorium der Deutschen Wirtschaft e.V., Landesgruppe Nordrhein-Westfalen, Düsseldorf 06255
Rationalisierungs-Kuratorium der Deutschen Wirtschaft e.V., Landesgruppe Rheinland-Pfalz, Mainz 06256
REFA-Verband für Arbeitsstudien und Betriebsorganisation e.V., Darmstadt 06261
Schmalenbach-Gesellschaft, Berlin 06277
Vereinigung der Versicherungs-Betriebswirte e.V., Köln 06402
Volks- und Betriebswirtschaftliche Vereinigung im Rheinisch-Westfälischen Industriegebiet e.V., Duisburg 06435
Wuppertaler Kreis e.V., Köln 06455

Greece
Ellinikon Kentron Paragogikotitos, Athinai 06516

Hong Kong
Asian Association of Management Organizations, Hong Kong 06588
Asian Pacific Federation of Human Resource Management, Hong Kong 06591

Hungary
Szervezési és Vezetési Tudományos Társaság, Budapest 06673

India
Asian Center for Organization, Research and Development, New Delhi 06721
Association of Management Development Institutions in Asia, Hyderabad 06734
Institute of Chartered Accountants of India, New Delhi 06825
National Productivity Council, New Delhi 06846

Indonesia
ASEAN Energy Management and Research Training Centre, Jakarta 06874

Ireland
Institute of Management Consultants in Ireland, Dublin 06963
Irish Management Institute, Dublin 06989
Irish Productivity Centre, Dublin 06996

Israel
The Israel Institute of Productivity, Tel Aviv 07078

Italy
Accademia degli Incamminati, Modigliana 07138
Associazione Italiana di Ricerca Operativa, Genova 07322
Associazione Italiana di Studio delle Relazioni Industriali, Roma 07326
Centro di Studi e Applicazioni di Organizzazione Aziendale della Produzione e dei Trasporti, Torino 07439
Centro Italiano di Studi Aziendali, Milano 07497
Comitato Nazionale Italiana per l'Organizzazione Scientifica, Roma 07618
Societá Italiana di Aziendologia, Milano 07840
Società Italiana per l'Organizzazione Internazionale, Roma 07969
Società per la Formazione la Ricerca e l'Addestramento per le Aziende e le Organizzazioni, Milano 08000

Japan
Asian Productivity Organization, Tokyo 08107
Association of Asian-Pacific Operational Research Societies, Tokyo 08112
Japan Techno-Economics Society, Tokyo 08146
Nippon Gyosei Gakkai, Tokyo 08302
Nippon Keiei Gakkai, Tokyo 08326
Nippon Noritsu Kyokai Shiryoshitsu, Tokyo 08350
Nippon Shogyo Gakkai, Tokyo 08380
Seisan Gijutsu Kenkyusho, Tokyo 08408

Kenya
Church Organization Research and Advisory Trust of Africa, Nairobi 08485

Luxembourg
European Business Associates On-line, Luxembourg 08584

Malaysia
ASEAN Institute of Forest Management, Kuala Lumpur 08630

Netherlands
International Academy of Management, 's-Gravenhage 08917
Nederlandse Vereniging voor Management, 's-Gravenhage 09025
Nederlandse Vereniging voor Personeelbeleid, Utrecht 09034
Nederlandse Vereniging voor Produktieleiding, 's-Gravenhage 09035

New Zealand
The New Zealand Institute of Management, Wellington 09159

Nigeria
Nigerian Institute of Management, Lagos 09239

Norway
Høyokoleutdannedes Forbund, Oslo 09269
Norske Siviløkonomers Forening, Oslo 09307

Pakistan
Institute of Cost and Management Accountants of Pakistan, Karachi 09366
Pakistan Council of Scientific and Industrial Research, Karachi 09384

Philippines
Asian Institute of Management, Manila 09498
Association of Deans of Southeast Asian Graduate Schools of Management, Manila 09509

Poland
Towarzystwo Naukowe Organizacji i Kierownictwa, Warszawa 09625

Portugal
Associação Portuguesa de Management, Lisboa 09647

Russia
Council for the Study of Productive Forces, Moskva 09861

Senegal
Centre Africain d'Etudes Supérieures en Gestion, Dakar 10023

South Africa
The South African Institute of Organization and Methods, Pretoria 10199

Spain
Asociación para el Progreso de la Dirección, Madrid 10279

Sweden
European Federation of Productivity Services, Stockholm 10455
Företagsekonomiska Föreningen, Stockholm 10461
Sveriges Rationaliseringsförbund, Stockholm 10594

Switzerland
Europäische Organisation für Qualität, Bern 10669
Fachgruppe für das Management im Bauwesen des SIA, Zürich 10692
Gesellschaft für Führungspraxis und Personalentwicklung, Zürich 10715
International Industrial Relations Association, Genève 10750
International Institute for Management Development, Lausanne 10751
International Project Management Association, Zürich 10754
Schweizerische Gesellschaft für Personalfragen, Zürich 10841
Schweizerische Management-Gesellschaft, Zürich 10871
Schweizerischer Verband für Betriebsorganisation und Fertigungstechnik, Winterthur 10909

Tanzania
Association of Management Training Institutions of Eastern and Southern Africa, Arusha 11047
Eastern and Southern African Management Institute, Arusha 11056

Thailand
Asian and Pacific Project for Labour Administration, Bangkok 11077

Trinidad and Tobago
Caribbean Industrial Research Institute, Tunapuna 11113

United Kingdom
Accountants Study Group of the EEC, London 11199
Association for Information Management, London 11267
Association of Principals of Colleges (Northern Ireland Branch), Lisburn 11379
Association of Scientific, Technical and Managerial Staffs, London 11388
Association of Teachers of Management, London 11396
Association of University Teachers in Accounting, Sheffield 11404
Chartered Institute of Public Finance and Accountancy, London 11765
Commonwealth Association for Public Administration and Management, London 11810
Commonwealth Consultative Group on Technology Management, London 11816
European Association for Personnel Management, London 11991
European Council for Industrial Marketing, Brixham 12021
European Group for Organization Studies, Plymouth 12030
Institute of Administrative Management, Orpington 12212
Institute of Business Administration, Wigan 12217
Institute of Directors, London 12232
Institute of Management, Corby 12253
Institute of Management Services, Enfield 12254
Institute of Materials Management, Corby 12257
Institute of Personnel Management, London 12266
International Management Centre from Buckingham, Buckingham 12375
Management Research Groups, Corby 12481
Management Systems Training Council, Birmingham 12482
Manpower Society, London 12489
Production Mangagement Action Group, Henley-on-Thames 12723
Society of Business Economists, Watford 13032

U.S.A.
Academy of Management, Ada 13296
Academy of Marketing Science, Coral Gables 13297
Administrative Management Society, Washington 13329
AMA International, New York 13372
American Accounting Association, Sarasota 13444
American Collegiate Retailing Association, Tallhassee 13743
American Institute of Management, Quincy 13887
American Management Association, New York 13925
Association for Systems Management, Cleveland 14339
Association for University Business and Economic Research, Indianapolis 14364
Association of Internal Management Consultants, East Bloomfield 14446
Association of Management Consultants, New York 14457
Association of Research Directors, Orangeburg 14491
Business Professionals of America, Columbus 14629
Career College Association, Washington 14645
Center for Management Development, New York 14684
Consortium for Graduate Study in Management, Saint Louis 14846
Council of Consulting Organizations, New York 14894
Direct Marketing Educational Foundation, New York 14974
Foundation for Accounting Education, New York 15119
Graduate Management Admission Council, Los Angeles 15169
Human Resource Certification Institute, Alexandria 15238
Human Resource Planning Society, New York 15239
Industrial Development Research Council, Norcross 15258
Industrial Relations Research Association, Madison 15260
The Institute of Management Sciences, Providence 15330
International Academy of Management – U.S. Branch, New York 15365
International Association for Personnel Women, Andover 15389
International Association of Career Consulting Firms, Arlington 15401
International Management Council, Omaha 15500
International Personnel Management Association, Alexandria 15511
National Association of Management and Technical Assistance Centers, Washington 15871
National Association of Supervisors of Business Education, Fort Worth 15909
National Business Education Association, Reston 15930
National Center for Housing Management, Washington 15945
National Institute of Management Counsellors, Great Neck 16094
National Records Management Council, New York 16135
Office Systems Research Association, Springfield 16248
Operation Enterprise, Hamilton 16252
Operations Research Society of America, Baltimore 16253
Organization Development Institute, Chesterland 16262
Project Management Institute, Drexel Hill 16362
Social Sciences Services and Resources, Wasco 16486
Society for Advancement of Management, Vinton 16491
Society for Human Resource Management, Alexandria 16535
Society for Information Management, Chicago 16540
Society for Range Management, Denver 16574
Society for the Preservation of American Business History, Williamsburg 16603
Society of Logistics Engineers, New Carrollton 16676

Venezuela
Consejo Zuliano de Planificación y Coordinación, Maracaibo 17004

Virgin Islands (U.S.)
Consortium of Caribbean Universities for Resource Management, Saint Thomas 17075

Yugoslavia
Jugoslovenski Zavod za Produktivnost Rada, Beograd 17096

Cancer Research s. Cell Biology, Cancer Research

Cardiology

Argentina
Sociedad Argentina de Cardiología, Buenos Aires 00160

Australia
The Cardiac Society of Australia and New Zealand, Sydney 00370

Austria
Österreichische Kardiologische Gesellschaft, Wien 00861

Belgium
Fondation Born-Bunge pour la Recherche, Wilrijk 01362
Société Belge de Cardiologie, Bruxelles 01427

Bulgaria
Bulgarian Society of Cardiology, Sofia 01757

Canada
British Columbia and Yukon Heart Foundation, Vancouver 01970
Canadian Cardiovascular Society, Westmount 02087
Canadian Diabetes Association, Toronto 02117
Canadian Heart Foundation, Ottawa 02155
Canadian Society of Cardiology Technologists, Winnipeg 02276
Canadian Society of Pulmonary and Cardiovascular Technology, Toronto 02294
Canadian Stroke Recovery Association, Don Mills 02304
Cardiology Technicians Associations of British Columbia, Vancouver 02321
Cardiology Technologists Association of Ontario, Downsview 02322
Fondation du Québec des Maladies du Coeur, Montréal 02466
Heart and Stroke Foundation of Alberta, Calgary 02498
International Society for Heart Research, Winnipeg 02557
Manitoba Association of Cardiology Technologists, Winnipeg 02604
Manitoba Heart Foundation, Winnipeg 02620
New Brunswick Heart Foundation, Saint John 02684
Nova Scotia Cardiology Technicians Association, Halifax 02726
Nova Scotia Heart Foundation, Halifax 02733
Saskatchewan Cardiology Technicians Association, Saskatoon 02886

Chile
Fundación Gildemeister, Santiago 03052
Sociedad Chilena de Cardiología y Cirugía Cardiovascular, Santiago 03068

China, Republic
International Academy of Chest Physicians and Surgeons of the American College of Chest Physicians, Republic of China Chapter, Taipei 03302
Republic of China Society of Cardiology, Taipei 03324

Colombia
Sociedad Colombina de Cardiología, Bogotá 03415

Costa Rica
Asociación de Cardiología, San José 03433

Czech Republic
Czech Society of Cardiology, Praha 03541

Denmark
Dansk Cardiologisk Selskab, Herlev 03597

Finland
Association of European Paediatric Cardiologists, Helsinki 03951
Suomen Kardiologinen Seura, Helsinki 04033

France
Société Française de Cardiologie, Paris 04789
Syndicat National des Médecins Français Spécialistes des Maladies du Coeur et des Vaisseaux, Paris 04954

Germany
Deutsche Gesellschaft für Herz- und Kreislaufforschung, Düsseldorf 05362

Greece
Elliniki Kardiologiki Etairia, Athinai 06507

Honduras
Sociedad Centroamericana de Cardiología, Tegucigalpa 06585

Hungary
Magyar Kardiologusok Társasága, Budapest 06641

Israel
Israel Heart Society, Jerusalem 07075

Italy
Associazione Italiana di Cardiostimolazione, Pisa 07300
Società Italiana di Cardiologia, Roma 07848

Japan
Nihon Junkanki Gakkai, Kyoto 08215

Korea, Republic
Asian-Pacific Society of Cardiology, Seoul 08522

Mexico
Sociedad Interamericana de Cardiología, México 08751
Sociedad Mexicana de Cardiología, México 08758

Netherlands
European Society of Cardiology, Rotterdam 08893
Nederlandse Vereniging voor Cardiologie, Nieuwegein 09017

Poland
Polskie Towarzystwo Kardiologiczne, Warszawa 09581

Portugal
Sociedade Portuguesa de Cardiologia, Lisboa 09694

Romania
Societatea de Cardiologie, Bucuresti 09757

Russia
Society of Cardiology, Moskva 09972

South Africa
Southern Africa Cardiac Society, Durban 10223

Switzerland
Schweizerische Gesellschaft für Kardiologie, Bern 10825
Société et Fédération Internationale de Cardiologie, Genève 10973

Turkey
Türk Kardiyoloji Derneği, Istanbul 11164

United Kingdom
British Cardiac Society, London 11537
The Chest, Heart and Stroke Association, London 11775
Northern Ireland Chest, Heart and Stroke Association, Belfast 12636
Society of Thoracic and Cardiovascular Surgeons of Great Britain and Ireland, Conventry 13068
Stroke Association, London 13103

U.S.A.
Academy of Veterinary Cardiology, Floral Park 13311
American Association of Cardiovascular and Pulmonary Rehabilitation, Middleton 13460
American College of Cardiology, Bethesda 13699
American College of Chest Physicians, Northbrook 13700
American Heart Association, Dallas 13844
American Society of Echocardiography, Raleigh 14120
American Society of Extra-Corporeal Technology, Reston 14122
Association of Physician's Assistants in Cardiovascular Surgery, Lynchburg 14483
Cardiovascular Credentialing International, Dayton 14643
Cardiovascular System Dynamics Society, Philadelphia 14644
Coronary Club, Cleveland 14863
Council on Arteriosclerosis of the American Heart Association, Dallas 14909
Heart Disease Research Foundation, Brooklyn 15199
International Atherosclerosis Society, Houston 15419
International Cardiology Foundation, Dallas 15427
International Society for Cardiovascular Surgery, Manchester 15527
Mended Hearts, Dallas 15718
Michael E. Debakey International Surgical Society, Houston 15726
National Heart Research Project, Baltimore 16075
National Society for Cardiovascular Technology, Fredericksburg 16154
Peruvian Heart Association, Wadsworth 16307
Society for Cardiac Angiography and Interventions, Breckenridge 16505

Venezuela
Sociedad Venezolana de Cardiología, Caracas 17017

Cartography

Australia
Australian Institute of Cartographers, Brisbane 00298

Austria
Internationale Coronelli-Gesellschaft für Globen- und Instrumentenkunde, Wien 00668

Belgium
European Centre for Ethnolinguistic Cartography, Bruxelles 01309
Société Belge de Photogrammétrie, de Télédétection et de Cartographie, Bruxelles 01440

Brazil
Sociedade Brasileira de Cartografia, Rio de Janeiro 01706

Canada
Association of Canadian Map Libraries, Ottawa 01924
Canadian Cartographic Association, Calgary 02088

China, People's Republic
Chinese Society of Geodesy, Photogrammetry and Cartography, Beijing 03191

France
Association de Recherches Universitaires Géographiques et Cartographiques, Paris 04161
Commission de la Carte Géologique du Monde, Paris 04422
Société Cartographique de France, Paris 04636
Société Française d'Etudes et de Réalisations Cartographiques, Paris 04852

Germany
Deutsche Geodätische Kommission, München 05310
Deutsche Gesellschaft für Kartographie e.V., Berlin 05369
GEO-KART, Oberkochen 05780

Hungary
Geodéziai és Kartográfiai Egyesület, Budapest 06605

Iran
National Cartographic Centre, Teheran 06901

Italy
Associazione Italiana di Cartografia, Napoli 07301

Netherlands
European Cartographic Institute, 's-Gravenhage 08875

New Zealand
New Zealand Cartographic Society, Thorndon 09139

Portugal
Centro de Estudos de História e Cartografia Antiga, Lisboa 09654

Russia
Commission for a Linguistic Atlas of Europe, Moskva 09812
Commission for International Tectonic Maps, Moskva 09819

South Africa
South African Society for Photogrammetry, Remote Sensing and Cartography, Newlands 10216

Sweden
Kartografiska Sällskapet, Gävle 10479

Switzerland
Schweizerische Gesellschaft für Kartographie, Zürich 10826

United Kingdom
British Cartographic Society, Lymington 11538
International Map Collector's Society, Flyford Flavell 12376
Remote Sensing Society, Nottingham 12749
Society of Cartographers, Glasgow 13033

U.S.A.
American Cartographic Association, Bethesda 13678
American Congress on Surveying and Mapping, Bethesda 13753
Western Association of Map Libraries, Clovis 16880

Cell Biology, Cancer Research

Argentina
Sociedad Argentina para el Estudio del Cáncer, Buenos Aires 00187

Australia
Australian Cancer Society, Sydney 00272

Austria
Gesellschaft für Manuelle Lymphdrainage nach Dr. Vodder, Walchsee 00626
Österreichische Ärztegesellschaft zur Bekämpfung der cystischen Fibrose, Wien 00720
Österreichische Gesellschaft für angewandte Zytologie, Graz 00756
Österreichische Krebshilfe – Österreichische Krebsgesellschaft, Wien 00863
Österreichische Krebshilfe Steiermark, Graz 00864

Belgium
Comité des Cancérologues de la Communauté Européenne, Bruxelles 01221
European Cancer Prevention Organizazion, Bruxelles 01306
Institut Européen d'Ecologie et de Cancérologie, Bruxelles 01383
Oeuvre Belge du Cancer, Bruxelles 01415

Brazil
Fundação Antonio Prudente, São Paulo 01650
Instituto Nacional do Cancer, Rio de Janeiro 01691

Canada
British Columbia Cancer Foundation, Vancouver 01973
Canadian Society for Cellular and Molecular Biology, Québec 02251
Canadian Society of Cytology, Hamilton 02281
Cancer Research Society, Montréal 02320
Children's Oncology Care of Ontario, Toronto 02345
Manitoba Cancer Treatment and Research Foundation, Winnipeg 02611
Ontario Cancer Treatment and Research Foundation, Toronto 02764

Chile
Sociedad Chilena de Cancerología, Santiago 03067

China, People's Republic
Chinese Anti-Cancer Association, Tian Jing 03157

Colombia
Comité Nacional de Lucha contra el Cáncer, Bogotá 03367
Fundación OFA para el Avance de las Ciencias Biomédicas, Bogotá 03375
Instituto Colombiano de Oncologia Pediatrica, Bogotá 03387
Instituto Nacional de Cancerología, Bogotá 03392
Liga Colombiana de Lucha Contra el Cancer, Bogotá 03398
Sociedad Colombiana de Cancerología, Bogotá 03405

Czech Republic
Czech Oncologic Society, Praha 03539

Denmark
Dansk Selskab for Cancerforskning, København 03674
Landsforeningen til Kraeftens Bekaempelse, København 03766
Nordic Society for Cell Biology, København 03779
Nordisk Forening for Celleforskning, København 03783

Finland
Suomen Syöpäyhdistys, Helsinki 04061

France
Association Française pour l'Etude du Cancer, Paris 04245
Ligue Nationale contre le Cancer, Paris 04598
Organisation Internationale de Recherche sur la Cellule, Paris 04616
Société Internationale pour la Lutte contre le Cancer du Sein, Paris 04884

Germany
Arbeitsgemeinschaft für Krebsbekämpfung des Landes Niedersachsen e.V., Hannover 05107
Bayerische Krebsgesellschaft e.V., München 05173
Deutsche Gesellschaft für Hämatologie und Onkologie e.V., München 05358
Deutsche Krebsgesellschaft e.V., Frankfurt 05461
Deutsches Krebsforschungszentrum, Heidelberg 05592
European Federation of Cytology Societies, Düsseldorf 05679
Hessische Krebsgesellschaft e.V., Marburg 05913
International Academy of Cytology, Freiburg 05984

India
Indian Cancer Society, Bombay 06781

Ireland
Irish Cancer Society, Dublin 06978

Italy
Associazione Italiana di Oncologia Medica, Milano 07317
Comitato Italiano per lo Studio del Dolore in Oncologia, Roma 07617
European Cell Biology Organization, Milano 07668
Lega Italiana per la Lotta contro i Tumori, Roma 07761
Società Italiana di Cancerologia, Milano 07847
Società Italiana di Citologia Clinica e Sociale, Roma 07859
Società Italiana di Oncologia Ginecologica, Roma 07920
Società Italiana per lo Studio della Cancerogenesi Ambientale ed Epidemiologia dei Tumori, Genova 07970

Japan
Asian-Pacific Organization for Cell Biology, Okayama 08101
National Cancer Center, Tokyo 08184
Nippon Gan Gakkai, Tokyo 08299

Mexico
Centro de Investigación y de Estudios Avanzados del Instituto Politécnico Nacional, México 08718

Netherlands
European Association for Cancer Education, Groningen 08851
European Calcified Tissue Society, Amsterdam 08874
International Society of Paediatric Oncology, 's-Hertogenbosch 08943
Vereniging Het Nederlands Kanker Instituut, Amsterdam 09085

Peru
Federación de Sociedades Latinoamericanas del Cáncer, Lima 09467

Russia
National Scientific Medical Society of Oncologists, St. Petersburg 09924

South Africa
Cancer Association of South Africa, Braamfontein 10127
Carbohydrate and Lipid Metabolism Research Group, Johannesburg 10128

Spain
Asociación Española contra el Cancer, Barcelona 10255

Sweden
Cancerfonden – Riksföreningen mot Cancer, Stockholm 10441
Scandinavian Forum for Lipid Research and Technology, Göteborg 10518

Switzerland
Institut Suisse de Recherches Expérimentales sur le Cancer, Epalinges 10732
Schweizerische Gesellschaft für cystische Fibrose, Spiegel 10815
Schweizerische Gesellschaft für Onkologie, Epalinges 10837
Schweizerische Krebsliga, Bern 10866
Union Internationale contre le Cancer, Genève 10993

Turkey
Türk Kanser Arastirma ve Savas Kurumu, Ankara 11163

United Kingdom
Association for International Cancer Research, Cameron-Markby 11268
Association of Head and Neck Oncologists of Great Britain, London 11350
British Association for Cancer Research, London 11489
British Association of Surgical Oncology, London 11525
British Society for Cell Biology, Cambridge 11651
British Society for Clinical Cytology, Manchester 11652
Cancer Research Campaign, London 11743
European Association for Cancer Research, Nottingham 11988
International Federation of Cell Biology, London 12355
Macmillan Cancer Fund, London 12475
National Association for the Relief of Pagets Disease, Manchester 12563

U.S.A.
American Association for Cancer Education, Houston 13470
American Association for Cancer Research, Philadelphia 13471
American Association for Study of Neoplastic Diseases, South Euclid 13493
American Cancer Society, Atlanta 13676
American College of Mohs' Micrographic Surgery and Cutaneous Oncology, Schaumburg 13715
American Industrial Health Council, Washington 13862
American Joint Committee on Cancer, Chicago 13906
American Society for Cell Biology, Bethesda 14049
American Society for Cytotechnology, Raleigh 14055
American Society for Therapeutic Radiology and Oncology, Philadelphia 14091
American Society of Clinical Oncology, Chicago 14109
American Society of Cytology, Philadelphia 14116
Association for Research of Childhood Cancer, Buffalo 14331
Association of Community Cancer Centers, Rockville 14418
Breast Cancer Advisory Center, Kensington 14614
Cancer Care, New York 14637
Cancer Control Society, Los Angeles 14638
Cancer Federation, Moreno Valley 14639
Cancer Guidance Institute, Pittsburgh 14640
Candlelighters Childhood Cancer Foundation, Bethesda 14641
Committee for Freedom of Choice in Medicine, Chula Vista 14794
DES Action USA, Oakland 14963
Foundation for Advancement in Cancer Therapy, New York 15120
National Cancer Center, Plainview 15931
National Leukemia Association, Garden City 16101
National Tumor Registrars Association, Mundelein 16174
Oncology Nursing Society, Pittsburgh 16249
R.A. Bloch Cancer Foundation, Kansas City 16383
Radiation Therapy Oncology Group, Philadelphia 16386
Skin Cancer Foundation, New York 16478
Society of Surgical Oncology, Palatine 16703
Spirit and Breath Association, Skokie 16743
United Cancer Council, Indianapolis 16808

Venezuela
Sociedad Venezolana de Oncología, Caracas 17028

Yugoslavia
Serbian Association against Cancer, Beograd 17123

Chemistry

Argentina
Asociación Química Argentina, Buenos Aires 00096
Comité Nacional de Cristalografía, Buenos Aires 00123

Australia
Asian Coordinating Group for Chemistry, Nedlands 00222
Asian Network for Analytical and Inorganic Chemistry, Perth 00223
Asian Pacific Federation of Clinical Biochemistry, Adelaide 00226
Australian Association of Clinical Biochemists, Maylands 00263
Australian Society of Cosmetic Chemists, Kingsgrove 00351
Royal Australian Chemical Institute, Parkville 00461
Society of Leather Technologists and Chemists, Botany 00494
Sydney University Chemical Engineering Association, Sydney 00500
Sydney University Chemical Society, Sydney 00501

Austria
Chemisch-Physikalische Gesellschaft in Wien, Wien 00577
Gesellschaft für Chemiewirtschaft, Wien 00615

Gesellschaft österreichischer Chemiker, Wien 00637
Österreichische Gesellschaft für Klinische Chemie, Wien 00802
Österreichische Gesellschaft für Mikrochemie und analytische Chemie, Wien 00810
Österreichische Tribologische Gesellschaft, Wien 00923
Österreichische Vereinigung der Zellstoff- und Papierchemiker und -techniker, Wien 00924
Verein Österreichischer Textilchemiker und Coloristen, Dornbirn 01002
Verein zur Förderung des physikalischen und chemischen Unterrichts, Wien 01005

Belgium

Branche Belge de la Société de Chimie Industrielle, Bruxelles 01171
Centre Européen des Silicones, Bruxelles 01186
Centre International d'Etudes du Lindane, Bruxelles 01200
Euro Chlor, Bruxelles 01265
European Association on Catalysis, Namur 01299
European Centre of Studies on Linear Alkylbenzene, Bruxelles 01314
European Chlorinated Solvents Association, Bruxelles 01316
European Committee of Organic Surfactants and their Intermediates, Bruxelles 01324
Société Royale de Chimie, Bruxelles 01481
Société Technique et Chimique de Sucrerie de Belgique, Bruxelles 01491
Vlaamse Chemische Vereniging, Gent 01507

Brazil

Associação Brasileira de Química, Rio de Janeiro 01593
Instituto Nacional de Pesquisas da Amazonia, Manaus 01688
Instituto Nacional de Tecnologia, Rio de Janeiro 01690
Sociedade de Farmácia e Química de São Paulo, São Paulo 01723

Bulgaria

Federation of European Biochemical Societies, Sofia 01767
Society of Bulgarian Chemists, Sofia 01778
Union of Chemistry and the Chemical Industry, Sofia 01788

Canada

Alberta Sulphur Research, Calgary 01849
Canadian Society for Color, Ottawa 02254
Chemical Industries Accident Prevention Association, Toronto 02341
Chemical Institute of Canada, Ottawa 02342
Council of Canadian University Chemistry Chairmen, Burnaby 02402
International Union of Pure and Applied Chemistry (Canadian National Committee), Edmonton 02562
Pulp and Paper Research Institute of Canada, Pointe Claire 02853
Society of Chemical Industry (Canadian Section), Mississauga 02941

Chile

Colegio de Quimico-Farmacéuticos de Chile, Santiago 03033
Sociedad Chilena de Química, Concepción 03082

China, People's Republic

Academy of Chemical Engineering, Beijing 03108
Chinese Chemical Society, Beijing 03167
Chinese Society of Chemical Engineering, Beijing 03186
Mukden Academy of Chemical Engineering, Liaoning 03211

China, Republic

Agricultural Chemistry Society of China, Taipei 03230
Asian Crystallographic Association, Taipei 03237
Association for Physics and Chemistry Education of the Republic of China, Taipei 03243
Chinese Chemical Society, Taipei 03265

Colombia

Sociedad Colombiana de Químicos e Ingenieros Químicos, Bogotá 03414

Cuba

Centro Nacional de Investigaciones Científicas, La Habana 03478

Czech Republic

Česká Společnost Chemická, Praha 03520
Česká Společnost Histo- a Cytochemnická, Praha 03524

Denmark

Jydsk Selskab for Fysik og Kemi, Arhus 03753
Kemisk Forening, København 03757

Finland

Finska Kemistsamfundet, Helsinki 03960
Suomalaisten Kemistien Seura, Helsinki 04002
Suomen Kemian Seura, Helsinki 04036

France

Association des Chimistes, Ingénieurs et Cadres des Industries Agricoles et Alimentaires, Paris 04173
Association Française des Ingénieurs, Chimistes et Techniciens des Industries du Cuir, Lyon 04232
Association pour la Recherche et le Développement en Informatique Chimique, Paris 04329
Centre de Coopération pour les Recherches Scientifiques Relatives au Tabac, Paris 04366
Division de Chimie Physique de la Société Française de Chimie, Paris 04466
European Association of Geochemistry, Paris 04486
Fédération Nationale des Associations de Chimie de France, Paris 04541
Groupe d'Etude et de Synthèse des Microstructures, Paris 04551
Groupe pour l'Avancement des Sciences Analytiques, Paris 04562
Société de Chimie Industrielle, Paris 04646
Société de Chimie Thérapeutique, Châtenay Malabry 04647
Société des Experts-Chimistes de France, Paris 04705
Société Française de Chimie, Paris 04791
Société Nationale des Sciences Naturelles et Mathématiques, Cherbourg 04902
Union des Physiciens, Paris 04975

Germany

Arbeitskreis Chemische Industrie, Köln 05136
Beilstein-Institut für Literatur der Organischen Chemie, Frankfurt 05182
Bundesverband der öffentlichen angestellten und vereidigten Chemiker e.V., Hamburg 05245
Deutsche Bunsen-Gesellschaft für Physikalische Chemie e.V., Frankfurt 05296
Deutscher Zentralausschuss für Chemie, Frankfurt 05565
Europäische Föderation für Chemie-Ingenieur-Wesen, Frankfurt 05652
Forschungsstelle für Acetylen, Dortmund 05737
Gesellschaft Deutscher Chemiker e.V., Frankfurt 05793
Gesellschaft für Informationsvermittlung und Technologieberatung, München 05831
Gesellschaft zur Förderung der Spektrochemie und angewandten Spektroskopie e.V., Dortmund 05884
Verein der Textilchemiker und Coloristen e.V., Heidelberg 06370
Verein für Gerberei-Chemie und -Technik e.V., München 06385

Greece

Enosis Ellinon Chimikon, Athinai 06519

Hungary

Magyar Kémikusok Egyesülete, Budapest 06643

India

Indian Association for the Cultivation of Science, Calcutta 06770
Indian Chemical Society, Calcutta 06783

Ireland

The Institute of Chemistry of Ireland, Dublin 06961

Italy

Associazione Italiana dei Chimici del Cuoio, Torino 07295
Associazione Italiana di Chimica Tessile e Coloristica, Milano 07302
Associazione Italiana di Ingegneria Chimica, Milano 07311
Centro di Studi Chimico-Fisici di Macromolecole Sintetiche e Naturali, Genova 07438
Consiglio Nazionale dei Chimici, Roma 07627
Istituto Italiano di Storia della Chimica, Roma 07740
Società Chimica Italiana, Roma 07774
Società Italiana di Biochimica, Perugia 07841

Japan

Denki Kagaku Kyokai, Tokyo 08119
Kobunshi Gakkai, Tokyo 08166
Nihon Aisotopu Kyokai, Tokyo 08186
Nihon Bunseki Kagaku-Kai, Tokyo 08189
Nippon Bitamin Gakkai, Kyoto 08281
Nippon Bunko Gakkai, Tokyo 08283
Nippon Butsuri-Kagaku Kenkyukai, Kyoto 08285
Nippon Kagakukai, Tokyo 08318
Nippon Nensho Gakkai, Tokyo 08345
Nippon Nogei Kagaku Kai, Tokyo 08348
Nippon Rikusui Gakkai, Tokyo 08357
Shokubai Gakkai, Tokyo 08422

Korea, Republic

Korean Chemical Society, Seoul 08530

Mexico

Centro de Investigación y de Estudios Avanzados del Instituto Politécnico Nacional, México 08718
Sociedad Química de México, México 08776

Moldova, Republic of

Mendeleev Chemical Society, Chişinău 08781

Netherlands

European Photochemistry Association, Amsterdam 08891
International Colour Association, Eindhoven 08927
International Commitee for Histochemistry and Cytochemistry, Leiden 08930
Koninklijke Nederlandse Chemische Vereniging, 's-Gravenhage 08955

New Zealand

New Zealand Fertiliser Manufacturers Research Association, Auckland 09146
New Zealand Institute of Chemistry, Wellington 09155

Norway

Norsk Kjemisk Selskap, Oslo 09323

Peru

Sociedad Química del Perú, Lima 09487

Philippines

Philippine Council of Chemists, Manila 09525

Poland

Polskie Towarzystwo Biochemiczne, Warszawa 09557
Polskie Towarzystwo Chemiczne, Warszawa 09559

Portugal

Sociedade Portuguesa de Química, Lisboa 09712

Romania

Societatea de Histochimie si Citochimie, Bucuresti 09766
Societatea de Stiinte Fizice si Chimice din Romania, Bucuresti 09783

Russia

D.I. Mendeleev Chemical Society, Moskva 09864
National Committee of Chemists, Moskva 09893
Programme Committee Surface Physics, Chemistry and Mechanics, Moskva 09946

Slovakia

Slovenská Chemická Spoločnost, Bratislava 10068

South Africa

Chemical Engineering Research Group, Durban 10129
Division of Energy Technology, Pretoria 10136
Natural Products Chemistry Research Group, Johannesburg 10157
Pollution Research Group, Durban 10162
Preclinical Diagnostic Chemistry Research Group of the South African Medical Research Council, Congella 10163
Research Group for Neurochemistry, Tygerberg 10169
The South African Chemical Institute, Yeoville 10186
Spectroscopic Society of South Africa, Pretoria 10229

Spain

Academia de Ciencias Exactas, Físicas, Químicas y Naturales de Zaragoza, Zaragoza 10239
Asociación Española de Químicos y Coloristas Textiles, Barcelona 10268
Asociación Nacional de Químicos de España, Madrid 10276
Asociación Química Española de la Industria del Cuero, Barcelona 10281
Real Sociedad Española de Física, Madrid 10361
Real Sociedad Española de Química, Madrid 10363

Sweden

Chemical Societies of the Nordic Countries, Stockholm 10444
Svenska Kemistsamfundet, Stockholm 10555

Switzerland

European Chemoreception Research Organization, Zürich 10681
Internationale Föderation der Vereine der Textilchemiker und Coloristen, Reinach 10737
Neue Schweizerische Chemische Gesellschaft, Basel 10774
Schweizerischer Chemiker-Verband, Binningen 10891
Schweizerischer Verein der Chemiker-Coloristen, Oftringen 10918
Schweizerische Vereinigung der Lack- und Farben-Chemiker, Frauenfeld 10936
Schweizerische Vereinigung diplomierter Chemiker HTL, Basel 10938
Verband der Kantonschemiker der Schweiz, Bern 10996

United Kingdom

Association of British Spectroscopists, Kent 11317
Association of Manufacturing Chemists, London 11364
British Electrophoresis Society, Aberdeen 11556
Chemicals Notation Association, Dartford 11771
Chemometrics Society, Sandwich 11772
Coordinating European Council for the Development of Performance Tests for Lubricants and Engine Fuels, London 11841
European Association of Organic Geochemists, Newcastle-upon-Tyne 11999
European Communities Chemistry Council, London 12013
European Communities Clinical Chemistry Committee, London 12014
International Association of Volcanology and Chemistry of the Earth's Interior JAVCEJ, Leeds 12336
International Federation of Societies of Cosmetic Chemists, Luton 12359
International Union of Pure and Applied Chemistry, Oxford 12404
Oil and Colour Chemists' Association, Wembley 12653
Phytochemical Society of Europe, Swansea 12699
Royal Society of Chemistry, London 12857
Science and Engineering Research Council, Swindon 12889
Society for the History of Alchemy and Chemistry, London 13003
Society of Chemical Industry, London 13034, 13035
Society of Cosmetic Scientists, Luton 13039
The Society of Dyers and Colourists, Bradford 13044
Ulster Chemists' Association, Belfast 13151

Uruguay

Asociación de Química y Farmacia del Uruguay, Montevideo 13245

U.S.A.

Adhesion Society, Cincinnati 13327
American Association for Aerosol Research, Cincinnati 13465
American Association for Clinical Chemistry, Washington 13474
American Association for Crystal Growth, Thousand Oaks 13476
American Association for Laboratory Accreditation, Gaithersburg 13483
American Association of Cereal Chemists, Saint Paul 13514
American Association of Textile Chemists and Colorists, Research Triangle Park 13592
American Chemical Society, Washington 13688
American Crystallographic Association, Buffalo 13774
American Institute of Chemical Engineers, New York 13878
American Institute of Chemists, Bethesda 13879
American Leather Chemists Association, Cincinnati 13913
American Microchemical Society, Princeton 13945
American Oil Chemists' Society, Champaign 13964
American Society for Neurochemistry, Palo Alto 14074
American Society of Brewing Chemists, Saint Paul 14103
AOAC International, Arlington 14230
Association for Chemoreception Sciences, Tallahassee 14277
Association of Analytical Chemists, Detroit 14391
Association of Consulting Chemists and Chemical Engineers, New York 14422
Association of Official Racing Chemists, Portland 14471
Association of Vitamin Chemists, Northville 14534
Catalysis Society (North America), West Chester 14648
Catecholamine Club, Lawrence 14649
Chemists' Club, New York 14726
Emulsion Polymers Institute, Bethlehem 15040
Federation of Analytical Chemistry and Spectroscopy Societies, Frederick 15082
Fiber Society, Princeton 15092
Geochemical Society, Washington 15152
Histamine Research Society of North America, Dallas 15215
Histochemical Society, Woods Hole 15216
Inter-Society Color Council, Lawrenceville 15578
National Registry in Clinical Chemistry, Washington 16136
New York Pigment Club, Clifton 16205
Society of Cosmetic Chemists, New York 16651
Society of Flavor Chemists, Hunt Valley 16665
Society of Rheology, New York 16700
Steel Structures Painting Council, Pittsburgh 16746
Sulphur Institute, Washington 16752

Vietnam

Vietnamese Association of Chemistry, Hanoi 17043

Yugoslavia

Srpsko Hemijsko Društvo, Beograd 17129

Cinematography

Austria

Österreichische Gesellschaft für Filmwissenschaft, Kommunikations- und Medienforschung, Wien 00779
SYNEMA – Gesellschaft für Film und Medien, Wien 00951

Belgium

Association Belge de Photographie et de Cinématographie, Bruxelles 01109
Centre International de Liaison des Ecoles de Cinéma et de Télévision, Bruxelles 01192
European Academy for Film and Television, Bruxelles 01267
Institut National de Cinématographie Scientifique de Belgique, Bruxelles 01390

Bulgaria

Union of Bulgarian Film Makers, Sofia 01785

Canada

Academy of Canadian Cinema, Toronto 01815
Canadian Film Institute, Ottawa 02134
Canadian Society of Cinematographers, Toronto 02279
Cinémathèque Québécoise, Montréal 02349

China, Republic

Chinese Film Critic's Association of China, Taipei 03267

France

Association des Auteurs de Films, Paris 04168
Association pour la Fondation Internationale du Cinéma et de la Communication Audiovisuelle, Cannes 04321
Centre International du Film pour l'Enfance et la Jeunesse, Paris 04380
Comité du Film Ethnographique, Paris 04399
Conseil International des Moyens du Film d'Enseignement, Paris 04457

Germany
Bundesverband Jugend und Film e.V., Frankfurt *05254*
Deutsche Akademie der Darstellenden Künste e.V., Frankfurt *05284*
Deutsche Film- und Fernsehakademie Berlin, Berlin *05304*
Deutsche Gesellschaft für Filmdokumentation, Wiesbaden *05349*
Deutsche Gesellschaft für Film- und Fernsehforschung, München *05350*
Deutsches Institut für Filmkunde e.V., Frankfurt *05584*
Filmbewertungsstelle Wiesbaden, Wiesbaden *05700*
Filmkritiker Kooperative, München *05701*
Institut für den Wissenschaftlichen Film, Göttingen *05956*
Institut für Film und Bild in Wissenschaft und Unterricht, Grünwald *05960*
Konferenz der Landesfilmdienste in der Bundesrepublik Deutschland e.V., Bonn *06082*

Hungary
Optikai, Akusztikai és Filmtechnikai Egyesület, Budapest *06667*

Iceland
Tónskáldafélag Islands, Reykjavik *06705*

Ireland
Espace Vidéo Européen, Dublin *06945*
Federation of Irish Film Societies, Dublin *06948*
Irish Film Society, Dublin *06985*

Italy
Associazione Italiana di Cinematografia Scientifica, Roma *07303*
Associazione Italiana per le Ricerche di Storia del Cinema, Roma *07342*
Associazione Tecnica Italiana per la Cinematografia, Roma *07407*
Centro Italiano Studi sull'Arte dello Spettacolo, Roma *07510*
Centro Sperimentale di Cinematografia, Roma *07545*
Centro Studi Cinematografici, Roma *07550*
Centro Studi Pietro Mancini, Cosenza *07585*
Cineteca Italiana Archivio Storico del Film, Milano *07601*
Ente Autonomo La Biennale de Venezia, Venezia *07649*
Federazione Italiana dei Cineclub, Roma *07690*

Netherlands
Stichting Verenigd Nederlands Filminstituut, Hilversum *09076*

Romania
Asociaţia Cineaştilor din Romania, Bucuresti *09739*

Russia
Union of Russian Filmmakers, Moskva *09988*

Spain
Asociación Española de Cine Científico, Madrid *10259*

Switzerland
Schweizerische Vereinigung für Filmkultur, Bern *10945*

United Kingdom
British Academy, London *11470*
British Academy of Film and Television Arts, London *11471*
British Federation of Film Societies, London *11560*
British Film Institute, London *11562*
British Kinematograph Sound and Television Society, London *11599*
British Universities Film and Video Council, London *11700*
Mercia Cinema Society, Birmingham *12519*
National Association for Film in Education, Bexleyheath *12558*

U.S.A.
Academy of Motion Picture Arts and Sciences, Beverly Hills *13298*
Academy of Science Fiction, Fantasy and Horror Films, Los Angeles *13307*
American Film Institute, Washington *13813*
Anthology Film Archives, New York *14228*
Black Filmmakers Hall of Fame, Oakland *14598*
Cinemists 63, Los Angeles *14752*
Consortium of College and University Media Centers, Ames *14848*
Dance Films Association, New York *14953*
International Council – National Academy of Television Arts and Sciences, New York *15455*
International Film Seminars, New York *15477*
Society for Cinema Studies, Chapel Hill *16507*
Society for the Furtherance and Study of Fantasy and Science Fiction, Madison *16595*
Society of Motion Picture and Television Art Directors, Studio City *16680*
Society of Motion Picture and Television Engineers, White Plains *16681*
University Film and Video Association, Los Angeles *16847*

Civil Engineering s. a. Architecture; Urban and Regional Planning

Argentina
Grupo Latinoamericano de R.I.L.E.M., Buenos Aires *00141*

Australia
Australian National Committee on Large Dams, Brisbane *00323*
Australian Road Research Board, Melbourne *00335*
Water Authorities Association of Victoria, Melbourne *00519*

Austria
Forschungsgesellschaft für das Verkehrs- und Straßenwesen im ÖIAV, Wien *00595*
Forschungsgesellschaft für Wohnen, Bauen und Planen, Wien *00597*
Österreichischer Stahlbauverband, Wien *00903*

Belgium
European Convention for Constructional Steelwork, Bruxelles *01332*
Union Belge des Géomètres-Experts Immobiliers, Bruxelles *01493*

Bulgaria
Union of Civil Engineering, Sofia *01789*
Union of Water Works, Sofia *01803*

Canada
Alberta Construction Association, Edmonton *01830*
Canadian Construction Association, Ottawa *02104*
Canadian Sheet Steel Building Institute, Willowdale *02249*
Canadian Society for Civil Engineering, Ottawa *02252*
Construction Management Institute, Willowdale *02392*
Housing and Urban Development Association of Nova Scotia, Halifax *02507*

Chile
Asociación Chilena para la Investigación y Desarrollo del Hormigón Estructural, Santiago *03016*
Colegio de Architectos de Chile, Santiago *03029*

China, People's Republic
Academy of Architectural Engineering, Beijing *03105*
China Civil Engineering Society, Beijing *03130*

China, Republic
Chinese Institute of Civil and Hydraulic Engineering, Taipei *03272*

Denmark
Dansk Geoteknisk Forening, Lyngby *03626*
Dansk Selskab for Bygningsstatik, Lyngby *03673*
Vej- og Byplanforeningen, København *03828*

Egypt
General Organization for Housing, Building and Planning Research, Cairo *03902*

France
Association Française de Recherches d'Essais sur les Matériaux et les Constructions, Paris *04224*
Association Internationale des Travaux en Souterrain, Bron *04280*
Commission Internationale des Grands Barrages, Paris *04425*
Office Général du Bâtiment et des Travaux Publics, Paris *04604*
Organisation Scientifique des Industries du Bâtiment, Paris *04617*
Réunion Internationale des Laboratoires d'Essais et de Recherche sur les Matériaux et les Constructions, Cachan *04622*

Germany
Arbeitsgemeinschaft für zeitgemässes Bauen e.V., Kiel *05114*
Arbeitsgemeinschaft Industriebau, Köln *05119*
Deutsche Gesellschaft für Bauingenieurwesen e.V., Karlsruhe *05327*
Deutsche Gesellschaft für Baukybernetik e.V, Holzminden *05328*
Deutsche Gesellschaft für Erd- und Grundbau e.V., Essen *05345*
Deutscher Ausschuss für Stahlbau, Köln *05500*
Deutscher Stahlbau-Verband, Köln *05541*
Forschungsgesellschaft für Strassen- und Verkehrswesen e.V., Köln *05720*
Forschungsinstitut für Wärmeschutz e.V., Gräfelfing *05732*
Fraunhofer-Gesellschaft zur Förderung der angewandten Forschung e.V., München *05759*
Gesellschaft des Bauwesens e.V., Eschborn *05792*
Hafenbautechnische Gesellschaft e.V., Hamburg *05904*
Institut für Bauforschung e.V., Hannover *05954*
Intitut für Baustoffprüfung und Fußbodenforschung, Troisdorf *06014*
Kuratorium für Kulturbauwesen, Hannover *06087*
Kuratorium für Technik und Bauwesen in der Landwirtschaft e.V., Darmstadt *06088*
Normenausschuss Akustik, Lärmminderung und Schwingungstechnik im DIN Deutsches Institut für Normung e.V., Berlin *06147*
Normenausschuss Bauwesen im DIN Deutsches Institut für Normung e.V., Berlin *06150*
Rationalisierungs-Gemeinschaft Bauwesen im RKW, Eschborn *06245*
Studiengemeinschaft für Fertigbau e.V., Wiesbaden *06297*
Studiengemeinschaft Holzleimbau e.V., Düsseldorf *06298*
VDI-Gesellschaft Bautechnik, Düsseldorf *06320*

Hungary
Epitéstudományi Egyesület, Budapest *06602*

Ireland
Building Materials Federation, Dublin *06935*
Institute of Architectural and Associated Technology, Dublin *06959*

Italy
Associazione Italiana di Genio Rurale, Padova *07309*
Associazione Italiana per la Promozione degli Studi e delle Ricerche per l'Edifizia, Milano *07336*
Associazione Italiana Tecnico-Economica del Cemento, Roma *07352*
Tecnocasa, L'Aquila *08033*

Japan
Doboku-Gakkai, Tokyo *08122*
Doshitsu Kogakkai, Tokyo *08123*
Nihon Kensetsu Kikaika Kyokai, Tokyo *08221*

Netherlands
European Asphalt Pavement Association, Breukelen *08850*
International Federation for Housing and Planning, 's-Gravenhage *08932*
Stichting Economisch Instituut voor de Bouwnijverheid, Amsterdam *09071*

New Zealand
The New Zealand National Society for Earthquake Engineering, Wellington *09171*

Norway
Norske Sivilingeniørers Forening, Oslo *09306*

Pakistan
Pakistan Concrete Institute, Karachi *09382*

Peru
Asociación de Ingenieros Civiles del Perú, Lima *09449*

Russia
Automobile and Road Building Society, Moskva *09807*

South Africa
The South African Institution of Civil Engineers, Yeoville *10203*
South African National Group of the International Society for Rock Mechanics, Marshalltown *10208*

Spain
International Association for Shell and Spatial Structures, Madrid *10328*
Sociedad Española de Mecánica del Suelo y Cimentaciones, Madrid *10389*

Sweden
Sveriges Civilingenjörsförbund, Stockholm *10588*

Switzerland
Fachgruppe für Brückenbau und Hochbau des SIA, Zürich *10691*
Fachgruppe für industrielles Bauen des SIA, Zürich *10694*
Fachgruppe für Untertagbau des SIA, Zürich *10696*
Internationale Vereinigung für Brückenbau und Hochbau (IVBH), Zürich *10742*
Schweizerischer Ingenieur- und Architekten-Verein, Zürich *10899*

United Kingdom
Association of Building Engineers, Northampton *11319*
British Tunnelling Society, London *11698*
Building Services Research and Information Association, Bracknell *11716*
The Chartered Institute of Building, Ascot *11761*
Chartered Institution of Building Services Engineers, London *11767*
Construction Industry Research and Information Association, London *11839*
Dry Stone Walling Association, Oswestry *11909*
Faculty of Building, Elstree *12050*
Institute of Concrete Technology, Beaconsfield *12225*
Institute of Construction Management, London *12226*
Institution of Civil Engineers, London *12294*
Institution of Highways and Transportation, London *12305*
Scottish Association for Building Education and Training, Paisley *12894*
Society for Earthquake and Civil Engineering Dynamics, London *12969*
Structural Fire Protection Association, London *13104*

U.S.A.
American Concrete Institute, Detroit *13748*
American Institute of Building Design, Shelton *13877*
American Institute of Constructors, Saint Petersburg *13880*
American Institute of Steel Construction, Chicago *13895*
American Society of Civil Engineers, Washington *14107*
Associated Schools of Construction, Peoria *14260*
Association of Asphalt Paving Technologists, Maplewood *14395*
Building Officials and Code Administrators International, Country Club Hills *14622*
Construction Specifications Institute, Alexandria *14850*
Expanded Shale Clay and Slate Institute, Salt Lake City *15069*
Federation of Societies for Coatings Technology, Blue Bell *15089*
International Bridge, Tunnel and Turnpike Association, Washington *15423*
International Conference of Building Officials, Whittier *15445*
The Masonry Society, Boulder *15698*
National Academy of Code Administration, Cincinnati *15771*
National Conference of States on Building Codes and Standards, Herndon *15972*
Reinforced Concrete Research Council, Urbana *16407*
Southern Building Code Congress – International, Birmingham *16724*
Underground Technology Research Council Education Committee, Brooklyn *16802*
United States Committee on Large Dams, Denver *16820*

Cocoa s. Coffee, Tea, Cocoa

Coffee, Tea, Cocoa

Italy
Centro Luigi Lavazza per gli Studi e Ricerche sul Caffe', Torino *07515*

Malawi
Tea Research Foundation (Central Africa), Mulanje *08625*

U.S.A.
American Cocoa Research Institute, McLean *13694*

Commerce

Canada
Société des Relations d'Affaires Hautes Etudes Commerciales, Montréal *02926*

Germany
Forschungsstelle für den Handel Berlin e.V., Berlin *05738*
Forschungsverband für den Handelsvertreter- und Handelsmaklerberuf, Köln *05740*
HWWA – Institut für Wirtschaftsforschung – Hamburg, Hamburg *05935*
Staats- und Handelspolitische Gesellschaft, Recklinghausen *06291*

Ireland
Irish Institute of Purchasing and Materials Management, Dublin *06988*

Italy
Centro per gli Studi sui Sistemi Distributivi e il Turismo, Milano *07521*

Russia
Society for Trade and Commerce, Moskva *09971*

United Kingdom
Institute of Purchasing and Supply, Ascot *12274*
Institute of Small Business, London *12282*
Institute of Supervisory Management, Lichfield *12285*

Communications

Australia
Inter-Union Commission on Frequency Allocations for Radio Astronomy and Space Science, Epping *00426*

Bulgaria
Union of Energetics, Sofia *01792*

Canada
Alberta Educational Communications Corporation, Edmonton *01832*
Canadian Institute for Studies in Telecommunications, Pierrefonds *02166*

Germany
Informationstechnische Gesellschaft im VDE, Frankfurt *05942*

Italy
Associazione Studi sull'Informazione, Milano *07405*
Istituto Internazionale delle Comunicazioni, Genova *07730*

Switzerland
International Telecommunication Union, Genève *10758*

United Kingdom
Independent Association of Telecommunications Users, London *12197*

U.S.A.
Association of College and University Telecommunications Administrators, Lexington *14415*

Communication Science

Austria
Kontext – Institut für Kommunikations- und Textanalysen, Wien *00696*
Mediacult, Internationales Forschungsinstitut für Medien, Kommunikation und kulturelle Entwicklung, Wien *00705*
Österreichische Gesellschaft für Filmwissenschaft, Kommunikations- und Medienforschung, Wien *00779*
Österreichische Gesellschaft für Kommunikationsfragen, Wien *00803*

Canada
Commonwealth Association for Education in Journalism and Communication, London *02364*

France

Association pour la Fondation Internationale du Cinéma et de la Communication Audiovisuelle, Cannes *04321*

Germany

Deutsche Gesellschaft für Kommunikationsforschung, München *05373*
Deutsche Gesellschaft für Publizistik- und Kommunikationswissenschaft e.V., Eichstätt *05413*
Deutsches Institut für angewandte Kommunikation und Projektförderung e.V., Bonn *05582*
Münchner Kreis, München *06125*

Italy

Accademia di Relazioni Pubbliche, Roma *07171*
Accademia Prenestina del Cimento di Musica, Lettere, Scienze, Arti Visive e Figurative, Palestrina *07230*
Centro Internazionale di Documentazione e Comunicazione, Roma *07467*
Istituto Italiano di Pubblicismo, Roma *07739*

Jamaica

Caribbean Institute of Mass Communications, Mona *08074*

Japan

Denshi Joho Tsushin Gakkai, Tokyo *08121*

Kenya

ACCE Institute for Communication Development and Research, Nairobi *08457*
African Council for Communication Education, Nairobi *08466*

Malaysia

Asian Institute for Development Communication, Kuala Lumpur *08634*

Saudi Arabia

Arab Satellite Communications Organization, Riyadh *09998*

Senegal

Association des Professionnelles Africaines de la Communication, Dakar *10020*

Singapore

Asian Mass Communication Research and Information Centre, Singapore *10035*

Switzerland

Schweizerische Gesellschaft für Kommunikations- und Medienwissenschaft, Bern *10828*

United Kingdom

Industrial Society, London *12202*

U.S.A.

Association for Communication Administration, Annadale *14281*
Association for Progressive Communications, San Francisco *14322*
Association for the Development of Religious Information Systems, Milwaukee *14352*
Association of Biomedical Communication Directors, Hershey *14400*
Council for Biomedical Communications Associations, Ann Arbor *14871*
Intermedia, New York *15361*
International Communications Association, Dallas *15444*
The Media Institute, Washington *15707*
World Communication Association, Westfield *16924*

Communism s. Socialism

Computer and Information Science, Data Processing

Argentina

Asociación Argentina de Bibliotecas y Centros de Información Científicos y Técnicos, Buenos Aires *00058*
Conference of Latin American Data-Processing Authorities, Buenos Aires *00125*

Armenia

Commission on Computer Technology, Yerevan *00204*
Council on Informatics, Yerevan *00206*

Australia

Australian Society for Operations Research, Melbourne *00343*

Austria

Gesellschaft für Input-Output-Analyse, Wien *00622*
Österreichische Computer-Gesellschaft, Wien *00735*
Österreichische Gesellschaft für Artificial Intelligence, Wien *00762*
Österreichische Gesellschaft für Dokumentation und Information, Wien *00774*
Österreichische Gesellschaft für Informatik, Linz *00795*
Österreichische Gesellschaft für Statistik, Wien *00834*

Bahrain

Bahrain Computer Society, Manama *01056*

Belgium

Advanced Informatics in Medicine in Europe, Bruxelles *01097*
ASAB-VEBI, Bruxelles *01103*
Association Internationale des Mathématiques et Calculateurs en Simulation, Liège *01141*
Cooperation for Open Systems Interconnection Networking in Europe, Bruxelles *01256*
European Association for Health Information and Libraries, Bruxelles *01276*
European Committee for Interoperable Systems, Bruxelles *01321*
European Committee on Computational Methods in Applied Sciences, Bruxelles *01325*
European Control Data Users Group, Diepenbeek *01331*

Canada

American Society for Information Science (Western Canada Chapter), Edmonton *01864*
Association for Computing Machinery, Verdun *01900*
Canadian Association for Information Science, Perth *02034*
Canadian Image Processing and Pattern Recognition Society, Toronto *02159*
Canadian Industrial Computer Society, Ottawa *02161*
Canadian Man-Computer Communications Society, Toronto *02193*
Canadian Society for the Computational Study of Intelligence, Toronto *02266*
Computer Communications Institute, Willowdale *02374*
Information Resource Management Association of Canada, Toronto *02517*
International Development Research Centre, Ottawa *02552*
Languages of Instruction Commission of Ontario, Toronto *02575*

Chile

Asociación de Informatica y Computación en Educación, Santiago *03018*

China, People's Republic

China Computer Federation, Beijing *03132*
Chinese Information Processing Society, Beijing *03173*
Inner Mongolian Academy of Social Sciences, Huhehot *03206*

China, Republic

Chinese Society for Electronic Data Processing, Taipei *03284*

Denmark

Dansk Dataforening, København *03599*
Dansk Føderation for Informationsbehandling og Virksomhedsstyring, København *03618*

Ecuador

Agence Latinoaméricaine d'Information, Quito *03841*

Egypt

Egyptian Association of Archives, Librarianship and Information Science, Cairo *03889*
National Information and Documentation Centre, Cairo *03910*

Finland

Nordic Council for Scientific Information, Espoo *03984*
Tietopalveluseura, Espoo *04077*

France

Association des Informaticiens de Langue Française, Paris *04188*
Association Internationale de Recherche en Informatique Toxicologique, Paris *04263*
Association Internationale des Ecoles des Sciences de l'Information, Paris *04270*
Association Internationale d'Information Scolaire, Universitaire et Professionnelle, Paris *04283*
Association Internationale Données pour le Développement, Marseille *04286*
Comité pour les Données Scientifiques et Technologiques, Paris *04420*
Committee on Data for Science and Technology, Paris *04431*
Conité Européen Lex Informatica Mercatoriaque, Montpellier *04451*
Conseil International pour l'Information Scientifique et Technique, Paris *04461*
Ecole Internationale d'Informatique de l'AFCET, Paris *04468*
Groupement Professionnel National de l'Informatique, Paris *04560*

Georgia

Scientific-Technical Council on Computer Technology, Mathematical Modelling, Automation of Scientific Research and Instrument Making, Tbilisi *05027*

Germany

Anwenderverband Deutscher Informationsverarbeiter e.V., Kiel *05077*
Berliner Arbeitskreis Information, Berlin *05186*
European Association for the Development of Databases in Education and Training, Berlin *05664*
European Association for Theoretical Computer Science, Paderborn *05665*
Forschungsvereinigung Programmiersprachen für Fertigungseinrichtungen e.V., Aachen *05752*
Fraunhofer-Gesellschaft zur Förderung der angewandten Forschung e.V., München *05759*
Gesellschaft für angewandte Informatik, Bad Homburg *05797*
Gesellschaft für Datenschutz und Datensicherung e.V., Bonn *05806*
Gesellschaft für Informatik e.V., Bonn *05829*
Gesellschaft für Informationsverarbeitung in der Landwirtschaft, Stuttgart *05830*
Gesellschaft für Mathematik und Datenverarbeitung, Sankt Augustin *05839*
Gesellschaft für Pädagogik und Information e.V., Paderborn *05848*
Gesellschaft für Programmierte Instruktion und Mediendidaktik e.V., Giessen *05852*
INFODAS, Köln *05940*
Informatica, Darmstadt *05941*
Normenausschuß Informationsverarbeitungssysteme im DIN Deutsches Institut für Normung e.V., Berlin *06179*

Hungary

Neumann János Számítógéptudományi Társaság, Budapest *06666*

India

Indian Association of Special Libraries and Information Centres, Calcutta *06775*
Operational Research Society of India, New Dehli *06847*

Iraq

Arab Gulf States Information Documentation Centre, Baghdad *06909*

Ireland

Irish Computer Society, Dublin *06980*

Israel

Israel Society of Special Libraries and Information Centres, Tel Aviv *07104*

Italy

Association of European Operational Research Societies, Bologna *07253*
Associazione Italiana di Documentazione e di Informazione, Roma *07307*
Associazione Italiana per l'Informatica et il Calcolo Automatico, Milano *07343*
European Coordinating Committee for Artificial Intelligence, Povo *07676*
ITALSIEL, Roma *07758*
Società Nazionale di Informatica delle Camere di Commercio per la Gestione dei Centri Elettronici Reteconnessi Valutazione Elaborazione Dati, Roma *07991*

Japan

Information Processing Society of Japan, Tokyo *08140*

Korea, Republic

Korean Library and Information Science Society, Seoul *08535*

Luxembourg

European Association of Information Services, Luxembourg *08583*

Macedonia

Sojuz na Društvata na Matematičarite i Informatičarite na Makedonija, Skopje *08610*

Morocco

African Network of Administrative Information, Tangiers *08799*

Netherlands

Centrum voor Wiskunde en Informatica, Amsterdam *08841*
Conference of European Computer User Associations, Bussum *08845*
European Association for Cognitive Ergonomics, Amsterdam *08852*
European Association for Microprocessing and Microprogramming, Apeldoorn *08858*
International Association for Statistical Computing, Voorburg *08921*
Nederlands Bureau voor Bibliotheekwezen en Informatieverzorging, 's-Gravenhage *08981*
Sectie Operationele Research, Amsterdam *09068*

New Zealand

Operational Research Society of New Zealand, Wellington *09184*

Nigeria

Africa Regional Centre for Information Science, Ibadan *09222*

Norway

Norsk Operasjonsanalyseforening, Oslo *09333*
Norsk Regnesentral, Oslo *09336*

Peru

Andean Technological Information System, Lima *09446*

Russia

Commission for Computer Technology, Moskva *09814*
Commission for the Use of Computers and for Raising the Qualifications of Computer Users, Moskva *09842*
Committee for Systems Analysis, Moskva *09847*
Co-ordination Committee for Computer Technology, Moskva *09855*
National Committee for the Collection and Assessment of Numerical Data in Science and Technology, Moskva *09889*

Slovakia

Slovenská Spoločnost pre Kybernetiku a Informatiku, Bratislava *10089*

South Africa

South African Council for Automation and Computation, Pretoria *10187*
South African Institute for Librarianship and Information Science, Pretoria *10193*

Spain

Sociedad de Estadística e Investigación Operativa, Madrid *10373*

Sweden

Svenska Operationsanalysföreningen, Stockholm *10571*
Svenska Samfundet för Informationsbehandling, Stockholm *10574*

Switzerland

Advisory Committee for the Coordination of Information Systems, Genève *10608*
Association for Computational Linguistics-Europe, Lugano *10625*
Digital Equipment Computer Users Society, Genève *10663*
European Association for Computer Graphics, Aire-la-Ville *10673*
European Association for Signal Processing, Lausanne *10674*
International Computing Centre, Genève *10735*
Schweizerische Informationskonferenz, Basel *10863*
Schweizerische Vereinigung für Datenverarbeitung, Zürich *10941*

Syria

Arab Centre for Information Studies on Population, Development and Construction, Damascus *11039*

Thailand

Agricultural Information Development Scheme, Bangkok *11064*

Tunisia

Arab Information Network for Terminology, Tunis *11130*

United Kingdom

Association for Literary and Linguistic Computing, Bangor *11271*
British Association for Information and Libraries Education and Research, Sheffield *11493*
The British Computer Society, London *11547*
British Robot Association, Birmingham *11641*
British Urban and Regional Information System Association, Winchester *11702*
Computer Arts Society, London *11834*
Construction Industry Computing Association, Cambridge *11838*
Institute of Data Processing Management, London *12231*
Institute of Information Scientists, London *12246*
Operational Research Society, Birmingham *12655*
Remote Sensing Society, Nottingham *12749*
Society for Computers and Law, Bristol *12967*
Society for the Study of Information Transfer, Beckenham *13017*

U.S.A.

Accountants Computer Users Technical Exchange, Indianapolis *13313*
Amdahl Users Group, Bellevue *13374*
American Association for Artificial Intelligence, Menlo Park *13468*
American Library Association, Chicago *13916*
American Medical Informatics Association, Bethesda *13937*
American Physicians Association of Computer Medicine, Pittsford *14002*
American Society for Information Science, Silver Spring *14068*
American Software Users Group, Chicago *14176*
ARMS Firms Users Association, Mount Clemens *14248*
Association for Computers and the Humanities, Provo *14284*
Association for Computing Machinery, New York *14285*
Association for the Development of Computer-Based Instructional Systems, Columbus *14350*
Association for Women in Computing, San Francisco *14366*
Association of Computer Users, Berkeley *14421*
Association of Data Communications Users, Bloomington *14425*
Association of Information and Dissemination Centers, Athens *14445*
Association of Minicomputers Users, Franklin *14466*
Association of the Institute for Certification of Computer Professionals, Des Plaines *14520*
Atex Newspaper Users Group, Chicago *14544*
Autoprep 5000 Users Group, South Windsor *14559*
BITNET, Princeton *14597*
Center for Computer-Assisted Legal Instruction, Minneapolis *14673*
Clinical Chemistry Data Communication Group, Minneapolis *14762*
Computer Aided Manufacturing International, Arlington *14817*
CUBE, Saint Clair Shores *14946*
Dyleague, Woodland Hills *14991*
Education Turnkey Systems, Falls Church *15029*
Federation of Government Information Processing Councils, Atlanta *15084*
Graphic Communications Association, Alexandria *15172*
Health Industry Business Communications Council, Phoenix *15193*
Independent Computer Consultants Association, Saint Louis *15252*
Institute for Certification of Computer Professionals, Des Plaines *15271*
International Association for Computer Information Systems, Eau Claire *15377*
International Association for Computing in Education, Eugene *15378*
Museum Computer Network, Pittsburgh *15759*
National Center for Automated Information Research, New York *15939*
National Center for Computer Crime Data, Santa Cruz *15941*

Robotics International of the Society of Manufacturing Engineers, Dearborn 16423
Society for Computer Simulation International, San Diego 16512
Society for Information Display, Playa Del Rey 16539
Special Interest Group for Computer Personnel Research, College Station 16734
Special Interest Group for Computer Uses in Education, College Station 16735
Special Interest Group on Artificial Intelligence, New York 16737
Special Interest Group on Biomedical Computing, Durham 16738
Special Interest Group on Information Retrieval, Blackburg 16739
Urban and Regional Information Systems Association, Washington 16851
VIM, Saint Paul 16862
Women's Computer Literacy Center, San Francisco 16919

Venezuela

Action Committee for the Establishment of a Latin American Network of Technological Information, Caracas 16964
Action for National Information Systems, Caracas 16965

Zimbabwe

African Regional Computer Confederation, Harare 17160

Corrosion s. Metallurgy

Criminology

Argentina

Sociedad Argentina de Criminología, Buenos Aires 00163

Canada

British Columbia Corrections Association, Vancouver 01974
Canadian Criminal Justice Association, Ottawa 02113
International Center for Comparative Criminology, Montréal 02543
Manitoba Society of Criminology, Winnipeg 02635
Nova Scotia Criminology and Corrections Association, Halifax 02730
Prince Edward Island Criminology and Corrections Association, Charlottetown 02831
Société de Criminologie du Québec, Montréal 02919

Denmark

Dansk Kriminalistforening, København 03640

France

Association Internationale des Amis de Vasile Stanciu, Paris 04264
Association Lyonnaise de Criminologie et Anthropologie Sociale, Lyon 04300
Comité Européen pour les Problèmes Criminels, Strasbourg 04402
Société de Médecine Légale et de Criminologie de France, Paris 04670
Société Internationale de Criminologie, Paris 04877

Germany

Neue Kriminologische Gesellschaft, Tübingen 06140

India

Indian Society of Criminology, Madras 06816

Iraq

Arab Bureau for Prevention of Crime, Baghdad 06906

Israel

Israel Society of Criminology, Jerusalem 07099

Italy

Società Italiana di Criminologia, Roma 07860
Società Lombarda di Criminologia, Milano 07981

Japan

Asia Crime Prevention Foundation, Tokyo 08094
Kyosei Kyokai Kyosei Toshokan, Tokyo 08177
Nippon Keiho Gakkai, Tokyo 08327

Saudi Arabia

Arab Security Studies and Training Center, Riyadh 09999

Sweden

Svenska Kriminalistföreningen, Stockholm 10557

United Kingdom

Citizens Protection Society, London 11788

U.S.A.

American Correctional Association, Laurel 13756
American Criminal Justice Association – Lambda Alpha Epsilon, Sacramento 13772
American Society of Criminology, Columbus 14115
Association of Forensic Document Examiners, Minneapolis 14440
National Center for Computer Crime Data, Santa Cruz 15941
National Council on Crime and Delinquency, San Francisco 16017
National Crime Prevention Institute, Louisville 16027
National Criminal Defense College, Macon 16028

Yugoslavia

Savez Udruženja za Krivično Pravo i Kriminolgiju Jugoslavije, Beograd 17122

Crop Husbandry

Belgium

European Crop Protection Association, Bruxelles 01336

Costa Rica

Central American, Mexican and Caribbean Network for Bean Research, San José 03439

Germany

Deutsche Phytomedizinische Gesellschaft e.V., Mainz 05489
Gemeinschaft zur Förderung der privaten deutschen Pflanzenzüchtung e.V., Bonn 05773

Japan

Nihon Sakumotsu Gakkai, Tokyo 08252
Nippon Shokubutsu-Byori Gakkai, Tokyo 08382

Mexico

Comisión Latinoamericana de Investigadores en Sorgo, México 08725

Netherlands

Centrum voor Plantenveredelings- en Reproduktieonderzoek, Wageningen 08840
European Association for Research on Plant Breeding, Dronten 08861
European Association for Research on Plant Breeding, Wageningen 08862

Peru

CIAT Andean Zone Network for Bean Research, Lima 09463

Spain

European Cooperative Research Network on Olives, Cordoba 10308

Uganda

Cotton Research Corporation, Kampala 11183

United Kingdom

Scottish Society for Crop Research, Dundee 12928

Virgin Islands (U.S.)

Caribbean Food Crops Society, Saint Croix 17073

Cultural History, History of Civilization

Andorra

Associació Cultural i Artística Els Esquirols, La Massana 00027

Austria

Commission Internationale d'Histoire du Sel, Innsbruck 00579
Österreichische Gesellschaft für Amerikastudien, Salzburg 00753
Österreichisches Ost- und Südosteuropa-Institut, Wien 00915

Barbados

Caribbean Conservation Association, Bridgetown 01080

Belgium

Association des Diplomés en Histoire de l'Art et Archéologie de l'Université Catholique de Louvain, Louvain-la-Neuve 01116
European Association for Burgundy Studies, Bruxelles 01274

Brazil

Sociedade Brasileira de Cultura, São Paulo 01707

Canada

American Osler Society, Hamilton 01862
Canadian Association for the Advancement of Netherlandic Studies, Windsor 02041
Canadian Railroad Historical Association, Saint Constant 02241
Canadian Rodeo Historical Association, Cochrane 02245
Canadian Science and Technology Historical Association, North York 02247
Canadian Society for the History and Philosophy of Science, Montréal 02267
International Cooperation in History of Technology Committee, Toronto 02547
Micmac Association of Cultural Studies, Sydney 02652
Ontario Electric Railway Historical Association, Scarborough 02775
Postal History Society of Canada, Ottawa 02825
Upper Canada Society for the History and Philosophy of Science and Technology, Hamilton 02974

Cuba

Sociedad Cubana de Historia de la Medicina, La Habana 03482

Denmark

Dansk Medicinsk-Historisk Selskab, København 03646
Skandinavisk Museumsforbund, Danish Section, Aalborg 03821

Egypt

The Egyptian Society for the Dissemination of Universal Culture and Knowledge, Cairo 03896

France

Arab World Institute, Paris 04136
Association pour la Recherche Interculturelle, Saint-Denis 04330
Commission Internationale d'Etudes Historiques Latinoaméricaines et des Caraïbes, Paris 04426
European Association for Chinese Studies, Paris 04481
International Association of Sanskrit Studies, Paris 04582
Société de l'Histoire de l'Art Français, Paris 04662
Société Française d'Histoire de la Médecine, Reims 04856

Germany

Deutsche Gesellschaft für Geschichte der Medizin, Naturwissenschaft und Technik e.V., München 05355
Frobenius-Gesellschaft e.V., Frankfurt 05765
Gesellschaft für Geschichte und Kultur e.V., Bonn 05825
Leiterkreis der Katholischen Akademien, Schwerte 06102
Moses Mendelssohn Zentrum für europäisch-jüdische Studien, Potsdam 06121
Ost-Akademie e.V., Lüneburg 06225
Verband Deutscher Kunsthistoriker e.V., Darmstadt 06353
Verein von Altertumsfreunden im Rheinlande, Bonn 06424
Zentralinstitut für Kunstgeschichte, München 06465

Greece

Etaireia Byzantinon Spoudon, Athinai 06524

Hong Kong

Royal Asiatic Society, Hong Kong Branch, Hong Kong 06595

India

Sri Aurobindo Centre, New Delhi 06862

Israel

Yad Izhak Ben-Zvi, Jerusalem 07118

Italy

Accademia delle Scienze e delle Arti degli Ardenti di Viterbo, Viterbo 07151
Accademia di Scienze, Lettere e Arti, Lucca 07173
Accademia Raffaello, Urbino 07231
Accademia Spoletina, Spoleto 07240
Associazione Nazionale per i Centri Storico-Artistici, Gubbio 07379
Centro di Studi e Ricerche di Museologia Agraria, Milano 07442
Centro di Studi per la Storia dell'Architettura, Roma 07450
Centro per la Storia della Tradizione Aristotelica Nel Veneto, Padova 07526
Centro Ricerche di Storia e Arte Bitontina, Bitonto 07537
Centro Studi Storici di Mestre, Mestre 07593
Istituto Italiano di Storia della Chimica, Roma 07740
Società Italiana di Storia della Medicina, Bologna 07952
Società Italiana per l'Archeologia e la Storia delle Arti, Napoli 07965
Società Tarquiniense d'Arte e Storia, Tarquinia 08021
Società Tiburtina di Storia e d'Arte, Tivoli 08023

Japan

Suifumeitokukai Shokokan, Ibaraki-ken 08427

Macedonia

Društvo na Istoričarite na Umetnosta od Makedonija, Skopje 08593

Mexico

Sociedad Mexicana de Historia de la Ciencia y la Tecnología, México 08765

Netherlands

Comité International d'Histoire de l'Art, Utrecht 08842
European Association for Japanese Studies, Leiden 08857
Koninklijk Oudheidkundig Genootschap Amsterdam, Amsterdam 08968

New Zealand

Commonwealth Heraldry Board, Auckland 09123

Pakistan

Lok Virsa, Islamabad 09373
Research Society of Pakistan, Lahore 09398

Peru

Sociedad Peruana de Historia de la Medicina, Lima 09484

Poland

Polskie Towarzystwo Historii Medycyny i Farmacji, Warszawa 09577
Stowarzyszenie Historyków Sztuki, Warszawa 09616

Russia

Commission for the History of Philology, Moskva 09836

Slovakia

Slovenská Spoločnost pre Dejiny Vied a Techniky pri SAV, Bratislava 10088

Slovenia

Slovensko Umetnostno Zgodovinsko Društvo, Ljubljana 10107

Spain

Comité International Permanent des Etudes Mycéniennes, Vitoria 10292

Sweden

Statens Kulturråd, Stockholm 10523

Switzerland

Association des Instituts d'Etudes Européennes, Genève 10621
Schweizerische Gesellschaft für Geschichte der Medizin und der Naturwissenschaften, Bern 10822
Schweizerische Gesellschaft Pro Technorama, Winterthur 10857

United Kingdom

Association of European Open Air Museums, Chichester 11344
British Archaeological Association, London 11481
British Postmark Society, Chorley 11634
British Vexillological Society, Beckenham 11706
The Costume Society, London 11846
Disinfected Mail Study Circle, London 11895
The Ephemera Society, London 11970
European Centre for Traditional and Regional Cultures, Llangollen 12006
Hertfordshire Local History Council, Hitchin 12164
Institute for Cultural Research, Tunbridge Wells 12207
Local Population Studies Society, London 12458
Medical Sciences Historical Society, Liverpool 12510
Printing Historical Society, London 12718
Scottish Society for the Preservation of Historical Machinery, Glasgow 12930
Scottish Society of the History of Medicine, Aberdeen 12931

U.S.A.

American Association for the History of Medicine, Boston 13500
American Institute for Patristic and Byzantine Studies, Kingston 13870
American Institute of the History of Pharmacy, Madison 13896
American Society for Ethnohistory, Chicago 14060
Buffalo Bill Memorial Association, Cody 14621
Forest History Society, Durham 15112
Hebrew Culture Foundation, New York 15200
Institute of Early American History and Culture, Williamsburg 15316
International Arthur Schnitzler Research Association, San Bernardino 15374
International Society for the History of Ideas, Philadelphia 15542
Lexington Group in Transportation History, Saint Cloud University 15666
Mid-Continent Railway Historical Society, North Freedom 15730
National Association for Outlaw and Lawman History, Killeen 15808
North American Society for Sport History, University Park 16223
Postal History Society, Pikesville 16341
Society for Creative Anachronism, Oakland 16513
Society for the History of Technology, Houghton 16598
Tennessee Folklore Society, Murfreesboro 16770

Venezuela

Sociedad Venezolana de Historia de la Medicina, Caracas 17024

Cybernetics

Austria

Forschungsgesellschaft für Psycho-Elektronik und Kybernetik, Wien 00596
Österreichische Studiengesellschaft für Kybernetik, Wien 00921

Azerbaijan

Scientific Council on Complex Problems (Cybernetics), Baku 01048

Belgium

Association Internationale de Cybernétique, Namur 01135

Czech Republic

Česká Společnost Kybernetiku a Informatiku, Praha 03525

France

Centre International de Cyto-Cybernétique, Paris 04375

Germany

Deutsche Gesellschaft für Baukybernetik e.V, Holzminden 05328

Greece

European Institute of Environmental Cybernetics, Athinai 06527

Japan

Tokei Kagaku Kenkyukai, Fukuoka 08435

Slovakia

Slovenská Spoločnost pre Kybernetiku a Informatiku, Bratislava 10089

U.S.A.

American Society for Cybernetics, Fairfax 14054

Dairy Science

Australia
Dairy Industry Association of Australia, Highett *00384*

Belgium
Société Nationale de Laiterie, Bruxelles *01467*

Denmark
Dansk Mejeringeniør Forening, Odense *03649*

Egypt
Egyptian Society of Dairy Science, Cairo *03897*

France
Centre Technique et de Promotion des Laitiers Sidérurgiques, Paris La Défense *04385*

Italy
Centro Lattiero Caseario di Assistenza e Sperimentazione Antonio Bizzozero, Parma *07513*

Netherlands
Genootschap ter Bevordering van Melkkunde, Rijswijk *08903*

Norway
Norske Meierifolks Landsforening, Oslo *09303*

United Kingdom
British Goat Society, Bovey Tracey *11569*
Society of Dairy Technology, Huntingdon *13042*

U.S.A.
Dairy Research Foundation, Oak Grove Village *14948*
Dairy Society International, Chambersburg *14949*
National Dairy Council, Rosemont *16029*

Data Processing s. Computer and Information Science, Data Processing

Decorative Arts s. Graphic and Decorative Arts, Design

Dentistry

Argentina
Asociación Odontológica Argentina, Buenos Aires *00091*

Australia
Australian Dental Association, Saint Leonards *00282*
Australian Dental Association, Northern Territory Branch, Darwin *00283*
Australian Dental Association, Queensland Branch, Bowen Hills *00284*
Australian Dental Association, South Australian Branch, Unley *00285*
Australian Dental Association, Tasmania Branch, New Town *00286*
Australian Dental Association, Victorian Branch, Toorak *00287*
Australian Dental Association, Western Australia Branch, West Perth *00288*
Australian Society of Endodontology, Subiaco *00352*
Australian Society of Orthodontists, Perth *00355*
Australian Society of Periodontology, Melbourne *00356*
Australian Society of Prosthodontists, Melbourne *00358*

Austria
Österreichische Dentistenkammer, Wien *00736*
Österreichische Gesellschaft für Zahn-, Mund- und Kieferheilkunde, Wien *00850*

Barbados
Caribbean Atlantic Regional Dental Association, Saint Michael *01078*

Belgium
Groupement International pour la Recherche Scientifique en Odontologie et en Stomatologie, Bruxelles *01371*
Groupement International pour la Recherche Scientifique en Stomatologie et Odontologie, Bruxelles *01372*

Bolivia
Confederación Boliviana de Odontólogos, La Paz *01542*

Canada
Alberta Dental Association, Edmonton *01831*
Association of Canadian Faculties of Dentistry, Edmonton *01922*
Canadian Academy of Periondontology, Toronto *02005*
Canadian Academy of Restorative Dentistry, London *02007*
Canadian Association for Dental Research, London *02029*
Canadian Dental Association, Ottawa *02114*
Canadian Dental Research Foundation, Ottawa *02115*
College of Dental Surgeons of British Columbia, Vancouver *02352*
College of Dental Surgeons of Saskatchewan, Saskatoon *02353*
Dental Association of Prince Edward Island, Charlottetown *02414*
Denturist Society of Alberta, Sherwood Park *02415*
Denturist Society of Nova Scotia, Halifax *02416*
Manitoba Dental Association, Winnipeg *02614*
National Dental Examining Board of Canada, Ottawa *02671*
New Brunswick Dental Society, Rothesay *02678*
Nova Scotia Dental Association, Halifax *02731*
Ontario Dental Association, Toronto *02773*
Royal College of Dental Surgeons of Canada, Toronto *02868*

Chile
Colegio de Dentistas de Chile, Concepción *03031*
Sociedad Odontológica de Concepción, Concepción *03103*

Colombia
Asociación Lationoamericana de Facultades de Odontologia, Medellín *03353*

Costa Rica
Academia Costarricense de Periodoncia, San José *03428*

Denmark
Dansk Odontologisk Selskab, Klampenborg *03657*
Dansk Tandlaegeforening, København *03695*
International Society for Prosthetics and Orthotics, København *03748*

Egypt
Cairo Odontological Society, Cairo *03884*
Egyptian Dental Federation, Cairo *03891*

Finland
Nordisk Odontologisk Förening, Helsinki *03986*
Scandinavian Dental Association, Turku *03990*
Suomen Hammaslääkäriliitto, Helsinki *04024*
Suomen Hammaslääkäriseura, Helsinki *04025*

France
Administration Universitaire Francophone et Européenne en Médecine et Odontologie, Paris *04123*
Association Dentaire Française, Paris *04155*
Association Internationale Francophone de Recherche Odontologique, Rennes *04287*
Association Internationale pour le Développement de l'Odonto-Stomatologie Tropicale, Bordeaux *04288*
Confédération Nationale des Syndicats Dentaires, Paris *04439*
Fédération des Chambres Syndicales des Chirurgiens Dentistes de la Région de Paris, Paris *04512*
Fédération des Syndicats Dentaires Libéraux, Paris *04518*
Groupement des Associations Dentaires Francophones, Paris *04553*
Ordre des Chirurgiens Dentistes, Paris *04608*
Ordre National des Chirurgiens Dentistes, Paris *04610*
Société Odontologique de Paris, Paris *04907*
Syndicat des Chirurgiens Dentistes de Paris, Paris *04924*
Union Nationale Patronale des Prothésistes Dentaire, Paris *04994*

Germany
Ärztekammer des Saarlandes, Abteilung Zahnärzte, Saarbrücken *05033*
Bayerische Landeszahnärztekammer, München *05175*
Bundeszahnärztekammer, Köln *05261*
Deutsche Arbeitsgemeinschaft für Paradontologie, Goslar *05292*
Deutsche Gesellschaft für Kieferorthopädie e.V., Würzburg *05370*
Deutsche Gesellschaft für Parodontologie, Hamburg *05396*
Deutsche Gesellschaft für Zahnärztliche Prothetik und Werkstoffkunde e.V., Hannover *05443*
Deutsche Gesellschaft für Zahnerhaltung, Berlin *05444*
Deutsche Gesellschaft für Zahn-, Mund- und Kieferheilkunde, Düsseldorf *05445*
Freier Verband Deutscher Zahnärzte e.V., Bonn *05761*
Kassenzahnärztliche Bundesvereinigung, Köln *06040*
Kassenzahnärztliche Vereinigung Berlin, Berlin *06041*
Kassenzahnärztliche Vereinigung für den Regierungsbezirk Freiburg, Freiburg *06042*
Kassenzahnärztliche Vereinigung für den Regierungsbezirk Karlsruhe, Mannheim *06043*
Kassenzahnärztliche Vereinigung für den Regierungsbezirk Stuttgart, Stuttgart *06044*
Kassenzahnärztliche Vereinigung für den Regierungsbezirk Tübingen, Tübingen *06045*
Kassenzahnärztliche Vereinigung Hamburg, Hamburg *06046*
Kassenzahnärztliche Vereinigung Hessen, Frankfurt *06047*
Kassenzahnärztliche Vereinigung im Lande Bremen, Bremen *06048*
Kassenzahnärztliche Vereinigung Koblenz-Trier, Koblenz *06049*
Kassenzahnärztliche Vereinigung Niedersachsen, Hannover *06050*
Kassenzahnärztliche Vereinigung Nordrhein, Düsseldorf *06051*
Kassenzahnärztliche Vereinigung Pfalz, Ludwigshafen *06052*
Kassenzahnärztliche Vereinigung Rheinhessen, Mainz *06053*
Kassenzahnärztliche Vereinigung Saarland, Saarbrücken *06054*
Kassenzahnärztliche Vereinigung Schleswig-Holstein, Kiel *06055*
Kassenzahnärztliche Vereinigung Westfalen-Lippe, Münster *06056*
Landeszahnärztekammer Baden-Württemberg, Stuttgart *06099*
Landeszahnärztekammer Hessen, Frankfurt *06100*
Landeszahnärztekammer Rheinland-Pfalz, Mainz *06101*
Normenausschuss Dental im DIN Deutsches Institut für Normung e.V., Pforzheim *06157*
Regionale Organisation der FDI für Europa (ERO), Köln *06262*
Saarländische Gesellschaft für zahnärztliche Fortbildung, Saarbrücken *06270*
Schleswig-Holsteinische Gesellschaft für Zahn-, Mund- und Kieferheilkunde, Lübeck *06276*
Vereinigung der Hochschullehrer für Zahn-, Mund- und Kieferheilkunde, Frankfurt *06398*
Westfälische Gesellschaft für Zahn-, Mund- und Kieferheilkunde, Münster *06440*
Zahnärztekammer Berlin, Berlin *06456*
Zahnärztekammer Bremen, Bremen *06457*
Zahnärztekammer Hamburg, Hamburg *06458*
Zahnärztekammer Niedersachsen, Hannover *06459*
Zahnärztekammer Nordrhein, Düsseldorf *06460*
Zahnärztekammer Schleswig-Holstein, Kiel *06461*
Zahnärztekammer Westfalen-Lippe, Münster *06462*

Greece
Etaireia Odontostomatologikis Ereunis, Athinai *06526*

Guatemala
Asociación de Ortodoncistas de Guatemala, Guatemala City *06559*
Asociacion Latinoamericana de Escuelas de Cirurgía Dental, Guatemala City *06560*

Hungary
Magyar Fögorvosok Egyesülete, Budapest *06628*

Iceland
Tannlaeknafélag Islands, Reykjavik *06703*

Iraq
Arab Dental Federation, Baghdad *06907*

Ireland
The Dental Council, Dublin *06941*
Irish Dental Association, Dublin *06981*

Israel
Israeli Dental Association, Tel Aviv *07076*

Italy
Accademia Europea Dentisti Implantologi, Milano *07180*
Associazione Italiana di Anestesia Odontostomatologica, Bologna *07298*
Associazione Medici Dentisti Italiani, Roma *07357*
Società Italiana di Odontoiatria Infantile, Roma *07918*
Società Italiana di Odontostomatologia e Chirurgia Maxillo-Facciale, L'Aquila *07919*

Japan
Koko Eisei Gakkai, Tokyo *08170*
Nihon Hozon Shika Gakkai, Tokyo *08208*
Nihon Koku Eisei Gakkai, Tokyo *08230*
Nihon Kyosei Shikagakkai, Tokyo *08240*
Nippon Hotetsu Shika Gakkai, Tokyo *08310*
Nippon Shika Hozon Gakkai, Tokyo *08375*
Nippon Shika Igakkai, Tokyo *08376*

Kenya
Kenya Dental Association, Nairobi *08503*

Luxembourg
Association des Médecins et Médecins-Dentistes du Grand-Duché de Luxembourg, Luxembourg *08575*

Malta
Dental Association of Malta, Saint Andrews *08659*

Netherlands
European College of Gerodontology, Nijmegen *08878*
Nederlandsche Maatschappij tot Bevordering der Tandheelkunde, Nieuwegein *08983*
Nederlandse Vereniging van Specialisten in den Dento-Maxillaire Orthopaedie, Nijmegen *09011*
Nederlandse Vereniging van Tandartsen, Maarssen *09012*
Nederlandse Vereniging voor Orthodontische Studie, Gorinchem *09031*

Norway
Den Norske Tannlegeforening, Oslo *09255*

Panama
Federación Odontológica de Centro América y Panamá, Panamá City *09419*

Portugal
Associação Portuguesa de Odontologia, Lisboa *09648*
Sindicato Nacional dos Odontologistas Portuguesas, Lisboa *09669*
Sindicato Nacional dos Protésicos Dentários, Lisboa *09672*

Singapore
Asian Pacific Dental Federation, Singapore *10039*

South Africa
Dental Association of South Africa, Houghton *10134*
Medical Research Council, Johannesburg *10153*
Oral and Dental Research Institute, Tygerberg *10159*
Prosthodontics Society of South Africa, Houghton *10166*

Spain
Association for Dental Education in Europe, Madrid *10282*
Consejo General de Colegios Oficiales de Odontólogos y Estomatólogos de España, Madrid *10299*

Sweden
Sveriges Tandläkarförbund, Stockholm *10596*

Switzerland
Arbeitsgemeinschaft für orale Implantologie, Zürich *10618*
Schweizerische Arbeitsgemeinschaft für Schul- und Jugendzahnpflege, Langnau *10797*
Schweizerischer Verband kantonal approbierter Zahnärzte, Zürich *10916*
Schweizerische Zahnärzte-Gesellschaft, Bern *10959*

Turkey
Istanbul Dişhekimleri Odasi, Istanbul *11146*

United Kingdom
American Dental Society of Europe, London *11225*
Anglo-Continental Dental Society, London *11231*
Association of Basic Science Teachers in Dentistry, Glasgow *11304*
Association of British Dental Surgery Assistants, Fleetwood *11309*
Association of Dental Hospitals of Great Britain and Northern Ireland, Birmingham *11335*
Bone and Tooth Society, Oxford *11458*
British Association of Orthodontists, Maidenhead *11514*
British Dental Association, London *11549*
British Dental Hygienists Association, London *11550*
British Endodontic Society, London *11557*
British Society for Oral Medicine, London *11660*
British Society for Restorative Dentistry, London *11665*
British Society for the Study of Orthodontics, London *11678*
British Society of Medical and Dental Hypnosis, London *11686*
Comité Dentaire de Liaison de la Communauté Européenne, London *11801*
Commonwealth Dental Association, London *11818*
Dentists' Liaison Committee for the EEC, London *11881*
European Dental Society, Pinner *12025*
European Organization for Caries Research, Leeds *12035*
European Orthodontic Society, London *12036*
Faculty of Dental Surgery, London *12051*
Federation of London Area Dental Committees, London *12066*
General Dental Council, London *12104*
General Dental Practitioner's Association, Manchester *12105*
International Association of Paediatric Dentistry, London *12332*
International Dental Federation, London *12348*
Royal College of Physicians and Surgeons of Glasgow, Glasgow *12786*
World Endodontics Confederation, London *13219*

Uruguay
Asociación Ondontológica Uruguaya, Montevideo *13246*

U.S.A.
Academy for Implants and Transplants, Springfield *13278*
Academy of Dental Materials, Charleston *13287*
Academy of Dentistry for the Handicapped, Chicago *13288*
Academy of Dentistry International, Washington *13289*
Academy of Denture Prosthetics, Bellevue *13290*
Academy of General Dentistry, Chicago *13292*
Academy of International Dental Studies, Boston *13294*
Academy of Operative Dentistry, Menomonie *13300*
Academy of Oral Dynamics, Frederick *13301*
American Academy for Plastics Research in Dentistry, Davison *13378*
American Academy of Dental Electrosurgery, New York *13387*
American Academy of Dental Group Practice, Phoenix *13388*
American Academy of Dental Practice Administration, Palatine *13389*
American Academy of Esthetic Dentistry, Chicago *13392*
American Academy of Fixed Prosthodontics, Dallas *13395*
American Academy of Gold Foil Operators, Noblesville *13398*
American Academy of Implant Dentistry, Chicago *13399*

American Academy of Implant Prosthodontics, Atlanta 13400
American Academy of Maxillofacial Prosthetics, Carmel 13404
American Academy of Oral and Maxillofacial Radiology, Aurora 13412
American Academy of Orofacial Pain, Lafayette 13414
American Academy of Orthodontics for the General Practitioner, Milwaukee 13415
American Academy of Pediatric Dentistry, Chicago 13420
American Academy of Periodontology, Chicago 13422
American Academy of Physiologic Dentistry, Naperville 13426
American Academy of Restorative Dentistry, Colorado Springs 13435
American Academy of the History of Dentistry, Chicago 13440
American Association for Dental Research, Washington 13477
American Association for Functional Orthodontics, Winchester 13478
American Association of Dental Examiners, Chigago 13527
American Association of Dental Schools, Washington 13528
American Association of Endodontists, Chicago 13531
American Association of Entrepreneurial Dentists, Tupelo 13533
American Association of Hospital Dentists, Chicago 13540
American Association of Oral and Maxillofacial Surgeons, Rosemont 13557
American Association of Orthodontists, Saint Louis 13558
American Association of Public Health Dentistry, Richmond 13576
American Association of Women Dentists, Chicago 13603
American Board of Dental Public Health, Gainesville 13631
American Board of Endodontics, Chicago 13633
American Board of Orthodontics, Saint Louis 13646
American Board of Pediatric Dentistry, Carmel 13650
American Board of Periodontology, Baltimore 13652
American Board of Prosthodontics, Atlanta 13659
American College of Dentists, Gaithersburg 13704
American College of Medical Group Administrators, Englewood 13714
American College of Prosthodontists, San Antonio 13733
American Dental Assistants Association, Chicago 13778
American Dental Association, Chicago 13779
American Dental Hygienists' Association, Chicago 13780
American Dental Society of Anesthesiology, Chicago 13781
American Endodontic Society, Fullerton 13797
American Equilibration Society, Morton Grove 13802
American Institute of Oral Biology, South Laguna 13891
American Orthodontic Society, Dallas 13970
American Prosthodontic Society, Chicago 14016
American Society for Dental Aesthetics, New York 14056
American Society for Geriatric Dentistry, Chicago 14062
American Society for the Study of Orthodontics, Oakland Gardens 14092
American Society of Dentistry for Children, Chicago 14118
American Society of Forensic Odontology, Chicago 14123
American Society of Master Dental Technologists, Oakland Gardens 14143
American Student Dental Association, Chicago 14184
Association of State and Territorial Dental Directors, Sacramento 14507
Christian Dental Society, Sumner 14745
Delta Dental Plans Association, Chicago 14958
Dental Assisting National Board, Chicago 14960
Dental Health International, Athens 14961
Federation of Orthodontic Associations, Milwaukee 15087
Federation of Prosthodontic Organizations, Chicago 15088
Flying Dentists Association, Dunwoody 15104
Holistic Dental Association, Oklahoma City 15224
Indian Dental Association U.S.A., Jamaica 15256
International Academy of Myodontics, Doylestown 15366
International Association for Dental Research, Washington 15379
International Association of Orthodontics, Oak Park 15415
International College of Dentists, Rockville 15438
International Congress of Oral Implantologists, Upper Montclair 15447
International Dental Health Foundation, Reston 15465
National Association of Dental Assistants, Falls Church 15848
National Association of Dental Laboratories, Alexandria 15849
National Board for Certification in Dental Technology, Alexandria 15922
National Board for Certification of Dental Laboratories, Alexandria 15923
National Dental Assistants Association, Washington 16030
National Dental Association, Washington 16031
National Dental Hygienists' Association, Washington 16032
National Denturist Association, Portland 16033
National Foundation of Dentistry for the Handicapped, Denver 16060
National Medical and Dental Association, Philadelphia 16106
North American Medical/Dental Association, Newport Beach 16218
Orthodontic Education and Research Foundation, Saint Louis 16268
Pierre Fauchard Academy, Justice 16319
Society for Occlusal Studies, Kansas City 16559

Venezuela
Sociedad Odontológica Zuliana de Prótesis Maracaibo, Maracaibo 17015

Vietnam
Vietnamese Association of Odonto-Maxillo-Facial Medicine, Hanoi 17055

Yugoslavia
Savez Društava Zubarskih Radnika Jugoslavije, Beograd 17115

Zimbabwe
Dental Association of Zimbabwe, Harare 17168

Depth Psychology

Italy
Centro Internazionale di Ipnosi Medica e Psicologica, Milano 07468

United Kingdom
Association for Applied Hypnosis, Grimsby 11259
British Hypnosis Research Association, Somersham 11576

U.S.A.
American Educational Studies Association, Cincinnati 13791

Dermatology

Argentina
Federación del Patronato del Enfermo de Lepra de la República Argentina, Buenos Aires 00135
Sociedad Argentina de Dermatología, Buenos Aires 00164
Sociedad Argentina de Leprología, Buenos Aires 00175
Sociedad de Dermatología y Sifilografía, Buenos Aires 00192

Australia
Australasian College of Dermatologists, Gladesville 00240

Austria
Österreichische Gesellschaft für Dermatologie und Venerologie, Graz 00773

Belgium
European Association against Pigmentary Dystrophy and its Syndromes, Bruxelles 01272
European Burns Association, Leuven 01304
Société Royale Belge de Dermatologie et de Vénérologie, Bruxelles 01472

Brazil
Sociedade Brasileira de Dermatologia, Rio de Janeiro 01708

Bulgaria
Bulgarian Dermatological Society, Sofia 01747

Canada
Canadian Dermatological Association, Vancouver 02116
Canadian Psoriasis Foundation, Ottawa 02230
Research, Education and Assistance for Canadians with Herpes, Toronto 02860

Czech Republic
Czech Dermatological Society, Praha 03538

Denmark
Dansk Dermatologisk Selskab, København 03601
Danske Dermato-Venerologers Organisation, Alborg 03603

El Salvador
Central American Dermatological Society, San Salvador 03922

France
Association des Dermatologistes et Syphiligraphes de Langue Française, Créteil 04175
Association des Léprologues de Langue Française, Paris 04191
European Federation of AIDS Research, Paris 04504
Société Française de Dermatologie et de Syphilographie, Paris 04799
Société Française de Médecine Esthétique, Courbevoie 04810
Syndicat National Français des Dermatologistes et Vénéréologistes, Paris 04969

Germany
Berufsverband der Deutschen Dermatologen e.V., Pforzheim 05196
Deutsche Dermatologische Gesellschaft, Kiel 05298
Münchner Dermatologische Gesellschaft, München 06123
Vereinigung Südwestdeutscher Dermatologen, Freiburg 06413

Guatemala
Sociedad Centroamericana de Dermatologia, Guatemala City 06573

Hong Kong
Asian Dermatological Association, Hong Kong 06589

Hungary
Magyar Dermatologiai Társulat, Budapest 06622

Israel
Israel Dermatological Society, Petah Tikva 07070

Italy
Società Italiana di Dermatologia e Sifilografia, Roma 07861

Japan
Nippon Hifuka Gakkai, Tokyo 08303
Nippon Rai Gakkai, Tokyo 08356

Kenya
African Association of Dermatology, Nairobi 08462

Netherlands
Nederlandse Vereniging vor Dermatologie en Venereologie, Breda 09044

Poland
Polskie Towarzystwo Dermatologiczne, Poznań 09560
Polskie Towarzystwo Dermatologiczne, Warszawa 09561

Portugal
Colégio Ibero-Latino-Americano de Dermatologia, Lisboa 09655
Sociedade Portuguesa de Dermatologia e Venereologia, Lisboa 09697

Romania
Societatea de Dermato-Venerologie, Bucuresti 09759

Russia
National Scientific Medical Society of Venereologists and Dermatologists, Moskva 09937

Spain
Academia Española de Dermatología y Sifilografía, Madrid 10242

Sweden
Svenska Dermatologiska Sällskapet, Stockholm 10541

Switzerland
Schweizerische Gesellschaft für Dermatologie und Venereologie, Lausanne 10816

United Kingdom
British Leprosy Relief Association, Colchester 11602
European Contact Dermatitis Society, Belfast 12018
International Federation of Anti-Leprosy Associations, London 12354
National Eczema Society, London 12586
Psoriasis Association, Northampton 12725
Saint John's Hospital Dermatological Society, London 12883

U.S.A.
American Academy of Dermatology, Schaumburg 13390
American Academy of Veterinary Dermatology, Phoenix 13441
American Board of Dermatology, Detroit 13632
American Dermatological Association, Iowa City 13782
American Leprosy Missions, Greenville 13915
American Osteopathic College of Dermatology, Atlanta 13983
American Society for Dermatologic Surgery, Schaumburg 14057
American Society of Dermatopathology, Baltimore 14119
American Society of Podiatric Dermatology, Bel Air 14159
Damien Dutton Society for Leprosy Aid, Bellmore 14951
Dermatology Foundation, Evanston 14962
Dystrophic Epidermolysis Bullosa Research Association of America, New York 14994
History of Dermatology Society, Philadelphia 15219
International Society of Dermatologic Surgery, Schaumburg 15549
International Society of Dermatology: Tropical, Geographic and Ecologic, Rochester 15550
Leonard Wood Memorial – American Leprosy Foundation, Rockville 15661
National Burn Information Exchange, Ann Arbor 15928
National Burn Victim Foundation, Orange 15929
National Psoriasis Foundation, Portland 16131
Noah Worcester Dermatological Society, Cincinnati 16207
North American Clinical Dermatologic Society, Jacksonville 16214
Pacific Dermatologic Association, Schaumburg 16280
PFB Project, Washington 16308
Psoriasis Research Association, San Carlos 16367
Psoriasis Research Institute, Palo Alto 16368
Society for Investigative Dermatology, Cleveland 16543
Society for Pediatric Dermatology, Ann Arbor 16561

Vietnam
Vietnamese Association of Dermatology, Hanoi 17044

Design
s. Graphic and Decorative Arts, Design

Developing Areas

Austria
Österreichische Forschungsstiftung für Entwicklungshilfe, Wien 00743
Österreichisches Lateinamerika-Institut, Wien 00913
Wiener Institut für Entwicklungsfragen und Zusammenarbeit, Wien 01019

Belgium
Association Internationale pour la Coopération et le Développement en Afrique Australe, Winksele 01143
Association pour l'Etude et l'Evaluation Epidémiologiques des Désastres dans les Pays en Voie de Développement, Bruxelles 01152
Centre International de Formation et de Recherche en Population et Développement en Association avec les Nations Unies, Louvain-la-Neuve 01191
Centre International d'Etudes, de Recherche et d'Action pour le Développemnt, Bruxelles 01199
European Association for Information on Local Development, Bruxelles 01277

Chile
Centro Interuniversitario de Desarollo, Santiago 03025

Colombia
Centro Internacional de Agricultura Tropical, Cali 03360
Corporación Latinoamericana de Investigación para el Desarrollo del Sector Rural y Zona Costeros, Bogotá 03372

Costa Rica
Instituto Centroamericano de Administración Pública, San José 03445

Egypt
Arab Center for Energy Studies, Cairo 03868

France
European Centre for Regional Development, Strasbourg 04495

Germany
Catholic Media Council, Aachen 05273
HWWA – Institut für Wirtschaftsforschung – Hamburg, Hamburg 05935
Institut für nationale und internationale Fleisch- und Ernährungswirtschaft, Heidelberg 05965

Grenada
Caribbean Agricultural and Rural Development, Advisory and Training Service, Saint George's 06550

Israel
Development Study Center, Rehovot 07052

Italy
Associazione per lo Sviluppo dell'Istruzione e della Formazione Professionale, Roma 07395
Centro Internazionale Ricerche sulle Strutture Ambientali Pio Manzu, Verucchio 07484
International Centre for Advanced Technical and Vocational Training, Torino 07709
Istituto per la Cooperazione Economica Internazionale e i Problemi di Sviluppo, Roma 07751

Ivory Coast
African Institute for Economic and Social Development, Abidjan 08057

Kenya
Committee of International Development Institutions on the Environment, Nairobi 08487

Malaysia
Association of Development Research and Training Institutes of Asia and the Pacific, Kuala Lumpur 08639

Netherlands
Stichting voor Wetenschappelijk Onderzoek van de Tropen, 's-Gravenhage 09079

Senegal
African Institute for Economic Development and Planning, Dakar 10011
African Network for Integrated Development, Dakar 10012
Association of African Women for Research and Development, Dakar 10021

Sweden
Dag Hammarskjöld Foundation, Uppsala 10449
Nordiska Afrikainstitutet, Uppsala 10507

Switzerland
Development Innovations and Networks, Genève 10662

United Kingdom
Agency for Cooperation and Research in Development, London 11212
Commonwealth Association for Development, London 11808

U.S.A.

Advisory Committee on Science and Technology for Development, New York *13335*
Association for the Advancement of Policy, Research and Development in the Third World, Washington *14344*
Association of Third World Affairs, Washington *14522*
Center for International Development and Environment, Washington *14683*
Commission for the Scientific and Technological Development of Central America and Panama, Washington *14784*
East-West Center Institute of Economic Development and Policy, Honolulu *15005*
Economic Development Institute, Washington *15010*
International Institute of Rural Reconstruction – U.S. Chapter, New York *15490*
Social Sciences Services and Resources, Wasco *16486*

Diabetes

Austria

Föderation der Internationalen Donausymposia über Diabetes mellitus, Wien *00593*
Österreichische Diabetikervereinigung, Wien *00737*

Belgium

International Diabetes Federation, Bruxelles *01395*

Canada

Juvenile Diabetes Foundation, Willowdale *02571*

Denmark

Landsforeningen for Sukkersyge, Odense *03765*

Germany

Deutsche Diabetes-Gesellschaft, Bad Oeynhausen *05299*
Deutscher Diabetiker-Verband e.V., Kaiserslautern *05508*
European Association for the Study of Diabetes, Düsseldorf *05666, 05667*

Hungary

Magyar Diabetes Társaság, Budapest *06623*

Italy

Associazione Italiana per la Difesa degli Interessi di Diabetici, Roma *07334*
Associazione per il Diabete Infantile e Giovanile, Roma *07387*
Società Italiana di Diabetologia, Torino *07862*
Società Italiana di Diabetologia e Endocrinologia Pediatrica, Milano *07863*

Japan

Nippon Tonyo-byo Gakkai, Tokyo *08393*

Netherlands

European Association of Diabetes Educators, Leiden *08864*

Portugal

Associação Protectora dos Diabéticos de Portugal, Lisboa *09650*

South Africa

Society for Endocrinology, Metabolism and Diabetes of Southern Africa, Sandton *10173*

Spain

Sociedad Española de Diabetes, Madrid *10385*

Switzerland

Schweizerische Diabetes-Gesellschaft, Zürich *10801*

Turkey

Türk Diabetler Cemiyeti, Istanbul *11153*

United Kingdom

British Diabetic Association, London *11551*

U.S.A.

American Association of Diabetes Educators, Chicago *13529*
American Diabetes Association, Alexandria *13785*
Joslin Diabetes Center, Boston *15626*
Juvenile Diabetes Foundation International, New York *15633*

Dietetics s. Nutrition

Documentation

Algeria

Centre National de Documentation et de Recherche en Pédagogie, Alger *00016*

Argentina

Centro Argentino de Información Científica y Tecnológica, Buenos Aires *00099*
Centro de Documentación Bibliotecológica, Bahía Blanca *00101*
Centro de Investigación Documentaria, Buenos Aires *00105*
Centro Nacional de Documentación e Información Educativa, Buenos Aires *00112*

Austria

Österreichische Gesellschaft für Dokumentation und Information, Wien *00774*

Belgium

Association Belge de Documentation, Bruxelles *01106*
Association Professionnelle des Bibliothécaires et Documentalistes, Bruxelles *01156*
Centre National de Documentation Scientifique et Technique, Bruxelles *01201, 01202*
INBEL, Institut Belge d'Information et de Documentation, Bruxelles *01374*

Bolivia

Centro Nacional de Documentación Científica y Tecnológica, La Paz *01536*

Brazil

Comissão Brasileira de Documentâcão Agricola, Brasília *01639*

Canada

Association of Wholly or Partially French Language Universities, Montréal *01948*
Association pour l'Avancement des Sciences et des Techniques de la Documentation, Montréal *01951*
Canadian Academy of Recording Arts and Sciences, Toronto *02006*
Canadian Association for Pastoral Education, Toronto *02038*
Canadian Classification Research Group, London *02096*
Canadian Institute for Historical Microreproductions, Ottawa *02163*
Société des Auteurs, Eecherchistes, Documentalistes et Compositeurs, Montréal *02923*

Chile

Centro Nacional de Información y Documentación, Santiago *03027*

Colombia

Grupo de Biblioteca y Documentación, Bogotá *03376*

Costa Rica

Inter-American Association of Agricultural Librarians and Documentalists, Turrialba *03449*

Dominican Republic

Servicio de Documentación y Biblioteca, Santo Domingo *03837*

Egypt

National Information and Documentation Centre, Cairo *03910*

France

Association de Documentation pour l'Industrie Nationale, Paris *04148*
Association des Diplômés de l'Ecole de Bibliothécaires-Documentalistes, Paris *04176*
Association des Professionnels de l'Information et de la Documentation, Paris *04198*
Association pour la Médiathèque Public, Cambrai *04326*
Bureau International de Documentation, Paris *04361*
Centre d'Etudes, de Documentation, d'Information et d'Action Sociales, Paris *04369*
Comité International pour l'Information et la Documentation en Sciences Sociales, Paris *04409*
International Association of Agricultural Information Specialists, Montpellier *04580*
Société d'Etudes et de Documentation Economiques, Industrielles et Sociales, Neuilly-sur-Seine *04732*
Union Française des Organismes de Documentation, Paris *04979*

Germany

Arbeitsgemeinschaft für juristisches Bibliotheks- und Dokumentationswesen, Augsburg *05105*
AWV-Fachausschuss Mikrofilm/Optische Informationssysteme, Eschborn *05160*
Berliner Arbeitskreis Information, Berlin *05186*
Deutsche Gesellschaft für Dokumentation e.V., Frankfurt *05341*
Deutsche Gesellschaft für Filmdokumentation, Wiesbaden *05349*
Deutsche Gesellschaft für Medizinische Informatik, Biometrie und Epidemiologie, Köln *05382*
Deutsches Institut für medizinische Dokumentation und Information, Köln *05586*
Gesellschaft für Bibliothekswesen und Dokumentation des Landbaues, Karlsruhe *05804*
Gesellschaft Information Bildung e.V., Frankfurt *05875*
Immuno, Heidelberg *05937*
Institut für nationale und internationale Fleisch- und Ernährungswirtschaft, Heidelberg *05965*
Internationales Institut für Traditionelle Musik e.V., Berlin *06001*
Normenausschuss Bibliotheks- und Dokumentationswesen im DIN Deutsches Institut für Normung e.V., Berlin *06152*
VDD-Berufsverband Dokumentation, Information, Kommunikation, Bonn *06316*
Zentralstelle für Agrardokumentation und -information, Bonn *06466*

Indonesia

Asosiasi Perpustakaan, Arsip dan Dokumentasi Indonesia, Jakarta *06885*

Italy

Associazione Internazionale Centro Studi di Storia e Documentazione delle Regioni, Reggio Emilia *07285*
Associazione Italiana di Documentazione e di Informazione, Roma *07307*
Centro di Documentazione, Pistoia *07420*
Centro di Documentazione Economica per Giornalisti, Roma *07421*
Centro di Documentazione Giornalistica, Roma *07424*
Centro di Studi Metodologici, Torino *07448*
Centro Internazionale di Documentazione e Comunicazione, Roma *07467*
Centro per la Documentazione Automatica, Milano *07523*
Documentation and Research Centre, Roma *07647*
International Centre for Advanced Technical and Vocational Training, Torino *07709*
Società Italiana per gli Archivi Sanitari Ospedalieri, Viareggio *07962*
SVP Italia, Milano *08030*

Japan

Gakujutsu Bunken Fukyu-Kai, Tokyo *08128*
Nihon Kagaku Gijutsu Joho Sentah, Tokyo *08216*
Nippon Dokumenesyon Kyokai, Tokyo *08296*

Morocco

Centre National de Documentation, Rabat *08805*

Netherlands

Federatie van Organisaties van Bibliotheek-, Informatie-, Dokumentatiewezen, 's-Gravenhage *08894*
Fédération Internationale d'Information et de Documentation, 's-Gravenhage *08895*
International Bureau of Fiscal Documentation, Amsterdam *08926*
Nederlandse Vereniging van Bibliothecarissen, Documentalisten en Literatuuronderzoekers, Schelluinen *09005*
Rijksbureau voor Kunsthistorische Documentatie, 's-Gravenhage *09066*

Paraguay

Comisión Paraguaya de Documentación e Información, Asunción *09429*

Portugal

Associação Portuguesa de Bibliotecários, Arquivistas e Documentalistas, Lisboa *09640*
Junta Nacional de Investigação Científica e Tecnológica, Lisboa *09663*

Senegal

Ecole de Bibliothécaires, Archivistes et Documentalistes, Dakar *10027*

Spain

Oficina Internacional de Información y Observación del Español, Madrid *10332*

Sweden

International Association of Music Libraries, Archives and Documentation Centres, Stockholm *10472*
Tekniska Litteratursällskapet, Stockholm *10601*

Switzerland

Centre Suisse de Documentation en Matière d'Enseignement et d'Education, Grand-Saconnex *10644*
Institut Romand de Recherches et de Documentation Pédagogiques, Neuchâtel *10731*
Schweizerische Vereinigung für Dokumentation, Thun *10942*

Thailand

Asian Institute of Technology Library and Regional Documentation Centre, Bangkok *11082*

Togo

Association Togolaise pour le Développement de la Documentation, des Bibliothèques, Archives et Musées, Lomé *11104*

Trinidad and Tobago

Caribbean Documentation Centre, Port of Spain *11111*

Tunisia

Association Tunisienne des Bibliothécaires, Documentalistes et Archivistes, Tunis *11138*

United Kingdom

British Record Society, London *11639*

U.S.A.

American College of Oral and Maxillofacial Surgeons, San Antonio *13723*
Use, Lanham *16853*

Venezuela

Asociación Interamericana de Bibliotecarios y Documentalistas Agrícolas, Filial Venezuela, Mérida *16969*

Yugoslavia

Jugoslovenski Centar za Tehničku i Naučnu Dokumentaciju, Beograd *17094*

Zaire

Association Zaïroise des Archivistes, Bibliothécaires et Documentalistes, Kinshasa *17141*

Ecology

Algeria

Centre National d'Etudes et de Recherches pour l'Aménagement du Territoire, Alger *00021*

Andorra

Associació per la Defensa de la Natura, Andorra la Vella *00028*

Argentina

Asociación Argentina de Ecología, Córdoba *00062*
Centro de Investigaciones de Recursos Naturales, Castelar *00107*

Australia

Australian Conservation Foundation, Fitzroy *00278*
Australian Council of National Trusts, Civic Square *00281*
Clean Air Society of Australia and New Zealand, Eastwood *00372*
Commission for the Conservation of Antarctic Marine Living Resources, Hobart *00373*
Ecological Society of Australia, Canberra *00387*
Field Naturalists' Club of Victoria, South Yarra *00396*
The Field Naturalists' Society of South Australia, Adelaide *00397*
Native Plants Preservation Society of Victoria, Toorak *00439*
Natural Resources Conservation League of Victoria, Springvale South *00440*
North Queensland Naturalists Club, Cairns *00441*
Plant Protection Society of Western Australia, Victoria Park *00446*
The Queensland Naturalists' Club, Brisbane *00449*
Western Australian Naturalists' Club, Nedlands *00524*
Western Australian Shell Club, Perth *00525*
Wildlife Conservation Society, Lavington *00526*
Wildlife Preservation Society of Australia, Sydney *00527*
Wildlife Preservation Society of Queensland, Brisbane *00528*

Austria

International Institute for Applied Systems Analysis, Laxenburg *00687*
Ökosoziales Forum, Wien *00718*
Österreichische Gesellschaft für Humanökologie, Wien *00792*
Österreichische Gesellschaft und Institut für Umweltschutz, Umwelttechnologie und Umweltwissenschaften, Wien *00852*
Österreichische Gesellschaft zur Föderung von Umweltschutz und Energieforschung, Wien *00856*
Österreichischer Arbeitsring für Lärmbekämpfung, Wien *00885*
Österreichischer Naturschutzbund, Salzburg *00898*
Österreichischer Verband für Strahlenschutz, Seibersdorf *00905*

Azerbaijan

Commission on Nature Conservation, Baku *01044*

Bahamas

Bahamas National Trust, Nassau *01052*

Barbados

Caribbean Conservation Association, Bridgetown *01080*

Belgium

Aire Méditerranéenne et Latinoaméricaine, Bruxelles *01098*
Bureau International de la Récupération, Bruxelles *01174*
Comité International de Recherche et d'Etude de Facteurs de l'Ambiance, Bruxelles *01229*
Committee on the Challenges of Modern Society, Bruxelles *01239*
European Association against Fibre Pollution, Bruxelles *01271*
European Bureau for Conservation and Development, Bruxelles *01302*
European Centre for Plastics in the Environment, Bruxelles *01311*
European Chemical Industry Ecology and Toxicology Centre, Bruxelles *01315*
European Environment Agency, Bruxelles *01341*
European Environmental Bureau, Bruxelles *01342*
European Environment Information and Observation Network, Bruxelles *01343*
Institut Européen d'Ecologie et de Cancérologie, Bruxelles *01383*
Koninklijke Vereniging voor Natuur- en Stedeschoon, Antwerpen *01411*
Société Internationale pour la Recherche sur les Maladies de Civilisation et sur l'Environnement, Bruxelles *01464*

Bermuda

Bermuda Audubon Society, Hamilton *01519*

Brazil

Associação Brasileira de Engenharia Sanitária e Ambiental, Rio de Janeiro *01588*
Instituto de Planejamento de Pernambuco, Recife *01668*
Instituto Nacional de Pesquisas da Amazonia, Manaus *01688*

Canada

British Columbia Pollution Control Association, West Vancouver *01987*
Canadian Acid Precipitation Foundation, Toronto *02011*
Canadian Association on Water Pollution Research and Control, Montréal *02076*
Canadian Coalition on Acid Rain, Toronto *02098*
Canadian Institute of Treated Wood, Ottawa *02185*
Canadian Society of Environmental Biologists, Toronto *02284*
Canadian Tinplate Recycling Council, Hamilton *02309*
Canadian Water Quality Association, Waterloo *02316*

Manitoba Environmental Council, Winnipeg 02615
Nature Conservancy of Canada, Toronto 02676
Pollution Control Association of Ontario, Aurora 02823
Pollution Probe Foundation, Toronto 02824
Recycling Council of Ontario, Toronto 02855
Société d'Animation du Jardin et de l'Institut Botaniques, Montréal 02917
Société de Protection des Plantes du Québec, Saint-Hyacinthe 02922
Sociéte pour vaincre la Pollution, Montréal 02935

Chile

Comité Nacional Pro Defensa de la Fauna y Flora, Santiago 03042

China, People's Republic

Chinese Society of Environmental Sciences, Beijing 03189
Ecological Society of China, Beijing 03197

China, Republic

Asian Ecological Society, Taichung 03238

Costa Rica

Tropical Science Center, San José 03452

Denmark

Danmarks Naturfredningsforening, København 03583
Dansk Økologisk Forening Oikos, København 03659
Det Danske Hedeselskab, Viborg 03705
Euroenviron, København 03719

Ethiopia

African Forum for Mathematical Ecology, Asmara 03936

Finland

Baltic Marine Environment Protection Commission, Helsinki 03952

France

Association Française pour la Protection des Eaux, Paris 04241
Association Internationale de Médecine et de Biologie de l'Environnement, Paris 04260
Association Nationale pour la Protection des Eaux, Paris 04313
Association pour la Prévention de la Pollution Atmosphérique, Paris 04327
Centre Naturopa, Strasbourg 04384
Conseil Mondial d'Ethique des Droits de l'Animal, Asnières 04462
European and Mediterranean Plant Protection Organization, Paris 04479
Fédération Française des Sociétés de Protection de la Nature, Paris 04524
Program on Man and the Biosphere, Paris 04621
Scientific Committee on Problems of the Environment, Paris 04623
Société Française d'Ecologie, Brunoy 04796
Société Française d'Ichtyologie, Paris 04860
Société Nationale de Protection de la Nature, Paris 04899
Société pour la Protection des Paysages, Sites et Monuments, Paris 04912
Universités Unies pour l'Environnement, Paris 04999

Georgia

Commission on Biosphere and Ecology Research, Tbilisi 05008

Germany

Aktionsgemeinschaft Natur- und Umweltschutz Baden-Württemberg, Stuttgart 05068
Arbeitsgemeinschaft beruflicher und ehrenamtlicher Naturschutz, Bonn 05081
Arbeitsgemeinschaft für Abfallwirtschaft, Köln 05100
Arbeitsgemeinschaft für Landschaftsentwicklung, Bonn 05108
Arbeitsgemeinschaft für Umweltfragen e.V., Bonn 05112
Arbeitskreis Chemische Industrie, Köln 05136
Badischer Landesverein für Naturkunde und Naturschutz e.V., Freiburg 05163
Bund für Umwelt und Naturschutz Deutschland e.V., Bonn 05266
Bund für Umwelt und Naturschutz Deutschland, Landesverband Hessen e.V., Mörfelden-Walldorf 05267
Bund Naturschutz in Bayern e.V., München 05269
Commission Internationale pour la Protection du Lac de Constance, München 05276
Dachverband wissenschaftlicher Gesellschaften der Agrar-, Forst-, Ernährungs-, Veterinär- und Umweltforschung e.V., Frankfurt 05282
Deutsche Gesellschaft für Gartenkunst und Landschaftspflege e.V., Karlsruhe 05353
Deutscher Naturschutzring e.V., Bundesverband für Umweltschutz, Bonn 05533
Deutscher Rat für Landespflege, Bonn 05538
Energie- und Umweltzentrum am Deister e.V., Springe-Eldagsen 05641
Forschungsgemeinschaft Eisenhüttenschlacken, Duisburg 05707
Forschungsgesellschaft Landschaftsentwicklung Landschaftsbau e.V., Bonn 05722
Fraunhofer-Gesellschaft zur Förderung der angewandten Forschung e.V., München 05759
Gesellschaft für Naturkunde in Württemberg e.V., Stuttgart 05843
Gesellschaft für Strahlen- und Umweltforschung, Ergersheim 05862
Gesellschaft für Technologiefolgenforschung e.V., Berlin 05863
Gruppe Ökologie, Hannover 05900
Institut für angewandte Verbraucherforschung e.V., Köln 05949
Institut für Energie- und Umweltforschung Heidelberg e.V., Heidelberg 05958
Institut für Europäische Umweltpolitik, Bonn 05959
Institut für gewerbliche Wasserwirtschaft und Luftreinhaltung e.V., Köln 05962
Institut für ökologische Forschung und Bildung e.V., Münster 05968
Internationale Kommission zum Schutze des Rheins gegen Verunreinigung, Koblenz 05993
Internationaler Rat für Umweltrecht, Bonn 05998
Katalyse-Umweltgruppe Köln e.V., Köln 06057
Kommission Reinhaltung der Luft im VDI und DIN, Düsseldorf 06079
Naturschutzbund Deutschland e.V., Bonn 06132
Öko-Institut, Institut für angewandte Ökologie e.V., Freiburg 06223
POLLICHIA e.V., Annweiler 06238
Rat von Sachverständigen für Umweltfragen, Wiesbaden 06258
VDI-Kommission Lärmminderung, Düsseldorf 06329
Vereinigung Deutscher Gewässerschutz e.V., Bonn 06404
Verein Naturschutzpark e.V., Niederhaverbeck 06422
Zentralstelle für Agrardokumentation und -information, Bonn 06466

Ghana

African Forestry and Wildlife Commission, Accra 06471

Guam

Association of South Pacific Environmental Institutions, Mangilao 06552

India

Asian Environmental Society, New Delhi 06722

Israel

Society for the Protection of Nature in Israel, Tel Aviv 07111

Italy

Associazione Italiana di Fisica Sanitaria e di Protezione contro le Radiazioni, Torino 07308
Associazione Micologica ed Ecologica Romana, Roma 07358
Centro di Ricerca Applicata e Documentazione, Udine 07431
Centro Internazionale di Ricerche sulle Strutture Ambiente Pio Manzù, Verucchio 07469
Ente Fauna Siciliana, Noto 07652
European Centre of Environmental Studies, Torino 07672
Federazione Nazionale pro Natura, Bologna 07698
International Juridical Organization for Environment and Development, Roma 07714
Società Ecologica Friulana, Campoformido 07798
Società Emiliana Pro Montibus et Silvis, Bologna 07800
Società Italiana per le Scienze Ambientali: Biometeorologia, Bioclimatologia ed Ecologia, Milano 07968
Società Italiana per lo Studio e l'Applicazione del Pirodiserbo, Piacenza 07974
Società Orticola Italiana, Firenze 07997

Japan

Nippon Seitai Gakkai, Sendai 08365

Kenya

African Society for Environmental Studies Programme, Nairobi 08473
Committee of International Development Institutions on the Environment, Nairobi 08487
Desert Locust Control Organization for Eastern Africa, Nairobi 08490
Earthwatch, Nairobi 08491
East African Wild Life Society, Nairobi 08498
Eastern Africa Environment Network, Nairobi 08500

Liechtenstein

Internationale Alpenschutzkommission, Vaduz 08570
Liechtensteinische Gesellschaft für Umweltschutz, Vaduz 08571

Malaysia

Asia-Pacific People's Environment Network, Penang 08637

Malta

Euro-Mediterranean Centre on Marine Contamination Hazards, Valletta 08661
Society for the Study and Conservation of Nature, Valletta 08671

Mauritania

Conseil Panafricain pour la Protection de l'Environnement et le Développement, Nouakchott 08672

Mexico

Asociación Latinoamericana de Ecodesarollo, Mexicali 08694
Environmental Training Network for Latin America and the Caribbean, México 08732
Instituto de Ecología, Xalapa 08735

Moldova, Republic of

Moldovan Society of Animal Protection, Chişinău 08784

Netherlands

Bond Heemschut, Amsterdam 08834
European Environmental Research Organization, Wageningen 08885
Koninklijke Nederlandsche Heidemaatschappij, Arnhem 08950
Stichting Natuur en Milieu, Utrecht 09073
Vereniging Natuurmonumenten, 's-Graveland 09086

New Zealand

Clean Air Society of Australia and New Zealand (N.Z. Branch), Wellington 09122
New Zealand Association of Soil Conservators, Blenheim 09136
New Zealand Ecological Society, Christchurch 09144

Nigeria

Ecological Society of Nigeria, Lagos 09224

Norway

Norsk Forening mot Støy, Oslo 09311

Philippines

Asian Recycling Association, Camarines Sur 09502

Russia

Commission for Ecology, Moskva 09817
Commission for Socio-ecological Research, Moskva 09828
Commission for the Protection of Natural Waters, Moskva 09839
Co-ordination Council for Scientific Problems Linked with Ecological Consequences of the Use of New Technological Systems, Moskva 09857
National Committee of the Scientific Committee for Problems of the Environment, Moskva 09903

St. Lucia

Caribbean Environmental Health Institute, Castries 09995

Senegal

Environmental Development in the Third World, Dakar 10028

South Africa

The Botanical Society of South Africa, Claremont 10126
National Association for Clean Air, Johannesburg 10155
Pollution Research Group, Durban 10162
Wildlife Society of Southern Africa, Linden 10236

Spain

Asociación para la Defensa de la Naturaleza, Madrid 10280

Sweden

Coalition Clean Baltic, Stockholm 10445
European Ecological Federation, Lund 10454
Oikos, Lund 10513
Svenska Naturskyddsföreningen, Stockholm 10568

Switzerland

Aqua Viva, Schweizerische Aktionsgemeinschaft zur Erhaltung der Flüsse und Seen, Bern 10617
Commission Internationale pour la Protection des Eaux du Léman contre la Pollution, Lausanne 10650
Cooperative Program for Monitoring and Evaluation of the Long-range Transmission of Air Pollutants in Europe, Genève 10658
Ecological and Toxicological Association of the Dyestuffs Manufacturing Industry, Basel 10664
Fachgruppe für Raumplanung und Umwelt des SIA, Zürich 10695
Internationale Vereinigung gegen den Lärm, Luzern 10746
Internationale Vereinigung wissenschaftlicher Fremdenverkehrsexperten, Sankt Gallen 10747
International Union for Conservation of Nature and Natural Resources, Gland 10759
Rheinaubund, Schweizerische Arbeitsgemeinschaft für Natur und Heimat, Schaffhausen 10783
Schweizer Heimatschutz, Zürich 10784
Schweizerische Gesellschaft für Umweltschutz, Zürich 10852
Schweizerische Liga gegen den Lärm, Luzern 10869
Schweizerischer Bund für Naturschutz, Basel 10890
Schweizerischer Werkbund, Zürich 10923
Schweizerische Vereinigung zum Schutz und zur Förderung des Berggebietes, Brig 10957
Verband Schweizerischer Abwasserfachleute, Zürich 11004

Thailand

Asian Society for Environmental Protection, Bangkok 11084
Environmental Sanitation Information Center, Bangkok 11091

Trinidad and Tobago

Caribbean Plant Protection Commission, Arima 11117

United Kingdom

Advisory Committee on Protection of the Sea, London 11207
The Association for the Protection of Rural Scotland, Edinburgh 11288
Association for the Reduction of Aircraft Noise, London 11289
Association of National Park Officers, Kendal 11371
Association of Public Analysts, London 11382
British Butterfly Conservation Society, Colchester 11536
The British Ecological Society, London 11554
British Naturalists' Association, Higham Ferrers 11617
British Wood Preserving and Damp-proofing Association, London 11709
Campaign for the Protection of Rural Wales, Welshpool 11741
Centre for Alternative Technology, Machynlleth 11753
Chiltern Society, Great Missenden 11776
The Civic Trust, London 11791
Commons, Open Spaces and Footpaths Preservation Society, Henley-on-Thames 11806
Council for Environmental Conservation, London 11851
Council for the Protection of Rural England, London 11855
Dartmoor Preservation Association, Plymouth 11878
Ecological Physics Research Group, Bedford 11925
Environmental Council, London 11969
Essex Wildlife Trust, Colchester 11978
European Association for the Science of Air Pollution, London 11993
European Centre for Pollution Research, London 12005
Fauna and Flora Preservation Society, London 12059
Filtration Society, Leicester 12075
The Game Conservancy Trust, Fordingbridge 12100
Gower Society, Swansea 12125
Gwent Wildlife Trust, Monmouth 12138
Institution of Environmental Health Officers, London 12301
Institution of Water and Environment Management, London 12318
International Association of Environmental Coordinators, Chalfont Saint Gills 12329
International Association on Water Quality, London 12337
International Petroleum Industry Environmental Conservation Association, London 12383
International Union of Air Pollution Prevention and Environmental Protection Associations, Brighton 12400
Marine Conservation Society, Kempley 12493
National Association of Principal Agricultural Education Officers, Beverley 12569
National Pure Water Association, Ellesmere 12603
National Society for Clean Air, Brighton 12607
National Trust for Places of Historic Interest or Natural Beauty, London 12614
National Trust for Scotland, Edinburgh 12615
Oxford Preservation Trust, Oxford 12668
River Thames Society, Richmond 12758
Royal Society for Nature Conservation, Lincoln 12850
Royal Society for the Protection of Birds, Sandy 12853
RSNC Wildlife Trusts Partnership, Lincoln 12877
Rural Preservation Association, Liverpool 12879
Scottish River Purification Boards Association, Glasgow 12926
Scottish Wildlife Trust, Edinburgh 12936
The Sempervivum Society, Burgess Hill 12940
Ulster Society for the Preservation of the Countryside, Holywood 13156
Vegan Society, Saint Leonards-on-Sea 13178
Water Research Centre, Marlow 13194

Uruguay

Association for the Study of Man Environment Relations, Montevideo 13249

U.S.A.

Air and Waste Management Association, Pittsburgh 13355
Alliance for Environmental Education, The Plains 13365
American Committee for International Conservation, Washington 13744
American Conservation Association, New York 13754
American Medical Fly Fishing Association, Lock Haven 13936
American Wood Preservers Institute, Vienna 14213
Animal Protection Institute of America, Sacramento 14221
Association for Arid Lands Studies, Lubbock 14267
Association of Applied Insect Ecologists, Sacramento 14393
Association of State and Interstate Water Pollution Control Administrators, Washington 14506
Center for International Development and Environment, Washington 14683
Center for Marine Conservation, Washington 14685
Citizens for a Better Environment, Chicago 14756
The Coastal Society, Gloucester 14767
Community Environmental Council, Santa Barbara 14811
Concern, Washington 14819
Conservation International, Washington 14845
The Cousteau Society, Chesapeake 14936
Crop Science Society of America, Madison 14942
Desert Botanical Garden, Phoenix 14964
Desert Fishes Council, Bishop 14965
Desert Protective Council, Valley Center 14966
Earthmind, Mariposa 14999
Ecological Society of America, Tempe 15006
Ecology Center, Berkeley 15007
EcoNet, San Francisco 15008
Elm Research Institute, Harrisville 15038
Environmental Defense Fund, New York 15048
Environmental Law Institute, Washington 15050
Environmental Technology Seminar, Bethpage 15052
Forum International: International Ecosystems University, Pleasant Hill 15117

Harvard Environmental Law Society, Cambridge 15189
International Association for Ecology, Aiken 15380
International Association for the Advancement of Earth and Environmental Sciences, Chicago 15392
International Bio-Environmental Foundation, Sherman Oaks 15420
International Bird Rescue Research Center, Berkeley 15421
International Center for the Solution of Environmental Problems, Houston 15431
International Society of Chemical Ecology, Tampa 15547
Interprofessional Council on Environmental Design, Washington 15577
National Association of Environmental Professionals, Washington 15856
National Council for Environmental Balance, Louisville 15987
National Fisheries Contaminant Research Center, Columbia 16052
National Wildlife Federation, Washington 16180
Natural Resources Defense Council, New York 16187
The Nature Conservancy, Arlington 16188
Open Space Institute, New York 16250
Outdoor Ethics Guild, Bucks Harbor 16276
Passaic River Coalition, Basking Ridge 16298
Sea Shepherd Conservation Society, Santa Monica 16455
Sierra Club, San Francisco 16471
Society for Vector Ecology, Santa Ana 16627
Soil and Water Conservation Society, Ankeny 16716
Solid Waste Association of North America, Silver Spring 16719
Thorne Ecological Institute, Boulder 16781
United New Conservationists, Campbell 16813
Water Environment Federation, Alexandria 16873
The Wilderness Society, Washington 16896
Wilderness Watch, Sturgeon Bay 16897
Wildlife Management Institute, Washington 16899
Wildlife Preservation Trust International, Philadelphia 16900
The Wildlife Society, Bethesda 16901

Venezuela
Centro Interamericano de Desarrollo Integras de Aguas y Tierra, Mérida 16987

Zambia
National Monuments Commission, Livingstone 17156
Wildlife Conservation Society of Zambia, Lusaka 17157

Zimbabwe
Wildlife Society of Zimbabwe, Harare 17183

Economics

Afghanistan
Asia Foundation, Kabul 00002

Algeria
Centre de Coordination des Etudes et des Recherches sur les Infrastructures et les Equipements, Alger 00009
Centre de Recherches en Economie Appliquée pour le Développement, Alger 00011

Argentina
Academia Nacional de Ciencias Económicas, Buenos Aires 00044
Centro de Estudios Económicos Sociales, Buenos Aires 00102
Centro de Investigaciones Económicas, Buenos Aires 00108
Colegio de Graduados en Ciencias Económicas, Buenos Aires 00116
Comisión Nacional Protectora de Bibliotecas Populares, Buenos Aires 00121

Armenia
Yerevan Academy of National Economy, Yerevan 00210

Australia
Committee for Economic Development of Australia, Melbourne 00374
Economic Society of Australia, Sydney 00388
Institute of Public Affairs, Jolimont 00418
Victorian Public Interest Research Group, Fitzroy 00517

Austria
Gemeinnütziger Verein zur Durchführung von Lehr- und Forschungsaufgaben an der Wirtschaftsuniversität Wien, Wien 00601
Gesellschaft für Chemiewirtschaft, Wien 00615
Institut für Österreichkunde, Wien 00658
International Institute for Applied Systems Analysis, Laxenburg 00687
Nationalökonomische Gesellschaft, Wien 00711
Österreichische Gesellschaft für angewandte Fremdenverkehrswissenschaft, Wien 00755
Österreichische Gesellschaft für Unternehmensgeschichte, Wien 00838
Österreichische Gesellschaft für Wirtschaftsraumforschung, Wien 00847
Österreichisches Institut für Wirtschaftsforschung, Wien 00912
Österreichisches Ost- und Südosteuropa-Institut, Wien 00915
Österreichische Tribologische Gesellschaft, Wien 00923
Verband Österreichischer Wirtschaftsakademiker, Wien 00975
Vereinigung der kooperativen Forschungsinstitute der Österreichischen Wirtschaft, Wien 00992
Volkswirtschaftliche Gesellschaft Österreich, Wien 01007
Wiener Institut für Entwicklungsfragen und Zusammenarbeit, Wien 01019
Wiener Institut für Internationale Wirtschaftsvergleiche, Wien 01020
Wiener Volkswirtschaftliche Gesellschaft, Wien 01029

Bangladesh
Bangladesh Economic Association, Dhaka 01068

Belgium
American Management Association / International, Bruxelles 01100
Association d'Instituts Européens de Conjoncture Economique, Louvain-la-Neuve 01119
Association Européenne des Centres Nationaux de Productivité, Bruxelles 01127
Centre International de Recherches et d'Information sur l'Economie Publique, Sociale et Coopérative, Liège 01193
Comité d'Etudes Economiques de l'Industrie du Gaz, Bruxelles 01224
Comité Economique et Social des Communautés Européennes, Bruxelles 01225
Comité Européen de Recherche et de Développement, Bruxelles 01227
Conseil Consultatif Economique et Social de l'Union Economique Benelux, Bruxelles 01250
European Economic Association, Leuven 01338
European Economic Research and Advisory Consortium, Bruxelles 01339
European Trade Union Institute, Bruxelles 01353

Brazil
Centro de Análise Conjuntura Econômica, Rio de Janeiro 01619
Fundação João Pineiro, Belo Horizonte 01657
Fundação Joaquim Nabuco, Recife 01658
Instituto Brasileiro de Economia, Rio de Janeiro 01664
Instituto de Planejamento de Pernambuco, Recife 01668

Bulgaria
Union of Economics, Sofia 01790

Burkina Faso
Centre of Economic and Social Studies and Experiments in Western Africa, Bobo-Dioulasso 01808

Canada
Canadian Association for Corporate Growth, Toronto 02028
Canadian Economics Association, Montréal 02120
Canadian Foundation for Economic Education, Toronto 02140
Canadian Society for Training and Development, Ottawa 02271
Central Ontario Industrial Relations Institute, Toronto 02325
Centre for Research on Latin America and the Caribbean, Toronto 02329
Conference Board of Canada, Ottawa 02377
Economic Council of Canada, Ottawa 02425
Economics Society of Alberta, Calgary 02426
New Brunswick Research and Productivity Council, Fredericton 02691
Ontario Educational Research Council, Ajax 02774
Professional Development Institute, Ottawa 02845
Sales Research Club, Toronto 02876

Chile
Comisíon Económica para América Latina y el Caribe, Santiago 03038
Corporación de Fomento de la Producción, Santiago 03048
Instituto Latinoamericano de Planificación Económica y Social, Santiago 03058

China, People's Republic
Chinese Society for EC Studies, Shanghai 03179
Inner Mongolian Academy of Social Sciences, Huhehot 03206

China, Republic
Association for Socio-Economic Development in China, Hsinchu 03244
Cooperative League of the Republic of China, Taipei 03293

Colombia
Centro de Estudios sobre Desarrollo Económico, Bogotá 03356
Consejo Nacional de Política Económica Planeación, Bogotá 03371
Sociedad Colombiana de Economistas, Bogotá 03407

Costa Rica
Junta Nacional de Planeamiento Económico, San José 03450

Croatia
Društvo Ekonomista Hrvatske, Zagreb 03456

Cuba
Sociedad Económica de Amigos del País, La Habana 03485

Cyprus
Cyprus Economic Association, Nicosia 03490

Czech Republic
Česká Společnost Ekonomická, Praha 03521

Denmark
Danmarks Jurist- og Økonomforbund, København 03580
Foreningen af Danske Civiløkonomer, København 03728
Nationaløkonomisk Forening, København 03774
Scandinavian Society for Economic and Social History, Odense 03806
Selskabet for Historie og Samfundsøkonomi, København 03814

Ecuador
Instituto Latinoamericano de Investigaciones Sociales, Quito 03854

Egypt
Egyptian Society of Political Economy, Statistics and Legislation, Cairo 03901

Finland
Ekonomiska Samfundet i Finland, Helsinki 03954
Kansantaloudellinen Yhdistys, Helsinki 03973
Kansantaloustieten Professorien ja Dosenttien Yhidistys, Jyväskylä 03974
Taloushistoriallinen Yhdistys, Jyväskylä 04070
Taloustieteellinen Seura, Helsinki 04071

France
Académie Commerciale Internationale, Paris 04084
Académie des Sciences Morales et Politiques, Paris 04109
Association de l'Economie des Institutions, Paris 04150
Association Française de Management, Paris 04220
Association Française de Science Economique, Toulouse 04226
Association Française des Historiens Economistes, Paris 04231
Association Française des Instituts de Recherche sur le Développement, Paris 04233
Association Française des Sociétés d'Etudes et de Conseils Exportatrices, Paris 04235
Association Internationale d'Histoire Economique, Paris 04282
Center for Studies, Research and Training in International Understanding and Cooperation, Paris 04364
Club of Rome, Paris 04390
Conseil International des Sciences Sociales, Paris 04459
Fondation pour la Recherche Sociale, Paris 04544
International Economic Association, Paris 04586
Mouvement Français pour la Qualité, Nanterre 04601
Société d'Economie et de Science Sociale, Paris 04651
Société d'Economie Politique, Paris 04652
Société d'Encouragement pour l'Industrie Nationale, Paris 04673
Société d'Etudes Economiques et Comptables, Paris 04730
Société d'Etudes et de Documentation Economiques, Industrielles et Sociales, Neuilly-sur-Seine 04732
Société d'Etudes Juridiques, Economiques et Fiscales, Paris 04745
Société d'Etudes pour le Développement Economique et Social, Paris 04750

Georgia
Commission for the Study of Production Forces and Natural Resources, Tbilisi 05007

Germany
Akademie der Arbeit in der Universität, Frankfurt 05046
Aktionsgemeinschaft Soziale Marktwirtschaft e.V., Tübingen 05069
Arbeitsgemeinschaft der wissenschaftlichen Institute des Handwerks der EG-Länder, München 05095
Arbeitsgemeinschaft deutscher wirtschaftswissenschaftlicher Forschungsinstitute e.V., München 05097
Arbeitsgemeinschaft Währungsethik, Köln 05129
Arnold-Bergstraesser-Institut für kulturwissenschaftliche Forschung e.V., Freiburg 05152
Bremer Ausschuss für Wirtschaftsforschung, Bremen 05222
Bremer Gesellschaft für Wirtschaftsforschung e.V., Bremen 05223
Bundesverband Katholischer Ingenieure und Wirtschaftler Deutschlands, Bonn 05255
Bundesvereinigung Logistik e.V., Bremen 05260
Center for International Research on Economic Tendency Surveys, München 05275
Deutsche Gesellschaft für Asienkunde e.V., Hamburg 05325
Deutsche Gesellschaft für Logistik e.V., Dortmund 05376
Deutsche Planungsgesellschaft EG Bonn, Bonn 05490
Deutsches Handwerksinstitut e.V., München 05579
Deutsches Institut für Wirtschaftsforschung, Berlin 05589
Deutsche Stiftung für internationale Entwicklung, Bonn 05601
Evangelische Sozialakademie Friedewald, Friedewald 05690
Forschungsgesellschaft für Agrarpolitik und Agrarsoziologie e.V., Bonn 05719
Fraunhofer-Gesellschaft zur Förderung der angewandten Forschung e.V., München 05759
Gesellschaft für Deutschlandforschung e.V., Berlin 05809
Gesellschaft für Dezentralisierte Energiewirtschaft e.V., Ludwigsburg 05810
Gesellschaft für Informationsvermittlung und Technologieberatung, München 05831
Gesellschaft für internationale Geldgeschichte, Frankfurt 05833
Gesellschaft für öffentliche Wirtschaft e.V., Berlin 05846
Gesellschaft für Sozial- und Wirtschaftsgeschichte, Heidelberg 05860
Gesellschaft für Unternehmensgeschichte e.V., Köln 05866
Gesellschaft für Wirtschaftskunde e.V., Siegen 05870
Gesellschaft für Wirtschafts- und Sozialwissenschaften, Mannheim 05871
Gesellschaft zur Erforschung des Markenwesens e.V., Wiesbaden 05879
HWWA – Institut für Wirtschaftsforschung – Hamburg, Hamburg 05935
Ifo Institut für Wirtschaftsforschung e.V., München 05936
Institut der deutschen Wirtschaft e.V., Köln 05943
Institut für Angewandte Wirtschaftsforschung, Tübingen 05950
Institut für angewandte Wirtschaftsforschung im Mittelstand, Düsseldorf 05951
Institut für Handwerkswirtschaft München, München 05963
Institut für Sozialforschung und Sozialwirtschaft e.V., Saarbrücken 05971
Institut für Wirtschaft und Gesellschaft Bonn e.V., Bonn 05977
Institut Neue Wirtschaft e.V., Hamburg 05981
Kuratorium der Deutschen Wirtschaft für Berufsbildung, Bonn 06085
Leiterkreis der Katholischen Akademien, Schwerte 06102
List Gesellschaft e.V., Bochum 06104
Normenausschuss Holzwirtschaft und Möbel im DIN Deutsches Institut für Normung e.V., Köln 06178
Normenausschuss Zeichnungswesen im DIN Deutsches Institut für Normung e.V., Berlin 06220
Osteuropa-Institut München, München 06226
Rheinisch-Westfälische Akademie der Wissenschaften, Düsseldorf 06265
Rheinisch-Westfälisches Institut für Wirtschaftsforschung, Essen 06267
Staats- und Wirtschaftspolitische Gesellschaft e.V., Hamburg 06292
Studiengruppe Unternehmer in der Gesellschaft, Königstein 06306
Südosteuropa-Gesellschaft e.V., München 06309
Verein für Kommunalwirtschaft und Kommunalpolitik e.V., Düsseldorf 06388
Vereinigung der unabhängigen freiberuflichen Versicherungs- und Wirtschaftsmathematiker in der Bundesrepublik Deutschland e.V., Grünwald 06401
Volks- und Betriebswirtschaftliche Vereinigung im Rheinisch-Westfälischen Industriegebiet e.V., Duisburg 06435
Wirtschaftsakademie für Lehrer e.V., Bad Harzburg 06443
Wirtschaftspolitischer Club Bonn e.V., Bonn 06445

Ghana
Economic Society of Ghana, Accra 06480

Guatemala
Central American Research Institute for Industry, Guatemala City 06564
Secretaria Permanente del Tratado General de Integración Económica Centroamericana, Guatemala City 06572

Guyana
Guyana Society, Georgetown 06578

India
Indian Economic Association, Dehli 06791
National Council of Applied Economic Research, New Delhi 06842

Iraq
Arab Industrial Development and Mining Organization, Baghdad 06911

Ireland
Irish Productivity Centre, Dublin 06996

Italy
Accademia Economico Agraria dei Georgofili, Firenze 07177
Accademia Internazionale per le Scienze Economiche, Sociali e Sanitarie, Roma 07192
Accademia Olimpica, Vicenza 07225
Accademia Simba, Roma 07238
Associazione Bresciana di Ricerche Economiche, Brescia 07262
Associazione Italiana per la Ricerca Industriale, Roma 07339
Associazione Nazionale Italiana Esperti Scientifici del Turismo, Roma 07377
Associazione per le Previsioni Econometriche, Bologna 07391
Centro di Documentazione Economica per Giornalisti, Roma 07421
Centro di Formazione e Studi per il Mezzogiorno, Roma 07428
Centro di Studi Aziendali e Amministrativi, Cremona 07436
Centro di Vita Europea, Roma 07458
Centro Europeo per il Progresso Economico e Sociale, Roma 07462
Centro Informazioni e Studi sulla Comunita' Europea, Milano 07465
Centro Internazionale di Studi e Documentazione sulle Comunità Europee, Milano 07473
Centro Internazionale Sonnenberg per l'Italia, Torino 07485
Centro Italiano di Ricerche e d'Informazione sull'Economia delle Imprese Pubbliche e di Pubblico Interesse, Milano 07494
Centro Italiano di Studi Sociali Economici e Giuridici, Roma 07502

Centro Italiano Ricerca e Informazione Economica, Roma *07505*
Centro Italiano Studi Politici Economici Sociali, Roma *07509*
Centro Italo-Nipponico di Studi Economici, Milano *07512*
Centro per gli Studi sui Mercati Esteri, Roma *07520*
Centro per la Statistica Aziendale, Firenze *07525*
Centro Ricerche Economiche ed Operative della Cooperazione, Napoli *07538*
Centro Ricerche Economiche Sociologiche e di Mercato nell'Edilizia, Roma *07539*
Centro Studi della Cooperazione nel Veneto, Vicenza *07551*
Centro Studi di Economia Applicata all'Ingegneria, Bari *07555*
Centro Studi di Economia Applicata all'Ingegneria, Napoli *07556*
Centro Studi di Estimo e di Economia Territoriale, Firenze *07557*
Centro Studi di Politica Economica, Torino *07559*
Centro Studi di Psicologia e Sociologia Applicate ad Indirizzo Adleriano, Roma *07560*
Centro Studi Economici, Roma *07566*
Centro Studi Economici e Sociali Giuseppe Toniolo, Pisa *07567*
Centro Studi Economici per l'Alta Italia, Milano *07568*
Centro Studi per il Mezzogiomo A. Ajon, Acireale *07580*
Centro Studi per la Programmazione Economica e Sociale, Roma *07582*
Centro Studi per la Valorizzazione delle Risorse del Mezzogiomo, Napoli *07583*
Ente Friulano di Economia Montana, Udine *07653*
Ente Studi Economici per la Calabria, Cosenza *07659*
European Association for Bioeconomic Studies, Milano *07663*
International Centre for Advanced Technical and Vocational Training, Torino *07709*
Istituto Cooperazione Economica Internazionale, Milano *07726*
Istituto Italo-Africano, Roma *07746*
Istituto per la Cooperazione Economica Internazionale e i Problemi di Sviluppo, Roma *07751*
Società Economica di Chiavari, Chiavari *07799*
Società Italiana Calcolo Ricerca Economica Operativa, Roma *07823*
Società Italiana degli Economisti, Genova *07824*
Società Italiana di Economia Demografia e Statistica, Roma *07865*
Società per la Formazione la Ricerca e l'Addestramento per le Aziende e le Organizzazioni, Milano *08000*
Società per la Matematica e l'Economia Applicate, Roma *08001*

Ivory Coast
African Institute for Economic and Social Development, Abidjan *08057*
Institut Africain pour le Développement Economique et Social, Abidjan *08065*

Jamaica
Association of Caribbean Economists, Kingston *08068*

Japan
Ajia Seikei Gakkai, Tokyo *08093*
Hokkaido Keizai Rengokai, Sapporo *08135*
Kansai Keizai Rengokai, Osaka *08154*
Keizai Chosa Kai Shiryoshitsu, Tokyo *08158*
Keizaigaku-shi Gakkai, Kobe *08159*
Keizai Tokei Kenkyukai, Tokyo *08161*
Kikai Shinko Kyokai Keizai Kenkyo-sho Shiryoshitsu, Tokyo *08165*
Kokusai Keizai Gakkai, Tokyo *08173*
Kyoto Daigaku Keizai Gakkai, Kyoto *08178*
Nihon Kasai Gakkai, Tokyo *08220*
Raten Amerika Kyokai Kenkyubu, Tokyo *08403*
Riron Keiryo Keizai Gakkai, Tokyo *08404*
Seito Kogyokai Jimubu Chosaka, Tokyo *08409*
Sekai Keizai Chosakai Shiryoshitsu, Tokyo *08410*
Shakai Keizaishi Gakkai, Tokyo *08415*
Tochi Seidoshi Gakkai, Tokyo *08432*
Tokyo Daigaku Keizai Gakkai, Tokyo *08438*

Kazakhstan
Council for Study of Productive Forces, Alma-Ata *08455*

Kenya
African Economic Research Consortium, Nairobi *08467*

Korea, Democratic People's Republic
Academy of Light Industry Science, Pyongyang *08511*

Korea, Republic
Korean Economic Association, Seoul *08531*

Kuwait
Arab Planning Institute, Kuwait *08546*

Macedonia
Sojuz na Ekonomistite na Makedonija, Skopje *08613*

Malaysia
Asia Foundation, Kuala Lumpur *08632*

Malta
The Economic Society of Malta, Paceville *08660*

Mexico
Center for Economic and Social Studies in the Third World, México *08709*
Centro de Estudios Económicos y Sociales del Tercer Mundo, México *08714*

Morocco
Centre d'Etudes, de Documentations et d'Informations Economiques et Sociales, Casablanca *08804*
Centre National de Documentation, Rabat *08805*
Société d'Etudes Economiques, Sociales et Statistiques du Maroc, Rabat *08809*

Netherlands
European Association of Law and Economics, Maastricht *08868*
Nationale Vereniging voor Economisch Onderwijs, Bergen *08978*
Nederlands Economisch Instituut, Rotterdam *08989*
Vereniging Het Nederlandsch Economisch-Historisch Archief, Amsterdam *09084*

Nigeria
Nigerian Economic Society, Ibadan *09236*

Norway
Sosialøkonomenes Forening, Oslo *09349*
Statsøkonomisk Forening, Oslo *09352*

Pakistan
Institute of Cost and Management Accountants of Pakistan, Karachi *09366*

Panama
Consejo de Economía Nacional, Panamá City *09416*

Paraguay
Centro Paraguayo de Estudios de Desarrollo Económico y Social, Asunción *09428*

Philippines
Asian Institute of Tourism, Manila *09499*

Poland
Polskie Towarzystwo Ekonomiczne, Warszawa *09562*

Portugal
Associação Portuguesa de Economistas, Lisboa *09641*

Romania
Asociaţia Generala a Economiştilor din Romania, Bucuresti *09741*

Russia
Agro-Industrial Society, Moskva *09799*
Association of Economic Scientific Institutions, Moskva *09801*
Commission for the Effective Use of Shales in the Russian Economy, Moskva *09833*
Commission for the Study of Productive Forces and Natural Resources, Moskva *09840*
Economics Society, Moskva *09865*
Municipal Economy Soiety, Moskva *09887*

Rwanda
African and Mauritian Institute of Statistics and Applied Economics, Kigali *09993*

Senegal
African Institute for Economic Development and Planning, Dakar *10011*
Council for the Development of Economic and Social Research in Africa, Dakar *10026*

Singapore
Economic Research Centre, Singapore *10046*
Institute of Southeast Asian Studies, Singapore *10049*

Slovakia
Slovenská Ekonomická Spoločnost, Bratislava *10070*

Slovenia
Zveza Ekonomistov Slovenije, Ljubljana *10114*

South Africa
Africa Institute of South Africa, Pretoria *10120*
Economic Society of South Africa, Pretoria *10137*

Spain
Real Sociedad Económica de Amigos del País de Tenerife, Tenerife *10360*

Sweden
Nationalekonomiska Föreningen, Stockholm *10503*
Studieförbundet Nämigsliv och Samhälle, Stockholm *10528*

Switzerland
Forschungsgemeinschaft für Nationalökonomie, Sankt Gallen *10706*
Internationale Vereinigung für Natürliche Wirtschaftsordnung, Aarau *10744*
Internationale Vereinigung wissenschaftlicher Fremdenverkehrsexperten, Sankt Gallen *10747*
International Industrial Relations Association, Genève *10750*
Schweizerische Franchising-Vereinigung, Zollikon *10807*
Schweizerische Gesellschaft für Statistik und Volkswirtschaft, Bem *10849*
Schweizerische Studiengruppe für Konsumentenfragen, Bem *10933*

Thailand
Asia Foundation, Bangkok *11075*

Trinidad and Tobago
Caribbean Information System for Economic and Social Planning, Port of Spain *11114*

Tunisia
Association of Arab Institutes and Centres for Economic and Social Research, Tunis *11137*

United Kingdom
Acton Society Trust, London *11203*
Association of Principals of Colleges (Northem Ireland Branch), Lisbum *11379*
Commonwealth Independent Centre for the Study of the South African Economy and International Finance, London *11822*
Economic and Social Science Research Association, London *11926*
Economic History Society, Heslington *11927*
The Economics and Business Education Association, Hassocks *11928*
Economic Study Association, London *11929*
Economists Advisory Group, London *11930*
European Association for Evolutionary Political Economy, Newcastle-upon-Tyne *11990*
The European Movement, London *12034*
Foundation for Business Responsibilities, London *12088*
Hotel, Catering and Institutional Management Associaton, London *12184*
Institute for Fiscal Studies, London *12208*
Institute of Chartered Accountants in England and Wales, London *12219*
Institute of Chartered Accountants of Scotland, Edinburgh *12220*
Institute of Economic Affairs, London *12233*
Institute of Training and Development, Marlow *12287*
Institute of Value Management, Buxton *12290*
International Institute of Social Economics, Hull *12370*
Irish Association for Cultural, Economic and Social Relations, Belfast *12410*
Liberal Industrial Relations Association, Darwen *12449*
Manchester Statistical Society, Manchester *12487*
National Institute of Economic and Social Research, London *12598*
Political and Economic Planning, London *12711*
Royal Economic Society, London *12799*
Royal Society for the Encouragement of Arts, Manufactures and Commerce, London *12851*
Scottish Economic Society, Glasgow *12904*

Uruguay
Centro de Nacionales y Comercio Internacional del Uruguay, Montevideo *13253*

U.S.A.
Academy of International Business, Detroit *13293*
AIESEC – United States, New York *13354*
Alternative Energy Resources Organization, Helena *13368*
American Agricultural Economics Association, Ames *13446*
American Assembly of Collegiate Schools of Business, Saint Louis *13462*
American Chamber of Commerce Researchers Association, Alexandria *13687*
American Council on Consumer Interests, Columbia *13765*
American Economic Association, Nashville *13789*
American Institute for Economic Research, Great Barrington *13868*
American Planning Association, Washington *14007*
American Production and Inventory Control Society, Falls Church *14015*
American Real Estate and Urban Economics Association, Bloomington *14026*
American Society for Training and Development, Alexandria *14093*
Association for Business Communication, Denton *14275*
Association for Comparative Economic Studies, Flushing *14282*
Association for Evolutionary Economics, Lincoln *14295*
Association for Social Economics, Ruston *14334*
Association for University Business and Economic Research, Indianapolis *14364*
Association of Business Officers of Preparatory Schools, Windsor *14406*
Atlantic Economic Society, Edwardsville *14547*
Business-Higher Education Forum, Washington *14627*
Business History Conference, Williamsburg *14628*
CDS International, New York *14660*
Center for Economic Conversion, Mountain View *14678*
Committee for Economic Development, Washington *14793*
Consumer Education Research Center, Orange *14853*
Corporate Data Exchange, New York *14864*
Decision Sciences Institute, Atlanta *14957*
East-West Center, Honolulu *15004*
Economic History Association, Washington *15011*
Environmental Design Research Association, Oklahoma City *15049*
Financial Management Association, Tampa *15093*
Global Options, San Francisco *15165*
History of Economics Society, Richmond *15220*
Industrial Development Research Council, Norcross *15258*
Industrial Relations Research Association, Madison *15260*
Inroads, Saint Louis *15264*
Institute for Econometric Research, Fort Lauderdale *15276*
Institute for Economic Analysis, Silver Springs *15277*
Institute for Food and Development Policy, San Francisco *15281*
Institute for Social Research, Ann Arbor *15296*
International Association for Research in Income and Wealth, New York *15391*
International Center for Research on Women, Washington *15430*
International Human Resources, Business and Legal Research Association, Washington *15487*
International Society for Business Education – United States Chapter, Reston *15526*
Joint Center for Political and Economic Studies, Washington *15615*
Junior Achievement, Colorado Springs *15631*
Labor Research Association, New York *15645*
Latin American Studies Association, Pittsburgh *15648*
League for Industrial Democracy, Washington *15657*
National Association for Industry-Education Cooperation, Buffalo *15805*
National Association for the Exchange of Industrial Resources, Galesburg *15819*
National Association of Business Education State Supervisors, Phoenix *15836*
National Association of Classroom Educators in Business Education, Boone *15838*
National Association of College and University Business Officers, Washington *15840*
National Association of State Development Agencies, Washington *15895*
National Association of Teacher Educators for Business Education, Columbia *15910*
National Bureau of Economic Research, Cambridge *15927*
National Center for Business and Economic Communication, Washington *15940*
National Congress for Community Economic Development, Washington *15978*
National Council for Urban Economic Development, Washington *15995*
National Council on Economic Education, New York *16018*
National Council on Employment Policy, Washington *16019*
National Industrial Zoning Committee, Columbus *16085*
National Schools Committee for Economic Education, Cos Cob *16149*
National Tax Association – Tax Institute of America, Columbus *16168*
Social Sciences Services and Resources, Wasco *16486*
Society for Business Ethics, Chicago *16503*
Society for Sedimentary Geology, Tulsa *16578*
Society for the Preservation of American Business History, Williamsburg *16603*
Society of Economic Geologists, Littleton *16655*

Venezuela
Centro Interamericano para el Desarrollo Regional, Maracaibo *16988*
Colegio de Economistas del Distrito Federal y Estado Miranda, Caracas *16993*
Federación Venezolana de Camaras y Asociaciones de Comercio y Producción, Caracas *17007*

Yugoslavia
Economists' Society of Serbia, Beograd *17092*
Savez Ekonomista Jugoslavije, Beograd *17116*
Yugoslav Economists' Association, Beograd *17137*

Education
s. a. Adult Education; Education of the Handicapped

Afghanistan
Asia Foundation, Kabul *00002*

Algeria
Centre National de Documentation et de Recherche en Pédagogie, Alger *00016*

Argentina
Asociación Latinoamericana de Centros de Educación, Rosario *00086*
Centro Nacional de Documentación e Información Educativa, Buenos Aires *00112*

Australia
Asian and Pacific Association for Social Work Education, Bundoora *00220*
Australian College of Education, Deakin *00274*
Australian Committee of Directors of Principals, Braddon *00277*
The Australian Council for Educational Research, Hawthorn *00279*
Australian Geography Teachers' Association, Adelaide *00292*
Australian Remedial Education Association, Camberwell *00332*
Australian Research Council, Canberra *00333*

Australian Science Teachers Association, Warradale 00337
Australian Vice-Chancellors' Committee, Canberra 00362
Commonwealth Council for Educational Administration, Armidale 00375
Geography Teachers' Association of New South Wales, Rozelle 00403
International Development Program of Australian Universities and Colleges, Canberra 00425
Primary English Teaching Association, Rozelle 00447
Queensland Institute for Educational Research, Saint Lucia 00448
Science Teachers Association of New South Wales, Rozelle 00488
Science Teachers Association of Queensland, Spring Hill 00489
Science Teachers' Association of Victoria, Parkville 00490
South Australian Science Teachers Association, Parkside 00497
Victorian Post-Secondary Education Commission, Hawthorn 00516

Austria

Arbeitsgemeinschaft der Musikerzieher Österreichs, Wien 00549
Arbeitsgemeinschaft für Deutschdidaktik, Klagenfurt 00550
Internationale Gesellschaft für Ingenieurpädagogik, Klagenfurt 00670
Österreichische Gesellschaft für Hochschuldidaktik, Wien 00790
Österreichische Gesellschaft für Sprachheilpädagogik, Wien 00833
Österreichische Pädagogische Gesellschaft, Wien 00876
Österreichische Rektorenkonferenz, Wien 00890
Österreichischer Fernschulverband, Wien 00892
Österreichischer Lehrerverband, Wien 00896
Österreichisches College: Collegegemeinschaft Wien, Wien 00908
Österreichisches Ost- und Südosteuropa-Institut, Wien 00915
Verband der Russischlehrer Österreichs, Wien 00962
Verband Österreichischer Bildungswerke, Wien 00967
Verein zur Förderung des physikalischen und chemischen Unterrichts, Wien 01005

Bahrain

Arab Regional Office of the United Schools International, Gudaibiya 01053

Bangladesh

Association of Universities of Bangladesh, Dhaka 01064

Barbados

Caribbean Examinations Council, Saint Michael 01081
Caribbean Network of Educational Innovation for Development, Saint Michael 01085

Belgium

Association Européenne F. Matthias Alexander, Bruxelles 01129
Association for European Training for Employees in Technology, Bruxelles 01132
Association for Teacher Education in Europe, Bruxelles 01133
Association Internationale des Métiers et Enseignements d'Art, Bruxelles 01142
Association Mondiale des Sciences de l'Education, Gent 01148
Board of Governors of the European Schools, Bruxelles 01170
Catholic International Education Office, Bruxelles 01179
Centre International de Liaison des Ecoles de Cinéma et de Télévision, Bruxelles 01192
Centre International d'Etudes de la Formation Religieuse, Bruxelles 01198
Centre of Promotion of Catholic Education in Europe, Bruxelles 01204
Coimbra Group, Bruxelles 01214
College of Europe, Brugge 01217
Committee on Higher Education in the European Community, Bruxelles 01238
Common Office for European Training, Mons 01240
Community Network for European Education and Training, Liège 01241
Comparative Education Society in Europe, Bruxelles 01242
Confédération d'Associations d'Ecoles Indépendantes de la Communauté Européenne, Bruxelles 01243
Conférence Internationale Permanente de Directeurs d'Instituts Universitaires pour la Formation de Traducteurs et d'Interprètes, Antwerpen 01248
Conseil des Recteurs des Institutions Universitaires Francophones de Belgique, Bruxelles 01251
Consortium of Institutions of Higher Education in Health and Rehabilitation in Europe, Gent 01255
Coordination et Promotion de l'Enseignement de la Réligion en Europe, Bruxelles 01257
Education Information Network in the European Community, Bruxelles 01260
European Association for Audiovisual Media Education, Bruxelles 01273
European Association for Institutions of Higher Education, Bruxelles 01278
European Association for the Teaching of Legal Theory, Bruxelles 01285
European Association of Architectural Education, Louvain-la-Neuve 01287
European Association of Education and Research in Public Relations, Gent 01290
European Association of Users of Satellites in Training and Education Programmes, Bruxelles 01297
European Capitals Universities Network, Bruxelles 01307
European Committee for Catholic Education, Bruxelles 01318
European Educational Association for News Distribution, Gerpinnes 01340
European Federation for the Education of Children of Occupational Travellers, Bruxelles 01346
European Finance Association, Bruxelles 01350
Fédération Nationale de l'Enseignement Moyen Catholique, Bruxelles 01359
Fondation Fernand Lazard, Bruxelles 01364
Fondation Francqui, Bruxelles 01365
International Educational and Cultural Association, Bruxelles 01396
International Secretariat for University Study of Education, Gent 01403
Jeunesse Intellectuelle, Bruxelles 01405
Office International de l'Enseignement Catholique, Bruxelles 01417
Vereniging Leraars Aardrijkskunde, Heverlee 01505

Bermuda

Amalgamated Bermuda Union of Teachers, Hamilton 01517
Bermuda Audubon Society, Hamilton 01519

Bolivia

Centro Nacional de Documentación e Información Educativa, La Paz 01537
Centro Pedagógico y Cultural de Portales, Cochabamba 01538
Consejo Nacional de Educación, La Paz 01544

Brazil

Asociación Latinoamericana de Escuelas de Trabajo Social, Saõ Luís 01583
Associação Brasileira de Educadores Lassalistas, São Carlos 01587
Associação Brasileira de Escolas Superiores Católicas, Brasília 01589
Associação de Educação Católica do Brasil, Brasília 01595
Associação de Ensino, Marília 01596
Associação de Ensino, Ribeirão Preto 01597
Associação de Ensino e Cultura Urubupungá, Pereira Barreto 01598
Associação Educacional Presidente Kennedy, Guarulhos 01600
Associação Itaquerense de Ensino, Itaquera 01602
Associação Prudentina de Educação e Cultura, Presidente Prudente 01608
Associação Tibiriçá de Educação, São Paulo 01610
Associação Universitaria Santa Ursula, Rio de Janeiro 01611
Association of Amazonian Universities, Belém 01613
Centro Regional de Pesquisas Educacionais do Sul, Porto Alegre 01638
Conselho Federal de Educação, Rio de Janeiro 01644
Fundação Educacional de Fortaleza, Fortaleza 01653
Instituto Brasileiro de Educâcão, Ciéncia e Cultura, Rio de Janeiro 01665
Instituto Nacional de Estudos e Pesquisas Educacionais, Brasília 01687
Organização Guarão de Ensino, Guaratinguetá 01699
Sociedade Civil de Educação São Marcos, São Paulo 01717
Sociedade de Cultura e Educação do Litoral Sul, Registro 01720
Sociedade de Ensino Piratininga, São Paulo 01722

Bulgaria

Society of Foreign Language Teachers, Sofia 01781

Burkina Faso

African and Malagasy Council on Higher Education, Ouagadougou 01804

Cameroon

Association des Institutions d'Enseignement Théologique en Afrique Centrale, Yaoundé 01811

Canada

Agency for Tele-Education in Canada, Toronto 01822
Alberta Educational Communications Corporation, Edmonton 01832
Alberta School Trustees' Association, Edmonton 01847
Alberta Teachers' Association, Edmonton 01850
American Association of Physics Teachers (Ontario), Kingston 01859
Association Canadienne d'Education de Langue Française, Sillery 01873
Association de l'Enseignement du Nouveau-Québec, Sainte-Foy 01876
Association des Collèges du Québec, Montréal 01879
Association des Diplômés de Polytechnique, Montréal 01880
Association des Enseignants Franco-Ontariens, Ottawa 01884
Association des Enseignants Francophones du Nouveau-Brunswick, Fredericton 01885
Association des Institutions de Niveaux Préscolaire et Elementaire du Québec, Montréal 01886
Association des Institutions d'Enseignement Secondaire, Montréal 01887
Association des Professeurs de Français de la Saskatchewan, Saskatoon 01890
Association des Professeurs de Français des Universités et Collèges Canadiens, Victoria 01891
Association for Media and Technology in Education, Toronto 01901
Association Française des Conseils Scolaires de l'Ontario, Ottawa 01907
Association Francophone Internationale des Directeurs d'Etablissements Scolaires, Montréal 01908
Association Internationale de Pédagogie Universitaire, Montréal 01909
Association of British Columbia School Superintendents, Osoyoos 01916
Association of Canadian Career Colleges, Brantford 01919
Association of Canadian Community Colleges, Toronto 01921
Association of Educators of Gifted, Talented and Creative Children in B.C., Coquitlam 01931
Association of Large School Boards in Ontario, Toronto 01934
Association of Nova Scotia Education Administrators, Sydney 01936
Association of Registrars of Universities and Colleges of Canada, Ottawa 01941
Association of University of New Brunswick Teachers, Fredericton 01947
British Columbia Council for Leadership in Educational Administration, Richmond 01975
British Columbia Parents in Crisis Society, Burnaby 01984
British Columbia Teachers' Federation, Vancouver 01992
Broadcast Education Association of Canada, Burnaby 01994
Canadian Association for Distance Education, Waterloo 02030
Canadian Association for Educational Psychology, London 02031
Canadian Association for the Study of Educational Administration, Halifax 02043
Canadian Association for University Continuing Education, Ottawa 02044
Canadian Association of Business Education Teachers, Medicine Hat 02047
Canadian Association of Deans of Education, Vancouver 02049
Canadian Association of Foundations of Education, Kingston 02050
Canadian Association of Immersion Teachers, Ottawa 02053
Canadian Association of Independent Schools, Toronto 02054
Canadian Association of University Development Officers, Ottawa 02071
Canadian Association of University Schools of Nursing, Ottawa 02072
Canadian Association of University Schools of Rehabilitation, Montréal 02073
Canadian Association of University Teachers, Ottawa 02074
Canadian Association of University Teachers of German, Edmonton 02075
Canadian Bureau for International Education, Ottawa 02084
Canadian College of Teachers, Edmonton 02100
Canadian Committee on Early Childhood, Edmonton 02101
Canadian Council of Teachers of English, Oakville 02108
Canadian Educational Researchers' Association, Ottawa 02121
Canadian Education Association, Toronto 02122
Canadian Federation of University Women, Nanaimo 02131
Canadian Foundation for Economic Education, Toronto 02140
Canadian Health Education Society, Toronto 02152
Canadian Society for Education through Art, Malton 02255
Canadian Society for the Study of Education, Ottawa 02268
Canadian Society for the Study of Higher Education, Ottawa 02269
Canadian Society for Training and Development, Ottawa 02271
Canadian Teachers' Federation, Ottawa 02305
Canadian Universities and Colleges Conference Officers Association, London 02310
Canadian University and College Counselling Association, Victoria 02311
Catholic Public Schools Inter-Society Committee, Prince George 02323
Centrale de l'Enseignement du Québec, Montréal 02324
Centre d'Animation, de Développement et de Recherche en Education, Montréal 02326
Centre Franco-Ontarien de Resources Pédagogiques, Ottawa 02330
Child Development Centre Society of Fort Saint John and District, Fort Saint John 02343
Commonwealth Association for Education in Journalism and Communication, London 02364
Commonwealth of Learning, Vancouver 02369
Conference of Alberta School Superintendents, Calgary 02378
Conseil International d'Education Mésologique, Québec 02384
Conseil Québecoise pour l'Enfance et la Jeunesse, Montréal 02385
Conseil Supérieur de l'Education, Sainte Foy 02386
Consortium-Distance Education Network, Sainte-Foy 02390
Consortium of Ontario Public Alternative Schools, Toronto 02391
Council of Parent Participation Pre-Schools in British Columbia, Burnaby 02404
Council of Private Technical Schools, Montréal 02406
Departments of Education Correspondence Schools Association (Canada), Barrhead 02417
Eastern Québec Teachers Association, Sainte Foy 02423
Eastern Townships Association of Teachers, Cookshire 02424
Education Relations Commission of Ontario, Toronto 02427
Estonian Teachers' League, Don Mills 02443
Fédération des Collèges d'Enseignement Général et Professionel, Montréal 02450
Fédération des Commissions Scolaires Catholiques du Québec, Sainte Foy 02451
Fédération Nationale des Enseignants et des Enseignantes du Québec, Montréal 02454
Federation of English Speaking Catholic Teachers, Montréal 02455
Federation of Independent School Associations in British Columbia, Vancouver 02456
Federation of Independent Schools in Canada, Edmonton 02457
Federation of Women Teachers' Associations of Ontario, Toronto 02460
Fédération Québécoise des Directeurs et Directrices d'Ecole, Anjou 02461
Greek Teachers Association, Toronto 02490
Harold Crabtree Foundation, Ottawa 02495
Independent Schools Association of British Columbia, Vancouver 02514
International Association of Master Penmen and Teachers of Handwriting, Ottawa 02542
International Council for Distance Education, Montréal 02549
International Development Education Committee of Ontario, Toronto 02550
International Development Education Resources Association, Vancouver 02551
Lakeshore Teachers Association, Dollard des Ormeaux 02574
League of Education Administrators, Directors and Superintendents, Warren 02585
Learning Resources Council, Edmonton 02587
Manitoba Association for Art Education, Winnipeg 02601
Manitoba Association for Multicultural Education, Winnipeg 02602
Manitoba Association of Mathematics Teachers, Winnipeg 02606
Manitoba Association of School Trustees, Winnipeg 02608
Manitoba Committee on Children and Youth, Winnipeg 02613
Manitoba Indian Education Association, Winnipeg 02622
Manitoba School Library Audio-Visual Association, Winnipeg 02632
Manitoba Social Science Teachers' Association, Winnipeg 02633
Manitoba Society for Training and Development, Winnipeg 02634
Manitoba Teachers' Society, Winnipeg 02638
Maritime Provinces Higher Education Commission, Fredericton 02640
New Brunswick Music Educators' Association, Quispamsis 02689
New Brunswick School Superintendents Association, Rexton 02693
New Brunswick Teachers' Federation, Fredericton 02696
Newfoundland Music Educators' Association, Corner Brook 02710
Newfoundland Teachers' Association, Saint John's 02712
North Island/Laurentian Teachers Union, Chomedey-Laval 02715
Northwest Territories Music Educators' Association, Fort Smith 02720
Northwest Territories Teachers' Association, Yellowknife 02721
Northwest Territories Teachers Association Home Economics Council, Inuvik 02722
Nova Scotia College Conference, Kingston 02728
Nova Scotia Music Educators' Association, Armdale 02739
Nova Scotia School Boards Association, Halifax 02744
Nova Scotia Teachers Union, Armdale 02747
Ontario Association for Continuing Education, Toronto 02754
Ontario Association for Curriculum Development, London 02755
Ontario Association of Alternative and Independent Schools, Toronto 02756
Ontario Association of Education Administrative Officials, Toronto 02758
Ontario Co-operative Education Association, Hamilton 02768
Ontario Council for Leadership in Educational Administration, Toronto 02769
Ontario Family Studies Home Economics Educators Association, Thorold 02776
Ontario Music Educators' Association, London 02786
Ontario Public School Boards' Association, North Bay 02791
Ontario Public School Teachers' Federation, Toronto 02792
Ontario Registered Music Teachers Association, Toronto 02793
Ontario School Counsellors' Association, Don Mills 02794
Ontario Secondary School Teachers' Federation, Toronto 02795
Ontario Teachers' Federation, Toronto 02801
Ontario Vocational Educational Association, Caledonia 02805
Polish Teachers' Association in Canada, Toronto 02822
Prince Edward Island Association of Community Schools, Belle River 02827
Prince Edward Island Physical Education Association, Charlottetown 02837
Prince Edward Island Teachers' Federation, Charlottetown 02841
Provincial Association of Catholic Teachers (Québec), Montréal 02848
Provincial Association of Protestant Teachers of Québec, Dollard des Ormeaux 02849
Public Affairs Council for Education, Sainte-Foy 02852
Quebec Association of Independent Schools, Beaconsfield 02854
Registered Music Teachers' Association, Winnipeg 02856
Rural Education and Development Association, Edmonton 02875
Saskatchewan Association of Teachers of German, Saskatoon 02884
Saskatchewan High School Principals' Group, Saskatoon 02890
Saskatchewan Registered Music Teachers' Association, Saskatoon 02898
Saskatchewan Teachers' Federation, Saskatoon 02905
Sex Information and Education Council of Canada, Toronto 02912
Société Nationale de Diffusion Educative et Culturelle, Montréal 02934
Society for Indian and Northern Education, Saskatoon 02939
Society of Christian Schools in British Columbia, Surrey 02942
Waldorf School Association of Alberta, Edmonton 02983
Western Québec Teachers Association, Hull 02993

Yukon Teachers' Association, Whitehorse *03004*

Chile

Academia Chilena de Ciencias Sociales, Politicas y Morales, Santiago *03009*
Centro de Perfeccionamiento, Experimentación e Investigaciones Pedagógicas, Santiago *03023*
Confederación de Educadores de América Latina, Santiago *03045*
Oficina Regional de Educación de la UNESCO para América Latina y el Caribe, Santiago *03062*
Sistema Nacional de Información en Educación, Santiago *03064*
Unión de Federaciones Universitárias Chilenas, Santiago *03104*

China, Republic

Association for Education through Art, Republic of China, Taichung Hsien *03242*
Association for Physics and Chemistry Education of the Republic of China, Taipei *03243*
Association of Child Education of the Republic of China, Taipei *03246*
China Education Society, Taipei *03254*
China International Education Research Association, Taipei *03255*
China Social Education Society, Taipei *03257*
Chinese Home Education Promotion Association, Taipei *03271*
Early Childhood Education Society of the Republic of China, Taipei *03295*
International Education Association of China, Taipei *03303*
National Audio-Visual Education Association of China, Taipei *03310*
Special Education Association of the Republic of China, Taipei *03330*

Colombia

Asociación de Universidades Confiadas a la Compañía de Jesús en América Latina, Bogotá *03350*
Confederación Interamericana de Educación Católica, Bogotá *03368*
Instituto Colombiano de Crédito Educativo y Estudios Técnicos en el Exterior, Bogotá *03382*
Instituto Colombiano de Pedagogía, Bogotá *03388*
Instituto Colombiano para el Fomento de la Educación Superior, Bogotá *03389*
Servicio Nacional de Aprendizaje, Bogotá *03401*

Costa Rica

Central American University Confederation, San José *03440*
Instituto Centroamericano de Extensión de la Cultura, San José *03446*

Croatia

Pedagoško-Književni zbor, Savez Pedagoških Društava Hrvatske, Zagreb *03465*

Czech Republic

European Centre for Leisure and Education, Praha *03543*
Kruh Moderních Filologu Při, Praha *03548*

Denmark

Association for World Education, Snedsted *03561*
Danmarks Laererforening, København *03581*
Danmarks Realskoleforening, Kolding *03585*
Dansk Historielaererforening, Vordingborg *03630*
Dansk Laererforeningen, København *03643*
Dansk Musikpaedagogisk Forening, København *03652*
Dansk Teknisk Laererforening, København *03696*
European Council of National Associations of Independent Schools, Slagelse *03724*
Foreningen af Danske Biologer., Birkerød *03727*
Foreningen af Geografilaerere ved de Gymnasiale Uddannelser, Ribe *03731*
Frankslaererforeningen, København *03738*
Gymnasieskolernes Laererforening, København *03740*
Gymnasieskolernes Tysklaererforening, Nykøbing *03741*
Matematiklaererforeningen, Birkerød *03771*
Nordic Federation for Medical Education, København *03778*
Paedagogisk Forening, København *03793*
Rektorkollegiet, København *03796*
Selskabet for Dansk Skolehistorie, Hellerup *03811*
Selskabet for Tekniske Uddanelsespørgsmal, København *03817*
Socialpaedagogernes Landsforbund, København *03822*
Teknisk Skoleforening, Odense *03826*

Ecuador

Asociación Ecuatoriana de Museos, Quito *03844*
Asociación Latinoamericana de Educación Radiofónica, Quito *03845*
Asociación Latinoamericana de Escuelas y Facultades de Enfermería, Quito *03846*

Egypt

Arab Council for Childhood and Development, Cairo *03870*
Association for Medical Education in the Middle East, Alexandria *03879*
Association of Medical Schools in the Middle East, Alexandria *03881*
National Centre for Educational Research, Cairo *03909*

Finland

Äidinkielen opettajain liitto, Helsinki *03948*
European Association for Research on Learning and Instruction, Turku *03956*
Finlands Svenska Lärarförbund, Helsinki *03959*
Kansantaloustieten Professorien ja Dosenttien Yhidistys, Jyväskylä *03974*
Lastentarhanopettajalütto, Helsinki *03981*
Opetusalan Ammattijarjestö, Helsinki *03987*
Suomen Kasvatusopillinen Yhdistys, Helsinki *04034*
Suomen Kasvatustieteellinen Seura, Oulu *04035*
Suomen Musiikinopettajain Liitto, Helsinki *04053*

France

Associated Schools Project in Education for International Cooperation and Peace, Paris *04143*
Association Catéchétique Nationale pour l'Audio Visuel, Paris *04146*
Association de l'Education Morale de la Jeunesse, Paris *04151*
Association de Réadaption Psychopédagogique et Scolaire, Paris *04158*
Association des Ecoles de Service Social de la Région Parisienne, Paris *04178*
Association des Professeurs de Langues Vivantes de l'Enseignement Public, Paris *04195*
Association des Professeurs de Mathématiques de l'Enseignement Publique, Paris *04196*
Association des Professeurs d'Histoire et de Géographie de l'Enseignement Public, Evry *04197*
Association Européenne de Recherches et d'Echanges Pédagogiques, Toulouse *04204*
Association Européenne des Enseignants, Section Française, Paris *04206*
Association Européenne pour l'Etude de l'Alimentation et du Développement de l'Enfant, Paris *04209*
Association Française de l'Ecole Paysanne, Paris *04219*
Association Française des Enseignants de Français, Paris *04228*
Association Francophone d'Education Comparée, Sèvres *04249*
Association Internationale des Ecoles des Sciences de l'Information, Paris *04270*
Association Internationale des Educateurs de Jeunes Inadaptés, Pau *04271*
Association Internationale des Professeurs de Langue et Littérature Russes, Paris *04274*
Association Internationale des Universités, Paris *04281*
Association Maria Montessori, Paris *04302*
Association Nationale des Educateurs de Jeunes Inadaptés, Paris *04310*
Association Nationale des Professeurs en Economie Sociale et Familiale, Paris *04311*
Association pour la Promotion de la Pédagogie Nouvelle, Paris *04328*
Association pour l'Enseignement de l'Assurance, Paris *04342*
Association Régionale d'Education Permanente, Paris *04345*
Association Régionale des Oeuvres Educatives et des Vacances de l'Education Nationale, Paris *04346*
Centre International de l'Enfance, Paris *04376*
Centre International Scolaire de Correspondance Sonore, Sainte-Savine *04381*
Comité d'Entente des Ecoles de Formation en Economie Sociale Familiale, Paris *04393*
Comité Européen pour l'Education des Enfants et Adolescents Précoces, Doués, Talentueux, Nîmes *04401*
Comité Français d'Education et d'Assistance de l'Enfance Déficiente, Paris *04403*
Comité National de l'Enseignement Libre, Paris *04413*
Comité National des Ecoles Françaises de Service Social, Paris *04415*
Comité Universitaire d'Information Pédagogique, Paris *04421*
Commission de l'Enseignement Supérieur en Biologie, Paris *04423*
Community of European Management Schools, Jouy-en-Josas *04432*
Conférence Internationale des Doyens des Facultés de Médecine d'Expression Française, Tours *04441*
Conférence Internationale des Facultés de Droit ayant en Commun l'Usage du Français, Aix-en-Provence *04442*
Conférence Internationale des Facultés, Instituts et Ecoles de Pharmacie d'Expression Française, Paris *04443*
Conférence Internationale des Formations d'Ingénieurs et Techniciens d'Expression Française, Villeurbanne *04444*
Conférence Internationale des Responsables des Universités et Instituts à Dominante Scientifique et Technique d'Expression Française, Talence *04446*
Dialogue et Coopération, Paris *04464*
Ecole Internationale de Bordeaux, Talence *04467*
Ecole Internationale d'Informatique de l'AFCET, Paris *04468*
Education for All Network, Paris *04469*
European Association of Establishments for Veterinary Education, Maisons-Alfort *04485*
European Council on Chiropractic Education, Poitiers *04500*
European Documentation and Information System for Education, Strasbourg *04502*
European Dyslexia Association – International Organization for Specific Learning Disabilities, Le Mesnil-Saint-Denis *04503*
Fédération Autonome de l'Education Nationale, Paris *04507*
Fédération de l'Education Nationale, Paris *04510*
Fédération Internationale des Associations d'Instituteurs, Paris *04531*
Fédération Internationale des Centres d'Entraînement aux Méthodes d'Education Active, Paris *04532*
Fédération Internationale des Mouvements d'Ecole Moderne, Cannes-La Bocca *04533*
Fédération Internationale des Professeurs de Français, Sèvres *04534*
Fédération Internationale des Professeurs de l'Enseignement Secondaire Officiel, Paris *04535*
Groupe de Recherche et pour l'Education et la Prospective, Paris *04548*
Institut International de Planification de l'Education, Paris *04573*
International Union for Health Education, Paris *04588*
Ligue Française de l'Enseignement et de l'Education Permanente, Paris *04594*
Ligue Française de l'Enseignement Oroleis de Paris, Paris *04595*
Ligue Internationale de l'Enseignement, de l'Education et de la Culture Populaire, Paris *04597*
Office National d'Information sur les Enseignements et les Professions, Paris *04606*
Société des Ecoles du Dimanche, Paris *04702*
Société des Professeurs de Dessin et d'Arts Plastiques de l'Enseignement Secondaire, Paris *04713*
Société Française de Pédagogie, Paris *04828*
Société Française des Professeurs de Russe, Paris *04845*
Syndicat des Enseignants, Paris *04926*
Syndicat Général de l'Education Nationale, Paris *04930*
Syndicat Général des Personnels de l'Education Nationale, Paris *04931*
Syndicat National de l'Enseignement Supérieur, Paris *04932*
Syndicat National de l'Intendance de l'Education Nationale, Paris *04934*
Syndicat National des Chefs d'Etablissements d'Enseignement Libre, Paris *04939*
Syndicat National des Collèges de la Région Parisienne, Paris *04942*
Syndicat National des Collèges et des Lycées, Paris *04943*
Syndicat National des Enseignements de Second Degré, Paris *04944*
Syndicat National des Instituteurs et Professeurs de Collège, Paris *04945*
Syndicat National des Lycées et Collèges, Paris *04946*
Syndicat National des Personnels de Direction de l'Enseignement Secondaire, Paris *04964*
Syndicat National des Professeurs des Ecoles Normales d'Instituteurs, Paris *04966*
Union des Physiciens, Paris *04975*
Union des Professeurs de Spéciales (Mathématiques et Sciences Physiques), Paris *04976*
Union Internationale pour la Liberté d'Enseignement, Paris *04990*

Germany

Akademie für Fernstudium und Weiterbildung Bad Harzburg, Bad Harzburg *05054*
Akademie Klausenhof, Hamminkeln *05062*
Akademie Remscheid für musische Bildung und Medienerziehung e.V., Remscheid *05064*
Aktion Bildungsinformation e.V., Stuttgart *05066*
Anthroposophische Gesellschaft in Deutschland, Stuttgart *05076*
Arbeitsgemeinschaft Bremer Schule e.V., Bremen *05082*
Arbeitsgemeinschaft der Ordenshochschulen, Vallendar *05088*
Arbeitsgemeinschaft der Verbände Gemeinnütziger Privatschulen in der Bundesrepublik, Köln *05094*
Arbeitsgemeinschaft Freier Schulen, Vereinigungen und Verbände gemeinnütziger Schulen in freier Trägerschaft, Berlin *05099*
Arbeitsgruppe für empirische Bildungsforschung e.V., Heidelberg *05133*
Arbeitskreis Bildung und Politik Rheinland, Remagen *05135*
Arbeitskreis deutscher Bildungsstätten e.V., Bonn *05137*
Arbeitskreis für Schulmusik und allgemeine Musikpädagogik e.V., Würzburg *05143*
Arbeitskreis katholischer Schulen in freier Trägerschaft in der Bundesrepublik Deutschland, Bonn *05146*
Arbeit und Leben, Düsseldorf *05149*
Association Européenne pour l'Enseignement de l'Architecture, Kassel *05154*
Bayerischer Lehrer- und Lehrerinnenverband, München *05178*
Bund der Freien Waldorfschulen e.V., Stuttgart *05225*
Bund Deutscher Kunsterzieher e.V., Hannover *05228*
Bundesakademie für musikalische Jugendbildung, Trossingen *05230*
Bundesarbeitsgemeinschaft Schule-Wirtschaft, Köln *05236*
Bundesverband Deutscher Privatschulen, Frankfurt *05251*
Bundesvereinigung Kulturelle Jugendbildung e.V., Remscheid *05258*
Bund katholischer Erzieher Deutschlands, Essen *05268*
Deutsche Gesellschaft für Erziehungswissenschaft, Berlin *05347*
Deutsche Gesellschaft für Sprechwissenschaft und Sprecherziehung e.V., Münster *05428*
Deutsche Gesellschaft für Suchtforschung und Suchttherapie e.V., Hamm *05429*
Deutsche Hauptstelle gegen die Suchtgefahren e.V., Hamm *05454*
Deutsche Montessori Gesellschaft, Würzburg *05472*
Deutscher Berufsverband der Sozialarbeiter und Sozialpädagogen e.V., Essen *05504*
Deutscher Hochschulverband, Bonn *05516*
Deutscher Lehrerverband Niedersachsen, Hannover *05526*
Deutscher Medizinischer Informationsdienst e.V., Frankfurt *05528*
Deutscher Tonkünstlerverband e.V., München *05543*
Deutscher Werkbund e.V., Frankfurt *05563*
Deutsches Handwerksinstitut e.V., München *05579*
Deutsches Institut für Internationale Pädagogische Forschung, Frankfurt *05585*
Dokumentationsring Pädagogik, Frankfurt *05634*
Europäische Akademie Bayern e.V., München *05646*
Europäische Akademie Berlin e.V., Berlin *05647*
Europäische Bildungs- und Aktionsgemeinschaft, Bonn *05649*
Europäische Bildungs- und Begegnungszentren, Idstein *05650*
Europäischer Erzieherbund e.V., Sektion Deutschland, Schwalbach *05655*
European Association for Research and Development in Higher Education, Berlin *05662*
European Association for Special Education, Stuttgart *05663*
European Association of Business and Management Teachers, Dortmund *05668*
European Baptist Theological Teachers' Conference, Hamburg *05672*
European Centre for the Development of Vocational Training, Berlin *05673*
Evangelischer Erziehungs-Verband e.V., Hannover *05689*
Fachverband Textilunterricht e.V., Münster *05697*
Freundeskreis Deutscher Auslandsschulen e.V., Bonn *05763*
Gesellschaft für Deutschlandforschung e.V., Berlin *05809*
Gesellschaft für Pädagogik und Information e.V., Paderborn *05848*
Gesellschaft für Programmierte Instruktion und Mediendidaktik e.V., Giessen *05852*
Gesellschaft für publizistische Bildungsarbeit e.V., Hagen *05854*
Gesellschaft Information Bildung e.V., Frankfurt *05875*
Gesellschaft zur Förderung Pädagogischer Forschung e.V., Frankfurt *05889*
Historische Kommision der Deutschen Gesellschaft für Erziehungswissenschaft, Hannover *05915*
Hochschullehrerbund e.V., Bonn *05928*
Institut für Angewandte Pädagogische Forschung e.V., Seelze *05948*
Institut für Neue Musik und Musikerziehung e.V., Darmstadt *05966*
Institut für Sozialarbeit und Sozialpädagogik, Frankfurt *05970*
Internationales Zentralinstitut für das Jugend- und Bildungsfernsehen, München *06002*
International Society for Group Activity in Education, Schriesheim *06010*
Intitut für Bildungsmedien e.V., Frankfurt *06015*
Katholische Erziehergemeinschaft in Bayern, München *06064*
Katholische Pädagogenarbeit Deutschlands, Bonn *06066*
Landesarbeitsgemeinschaft Jugend und Literatur NRW e.V., Pulheim-Brauweiler *06095*
Outward Bound – Deutsche Gesellschaft für Europäische Erziehung e.V., München *06228*
Pädagogisches Zentrum, Berlin *06230*
Philosophischer Fakultätentag, Saarbrücken *06235*
Saarländischer Gymnasiallehrerverband e.V., Losheim *06271*
Verband Bildung und Erziehung e.V., Bonn *06332*
Verband der Geschichtslehrer Deutschlands, Bokholt-Hanredder *06339*
Verband Deutscher Lehrer im Ausland e.V., Husum *06355*
Verband deutscher Musikschulen e.V., Bonn *06356*
Verband Deutscher Realschullehrer im Deutschen Beamtenbund, Viersen *06358*
Verband Deutscher Schulgeographen e.V., Burgwedel *06360*
Verband Deutscher Schulmusiker e.V., Mainz *06361*
Vereinigung deutscher Landerziehungsheime, Berlin *06405*
Verein katholischer deutscher Lehrerinnen e.V., Essen *06421*
Weltbund für Erneuerung der Erziehung, Deutschsprachige Sektion, Heidelberg *06437*
Zentrale für Fallstudien e.V., Erftstadt *06464*

Ghana

All Africa Teachers' Organization, Accra *06475*
Association of African Universities, Accra *06477*
West African Examinations Council, Accra *06493*

Gibraltar

Gibraltar Teachers' Association, Gibraltar *06497*

Greece

Omospondia Didaskaliki Ellados, Athinai *06539*
Syllogos pros Diadosin ton Hellenikon Grammaton, Athinai *06546*

Grenada

Caribbean Association of Catholic Teachers, Saint George's *06551*

Hong Kong

International Council of Associations for Science Education, Hong Kong *06594*

Hungary

European Distance Education Network, Budapest *06603*

Iceland
Félag Enskukennara á Islandi, Reykjavik *06682*
Félag Menntaskólakennara, Reykjavik *06686*
Kennarasamband Islands, Reykjavik *06693*

India
Association for Engineering Education in South and Central Asia, Mysore *06730*
Association for Medical Education in South-East Asia, New Delhi *06731*
Bal Bhavan Society, Calcutta *06737*
Bal Bhavan Society, New Delhi *06738*
Hyderabad Educational Conference, Hyderabad *06765, 06766*
Indian Psychometric and Educational Research Association, Patna *06810*
National Council of Educational Research and Training, New Dehli *06843*
United Schools International, New Dehli *06868*

Indonesia
ASEAN Council of Teachers, Jakarta *06873*
Asian Physics Education Network, Jakarta *06881*
Cooperative Program in Technological Research and Higher Education in Southeast Asia and the Pacific, Jakarta *06887*

Iran
Sazemane Pachuhesh va Barnamerizi Amuzeshi, Teheran *06903*

Iraq
Arab Federation for Technical Education, Baghdad *06908*
Teachers' Society, Baghdad *06920*

Ireland
Association of Irish Headmistresses, Dun Laoghaire *06927*
Association of Secondary Teachers, Ireland, Dublin *06931*
Church Education Society for Ireland, Dublin *06936*
Irish Association of Curriculum Development, Dublin *06974*
Irish Federation of University Teachers, Dublin *06984*
Irish National Teachers' Organisation, Dublin *06994*
Irish Science Teachers Association, Dublin *06999*

Italy
Accademia degli Incolti, Roma *07139*
Accademia degli Ottimi, Roma *07140*
Accademia Internazionale d'Arte Moderna, Roma *07189*
Accademia Internazionale della Tavola Rotonda, Milano *07190*
Accademia Italiana di Stenografia e di Dattilografia Giuseppe Aliprandi, Firenze *07195*
Association of Advisers on Education in International Religious Congregations, Roma *07251*
Associazione Centri di Orientamento Scolastico Professionale e Sociale, Roma *07264*
Associazione Educatrice Italiana, Roma *07275*
Associazione Italiana degli Insegnanti di Geografia, Trieste *07293*
Associazione Nazionale per il Progresso della Scuola Italiana, Roma *07380*
Associazione Nazionale per la Scuola Italiana, Roma *07381*
Associazione Pedagogica Italiana, Chiusi Città *07386*
Centre for Coordination of Research of the International Federation of Catholic Universities, Roma *07413*
Centro Europeo per il Progresso della Scuola, Roma *07461*
Centro Internazionale per l'Avanzamento della Ricerca e dell'Educazione, Vigliano Biellese *07479*
Centro Internazionale per l'Educazione Artistica della Fondazione Giorgio Cini, Venezia *07481*
Centro Internazionale Sonnenberg per l'Italia, Torino *07485*
Centro Isec, Iniziative per Studi e Convegni, Roma *07489*
Centro Psico-Pedagogico Didattico, Bologna *07531*
Centro Ricerche Didattiche Ugo Morin, Paderno del Grappa *07536*
Centro Studi ed Esperienze Scout Baden Powell, Roma *07569*
Centro Studi e di Educazione Civica Enrico Mattei, Torino *07570*
Centro Studi e Iniziative Pier Santi Mattarella, Trapani *07572*
Centro Studi Filippo e Marta Larizza per la Formazione Permanente degli Educatori e per la Prevenzione del Disadattamento Giovanile, Bassano del Grappa *07575*
Centro Studi Mario Mazza, Genova *07576*
Centro Studi Mutualistici Emancipazione e Partecipazione, Roma *07577*
Centro Studi per il Progresso della Educazione Sanitaria e del Diritto Sanitario, Roma *07581*
Centro Studi Politici e Sociali Alcide de Gasperi, Cosenza *07586*
Comitato Italiano per l'Educazione Sanitaria, Perugia *07615*
Community of Mediterranean Universities, Bari *07625*
Ente Eugenio e Claudio Faina per l'Istruzione Professionale Agraria, Roma *07651*
International Centre for Advanced Technical and Vocational Training, Torino *07709*
International League of Esperantist Teachers, Massa *07716*
Sindacato Autonomo Scuola Media Italiana, Roma *07765*
Sindacato Nazionale Istruzione Artistica, Roma *07767*
Società Italiana per l'Educazione Musicale, Milano *07967*
Unione Cattolica Italiana Insegnanti Medi, Roma *08042*
Union Mondiale des Enseignants Catholiques, Roma *08053*

Jamaica
Caribbean Council of Legal Education, Kingston *08071*

Japan
Asian Pacific University Presidents Conference, Tokyo *08103*
Kyoikushi Gakkai, Tokyo *08175*
Kyoiku Tetsugakkai, Tokyo *08176*
Nihon Kokusai Kyoiku Kyokai, Tokyo *08232*
Nihon Kyoiku Gakkai, Tokyo *08237*
Nihon Kyoiku-shakai Gakkai, Tokyo *08238*
Nihon Kyoiku Shinri Gakkai, Tokyo *08239*
Nippon Kagaku Kyoiku Gakukai, Tokyo *08319*
Nippon Sugaku Kyoiku Gakkai, Tokyo *08387*

Jordan
Association of Arab Universities, Amman *08451*

Kenya
African Association for Correspondence Education, Nairobi *08460*
African Commission on Mathematics Education, Nairobi *08465*
African Council for Communication Education, Nairobi *08466*
Association for Teacher Education in Africa, Nairobi *08478*
Association of International Schools in Africa, Nairobi *08480*
Basic Education Resource Centre, Nairobi *08483*
Commonwealth Association of Polytechnics in Africa, Nairobi *08488*

Korea, Republic
Asian Association of Open Universities, Seoul *08518*

Kuwait
Educational Innovation Programme for Development in the Arab States, Kuwait *08547*

Luxembourg
Association des Professeurs de l'Enseignement Secondaire et Supérieur du Grand-Duché de Luxembourg, Luxembourg *08576*
Association Internationale d'Orientation Scolaire et Professionelle, Strassen *08578*
Commission Grand-Ducale d'Instruction, Luxembourg *08581*
Syndicat National des Enseignants, Luxembourg *08590*

Malaysia
Asia Foundation, Kuala Lumpur *08632*
Association for Medical Education in the Western Pacific Region, Kuala Lumpur *08638*

Malta
Graduate Teachers' Association, Paceville *08664*
Malta Union of Teachers, Valletta *08667*

Mexico
Asociación Nacional de Universidades e Institutos de Enseñanza Superior, México *08705*
Centro de Estudios Educativos, México *08715*
Centro Regional de Educación de Adultos y Alfabetización Funcional para América Latina, Pátzcuaro *08722*
El Colegio Nacional, México *08724*
Confederación de Educadores Americanos, México *08727*
Departamento de Educación Audiovisual, México *08731*
Sociedad de Educación, México *08747*
Sociedad Mexicana de Estudios Psico-Pedagógicos, México *08760*

Morocco
Association of African Faculties of Agriculture, Rabat *08803*
Conférence Internationale des Directeurs et Doyens des Etablissements d'Enseignement Supérieur et Facultés d'Expression Française des Sciences de l'Agriculture et de l'Alimentation, Rabat *08807*

Netherlands
Algemene Bond van Onderwijzend Personeel, Amsterdam *08820*
Association Montessori Internationale, Amsterdam *08825*
Association of European Correspondence Schools, Arnhem *08826*
Association of European Schools of Planning, Nijmegen *08827*
Association of Schools of Public Health in the European Region, Maastricht *08829*
Centrum voor de Studie van het Onderwijs in Ontwikkelingslanden, 's-Gravenhage *08839*
Commission of Socialist Teachers of the European Community, Zutphen *08843*
European Association for Cancer Education, Groningen *08851*
European Association for International Education, Amsterdam *08856*
European Association of Diabetes Educators, Leiden *08864*
European Association of Distance Teaching Universities, Heerlen *08865*
European Association of Teachers, Beek-Ubbergen *08871*
European Development Education Curriculum Network, Alkmaar *08883*
Instituut voor Onderwijskundige Dienstverlening, Nijmegen *08913*
The International Association for the Evaluation of Educational Achievement, 's-Gravenhage *08922*
International Council for Children's Play, Groningen *08931*
International Federation of Free Teachers' Unions, Amsterdam *08933*
Nationale Vereniging voor Economisch Onderwijs, Bergen *08978*
Nederlandse Organisatie voor Internationale Samenwerking in het Hoger Onderwijs, 's-Gravenhage *08997*
Nederlandse Vereniging van Pedagogen, Onderwijskundigen en Andragologen, Utrecht *09009*
Nederlandse Vereniging van Wiskundeleraren, 's-Gravenhage *09013*
Nederlandse Vereniging voor Medisch Onderwijs, Utrecht *09026*
Nederlands Genootschap van Leraren, Dordrecht *09048*
Protestants-Christelijke Onderwijsvakorganisatie, 's-Gravenhage *09062*
Vereniging van Katholieke Leraren Sint-Bonaventura, Hengelo *09091*
Vereniging voor Filosofie-Onderwijs, Zoetermeer *09099*

New Zealand
Association for Engineering Education in Southeast Asia and the Pacific, Christchurch *09118*
New Zealand Council for Educational Research, Wellington *09141*

Nigeria
Accrediting Council for Theological Education in Africa, Kaduna *09209*
Nigeria Educational Research Council, Lagos *09233*

Norway
Landslaget Musikk i Skolen, Oslo *09277*
Norske Musikklaereres Landsforbund, Kristiansand *09304*
Norsk Laererlag, Oslo *09325*
Organization of Nordic Teachers Associations, Oslo *09342*
The Teachers' Union, Oslo *09354*

Pakistan
All-Pakistan Educational Conference, Karachi *09357*
Punjab Bureau of Education, Lahore *09393*

Panama
Educational and Development Foundation of the Latin American Confederation of Credit Unions, Panamá City *09418*

Papua New Guinea
Papua New Guinea Teachers' Association, Boroko *09423*

Philippines
ASEAN Training Centre for Preventive Drug Education, Manila *09492*
Asian Association for Biology Education, Manila *09494*
Asian Association of Agricultural Colleges and Universities, Laguna *09495*
Asian Confederation of Teachers, Manila *09496*
Asian Regional Training and Development Organization, Manila *09503*
Asia Pacific Physics Teachers and Educators Association, Manila *09507*
Association of Deans of Southeast Asian Graduate Schools of Management, Manila *09509*
Colombo Plan Staff College for Technician Education, Manila *09513*

Portugal
Associação das Universidades de Lingua Portuguesa, Lisboa *09637*
Association of Paediatric Education in Europe, Lisboa *09652*
Conselho de Reitores das Universidades Portuguesas, Lisboa *09656*
Instituto Açoriano de Cultura, Angra do Heroismo *09658*
Instituto de Coímbra, Coímbra *09659*
Sindicato Nacional dos Professores, Lisboa *09671*
Sociedade Portuguesa de Educação Médica, Lisboa *09698*

Puerto Rico
Asociación de Maestros de Puerto Rico, Hato Rey *09720*
Caribbean Hospitality Training Centre, Santurce *09728*

Romania
European Centre for Higher Education, Bucuresti *09750*

Russia
Commission for Links with US Research Establishments in the Use of Technology and New Communications Technologies in Education, Moskva *09821*
Commission for Work with Young People, Moskva *09844*
Council for Links between the Academy of Sciences and Higher Education, Moskva *09859*
Russian Academy of Pedagogical Sciences, Moskva *09953*

Saudi Arabia
Arab Bureau of Education for the Gulf States, Riyadh *09997*
Center for Research in Islamic Education, Mecca *10001*

Senegal
Conférence des Facultés et Ecoles de Médecine d'Afrique d'Expression Française, Dakar *10025*

Singapore
Association for Theological Education in South East Asia, Singapore *10043*

Slovakia
Slovenská Pedagogická Spoločnost, Prešov *10083*

Slovenia
Zveza Pedagoskih Društev Slovenije, Ljubljana *10116*

South Africa
South African Association for Technical and Vocational Education, Arcadia *10180*
South African Institute of Race Relations, Braamfontein *10202*

Spain
Asociación Iberoamericana de Educación Superior a Distancia, Madrid *10273*
Association for Dental Education in Europe, Madrid *10282*
Conferencia de Rectores de las Universidades del Estado, Córdoba *10294*
Consejo de Universidades, Madrid *10295*
European Centre for Professional Training in Environment and Tourism, Madrid *10307*
Federación Española de Religiosos de Enseñanza, Madrid *10310*
Organización de Estados Iberoamericanos para la Educación, la Ciencia y la Cultura, Madrid *10333*

Sri Lanka
National Education Society of Ceylon, Colombo *10417*

Sudan
Association des Professeurs de Français en Afrique, Khartoum *10425*

Sweden
Asociación Nórdica de Intercambio en Educación Popular con América Latina, Kungälv *10436*
Council of Nordic Teachers' Associations, Stockholm *10448*
Geografiläramas Riksförening, Lund *10464*
International Association for the Evaluation of Educational Achievement, Stockholm *10470*
Kungliga Gustav Adolfs Akademien, Uppsala *10484*
Lärarförbundet, Stockholm *10494*
Läramas Riksförbund, Stockholm *10495*
Riksföreningen för Lärarna i Moderna Språk, Hisings Kärra *10515*
Skolledarförbundet, Stockholm *10521*
Svenska Akademiska Rektorskonferensen, Uppsala *10531*
Svenska Yrketsutbildningsföreningen, Ludvika *10577*

Switzerland
Association des Ecoles Internationales, Genève *10620*
Association des Sociétés Suisses des Professeurs de Langues Vivantes, Luzern *10622*
Bureau International d'Education, Genève *10639*
Campagne d'Education Civique Européenne, Genève *10641*
Centre Suisse de Documentation en Matière d'Enseignement et d'Education, Grand-Saconnex *10644*
Christlicher Lehrer- und Erzieherverein der Schweiz, Horw *10645*
Dachverband Schweizerischer Lehrerinnen und Lehrer, Zürich *10659*
Europäische Rektorenkonferenz, Genève *10670*
European Centre for Insurance Education and Training, Sankt Gallen *10680*
Fédération des Ecoles Privées de la Suisse Romande, Rolle *10699*
Fédération Suisse des Ecoles Privées, Genève *10704*
Gesellschaft schweizerischer Zeichenlehrer, Einsiedeln *10722*
Institut Romand de Recherches et de Documentation Pédagogiques, Neuchâtel *10731*
Internationale Union Demokratisch-Sozialistischer Erzieher, Wildegg *10741*
Konferenz Schweizerischer Gymnasialrektoren, Nyon *10761*
Schweizerische Arbeitsgemeinschaft für akademische Berufs- und Studienberatung, Luzern *10792*
Schweizerische Arbeitsgemeinschaft für Jugendmusik und Musikerziehung, Neuheim *10794*
Schweizerische Gesellschaft für Bildungs- und Erziehungsfragen, Langnau *10814*
Schweizerische Gesellschaft für Lehrerinnen- und Lehrerbildung, Bern *10830*
Schweizerische Heilpädagogische Gesellschaft, Bern *10860*
Schweizerische Hochschulrektoren-Konferenz, Bern *10862*
Schweizerischer Ballettlehrer-Verband, Hombrechtikon *10886*
Schweizerischer Berufsverband diplomierter Sozialarbeiterinnen und Sozialpädagoginnen, Bern *10887*
Schweizerischer Musikpädagogischer Verband, Zürich *10900*
Schweizerischer Verband für Fernunterricht, Kreuzlingen *10912*
Schweizerischer Verband für Sport in der Schule, Bern *10915*
Schweizerische Zentralstelle für Hochschulwesen, Bern *10960*
Société des Professeurs d'Allemand en Suisse Romande et Italienne, Genève *10970*
Société Pédagogique de la Suisse Romande, Carouge *10981*
Société Suisse des Professeurs de Français, Chancy *10986*
Société Suisse des Professeurs de Musique de l'Enseignement Secondaire, Emmenbrücke *10987*
Société Suisse pour la Recherche en Education, Carouge *10989*
Verband der Heilpädagogischen Ausbildungsinstitute der Schweiz, Luzern *10995*

Verband Jüdischer Lehrer und Kantoren der Schweiz, Zürich *11002*
Verband Musikschulen Schweiz, Liestal *11003*
Verein der Freien Pädagogischen Akademie, Hedingen *11009*
Vereinigung Schweizerischer Naturwissenschaftslehrer, Neerach *11020*
Verein katholischer Lehrerinnen der Schweiz, Hochdorf *11023*
Verein Schweizerischer Geschichtslehrer, Riehen *11025*
World Association for the School as an Instrument of Peace, Genève *11031*
World Confederation of Organizations of the Teaching Profession, Morges *11032*

Syria

Arab States Regional Broadcasting Training Centre, Damascus *11043*

Tanzania

East Africa Association for Theological Education by Extension, Njombe *11049*
East, Central and Southern African College of Nursing, Arusha *11053*
Eastern and Southern African Universities Research Programme, Dar es Salaam *11058*

Thailand

Asia and Pacific Programme of Educational Innovation for Development, Bangkok *11074*
Asia Foundation, Bangkok *11075*
Asian Centre of Educational Innovation for Development, Bangkok *11079*
Asia-Pacific Programme of Education for All, Bangkok *11087*
Association of Southeast Asian Institutions of Higher Learning, Bangkok *11089*
Rectors' Conference, Bangkok *11096*

Togo

African Association of Education for Development, Lomé *11103*

Tunisia

Arab League Educational, Cultural and Scientific Organization, Tunis *11131*

Uganda

Association for Teacher Education in Africa, Kampala *11181*
Association of Eastern and Southern African Universities, Kampala *11182*
East African School of Librarianship, Kampala *11185*

United Kingdom

Academia Europaea, London *11197*
Advisory Centre for Education, London *11206*
Association for Child Psychology and Psychiatry, London *11262*
Association for Educational and Training Technology, Axminster *11263*
Association for Language Learning, Rugby *11269*
Association for Medical Education in Europe, Dundee *11273*
Association for Recurrent Education, Sheffield *11278*
Association for Religious Education, Cumnor Hill *11279*
The Association for Science Education, Hatfield *11282*
Association for Teaching Psychology, Leicester *11286*
Association for the Reform of the Latin Teaching, Woking *11290*
Association for the Teaching of the Social Sciences, Saint Albans *11295*
Association of Beauty Teachers, Southampton *11305*
Association of Blind and Partially-Sights Teachers and Students, London *11306*
Association of British Correspondence Colleges, London *11308*
Association of Christian Teachers, London *11323*
Association of Colleges for Further and Higher Education, Swindon *11326*
Association of Commonwealth Teachers, London *11328*
Association of Commonwealth Universities, London *11329*
Association of Directors of Education in Scotland, Dundee *11336*
Association of Educational Psychologists, Durham *11341*
Association of Governing Bodies of Girls Public Schools, Petersfield *11347*
Association of Governing Bodies of Public Schools (Boys), Petersfield *11348*
Association of Headmistresses of Preparatory Schools, London *11351*
Association of Heads of Independent Schools, Windor *11352*
Association of Institute and School of Education In-Service Tutors, Durham *11358*
The Association of Law Teachers, Leicester *11360*
Association of Lecturers in Colleges of Education in Scotland, Beech Grove *11362*
Association of Polytechnic Teachers, Southsea *11377*
Association of Principals of Colleges, Broxborn *11378*
Association of Principals of Colleges (Northern Ireland Branch), Lisburn *11379*
Association of Recognised English Language Services, London *11383*
Association of Religious in Education, London *11384*
Association of Schools of Public Health in the European Region, Bristol *11387*
Association of Teachers of Management, London *11396*
Association of Teachers of Mathematics, Derby *11397*
Association of University Teachers (Scotland), Glasgow *11405*
Association of Veterinary Teachers and Research Workers, Edinburgh *11407*
Association of Vice-Principals in Colleges, Maidenhead *11408*
Association of Voluntary Aided Secondary Schools, Beckenham *11409*
Benslow Music Trust, Hitchin *11434*
Boarding Schools Association, Reigate *11457*
British Educational Management and Administration Society, Oxford *11555*
British Schools Exploring Society, London *11643*
Campaign for the Advancement of State Education, London *11740*
Central Bureau for Educational Visits and Exchanges, London *11749*
Choir Schools Association, Wells *11781*
Christian Education Movement, Derby *11783*
City and Guilds of London Institute, London *11789*
The College of Preceptors, Epping *11797*
Committee of Directors of Polytechnics, London *11804*
Commonwealth Association of Architects – Education Committee, London *11811*
Commonwealth Association of Science and Mathematics Educators, London *11812*
Commonwealth Education Liaison Committee, London *11819*
Commonwealth Education Programme, London *11820*
Commonwealth Industrial Training and Experience Programme, London *11823*
Commonwealth Legal Education Association, London *11826*
Commonwealth Secretariat, Education Programme, Human Resource, Development Group, London *11830*
Council for Education in World Citizenship, London *11850*
Council for Environmental Education, Reading *11852*
Council for the Accreditation of Correspondence Colleges, London *11853*
Council of Legal Education, London *11857*
County Education Officers' Society, Winchester *11860*
County Emergency Planning Officers Society, Worlester *11861*
The Economics and Business Education Association, Hassocks *11928*
Educational Centres Association, London *11941*
Educational Development Association, Wisbech *11942*
Educational Drama Association, Birmingham *11943*
The Educational Institute of Design, Craft and Technology, Melton Mowbray *11944*
Educational Institute of Scotland, Edinburgh *11945*
Educational Television Association, York *11946*
Esperanto Teachers Association, Leicester *11974*
European Council of International Schools, Petersfield *12022*
European Design Education Network, Glasgow *12026*
Federal Trust for Education and Research, London *12060*
General Studies Association, York *12107*
The General Teaching Council for Scotland, Edinburgh *12108*
Girl's Schools Association, Oxford *12115*
Greater London Federation of Parent-Teacher Associations, London *12126*
Head Teachers' Association of Scotland, Inverness *12153*
Higher Education Funding Council for England, Bristol *12168*
History of Education Society, Leicester *12179*
Honours Graduate Teachers' Association, Ayr *12181*
Incorporated Association of Preparatory Schools, London *12194*
Independent Schools Joint Committee, Petersfield *12198*
Institute of Training and Development, Marlow *12287*
International Association of Teachers of English as a Foreign Language, Whitstable *12333*
International Commission on Physics Education, Malvern *12343*
International Federation of Surgical Colleges, Edinburgh *12361*
International Study Group for Mathematics Learning, London *12396*
Isotype Institute, London *12414*
Jersey Association of the National Association of Schoolmasters, Saint Saviour's *12417*
The Joint Association of Classical Teachers, London *12422*
League for the Exchange of Commonwealth Teachers, London *12443*
London Association of Science Teachers, London *12461*
Mathematical Instruction Subcommittee, London *12499*
Medau Society of Great Britain and Northern Ireland, London *12500*
The Montessori Society A.M.I. (UK), London *12536*
Museum Training Institute, Bradford *12546*
Music Advisers' National Association, Warwick *12547*
Music Masters' and Mistresses' Association, Guildford *12549*
National Association for Environmental Education, Walsall *12557*
National Association for Film in Education, Bexleyheath *12558*
National Association for Multiracial Education, Slough *12559*
National Association for Outdoor Education, Sheffield *12560*
National Association for Road Safety Instruction in Schools, London *12561*
National Association for the Teaching of English, Sheffield *12564*
National Association of Head Teachers, Haywards Heath *12567*
National Association of Schoolmasters, Birmingham *12570*
National Association of Teachers in Further and Higher Education, London *12571*
National Associations of Teachers of Home Economics and Technology, London *12573*
National Campaign for Nursery Education, London *12578*
National Christian Education Council, Redhill *12580*
National Conference of University Professors, Uxbridge *12581*
National Council for Educational Technology, Coventry *12582*
National Council of Jewish Religious Day Schools, London *12584*
National Education Association, Tewin Wood *12587*
National Foundation for Educational Research in England and Wales, Slough *12591*
National Society (Church of England), London *12606*
National Society for Education in Art and Design, Corsham *12608*
National Strict Baptist Sunday School Association, Bishops Stortford *12612*
National Union of Teachers, London *12616*
Northern Ireland Women Teacher's Association, Belfast *12639*
Organisation of Teachers of Transport Studies, Southall *12660*
Philosophy of Education Society of Great Britain, Coleraine *12695*
Polish Educational Society, London *12708*
Professional Association of Teachers, Derby *12724*
Right to a Comprehensive Education, London *12757*
Royal College of Physicians of London, London *12788*
School Natural Science Society, Gillingham *12888*
Scottish Association of Geography Teachers, Hamilton *12898*
Scottish Association of Local Government to Educational Psychologists, Glasgow *12899*
Scottish Secondary Teachers' Association, Edinburgh *12927*
Socialist Educational Association, Stockport *12956*
Society for Education in Film and Television, London *12970*
Society for Promotion of Educational Reform through Teacher Training, London *12988*
Society for Research in the Psychology of Music and Music Education, Dagenham *12993*
Society for Research into Higher Education, London *12994*
Society for the Advancement of Games and Simulations in Education and Training, Loughborough *12999*
Society for the Health Education, London *13002*
Society for the Protection of Science and Learning, London *13011*
Society of Assistants Teaching in Preparatory Schools, Altrincham *13029*
Society of Indexers, London *13052*
Society of Public Teachers of Law, Buckingham *13062*
Society of Teachers in Business Education, Sheffield *13065*
Society of Teachers of Speech and Drama, Canterbury *13066*
Sound Learning Society, Rickmansworth *13076*
South Place Ethical Society, London *13082*
Ulster Teachers' Union, Belfast *13158*
Union of Educational Institutions, Birmingham *13160*
Union of Lancashire and Cheshire Institutes, Manchester *13161*
United Kingdom Reading Association, Widnes *13167*
United World Colleges, London *13173*
University Association for Contemporary European Studies, London *13174*
Welsh Federation of Head Teachers' Associations, Swansea *13200*
Workers' Educational Association, London *13215*
World Education Fellowship, London *13218*
World Federation for Medical Education, Edinburgh *13220*
World-wide Education Service of the Bell Educational Trust, London *13227*

Uruguay

Consejo Nacional de Educación, Centro Nacional de Información y Documentación, Montevideo *13256*

U.S.A.

Academy for Educational Development, Washington *13277*
Academy of Criminal Justice Sciences, Highland Heights *13286*
Academy of Security Educators and Trainees, Berryville *13309*
Accordion Teacher's Guild, Kansas City *13312*
Aerospace Education Foundation, Arlington *13337*
Agency for Instructional Technology, Bloomington *13349*
ALI-ABA Committee on Continuing Professional Education, Philadelphia *13363*
American Academy of Teachers of Singing, New York *13439*
American Association for Agricultural Education, Ithaca *13466*
American Association for Career Education, Hermosa Beach *13472*
American Association for Higher Education, Washington *13482*
American Association for Paralegal Education, Overland *13488*
American Association of Christian Schools, Fairfax *13518*
American Association of Colleges for Teacher Education, Washington *13519*
American Association of Community and Junior Colleges, Washington *13524*
American Association of Diabetes Educators, Chicago *13529*
American Association of Housing Educators, College Station *13542*
American Association of Philosophy Teachers, Oklahoma City *13563*
American Association of Physics Teachers, College Park *13567*
American Association of Professors of Yiddish, Flushing *13572*
American Association of School Personnel Administrators, Sacramento *13580*
American Association of Sex Educators, Counselors and Therapists, Chicago *13581*
American Association of Teachers of Arabic, Provo *13585*
American Association of Teachers of Esperanto, Santa Barbara *13586*
American Association of Teachers of French, Champaign *13587*
American Association of Teachers of German, Cherry Hill *13588*
American Association of Teachers of Italian, Columbus *13589*
American Association of Teachers of Spanish and Portuguese, Mississippi State *13590*
American Association of Teachers of Turkish, Princeton *13591*
American Association of University Women Educational Foundation, Washington *13596*
American Association of Women in Community and Junior Colleges, Middletown *13604*
American Board of Funeral Service Education, Cumberland Center *13634*
American Board of Master Educators, Nashville *13637*
American Board of Vocational Experts, Topeka *13666*
American Classical League, Oxford *13691*
American College Testing, Iowa City *13742*
American Correctional Association, Laurel *13756*
American Council for Construction Education, Monroe *13759*
American Council on Education, Washington *13766*
American Council on Rural Special Education, Bellingham *13768*
American Council on Schools and Colleges, Tampa *13769*
American Council on the Teaching of Foreign Languages, Yonkers *13770*
American Counseling Association, Alexandria *13771*
American Educational Research Association, Washington *13790*
American Education Association, Center Moriches *13792*
American Forum for Global Education, New York *13821*
American Friends of Cambridge University, Arlington *13827*
American Hungarian Educators' Association, Silver Spring *13855*
American Institute for Foreign Study, Greenwich *13869*
American Mathematical Association of Two Year Colleges, Flint *13928*
American Medical Association Education and Research Foundation, Chicago *13934*
American Montessori Society, New York *13947*
American Nepal Education Foundation, Tillamook *13957*
American Orthopsychiatric Association, New York *13974*
American Osteopathic College of Radiology, Milan *13986*
American Schools Association, Atlanta *14037*
American Society for Healthcare Education and Training of the American Hospital Association, Chicago *14064*
American Society for Training and Development, Alexandria *14093*
American String Teachers Association, Dallas *14183*
American Vocational Education Personnel Development Association, Stillwater *14207*
Armenian Educational Foundation, Glendale *14246*
Arts International Program of the Institute of International Education, New York *14251*
Associated Colleges of the Midwest, Chicago *14258*
Association for Canadian Studies in the United States, Washington *14276*
Association for Childhood Education International, Wheaton *14279*
Association for Counselor Education and Supervision, Alexandria *14287*
Association for Educational Communications and Technology, Washington *14290*
Association for Education of Teachers in Science, Auburn *14293*
Association for General and Liberal Studies, Columbus *14298*
Association for Humanistic Education, Laramie *14303*
Association for Humanistic Education and Development, Alexandria *14304*
Association for Institutional Research, Tallahassee *14306*
Association for International Practical Training, Columbia *14308*
Association for Religious and Value Issues in Counseling, Alexandria *14326*
Association for Specialists in Group Work, Alexandria *14335*
Association for Supervision and Curriculum Development, Alexandria *14336*
Association for Surgical Education, Salt Lake City *14337*
Association for the Advancement of Health Education, Reston *14342*
Association for the Advancement of International Education, New Wilmington *14343*
Association for Theatre in Higher Education, Evansville *14346*
Association for the Behavioral Sciences and Medical Education, McLean *14347*
Association for the Development of Human Potential, Spokane *14351*
Association for the Study of Higher Education, College Station *14358*
Association of Advanced Rabbinical and Talmudic Schools, New York *14376*

Association of American Colleges, Washington *14378*
Association of American Schools in South America, Miami *14386*
Association of Black Women in Higher Education, Albany *14404*
Association of Catholic Colleges and Universities, Washington *14408*
Association of Chiropractic Colleges, San Lorenzo *14411*
Association of Community College Trustees, Washington *14419*
Association of Community Tribal Schools, Sisseton *14420*
Association of Continuing Legal Education Administrators, Chicago *14423*
Association of Episcopal Colleges, New York *14435*
Association of Graduate Liberal Studies Programs, Chicago *14441*
Association of Graduate Schools in Association of American Universities, Washington *14442*
Association of Independent Colleges and Schools, Washington *14444*
Association of Jesuit Colleges and Universities, Washington *14450*
Association of Lutheran College Faculties, Minneapolis *14454*
Association of Lutheran Secondary Schools, Saint Lucas *14455*
Association of Medical Education and Research in Substance Abuse, Providence *14459*
Association of Mercy Colleges, Erie *14463*
Association of Minority Health Professions Schools, Washington *14467*
Association of Naval R.O.T.C. Colleges and Universities, Lafayette *14470*
Association of Orthodox Jewish Teachers, Brooklyn *14476*
Association of Overseas Educators, Indiana *14478*
Association of Presbyterian Colleges and Universities, Louisville *14484*
Association of School Business Officials International, Reston *14494*
Association of Seventh-Day Adventist Educators, Silver Spring *14501*
Association of Southern Baptist Colleges and Schools, Nashville *14503*
Association of Teacher Educators, Reston *14513*
Association of Teachers of Japanese, Middlebury *14514*
Association of Teachers of Technical Writing, Orlando *14517*
Association of the Health Occupations Teacher Educators, Raleigh *14519*
Association of Theological Schools, Pittsburgh *14521*
Association on Boarding Schools, Boston *14536*
Audubon Naturalist Society of the Central Atlantic States, Chevy Chase *14549*
AVKO Educational Research Foundation, Birch Run *14567*
Basque Educational Organization, San Francisco *14570*
Belgian American Educational Foundation, New Haven *14576*
Better Education thru Simplified Spelling, Detroit *14581*
Black Women's Educational Alliance, Philadelphia *14602*
B'Nai B'Rith International Commission on Continuing Jewish Education, Washington *14604*
Brandeis-Bardin Institute, Brandeis *14613*
British Schools and Universities Foundation, New York *14616*
Business-Higher Education Forum, Washington *14627*
Business Professionals of America, Columbus *14629*
Carnegie Foundation for the Advancement of Teaching, Princeton *14646*
Catholic Audio-Visual Educators Association, Pittsburgh *14651*
Catholic Biblical Association of America, Washington *14652*
Center for Computer-Assisted Legal Instruction, Minneapolis *14673*
Center for Death Education and Research, Minneapolis *14675*
Center for Early Adolescence, Carrboro *14677*
Center for Global Education, Minneapolis *14681*
Center for Sutton Movement Writing, Newport Beach *14705*
Center for the Study of Parent Involvement, Orinda *14707*
Central Organization for Jewish Education, Brooklyn *14715*
Ceramic Educational Council, Westerville *14718*
Character Education Institute, San Antonio *14720*
Child Care Employee Project, Oakland *14731*
Chinese Language Teachers Association, Kalamazoo *14740*
Christian College Coalition, Washington *14744*
Christian Schools International, Grand Rapids *14748*
Cities in Schools, Alexandria *14755*
Coalition for the Advancement in Jewish Education, New York *14765*
The College Board, New York *14771*
College Reading and Learning Association, Salem *14778*
College-University Resource Institute, Washington *14780*
Commission on Gay/Lesbian Issues in Social Work Education, Alexandria *14786*
Committee on Allied Health Education and Accreditation, Chicago *14799*
Committee on Continuing Education for School Personnel, Union *14801*
Committee on Institutional Cooperation, Champaign *14803*
Committee on the Role and Status of Women in Educational Research and Development, Hayward *14806*
Committee on the Teaching of Science, Kalamazoo *14807*
Community College Association for Instruction and Technology, Roswell *14809*
Comparative and International Education Society, Buffalo *14816*
Conduit, Iowa City *14820*
Conference on Christianity and Literature, Saint Petersburg *14836*
Conference on College Composition and Communication, Urbana *14837*
Conference on English Education, Urbana *14839*
Consortium for International Studies Education, Columbus *14847*
Consumer Education Research Center, Orange *14853*
Correctional Education Association, Laurel *14865*
Council for Advancement and Support of Education, Washington *14867*
Council for American Private Education, Washington *14869*
Council for Basic Education, Washington *14870*
Council for Educational Development and Research, Washington *14873*
Council for Indian Education, Billings *14876*
Council for Interinstitutional Leadership, Leawood *14877*
Council for Jewish Education, New York *14878*
Council for Religion in Independent Schools, Washington *14882*
Council for Research in Music Education, Urbana *14883*
Council of 1890 College Presidents, Langston *14888*
Council of Administrators of Special Education, Albuquerque *14889*
Council of Chief State School Officers, Washington *14891*
Council of Graduate Schools, Washington *14896*
Council of Mennonite Colleges, Hillsboro *14899*
Council of Scientific Society Presidents, Washington *14902*
Council of the Great City Schools, Washington *14905*
Council of Undergraduate Research, Asheville *14906*
Council on Accreditation of Nurse Anesthesia Educational Programs/Schools, Park Ridge *14907*
Council on Anthropology and Education, Washington *14908*
Council on Chiropractic Education, West Des Moines *14910*
Council on Electrolysis Education, Memphis *14917*
Council on Postsecondary Accreditation, Washington *14930*
Country Day School Headmasters Association of the U.S., Charlotte *14935*
Designs for Change, Chicago *14969*
Dietary Managers Association, Lombard *14973*
Direct Marketing Educational Foundation, New York *14974*
Distributive Education Clubs of America, Reston *14978*
Educational Center for Applied Ekistics, Atlanta *15017*
Educational Foundation for the Fashion Industries, New York *15019*
Educational Leadership Institute, Shorewood *15020*
Educational Planning, Memphis *15021*
Educational Planning Institute, New York *15022*
Educational Records Bureau, New York *15023*
Educational Research Analysts, Longview *15024*
Educational Research Service, Arlington *15025*
Educational Testing Service, Princeton *15026*
Education Commission of the States, Denver *15027*
Education Development Center, Newton *15028*
Education Writers Association, Washington *15030*
Evangelical Lutheran Education Association, North Hollywood *15066*
Exodus Trust, San Francisco *15068*
Federated Council of Beth Jacob Schools, New York *15078*
Federation for Unified Science Education, Columbus *15079*
Foundation for Accounting Education, New York *15119*
Foundation for Educational Futures, Charlotte *15123*
Foundation for Interior Design Education Research, Grand Rapids *15126*
Foundation for International Cooperation, Park Ridge *15127*
General Educational Development Institute, Seattle *15149*
German-American Information and Education Association, Burke *15156*
Global Education Associates, New York *15163*
Global Learning, Union *15164*
Great Lakes Colleges Association, Ann Arbor *15174*
Headmasters Association, Washington *15192*
Higher Education Consortium for Urban Affairs, Saint Paul *15206*
Higher Education Panel, Washington *15207*
Higher Education Resource Services, Wellesley *15208*
History of Education Society, Eugene *15221*
Home Economics Education Association, Gainesville *15226*
Horace Mann League of the U.S.A., Summit *15228*
Independent Educational Services, Princeton *15253*
Institute for Childhood Resources, San Francisco *15272*
Institute for Development of Educational Activities, Dayton *15275*
Institute for Educational Leadership, Washington *15278*
Institute for Fluitronics Education, Elm Grove *15280*
Institute for Public Relations Research and Education, Sarasota *15291*
Institute for Responsive Education, Boston *15295*
Institute of Electrology Educators, Bellingham *15318*
Institute of International Education, New York *15326*
Inter-American College Association, Pomona *15350*
Inter-American Council for Education, Science and Culture, Washington *15352*
Intercollegiate Broadcasting System, Vails Gate *15357*
International Association for Computer Information Systems, Eau Claire *15377*
International Association for Computing in Education, Eugene *15378*
International Association for Learning Laboratories, Philadelphia *15386*
International Association for the Study of Cooperation in Education, Santa Cruz *15395*
International Association of Counseling Services, Alexandria *15403*
International Association of Educational Peace Officers, Houston *15405*
International Association of Educators for World Peace, Huntsville *15406*
International Association of Jazz Educators, Manhattan *15409*
International Commission on Physics Education, Columbus *15441*
International Council on Education for Teaching, Arlington *15460*
International Education Research Foundation, Los Angeles *15468*
International Institute of Rural Reconstruction – U.S. Chapter, New York *15490*
International Organization for the Education of the Hearing Impaired, Washington *15508*
International Reading Association, Newark *15518*
International Schools Services, Princeton *15521*
International Society for Intercultural Education, Training and Research, Washington *15534*
International Tele-Education, Philadelphia *15562*
International Textile and Apparel Association, Monument *15564*
Jesuit Secondary Education Association, Washington *15601*
Jewish Education Service of North America, New York *15605*
Jewish Educators Assembly, New York *15606*
Jewish Teachers Association – Morim, New York *15610*
John Dewey Society, Indianapolis *15612*
Joint Review Committee on Educational Programs for the EMT-Paramedic, Euless *15623*
Joint Review Committee on Education in Diagnostic Medical Sonography, Chicago *15624*
Joint Review Committee on Education in Radiologic Technology, Chicago *15625*
Keyboard Teachers Association International, Westbury *15637*
Lincoln Institute for Research and Education, Washington *15673*
Luso-American Education Foundation, Oakland *15679*
Lutheran Educational Conference of North America, Washington *15681*
Lutheran Education Association, River Forest *15682*
Middle East Institute, Washington *15732*
Middle East Studies Association of North America, Tucson *15735*
Middle States Association of Colleges and Schools, Philadelphia *15736*
Moby Dick Academy, Ocean Park *15748*
The Monroe Institute, Faber *15753*
Museum Education Roundtable, Washington *15760*
Music Educators National Conference, Reston *15763*
Music Teachers National Association, Cincinnati *15765*
National Academy of Education, Stanford *15773*
National Alliance for Safe Schools, Bethesda *15783*
National Alliance of Black School Educators, Washington *15784*
National Art Education Association, Reston *15790*
National Association for Bilingual Education, Washington *15794*
National Association for Core Curriculum, Kent *15797*
National Association for Equal Opportunity in Higher Education, Washington *15798*
National Association for Humane and Environmental Education, East Haddam *15803*
National Association for Humanities Education, Saint Joseph *15804*
National Association for Legal Support of Alternative Schools, Santa Fe *15807*
National Association for Research in Science Teaching, Manhattan *15812*
National Association for the Advancement of Black Americans in Vocational Education, Detroit *15816*
National Association for the Education of Young Children, Washington *15818*
National Association for the Exchange of Industrial Resources, Galesburg *15819*
National Association for Women in Education, Washington *15823*
National Association for Year-Round Education, San Diego *15824*
National Association of Academic Advisors for Athletics, Charlottesville *15825*
National Association of Advisers for the Health Professions, Champaign *15827*
National Association of Biology Teachers, Reston *15831*
National Association of Blind Teachers, Washington *15833*
National Association of Business Education State Supervisors, Phoenix *15836*
National Association of Classroom Educators in Business Education, Boone *15838*
National Association of College Deans, Registrars and Admissions Officers, Albany *15841*
National Association of Colleges and Teachers of Agriculture, Urbana *15842*
National Association of Educational Buyers, Hauppauge *15852*
National Association of Elementary School Principals, Alexandria *15853*
National Association of Episcopal Schools, New York *15857*
National Association of Extension 4-H Agents, Athens *15858*
National Association of Geology Teachers, Bellingham *15861*
National Association of Hebrew Day School Administrators, Brooklyn *15864*
National Association of Hebrew Day School PTA'S, New York *15865*
National Association of Independent Schools, Boston *15868*
National Association of Principals of Schools for Girls, Hendersonville *15879*
National Association of Private Schools for Exceptional Children, Washington *15880*
National Association of Professional Educators, Washington *15881*
National Association of School Safety and Law Enforcement Officers, Upper Marlboro *15885*
National Association of Schools and Colleges of the United Methodist Church, Nashville *15886*
National Association of Schools of Art and Design, Reston *15887*
National Association of Schools of Music, Reston *15888*
National Association of Schools of Theatre, Reston *15890*
National Association of Secondary School Principals, Reston *15892*
National Association of State Boards of Education, Alexandria *15894*
National Association of State Directors of Special Education, Alexandria *15896*
National Association of State Directors of Teacher Education and Certification, Seattle *15897*
National Association of State Educational Media Professionals, Nashville *15899*
National Association of State Supervisors and Directors of Secondary Education, Pierre *15904*
National Association of State Supervisors of Vocational Home Economics, Austin *15905*
National Association of State Universities and Land-Grant Colleges, Washington *15907*
National Association of Teacher Educators for Business Education, Columbia *15910*
National Association of Teachers of Singing, Jacksonville *15911*
National Association of Test Directors, Englewood *15912*
National Association of University Women, Tallahassee *15913*
National Association of Vocational Education Special Needs Personnel, McKeesport *15914*
National Black MBA Association, Chicago *15919*
National Catholic Educational Association, Washington *15935*
National Center for Disability Services, Albertson *15942*
National Center for Fair and Open Testing, Cambridge *15943*
National Coalition for Public Education and Religious Liberty, Washington *15954*
National Coalition for Women and Girls in Education, Washington *15955*
National Coalition of Alternative Community Schools, Summertown *15956*
National Collegiate Conference Association, Albany *15959*
National Commission for Cooperative Education, Boston *15960*
National Committee for the Furtherance of Jewish Education, Brooklyn *15964*
National Community Education Association, Alexandria *15966*
National Conference of Yeshiva Principals, New York *15973*
National Conference on Research in English, Amherst *15975*
National Consortium of Arts and Letters for Historically Black Colleges and Universities, Washington *15982*
National Council for Accreditation of Teacher Education, Washington *15984*
National Council for Black Studies, Columbus *15985*
National Council for Geographic Education, Indiana *15988*
National Council for Research and Planning, Radford *15990*
National Council for Torah Education, New York *15994*
National Council of BIA Educators, Albuquerque *15997*
National Council of State Directors of Community and Junior Colleges, Lansing *16004*
National Council of State Education Associations, Washington *16005*
National Council of State Supervisors of Foreign Languages, Little Rock *16007*
National Council of State Supervisors of Music, Columbus *16008*
National Council of Teachers of English, Urbana *16010*
National Council of Teachers of Mathematics, Reston *16011*
National Council of Urban Education Associations, Washington *16014*
National Council on Economic Education, New York *16018*
National Council on Measurement in Education, Washington *16023*
National Council on Religion and Public Education, Ames *16026*
National Earth Science Teachers Association, Fargo *16037*
National Education Association, Washington *16039*
National Federation of Abstracting and Information Services, Philadelphia *16045*
National Federation of Modern Language Teachers Associations, Omaha *16047*
National Foundation for Children's Hearing Education and Research, Yonkers *16058*
National Fund for Medical Education, Boston *16061*

National Guild of Community Schools of the Arts, Englewood 16067
National Guild of Piano Teachers, Austin 16068
National Home Study Council, Washington 16076
National Humane Education Society, Leesburg 16079
National Identification Program for the Advancement of Women in Higher Education Administration, Washington 16082
National Indian Education Association, Washington 16084
National Information Center for Educational Media, Albuquerque 16086
National Medical Fellowships, New York 16108
National Middle School Association, Columbus 16110
National Organization for Continuing Education of Roman Catholic Clergy, Chicago 16117
National Organization on Legal Problems of Education, Topeka 16120
National Reading Conference, Chicago 16134
National Resident Matching Program, Washington 16140
National Rural and Small Schools Consortium, Bellingham 16142
National Rural Education Association, Fort Collins 16144
National School Boards Association, Alexandria 16147
National School Public Relations Association, Arlington 16148
National Schools Committee for Economic Education, Cos Cob 16149
National Science Supervisors Association, Glastonbury 16150
National Science Teachers Association, Washington 16151
National Society for Experiential Education, Raleigh 16155
National Society for Performance and Instruction, Washington 16156
National Society for the Study of Education, Chicago 16158
National Study of School Evaluation, Bloomington 16167
National Vocational Technical Educational Foundation, Washington 16178
National Writing Project, Berkeley 16184
Nation's Report Card – National Assembly of Educational Progress, Princeton 16185
Near East College Association, New York 16190
New England Association of Schools and Colleges, Winchester 16196
North American Association for Environmental Education, Washington 16211
North American Association of Professors of Christian Education, Elgin 16212
North American Association of Summer Sessions, Saint Louis 16213
North American Simulation and Gaming Association, Indianapolis 16221
North Central Association of Colleges and Schools, Tempe 16228
North Central Conference on Summer Schools, River Falls 16229
Northeast Conference on the Teaching of Foreign Languages, South Burlington 16230
Northwest Association of Schools and Colleges, Boise 16232
Ocean Education Project, Ararat 16240
Oceanic Educational Foundation, Falls Church 16241
Office for Advancement of Public Black Colleges of the National Association of State Universities and Land Grant Colleges, Washington 16245
Office of International Education of the American Council on Education, Washington 16246
Organizational Behavior Teaching Society, Norman 16261
Organization of American Kodaly Educators, Thibodaux 16266
Organization of Professional Acting Coaches and Teachers, Studio City 16267
Orthodontic Education and Research Foundation, Saint Louis 16268
Outdoor Education Association, Denville 16275
Overseas Education Association, Washington 16278
Parkinson's Educational Program – U.S.A., Newport Beach 16297
People United for Rural Education, Alden 16302
Play Schools Association, New York 16327
Probe Ministries International, Richardson 16355
Professors of Curriculum, Orlando 16359
Project on Equal Education Rights, New York 16363
Public Leadership Education Network, Washington 16376
Radix Teachers Association, Dallas 16391
Regional Education Board of the Christian Brothers, Romeoville 16401
Research Association of Minority Professors, Prairie View 16413
School Management Study Group, Salt Lake City 16442
School Science and Mathematics Association, Bowling Green 16443
Science Service, Washington 16445
Secondary School Admission Test Board, Princeton 16456
Sex Information and Education Council of the U.S., New York 16464
Smithsonian Institution, Washington 16481
Social Science Education Consortium, Boulder 16484
Société des Professeurs Français et Francophones en Amérique, New York 16488
Society for Applied Learning Technology, Warrenton 16495
Society for Educational Reconstruction, New York 16518
Society for French American Cultural Services and Educational Aid, New York 16524
Society for History Education, Long Beach 16531
Society for Music Teacher Education, Reston 16553
Society for Research in Child Development, Chicago 16576
Society for South India Studies, Berkeley 16581
Society for the Advancement of Education, Valley Stream 16588
Society for Values on Higher Education, Washington 16625
Society of American Law Teachers, New York 16636
Society of Ethnic and Special Studies, Edwardsville 16658
Society of Professors of Education, Knoxville 16696
Society of Teachers in Education of Professional Photography, West Carrollton 16705
Solomon Schecher Day School Association, New York 16720
Southern Association of Colleges and Schools, Decatur 16723
Student National Medical Association, Washington 16749
Swedish Women's Educational Association International, La Jolla 16760
Teachers of English to Speakers of Other Languages, Alexandria 16766
Theatre in Education, Old Lyme 16774
Theodor Herzl Institute, New York 16777
Tripoli Rocketry Association, Kenner 16792
Union Institute, Cincinnati 16804
United Board for Christian Higher Education in Asia, New York 16807
United Cerebral Palsy Research and Educational Foundation, New York 16810
United States Institute for Theatre Technology, New York 16825
United Synagogue Commission on Jewish Education, New York 16836
University and College Labor Education Association, New Brunswick 16842
University Consortium for Instructional Development and Technology, Athens 16844
University Council for Educational Administration, University Park 16846
Vocational Instructional Materials Section, Columbia 16867
Western College Association, Oakland 16882
Wilderness Education Association, Saranac Lake 16894
Women and Mathematics Education, South Hadley 16908
Women Educators, Toledo 16909
Women's College Coalition, Washington 16918
World Education Fellowship, United States Section, New Rochelle 16925

Venezuela

Caribbean Institute for Social Formation, Caracas 16980
Consejo Venezolano del Niño, Caracas 17003

Virgin Islands (U.S.)

Consortium of Caribbean Universities for Resource Management, Saint Thomas 17075

Yugoslavia

Pedagogical Society of Yugoslavia, Beograd 17106
Savez Društava za Strane Jezike i Knjizevnosti Jugoslavije, Beograd 17114
Savez Pedagoških Društava Jugoslavije, Beograd 17119
Savez Pedagoških Društava Srbije, Beograd 17120
Sindikat Radnicka Drustvenih Delatnosti Jugoslavije, Beograd 17124
Udruženje Univerzitetskih Nastavnika i Van-Univerzitetskih Naučnik Radnika, Beograd 17131
Zajednica Universitéta Jugoslavije, Beograd 17138

Zaire

African Bureau of Educational Sciences, Kinshasa 17139
Association des Institutions d'Enseignement Théologique en Afrique Occidentale, Kinshasa 17140
Centrale des Enseignants Zairois, Kinshasa 17143
Conférence des Recteurs des Universités Francophones d'Afrique, Kinshasa 17145

Zambia

Association of Medical Schools in Africa, Lusaka 17151

Zimbabwe

Zimbabwe Association for Science Education, Harare 17185

Education of the Handicapped

Austria

Österreichische Gesellschaft für Sprachheilpädagogik, Wien 00833

Belgium

Groupe Belge d'Etude de l'Arriération Mentale, Bruxelles 01368

Canada

Atlantic Conference on Learning Disabilities, Dartmouth 01958
Canadian National Institute for the Blind, Toronto 02206
Learning Disabilities Association of Canada, Ottawa 02586
New Brunswick Speech and Hearing Association, Fredricton 02695
Ontario Deafness Research Foundation, Toronto 02772
World Federation for Mental Health, Vancouver 02998

France

Société d'Etudes et de Soins pour les Enfants Paralysés, Paris 04736

Germany

Berufsverband Deutscher Hörgeschädigtenpädagogen, Hamburg 05202
Bundesverband Hilfe für das autistische Kind e.V., Hamburg 05253
Bundesverband Legasthenie e.V., Hannover 05256
Bundesvereinigung Lebenshilfe für geistig Behinderte e.V., Marburg 05259
Deutsche Gesellschaft zur Förderung der Gehörlosen und Schwerhörigen e.V., München 05448
Gesellschaft für wissenschaftliche Gesprächspsychotherapie e.V., Köln 05872
Lernen Fördern – Bundesverband zur Förderung Lernbehinderter e.V., Köln 06103

Italy

Associazione La Nostra Famiglia, Ponte Lambro 07354
Associazione Nazionale Famiglie di Fanciulli e Adulti Subnormali, Roma 07371

United Kingdom

Association for Independent Disabled Self-Sufficiency, Bath 11265
Association for the Education and Welfare of the Visually Handicapped, Droitwich Spa 11287
Association of Blind and Partially-Sights Teachers and Students, London 11306
Association of Workers for Maladjusted Children, Kettering 11410
British Association of Teachers of the Deaf, Cheadle Hulme 11526
British Institute for Brain Injured Children, Bridgwater 11579
British Institute of Learning Disabilities, Kidderminster 11584
British Society for the Study of Mental Subnormality, Cranage 11677
Disabled Living Foundation, London 11894
Dyslexia Institute, Staines 11914
Hyper Active Children's Support Group, Chichester 12190
National Association of Teachers of the Mentally Handicapped, Aylesbury 12572

U.S.A.

Alexander Graham Bell Association for the Deaf, Washington 13361
American Association on Mental Retardation, Washington 13608
Association for Education and Rehabilitation of the Blind and Visually Impaired, Alexandria 14291
Conference of Educational Administrators Serving the Deaf, Tucson 14830
Convention of American Instructors of the Deaf, Rochester 14857
Council for Children with Behavioral Disorders, Reston 14872
Council on Education of the Deaf, Washington 14916
Division for Early Childhood, Reston 14979
Division on Mental Retardation of the Council for Exceptional Children, Athens 14981
Foundation for Exceptional Children, Reston 15124
Foundation for Science and the Handicapped, Morgantown 15132
Model Secondary School for the Deaf, Washington 15749
National Captioning Institute, Falls Church 15932
National Hearing Aid Society, Livonia 16073

Electrical Engineering

Argentina

Asociación Electrotécnica Argentina, Buenos Aires 00079, 00080

Australia

The Australian Institute of Refrigeration, Air Conditioning and Heating, Parkville 00314
Telecommunication Society of Australia, Melbourne 00509

Austria

Österreichischer Verband für Elektrotechnik, Wien 00904

Belgium

Association Belge de l'Eclairage, Bruxelles 01108
Comité Européen de Normalisation Electrotechnique, Bruxelles 01226
Société Royale Belge des Electriciens, Bruxelles 01478
Union Radio-Scientifique Internationale, Bruxelles 01500

Bulgaria

Union of Electronics, Electrical Engineering and Communications, Sofia 01791
Union of Energetics, Sofia 01792

Canada

Canadian Electrical Association, Montréal 02123
Canadian Society for Electrical and Computer Engineering, Ottawa 02256
Institute of Electrical and Electronics Engineers, Thornhill 02529

China, People's Republic

Beijing Academy of Hydroelectrical Engineering Design, Beijing 03119
Changchun Academy of Electrical Power Engineering Design, Chilin 03123
Chinese Electrotechnical Society, Beijing 03168
Shanghai Academy of Hydroelectrical Engineering Design, Shanghai 03215
Sian Academy of Electric Power Design, Shanhsi 03220

Denmark

Dansk Køleforening, Lyngby 03638
Elektroteknisk Forening, Hellerup 03717
Lysteknisk Selskab, Stenløse 03769

France

Association Française de l'Eclairage, Paris 04218
Comité Français de l'Electricité, Paris-La Défense 04404
Conférence Internationale des Grands Réseaux Electriques à Haute Tension, Paris 04445
Union Internationale d'Electrothermie, Paris-La Défense 04983
Union Technique de l'Electricité, Paris-La Défense 04998

Germany

Deutsche Elektrotechnische Kommission im DIN und VDE, Frankfurt 05301
Deutsche Lichttechnische Gesellschaft e.V., Berlin 05464
Dokumentationsring Elektrotechnik, Erlangen 05633
Forschungsgemeinschaft für Hochspannungs- und Hochstromtechnik e.V., Mannheim 05709
Forschungsvereinigung Elektrotechnik beim ZVEI e.V., Frankfurt 05746
Gesellschaft für Elektrische Hochleistungsprüfungen, Frankfurt 05812
Gesellschaft für Informationsvermittlung und Technologieberatung, München 05831
Informationstechnische Gesellschaft im VDE, Frankfurt 05942
Normenausschuss Heiz-, Koch- und Wärmegeräte im DIN Deutsches Institut für Normung e.V., Frankfurt 06176
Normenausschuss Lichttechnik im DIN Deutsches Institut für Normung e.V., Berlin 06189
Verband Deutscher Elektrotechniker e.V., Frankfurt 06351

Hungary

Hiradástechnikai Tudományos Egyesület, Budapest 06607
Magyar Elektrotechnikai Egyesület, Budapest 06624

Ireland

Electro-Technical Council of Ireland, Dublin 06943
Institution of Electrical Engineers (Irish Branch), Dun Laoghaire 06967

Italy

Comitato Elettrotecnico Italiano, Milano 07613
Istituto Elettrotecnico Nazionale Galileo Ferraris, Torino 07728

Japan

Denki Tsushin Kyokai, Tokyo 08120
Handotai Kenkyu Shinkokai Handotai Kenkyusho Toshoshitsu, Sendai 08129
Nihon Reito Kyokai, Tokyo 08248
Shomei Gakkai, Tokyo 08424

Pakistan

Institution of Electrical Engineers, Lahore 09368

Peru

Asociación Electrotécnica Peruana, Lima 09450

Russia

Power and Electrical Power Engineering Society, St. Petersburg 09944
Radio Engineering, Electronics and Telecommunications Society, Moskva 09948

South Africa

South African Council for Automation and Computation, Pretoria 10187
The South African National Committee on Illumination, Pretoria 10207

Spain

Asociación Electrotécnica Española, Barcelona 10253

Sweden

Svenska Föreningen för Ljuskultur, Stockholm 10543

Switzerland

Commission Electrotechnique Internationale, Genève 10649
Schweizerischer Elektrotechnischer Verein, Zürich 10892

United Kingdom

Battery Vehicle Society, Peacehaven 11428
Electric Vehicle Development Group, London 11952
ERA Technology, Leatherhead 11971
Institution of Electrical Engineers, London 12296
Institution of Electronics and Electrical Incorporated Engineers, London 12298
Institution of Lighting Engineers, Rugby 12306
Institution of Post Office Electrical Engineers, London 12315
International Special Committee on Radio Interference, London 12395
Radio Society of Great Britain, Potters Bar 12737
The Royal Television Society, London 12868
Visible Record and Minicomputer Society, Croydon 13190
Wireless Preservation Society, Shanklin 13211

U.S.A.
Aerospace Electrical Society, Anaheim 13338
Antenna Measurement Techniques Association, Atlanta 14227
Association of Federal Communications Consulting Engineers, Washington 14436
Edison Electric Institute, Washington 15015
Electricity Consumers Resource Council, Washington 15033
Electric Vehicle Association, Cupertino 15034
IEEE Control Systems Society, New York 15245
IEEE Ultrasonics, Ferroelectrical and Frequency Control Society, Orlando 15247
Illuminating Engineering Society of North America, New York 15248
The Institute of Electrical and Electronics Engineers, New York 15317
International Kirlian Research Association, Brooklyn 15495
North American Electric Reliability Council, Princeton 16217
North American Society of Pacing and Electrophysiology, Newton Upper Falls 16225
Radio Technical Commission for Aeronautics, Washington 16389
Signal Processing Society, New York 16474
Society of Broadcast Engineers, Indianapolis 16644
United States National Committee of the International Union of Radio Science, Washington 16829
Women in Broadcast Technology, Berkeley 16911

Venezuela
Asociacíon Venezolana de Ingeniería Eléctrica y Mecánica, Caracas 16975

Electrochemistry

India
The Electrochemical Society of India, Bangalore 06757

Japan
Nihon Kandenchi Kogyokai, Tokyo 08219

United Kingdom
Scottish Electrophysiological Society, Saint Andrews 12905

U.S.A.
Electrochemical Society, Pennington 15035

Electronic Engineering

Australia
Institution of Radio and Electronics Engineers – Australia, Sydney 00422

Austria
Forschungsgesellschaft für Psycho-Elektronik und Kybernetik, Wien 00596

Belgium
European Power Electronics and Drives Association, Bruxelles 01351
Société Belge de Microscopie Electronique, Liège 01435

Canada
Institute of Electrical and Electronics Engineers, Thornhill 02529

China, People's Republic
Chinese Society of Electronics, Beijing 03187

France
Organisation Européenne pour l'Equipement de l'Aviation Civile, Paris 04613
Société Française de Microscopie Electronique, Ivry-sur-Seine 04818

Germany
CENELEC Electronic Components Committee, Frankfurt 05274
Deutsche Gesellschaft für Chronometrie e.V., Stuttgart 05337
Electronic Components Quality Assurance Committee, Frankfurt 05638
Elektrotechnischer Verein Berlin e.V., Berlin 05639
Fraunhofer-Gesellschaft zur Förderung der angewandten Forschung e.V., München 05759
Gesellschaft für Informationsvermittlung und Technologieberatung, München 05831
GES-Gesellschaft für elektronische Systemforschung, Allensbach 05891
VDE/VDI-Gesellschaft Mikroelektronik, Frankfurt 06318

Italy
Automazione Energia Informazione, Milano 07411
ITALSIEL, Roma 07758
Società Italiana per la Robotica Industriale, Milano 07966
Società Nazionale di Informatica delle Camere di Commercio per la Gestione dei Centri Elettronici Reteconnessi Valutazione Elaborazione Dati, Roma 07991

Japan
Denshi Joho Tsushin Gakkai, Tokyo 08121
Nihon Denshi Kogyo Shinko Kyokai, Tokyo 08192

Mexico
Centro de Investigación y de Estudios Avanzados del Instituto Politécnico Nacional, México 08718

New Zealand
New Zealand Computer Society, Wellington 09140
New Zealand Electronics Institute, Auckland 09145
New Zealand Society for Electron Microscopy, Auckland 09174

Russia
Commission for Computer Technology, Novosibirsk 09815
Radio Engineering, Electronics and Telecommunications Society, Moskva 09948

Switzerland
Fachgruppe für Elektronik des STV, Lausanne 10693
Schweizerische Gesellschaft für Mikrotechnik, Zürich 10834

United Kingdom
European Airlines Electronic Committee, London 11985
Institution of Electrical Engineers, London 12296
Institution of Electronics, Rochdale 12297
Institution of Electronics and Electrical Incorporated Engineers, London 12298
International Society for Microelectronics, Harbertonford 12389
Visible Record and Minicomputer Society, Croydon 13190

U.S.A.
Aircraft Electronics Association, Independence 13356
American Electronics Association, Santa Clara 13795
American Microscopical Society, Chestertown 13946
Association for Intelligent Systems Technology, Syracuse 14307
Association of Old Crows, Alexandria 14473
Bioelectromagnetics Society, Frederick 14586
The Institute of Electrical and Electronics Engineers, New York 15317
Insulated Cable Engineers Association, South Yarmouth 15345
International Association of Knowledge Engineers, Gaithersburg 15410
International Electronics Packaging Society, Wheaton 15469
International Society for Hybrid Microelectronics, Reston 15533
Microscopy Society of America, Woods Hole 15729
National Engineering Consortium, Chicago 16042
Robotics International of the Society of Manufacturing Engineers, Dearborn 16423

Vietnam
Vietnamese Association of Radio-Electronics, Hanoi 17063

Endocrinology

Argentina
Servicio de Endocrinología y Metabolismo, Buenos Aires 00152
Sociedad Argentina de Endocrinología y Metabolismo, Buenos Aires 00165

Australia
Endocrine Society of Australia, Camperdown 00389

Bulgaria
Society of Bulgarian Anatomists, Histologists and Embryologists, Sofia 01777

Canada
Canadian Society for Endocrinology and Metabolism, Montréal 02257
Thyroid Foundation of Canada, Kingston 02963
Turner's Syndrome Society, Downsview 02971

Denmark
Committee of the Acta-Endocrinologica Countries, Glostrup 03574

France
Société Française d'Endocrinologie, Paris 04822
Syndicat National des Médecins Spécialistes de l'Endocrinologie et de la Nutrition, Paris 04960

Germany
Deutsche Gesellschaft für Endokrinologie, Berlin 05344

Hong Kong
Asia and Oceania Society for Comparative Endocrinology, Hong Kong 06587

Indonesia
ASEAN Federation of Endocrine Societies, Jakarta 06875

Israel
Coordination Office of Paediatric Endocrine Societies, Petah Tikva 07051

Italy
Società Italiana di Endocrinologia, Roma 07868

Mexico
Sociedad Mexicana de Nutrición y Endocrinología, México 08769

Poland
Polskie Towarzystwo Endokrynologiczne, Warszawa 09563

Romania
Societatea de Endocrinologie, Bucuresti 09760

Russia
National Scientific Medical Society of Endocrinologists, Moskva 09913

South Africa
Society for Endocrinology, Metabolism and Diabetes of Southern Africa, Sandton 10173

United Kingdom
Europeam Federation of Endocrine Societies, London 11982
European Society for Comparative Endocrinology, Reading 12039
Society for Endocrinology, Bristol 12971

U.S.A.
American Thyroid Association, Washington 14195
Endocrine Society, Bethesda 15041
Women's Caucus of the Endocrine Society, Baltimore 16916

Energy

Australia
Australian and New Zealand Solar Energy Society, Caulfield East 00260
Australian Institute of Energy, Wahroonga 00300

Austria
International Atomic Energy Agency, Wien 00663
Österreichische Gesellschaft zur Föderung von Umweltschutz und Energieforschung, Wien 00856

Belgium
Benelux Association of Energy Economists, Bruxelles 01166
European Carbon Black Centre, Bruxelles 01308

Bolivia
Asociación de Ingenieros y Geólogos de Yacimientos Petrolíferos Fiscales Bolivianos, La Paz 01533
Instituto Boliviano del Petróleo, La Paz 01546

Bulgaria
Union of Electronics, Electrical Engineering and Communications, Sofia 01791

Canada
Biomass Energy Institute, Winnipeg 01966
Canadian Energy Research Institute, Calgary 02124
Canadian Gas Association, Don Mills 02143
Canadian National Committee of World Energy Conference, Ottawa 02205
Canadian Wind Engineering Association, London 02318
Energy Probe Research Foundation, Toronto 02434

Chile
Colegio de Architectos de Chile, Santiago 03029

China, People's Republic
Chinese Society of High Energy Physics, Beijing 03192

Denmark
Dansk Ingeniørforening, København 03635

France
Agence Internationale de l'Energie, Paris 04126
Association Technique de l'Industrie du Gaz en France, Paris 04353
Atlantic Gas Research Rxchange, La Plaine-Saint-Denis 04360
Institut Français de l'Energie, Paris 04568
Société Française d'Energie Nucléaire, Paris 04824

Germany
Arbeitsgemeinschaft für sparsamen und umweltfreundlichen Energieverbrauch e.V., Hamburg 05111
Deutsche Gesellschaft für Sonnenenergie e.V., München 05420
Deutscher Verein des Gas- und Wasserfaches e.V., Eschborn 05556
Deutsches Nationales Komitee des Weltenergierats, Düsseldorf 05595
Energietechnische Gesellschaft im VDE, Frankfurt 05640
Energie- und Umweltzentrum am Deister e.V., Springe-Eldagsen 05641
Fraunhofer-Gesellschaft zur Förderung der angewandten Forschung e.V., München 05759
Gesellschaft für Dezentralisierte Energiewirtschaft e.V., Ludwigsburg 05810
Gesellschaft für Informationsvermittlung und Technologieberatung, München 05831
Gesellschaft für praktische Energiekunde e.V., München 05851
Gesellschaft für Rationelle Energieverwendung e.V., Berlin 05856
Institut für Energie- und Umweltforschung Heidelberg e.V., Heidelberg 05958
Technische Vereinigung der Großkraftwerksbetreiber e.V., Essen 06312
VDI-Gesellschaft Energietechnik, Düsseldorf 06321
VGB-Forschungsstiftung, Essen 06432

Hungary
Energiagazdálkodási Tudományos Egyesület, Budapest 06600

Italy
Centro per gli Studi e le Applicazioni delle Risorse Energetiche, Rovigo 07519
International Centre for Advanced Technical and Vocational Training, Torino 07709
International Juridical Organization for Environment and Development, Roma 07714
International Solar Energy Society, Sezione Italiana, Napoli 07722
Tecneco, Fano 08032

Jamaica
Caribbean Energy Information System, Kingston 08072

Japan
Nihon Gas Kyokai Chosabu Chosaka, Tokyo 08201
Nippon Denki Kyokai Chosabu Chosaka, Tokyo 08293
Nippon Netsushori Gijutsu Kyokai, Tokyo 08347

Mexico
Centro de Investigación y de Estudios Avanzados del Instituto Politécnico Nacional, México 08718

New Zealand
Institute of Energy (New Zealand Section), Wellington 09128

Russia
Consultative Working Group for the Preparation of New Questions of Longterm Prospects of Development of Energy, Moskva 09854
Permanent Commission of Gosplan, Moskva 09941
Petroleum and Gas Society, Moskva 09942
Power and Electrical Power Engineering Society, St. Petersburg 09944

Sweden
Statens Råd för Byggnadsforskning, Stockholm 10524

Switzerland
Schweizerische Energie-Stiftung, Zürich 10803
Schweizerische Lichttechnische Gesellschaft, Bern 10868
Schweizerischer Verein des Gas- und Wasserfaches, Zürich 10919

Thailand
ASEAN Sub Committee on Non-Conventional Energy Research, Bangkok 11071

United Kingdom
British Wind Energy Association, Milton Keynes 11708
Engineering Equipment and Materials Users' Association, London 11959
European Association for the Conservation of Energy, London 11992
Institute of Energy, London 12235
Institution of Gas Engineers, London 12304

U.S.A.
Alliance to Save Energy, Washington 13366
Alternative Energy Resources Organization, Helena 13368
Alternative Sources of Energy, Milaca 13369
American Association of Blacks in Energy, Washington 13510
American Coucil for an Energy Efficient Economy, Washington 13758
American Petroleum Institute, Washington 13995
Americans for Energy Independence, Washington 14040
American Society of Gas Engineers, Independence 14124
American Solar Energy Society, Boulder 14177
American Wind Energy Association, Washington 14211
Association of Energy Engineers, Atlanta 14431
Association of Professional Energy Managers, Sacramento 14486
Biomass Energy Research Association, Washington 14592
Calorimetry Conference, Provo 14633
Center for Energy Policy and Research, Old Westbury 14679
Consumer Energy Council of America Research Foundation, Washington 14854
Critical Mass Energy Project of Public Citizen, Washington 14941
Energy Research Institute, Naples 15043
Fluid Power Society, Milwaukee 15102
Gas Research Institute, Chicago 15147
Institute of Gas Technology, Chicago 15322
International Association for Energy Economics, Cleveland 15381
International Association for Hydrogen Energy, Coral Gables 15385
International Microwave Power Institute, Clifton 15502
National Association of Power Engineers, Chicopee 15878
National Center for Appropriate Technology, Butte 15938
National Conference on Fluid Power, Milwaukee 15974
National Energy Management Institute, Alexandria 16040
National Energy Resources Organization, Washington 16041
National Hydropower Association, Washington 16081

National Old Timers' Association of the Energy Industry, Mineola 16113
The Network, Cleveland 16192
North American Thermal Analysis Society, Buffalo 16226
Oak Ridge Associated Universities, Oak Ridge 16239
Scientists and Engineers for Secure Energy, New York 16447
Society of Professional Well Log Analysts, Houston 16694
Solartherm, Silver Spring 16718
United States Energy Association, Washington 16822

Venezuela

Center for OPEC Studies, Caracas 16981

Engineering

Algeria

Arab Scientific and Technical Information Network, Algiers 00008

Argentina

Asociación Argentina del Frío, Buenos Aires 00068
Centro Argentino de Información Científica y Tecnológica, Buenos Aires 00099
Centro Argentino de Ingenieros, Buenos Aires 00100
Consejo Nacional de Investigaciones Científicas y Técnicas, Buenos Aires 00130
Instituto Nacional de Tecnologia Industrial, Buenos Aires 00145

Australia

Agricultural Engineering Society, Australia, Parkville 00213
Asian Pacific Confederation of Chemical Engineering, Barton 00224
Australian Academy of Technological Sciences and Engineering, Parkville 00251
Australian Federation for Medical and Biological Engineering, Parkville 00291
Australian Geomechanics Society, Barton 00293
Australian Institute of Food Science and Technology, Pymble 00301
Commonwealth Scientific and Industrial Research Organisation, Dickson 00377
Council of Australian Food Technology Associations, North Sydney 00383
Federation of Australian Scientific and Technological Societies, Canberra 00394
Food Technology Association of New South Wales, Sydney 00398
Food Technology Association of Western Australia, Perth 00401
Horological Guild of Australasia, Sydney 00409
Institute of Photographic Technology, Melbourne 00417
Institution of Surveyors – Australia, Canberra 00420
Institution of Engineers – Australia, Barton 00421
Royal Melbourne Institute of Technology, Melbourne 00472
Society of Leather Technologists and Chemists, Botany 00494
Sydney University Chemical Engineering Association, Sydney 00500
University of New South Wales Chemical Engineering Association, Kensington 00514

Austria

Commission Internationale de l'Eclairage, Wien 00578
Internationale Gesellschaft für Ingenieurpädagogik, Klagenfurt 00670
International Federation of Automatic Control, Laxenburg 00686
International Institute for Applied Systems Analysis, Laxenburg 00687
Mikrographische Gesellschaft, Wien 00707
Österreichische Gesellschaft für Schweisstechnik, Wien 00829
Österreichische Gesellschaft für Vakuumtechnik, Wien 00841
Österreichische Gesellschaft für Warenkunde und Technologie, Wien 00844
Österreichische Gesellschaft und Institut für Umweltschutz, Umwelttechnologie und Umweltwissenschaften, Wien 00852
Österreichische Tribologische Gesellschaft, Wien 00923
Österreichische Vereinigung der Zellstoff- und Papierchemiker und -techniker, Wien 00924
Verein Forschung für das graphische Gewerbe, Wien 00980
Verein Österreichischer Lebensmittel- und Biotechnologen, Wien 01000
Verein österreichischer Ledertechniker, Wien 01001

Azerbaijan

Commission for the Study of Productive Forces and Natural Resources, Baku 01041
Council on Exploitation of Sciemtific Equipment, Baku 01046
Science Production Association Biotech, Baku 01047

Bahrain

Bahrain Society of Engineers, Manama 01059

Bangladesh

Bangladesh Council of Scientific and Industrial Research, Dhaka 01067

Barbados

Caribbean Appropriate Technology Centre, Bridgetown 01076

Belgium

ASM International-European Council, Bruxelles 01104
Association Européenne Camac, Louvain-la-Neuve 01121
Association for European Training for Employees in Technology, Bruxelles 01132
Biotechnology Research for Innovation, Development and Growth in Europe, Bruxelles 01168
Brite-EuRam II, Bruxelles 01172
Bureau International Technique de l'ABS, Bruxelles 01175
Bureau International Technique des Polyesters Insaturés, Bruxelles 01176
Bureau International Technique du Spathfluor, Bruxelles 01177
Centre International d'Etude de Tantale et de Niobium, Bruxelles 01197
Collège Européen des Technologies, Arlon 01215
Eureka Organization, Bruxelles 01262
European Association for the Study of Safety Problems in the Production and Use of Propellant Powders, Bruxelles 01284
European Committee for EC Agricultural Engineers, Bruxelles 01319
European Consulting Engineering Network, Bruxelles 01329
European Cooperation in the Field of Scientific and Technical Research, Bruxelles 01333
Institut Belge de la Soudure, Bruxelles 01378
Institut Belge de Régulation et d'Automatisme, Bruxelles 01380
Institut Européen des Armes de Chasse et de Sport, Liège 01384
Koninklijke Academie voor Wetenschappen, Letteren en Schone Kunsten van België, Bruxelles 01409
Société Belge d'Ergologie, Bruxelles 01446
Société Belge de Vacuologie et de Vacuotechnique, Bruxelles 01449
Société Européenne pour la Formation des Ingénieurs, Bruxelles 01460
Société Technique et Chimique de Sucrerie de Belgique, Bruxelles 01491

Bermuda

Bermuda Technical Society, Hamilton 01524

Brazil

Asociación Interamericana de Ingenieria Sanitaria y Ambiental, Saõ Paulo 01582
Associação Brasileira de Mecânica dos Solos, São Paulo 01592
Conselho Nacional de Desenvolvimento Científico e Tecnológico, Brasília 01645
Fundação Instituto Tecnológico do Estado de Pernambuco, Recife 01656
Instituto de Engenharia de São Paulo, São Paulo 01667
Instituto Nacional de Pesquisas da Amazonia, Manaus 01688
Instituto Nacional de Tecnologia, Rio de Janeiro 01690
Sociedade de Engenharia do Rio Grande do Sul, Pôrto Alegre 01721

Bulgaria

Central Council of Scientific and Technical Unions, Sofia 01766
National Scientific Society of Biomedical Physics and Engineering, Sofia 01769
Research Association on Fundamental Problems of Technical Sciences, Sofia 01771
Union of Energetics, Sofia 01792
Union of Mechanical Engineering, Sofia 01795
Union of the Food Industry, Sofia 01801

Burkina Faso

Association des Techniciens et Ingénieurs Sanitaires Africains, Ouagadougou 01805

Canada

Assocation of Canadian Industrial Designers, Toronto 01872
British Columbia Research Council, Vancouver 01988
Canadian Academy of Engineering, Ottawa 02003
Canadian Advanced Technology Association, Ottawa 02014
Canadian Council of Professional Engineers, Ottawa 02107
Canadian Drilling Research Association, Calgary 02119
Canadian Flexible Packaging Institute, Toronto 02136
Canadian Geotechnical Society, Rexdale 02147
Canadian Industrial Arts Association, Burnaby 02160
Canadian Medical and Biological Engineering Society, Ottawa 02196
Canadian Science and Technology Historical Association, North York 02247
Canadian Society for Mechanical Engineering, Ottawa 02262
Canadian Society of Engineering Management, Ottawa 02283
Canadian Technion Society, Montréal 02306
Council of Regents for Colleges of Applied Arts and Technology, Toronto 02407
Eastern Newfoundland Engineering Society, St. John's 02422
Engineering Institute of Canada, Ottawa 02435
Foundation for Independent Research on Technology and Health, Goodwood 02473
Human Factors Association of Canada, Mississauga 02508
Hungarian Canadian Engineers' Association, Don Mills 02510
Institut d'Horlogerie du Canada, Montréal 02523
International Cooperation in History of Technology Committee, Toronto 02547
International Federation for Medical and Biological Engineering, Ottawa 02553
Mechanics' Institute of Montréal, Montréal 02642
National Association of Women in Construction, Edmonton 02669
Natural Sciences and Engineering Research Council of Canada, Ottawa 02675
Ontario Society for Cable Television Engineering, Mississauga 02796
Pulp and Paper Research Institute of Canada, Pointe Claire 02853
Research and Productivity Council, Fredericton 02859
Science Council of Canada, Ottawa 02907
Society for Cable Television Engineering, Mississauga 02938
Society of Ukrainian Engineers and Associates in Canada, Toronto 02948
Truss Plate Institute of Canada, Concord 02970
Welding Institute of Canada, Oakville 02985
York Technology Association, Markham 03000

Chile

Appropriate Technology Study Centre for Latin America, Valparaiso 03013
Caribbean Technical Cooperation Network on Upper Watershed Management, Santiago 03021
Colegio de Architectos de Chile, Santiago 03029
Colegio de Ingenieros de Chile, Concepción 03032
Colegio de Técnicos, Concepción 03035
Comisíon Nacional de Investigación Científica y Tecnológica, Santiago 03039
Instituto de Ingenieros de Chile, Santiago 03055

China, People's Republic

Academy of Chemical Engineering, Beijing 03108
Academy of Hydrotechnology, Beijing 03110
Aerodynamics Research Society, Beijing 03114
Chemical Industry and Engineering Society of China, Beijing 03125
China Academy of Railway Sciences, Beijing 03126
China Association for Science and Technology, Beijing 03128
Chinese Association of Automation, Beijing 03162
The Chinese Association of Fire Protection, Beijing 03163
Chinese Engineering Graphics Society, Wuhan 03169
Chinese Light Industry Society, Beijing 03174
Chinese Mechanical Engineering Society, Beijing 03176
Chinese Mechanics Society, Beijing 03177
Chinese Society of Chemical Engineering, Beijing 03186
Chinese Society of Engineering Thermophysics, Beijing 03188
Mukden Academy of Chemical Engineering, Liaoning 03211
Society of Autmotive Engineering of China, Beijing 03221

China, Republic

The Chinese Institute of Engineers, Taipei 03273

Colombia

Instituto Colombiano de Crédito Educativo y Estudios Técnicos en el Exterior, Bogotá 03382
Instituto Colombiano de Normas Técnicas, Bogotá 03386
Sociedad Antioqueña de Ingenieros y Architectos, Medellín 03403
Sociedad Colombiana de Químicos e Ingenieros Químicos, Bogotá 03414

Congo

Conseil National de la Recherche Scientifique et Technique, Brazzaville 03425

Croatia

European Association for Earthquake Engineering, Zagreb 03458

Cuba

Sociedad Cubana de Historia de la Ciencia y de la Técnica, La Habana 03481
Sociedad Cubana de Ingenieros, La Habana 03483

Cyprus

Cyprus Joint Technical Council, Nicosia 03492

Czech Republic

Česká Vědeckothechnická Společnost, Praha 03536

Denmark

Akademiet for de tekniske Videnskaber, Lyngby 03557
Brandteknisk Selskab, København 03569
Dansk Astronautisk Forening, København 03591
Dansk Automationsselskab, Københan 03592
Dansk Medikoteknisk Selskab, Holte 03648
Dansk Mejeringeniør Forening, Odense 03649
Dansk Selskab for Opvarmnings- og Ventilationsteknik, København 03685
Dansk Svejseteknisk Landsforening, Broendby 03694
danvak VVS Teknisk Forening, Lyngby 03699
FORCE Instituttene, Glostrup 03726
Militaerteknisk Forening, København 03773
Polymerteknisk Selskab, København 03794
Polyteknisk Forening, Lyngby 03795
Selskabet for Tekniske Uddanelsespørgsmal, København 03817
Teknisk Skoleforening, Odense 03826

Ecuador

Centro Andino de Tecnología Rural, Loja 03848

Egypt

Academy of Scientific Research and Technology, Cairo 03859
Egyptian Society of Engineers, Cairo 03898

Finland

Skandinaviska Simuleringssällskapet, Espoo 03991
Suomen Avaruustutkimusseura, Helsinki 04009
Suomen Hitsausteknillinen Yhdistys, Helsinki 04028
Suomen Ilmailuliitto, Helsinki 04030
Suomen Tekstiiliteknillinen Liitto, Tampere 04063
Svenska Tekniska Vetenskapsakademien i Finland, Helsinki 04068
Tekniikan Akateemisten Liitto, Helsinki 04072
Teknillisten Tieteiden Akatemia, Helsinki 04074
Tekniska Föreningen i Finland, Helsinki 04075

France

African Technical Association, Paris 04125
Association d'Etudes Techniques des Industries de l'Estampage et de la Forge, Paris 04201
Association Européenne pour l'Administration de la Recherche Industrielle, Paris 04208
Association Française des Ingénieurs, Chimistes et Techniciens des Industries du Cuir, Lyon 04232
Association Française du Froid, Paris 04239
Association Générale des Hygiénistes et Techniciens Municipaux, Paris 04253
Association Internationale des Techniciens Biologistes de Langue Française, Le Perreux 04279
Association Internationale d'Irradiation Industrielle, Charbonnières-les-Bains 04284
Association Nationale de la Recherche Technique, Paris 04305
Association Technique de la Réfrigération et de l'Equipement Ménager, Paris 04351
Association Technique de l'Industrie Papetière, Paris 04354
Association Technique Maritime et Aéronautique, Paris 04355
Centre de Perfectionnement des Industries Textiles Rhône-Alpes, Lyon 04367
Collège International pour l'Etude Scientifique des Techniques de Production Mécanique, Paris 04391
Comité National des Conseillers de l'Enseignement Technique, Paris 04414
Conférence Internationale des Formations d'Ingénieurs et Techniciens d'Expression Française, Villeurbanne 04444
Conférence Internationale des Responsables des Universités et Instituts à Dominante Scientifique et Technique d'Expression Française, Talence 04446
Eurocoast, Marseille 04474
European Association of Contract Research Organizations, Plaisir 04484
European Association of Marine Sciences and Techniques, Talence 04487
European Consortium for Ocean Drilling, Strasbourg 04498
France Intec, Paris 04547
Société d'Ergonomie de Langue Française, Caen 04690
Société d'Etudes Financières et Meunières, Paris 04738
Société d'Etudes Minières, Industrielles et Financières, Paris 04748
Société d'Etudes Techniques, Paris 04755
Société Française de Chronométrie et de Microtechnique, Besançon 04795
Société Française des Thermiciens, Paris 04846
Société Française du Vide, Paris 04866
Société Hydrotechnique de France, Paris 04874
Société Technique d'Etudes Mécaniques et d'Outillage, Paris 04920

Georgia

Council on the History of Natural Sciences and Technology, Tbilisi 05010
Scientific-Technical Council on Computer Technology, Mathematical Modelling, Automation of Scientific Research and Instrument Making, Tbilisi 05027

Germany

Akademie für das Grafische Gewerbe, München 05053
Arbeitsausschuß Wälzlager im DIN Deutsches Institut für Normung e.V., Köln 05078
Arbeitsgemeinschaft industrieller Forschungsvereinigungen „Otto von Guericke" e.V., Köln 05120
Arbeitswissenschaft im Landbau e.V., Stuttgart 05148
Battelle-Institut e.V., Frankfurt 05166
Bundesverband für den Selbstschutz, Bonn 05252
Bundesverband Katholischer Ingenieure und Wirtschaftler Deutschlands, Bonn 05255
Bund Technischer Experten e.V., Bremen 05270
Carl-Cranz-Gesellschaft e.V., Weßling 05271
Deutsche Arbeitsgemeinschaft Vakuum, Jülich 05294
Deutsche Forschungsgesellschaft für Oberflächenbehandlung e.V., Düsseldorf 05306
Deutsche Gesellschaft für Biomedizinische Technik e.V., Berlin 05331
Deutsche Gesellschaft für Chemisches Apparatewesen, Chemische Technik und Biotechnologie e.V., Frankfurt 05334

Deutsche Gesellschaft für Chronometrie e.V., Stuttgart *05337*
Deutsche Gesellschaft für Forschung im Graphischen Gewerbe, München *05351*
Deutsche Gesellschaft für Galvano- und Oberflächentechnik e.V., Düsseldorf *05352*
Deutsche Gesellschaft für Geschichte der Medizin, Naturwissenschaft und Technik e.V., München *05355*
Deutsche Gesellschaft für Technische Zusammenarbeit, Eschborn *05430*
Deutsche Gesellschaft für Zerstörungsfreie Prüfung e.V., Berlin *05446*
Deutsche Glastechnische Gesellschaft e.V., Frankfurt *05449*
Deutsche Kommission für Ingenieurausbildung, Düsseldorf *05460*
Deutscher Dampfkesselausschuss, Essen *05507*
Deutscher Erfinderring e.V., Nürnberg *05510*
Deutscher Verband für Schweisstechnik e.V., Düsseldorf *05551*
Deutscher Verband Technischer Assistenten in der Medizin e.V., Hamburg *05554*
Deutscher Verband technisch-wissenschaftlicher Vereine, Düsseldorf *05555*
Deutsches Handwerksinstitut e.V., München *05579*
Deutsches Komitee Instandhaltung e.V., Düsseldorf *05591*
Deutsche Vereinigung für Verbrennungsforschung e.V., Essen *05621*
Europäische Föderation für Chemie-Ingenieur-Wesen, Frankfurt *05652*
Europäischer Verband für Produktivitätsförderung, Hamburg *05656*
European Committee for Future Accelerators, Aachen *05675*
Fördergemeinschaft für das Süddeutsche Kunststoff-Zentrum e.V., Würzburg *05703*
FOGRA Forschungsgesellschaft Druck e.V., München *05705*
Forschungsgemeinschaft Eisenhüttenschlacken, Duisburg *05707*
Forschungsgemeinschaft Industrieofenbau e.V., Frankfurt *05712*
Forschungsgemeinschaft Werkzeuge und Werkstoffe e.V., Remscheid *05716*
Forschungsgesellschaft Stahlverformung e.V., Hagen *05723*
Forschungs-Gesellschaft Verfahrenstechnik e.V., Düsseldorf *05725*
Forschungsvereinigung Antriebstechnik e.V., Frankfurt *05741*
Forschungsvereinigung Feinmechanik und Optik e.V., Köln *05747*
Forschungsvereinigung für angewandte Schloß-, Beschlag- und präventive Sicherheitstechnik e.V., Velbert *05748*
Fraunhofer-Gesellschaft zur Förderung der angewandten Forschung e.V., München *05759*
Frontinus-Gesellschaft e.V., Bergisch Gladbach *05766*
Georg-Agricola-Gesellschaft zur Förderung der Geschichte der Naturwissenschaften und der Technik e.V., Düsseldorf *05782*
Gesellschaft für Angewandte Mathematik und Mechanik, Hamburg *05798*
Gesellschaft für die Geschichte und Bibliographie des Brauwesens e.V., Berlin *05811*
Gesellschaft für Technologiefolgenforschung e.V., Berlin *05863*
Gesellschaft für Tribologie e.V., Moers *05864*
Gesundheitstechnische Gesellschaft e.V., Berlin *05893*
GKSS-Forschungszentrum Geesthacht, Geesthacht *05895*
Gutenberg-Gesellschaft, Mainz *05902*
GVC/VDI-Gesellschaft Verfahrenstechnik und Chemieingenieurwesen, Düsseldorf *05903*
Hahn-Schickard-Gesellschaft für angewandte Forschung e.V., Stuttgart *05905*
Hannoversches Forschungsinstitut für Fertigungsfragen e.V., Hannover *05908*
Hüttentechnische Vereinigung der Deutschen Glasindustrie e.V., Frankfurt *05932*
Industrie-Gemeinschaft Aerosole e.V., Frankfurt *05939*
Institut für Europäische Umweltpolitik, Bonn *05959*
Institut für technische Weiterbildung Berlin e.V., Berlin *05974*
International Commission of Sugar Technology, Rain *05986*
Kuratorium für Forschung und Technik der Zellstoff- und Papierindustrie, Bonn *06086*
Kuratorium für Technik und Bauwesen in der Landwirtschaft e.V., Darmstadt *06088*
Max-Eyth-Gesellschaft für Agrartechnik e.V., Darmstadt *06113*
Normenausschuss Armaturen im DIN Deutsches Institut für Normung e.V., Köln *06149*
Normenausschuss Chemischer Apparatebau im DIN Deutsches Institut für Normung e.V., Köln *06156*
Normenausschuss Dichtungen im DIN Deutsches Institut für Normung e.V., Köln *06158*
Normenausschuss Druckgasanlagen im DIN Deutsches Institut für Normung e.V., Berlin *06159*
Normenausschuss Druck- und Reproduktionstechnik im DIN Deutsches Institut für Normung e.V., Berlin *06160*
Normenausschuss Eisen-, Blech- und Metallwaren im DIN Deutsches Institut für Normung e.V., Düsseldorf *06162*
Normenausschuss Erdöl- und Erdgasgewinnung im DIN Deutsches Institut für Normung e.V., Köln *06164*
Normenausschuss Fahrräder im DIN Deutsches Institut für Normung e.V., Köln *06166*
Normenausschuss Feinmechanik und Optik im DIN Deutsches Institut für Normung e.V., Pforzheim *06168*
Normenausschuss Feuerwehrwesen im DIN Deutsches Institut für Normung e.V., Berlin *06169*
Normenausschuss Gastechnik im DIN Deutsches Institut für Normung e.V., Eschborn *06170*
Normenausschuss Gleitlager im DIN Deutsches Insitut für Normung e.V., Köln *06172*
Normenausschuss Graphische Symbole im DIN Deutsches Institut für Normung e.V., Berlin *06173*
Normenausschuss Heiz- und Raumlufttechnik im DIN Deutsches Institut für Normung e.V., Berlin *06177*
Normenausschuss Instandhaltung im DIN Deutsches Institut für Normung e.V., Köln *06180*
Normenausschuss Kältetechnik im DIN Deutsches Institut für Normung e.V., Köln *06181*
Normenausschuss Kautschuktechnik im DIN Deutsches Institut für Normung e.V., Frankfurt *06182*
Normenausschuss Kommunale Technik im DIN Deutsches Institut für Normung e.V., Berlin *06184*
Normenausschuss Kraftfahrzeuge im DIN Deutsches Institut für Normung e.V., Frankfurt *06185*
Normenausschuss Laborgeräte und Laboreinrichtungen im DIN Deutsches Institut für Normung e.V., Frankfurt *06187*
Normenausschuss Mechanische Verbindungselemente im DIN Deutsches Institut für Normung e.V., Köln *06192*
Normenausschuss Rohre, Rohrverbindungen und Rohrleitungen im DIN Deutsches Institut für Normung e.V., Köln *06199*
Normenausschuss Rundstahlketten im DIN Deutsches Institut für Normung e.V., Köln *06200*
Normenausschuss Schienenfahrzeuge im DIN Deutsches Institut für Normung e.V., Kassel *06201*
Normenausschuss Schmiedetechnik im DIN Deutsches Institut für Normung e.V., Hagen *06202*
Normenausschuss Schweisstechnik im DIN Deutsches Institut für Normung e.V., Berlin *06204*
Normenausschuss Siebböden und Kornmessung im DIN Deutsches Institut für Normung e.V., Berlin *06205*
Normenausschuss Sport- und Freizeitgerät im DIN Deutsches Institut für Normung e.V., Köln *06206*
Normenausschuss Stahldraht und Stahldrahterzeugnisse im DIN Deutsches Institut für Normung e.V., Köln *06207*
Normenausschuss Transportkette im DIN Deutsches Institut für Normung e.V., Berlin *06210*
Normenausschuss Überwachungsbedürftige Anlagen im DIN Deutsches Institut für Normung e.V., Köln *06211*
Normenausschuss Uhren und Schmuck im DIN Deutsches Institut für Normung e.V., Pforzheim *06212*
Normenausschuss Vakuumtechnik im DIN Deutsches Institut für Normung e.V., Köln *06213*
Normenausschuss Waagenbau im DIN Deutsches Institut für Normung e.V., Berlin *06215*
Normenausschuss Wärmebehandlungstechnik metallischer Werkstoffe im DIN Deutsches Institut für Normung e.V., Köln *06216*
Normenausschuss Werkzeuge und Spannzeuge im DIN Deutsches Institut für Normung e.V., Köln *06218*
Normenausschuss Werkzeugmaschinen im DIN Deutsches Institut für Normung e.V., Frankfurt *06219*
Rheinisch-Westfälische Akademie der Wissenschaften, Düsseldorf *06265*
Schiffbautechnische Gesellschaft e.V., Hamburg *06275*
Studiengesellschaft Stahlanwendung e.V., Düsseldorf *06303*
Studiengruppe Entwicklung Technischer Hilfsmittel für Behinderte, Heidelberg *06304*
Technische Akademie Wuppertal e.V., Wuppertal *06310*
VDI-Gesellschaft Agrartechnik, Düsseldorf *06319*
VDI-Gesellschaft Entwicklung Konstruktion Vertrieb, Düsseldorf *06322*
VDI-Gesellschaft Fördertechnik Materialfluss Logistik, Düsseldorf *06324*
VDI-Gesellschaft Kunststofftechnik, Düsseldorf *06325*
VDI-Gesellschaft Produktionstechnik, Düsseldorf *06326*
VDI-Gesellschaft Technische Gebäudeausrüstung, Düsseldorf *06327*
VDI-Kommission Lärmminderung, Düsseldorf *06329*
VDI/VDE-Gesellschaft Mess- und Automatisierungstechnik, Düsseldorf *06330*
VDI/VDE-Gesellschaft Mikro- und Feinwerktechnik, Düsseldorf *06331*
Verband der Dozenten an Deutschen Ingenieurschulen, Mainz *06336*
Verband der Technischen Überwachungs-Vereine e.V., Essen *06343*
Verein Deutscher Emailfachleute e.V., Hagen *06375*
Verein Deutscher Ingenieure, Düsseldorf *06377*
Verein für Gerberei-Chemie und -Technik e.V., München *06385*

Ghana

Council for Scientific and Industrial Research, Accra *06479*

Greece

Epimelitirion Technikon tis Ellados, Athinai *06522*
Panellinios Enosis Technikon, Athinai *06541*

Guatemala

Center for Meso American Studies on Appropriate Technology, Guatemala City *06563*
Colegio de Ingenieros de Guatemala, Guatemala City *06565*
Escuela Regional de Ingeniería Sanitaria, Guatemala City *06567*
Instituto Centroamericano de Investigación y Tecnología, Guatemala City *06569*

Hungary

Faipari Tudományos Egyesület, Budapest *06604*
Gépipari Tudományos Egyesület, Budapest *06606*
International Measurement Confederation, Budapest *06608*
Magyar Asztonautikai Egyesület, Budapest *06617*
Magyar Iparjogvédelmi Egyesület, Budapest *06637*
Méréstechnikai és Automatizálási Tudományos Egyesület, Budapest *06664*
Müszaki és Természettudományi Egyesületek Szövetségi Kamarája, Budapest *06665*
Papir- és Nyomdaipari Müszaki Egyesület, Budapest *06672*
Szilikátipari Tudományos Egyesület, Budapest *06674*
Textilipari Müszaki és Tudományos Egyesület, Budapest *06675*

Iceland

Verkfraedingafélag Islands, Reykjavik *06706*

India

Asian and Pacific Centre for Transfer of Technology, Bangalore *06720*
Association for Engineering Education in South and Central Asia, Mysore *06730*
Committee on Science and Technology in Developing Countries, Madras *06752*
Council of Scientific and Industrial Research, New Delhi *06753*
Indian Jute Industries' Research Association, Calcutta *06797*
Indian National Academy of Engineering, New Delhi *06803*
Indian Science Congress Association, Calcutta *06813*
Research Designs and Standards Organization, Lucknow *06853*
Silk and Art Silk Mills' Research Association, Bombay *06857*
The Theosophical Society, Madras *06865*

Indonesia

Akademi Teknologi Kulit, Jogjakarta *06872*
Cooperative Program in Technological Research and Higher Education in Southeast Asia and the Pacific, Jakarta *06887*
Persatuan Insinyur Indonesia, Jakarta Pusat *06893*

Ireland

Association of Consulting Engineers of Ireland, Dublin *06926*
Automobile, General Engineering and Mechanical Operations Union, Dublin *06933*
Engineering and Scientific Association of Ireland, Mount Merrion *06944*
Institute of Industrial Engineers, Dublin *06962*
Institution of Civil Engineers (Republic of Ireland), Killiney *06966*
Institution of Engineers of Ireland, Dublin *06968*
Irish Welding Association, Dublin *07005*

Israel

Technion Research and Development Foundation Ltd, Haifa *07114*

Italy

Associazione Criogenica Italiana, Genova *07265*
Associazione Italiana di Ingegneria Chimica, Milano *07311*
Associazione Italiana di Ingegneria Medica e Biologica, Napoli *07312*
Associazione Italiana di Strumentisti, Milano *07325*
Associazione Italiana per l'Analisi delle Sollecitazioni, Trieste *07335*
Associazione Italiana per la Ricerca Industriale, Roma *07339*
Associazione Nazionale di Ingegneria Sanitaria, Roma *07367*
Associazione Nazionale di Meccanica, Milano *07368*
Associazione Nazionale Industria Meccanica Varia ed Affine, Milano *07373*
Associazione Nazionale Ingegneri ed Architetti Italiani, Roma *07374*
Associazione Nazionale Italiana per l'Automazione, Milano *07378*
Associazione Tecnica Italiana per la Cellulosa e la Carta, Milano *07406*
Centro Internazionale di Scienze Meccaniche, Udine *07470*
Centro Provinciale Impiego Combinato Tecniche Agricole, Bologna *07530*
Centro Studi di Economia Applicata all'Ingegneria, Bari *07555*
Centro Studi di Economia Applicata all'Ingegneria, Napoli *07556*
Centro Studi e Applicazioni in Tecnologie Avanzate, Bari *07564*
Collegio dei Tecnici dell' Acciaio, Milano *07608*
Committee for the European Development of Science and Technology, Roma *07623*
European Safeguards Research and Development Association, Ispra *07681*
Federazione delle Associazioni Scientifiche e Tecniche, Milano *07686*
International Association of Engineering Geology, Sezione Italiana, Bari *07707*
International Centre for Advanced Technical and Vocational Training, Torino *07709*
International Centre for Mechanical Sciences, Udine *07710*
Istituto Italiano della Saldatura, Genova *07736*
ITALCONSULT, Roma *07756*
Società Dauna di Cultura, Foggia *07778*
Società degli Ingegneri e degli Architetti in Torino, Torino *07780*

Jamaica

Caribbean Council of Engineering Organizations, Kingston *08070*

Japan

Asian Fluid Mechanics Committee, Tokyo *08099*
Kami Parupu Gijutsu Kyokai, Tokyo *08153*
Kansai Zosen Kyokai, Osaka *08155*
Kogyo Kayaku Kyokai, Tokyo *08168*
Kuki-Chowa Eisei Kogakkai, Tokyo *08174*
Nihon Aisotopu Kyokai, Tokyo *08186*
Nihon Kikai Gakkai, Tokyo *08222*
Nihon Yosetsu Kyokai, Tokyo *08274*
Nihon Zosen Gakkai, Tokyo *08279*
Nippon Jidoseigyo Kyokai, Kyoto *08314*
Nippon Kogakukai, Tokyo *08336*
Nippon Koku Uchu Gakkai, Tokyo *08340*
Nippon Nensho Gakkai, Tokyo *08345*
Nippon Nensho Kenkyukai, Kyoto *08346*
Nippon Shashin Gakkai, Tokyo *08371*
Nogyokikai Gakkai, Tokyo *08400*
San-yo Gijutsu Shinkokai, Okayama-ken *08406*
Sekiyu Gijutsu Kyokai, Tokyo *08411*
Sen-i Gakkai, Tokyo *08413*
Yosetsu Gakkai, Tokyo *08446*

Kenya

African Centre for Technology Studies, Nairobi *08464*
African Institute for Higher Technical Training and Research, Nairobi *08469*
African Network of Scientific and Technological Institutions, Nairobi *08471*
East African Engineering Consultants, Nairobi *08494*
East African Industrial Research Organization, Nairobi *08495*

Korea, Democratic People's Republic

Academy of Light Industry Science, Pyongyang *08511*
Academy of Railway Sciences, Pyongyang *08513*

Luxembourg

European Association for the Transfer of Technologies, Innovations and Industrial Information, Luxembourg *08582*

Macedonia

Sojuz na Inženeri i Tehničari na Makedonija, Skopje *08614*

Mauritius

Société de Technologie Agricole et Sucrière de l'Ile Maurice, Réduit *08676*

Mexico

Asociación de Ingenieros y Arquitectos de México, México *08692*
Centro Científico y Técnico Francés en México, México *08713*
Centro Nacional de Ciencias y Tecnologías Marinas, Veracruz *08721*
Consejo Nacional de Ciencia y Tecnologia, México *08729*

Netherlands

European Committee for the Advancement of Thermal Sciences and Heat Transfer, Delft *08880*
International Association for Hydraulic Research, Delft *08918*
International Technical and Scientific Organization for Soaring Flight, Wessling *08945*
Koninklijke Nederlandse Vereniging voor Luchtvaart, 's-Gravenhage *08960*
Koninklijk Instituut van Ingenieurs, 's-Gravenhage *08963*
Nederlandse Organisatie voor Wetenschappelijk Onderzoek, 's-Gravenhage *08998*
Stichting voor de Technische Wetenschappen, Utrecht *09077*

New Zealand

Association for Engineering Education in Southeast Asia and the Pacific, Christchurch *09118*
Institution of Professional Engineers New Zealand, Wellington *09129*
New Zealand Dairy Technology Society, Waitoa *09142*
New Zealand Institute of Food Science and Technology, Christchurch *09156*
New Zealand Institute of Surveyors, Wellington *09161*
New Zealand Society of Dairy Science and Technology, Edgecumbe *09178*

Nigeria

African Regional Centre for Engineering Design and Manufacturing, Ibadan *09219*
Organization of African Unity – Scientific, Technical and Research Commission, Lagos *09244*

Norway

Den Polytekniske Forening, Oslo *09257*
Norges Tekniske Vitenskapsakademi, Trondheim *09288*
Norges Teknisk-Naturvitenskapelige Forskningsråd, Oslo *09289*
Norsk Astronautisk Forening, Oslo *09292*
Selskapet for Lyskultur, Sandvika *09346*
Stiftelsen for Industriell og Teknisk Forskning ved Norges Tekniske Høgskole, Trondheim *09353*

Pakistan

Institution of Engineers, Pakistan, Lahore *09369*
Pakistan Concrete Institute, Karachi *09382*

Paraguay

Servicio Técnico Interamericano de Cooperación Agrícola, Asunción 09433
Unión Sudamericana de Asociaciones de Ingenieros, Asunción 09436

Peru

Centro Latinoamericana de Estudios y Difusión de la Construcción en Tierra, Lima 09461
Consejo Andino de Cienccia y Tecnología, Lima 09465
Instituto Peruano de Ingenieros Mecánicos, Lima 09471
Sociedad de Ingenieros del Perú, Lima 09475

Philippines

Asian Alliance of Appropriate Technology Practitioners, Manila 09493
Asia Pacific Grouping of Consulting Engineers, Manila 09506
United Technological Organizations of the Philippines, Manila 09538

Poland

Polskie Towarzystwo Mechaniki Teoretycznej i Stosowanej, Warszawa 09586
Polski Komitet Pomiarów Automatyki, Warszawa 09611

Portugal

Associação Técnica da Indústria do Cimento, Lisboa 09651
Ordem dos Engenheiros, Lisboa 09664
Sociedade de Estudos Técnicos SARL-SETEC, Lisboa 09681

Russia

Biological Engineering Society, Moskva 09809
Civil Engineering Society, Moskva 09810
Commission for Scientific and Technical Co-operation of the Academy of Sciences and Organizations of Moscow Oblast, Moskva 09827
Commission for Synchroton Irradiation, Moskva 09830
Committee for the Coordination of Construction of Scientific Apparatus and Antomation of Research Work, Moskva 09848
Mapping and Prospecting Engineering Society, Moskva 09879
Mechanical Engineering Society, Moskva 09881
Medical Engineering Society, Moskva 09882
National Committee of History and Philosophy of Natural Science and Technology, Moskva 09896
National Medical and Technical Scientific Society, Moskva 09909
Press and Publishing Engineering Society, Moskva 09945
Programme Committee Surface Physics, Chemistry and Mechanics, Moskva 09946
Russian Pugwash Committee, Moskva 09964
Scientific-Technical Association, St. Petersburg 09968
Shipbuilding Engineering Society, St. Petersburg 09969

Senegal

African Regional Centre for Technology, Dakar 10013

Singapore

Asian Network for Industrial Technology Information and Extension, Singapore 10037

Slovenia

Association of Engineers and Technicians of Slovenia, Ljubljana 10097
Društvo za Medicinsku i Biološku Tehniku, Ljubljana 10100
Raziskovalna Skupnost Slovenije, Ljubljana 10103
Zveza Društev za Varilno Tehniko Slovenije, Ljubljana 10113

South Africa

The Associated Scientific and Technical Societies of South Africa, Marshalltown 10123
Chemical Engineering Research Group, Durban 10129
CSIR, Pretoria 10133
Division of Energy Technology, Pretoria 10136
Pollution Research Group, Durban 10162
South African Filtration Society, Hillcrest 10190
South African Institute of Aeronautical Engineers, Sunnyside 10195
South African Institute of Assayers and Analysts, Mashalltown 10196
South African Institute of Printing, Cape Town 10201

Spain

Asociación de Investigación Técnica de la Industria Papelera Española, Madrid 10248
Asociación Española de Técnicos de Cerveza y Malta, Madrid 10269
Colegio Oficial de Ingenieros Industriales de Cataluña, Barcelona 10291
Instituto de la Engeniería de España, Madrid 10324
International Towing Tank Conference, Madrid 10331

Sweden

Ingenjörsvetenskapsakademien, Stockholm 10469
Kungliga Fysiografiska Sällskapet i Lund, Lund 10483
Styrelsen för Teknisk Utveckling, Stockholm 10529
Svenska Färgeritekniska Riksförbundet, Göteborg 10542
Svenska Föreningen för Medicinsk Teknik och Fysik, Umea 10544
Svenska Geotekniska Föreningen, Linköping 10552
Svenska Kyltekniska Föreningen, Järfälla 10558
Svenska Livsmedelstekniska Föreningen, Stockholm 10563
Svetstekniska Föreningen, Stockholm 10599
Teknikkonsulterna, Stockholm 10600
Tekniska Samfundet i Göteborg, Göteborg 10602

Switzerland

Alliance Graphique Internationale, Zürich 10615
Conférence des Directeurs des Ecoles Techniques Supérieures de Suisse, Winterthur 10653
Consultative Council for Postal Studies, Bern 10657
Fachgruppe Carosserie- und Fahrzeugtechnik des STV, Sissach 10687
Fachgruppe für Arbeiten im Ausland des SIA, Zürich 10688
Fachgruppe für Betriebstechnik des STV, Winterthur 10690
Fachgruppe für Verfahrenstechnik des SIA, Basel 10697
Organisation Internationale de Protection Civile, Petit-Lancy 10776
Schweizerische Akademie der Technischen Wissenschaften, Zürich 10790
Schweizerische Gesellschaft für Automatik, Bern 10812
Schweizerische Gesellschaft für Feintechnik, Zürich 10820
Schweizerische Gesellschaft für Lebensmittel-Wissenschaft und -Technologie, Wädenswil 10829
Schweizerische Gesellschaft für Oberflächentechnik, Bettlach 10836
Schweizerische Gesellschaft für Vakuum-Physik und -Technik, Zürich 10854
Schweizerische Gesellschaft Pro Technorama, Winterthur 10857
Schweizerische Lichttechnische Gesellschaft, Bern 10868
Schweizerischer Fachverband der diplomierten medizinischen Laborantinnen und Laboranten, Sankt Gallen 10893
Schweizerischer Technischer Verband, Zürich 10905
Schweizerischer Verband für die Wärmebehandlung der Werkstoffe, Winterthur 10911
Schweizerischer Verband für Landtechnik, Riniken 10914
Schweizerischer Verein für Schweisstechnik, Basel 10920
Schweizerischer Verein für Vermessung und Kulturtechnik, Solothurn 10921
Schweizerische Vereinigung für Gesundheitstechnik, Zürich 10946
Verband Schweizerischer Abwasserfachleute, Zürich 11004

Tanzania

Tanzania Commission for Science and Technology, Dar es Salaam 11059

Thailand

ASEAN Federation of Engineering Organizations, Bangkok 11067
Asian Disaster Preparedness Center, Bangkok 11080
Asian Geotechnical Engineering Information Center, Bangkok 11081
Asian Institute of Technology Library and Regional Documentation Centre, Bangkok 11082
Southeast Asian Society of Soil Engineering, Bangkok 11100

Trinidad and Tobago

Caribbean Council for Science and Technology, Port of Spain 11110
Sugar Technologist's Association of Trinidad and Tobago, Port of Spain 11123

Turkey

Türk Mühendis ve Mimar Odalari Birligi, Ankara 11167
Türk Otomatik Kontrol Kurumu, Istanbul 11171

Ukraine

Scientific and Technical Societies National Headquarters, Kiev 11190

United Kingdom

Antiquarian Horological Society, Wadhurst 11238
Association for Medical Physics Technology, Ickenham 11274
Association of Advisers in Design and Technical Studies, Camberley 11298
Association of Independent Research and Technology Organisations, Cambridge 11357
Association of Marine Engineering Schools, Liverpool 11365
Association of Municipal Engineers, London 11368
Association of Noise Consultants, Guilden Morden 11372
Association of Polytechnic Teachers, Southsea 11377
Association of Principals of Colleges (Northern Ireland Branch), Lisburn 11379
Association of Professional Scientists and Technologists, London 11380
Association of Scientific, Technical and Managerial Staffs, London 11388
Association of Teachers of Printing and Allied Subjects, Liverpool 11398
Audio Engineering Society, British Section, Sevenoaks 11416
BHRA Fluid Engineering, Bedford 11437
Biological Engineering Society, London 11443
BMT Cortec, Wallsend 11455
BNF Metals Technology Centre, Wantage 11456
BRF International, Nutfield 11467
British Association for Brazing and Soldering, Cambridge 11487
British Brush Manufacturers Research Association, London 11534
British Compressed Air Society, London 11546
The British Cryogenics Council, London 11548
British Geotechnical Society, London 11567
British Horological Institute, Upton 11573
The British Institute of Cleaning Science, Northampton 11580
British Internal Combustion Engine Research Institute, Slough 11593
British Society for Strain Measurement, Newcastle-upon-Tyne 11668
Brunel Society, Bristol 11714
Centre for Alternative Technology, Machynlleth 11753
Confederation of British Industry, London 11837
Design and Industries Association, London 11883
The Educational Institute of Design, Craft and Technology, Melton Mowbray 11944
Engineering Committee on Oceanic Resources, London 11957
The Engineering Council, London 11958
Ergonomics Society, Loughborough 11972
European Committee for Civil Engineers, London 12012
Fabric Care Research Association, Harrogate 12046
Fire Protection Association, London 12076
Furniture Industry Research Association, Stevenage 12096
HATRA, Nottingham 12149
High Pressure Technology Association, Leeds 12169
Industrial Unit of Tribology, Leeds 12203
Institute of Engineers and Technicians, London 12236
Institute of Food Science and Technology of the United Kingdom, London 12239
Institute of Hospital Engineering, Portsmouth 12243
Institute of Inventors, London 12247
Institute of Marine Engineers, London 12255
Institute of Measurement and Control, London 12259
Institute of Packaging, Melton Mowbray 12265
Institute of Printing, Tunbridge Wells 12270
Institute of Refrigeration, Carshalton 12277
Institute of Science Technology, Lichfield 12279
Institute of Scientific and Technical Communicators, London 12280
Institute of Sheet Metal Engineering, Birmingham 12281
Institution of Agricultural Engineers, Bedford 12292
The Institution of Chemical Engineers, Rugby 12293
Institution of Engineers amd Shipbuilders in Scotland, Glasgow 12300
Institution of Fire Engineers, Leicester 12303
Institution of Gas Engineers, London 12304
Institution of Mechanical Engineers, London 12307
Institution of Mechanical Incorporated Engineers, London 12308
Institution of Mining Engineers, Doncaster 12310
Institution of Plant Engineers, London 12313
Institution of Polish Engineers in Great Britain, London 12314
Institution of Railway Signal Engineers, Dawlish 12316
Institution of Structural Engineers, London 12317
Intermediate Technology Development Group, Rugby 12320
International Enamellers Institute, Ripley near Derby 12349
International Federation of Airworthiness, Ruislip 12353
International Hydrofoil and Multihull Society, Great Yarmouth 12367
International Hydrofoil Society, London 12368
International Society for Soil Mechanics and Foundation Engineering, Cambridge 12390
Lambeg Industrial Research Association, Lisburn 12436
National Foundry and Engineering Training Association, Glasgow 12592
The Newcomen Society, London 12625
Parliamentary and Scientific Committee, London 12679
Pira International, Leatherhead 12701
Printing Historical Society, London 12718
Production Engineering Research Association, Melton Mowbray 12722
Research and Development Society, London 12751
Royal Academy of Engineering, London 12763
The Royal Institution of Naval Architects, London 12823
Safety and Reliability Society, Southport 12881
SATRA Footwear Technology Centre, Kettering 12886
Science and Engineering Research Council, Swindon 12889
Scottish Society for the Preservation of Historical Machinery, Glasgow 12930
Sheffield Metallurgical and Engineering Association, Rotherham 12948
Sira, Chislehurst 12953
Society for Underwater Technology, London 13021
Society of Consulting Marine Engineers and Ship Surveyors, London 13038
The Society of Engineers, Colchester 13045
Society of Licensed Aircraft Engineers and Technologists, Kingston-upon-Thames 13055
Society of Professional Engineers, London 13059
Society of Technical Analysts, Cambridge 13067
South Wales Institute of Engineers, Cardiff 13084
Spring Research and Manufacturers Association, Sheffield 13089
Tensor Society of Great Britain, Surbiton 13119
TWI, Cambridge 13147
UMIST Association, Manchester 13159

Uruguay

Academia Nacional de Ingeniería, Montevideo 13239
Asociación de Ingenieros del Uruguay, Montevideo 13244
Consejo Nacional de Investigaciones Científicas y Técnicas, Montevideo 13258
Instituto Uruguayo de Normas Técnicas, Montevideo 13262

U.S.A.

AACE International, Morgantown 13273
Abrasive Engineering Society, Butler 13276
Academy of Security Educators and Trainees, Berryville 13309
Accreditation Board for Engineering and Technology, New York 13315
ACEC Research and Management Foundation, Washington 13320
Advanced Technology Alert System, New York 13332
American Academy of Environmental Engineers, Annapolis 13391
American Academy of Mechanics, La Jolla 13405
American Association of Engineering Societies, Washington 13532
American Automatic Control Council, Evanston 13612
American Canal Society, York 13675
American Consulting Engineers Council, Washington 13755
American Council of the International Institute of Welding, Miami 13764
American Engineering Model Society, Boston 13798
American Indian Council of Architects and Engineers, Tigard 13858
American Indian Science and Engineering Society, Boulder 13861
American Institute of Aeronautics and Astronautics – Technical Information Division, New York 13873
American Institute of Plant Engineers, Cincinnati 13894
American Institute of Timber Construction, Vancouver 13897
American Paper Institute, New York 13990
American Society for Artificial Internal Organs, Boca Raton 14046
American Society for Engineering Management, Annapolis 14059
American Society for Hospital Engineering, Chicago 14067
American Society for Medical Technology, Bethesda 14072
American Society for Technion-Israel Institute of Technology, New York 14088
American Society of Agricultural Engineers, Saint Joseph 14097
American Society of Body and Design Engineers, Redford 14102
American Society of Certified Engineering Technicians, El Paso 14105
American Society of Danish Engineers, Larchmont 14117
American Society of Mechanical Engineers, New York 14145
American Society of Naval Engineers, Alexandria 14147
American Society of Plumbing Engineers, Westlake 14158
American Society of Sanitary Engineering, Bay Village 14166
American Society of Swedish Engineers, New York 14169
American Technical Education Association, Wahpeton 14188
American Underground-Space Association, Minneapolis 14200
Annular Bearing Engineers Committee, Washington 14223
APEC, Dayton 14232
ASME International Gas Turbine Institute, Atlanta 14254
Association for Finishing Processes of the Society of Manufacturing Engineers, Dearborn 14297
Association for Science, Technology and Innovation, Arlington 14332
Association of Chairmen of Departments of Mechanics, Gainesville 14410
Association of Environmental Engineering Professors, Houghton 14434
Association of Firearm and Tool Mark Examiners, La Jolla 14438
Association of Iron and Steel Engineers, Pittsburgh 14449
Association of Muslim Scientists and Engineers, Plainfield 14468
Association of Science-Technology Centers, Washington 14500
Association of Surgical Technologists, Englewood 14511
ATAS Technology Assessment Network, New York 14543
Audio Engineering Society, New York 14548
Ball Manufacturers Engineers Committee, Washington 14569
Biomedical Engineering Society, Culver City 14593
Career College Association, Washington 14645
Coastal Engineering Research Council, Charleston 14766
Commission for the Scientific and Technological Development of Central America and Panama, Washington 14784
Commission on Professionals in Science and Technology, Washington 14790
Cooling Tower Institute, Houston 14858
Coordinating Research Council, Atlanta 14861
Council of Engineering and Scientific Society Executives, Washington 14895
Cryogenic Engineering Conference, Moffet Field 14944
Ductile Iron Pipe Research Association, Birmingham 14988
Earthquake Engineering Research Institute, Oakland 15000
Emulsion Polymers Institute, Bethlehem 15040
Engineering Manpower Commission, Washington 15044
Environic Foundation International, South Bend 15047

Environmental Design Research Association, Oklahoma City *15049*
Experiments in Art and Technology, Berkeley Heights *15071*
Federation of Societies for Coatings Technology, Blue Bell *15089*
Fiber Society, Princeton *15092*
General Society of Mechanics and Tradesmen, New York *15150*
Goudy Society, New York *15167*
Graphic Communications Association, Alexandria *15172*
Gravure Association of America, Rochester *15173*
Human Factors Society, Santa Monica *15235*
IEEE Engineering in Medicine and Biology Society, New York *15246*
Illuminating Engineering Society of North America, New York *15248*
Industrial Research Institute, Washington *15261*
Institute for the Advancement of Engineering, Woodland Hills *15299*
Institute of Gas Technology, Chicago *15322*
Institute of Industrial Engineers, Norcross *15325*
Insulated Cable Engineers Association, South Yarmouth *15345*
Inter-American Association of Sanitary Engineering and Environmental Sciences, Brookeville *15348*
International Association for Learning Laboratories, Philadelphia *15386*
International Conference on Mechanics in Medicine and Biology, Ann Arbor *15446*
International Council for Pressure Vessel Technology, Milwaukee *15453*
International Microwave Power Institute, Clifton *15502*
International Society for Clinical Laboratory Technology, Saint Louis *15529*
International Society for Terrain-Vehicle Systems, Hanover *15541*
International Test and Evaluation Association, Fairfax *15563*
Laser Institute of America, Orlando *15647*
Leonardo – The International Society for the Arts, Sciences and Technology, San Francisco *15660*
Marine Technology Society, Washington *15691*
Missile, Space and Range Pioneers, Patrick Air Force Base *15746*
National Academy of Engineering, Washington *15774*
National Academy of Recording Arts and Sciences, Burbank *15777*
National Accrediting Agency for Clinical Laboratory Sciences, Chicago *15780*
National Association of County Engineers, Washington *15846*
National Association of Industrial and Technical Teacher Educators, Knoxville *15869*
National Certification Agency for Medical Lab Personnel, Bethesda *15951*
National Council of Examiners for Engineering and Surveying, Clemson *15998*
National Council of the Paper Industry for Air and Stream Improvement, New York *16012*
National Information Service for Earthquake Engineering, Richmond *16087*
National Institute for Certification in Engineering Technologies, Alexandria *16089*
National Institute for Rehabilitation Engineering, Hewitt *16092*
National Institute of Ceramic Engineers, Westerville *16093*
National Research Council, Washington *16138*
National Society of Professional Engineers, Alexandria *16159*
National Technical Association, Washington *16169*
National Technical Services Association, Alexandria *16170*
Numerical Control Society/AIM Tech, Research Triangle Park *16235*
Pacific Rocket Society, Oxnard *16281*
Plastics Institute of America, Fairfield *16325*
Precision Measurements Association, Los Angeles *16350*
Print Council of America, Baltimore *16354*
Program for Appropriate Technology in Health, Seattle *16360*
Research and Engineering Council of the Graphic Arts Industry, Chadds Ford *16412*
Research Council on Structural Connections, Chicago *16414*
Roller Bearing Engineers Committee, Washington *16428*
SAE International, Warrendale *16433*
Safe Association, Yoncalla *16434*
SFI Foundation, Poway *16465*
Society for Biomaterials, Minneapolis *16502*
Society for Experimental Mechanics, Bethel *16523*
Society for Technical Communication, Arlington *16584*
Society for the History of Technology, Houghton *16598*
Society of Allied Weight Engineers, Chula Vista *16632*
Society of Cable Television Engineers, Exton *16645*
Society of Carbide and Tool Engineers, Metals Park *16646*
Society of Engineering Illustrators, Madison Heights *16656*
Society of Engineering Science, Charlottesville *16657*
Society of Flight Test Engineers, Lancaster *16666*
Society of Manufacturing Engineers, Dearborn *16677*
Society of Naval Architects and Marine Engineers, Jersey City *16683*
Society of Plastics Engineers, Brookfield *16692*
Society of Professional Well Log Analysts, Houston *16694*
Society of Turkish Architects, Engineers and Scientists in America, New York *16709*
Society of Women Engineers, New York *16713*
Structural Stability Research Council, Bethlehem *16748*
Sulphur Institute, Washington *16752*
Technical Association of the Graphic Arts, Rochester *16767*
Technology Transfer Society, Indianapolis *16769*
Tripoli Rocketry Association, Kenner *16792*
Ukrainian Engineers' Society of America, Cliffside Park *16797*
United Engineering Trustees, New York *16811*
United States Institute for Theatre Technology, New York *16825*
United States National Committee on Theoretical and Applied Mechanics, Washington *16830*
United States National Society for the International Society of Soil Mechanics and Foundation Engineering, Raleigh *16831*
Water Environment Federation, Alexandria *16873*
Welding Research Council, New York *16878*

Venezuela

Asociacíon Venezolana de Ingeniería Eléctrica y Mecánica, Caracas *16975*
Asociación Venezolana de Ingeniería Sanitaria y Ambiental Caracas, Caracas *16976*
Consejo Nacional de Investigaciones Científicas y Tecnológicas, Caracas *17001*

Vietnam

Vietnamese Association of Engineering, Hanoi *17045*
Vietnam Union of Sientific and Technical Associations, Hanoi *17071*

Yugoslavia

Jugoslovenski Centar za Tehničku i Naučnu Dokumentaciju, Beograd *17094*
Jugoslovenski Savez za Zavarivanje, Beograd *17095*
Jugoslovensko Društvo za Mehaniku, Beograd *17099*
Nikola Tesla Association of Societies for Promotion of Technical Sciences in Yugoslavia, Beograd *17105*
Union of Engineers and Technicians of Serbia, Beograd *17133*
Union of Engineers and Technicians of Yugoslavia, Beograd *17134*

Zambia

The Engineering Institution of Zambia, Lusaka *17152*

Entomology

Argentina

Sociedad Entomológica Argentina, La Plata *00197*

Australia

The Australian Entomological Society, Burnley *00289*
Council for International Congresses of Entomology, Canberra *00381*
Entomological Society of New South Wales, Sydney *00391*
Entomological Society of Queensland, Brisbane *00392*

Austria

Arbeitsgemeinschaft Österreichischer Entomologen, Wien *00560*
Österreichische Entomologische Gesellschaft, Graz *00738*

Belgium

Société Royale Belge d'Entomologie, Bruxelles *01476*

Brazil

Sociedade Brasileira de Entomologia, São Paulo *01709*

Canada

Council for International Congresses of Dipterology, Edmonton *02400*
Entomological Society of British Columbia, Victoria *02436*
Entomological Society of Canada, Ottawa *02437*
Entomological Society of Manitoba, Winnipeg *02438*
Entomological Society of Ontario, Guelph *02439*
Entomological Society of Saskatchewan, Indian Head *02440*
Societe d'Entomologie du Quebec, Sainte-Foy *02921*

Chile

Sociedad Chilena de Entomología, Santiago *03070*

China, People's Republic

Entomological Society of China, Beijing *03198*

Czech Republic

Česká Společnost Entomologická, Praha *03522*

Denmark

Entomologisk Forening, København *03718*

Egypt

Société Entomologique d'Egypte, Cairo *03914*

Finland

Suomen Hyönteistieteellinen Seura, Helsinki *04029*

France

Société des Lépidopteristes Français, Paris *04708*
Société Entomologique de France, Paris *04777*

Germany

Deutsche Gesellschaft für allgemeine und angewandte Entomologie e.V., Dossenheim *05317*
Münchner Entomologische Gesellschaft e.V., München *06124*

Hungary

Magyar Rovartani Társaság, Budapest *06656*

Italy

Accademia Nazionale di Entomologia, Firenze *07218*
Accademia Nazionale Italiana di Entomologia, Firenze *07223*
Apimondia, Roma *07248*
Associazione Romana di Entomologia, Roma *07400*
Società Entomologica Italiana, Genova *07801*

Japan

Nippon Kontyu Gakkai, Tokyo *08341*
Nippon Oyo-Dobutsu-Konchu Gakkai, Tokyo *08355*

Kenya

African Association of Insect Scientists, Nairobi *08463*

Mali

International African Migratory Locust Organisation, Bamako *08656*

Mexico

Sociedad Mexicana de Entomología, Chapingo *08759*

Netherlands

Nederlandse Entomologische Vereniging, Amsterdam *08992*

New Zealand

Entomological Society of New Zealand, Upper Hutt *09124*
New Zealand Society for Parasitology, Hamilton *09176*

Nigeria

Entomological Society of Nigeria, Zaria *09226*

Peru

Sociedad Entomológica del Perú, Lima *09476*

Poland

Polskie Towarzystwo Entomologiczne, Warszawa *09564*

South Africa

Entomological Society of Southern Africa, Pretoria *10140*

Switzerland

Schweizerische Entomologische Gesellschaft, Zürich *10804*

United Kingdom

Amateur Entomologists' Society, Feltham *11223*
C.A.B. International Institute of Entomology, London *11732*
Royal Entomological Society of London, London *12800*

U.S.A.

American Entomological Society, Philadelphia *13799*
American Mosquito Control Association, Lake Charles *13948*
Association of Applied Insect Ecologists, Sacramento *14393*
Entomological Society of America, Lanham *15046*

Venezuela

Sociedad Venezolana de Entomología, Maracay *17022*

Zambia

International Red Locust Control Organisation for Central and Southern Africa, Ndola *17153*

Environmental Protection s. Ecology

Ethnology s. a. Folklore

Algeria

Centre de Recherches Anthropologiques, Préhistoriques et Ethnographiques, Alger *00010*

Argentina

Asociación Argentina de Estudios Americanos, Buenos Aires *00063*
Departamento de Estudios Etnográficos y Coloniales, Santa Fé *00131*

Armenia

Interdepartmental Council for Study of Ethnic Problems, Yerevan *00208*

Australia

Australian Institute of Aboriginal Studies, Canberra City *00295*

Austria

Akademische Arbeitsgemeinschaft für Volkskunde, Wien *00542*
Anthropologische Gesellschaft in Wien, Wien *00544*
Arbeitskreis für Tibetische und Buddhistische Studien, Wien *00567*
Gesellschaft für Ost- und Südostkunde, Linz *00628*
Gesellschaft zur Erforschung slawischer Sprachen und Kulturen, Graz *00640*
Gesellschaft zur Förderung von Nordamerika-Studien an der Universität Wien, Wien *00644*
Niederösterreichisches Bildungs- und Heimatwerk, Arbeitsgemeinschaft für Volkskunde, Wien *00716*
Österreichische Ethnologische Gesellschaft, Wien *00739*
Österreichische Ethnomedizinische Gesellschaft, Wien *00740*
Österreichische Kulturgemeinschaft, Wien *00865*
Österreichische Orient-Gesellschaft Hammer-Purgstall, Wien *00875*
Österreichischer Fachverband für Volkskunde, Graz *00891*
Österreichisches Lateinamerika-Institut, Wien *00913*
Orientalische Gesellschaft, Wien *00930*
Pro Austria, Wien *00936*
Salzburger Kulturvereinigung, Salzburg *00942*
SIGMA – Salzburger Gesellschaft für Semiologie, Salzburg *00943*
Verein Freunde der Völkerkunde, Wien *00982*
Verein für Volkskunde in Wien, Wien *00988*

Belgium

Association Européenne de l'Ethnie Française, Bruxelles *01123*
Association Européenne pour l'Analyse Transculturelle de Groupe, Bruxelles *01130*
Fondation Egyptologique Reine Elisabeth, Bruxelles *01363*

Brazil

Centro de Pesquisas Folclóricas, Rio de Janeiro *01629*
Comissão Nacional de Folclore, Rio de Janeiro *01640*
Coordenação de Folclore e Cultura Popular, Rio de Janeiro *01647*
Fundação Nacional do Indio, Brasília *01660*
Instituto Archeológico, Histórico e Geográfico Pernambucano, Recife *01663*
Instituto Histórico, Geográfico e Etnográfico Paranaense, Curitiba *01686*

Canada

American Institute of Iranian Studies, Toronto *01861*
Canadian Association for Scottish Studies, Guelph *02040*
Canadian Association of African Studies, Toronto *02045*
Canadian Ethnic Studies Association, Toronto *02127*
Canadian Institute of Ukrainian Studies, Edmonton *02186*
Glenbow Alberta Institute, Calgary *02486*

China, People's Republic

Inner Mongolian Academy of Social Sciences, Huhehot *03206*

China, Republic

China Academy, Hwa Kang *03252*
The China Society, Taipei *03258*
Chinese Association for Folklore, Taipei *03260*
Ethnological Society of China, Taipei *03296*

Colombia

Comité Nacional del Consejo Internacional de Museos, Bogotá *03366*
Junta Nacional de Folclore, Bogotá *03397*

Cyprus

Society of Cypriot Studies, Nicosia *03499*

Czech Republic

Česká Národopisná Spolecnost, Praha *03511*
Česká Orientalistická Společnost, Praha *03513*

Denmark

Det Danske Afrika Selskab, Hellerup *03703*
Orientalsk Samfund, København *03792*

Egypt

The Egyptian Society for the Dissemination of Universal Culture and Knowledge, Cairo *03896*
Institut d'Egypte, Cairo *03906*

Finland

International Society for Folk-Narrative Research, Turku *03970*
Kalevalaseura, Helsinki *03972*
Museovirasto, Helsinki *03983*
Suomen Itämainen Seura, Helsingin Yliopisto *04031*
Suomen Muinaismuistoyhdistys, Helsinki *04051*

France

Asociata Internationalà de Studi Romåne, Paris *04140*
Association des Etudes Tsiganes, Paris *04182*
Association des Langues et Civilisations, Paris *04190*
Association Française des Arabisants, Paris *04225*
Association Française d'Etudes Américaines, Paris *04237*
Association Internationale des Etudes de l'Asie du Sud-Est, Paris *04272*
Comité International Rom, Pierrefitte *04410*

Société Africaine de Culture, Paris 04626
Société Asiatique de Paris, Paris 04631
Société des Africanistes, Paris 04691
Société des Américanistes, Paris 04693
Société des Amis du Musée de l'Homme, Paris 04698
Société des Océanistes, Paris 04711
Société d'Ethnographie de Paris, Paris 04722
Société d'Ethnologie Française, Paris 04723
Société d'Etudes et de Recherches pour la Connaissance de l'Homme, Paris 04735
Société d'Etudes Folkloriques du Centre-Ouest, Saint-Jean d'Angely 04739
Société d'Etudes Hispaniques et de Diffusion de la Culture Française à l'Etranger, Périgueux 04740
Société d'Etudes Italiennes, Paris 04742
Société d'Etudes Linguistiques et Anthropologiques de France, Paris 04746
Union Internationale des Etudes Orientales et Asiatiques, Saint-Maur 04986

French Polynesia

Société des Etudes Océaniennes, Papeete 05003

Germany

Arbeitskreis Ethnomedizin, Hamburg 05138
Arnold-Bergstraesser-Institut für kulturwissenschaftliche Forschung e.V., Freiburg 05152
Berliner Gesellschaft für Anthropologie, Ethnologie und Urgeschichte, Berlin 05187
Deutsche Gesellschaft für Amerikastudien e.V., Nürnberg 05318
Deutsche Gesellschaft für Asienkunde e.V., Hamburg 05325
Deutsche Gesellschaft für Osteuropakunde e.V., Berlin 05394
Deutsche Gesellschaft für Völkerkunde e.V., Freiburg 05437
Deutsche Gesellschaft für Volkskunde e.V., Göttingen 05439
Deutsche Morgenländische Gesellschaft e.V., Heidelberg 05473
Deutsches Orient-Institut, Hamburg 05598
Frobenius-Gesellschaft e.V., Frankfurt 05765
Geographische Gesellschaft Bremen, Bremen 05777
Gesellschaft für bedrohte Völker e.V., Göttingen 05803
Gesellschaft für Erd- und Völkerkunde, Bonn 05817
Gesellschaft für Erd- und Völkerkunde zu Stuttgart e.V., Stuttgart 05818
Gesellschaft für Natur- und Völkerkunde Ostasiens e.V., Hamburg 05844
Nah- und Mittelost-Verein e.V., Hamburg 06126
Naturhistorische Gesellschaft Nürnberg e.V., Nürnberg 06130
Osteuropa-Institut München, München 06226
Rheinische Vereinigung für Volkskunde, Bonn 06264
Sorbisches Institut e.V., Bautzen 06282
Südosteuropa-Gesellschaft e.V., München 06309
Volkskundliche Kommission für Westfalen, Münster 06434

Greece

Elliniki Laographiki Etaireia, Athinai 06509
Istoriki kai Ethnologiki Etaireia tis Ellados, Athinai 06537

Hong Kong

Royal Asiatic Society, Hong Kong Branch, Hong Kong 06595

Hungary

Magyar Néprajzi Társasag, Budapest 06647

India

Asiatic Society, Calcutta 06728
Asiatic Society of Bombay, Bombay 06729
Ethnographic and Folk Culture Society, Lucknow 06758
International Academy of Indian Culture, New Delhi 06826
Jammu and Kashmir Academy of Art, Culture and Languages, Srinagar 06830

Ireland

Folklore of Ireland Society, Dublin 06949

Israel

Association Internationale des Etudes Arméniennes, Jerusalem 07045
Israel Association for Asian Studies, Jerusalem 07062
Israel Oriental Society, Jerusalem 07084

Italy

Accademia di Romania, Roma 07172
Accademia Polacca delle Scienze, Roma 07226
Associazione degli Africanisti Italiani, Pavia 07266
Associazione Italiana di Cultura Classica, Firenze 07304
Associazione Italiana Studi Americanisti, Genova 07351
Associazione per la Conservazione delle Tradizioni Popolari, Palermo 07389
Associazione Sole Italico, Sulmona 07404
Centro Camuno di Studi Preistorici ed Etnologici, Capo di Ponte 07417
Centro di Studi Storici ed Etnografici del Piceno, Ascoli Piceno 07455
Centro Internazionale di Studi Sardi, Cagliari 07475
Centro Polesano di Studi Storici, Archeologici ed Etnografici, Rovigo 07529
Centro Ricerche di Storia e Arte Bitontina, Bitonto 07537
Centro Studi Zingari, Roma 07597
International Association of Biblicists and Orientalists, Ravenna 07706
Istituto Italiano per il Medio ed Estremo Oriente, Roma 07742
Istituto Italo-Africano, Roma 07746
Istituto per l'Oriente, Roma 07754
Società Italiana di Antropologia ed Etnologia, Firenze 07838
Società Italiana per l'Antropologia e la Etnologia, Firenze 07964
Union Generela di Ladins dla Dolomites, Ortisei-Urtijëi 08051
Union Ladins Val Badia, Pedraces 08052

Ivory Coast

Centre des Sciences Humaines, Abidjan 08063

Jamaica

Institute of Jamaica, Kingston 08080

Japan

Itaria Gakkai, Kyoto 08145
Nippon Afurika Gakkai, Tokyo 08280
Nippon Dokyo Gakkai, Tokyo 08297
Nippon Orient Gakkai, Tokyo 08354
Todai Chugoku Gakkai, Tokyo 08433
Toyo Gakujutsu Kyôkai, Tokyo 08440

Korea, Republic

Korea Branch of the Royal Asiatic Society, Seoul 08527
Korean Association of Sinology, Seoul 08529

Malaysia

Malaysian Branch of the Royal Asiatic Society, Kuala Lumpur 08646

Malta

Ghaqda Tal Folklor, Sliema 08663

Mexico

Instituto Indigenista Interamericano, México 08736

Netherlands

Association Internationale d'Etudes Occitanes, Oegstgeest 08824
Koninklijk Instituut voor Taal-, Land- en Volkenkunde, Leiden 08964

New Caledonia

Société des Etudes Mélanésiennes, Nouméa 09114

New Zealand

Polynesian Society, Auckland 09189

Pakistan

Institute of Islamic Culture, Lahore 09367
Lok Virsa, Islamabad 09373

Paraguay

Asciación Indeginista del Paraguay, Asunción 09425

Poland

Polskie Towarzystwo Ludoznawcze, Wroclaw 09584
Polskie Towarzystwo Orientalistyczne, Warszawa 09595

Portugal

Sociedade Portuguesa de Antropologia e Etnologia, Porto 09691

Puerto Rico

Caribbean Studies Association, Rio Piedras 09729

Romania

International Association of South-East European Studies, Bucuresti 09752

Russia

Association of Orientalists, Moskva 09803
Association of Sinologists, Moskva 09805

Singapore

The China Society, Singapore 10044

Slovakia

Slovenská Národopisná Spoločnost, Bratislava 10080
Slovenská Orientalistická Spoločnost, Bratislava 10081

South Africa

Africa Institute of South Africa, Pretoria 10120

Spain

Asociación Española de Orientalistas, Madrid 10265
Instituto de Estudios Africanos, Madrid 10319
Real Sociedad Económica de Amigos del País de Tenerife, Tenerife 10360
Sociedad de Ciencias, Letras y Artes El Museo Canario, Las Palmas 10372

Sri Lanka

Royal Asiatic Society of Sri Lanka, Colombo 10418

Switzerland

Geographisch-Ethnologische Gesellschaft Basel, Basel 10709
Schweizerische Afrika-Gesellschaft, Bern 10786
Schweizerische Asiengesellschaft, Zürich 10798
Schweizerische Ethnologische Gesellschaft, Bern 10806
Schweizerische Gesellschaft für Volkskunde, Basel 10856
Schweizerisches Institut für Volkskunde, Basel 10928
Société Suisse des Américanistes, Genève 10985

Tanzania

The Tanzania Society, Dar es Salaam 11062

Thailand

The Siam Society, Bangkok 11099

Tunisia

Comité International d'Etudes Morisques, Zaghouan 11139
Institut National d'Archéologie et d'Art, Tunis 11142

Turkey

Milletlerarasi Sark Tetkikleri Cemiyeti, Istanbul 11148
Türk Halk Bilgisi Derneği, Istanbul 11158

United Kingdom

African Studies Association of the United Kingdom, London 11211
An Comunn Gaidhealach, Inverness 11229
Arab Research Centre, London 11240
British Association for American Studies, Newcastle-upon-Tyne 11485
China Society, London 11778
The Egypt Exploration Society, London 11948
Folklore Society, London 12082
Gypsy Lore Society, Sandon near Stafford 12139
Hispanic and Luso-Brazilian Council, London 12170
Institution of Polish Engineers in Great Britain, London 12314
International African Institute, London 12321
International Association for Scandinavian Studies, Norwich 12325
Maghreb Studies Association, London 12476
National Association for Soviet and East European Studies, Milton Keynes 12562
The Royal African Society, London 12766
The Royal Asiatic Society, London 12770
Royal Celtic Society, Edinburgh 12776
Royal Commonwealth Society, London 12796
Royal Society for Asian Affairs, London 12849
Scottish Society for Northern Studies, Edinburgh 12929
Society for Cooperation in Russian and Soviet Studies, London 12968
Society for Folk Life Studies, Hollywood 12974
Society for Latin American Studies, York 12977
Society for Libyan Studies, London 12978
The Society for South Asian Studies, London 12997
Ulster Folklife Society, Holywood 13152

U.S.A.

Aboriginal Research Club, Dearborn 13275
Afghanistan Studies Association, Omaha 13340
African Heritage Center for African Dance and Music, Washington 13344
African Heritage Studies Association, Flushing 13345
African Studies Association, Atlanta 13346
Afro-Asian Center, Saugerties 13348
American Academy for Jewish Research, New York 13377
American Conference for Irish Studies, Fort Wayne 13749
American Ethnological Society, Lancaster 13803
American Folklore Society, Harrisburg 13817
American Indian Lore Association, Logan 13860
American Institute of Indian Studies, Chicago 13883
American Institute of Iranian Studies, Columbus 13884
American Institute of Islamic Studies, Denver 13885
American Oriental Society, Ann Arbor 13968
American Research Institute in Turkey, Philadelphia 14029
American Siam Society, Santa Monica 14041
American Society for Ethnohistory, Chicago 14060
American Swedish Institute, Minneapolis 14187
Amerind Foundation, Dragoon 14214
The Asia Society, New York 14253
Association for Asian Studies, Ann Arbor 14268
Association for Jewish Demography and Statistics – American Branch, Los Angeles 14310
Association for Korean Studies, Rancho Palos Verdes 14312
Association for the Advancement of Baltic Studies, Hackettstown 14341
Association for the Study of Afro-American Life and History, Washington 14356
Association of Arab-American University Graduates, Belmont 14394
Association of Caribbean Studies, Lexington 14407
Bet Nahrain, Modesto 14580
Center for Austrian Studies, Minneapolis 14669
Center for Chinese Research Materials, Oakton 14671
Center for Migration Studies of New York, Staten Island 14689
China Institute in America, New York 14734
Chinese Historical Society of America, San Francisco 14739
Estonian Learned Society of America, New York 15060
Foundation for Mideast Communication, Beverly Hills 15130
Gypsy Lore Society, Cheverly 15183
Hispanic Institute, New York 15212
Independent Scholars of Asia, Berkeley 15255
Institute of Chinese Culture, New York 15313
Institute of Lithuanian Studies, Chicago 15329
Jugoslavia Study Group, Wausau 15630
Middle East Research and Information Project, Washington 15734
Mongolia Society, Bloomington 15752
National Association for Armenian Studies and Research, Belmont 15793
National Association for Ethnic Studies, Tempe 15799
Nepal Studies Association, Oakland 16191
North American Conference on British Studies, Seattle 16215
Scandinavian Seminar, Amherst 16441
Society for Ethnomusicology, Bloomington 16520
Society for Iranian Studies, New York 16544
Society for Slovene Studies, Bloomington 16579
Society for the Advancement of Scandinavian Study, Provo 16590
Society of Ethnic and Special Studies, Edwardsville 16658
Tibet Society, Bloomington 16782
Ukrainian Institute of America, New York 16798
YIVO Institute for Jewish Research, New York 16941

Venezuela

Asociación Cultural Humboldt, Caracas 16966
Centro de Estudios Venezolanos Indigenas, Caracas 16982

Family Planning

Canada

Association du Planning Familial des Sept-Iles, Sept-Iles 01896
Cumberland County Family Planning Association, Amherst 02411
Family Planning Resource Team, Sydney 02447
Fédération du Québec pour le Planning des Naissances, Montréal 02453
International Childbirth Education Association, Calgary 02545
Serena British Columbia, Surrey 02910
Serena Canada, Ottawa 02911

Dominican Republic

Asociación Dominicana pro Bienestar de la Familia, Santo Domingo 03832

Germany

Deutsches Institut für Vormundschaftswesen, Heidelberg 05588

Ireland

Irish Family Planning Association, Dublin 06983

Italy

Centro di Informazioni Sterilizzazione e Aborto, Milano 07429
Centro di Studi e di Ricerche per la Psicoterapia della Coppia e della Famiglia, Roma 07440
Centro Internazionale Studi Famiglia, Milano 07486
Centro per la Riforma del Diritto di Famiglia, Milano 07524
Famiglia e Libertà, Roma 07685
Unione Consultori Italiani Prematrimoniali e Matrimoniali, Bologna 08043

Kenya

Centre for African Family Studies, Nairobi 08484

United Kingdom

The Family Planning Association, London 12056

Uruguay

Centro de Investigaciones y Estudios Familiares, Montevideo 13252

U.S.A.

Family Health International, Durham 15075
International Federation for Family Life Promotion, Washington 15473
National Council on Family Relations, Minneapolis 16020
Planned Parenthood Federation of America, New York 16324
Society of Teachers of Family Medicine, Kansas City 16706

Vietnam

Vietnamese Association of Obstetrics, Gynaecology and Family Planning, Hanoi 17054

Finance

Argentina

Instituto para la Integración de América Latina, Buenos Aires 00147

Australia

Australian Institute of Credit Management, Artarmon 00299
Australian Institute of Valuers and Land Economists, Deakin 00316
Commonwealth Institute of Valuers, Sydney 00376

Austria

Österreichische Bankwissenschaftliche Gesellschaft an der Wirtschaftsuniversität Wien, Wien 00729
Österreichisches Forschungsinstitut für Sparkassenwesen, Wien 00909

Belgium
Association Belge des Analystes Financiers, Bruxelles *01111*
European Finance Association, Bruxelles *01350*

Benin
African Union for the Management of Development Banks, Cotonou *01515*

Canada
Canadian Credit Institute Educational Foundation, Mississauga *02112*
Canadian Institute of Cost Reduction, Ottawa *02170*
Canadian Institute of Credit and Financial Management, Mississauga *02171*
Canadian Institute of Financial Planning, Toronto *02173*
Institute of Chartered Accountants of Ontario, Toronto *02528*
Toronto Society of Financial Analysts, Toronto *02967*
Winnipeg Society of Financial Analysts, Winnipeg *02996*

China, Republic
Chinese Society of Budgetary Management, Taipei *03286*
Finance Association of China, Taipei *03297*

Denmark
International Banking Research Institute, København *03745*

France
Centre for the Advancement and Study of the European Currency, Lyon *04373*
Institut des Actuaires Français, Paris *04565*
Société d'Etudes Financières et Meunières, Paris *04738*
Société d'Etudes Minières, Industrielles et Financières, Paris *04748*
Société Financière Européenne, Paris *04780*

Germany
Bankakademie e.V., Frankfurt *05165*
Deutsches wissenschaftliches Steuerinstitut der Steuerberater und Steuerbevollmächtigten e.V., Bonn *05605*
Deutsche Vereinigung für Finanzanalyse und Anlageberatung e.V., Dreieich *05613*
Deutsche Vereinigung für Internationales Steuerrecht, Köln *05615*
Deutsche Vereinigung für internationales Steuerrecht im Verband der Fiscal Association, Bayerische Sektion e.V., München *05616*
Europäische Vereinigung für Eigentumsbildung, Hermann-Lindrath-Gesellschaft e.V., Hannover *05658*
Gesellschaft für Finanzwirtschaft in der Unternehmensführung e.V., Frankfurt *05822*
Gesellschaft für internationale Geldgeschichte, Frankfurt *05833*
Gesellschaft für Wohnungsrecht und Wohnungswirtschaft Köln e.V., Köln *05874*
Gesellschaft zur Förderung der finanzwissenschaftlichen Forschung e.V., Köln *05881*
Gesellschaft zur Förderung der Wissenschaftlichen Forschung über das Spar- und Girowesen e.V., Bonn *05885*
Institut Finanzen und Steuern e.V., Bonn *05945*
Institut für bankhistorische Forschung e.V., Frankfurt *05953*
Institut für Städtebau, Wohnungswirtschaft und Bausparwesen e.V., Bonn *05972*
Internationales Institut für Öffentliche Finanzen, Saarbrücken *06000*
International Institute of Public Finance, Saarbrücken *06008*
Vereinigung für Bankbetriebsorganisation e.V., Frankfurt *06409*

Ireland
Institute of Taxation in Ireland, Dublin *06965*

Italy
Associazione Nazionale per lo Studio dei Problemi del Credito, Roma *07382*
Associazione per lo Sviluppo degli Studi di Banca e Borsa, Milano *07393*
Centro Italiano di Studi Finanziari, Roma *07501*
International Centre for Advanced Technical and Vocational Training, Torino *07709*

Japan
Asian Council of Securities Analysts, Tokyo *08098*
Nihon Zaisei Gakkai, Tokyo *08277*

Mexico
Centre for Latin American Monetary Studies, México *08712*
Centro de Estudios Monetarios Latinoamericanos, México *08716*

Netherlands
Société Universitaire Européenne de Recherches Financières, Tilburg *09069*

Senegal
African Centre for Monetary Studies, Dakar *10009*

Singapore
Asian-Pacific Tax and Investment Research Centre, Singapore *10042*

South Africa
Institute of Bankers in South Africa, Johannesburg *10148*

United Kingdom
Association of Health Service Treasurers, Warwick *11354*
The Chartered Institute of Bankers, London *11760*
Chartered Institute of Management Accountants, London *11762*
Chartered Institute of Public Finance and Accountancy, London *11765*
Institute of Bankers in Scotland, Edinburgh *12214*
Institute of Credit Management, Stamford *12230*
Institute of Investment Management and Research, Bromley *12248*

U.S.A.
American Association for Budget and Program Analysis, Falls Church *13469*
American Council of Independent Laboratories, Washington *13762*
American Finance Association, New York *13814*
Association for Investment Management and Research, Charlottesville *14309*
Eastern Finance Association, Statesboro *15003*
Institute of Certified Financial Planners, Denver *15312*
Institute of Financial Education, Chicago *15320*
International Society of Parametric Analysts, Germantown *15553*
National Economic Association, Ann Arbor *16038*
National Grants Management Association, Washington *16065*
National Tax Association – Tax Institute of America, Columbus *16168*
New York Financial Writers' Association, Syosset *16203*
Society of Cost Estimating and Analysis, Alexandria *16652*

Fine Arts
s. a. Graphic and Decorative Arts, Design

Austria
Adalbert Stifter-Gesellschaft, Wien *00531*

Belgium
Centre International d'Etude de la Peinture Médiévale des Bassins de l'Exant et de la Meuse, Bruxelles *01196*
Koninklijke Academie voor Wetenschappen, Letteren en Schone Kunsten van België, Bruxelles *01409*
Société Royale des Beaux-Arts, Bruxelles *01485*

Canada
Canadian Art Museums Directors' Organization, Victoria *02024*
Canadian Society of Painters in Watercolour, Toronto *02290*
Council of Regents for Colleges of Applied Arts and Technology, Toronto *02407*

Denmark
Danske Kunsthåndvårkeres Landssammenslutning, København *03609*

France
Société des Amis du Louvre, Paris *04697*

Germany
Akademie der Künste, Berlin *05048*
Bundesverband Bildender Künstler Bundesrepublik Deutschland e.V., Bonn *05241*
Ernst Barlach Gesellschaft e.V., Hamburg *05642*
Gesellschaft für Goldschmiedekunst, Hanau *05826*
Gesellschaft zur Förderung Frankfurter Malerei des 19. und 20. Jahnhunderts e.V., Frankfurt *05888*
Peter-Schwingen-Gesellschaft e.V., Bonn *06233*
Verband der Gemeinschaften der Künstlerinnen und Kunstfreunde e.V., Wuppertal *06338*
Wilhelm-Busch-Gesellschaft e.V., Hannover *06442*

Hungary
Magyar Régészeti és Művészettörténeti Társulat, Budapest *06655*

Israel
Wilfrid Israel House for Oriental Art and Studies, Hazorea *07117*

Italy
Academia Gentium Pro Pace, Roma *07123*
Accademia Raffaello, Urbino *07231*
Società per le Belle Arti ed Esposizione Permanente, Milano *08002*
Società Promotrice di Belle Arti, Napoli *08005*

Macedonia
Društvo na Likovnite Umetnici na Makedonija, Skopje *08595*

Norway
Norske Billedkunstnere, Oslo *09297*

Spain
Instituto Amatller de Arte Hispánico, Barcelona *10316*

Tunisia
Union Nationale des Arts Plastiques, Tunis *11143*

United Kingdom
British Artists in Glass, Brampton *11482*
British Origami Society, Birmingham *11624*
British Society of Painters in Oil, Pastels and Acrylic, Ilkley *11687*
Charles Rennie Mackintosh Society, Glasgow *11758*
Federation of British Artists, London *12063*
Hesketh Hubbard Art Society, London *12166*
Marquetry Society, Saint Albans *12497*
Pastel Society, London *12681*
Royal Fine Art Commission for Scotland, Edinburgh *12804*
Royal Glasgow Institute of the Fine Arts, Glasgow *12807*
Royal Institute of Oil Painters, London *12815*
Royal Institute of Painters in Water Colours, London *12816*
Royal Society of British Artists, London *12855*
Royal Society of British Sculptors, London *12856*
Royal Society of Marine Artists, London *12861*
Royal Society of Portrait Painters, London *12863*
Royal Watercolour Society, London *12873*
Society of Miniaturists, Ilkley *13057*
Society of Wildlife Artists, London *13070*
Society of Women Artists, London *13071*
Turner Society, London *13145*
United Society of Artists, Alderney *13172*

U.S.A.
Catholic Fine Arts Society, Rockville Centre *14654*
Drawing Society, New York *14985*
National Sculpture Society, New York *16152*

Venezuela
Sociedad Amigos del Museo de Bellas Artes, Caracas *17008*

Vietnam
Vietnamese Fine Arts Association, Hanoi *17068*

Fisheries

Argentina
Instituto Nacional de Investigación y Desarrollo Pesquero, Mar del Plata *00144*
Instituto Nacional de Tecnologia Industrial, Buenos Aires *00145*

Canada
Conseil des Recherches en Pêche et en Agro-Alimentaire du Québec, Québec *02382*
Fish and Seafood Association of Ontario, Toronto *02464*
Fisheries Association of British Columbia, Vancouver *02465*

China, People's Republic
China Society of Fisheries, Beijing *03134*

France
Société Française d'Ichtyologie, Paris *04860*

India
Inland Fisheries Society of India, Barrackpore *06824*

Italy
Aquatic Sciences and Fisheries Information System, Roma *07249*
Centro Italiano di Studi e Programmazioni per la Pesca, Roma *07499*
International Centre for Advanced Technical and Vocational Training, Torino *07709*

Japan
Hyomen Gijutsu Kyokai, Tokyo *08138*
Nippon Suisan Gakkai, Tokyo *08388*

Norway
Norges Fiskeriforskningsråd, Trondheim *09282*

Philippines
Asian Fisheries Society, Manila *09497*

United Kingdom
European Association of Fisheries Economists, Southsea *11995*
Institute of Fisheries Management, Holmer *12237*
National Association of Principal Agricultural Education Officers, Beverley *12569*

U.S.A.
American Fisheries Society, Bethesda *13816*
Gulf and Caribbean Fisheries Institute, Charleston *15182*
Inter-American Tropical Tuna Comission, La Jolla *15355*
National Fisheries Contaminant Research Center, Columbia *16052*
National Fisheries Education and Research Foundation, Arlington *16053*
National Shellfisheries Association, Southampton *16153*
Society for the Protection of Old Fishes, Seattle *16607*
Trout Unlimited, Vienna *16793*

Folklore
s. a. Ethnology

Austria
Akademische Arbeitsgemeinschaft für Volkskunde, Wien *00542*

Belgium
Institut Archéologique du Luxembourg, Arlon *01375*

Brazil
Comissão Nacional de Folclore, Rio de Janeiro *01640*

Canada
Canadian Folk Arts Council, Toronto *02137*
Canadian Folk Music Society, Calgary *02138*
The Lithuanian Folk Art Institute, Toronto *02594*

Finland
Kalevalaseura, Helsinki *03972*
Suomalaisen Kirjallisuuden Seura, Helsinki *04001*

Germany
Europäische Märchengesellschaft e.V., Rheine *05654*

India
Indian Folklore Society, Calcutta *06792*

Italy
Associazione di Cultura Lao Silesu, Iglesias *07271*
Associazione per la Conservazione delle Tradizioni Popolari, Palermo *07389*
Centro Internazionale di Studi Sardi, Cagliari *07475*
Società Istriana di Archeologia e Storia Patria, Trieste *07820*

Japan
Minzokugaku Shinkokai, Tokyo *08182*

Macedonia
Združenie na Folkloristite na Makedonija, Skopje *08619*

Pakistan
Lok Virsa, Islamabad *09373*

Switzerland
Schweizerische Trachtenvereinigung, Burgdorf *10935*

United Kingdom
British Caspian Trust, Droitwich *11539*

U.S.A.
American Historical Society of Germans from Russia, Lincoln *13847*
Pioneer America Society, Waukesha *16321*

Food
s. a. Nutrition

Argentina
Instituto Nacional de Investigación y Desarrollo Pesquero, Mar del Plata *00144*
Instituto Nacional de Tecnologia Industrial, Buenos Aires *00145*

Australia
Asia Pacific Food Analysis Network, Canberra *00228*
Australian Institute of Food Science and Technology, Pymble *00301*
Council of Australian Food Technology Associations, North Sydney *00383*

Austria
Asian Regional Cooperative Project on Food Irradiation, Wien *00570*
Österreichische Gesellschaft für Ernährungsforschung, Wien *00778*
Verein Österreichischer Lebensmittel- und Biotechnologen, Wien *01000*

Belgium
European Alliance of Safe Meat, Bruxelles *01269*

Brazil
Centro Nacional de Pesquisa de Arroz e Feijão, Goiânia *01632*
Centro Nacional de Pesquisa de Mandioca e Fruticultura, Cruz das Almas *01633*
Centro Nacional de Pesquisa de Soja, Londrina *01636*

Bulgaria
Bulgarian Nutrition Society, Sofia *01752*
Union of the Food Industry, Sofia *01801*

Canada
Association des Professeurs en Alimentation du Québec, Montréal *01892*
Canadian Institute of Food Science and Technology, Ottawa *02174*
Canadian Spice Association, Montréal *02300*
Fish and Seafood Association of Ontario, Toronto *02464*
Food Products Accident Prevention Association, Toronto *02470*
International Development Research Centre, Ottawa *02552*
New Brunswick Dietetic Association, Saint John *02679*
Sugar Industry Technologists, Sainte-Thérèse de Blainville *02962*

China, Republic
Asian Food Council, Taipei *03239*

Denmark
Danske Veterinårhygiejnikeres Organisation, Slagense 03614
Levnedsmiddelselskabet, København 03768

Germany
Bund für Lebensmittelrecht und Lebensmittelkunde e.V., Bonn 05265
Förderungsgemeinschaft der Kartoffelwirtschaft e.V., Munster 05704
Forschungsgemeinschaft für Verpackungs- und Lebensmitteltechnik e.V., München 05711
Institut für nationale und internationale Fleisch- und Ernährungswirtschaft, Heidelberg 05965
Normenausschuss Lebensmittel und Landwirtschaftliche Produkte im DIN Deutsches Institut für Normung e.V., Berlin 06188
Verein Deutscher Zuckertechniker, Schladen 06378

Guatemala
Instituto de Nutrición de Centro América y Panamá, Guatemala City 06570

Hungary
Magyar Elemezésipari Tudományos Egyesület, Budapest 06625

Iceland
Rannsóknastofnun Fiskidnadarins, Reykjavik 06697

Ireland
International Union of Food Science and Technology, Dublin 06970

Italy
Associazione Italiana di Dietetica e Nutrizione Clinica, Roma 07305
Centro Appenninico del Terminillo Carlo Jucci, Rieti 07415
Centro Informazione Farine e Pane, Roma 07464
Codex Coordinating Committee for Africa, Roma 07606
Società Italiana di Nutrizione Umana, Roma 07917

Jamaica
Caribbean Food and Nutrition Institute, Kingston 08073
Jamaican Association of Sugar Technologists, Mandeville 08084

Japan
Nihon Denpun Gakkai, Ibaraki 08191
Nippon Bitamin Gakkai, Kyoto 08281
The Society of Fermentation and Bioengineering, Japan, Osaka 08425

Malaysia
ASEAN Food Handling Bureau, Kuala Lumpur 08628

Mexico
Sociedad Mexicana de Nutrición y Endocrinología, México 08769

Morocco
Conférence Internationale des Directeurs et Doyens des Etablissements d'Enseignement Supérieur et Facultés d'Expression Française des Sciences de l'Agriculture et de l'Alimentation, Rabat 08807

Netherlands
European Association for Potato Research, Wageningen 08860
Vereniging voor Zuivelindustrie en Melkhygiëne, 's-Gravenhage 09107

New Zealand
New Zealand Institute of Food Science and Technology, Christchurch 09156
New Zealand Society of Dairy Science and Technology, Edgecumbe 09178
Nutrition Society of New Zealand, Dunedin 09183

Nigeria
African Council of Food and Nutrition Sciences, Harare 09214
Nutrition Society of Nigeria, Ile-Ife 09243

Philippines
Philippine Association of Nutrition, Manila 09523
Society for the Advancement of the Vegetable Industry, Laguna 09537

Russia
Society of the Food Industry, Moskva 09980

Senegal
Comité Scientifique Inter-Africain Post-Récolte, Dakar 10024

Slovakia
Slovenská Spoločnost pre Vedy Polnohospodarske, Lesnické a Potravinárské, Ivanka pri Dunaji 10091

South Africa
The Nutrition Society of Southern Africa, Medunsa 10158
The South African Association for Food Science and Technology, Edenvale 10179

Spain
Sociedad Española de Bromatología, Madrid 10379

Sweden
International Society for Fat Research, Karlshamn 10477
Svenska Livsmedelstekniska Föreningen, Stockholm 10563

Switzerland
Schweizerische Gesellschaft für Ernährungsforschung, Basel 10818
Schweizerische Gesellschaft für Lebensmittel-Wissenschaft und -Technologie, Wädenswil 10829
Schweizerische Vereinigung für Ernährung, Zollikofen 10943

Tunisia
Association Africaine de Microbiologie et d'Hygiène Alimentaire, Sousse 11134

United Kingdom
Agricultural and Food Research Council, Swindon 11213
Association of Public Analysts, London 11382
British Food Manufacturing Industries Research Association, Leatherhead 11564
British Society for the Promotion of Vegetable Research, Wellesbourne 11675
British Society of Flavourists, Haverhill 11683
Campden Food and Drink Research Association, Chipping Campden 11742
Farm and Food Society, London 12057
Food Education Society, London 12083
Institute of Food Science and Technology of the United Kingdom, London 12239
Institute of Refrigeration, Carshalton 12277
International Food Information Service, Reading 12364
Nutrition Society, London 12651

U.S.A.
American Association of Feed Microscopists, Sacramento 13535
American Board of Nutrition, Bethesda 13641
American Dairy Science Association, Champaign 13775
American Meat Science Association, Chicago 13931
American Peanut Research and Education Society, Stillwater 13993
American Pomological Society, University Park 14011
American Society for Clinical Nutrition, Bethesda 14051
American Society of Bakery Engineers, Chicago 14101
American Society of Sugar Beet Technologists, Denver 14168
Association of Food and Drug Officials, York 14439
Association of Vitamin Chemists, Northville 14534
Center for Hospitality Research and Service, Blacksburg 14682
Distillers Feed Research Council, Des Moines 14977
Food and Nutrition Board, Washington 15108
Food Distribution Research Society, Newark 15109
Institute of Food Technologists, Chicago 15321
National Dry Bean Council, Washington 16036
Potato Association of America, Hancock 16344
Price-Pottenger Nutrition Foundation, La Mesa 16352
Society of Soft Drink Technologists, Hartfield 16701

Forensic Medicine

Australia
Association of Australasian and Pacific Area Police Medical Officers, Melbourne 00229

Austria
Österreichische Gesellschaft für gerichtliche Medizin, Wien 00784

France
Société de Médecine Légale et de Criminologie de France, Paris 04670

Germany
Deutsche Gesellschaft für Rechtsmedizin, Köln 05416

Italy
Consiglio Nazionale Forense, Roma 07629

Russia
Mational Scientific Medical Society of Forensic Medical Officers, Moskva 09880

United Kingdom
Association of British Forensic Specialists, Moseley 11311
Association of Police Surgeons, Northampton 11376
British Academy of Forensic Sciences, London 11472
British Association in Forensic Medicine, Leeds 11498

U.S.A.
American Society of Forensic Odontology, Chicago 14123
American Society of Law and Medicine, Boston 14140
Association of Forensic Document Examiners, Minneapolis 14440
International Reference Organization in Forensic Medicine and Sciences, Wichita 15519
Milton Helpern Institute of Forensic Medicine, New York 15740

Vietnam
Vietnamese Association of Forensic Scientists, Hanoi 17046

Forestry

Algeria
Centre National de Recherches et Expérimentations Forestières, Alger 00018

Austria
Internationaler Verband Forstlicher Forschungsanstalten, Wien 00677
Österreichische Gesellschaft für Holzforschung, Wien 00791

Brazil
Instituto Nacional de Pesquisas da Amazonia, Manaus 01688

Bulgaria
Union of Forest Engineering, Sofia 01793

Canada
Alberta Forest Development Research Trust Fund, Spruce Grove 01833
Association de Securité des Exploitations Forestières du Québec, Québec 01881
Association de Securité des Industriels Forestiers du Québec, Québec 01882
Association of University Forestry Schools of Canada, Thunder Bay 01946
Canadian Forestry Association, Ottawa 02139
Canadian Institute of Forestry, Ottawa 02175
Canadian Institute of Treated Wood, Ottawa 02185
Forest Products Accident Prevention Association, North Bay 02471
Forestry Canada – Ontario Region, Sault Sainte Marie 02472
Interior Forest Labour Relations Association, Kelowna 02536
Newfoundland Forest Protection Association, Corner Brook 02704
Woodworkers Accident Prevention Association, Toronto 02997

China, People's Republic
Chinese Academy of Forestry, Beijing 03138
Chinese Society of Forestry, Beijing 03190
Heilongjiang Academy of Forestry Sciences, Harbin 03201

China, Republic
Chinese Forestry Association, Taipei 03268

Congo
Centre Technique Forestier Tropical, Pointe-Noire 03424

Costa Rica
Tropical Science Center, San José 03452

Cuba
Centro de Investigación Forestal, La Habana 03476

Denmark
Dansk Dendrologisk Forening, Hørsholm 03600
Danske Forstkandidaters Forening, Klampenborg 03605
Dansk Skovforening, Frederiksberg 03690
Det Danske Hedeselskab, Viborg 03705

Finland
Suomen Metsätieteellinen Seura, Helsinki 04050

Germany
AID Auswertungs- und Informationsdienst für Ernährung, Landwirtschaft und Forsten e.V., Bonn 05045
Arbeitsgemeinschaft wildbiologischer und jagdkundlicher Forschungsstätten, Bonn 05130
Bayerischer Holzwirtschaftsrat, München 05176
Dachverband wissenschaftlicher Gesellschaften der Agrar-, Forst-, Ernährungs-, Veterinär- und Umweltforschung e.V., Frankfurt 05282
Deutsche Dendrologische Gesellschaft, Trier 05297
Deutsche Gesellschaft für Holzforschung e.V., München 05363
Deutscher Forstverein e.V., Frankfurt 05511
Deutscher Forstwirtschaftsrat e.V., Rheinbach 05512
Deutscher Holzwirtschaftsrat, Wiesbaden 05517
Deutscher Verband Forstlicher Forschungsanstalten, Trippstadt 05546
Kuratorium für Waldarbeit und Forsttechnik e.V., Groß-Umstadt 06089
Verein für Forstliche Standortskunde und Forstpflanzenzüchtung e.V., Freiburg 06384

Ghana
African Forestry and Wildlife Commission, Accra 06471

Greece
Omospondia Panellinios Syndesmon Dasoponon, Athinai 06540

Hungary
Országos Erdészeti Egyesület, Budapest 06669

Iceland
Skógraektarfélag Islands, Reykjavik 06698

Ireland
Irish Timber Growers Association, Dublin 07003
Society of Irish Foresters, Dublin 07033

Italy
Accademia Italiana di Scienze Forestali, Firenze 07194
Associazione Forestale Italiana, Roma 07277
Committee on Forest Development in the Tropics, Roma 07624
Federazione Italiana Dottori in Agraria e Forestali, Roma 07692
International Poplar Commission, Roma 07717
Società Orticola Italiana, Firenze 07997

Ivory Coast
Centre Technique Forestier Tropical, Abidjan 08064

Japan
Nippon Mokuzai Gakkai, Tokyo 08344
Nippon Ringakukai, Tokyo 08358
Sapporo Norin Gakkai, Sapporo 08407

Korea, Democratic People's Republic
Academy of Forestry Science, Pyongyang 08510

Korea, Republic
Korean Forestry Society, Kyonggido 08532

Macedonia
Sojuz na Inženeri i Tehničari po Sumarstvo i Industrija za Prerabotka na Drvo na Makedonija, Skopje 08615

Malaysia
ASEAN Institute of Forest Management, Kuala Lumpur 08630

Netherlands
International Association of Wood Anatomists, Utrecht 08925
Koninklijke Nederlandse Bosbouw Vereniging, Arnhem 08953
Nederlandse Dendrologische Vereniging, Boskoop 08990

New Zealand
New Zealand Institute of Forestry, Christchurch 09157

Nigeria
Forestry Association of Nigeria, Ibadan 09228

Norway
Det Norske Skogselskap, Oslo 09263
Skogbrukets og Skogindustrienes Forskningsforening, Aas 09348

Poland
Polskie Towarzystwo Lesne, Warszawa 09583

Romania
Academia de Stiinte Agricole si Silvice, Bucuresti 09734

Russia
Paper and Wood-Working Society, Moskva 09940
Society of the Timber and Forestry Industry, Moskva 09982

Slovakia
Slovenská Spoločnost pre Vedy Polnohospodarske, Lesnické a Potravinárské, Ivanka pri Dunaji 10091

South Africa
Southern African Institute of Forestry, Pretoria 10225
The Tree Society of Southern Africa, Johannesburg 10232

Sweden
International Research Group on Wood Preservation, Stockholm 10476
Kungliga Skogs- och Lantbruksakademien, Stockholm 10489
Skogs- och Jordbrukets Forskningsråd, Stockholm 10520

Switzerland
Schweizerischer Forstverein, Zürich 10896
Waldwirtschaftsverband Schweiz, Solothurn 11029

Syria
Arab Institute for Forestry and Ranges, Lattakia 11042

Tanzania
East African Agriculture and Forestry Research Organization, Arusha 11051

Thailand
Asia-Pacific Forestry Commission, Bangkok 11085

Trinidad and Tobago
Eastern Caribbean Institute of Agriculture and Forestry, Arima 11120

United Kingdom
Arboricultural Association, Romsey 11241
Association of National Park Officers, Kendal 11371
CAB International Forestry Bureau, Wallingford 11731
Commonwealth Forestry Association, Oxford 11821
Institute of Chartered Foresters, Edinburgh 12221
Institute of Wood Science, High Wycombe 12291
National Association of Principal Agricultural Education Officers, Beverley 12569
Royal Forestry Society of England, Wales and Northern Ireland, Tring 12805
Royal Scottish Forestry Society, Dunfermline 12845
Scottish Wildlife Trust, Edinburgh 12936

Standing Committee on Commonwealth Forestry, Edinburgh *13094*
Timber Research and Development Association, High Wycombe *13133*

U.S.A.

American Boxwood Society, Boyce *13668*
American Forestry Association, Washington *13820*
American Wood-Preservers' Association, Woodstock *14212*
Forest History Society, Durham *15112*
Forest Products Research Society, Madison *15113*
Hardwood Research Council, Memphis *15187*
International Tree Crops Institute U.S.A., Davis *15567*
National Council of Forestry Association Executives, Augusta *15999*
Society of American Foresters, Washington *16634*
Society of Wood Science and Technology, Madison *16714*
Tree-Ring Society, Tucson *16791*

Vietnam

Vietnamese Association of Forestry, Hanoi *17047*

Futurology

Canada

Canadian Association for Future Studies, Montréal *02032*

China, People's Republic

Chinese Society for Future Studies, Beijing *03180*

France

International Association Futuribles, Paris *04579*

Italy

World Future Studies Federation, Roma *08055*

Switzerland

Schweizerische Vereinigung für Zukunftsforschung, Horgen *10955*

U.S.A.

American Teilhard Association, Syosset *14189*
Earthrise, Tulsa *15001*
Institute for the Future, Menlo Park *15303*
World Future Society, Bethesda *16927*

Gardening s. Horticulture

Gastroenterology

Argentina

Sociedad Argentina de Gastroenterología, Buenos Aires *00170*

Australia

Asian Pan-Pacific Society for Paediatric Gastroenterology and Nutrition, East Perth *00227*

Austria

Österreichische Gesellschaft für Gastroenterologie und Hepatologie, Wien *00781*

Belgium

Société Belge de Gastro-Entérologie, Bruxelles *01429*

Bulgaria

Société Bulgare de Gastroentérologie, Sofia *01776*

Canada

Canadian Association of Gastroenterology, Montréal *02051*

Chile

Sociedad Chilena de Gastroenterología, Santiago *03072*

Czech Republic

Gastroenterologická Společnost, Praha *03544*

Denmark

Dansk Gastroenterologisk Selskab, Hvidovre *03623*

Finland

Suomen Gastroenterologiayhdistys, Helsinki *04020*

France

Société Nationale Française de Gastro-Entérologie, Paris *04905*

Germany

Deutsche Gesellschaft für Phlebologie, Norderney *05401*

Hungary

Magyar Gastroenterologiai Társaság, Budapest *06630*

Israel

Israel Society for Gastroenterology, Tel Aviv *07092*

Italy

Società Italiana di Gastroenterologia, Bologna *07879*

Japan

Nihon Shokaki Naishikyo Gakkai, Tokyo *08263*
Nippon Shokaki-byo Gakkai, Tokyo *08381*

Mexico

Asociación Interamericana de Gastroentereología, México *08693*

Netherlands

European Association for Gastroenterology and Endoscopy, Amsterdam *08853*
Nederlandse Vereniging voor Gastro-Enterologie, Haarlem *09018*

Norway

Norsk Gastroenterologisk Selskap, Oslo *09315*

Portugal

Sociedade Portuguesa de Gastroenterologia, Lisboa *09701*

Romania

Societatea de Gastroenterologie, Bucuresti *09764*

Russia

National Scientific Medical Society of Gastroenterologists, Moskva *09914*

Sweden

Svensk Förening för Gastroenterologi och Gastrointestinal Endoskopi, Helsingborg *10581*

Turkey

Türk Gastroenterologii Derneği, Ankara *11157*

United Kingdom

Association of National European and Mediterranean Societies of Gastroenterology, London *11369*
British Society of Gastroenterology, London *11684*
Ileostomy Association of Great Britain and Ireland, Mansfield *12191*

U.S.A.

American Celiac Society, West Orange *13682*
American College of Gastroenterology, Arlington *13708*
American Gastroenterological Association, Bethesda *13829*
American Osteopathic College of Proctology, Union *13985*
American Pancreatic Association, Sepulveda *13989*
American Society for Gastrointestinal Endoscopy, Manchester *14061*
Bockus International Society of Gastroenterology, La Jolla *14605*
National Digestive Diseases Information Clearinghouse, Bethesda *16034*
North American Society for Pediatric Gastroenterology and Nutrition, Denver *16222*
Society of Gastroenterology Nurses and Associates, Rochester *16668*
United Ostomy Association, Irvine *16814*

Genealogy, Heraldry

Australia

Society of Australian Genealogists, Sydney *00492*

Austria

Heraldisch-Genealogische Gesellschaft Adler, Wien *00649*

Belgium

Institut Archéologique du Luxembourg, Arlon *01375*
Office Généalogique et Héraldique de Belgique, Bruxelles *01416*
Vlaamse Vereniging voor Familiekunde, Merksen *01510*

Brazil

Instituto Genealógico Brasileiro, São Paulo *01671*

Canada

Alberta Genealogical Society, Edmonton *01834*
British Columbia Genealogical Society, Richmond *01978*
Genealogical Association of Nova Scotia, Halifax *02479*
Genealogical Institute of the Maritimes, Halifax *02480*
Heraldry Society of Canada, Ottawa *02499*
Manitoba Genealogical Society, Winnipeg *02617*
New Brunswick Genealogical Society, Fredericton *02681*
The Ontario Genealogical Society, Toronto *02778*
Royal Nova Scotia Historical Society, Halifax *02870*
Saskatchewan Genealogical Society, Regina *02888*
Société de Généalogie de Québec, Québec *02920*
Société Généalogique Canadienne-Française, Montréal *02928*

Denmark

Bureau Permanent des Congrès Internationaux des Sciences Généalogique et Héraldique, København *03570*
Samfundet for Dansk Genealogi og Personalhistorie, Holte *03797*
Societas Heraldica Scandinavica, Bagsvaerd *03823*

Finland

Suomen Heraldinen Seura, Helsinki *04026*
Suomen Sukututkimusseura, Helsinki *04060*

France

Fédération des Sociétés Françaises de Généalogie, d'Héraldique et de Sigillographie, Versailles *04515*
Société Française d'Héraldique et de Sigillographie, Paris *04855*

Germany

Bayerischer Landesverein für Familienkunde e.V., München *05177*
Deutsche Arbeitsgemeinschaft genealogischer Verbände e.V., Brühl *05293*
Genealogische Gesellschaft Hamburg e.V., Hamburg *05775*
Gesellschaft für Familienkunde in Franken e.V., Nürnberg *05821*
Verein für Familienforschung in Ost- und Westpreussen e.V., Hamburg *06382*
Verein für Familien- und Wappenkunde in Württemberg und Baden e.V., Stuttgart *06383*
Westdeutsche Gesellschaft für Familienkunde e.V., Köln *06438*

Ireland

Genealogical Office, Dublin *06952*

Japan

Keizu Kyokai Toshokan, Tokyo *08162*

Netherlands

Centraal Bureau voor Genealogie, 's-Gravenhage *08838*
Koninklijk Nederlandsch Genootschap voor Geslacht- en Wapenkunde, 's-Gravenhage *08966*
Nederlandse Genealogische Vereniging, Amsterdam *08993*

Norway

Norsk Heraldisk Forening, Oslo *09319*
Norsk Slektshistorisk Forening, Oslo *09339*

South Africa

Genealogical Society of South Africa, Kelvin *10143*
Heraldry Society of Southern Africa, Claremont *10145*

Switzerland

Académie Internationale d'Héraldique, Genève *10607*
Schweizerische Gesellschaft für Familienforschung, Basel *10819*
Société Suisse d'Héraldique, Genève *10988*

United Kingdom

Association of Genealogists and Record Agents, Horsham *11345*
Federation of Family History Societies, Birmingham *12065*
The Harleian Society, London *12147*
The Heraldry Society, London *12161*
Heraldry Society of Scotland, Edinburgh *12162*
Institute of Heraldic and Genealogical Studies, Canterbury *12242*
Manorial Society of Great Britain, London *12488*
Royal Institution of Cornwall, Truro *12821*
Scots Ancestry Research Society, Edinburgh *12892*
Scottish Genealogy Society, Edinburgh *12910*
Scottish Record Society, Glasgow *12922*
Society of Genealogists, London *13047*
The Stewart Society, Edinburgh *13102*
The Ulster Genealogical and Historical Guild, Belfast *13153*

U.S.A.

Afro-American Historical and Genealogical Society, Washington *13347*
American-Canadian Genealogical Society, Manchester *13674*
American College of Heraldry, Tuscaloosa *13711*
American-French Genealogical Society, Pawtucket *13826*
American Historical Society of Germans from Russia, Lincoln *13847*
American Society of Genealogists, Columbus *14125*
Ark-La-Tex Genealogical Association, Shreveport *14242*
Association of Professional Genealogists, Washington *14487*
Association of the German Nobility in North America, Benicia *14518*
Augustan Society, Torrance *14552*
Black Resources Information Coordinating Services, Tallahassee *14601*
International Society for British Genealogy and Family History, Salt Lake City *15524*
National Genealogical Society, Arlington *16062*
New England Historic Genealogical Society, Boston *16197*
New York Genealogical and Biographical Society, New York *16204*
Polish Genealogical Society, Chicago *16334*
Southern Society of Genealogists, Piedmont *16728*
Vesterheim Genealogical Center, Madison *16858*

Genetics

Austria

Austrotransplant – Österreichische Gesellschaft für Transplantation, Transfusion und Genetik, Wien *00572*
Gregor Mendel-Gesellschaft Wien, Wien *00646*

Brazil

Sociedade Brasileira de Genética, Ribeirão Preto *01711*

Canada

Canadian College of Medical Geneticists, Calgary *02099*
Genetics Society of Canada, Ottawa *02481*

Chile

Sociedad de Genética de Chile, Santiago *03093*

China, People's Republic

Genetics Society of China, Beijing *03199*

France

Equipe de Recherche Associée au C.N.R.S.-Laboratoire de Génétique, Talence *04470*
Groupe de Recherche Génétique Epidémiologique, Paris *04549*
Société Française de Génétique, Orléans *04801*

Georgia

Georgian Society of Genetics and Selectionists, Tbilisi *05022*

Germany

Gesellschaft für Anthropologie und Humangenetik, Bremen *05800*

India

Indian Society of Genetics and Plant Breeding, New Dehli *06818*

Israel

The Genetics Society of Israel, Neve Ya'ar *07054*

Italy

Associazione Genetica Italiana, Padova *07278*
Centro Ricerche Applicazione Bioritmo, Roma *07533*
International Board for Plant Genetic Resources, Roma *07708*
International Society for Twin Studies, Roma *07720*
Società Internazionale di Studi Gemellari, Roma *07818*
Società Italiana di Genetica Agraria, Foggia *07880*

Japan

Nihon Iden Gakkai, Shizuoka *08209*
Nippon Jinrui Iden Gakkai, Tokyo *08316*

Kyrgyzstan

Kyrgyz Genetics and Selection Society, Bishkek *08550*

Mexico

Sociedad Mexicana de Eugenesia, México *08761*

Moldova, Republic of

Society of Genetics of Moldova, Chişinău *08791*

Netherlands

European Association for Research on Plant Breeding, Wageningen *08862*
Nederlandse Genetische Vereniging, Leiden *08994*
Vereniging Het Nederlands Kanker Instituut, Amsterdam *09085*

New Zealand

New Zealand Genetical Society, Christchurch *09147*

Nigeria

Africa Genetics Association, Benin City *09210*
Genetics Society of Nigeria, Ibadan *09229*

Peru

Sociedad Peruana de Eugenesia, Lima *09483*

Russia

National Scientific Medical Society of Medical Geneticists, Moskva *09919*
Russian Society of Genetics and Breeders, Moskva *09965*

South Africa

South African Genetic Society, Stellenbosch *10191*

Sweden

European Drosophila Centre, Umea *10453*
Nordisk Genetikerforening, Stockholm *10512*

United Kingdom

C.A.B. International Division of Animal Production, Wallingford *11730*
European Environmental Mutagen Society, Edinburgh *12028*
Galton Institute, London *12099*
The Genetical Society of Great Britain, Edinburgh *12109*
Institute for Animal Health Pirbright, Pirbright Woking *12205*
International Genetics Federation, Cambridge *12365*

U.S.A.

American Genetic Association, Buckeystown *13830*
American Society of Human Genetics, Bethesda *14133*
Behavior Genetics Association, Saint Louis *14575*
Genetics Society of America, Bethesda *15151*
National Council on Gene Resources, Berkeley *16021*

Geography

Argentina
Academia Nacional de Geografía, Buenos Aires *00049*
Sociedad Argentina de Estudios Geográficos, Buenos Aires *00167*

Australia
Australian and New Zealand Association for Canadian Studies, North Ryde *00256*
Australian Geography Teachers' Association, Adelaide *00292*
Geographical Society of New South Wales, Gladesville *00402*
Geography Teachers' Association of New South Wales, Rozelle *00403*
Institute of Australian Geographers, Canberra *00412*
Royal Geographical Society of Queensland, Fortitude Valley *00468*
University Geographical Society, Nedlands *00513*

Austria
Institut für Österreichkunde, Wien *00658*
Österreichische Geographische Gesellschaft, Wien *00745*
Österreichische Geographische Gesellschaft, Zweigverein Innsbruck, Innsbruck *00746*
Österreichisches Ost- und Südosteuropa-Institut, Wien *00915*
Vereinigung Burgenländischer Geographen, Stegersbach *00990*

Belgium
Belgische Vereniging voor Aardrijkskundige Studies, Heverlee *01161*
Société Royale Belge de Géographie, Bruxelles *01473*
Union Royale Belge pour les Pays d'Outre-Mer et l'Europe Unie, Bruxelles *01501*
Vereniging Leraars Aardrijkskunde, Heverlee *01505*

Benin
African Geographers' Association, Cotonou *01514*

Bolivia
Sociedad de Estudios Geográficos e Históricos, Santa Cruz de la Sierra *01551*
Sociedad Geográfica de La Paz, La Paz *01553*
Sociedad Geográfica Sucre, Sucre *01554*
Sociedad Geográfica y de Historia Potosío, Potosí *01555*

Brazil
Associaçcão de Geografia Teorética, Rio Claro *01612*
Centro de Pesquisas de Geografia do Brasil, Rio de Janeiro *01628*
Fundação Instituto Brasileiro de Geografia e Estatística, Rio de Janeiro *01655*
Instituto Archeológico, Histórico e Geográfico Pernambucano, Recife *01663*
Instituto do Ceará, Fortaleza *01669*
Instituto Geográfico e Histórico da Bahia, Salvador *01672*
Instituto Geográfico e Histórico do Amazonas, Manaus *01673*
Instituto Histórico e Geográfico Brasileiro, Rio de Janeiro *01675*
Instituto Histórico e Geográfico de Goiás, Goiânia *01676*
Instituto Histórico e Geográfico de Santa Catarina, Florianópolis *01677*
Instituto Histórico e Geográfico de Santos, Santos *01678*
Instituto Histórico e Geográfico de São Paulo, São Paulo *01679*
Instituto Histórico e Geográfico de Sergipe, Aracajú *01680*
Instituto Histórico e Geográfico do Maranhão, São Luís *01681*
Instituto Histórico e Geográfico do Pará, Belém *01682*
Instituto Histórico e Geográfico do Rio Grande do Norte, Natal *01683*
Instituto Histórico e Geográfico do Rio Grande do Sul, Porto Alegre *01684*
Instituto Histórico e Geográfico Paraibano, João Pessoa *01685*
Instituto Histórico, Geográfico e Etnográfico Paranaense, Curitiba *01686*
Istituto Histórico e Geográfico do Espirito Santo, Vitória *01697*
Sociedade Brasileira de Geografia, Rio de Janeiro *01712*
Sociedade Geográfica Brasileira, São Paulo *01726*

Bulgaria
Bulgarian Geographical Society, Sofia *01748*

Canada
Arctic Institute of North America, Calgary *01871*
Canadian Association of Geographers, Montréal *02052*
Canadian National Committee for The International Geographical Union, Vanier *02204*
Canadian Quaternary Association, Waterloo *02239*
Commonwealth Geographical Bureau, Montréal *02368*
Huntington Society of Canada, Cambridge *02511*
Royal Canadian Geographical Society, Vanier *02865*

Chile
Comité Nacional de Geografía, Geodesía y Geofísica, Santiago *03041*
Sociedad Chilena de Historia y Geografía, Santiago *03076*

China, People's Republic
Geographical Society of China, Beijing *03200*

China, Republic
The Geographical Society of China, Taipei *03298*

Colombia
Sociedad Geográfica de Colombia, Bogotá *03419*

Costa Rica
Organización de Estudios Tropicales, San José *03451*

Croatia
Croatiam Geographic Society, Zagreb *03453*

Cyprus
Cyprus Geographical Association, Nicosia *03491*

Czech Republic
Česká Geografická Společnost, Praha *03506*

Denmark
Arktisk Institut, København *03560*
Det Grønlandske Selskab, Charlottenlund *03709*
Foreningen af Geografilaerere ved de Gymnasiale Uddannelser, Ribe *03731*
Kongelige Danske Geografiske Selskab, København *03759*

Finland
Suomen Maantieteellinen Seura, Helsingin Yliopisto *04047*

France
Académie des Sciences d'Outre-Mer, Paris *04105*
Association de Recherches Universitaires Géographiques et Cartographiques, Paris *04161*
Association des Géographes de l'Est, Nancy *04184*
Association des Géographes Français, Paris *04185*
Association des Professeurs d'Histoire et de Géographie de l'Enseignement Public, Evry *04197*
Comité National Français de Géographie, Paris *04417*
Comité National Français des Recherches Antarctiques, Paris *04419*
Conseil International des Sciences Sociales, Paris *04459*
European Committee on Ocean and Polar Sciences, Strasbourg *04497*
European Council on African Studies, Strasbourg *04499*
Expéditions Polaires Françaises, Paris *04505*
Groupe Rhône-Alpes de Recherche et d'Etudes en Gestion, Lyon *04563*
Institut Géographique National, Paris *04570*
Société de Biogéographie, Paris *04642*
Société de Géographie, Paris *04654*
Société de Géographie Commerciale de Bordeaux, Bordeaux *04655*
Société de Géographie Commerciale de Paris, Paris *04656*
Société de Géographie de Lille, Lille *04657*
Société de Géographie de Lyon, Lyon *04658*
Société de Géographie de Toulouse, Toulouse *04659*
Société des Amis de la Revue de Géographie de Lyon, Lyon *04694*
Société des Explorateurs et des Voyageurs Français, Paris *04706*
Société Française de Géographie Economique, Paris *04802*
Société Longuédocienne de Géographie, Montpellier *04889*

Georgia
Georgian Geographical Society, Tbilisi *05015*

Germany
Arbeitsgemeinschaft für Osteuropaforschung, Tübingen *05110*
Arbeitskreis für Medizinische Geographie im Zentralverband der deutschen Geographen, Heidelberg *05141*
Deutsche Gesellschaft für Polarforschung, Bremerhaven *05409*
Deutscher Verband für Angewandte Geographie e.V., Hamburg *05547*
Fränkische Geographische Gesellschaft e.V., Erlangen *05757*
Frankfurter Geographische Gesellschaft e.V., Frankfurt *05758*
Geographische Gesellschaft, München *05776*
Geographische Gesellschaft Bremen, Bremen *05777*
Geographische Gesellschaft in Hamburg e.V., Hamburg *05778*
Geographische Gesellschaft zu Hannover, Hannover *05779*
Gesellschaft für Erdkunde zu Berlin, Berlin *05815*
Gesellschaft für Erdkunde zu Köln e.V., Köln *05816*
Gesellschaft für Erd- und Völkerkunde, Bonn *05817*
Gesellschaft für Erd- und Völkerkunde zu Stuttgart e.V., Stuttgart *05818*
Gesellschaft für Geographie und Geologie Bochum, Bochum *05824*
Institut für Länderkunde e.V., Leipzig *05964*
International Geographical Union, Bonn *06007*
Verband Deutscher Hochschullehrer der Geographie, Heidelberg *06352*
Verband Deutscher Schulgeographen e.V., Burgwedel *06360*
Zentralausschuss für Deutsche Landeskunde e.V., Frankfurt *06463*
Zentralverband der Deutschen Geographen, Heidelberg *06468*

Ghana
Ghana Geographical Association, Accra *06485*

Greece
Elliniki Geografiki Etaireia, Athinai *06506*

Guatemala
Academia de Geografía e Historia de Guatemala, Guatemala City *06554*

Honduras
Academia Hondureña de Geografía e Historia, Tegucigalpa *06581*
Sociedad de Geografía e Historia de Honduras, Tegucigalpa *06586*

Hungary
Magyar Földrajzi Társaság, Budapest *06629*

India
Geographical Society of India, Calcutta *06760*
Indian Geographical Society, Madras *06793*
National Geographical Society of India, Varanasi *06844*
Survey of India, Dehra Dun *06863*

Ireland
The Geographical Society of Ireland, Galway *06953*

Israel
Israel Geographical Society, Jerusalem *07072*

Italy
Accademia Lunigianese di Scienze Giovanni Capellini, La Spezia *07202*
Associazione dei Geografi Italiani, Roma *07270*
Associazione Italiana degli Insegnanti di Geografia, Trieste *07293*
Centro di Studio e Documentazione sul Vietnam e il Terzo Mondo, Milano *07449*
Centro Studi e Ricerche per la Conoscenza della Liguria Attraverso le Testimonianze dei Viaggiatori Stranieri, Genova *07573*
Centro Studi Nord e Sud, Napoli *07578*
Comitato dei Geografi Italiani, Roma *07612*
Istituto Geografico Militare, Firenze *07729*
Società di Studi Geografici, Firenze *07794*
Società di Studi Romagnoli, Cesena *07795*
Società Geografica Italiana, Roma *07812*
Società Istriana di Archeologia e Storia Patria, Trieste *07820*

Jamaica
Jamaican Geographical Society, Kingston *08086*

Japan
Jinbun Chiri Gakkai, Kyoto *08148*
Keizai Chiri Gakkai, Tokyo *08157*
Nippon Chiri Gakkai, Tokyo *08289*
Shigaku Kenkyukai, Kyoto *08418*
Tokyo Chigaku Kyokai, Tokyo *08437*

Korea, Republic
Korean Geographical Society, Seoul *08533*

Kyrgyzstan
Kyrgyz Geographical Society, Bishkek *08551*

Macedonia
Geografsko Društvo na Makedonija, Skopje *08602*

Malta
Malta Geographical Society, Gzira *08665*

Mauritius
Centre for Documentation, Research and Training on the Islands of the South West Indian Ocean, Rose Hill *08673*

Mexico
Academia Nacional de Historia y Geografía, México *08689*
Center of Coordination and Diffusion of Latin American Studies, México *08710*
Instituto Panamericano de Geografía e Historia, México *08740*
Sociedad Mexicana de Geografía y Estadística, México *08764*
Sociedad Nuevoleonesa de Historia, Geografía y Estadística, Monterrey *08775*

Moldova, Republic of
Geographic Society of Moldova, Chişinău *08779*

Monaco
Comité Arctique International, Monte Carlo *08795*

Morocco
Association Nationale des Géographes Marocains, Rabat *08802*
Comité National de Géographie du Maroc, Rabat *08806*

Netherlands
Koninklijk Nederlands Aardrijkskundige Genootschap, Amsterdam *08965*

New Zealand
New Zealand Geographical Society, Christchurch *09148*
Ross Dependency Research Committee, Wellington *09192*

Nigeria
Nigerian Geographical Association, Ibadan *09237*

Norway
Det Norske Geografiske Selskap, Oslo *09259*
Norsk Samfunnsgeografisk Forening, Dragvoll *09337*

Peru
Instituto Geográfico Nacional, Lima *09470*
Sociedad Geográfica de Lima, Lima *09477*

Poland
Polskie Towarzystwo Geograficzne, Warszawa *09571*

Portugal
Sociedade de Geografia de Lisboa, Lisboa *09682*

Romania
Societatea de Stiinte Geografice din Romania, Bucuresti *09784*

Russia
Commission for the Study of the Arctic, Moskva *09841*
Russian Geographical Society, St. Petersburg *09957*
Tropical Committee, Moskva *09986*

Saudi Arabia
King Abdul Aziz Research Centre, Riyadh *10004*

Slovakia
Slovenská Geografická Spoločnost, Bratislava *10072*

Slovenia
Zveza Geografskih Društev Slovenije, Ljubljana *10115*

South Africa
Africa Institute of South Africa, Pretoria *10120*
Climatology Research Group, Johannesburg *10131*
South African Geographical Society, Wits *10192*

Spain
Instituto Geográfico Nacional, Madrid *10325*
Real Sociedad Geográfica, Madrid *10365*

Sri Lanka
Ceylon Geographical Society, Colombo *10410*

Sweden
Geografilärarnas Riksförening, Lund *10464*
Geografiska Förbundet, Stockholm *10465*
Svenska Sällskapet för Antropologi och Geografi, Stockholm *10573*

Switzerland
Geographische Gesellschaft Bern, Bern *10707*
Geographisch-Ethnographische Gesellschaft Zürich, Zürich *10708*
Geographisch-Ethnologische Gesellschaft Basel, Basel *10709*
Schweizerische Stiftung für Alpine Forschungen, Zürich *10931*
Société de Géographie de Genève, Genève *10968*
Verband der Schweizer Geographen, Zürich *10998*

United Kingdom
Council for British Geography, London *11848*
Council of British Geography, London *11856*
The Geographical Association, Sheffield *12110*
The Hakluyt Society, London *12141*
Hispanic and Luso-Brazilian Council, London *12170*
Institute of British Geographers, London *12216*
Manchester Geographical Society, Manchester *12483*
Palestine Exploration Fund, London *12676*
Remote Sensing Society, Nottingham *12749*
Royal Geographical Society, London *12806*
Royal Scottish Geographical Society, Glasgow *12846*
Scientific Committee on Antarctic Research, Cambridge *12890*
Scientific Exploration Society, Marlborough *12891*
Scottish Association of Geography Teachers, Hamilton *12898*
Society for Libyan Studies, London *12978*
Trans-Antarctic Association, Cambridge *13139*

Uruguay
Instituto Histórico y Geográfico del Uruguay, Montevideo *13261*

U.S.A.
American Geographical Society, New York *13831*
American Polar Society, Columbus *14009*
Antarctica and Southern Ocean Coalition, Washington *14226*
Association of American Geographers, Washington *14379*
Environmental Design Research Association, Oklahoma City *15049*
The Explorers Club, New York *15072*
Human Relations Area Files, New Haven *15237*
International Association for Great Lakes Research, Ann Arbor *15382*

National Council for Geographic Education, Indiana *15988*
National Geographic Society, Washington *16063*
South American Explorers Club, Ithaca *16722*
United States Antarctic Research Program, Washington *16815*

Venezuela
Asociación Cultural Humboldt, Caracas *16966*

Vietnam
Vietnamese Association of Geography, Hanoi *17048*

Yugoslavia
Srpsko Geografsko Društvo, Beograd *17127*

Zimbabwe
Geographical Association of Zimbabwe, Harare *17170*

Geology

Albania
Shoqata e Gjeologeve te Shqiperise, Tirana *00007*

Algeria
Centre National de Recherches et d'Application des Géosciences, Alger *00017*

Angola
Direcção Provincial dos Servicos de Geologia e Minas de Angola, Luanda *00033*

Argentina
Asociación Argentina de Geofísicos y Geodestas, Buenos Aires *00065*
Asociación Geológica Argentina, Buenos Aires *00082*
Centro Argentino de Espeleología, Buenos Aires *00098*
Comité Nacional de Cristalografía, Buenos Aires *00123*
Fundación Miguel Lillo, San Miguel de Tucumán *00140*

Australia
Australian Geomechanics Society, Barton *00293*
Geological Society of Australia, Sydney *00404*
Geological Survey of New South Wales, Sydney *00405*
Geological Survey of Western Australia, East Perth *00406*
Tasmanian Geological Survey, Rosny *00505*

Austria
Gesellschaft der Geologie- und Bergbaustudenten in Österreich, Wien *00610*
Österreichische Bodenkundliche Gesellschaft, Wien *00733*
Österreichische Geologische Gesellschaft, Wien *00747*
Österreichische Gesellschaft für Erdölwissenschaften, Wien *00777*
Österreichische Himalaya-Gesellschaft, Wien *00859*
Österreichischer Alpenverein, Wissenschaftlicher Unterausschuss, Innsbruck *00881*
Vereinigung für Angewandte Lagerstättenforschung, Leoben *00993*
Vereinigung für Hydrogeologische Forschungen in Graz, Graz *00994*

Barbados
Barbados Museum and Historical Society, Bridgetown *01074*

Belgium
European Centre for Geodynamics and Seismology, Bruxelles *01310*
Société Belge de Géologie, Bruxelles *01430*
Société Géologique de Belgique, Liège *01461*

Bolivia
Asociación de Ingenieros y Geólogos de Yacimientos Petrolíferos Fiscales Bolivianos, La Paz *01533*
Instituto Boliviano del Petróleo, La Paz *01546*
Sociedad Geológica Boliviana, La Paz *01556*

Brazil
Sociedade Brasileira de Geologia, São Paulo *01713*

Bulgaria
Bulgarian Geological Society, Sofia *01749*
Carpathian Balkan Geological Association, Sofia *01765*
Union of Mining Engineering, Geology and Metallurgy, Sofia *01796*

Canada
Alberta Research Council, Edmonton *01845*
Association of Exploration Geochemists, Rexdale *01932*
Association Professionnelle des Géologues et des Géophysiciens du Québec, Montréal *01952*
Canadian Geoscience Council, Waterloo *02146*
Canadian Quaternary Association, Waterloo *02239*
Canadian Society of Petroleum Geologists, Calgary *02291*
Geological Association of Canada, Saint John's *02482*
Saskatchewan Geological Society, Regina *02889*

Chile
Corporación para el Desarrollo de la Ciencia, Santiago *03049*
Servicio Nacional de Geología y Minería, Santiago *03063*
Sociedad Geológica de Chile, Santiago *03097*

China, People's Republic
Academy of Petroleum Research, Beijing *03112*
Chinese Academy of Geological Sciences, Beijing *03139*
Chinese Geological Society, Beijng *03170*

China, Republic
Geological Society of China, Taipei *03299*

Colombia
Instituto de Ciencias Naturales, Bogotá *03391*
Instituto Nacional de Investigaciones Geológico-Mineras, Bogotá *03393*

Costa Rica
Centro de Estudios Médicos Ricardo Moreno Cañas, San José *03441*

Czech Republic
Česká Geologická Společnost, Praha *03507*
International Association of Planetology, Praha *03545*
International Association on the Genesis of Ore Deposits, Jeseník *03546*

Denmark
Collegium Palynologicum Scandinavicum, Aarhus *03573*
Dansk Geologisk Forening, København *03625*

Ecuador
Dirección General de Geología y Minas, Quito *03852*

El Salvador
Centro de Estudios e Investigaciones Geotécnicas, San Salvador *03925*

Ethiopia
African Mountain Association, Asmara *03937*

Finland
Geologian Tutkimuskeskus, Espoo *03963*
Geologiliitto, Helsinki *03964*
International Peat Society, Jyskä *03969*
Suomen Geologinen Seura, Espoo *04022*

France
Association des Palynologues de Langue Française, Strasbourg *04192*
Association Française pour l'Etude du Quaternaire, Caen *04246*
Association Nationale pour l'Etude de la Neige et des Avalanches, Grenoble *04316*
Association of African Geological Surveys, Orléans *04317*
Eurocoast, Marseille *04474*
EUROLAT – European Network for Studies on Laterites and Tropical Environment, Strasbourg *04475*
International Association of Engineering Geology, Paris *04581*
Société des Lettres, Sciences et Arts de la Haute-Auvergne, Aurillac *04709*
Société Géologique de France, Paris *04868*
Société Géologique et Minéralogique de Bretagne, Rennes *04869*
Société Parisienne d'Etudes et de Recherches Foncières, Paris *04909*
Union Géodésique et Géophysique Internationale, Toulouse *04982*

Georgia
Georgian Geological Society, Tbilisi *05016*
Georgian National Speleological Society, Tbilisi *05019*

Germany
Deutsche Geologische Gesellschaft, Hannover *05311*
Deutsche Gesellschaft für Moor- und Torfkunde, Hannover *05385*
Deutsche Gesellschaft für Polarforschung, Bremerhaven *05409*
Deutsche Quartärvereinigung, Hannover *05493*
Geologische Vereinigung e.V., Mendig *05781*
Gesellschaft für Geographie und Geologie Bochum, Bochum *05824*
Verband der deutschen Höhlen- und Karstforscher e.V., Leinfelden-Echterdingen *06335*
Vereinigung der Freunde der Mineralogie und Geologie e.V., Heidelberg *06397*

Ghana
Geological Survey of Ghana, Accra *06481*

Guatemala
Instituto Nacional de Sismología, Vulcanología, Meteorología e Hidrología, Guatemala *06571*

Hungary
Magyarhoni Földtani Társulat, Budapest *06635*

India
Geological, Mining and Metallurgical Society of India, Calcutta *06761*
Geological Survey of India, Calcutta *06762*
Indian Geologists' Association, Chandigarh *06794*

Indonesia
Asia Soil Conservation Network for the Humid Tropics, Jakarta *06884*

Iraq
Association of Arab Geologists, Baghdad *06916*

Ireland
Irish Association for Economic Geology, Dublin *06972*

Israel
Israel Geological Society, Jerusalem *07073*
The Natural Resources Research Organization, Jerusalem *07109*

Italy
Accademia Valdarnese del Poggio, Montevarchi *07246*
Associazione Geo-Archeologica Italiana, Roma *07279*
Comitato Glaciologico Italiano, Torino *07614*
Earthnet Programme Office, Frascati *07648*
European Association of Remote Sensing Laboratories, Napoli *07665*
International Association of Engineering Geology, Sezione Italiana, Bari *07707*
Istituto Internazionale di Vulcanologia, Catania *07733*
Società Geologica Italiana, Roma *07813*
Società Italiana della Scienza del Suolo, Firenze *07826*

Jamaica
Geological Society of Jamaica, Kinston *08079*

Japan
Chigaku Dantai Kenkyu-Kai, Tokyo *08116*
East and Southeast Asia Federation of Soil Science Societies, Tokyo *08124*
Jishin Gakkai, Tokyo *08149*
Nippon Chishitsu Gakkai, Tokyo *08290*
Nippon Dai-Yonki Gakkai, Tokyo *08292*
Nippon Ganseki Kobutsu Kosho Gakkai, Sendai *08300*
Tokyo Chigaku Kyokai, Tokyo *08437*

Kenya
Desert Locust Control Organization for Eastern Africa, Nairobi *08489*

Kyrgyzstan
Kyrgyz Commission on Earthquake Forecasting, Bishkek *08549*

Liberia
Geological, Mining and Metallurgical Society of Liberia, Monrovia *08563*

Macedonia
Makedonsko Geološko Društvo, Skopje *08604*

Malaysia
Asian Wetland Bureau, Kuala Lumpur *08635*
Geological Survey of Malaysia, Ipoh *08642*
Geological Survey of Malaysia, Kuching *08643*

Mexico
Asociación Mexicana de Geólogos Petroleros, México *08701*
Sociedad Geológica Mexicana, México *08750*

Netherlands
European Association of Petroleum Geoscientists and Engineers, Zeist *08869*
Koninklijk Nederlands Geologisch Mijnbouwkundig Genootschap, Haarlem *08967*

New Zealand
Geological Society of New Zealand, Lower Hutt *09126*
Waikato Geological and Lapidary Society, Hamilton *09199*

Nigeria
Geological Survey of Nigeria, Kaduna South *09230*

Norway
Norges Geologiske Undersökelse, Trondheim *09283*
Norsk Geologisk Forening, Trondheim *09317*

Peru
Sociedad Geológica del Perú, Lima *09478*

Poland
Polskie Towarzystwo Geologiczne, Kraków *09572*

Portugal
International Society for Rock Mechanics, Lisboa *09662*
Sociedade Geológica de Portugal, Lisboa *09685*

Romania
Comitetul National al Geologilor din Romania, Bucuresti *09748*
Societatea de Stiinte Geologice din Romania, Bucuresti *09785*

Russia
Commission for the History of Geological Knowledge and Geological Study of the Russian Federation, Moskva *09835*
Committee of the International Programme of Geological Correlation, Moskva *09851*
Committee on Petrography, Moskva *09853*
Geological Society, Moskva *09870*
Geomorphological Commission, Moskva *09871*
National Committee of Geologists, Moskva *09895*
Soil Science Society, Moskva *09984*

Slovakia
Slovenská Geologická Spoločnost, Bratislava *10073*

South Africa
The Geological Society of South Africa, Linden *10144*

Spain
International Congress of Carboniferous Stratigraphy and Geology, Madrid *10329*

Suriname
Geologisch Mijnbouwkundige Dienst, Paramaribo *10431*

Sweden
Geologiska Föreningen, Uppsala *10466*
Svenska Geotekniska Föreningen, Linköping *10552*
Svenska Nationalkommittén för Geologi, Uppsala *10566*
Sveriges Geologiska Undersökning, Uppsala *10591*

Switzerland
European Crystallographic Committee, Genève *10683*
Société Géologique Suisse, Fribourg *10975*
Vereinigung Schweizerischer Petroleumgeologen und -Ingenieure, Gelterkinden *11021*

Syria
Arab Center for the Studies of Arid Zones and Dry Lands, Damascus *11038*

Thailand
Association of Geoscientists for International Development, Bangkok *11088*

Turkey
Jeoloji Mühendisleri Odasi, Ankara *11147*

United Kingdom
Association of European Geological Societies, London *11342*
British Geological Survey, Keyworth *11565*
British Geomorphological Research Group, London *11566*
The British Society of Dowsers, Ashford *11682*
Dryland Professional Network, London *11908*
EUROMINERALS – Confederation of Learned/Engineering Societies in the Mineral Industry, London *11981*
The Geological Society, London *12111*
Geologists' Association, London *12112*
Hertfordshire Natural History Society and Field Club, Potters Bar *12165*
International Association of Volcanology and Chemistry of the Earth's Interior JAVCEJ, Leeds *12336*
International Glaciological Society, Cambridge *12366*
International Seismological Centre, Newbury *12387*
International Union for Quaternary Research, Egham *12399*
Petroleum Exploration Society of Great Britain, London *12691*
Society for Libyan Studies, London *12978*
United Earth Sciences Exploration Group, Leigh-on-Sea *13163*
Yorkshire Geological Society, Cottingham *13233*

U.S.A.
American Association for Crystal Growth, Thousand Oaks *13476*
American Association of Petroleum Geologists, Tulsa *13562*
American Crystallographic Association, Buffalo *13774*
American Geological Institute, Alexandria *13832*
American Spelean History Association, Altoona *14179*
Association for Mexican Cave Studies, Austin *14318*
Association for Women Geoscientists, Saint Paul *14365*
Association of American State Geologists, Champaign *14388*
Association of Earth Science Editors, Lawrence *14430*
Association of Engineering Geologists, Sudbury *14433*
Carnegie Institution of Washington, Washington *14647*
Flat Earth Research Society International, Lancaster *15099*
Geochemical Society, Washington *15152*
Geological Society of America, Boulder *15153*
Geothermal Resources Council, Davis *15155*
International Association for Mathematical Geology, Reston *15387*
International Commission on the History of the Geological Sciences, Cambridge *15443*
Jesuit Seismological Association, Weston *15602*
National Association of Geology Teachers, Bellingham *15861*
National Earth Science Teachers Association, Fargo *16037*
Seismological Society of America, El Cerrito *16460*
Society of Economic Geologists, Littleton *16655*
Society of Independent Professional Earth Scientists, Dallas *16673*

Society of Professional Well Log Analysts, Houston *16694*
Stonehenge Study Group, Santa Barbara *16747*
Yellowstone-Bighorn Research Association, Billings *16940*

Venezuela
Asociación Venezolana de Geología, Minería y Petróleo, Caracas *16974*

Vietnam
Vietnamese Association of Geology, Hanoi *17049*

Virgin Islands (U.S.)
Caribbean Natural Resources Institute, Saint Croix *17074*

Yugoslavia
Srpsko Geološko Društvo, Beograd *17128*

Zimbabwe
Geological Society of Zimbabwe, Harare *17171*
Mennel Society, Harare *17175*

Geomorphology

Finland
Suomen Geoteknillinen Yhdistys, Espoo *04023*

France
Association Française pour l'Etude du Sol, Plaisir *04247*

U.S.A.
International Erosion Control Association, Steamboat Springs *15472*

Geophysics

Algeria
Centre National d'Astronomie, d'Astrophysique et de Géophysique, Alger *00015*
Centre National de Recherche sur les Zones Arides, Alger *00019*

Argentina
Consejo Federal de Inversiones, Buenos Aires *00127*

Australia
Royal Meteorological Society, Australian Branch, Parkville *00473*

Austria
Österreichische Gesellschaft für Meteorologie, Wien *00809*
Sonnblick-Verein, Wien *00946*

Barbados
Caribbean Institute for Meteorology and Hydrology, Bridgetown *01082*

Belgium
Société Royale Belge d'Astronomie, de Météorologie et de Physique du Globe, Bruxelles *01470*

Canada
Canadian Meteorological and Oceanographic Society, Newmarket *02199*
International Society of Biometeorology, Sainte Anne de Bellevue *02559*

Chile
Asociación Chilena de Sismología e Ingeniería Antisísmica, Santiago *03015*

China, People's Republic
Chinese Geophysical Socitey, Beijing *03171*

China, Republic
The Meteorological Society of the Republic of China, Taipei *03307*

Czech Republic
Česká Meteorologická Společnost, Praha *03510*

Denmark
Dansk Geofysisk Forening, København *03624*

Finland
Geofysiikan Seura, Espoo *03962*

France
Comité National Français de Géodesie et Géophysique, Paris *04416*
Groupe des Méthodes Pluridisciplinaires Contribuant à l'Archéologie, Rennes *04550*
Société de Recherches Géophysiques, Paris *04687*
Société Française d'Hydrologie et de Climatologie Médicales, Paris *04858*
Société Météorologique de France, Boulogne-Billancourt *04896*
Union Géodésique et Géophysique Internationale, Toulouse *04982*

Germany
Deutsche Geophysikalische Gesellschaft e.V., Münster *05312*
Deutsche Gesellschaft für Polarforschung, Bremerhaven *05409*
Deutsche Meteorologische Gesellschaft e.V., Traben-Trarbach *05469*
Deutsche Meteorologische Gesellschaft e.V., Zweigverein Hamburg, Hamburg *05470*
Forschungsgemeinschaft Angewandte Geophysik e.V., Hannover *05706*

Guatemala
Instituto Nacional de Sismología, Vulcanología, Meteorología e Hidrología, Guatemala *06571*

Hungary
Magyar Geofizikusok Egyesülete, Budapest *06631*
Magyar Meteorológiai Társaság, Budapest *06645*

India
Indian Society of Earthquake Technology, Roorkee *06817*

Italy
Associazione Geofisica Italiana, Roma *07280*
Società Italiana di Geofisica e Meteorologia, Genova *07881*
Società Italiana di Meteorologia Applicata, Roma *07905*
Società Italiana per le Scienze Ambientali: Biometeorologia, Bioclimatologia ed Ecologia, Milano *07968*
Société de la Flore Valdôtaine, Aosta *08027*

Japan
Japan Weather Association, Tokyo *08147*
Nippon Kazan Gakkai, Tokyo *08325*
Nippon Kisho Gakkai, Tokyo *08335*
Nippon Nogyo-Kisho Gakkai, Tokyo *08349*
Tokyo Chigaku Kyokai, Tokyo *08437*
Zisin Gakkai, Tokyo *08448*

Kenya
Climate Network Africa, Nairobi *08486*

Netherlands
European Association of Exploration Geophysicists, Zeist *08867*

New Zealand
Meteorological Society of New Zealand, Wellington *09131*
New Zealand Geophysical Society, Wellington *09149*

Niger
Centre Régional de Formation et d'Application en Agrométéorologie et Hydrologie Opérationnelle, Niamey *09208*

Norway
Den Geofysiske Kommisjon, Oslo *09250*
Norges Geotekniske Institutt, Oslo *09284*
Norsk Geofysisk Forening, Trondheim *09316*
Norsk Meteorologforening, Oslo *09330*

Philippines
ESCAP/WMO Typhoon Committee, Manila *09519*

Portugal
Sociedade de Estudos Açoreanos Afonso Chaves, Ponta Delgada *09680*

Russia
Committee of the International Programme „Litosphere", Moskva *09850*

Sweden
Nordic Association of Applied Geophysics, Luleå *10506*
Nordisk Förening för Tillämpad Geofysik, Stockholm *10511*
Svenska Geofysiska Föreningen, Norrköping *10551*

Switzerland
Organisation Météorologique Mondiale, Genève *10777*

Trinidad and Tobago
Caribbean Meteorological Organization, Port of Spain *11116*

United Kingdom
Association of British Climatologists, London *11307*
European Centre for Medium-Range Weather Forecasts, Reading *12004*
International Commission on Polar Meteorology, Cambridge *12344*
Royal Meteorological Society, Reading *12832*
Scientific Committee on Antarctic Research, Cambridge *12890*
Society for Earthquake and Civil Engineering Dynamics, London *12969*

U.S.A.
American Geophysical Union, Washington *13833*
American Meteorological Society, Boston *13943*
American Meteor Society, Geneseo *13944*
Earthquake Engineering Research Institute, Oakland *15000*
Geochemical Society, Washington *15152*
Meteoritical Society, Piscataway *15723*
National Information Service for Earthquake Engineering, Richmond *16087*
Seismological Society of America, El Cerrito *16460*
Society of Exploration Geophysicists, Tulsa *16662*
University Corporation for Atmospheric Research, Boulder *16845*
Weather Modification Association, Fresno *16876*

Geriatrics

Argentina
Sociedad Argentina de Gerontología y Geriatría, Buenos Aires *00171*

Australia
Australian Association of Gerontology, Parkville *00264*

Austria
Österreichische Gesellschaft für Geriatrie und Gerontologie, Wien *00783*
Österreichische Gesellschaft für Nephrologie, Linz *00812*
Tiroler Gesellschaft zur Förderung der Alterswisschaft und des Seniorenstudiums an der Universität Innsbruck, Innsbruck *00954*

Canada
Alberta Association on Gerontology, Edmonton *01828*
Alzheimer Society of Manitoba, Winnipeg *01857*
Canadian Geriatrics Research Society, Toronto *02148*
Canadian Institute of Religion and Gerontology, Toronto *02182*
G.A. Frecker Association on Gerontology, St. John's *02476*
Gerontology Association of Nova Scotia, Wolfville *02484*
International Federation on Ageing, Montréal *02554*
Manitoba Association on Gerontology, Winnipeg *02610*
New Brunswick Gerontology Association, Saint John *02682*
Ontario Gerontology Association, Waterloo *02779*
Ontario Psychogeriatric Association, Kingston *02789*
Saskatchewan Association on Gerontology, Saskatoon *02885*

Chile
Sociedad Chilena de Gerontología, Santiago *03073*

Denmark
Dansk Gerontologisk Selskab, Naestved *03627*

France
Association de Gérontologie du 13e, Paris *04149*

Germany
Deutsche Gesellschaft für Gerontologie, Lübeck *05354*
Deutsches Zentrum für Altersfragen e.V., Berlin *05609*

Ghana
African Gerontological Society, Legon *06472*

Hungary
Magyar Gerontológiai Társaság, Budapest *06632*

Italy
Società Italiana di Gerontologia e Geriatria, Firenze *07882*
Società Italiana Medici e Operatori Geriatrici, Firenze *07959*

Japan
Nihon Ronen Igakukai, Tokyo *08251*

Mexico
International Association of Gerontology, México *08741*

Romania
Societatea de Gerontologie, Bucuresti *09765*

Switzerland
Schweizerische Gesellschaft für Gerontologie, Bern *10821*

Ukraine
Gerontology and Geriatrics Society, Kiev *11189*

United Kingdom
British Geriatrics Society, London *11568*
International Federation on Aging – European Office, London *12362*

U.S.A.
American Aging Association, Omaha *13445*
American Association for Geriatric Psychiatry, Greenbelt *13480*
American Geriatrics Society, New York *13834*
American Longevity Association, Torrance *13921*
American Society for Geriatric Dentistry, Chicago *14062*
American Society of Nephrology, Washington *14148*
Association for Gerontology in Higher Education, Washington *14299*
Gerontological Society of America, Washington *15158*
National Association for Human Development, Washington *15802*
National Association of Nutrition and Aging Services Programs, Grand Rapids *15875*
National Association of State Units on Aging, Washington *15906*
National Center on Arts and the Aging, Washington *15949*
National Geriatrics Society, Eustis *16064*

Graphic and Decorative Arts, Design

Austria
Institut für Soziales Design – Entwicklung und Forschung, Wien *00660*

Canada
Association of Canadian Industrial Designers, Toronto *01872*
Association of Canadian Industrial Designers, Rexdale *01923*
Canadian Academy of Medical Illustrators, Toronto *02004*
Canadian Graphic Arts Institute, Toronto *02149*
Canadian Society of Children's Authors, Illustrators and Performers, Toronto *02277*

Germany
AW produktplanung, Essen *05159*
Design Zentrum Nordrhein-Westfalen e.V., Essen *05283*
Deutscher Beton-Verein e.V., Wiesbaden *05505*
Rat für Formgebung, Frankfurt *06244*

India
National Institute of Design, Ahmedabad *06845*

Ireland
Society of Designers in Ireland, Dublin *07032*

Italy
Accademia di Costume e di Moda, Libero Istituto di Studi Superiori di Belle Arti, Roma *07166*
Centro di Studi Grafici, Milano *07447*

Netherlands
Nederlandsche Vereniging voor Druk- en Boekkunst, Haarlem *08984*

South Africa
South African Institute of Printing, Cape Town *10201*

Sweden
Svenska Psykoanalytiska Föreningen, Stockholm *10572*

United Kingdom
Association of Advisers in Design and Technical Studies, Camberley *11298*
Bookplate Society, London *11459*
Chartered Society of Designers, London *11769*
Commemorative Collectors Society, Nottingham *11802*
Design Council, London *11884*
The Design Council – Scotland, Glasgow *11885*
The Educational Institute of Design, Craft and Technology, Melton Mowbray *11944*
Faculty of Royal Designers for Industry, London *12053*
Institution of Engineering Designers, Westbury *12299*
Society of Architectural and Industrial Illustrators, Stroud *13026*
Society of Designer Craftsmen, London *13043*

U.S.A.
American Center for Design, Chicago *13683*
American Design Drafting Association, Rockville *13784*
American Institute of Graphic Arts, New York *13881*
Fostoria Glass Society of America, Moundsville *15118*
Foundation for Interior Design Education Research, Grand Rapids *15126*
Glass Art Society, Seattle *15161*
Interior Design Educators Council, Irvine *15360*
International Design Conference in Aspen, Aspen *15466*
International Graphic Arts Education Association, Pittsburgh *15481*
National Academy of Design, New York *15772*
National Association of Schools of Art and Design, Reston *15887*
Society for Calligraphy, Los Angeles *16504*
Technical Association of the Graphic Arts, Rochester *16767*
University and College Designers Association, Walkerton *16841*

Graphology

Denmark
Dansk Grafologisk Selskab, Holte *03628*

France
Société Française de Graphologie, Paris *04803*

Germany
Berufsverband Geprüfter Graphologen/Psychologen e.V., München *05206*
Deutsche Graphologische Vereinigung e.V., Schwerte *05450*

Italy
Associazione Grafologica Italiana, Ancona *07283*
Associazione Nazionale dei Periti Grafici a Base Psicologica, Milano *07364*
Società Internazionale di Psicologia della Scrittura, Milano *07816*

Switzerland
Europäische Gesellschaft für Schriftpsychologie und Schriftexpertise, Zürich *10668*
Schweizerische Graphologische Berufsvereinigung, Erlenbach *10858*

U.S.A.
American Association of Handwriting Analysts, Hinsdale *13538*
Association of Forensic Document Examiners, Minneapolis *14440*
Handwriting Analysts – International, Davenport *15186*
International Graphoanalysis Society, Chicago *15482*

Gynecology

Argentina
Sociedad Argentina para el Estudio de la Esterilidad, Buenos Aires 00186

Australia
The Australian College of Obstetricians and Gynaecologists, Melbourne 00275

Austria
Österreichische Gesellschaft für Gynäkologie und Geburtshilfe, Wien 00788
Österreichische Gesellschaft für Perinatale Medizin, Wien 00821
Österreichische Gesellschaft zum Studium der Sterilität und Fertilität, Wien 00853

Belgium
Société Royale Belge de Gynécologie et d'Obstétrique, Bruxelles 01474

Bulgaria
The Bulgarian Gynecological and Obstetrical Society, Sofia 01750

Canada
Society of Obstetricians and Gynaecologists of Canada, Toronto 02946

Chile
Sociedad Chilena de Obstetricia y Ginecología, Santiago 03078

China, Republic
The Association of Obstetrics and Gynecology of the Republic of China, Taipei 03247

Colombia
Sociedad Colombiana de Obstetricia y Ginecologia, Bogotá 03410

Costa Rica
Asociación de Obstetricia y Ginecología, San José 03435

Denmark
Danske Fødsels- og Kvindelaegers Organisation, Søborg 03604
Dansk Selskab for Obstetrik og Gynaekologi, København 03680

El Salvador
Sociedad de Ginecología y Obstetricia de El Salvador, San Salvador 03931

France
Fédération des Gynécologues et Obstétriciens de Langue Française, Paris 04513
Société d'Obstétrique et de Gynécologie de Marseille, Marseille 04769
Société d'Obstétrique et de Gynécologie de Toulouse, Toulouse 04770
Société Française de Gynécologie, Paris 04804
Société Internationale de Psycho-Prophylaxie Obstétricale, Paris 04880

Germany
Deutsche Gesellschaft für Gynäkologie und Geburtshilfe, Amberg 05357
Deutsche Gesellschaft für Perinatale Medizin, Berlin 05398
Oberrheinische Gesellschaft für Geburtshilfe und Gynäkologie, Tübingen 06222

Italy
Società Internazionale di Studi Gemellari, Roma 07818
Società Italiana della Continenza, Roma 07825
Società Italiana di Ginecologia Pediatrica, Roma 07883
Società Italiana di Oncologia Ginecologica, Roma 07920
Società Italiana di Ostetricia e Ginecologia, Roma 07922
Società Italiana per lo Studio della Fertilità e della Sterilità, Roma 07971
Società Laziale - Abruzzese Marchigiana Molisana di Ostetricia e Ginecologia, Roma 07978

Japan
Nippon Sanka-Fujinka Gakkai, Tokyo 08362

Mexico
Asociación Mexicana de Ginecologia y Obstetricia, México 08702

Netherlands
Caribbean Institute of Perinatology, Groningen 08836
European Association for Gynaecologists and Obstetricians, Nijmegen 08855
European College of Obstetrics and Gynaecology, Rotterdam 08879
Nederlandse Vereniging voor Obstetrie en Gynaecologie, Utrecht 09030

Poland
Polskie Towarzystwo Ginekologiczne, Warszawa 09573

Romania
Societatea de Obstetrică si Ginecologie, Bucuresti 09774

Russia
National Scientific Medical Society of Obstetricians and Gynaecologists, Moskva 09923

Singapore
Asia and Oceania Federation of Obstetrics and Gynecology, Singapore 10034

Switzerland
Schweizerische Gesellschaft für Gynäkologie und Geburtshilfe, Lausanne 10823

United Kingdom
Blair Bell Research Society, Leicester 11453
Edinburgh Obstetrical Society, Edinburgh 11937
Glasgow Obstetrical and Gynaecological Society, Glasgow 12120
International Federation of Gynecology and Obstetrics, London 12356
Royal College of Midwives, London 12780
Royal College of Obstetricians and Gynaecologists, London 12783

U.S.A.
American Association of Gynecological Laparoscopists, Santa Fe Springs 13537
American Association of Pro Life Obstetricians and Gynecologists, Elm Grove 13573
American Board of Obstetrics and Gynecology, Seattle 13642
American College of Obstetricians and Gynecologists, Washington 13722
American College of Osteopathic Obstetricians and Gynecologists, Pontiac 13726
American Fertility Society, Birmingham 13811
American Gynecological and Obstetrical Society, Salt Lake City 13838
American Society for Colposcopy and Cervical Pathology, Washington 14052
American Society for Psychoprophylaxis in Obstetrics, Washington 14083
Association for Childbirth at Home, International, Glendale 14278
Association of Professors of Gynecology and Obstetrics, Washington 14489
Childbirth without Pain Education Association, Detroit 14730
Council on Resident Education in Obstetrics and Gynecology, Washington 14931
Endometriosis Association, Milwaukee 15042
International Childbirth Education Association, Minneapolis 15434
Maternity Center Association, New York 15702
National Perinatal Association, Tampa 16124
Read Natural Childbirth Foundation, San Rafael 16398
Society for Gynecologic Investigation, Washington 16528
Society for Obstetric Anesthesia and Perinatology, Houston 16558

Venezuela
Sociedad de Obstetricia y Ginecología de Venezuela, Caracas 17011

Vietnam
Vietnamese Association of Obstetrics, Gynaecology and Family Planning, Hanoi 17054

Heat Engineering s. Air Conditioning

Hematology

Argentina
Sociedad Argentina de Hematología, Buenos Aires 00172

Austria
Österreichische Hämophilie-Gesellschaft, Wien 00858

Canada
Canadian Hematology Society, Edmonton 02156
Canadian Sickle Cell Society, Toronto 02250

Chile
Sociedad Chilena de Hematología, Santiago 03074

Denmark
Dansk Haematologisk Selskab, Hellerup 03629

France
Association de Transfusion Sanguine, Paris 04199
Association Française des Hémophiles, Paris 04230
Société Française d'Angéiologie, Neuilly-sur-Seine 04784
Société Française d'Hématologie, Paris 04854
Société Internationale de Transfusion Sanguine, Les Ulis 04882
Société Nationale de Transfusion Sanguine, Paris 04903
Union Syndicale Nationale des Angiologues, Paris 04996

Germany
Deutsche Gesellschaft für Angiologie, Esslingen 05322
Deutsche Gesellschaft für Bluttransfusion und Immunhämatologie e.V., Frankfurt 05333
Deutsche Gesellschaft für Hämatologie und Onkologie e.V., München 05358
Deutsche Hämophilieberatung, Marl-Hüls 05452
Deutsche Hämophiliegesellschaft zur Bekämpfung von Blutungskrankheiten e.V., Hamburg 05453
Gesellschaft zur Förderung der Erforschung der Zuckerkrankheit e.V., Düsseldorf 05880

Hungary
Magyar Angiologiai Társaság, Budapest 06616

Israel
Israel Society for Hematology and Blood Transfusion, Jerusalem 07093

Italy
Associazione Dietetica Italiana, Roma 07273
Associazione Italiana di Dietetica e Nutrizione Clinica, Roma 07305
European Dialysis and Transplant Association - European Renal Association, Parma 07678
Società Italiana della Trasfusione del Sangue, Torino 07827
Società Italiana di Angiologia, Pisa 07837
Società Italiana di Ematologia, Milano 07867
Società Italiana di Immunoematologia e Trasfusione del Sangue, Roma 07886

Jamaica
Medical Research Council Laboratories, Kingston 08088

Japan
Nippon Ketsueki Gakkai, Kyoto 08331
Nippon Yuketsu Gakkai, Tokyo 08397

Norway
Association for Nordic Dialysis and Transplant Personnel, Trondheim 09248

Poland
Polskie Towarzystwo Hematologiczne, Wroclaw 09575

Russia
National Scientific Medical Society of Haematologists and Transfusiologists, Moskva 09915

South Africa
South African Dietetics and Home Economics Association, Stellenbosch 10189

Switzerland
Schweizerische Hämophilie-Gesellschaft, Zürich 10859

United Kingdom
British Committee for Standards in Haematology, London 11545
British Society for Haematology, London 11656
Haemophilia Society, London 12140

U.S.A.
American Blood Commission, Chicago 13624
American Celiac Society, West Orange 13682
American College of Nutrition, Wilmington 13721
American Council of Applied Clinical Nutrition, Florissant 13761
American Society for Parenteral and Enteral Nutrition, Silver Spring 14076
American Society of Hematology, Washington 14131
Community Nutrition Institute, Washington 14814
International Association for Comparative Research on Leukemia and Related Diseases, Columbus 15376
International Society of Hematology, Rochester 15552
International Society on Thrombosis and Hemostasis, Chapel Hill 15557
Iron Overload Diseases Association, North Palm Beach 15586
Leukemia Society of America, New York 15664
Society for the Study of Blood, Brooklyn 16613
World Hemophilia Aids Center, Pasadena 16928

Venezuela
Sociedad Venezolana de Angiologia, Caracas 17016
Sociedad Venezolana de Hematología, Caracas 17023

Heraldry s. Genealogy, Heraldry

History

Algeria
Société Historique Algérienne, Alger 00024

Andorra
Comitè Andorrà de Ciències Històriques, Andorra la Vella 00030

Argentina
Academia Nacional de la Historia, Buenos Aires 00050
Departamento de Estudios Históricos Navales, Buenos Aires 00132

Australia
Anthropological Society of South Australia, Adelaide 00216
Australasian Association for the History, Philosophy and Social Studies of Science, Sydney 00237
Australian and New Zealand Association for Canadian Studies, North Ryde 00256
Australian and New Zealand Association for Medieval and Renaissance Studies, Sydney 00257
Australian Association for the Study of Religions, Saint Lucia 00261
Field Naturalists' Club of Victoria, South Yarra 00396
The Field Naturalists' Society of South Australia, Adelaide 00397
Gosford District Historical Research and Heritage Association, Hardy's Bay 00407
Indo-Pacific Prehistory Association, Canberra 00410
Linnean Society of New South Wales, Milsons Point 00429
North Queensland Naturalists Club, Cairns 00441
Oaks Historical Society, The Oaks 00442
The Queensland Naturalists' Club, Brisbane 00449
Royal Australian Historical Society, Sydney 00464
The Royal Historical Society of Queensland, Brisbane North Quay 00469
Royal Historical Society of Victoria, Melbourne 00470
Royal Western Australian Historical Society, Nedlands 00483
Tasmanian Historical Research Association, Sandy Bay 00506
Western Australian Naturalists' Club, Nedlands 00524

Austria
Anthropologische Gesellschaft in Wien, Wien 00544
Arbeitsgemeinschaft für Historische Sozialkunde, Wien 00551
Deutscher Rechtshistorikertag, Graz 00580
Dokumentationsarchiv des österreichischen Widerstandes, Wien 00582
Eranos Vindobonensis, Wien 00584
Geschichtsverein für Kärnten, Klagenfurt 00602
Gesellschaft für die Geschichte des Protestantismus in Österreich, Wien 00617
Gesellschaft für Geschichte der Neuzeit, Salzburg 00620
Gesellschaft für österreichische Kulturgeschichte, Eisenstadt 00627
Gesellschaft für Photographie und Geschichte, Wien 00630
Gesellschaft für Salzburger Landeskunde, Salzburg 00632
Historische Landeskommission für Steiermark, Graz 00650
Historischer Verein für Steiermark, Graz 00651
Institut für Österreichkunde, Wien 00658
Internationale Tagung der Historiker der Arbeiterbewegung, Wien 00684
Kommission für Neuere Geschichte Österreichs, Salzburg 00695
Kulturgeschichtliche Gesellschaft am Landesmuseum Joanneum, Graz 00698
Kunsthistorische Gesellschaft, Wien 00699
Kunsthistorische Gesellschaft an der Universität Graz, Graz 00700
Montanhistorischer Verein für Österreich, Leoben-Donawitz 00708
Österreichische Byzantinische Gesellschaft, Wien 00734
Österreichische Gesellschaft für Amerikastudien, Salzburg 00753
Österreichische Gesellschaft für Geschichte der Pharmazie, Wien 00785
Österreichische Gesellschaft für Neugriechische Studien, Wien 00813
Österreichische Gesellschaft für Unternehmensgeschichte, Wien 00838
Österreichische Gesellschaft für Ur- und Frühgeschichte, Wien 00840
Österreichische Gesellschaft für Wissenschaftsgeschichte, Wien 00849
Österreichische Gesellschaft für Zeitgeschichte, Wien 00851
Österreichische Gesellschaft zur Erforschung des 18. Jahrhunderts, Wien 00855
Österreichischer Arbeitskreis für Stadtgeschichtsforschung, Linz 00884
Österreichischer Burgenlandbund, Arbeitsgemeinschaft für Burgenländische Geschichte und Persönlichkeiten-Deutschtum in Ungarn, Eisenstadt 00888
Österreichischer Burgenverein, Grieskirchen 00889
Österreichischer Kunsthistorikerverband, Wien 00895
Österreichisches Ost- und Südosteuropa-Institut, Wien 00915
SIGMA - Salzburger Gesellschaft für Semiologie, Salzburg 00943
Verband der Österreicher zur Wahrung der Geschichte Österreichs, Wien 00960
Verband Österreichischer Geschichtsvereine, Wien 00968
Verein der Freunde des Radwerkes IV in Vordernberg, Vordernberg 00977
Verein für Geschichte der Arbeiterbewegung, Wien 00983
Verein für Geschichte der Stadt Wien, Wien 00984
Verein für Landeskunde von Niederösterreich, Wien 00985
Verein Tiroler Landesmuseum Ferdinandeum, Innsbruck 01003
Vorarlberger Landesmuseumsverein, Dornbirn 01008
Waldviertler Heimatbund, Wissenschaftliche Sektion, Horn 01009
Wiener Humanistische Gesellschaft, Wien 01018

Bahamas
Association of Caribbean Historians, Nassau 01050
Bahamas Historical Society, Nassau 01051

Bahrain
Bahrein Historical and Archaeological Society, Manama 01062

Barbados
Barbados Museum and Historical Society, Bridgetown 01074

Belgium

Association des Diplomés en Histoire de l'Art et Archéologie de l'Université Catholique de Louvain, Louvain-la-Neuve *01116*
Centre International de Recherches Glyptographiques, Braine-le-Château *01194*
Centre International des Langues, Littératures et Traditions d'Afrique au Service du Développement, Louvain-la-Neuve *01195*
Cercle Benelux d'Histoire de la Pharmacie, Kortrijk *01207*
Comité Belge d'Histoire des Sciences, Bruxelles *01218*
Comité International d'Etude des Géants Processionnels, Bruxelles *01231*
Commission of the European Communities Liaison Committee of Historians, Louvain-la-Neuve *01236*
European Association of Historical Associations, Louvain-la-Neuve *01293*
Institut Archéologique du Luxembourg, Arlon *01375*
Institut Archéologique Liégeois, Liège *01376*
Institut Belge des Hautes Etudes Chinoises, Bruxelles *01381*
Institut Historique Belge de Rome, Bruxelles *01386*
Koninklijke Academie voor Wetenschappen, Letteren en Schone Kunsten van België, Bruxelles *01409*
Société Archéologique de Namur, Namur *01421*
Société Belge d'Etudes Byzantines, Bruxelles *01448*
Société Belge d'Histoire des Hôpitaux, Bruxelles *01450*
Société des Bollandistes, Bruxelles *01456*
Société d'Etudes Latines de Bruxelles, Tournai *01458*
Société Royale Belge d'Anthropologie et de Préhistoire, Bruxelles *01469*
Société Royale des Amis du Musée Royal de l'Armée et d'Histoire Militaire, Bruxelles *01484*
Union Internationale des Sciences Préhistoriques et Protohistoriques, Gent *01497*

Bermuda

Bermuda Historical Society, Hamilton *01520*
Saint George's Historical Society, Saint George *01527*

Bolivia

Academia Nacional de la Historia, La Paz *01530*
Sociedad de Estudios Geográficos e Históricos, Santa Cruz de la Sierra *01551*
Sociedad Geográfica y de Historia Potosío, Potosí *01555*

Brazil

Centro de Estudios de Demografía Histórica de América Latina, São Paulo *01623*
Fundação João Pineiro, Belo Horizonte *01657*
Instituto Archeológico, Histórico e Geográfico Pernambucano, Recife *01663*
Instituto do Ceará, Fortaleza *01669*
Instituto Geográfico e Histórico da Bahia, Salvador *01672*
Instituto Geográfico e Histórico do Amazonas, Manaus *01673*
Instituto Histórico de Alagoas, Maceío *01674*
Instituto Histórico e Geográfico Brasileiro, Rio de Janeiro *01675*
Instituto Histórico e Geográfico de Goiás, Goiânia *01676*
Instituto Histórico e Geográfico de Santa Catarina, Florianópolis *01677*
Instituto Histórico e Geográfico de Santos, Santos *01678*
Instituto Histórico e Geográfico de São Paulo, São Paulo *01679*
Instituto Histórico e Geográfico de Sergipe, Aracajú *01680*
Instituto Histórico e Geográfico do Maranhão, São Luís *01681*
Instituto Histórico e Geográfico do Pará, Belém *01682*
Instituto Histórico e Geográfico do Rio Grande do Norte, Natal *01683*
Instituto Histórico e Geográfico do Rio Grande do Sul, Porto Alegre *01684*
Instituto Histórico e Geográfico Paraíbano, João Pessoa *01685*
Instituto Histórico, Geográfico e Etnográfico Paranaense, Curitiba *01686*
Istituto Histórico e Geográfico do Espirito Santo, Vitória *01697*
Secretaria do Patrimônio Histórico e Artístico Nacional, Rio de Janeiro *01702*

Bulgaria

Bulgarian Historical Society, Sofia *01751*

Canada

Alberta Historical Resources Foundation, Calgary *01837*
Association for 18th Century Studies, Hamilton *01898*
Association for Canadian Studies, Montréal *01899*
Association of Canadian Universities for Northern Studies, Ottawa *01926*
British Columbia Historical Association, Nanaimo *01979*
Canadian Association for American Studies, Burnaby *02026*
Canadian Aviation Historical Society, Willowdale *02080*
Canadian Catholic Historical Association (English Section), London *02089*
Canadian Historical Association, Ottawa *02157*
Canadian Oral History Association, Ottawa *02214*
Canadian Polish Research Institute, Toronto *02228*
Canadian Railroad Historical Association, Saint Constant *02241*
Canadian Society for Italian Studies, Ottawa *02261*
Canadian Society of Church History, Halifax *02278*
Canadian Society of Military Medals and Insignia, Thorold *02287*
Centre for Monarchial Studies, Toronto *02328*
Champlain Society, Toronto *02339*
German-Canadian Historical Association, Montréal *02483*
Glenbow Alberta Institute, Calgary *02486*
Historical and Scientific Society of Manitoba, Winnipeg *02501*
Historical Society of Alberta, Calgary *02502*
Historical Society of Mecklenburg Upper Canada, Toronto *02503*
Historical Society of the Gatineau, Old Chelsea *02504*
Institut d'Histoire de l'Amérique Française, Montréal *02522*
MacBride Museum Society, Whitehorse *02597*
Manitoba Historical Society, Winnipeg *02621*
Miramichi Historical Society, Newcastle *02659*
Multicultural History Society of Ontario, Toronto *02663*
New Brunswick Historical Society, Saint John *02685*
Newfoundland Historical Society, Saint John's *02705*
Ontario Black History Society, Willowdale *02763*
Ontario Historical Society, Willowdale *02782*
Royal Nova Scotia Historical Society, Halifax *02870*
Société Historique Acadienne, Moncton *02929*
Société Historique de la Saskatchewan, Regina *02930*
Société Historique de Québec, Québec *02931*
Société Historique du Saguenay, Chicoutimi *02932*
Vancouver Museum Association, Vancouver *02976*
Victorian Studies Association of Western Canada, Edmonton *02981*
Waterloo Historical Society, Kitchener *02984*
Yukon Historical and Museums Association, Whitehorse *03003*

Chile

Academia Chilena de la Historia, Santiago *03010*
Sociedad Chilena de Historia y Geografía, Santiago *03076*

China, People's Republic

Chinese Society of History of Science, Bejing *03193*
Inner Mongolian Academy of Social Sciences, Huhehot *03206*

China, Republic

Academia Historica, Taipei *03226, 03227*
Chinese Historical Association, Taipei *03270*
Chinese Medical History Association, Taipei *03278*
Historical Research Commission of Taiwan, Taipei *03301*
Society of the Chinese Borders History and Languages, Taipei *03329*

Colombia

Academia Antioqueña de Historia, Medellín *03334*
Academia Boyacense de Historia, Tunja *03335*
Academia Colombiana de Historia, Bogotá *03337*
Academia de Historia de Cartagena de Indias, Cartagena *03340*
Academia de Historia del Norte de Santander, Cúcuta *03341*
Instituto Caro y Cuervo, Bogotá *03377*
Instituto Colombiano de Cultura Hispánica, Bogotá *03384*

Costa Rica

Academia de Geografía e Historia de Costa Rica, San José *03429*

Cuba

Centro Asturiano, La Habana *03472*
Centro de Estudios de Historia y Organización de la Ciencia Carlos J. Finlay, La Habana *03473*

Cyprus

Cyprus Research Centre, Nicosia *03496*
Society of Cypriot Studies, Nicosia *03499*

Czech Republic

Česká Historická Společnost, Praha *03508*
Matice Moravská, Brno *03551*

Denmark

Chakoten Dansk Militaershistorisk Selskab, København *03572*
Danske Komité for Historikernes Internationale Samarbejde, København *03608*
Dansk Farmacihistorisk Selskab, København *03617*
Dansk Historielaererforening, Vordingborg *03630*
Dansk Lokalhistorisk Forening, Bagsvaerd *03644*
Dansk Naturhistorisk Forening, København *03653*
Dansk Pressehistorisk Selskab, København *03665*
Dansk Selskab for Oldtids- og Middelalderforskning, København *03681*
Den Danske Historiske Forening, København *03702*
Det Danske Sprog- og Litteraturselskab, København *03708*
Det Kongelige Nordiske Oldskriftselskab, København *03711*
Historisk Samfund for Sønderjylland, Abenrå *03742*
Historisk-Topografisk Selskab, Frederiksberg *03743*
Jysk Selskab for Historie, Arhus *03756*
Marinehistorisk Selskab, København *03770*
Naturhistorisk Forening for Jylland, Arhus *03775*
Naturhistorisk Forening for Nordsjaelland, Hillerød *03776*
Scandinavian Society for Economic and Social History, Odense *03806*
Selskabet for Danmarks Kirkehistorie, København *03809*
Selskabet for Dansk Kulturhistorie, København *03810*
Selskabet for Dansk Skolehistorie, Hellerup *03811*
Selskabet for Dansk Teaterhistorie, København *03812*
Selskabet for Historie og Samfundsøkonomi, København *03814*
Selskabet for Københavns Historie, København *03815*
Selskabet til Udgivelse af Danske Mindesmaerker, Nivaa *03818*

Dominican Republic

Academia Dominicana de la Historia, Santo Domingo *03829*

Ecuador

Centro de Investigaciones Históricas, Guayaquil *03849*

Egypt

Hellenic Society of Ptolemaic Egypt, Alexandria *03903*

El Salvador

Academia Salvadoreña de la Historia, San Salvador *03917*

Finland

European Association for Environmental History, Helsinki *03955*
Historian Ystäväin Liitto, Tampere *03965*
Kalevalaseura, Helsinki *03972*
Klassillis-Filologinen Yhdistys, Helsingin Yliopisto *03978*
Museovirasto, Helsinki *03983*
Suomen Historiallinen Seura, Helsinki *04027*
Suomen Kirkkohistoriallinen Seura, Helsingin Yliopisto *04040*
Suomen Muinaismuistoyhdistys, Helsinki *04051*
Taloushistoriallinen Yhdistys, Jyväskylä *04070*
Turun Historiallinen Yhdistys, Turku *04078*

France

Académie des Belles-Lettres, Sciences et Arts de La Rochelle, La Rochelle *04094*
Académie des Sciences, Agriculture, Arts et Belles-Lettres d'Aix, Aix-en-Provence *04099*
Académie des Sciences, Arts et Belles-Lettres de Dijon, Dijon *04100*
Académie des Sciences d'Outre-Mer, Paris *04105*
Académie Internationale de Science Politique et d'Histoire Constitutionnelle, Paris *04113*
Académie Nationale de Metz, Metz *04119*
Amici Thomae Mori, Angers *04132*
Asociación de Historadores Latinoamericanos y del Caribe, Paris *04139*
Association des Professeurs d'Histoire et de Géographie de l'Enseignement Public, Evry *04197*
Association Française des Historiens Economistes, Paris *04231*
Association Internationale d'Histoire Economique, Paris *04282*
Association Marc Bloch, Paris *04301*
Centre d'Etudes, de Documentation, d'Information et d'Action Sociales, Paris *04369*
Centre d'Histoire Militaire et d'Etudes de Défense Nationale, Montpellier *04370*
Centre International d'Etude des Textiles Anciens, Lyon *04377*
Comité Historique du Centre-Est, Lyon *04405*
Comité International des Sciences Historiques, Paris *04408*
Commission Internationale d'Histoire des Mouvements Sociaux et des Structures Sociales, Paris *04427*
Demeure Historique, Paris *04463*
Etudes Préhistoriques, Sainte-Foy-lès-Lyon *04472*
European Association for American Studies, Lyon *04480*
European Association of Museums of the History of Medical Sciences, Denicé *04488*
Fédération d'Associations et Groupements pour les Etudes Corses, Bastia *04509*
Fédération des Sociétes Historiques et Archéologiques de Paris et de l'Ile de France, Paris *04516*
Féderation des Sociétés Savantes de la Charente-Maritime, La Rochelle *04517*
Fédération Historique de Provence, Marseille *04528*
Institut des Sciences Historiques, Paris *04566*
Institut Français d'Histoire Sociale, Paris *04569*
International Academy of the History of Science, Paris *04578*
Paris et son Histoire, Paris *04618*
Le Pays Bas-Normand, Flers *04619*
Société Archéologique et Historique du Limousin, Limoges *04630*
Société d'Archéologie et d'Histoire de la Manche, Saint-Lo *04640*
Société d'Archéologie et d'Histoire de l'Aunis, La Rochelle *04641*
Société de Démographie Historique, Paris *04653*
Société de l'Histoire de France, Paris *04661*
Société de l'Histoire de l'Art Français, Paris *04662*
Société de l'Histoire du Protestantisme Français, Paris *04663*
Société d'Emulation Historique et Littéraire d'Abbéville, Abbéville *04671*
Société de Mythologie Française, Paris *04672*
Société de Recherches et d'Etudes Historiques Corses, Ajaccio *04686*
Société des Amis du Musée de l'Homme, Paris *04698*
Société des Lettres, Sciences et Arts de la Haute-Auvergne, Aurillac *04709*
Société des Sciences Historiques et Naturelles de la Corse, Bastia *04716*
Société de Statistique, d'Histoire et d'Archéologie de Marseille et de Provence, Marseille *04720*
Société d'Etude du Dix-Septième Siècle, Paris *04727*
Société d'Etudes Ardennaises, Charleville-Mézières *04728*
Société d'Etudes Historiques, Paris *04741*
Société d'Etudes Juives, Paris *04744*
Société d'Etudes Médiévales, Poitiers *04747*
Société d'Etudes Robespierristes, Paris *04752*
Société d'Etudes Romantiques, Clermont-Ferrand *04753*
Société d'Histoire de Bordeaux, Bordeaux *04756*
Société d'Histoire de la Médecine Hébraïque, Paris *04757*
Société d'Histoire de la Pharmacie, Paris *04758*
Société d'Histoire du Droit, Paris *04759*
Société d'Histoire du Droit Normand, Caen *04760*
Société d'Histoire du Théâtre, Paris *04761*
Société d'Histoire et d'Archéologie de Bretagne, Rennes *04762*
Société d'Histoire et d'Archéologie de la Lorraine, Saint Julien les Metz *04763*
Société d'Histoire Générale et d'Histoire Diplomatique, Paris *04765*
Société d'Histoire Moderne, Paris *04766*
Société d'Histoire Religieuse de la France, Paris *04767*
Société Française d'Egyptologie, Paris *04805*
Société Française d'Etude du Dix-Huitième Siècle, Pau *04850*
Société Française d'Histoire d'Outre-Mer, Paris *04857*
Société Historique, Archéologique et Littéraire de Lyon, Lyon *04870*
Société Historique de la Province de Maine, Le Mans *04871*
Société Historique du Bas-Limousin, Tulle *04872*
Société Historique et Archéologique du Périgord, Périgueux *04873*
Société Saint-Simon, Sceaux *04916*
Société Savoisienne d'Histoire et d'Archéologie, Chambéry *04917*

Georgia

Council on the History of Natural Sciences and Technology, Tbilisi *05010*
Georgian History Society, Tbilisi *05017*

Germany

Aachener Geschichtsverein, Aachen *05028*
Arbeitsgemeinschaft Historischer Kommissionen und Landesgeschichtlicher Institute e.V., Marburg *05118*
Arbeo-Gesellschaft e.V., Bachenhausen *05150*
Baltische Gesellschaft in Deutschland e. V., München *05164*
Bergischer Geschichtsverein e. V., Wuppertal *05184*
Berliner Gesellschaft für Anthropologie, Ethnologie und Urgeschichte, Berlin *05187*
Deutsche Morgenländische Gesellschaft e.V., Heidelberg *05473*
Deutsche Orient-Gesellschaft e.V., Berlin *05480*
Deutsches Orient-Institut, Hamburg *05598*
Faust-Gesellschaft, Mühlacker *05698*
Friedrich-Ebert-Stiftung e.V., Bonn *05764*
Frobenius-Gesellschaft e.V., Frankfurt *05765*
Fuldaer Geschichtsverein e.V., Fulda *05767*
Germania Judaica, Köln *05786*
Gesamtverein der deutschen Geschichts- und Altertumsvereine, Köln *05787*
Gesellschaft für Agrargeschichte, Stuttgart *05796*
Gesellschaft für Deutsche Postgeschichte, Frankfurt *05807*
Gesellschaft für Geistesgeschichte e.V., Erlangen *05823*
Gesellschaft für Geschichte und Kultur e.V., Bonn *05825*
Gesellschaft für Historische Waffen- und Kostümkunde, Hannover *05827*
Gesellschaft für Mittelrheinische Kirchengeschichte, Koblenz *05841*
Gesellschaft für Regionalforschung e.V., Bonn *05858*
Gesellschaft für Sozial- und Wirtschaftsgeschichte, Heidelberg *05860*
Gesellschaft für Unternehmensgeschichte e.V., Köln *05866*
Göttinger Arbeitskreis, Göttingen *05898*
Gutenberg-Gesellschaft, Mainz *05902*
Harzverein für Geschichte und Altertumskunde, Goslar *05910*
Historische Kommision der Deutschen Gesellschaft für Erziehungswissenschaft, Hannover *05915*
Historische Kommission des Börsenvereins des Deutschen Buchhandels, Frankfurt *05916*
Historische Kommission zu Berlin, Berlin *05917*
Historische Kommission zur Erforschung des Pietismus an der Universität Münster, Münster *05918*
Historischer Verein Bamberg, Bamberg *05919*
Historischer Verein der Pfalz e.V., Speyer *05920*
Historischer Verein Dillingen an der Donau, Dillingen *05921*
Historischer Verein für die Saargegend e.V., Saarbrücken *05922*
Historischer Verein für Hessen, Darmstadt *05923*

Ghana

Gibraltar

Greece

Guatemala

Honduras

Hong Kong

Hungary

Iceland

India

Indonesia

Iran

Iraq

Ireland

Israel

Italy

Jamaica

Japan

Kenya

Korea, Republic

Liechtenstein

Macedonia

Malawi

Malaysia

Malta

Mauritius

Mexico

Monaco

Netherlands

New Zealand

Niger

Nigeria

Norway

Oman

Pakistan

Panama

Paraguay

Peru

Philippines

Poland

Portugal

Centro de Estudos de História e Cartografia Antiga, Lisboa 09654
Instituto Português de Arqueologia, História e Etnografia, Lisboa 09661
Sociedade Histórica da Independência de Portugal, Lisboa 09686

Puerto Rico
Academia Puertorriqueña de la Historia, Santurce 09718

Réunion
Association Historique Internationale de l'Océan Indien, Sainte-Clotilde 09733

Romania
Societatea de Istorie Medicinei, Bucuresti 09768

Russia
National Committee of History and Philosophy of Natural Science and Technology, Moskva 09896
National Committee of Russian Historians, Moskva 09898
National Scientific Medical Society of History of Medicine, Moskva 09916
Russian Palestine Society, Moskva 09961

Saudi Arabia
King Abdul Aziz Research Centre, Riyadh 10004
Society of Esaff Alkhairia, Mecca 10006

Sierra Leone
Historical Society of Sierra Leone, Freetown 10030

Slovakia
Slovenská Historická Spoločnost, Bratislava 10074

Slovenia
Zveza Zgodovinskih Društev Slovenije, Ljubljana 10117

South Africa
Federasie van Afrikaanse Kultuurvereniginge, Auckland Park 10141
Van Riebeeck Society, Cape Town 10233

Spain
Academia Iberoamericana y Filipina de Historia Postal, Madrid 10243
Institut d'Estudis Catalans, Barcelona 10315
Instituto de Estudios Asturianos, Oviedo 10320
Instituto de Historia y Cultura Naval, Madrid 10323
Real Academia de Bellas Artes y Ciencias Históricas de Toledo, Toledo 10341
Real Academia de la Historia, Madrid 10348
Real Sociedad Bascongada de los Amigos del Pais, San Sebastian 10359
Real Sociedad Española de Historia Natural, Madrid 10362
Real Sociedad Vascongada de los Amigos del País, San Sebastián 10367
Servicio de Investigación Prehistórica y Museo de Prehistoria, Valencia 10369
Sociedad Española de Estudios Clásicos, Madrid 10386

Sri Lanka
The Classical Association of Ceylon, Colombo 10415
Royal Asiatic Society of Sri Lanka, Colombo 10418

Suriname
Stichting Cultureel Centrum Suriname, Paramaribo 10432

Sweden
Association of Nordic Paper Historians, Stockholm 10437
Carl Johans Förbundet, Uppsala 10442
Karolinska Förbundet, Stockholm 10478
Konsthistoriska Sällskapet, Stockholm 10481
Kungliga Samfundet för Utgivande av Handskrifter rörande Skandinaviens Historia, Stockholm 10488
Kungliga Vitterhets Historie och Antikvitets Akademien, Stockholm 10493
Svenska Historiska Föreningen, Stockholm 10554
Svenska Klassikerförbundet, Stockholm 10556
Svenska Kyrkohistoriska Föreningen, Uppsala 10559
Svensk Flyghistorisk Förening, Stockholm 10578

Switzerland
Allgemeine Geschichtforschende Gesellschaft der Schweiz, Bern 10614
Antiquarische Gesellschaft in Zürich, Zürich 10616
Association d'Histoire et de Science Politique, Wabern 10623
Association Internationale pour l'Histoire Contemporaine de l'Europe, Genève 10629
Fédération Internationale des Associations d'Etudes Classiques, Bellevue 10701
Gesellschaft für Schweizerische Kunstgeschichte, Bern 10716
Historisch-Antiquarischer Verein Heiden, Heiden 10726
Historischer Verein des Kantons Bern, Bern 10727
Historische und Antiquarische Gesellschaft zu Basel, Basel 10728
Opera Svizzera dei Monumenti d'Arte, Locarno 10775
Schweizerische Gesellschaft für orientalische Altertumswissenschaft, Genève 10838
Schweizerische Vereinigung für Altertumswissenschaft, Basel 10939
Società Storica Locarnese, Locarno 10964
Société de Physique et d'Histoire Naturelle de Genève, Genève 10969
Société d'Histoire de la Suisse Romande, Lausanne 10971
Société d'Histoire et d'Archéologie, Genève 10972
Société Vaudoise d'Histoire et d'Archéologie, Chavannes-près-Renens 10991
Vereinigung für Walsertum, Brig 11015
Verein Schweizerischer Geschichtslehrer, Riehen 11025

Tanzania
The Tanzania Society, Dar es Salaam 11062

Trinidad and Tobago
Historical Society of Trinidad and Tobago, Port of Spain 11121

Tunisia
Arab Committee for Ottoman Studies, Zaghouan 11128

Turkey
Türk Tarih Kurumu, Ankara 11175

Uganda
The Uganda Society, Kampala 11188

United Kingdom
American Civil War Round Table, United Kingdom, Richmond 11224
Anglesey Antiquarian Society and Field Club, Menai Bridge 11230
Antiquarian Horological Society, Wadhurst 11238
Ashmolean Natural History Society of Oxfordshire, Oxford 11255
Association of European Latin Americanist Historians, Liverpool 11343
Association of Genealogists and Record Agents, Horsham 11345
Association of Northumberland Local History Societies, Newcastle-upon-Tyne 11373
Association of School Natural History Societies, Lancing 11386
Ayrshire Archaeological and Natural History Society, Ayr 11419
The Baptist Historical Society, Oxford 11422
Bedfordshire Historical Record Society, Bedford 11431
Belfast Natural History and Philosophical Society, Belfast 11432
Birmingham Natural History Society, Birmingham 11450
Birmingham Transport Historical Group, Folkestone 11451
Blackcountry Society, Kingswinford 11452
Borthwick Institute of Historical Research, York 11462
Brewery History Society, Sutton 11466
Bristol and Gloucestershire Archaeological Society, Gloucester 11468
British Agricultural History Society, Reading 11474
British Association for American Studies, Newcastle-upon-Tyne 11485
British Association for Canadian Studies, London 11488
British Association for the History of Religions, Cardiff 11496
British Records Association, London 11638
British Record Society, London 11639
British Society for Eighteenth Century Studies, Aberdeen 11654
British Society for the History of Mathematics, Milton Keynes 11670
British Society for the History of Medicine, London 11671
British Society for the History of Pharmacy, Edinburgh 11672
The British Society for the History of Science, Faringdon 11673
British Universities Association of Slavists, Reading 11699
Buteshire Natural History Society, Bute 11718
Caernarvonshire Historical Society, Caernarfon 11733
Cambridge Antiquarian Society, Cambridge 11735
Canterbury and York Society, Twickenham 11744
Catholic Record Society, London 11748
Cinema Theatre Association, Teddington 11786
Confederate Historical Society, Leigh-on-Sea 11836
Cornish Methodist Historical Association, Carnon Downs 11842
County Museum Society, Kidderminster 11862
Cross and Cockade, Great Britain Society of World Ware One Aero Historians, Farnborough 11869
Crown Imperial Society, Southall 11870
Cumberland and Westmorland Antiquarian and Archaeological Society, Kendal 11871
Cymdeithas Hanes Sir Ddinbych, Denbigh 11874
Devon and Cornwall Record Society, Exeter 11888
Dorothy L. Sayers Society, Hurstpierpoint 11901
Dorset Natural History and Archaeological Society, Dorchester 11902
Dorset Record Society, Dorchester 11903
The Dugdale Society, Stratford-upon-Avon 11910
Dumfriesshire and Galloway Natural History and Antiquarian Society, Dumfries 11911
Durham County Local History Society, Durham 11913
East London History Society, London 11918
East Lothian Antiquarian and Field Naturalists' Society, Dunbar 11919
Ecclesiastical History Society, Glasgow 11923
Economic History Society, Heslington 11927
Edmonton Hundred Historical Society, London 11940
Edwardian Studies Association, Dagenham 11947
Eighteen Nineties Society, London 11949
English Westerners Society, London 11968
Essex Archaeological and Historical Congress, Stratford Saint Mary 11976
Essex Society for Archaeology and History, Colchester 11977
The European-Atlantic Movement, Exeter 12003
Federation for Ulster Local Studies, Belfast 12061
Federation of Old Cornwall Societies, Launceston 12067
Flag Institute, Chester 12078
Flintshire Historical Society, Clwyd 12079
Fort Cumberland and Portsmouth Militaria Society, Portsmouth 12086
Friends Historical Society, London 12093
Furniture History Society, London 12095
Glamorgan History Society, Cardiff 12116
The Greek Institute, London 12128
Group for the Study of Irish Historic Settlement, Newtonabbey 12131
The Hakluyt Society, London 12141
Hampshire Field Club and Archaeological Society, Winchester 12144
Hertfordshire Local History Council, Hitchin 12163
Hispanic and Luso-Brazilian Council, London 12170
Historical Association, London 12172
The Historical Metallurgy Society, Swansea 12173
Historical Society of the Church in Wales, Carmarthen 12175
Historical Society of the Methodist Church in Wales, Pwllheli 12176
Historic Breechloading Smallarms Association, London 12177
Historic Society of Lancashire and Cheshire, Liverpool 12178
History of Education Society, Leicester 12179
Huguenot Society of Great Britain and Ireland, London 12186
Huntingdonshire Local History Society, Thurleigh 12189
Institute of Contemporary History and Wiener Library, London 12228
Institute of Jewish Affairs, London 12249
International Commission for Orders of Chivalry, Edinburgh 12341
International Commission for the History of Representative and Parliamentary Institutions, Brighton 12342
Isle of Man Natural History and Antiquarian Society, Douglas 12412
Isle of Wight Natural History and Archaeological Society, Shanklin 12413
Jewish Historical Society of England, London 12418
The Joint Association of Classical Teachers, London 12422
Lancashire Parish Register Society, Hemel Hempstead 12439
Leicestershire Archaeological and Historical Society, Leicester 12445
Leicestershire Local History Council, Leicester 12446
Lincoln Record Society, Lincoln 12452
Linnean Society of London, London 12454
Literary and Philosophical Society of Newcastle upon Tyne, Newcastle-upon-Tyne 12457
London and Middlesex Archaeological Society, London 12460
London Medieval Society, London 12463
London Natural History Society, Frinton-on-Sea 12465
London Record Society, London 12467
London Topographical Society, Richmond 12470
Manorial Society of Great Britain, London 12488
Medieval Combat Society, London 12516
Medieval Settlement Research Group, Taunton 12517
Military Historical Society, Bromley 12527
Moray Society, Elgin 12538
Napoleonic Society, London 12552
Natural History Society of Northumbria, Newcastle-upon-Tyne 12620
Navy Records Society, London 12622
The Newcomen Society, London 12625
Norfolk Record Society, Norwich 12630
Northamptonshire Natural History Society and Field Club, Northampton 12632
Northamptonshire Record Society, Northampton 12633
North Staffordshire Field Club, Stoke-on-Trent 12642
North Western Society for Industrial Archaeology and History, Liverpool 12647
Nottinghamshire Local History Association, Nottingham 12650
Oral History Society, Colchester 12656
Orders and Medals Research Society, Croydon 12657
Oxfordshire Architectural and Historical Society, Oxford 12669
Oxfordshire Record Society, Oxford 12670
Palestine Exploration Fund, London 12676
Peak District Mines Historical Society, Nottingham 12683
Pembrokeshire Historical Society, Scloddau Fishguard 12686
Pendragon Society, Bristol 12687
Pipe Roll Society, London 12700
Polish Underground Movement (1939-1945) Study Trust, London 12710
Prehistoric Society, Swindon 12715
Presbyterian Church of Wales Historical Society, Aberystwyth 12716
Psywar Society, Kettering 12727
Radnorshire Society, Llandrindod Wells 12739
The Ray Society, London 12745
Richard III Society, London 12755
Royal Commission for the Exhibition of 1851, London 12794
Royal Commission on Historical Manuscripts, London 12795
Royal Historical Society, London 12809
Royal Institution of Cornwall, Truro 12821
Royal Stuart Society, Huntingdon 12867
Saint Albans and Hertfordshire Architectural and Archaeological Society, Saint Albans 12882
Scottish Catholic Historical Association, Glasgow 12902
Scottish Church History Society, Edinburgh 12903
Scottish History Society, Edinburgh 12911
Scottish Record Society, Glasgow 12922
Scottish Society for Northern Studies, Edinburgh 12929
Scottish Tartans Society, Pitlochry 12932
Selborne Society, Oxted 12938
Selden Society, London 12939
Seventeenhundredandfourtyfive and National Military History Society, Edinburgh 12941
Shropshire Archaeological and Parish Register Society, Much Wenlock 12951
Social History Society of the United Kingdom, Lancaster 12955
La Société Guernesiaise, Saint Peter Port 12961
Society for African Church History, Aberdeen 12963
Society for Army Historical Research, London 12965
Society for Libyan Studies, London 12978
Society for Lincolnshire History and Archaeology, Lincoln 12979
Society for Nautical Research, London 12984
The Society for Renaissance Studies, London 12992
The Society for South Asian Studies, London 12997
Society for the History of Alchemy and Chemistry, London 13003
Society for the History of Natural History, London 13004
Society for the Promotion of Hellenic Studies, London 13005
Society for the Promotion of Roman Studies, London 13008
Society for the Social History of Medicine, Manchester 13012
Society for the Study of Labour History, London 13018
Society of Antiquaries of Scotland, Edinburgh 13024
Society of Architectural Historians of Great Britain, London 13027
Society of Cirplanologists, Hyde 13037
Society of Writers to Her Majesty's Signet, Edinburgh 13072
Somerset Archaeological and Natural History Society, Taunton 13074
Somerset Record Society, Taunton 13075
Southern Skirmish Association, London 13080
South Staffordshire Archaeological and Historical Society, Walsall 13083
Staffordshire Parish Registers Society, Penn 13091
Staffordshire Record Society, Stafford 13092
Stair Society, Edinburgh 13093
Standing International Committee for Mycenaean Studies, Cambridge 13098
Suffolk Institute of Archaeology and History, Ipswich 13106
Suffolk Records Society, Ipswich 13107
Surrey Record Society, Guildford 13112
Sussex Archaeological Society, Lewes 13113
Thomas Paine Society, Selsey 13127
Thoresby Society, Leeds 13130
Thoroton Society of Nottinghamshire, Nottingham 13131
Trevithick Society, Truro 13142
The Ulster Genealogical and Historical Guild, Belfast 13153
Ulster Historical Foundation, Belfast 13154
Ulster Society for Irish Historical Studies, Belfast 13155
Ulster Society in London, London 13157
Unitarian Historical Society, Edinburgh 13162
United Reformed Church History Society, London 13169
University Association for Contemporary European Studies, London 13174
Victorian Military Society, Northampton 13183
Viking Society for Northern Research, London 13185
The Walpole Society, London 13193
Welsh Baptist Historical Society, Llanelli 13199
Wesley Historical Society, Birmingham 13205
William Morris Society, London 13208
Wiltshire Archaeological and Natural History Society, Devizes 13209
Wiltshire Record Society, Trowbridge 13210
Woolhope Naturalists' Field Club, Hereford 13213

Uruguay
Instituto Histórico y Geográfico del Uruguay, Montevideo 13261

U.S.A.
Aaron Burr Association, Annandale 13274
Academy of American Franciscan History, Berkeley 13281
Adirondack Historical Association, Blue Mountain Lake 13328
African American Museum, Cleveland 13343
Afro-American Historical and Genealogical Society, Washington 13347
Agricultural History Society, Washington 13351
American Academy for Jewish Research, New York 13377
American Academy of the History of Dentistry, Chicago 13440
American Antiquarian Society, Worcester 13454
American Association for State and Local History, Nashville 13492
American Baptist Historical Society, Rochester 13616
American Canal Society, York 13675

Vatican City

Venezuela

Yugoslavia

Zaire

Zimbabwe

History of Civilization s. Cultural History, History of Civilization

Home Economics

Canada

China, Republic

Denmark

Finland

France

Germany

Korea, Republic

South Africa

Switzerland

United Kingdom

U.S.A.

Homeopathy

France

Germany

Italy

Netherlands

Norway

United Kingdom

U.S.A.

Horticulture

Australia

Austria

Canada

China, Republic

Denmark

Egypt

Finland

France

Germany

India

Ireland

Italy

Japan

Morocco

Netherlands

International Society for Horticultural Science, Wageningen *08940*
Koninklijke Maatschappij Tuinbouw en Plantkunde, 's-Gravenhage *08948*
Nederlandse Tuinbouwraad, 's-Gravenhage *09003*

New Zealand

New Zealand Society for Horticultural Science, Havelock North *09175*

Norway

Det Norske Hageselskap, Oslo *09260*

Philippines

Asian-Pacific Weed Science Society, Laguna *09501*

South Africa

The Botanical Society of South Africa, Claremont *10126*
Organic Soil Association of South Africa, Parklands *10160*
Pretoria Horticultural Society, Pretoria *10164*

Spain

Sociedad Española de Horticultura, Madrid *10388*

Sweden

Bergianska Stiftelsen, Stockholm *10439*

United Kingdom

British Association of Landscape Industries, Keighley *11510*
CAB International Bureau of Horticulture and Plantation Crops, Wallingford *11722*
C.A.B. International Bureau of Horticulture and Plantation Crops, Wallingford *11723*
Good Gardeners' Association, Churcham *12123*
Hardy Plant Society, West Ewell *12146*
Henry Doubleday Research Association, Coventry *12158*
The Landscape Institute, London *12440*
National Association of Principal Agricultural Education Officers, Beverley *12569*
Northern Horticultural Society, Harrogate *12634*
Processors and Growers Research Organisation, Peterborough *12720*
Royal Caledonian Horticultural Society, Edinburgh *12774*
Royal Horticultural Society, London *12810*
Royal Jersey Agricultural and Horticultural Society, Saint Helier *12826*
Saintpaulia and Houseplant Society, Ilford *12884*
Wellesbourne Vegetable Research Association, Warwick *13197*

U.S.A.

American Horticultural Society, Alexandria *13852*
American Horticultural Therapy Association, Gaithersburg *13853*
American Pomological Society, University Park *14011*
American Society for Horticultural Science, Alexandria *14066*
Dynamics International Gardening Association, Asheboro *14992*
Horticultural Research Institute, Washington *15230*
Hydroponic Society of America, Concord *15243*

Yugoslavia

Jugoslovensko Naučno Voársko Društvo, Čačak *17101*

Humanities, general

Australia

Australian Academy of the Humanities, Canberra *00252*

Austria

Innsbrucker Gesellschaft zur Pflege der Geisteswissenschaften, Innsbruck *00654*
Österreichische Humanistische Gesellschaft für die Steiermark, Graz *00860*

Belgium

Koninklijke Academie voor Wetenschappen, Letteren en Schone Kunsten van België, Bruxelles *01409*

Bermuda

The Bermuda National Trust, Hamilton *01522*
Royal Commonwealth Society, Bermuda Branch, Devonshire *01526*

Canada

Canadian Federation for the Humanities, Ottawa *02128*
Institute for the Encyclopedia of U.R.A.M., Toronto *02526*
Social Sciences and Humanities Research Council of Canada, Ottawa *02915*

Denmark

Nordisk Institut for Asienstudier, København *03785*

France

Centre National de la Recherche Scientifique, Paris *04382*
Conseil International de la Philosophie et des Sciences Humaines, Paris *04455*

Germany

Anthroposophische Gesellschaft in Deutschland, Stuttgart *05076*
Deutscher Wissenschaftler Verband, Goslar *05564*
Gesellschaft für Geistesgeschichte e.V., Erlangen *05823*
Joachim Jungius-Gesellschaft der Wissenschaften e.V., Hamburg *06016*
Rheinisch-Westfälische Akademie der Wissenschaften, Düsseldorf *06265*
Stifterverband für die Deutsche Wissenschaft, Essen *06295*

Iran

Philosophy and Humanities Society, Teheran *06902*

Ireland

Royal Irish Academy, Dublin *07028*

Israel

The Israel Academy of Sciences and Humanities, Jerusalem *07060*

Italy

Accademia Ambrosiana Medici Umanisti e Scrittori, Milano *07127*
Accademia Cosentina, Cosenza *07133*
Accademia Culturale d'Europa, Bassano Romano *07134*
Accademia Culturale di Rapallo, Rapallo *07135*
Accademia Lancisiana di Roma, Roma *07198*
Accademia Polacca delle Scienze, Roma *07226*
Centro di Studi Metodologici, Torino *07448*
Centro Internazionale di Studi Umanistici, Roma *07476*

Mexico

Centro de Información Científica y Humanística, México *08717*

Netherlands

Nederlandse Organisatie voor Wetenschappelijk Onderzoek, 's-Gravenhage *08998*

Russia

Commission for the Co-ordination of Co-operation of Humanities Institutions of the Russian Academy of Sciences with UNESCO, Moskva *09831*

Senegal

Association des Facultés ou Etablissements de Lettres et Sciences Humaines des Universités d'Expression Française, Dakar *10019*

South Africa

Human Sciences Research Council, Pretoria *10147*

Sweden

Humanistisk-Samhällsvetenskapliga Forskningsradet, Stockholm *10468*
Kungliga Humanistiska Vetenskaps-Samfundet i Uppsala, Uppsala *10485*

Switzerland

Institut National Genevois, Genève *10730*
Schweizerische Akademie der Sozial- und Geisteswissenschaften, Bern *10789*
Stiftung für Humanwissenschaftliche Grundlagenforschung, Zürich *10992*

U.S.A.

Association for Humanistic Education, Laramie *14303*
Catholic Commission on Intellectual and Cultural Affairs, Philadelphia *14653*
Center for Advanced Study in the Behavioral Sciences, Stanford *14663*
Chinese Culture Association, Palo Alto *14737*
Latin American Studies Association, Pittsburgh *15648*
National Association for Humanities Education, Saint Joseph *15804*
National Humanities Center, Research Triangle Park *16080*
Rhetoric Society of America, Norman *16420*
SIGMA Xi – The Scientific Research Society, Research Triangle Park *16472*
Smithsonian Institution, Washington *16481*
Society for the Humanities, Ithaca *16600*
Southern Humanities Conference, Chattanooga *16726*
Women's Classical Caucus, Rye *16917*

Venezuela

Consejo de Desarrollo Científico y Humanístico, Caracas *17000*

Hydrology

Australia

Australian Society for Limnology, Abbotsford *00340*

Azerbaijan

Commission on Mountain Mud Flows, Baku *01043*
Commission on the Caspian Sea, Baku *01045*

Barbados

Caribbean Institute for Meteorology and Hydrology, Bridgetown *01082*

China, People's Republic

Academy of Hydrotechnology, Beijing *03110*
Chinese Society for Oceanology and Limnology, Quingdao *03181*

France

Société Hydrotechnique de France, Paris *04874*

Germany

Deutsche Gesellschaft für Hydrokultur e.V., Herten *05365*
Deutscher Verband für Wasserwirtschaft und Kulturbau e.V., Bonn *05552*

Hungary

Magyar Hidrológiai Társaság, Budapest *06634*

India

Asian Regional Coordinating Committee on Hydrology, Roorkee *06725*

Italy

Società Internazionale di Tecnica Idrotermale, Castellamare di Stabia *07819*

Japan

Nippon Rikusui Gakkai, Tokyo *08357*

Netherlands

Committee on Water Research, Delft *08844*

New Zealand

New Zealand Limnological Society, Rotorua *09165*

Niger

Centre Régional de Formation et d'Application en Agrométéorologie et Hydrologie Opérationnelle, Niamey *09208*

South Africa

Southern African Society of Aquatic Scientists, Rondebosch *10228*
Water Research Commission, Pretoria *10234*
Water Systems Research Group, Johannesburg *10235*

United Kingdom

The British Society of Dowsers, Ashford *11682*
Freshwater Biological Association, Ambleside *12092*
Institute of Hydrology, Wallingford *12245*
Institution of Water and Environment Management, London *12318*
Pure Water Preservation Society, Woking *12730*
River Thames Society, Richmond *12758*

U.S.A.

American Society of Limnology and Oceanography, Gloucester Point *14141*

Hygiene s. a. Public Health

Argentina

Sociedad Argentina de Medicina Social, Buenos Aires *00177*

Austria

Österreichische Gesellschaft für Arbeitsmedizin, Hall *00758*
Österreichische Gesellschaft für Hygiene, Mikrobiologie und Präventivmedizin, Wien *00793*
Österreichische Gesellschaft für Psychische Hygiene, Wien *00825*
Österreichische wissenschaftliche Gesellschaft für Prophylaktische Medizin und Sozialhygiene, Wien *00929*

Belgium

Association Belge d'Hygiène et de Médecine Sociale, Bruxelles *01112*
Collège Européen d'Hygiène et de Médecine Naturelles, Enghien *01216*
European Association for the Promotion of the Hand Hygiene, Bruxelles *01283*

Brazil

Associação Prudentina de Educação e Cultura, Presidente Prudente *01608*

Canada

Association des Médecins de Travail du Québec, Montréal *01889*
Canadian Association on Water Pollution Research and Control, Montréal *02076*
Canadian Natural Hygiene Society, Toronto *02207*

China, Republic

Chinese National Association for Mental Hygiene, Taipei *03280*

Denmark

Danske Veterinårhygiejnikeres Organisation, Slagense *03614*
Dansk Industrimedicinsk Selskab, Charlottenlund *03634*
Dansk Selskab for Social Medicin, Odense *03687*

Egypt

Egyptian Society of Medicine and Tropical Hygiene, Alexandria *03900*

France

Association Générale des Hygiénistes et Techniciens Municipaux, Paris *04253*
Congrès International de Médecine Légale et de Médecine Sociale de Langue Française, Paris *04450*
Groupement Médical d'Etudes sur l'Alcoolisme, Nantes *04557*
Ligue Française d'Hygiène Mentale, Paris *04596*
Société de Médecine et d'Hygiène du Travail, Paris *04669*
Société d'Hygiène International, Paris *04768*
Société Française de Médecine Préventive et Sociale, Paris *04813*
Société Française d'Hygiène, de Médecine Sociale et de Génie Sanitaire, Vandoeuvre-lès-Nancy *04859*
Société Scientifique d'Hygiène Alimentaire, Paris *04919*
Syndicat National Professionnel des Médecins du Travail, Toulouse *04970*

Germany

Bayerische Akademie für Arbeits- und Sozialmedizin, München *05170*
Deutsche Gesellschaft für Arbeitsmedizin e.V., München *05324*
Deutsche Gesellschaft für Hygiene und Mikrobiologie e.V., Heidelberg *05366*
Deutsche Gesellschaft für Sozialmedizin und Prävention e.V., Bochum *05422*
Deutsche Gesellschaft für Wohnmedizin und Bauhygiene e.V., Spöck-Stutensee *05442*
Gesellschaft zur Förderung der Lufthygiene und Silikoseforschung e.V., Düsseldorf *05882*
Verein für Wasser-, Boden- und Lufthygiene e.V., Berlin *06392*

Israel

Industrial Medical Association, Haifa *07057*

Italy

Accademia Internazionale di Medicina Legale e di Medicina Sociale, Roma *07191*
Accademia Internazionale per le Scienze Economiche, Sociali e Sanitarie, Roma *07192*
Associazione Italiana per L'Educazione Sanitaria, Perugia *07341*
Centro Italiano di Sessuologia, Roma *07495*
Società Italiana di Medicina del Lavoro e di Igiene Industriale, Pavia *07893*
Società Italiana di Medicina e Igiene della Scuola, Milano *07895*
Società Italiana di Medicina Preventiva e Sociale, Genova *07900*
Società Italiana di Medicina Sociale, Torino *07902*
Società Italiana pro Deontologia Sanitaria, Milano *07975*
Società Laziale Abruzzese di Medicina del Lavoro, Roma *07977*

Japan

Nihon Eisei Gakkai, Tokyo *08197*

Mexico

Instituto Nacional de Higiene, México *08738*

Netherlands

Nederlandse Vereniging voor Algemene Gezondheidszorg, Rotterdam *09015*

Peru

Liga Nacional de Higiene y Profilaxia Social, Lima *09473*

Poland

Polskie Towarzystwo Higieny Psychicznej, Warszawa *09576*

Romania

Societatea de Igiena si Sànàtate Publica, Bucuresti *09767*

Russia

National Scientific Medical Society of Hygienists, Moskva *09917*

Sweden

Föreningen för Vattenhygien, Vällingby *10459*

Switzerland

Schweizerische Gesellschaft für Sozial- und Präventivmedizin, Bern *10847*
Verein zur Förderung der Wasser- und Lufthygiene, Zürich *11027*

United Kingdom

British Natural Hygiene Society, Frinton-on-Sea *11616*
British Occupational Hygiene Society, London *11623*
European Federation of Associations of Industrial Safety and Medical Officers, Leicester *12029*
Institute of Occupational Medicine, Edinburgh *12264*
International Association on Water Quality, London *12337*
International Conference on Social Science and Medicine, Brighton *12346*
The Royal Institute of Public Health and Hygiene, London *12819*
Royal Society of Tropical Medicine and Hygiene, London *12864*
Society of Occupational Medicine, London *13058*

Uruguay

Consejo Nacional de Higiene, Montevideo *13257*

U.S.A.

American Board of Preventive Medicine, Dayton *13657*
American College of Preventive Medicine, Washington *13732*
American Conference of Governmental Industrial Hygienists, Cincinnati *13751*
American Dental Hygienists' Association, Chicago *13780*
American Industrial Hygiene Association, Akron *13863*
American Natural Hygiene Society, Tampa *13955*
American Society of Sanitary Engineering, Bay Village *14166*
American Society of Tropical Medicine and Hygiene, McLean *14171*
Association of Teachers of Preventive Medicine, Washington *14516*
Inter-American Association of Sanitary Engineering and Environmental Sciences, Brookeville *15348*
International Academy of Nutrition and Preventive Medicine, Asheville *15367*

International Christian Leprosy Mission, Portland 15437
International Society for Burn Injuries, Denver 15525
National Dental Hygienists' Association, Washington 16032
Water Quality Research Council, Lisle 16874

Venezuela

Sociedad Venezolana de Medicina del Trabajo y Deportes, Caracas 17026

Vietnam

Vietnamese Association of Prophylactic Hygiene, Hanoi 17062

Immunology

Argentina

Asociación Argentina de Alergía e Inmunología, Buenos Aires 00055
Sociedad Argentina de Alergia e Inmunopatología, Buenos Aires 00153

Australia

Australian Society for Parasitology, Canberra 00344
International Association of Trichologists, Sydney 00424

Belgium

Académie Européenne d'Allergologie et d'Immunologie Clinique, Bruxelles 01090
Association Européenne contre les Maladies à Virus, Bruxelles 01122
Societé Belge d'Allergologie et d'immunologie Clinique, Gent 01423

Canada

Addiction Research Foundation, Toronto 01819
Allergy Association of Calgary, Calgary 01854
Allergy Foundation of Canada, Saskatoon 01855
Allergy Information Association, Etobicoke 01856
Canadian Association for Clinical Micro and Infectious Diseases, Willowdale 02027
Canadian Society for Immunology, Toronto 02259

Denmark

Dansk Selskab for Allergologi og Immunologi, København 03671
European Academy of Allergology and Clinical Immunology, København 03720

Ecuador

Sociedad Ecuatoriana de Alergía y Ciencias Afinas, Quito 03855

Finland

Suomen Allergologi- ja Immunologiyhdistys, Helsinki 04006

France

Groupe de Recherche Génétique Epidémiologique, Paris 04549
Société Française d'Allergologie, Paris 04782
Syndicat National des Allergologistes Français, Tours 04936

Germany

Arbeitsgemeinschaft Allergiekrankes Kind, Herborn 05080
Deutsche Gesellschaft für Allergieforschung, Tübingen 05315
Deutsche Gesellschaft für Allergie- und Immunitätsforschung, Bochum 05316
Deutsche Vereinigung zur Bekämpfung der Viruskrankheiten e.V., München 05622
Gesellschaft für Immunologie e.V., Marburg 05828
Gesellschaft zur Förderung der Lufthygiene und Silikoseforschung e.V., Düsseldorf 05882
Immuno, Heidelberg 05937

Hungary

Magyar Allergologiai és Klinikai Immunológiai Társaság, Kékestetö 06613
Magyar Immunológiai Társaság, Budapest 06636

Israel

Israel Society of Allergology, Tel Aviv 07096

Italy

Associazione Italiana di Immuno-Oncologia Clinico-Pratica, Mantova 07310
European Academy of Allergology and Clinical Immunology, Roma 07662
International Society for the Study of Infectious and Parasitic Diseases, Torino 07719
Società Italiana di Allergologia e Immunologia Clinica, Roma 07831
Società Italiana di Immunoematologia e Trasfusione del Sangue, Roma 07886
Società Italiana di Immunologia e di Immunopatologia, Bari 07887

Mexico

Sociedad Latinoamericana de Alergología, México 08752

Netherlands

Vereniging Het Nederlands Kanker Instituut, Amsterdam 09085

New Zealand

New Zealand Society for Parasitology, Hamilton 09176

Poland

Polskie Towarzystwo Immunologiczne, Lódź 09579

Portugal

Sociedade Portuguesa de Alergologia e Imunologia Clínica, Lisboa 09689

Russia

National Immunological Society, Moskva 09908

Senegal

Africa AIDS Research Network, Dakar 10007

Turkey

Türk Tüberküloz Cemiyeti, Istanbul 11176

United Kingdom

Action Against Allergy, London 11200
British Society for Allergy and Clinical Immunology, Manchester 11649
The British Society for Immunology, East Grinstead 11657
British Society for the Study of Infection, London 11676
Institute for Animal Health Pirbright, Pirbright Woking 12205
Institute of Trichologists, London 12289
International Filariasis Association, London 12363
Midlands Asthma and Allergy Research Association, Derby 12525
National Society for Research into Allergy, Hinckley 12610
Society for Environmental Therapy, Manchester 12972

U.S.A.

American Academy of Allergy and Immunology, Milwaukee 13381
American Academy of Otolaryngic Allergy, Silver Spring 13419
American Allergy Association, Menlo Park 13448
American Association of Immunologists, Bethesda 13543
American Board of Allergy and Immunology, Philadelphia 13627
American College of Allergy and Immunology, Palatine 13697
American Dermatologic Society of Allergy and Immunology, Rochester 13783
American Foundation for Aids Research, Los Angeles 13822
American Osteopathic College of Allergy and Immunology, Scottsdale 13981
Haitian Coalition on Aids, Brooklyn 15185
Histamine Research Society of North America, Dallas 15215
Infectious Diseases Society of America, New Haven 15262
International Aids Prospective Epidemiology Network, Chicago 15370
International Association of Allergology and Clinical Immunology, Milwaukee 15397
International Correspondence Society of Allergists, Shawnee Mission 15450
Joint Council of Allergy and Immunology, Arlington Heights 15618
San Francisco Aids Foundation, San Francisco 16438
Society for Epidemiologic Research, Fort Collins 16519

Vietnam

Vietnamese Association of Anti-Contagious Diseases, Hanoi 17039

Information Science s. Computer and Information Science, Data Processing

Insurance

Australia

The Institute of Actuaries of Australia, Saint Ives 00411

Austria

Österreichische Gesellschaft für Versicherungsfachwissen, Wien 00842

Belgium

Association Actuarielle Internationale, Bruxelles 01105
Association Royale des Actuaires Belges, Bruxelles 01157

Canada

Insurers' Advisory Organization, Toronto 02534
Reinsurance Research Council, Toronto 02858

China, Republic

Actuarial Institute of the Republic of China, Taipei 03229

Denmark

Den Danske Aktuarforening, Lyngby 03700

Finland

Suomen Aktuaariyhdistys, Helsinki 04005

France

Association des Actuaires Diplomés de l'Institut de Science Financière et d'Assurances, Paris 04162
Association pour l'Enseignement de l'Assurance, Paris 04342
Institut des Actuaires Français, Paris 04565

Germany

Deutsche Gesellschaft für Versicherungsmathematik e.V., Bonn 05436
Deutscher Verein für Versicherungswissenschaft e.V., Berlin 05560
Gesellschaft für Versicherungswissenschaft und -gestaltung e.V., Köln 05868
Institut für Schadenverhütung und Schadenforschung der öffentlich-rechtlichen Versicherer e.V., Kiel 05969
Verein für Versicherungs-Wissenschaft und -Praxis Nordhessen e.V., Kassel 06391
Vereinigung der unabhängigen freiberuflichen Versicherungs- und Wirtschaftsmathematiker in der Bundesrepublik Deutschland e.V., Grünwald 06401
Vereinigung der Versicherungs-Betriebswirte e.V., Köln 06402
Verein zur Förderung der Versicherungswissenschaft in Hamburg e.V., Hamburg 06427
Verein zur Förderung der Versicherungswissenschaft in München e.V., München 06428
Versicherungswissenschaftlicher und Versicherungswirtschaftlicher Fördererverein e.V., Nürnberg 06429
Versicherungswissenschaftlicher Verein in Hamburg e.V., Hamburg 06430

Iceland

Félag Islenskra Tryggingastaerdfraedinga, Reykjavik 06685

Ireland

Insurance Institute of Ireland, Dublin 06969

Italy

Centro Italiano Ricerche e Studi Assicurativi, Roma 07506
Centro Studi Assicurativi Piero Sacerdoti, Milano 07549
Istituto Italiano degli Attuari, Roma 07735
Istituto Nazionale delle Assicurazioni, Roma 07747
Ordine Nazionale degli Attuari, Roma 07762

Japan

Nihon Hoken Gakkai, Tokyo 08206

Netherlands

Actuarieel Genootschap, Amsterdam 08818

Norway

Den Norske Aktuarforening, Oslo 09251

Portugal

Instituto dos Actuarios Portugueses, Lisboa 09660

Sweden

Svenska Aktuarieföreningen, Stockholm 10532
Svenska Försäkringsföreningen, Stockholm 10546

Switzerland

Europäisches Zentrum für die Bildung im Versicherungswesen, Sankt Gallen 10671
European Centre for Insurance Education and Training, Sankt Gallen 10680
International Association for the Study of Insurance Economics, Genève 10733
Schweizerische Vereinigung der Versicherungsmathematiker, Zürich 10937
Vereinigung für Berufsbildung der Schweizerischen Versicherungswirtschaft, Bern 11013

Turkey

Türkiye Aktüerler Cemiyeti, Istanbul 11160

United Kingdom

Association of Consulting Actuaries, London 11332
Chartered Insurance Institute, London 11768
Faculty of Actuaries in Scotland, Edinburgh 12047
Institute of Actuaries, London 12211

U.S.A.

Actuarial Studies in Non-Life Insurance, Philadelphia 13323
American Academy of Actuaries, Washington 13379
American Academy of Insurance Medicine, Springfield 13401
American Council of Independent Laboratories, Washington 13762
American Institute for Chartered Property Casuality Underwriters, Malvern 13865
American Risk and Insurance Association, Sacramento 14032
Association of Defense Trial Attorneys, Peoria 14426
Association of Life Insurance Counsel, Newark 14453
Conference of Consulting Actuaries, Schaumburg 14829
Federation of Insurance and Corporate Counsel, Walpole 15085
Insurance Information Institute, New York 15346
International Association of Defense Counsel, Chicago 15404
International Insurance Society, Tuscaloosa 15491
Life Insurance Marketing and Research Association, Hartford 15671
Society of Actuaries, Schaumburg 16631
Society of Insurance Research, Marietta 16674

Internal Medicine

Argentina

Sociedad de Medicina Interna de Buenos Aires, Buenos Aires 00193
Sociedad de Medicina Interna de Córdoba, Córdoba 00194

Austria

Gesellschaft für Innere Medizin in Wien, Wien 00621
Österreichische Gesellschaft für Innere Medizin, Wien 00796
Österreichische Gesellschaft für internistische Intensivmedizin, Wien 00797
Wiener Gesellschaft für Innere Medizin, Wien 01015

Belgium

Société Belge de Médecine Interne, Bruxelles 01433

Colombia

Instituto Nacional de Cancerología, Bogotá 03392

Costa Rica

Asociación de Medicina Interna, San José 03434

Denmark

Association of European Cancer Leagues, København 03563
Dansk Nefrologisk Selskab, Århus 03654
Dansk Selskab for Intern Medicin, Gentofte 03676

Germany

Deutsche Gesellschaft für Innere Medizin, Wiesbaden 05367
Deutsche Gesellschaft für Internistische Intensivmedizin, Hamburg 05368
Deutsche Gesellschaft für Verdauungs- und Stoffwechselkrankheiten, Frankfurt 05434
Deutsche Medizinische Arbeitsgemeinschaft für Herd- und Regulationsforschung e.V., Düren 05467
Nordwestdeutsche Gesellschaft für innere Medizin, Hamburg 06145
Saarländisch-Pfälzische Internistengesellschaft e.V., Ludwigshafen 06272

Italy

Società Italiana di Medicina Interna, Roma 07898

Japan

Asian and Oceanian Thyroid Association, Nagasaki 08095
Asian and Pacific Federation of Organizations for Cancer Research and Control, Tokyo 08096
Asian Pacific Society of Respirology, Tokyo 08102

Korea, Republic

Asian Pacific Association for the Study of the Liver, Seoul 08521

Netherlands

European Association for the Study of the Liver, Amsterdam 08863
Nederlandsche Internisten Vereeniging, Utrecht 08982

Pakistan

Pakistam Anti Tuberculosis Association, Karachi 09377

Poland

Towarzystwo Internistów Polskich, Wroclaw 09621

Romania

Societatea de Medicinà Internà, Bucuresti 09770

South Africa

African Association for the Study of Liver Diseases, Cape Town 10121
Diffuse Obstructive Pulmonary Syndrome Research Group, Tygerberg 10135

Switzerland

International Society of Internal Medicine, Langenthal 10757
Schweizerische Gesellschaft für Innere Medizin, Langenthal 10824
Société Internationale de Médecine Interne, Langenthal 10977

United Kingdom

British Microcirculation Society, London 11608
European Association of Internal Medicine, Brighton 11996, 11997
European Atherosclerosis Society, London 12002

U.S.A.

American Association for the Study of Liver Diseases, Thorofare 13502
American Board of Internal Medicine, Philadelphia 13636
American Society of Internal Medicine, Washington 14136
American Thoracic Society, New York 14194
Society for the Study of Breast Disease, Dallas 16614
Wilson's Disease Association, Washington 16906

Vietnam

Vietnamese Association of Internal Medicine, Hanoi 17050
Vietnamese Association of Serum and Blood Transfusion, Hanoi 17065

Zimbabwe

African Organization for Research and Training in Cancer, Harare 17159

International Relations

Austria
Wiener Institut für Entwicklungsfragen und Zusammenarbeit, Wien 01019

Canada
Canadian Council for European Affairs, Saskatoon 02105

China, Republic
The Asia Foundation, Taipei 03235

Denmark
Danske Komité for Historikernes Internationale Samarbejde, København 03608
Mellemfolkeligt Samvirke, København 03772

Egypt
The African Society, Cairo 03861
Institut d'Egypte, Cairo 03906

Finland
International Law Association: Finnish Branch, Helsinki 03968

France
Bureau International de Liaison et de Documentation, Paris 04362

Germany
Deutsche Stiftung für internationale Entwicklung, Bonn 05601
Deutsches Übersee-Institut, Hamburg 05604
Deutsch-Pazifische Gesellschaft e.V., München 05630
HWWA – Institut für Wirtschaftsforschung – Hamburg, Hamburg 05935
Institut für Auslandsbeziehungen, Stuttgart 05952
Inter Nationes e.V., Bonn 06012
Rheinisch-Westfälische Auslandsgesellschaft e.V., Dortmund 06266
Südosteuropa-Gesellschaft e.V., München 06309
Walter Eucken Institut e.V., Freiburg 06436

Greece
Hellenic Institute of International and Foreign Law, Athinai 06532

Iceland
Danske Selskab i Reykjavik, Reykjavik 06680

Indonesia
The Asia Foundation, Jakarta 06879

Italy
Associazione Italiana per lo Sviluppo Internazionale, Roma 07346
Centro per gli Studi sui Mercati Esteri, Roma 07520
Istituto Cooperazione Economica Internazionale, Milano 07726
Istituto di Sociologia Internazionale di Gorizia, Gorizia 07727
Istituto per la Cooperazione Economica Internazionale e i Problemi di Sviluppo, Roma 07751
Società Europea di Cultura, Venezia 07802
Società Internazionale di Diritto Penale Militare e di Diritto della Guerra, Gruppo Italiano, Roma 07815
Society for International Development, Roma 08028

Japan
Africa Kyokai, Tokyo 08090
Ajia Chosakai, Tokyo 08091
Ajiakurabu, Tokyo 08092
Kajima Heiwa Kenkyujo, Tokyo 08152
Nihon Kokusai Kyoiku Kyokai, Tokyo 08232

Korea, Republic
Asia Foundation, Seoul 08517
International Cultural Society of Korea, Seoul 08526

Laos
Laos-China Association, Vientiane 08553
Laos-Soviet Association, Vientiane 08554
Laos-Viet Nam Association, Vientiane 08555

Netherlands
Nederlands Instituut voor Internationale Betrekkingen Clingendael, 's-Gravenhage 09053
Nederlands-Zuidafrikaanse Vereniging, Amsterdam 09057

Pakistan
Asia Foundation, Islamabad 09360

Romania
Asociatia de Drept International si Relatii Internationale din Romania, Bucuresti 09740

Singapore
Singapore Institute of International Affairs, Singapore 10054

South Africa
Africa Institute of South Africa, Pretoria 10120

Sweden
Centre for the Study of International Relations, Stockholm 10443
International Law Association, Swedish Branch, Stockholm 10475
Nordiska Samfundet för Latinamerika-forskning, Stockholm 10510

Switzerland
Pro Helvetia, Zürich 10781

United Kingdom
British Association of the Experiment in International Living, Malvern 11527
Franco-British Society, London 12091

U.S.A.
American Association for Study of the United States in World Affairs, Annandale 13494
Committee on Changing International Realities, Washington 14800
Harry S. Truman Library Institute for National and International Affairs, Independence 15188
Middle East Studies Association of North America, Tucson 15735

Journalism s. a. Literature

Austria
Österreichische Gesellschaft für Kommunikationsfragen, Wien 00803

Canada
Association of Directors of Journalism Programs in Canadian Universities, London 01930
Canadian Theatre Critics Association, Toronto 02307
Newspaper Research Centre, Toronto 02714

Denmark
Dansk Pressehistorisk Selskab, København 03665
Sammenslutningen af Medieforskere i Danmark, Alborg 03803

Germany
Akademie für Publizistik in Hamburg e.V., Hamburg 05059
Catholic Media Council, Aachen 05273
Deutsche Gesellschaft für Publizistik- und Kommunikationswissenschaft e.V., Eichstätt 05413
Deutsche Studiengesellschaft für Publizistik, Stuttgart 05603
Deutsche Zeitungswissenschaftliche Vereinigung e.V., München 05627
Gemeinschaftswerk der Evangelischen Publizistik e.V., Frankfurt 05772
Gesellschaft für publizistische Bildungsarbeit e.V., Hagen 05854
INCA-FIEJ Research Association, Darmstadt 05938
Institut für Zeitungsforschung, Dortmund 05979
Kollegium der Medizinjournalisten, Oberaudorf 06074
Verband Deutscher Agrarjournalisten e.V., Bonn 06348

Italy
Accademia di Paestum Eremo Italico, Mercato San Severino 07170
Centro di Demodossalogia, Roma 07419
Centro di Documentazione Giornalistica, Roma 07424
Centro Emilia-Romagna per la Storia del Giornalismo, Bologna 07459
Centro Sperimentale Italiano di Giornalismo, Milano 07546
Centro Studi Pietro Mancini, Cosenza 07585
Istituto Italiano di Pubblicismo, Roma 07739

Japan
Nippon Shimbun Gakkai, Tokyo 08377

United Kingdom
Critics' Guild, London 11868
Historical Newspaper Service, New Barnet 12174
International Association for Mass Communication Research, Leicester 12324

U.S.A.
Academy of Television Arts and Sciences, North Hollywood 13310
Accrediting Council on Education in Journalism and Mass Communications, Lawrence 13319
American Press Institute, Reston 14013
Association for Education in Journalism and Mass Communication, Columbia 14292
Association of Schools of Journalism and Mass Communication, Columbia 14497
Broadcast Education Association, Washington 14617
College Press Service, Orlando 14777
Editorial Projects in Education, Washington 15016
International Communications Association, Dallas 15444
Journalism Association of Community Colleges, San Pablo 15627
Journalism Education Association, Manhattan 15628
Music Critics Association, Westfield 15762
National Association of Science Writers, Greenlawn 15891
New York Drama Critics Circle, New York 16202
New York Financial Writers' Association, Syosset 16203
Outer Critics Circle, New York 16277
Research Society for Victorian Periodicals, Edwardsville 16417
Washington Journalism Center, Washington 16872

Labour Movement s. Socialism

Language and Languages s. Linguistics

Laryngology s. Otorhinolaryngology

Law

Argentina
Academia Nacional de Derecho y Ciencias Sociales, Buenos Aires 00047
Academia Nacional de Derecho y Ciencias Sociales (Córdoba), Córdoba 00048
American Association of Jurists, Buenos Aires 00052
Colegio de Abogados de la Ciudad de Buenos Aires, Buenos Aires 00115

Australia
Australian Bar Association, Melbourne 00269
Australian Law Librarians' Group, Melbourne 00317
Law Council of Australia, Braddon 00427
Law Society of New South Wales, Sydney 00428
Medico Legal Society of New South Wales, Saint Leonards 00433

Austria
Deutscher Rechtshistorikertag, Graz 00580
Gesellschaft für Arbeitsrecht und Sozialrecht, Linz 00613
Gesellschaft für das Recht der Ostkirchen, Wien 00616
Gesellschaft zum Studium und zur Erneuerung der Struktur der Rechtsordnung, Wien 00639
International Law Association, Österreichischer Zweigverein, Wien 00688
Kärntner Juristische Gesellschaft, Klagenfurt 00693
Österreichische Gesellschaft für Agrar- und Umweltrecht, Wien 00750
Österreichische Gesellschaft für Arbeitsrecht und Sozialrecht, Linz-Auhof 00759
Österreichische Gesellschaft für Gesetzgebungslehre, Salzburg 00786
Österreichische Gesellschaft für Kirchenrecht, Wien 00801
Österreichische Gesellschaft für Rechtsvergleichung, Wien 00827
Österreichischer Juristentag, Wien 00894
Österreichisches Ost- und Südosteuropa-Institut, Wien 00915
Salzburger Institut für juristische Information und Fortbildung, Salzburg 00940
Salzburger Juristische Gesellschaft, Salzburg 00941
Tiroler Juristische Gesellschaft, Innsbruck 00955
Wiener Juristische Gesellschaft, Wien 01021
Wiener Rechtsgeschichtliche Gesellschaft, Wien 01026

Bahrain
Bahrain Bar Society, Manama 01055

Barbados
Caribbean Law Institute, Bridgetown 01083

Belgium
Association Belge de Droit Rural, Bruxelles 01107
Association Euro-Arabe de Juristes, Bruxelles 01120
Association Européenne des Barreaux des Courts Suprêmes, Grimbergen 01125
Association Européenne pour le Droit Bancaire et Financier, Bruxelles 01131
Association Internationale de Droit Economique, Louvain-la-Neuve 01136
Association Internationale des Juristes Démocrates, Bruxelles 01139
Belgisch Instituut voor Arbeidsverhoudingen, Leuven 01165
CELEX, Bruxelles 01181
Conférence Diplomatique de Droit Maritime International, Bruxelles 01247
European Association for the Teaching of Legal Theory, Bruxelles 01285
European Consumer Law Group, Louvain-la-Neuve 01330
European Council for Rural Law, Gent 01334
Institut Belge de Droit Comparé, Bruxelles 01377
Institut Royal des Relations Internationales, Bruxelles 01392
Koninklijke Academie voor Wetenschappen, Letteren en Schone Kunsten van België, Bruxelles 01409
Société Internationale de Droit Pénal Militaire et de Droit de la Guerre, Bruxelles 01463

Brazil
Associação dos Advogados, São Paulo 01599
Association of Latin American Lawyers for the Defense of Human Rights, São Paulo 01614
Instituto dos Advogados Brasileiros, Rio de Janeiro 01670

Bulgaria
Bulgarian Association of Penal Law, Sofia 01743

Canada
Alberta Law Foundation, Calgary 01838
Arbitrators' Institute of Canada, Toronto 01869
Canadian Bar Association, Ottawa 02081
Canadian Canon Law Society, Ottawa 02086
Canadian Institute for Environmental Law and Policy, Toronto 02162
Canadian Institute for the Administration of Justice, Montréal 02167
Canadian Law and Society Association, Downsview 02187
Canadian Law Information Council, Ottawa 02188
Canadian Maritime Law Association, Montréal 02194
Canadian Petroleum Law Foundation, Calgary 02222
Commonwealth Association of Legislative Counsel, Edmonton 02365
Computer Law Association, Toronto 02375
Council of Canadian Law Deans, Ottawa 02401
County of York Law Association, Toronto 02410
Estonian Jurists' Association, Mississauga 02442
Inter-American Commercial Arbitration Commission (Canadian Section), Ottawa 02535
Law Foundation of Nova Scotia, Halifax 02578
Law Society of Manitoba, Winnipeg 02579
Law Society of Newfoundland, Saint John's 02580
Law Society of Saskatchewan, Regina 02581
Law Society of the Northwest Territories, Yellowknife 02582
Law Society of Upper Canada, Toronto 02583
Northwest Territories Law Foundation, Yellowknife 02718
Nova Scotia Barristers' Society, Halifax 02725
West Coast Environmental Law Association, Vancouver 02986
West Coast Environmental Law Research Foundation, Vancouver 02987

Chile
Colegio de Abogados, Concepción 03028

China, People's Republic
China Law Society, Beijing 03133

China, Republic
Chinese Society of International Law, Taipei 03287
National Bar Association, Taipei 03311
National Tax Research Association of China, Taipei 03315
Society of Chinese Constitutional Law, Taipei 03328

Colombia
Academia Colombiana de Jurisprudencia, Bogotá 03338
Instituto Nacional de Medicina Legal, Bogotá 03394
Sociedad Jurídica de la Universidad Nacional, Bogotá 03420

Costa Rica
Instituto Latinoamericano de las Naciones Unidas para la Prevención del Delito y Tratamiento del Delincuente, San José 03448

Croatia
League of Jurists' Associations of Croatia, Zagreb 03463

Denmark
Danmarks Jurist- og Økonomforbund, København 03580
Dansk Forening for Europaret, København 03619
Dansk Forening for Retssociologi, København 03621
Dansk Skattevidenskabelig Forening, København 03689
International Law Association: Danish Branch, København 03747
Juridisk Forening, København 03752
Nordisk Forening for Rettssociologi, København 03784

Egypt
Arab Commission for International Law, Cairo 03869
Egyptian Society of International Law, Cairo 03899
Egyptian Society of Political Economy, Statistics and Legislation, Cairo 03901

Fiji
The Fiji Law Society, Suva 03946

Finland
International Law Association: Finnish Branch, Helsinki 03968
Juridiska Föreningen i Finland, Helsinki 03971
Suomalainen Lakimiesyhdistys, Helsinki 03998
Suomen Lainopillinen Yhdistys, Helsinki 04044
Suomen Lakimiesliitto, Helsinki 04045

France
Académie des Sciences d'Outre-Mer, Paris 04105
Académie des Sciences Morales et Politiques, Paris 04109
Association Internationale de Droit Pénal, Pau 04256
Association Internationale des Sciences Juridiques, Paris 04276
Association Nationale des Docteurs en Droit, Paris 04309
Association pour le Développement du Droit Mondial, Paris 04341
Centre Français de Droit Comparé, Paris 04374
Comité d'Etudes Fiscales et Contentieuses, Paris 04397
Comité Européen pour les Problèmes Criminels, Strasbourg 04402
Conférence Internationale des Facultés de Droit ayant en Commun l'Usage du Français, Aix-en-Provence 04442
Conseil International des Sciences Sociales, Paris 04459
European Council on Environmental Law, Strasbourg 04501

Institut International de Droit d'Expression et d'Inspiration Françaises, Paris *04572*
Institut International des Droits de l'Homme, Strasbourg *04574*
Société de Législation Comparée, Paris *04660*
Société d'Etudes Economiques et Comptables, Paris *04730*
Société d'Etudes et de Contrôles Juridiques, Paris *04731*
Société d'Etudes Juridiques, Economiques et Fiscales, Paris *04745*
Société d'Histoire du Droit, Paris *04759*
Société d'Histoire du Droit Normand, Caen *04760*
Société Française de Droit Aérien et Spatial, Paris *04800*
Société Française d'Etudes Juridiques, Paris *04853*
Société Française pour le Droit International, Strasbourg *04867*
Société Juridique et Fiscale de France, Levallois Perret *04887*

Gabon

African Women Jurists Federation, Libreville *05004*

Gambia

African Centre for Democracy and Human Rights Studies, Banjul *05005*

Germany

Akademie der Arbeit in der Universität, Frankfurt *05046*
Arbeitsgemeinschaft für juristisches Bibliotheks- und Dokumentationswesen, Augsburg *05105*
Bund für Lebensmittelrecht und Lebensmittelkunde e.V., Bonn *05265*
Cusanus-Institut, Trier *05280*
Deutsche Akademie für Verkehrswissenschaft e.V., Hamburg *05291*
Deutsche Gesellschaft für Agrarrecht, Bonn *05314*
Deutsche Gesellschaft für Baurecht e.V., Frankfurt *05329*
Deutsche Gesellschaft für Völkerrecht, Bonn *05438*
Deutscher Arbeitsgerichtsverband e.V., Köln *05498*
Deutsche Richterakademie, Trier *05518*
Deutscher Juristen-Fakultätentag, Würzburg *05519*
Deutscher Juristentag e.V., Bonn *05520*
Deutscher Verein für Internationales Seerecht e.V., Hamburg *05557*
Deutsches Anwaltsinstitut e.V., Bochum *05567*
Deutsches Handwerksinstitut e.V., München *05579*
Deutsche Vereinigung für gewerblichen Rechtsschutz und Urheberrecht e.V., Köln *05614*
European Association for Chinese Law, Aachen *05661*
Gesellschaft für Datenschutz und Datensicherung e.V., Bonn *05806*
Gesellschaft für Deutschlandforschung e.V., Berlin *05809*
Gesellschaft für Rechtsvergleichung e.V., Göttingen *05857*
Gesellschaft für Wohnungsrecht und Wohnungswirtschaft Köln e.V., Köln *05874*
Hamburger Gesellschaft für Völkerrecht und Auswärtige Politik e.V., Hamburg *05907*
Institut für Deutsches und Internationales Baurecht e.V., Frankfurt *05957*
Institut für Urheber- und Medienrecht e.V., München *05976*
International Copyright Society, München *05987*
Internationaler Rat für Umweltrecht, Bonn *05998*
Internationales Dokumentations- und Studienzentrum für Jugendkonflikte, Wuppertal *05999*
Internationale Vereinigung für Rechts- und Sozialphilosophie e.V., Göttingen *06004*
Internationale Vereinigung von Versicherungsjuristen (A.I.D.A.), Deutsche Landesgruppe, Berlin *06006*
Johannes-Althusius-Gesellschaft e.V., Münster *06017*
Katholische Juristenarbeit Deutschlands, Bonn *06065*
Rechts- und Staatswissenschaftliche Gesellschaft, Recklinghausen *06259*
Rechts- und Staatswissenschaftliche Vereinigung Düsseldorf e.V., Düsseldorf *06260*
Studienkreis für Presserecht und Pressefreiheit, Stuttgart *06307*
Wissenschaftliche Gesellschaft für Europarecht, Trier *06447*

Ghana

Ghana Bar Association, Accra *06484*

Greece

Hellenic Institute of International and Foreign Law, Athinai *06532*

Hungary

Magyar Jogász Egylet, Budapest *06640*

India

Asian-African Legal Consultative Committee, New Dehli *06719*
Bar Association of India, New Dehli *06739*
Indian Law Institute, New Delhi *06798*

Indonesia

ASEAN Law Association, Jakarta *06876*

Ireland

Association of Irish Jurists, Dublin *06928*
Honorable Society of King's Inns, Dublin *06955*
Incorporated Law Society of Ireland, Dublin *06956*
Irish Maritime Law Association, Dublin *06991*

Israel

Israel Bar Association, Tel Aviv *07068*

Italy

Associazione Forense Italiana, Roma *07276*
Associazione Internazionale di Diritto Nucleare, Roma *07287*
Associazione Internazionale Giuristi Italia-USA, Roma *07289*
Associazione Italiana dei Giuristi Europei, Roma *07296*
Associazione Italiana di Diritto Marittimo, Roma *07306*
Associazione Italiana Giuristi Democratici, Roma *07332*
Centro di Informazioni Sterilizzazione e Aborto, Milano *07429*
Centro di Studi Aziendali e Amministrativi, Cremona *07436*
Centro Internazionale Magistrati Luigi Severini, Perugia *07477*
Centro Internazionale per l'Iniziativa Giuridica, Roma *07482*
Centro Italiano di Studi di Diritto dell'Energia, Roma *07498*
Centro Italiano di Studi Sociali Economici e Giuridici, Roma *07502*
Centro per la Documentazione Automatica, Milano *07523*
Centro Studi di Diritto del Lavoro, Milano *07552*
Centro Studi di Diritto Fluviale e della Navigazione Interna, Venezia *07553*
Centro Studi di Diritto Sportivo, Vicenza *07554*
Centro Studi Diritto Comunitario, Roma *07562*
Circolo Giuridico Italiano, Roma *07603*
International Association for the Study of Canon Law, Roma *07703*
International Association for Water Law, Roma *07705*
International Juridical Organization for Environment and Development, Roma *07714*
International Law Association, Sezione Italiana, Roma *07715*
Istituto Internazionale di Diritto Umanitario, San Remo *07731*
Istituto Internazionale per l'Unificazione del Diritto Privato, Roma *07734*
Società Internazionale di Diritto Penale Militare e di Diritto della Guerra, Gruppo Italiano, Roma *07815*
Società Italiana di Filosofia Giuridica e Politica, Roma *07873*
Società Sassarese per le Scienze Giuridiche, Sassari *08011*
Tecnocentro Italiano, Milano *08034*
Unione Giuristi Cattolici Italiani, Roma *08046*

Jamaica

Caribbean Council of Legal Education, Kingston *08071*

Japan

Asian Patent Attorneys Association, Tokyo *08105*
Daini-Tokyo Bengoshikai Toshokan, Tokyo *08118*
Hikaku-ho Gakkai, Tokyo *08131*
Hogaku Kyokai, Tokyo *08134*
Hosei-shi Gakkai, Tokyo *08136*
Hosokai, Tokyo *08137*
Keizai-ho Gakkai, Tokyo *08160*
Kokusaiho Gakkai, Tokyo *08172*
Minji Soshoho Gakkai, Tokyo *08181*
Nichibei Hogakkai, Tokyo *08185*
Nihon Koho Gakkai, Tokyo *08227*
Nihon Kuho Gakkai, Tokyo *08235*
Nihon Zeiho Gakkai, Kyoto *08278*
Nippon Hoi Gakkai, Tokyo *08306*
Nippon Hoshakai Gakkai, Tokyo *08307*
Nippon Kaiho Gakkai, Tokyo *08321*
Nippon Shiho Gakkai, Tokyo *08373*
Raten Amerika Kyokai Kenkyubu, Tokyo *08403*
Tokyo Bengoshikai Toshokan, Tokyo *08436*

Korea, Republic

Asia-Pacific Lawyers Association, Seoul *08525*

Lebanon

Association Libanaise des Sciences Juridiques, Beirut *08560*

Macedonia

Sojuz na Združenijata na Pravnicite na Makedonija, Skopje *08616*

Malaysia

Asia Pacific Forum on Women, Law and Development, Kuala Lumpur *08636*

Mexico

Academia Mexicana de Jurisprudencia y Legislación, México *08685*
Asociación Mexicana de Administración Científica, México *08698*
Barra Mexicana-Colegio de Abogados, México *08708*

Netherlands

European Association of Law and Economics, Maastricht *08868*
European Company Lawyers Association, 's-Gravenhage *08882*
Hague Academy of International Law, 's-Gravenhage *08907*
International Bureau of Fiscal Documentation, Amsterdam *08926*
Nederlandse Vereniging voor Internationaal Recht, 's-Gravenhage *09021*
Vereniging voor Agrarisch Recht, Wageningen *09094*
Vereniging voor Arbeidsrecht, Woerden *09095*
Vereniging voor Bouwrecht, 's-Gravenhage *09096*
Vereniging voor Wijsbegeerte van het Recht, Amstelveen *09106*
Volkenrechtelijk Instituut, Utrecht *09108*

New Zealand

New Zealand Law Society, Wellington *09162*

Nigeria

Nigerian Bar Association, Warri *09235*

Norway

Norsk Forening for Internasjonal Rett, Oslo *09309*
Norsk Forsikringsjuridisk Forening, Oslo *09312*

Peru

Andean Commission of Jurists, Lima *09443*

Romania

Asociatia de Drept International si Relatii Internationale din Romania, Bucuresti *09740*
Asociatia Juristilor din Romania, Bucuresti *09742*

Russia

Association of International Law, Moskva *09802*

Slovakia

Slovenská Spoločnost Medzinárodné Právo, Bratislava *10087*

Slovenia

Society of Jurists of Slovenia, Ljubljana *10109*

Spain

Asociación Latinoamericana de Derecho Aeronáutico y Espacial, Madrid *10275*
Centro de Estudios Constitucionales, Madrid *10287*
Colegio de Abogados de Barcelona, Barcelona *10288*
Colegio Notarial, Barcelona *10290*
Real Academia de Jurisprudencia y Legislación, Madrid *10347*

Sweden

International Law Association, Swedish Branch, Stockholm *10475*

Switzerland

Aktion Freiheitliche Bodenordnung, Zürich *10611*
Association for the International Collective Management of Audiovisual Works, Genève *10626*
Association of International Consultants on Human Rights, Genève *10630*
Association Suisse de Droit Aérien et Spatial, Genève *10632*
Budapest Union for the International Recognition of the Deposit of Microorganisms for the Purposes of Patent Procedure, Genève *10636*
Centre for the Independence of Judges and Lawyers, Genève *10643*
Internationale Vereinigung für gewerblichen Rechtsschutz, Zürich *10743*
Schweizerische Gesellschaft für die Rechte der Urheber musikalischer Werke, Zürich *10817*
Schweizerische Gesellschaft für Versicherungsrecht, Zürich *10855*
Schweizerische Kriminalistische Gesellschaft, Giubiasco *10867*
Schweizerischer Anwaltsverband, Bern *10884*
Schweizerischer Notaren-Verband, Bern *10901*
Schweizerischer Verband für Frauenrechte, Lausanne *10913*
Schweizerische Vereinigung für Internationales Recht, Zürich *10947*
Schweizerische Vereinigung für Schiedsgerichtsbarkeit, ZZürich *10950*
Schweizerische Vereinigung für Steuerrecht, Zürich *10952*
Schweizerische Vereinigung für Urheber- und Medienrecht, Zürich *10953*
Société Suisse de Juristes, Lausanne *10984*
Union Internationale pour la Protection des Obtentions Végétales, Genève *10994*
Vereinigung für Rechsstaat und Individualrechte, Solothurn *11014*
Wissenschaftliche Vereinigung zur Pflege des Wirtschafts- und Konsumentenschutzrechts, Zürich *11030*
World Association for the School as an Instrument of Peace, Genève *11031*
World Intellectual Property Organization, Genève *11034*

Tanzania

African Participatory Research Network, Dar es Salaam *11046*

Turkey

Türk Hukuk Kurumu, Ankara *11159*

United Kingdom

Abortion Law Reform Association, London *11196*
African Society of International and Comparative Law, London *11210*
Association of International Accountants, Gateshead *11359*
The Association of Law Teachers, Leicester *11360*
Bar Association for Commerce, Finance and Industry, London *11423*
British Institute of International and Comparative Law, London *11583*
British Insurance Law Association, London *11591*
Chartered Association of Certified Accountants, London *11759*
Chartered Institute of Patent Agents, London *11764*
Commonwealth Commercial Crime Unit, London *11815*
Commonwealth Lawyers' Association, London *11825*
Commonwealth Legal Education Association, London *11826*
Council of Legal Education, London *11857*
Cowethas Lacha Sten Cernow, Gunnislake *11866*
Electoral Reform Society of Great Britain and Ireland, London *11950*
European Air Law Association, London *11984*
Faculty of Advocates, Edinburgh *12048*
The Forensic Science Society, Harrogate *12085*
General Council of the Bar, London *12103*
International Bar Association, London *12338*
International Esperanto-Association of Jurists, Wembley *12350*
International Law Association, London *12373*
The Josephine Butler Society, Hatfield *12423*
Jury System Reform Society, Beckenham *12425*
Law Society, London *12441*
Law Society of Scotland, Edinburgh *12442*
Mansfield Law Club, London *12490*
Medico-Legal Society, London *12515*
Minority Rights Group, London *12533*
Northern Ireland Civil Rights Association, Belfast *12637*
Procurators Fiscal Society, Dumbarton *12721*
The Royal Faculty of Procurators in Glasgow, Glasgow *12802*
Scottish Rights of Way Society, Edinburgh *12925*
Selden Society, London *12939*
Sexual Law Reform Society, London *12942*
Society for Computers and Law, Bristol *12967*
Society of Public Teachers of Law, Buckingham *13062*
Society of Writers to Her Majesty's Signet, Edinburgh *13072*
Stair Society, Edinburgh *13093*
Statute Law Society, London *13099*
United Kingdom Association for European Law, London *13164*

U.S.A.

Academy of Criminal Justice Sciences, Highland Heights *13286*
Academy of Legal Studies in Business, Oxford *13295*
American Academy of Forensic Sciences, Colorado Springs *13396*
American Academy of Matrimonial Lawyers, Chicago *13403*
American Academy of Psychiatry and the Law, Baltimore *13431*
American Agricultural Law Association, Fayetteville *13447*
American Arbitration Association, New York *13456*
American Association for the Comparative Study of Law, New York *13499*
American Association of Medico-Legal Consultants, Philadelphia *13551*
American Bar Association, Chicago *13617*
American Criminal Justice Association – Lambda Alpha Epsilon, Sacramento *13772*
American Foreign Law Association, New York *13818*
American Forensic Association, River Falls *13819*
American Intellectual Property Law Association, Arlington *13900*
American Judges Association, Williamsburg *13907*
American Judicature Society, Chicago *13908*
American Law Institute, Philadelphia *13912*
American Legal Studies Association, Boston *13914*
American Orthopsychiatric Association, New York *13974*
American Psychology-Law Society, Worcester *14020*
American Society for Legal History, University *14070*
American Society for Pharmacy Law, Vienna *14078*
American Society of International Law, Washington *14137*
American Society of Law and Medicine, Boston *14140*
American Society of Notaries, Washington *14150*
American Society of Questioned Document Examiners, Rockville *14162*
American Society of Trial Consultants, Towson *14170*
Association for Assessment in Counseling, Alexandria *14269*
Association Henri Capitant, Baton Rouge *14371*
Association of American Jurists, Berkeley *14381*
Association of American Law Schools, Washington *14382*
Association of Continuing Legal Education Administrators, Chicago *14423*
Association of Legal Administrators, Vernon Hills *14452*
Association of Trial Lawyers of America, Washington *14523*
Association of United States Members of the International Institute of Space Law, Dumfries *14524*
Bay Area Physicians for Human Rights, San Francisco *14571*
Canon Law Society of America, Washington *14642*
Center for Computer-Assisted Legal Instruction, Minneapolis *14673*
Center for Oceans Law and Policy, Charlottesville *14691*
Center for Philosophy, Law, Citizenship, Farmingdale *14692*
Center for Studies in Criminal Justice, Chicago *14703*
Commission on Mental and Physical Disability Law, Washington *14787*
Commission on Pastoral Research, Columbia *14788*
Computer Law Association, Fairfax *14818*
Conference of Chief Justices, Williamsburg *14828*
Conference on Consumer Finance Law, Chicago *14838*
Copyright Society of the U.S.A., New York *14862*
Council on Legal Education Opportunity, Washington *14922*
Earl Warren Legal Training Program, New York *14996*
Educators' ad hoc Committee on Copyright Law, Alexandria *15031*

Environmental Law Institute, Washington 15050
Federal Bar Association, Washington 15077
Independent Association of Questioned Document Examiners, Red Oak 15250
Institute for the Development of Indian Law, Oklahoma City 15302
Institute of Judicial Administration, New York 15327
Inter-American Bar Association, Washington 15349
Inter-American Commercial Arbitration Commission, Washington 15351
International Academy of Trial Lawyers, San Jose 15369
International Association of Campus Law Enforcement Administrators, Hartford 15400
International Association of Defense Counsel, Chicago 15404
International Center for Law in Development, New York 15429
International Criminal Law Commission, Santa Barbara 15463
International Human Resources, Business and Legal Research Association, Washington 15487
International Law Institute, Washington 15496
International Society for Labor Law and Social Security – U.S. National Branch, New York 15535
Japanese American Society for Legal Studies, Seattle 15594
Jewish Lawyers Guild, New York 15607
Joint Committee of the States to Study Alcoholic Beverage Laws, Alexandria 15617
Law and Society Association, Amherst 15653
Law of the Sea Institute, Honolulu 15654
Law School Admission Council – Law School Admission Services, Newtown 15655
Lawyers' Committee for Civil Rights under Law, Washington 15656
National Association of Bar Executives, Chicago 15830
National Association of Bond Lawyers, Hinsdale 15835
National Association of College and University Attorneys, Washington 15839
National Association of Legal Assistants, Tulsa 15870
National Association of School Safety and Law Enforcement Officers, Upper Marlboro 15885
National Association of Women Lawyers, Chicago 15915
National Catholic Conference for Interracial Justice, Washington 15934
National Catholic Forensic League, Milford 15936
National Center for Automated Information Research, New York 15939
National Center for Law and Deafness, Washington 15946
National Center for State Courts, Williamsburg 15947
National Center for Youth Law, San Francisco 15948
National Center on Institutions and Alternatives, Alexandria 15950
National College of District Attorneys, Houston 15957
National Conference of Bankruptcy Judges, Rockville 15967
National Conference of Black Lawyers, New York 15968
National Conference of Commissioners on Uniform State Laws, Chicago 15969
National Consumer Law Center, Boston 15983
National Council for Labor Reform, Chicago 15989
National Council of Intellectual Property Law Associations, Arlington 16001
National Council of Juvenile and Family Court Judges, Reno 16002
National Criminal Defense College, Macon 16028
National District Attorneys Association, Alexandria 16035
National Forensic League, Ripon 16055
National Health Lawyers Association, Washington 16072
National Immigration Law Center, Los Angeles 16083
National Judicial College, Reno 16099
National Organization on Legal Problems of Education, Topeka 16120
National Resource Center for Consumers of Legal Services, Gloucester 16141
National Tax Association – Tax Institute of America, Columbus 16168
National Urban League, New York 16176
Philip C. Jessup International Law Moot Court Competition, Washington 16310
Practising Law Institute, New York 16348
Public Law Education Institute, Washington 16375
Scribes, Winston-Salem 16450
Search Group, Sacramento 16454
Section on Women in Legal Education of the AALS, Washington 16458
Society of American Law Teachers, New York 16636
Society of Medical Jurisprudence, New York 16679
Southwestern Legal Foundation, Richardson 16730
Tax Analysts, Arlington 16764

Venezuela

Colegio de Abogados del Distrito Federal, Caracas 16992

Vietnam

Vietnamese Association of Forensic Scientists, Hanoi 17046

Yugoslavia

Association of Jurists of Serbia, Beograd 17077
Savez Udruženja Pravnika Jugoslavije, Beograd 17121
Savez Udruženja za Krivično Pravo i Kriminolgiju Jugoslavije, Beograd 17122
Union of Jurists' Associations of Yugoslavia, Beograd 17135

Leisure s. Travel and Tourism

Librarianship and Book Science

Albania

Council of Libraries, Tirana 00004

Antigua and Barbuda

Library Association of Antigua and Barbuda, Saint John's 00036

Argentina

Asociación Argentina de Bibliotecas y Centros de Información Científicis y Tecnical, Buenos Aires 00057
Asociación Argentina de Bibliotecas y Centros de Información Científicos y Técnicos, Buenos Aires 00058
Asociación Bernardino Rivadavia, Bahía Blanca 00073
Asociación de Bibliotecarios Graduados de la República Argentina, Buenos Aires 00077
Centro de Documentación Bibliotecológica, Bahía Blanca 00101

Australia

Australian Law Librarians' Group, Melbourne 00317
Australian Library and Information Association, Canberra 00318
Australian School Library Association, Alderley 00336
Australian Society of Indexers, Melbourne 00354
Bibliographical Society of Australia and New Zealand, Canberra 00366
School Library Association of the Northern Territory, Darwin 00487

Austria

Gesellschaft der Freunde der Österreichischen Nationalbibliothek, Wien 00608
Österreichisches Institut für Bibliographie, Wien 00911
Verband Österreichischer Volksbüchereien und Volksbibliothekare, Wien 00973
Vereinigung Österreichischer Bibliothekarinnen und Bibliothekare, Innsbruck 00997
Wiener Bibliophilen-Gesellschaft, Wien 01013

Bangladesh

Library Association of Bangladesh, Dhaka 01069

Barbados

Caribbean Association of Law Librarians, Cave Hill 01077
Library Association of Barbados, Bridgetown 01087

Belgium

Archives et Bibliothèques de Belgique, Bruxelles 01102
Association des Bibliothécaires et du Personnel des Bibliothèques des Ministères de Belgique, Bruxelles 01115
Association pour la Promotion des Publications Scientifiques, Bruxelles 01150
Association Professionnelle des Bibliothécaires et Documentalistes, Bruxelles 01156
Commission Belge de Bibliographie, Bruxelles 01234
European Association for Health Information and Libraries, Bruxelles 01276
Société Royale des Bibliophiles et Iconophiles de Belgique, Bruxelles 01486
Vereniging der Antwerpsche Bibliophielen, Antwerpen 01504
Vereniging van Religieus-Wetenschappelijke Bibliothecarissen, Sint Truiden 01506

Belize

National Library Service, Belize City 01513

Brazil

Associação Brasileira de Imprensa, Rio de Janairo 01591
Associação Paulista de Bibliotecários, São Paulo 01605
Associação Pernambucana de Bibliotecarios, Recife 01607
Associação Rio-Grandense de Bibliotecarios, Pôrto Alegre 01609
Conselho Federal de Biblioteconomia, Brasília 01643
Federação Brasileira de Associaeões de Bibliotecarios, São Paulo 01649
Instituto Nacional do Livro, Brasília 01692
Instituto Nacional do Livro, Rio de Janeiro 01693

Bulgaria

Bulgarian Union of Public Libraries, Sofia 01764

Canada

Alberta Association of College Librarians, Calgary 01824
Alberta Association of Library Technicians, Edmonton 01825
Association of Canadian Map Libraries, Ottawa 01924
Association of Parliamentary Librarians in Canada, Ottawa 01938
Association of Small Public Libraries of Ontario, Tillsonburg 01944
Atlantic Provinces Library Association, Halifax 01962
Bibliographical Society of Canada, Toronto 01964
British Columbia Library Trustees' Association, Victoria 01980
British Columbia Teacher-Librarians' Association, Vancouver 01991
Canadian Association of Law Libraries, North York 02056
Canadian Association of Library Schools, Halifax 02057
Canadian Association of Music Libraries, Ottawa 02059
Canadian Association of Research Libraries, Toronto 02067
Canadian Association of Toy Libraries and Parent Resources Centres, Ottawa 02070
Canadian Children's Book Centre, Toronto 02094
Canadian Council of Library Schools, Vancouver 02106
Canadian Health Libraries Association, Ottawa 02153
Canadian Library Association, Ottawa 02189
Church Library Association, Aylmer 02348
Congregational Libraries Association of British Columbia, Victoria 02381
Council of Prairie University Libraries, Calgary 02405
Health Libraries Association of British Columbia, Vancouver 02497
Indexing and Abstracting Society of Canada, Toronto 02515
Institute of Professional Libraries of Ontario, Toronto 02530
Library Association of Alberta, Edmonton 02589
Library Association of Ottawa-Hull, Ottawa 02590
Library Science Alumni Association, Toronto 02591
Library Technicians Association of British Columbia, Vancouver 02592
Manitoba Association of Library Technicians, Winnipeg 02605
Manitoba Health Libraries Association, Winnipeg 02618
Manitoba Library Association, Winnipeg 02625
Manitoba Library Trustee Association, Neepawa 02626
Manitoba School Library Audio-Visual Association, Winnipeg 02632
Northwest Territories Library Association, Yellowknife 02719
Nova Scotia Library Association, Halifax 02735
Ontario Association of Library Technician Instructors, Welland 02759
Ontario Association of Library Technicians, Oakville 02760
Ontario Library Association Literacy Guild, Toronto 02784
Polish Canadian Librarians Association, Toronto 02821
Saskatchewan Association of Library Technicians, Saskatoon 02880
Saskatchewan Library Association, Regina 02893
West Coast Library Association, Corner Brook 02988

Chile

Asociación de Bibliotecarios de Chile, Santiago 03017
Colegio de Bibliotecarios de Chile, Santiago 03030

China, People's Republic

Chinese Society of Library Science, Beijing 03194

China, Republic

The Library Association of China, Taipei 03305

Colombia

Asociación Colombiana de Bibliotecarios, Bogotá 03343
Asociación de Egresados de la Escuela Interamericana de Bibliotecología de la Universidad de Antioquia, Medellín 03349
Centro Regional para el Fomento del Libro en América Latina, Bogotá 03362
Colegio de Bibliotecarios Colombianos, Bogotá 03364
Consejo Nacional de Archivos, Bogotá 03370
Grupo de Biblioteca y Documentación, Bogotá 03376

Congo

Bureau des Archives et Bibliothèques de Brazzaville, Brazzaville 03423

Costa Rica

Asociación Costarricense de Bibliotecarios, San José 03430
Inter-American Association of Agricultural Librarians and Documentalists, Turrialba 03449

Croatia

Hrvatsko Bibliotekarsko Društvo, Zagreb 03460

Cuba

Asociación Cubana de Bibliotecarios, La Habana 03470

Cyprus

Library Association of Cyprus, Nicosia 03497

Denmark

Danmarks Biblioteksforening, Ballerup 03576
Danmarks Forskningsbiblioteksforening, Lyngby 03578
Danmarks Skolebibliotekarforening, Viby 03586
Danmarks Skolebiblioteksforening, København 03587
Dansk Exlibris Selskab, Brœnshœj 03615
Dansk Musikbiblioteksforening, København 03651
Foreningen af Medarbejdere ved Danmarks Forskningsbiblioteker, Lyngby 03733
International Society of Libraries and Museums of the Performing Arts, København 03749
Sammenslutningen af Danmarks Forskningsbiblioteker, Lyngby 03800
Scandinavian Library Center, Ballerup 03805

Dominican Republic

Asociación Dominicana de Bibliotecarios, Santo Domingo 03831
Grupo Bibliográfico Nacional de la República Dominicana, Santo Domingo 03836
Servicio de Documentación y Biblioteca, Santo Domingo 03837

Ecuador

Asociación Ecuatoriana de Bibliotecarios, Quito 03843

Egypt

Egyptian Association of Archives, Librarianship and Information Science, Cairo 03889
Egyptian School Library Association, Cairo 03895

El Salvador

Asociación de Bibliotecarios de El Salvador, San Salvador 03919

Ethiopia

Ethiopian Library Association, Addis Ababa 03943

Finland

Kirjastonhoitajaliitto, Helsinki 03977
Nordic Council for Scientific Information, Espoo 03984
Suomen Kirjastoseura, Helsinki 04039
Suomen Tieteellinen Kirjastoseura, Helsinki 04064
Tietohuollon Neuvottelukunta, Helsinki 04076

France

Amicale des Directeurs de Bibliothèques Universitaires, Paris 04130
Association des Bibliothécaires Français, Paris 04169
Association des Bibliothèques de Judaïca et Hébraïca en Europe, Paris 04170
Association des Bibliothèques Ecclésiastiques de France, Paris 04171
Association des Conservateurs de Bibliothèques, Paris 04174
Association des Diplômés de l'Ecole de Bibliothécaires-Documentalistes, Paris 04176
Association des Professionnels de l'Information et de la Documentation, Paris 04198
Association Nationale des Bibliothécaires, Paris 04307
Association pour la Médiathèque Public, Cambrai 04326
Commission Internationale de Bibliographie, Paris 04424
Fédération des Amicales des Documentalistes et Bibliothcaires de l'Education Nationale, Paris 04511
Sociéte des Bibliophiles de Guyenne, Bordeaux 04701
Société Internationale de Bibliographie Classique, Paris 04875

Germany

Arbeitsgemeinschaft der Archive und Bibliotheken in der evangelischen Kirche, Nürnberg 05083
Arbeitsgemeinschaft der kirchlichen Büchereiverbände, Bonn 05087
Arbeitsgemeinschaft der Parlaments- und Behördenbibliotheken, München 05089
Arbeitsgemeinschaft der Regionalbibliotheken, Hannover 05090
Arbeitsgemeinschaft der Spezialbibliotheken e.V., Leverkusen 05093
Arbeitsgemeinschaft für juristisches Bibliotheks- und Dokumentationswesen, Augsburg 05105
Arbeitsgemeinschaft für medizinisches Bibliothekswesen, Mannheim 05109
Arbeitsgemeinschaft katholisch-theologischer Bibliotheken, Paderborn 05122
Beta Beta Delta, Stuttgart 05207
Bundesarbeitsgemeinschaft der katholisch-kirchlichen Büchereiarbeit, Bonn 05233
Conseil International des Associations de Bibliothèques de Théologie, Köln 05279
Deutsche Exlibris Gesellschaft e.V., Konstanz 05302
Deutscher Bibliotheksverband e.V., Berlin 05506
Deutscher Verband Evangelischer Büchereien e.V., Göttingen 05544
Deutsches Bibliotheksinstitut, Berlin 05570
Gesellschaft der Bibliophilen e.V., Höchberg 05790
Gesellschaft für Bibliothekswesen und Dokumentation des Landbaues, Karlsruhe 05804
Gesellschaft für Klassifikation e.V., Aachen 05835
Historische Kommission des Börsenvereins des Deutschen Buchhandels, Frankfurt 05916
Internationale Arbeitsgemeinschaft der Archiv-, Bibliotheks- und Graphikrestauratoren, Marburg 05989
Internationale Vereinigung der Musikbibliotheken, Musikarchive und Musikdokumentationszentren, Gruppe Bundesrepublik Deutschland, Berlin 06003
Normenausschuss Bibliotheks- und Dokumentationswesen im DIN Deutsches Institut für Normung e.V., Berlin 06152
Verband der Bibliotheken des Landes Nordrhein-Westfalen e.V., Witten 06334
Verband deutscher Werkbibliotheken e.V., Leverkusen 06363
Verein der Bibliothekare an Öffentlichen Bibliotheken e.V., Reutlingen 06368
Verein der Diplom-Bibliothekare an wissenschaftlichen Bibliotheken e.V., Regensburg 06369

Verein Deutscher Bibliothekare e.V., Mainz *06373*
Württembergische Bibliotheksgesellschaft, Vereinigung der Freunde der Landesbibliothek, Stuttgart *06454*

Ghana

Ghana Library Association, Accra *06486*
Ghana Library Board, Accra *06487*

Greece

Enosis Ellinon Bibliothekarion, Athinai *06518*

Guatemala

Asociación Bibliotecológica Guatemalteca, Guatemala City *06557*

Guyana

Guyana Library Association, Georgetown *06577*

Honduras

Asociación de Bibliotecarios y Archiveros de Honduras, Tegucigalpa *06582*

Hong Kong

Hong Kong Library Association, Hong Kong *06593*

Hungary

Magyar Könyvtárosok Egyesülete, Budapest *06644*

Iceland

Bókavardafélag Islands, Reykjavik *06678*
Félag Bókavarda í Rannsókarbókasöfnum, Reykjavik *06681*

India

Delhi Library Association, Dehli *06755*
Federation of Indian Library Associations, Patiala *06759*
Indian Association of Academic Librarians, New Delhi *06771*
Indian Association of Special Libraries and Information Centres, Calcutta *06775*
Indian Association of Teachers of Library Science, Nagpur *06777*
Indian College Library Association, Hyderabad *06784*
Indian Library Association, Delhi *06799*
Tripura Library Association, Agartala *06866*
Uttar Pradesh (India) Library Association, Lucknow *06869*

Indonesia

Asosiasi Perpustakaan, Arsip dan Dokumentasi Indonesia, Jakarta *06885*
Himpunam Pustakawan Chusus Indonesia, Jakarta *06888*
Ikatan Pustakawan Indonesia, Jakarta *06890*

Ireland

Library Association of Ireland, Dublin *07007*

Israel

Israel Library Association, Tel Aviv *07080*
Israel Society of Special Libraries and Information Centres, Tel Aviv *07104*

Italy

Agricultural Libraries Network, Roma *07247*
Arbeitsgemeinschaft der Kunstbibliotheken, Roma *07250*
Associazione Italiana Biblioteche, Roma *07291*
Centro Bibliografica Francescano, Benevento *07416*

Jamaica

Commonwealth Library Association, Mandeville *08078*
Jamaica Library Association, Kingston *08083*

Japan

Nihon Toshokan Kyokai, Tokyo *08271*
Nippon Toshokan Gakkai, Tokyo *08394*

Jordan

Jordan Library Association, Amman *08452*

Kenya

East African Library Association, Nairobi *08496*
Kenya Library Association, Nairobi *08504*
Standing Conference of African University Libraries, Nairobi *08507*

Korea, Democratic People's Republic

Library Association of the Democratic People's Republic of Korea, Pyongyang *08516*

Korea, Republic

Korean Library and Information Science Society, Seoul *08535*
Korean Library Association, Seoul *08536*
Korean Micro-Library Association, Seoul *08538*

Latvia

Council on Libraries, Riga *08556*

Lebanon

Lebanese Library Association, Beirut *08562*

Macedonia

Sojuz na Društvata na Bibliotekarite na Makedonija, Skopje *08608*

Malawi

Malawi Library Association, Zomba *08623*

Malaysia

Malaysian Library Association, Kuala Lumpur *08649*

Malta

Ghaqda Bibljotekarji, Floriana *08662*
Professional Librarians of Malta Association, Paceville *08670*

Mexico

Asociación Mexicana de Bibliotecarios, AC, México *08699*
Consejo Interamericano de Archiveros, México *08728*
Sociedad Mexicana de Bibliografía, México *08757*

Netherlands

Federatie van Organisaties van Bibliotheek-, Informatie-, Dokumentatiewezen, 's-Gravenhage *08894*
International Federation of Library Associations and Institutions, 's-Gravenhage *08934*
Nederlands Bibliotheek en Lektuur Centrum, 's-Gravenhage *08980*
Nederlands Bureau voor Bibliotheekwezen en Informatieverzorging, 's-Gravenhage *08981*
Nederlandsche Vereniging voor Druk- en Boekkunst, Haarlem *08984*
Nederlandse Vereniging van Bibliothecarissen, Documentalisten en Literatuuronderzoekers, Schelluinen *09005*
Samenwerkingsverband van de Universiteitsbibliotheken, de Koninklijke Bibliotheek en de Bibliotheek van de Koninklijke Nederlandse Akademie van Wetenschappen, Amsterdam *09067*
Vereniging voor het Theologisch Bibliothecariaat, Nijmegen *09100*

New Zealand

New Zealand Book Council, Wellington *09138*
New Zealand Library Association, Wellington *09164*

Nicaragua

Asociación de Bibliotecas Universitarias y Especializadas de Nicaragua, León *09203*
Asociación Nicaragüense de Bibliotecarios, Managua *09204*

Nigeria

Nigerian Library Association, Lagos *09240*

Norway

Kommunale Bibliotekarbeiderers Forening, Kolbotn *09274*
Norsk Bibliotekforening, Oslo *09293*
Norske Fagbibliotek Forening, Oslo *09298*
Riksbibliotektjenesten, Oslo *09344*

Pakistan

Federal Library Association, Rawalpindi *09363*
Mehran Library Association, Hyderabad *09374*
National Book Council of Pakistan, Islamabad *09375*
Pakistan Library Association, Karachi *09387*
Sind Library Association, Hyderabad *09402*
Society for the Promotion and Improvement of Libraries, Karachi *09403*

Panama

Asociación de Bibliotecarios Graduados del Istmo de Panamá, Panamá City *09409*
Asociación Panameña de Bibliotecarios, Panamá City *09410*

Papua New Guinea

Papua New Guinea Library Association, Boroko *09421*

Paraguay

Asociación de Bibliotecarios del Paraguay, Asunción *09426*
Asociación de Bibliotecarios Universitarios del Paraguay, Asunción *09427*

Peru

Agrupación de Bibliotecas para la Información Socio-Económica, Lima *09442*
Asociación Peruana de Bibliotecarios, Lima *09456*

Philippines

Association of Special Libraries of the Philippines, Manila *09512*
Philippine Library Association, Manila *09528*

Poland

Stowarzyszenie Bibliotekarzy Polskich, Warszawa *09615*

Portugal

Associação Portuguesa de Bibliotecários, Arquivistas e Documentalistas, Lisboa *09640*

Puerto Rico

Association of Caribbean University, Research and Institutional Libraries, San Juan *09723*

Romania

Asociatia Bibliotecarilor din Romania, Bucuresti *09738*

Senegal

African Standing Conference on Bibliographic Control, Dakar *10014*
Association des Bibliothèques de l'Enseignement Supérieure de l'Afrique de l'Ouest Francophone, Dakar *10017*
Ecole de Bibliothécaires, Archivistes et Documentalistes, Dakar *10027*

Sierra Leone

Sierra Leone Association of Archivists, Librarians and Information Scientists, Freetown *10031*

Singapore

Library Association of Singapore, Singapore *10051*

Slovakia

Zväz Slovenských Knihovníkov a Informatikov, Bratislava *10095*

Slovenia

Zveza Bibliotekarskih Društev Slovenije, Ljubljana *10111*

South Africa

African Library Association of South Africa, Pietersburg *10122*
South African Institute for Librarianship and Information Science, Pretoria *10193*

Spain

Asociación Española de Bibliotecarios, Archiveros, Museologos y Documentalistas, Madrid *10257*

Sri Lanka

Sri Lanka Library Association, Colombo *10420*

Sudan

Sudan Library Association, Khartoum *10429*

Swaziland

Swaziland Library Association, Mbabane *10435*

Sweden

International Association of Music Libraries, Archives and Documentation Centres, Stockholm *10472*
Svenska Bibliotekariesamfundet, Borås *10539*
Svenska Folkbibliotekarie Förbundet, Nacka *10547*
Sveriges Allmänna Biblioteksförening, Stockholm *10585*
Sveriges Vetenskapliga Specialbiblioteks Förening, Stockholm *10597*

Switzerland

Association des Bibliothèques Internationales, Genève *10619*
Schweizerische Bibliophilen-Gesellschaft, Zürich *10799*
Vereinigung Schweizerischer Bibliothekare, Bern *11017*

Tanzania

Tanzania Library Association, Dar es Salaam *11060*

Thailand

Thai Library Association, Bangkok *11102*

Trinidad and Tobago

Library Association of Trinidad and Tobago, Port of Spain *11122*

Tunisia

Association Tunisienne des Bibliothécaires, Documentalistes et Archivistes, Tunis *11138*

Turkey

Türk Kütüphaneciler Derneği, Ankara *11165*

Uganda

East African School of Librarianship, Kampala *11185*
Uganda Library Association, Kampala *11187*

United Kingdom

Art Libraries Society of the United Kingdom and Ireland, Bromsgrove *11251*
Aslib, The Association for Information Management, London *11256*
Association of British Theological and Philosophical Libraries, Cambridge *11318*
Association of Independent Libraries, Manchester *11356*
Association of London Chief Librarians, Barking *11363*
The Bibliographical Society, London *11438*
Birmingham Bibliographical Society, Birmingham *11448*
Bliss Classification Association, Cambridge *11454*
Bookplate Society, London *11459*
British and Irish Association of Law Librarians, Birmingham *11479*
British Association for Information and Libraries Education and Research, Sheffield *11493*
British Association of Picture Libraries, London *11518*
Cambridge Bibliographical Society, Cambridge *11736*
Circle of State Librarians, London *11787*
Edinburgh Bibliographical Society, Edinburgh *11932*
Friends of The National Libraries, London *12094*
International Association for Esperanto in Libraries, London *12323*
International Association of Music Libraries, Archives and Documentation Centres (United Kingdom Branch), Exeter *12331*
International Association of Technological University Libraries, Edinburgh *12334*
Library Association, London *12450*
The Library Association – Personnel, Training and Education Group, London *12451*
Marine Librarians Association, London *12494*
Play Matters – The National Association of Toy and Leisure Libraries, London *12704*
The Private Libraries Association, Pinner *12719*
Scottish Library Association, Motherwell *12915*
Society of County Librarians, Worcester *13040*
Society of Scribes and Illuminators, London *13064*
The Standing Conference of National and University Libraries, London *13096*
Standing Conference on Library Materials on Africa, Milton Keynes *13097*

Uruguay

Asociacíon de Bibliotecarios del Uruguay, Montevideo *13242*
Asociación de Bibliotecólogos del Uruguay, Montevideo *13243*
Library Association of Uruguay, Montevideo *13263*

U.S.A.

American Association of Law Libraries, Chicago *13547*
American Association of School Librarians, Chicago *13579*
American Film and Video Association, Niles *13812*
American Library Association, Chicago *13916*
American Library Trustee Association, Chicago *13917*
American Merchant Marine Library Association, New York *13942*
American Printing History Association, New York *14014*
American Society of Indexers, Port Aransas *14135*
American Theological Library Association, Evanston *14192*
Art Libraries Society/North America, Tucson *14250*
Association for Documentary Editing, Carbondale *14289*
Association for Library and Information Science Education, Raleigh *14313*
Association for Library Collections and Technical Services, Chicago *14314*
Association for Library Service to Children, Chicago *14315*
Association for Population/Family Planning Libraries and Information Centers International, New York *14319*
Association for Recorded Sound Collections, Silver Spring *14325*
Association of Academic Health Sciences Library Directors, Houston *14373*
Association of Christian Librarians, Cedarville *14412*
Association of College and Research Libraries, Chicago *14414*
Association of Jewish Libraries, New York *14451*
Association of Research Libraries, Washington *14492*
Association of Specialized and Cooperative Library Agencies, Chicago *14505*
Association of Visual Science Librarians, Portland *14533*
Bibliographical Society of America, New York *14584*
Bibliographical Society of the University of Virginia, Charlottesville *14585*
Catholic Library Association, Haverford *14656*
Center for Chinese Research Materials, Oakton *14671*
Center for Research Libraries, Chicago *14696*
Chief Officers of State Library Agencies, Lexington *14729*
Chinese-American Librarians Association, Sacramento *14736*
Church and Synagogue Library Association, Portland *14751*
Committee on Research Materials on Southeast Asia, Ann Arbor *14805*
Continuing Library Education Network and Exchange Round Table, Chicago *14856*
Council of National Library and Information Associations, Washington *14900*
Council of Planning Librarians, Chicago *14901*
Council on Interracial Books for Children, New York *14921*
Council on Library-Media Technical Assistants, Cleveland *14923*
Council on Library Resources, Washington *14924*
Dictionary Society of North America, Cleveland *14972*
Esperanto Librarians International, Naperville *15058*
Ethnic Materials and Information Exchange Round Table, Chicago *15062*
Herbert Hoover Presidential Library Association, West Branch *15203*
Independent Research Libraries Association, Washington *15254*
International Association of Music Libraries – United States Branch, New York *15412*
International Association of School Librarianship, Kalamazoo *15417*
International Council of Library Association Executives, Sacramento *15456*
International Survey Library Association, Storrs *15559*
Italian American Librarians Caucus, New York *15587*
Jewish Librarians Task Force, Franklin Square *15608*
Library Administration and Management Association, Chicago *15668*
Library and Information Technology Association, Chicago *15669*
Library Public Relations Council, East Brunswick *15670*
Lutheran Church Library Association, Minneapolis *15680*

Medical Library Association, Chicago *15710*
Middle East Librarians' Association, Santa Barbara *15733*
Music Library Association, Canton *15764*
National Librarians Association, Alma *16102*
Northern Libraries Colloquy, Anchorage *16231*
Office of Management Services, Washington *16247*
Public Library Association, Chicago *16377*
Reference and Adult Services Division (of ALA), Chicago *16399*
Reforma: National Association to Promote Library Services to the Spanish-Speaking, New York *16400*
Research Libraries Group, Mountain View *16416*
Seminar on the Acquisition of Latin American Library Materials, Albuquerque *16461*
Social Responsibilities Round Table, Chicago *16483*
Society of American Archivists, Chicago *16633*
Special Libraries Association, Washington *16740*
Theatre Library Association, New York *16775*
Universal Serials and Book Exchange, Cleveland *16837*
Urban Libraries Council, State College *16852*
Western Association of Map Libraries, Clovis *16880*
Zionist Archives and Library of World Zionist Organization – American Section, New York *16945*

Venezuela

Asociación Interamericana de Bibliotecarios y Documentalistas Agrícolas, Filial Venezuela, Mérida *16969*

Yugoslavia

Društvo Bibliotekara Vojvodine, Novi Sad *17083*
Savez Bibliotečkih Radnika Srbije, Beograd *17109*

Zaire

Association Zaïroise des Archivistes, Bibliothécaires et Documentalistes, Kinshasa *17141*

Zambia

Zambia Library Association, Lusaka *17158*

Zimbabwe

Zimbabwe Library Association, Harare *17186*

Linguistics s. a. Literature

Algeria

Centre National de Traduction et de Terminologie Arabe, Alger *00020*

Argentina

Asociación Dante Alighieri, Buenos Aires *00076*

Australia

Australasian Universities Language and Literature Association, Sydney *00248*
English Association – Sydney Branch, Rozelle *00390*

Austria

Arbeitsgemeinschaft für Deutschdidaktik, Klagenfurt *00550*
Arbeitskreis der Wiener Altgermanisten, Wien *00564*
Arbeitskreis der Wiener Skandinavisten, Wien *00565*
Austria Esperanto Federacio, Wien *00571*
Eranos Vindobonensis, Wien *00584*
Gesellschaft für Klassische Philologie in Innsbruck, Innsbruck *00623*
Gesellschaft zur Erforschung slawischer Sprachen und Kulturen, Graz *00640*
Gesellschaft zur Förderung Slawistischer Studien, Wien *00643*
Innsbrucker Germanistische Arbeitsgemeinschaft, Innsbruck *00653*
Innsbrucker Sprachwissenschaftliche Gesellschaft, Innsbruck *00655*
Institut für Österreichkunde, Wien *00658*
Klagenfurter Sprachwissenschaftliche Gesellschaft, Klagenfurt *00694*
Kontext – Institut für Kommunikations- und Textanalysen, Wien *00696*
Österreichische Gesellschaft für Neugriechische Studien, Wien *00813*
Österreichische Gesellschaft für Semiotik, Wien *00830*
Österreichische Orient-Gesellschaft Hammer-Purgstall, Wien *00875*
Orientalische Gesellschaft, Wien *00930*
Societas Linguistica Europaea, Wien *00945*
Verband der österreichischen Neuphilologen, Wien *00961*
Verband der Russischlehrer Österreichs, Wien *00962*
Verein der Freunde der im Mittelalter von Österreich aus besiedelten Sprachinseln, Wien *00976*
Verein Muttersprache, Lang-Enzersdorf *00999*
Wiener Humanistische Gesellschaft, Wien *01018*
Wiener Sprachgesellschaft, Wien *01028*

Belgium

Académie Royale de Langue et de Littérature Françaises, Bruxelles *01093*
Association Internationale pour l'Utilisation des Langues Régionales à l'Ecole, Liège *01145*
Belgische Vereniging voor Toegepaste Linguistiek, Bruxelles *01162*
Bureau International d'Audiophonologic, Bruxelles *01173*
Centre International des Langues, Littératures et Traditions d'Afrique au Service du Développement, Louvain-la-Neuve *01195*
Chambre Belge des Traducteurs, Interprètes et Philologues, Bruxelles *01210*
Conférence Internationale Permanente de Directeurs d'Instituts Universitaires pour la Formation de Traducteurs et d'Interprètes, Antwerpen *01248*
Europa Esperanto-Centro, Brugge *01266*
European Bureau of Lesser-Used Languages, Bruxelles *01303*
Institut Belge des Hautes Etudes Chinoises, Bruxelles *01381*
International Committee of Dialectologists, Leuven *01393*
International Council of Onomastic Sciences, Leuven *01394*
Koninklijke Academie voor Nederlandse Taal- en Letterkunde, Gent *01408*
Koninklijke Academie voor Wetenschappen, Letteren en Schone Kunsten van België, Bruxelles *01409*
Société de Langue et de Littérature Wallonnes, Liège *01454*
Société d'Etudes Latines de Bruxelles, Tournai *01458*

Brazil

Sociedade Brasileira de Romanistas, Rio de Janeiro *01716*

Bulgaria

Society of Foreign Language Teachers, Sofia *01781*
Union of Translators' in Bulgaria, Sofia *01802*

Canada

Académie des Lettres du Québec, Montréal *01814*
American Dialect Society, London *01860*
Association Canadienne d'Education de Langue Française, Sillery *01873*
Association des Professeurs de Français de la Saskatchewan, Saskatoon *01890*
Association des Professeurs de Français des Universités et Collèges Canadiens, Victoria *01891*
Association for the Advancement of Scandinavian Studies in Canada, Guelph *01904*
Association of Wholly or Partially French Language Universities, Montréal *01948*
Canadian Association for the Advancement of Netherlandic Studies, Windsor *02041*
Canadian Association of Latin American Studies, Vancouver *02055*
Canadian Association of University Teachers of German, Edmonton *02075*
Canadian Council of Teachers of English, Oakville *02108*
Canadian Esperanto Association, Montréal *02125*
Canadian Esperanto Youth, Montréal *02126*
Canadian Linguistic Association, Toronto *02191*
Canadian Metric Association, Fonthill *02200*
Canadian Semiotic Association, Kingston *02248*
Estonian Teachers' League, Don Mills *02443*
International Center for Research on Bilingualism, Sainte Foy *02544*
Manitoba Association for the Promotion of Ancestral Languages, Winnipeg *02603*
Polish Teachers' Association in Canada, Toronto *02822*
Saskatchewan Association of Teachers of German, Saskatoon *02884*

Chile

Academia Chilena de la Lengua, Santiago *03011*
Asociación de Lingüística y Filología de América Latina, Santiago *03019*
Sociedad Chilena de Lingüística, Santiago *03077*

China, People's Republic

Inner Mongolian Academy of Social Sciences, Huhehot *03206*

China, Republic

Chinese Language Society, Taipei *03275*
National Institute for Compilation and Translation, Taipei *03312*
Society of the Chinese Borders History and Languages, Taipei *03329*

Colombia

Academia Colombiana de la Lengua, Bogotá *03339*
Instituto Caro y Cuervo, Bogotá *03377*

Congo

Institut Africain, Mouyondzi *03426*

Costa Rica

Academia Costarricense de la Lengua, San José *03427*

Croatia

Nakladni Zavod Matice Hrvatske, Zagreb *03464*

Cuba

Academia Cubana de la Lengua, La Habana *03468*

Cyprus

Society of Cypriot Studies, Nicosia *03499*

Denmark

Dansk Laererforeningen, København *03643*
Dansk Selskab for Logopaedi og Foniatri, København *03677*
Dansk Sprogvaern, Vaerløse *03693*
Det Danske Sprog- og Litteraturselskab, København *03708*
Filologisk-Historiske Samfund, København *03725*
Frankslaererforeningen, København *03738*
Gymnasieskolernes Tysklaererforening, Nykøbing *03741*
Selskab for Nordisk Filologi, København *03820*

Dominican Republic

Academia Dominicana de la Lengua, Santo Domingo *03830*

Egypt

Academy of the Arabic Language, Cairo *03860*

Finland

Äidinkielen opettajain liitto, Helsinki *03948*
Kalevalaseura, Helsinki *03972*
Kotikielen Seura, Helsingin Yliopisto *03979*
Suomalaisen Kirjallisuuden Seura, Helsinki *04001*
Suomalais-Ugrilainen Seura, Helsinki *04003*
Suomen Kielen Seura, Turku *04037*
Uusfilologinen Yhdistys, Helsingin Yliopisto *04081*

France

Académie Française, Paris *04111*
Association des Langues et Civilisations, Paris *04190*
Association des Professeurs de Langues Vivantes de l'Enseignement Public, Paris *04195*
Association for the Expansion of International Roles of the Languages of Continental Europe, Paris *04210*
Association Française des Enseignants de Français, Paris *04228*
Association Française des Professeurs de Langues Vivantes, Paris *04234*
Association Internationale d'Epigraphie Grecque et Latine, Paris *04262*
Association Internationale des Etudes Françaises, Paris *04273*
Association Internationale des Professeurs de Langue et Littérature Russes, Paris *04274*
Association Linguistique Franco-Européenne, Paris *04298*
Association Philotechnique, Paris *04320*
Association pour le Développement de la Traduction Automatique et de Linguistique Appliqué, Paris *04336*
Centre International d'Etudes Latines, Paris *04378*
Comité International de Paléographie Latine, Paris *04407*
Conseil International de la Langue Française, Paris *04453*
Esperanto Academy, Paris *04471*
Fédération Internationale des Professeurs de Français, Sèvres *04534*
Groupe Leibniz, Grenoble *04552*
Groupe Phonétique de Paris, Paris *04561*
International Association of Sanskrit Studies, Paris *04582*
Organisation de la Jeunesse Esperantiste Française, Paris *04612*
Société de Linguistique de Paris, Paris *04664*
Société de Linguistique Romane, Strasbourg *04665*
Société d'Etudes Folkloriques du Centre-Ouest, Saint-Jean d'Angely *04739*
Société d'Etudes Linguistiques et Anthropologiques de France, Paris *04746*
Société Française des Professeurs de Russe, Paris *04845*
Union des Travailleurs Espérantistes des Pays de Langue Française, Paris *04977*
Union Française pour l'Espéranto, Paris *04980*

Georgia

Amateur Society of Basque Language, Tbilisi *05006*
Council on Co-ordinating Scientific Studies of the Georgian Language, Tbilisi *05009*

Germany

Arbeitskreis Rhetorik in Wirtschaft, Politik und Verwaltung, Bonn *05147*
Bund für Deutsche Schrift und Sprache, Ahlhorn *05263*
Deutsche Akademie für Sprache und Dichtung e.V., Darmstadt *05289*
Deutsche Gesellschaft für Sprachwissenschaft, Passau *05427*
Deutsche Gesellschaft für Sprechwissenschaft und Sprecherziehung e.V., Münster *05428*
Deutsche Morgenländische Gesellschaft e.V., Heidelberg *05473*
Deutscher Altphilologen-Verband, Puchheim *05497*
Deutscher Germanistenverband, Aachen *05513*
Deutscher Philologen-Verband e.V., Unterhaching *05536*
Fachverband Moderne Fremdsprachen, Augsburg *05695*
Germana Esperanto Asocio r.a., Bonn *05785*
Gesellschaft für deutsche Sprache e.V., Wiesbaden *05808*
Gesellschaft für Interkulturelle Germanistik e.V., Bayreuth *05832*
Gesellschaft für internationale Sprache e.V., Reinbek *05834*
Hessischer Philologen-Verband, Wiesbaden *05914*
Sprachverband Deutsch für ausländische Arbeitnehmer e.V., Mainz *06290*
Ständiger Ausschuss für Geographische Namen, Frankfurt *06294*
Südosteuropa-Gesellschaft e.V., München *06309*
Verein Nordfriesisches Institut e.V., Bredstedt *06423*

Greece

Association of Arts and Letters, Athinai *06500*
Hetaireia Hellenon Philologon, Athinai *06534*

Guatemala

Academia de la Lengua Maya Quiché, Quezaltenango *06555*

Hungary

Magyar Nyelvtudományi Társaság, Budapest *06649*

Iceland

Félag Enskukennara á Islandi, Reykjavik *06682*

India

International Tamil League, Madras *06828*
Jammu and Kashmir Academy of Art, Culture and Languages, Srinagar *06830*
Linguistic Society of India, Poona *06833*
Sanskrit Academy, Madras *06856*
Tamil Association, Thanjavur *06864*

Indonesia

Pusat Pembinaan dan Pengembangan Bahasa, Jakarta *06894*

Ireland

Conradh na Gaeilge, Dublin *06937*
Institiuid Teangeolaiochta Eireann, Dublin *06957*

Israel

Academy of the Hebrew Language, Jerusalem *07040*

Italy

Accademia della Crusca, Firenze *07147*
Accademia di Agricoltura Scienze e Lettere, Verona *07154*
Accademia Ligustica de Belle Arti, Genova *07201*
Associazione Italiana degli Slavisti, Pisa *07294*
Associazione Italiana di Anglistica, Pisa *07299*
Associazione Italiana di Studi Semiotici, Torino *07327*
Centro di Studi Filologici e Linguistici Siciliani, Palermo *07445*
Centro Esperantista Contro l'Imperialismo Linguistico, Milano *07460*
Centro per lo Studio dei Dialetti Veneti dell'Istria, Trieste *07527*
Europa Club, Sezione Italiana, Verona *07661*
Gruppo Italiano di Linguistica Applicata, Pisa *07702*
International League of Esperantist Teachers, Massa *07716*
Istituto Italiano di Studi Germanici, Roma *07741*
Società di Linguistica Italiana, Roma *07784*
Società Filologica Friulana G.I. Ascoli, Udine *07806*
Società Filologica Romana, Roma *07807*
Società Istriana di Archeologia e Storia Patria, Trieste *07820*
Società Italiana di Glottologia, Pisa *07885*
Union Ladins Val Badia, Pedraces *08052*

Jamaica

Association for Commonwealth Language and Literature Studies, Kingston *08067*

Japan

East-West Sign Language Association, Tokyo *08126*
Hanshin Doitsubungakukai, Kobe *08130*
Kokugo Gakkai, Tokyo *08171*
Nihon Chugoku Gakkai, Tokyo *08190*
Nihon Eibungakkai, Tokyo *08196*
Nihon Esperanto Gakkai, Tokyo *08198*
Nihon Furansu-go Furansu-bun-gaku-kai, Tokyo *08199*
Nihon Gengogakkai, Tokyo *08202*
Nippon Rosiya Bungakkai, Tokyo *08360*
Toho Gakkai, Tokyo *08434*

Malaysia

Asian Association of National Languages, Kuala Lumpur *08633*
Tamil Language Society, Kuala Lumpur *08655*

Mexico

Academia Mexicana de la Lengua, México *08687*

Netherlands

Center for Research and Documentation on the World Language Problem, Rotterdam *08837*
Fryske Akademy, Leeuwarden *08899*
Institute for Esperanto in Commerce and Industry, Valkenswaard *08911*
Koninklijk Instituut voor Taal-, Land- en Volkenkunde, Leiden *08964*
Kristaina Esperantista Ligo Internacia, Alphen *08969*
Maatschappij der Nederlandse Letterkunde, Leiden *08974*
Nederlandse Vereniging voor Toegepaste Taalwetenschap, Utrecht *09039*
Permanent International Committee of Linguists, Leiden *09060*
Universal Esperanto Association, Rotterdam *09083*

Nicaragua

Academia Nicaragüense de la Lengua, Managua *09202*

Niger

Centre for Linguistic and Historical Studies by Oral Tradition, Niamey *09207*

Norway

Landslaget for Språklig Samling, Oslo *09276*

Norske Akademi for Sprog og Litteratur, Oslo 09295

Pakistan

Baluchi Academy, Quetta 09361
Iqbal Academy, Lahore 09370
Pashto Academy, Peshawar 09392
Punjabi Adabi Academy, Lahore 09394
Urdu Academy, Bahawalpur 09404

Panama

Academia Panameña de la Lengua, Panamá City 09407

Paraguay

Academia de la Lengua y Cultura Guaraní, Asunción 09424

Poland

Polskie Towarzystwo Filologiczne, Warszawa 09566
Polskie Towarzystwo Jezykoznawcze, Kraków 09580
Polskie Towarzystwo Neofilologiczne, Poznan 09593
Polskie Towarzystwo Semiotyczne, Warszawa 09604
Towarzystwo Milosnikow Jezyka Polskiego, Kraków 09624

Portugal

Sociedade de Lingua Portuguesa, Lisboa 09683

Puerto Rico

Academia Puertorriqueña de la Lengua Española, San Juan 09719

Romania

Societatea de Stiinte Filologice din Romania, Bucuresti 09782
Societatea Romana de Linguistica, Bucuresti 09792

Russia

Commission for a Linguistic Atlas of Europe, Moskva 09812
Commission for Philology and Phonetics, Moskva 09826
Commission for the History of Philology, Moskva 09836
National Committee of Finno-Ugric Philologists, Moskva 09894
National Committee of Slavonic Philologists, Moskva 09899
National Committee of Turkish Philologists, Moskva 09904
Russian Linguistics Society, Moskva 09958

Singapore

Chinese Language and Research Centre, Singapore 10045

Slovakia

Kružok Modemých Filológov, Bratislava 10058
Slovenská Jazykovedná Spoločnost, Bratislava 10075
Slovenská Jednota Klassických Filológov, Bratislava 10076

Slovenia

Slavistično Društvo Slovenije, Ljubljana 10104
Slovenska Matica, Ljubljana 10106

South Africa

Classical Association of South Africa, Stellenbosch 10130
English Academy of Southern Africa, Wits 10139
Federasie van Rapportryerskorpse, Auckland Park 10142

Spain

Institución Fernando el Católico, Zaragoza 10312
Instituto de Filologia, Madrid 10322
Oficina Internacional de Información y Observación del Español, Madrid 10332
Real Academia de la Lengua Vasca, Bilbao 10349
Real Academia Española, Madrid 10353
Seminario de Filología Vasca Julio de Urquijo, San Sebastián 10368

Sri Lanka

Royal Asiatic Society of Sri Lanka, Colombo 10418

Sudan

Association des Professeurs de Français en Afrique, Khartoum 10425

Sweden

Riksföreningen för Lärarna i Moderna Språk, Hisings Kärra 10515
Svenska Akademien, Stockholm 10530
Svenska Klassikerförbundet, Stockholm 10556

Switzerland

Association des Sociétés Suisses des Professeurs de Langues Vivantes, Luzern 10622
Bund für vereinfachte rechtschreibung, Zürich 10637
Collegium Romanicum, Lausanne 10648
Deutschschweizerischer Sprachverein, Bremgarten 10660
Fédération Internationale des Associations d'Etudes Classiques, Bellevue 10701
Gesellschaft für deutsche Sprache und Literatur in Zürich, Zürich 10713
Schweizerische Akademische Gesellschaft der Anglisten, Genève 10791
Schweizerische Esperanto-Gesellschaft, Kriens 10805
Schweizerische Gesellschaft für Skandinavische Studien, Zürich 10846
Schweizerischer Altphilologen-Verband, Basel 10882
Schweizerischer Anglistenverband, Lugano 10883
Schweizerischer Romanistenverband, Luzern 10903
Schweizerische Sprachwissenschaftliche Gesellschaft, Basel 10930
Società Retorumantscha, Chur 10963
Société des Professeurs d'Allemand en Suisse Romande et Italienne, Genève 10970
Société Suisse des Professeurs de Français, Chancy 10986
Verein Schweizerdeutsch, Lachen 11024

Syria

L'Académie Arabe de Damas, Damascus 11036

Tanzania

Eastern African Centre for Research on Oral Traditions and African National Languages, Zanzibar 11054

Thailand

Thai-Bhara Cultural Lodge and Swami Satyananda Puri Foundation, Bangkok 11101

Turkey

Türk Dil Kurumu, Ankara 11154

United Kingdom

Academi Gymreig, Cardiff 11198
An Comunn Gaidhealach, Inverness 11229
Association for Language Learning, Rugby 11269
Association for the Reform of the Latin Teaching, Woking 11290
Association of Recognised English Language Services, London 11383
Basic English Foundation, London 11426
British Association for Applied Linguistics, Clevedon 11486
British Association of Academic Phoneticians, Glasgow 11499
British Association of Blind Esperantists, Middlesborough 11502
British Institute of Persian Studies, London 11586
British Interlingua Society, Sheffield 11592
British Universities Association of Slavists, Reading 11699
Cymdeithas yr Iaith Gymraeg, Aberystwyth 11875
English Association, London 11960
English Place-Name Society, Nottingham 11964
English Speaking Board (International), Southport 11965
English-Speaking Board (International), Southport 11966
English Spelling Association, London 11967
Esperanto-Asocio de Britio, London 11973
Esperanto Teachers Association, Leicester 11974
Fellowship of British Christian Esperantists, Hornchurch 12071
Gaelic League of Scotland, Glasgow 12097
The Greek Institute, London 12128
Institute of Linguists, London 12251
Institute of Translation and Interpreting, London 12288
International Association for Esperanto in Libraries, London 12323
International Association of Applied Linguistics, Edinburgh 12327
International Association of Teachers of English as a Foreign Language, Whitstable 12333
International Association of University Professors of English, Marazion 12335
International Federation for Modern Languages and Literatures, Belfast 12351
International Language Society of Great Britain, Barnsley 12371
International Language Union, Harrogate 12372
International Phonetic Association, Leeds 12384
Lakeland Dialect Society, Carlisle 12435
Lancashire Dialect Society, Stockport 12438
Linguistics Association of Great Britain, Glasgow 12453
London Welsh Association, London 12473
The Manx Gaelic Society, Douglas 12491
Names Society, Thames Ditton 12550
National Association for the Teaching of English, Sheffield 12564
National Association of Language Advisers, Rugby 12568
The Philological Society, London 12693
Quaker Esperanto Society, Morecambe 12732
Queen's English Society, Guildford 12733
Scottish Language Society, Inverness 12914
Scottish National Dictionary Association, Edinburgh 12916
Simplified Speeling Society, Beckenham 12952
Society for Italic Handwriting, London 12976
Society for the Study of Medieval Languages and Literature, Oxford 13019
Society of Teachers of Speech and Drama, Canterbury 13066
Yorkshire Dialect Society, Leeds 13232

Uruguay

Academia Nacional de Letras, Montevideo 13240

U.S.A.

American Association for Applied Linguistics, Oklahoma City 13467
American Association for Chinese Studies, Columbus 13473
American Association of Language Specialists, Washington 13546
American Association of Phonetic Sciences, Gainesville 13564
American Association of Professors of Yiddish, Flushing 13572
American Association of Teachers of Arabic, Provo 13585
American Association of Teachers of Esperanto, Santa Barbara 13586
American Association of Teachers of French, Champaign 13587
American Association of Teachers of German, Cherry Hill 13588
American Association of Teachers of Italian, Columbus 13589
American Association of Teachers of Spanish and Portuguese, Mississippi State 13590
American Association of Teachers of Turkish, Princeton 13591
American Catholic Esperanto Society, La Jolla 13679
American Classical League, Oxford 13691
American Council on the Teaching of Foreign Languages, Yonkers 13770
American Dialect Society, Jacksonville 13786
American Institute for Contemporary German Studies, Washington 13867
American Name Society, New York 13952
American Oriental Society, Ann Arbor 13968
American Society of Geolinguistics, Fairfield 14126
Association for Computational Linguistics, Morristown 14283
Association for the Study of Jewish Languages, Oakland Garden 14359
Association of Departments of English, New York 14427
Association of Departments of Foreign Languages, New York 14428
Association of Teachers of Japanese, Middlebury 14514
Basque Educational Organization, San Francisco 14570
Bet Nahrain, Modesto 14580
Better Education thru Simplified Spelling, Detroit 14581
Center for Applied Linguistics, Washington 14665
Center for Arab-Islamic Studies, Brattleboro 14667
Chinese-English Translation Assistance Group, Kensington 14738
Chinese Language Teachers Association, Kalamazoo 14740
College English Association, Rochester 14772
College Language Association, Atlanta 14773
Committee for the Implementation of the Standardized Yiddish Orthography, New York 14796
Conference on English Education, Urbana 14839
Conseil International d'Etudes Francophones, Upper Montclair 14844
Cross-Examination Debate Association, Northridge 14943
Esperantic Studies Foundation, Washington 15056
Esperanto League for North America, El Cerrito 15057
German-American Information and Education Association, Burke 15156
Hispanic Society of America, New York 15213
Histadruth Ivrith of America, New York 15214
Institute of General Semantics, Englewood 15323
Intercultural Development Research Association, San Antonio 15358
International Association for Learning Laboratories, Philadelphia 15386
International Catholic Esperanto Association, Champaign 15428
International Christian Esperanto Association, Levittown 15436
International Maledicta Society, Santa Rosa 15499
International Society for General Semantics, Concord 15530
International Society for Humor Studies, Tempe 15532
Jargon Society, Winston-Salem 15595
Junior Classical League, Oxford 15632
Linguistic Society of America, Washington 15674
Modern Language Association of America, New York 15750
National Association for Bilingual Education, Washington 15794
National Association of Professors of Hebrew, Madison 15882
National Assoiation of Self-Instructional Language Programs, Philadelphia 15917
National Committee for Latin and Greek, Cypress 15962
National Conference on Research in English, Amherst 15975
National Council of State Supervisors of Foreign Languages, Little Rock 16007
National Council of Teachers of English, Urbana 16010
National Federation of Modern Language Teachers Associations, Omaha 16047
National Forensic Association, Westerville 16054
Nieman Foundation, Cambridge 16206
Northeast Conference on the Teaching of Foreign Languages, South Burlington 16230
Reading is Fundamental, Washington 16396
Reading Reform Foundation, Tacoma 16397
Religious Speech Communication Association, Upland 16409
Semiotic Society of America, Davis 16462
Société des Professeurs Français et Francophones en Amérique, New York 16488
Society for New Language Study, Denver 16556
Society for Slovene Studies, Bloomington 16579
Society for the Preservation of English Language and Literature, Waleska 16604
Society of Basque Studies in America, Brooklyn 16640
Society of Federal Linguists, Washington 16663
Southern States Communication Association, Hattiesburg 16729
Speech Communication Association, Annandale 16741
Teachers of English to Speakers of Other Languages, Alexandria 16766
United States Branch of the International Committee for the Defense of the Breton Language, Jenkintown 16816
Women's Caucus for the Modern Languages, Baton Rouge 16915

Venezuela

Academia Venezolana de la Lengua, Caracas 16963
Asociación de Linguistica y Filologia de América Latina, Caracas 16968
Comité para la Defensa de las Lenguas Indígenas de América Latina y el Caribe, Caracas 16999

Yugoslavia

Akademia e Shkencave dhe e Arteve e Kosovës, Priština 17076
Društvo za Srpski Jezik i Književnost, Beograd 17090
Društvo za Srpskohrvatski Jezik i Književnost, Beograd 17091
Matica Srpska, Novi Sad 17104
Savez Društava za Strane Jezike i Knjizevnosti Jugoslavije, Beograd 17114

Literature s. a. Journalism; Linguistics

Albania

Lidhjy e Shkrimtareve dhe e Artisteve te Shqiperise, Tirana 00005
PEN Centre of Albania, Tirana 00006

Algeria

Union des Ecrivains Algériens, Alger 00025

Andorra

Cercle de les Arts i de les Lletres, Escaldes-Engordany 00029

Angola

União dos Escritores Angolanos, Luanda 00034

Argentina

Academia Argentina de Letras, Buenos Aires 00038
Asociación Argentina de la Cultura Inglesa, Buenos Aires 00067
Asociación Dante Alighieri, Buenos Aires 00076
Asociación Interamericana de Escritores, Buenos Aires 00084
Centro Lincoln, Buenos Aires 00111
Congreso Internacional de Americanistas, La Plata 00126
Instituto de Literatura, La Plata 00143
PEN Club Argentino, Buenos Aires 00151
Sociedad Argentina de Autores y Compositores de Música, Buenos Aires 00156
Sociedad Argentina de Escritores, Buenos Aires 00166
Sociedad General de Autores de la Argentina, Buenos Aires 00198
Sociedad Hebraica Argentina, Buenos Aires 00199

Australia

Australasian Universities Language and Literature Association, Sydney 00248
Australian and New Zealand Association for Medieval and Renaissance Studies, Sydney 00257
Australian Society of Authors, Strawberry Hills 00349
Fellowship of Australian Writers, Sydney 00395
PEN International (Sydney Centre), Woollahra 00445

Austria

Adalbert Stifter-Gesellschaft, Wien 00531
Arbeitsgemeinschaft für Deutschdidaktik, Klagenfurt 00550
Arbeitskreis der Wiener Altgermanisten, Wien 00564
Arbeitskreis der Wiener Skandinavisten, Wien 00565
Arthur Schnitzler-Institut, Wien 00569
Autorenkreis Linz, Linz 00573
Dokumentationsstelle für neuere österreichische Literatur, Wien 00583
Eranos Vindobonensis, Wien 00584
Friedrich Hebbel-Gesellschaft, Wien 00599
Gesellschaft für Klassische Philologie in Innsbruck, Innsbruck 00623
Gesellschaft zur Förderung Slawistischer Studien, Wien 00643
Grillparzer-Gesellschaft, Wien 00647
Innsbrucker Germanistische Arbeitsgemeinschaft, Innsbruck 00653
Institut für Österreichkunde, Wien 00658
Internationale Nestroy-Gesellschaft, Wien 00675
Internationales Institut für Jugendliteratur und Leseforschung, Wien 00681
Johann-Joseph-Fux-Gesellschaft, Graz 00691
Österreichische Exlibris-Gesellschaft, Wien 00741
Österreichische Gesellschaft für Amerikastudien, Salzburg 00753
Österreichische Gesellschaft für Literatur, Wien 00805
Österreichischer PEN-Club, Wien 00900
Österreichischer Schriftstellerverband, Wien 00901
Rudolf Kassner-Gesellschaft, Wien 00937
SIGMA – Salzburger Gesellschaft für Semiologie, Salzburg 00943
Verband der österreichischen Neuphilologen, Wien 00961
Verein der Mundartfreunde Österreichs, Wien 00978
Verein Wiener Frauenverlag, Wien 01004
Walter Buchebner Gesellschaft, Mürzzuschlag 01010
Wiener Goethe-Verein, Wien 01017
Wiener Humanistische Gesellschaft, Wien 01018

Bahrain

Bahrain Writers and Literators Association, Manama 01061

Bangladesh

Society of Arts, Literature and Welfare, Chittagong 01070

Belgium

Académie Européenne des Ecrivains Publics, Bruxelles 01091
Académie Royale de Langue et de Littérature Françaises, Bruxelles 01093
Académie Royale des Sciences, des Lettres et des Beaux-Arts de Belgique, Bruxelles 01095
Association des Ecrivains Belges de Langue Française, Bruxelles 01117
Centre International des Langues, Littératures et Traditions d'Afrique au Service du Développement, Louvain-la-Neuve 01195
Chambre Belge des Traducteurs, Interprètes et Philologues, Bruxelles 01210
European Association for the Promotion of Poetry, Leuven 01282
Icon, Leuven 01373
International PEN Club, Flemish Centre, Dilbeek 01400
International PEN Club, French Speaking Branch, Kraainem 01401
Koninklijke Academie voor Nederlandse Taal- en Letterkunde, Gent 01408
Koninklijke Academie voor Wetenschappen, Letteren en Schone Kunsten van België, Bruxelles 01409
Société Belge des Auteurs, Compositeurs et Editeurs, Bruxelles 01447
Société de Langue et de Littérature Wallonnes, Liège 01454
Société des Auteurs et Compositeurs Dramatiques, Bruxelles 01455
Union Belge des Journalistes et Ecrivains du Tourisme, Vilvoorde 01494

Bolivia

PEN Club de Bolivia-Centro Internacional de Escritores, La Paz 01547

Brazil

Academia Alagoana de Letras, Maceió 01560
Academia Amazonense de Letras, Manaus 01561
Academia Brasileira de Letras, Rio de Janeiro 01564
Academia Cachoeirense de Letras, Cachoeira de Itapemerim 01565
Academia Campinense de Letras, Campinas 01566
Academia Catarinense de Letras, Florianópolis 01567
Academia Cearense de Letras, Fortaleza 01568
Academia de Letras, Jõao Pessoa 01569
Academia de Letras da Bahia, Salvador 01570
Academia de Letras de Piauí, Teresina 01571
Academia de Letras e Artes do Planalto, Luziânia 01572
Academia Matogrossense de Letras, Cuiabá 01574
Academia Miniera de Letras, Belo Horizonte 01575
Academia Paraibana de Letras, João Pessoa 01578
Academia Paulista de Letras, São Paulo 01579
Academia Pernambucana de Letras, Recife 01580
Academia Riograndense de Letras, Pôrto Alegre 01581
Centro de Ciências, Letras e Artes, Campinas 01622
Fundação Moinho Santista, São Paulo 01659
Grémio Literario Carlos Ferreira, Amparo 01661
Grémio Literario e Comercial Portugués, Belém 01662
Instituto Archeológico, Histórico e Geográfico Pernambucano, Recife 01663
International Communication Agency, São Paulo 01694
PEN Clube do Brasil-Associação Universal de Escritores, Rio de Janeiro 01700

Bulgaria

Union of Bulgarian Writers, Sofia 01787

Canada

Académie des Lettres du Québec, Montréal 01814
African Literature Association, Edmonton 01821
Association for the Advancement of Scandinavian Studies in Canada, Guelph 01904
Association of Canadian Universities for Northern Studies, Ottawa 01926
Canadian Association for the Advancement of Netherlandic Studies, Windsor 02041
Canadian Association of Deans of Arts and Sciences, Charlottetown 02048
Canadian Association of Latin American Studies, Vancouver 02055
Canadian Authors Association, Ottawa 02079
Canadian Metric Association, Fonthill 02200
Canadian Research Society for Children's Literature, London 02244
Canadian Society of Children's Authors, Illustrators and Performers, Toronto 02277
Canadian Writers' Foundation, Ottawa 02319
Composers, Authors and Publishers Association of Canada, Toronto 02373
G.K. Chesterton Society, Saskatoon 02485
International Arthurian Society – North American Branch, Halifax 02538
Jane Austen Society of North America, North Vancouver 02567, 02568
Laubach Literacy of Canada (Prince Edward Island), Charlottetown 02577
League of Canadian Poets, Toronto 02584
The Malraux Society, Edmonton 02598
Ontario Library Association Literacy Guild, Toronto 02784
PEN International, Centre Québécois, Montréal 02812
Société des Auteurs, Eecherchistes, Documentalistes et Compositeurs, Montréal 02923
Société des Ecrivains Canadiens, Québec 02924
Society of Composers, Authors and Music Publishers of Canada, Don Mills 02944
Writers Development Trust, Toronto 02999

China, People's Republic

Inner Mongolian Academy of Social Sciences, Huhehot 03206

China, Republic

China National Association of Literature and the Arts, Taipei 03256
Chinese Center, International PEN, Taipei 03264
Chinese Women Writer's Association, Taipei 03290
National Young Writers Association of China, Taipei 03316
Playwriters Association of the Republic of China, Taipei 03320

Colombia

Instituto Caro y Cuervo, Bogotá 03377
Instituto Colombiano de Cultura, Bogotá 03383
Instituto Colombiano de Cultura Hispánica, Bogotá 03384
PEN Club de Colombia, Bogotá 03399

Croatia

Nakladni Zavod Matice Hrvatske, Zagreb 03464

Cuba

Centro Asturiano, La Habana 03472
Unión de Escritores y Artistas de Cuba, La Habana 03486

Czech Republic

Literárněvědná Společnost, Praha 03549
Matice Moravská, Brno 03551

Denmark

Dansk Forfatterforening, København 03622
Det Danske Shakespeare Selskab, København 03707
Det Danske Sprog- og Litteraturselskab, København 03708
DTL – Dansk Forening for Information og Dokumentation, Lyngby 03716
Filologisk-Historiske Samfund, København 03725
Islandske Litteratursamfund i København, Hillerød 03751
Samfund til Udgivelse af Gammel Nordisk Litteratur, Valby 03799
Selskab for Nordisk Filologi, København 03820

Ecuador

Academia Ecuatoriana de la Lengua, Quito 03839
Casa de la Cultura Ecuatoriana, Quito 03847

Egypt

Afro-Asian Writers' Permanent Bureau, Cairo 03864
Atelier, Alexandria 03882
High Council of Arts and Literature, Cairo 03904
High Council of Culture, Cairo 03905

Ethiopia

African Association of Science Editors, Addis Ababa 03935

Finland

Äidinkielen opettajain liitto, Helsinki 03948
Finlands Svenska Författareförening, Helsinki 03958
International Society for Folk-Narrative Research, Turku 03970
Kirjallisuudentutkijain Seura, Helsinki 03976
Klassillis-Filologinen Yhdistys, Helsingin Yliopisto 03978
Kotikielen Seura, Helsingin Yliopisto 03979
Suomalainen Tiedeakatemia, Helsinki 04000
Suomalaisen Kirjallisuuden Seura, Helsinki 04001
Suomen Kirjailijaliitto, Helsinki 04038
Svenska Litteratursällskapet i Finland, Helsinki 04067
Uusfilologinen Yhdistys, Helsingin Yliopisto 04081

France

Académie des Belles-Lettres, Sciences et Arts de La Rochelle, La Rochelle 04094
Académie des Inscriptions et Belles-Lettres, Paris 04095
Académie des Jeux Floraux, Toulouse 04096
Académie des Lettres, Sciences et Arts d'Amiens, Amiens 04097
Académie des Sciences, Agriculture, Arts et Belles-Lettres d'Aix, Aix-en-Provence 04099
Académie des Sciences, Arts et Belles-Lettres de Dijon, Dijon 04100
Académie des Sciences, Belles-Lettres et Arts de Clermont, Clermont-Ferrand 04101
Académie des Sciences, Belles-Lettres et Arts de Lyon, Lyon 04102
Académie des Sciences, Belles-Lettres et Arts de Rouen, Rouen 04103
Académie des Sciences, Belles-Lettres et Arts de Savoie, Chambéry 04104
Académie des Sciences et Lettres de Montpellier, Montpellier 04106
Académie des Sciences, Lettres et Arts d'Arras, Arras 04107
Académie des Sciences, Lettres et Arts de Marseille, Marseille 04108
Académie Française, Paris 04111
Académie Goncourt, Paris 04112
Académie Mallarmé, Paris 04114
Académie Montaigne, Sillé-le-Guillaume 04115
Académie Nationale des Sciences, Belles-Lettres et Arts de Bordeaux, Bordeaux 04120
Amici Thomae Mori, Angers 04132
Amis de Guy de Maupassant, Paris 04133
Amis de Rimbaud, Paris 04134
Association des Amis d'Alfred de Vigny, Paris 04163
Association des Amis de Miguel Angel Asturias, Paris 04164
Association des Ecrivains Combattants, Paris 04179
Association des Ecrivains de Langue Française, Paris 04180
Association Guillaume Budé, Paris 04255
Association Internationale de Littérature Comparée, Paris 04259
Association Internationale des Critiques Littéraires, Paris 04267
Association Internationale des Etudes Françaises, Paris 04273
Association Internationale des Professeurs de Langue et Littérature Russes, Paris 04274
Association Littéraire et Artistique Internationale, Paris 04299
Confédération Internationale des Sociétés d'Auteurs et Compositeurs, Paris 04438
Dickens Fellowship, Boulogne-sur-Mer 04465
European Association for American Studies, Lyon 04480
Fondation Saint-John Perse, Aix-en-Provence 04546
International Association of Sanskrit Studies, Paris 04582
Jeunesses Littéraires de France, Paris 04590
P.E.N. Maison Internationale, Paris 04620
Société Anatole France, Paris 04627
Société d'Agriculture, Sciences, Belles-Lettres et Arts d'Orléans, Orléans 04638
Société d'Emulation Historique et Littéraire d'Abbéville, Abbéville 04671
Société des Amis de Marcel Proust et des Amis d'Illiers-Combray, Paris 04695
Société des Anciens Textes Français, Paris 04699
Société des Auteurs et Compositeurs Dramatiques, Paris 04700
Société des Etudes Latines, Paris 04703
Société des Gens de Lettres de France, Paris 04707
Société des Lettres, Sciences et Arts de la Haute-Auvergne, Aurillac 04709
Société des Poètes Français, Paris 04712
Société des Sciences, Arts et Belles-Lettres de Bayeux, Bayeux 04714
Société d'Etude du Dix-Septième Siècle, Paris 04727
Société d'Etudes Dantesques, Nice 04729
Société d'Etudes Juives, Paris 04744
Société d'Etudes Romantiques, Clermont-Ferrand 04753
Société d'Histoire et d'Archéologie Le Vieux Montmartre, Paris 04764
Société Française d'Etude du Dix-Huitième Siècle, Pau 04850
Société Historique, Archéologique et Littéraire de Lyon, Lyon 04870
Société Internationale des Amis de Montaigne, Paris 04881
Société Italienne des Auteurs et Editeurs, Paris 04885
Société Racinienne, Neuilly-sur-Seine 04915
Syndicat des Ecrivains, Paris 04925
Syndicat National des Auteurs et Compositeurs de Musique, Paris 04938
Union Générale des Auteurs et Musiciens Professionnels, Paris 04981

Germany

Arbeitsgemeinschaft wissenschaftliche Literatur e.V., Frankfurt 05131
Arbeo-Gesellschaft e.V., Bachenhausen 05150
Bundesverband der Friedrich-Bödecker-Kreise e.V., Mainz 05244
Bund für Deutsche Schrift und Sprache, Ahlhorn 05263
Deutsche Akademie für Kinder- und Jugendliteratur e.V., Würzburg 05286
Deutsche Akademie für Sprache und Dichtung e.V., Darmstadt 05289
Deutsche Morgenländische Gesellschaft e.V., Heidelberg 05473
Deutscher Altphilologen-Verband, Puchheim 05497
Deutscher Autoren-Verband e.V., Hannover 05501
Deutscher Germanistenverband, Aachen 05513
Deutscher Philologen-Verband e.V., Unterhaching 05536
Deutsche Schillergesellschaft e.V., Marbach 05571
Deutsche Shakespeare-Gesellschaft, Weimar 05577
Deutsche Shakespeare-Gesellschaft West e.V., Bochum 05578
Deutsches Jugendinstitut e.V., München 05590
Eichendorff-Gesellschaft e.V., Ratingen 05637
Ernst Barlach Gesellschaft e.V., Hamburg 05642
E.T.A. Hoffmann-Gesellschaft e.V., Bamberg 05644
Europäische Märchengesellschaft e.V., Rheine 05654
European Authors' Association Die Kogge, Minden 05671
Freier Deutscher Autorenverband e.V., München 05760
Gesellschaft für Interkulturelle Germanistik e.V., Bayreuth 05832
Gesellschaft für Literatur in Nordrhein-Westfalen e.V., Münster 05837
Goethe-Gesellschaft in Weimar e.V., Weimar 05897
Gustav Freytag Gesellschaft e.V., Ratingen 05901
Hamburger Autorenvereinigung e.V., Hamburg 05906
Hölderlin-Gesellschaft, Tübingen 05930
Interessengemeinschaft deutschsprachiger Autoren, Eutin 05982
Landesarbeitsgemeinschaft Jugend und Literatur NRW e.V., Pulheim-Brauweiler 06095
Literarischer Verein in Stuttgart e.V., Stuttgart 06105
Literarisches Colloquium, Berlin 06106
Mommsen-Gesellschaft, Kiel 06119
PEN Zentrum Bundesrepublik Deutschland, Darmstadt 06232
Verein Nordfriesisches Institut e.V., Bredstedt 06423
Wilhelm-Busch-Gesellschaft e.V., Hannover 06442

Ghana

Ghana Association of Writers, Accra 06483

Greece

Association of Arts and Letters, Athinai 06500
Etaireia Byzantinon kai Metabyzantinon Meleton, Athinai 06523
Hetaireia Hellenon Philologon, Athinai 06534
Hetairia Hellinon Logotechnon, Athinai 06535
Hetairia Hellinon Thetricon Syngrapheon, Athinai 06536
PEN Centre, Athinai 06543
Women's Literary Society, Athinai 06548

Hungary

Magyar Irodalomtörténeti Társaság, Budapest 06638
Magyar Irók Szövetsége, Budapest 06639
Magyar PEN Club, Budapest 06652

Iceland

International PEN Centre, Reykjavik 06690
Islenzka bókmennatafélag, Reykjavik 06691

India

Jammu and Kashmir Academy of Art, Culture and Languages, Srinagar 06830
Madras Literary Society and Auxiliary of the Royal Asiatic Society, Madras 06834
PEN All-India Centre, Bombay 06850
Sahitya Akademi, New Dehli 06854

Indonesia

Pusat Pembinaan dan Pengembangan Bahasa, Jakarta 06894

Ireland

Apothecaries' Hall, Dublin 06923
Authors' Guild of Ireland, Dublin 06932
The Bram Stoker Society, Dublin 06934
Irish Academy of Letters, Dublin 06971
Irish PEN, Bray 06995

Israel

Association of Religious Writers, Jerusalem 07047
Hebrew Writers Association in Israel, Tel Aviv 07055
Society of Authors, Composers and Music Publishers in Israel, Tel Aviv 07112
Verband deutschsprachiger Schriftsteller in Israel, Tel Aviv 07116

Italy

Academia Gentium Pro Pace, Roma 07123
Academia Petrarca di Lettere, Arti e Scienze, Arezzo 07124
Accademia degli Euteleti, San Miniato 07137
Accademia degli Incamminati, Modigliana 07138
Accademia dei Filopatridi (Rubiconia), Savignano sul Rubicone 07144
Accademia di Paestum Eremo Italico, Mercato San Severino 07170
Accademia di Scienze, Lettere e Arti, Udine 07174
Accademia di Scienze, Lettere e Belle Arti degli Zelanti e dei Dafnici, Acireale 07175
Accademia di Scienze, Lettere ed Arti, Palermo 07176
Accademia Fulginia di Arti, Lettere, Scienze, Foligno 07184
Accademia Gli Amici Dei Sacri Lari, Bergamo 07187
Accademia Il Tetradramma, Roma 07188
Accademia Letteraria Italiana, Arcadia, Roma 07199
Accademia Ligure di Scienze e Lettere, Genova 07200
Accademia Nazionale di Scienze, Lettere e Arti, Modena 07222
Accademia Nazionale Virgiliana di Scienze, Lettere ed Arti di Mantova, Mantova 07224
Accademia Pomposiana, Codigoro 07227
Accademia Pontaniana, Napoli 07228
Accademia Prenestina del Cimento di Musica, Lettere, Scienze, Arti Visive e Figurative, Palestrina 07230
Accademia Romana di Cultura, Roma 07232
Accademia Roveretana degli Agiati, Rovereto 07234
Accademia Salentina di Lettere ed Arti, Lecce 07235
Accademia Scientifica, Letteraria, Artistica del Frignano Lo Scoltenna, Pievepelago 07236
Accademia Senese degli Intronati, Siena 07237
Accademia Tedesca Villa Massimo, Roma 07241
Accademia Toscana di Scienze e Lettere La Colombaria, Firenze 07243

Accademia Universale Citta' Eterna, Roma 07244
Accademia Universale Guglielmo Marconi, Roma 07245
Associazione dei Critici Letterari Italiani, Roma 07269
Associazione di Cultura Lao Silesu, Iglesias 07271
Associazione Internazionale di Poesia, Roma 07288
Associazione Italiana degli Slavisti, Pisa 07294
Associazione Siciliana per le Lettere e le Arti, Palermo 07402
Centro d'Arte e di Cultura, Bologna 07418
Centro Europeo per la Diffusione della Cultura, Roma 07463
Centro Studi di Poesia e di Storia delle Poetiche, Roma 07558
Centro Studi Russia Cristiana, Milano 07591
Centro Thomas Mann, Roma 07600
Ente Nazionale Francesco Petrarca, Padova 07655
Istituto Italiano di Studi Germanici, Roma 07741
Keats-Shelley Memorial Association, Roma 07759
P.E.N. International Centre, Roma 07763
Sindacato Nazionale Autori Drammatici, Roma 07766
Sindacato Nazionale Scrittori, Roma 07768
Società Dante Alighieri, Roma 07776
Società Dantesca Italiana, Firenze 07777
Società Dauna di Cultura, Foggia 07778
Società di Letture e Conversazioni Scientifiche, Genova 07783
Società Ecologica Friulana, Campoformido 07798
Società Filologica Friulana G.I. Ascoli, Udine 07806
Società Filologica Romana, Roma 07807
Società Internazionale di Psicologia della Scrittura, Milano 07816
Società Letteraria, Verona 07979
Società Nazionale di Scienze, Lettere ed Arti, Ex Società Reale, Napoli 07992
Società Torricelliana di Scienze e Lettere, Faenza 08024
Union Ladins Val Badia, Pedraces 08052

Jamaica

Association for Commonwealth Language and Literature Studies, Kingston 08067
International P.E.N. Club, Kingston 08081

Japan

Hanshin Doitsubungakukai, Kobe 08130
Nihon Chugoku Gakkai, Tokyo 08190
Nihon Dokubun Gakkai, Tokyo 08194
Nihon Eibungakkai, Tokyo 08196
Nihon Furansu-go Furansu-bun-gaku-kai, Tokyo 08199
Nihon PEN Kurabu, Tokyo 08247
Nippon Bungaku Kyokai, Tokyo 08282
Nippon Hikaku Bungakukai, Tokyo 08304
Nippon Rosiya Bungakkai, Tokyo 08360
Nippon Seiyo Koten Gakkai, Kyoto 08366
Toho Gakkai, Tokyo 08434
Waka Bungakkai, Tokyo 08444

Kenya

African Training Centre for Literacy and Adult Education, Nairobi 08474

Kuwait

Arab Center for Medical Literature, Kuwait 08544

Liberia

Society of Liberian Authors, Monrovia 08565

Lithuania

PEN Centre of Lithuania, Vilnius 08574

Macedonia

Društvo na Literaturnite Preveduvači na Makedonija, Skopje 08596
Društvo na Pisatelite na Makedonija, Skopje 08598
Sojuz na Društvata za Makedonski Jazik i Literatura, Skopje 08612

Mexico

Anuarios de Filosofía y Letras, México 08691
Centro Vasco, México 08723
International Communication Agency, Guadalajara 08742
International Communication Agency, México 08743
International Communication Agency, Monterrey 08744

Moldova, Republic of

PEN Centre of Moldova, Chişinău 08787

Netherlands

Auteursunie, Amsterdam 08830
Esperanto Writers' Association, Amsterdam 08847
European Association for Grey Literature Exploitation, 's-Gravenhage 08854
Maatschappij der Nederlandse Letterkunde, Leiden 08974
Nederlands Bibliotheek en Lektuur Centrum, 's-Gravenhage 08980
Nederlandse Vereniging van Bibliothecarissen, Documentalisten en Literatuuronderzoekers, Schelluinen 09005
Netherlands Centre of the International PEN, Maastricht 09059

New Zealand

P.E.N. New Zealand, Auckland 09187

Norway

Det Norske Samlaget, Oslo 09262
Norske Akademi for Sprog og Litteratur, Oslo 09295
Norske Forfatterforening, Oslo 09299
Norsk P.E.N., Oslo 09334

Pakistan

Academy of Letters, Islamabad 09356
Asia Foundation, Islamabad 09360
Baluchi Academy, Quetta 09361
Idarah-i-Yadgar-i-Ghalib, Karachi 09365
Institute of Islamic Culture, Lahore 09367
Iqbal Academy, Lahore 09370
National Book Council of Pakistan, Islamabad 09375
Pakistan Academy of Letters, Islamabad 09378
Pakistan Board for Advancement of Literature, Lahore 09381
Pashto Academy, Peshawar 09392
Punjabi Adabi Academy, Lahore 09394
Punjab Text Board, Lahore 09395
Sindhi Adabi Board, Tamshoro 09401
Urdu Academy, Bahawalpur 09404

Paraguay

Academia de la Lengua y Cultura Guaraní, Asunción 09424

Peru

Asociación Nacional de Escritores y Artistas, Lima 09453
Centro del PEN Internacional, Lima 09460

Poland

Polskie Towarzystwo Filologiczne, Warszawa 09566
Polskie Towarzystwo Neofilologiczne, Poznan 09593
Polski Klub Literacki P.E.N., Warszawa 09610
Stowarzyszenie Autorów ZAIKS, Warszawa 09614
Towarzystwo Literackie im. Adama Mickiewicza, Warszawa 09622

Portugal

Associação Portuguesa de Escritores, Lisboa 09642
Sociedade Martins Sarmento, Guimarães 09687
Sociedade Portuguesa de Autores, Lisboa 09692

Puerto Rico

Ateneo de Ponce, Ponce 09725
Ateneo Puertorriqueño, San Juan 09726
Congreso de Poesía de Puerto Rico, Mayaguez 09730
Sociedad Puertorriqueña de Escritores, San Juan 09732

Romania

Societatea de Stiinte Filologice din Romania, Bucuresti 09782
Uniunea Scriitorilor din Romania, Bucuresti 09798

Russia

Commission for Philology and Phonetics, Moskva 09826
Commission for the History of Philology, Moskva 09836
National Committee of Finno-Ugric Philologists, Moskva 09894
National Committee of Slavonic Philologists, Moskva 09899
National Committee of Turkish Philologists, Moskva 09904
Pushkin Commission, St. Petersburg 09947
Russian Association for Comparative Literature, Moskva 09955
Russian PEN Centre, Moskva 09962
Union of Russian Writers, Moskva 09989

Saudi Arabia

Higher Council for Promotion of Arts and Letters, Riyadh 10003
King Abdul Aziz Research Centre, Riyadh 10004

Slovakia

Kružok Moderných Filológov, Bratislava 10058
Slovenská Jednota Klassických Filológov, Bratislava 10076
Slovenská Literárnovedná Spoločnost, Bratislava 10078

Slovenia

Slavistično Društvo Slovenije, Ljubljana 10104
Slovenska Matica, Ljubljana 10106

South Africa

Classical Association of South Africa, Stellenbosch 10130
Federasie van Rapportryerskorpse, Auckland Park 10142
The Science Writers' Association of South Africa, Johannesburg 10172
South African PEN Centre, Claremont 10214

Spain

Academia de Buenas Letras de Barcelona, Barcelona 10238
Asociación de Escritores y Artistas Españoles, Madrid 10247
Ateneo Barcelonés, Barcelona 10283
Ateneo Científico, Literario y Artístico, Madrid 10284
Ateneo Científico, Literario y Artístico, Mahón 10285
Consejo General de Colegios Oficiales de Doctores y Licenciados en Filosofía y Letras y en Ciencias, Madrid 10297
Institución Fernando el Católico, Zaragoza 10312
Instituto Aula de Mediterráneo, Valencia 10317
Instituto de Filologia, Madrid 10322
PEN Club, Madrid 10336
Real Academia de Córdoba de Ciencias, Bellas Letras y Nobles Artes, Córdoba 10345
Real Academia Española, Madrid 10353
Real Academia Sevillana de Buenas Letras, Sevilla 10357
Seminario de Filología Vasca Julio de Urquijo, San Sebastián 10368
Sociedad de Ciencias, Letras y Artes El Museo Canario, Las Palmas 10372
Sociedad General de Autores de España, Madrid 10398

Sri Lanka

Royal Asiatic Society of Sri Lanka, Colombo 10418

Suriname

Stichting Cultureel Centrum Suriname, Paramaribo 10432

Sweden

Kungliga Vitterhets Historie och Antikvitets Akademien, Stockholm 10493
Litteraturfrämjandet, Stockholm 10497
Pennklubben, Stockholm 10514
Samfundet De Nio, Stockholm 10517
Strindbergssällskapet, Kungsängen 10527
Svenska Akademien, Stockholm 10530
Svenska Fornskriftsällskapet, Stockholm 10549
Svenska Litteratursällskapet, Uppsala 10562
Svenska Österbottens Litteraturförening, Krylbo 10569
Sveriges Författarförbund, Stockholm 10589
Tekniska Litteratursällskapet, Stockholm 10601

Switzerland

Collegium Romanicum, Lausanne 10648
Deutschschweizerisches PEN-Zentrum, Bern 10661
Fédération Internationale des Associations d'Etudes Classiques, Bellevue 10701
Gesellschaft für deutsche Sprache und Literatur in Zürich, Zürich 10713
Gruppe Olten, Schweizer Autorengruppe, Tägerwilen 10724
International Board on Books for Young People, Basel 10734
PEN Club de Suisse Romande, Carouge 10778
Schweizerische Akademische Gesellschaft der Anglisten, Genève 10791
Schweizerische Asiengesellschaft, Zürich 10798
Schweizerische Gesellschaft für Skandinavische Studien, Zürich 10846
Schweizerischer Altphilologen-Verband, Basel 10882
Schweizerischer Anglistenverband, Lugano 10883
Schweizerischer Romanistenverband, Luzern 10903
Schweizerischer Schriftstellerinnen- und Schriftsteller-Verband, Zürich 10904
Schweizerische Schillerstiftung, Zumikon 10925
Société Jean-Jacques Rousseau, Genève 10978

Syria

L'Académie Arabe de Damas, Damascus 11036

Tanzania

East African Literature Bureau, Dar es Salaam 11052

Thailand

Thai-Bhara Cultural Lodge and Swami Satyananda Puri Foundation, Bangkok 11101

Turkey

P.E.N. Yazarlar Derneği, Istanbul 11149

Uganda

The Uganda Society, Kampala 11188

United Kingdom

Academi Gymreig, Cardiff 11198
Anglo-Norman Text Society, London 11233
Arnold Bennett Literary Society, Stoke-on-Trent 11249
Association for Scottish Literary Studies, Aberdeen 11283
Association of British Science Writers, London 11316
Association of Hispanists of Great Britain and Ireland, London 11355
The Bantock Society, London 11421
Birmingham and Midland Institute, Birmingham 11446
Book Trust, London 11460
British Association for American Studies, Newcastle-upon-Tyne 11485
British Institute of Persian Studies, London 11586
British Science Fiction Association, Wantage 11644
British Society for Eighteenth Century Studies, Aberdeen 11654
British Universities Association of Slavists, Reading 11699
The Brontë Society, Keighley 11712
Browning Society of London, London 11713
The Burns Federation, Kilmarnock 11717
Byron Society, London 11719
Carlyle Society of Edinburgh, Edinburgh 11746
The Charles Lamb Society, Richmond 11757
Chester Society of Natural Science, Literature and Art, Chester 11774
The Classical Association, Cambridge 11792
Critics' Circle, London 11867
The Devonshire Association for the Advancement of Science, Literature and Art, Exeter 11890
Dickens Fellowship, London 11891
Diplomatic and Commonwealth Writers Association of Britain, London 11892
Dorothy L. Sayers Society, Hurstpierpoint 11901
Early English Text Society, Oxford 11915
Edinburgh Sir Walter Scott Club, Edinburgh 11939
English Association, London 11960
English Goethe Society, London 11962
Fellowship of Christian Writers, London 12072
The Francis Bacon Society, London 12090
George Eliot Fellowship, Coventry 12113
The Greek Institute, London 12128
Guild of Travel Writers, Reading 12137
Henry Williamson Society, Freshwater 12159
Henty Society, Cheltenham 12160
H.G. Wells Society, Moulton Park 12167
Hispanic and Luso-Brazilian Council, London 12170
Historical Newspaper Service, New Barnet 12174
Honourable Society of Cymmrodorion, London 12180
Horatian Society, London 12182
Housman Society, Bromsgrove 12185
International Federation for Modern Languages and Literatures, Belfast 12351
International P.E.N. Writers Association, London 12382
Irish Heritage, London 12411
Jane Austen Society, Alton 12415
Johnson Society, Lichfield 12420
Johnson Society of London, Hindhead 12421
Keats-Shelley Memorial Association, Leamington Spa 12426
Kent and Sussex Poetry Society, Hadlow 12427
The Kipling Society, Haslemere 12431
Leeds Philosophical and Literary Society, Leeds 12444
Lewis Carroll Society, Chislehurst Common 12448
Literary and Philosophical Society of Liverpool, Wallasey 12456
Literary and Philosophical Society of Newcastle upon Tyne, Newcastle-upon-Tyne 12457
Malone Society, Oxford 12480
Manchester Literary and Philosophical Society, Manchester 12484
Mervyn Peake Society, London 12523
Modern Humanities Research Association, London 12535
Nicholas Roerich Society, London 12627
Pali Text Society, London 12677
P.E.N. English Centre, London 12688
Poetry Society, London 12706
Poets' Theatre Guild, South Croydon 12707
The Powys Society, Kilmersdon 12714
Pushkin Club, London 12731
Queen's English Society, Guildford 12733
Richard Jefferies Society, Swindon 12756
Romantic Novelists' Association, Richmond 12759
The Royal Society of Edinburgh, Edinburgh 12858
Royal Society of Literature of the United Kingdom, London 12860
Scottish Association of Writers, Helensburgh 12900
Scottish Gaelic Texts Society, Glasgow 12909
Scottish Society for Northern Studies, Edinburgh 12929
Scottish Text Society, Edinburgh 12934
Shakespeare Authorship Society, London 12943
Shakespeare Birthplace Trust, Stratford-upon-Avon 12944
Shakespeare Institue, Stratford-upon-Avon 12945
Shakespeare Reading Society, London 12946
Shaviana, Dagenham 12947
Sherlock Holmes Society of London, Southsea 12949
Society for the Promotion of Hellenic Studies, London 13005
Society for the Promotion of Roman Studies, London 13008
Society for the Study of Medieval Languages and Literature, Oxford 13019
Society of Authors, London 13030
Sunday Shakespeare Society, East Grinstead 13108
The Tennyson Society, Lincoln 13118
Tolkien Society, Hove 13136
United Society for Christian Literature, Guildford 13170
Viking Society for Northern Research, London 13185
Virgil Society, London 13189
Welsh Library Association, Aberystwyth 13203
Wesley Historical Society, Birmingham 13205
William Morris Society, London 13208
Writers' Guild of Great Britain, London 13228

Uruguay

Asociación Uruguaya de Escritores, Montevideo 13248

U.S.A.

Academy of American Poets, New York 13282
American Academy and Institute of Arts and Letters, New York 13375
American Boccaccio Association, Saint Louis 13667
American Comparative Literature Association, Provo 13747
American Hobbit Association, Cincinnati 13848
American Holistic Medical Association, Raleigh 13849
American Literary Translators Association, Richardson 13919
American Medical Writers' Association, Bethesda 13940
American Oriental Society, Ann Arbor 13968
American Philological Association, Worcester 13997
American Society for Eighteenth-Century Studies, Logan 14058
American Studies Association, College Park 14185
American Tolkien Society, Highland 14197
AOIP, New York 14231
Armenian Literary Society, Demarest 14247

Asian Literature Division (of MLA), Bloomington 14252
Associated Business Writers of America, Aurora 14257
Association for the Study of Dada and Surrealism, Los Angeles 14357
Athenaeum of Philadelphia, Philadelphia 14545
Augustan Reprint Society, New York 14551
August Derleth Society, Uncasville 14553
Authors Guild, New York 14554
Authors League of America, New York 14555
Aviation/Space Writers Association, Columbus 14564
Baker Street Irregulars, Norwood 14568
The Bernard Shaw Society, New York 14578
Bertrand Russell Society, Skokie 14579
Bloomsday Club, New Brunswick 14603
Book Industry Study Group, New York 14607
Bram Stoker Memorial Association, New York 14612
Browning Institute, New York 14618
Burns Society of the City of New York, Smithtown 14625
Business Council for Effective Literacy, New York 14626
Byron Society, Collingswood 14630
Chautauqua Literary and Scientific Circle, Chautauqua 14723
Children's Literature Association, Battle Creek 14733
Christian Literacy Associates, Pittsburgh 14746
Christopher Morley Knothole Association, Roslyn 14749
Confederate Memorial Literary Society, Richmond 14821
Conference for Chinese Oral and Performing Literature, Hanover 14824
Conference on Christianity and Literature, Saint Petersburg 14836
Construction Writers Association, Aldie 14851
Contact Literacy Center, Linoln 14855
Council on National Literatures, Whitestone 14926
Dante Society of America, Cambridge 14955
D.H. Lawrence Society of North America, San Marcos 14970
Dickens Society, Worcester 14971
Edgar Allan Poe Society of Baltimore, Baltimore 15013
Education Writers Association, Washington 15030
Elizabeth Linington Society, Glassboro 15037
English Institute, Cambridge 15045
Eugene O'Neill Society, Kingston 15064
Evelyn Waugh Society, Garden City 15067
Ezra Pound Society, Orono 15074
F. Marion Crawford Memorial Sociey, Santa Clara 15107
Francis Bacon Foundation, Claremont 15135
Friends of George Sand, Hempstead 15143
Goethe Society of North America, Irvine 15166
Haiku Society of America, New York 15184
Harvey Society, New York 15190
Hemingway Society, Grand Forks 15202
Hispanic Society of America, New York 15213
Horatio Alger Society, Lansing 15229
Ibsen Society of America, Brooklyn 15244
Institute for Twenty-First Century Studies, Arlington 15307
Intermedia, New York 15361
International Arthur Schnitzler Research Association, San Bernardino 15374
International Brecht Society, Athens 15422
International Courtly Literature Society, East Lansing 15461
International Dostoevsky Society, Kensington 15467
International Galdos Association, Davis 15480
International Imagery Association, Bronx 15488
International John Steinbeck Society, Muncie 15493
International Society for Humor Studies, Tempe 15532
International Society for the Study of Expressionism, Cincinnati 15543
International Visual Literacy Association, Blacksburg 15571
International Wizard of Oz Club, Kinderhook 15574
Irish American Cultural Institute, Saint Paul 15585
Jack London Research Center, Glen Ellen 15589
James Branch Cabell Society, Alstead 15591
James Joyce Society, New York 15592
James Willard Schultz Society, New Bern 15593
Jesse Stuart Foundation, Ashland 15599
The Johnsonians, New Haven 15614
Kafka Society of America, Philadelphia 15634
Kate Greenaway Society, Norwood 15635
Keats-Shelley Association of America, New York 15636
Kipling Society of North America, Rockford 15638
Laubach Literacy Action, Syracuse 15649
Laubach Literacy International, Syracuse 15650
Laura Ingalls Wilder Memorial Society, Peplin 15652
Lessing Society, Cincinnati 15663
Lewis Carroll Society of North America, Silver Spring 15665
Literacy and Evangelism International, Tulsa 15675
Literacy Volunteers of America, Syracuse 15676
Longfellow Society of Sudbury and Wayside Inn, Sherborn 15677
Louisa May Alcott Memorial Association, Concord 15678
Mark Twain Association, New York 15694
Mark Twain Research Foundation, Perry 15695
Marlowe Society of America, Lubbock 15696
Marquandia Society, Williamsburg 15697
Medieval Academy of America, Cambridge 15714
Melville Society, Arlington 15716
Mencken Society, Baltimore 15717
Milton Society of America, Pittsburgh 15742
Modern Poetry Association, Chicago 15751
Mythopoeic Society, Altadena 15768
Nathaniel Hawthorne Society, Brunswick 15769
National Association for Applied Arts, Science and Education, Milwaukee 15792
National Association of Home and Workshop Writers, New York 15866
National Association of Professors of Hebrew, Madison 15882
National Coalition for Literacy, Washington 15953
National Federation of State Poetry Societies, Ogden 16049
National League of American PEN Women, Washington 16100
National Writers Club, Aurora 16183
New York Browning Society, New York 16200
New York C.S. Lewis Society, Jamaica 16201
Nockian Society, Fort Mitchell 16208
North American Dostoevsky Society, Lexington 16216
Outer Critics Circle, New York 16277
Panamerican Cultural Circle, Verona 16291
PEN American Center, New York 16301
P.G. Wodehouse Society, Santa Monica 16309
Philip Jose Farmer Society, Champaign 16311
Pirandello Society of America, Jamaica 16322
Poe Foundation, Richmond 16329
Poe Studies Association, Monmouth 16330
Poetry Society of America, New York 16331
Powys Society of North America, Valparaiso 16346
Proust Research Association, Lawrence 16365
Ralph Waldo Emerson Memorial Association, Boston 16393
R. Austin Freeman Society, Lexington 16395
Renaissance English Text Society, Amherst 16410
Research Society for Victorian Periodicals, Edwardsville 16417
Robinson Jeffers Committee, Long Beach 16422
Science Fiction Research Association, Westminster 16444
Scribes, Winston-Salem 16450
Seingalt Society, Salt Lake City 16459
Shakespeare Association of America, Dallas 16466
Shakespeare Data Bank, Evanston 16467
Shakespeare Society of America, West Hollywood 16468
Sherwood Anderson Society, Blacksburg 16469
Society for Slovene Studies, Bloomington 16579
Society for the Furtherance and Study of Fantasy and Science Fiction, Madison 16595
Society for the Preservation of English Language and Literature, Waleska 16604
Society for the Study of Southern Literature, New Orleans 16622
Society of Biblical Literature, Decatur 16642
Spenser Society, Seattle 16742
Thomas Wolfe Society, Richmond 16778
Thoreau Lyceum, Concord 16779
Thoreau Society, Concord 16780
Vachel Lindsay Association, Springfield 16854
Virginia Woolf Society, Washington 16864
Vladimir Nabokov Society, Lawrence 16865
Western Literature Association, Logan 16884
William Morris Society in the United States, Washington 16904
Wordsworth-Coleridge Association, Ann Arbor 16920
W.T. Bandy Center for Baudelaire Studies, Nashville 16938
Yuki Teikei Haiku Society of the United States and Canada, San Jose 16943
Zane Grey's West Society, Fort Wayne 16944

Venezuela
Asociación Cultural Humboldt, Caracas 16966
Asociación de Linguistica y Filologia de América Latina, Caracas 16968
Asociación Nacional de Escritores Venezolanos, Caracas 16970

Vietnam
Vietnamese Writers' Association, Hanoi 17070
Writers and Artists Union, Hanoi 17072

Yugoslavia
Akademia e Shkencave dhe e Arteve e Kosovës, Priština 17076
Društvo za Srpski Jezik i Književnost, Beograd 17090
Društvo za Srpskohrvatski Jezik i Književnost, Beograd 17091
Matica Srpska, Novi Sad 17104
Savez Društava za Strane Jezike i Knjizevnosti Jugoslavije, Beograd 17114

Zambia
Africa Literature Centre, Kitwe 17149

Zimbabwe
The Literature Bureau, Zimbabwe, Harare 17173
PEN Centre of Zimbabwe, Harare 17177
Zimbabwe Writers Union, Harare 17190

Logic

Belgium
Centre National de Recherches de Logique, Bruxelles 01203

Italy
Centro Superiore di Logica e Scienze Comparate, Bologna 07598

United Kingdom
Institute of Logistics, Corby 12252

U.S.A.
Association for Symbolic Logic, Urbana 14338

Logopedy

Austria
Österreichische Gesellschaft für Logopädie, Phoniatrie und Pädoaudiologie, Wien 00806

Belgium
Societas Logopedica Latina, Leuven 01420

Canada
British Columbia Association of Speech, Vancouver 01972
Canadian Hearing Society Foundation, Toronto 02154
Canadian Speech and Hearing Society, Edmonton 02299
Corporation Professionnelle des Orthophonistes et Audiologistes du Québec, Montréal 02397
Manitoba Speech and Hearing Association, Winnipeg 02637
Newfoundland Speech and Hearing Association, Saint John's 02711
The Ontario Speech and Hearing Association, Toronto 02800
Prince Edward Island Speech and Hearing Association, Charlottetown 02840
Saskatchewan Speech and Hearing Association, Regina 02904
Speech and Hearing Association of Alberta, Edmonton 02950
Speech and Hearing Association of Nova Scotia, Halifax 02951, 02952
Speech Foundation of Ontario, Toronto 02953

Denmark
Audiologopaedisk Forening, Kokkedal 03565
Dansk Selskab for Logopaedi og Foniatri, København 03677

France
Association pour la Rééducation de la Parole et du Langage Oral et Ecrit et de la Voix, Paris 04331

Germany
Deutsche Gesellschaft für Sprachheilpädagogik e.V., Berlin 05425
Deutsche Gesellschaft für Sprach- und Stimmheilkunde, Münster 05426

Italy
Associazione Italiana tra Foniatri e Logopedisti, Padova 07353

Norway
Norsk Logopedlag, Trondheim 09326

South Africa
South African Speech and Hearing Association, Braamfontein 10220

Spain
Asociación Española de Logopedia, Foniatria y Audiología, Barcelona 10263

United Kingdom
Action for Dysphasic Adults, London 11201
Association for All Speech Impaired Children, London 11258
Association for Brain Damaged Children, Coventry 11260
British Dyslexia Association, Reading 11553

U.S.A.
Academy of Aphasia, Los Angeles 13283
American Speech-Language-Hearing Association, Rockville 14178
National Association for Hearing and Speech Action, Rockville 15801
National Student Speech Language Hearing Association, Rockville 16166

Machine Engineering s. a. Automotive Engineering

China, People's Republic
Chinese Hydraulic Engineering Society, Beijing 03172

Czech Republic
Česká Společnost pro Mechaniku, Praha 03528

France
European Aggregate Association, Paris 04478

Germany
Deutsche Gesellschaft für Wirtschaftliche Fertigung und Sicherheitstechnik e.V., Kaarst 05441
Forschungsgesellschaft Druckmaschinen e.V., Frankfurt 05718
Forschungskuratorium Maschinenbau e.V., Frankfurt 05734
Normenausschuss Maschinenbau im DIN Deutsches Institut für Normung e.V., Frankfurt 06190
Normenausschuss Textil und Textilmaschinen im DIN Deutsches Institut für Normung e.V., Berlin 06209
Normenausschuss Werkzeugmaschinen im DIN Deutsches Institut für Normung e.V., Frankfurt 06219

Italy
Associazione Nazionale Disegno di Macchine, Palermo 07369

Japan
Kikai Shinko Kyokai Keizai Kenkyo-sho Shiryoshitsu, Tokyo 08165
Nihon Bearingu Kogyokai, Tokyo 08188
Nihon Sen'i Kikai Gakkai, Osaka 08258
Nippon Koku Gakkai, Tokyo 08338

Russia
I.I. Polzunov Science Production Association for Research and Design of Power Equipment, St. Petersburg 09874
National Committee on Automatic Control, Moskva 09905
National Committee on Theoretical and Applied Mechanics, Moskva 09907
Society of Light Industry, Moskva 09974
Society of the Instrument Building Industry, Moskva 09981

United Kingdom
British Hydromechanics Research Association, Bedford 11575
Industrial Locomotive Society, Woking 12200

Management s. Business Administration, Management

Marine Sciences s. Oceanography, Marine Sciences

Marketing s. a. Advertising

Argentina
Federación Lanera Argentina, Buenos Aires 00137

Austria
Verband der Marktforscher Österreichs, Wien 00959

Canada
Canadian Association of Marketing Research Organizations, Toronto 02058
Professional Marketing Research Society, Toronto 02847

Germany
Bundesvereinigung Logistik e.V., Bremen 05260
Deutsche Gesellschaft für Logistik e.V., Dortmund 05376
Deutsches Übersee-Institut, Hamburg 05604
Fördergemeinschaft für Absatz- und Werbeforschung, Frankfurt 05702
Gesellschaft für Konsum-, Markt- und Absatzforschung, Nürnberg 05836
GFM-GETAS Gesellschaft für Marketing-, Kommunikations- und Sozialforschung, Hamburg 05894
HWWA – Institut für Wirtschaftsforschung – Hamburg, Hamburg 05935
Institut für Angewandte Wirtschaftsforschung, Tübingen 05950
Institut für angewandte Wirtschaftsforschung im Mittelstand, Düsseldorf 05951
Rheinisch-Westfälisches Institut für Wirtschaftsforschung, Essen 06267
Vereinigung Getreide-, Markt- und Ernährungsforschung e.V., Bonn 06410

Italy
Associazione Italiana per gli Studi di Marketing, Milano 07333
Centro per gli Studi sui Mercati Esteri, Roma 07520
Centro Ricerche Economiche Sociologiche e di Mercato nell'Edilizia, Roma 07539
International Centre for Advanced Technical and Vocational Training, Torino 07709

Netherlands
European Federation of Associations of Market Research Organizations, Amsterdam 08886
European Society for Opinion and Marketing Research, Amsterdam 08892
Nederlands Instituut voor Marketing, Amsterdam 09054

South Africa
South African Market Research Association, Johannesburg 10204

Switzerland
Groupement Romand du Marketing, Lausanne 10723
Schweizerische Gesellschaft für Marketing, Zürich 10832
Verband Schweizerischer Marketing- und Sozialforscher, Hergiswil 11007

United Kingdom

U.S.A.

Mass Media

Austria

Belgium

Canada

Germany

Netherlands

Poland

Russia

United Kingdom

U.S.A.

Materials Science

Argentina

Australia

Belgium

Bosnia and Herzegowina

Canada

China, People's Republic

China, Republic

Denmark

Finland

France

Germany

Hungary

India

Ireland

Israel

Italy

Japan

Libya

Malaysia

Netherlands

New Zealand

Norway

Philippines

Portugal

South Africa

Spain

Sri Lanka

Switzerland

Syria

United Kingdom

U.S.A.

Mathematics

Argentina

Australia

Austria

Österreichische Mathematische Gesellschaft, Wien *00867*

Azerbaijan

Azerbaijan Mathematics Society, Baku *01037*

Belgium

Association Internationale des Mathématiques et Calculateurs en Simulation, Liège *01141*
Société Mathématique de Belgique, Bruxelles *01466*

Brazil

International Mathematical Union, Rio de Janeiro *01695*
Sociedade Paranaense de Matemática, Curitiba *01728*

Bulgaria

Scientific Information Centre at the Bulgarian Academy of Sciences, Sofia *01775*
Union of Bulgarian Mathematicians, Sofia *01786*

Canada

Association Mathématique du Québec, L'Assomption *01911*
Canadian Mathematical Society, Ottawa *02195*
Manitoba Association of Mathematics Teachers, Winnipeg *02606*

Chile

Asociación Latinoamericana de Biomatemática, Santiago *03020*

China, People's Republic

Chinese Mathematical Society, Beijing *03175*

China, Republic

Chinese Mathematical Society, Taipei *03276*

Colombia

Sociedad Colombiana de Matemáticas, Bogotá *03409*

Czech Republic

Jednota Českých Matematikû a Fysikû, Praha *03547*

Denmark

Dansk Mathematisk Forening, København *03645*
Matematiklaererforeningen, Birkerœd *03771*

Ethiopia

African Forum for Mathematical Ecology, Asmara *03936*

Finland

European Consortium for Mathematics in Industry, Lahti *03957*
Suomen Matemaattinen Yhdistys, Helsingin Yliopisto *04049*

France

Association des Professeurs de Mathématiques de l'Enseignement Publique, Paris *04196*
Comité National Français de Mathématiciens, Paris *04418*
Société Internationale de Biologie Mathématique, Antony *04876*
Société Mathématique de France, Paris *04890*
Société Nationale des Sciences Naturelles et Mathématiques, Cherbourg *04902*

Georgia

Scientific-Technical Council on Computer Technology, Mathematical Modelling, Automation of Scientific Research and Instrument Making, Tbilisi *05027*

Germany

Berliner Mathematische Gesellschaft e.V., Berlin *05188*
Deutsche Mathematiker-Vereinigung e.V., Freiburg *05466*
Gesellschaft für Angewandte Mathematik und Mechanik, Hamburg *05798*
Gesellschaft für Mathematik und Datenverarbeitung, Sankt Augustin *05839*
Gesellschaft für mathematische Forschung e.V., Oberwolfach *05840*
Mathematische Gesellschaft in Hamburg, Hamburg *06111*
Mathematisch-Naturwissenschaftlicher Fakultätentag, Hamburg *06112*
Vereinigung der unabhängigen freiberuflichen Versicherungs- und Wirtschaftsmathematiker in der Bundesrepublik Deutschland e.V., Grünwald *06401*

Greece

Elliniki Mathimatiki Eteria, Athinai *06510*

Hungary

Bolyai János Matematikai Társulat, Budapest *06597*

India

Allahabad Mathematical Society, Allahabad *06713*
Bharata Ganita Parisad, Lucknow *06741*
Calcutta Mathematical Society, Calcutta *06750*
Indian Association for the Cultivation of Science, Calcutta *06770*
Indian Mathematical Society, New Delhi *06800*

Israel

Israel Mathematical Union, Beer Sheva *07081*

Italy

Società Italiana di Scienze Fisiche e Matematiche Mathesis, Roma *07945*
Società per la Matematica e l'Economia Applicate, Roma *08001*
Unione Matematica Italiana, Bologna *08048*

Japan

Nippon Sugaku Kai, Tokyo *08386*
Nippon Sugaku Kyoiku Gakkai, Tokyo *08387*
Tensor Society, Chigasaki *08428*

Kenya

African Commission on Mathematics Education, Nairobi *08465*

Macedonia

Sojuz na Društvata na Matematičarite i Informatičarite na Makedonija, Skopje *08610*

Mexico

Centro de Investigación y de Estudios Avanzados del Instituto Politécnico Nacional, México *08718*
Centro Nacional de Cálcula, México *08720*
Sociedad Matemática Mexicana, México *08753*

Netherlands

Bernoulli Society for Mathematical Statistics and Probability, Voorburg *08833*
Centrum voor Wiskunde en Informatica, Amsterdam *08841*
Nederlandse Vereniging van Wiskundeleraren, 's-Gravenhage *09013*
Wiskundig Genootschap, Amsterdam *09110*

New Zealand

New Zealand Mathematical Society, Dunedin *09168*

Nigeria

African Mathematical Union, Ibadan *09217*

Norway

Norsk Matematisk Forening, Oslo *09328*

Poland

Polskie Towarzystwo Matematyczne, Warszawa *09585*

Romania

Latin Language Mathematicians' Group, Bucuresti *09753*
Societatea de Stiinte Matematice din Romania, Bucuresti *09786*

Russia

National Committee of Mathematicians, Moskva *09897*
Rersearch Council for Applied Mathematics, Moskva *09949*

Slovakia

Jednota Slovenských Matematikov a Fyzikov, Bratislava *10057*

Slovenia

Društvo Matematikov, Fizikov i Astronomov Slovenije, Ljubljana *10098*

South Africa

The South African Mathematical Society, Pretoria *10205*

Spain

Academia de Ciencias Exactas, Físicas, Químicas y Naturales de Zaragoza, Zaragoza *10239*
Real Sociedad Matemática Española, Madrid *10366*

Sweden

Lunds Matematiska Sällskap, Lund *10499*
Matematiska Föreningen, Uppsala *10500*
Svenska Matematikersamfundet, Umea *10564*

Switzerland

Schweizerische Mathematische Gesellschaft, Zürich *10872*

Tunisia

Conseil International sur les Mathématiques dans les Pays en Voie de Développement, Tunis *11141*

Turkey

Türk Sirfi ve Tatbiki Matematik Derneği, Istanbul *11173*

United Kingdom

Association of Teachers of Mathematics, Derby *11397*
British Society for the History of Mathematics, Milton Keynes *11670*
Comité de Liaison des Géomètres-Experts Européens, London *11800*
Commonwealth Association of Science and Mathematics Educators, London *11812*
Edinburgh Mathematical Society, Edinburgh *11935*
Glasgow Mathematical Association, Glasgow *12119*
Institute of Mathematics and its Applications, Southend-on-Sea *12258*
International Study Group for Mathematics Learning, London *12396*
London Mathematical Society, London *12462*
Mathematical Association, Leicester *12498*
Mathematical Instruction Subcommittee, London *12499*
Science and Engineering Research Council, Swindon *12889*
Time Series Analysis and Forecasting Society, Nottingham *13134*

U.S.A.

American Mathematical Association of Two Year Colleges, Flint *13928*
American Mathematical Society, Providence *13929*
Association for Women in Mathematics, College Park *14367*
Association of State Supervisors of Mathematics, Little Rock *14510*
Conference Board of the Mathematical Sciences, Washington *14823*
Dozenal Society of America, Garden City *14983*
Econometric Society, Evanston *15009*
Industrial Mathematics Society, Roseville *15259*
The Madison Project, New Brunswick *15685*
Mathematical Association of America, Washington *15703*
Math/Science Network, Oakland *15704*
National Council of Supervisors of Mathematics, Golden *16009*
National Council of Teachers of Mathematics, Reston *16011*
School Science and Mathematics Association, Bowling Green *16443*
Society for Industrial and Applied Mathematics, Philadelphia *16536*
Society for Risk Analysis, McLean *16577*
Special Interest Group for Symbolic and Algebraic Manipulation, Kent *16736*
Study Group for Mathematical Learning, Highland Park *16750*
United States Metric Association, Los Angeles *16826*
Women and Mathematics Education, South Hadley *16908*

Venezuela

Academia de Ciencias Físicas, Matemáticas y Naturales, Caracas *16958*

Vietnam

Vietnamese Association of Mathematics, Hanoi *17051*

Yugoslavia

Association of the Mathematicians', Physicists' and Astronomers' Societies of Yugoslavia, Priština *17079*
Društvo Matematičara Srbije, Beograd *17087*
Savez Društava Matematičara, Fizičara i Astronoma Jugoslavije, Novi Sad *17112*

Medicine

Algeria

Union Médicale Algérienne, Alger *00026*

Argentina

Academia de Ciencias Médicas, Córdoba *00039*
Academia Nacional de Medicina, Buenos Aires *00051*
Asociación Internacional de Hidatidología, Buenos Aires *00085*
Asociación Médica Argentina, Buenos Aires *00090*
Círculo Médico de Córdoba, Córdoba *00113*
Círculo Médico de Rosario, Rosario *00114*
Confederación Sudamericana de Medicina del Deporte, Buenos Aires *00124*
Sociedad Argentina de Investigación Clínica, Buenos Aires *00173*

Australia

Association of Medical Directors of the Australian Pharmaceutical Industry, West Ryde *00230*
Australasian College of Physical Scientists and Engineers in Medicine, Melbourne *00241*
Australian Federation for Medical and Biological Engineering, Parkville *00291*
Australian Institute of Homoeopathy, Roseville *00302*
Australian Medical Association, Barton *00322*
Australian Postgraduate Federation in Medicine, Camperdown *00330*
Australian Society for Medical Research, Sydney *00341*
International Association of Trichologists, Sydney *00424*
Medical Foundation, Sydney *00431*
Medical Society of Victoria, Parkville *00432*
National Health and Medical Research Council, Camberra *00438*
Royal Australasian College of Physicians, Sydney *00456*
Royal Australian College of General Practitioners, Sydney *00462*
Royal College of Nursing – Australia, Melbourne *00466*
Sydney University Medical Society, Sydney *00502*
Victorian Society of Pathology and Experimental Medicine, Parkville *00518*

Austria

Ärztegemeinschaft im Katholischen Akademieverband der Erzdiözese Wien, Wien *00532*
Ärztegesellschaft Innsbruck, Innsbruck *00533*
Ärztekammer für Kärnten, Klagenfurt *00534*
Ärztekammer für Niederösterreich, Wien *00535*
Ärztekammer für Wien, Wien *00536*
Akademie für Allgemeinmedizin, Graz *00540*
Arbeitsgemeinschaft für theoretische und klinische Leistungsmedizin der Hochschullehrer Österreichs, Graz *00557*
Arbeitsgemeinschaft zur Erforschung der Ärztlichen Allgemeinpraxis, Brunn *00563*
Gesellschaft der Ärzte in Vorarlberg, Dornbirn *00603*
Gesellschaft der Ärzte in Wien, Wien *00604*
Gesellschaft zur Errichtung der Akademie für Allgemeinmedizin, Graz *00641*
Internationale Paracelsus-Gesellschaft, Salzburg *00676*
Medizinische Gesellschaft für Oberösterreich, Linz *00706*
Naturwissenschaftlich-medizinischer Verein in Innsbruck, Innsbruck *00714*
Naturwissenschaftlich-Medizinische Vereinigung in Salzburg, Salzburg *00715*
Österreichische Ärztekammer, Wien *00721*
Österreichische Ethnomedizinische Gesellschaft, Wien *00740*
Österreichische Gesellschaft für Akupunktur und Aurikulotherapie, Wien *00751*
Österreichische Gesellschaft für Allgemeinmedizin, Klagenfurt *00752*
Österreichische Gesellschaft für Angiologie, Wien *00757*
Österreichische Gesellschaft für Arbeitsmedizin, Hall *00758*
Österreichische Gesellschaft für Biomedizinische Technik, Wien *00766*
Österreichische Gesellschaft für Hygiene, Mikrobiologie und Präventivmedizin, Wien *00793*
Österreichische Gesellschaft für Laboratoriumsmedizin, Wien *00804*
Österreichische Gesellschaft für Medizinsoziologie, Wien *00808*
Österreichische Gesellschaft für Nephrologie, Linz *00812*
Österreichische Gesellschaft zur Förderung medizin-meteorologischer Forschung in Österreich, Wien *00857*
Österreichische wissenschaftliche Gesellschaft für Prophylaktische Medizin und Sozialhygiene, Wien *00929*
Salzburger Ärztegesellschaft, Salzburg *00938*
Verband Österreichischer Kurärzte, Wien *00970*
Verband Österreichischer Sportärzte, Wien *00972*
Vereinigung Österreichischer Ärzte, Wien *00996*
Wiener Medizinische Akademie für Ärztliche Fortbildung und Forschung, Wien *01024*
Wissenschaftliche Ärztegesellschaft Innsbruck, Innsbruck *01030*
Wissenschaftliche Arbeitsgemeinschaft für Leibeserziehung und Sportmedizin in Innsbruck, Innsbruck *01031*
Wissenschaftliche Gesellschaft der Ärzte in der Steiermark, Graz *01032*

Bahrain

Bahrain Medical Society, Manama *01058*

Barbados

Barbados Association of Medical Practitioners, Saint Michael *01072*

Belgium

Académie Royale de Médecine de Belgique, Bruxelles *01094*
Académie Royale des Sciences d'Outre-Mer, Bruxelles *01096*
Alzheimer Europe, Bruxelles *01099*
Association des Sociétés Scientifiques Médicales Belges, Bruxelles *01118*
Association for the Epidemiological Study and Assessment of Disasters in Developing Countries, Bruxelles *01134*
Association Médicale Européenne, Bruxelles *01147*
Benelux Phlebology Society, Bruxelles *01167*
Bureau International d'Audiophonologie, Bruxelles *01173*
Centre for Research on the Epidemiology of Disasters, Bruxelles *01189*
Collège Européen d'Hygiène et de Médecine Naturelles, Enghien *01216*
Comité d'Associations Européennes de Médecins Catholiques, Bruxelles *01219*
Comité International de Médecine Militaire, Liège *01228*
Conseil National de l'Ordre des Médecins, Bruxelles *01253*
European Association of Podologists, Bruxelles *01295*
European Committee for Treatment and Research in Multiple Sclerosis, Melsbroek *01323*
European Council of Integrated Medicine, Bruxelles *01335*
Fédération Belge des Chambres Syndicales de Médecins, Braine l'Allend *01355*
Fédération Européenne de Médecine Physique et Réadaption, Gent *01357*
Fondation Médicale Reine Elisabeth, Bruxelles *01366*
Groupement Belge des Omnipraticiens, Bruxelles *01369*
Groupement des Unions Professionnelles Belges de Médecins Spécialistes, Bruxelles *01370*
Koninklijke Academie voor Geneeskunde van België, Bruxelles *01407*
Société Belge d'EMG, Bruxelles *01434*
Société Internationale pour la Recherche sur les Maladies de Civilisation et sur l'Environnement, Bruxelles *01464*
Société Royale des Sciences Médicales et Naturelles de Bruxelles, Bruxelles *01488*
Union Européenne des Médecins Spécialistes, Bruxelles *01496*
Union Internationale des Services Médicaux des Chemins de Fer, Bruxelles *01498*

Bermuda

Bermuda Medical Society, Paget *01521*

Bolivia

Ateneo de Medicina de Sucre, Sucre *01534*

Brazil

Academia de Medicina de São Paulo, São Paulo *01573*

Academia Nacional de Medicina, Rio de Janeiro 01577
Associação Bahiana de Medicina, Salvador 01586
Associação Médica Brasileira, São Paulo 01603
Associação Pananericana de Medicina Social, Rio de Janeiro 01604
Associação Paulista de Medicina, São Paulo 01606
Centro de Biomédica de Campina Grande, Campina Grande 01621
Confederación Panamericana de Medicina Deportiva, Porto Alegre 01641
Sociedade de Medicina de Alagoas, Maceió 01724

Bulgaria

Medical Academy, Sofia 01768
Society of Bulgarian Anatomists, Histologists and Embryologists, Sofia 01777
Society of Sport Medicine, Sofia 01782
Union of Scientific Medical Societies in Bulgaria, Sofia 01798

Burkina Faso

Centre Muraz, Bobo-Dioulasso 01807
Coordination and Cooperation Organization for the Control of the Major Endemioc Diseases, Bobo-Dioulasso 01809

Canada

Academy of Medicine, Toronto 01816
Acupuncture Foundation of Canada, Markham 01818
Alberta Heritage Foundation for Medical Research, Edmonton 01836
Alberta Medical Council, Calgary 01840
American Osler Society, Hamilton 01862
Association des Médecins de Langue Française du Canada, Montréal 01888
Association du Syndrôme de Turner du Québec, Boucherville 01897
Association Médicale du Québec, Montréal 01912
Association Mondiale des Médecins Francophones, Ottawa 01913
Association of Canadian Medical Colleges, Ottawa 01925
Banting Research Foundation, Toronto 01963
British Columbia Medical Association, Vancouver 01982
British Columbia Parkinsons's Disease Association, Vancouver 01985
Canadian Academy of Medical Illustrators, Toronto 02004
Canadian College of Medical Geneticists, Calgary 02099
Canadian Medical and Biological Engieneering Society, Ottawa 02196
Canadian Medical Association, Ottawa 02197
Canadian Osteogenesis Imperfecta Society, Toronto 02216
Canadian Osteopathic Aid Society, London 02217
Canadian Osteopathic Association, London 02218
Canadian Paraplegic Association, Toronto 02220
Canadian Society for Clinical Investigation, Montréal 02253
Canadian Society of Electroencephalographers, Electromyographers and Clinical Neurophysiologists, Sherbrooke 02282
Centre de Bioéthique, Montréal 02327
Christian Medical Society, Islington 02347
Civil Aviation Medical Association, Ottawa 02350
Clinical Research Society of Toronto, Toronto 02351
College of Family Physicians of Canada, Willowdale 02354
College of Physicians and Surgeons of New Brunswick, Saint John 02358
Collegium Internationale Allergologicum, Hamilton 02362
Defence Medical Association of Canada, Ottawa 02413
Dystonia Medical Research Foundation, Vancouver 02419
Federation of Medical Women of Canada, Ottawa 02458
Hamilton Academy of Medicine, Hamilton 02492
Harold Crabtree Foundation, Ottawa 02495
International Federation for Medical and Biological Engineering, Ottawa 02553
Manitoba Medical Association, Winnipeg 02628
Medical Council of Canada, Ottawa 02643
Medical Council of New Brunswick, Saint John 02644
Medical Council of Prince Edward Island, Charlottetown 02645
Medical Research Council of Canada, Ottawa 02646
Medical Society of Nova Scotia, Halifax 02647
Medical Society of Prince Edward Island, Charlottetown 02648
Medical Staff Council, Winnipeg 02649
Medical Symposium Association of Canada, Edmonton 02650
New Brunswick Medical Society, Fredericton 02688
Newfoundland Medical Association, Saint John's 02707
Newfoundland Medical Board, Saint John's 02708
Newfoundland Medical Council, Saint John's 02709
Nova Scotia Medical Council, Halifax 02737
Occupational Medical Association of Canada, Toronto 02752
Prince Edward Island Medical Society, Charlottetown 02835
Provincial Medical Board of Nova Scotia, Halifax 02850
Saskatchewan Amyotrophic Lateral Sclerosis Society Foundation, Radville 02878
Saskatchewan Medical Association, Saskatoon 02895
Saskatchewan Society of Medical Laboratory Technologists, Saskatoon 02901
Sport Medicine Council of Canada, Vanier 02957

Chile

Academia Chilena de Medicina, Santiago 03012
Sociedad Médica de Concepción, Concepción 03098
Sociedad Médica de Santiago, Santiago 03099
Sociedad Médica de Valparaíso, Valparaíso 03100

China, People's Republic

China Academy of Traditional Chinese Medicine, Beijing 03127
China Association of Traditional Chinese Medicine, Beijing 03129
Chinese Abacus Association, Beijing 03135
Chinese Academy of Medical Sciences, Beijing 03140
Chinese Association of Integrated Traditional and Western Medicine, Beijing 03164
Chinese Medical Society, Beijing 03178

China, Republic

Chinese Medical Association, Taipei 03277
Chinese Medical Woman's Association, Taipei 03279

Colombia

Academia Nacional de Medicina, Bogotá 03342
Asociación Colombiana de Facultades de Medicina, Bogotá 03344
Instituto Central de Medicina Legal, Bogotá 03378
Instituto Nacional de Medicina Legal, Bogotá 03394

Costa Rica

Confederación Centroamericana de Medicina del Deporte, Heredia 03443

Croatia

Croatian Medical Association, Zagreb 03454

Cuba

Centro Nacional de Información de Ciencias Médicas, La Habana 03477
Sociedad Cubana de Historia de la Medicina, La Habana 03482

Cyprus

Association of Sports Medicine of the Balkan, Nicosia 03487
Pancyprian Medical Association, Nicosia 03498

Czech Republic

Česká Lékařská Společnost J.E. Purkyně, Praha 03509

Denmark

Danske Interne Medicineres Organisation, København 03607
Dansk Industrimedicinsk Selskab, Charlottenlund 03634
Dansk Medicinsk-Historisk Selskab, København 03646
Dansk Medicinsk Selskab, København 03647
Dansk Medikoteknisk Selskab, Holte 03648
Dansk Selskab for Almen Medicin, Greve 03672
Det Medicinske Selskab i København, København 03714
Foreningen af Speciallåger, København 03734
Landsforeningen for Polio-, Trafik- og Ulykkesskadede, Hellerup 03764
Nordic Federation for Medical Education, København 03778

Dominican Republic

Asociación Médica de Santiago, Santiago de los Caballeros 03833
Asociación Médica Dominicana, Santo Domingo 03834

Ecuador

Academia Ecuatoriana de Medicina, Quito 03840
Centro Médico Federal del Azuay, Cuenca 03850
Federación Nacional de Médicos del Ecuador, Quito 03853

Egypt

Alexandria Medical Association, Alexandria 03866
Arab Scientific Advisory Committee for Blood Transfer, Cairo 03876
Association for Medical Education in the Middle East, Alexandria 03879
Association of Medical Schools in the Middle East, Alexandria 03881
Egyptian Medical Association, Cairo 03893

El Salvador

Colegio Médico de El Salvador, San Salvador 03927

Ethiopia

Ethiopian Medical Association, Addis Ababa 03944

Finland

Finska Läkaresällskapet, Helsinki 03961
Lääketieteellisen Fysiikan ja Tekniikan Yhdistys, Tampere 03980
Suomalainen Lääkäriseura Duodecim, Helsinki 03997

France

Académie Nationale de Médecine, Paris 04118
Administration Universitaire Francophone et Européenne en Médecine et Odontologie, Paris 04123
Association de Médecine Rurale, Paris 04152
Association de Médecine Urbaine, Paris 04153
Association des Centres Médicaux de la Seine, Paris 04172
Association des Epidémiologists de Langue Française, Le Kremlin-Bicêtre 04181
Association Française des Femmes Médecins, Paris 04229
Association Française Inter Médicale, Paris 04240
Association Générale des Médecins de France, Paris 04254
Association Internationale de Médecine et de Biologie de l'Environnement, Paris 04260
Association Interprofessionnelle des Centres Médicaux et Sociaux de la Région Parisienne, Paris 04295
Association Médico-Sociale Protestante de Langue Française, Alfortville 04303
Association pour le Développement des Relations Médicales entre la France et les Pays Etrangers, Paris 04339
Confédération des Syndicats Médicaux Français, Paris 04435
Confédération Internationale des Associations de Médecines Alternatives Naturelles, Pierrelatte 04437
Conférence Internationale des Doyens des Facultés de Médecine d'Expression Française, Tours 04441
European Association of Museums of the History of Medical Sciences, Denicé 04488
Fédération des Médecins de France, Paris 04514
Fédération Hospitalière de France, Paris 04529
Groupement des Bureaux Médicaux, Paris 04554
Groupement d'Etudes et de Réalisations Médicales, Paris 04555
Groupement Médical Saint-Augustin, Paris 04558
Médecins sans Frontières, Paris 04600
Ordre National des Médecins, Paris 04611
Société de Médecine, Chirurgie et Pharmacie de Toulouse, Toulouse 04666
Société de Médecine de Strasbourg, Strasbourg 04667
Société de Médecine et de Chirurgie de Bordeaux, Bordeaux 04668
Société de Médecine Légale et de Criminologie de France, Paris 04670
Société de Neurophysiologie Clinique de Langue Française, Paris 04675
Société de Réanimation de Langue Française, Caen 04685
Société des Médecins-Chefs des Compagnies Européennes d'Aviation, Paris 04710
Société Française de Médecine Aérospatiale, Bretigny 04807
Société Française de Médecine du Sport, Paris 04808
Société Française de Médecine du Trafic, Paris 04809
Société Française de Médecine Générale, Paris 04811
Société Française de Phlébologie, Paris 04831
Société Française d'Histoire de la Médecine, Reims 04856
Société Française d'Hygiène, de Médecine Sociale et de Génie Sanitaire, Vandoeuvre-lès-Nancy 04859
Société Médicale des Hôpitaux de Paris, Paris 04891
Société Médico-Psychologique, Boulogne 04895
Syndicat National des Médecins Biologistes, Paris 04949
Syndicat National des Médecins de Groupe, Paris 04950
Syndicat National des Médecins des Hôpitaux Publics, Paris 04951
Syndicat National des Médecins du Sport, Paris 04952
Union des Biologistes de France, Paris 04973
Union Internationale de Phlébologie, Paris 04984
Union Nationale des Médecins Spécialistes Confédérés, Paris 04992
World Medical Association, Ferney-Voltaire 05002

Georgia

Georgian Bio-Mecico-Technical Society, Tbilisi 05012

Germany

Ärztekammer Berlin, Berlin 05030
Ärztekammer Bremen, Bremen 05031
Ärztekammer des Saarlandes, Saarbrücken 05032
Ärztekammer Frankfurt, Frankfurt 05034
Ärztekammer Hamburg, Hamburg 05035
Ärztekammer Land Brandenburg, Cottbus 05036
Ärztekammer Mecklenburg-Vorpommern, Rostock 05037
Ärztekammer Niedersachsen, Hannover 05038
Ärztekammer Nordrhein, Düsseldorf 05039
Ärztekammer Sachsen-Anhalt, Magdeburg 05040
Ärztekammer Schleswig-Holstein, Bad Segeberg 05041
Ärztekammer Westfalen-Lippe, Münster 05042
Arbeitskreis Ethnomedizin, Hamburg 05138
Baden-Württembergischer Sportärzteverband, Bezirksgruppe Süd-Baden, Freiburg 05162
Bayerische Akademie für Arbeits- und Sozialmedizin, München 05170
Bayerische Landesärztekammer, München 05174
Bayerischer Sportärzteverband e.V., München 05180
Berliner Medizinische Gesellschaft, Berlin 05189
Berliner Sportärztebund e.V., Berlin 05191
Berufsverband der Praktischen Ärzte und Ärzte für Allgemeinmedizin Deutschlands e.V., Köln 05201
Bezirksärztekammer Koblenz, Koblenz 05210
Bezirksärztekammer Nordwürttemberg, Stuttgart 05211
Bezirksärztekammer Pfalz, Neustadt 05212
Bezirksärztekammer Trier, Trier 05213
Bezirkszahnärztekammer Koblenz, Koblenz 05214
Bezirkszahnärztekammer Pfalz, Ludwigshafen 05215
Bezirkszahnärztekammer Rheinhessen, Mainz 05216
Bezirkszahnärztekammer Trier, Trier 05217
Bremer Sportärztebund, Bremen 05224
Bundesärztekammer, Köln 05229
Bundesverband der Vertrauens- und Rentenversicherungsärzte, Aurich 05247
Bundesverband Deutscher Ärzte für Naturheilverfahren e.V., München 05249
Deutsche Akademie der Naturforscher Leopoldina, Halle 05285
Deutsche Akademie für medizinische Fortbildung, Kassel 05287
Deutsche EEG-Gesellschaft, Berlin 05300
Deutsche Gesellschaft für Aesthetische Medizin, Berlin 05313
Deutsche Gesellschaft für Anästhesiologie und Intensivmedizin, Nürnberg 05319
Deutsche Gesellschaft für Biomedizinische Technik e.V., Berlin 05331
Deutsche Gesellschaft für Geschichte der Medizin, Naturwissenschaft und Technik e.V., München 05355
Deutsche Gesellschaft für Laboratoriumsmedizin e.V., Düsseldorf 05374
Deutsche Gesellschaft für Luft- und Raumfahrtmedizin e.V., Ulm 05378
Deutsche Gesellschaft für Manuelle Medizin e.V., Boppard 05380
Deutsche Gesellschaft für Medizinische Informatik, Biometrie und Epidemiologie, Köln 05382
Deutsche Gesellschaft für Parasitologie e.V., Marburg 05395
Deutsche Gesellschaft für Suchtforschung und Suchttherapie e.V., Hamm 05429
Deutsche Hauptstelle gegen die Suchtgefahren e.V., Hamm 05454
Deutsche Multiple Sklerose Gesellschaft Bundesverband e.V., Hannover 05475
Deutscher Ärztinnenbund e.V., Köln 05494
Deutscher Kassenarztverband e.V., Gross-Gerau 05522
Deutscher Naturheilbund e.V., Crailsheim 05531
Deutscher Sportärztebund, Heidelberg 05539
Deutscher Verband Technischer Assistenten in der Medizin e.V., Hamburg 05554
Deutsches Institut für Ärztliche Mission, Tübingen 05581
Deutsches Institut für medizinische Dokumentation und Information, Köln 05586
Fachverband Deutscher Heilpraktiker e.V., Bonn 05694
Gesellschaft anthroposophischer Ärzte, Stuttgart 05788
Gesellschaft der Ärzte für Erfahrungsheilkunde e.V., Heidelberg 05789
Gesellschaft Deutscher Naturforscher und Ärzte e.V., Leverkusen 05795
Gesellschaft für Epilepsieforschung e.V., Bielefeld 05814
Gesellschaft für Informationsvermittlung und Technologieberatung, München 05831
Gesellschaft für sozialwissenschaftliche Forschung in der Medizin, Freiburg 05861
Immuno, Heidelberg 05937
Kassenärztliche Bundesvereinigung, Köln 06021
Kassenärztliche Vereinigung Bayerns, München 06022
Kassenärztliche Vereinigung Berlin, Berlin 06023
Kassenärztliche Vereinigung Bremen, Bremen 06024
Kassenärztliche Vereinigung Hamburg, Hamburg 06025
Kassenärztliche Vereinigung Hessen, Frankfurt 06026
Kassenärztliche Vereinigung Koblenz, Koblenz 06027
Kassenärztliche Vereinigung Niedersachsen, Hannover 06028
Kassenärztliche Vereinigung Nordbaden, Karlsruhe 06029
Kassenärztliche Vereinigung Nordrhein, Düsseldorf 06030
Kassenärztliche Vereinigung Nord-Württemberg, Stuttgart 06031
Kassenärztliche Vereinigung Pfalz, Neustadt 06032
Kassenärztliche Vereinigung Rheinhessen, Mainz 06033
Kassenärztliche Vereinigung Saarland, Saarbrücken 06034
Kassenärztliche Vereinigung Schleswig-Holstein, Bad Segeberg 06035
Kassenärztliche Vereinigung Südbaden, Freiburg 06036
Kassenärztliche Vereinigung Südwürttemberg, Tübingen 06037
Kassenärztliche Vereinigung Trier, Trier 06038
Kassenärztliche Vereinigung Westfalen-Lippe, Dortmund 06039
Katholische Ärztearbeit Deutschlands, Bonn 06058
Klinische Forschungsgruppe für Multiple Sklerose, Würzburg 06071
Klinische Forschungsgruppe für Reproduktionsmedizin, Münster 06072
Kollegium der Medizinjournalisten, Oberaudorf 06074
Kongreßgesellschaft für ärztliche Fortbildung e.V., Berlin 06083
Landesärztekammer Hessen, Frankfurt 06092
Landesärztekammer Rheinland-Pfalz, Mainz 06093
Landesärztekammer Thüringen, Jena 06094
Marburger Bund, Köln 06109
MEDICA, Deutsche Gesellschaft zur Förderung der Medizinischen Diagnostik e.V., Stuttgart 06115
Medical Women's International Association, Köln 06116
Medizinischer Fakultätentag der Bundesrepublik Deutschland, Münster 06117

NAV-Virchowbund, Köln 06137
Nephrologischer Arbeitskreis Saar-Pfalz-Mosel e.V., Kaiserslautern 06138
Nordwestdeutsche Gesellschaft für ärztliche Fortbildung e.V., Steinburg 06144
Normenausschuss Medizin im DIN Deutsches Institut für Normung e.V., Berlin 06193
Physikalisch-Medizinische Sozietät zu Erlangen, Erlangen 06236
Sächsische Landesärztekammer, Dresden 06274
Sportärztebund Hamburg, Hamburg 06284
Sportärztebund Hessen, Frankfurt 06285
Sportärztebund Niedersachsen, Göttingen 06286
Sportärztebund Nordrhein, Leverkusen 06287
Sportärztebund Rheinland-Pfalz, Kaiserslautern 06288
Sportärzteverband Schleswig-Holstein, Kiel 06289
Unabhängiger Ärzteverband Deutschlands e.V., Köln 06315
Verband der Betriebs- und Werksärzte e.V., Karlsruhe 06333
Verband der leitenden Krankenhausärzte Deutschlands e.V., Düsseldorf 06341
Verband Deutscher Badeärzte e.V., Bad Oeynhausen 06349
Vereinigung der Ärzte der Medizinaluntersuchungsämter, Osnabrück 06395
Vereinigung der Praktischen und Allgemeinärzte Bayerns e.V., München 06400
Vereinigung deutscher Ärzte, Frankfurt 06403
Vereinigung Südwestdeutscher Radiologen und Nuklearmediziner, Sindelfingen 06415
Westdeutscher Medizinischer Fakultätentag, Erlangen 06439

Greece

Syllogos Iatrikos Panellinios, Athinai 06545

Guatemala

Academia de Ciencias Médicas, Físicas y Naturales de Guatemala, Guatemala City 06553

Honduras

Asociación de Facultades de Medicina, Tegucigalpa 06583

Hungary

Korányi Sandor Társaság, Budapest 06610
Magyar Altalános Orvosok Tudományos Egyesülete, Budapest 06614
Magyar Orvostudományi Társaságok és Egyesületek Szövetsége, Budapest 06650

Iceland

Loeknafélag Islands, Reykjavik 06694

India

Association for Medical Education in South-East Asia, New Delhi 06731
Bombay Medical Union, Bombay 06746
Indian Council of Medical Research, New Dehli 06787
The Indian Medical Association, New Dehli 06801
Indian Science Congress Association, Calcutta 06813
Medical Council of India, New Dehli 06835

Indonesia

Ikatan Dokter Indonesia, Jakarta 06889

Iran

Medical Nomenclature Society of Iran, Isfahan 06900
Society of Iranian Clinicians, Teheran 06904

Iraq

Iraqi Medical Society, Baghdad 06918

Ireland

European Alliance of Muscular Dystrophy Associations, Dublin 06946
Irish Medical Association, Dublin 06992
The Medical Council, Dublin 07010
Medical Research Council of Ireland, Dublin 07011
Medical Union, Dublin 07012
Royal Academy of Medicine, Dublin 07021
Royal College of Physicians of Ireland, Dublin 07022

Israel

International Society of Computerized and Quantitative EMG, Jerusalem 07058
Israel Association of General Practitioners, Hof Ha-Karmel 07065
Israel Medical Association, Tel Aviv 07082
Israel Society for Experimental Biology and Medicine, Rehovoth 07091
Israel Society of Clinical Neurophysiology, Jerusalem 07098

Italy

Academia Gentium Pro Pace, Roma 07123
Accademia delle Scienze Mediche di Palermo, Palermo 07152
Accademia di Medicina di Torino, Torino 07169
Accademia Internazionale di Medicina Legale e di Medicina Sociale, Roma 07191
Accademia Medica, Genova 07203
Accademia Medica di Roma, Roma 07204
Accademia Medica Lombarda, Milano 07205
Accademia Medica Pistoiese Filippo Pacini, Pistoia 07206
Accademia Medico-Chirurgica del Piceno, Ancona 07207
Accademia Medico-Fisica Fiorentina, Firenze 07208
Accademia Pratese di Medicina e Scienze, Prato 07229
Accademia Romana di Scienze Mediche e Biologiche, Roma 07233
Associazione Italiana di Ingegneria Medica e Biologica, Napoli 07312
Associazione Italiana di Medicina Aeronautica e Spaziale, Roma 07313
Associazione Italiana di Medicina dell'Assicurazione Vita, Roma 07314
Centro di Informazioni Sterilizzazione e Aborto, Milano 07429
Centro Internazionale di Ipnosi Medica e Psicologica, Milano 07468
Centro Italiano di Sessuologia, Roma 07495
Centro Italiano per lo Studio e lo Sviluppo dell'Agopuntura Moderna e dell'Altra Medicina, Vicenza 07503
Centro Ricerche Applicazione Bioritmo, Roma 07533
Centro Studi Problemi Medici, Milano 07588
European Association of Senior Hospital Physicians, Udine 07666
European Association of Social Medicine, Torino 07667
Federazione Medico-Sportiva Italiana, Roma 07693
Federazione Nazionale degli Ordini dei Medici Chirurghi e Odontoiatri, Roma 07695
International Society for Medical and Psychological Hypnosis, Milano 07718
International Society for Twin Studies, Roma 07720
Società Italiana di Liposcultura, Roma 07891
Società Italiana di Medicina del Lavoro e di Igiene Industriale, Pavia 07893
Società Italiana di Medicina e Igiene della Scuola, Milano 07895
Società Italiana di Medicina Legale e delle Assicurazioni, Roma 07899
Società Italiana di Parassitologia, Roma 07926
Società Italiana di Senologia, Roma 07947
Società Italiana di Storia della Medicina, Bologna 07952
Società Italiana per gli Archivi Sanitari Ospedalieri, Viareggio 07962
Società Jonico-Salentina de Medicina e Chirurgia, Taranto 07976

Ivory Coast

African Union of Sports Medicine, Algiers 08060

Jamaica

Commonwealth Caribbean Medical Research Council, Kingston 08077
Medical Association of Jamaica, Kingston 08087

Japan

Asian Parasite Control Organization, Tokyo 08104
Association of Medical Doctors for Asia, Okayana 08113
Nihon Arerugi Gakkai, Tokyo 08187
Nihon Hoken Igakkai, Tokyo 08207
Nihon Ishi-Kai, Tokyo 08212
Nihon Kakuigakukai, Tokyo 08218
Nihon Koko Geka Gakkai, Tokyo 08229
Nihon Kotsu Igakkai, Tokyo 08233
Nihon Naika Gakkai, Tokyo 08243
Nihon Nettai Igakkai, Nagasaki 08244
Nihon Tairyoku Igakkai, Tokyo 08267
Nihon Uirusu Gakkai, Tokyo 08272
Nippon Bitamin Gakkai, Kyoto 08281
Nippon Hoi Gakkai, Tokyo 08306
Nippon Junkatsu Gakkai, Tokyo 08317
Nippon Kango Kyokai, Tokyo 08324
Toa Igaku Kyokai, Tokyo 08430

Kenya

African Medical and Research Foundation, Nairobi 08470

Korea, Democratic People's Republic

Academy of Medical Sciences, Pyongyang 08512

Korea, Republic

Asian Federation of Catholic Medical Associations, Seoul 08519
Korean Medical Association, Seoul 08537

Kuwait

Arab Medical Information Network, Kuwait 08545

Latvia

Latvian Academy of Medicine, Riga 08557

Liechtenstein

Liechtensteinischer Ärzteverein, Vaduz 08572

Luxembourg

Association des Médecins et Médecins-Dentistes du Grand-Duché de Luxembourg, Luxembourg 08575
Association Européenne de Méthodes Médicales Nouvelles, Luxembourg 08577
Société des Sciences Médicales du Grand-Duché de Luxembourg, Luxembourg 08588

Macedonia

Makedonsko Lekarsko Društvo, Skopje 08605

Malaysia

Association for Medical Education in the Western Pacific Region, Kuala Lumpur 08638
Malaysian Medical Association, Kuala Lumpur 08650

Malta

Medical Association of Malta, Paceville 08668

Mexico

Academia Nacional de Medicina, México 08690
Asociación Médica del Hospital Beistegui, México 08696
Asociación Médica Franco-Mexicana, México 08697
Asociación Mexicana de Facultades y Escuelas de Medicina, San Luis de Potosí 08700
Sociedad Médica del Hospital General, México 08754

Myanmar

Burma Medical Research Council, Rangoon 08811

Netherlands

Genootschap ter Bevordering van Natuur-, Genees- en Heelkunde, Amsterdam 08904
International Huntington Association, Harfsen 08937
Koninklijke Nederlandsche Maatschappij tot Bevordering der Geneeskunst, Utrecht 08951
Landelijke Huisartsen Vereniging, Utrecht 08970
Landelijke Specialisten Vereniging, Utrecht 08971
Nederlandsche Vereniging voor Levensverzekeringgeneeskunde, Rotterdam 08985
Nederlandse Bond voor Natuurgeneeswijze, Amsterdam 08988
Nederlandse Organisatie voor Wetenschappelijk Onderzoek, 's-Gravenhage 08998
Nederlandse Vereniging van Laboratoriumartsen, Leeuwarden 09007
Nederlandse Vereniging voor Medisch Onderwijs, Utrecht 09026
Nederlandse Werkgroep van Praktizijns in de Natuurlijke Geneeskunst, Leusden 09045
Nederlands Huisartsen Genootschap, Utrecht 09052
Vereniging van Medische Analisten, Utrecht 09092

New Zealand

Auckland Medical Research Foundation, Auckland 09119
Canterbury Medical Research Foundation, Christchurch 09121
Hawke's Bay Medical Research Foundation, Napier 09127
Medical Research Council of New Zealand, Auckland 09130
New Zealand Medical Association, Wellington 09169
Palmerston North Medical Research Foundation, Palmerston North 09186
Wellington Medical Research Foundation, Wellington 09200

Nigeria

The Medical and Dental Council of Nigeria, Lagos 09232

Norway

Den Norske Laegeforening, Lysaker 09253
Det Norske Medicinske Selskab, Oslo 09261
Medicinske Selskap i Bergen, Bergen 09278

Pakistan

Pakistan Medical Association, Karachi 09388
Pakistan Medical Research Council, Karachi 09389

Peru

Academia Nacional de Medicina, Lima 09439
Asociación Médica Peruana Daniel A. Carrión, Lima 09452
Federación Médica Peruana, Lima 09468
Sociedad Peruana de Historia de la Medicina, Lima 09484

Philippines

Confederation of Medical Associations in Asia and Oceania, Manila 09514
Philippine Medical Association, Quezon City 09529
Philippine Society of Parasitology, Manila 09532

Poland

Polskie Towarzystwo Historii Medycyny i Farmacji, Warszawa 09577
Polskie Towarzystwo Lekarski, Warszawa 09582
Polskie Towarzystwo Medycyny Pracy, Sosnowiec 09587
Zrzeszenie Polskich Towarzystw Medycznych, Warszawa 09631

Portugal

Ordem dos Médicos, Lisboa 09665
Sociedade das Ciências Médicas de Lisboa, Lisboa 09678
Sociedade Portuguesa de Hemorreologia e Microcirculaçã0, Lisboa 09702

Romania

Academia de Stiinte Medicale, Bucuresti 09735
Balkan Medical Union, Bucuresti 09747
Entente Médicale Méditerranéenne, Bucuresti 09749
Romanian Medical Association, Bucuresti 09754
Societatea de Medicinà Generalà, Bucuresti 09769
Societatea de Medici si Naturalisti, Iasi 09772
Societatea Nationala de Medicina Generala din Romania, Bucuresti 09789

Russia

Council of Scientific Medical Societies, Moskva 09862
National Medical and Technical Scientific Society, Moskva 09909
National Scientific Medical Society of History of Medicine, Moskva 09916
National Scientific Medical Society of Infectionists, Moskva 09918
National Scientific Medical Society of Nephrologists, Moskva 09920
National Scientific Medical Society of Phtisiologists, Moskva 09927
National Scientific Medical Society of Physical Therapists and Health-Resort Physicians, Moskva 09928
National Scientific Medical Society of Physicians-Analysts, Moskva 09929
National Scientific Medical Society of Physicians in Curative Physical Culture and Sports Medicine, Moskva 09930
Russian Academy of Medical Sciences, Moskva 09952

San Marino

European Centre for Disaster Medicine, San Marino 09996

Senegal

Conférence des Facultés et Ecoles de Médecine d'Afrique d'Expression Française, Dakar 10025

Singapore

Academy of Medicine, Singapore, Singapore 10033
Singapore Medical Association, Singapore 10055

Slovakia

Slovenská Lekárska Společnost, Bratislava 10077
Slovenská Parazitologická Spoločnost, Košice 10082

South Africa

Carbohydrate and Lipid Metabolism Research Group, Johannesburg 10128
The College of Medicine of South Africa, Rondebosch 10132
The Medical Association of South Africa, Lynnwood 10151
Medical Graduates' Association of the University of the Witwatersrand, Johannesburg 10152
The South African Association for Food Science and Technology, Edenvale 10179
The South African Association of Physicists in Medicine and Biology, Pretoria 10183
The South African Institute for Medical Research, Johannesburg 10194
The South African Medical Research Council, Tygerberg 10206
South African National Multiple Sclerosis Society, Walmer 10209

Spain

Academia de Ciencias Médicas de Bilbao, Bilbao 10240
Académia de Ciències Mèdiques de Catalunya i de Baleares, Barcelona 10241
Academia Médico-Quirúrgica Española, Madrid 10244
Asociación de Medicina Aeronáutica y Espacial, Barcelona 10250
Asociación Española de la Lucha contra la Poliomielitis, Madrid 10262
European Association of Football Teams Physicians, Barcelona 10305
Organización Médica Colegial-Consejo General de Colegios Médicos de España, Madrid 10334
Real Academia de Medicina de Sevilla, Sevilla 10350
Real Academia de Medicina y Cirugía de Palma de Mallorca, Palma de Mallorca 10351
Real Academia Nacional de Medicina, Madrid 10356

Sri Lanka

Sri Lanka Medical Association, Colombo 10421

Sweden

Göteborgs Läkarförening, Göteborg 10467
International Association of Medical Laboratory Technologists, Stockholm 10471
Kungliga Fysiografiska Sällskapet i Lund, Lund 10483
Medicinska Forskningsrådet, Stockholm 10501
Svenska Föreningen för Medicinsk Teknik och Fysik, Umea 10544
Svenska Läkaresällskapet, Stockholm 10560
Sveriges Praktiserande Läkares Förening, Stockholm 10593

Switzerland

Association Olympique Internationale pour la Recherche Médico-Sportive, Lausanne 10631
Conseil des Organisations Internationales des Sciences Médicales, Genève 10655
Gesellschaft schweizerischer Amtsärzte, Bern 10719
Kommission Sportmed des Schweizerischen Landesverbandes für Sport, Bern 10760
Permanent Working Group of European Junior Hospital Doctors, Bern 10779
Schweizerische Ärztegesellschaft für Manuelle Medizin, Zürich 10785
Schweizerische Akademie der medizinischen Wissenschaften, Basel 10787
Schweizerische Gesellschaft für Geschichte der Medizin und der Naturwissenschaften, Bern 10822
Schweizerischer Fachverband der diplomierten medizinischen Laborantinnen und Laboranten, Sankt Gallen 10893
Société Médicale de la Suisse Romande, Sion 10979
Verband deutschschweizerischer Ärzte-Gesellschaften, Tuggen 11000
Verband Schweizerischer Assistenz- und Oberärzte, Bern 11005

Syria

Arab Council for Medical Specialization, Damascus 11040
Arab Federation for the Organs of the Deaf, Damascus 11041

Tanzania

Confederation of African Medical Associations and Societies, Dar es Salaam 11048
East Africa Medical Research Council, Arusha 11050
East, Central and Southern African College of Nursing, Arusha 11053
Tanzania Medical Association, Dar es Salaam 11061

Thailand

Medical Association of Thailand, Bangkok 11093

Trinidad and Tobago

Caribbean Sports Medicine Association, Port of Spain 11119

Tunisia

Arab Federation of Sports Medicine, Tunis 11129
Arab Medical Union, Tunis 11132

Turkey

Danubian League against Thrombosis and Hemorrhagic Disorders, Istanbul 11144

Ukraine

Ukrainian Academy of Medical Sciences, Kiev 11192

United Kingdom

Action in International Medicine, London 11202
Anglo-European College of Chiropractic, Bournemouth 11232
Association for Medical Deans in Europe, Belfast 11272
Association for Medical Education in Europe, Dundee 11273
Association for Spina Bifida and Hydrocephalus, Peterborough 11284
Association for the Study of Medical Education, Dundee 11293
Association of Crematorium Medical Referees, Westcliff-on-Sea 11334
Association of District Medical Officers, Bath 11340
Association of Health Care Information and Medical Records Officers, Hopton 11353
Association of Researchers in Medicine and Science, London 11385
Association to Combat Huntington's Chorea, Hinckley 11412
Assurance Medical Society, London 11413
British Acupuncture Association and Register, London 11473
British Association for Immediate Care, Ipswich 11492
British Association of Sport Medicine, New Malden 11524
British Institute of Musculoskeletal Medicine, Northwood 11585
British Medical Association, London 11607
British Migraine Association, West Byfeet 11609
British Naturopathic and Osteopathic Association, London 11618
British Postgraduate Medical Federation, London 11633
British Society of Medical and Dental Hypnosis, London 11686
Commonwealth Medical Association, London 11827
Disinfected Mail Study Circle, London 11895
Edinburgh Medical Missionary Society, Edinburgh 11936
European Association of Science Editors, London 12000
European Confederation of Huntington Associations, Hinckley 12015
European Society for Clinical Investigation, London 12038
European Underseas Bio-Medical Society, Aberdeen 12042
The Fellowship for Freedom in Medicine, Storrington 12070
Fellowship of Postgraduate Medicine, London 12073
General Council and Register of Osteopaths, Reading 12102
General Medical Council, London 12106
Guild of Catholic Doctors, Bristol 12132
Guild of Health, London 12135
The Harveian Society of London, London 12148
Hospitals Consultants and Specialists Association, Overton 12183
Hunterian Society, London 12188
Incorporated Society of Registered Naturopaths, Edinburgh 12196
Institute of Medical Laboratory Sciences, London 12261
Institute of Psycho-Sexual Medicine, London 12273
Institute of Sports Medicine, London 12283
Institute of Trichologists, London 12289
International Conference on Social Science and Medicine, Brighton 12346
International Federation of Multiple Sclerosis Societies, London 12357
International Federation of Sportive Medicine, Slough 12360
International Filariasis Association, London 12363
Manchester Medical Society, Manchester 12485
The Medical Defence Union, London 12503
Medical Practices Committee, London 12505
Medical Practitioners' Union, London 12506
Medical Protection Society, London 12507
Medical Research Council, London 12508
Medical Research Society, London 12509
Medical Sciences Historical Society, Liverpool 12510
The Medical Society of London, London 12513
Medical Women's Federation, London 12514
Medico-Legal Society, London 12515
The Multiple Sclerosis Society of Great Britain and Northern Ireland, London 12541
Muscular Dystrophy Group of Great Britain & Northern Ireland, London 12542
National Back Pain Association, Teddington 12576
National Institute for Medical Research, London 12594
National Institute of Medical Herbalists, Exeter 12600
Northern Ireland Polio Fellowship, Belfast 12638
Osteopathic Medical Association, London 12665
Overseas Doctors Association in the UK, Manchester 12666
Parkinson's Disease Society of the United Kingdom, London 12678
Polish Medical Association, London 12709
Renal Association, Oxford 12750
Royal College of General Practitioners, London 12779
Royal College of Nursing of the United Kingdom, London 12782
Royal College of Physicians and Surgeons of Glasgow, Glasgow 12786
Royal College of Physicians of London, London 12788
Royal Medical Society, Edinburgh 12830
Royal Medico-Chirurgical Society of Glasgow, Glasgow 12831
The Royal Society of Medicine, London 12862
Scottish Society of the History of Medicine, Aberdeen 12931
Scottish Teachers' Nursing Association, Glasgow 12933
Society for Back Pain Research, London 12966
Society for Social Medicine, London 12996
Society for the Social History of Medicine, Manchester 13012
Society for the Study of Addiction, Horsham 13013
Society for the Study of Fertility, Cambridge 13014
Society for the Study of Inborn Errors of Metabolism, Manchester 13016
Society of Occupational Medicine, London 13058
Vasectomy Advancement Society of Great Britain, London 13177
World Federation for Medical Education, Edinburgh 13220

Uruguay

Academia Nacional de Medicina del Uruguay, Montevideo 13241

U.S.A.

Accreditation Association for Ambulatory Health Care, Skokie 13314
Accrediting Bureau of Health Education Schools, Elkhart 13316
Acupuncture International Association, Saint Louis 13324
Acupuncture Research Institute, Monterey Park 13325
Addiction Research and Treatment Corporation, Brooklyn 13326
Adrenal Metabolic Research Society of the Hypoglycemia Foundation, Troy 13330
Aerospace Medical Association, Alexandria 13339
Aid for International Medicine, Rocklamd 13353
Airline Medical Directors Association, Chicago 13357
Alcor Life Extension Foundation, Riverside 13360
Alzheimer's Association, Chicago 13370
Alzheimer's Disease International, Chicago 13371
American Academy for Cerebral Palsy and Developmental Medicine, Richmond 13376
American Academy of Family Physicians, Kansas City 13394
American Academy of Insurance Medicine, Springfield 13401
American Academy of Legal and Industrial Medicine, New York 13402
American Academy of Medical Administrators, Southfield 13406
American Academy of Nursing, Kansas City 13409
American Academy of Osteopathy, Newark 13418
American Academy of Physical Medicine and Rehabilitation, Chicago 13424
American Academy of Sanitarians, Miami 13437
American Academy of Sports Physicians, Northridge 13438
American Anorexia/Bulimia Association, New York 13452
American Association for Clinical Chemistry, Washington 13474
American Association for Medical Transcription, Modesto 13486
American Association for the Advancement of Automotive Medicine, Des Plaines 13496
American Association for the History of Medicine, Boston 13500
American Association for the Study of Headache, West Deptford 13501
American Association of Certified Allergists, Palatine 13515
American Association of Colleges of Nursing, Washington 13520
American Association of Critical-Care Nurses, Aliso Viejo 13526
American Association of Medical Assistants, Chicago 13548
American Association of Medical Milk Commissions, Los Angeles 13549
American Association of Medical Society Executives, Chicago 13550
American Association on Mental Retardation, Washington 13608
American Behcet's Association, Minneapolis 13620
American Black Chiropractors Association, Saint Louis 13623
American Board of Bioanalysis, Saint Louis 13629
American Board of Preventive Medicine, Dayton 13657
American Bureau for Medical Advancement in China, New York 13672
American College of Emergency Physicians, Dallas 13705
American College of General Practitioners in Osteopathic Medicine and Surgery, Arlington Heights 13709
American College of Healthcare Executives, Chicago 13710
American College of Medical Group Administrators, Englewood 13714
American College of Nurse-Midwives, Washington 13720
American College of Preventive Medicine, Washington 13732
American College of Sports Medicine, Indianapolis 13737
American Cryonics Society, Cupertino 13773
American Electrology Association, Trumbull 13794
American Federation for Clinical Research, Thorofare 13806
American Group Practice Association, Alexandria 13835
American Healing Association, Glendale 13840
American Health Information Management Association, Chicago 13841
American Holistic Medical Foundation, Raleigh 13850
American Hypnotists' Association, San Francisco 13857
American Institute of Ultrasound in Medicine, Rockville 13898
American Medical Association, Chicago 13933
American Medical Association Education and Research Foundation, Chicago 13934
American Medical Fly Fishing Association, Lock Haven 13936
American Medical Informatics Association, Bethesda 13937
American Medical Technologists, Park Ridge 13938
American Medical Women's Association, Alexandria 13939
American Medical Writers' Association, Bethesda 13940
American Naprapathic Association, Chicago 13953
American Osteopathic Academy of Sports Medicine, Middleton 13979
American Physicians Association of Computer Medicine, Pittsford 14002
American Society for Clinical Investigation, Thorofare 14050
American Society for Healthcare Human Resources Administration, Chicago 14065
American Society for Hospital Engineering, Chicago 14067
American Society for Medical Technology, Bethesda 14072
American Society of Addiction Medicine, Washington 14096
American Society of Clinical Hypnosis, Des Plaines 14108
American Society of Contemporary Medicine and Surgery, Chicago 14113
American Society of Law and Medicine, Boston 14140
American Society of Nephrology, Washington 14148
ANAD – National Association of Anorexia Nervosa and Associated Disorders, Highland Park 14215
Association for Faculty in the Medical Humanities, McLean 14296
Association for Hospital Medical Education, Washington 14302
Association for Practitioners in Infection Control, Mundelein 14321
Association of Air Medical Services, Pasadena 14377
Association of American Medical Colleges, Washington 14383
Association of Chiropractic Colleges, San Lorenzo 14411
Association of Clinical Scientists, Farmington 14413
Association of Medical Education and Research in Substance Abuse, Providence 14459
Association of Professors of Medicine, Washington 14490
Asthma and Allergy Foundation of America, Washington 14538
Biomedical Engineering Society, Culver City 14593
Catholic Medical Mission Board, New York 14657
Center for Advanced Study in the Behavioral Sciences, Stanford 14663
Center for Research in Ambulatory Health Care Administration, Englewood 14695
Central Society for Clinical Research, Chicago 14716
China Medical Board of New York, New York 14735
Chronic Fatigue Syndrome Society, Portland 14750
City of Hope, Duarte 14757
Commission on Professional and Hospital Activities, Ann Arbor 14789
Committee for the Promotion of Medical Research, Yonkers 14797
Council for Medical Affairs, Chicago 14879
Council of Community Blood Centers, Washington 14893
Council of Teaching Hospitals, Washington 14904
Council on Chiropractic Education, West Des Moines 14910
Council on Medical Education of the American Medical Association, Chicago 14925
Cranial Academy, Indianapolis 14937
Drug and Alcohol Nursing Association, Baltimore 14986
ECRI, Plymouth Meeting 15012
Educational Commission for Foreign Medical Graduates, Philadelphia 15018
Federation of State Medical Boards of the United States, Fort Worth 15090
First – Foundation for Ichthyosis and Related Skin Types, Raleigh 15097
Flower Essence Society, Nevada City 15101
Flying Chiropractors Association, Philadelphia 15103
The Forum for Health Care Planning, Washington 15115
Forum for Medical Affairs, Lake Success 15116
Foundation for the Advancement of Chiropractic Tenets and Science, Arlington 15133
Foundation for the Support of International Medical Training, Lewiston 15134
Gerson Institute, Bonita 15159
Graduates of Italian Medical Schools, Bronx 15171
Harvey Society, New York 15190
Herpes Resource Center – American Social Health Association, Research Triangle Park 15204
Hospital Research and Educational Trust, Chicago 15231
IEEE Engineering in Medicine and Biology Society, New York 15246
Independent Citizens Research Foundation for the Study of Degenerative Diseases, Ardsley 15251
Indians Into Medicine, Grand Forks 15257
Institute for Advanced Research in Asian Science and Medicine, Garden City 15265
Institute for Hospital Clinical Nursing Education, Chicago 15283
Institute for the Advancement of Human Behavior, Stanford 15300
Institute of Electrology Educators, Bellingham 15318
Institute of Medicine, Washington 15332
Interchurch Medical Assistance, New Windsor 15356
International Academy of Behavioral Medicine, Dallas 15363
International Academy of Nutrition and Preventive Medicine, Asheville 15367
International Association for Healthcare Security and Safety, Lombard 15383
International Association of Coroners and Medical Examiners, Peoria 15402
International Bundle Branch Block Association, Los Angeles 15426
International Conference on Mechanics in Medicine and Biology, Ann Arbor 15446
International Epidemiological Association, Los Angeles 15471
International Hospital Federation, Chicago 15485
International Joseph Diseases Foundation, Livermore 15494
International Professional Surrogates Association, Los Angeles 15516
International Rescue and Emergency Care Association, Covington 15520
International Society for Artificial Organs, Cleveland 15522
International Society for Clinical Laboratory Technology, Saint Louis 15529
International Society of Reproductive Medicine, Rancho Mirage 15555
Interstate Postgraduate Medical Association of North America, Madison 15581
Joint Commission on Accreditation of Healthcare Organizations, Oak Brook Terrace 15616
Joint Review Committee on Educational Programs for the EMT-Paramedic, Euless 15623
Joint Review Committee on Education in Diagnostic Medical Sonography, Chicago 15624
Liaison Committee on Medical Education, Chicago 15667
Medical Seminars International, Chatsworth 15712
Medical Society of the United States and Mexico, Phoenix 15713
Mental Research Institute, Palo Alto 15720
Microbeam Analysis Society, Deerfield Beach 15727
Mission Doctors Association, Los Angeles 15747
National Accrediting Agency for Clinical Laboratory Sciences, Chicago 15780
National Alopecia Areata Foundation, San Rafael 15786
National Anorexic Aid Society, Columbus 15787
National Association for Ambulatory Care, Grand Rapids 15791
National Association for Biomedical Research, Washington 15795
National Association for Practical Nurse Education and Service, Silver Spring 15811
National Association for Search and Rescue, Fairfax 15814
National Association of Advisers for the Health Professions, Champaign 15827
National Association of Disability Examiners, Frankfort 15850
National Association of Emergency Medical Technicians, Kansas City 15854
National Association of Medical Examiners, Saint Louis 15873
National Board of Medical Examiners, Philadelphia 15926
National Certification Agency for Medical Lab Personnel, Bethesda 15951
National Committee for Clinical Laboratory Standards, Villanova 15961
National Committee on the Treatment of Intractable Pain, Washington 15965
National Foundation for Non-Invasive Diagnostics, Princeton 16059
National Fund for Medical Education, Boston 16061
National Hormone and Pituitary Program, Rockville 16077
National Institute for Burn Medicine, Ann Arbor 16088
National Jewish Center for Immunology and Respiratory Medicine, Denver 16098
National Medical and Dental Association, Philadelphia 16106
National Medical Association, Washington 16107
National Medical Fellowships, New York 16108
National Multiple Sclerosis Society, New York 16111

National Organization for Rare Disorders, New Fairfield 16118
National Resident Matching Program, Washington 16140
National Student Nurses' Association, New York 16165
North American Academy of Muscoloskeletal Medicine, Middleton 16210
North American Medical/Dental Association, Newport Beach 16218
North American Primary Care Research Group, Richmond 16220
Orton Dyslexia Society, Baltimore 16272
Pan-American Medical Association, New York 16293
Parker Chiropractic Resource Foundation, Fort Worth 16296
Planned Parenthood Federation of America, New York 16324
Practical Allergy Research Foundation, Buffalo 16347
Public Responsibility in Medicine and Research, Boston 16379
The Radiance Technique Association International, Saint Petersburg 16384
Registered Medical Assistants of American Medical Technologists, Park Ridge 16404
Research Discussion Group, Morgantown 16415
Society for Academic Emergency Medicine, Lansing 16489
Society for Biomaterials, Minneapolis 16502
Society for Clinical and Experimental Hypnosis, Liverpool 16508
Society for Clinical Trials, Baltimore 16509
Society for Experimental Biology and Medicine, New York 16522
Society for Psychophysiological Research, Dallas 16572
Society for the Advancement of Ambulatory Care, Washington 16585
Society for the Study of Reproduction, Champaign 16619
Society of Critical Care Medicine, Anaheim 16653
Society of Diagnostics Medical Sonographers, Dallas 16654
Society of Medical Consultants to the Armed Forces, Bethesda 16678
Society of Medical Jurisprudence, New York 16679
Society of Prospective Medicine, Indianapolis 16697
Society of Teachers of Family Medicine, Kansas City 16706
Student National Medical Association, Washington 16749
Tissue Culture Association, Columbia 16784
Touch for Health Foundation, Pasadena 16786
Transplantation Society, Columbus 16788
Ukrainian Medical Association of North America, Chicago 16799
Undersea and Hyperbaric Medical Society, Bethesda 16803
United Cerebral Palsy Associations, Washington 16809
United Cerebral Palsy Research and Educational Foundation, New York 16810
Virchow-Pirquet Medical Society, New York 16863
Wilderness Medical Society, Indiamapolis 16895
Women's Auxiliary of the ICA, Muskegon 16912
Wound, Ostomy and Continence Nurses – An Association of E.T. Nurses, Costa Mesa 16937

Venezuela
Academia Nacional de Medicina, Caracas 16962
Asociación Venezolana de Facultades de Medicina, Caracas 16973
Colegio de Médicos del Distrito Federal, Caracas 16995
Colegio de Médicos del Estado Anzoátegui, Barcelona 16996
Colegio de Médicos del Estado Mérida, Mérida 16997
Colegio de Médicos del Estado Miranda, Caracas 16998
Federación Médica Venezolana, El Bosque 17005
Federación Panamericana de Asociaciones de Facultades y Escuelas de Medicina, Caracas 17006
Sociedad Médica, Bolivar 17014
Sociedad Venezolana de Historia de la Medicina, Caracas 17024
Sociedad Venezolana de Medicina del Trabajo y Deportes, Caracas 17026

Yugoslavia
Društvo Lekara Vojvodine, Novi Sad 17085
Društvo Ljekara Crne Gore, Podgorica 17086
Savez Lekarskih Društava Jugoslavije, Beograd 17117
Srpsko Lekarsko Društvo, Beograd 17130

Zambia
Association of Medical Schools in Africa, Lusaka 17151

Zimbabwe
Zimbabwe Medical Association, Harare 17187

Metallurgy

Australia
Asian-Pacific Corrosion Control Organization, Sydney 00225
Australasian Corrosion Association, Clayton 00243
The Australasian Institute of Mining and Metallurgy, Parkville 00244
Australian Institute of Mining and Metallurgy, Parkville 00307
Australian Zinc Development Association, Melbourne 00364
Institute of Metals and Materials Australasia, Parkville 00415

Austria
Österreichisches Giesserei-Institut, Verein für Praktische Giessereiforschung, Leoben 00910
Technisch-wissenschaftlicher Verein Eisenhütte Österreich, Leoben 00953

Belgium
Cadmium Pigments Association, Bruxelles 01178
Centre Belge d'Etude de la Corrosion, Bruxelles 01184
Comité d'Etude de la Corrosion et de la Protection des Canalisations, Liège 01222
Comité National Belge de l'Organisation Scientifique, Bruxelles 01232
C.R.I.F., Zwijnaarde 01258
European Association of Metals, Bruxelles 01294
European Coil Coating Association, Bruxelles 01317
Institut International du Fer et de l'Acier, Bruxelles 01389
Société Beneluxienne de Métallurgie, Bruxelles 01452

Brazil
Centro de Aperfeiçoamento e Especialização Médica, Rio de Janeiro 01620

Bulgaria
Union of Mining Engineering, Geology and Metallurgy, Sofia 01796

Canada
American Society for Metals, Winnipeg 01865
Canadian Institute of Mining and Metallurgy, Montréal 02180
Canadian Steel Environmental Association, Toronto 02302
Canadian Steel Industry Research Association, Toronto 02303
Industrial Accident Prevention Association, Toronto 02516
Reinforcing Steel Institute of Ontario, Willowdale 02857
Steel Castings Institute of Canada, Ottawa 02961

Chile
Centro de Investigación Minera y Metalúrgica, Santiago 03022

China, People's Republic
Academy of Non-Ferrous Metallurgical Design, Beijing 03111
The Chinese Society of Metals, Beijing 03195

China, Republic
Chinese Foundrymen Association, Kaohsiung 03269
Chinese Institute of Mining and Metallurgical Engineers, Taipei 03274

Czech Republic
Védecká Společnost pro Nauku o Kovech, Brno 03555

Denmark
Dansk Metallurgisk Selskab, Lyngby 03650

Finland
Suomen Korroosioyhdists Sky, Helsinki 04043

France
Association Technique de la Fonderie, Paris 04350
Association Technique de la Sidérurgie Française, Paris la Défense 04352
Fédération Métallurgique Française, Paris 04539
Société Française de Métallurgie et de Matériaux, Paris 04816

Germany
Akademischer Verein Hütte, Berlin 05065
Arbeitsgemeinschaft Korrosion e.V., Frankfurt 05123
Deutsche Gesellschaft für Materialkunde e.V., Oberursel 05381
Europäische Föderation Korrosion, Frankfurt 05653
European Aluminium Association, Düsseldorf 05659
European Aluminium Foil Association, Frankfurt 05660
European Federation of Corrosion, Frankfurt 05678
Fachverband Pulvermetallurgie, Hagen 05696
Forschungsgemeinschaft Eisenhüttenschlacken, Duisburg 05707
Forschungsvereinigung Schweißen und Schneiden e.V., Düsseldorf 05753
Gemeinschaftsausschuss Kaltformgebung e.V., Düsseldorf 05771
Gesellschaft Deutscher Metallhütten- und Bergleute e.V., Clausthal-Zellerfeld 05794
Normenausschuß Eisen und Stahl im DIN Deutsches Institut für Normung e.V., Düsseldorf 06163
Normenausschuss Giessereiwesen im DIN Deutsches Institut für Normung e.V., Köln 06171
Normenausschuss Nichteisenmetalle im DIN Deutsches Institut für Normung e.V., Köln 06194
Normenausschuss Pulvermetallurgie im DIN Deutsches Institut für Normung e.V., Köln 06198
Normenausschuss Wärmebehandlungstechnik metallischer Werkstoffe im DIN Deutsches Institut für Normung e.V., Köln 06216
Stifterverband Metalle, Düsseldorf 06296
VDEh-Gesellschaft zur Förderung der Eisenforschung, Düsseldorf 06317
Verein Deutscher Eisenhüttenleute, Düsseldorf 06374
Verein Deutscher Giessereifachleute e.V., Düsseldorf 06376
Verein für das Forschungsinstitut für Edelmetalle und Metallchemie e.V., Schwäbisch Gmünd 06381
Verein zur Förderung der Gießerei-Industrie e.V., Düsseldorf 06426

Hungary
Országos Magyar Bányászati és Kohászati Egyesület, Budapest 06670

India
Geological, Mining and Metallurgical Society of India, Calcutta 06761
Indian Institute of Metals, Calcutta 06796

Italy
Associazione Italiana di Metallurgia, Milano 07315
Centro Sperimentale Metallurgico, Roma 07547
Collegio dei Tecnici dell' Acciaio, Milano 07608
Commissione di Studio dei Fenomeni di Corrosione Elettrolitica, Bologna 07621

Japan
Keikinzoku Gakkai, Tokyo 08156
Nihon Do Senta, Tokyo 08195
Nippon Imono Kyokai, Tokyo 08312
Nippon Kinzoku Gakkai, Sendai 08333
Nippon Tekko Kyokai, Tokyo 08392

Liberia
Geological, Mining and Metallurgical Society of Liberia, Monrovia 08563

Netherlands
Nederlandse Vereniging van Gieterijtechnici, Zoetermeer 09006

Norway
Norsk Korrosjonsteknisk Forening, Oslo 09324
Norsk Metallurgisk Selskap, Oslo 09329

Portugal
Associação Portuguesa de Fundição, Porto 09646

Russia
Ferrous Metallurgy Society, Moskva 09868
Society of Non-Ferrous Metallurgy, Moskva 09976

South Africa
The South African Institute of Mining and Metallurgy, Marshalltown 10198

Switzerland
Internationales Komitee Giessereitechnischer Vereinigungen, Zürich 10740

United Kingdom
The Birmingham Metallurgical Association, Birmingham 11449
BNF Metals Technology Centre, Wantage 11456
British Association for Brazing and Soldering, Cambridge 11487
British Non-Ferrous Metals Research Association, Wantage 11619
Castings Technology International, Sheffield 11747
Drop Forging Research Association, Sheffield 11907
Institute of Corrosion, Leighton Buzzard 12229
The Institute of Metals, London 12262
Institution of Corrosion Science and Technology, Leighton Buzzard 12295
Institution of Mining and Metallurgy, London 12309
International Lead and Zinc Study Group, London 12374
London Metallurgical Society, Wallington 12464
Metals Society, London 12524
National Council of Corrosion Societies, London 12583
National Foundry and Engineering Training Association, Glasgow 12592
Sheffield Metallurgical and Engineering Association, Rotherham 12948
Spring Research and Manufacturers Association, Sheffield 13089
The Steel Construction Institute, Ascot 13100
World Bureau of Metal Statistics, London 13217
Zinc Development Association, London 13235

U.S.A.
American Council of the International Institute of Welding, Miami 13764
American Electroplaters and Surface Finishers Society, Orlando 13796
American Foundrymen's Society, Des Plaines 13824
American Iron and Steel Institute, Washington 13903
American Welding Society, Miami 14210
ASM International, Materials Park 14255
Association of Iron and Steel Engineers, Pittsburgh 14449
Ductile Iron Society, North Olmsted 14989
Extractive Metallurgy Institute, Point Roberts 15073
International Copper Association, New York 15449
International Lead Zinc Research Organization, Research Triangle Park 15497
International Magnesium Association, McLean 15498
International Metallographic Society, Columbus 15501
International Precious Metals Institute, Allentown 15515
Materials Properties Council, New York 15701
Metal Treating Institute, Neptune Beach 15721
Minerals, Metals and Materials Society, Warrendale 15744
Mining and Metallurgical Society of America, Larkspur 15745
National Association of Corrosion Engineers, Houston 15845
National Institute of Steel Detailing, Anaheim 16097
Steel Founders' Society of America, Des Plaines 16745
Steel Structures Painting Council, Pittsburgh 16746
Tin Research Institute, Columbus 16783

Vietnam
Vietnamese Associaton of Cast and Metallurgy, Hanoi 17067

Meteorology s. Geophysics

Microbiology s. a. Immunology

Argentina
Departamento de Microbiología, Castelar 00133

Australia
Australian Society for Microbiology, Parkville 00342
Australian Society for Parasitology, Canberra 00345

Austria
Österreichische Gesellschaft für Hygiene, Mikrobiologie und Präventivmedizin, Wien 00793

Brazil
Centro Nacional de Informação Científica em Microbiologia, Rio de Janeiro 01631
Sociedade Brasileira de Microbiologia, Rio de Janeiro 01715

Bulgaria
Bulgarian Society for Microbiology, Sofia 01755

Canada
Canadian Society of Microbiologists, Ottawa 02286

Chile
Asociación Chilena de Microbiología, Santiago 03014
Federación Latinoamericana de Parasitología, Santiago 03051
Instituto de Salud Publica de Chile, Santiago 03057
Sociedad Chilena de Parasitología, Santiago 03079

Colombia
Asociación Latinoamericana para Microbiologa, Bogotá 03352

Czech Republic
Česká Spolecnost Mikrobiologická, Praha 03526

Denmark
Danmarks Mikrobiologiske Selskab, København 03582
Dansk Selskab for Patologi, København 03686

Finland
Societas Biochemica, Biophysica et Microbiologica Fenniae, Helsingin Yliopisto 03992

France
International Union of Microbiological Societies, Strasbourg 04589
Société Française de Microbiologie, Paris 04817

Germany
Deutsche Gesellschaft für Hygiene und Mikrobiologie e.V., Heidelberg 05366
Deutsche Vereinigung zur Bekämpfung der Viruskrankheiten e.V., München 05622

Greece
Elliniki Microbiologiki Etaireia, Athinai 06511

Hungary
Magyar Mikrobiológiai Társaság, Budapest 06646

India
Indian Association of Parasitologists, Calcutta 06774

Iran
Iranian Society of Microbiology, Teheran 06899

Israel
Israel Society for Microbiology, Jerusalem 07094

Italy
Associazione Italiana di Microbiologia Applicata, Milano 07316
International Society for the Study of Infectious and Parasitic Diseases, Torino 07719
Società Italiana di Microangiologia e Microcorcolazione, Pisa 07906
Società Italiana di Microbiologia, Catania 07907

Japan
Nippon Kisei-chu Gakkai, Tokyo 08334

Mexico
Asociación Mexicana de Profesores de Microbiología y Parasitología en Escuelas de Medicina, Guadalajara 08704
Federación Latinoamericana de Parasitología, México 08734

Moldova, Republic of
Microbiological Society of Moldova, Chişinău *08782*

Netherlands
Nederlandse Vereniging voor Microbiologie, Bilthoven *09027*
Nederlandse Vereniging voor Parasitologie, Nijmegen *09032*

New Zealand
New Zealand Leather and Shoe Research Association, Palmerston North *09163*
New Zealand Microbiological Society, Dunedin *09170*
New Zealand Society for Parasitology, Hamilton *09176*

Nigeria
African Regional Network for Microbiology, Okigwi *09220*
Nigerian Society for Microbiology, Ibadan *09241*

Norway
Den Norske Mikrobionomforening, Oslo *09254*
Norsk Forening for Mikrobiologi, Aas *09310*

Paraguay
Instituto Nacional de Parasitología, Asunción *09431*

Poland
Polskie Towarzystwo Mikrobiologów, Warszawa *09588*
Polskie Towarzystwo Parazytologiczne, Wroclaw *09598*

Romania
Societatea de Microbiologie, Bucuresti *09773*

Russia
Microbiological Society, Moskva *09883*

Spain
Sociedad Española de Microbiología, Madrid *10391*

Sweden
Svenska Föreningen för Mikrobiologi, Lund *10545*

Switzerland
Schweizerische Gesellschaft für Mikrobiologie, Lausanne *10833*

Tunisia
Association Africaine de Microbiologie et d'Hygiène Alimentaire, Sousse *11134*

Turkey
Türk Mikrobiyoloji Cemiyeti, Istanbul *11166*

United Kingdom
British Society for Parasitology, Reading *11661*
British Society for Plant Pathology, Leeds *11663*
European Collection of Animal Cell Cultures, Salisbury *12009*
European Society of Nematologists, Invergowrie *12040*
Society for Applied Bacteriology, Newton Abbot *12964*
Society for General Microbiology, Reading *12975*

U.S.A.
American Academy of Microbiology, Washington *13407*
American Society for Microbiology, Washington *14073*
American Society of Parasitologists, El Paso *14152*
Foundation for Microbiology, New York *15129*
International Organization for Mycoplasmology, Seattle *15507*
National Hormone and Pituitary Program, Rockville *16077*
National Pest Control Association, Dunn Loring *16125*
Society for Industrial Microbiology, Arlington *16538*

Military Science

Belgium
Comité International de Médecine Militaire, Liège *01228*
Société Royale des Amis du Musée Royal de l'Armée et d'Histoire Militaire, Bruxelles *01484*

Canada
Canadian Society of Military Medals and Insignia, Thorold *02287*
Conference of Defence Associations Institute, Ottawa *02379*
Institut Militaire de Québec, Québec *02533*
Organization of Military Museums of Canada, Ottawa *02807*
Royal Canadian Military Institute, Toronto *02867*

Denmark
Chakoten Dansk Militaershistorisk Selskab, København *03572*
Det Krigsvidenskabelige Selskab, Frederiksberg *03712*
Marinehistorisk Selskab, København *03770*
Militaerteknisk Forening, København *03773*

France
Centre d'Histoire Militaire et d'Etudes de Défense Nationale, Montpellier *04370*
Union Fédérative des Sociétés d'Education Physique et de Préparation Militaire, Paris *04978*

Germany
Arbeitskreis für Wehrforschung, Stuttgart *05144*
Deutsche Gesellschaft für Heereskunde e.V., Beckum *05360*
Deutsche Gesellschaft für Wehrtechnik e.V., Bonn *05440*
Gesellschaft für Historische Waffen- und Kostümkunde, Hannover *05827*

Ireland
The Military History Society of Ireland, Dublin *07014*

Japan
Koeki Jigyo Gakkai, Tokyo *08167*

Peru
Centro de Estudios Histórico-Militares del Perú, Lima *09458*

Sweden
Kungliga Krigsvetenskapsakademien, Stockholm *10486*
Nordiska Samarbetskommittén för Internationell Politik, inklusive Konfliktoch Fredsforskning, Stockholm *10509*

United Kingdom
Arms and Armour Society of Great Britain, Staines *11248*
British Model Soldier Society, London *11610*
Commonwealth Defence Science Organisation, London *11817*
EUCLID, London *11980*
Fort Cumberland and Portsmouth Militaria Society, Portsmouth *12086*
Fortress Study Group, Newport *12087*
Historic Breechloading Smallarms Association, London *12177*
London Wargames Section, Barnet *12472*
Military Heraldry Society, Southall *12526*
Military Historical Society, Bromley *12527*
Napoleonic Association, Knaresborough *12551*
Orders and Medals Research Society, Croydon *12657*
Seventeenhundredandfourtyfive and National Military History Society, Edinburgh *12941*
Society for Army Historical Research, London *12965*
Victorian Military Society, Northampton *13183*

U.S.A.
Armed Forces Civilian Instructors Association, Sa Ysidro *14243*
Association of Military Colleges and Schools of the United States, Alexandria *14464*
Center for War/Peace Studies, New York *14712*
Company of Military Historians, Westbrook *14815*
Firearms Research and Identification Association, Rowland Heights *15096*
George C. Marshall Foundation, Lexington *15154*
Institute for Defense Analyses, Alexandria *15274*
Military Operations Research Society, Alexandria *15739*

Mineralogy

Australia
Australian Clay Minerals Society, Canberra *00273*
Australian Geomechanics Society, Barton *00293*

Austria
Österreichische Mineralogische Gesellschaft, Wien *00869*

Belgium
Bisnuth Institute, Grimbergen *01169*

Canada
Canadian Gemmological Association, Toronto *02144*
Lapidary, Rock and Mineral Society of British Columbia, Delta *02576*
Mineralogical Association of Canada, Ottawa *02655*
Nova Scotia Mineral and Gem Society, Dartmouth *02738*

Chile
Centro de Investigación Minera y Metalúrgica, Santiago *03022*

Czech Republic
Česká Geologická Společnost, Praha *03507*

Finland
Suomen Gemmologinen Seura, Helsinki *04021*

France
Association Française de Gemmologie, Paris *04216*
European Committee for the Study of Salt, Paris *04496*
Société Française de Minéralogie et de Cristallographie, Paris *04819*
Société Géologique et Minéralogique de Bretagne, Rennes *04869*

Germany
Deutsche Gemmologische Gesellschaft e.V., Idar-Oberstein *05309*
Deutsche Mineralogische Gesellschaft e.V., Münster *05471*
Gemologisches Institut, Idar-Oberstein *05774*
International Mineralogical Association, Marburg *06009*
Vereinigung der Freunde der Mineralogie und Geologie e.V., Heidelberg *06397*

India
Mineralogical Society of India, Mysore *06836*

Israel
Israel Crystallographic Society, Haifa *07069*

Italy
Associazione Nazionale Ingegneri Minerari, Roma *07375*
Società degli Amici del Museo Civico di Storia Naturale Giacomo Doria, Genova *07779*
Società Italiana di Mineralogia e Petrologia, Milano *07908*

Japan
Nihon Kobutsu Gakkai, Tokyo *08226*
Nippon Ganseki Kobutsu Kosho Gakkai, Sendai *08300*
Nippon Kessho Gakkai, Tokyo *08330*

Poland
Polskie Towarzystwo Mineralogiczne, Kraków *09590*

Russia
Russian Mineralogical Society, St. Petersburg *09959*

South Africa
Mintek, Randburg *10154*
South African Crystallographic Society, Johannesburg *10188*

Spain
Asociación Española de Gemología, Barcelona *10261*

Sri Lanka
Ceylon Gemmologists Association, Colombo *10409*

Sweden
Svenska Nationalkommittén för Kristallografi, Stockholm *10567*
Sveriges Gemmologiska Riksförening, Stockholm *10590*

Tanzania
Eastern and Southern African Mineral Resources Development Centre, Dodoma *11057*

United Kingdom
British Association of Crystal Growth, Towcester *11505*
British Zeolite Association, London *11710*
Gemmological Association and Gem Testing Laboratory of Great Britain, London *12101*
International Union of Crystallography, Chester *12401*
Mineralogical Society of Great Britain and Ireland, London *12530*

U.S.A.
American Federation of Mineralogical Societies, Oklahoma City *13809*
Clay Minerals Society, Boulder *14761*
Geochemical Society, Washington *15152*
Mineralogical Society of America, Washington *15743*
Minerals, Metals and Materials Society, Warrendale *15744*
Salt Institute, Alexandria *16436*
Society for Sedimentary Geology, Tulsa *16578*

Mining

Angola
Direcção Provincial dos Servicos de Geologia e Minas de Angola, Luanda *00033*

Australia
The Australasian Institute of Mining and Metallurgy, Parkville *00244*
Australian Geomechanics Society, Barton *00293*
Australian Institute of Mining and Metallurgy, Parkville *00307*
Australian Petroleum Exploration Association, Sydney *00327*

Austria
Gesellschaft der Geologie- und Bergbaustudenten in Österreich, Wien *00610*
Montanhistorischer Verein für Österreich, Leoben-Donawitz *00708*
Technisch-Wissenschaftlicher Verein Bergmännischer Verband Österreichs, Leoben *00952*

Azerbaijan
Commision on Mining, Baku *01040*

Belgium
Comité d'Etude des Producteurs de Charbon d'Europe Occidentale, Bruxelles *01223*

Bulgaria
Union of Mining Engineering, Geology and Metallurgy, Sofia *01796*

Canada
Canadian Institute of Mining and Metallurgy, Montréal *02180*
Mining Association of British Columbia, Vancouver *02656*
Mining Association of Manitoba, Winnipeg *02657*
Mining Society of Nova Scotia, Glace Bay *02658*

Chile
Centro de Investigación Minera y Metalúrgica, Santiago *03022*
Servicio Nacional de Geología y Minería, Santiago *03063*
Sociedad Nacional de Minería, Santiago *03102*

China, People's Republic
Beijing Academy of Coal Mine Design, Beijing *03118*
China Coal Society, Beijing *03131*
Chinese Academy of Coal Mining Sciences, Beijing *03137*

China, Republic
Chinese Institute of Mining and Metallurgical Engineers, Taipei *03274*

Colombia
Instituto Nacional de Investigaciones Geológico-Mineras, Bogotá *03393*

Ecuador
Dirección General de Geología y Minas, Quito *03852*

Finland
Suomen Geoteknillinen Yhdistys, Espoo *04023*

France
Société d'Etudes Minières, Industrielles et Financières, Paris *04748*

Germany
Bergbau-Forschung, Essen *05183*
Deutscher Markscheider-Verein e.V., Herne *05527*
Gesellschaft Deutscher Metallhütten- und Bergleute e.V., Clausthal-Zellerfeld *05794*
Normenausschuss Bergbau im DIN Deutsches Institut für Normung e.V., Essen *06151*
Normenausschuss Erdöl- und Erdgasgewinnung im DIN Deutsches Institut für Normung e.V., Köln *06164*
Versuchsgrubengesellschaft, Dortmund *06431*

Hungary
Országos Magyar Bányászati és Kohászati Egyesület, Budapest *06670*

India
Geological, Mining and Metallurgical Society of India, Calcutta *06761*

Iraq
Arab Industrial Development and Mining Organization, Baghdad *06911*

Ireland
Irish Mining and Quarrying Society, Dublin *06993*

Italy
Associazione Geotecnica Italiana, Roma *07281*

Ivory Coast
Société pour le Développement Minier de la Côte d'Ivoire, Abidjan *08066*

Liberia
Geological, Mining and Metallurgical Society of Liberia, Monrovia *08563*

Netherlands
Koninklijk Nederlands Geologisch Mijnbouwkundig Genootschap, Haarlem *08967*
Mijnbouwkundige Vereeniging, Delft *08976*

Norway
Norsk Geoteknisk Forening, Oslo *09318*

Peru
Sociedad Nacional de Minería, Lima *09481*

Russia
Mining Engineering Society, Moskva *09884*

South Africa
The South African Institute of Mining and Metallurgy, Marshalltown *10198*
South African National Group of the International Society for Rock Mechanics, Marshalltown *10208*

Suriname
Geologisch Mijnbouwkundige Dienst, Paramaribo *10431*

Sweden
Svenska Bergmannaföreningen, Stockholm *10538*

United Kingdom
Cornish Mining Development Association, Camborne *11843*
Institute of Quarrying, Nottingham *12275*
Institution of Mining and Metallurgy, London *12309*
Institution of Mining Engineers, Doncaster *12310*
International Lead and Zinc Study Group, London *12374*
Minerals Engineering Society, Burton-on-Trent *12531*
Mining Institute of Scotland, Edinburgh *12532*
Peak District Mines Historical Society, Nottingham *12683*

U.S.A.
American Institute of Mining, Metallurgical and Petroleum Engineers, New York *13888*
Mining and Metallurgical Society of America, Larkspur *15745*
Perlite Institute, Staten Island *16305*
Society for Mining, Metallurgy and Exploration, Littleton *16552*

Venezuela
Asociación Venezolana de Geología, Minería y Petróleo, Caracas *16974*
Sociedad Venezolana de Ingenieros de Minas y Metalúrgicos, Caracas *17025*

Musicology

Algeria
El-Djazairia El-Mossilia, Alger 00022

Argentina
Sociedad Argentina de Autores y Compositores de Música, Buenos Aires 00156

Australia
Musicological Society of Australia, Canberra 00436

Austria
Anton-Bruckner-Institut Linz, Linz 00547
Arbeitsgemeinschaft der Musikerzieher Österreichs, Wien 00549
Franz Schmidt-Gesellschaft, Wien 00598
Gebrüder Schrammel-Gesellschaft, Wien 00600
Gesellschaft der Musikfreunde in Wien, Wien 00612
Gesellschaft zur Herausgabe von Denkmälern der Tonkunst in Österreich, Wien 00645
Institut für Österreichische Musikdokumentation, Wien 00657
Internationale Albrechtsberger-Gesellschaft, Klosteneuburg 00665
Internationale Bruckner-Gesellschaft, Wien 00666
Internationale Chopin-Gesellschaft in Wien, Wien 00667
Internationale Gesellschaft für Jazzforschung, Graz 00671
Internationale Gesellschaft zur Erforschung und Förderung der Blasmusik, Graz 00672
Internationale Gustav Mahler Gesellschaft, Wien 00673
Internationale Hugo Wolf-Gesellschaft, Wien 00674
Internationale Schönberg-Gesellschaft, Mödling 00678
Internationales Musikzentrum, Wien 00682
Internationale Stiftung Mozarteum, Salzburg 00683
Johann Strauß-Gesellschaft, Wien 00692
Mediacult, Internationales Forschungsinstitut für Medien, Kommunikation und kulturelle Entwicklung, Wien 00705
Österreichische Gesellschaft für Musik, Wien 00811
Österreichisches Volksliedwerk, Wien 00922
Verband der Geistig Schaffenden Österreichs, Wien 00958
Walter Buchebner Gesellschaft, Mürzzuschlag 01010
Wiener Beethoven-Gesellschaft, Wien 01012
Wiener Konzerthausgesellschaft, Wien 01022
Wiener Schubertbund, Wien 01027

Belgium
Koninklijke Academie voor Wetenschappen, Letteren en Schone Kunsten van België, Bruxelles 01409
Société Belge de Musicologie, Bruxelles 01436
Société Belge des Auteurs, Compositeurs et Editeurs, Bruxelles 01447
Société des Auteurs et Compositeurs Dramatiques, Bruxelles 01455

Bulgaria
Union of Bulgarian Composers, Sofia 01784

Canada
Canadian Aldeburgh Foundation, Toronto 02019
Canadian Association of Music Libraries, Ottawa 02059
Canadian Bureau for the Advancement of Music, Toronto 02085
Canadian Folk Music Society, Calgary 02138
Canadian Music Centre, Toronto 02202
Canadian Music Council, Ottawa 02203
Canadian University Music Society, Windsor 02312
Composers, Authors and Publishers Association of Canada, Toronto 02373
New Brunswick Music Educators' Association, Quispamsis 02689
Newfoundland Music Educators' Association, Corner Brook 02710
Northwest Territories Music Educators' Association, Fort Smith 02720
Nova Scotia Music Educators' Association, Armdale 02739
Ontario College of Percussion, Toronto 02766
Ontario Music Educators' Association, London 02786
Ontario Registered Music Teachers Association, Toronto 02793
Registered Music Teachers' Association, Winnipeg 02856
Saskatchewan Registered Music Teachers' Association, Saskatoon 02898
Société des Auteurs, Eecherchistes, Documentalistes et Compositeurs, Montréal 02923
Society of Composers, Authors and Music Publishers of Canada, Don Mills 02944
Western Board of Music, Edmonton 02989

China, Republic
Chinese Classical Music Association, Taipei 03266
National Music Council of China, Taipei 03313

Czech Republic
Společnost Antonína Dvořáka, Praha 03554

Denmark
Dansk Komponistforening, København 03639
Dansk Musikbiblioteksforening, København 03651
Dansk Musikpaedagogisk Forening, København 03652
Dansk Selskab for Musikforskning, København 03679
Det Danske Orgelselskab, Vanløse 03706
Samfundet til Udgivelse af Dansk Musik, København 03798

Egypt
Atelier, Alexandria 03882
Institute of Arab Music, Alexandria 03907
Institute of Arab Music, Cairo 03908

Finland
Suomen Musiikinopettajain Liitto, Helsinki 04053
Suomen Säveltäjät, Helsinki 04058
Turun Soitannollinen Seura, Turku 04079

France
Académie Nationale de Danse, Paris 04117
Arab Music Rostrum, Paris 04135
Asian Music Rostrum, Paris 04138
Association Européenne des Conservatoires, Académies de Musique et Musikhochschulen, Angers 04205
Association Française pour la Recherche et la Création Musicales, Paris 04242
Confédération Internationale des Sociétés d'Auteurs et Compositeurs, Paris 04438
Conseil International de la Danse, Paris 04452
Conseil International de la Musique, Paris 04454
Jeunesses Musicales de France, Paris 04591
Société des Auteurs et Compositeurs Dramatiques, Paris 04700
Société Française de Composition, Paris 04797
Société Française de Musicologie, Paris 04820
Société J. S. Bach, Paris 04886
Syndicat National des Auteurs et Compositeurs de Musique, Paris 04938
Union Générale des Auteurs et Musiciens Professionnels, Paris 04981

Germany
Allgemeiner Cäcilien-Verband für Deutschland, Regensburg 05073
Arbeitskreis für Schulmusik und allgemeine Musikpädagogik e.V., Würzburg 05143
Bach-Verein Köln e.V., Köln 05161
Bundesakademie für musikalische Jugendbildung, Trossingen 05230
Deutsche Jazz-Föderation e.V., Frankfurt 05456
Deutsche Mozart-Gesellschaft e.V., Augsburg 05474
Deutsche Phono-Akademie e.V., Hamburg 05486
Deutscher Komponisten-Verband e.V., Berlin 05524
Deutscher Musikrat e.V., Nationalkomitee der Bundesrepublik Deutschland im Internationalen Musikrat, Bonn 05530
Deutscher Tonkünstlerverband e.V., München 05543
Georg-Friedrich-Händel-Gesellschaft e.V., Halle 05783
Gesellschaft der Musik- und Kunstfreunde Heidelberg e.V., Heidelberg 05791
Gesellschaft für Musikforschung e.V., Kassel 05842
Gesellschaft für Neue Musik e.V., Frankfurt 05845
Institut für Neue Musik und Musikerziehung e.V., Darmstadt 05966
Internationale Heinrich Schütz-Gesellschaft e.V., Kassel 05992
Internationaler Arbeitskreis für Musik e.V., Kassel 05994
Internationales Institut für Traditionelle Musik e.V., Berlin 06001
Neue Bachgesellschaft e.V., Leipzig 06139
Orchester-Akademie des Berliner Philharmonischen Orchesters e.V., Berlin 06224
Richard-Wagner-Verband Bayreuth e.V., Bayreuth 06269
Verband deutscher Musikschulen e.V., Bonn 06356
Verband Deutscher Schulmusiker e.V., Mainz 06361
Verein zur Förderung der deutschen Tanz- und Unterhaltungsmusik e.V., Bonn 06425

Greece
Enosis Hellinon Mousourgon, Athinai 06521

Guatemala
Sociedad Pro-Arte Musical, Guatemala City 06574

Hungary
Liszt Ferenc Társaság, Budapest 06611
Magyar Zenei Tanács, Budapest 06663
Országos Magyar Cecília Társulat, Budapest 06671

Iceland
Tónlistarfélagid, Reykjavik 06704

India
Indian Musicological Society, Baroda 06802
Sangeet Natak Akademi, New Delhi 06855

Iraq
Arab Academy of Music, Baghdad 06905

Ireland
Association of Irish Musical Societies, Dublin 06929
The Music Association of Ireland, Dublin 07015
The Royal Irish Academy of Music, Dublin 07029

Israel
Arthur Rubinstein International Music Society, Tel Aviv 07043
Israel Music Institute, Tel Aviv 07083
Society of Authors, Composers and Music Publishers in Israel, Tel Aviv 07112

Italy
Accademia Corale Stefano Tempia, Torino 07132
Accademia Filarmonica di Bologna (Reale), Bologna 07181
Accademia Filarmonica di Verona, Verona 07182
Accademia Filarmonica Romana, Roma 07183
Accademia Musicale Chigiana, Siena 07209
Accademia Musicale Ottorino Respighi, Roma 07210
Accademia Nazionale di Santa Cecilia, Roma 07221
Accademia Prenestina del Cimento di Musica, Lettere, Scienze, Arti Visive e Figurative, Palestrina 07230
Accademia Tedesca Villa Massimo, Roma 07241
Associazione Alessandro Scarlatti, Napoli 07254
Associazione Italiana Santa Cecilia per la Musica Sacra, Roma 07347
Associazione Nazionale dei Musei di Enti Locali e Istituzionale, Padova 07362
Centro di Musicologia Walter Stauffer, Cremona 07430
Centro Internazionale Studi Musicali, Roma 07487
Centro Italiano di Musica Antica, Roma 07492
Centro Ricerche di Storia e Arte Bitontina, Bitonto 07537
Centro Rossiniano di Studi, Pesaro 07544
Collegium Musicum di Latina, Latina 07611
Consociatio Internationalis Musicae Sacrae, Roma 07630
Ente Autonomo La Biennale de Venezia, Venezia 07649
Istituto Italiano per la Storia della Musica, Roma 07744
Società Filarmonica di Trento, Trento 07805
Società Italiana di Musicologia, Bologna 07909
Società Italiana per l'Educazione Musicale, Milano 07967

Ivory Coast
African Music Rostrum, Abidjan 08058

Japan
Nippon Ongaku Gakkai, Tokyo 08352
Toyo Ongaku Gakkai, Tokyo 08441

Korea, Republic
Music Association of Korea, Seoul 08541

Macao
Circulo de Cultura Musical, Macao 08591

Macedonia
Društvo na Kompozitorite na Makedonija, Skopje 08594

Mexico
Academia de Música, Guanajuato 08683
Academia de Música, Hermosillo 08684

Morocco
Association des Amateurs de la Musique Andalouse, Casablanca 08801

Netherlands
Koninklijke Nederlandse Toonkunstenaars-vereniging, Amsterdam 08959
Maatschappij tot Bevordering der Toonkunst, Amsterdam 08975
Nederlandse Sint-Gregorius Vereniging ter Bevordering van Liturgische Muziek, Utrecht 09001
Nederlandse Toonkunstenaarsraad, Amsterdam 09002
Vereniging voor Nederlandse Muziekgeschiedenis, Utrecht 09102
Wagnervereeniging, Amsterdam 09109

Norway
Landslaget Musikk i Skolen, Oslo 09277
Norske Musikklaereres Landsforbund, Kristiansand 09304
Norsk Musikkinformasjon, Oslo 09331

Poland
Towarzystwo imienia Fryderyka Chopina, Warszawa 09620

Russia
Russian Union of Composers, Moskva 09966

Slovenia
Društvo Slovenskih Skladateljev, Ljubljana 10099

Spain
Comité Nacional Español del Consejo Internacional de la Música, Madrid 10293
Fundación Juan March, Madrid 10311

Sweden
Föreningen Svenska Tonsättare, Stockholm 10460
Fylkingen, Stockholm 10463
International Association of Music Libraries, Archives and Documentation Centres, Stockholm 10472
Kungliga Musikaliska Akademien, Stockholm 10487
Musikaliska Konstföreningen, Stockholm 10502
Svenska Samfundet för Musikforskning, Stockholm 10575

Switzerland
Eidgenössischer Musikverband, Dagmersellen 10666
European Association of Music Conservatories, Academies and High Schools, Bern 10678
European Association of Music Festivals, Genève 10679
Fondation Hindemith, Vevey 10705
Gesellschaft der Freunde alter Musikinstrumente, Binningen 10710
Internationale Gesellschaft für Musikwissenschaft, Basel 10738
Schweizerische Arbeitsgemeinschaft für Jugendmusik und Musikerziehung, Neuheim 10794
Schweizerische Gesellschaft für die Rechte der Urheber musikalischer Werke, Zürich 10817
Schweizerische Musikforschende Gesellschaft, Basel 10874
Schweizerischer Musikpädagogischer Verband, Zürich 10900
Schweizer Musikrat, Aarau 10962
Société Suisse des Professeurs de Musique de l'Enseignement Secondaire, Emmenbrücke 10987
Verband Musikschulen Schweiz, Liestal 11003

Trinidad and Tobago
Trinidad Music Association, Port of Spain 11126

United Kingdom
Alkan Society, Salisbury 11221
Association for British Music, Boston 11261
Barbirolli Society, Retford 11424
Benesh Institute of Choreology, London 11433
Benslow Music Trust, Hitchin 11434
British Federation of Festivals, Macclesfield 11559
British Jazz Society, Twickenham 11597
British Library National Sound Archive, London 11603
British Music Society, Upminster 11613
British Society for Electronic Music, London 11655
Choir Schools Association, Wells 11781
The Chopin Society, Wembley Park 11782
Church Music Association, London 11784
Cinema Organ Society, Cambridge 11785
Commonwealth Music Association, London 11828
The Composers' Guild of Great Britain, London 11833
Delius Society, Chatham 11879
Dolmetsch Foundation, Chichester 11898
The Donizetti Society, London 11900
Edinburgh Festival Society, Edinburgh 11933
The Edinburgh Highland Reel and Strathspey Society, Edinburgh 11934
Edinburgh Royal Choral Union, Edinburgh 11938
Elgar Society, Woldingham 11955
Elgar Society (London), Northwood 11956
The English Folk Dance and Song Society, London 11961
Fair Organ Preservation Society, Norwich 12055
The Galpin Society for the Study of Musical Instruments, Leicester 12098
Gilbert and Sullivan Society of Edinburgh, Edinburgh 12114
Glenn Miller Society, London 12121
Guild of Church Musicians, Blechingley 12133
Halle Concerts Society, Manchester 12143
Havergal Brian Society, Watford 12150
Incorporated Association of Organists, Redditch 12193
Incorporated Society of Musicians, London 12195
Irish Heritage, London 12411
Jazz Centre Society, London 12416
Johann Strauss Society of Great Britain, Caterham 12419
The Liszt Society, London 12455
London Orchestral Association, London 12466
Lute Society, Richmond 12474
The Morris Ring, Ipswich 12539
Music Advisers' National Association, Warwick 12547
Musical Box Society of Great Britain, Cambridge 12548
Music Masters' and Mistresses' Association, Guildford 12549
Peter Warlock Society, London 12690
Plainsong and Mediaeval Music Society, Cambridge 12702
Royal Academy of Music, London 12764
Royal Choral Society, London 12777
Royal College of Music, London 12781
The Royal College of Organists, London 12784
Royal Liverpool Philharmonic Society, Liverpool 12828
The Royal Musical Association, Oxford 12834
The Royal Philharmonic Society, London 12839
Scottish National Orchestra Society, Glasgow 12918
Society for Research in the Psychology of Music and Music Education, Dagenham 12993
Society for the Promotion of New Music, London 13006
Society of Authors, London 13030
Spohr Society of Great Britain, Sheffield 13087
Theatre Organ Preservation Society, Slough 13123
Thomas Tallis Society, London 13128
Vintage Light Music Society, West Wickham 13186
Viola da Gamba Society, London 13187
Viola d'Amore Society, Chorleywood 13188
Vivaldi Society, London 13191
Wagner Society, London 13192

Welsh Folk Dance Society, Llanfair Caereinion 13201
Welsh Folk Song Society, Aberystwyth 13202
Wilhelm Furtwängler Society, London 13207
Workers' Music Association, London 13216

U.S.A.

Accordion Teacher's Guild, Kansas City 13312
American Academy of Teachers of Singing, New York 13439
American Bach Foundation, Washington 13614
American Beethoven Society, San Jose 13618
American College of Musicians, Austin 13716
American Harp Society, Hollywood 13839
American Institute for Verdi Studies, New York 13871
American Institute of Musical Studies, Dallas 13889
American Liszt Society, Rochester 13918
American Matthay Association, Pittsburgh 13930
American Musical Instrument Society, Vermillion 13949
American Music Conference, Carlsbad 13950
The American Musicological Society, Philadelphia 13951
American String Teachers Association, Dallas 14183
American Theatre Organ Society, Sacramento 14191
Association for Technology in Music Instruction, Eugene 14340
Bernard Herrmann Society, North Hollywood 14577
Bruckner Society of America, Iowa City 14620
Catgut Acoustical Society, Montclair 14650
Charles Ives Society, Bloomington 14721
Chinese Musical and Theatrical Association, New York 14741
Chinese Music Society of North America, Woodridge 14742
College Music Society, Missoula 14774
Council for Research in Music Education, Urbana 14883
Dalcroze Society of America, Seattle 14950
Django Reinhardt Society, Middletown 14982
The Duke Ellington Society, New York 14990
Electronic Music Consortium, Columbus 15036
Ernest Bloch Society, Gualala 15054
Gilbert and Sullivan Society, Brooklyn 15160
Glazounov Society, Sonoma 15162
Institute for Studies in American Music, Brooklyn 15298
Institute of the American Musical, Los Angeles 15340
Inter-American Music Council, Washington 15353
International Alban Berg Society, New York 15371
International Association of Jazz Educators, Manhattan 15409
The International Council for Traditional Music, New York 15454
International Percy Grainger Society, White Plains 15510
Jack Point Preservation Society, New Ellenton 15590
Keyboard Teachers Association International, Westbury 15637
Kurt Weill Foundation for Music, New York 15641
Leschetizky Association, New York 15662
Massenet Society, Fort Lee 15699
Max Steiner Memorial Society, Los Angeles 15706
Medtner Society, United States of America, Chicago 15715
Music and Arts Society of America, Washington 15761
Music Critics Association, Westfield 15762
Music Educators National Conference, Reston 15763
Music Teachers National Association, Cincinnati 15765
National Academy of Recording Arts and Sciences, Burbank 15777
National Association of College Wind and Percussion Instructors, Kirksville 15843
National Association of Music Executives in State Universities, Tucson 15874
National Association of Pastoral Musicians, Washington 15877
National Association of Schools of Music, Reston 15888
National Association of Teachers of Singing, Jacksonville 15911
National Black Music Caucus of the Music Educators National Conference, Ann Arbor 15920
National Council of State Supervisors of Music, Columbus 16008
National Federation Interscholastic Music Association, Kansas City 16044
National Guild of Community Schools of the Arts, Englewood 16067
National Guild of Piano Teachers, Austin 16068
National League of American PEN Women, Washington 16100
National Music Council, Englewood 16112
National Opera Association, Shreveport 16114
National Orchestral Association, New York 16116
National Piano Foundation, Dallas 16128
Opera America, Washington 16251
Organ Historical Society, Richmond 16260
Organization of American Kodaly Educators, Thibodaux 16266
Percussive Arts Society, Lawton 16303
Sir Thomas Beecham Society, Redondo Beach 16476
Society for Asian Music, Ithaca 16500
Society for Ethnomusicology, Bloomington 16520
Society for General Music, Reston 16526
Society for Music Teacher Education, Reston 16553
Society of Composers, New York 16650
Sonneck Society, Canton 16721
Suzuki Association of the Americas, Boulder 16757
Wagner Society of New York, New York 16870
Willem Mengelberg Society, Manitowoc 16902

Zambia

Africa Association for Liturgy, Music and Arts, Lusaka 17148

Mycology s. Botany, Specific

Natural History s. Natural Sciences, general

Natural Resources

Argentina

Comisión Nacional de Investigaciones Espaciales, Buenos Aires 00119

Australia

Australian Conservation Foundation, Fitzroy 00278

Canada

Alberta Sulphur Research, Calgary 01849
Atlantic Petroleum Association, Halifax 01959
Canadian Arctic Resources Committee, Ottawa 02023

Natural Sciences, general

Algeria

Arab Scientific and Technical Information Network, Algiers 00008

Argentina

Academia Nacional de Ciencias de Córdoba, Córdoba 00043
Academia Nacional de Ciencias Exactas, Físicas y Naturales, Buenos Aires 00045
Asociación Argentina de Ciencias Naturales, Buenos Aires 00060
Asociación Argentina para el Progreso de las Ciencias, Buenos Aires 00072
Asociación de Ciencias Naturales del Litoral, Santo Tomé 00078

Australia

Australian Research Grants Committee, Woden 00334
Field Naturalists' Club of Victoria, South Yarra 00396
The Field Naturalists' Society of South Australia, Adelaide 00397
Linnean Society of New South Wales, Milsons Point 00429
North Queensland Naturalists Club, Cairns 00441
The Queensland Naturalists' Club, Brisbane 00449
Western Australian Naturalists' Club, Nedlands 00524

Austria

Internationales Institut für den Frieden, Wien 00680
Naturwissenschaftlicher Verein für Kärnten, Klagenfurt 00712
Naturwissenschaftlicher Verein für Steiermark, Graz 00713
Naturwissenschaftlich-medizinischer Verein in Innsbruck, Innsbruck 00714
Österreichische Gesellschaft für China-Forschung, Wien 00768
Österreichische Gesellschaft für Wissenschaftsgeschichte, Wien 00849
Österreichische Gesellschaft und Institut für Umweltschutz, Umwelttechnologie und Umweltwissenschaften, Wien 00852
Verein Tiroler Landesmuseum Ferdinandeum, Innsbruck 01003
Verein zur Verbreitung naturwissenschaftlicher Kenntnisse, Wien 01006

Azerbaijan

Commission for the Study of Productive Forces and Natural Resources, Baku 01041
Terminological Committee, Baku 01049

Belgium

Académie Royale des Sciences d'Outre-Mer, Bruxelles 01096
Koninklijke Academie voor Wetenschappen, Letteren en Schone Kunsten van België, Bruxelles 01409
Les Naturalistes Belges, Bruxelles 01414
Société Royale des Sciences Médicales et Naturelles de Bruxelles, Bruxelles 01488

Bulgaria

Bulgarian Society of Natural History, Sofia 01759
Scientific Information Centre at the Bulgarian Academy of Sciences, Sofia 01775

Canada

Association Canadienne-Française pour l'Avancement des Sciences, Montréal 01874
Cercles des Jeunes Naturalistes, Montréal 02332
Manitoba Naturalists Society, Winnipeg 02629
Natural History Society of Prince Edward Island, Charlottetown 02674
Natural Sciences and Engineering Research Council of Canada, Ottawa 02675
New Brunswick Federation of Naturalists, Saint John 02680
Vancouver Natural History Society, Vancouver 02977

Chile

Instituto de Estudios y Publicaciones Juan Molina, Santiago 03054
Sociedad Chilena de Historia Natural, Santiago 03075

Colombia

Academia Colombiana de Ciencias Exactas, Físicas y Naturales, Bogotá 03336
Instituto de Ciencias Naturales, Bogotá 03391
Sociedad de Ciencias Naturales Caldas, Medellín 03417

Croatia

Hrvatsko Prirodoslovno Društvo, Zagreb 03462

Cuba

Centro de Estudios de Historia y Organización de la Ciencia Carlos J. Finlay, La Habana 03473
Sociedad Cubana de Historia de la Ciencia y de la Técnica, La Habana 03481

Denmark

Danmarks Naturvidenskabelige Samfund, Roskilde 03584
Dansk Naturhistorisk Forening, København 03653
Jysk Forening for Naturvidenskab, Arhus 03755
Naturhistorisk Forening for Jylland, Arhus 03775
Naturhistorisk Forening for Nordsjaelland, Hillerød 03776
Selskabet for Naturlaerens Udbredelse, København 03816

Ecuador

Asociación Ecuatoriana de Museos, Quito 03844

Finland

European Association for Environmental History, Helsinki 03955

France

Fédération Française de Sociétés des Sciences Naturelles, Paris 04521
Fédération Française des Sociétés de Sciences Naturelles, Paris 04525
Naturalistes Parisiens, Paris 04603
Société des Sciences Historiques et Naturelles de la Corse, Bastia 04716
Société des Sciences Naturelles de Bourgogne, Dijon 04717
Société des Sciences Physiques et Naturelles de Bordeaux, Talence 04718
Société Linnéenne de Provence, Marseille 04888
Société Nationale des Sciences Naturelles et Mathématiques, Cherbourg 04902

Germany

Battelle-Institut e.V., Frankfurt 05166
Deutsche Akademie der Naturforscher Leopoldina, Halle 05285
Deutsche Gesellschaft für Elektronenmikroskopie e.V., Berlin 05343
Deutsche Gesellschaft für Geschichte der Medizin, Naturwissenschaft und Technik e.V., München 05355
Deutscher Naturkundeverein e.V., Stuttgart 05532
Evangelische Studiengemeinschaft e.V., Heidelberg 05691
Fraunhofer-Gesellschaft zur Förderung der angewandten Forschung e.V., München 05759
Georg-Agricola-Gesellschaft zur Förderung der Geschichte der Naturwissenschaften und der Technik e.V., Düsseldorf 05782
Gesellschaft Deutscher Naturforscher und Ärzte e.V., Leverkusen 05795
Gesellschaft für Strahlen- und Umweltforschung, Ergersheim 05862
Joachim Jungius-Gesellschaft der Wissenschaften e.V., Hamburg 06016
Mathematisch-Naturwissenschaftlicher Fakultätentag, Hamburg 06112
Naturforschende Gesellschaft Bamberg e.V., Viereth-Trunstadt 06127
Naturforschende Gesellschaft Freiburg, Freiburg 06128
Naturhistorische Gesellschaft Hannover, Hannover 06129
Naturhistorische Gesellschaft Nürnberg e.V., Nürnberg 06130
Naturhistorischer Verein der Rheinlande und Westfalens, Bonn 06131
Naturwissenschaftlicher und Historischer Verein für das Land Lippe e.V., Detmold 06133
Naturwissenschaftlicher Verein für das Fürstentum Lüneburg von 1851 e.V., Lüneburg 06134
Naturwissenschaftlicher Verein in Hamburg, Hamburg 06135
Naturwissenschaftlicher Verein zu Bremen, Bremen 06136
POLLICHIA e.V., Annweiler 06238
Rheinische Naturforschende Gesellschaft, Mainz 06263
Senckenbergische Naturforschende Gesellschaft, Frankfurt 06280

Guatemala

Academia de Ciencias Médicas, Físicas y Naturales de Guatemala, Guatemala City 06553
Asociación Centroamericana de Historia Natural, Guatemala City 06558

Haiti

Conseil National des Recherches Scientifiques, Port-au-Prince 06579

Hungary

Müszaki és Természettudományi Egyesületek Szövetségi Kamarája, Budapest 06665
Tudományos Ismeretterjesztő Társulat, Budapest 06676

Iceland

Surtseyjarfélagid, Reykjavik 06701

India

Bengal Natural History Society, Darjeeling 06740
Bombay Natural History Society, Bombay 06747

Indonesia

Perkumpulan Penggemar Alam di Indonesia, Bogor 06892

Ireland

Irish Science Teachers Association, Dublin 06999

Italy

Accademia Gioenia di Scienze Naturali, Catania 07186
Accademia Nazionale delle Scienze, detta dei XL, Roma 07213
Accademia Olimpica, Vicenza 07225
Associazione Campana degli Insegnanti di Scienze Naturali, Napoli 07263
Centro Studi Ricerche Ligabue, Venezia 07590
Società dei Naturalisti in Napoli, Napoli 07781
Società di Scienze Naturali del Trentino, Trento 07789
Società Italiana di Scienze Naturali, Milano 07946
Società Naturalisti Veronesi F. Zorzi, Verona 07990
Società Toscana di Scienze Naturali, Pisa 08026
Unione Bolognese Naturalisti, Bologna 08041

Japan

Nihon Gakujutsu Shinko-kai, Tokyo 08200

Kenya

African Academy of Sciences, Nairobi 08458
African Network of Scientific and Technological Institutions, Nairobi 08471
Association of Faculties of Science in African Universities, Nairobi 08479
East Africa Natural History Society, Nairobi 08493
East African Natural History Society Nairobi, Nairobi 08497

Luxembourg

Société des Naturalistes Luxembourgeois, Luxembourg 08587

Malaysia

Malayian Nature Society, Kuala Lumpur 08644

Mexico

Centro Científico y Técnico Francés en México, México 08713
Consejo Nacional de Ciencia y Tecnología, México 08729
Sociedad Mexicana de Historia Natural, México 08766

Moldova, Republic of

Teriological Society of Moldova, Chişinău 08793

Morocco

Société des Sciences Naturelles et Physiques du Maroc, Rabat 08808

Netherlands

Koninklijke Maatschappij voor Natuurkunde onder de Zinspreuk Diligentia, 's-Gravenhage 08949
Koninklijke Nederlandse Natuurhistorische Vereniging, Utrecht 08958
Natuurhistorisch Genootschap in Limburg, Maastricht 08979
Nederlandse Organisatie voor Wetenschappelijk Onderzoek, 's-Gravenhage 08998

Norway

Joint Committee of the Nordic Natural Science Research Councils, Oslo 09272
Norges Teknisk-Naturvitenskapelige Forskningsråd, Oslo 09289
Norske Naturhistoriske Museers Landsforbund, Tromsø 09305
Scandinavian Society for Electron Microscopy, Oslo 09345

Peru

Academia Nacional de Ciencias Exactas, Físicas y Naturales de Lima, Lima 09438

Poland

Polskie Towarzystwo Przyrodników im. Kopernika, Warszawa 09601

Portugal

Sociedade de Estudos Açoreanos Afonso Chaves, Ponta Delgada 09680
Sociedade Portuguesa de Ciências Naturais, Lisboa 09695

Russia

Commission for Co-ordination of Research in State Nature Reserves, Moskva 09816

Moscow Society of Naturalists, Moskva 09886
National Committee of History and Philosophy of Natural Science and Technology, Moskva 09896

Slovenia

Raziskovalna Skupnost Slovenije, Ljubljana 10103
Society for Natural Sciences of Slovenia, Ljubljana 10108

Spain

Academia de Ciencias Exactas, Físicas, Químicas y Naturales de Zaragoza, Zaragoza 10239
Real Academia de Ciencias Exactas, Físicas y Naturales, Madrid 10342
Real Sociedad Española de Historia Natural, Madrid 10362
Sociedad de Ciencias Aranzadi, San Sebastián 10371

Sweden

Naturvetenskapliga Forskningsrådet, Stockholm 10504
Svenska Linné-Sällskapet, Uppsala 10561
Wenner-Gren Center Foundation for Scientific Research, Stockholm 10605

Switzerland

Institut National Genevois, Genève 10730
Naturforschende Gesellschaft des Kantons Glarus, Glarus 10766
Naturforschende Gesellschaft in Basel, Riehen 10767
Naturforschende Gesellschaft in Bern, Bern 10768
Naturforschende Gesellschaft in Zürich, Forch 10769
Naturforschende Gesellschaft Luzern, Luzern 10770
Naturforschende Gesellschaft Schaffhausen, Schaffhausen 10771
Naturforschende Gesellschaft Solothurn, Brugglen 10772
Naturwissenschaftliche Gesellschaft Winterthur, Wiesendangen 10773
Schweizerische Akademie der Naturwissenschaften, Bern 10788
Società Ticinese di Scienze Naturali, Lugano 10965
Société de Physique et d'Histoire Naturelle de Genève, Genève 10969
Société Fribourgeoise des Sciences Naturelles, Fribourg 10974
Société Neuchâteloise des Sciences Naturelles, Neuchâtel 10980
Société Vaudoise des Sciences Naturelles, Lausanne 10990
Vereinigung Schweizerischer Naturwissenschaftslehrer, Neerach 11020

Trinidad and Tobago

Caribbean Council for Science and Technology, Port of Spain 11110

United Kingdom

Armagh Field Naturalists Society, Armagh 11247
Ashmolean Natural History Society of Oxfordshire, Oxford 11255
Association of School Natural History Societies, Lancing 11386
Ayrshire Archaeological and Natural History Society, Ayr 11419
Belfast Natural History and Philosophical Society, Belfast 11432
Berwickshire Naturalists' Club, Eyemouth 11436
Birmingham Natural History Society, Birmingham 11450
The British Society for the History of Science, Faringdon 11673
Buteshire Natural History Society, Bute 11718
Cambridge Philosophical Society, Cambridge 11737
Chester Society of Natural Science, Literatur and Art, Chester 11774
Dorset Natural History and Archaeological Society, Dorchester 11902
East Lothian Antiquarian and Field Naturalists' Society, Dunbar 11919
Field Studies Council, Shrewsburry 12074
Hampshire Field Club and Archaeological Society, Winchester 12144
Institution of Environmental Sciences, London 12302
Isle of Man Natural History and Antiquarian Society, Douglas 12412
Isle of Wight Natural History and Archaeological Society, Shanklin 12413
Linnean Society of London, London 12454
London Natural History Society, Frinton-on-Sea 12465
Natural Environment Research Council, Swindon 12619
Natural History Society of Northumbria, Newcastle-upon-Tyne 12620
Northamptonshire Natural History Society and Field Club, Northampton 12632
North Staffordshire Field Club, Stoke-on-Trent 12642
North Western Naturalists' Union, Stockport 12646
Nottingham and Nottinghamshire Field Club, Nottingham 12649
Perthshire Society of Natural Science, Perth 12689
Quekett Microscopical Club, London 12734
The Ray Society, London 12745
Royal Microscopical Society, Oxford 12833
The Royal Society, London 12848
School Natural Science Society, Gillingham 12888
Scottish Field Studies Association, Blairgowrie 12907
Selborne Society, Oxted 12938
La Société Guernesiaise, Saint Peter Port 12961
Société Jersiaise, Saint Helier 12962
Society for the History of Natural History, London 13004
Somerset Archaeological and Natural History Society, Taunton 13074
Wiltshire Archaeological and Natural History Society, Devizes 13209
Woolhope Naturalists' Field Club, Hereford 13213

U.S.A.

Academy of Natural Sciences, Philadelphia 13299
American Nature Study Society, Homer 13956
American Quaternary Association, Amherst 14023
American Society of Ichthyologists and Herpetologists, Carbondale 14134
Annual Reviews, Palo Alto 14222
Applied Technology Council, Redwood City 14233
Association for Science, Technology and Innovation, Arlington 14332
Audubon Naturalist Society of the Central Atlantic States, Chevy Chase 14549
Bio-Integral Resource Center, Berkeley 14587
Foundation for Field Research, Alpine 15125
Institute of Environmental Sciences, Mount Prospect 15319
John Burroughs Association, New York 15611
National Association for Interpretation, Fort Collins 15806
SIGMA Xi – The Scientific Research Society, Research Triangle Park 16472
Smithsonian Institution, Washington 16481
Society for the Promotion of Science and Scholarship, Palo Alto 16606
Western Society of Naturalists, Moss Landing 16886
XPLOR International, Palos Verdes Estates 16939
Yosemite Association, Yosemite National Park 16942

Venezuela

Academia de Ciencias Físicas, Matemáticas y Naturales, Caracas 16958
Sociedad de Ciencias Naturales La Salle, Caracas 17010
Sociedad Venezolana de Ciencias Naturales, Caracas 17018

Yugoslavia

Akademia e Shkencave dhe e Arteve e Kosovës, Priština 17076

Zimbabwe

Lowveld Natural History Branch, Wildlife Society of Zimbabwe, Chiredzi 17174

Navigation

Argentina

Centro de Navegación Transatlántica, Buenos Aires 00110
Departamento de Estudios Históricos Navales, Buenos Aires 00132

Australia

Australian Institute of Navigation, Sydney 00308

Canada

Canadian Nautical Research Society, Ottawa 02208

Chile

Liga Marítima de Chile, Valparaíso 03061

Croatia

Drustvo za Proučavanje i Unapredenje Pomorstva, Rijeka 03457

Egypt

Association of African Maritime Training Institutes, Alexandria 03880

France

Académie de Marine, Paris 04090

Germany

Deutsche Gesellschaft für Ortung und Navigation e.V., Düsseldorf 05393
Deutscher Nautischer Verein von 1868 e.V., Hamburg 05534
Forschungszentrum des Deutschen Schiffbaues e.V., Hamburg 05756
Verband Deutscher Schiffahrts-Sachverständiger e.V., Hamburg 06359

Ireland

Maritime Institute of Ireland, Dun Laoghaire 07008

Italy

Accademia Nazionale di Marina Mercantile, Genova 07219
Associazione Italiana di Tecnica Navale, Genova 07328
Centro per gli Studi di Tecnica Navale, Genova 07518
Centro Studi di Diritto Fluviale e della Navigazione Interna, Venezia 07553
Ente di Unificazione Navale, Genova 07650

Japan

Nippon Kokai Gakkai, Tokyo 08337

Poland

Polskie Towarzystwo Nautologiczne, Gdynia 09592

United Kingdom

International Association of Institutes of Navigation, London 12330
The Nautical Institute, London 12621
Royal Institute of Navigation, London 12814
Royal Naval Bird Watching Society, Southsea 12835
Society for Nautical Research, London 12984

U.S.A.

American Society of Naval Engineers, Alexandria 14147
Association of Naval R.O.T.C. Colleges and Universities, Lafayette 14470
Great Lakes Commission, Ann Arbor 15175
Great Lakes Maritime Institute, Detroit 15177
Institute of Navigation, Alexandria 15334
International Omega Association, Arlington 15506
Mystic Seaport Museum, Mystic 15767
National Association of Marine Surveyors, Chesapeake 15872
Permanent International Association of Navigation Congresses – United States Section, Washington 16306
Whaling Museum Society, Cold Spring Harbor 16890
Wild Goose Association, The Plains 16898

Neurology

Argentina

Sociedad Argentina de Ciencias Neurológicas, Psiquiátricas y Neuroquirúrgicas, Buenos Aires 00162
Sociedad Argentina de Neurología, Psiquiatría y Neurocirugía, Buenos Aires 00178

Australia

Asian-Australasian Society of Neurological Surgeons, Brisbane 00221
Australian Association of Neurologists, Melbourne 00265

Austria

Arbeitsgemeinschaft für Neurobiologie, Wien 00554
Gesellschaft für biologische und psychosomatische Medizin, Wien 00614
Gesellschaft Österreichischer Nervenärzte und Psychiater, Wien 00638
Österreichische Multiple Sklerosis Gesellschaft, Wien 00870
Verein für Psychiatrie und Neurologie, Wien 00986

Belgium

Belgische Wetenschappelijke Vereniging voor Neurochirurgie, Brugge 01164
European Brain Injury Society, Bruxelles 01301
European Federation of Child Neurology Societies, Bruxelles 01348
Fondation Born-Bunge pour la Recherche, Wilrijk 01362
Société Belge d'Electroencéphalographie et de Neurophysiologie Clinique, Ottignies 01431
Société Belge de Neurologie, Bruxelles 01437

Bulgaria

Bulgarian Society of Electroencephalography, Electromyography and Clinical Neurophysiology, Sofia 01758
Scientific Association of the Bulgarian Neurologists, Sofia 01774

Canada

Amyotrophic Lateral Sclerosis Society of Canada, Toronto 01867
Association de Paralysie Cérébrale du Québec, Québec 01877
Atlantic Cerebral Palsy Association, Fredericton 01957
British Columbia Parkinsons's Disease Association, Vancouver 01985
The Canadian Neurological Society, Montréal 02209
Canadian Neurosurgical Society, Toronto 02210
Canadian Society of Clinical Neurophysiologists, Ottawa 02280
Canadian Society of Electroencephalographers, Electromyographers and Clinical Neurophysiologists, Sherbrooke 02282
Cerebral Palsy Association in Alberta, Calgary 02333
Cerebral Palsy Association of Manitoba, Winnipeg 02334
Cerebral Palsy Association of Newfoundland, St. John's 02335
Cerebral Palsy Association of Nova Scotia, Halifax 02336
Cerebral Palsy Association of Prince Edward Island, Charlottetown 02337
Cerebral Palsy Association of Prince George and District, Prince George 02338
Children's Rehabilitation and Cerebral Palsy Association, Vancouver 02346
Manitoba Cerebral Palsy Association, Winnipeg 02612
Myasthenia Gravis Foundation of British Columbia, Vancouver 02666
Nanaimo Neurological and Cerebral Palsy Association, Nanaimo 02667
North Okanagan Neurological Association, Vernon 02716
Parkinson Foundation of Canada, Toronto 02811
Prince Edward Island Cerebral Palsy Association, Charlottetown 02830

Colombia

Fundación OFA para el Avance de las Ciencias Biomédicas, Bogotá 03375

Denmark

Danske Nervelaegers Organisation, Odense 03611
Dansk Epilepsiforening, Hvidovre 03612
Dansk Neurologisk Selskab, Hellerup 03655

France

Club Européen d'Histoire de la Neurologie, Lyon 04389
Congrès de Psychiatrie et de Neurologie de Langue Française, Limoges 04448
Société de Neurophysiologie Clinique de Langue Française, Paris 04675
Société d'Oto-Neuro-Ophthalmologie du Sud-Est de la France, Marseille 04775
Société Française de Neurologie, Paris 04825

Germany

Berufsverband Deutscher Nervenärzte e.V., Frankfurt 05203
Deutsche EEG-Gesellschaft, Berlin 05300
Deutsche Gesellschaft für Neurologie, Würzburg 05388
Deutsche Gesellschaft für Neuropathologie und Neuroanatomie e.V., München 05389
Deutsche Gesellschaft für Psychiatrie und Nervenheilkunde e.V., Köln 05410
Deutsche Gesellschaft für Psychosomatische Medizin e.V., München 05412
Deutsche Neurovegetative Gesellschaft, Düsseldorf 05476
Vereinigung Deutscher Neuropathologen und Neuroanatomen, Giessen 06406

India

Indian Brain Research Association, Calcutta 06780

Israel

International Society of Computerized and Quantitative EMG, Jerusalem 07058
Israeli Neurological Association, Tel Hashomer 07077
Israel Society of Clinical Neurophysiology, Jerusalem 07098

Italy

Società Italiana Attività Nervosa Superiore, Milano 07822
Società Italiana di Elettroencefalografia e Neurofisiologia, Bologna 07866
Società Italiana di Neurologia, Roma 07911
Società Italiana di Neuropediatria, Siena 07912
Società Italiana di Neurosonologia, Bologna 07915
Società Italiana di Oto-Neuro-Oftalmologia, Bologna 07923

Japan

Nihon Seishin Shinkei Gakkai, Tokyo 08257
Nippon No-Shinkei Gek Gakkai, Tokyo 08351
Nippon Shinkei Gakkai, Tokyo 08378
Nippon Shinkeikagaku Gakkai, Tokyo 08379
Nippon Teiinoshujutsu Kenkyukai, Tokyo 08391

Mexico

Sociedad Mexicana de Neurologia y Psiquiatria, México 08768

Netherlands

Nederlandse Vereniging voor Neurologie, Utrecht 09029

Nigeria

Pan African Association of Neurological Sciences, Lagos 09245

Norway

European Brain and Behaviour Society, Oslo 09265

Philippines

ASEAN Neurological Society, Manila 09491

Poland

Polskie Towarzystwo Neurologiczne, Konstancin-Jeziorna 09594

Portugal

Sociedade Portuguesa de Neurologia e Psiquiatria, Lisboa 09705

Romania

Societatea Romana de Neurochirurgie, Bucuresti 09793
Societatea Romana de Sprijin a Virstnicilor Suferinzi de Afectiuni de tip Alzheimer, Bucuresti 09795

South Africa

Research Group for Neurochemistry, Tygerberg 10169

Switzerland

Schweizerische Neurologische Gesellschaft, Aarau 10875

Turkey

Türk Nöro-Psikiyatri Derněi, Istanbul 11168

United Kingdom

Association of British Neurologists, London 11313
Electroencephalographic Society, Birmingham 11953
Parkinson's Disease Society of the United Kingdom, London 12678
Society for Psychosomatic Research, London 12990
Society of British Neurological Surgeons, Glasgow 13031

U.S.A.

American Academy for Cerebral Palsy and Developmental Medicine, Richmond 13376
American Academy of Neurological Surgery, Philadelphia 13408
American Association for the Study of Headache, West Deptford 13501
American Association of Electrodiagnostic Medicine, Rochester 13530
American Association of Neurological Surgeons, Park Ridge 13555

American Association of Neuropathologists, Salt Lake City *13556*
American Board of Neurological Surgery, Houston *13639*
American Board of Psychiatry and Neurology, Deerfield *13660*
American Brain Tumor Association, Chicago *13669*
American College of Neuropsychiatrists, Farmington Hills *13717*
American Electroencephalographic Society, Bloomfield *13793*
American Medical Electroencephalographic Association, Elm Grove *13935*
American Neurological Association, Minneapolis *13958*
American Parkinson Disease Association, Staten Island *13991*
American Society for Neurochemistry, Palo Alto *14074*
American Society for Stereotactic and Functional Neurosurgery, Houston *14086*
American Society of Electroneurodiagnostic Technologists, Carroll *14121*
Association for Research in Nervous and Mental Disease, New York *14329*
Brain Information Service, Los Angeles *14611*
Cajal Club, Denver *14631*
Child Neurology Society, Saint Paul *14732*
Congress of Neurological Surgeons, Cincinnati *14842*
Dystonia Medical Research Foundation, Beverly Hills *14993*
Huntington's Disease Society of America, New York *15241*
Microneurography Society, Richmond *15728*
Neurosurgical Society of America, Cleveland *16193*
Society for Neuroscience, Washington *16555*
Society of Neurological Surgeons, Boston *16685*
Society of Neurosurgical Anesthesia and Critical Care, Richmond *16686*
Tourette Syndrome Association, Bayside *16787*
World Society for Stereotactic and Functional Neurosurgery, Houston *16932*

Venezuela
Sociedad Venezolana de Psiquiatría y Neurología, Caracas *17030*

Vietnam
Vietnamese Association of Neurology, Psychiatry and Neurosurgery, Hanoi *17053*

Nuclear Medicine

Argentina
Asociación Argentina de Biología y Medicina Nuclear, Buenos Aires *00059*
Asociación Latinoamericana de Sociedades de Biología y Medicina Nuclear, Buenos Aires *00087*
Centro de Medicina Nuclear, Buenos Aires *00109*
Sociedad Argentina de Medicina Nuclear, Buenos Aires *00176*

Australia
Australian and New Zealand Society of Nuclear Medicine, Darlinghurst *00259*

Austria
Österreichische Gesellschaft für Nuclearmedizin, Wien *00816*
Österreichische Röntgengesellschaft – Gesellschaft für medizinische Radiologie und Nuklearmedizin, Wien *00899*

Canada
Canadian Association of Radiologists, Montréal *02066*
Manitoba Association of Medical Radiation Technologists, Winnipeg *02607*
Newfoundland Association of Medical Radiation Technologists, Saint John's *02701*
Nova Scotia Society of Medical Radiation Technologists, Armdale *02745*
Prince Edward Island Association of Medical Radiation Technologists, Montague *02828*
Saskatchewan Association of Medical Radiation Technologists, Saskatoon *02881*

China, Republic
Asia and Oceania Federation of Nuclear Medicine and Biology, Taipei *03233*

Colombia
Instituto Nacional de Cancerología, Bogotá *03392*

Germany
Berufsverband der Deutschen Radiologen und Nuklearmediziner e.V., München *05198*
Deutsche Akademie für Nuklearmedizin, Hannover *05288*
Deutsche Gesellschaft für Nuklearmedizin, Bonn *05391*
Deutsche Röntgengesellschaft, Neu-Isenburg *05535*
Vereinigung Südwestdeutscher Radiologen und Nuklearmediziner, Sindelfingen *06415*

Italy
Associazione Italiana di Radiologia e Medicina Nucleare, Roma *07321*
Società Italiana di Radiologia Medica e di Medicina Nucleare, Milano *07941*

Netherlands
Nederlandse Vereniging van Radiologische Laboranten, Utrecht *09010*

Spain
Sociedad Española de Radiología Medica, Madrid *10396*

United Kingdom
European Association of Nuclear Medicine, London *11998*

U.S.A.
American Board of Nuclear Medicine, Los Angeles *13640*
American College of Nuclear Physicians, Washington *13719*
Foundation for the Support of International Medical Training, Lewiston *15134*
Society of Nuclear Medicine, New York *16687*

Nuclear Research

Algeria
Centre des Sciences et de la Technologie Nucléaires, Alger *00014*

Argentina
Comisión Nacional de Energía Atómica, Buenos Aires *00118*

Armenia
Council on Use of Atomic Energy and Technology, Yerevan *00207*

Australia
Australian Atomic Energy Commission Research Establishment, Sutherland *00268*
Australian Institute of Nuclear Science and Engineering, Menai *00309*

Austria
International Atomic Energy Agency, Wien *00663*

Belgium
Central Bureau for Nuclear Measurements, Geel *01183*
Euratom Scientific and Technical Committee, Bruxelles *01261*

Bolivia
Comisión Bolivana de Energía Nuclear, La Paz *01540*

Canada
Canadian Coalition for Nuclear Responsibility, Montréal *02097*
Canadian Nuclear Association, Toronto *02211*

Chile
Comisión Chilena de Energía Nuclear, Santiago *03037*

China, Republic
Atomic Energy Council, Taipei *03250*

Costa Rica
Comisión de Energía Atómica de Costa Rica, San José *03442*

Denmark
Danatom, Haslev *03575*
Dansk Kerneteknisk Selskab, København *03636*
Nordisk Institut for Teoretisk Atomfysik, København *03786*

Ecuador
Comisión Ecuatoriana de Energía Atómica, Quito *03851*

Finland
Suomen Atomiteknillinen Seura, Helsinki *04007*

France
European Atomic Energy Society, Paris *04491*
Société Française d'Energie Nucléaire, Paris *04824*

Germany
Arbeitsgemeinschaft Versuchsreaktor, Düsseldorf *05128*
Deutsches Atomforum e.V., Bonn *05569*
European Fast Reactor Association, Eggenstein *05677*
Gesellschaft für Anlagen- und Reaktorsicherheit, Köln *05799*
Normenausschuss Kerntechnik im DIN Deutsches Institut für Normung e.V., Berlin *06183*

Greece
Helliniki Epitropi Atomikis Energhias, Athinai *06533*

India
Indian Society for Nuclear Techniques in Agriculture and Biology, New Delhi *06814*

Israel
Israel Atomic Energy Commission, Tel Aviv *07067*

Italy
Associazione Nazionale di Ingegneria Nucleare, Roma *07366*
Centro Italiano di Studi di Diritto dell'Energia, Roma *07498*
Ente per le Nuove Tecnologie, l'Energia e l'Ambiente, Roma *07658*
Forum Italiano dell'Energia Nucleare, Roma *07700*
Società Nucleare Italiana, Firenze *07993*
Società Ricerche Impianti Nucleari, Vercelli *08008*

Japan
Nihon Genshiryoku Gakkai, Tokyo *08203*

Mexico
Sociedad Nuclear Mexicana, México *08774*

Netherlands
FOM-Instituut voor Atoom- en Molecuulfysica, Amsterdam *08896*
FOM-Instituut voor Plasmafysica Rijnhuizen, Nieuwegein *08897*

Pakistan
Pakistan Atomic Energy Commission, Islamabad *09380*

Peru
Junta de Control de Energía Atómica, Lima *09472*

Philippines
Philippine Atomic Energy Commission, Quezon City *09524*
Radioisotope Society of the Philippines, Quezon City *09535*

Russia
Commission for Atomic Energy, Moskva *09813*

Switzerland
European Nuclear Society, Bern *10685*
Laboratoire Européen pour la Physique des Particules, Genève *10764*
Schweizerische Gesellschaft der Kernfachleute, Bern *10808*
Schweizerische Vereinigung für Atomenergie, Bern *10940*

Thailand
Office of the Atomic Energy Commission for Peace, Bangkok *11095*

Tunisia
Arab Atomic Energy Agency, Tunis *11127*

United Kingdom
British Nuclear Energy Society, London *11620*
Institution of Nuclear Engineers, London *12311*
Science and Engineering Research Council, Swindon *12889*

Uruguay
Comisión Nacional de Energía Atómica, Montevideo *13255*

U.S.A.
American Nuclear Energy Council, Washington *13959*
American Nuclear Society, La Grange Park *13960*
Institute of Nuclear Materials Management, Northbrook *15336*
United States Council for Energy Awareness, Washington *16821*

Numismatics

Argentina
Instituto Bonaerense de Numismática y Antigüedades, Buenos Aires *00142*

Australia
Australian Numismatic Society, Sydney *00324*

Austria
Österreichische Numismatische Gesellschaft, Wien *00872*

Belgium
Cercle d'Etudes Numismatiques, Bruxelles *01208*
Institut Archéologique du Luxembourg, Arlon *01375*

Canada
Canadian Numismatic Association, Barrie *02212*
Numismatic Educational Services Association, Barrie *02750*
Société d'Archéologie et de Numismatique de Montréal, Montréal *02918*

Croatia
Hrvatsko Numizmatičko Društvo, Zagreb *03461*

Cyprus
Cyprus Numismatic Society, Nicosia *03493*

Denmark
Dansk Numismatisk Forening, Taastrup *03656*
Nordisk Numismatisk Union, København *03790*

Finland
Museovirasto, Helsinki *03983*

France
Société Française de Numismatique, Paris *04826*

Germany
Deutsche Numismatische Gesellschaft, Speyer *05477*
Numismatische Kommission der Länder in der Bundesrepublik Deutschland, Berlin *06221*

Greece
Elliniki Nomismatiki Etaireia, Athinai *06512*

Hungary
Magyar Numizmatikai Társulat, Budapest *06648*

Italy
Accademia Italiana di Studi Filatelici e Numismatici, Reggio Emilia *07197*
Istituto Italiano di Numismatica, Roma *07737*
Società Istriana di Archeologia e Storia Patria, Trieste *07820*

Netherlands
Vereniging voor Penningkunst, Alkmaar *09103*

Poland
Polskie Towarzystwo Archeologiczne i Numizmatyczne, Warszawa *09553*

Portugal
Sociedade Portuguesa de Numismática, Porto *09706*

Romania
Societatea Numismatica Romana, Bucuresti *09791*

South Africa
The South African Numismatic Society, Cape Town *10211*

Spain
Asociación Numismática Española, Barcelona *10278*
Sociedad Ibero-Americana de Estudios Numismáticos, Madrid *10399*

Switzerland
Schweizerische Numismatische Gesellschaft, Bern *10877*

United Kingdom
British Association of Numismatic Societies, London *11511*
British Numismatic Society, Pontyclun *11621*
International Numismatic Commission, London *12379*
Oriental Numismatic Society, Reading *12662*
Royal Numismatic Society, London *12837*

U.S.A.
American Numismatic Association, Colorado Springs *13961*
American Numismatic Society, New York *13962*
International Numismatic Society Authentication Bureau, Philadelphia *15505*
Maximilian Numismatic and Historical Society, Tekonsha *15705*
Numismatic Literary Guild, Glen Rock *16236*
Professional Numismatists Guild, Van Nuys *16357*
Society for International Numismatics, Santa Monica *16541*
Society of Philatelists and Numismatists, Montebello *16690*

Nutrition
s. a. Food

Australia
Asian Pan-Pacific Society for Paediatric Gastroenterology and Nutrition, East Perth *00227*
Australian Institute of Food Science and Technology, Pymble *00301*
Dietitians Association of Australia, Canberra *00385*
Food Technology Association of New South Wales, Sydney *00398*
Food Technology Association of Queensland, Brisbane *00399*
Food Technology Association of Tasmania, Hobart *00400*
Food Technology Association of Western Australia, Perth *00401*

Austria
Arbeitsgemeinschaft für klinische Ernährung, Wien *00553*

Brazil
Instituto Nacional de Pesquisas da Amazonia, Manaus *01688*

Canada
Alberta Registered Dietitians Association, Calgary *01844*
Brewing and Malting Barley Research Institute, Winnipeg *01969*
Canadian Dietetic Association, Toronto *02118*
Canadian Society for Nutritional Sciences, Halifax *02264*
Grain, Feed and Fertilizer Accident Prevention Association, Toronto *02488*
Human Nutrition Research Council of Ontario, Stittsville *02509*
International Development Research Centre, Ottawa *02552*
Newfoundland Dietetic Association, Saint John's *02703*
Nova Scotia Dietetic Association, Halifax *02732*
Prince Edward Island Dietetic Association, Charlottetown *02832*
Saskatchewan Dietetic Association, Regina *02887*

Finland
Agronomiliitto, Helsinki *03949*

France
American European Dietetic Association, Montesson *04129*
Association Européenne pour l'Etude de l'Alimentation et du Développement de l'Enfant, Paris *04209*
Société de Nutrition et de Diététique de Langue Française, Paris *04677*
Syndicat National des Médecins Spécialistes de l'Endocrinologie et de la Nutrition, Paris *04960*

Germany

AID Auswertungs- und Informationsdienst für Ernährung, Landwirtschaft und Forsten e.V., Bonn *05045*
Dachverband wissenschaftlicher Gesellschaften der Agrar-, Forst-, Ernährungs-, Veterinär- und Umweltforschung e.V., Frankfurt *05282*
Deutsche Gesellschaft für Ernährung e.V., Frankfurt *05346*
Förderungsgemeinschaft der Kartoffelwirtschaft e.V., Munster *05704*
Gesellschaft für Ernährungsbiologie e.V., München *05819*
Institut für nationale und internationale Fleisch- und Ernährungswirtschaft, Heidelberg *05965*
International Commission for Uniform Methods of Sugar Analysis, Braunschweig *05985*
Margarine-Institut für gesunde Ernährung, Hamburg *06110*
Vereinigung Getreide-, Markt- und Ernährungsforschung e.V., Bonn *06410*

Italy

Associazione Italiana di Tecnologia Alimentare, Milano *07329*
Centro di Studi e Ricerche sulla Nutrizione e Sugli Alimenti, Parma *07443*
Centro Nazionale di Dietobiologia ed Igiene della Alimentazione, Milano *07516*
Società Italiana di Nutrizione Umana, Roma *07917*

Jamaica

Caribbean Food and Nutrition Institute, Kingston *08073*

Japan

Nippon Bitamin Gakkai, Kyoto *08281*

Netherlands

European Federation of the Associations of Dieticians, Oss *08888*

New Zealand

New Zealand Dietetic Association, Wellington *09143*

Nigeria

African Council of Food and Nutrition Sciences, Harare *09214*

Portugal

Sociedade Portuguesa de Higiene Alimentar, Lisboa *09703*

South Africa

Organic Soil Association of South Africa, Parklands *10160*

Spain

Sociedad Española de Patología Digestiva y de la Nutrición, Madrid *10394*

Switzerland

Schweizerischer Verband der Ingenieur-Agronomen und der Lebensmittel-Ingenieure, Zollikofen *10907*

Trinidad and Tobago

Caribbean Association of Nutritionists and Dieticians, West Moorings *11109*

United Kingdom

The British Dietetic Association, Birmingham *11552*
British Nutrition Foundation, London *11622*
CAB International Bureau of Nutrition, Wallingford *11724*
Gooseberry Society, Barnet *12124*
Institute of Food Research, Reading *12238*
International Union of Nutritional Sciences, London *12403*

U.S.A.

American Celiac Society, West Orange *13682*
American Dietetic Association, Chicago *13787*
American Institute of Nutrition, Bethesda *13890*
Dietary Managers Association, Lombard *14973*
Food and Nutrition Board, Washington *15108*
Institute for Food and Development Policy, San Francisco *15281*
National Association of Nutrition and Aging Services Programs, Grand Rapids *15875*
Nutrition Institute of America, New York *16237*
Society for Nutrition Education, Minneapolis *16557*

Zambia

National Food and Nutrition Commission, Lusaka *17155*

Occultism s. Parapsychology

Oceanography, Marine Sciences

Algeria

Centre de Recherches Océanographiques et des Pêches, Alger *00012*

Argentina

Centro de Investigación de Biologia Marina, San Martín *00104*
Instituto Nacional de Investigación y Desarrollo Pesquero, Mar del Plata *00144*

Australia

Australian Institute of Marine Science, Townsville *00306*

Austria

Haus des Meeres Vivarium Wien, Wien *00648*

Canada

Canadian Meteorological and Oceanographic Society, Newmarket *02199*

Chile

Comité Oceanográfico Nacional, Valparaíso *03043*
Instituto Oceanográfico de Valparaíso, Valparaíso *03059*

China, People's Republic

Chinese Society for Oceanology and Limnology, Quingdao *03181*

Croatia

Drustvo za Proučavanje i Unapredenje Pomorstva, Rijeka *03457*

Denmark

International Council for the Exploration of the Sea, København *03746*
Nordisk Kollegium for Fysisk Oceanografi, København *03787*

Finland

Baltic Marine Environment Protection Commission, Helsinki *03952*
Geofysiikan Seura, Espoo *03962*
Nordisk Kollegium for Fysisk Oceanografi, Helsinki *03985*

France

Association Internationale d'Océanographie Médicale, Nice *04285*
Commission Océanographique Intergouvernementale, Paris *04428*
Committee for European Marine Biology Symposia, Brest *04429*
European Association of Marine Sciences and Techniques, Talence *04487*
European Committee on Ocean and Polar Sciences, Strasbourg *04497*
Fédération Française d'Etudes et de Sports Sous Marins, Marseille *04526*
Joint Committee on Climatic Changes and the Ocean, Paris *04592*
Société d'Océanographie de France, Paris *04771*
Union des Océanographes de France, Paris *04974*

Germany

Conference of Baltic Oceanographers, Kiel *05278*
Deutsche Meteorologische Gesellschaft e.V., Traben-Trarbach *05469*
Deutsche Meteorologische Gesellschaft e.V., Zweigverein Hamburg, Hamburg *05470*
Deutsche Wissenschaftliche Kommission für Meeresforschung, Hamburg *05626*

Ghana

Accra Regional Maritime Academy, Accra *06469*

Italy

Società Italiana di Biologia Marina, Livorno *07843*
Tecnomare, Venezia *08036*

Ivory Coast

Académie Régionale des Sciences et Techniques de la Mer d'Abidjan, Abidjan *08056*
Centre de Recherches Océanographiques, Abidjan *08061*

Japan

Nippon Kaiyo Gakkai, Tokyo *08323*

Kenya

East Asian Seas Action Plan, Nairobi *08499*

Malta

Euro-Mediterranean Centre on Marine Contamination Hazards, Valletta *08661*

Monaco

Commission Internationale pour l'Exploration Scientifique de la Mer Méditerranée, Monaco *08796*

New Zealand

New Zealand Marine Sciences Society, Wellington *09167*

Norway

Norske Havforskeres Forening, Nordnes *09301*

Philippines

Association of Southeast Asian Marine Scientists, Manila *09511*

Puerto Rico

Association of Marine Laboratories of the Caribbean, Mayaguez *09724*

Russia

CMEA Coordination Centre for Study of World Oceans Development of Techniques for Exploration and Utilization of Resources, Moskva *09811*
Commission for the World's Oceans, Moskva *09843*
National Committee of the Pacific Ocean Research Association, Moskva *09902*

South Africa

Transvaal Underwater Research Group, Johannesburg *10231*

Sweden

Baltic Marine Biologists, Lysekil *10438*
Conference of the Baltic Oceanographers, Norrköping *10447*

United Kingdom

Advisory Committee on Protection of the Sea, London *11207*
British Marine Aquarist Association, Hull *11605*
Challenger Society for Marine Science, Godalming *11756*
International Glaciological Society, Cambridge *12366*
International Hydrofoil Society, London *12368*
Marine Biological Association of the United Kingdom, Plymouth *12492*
Maritime Trust, London *12495*
The Scottish Association for Marine Sciemce, Oban *12895*

U.S.A.

American Society of Limnology and Oceanography, Gloucester Point *14141*
Aquatic Research Institute, Hayward *14235*
Center for Oceans Law and Policy, Charlottesville *14691*
Coastal Engineering Research Council, Charleston *14766*
Estuarine Research Federation, Crownsville *15061*
International Association for the Physical Sciences of the Ocean, Del Mar *15393*
International Tsunami Information Center, Honolulu *15568*
Marine Technology Society, Washington *15691*
Oceanic Society, Washington *16242*
Oceanic Society Expeditions, San Francisco *16243*
Scientific Committee on Oceanic Research, Baltimore *16446*
Sea Grant Association, La Jolla *16452*
United States Antarctic Research Program, Washington *16815*
World Aquaculture Society, Baton Rouge *16921*

Ophthalmology

Argentina

Sociedad Argentina de Oftalmología, Buenos Aires *00179*

Australia

The Contact Lens Society of Australia, Sydney *00379*
Opticians and Optometrists Association of New South Wales, Sydney *00443*
Royal Australian College of Ophthalmologists, Sydney *00463*

Austria

Österreichische Ophthalmologische Gesellschaft, Wien *00873*

Belgium

European Centre of Ophthalmology, Bruxelles *01313*
European Community University Professors in Ophthalmology, Gent *01328*
Société Belge d'Ophthalmologie, Bruxelles *01451*
Union Professionnelle Belge des Médécins Ophthalmologistes, Bruxelles *01499*

Canada

Association of Schools of Optometry of Canada, Waterloo *01943*
Canadian Association of Optometrists, Ottawa *02060*
Canadian Orthoptic Society, Montréal *02215*
E.A. Baker Foundation for Prevention of Blindness, Toronto *02420*
Nova Scotia Association of Optometrists, Halifax *02723*
Operation Eyesight Universal, Calgary *02806*
RP Eye Research Foundation, Toronto *02872*
Société Québécoise d'Assainissement des Eaux, Montréal *02936*

China, Republic

Ophthalmological Society of the Republic of China, Taipei *03317*

Colombia

Sociedad Americana de Oftalmologia y Optometria, Bogotá *03402*

Cyprus

Cyprus Ophthalmological Society, Larnaca *03494*

Czech Republic

Česká Oftalmologická Společnost, Praha *03512*

Denmark

Dansk Oftalmologisk Selskab, København *03658*

Egypt

Ophthalmological Society of Egypt, Cairo *03912*

France

Academia Ophthalmologica Internationalis, Albi *04083*
Association of European Schools and Colleges of Optometry, Bures-sur-Yvette *04318*
Organisation Internationale contre le Trachome, Créteil *04614*
Société d'Ophtalmologie de l'Est de la France, Nancy *04772*
Société d'Ophtalmologie de Lyon, Lyon *04773*
Société d'Ophtalmologie de Paris, Paris *04774*
Société d'Oto-Neuro-Ophthalmologie du Sud-Est de la France, Marseille *04775*
Société Française d'Ophthalmologie, Paris *04861*

Germany

Berufsverband der Augenärzte Deutschlands e.V., Düsseldorf *05193*
Deutsche Ophthalmologische Gesellschaft Heidelberg, Heidelberg *05478*
Vereinigung Bayerischer Augenärzte, München *06394*
Wissenschaftliche Vereinigung für Augenoptik und Optometrie e.V., Mainz *06450*

Hong Kong

Asian Foundation for the Prevention of Blindness, Hong Kong *06590*

India

Indian Optometric Association, New Delhi *06805*

Iran

Association of Ophthalmists, Teheran *06897*

Ireland

Association of Ophthalmic Opticians of Ireland, Dublin *06930*

Italy

Associazione Professionale Italiana Medici Oculisti, Roma *07398*
Società Italiana di Oto-Neuro-Oftalmologia, Bologna *07923*
Società Oftalmologica Italiana, Bologna *07994*

Japan

Asia Pacific Academy of Ophthalmology, Tokyo *08111*
Ken-i Kai Ganka Toshokan, Tokyo *08164*

Kenya

Ophthalmological Society of East Africa, Nairobi *08506*

Mexico

Sociedad de Oftalmología del Hospital de Oftalmológico de Nuestra Señora de la Luz, México *08749*
Sociedad Médica del Hospital Oftalmológico de Nuestra Señora de la Luz, México *08755*

Netherlands

International Federation of Ophthalmological Societies, Nijmegen *08935*
Nederlands Oogheelkundig Gezelschap, Maastricht *09055*

Nicaragua

Sociedad de Oftalmología Nicaragüense, Managua *09205*

Panama

Asociación Panamericana de Oftalmología, Panamá City *09411*

Portugal

Sociedade Portuguesa de Oftalmologia, Lisboa *09708*

Romania

Societatea de Oftalmologie, Bucuresti *09775*

Russia

National Ophthalmological Society, Moskva *09910*

Spain

Sociedad Oftalmológica Hispano-Americano, Madrid *10401*

Sweden

Svenska Ogonläkareföreningen, Stockholm *10570*

Switzerland

Schweizerische Ophthalmologische Gesellschaft, Horgen *10878*

Turkey

Türk Oftalmoloji Derneği, Istanbul *11169*

United Kingdom

British Orthoptic Society, London *11628*
British Retinitis Pigmentosa Society, Buckingham *11640*
College of Ophthalmologists, London *11796*
European Contact Lens Society of Ophthalmologists, London *12019*
Medical Contact Lens Association, London *12502*

U.S.A.

American Academy of Ophthalmology, San Francisco *13410*
American Academy of Optometry, Bethesda *13411*
American Academy of Otolaryngic Allergy, Silver Spring *13419*
American Association for Pediatric Ophthalmology and Strabismus, San Francisco *13489*
American Board of Ophthalmology, Bala Cynwyd *13643*
American Diopter and Decibel Society, Pittsburgh *13788*
American Foundation for Vision Awareness, Saint Louis *13823*
American Ophthalmological Society, Durham *13965*
American Optometric Association, Saint Louis *13966*
American Orthoptic Council, Madison *13975*

American Society of Contemporary Ophthalmology, Chicago *14114*
American Society of Veterinary Ophthalmology, Stillwater *14172*
Armed Forces Optometric Society, Harrison *14245*
Association for Macular Diseases, New York *14317*
Association for Research in Vision and Ophthalmology, Bethesda *14330*
Association of Optometric Educators, Tahlequah *14474*
Association of Schools and Colleges of Optometry, Rockville *14495*
Association of University Professors of Ophthalmology, San Francisco *14528*
Better Vision Institute, Rosslyn *14583*
College of Optometrists in Vision Development, Chula Vista *14776*
Council on Clinical Optometric Care, Saint Louis *14913*
Dana Center for Preventive Ophthalmology, Baltimore *14952*
International Association of Boards of Examiners in Optometry, Bethesda *15398*
International Association of Ocular Surgeons, Chicago *15413*
International Association of Optometric Executives, Bismarck *15414*
International Myopia Prevention Association, Ligonier *15503*
International Society on Metabolic Eye Disease, New York *15556*
Joint Review Committee for the Ophthalmic Medical Personnel, Saint Paul *15622*
National Association of Optometrists and Opticians, Cleveland *15876*
National Optometric Association, Columbus *16115*
Ophthalmic Research Institute, Bethesda *16254*
Optometric Historical Society, Saint Louis *16257*
Osteopathic College of Ophthalmology and Otorhinolaryngology, Dayton *16273*
Pan-American Association of Ophthalmology, Arlington *16289*
Research to Prevent Blindness, New York *16419*

Venezuela

Sociedad Venezolana de Oftalmología, Caracas *17027*

Vietnam

Vietnamese Association of Ophthalmology, Hanoi *17056*

Optics

Australia

Australian Optometrical Association, Carlton South *00325*
Opticians and Optometrists Association of New South Wales, Sydney *00443*

Austria

Österreichische Gesellschaft für Elektronenmikroskopie, Graz *00776*

Canada

Microscopical Society of Canada, Toronto *02653*
Nova Scotia Association of Optometrists, Halifax *02723*

Denmark

Dansk Selskab for Optometri, Vedbaek *03684*

France

Société Française d'Optique Physiologique, Paris *04862*

Germany

Arbeitsgemeinschaft für Elektronenoptik e.V., Karlsruhe *05103*
Deutsche Gesellschaft für angewandte Optik e.V., Jena *05321*
Deutsches Optisches Komitee, München *05597*
Forschungsvereinigung Feinmechanik und Optik e.V., Köln *05747*
Normenausschuss Feinmechanik und Optik im DIN Deutsches Institut für Normung e.V., Pforzheim *06168*
Vereinigung zur Förderung der technischen Optik e.V., Wetzlar *06418*
Wissenschaftliche Vereinigung für Augenoptik und Optometrie e.V., Mainz *06450*

Hungary

Optikai, Akusztikai és Filmtechnikai Egyesület, Budapest *06667*

India

Optical Society of India, Calcutta *06848*

Ireland

Association of Ophthalmic Opticians of Ireland, Dublin *06930*

Italy

Associazione Ottica Italiana, Firenze *07385*
Istituto Nazionale di Ottica, Firenze *07748*

New Zealand

New Zealand Society for Electron Microscopy, Auckland *09174*

South Africa

The Electron Microscopy Society of Southern Africa, Durban *10138*
South African Optometric Association, Pretoria *10212*

Spain

Sociedad Española de Optica, Madrid *10392*

Switzerland

Verein zur Förderung der Augenoptik, Zürich *11026*

United Kingdom

Association of British Dispensing Opticians, London *11310*
British College of Optometrists, London *11544*
European Council of Optics and Optometry, London *12023*

U.S.A.

Air National Guard Optometric Society, Birmingham *13358*
American Board of Opticianry, Fairfax *13644*
International Congress on High-Speed Photography and Photonics, Janesville *15448*
International Federation of Societies for Electron Microscopy, Berkeley *15474*
Microscopy Society of America, Woods Hole *15729*
National Academy of Opticianry, Bowie *15775*
National Association of Optometrists and Opticians, Cleveland *15876*
National Board of Examiners in Optometry, Washington *15925*
Optical Society of America, Washington *16255*
Opticians Association of America, Fairfax *16256*

Ornithology

Argentina

Asociación Ornitológica del Plata, Buenos Aires *00092*

Australia

Australian Bird Study Association, Sydney South *00271*
Bird Observers Club of Australia, Nunawading *00368*
Royal Australasian Ornithologists Union, Moonee Ponds *00459*
South Australian Ornithological Association, Adelaide *00496*

Austria

Österreichische Gesellschaft für Vogelkunde, Wien *00843*

Belgium

Koninklijke Vereniging voor Vogel- en Natuurstudie de Wielewaal, Turnhout *01412*

Canada

Jack Miner Migratory Bird Foundation, Kingsville *02566*

Denmark

Bird Strike Committee Europe, København *03568*
Dansk Ornithologisk Forening, København *03660*

Finland

Suomen Lintutieteellinen Yhdistys, Helsinki *04046*

France

Ligue pour la Protection des Oiseaux, Rochefort *04599*
Société d'Etudes Ornithologiques, Brunoy *04749*
Société Ornithologique de France, Paris *04908*

Germany

Deutsche Ornithologen-Gesellschaft e.V., Radolfzell *05481*
Gesellschaft Rheinischer Ornithologen e.V., Düsseldorf *05876*
Verband deutscher Waldvogelpfleger und Vogelschützer, Mainz *06362*

Gibraltar

Gibraltar Ornithological and Natural History Society, Gibraltar *06495*

Ireland

Irish Wildbird Conservancy, Monkstown *07006*

Italy

Società Ornitologica Italiana, Ravenna *07995*
Società Ornitologica Reggiana, Reggio Emilia *07996*

Japan

East Asian Bird Protection Society, Tokyo *08125*
Nippon Chô Gakkai, Tokyo *08291*

Luxembourg

Ligue Luxembourgeoise pour la Protection de la Nature et des Oiseaux, Luxembourg *08586*

Malta

The Ornithological Society, Valletta *08669*

Moldova, Republic of

Ornithological Society of Moldova, Chişinău *08786*

Netherlands

Nederlandse Ornithologische Unie, Odijk *08999*

New Zealand

Ornithological Society of New Zealand, Wellington *09185*

Russia

Society of Ornithologists, Moskva *09977*

South Africa

Southern African Ornithological Society, Johannesburg *10227*

Sweden

Esperantist Ornithologists' Association, Nynäshamn *10450*

Switzerland

Ala – Schweizerische Gesellschaft für Vogelkunde und Vogelschutz, Sempach *10612*

United Kingdom

The Avicultural Society, Ascot *11417*
Bird Life International, Cambridge *11445*
British Ornithologists' Club, Oakham *11625*
British Ornithologists' Union, Tring *11626*
British Trust for Ornithology, Thetford *11697*
Hawk and Owl Trust, London *12152*
Hertfordshire Natural History Society and Field Club, Potters Bar *12165*
International Waterfowl and Wetlands Research Bureau, Slimbridge *12405*
Ornithological Society of the Middle East, Sandy *12663*
Parrot Society, Bedford *12680*
Royal Naval Bird Watching Society, Southsea *12835*
Royal Society for the Protection of Birds, Sandy *12853*
Scottish Ornithologists' Club, Edinburgh *12919*
Seabird Group, Sandy *12937*
World's Poultry Science Association, United Kingdom Branch, Aylesbury *13224*

U.S.A.

American Birding Association, Colorado Springs *13622*
American Ornithologists' Union, Washington *13969*
Association of Field Ornithologists, South Natick *14437*
Avicultural Society of America, Yorba Linda *14566*
Cooper Ornithological Society, Los Angeles *14859*
Eastern Bird Banding Association, Hopewell *15002*
Inland Bird Banding Association, Wisner *15263*
International Bird Rescue Research Center, Berkeley *15421*
Nuttall Ornithological Club, Cambridge *16238*
Pacific Seabird Group, Aiken *16283*
Soiety for Northwestern Vertebrate Biology, Olympia *16715*
Western Bird Banding Association, Bakersfield *16881*
Whooping Crane Conservation Association, Lafayette *16892*
Wilson Ornithological Society, Ann Arbor *16905*
World's Poultry Science Association – U.S.A. Branch, Hyattsville *16934*

Zimbabwe

Ornithological Association of Zimbabwe, Harare *17176*

Orthopedics

Argentina

Asociación Argentina de Ortopedía y Traumatología, Buenos Aires *00070*

Australia

Australian Orthopaedic Association, Sydney *00326*

Austria

Forschungsgemeinschaft für Erkrankungen des Bewegungsapparates, Wien *00594*
Österreichische Gesellschaft für Orthopädie und Orthopädische Chirurgie, Wien *00818*

Belgium

Chambre Belge des Pédicures Médicaux, Bruxelles *01209*

Canada

Alberta Chiropractic Association, Edmonton *01829*
British Columbia Association of Podiatry, Vancouver *01971*
Canadian Podiatry Association, Ottawa *02227*
Canadian Society of Orthopaedic Technologists, Agincourt *02288*
Nova Scotia Chiropractic Association, Antigonish *02727*
Ontario Chiropractic Association, Toronto *02765*
Ontario Podiatry Association, Toronto *02788*

Denmark

Dansk Ortopaedisk Selskab, København *03661*

France

Groupement National d'Etude des Médecins du Bâtiment et des Travaux Publics, Paris *04559*
Société Française de Chirurgie Orthopédique et Traumatologique, Paris *04792*
Société Française de Médecine Orthopédique et Thérapeutique Manuelle, Paris *04812*
Société Française d'Orthopédie, Paris *04863*
Syndicat des Spécialistes Français en Orthopédie Dento-Faciale, Paris *04928*
Syndicat National de l'Orthopédie Française, Paris *04935*

Germany

Berliner Orthopädische Gesellschaft e.V., Berlin *05190*
Berufsverband der Ärzte für Orthopädie e.V., Karlsruhe *05192*
Bundesarbeitsgemeinschaft zur Förderung haltungsgefährdeter Kinder und Jugendlicher e.V., Mainz *05237*
Deutsche Gesellschaft für Orthopädie und Traumatologie e.V., Frankfurt *05392*
Vereinigung Süddeutscher Orthopäden e.V., Baden-Baden *06412*

Israel

Society of Orthopedic Surgeons of the Israel Medical Association, Tel Aviv *07113*

Italy

Società Italiana di Medicina del Traffico, Roma *07894*
Società Italiana di Ortopedia e Traumatologia, Roma *07921*
Società Piemontese, Ligure, Lombarda di Ortopedia e Traumatologia, Genova *08003*

Japan

Nippon Seikei Geka Gakkai, Tokyo *08363*

Netherlands

Nederlandse Vereniging van Specialisten in den Dento-Maxillaire Orthopaedie, Nijmegen *09011*
Nederlanedse Orthopaedische Vereniging, Nijmegen *09058*

Peru

Sociedad Peruana de Ortopedia y Traumatología, Lima *09485*

Poland

Polskie Towarzystwo Ortopedyczne i Traumatologiczne, Warszawa *09596*

Portugal

Sociedade Portuguesa de Ortopedia e Traumatologia, Lisboa *09709*

Romania

Societatea Română de Ortopedie şi Traumatologie, Bucuresti *09794*

Russia

National Scientific Medical Society of Traumatic Surgeons and Orthopaedists, Moskva *09936*

South Africa

South African Orthopaedic Association, Durban *10213*

Sweden

Scandinavian Orthopaedic Association, Stockholm *10519*

Switzerland

Schweizerische Gesellschaft für Orthopädie, Bern *10839*

Turkey

Türk Ortopedi ve Travmatoloji Dernegi, Istanbul *11170*

United Kingdom

British Orthopaedic Association, London *11627*
Chiropractic Advancement Association, Thames Ditton *11780*
NICOD, Belfast *12628*

U.S.A.

American Academy of Gnathologic Orthopedics, Richmond *13397*
American Academy of Orthopaedic Surgeons, Rosemont *13416*
American Academy of Podiatric Sports Medicine, Potomac *13427*
American Association of Certified Orthoptists, Houston *13516*
American Association of Colleges of Podiatric Medicine, Rockville *13523*
American Association of Hospital Podiatrists, Brooklyn *13541*
American Association of Orthopedic Medicine, Lewiston *13560*
American Board of Orthopedic Surgery, Chapel Hill *13647*
American Board of Podiatric Orthopedics, Redlands *13655*
American Board of Podiatric Surgery, San Francisco *13656*
American Chiropractic Association, Arlington *13690*
American College for Continuing Education, Union *13695*
American College of Chiropractic Orthopedists, El Centro *13701*
American College of Foot Orthopedists, Redlands *13706*
American College of Foot Surgeons, Park Ridge *13707*
American College of Podopediatrics, Cleveland *13731*
American Fracture Association, Bloomington *13825*
American Orthopaedic Association, Rosemont *13971*
American Orthopaedic Foot and Ankle Society, Rosemont *13972*
American Orthopaedic Society for Sports Medicine, Des Plaines *13973*
American Osteopathic Academy of Addictionology, Cherry Hill *13977*
American Podiatric Medical Association, Bethesda *14008*
Association of Bone and Joint Surgeons, Rosemont *14405*
Clinical Orthopedic Society, Bloomington *14763*
Conference of Podiatry Executives, Columbus *14831*
Council on Chiropractic Orthopedics, Provo *14911*
Foundation for Chiropractic Education and Research, Arlington *15122*
International Chiropractors Association, Arlington *15435*
National College of Foot Surgeons, Woodland Hills *15958*
National Podriatic Medical Association, Chicago *16130*

Orthopedic Research Society, Rosemont 16271
Precision Chiropractic Research Society, Brea 16349
Sacro Occipital Research Society International, Prairie Village 16432
Scoliosis Research Society, Rosemont 16449

Venezuela

Sociedad Venezolana de Cirurgía Ortopédica y Traumatología, Caracas 17021

Otorhinolaryngology

Austria

Österreichische Gesellschaft für Hals-, Nasen- und Ohrenheilkunde, Kopf- und Halschirurgie, Wien 00789
Österreichische Gesellschaft für Logopädie, Phoniatrie und Pädoaudiologie, Wien 00806
Wiener Gesellschaft der Hals-Nasen-Ohren-Ärzte, Wien 01014

Belgium

Belgische Neus-, Keel- en Oorheelkundige Vereniging, Brugge 01160
European Association of Hearing Aid Audiologists, Fleurus 01292
International Rhinologic Society, Bruxelles 01402

Canada

Canadian Society of Otolaryngology, Head and Neck Surgery, Islington 02289

Chile

Sociedad de Otorinolaringología de Valparaíso, Valparaíso 03095

China, Republic

The Otolaryngological Society of the Republic of China, Taipei 03318

Czech Republic

Czech Otolaryngological Society, Praha 03540

Denmark

Dansk Oto-laryngologisk Selskab, Hellerup 03662

France

European Audio Phonological Centers Association, Saint-Etienne 04492
Societas Oto-Rhino-Laryngologia Latina, Montpellier 04624
Société d'Oto-Neuro-Ophthalmologie du Sud-Est de la France, Marseille 04775
Société Française de Phoniatrie, Paris 04832
Société Française d'Oto-Rhino-Laryngologie et de Pathologie Cervico-Faciale, Paris 04864
Syndicat National des Médecins Spécialisés en Phoniatrie, Saint-Denis 04959
Syndicat National des Oto-Rhino-Laryngologistes Français, Paris 04962

Germany

Berufsverband der Berliner Hals-, Nasen-, Ohren-Ärzte, Berlin 05194
Confederation of European Laryngectomees, Bebra 05277
Deutsche Gesellschaft für Hals-Nasen-Ohren-Heilkunde, Kopf- und Hals-Chirurgie, Bonn 05359
Deutscher Berufsverband der Hals-Nasen-Ohrenärzte e.V., Neumünster 05503
Nordwestdeutsche Vereinigung der Hals-Nasen-Ohrenärzte, Kiel 06146
Schleswig-Holsteinische Gesellschaft für Zahn-, Mund- und Kieferheilkunde, Lübeck 06276
Vereinigung Südwestdeutscher HNO-Ärzte, Köln 06414
Vereinigung Westdeutscher Hals-, Nasen- und Ohrenärzte, Köln 06416

Iceland

Félag Háls-, Nef- og Eyrnalaekna, Reykjavik 06683

India

Asia-Oceania Association of Otolaryngological Societies, Bombay 06726

Indonesia

Asian Society of Oto-Rhino-Laryngology, Jakarta 06882

Israel

Oto-Laryngological Society of Israel, Tel Aviv 07110

Italy

Associazione Otologica Ospedaliera Italiano, Roma 07384
Società Italiana di Audiologia e Foniatria, Roma 07839
Società Italiana di Foniatria, Catania 07876
Società Italiana di Laringologia, Otologia, Rinologia e Patologia Cervico-Facciale, Roma 07890
Società Italiana di Oto-Neuro-Oftalmologia, Bologna 07923
Società Italiana di Otorinolaringologia e Chirurgia Cervico Facciale, Roma 07924
Società Italiana di Otorinolaringologia Pediatrica, Roma 07925

Japan

Nippon Jibi-Inkoka Gakkai, Tokyo 08313
Shika Kiso Igakkai, Tokyo 08419

Netherlands

Nederlandse Vereniging voor Mondziekten en Kaakchirurgie, 's-Gravenhage 09028

Poland

Polskie Towarzystwo Otolaryngologiczne, Warszawa 09597

Portugal

Sociedade Portuguesa de Otorrinolaringologia e Bronco-Esofagologia, Lisboa 09710

Romania

Societatea de Oto-Rino-Laringologie, Bucuresti 09776

Russia

National Scientific Medical Society of Oto-Rhino-Laryngologists, Moskva 09925

South Africa

South African Society of Otorhinolaryngology, Bellville 10218

Spain

Asociación Española de Logopedia, Foniatria y Audiología, Barcelona 10263
Sociedad Española de Otorinolaringología y Patología Cévico-Facial, Madrid 10393

Sweden

Svensk Otolaryngologisk Förening, Stockholm 10584

Switzerland

Schweizerische Gesellschaft für Oto-Rhino-Laryngologie, Hals- und Gesichtschirurgie, Bern 10840
Société Romande d'Audiophonologie et de Pathologie du Langage, Delémont 10982

Thailand

ASEAN Otorhinolaryngological Head and Neck Federation, Bangkok 11070

Turkey

Türk Oto-Rino-Laringoloji Cemiyeti, Istanbul 11172

United Kingdom

British Association of Oral and Maxillo-Facial Surgeons, London 11513
British Association of Otolaryngologists, London 11515
British Kidney Patient Association, Bordon 11598
British Society of Audiology, Reading 11681
British Tinnitus Association, London 11695
International Society of Audiology, London 12391

U.S.A.

Academy of Dispensing Audiologists, Columbia 13291
Academy of Rehabilitative Audiology, Akron 13305
American Academy of Oral Medicine, Houston 13413
American Academy of Otolaryngic Allergy, Silver Spring 13419
American Auditory Society, Dallas 13611
American Board of Otolaryngology, Houston 13648
American Broncho-Esophagological Association, Chicago 13670
American Cleft Palate-Craniofacial Association, Pittsburgh 13692
American Laryngological Association, Rochester 13910
American Laryngological, Rhinological and Otological Society, East Greenville 13911
American Otological Society, Maywood 13988
American Rhinologic Society, Brooklyn 14031
American Society for Head and Neck Surgery, Baltimore 14063
American Tinnitus Association, Portland 14196
Better Hearing Institute, Washington 14582
Deafness Research Foundation, New York 14956
Hear Center, Pasadena 15198
House Ear Institute, Los Angeles 15232
International Association of Laryngectomees, Atlanta 15411
International Bronchoesophagological Society, Scottsdale 15424
National Center for Law and Deafness, Washington 15946
National Foundation for Children's Hearing Education and Research, Yonkers 16058
National Hearing Conservation Association, Des Moines 16074
Osteopathic College of Ophthalmology and Otorhinolaryngology, Dayton 16273
Otosclerosis Study Group, Tulsa 16274

Venezuela

Sociedad Venezolana de Otorinolaringología, Caracas 17029

Vietnam

Vietnamese Association of Oto-Rhino-Laryngology, Hanoi 17057

Paleontology

Argentina

Asociación Paleontológica Argentina, Buenos Aires 00094

Austria

Österreichische Paläontologische Gesellschaft, Wien 00877

Brazil

Asociación Latinoamericana de Paleobotánica y Palinología, Porto Alegre 01584

China, People's Republic

Palaeontological Society of China, Nanjing 03213

France

Association Internationale pour l'Etude de la Paléontologie Humaine, Paris 04292

Germany

Paläontologische Gesellschaft, Frankfurt 06231

Italy

Accademia Valdarnese del Poggio, Montevarchi 07246
Centro Studi Ricerche Ligabue, Venezia 07590
Istituto Italiano di Paleontologia Umana, Roma 07738
Società degli Amici del Museo Civico di Storia Naturale Giacomo Doria, Genova 07779
Società Italiana di Biogeografia, Siena 07842

Japan

International Palaeontological Association, Sapporo 08144
Nippon Koseibutsu Gakkai, Tokyo 08342

Malta

Society for the Study and Conservation of Nature, Valletta 08671

Russia

Russian Palaeontological Society, St. Petersburg 09960

Sri Lanka

Ceylon Palaeological Society, Colombo 10413

Switzerland

Schweizerische Paläontologische Gesellschaft, Genève 10879

United Kingdom

Moray Society, Elgin 12538
Palaeontographical Society, Nottingham 12674
Palaeontological Association, Cardiff 12675

U.S.A.

Paleontological Research Institution, Ithaca 16286
Paleontological Society, Socorro 16287
Society for Sedimentary Geology, Tulsa 16578
Society of Vertebrate Paleontology, Lincoln 16712

Parapsychology

Austria

Österreichische Gesellschaft für Parapsychologie, Wien 00819

Canada

Canadian Association of Parapsychologists, Ottawa 02062
International Society for Research in Palmistry, Westmount 02558

France

European Association for the Study of Dreans, Ile-Saint-Denis 04483

India

The Theosophical Society, Madras 06865

Italy

Associazione Italiana Scientifica di Metapsichica, Milano 07348
Centro Italiano di Parapsicologia, Napoli 07493
Centro Ricerche Metapsichiche e Psicofoniche, Fermo 07540

Japan

International Association for Religion and Parapsychology, Tokyo 08141

United Kingdom

Faculty of Astrological Studies, London 12049
Society for Psychical Research, London 12989

U.S.A.

American Federation of Astrologers, Tempe 13808
Association for Research and Enlightenment, Virginia Beach 14328
International Association for Near-Death Studies, Philadelphia 15388
International Society for Astrological Research, Los Angeles 15523
Parapsychological Association, Research Triangle Park 16295
Professional Psychics United, Oak Park 16358

Parasitology s. Microbiology

Pathology

Argentina

Asociación Internacional de Hidatidología, Buenos Aires 00085
Sociedad Argentina de Anatomía Normal y Patológica, Buenos Aires 00154
Sociedad Argentina de Patología, Buenos Aires 00180

Australia

Royal College of Pathologists of Australasia, Surrey Hills 00467
Victorian Society of Pathology and Experimental Medicine, Parkville 00518

Austria

Österreichische Arbeitsgemeinschaft für morphologische und funktionelle Atherosklerosforschung, Wien 00724
Österreichische Gesellschaft für Neuropathologie, Wien 00815
Österreichische Gesellschaft für Pathologie, Wien 00820
Österreichische Gesellschaft zur Bekämpfung der Cystischen Fibrose, Wien 00854

Belgium

Association pour les Etudes et Recherches de Zoologie Appliquée et de Phytopathologie, Louvain-la-Neuve 01151
International Institute of Cellular and Molecular Pathology, Bruxelles 01398
Ligue Nationale Belge contre l'Epilepsie, Bruxelles 01413

Canada

Amyotrophic Lateral Sclerosis Society of British Columbia, Vancouver 01866
Association du Syndrôme de Turner du Québec, Boucherville 01897
British Columbia Epilepsy Society, Vancouver 01977
Canadian Association of Pathologists, Kingston 02063
Canadian Osteogenesis Imperfecta Society, Toronto 02216
Canadian Osteopathic Aid Society, London 02217
Canadian Osteopathic Association, London 02218
Canadian Paraplegic Association, Toronto 02220
Canadian Society of Electroencephalographers, Electromyographers and Clinical Neurophysiologists, Sherbrooke 02282
Epilepsy Prince Edward Island, Richmond 02441
Manitoba Epilepsy Association, Winnipeg 02616
New Brunswick Association of Pathologists, Saint John 02677
Ontario Osteopathic Association, Toronto 02787
Osteoporosis Society of Canada, Toronto 02808
Saskatchewan Amyotrophic Lateral Sclerosis Society Foundation, Radville 02878
Saskatchewan Association of Pathologists, Regina 02882
Saskatchewan Society of Osteopathic Physicians, Regina 02903
Spina Bifida and Hydrocephalus Association of Ontario, Toronto 02954
Spina Bifida Association of British Columbia, Surrey 02955
Spina Bifida Association of Canada, Winnipeg 02956
Western Cleft Lip and Palate Association, Burnaby 02992

Chile

Sociedad Chilena de Patología de la Adaptación y del Mesenquima, Santiago 03080
Sociedad de Anatomía Normal y Patológica de Chile, Santiago 03087

Colombia

Instituto Nacional de Cancerología, Bogotá 03392
Sociedad Colombiana de Patología, Cali 03411

Denmark

Dansk Selskab for Akupunktur, København 03670
Dansk Selskab for Patologi, København 03686
European Association for Haematopathology, Aarhus 03722

France

Ligue Française contre la Sclérose en Plaques, Paris 04593
Société de Pathologie Comparée, Paris 04678
Société de Pathologie Exotique, Paris 04679
Société Française de Néonatologie, Paris 04823
Société Française de Pathologie Respiratoire, Paris 04827
Syndicat National des Médecins Anatomo-Cyto-Pathologistes Français, Paris 04948

Germany

Deutsche Gesellschaft für die Bekämpfung der Muskelkrankheiten e.V., Freiburg 05339
Deutsche Gesellschaft für Pathologie, Frankfurt 05397
Deutsche Gesellschaft für Phlebologie, Norderney 05401
Deutsche Multiple Sklerose Gesellschaft Bundesverband e.V., Hannover 05475
Deutsche Sektion der Internationalen Liga gegen Epilepsie, Bielefeld 05573
Internationale Akademie für Pathologie, Deutsche Abteilung e.V., Bonn 05988
Vereinigung Deutscher Neuropathologen und Neuroanatomen, Giessen 06406

Hungary

Magyar Elettani Társaság, Budapest 06626
Magyar Pathológusok Társasága, Budapest 06651

Ireland

Irish Epilepsy Association, Dublin 06982

Israel
Israel Society of Pathologists, Petan Tiqva *07101*

Italy
Associazione Italiana per lo Studio del Dolore, Firenze *07344*
Associazione Italiana Sclerosi Multipla, Roma *07349*
International Study Group for Steroid Hormones, Roma *07723*
Società Europea di Patologia, Milano *07803*
Societá Italiana della Continenza, Roma *07825*
Società Italiana di Anatomia Patologica, Messina *07833*
Società Italiana di Andrologia, Pisa *07834*
Società Italiana di Immunologia e di Immunopatologia, Bari *07887*
Società Italiana di Patologia, Milano *07927*
Società Italiana di Patologia Vascolare, Roma *07931*
Società Italiana per lo Studio dell'Arteriosclerosi, Bari *07972*
Società Laziale Abruzzese di Medicina del Lavoro, Roma *07977*

Japan
Nihon Rinsho Byori Gakkai, Tokyo *08250*
Nippon Byori Gakkai, Tokyo *08286*
World Association of Societies of Pathology (Anatomic and Clinical), Tokyo *08445*

Netherlands
Nederlandse Patholoog Anatomen Vereniging, Maastricht *09000*
Nederlandse Vereniging voor Pathologie, Rotterdam *09033*

Poland
Polskie Towarzystwo Patologów, Lódź *09599*

Romania
Societatea de Ftiziologie, Bucuresti *09763*
Societatea de Patologie Infectioasà, Bucuresti *09777*

Russia
National Scientific Medical Society of Neuropathologists and Psychiatrists, Moskva *09921*
Scientific Medical Society of Anatomists-Pathologists, Moskva *09967*

South Africa
Carbohydrate and Lipid Metabolism Research Group, Johannesburg *10128*
Preclinical Diagnostic Chemistry Research Group of the South African Medical Research Council, Congella *10163*
South African National Multiple Sclerosis Society, Walmer *10209*

Spain
Asociación Española de Biopatología Clínica, Madrid *10258*
Sociedad Española de Patología Digestiva y de la Nutrición, Madrid *10394*

Switzerland
European Cytoskeletal Club, Genève *10684*
Schweizerische Gesellschaft für cystische Fibrose, Spiegel *10815*
Schweizerische Gesellschaft für Muskelkranke, Zürich *10835*
Schweizerische Liga gegen Epilepsie, Zürich *10870*
Schweizerische Multiple Sklerose Gesellschaft, Zürich *10873*

United Kingdom
Association for Research into Restricted Growth, Witham *11281*
Association of Clinical Pathologists, Brighton *11325*
British Epilepsy Association, Leeds *11558*
General Council and Register of Osteopaths, Reading *12102*
Incorporated Society of Registered Naturopaths, Edinburgh *12196*
International Medical Society of Paraplegia, Aylesbury *12377*
International Society of Neuropathology, London *12393*
The Multiple Sclerosis Society of Great Britain and Northern Ireland, London *12541*
Muscular Dystrophy Group of Great Britain & Northern Ireland, London *12542*
National Society for Epilepsy, Chalfont Saint Peter *12609*
Osteopathic Association of Great Britain, London *12664*
Osteopathic Medical Association, London *12665*
Pathological Society of Great Britain and Ireland, London *12682*
Royal College of Pathologists, London *12785*
Scottish Epilepsy Association, Glasgow *12906*
Society for Research into Hydrocephalus and Spina Bifida, Manchester *12995*

Uruguay
Sociedad Uruguaya de Patología Clínica, Montevideo *13270*

U.S.A.
American Academy of Osteopathy, Newark *13418*
American Association of Avian Pathologists, Kennett Sq *13507*
American Association of Colleges of Osteopathic Medicine, Rockville *13521*
American Association of Neuropathologists, Salt Lake City *13556*
American Association of Pathologists, Bethesda *13561*
American Board of Oral Pathology, Tampa *13645*
American Board of Pathology, Tampa *13649*
American College of General Practitioners in Osteopathic Medicine and Surgery, Arlington Heights *13709*
American College of Osteopathic Internists, Washington *13725*
American College of Osteopathic Pediatricians, Trenton *13727*
American College of Osteopathic Surgeons, Alexandria *13728*
American Epidemiological Society, Atlanta *13800*
American Epilepsy Society, Hartford *13801*
American Hearing Research Foundation, Chicago *13843*
American Osteopathic Academy of Sclerotherapy, Wilmington *13978*
American Osteopathic Association, Chicago *13980*
American Osteopathic College of Anesthesiologists, Independence *13982*
American Osteopathic College of Pathologists, Pembroke Pines *13984*
American Osteopathic College of Proctology, Union *13985*
American Osteopathic College of Rehabilitation Medicine, Des Plaines *13987*
American Society of Clinical Pathologists, Chicago *14110*
American Society of Dermatopathology, Baltimore *14119*
American Society of Psychopathology of Expression, Brookline *14161*
American Spinal Injury Association, Chicago *14180*
Armed Forces Institute of Pathology, Washington *14244*
Association of Osteopathic State Executive Directors, Sacramento *14477*
Association of Pathology Chairmen, Bethesda *14480*
Auxiliary of the American Osteopathic Association, Chicago *14560*
Bureau of Professional Education of the American Osteopathic Association, Chicago *14623*
College of American Pathologists, Northfield *14775*
Institute for Gravitational Strain Pathology, Rangeley *15282*
International Academy of Pathology, Tucson *15368*
Intersociety Committee on Pathology Information, Bethesda *15579*
National Foundation for Brain Research, Washington *16057*
National Multiple Sclerosis Society, New York *16111*
National Osteopathic Guild Association, Chicago *16123*
Society for Leukocyte Biology, Augusta *16547*
Society for Occupational and Environmental Health, McLean *16560*

Vietnam
Vietnamese Asscoiation of Morphology, Hanoi *17035*
Vietnamese Association of Acupuncture, Hanoi *17036*

Pediatrics

Argentina
Asociación pala la Lucha la Parálisis Infantil, Buenos Aires *00093*
Sociedad Argentina de Pediatría, Buenos Aires *00181*

Australia
Asian Pan-Pacific Society for Paediatric Gastroenterology and Nutrition, East Perth *00227*
Australian College of Paediatrics, Parkville *00276*
Paediatric Society of Victoria, Parkville *00444*

Austria
Österreichische Gesellschaft für Kinder- und Jugendheilkunde, Wien *00800*

Belgium
Association Professionnelle Belge des Pédiatres, Bruxelles *01155*
Confédération Européenne des Syndicats Nationaux et Associations Professionnelles de Pédiatres, Bruxelles *01244*
Société Belge de Pédiatrie, Bruxelles *01438*

Benin
Association des Pédiatres d'Afrique Noire Francophone, Cotonou *01516*

Bolivia
Sociedad de Pediatría de Cochabamba, Cochabamba *01552*

Brazil
Sociedade de Pediatria da Bahia, Salvador *01725*

Canada
Canadian Association of Paediatric Surgeons, London *02061*
Canadian Foundation for the Study of Infant Deaths, Toronto *02142*
Canadian Institute of Child Health, Ottawa *02169*
Canadian Paediatric Society, Ottawa *02219*
Children's Oncology Care of Ontario, Toronto *02345*
Children's Rehabilitation and Cerebral Palsy Association, Vancouver *02346*
Garrod Association of Canada, Halifax *02478*
Hospital for Sick Children Foundation, Toronto *02505*

Chile
Sociedad Chilena de Pediatría, Santiago *03081*
Sociedad de Pediatría de Valparaíso, Valparaíso *03096*

China, Republic
The Chinese Taipei Pediatric Association, Taipei *03289*

Colombia
Instituto Colombiano de Oncologia Pediatrica, Bogotá *03387*
Sociedad Colombiana de Pediatría, Bogotá *03412*
Sociedad de Pediatría y Puericultura del Atlántico, Barranquilla *03418*

Costa Rica
Asociación Costarricense de Pediatría, San José *03432*

Czech Republic
Česká Pediatrická Společnost, Praha *03514*

Denmark
Dansk Pediatrisk Selskab, Virum *03663*

Ecuador
Sociedad Ecuatoriana de Pediatría, Guayaquil *03857*

El Salvador
Central American Paediatric Society, San Salvador *03923*

France
Association des Pédiatres de Langue Française, Clamart *04193*
Association Internationale de Pédiatrie, Paris *04261*
Cercle d'Etudes Pédiatriques, Paris *04387*
Société d'Etudes et de Soins pour les Enfants Paralysés, Paris *04736*
Société Française de Néonatologie, Paris *04823*
Société Française de Pédiatrie, Paris *04829*
Société Provençale de Pédiatrie, Marseille *04913*
Syndicat National des Pédiatres Français, Clermont-Ferrand *04963*

Germany
Berufsverband der Kinderärzte Deutschlands e.V., Köln *05200*
Deutsche Gesellschaft für Kinderheilkunde, Hannover *05371*
Deutsche Gesellschaft für Sozialpädiatrie e.V., München *05423*

Greece
Elliniki Paidiatriki Etairia, Athinai *06513*
Union of Middle East Mediterranean Paediatric Societies, Athinai *06547*

Guatemala
Asociación Pediátrica de Guatemala, Guatemala City *06561*

Israel
Clinical Paediatric Club of Israel, Zerifin *07050*
Coordination Office of Paediatric Endocrine Societies, Petah Tikva *07051*
Israel Pediatric Association, Petach Tikva *07085*

Italy
Associazione per il Diabete Infantile e Giovanile, Roma *07387*
Società Italiana di Chirurgia Pediatrica, Roma *07856*
Società Italiana di Ginecologia Pediatrica, Roma *07883*
Società Italiana di Neuropediatria, Siena *07912*
Società Italiana di Nipiologia, Roma *07916*
Società Italiana di Otorinolaringologia Pediatrica, Roma *07925*
Società Italiana di Pediatria, Roma *07932*

Japan
Nihon Shoni Geka Gakkai, Tokyo *08264*
Nihon Shonika Gakkai, Tokyo *08265*

Mexico
Sociedad Mexicana de Pediatría, México *08770*

Netherlands
International Society of Paediatric Oncology, 's-Hertogenbosch *08943*
Nederlandse Vereniging voor Kindergeneeskunde, Amsterdam *09022*

Paraguay
Sociedad de Pediatría y Puericultura del Paraguay, Asunción *09435*

Peru
Sociedad Latinoamericana de Investigación Pediátrica, Lima *09479*

Philippines
Association of Pediatric Societies of the Southeast Asian Region, Manila *09510*
Philippine Paediatric Society, Manila *09530*

Poland
Polskie Towarzystwo Pediatryczne, Warszawa *09600*

Portugal
Sociedade Portuguesa de Pediatria, Lisboa *09711*

Romania
Societatea de Pediatrie, Bucuresti *09778*

Russia
National Scientific Medical Society of Paediatriciams, Moskva *09926*

Sierra Leone
African Paediatric Club, Kenema *10029*

Spain
Asociación Española de Pediatría, Madrid *10266*
Sociedad de Pediatría de Madrid y Castilla La Mancha, Madrid *10374*
Sociedad de Pediatría de Madrid y Región Centro, Madrid *10375*

Sweden
European Association of Perinatal Medicine, Uppsala *10452*

Switzerland
European Society for Pediatric Nephrology, Zürich *10686*

United Kingdom
Association for Brain Damaged Children, Coventry *11260*
Association of British Paediatric Nurses, London *11314*
British Association of Paediatric Nephrology, Birmingham *11516*
British Association of Paediatric Surgeons, Edinburgh *11517*
British Paediatric Association, London *11629*
Foundation for the Study of Infant Deaths, London *12089*
Paediatric Welfare and Research Association, Macclesfield *12672*

Uruguay
Sociedad Uruguaya de Pediatría, Montevideo *13271*

U.S.A.
ALSAC – Saint Jude Children's Research Hospital, Memphis *13367*
Ambulatory Pediatric Association, McLean *13373*
American Academy of Pediatrics, Elk Grove Village *13421*
American Board of Pediatrics, Chapel Hill *13651*
American College of Osteopathic Pediatricians, Trenton *13727*
American Orthopsychiatric Association, New York *13974*
American Pediatric Society, Elk Grove Village *13994*
Association for Research of Childhood Cancer, Buffalo *14331*
Association for the Care of Children's Health, Bethesda *14349*
Association of Medical School Pediatric Department Chairmen, Salt Lake City *14461*
Child Neurology Society, Saint Paul *14732*
Family Health International, Durham *15075*
International Society for Pediatric Neurosurgery, Salt Lake City *15537*
National Association of Children's Hospitals and Related Institutions, Alexandria *15837*
National Council of Guilds for Infant Survival, Davenport *16000*
North American Society for Pediatric Gastroenterology and Nutrition, Denver *16222*
Society for Adolescent Medicine, Independence *16490*
Society for Pediatric Dermatology, Ann Arbor *16561*
Society for Pediatric Radiology, Oak Brook *16563*
Society for Pediatric Research, Elk Grove Village *16564*
Society for Pediatric Urology, Seattle *16565*

Venezuela
Sociedad Venezolana de Puericultura y Pediatría, Caracas *17031*

Vietnam
Vietnamese Association of Paediatrics, Hanoi *17058*

Performing Arts, Theater

Austria
Gesellschaft für Forschungen zur Aufführungspraxis, Graz *00618*
Internationale Nestroy-Gesellschaft, Wien *00675*
Mediacult, Internationales Forschungsinstitut für Medien, Kommunikation und kulturelle Entwicklung, Wien *00705*
Wiener Gesellschaft für Theaterforschung, Wien *01016*

Brazil
Sociedade Brasileira de Autores Teatrais, Rio de Janeiro *01704*

Canada
Association Internationale du Théâtre pour l'Enfance et la Jeunesse, Montréal *01910*
British Columbia Drama Association, Victoria *01976*
Canadian Society of Children's Authors, Illustrators and Performers, Toronto *02277*
Canadian Theatre Critics Association, Toronto *02307*
International Theatre Institute, Toronto *02561*
Newfoundland and Labrador Drama Society, Saint John's *02699*

Denmark
Association Internationale du Théâtre pour l'Enfance et la Jeunesse, København *03562*

International Society of Libraries and Museums of the Performing Arts, København 03749
Selskabet for Dansk Teaterhistorie, København 03812

Finland
Suomen Näytelmäkirjailijaliitto, Helsinki 04054

France
Association Internationale des Critiques de Théâtre, Paris 04266
Conseil International de la Danse, Paris 04452
Institut International du Théâtre, Paris 04576
Société des Auteurs et Compositeurs Dramatiques, Paris 04700
Société d'Histoire du Théâtre, Paris 04761

Germany
Deutsche Akademie der Darstellenden Künste e.V., Frankfurt 05284
Dramatiker-Union e.V., Berlin 05635

Hungary
Magyar Szinházi Intézet, Budapest 06658

Italy
Accademia Artistica Internazionale Pinocchio d'Oro, Napoli 07130
Accademia degli Sbalzati, Sansepolcro 07141
Accademia dei Filodrammatici, Milano 07143
Accademia Nazionale di Arte Drammatica Silvio d'Amico, Roma 07215
Accademia Nazionale di Danza, Roma 07217
Associazione Nazionale Esercenti Teatri, Roma 07370
Associazione per la Conservazione delle Tradizioni Popolari, Palermo 07389
Centro di Ricerca per il Teatro, Milano 07434
Centro di Studi sul Teatro Medioevale e Rinascimentale, Viterbo 07457
Centro Italiano Studi sull'Arte dello Spettacolo, Roma 07510
Centro Studi Pietro Mancini, Cosenza 07585
Ente Autonomo La Biennale de Venezia, Venezia 07649

Japan
Nippon Engeki Gakkai, Tokyo 08298

Mexico
Academia de Arte Dramática, Hermosillo 08677
Academia de Dramática, Guanajuato 08681

Netherlands
Theater Instituut Nederland, Amsterdam 09081

Peru
Asociación de Artistas Aficionados, Lima 09447

Russia
International Confederation of Theatre, Moskva 09876
Theatre Union of the Russian Federation, Moskva 09985

Spain
Institut del Teatre, Barcelona 10314

Switzerland
Schweizerische Gesellschaft für Theaterkultur, Bonstetten 10850
Schweizerischer Ballettlehrer-Verband, Hombrechtikon 10886

United Kingdom
British Ballet Organization, London 11532
The British Puppet and Model Theatre Guild, Yeading 11637
British Theatre Association, London 11693
British Theatre Institute, London 11694
Council for Dance Education and Training, London 11849
Dolmetsch Historical Dance Society, Southampton 11899
Drama Association of Wales, Cardiff 11905
Drama Board Association, Chilehurst 11906
Educational Drama Association, Birmingham 11943
International Council of Kinetography, London 12347
International Federation for Theatre Research, Lancaster 12352
Irish Heritage, London 12411
Marlowe Society, London 12496
National Association of Drama Advisers, Romford 12566
National Operatic and Dramatic Association, London 12602
Radius, London 12738
Royal Academy of Dancing, London 12761
Royal Academy of Dramatic Art, London 12762
The Society for Theatre Research, London 13000
Society of Teachers of Speech and Drama, Canterbury 13066
Theatre Arts Society, London 13122

U.S.A.
American Alliance for Theatre and Education, Tempe 13450
American Association of Community Theatre, Muncie 13525
American Society for Theatre Research, Kingston 14090
American Theatre Critics Association, Nashville 14190
Burlesque Historical Society, Helendale 14624
Cecchetti Council of America, Ann Arbor 14661
Chinese Musical and Theatrical Association, New York 14741
Congress on Research in Dance, Brockport 14843
Drama Tree, New York 14984
Friars Club, New York 15139
Institute for Advanced Studies in the Theatre Arts, New York 15266
Institute of Outdoor Drama, Chapel Hill 15337
International Thespian Society, Cincinnati 15565
Lincoln Center for the Performing Arts, New York 15672
National Association of Dramatic and Speech Arts, Blacksburg 15851
National Theatre Institute, Waterford 16172
New York Drama Critics Circle, New York 16202
Outer Critics Circle, New York 16277
Theatre Guild, New York 16772
Theatre Historical Society, Elmhurst 16773
Theatre in Education, Old Lyme 16774
Theatre Library Association, New York 16775
United States Institute for Theatre Technology, New York 16825

Petrochemistry

Azerbaijan
Azerbaijan Petroleum Academy, Baku 01038

Canada
Atlantic Petroleum Association, Halifax 01959
Independent Petroleum Association of Canada, Calgary 02513
Mineralogical Association of Canada, Ottawa 02655

Iraq
Arab Petroleum Training Institute, Baghdad 06914

Italy
Società Italiana di Mineralogia e Petrologia, Milano 07908

Japan
Nippon Ganseki Kobutsu Kosho Gakkai, Sendai 08300
Nippon Yukagaku Kyokai, Tokyo 08396
Sekiyu Remmei Kohobu Shiryoka, Tokyo 08412

Sweden
Kommittén för Petrokemisk Forskning och Utveckling, Stockholm 10480

United Kingdom
Institute of Petroleum, London 12267

U.S.A.
Society of Petroleum Engineers, Richardson 16689

Pharmacology

Argentina
Asociación Argentina de Farmacia y Bioquímica Industrial, Buenos Aires 00064
Sociedad Argentina de Farmacología y Terapéutica, Buenos Aires 00168

Australia
Association of Medical Directors of the Australian Pharmaceutical Industry, West Ryde 00230
Australasian Pharmaceutical Science Association, Saint Lucia 00245
Australian Physiological and Pharmacological Society, Melbourne 00328

Austria
Österreichische Gesellschaft für Geschichte der Pharmazie, Wien 00785
Österreichische Pharmazeutische Gesellschaft, Wien 00878

Barbados
Barbados Pharmaceutical Society, Bridgetown 01075

Belgium
Cercle Benelux d'Histoire de la Pharmacie, Kortrijk 01207
Comité International de Médecine Militaire, Liège 01228
Société Belge de Physiologie et de Pharmacologie, Bruxelles 01441

Brazil
Academia Nacional de Farmacia, Rio de Janeiro 01576
Associação Brasileira de Farmacêuticos, Rio de Janeiro 01590
Secção de Farmacia Galénica, Curitiba 01701
Sociedade de Farmácia e Química de São Paulo, São Paulo 01723

Bulgaria
Bulgarian Scientific Pharmaceutical Association, Sofia 01754

Canada
Association of Deans of Pharmacy of Canada, Toronto 01929
Association of Faculties of Pharmacy of Canada, Vancouver 01933
British Columbia Pharmacists' Society, Vancouver 01986
Canadian Academy of the History of Pharmacy, Toronto 02009
Canadian Conference on Continuing Education in Pharmacy, Saint John 02103
Canadian Foundation for the Advancement of Pharmacy, Toronto 02141
Canadian Pharmaceutical Association, Ottawa 02223
Nova Scotia Pharmaceutical Society, Halifax 02741
Pharmacological Society of Canada, Vancouver 02813
Pharmacy Association of Nova Scotia, Halifax 02814
Saskatchewan Pharmaceutical Association, Regina 02896

Chile
Colegio de Quimico-Farmacéuticos de Chile, Santiago 03033
Colegio de Químicos Farmacéuticos, Concepción 03034
Colegio Farmacéutico de Chile, Concepción 03036

Croatia
Croatian Pharmaceutical Society, Zagreb 03455

Czech Republic
Česká Farmaceutická Společnost, Praha 03505

Denmark
Danmarks Farmaceutiske Selskab, København 03577
Dansk Farmaceutforening, København 03616
Dansk Farmacihistorisk Selskab, København 03617
European College of Neuropsychopharmacology, Hillerød 03723

Ecuador
Sociedad Latinoamericana de Farmacología, Quito 03858

Egypt
Arab Higher Committee for Pharmacological Affairs, Cairo 03871

Finland
Suomen Farmaseuttinen Yhdistys, Helsingin Yliopisto 04014
Suomen Farmasialiitto, Helsinki 04015

France
Académie de Pharmacie de Paris, Paris 04092
Association des Internes et Anciens Internes en Pharmacie des Hôpitaux de Nancy, Nancy 04189
Association Nationale des Cours Professionnels pour les Préparateurs en Pharmacie, Paris 04308
Comité d'Education Sanitaire et Sociale de la Pharmacie Française, Paris 04392
Conférence Internationale des Facultés, Instituts et Ecoles de Pharmacie d'Expression Française, Paris 04443
Fédération des Syndicats Pharmaceutiques de France, Paris 04519
Société de Médecine, Chirurgie et Pharmacie de Toulouse, Toulouse 04666
Société de Pharmacie de Bordeaux, Bordeaux 04680
Société de Pharmacie de Lyon, Lyon 04681
Société de Pharmacie de Marseille, Marseille 04682
Société de Recherches Pharmaceutiques et Cientifiques, Paris 04688
Société Française de Phytiatrie et de Phytopharmacie, Versailles 04837
Société Française de Sciences et Techniques Pharmaceutiques, Paris 04841

Germany
Deutsche Gesellschaft für Pharmakologie und Toxikologie, Darmstadt 05400
Deutsche Pharmakologische Gesellschaft e.V., Wuppertal 05484
Deutsche Pharmazeutische Gesellschaft e.V., Eschborn 05485
Gesellschaft für Arzneipflanzenforschung e.V., Kleinrinderfeld 05802
Internationale Gesellschaft für Geschichte der Pharmazie, Bremen 05991
Medizinisch Pharmazeutische Studiengesellschaft e.V., Bonn 06118
Pharma-Dokumentationsring e.V., Wuppertal 06234

Ghana
Pharmaceutical Society of Ghana, Accra 06492

Greece
Elliniki Pharmakeutiki Etaireia, Athinai 06514

Hungary
Magyar Farmakológiai Társaság, Budapest 06627
Magyar Gyógyszerészeti Társaság, Budapest 06633

India
Indian Pharmaceutical Association, Bombay 06806
Pharmacy Council of India, New Dehli 06851

Ireland
Pharmaceutical Society of Ireland, Dublin 07019

Italy
Accademia Italiana di Storia della Farmacia, Pisa 07196
Associazione dei Biologi delle Facoltá di Farmacia, Padova 07268
Centro Ricerche Cosmetologiche, Roma 07535
Collegium Internationale Neuro-Psychopharmacologicum, Milano 07610
Latin-Mediterranean Society of Pharmacy, Bologna 07760
Società di Scienze Farmacologiche Applicate, Milano 07788
Società Farmaceutica del Mediterraneo Latino, Messina 07804
Società Italiana di Farmacologia, Milano 07871
Società Italiana di Farmacologia Clinica, Pisa 07872
Società Italiana di Scienze Farmaceutiche, Milano 07944
Tecnofarmaci, Pomezia 08035
Unione Tecnica Italiana Farmacisti, Genova 08050

Jamaica
Caribbean Regional Drug Testing Laboratory, Kingston 08075

Japan
Nihon Yakuri Gakkai, Tokyo 08273
Nippon Bitamin Gakkai, Kyoto 08281
Nippon Syoyakugakkai, Tokyo 08389
Nippon Yakugaku-Kai, Tokyo 08395

Korea, Republic
The Korean Society of Pharmacology, Seoul 08540

Macedonia
Farmacevtsko Društvo na Makedonija, Skopje 08601

Malaysia
Action for Rational Drugs in Asia, Penang 08626

Netherlands
European Committee on Radiopharmaceuticals, Rotterdam 08881
Koninklijke Nederlandse Maatschappij ter Bevordering der Pharmacie, 's-Gravenhage 08956
Pharma-Dokumentationsring e.V., Oss 09061

Panama
Central American Society of Pharmacology, Panamá City 09413

Philippines
Philippine Pharmaceutical Association, Manila 09531

Poland
Polskie Towarzystwo Farmaceutyczne, Warszawa 09565
Polskie Towarzystwo Historii Medycyny i Farmacji, Warszawa 09577

Portugal
Sociedade Farmacêutica Lusitana, Lisboa 09684

Romania
Societatea de Farmacie, Bucuresti 09761
Societatea de Stiinte Farmaceutice, Bucuresti 09781

Russia
National Pharmaceutical Society, Moskva 09911
Russian Pharmacological Society, Moskva 09963

South Africa
Society of Physiologists, Biochemists and Pharmacologists, Pretoria 10175

Spain
Consejo General de Colegios Oficiales de Farmacéuticos, Madrid 10298
Instituto de Farmacología Española, Madrid 10321
Real Academia de Farmacia, Madrid 10346

Switzerland
Gesellschaft für Arzneipflanzenforschung e.V., Zürich 10711
Schweizerischer Apotheker-Verein, Bern-Liebefeld 10885

Turkey
Türk Eczacilari Birliği, Istanbul 11155

United Kingdom
Association for Veterinary Clinical Pharmacology and Therapeutics, Dorking 11296
Association of Medical Advisers to the Pharmaceutical Industry, London 11367
British Association for Psychopharmacology, London 11494
British Pharmacological Society, London 11630
British Society for the History of Pharmacy, Edinburgh 11672
Commonwealth Pharmaceutical Association, London 11829
The Company Chemists' Association, Nottingham 11832
Pharmaceutical Society of Northern Ireland, Belfast 12692
Royal Pharmaceutical Society of Great Britain, London 12838
Scottish Pharmaceutical Federation, Glasgow 12920
Society for Environmental Therapy, Manchester 12972
Society for Medicine Research, London 12981

Uruguay
Asociación de Química y Farmacia del Uruguay, Montevideo 13245

U.S.A.
Academy of Pharmacy Practice and Management, Washington 13302
American Association of Colleges of Pharmacy, Alexandria 13522
American Association of Homeopathic Pharmacists, Norwood 13539
American College of Apothecaries, Alexandria 13698

American College of Clinical Pharmacology, Wayne *13702*
American College of Neuropsychopharmacology, Nashville *13718*
American Council on Pharmaceutical Education, Chicago *13767*
American Institute of the History of Pharmacy, Madison *13896*
American Pharmaceutical Association, Washington *13996*
American Society for Pharmacology and Experimental Therapeutics, Bethesda *14077*
American Society for Pharmacy Law, Vienna *14078*
American Society of Consultant Pharmacists, Alexandria *14112*
American Society of Hospital Pharmacists, Bethesda *14132*
American Society of Pharmacognosy, Monroe *14153*
Auxiliary to the American Pharmaceutical Association, Pittsburgh *14561*
Behavioral Pharmacology Society, Atlanta *14573*
Drug Information Association, Maple Glen *14987*
International Aloe Science Council, Austin *15372*
Jewish Pharmaceutical Society of America, Brooklyn *15609*
Medical Letter, New Rochelle *15709*
National Association of Boards of Pharmacy, Park Ridge *15834*
National Catholic Pharmacists Guild of the United States, Saint Louis *15937*
National Council of State Pharmacy Executives, Chapel Hill *16006*
National Pharmaceutical Association, Washington *16126*
United States Pharmacopeial Convention, Rockville *16832*

Venezuela
Colegio de Farmacéuticos del Distrito Federal y Estado Miranda, Caracas *16994*
Sociedad Latinoamericana de Farmacología, Caracas *17013*

Vietnam
Vietnamese Association of Pharmacy, Hanoi *17059*

Yugoslavia
Association of Pharmacists of Serbia, Beograd *17078*

Zimbabwe
Pharmaceutical Society of Zimbabwe, Harare *17178*

Philosophy

Argentina
Academia Nacional de Ciencias Morales y Políticas, Buenos Aires *00046*

Australia
Australasian Association for the History, Philosophy and Social Studies of Science, Sydney *00237*
Australasian Association of Philosophy, Melbourne *00238*

Austria
Anthroposophische Gesellschaft in Österreich, Wien *00545*
Gesellschaft für Ganzheitsforschung, Wien *00619*
Internationale Paracelsus-Gesellschaft, Salzburg *00676*
Österreichische Gesellschaft für Gruppendynamik und Organisationsberatung, Klagenfurt *00787*
Österreichische Gesellschaft für Kinderphilosophie, Graz *00799*
Österreichische Gesellschaft für Philosophie, Salzburg *00822*
Österreichische Ludwig-Wittgenstein-Gesellschaft, Kirchberg *00866*
Philosophische Gesellschaft an der Universität Graz, Graz *00931*
Philosophische Gesellschaft Innsbruck, Innsbruck *00932*
Philosophische Gesellschaft in Salzburg, Salzburg *00933*
Philosophische Gesellschaft Klagenfurt, Klagenfurt *00934*
Philosophische Gesellschaft Wien, Wien *00935*
Vereinigung der Humanistischen Gesellschaften Österreichs, Graz *00991*
Vereinigung für wissenschaftliche Grundlagenforschung, Graz *00995*

Belgium
Centre Européen pour l'Etude de l'Argumentation, Bruxelles *01187*
European Association of Centers of Medical Ethics, Bruxelles *01288*
European Business Ethics Network, Bruxelles *01305*
Institut Belge des Hautes Etudes Chinoises, Bruxelles *01381*
Koninklijke Academie voor Wetenschappen, Letteren en Schone Kunsten van België, Bruxelles *01409*
Société Belge de Logique et de Philosophie des Sciences, Bruxelles *01432*
Société Belge de Philosophie, Bruxelles *01439*
Société Internationale pour l'Etude de la Philosophie Médiévale, Louvain-la-Neuve *01465*
Société Philosophique de Louvain, Louvain-la-Neuve *01468*

Brazil
Associação Católica Interamericana de Filosofía, Rio de Janeiro *01594*
Sociedade Brasileira de Filosofia, Rio de Janeiro *01710*

Bulgaria
Bulgarian Philosophical Society, Sofia *01753*

Canada
Anthroposophical Society in Canada, Toronto *01868*
Association of Concern for Ultimate Reality and Meaning, Toronto *01928*
Canadian Philosophical Association, Ottawa *02224*
Canadian Society for the History and Philosophy of Science, Montréal *02267*
Society of Philosophy and Social Science, Kingston *02947*
Upper Canada Society for the History and Philosophy of Science and Technology, Hamilton *02974*

China, People's Republic
Inner Mongolian Academy of Social Sciences, Huhehot *03206*

China, Republic
Confucius-Mencius Society of the Republic of China, Taipei *03292*
World Wide Ethical Society, Taipei *03333*

Denmark
Selskabet for Filosofi og Psykologi, København *03813*

Egypt
Afro-Asian Philosophy Association, Cairo *03863*

Finland
Suomen Filosofinen Yhdistys, Helsingin Yliopisto *04016*

France
Académie des Sciences Morales et Politiques, Paris *04109*
Association Anthroposophique en France, Paris *04145*
Conseil International de la Philosophie et des Sciences Humaines, Paris *04455*
Fondation pour la Science – Centre International de Synthèse, Paris *04545*
Société de Philosophie de Toulouse, Toulouse *04683*
Société d'Etude du Dix-Septième Siècle, Paris *04727*
Société d'Etudes Romantiques, Clermont-Ferrand *04753*
Société Française de Philosophie, Paris *04830*

Georgia
Georgian Philosophy Centre, Tbilisi *05020*

Germany
Albertus-Magnus-Institut, Bonn *05070*
Allgemeine Gesellschaft für Philosophie in Deutschland e.V., Giessen *05072*
Anthroposophische Gesellschaft in Deutschland, Stuttgart *05076*
Association Internationale des Professeurs de Philosophie, Minden *05155*
Cusanus-Institut, Trier *05280*
Evangelische Studiengemeinschaft e.V., Heidelberg *05691*
Gemeinnütziger Verein zur Förderung von Philosophie und Theologie e.V., Bornheim *05769*
Gottfried-Wilhelm-Leibniz-Gesellschaft e.V., Hannover *05899*
Internationale Vereinigung für Rechts- und Sozialphilosophie e.V., Göttingen *06004*
Kant-Gesellschaft, Bonn *06020*
Philosophischer Fakultätentag, Saarbrücken *06235*
Rheinisch-Westfälische Akademie der Wissenschaften, Düsseldorf *06265*
Schopenhauer-Gesellschaft e.V., Bonn *06278*

India
Mythic Society, Bangalore *06839*
The Theosophical Society, Madras *06865*
United Lodge of Theosophists, Bombay *06867*

Iran
Philosophy and Humanities Society, Teheran *06902*

Ireland
University Philosophical Society, Dublin *07037*

Israel
Israel Society of Logic and Philosophy of Science, Jerusalem *07100*
Jerusalem Philosophical Society, Jerusalem *07106*

Italy
Academia Gentium Pro Pace, Roma *07123*
Accademia Olimpica, Vicenza *07225*
Associazione Nazionale del Libero Pensiero Giordano Bruno, Roma-Prati *07365*
Associazione Nazionale Filosofia Arti Scienze, Bologna *07372*
Associazione Piemontese di Studi Filosofici, Biella *07397*
Associazione Teologica Italiana per lo Studio della Morale, Novara *07408*
Centro di Studi Bonaventuriani, Bagnoregio *07437*
Centro di Studi Filosofici di Gallarate, Padova *07446*
Centro Internazionale di Studi Rosminiani, Stresa *07474*
Centro Internazionale di Studi Umanistici, Roma *07476*
Centro per la Storia della Tradizione Aristotelica Nel Veneto, Padova *07526*
Circolo Filosofico di Studi Tomistici, Roma *07602*
Società Filosofica Calabrese, Palmi *07808*
Società Filosofica Italiana, Milano *07809*
Società Filosofica Romana, Roma *07810*
Società Italiana di Filosofia Giuridica e Politica, Roma *07873*

Japan
Bigaku-Kai, Tokyo *08114*
Chusei Tetsugakkai, Tokyo *08117*
Hiroshima Tetsugakkai, Hiroshima-ken *08133*
Kagaku Kisoron Gakkai, Tokyo *08150*
Kyoiku Tetsugakkai, Tokyo *08176*
Kyoto Tetsugakkai, Kyoto *08179*
Moralogy Kenkyusho, Chiba *08183*
Nihon Indogaku Bukkyogaku Kai, Tokyo *08211*
Nihon Rinrigakukai, Tokyo *08249*
Nihon Tetsugakkai, Tokyo *08269*
Nippon Hotetsu Gakkai, Kyoto *08309*
Nippon Seiyo Koten Gakkai, Kyoto *08366*
Tetsugaku-kai, Tokyo *08429*
Toho Gakkai, Tokyo *08434*

Macedonia
Društvo za Filozofija, Sociologija i Politikologija na Makedonija, Skopje *08599*

Mexico
Anuarios de Filosofía y Letras, México *08691*

Netherlands
Algemene Nederlandse Vereniging voor Wijsbegeerte, Uithoorn *08821*
Bataafsch Genootschap der Proefondervindelijke Wijsbegeerte, Rotterdam *08831*
Genootschap voor Wetenschappelijke Filosofie, Paterswolde *08905*
International Humanist and Ethical Union, Utrecht *08936*
Nederlandse Vereniging voor Logica en Wijsbegeerte der Exacte Wetenschappen, Utrecht *09023*
Nederlands Filosofisch Genootschap, Zeist *09047*
Vereniging voor Calvinistische Wijsbegeerte, Amsterdam *09097*
Vereniging voor Filosofie-Onderwijs, Zoetermeer *09099*
Vereniging voor Wijsbegeerte te s'-Gravenhage, 's-Gravenhage *09105*
Vereniging voor Wijsbegeerte van het Recht, Amstelveen *09106*

Nicaragua
Academia Nacional de Filosofía, Managua *09201*

Pakistan
Pakistan Philosophical Congress, Lahore *09391*
Shah Waliullah Academy, Hyderabad *09400*

Philippines
Philosophical Association of the Philippines, Manila *09534*

Poland
Polskie Towarzystwo Filozoficzne, Warszawa *09567*

Russia
National Committee of History and Philosophy of Natural Science and Technology, Moskva *09896*
Philosophy Society, Moskva *09943*

Slovakia
Slovenské Filozofické Združenie, Bratislava *10093*

Spain
Consejo General de Colegios Oficiales de Doctores y Licenciados en Filosofía y Letras y en Ciencias, Madrid *10297*
Real Academia de Ciencias Morales y Políticas, Madrid *10343*
Sociedad Española de Filosofía, Madrid *10387*

Sri Lanka
Ceylon Humanist Society, Jaffna *10411*

Sudan
Philosophical Society, Khartoum *10428*

Switzerland
Allgemeine Anthroposophische Gesellschaft, Dornach *10613*
Fédération Internationale des Sociétés de Philosophie, Fribourg *10703*
Schweizerische Asiengesellschaft, Zürich *10798*
Schweizerische Gesellschaft für Logik und Philosophie der Wissenschaften, Bern *10831*
Schweizerische Philosophische Gesellschaft, Sankt Gallen *10880*
Société Jean-Jacques Rousseau, Genève *10978*

Thailand
Thai-Bhara Cultural Lodge and Swami Satyananda Puri Foundation, Bangkok *11101*

Turkey
Yeni Felsefe Cemiyeti, Istanbul *11179*

United Kingdom
Anthroposophical Society in Great Britain, London *11237*
Aristotelian Society, London *11246*
Belfast Natural History and Philosophical Society, Belfast *11432*
Bertrand Russell Society, Manchester *11435*
British Humanist Association, London *11574*
The British Society for Phenomenology, Cardiff *11662*
British Society for the Philosophy of Science, London *11674*
British Society of Aesthetics, Nottingham *11679*
Cambridge Philosophical Society, Cambridge *11737*
Leeds Philosophical and Literary Society, Leeds *12444*
Literary and Philosophical Society of Liverpool, Wallasey *12456*
Literary and Philosophical Society of Newcastle upon Tyne, Newcastle-upon-Tyne *12457*
Manchester Literary and Philosophical Society, Manchester *12484*
Mind Association, Saint Andrews *12528*
National Secular Society, London *12605*
The Philosophical Society of England, Brighton *12694*
Philosophy of Education Society of Great Britain, Coleraine *12695*
Radical Philosophy Society, Bristol *12735*
Research into Lost Knowledge Organisation, Orpington *12753*
The Royal Institute of Philosophy, London *12817*
Royal Philosophical Society of Glasgow, Glasgow *12840*
Society for the Promotion of Principles, Birmingham *13007*
The Society of Metaphysicians, Hastings *13056*
South Place Ethical Society, London *13082*
The Swedenborg Society, London *13115*
Theosophical Society in England, London *13124*
Verulam Institute, Chichester *13180*
The Victoria Institute or Philosophical Society of Great Britain, London *13182*
World Union of Pythagorean Organizations, Lutton *13225*
Yorkshire Philosophical Society, York *13234*

U.S.A.
American Association of Philosophy Teachers, Oklahoma City *13563*
American Catholic Philosophical Association, Washington *13681*
American Mensa, Brooklyn *13941*
American Philosophical Association, Newark *13998*
American Society for Eighteenth-Century Studies, Logan *14058*
American Society for Political and Legal Philosophy, Norton *14081*
American Society for Value Inquiry, Raleigh *14094*
Bertrand Russell Society, Skokie *14579*
Center for Philosophy, Law, Citizenship, Farmingdale *14692*
Center for Process Studies, Claremont *14693*
Charles S. Peirce Society, Indianapolis *14722*
Council for Philosophical Studies, New Haven *14881*
Council for Research in Values and Philosophy, Washington *14884*
Foreign Services Research Institute, Washington *15111*
Hegel Society of America, Villanova *15201*
Himalayan International Institute of Yoga Science and Philosophy of the U.S.A., Honesdale *15211*
Institute for Philosophy and Public Policy, College Park *15288*
Institute for Policy Studies, Washington *15289*
International Association for Philosophy of Law and Social Philosophy – American Section, Gainesville *15390*
International New Thought Alliance, Mesa *15504*
International Phenomenological Society, Providence *15512*
International Society for Neoplatonic Studies, Norfolk *15536*
International Society for Philosophical Enquiry, Hudson *15538*
International Society for the History of Ideas, Philadelphia *15542*
Intertel, Tulsa *15582*
Jesuit Philosophical Association of the United States and Canada, Chestnut Hill *15600*
Mandala Society, Del Mar *15688*
Medieval Academy of America, Cambridge *15714*
Metaphysical Society of America, Huntsville *15722*
Philosophical Research Society, Los Angeles *16312*
Philosophic Society for the Study of Sport, Union *16313*
Philosophy of Science Association, East Lansing *16314*
Society for Ancient Greek Philosophy, Binghamton *16493*
Society for Asian and Comparative Philosophy, Oneonta *16498*
Society for Natural Philosophy, Lexington *16554*
Society for Phenomenology and Existential Philosophy, Carbondale *16567*
Society for Philosophy and Public Affairs, Lexington *16568*
Society for the Philosophy of Religion, Athens *16569*
Society for the Advancement of American Philosophy, Seattle *16586*
Society for the Philosophical Study of Marxism, Bridgeport *16602*
Society for the Scientific Study of Religion, West Lafayette *16610*
Society for the Study of Process Philosophies, Portland *16618*
Society for Women in Philosophy – Eastern Division, Scranton *16629*
Society for Women in Philosophy – Pacific Division, San Jose *16630*
Society of Christian Ethics, Boston *16649*

Yugoslavia
Filozofsko Društvo Srbije, Beograd *17093*
Jugoslovensko Udruženje za Filozofiju, Beograd *17102*

Photogrammetry s. Surveying, Photogrammetry

Photography

Australia
Institute of Photographic Technology, Melbourne 00417

Austria
Gesellschaft für Photographie und Geschichte, Wien 00630

Belgium
Association Belge de Photographie et de Cinématographie, Bruxelles 01109

Canada
National Association for Photographic Art, Scarborough 02668
The Photographic Historical Society of Canada, Toronto 02816

Egypt
Atelier, Alexandria 03882

France
Société Française de Photographie et Cinématographie, Paris 04834

Germany
Deutsche Gesellschaft für Photographie e.V., Köln 05403
Normenausschuss Bild und Film im DIN Deutsches Institut für Normung e.V., Berlin 06153

Italy
Centro Studi e Archivio della Comunicazione, Università di Parma, Parma 07565

United Kingdom
British Institute of Professional Photography, Ware 11588
The Royal Photographic Society, Bath 12841

U.S.A.
Biological Photographic Association, Chapel Hill 14590
International Center of Photography, New York 15433
International Congress on High-Speed Photography and Photonics, Janesville 15448
National Photography Instructors Association, Eagle Rock 16127
Photographic Art and Science Foundation, Des Plaines 16316
Society for Photographic Education, Dallas 16570
Society of Imaging Science and Technology, Springfield 16672
Society of Teachers in Education of Professional Photography, West Carrollton 16705
University Photographers Association of America, Provo 16848

Vietnam
Vietnamese Photographers Association, Hanoi 17069

Physical Therapy

Austria
Österreichische Gesellschaft für Akupunktur und Aurikulotherapie, Wien 00751
Österreichische Gesellschaft für Physikalische Medizin und Rehabilitation, Wien 00823
Verband der diplomierten Physiotherapeuten Österreichs, Wien 00957

Belgium
Fédération Européenne de Médecine Physique et Réadaption, Gent 01357

Brazil
Associação Prudentina de Educação e Cultura, Presidente Prudente 01608

Bulgaria
Society of Sport Medicine, Sofia 01782

Canada
Canadian Academy of Sport Medicine, Vanier City 02008
Canadian Association of Physical Medicine and Rehabilitation, London 02064
Canadian Athletic Therapists Association, Alorchester 02078
Physiotherapy Association of British Columbia, Burnaby 02818
Physiotherapy Foundation of Canada, London 02819

Chile
Confederación Latinoamericano de Fisioterapia y Kinesiología, Santiago 03046

Colombia
Asociación Colombiana de Fisioterapía, Bogotá 03345

Finland
Societas Medicinae Physicalis et Rehabilitationis Fenniae, Helsinki 03994

France
Association Européenne de Thermographie, Strasbourg 04207
Société Française de Médecine du Sport, Paris 04808

Germany
Deutsche Gesellschaft für Physikalische Medizin und Rehabilitation, Hannover 05404
Deutscher Sportärztebund, Heidelberg 05539

Hungary
Magyar Balneológiai Egyesület, Budapest 06618

Indonesia
Asian Confederation of Physical Therapy, Jakarta 06880

Italy
Associazione Medica Italiana di Idroclimatologia, Talassologia e Terapia Fisica, Roma 07355
Centro Italiano per lo Studio e lo Sviluppo dell'Agopuntura Moderna e dell'Altra Medicina, Vicenza 07503
Società Italiana di Ginnastica Medica, Medicina Fisica e Riabilitazione, Milano 07884
Unione Italiana Lotta alla Distrofia Muscolare, Padova 08047

Netherlands
Nederlandse Vereniging van Artsen voor Revalidatie en Physische Geneeskunde, Groningen 09004

Poland
Polskie Towarzystwo Balneologii, Bioklimatologii i Medycyny Fizykalnej, Poznan 09556

Romania
Societatea de Balneologie, Bucuresti 09756
Societatea de Medicinà Sportiva, Bucuresti 09771

Russia
National Scientific Medical Society of Physical Therapists and Health-Resort Physicians, Moskva 09928

Switzerland
Schweizerische Gesellschaft für Balneologie und Bioklimatologie, Sankt Moritz 10813

United Kingdom
British Acupuncture Association and Register, London 11473

U.S.A.
Acupuncture International Association, Saint Louis 13324
Acupuncture Research Institute, Monterey Park 13325
American Clinical and Climatological Association, Charleston 13693
American College of Sports Medicine, Indianapolis 13737
American Society of Hand Therapists, Garner 14129
Association of Professional Baseball Physicians, Minneapolis 14485
International Council for Health, Physical Education and Recreation, Reston 15452
Rolf Institute, Boulder 16427
United States Physical Therapy Association, Arlington Heights 16833

Physics

Argentina
Academia Nacional de Ciencias Exactas, Físicas y Naturales, Buenos Aires 00045
Asociación Física Argentina, Villa Elisa 00081
Comité Nacional de Cristalografía, Buenos Aires 00123

Australia
Australian Institute of Physics, Parkville 00310

Austria
Chemisch-Physikalische Gesellschaft in Wien, Wien 00577
Mathematisch-Physikalische Gesellschaft in Innsbruck, Innsbruck 00704
Österreichische Physikalische Gesellschaft, Graz 00879
Österreichische Tribologische Gesellschaft, Wien 00923
Verein zur Förderung des physikalischen und chemischen Unterrichts, Wien 01005

Azerbaijan
Azerbaijan Physical Society, Baku 01039

Belgium
Société Belge de Physique, Bruxelles 01442
Société Belge de Vacuologie et de Vacuotechnique, Bruxelles 01449

Brazil
Centro Brasileiro de Pesquisas Fisicas, Rio de Janeiro 01617
Centro Latino Americano de Física, Rio de Janeiro 01630
Latin American Centre for Physics, Rio de Janeiro 01698

Bulgaria
National Scientific Society of Biomedical Physics and Engineering, Sofia 01769
Society of Bulgarian Physicists, Sofia 01779

Canada
American Association of Physics Teachers (Ontario), Kingston 01859
Canadian Association of Physicists, Ottawa 02065
Electrolysis Association of Canada, Toronto 02429
International Union of Pure and Applied Physics, Québec 02563
Pulp and Paper Research Institute of Canada, Pointe Claire 02853
Spectroscopy Society of Canada, Ottawa 02949

Chile
Sociedad Chilena de Física, Santiago 03071

China, People's Republic
Chinese Academy of Meteorological Sciences, Beijing 03141

China, Republic
Association for Physics and Chemistry Education of the Republic of China, Taipei 03243
Physical Society of China, Taipei 03319

Colombia
Academia Colombiana de Ciencias Exactas, Físicas y Naturales, Bogotá 03336

Cuba
Centro Nacional de Investigaciones Científicas, La Habana 03478

Czech Republic
Jednota Českých Matematiků a Fysiků, Praha 03547
Spektroskopická Společnost Jana Marca Marci, Praha 03553

Denmark
Fysisk Forening, København 03739
Jydsk Selskab for Fysik og Kemi, Arhus 03753
Nordisk Institut for Teoretisk Atomfysik, København 03786

Finland
Geofysiikan Seura, Espoo 03962
Suomen Fyysikkoseura, Helsingin Yliopisto 04019

France
Association Francophone de Spectrométrie des Masses Solides, Lannion 04250
European Centre for Advanced Studies in Thermodynamics, Paris 04494
Groupe d'Etude et de Synthèse des Microstructures, Paris 04551
Groupe pour l'Avancement des Sciences Analytiques, Paris 04562
Joint Committee on Climatic Changes and the Ocean, Paris 04592
Société des Sciences Physiques et Naturelles de Bordeaux, Talence 04718
Société Française d'Acoustique, Paris 04781
Société Française de Physique, Paris 04836
Société Française des Physiciens d'Hôpital, Paris 04844
Société Nationale des Sciences Naturelles et Mathématiques, Cherbourg 04902
Union des Physiciens, Paris 04975

Germany
Deutsche Arbeitsgemeinschaft Vakuum, Jülich 05294
Deutsche Gesellschaft für Lichtforschung, Hanau 05375
Deutsche Physikalische Gesellschaft e.V., Bad Honnef 05487
Deutsches Elektronen-Synchrotron, Hamburg 05575
Ernst-Mach-Institut, Freiburg 05643
Gesellschaft zur Förderung der Spektrochemie und angewandten Spektroskopie e.V., Dortmund 05884
Physikalisch-Medizinische Sozietät zu Erlangen, Erlangen 06236
Verband Deutscher Physikalischer Gesellschaften, Stuttgart 06357

Greece
Balkan Physical Union, Thessaloniki 06502
Enosis Ellinon Physikon, Athinai 06520

Guatemala
Academia de Ciencias Médicas, Físicas y Naturales de Guatemala, Guatemala City 06553

Hungary
Eötvös Loránd Fizikai Társulat, Budapest 06601
Magyar Biofizikai Társaság, Budapest 06619

India
Indian Association for the Cultivation of Science, Calcutta 06770

Indonesia
Asian Physics Education Network, Jakarta 06881

Israel
Israel Physical Society, Haifa 07086

Italy
Academia Gentium Pro Pace, Roma 07123
European Society for the Study of Ultrasonics, Roma 07683
International Centre for Theoretical Physics, Trieste 07711
Società Italiana di Fisica, Bologna 07874
Società Italiana di Reologia, Napoli 07942
Società Italiana di Scienze Fisiche e Matematiche Mathesis, Roma 07945

Japan
Kobunshi Gakkai, Tokyo 08166
Nihon Shinku Kyokai, Tokyo 08261
Nippon Bunko Gakkai, Tokyo 08283
Nippon Butsuri Gakkai, Tokyo 08284
Nippon Butsuri-Kagaku Kenkyukai, Kyoto 08285
Nippon Chikyu Denki Ziki Gakkai, Tokyo 08288
Nippon Denshi Kenbikyo Gakkai, Tokyo 08294
Nippon Rikusui Gakkai, Tokyo 08357
Oyo-buturi Gakkai, Tokyo 08402

Malaysia
ASEAN Institute for Physics, Kuala Lumpur 08629

Mexico
Centro de Investigación y de Estudios Avanzados del Instituto Politécnico Nacional, México 08718

Moldova, Republic of
Physical Society of Moldova, Chişinău 08788

Morocco
Société des Sciences Naturelles et Physiques du Maroc, Rabat 08808

Netherlands
FOM-Instituut voor Atoom- en Molecuulfysica, Amsterdam 08896
Stichting voor Fundamenteel Onderzoek der Materie, Utrecht 09078

New Zealand
New Zealand Institute of Physics, Auckland 09160

Norway
Fysikkforeningen, Oslo 09267
Norsk Fysisk Selskap, Kjeller 09314

Peru
Academia Nacional de Ciencias Exactas, Físicas y Naturales de Lima, Lima 09438

Philippines
Asia Pacific Physics Teachers and Educators Association, Manila 09507

Poland
Polskie Towarzystwo Fizyczne, Warszawa 09569
Polskie Towarzystwo Mechaniki Teoretycznej i Stosowanej, Warszawa 09586

Romania
Societatea de Stiinte Fizice si Chimice din Romania, Bucuresti 09783

Russia
Commission for Mechanics and Physics of Polymers, Moskva 09823
Commission for Nuclear Physics, Moskva 09825
National Committee for Thermal Analysis, Moskva 09890
Programme Committee Surface Physics, Chemistry and Mechanics, Moskva 09946

Singapore
Institute of Physics, Singapore, Singapore 10048

Slovakia
Jednota Slovenských Matematikov a Fyzikov, Bratislava 10057
Slovenská Spoločnost pre Mechaniku, Bratislava 10090

Slovenia
Društvo Matematikov, Fizikov i Astronomov Slovenije, Ljubljana 10098

South Africa
The South African Institute of Mining and Metallurgy, Marshalltown 10198
The South African Institute of Physics, Faure 10200

Spain
Academia de Ciencias Exactas, Físicas, Químicas y Naturales de Zaragoza, Zaragoza 10239
European Bioelectromagnetics Association, Madrid 10306
Instituto Nacional de Meteorología, Madrid 10327
Real Academia de Ciencias Exactas, Físicas y Naturales, Madrid 10342
Real Sociedad Española de Física, Madrid 10361

Sweden
Svenska Fysikersamfundet, Uppsala 10550
Svensk Förening for Radiofysik, Danderyd 10582

Switzerland
Centre Européen de Réflexion et d'Etude en Thermodynamique, Lausanne 10642
International Society for General Relativity and Gravitation, Bern 10756
Physikalische Gesellschaft Zürich, Zürich 10780
Schweizerische Gesellschaft für Vakuum-Physik und -Technik, Zürich 10854
Schweizerische Physikalische Gesellschaft, Genève 10881
Société de Physique et d'Histoire Naturelle de Genève, Genève 10969

Turkey
Türk Fizik Derneği, Istanbul 11156

United Kingdom

Association for Medical Physics Technology, Ickenham 11274
Association of British Spectroscopists, Kent 11317
The British Cryogenics Council, London 11548
Direct Investigation Group on Aerial Phenomena, Bury 11893
Ecological Physics Research Group, Bedford 11925
European Mechanics Council, Cambridge 12033
Institute of Electrolysis, Manchester 12234
Institute of Logistics, Corby 12252
Institute of Physics, London 12268
International Commission on Physics Education, Malvern 12343
Royal Physical Society of Edinburgh, Edinburgh 12842
Science and Engineering Research Council, Swindon 12889
UV Spectrometry Group, Cambridge 13176

U.S.A.

American Association for Crystal Growth, Thousand Oaks 13476
American Association of Physics Teachers, College Park 13567
American Carbon Society, Saint Marys 13677
American College of International Physicians, Washington 13712
American Crystallographic Association, Buffalo 13774
American Institute of Physics, New York 13893
American Physical Society, New York 14000
American Vacuum Society, New York 14202
Carnegie Institution of Washington, Washington 14647
Center for Short Lived Phenomena, Cambridge 14700
Coblentz Society, Norwalk 14768
Cryogenic Society of America, Oak Park 14945
Federation of Analytical Chemistry and Spectroscopy Societies, Frederick 15082
Fiber Society, Princeton 15092
Health Physics Society, McLean 15195
International Association for the Properties of Water and Steam, Palo Alto 15394
International Commission on Physics Education, Columbus 15441
Microbeam Analysis Society, Deerfield Beach 15727
Society for Applied Spectroscopy, Frederick 16496
Society for Magnetic Resonance Imaging, Chicago 16550
Society of Rheology, New York 16700
United States National Committee on Theoretical and Applied Mechanics, Washington 16830
Universities Research Association, Washington 16839

Uzbekistan

Scientific Industrial Association „Solar Physics", Tashkent 16947

Venezuela

Academia de Ciencias Físicas, Matemáticas y Naturales, Caracas 16958

Vietnam

Vietnamese Association of Physics, Hanoi 17060

Yugoslavia

Association of the Mathematicians', Physicists' and Astronomers' Societies of Yugoslavia, Priština 17079
Savez Društava Matematičara, Fizičara i Astronoma Jugoslavije, Novi Sad 17112

Physiology

Argentina

Asociación Internacional de Hidatidología, Buenos Aires 00085
Sociedad Argentina de Ciencias Fisiológicas, Buenos Aires 00161

Australia

Australian Physiological and Pharmacological Society, Melbourne 00328

Austria

Österreichische Arbeitsgemeinschaft für morphologische und funktionelle Atheroskleroseforschung, Wien 00724
Österreichische Gesellschaft für angewandte Zytologie, Graz 00756
Österreichische Gesellschaft für Balneologie und medizinische Klimatologie, Wien 00765
Österreichische Gesellschaft für Elektroencephalographie und klinische Neurophysiologie, Innsbruck 00775
Österreichische Gesellschaft für Physikalische Medizin und Rehabilitation, Wien 00823
Österreichische Gesellschaft zur Bekämpfung der Cystischen Fibrose, Wien 00854
Österreichische Physiologische Gesellschaft, Wien 00880

Belgium

Société Belge d'Electroencéphalographie et de Neurophysiologie Clinique, Ottignies 01431
Société Belge de Physiologie et de Pharmacologie, Bruxelles 01441
Société des Physiciens des Hôpitaux d'Expression Française, Liège 01457
Société Royale Belge de Médecine Physique et de Réhabilitation, Bruxelles 01475

Bulgaria

Bulgarian Society of Electroencephalography, Electromyography and Clinical Neurophysiology, Sofia 01758
Bulgarian Society of Physiological Sciences, Sofia 01761

Canada

Canadian Physiological Society, Kingston 02225
College of Physicians and Surgeons of Alberta, Edmonton 02355
College of Physicians and Surgeons of British Columbia, Vancouver 02356
College of Physicians and Surgeons of Manitoba, Winnipeg 02357
College of Physicians and Surgeons of Ontario, Toronto 02359
College of Physicians and Surgeons of Saskatchewan, Saskatoon 02360
College of Psychologists of New Brunswick, Fredericton 02361
Corporation Professionnelle des Médecins du Québec, Montréal 02395
Cystic Fibrosis Foundation of Alberta, Edmonton 02412
Human Factors Association of Canada, Mississauga 02508
International Society of Electrophysiological Kinesiology, Montréal 02560
Royal College of Physicians and Surgeons of Canada, Ottawa 02869
Saskatchewan Society of Osteopathic Physicians, Regina 02903

Chile

Latin American Association of Physiological Sciences, Santiago 03060

China, Republic

Chinese Physiological Society, Taipei 03282

Denmark

Sektionen for Klinisk Neurofysiologi, Gentofte 03808

Finland

Lääketieteellisen Fysiikan ja Tekniikan Yhdistys, Tampere 03980
Societas Medicinae Physicalis et Rehabilitationis Fenniae, Helsinki 03994
Suomen Fysiologiyhdistys, Oulu 04017
Suomen Kliinisen Neurofysiologian Yhdistys, Helsinki 04042

France

Association des Physiologistes, Aubière 04194
Société de Neurophysiologie Clinique de Langue Française, Paris 04675
Société d'Ergonomie de Langue Française, Caen 04690
Société Française de Médecine Aérospatiale, Bretigny 04807
Société Française de Néonatologie, Paris 04823
Société Française d'Optique Physiologique, Paris 04862
Syndicat National des Médecins Phlébologues Français, Paris 04957

Georgia

Georgian Society of Physiologists, Tbilisi 05025

Germany

Arbeitsgemeinschaft Spina bifida und Hydrocephalus e.V., Dortmund 05126
Deutsche Gesellschaft für Phlebologie, Nordemey 05401
Deutsche Gesellschaft für Physikalische Medizin und Rehabilitation, Hannover 05404
Deutsche Physiologische Gesellschaft e.V., Heidelberg 05488

Hungary

Magyar Elettani Társaság, Budapest 06626

Ireland

Cystic Fibrosis Association of Ireland, Dublin 06940

Israel

International Society of Computerized and Quantitative EMG, Jerusalem 07058
Israel Association for Physical Medicine and Rheumatology, Tel Aviv 07063
Israel Society of Clinical Neurophysiology, Jerusalem 07098

Italy

Accademia Medico-Fisica Fiorentina, Firenze 07208
European Society for Clinical Respiratory Physiology, Milano 07682
Società Italiana di Elettroencefalografia e Neurofisiologia, Bologna 07866
Società Italiana di Ergonomia Stomatologica, Milano 07870
Società Italiana di Fisiologia, Firenze 07875
Società Italiana di Medicina Estetica, Roma 07896
Società Italiana di Medicina Fisica e Riabilitazione, Moncalieri 07897
Società Italiana di Patologia Vascolare, Roma 07931

Japan

Nippon Seiri Gakkai, Tokyo 08364

Netherlands

Nederlandse Vereniging van Artsen voor Revalidatie en Physische Geneeskunde, Groningen 09004

New Zealand

Physiological Society of New Zealand, Palmerston North 09188

Norway

Norsk Fysiologisk Forening, Oslo 09313

Pakistan

Defence Housing Society, Karachi 09362

Poland

Polskie Towarzystwo Balneologii, Bioklimatologii i Medycyny Fizykalnej, Poznan 09556
Polskie Towarzystwo Fizjologiczne, Lublin 09568
Polskie Towarzystwo Fizyki Medycznej, Warszawa 09570

Romania

Societatea de Fiziologie, Bucuresti 09762

Russia

International Society for Pathophysiology, Moskva 09877
I.P. Pavlov Physiological Society, Moskva 09878
National Scientific Medical Society of Anatomists, Histologists and Embryologists, Moskva 09912

South Africa

Physiological and Biochemical Society of Southern Africa, Pretoria 10161
Society of Physiologists, Biochemists and Pharmacologists, Pretoria 10175
The South African Association of Physicists in Medicine and Biology, Pretoria 10183

Spain

Sociedad Española de Ciencias Fisiológicas, Madrid 10381

Sweden

Svenska Föreningen för Medicinsk Teknik och Fysik, Umea 10544

Switzerland

Schweizerische Gesellschaft für Muskelkranke, Zürich 10835
Schweizerische Gesellschaft für Physikalische Medizin und Rehabilitation, Zürich 10842

United Kingdom

Association for the Study of Obesity, London 11294
Association of Head and Neck Oncologists of Great Britain, London 11350
British Microcirculation Society, London 11608
Brittle Bone Society, Dundee 11711
Central Sterilising Club, Kingston-upon-Thames 11752
Cystic Fibrosis Research Trust, Bromley 11876
Electro Physiological Technologists' Association, London 11954
Ergonomics Society, Loughborough 11972
Neonatal Society, Oxford 12623
The Physiological Society, Oxford 12698
Royal College of Physicians of Edinburgh, Edinburgh 12787

U.S.A.

American Academy of Family Physicians, Kansas City 13394
American Academy of Physician Assistants, Alexandria 13425
American Association of Physicists in Medicine, New York 13566
American Association of Public Health Physicians, Madison 13577
American Board of Physical Medicine and Rehabilitation, Rochester 13653
American College of Nuclear Physicians, Washington 13719
American College of Orgonomy, Princeton 13724
American College of Physicians, Philadelphia 13729
American Physicians Fellowship for Medicine in Israel, Brookline 14003
American Physiological Society, Bethesda 14005
American Spinal Injury Association, Chicago 14180
Association for Applied Psychophysiology and Biofeedback, Wheat Ridge 14266
Association of American Indian Physicians, Oklahoma City 14380
Association of American Physicians, Indianapolis 14384
Association of American Physicians and Surgeons, Tucson 14385
Association of Philippine Physicians in America, Burbank 14481
Association of Physician Assistant Programs, Alexandria 14482
Association of Professional Baseball Physicians, Minneapolis 14485
Bay Area Physicians for Human Rights, San Francisco 14571
Flying Physicians Association, Kansas City 15105
Human Factors Society, Santa Monica 15235
International Human Powered Vehicle Association, Indianapolis 15486
Mental Research Institute, Palo Alto 15720
MTM Association for Standards and Research, Park Ridge 15757
National Federation of Catholic Physicians Guilds, Elm Grove 16046
North American Society of Pacing and Electrophysiology, Newton Upper Falls 16225
Physicians Forum, Chicago 16317
Society for Leukocyte Biology, Augusta 16547
Society for Psychophysiological Research, Dallas 16572
Society of General Physiologists, Woods Hole 16669
Turkish American Physicians Association, Smithtown 16795

Vietnam

Vietnamese Association of Physiology, Hanoi 17061

Yugoslavia

Jugoslovensko Društvo za Fiziologiju, Novi Sad 17098

Pneumonic Disease s. Pulmonary Disease

Political Science

Argentina

Academia Nacional de Ciencias Morales y Políticas, Buenos Aires 00046
Instituto para la Integración de América Latina, Buenos Aires 00147

Australia

Australasian Political Studies Association, Canberra 00247
Australian Fabian Society, Melbourne 00290
Australian Institute of International Affairs, Canberra 00304
Australian Institute of Political Science, Sydney 00311

Austria

Arbeitsgemeinschaft für Wissenschaft und Politik, Innsbruck 00559
Diplomatische Akademie Wien, Wien 00581
Europäisches Zentrum für Wohlfahrtspolitik und Sozialforschung, Wien 00585
Gesellschaft für politische Aufklärung, Wien 00631
Institut für Österreichkunde, Wien 00658
Internationales Institut für den Frieden, Wien 00680
Österreichische Gesellschaft für Aussenpolitik und Internationale Beziehungen, Wien 00763
Österreichische Gesellschaft für Friedensforschung, Wien 00780
Österreichische Gesellschaft für Politikwissenschaft, Wien 00824
Österreichische Gesellschaft für Wirtschaftspolitik, Wien 00846
Österreichisches Lateinamerika-Institut, Wien 00913
Österreichisches Ost- und Südosteuropa-Institut, Wien 00915
Österreichische Vereinigung für politische Wissenschaften, Wien 00925
Verein für Sozial- und Wirtschaftspolitik, Wien 00987
Wiener Institut für Entwicklungsfragen und Zusammenarbeit, Wien 01019

Barbados

Caribbean Policy Development Centre, Bridgetown 01086

Belgium

Académie Royale des Sciences d'Outre-Mer, Bruxelles 01096
European Trade Union Institute, Bruxelles 01353
Institut Royal des Relations Internationales, Bruxelles 01392
Société Royale d'Economie Politique de Belgique, Charleroi 01482

Brazil

Fundação João Pineiro, Belo Horizonte 01657
Sociedade Brasileira de Instrução, Botafogo 01714

Canada

Canadian Association for the Advancement of Netherlandic Studies, Windsor 02041
Canadian Institute for Environmental Law and Policy, Toronto 02162
Canadian Institute for Middle East Research, Calgary 02164
Canadian Institute of International Affairs, Toronto 02177
Canadian Political Science Association, Ottawa 02229
Centre for Research on Latin America and the Caribbean, Toronto 02329
Committee on Atlantic Studies, Ottawa 02363
Comparative Federation and Federalism Research Committee, Windsor 02372
Institute for Research on Public Policy, Halifax 02525

Chile

Academia Chilena de Ciencias Sociales, Politicas y Morales, Santiago 03009

China, Republic

Chinese National Foreign Relations Association, Taipei 03281

Costa Rica

Instituto Centroamericano de Administración Pública, San José 03445

Czech Republic

Česká Společnost pro Politické Vědy, Praha 03529

Denmark

Det Udenrigspolitiske Selskab, København 03715

Ecuador

Instituto Latinoamericano de Investigaciones Sociales, Quito 03854

Egypt

Egyptian Society of Political Economy, Statistics and Legislation, Cairo 03901

Finland

Valtiotieteellinen Yhdistys, Helsingin Yliopisto 04082

France

Académie des Sciences Morales et Politiques, Paris 04109
Académie Diplomatique Internationale, Paris 04110
Académie Internationale de Science Politique et d'Histoire Constitutionnelle, Paris 04113
Association Française de Science Politique, Paris 04227
Centre d'Etudes, de Documentation, d'Information et d'Action Sociales, Paris 04369
Club of Rome, Paris 04390
Conseil International des Sciences Sociales, Paris 04459
Fondation Nationale des Sciences Politiques, Paris 04543
Société d'Economie Politique, Paris 04652
Société d'Etudes Jaurésiennes, Paris 04743
Société d'Histoire Générale et d'Histoire Diplomatique, Paris 04765

Gambia

African Centre for Democracy and Human Rights Studies, Banjul 05005

Germany

Akademie für Politische Bildung, Tutzing 05058
Akademie Kontakte der Kontinente, Bonn 05063
Arbeitsgemeinschaft für betriebliche Altersversorgung e.V., Heidelberg 05102
Arbeitsgemeinschaft für Jugendhilfe, Bonn 05104
Arbeitskreis Bildung und Politik Rheinland, Remagen 05135
Arbeitskreis für Ost-West-Fragen, Vlotho 05142
Arbeit und Leben, Düsseldorf 05149
Arnold-Bergstraesser-Institut für kulturwissenschaftliche Forschung e.V., Freiburg 05152
Bildungswerk der Konrad-Adenauer-Stiftung, Politische Akademie der Konrad-Adenauer-Stiftung, Wesseling 05219
Deutsche Gesellschaft für Asienkunde e.V., Hamburg 05325
Deutsche Gesellschaft für Auswärtige Politik e.V., Bonn 05326
Deutsche Gesellschaft für die Vereinten Nationen e.V., Bonn 05340
Deutsche Parlamentarische Gesellschaft e.V., Bonn 05482
Deutscher Politologen-Verband e.V., Bonn 05537
Deutsches Forum für Entwicklungspolitik, Bonn 05576
Deutsches Orient-Institut, Hamburg 05598
Deutsche Vereinigung für Parlamentsfragen e.V., Bonn 05617
Deutsche Vereinigung für Politische Wissenschaft, Darmstadt 05618
Europäische Akademie Otzenhausen e.V., Nonnweiler 05648
Forschungsgesellschaft für Agrarpolitik und Agrarsoziologie e.V., Bonn 05719
Forschungsinstitut für Gesellschaftspolitik und beratende Sozialwissenschaft e.V., Göttingen 05729
Georg-von-Vollmar-Akademie e.V., München 05784
Gesellschaft für bedrohte Völker e.V., Göttingen 05803
Gesellschaft für Deutschlandforschung e.V., Berlin 05809
Gesellschaft für rationale Verkehrspolitik e.V., Düsseldorf 05855
Gesellschaft für Übernationale Zusammenarbeit e.V., Bonn 05865
Gesellschaft zum Studium strukturpolitischer Fragen e.V., Bonn 05878
Gesundheitspolitische Gesellschaft e.V., Kiel 05892
Hamburger Gesellschaft für Völkerrecht und Auswärtige Politik e.V., Hamburg 05907
Humanistische Union e.V., München 05933
Institut für Wirtschaft und Gesellschaft Bonn e.V., Bonn 05977
Interparlamentarische Arbeitsgemeinschaft, Bonn 06013
Kommission für Geschichte des Parlamentarismus und der politischen Parteien e.V., Bonn 06077
Kommunalpolitische Vereinigung der CDU und CSU Deutschland, Bonn 06080
Kulturpolitische Gesellschaft e.V., Hagen 06084
Leiterkreis der Katholischen Akademien, Schwerte 06102
Ost-Akademie e.V., Lüneburg 06225
Otto A. Friedrich-Kuratorium für Grundlagenforschung zur Eigentumspolitik, Köln 06227
Politische Akademie Biggesee, Attendorn-Neulisternohl 06237
Rechts- und Staatswissenschaftliche Gesellschaft, Recklinghausen 06259
Rechts- und Staatswissenschaftliche Vereinigung Düsseldorf e.V., Düsseldorf 06260
Staats- und Handelspolitische Gesellschaft, Recklinghausen 06291
Staats- und Wirtschaftspolitische Gesellschaft e.V., Hamburg 06292
Verein für Kommunalwirtschaft und Kommunalpolitik e.V., Düsseldorf 06388
Verein für Kommunalwissenschaften e.V., Berlin 06389
Wirtschaftspolitische Gesellschaft von 1947, Frankfurt 06444
Wirtschaftspolitischer Club Bonn e.V., Bonn 06445

Guyana

Guyana Institute of International Affairs, Georgetown 06576

India

Indian Council of World Affairs, New Delhi 06789
Indian Political Science Association, Madras 06808

Ireland

Irish Association of Civil Liberty, Dublin 06973

Israel

Israel Political Science Association, Haifa 07087

Italy

Accademia Simba, Roma 07238
Associazione Italiana di Scienze Politiche e Sociali, Roma 07323
Associazione per lo Studio del Problema Mondiale dei Rifugiati, Sezione Italiana, Roma 07392
Centro di Documentazione e di Iniziativa Politica, Roma 07422
Centro di Studi Sociali e Politici Lorenzo Milani, Arezzo 07454
Centro di Vita Europea, Roma 07458
Centro Informazioni e Studi sulla Comunita' Europea, Milano 07465
Centro Internazionale di Studi Rosminiani, Stresa 07474
Centro Italiano di Studi Europei, Roma 07500
Centro Romano per lo Studio dei Problemi di Attualita' Sociale, Roma 07543
Centro Studi di Politica Economica, Torino 07559
Centro Studi Parlamentari, Roma 07579
Centro Studi per il Mezzogiorno A. Ajon, Acireale 07580
Centro Studi per la Valorizzazione delle Risorse del Mezzogiorno, Napoli 07583
Centro Studi Politici e Sociali Alcide de Gasperi, Cosenza 07586
Centro Studi Politico-Sociali Achille Grandi, Milano 07587
Centro Studi sulla Resistenza, Urbino 07595
Club Turati, Milano 07605
European Consortium for Church and State Research, Firenze 07675
Istituto di Sociologia Internazionale di Gorizia, Gorizia 07727
Istituto Italo-Africano, Roma 07746
Istituto per gli Studi di Politica Internazionale, Milano 07750
Società Ecologica Friulana, Campoformido 07798
Società Italiana di Filosofia Giuridica e Politica, Roma 07873
Società Mazziniana Pensiero e Azione, Roma 07982
Tecnocentro Italiano, Milano 08034

Japan

Ajia Chosakai, Tokyo 08091
Ajia Seikei Gakkai, Tokyo 08093
Kokka Gakkai, Tokyo 08169
Nippon Keizai Seisaku Gakkai, Tokyo 08328
Nippon Kokusai Seiji Gakkai, Tokyo 08339
Nippon Seizi Gakkai, Tokyo 08368
Okinawa Kyokai Chosa Engoka, Tokyo 08401
Shakai Seisaku Gakkai, Tokyo 08416

Luxembourg

European Centre for Parliamentary Research and Documentation, Luxembourg 08585

Macedonia

Društvo za Filozofija, Sociologija i Politikologija na Makedonija, Skopje 08599

Netherlands

Europe 2000, Son en Breugel 08849
European Centre for Development Policy Management, Maastricht 08876
Vereniging voor de Staathuishoudkunde, Delft 09098

New Zealand

New Zealand Institute of International Affairs, Wellington 09158

Nigeria

Africa Leadership Forum, Abeokuta 09211
African Association of Political Science, Lagos 09212
Nigerian Institute of International Affairs, Lagos 09238

Norway

Committee on Conceptual and Terminological Analysis, Oslo 09249
International Political Science Association, Oslo 09271

Pakistan

Pakistan Institute of International Affairs, Karachi 09386

Romania

Academia de Stiinte Sociale si Politice, Bucuresti 09736
Asociatia de Drept International si Relatii Internationale din Romania, Bucuresti 09740
Asociatia Romana de Ştiinte Politice, Bucuresti 09746

Russia

Association of Political Sciences, Moskva 09804

Singapore

Asian-Pacific Political Science Association, Singapore 10040
Institute of Southeast Asian Studies, Singapore 10049

South Africa

Africa Institute of South Africa, Pretoria 10120
The South African Institute of International Affairs, Braamfontein 10197
South African Institute of Race Relations, Braamfontein 10202

Spain

Centro de Estudios Constitucionales, Madrid 10287
Real Academia de Ciencias Morales y Políticas, Madrid 10343

Sri Lanka

Ceylon Institute of World Affairs, Colombo 10412

Sweden

Nordiska Samarbetskommittén för Internationell Politik, inklusive Konfliktoch Fredsforskning, Stockholm 10509
Statsvetenskapliga Förbundet, Stockholm 10526
Utrikespolitiska Institutet, Stockholm 10603

Switzerland

Association d'Histoire et de Science Politique, Wabern 10623
Association Suisse de Science Politique, Genève 10633
Bureau International de la Paix, Genève 10640
Schweizerische Gesellschaft für Aussenpolitik, Lenzburg 10811
Schweizerische Vereinigung für Sozialpolitik, Zürich 10951

United Kingdom

Acton Society Trust, London 11203
Association for the Study of German Politics, Sheffield 11292
British Institute of Human Rights, London 11581
British International Studies Association, Bailrigg 11594
China Policy Study Group, London 11777
European Association for Evolutionary Political Economy, Newcastle-upon-Tyne 11990
European Consortium for Political Research, Colchester 12016, 12017
Fabian Society, London 12045
Foreign Affairs Circle, Richmond 12084
Hansard Society for Parliamentary Government, London 12145
Manchester Statistical Society, Manchester 12487
Political and Economic Planning, London 12711
Political Studies Association of the United Kingdom, Belfast 12712
Politics Association, London 12713
Pugwash Conferences on Science and World Affairs, London 12729
Royal Institute of International Affairs, London 12813
Royal Institute of Public Administration, London 12818
William Morris Society, London 13208

U.S.A.

Academy of Political Science, New York 13303
The Africa Fund, New York 13341
African-American Institute, New York 13342
American Academy of Political and Social Science, Philadelphia 13429
American Assembly, New York 13461
American Institute of Parliamentarians, Fort Wayne 13892
American Peace Society, Washington 13992
American Political Science Association, Washington 14010
American Society for Eighteenth-Century Studies, Logan 14058
American Society for Political and Legal Philosophy, Norton 14081
The Asia Society, New York 14253
Association for the Advancement of Policy, Research and Development in the Third World, Washington 14344
Association of Arab-American University Graduates, Belmont 14394
Atlantic Council of the United States, Washington 14546
Canadian-American Committee, Washington 14635
Center for Philosophy, Law, Citizenship, Farmingdale 14692
Center for Public Justice, Washington 14694
Center for the Study of Human Rights, New York 14706
Center for the Study of the Presidency, New York 14708
Center for War, Peace and the News Media, New York 14711
Committee on Political Education, AFL-CIO, Washington 14804
Community for Religious Research and Education, Loretto 14812
Conference Group on French Politics and Society, Cambridge 14825
Conference Group on German Politics, Durham 14826
Conference Group on Italian Politics and Society, Chicago 14827
Consortium on Peace Research, Education and Development, Fairfax 14849
Council on Governmental Relations, Washington 14918
East-West Center, Honolulu 15004
East-West Center Institute of Economic Development and Policy, Honolulu 15005
European Community Studies Association, Cleveland 15065
Foreign Policy Association, New York 15110
Foreign Services Research Institute, Washington 15111
Global Options, San Francisco 15165
Governmental Research Association, Birmingham 15168
Harry S. Truman Library Institute for National and International Affairs, Independence 15188
Industrial Relations Research Association, Madison 15260
Institute for Mediterranean Affairs, New York 15286
Institute for Palestine Studies, Washington 15287
Institute for Philosophy and Public Policy, College Park 15288
Institute for Social Research, Ann Arbor 15296
Institute of Public Administration, New York 15338
International Association of Educators for World Peace, Huntsville 15406
International Institute of Rural Reconstruction – U.S. Chapter, New York 15490
Inter-University Consortium for Political and Social Research, Ann Arbor 15583
Joint Center for Political and Economic Studies, Washington 15615
Labor Policy Association, Washington 15644
Latin American Studies Association, Pittsburgh 15648
League for Industrial Democracy, Washington 15657
Lentz Peace Research Laboratory, Saint Louis 15658
Lincoln Institute for Research and Education, Washington 15673
Middle East Research and Information Project, Washington 15734
National Council for Labor Reform, Chicago 15989
National Council on Employment Policy, Washington 16019
National Institute for Public Policy, Fairfax 16091
Pacific Studies Center, Mountain View 16284
Policy Studies Organization, Urbana 16332
Society for Philosophy and Public Affairs, Lexington 16568
Ukrainian Political Science Association in the United States, Philadelphia 16800
United Nations Committee on the Peaceful Uses of Outer Space, New York 16812
Walter Bagehot Research Council on National Sovereignty, Whitestone 16871
Women's Caucus for Political Science, Atlanta 16914

Venezuela

Academia de Ciencias Políticas y Sociales, Caracas 16959
Asociación Cultural Humboldt, Caracas 16966

Preservation of Historical Monuments, Restoration

Argentina

Comisión Nacional de Museos y de Monumentos y Lugares Históricos, Buenos Aires 00120

Australia

Australian Council of National Trusts, Civic Square 00281

Austria

Österreichische Gesellschaft für Denkmal- und Ortsbildpflege, Wien 00772
Verein Montandenkmal Altböckstein, Leoben 00998

Belgium

Association Royale des Demeures Historiques de Belgique, Bruxelles 01158

Canada

Conservation and Development Association of Saskatchewan, Canora 02387
Conservation Council of New Brunswick, Fredericton 02388
Conservation Council of Ontario, Toronto 02389

Finland

Museovirasto, Helsinki 03983
Suomen Egyptologinen Seura, Helsinki 04011

France

Association Nationale pour la Protection des Villes d'Art, Paris 04314
Conseil International des Monuments et des Sites, Paris 04456

Germany

Deutsches Nationalkomitee für Denkmalschutz, Bonn 05596
Schutzgemeinschaft Alt Bamberg e.V., Bamberg 06279
Vereinigung der Landesdenkmalpfleger in der Bundesrepublik Deutschland, Hannover 06399

Italy

Associazione Giacomo Boni per la Difesa dei Monumenti di Roma Antica, Roma 07282
Associazione per Imola Storico-Artistica, Imola 07388
European Centre for Training Craftsmen in the Conservation of the Architectural Heritage, Venezia 07671
Italia Nostra – Associazione Nazionale per la Tutela del Patrimonio Storico Artistico e Naturale della Nazione, Roma 07757

Macedonia

Republički Zavod za Zaštita na Spomenicite na Kulturata, Skopje 08606

Netherlands

Europa Nostra/IBI, 's-Gravenhage 08848
Monumentenraad, Zeist 08977
Raad voor het Cultuurbeheer, Rijswijk 09064

Norway

Foreningen til Norske Fortidsminnesmerkers Bevaring, Oslo 09266

Panama

Comisión Nacional de Arqueología y Monumentos Históricos, Panamá City 09415

Peru

Centro de Investigación y Restauración de Bienes Monumentales del Instituto Nacional de Cultura, Lima 09459

Poland

Stowarzyszenie Historyków Sztuki, Warszawa 09616
Towarzystwo Milosników Historii i Zabytków Krakowa, Kraków 09623

Spain

Asociación Española de Amigos de los Castillos, Madrid 10256

United Kingdom

Ancient Monuments Society, London 11228
Association for Studies in the Conservation of Historic Buildings, London 11285
Chiltern Society, Great Missenden 11776
Council for the Care of Churches, London 11854
International Institute for Conservation of Historic and Artistic Works, London 12369
Monumental Brass Society, London 12537
National Trust for Places of Historic Interest or Natural Beauty, London 12614
National Trust for Scotland, Edinburgh 12615
Oxford Preservation Trust, Oxford 12668
Scottish Society for the Preservation of Historical Machinery, Glasgow 12930
Shakespeare Birthplace Trust, Stratford-upon-Avon 12944
La Société Guernesiaise, Saint Peter Port 12961
Society for the Protection of Ancient Buildings, London 13010
Twentieth Century Society, London 13146
Ulster Architectural Heritage Society, Belfast 13150
United Kingdom Institute for Conservation of Historic and Artistic Works, London 13166

U.S.A.

Association for the Preservation of Virginia Antiquities, Richmond 14353
Edison Birthplace Association, Milan 15014
Friends of Cast Iron Architecture, New York 15142
National Society for the Preservation of Covered Bridges, Worcester 16157
Society for the Preservation of New England Antiquities, Boston 16605

Zambia

National Monuments Commission, Livingstone 17156

Promotion of Peace

Austria

Österreichische Gesellschaft für Friedensforschung, Wien 00780
Österreichisches Studienzentrum für Frieden und Konfliktforschung, Burg Schlaining 00918

Canada

Canadian Peace Research and Education Association, Brandon 02221
Science for Peace, Toronto 02908

Italy

Centro Internazionale della Pace, Torino 07466

New Zealand

Asian Peace Research Association, Christchurch 09117

Nigeria

African Peace Research Institute, Lagos 09218

Russia

Committee for Russian Scientists in Defence of Peace and against Nuclear War, Moskva 09846

Switzerland

Bureau International de la Paix, Genève 10640

U.S.A.

Association of American Jurists, Berkeley 14381
Center for the Study of Human Rights, New York 14706
Center for War/Peace Studies, New York 14712
Coalition for International Cooperation and Peace, New York 14764
Commission to Study the Organization of Peace, Washington 14791
Council on Peace Research in History, Detroit 14928
International Association of Educators for World Peace, Huntsville 15406
Lentz Peace Research Laboratory, Saint Louis 15658
National Research Council on Peace Strategy, New York 16139
Peace Science Society, Binghamton 16300

Protection s. Safety and Protection, Safety Engineering

Psychiatry

Argentina

Asociación Argentina para el Estudio Científico de la Deficiencia Mental, Buenos Aires 00071
Sociedad Argentina de Ciencias Neurológicas, Psiquiátricas y Neuroquirúrgicas, Buenos Aires 00162
Sociedad Argentina de Neurología, Psiquiatría y Neurocirugía, Buenos Aires 00178
Sociedad de Psicología Médica, Psicoanálisis y Medicina Psicosomática, Buenos Aires 00195

Australia

Institute of Mental Health Research and Postgraduate Training, Parkville 00414
The Royal Australian and New-Zealand College of Psychiatrists, Carlton 00460
Western Australian Mental Health Association, Subiaco 00523

Austria

Gesellschaft für biologische und psychosomatische Medizin, Wien 00614
Gesellschaft Österreichischer Nervenärzte und Psychiater, Wien 00638
Internationale Vereinigung für Selbstmordprophylaxe, Wien 00685
Österreichische Arbeitsgemeinschaft für Neuropsychiatrie und Psychologie des Kindes- und Jugendalters und verwandter Berufe, Wien 00725
Verein für Psychiatrie und Neurologie, Wien 00986

Belgium

Commission Mixte sur les Aspects Internationaux de l'Arriération Mentale, Bruxelles 01235
Société Royale de Médecine Mentale de Belgique, Bruxelles 01483

Canada

Alberta Psychiatric Association, Edmonton 01842
Association des Psychiatres du Québec, Montréal 01893
Canadian Psychiatric Association, Ottawa 02231
Canadian Psychiatric Research Foundation, Toronto 02232
Canadian Schizophrenia Foundation, Burnaby 02246
New Brunswick Psychiatric Association, Saint John 02690
Prince Edward Island Psychiatric Association, Charlottetown 02838
Saskatchewan Psychiatric Association, Saskatoon 02897

Colombia

Sociedad Colombiana de Psiquiatría, Bogotá 03413

Croatia

Union of the Societies of Engineers and Technicians of Croatia, Zagreb 03467

Denmark

Dansk Selskab for Oligofreniforskning, København 03682

France

Association Française de Psychiatrie, Ville d'Avray 04223
Association pour le Développement des Etudes Biologiques en Psychiatrie, Paris 04338
Congrès de Psychiatrie et de Neurologie de Langue Française, Limoges 04448
Fédération Caribe de Santé Mentale, Fort de France 04508
Société d'Etudes Psychiques, Nancy 04751
Société Française de Médecine Psychosomatique, Paris 04814
Société Internationale de Psychopathologie de l'Expression, Paris 04879
Société Médico-Psychologique, Boulogne 04895
Syndicat des Psychiatres Français, Ville d'Avray 04927
Syndicat National des Psychiatres des Hôpitaux, Paris 04967
Union Internationale des Sociétés d'Aide à la Santé Mentale, Bordeaux 04989

Germany

Aktion Psychisch Kranke, Bonn 05067
Deutsche Gesellschaft für Dynamische Psychiatrie, München 05342
Deutsche Gesellschaft für Kinder- und Jugendpsychiatrie e.V., Marburg 05372
Deutsche Gesellschaft für Psychiatrie und Nervenheilkunde e.V., Köln 05410
Deutsche Gesellschaft für Psychosomatische Medizin e.V., München 05412
Deutsche Gesellschaft für Soziale Psychiatrie e.V., Köln 05421

Iceland

Gedverndarfélag Islands, Reykjavik 06687

Ireland

Mental Health Association of Ireland, Dublin 07013

Italy

Associazione Italiana di Terapia Occupazionale, Roma 07330
Associazione Italiana di Terapie Psicologiche, Milano 07331
Centro di Studi e di Ricerche per la Psicoterapia della Coppia e della Famiglia, Roma 07440
Centro Italiano per lo Studio e lo Sviluppo della Psicoterapia e dell'Autogenes Training, Padova 07504
Centro Studi di Psicoterapia e Psicologia Clinica, Genova 07561
Società Italiana di Medicina Psicosomatica, Roma 07901
Società Italiana di Neuropsichiatria Infantile, Napoli 07913
Società Italiana di Psichiatria, Milano 07934
Società Italiana di Psichiatria Biologica, Napoli 07935

Japan

Nihon Seishin Shinkei Gakkai, Tokyo 08257

Luxembourg

Association of European Psychiatrists, Luxembourg 08580

Malaysia

ASEAN Federation for Psychiatric and Mental Health, Kuala Lumpur 08627

Mexico

Sociedad Mexicana de Neurologia y Psiquiatria, México 08768

Netherlands

Nederlandse Vereniging voor Psychiatrie, Utrecht 09036

Nicaragua

Sociedad Nicaragüense de Psiquiatría y Psicología, Managua 09206

Poland

Polskie Towarzystwo Psychiatryczne, Warszawa 09602

Portugal

Sociedade Portuguesa de Neurologia e Psiquiatria, Lisboa 09705

Romania

Asociaţia Psihiatricǎ Românǎ, Bucuresti 09744

Russia

National Scientific Medical Society of Neuropathologists and Psychiatrists, Moskva 09921

Spain

Asociación Española para el Estudio Científico del Retraso Mental, Madrid 10271
Sociedad Española de Medicina Psicosomática y Psicoterapía, Barcelona 10390

Switzerland

Schweizerisches Nationalkomitee für geistige Gesundheit, Zürich 10929
World Psychiatric Association, Prilly 11035

Trinidad and Tobago

Caribbean Federation for Mental Health, Port of Spain 11112

Turkey

Türk Nöro-Psikiyatri Derněi, Istanbul 11168

United Kingdom

Association for Child Psychology and Psychiatry, London 11262
Association for Professionals in Services for Adolescents, Edinburgh 11276
Association of Psychiatrists in Training, Edinburgh 11381
British Association of Social Psychiatry, Kew 11522
British Society for the Study of Mental Subnormality, Cranage 11677
MIND – National Association for Mental Health, London 12529
National Schizophrenia Fellowship, Surbiton 12604
Northern Ireland Association for Mental Health, Belfast 12635
Psychiatric Rehabilitation Association, London 12726
The Royal College of Psychiatrists, London 12789
Schizophrenia Association of Great Britain, Bangor 12887
Scottish Association for Mental Health, Edinburgh 12896

U.S.A.

American Academy of Child and Adolescent Psychiatry, Washington 13383
American Academy of Clinical Psychiatrists, San Diego 13384
American Academy of Psychiatrists in Alcoholism and Addictions, Greenbelt 13430
American Academy of Psychiatry and the Law, Baltimore 13431
American Association for Geriatric Psychiatry, Greenbelt 13480
American Association for Social Psychiatry, Washington 13491
American Association of Chairmen of Departments of Psychiatry, Little Rock 13517
American Association of Mental Health Professionals in Corrections, Sacramento 13552
American Association of Orthomolecular Medicine, Boca Raton 13559
American Association of Psychiatric Administrators, Atlanta 13574
American Association of Psychiatric Services for Children, Rochester 13575
American Association of Suicidology, Denver 13584
American Board of Psychiatry and Neurology, Deerfield 13660
American College of Neuropsychiatrists, Farmington Hills 13717
American College of Psychiatrists, Greenbelt 13734
American Institute of Life Threatening Illness and Loss, New York 13886
American Orthopsychiatric Association, New York 13974
American Parkinson Disease Association, Staten Island 13991
American Psychiatric Association, Washington 14017
American Psychoanalytic Association, New York 14018
American Psychopathological Association, Pittsburgh 14021
American Schizophrenia Association, Boca Raton 14035
American Society for Adolescent Psychiatry, Bethesda 14043
Association for Advancement of Behavior Therapy, New York 14263
Association for Research in Nervous and Mental Disease, New York 14329
Association of Mental Health Administrators, Northbrook 14462
Black Psychiatrists of America, Oakland 14600
Group for the Advancement of Psychiatry, Dallas 15179
Institute for Labor and Mental Health, Oakland 15285
Institute on Hospital and Community Psychiatry, Washington 15344
International Institute for Bioenergetic Analysis, New York 15489
International Transactional Analysis Association, San Francisco 15566
Laughter Therapy, Los Angeles 15651
Mental Health Materials Center, Bronxville 15719
Metropolitan College Mental Health Association, New York 15725
Milton H. Erickson Foundation, Phoenix 15741
National Association for Rural Mental Health, Kansas City 15813
National Association of State Mental Health Program Directors, Alexandria 15900
National Association of State Mental Retardation Program Directors, Alexandria 15901
National Consortium for Child Mental Health Services, Washington 15981
National Guild of Catholic Psychiatrists, Ellicott City 16066
National Mental Health Association, Alexandria 16109
Postgraduate Center for Mental Health, New York 16342
Sleep Research Society, Boston 16479
Social Psychiatry Research Institute, New York 16482
Society for Life History Research, Saint Louis 16548
Society for Psychophysiological Research, Dallas 16572
Society of Biological Psychiatry, Dallas 16643
Society of Professors of Child and Adolescent Psychiatry, Washington 16695
Tourette Syndrome Association, Bayside 16787

Venezuela

Asociación Psiquiátrica de la América Latina, Caracas 16971
Sociedad Venezolana de Psiquiatría y Neurología, Caracas 17030

Vietnam

Vietnamese Association of Neurology, Psychiatry and Neurosurgery, Hanoi 17053

Psychoanalysis

Argentina

Asociación Psicoanalitica Argentina, Buenos Aires 00095

Austria

Gesellschaft für Logotherapie und Existenzanalyse, Wien 00625
Innsbrucker Arbeitskreis für Psychoanalyse, Psychoanalytisches Forschungs- und Ausbildungsinstitut, Innsbruck 00652
Österreichische Gesellschaft für Autogenes Training und allgemeine Psychotherapie, Wien 00764
Österreichische Studiengesellschaft für Kinderpsychoanalyse, Salzburg 00920
Salzburger Arbeitskreis für Psychoanalyse, Salzburg 00939
Sigmund Freud-Gesellschaft, Wien 00944
Wiener Arbeitskreis für Psychoanalyse, Wien 01011
Wiener Psychoanalytische Vereinigung, Wien 01025

Belgium

Société Belge de Psychoanalyse, Loverval 01444

Canada

Canadian Institute of Hypnotism, Sainte Anne de Bellevue 02176
Canadian Psychoanalytic Society, Montréal 02233
Toronto Psychoanalytic Society, Toronto 02966

Denmark

Dansk Psykoanalytisk Selskab, København 03666

France

Association Psychoanalytique de France, Paris 04344
Congrès des Psychoanalystes de Langues Romanes, Paris 04449
Société Psychoanalytique de Paris, Paris 04914

Germany

Deutsche Gesellschaft für Analytische Psychologie e.V., Berlin 05320
Deutsche Psychoanalytische Gesellschaft e.V., München 05491

India

Indian Psychoanalytical Society, Calcutta 06809

Ireland

Irish Psychoanalytical Association, Dublin *06997*

Italy

Associazione Italiana per lo Studio della Psicologia Analitica, Roma *07345*
Associazione Medica Italiana per lo Studio della Ipnosi, Milano *07356*
Associazione Psicanalitica Italiana, Milano *07399*
International Society of Theoretical and Experimental Hypnosis, Milano *07721*

Netherlands

Nederlands Psychoanalytisch Genootschap, Utrecht *09056*

Spain

Sociedad Española de Psicoanàlisis, Barcelona *10395*

Sweden

Svenska Psykoanalytiska Föreningen, Stockholm *10572*

Switzerland

CIP International Study and Research Center in Psychosynteresis, Lausanne *10646*
European Association for Transactional Analysis, Genève *10676*

United Kingdom

British Psychoanalytical Society, London *11635*
Institute of Psycho-Analysis, London *12272*
International Psychoanalytical Association, London *12385*

U.S.A.

American Academy of Psychoanalysis, New York *13432*
American College of Psychoanalysts, Oakland *13735*
Association for Advancement of Psychoanalysis, New York *14264*
Association for Behavior Analysis, Kalamazoo *14271*
Association for Child Psychoanalysis, Great Falls *14280*
Association for Group Psychoanalysis and Process, New York *14301*
Association for Psychoanalytic Medicine, New York *14323*
Association to Advance Ethical Hypnosis, Cuyahoga Falls *14537*
C.G. Jung Foundation for Analytical Psychology, New York *14719*
National Association for the Advancement of Psychoanalysis and The American Boards for Accreditation and Certification, New York *15817*
National Psychological Association for Psychoanalysis, New York *16132*
Society for the Advancement of Behavior Analysis, Kalamazoo *16587*

Psychology s. a. Behavioral Sciences; Depth Psychology; Psychoanalysis

Argentina

Sociedad Argentina de Psicología, Buenos Aires *00182*
Sociedad de Psicología Médica, Psicoanálisis y Medicina Psicosomática, Buenos Aires *00195*

Australia

Australian Institute of Industrial Psychology, Sydney *00303*
Australian Psychological Society, Parkville *00331*
Sydney University Psychological Society, Sydney *00503*

Austria

Arbeitsgemeinschaft für Präventivpsychologie, Wien *00555*
Arbeitsgemeinschaft für Psychotechnik in Österreich, Wien *00556*
Arbeitsgemeinschaft für Verhaltensmodifikation, Salzburg *00558*
Arbeitsgemeinschaft Personenzentrierte Psychotherapie und Gesprächsführung, Wien *00562*
Berufsverband Österreichischer Psychologen, Wien *00575*
Gesellschaft für Kulturpsychologie, Salzburg *00624*
Internationale Vereinigung für Selbstmordprophylaxe, Wien *00685*
Österreichische Ärztegesellschaft für Psychotherapie, Wien *00719*
Österreichische Arbeitsgemeinschaft für Neuropsychiatrie und Psychologie des Kindes- und Jugendalters und verwandter Berufe, Wien *00725*
Österreichische Arbeitskreise für Tiefenpsychologie, Salzburg *00728*
Österreichische Gesellschaft für ärztliche Hypnose und autogenes Training, Wien *00749*
Österreichische Gesellschaft für Bionome Psychotherapie, Wien *00767*
Österreichischer Verein für Individualpsychologie, Wien *00906*
Sigmund Freud-Gesellschaft, Wien *00944*
Steirische Gesellschaft für Psychologie, Graz *00950*

Belgium

Association de l'Europe Occidentale pour la Psychologie Aéronautique, Bruxelles *01114*
European Association of Experimental Social Psychology, Louvain-la-Neuve *01291*
European Down's Syndrome Association, Verviers *01337*
Société Belge de Psychologie, Bruxelles *01445*

Bulgaria

Society of Bulgarian Psychologists, Sofia *01780*

Canada

Association for Social Psychology, Toronto *01902*
Association of Psychologists of Nova Scotia, Millstream *01939*
Calgary Society for the Treatment of Autism, Calgary *01998*
Canadian Association for Educational Psychology, London *02031*
Canadian Group Psychotherapy Association, Montréal *02150*
Canadian Hypnotherapy Association, Montréal *02158*
Canadian Psychological Association, Old Chelsea *02234*
Canadian Society for Psychomotor Learning and Sport Psychology, Kingston *02265*
Corporation Professionnelle des Psychologues du Québec, Montréal *02398*
Human Factors Association of Canada, Mississauga *02508*
Institut de Recherches Psychologiques, Montréal *02521*
International Association for Cross-Cultural Psychology, Kingston *02539*
International Association of Group Psychotherapy, Montréal *02541*
Ontario Group Psychotherapy Association, Etobicoke *02780*
Psychological Society of Saskatchewan, Regina *02851*

China, Republic

Chinese Association of Psychological Testing, Taipei *03262*
Chinese Psychological Association, Taipei *03283*

Czech Republic

Česká Psychologická Společnost, Praha *03515*

Denmark

Dansk Psykoanalytisk Selskab, København *03666*
Dansk Psykolog Forening, Valby *03667*
Selskabet for Filosofi og Psykologi, København *03813*

Egypt

Egyptian Association for Psychological Studies, Cairo *03888*

Finland

Suomen Psykologiliitto, Helsinki *04056*

France

Association de Psychologie du Travail de Langue Française, Versailles *04156*
Association de Psychologie Scientifique de Langue Française, Paris *04157*
Association Internationale pour l'Etude du Comportement des Conuducteurs, Paris *04293*
Conseil International des Sciences Sociales, Paris *04459*
Institut Français d'Analyse de Groupe et de Psychodrame, Paris *04567*
Société de Neuropsychologie de Langue Française, Paris *04676*
Société de Psychologie Médicale de Langue Française, Paris *04684*
Société d'Ergonomie de Langue Française, Caen *04690*
Société d'Etude de Psychodrame Pratique et Théorique, Paris *04726*
Société Française de Psychologie, Paris *04838*
Société Française d'Etudes des Phénomènes Psychiques, Paris *04851*
Société Médico-Psychologique, Boulogne *04895*

Georgia

Georgian Society of Psychologists, Tbilisi *05026*

Germany

Berufsverband Deutscher Psychologen e.V., Bonn *05204*
Berufsverband Geprüfter Graphologen/Psychologen e.V., München *05206*
Bundesverband Legasthenie e.V., Hannover *05256*
Dachverband Psychosozialer Hilfsvereinigungen e.V., Bonn *05281*
Deutsche Gesellschaft für Analytische Psychologie e.V., Berlin *05320*
Deutsche Gesellschaft für Psychologie e.V., Münster *05411*
Deutsche Gesellschaft für Psychosomatische Medizin e.V., München *05412*
Deutsche Gesellschaft für Sexualforschung e.V., Hamburg *05419*
Deutsche Gesellschaft für Suchtforschung und Suchttherapie e.V., Hamm *05429*
Deutsche Hauptstelle gegen die Suchtgefahren e.V., Hamm *05454*
Deutsche Psychoanalytische Vereinigung e.V., Berlin *05492*
Gesellschaft für Psychotherapie, Psychosomatik und Medizinische Psychologie e.V., Lipzig *05853*
Psychobiologische Gesellschaft, Mülheim *06242*

Hungary

Magyar Pszichológiai Társaság, Budapest *06653*

Italy

Associazione di Ricerca e Interventi Psicosociali e Psicoterapeutici, Molinetto di Mazzano *07274*
Associazione Italiana di Psicologia dello Sport, Ferrara *07319*
Associazione Italiana per la Psicologia Umanistica e Transpersonale, Roma *07337*
Associazione Italiana per lo Studio della Psicologia Analitica, Roma *07345*
Associazione Italiana Socioanalisi Individuale, Milano *07350*
Associazione Nazionale dei Periti Grafici a Base Psicologica, Milano *07364*
Centro di Demodossalogia, Roma *07419*
Centro di Ricerche Biopsichiche di Padova, Padova *07435*
Centro Internazionale di Ipnosi Medica e Psicologica, Milano *07468*
Centro Internazionale Studi Umanistici Scientifici Psicologici, Roma *07488*
Centro Italiano di Sessuologia, Roma *07495*
Centro Psico-Pedagogico Didattico, Bologna *07531*
Centro Studi di Psicologia e Sociologia Applicate ad Indirizzo Adleriano, Roma *07560*
Centro Studi Wilhelm Reich, Napoli *07596*
International Society for Medical and Psychological Hypnosis, Milano *07718*
Sezione di Training Autogeno del Centro Internazionale de Ipnosi, Milano *07764*
Società Internazionale di Psicologia della Scrittura, Milano *07816*
Società Italiana di Psicologia, Roma *07936*
Società Italiana di Psicologia Individuale, Milano *07937*
Società Italiana di Psicologia Scientifica, Palermo *07938*
Unione Italiana Lotta alla Distrofia Muscolare, Padova *08047*

Japan

Asian Society for Sport Psychology, Tokyo *08110*
International Association for Religion and Parapsychology, Tokyo *08141*
Nihon Kyoiku Shinri Gakkai, Tokyo *08239*
Nihon Oyo Shinri-gakkai, Tokyo *08246*
Nihon Shakai Shinri Gakkai, Tokyo *08260*
Nihon Shinrigakkai, Tokyo *08262*

Korea, Republic

Korean Psychological Association, Seoul *08539*

Macedonia

Društvo za Nauka i Umetnost, Bitola *08600*

Mexico

Sociedad Mexicana de Estudios Psico-Pedagógicos, México *08760*

Netherlands

European Association of Experimental Social Psychology, Amsterdam *08866*
International Association of Applied Psychology, Nijmegen *08924*
Nederlandse Vereniging voor Psychotherapie, Utrecht *09037*
Studievereniging voor Psychical Research, Voorburg *09080*

New Zealand

New Zealand Psychological Society, Wellington *09173*

Nicaragua

Sociedad Nicaragüense de Psiquiatría y Psicología, Managua *09206*

Poland

Polskie Towarzystwo Psychologiczne, Warszawa *09603*

Puerto Rico

Asociación de Psicólogos de Puerto Rico, Rio Piedras *09721*

Romania

Asociatia Psihologilor din Romania, Bucuresti *09745*

Russia

Society of Psychologists, Moskva *09979*

Slovakia

Slovenská Psychologická Spoločnost, Bratislava *10085*

South Africa

Psychological Association of South Africa, Pretoria *10167*

Sweden

Svenska Psykoanalytiska Föreningen, Stockholm *10572*

Switzerland

Institut International de Psychologie et de Psychothérapie Charles Baudouin, Coppet *10729*
International Federation for Psychotherapy, Bern *10749*
Schweizerische Gesellschaft für Lehrerinnen- und Lehrerbildung, Bern *10830*
Schweizerische Gesellschaft für Psychologie, Lausanne *10844*
Schweizerischer Berufsverband für Angewandte Psychologie, Forch *10888*
Schweizerische Stiftung für Angewandte Psychologie, Fribourg *10932*
Vereinigung Schweizerischer Kinder- und Jugendpsychologen, Solothurn *11019*

United Kingdom

Association for Child Psychology and Psychiatry, London *11262*
Association for Group and Individual Psychotherapy, London *11264*
Association for Teaching Psychology, Leicester *11286*
Association of Educational Psychologists, Durham *11341*
Balint Society, London *11420*
British Association of Psychotherapists, London *11520*
British Institute of Practical Psychology, London *11587*
British Psychological Society, Leicester *11636*
British Society for the Study of Mental Subnormality, Cranage *11677*
Clinical Theology Association, Oxford *11794*
Ergonomics Society, Loughborough *11972*
European Psycho-Analytical Federation, London *12037*
Experimental Psychology Society, Egham *12044*
Guild of Pastoral Psychology, London *12136*
National Council of Psycho-Therapists, London *12585*
National Institute of Industrial Psychology, Slough *12599*
Scottish Association of Local Government to Educational Psychologists, Glasgow *12899*
Society for Multivariate Experimental Psychology, Upton Park *12983*
Society for Research in the Psychology of Music and Music Education, Dagenham *12993*
Society for the Study of Normal Psychology, London *13020*
The Tavistock Institute of Medical Psychology, London *13117*
Theosophical Society in England, London *13124*

U.S.A.

Academy of Psychosomatic Medicine, Chicago *13304*
Academy of Religion and Psychical Research, Bloomfield *13306*
Academy of Scientific Hypnotherapy, San Diego *13308*
Alfred Adler Institute, New York *13362*
American Academy of Crisis Interveners, Louisville *13386*
American Academy on Mental Retardation, Omaha *13443*
American Association for Correctional Psychology, Institute *13475*
American Association of Laban Movement Analysts, New York *13545*
American Association of Mental Health Professionals in Corrections, Sacramento *13552*
American Association on Mental Retardation, Washington *13608*
American Board of Professional Psychology, Columbia *13658*
American Board of Psychological Hypnosis, Chicago *13661*
American Guild of Hypnotherapists, Omaha *13837*
American Hypnotists' Association, San Francisco *13857*
American Orthopsychiatric Association, New York *13974*
American Psychological Association, Washington *14019*
American Psychology-Law Society, Worcester *14020*
American Society for Psychical Research, New York *14082*
American Society of Clinical Hypnosis, Des Plaines *14108*
American Society of Physoanalytic Physicians, Rockville *14154*
American Society of Psychopathology of Expression, Brookline *14161*
Arica Institute, New York *14241*
Association for Advancement of Psychology, Colorado Springs *14265*
Association for Astrological Psychology, San Rafael *14270*
Association for Birth Psychology, New York *14273*
Association for Creative Change, Minneapolis *14288*
Association for Group Psychoanalysis and Process, New York *14301*
Association for Humanistic Psychology, San Francisco *14305*
Association for the Development of Human Potential, Spokane *14351*
Association for Transpersonal Psychology, Stanford *14362*
Association for Women in Psychology, Philadelphia *14368*
Association of Aviation Psychologists, Colorado Springs *14397*
Association of Black Psychologists, Washington *14402*
Association of Management, Grafton *14456*
Astro-Psychology Institute, San Francisco *14542*
Autism Services Center, Huntington *14556*
Autism Society of America, Silver Spring *14557*
Behavioral Research Council, Great Barrington *14574*
Center for Responsive Psychology, Brooklyn *14697*
C.G. Jung Foundation for Analytical Psychology, New York *14719*
Christian Association for Psychological Studies, Temecula *14743*
Cognitive Science Society, Pittsburgh *14769*
Committee for the Scientific Investigation of Claims of the Paranormal, Buffalo *14798*
Division of Applied Experimental and Engineering Psychologists, Savoy *14980*
Esalen Institute, Big Sur *15055*
Federation of Behavioral, Psychological and Cognitive Sciences, Washington *15083*
Federation of Trainers and Training Programs in Psychodrama, Dallas *15091*
The Forum, Santa Fe *15114*
Group for the Use of Psychology in History, Springfield *15180*
Haunt Hunters, Chesterfield *15191*
Himalayan International Institute of Yoga Science and Philosophy of the U.S.A., Honesdale *15211*
Human Factors Society, Santa Monica *15235*
Human Relations Area Files, New Haven *15237*
Huxley Institute for Biosocial Research, Boca Raton *15242*
Institute for Psychohistory, New York *15290*
Institute for Rational-Emotive Therapy, New York *15292*
Institute for Reality Therapy, Canoga Park *15293*
Institute for Research in Hypnosis and Psychotherapy, New York *15294*

Institute for Social Research, Ann Arbor *15296*
Institute of Cultural Affairs, Chicago *15315*
International Council of Psychologists, Hopkinton *15458*
International Imagery Association, Bronx *15488*
International Organization for the Study of Group Tensions, New York *15509*
International Psychohistorical Association, New York *15517*
International Society for Research on Aggression, Chicago *15540*
Jean Piaget Society, Buffalo *15596*
Jefferson Center for Character Education, Pasadena *15597*
Laban/Bartenieff Institute of Movement Studies, New York *15642*
Laughter Therapy, Los Angeles *15651*
Lentz Peace Research Laboratory, Saint Louis *15658*
Milton H. Erickson Foundation, Phoenix *15741*
National Association of School Psychologists, Silver Spring *15884*
National Character Laboratory, El Paso *15952*
North American Society of Adlerian Psychology, Chicago *16224*
Organizational Behavior Teaching Society, Norman *16261*
Psi Chi – The National Honor Society in Psychology, Chattanooga *16366*
Psychic Science International Special Interest Group, Huber Heights *16369*
Psychologists Interested in Religious Issues, Chestnut Hill *16370*
Psychology Society, New York *16371*
Psychometric Society, Champaign *16372*
Psychonomic Society, San Jose *16373*
Radix Institute, Granbury *16390*
Sigmund Freud Archives, Roslyn *16473*
Society for Clinical and Experimental Hypnosis, Liverpool *16508*
Society for Pediatric Psychology, Gainesville *16562*
Society for Personality Assessment, Saint Petersburg *16566*
Society for Psychophysiological Research, Dallas *16572*
Society for the Psychological Study of Social Issues, Ann Arbor *16608*
Society for the Study of Male Psychology and Physiology, Montpellier *16617*
Society of Experimental Psychologists, University Park *16660*
Society of Multivariate Experimental Psychology, Charlottesville *16682*
Wilbur Hot Springs Health Sanctuary, Wilbur Springs *16893*

Yugoslavia
Društvo Psihologa Srbije, Beograd *17088*

Psychotherapy s. Therapeutics

Public Administration

Australia
Commonwealth Council for Educational Administration, Armidale *00375*
Institute of Municipal Management, South Melbourne *00416*

Belgium
European Centre for Strategic Management of Universities, Bruxelles *01312*
Institut International des Sciences Administratives, Bruxelles *01388*

Brazil
Academia Brasileira de Ciência da Administração, Rio de Janeiro *01562*
Centro de Estudos e Pesquisas em Administração, Porto Alegre *01624*
Fundação Centro de Pesquisas e Estudos, Salvador *01651*
Fundação João Pineiro, Belo Horizonte *01657*

Canada
Administrative Sciences Association of Canada, Montréal *01820*
Association Française des Conseils Scolaires de l'Ontario, Ottawa *01907*
Association of Large School Boards in Ontario, Toronto *01934*
Association of School Business Officials of Saskatchewan, Yorkton *01942*
Canadian Federation of Deans of Management and Administrative Studies, Ottawa *02130*
Canadian University and College Counselling Association, Victoria *02311*
Fédération des Commissions Scolaires Catholiques du Québec, Sainte Foy *02451*
Fédération Québécoise des Directeurs et Directrices d'Ecole, Anjou *02461*
Institute of Public Administration of Canada, Toronto *02531*
Nova Scotia School Boards Association, Halifax *02744*
Ontario Association of Education Administrative Officials, Toronto *02758*
Professional Institute of the Public Service of Canada, Ottawa *02846*

Costa Rica
Central American Institute of Public Administration, San José *03437*
Instituto Centroamericano de Administración Pública, San José *03445*

Ethiopia
African Association for Public Administration and Management, Addis Ababa *03934*

France
Association des Hautes Etudes Hospitalières, Paris *04186*
Association Nationale des Assistants de Service Social, Paris *04306*
Institut International d'Administration Publique, Paris *04571*

Germany
Bundes-Arbeitsgemeinschaft Akademischer Räte in der Bundesrepublik, München *05232*
Deutsche Sektion des Internationalen Instituts für Verwaltungswissenschaften, Bonn *05574*
Gesellschaft für Deutsche Postgeschichte, Frankfurt *05807*
Gesellschaft für Organisation e.V., Giessen *05847*

Hungary
Szervezési és Vezetési Tudományos Társaság, Budapest *06673*

Iran
Sazemane Pachuhesh va Barnamerizi Amuzeshi, Teheran *06903*

Jordan
Arab Administrative Development Organization, Amman *08449*

Morocco
African Training and Research Centre in Administration for Development, Tangiers *08800*

Philippines
Eastern Regional Organization for Public Administration, Manila *09518*

United Kingdom
Association for Research in the Voluntary and Community Sector, Wivenhoe *11280*
Commonwealth Association for Public Administration and Management, London *11810*
Institute of Chartered Secretaries and Administrators, London *12222*
Social Policy Association, Hull *12959*

U.S.A.
American Academy of Podiatry Administration, West Hartford *13428*
American Society for Public Administration, Washington *14084*
Council of Planning Librarians, Chicago *14901*
Federation of Government Information Processing Councils, Atlanta *15084*
Governmental Research Association, Birmingham *15168*
Institute of Public Administration, New York *15338*
International Cost Engineering Council, Morgantown *15451*
National Academy of Public Administration, Washington *15776*
National Association of Schools of Public Affairs and Administration, Washington *15889*
National Records Management Council, New York *16135*
Public Service Research Council, Reston *16380*
Social Sciences Services and Resources, Wasco *16486*

Venezuela
Centro Interamericano para el Desarrollo Regional, Maracaibo *16988*
Centro Latinoamericano de Adminstración para el Desarrollo, Caracas *16989*

Public Health s. a. Hygiene

Australia
The Australian Council for Health, Physical Education and Recreation, Hindmarsh *00280*
National Health and Medical Research Council, Camberra *00438*

Austria
Österreichische Arbeitsgemeinschaft für Volksgesundheit, Wien *00727*

Belgium
Association pour l'Etude et l'Evaluation Epidémiologiques des Désastres dans les Pays en Voie de Développement, Bruxelles *01152*
Consortium of Institutions of Higher Education in Health and Rehabilitation in Europe, Gent *01255*
Société Internationale pour la Recherche sur les Maladies de Civilisation et sur l'Environnement, Bruxelles *01464*

Bermuda
Bermuda Tuberculosis, Cancer and Health Association, Hamilton *01525*

Bolivia
Sociedad Boliviana de Salud Pública, La Paz *01550*

Burkina Faso
Centre d'Epidémiologie, Statistique et Information Sanitaire, Bobo-Dioulasso *01806*

Canada
Alberta Health Records Association, Edmonton *01835*
Alberta Public Health Association, Olds *01843*
Association Paritaire de Prévention pour la Santé et la Securité du Travail, Montréal *01950*
Canadian Active Health Foundation, Vancouver *02013*
Canadian Association for Health Physical Education and Recreation, Vanier *02033*
Canadian Health Care Material Management Association, High River *02151*
Canadian Health Education Society, Toronto *02152*
Canadian Health Libraries Association, Ottawa *02153*
Canadian Institute of Stress, Toronto *02183*
Canadian Medical Protective Association, Ottawa *02198*
Canadian Public Health Association, Ottawa *02235*
Canadian Society for International Health, Ottawa *02260*
E.A. Baker Foundation for Prevention of Blindness, Toronto *02420*
Foundation for Independent Research on Technology and Health, Goodwood *02473*
Foundation of the Canadian College of Health Service Executives, Ottawa *02474*
Health Labour Relations Association of British Columbia, Vancouver *02496*
Health Libraries Association of British Columbia, Vancouver *02497*
Hospital Medical Records Institute, Don Mills *02506*
International Development Research Centre, Ottawa *02552*
Manitoba Health Libraries Association, Winnipeg *02618*
Manitoba Health Records Association, Winnipeg *02619*
Manitoba Paramedical Association, Winnipeg *02630*
New Brunswick Health Records Association, Saint John *02683*
Occupational Health and Safety Council of Prince Edward Island, Charlottetown *02751*
Ontario Council of Health, Toronto *02770*
Ontario Health Record Association, Harrow *02781*
Yukon Tuberculosis and Health Association, Whitehorse *03005*

Chile
Sociedad Chilena de Sanidad, Santiago *03083*

China, Republic
School Health Association of the Republic of China, Taipei *03325*

Colombia
Centro de Educación en Administración de Salud, Bogotá *03355*
Instituto Nacional de Salud, Bogotá *03395*

Congo
Association for Health Information and Libraries in Africa, Brazzaville *03422*

Denmark
European Advisory Committee on Health Research, København *03721*

El Salvador
Central American Public Health Council, San Salvador *03924*
Sociedad Médica de Salud Pública, San Salvador *03932*

France
Association des Hautes Etudes Hospitalières, Paris *04186*
Fédération Caribe de Santé Mentale, Fort de France *04508*
Médecins sans Frontières, Paris *04600*
Société Française de Santé Publique, Vandoeuvre lès Nancy *04840*
Société Française de Sexologie Clinique, Paris *04842*
Société Française d'Hygiène, de Médecine Sociale et de Génie Sanitaire, Vandoeuvre-lès-Nancy *04859*

Germany
Akademie für öffentliches Gesundheitswesen in Düsseldorf, Düsseldorf *05056*
Arbeitsgemeinschaft der Sozialdemokraten im Gesundheitswesen, Bonn *05092*
Arbeitskreis Gesundheitskunde e.V., Sankt Georgen *05145*
Bundesvereinigung für Gesundheit e.V., Bonn *05257*
Deutsche Gesellschaft für das Badewesen e.V., Essen *05338*
Deutsche Gesellschaft für Gesundheitsvorsorge e.V., Leverkusen *05356*
Deutsche Gesellschaft für Suchtforschung und Suchttherapie e.V., Hamm *05429*
Deutsche Hauptstelle gegen die Suchtgefahren e.V., Hamm *05454*
Deutscher Arbeitsring für Lärmbekämpfung e.V., Düsseldorf *05499*
Deutscher Bäderverband e.V., Bonn *05502*
Deutscher Medizinischer Informationsdienst e.V., Frankfurt *05528*
Deutsche Zentrale für Volksgesundheitspflege e.V., Frankfurt *05628*
Fraunhofer-Gesellschaft zur Förderung der angewandten Forschung e.V., München *05759*
Gesundheitstechnische Gesellschaft e.V., Berlin *05893*
Hartmannbund, Bonn *05909*
Kneipp-Bund e.V., Bad Wörishofen *06073*
Landesärztekammer Baden-Württemberg, Stuttgart *06091*

India
Indian Public Health Association, Calcutta *06811*

Ireland
European Healthcare Management Association, Dublin *06947*
Health Research Board, Dublin *06954*

Italy
Associazione Italiana di Fisica Sanitaria e di Protezione contro le Radiazioni, Torino *07308*
Centro Studi per il Progresso della Educazione Sanitaria e del Diritto Sanitario, Roma *07581*
Comitato Italiano per l'Educazione Sanitaria, Perugia *07615*
European Centre for Research and Development in Primary Health Care, Perugia *07669*
Società di Medicina Legale e delle Assicurazioni, Roma *07785*
Società Italiana di Medicina Preventiva e Sociale, Genova *07900*
Società Italiana per lo Studio della Cancerogenesi Ambientale ed Epidemiologia dei Tumori, Genova *07970*

Japan
Asian Health Institute, Aichi *08100*
Nippon Koshu-Eisei Kyokai, Tokyo *08343*

Mexico
Sociedad Mexicana de Salud Pública, México *08771*

Netherlands
Association of Schools of Public Health in the European Region, Maastricht *08829*
Genootschap ter Bevordering von Natuur-, Genees- en Heelkunde, Amsterdam *08904*
Nederlandse Vereniging voor Algemene Gezondheidszorg, Rotterdam *09015*

Paraguay
Servicio Cooperativo Interamericano de Salud Pública, Asunción *09432*

Romania
Societatea de Igiena si Sànàtate Publica, Bucuresti *09767*

Singapore
International Commission on Occupational Health, Singapore *10050*

South Africa
Organic Soil Association of South Africa, Parklands *10160*

Switzerland
Ärztekommission für Rettungswesen SRK, Bern *10609*
Commission Médicale Chrétienne, Genève *10651*
Schweizerisches Institut für das Gesundheitswesen, Aarau *10926*
Volksgesundheit Schweiz, Zürich *11028*
World Health Organization, Genève *11033*

Thailand
ASEAN Institute for Health Development, Nakhon Pathom *11069*
Asian Association of Occupational Health, Bangkok *11078*

United Kingdom
Appropriate Health Resources and Technologies Action Group, London *11239*
Association of Community Health Councils for England and Wales, London *11330*
Association of County Public Health Officers, Malmesbury *11333*
Association of General Practitioner Hospitals, Lichfield *11346*
Association of Health Care Information and Medical Records Officers, Hopton *11353*
Association of Health Service Treasurers, Warwick *11354*
Association of National Health Service Officers, Wareham *11370*
Association of Schools of Public Health in the European Region, Bristol *11387*
Association of Scottish Local Health Councils, Edinburgh *11389*
British Fluoridation Society, Alderley Edge *11563*
British Naturopathic and Osteopathic Association, London *11618*
The British Society of Dowsers, Ashford *11682*
Disinfected Mail Study Circle, London *11895*
Group and Association of County Medical Officers of Health of England and Wales, Stafford *12130*
Health Service Social Worker Group, Barnet *12154*
Hyper Active Children's Support Group, Chichester *12190*
Institute for Consumer Ergonomics, Loughborough *12206*
Institute of Health Education, Hale Barns *12241*
Institution of Occupational Safety and Health, Leicester *12312*
Medical Officers of Schools Association, London *12504*
National Campaign for Nursery Education, London *12578*
Organic Federation, Malvern *12658*
Royal Environmental Health Institute of Scotland, Glasgow *12801*
The Royal Institute of Public Health and Hygiene, London *12819*
Royal Sanitary Association of Scotland, Glasgow *12843*
The Royal Society of Health, London *12859*
Socialist Health Association, London *12957*
Society for the Health Education, London *13002*
Society of Health and Beauty Therapists, London *13049*
The Society of Public Health, Huddersfield *13061*

U.S.A.
Accrediting Commission on Education for Health Services Administration, Arlington *13317*

Public Relations s. Advertising

Pulmonary Disease

Argentina

Austria

Belgium

Bermuda

Canada

Chile

China, People's Republic

Cuba

Czech Republic

Denmark

France

Germany

Italy

Japan

Mexico

Peru

Slovakia

South Africa

Switzerland

United Kingdom

Uruguay

U.S.A.

Venezuela

Vietnam

Radiology s. a. X-Ray Technology; Nuclear Medicine

Argentina

Australia

Austria

Belgium

Canada

China, Republic

Colombia

Cuba

Czech Republic

Denmark

El Salvador

Finland

France

Germany

Hungary

Israel

Italy

Japan

Korea, Republic

Luxembourg

Mexico

Netherlands

Norway

Poland
Polskie Lekarskie Towarzystwo Radiologiczne, Warszawa *09550*

Portugal
Associação dos Técnicos e Auxiliares de Radiologia de Portugal, Lisboa *09639*

Romania
Societatea de Radiologie, Bucuresti *09779*

Russia
National Scientific Medical Society of Roentgenologists and Radiologists, Moskva *09932*

Senegal
Association de Radiologie d'Afrique Francophone, Dakar *10016*

Spain
Asociación Española de Técnicos de Radiología, Madrid *10270*
Sociedad Española de Radiología Medica, Madrid *10396*

Switzerland
European Association for the Advancement of Radiation Curing by UV, EB and Laser Beams, Fribourg *10675*

United Kingdom
Association for Radiation Research, Northwood *11277*
Association of University Radiation Protection Officers, Cardiff *11402*
British Institute of Radiology, London *11589*
British Veterinary Radiology Association, Hatfield *11704*
The College of Radiographers, London *11798*
International Society of Radiographers and Radiological Technicians, Cardiff *12394*
Medical Society for the Study of Radiesthesia, Bournemouth *12511*
Radionic Association, Guildford *12736*
Royal College of Radiologists, London *12790*
Society for Radiological Protection, Didcot *12991*

Uruguay
Gremial Uruguaya de Médicos Radiólogos, Montevideo *13259*
Sociedad de Radiología del Uruguay, Montevideo *13268*

U.S.A.
American Academy of Oral and Maxillofacial Radiology, Aurora *13412*
American Board of Radiology, Troy *13662*
American College of Podiatric Radiologists, Miami Beach *13730*
American College of Veterinary Radiology, Glencoe *13741*
American Radium Society, Philadelphia *14025*
American Registry of Radiologic Technologists, Mendota Heights *14027*
American Society for Laser Medicine and Surgery, Wausau *14069*
American Society for Therapeutic Radiology and Oncology, Philadelphia *14091*
American Society of Neuroradiology, Oak Brook *14149*
American Society of Radiologic Technologists, Albuquerque *14163*
Association of University Radiologists, Reston *14530*
International Commission on Radiation Units and Measurements, Bethesda *15442*
Joint Review Committee on Education in Radiologic Technology, Chicago *15625*
National Council on Radiation Protection and Measurements, Bethesda *16025*
Radiation Research Society, Reston *16385*
Radiation Therapy Oncology Group, Philadelphia *16386*
Radiological Society of North America, Oak Brook *16388*
Society for Pediatric Radiology, Oak Brook *16563*
Society of Cardiovascular and Interventional Radiology, Fairfax *16647*

Venezuela
Sociedad Venezolana de Radiología, Caracas *17032*

Vietnam
Vietnamese Association of Radiology, Hanoi *17064*

Recreation s. Travel and Tourism

Refrigeration s. Air Conditioning

Regional Planning s. Urban and Regional Planning

Rehabilitation

Austria
Österreichische Arbeitsgemeinschaft für Rehabilitation, Wien *00726*
Österreichische Gesellschaft für Physikalische Medizin und Rehabilitation, Wien *00823*
Österreichische Gesellschaft für Sprachheilpädagogik, Wien *00833*

Belgium
Consortium of Institutions of Higher Education in Health and Rehabilitation in Europe, Gent *01255*
Fédération Européenne de Médecine Physique et Réadaption, Gent *01357*
International League of Societies for Persons with Mental Handicap, Bruxelles *01399*
Société Royale Belge de Médecine Physique et de Réhabilitation, Bruxelles *01475*

Canada
Alberta Association of Rehabilitation Centres, Calgary *01826*
Association des Services de Réhabilitation Sociale du Québec, Montréal *01894*
Canadian Association of University Schools of Rehabilitation, Montréal *02073*
Children's Rehabilitation and Cerebral Palsy Association, Vancouver *02346*
Island Association of Rehabilitation Workshops, Charlottetown *02565*
Kinsmen Rehabilitation Foundation of British Columbia, Vancouver *02573*

Denmark
Dansk Reumatologisk Selskab, København *03669*
Landsforeningen for Polio-, Trafik- og Ulykkesskadede, Hellerup *03764*

Ethiopia
All Africa Leprosy and Rehabilitation Training Center, Addis Ababa *03940*

Finland
Societas Medicinae Physicalis et Rehabilitationis Fenniae, Helsinki *03994*

France
Association Francophone Internationale des Groupes d'Animation de la Paraplégie, Brie-Comte-Robert *04251*
Association Nationale pour la Réhabilitation Professionnelle par le Travail Protégé, Paris *04315*
Société Nationale Française de Rééducation et Réadaption Fonctionnelles, Paris *04906*

Germany
Bundesarbeitsgemeinschaft für Rehabilitation, Frankfurt *05234*
Bundesarbeitsgemeinschaft zur Förderung haltungsgefährdeter Kinder und Jugendlicher e.V., Mainz *05237*
Dachverband Psychosozialer Hilfsvereinigungen e.V., Bonn *05281*
Deutsche Gesellschaft für Physikalische Medizin und Rehabilitation, Hannover *05404*
Deutsche Gesellschaft zur Förderung der Gehörlosen und Schwerhörigen e.V., München *05448*
Deutsche Ileostomie-Kolostomie-Urostomie-Vereinigung e.V., Freising *05455*
Deutsche Vereinigung für die Rehabilitation Behinderter e.V., Heidelberg *05612*
Lernen Fördern – Bundesverband zur Förderung Lernbehinderter e.V., Köln *06103*

Italy
Società Italiana di Ginnastica Medica, Medicina Fisica e Riabilitazione, Milano *07884*
Società Italiana di Medicina Fisica e Riabilitazione, Moncalieri *07897*

Netherlands
Nederlandse Vereniging van Artsen voor Revalidatie en Physische Geneeskunde, Groningen *09004*

Netherlands Antilles
Caribbean Association on Mental Retardation and other Developmental Disabilities, Jong Bloed *09112*

Portugal
Sociedade Portuguesa de Medicina Fisica e Reabilitação, Lisboa *09704*

St. Lucia
Caribbean Association for Rehabilitation Therapists, Castries *09994*

Spain
Center of Information and Research SIIS, San Sebastián *10286*
Sociedad Española de Rehabilitación y Medicina Fisica, Madrid *10397*

Switzerland
Schweizerische Arbeitsgemeinschaft für Rehabilitation, Basel *10796*
Schweizerische Heilpädagogische Gesellschaft, Bern *10860*
Verband der Heilpädagogischen Ausbildungsinstitute der Schweiz, Luzern *10995*

Turkey
Fizik Tedavi ve Rehabilitasyon Derneği, Istanbul *11145*

United Kingdom
Association of Disabled Professionals, Horbury *11339*
Commonwealth Association for Mental Handicap and Developmental Disabilities, Sheffield *11809*
Hearing Concern, London *12155*
International Cerebral Palsy Society, London *12340*
National Autistic Society, London *12575*
Northern Ireland Chest, Heart and Stroke Association, Belfast *12636*
Psychiatric Rehabilitation Association, London *12726*
Royal Association for Disability and Rehabilitation, London *12771*
Society of Hearing Aid Audiologists, South Croydon *13050*
The Spastics Society, London *13086*

U.S.A.
American Academy of Orthotists and Prosthetists, Alexandria *13417*
American Academy of Physical Medicine and Rehabilitation, Chicago *13424*
American Association of Cardiovascular and Pulmonary Rehabilitation, Middleton *13460*
American Association for Rehabilitation Therapy, North Little Rock *13490*
American Association of Electrodiagnostic Medicine, Rochester *13530*
American Board for Certification in Orthotics and Prosthetics, Alexandria *13625*
American Board of Physical Medicine and Rehabilitation, Rochester *13653*
American Congress of Rehabilitation Medicine, Chicago *13752*
American Deafness and Rehabilitation Association, Little Rock *13777*
American Horticultural Therapy Association, Gaithersburg *13853*
American Kinesiotherapy Association, Dayton *13909*
American Orthotic and Prosthetic Association, Alexandria *13976*
American Osteopathic College of Rehabilitation Medicine, Des Plaines *13987*
American Rehabilitation Counseling Association, Alexandria *14028*
Association of Academic Physiatrists, Indianapolis *14374*
Association of Medical Rehabilitation Administrators, San Diego *14460*
Commission on Accreditation of Rehabilitation Facilities, Tucson *14785*
Disability Rights Center, Washington *14976*
Jewish Braille Institute of America, New York *15604*
National Association of Rehabilitation Facilities, Washington *15883*
National Association on Drug Abuse Problems, New York *15916*
National Center for Disability Services, Albertson *15942*
National Institute for Rehabilitation Engineering, Hewitt *16092*
National Orthotic and Prosthetic Research Institute, Lenox Hill *16122*
National Rehabilitation Association, Alexandria *16137*
Odyssey Institute Corporation, Westport *16244*
Phoenix Society for Burn Survivors, Levittown *16315*
Project Magic, Topeka *16361*
Rehabilitation Information Round Table, Alexandria *16405*
Rehabilitation International, New York *16406*
Section for Rehabilitation Hospitals and Programs, Chicago *16457*
Sister Kenny Institute, Minneapolis *16477*
W.E. Upjohn Institute for Employment Research, Kalamazoo *16889*
World Organization for Human Potential, Philadelphia *16929*
World Rehabilitation Fund, New York *16931*

Zimbabwe
African Rehabilitation Institute, Harare *17161*

Religions and Theology

Argentina
Junta de Historia Eclesiástica Argentina, Buenos Aires *00148*

Australia
Australian Association for the Study of Religions, Saint Lucia *00261*

Austria
Arbeitskreis für Tibetische und Buddhistische Studien, Wien *00567*
Evangelische Akademie in Wien, Wien *00592*
Gesellschaft für das Recht der Ostkirchen, Wien *00616*
Gesellschaft für die Geschichte des Protestantismus in Österreich, Wien *00617*
Internationale Paracelsus-Gesellschaft, Salzburg *00676*
Österreichische Bibelgesellschaft, Wien *00730*
Österreichische Gesellschaft für Kirchenrecht, Wien *00801*
Österreichische Gesellschaft für Religionswissenschaft, Salzburg *00828*

Bahrain
The Islamic Association, Manama *01063*

Belgium
Association Internationale des Etudes Coptes, Louvain-la-Neuve *01138*
Centre International d'Etudes de la Formation Religieuse, Bruxelles *01198*
Centre pour l'Etude des Problèmes du Monde Musulman Contemporain, Bruxelles *01205*
Coordination et Promotion de l'Enseignement de la Religion en Europe, Bruxelles *01257*
Fédération Internationale des Instituts de Recherches Socio-Religieuses, Louvain-la-Neuve *01358*
Institut Belge des Hautes Etudes Chinoises, Bruxelles *01381*
Koninklijke Academie voor Wetenschappen, Letteren en Schone Kunsten van België, Bruxelles *01409*
Office International de l'Enseignement Catholique, Bruxelles *01417*
Ruusbroecgenootschap, Antwerpen *01419*
Société des Bollandistes, Bruxelles *01456*

Cameroon
Association des Institutions d'Enseignement Théologique en Afrique Centrale, Yaoundé *01811*
Association Oecuménique des Théologiens Africains, Yaoundé *01812*

Canada
Association for the Advancement of Christian Scholarship, Toronto *01903*
Association of Canadian Bible Colleges, Kitchener *01918*
Canadian Canon Law Society, Ottawa *02086*
Canadian Catholic Historical Association (English Section), London *02089*
Canadian Institute of Religion and Gerontology, Toronto *02182*
Canadian Society of Biblical Studies, Calgary *02275*
Canadian Society of Church History, Halifax *02278*
Conférence Religieuse Canadienne, Ottawa *02380*
Council on Homosexuality and Religion, Winnipeg *02409*
International Organization for Septuagint and Cognate Studies, Toronto *02555*

China, Republic
Chinese Buddhist Association, Taipei *03263*

Colombia
Comisión Episcopal Latino Americano, Bogotá *03365*
Unión Parroquial del Sur, Bogotá *03421*

Denmark
Det Danske Bibelselskab, København *03704*
Kirkeligt Centrum, Hellerup *03758*
Selskabet for Danmarks Kirkehistorie, København *03809*
Teologisk Forening ved Københavns Universitet, København *03827*

Finland
Suomalainen Teologinen Kirjallisuusseura, Helsinki *03999*
Suomen Eksegeettinen Seura, Helsinki *04012*
Suomen Kirkkohistoriallinen Seura, Helsingin Yliopisto *04040*

France
Amici Thomae Mori, Angers *04132*
Asocio de Studato Internacia pri Spiritaj kaj Teologiaj Instruoj, Bourges *04141*
Association de Recherche et d'Etudes Catéchétiques, Paris *04159*
Association Episcopale Catéchistique, Paris *04203*
Centre Européen de Recherches sur les Congrégations et Ordres Religieux, Saint-Etienne *04372*
Société de l'Histoire du Protestantisme Français, Paris *04663*
Société des Etudes Renaniennes, Paris *04704*
Société d'Etudes Juives, Paris *04744*
Société Théosophique de France, Paris *04921*

Germany
Akademie der Diözese Rottenburg-Stuttgart, Stuttgart *05047*
Albertus-Magnus-Institut, Bonn *05070*
Arbeitsgemeinschaft der Ordenshochschulen, Vallendar *05088*
Arbeitskreis katholischer Schulen in freier Trägerschaft in der Bundesrepublik Deutschland, Bonn *05146*
Cusanus-Institut, Trier *05280*
Deutsche Gesellschaft für Missionswissenschaft, Heidelberg *05384*
Deutsche Paul-Tillich-Gesellschaft e.V., Göttingen *05483*
Deutsche Religionsgeschichtliche Studiengesellschaft, Saarbrücken *05509*
Deutsche Vereinigung für Religionsgeschichte, Hannover *05619*
Evangelische Sozialakademie Friedewald, Friedewald *05690*
Evangelische Studiengemeinschaft e.V., Heidelberg *05691*
Forschungsgruppe für Anthropologie und Religionsgeschichte e.V., Saarbrücken *05726*
Gemeinnütziger Verein zur Förderung von Philosophie und Theologie e.V., Bornheim *05769*
Gemeinschaftswerk der Evangelischen Publizistik e.V., Frankfurt *05772*
Gesellschaft für Mittelrheinische Kirchengeschichte, Koblenz *05841*
Gesellschaft zur Herausgabe des Corpus Catholicorum, Münster *05890*
Historische Kommission zur Erforschung des Pietismus an der Universität Münster, Münster *05918*
Leiterkreis der Katholischen Akademien, Schwerte *06102*
Rabanus-Maurus-Akademie, Frankfurt *06243*
Thomas-Morus-Akademie Bensberg, Bergisch Gladbach *06313*
Verein für bayerische Kirchengeschichte, Nürnberg *06379*
Verein für Westfälische Kirchengeschichte, Bielefeld *06393*
Wissenschaftliche Gesellschaft für Theologie e.V., Tübingen *06448*

India
Asia Theological Association, Bangalore *06727*
Islamic Research Association, Bombay *06829*
The Theosophical Society, Madras *06865*
United Lodge of Theosophists, Bombay *06867*

Ireland
Church Education Society for Ireland, Dublin *06936*
Theosophical Society in Ireland, Dublin *07036*

Israel
Ecumenical Institute for Theological Research, Jerusalem *07053*
Israel Society for Biblical Research, Jerusalem *07090*
Yeshivat Dvar Yerushalayim, Jerusalem *07119*

Italy
Academia Cardinalis Bessarionis, Roma *07122*
Academia Gentium Pro Pace, Roma *07123*
Accademia Gli Amici Dei Sacri Lari, Bergamo *07187*
Associazione Internazionale per lo Studio del Diritto Canonico, Roma *07290*
Associazione per lo Sviluppo delle Scienze Religiose in Italia, Bologna *07394*
Centro di Ricerca e di Studio sul Movimento dei Disciplinati, Perugia *07432*
Centro di Studi Bonaventuriani, Bagnoregio *07437*
Centro Internazionale di Studi Rosminiani, Stresa *07474*
Centro Ricerche Socio-Religiose, Napoli *07541*
Centro Studi Russia Cristiana, Milano *07591*
Centro Studi Santa Veronica Giuliani, Città di Castello *07592*
Circolo Filosofico di Studi Tomistici, Roma *07602*
Documentation and Research Centre, Roma *07647*
International Association for the Study of Canon Law, Roma *07703*
International Association of Biblicists and Orientalists, Ravenna *07706*
Società di Studi Celestiniani, Isernia *07793*
Società Internazionale di Studi Francescani, Assisi *07817*
Società Teosofica in Italia, Trieste *08022*
Studio Teologico Accademico Bolognese, Bologna *08029*
Unione Giuristi Cattolici Italiani, Roma *08046*

Japan
Nihon Indogaku Bukkyogaku Kai, Tokyo *08211*
Nihon Kirisutokyo Gakkai, Tokyo *08225*
Nihon Shukyo Gakkai, Tokyo *08266*
Nippon Dokyo Gakkai, Tokyo *08297*
Rissho Koseikai, Tokyo *08405*
Shinto Gakkai, Tokyo *08420*
Shinto Shukyo Gakkai, Tokyo *08421*

Kenya
Association of Theological Institutions of Eastern Africa, Nairobi *08482*

Lebanon
Association of Theological Institutes in the Middle East, Beirut *08561*

Mexico
Commission of Studies for Latin American Church History, México *08726*

Netherlands
Katholiek Documentatie Centrum, Nijmegen *08946*

Nigeria
Ecumenical Association of Third World Theologians, Port Harcourt *09225*

Norway
Kirkehistorisk Samfunn, Sandvika *09273*

Pakistan
Jamiyat-ul-Falah, Karachi *09371*
Karachi Theosophical Society, Karachi *09372*

Philippines
Conference of Asian-Pacific Pastoral Institutes, Manila *09515*
East Asian Pastoral Institute, Manila *09517*

Spain
Federación Española de Religiosos de Enseñanza, Madrid *10310*

Sri Lanka
Buddhist Academy of Ceylon, Colombo *10408*
Maha Bodhi Society of Ceylon, Colombo *10416*
Theosophical Society of Ceylon, Colombo *10422*

Sweden
Svenska Kyrkohistoriska Föreningen, Uppsala *10559*

Switzerland
Association Européenne Francophone pour les Etudes Bahá'íes, Bern *10624*
Ecumenical Institute Bossey, Céligny *10665*
Schweizerische Asiengesellschaft, Zürich *10798*
Schweizerische katholische Arbeitsgemeinschaft für Ausländerfragen, Luzern *10864*
Schweizerische Theologische Gesellschaft, Bern *10934*

Tanzania
East Africa Association for Theological Education by Extension, Njombe *11049*

Trinidad and Tobago
Theosophical Society of Trinidad, Guaico *11124*

United Kingdom
The Aetherius Society, London *11209*
Alcuin Club, Runcorn *11220*
Association for Latin Liturgy, Chard *11270*
Association for Religious Education, Cumnor Hill *11279*
Association of Religious in Education, London *11384*
British and Foreign Bible Society, Swindon *11478*
British Association for the History of Religions, Cardiff *11496*
British Association for the Study of Religions, Cardiff *11497*
The Buddhist Society, London *11715*
Clinical Theology Association, Oxford *11794*
Cornish Methodist Historical Association, Carnon Downs *11842*
Ecclesiastical History Society, Glasgow *11923*
Ecclesiological Society, London *11924*
Henry Bradshaw Society, Norwich *12157*
Historical Society of the Church in Wales, Carmarthen *12175*
Historical Society of the Methodist Church in Wales, Pwllheli *12176*
Intercontinental Church Society, London *12319*
International African Institute, London *12321*
International Organization for the Study of the Old Testament, Cambridge *12380*
Maha Bodhi Society of Sri Lanka (U.K.), London *12477*
Modern Churchpeople's Union, London *12534*
National Bible Society of Scotland, Edinburgh *12577*
National Council of Jewish Religious Day Schools, London *12584*
National Society (Church of England), London *12606*
National Strict Baptist Sunday School Association, Bishops Stortford *12612*
Odinic Rite, London *12652*
Oxford Centre for Islamic Studies, Oxford *12667*
Presbyterian Church of Wales Historical Society, Aberystwyth *12716*
Religious Society of Friends, London *12748*
Research into Lost Knowledge Organisation, Orpington *12753*
Scottish Catholic Historical Association, Glasgow *12902*
Scottish Church History Society, Edinburgh *12903*
Scottish Reformation Society, Edinburgh *12924*
Society for African Church History, Aberdeen *12963*
Society for New Testament Study, Lancaster *12985*
Society for Promoting Christian Knowledge, London *12987*
Society of Cirplanologists, Hyde *13037*
Society of King Charles the Martyr, London *13054*
The Swedenborg Society, London *13115*
Theosophical Society in England, London *13124*
Theosophical Society in Europe, Camberley *13125*
Theosophical Society in Scotland, Edinburgh *13126*
Trinitarian Bible Society, London *13143*
United Reformed Church History Society, London *13169*
United Society for the Propagation of the Gospel, London *13171*
Welsh Baptist Historical Society, Llanelli *13199*

U.S.A.
Academy of American Franciscan History, Berkeley *13281*
Academy of Religion and Psychical Research, Bloomfield *13306*
American Academy for Jewish Research, New York *13377*
American Academy of Religion, Syracuse *13434*
American Association of Bible Colleges, Fayetteville *13509*
American Baptist Historical Society, Rochester *13616*
American Biblical Encyclopedia Society, Monsey *13621*
American Institute of Islamic Studies, Denver *13885*
American Teilhard Association, Syosset *14189*
American Theological Library Association, Evanston *14192*
American Theological Society – Midwest Division, Chicago *14193*
Association for Jewish Studies, Cambridge *14311*
Association for Practical Theology, Saint Louis *14320*
Association for the Sociology of Religion, Lancaster *14355*
Association of Advanced Rabbinical and Talmudic Schools, New York *14376*
Association of Disciples for Theological Discussion, Saint Louis *14429*
Association of Episcopal Colleges, New York *14435*
Association of Jesuit Colleges and Universities, Washington *14450*
Association of Orthodox Jewish Scientists, New York *14475*
Association of Theological Schools, Pittsburgh *14521*
Boston Theological Institute, Newton Centre *14609*
Cambodian Buddhist Society, Silver Spring *14634*
Canon Law Society of America, Washington *14642*
Catholic Biblical Association of America, Washington *14652*
Catholic Theological Society of America, Philadelphia *14658*
Center for Applied Research in the Apostolate, Washington *14666*
Christian Research Institute International, San Juan Capistrano *14747*
College Theology Society, San Diego *14779*
Community for Religious Research and Education, Loretto *14812*
Conference on Christianity and Literature, Saint Petersburg *14836*
Council for Religion in Independent Schools, Washington *14882*
Council of Societies for the Study of Religion, Valparaiso *14903*
Friends Historical Association, Haverford *15140*
Historical Committee of the Mennonite Church, Goshen *15217*
Institute for Advanced Studies of World Religions, Carmel *15267*
Institute for Biblical Research, Wheaton *15270*
Institute for Theological Encounter with Science and Technology, Saint Louis *15304*
International Association of Buddhist Studies, Berkeley *15399*
Literacy and Evangelism International, Tulsa *15675*
Lutheran Historical Conference, Saint Louis *15683*
Mormon History Association, Provo *15755*
National Association of Baptist Professors of Religion, Macon *15829*
National Coalition for Public Education and Religious Liberty, Washington *15954*
National Council for Torah Education, New York *15994*
National Council on Religion and Public Education, Ames *16026*
National Ramah Commission, New York *16133*
North American Academy of Ecumenists, Albuquerque *16209*
Orthodox Theological Society in America, Brookline *16269*
Psychologists Interested in Religious Issues, Chestnut Hill *16370*
Religious Research Association, Washington *16408*
Society for Humanistic Judaism, Farmington Hills *16534*
Society for Philosophy of Religion, Athens *16569*
Society for the Arts, Religion and Contemporary Culture, Boston *16592*
Society for the Scientific Study of Religion, West Lafayette *16610*
Society of Biblical Literature, Decatur *16642*

Vatican City
Accademia Romana di S. Tommaso d'Aquino e di Religione Cattolica, Roma *16949*
Collegium Cultorum Martyrum, Roma *16950*
Commission Théologique Internationale, Roma *16951*
Pontificia Accademia dell'Immacolata, Roma *16953*
Pontificia Accademia Mariana Internationalis, Roma *16954*
Pontificia Accademia Teologica Romana, Roma *16956*

Zaire
Association des Institutions d'Enseignement Théologique en Afrique Occidentale, Kinshasa *17140*

Zambia
Africa Association for Liturgy, Music and Arts, Lusaka *17148*

Zimbabwe
Conference of African Theological Institutions, Harare *17166*
Ecumenical Documentation and Information Centre for Eastern and Southern Africa, Harare *17169*

Restoration s. Preservation of Historical Monuments, Restoration

Rheumatology

Belgium
Société Royale Belge de Rheumatologie, Bruxelles *01477*

Bulgaria
Bulgarian Society of Cardiology, Sofia *01757*

Canada
Asthma Association of Canada, Toronto *01956*

Denmark
Dansk Reumatologisk Selskab, København *03669*

France
Syndicat National des Médecins Rhumatologues, Antony *04958*

Germany
Deutsche Gesellschaft für Rheumatologie e.V., Bad Bramstedt *05417*
Deutsche Rheuma-Liga Bundesverband e.V., Bonn *05515*

Indonesia
Asia Pacific League against Rheumatism, Semarang *06883*

Israel
Israel Association for Physical Medicine and Rheumatology, Tel Aviv *07063*

Italy
Società Italiana di Reumatologia, Bari *07943*

Portugal
Sociedade Portuguesa de Reumatologia, Lisboa *09714*

Russia
National Scientific Medical Society of Rheumatologists, Moskva *09931*

South Africa
The South African Rheumatism and Arthritis Association, Johannesburg *10215*

Switzerland
Schweizerische Gesellschaft für Rheumatologie, Zürich *10845*
Schweizerische Rheumaliga, Zürich *10898*

United Kingdom
The Arthritis and Rheumatism Council, Chesterfield *11250*
British Society for Rheumatology, London *11666*

U.S.A.
American College of Rheumatology, Atlanta *13736*

Rhinology s. Otorhinolaryngology

Roentgenology s. X-Ray Technology

Safety and Protection, Safety Engineering

Argentina
Sociedad Argentina de Socorros, Esperanza *00185*

Belgium
Association Nationale pour la Protection contre l'Incendie, Ottignies *01149*

Canada
Accident Prevention Association of Manitoba, Winnipeg *01817*
Alberta Safety Council, Edmonton *01846*
Association de Securité des Exploitations Forestières du Québec, Québec *01881*
Association de Securité des Industriels Forestiers du Québec, Québec *01882*
Association de Securité des Pâtes et Papiers du Québec, Québec *01883*
Association of Workers' Compensation Boards of Canada, Scarborough *01949*
Association Paritaire de Prévention pour la Santé et la Securité du Travail, Montréal *01950*
Canada Safety Council, Ottawa *02001*
Canadian Alarm and Security Association, Willowdale *02018*
Canadian Fire Safety Association, Willowdale *02135*
Canadian Radiation Protection Association, Ottawa *02240*
Canadian Society of Safety Engineering, Mississauga *02295*
Ceramics and Stone Accident Prevention Association, Toronto *02331*
Chemical Industries Accident Prevention Association, Toronto *02341*
Construction Safety Association of Ontaria, Toronto *02393*
Fire Fighters Burn Treatment Society, Edmonton *02462*
Fire Prevention Canada Association, Ottawa *02463*
Food Products Accident Prevention Association, Toronto *02470*
Forest Products Accident Prevention Association, North Bay *02471*
Grain, Feed and Fertilizer Accident Prevention Association, Toronto *02488*
Ligue de Sécurite du Québec, Montréal *02593*
Manitoba Safety Council, Winnipeg *02631*
Motor Vehicle Safety Association, Mississauga *02661*
New Brunswick Safety Council, Fredericton *02692*
Nova Scotia Safety Council, Halifax *02743*
Occupational Health and Safety Council of Prince Edward Island, Charlottetown *02751*
Ottawa-Carleton Safety Council, Ottawa *02809*
Printing Trades Accident Prevention Association, Toronto *02844*
Saskatchewan Safety Council, Regina *02900*
Vancouver Safety Council, Vancouver *02978*
Woodworkers Accident Prevention Association, Toronto *02997*

Denmark
Brandteknisk Selskab, København *03569*

Finland
Säteilyturvakeskus, Helsinki *03989*
Suomen Palontorjuntaliitto, Helsinki *04055*

France
Association des Industriels de France contre les Accidents du Travail, Paris *04187*
Association Française de Prévention des Accidents de Travail et Incendie, Paris *04222*
Association Interprofessionnelle de France, Seclin *04294*
Comité International de l'AISS pour la Prévention des Risques Professionnels du Bâtiment et des Travaux Publics, Boulogne-Billancourt *04406*

Germany
Deutsche Gesellschaft für Wirtschaftliche Fertigung und Sicherheitstechnik e.V., Kaarst *05441*
Deutscher Verkehrssicherheitsrat e.V., Bonn *05561*
Gesellschaft für Anlagen- und Reaktorsicherheit, Köln *05799*
Gesellschaft für Sicherheitswissenschaft e.V., Wuppertal *05859*
Normenausschuss Feuerwehrwesen im DIN Deutsches Institut für Normung e.V., Berlin *06169*

Normenausschuss Persönliche Schutzausrüstung und Sicherheitskennzeichnung im DIN Deutsches Insitut für Normung e.V., Berlin *06196*
Verband der Technischen Überwachungs-Vereine e.V., Essen *06343*
Vereinigung zur Förderung des Deutschen Brandschutzes e.V., Altenberge *06419*

Ireland
National Safety Council, Dublin *07017*

Netherlands
Nederlandse Vereniging voor Veiligheidskunde, Amsterdam *09042*

Sweden
Brandförsvarsföreningen, Stockholm *10440*

United Kingdom
Association for Petroleum and Explosives Administration, Huntington *11275*
British Safety Council, London *11642*
European Federation of Associations of Industrial Safety and Medical Officers, Leicester *12029*
Industrial Fire Protection Association of Great Britain, London *12199*
North West Regional Association of Industrial Safety Groups, Preston *12648*
Royal Society for the Prevention of Accidents, Birmingham *12852*
Safety and Reliability Society, Southport *12881*

U.S.A.
American Academy of Safety Education, Warrensburg *13436*
American Security Council, Boston *14038*
American Security Council Foundation, Boston *14039*
American Society of Safety Engineers, Des Plaines *14165*
Association of Safety Council Executives, Gainesville *14493*
Aviation Safety Institute, Worthington *14563*
Center for Auto Safety, Washington *14670*
Center for Safety in the Arts, New York *14698*
Flight Safety Foundation, Arlington *15100*
Highway Users Federation for Safety and Mobility, Washington *15210*
Inter-American Safety Council, Englewood *15354*
International Rescue and Emergency Care Association, Covington *15520*
International Society of Air Safety Investigators, Sterling *15545*
Motorcycle Safety Foundation, Irvine *15756*
National Fire Protection Association, Quincy *16051*
National Institute for Farm Safety, Columbia *16090*
National Safety Council, Itasca *16145*
National Safety Management Society, Weaverville *16146*
Safety Equipment Institute, Arlington *16435*
Society of Fire Protection Engineers, Boston *16664*

Science, general

Afghanistan
Afghanistan Academy of Sciences, Kabul *00001*

Albania
Albanian Academy of Sciences, Tirana *00003*

Andorra
Societat Andorrana de Ciències, Andorra la Vella *00032*

Argentina
Academia Nacional de Ciencias de Buenos Aires, Buenos Aires *00042*
Asociación Científica Argentino-Alemana, Buenos Aires *00075*
Centro Argentino de Información Científica y Tecnológica, Buenos Aires *00099*
Comisión de Investigaciones Científicas de la Provincia de Buenos Aires, La Plata *00117*
Consejo Internacional de Administración Científica, Buenos Aires *00128*
Consejo Nacional de Investigaciones Científicas y Técnicas, Buenos Aires *00130*
Organización de Universidades Católicas de América Latina, Buenos Aires *00150*
Sociedad Científica Argentina, Buenos Aires *00189*

Armenia
Armenian Academy of Sciences, Yerewan *00203*

Australia
Australian Academy of Science, Canberra *00250*
Australian and New Zealand Association for the Advancement of Science, Sydney *00258*
Australian Science Teachers Association, Warradale *00337*
Commonwealth Scientific and Industrial Research Organisation, Dickson *00377*
Federation of Australian Scientific and Technological Societies, Canberra *00394*
Royal Society of Canberra, Canberra *00475*
Royal Society of New South Wales, North Ryde *00476*
Royal Society of Queensland, Saint Lucia *00477*
Royal Society of South Australia, Adelaide *00478*
Royal Society of Tasmania, Hobart *00479*
Royal Society of Victoria, Melbourne *00480*
Royal Society of Western Australia, Perth *00481*
Science Teachers Association of New South Wales, Rozelle *00488*
Science Teachers Association of Queensland, Spring Hill *00489*
Science Teachers' Association of Victoria, Parkville *00490*
South Australian Science Teachers Association, Parkside *00497*

Austria
Arbeitsgemeinschaft für Wissenschaft und Politik, Innsbruck *00559*
Institut für Grenzgebiete der Wissenschaft, Innsbruck *00656*
Institut für Wissenschaft und Kunst, Wien *00661*
Internationales Forschungszentrum für Grundfragen der Wissenschaften, Salzburg *00679*
Ludwig Boltzmann-Gesellschaft, Österreichische Vereinigung zur Förderung der wissenschaftlichen Forschung, Wien *00703*
Österreichische Akademie der Wissenschaften, Wien *00723*
Österreichische Forschungsgemeinschaft, Wien *00742*
Verband der Akademikerinnen Österreichs, Wien *00956*
Verband der wissenschaftlichen Gesellschaften Österreichs, Wien *00963*
Vereinigung für wissenschaftliche Grundlagenforschung, Graz *00995*
Vorarlberger Landesmuseumsverein, Dornbirn *01008*
Wiener Kulturkreis, Wien *01023*

Azerbaijan
Azerbaijan Academy of Sciences, Baku *01036*
Commission on International Scientific Contacts, Baku *01042*

Bangladesh
Bangla Academy, Dhaka *01065*
Bangladesh Academy of Sciences, Dhaka *01066*
Bangladesh Council of Scientific and Industrial Research, Dhaka *01067*

Belarus
Belarussian Academy of Sciences, Minsk *01088*

Belgium
Académie Royale des Sciences, des Lettres et des Beaux-Arts de Belgique, Bruxelles *01095*
Centre International de Documentation Marguerite Yourcenar, Bruxelles *01190*
Comité Belge d'Histoire des Sciences, Bruxelles *01218*
Committee on the Challenges of Modern Society, Bruxelles *01239*
Conférence des Recteurs des Universités Belges, Bruxelles *01246*
European Community Network of the National Academic Recognition Information Centres, Bruxelles *01327*
European Cooperation in the Field of Scientific and Technical Research, Bruxelles *01333*
Fédération Belge des Sociétés Scientifiques, Bruxelles *01356*
Fondation Universitaire, Bruxelles *01367*
Société Royale des Sciences de Liège, Liège *01487*
Société Scientifique de Bruxelles, Namur *01490*
Union Académique Internationale, Bruxelles *01492*

Bolivia
Academia Boliviana, La Paz *01528*
Academia Nacional de Ciencias de Bolivia, La Paz *01529*
Centro Intelectual Galindo, La Paz *01535*
Comisión de Planeamiento y Coordinación de las Universidades Bolvianas, Cochabamba *01541*

Botswana
The Botswana Society, Gaborone *01559*

Brazil
Academia Brasileira de Ciências, Rio de Janeiro *01563*
Centro Académico Hugo Simas, Curitiba *01615*
Centro Brasileiro de Estudos, Campinas *01616*
Centro Cultural de Botucatu, Botucatu *01618*
Centro Nacional de Pesquisa de Seringueira, Manaus *01635*
Fundação Getulio Vargas, Rio de Janeiro *01654*
Fundação Moinho Santista, São Paulo *01659*
Sociedade Propagadora Esdeva, Juiz de Fora *01729*
Sociedade Visconde de São Leopoldo, Santos *01730*
Unidade de Pesquisa de Ambito Estadual em Barreiras, Barreiras *01731*
Unidade de Pesquisa de Ambito Estadual em Campos, Campos *01732*
Unidade de Pesquisa de Ambito Estadual em Corumba, Corumba *01733*
Unidade de Pesquisa de Ambito Estadual em Dourados, Dourados *01734*
Unidade de Pesquisa de Ambito Estadual em Itaguai, Nova Iguaçu *01735*
Unidade de Pesquisa de Ambito Estadual em Itapirema, Goiânia *01736*
Unidade de Pesquisa de Ambito Estadual em Manaus, Manaus *01737*
Unidade de Pesquisa de Ambito Estadual em Ponta Grossa, Ponta Grossa *01738*
Unidade de Pesquisa de Ambito Estadual em Porto Velho, Porto Velho *01739*
Unidade de Pesquisa de Ambito Estadual em Teresina, Teresina *01740*

Bulgaria
Bulgarian Academy of Sciences, Sofia *01742*
Central Council of Scientific and Technical Unions, Sofia *01766*
Union of Scientific Workers in Bulgaria, Sofia *01799*

Canada
Association Canadienne-Française pour l'Avancement des Sciences, Montréal *01875*
Association des Universités Partiellement ou Entièrement de Langue Française, Montréal *01895*
Association for the Advancement of Science in Canada, Ottawa *01905*
Association of Canadian College and University Information Bureaus, Halifax *01920*
Association of Universities and Colleges of Canada, Ottawa *01945*
Association of Wholly or Partially French Language Universities, Montréal *01948*
Atlantic Provinces Council on the Sciences, St. John's *01961*
Birks Family Foundation, Montréal *01967*
Bobechko Foundation, Toronto *01968*
British Columbia Research Council, Vancouver *01988*
Canadian Association for Latin American and Caribbean Studies, Ottawa *02036*
Canadian Association for University Continuing Education, Ottawa *02044*
Canadian Association of Deans of Arts and Sciences, Charlottetown *02048*
Canadian Society for the Weizmann Institute of Science, Downsview *02270*
Charles H. Ivey Foundation, Willowdale *02340*
Confederation of Alberta Faculty Associations, Edmonton *02376*
Conseil des Universités du Québec, Sainte Foy *02383*
Council of Ontario Universities, Toronto *02403*
Council of Western Canadian University Presidents, Winnipeg *02408*
Donner Canadian Foundation, Toronto *02418*
E.W. Bickle Foundation, Toronto *02444*
Fédération des Associations de Professeurs des Universités du Québec, Montréal *02449*
Fédération des Professionnelles et Professionnels de Cégeps et de Collèges, Montréal *02452*
Federation of New Brunswick Faculty Associations, Fredericton *02459*
Fondation J. Armand Bombardier, Valcourt *02467*
Fondation Justine Lacoste-Beaubien, Montréal *02468*
Fondation Lionel Groulx, Outremont *02469*
Fraser Institute, Vancouver *02475*
Gairdner Foundation, Willowdale *02477*
Goodwin's Foundation, Ottawa *02487*
Hamber Foundation, Vancouver *02491*
Hamilton Foundation, Hamilton *02493*
H.G. Bertram Foundation, Hamilton *02500*
Historical and Scientific Society of Manitoba, Winnipeg *02501*
Institut Canadien de Québec, Québec *02520*
Inter-University Council on Academic Exchanges with the USSR and Eastern Europe, Waterloo *02564*
J.P. Bickell Foundation, Toronto *02570*
J.W. MacConnell Foundation, Montréal *02572*
Leon and Thea Koerner Foundation, Vancouver *02588*
Max Bell Foundation, Toronto *02641*
Molson Family Foundation, Montréal *02660*
M.S.I. Foundation, Edmonton *02662*
National Research Council of Canada, Ottawa *02673*
Nova Scotia Confederation of University Faculty Associations, Halifax *02729*
Nova Scotian Institute of Science, Halifax *02740*
Nova Scotia Research Foundation Corporation, Dartmouth *02742*
Ontario Council on University Affairs, Toronto *02771*
Richard Ivey Foundation, London *02861*
Royal Canadian Institute, Toronto *02866*
Royal Society of Canada, Ottawa *02871*
R. Samuel McLaughlin Foundation, Toronto *02873*
Sandford Fleming Foundation, Waterloo *02877*
Saskatchewan Research Council, Saskatoon *02899*
Science Council of Canada, Ottawa *02907*
Sir Joseph Flavelle Foundation, Toronto *02913*
Victoria Foundation, Victoria *02979*
W. Garfield Weston Foundation, Toronto *02994*
Winnipeg Foundation, Winnipeg *02995*

Chile
Academia Chilena de Ciencias, Santiago *03008*
Consejo de Rectores de Universidades Chilenas, Santiago *03047*
Instituto de Chile, Santiago *03053*
Sociedad Científica Chilena Claudio Gay, Santiago *03085*
Sociedad Científica de Chile, Santiago *03086*

China, People's Republic
China Association for Science and Technology, Beijing *03128*
Chinese Academy of Sciences, Beijing *03142*
Chinese Academy of Sciences – Anhwei Branch, Anhui *03143*
Chinese Academy of Sciences – Central South Branch, Kuangtung *03144*
Chinese Academy of Sciences – Chekiang Branch, Chechiang *03145*
Chinese Academy of Sciences – East China Branch, Shanghai *03146*
Chinese Academy of Sciences – Fukien Branch, Fuchien *03147*
Chinese Academy of Sciences – Hopeh Branch, Hopei *03148*
Chinese Academy of Sciences – Kiangsu Branch, Chiangsu *03149*
Chinese Academy of Sciences – Kirin Branch, Chilin *03150*
Chinese Academy of Sciences – Northwestern Branch, Shanhsi *03151*
Chinese Academy of Sciences – Shansi Branch, Shanhsi *03152*
Chinese Academy of Sciences – Shantung Branch, Shantung *03153*
Chinese Academy of Sciences – Sinkiang Branch, Hsinchiang *03154*

China, Republic
Academia Sinica, Taipei *03228*
Chinese Association for the Advancement of Science, Taipei *03261*
Chinese Youth Academic Research Association, Taipei *03291*
National Science Council, Taipei *03314*
Television Academy of Arts and Sciences of the Republic of China, Taipei *03332*

Colombia
Asociación Colombiana de Sociedades Científicas, Bogotá *03347*
Asociación Colombiana de Universidades, Bogotá *03348*
Fondo Colombiano de Investigaciones Científicas y Proyectos Especiales Francisco José de Caldas, Bogotá *03374*
IUS Federación Universitaria Nacional, Bogotá *03396*

Congo
Conseil National de la Recherche Scientifique et Technique, Brazzaville *03425*

Costa Rica
Consejo Superior Universitario Centroemericano, San José *03444*

Croatia
Hrvatska Akademija Znanosti i Umjetnosti, Zagreb *03459*

Cuba
Academia de Ciencias de Cuba, La Habana *03469*
Centro de Estudios de Historia y Organización de la Ciencia Carlos J. Finlay, La Habana *03473*

Czech Republic
Česká Akademie Věd, Praha *03500*
Česká Vědeckothechnická Společnost, Praha *03536*

Denmark
Akademikernes Centralorganisation, København *03558*
Det Kongelige Danske Videnskabernes Selskab, København *03710*
Det Laerde Selskab i Arhus, Arhus *03713*
Nordforsk, København *03777*

Egypt
Academy of Scientific Research and Technology, Cairo *03859*
National Research Centre, Cairo *03911*

El Salvador
Academia Salvadoreña, San Salvador *03916*

Estonia
Academy of Sciences, Tallinn *03933*

Fiji
Fiji Society, Suva *03947*

Finland
Societas Scientiarum Fennica, Helsinki *03996*
Suomen Akatemia, Helsinki *04004*
Tietohuollon Neuvottelukunta, Helsinki *04076*

France
Académie d'Arles, Arles *04087*
Académie de la Réunion, Saint-Denis *04089*
Académie de Nîmes, Nîmes *04091*
Académie des Jeux Floraux, Toulouse *04096*
Académie des Lettres, Sciences et Arts d'Amiens, Amiens *04097*
Académie des Sciences, Paris *04098*
Académie des Sciences, Agriculture, Arts et Belles-Lettres d'Aix, Aix-en-Provence *04099*
Académie des Sciences, Arts et Belles-Lettres de Dijon, Dijon *04100*
Académie des Sciences, Belles-Lettres et Arts de Clermont, Clermont-Ferrand *04101*
Académie des Sciences, Belles-Lettres et Arts de Lyon, Lyon *04102*
Académie des Sciences, Belles-Lettres et Arts de Rouen, Rouen *04103*
Académie des Sciences, Belles-Lettres et Arts de Savoie, Chambéry *04104*
Académie des Sciences et Lettres de Montpellier, Montpellier *04106*
Académie des Sciences, Lettres et Arts d'Arras, Arras *04107*
Académie des Sciences, Lettres et Arts de Marseille, Marseille *04108*
Académie Nationale de Metz, Metz *04119*
Académie Nationale des Sciences, Belles-Lettres et Arts de Bordeaux, Bordeaux *04120*
Académie Polonaise des Sciences, Paris *04121*
Association des Françaises Diplômées des Universités, Paris *04183*

Association d'Etudes pour l'Expansion de la Recherche Scientifique, Paris 04200
Association Internationale des Professeurs et Maîtres de Conférences des Universités, Nancy 04275
Association Internationale pour le Développement des Universités Internationales et Mondiales, Aulnay-sous-Bois 04290
Association pour le Développement de l'Enseignement et des Recherches Scientifiques auprès des Universités de la Région Parisienne, Paris 04337
Association pour l'Innovation Scientifique, Paris 04343
Association Universitaire pour la Diffusion Internationale de la Recherche, Paris 04357
Chambre Syndicale des Sociétés d'Etudes et de Conseils, Paris 04388
Confédération des Sociétés Scientifiques Françaises, Paris 04434
Conférence Internationale des Responsables des Universités et Instituts à Dominante Scientifique et Technique d'Expression Française, Talence 04446
Conseil International des Unions Scientifiques, Paris 04460
European Academic and Research Network, Orsay 04476
Fédération Internationale des Universités Catholiques, Paris 04536
Fondation pour la Science – Centre International de Synthèse, Paris 04545
Institut de France, Paris 04564
International Academy of the History of Science, Paris 04578
Mouvement Universel de la Responsabilité Scientifique, Paris 04602
Société d'Agriculture, Sciences, Belles-Lettres et Arts d'Orléans, Orléans 04638
Société des Sciences, Arts et Belles-Lettres de Bayeux, Bayeux 04714
Société des Sciences et Arts, Saint-Denis 04715
Société d'Etudes Scientifiques et de Recherches, Herblay 04754
Société Nationale Académique de Cherbourg, Cherbourg 04898
Société Parisienne d'Etudes Spéciales, Paris 04910
Société Scientifique de Bretagne, Rennes 04918
World Federation of Scientific Workers, Montreuil 05001

Georgia

Council on the History of Natural Sciences and Technology, Tbilisi 05010
Georgian Academy of Sciences, Tbilisi 05011

Germany

Akademie der Wissenschaften in Göttingen, Göttingen 05049
Akademie der Wissenschaften und der Literatur zu Mainz, Mainz 05050
Akademie der Wissenschaften zu Berlin, Berlin 05051
Arbeitsgemeinschaft der Grossforschungs-Einrichtungen, Bonn 05086
Arbeitsgemeinschaft Deutsche Lateinamerika-Forschung, Eichstätt 05096
Bayerische Akademie der Wissenschaften, München 05169
Berlin-Brandenburgische Akademie der Wissenschaften, Berlin 05185
Braunschweigische Wissenschaftliche Gesellschaft, Braunschweig 05221
Bund Freiheit der Wissenschaft e.V., Bonn 05262
Carl Duisberg Gesellschaft e.V., Köln 05272
Deutsche Forschungsgemeinschaft, Bonn 05305
Deutsche MERU Gesellschaft, Bissendorf 05468
Deutscher Akademikerinnenbund e.V., Nürnberg 05495
Deutscher Akademischer Austauschdienst, Bonn 05496
Evangelische Akademikerschaft in Deutschland, Stuttgart 05688
Forschungsgruppe Köln, Köln 05727
Gemeinschaft katholischer Studierender und Akademiker, Viernheim 05770
Gesellschaft für Wissenschaft und Leben im Rheinisch-Westfälischen Industriegebiet e.V., Essen 05873
Gesellschaft zur Förderung der wissenschaftlichen Zusammenarbeit mit der Universität Tel-Aviv, Bonn 05886
Görres-Gesellschaft zur Pflege der Wissenschaft, Köln 05896
Heidelberger Akademie der Wissenschaften, Heidelberg 05911
Humboldt-Gesellschaft für Wissenschaft, Kunst und Bildung e.V., Mannheim 05934
Institut für angewandte Arbeitswissenschaft e.V., Köln 05946
Institut für den Wissenschaftlichen Film, Göttingen 05956
Institut für Wissenschaftliche Zusammenarbeit, Tübingen 05978
Internationaler Arbeitskreis Sonnenberg, Braunschweig 05995
Katholische Akademikerarbeit Deutschlands, Bonn 06061
Katholischer Akademikerverband, Bonn 06067
Katholischer Akademischer Ausländer-Dienst, Bonn 06068
Konferenz der deutschen Akademien der Wissenschaften, Mainz 06081
Max-Planck-Gesellschaft zur Förderung der Wissenschaften e.V., München 06114
Sächsische Akademie der Wissenschaften zu Leipzig, Leipzig 06273
Ständiger Arbeitsausschuss für die Tagungen der Nobelpreisträger in Lindau, Lindau 06293
Stifterverband für die Deutsche Wissenschaft, Essen 06295
Verband Hochschule und Wissenschaft im Deutschen Beamtenbund, Bonn 06365
Wissenschaftliche Gesellschaft an der Johann Wolfgang Goethe Universität Frankfurt am Main, Frankfurt 06446
Wissenschaftsrat, Köln 06451
Wissenschaftszentrum Berlin, Berlin 06452
Wittheit zu Bremen e.V., Bremen 06453

Ghana

African Union for Scientific Development, Accra 06474
Council for Scientific and Industrial Research, Accra 06479
Ghana Academy of Arts and Sciences, Accra 06482
Ghana Science Association, Accra 06489
West African Science Association, Accra 06494

Greece

Akadimia Athinon, Athinai 06498

Guatemala

Academia Guatemalteca de la Lengua, Guatemala City 06556
Federación de Universidades Privadas de América Central, Guatemala City 06568

Honduras

Academia Hondureña, Tegucigalpa 06580
Instituto Hondureño de Cultura Interamericana, Tegucigalpa 06584

Hong Kong

International Council of Associations for Science Education, Hong Kong 06594

Hungary

Magyar Tudományos Akadémia, Budapest 06661

Iceland

Rannsóknaráð Ríkisins, Reykjavík 06696
Vísindafélag Islands, Reykjavík 06707

India

Association of Indian Universities, New Delhi 06733
Committee on Science and Technology in Developing Countries, Madras 06752
Council of Scientific and Industrial Research, New Delhi 06753
Indian Academy of Sciences, Bangalore 06767
Indian National Science Academy, New Delhi 06804
Indian Science Congress Association, Calcutta 06813
National Academy of Sciences, Allahabad 06841
Rajasthan Academy of Science, Pilani 06852
Society of Young Scientists, New Dehli 06859

Iraq

Iraq Academy, Baghdad 06917

Ireland

Engineering and Scientific Association of Ireland, Mount Merrion 06944
Royal Dublin Society, Dublin 07024
Royal Irish Academy, Dublin 07028

Israel

Academic Circle of Tel Aviv, Tel Aviv 07039
Association for the Advancement of Science in Israel, Ramat-Gan 07044
The Israel Academy of Sciences and Humanities, Jerusalem 07060
National Council for Research and Development, Jerusalem 07108

Italy

Academia Petrarca di Lettere, Arti e Scienze, Arezzo 07124
Accademia Americana, Roma 07128
Accademia degli Abruzzi per le Scienze e le Arti, Chieti 07136
Accademia degli Euteleti, San Miniato 07137
Accademia degli Incamminati, Modigliana 07138
Accademia dei Filedoni, Perugia 07142
Accademia dei Filopatridi (Rubiconia), Savignano sul Rubicone 07144
Accademia dei Sepolti, Volterra 07146
Accademia della Crusca, Firenze 07147
Accademia delle Scienze dell'Istituto di Bologna, Bologna 07148
Accademia delle Scienze di Ferrara, Ferrara 07149
Accademia delle Scienze di Torino, Torino 07150
Accademia di Agricoltura Scienze e Lettere, Verona 07154
Accademia di Danimarca, Roma 07167
Accademia di Francia, Roma 07168
Accademia di Paestum Eremo Italico, Mercato San Severino 07170
Accademia di Romania, Roma 07172
Accademia di Scienze, Lettere e Arti, Udine 07174
Accademia di Scienze, Lettere e Belle Arti degli Zelanti e dei Dafnici, Acireale 07175
Accademia di Scienze, Lettere ed Arti, Palermo 07176
Accademia Etrusca, Cortona-Arezzo 07178
Accademia Fulginia di Arti, Lettere, Scienze, Foligno 07184
Accademia Il Tetradramma, Roma 07188
Accademia Internazionale della Tavola Rotonda, Milano 07190
Accademia Ligure di Scienze e Lettere, Genova 07200
Accademia Lunigianese di Scienze Giovanni Capellini, La Spezia 07202
Accademia Nazionale dei Lincei, Roma 07211
Accademia Nazionale dei Sartori, Roma 07212
Accademia Nazionale di Scienze, Lettere e Arti, Modena 07222
Accademia Nazionale Virgiliana di Scienze, Lettere ed Arti di Mantova, Mantova 07224
Accademia Polacca delle Scienze, Roma 07226
Accademia Pontaniana, Napoli 07228
Accademia Pratese di Medicina e Scienze, Prato 07229
Accademia Romana di Cultura, Roma 07232
Accademia Roveretana degli Agiati, Rovereto 07234
Accademia Tiberina, Roma 07242
Accademia Toscana di Scienze e Lettere La Colombaria, Firenze 07243
Accademia Universale Guglielmo Marconi, Roma 07245
Associazione Nazionale Filosofia Arti Scienze, Bologna 07372
Committee for the European Development of Science and Technology, Roma 07623
Consiglio Nazionale delle Ricerche, Roma 07628
European Network of Scientific Information Referral Centres, Frascati 07680
Federazione delle Associazioni Scientifiche e Tecniche, Milano 07686
Società Adriatica di Scienze, Trieste 07769
Società di Letture e Conversazioni Scientifiche, Genova 07783
Società Nazionale di Scienze, Lettere ed Arti, Ex Società Reale, Napoli 07992
Società Torricelliana di Scienze e Lettere, Faenza 08024
Unione della Legion d'Oro, Roma 08044

Jamaica

Scientific Research Council, Kingston 08089

Japan

Manyo Gakkai, Osaka 08180
Nihon Myakkan Gakkai, Tokyo 08242
Nihon Sangyo Eisei Gakkai, Tokyo 08253
Nippon Rodo-ho Gakkai, Tokyo 08359
Nippon Saikingakkai, Tokyo 08361
Nogyo-Doboku Gakkai, Tokyo 08398
San-yo Gijutsu Shinkokai, Okayama-ken 08406
Toa Kumo Gakkai, Osaka 08431

Jordan

Jordan Research Council, Amman 08453
Royal Scientific Society, Amman 08454

Kazakhstan

Kazakh Academy of Sciences, Alma-Ata 08456

Kenya

East African Academy, Nairobi 08492
Kenya National Academy of Sciences, Nairobi 08505

Korea, Democratic People's Republic

Academy of Sciences, Pyongyang 08514

Korea, Republic

National Academy of Sciences, Seoul 08543

Kyrgyzstan

Kyrgyz Academy of Sciences, Bishkek 08548

Latvia

Latvian Academy of Sciences, Riga 08558
Terminological Commission, Riga 08559

Libya

Intellectual Society of Libya, Tripoli 08568

Lithuania

Lithuanian Academy of Sciences, Vilnius 08573

Macedonia

Makedonska Akademija na Naukite i Umetnostite, Skopje 08603

Madagascar

Académie Malgache, Antananarivo 08620

Malawi

Society of Malawi, Blantyre 08624

Malaysia

Malaysian Scientific Association, Kuala Lumpur 08652

Mauritius

Royal Society of Arts and Sciences of Mauritius, Réduit 08674

Mexico

Academia de la Investigación Científica, México 08682
Academia Nacional de Ciencias, México 08688
Asociación Nacional de Universidades e Institutos de Enseñanza Superior, México 08705
Unión de Universidades de América Latina, México 08777

Moldova, Republic of

Moldovan Academy of Sciences, Chişinău 08783

Mongolia

Academy of Sciences, Ulan Bator 08798

Myanmar

Burma Research Society, Rangoon 08812
Central Research Organization, Rangoon 08813

Namibia

South West Africa Scientific Society, Windhoek 08816

Nepal

Royal Nepal Academy, Kathmandu 08817

Netherlands

Bataafsch Genootschap der Proefondervindelijke Wijsbegeerte, Rotterdam 08831
Coordinating Committee for Intercontinental Research Networking, Amsterdam 08846
Hollandsche Maatschappij der Wetenschappen, Haarlem 08910
International Technical and Scientific Organization for Soaring Flight, Wessling 08945
Koninklijke Nederlandse Akademie van Wetenschappen, Amsterdam 08952

New Zealand

Foundation for Research, Science and Technology, Wellington 09125
New Zealand Association of Scientists, Wellington 09135
Royal Society of New Zealand, Wellington 09196

Nigeria

Nigerian Academy of Science, Akoka 09234
Organization of African Unity – Scientific, Technical and Research Commission, Lagos 09244

Norway

Det Kongelige Norske Videnskabers Selskab, Trondheim 09258
Det Norske Videnskaps-Akademi, Oslo 09264
Hovedkomiteen for Norsk Forskning, Oslo 09268
Norges Almenvitenskapelige Forskningsråd, Oslo 09281
Selskapet til Vitenskapenes Fremme, Bergen 09347

Pakistan

National Science Council of Pakistan, Islamabad 09376
Pakistan Academy of Sciences, Islamabad 09379
Pakistan Council of Scientific and Industrial Research, Karachi 09384
Quaid-i-Azam Academy, Karachi 09397
Scientific Society of Pakistan, Karachi 09399

Panama

Academia Nacional de Ciencias de Panamá, Panamá City 09405
Centro para el Desarrollo de la Capacidad Nacional de Investigación, Panamá City 09414
Consejo Nacional de Ciencia, Panamá City 09417

Papua New Guinea

Papua New Guinea Scientific Society, Boroko 09422

Paraguay

Sociedad Científica del Paraguay, Asunción 09434

Peru

Consejo Andino de Cienccia y Tecnología, Lima 09465
Consejo Nacional de la Universidad Peruana, Lima 09466

Philippines

Academia Filipina, Manila 09488
National Research Council of the Philippines, Manila 09522

Poland

Białostockie Towarzystwo Naukowe, Bialystok 09539
Bydgoskie Towarzystwo Naukowe, Bydgoszcz 09540
Gdánskie Towarzystwo Naukowe, Gdánsk 09543
Lódzkie Towarzystwo Naukowe, Lódź 09545
Lubelskie Towarzystwo Naukowe, Lublin 09546
Lubuskie Towarzystwo Naukowe, Zielona Góra 09547
Opolskie Towarzystwo Przyjaciót Nauk, Opole 09548
Polska Akademia Nauk, Warszawa 09549
Poznánskie Towarzystwo Przyjaciól Nauk, Poznan 09612
Szczecinskie Towarzystwo Naukowe, Szczecin 09617
Towarzystwo Naukowe Płockie, Płock 09626
Towarzystwo Naukowe w Toruniu, Torun 09627
Towarzystwo Przyjaciól Nauk w Przymyslu, Przemysl 09628
Wroclawskie Towarzystwo Naukowe, Wroclaw 09630

Portugal

Academia das Ciências de Lisboa, Lisboa 09633
Fundação Calouste Gulbenkian, Lisboa 09657
Sociedade Científica da Universidade Católica Portuguesa, Lisboa 09677

Puerto Rico

Association of Caribbean Universities and Research Institutes, San Juan 09722

Romania

Academia Romana, Bucuresti 09737
Asociatia Oamenilor de Stiinta din Romania, Bucuresti 09743
European Committee for Scientific and Cultural Relations with Romania, Bucuresti 09751

Russia
Commission for International Scientific Links, Moskva 09818
Commission for Scientific and Technical Co-operation of the Academy of Sciences and Organizations of Moscow Oblast, Moskva 09827
Commission for the Development of Scientific Co-operation with Great Britain, Moskva 09832
Commission for the Processing of the Scientific Legacy of Academician V.I. Vernadsky, Moskva 09837
Commission for the Prospects of Development of Science in the Russian Federation, Moskva 09838
Co-ordination Council for Information on Achievements, Moskva 09856
Group of the Far Eastern Division, Moskva 09872
Moscow Science-Production Association, Moskva 09885
National Committee of the International Council of Scientific Unions, Moskva 09900
Russian Academy of Sciences, Moskva 09954
Union of Scientific and Learned Societies, Moskva 09990
Znanie, Moskva 09992

Sierra Leone
Sierra Leone Science Association, Freetown 10032

Singapore
Singapore Association for the Advancement of Science, Singapore 10052
Singapore National Academy of Science, Singapore 10056

Slovakia
Organizačné Stredisko Vedeckých Spoločností pri SAV, Bratislava 10059
Slovenská Akadémia Vied, Bratislava 10060

Slovenia
Prirodoslovno Društvo Slovenije v Ljubljani, Ljubljana 10102
Slovenska Akademija Znanosti in Umetnosti, Ljubljana 10105

South Africa
The Associated Scientific and Technical Societies of South Africa, Marshalltown 10123
CSIR, Pretoria 10133
Joint Council of Scientific Societies, Yeoville 10150
The National Association of Scientists, Pretoria 10156
Royal Society of South Africa, Cape Town 10171
South African Academy of Science and Arts, Pretoria 10177
Southern African Association for the Advancement of Science, Arcadia 10224

Spain
Asociación de Personal Investigador del CSIC, Madrid 10252
Asociación Española para el Progreso de las Ciencias, Madrid 10272
Instituto de España, Madrid 10318
Organización de Estados Iberoamericanos para la Educación, la Ciencia y la Cultura, Madrid 10333
Real Academia de Ciencias Exactas, Físicas y Naturales, Madrid 10342
Real Academia de Ciencias y Artes de Barcelona, Barcelona 10344
Real Academia Gallega, La Coruña 10354
Real Academia Hispano-Americana, Cádiz 10355

Sri Lanka
Sri Lanka Association for the Advancement of Science, Colombo 10419

Sudan
National Council for Research, Khartoum 10427

Suriname
Stichting voor Wetenschappelijk Ondersoek van de Tropen, Paramaribo 10433

Sweden
European Incoherent Scatter Scientific Association, Kiruna 10456
Forskningsrådsnämnden, Stockholm 10462
Kungliga Fysiografiska Sällskapet i Lund, Lund 10483
Kungliga Vetenskapsakademien, Stockholm 10490
Kungliga Vetenskaps- och Vitterhets-Samhället i Göteborg, Göteborg 10491
Kungliga Vetenskaps-Societeten i Uppsala, Uppsala 10492
Nobelstiftelsen, Stockholm 10505
SACO Sveriges Akademikers Centralorganisation, Stockholm 10516
Svenska Linné-Sällskapet, Uppsala 10561
Vetenskapliga Bibliotekens Tjänstemannaförening, Nacka 10604

Switzerland
Commission pour l'Encouragement de la Recherche Scientifique, Bern 10652
European Association of Development Research and Training, Genève 10677
Fédération Internationale des Femmes Diplômées des Universités, Genève 10702
Institut National Genevois, Genève 10730
Mouvement International des Intellectuels Catholiques/MIIC-Pax Romana, Genève 10765
Renaissance, Schweizerischer Verband katholischer Akademiker-Gesellschaften, Luzern 10782
Schweizerischer Verband der Akademikerinnen, Chur 10906
Schweizerischer Wissenschaftsrat, Bern 10924
Société Académique, Neuchâtel 10966

Tajikistan
Tajik Academy of Sciences, Dushanbe 11045

Tanzania
Tanzania Commission for Science and Technology, Dar es Salaam 11059

Thailand
Applied Scientific Research Corporation of Thailand, Bangkok 11066
National Culture Commission, Bangkok 11094
Royal Institute, Bangkok 11097
Science Society of Thailand, Bangkok 11098

Tunisia
Arab League Educational, Cultural and Scientific Organization, Tunis 11131

Turkmenistan
Turkmen Academy of Sciences, Ashkhabad 11180

Ukraine
Scientific and Technical Societies National Headquarters, Kiev 11190
Ukrainian Academy of Sciences, Kiev 11195

United Kingdom
The Association for Science Education, Hatfield 11282
Association of British Science Writers, London 11316
Association of Principals of Colleges (Northern Ireland Branch), Lisburn 11379
Association of Professional Scientists and Technologists, London 11380
Association of Scientific, Technical and Managerial Staffs, London 11388
British American Scientific Research Association, Watford 11476
British Association for the Advancement of Science, London 11495
British Association of Friends of Museums, Manchester 11506
British Federation of Women Graduates, London 11561
British Society for Middle Eastern Studies, Cambridge 11658
Committee of Vice-Chancellors and Principals of the Universities of the United Kingdom, London 11805
Commonwealth Association of Science and Mathematics Educators, London 11812
Commonwealth Institute, London 11824
Commonwealth Trust, London 11831
The Devonshire Association for the Advancement of Science, Literature and Art, Exeter 11890
European Association of Science Editors, London 12000
Institute for Scientific Information, Uxbridge 12209
Institute of Scientific and Technical Communicators, London 12280
International Union of Independent Laboratories, Elstree 12402
Kilvert Society, Worcester 12430
Parliamentary and Scientific Committee, London 12679
The Plymouth Athenaeum, Plymouth 12705
Pugwash Conferences on Science and World Affairs, London 12729
Research and Development Society, London 12751
Research Defence Society, London 12752
Royal Institution of Cornwall, Truro 12821
Royal Institution of Great Britain, London 12822
Royal Institution of South Wales, Swansea 12824
Royal Microscopical Society, Oxford 12833
Royal Scottish Academy, Edinburgh 12844
The Royal Society of Edinburgh, Edinburgh 12858
Royal Ulster Academy, Lurgan 12870
Royal Ulster Academy Association, Lambeg 12871
Socialist International Research Council, London 12958
Society for the Protection of Science and Learning, London 13011
South-Eastern Union of Scientific Societies, Shoreham-by-Sea 13078
Tensor Society of Great Britain, Surbiton 13119
Theosophical Society in England, London 13124
United Kingdom Science Park Association, Birmingham 13168

Uruguay
Consejo Nacional de Investigaciones Científicas y Técnicas, Montevideo 13258

U.S.A.
Academy for Interscience Methodology, Hinsdale 13279
Academy of Applied Science, Concord 13284
Academy of Arts and Sciences of the Americas, Miami 13285
American Academy of Arts and Sciences, Cambridge 13382
American Association for the Advancement of Science, Washington 13497
American Association for the Advancement of Slavic Studies, Stanford 13498
American Association of Meta-Science, Huntsville 13553
American Association of Presidents of Independent Colleges and Universities, Malibu 13570
American Association of State Colleges and Universities, Washington 13583
American Committee for the Weizmann Institute of Science, New York 13745
American Indian Science and Engineering Society, Boulder 13861
American Philosophical Society, Philadelphia 13999
American Society for Eighteenth-Century Studies, Logan 14058
Associated Universities, Washington 14261
Association for Education of Teachers in Science, Auburn 14293
Association for Women in Science, Washington 14369
Association of American Universities, Washington 14389
Association of International Colleges and Universities, Independence 14447
Association of Jesuit Colleges and Universities, Washington 14450
Association of Muslim Scientists and Engineers, Plainfield 14468
Association of Orthodox Jewish Scientists, New York 14475
Association of Science Museum Directors, Chicago 14499
Association of Science-Technology Centers, Washington 14500
Association of University Related Research Parks, Tempe 14531
Association of University Summer Sessions, Bloomington 14532
Center for Advanced Study in the Behavioral Sciences, Stanford 14663
Center for Field Research, Watertown 14680
Center for Science in the Public Interest, Washington 14699
Conference Board of Associated Research Councils, New York 14822
Council for Elementary Science International, Westport 14874
Council for the Advancement of Science Writing, Greenlawn 14886
Council of Colleges of Arts and Sciences, Columbus 14892
Council of Engineering and Scientific Society Executives, Washington 14895
Council of Scientific Society Presidents, Washington 14902
Creation Research Society, Kansas City 14938
Czechoslovak Society of Arts and Sciences, Sherman Oaks 14947
Essentia, Boulder 15059
Federation of American Scientists, Washington 15080
History of Science Society, Worcester 15222
Institute for Advanced Research in Asian Science and Medicine, Garden City 15265
Institute for Policy Studies, Washington 15289
Institute for the Study of Human Knowledge, Los Altos 15305
Institute of World Affairs, Salisbury 15343
International Academy at Santa Barbara, Santa Barbara 15362
International Society for the Systems Sciences, Pocatello 15544
International Studies Association, Provo 15558
Jewish Academy of Arts and Sciences, New York 15603
Leonardo – The International Society for the Arts, Sciences and Technology, San Francisco 15660
Moody Institute of Science, Chicago 15754
National Academic Advising Association, Manhattan 15770
National Association of Academies of Science, Columbia 15826
National Association of Independent Colleges and Universities, Washington 15867
National Association of Science Writers, Greenlawn 15891
National Association of State Universities and Land-Grant Colleges, Washington 15907
National Conference on the Advancement of Research, San Antonio 15976
National Congress of Inventors Organizations, Rheem Valley 15979
National Council of University Research Administrators, Washington 16013
National Institute of Science, Nashville 16095
National Research Council, Washington 16138
New York Academy of Sciences, New York 16199
Pacific Science Association, Honolulu 16282
Pattern Recognition Society, Washington 16299
Polish Institute of Arts and Sciences in America, New York 16335
Rare Earth Research Conference, Lexington 16394
Shevchenko Scientific Society, New York 16470
Society of Jewish Science, Plainview 16675
Society of Research Administrators, Chicago 16699
Ukrainian Academy of Arts and Sciences in the U.S., New York 16796
Union Institute, Cincinnati 16804
United States Federation of Scholars and Scientists, Fullerton 16824
United States Metric Association, Los Angeles 16826

Uzbekistan
Uzbek Academy of Sciences, Tashkent 16948

Vatican City
Pontificia Academia Scientiarum, Roma 16952

Venezuela
Asociación Venezolana para el Avance de la Ciencia, Caracas 16977
Consejo de Desarrollo Científico y Humanístico, Caracas 17000
Consejo Nacional de Investigaciones Científicas y Tecnológicas, Caracas 17001

Vietnam
Vietnam Union of Sientific and Technical Associations, Hanoi 17071

Yugoslavia
Crnogorska Akademija Nauka i Umjetnosti, Podgorica 17080
Srpska Akademija Nauka i Umetnosti, Beograd 17125

Zambia
National Council for Scientific Research, Lusaka 17154

Zimbabwe
Scientific Council of Zimbabwe, Harare 17180
Zimbabwe Association for Science Education, Harare 17185
Zimbabwe Scientific Association, Harare 17188

Socialism

Australia
Australian Fabian Society, Melbourne 00290

Austria
Internationale Tagung der Historiker der Arbeiterbewegung, Wien 00684
Verein für Geschichte der Arbeiterbewegung, Wien 00983

Belgium
European Trade Union Institute, Bruxelles 01353

Canada
Société des Etudes Socialistes, Winnipeg 02925

Italy
Centro Ligure di Storia Sociale, Genova 07514

U.S.A.
Center for Socialist History, Berkeley 14701
Global Options, San Francisco 15165

Social Sciences

Argentina
Asociación Iberoamericana de Estudio de los Problemas del Alcohol y la Droga, Córdoba 00083
Center for Integrated Social Development, Rosario 00097

Australia
Asian and Pacific Association for Social Work Education, Bundoora 00220

Austria
AGEMUS Arbeitsgemeinschaft Evolution, Menschheitszukunft und Sinnfragen, Wien 00538
Akademie für Sozialarbeit der Stadt Wien, Wien 00541
European Centre for Social Welfare Policy and Research, Wien 00588
European Cooperation in Social Science Information and Documentation, Wien 00590
European Coordination Centre for Research and Documentation in Social Sciences, Wien 00591
Institut für Sozialdienste, Bregenz 00659
International Council on Social Welfare, Wien 00664
Internationales Institut für den Frieden, Wien 00680

Belgium
Association Centrale des Assistants Sociaux, Bruxelles 01113
Catholic Office for Information on European Problems, Bruxelles 01180
Centre for Research on European Women, Bruxelles 01188
Centre International de Formation et de Recherche en Population et Développement en Association avec les Nations Unies, Louvain-la-Neuve 01191
Institut Européen Interuniversitaire de l'Action Sociale, Marcinelle 01385

Brazil
Asociación Latinoamericana de Escuelas de Trabajo Social, São Luís 01583
Fundação João Pineiro, Belo Horizonte 01657

Burkina Faso
Centre of Economic and Social Studies and Experiments in Western Africa, Bobo-Dioulasso 01808

Canada
Centre for Research on Latin America and the Caribbean, Toronto 02329
Fédération des Affaires Sociales, Montréal 02448
International Development Research Centre, Ottawa 02552
Manitoba Social Science Teachers' Association, Winnipeg 02633
Nova Scotia Association of Social Workers, Dartmouth 02724
Ontario Teachers' Superannuation Commission, North York 02802
Social Sciences and Humanities Research Council of Canada, Ottawa 02915
Society of Philosophy and Social Science, Kingston 02947

China, People's Republic

Chinese Academy of Social Sciences, Beijing *03155*

Costa Rica

Central American Institute of Social Studies, San José *03438*

Denmark

Nordisk Institut for Asienstudier, København *03785*

Egypt

Centre for Social Research and Documentation for the Arab Region, Cairo *03885*

France

Association Internationale des Démographes de Langue Française, Paris *04268*
Association Médico-Sociale Protestante de Langue Française, Alfortville *04303*
Center for Studies, Research and Training in International Understanding and Cooperation, Paris *04364*
Club of Rome, Paris *04390*
Committee for International Cooperation in National Research in Demography, Paris *04430*
European Association for American Studies, Lyon *04480*
Société d'Ethnographie de Paris, Paris *04722*

Germany

Akademie gemeinnütziger Wissenschaften zu Erfurt e.V., Erfurt *05061*
Arbeitsgemeinschaft für betriebliche Altersversorgung e.V., Heidelberg *05102*
Arbeitsgemeinschaft Sozialwissenschaftlicher Institute e.V., Bonn *05125*
Deutsches Zentralinstitut für soziale Fragen, Berlin *05607*
Deutsches Zentrum für Altersfragen e.V., Berlin *05609*
Gesellschaft für Sozial- und Wirtschaftsgeschichte, Heidelberg *05860*
Gesellschaft Sozialwissenschaftlicher Infrastruktureinrichtungen e.V., Mannheim *05877*
GFM-GETAS Gesellschaft für Marketing-, Kommunikations- und Sozialforschung, Hamburg *05894*
Thomas-Morus-Akademie Bensberg, Bergisch Gladbach *06313*

Greece

Athens Centre of Ekistics, Athinai *06501*

Hungary

Magyar Gerontológiai Társaság, Budapest *06632*

India

Asian Network of Human Resource Development Planning Institutes, New Delhi *06724*
Association of Asian Social Science Research Councils, New Delhi *06732*

Indonesia

ASEAN Population Coordination Unit, Jakarta *06877*
ASEAN Population Information Network, Jakarta *06878*

Ireland

Irish Association of Social Workers, Dublin *06976*
Royal Irish Academy, Dublin *07028*

Israel

Israel Gerontological Society, Tel Aviv *07074*

Italy

Accademia Internazionale per le Scienze Economiche, Sociali e Sanitarie, Roma *07192*
Accademia Polacca delle Scienze, Roma *07226*
Accademia Simba, Roma *07238*
Centro Internazionale per le Communicazione Sociali, Roma *07480*
Centro Studi e Ricerche sui Rapporti Umani, Roma *07574*
European Council for Social Research on Latin America, Roma *07677*

Ivory Coast

African Institute for Economic and Social Development, Abidjan *08057*

Libya

African Centre for Applied Research and Training in Social Development, Tripoli *08566*

Malaysia

Asia Pacific Forum on Women, Law and Development, Kuala Lumpur *08636*

Mexico

Center for Economic and Social Studies in the Third World, México *08709*
Center of Coordination and Diffusion of Latin American Studies, México *08710*

Netherlands

European Association for Population Studies, 's-Gravenhage *08859*
European Centre for Work and Society, Maastricht *08877*
Fryske Akademy, Leeuwarden *08899*
Instituut voor Toegepaste Sociale Wetenschapen, Nijmegen *08915*
Nederlandse Organisatie voor Wetenschappelijk Onderzoek, 's-Gravenhage *08998*

New Zealand

Population Association of New Zealand, Wellington *09190*

Nigeria

African Curriculum Organization, Ibadan *09215*

Norway

International Federation of Social Workers, Oslo *09270*

Pakistan

Hamdard Foundation, Karachi *09364*

Peru

Andean Institute for Population Studies and Development, Lima *09444*
Andean Institute of Social Studies, Lima *09445*

Philippines

Asian Women's Research and Action Network, Manila *09505*

Russia

Commission for Links in Social Science with the American Council of Learned Societies, Moskva *09820*
Commission for the European (Vienna) Centre for Co-ordination of Research and Documentation in Social Sciences, Moskva *09834*
Committee for UNESCO Programme „Man and Biosphere", Moskva *09849*
Council for International Co-operation in Social Sciences, Moskva *09858*

Senegal

African Council for Social and Human Sciences, Dakar *10010*

Sri Lanka

Royal Asiatic Society of Sri Lanka, Colombo *10418*

Sweden

European Association of Labour Economists, Stockholm *10451*
Humanistisk-Samhällsvetenskapliga Forskningsradet, Stockholm *10468*

Switzerland

Internationale Vereinigung wissenschaftlicher Fremdenverkehrsexperten, Sankt Gallen *10747*
Schweizerische Akademie der Sozial- und Geisteswissenschaften, Bern *10789*
Schweizerische Asiengesellschaft, Zürich *10798*
Schweizerischer Verband von Fachleuten für Alkoholgefährdeten- und Suchtkrankenhilfe, Zürich *10917*
Verband Schweizerischer Marketing- und Sozialforscher, Hergiswil *11007*

Syria

Arab Centre for Information Studies on Population, Development and Construction, Damascus *11039*

Thailand

Asia-Pacific Information Network in Social Sciences, Bangkok *11086*

Trinidad and Tobago

Caribbean Association for Feminist Research and Action, Tunapuna *11108*
Caribbean Information System for Economic and Social Planning, Port of Spain *11114*

Tunisia

Association of Arab Institutes and Centres for Economic and Social Research, Tunis *11137*

United Kingdom

Association for Child Psychology and Psychiatry, London *11262*
Centre for Iberian Studies, Keele *11754*
National Institute of Economic and Social Research, London *12598*
Standing Conference of Arts and Social Sciences, Cardiff *13095*
Theosophical Society in England, London *13124*

U.S.A.

American Association on Mental Retardation, Washington *13608*
American Council of Learned Societies, New York *13763*
American Orthopsychiatric Association, New York *13974*
Caucus for a New Political Science, Boston *14659*
Commission on Gay/Lesbian Issues in Social Work Education, Alexandria *14786*
Council for European Studies, New York *14875*
Institute for Policy Studies, Washington *15289*
Lentz Peace Research Laboratory, Saint Louis *15658*
National Association of Alcoholism and Drug Abuse Counselors, Arlington *15828*
New World Foundation, New York *16198*
Research Society on Alcoholism, Austin *16418*
Social Science Education Consortium, Boulder *16484*
Social Science Research Council, New York *16485*
Society for Social Studies of Science, Baton Rouge *16580*

Venezuela

Caribbean Institute for Social Formation, Caracas *16980*

Yugoslavia

Akademia e Shkencave dhe e Arteve e Kosovës, Priština *17076*

Zaire

Centre for the Coordination of Research and Documentation in Social Science for Sub-Saharan Africa, Kinshasa *17144*

Sociology

Argentina

Academia Nacional de Derecho y Ciencias Sociales, Buenos Aires *00047*
Academia Nacional de Derecho y Ciencias Sociales (Córdoba), Córdoba *00048*
Asociación Latinoamericana de Sociología, Córdoba *00088*
Centro de Estudios Económicos Sociales, Buenos Aires *00102*
Sociedad Argentina de Sociología, Córdoba *00184*

Australia

Academy of the Social Sciences in Australia, Canberra *00211*
Australasian Association for the History, Philosophy and Social Studies of Science, Sydney *00237*
Australian Association of Social Workers, North Richmond *00267*
Australian Sociological Association, Bundoora *00360*
Society for Social Responsibility in Science (A.C.T.), Canberra *00491*
Sociological Association of Australia and New Zealand, Bathurst *00495*

Austria

Akademisch-soziale Arbeitsgemeinschaft Österreichs, Wien *00543*
Arbeitsgemeinschaft für Historische Sozialkunde, Wien *00551*
Arbeitsgemeinschaft für interdisziplinäre angewandte Sozialforschung, Wien *00552*
Europäisches Zentrum für Wohlfahrtspolitik und Sozialforschung, Wien *00585*
Gesellschaft für Soziologie an der Universität Graz, Graz *00633*
International Association of Schools of Social Work, Wien *00662*
Mediacult, Internationales Forschungsinstitut für Medien, Kommunikation und kulturelle Entwicklung, Wien *00705*
Österreichische Gesellschaft für China-Forschung, Wien *00768*
Österreichische Gesellschaft für Medizinsoziologie, Wien *00808*
Österreichische Gesellschaft für Sexualforschung, Wien *00831*
Österreichische Gesellschaft für Soziologie, Wien *00832*
Österreichische Gesellschaft für Wirtschaftssoziologie, Wien *00848*
Österreichischer Arbeitskreis für Soziologie des Sports und der Leibeserziehung, Wien *00883*
Österreichisches Lateinamerika-Institut, Wien *00913*
Sozialwissenschaftliche Arbeitsgemeinschaft, Wien *00947*
Sozialwissenschaftliche Studiengesellschaft, Wien *00948*
Verein für Geschichte der Arbeiterbewegung, Wien *00983*

Bahrain

Bahrain Society of Sociologists, Manama *01060*

Belgium

Association Internationale pour le Progrès Social, Bruxelles *01144*
Comité Economique et Social des Communautés Européennes, Bruxelles *01225*
Comité National pour l'Etude et la Prévention de l'Alcoolisme et des Autres Toxicomanies, Bruxelles *01233*
Commission Permanente des Groupements Professionnels d'Assistants Sociaux, Bruxelles *01237*
Conseil Consultatif Economique et Social de l'Union Economique Benelux, Bruxelles *01250*
European Trade Union Institute, Bruxelles *01353*
Fédération Internationale des Instituts de Recherches Socio-Religieuses, Louvain-la-Neuve *01358*
Institut Européen Interuniversitaire de l'Action Sociale, Marcinelle *01385*
International Institute for Organizational and Social Development, Leuven *01397*
International Union for the Scientific Study of Population, Liège *01404*
Union des Associations d'Assistants Sociaux Francophones, Bruxelles *01495*

Brazil

Conselho Nacional Serviço Social, Rio de Janeiro *01646*
Fundação de Estudos Sociais do Paraná, Curitiba *01652*
Fundação Joaquim Nabuco, Recife *01658*
Instituto Archeológico, Histórico e Geográfico Pernambucano, Recife *01663*
Instituto de Planejamento de Pernambuco, Recife *01668*
Sociedade Brasileira de Instrução, Botafogo *01714*

Bulgaria

Bulgarian Sociological Society, Sofia *01762*
Scientific Information Centre at the Bulgarian Academy of Sciences, Sofia *01775*

Canada

Alberta Association of Social Workers, Edmonton *01827*
Association for Canadian Studies, Montréal *01899*
Association of Canadian Universities for Northern Studies, Ottawa *01926*
Canadian Association for the Advancement of Netherlandic Studies, Windsor *02041*
Canadian Association for the Social Studies, Fredericton *02042*
Canadian Association of Social Workers, Ottawa *02068*
Canadian Council on Social Development, Ottawa *02111*
Canadian Institute for Middle East Research, Calgary *02164*
Canadian Polish Research Institute, Toronto *02228*
Canadian Research Institute for the Advancement of Women, Ottawa *02242*
Canadian Sociology and Anthropology Association, Montréal *02298*
Corporation Professionnelle des Travailleurs Sociaux du Québec, Montréal *02399*
Elizabeth Fry Society of Alberta (Edmonton), Edmonton *02430*
Elizabeth Fry Society of Manitoba, Winnipeg *02431*
Elizabeth Fry Society of New Brunswick, Moncton *02432*
Elizabeth Fry Society of Saskatchewan, Saskatoon *02433*
Fédération des Affaires Sociales, Montréal *02448*
Manitoba Association of Social Workers, Winnipeg *02609*
Manitoba Committee on Children and Youth, Winnipeg *02613*
Manitoba Institute of Registered Social Workers, Winnipeg *02624*
Migraine Foundation, Toronto *02654*
Newfoundland Association of Social Workers, Saint John's *02702*
Ontario Public Interest Research Group, Ottawa *02790*
Prince Edward Island Association of Social Workers, Charlottetown *02829*
Rural Development Council of Prince Edward Island, Charlottetown *02874*
Saskatchewan Association of Social Workers, Regina *02883*
Social Science Federation of Canada, Ottawa *02914*
Youth Science Foundation, Ottawa *03002*

Chile

Academia Chilena de Ciencias Sociales, Politicas y Morales, Santiago *03009*
Instituto Latinoamericano de Planificación Económica y Social, Santiago *03058*

China, People's Republic

Inner Mongolian Academy of Social Sciences, Huhehot *03206*

China, Republic

Association for Socio-Economic Development in China, Hsinchu *03244*
China Social Education Society, Taipei *03257*
Population Association of China, Taipei *03321*

Colombia

Centro de Investigación y Acción Social, Bogotá *03357*
Centro Interamericano de Vivienda y Planeamiento, Bogotá *03359*
Instituto Colombiano de Bienestar Familiar, Bogotá *03381*

Cyprus

Cyprus Research Centre, Nicosia *03496*

Czech Republic

Masarykova Česká Sociologická Společnost, Praha *03550*

Denmark

Dansk Forening for Retssociologi, København *03621*
Dansk Socialradgiverforening, København *03691*
Dansk Sociologisk Selskab, København *03692*
Nordisk Forening for Rettssociologi, København *03784*
Paedagogisk Forening, København *03793*
Scandinavian Society for Economic and Social History, Odense *03806*
Scandinavian Sociological Association, København *03807*
Selskab for Arbejdsmiljø, København *03819*
Socialpaedagogernes Landsforbund, København *03822*

Ecuador

Instituto Latinoamericano de Investigaciones Sociales, Quito *03854*

Ethiopia

African Training and Research Centre for Women, Addis Ababa *03939*
Association for Social Work Education in Africa, Addis Ababa *03941*

France

Association des Ecoles de Service Social de la Région Parisienne, Paris *04178*
Association Française d'Etude des Relations Professionnelles, Fresnes *04236*
Association Internationale des Sociologues de Langue Française, Toulouse *04277*
Association Interprofessionnelle des Centres Médicaux et Sociaux de la Région Parisienne, Paris *04295*

Association Nationale des Professeurs en Economie Sociale et Familiale, Paris *04311*
Association Régionale d'Informations Sociales, Paris *04347*
Centre d'Etudes, de Documentation, d'Information et d'Action Sociales, Paris *04369*
Centre National de la Recherche Scientifique, Paris *04382*
Comité d'Entente des Ecoles de Formation en Economie Sociale Familiale, Paris *04393*
Comité d'Etudes pour un Nouveau Contrat Social, Paris *04398*
Comité International pour l'Information et la Documentation en Sciences Sociales, Paris *04409*
Comité National des Ecoles Françaises de Service Social, Paris *04415*
Commission Internationale d'Histoire des Mouvements Sociaux et des Structures Sociales, Paris *04427*
Fondation pour la Recherche Sociale, Paris *04544*
Institut Français d'Analyse de Groupe et de Psychodrame, Paris *04567*
Société d'Economie et de Science Sociale, Paris *04651*
Société d'Etudes et de Documentation Economiques, Industrielles et Sociales, Neuilly-sur-Seine *04732*
Société d'Etudes et de Recherches en Sciences Sociales, Paris *04734*
Société d'Etudes Juives, Paris *04744*
Société d'Etudes pour le Développement Economique et Social, Paris *04750*
Société Française de Sociologie, Paris *04843*

Germany
Agrarsoziale Gesellschaft e.V., Göttingen *05044*
Akademie der Arbeit in der Universität, Frankfurt *05046*
Akademie für Arbeit und Sozialwesen, Saarbrücken *05052*
Arbeitsgemeinschaft für angewandte Sozialforschung, München *05101*
Arbeit und Leben, Düsseldorf *05149*
Arnold-Bergstraesser-Institut für kulturwissenschaftliche Forschung e.V., Freiburg *05152*
Berufsverband Deutscher Soziologen e.V., Bielefeld *05205*
Deutsche Gesellschaft für Bevölkerungswissenschaft e.V., Wiesbaden *05330*
Deutsche Gesellschaft für Medizinische Soziologie, Ulm *05383*
Deutsche Gesellschaft für Publizistik- und Kommunikationswissenschaft e.V., Eichstätt *05413*
Deutsche Gesellschaft für Soziologie, Mannheim *05424*
Deutsche Gesellschaft für Suchtforschung und Suchttherapie e.V., Hamm *05429*
Deutsche Hauptstelle gegen die Suchtgefahren e.V., Hamm *05454*
Deutscher Berufsverband der Sozialarbeiter und Sozialpädagogen e.V., Essen *05504*
Europäische Vereinigung für Eigentumsbildung, Hermann-Lindrath-Gesellschaft e.V., Hannover *05658*
Evangelische Sozialakademie Friedewald, Friedewald *05690*
Ferdinand Tönnies-Gesellschaft e.V., Kiel *05699*
Forschungsgesellschaft für Agrarpolitik und Agrarsoziologie e.V., Bonn *05719*
Forschungsinstitut für Gesellschaftspolitik und beratende Sozialwissenschaft e.V., Göttingen *05729*
Gesellschaft für bedrohte Völker e.V., Göttingen *05803*
Gesellschaft für empirische soziologische Forschung e.V., Nürnberg *05813*
Gesellschaft für Regionalforschung e.V., Bonn *05858*
Gesellschaft für sozialwissenschaftliche Forschung in der Medizin, Freiburg *05861*
Gesellschaft für Technologiefolgenforschung e.V., Berlin *05863*
Gesellschaft für Wirtschafts- und Sozialwissenschaften, Mannheim *05871*
Gesellschaft für Wissenschaft und Leben im Rheinisch-Westfälischen Industriegebiet e.V., Essen *05873*
Institut der deutschen Wirtschaft e.V., Köln *05943*
Institut für angewandte Verbraucherforschung e.V., Köln *05949*
Institut für Gesellschaftswissenschaften Walberberg e.V., Bonn *05961*
Institut für Sozialarbeit und Sozialpädagogik, Frankfurt *05970*
Institut für Sozialforschung und Sozialwirtschaft e.V., Saarbrücken *05971*
Internationale Vereinigung für Rechts- und Sozialphilosophie e.V., Göttingen *06004*
Katholische Erziehergemeinschaft in Bayern, München *06064*
Pro Familia, Frankfurt *06240*
Sozialakademie Dortmund, Dortmund *06283*
Studiengruppe für Sozialforschung e.V., Marquartstein *06305*

Ghana
Ghana Sociological Association, Accra *06490*

Greece
World Society for Ekistics, Athinai *06549*

India
Indian Council of Social Science Research, New Dehli *06788*
The Theosophical Society, Madras *06865*

Ireland
Statistical and Social Inquiry Society of Ireland, Dublin *07035*

Italy
Accademia di Scienze, Lettere e Arti, Udine *07174*
Accademia Universale Citta' Eterna, Roma *07244*
Associazione di Ricerca e Interventi Psicosociali e Psicoterapeutici, Molinetto di Mazzano *07274*
Associazione Italiana di Scienze Politiche e Sociali, Roma *07323*
Associazione Italiana di Sociologia, Roma *07324*
Associazione Italiana Socioanalisi Individuale, Milano *07350*
Associazione Nazionale Assistenti Sociali, Roma *07360*
Associazione per lo Studio del Problema Mondiale dei Rifugiati, Sezione Italiana, Roma *07392*
Associazione Sociologi Lucani, Lauria Inferiore *07403*
Centro di Studi Sociali e Politici Lorenzo Milani, Arezzo *07454*
Centro Europeo per il Progresso Economico e Sociale, Roma *07462*
Centro Italiano di Studi Sociali Economici e Giuridici, Roma *07502*
Centro Ligure di Storia Sociale, Genova *07514*
Centro Regionale di Studi Sociali V.G. Galati, Lamezia Terme *07532*
Centro Ricerche Economiche Sociologiche e di Mercato nell'Edilizia, Roma *07539*
Centro Ricerche Socio-Religiose, Napoli *07541*
Centro Romano per lo Studio dei Problemi di Attualita' Sociale, Roma *07543*
Centro Studi Mutualistici Emancipazione e Partecipazione, Roma *07577*
Centro Studi Nord e Sud, Napoli *07578*
Centro Studi per la Programmazione Economica e Sociale, Roma *07582*
Centro Studi Politici e Sociali Alcide de Gasperi, Cosenza *07586*
Centro Studi Politico-Sociali Achille Grandi, Milano *07587*
Comitato Italiano per lo Studio dei Problemi della Popolazione, Roma *07616*
Famiglia e Libertà, Roma *07685*
International Centre for Advanced Technical and Vocational Training, Torino *07709*
Istituto di Sociologia Internazionale di Gorizia, Gorizia *07727*
Società di Storia Patria di Terra di Lavoro, Caserta *07790*
Società Italiana di Ergonomia, Milano *07869*
Società Italiana di Sociologà, Roma *07950*
Unione Associazioni Regionali, Roma *08040*

Ivory Coast
Centre des Sciences Humaines, Abidjan *08063*
Institut Africain pour le Développement Economique et Social, Abidjan *08065*

Japan
Asian Population and Development Association, Tokyo *08106*
Keizaigaku-shi Gakkai, Kobe *08159*
Kokka Gakkai, Tokyo *08169*
Nihon Kyoiku-shakai Gakkai, Tokyo *08238*
Nippon Hoshakai Gakkai, Tokyo *08307*
Nippon Shakai Gakkai, Tokyo *08370*

Korea, Democratic People's Republic
Academy of Social Sciences, Pyongyang *08515*

Macedonia
Društvo za Filozofija, Sociologija i Politikologija na Makedonija, Skopje *08599*

Malaysia
Asia Foundation, Kuala Lumpur *08632*

Mexico
Asociación Latinoamericana de Sociología, México *08695*
Centro de Estudios Económicos y Sociales del Tercer Mundo, México *08714*

Moldova, Republic of
Moldovan Sociological Association, Beltsi *08785*

Morocco
Centre d'Etudes, de Documentations et d'Informations Economiques et Sociales, Casablanca *08804*
Centre National de Documentation, Rabat *08805*
Société d'Etudes Economiques, Sociales et Statistiques du Maroc, Rabat *08809*

Netherlands
Katholiek Documentatie Centrum, Nijmegen *08946*
Research Group for European Migration Problems, 's-Gravenhage *09065*

Panama
Asociación Centroamericana de Sociología, Panamá City *09408*

Paraguay
Centro Paraguayo de Estudios de Desarrollo Económico y Social, Asunción *09428*

Poland
Polskie Towarzystwo Socjologiczne, Warszawa *09605*

Portugal
Sindicato Nacional dos Professionais de Serviço Social, Lisboa *09670*
Sociedade Portuguesa Veterinária de Estudos Sociológicos, Lisboa *09716*

Romania
Academia de Stiinte Sociale si Politice, Bucuresti *09736*

Russia
Sociological Association, Moskva *09983*

Senegal
Council for the Development of Economic and Social Research in Africa, Dakar *10026*

Singapore
Institute of Southeast Asian Studies, Singapore *10049*

Slovakia
Slovenská Sociologická Spoločnost, Bratislava *10086*

South Africa
Human Sciences Research Council, Pretoria *10147*
South African Institute of Race Relations, Braamfontein *10202*

Spain
Consejo General de Colegios Oficiales de Diplomados en Trabajo Social, Madrid *10296*
International Sociological Association, Madrid *10330*

Sweden
Committee on Family Research, Uppsala *10446*
Sveriges Socianomer Riksförbund, Stockholm *10595*

Switzerland
Association Internationale de la Securité Sociale, Genève *10627*
Heimverband Schweiz, Zürich *10725*
Schweizerische Gesellschaft für Soziologie, Zürich *10848*
Schweizerischer Berufsverband diplomierter Sozialarbeiterinnen und Sozialpädagoginnen, Bern *10887*
Schweizerischer Fachverband Sozialdienst in Spitälern, Bern *10895*
Vereinigung Umwelt und Bevölkerung, Zollikofen *11022*

Thailand
Asia Foundation, Bangkok *11075*

Tunisia
Association Maghrébine des Etudes de la Population, Tunis *11136*

United Kingdom
Alternative Society, Kidlington *11222*
Association for the Teaching of the Social Sciences, Saint Albans *11295*
Association of Community Workers, Newcastle-upon-Tyne *11331*
Association of Directors of Social Services, Stockport *11337*
Association of Directors of Social Work, Melrose *11338*
Association of Social Anthropologists of the Commonwealth, London *11392*
Association of Social Research Organisation, London *11393*
British Association of Social Workers, Birmingham *11523*
British Society for Social Responsibility in Science, London *11667*
The British Sociological Association, Durham *11691*
Economic and Social Science Research Association, London *11926*
Haldane Society, London *12142*
Health Service Social Worker Group, Barnet *12154*
Institute of Community Studies, London *12224*
Institute of Race Relations, London *12276*
Irish Association for Cultural, Economic and Social Relations, Belfast *12410*
National Institute for Social Work, London *12595*
Royal London Aid Society, London *12829*
Shetland Council of Social Service, Lerwick *12950*
Social Research Association, London *12960*
Society for the Study of Labour History, London *13018*

U.S.A.
American Academy of Political and Social Science, Philadelphia *13429*
American Sociological Association, Washington *14175*
The Aspen Institute, Queenstown *14256*
Association for Evolutionary Economics, Lincoln *14295*
Association for the Sociological Study of Jewry, New London *14354*
Association for the Sociology of Religion, Lancaster *14355*
Association of Black Sociologists, Washington *14403*
Association of Muslim Social Scientists, Herndon *14469*
Association of Social and Behavioral Scientists, Washington *14502*
Center for Community Change, Washington *14672*
Center for Social Research and Education, San Francisco *14702*
Center for Urban Black Studies, Berkeley *14710*
Community Development Society, Milwaukee *14810*
Community for Religious Research and Education, Loretto *14812*
Council of International Programs, Washington *14897*
Council on Social Work Education, Alexandria *14932*
Employee Benefit Research Institute, Washington *15039*
Foreign Services Research Institute, Washington *15111*
Global Options, San Francisco *15165*
Human Relations Area Files, New Haven *15237*
Human Resources Research Organization, Alexandria *15240*
Industrial Relations Research Association, Madison *15260*
Institute for Food and Development Policy, San Francisco *15281*
Institute for Policy Studies, Washington *15289*
Institute for Social Research, Ann Arbor *15296*
Institute of Cultural Affairs, Chicago *15315*
International Center for Research on Women, Washington *15430*
International Federation for Family Life Promotion, Washington *15473*
International Institute of Rural Reconstruction – U.S. Chapter, New York *15490*
Inter-University Consortium for Political and Social Research, Ann Arbor *15583*
Jefferson Center for Character Education, Pasadena *15597*
Law and Society Association, Amherst *15653*
National Center on Institutions and Alternatives, Alexandria *15950*
National Congress for Community Economic Development, Washington *15978*
National Council for the Social Studies, Washington *15992*
National Council on Family Relations, Minneapolis *16020*
National Council on Public History, Indianapolis *16024*
National Federation of Societies for Clinical Social Work, Arlington *16048*
National Institute of Social Sciences, New York *16096*
Planned Parenthood Federation of America, New York *16324*
Population Association of America, Washington *16337*
Population Council, New York *16338*
Population Crisis Committee, Washington *16339*
Population Reference Bureau, Washington *16340*
Regional Institute of Social Welfare Research, Athens *16402*
Religious Research Association, Washington *16408*
Rural Sociological Society, Bozeman *16430*
Social Science Education Consortium, Boulder *16484*
Social Sciences Services and Resources, Wasco *16486*
Social Welfare History Group, East Lansing *16487*
Society for Hospital Social Work Directors, Chicago *16532*
Society for the Comparative Study of Society and History, Ann Arbor *16593*
Society for the Psychological Study of Social Issues, Ann Arbor *16608*
Society for the Study of Social Problems, Knoxville *16621*
Society for the Study of Symbolic Interaction, Saint Petersburg *16623*
Southern Regional Council, Atlanta *16727*
Subterranean Sociological Association, Ypsilanti *16751*
Unitarian Universalist Association of Congregations, Washington *16805*
World Population Society, Washington *16930*

Venezuela
Academia de Ciencias Políticas y Sociales, Caracas *16959*

Yugoslavia
Jugoslovensko Udruženje za Sociologiju, Beograd *17103*

Soil Science s. Agriculture

Space Technology s. Aeronautics, Aviation, Space Technology

Speleology

Austria
Landesverein für Höhlenkunde in Wien und Niederösterreich, Wien *00702*
Verband Österreichischer Höhlenforscher, Wien *00969*

Canada
British Columbia Speleological Federation, Gold River *01990*

France
Fédération Française de Spéléologie, Paris *04522*

Germany
Verband der deutschen Höhlen- und Karstforscher e.V., Leinfelden-Echterdingen *06335*

Greece
Elliniki Spilaiologiki Etaireia, Athinai *06515*

Hungary
Magyar Karszt- és Barlangkutató Társulat, Budapest *06642*

Italy
Circolo Speleologico Romano, Roma *07604*
Società Speleologica Italiana, Milano *08015*

Monaco
Association Monégasque de Préhistoire, Monaco *08794*

Peru
Sociedad Peruana de Espeleología, Lima *09482*

Portugal
Sociedade Portuguesa de Espeleologia, Lisboa *09700*

Slovenia

Jamarska zveza Slovenije, Ljubljana *10101*

United Kingdom

British Cave Research Association, Bristol *11540*
National Caving Association, Derby *12579*
University of Bristol Spelaeological Society, Bristol *13175*

U.S.A.

National Speleological Society, Huntsville *16163*

Sports

Australia

The Australian Council for Health, Physical Education and Recreation, Hindmarsh *00280*

Austria

Österreichischer Arbeitskreis für Soziologie des Sports und der Leibeserziehung, Wien *00883*
Österreichischer Sportlehrerverband, Wien *00902*
Österreichische Sportwissenschaftliche Gesellschaft, Salzburg *00916*
Sportwissenschaftliche Gesellschaft der Universität Graz, Graz *00949*
Wissenschaftliche Arbeitsgemeinschaft für Leibeserziehung und Sportmedizin in Innsbruck, Innsbruck *01031*
Wissenschaftliche Gesellschaft für Sport und Leibeserziehung am Institut für Sportwissenschaften der Universität Salzburg, Salzburg *01033*

Belgium

Association Internationale des Ecoles Supérieures d'Education Physique, Liège *01137*
European Association for Research into Adapted Physical Activity, Bruxelles *01280*
Fédération Belge d'Education Physique, Bruxelles *01354*

Canada

American and Canadian Underwater Certifications, Burlington *01858*
Canadian Association of Sport Sciences, Gloucester *02069*
Canadian Council of University Physical Education Administrators, Edmonton *02110*
Physical Education Council of the New Brunswick Teachers Association, Moncton *02817*
Prince Edward Island Physical Education Association, Charlottetown *02837*
Sport Medicine Council of Canada, Vanier *02957*

Denmark

Dansk Idraetslaererforening, Nyborg *03633*
International Association of Physical Education and Sports for Girls and Women, Viborg *03744*

Finland

International Council of Sport Science and Physical Education, Jyväskylä *03967*

France

Association du Sport Scolaire et Universitaire, Paris *04202*
Fédération Française d'Education Physique et de Gymnastique Volontaire, Paris *04520*
Fédération Internationale Catholique d'Education Physique et Sportive, Paris *04530*
Syndicat National des Professeurs d'Arts Martiaux, Paris *04965*
Union Fédérative des Sociétés d'Education Physique et de Préparation Militaire, Paris *04978*

Germany

Arbeitsgemeinschaft der Direktoren der Institute für Leibesübungen an Universitäten und Hochschulen der Bundesrepublik Deutschland, Karlsruhe *05085*
Bundesausschuss für Wissenschaft und Bildung des Deutschen Sportbundes, Frankfurt *05240*
Bundesverband Deutscher Leibeserzieher, Stuttgart *05250*
Deutscher Sportlehrerverband e.V., Wetzlar *05540*
Deutscher Verband für das Skilehrwesen e.V., Oberstdorf *05548*
Deutsche Vereinigung für Sportwissenschaft, Hamburg *05620*
Normenausschuss Sport- und Freizeitgerät im DIN Deutsches Institut für Normung e.V., Köln *06206*

Italy

Associazione Italiana di Psicologia dello Sport, Ferrara *07319*
Centro di Studi e Documentazione delle Ricerche sulla Didattica dell'Educazione Fisica e dello Sport, Napoli *07441*
Centro di Studi per l'Educazione Fisica, Bologna *07451*
Centro Studi di Diritto Sportivo, Vicenza *07554*
Federazione Medico-Sportiva Italiana, Roma *07693*
Federazione Nazionale Insegnanti Educazione Fisica, Roma *07697*

Japan

Asian Society for Adapted Physical Education and Exercise, Tsukuba *08109*
Nippon Taiiku Gakkai, Tokyo *08390*

Netherlands

Koninklijke Vereniging van Leraren en Onderwijzers in de Lichamelijke Opvoeding, Zeist *08961*

Romania

Societatea de Medicinà Sportiva, Bucuresti *09771*

Sweden

Svenska Gymnastikläraresällskapet, Vällingby *10553*

Switzerland

Schweizerischer Verband für Sport in der Schule, Bern *10915*
Vereinigung der Gymnastiklehrer, Zürich *11011*

United Kingdom

Association of Ski Schools in Great Britain, Grantown-on-Spey *11391*
British Association of Advisers and Lecturers in Pysical Education, Exmouth *11500*
Central Council of Physical Recreation, London *11750*
Fédération Internationale d'Education Physique, Cheltenham *12062*
National Federation of Sailing Schools, Lymington *12589*
Physical Education Association of Great Britain and Northern Ireland, Birmingham *12697*
Rugby Football Schools Union, Twickenham *12878*
Scottish Association of Advisers in Physical Education, Dumbarton *12897*
Scottish Physical Education Association, Glasgow *12921*
The Sports Council, London *13088*

U.S.A.

American Academy of Physical Education, Tallahassee *13423*
American Academy of Sports Physicians, Northridge *13438*
American Alliance for Health, Physical Education, Recreation and Dance, Reston *13449*
American Association of Laban Movement Analysts, New York *13545*
American Center for the Alexander Technique, New York *13684*
American Osteopathic Academy of Sports Medicine, Middleton *13979*
American Sports Education Institute, North Palm Beach *14181*
Association for Research, Administration, Professional Councils and Societies, Reston *14327*
The Association of Higher Education Facilities Officers, Alexandria *14443*
International Association of Physical Education and Sports for Girls and Women, Mankato *15416*
Laban/Bartenieff Institute of Movement Studies, New York *15642*
National Association for Physical Education in Higher Education, San Jose *15809*
National Association for Sport and Physical Education, Reston *15815*
National Association of Academic Advisors for Athletics, Charlottesville *15825*
National Council of Athletic Training, Reston *15996*
North American Society for Sport History, University Park *16223*
Philosophic Society for the Study of Sport, Union *16313*
Professional Football Researches Association, North Huntington *16356*
Society of State Directors of Health, Physical Education and Recreation, Kensington *16702*
United States Sports Academy, Daphne *16835*

Standardization

Argentina

Comisión Panamericana de Normas Técnicas, Buenos Aires *00122*

Australia

Standards Association of Australia, North Sydney *00498*

Austria

Österreichisches Normungsinstitut, Wien *00914*

Belgium

Association pour l' Introduction de la Nomenclature Biologique Nouvelle, Kalmthout *01154*
CEN/CENELEC, Bruxelles *01182*
Comité Européen de Normalisation Electrotechnique, Bruxelles *01226*
Comité International de Standardisation en Biologie Humaine, Bruxelles *01230*
European Association of Classification Societies, Antwerpen *01289*
European Committee for Electrotechnical Standardization, Bruxelles *01320*
European Committee for Iron and Steel Standardization, Bruxelles *01322*
Institut Belge de Normalisation, Bruxelles *01379*

Canada

Canadian Classification Research Group, London *02096*
Canadian General Standards Board, Hull *02145*
Canadian Standards Association, Rexdale *02301*
Canadian Welding Bureau, Mississauga *02317*
Indexing and Abstracting Society of Canada, Toronto *02515*
Multiple Dwelling Standards Association, Willowdale *02664*
Ontario Association of Property Standards Officers, Tillsonburg *02761*
Standards Council of Canada, Ottawa *02958*
Standards Engineering Society (Canadian Region), Nepean *02959*

Colombia

Instituto Colombiano de Normas Técnicas, Bogotá *03386*

Egypt

Arab Organization for Standardization and Metrology, Cairo *03875*

Finland

Suomen Standardisoimisliitto, Helsinki *04059*

France

Association Française de Normalisation, Paris La Défence *04221*
Bureau International des Poids et Mesures, Sèvres *04363*
Centre de Coopération pour les Recherches Scientifiques Relatives au Tabac, Paris *04366*
Organisation Internationale de Métrologie Légale, Paris *04615*

Germany

Arbeitsausschuß Wälzlager im DIN Deutsches Institut für Normung e.V., Köln *05078*
Arbeitswissenschaft im Landbau e.V., Stuttgart *05148*
Ausschuss Normenpraxis im DIN Deutsches Institut für Normung e.V., Berlin *05158*
Deutsche Elektrotechnische Kommission im DIN und VDE, Frankfurt *05301*
Deutscher Beton-Verein e.V., Wiesbaden *05505*
DIN Deutsches Institut für Normung e.V., Berlin *05632*
European Collaboration on Measurement Standards, Braunschweig *05674*
Hannoversches Forschungsinstitut für Fertigungsfragen e.V., Hannover *05908*
Institut für Neue Technische Form, Darmstadt *05967*
Normenausschuss Akustik, Lärmminderung und Schwingungstechnik im DIN Deutsches Institut für Normung e.V., Berlin *06147*
Normenausschuss Anstrichstoffe und ähnliche Beschichtungsstoffe im DIN Deutsches Institut für Normung e.V., Berlin *06148*
Normenausschuss Armaturen im DIN Deutsches Institut für Normung e.V., Köln *06149*
Normenausschuss Bauwesen im DIN Deutsches Institut für Normung e.V., Berlin *06150*
Normenausschuss Bergbau im DIN Deutsches Institut für Normung e.V., Essen *06151*
Normenausschuss Bibliotheks- und Dokumentationswesen im DIN Deutsches Institut für Normung e.V., Berlin *06152*
Normenausschuss Bild und Film im DIN Deutsches Institut für Normung e.V., Berlin *06153*
Normenausschuss Bühnentechnik in Theatern und Mehrzweckhallen im DIN Deutsches Institut für Normung e.V., Berlin *06154*
Normenausschuss Bürowesen im DIN Deutsches Institut für Normung e.V., Berlin *06155*
Normenausschuss Chemischer Apparatebau im DIN Deutsches Institut für Normung e.V., Köln *06156*
Normenausschuss Dental im DIN Deutsches Institut für Normung e.V., Pforzheim *06157*
Normenausschuss Dichtungen im DIN Deutsches Institut für Normung e.V., Köln *06158*
Normenausschuss Druckgasanlagen im DIN Deutsches Institut für Normung e.V., Berlin *06159*
Normenausschuss Druck- und Reproduktionstechnik im DIN Deutsches Institut für Normung e.V., Berlin *06160*
Normenausschuss Einheiten und Formelgrössen im DIN Deutsches Institut für Normung e.V., Berlin *06161*
Normenausschuss Eisen-, Blech- und Metallwaren im DIN Deutsches Institut für Normung e.V., Düsseldorf *06162*
Normenausschuß Eisen und Stahl im DIN Deutsches Institut für Normung e.V., Düsseldorf *06163*
Normenausschuss Erdöl- und Erdgasgewinnung im DIN Deutsches Institut für Normung e.V., Köln *06164*
Normenausschuss Ergonomie im DIN Deutsches Institut für Normung e.V., Berlin *06165*
Normenausschuss Fahrräder im DIN Deutsches Institut für Normung e.V., Köln *06166*
Normenausschuss Farbe im DIN Deutsches Institut für Normung e.V., Berlin *06167*
Normenausschuss Feinmechanik und Optik im DIN Deutsches Institut für Normung e.V., Pforzheim *06168*
Normenausschuss Feuerwehrwesen im DIN Deutsches Institut für Normung e.V., Berlin *06169*
Normenausschuss Gastechnik im DIN Deutsches Institut für Normung e.V., Eschborn *06170*
Normenausschuss Giessereiwesen im DIN Deutsches Institut für Normung e.V., Köln *06171*
Normenausschuss Gleitlager im DIN Deutsches Insitut für Normung e.V., Köln *06172*
Normenausschuss Graphische Symbole im DIN Deutsches Institut für Normung e.V., Berlin *06173*
Normenausschuss Grundlagen der Normung im DIN Deutsches Institut für Normung e.V., Berlin *06174*
Normenausschuss Hauswirtschaft im DIN Deutsches Institut für Normung e.V., Berlin *06175*
Normenausschuss Heiz-, Koch- und Wärmegeräte im DIN Deutsches Institut für Normung e.V., Frankfurt *06176*
Normenausschuss Heiz- und Raumlufttechnik im DIN Deutsches Institut für Normung e.V., Berlin *06177*
Normenausschuss Holzwirtschaft und Möbel im DIN Deutsches Institut für Normung e.V., Köln *06178*
Normenausschuß Informationsverarbeitungssyteme im DIN Deutsches Institut für Normung e.V., Berlin *06179*
Normenausschuss Instandhaltung im DIN Deutsches Institut für Normung e.V., Köln *06180*
Normenausschuss Kältetechnik im DIN Deutsches Institut für Normung e.V., Köln *06181*
Normenausschuss Kautschuktechnik im DIN Deutsches Institut für Normung e.V., Frankfurt *06182*
Normenausschuss Kerntechnik im DIN Deutsches Institut für Normung e.V., Berlin *06183*
Normenausschuss Kommunale Technik im DIN Deutsches Institut für Normung e.V., Berlin *06184*
Normenausschuss Kraftfahrzeuge im DIN Deutsches Institut für Normung e.V., Frankfurt *06185*
Normenausschuss Kunststoffe im DIN Deutsches Institut für Normung e.V., Berlin *06186*
Normenausschuss Laborgeräte und Laboreinrichtungen im DIN Deutsches Institut für Normung e.V., Frankfurt *06187*
Normenausschuss Lebensmittel und Landwirtschaftliche Produkte im DIN Deutsches Institut für Normung e.V., Berlin *06188*
Normenausschuss Lichttechnik im DIN Deutsches Institut für Normung e.V., Berlin *06189*
Normenausschuss Maschinenbau im DIN Deutsches Institut für Normung e.V., Frankfurt *06190*
Normenausschuss Materialprüfung im DIN Deutsches Institut für Normung e.V., Berlin *06191*
Normenausschuss Mechanische Verbindungselemente im DIN Deutsches Institut für Normung e.V., Köln *06192*
Normenausschuss Medizin im DIN Deutsches Institut für Normung e.V., Berlin *06193*
Normenausschuss Nichteisenmetalle im DIN Deutsches Institut für Normung e.V., Köln *06194*
Normenausschuss Papier und Pappe im DIN Deutsches Institut für Normung e.V., Berlin *06195*
Normenausschuss Persönliche Schutzausrüstung und Sicherheitskennzeichnung im DIN Deutsches Insitut für Normung e.V., Berlin *06196*
Normenausschuss Pigmente und Füllstoffe im DIN Deutsches Institut für Normung e.V., Berlin *06197*
Normenausschuss Pulvermetallurgie im DIN Deutsches Institut für Normung e.V., Köln *06198*
Normenausschuss Rohre, Rohrverbindungen und Rohrleitungen im DIN Deutsches Institut für Normung e.V., Köln *06199*
Normenausschuss Rundstahlketten im DIN Deutsches Institut für Normung e.V., Köln *06200*
Normenausschuss Schienenfahrzeuge im DIN Deutsches Institut für Normung e.V., Kassel *06201*
Normenausschuss Schmiedetechnik im DIN Deutsches Institut für Normung e.V., Hagen *06202*
Normenausschuss Schmuck im DIN Deutsches Insitut für Normung e.V., Pforzheim *06203*
Normenausschuss Schweisstechnik im DIN Deutsches Institut für Normung e.V., Berlin *06204*
Normenausschuss Siebböden und Kornmessung im DIN Deutsches Institut für Normung e.V., Berlin *06205*
Normenausschuss Sport- und Freizeitgerät im DIN Deutsches Institut für Normung e.V., Köln *06206*
Normenausschuss Stahldraht und Stahldrahterzeugnisse im DIN Deutsches Institut für Normung e.V., Köln *06207*
Normenausschuss Terminologie im DIN Deutsches Institut für Normung e.V., Berlin *06208*
Normenausschuss Textil und Textilmaschinen im DIN Deutsches Institut für Normung e.V., Berlin *06209*
Normenausschuss Transportkette im DIN Deutsches Institut für Normung e.V., Berlin *06210*
Normenausschuss Überwachungsbedürftige Anlagen im DIN Deutsches Institut für Normung e.V., Köln *06211*
Normenausschuss Uhren und Schmuck im DIN Deutsches Institut für Normung e.V., Pforzheim *06212*
Normenausschuss Vakuumtechnik im DIN Deutsches Institut für Normung e.V., Köln *06213*
Normenausschuss Verpackungswesen im DIN Deutsches Institut für Normung e.V., Berlin *06214*
Normenausschuss Waagenbau im DIN Deutsches Institut für Normung e.V., Berlin *06215*
Normenausschuss Wärmebehandlungstechnik metallischer Werkstoffe im DIN Deutsches Institut für Normung e.V., Köln *06216*
Normenausschuss Wasserwesen im DIN Deutsches Institut für Normung e.V., Berlin *06217*
Normenausschuss Werkzeuge und Spannzeuge im DIN Deutsches Institut für Normung e.V., Köln *06218*
Normenausschuss Werkzeugmaschinen im DIN Deutsches Institut für Normung e.V., Frankfurt *06219*
Normenausschuss Zeichnungswesen im DIN Deutsches Institut für Normung e.V., Berlin *06220*

Greece

Ellinikos Organismos Tupopoiesseos, Athinai *06517*

India

Indian Standards Institution, New Dehli *06822*
Research Designs and Standards Organization, Lucknow *06853*

Indonesia
Jajasan Dana Normalisasi Indonesia, Bandung *06891*
Yayasan Dana Normalisasi Indonesia, Bandung *06895*

Iran
Medical Nomenclature Society of Iran, Isfahan *06900*

Italy
Ente Nazionale Italiano di Unificazione, Milano *07656*

Japan
Nihon Kikaku Kyokai Gaikoku Kikaku Raiburari, Tokyo *08223*

Kenya
African Regional Organization for Standardization, Nairobi *08472*

Norway
Norges Standardiseringsforbund, Oslo *09287*
Standardiseringsforeningen, Oslo *09350*

Russia
Council for Metrological Provision and Standardization, St. Petersburg *09860*
Institute for Standardization, Moskva *09875*

Sweden
Standardiseringskommissionen i Sverige, Stockholm *10522*

Switzerland
International Organization for Standardization, Genève *10753*
Schweizerische Normen-Vereinigung, Zürich *10876*

Turkey
Türk Standartlari Enstitüsü, Ankara *11174*

United Kingdom
British Measures Group, Teddington *11606*
British Standards Institution, London *11692*
The Dozenal Society of Great Britain, Moulsford *11904*
European Council for Clinical and Laboratory Standardization, Coventry *12020*
International Anatomical Nomenclature Committee, London *12322*
International Association of Biological Standardization, Hassocks *12328*
International Commission on Zoological Nomenclature, London *12345*
Society of Indexers, London *13052*

Uruguay
Instituto Uruguayo de Normas Técnicas, Montevideo *13262*

U.S.A.
American National Standards Institute, New York *13954*
American Society for Testing and Materials, Philadelphia *14089*
Dozenal Society of America, Garden City *14983*
International Association for the Properties of Water and Steam, Palo Alto *15394*
Joint Industry Council, Champaign *15619*
Manufacturers Standardization Society of the Valve and Fittings Industry, Vienna *15689*
National Conference of Regulatory Utility Commission Engineers, Nashville *15970*
National Conference of Standards Laboratories, Boulder *15971*
National Conference of States on Building Codes and Standards, Herndon *15972*
National Conference on Weights and Measures, Gaithersburg *15977*
National Standard Plumbing Code Committee, Falls Church *16164*
Society of American Value Engineers, Northbrook *16638*
Standards Engineering Society, Dayton *16744*

Yugoslavia
Jugoslovenski Zavod za Standardizaciju, Beograd *17097*

Zimbabwe
Standards Association of Central Africa, Harare *17181*

Statistics

Australia
Statistical Society of Australia, Canberra *00499*

Austria
Österreichische Gesellschaft für Statistik, Wien *00834*
Österreichische Statistische Gesellschaft, Wien *00919*

Belgium
Conseil Supérieur de Statistique, Bruxelles *01254*

Brazil
Fundação Instituto Brasileiro de Geografia e Estatística, Rio de Janeiro *01655*

Canada
Statistical Society of Canada, Hamilton *02960*

Chile
Centro Interamericano de Enseñanza de Estadística, Santiago *03024*
Centro Latinoamericano de Demografía, Santiago *03026*
Comisíon Económica para América Latina y el Caribe, Santiago *03038*

China, Republic
Chinese Statistical Association, Taipei *03288*

Colombia
Departamento Administrativo Nacional de Estadística, Bogotá *03373*

Denmark
Dansk Selskab for Teoretisk Statistik, København *03688*
Nordisk Statistisk Sekretariat, København *03791*

Egypt
Cairo Demographic Centre, Cairo *03883*
Egyptian Society of Political Economy, Statistics and Legislation, Cairo *03901*

El Salvador
Dirección General de Estadística y Censos, San Salvador *03928*

Finland
Suomen Tilastoseura, Helsinki *04065*
Suomen Väestötieteen Yhdistys, Helsinki *04066*

France
Association Internationale des Statisticiens d'Enquêtes, Paris *04278*
Centre Européen de Formation des Statisticiens Economistes des Pays en Voie de Développement, Malakoff *04371*
Société de Statistique de Paris, Paris *04719*
Société de Statistique, d'Histoire et d'Archéologie de Marseille et de Provence, Marseille *04720*

Germany
Arbeitsgemeinschaft Allensbach e.V., Allensbach *05079*
Deutsche Gesellschaft für Bevölkerungswissenschaft e.V., Wiesbaden *05330*
Deutsche Gesellschaft für Medizinische Informatik, Biometrie und Epidemiologie, Köln *05382*
Deutsche Statistische Gesellschaft, Konstanz *05599*

Ghana
African Commission on Agricultural Statistics, Accra *06470*

Iceland
Félag Islenskra Tryggingastaerdfraedinga, Reykjavik *06685*

India
Calcutta Statistical Association, Calcutta *06751*

Iraq
Arab Institute for Training and Research in Statistics, Baghdad *06912*

Ireland
Institute of Chartered Accountants in Ireland, Dublin *06960*
Statistical and Social Inquiry Society of Ireland, Dublin *07035*

Italy
Associazione degli Statistici, Udine *07267*
Centro di Documentazione Economica per Giornalisti, Roma *07421*
Centro di Documentazione Statistica Internazionale, Bologna *07425*
Centro Italiano di Biostatistica, Roma *07491*
Società Italiana di Economia Demografia e Statistica, Roma *07865*
Società Italiana di Statistica, Roma *07951*
SVP Italia, Milano *08030*

Japan
Keizai Tokei Kenkyukai, Tokyo *08161*
Nihon Tokei Gakkai, Tokyo *08270*
Tokei Kagaku Kenkyukai, Fukuoka *08435*

Mexico
Sociedad Mexicana de Geografía y Estadística, México *08764*
Sociedad Nuevoleonesa de Historia, Geografía y Estadística, Monterrey *08775*

Morocco
Société d'Etudes Economiques, Sociales et Statistiques du Maroc, Rabat *08809*

Netherlands
Bernoulli Society for Mathematical Statistics and Probability, Voorburg *08833*
International Association for Official Statistics, Voorburg *08920*
International Association for Statistical Computing, Voorburg *08921*
International Statistical Institute, Voorburg *08944*
Vereniging voor Statistiek, Heiloo *09104*

New Zealand
New Zealand Statistical Association, Wellington *09181*

Nigeria
African Statistical Association, Lagos *09221*

Norway
Den Norske Aktuarforening, Oslo *09251*
Statistics Norway, Oslo *09351*

Panama
Inter-American Statistical Institute, Panamá City *09420*

Rwanda
African and Mauritian Institute of Statistics and Applied Economics, Kigali *09993*

Slovakia
Slovenská Demografická a Štatistická Spoločnosť, Bratislava *10069*

South Africa
The South African Statistical Association, Sunnyside *10221*

Spain
Instituto Nacional de Estadística, Madrid *10326*
Sociedad de Estadística e Investigación Operativa, Madrid *10373*

Sweden
Statistika Föreningen, Stockholm *10525*

Switzerland
Conférence des Statisticiens Européens, Genève *10654*
Schweizerische Gesellschaft für Statistik und Volkswirtschaft, Bern *10849*

Tanzania
Eastern Africa Statistical Training Centre, Dar es Salaam *11055*

Thailand
Asia and Pacific Commission on Agricultural Statistics, Bangkok *11072*

United Kingdom
Association of Track and Field Statisticians, Salisbury *11399*
British Universities Industrial Relations Association, Manchester *11701*
Institute of Population Registration, London *12269*
Institute of Statisticians, Preston *12284*
Organisation of Professional Users of Statistics, Esher *12659*
Royal Statistical Society, London *12866*
World Bureau of Metal Statistics, London *13217*

Uruguay
Centro de Nacionales y Comercio Internacional del Uruguay, Montevideo *13253*

U.S.A.
American Statistical Association, Alexandria *14182*
Association for Jewish Demography and Statistics – American Branch, Los Angeles *14310*
Center for Advanced Study in the Behavioral Sciences, Stanford *14663*
Econometric Society, Evanston *15009*
Institute for Econometric Research, Fort Lauderdale *15276*
Institute of Mathematical Statistics, Hayward *15331*
Tax Analysts, Arlington *16764*

Stomatology

Austria
Österreichische Gesellschaft für Zahn-, Mund- und Kieferheilkunde, Wien *00850*

Belgium
Groupement International pour la Recherche Scientifique en Odontologie et en Stomatologie, Bruxelles *01371*
Groupement International pour la Recherche Scientifique en Stomatologie et Odontologie, Bruxelles *01372*
Société Royale Belge de Stomatologie et Chirurgie Maxillo-Faciale, Gent *01479*

Czech Republic
Česká Stomatologická Společnost, Praha *03532*

France
Association Française pour le Développement de la Stomatologie, Paris *04243*
Association Internationale pour le Développement de l'Odonto-Stomatologie Tropicale, Bordeaux *04288*
Association pour le Développement de la Stomatologie, Paris *04335*
Association Stomatologique Internationale, Paris *04349*
Société de Stomatologie de France, Paris *04721*
Syndicat National des Médecins Spécialistes en Stomatologie et Chirurgie Maxillo-Faciale, Paris *04961*

Greece
Etaireia Odontostomatologikis Ereunis, Athinai *06526*

Italy
Associazione Medici Dentisti Italiani, Roma *07357*
Società Italiana di Ergonomia Stomatologica, Milano *07870*
Società Italiana di Odontostomatologia e Chirurgia Maxillo-Facciale, L'Aquila *07919*

Japan
Nihon Kokukai Gakkai, Tokyo *08231*

Nigeria
The Medical and Dental Council of Nigeria, Lagos *09232*

Peru
Academia de Estomatología del Perú, Lima *09437*

Poland
Polskie Towarzystwo Stomatologiczne, Lódź *09606*

Portugal
Sociedade Portuguese de Estomatologia, Lisboa *09717*

Romania
Societatea de Stomatologie, Bucuresti *09787*

Russia
National Scientific Medical Society of Stomatologists, Moskva *09933*

Spain
Consejo General de Colegios Oficiales de Odontólogos y Estomatólogos de España, Madrid *10299*

Surgery

Argentina
Academia Argentina de Cirugía, Buenos Aires *00037*
Asociación Argentina de Cirugía, Buenos Aires *00061*
Sociedad Argentina de Ciencias Neurológicas, Psiquiátricas y Neuroquirúrgicas, Buenos Aires *00162*
Sociedad Argentina de Investigación Clínica, Buenos Aires *00173*
Sociedad Argentina de Neurología, Psiquiatría y Neurocirugía, Buenos Aires *00178*
Sociedad de Cirugía de Buenos Aires, Buenos Aires *00191*

Australia
Asian-Australasian Society of Neurological Surgeons, Brisbane *00221*
Royal Australasian College of Dental Surgeons, Sydney *00455*
Royal Australasian College of Surgeons, Melbourne *00458*

Austria
Austrotransplant – Österreichische Gesellschaft für Transplantation, Transfusion und Genetik, Wien *00572*
European Bone Marrow Transplant Group, Wien *00587*
Gesellschaft der Chirurgen in Wien, Wien *00605*
Österreichische Gesellschaft für Chirurgie, Wien *00769*
Österreichische Gesellschaft für Chirurgische Forschung, Wien *00770*
Österreichische Gesellschaft für Gefässchirurgie, Salzburg *00782*
Österreichische Gesellschaft für Hals-, Nasen- und Ohrenheilkunde, Kopf- und Halschirurgie, Wien *00789*
Österreichische Gesellschaft für Kinderchirurgie, Graz *00798*
Österreichische Gesellschaft für Neurochirurgie, Wien *00814*
Österreichische Gesellschaft für Orthopädie und Orthopädische Chirurgie, Wien *00818*
Österreichische Gesellschaft für Unfallchirurgie, Wien *00837*

Belgium
Belgische Wetenschappelijke Vereniging voor Neurochirurgie, Brugge *01164*
Société Belge de Chirurgie Orthopédique et de Traumatologie, Hony-Esneux *01428*
Société Internationale de Chirurgie Orthopédique et de Traumatologie, Bruxelles *01462*
Société Royale Belge de Chirurgie, Bruxelles *01471*
Société Royale Belge de Stomatologie et Chirurgie Maxillo-Faciale, Gent *01479*

Bolivia
Sociedad Boliviana de Cirugía, La Paz *01549*

Bulgaria
Bulgarian Society of Neurosurgery, Sofia *01760*

Canada
Association des Chirurgiens Généraux de la Province de Québec, Montréal *01878*
Canadian Academy of Urological Surgeons, Vancouver *02010*
Canadian Association of Paediatric Surgeons, London *02061*
Canadian Chiropractic Association, Toronto *02095*
Canadian Institute of Facial Plastic Surgery, Toronto *02172*
Canadian Neurosurgical Society, Toronto *02210*
Canadian Society of Otolaryngology, Head and Neck Surgery, Islington *02289*
Canadian Society of Plastic Surgeons, Willowdale *02293*
College of Physicians and Surgeons of Alberta, Edmonton *02355*
College of Physicians and Surgeons of British Columbia, Vancouver *02356*
College of Physicians and Surgeons of Manitoba, Winnipeg *02357*
College of Physicians and Surgeons of New Brunswick, Saint John *02358*
College of Physicians and Surgeons of Ontario, Toronto *02359*
College of Physicians and Surgeons of Saskatchewan, Saskatoon *02360*
Royal College of Physicians and Surgeons of Canada, Ottawa *02869*

Chile
Fundación Gildemeister, Santiago *03052*
Sociedad Chilena de Cirugía Plástica y Reparadora, Santiago *03069*

Sociedad de Cirujanos de Chile, Santiago 03092

China, Republic

Surgical Association of the Republic of China, Taipei 03331

Colombia

Colegio Colombiano de Cirujanos, Bogotá 03363
Instituto Nacional de Cancerología, Bogotá 03392
Sociedad Colombiana de Cirugía, Bogotá 03406

Costa Rica

Asociación Costarricense de Cirurgíca, San José 03431

Czech Republic

Czech Surgical Society, Praha 03542

Denmark

Dansk Kirurgisk Selskab, Hørsholm 03637
Nordisk Neurokirurgisk Forening, Arhus 03789

Egypt

Eastern Mediterranean Hand Society, Cairo 03886

Finland

Suomen Kirurgiyhdistys, Helsinki 04041

France

Académie de Chirurgie, Paris 04088
Académie Nationale de Chirurgie Dentaire, Paris 04116
Association Française de Chirurgie, Paris 04213
European Association of Plastic Surgeons, Paris 04489
Fédération des Chambres Syndicales des Chirurgiens Dentistes de la Région de Paris, Paris 04512
Ordre des Chirurgiens Dentistes, Paris 04608
Ordre National des Chirurgiens Dentistes, Paris 04610
Société de Chirurgie de Marseille, Marseille 04648
Société de Chirurgie de Toulouse, Toulouse 04649
Société de Chirurgie Thoracique et Cardio-Vasculaire de Langue Française, Le Plessis Robinson 04650
Société de Médecine, Chirurgie et Pharmacie de Toulouse, Toulouse 04666
Société de Médecine et de Chirurgie de Bordeaux, Bordeaux 04668
Société de Neurochirurgie de Langue Française, Lyon 04674
Société de Stomatologie de France, Paris 04721
Société de Transplantation, Paris 04725
Société Française de Chirurgie Orthopédique et Traumatologique, Paris 04792
Société Française de Chirurgie Pédiatrique, Lyon 04793
Société Française de Chirurgie Plastique et Reconstructive, Paris 04794
Société Française de Médecine Esthétique, Courbevoie 04810
Société Internationale de Podologie Médico-Chirurgicale, Cannes-la-Bocca 04878
Société Médico-Chirurgicale des Hôpitaux et Formations Sanitaires des Armées, Paris 04893
Société Médico-Chirurgicale des Hôpitaux Libres, Paris 04894
Syndicat des Chirurgiens Dentistes de Paris, Paris 04924
Syndicat National des Chirurgiens Français, Paris 04940
Syndicat National des Chirurgiens Plasticiens, Paris 04941

Germany

Arbeitsgemeinschaft für Kieferchirurgie innerhalb der Deutschen Gesellschaft für Zahn-, Mund- und Kieferheilkunde, Kiel 05106
Berufsverband der Deutschen Chirurgen, Hamburg 05195
Bundesverband Deutscher Ärzte für Mund-Kiefer-Gesichtschirurgie e.V., Hamburg 05248
Deutsche Gesellschaft für Chirurgie, Bonn 05335
Deutsche Gesellschaft für Hals-Nasen-Ohren-Heilkunde, Kopf- und Hals-Chirurgie, Bonn 05359
Deutsche Gesellschaft für Mund-, Kiefer- und Gesichtschirurgie, Bonn 05386
Deutsche Gesellschaft für Neurochirurgie, Essen 05387
Deutsche Gesellschaft für Plastische und Wiederherstellende Chirurgie e.V., Rotenburg 05406
Deutsche Gesellschaft für Thorax-, Herz- und Gefässchirurgie, Bad Nauheim 05431
Deutsche Gesellschaft für Unfallheilkunde e.V., Frankfurt 05432
Deutschsprachige Arbeitsgemeinschaft für Handchirurgie, Hamburg 05631
European Association of Neurosurgical Societies, Berlin 05669
Fachschaft Berliner Chirurgen e.V., Berlin 05693
Vereinigung der Bayerischen Chirurgen e.V., Altötting 06396

Greece

Elliniki Cheirourgiki Etaireia, Athinai 06505

Hong Kong

Asian Surgical Association, Hong Kong 06592

Hungary

Magyar Sebész Társaság, Budapest 06657

Iceland

Skurdlaeknafélag Islands, Reykjavik 06699

India

Association of Surgeons of India, Madras 06735

Ireland

Royal College of Surgeons in Ireland, Dublin 07023

Israel

Israel Association of Plastic Surgeons, Ramat-Gan 07066
Society of Orthopedic Surgeons of the Israel Medical Association, Tel Aviv 07113

Italy

Accademia Anatomico-Chirurgica, Perugia 07129
Accademia Medica Pistoiese Filippo Pacini, Pistoia 07206
Accademia Medico-Chirurgica del Piceno, Ancona 07207
European Dialysis and Transplant Association – European Renal Association, Parma 07678
Società Italiana di Chirurgia, Roma 07850
Società Italiana di Chirurgia Cardiaca e Vascolare, Roma 07851
Società Italiana di Chirurgia Clinica, Roma 07852
Società Italiana di Chirurgia della Mano, Firenze 07853
Società Italiana di Chirurgia d'Urgenza, di Pronto Soccorso e di Terapia Intensiva Chirurgica, Milano 07854
Società Italiana di Chirurgia Estetica, Roma 07855
Società Italiana di Chirurgia Pediatrica, Roma 07856
Società Italiana di Chirurgia Plastica, Verona 07857
Società Italiana di Chirurgia Toracica, Roma 07858
Società Italiana di Ingegneria, Aerofotogrammetria e Topografia, Roma 07888
Società Italiana di Medicina del Traffico, Roma 07894
Società Italiana di Neurochirurgia, Ancona 07910
Società Italiana di Odontostomatologia e Chirurgia Maxillo-Facciale, L'Aquila 07919
Società Italiana Medico-Chirurgica di Pronto Soccorso, Bologna 07960
Società Italiana Organi Artificiali, Roma 07961
Società Jonico-Salentina de Medicina e Chirurgia, Taranto 07976
Società Medica Chirurgica di Bologna, Bologna 07983
Società Medico-Chirugica, Bari 07984
Società Medico-Chirurgica di Ferrara, Ferrara 07985
Società Medico-Chirurgica di Modena, Modena 07986
Società Napoletana di Chirurgia, Napoli 07988
Società Romana di Chirurgia, Roma 08009

Japan

Nihon Koko Geka Gakkai, Tokyo 08229
Nihon Kyobu Geka Gakkai, Tokyo 08236
Nihon Noshinkei Geka Gakkai, Tokyo 08245
Nippon Geka Gakkai, Tokyo 08301
Nippon No-Shinkei Gek Gakkai, Tokyo 08351

Kenya

Association of Surgeons of East Africa, Nairobi 08481

Mexico

Asociación Panamericana de Cirurgía Pediátrica, México 08706

Netherlands

Nederlandse Vereniging van Neurochirurgen, Rotterdam 09008
Nederlandse Vereniging voor Heelkunde, Utrecht 09020
Nederlandse Vereniging voor Mondziekten en Kaakchirurgie, 's-Gravenhage 09028
Nederlandse Vereniging voor Thoraxchirurgie, Groningen 09038
World Federation of Neurosurgical Societies, Nijmegen 09111

Norway

Norsk Kirurgisk Forening, Lysaker 09322

Pakistan

Defence Housing Society, Karachi 09362

Peru

Academia Peruana de Cirurgía, Lima 09440

Poland

Towarzystwo Chirurgów Polskich, Warszawa 09619

Romania

Societatea de Chirurgie, Bucuresti 09758
Societatea Romana de Neurochirurgie, Bucuresti 09793

Russia

National Scientific Medical Society of Neurosurgeons, Moskva 09922
National Scientific Medical Society of Surgeons, Moskva 09934
National Scientific Medical Society of Traumatic Surgeons and Orthopaedists, Moskva 09936
National Scientific Society of Urological Surgeons, Moskva 09939

Singapore

Asian-Pacific Association for Laser Medicine and Surgery, Singapore 10038
Asian Pacific Section of the International Confederation for Plastic and Reconstructive Surgery, Singapore 10041

South Africa

Association of Surgeons of South Africa, Johannesburg 10124

Spain

Academia Médico-Quirúrgica Española, Madrid 10244
Asociación Española de Cirujanos, Madrid 10260
Real Academia de Medicina y Cirugía de Palma de Mallorca, Palma de Mallorca 10351
Sociedad Española de Cirugía Oral y Maxillofacial, Madrid 10382
Sociedad Española de Cirugía Ortopédica y Traumatología, Madrid 10383
Sociedad Española de Cirugía Plástica, Reparadora y Estética, Madrid 10384
Sociedad Luso-Española de Neurocirugía, Madrid 10400
World Federation of Associations of Pediatric Surgeons, Barcelona 10404

Sweden

Svensk Kirurgisk Förening, Uppsala 10583

Switzerland

European Association for Cardio-Thoracic Surgery, Zürich 10672
European Council of Coloproctology, Genève 10682
Schweizerische Gesellschaft für Oto-Rhino-Laryngologie, Hals- und Gesichtschirurgie, Bern 10840
Société Internationale de Chirurgie, Reinach 10976
Société Suisse de Chirurgie, Genève 10983

Thailand

ASEAN Federation of Plastic Surgeons, Bangkok 11068

Turkey

Türk Cerrahi Cemiyeti, Istanbul 11152

United Kingdom

Association of British Dental Surgery Assistants, Fleetwood 11309
Association of Surgeons of Great Britain and Ireland, London 11394
British Association for Accident and Emergency Medicine, London 11484
British Association of Clinical Anatomists, Manchester 11503
British Association of Cosmetic Surgeons, London 11504
British Association of Hair Transplant Surgeons, Birmingham 11507
British Association of Oral and Maxillo-Facial Surgeons, London 11513
British Association of Paediatric Surgeons, Edinburgh 11517
British Association of Plastic Surgeons, London 11519
British Association of Surgical Oncology, London 11525
British Association of Urological Surgeons, London 11528
British Institute of Surgical Technologists, Sevenoaks 11590
British Society for Surgery of the Hand, London 11669
European Academy of Facial Surgery, London 11983
European Association for Cranio-Maxillo-Facial Surgery, London 11989
Faculty of Dental Surgery, London 12051
International Federation of Surgical Colleges, Edinburgh 12361
National Society for Transplant Surgery, Cardiff 12611
Royal College of Physicians and Surgeons of Glasgow, Glasgow 12786
Royal College of Surgeons of Edinburgh, Edinburgh 12791
The Royal College of Surgeons of England, London 12792
Royal Medico-Chirurgical Society of Glasgow, Glasgow 12831
Society of British Neurological Surgeons, Glasgow 13031
Society of Thoracic and Cardiovascular Surgeons of Great Britain and Ireland, Conventry 13068
Surgical Research Society, Glasgow 13109
Transplantation Society, London 13140

Uruguay

Ateneo de Clínica Quirúrgica, Montevideo 13250
Sociedad de Cirugía del Uruguay, Montevideo 13267

U.S.A.

Academy of Ambulatory Foot Surgery, Tuscaloosa 13280
American Academy of Cosmetic Surgery, Arcadia 13385
American Academy of Dental Electrosurgery, New York 13387
American Academy of Facial Plastic and Reconstructive Surgery, Washington 13393
American Academy of Neurological Surgery, Philadelphia 13408
American Academy of Orthopaedic Surgeons, Rosemont 13416
American Association for Hand Surgery, Chicago 13481
American Association for Thoracic Surgery, Manchester 13503
American Association of Genito-Urinary Surgeons, Houston 13536
American Association of Neurological Surgeons, Park Ridge 13555
American Association of Plastic Surgeons, La Jolla 13568
American Association of Railway Surgeons, Daleville 13578
American Association of Tissue Banks, McLean 13593
American Board of Abdominal Surgery, Melrose 13626
American Board of Colon and Rectal Surgery, Taylor 13630
American Board of Neurological Surgery, Houston 13639
American Board of Orthopedic Surgery, Chapel Hill 13647
American Board of Plastic Surgery, Philadelphia 13654
American Board of Podiatric Surgery, San Francisco 13656
American Board of Surgery, Philadelphia 13663
American Board of Thoracic Surgery, Evanston 13664
American Clinical and Climatological Association, Charleston 13693
American College of Cryosurgery, Schaumburg 13703
American College of Foot Surgeons, Park Ridge 13707
American College of General Practitioners in Osteopathic Medicine and Surgery, Arlington Heights 13709
American College of Mohs' Micrographic Surgery and Cutaneous Oncology, Schaumburg 13715
American College of Osteopathic Surgeons, Alexandria 13728
American College of Surgeons, Chicago 13738
American Fracture Association, Bloomington 13825
American Laryngological Association, Rochester 13910
American Society for Aesthetic Plastic Surgery, Long Beach 14045
American Society for Artificial Internal Organs, Boca Raton 14046
American Society for Dermatologic Surgery, Schaumburg 14057
American Society for Head and Neck Surgery, Baltimore 14063
American Society for Laser Medicine and Surgery, Wausau 14069
American Society for Surgery of the Hand, Aurora 14087
American Society of Abdominal Surgery, Melrose 14095
American Society of Cataract and Refractive Surgery, Fairfax 14104
American Society of Colon and Rectal Surgeons, Palatine 14111
American Society of Contemporary Medicine and Surgery, Chicago 14113
American Society of Maxillofacial Surgeons, Arlington Heights 14144
American Society of Plastic and Reconstructive Surgeons, Arlington Heights 14157
American Surgical Association, Chapel Hill 14186
Association for Academic Surgery, Minneapolis 14262
Association for Surgical Education, Salt Lake City 14337
Association of American Physicians and Surgeons, Tucson 14385
Association of Bone and Joint Surgeons, Rosemont 14405
Association of Military Surgeons of the U.S., Bethesda 14465
Association of Physician's Assistants in Cardiovascular Surgery, Lynchburg 14483
Association of Surgical Technologists, Englewood 14511
Collegium Internationale Chirurgiae Digestivae, Milwaukee 14781
Congress of Neurological Surgeons, Cincinnati 14842
International Academy of Chest Physicians and Surgeons, Northbrook 15364
International Association of Ocular Surgeons, Chicago 15413
International College of Surgeons, Chicago 15439
International Society for Cardiovascular Surgery, Manchester 15527
International Society for Pediatric Neurosurgery, Salt Lake City 15537
International Society of Dermatologic Surgery, Schaumburg 15549
Lyman A. Brewer International Surgical Society, Los Angeles 15684
National College of Foot Surgeons, Woodland Hills 15958
Neurosurgical Society of America, Cleveland 16193
Pan-Pacific Surgical Association, Honolulu 16294
Plastic Surgery Research Council, Chapel Hill 16326
Society for Surgery of the Alimentary Tract, New York 16583
Society for Vascular Surgery, Manchester 16626
Society of Head and Neck Surgeons, Arlington 16671
Society of Neurological Surgeons, Boston 16685
Society of Pelvic Surgeons, New York 16688
Society of Philippine Surgeons of America, Roanoke 16691
Society of Thoracic Surgeons, Chicago 16707
Society of United States Air Force Flight Surgeons, Brooks Air Force Base 16710
Society of University Surgeons, New Haven 16711
Western Surgical Association, Rochester 16887
World Society for Stereotactic and Functional Neurosurgery, Houston 16932

Venezuela

Sociedad Venezolana de Cirugía, Caracas 17019
Sociedad Venezolana de Cirugía Plástica y Reconstrucción, Caracas 17020
Sociedad Venezolana de Cirugía Ortopédica y Traumatología, Caracas 17021

Vietnam

Vietnamese Association of Neurology, Psychiatry and Neurosurgery, Hanoi 17053
Vietnamese Association of Surgery, Hanoi 17066

Surveying, Photogrammetry s. a. Cartography

Argentina
Asociación Argentina de Geofísicos y Geodestas, Buenos Aires *00065*

Australia
Australian Institute of Quantity Surveyors, Deakin *00312*

Austria
Österreichische Kommission für Internationale Erdmessung, Wien *00862*
Österreichischer Verein für Vermessungswesen und Photogrammetrie, Wien *00907*

Belgium
Société Belge de Photogrammétrie, de Télédétection et de Cartographie, Bruxelles *01440*
Union Belge des Géomètres-Experts Immobiliers, Bruxelles *01493*

Bulgaria
Union of Geodesy and Cartography, Sofia *01794*

Canada
Canadian Institute of Surveying, Ottawa *02184*

Chile
Comité Nacional de Geografía, Geodesía y Geofísica, Santiago *03041*

China, People's Republic
Chinese Society of Geodesy, Photogrammetry and Cartography, Beijing *03191*

Colombia
Centro Interamericano de Fotointerpretación, Bogotá *03358*
Servicio Interamericana de Geodesía, Bogotá *03400*

Cyprus
Cyprus Photogrammetric Society, Nicosia *03495*

Czech Republic
Česká Vědecko-Technická Společnost pro Geodezii a Kartografii, Praha *03535*

Denmark
Dansk Selskab for Fotogrammetri og Landmåling, Ålborg *03675*

France
Association Internationale de Géodésie, Paris *04257*
Comité National Français de Géodesie et Géophysique, Paris *04416*
Ordre des Géomètres-Experts, Paris *04609*
Société Française de Photogrammétrie et de Télédétection, Saint-Mandé *04833*

Germany
Bayerische Kommission für die Internationale Erdmessung, München *05172*
Deutsche Geodätische Kommission, München *05310*
Deutsche Gesellschaft für Photogrammetrie und Fernerkundung, Neubiberg *05402*
Deutscher Verein für Vermessungswesen e.V., Heidelberg *05559*
GEO-KART, Oberkochen *05780*
Institut für Angewandte Geodäsie, Frankfurt *05947*

Hungary
Geodéziai és Kartográfiai Egyesület, Budapest *06605*

Ireland
Royal Institution of Chartered Surveyors, Dublin *07027*

Italy
Società di Fotogrammetria e Topografia, Milano *07782*
Società Italiana di Fotogrammetria e Topografia, Milano *07878*

Japan
Asian Association on Remote Sensing, Tokyo *08097*
Nippon Shashin Sokuryo Gakkai, Tokyo *08372*
Nippon Sokuchi Gakkai, Ibaraki *08385*

Malawi
Geological Survey of Malawi, Zomba *08622*

Malaysia
Geological Survey of Malaysia, Ipoh *08642*

Netherlands
Nederlandse Vereniging voor Geodesie, Apeldoorn *09019*

Portugal
Associação Portuguesa de Fotogrametria, Lisboa *09645*

South Africa
Aerial Survey, Photogrammetric and Remote Sensing Research Group, Durban *10118*

Switzerland
Fachgruppe für Vermessung und Kulturtechnik (Deutsch-Schweiz) des STV, Weiningen *10698*
Schweizerischer Verein für Vermessung und Kulturtechnik, Solothurn *10921*
Verband Schweizerischer Vermessungs-Techniker, Burgdorf *11008*

United Kingdom
Architects and Surveyors Institute, Chippenham *11242*
Association of British Geodesists, Dagenham *11312*
Commonwealth Association on Surveying and Land Economy, London *11813*
London Subterranean Survey Association, London *12469*
The Photogrammetric Society, London *12696*
The Royal Institution of Chartered Surveyors, London *12820*

U.S.A.
American Association for Geodetic Surveying, Bethesda *13479*
American Congress on Surveying and Mapping, Bethesda *13753*
American Society for Photogrammetry and Remote Sensing, Bethesda *14080*
Management Association for Private Photogrammetric Surveyors, Reston *15687*
National Association of Marine Surveyors, Chesapeake *15872*
Surveyors Historical Society, Lansing *16756*

Zimbabwe
Survey Institute of Zimbabwe, Harare *17182*

Tea s. Coffee, Tea, Cocoa

Textiles

Argentina
Instituto Nacional de Tecnologia Industrial, Buenos Aires *00145*

Australia
Australian Road Research Board, Melbourne *00335*
Australian Wool Corporation, Parkville *00363*
Textile Society of Australia, Melbourne *00510*

Belgium
Association Internationale des Laboratoires Textiles Lainiers, Bruxelles *01140*
European Association for Textile Polyolefins, Bruxelles *01281*

Bulgaria
Union of Textiles, Clothing and Leather, Sofia *01800*

Cameroon
Compagnie Française pour le Développement des Fibres Textiles, Douala *01813*

Canada
Canadian Apparel Manufacturers Institute, Ottawa *02022*
Institute of Textile Science, Montréal *02532*

China, People's Republic
Shanghai Textile Engineering Society, Shanghai *03216*
Textile Academy, Beijing *03222*
Wuhsi Academy of Textile Engineering, Wuhsi *03223*

France
Association Interprofessionnelle pour la Formation Permanente dans le Commerce Textile, Paris *04296*
Centre de Perfectionnement des Industries Textiles Rhône-Alpes, Lyon *04367*
Centre International d'Etude des Textiles Anciens, Lyon *04377*

Germany
Deutsches Textilforschungszentrum Nord-West e.V., Krefeld *05600*
Deutsches Wollforschungsinstitut, Aachen *05606*
Fachinformationszentrum Technik e.V., Frankfurt *05692*
Fachverband Textilunterricht e.V., Münster *05697*
Forschungskuratorium Gesamttextil, Eschborn *05733*
Institut für Chemiefasern, Denkendorf *05955*
Institut für Textil- und Faserforschung Stuttgart, Denkendorf *05975*
Normenausschuss Textil und Textilmaschinen im DIN Deutsches Institut für Normung e.V., Berlin *06209*

Hungary
Textilipari Müszaki és Tudományos Egyesület, Budapest *06675*

India
Ahmedabad Textile Industry's Research Association, Ahmadabad *06712*
Bombay Textile Research Association, Bombay *06748*
South India Textile Research Association, Coimbatore *06861*

Ireland
Irish Textiles Federation, Dublin *07002*

Italy
Associazione Italiana di Chimica Tessile e Coloristica, Milano *07302*
Associazione per lo Sviluppo di Studi e Ricerche nell'Industria Tessile Laniera Oreste Rivetti, Biella *07396*
Tecnotessile, Prato *08037*

Norway
Norsk Tekstil Teknisk Forbund, Bergen *09340*

Spain
Asociación Española de Químicos y Coloristas Textiles, Barcelona *10268*

Switzerland
Internationale Föderation der Vereine der Textilchemiker und Coloristen, Reinach *10737*

United Kingdom
International Wool Secretariat, London *12406*
The Textile Institute, Manchester *13120*
Textile Research Council, Nottingham *13121*

U.S.A.
American Association for Textile Technology, Gastonia *13495*
American Association of Textile Chemists and Colorists, Research Triangle Park *13592*
Costume Society of America, Earleville *14866*
Educational Foundation for the Fashion Industries, New York *15019*
Institute of Textile Technology, Charlottesville *15339*
International Textile and Apparel Association, Monument *15564*
National Council for Textile Education, Charlottesville *15991*

Theater s. Performing Arts, Theater

Theology s. Religions and Theology

Therapeutics s. a. Psychiatry; Rehabilitation; Psychology

Argentina
Sociedad Argentina de Farmacología y Terapéutica, Buenos Aires *00168*

Australia
Australian Association of Occupational Therapists, Alphington *00266*
Australian Physiotherapy Association, Concord *00329*
Australian Society of Clinical Hypnotherapy, Eastwood *00350*

Austria
Ärztliche Gesellschaft für Physiotherapie, Österreichischer Kneippärztebund, Neunkirchen *00537*
Arbeitsgemeinschaft Personenzentrierte Psychotherapie und Gesprächsführung, Wien *00562*
Gesellschaft für Logotherapie und Existenzanalyse, Wien *00625*
Österreichische Ärztegesellschaft für Psychotherapie, Wien *00719*
Österreichische ärztliche Gesellschaft für medizinisches und technisches Ozon, Wien *00722*
Österreichische Gesellschaft für Anästhesiologie, Reanimation und Intensivtherapie, Wien *00754*
Österreichische Gesellschaft für Autogenes Training und allgemeine Psychotherapie, Wien *00764*
Österreichische Gesellschaft für Bionome Psychotherapie, Wien *00767*
Österreichische medizinische Gesellschaft für Neuraltherapie nach Huneke-Regulationsforschung, Graz *00868*
Österreichischer Arbeitskreis für Gruppentherapie und Gruppendynamik, Wien *00882*

Belgium
European Family Therapy Association, Bruxelles *01345*
Vlaams Kinesitherapeuten Verbond, Poperinge *01511*

Canada
Association of Occupational Therapists of Manitoba, Winnipeg *01937*
British Columbia Society of Occupational Therapists, Burnaby *01989*
Canadian Group Psychotherapy Association, Montréal *02150*
Corporation Professionnelle des Ergothèrapeutes du Québec, Montréal *02394*
Corporation Professionnelle des Orthophonistes et Audiologistes du Québec, Montréal *02397*
International Association of Group Psychotherapy, Montréal *02541*
Manitoba Society of Occupational Therapists, Winnipeg *02636*
New Brunswick Society of Occupational Therapists, Fredericton *02694*
Newfoundland and Labrador Association of Occupational Therapists, Saint John's *02698*
Nova Scotia Society of Occupational Therapists, Halifax *02746*
Ontario Athletic Therapists Association, Downsview *02762*
Ontario Group Psychotherapy Association, Etobicoke *02780*
Ontario Society of Occupational Therapists, Toronto *02799*
Pacific Coast Family Therapy Training Association, Vancouver *02810*
Prince Edward Island Society of Occupational Therapists, Charlottetown *02839*
Saskatchewan Society of Occupational Therapists, Regina *02902*

China, Republic
China Spiritual Therapy Study Association, Taipei *03259*

Denmark
Danske Fysioterapeuter, København *03606*
Landsforeningen af Foldterapeuter, København *03763*

Finland
Suomen Fysioterapeuttiliitto, Helsinki *04018*

France
Confédération Européenne pour la Thérapie Physique, Paris *04436*
European Brachytherapy Group, Villejuif *04493*
Institut Français d'Analyse de Groupe et de Psychodrame, Paris *04567*
Société de Recherches Psychothérapiques de Langue Française, Versailles *04689*
Société Française de Médecine Orthopédique et Thérapeutique Manuelle, Paris *04812*
Société Française de Mesothérapie, Paris *04815*
Société Française de Thérapeutique et de Pharmacologe Clinique, Paris *04848*
Syndicat National des Médecins Ostéothérapeutes Français, Nancy *04956*
Union Internationale Thérapeutique, Paris *04991*

Germany
Ärztliche Gesellschaft für Physiotherapie, Kneippärztebund e.V., Bad Münstereifel *05043*
Allgemeine Ärztliche Gesellschaft für Psychotherapie, Düsseldorf *05071*
Bundesverband Hilfe für das autistische Kind e.V., Hamburg *05253*
Dachverband Psychosozialer Hilfsvereinigungen e.V., Bonn *05281*
Deutsche Gesellschaft für Poesie- und Bibliotherapie e.V., Köln *05408*
Deutsche Gesellschaft für Verhaltenstherapie e.V., Tübingen *05435*
Deutsche Gruppenpsychotherapeutische Gesellschaft e.V., Berlin *05451*
Deutscher Verband für Physiotherapie, Köln *05550*
Gesellschaft für prä- und postoperative Tumortherapie e.V., Undenheim *05850*
Gesellschaft für Psychotherapie, Psychosomatik und Medizinische Psychologie e.V., Lipzig *05853*
Verband Physikalische Therapie, Hamburg *06367*

Italy
Associazione Medica Italiana di Idroclimatologia, Talassologia e Terapia Fisica, Roma *07355*
Centro di Ricerche Biopsichiche di Padova, Padova *07435*
European Society of Hypnosis in Psychotherapy and Psychosomatic Medicine-Italian Constituent Society, Verona *07684*
Istituto Italiano Studi di Ipnosi Clinica e Psicoterapia, Verona *07745*
Società Italiana di Chemioterapia, Pavia *07849*
Società Italiana di Medicina Subacquea ed Iperbarica, Napoli *07903*
Società Italiana di Mesoterapia, Roma *07904*
Società Italiana di Psicosintesi Terapeutica, Firenze *07939*
Società Italiana di Psicoterapia Analitica Immaginativa, Cremona *07940*
Società Italiana di Terapia Familiare, Roma *07954*
Società Italiana Medica del Training Autogeno, Bologna *07958*

Luxembourg
Association Luxembourgeoise des Kinésithérapeutes Diplomés, Luxembourg *08579*

Netherlands
Nederlandse Vereniging voor Psychotherapie, Utrecht *09037*
Nederlands Genootschap voor Fysiotherapie, Amersfoort *09050*

Norway
Norske Fysioterapeuters Forbund, Oslo *09300*

Portugal
Associação Portuguesa de Fisioterapeutas, Lisboa *09644*

Romania
Societatea de Anestezie si Terapie Intensiva, Bucuresti *09755*

Russia
National Scientific Medical Soiety of Therapists, Moskva *09938*

St. Lucia
Caribbean Association for Rehabilitation Therapists, Castries *09994*

South Africa
The South African Society of Physiotherapy, Johannesburg *10219*

Spain
Asociación Española de Psicoterapia Analítica, Madrid *10267*
Sociedad Española de Medicina Psicosomática y Psicoterapia, Barcelona *10390*

Sweden

Förbundet Sveriges Arbetsterapeuter, Nacka *10458*
Legitimerade Sjukgymnasters Riksförbund, Stockholm *10496*

Switzerland

Gesellschaft für Arzneipflanzenforschung e.V., Zürich *10711*
Institut International de Psychologie et de Psychothérapie Charles Baudouin, Coppet *10729*
International Federation for Psychotherapy, Bern *10749*
Schweizerischer Physiotherapeuten-Verband, Sempach-Stadt *10902*
Verband Schweizerischer Ergotherapeuten, Zürich *11006*

Thailand

Asian-Pacific Federation of Therapeutic Communities, Bangkok *11083*

Turkey

Fizik Tedavi ve Rehabilitasyon Derneği, Istanbul *11145*

United Kingdom

Association for Group and Individual Psychotherapy, London *11264*
Association of Child Psychotherapists, London *11322*
Association of Swimming Therapy, Shrewsbury *11395*
British Association for Counselling, Rugby *11491*
British Association of Art Therapists, Brighton *11501*
British Association of Occupational Therapists, London *11512*
British Association of Psychotherapists, London *11520*
British Hypnotherapy Association, London *11577*
British Institute of Industrial Therapy, Southampton *11582*
British Society for Antimicrobial Chemotherapy, Birmingham *11650*
British Society for Music Therapy, East Barnet *11659*
British Society of Hypnotherapists, London *11685*
The Chartered Society of Physiotherapy, London *11770*
European Association for Behaviour Therapy, Harrow *11987*
International Federation of Practitioners of Natural Therapeutics, London *12358*
National Council of Psycho-Therapists, London *12585*
Research Society for Natural Therapeutics, Bournemouth *12754*
Society of Health and Beauty Therapists, London *13049*
Society of Remedial Gymnasts and Recreational Therapies, Cardiff *13063*
World Medical Tennis Association, Poole *13222*

U.S.A.

Academy of Scientific Hypnotherapy, San Diego *13308*
Ackerman Institute for Family Therapy, New York *13321*
American Academy of Psychotherapists, Decatur *13433*
American Apitherapy Society, Red Bank *13455*
American Art Therapy Association, Mundelein *13459*
American Association for Marriage and Family Therapy, Washington *13485*
American Association for Music Therapy, Valley Forge *13487*
American Association of Behavioral Therapists, Roswell *13508*
American Association of Professional Hypnotherapists, Boones Mill *13571*
American Dance Therapy Association, Columbia *13776*
American Family Therapy Association, Washington *13804*
American Group Psychotherapy Association, New York *13836*
American Horticultural Therapy Association, Gaithersburg *13853*
American Massage Therapy Association, Chicago *13927*
American Occupational Therapy Association, Rockville *13963*
American Osteopathic Academy of Sclerotherapy, Wilmington *13978*
American Physical Therapy Association, Alexandria *14001*
American Society of Group Psychotherapy and Psychodrama, McLean *14128*
Association for the Advancement of Psychotherapy, New York *14345*
The Bridge, New York *14615*
Circulo de Radioterapeutas Ibero-Latinoamericanos, Houston *14753*
Community Guidance Service, New York *14813*
Council on Chiropractic Physiological Therapeutics, Idaho Falls *14912*
Federation of Trainers and Training Programs in Psychodrama, Dallas *15091*
Foundation for Advancement in Cancer Therapy, New York *15120*
Institute for Expressive Analysis, New York *15279*
Institute for Rational-Emotive Therapy, New York *15292*
Joint Review Committee for Respiratory Therapy Education, Euless *15621*
National Association for Poetry Therapy, Potomac *15810*
National Board for Respiratory Care, Lenexa *15924*
National Therapeutic Recreation Society, Arlington *16173*
Rabbinic Center for Research and Counseling, Westfield *16382*
Radiation Therapy Oncology Group, Philadelphia *16386*
Society for Clinical and Experimental Hypnosis, Liverpool *16508*

Tobacco

France

Centre de Coopération pour les Recherches Scientifiques Relatives au Tabac, Paris *04366*

Topography s. Surveying, Photogrammetry

Tourism s. Travel and Tourism

Toxicology

Belgium

Association Européenne des Centres de Lutte contre les Poisons, Bruxelles *01126*
European Chemical Industry Ecology and Toxicology Centre, Bruxelles *01315*

Canada

Canadian Centre for Toxicology, Guelph *02091*

France

Association Internationale de Recherche en Informatique Toxicologique, Paris *04263*
Association pour le Développement de la Recherche en Toxicologie Expérimentale, Paris *04334*
Comité Européen Permanent de Recherches pour la Protection des Populations contre les Risques d'Intoxication à Long Terme, Paris *04400*
Société de Médecine Légale et de Criminologie de France, Paris *04670*
Société Française de Toxicologie, Paris *04849*

Germany

Deutsche Gesellschaft für Pharmakologie und Toxikologie, Darmstadt *05400*
Gesellschaft zur Förderung der Lufthygiene und Silikoseforschung e.V., Düsseldorf *05882*

Italy

Centro Italiano di Solidarieta', Roma *07496*
Società Italiana di Tossicologia, Roma *07955*

Russia

National Scientific Medical Society of Toxicologists, St. Petersburg *09935*

Switzerland

Conseil International sur les Problèmes de l'Alcoolisme et des Toxicomanies, Lausanne *10656*
Ecological and Toxicological Association of the Dyestuffs Manufacturing Industry, Basel *10664*

United Kingdom

Society for Environmental Therapy, Manchester *12972*

U.S.A.

American Association of Poison Control Centers, Tucson *13569*
American Board of Medical Toxicology, Salt Lake City *13638*
American Physicians Poetry Association, Southampton *14004*
Chemical Industry Institute of Toxicology, Research Triangle Park *14724*
International Association of Forensic Toxicologists, Newport Beach *15407*
Society of Toxicology, Washington *16708*
Teratology Society, Bethesda *16771*

Trade s. Commerce

Transport and Traffic

Australia

Australian Road Research Board, Melbourne *00335*
Chartered Institute of Transport in Australia, Sydney *00371*
Institute of Transport, Beverly Hills *00419*

Austria

Forschungsgesellschaft für das Verkehrs- und Straßenwesen im ÖIAV, Wien *00595*
Kuratorium für Verkehrssicherheit, Wien *00701*
Österreichische Gesellschaft für Strassenwesen, Wien *00835*
Österreichische Verkehrswissenschaftliche Gesellschaft, Wien *00926*

Belgium

Dedicated Road Infrastructure for Vehicle Safety in Europe, Bruxelles *01259*

Bulgaria

Scientific and Technical Union of Transport, Sofia *01773*

Canada

Canadian Urban Transit Association, Toronto *02313*
Traffic Injury Research Foundation of Canada, Ottawa *02968*
Transport 2000 Canada, Ottawa *02969*

China, People's Republic

Academy of Highway Sciences, Beijing *03109*

Denmark

Marinehistorisk Selskab, København *03770*

Egypt

Arab Institute of Navigation, Alexandria *03872*
Arab Maritime Transport Academy, Alexandria *03874*

France

Association pour le Développement des Techniques de Transport, d'Environnement et de Circulation, Paris *04340*
Conference on the Development and the Planning of Urban Transport in Developing Countries, La Défense *04447*
Société d'Etudes Ferroviaires, Paris *04737*

Germany

Deutsche Akademie für Verkehrswissenschaft e.V., Hamburg *05291*
Deutscher Verkehrssicherheitsrat e.V., Bonn *05561*
Deutsche Straßenliga e.V., Bonn *05602*
Deutsche Verkehrswissenschaftliche Gesellschaft e.V., Bergisch Gladbach *05623*
Forschungsgesellschaft für Strassen- und Verkehrswesen e.V., Köln *05720*
Freie Vereinigung von Fachleuten öffentlicher Verkehrsbetriebe, Gelsenkirchen *05762*
Gesellschaft für Ursachenforschung bei Verkehrsunfällen e.V., Köln *05867*
Normenausschuss Transportkette im DIN Deutsches Institut für Normung e.V., Berlin *06210*
SNV Studiengesellschaft Verkehr, Hamburg *06281*
Studiengesellschaft für den kombinierten Verkehr e.V., Frankfurt *06299*
Studiengesellschaft für unterirdische Verkehrsanlagen e.V., Köln *06302*
Verein für Binnenschiffahrt und Wasserstraßen e.V., Duisburg *06380*
Wissenschaftlicher Verein für Verkehrswesen e.V., Dortmund *06449*

Guatemala

Corporación Centroamericana de Servicios de Navegación Aéreal, Guatemala City *06566*

Hungary

Közlekedéstudományi Egyesület, Budapest *06609*

Italy

Centro di Studi e Applicazioni di Organizzazione Aziendale della Produzione e dei Trasporti, Torino *07439*
Centro Italiano Ricerche e Studi Trasporto Aereo, Roma *07507*
Centro Italiano Studi Containers, Genova *07508*
Collegio degli Ingegneri Ferroviari Italiani, Roma *07607*

Japan

International Association of Traffic and Safety Sciences, Tokyo *08142*
Nihon Kotsu Kyokai Toshokan, Tokyo *08234*
Un-yu Chosakyoku Johobu Toshoshitsu, Tokyo *08443*

Netherlands

Nederlandse Vereniging voor Zeegeschiedenis, Leiden *09043*

Norway

Nordic Road Safety Council, Oslo *09279*
Norsk Senter for Samferdselsforskning, Oslo *09338*

Portugal

Prevenção Rodoviária Portuguesa, Lisboa *09666*

Russia

Society for Railway Transport, Moskva *09970*

Spain

Instituto de Historia y Cultura Naval, Madrid *10323*

Switzerland

Aktion 100, Verein zur Förderung der Sicherheit im Strassenverkehr, Wabern *10610*
Schweizerischer Verkehrssicherheitsrat, Bern *10922*
Schweizerische Verkehrswirtschaftliche Gesellschaft, Sankt Gallen *10958*

United Kingdom

Barge and Canal Development Association, Wakefield *11425*
Birmingham Transport Historical Group, Folkestone *11451*
Branch Line Society, Huddersfield *11465*
British Trolleybus Society, Reading *11696*
Cambridge Refrigeration Technology, Cambridge *11738*
The Chartered Institute of Transport, London *11766*
County Surveyors Society, Northampton *11864*
Electric Railway Society, Sutton Coldfield *11951*
Great Western Society, Didcot *12127*
London Underground Railway Society, Morden *12471*
Narrow Gauge Railway Society, Huddersfield *12553*
National Association for Road Safety Instruction in Schools, London *12561*
National Trolleybus Association, London *12613*
National Waterways Transport Association, London *12617*
Omnibus Society, London *12654*
Organisation of Teachers of Transport Studies, Southall *12660*
Railway Club, London *12740*
Railway Correspondence and Travel Society, London *12741*
Railway Development Association, Purley *12742*
Scottish Inland Waterways Association, Edinburgh *12913*
Scottish Tramway Museum Society, Glasgow *12935*
Stephenson Locomotive Society, London *13101*
Tramway Museum Society, Matlock *13138*
Transport Ticket Society, Luton *13141*

U.S.A.

Advanced Transit Association, Fairfax *13333*
Committee for Better Transit, Long Island City *14792*
Electric Auto Association, Belmont *15032*
Institute of Transportation Engineers, Washington *15342*
Lexington Group in Transportation History, Saint Cloud University *15666*
Safe Association, Yoncalla *16434*
Society of Freight Car Historians, Monrovia *16667*
Transportation Alternatives, New York *16789*
Transportation Research Board, Washington *16790*

Traumatology

Argentina

Asociación Argentina de Ortopedía y Traumatología, Buenos Aires *00070*

Belgium

Société Internationale de Chirurgie Orthopédique et de Traumatologie, Bruxelles *01462*

France

Groupement National d'Etude des Médecins du Bâtiment et des Travaux Publics, Paris *04559*
International Research Council on Biokinetics of Impacts, Bron *04587*
Société Française de Chirurgie Orthopédique et Traumatologique, Paris *04792*

Hungary

Magyar Traumatológus Társaság, Budapest *06660*

Italy

Società di Ortopedia e Traumatologia dell'Istituto Meridionale ed Insulare, Napoli *07787*
Società Italiana di Medicina del Traffico, Roma *07894*
Società Italiana di Ortopedia e Traumatologia, Roma *07921*
Società Piemontese, Ligure, Lombarda di Ortopedia e Traumatologia, Genova *08003*

Peru

Sociedad Peruana de Ortopedia y Traumatología, Lima *09485*

Poland

Polskie Towarzystwo Ortopedyczne i Traumatologiczne, Warszawa *09596*

Portugal

Sociedade Portuguesa de Ortopedia e Traumatologia, Lisboa *09709*

Romania

Societatea Română de Ortopedie şi Traumatologie, Bucuresti *09794*

Turkey

Türk Ortopedi ve Travmatoloji Derneği, Istanbul *11170*

U.S.A.

American Burn Association, Baltimore *13673*
American Trauma Society, Upper Marlboro *14198*

Venezuela

Sociedad Venezolana de Cirurgía Ortopédica y Traumatología, Caracas *17021*

Travel and Tourism

Italy

Accademia Euro-Afro-Asiatica del Turismo, Catania *07179*
Centro di Demodossalogia, Roma *07419*
Centro per gli Studi sui Sistemi Distributivi e il Turismo, Milano *07521*
Centro Studi e Ricerche per la Conoscenza della Liguria Attraverso le Testimonianze dei Viaggiatori Stranieri, Genova *07573*
International Centre for Advanced Technical and Vocational Training, Torino *07709*

United Kingdom

British Association of the Experiment in International Living, Malvern *11527*
Institute of Leisure and Amenity Management, Reading *12250*
Leisure Studies Association, Edinburgh *12447*
Scottish Recreational Land Association, Edingburgh *12923*

U.S.A.

Southwest Parks and Monuments Association, Tucson *16731*

Tropical Medicine

Austria
Österreichische Gesellschaft für Tropenmedizin und Parasitologie, Wien *00836*

Belgium
Association pour l'Etude et l'Evaluation Epidémiologiques des Désastres dans les Pays en Voie de Développement, Bruxelles *01152*
Belgische Vereniging voor Tropische Geneeskunde, Antwerpen *01163*

Egypt
Egyptian Society of Medicine and Tropical Hygiene, Alexandria *03900*

France
Société de Pathologie Exotique, Paris *04679*

Germany
Deutsche Tropenmedizinische Gesellschaft e.V., Frankfurt *05611*

Hungary
Council of Directors of Institutes of Tropical Medicine in Europe, Budapest *06599*

Netherlands
Nederlandse Vereniging voor Tropische Geneeskunde, 's-Gravenhage *09040*

Switzerland
Schweizerische Gesellschaft für Tropenmedizin und Parasitologie, Zürich *10851*

United Kingdom
Royal Society of Tropical Medicine and Hygiene, London *12864*

U.S.A.
American Society of Tropical Medicine and Hygiene, McLean *14171*

Tuberculosis s. Pulmonary Disease

Urban and Regional Planning s. a. Architecture; Civil Engineering

Argentina
Centro de Estudios Urbanos y Regionales, Buenos Aires *00103*

Australia
Australian Institute of Urban Studies, Canberra *00315*
Metropolitan Research Trust, Canberra City *00434*
Royal Australian Planning Institute, Hawthorn *00465*
Town and Country Planning Association of Victoria, Melbourne *00511*

Austria
Arbeitskreis für neue Methoden in der Regionalforschung, Wien *00566*
Forschungsgesellschaft für Wohnen, Bauen und Planen, Wien *00597*
Österreichische Gesellschaft für Raumforschung und Raumplanung, Wien *00826*
Österreichische Raumordnungskonferenz, Wien *00887*

Belgium
Association Européenne des Institutions d'Aménagement Rural, Bruxelles *01128*
Conferentie voor Regionale Ontwikkeling in Noord-West-Europa, Brugge *01249*
European Association for Country Planning Institutions, Bruxelles *01275*
Koninklijke Vereniging voor Natuur- en Stedeschoon, Antwerpen *01411*

Bolivia
Amigos de la Ciudad, La Paz *01531*

Brazil
Fundação João Pineiro, Belo Horizonte *01657*

Canada
Association of Canadian University Planning Programs, Halifax *01927*
Atlantic Planners Institute, Halifax *01960*
Canadian Institute of Planners, Ottawa *02181*
Commonwealth Association of Planners, Ottawa *02367*
Community Planning Association of Canada, Regina *02370*
Community Planning Association of Nova Scotia, Dartmouth *02371*
Housing and Urban Development Association of Nova Scotia, Halifax *02507*
Ontario Community Development Society, Guelph *02767*
Planning Institute of British Columbia, Vancouver *02820*
Urban Development Institute of Ontario, Toronto *02975*

China, Republic
International House Association, Taipei Chapter, Taipei *03304*

Colombia
Centro Interamericano de Vivienda y Planeamiento, Bogotá *03359*

Denmark
Byggecentrum, Hørsholm *03571*
Dansk Byplanlaboratorium, København *03596*
Nordisk Byggedag, København *03781*

Egypt
General Organization for Housing, Building and Planning Research, Cairo *03902*

France
Société Française des Urbanistes, Paris *04847*

Germany
Akademie für Raumforschung und Landesplanung, Hannover *05060*
Deutsche Akademie für Städtebau und Landesplanung e.V., München *05290*
Deutscher Verband für Wohnungswesen, Städtebau und Raumplanung e.V., Bonn *05553*
Deutsches Institut für Urbanistik, Berlin *05587*
Gesellschaft für Wohnungsrecht und Wohnungswirtschaft Köln e.V., Köln *05874*
Institut für Städtebau, Wohnungswirtschaft und Bausparwesen e.V., Bonn *05972*
Vereinigung Stadt-, Regional- und Landesplanung e.V., Bochum *06411*

Greece
World Society for Ekistics, Athinai *06549*

Hungary
Magyar Urbanisztikai Társaság, Budapest *06662*

India
Eastern Regional Organization for Planning and Housing, New Delhi *06756*

Italy
Associazione Nazionale degli Urbanisti, Treviso *07361*
Centro Nazionale di Studi Urbanistici, Roma *07517*
Centro Ricerche Urbanistiche e di Progettazione, Milano *07542*
Istituto Nazionale di Urbanistica, Roma *07749*

Jamaica
Commonwealth Association of Planners, Kingston *08076*
Jamaica National Trust Commission, Kingston *08085*

Japan
Toshi Kaihatsu Kyokai Joho Sabisu Senta Shiryoshitsu, Tokyo *08439*

Netherlands
Bond van Nederlandse Stedebouwkundigen, Amsterdam *08835*
International Federation for Housing and Planning, 's-Gravenhage *08932*
International Society of City and Regional Planners, 's-Gravenhage *08942*
Stichting Centrale Raad voor de Academies van Bouwkunst, Amsterdam *09070*

Norway
Norsk Institutt for By- og Regionforskning, Oslo *09321*

Pakistan
Pakistan Council of Architects and Town Planners, Karachi *09383*

Philippines
ASEAN Association for Planning and Housing, Manila *09490*

Poland
Towarzystwo Urbanistów Polskich, Warszawa *09629*

Russia
Commission for Management of the Development of Cities, Moskva *09822*

Saudi Arabia
Arab Urban Development Institute, Riyadh *10000*

Spain
Federación de Urbanismo y de la Vivienda, Madrid *10309*

Sweden
Nordiska Institutet för Samhällsplanering, Stockholm *10508*

Switzerland
Fachgruppe für Raumplanung und Umwelt des SIA, Zürich *10695*
Schweizerische Vereinigung für Landesplanung, Bern *10949*

United Kingdom
District Planning Officers Society, Crowborough *11897*
European Council of Town Planners, London *12024*
Institute of Municipal Building Management, Colwyn Bay *12263*
London Society, London *12468*
National Housing and Town Planning Council, London *12593*
Regional Studies Association, London *12747*
Royal Town Planning Institute, London *12869*
Scottish National Housing and Town Planning Council, Troon *12917*
Society of Town Planning Technicians, London *13069*
Town and Country Planning Association, London *13137*

U.S.A.
Association of Collegiate Schools of Planning, Madison *14417*
Center for Design Planning, Pensacola *14676*
Center for Urban and Regional Studies, Chapel Hill *14709*
Community Associations Institute, Alexandria *14808*
Council of Planning Librarians, Chicago *14901*
Environmental Design Research Association, Oklahoma City *15049*
Metropolitan Association of Urban Designers and Environmental Planners, New York *15724*
National Association of County Planners, Washington *15847*
National Council for Urban Economic Development, Washington *15995*
Regional Science Association, Urbana *16403*
Small Towns Institute, Ellensburg *16480*
ULI – The Urban Land Institute, Washington *16801*
Urban Affairs Association, Newark *16850*

Venezuela
Consejo Zuliano de Planificación y Coordinación, Maracaibo *17004*

Yugoslavia
Urbanisticki Savez Jugoslavije, Novi Sad *17136*

Urology

Austria
Österreichische Gesellschaft für Urologie, Wien *00839*

Canada
Canadian Academy of Urological Surgeons, Vancouver *02010*
Canadian Urological Association, Halifax *02314*

France
European Association of Urology, Paris *04490*
Société Française d'Urologie, Paris *04865*
Société Internationale d'Urologie, Paris *04883*

Germany
Berufsverband der Deutschen Fachärzte für Urologie, Hamburg *05197*
Berufsverband der Deutschen Urologen e.V., Dorfen *05199*
Deutsche Gesellschaft für Urologie, Hannover *05433*
Nordrhein-Westfälische Gesellschaft für Urologie, Osnabrück *06143*

Israel
Israeli Urological Association, Tel Aviv *07079*

Italy
Società Italiana di Urodinamica, Bologna *07956*
Società Italiana di Urologia, Roma *07957*

Japan
Nippon Hinyoki-ka Gakkai, Tokyo *08305*

Netherlands
Nederlandse Vereniging voor Urologie, Utrecht *09041*

Poland
Polskie Towarzystwo Urologiczne, Katowice *09607*

Russia
National Scientific Society of Urological Surgeons, Moskva *09939*

Turkey
Türk Uroloji Dernegi, Istanbul *11177*

United Kingdom
British Association of Urological Surgeons, London *11528*

U.S.A.
American Association of Genito-Urinary Surgeons, Houston *13536*
American Board of Urology, Bingham Farms *13665*
American Fertility Society, Birmingham *13811*
American Urological Association, Baltimore *14201*
Society for Pediatric Urology, Seattle *16565*

Venezuela
Sociedad Venezolana de Urología, Caracas *17034*

Venereology

Australia
Australasian College of Venereologists, NSW 2000 *00242*

Belgium
Société Royale Belge de Dermatologie et de Vénérologie, Bruxelles *01472*

Canada
Canadian Fertility and Andrology Society, Montréal *02132*

Denmark
Danske Dermato-Venerologers Organisation, Alborg *03603*

France
Association des Dermatologistes et Syphiligraphes de Langue Française, Créteil *04175*
Société Française de Dermatologie et de Syphiligraphie, Paris *04799*
Syndicat National Français des Dermatologistes et Vénéréologistes, Paris *04969*

Italy
Società Italiana di Dermatologia e Sifilografia, Roma *07861*
Società Italiana di Sessuologia Clinica, Roma *07948*
Società Italiana di Sessuologia Medica, Roma *07949*
Società Italiana per lo Studio della Fertilità e della Sterilità, Roma *07971*

Netherlands
Nederlandse Vereniging vor Dermatologie en Venereologie, Breda *09044*

Poland
Polskie Towarzystwo Dermatologiczne, Poznań *09560*

Portugal
Sociedade Portuguesa de Dermatologia e Venereologia, Lisboa *09697*

Romania
Societatea de Dermato-Venerologie, Bucuresti *09759*

Russia
National Scientific Medical Society of Venereologists and Dermatologists, Moskva *09937*

Spain
Academia Española de Dermatología y Sifilografía, Madrid *10242*

Sweden
Svenska Dermatologiska Sällskapet, Stockholm *10541*

Switzerland
Schweizerische Gesellschaft für Dermatologie und Venereologie, Lausanne *10816*

United Kingdom
Institute of Technicians in Venereology, Orsett *12286*
International Union Against the Venereal Diseases and the Treponematoses, London *12398*
Medical Society for the Study of Venereal Diseases, London *12512*

U.S.A.
American Venereal Disease Association, Baltimore *14203*

Ventilation s. Air Conditioning

Veterinary Medicine

Argentina
Academia Nacional de Agronomía y Veterinaria, Buenos Aires *00040*
Sociedad Rural Argentina, Buenos Aires *00201*

Australia
Australian Veterinary Association, Artarmon *00361*
Commonwealth Veterinary Association, Scotsburn *00378*
Sydney University Veterinary Society, Sydney *00504*

Austria
Berufsverband freiberuflich tätiger Tierärzte Österreichs, Irdning *00574*
Bundeskammer der Tierärzte Österreichs, Wien *00576*
Österreichische Gesellschaft der Tierärzte, Wien *00748*

Belgium
European Association of Veterinary Anatomists, Gent *01298*
European Federation of Animal Health, Bruxelles *01347*
Union Syndicale Vétérinaire Belge, Bruxelles *01503*

Brazil
Associação Prudentina de Educação e Cultura, Presidente Prudente *01608*
Empresa Brasileira de Pesquisa Agropecuária, Campo Grande *01648*

Canada
Alberta Veterinary Medical Association, Edmonton *01852*
Canadian Association for Laboratory Animal Science, Edmonton *02035*
Canadian Veterinary Medical Association, Ottawa *02315*
Corporation Professionnelle des Médecins Vétérinaires du Québec, Saint Hyacinthe *02396*
Manitoba Veterinary Medical Association, Winnipeg *02639*
New Brunswick Veterinary Medical Association, Fredericton *02697*
Newfoundland and Labrador Veterinary Medical Association, Mount Pearl *02700*
Nova Scotia Veterinary Medical Association, Kentville *02749*
Ontario Veterinary Association, Guelph *02804*
Prince Edward Island Veterinary Medical Association, Charlottetown *02842*
Saskatchewan Veterinary Medical Association, Saskatoon *02906*

Chile
Comité Chileno Veterinario de Zootecnia, Santiago *03040*
Sociedad de Medicina Veterinaria de Chile, Santiago *03094*

China, People's Republic
Chinese Association of Animal Science and Veterinary Medicine, Beijing *03161*

China, Republic
Association of Animal Husbandry and Veterinary Medicine of Taiwan, Nantou Hsien *03245*

Denmark
Danske Veterinårhygiejnikeres Organisation, Slagense *03614*
Dansk Veterinårhistorisk Samfund, Egtved *03698*
Den Danske Dyrlaegeforening, Vanlöse *03701*
Sammenslutningen af Praktiserende Dyrlaeger, Vanløse *03804*

Finland
International Council for Laboratory Animal Science, Kuopio *03966*
Suomen Eläinlääkäriliitto, Helsinki *04013*

France
Académie Vétérinaire de France, Paris *04122*
Association Centrale des Vétérinaires, Paris *04147*
Association Mondiale des Vétérinaires Microbiologistes, Immunologistes et Spécialistes des Maladies Infectieuses, Maisons Alfort *04304*
European Association of Establishments for Veterinary Education, Maisons-Alfort *04485*
Société Vétérinaire Pratique de France, Paris *04922*
Syndicat des Vétérinaires de la Région Parisienne, Paris *04929*
Syndicat National des Vétérinaires Français, Paris *04968*

Germany
Bundesverband der beamteten Tierärzte, Heinsberg *05242*
Dachverband wissenschaftlicher Gesellschaften der Agrar-, Forst-, Ernährungs-, Veterinär- und Umweltforschung e.V., Frankfurt *05282*
Deutsche Gesellschaft für Parasitologie e.V., Marburg *05395*
Deutsche Tierärzteschaft e.V., Bonn *05610*
Deutsche Veterinärmedizinische Gesellschaft e.V., Giessen *05624*
Gesellschaft für Ernährungsphysiologie, Frankfurt *05820*
Münchener Tierärztliche Gesellschaft, München *06122*
Verband der Gemeinde-Tierärzte Baden-Württembergs, Baden-Baden *06337*
Verband der Tierheilpraktiker e.V., Meitingen *06344*

Greece
Elliniki Ktiniatriki Eteria, Athinai *06508*

Hungary
Országos Állategészségügyi Intézet, Budapest *06668*

India
Helminthological Society of India, Mathura *06764*

Iraq
Arab Union of Veterinary Surgeons, Baghdad *06915*

Ireland
Irish Veterinary Association, Dublin *07004*
Veterinary Council, Dublin *07038*

Israel
Israel Veterinary Medical Association, Tel Aviv *07105*

Italy
European Commission for the Control of Foot-and-Mouth Disease, Roma *07673*
Federazione Nazionale degli Ordini dei Veterinari Italiani, Roma *07696*
International Association for Veterinary Homeopathy, Milano *07704*
Società Italiana delle Scienze Veterinarie, Brescia *07828*
Società Italiana di Buiatria, Torino *07846*
Società Italiana di Ippologia, Roma *07889*
Società Italiana di Patologia Aviare, Perugia *07928*
Società Italiana di Patologia ed Allevamento dei Suini, Parma *07929*
Società Italiana di Patologia e di Allevamento degli Ovini e dei Caprini, Catania *07930*

Jamaica
British Caribbean Veterinary Association, Kingston *08069*

Japan
Nihon Dobutsu Shinri Gakkai, Ibaraki *08193*
Nihon Jui Gakkai, Tokyo *08214*

Kenya
African Trypanotolerant Livestock Network, Nairobi *08475*

Macedonia
Sojuz na Društvata na Veterinarnite Lekari i Tehničari na Makedonija, Skopje *08611*

Netherlands
European Association of State Veterinary Officers, Rotterdam *08870*
European Association of Veterinary Pharmacology and Toxicology, Utrecht *08872*
Koninklijke Nederlandse Maatschappij voor Diergeneeskunde, Utrecht *08957*

New Zealand
New Zealand Veterinary Association, Wellington *09182*
Veterinary Services Council, Wellington *09198*

Nigeria
Nigerian Veterinary Medical Association, Vom *09242*

Norway
Den Norske Veterinaerforening, Oslo *09256*

Peru
Centro Nacional de Patologia Animal, Lima *09462*

Philippines
Philippine Veterinary Medical Association, Quezon City *09533*

Poland
International Commission on Trichinellosis, Poznan *09544*
Polskie Towarzystwo Nauk Weterynaryjnych, Warszawa *09591*

Portugal
Sociedade Portuguesa de Ciências Veterinárias, Lisboa *09696*
Sociedade Portuguesa Veterinária de Anatomia Comparativa, Lisboa *09715*
Sociedade Portuguesa Veterinária de Estudos Sociológicos, Lisboa *09716*

Senegal
Association des Etablissements d'Enseignement Vétérinaire Totalement ou Partiellement de Langue Française, Dakar *10018*

South Africa
South African Veterinary Association, Monument Park *10222*

Spain
Asociación del Cuerpo Nacional Veterinario, Madrid *10249*
Asociación Internacional Veterinaria de Producción Animal, Madrid *10274*
Asociación Nacional de Veterinarios Titulares, Madrid *10277*
Consejo General de Colegios Veterinarios de España, Madrid *10300*
European Association of Coleopterology, Barcelona *10304*
World Veterinary Association, Madrid *10405*

Sweden
Sveriges Veterinärförbund, Stockholm *10598*

Switzerland
Gesellschaft Schweizerischer Tierärzte, Bern *10721*
Internationale Veterinäranatomische Nomenklatur-Kommission, Zürich *10748*
Schweizerische Vereinigung für Kleintiermedizin, Bern *10948*

Tanzania
Tanzania Veterinary Association, Morogoro *11063*

Turkey
Türk Veteriner Hekimleri Demegi, Ankara *11178*

United Kingdom
Animal Diseases Research Association, Edinburgh *11235*
Animal Health Trust, Newmarket *11236*
Association for Veterinary Clinical Pharmacology and Therapeutics, Dorking *11296*
Association of Meat Inspectors, Barnet *11366*
Association of Veterinary Anaesthetists of Great Britain and Ireland, Bury Saint Edmunds *11406*
Association of Veterinary Teachers and Research Workers, Edinburgh *11407*
British Association of Homoeopathic Veterinary Surgeons, Stanford-in-the-Vale *11509*
British Laboratory Animals Veterinary Association, Salisbury *11600*
British Small Animal Veterinary Association, Cheltenham *11646*
British Veterinary Association, London *11703*
British Veterinary Radiology Association, Hatfield *11704*
British Veterinary Zoological Society, Studham *11705*
CAB International Division of Animal Health and Medical Parasitology, Wallingford *11729*
Federation of European Veterinaries in Industry and Research, Hounslow *12064*
Federation of Veterinarians of the EEC, London *12068*
Institute for Animal Health, Newbury *12204*
International Society for Applied Ethology, Edinburgh *12388*
National Anti-Vivisection Society, London *12555*
Royal College of Veterinary Surgeons, London *12793*
Veterinary History Society, London *13181*
World Veterinary Poultry Association, Huntingdon *13226*

U.S.A.
Academy of Veterinary Cardiology, Floral Park *13311*
American Academy of Veterinary Dermatology, Phoenix *13441*
American Academy of Veterinary Nutrition, Athens *13442*
American Animal Hospital Association, Denver *13451*
American Association for Accreditation of Laboratory Animal Care, Rockville *13463*
American Association for Laboratory Animal Science, Cordova *13484*
American Association of Avian Pathologists, Kennett Sq *13507*
American Association of Bovine Practitioners, West Lafayette *13513*
American Association of Equine Practitioners, Lexington *13534*
American Association of Industrial Veterinarians, Columbia *13544*
American Association of Small Ruminant Practitioners, Ithaca *13582*
American Association of Veterinary Anatomists, Auburn *13598*
American Association of Veterinary Laboratory Diagnosticians, Columbia *13599*
American Association of Veterinary Parasitologists, Kalamazoo *13600*
American Association of Veterinary State Boards, Jefferson City *13601*
American Association of Wildlife Veterinarians, Fort Collins *13602*
American Association of Zoo Veterinarians, Philadelphia *13607*
American College of Laboratory Animal Medicine, Hershey *13713*
American College of Veterinary Internal Medicine, Blacksburg *13739*
American College of Veterinary Pathologists, West Deptford *13740*
American College of Veterinary Radiology, Glencoe *13741*
American Fertility Society, Birmingham *13811*
American Society of Veterinary Ophthalmology, Stillwater *14172*
American Society of Veterinary Physiologists and Pharmacologists, Mississippi State *14173*
American Veterinary Medical Association, Schaumburg *14204*
American Veterinary Society of Animal Behavior, Ithaca *14205*
Animal Medical Center, New York *14219*
Animal Nutrition Research Council, Chantilly *14220*
Association for Equine Sports Medicine, Santa Barbara *14294*
Association for Gnotobiotics, East Aurora *14300*
Association for Women Veterinarians, Union City *14370*
Association of American Veterinary Medical Colleges, Washington *14390*
Association of Avian Veterinarians, Boca Raton *14396*
Beef Improvement Federation, Athens *14572*
Conference of Public Health Veterinarians, Arlington *14833*
Conference of Research Workers in Animal Diseases, Fort Collins *14834*
Flying Veterinarians Association, Columbia *15106*
Foundation for Biomedical Research, Washington *15121*
Institute of Laboratory Animal Resources, Washington *15328*
International Association for Aquatic Animal Medicine, San Leandro *15375*
International Embryo Transfer Society, Champaign *15470*
International Veterinary Acupuncture Society, Chester Springs *15570*
Laboratory Animal Management Association, Silver Spring *15643*
National Association for Veterinary Acupuncture, Fullerton *15822*
National Association of Federal Veterinarians, Washington *15859*
National Association of State Public Health Veterinarians, Columbus *15903*
National Mastitis Council, Arlington *16104*
Orthopedic Foundation for Animals, Columbia *16270*
Scientists Center for Aminal Welfare, Bethesda *16448*
Society for Invertebrate Pathology, Bethesda *16542*
Society for Theriogenology, Hastings *16609*
Veterinary Cancer Society, Rochester *16859*
Veterinary Orthopaedic Society, Salt Lake City *16860*
Western Veterinary Conference, Las Vegas *16888*
World Association of Veterinary Anatomists, West Lafayette *16923*

Yugoslavia
Društvo Veterinara Srbije, Beograd *17089*
Savez Društava Veterinara Jugoslavije, Beograd *17113*

Zimbabwe
Zimbabwe Veterinary Association, Harare *17189*

Vocational Training s. Adult Education

Water Resources

Australia
Safe Water Association of New South Wales, Sydney *00486*
Water Research Foundation of Australia, Kingsford *00520*

Brazil
Instituto Nacional de Pesquisas Hidroviarias, Rio de Janeiro *01689*

Canada
Association Québécoise des Techniques de l'Eau, Montréal *01955*
Canadian Association on Water Pollution Research and Control, Montréal *02076*
Canadian Water Quality Association, Waterloo *02316*
Northwestern Ontario Water and Waste Conference, Thunder Bay *02717*
Western Canada Water and Wastewater Association, Calgary *02990*

Denmark
Dansk Vandteknisk Forening, Viby *03697*

France
Association Française pour l'Etude des Eaux, Limoges *04244*
Intergovernmental Council for the International Hydrological Programme, Paris *04577*
Société Française d'Hydrologie et de Climatologie Médicales, Paris *04858*

Germany
Abwassertechnische Vereinigung e.V., Sankt Augustin *05029*
Deutscher Verband für Wasserwirtschaft und Kulturbau e.V., Bonn *05552*
Deutscher Verein des Gas- und Wasserfaches e.V., Eschborn *05556*
Institut für gewerbliche Wasserwirtschaft und Luftreinhaltung e.V., Köln *05962*
Normenausschuss Wasserwesen im DIN Deutsches Institut für Normung e.V., Berlin *06217*

Guatemala
Instituto Nacional de Sismología, Vulcanología, Meteorología e Hidrología, Guatemala *06571*

Hungary
Magyar Hidrológiai Társaság, Budapest *06634*

India
Indian Association of Geohydrologists, Calcutta *06773*

Italy
Associazione Idrotecnica Italiana, Roma *07284*
Centro Internazionale per gli Studi sulla Irrigazione, Verona *07478*
Società Internazionale di Tecnica Idrotermale, Castellamare di Stabia *07819*
Tecneco, Fano *08032*
Tecnomare, Venezia *08036*

Japan
Nippon Kaisui Gakkai, Tokyo *08322*

Kenya
African Water Network, Nairobi *08476*

Monaco
Organisation Hydrographique Internationale, Monte Carlo *08797*

Netherlands
Nederlandse Vereniging voor Afvalwaterbehandeling en Waterkwaliteitsbeheer, Rijswijk *09014*

New Zealand
New Zealand Hydrological Society, Wellington *09152*

Russia
Water Management Society, Moskva *09991*

South Africa
Water Systems Research Group, Johannesburg *10235*

Switzerland
Schweizerischer Verein des Gas- und Wasserfaches, Zürich *10919*
Verband Schweizerischer Abwasserfachleute, Zürich *11004*

Thailand
Committee for Co-ordination of Investigations of the Lower Mekong Basin, Bangkok *11090*

United Kingdom
Estuarine and Brackish-Water Sciences Association, Cambridge *11979*
International Association on Water Quality, London *12337*
Water Research Centre, Marlow *13194*

U.S.A.
American Cetacean Society, San Pedro *13686*
American Water Resources Association, Bethesda *14209*
Associated Laboratories, Palatine *14259*
Colorado River Association, Los Angeles *14783*
Great Lakes Commission, Ann Arbor *15175*
International Association of Theoretical and Applied Limnology, Tuscaloosa *15418*
International Water Resources Association, Urbana *15572*
Interstate Conference on Water Policy, Washington *15580*
National Hydropower Association, Washington *16081*
National Water Resources Association, Arlington *16179*
Rural Community Assistance Program, Leesburg *16429*
Trout Unlimited, Vienna *16793*
United States Committee on Irrigation and Drainage, Denver *16819*
Universities Council on Water Resources, Carbondale *16838*
Water Quality Research Council, Lisle *16874*
Water Resources Congress, Arlington *16875*

Wines and Wine Making

Canada
Grape and Wine Institute of British Columbia, Kelowna *02489*

Italy
Ente per la Valorizzazione dei Vini Astigiani, Asti *07657*

X-Ray Technology
s. a. Radiology

Austria
Österreichische Röntgengesellschaft – Gesellschaft für medizinische Radiologie und Nuklearmedizin, Wien 00899

Denmark
Danske Radiologers Organisation, Sønderborg 03613

Germany
Bayerische Röntgengesellschaft, Fürth 05179
Deutsche Röntgengesellschaft, Neu-Isenburg 05535
Vereinigung Südwestdeutscher Radiologen und Nuklearmediziner, Sindelfingen 06415

Iceland
Félag Islenskra Röntgenlaekna, Reykjavik 06684

Russia
National Scientific Medical Society of Roentgenologists and Radiologists, Moskva 09932

U.S.A.
American Roentgen Ray Society, Reston 14034
Council on Diagnostic Imaging, Ashtabula 14914

Zoology
s. a. Ornithology; Entomology

Argentina
Fundación Miguel Lillo, San Miguel de Tucumán 00140

Australia
Australian Mammal Society, Lyneham 00319
Australian Society for Fish Biology, Queenscliff 00339
Australian Society of Herpetologists, Lyneham 00353
Malacological Society of Australia, Melbourne 00430
Royal Zoological Society of New South Wales, Mosman 00484
Royal Zoological Society of South Australia, Adelaide 00485
World's Poultry Science Association, Australian Branch, Seven Hills 00529
Zoological Board of Victoria, Parkville 00530

Austria
Zoologisch-Botanische Gesellschaft in Österreich, Wien 01035

Bangladesh
Zoological Society of Bangladesh, Dhaka 01071

Barbados
Barbados Museum and Historical Society, Bridgetown 01074

Belgium
Association pour les Etudes et Recherches de Zoologie Appliquée et de Phytopathologie, Louvain-la-Neuve 01151
Koninklijke Maatschappij voor Dierkunde van Antwerpen, Antwerpen 01410
Société Royale Zoologique de Belgique, Bruxelles 01489

Brazil
Congresso da América do Sul da Zoologia, São Paulo 01642

Burkina Faso
Organization for Co-ordination and Co-operation in the Control of Major Endemic Diseases, Bobo-Dioulasso 01810

Canada
Calgary Zoological Society, Calgary 01999
Canadian Amphibian and Reptile Conservation Society, Mississauga 02020
Canadian Society of Animal Science, Ottawa 02274
Canadian Society of Zoologists, Toronto 02297
Metropolitan Toronto Zoological Society, West Hill 02651
Société Zoologique de Québec, Charlesbourg 02937
Zoological Society of Manitoba, Winnipeg 03006

Chile
Comité Chileno Veterinario de Zootecnia, Santiago 03040

China, Republic
Association of Animal Husbandry and Veterinary Medicine of Taiwan, Nantou Hsien 03245
Malacological Society of China, Taipei 03306

Colombia
Instituto de Ciencias Naturales, Bogotá 03391

Czech Republic
Česká Společnost Zoologická, Praha 03531

Denmark
Lepidopterologisk Forening, Holte 03767

Finland
Societas pro Fauna et Flora Fennica, Helsingin Yliopisto 03995

France
Société d'Ethnozoologie et d'Ethnobotanique, Paris 04724
Société Française de Malacologie, Paris 04806
Société Française d'Ichtyologie, Paris 04860
Société Zoologique de France, Paris 04923

Georgia
Georgian Society of Helminthologists, Tbilisi 05023

Germany
Deutsche Gesellschaft für Herpetologie und Terrarienkunde e.V., Rheinbach 05361
Deutsche Gesellschaft für Parasitologie e.V., Marburg 05395
Deutsche Gesellschaft für Säugetierkunde e.V., Tübingen 05418
Deutsche Gesellschaft für Züchtungskunde e.V., Bonn 05447
Deutsche Malakozoologische Gesellschaft, Frankfurt 05465
Deutsche Zoologische Gesellschaft e.V., Bonn 05629
Verband Deutscher Zoodirektoren e.V., Heidelberg 06364

India
The Academy of Zoology, Agra 06708
Indian Association of Systematic Zoologists, Calcutta 06776
Zoological Society of Calcutta, Calcutta 06870
Zoological Society of India, Calcutta 06871

Ireland
Royal Zoological Society of Ireland, Dublin 07031

Israel
Israel Association for Applied Animal Genetics, Haifa 07061
Zoological Society of Israel, Jerusalem 07120

Italy
Associazione Scientifica di Produzione Animale, Napoli 07401
Federazione Europea di Zootecnia, Roma 07688
Società degli Amici del Museo Civico di Storia Naturale Giacomo Doria, Genova 07779
Società Italiana di Malacologia, Milano 07892
Società Italiana per il Progresso della Zootecnia, Milano 07963
Unione Erpetologica Italiana, Roma 08045

Ivory Coast
Centre de Recherches Zootechniques, Bouaké 08062

Japan
Nihon Kairui Gakkai, Tokyo 08217
Nippon Chikusan Gakkai, Tokyo 08287
Nippon Dobutsu Gakkai, Tokyo 08295
Nippon Oyo-Dobutsu-Konchu Gakkai, Tokyo 08355

Kenya
African Elephant and Rhino Specialist Group, Nairobi 08468

Korea, Democratic People's Republic
Academy of Fisheries, Sinpo City 08509

Malaysia
Malaysian Zoological Society, Kuala Lumpur 08654

Moldova, Republic of
Entomological Society of Moldova, Chişinău 08778
Protozoological Society of Moldova, Chişinău 08789

Netherlands
European Endangered Species Programme, Amsterdam 08884
European Federation of Branches of the World's Poultry Science Association, Beekbergen 08887
Nederlandse Dierkundige Vereniging, Wageningen 08991
Nederlandse Zootechnische Vereniging, Wassenaar 09046
Stichting Koninklijk Zoölogisch Genootschap Natura Artis Magistra, Amsterdam 09072
Unitas Malacologica, Leiden 09082

New Zealand
New Zealand Society for Parasitology, Hamilton 09176

Nigeria
Fisheries Society of Nigeria, Lagos 09227

Peru
Instituto de Zoonosis e Investigación Pecuaria, Lima 09469

Poland
Polskie Towarzystwo Zoologiczne, Wroclaw 09608
Polskie Towarzystwo Zootechniczne, Warszawa 09609

Portugal
Sociedade Portuguesa de Especialistas de Pequenos Animales, Lisboa 09699

Russia
Entomological Society, St. Petersburg 09866
Society of Mammalogists, Moskva 09975
Society of Protozoologists, St. Petersburg 09978

Slovakia
Slovenská Entonologická Spoločnost, Bratislava 10071
Slovenská Zoologická Spoločnost, Bratislava 10092

South Africa
Herpetological Association of Africa, Port Elizabeth 10146
Zoological Society of Southern Africa, Port Elizabeth 10237

Spain
Patronato de Biología Animal, Madrid 10335
Sociedad Veterinaria de Zootecnica de España, Madrid 10402

Switzerland
Gesellschaft für Versuchstierkunde, Füllinsdorf 10717
Schweizerische Zoologische Gesellschaft, Basel 10961

United Kingdom
British Arachnological Society, Banff 11480
British Herpetological Society, London 11571
British Ichthyological Society, Welwyn Garden City 11578
British Morgan Horse Society, London 11611
British Shell Collectors Club, New Maldon 11645
British Veterinary Zoological Society, Studham 11705
Conchological Society of Great Britain and Ireland, Ilford 11835
Delphinium Society, Farnham 11880
European Association for Aquatic Mammals, Dunstable 11986
European Cetacean Society, Oxford 12007
The Federation of Zoological Gardens of Great Britain and Ireland, London 12069
Fisheries Society of the British Isles, Cambridge 12077
International Bee Research Association, Cardiff 12339
International Commission on Zoological Nomenclature, London 12345
Laboratory Animal Science Association, London 12432
Lancashire and Cheshire Fauna Society, Manchester 12437
Loch Ness Phenomena Investigation Bureau, Drumnadrochit 12459
The Malacological Society of London, Canterbury 12478
National Zoological Association of Great Britain, Bury Saint Edmunds 12618
The North of England Zoological Society, Chester 12640
The Royal Zoological Society of Scotland, Edinburgh 12876
Society of Protozoologists, British Section, London 13060
Wildlife Sound Recording Society, Farnham 13206
World Rabbit Science Association, Cheltenham 13223
Zoological Society of Glasgow and West of Scotland, Glasgow 13236
The Zoological Society of London, London 13237
Zoological Society of Northern Ireland, Newtownards 13238

Uruguay
Sociedad Malacológica del Uruguay, Montevideo 13269
Sociedad Zoológica del Uruguay, Montevideo 13272

U.S.A.
American Association of Zoo Keepers, Topeka 13605
American Association of Zoological Parks and Aquariums, Wheeling 13606
American Littoral Society, Highlands 13920
American Malacological Union, North Myrtle Beach 13924
American Society of Animal Science, Campaign 14100
American Society of Ichthyologists and Herpetologists, Carbondale 14134
American Society of Mammalogists, Provo 14142
American Society of Zoologists, Chicago 14174
Animal Behavior Society, Denver 14218
Desert Tortoise Council, Palm Desert 14967
Friends of the Sea Lion Marine Mammal Center, Laguna Beach 15145
Herpetologists' League, Austin 15205
Human/Dolphin Foundation, Malibu 15234
Inter-American Tropical Tuna Comission, La Jolla 15355
International Society of Cryptozoology, Tucson 15548
Poultry Science Association, Champaign 16345
Simian Society of America, Saint Louis 16475
Society for Experimental and Descriptive Malacology, Ann Arbor 16521
Society for the Study of Amphibians and Reptiles, Oxford 16612
Society of Nematologists, Beetsville 16684
Society of Protozoologists, New York 16698
Soiety for Northwestern Vertebrate Biology, Olympia 16715
Tarleton Foundation, San Francisco 16763
The Trumpeter Swan Society, Maple Plain 16794
Western Society of Malacologists, Santa Barbara 16885

Publications Index

Register der periodischen Publikationen

003-Euskararen Lekukoak *10349*
AAACE Newsletter *13464*
AAA Newsletter *13380*
AAASA Newsletter *03942*
AABC Newsletter *13509*
AABS Newsletter *14341*
AACE International: Directory of Members *13273*
AACE International: Transactions *13273*
Aachener Beiträge für Baugeschichte und Heimatkunst *05028*
AACN Clinical Issues in Critical Care Nursing *13526*
AACP News *13522*
AACT Directory of Community Theatres in the United States *13525*
AACTE Policy Papers *13519*
AAEA Journal *04164*
AAEA Newsletter *04164*
AAEE Annual Courses *13530*
AAEE Annual Meetings *13530*
AAEE Case Reports *13530*
AAEE Minimonographs *13530*
AA Files *11245*
AAFP Reporter *13394*
AAHA Trends *13451*
AAHE Bulletin *13482*
AAHSLD News *14373*
AAIE Bulletin *14393*
AAI Green Book *06924*
AAI Newsletter *06924*
AAIV Highlights *13544*
AALS Newsletter *14267*
AAMO Newsletter *06588*
AAMTI Newsletter *03880*
AAO Journal *13418*
AAOM Membership Directory *13560*
AAOM News *13560*
AAPA Bulletin *13425*
AAPAM Newsletter *03934*
AAPH Bulletin *09490*
AAPH Data Resouces *09490*
AAP News *13421*
AAP Newsletter *13433, 13561*
AAPO Newsletter *13428*
AAPS News *14385*
AAPS Newsletter *09212*
AAPT News *13563*
Aarbøger for Nordisk Oldkyndighed og Historie *03711*
De Aardrijkskunde *01505*
Aaron Burr Association: Newsletter *13274*
AARU Bulletin *08451*
AASE Newsletter *03935*
AASL Presidential Hotline *13579*
AASSREC Panorama Newsletter *06732*
AATCC Technical Manual *13592*
AATT Newsletter *13591*
AAU Newsletter *06477*
AAVSO Circular *13597*
AAVSO Report *13597*
AAWCJC Journal *13604*
AAWCJC Quarterly *13604*
AAZV Newsletter *13607*
ABA Journal *13617*
A.B.C.A. Newsletter *01915*
ABC Bulletin *13876*
ABC Infos *04174*
ABC's of Study in Japan *08232*
ABD Flash *01106*
Abdominal Surgery *14095*
Abdruck von durchzuführenden Weiterbildungs- und Fortbildungskursen *05539*
Aberdeen-Angus Herdbook *11194*
Aberdeen-Angus Review *11194*
Abhandlungen der Zoologisch-Botanischen Gesellschaft *01035*
Abhandlungen des Naturwissenschaftlichen Vereins zu Bremen *06136*
ABH Bulletin *14348*
ABHES News *13316*
ABLA-Papers *01162*
ABN Newsletter *10008*
About your Medicines *16832*
Absatzwirtschaft: Zeitschrift für Marketing *06322*
ABS Newsletter *14403*
Abstract Journal in Earthquake Engineering *16087*
Abstracts of AIT Reports and Publications on Energy *11082*
Abstracts of Communications *10686*
Abstracts of Papers Presented to the American Mathematical Society *13929*
Abstracts of Research in Pastoral Care and Counseling *14788*
Abstracts on Cassava *03360*
Abstracts on Hygiene and Communicable Diseases *11720*
Abstracts on Phaseolus Vulgaris *03360*
ABYDAP Informa *09448*
ACA Bulletin *01917*
ACA Conference Proceedings *00243*
Academia Amazonense de Letras: Revista *01561*
Academia Argentina de Letras: Boletín *00038*
Academia Boliviana: Revista *01528*
Academia Brasileira de Letras: Revista *01564*
Academia Campinense de Letras: Publicações *01566*
Academia Catarinense de Letras: Revista *01567*
Academia Cearense de Letras: Revista *01568*
Academia Chilena de Bellas Artes: Boletín *03007*
Academia Chilena de Ciencias: Boletín *03008*
Academia Chilena de Ciencias Sociales, Politicas y Morales: Boletín *03009*
Academia Chilena de Medicina: Boletín *03012*
Academia Colombiana de Ciencias Exactas, Físicas y Naturales: Revista *03336*
Academia Colombiana de Jurisprudencia: Anuario *03338*
Academia Colombiana de Jurisprudencia: Revista *03338*
Academia Colombiana de la Lengua: Anuario *03339*
Academia Colombiana de la Lengua: Boletín *03339*
Academia Costarricense de la Lengua: Boletín *03427*
Academia das Ciências de Lisboa: Anuário Académico *09633*
Academia das Ciências de Lisboa: Boletim *09633*
Academia de Buenas Letras de Barcelona: Boletín *10238*
Academia de Buenas Letras de Barcelona: Memorias *10238*
Academia de Ciencias Exactas, Físicas, Químicas y Naturales de Zaragoza: Revista *10239*
Academia de Ciencias Médicas, Físicas y Naturales de Guatemala: Annals *06553*
Academia de Ciencias Políticas y Sociales: Boletín *16959*
Academia de Estomatología del Perú: Actualidades Académicas *09437*
Academia de Estomatología del Perú: Boletín *09437*
Academia de Geografía e Historia de Costa Rica: Anales *03429*
Academia de Historia del Zulia: Boletín *16960*
Academia de la Investigación Científica: Ciencia *08682*
Academia de la Lengua y Cultura Guaraní: Revista *09424*
Academia de Letras da Bahia: Revista *01570*
Academia de Letras de Piauí: Revista *01571*
Academia de Stiinte Medicale: Buletinul *09735*
Academiae Analecta: Mededelingen van de K.A.W.L.S.K. Klasse der Letteren *01409*
Academiae Analecta: Mededelingen van de K.A.W.L.S.K. Klasse der Schone Kunsten *01409*
Academiae Analecta: Mededelingen van de K.A.W.L.S.K. Klasse der Wetenschappen *01409*
Academia Ecuatoriana de la Lengua: Memorias *03839*
Academia Gentium Pro Pace: Annuario *07123*
Academia Guatemalteca de la Lengua: Boletín *06556*
Academia Hondureña: Boletín *06580*
Academia Hondureña de Geografía e Historia: Revista *06581*
Academia Iberoamericana y Filipina de Historia Postal: Boletín *10243*
Academia Matogrossense de Letras: Revista *01574*
Academia Médico-Quirúrgica Española: Anales *10244*
Academia Mexicana de la Historia: Memorias *08686*
Academia Nacional de Agronomía y Veterinaría: Anales *00040*
Academia Nacional de Belas-Artes: Boletim *09634*
Academia Nacional de Ciencias de Bolivia: Publicaciones *01529*
Academia Nacional de Ciencias de Bolivia: Revista *01529*
Academia Nacional de Ciencias de Córdoba: Actas *00043*
Academia Nacional de Ciencias de Córdoba: Boletín *00043*
Academia Nacional de Ciencias de Córdoba: Miscelanea *00043*
Academia Nacional de Ciencias Económicas: Anales *00044*
Academia Nacional de Ciencias Exactas, Físicas y Naturales: Anales *00045*
Academia Nacional de Ciencias Exactas, Físicas y Naturales de Lima: Actas *09438*
Academia Nacional de Ciencias: Memorias *08688*
Academia Nacional de Ciencias: Revista *08688*
Academia Nacional de Derecho y Ciencias Sociales: Anales *00047*
Academia Nacional de Derecho y Ciencias Sociales (Córdoba): Anales *00048*
Academia Nacional de Farmacia: Boletim *01576*
Academia Nacional de Geografía: Anales *00049*
Academia Nacional de Historia y Geografía: Revista *08689*
Academia Nacional de la Historia: Anuario *16961*
Academia Nacional de la Historia: Boletín *00050, 16961*
Academia Nacional de la Historia: Memorias *16961*
Academia Nacional de Medicina: Boletim *01577*
Academia Nacional de Medicina: Boletín *00051*
Academia Nacional de Medicina del Uruguay: Boletín *13241*
Academia Panameña de la Historia: Boletín *09406*
Academia Panameña de la Lengua: Boletín *09407*
Academia Paraibana de Letras: Discursos e Ensaios *01578*
Academia Paraibana de Letras: Revista *01578*
Academia Pernambucana de Letras: Revista *01580*
Academia Peruana de Cirurgía: Revista *09440*
Academia Peruana de la Lengua: Boletín *09441*
Academia Petrarca di Lettere, Arti e Scienze: Atti e Memorie *07124*
Academia Portuguesa da História: Boletim *09635*
Academia Puertorriqueña de la Lengua Española: Boletín *09719*
Academia Riograndense de Letras: Revista *01581*
Academia Salvadoreña de la Historia: Boletín *03917*
Academia Venezolana de la Lengua: Boletín *16963*
Academic Athletic Journal *15825*
Academic Position Report *14489*
Académie de la Réunion: Bulletin *04089*
Académie de Nîmes: Bulletin Trimestriel *04091*
Académie de Nîmes: Mémoires *04091*
Académie des Beaux-Arts: Annuaire *04093*
Académie des Inscriptions et Belles-Lettres: Mémoires *04095*
Académie des Sciences, Agriculture, Arts et Belles-Lettres d'Aix: Bulletin *04099*
Académie des Sciences, Arts et Belles-Lettres de Dijon: Mémoires *04100*
Académie des Sciences, Lettres et Arts d'Arras: Mémoires *04107*
Académie Malgache: Bulletin *08620*
Académie Malgache: Mémoires *08620*
Académie Nationale de Chirurgie Dentaire: Bulletin *04116*
Académie Nationale de Médecine: Bulletin *04118*
Académie Polonaise des Sciences: Conférences *04121*
Académie Royale de Langue et de Littérature Françaises: Annuaire *01093*
Académie Royale de Langue et de Littérature Françaises: Bulletin *01093*
Académie Royale de Médecine de Belgique: Monthly Bulletin *01094*
Académie Royale des Sciences d'Outre-Mer: Bulletin des Séances *01096*
Académie Vétérinaire de France: Bulletin *04122*
Academy for Educational Development: Academy News *13277*
Academy for Educational Development: Newsletter *13277*
The Academy Forum *13432*
Academy of Ambulatory Foot Surgery: Directory *13280*
Academy of Ambulatory Foot Surgery: Newsletter *13280*
Academy of American Poets: Booklist *13282*
Academy of Dentistry for the Handicapped: Membership Referral Roster *13288*
Academy of International Business: Membership Directory *13293*
Academy of International Business: Newsletter *13293*
Academy of Legal Studies in Business: Newsletter *13295*
Academy of Management Executive *13296*
Academy of Management Journal *13296*
Academy of Management Newsletter *13296*
Academy of Management Proceedings *13296*
Academy of Management Review *13296*
Academy of Marketing Science *13297*
Academy of Marketing Science News *13297*
Academy of Medicine: Bulletin *01816*
Academy of Natural Sciences: Academy News *13299*
Academy of Operative Dentistry: Membership Roster *13300*
Academy of Political Science: Proceedings *13303*
Academy of Rehabilitative Audiology: Membership Directory *13305*
Academy of Science Fiction, Fantasy and Horror Films: Newsletter *13307*
Academy of Sciences: Journal *08514*
Academy of Scientific Hypnotherapy: Bulletin *13308*
Academy of Television Arts and Sciences: Debut *13310*
Academy of the Arabic Language: Councils and Conferences Proceedings *03860*
Academy of the Arabic Language: Review of the Academy *03860*
Academy of the Social Sciences in Australia: Newsletter *00211*
Academy of the Social Sciences in Australia: Report *00211*
Academy of Veterinary Cardiology: Membership Directory *13311*
Academy of Veterinary Cardiology: Newsletter *13311*
The Academy Papers *13423*
Academy Players Directory *13298*
Academy PM&R News *13424*
ACA Monographs *13774*
ACA Newsletter *13774*
ACA Program and Abstracts *13774*
ACARTSOD Newsletter *08566*
ACAT News *13684*
ACA Transactions *13774*
ACA Update *13760*
Accademia Cosentina: Atti *07133*
Accademia delle Scienze Mediche di Palermo: Atti *07152*
Accademia di Agricoltura Scienze e Lettere: Atti e Memorie *07154*
Accademia Etrusca: L'Annuario *07178*
Accademia Etrusca: Note e Documenti *07178*
Accademia Gioenia di Scienze Naturali: Atti *07186*
Accademia Letteraria Italiana, Arcadia: Atti e Memorie: Serie Terza *07199*
Accademia Medica Pistoiese Filippo Pacini: Bollettino *07206*
Accademia Nazionale dei Lincei: Annuario *07211*
Accademia Nazionale dei Lincei: Celebrazioni lincee *07211*
Accademia Nazionale delle Scienze, detta dei XL: Annali *07213*
Accademia Nazionale di Agricoltura: Annali *07214*
Accademia Nazionale di Scienze, Lettere e Arti: Atti e Memorie *07222*
Accademia Nazionale Virgiliana di Scienze, Lettere ed Arti di Mantova: Atti e Memorie *07224*
Accademia Polacca delle Scienze: Conferenze *07226*
Accademia Pontaniana: Quaderni *07228*
Accademia Roveretana degli Agiati: Atti *07234*
Accademia Toscana di Scienze e Lettere La Colombaria: Atti e Memorie *07243*
ACCH Network *14349*
ACCH News *14349*
Acciaio *07511*
ACCI Conference Proceedings *13765*
Accident Analysis and Prevention *13496*
Accident Prevention *02516*
ACCI Newsletter *13765*
ACCIS Newsletter *10608*
Accordion Teacher's Guild Newsletter *13312*
The Account *13313*
Accountancy *12219*
Accountants Computer Users Technical Exchange: Membership Directory *13313*
Accountants Digest *12219*
Accountants' Guide *11759*
The Accountant's Magazine *12220*
Accounting and Business Research *12219*
Accounting and Finance *00212*
Accounting Review *13444*
Accounts of Chemical Research *13688*
ACCRA Newsletter *13687*
Accreditation *14930*
Accreditation Yearbook *13315*
Accredited Journalism and Mass Communications Education *13319*
ACCT-O-Line *14419*
ACE-Bulletin *11206*
ACEI Exchange *14279*
ACE News *00274*
ACE Newsletter *01101*
ACEP News *13705*
ACES Bulletin *14282*
ACHPER National Journal *00280*
ACI Materials Journal *13748*
ACISN: Bollettino Periodico delle Attività *07263*
ACI Structural Journal *13748*
ACJS Employment Bulletin *13286*
ACJS Membership Directory *13286*
ACJS Today *13286*
ACLALS Newsletter *08067*
ACME Survey of Key Management Information *14894*
ACM Guide to Computing Literature (Guide) *14285*
ACMS Research Report *10009*
ACM Transactions on Computer Systems (TOCS) *14285*
ACM Transactions on Database Systems (TODS) *14285*
ACM Transactions on Graphics (TOG) *14285*
ACM Transactions on Mathematical Software (TOMS) *14285*
ACM Transactions on Office Information Systems (TOOIS) *14285*
ACM Transactions on Programming Languages & Systems (TOPLAS) *14285*
ACNM Membership Directory *13720*
ACOOG Newsletter *13726*
ACOPS News *11207*
ACOPS Yearbook *11207*
Açoreana *09680*
Acoustics Australia *00253*
ACP Newsletter *13734*
Acquisition du language et pathologie *04746*
The Acquisition of Film Rights in Literary Properties *00349*
ACRES Membership Newsletter *13768*
ACR-Info *00992*
ACSA News *14416*
ACSM Membership Directory *13737*
ACS Newsletter/Bulletin de l'AEC *01899*
Acta Acustica *03113, 04781*
Acta Agriculturae Scandinavica *10489*
Acta Agrobotanica *09558*
Acta Agronomica *06661*
Acta Alimentaria *06661*
Acta Amazonica *01688*
Acta Anaesthesiologica Belgica *01424*
Acta Anaesthesiologica Scandinavica *09280*
Acta Anaesthesiologica Sinica *03326*
Acta Anaesthisiologica Hellenica *06503*
Acta Anatomica Nipponica *08320*
Acta Anesthesiologica Italia *07835*
Acta Antiqua *06661*
Acta Archaeologica *06661*
Acta Archaeologica Lodziensia *09545*
Acta Arctica *03560*
Acta Asiatica *08434*
Acta Astronomica Sinica *03165*
Acta Astrophysica Sinica *03165*
Acta Automatica Sinica *03162*
Acta Biochimica et Biphysica *06661*
Acta Biologiae et Medicinae Experimentalis *17076*
Acta Biologica *06661, 09543*
Acta Botanica *06661, 08735*
Acta Botanica Fennica *03993, 03995*
Acta Botanica Gallica *04635*
Acta Botanica Neerlandica *08954*
Acta Botanica Sinica *03121*
Acta Cardiologica Belgica *01427*
Acta Chemica Scandinavica *10555*
Acta Chimica *06661*
Acta Chimica Sinica *03167*
Acta Chirurgica *06661*
Acta Chirurgica Austriaca *00769*
Acta Chirurgica Belgica *01471*
Acta Científica Venezolana: Multidisciplinary Scientific Journal *16977*
Acta Classica *10130*
Acta Crystallographica, Section B *12401*
Acta Crystallographica, Section C *12401*
Acta Cystallographica, Section A *12401*
Acta Cytologica *00756, 05984, 14116*
Acta Demographica *05330*
Acta Endoscopica *04892*
Acta Entomologica *03198*
Acta Entomologica Bohemoslovaca *03500*
Acta Ethnographica *06661*
Acta Forestalia Fennica *04050*
Acta Gastroenterológica Latinoamericana *00170*
Acta Geneticae Medicae et Gemellologiae/Twin Research *07720*
Acta Genetica Medicae et Gemellologiae: Twin Research *07818*
Acta Genetica Sinica *03199*
Acta Geodaetica, Geophysica et Montanistica *06661*
Acta Geodetica et Cartographica Sinica *03191*
Acta Geographica *04654*
Acta Geographica Lodziensia *09545*
Acta Geologia *03170*
Acta Geologica *06661*
Acta Geologica Lilloana *00140*
Acta Geophysica Sinica *03171*
Acta Gepographica Sinica *03200*
Acta Haematologica Japonica: International Journal of Hematology *08331*
Acta Historiae Artium *06661*
Acta Historica *06661*
Acta Historica Leopoldina: Abhandlungen aus dem Archiv für Geschichte der Naturforschung und Medizin der Deutschen Akademie der Naturforscher Leopoldina *05285*
Acta Horticulturae *08940*
Acta Juridica *06661*
Acta Leprologica *04191*
Acta Linguistica *06661*
Acta Litteraria *06661*
Acta Mathematica *06661, 10490*
Acta Medica *06661*
Acta medica austriaca *00861*
Acta Medicinae Legalis et Medicinae Socialis *07191*
Acta Microbiologica *06661*
Acta Microbiologica et Immunologica Hungarica *06636*
Acta Microbiologica Polonica *09588*
Acta Morphologica *06661*

Acta Mozartiana 05474
Acta Museorum Agriculturae Pragae 12326
Acta Museorum Italicorum Agricultare: Rivista di Storia dell'Agricoltura 07442
Acta Musicologica 10738
Acta Mycologica Sinica 03121
Acta Myologica 09558
Acta Obstetrica et Gynaecologica Japonica 08362
Acta Oeconomica 06661
Acta Oncologica Brasileira 01650
Acta Ophthalmalogica Scandinavica 03658
Acta Orientalia 03792, 06661
Acta O.R.L. Italica 07890, 07924
ACTA Orthopaedica Belgica 01428
Acta Orthopaedica et Traumatologica Turcica 11170
Acta Otorhinolaryngologica Italica 07384
Acta Otorinolaringologica Española 10393
Acta Paediatrica 06661
Acta Paediatrica Japonica 08265
Acta Palaeontologica Sinica 03213
Acta Pediatrica Sinica 03289
Acta Pharmaceutica Hungarica 06633
Acta Philologica Aenipotana 00623
Acta Philosophica Fennica 04016
Acta Physica 06661
Acta Physica Polonica ser. A: Europhysics Journal 09569
Acta Physica Polonica ser. B: Europhysics Journal 09569
Acta Physica Slovaca 10060
Acta Physiologica 06661
Acta Physiologica, Pharmacologica et Therapeutica Latino-Americana 03060
Acta Phytoecologica et Geobotanica Sinica 03121
Acta Phytogeographica Suecica 10576
Acta Phytopathologica 06661
Acta Phytotaxonimica Sinica 03121
Acta Phytotaxonomica et Geobotanica 08423
Acta Poloniae Pharmaceutica 09565
Acta Polytechnica Scandinavica 04074
Acta Psychiatrica Belgica 01483
Acta Reumatológica Portuguesa 09714
Acta Sagittariana 05992
Actas Dermosifiliográficas 10242
Acta Societatis Botanicorum Poloniae 09558
Acta Supplementa 08828
Acta Technica 03500, 06661
Acta Technica Gedanensia 09543
Acta Urologica Italica 07957
Acta Venezolana de ORL 17029
Acta Veterinaria 06661
Acta Virologica 03500, 10060
Acta Zoologica 06661, 08735, 10490
Acta Zoologica Fennica 03993, 03995, 04029
Acta Zoologica Lilloana 00140
Acta Zootaxonomia Sinica 03198
ACTEA Bulletin 09209
ACTEA Tools and Studies 09209
ACTFL Foreign Language Education Series 13770
Action for Dysphasic Adults: Newsletter 11201
Activities of the Federation of Polish Medical Societies 09631
ACTivity 13742
Actualidades Bibliotecológicas 13242, 13243
L'Actualité Chimique 04791
L'Actualité Chimique et Analysis 04646
Actualité Combustibles Energie 04568
The Actuarial Society of Finland Working Papers 04005
Actuarial Update 13379
De Actuaris 08818
The Actuary 16631
A.C.U. Bulletin of Current Documentation 11329
Acupuncture Foundation of Canada: Newsletter 01818
Acupuncture International Association: Bulletin 13324
ACURIL Newsletter 09723
ACUTA News 14415
ADA Feedback 13291
Adbiyat 09378
Addarah 10004
Addiction Research Foundation: The Journal 01819
ADE Bulletin 04209, 14427
Der Adelsbote 14518
ADFL Bulletin 14428
Adhesion Society: Newsletter 13327
ADHILAC Informa 04139
ADI Global Perspective 13371
ADIPA Newsletter 08639
Adirondack Historical Association: Guide Line 13328
ADLAF-Info 05096
Adler: Zeitschrift für Genealogie und Heraldik 00649
Administration: Journal 06964
Administration Publique 01382
Administration Universitaire Francophone et Européenne en Médecine et Odontologie: Report 04123
Administrative Law Review 13617
Administrative Minutes 16853
Administrative Sciences Association of Canada: Newsletter 01820
Administrative Sciences Association of Canada: Proceedings 01820
Administrator 12222
The Administrator 13518
Admissions Requirements at U.S. and Canadian Dental Schools 13528
ADM Newsletter 13287
ADSA Directory 13781
Adult Christian Education Foundation: Annual Report 13331
Adult Education 13464
Adult Education and Development 05562
Adult Education in Finland 03975
Adult Learner 15149
Adults Learning 12596
Adunanze straordinarie per il conferimento dei premi A. Feltrinelli 07211
Advance 14265, 15122
Advanced Ceramic Materials 13685
Advanced Lighter-Than-Air Review 13359
Advanced Materials and Processes 14255
Advances in Cement Research 12294
Advances in Cryogenic Engineering 14944
Advances in Physiology Education 14005
Advances in Space Research 04411
Advances in Thanatology 13886
Advice for the Patient 16832
Advisor 14419
Advisory Notes 09227
The Advocate 01827
Advocate 14523
Advocate for Education of the Deaf 14830
AEDA News 04129
AEDEC Newsletter 04481
AEEMA Newsletter 01273
AEESEAP Journal of Engineering Education 09118
AEESEAP Newsletter 09118
Äidinkielen opettajain liitto: Yearbook 03948
Aerospace Market Outlook 16768
Årbok for Foreningen til norske Fortidsminnesmerkers Bevaring 09266
AERC Newsletter 08467
Aeromedical Journal 14377
Aeronautical Journal 12765
Aeronautical Society of South Africa: Journal 10119
Aeronautica Meridiana 10195
L'Aéronautique et l'Astronautique 04144
Aerosol Science and Technology 13465
Aerospace 12765
Aerospace America 13872
Aerospace and Electronic Systems Magazine 15317
Aerospace Education Foundation: Newsletter 13337
Aerospace Electrical Society: News & Views 13338
Aerospace Engineering 16433
Aerospace Journal 14107
Aerospace Medical Association: Membership Directory 13339
The Aerostat 11533
Aerostation 14398
AERO Sun-Times 13368
L'Aerotecnica-Missili e Spazio 07297
AER Report 14291
Ärzteblatt Baden-Württemberg 05211
Ärzteblatt Baden-Württemberg, (ÄBW): Offizielles Organ der Landesärztekammer Baden-Württemberg 06091
Ärzteblatt Rheinland-Pfalz 06093
Ärztin: Mitteilungsblatt des Deutschen Ärztinnenbundes 05494
AES-Magazine 13276
AESOP News 08827
AESOP Papers 08827
Aesthetics 08114
Aesthetic Surgery 14045
AESTM Newsletter 04487
AETFAT Bulletin 08621
AETFAT Index 08621
AFAA Newsletter 08803
AFA Newsletter 13807
AFCR News 11213
Afford Ability 10147
Afghanistan Studies Journal 13340
Afnetan 09223
AFOD Newsletter 11041
Africa 11142
Africa 2001 Dialogue with the Future 10147
Africa and Africa Bibliography 12321
Africa Development 10026
The Africa Fund: Annual Report 13341
Africa Insight 10120
Africa Management Development Forum 11056
Africa Media Review 08466
Africana 10507
African Affairs 11211, 12766
African-American Institute: Annual Report 13342
African-American Institute: Bulletin 13342
African American Museum Newsletter 13343
African Bureau of Educational Sciences: Bulletin d'Information 17139
African Centre Human Rights Newsletter 05005
African Entomology 10140
Africa Newsletter 03861
African Geology 04317
African Heritage Studies Association: Newsletter 13345
African Herp News 10146
African Journal of Agricultural Sciences 03942
The African Journal of Ecology 08498
African Journal of Political Economy 09212
African Journal of Science and Technology 08471
African Literature Association: Annual Selected Conference Paper 01821
African Mountains and Highlands Newsletter 03937
African Regional Network for Microbiology: Newsletter 09220
African Rehabilitation Journal 17161
African Research and Documentation 13097
African Review 09212
African Small Ruminant Research Network: Newsletter 03938
African Social Studies Forum 08473
African Soils 09244
African Studies 03861
African Studies Association: Issue 13346
African Studies Review 13346
African Wildlife 10236
Africa Palm 08059
Africa Report 13342
AFRICOM Newsletter 08466
Afrika Mathematika 09217
Afro-American Historical and Genealogical Society: Journal 13347
Afro-American Historical and Genealogical Society: Newsletter 13347
AGA News 13829
Agave 14964
Agbiotech News and Information 11720
AgBiotech News and Information 11730
AGD Impact 13292
AGE 13445
Age and Ageing 11568
AGE Current Awareness Service 11081, 11082
AGEMUS Nachrichten Wien: Internes Informationsorgan der Arbeitsgemeinschaft 00538
Agence Spatiale Européenne: Rapport Annuel 04127
Agency Issues 15511
L'Agenda del Giornalista 07424
Agenda dell'Archeologia Italiana 07423
Agenda du Musicien: Musikerkalender 10900
AGE News 11081, 11082
AGENPP Press: Agenzia di Stampa Quotidiana 07123
Agenzia Giornalistica SAFJ-Press 07510
AGE Refdex 11082
Agest for Japan SAE Revlon 12540
Agest for SAE Publications 12540
Aggiornamenti di Psicoterapia e Psicologia Clinica 07561
Aggiornamenti in Ostetricia e Ginecologia 07978
AGHE Exchange Newsletter 14299
AGID News 11088
Aging International 12362
A Glimpse of the Associated Schools Project 04143
AGMANZ News 09116
Agopuntura Moderna, Scienze dell'Uomo Totale: Rivista di Bioterapia e Psicobiofisica 07503
Agora 15646
Agraria dei Georgofili 07177
Agrarische Rundschau 00718, 00750
Agrarrecht: Zeitschrift für das gesamte Recht der Landwirtschaft, der Agrarmärkte und des ländlichen Raumes 05314
Agrarsoziale Gesellschaft e.V.: Geschäfts- und Arbeitsbericht 05044
Agrarsoziale Gesellschaft e.V.: Kleine Reihe 05044
Agrarsoziale Gesellschaft e.V.: Materialsammlung 05044
Agrarspectrum 05282
Agrarspectrum: Schriftenreihe 05282
Agrárvilág 06612
Agressive Behavior 15540
Agricolas 01639
Agricultural and Economic Report 11162
Agricultural and Food Research Council: Annual Report 11213
Agricultural Communicators in Education: Newsletter 13350
The Agricultural Engineer incorporating Soil & Water 12292
Agricultural Engineering Abstracts 11720
Agricultural Engineering Australia 00213
Agricultural Engineering Magazine 14097
Agricultural History 13351
Agricultural History Review 11474
Agricultural History Society: Symposium Proceedings 13351
Agricultural Information Development Bulletin 11064
Agricultural Law Update 13447
Agricultural Literature 03501
Agricultural Manpower 11648
Agricultural Research Council of Zimbabwe: Annual Report 17162
Agricultural Science 00296
Agricultural Society of Trinidad and Tobago: Journal 11105
Agricultura Técnica 03056
Agriculture Association of China: Journal 03232
Agriculture Digest 16436
Agrindex 07248
Agritrop Tropical and Subtropical 04365
Agroanalysis 01664
Agroforestry Abstracts 11720
Agrohemija 17100
Agrokhimiya: Agrochemistry 09954
Agronomica Tropical 16991
Agronomy Abstracts 14098
Agronomy Journal 14098, 14942
Agronomy News 14098
Agrotecnia de Cuba 03474
Aguiaine 04739
AHANA 06862
AHA Perspectives: Newsletter 13846
AHEA Action 13851
AHI News 08100
AHP Perspective 14305
AHZ: Allgemeine homöopathische Zeitung 05566
AIAA Journal 13872
AIA Bulletin 11266
AIA Newsletter 04137
AIBDA Actualidades 03449
AIB Notizie 07291
AICCP Newsletter 14520
AICS Bulletin 15793
AIDCOM Information 08634
AIDS Action 11239
AIDS Bulletin 10206
Aids/HIV Treatment Directory 13822
Aids Newsletter 11720
AID Verbraucherdienst: Zeitschrift für Fach-, Lehr- und Beratungskräfte im Bereich Ernährung 05045
AIESEC-U.S. Annual Report 13354
AIGA Journal of Graphic Design 13881
Aikakauskirja 04051
AIMS Bulletin 13889
AIN News Bulletin 03872
AINSE Annual Report 00309
AIPE Facilities Management, Operations and Engineering 13894
AIPE Newsline 13894
AIPLA Bulletin 13900
AIPLA Journal 13900
AIPPI Newsletter 10743
Airfields Environment Federation: Newsletter 11218
Air National Guard Optometric Society: Newsletter 13358
Air Sports International 04506
AISA Newsletter 08480
AITIA Magazine 14692
AIT Newsbrief 09499
AIUM Newsletter 13898
Ajatus-yearbook 04016
AJDC: American Journal of Diseases of Children 13933
Ajia Jiho: Monthly Asia Review 08091
Ajia Kenkyu 08093
AJL Newsletter 14451
AJS Newsletter 14311
AJS Review 14311
Akademia e Shkencave dhe e Arteve e Kosovës: Bibliografia/Bibliografija 17076
Akademie der Arbeit in der Universität: Mitteilungen 05046
Akademie der Wissenschaften in Göttingen: Abhandlungen: I. Philologisch-Historische Klasse, II. Mathematisch-Physikalische Klasse 05049
Akademie der Wissenschaften in Göttingen: Jahrbuch 05049
Akademie der Wissenschaften in Göttingen: Nachrichten: I. Philologisch-Historische Klasse, II. Mathematisch-Physikalische Klasse 05049
Akademie der Wissenschaften und der Literatur zu Mainz: Abhandlungen: Geistes- und Sozialwissenschaftliche Klasse 05050
Akademie der Wissenschaften und der Literatur zu Mainz: Abhandlungen: Klasse der Literatur 05050
Akademie der Wissenschaften und der Literatur zu Mainz: Abhandlungen: Mathematisch-Naturwissenschaftliche Klasse 05050
Akademie der Wissenschaften und der Literatur zu Mainz: Jahrbuch 05050
Akademie der Wissenschaften und der Literatur zu Mainz: Veröffentlichungen: Orientalische Kommission 05050
Akademie für öffentliches Gesundheitswesen in Düsseldorf: Blickpunkt 05056
Akhbar-ut-Tib 09364
Akroterion 10130
Aktiv 05475
Aktuelle Gerontologie 00783
Aktuelle Gespräche 05682
Akuserstvo i Ginekologija 01750
Akusticheskii Zhurnal: Acoustics Journal 09954
Akzente, Profile, Innovationen 05120
ALA Bulletin 01821
ALA Handbook of Organization and Membership Directory 13916
ALAI Servicio Mensual de Información y Documentación 03841
ALA News 14452
ALA Newsletter 06876
Al-Arabiyya 13585
A Lavoura 01727
Alberta Construction 01830
Alberta Construction Association: Membership Roster 01830
Alberta Sulphur Research: Bulletin 01849
Alberta Veterinary Medical Association: Newsletter 01852
Albion 16215
Album des Normes de l'Automobile 04997
ALCTS Newsletter 14314
Alcuin Club: Collections 11220
ALECSO Newsletter 11131
Alectoris 06495
Alert 02691
ALERT 03940
Alert 15692
The Alexandria Medical Journal 03866
Alfred Adler Institute: Bulletin 13362
Alfred Adler Institute: Newsletter 13362
Algar 09700
Algebra i Analiz: Algebra and Analysis 09954
Algérie Médicale 00026
ALI-ABA CLR Review 13363
ALI-ABA Course Materials Journal 13363
ALI-ABA Reporter 13363
Al-Ilm 09357
ALI Reporter 13912
Alkan Society: Bulletin 11221
All Africa Teachers' Organization Newsletter 06475
Allahabad Mathematical Society: Bulletin 06713
All Differentiation 12392
Allensbacher Almanach 05079
Allergo Journal 05316
Allergologia e Imunologia Clinica: Boletim 09689
Allergy: European Journal of Allergy and Clinical Immunology 03720
Allergy Alert 01855
Allergy & Clinical Immunology News 15397
The Allergy Letters 15450
Allergy Quarterly 01856
Allgemeines Statistisches Archiv 05599
Alliance for Environmental Education: Annual Report 13365
Alliance to Save Energy: Annual Report 13366
Alliance Update 13366
Allied Health Education Directory 14925
Allied Health Trends 14496
Al-Lissan Al-Arabi Magazine 11131
Al-Ma'arif 09367
Al-Mohandes 01059
Al-Moukhtarat 04136
Aloe: Journal of the Succulent Society of South Africa 10230
Al-Omran Al-Arabi 11044
Al-Qantara 10322
ALSAC News 13367
ALS News 01867
Alta Frequenza 07686
Alta Frequenza: Rivista di Elettronica 07411
Általános Földtani Szemle: General Geological Review 06635
Alta Musica: Alta Bücher 00672
Altenhilfe: Beispiele, Informationen, Meinungen 05609
Altern und Entwicklung: Aging and Development 05050
Altpreussische Geschlechterkunde 06382
Alumnae Newsletter 16172
Alwali 09400
Alzheimer Europe Annual Newsletter 01099
Alzheimer's Association Newsletter 13370
Alzheimer Society of Manitoba: Newsletter 01857
AMA Annual Membership Roster & International Buyers' Guide 13926
A Magyar Nyelvtudományi Társaság Kiadványai 06649
A Matematika Tanitása 06597
Amateur Entomologists' Society: Bulletin: Wants & Exchange List 11223
AMATYC News 13928
AMATYC Review 13928
Ambiente 07559
Ambio 10490
Ambix 13003
Ambulatory Pediatric Association: Membership Directory 13373
Ambulatory Pediatric Association: Newsletter 13373
Amdahl Users Group: Directory 13374
Ameghiniana 00094
America 00131
América Indígena 08736
American Academy and Institute of Arts and Letters: Proceedings 13375
American Academy and Institute of Arts and Letters: Yearbook 13375
American Academy for Cerebral Palsy and Developmental Medicine: Newsletter 13376

American Academy for Jewish Research: Monograph Series 13377
American Academy of Advertising: Proceedings of the Conference 13380
American Academy of Allergy and Immunology Abstract Book 13381
American Academy of Allergy and Immunology: News & Notes 13381
American Academy of Arts and Sciences: Bulletin 13382
American Academy of Child and Adolescent Psychiatry: Membership Directory 13383
American Academy of Child and Adolescent Psychiatry: Newsletter 13383
American Academy of Cosmetic Surgery: Membership Roster 13385
American Academy of Cosmetic Surgery: Newsletter 13385
American Academy of Dental Electrosurgery: Current Events 13387
American Academy of Dental Group Practice: Directory 13388
American Academy of Dental Group Practice: Newsletter 13388
American Academy of Dental Practice Administration: Essay Tapes 13389
American Academy of Dental Practice Administration: Proceedings 13389
American Academy of Dermatology: Bulletin 13390
American Academy of Fixed Prosthodontics: Newsletter 13395
American Academy of Forensic Sciences: Membership Directory 13396
American Academy of Forensic Sciences: Newsletter 13396
American Academy of Gnathologic Orthopedics: Journal 13397
American Academy of Gnathologic Orthopedics: Membership Roster 13397
American Academy of Gold Foil Operators: Roster 13398
American Academy of Implant Dentistry: Newsletter 13399
American Academy of Implant Prosthodontics: Membership Directory 13400
American Academy of Implant Prosthodontics: Newsletter 13400
American Academy of Insurance Medicine: Transactions 13401
American Academy of Matrimonial Lawyers: Newsletter 13403
American Academy of Mechanics: Directory 13405
American Academy of Medical Administrators: Executive 13406
American Academy of Ophthalmology: Directory 13410
American Academy of Oral and Maxillofacial Radiology: Roster of Membership 13412
American Academy of Oral Medicine: Newsletter 13413
American Academy of Orthodontics for the General Practitioner: Academic Calendar 13415
American Academy of Orthopaedic Surgeons: Bulletin 13416
American Academy of Otolaryngic Allergy: Directory 13419
American Academy of Otolaryngic Allergy: Newsletter 13419
American Academy of Otolaryngic Allergy: Transactions 13419
American Academy of Pediatric Dentistry: Newsletter 13420
American Academy of Periodontology: Newsletter 13422
American Academy of Periodontology: Roster of Members 13422
American Academy of Physician Assistants: Newsletter 13425
American Academy of Podiatric Sports Medicine: Membership Directory 13427
American Academy of Podiatric Sports Medicine: Newsletter 13427
American Academy of Political and Social Science: The Annals 13429
American Academy of Psychiatrists in Alcoholism and Addictions: Newsletter 13430
American Academy of Psychiatry and the Law: Newsletter 13431
American Academy of Psychotherapists: Directory 13433
American Academy of Religion: Academy Series 13434
American Academy of Religion: Directory of Position-Holders 13434
American Academy of Religion: Journal 13434
American Academy of Religion: Monograph Series 13434
American Academy of Restorative Dentistry: Roster 13435
American Academy of Sanitarians: Newsletter 13437
American Academy of Sanitarians: Roster of Diplomates 13437
American Academy of Sports Physicians: Newsletter 13438
American Academy of Veterinary Dermatology: Membership Directory 13441
American Academy of Veterinary Dermatology: Newsletter 13441
American Academy of Veterinary Nutrition: Membership Directory 13442
American Academy on Mental Retardation: Newsletter 13443
American Accounting Association: Newsletter 13444
American Agricultural Economics Association: Choices 13446
American Agricultural Economics Association: Newsletter 13446
American Agricultural Law Association Membership Directory 13447
American Alliance for Health, Physical Education, Recreation and Dance: Update 13449
American Animal Hospital Association: Proceedings 13451
American Annals of the Deaf 14830, 14857
American Anorexia/Bulimia Association: Newsletter 13452
American Anthropologist 13453, 15148
American Antiquity 16492
American Arbitration Association: Newsletter 13456
American Architectural Foundation: Forum 13457
American Archivist 16633
American Art Therapy Association: Journal 13459
American Art Therapy Association: Newsletter 13459
American Art Therapy Association: Proceedings of Annual Conference 13459
American Assembly: Annual Report 13461
American Assembly of Collegiate Schools of Business: Newsline 13462
American Association for Accreditation of Laboratory Animal Care: Communique 13463
American Association for Adult and Continuing Education: Membership Directory 13464
American Association for Aerosol Research: Newsletter 13465
American Association for Agricultural Education: Directory 13466
American Association for Agricultural Education: Journal 13466
American Association for Agricultural Education: Summaries of Studies 13466
American Association for Applied Linguistics: Newsletter 13467
American Association for Artificial Intelligence: AI Magazine 13468
American Association for Artificial Intelligence: Conference Proceedings 13468
American Association for Budget and Program Analysis: Newsletter 13469
American Association for Cancer Education: Directory of Members 13470
American Association for Cancer Research: Directory 13471
American Association for Cancer Research: Proceedings 13471
American Association for Career Education: Newsletter 13472
American Association for Chinese Studies: Directory of Members 13473
American Association for Chinese Studies: Newsletter 13473
American Association for Clinical Chemistry: Annual Report 13474
American Association for Correctional Psychology: Newsletter 13475
American Association for Crystal Growth: Membership Directory 13476
American Association for Crystal Growth: Newsletter 13476
American Association for Dental Research: Around 13477
American Association for Dental Research: Membership Directory 13477
American Association for Geriatric Psychiatry: Membership Directory 13480
American Association for Geriatric Psychiatry: Newsletter 13480
American Association for Laboratory Accreditation: Directory of Accredited Laboratories 13483
American Association for Laboratory Accreditation: Newsletter 13483
American Association for Laboratory Animal Science: Bulletin 13484
American Association for Marriage and Family Therapy: Journal 13485
American Association for Marriage and Family Therapy: Membership Directory 13485
American Association for Marriage and Family Therapy: Newspaper 13485
American Association for Medical Transcription: Journal 13486
American Association for Music Therapy: International Newsletter 13487
American Association for Pediatric Ophthalmology and Strabismus: Journal 13489
American Association for Pediatric Ophthalmology and Strabismus: Membership Directory 13489
American Association for Rehabilitation Therapy: Directory 13490
American Association for Rehabilitation Therapy: Journal 13490
American Association for Rehabilitation Therapy: Newsletter 13490
American Association for Social Psychiatry: Newsletter 13491
American Association for Textile Technology: Annual Conference Proceedings 13495
American Association for Textile Technology: Membership Directory 13495
American Association for Textile Technology: Newsletter 13495
American Association for the Advancement of Science: Handbook 13497
American Association for the Advancement of Slavic Studies: Directory of Members 13498
American Association for the Advancement of Slavic Studies: Newsletter 13498
American Association for the History of Medicine: Membership Directory 13500
American Association for the History of Medicine: Newsletter 13500
American Association for the Study of Liver Diseases: Journal 13502
American Association for the Study of Liver Diseases: Newsletter 13502
American Association for World Health: Newsletter 13505
American Association of Blacks in Energy: Bulletin 13510
American Association of Blood Banks: News Briefs 13511
American Association of Botanical Gardens and Arboreta: Newsletter 13512
American Association of Botanical Gardens and Arboreta: Proceedings 13512
American Association of Bovine Practitioners: Directory 13513
American Association of Bovine Practitioners: Newsletter 13513
American Association of Bovine Practitioners: Proceeding of Annual Meeting 13513
American Association of Certified Orthoptists: Directory 13516
American Association of Chairmen of Departments of Psychiatry: Membership List 13517
American Association of Christian Schools: Directory 13518
American Association of Christian Schools: Newsletter 13518
American Association of Colleges of Nursing: Annual Report 13520
American Association of Colleges of Nursing: Newsletter 13520
American Association of Colleges of Osteopathic Medicine: Annual Organizational Guide 13521
American Association of Colleges of Osteopathic Medicine: Annual Statistical Report 13521
American Association of Colleges of Podiatric Medicine: Newsletter 13523
American Association of Dental Examiners: Bulletin 13527
American Association of Dental Examiners: Proceedings 13527
American Association of Dental Schools: Directory of Institutional Members 13528
American Association of Diabetes Educators: Newsletter 13529
American Association of Endodontists: Communique 13531
American Association of Endodontists: Journal 13531
American Association of Endodontists: Membership Roster 13531
American Association of Equine Practitioners: Directory 13534
American Association of Equine Practitioners: Newsletter 13534
American Association of Equine Practitioners: Proceedings 13534
American Association of Feed Microscopists: Newsletter 13535
American Association of Gynecological Laparoscopists: News-Scope 13537
American Association of Handwriting Analysts: Annals 13538
American Association of Handwriting Analysts: Directory 13538
American Association of Handwriting Analysts: Newsletter 13538
American Association of Hospital Podiatrists: Newsletter 13541
American Association of Housing Educators: Newsletter 13542
American Association of Housing Educators: Proceedings 13542
American Association of Industrial Veterinarians: Directory 13544
American Association of Laban Movement Analysts: Directory 13545
American Association of Laban Movement Analysts: Newsletter 13545
American Association of Language Specialists: Yearbook 13546
American Association of Medical Society Executives: Hotline 13550
American Association of Museums: Aviso 13554
American Association of Neurological Surgeons: Bulletin 13555
American Association of Neuropathologists: Roster of Members 13556
American Association of Oral and Maxillofacial Surgeons: Forum 13557
American Association of Orthodontists: Membership Directory 13558
American Association of Petroleum Geologists: Bulletin 13562
American Association of Phonetic Sciences: Newsletter 13564
American Association of Poison Control Centers: Annual Report 13569
American Association of Poison Control Centers: Membership Directory 13569
American Association of Pro Life Obstetricians and Gynecologists: Newsletter 13573
American Association of Psychiatric Administrators: Journal 13574
American Association of Psychiatric Administrators: List of Members 13574
American Association of Psychiatric Administrators: Newsletter 13574
American Association of Psychiatric Services for Children: Membership Directory 13575
American Association of Psychiatric Services for Children: Newsletter 13575
American Association of Public Health Dentistry: Communique 13576
American Association of Public Health Physicians: Bulletin 13577
American Association of Railway Surgeons: Newsletter 13578
American Association of School Personnel Administrators: Bulletin 13580
American Association of School Personnel Administrators: Membership Roster 13580
American Association of School Personnel Administrators: Report 13580
American Association of State Colleges and Universities: Membership List 13583
American Association of Suicidology: Newslink 13584
American Association of Teachers of German: Newsletter 13588
American Association of Teachers of German: Rundbrief 13588
American Association of Teachers of Italian: Newsletter 13589
American Association of Teachers of Spanish and Portuguese: Directory 13590
American Association of Tissue Banks: Membership Directory 13593
American Association of Tissue Banks: Newsletter 13593
American Association of University Professors: Academe 13594
American Association of University Women: Action Alert 13595
American Association of Variable Star Observers: Bulletin 13597
American Association of Veterinary Anatomists: Directory 13598
American Association of Veterinary Anatomists: Newsletter 13598
American Association of Veterinary Laboratory Diagnosticians: Newsletter 13599
American Association of Veterinary Parasitologists: Directory 13600
American Association of Veterinary Parasitologists: Newsletter 13600
American Association of Veterinary Parasitologists: Proceedings 13600
American Association of Wildlife Veterinarians: Membership Directory 13602
American Association of Wildlife Veterinarians: Newsletter 13602
American Association of Women Dentists: Chronicle 13603
American Association of Women Dentists: Newsletter 13603
American Association of Zoological Parks and Aquariums: Annual Conference Proceedings 13606
American Association of Zoo Veterinarians: Conference Proceedings 13607
American Association of Zoo Veterinarians: Membership Directory 13607
American Astronautical Society: Newsletter 13609
American Astronomical Society: Bulletin 13610
American Automatic Control Council: Newsletter 13612
American Aviation Historical Society: Catalog 13613
American Aviation Historical Society: Index 13613
American Aviation Historical Society: Journal 13613
American Aviation Historical Society: Newsletter 13613
American Bamboo Society: Newsletter 13615
American Baptist Quarterly 13616
American Bar Association: Letter 13617
American Behcet's Association: Newsletter 13620
American Bibliography of Slavic & East European Studies 13498
The American Biology Teacher 15831
American Birding Association: Directory 13622
American Blood Commission: Report 13624
American Board of Dental Public Health: Newsletter 13631
American Board of Dermatology: Booklet of Information 13632
American Board of Endodontics: Membership Roster 13633
American Board of Medical Toxicology Journal 13638
American Board of Obstetrics and Gynecology: Bulletin 13642
American Board of Orthodontics Directory 13646
American Board of Otolaryngology: Newsletter 13648
The American Board of Pathology 13649
American Board of Pathology: Information Booklet 13649
American Board of Podiatric Orthopedics: Newsletter 13655
American Board of Preventive Medicine: Bulletin 13657
American Board of Professional Psychology: Newsletter 13658
American Board of Surgery: Booklet of Information 13663
American Broncho-Esophagological Association: Transactions 13670
American Bureau for Medical Advancement in China: Annual Report 13672
American Business Law Journal 13295
American Canal Guides 13675
American Cancer Society: Annual Report 13676
American Cartographer 13753
American Catholic Esperanto Society: Annual Report of the Representative 13679
American Celiac Society: Newsletter 13682
American Chamber of Commerce Researchers Association: Membership Directory 13687
American Chiropractic Association: Membership Directory 13690
American Civil War Round Table, United Kingdom: Newsletter 11224
American Classical Studies 13997
American Clinical and Climatological Association: Transactions 13693
American College Health Association: Action Newsletter 13696
American College of Allergy and Immunology: Membership Directory 13697
American College of Allergy and Immunology: Newsletter 13697
American College of Apothecaries: Newsletter 13698
American College of Chest Physicians: Bulletin 13700
American College of Chest Physicians: Membership Directory 13700
American College of Dentists: Journal 13704
American College of Dentists: News and Views 13704
American College of Foot Orthopedists: Newsletter 13706
American College of General Practitioners in Osteopathic Medicine and Surgery: Membership Directory 13709
American College of General Practitioners in Osteopathic Medicine and Surgery: Newsletter 13709
American College of International Physicians: Bulletin 13712
American College of Laboratory Animal Medicine: Membership Directory 13713
American College of Laboratory Animal Medicine: Newsletter 13713
American College of Mohs' Micrographic Surgery and Cutaneous Oncology: Bulletin 13715
American College of Neuropsychiatrists: Directory 13717
American College of Neuropsychiatrists: Journal 13717
American College of Neuropsychopharmacology: Roster 13718
American College of Nuclear Physicians: Directory 13719
American College of Nutrition: Journal 13721
American College of Nutrition: Newsletter 13721

American College of Obstetricians and Gynecologists: Newsletter *13722*
American College of Oral and Maxillofacial Surgeons: Newsletter *13723*
American College of Osteopathic Internists: Directory *13725*
American College of Osteopathic Internists: Newsletter *13725*
American College of Osteopathic Obstetricians and Gynecologists: Directory of Members *13726*
American College of Osteopathic Pediatricians: Membership Directory *13727*
American College of Osteopathic Pediatricians: Newsletter *13727*
American College of Osteopathic Surgeons: Membership Directory and By-laws *13728*
American College of Osteopathic Surgeons: News *13728*
American College of Physicians: Directory *13729*
American College of Podiatric Radiologists: Newsletter *13730*
American College of Podopediatrics: Newsletter *13731*
American College of Preventive Medicine: Membership Roster *13732*
American College of Preventive Medicine: Newsletter *13732*
American College of Prosthodontists: Newsletter *13733*
American College of Psychoanalysts: Bulletin *13735*
American College of Surgeons: Bulletin *13738*
American College of Surgeons: Yearbook *13738*
American College of Veterinary Pathologists: Membership List *13740*
American College of Veterinary Pathologists: Proceedings *13740*
American Collegiate Retailing Association: Directory of Members *13743*
American Collegiate Retailing Association: Newsletter *13743*
American Committee to Advance the Study of Petroglyphs and Pictographs: Membership Directory *13746*
American Committee to Advance the Study of Petroglyphs and Pictographs: Newsletter *13746*
American Communities Tomorrow *16486*
American Comparative Literature Association: Newsletter *13747*
American Concrete Institute: Concrete Abstracts *13748*
American Conference for Irish Studies: Newsletter *13749*
American Conference of Academic Deans: Proceedings *13750*
American Congress of Rehabilitation Medicine: Membership Directory *13752*
American Congress on Surveying and Mapping: Bulletin *13753*
American Congress on Surveying and Mapping: News *13753*
American Consulting Engineers Council: Membership Directory *13755*
American Coucil for an Energy Efficient Economy: Conference Proceedings *13758*
American Council for Construction Education: Annual Report *13759*
American Council of Independent Laboratories: Directory *13762*
American Council of Independent Laboratories: Newsletter *13762*
American Council of Learned Societies: Annual Report *13763*
American Council of Learned Societies: Newsletter *13763*
American Cryonics *13773*
American Cryonics News *13773*
American Dance Therapy Association: Newsletter *13776*
American Deafness and Rehabilitation Association: Newsletter *13777*
American Dental Association: Directory *13779*
American Dental Association: Journal *13779*
American Dental Association: News *13779*
American Dental Hygienists' Association: Access *13780*
American Dialect Society: Newsletter *13786*
American Dietetic Association: Journal *13787*
American Diopter and Decibel Society: Directory *13788*
American Diopter and Decibel Society: Seminar Transactions of Meetings *13788*
American Economic Association: Papers and Proceeding *13789*
American Economic Review *13789*
American Educational Research Journal *13790*
American Educational Studies Association: Newsletter *13791*
American Education Association: Newsletter *13792*
American Electrology Association: Roster *13794*
American Electronics Association: Directory *13795*
American Electronics Association: Update *13795*
American Endodontic Society: Hotline *13797*
American Endodontic Society: Newsletter *13797*
American Engineering Model Society: Annual Seminar Papers *13798*
American Entomological Society: Memoirs *13799*
American Entomological Society: Transactions *13799*
American Equilibration Society: Newsletter *13802*
American Ethnological Society: Monograph Series *13803*
American Ethnologist *13453, 13803*
American Family Physician *13394*
American Family Therapy Association: Membership Directory *13804*
American Family Therapy Association: Newsletter *13804*
American Federation of Mineralogical Societies: Newsletter *13809*
American Fern Journal *13810*
American Film *13813*
American Film and Video Association: Bulletin *13812*
American Film and Video Association: Evaluations *13812*
American Folklore Society Newsletter *13817*
American Foreign Law Association: Newsletter *13818*
American Forensic Association: Journal *13819*
American Forensic Association: Newsletter *13819*
American Forests Magazine *13820*
American Forum for Global Education: Access *13821*
American Fracture Association: Directory *13825*
American Friends of Cambridge University: Annual Report *13827*
American Geographical Society: Focus *13831*
American Geriatrics Society: Journal *13834*
American Geriatrics Society: Newsletter *13834*
American Group Practice Association: Directory *13835*
American Group Practice Association: Executive News Service *13835*
American Group Psychotherapy Association: Membership Directory *13836*
American Group Psychotherapy Association: Newsletter *13836*
American Gynecological and Obstetrical Society: Transactions *13838*
American Harp Journal *13839*
American Healing Association: Newsletter *13840*
American Health Information Management Association: Journal *13841*
American Hearing Research Foundation: Newsletter *13843*
American Heart Association: Circulation *13844*
American Heart Association: Circulation Research *13844*
American Heart Association: Current Concepts *13844*
American Herb Association: Newsletter *13845*
American Historical Review *13846*
American Historical Society of Germans from Russia: Journal *13847*
American Historical Society of Germans from Russia: Newsletter *13847*
American Horticulturist *13852*
American Hospital Formulary Service *14132*
American Indian Journal *15302*
American Indian Science and Engineering Society: Annual Report *13861*
American Industrial Health Council: Newsletter *13862*
American Industrial Hygiene Association: Directory *13863*
American Industrial Hygiene Association: Journal *13863*
American Institute for Archaeological Research: Newsletter *13864*
American Institute for Conservation of Historic and Artistic Works: Journal *13866*
American Institute for Conservation of Historic and Artistic Works: Newsletter *13866*
American Institute for Economic Research: Research Report *13868*
American Institute of Architects: Memo *13874*
American Institute of Biological Sciences: Forum *13875*
American Institute of Biological Sciences: Membership Directory *13875*
American Institute of Building Design: Bulletin *13877*
American Institute of Building Design: Roster of Members *13877*
American Institute of Chemists: Professional Directory *13879*
American Institute of Constructors: Newsletter *13880*
American Institute of Constructors: Register *13880*
American Institute of Homeopathy: Newsletter *13882*
American Institute of Indian Studies: Annual Report *13883*
American Institute of Iranian Studies: Newsletter *13884*
American Institute of Islamic Studies: Bibliographic Series *13885*
American Institute of Management: The Institute Bulletin *13887*
American Institute of Ultrasound in Medicine: Membership Directory *13898*
American Italian Historical Association: Newsletter *13904*
American Jewish History *13905*
American Journal of Addictions *13430*
American Journal of Agricultural Economics *13446*
American Journal of Anatomy *13506*
American Journal of Archaeology *14237*
American Journal of Botany *14610*
American Journal of Chinese Medicine *15265*
American Journal of Clinical Hypnosis *14108*
The American Journal of Clinical Nutrition *14051*
American Journal of Clinical Pathology *14110*
The American Journal of Comparative Law *13499*
American Journal of Comparative Law *13818*
American Journal of Cosmetic Surgery *13385*
American Journal of Critical Care *13526*
American Journal of Dance Therapy *13776*
American Journal of EEG Technology *14121*
American Journal of Forensic Medicine and Pathology *15873*
American Journal of Gastroenterology *13708*
American Journal of Hospital Pharmacy *14132*
American Journal of Human Genetics *14133*
American Journal of Hypnotherapy *13837*
American Journal of Infection Control *14321*
American Journal of International Law *14137*
American Journal of Islamic Social Sciences *14469*
American Journal of Law and Medicine *14140*
American Journal of Mental Deficiency *13608*
American Journal of Neuroradiology *14149*
American Journal of Occupational Therapy *13963*
American Journal of Orthodontics and Dentofacial Orthopedics *13558*
American Journal of Orthopsychiatry *13974*
American Journal of Pathology *13561*
American Journal of Pharmaceutical Education *13522*
American Journal of Physical Anthropology *13565*
American Journal of Physics *13567*
American Journal of Physiology *14005*
American Journal of Physiology: Cell Physiology *14005*
American Journal of Physiology: Endocrinology and Metabolism *14005*
American Journal of Physiology: Gastrointestinal and Liver Physiology *14005*
American Journal of Physiology: Heart and Circulatory Physiology *14005*
American Journal of Physiology: Lung Cellular and Molecular Physiology *14005*
American Journal of Physiology: Regulatory, Integerative and Comparative Physiology *14005*
American Journal of Physiology: Renal, Fluid and Electrolyte Physiology *14005*
American Journal of Preventive Medicine *13732, 14516*
American Journal of Psychiatry *14017*
The American Journal of Psychoanalysis *14264*
American Journal of Psychotherapy *14345*
American Journal of Public Health *14022*
American Journal of Respiratory Cell and Molecular Biology *14194*
American Journal of Roentgenology *14034*
American Journal of Semiotics *16462*
The American Journal of Social Psychiatry *13491*
American Journal of Sports Medicine *13973*
American Journal of Surgery *15684*
American Journal of Tropical Medicine and Hygiene *14171*
American Journal of Veterinary Research *14204*
American Kinesiotherapy Association: Journal *13909*
American Laryngological Association: Transactions *13910*
American Libraries *13916*
American Library Association: Booklist *13916*
American Library Association: Choice *13916*
American Library Trustee Association: Newsletter *13917*
American Literary Translators Association: Newsletter *13919*
American Malacological Bulletin *13924*
American Malacological Union: Membership List *13924*
American Malacological Union: Newsletter *13924*
American Massage Therapy Association: Yearbook and Registry *13927*
American Mathematical *15703*
American Mathematical Association of Two Year Colleges: Newsletter *13928*
American Mathematical Society: Memoirs *13929*
American Mathematical Society: Notices *13929*
American Mathematical Society: Transactions *13929*
American Medical Fly Fishing Association: Newsletter *13936*
American Medical Informatics Association: Proceedings *13937*
American Medical News *13933*
American Medical Technologists: Journal *13938*
American Medical Women's Association: Journal *13939*
American Medical Women's Association: Quarterly Newsletter *13939*
American Meteorological Society: Bulletin *13943*
American Meteor Society: Annual Report *13944*
American Mineralogist *15743*
American Montessori Society: Annual Report *13947*
American Music *16721*
American Musical Instrument Society: Journal *13949*
American Musical Instrument Society: Membership Directory *13949*
American Musical Instrument Society: Newsletter *13949*
The American Musicological Society: Directory *13951*
The American Musicological Society: Newsletter *13951*
American Music Teacher Magazine *15765*
American Naprapathic Association: Directory *13953*
American National Standards Institute: Catalog of Standards *13954*
American Nature Study Society: News *13956*
American Neurological Association: Abstract Program *13958*
American Notary *14150*
American Oil Chemists' Society: Journal: JAOCS *13964*
American Optometric Association: Journal *13966*
American Optometric Association: News *13966*
American Orchid Society: Bulletin *13967*
American Oriental Series *13968*
American Oriental Series Essays *13968*
American Oriental Society: Journal *13968*
American Orthodontic Society: Newsletter *13970*
American Orthoptic Journal *13516, 13975*
American Orthotic and Prosthetic Association: Almanac *13976*
American Osteopathic Association: Yearbook & Directory *13980*
American Osteopathic College of Allergy and Immunology: Newsletter *13981*
American Osteopathic College of Anesthesiologists: Membership Directory *13982*
American Osteopathic College of Anesthesiologists: Newsletter *13982*
American Osteopathic College of Dermatology: Directory *13983*
American Osteopathic College of Dermatology: Newsletter *13983*
American Osteopathic College of Pathologists Directory *13984*
American Osteopathic College of Rehabilitation Medicine: Annual Directory *13987*
American Osteopathic College of Rehabilitation Medicine: Newsletter *13987*
American Otological Society: Transactions *13988*
American Paper Institute: Report *13990*
American Parkinson Disease Association: Annual Report *13991*
American Parkinson Disease Association: Newsletter *13991*
American Peanut Research and Education Society: Proceedings *13993*
American Pediatric Society: Program and Abstracts of Annual Meeting *13994*
American Petroleum Institute: Directory *13995*
American Pharmaceutical Association: Academy Reporter *13996*
American Pharmacy *13996*
American Philological Association: Newsletter *13997*
American Philological Association: Textbook Series *13997*
American Philological Association: Transactions *13997*
American Philosophical Society: Proceedings *13999*
American Philosophical Society: Transactions *13999*
American Physical Society: Bulletin *14000*
American Physical Society: Membership Directory *14000*
American Physicians Fellowship for Medicine in Israel: News *14003*
American Physicians Poetry Association: Membership List *14004*
American Physicians Poetry Association: Newsletter *14004*
American Podiatric Medical Association: Journal *14008*
American Podiatric Medical Association: News *14008*
The American Political Science Review *14010*
American Poultry Historical Society: Newsletter *14012*
American Press Institute: Bulletin *14013*
American Production and Inventory Control Society: Journal *14015*
American Professional Constructor *13880*
American Psychiatric Press Review of Psychiatry *14017*
American Psychoanalytic Association: Journal *14018*
American Psychoanalytic Association: Newsletter *14018*
American Psychologist *14019*
American Psychology-Law Society: Newsletter *14020*
American Psychopathological Association: Proceedings of Annual Meeting *14021*
American Quarterly *14185*
American Quaternary Association: Newsletter *14023*
American Radium Society: Membership Directory *14025*
American Real Estate and Urban Economics Association: Journal *14026*
American Real Estate and Urban Economics Association: Newsletter *14026*
American Rehabilitation Counseling Association: Newsletter *14028*
American Research Institute in Turkey: Newsletter *14029*
American Review of Canadian Studies *14276*
American Review of Respiratory Disease *14194*
American Review of Respiratory Diseases *13923*
American Revolution Round Table: Newsletter *14030*
American Rhinologic Society: Membership Directory *14031*
American Rhinologic Society: Newsletter *14031*
American Roentgen Ray Society: Membership Directory *14034*
The American School Board Journal *16147*
American Scientist *16472*
Americans for Energy Independence: Newsletter *14040*
American Society for Adolescent Psychiatry: Newsletter *14043*
American Society for Advancement of Anesthesia in Dentistry: Proceedings *14044*
American Society for Artificial Internal Organs: Abstracts *14046*
American Society for Artificial Internal Organs: Author Index *14046*
American Society for Artificial Internal Organs: Transactions *14046*
American Society for Bone and Mineral Research: Membership Roster *14048*
American Society for Bone and Mineral Research: Newsletter *14048*
American Society for Cell Biology: Newsletter *14049*
American Society for Cybernetics: Conference Proceedings *14054*
American Society for Cybernetics: Newsletter *14054*
American Society for Cytotechnology: Newsletter *14055*
American Society for Dental Aesthetics: Newsletter *14056*
American Society for Dermatologic Surgery: Roster *14057*
American Society for Engineering Management: Newsletter *14059*
American Society for Engineering Management: Proceedings of Annual Meeting *14059*

American Society for Geriatric Dentistry: Newsletter 14062
American Society for Information Science: Bulletin 14068
American Society for Information Science: Journal 14068
American Society for Legal History: Newsletter 14070
American Society for Neurochemistry: Membership Directory 14074
American Society for Neurochemistry: Newsletter 14074
American Society for Neurochemistry: Transactions 14074
American Society for Photobiology: Directory and Constitution 14079
American Society for Photobiology: Newsletter 14079
American Society for Psychoprophylaxis in Obstetrics: Annual Directory 14083
American Society for Technion-Israel Institute of Technology: Update Newsletter 14088
American Society for Theatre Research: Newsletter 14090
American Society for Training and Development: Journal 14093
American Society for Value Inquiry: Newsletter 14094
American Society of Addiction Medicine: Newsletter 14096
American Society of Agricultural Engineers: Newsletter 14097
American Society of Animal Science: Combined Abstracts 14100
American Society of Bakery Engineers: Letter 14101
American Society of Bakery Engineers: Proceedings 14101
American Society of Body and Design Engineers: Directory 14102
American Society of Body and Design Engineers: Newsletter 14102
American Society of Body and Design Engineers: Proceedings 14102
American Society of Certified Engineering Technicians: Annual Report 14105
American Society of Civil Engineers: Professional Journal 14107
American Society of Clinical Hypnosis: Directory 14108
American Society of Clinical Hypnosis: News Letter 14108
American Society of Clinical Oncology: Directory 14109
American Society of Consultant Pharmacists: Update Newsletter 14112
American Society of Danish Engineers: Newsletter 14117
American Society of Forensic Odontology: Membership Directory 14123
American Society of Forensic Odontology: Newsletter 14123
American Society of Gas Engineers: Digest 14124
American Society of Hand Therapists: Newsletter 14129
American Society of Heating, Refrigerating and Air-Conditioning Engineers: Handbook 14130
American Society of Heating, Refrigerating and Air-Conditioning Engineers: Journal 14130
American Society of Heating, Refrigerating and Air-Conditioning Engineers: Transactions 14130
American Society of Hematology: Directory of Members 14131
American Society of Hematology: Meeting Program 14131
American Society of International Law: Newsletter 14137
American Society of Irrigation Consultants: Bulletin 14138
American Society of Irrigation Consultants: Membership Roster 14138
American Society of Nephrology: Abstracts and Program 14148
American Society of Nephrology: Membership Directory 14148
American Society of Neuroradiology: Membership Roster 14149
American Society of Papyrologists: Bulletin 14151
American Society of Parasitologists: Newsletter 14152
American Society of Pharmacognosy: Newsletter 14153
American Society of Physoanalytic Physicians: Newsletter 14154
American Society of Physoanalytic Physicians: Proceedings 14154
American Society of Plant Physiologists: Newsletter 14155
American Society of Podiatric Dermatology: Newsletter 14159
American Society of Primatologists: Bulletin 14160
American Society of Radiologic Technologists: Journal 14163
American Society of Safety Engineers: Conference Proceedings 14165
American Society of Swedish Engineers: Membership Directory 14169
American Society of Trial Consultants: Directory 14170
American Society of Veterinary Ophthalmology: Newsletter 14172
American Sociological Association: Directory of Departments 14175
American Sociological Association: Directory of Members 14175
American Sociological Association: Footnotes 14175
American Sociological Review 14175
American Speech 13786
American Statistical Association: Proceedings 14182
The American Statistician 14182
American String Teacher 14183
American Student Dental Association: News 14184
American Studies Association Newsletter 14185
American Surgical Association: Transactions 14186
American Theatre Critics Association: Newsletter 14190
American Theological Library Association: Newsletter 14192
American Theological Library Association: Summary of Proceedings of the Annual Conference 14192
American Theological Society – Midwest Division: Membership Directory 14193
American Type Culture Collection: Newsletter 14199
American Urological Association: Membership Roster 14201
American Veterinary Medical Association: Journal 14204
American Vocational Association: Update 14206
American Water Resources Association: Proceedings 14209
American Zoologist 14174
The Americas 13281
Amerika Ho: Law in the United States 08185
Amerikastudien/American Studies: Vierteljahresschrift 05318
AMFI Industry News 14562
AMFI Industry Survey Report 14562
Amicus Journal 16187
AMMF Bulletin 01913
AMMLA Annual Report 13942
AMM Newsletter 01935
Amon Hen 13136
AMS Insights 13329
AMS Newsletter 13943
AMSPDC Membership List 14461
AMSTAT News 14182
AMT Directory 16404
AMT Events 13938
AMTIESA Bulletin 11047
Der Amtsvormund 05588
A Musical Guide for Holland 09002
AMWA Freelance Directory of Medical Communication Services 13940
AMWA Journal 13940
AMWA Membership Directory 13940
Anaesthesia 11300
Anaesthesia and Intensive Care 00347
Anästhesiologie und Intensivmedizin 05319
Der Anaesthesist 00754
Anaesthetic Research Society: Proceedings 11226
ANAI Index 08799
Anais 09635
Anais Brasileiros de Dermatologia 01708
Anais da Academia Brasileira de Ciências 01563
Anais da Real Sociedade Arqueológica Lusitana 09668
Anais de Farmácia e Química 01723
Anakainosis 01903
Analecta Bollandiana 01456
Analecta Hibernica 06990
Analecta Romana Instituti Danici 07167
Anales Cervantinos 10322
Anales de Bromatologia 10379
Anales de Física 10361
Anales de Invemar 03374
Anales de la Academia de Geografía e Historia 06554
Anales de Microbiologia 03014
Anales de Quimica 10363
Anales/Societas 03009
Análise & Conjuntura 01657
Analyse 09092
L'Analyse Numérique et la Théorie de l'Approximation 09737
Analyses de la SEDEIS 04732
Analytical Abstracts 12857
Analytical Chemistry 13688
Analytical Sciences 08189
Analytische Psychologie 05320
Anatomia, Histologia, Embryologia 16923
Anatomia, Histologia, Embryologia: Journal of Veterinary Medicine 01298
Anatomical News 13506
Anatomical Record 13506
Anatomischer Anzeiger: Annals of Anatomy 05074
Ancient Coins in North American Collections 13962
Ancient Monuments Society: Transactions 11228
Ancient Skies 14216
ANCOLD Bulletin 00323
Andean Newsletter 09443
Andrologia 07834
Andvari 06695
Anesteziologie a neodkladná péče: Anaesthesiology and Emergency Care 03509
Anesteziologie a neodkladná péče: Anaesthesiology and Emergency Medicine 03517
Anesthesia and Analgesia 15373
Anesthesia File 14954
Anesthesia Progress 13781
Anesthesiology 14099
An Gael 14217
Angéiologie 04784
Angewandte Arbeitswissenschaft 05946
Angewandte Botanik 06408
Angewandte Chemie 05793
Angewandte Chemie International Edition 05793
Angewandte Graphologie und Charakterkunde 05206
Angewandte Semiotik 00830
Angewandte Sozialforschung 00552
Angla 06483
Animal Behavior 14218
Animal Behavior Society: Newsletter 14218
Animal Behaviour 11291
Animal Breeding Abstracts 11720, 11730
Animal Diseases Research Association: Association Report and Accounts 11235
Animal Health Focus 01347
Animal Health Trust: Annual Report 11236
Animal Keeper's Forum 13605
Animal Learning and Behavior 16373
Animal Production 11680
Animal Production Review 09609
Animal Production Review Applied Sciences Report 09609
Animato 11003
An Leabharlann: The Irish Library 07007
Annalas da la Società Retorumantscha 10963
Annales Academiae Scientiarum Fennicae 04000
Annales Botanici Fennici 03993, 03995
Annales Chirurgiae & Gynaecologiae 04041
Annales de chirurgie 04650
Annales d'Economie Politique 04652
Annales de démographie historique 04653
Annales de Dermatologie 04175
Annales de la Société Archéologique de Namur (ASAN) 01421
Annales de la Société Jean-Jacques Rousseau 10978
Annales de l'Institut Archéologique du Luxembourg 01375
Annales de Médecine Interne 04891
Annales d'Endocrinologie 04822
Annales de Psychothérapie 04689
Annales de Radiologie 04778
Annales des Falsifications de l'Expertise Chimique et Toxicologique 04705
Les Annales du CNREF 00018
Annales Entomologici Fennici 04029
Annales Francaises d'Anesthésie et de Réanimation 04783
Annales Historiques de la Révolution Française 04752
Annales Juris Aquarum 07705
Annales Médico-Psychologiques 04895
Annales Silesiae 09630
Annales Societatis Geologorum Poloniae: Rocznik Polskiego Towarzystwa Geologicznego 09572
Annales Zoologici Fennici 03993, 03995
Annali dell'Accademia di Agricoltura di Torino 07153
Annali dell'Accademia Italiana di Scienze Forestali 07194
Annali della Facoltà di Medicina e Chirurgia dell'Università degli Studi di Perugia e Atti dell'Accademia Anatomico-Chirurgica 07129
Annali dell'Athenaeum 07123
Annali dell'Istituto Italiano di Numismatica 07737
Annali di Architettura 07472
Annali di Chimica 07774
Annals and Bulletin 12792
Annals of Adolescent Psychiatry 14043
Annals of Applied Biology 11301
Annals of Behavioral Medicine 16641
Annals of Biomedical Engineering 14593
Annals of Clinical Biochemistry 11324
Annals of Clinical Laboratory Science 14413
Annals of Clinical Psychiatry 13384
Annals of Emergency Medicine 13705, 16489
Annals of Glaciology: Conference Proceedings 12366
Annals of Human Biology 13015
Annals of IMA Academy of Medical Specialities 06801
Annals of Internal Medicine 13729
Annals of Medicine 03997
Annals of Occupational Hygiene 11623
Annals of Ophthalmology 14114
The Annals of Probability 15331
Annals of Public and Cooperative Economics 01193
The Annals of Statistics 15331
Annals of Surgery 14186
Annals of the Academy 10033
Annals of the History of Hungarian Geology: Földtani Tudománytörtáneti Évkönyv 06635
Annals of the Japan Association for Philosophy of Science 08150
Annals of the Japan Association of Economic Geographers 08157
Annals of the Phytopathological Society of Japan 08382
Annals of Thoracic Surgery 16707
Annals RCPSC 02869
L'Année Biologique 04521, 04525
Année philologique 10701
L'Année Philologique: Bibliographie Critique et Analytique de l'Antiquité Gréco-Latine 04875
Année Politique Suisse 10633
Ann. Haemat. 05358
Annuaire Africain des Sciences de l'Education 17139
Annuaire de l'Académie 01095
Annuaire de l'Académie Française 04111
Annuaire de l'Afrique du Nord 04382
Annuaire de l'AISLF 04277
Annuaire des Associations de Bibliothécaires et Documentalistes 04174
Annuaire des Cégeps 02450
Annuaire des Pays de l'Océan Indien 04382
Annuaire des Psychiatres Français 04927
Annuaire Français de Droit International 04382
Annuaire Roumain d'Anthropologie 09737
Annuaire Suisse de Science Politique 10633
Annual Bibliography of English Language and Literature 12535
Annual Book of ASTM Standards 14089
Annual Conference of Fellmongers and Hide Processors 09163
Annual Conference of Leather Technicians 09163
Annual Developmental Mathematics Committee Report 13928
Annual Directory of Jesuit High Schools and Universities 15601
Annual Directory of Registered Technologists 14027
Annual Index of Motion Picture Credits 13298
Annual Journal of International Affairs 06576
Annual National Compendium of Animal Rabies Control 15903
Annual of Pathological Autopsy Cases in Japan 08286
Annual of the Best Advertising, Graphics, Television & Editorial Design 11886
Annual Proceedings 02440, 09391, 10123, 14161, 14220
Annual Register of Pharmaceutical Chemists 12838
The Annual Report of Educational Psychology in Japan 08239
Annual Report of The Medical Defence Union 12503
Annual Review of Anthropology 14222
Annual Review of Astronomy & Astrophysics 14222
Annual Review of Biochemistry 14222
Annual Review of Biophysics & Biophysical Chemistry 14222
Annual Review of Cell Biology 14222
Annual Review of Computer Science 14222
Annual Review of Earth & Planetary Sciences 14222
Annual Review of Ecology & Systematics 14222
Annual Review of Energy 14222
Annual Review of Entomology 14222, 15046
Annual Review of Environmental Education 11852
Annual Review of Fluid Mechanics 14222
Annual Review of Genetics 14222
Annual Review of Immunology 14222
Annual Review of Materials Science 14222
Annual Review of Medicine 14222
Annual Review of Microbiology 14222
Annual Review of Neuroscience 14222
Annual Review of Nuclear & Particle Science 14222
Annual Review of Nutrition 14222
Annual Review of Pharmacology & Toxicology 14222
Annual Review of Physical Chemistry 14222
Annual Review of Physiology 14222
Annual Review of Phytopathology 14222
Annual Review of Plant Physiology & Plant Molecular Biology 14222
Annual Review of Psychology 14222
Annual Review of Public Health 14222
Annual Review of Sociology 14222
Annuario ATI 07409
Annuario dell'Accademia Nazionale dei XL 07213
Annuario dell'Accademia Nazionale di San Luca 07220
Annuario dell'Accademia Nazionale di Santa Cecilia 07221
Annuario dell'Archidiocesi di Napoli 07541
Annuario dell'Istituto Storico Italiano per l'Età moderna e contemporanea 07755
An Ounce of Prevention 16697
ANS Museum Notes 13962
ANS Numismatic Studies 13962
ANSTI Newsletter 08471
Antarctic Journal of the United States 16815
Antenna 12800
Antenna Measurement Techniques Association: Call for Papers 14227
Antenna Measurement Techniques Association: Newsletter 14227
Anthropological Index 12768
Anthropological Survey of India: Annual Report 06717
Anthropological Survey of India: Memoir 06717
Anthropological Survey of India: Newsletter 06717
Anthropological Survey of India: Occasional Publication 06717
Anthropologie et Préhistoire 01469
Anthropologische Forschungen 00544
Anthropologists in India 06769
Anthropology and Education 14908
Anthropology News: Proceedings 00217
Anthropology Newsletter 13453
Anthropology Today 12768
Anthropos: International Review of Ethnology and Linguistics Internationale Zeitschrift für Völker- und Sprachenkunde 05075
Antike Kunst 11010
Antinicrobial Agents and Chemotherapy 14073
Antiquarian Horology 11238
Antiquaries Journal 13022
Antiquités Africaines 04382
Antitrust 13617
Antitrust Law Journal 13617
Antropologia Contemporanea 08039
Antropologia Culturale: Proceedings of Seminars 07490
ANTS Publications 11233
An t Ultach 06937
Anual do Centro de Pesquisa Agropecuária dos Cerrados 01625
Anuario Bibliográfico Colombiano 03377
Anuario de Eusko-Folklore Aranzadiana Orria 10367
Anuario de la Minería 03063
Anuario del Teatro Argentino 00139
Anuario de Sociologia y Psicologia Juridicas 10288
Anuario Estadística de España 10326
Anuario Estadistico de America Latina y el Caribe: Statistical Yearbook for Latin America and the Caribbean 03038
Anuario Geográfico del Perú 09477
Anuario Indigenista 08736
Anuario Meteorologico 03361
An Up-to-Date in Pollenosis and Drug Allergy 00153
Anzeiger der mathemathisch-naturwissenschaftlichen Klasse 00723
Anzeiger der philosophisch-historischen Klasse 00723
AORTIC Bulletin 17159
APAA News 08105
APA Monitor 14019
APA Newsletter 13998
APASWE Newsletter 00220
APCChE Newsletter 00224
APDF Newsletter 10039
APEA Journal 00327
APEC: Journal 14232
APEC: Membership Directory 14232
APEE Bulletin 09652
Apevine 14221
APF Bulletin 03116
APFCB News 00226
APFHRM Newsletter 06591
APG Quarterly 14487
APHA Auxiliary Newsletter 14561
APHA Newsletter 14014
Apiacta 07248
APIC Journal 10406
Apicultural Abstracts 12339
APINESS Newsletter 11086
API Report 13995
APLA Bulletin 01962
APLIC Bulletin: Bulletin ABPAC 01938
APLIC Communicator 14319
Aplikace matematiky 03500
APPA Newsletter 14443
APPA Quarterly 14481
APPEN Features Service 11086
APP-Info: Vereinszeitschrift der Arge für Präventivpsychologie 00555
APPITA Journal 00508
Applications et Transferts 04746
Applied Acoustics 03113
Applied and Environmental Microbiology 14073
Applied Entomology and Zoology 08355
Applied Ergonomics 11972
Applied Experimental and Engineering Psychology 14980
Applied Industrial Hygiene 13751
Applied Linguistics 11486, 13467
Applied Market Research: A Journal for Practitioniers 15692

Applied Mathematics Notes 02195
Applied Mechanics 14145
Applied Mechanics Review 14145
Applied Neurophysiology 14086, 16932
Applied Optics 16255
Applied Spectroscopy 16496
APPM Update 13302
Apprentissage et Socialisation en Piste 02385
Approaching Ontario's Past 02782
Appropriate Technology 12320
APPSA Newsletter 10040
APRI Journal 09218
APSA Newsletter 00247
APSC Newsletter 08522
APWSS Newsletter 09501
Aquaforum 07705
Aquariculture 14235
Aquatica 14235
Aquatic Plant Management Society: Newsletter 14234
Aquilo 03988
Arab Archives Journal 11133
Arab Biosciences Newsletter 08450
Arab Culture Magazine 11131
Arab Gulf Journal of Scientific Research 09997
Arab Historian 06910
Arabic Training 06914
Arab Journal for Technical Education 06908
Arab Journal of Administration 08449
Arab Manuscripts Magazine 11131
Arab Maritime Transport Academy: News Bulletin 03874
Arab Music Magazine 06905
Arab Paper 11240
The Arab Researcher 11240
Arabs before Islam 06917
Arab Science Abstracts 03910
Arab Studies 14394
Arab Technical Education Newsletter 06908
ARA Membership Directory 13736
ARA Newsletter 09502
A Range of Information Publications 11857
Aranzadiana 10371
ARA Scientific Program 13736
Arbeiten zur Geschichte des Pietismus 05918
Arbeiten zur Rechtsvergleichung 05857
Arbeitsbehelfe 00884
Arbeitsberichte 05347
Arbeitsgemeinschaft der Spezialbibliotheken e.V.: Report 05093
Arbeitsgemeinschaft für zeitgemässes Bauen e.V.: Mitteilungsblätter 05114
Arbeitsgemeinschaft Historischer Kommissionen und Landesgeschichtlicher Institute e.V.: Mitteilungsblatt 05118
Arbeitsgemeinschaft katholisch-theologischer Bibliotheken: Mitteilungsblatt 05122
Arbeitsgruppe Asphaltstrassen: Schriftenreihe 05720
Arbeitsgruppe Betonstrassen: Schriftenreihe 05720
Arbeitsgruppe Erd- und Grundbau 05720
Arbeitsgruppe für empirische Bildungsforschung e.V.: Übersicht über die bisherigen Arbeiten 05133
Arbeitsgruppe Mineralstoffe im Strassenbau 05720
Arbeitshefte der Arbeitsgemeinschaft der Parlaments- und Behördenbibliotheken 05089
Arbeitsschriften Forschung und Entwicklung in der Wirtschaft 06295
Arbetsterapeuten 10458
ARBIDO 10942
ARBIDO-B; ARBIDO-R 11017
ARBIDO-Bulletin 11016
ARBIDO-Revue 11016
Arbitration Journal 13456
Arbitration Times 13456
Årbog for Dansk Skolehistorie: Yearbook on History of Danish Schools (Education) 03811
Arboretum and Botanical Garden Bulletin 13512
Arboricultural Journal: The International Journal of Urban Forestry 11241
Arcanes 04123
ARCCOH Newsletter 06725
ARCEDEM Bulletin 09219
Archaeologia 13022
Archaeologia Aeliana 13023
Archaeologia Cambrensis 11734
Archaeologia Cantiana 12428
Archaeológiai Ertesítő 06655
Archaeologia Japonica 08228
Archaeologia Zambiana Newsletter 17156
Archaeological Conservancy: Newsletter 14236
The Archaeological Journal 12769
Archäologie der Schweiz, Archéologie Suisse, Archeologia Svizzera 10853
Archäologie Österreichs 00840
Archaeologiki Ephimeris 06499
Archäologische Berichte aus dem Yemen 05568
Archäologische Berichte aus Iran 05568
Archäologische Bibliographie 05568
Archäologischer Anzeiger 05568
Archaeology 03155, 14237
Archaeology in Britain 11847
Archaeology in New Zealand 09133
Archaeonautica 04382
Archeia tis Pharmakeftikis 06514
Archeografo Triestino 07786
Archeologia 07701
Archeologické rozhledy 03500
Archeologie in Limburg 08972
Architect and Surveyor 12192
Architectes 04607
Architectural History 13027
Architectural Journal 03115
Architectural Knowledge 03115
Architecture 13874
Architecture and Society 01783
Architecture (SA) 10149
Architectuur/Bouwen 08947
Der Architekt 05226
Architettura Navale agli Inizi del '600 07573
Archiva Ecclesiae 07260
Der Archivar 06372
Archiv der Mathematik 05840
Archiv der Pharmazie 05485
Archiv des Badewesens 05338
Archiv des Historischen Vereins des Kantons Bern 10727
Archive of Biological Science 17126
Archive of Fishery and Marine Research: Archiv für Fischerei und Meeresforschung 05626
Archives 09747, 11638
Archives and Otolaryngology 14063
Archives and the User 11638
Archives de l'Art Français 04662
Archives des Maladies du Coeur et des Vaisseaux 04789
Archives des Sciences 10969
Archives d'Histoire de l'Art: Zbornik za Umetnostno Zgodovino 10107
Archives et Bibliothèques de Belgique: Archief- en Bibliotheekwezen in Belgie 01102
Archives Françaises de Pédiatrie 04829
Archives Héraldiques Suisses 10988
Archives Historiques 04762
Archives Internationales d'Histoire des Sciences 04578
Archives of American Art Journal 14240
Archives of Andrology 07834
Archives of Dermatology 13933
Archives of Emergency Medicine 11484
Archives of Environmental Health 10159
Archives of General Psychiatry 13933
Archives of Internal Medicine 13933
Archives of Natural History 13004
Archives of Neurology 13933
Archives of Ophthalmology 13933
Archives of Otolaryngology, Head & Neck Surgery 13933
Archives of Pathology & Laboratory Medicine 13933
Archives of Physical Medicine and Rehabilitation 13424, 13752
Archives of Surgery 13933
Archives of the Foundation of Thanatology 13886
Archives of the TSC 11164
Archives of Virology 04589
Archives scientifiques 08061
Archive und Archivare in der Bundesrepublik Deutschland, Österreich und der Schweiz 06372
Archiv für deutsche Postgeschichte 05807
Archiv für Geschichte des Buchwesens 05916
Archiv für Geschichte und Kultur 05825
Archiv für Geschichte von Oberfranken 05924
Archiv für Hessische Geschichte und Altertumskunde 05923
Archiv für Kommunalwissenschaften 05587, 06389
Archiv für Mittelrheinische Kirchengeschichte 05841
Archiv für Sozialgeschichte 05764
Archiv für Völkerkunde 00982
Archivi e Cultura 07359
Archivi e Cultura: Organo Culturale dell Centro di Ricerca Pergamene Medievali e Protocolli Notarili 07433
Archivio della Società Romana di Storia Patria 08010
Archivio di Chirurgia Toracica e Cardiovascolare 07851
Archivio di Filosofia 07476
Archivio per l'Antropologia e la Etnologia 07964
Archivio per l'Antropologia e l'Etnologia 07838
Archivio Storico di Terra di Lavoro 07790
Archivio Storico Italiano 07637
Archivio Storico Lombardo 08018
Archivio Storico Messinese 07987
Archivio Storico per la Sicilia Orientale 07792
Archivio Storico per le Province Napoletane 07989
Archivio Storico per le Province Parmensi 07641
Archivio Storico Pugliese 07791
Archivio Storico Siciliano 08013
Archivio Storico Siracusano 08014
Archivio Veneto 07642
Archivo de Prehistoria Levantina 10369
Archivo Español de Arte 10303
Archiv orientální 03500
Archivos Argentinos de Pediatría 00181
Archivos de Biología y Medicina Experimentales 03089
Archivos de Oftalmología de Buenos Aires 00179
Archivos de Pediatría del Uruguay 13271
Archiwum Ethnograficzne: Ethnographic Archives 09584
Archiwum Historii Medycyny 09577
Arch Notes 02753
Arc' News 11790
ARCSS Newsletter 03885
Arctique 04746
Arctos: Acta philologica Fennica 03978
Ardea 08999
Area 12216
A.R.E. Bulletin 11384
ARF Transcript Proceedings 13334
Argonauta 02208
Argus 13410
Arheološki vestnik: Acta Archaeologica 10105
Arhitectura 09796
Arhitekturata i zizenata sreda na coveka 01770
Arhiv za Farmaciju 17078
Arica Newsletter 14241
ARI News 11239
Aristotelian Society: Monographs 11246
Aristotelian Society: Proceedings 11246
Arithmetic Teacher 16011
Arkeologia Suomessa: Archaeology in Finland 03983
Ark File 12876
Arkheologiya: Archaeology 11193
Arkhimedes 04019, 04049
Arkhiv Anatomii, Gistologii i Embriologii: Anatomy, Histology and Embryology Archive 09952
Arkhiv Patologii i Meditsiny: Pathology and Medicine Archive 09952
Arkisto 03950
Arkitekten 03602
Arkitektnytt 09296
Arkitekttidningen 10535
Arkitektur DK 03602
Arkitídindi 06677
Arkiv för Matematik 10490
Arkivforeningens Seminarrapporter 03559
Arkiv, samhälle och forskning: Svenska arkiv samfundets skriftserie 10536
ARK Newsletter 12704
ARL Annual Solary Sunvey 14492
ARLIS/NA Update 14250
ARL Newsletter 14492
ARL Statistics 14492
Arma 10145
Armed Forces Institute of Pathology: Letter 14244
Armed Forces Optometric Society: Newsletter 14245
Armed Forces Optometric Society: Roster 14245
Armenian Academy of Sciences: Doklady: Report 00203
Armenian Academy of Sciences: Izvestiya: Bulletin 00203
The Armiger's News 13711
Arms and Armour Society of Great Britain: Journal 11248
ARMS Firms Users Association: Newsletter 14248
ARPR Bulletin 13306
Arqueólogia e História 09638
Arquivo de Anatomia e Antropologia 09673
Arriba 14795
ARR Research Reports 00335
ARSO Newsletter 08472
Arta 09797
Art & Archaeology Technical Abstracts 12369
The Art Bulletin 14770
Art Documentation 14250
ARTDO Journal 09503
ARTDO Report 09503
Art Education 15790
Artefact 00218
Art Enfantin 04533
Arteriosclerosis 13844, 14909
Art Exhibition Catalogs 14734
Art Galleries and Museums Association of New Zealand: Newsletter 09116
Arthritis and Rheumatism 13736
Arthur Rubinstein International Music Society: Bulletin 07043
Artifacts 15269
Art Journal 06744, 14770
Art-Journal of the Professional Artist 04265
Art Libraries Journal 11251
Art Libraries Society of the United Kingdom and Ireland: News-sheet 11251
Artlook 13079
The Art Magazine 12063
Art Matters 06925
ART Network Bulletin 08108
The Art of Translation: Collection of Theoretical Articles 01802
Art publications 10586
Arts Association of Namibia: Newsletter 08814
Arts Council: Annual Report 06925
Arts Council of Great Britain: Report 11253
Arts Council of Northern Ireland: Annual Report 11254
Arts Fines 09191
Arts News 06714
Artur 03602
Artviews 02982
ARVAC Bulletin 11280
Arv og Eje 03641
ARX: Burgen und Schlösser in Bayern, Österreich und Südtirol 00889
Der Arzt im Krankenhaus und im Gesundheitswesen: Monatsschrift des Marburger Bundes 06109
Arzt und Krankenhaus 06341
ASA News 13346
ASA Newsletter 01863, 11392, 13340, 14099
ASAPE Newsletter 08109
ASAS Handbook and Membership Directory 14100
ASA Studies 11392
ASB-Aktuell 05153
ASbH-Brief 05126
ASBO Accents 14494
ASCA Report 14053
ASCD Update 14336
ASCD Yearbook 14336
ASCE News 14107
ASC Newsletter 14512
ASDA SVLR 10632
ASDC Newsletter 14118
ASEAN Economic Bulletin 10049
ASEAN Food Handling Newsletter 08628
ASEAN Food Handling Project Review 08628
ASEAN Food Journal 08628
ASEAN Law Journal 06876
ASEAN Mental Health Bulletin 08627
ASEP Newsletter 11084
ASE Primary Science 11282
ASET Membership 13309
ASET Newsletter 14121
ASGCA Newsletter 14127
ASHA 14178
ASHA Monograph 14178
ASHA Report 14178
ASHS Newsletter 14066
ASIA Bulletin 14180
Asia Journal of Theology 10043
Asian-Australian Journal of Animal Science 03236
Asian Fisheries Science 09497
Asian Journal of Communication 10035
Asian Journal of Crime Prevention and Criminal Justice 08094
Asian Journal of Surgery 10407
The Asian Manager 09498
Asian Mass Communication Bulletin 10035
Asian Medical Journal 08212
Asian Music 16500
Asian Network for Industrial Technology Information and Extension: Newsletter 10037
Asian NGO Coalition for Agrarian Reform and Rural Development: Information Notes 09500
Asian Pacific Section of IPRS Newsletter 10041
Asian-Pacific Tax and Investment Bulletin 10042
Asian-Pacific Weed Science Society: Proceedings of Conferences 09501
Asian Religious Studies Information 15267
Asian Review of Public Administration 09518
Asian Social Scientists 06732
Asian Studies Newsletter 14268
Asia-Oceania Journal of Obstetrics and Gynecology 10034
Asia-Pacific Environment Newsletter 11086
Asia-Pacific Journal of Operational Research 08112
Asia-Pacific Journal of Physics Education 06881
Asia-Pacific Physics News 06881
Asia-Pacific Report 15005
Asia-Pacific Tech Monitor 06720
Asia Quarterly 08091
The Asia Society: Annual Report 14253
The Asia Society: Newsletter 14253
Asiatic Society: Journal 06728
Asiatic Society: Monthly Bulletin 06728
Asiatic Society of Bombay: Journal 06729
Asiatic Society: Year Book 06728
Asiatische Studien/Etudes Asiatiques 10798
ASIC Newsletter 14138
ASIDIC Newsletter 14445
Asien: Deutsche Zeitschrift für Politik, Wirtschaft und Kultur 05325
ASI Journal 11242
ASI Newsletter 14135
ASI & PROPAG Bulletin: Informative Pamphlets 06570
Asistilo 04141
Aslib Booklist 11256
ASLIB Book List 11267
ASLIB Information 11267
Aslib Informations 11256
Aslib Proceedings 11256
ASLIB Proceedings 11267
Aslib, The Association for Information Management: Program 11256
ASLP Bulletin 09512
ASLP Newsletter 09512
ASM-Bulletin 05069
ASME Membership Directory 14468
ASM News 14073, 14255
Asociación Archivística Argentina: Boletín 00053
Asociación Argentina de Astronomía: Boletín 00056
Asociación Argentina de Cirugía: Anuario 00061
Asociación Argentina de Cirugía: Boletín 00061
Asociación Argentina de Ecología: Bulletin 00062
Asociación Argentina de Geofísicos y Geodestas: Boletín 00065
Asociación Argentina de la Ciencia del Suelo: Boletín 00066
Asociación Argentina de Micología: Revista 00069
Asociación Colombiana de Bibliotecarios: Boletín 03343
Asociación Cultural Humboldt: Boletín 16966
Asociación de Bibliotecarios Graduados de la República Argentina: Boletín Informativo 00077
Asociación de Bibliotecarios Graduados de la República Argentina: Trabajos de Congresos 00077
Asociación Ecuatoriana de Museos: Boletín 03844
Asociación Española para el Progreso de las Ciencias: Las Ciencias 10272
Asociación Física Argentina: Revista 00081
Asociación Médica Argentina: Revista 00090
Asociación Mexicana de Bibliotecarios, AC: Memorias 08699
Asociación Odontológica Argentina: Revista 00091
Asociación Química Argentina: Anales 00096
Asociación Química Argentina: Boletín 00096
ASOR Bulletin 00343
ASPBAE Courier Service 10407
ASPBAE News 10407
Aspects of Educational Technology 11263
Aspects of Standardization 09875
Aspen Institute Monograph 14256
ASPEN Newsletter 06881
Asphalt Paving Technology 14395
Asphalt Technology 12213
ASPL Membership Directory 14078
ASPR Newsletter 14082
The Assembly Communique 15283
A.S.S.E. Year Book 14166
Assicurazioni: Bimestrale di Diritto, Economia e Finanza delle Assicurazioni Private 07747
Associação Portuguesa de Bibliotecários, Arquivistas e Documentalistas: Noticia 09640
Associação Portuguesa de Fundição: Anuario 09646
The Associate 13616
Associated Schools of Construction: Annual Report 14260
Associated Schools of Construction: Proceedings of the Annual Meeting 14260
The Associates Digest 13887
Association Belge de Photographie et de Cinématographie: Informations 01109
Association Belge de Radioprotection: Annales: Annalen 01110
Association Canadienne-Française pour l'Avancement des Sciences: Annales 01875
Association des Amis d'Alfred de Vigny: Bulletin 04163
Association des Anatomistes: Bulletin 04165
Association des Bibliothécaires Français: Bulletin d'Informations 04169
Association des Bibliothèques Ecclésiastiques de France: Bulletin 04171
Association des Diplômés de l'Ecole de Bibliothécaires-Documentalistes: Bulletin d'Information 04176
Association des Ecrivains Belges de Langue Française: Nos Lettres 01117
Association des Hautes Etudes Hospitalières: Magazine 04186
Association des Médecins de Travail du Québec: Bulletin 01889
Association d'Instituts Européens de Conjoncture Economique: Report 01119
Association Européenne des Affaires Internationales: Newsletter 01124
Association Européenne des Centres de Lutte contre les Poisons: Newsletter 01126
Association Européenne des Centres de Lutte contre les Poisons: Reports of Meetings 01126
Association Européenne pour l'Administration de la Recherche Industrielle: Conference Papers 04208
Association for Advancement of Behavior Therapy: Membership Directory 14263

Association for All Speech Impaired Children: Annual Report 11258
Association for All Speech Impaired Children: Newsletter 11258
Association for Applied Psychophysiology and Biofeedback: Proceedings of the Annual Meeting 14266
Association for Arid Lands Studies: Abstracts 14267
Association for Arid Lands Studies: Forum 14267
Association for Asian Studies: Monograph Series 14268
Association for Assessment in Counseling: Newsnotes 14269
Association for Behavior Analysis: Membership Directory 14271
Association for Behavior Analysis: Newsletter 14271
Association for Behavior Analysis: Program Book 14271
Association for Child Psychoanalysis: Membership Roster 14280
Association for Child Psychoanalysis: Newsletter 14280
Association for Communication Administration: Bulletin 14281
Association for Computers and the Humanities: Newsletter 14284
Association for Counselor Education and Supervision: Newsletter 14287
Association for Creative Change: Newsletter 14288
Association for Documentary Editing: Directory 14289
Association for Education in Journalism and Mass Communication: Newsletter 14292
Association for General and Liberal Studies: Directory of Members 14298
Association for General and Liberal Studies: Newsletter 14298
Association for General and Liberal Studies: Perspectives 14298
Association for Gnotobiotics: Membership Directory 14300
Association for Gnotobiotics: Newsletter 14300
Association for Hospital Medical Education: Membership Directory 14302
Association for Hospital Medical Education: Newsletter 14302
Association for Humanistic Education: Celebrations 14303
Association for Humanistic Education: Membership Roster 14303
Association for Institutional Research: Directory 14306
Association for Institutional Research: Journal 14306
Association for Institutional Research: Newsletter 14306
Association for Institutional Research: Professional File 14306
Association for Library Service to Children: Newsletter 14315
Association for Living Historical Farms and Agricultural Museums: Convention Proceedings 14316
Association for Living Historical Farms and Agricultural Museums: Membership List 14316
Association for Mexican Cave Studies: Bulletin 14318
Association for Mexican Cave Studies: Newsletter 14318
Association for Petroleum and Explosives Administration: The Bulletin 11275
Association for Population/Family Planning Libraries and Information Centers International: Proceedings of Annual Conference 14319
Association for Psychoanalytic Medicine: Bulletin 14323
Association for Psychoanalytic Medicine: Roster 14323
Association for Quality and Participation: Circle Report 14324
Association for Quality and Participation: Conference Proceedings 14324
Association for Recorded Sound Collections: Journal 14325
Association for Recorded Sound Collections: Membership Directory 14325
Association for Recorded Sound Collections: Newsletter 14325
Association for Research in Nervous and Mental Disease: Proceedings 14329
Association for Research of Childhood Cancer: Newsletter 14331
Association for Science, Technology and Innovation: Newsletter 14332
Association for Social Anthropology in Oceania: Newsletter 14333
Association for Social Economics: Forum 14334
Association for Specialists in Group Work: Newsletter 14335
Association for Studies in the Conservation of Historic Buildings: Newsletter 11285
Association for Studies in the Conservation of Historic Buildings: Transactions 11285
Association for the Advancement of Christian Scholarship: Perspective 01903
Association for the Advancement of Scandinavian Studies in Canada: Newsbulletin 01904
Association for the Bibliography of History: Membership Directory 14348
The Association for the Protection of Rural Scotland: Annual Report 11288
The Association for the Protection of Rural Scotland: Newsletter 11288
Association for the Sociological Study of Jewry: Journal 14354
Association for the Sociological Study of Jewry: Newsletter 14354
Association for the Study of Medical Education: Annual Report 11293
Association for the World University: Newsletter 14361
Association for Transpersonal Psychology: Listing of Professional Members 14362
Association for Transpersonal Psychology: Newsletter 14362
Association for University Business and Economic Research: Bibliography of Publications 14364
Association for University Business and Economic Research: Membership Directory 14364
Association for University Business and Economic Research: Newsletter 14364
Association for Women in Computing: Conference Proceedings 14366
Association for Women in Computing: Directory 14366
Association for Women in Computing: Newsletter 14366
Association for Women in Mathematics: Newsletter 14367
Association for Women in Psychology: Newsletter 14368
Association for Women in Science: Newsletter 14369
Association Française des Conseils Scolaires de l'Ontario: Cahier des actualités 01907
Association Française d'Observateurs d'Etoiles Variables: Bulletin 04238
Association Française pour l'Etude des Eaux: Bulletin 04244
Association Française pour l'Etude du Quaternaire: Quaternaire: Bulletin de l'Association Française pour l'Etude du Quaternaire 04246
Association Générale des Médecins de France: Bulletin 04254
Association in Scotland to Research into Astronautics: Asgard 11297
Association Internationale des Critiques de Théâtre: Prospectus 04266
Association Internationale des Etudes de l'Asie du Sud-Est: Lettre d'Information 04272
Association Internationale des Etudes Françaises: Cahiers 04273
Association Internationale des Universités: Bulletin 04281
Association Internationale d'Information Scolaire, Universitaire et Professionnelle: Bibliographical Bulletin 04283
Association Internationale d'Irradiation Industrielle: Newsletter 04284
Association Internationale pour le Développement des Universités Internationales et Mondiales: Memoranda 04290
Association Internationale pour le Progrès Social: Bulletin d'Information 01144
Association of Academic Health Sciences Library Directors: Annual Report 14373
Association of Advanced Rabbinical and Talmudic Schools: Annual Handbook 14376
Association of Air Medical Services: Membership Directory 14377
Association of American Geographers: Annals 14379
Association of American Geographers: Newsletter 14379
Association of American Indian Physicians: Newsletter 14380
Association of American Jurists Newsletter 14381
Association of American Law Schools: Newsletter 14382
Association of American Medical Colleges: Curriculum Directory 14383
Association of American State Geologists: Journal 14388
Association of Ancient Historians: Newsletter 14392
Association of Arab-American University Graduates: Newsletter 14394
Association of Black Sociologists: Roster of Membership 14403
Association of British Paediatric Nurses: Newssheet 11314
Association of British Theological and Philosophical Libraries: Bulletin 11318
Association of Canadian Archivists: Archivaria 01917
Association of Canadian Community Colleges: Bulletin 01921
Association of Canadian Community Colleges: Communiqué 01921
Association of Canadian Community Colleges: Newsletter 01921
Association of Canadian Medical Colleges: Forum 01925
Association of Caribbean Historians Bulletin 01050
Association of Caribbean Universities and Research Institutes: Newsletter 09722
Association of Chairmen of Departments of Mechanics: Newsletter 14410
Association of Clinical Biochemists: News Sheet 11324
Association of College and Research Libraries: Choice 14414
Association of College and University Telecommunications Administrators: Membership Directory 14415
Association of Collegiate Schools of Planning: Directory 14417
Association of Commonwealth Archivists and Records Managers Newsletter 11327
Association of Data Communications Users: Membership Mailing 14425
Association of Data Communications Users: Newsletter 14425
Association of Defense Trial Attorneys: Membership Roster 14426
Association of Defense Trial Attorneys: Newsletter 14426
Association of Disabled Professionals: House Bulletin 11339
Association of Disciples for Theological Discussion: Papers 14429
Association of Earth Science Editors: Membership Directory 14430
Association of Educational Psychologists: Journal 11341
Association of Energy Engineers: Directory 14431
Association of Energy Engineers: Newsletter 14431
Association of Engineers and Architects in Israel: Bulletin 07046
Association of Environmental Engineering Professors: Membership List 14434
Association of Environmental Engineering Professors: Newsletter 14434
Association of Firearm and Tool Mark Examiners: Journal 14438
Association of Food and Drug Officials: Bulletin 14439
Association of Food and Drug Officials: News and Views 14439
Association of Graduate Liberal Studies Programs: Newsletter 14441
Association of Graduate Schools in Association of American Universities: Newsletter 14442
Association of Independent Colleges and Schools: Compass 14444
Association of Independent Libraries: Directory 11356
Association of Independent Libraries: Newsletter 11356
Association of International Colleges and Universities: Directory 14447
Association of Life Insurance Counsel: Membership Directory 14453
Association of Life Insurance Counsel: Papers 14453
Association of Life Insurance Counsel: Proceedings 14453
Association of Marine Laboratories of the Caribbean: Newsletter 09724
Association of Medical Rehabilitation Administrators: Directory 14460
Association of Medical Rehabilitation Administrators: Quarterly Bulletin 14460
Association of Mental Health Administrators: Journal 14462
Association of Mental Health Administrators: Newsletter 14462
Association of Military Colleges and Schools of the United States: Membership List 14464
Association of Military Colleges and Schools of the United States: Newsletter 14464
Association of Muslim Scientists and Engineers: Newsletter 14468
Association of Muslim Scientists and Engineers: Proceedings of Conference 14468
Association of Muslim Social Scientists: Bulletin 14469
Association of Muslim Social Scientists: Newsletter 14469
Association of Muslim Social Scientists: Proceedings of Conferences 14469
Association of Official Seed Analysts: Newsletter 14472
Association of Orthodox Jewish Scientists: Membership Directory 14475
Association of Orthodox Jewish Scientists: Newsletter 14475
Association of Orthodox Jewish Scientists: Proceedings 14475
Association of Orthodox Jewish Teachers: Journal 14476
Association of Orthodox Jewish Teachers: Newspaper 14476
Association of Osteopathic State Executive Directors: Newsletter 14477
Association of Overseas Educators: Directory 14478
Association of Overseas Educators: News 14478
Association of Pacific Systematists Newsletter 14479
Association of Pathology Chairmen: Newsletter 14480
Association of Philippine Physicians in America: Newsletter 14481
Association of Physician Assistant Programs: Conference Proceedings 14482
Association of Physician Assistant Programs: Newsletter 14482
Association of Polytechnic Teachers: Bulletin 11377
Association of Professional Energy Managers: Membership Directory 14486
Association of Professors of Gynecology and Obstetrics: Membership Directory 14489
Association of Professors of Gynecology and Obstetrics: Newsletter 14489
Association of Public Analysts: Journal 11382
Association of Research Directors: Membership List 14491
Association of Researchers in Medicine and Science: Newsletter 11385
Association of Roman Ceramic Archaeologists: Kongressacta 08828
Association of Schools and Colleges of Optometry: Proceedings 14495
Association of Schools of Allied Health Professions: Directory 14496
Association of Social Anthropologists of the Commonwealth: Annals 11392
Association of Social Anthropologists of the Commonwealth: Monograph Series 11392
Association of State and Interstate Water Pollution Control Administrators: Annual Report 14506
Association of State and Interstate Water Pollution Control Administrators: Membership Directory 14506
Association of State and Territorial Directors of Public Health Education: Conference Call 14508
Association of State and Territorial Directors of Public Health Education: Proceedings 14508
Association of State and Territorial Health Officials: Conference Proceedings 14509
Association of State and Territorial Health Officials: Membership Directory 14509
Association of State and Territorial Health Officials: Newsletter 14509
Association of Surgeons of East Africa: Proceedings 08481
Association of Swimming Therapy: Report 11395
Association of Teachers of Preventive Medicine: Newsletter 14516
Association of the German Nobility in North America: Review 14518
Association of the Institute for Certification of Computer Professionals: Bulletin 14520
Association of Third World Affairs: Monographs 14522
Association of Trial Lawyers of America: Directory 14523
Association of University Anesthesiologists: Directory 14526
Association of University Architects: Newsletter 14527
Association of University Professors of Ophthalmology: Bulletin 14528
Association of University Professors of Ophthalmology: Membership Directory 14528
Association of Vice-Principals in Colleges: Conference Reports 11408
Association of Visual Science Librarians: Membership List 14533
Association pour la Fondation Internationale du Cinéma et de la Communication Audiovisuelle: Script 04321
Association pour l'Enseignement de l'Assurance: La Lettre d'Information 04342
Association Professionnelle des Bibliothécaires et Documentalistes: Bloc-Notes 01156
Association Royale des Actuaires Belges: Bulletin 01157
Association to Combat Huntington's Chorea: Newsletter 11412
Associazione Archeologica Centumcellae: Bollettino d'Informazioni 07258
Associazione Archivistica Ecclesiastica: Notiziario: Organo di Collegamento 07260
Associazione degli Africanisti Italiani: Bollettino 07266
Associazione Internazionale di Archeologia Classica: Annuario 07286
Associazione Italiana di Oncologia Medica: Abstracts Annual Meeting 07317
Associazione Italiana di Oncologia Medica: Association News 07317
Associazione Sole Italico: Bollettino Informativo 07404
Associazione Tecnica Italiana per la Cinematografia: Notiziario 07407
Associazione Termotecnica Italiana: Atti dei Congressi Nazionali 07409
ASSO Newsletter 14092
ASSP Magazine 15317
Assurance Medical Society: Transactions 11413
ASTC Newsletter 14500
ASTC Staff Directory 14500
Astérisque 04890
Asthma News 12574
ASTIN Bulletin 01105, 09209
ASTIR 06931
Astrological Review 14539
Astronautyka: Astronautics 09554
ASTRO Newsletter 14091
Astronomical Bulletin 08573
Astronomical Circular 03165
Astronomical Data 00233
The Astronomical Herold 08268
Astronomical Journal 13610
The Astronomical Journal 13893
Astronomical League: Convention Proceedings 14540
Astronomical Society of Australia: Proceedings 00231
Astronomical Society of South Australia: The Bulletin 00233
Astronomical Society of Southern Africa: Handbook 10125
Astronomical Society of Southern Africa: Notes 10125
Astronomical Society of Tasmania: Bulletin 00234
Astronomical Society of Victoria: Yearbook 00235
Astronomical Society of Western Australia: Journal 00236
Astronomicheskii Vestnik: Astronomical Herald 09954
Astronomicheskii Zhurnal: Astronomy Journal 09954
L'Astronomie 04633
Astronomische Gesellschaft e.V.: Mitteilungen 05157
Astronomisk Tidskrift 10537
Astronomisk Tidsskrift 03564
Astrophysical Journal 13610
ASV Newsletter 00235
The ATA Magazine 01850
The ATA News 01850
ATA News 06727
ATAS Bulletin 13332
ATEA Journal 14188
ATEE News 01133
Atemwegs- und Lungenkrankheiten 05838
Ateneo de Clínica Quirúrgica: Anales 13250
Atex Newspaper Users Group: Membership Directory 14544
ATFS Bulletin 11399
Athenaeum Annotations 14545
Athenaeum Architectural Archives 14545
Athenische Mitteilungen 05568
ATIC Magazine 09651
ATIRA Communications on Textiles 06712
ATJ Newsletter 14514
Atlantic Community News 14546
The Atlantic Community Quarterly 14546
Atlantic Council of the United States: Issues and Options 14546
Atlantic Economic Journal 14547
The Atlantic Planners Pen 01960
Atlas of Sweden 10573
Atlas Polskich Strojów Ludowych: Atlas of Polish Folk Costumes 09584
Atmosphere-Ocean 02199
Atmospheric Optics 16255
Atmospheric Science 03307
Atom-Informationen 05569
Atomnaya Energiya: Atomic Energy 09954
A Touch of Class 09727
ATRA Updates 13333
ATRCW Update 03939
Atrofizika: Astrophysics 00203
ATS Hotline 14198
Atti Congresso Internazionale di Studi Sardi 07475
Atti dei Convegni Lincei 07211
Atti dei Convegni Musicologici Annuali 07210
Atti dei Convegni Nazionale di Geotecnica 07281
Atti del Convegno Nazionale di Medicina dell'Assicurazione Vita 07314
Atti del Convegno Siciliano di Ecologia 07652
Atti della Accademia Ligure di Scienze e Lettere 07200
Atti della Accademia Nazionale di Marina Mercantile 07219
Atti dell'Accademia delle Scienze di Torino: Classe di Scienze fisiche 07150
Atti dell'Accademia delle Scienze di Torino: Classe di Scienze morali 07150
Atti dell'Accademia di Scienze, Lettere e Arti 07174
Atti dell'Accademia di Scienze, Lettere ed Arti di Palermo 07176

Atti dell'Accademia e Supplementi 07149
Atti dell'Accademia Lancisiana di Roma 07198
Atti dell'Accademia Medica Lombarda 07205
Atti dell'Accademia Pontaniana 07228
Atti della Società Economica di Chiavari 07799
Atti della Società Italiana di Buiatria 07846
Atti della Società Italiana di Scienze Naturali e del Museo Civico di Storia Naturale di Milano 07946
Atti della Società Toscana di Scienze Naturali 08026
Atti dell'Associazione Genetica Italiana 07278
Atti e Memorie della Deputazione di Storia Patria per le Marche 07639
Atti e Memorie della Deputazione di Storia Patria per le Province di Romagna 07640
Atti e Memorie della Deputazione Provinciale Ferrarese di Storia Patria 07644
Atti e Memorie della Società Tiburtina di Storia e d'Arte 08023
Atti e Memorie della S.O.T.I.M.I. 07787
Atti e Rassegna Tecnica della Società degli Ingegneri e degli Architetti in Torino 07780
Atti Sociali 07865
Attualità Grafologica 07283
AUA Today 14201
AUCBM Journal 11044
Auckland Medical Research Foundation: Annual Report 09119
Audecibel 16073
Audio Engineering Society, British Section: Journal 11416
Das audiovisuelle Archiv 00548
Audubon Naturalist News 14549
Der Aufschluss 06397
Der Augenarzt 05193
The Augustan 14552
The Augustan Age 16856
Augustana Historical Society: Newsletter 14550
August Derleth Society: Newsletter 14553
Aula Medica 07957
AUMLA 00248
Aurora: Jahrbuch der Eichendorff-Gesellschaft 05637
Aurora-Buchreihe 05637
AURPO Newsletter 11402
Ausbildung und Beratung in Land- und Hauswirtschaft: Monatsschrift für Lehr- und Beratungskräfte 05045
Ausgewählte neue Literatur zur Entwicklungspolitik 00743
Ausländische Aktiengesetze 05857
Außerschulische Bildung 05137
Die Aussprache 01029
The Australasian Ceramic Society: Journal 00239
Australasian Geomechanics Computing Newsletter 00293
The Australasian Journal of Dermatology 00240
Australasian Journal of Philosophy 00238
Australasian Pharmaceutical Science Association: Newsletter 00245
Australasian Radiology 00457
Australian Aboriginal Studies 00295
Australian Academic Research Libraries 00318
Australian Academy of Science: Yearbook 00250
Australian Academy of Technological Sciences and Engineering: Annual Report 00251
Australian Academy of Technological Sciences and Engineering: Handbook 00251
Australian Academy of Technological Sciences and Engineering: Symposia Series 00251
Australian Academy of the Humanities: Proceedings 00252
Australian Advances in Veterinary Science 00361
Australian and New Zealand Journal of Medicine 00456
Australian and New Zealand Journal of Ophthalmology 00463
The Australian and New Zealand Journal of Psychiatry 00460
The Australian and New Zealand Journal of Sociology 00495
Australian and New Zealand Journal of Surgery 00458
Australian Association of Adult and Community Education: Newsletter 00262
Australian Association of Clinical Biochemists – New Zealand Branch: Newsletter 09120
Australian Association of Gerontology: Proceedings 00264
Australian Atomic Energy Commission Research Establishment: Annual Report 00268
The Australian Author 00349
Australian Biochemical Society: Proceedings 00270
The Australian Bird Watcher 00368
Australian Book Contracts 00349
Australian Clay Minerals Society: Proceedings National Conference 00273
Australian Committee of Directors of Principals: Newsletter 00277
Australian Conservation Foundation: Newsletter 00278
The Australian Council for Educational Research: Newsletter 00279
Australian Dental Journal 00282
Australian Education Index 00279
Australian Education Review 00279
Australian Entomological Magazine 00392
The Australian Fluoridation News 00486
Australian Forest Research 00377
Australian Geographer 00402
Australian Geographical Studies 00412
Australian Geomechanics 00293
Australian Geomechanics News 00421
Australian Institute of Aboriginal Studies: Bibliographies 00295
Australian Institute of Aboriginal Studies: Manuals 00295
Australian Institute of Aboriginal Studies: Newsletter 00295
Australian Institute of Energy: News Journal 00300
Australian Institute of Homoeopathy: Newsletter 00302
Australian Institute of Marine Science: Annual Report: Projected Research Activities 00306
Australian Institute of Mining and Metallurgy: Bulletin 00307
Australian Institute of Mining and Metallurgy: Proceedings 00307
Australian Institute of Urban Studies 00315
Australian Journal of Adult Education 00262
Australian Journal of Agricultural Economics 00255
Australian Journal of Agricultural Research 00377
Australian Journal of Biological Sciences 00377
Australian Journal of Botany 00377
Australian Journal of Chemistry 00377
The Australian Journal of Clinical Hypnotherapy and Hypnosis 00350
Australian Journal of Dairy Technology 00384
Australian Journal of Ecology 00387
Australian Journal of Education 00279
Australian Journal of Experimental Agriculture and Animal Husbandry 00377
Australian Journal of Marine and Freshwater Research 00377
Australian Journal of Physics 00377
Australian Journal of Plant Physiology 00377
Australian Journal of Psychology 00331
Australian Journal of Science and Medicine in Sport 00280
Australian Journal of Soil Research 00377
Australian Journal of Statistics 00499
Australian Journal of Zoology 00377
Australian Law Librarians' Group Newsletter 00317
Australian Law News 00427
Australian Library Journal 00318
Australian Mammalogy 00319
The Australian Mathematical Society: Bulletin 00320
The Australian Mathematical Society: Journal: Series A 00320
The Australian Mathematical Society: Journal: Series B 00320
Australian Mathematics Competition for the Westpac Awards: Solutions and Statistics 00321
Australian Medicine 00322
Australian Microbiologist 00342
Australian Naturalist 00524
Australian Numismatic Society: Report 00324
Australian Occupational Therapy Journal 00266
Australian Outlook 00304
Australian Paediatric Journal 00276
The Australian Physicist 00310
Australian Physiological and Pharmacological Society: Proceedings 00328
Australian Planner 00465
Australian Psychologist 00331
The Australian Quarterly 00311
Australian Refrigeration, Air Conditioning and Heating: AIRAH Journal 00314
Australian Religion Studies Review 00261
Australian Road Research Board: Proceedings of Biennial Conference 00335
Australian Road Research in Progress 00335
Australian Road Research Journal 00335
Australian Shell News 00430
Australian Society for Biochemistry and Molecular Biology: Proceedings 00338
Australian Society for Fish Biology: Newsletter 00339
Australian Society for Limnology: Bulletin 00340
Australian Society for Limnology: Newsletter 00340
Australian Society for Limnology: Special Publications 00340
Australian Society for Medical Research: Proceedings 00341
Australian Society for Parasitology: Abstracts of Annual Conference 00345
Australian Society of Animal Production: Proceedings 00348
Australian Society of Herpetologists: Newsletter 00353
Australian Society of Indexers: Newsletter 00354
Australian Special Libraries News 00318
The Australian Standard 00498
Australian Transport 00371
Australian Urban Studies 00315
Australian Veterinary Journal 00361
Australian Vice-Chancellors' Committee: Information Summaries 00362
Australian Vice-Chancellors' Committee: Occasional Papers 00362
Australian Wildlife Newsletter 00527
Australian Wildlife Research 00377
Australian Wool Corporation: Annual Report 00363
Australian Zoologist 00484
Australind 00513
Austria-Esperanto-Revuo 00571
Austrian History Yearbook 16501
The Author 13030
The Authors Guild Bulletin 14554, 14555
Authorship 14257, 16183
Autismus 05253
Automatica 00686
Automation Panorama 03162
Automatique 01380
Automazione e Strumentazione 07686
Automobile Abstracts 12540
Automotive Data Bibliography 14558
Automotive Engineering 12307, 16433
Automotive Market Research Council: Newsletter 14558
Autoprep 5000 Users Group: Newsletter 14559
Autoren lesen vor Schülern – Autoren sprechen mit Schülern 05244
The Autz 13969
Auvimages 04146
Auxiliary of the American Osteopathic Association: Annual Report 14560
Auxiliary of the American Osteopathic Association: Newsletter 14560
Auxiliary of the American Osteopathic Association: Record 14560
Auxiliary of the American Osteopathic Association: Roster of Affiliates 14560
Auxiliary to the American Pharmaceutical Association: Journal 14561
Avances Técnicos 03361
Avance y Perspectiva 08718
AVA News 14200
Avaruusluotain 04009
AVA Yearbook 00361
Avian Diseases 13507
Avian Pathology 13226
Aviation, Space and Envivonmental Medicine 13339
Aviation/Space Writers Association: Newsletter 14564
Aviation/Space Writers Association: Yearbook and Directory 14564
Aviation Technician Education Council: Newsletter 14565
Avicultural Bulletin 14566
Avionics News 13356
AV Newsletter 03310
Avtomaticheskaya Svarka: Automatic Welding 11193
Avtomatika: Automation 11193
Avtomekhanika i Telemekhanika: Automation and Telemechanics 09954
Avtometriya: Autometry 09954
AWEA Wind Energy 14211
AWIS Legislative Update 14369
AWV Bulletin 14370
AWV-Informationen 05113, 05160
A Year Between 11749
Azerbaidzhanskii Khimicheskii Zhurnal: Azerbaijan Chemical Journal 01036
Azerbaijan Academy of Sciences: Doklady: Report 01036
Azerbaijan Academy of Sciences: Izvestiya: Bulletin 01036
Aziya i Afrika Segodnya: Asia and Africa Today 09954
Az-Zubair 09404

BAAS Pamphlets in American Studies 11485
Bach-Jahrbuch 06139
BACIE Journal 11490
Backsight 16756
Baconiana 12090
Bacteriologia, Virusologia, Parazitologia si Epidemiologia 09773
Badania Fizjograficzne: Seria A, B, C 09612
Badania z Dziejów Spolecznych i Gospodarczych 09612
Il Bagatto 07330
Baghdader Mitteilungen 05568
BAG-Magazin 05235
Bahamas Historical Society: Journal 01051
Bahamas National Trust: Currents 01052
The Bahamas Naturalist 01052
Bahasa dan Sastra 06894
The Baker Street Journal 14568
Balance 11551
Balance Wheel for Accreditation 14930
Balgarski folklor 01770
Balkan Medical Union: Annuaire 09747
Balkan Medical Union: Bulletin 09747
The Baltic Marine Biologists Publications 10438
Bangla Academy: Journal 01065
Bangladesh Journal of Zoology 01071
Bank-Archiv 00729
Bankhistorisches Archiv 05953
Banking World 11760
Bankwissenschaftliche Schriftenreihe 00729
La Banque des Mots 04453
Bányászat: Mining 06670
BAPHRON 14571
Baptist Quarterly 11422
Barbados Association of Medical Practitioners: Newsletter 01072
Barbados Astronomical Society: Journal 01073
Barbados Museum and Historical Society: Newsletter 01074
Barbirolli Society: Newsletter 11424
BARB News and Views 11529
Barge and Canal Development Association: Report 11425
Bar Leader 15830
Barley Yellow Dwarf Newsletter 08719
Barra Mexicana-Colegio de Abogados: El Foro 08708
BAR-REHA-INFO 05234
Barrister 13617
Basic Aspects of Glaucoma Research 05050
The Basic Biorhythm Story 00367
Basic Education 14870
Basic Education: Issues, Answers and Facts 14870
BASICS 11492
Basler Beiträge zur Geographie 10709
Basler Zeitschrift für Geschichte und Altertumskunde 10728
BASRA 11476
BASW News 11523
Battelle-Information 05166
Bauen mit Kopf 05328
Bauingenieur 06320
The Baum Bugle 15574
BAW-Monatsbericht 05222
Bayerische Akademie der Wissenschaften: Abhandlungen und Sitzungsbericht: Mathematisch-Naturwissenschaftliche Klasse 05169
Bayerische Akademie der Wissenschaften: Abhandlungen und Sitzungsbericht: Philosophisch-Historische Klasse 05169
Bayerische Akademie der Wissenschaften: Jahrbuch 05169
Bayerische Krebsgesellschaft e.V.: Rundschreiben 05173
Bayerischer Landesverein für Familienkunde e.V.: Blätter 05177
Bayerischer Volkshochschulverband e.V.: Das Forum: Zeitschrift der Volkshochschulen in Bayern 05181
Bayerisches Ärzteblatt 05174
Bayerische Schule 05178
Bayerisches Zahnärzteblatt (BZB) 05175
B.C.A.P. Journal 01971
BC News 11429
B.C. Notizie 07417
BDK-Mitteilungen 05228
The Beacon 14208
Béaloideas: Journal 06949
Beautiful China Pictorial Monthly 03252
Bedrijfsvoering 09035
Beef Improvement Federation: Update 14572
De Beeldenaar 09103
The Beethoven Newsletter 13618
Bee World 12339
Behavioral Disorders 14872
Behavioral Medicine Abstracts 16641
Behavioral Neuroscience 14019
The Behavior Analyst 16587
Behavior Genetics 14575
Behavior Research Methods 16373
Behavior Science Research: HRAF Journal ot Comparative Studies 15237
The Behavior Therapist 14263
Behavior Therapy 14263
Beihefte zu den Berichten der Naturhistorischen Gesellschaft Hannover 06129
Beilstein: Handbuch der Organischen Chemie 05182
Beiträge zur Allgemeinen und Vergleichenden Archäologie 05568
Beiträge zur Gerichtlichen Medizin 00784
Beiträge zur historischen Sozialkunde 00551
Beiträge zur Jazzforschung: Studies in Jazz Research 00671
Beiträge zur Landeskunde von Oberösterreich: I. Historische Reihe 00717
Beiträge zur Landeskunde von Oberösterreich: II. Naturwissenschaftliche Reihe 00717
Beiträge zur Lehrerbildung 10830
Beiträge zur nordischen Philologie 10846
Beiträge zur Physik der Atmosphäre 05469
Beiträge zur Sexualforschung 05419
Beiträge zur Sprachinselforschung 00976
Beiträge zur Wertpapieranalyse 05613
Békési Élet: Békés Life 06676
Belarussian Academy of Sciences: Doklady: Report 01088
Belas Artes: Revista-Boletim 09634
Belfast Natural History and Philosophical Society: Proceedings and Reports 11432
Belgeler 11175
Belgian Journal of Zoology 01489
Belgisch Brandtijdschrift 01149
Belgische Vereniging voor Tropische Geneeskunde: Annales 01163
Beliefs and Policies 16147
Belleten 11175
Ben Bowen Thomas Lecture 12645
Benesh Institute of Choreology: Newsletter 11433
Bengal Natural History Society: Journal 06740
Bensberger Manuskripte 06313
Bensberger Protokolle 06313
Benslow Music Trust: Course Brochure 11434
Bergbau-Verzeichnis 06151
Bergsmannen med JKA 10538
Berg- und Hüttenmännische Monatshefte 00952, 00953
Bericht 05643, 10879
Bericht des Historischen Vereins Bamberg 05919
Berichte 05171
Berichte: Reports 00921
Berichte der Bunsen-Gesellschaft für Physikalische Chemie: An International Journal of Physical Chemistry 05296
Berichte der Naturforschenden Gesellschaft Bamberg e.V. 06127
Berichte der Naturforschenden Gesellschaft Freiburg 06128
Berichte der Naturhistorischen Gesellschaft Hannover 06129
Berichte der naturwissenschaftlich-medizinischen Vereinigung Salzburg 00715
Berichte der ÖGKC 00802
Berichte der Römisch-Germanischen-Kommission 05568
Berichte Naturwissenschaftlich-medizinischer Verein in Innsbruck 00714
Berichte über GfS-Sommer-Symposien 05859
Berichte zur deutschen Landeskunde 06463
Berita Akitek 08648
Berita PPM 08649
Berliner Ärzte 05030
Berliner Arbeitskreis Information: Mitteilungen 05186
Berliner Gesellschaft für Anthropologie, Ethnologie und Urgeschichte: Mitteilungen 05187
Berliner Mathematische Gesellschaft e.V.: Sitzungsberichte 05188
Berliner Winckelmannsprogramm 05151
Bermuda Audubon Society: Newsletter 01519
Bermuda Books Out of Print 01520
The Bermuda National Trust: Newsletter 01522
Bermuda Sky Watch 01518
Berner Geographische Mitteilungen 10707
Berner Zeitschrift für Geschichte und Heimatkunde 10727
Bernheim Newsletter 07745
Bernische Botanische Gesellschaft: Sitzungsberichte 10635
Berufsverband der Ärzte für Orthopädie e.V.: Informationen 05192
Berufsverband der Deutschen Radiologen und Nuklearmediziner e.V.: Mitglieder-Info 05198
Bessarione: Atti Convegni sul Paleocristiano 07122
Best Books in the Social Sciences 16486
Beszámoló 06642
Beth Mikra 07090
Bet-Nahrain Journal 14580
Betoni 04010
Betriebliche Altersversorgung 05102
Betriebliches Vorschlagswesen 05583
Better Crops International 16343
Better Crops with Plant Food 16343
Better Vision Institute: Newsletter 14583
Beweging 09097
BFFS Handbook 11560
BFFS Register 11560
BFI Film and Television Handbook 11562
BFWG News 11561
Bharata Itihasa Samshodhaka Mandala: Journal 06742
BHRS Publications 11431
Bi'af 07095
Bialostoczyzna 09539

Bianco e Nero 07545
BIAS Bulletin 11469
BIAS Journal 11469
Biblia Revuo 07706
Bibliografía Agrícola Chilena 03056
Bibliografia Argentina de Artes y Letras 00139
Bibliografía Folclórica 01647
Bibliografia Geografica della Regione Italiana 07812
Bibliografia Italiana sull'Informazione 07739
Bibliografía Latinoamericana 08717
Bibliografie Doctorale Scripties Theologie 09100
Bibliographica Belgica 01234
Bibliographical Society of America: Papers 14584
Bibliographical Society of Australia and New Zealand: Bulletin 00366
Bibliographical Society of Canada: Bulletin 01964
Bibliographical Society of Canada: Papers/Cahiers 01964
Bibliographie Annuelle de l'Histoire de France 04382
Bibliographie der Hochschulrektorenkonferenz 05929
Bibliographie der Wirtschaftspresse: Documentation of selected articles from periodicals 05935
Bibliographie genevoise 10972
Bibliographie Lorraine 04119
Bibliographie Pädagogik: Bibliographie Pédagogique 05634
Bibliographie Papyrologique 01363
Bibliographie Pédagogique Suisse 10644
Bibliographies in American Music 14774
Bibliographie zur Kunstgeschichtlichen Literatur in Ost-, Mittelost- und Südosteuropäischen Zeitschriften 06465
Bibliography and Index of Geology 13832
Bibliography of Asian Studies 14268
Bibliography of Educational Theses in Australia 00279
Bibliography of Fossil Vertebrates 16712
Bibliography of Governmental Research 15168
Bibliography of Latin American and Caribbean Bibliographies 16461
Bibliography of Library Science 08394
Bibliography of Seismology 12387
Bibliography of Systematic Mycology 11720, 12378
Biblioteca Clásica de la Medicina Española 10356
Biblioteca José Jerónimo Triana 03391
Bibliotecología y Documentación 00077
Biblioteka 10105
Biblioteka Populamonaukowa: Library of Popular Science 09584
Biblioteka Przemyska 09628
Bibliotekar 17109
Bibliotekariesamfundet meddelar 10539
Bibliotekarska iskra 08608
Bibliotekarz 09615
Biblioteksbladet 10585
Biblioteksvejviser: Library Directory 03576
Bibliotheek en Samenleving 08980
Bibliothek: Editionsreihe 06105
Bibliotheka Indica 06728
Bibliotheksdienst 05570
Bibliotheksinfo 05570
Bibliothèque Archéologique 11142
Bibliothèque de la SELAF 04746
Bibliothèque de travail 04533
Bibliothikes kai Plicophocis: Bibliothèques et information 06518
BIBLOS: Mitteilungen der Vereinigung Österreichischer Bibliothekarinnen und Bibliothekare 00997
Biblos: Österreichische Zeitschrift für Buch- und Bibliothekswesen, Dokumentation, Bibliographie und Bibliophilie 00608
BIBRA Bulletin 11439
BIBRA Toxicity Profiles 11439
Bienal de Arquitectura 03029
Bigaku 08114
The Big Paper 11885
Bihar Research Society: Journal 06743
Bijdragen en Mededelingen betreffende de Geschiedenis der Nederlanden 09051
Bijdragen tot de Taal-, Land- en Volkenkunde 08964
Bijutsu-shi Gakkai: Journal 08115
BILA Bulletin 11591
Bildmessung und Luftbildwesen 05402
Bildungsarbeit in der Zweitsprache Deutsch: Konzepte und Materialien 06290
Bildungsforschung und Bildungspraxis: Education et Recherche 10989
Bildung und Wissenschaft 06012
Bilingual Books and Document Series 14674
Billedkunstneren 09297
Bilten 17078
Bilten: Astronomy 17079
Bilten: Bulletin 08594, 08601
Binnenschiffahrt 05904
Binnenschiffahrt: Zeitschrift für Binnenschiffahrt und Wasserstraßen (ZfB) 06380
Biocenoses 00013
The Biochemical Journal 11440
Biochemical Review 06858
Biochemical Society Symposia 11440
Biochemical Society Transactions 11440
The Biochemist 11440
Biochemistry 13688
Biochimie 04645
Bioconjugate Chemistry 13688
Biocontrol News and Information 11720
Biodeterioration Abstracts 11720
Bioelectromagnetics 14586
Bioenergetic Analysis 15489
BioEnergy Update 14592
Biofag 03727
Biofizika: Biophysics 09954
Biogeographia 07842
Biographical Series 13154
Bioimaging 12268
Biokémia 06620
Biokhimiya: Biochemistry 09954
Biologen in unserer Zeit 06350
Biologia Plantarum 03500
Biological and Cultural Tests for Control of Plant Diseases 14006
Biological Engineering Society: Proceedings of Conferences and Symposia 11443
Biological Journal 12454
Biological Pharmaceutical Bulletin 08395
Biological Photographic Association: Journal 14590
Biological Psychiatry 16643
Biological Research 03093
Biological Reviews 11737
Biologicheskii Zhurnal Armenii: Biological Journal of Armenia 00203
Biologicheski Membrany: Biological Membranes 09954
Biologické listy 03500
Biologiske Skrifter 03710
Biologist 12215
Biologiya Morya: Biology of the Sea 09954
Biologue 14592
Biology 01775
Biology in Action 00250
Biology International 04988
Biology of Reproduction 16619
Biology of the Cell 04818
Biomechanical Engineering 14145
Biomedical Engineering 09909
Biomedical Engineering Society: Bulletin 14593
Biomedizinische Technik: Biomedical Engineering 05331
Biometric Bulletin 14595
Biometrics 14594, 14595
Biona Report 05050
Bionieuws 08954, 09027
Bioorganicheskaya Khimiya: Bioorganic Chemistry 09954
Biopolimery i Kletki: Biopolymers and Cells 11193
BioScience 13875
Bioscience, Biotechnology and Biochemistry 08348
Biotechnologie – Verfahren, Anlagen, Apparate 05334
Le Biotechnologiste 04993
Biotechnology Progress 13688
Bioware: Zeitschrift für Biologie und Warenlehre 00844
Birding 13622
Birding in Southern Africa 10227
Bird Life International: Annual Report 11445
Bird NL 12165
The Bird Observer 00368
Birds 12853
Bird's Eye View 08669
Bird Study 11697
Birmingham and Warwickshire Archaeological Society Transactions 11447
Birth Notes 14278
Birth Psychology Bulletin 14273
Biuletyn Numizmatyczny 09553
Biuletyn Peryglacjalny 09545
Biuletyn Polskiego Towarzystwa Jezykoznawczego: Bulletin de la Société Polonaise de Linguistique 09580
Biuletyn ZIH 09632
BJF-magazin 05254
BKSTS News 11599
The Blackcountry Man Quarterly Magazine 11452
Black History Kit 14356
Black Lechwe 17157
Black Resources Information Coordinating Services: Newsletter 14601
Blätter für Deutsche Landesgeschichte 05787
Blätter für Fränkische Familienkunde 05821
Blätter für Heimatkunde 00651
Blanket Statements 14024
Blauwekamer/Profiel 08835
Bleter far Geszichte 09632
blickpunkt schule 05914
The Blind Teacher 15833
Bliss Classification Bulletin 11454
BLL-Schriftenreihe 05265
Blood 14131
Blood Transfusion and Immunhematology 04903
Blueline 14430
Blut 05333
Blyttia 09294
B.M.I. Magazine 11446
BMT Abstracts 11455
BNF Abstracts 11456
BNF Bulletin 11622
Boarding School 11457
Boarding School Life 14536
Boarding Schools Association: Occasional Papers 11457
Boarding Schools Directory 14536
BOCA Magazine 14622
Boccaccio Newsletter 13667
Bocina de los Andes 03842
Boden-Wasser-Luft 11027
BodyCast 02288
Body Engineering Journal 14102
Böcksteiner Montana 00998
Bőr és Cipőtechnika: Leather and Shoemaking 06596
Bogens Verden: Periodical on Culture and Librarianship 03576
Bohászat: Metallurgy 06670
Boletim Agrometeorológico 01633, 01648
Boletim da Comissão Nacional de Folclore 01640
Boletim da Sociedade Paranaense de Matemática 01728
Boletim da Sociedade Portuguesa de Cardiologia 09694
Boletim da Sociedade Portuguesa de Educação Médica 09698
Boletim da Sociedade Portuguesa de Hemorreologia e Microcirculação 09702
Boletim da Sociedade Portuguesa de Química 09712
Boletim de Geografia Teorética 01612
Boletim de Pesquisa 01625, 01627, 01633, 01635, 01648
Boletim Informativo 01578, 01633, 09666, 09714
Boletim Informativo Nacional 09681
Boletim Mensal de Estatistica 09673
Boletim Técnico 01656
Boletim Técnico 01738
Boletin ASEIBI 03349
Boletín Bibliografico 03062
Boletín Bibliográfico 03386
Boletín CEHILA 08726
Boletín de Antropología Americana 08740
Boletín de Astronomia 10371
Boletín de Bellas Artes y Temas de Estetica y Arte 10339
Boletin de Educación 03062
Boletin de Entomologia Venezolana 17022
Boletín de Estudios Históricos sobre San Sebastián 10367
Boletín de Hidatidosis 00085
Boletín de Historia y Antigüedades 03337
Boletín de Información 10298
Boletín de Información Cientifica 08719
Boletín de la Academia 03010
Boletín de la Academia Chilena de la Lengua 03011
Boletín de la Academia de Ciencias Fı̇sicas, Matemáticas y Naturales 16958
Boletín de la Academia Nacional de Letras 13240
Boletin de la ANABAD 10257
Boletin de la Asociación Española de Orientalistas 10265
Boletín de la Asociación Latinoamericana de Paleobotánico y Palinología 01584
Boletín de la Cofradía Vasca de Gastronomía 10367
Boletín de la Comisión Andina de Juristas 09443
Boletín de la Federación Médica Peruana 09468
Boletín de la FERE 10310
Boletín de la Real Sociedad Bascongada de los Amigos del Pais 10359
Boletín de la Sociedad Chilena de Química 03082
Boletín de la Sociedad Española de Mecanica del Suelo y Cimentaciones 10389
Boletín de la Sociedad Geográfica de Lima 09477
Boletín de la Sociedad Geológica Mexicana 08750
Boletín del Centro de Historia del Táchira 16983
Boletín del Instituto Boliviano del Petróleo 01546
Boletín de Mercado 03101
Boletín Demográfico 03026
Boletín Demográfico 03038
Boletín de Reseñas 03474
Boletín desde Colombia 03382
Boletin do Instituto de Botánica 01666
Boletín Económico 03101
Boletines Climaticos 00107
Boletín FEPAFEM: PAFAMS Bulletin 17006
Boletín Formativo e Informativo 10334
Boletín Historial 03340
Boletín Informativo 00073, 00116, 00134, 00144, 01529, 03386, 09470, 10311, 16988
Boletín Informativo ALADEFE 03846
Boletín Informativo ALER 03845
Boletín Informativo CINDA 03025
Boletín Informativo del CLAD 16989
Boletín Informativo de Pastos Tropicales 03360
Boletin Informativo SAIT 09446
Boletín Lanero 00137
Boletín Mensual de Estadística 10326
Boletín Meteorológico 10327
Boletín Minero 03102
Boletín RED 06563
Boletín: Sección Biológica 10362
Boletín: Sección Geológica 10362
Boletín SECYT-CONICET 00130
Boletín SMSR 08772
Boletín Técnico 03361, 09448
Boletín de Medicamentos y Terapéutica: Boletín Informativo de Ascofame 03344
Boletín Informativo de AIBDA 03449
Bolletin ADHILAC 04139
Bolletino 08021
Bolletino AIB 07291
Bolletino delle Sedute 07186
Bollettino A.I.R.P. 07318
Bollettino AMDI: Annuario degli Atti Ufficiali dell'Associazione 07357
Bollettino Bibliografico 07172
Bollettino Ceciliano 07347
Bollettino Charitas: Di Spiritualità 07474
Bollettino dei Classici 07211
Bollettino del Centro Camuno di Studi Preistorici 07417
Bollettino del Centro di Studi Bonaventuriani, Doctor Seraphicus 07437
Bollettino del Centro di Studi Filologici e Linguistici Siciliani 07445
Bollettino del Centro per lo Studio dei Dialetti Veneti dell'Istria 07527
Bollettino del Centro Rossiniano di Studi 07544
Bollettino del Centro Studi Archaeologici 07548
Bollettino del C.S.A.R.E. 07519
Bollettino della As.Pe.I. 07386
Bollettino della Bibliografia Forestale Italiana 07194
Bollettino dell'Accademia degli Euteleti 07137
Bollettino della Deputazione Abruzzese di Storia Patria 07632
Bollettino della Deputazione di Storia Patria per l'Umbria 07643
Bollettino della SIOI 07918
Bollettino della S.L.I. 07784
Bollettino della Società Adriatica di Scienze 07769
Bollettino della Società dei Naturalisti in Napoli 07781
Bollettino della Società Filosofica Italiana 07809
Bollettino della Società Geografica Italiana 07812
Bollettino della Società Italiana della Scienza del Suolo: Nuova Serie 07826
Bollettino della Società Italiana di Fotogrammetria e Topografia 07878
Bollettino della Società Jonico Salentina di Medicina e Chirurgia 07976
Bollettino della Società Letteraria di Verona 07979
Bollettino dell'Associazione 07342, 07379
Bollettino dell'Associazione Italiana di Cartografia 07301
Bollettino dell'Associazione Italiana di Chimica Tessile e Coloristica 07302
Bollettino dell'Associazione Italiana per lo Studio del Dolore 07344
Bollettino dell'Associazione Romana di Entomologia 07400
Bollettino della SSSAA 07999
Bollettino delle Scienze Mediche 07983
Bollettino di Audiologia e Foniatria 07839
Bollettino di Geodesia e Scienze Affini 07729
Bollettino di Informazione ANIAI 07374
Bollettino di Informazioni 07498
Bollettino di Psichiatria Biologica 07935
Bollettino ed Atti dell'Accademia Medica di Roma 07204
Bollettino Geofisico 07280
Bollettino Lavoro 07893
Bollettino Malacologico 07892
Bollettino Società Italiana di Biologia Sperimentale 07844
Bollettino SSL 10964
Bollettino Storico Bibliografico Subalpino 07646
Bollettino Storico della Basilicata 07635
Bollettino Storico della Città di Foligno 07184
Bollettino Storico per la Provincia di Novara 08019
Bollettino Storico Pisano 08020
Bollettino Storico Reggiano 07645
Bollettino Tecnico 07513
Bollettino Ufficiale del CNR 07628
Bolsilibros 03334
Bond Management Review 00365
Bone Tumor Registration in Japan 08184
Bonner Jahrbücher 06424
Bookbird 00681
Bookbird: World of Children's Books 10734
Book List 14545
Bookplate Society: Journal 11459
Bookplate Society: Newsletter 11459
Book Review Journal 16444
Books and Articles on Oriental Subjects published in Japan 08434
Booster 14181
Border Health Journal 16827
Borderline Science Series: Esoteric Series 13056
Børn & Bøger 03587
Borsodi Szemle: Borsodi Review 06676
Bossey Newsletter 10665
Bostonian Society: Proceedings 14608
Boston Theological Institute: Catalog 14609
Boston Theological Institute: Journal 14609
Boston Theological Institute: Newsletter 14609
Botanica 10491
Botanica Acta: Berichte der Deutschen Botanischen Gesellschaft 05295
Botanica Helvetica 10800
Botanical Journal 12454
Botanical Society of Scotland 11463
Botanical Survey of India: Bulletin 06749
Botanical Survey of India: Records and Reports 06749
Botanicheskii Zhurnal: Journal of Botany 09954
Botime të Veçanta/Posebna Izdanja: Monographs 17076
Botswana Notes and Records 01559
Bouw Werk: De Bouw in Feiten, Cijfers en Analyses 09071
Bovine Practitioner 13513
Boxwood Bulletin 13668
BPA Quarterly 14600
Bradlow Series 10197
The Brain and the Integration of Sciences 07414
Brain News 06780
Branch Line News 11465
Brand Book 11968
Brandeis-Bardin Institute: News 14613
Brandförsvar 10440
Breaking Chains 11196
Breakthrough 15163, 15753
Breakthroughs 16565
Brecht Yearbook 15422
Breeders Lists 12744
Bremer Zeitschrift für Wirtschaftspolitik 05222
Brics Bracs 14601
The Bridge 13020
The Bridge Newsletter 09517
The Brief 13617
Brief Bibliography Series 14299
Briefings 11295
BRIO 12331
Bristol and Gloucestershire Archaeological Society: Transactions 11468
La Brita Esperantisto 11973
Britannia 13008
Britannia Monographs 13008
British Academy: Annual Report 11470
British Academy: Proceedings 11470
British and Irish Association of Law Librarians: Newsletter 11479
British Arachnological Society: Bulletin 11480
British Arachnological Society: Newsletter 11480
British Archaeological Association: Conference Transactions 11481
British Archaeological Association: Journal 11481
British Archaeological Bibliography 11847
British Archaeological News 11847
British Association for Applied Linguistics: Newsletter 11486
British Association for Counselling: Newsletter 11491
British Association for the History of Religions: Bulletin 11496
British Association for the Study of Religions: Bulletin 11497
British Association of Clinical Anatomists: Proceedings of Meetings 11503
British Association of Friends of Museums: Newsletter 11506
British Association of Friends of Museums: Yearbook 11506
British Association of Homoeopathic Veterinary Surgeons: Newsletter 11509
British Association of Landscape Industries: Newsletter 11510
British Association of Teachers of the Deaf: Journal: Teacher of the Deaf 11526
British Astronomical Association Handbook 11530
British Aviation Preservation Council: Update 11531
British Butterfly Conservation Society: News Bulletin 11536
British Cartographic Society: Newsletter 11538
British Caspian Trust News-Letter 11539
The British Chiropractic Handbook 11780
British Columbia Genealogical Society: Newsletter 01978
British Columbia Generalogist 01978
British Columbia Parents in Crisis Society: Newsletter 01984
British Columbia Research Council: Annual Report 01988

British Columbia Research Council: List of Publications *01988*
British Corrosion Journal *12262, 12524*
British Cryogenics Council Newsletter *11548*
British Dental Journal *11549*
The British Dental Surgery Assistant *11309*
The British Ecological Society: Symposium *11554*
British Film Institute: Directions *11562*
British Food Manufacturing Industries Research Association: Abstracts *11564*
British Goat Society: The Monthly Journal *11569*
British Goat Society: The Year Book *11569*
British Heart Journal *11537*
British Herpetological Society: Bulletin *11571*
The British Homeopathic Journal *12052*
British Institute for Brain Injured Children: Newsletter *11579*
British Institute of Surgical Technologists: Journal *11590*
The British Journal for the History of Science *11673*
The British Journal for the Philosophy of Science *11674*
The British Journal of Acupuncture *11473*
British Journal of Aesthetics *11679*
British Journal of Audiology *11681*
British Journal of Biomedical Science *12261*
British Journal of Cancer *11743*
British Journal of Chemical Pharmacology *11630*
British Journal of Clinical Psychology *11636*
British Journal of Developmental Psychology *11636*
British Journal of Educational Psychology *11636*
British Journal of Educational Technology *12582*
The British Journal of Haematology *11656*
British Journal of Mathematical and Statistical Psychology *11636*
British Journal of Medical Psychology *11636*
British Journal of Mental Subnormality *11677*
British Journal of Middle Eastern Studies *11658*
The British Journal of Oral & Maxillo-Facial Surgery *11513*
British Journal of Orthodontics *11678*
British Journal of Pharmacology *11630*
British Journal of Physical Education *12697*
British Journal of Plastic Surgery *11519*
British Journal of Psychiatry *12789*
British Journal of Psychology *11636*
The British Journal of Radiology *11589*
The British Journal of Religions Education *11783*
British Journal of Rheumatology *11666*
British Journal of Social Psychology *11636*
British Journal of Social Work *11523*
The British Journal of The Nutrition Society *12651*
British Journal of Visual Impairment *11287*
British Lichen Society Bulletin *11604*
British Linguistic Newsletter *12453*
British Medical Journal *11607*
British Migraine Association: Newsletter *11609*
British Model Soldier Society: The Bulletin *11610*
British Museum Magazine *11612*
British Music Society: Journal *11613*
British Music Society: Newsletter *11613*
British Naturalist *11617*
British Naturopathic Journal *11618*
British Numismatic Journal *11621*
British Origami *11624*
British Orthoptic Journal *11628*
British Osteopathic Journal *12664*
British Postmark Society: Bulletin *11634*
British Psycho-Analytical Society AR *12272*
British Psycho-Analytical Society Bulletin *12272*
British Psycho-Analytical Society Roster *12272*
The British Puppet and Model Theatre Guild: Magazine *11637*
The British Puppet and Model Theatre Guild: Newsletter *11637*
British Science Fiction Association: Focus: New Writers Forum *11644*
British Society for Cell Biology: Newsletter *11651*
British Society for Eighteenth Century Studies: Bulletin *11654*
British Society for Eighteenth Century Studies: Journal *11654*
British Society for Middle Eastern Studies: Newsletter *11658*
The British Society for Phenomenology: Journal *11662*
The British Society for the History of Mathematics Newsletter *11670*
The British Society for the History of Science: Monograph Series *11673*
The British Society for the History of Science: Newsletter *11673*
British Society for the Promotion of Vegetable Research: Annual Report *11675*
British Society for the Study of Mental Subnormality: Newsletter *11677*
The British Society of Dowsers: Journal *11682*
British Society of Flavourists: Newsletter *11683*
British Society of Scientific Glassblowers: Journal *11689*
British Standards Institution: Annual Report *11692*
British Universities' Guide to Graduate Study *11805*
British Urban and Regional Information System Association: Newsletter *11702*
British Wood Preserving and Damp-proofing Association: Information Leaflets *11709*
British Wood Preserving and Damp-proofing Association: News Sheet *11709*
Brittle Bone Society: Newsletter *11711*
Broadcast *15704*
Broadsheet *00366, 11789*
Broadsheet Digest *11850*
Broadside: Newsletter of the TLA *16775*
Bromatologia i Chemia Toksykologiczna *09565*
Bronnen en Studies *08946*
The Brontë Society: Transactions *11712*
BRPS Newsletter *11640*
Bruckner-Jahrbuch *00547*
Brücken *06266*
Brunel Lectures *12050*
Brunonia *00377*
The Bryologist *13671*
Brytpunkt *10468*
BSBI Abstracts *11464*
BSCC Newsletter *11645*
BSI News *11692*
BSI Newsletter *11657*
BSMT Bulletin *11659*
BTA Newsletter *11695*
BTO News *11697*
BTRA Bulletin *06748*
BTRA Cleanings *06748*
BTRA Current Textile Literature *06748*
BTRA Scan *06748*
Buch und Bibliothek (BuB) *06368*
Buddhist Text Information *15267*
Buffalo Bill Memorial Association: Newsletter *14621*
The Building Economist *00312*
Building Engineer *11319*
Building Services *11767*
Building Services Engineering Research and Technology *11767*
Building Standards *15445*
Buletin de fizica si chimie *09783*
Buletinul informativ al Academiei de Stiinte Agricole si Silvice *09734*
Buletinul SSF *09782*
Buletinul S.S.G.: Bulletin de la Société Roumaine de Géographie *09784*
Bulletin AAI *01105*
Bulletin Analytique de Documentation *04543*
Bulletin and Clinical Review of Burn Injuries *15525*
Bulletin A.P.D.P. *09650*
Bulletin ATMA *04355*
Bulletin Baudelairien *16938*
Bulletin Catalogue of the Inventions of BAS *01775*
Bulletin České Společnosti Mikrobiologicke *03526*
Bulletin CIA *04585*
Bulletin d'Archéologie Sud-Est Européenne *09752*
Bulletin de l'Académie des Sciences Agricoles et Forestières *09734*
Bulletin de l'Académie des Sciences et Lettres de Montpellier *04106*
Bulletin de la Classe des Lettres *01095*
Bulletin de la Classe des Sciences *01095*
Bulletin de l'AET *04207*
Bulletin de l'AIEO *08824*
Bulletin de l'AIESEE *09752*
Bulletin de l'AISLF *04277*
Bulletin de l'ALAI *04299*
Bulletin de l'AMQ *01911*
Bulletin de la SAJIB *02917*
Bulletin de la Section des Naturalistes Enseignants *08808*
Bulletin de la SFG *04801*
Bulletin de la Société Archéologique et Historique du Limousin *04630*
Bulletin de la Société Chimique de France *04791*
Bulletin de la Société de l'Histoire de l'Art Français *04662*
Bulletin de la Société de Linguistique de Paris *04664*
Bulletin de la Société de Mythologie Française *04672*
Bulletin de la Societé de Pathologie Exotique *04679*
Bulletin de la Société des Naturalistes Luxembourgeois *08587*
Bulletin de la Société des Sciences et des Lettres de Łódź *09545*
Bulletin de la Société Française de Philosophie *04830*
Bulletin de la Société Fribourgeoise des Sciences Naturelles *10974*
Bulletin de la Société Géologique de France *04868*
Bulletin de la Société Historique, Archéologique et Littéraire de Lyon *04870*
Bulletin de la Société Historique et Archéologique du Périgord *04873*
Bulletin de la Société Vaudoise des Sciences Naturelles *10990*
Bulletin de la Société Vétérinaire Pratique de France *04922*
Bulletin de l'Association de Géographes Français *04185*
Bulletin de l'Association Professionnelle des Géologues et des Géophysiciens du Québec *01952*
Bulletin de l'Association Suisse de Science Politique *10633*
Bulletin de Liaison *04813, 04824*
Bulletin de Liaison de la SELF *04690*
Bulletin de Liaison et d'Information *10629*
Bulletin de l'I.D.E.F. *04572*
Bulletin de l'Institut Archéologique Liègeois *01376*
Bulletin de l'Institut des Actuaires Français *04565*
Bulletin de l'Institut Historique Belge de Rome *07121*
Bulletin de l'OIEC *01417*
Bulletin de l'Ordre des Médecins *04611*
Bulletin de l'Organisation Internationale de Métrologie Légale *04615*
Bulletin de Médecine Légale et de Toxicologie Médicale *01126*
Bulletin de Philosophie Médievale *01465*
Bulletin der Aktion *10610*
Bulletin der Wiener Psychoanalytischen Vereinigung *01025*
Bulletin des Amis de Gustave Courbet *04297*
Le Bulletin des Auteurs *04938*
Bulletin des Epizooties en Afrique *09244*
Bulletin des Géosciences en Afrique de l'Ouest *11088*
Bulletin des Sociétés Chimiques Belges *01507*
Bulletin des Travaux de la Société de Pharmacie de Lyon *04681*
Bulletin dh *04653*
Bulletin d'Information AIPPHi *05155*
Bulletin d'Information de l'Institut Géographique National *04570*
Bulletin d'Information du CDS *04359*
Bulletin d'Information GAMS *04562*
Bulletin d'Information Professionelle, BIP *01113*
Bulletin du Bureau International d'Education: Bulletin of the International Bureau of Education *10639*
Bulletin du Cancer *04245, 04844*
Bulletin du CIETA *04377*
Bulletin du Conseil National: Tijdschrift Nationale Raad *01253*
Bulletin du CORESTA *04366*
Bulletin du GADEF *04553*
Bulletin du GIRSO *01372*
Bulletin du Musée d'Anthropologie Préhistorique de Monaco *08794*
Bulletin ESA *04127*
Bulletin et Mémoires de la S.M.F. *04890*
Bulletin for International Fiscal Documentation *08926*
Bulletin Historique et Littéraire *04663*
Bulletin IBION *08673*
Bulletin ICTM *15454*
Bulletin Interafricain d'Informations Phytosanitaires *09244*
Bulletin International de la F.I.P.E.S.O. *04535*
Bulletin International Starine *03459*
Bulletin IRP *02521*
Bulletin Mathématique *09786*
Bulletin Mensuel SSPM: Mitteilungsblatt *10900*
Bulletin Monumental *04786*
Bulletin OCCGE Information *01810*
Bulletin of African Insect Science *08463*
Bulletin of Agricultural and Horticultural Sciences *09153*
Bulletin of Alloy Phase Diagrams *14255*
Bulletin of American Paleontology *16286*
Bulletin of Arab Publications *11131*
Bulletin of Biology *03121*
Bulletin of Botany *03121*
Bulletin of Canadian Petroleum Geology *02291*
Bulletin of Canadian Studies *11488*
Bulletin of Classification Society of India *06777*
Bulletin of Documentation *17094*
Bulletin of Educational Statistics for the Arab States *11131*
Bulletin of Entomological Research *11720, 11732*
Bulletin of Epizootic Diseases of Africa *09244*
Bulletin of Hospital Dental Practice *13540*
Bulletin of IAPESGW *03744*
Bulletin of Informatics and Cybernetics *08435*
Bulletin of Information for Candidates *13644*
Bulletin of Latin American Research *12977*
Bulletin of Legal Developments *11583*
Bulletin of Malacology *03306*
Bulletin of Maritime Transport Information Analysis *03874*
Bulletin of Materials Science *06767*
Bulletin of News *09231*
Bulletin of Physical Education: Safe Practice in Physical Education *11500*
Bulletin of Structural Integration *16427*
The Bulletin of the AAS *13611*
Bulletin of the American Academy of Psychiatry and the Law *13431*
Bulletin of the American Mathematical Society *13929*
The Bulletin of the Association for Business Communication *14275*
Bulletin of the Association of Engineering Geologists *14433*
Bulletin of the Association of Pediatric Societies of the Southeast Asian Region *09510*
Bulletin of the Astronomical Society of India *06736*
Bulletin of the Australian Psychological Society *00331*
Bulletin of the British Bryological Society *11535*
Bulletin of the British Ornithologists' Club *11625*
The Bulletin of the Calcutta Mathematical Society *06750*
Bulletin of the Canadian Nuclear Society *02211*
Bulletin of the Ceramic Society of Japan *08369*
Bulletin of the Chinese Geophysical Society *03171*
Bulletin of the Czech Astronomical Institute *03500*
Bulletin of the EATCS *05665*
Bulletin of the Geological Society of Denmark *03625*
Bulletin of the Hellenic Veterinary Medical Society *06508*
Bulletin of the History of Dentistry *13440*
Bulletin of the History of Medicine *13500*
Bulletin of the Hungarian Biophysical Society *06619*
Bulletin of the ICDE *02549*
Bulletin of the Indian Geologists' Association *06794*
Bulletin of the Indian Society of Soil Science *06820*
Bulletin of the International Association of Engineering Geology *07707*
Bulletin of the International Seismological Centre *12387*
Bulletin of the IPS *03969*
Bulletin of the Kansai Society of Naval Architects *08155*
Bulletin of the Meteorological Society of the Republic of China *03307*
Bulletin of the Pan American Health Organization *13585, 16292*
Bulletin of the Printing Historial Society *12718*
Bulletin of the Society for Near Eastern Studies in Japan *08354*
Bulletin of the Society for Renaissance Studies *12992*
Bulletin of The Society for Vector Ecology *16627*
Bulletin of the Soiety of Cartographers *13033*
Bulletin of the TBC *16785*
Bulletin of the World Health Organization *11033*
Bulletin of the WVA *10405*
Bulletin of Volcanic Eruptions *08325*
Bulletin of Zoological Nomenclature *12345*
Bulletin on Continuing Medical Education *06801*
Bulletin Paris et son Histoire *04618*
Bulletin SBE *10889*
Bulletin Scientifique de Bourgogne *04717*
Bulletin Spektroskopické Jana Marca Marci *03553*
Bulletin S.R.L.F. *04685*
Bullettino Senese di Storia Patria *07237*
Bullettino Storico Pistoiese *08004*
Bulteno *13586*
The Bulwark *12924*
Bùnadarrit *06679*
Bunseki *08189*
Bunseki Kagaku *08189*
Das bunte Blatt *05202*
Bureau Bulletin *17173*
Bureau de Statistiques Universitaires: Dienst voor Universitaire Statistiek *01367*
Bureau International d'Audiophonologie: Rapport Annuel: Recommandations *01173*
Bureau International de Liaison et de Documentation: Documents *04362*
Buried History *00297*
Burma Research Society: Journal *08812*
Burn Care Services in North America *13673*
Business Accounting for Lawyers Newsletter *16348*
Business Administrator *12217*
Business and Management Education Funding Alert *13462*
The Business Economist *13032*
Business Education Forum *15930*
Business Education Guide *12901*
Business History Conference: Conference Newsletter *14628*
Business History Conference: List of Members *14628*
Business Ideas Letter *12282*
Business Lawyer *13617*
Business News Index *12540*
Business Officer *15840*
Business Today *09727*
Business Tools *14421*
Busqueda *02552*
Butlletí Andorrà de Ciències Històriques *00030*
Butlleti Arqueologic *10358*
Butlleti de la Societat Arqueològica Lul-liana *10403*
Butsuri *08284*
BVRA Abstracts *11704*
Bygg *09306*
Byggekunst *09296*
Bygningsstatiske Meddelelser *03673*
Byline: AAMA *13548*
Byplan *03596*
The Byron Society Journal *11719*
Byulleten Eksperimentalnoi Biologii i Meditsiny: Bulletin of Experimental Biology and Medicine *09952*
Byulleten Moskovskogo Obshchestva Ispytatelei Prirody *09886*
Byulleten Sibirskogo Otdeleniya Rossiiskoi AMN: Bulletin of the Siberian Division of the Russian Academy of Medical Sciences *09952*
Byzantinoslavica *03500*
Byzantion: Revue Internationale des Etudes Byzantines *01448*

CAA Bulletin *02046*
CA-A Cancer Journal for Clinicians *13676*
Caderno de Conjuntura *01714*
Cadernos de Biblioteconomia, Arquivistica e Documentagâo *09640*
Cadernos de Estudos Sociais *01658*
CAD Newsletter *11808*
CAEJC Newsletter *02364*
Caernarvonshire Historical Society: Transactions *11733*
CAETA Newsletter *17165*
CAFRA News *11108*
CAFS Newsletter *08484*
Cagrindex *11115*
CAGRIS Newsletter *11115*
Cahiers d'Anthropologie et Biométrie Humaine *04644*
Cahiers de Civilisation Médiévale *04747*
Cahiers de Généalogie Protestante *04663*
Les Cahiers de l'Académie *01814*
Cahiers de la Documentation *01106*
Cahiers de la Fondation Universitaire: Cahiers van de Universitaire Stichting *01367*
Cahiers de la Société Historique Acadienne *02929*
Cahiers de Linguistique Théoretique et Appliquée *09737*
Cahiers de Micropaléontologie *04382*
Les Cahiers de Montpellier: Forces Armées et Politiques de Défense *04370*
Cahiers de Nutrition et de Diététique *04677*
Cahiers des Arts et Traditions Populaires *11142*
Cahiers de Sexologie Clinique *04842*
Cahiers des Naturalistes: Bulletin des Naturalistes Parisiens *04603*
Les Cahiers du CREAD *00011*
Cahiers du MURS *04602*
Les Cahiers du Psychologue Québecois *02398*
Cahiers du Seminaire d'Econométrie *04382*
Cahiers Ferdinand de Saussure *10930*
Les Cahiers Français de l'Electricité *04404*
Cahiers Ligures de Préhistoire et de Protohistoire *07732*
Les Cahiers Lorrains *04763*
Cahiers Saint-John Perse *04546*
C.A.H.S. Journal *02080*
CAI Informa *00100*
CAIS Newsletter: Nouvelles ACSI *02034*
Cajal Club: Proceedings *14631*
Cajanus *08073*
Calcutta Mathematical Society: News Bulletin *06750*
Calcutta Statistical Association Bulletin *06751*
Caldasía: Boletín *03391*
Calendari del Pagés *10313*
Calendar of Congresses of Medical Sciences *10655*
California Military Monitor *16284*
CALL Newsletter *02056*
Calypso Log *14936*
CA Magazine *02168*

Cambridge Antiquarian Society: Proceedings 11735
Cambridge Bibliographical Society: Monographs 11736
Cambridge Bibliographical Society: Transactions 11736
Cambridge Refrigeration Technology: Newsletter 11738
Camden 12809
CAMHDD Newsletter 11809
CAML Newsletter 02059
CAM Newsletter 02366
The Campaigner and Animals Defender 12555
Campus Law Enforcement Journal 15400
CAMRODD Bulletin 09112
Canada Council: Annual Report 02000
Canada Safety Council: Council Update 02001
Canada's Ethnic Groups Series 02157
Canadian Academy of Engineering: Newsbrief 02003
Canadian Academy of Recording Arts and Sciences: Newsletter 02006
Canadian Accounting Education and Research News 02002
Canadian Acoustics/Acoustique Canadienne 02012
Canadian Aeronautics and Space Journal 02016
Canadiana Germanica 02483
Canadian Agricultural Engineering Journal 02272
Canadian Association for American Studies: Newsletter 02026
Canadian Association for Information Science: Proceedings of the Annual Conference: Procesverbaux de la conférence 02034
Canadian Association for Laboratory Animal Science: Newsletter 02035
Canadian Association for Latin American and Caribbean Studies: Newsletter 02036
Canadian Association of African Studies: Newsletter 02045
Canadian Association of Geographers: The Directory 02052
Canadian Association of Latin American Studies: Newsletter 02055
Canadian Association of Physical Medicine and Rehabilitation: Newsletter 02064
Canadian Association of University Teachers of German: Seminar 02075
Canadian Author & Bookman 02079
Canadian Bar Review 02081
Canadian Biochemical Society: Bulletin 02082
Canadian Bureau for International Education: Annual Report 02084
Canadian Canon Law Society: Newsletter/Bulletin de Nouvelles 02086
Canadian Ceramics 02092
Canadian Chemical News 02342
The Canadian Children's Film and Video Directory 02344
The Canadian Composer 02373, 02944
Canadian Dental Association Journal 02114
Canadian Electrical and Computer Engineering Journal 02435
Canadian Electrical Association: Bulletin 02123
Canadian Electrical Association: Research Reports 02123
Canadian Electrical Engineering Journal 02256
Canadian Entomologist 02437
Canadian Ethnic Studies 02127
Canadian Ethnic Studies Association: Proceedings 02127
Canadian Family Physician 02354
Canadian Federation for the Humanities: Bulletin 02128
Canadian Federation of Biological Societies: Programme & Proceedings 02129
Canadian Gemmologist 02144
The Canadian Geographer 02052
Canadian Geotechnical Journal 02435
Canadian Institute of Chartered Accountants: Members Directory 02168
Canadian Institute of International Affairs: Behind the Headlines 02177
Canadian Institute of International Affairs: International Journal 02177
Canadian Institute of Mining and Metallurgy: Directory 02180
Canadian Issues 01899
Canadian Journal of Administrative Sciences 01820
Canadian Journal of African Studies 02045
Canadian Journal of Agricultural Economics 02017
Canadian Journal of Animal Science 01823, 02274
Canadian Journal of Applied Sport Sciences 02069
Canadian Journal of Behavioural Sience 02234
The Canadian Journal of Chemical Engineering 02342
The Canadian Journal of Civil Engineering 02252
Canadian Journal of Civil Engineering 02435
Canadian Journal of Comparative Medicine 02315
Canadian Journal of Criminology: Revue canadienne de criminologie 02113
Canadian Journal of Economics 02120
Canadian Journal of Education: Revue canadienne de l'éducation 02268
Canadian Journal of Higher Education 02269
Canadian Journal of Information Science: Revue canadienne des sciences de l'information 02034
The Canadian Journal of Latin American and Caribbean Studies 02036
Canadian Journal of Latin American Studies 02055
The Canadian Journal of Linguistics 02191
Canadian Journal of Netherlandic Studies 02041
Canadian Journal of Neurological Science 02210
The Canadian Journal of Nursing/Revue canadienne de recherche en sciences infirmières 02072
The Canadian Journal of Optometry 02060
Canadian Journal of Physiology and Pharmacology 02813
Canadian Journal of Plant Pathology 02226
Canadian Journal of Plant Science 01823
The Canadian Journal of Political Science: Revue canadienne de science politique 02229
The Canadian Journal of Psychiatry 02231
Canadian Journal of Psychology 02234
Canadian Journal of Public Health 02235
Canadian Journal of Soil Science 01823
Canadian Journal of Spectroscopy 02949
Canadian Journal of Statistics 02960
Canadian Journal of Surgery: Journal canadien de chirurgie 02197
Canadian Library Journal 02189
Canadian Lung Association Bulletin 02192
Canadian Manager Magazine 02178
Canadian Maritime Bibliography 02208
Canadian Materials 02189
Canadian Mathematical Bulletin 02195
The Canadian Mineralogist 02655
Canadian Music Centre: Acquisitions 02202
The Canadian Numismatic Journal 02212
Canadian Operational Research Society: Bulletin 02213
Canadian Paediatric Society: News Bulletin 02219
Canadian Papers in Peace Studies 02908
Canadian Pharmaceutical Journal 02223
Canadian Philosophical Association: Bulletin 02224
Canadian Polish Research Institute: Studies 02228
Canadian Political Science Association: The Bulletin: Le Bulletin 02229
Canadian Psoriasis Foundation Newsletter 02230
Canadian Psychological Association: Highlights 02234
Canadian Psychology 02234
Canadian Radiation Protection Association Bulletin 02240
Canadian Review of American Studies 02026
Canadian Review of Art Education Research 02255
Canadian Review of Sociology and Anthropology 02298
Canadian Society for Cellular and Molecular Biology: Bulletin 02251
Canadian Society for Education through Art: Newsletter 02255
Canadian Society for Education through Art: The Journal 02255
Canadian Society for Immunology: Bulletin 02259
Canadian Society of Biblical Studies: Society Bulletin 02275
Canadian Society of Cytology Bulletin 02281
Canadian Society of Microbiologists: Programme & Abstracts 02286
Canadian Society of Military Medals and Insignia: Journal 02287
Canadian Society of Military Medals and Insignia: Newsletter 02287
Canadian Society of Orthopaedic Technologists: NewsCast 02288
Canadian Studies Update 14276
Canadian Textile Journal 02532
Canadian University Music Review: Revue de musique des universités canadiennes 02312
Canadian Urban Transit Association: Annual Report 02313
Canadian Veterinary Journal 02315
CANASA Newsletter 02018
Cancer Book House List 14638
Cancer Facts and Figures 13676
Cancer Federation: Newsletter 14639
Cancer Forum 00272
Cancer in Ontario 02764
Cancer News 13676
Cancer Nursing News 13676
Cancer Research 13471
Cancer Research Campaign: Annual Report 11743
Cancer Research News 01973
Canon Law Society of America: Convention Proceedings 14642
Canon Law Society of America: Membership Directory 14642
Canon Law Society of America: Newsletter 14642
CANQA Newsletter/Bulletin 02239
CANSCAIP News 02277
Canu Gwerin: Folk Song 13202
Capacity Report 13990
CAPA Journal of Technical Education and Training 08488
CAPA Newsletter 08488
CAPFSA Reporter 10206
Capital Comments 14444
La Capitale 07463
Capital Outlook 15963
The Capitol Dome 16817
CAP News 08076
CAPPI Newsletter 09515
CAP Today 14775
Cara Seminary Directory 14666
Cara Seminary Forum 14666
Carbide and Tool Journal 16646
CARCAE Newsletter 00035
Cardichte's Guide 15925
Cardiologia 07848
Cardiologia Hungarica 06641
Cardiostimolazione 07300
Cardiovascolar Research 11537
Cardiovascular Nursing 13844
Career Brochure 12801
The Career Development 13771
Career Directory 15811
Career Information Bulletin 13565
Careers in Finance 15093
Careers Leaflets 12901
Career Teacher 12570
Career Training 14645
Care in the Home 12852
Care on the Road 12852
CARF Newsletter 02015
Caribbean Agro-Economic Society Journal 11107
Caribbean Agro-Economic Society Newsletter 11107
Caribbean Biotechnology Newsletter 11113
Caribbean Conservation News 01080
Caribbean Educational Bulletin 09722
CARICAD Newsletter 01079
Caries Research 12035
CARIMAC Newsletter 08074
Carinthia I: Zeitschrift für geschichtliche Landeskunde von Kärnten 00602
Carinthia II: Naturwissenschaftliche Beiträge zur Heimatkunde Kärntens 00712
CARIRI Technochat 11113
CARISPLAN Abstracts 11114
Carl Duisberg Forum 05272
Carnegie Institution of Washington: Year Book 14647
CARNEID Newsletter 01085
Carta de AUSJAL 03350
Carta de Colciencias 03374
Carta del CCYDEL 08710
Carta Geológica de Chile 03063
Carta Hidrogeológica de Chile 03063
Carta Informativa 03382, 09414
Carta Informativa de la Secretaría General 03440
Carta Magnética de Chile 03063
Carta Pediátrica 03432
Cartogram Newsletter 09139
Cartographica 02088
The Cartographic Journal 11538
Cartography 00298
CASAS Newsletter 08641
CASE in Point 14889
Casemate 12087
Časopis lékařu českých: Czech Medical Journal 03509
Časopis Matice Moravské 03508
Časopis Matice moravské 03551
Časopis pro mineralogii a geologii 03500
Časopis pro pěestování matematiky 03500
Cassava Newsletter 03360
Casteel Magazine 16745
Casting 03269
Casting Plant and Technology: CP+T International 06376
Castings Technology International: Annual Conference Proceedings 11747
Catalogo 10343
Catálogo Brasileiro de Engenharia Sanitária e Ambiental – CABES 01588
Catalogo Ilustrado de las Plantas de Cundinamarca 03391
Catálogos Bibliográficos 00156
Catalogs of Publications and Visual Aids 16051
Catalogue des Coléoptères de Belgique 01476
Catalogue of Hellenic Standards 06517
Catalogue of Norwegian Standards 09287
Cataloguing Australia 00318
Catalysis Society (North America): Newsletter 14648
Cathedra: History of Palestine Studies 07118
Catholic Audio-Visual Educators Association: Newsletter 14651
The Catholic Biblical Quarterly 14652
Catholic Fine Arts Society: Membership List 14654
Catholic Fine Arts Society: Newsletter 14654
Catholic Health World: Official Assoc. Newspaper 14655
Catholic Historical Review 13680
Catholic Library Association: Handbook and Membership Directory 14656
Catholic Library World 14656
Catholic Media Council: Information Bulletin 05273
Catholic Periodical and Literature Index 14656
The Catholic Pharmacist 15937
Catholic Theological Society of America: Proceedings 14658
CAUSN Newsletter/Bulletin d'information 02072
Caveat 00428
Caveat Emptor 14853
CAVE Evaluation 14651
Caves and Caving 11540
Cave Science: The Transactions of the B.C.R.A. 11540
c+b: Chemie und Biologie 11020
CBA/ABC Bulletin 02083
CBE Environmental Review 14756
CBE Views 14890
CBR News 11239
CCAMLR Newsletter 00373
CCAMLR Statistical Bulletin 00373
CCA News 14645
CCA Newsletter 14661
CCAS Newsletter 14892
CCBC Newsletter 14893
CCBD Newsletter 14872
CCDCG Newsletter 14762
CCEA Newsletter 00375
CCICA Annual 14653
CCST Newsletter 11110
CCTE Newsletter 02108
CCWHP/CGWH Newsletter 14860
CDC Acquisitions Lists 03883
CDC Newsletter 03883
CDT Yearbook and Directory 11944
The CEA Critic 14772
The CEA Forum 14772
CEA Handbook 02122
CEA Newsletter: Nouvelles 02122
CEC-M Report 14981
CEDAM International: Bulletin 14662
CED Approved Programs 14916
C.E.D.I.E.S. Informations 08804
Ce Fastu? 07806
Cégepropos 02450
Cell Biology International Reports 12355
Cellulose Chemistry and Technology 09737
Cemento 07352
Cements Research Progress 13685
CEMLA Bulletin 08712
CEN/CENELEC Review 01182
CENICAFE 03361
Censimento dei Codici Petrarcheschi 07655
Center for Advanced Study in the Behavioral Sciences: Annual Report 14663
Center for American Archaeology Annual Report 14664
Center for American Archaeology Newsletter 14664
Center for Chinese Research Materials: Newsletter 14671
Center for Cuban Studies: Newsletter 14674
Center for Early Adolescence: Common Focus 14677
Center for Marine Conservation: Report 14685
Center for Medieval and Early Renaissance Studies: Acta 14687
Center for Medieval and Early Renaissance Studies: Proceedings of Conference 14687
Center for Medieval and Renaissance Studies: Nouvelles 14688
Center for Migration Studies of New York: Newsletter 14689
Center for Neo-Hellenic Studies: Bulletin 14690
Center for Process Studies Newsletter 14693
Center for Reseach Libraries Directory 14696
Center for Research Libraries: Annual Report 14696
Center for Research Libraries: Focus 14696
Center for Research Libraries: Handbook 14696
Center for Safety in the Arts: Newsletter 14698
Center for the Study of Human Rights: Annual Report 14706
Center for the Study of Human Rights: Newsletter 14706
Center for the Study of the Presidency: Proceedings 14708
Center for Urban Black Studies: Newsletter 14710
Center for War/Peace Studies: Global Report: Newsletter 14712
Center for War/Peace Studies: Special Studies 14712
Centergram 14713
Center House Bulletin 14708
Centerpoint Newsmagazine 03240
The Center Scope 14219
Centraal Bureau voor Genealogie: Jaarboek 08838
Centraal Bureau voor Genealogie: Mededelingen 08838
Central African Specifications and Codes of Practice 17181
Central America Education Project 15165
Central American Journal of Public Administration 03445
Central European Journal of Public Health 03509
Central Mediterranean Naturalist 08671
Central States Anthropology Society Bulletin 14717
Centre d'Animation, de Développement et de Recherche en Education: Prospectives 02326
Centre de Recherches en Economie Appliquée pour le Développement: Bulletin Bibliographique 00011
Centre for the Study of International Relations: Review 10443
Centre International de l'Enfance: Bulletins bibliographiques 04376
Centre International d'Etude de Tantale et de Niobium: Bulletin 01197
Centre Suisse de Documentation en Matière d'Enseignement et d'Education: Bulletin: Mitteilungen 10644
Centro de Información Científica y Humanística: Periodica: Indice de Revistas Latinoamericanas en Ciencias 08717
Centro de Información e Divulgación Agropecuario: Informaciòn Express 03474
Centro de Perfeccionamiento, Experimentación e Investigaciones Pedagógicas: Revista de Estudios 03023
Centro de Pesquisa Agropecuária dos Cerrados: Documentos 01625
Centro de Pesquisa Agropecuária do Trópico-Árido: Annual Research Report 01626
Centro de Pesquisa Agropecuária do Trópico-Árido: Research Bulletin 01626
Centro de Pesquisa Agropecuária do Trópico Umido: Documentos 01627
Centro di Studi per la Storia dell'Architettura: Bollettino 07450
Centro di Vita Europea: Anni Nuovi 07458
Centro Latinoamericano de Adminstración para el Desarrollo: Boletín de Resumenes 16989
Centro Latino Americano de Física: Noticia 01630
Centro Nacional de Documentación Científica y Tecnológica: Actualidades 01536
Centro Nacional de Pesquisa de Mandioca e Fruticultura: Documentos 01633
Centro Nacional de Pesquisa de Milho e Sorgo: Documentos 01634
Centro per la Statistica Aziendale: Circolare 07525
Centro per la Statistica Aziendale: Index 07525
Centro Ricerche Biopsichiche 07435
Centro Ricerche Economiche Sociologiche e di Mercato nell'Edilizia: Bulletins 07539
Centro Ricerche Economiche Sociologiche e di Mercato nell'Edilizia: Cahiers 07539
Centro Studi ed Esperienze Scout Baden Powell: Esperienze e Progetti 07569
Centro Studi Politici e Sociali Alcide de Gasperi: Prospettive 07586
Centrum voor de Studie van het Onderwijs in Ontwikkelingslanden: Verhandelingen 08839
Centrum voor Plantenveredelings- en Reproduktieonderzoek: Annual Report 08840
Centrum voor Wiskunde en Informatica: Jaarverslagen: Annual Report 08841
Centrum voor Wiskunde en Informatica: Report Series 08841
Cepal Review 03038
CEPCEO Report 01223
Cepsa-Informa 07560
Ceramic Abstracts 13685, 16093
Ceramic Bulletin 13685, 16093
Ceramic Engineering and Science Proceedings 13685, 16093
Ceramic Forum International: Berichte 05459
Ceramic Source 16093
Céramique 10606
CERAM Research News 11755
CERAM Research Progress 11755

CERDAS Liaison 17144
Cereal Chemistry 13514
Cereal Foods World 13514
CERN Courier/Courrier 10764
Certified Accountants 11759
Certified Engineering Technician Magazine 14105
Certified Occupancy Specialists 15945
CESA Bulletin 02127
CES Bulletin 01225
CESE Newsletter 01242
Ceshtje te folklorit shqiptar: Problems of the Albanian Folklore 00003
Ceshtje te gramatikes se shqipes se sotme: Problems of the Grammar of the Modern Albanian 00003
CESI Directory 14874
CESI News 14874
CESI Source Books 14874
Česká dermatologie 03538
Česká fysiologie 03500
Česká Geografická Společnost: Sborník: Journal 03506
Česká literatura 03500
Česká Meteorologická Společnost: Bulletin for Members 03510
Česká mykologie 03500
Česká Mykologie 03534
Česká Mykologie: Czech Mycology 03534
Česká otolaryngologie 03540
Česká psychologie 03500
Česká revmatologie: Czech Rheumatology 03509
Česká rusistika 03500
Česká Společnost Anesteziologie a Resuscitace: Information Bulletin 03517
Česká Společnost Biochemická: Bulletin 03518
Česká Společnost Bioklimatologická: Zypravodaj 03519
Česká Společnost Fyziologie a Patologie Dýchání: Bulletin 03523
Česká Společnost Histo- a Cytochemnická: Proceedings 03524
Česká Společnost pro Mechaniku: Bulletin 03528
Česká Společnost pro Vědy Zemědělské, Lesnické, Veterinární a Potravinářské: Informační Zpravodaj 03530
Česká stomatologie 03532
Česko-slovenská dermatologie: Czecho-Slovak Dermatology 03509
Česko-slovenská epidemiologie, mikrobiologie, immunologie: Czecho-Slovak Epidemiology, Microbiology, Immunology 03509
Česko-Slovenská Farmacie 03505
Česko-slovenská farmacie: Czecho-Slovak Pharmacy 03509
Československá fyziologie: Czecho-Slovak Physiology 03509
Česko-slovenská gastroenterologie a výživa: Czecho-Slovak Gastroenterology and Nutrition 03509
Česko-Slovenská gynekologie: Czecho-Slovak Gynaecology 03509
Česko-Slovenská hygiena: Czecho-Slovak Hygiene 03509
Česko-slovenská neurologie a neurochirurgie: Czecho-Slovak Neurology and Neurosurgery 03509
Československá oftalmologie 03512
Česko-Slovenská oftalmologie: Czecho-Slovak Ophthalmology 03509
Česko-Slovenská otolaryngologie a foniatrie: Czecho-Slovak Otorhinolaryngology and Phoniatrics 03509
Česko-Slovenská patologie s přílohou Soudní lékařství: Czecho-Slovak Pathology with Supplement on Forensic Medicine 03509
Československá pediatrie 03514
Česko-Slovenská pediatrie: Czecho-Slovak Paediatrics 03509
Česko-Slovenská psychiatrie: Czecho-Slovak Psychiatry 03509
Československá radiologie 03516
Česko-Slovenská radiologie: Czecho-Slovak Radiology 03509
Česko-Slovenská stomatologie: Czecho-Slovak Stomatology 03509
Český časopis historický 03500
Český časopis pro fyziku 03500
Český lid 03500
The Ceylon Geographer 10410
Ceylon Humanist Society: Journal 10411
Ceylon Library Review 10420
The Ceylon Medical Journal 10421
CFBS Newsletter 02129
cf-Bulletin 10815
CFCS Newsletter 17073
CF News 11876
CFR-Gazette 10446
CGA News 02143
CGB Newsletter 02368
C.G. Jung Foundation for Analytical Psychology: Annual Report 14719
CGMW Bulletin 04422
Chadburn 15176
Chadshot ACUM 07112
Chakoten 03572
The Challenge 09377
Challenge 14639
Change 13482
Charadrius: Zeitschrift für Vogelkunde, Vogelschutz und Naturschutz in Nordrhein-Westfalen 05876
The Charles Lamb Society Bulletin 11757
Charles Rennie Mackintosh Society: Newsletter 11758
Charles S. Peirce Society: Transactions 14722
Chartered Association of Certified Accountants: List of Members 11759
Chartered Builder 11761
The Chartered Institute of Building: Yearbook & Directory of Members 11761
The Chartered Institute of Transport: Proceedings 11766
Chartered Quantity Surveyor 12820
Chartered Society of Designers: Newsletter 11769
Chartered Surveyor 12820
Chatham House Papers 12813
CHC News 11330
Check Mark 02528
Checkpoints 16402
Chelovek: Man 09954
Chelys 13187
Chemdata 10133
Chemical Business Bulletin 12857
Chemical Business Newsbase 12857
Chemical Business Update 12857
Chemical Education 08530
Chemical Engineer 12293
Chemical Engineer Diary & Institution News 12293
Chemical Engineering Abstracts 12857
Chemical Engineering and Technology 05903
Chemical Engineering in Australia 00421
Chemical & Engineering News 13688
Chemical Engineering Progress 13878
Chemical Engineering Research & Design: Transactions of The Institution of Chemical Engineers 12293
Chemical Marketing and Management 14725
Chemical Pharmaceutical Bulletin 08395
Chemical Processing SA 10186
Chemical Research in Toxicology 13688
Chemical Review 17129
Chemical Reviews 13688
Chemica Scripta 10490
Chemické listy 03500
Chemické Listy 03520
Chemické zvesti: Chemical Papers 10068
Chemie-Ingenieur-Technik 05334, 05793, 05903
Chemie in unserer Zeit 05793
Chemie Magazine 01507
Chemische Berichte 05793
Chemisch-Physikalische Gesellschaft in Wien: Bulletin 00577
The Chemist 13879
Chemistry 01775, 03265
Chemistry and Industry 13035, 13035
Chemistry in Australia 00461
Chemistry in Britain 12857
Chemistry in New Zealand 09155
Chemistry International 12404
Chemistry Letters 08318
Chemistry of Life 03166
Chemistry of Materials 13688
Chemists' Club: Newsletter 14726
CHEMTECH 13688
ChemWorld 08530
Chesapeake and Ohio Historical Magazine 14728
CHES Newsletter 02152
Chest 13700
The Chester Archaeological Society: Journal 11773
Chester Zoo Life 12640
Chigiana: Rivista annuale di studi musicologici 07209
Chikyu-kagaku: Earth Science 08116
Childbirth without Pain Education Association: Memo 14730
Child Development 16576
Child Development Abstracts & Bibliography 16576
Child Health 02169
Childhood Education 14279
Children and Animals 15803
Children's Book News 02094
Children's Health Care 14349
Children's Literature: An International Journal 14733
Children's Literature Association: Journal 14733
Children's Literature Association: Proceedings 14733
Children's Television 02344
Child's Brain Nervous System 15537
Chile Economic Report 03048
Chiltern News 11776
Chimia 10891
La Chimica e l'Industria: Quaderni dell'Ingegnere Chimico Italiano 07774
Chimica nella Scuola 07774
Chimie Nouvelle 01481
Chimika Chronika 06519
China Civil Engineering Journal 03130
China Institute in America: Annual Report 14734
China Institute in America: Bulletin 14734
China Journal of TCM 03129
China Law Reporter 13617
China Medical Board of New York: Annual Report 14735
China Railway Sciences 03126
China-Report 00768
The China Society: Journal 03258, 10044
Chinese Academy of Geological Sciences: Bulletin 03139
Chinese Acupuncture and Moxibustion 03127
Chinese AEC Bulletin 03250
Chinese America: History and Perspectives 14739
Chinese-American Librarians Association: Newsletter 14736
Chinese Biochemical Journal 03166
Chinese Chemical Society: Journal 03265
Chinese Culture 03252
Chinese Culture Association: Journal 14737
Chinese Culture Association: Newsletter 14737
Chinese-English Translation Assistance Group: Bulletin 14738
Chinese Historical Society of America: Bulletin 14739
Chinese Journal of Acoustics 03113
Chinese Journal of Cardiology: Acta Cardiologica Sinica 03324
Chinese Journal of Materials Science 03285
Chinese Journal of Oceanology and Limnology 03181
Chinese Journal of Physics 03319
Chinese Journal of Polymer Science 03167
Chinese Journal of Psychology 03283
Chinese Journal of Radiology 03323
Chinese Language Monthly 03275
Chinese Language Teachers Association: Journal 14740
Chinese Medical Journal 03178, 03277
Chinese Music 14742
The Chinese PEN 03264
Chinese Statistical Journal 03288
Chinook 02199
CHINOPERL Papers 14824
Chiribotan 08217
Chirihak 08533
Chiron 05568
Chiron: Mitteilungen der Kommission für Alte Geschichte und Epigraphik des Deutschen Archäologischen Instituts 06075
Chiropractic Advancement Association: News Bulletin 11780
Der Chirurg BDC: Informationen des Berufsverbandes der Deutschen Chirurgen 05195
Chirurgia Narządów Ruchu i Ortopedia Polska 09596
Chirurgie 04088, 09758
Le Chirurgien-Dentiste de France 04439
Le Chirurgien-Dentiste de Paris 04924
Chirurgie Pédiatrique 04793
The Chopin Society: Newsletter 11782
Chord and Discord 14620
The Choreologist 11433
CHRIE Communique 14920
Christian College News 14744
Christian Education Journal 16212
Christian Home and School 14748
Christianity and Literature 14836
Christian Librarian 14412
Christian Schools International: Directory 14748
Christ und Bildung 06064
Chronica Horticulturae 08940
The Chronicle 13274
Chronique d'Egypte 01363
La Chronique de l'AUFEMO 04123
Chronique UGGI 04982
Chronobiologia 15528
Church and King 13054
Church and Synagogue Libraries 14751
Church Education Society for Ireland: Annual Report 06936
Church History 14106
Churchscape 11854
CIAT International 03360
CIAT Report 03360
CIC Directory of Minority Ph.D. Candidates and Recipients 14803
CICRED Bulletin 04430
CIE 00578
Le Ciel 01422
Ciel et Espace 04212
Ciel et Terre 01470
Ciencia del Suelo 00066
Ciencia e Investigación 00072
Ciencia, Tecnología y Desarrollo 03374
Ciencia y Técnica en la Agricultura 03474
CIE Newsletter 03273
CIFST Journal 02174
C.I.I. Journal 11768
CIIT Activities 14724
CIJL Bulletin 10643
CILECT Newsletter 01192
CIM Bulletin 02180
CIMMYT Aujourd'hui 08719
CIMMYT Economics Paper 08719
CIMMYT Economics Working Paper 08719
CIMMYT EN 08719
CIMMYT hechos y tendencias mundiales relacionadas con trigo 08719
CIMMYT hechos y tendencias mundiales relacionadas con maíz 08719
CIMMYT Hoy 08719
CIMMYT IN 08719
CIMMYT Réalité et tendances-le mais dans le monde: le potentiel maisicole de l'Afrique subsaharienne 08719
CIMMYT Research Report 08719
CIMMYT Today 08719
CIMMYT Wheat Special Report 08719
CIMMYT World Maze Facts and Trends 08719
CIMMYT World Wheat Facts and Trends 08719
Cina 07742
Cinema Journal 16507
Cinema Organ 11785
Cinema Technology 11599
Cinema Theatre Association: Bulletin 11786
Cinésiologie: La Revue Internationale des Médecins du Sport 04952
CIOMS Organization and Activities/Directory of Members 10655
CIPA 11764
CIPAC Proceedings 11795
Circuits and Devices Magazine 15317
Circular Técnica 01625, 01627, 01633
Circular Técnica 01634, 01635
Circular Técnica 01648
Circulo Poético 16291
Circulo: Revista de Cultura 16291
CIRP Annals 04391
Cirúgia del Uruguay 13267
Cirugía Plástica Iberolatinoamericana 10384
CISCo-News: Notiziario 07508
CISM Journal 02184
Citologia Informazioni 07859
Città Eterna 07244
City and Guilds of London Institute: Handbook 11789
City and Guilds of London Institute: Report and Accounts 11789
Civil and Hydraulic Engineering 03272
Civil Engineering 10203, 11437, 14107
Civil Engineering Hydraulics Abstracts 11575
Civil Engineering in Japan 08122
Civil Engineering Transactions 00421
Civilingenjören 10588
Civil War Byline 14758
Civil War Round Table Digest 14759
Civitas Pacis 07466
CLAH Newsletter 14840
CLANews 11825
Clase: Citas Latinoamericanas en Sociología en Ciencias y Humanidades 08717
Clásicos Venezolanos 16963
Classical and Quantum Gravity 12268
The Classical Outlook 13691
Classical Quarterly 11792
Classical Resources Series 13434
Classical Review 11792
Classic America 16861
Clay Minerals 12530
Clay Minerals Society: Newsletter 14761
Clays and Clay Minerals 14761
Clean Air 09122, 12607
Clean Air Journal 10155
Clean Slate 11753
Cleft Palate Journal 13692
Clima 00068
Climatic Bulletin 06571
Climatic Data 01082
Climatological Bulletin 02199
Climax 00078
Clinica: Rivista Internazionale di Psichiatria 07399
Clinical Allergy 11649
Clinical and Experimental Dermatology 12883
Clinical and Experimental Neurology 00265
The Clinical Biochemist Newsletter 00263
The Clinical Biochemist Review 00263
Clinical Chemistry Journal 13474
Clinical Chemistry News 13474
Clinical Chemistry Reference Edition 13474
Clinical Consult Newsletter 14112
Clinical Diabetes 13785
Clinical EEG 13935
Clinical & Experimental Immunology 11657
Clinical & Experimental Optometry 00325
Clinical Justice and Behavior 13475
Clinical Laboratory Science 14072
Clinical Management in Physical Therapy 14001
Clinical Microbiology Review 14073
Clinical Neurology and Neurosurgery 09029
Clinical Neurology and Neurosurgery: CNN 09008
Clinical Neurosurgery 14842
Clinical Oncology 12790
Clinical Orthopedics and Related Research 14405
Clinical Orthopedic Society: Directory 14763
Clinical Pharmacy 14132
Clinical Preventive Dentistry 10159
Clinical Psychiatry Quarterly 13384
Clinical Radiology 12790
Clinical Report on Aging 13834
Clinical Reports of Allergy and Immunology 00153
Clinical Research Journal 13806
Clinical Science 11440
Clinical Series 16166
Clinical Social Work Journal 16048
Clinical Staging of Lung Cancer 08184
Clinical Standards Digest 15616
Clinical Theology Association: Contact 11794
La Clinica Termale 07355
Clinics in Developmental Medicine 13086
Clio 03829
Clues 13847
CMA Journal: Journal de l'Association médicale canadienne 02197
CMC Contact Newsletter 10651
CMU Bulletin 07625
CNA Annual International Conference Proceedings 02211
CNL/Review of Books 14926
CNL/World Report 14926
Coalition for International Cooperation and Peace: Conference Reports 14764
Coalition for International Cooperation and Peace: Newsletter 14764
Coastal Engineering in Japan 08122
Coastal Engineering Research Council: Proceedings 14766
Coastal Reporter 13920
The Coastal Society: Bulletin 14767
The Coastal Society: Directory 14767
Coating Regulations and Environmental Issues 12673
The Coat of Arms 12161
Coblentz Society Mailings 14768
COCTA News 09249
CODATA Bulletin 04420
CODATA Newsletter 04420
CODEFF Actual 03042
CODEFF Informa 03042
Code of Conduct 12103
CODESRIA Bulletin 10026
Coffin Corner 16356
COHEHRE Newsletter 01255
Coin Hoards 12837
The Cold Facts 14945
Cold Regions Journal 14107
Colegio de Abogados de la Ciudad de Buenos Aires: Revista 00115, 00115
Colegio de Architectos de Chile: Boletín 03029
Colfar 16994
Collage: Cultural Enrichment and Older Adults 15949
Collection d'Etudes Latines 04703
Collection linguistique 04664
Collection of Czech Chemical Publications 03500
Collections for a History of Staffordshire 13092
Collective Bargaining Newsletter 13594
College and Research Libraries 14414
College and Research Libraries News 14414
College Art Association: Newsletter 14770
The College Board: Membership Directory 14771
College Board News 14771
The College Board: Proceedings 14771
College Board Review 14771
College Canada 01921
College Composition and Communication 14837
College English 16010
College Mathematics Journal 15703
College Music Society: Symposium 14774
College of American Pathologists: Directory 14775
College of American Pathologists: Newsletter 14775
College of Europe: Information 01217
College of Maritime Studies and Technology Newsletter 03874
College of Optometrists in Vision Development: Newsletter 14776
College Press Service 14777
College Review 13714
Colleges and Schools of Pharmacy, Accredited Professional Degree Programs 13767
College Times 14771
Colombia, Ciencia y Tecnología 03374
Colombo Plan Staff College for Technician Education Newsletter 09513
Colorado River Association: Newsletter 14783
Colour Index 13044
The Colposcopist 14052
The Columns 14727
Combinatorica 06597
Comecon Data 01020
Comércio Externo 09673
Comet 12673
Comisión de Investigaciones Científicas de la Provincia de Buenos Aires: Monografías 00117
Comisión Nacional de Museos y de Monumentos y Lugares Históricos: Boletín 00120

Comisión Nacional Protectora de Bibliotecas Populares: Boletín 00121
Comisíon Nacional de Investigación Científica y Tecnológica: Panorama Científico 03039
Comité International des Sciences Historiques: Bulletin d'Information 04408
Comité National Français des Recherches Antarctiques: Report 04419
Comité pour les Données Scientifiques et Technologiques: Conference Proceedings 04420
COMLA Newsletter 08078
Com-line 15748
COMM: Community Work and Communication 01385
Commemorative Collectors Society: Collecting Commemorabilia 11802
Commemorative Collectors Society: Newsletter 11802
Commentarii Mathematici Helvetici 10872
Commentationes Mathematicae 09585
Commission Belge de Bibliographie: Bulletin 01234
Commission Internationale d'Etudes Historiques Latinoaméricaines et des Caraïbes: Newsletter 04426
Commission on Accreditation of Rehabilitation Facilities: Report 14785
Committee on Institutional Cooperation: Biennial Report 14803
Common Sense Pest Control 14587
Commonwealth Education 11826
Commonwealth Education News 11820, 11830
Commonwealth Forestry Association: Review 11821
Commonwealth Heraldry Bulletin 09123
The Commonwealth Lawyer 11825
Commonwealth Trust: Newsletter 11831
Commonwealth Universities Yearbook 11329
Communication 12575, 16850
Communication Careers 14281
Communication Education 16741
Communicationes 01148
Communication: Journalism Education Today 15628
Communication Monographs 16741
Communications 02846, 04318, 08825, 15422
Communications and Acta Electrografica 15495
Communications et Mémoires 04090
Communications Law 13617
Communications Magazine 15317
Communications of the ACM 14285
Communication Theory 15444
Communication Yearbook 15444
The Communicator 12280
Communicator 13389
The Communicator 15481
Communicator 15853
Communiqué Bibliographique 01809
Community Associations Institute: Common Ground 14808
Community Associations Institute: News 14808
Community Associations Institute: Report 14808
Community College Journal for Research and Planning 15990
Community Dentistry and Oral Epidemiology 10159
Community Education Journal 15966
Community Education Today 15966
Community Environmental Council: Newsletter 14811
Community Health Guides 15844
Community Health Listing 15844
Community Management Initiative Reports 15901
Community Network 13137
Community, Technical and Junior College Journal 13524
Community, Technical and Junior College Times 13524
Community Work 11331
COMNET Newsletter 01241
Company Secretarial Practice 12222
Comparative and International Education Society: Directory 14816
Comparative and International Education Society: Newsletter 14816
Comparative Education Review 14816
Comparative Federation and Federalism Research Committee: Newsletter 02372
Comparative Immunology Microbiology and infection disease 04304
Comparative Labor Law Journal 15535
Comparative Psychology and Behavior 14019
Comparative Studies in Society and History 16593
Compensation and Benefits Review 13925
Compensation Medicine Newsletter 13402
Comp Flash 13372, 13925
Compleat Anachronist 16513
The Compleat Lawyer 13617
The Composers' Guild of Great Britain: Compass News 11833
Comprehensive Bibliography in Neuroanesthesia 16686
Comprehensive Education 12757
Comprehensive Psychiatry 14021
Comprehensive Therapy 14113
Comprendre 07802
Compte rendus de la Société de Biologie 04643
Comptes Rendus 04864
Comptes rendus de l'Académie d'Agriculture de France 04085
Comptes-Rendus de l'Académie des Sciences 04098
Les comptes rendus de l'Académie des Sciences 04564
Comptes Rendus des Travaux 01482
Computational Linguistics 14283
Computer Aided Process Control 11437
Computer Applications in Power Magazine 15317
Computer Applications Program Catalog 16087
Computer Bulletin 11547
Computer Crime Chronicles 15941
Computer Crime Law Reporter 15941
Computer & Geosciences 15387
Computer Graphics and Applications Magazine 15317
Computer Journal 11547
Computer Law Association: Membership Directory 14818
Computer Law Association Newsletter 14818
Computer Magazine 15317
Computer Newssheet 11838
Computer Organ Inform 01103
Computers and the Humanities 14284
Computers & Control Abstracts 12296
Computers in Genealogy 13047
Computers & Law 12967
Computing, Communications, Media and Socio-Technology-Trend Monitor 11267
Computing Control Engineering 12296
Computing in Civil Engineering Journal 14107
Computing Reviews 14285
Computing Surveys 14285
Comunicado Técnico 01625, 01633, 01648
La Comunità Internazionale 07969
The Concord Saunterer 16779, 16780
Concrete International: Design & Construction 13748
The Concrete Yearbook 12294
Condition Monitor 11437
Condizionamento dell'Aria 07292
Condor 14859
Conduit 11735
Confederate Memorial Literary Society: Journal 14821
Conference Group on German Politics: Directory of Current Research 14826
Conference Group on German Politics: Newsletter 14826
Conference Group on Italian Politics and Society: Membership List 14827
Conference Group on Italian Politics and Society: Newsletter 14827
Conférence Internationale des Grands Réseaux Electriques à Haute Tension: Proceedings of the biennial Sessions 04445
Conference Model Guidelines 16150
Conference of Consulting Actuaries: Proceedings 14829
Conference of Educational Administrators Serving the Deaf: Newsletter 14830
Conference of Public Health Laboratorians: Newsletter 14832
Conference of Public Health Veterinarians: Newsletter 14833
Conference on Consumer Finance Law: Quarterly Report 14838
Conference Software Notes 16853
Conferencias y Estudios de Historia y Organización de la Ciencia 03473
Conferentie voor Regionale Ontwikkeling in Noord-West-Europa: Proceedings of Seminars Organized 01249
Conflict Management and Peace Science 16300
Confrontation 15607
Confucius-Mencius Monthly 03292
La Congiuntura in Toscana 07525
Congrès Archéologique de France 04786
Congrès de Psychiatrie et de Neurologie de Langue Française: Annual Report 04448
Congress and the Presidency 16817
Congress of Lung Association Staff: Membership Directory 14841
Congress of Lung Association Staff: Newsletter 14841
Congress of Neurological Surgeons: Newsletter 14842
Connect 13147
Connection 12575
Conquest 12752
Conseil Consultatif Economique et Social de l'Union Economique Benelux: Report of Council Meetings 01250
Conseil-Education: Edu Council 02386
Conseil International pour l'Information Scientifique et Technique: Forum 04461
Conservation News 13166
The Conservator 13166
The Consort 11898
Construction 02104
Construction Computing 11761
Construction Industry Computing Association: Evaluation Reports 11838
Construction Industry Software Selector 11838
Construction Journal 14107
Construction Management-Focus 12226
The Construction Specifier 14850
Construction Today 12294
Construction Writers Association: Newsletter 14851
Constructive Triangle 13947
Construire Ensemble 01808
The Consultant 12183
The Consultant Pharmacist Journal 14112
Contact Bulletin 01303
Contemporary Accounting Research 02002
Contemporary Biology 17126
Contemporary Psychology 14019
Contemporary Sexuality Newsletter 13581
Contemporary Sociology: A Journal of Reviews 14175
Contemporary Southeast Asia 10049
Contemporary Trends in Education 03909
Contingencies 13379
Continuing Pharmaceutical Education, Approved Providers 13767
Continuum 10206
Contra Watch 15165
Contribuciones INIDEP 00144
Contributi del Centro Linceo Interdisciplinare B. Segre 07211
Contributions à l'Etude des Primitifs Flamands 01196
Control Column 11208
Control Systems Magazine 15245, 15317
Convention of American Instructors of the Deaf: Newsletter 14857
Convergence 10765
Convivium: Revista de Investigação e Cultura 01707
Convorbiri literare 09798
Coolia: Journal 08996
Co-op Education Undergraduate Program Directory 15960
Cooperation and Conflict: Nordic Journal of International Politics 10509
Cooperative Learning 15395
Cooperative Research Report 03746
Coordination 10731
Coordination and Cooperation Organization for the Control of the Major Endemioc Diseases: Bulletin Mensuel d'Informations 01809
Coordination of Outer Space Research 16812
Copeia 14134
Copyright 11034
Copyright Society of the U.S.A.: Journal 14862
Corelle 00271
The Core Teacher 15797
Corhealth 13757
CORMOSEA Bulletin 14805
Cornish Archaeology 11844
Corporación de Fomento de la Producción: Annual Report 03048
Corporate Plan 11213, 12508
Corporation Professionnelle des Médecins du Québec: Bulletin 02395
Le Corps Médical 08575
Corpus de la Peinture des Anciens Pays-Bas Méridionaux au 15e Siècle 01196
Correctional Education Association: Newsletter 14865
Correctional Education Association: Yearbook 14865
Corrections Today 13756
Corrective and Social Psychiatry and Journal of Behavior Technology Methods and Therapy 13552
Correio da Unesco 01654
Correio do IBECC 01665
Corriere Africano: Mensile di Relazioni Africa-Europa-Medio Oriente 07238
Corriere del Friuli 07798
Corriere di Roma 07510
Corrosão & Protecão Boletim Informativo 01690
Corrosion 15845
Corrosion Abstracts 15845
Corrosion Australasia 00243
Corrosion Science 12229, 12295
Cosmetic Science in Australia 00351
Cosmic Voice 11209
Cosmos 01518
COSPAR Directory of Organization and Members 04411
COSPAR Information Bulletin 04411
COSTED News 06752
Cost Engineering 13273
Cost Engineers Notebook 13273
Cost & Management 02945
Costruzioni Metalliche: Organo Ufficiale del CTA 07608
Costume 11846
Costumes et Coutures 10935
COTH Report 14904
Cotton Industry 06675
Council for Agricultural Science and Technology: Comments 14868
Council for Agricultural Science and Technology: Papers 14868
Council for Agricultural Science and Technology: Reports 14868
Council for Agricultural Science and Technology: Special Publications 14868
Council for Educational Development and Research: Directory 14873
Council for Educational Development and Research: Newsletter 14873
Council for Education in World Citizenship: Newsletter 11850
Council for Elementary Science International: Monograph 14874
Council for Environmental Education: Newsheet 11852
Council for Interinstitutional Leadership: Consortium Directory 14877
Council for Interinstitutional Leadership: Newsletter 14877
Council for Jewish Education: Membership Directory 14878
Council for Religion in Independent Schools: Newsletter 14882
Council for Research in Music Education: Bulletin 14883
Council for Scientific and Industrial Research: Annual Report 06479
Council for the Accreditation of Correspondence Colleges: Information Leaflet 11853
Council for the Protection of Rural England: Membership Magazine 11855
Council for the Protection of Rural England: Report 11855
Council for Tobacco Research – U.S.A.: Report 14887
Council of Administrators of Special Education: Newsletter 14889
Council of Agriculture: General Report 03294
Council of Chief State School Officers: Membership List 14891
Council of Colleges of Arts and Sciences: Membership Directory 14892
Council of Consulting Organizations: Directory of Members 14894
Council of Engineering and Scientific Society Executives: Yearbook 14895
Council of Graduate Schools: Brochure 14896
Council of Graduate Schools: Newsletter 14896
Council of International Programs: Newsletter 14897
Council of National Library and Information Associations: Roster 14900
Council of National Library and Information Associations: Update 14900
Council of Planning Librarians: Bibliographies 14901
Council of Planning Librarians: Membership Directory 14901
Council on Arteriosclerosis of the American Heart Association: Newsletter 14909
Council on Chiropractic Orthopedics: Directory 14911
Council on Education of the Deaf: Newsletter 14916
Council on Interracial Books for Children: Bulletin 14921
Council on Library-Media Technical Assistants: Membership Directory and Data Book 14923
Council on Library-Media Technical Assistants: Newsletter 14923
Council on Library Resources: Annual Report 14924
Council on Tall Buildings and Urban Habitat: Brochure 14933
Council on Technology Teacher Education: Membership Directory 14934
Council on Technology Teacher Education: Newsletter 14934
Counseling and Values 13771, 14326
Counseling Services 15403
Counselling 11491
The Counsellor 02393
Counselor Education and Supervision 13771, 14287
Counterpoint 15896
Country-Side 11617
The Countryside Recorder 13156
County Louth Archaeological and Historical Journal 06939
The Courier 01974
Le Courier de la Nature 04899
Courier de la SEP 04593
Le Courier de l'ASSITEB 04279
Courrier CERN / CERN Courier 10764
Le Courrier de la Corée 08526
Court Call 14170
Court Review 13907
Covered Bridge Topics 16157
COWAR Bulletin 08844
CPA Newsletter 11829
CPL Newsletter 14901
CP Reporter 02333
Cranial Academy: Directory 14937
Cranial Academy: News Letter 14937
Craniofacial: Cleft Palate Bibliography 13692
CRE-action 10670
Creation Research Society Quarterly 14938
Creative Communicator Magazine 13683
The Credit and Financial Journal 02171
Credit Management 12230
Credit Review 00299
CREFFT 13198
CREW Report 01188
Crime and Delinquency 16017
Crime and Social Justice 15165
Criminalist's Sourcebook 15519
Criminal Justice 13617
Criminal Justice Newsletter 16017
Criminologie 02543
The Criminologist Newsletter 14115
Criminology 14115
Crisis: Journal of Crisis Intervention and Suicide Prevention 00685
Criteria and Procedures 15788
Critical Care Medicine 16653
Critical Care Nurse 13526
Critical Care: State of the Art 16653
Critical Mass Energy Bulletin 14941
Critical Review of Books 13434
Critical Studies In Mass Communication 16741
Critique 11267
Cronache Ercolanesi 07483
Cronache Farmaceutiche 07944
Crop Physiology Abstracts 11720
Crop Science 14942
Cross and Cockade, Great Britain Society of World Ware One Aero Historians: Journal 11869
Crosscurrent 12606
Crossfire 11224
Crossover 08074
Cross Reference (The Hospital Manager) 13854
Cryobiology: International Journal of Low Temperature Biology and Medicine 16514
Cryonics 13360
Cryptozoology 15548
CSA + Le Consommateur 02301
CSAS Newsletter 02274
CSA + The Consumer 02301
CSC Arquitectos-Q 10302
CSD/SWO Bulletin 10147
CSIR Annual Report 10133
CSIR Handbook 06479
CSIR News 06753
CSLA Bulletin 02285
CSME Transactions 02262
CSM Newsletter 02286
CSOA: Computational Statistics and Data Analysis 08921
CSPWC Newsletter 02290
CSSE News: Nouvelles SCEE 02268
CSSR Bulletin 14903
CTI News 14858
Cuadernos de Arquitectura y Urbanismo 10289
Cuadernos de Folclore 01647
Cuadernos de Investigación 03440
Cuadernos de la Cepal 03038
Cuadernos del CREFAL 08711, 08722
Cuadernos del Ilpes 03038
Cuadernos de Lingüística 03019
Cuba Update 14674
Cube 16595
CUBE: Newsletter 14946
CUFC Leader 14848
Cultura e Natura 07414
Cultural Anthropology 16515
Cultural Experts Directory 15820
Culture 01775
Culture et Cité 01361
Cumberland and Westmorland Antiquarian and Archaeological Society: Transactions 11871
Cuoio, Pelli e Materie Concianti 07295
Current Advances in Clinical Chemistry 11324
Current Affairs Broadsheet 11850
Current Aids Literature 11720
Current Antarctic Literature 16815
Current Anthropology 16879
Current Awareness Bulletin 03874, 11111, 11256, 11267
Current Awareness Service 11584, 12780
Current Bibliography on Science and Technology: Kagaku Gijutsu Bunken Sokuho 08216
Current Biotechnology Abstracts 12857
Current Catalog Proof Sheets 15710
Current Index to Statistics: Applications, Methods and Theory 14182
Current Issues in Anthropology Newsletter 14717
Current Literature on Science of Science 06753
Current Mathematical Publications 13929
Current Physics Index 13893
Current Published information on Standardization 06822
Current Science 06767
Current Sociology 04409
Current Sociology/Sociologie Contemporaine 10330
Current Topics in Neuropathology 00815
Current World Leaders 15362
Customized Current Information Service 16087
Cutis 16214
CVA News 00378
CWEA Newsletter 02318

CWI Monographs 08841
CWI Syllabus 08841
CWI Tracts 08841
Cyberline 16862
Cybernetic 14054
Cybernetica 01135
Cybernetics and Systems: An International Journal 00921
Cybium: Annuaire des Ichtyologistes Français 04860
Cyfres Cyfieith-iadau'r Academi 11198
Cymdeithas Hanes Sir Ddinbych: Transactions 11874
Cyprus Research Centre: Publications 03496
Cystic Fibrosis Association of Ireland: Newsletter 06940
Cystic Fibrosis Research Trust: Directory 11876
Cystic Fibrosis Research Trust: Information Leaflets 11876
The Cytotechnologist's Bulletin 14116
Czasopismo Geograficzne 09571
Czasopismo Stomatologiczne 09606
Czech Journal of Physics 03500
Czech Mathematical Journal 03500
Czechoslovak Society of Arts and Sciences: Bulletin 14947
Czechoslovak Society of Arts and Sciences: Zpravy 14947
Czech Otolaryngological Society: Proceedings 03540

Dacia: Revue d'Archéologie et d'Histoire Ancienne 09736
Dada-Surrealism 14357
DADOS 01714
Daedalus 13382
Daffodils 12810
Dairy Council Digest 16029
Dairying-in-India 06790
Dairy Research Digest 14948
Dairy Science Abstracts 11720, 11727
Dairy Society International: Bulletin 14949
Dairy Society International: Report to Members 14949
Damaszener Mitteilungen 05568
Damien-Dutton Call 14951
Dance on Camera News 14953
Dance Research Journal 14843
The Dancer Magazine 11532
Dancing Gazette 12761
Danish Dairy and Food Industry 03649
Danish Medical Bulletin 03647
Danish Medical Historical Yearbook 03646
Danish Yearbook of Musicology 03679
Dansjaarboek 09081
Dansk Audiologopaedi: Journal of the Danish Speech and Hearing Association 03565
Dansk Cardiologisk Selskab: Newsletter 03597
Dansk Dendrologisk Årsskrift 03600
Danske Fysioterapeuter 03606
Danske Komponister af i Dag: Danish Composers of Today – Dänische Komponisten von heute 03639
Dansk Kunsthåndverk 03609
Dansk Metallurgisk Selskabs Arbog 03650
Dansk Naturhistorisk Forening Arsskrift 03653
Dansk Oftalmologisk Selskab: Transactions 03658
Dansk Ornitologisk Forenings Tidsskrift 03660
Dansk Skovforenings Tidsskrift and Skoven 03690
Dansk Veterinårhistorisk Årbog 03698
Dansk Veterinaertidsskrift (Danish Veterinary Journal) 03701
Dante Studies 14955
Darmstädter Archivschriften 05923
Dartmoor Diary 11877
Darvoobrabotvasta i mebelna Promislenost 01793
Data Acquisitions Catalog 15559
Data Base Reports 15346
Data Book 14158
Data for Development Newletters 04286
DATA posten 03599
Data Sheets on Quarantine Organismes 04479
Date 16765
Dawn 15211
DAWN: DAWN Plus 11905
Dawson & Hind 01935
Daxue Huaxue 03167
DBS Pressedienst 05504
DBV-INFO 05506
The DEC Communicator 14979
Decheniana 06131
Decheniana-Beihefte 06131
Decision Line 14957
Decision Sciences 14957
Defektokopiya: Defectoscopy 09954
Defektologiya: Defectology 09953
DeGarmo Lectures 16696
Dějiny věd a techniky 03500
Dela: Opera 10105
Il Delfino 07496
Delta 09569
Delta Dental Plans Association: Newsletter 14958
Delta Dictum 14958
Deltion: Bulletin 03499
La Demeure Historique 04463
Democrazia e Diritto: Trimestrale di Diritto e Giurisprudenza 07332
Demography 16337
Denki Kagaku 08119
Denkmalschutz-Informationen 05596
Den Norske Tannlegeforenings Tidende 09255
Denpun Kagaku: Journal for Starch and Its Related Carbohydrates and Enzymes 08191
Dental Abstracts 13779
The Dental Assistant 13778
Dental Hygiene 13780
Dental Materials 13287
Dental Radiology 08374
Dental Student Handbook 14184
Dental World 16319
Dentistry 14184
Dentistry in Japan 08376
The Dentists Register 12104
La Dépèche Technique 04968
La Dépèche Vétérinaire 04968
Deputazione di Storia Patria per le Antiche Province Modenesi: Atti e Memorie 07638
Derbyshire Archaeological Journal 11882
Derecho Privado y Constitución 10287
der kinderarzt: Zeitschrift für Kinderheilkunde und Jugendmedizin 05200
Dermatologia i Venerologia 01747
Dermato-Venerologia 09759
Derm-Dialogue 13441
DES Action Voice 14963
Desafio 01582
Desarrollo Rural en las Américas 03447
Descent 00492
Description de langues et monographies ethnolinguistiques 04746
Descriptions of Pathogenic Fungi & Bacteria 11720
Desempenho da Economia de Pernambuco 01668
Desert Locus Situation Report 08490
Desert Tortoise Council: Proceedings of Symposium 14967
Design 11884
Design and Drafting News 13784
Design and Test of Computers Magazine 15317
Design & Art Direction 11886
Design/Build Digest 14968
Design Concerns: Student Colloquium Papers 06845
Designer 16841
The Designer-Craftsman 13043
Design Folio 06845
Designing 11884, 11885
Designing and Making 11944
Design Report 06244
Design Research News 15049
Design Review 11769
DESY Journal 05575
Det Danske Bibelselskabs Årbog 03704
Det Kongelige Norske Videnskabers Selskab: Skrifter: Scientific Papers 09258
Det Paedagogiske Selskab: Dansk Paedagogisk Tidsskrift 03793
Detritus 17175
Deutsche Akademie der Naturforscher Leopoldina: Informationen 05285
Deutsche Akademie für Städtebau und Landesplanung e.V.: Mitteilungen 05290
Deutsche Bauzeitung 05227
Deutsche Dendrologische Gesellschaft: Mitteilungen 05297
Der Deutsche Dermatologe 05196
Deutsche Exlibris Gesellschaft e.V.: Mitteilungen 05302
Deutsche Forschungsgemeinschaft: Forschung 05305
Deutsche Geophysikalische Gesellschaft e.V.: Mitteilungen 05312
Deutsche Gesellschaft für Erziehungswissenschaft: Tagungsberichte 05347
Deutsche Gesellschaft für Geschichte der Medizin, Naturwissenschaft und Technik e.V.: Nachrichtenblatt 05355
Deutsche Gesellschaft für Holzforschung e.V.: Merkhefte 05363
Deutsche Gesellschaft für Holzforschung e.V.: Mitteilungshefte 05363
Deutsche Gesellschaft für Psychiatrie und Nervenheilkunde e.V.: Spektrum 05410
Deutsche Gesellschaft für Publizistik- und Kommunikationswissenschaft e.V.: Aviso: Informationsdienst der Deutschen Gesellschaft für Publizistik- und Kommunikationswissenschaft 05413
Deutsche Gesellschaft für Qualitätsforschung (Pflanzliche Nahrungsmittel) e.V.: Proceedings der Jahreskongresse 05415
Deutsche Gesellschaft für Sprachwissenschaft: Bulletin 05427
Deutsche Gesellschaft für Technische Zusammenarbeit: Akzente 05430
Deutsche Gesellschaft für Versicherungsmathematik e.V.: Blätter 05436
Deutsche in der ehemaligen Sowjetunion 05898
Deutsche Keramische Gesellschaft e.V.: Fachausschussberichte 05459
Deutsche Krebsgesellschaft e.V.: Mitteilungen 05461
Deutsche Kunst und Denkmalpflege 06399
Deutsche Landschaften: Landeskundliche Erläuterungen zur Topographischen Karte 1:50000 06463
Deutsche Mathematiker-Vereinigung e.V.: Jahresbericht 05466
Deutscher Arbeitsgerichtsverband e.V.: Mitteilungen 05498
Deutscher Forstverein e.V.: Tagungsbericht 05511
Deutscher Komponisten-Verband e.V.: Informationen 05524
Deutscher Verband Evangelischer Büchereien e.V.: Buchauswahl 05544
Deutscher Volkshochschul-Verband e.V.: Agenda 05562
Deutsches Ärzteblatt: Publikationsorgan der deutschen Ärzteschaft 05229
Deutsches Archäologisches Institut: Jahrbuch 05568
Deutsches Archiv für Erforschung des Mittelalters 06120
Die Deutsche Schrift: Vierteljahreshefte zur Förderung der deutschen Sprache und Schrift 05263
Deutsche Sektion der Internationalen Liga gegen Epilepsie: Rundbrief 05573
Deutsche Shakespeare-Gesellschaft West e.V.: Jahrbuch 05578
Deutsches Institut für Internationale Pädagogische Forschung: Mitteilungen und Nachrichten 05585
Deutsches Jugendinstitut e.V.: Diskurs: Studien zu Kindheit, Jugend, Familie und Gesellschaft 05590
Deutsches Orient-Institut: Mitteilungen 05598
Deutsches Tierärzteblatt 05610
Deutsche Studien 06225
Deutsche zahnärztliche Zeitschrift 05106
Deutsche Zahnärztliche Zeitschrift: DZZ 05445
Deutsche Zeitschrift für Akupunktur 00751
Deutsche Zeitschrift für Sportmedizin 05539
Deutsche Zoologische Gesellschaft e.V.: Verhandlungen 05629
Deutschland-Journal 06292
Deutsch lernen: Zeitschrift für den Sprachunterricht mit ausländischen Arbeitnehmern 06290
Deutsch-Pazifische Gesellschaft e.V.: Bulletins: Informationshefte 05630
Development 08028
Developmental Biology 16516
Developmental Medicine + Child Neurology 13086
Developmental Psychology 14019
Development Dialogue 10449
Development Hotline 08028
Development Reports 15281
Developments in Industrial Microbiology 16538
Development Trends 16801
DF-Revy 03578, 03733, 03800
DFVLR-Forschungsberichte 05307
DFVLR-Mitteilungen 05307
DFVLR-Nachrichten 05307
DFWR-Dreijahresbericht 05512
DGC-Mitteilungen 05337
DGD-Newsletter 05341
DGfH-Nachrichten 05363
DGPh Intern 05403
DGP-Nachrichten 05396
dgv informationen 05439
Dhan Shaliker Desh 01065
Diabetes 13785
Diabetes aktuell / Hallo Du auch 05508
Diabetes Care 13785
Diabetes Contents 11551
The Diabetes Educator 13529
Diabetic Medicine 11551
Diabetologia 05667
Diabetologia Hungarica 06623
Diabetologie-Informationen 05299
Diagnostic and Interventional Radiology 04839
DiA-kunta 04007
Dialectes de Wallonie 01454
Dialektologjia Shqiptare: Albanian Dialectology 00003
dialog 06351
Dialogue 02224
Dialogue on Diarrhoea 11239
Dialogue Series 09238
Dialogues et Cultures 04534
Dia Regno: Divine Kingdom 15436
Diary of Agricultural Show Dates 12767
Diccionario Geográfico del Perú 09477
The Dickensian 11891
Dickens Quarterly 14971
Dictionary of Fire Technology 12303
die Drei: Zeitschrift für Anthroposophie 05076
Dieren 09072
Dietary Managers Association: Issues 14973
Dietitians Association of Australia: Annual Conference Proceedings 00385
Differentation 15551
Differentsialnye Uravneniya: Differential Equations 01088
Difv-Materialien 05587
Digesto Familiar 13252
The Digging Stick 10178
DIK-forum 10547, 10587, 10592, 10604
Dilmun 01062
Dimensione Europa: Periodico d'Informazioni e Studi sulla Comunità Europea 07465
DIN-Mitteilungen 06156, 06179, 06181, 06199, 06211
DIN-Mitteilungen + electronorm: Zentralorgan der deutschen Normung 06174
DIN-Mitteilungen + elektronorm 05301, 05632
DIN-Mitteilungen + elektronorm: Zentralorgan der deutschen Normung 06193, 06220
Dinteria 08816
Diogenes 04455
Diplomatic History 16529
Diplomatische Akademie Wien: Jahrbuch 00581
Diptycha 06523
Diptychon Parafylla 06523
Direcção Provincial dos Servicos de Geologia e Minas de Angola: Boletim 00033
Direcção Provincial dos Servicos de Geologia e Minas de Angola: Memória 00033
The Director 12232
Directorio de la Educación Superior en Colombio 03389
Directory for Invertebrate Pathology 16542
Directory for the Nursery Industry and Related Associations 15230
Directory Iron and Steel Plants 14449
Directory of Accredited Cosmetology Schools 15781
Directory of Accredited Noncollegiate Continuing Education Programs 13318
Directory of Certified Hearing Aid Audiologists 16073
Directory of Certified Rolfers and Movement Teachers 16427
Directory of Climatologists 11307
Directory of College Mental Health Services in Greater New York City 15725
Directory of College Transfer Information 14037
Directory of Demographic Research Centres 04430
Directory of Development Agencies and Officials 15895
Directory of Digestive Diseases Organizations 16034
Directory of Diplomates 13655
Directory of Educational Services and Programs 14904
Directory of Environmental Education Resources 14685
Directory of Episcopal Church Schools 15857
Directory of Ethiopian Libraries 03943
Directory of Fellows of the American Academy of Microbiology 13407
Directory of Former Participants 14897
Directory of Graduate and Undergraduate Programs and Courses in Middle East Studies in the U.S., Canada and Abroad 15735
Directory of Healers and Counselors 13840
Directory of Homeopathic Physicians in the U.S. 15944
Directory of Irish Family History Research 13153
Directory of Laboratories 00437
Directory of Law Libraries 13547
Directory of Libraries in Papua New Guinea 09421
Directory of London Public Libraries 11363
Directory of Marine Education Resources 14685
Directory of Mathematicians from Dev. Countries 07711
Directory of Membership APMS 14234
Directory of Music Faculties in Colleges and Universities, U.S. and Canada 14774
Directory of Nationally Certified Teachers 15765
Directory of Network Organizations 16429
Directory of Osteopathic Publications 14477
Directory of Pathology Training Programs 15579
Directory of Peace Studies Programs 14849
Directory of Physicists from Dev. Countries 07711
Directory of Planetaria and Planetarians 15513
Directory of Planning Libraries 14901
Directory of Polar and Regions Library Resources 16231
Directory of Private Schools 14869
Directory of Professional Archaeoloists 16693
Directory of Professional Engineers in Private Practice 16159
Directory of Programs in Societ & East European Studies 13498
Directory of Psychosocial Policy and Programs 14349
Directory of Radio-TV-Film Programs 14281
Directory of Residency Training Programs 14925
Directory of Southern Baptist Colleges and Schools 14503
Directory of Specialty Referral Centers 14198
Directory of State and Local Health Planning Agencies 13842
Directory of State Education Agencies 14891
Directory of Steel Foundries 16745
Directory of Suppliers 15243
Directory of Theatre Programs 14281
Directory of University Urban Programs 16850
Directory of Visual Anthropologists 16628
Directory of Wargame Section 12551
Directory of Year-Round Schools 15824
Directory Services for Victims of Crime: Répertoire Services aux victimes d'actes criminels 02113
Diritto e Pratica nell'Assicurazione 07549
Disability Law Reporter 14787
Disability Now 13086
Disaster Preparedness in the Americas 16292
Discharge Planning Update 13854, 16532
Discorsi e Immagini 07427
Discours de réception des académiciens 04111
The Discourses of the Academy 09234
Discovery 02977, 14353
Discovery and Innovations 08458
Discussiones Palaeontologicae: Őslénytani Viták 06635
Disease in Childhood 11629
Diseases of the Colon and Rectum 14111
Disegno di Macchine 07369
Dişhekimliğinde Klinik 11146
Diskretnaya Matematika: Discrete Mathematics 09954
Dissertation Series 15829
Distillers Feed Research Council: Proceedings 14977
Distributed Systems Engineering 12268
Distribution Maps of Pests 11720, 11732
Distribution Maps of Plant Diseases 11720
Distributive Education Clubs of America: Guide 14978
Distributive Education Clubs of America: Newsletter 14978
District Memoirs 08642
District Planning Officers Society: Newsletter 11897
Distrofia Muscolare 08047
Division of Applied Experimental and Engineering Psychologists: Newsletter 14980
Dix Huitième Siècle 04850
DKIN-Empfehlungen 05591
DLF Hamilton Index 11894
DLG-Mitteilungen 05463
DNR-Kurier 05533
The D.O. 13980
Doble Cruz 00149
Doboku-Gakkai: Journal 08122
Doboku-Gakkai: Proceedings 08122
Dobutsu-shinrigaku-kenkyu: The Japanese Journal of Animal Psychology 08193
DOCPAL Resúmenes sobre Población en América Latina 03038
DOCPAL Resúmentes sobre Población en América Latina 03026
Doctoral Dissertations on Asia 14268
Doctor and Clinic Directory 14638
Documentación Bibliotecológica 00101
Documentaliste – Sciences de l'Information 04198
Documentário Sonoro 01647
Documentary Editing 14289
Documentation 16524
Documentation et Bibliothèques 01951
Documentation Notes 06853
Documento de trabajo de Programa de Economía del CIMMYT 08719
Documentos Medievais Portugueses 09635
Documents de Recherches en Médecine Générale 04811
Documents of American Architecture 07571
Documents scientifiques du CRO 08061
Do Ihges 01697
Doitsu Bungaku: German Literature 08194
Doitsu-Bungaku-Ronko: Forschungsberichte zur Germanistik 08130
Doklady Akademii Nauk Rossiiskoi: Proceedings of the Russian Academy of Sciences 09954
Dokumentation Bibliographie Jugendhilfe 05590
Dokumentation der Gesetze und Verordnungen Osteuropas (DGVO) 00915

Dokumentation für Bodenmechanik, Grundbau, Felsmechanik, Ingenieurgeologie 05345
Dokumentationsarchiv des österreichischen Widerstandes: Jahrbuch 00582
Dokumentation Straße: Kurzauszüge aus dem Schrifttum über das Straßenwesen 05720
Dokumente und Schriften der Europäischen Akademie Otzenhausen 05648
Dollars and Cents of Shopping Centers 16801
Dolmetsch Foundation: Bulletin Newsletter 11898
Dolphin Log 14936
The Donizetti Society: Journal 11900
Dopovidi Akademii Nauk Ukrainy: Report of the Ukrainian Academy of Sciences 11193
Dorothy L. Sayers Society: Annual Seminar Proceedings 11901
Dorothy L. Sayers Society: Bulletin 11901
Dorset Monographs 11902
Dostoevsky Studies 15467
Il Dottore in Scienze Agrarie e Forestali 07692
The Double Ressure 12160
The Double Treasure 12162
The Dozenal Journal 11904
Dozzle 14958
DP International 12231
DP International Quarterly Journal 12231
Drama 11693
Dramatics Magazine 15565
The Dramatist Guild Quarterly 14555
Drawing 14985
DRG, Dicziunari Rumantsch Grischun 10963
Drilling News 11437
Drita: Journal 00005
Le Droit d'Auteur 11034
Droit et Economie 04309
Droit Médical 04670
Drop Forging Research Association: Annual Report 11907
Drop Forging Research Association: Newsletter 11907
Drug Information Association: Journal 14987
Drug Information Association: Membership Directory 14987
Drug Information for the Health Care Professional 16832
Drug Metabolism and Disposition 14077
Drugs under Experimental and Clinical Research 07872
Društvo za Filozofija, Sociologija i Politikologija na Makedonija: Zbornik: Collected Papers 08599
DSFL Meddelelser 03675
DTL-Nyt 03696
DTP Publishing Commentary 12701
Dublin Historical Record 07018
Ductile Iron News 14989
Ductile Iron Society: Membership Directory 14989
Ductile Technical Notes 14989
Duodecimal Bulletin 14983
Durham Archaeological Journal 11244
DVR-report 05561
dvs-Informationen 05620
DWI-Report 05606
DWV-Mitteilungen 05564
Dydaktyka Matematyki 09585
Dyleague: Proceedings 14991
Dynamic Psychotherapy Journal 16342
Dynamic Systems, Measurement and Control 14145
dynamik im handel 05645
Dynamische Psychiatrie: Internationale Zeitschrift für Psychiatrie und Psychoanalyse 05451
Dyslexia Contact 11553
Dyslexia Review 11914
Dystonia Medical Research Foundation: Newsletter 14993

EAAE Newsletter 01286
EAAE News Sheet 01287
The EAA Experimenter 15070
EAAM Newsletter 11986
EAA Newsletter 01268
EACE Newsletter 08852
EACL Information Bulletin 05661
Eacronews 04484
EACROTANAL Newsletter 11054
EADI Newsletter 10677
EADTU-News 08865
EAIE Newsletter 08856
EALE Newsletter 10451
EAMDA Bulletin 06946
EAMDA Newsletter 06946
EANHS Bulletin 08497
EAPM Newsletter 11991
EAPS Newsletter 08859
Ear and Hearing 13611
Early China 16615
Early Days 00483
Early Human Development 12623
Early Law Reports 12939
Early Settlers Association of the Western Reserve: Annals 14997
Early Settlers Association of the Western Reserve: Roster 14997
Early Sites Research Society: Bulletin 14998
Early Sites Research Society: Newsletter 14998
EAROPH News and Notes 06756
EARSeL News 07665
Earth in Space 13833
Earthlines 11852
Earthmind: Newsletter 14999
Earth Observation Quarterly 07648
Earthquake Engineering Research Center Reports 16087
Earthquake Spectra 15000
Earth Science 13832
EASE Circular Letter 05663
EASL Bulletin 11185
EASL Newsletter 11185
East and West 07742
East Asian Pastoral Review 09517
The Eastern Librarian 01069
East European Jewish Affairs 12249
East Lothian Antiquarian and Field Naturalists' Society: Transactions 11919
East of England Journal 11921
East-West European Economic Interaction/Workshop Papers 01020
EBBS Newsletter 09265
EBM Review 05668
E.B.S.A. Bulletin 11979
ECA Newsletter 01121
Ecclesiological Society: Papers 11924
Ecclesiology Today 11924
ECFMG Annual Report 15018
ECFMG Information Booklet 15018
Echo 11032
Echo aus Deutschland 05272
Echos Phytosanitaires 02922
Eclogae Geologicae Helvetiae 10975
Eco Allert 02388
Ecography 10513
Ecole e Paix 11031
Ecole et Vie: Bulletin dınformation syndical, pédagogique et culturel 08590
Ecological Entomology 12800
Ecological Monographs 15006
Ecological Research 08365
Ecological Society of America: Bulletin 15006
Ecology 15006
Ecology Center: Newsletter 15007
Ecology International 15380
Econometrica 15009
Economia Pubblica 07494
Economic Affairs 12233
Economic and Business History 14628
Economic Botany 16517
Economic Bulletin of Ghana 06480
Economic Bulletins 08642
Economic Council of Canada: Au Courant 02425
Economic Education Bulletin 13868
Economic Education Update 16018
Economic Entomology 02436
Economic Geology 16655
The Economic History Review 11927
Economic Journal 12799
Economic Notes 15645
Economic Outlook USA 15296
Economic Papers 00388
The Economic Record 00388
Economics 00210
Economics and Business Education 11928
Economics and Law 01775
Economic Situation Report 11837
Economic Survey of Latin America and the Caribbean 03038
Economic Systems 06226
Economic Systems Research 00622
Economic Trends 15231
Economie Familiale-Home Economics-Hauswirtschaft 04537
Economie Rurale 04798
Economisch- en sociaal-historisch Jaarboek 09084
Economisch Statistische Berichten 08989
Economistul 09741
ECOR Newsletter 11957
Ecosphere 15117
ECPR News 12017
ECSSID Bulletin 00590
EDF Letter 15048
Edicije DSS 10099
Ediciones Academia 00048
Edinburgh Bibliographical Society: Transactions 11932
Edinburgh Festival Society: Festival Preliminary Brochure 11933
Edinburgh Festival Society: Festival Programme Brochure 11933
Edinburgh Mathematical Society: Proceedings 11935
Educación Superior y Desarrollo 03389
Educadores 10310
L'Educateur 04533, 10981
Educational Administration Abstracts 16846
Educational and Training Technology International 11263
Educational Communication & Technology Journal 14290
Educational Evaluation and Policy Analysis 13790
Educational Film Locator 14848
The Educational Guide to NTS 12615
Educational Information Bulletin 03909
Educational Innovation and Education 10639
Educational Leadership 14336
Educational Management and Administration 11555
Educational Measurement: Issues and Practice 16023
Educational Planning 15021
Educational Planning: Directory 15021
Educational Record 13766
Educational Records Bureau: Catalog of Programs and Services 15023
Educational Records Bureau: Newsletter 15023
Educational Records Bureau: Prospectus 15023
Educational Research 12591
Educational Research Analysts: Newsletter 15024
Educational Researcher 13790
Educational Research News 12591
Educational Research Service: Bulletin 15025
Educational Series 13154
Educational Statistics Higher Education 09393
Educational Statistics (Schools) 09393
Educational Studies 13791
Education and Training in Mental Retardation 14981
Education and Training Programmes for Information Personnel: ET Newsletter 08895
Education Canada 02122
Education Comparée 04249
Education Development Center: Annual Report 15028
Education européenne 04206
Education in Chemistry 12857
Education in Eastern Africa 11181
Education in Science 11282
Education of the Masses 06913
Education of the Visually Handicapped 14291
Education permanente 10944
The Education Reporter 15030
Education Review 12616
Education Today 06994, 11797
Education USA 16148
Education Week 15016
Education Writers Association: Membership Directory 15030
The Educator 13309, 13855
Educazione Fisica e Sport nella Scuola 07451
Educazione Sanitaria e Medicina Preventiva 07615
EEG/EMG 00775
The EEG Journal 11953
E.E.M.U.A. Publications 11959
EERI Newsletter 15000
Eesti Loodus: Nature of Estonia 03933
EFB-Newsletter 05651
EFCE Newsletter 05652
EFC Newsletter 05653, 05678
Effluent 01955
Egan 10367
EGCS Herdbook & Type & Production Register 11963
Egészség: Health 06676
Egretta: Vogelkundliche Nachrichten aus Österreich 00843
EGS-Bulletin 10668
L'Egypte Contemporaine 03901
Egyptian Dental Journal 03891
Egyptian Journal of Botany 03890
Egyptian Journal of Dairy Science 03897
Egyptian Medical Association: Journal 03893
Eichholzbrief 05219
e & i Elektrotechnik und Informationstechnik 00904
Eighteenth-Century Studies 14058
einblick: Zeitschrift des Deutschen Krebsforschungszentrums 05592
Eire-Ireland 15585
Eiseikagaku 08395
Eiszeitalter und Gegenwart 05493
EJNM Journal 11998
Ekistical Education 15017
Ekologiya: Ecology 09954
Ekonomicko-matematicky obzor 03500
Ekonomika i Matematicheskie Metody: Economics and Mathematical Methods 09954
Ekonomika i Organizatsiya Promyshlennogo Proizvodstva: Economics amd the Organisation of Industrial Production 09954
Ekonomika preduzeća 17092
Ekonomika Ukrainy: Economy of the Ukraine 11193
Ekonomisk Debatt 10503
Ekonomist 17116, 17137
Ekonomista 09562
Eksperimentalnaya Onkologyia: Experimental Oncology 11193
elaphe 05361
Electra 04445
Electrical & Electronics Abstracts 12296
Electrical Insulation Magazine 15317
Electrical Report 15015
Electric Auto Association: News 15032
Electricity Consumers Resource Council: Report 15033
Electricity Supply and Demand 16217
Electric Railway Society: Journal 11951
Electric Vehicle Developments 11952
Electrochemical Society: Directory 15035
Electrochemical Society: Journal 15035
Electrolysis World 13794
Electron Device Letters 15317
The Electronic Author 13030
Electronics Education 12296
Electronics Letters 12296
The Electron Microscopy Society of Southern Africa: Proceedings 10138
Electronnoe Modelirovanie: Electronic Modelling 11193
Electrotechnology 12298
Elektrichestvo: Electricity 09954
Elektrik Mühendisliği: Electrical Engineering 11167
Elektro 09306
Elektrokhimiya: Electrochemistry 09954
Elektronenmikroskopie 05343
Elektronnaya Obrabotka Materialov: Electronic Processing of Materials 08783
Elektronnoe Modelirovanie: Electronic Modelling 09954
Elektrotechnika 06624
Elektrotechnische Zeitschrift 05301, 05640
Élelmezési Ipar 06625
Elementary School Guidance and Counseling 13771
Elemente der Mathematik 10872
Élet és Tudomány: Life and Science 06676
L'Elettrotecnica 07411, 07686
ELI Associates Newsletter 15050
Ellinika: Classical studies 06525
Elliniki Geografiki Etaireia: Bulletin 06506
Elliniki Mathimatiki Eteria: Bulletin 06510
El-Mokhtar 03870
Elm Research Institute: Press Release 15038
ELNA-Adresaro 15057
ELNA Newsletter 15057
Emergency Update 14221
Emerita 10322
EMIERT Bulletin 15062
Emmy Directory 13310
Emmy Magazine 13310
E.M. News 09174
Empirica: Austrian Economic Papers 00912
Empirica – Austrian Economic Papers 00711
Employee Benefit Notes 15039
Employee Benefit Research Institute: Issue Briefs 15039
Employee Benefit Research Institute: Proceedings 15039
Employer News 02802
Empresa Brasileira de Pesquisa Agropecuária: Documentos 01648
EMS News 14198
En bok för alla 10497
Enciclopedia Araldica 07799
Encomia: Bibliographical Bulletin of the International Courtly Literature Society 15461
Encore 15851
Endocrine-Related Cancer 12971
Endocrine Review 15041
Endocrine Society: Newsletter 15041
Endocrine Society of Australia: Proceedings 00389
Endocrinology 15041
End of Year Journal 16356
Endokrinologie-Informationen 05344
Endokrynologia Polska 09563
Endometriosis Association: Newsletter 15042
Endüstri Mühendisliği: Industrial Engineering 11167
Energetika 01792
L'Energia Elettrica 07411
Energia és Atomtechnika 06600
Energiagazdálkodás 06600
Energie Alternative: Bimestrale della Sezione Italiana 07722
Energie + Umwelt 10803
Energiteknik 10529
Energiya: Energy 09954
Energy & Fuels 13688
Energy Journal 14107
The Energy Journal 15381
Energy Management 06846
Energy Mechnics Journal 14107
Energy News 13510
Energy Research Institute: Directory 15043
Energy Research Institute: Report 15043
Energy Resources Technology 14145
Energy Review 15362
Energy Statistics 15322
Energy Technology 10529
Energy World 12235
Energy World Yearbook 12235
L'Enfant en Milieu Tropical: Children's in the Tropics 04376
ENFO Newsletter 11082
Engagement: Zeitschrift für Erziehung und Schule 05146
Engei Gakkai: Journal 08127
Engineering 11884
Engineering and Scientific Association of Ireland: Annual Report 06944
Engineering and Technology Degrees 13532, 15044
Engineering and Technology Enrollments 13532
The Engineering Council: Newsletter 11958
Engineering Design Education & Training 11885
The Engineering Designer 12299
The Engineering Economist 15325
Engineering for Gas Turbines and Power 14145
Engineering for Industry 14145
Engineering Geological Review: Mérnökgeológiai-Környezetföldtani Szemle 06635
Engineering in Medicine and Biology Magazine 15246, 15317
The Engineering Institution of Zambia: Journal 17152
Engineering Journal 03273, 13895
Engineering Management International 14059
Engineering Management Journal 12296
Engineering Management Review 15317
Engineering Materials 12307
Engineering Materials and Technology 14145
Engineering News 12307
Engineering Science and Education Journal 12296
Engineering Services Management 11716
Engineering Times 16159
Engineers Australia 00421
English 11960
English Academy Review 10139
English Association: Essays and Studies 11960
English Dance and Song 11961
English Education 14839
English in Education 12564
English Journal 16010
English Quarterly 02108
English Studies at Swiss Universities 10791
English Teaching Newsletter 03023
Enhanced Oil-Recovery Field Reports 16689
Enimerossi 06510
Enjeux 04221
Enrolled Actuaries Report 13379
L'Enseignement du Russe 04845
L'Enseignement Public 04510
Enterprise Europe Bulletin 12034
The Entertainment and Sports Lawyer 13617
Entomologia Experimentalis et Applicata 08992
Entomological News 13799
Entomological Society of America: Annals 15046
Entomological Society of America: Bulletin 15046
Entomological Society of Canada: Bulletin 02437
Entomological Society of Canada: Memoirs 02437
Entomological Society of Manitoba: Proceedings of the Society 02438
Entomological Society of Queensland: News Bulletin 00392
Entomological Society of Saskatchewan: Newsletter 02440
Entomological Society of Southern Africa: Memoirs 10140
Entomologicheskoe Obozrenie: Entomological Survey 09954
Entomologische Berichte 08992
Entomologiske Meddelelser 03718
The Entomologist 12800
Entomophaga 10752
Entrepreneurial News 13533
Entwicklung und Zusammenarbeit 05601
Environmental and Molecular Mutagenesis 15051
Environmental Bulletin 15497
The Environmental Communicator 16211
Environmental Defense Fund: Annual Report 15048
Environmental Education 12557
Environmental Engineer 13391
Environmental Engineering Selection Guide 13391
Environmental Entomology 15046
Environmental Health Journal 12301
Environmental Health News 12301
Environmental Health Scotland 12801
The Environmentalist 12302, 15052
Environmental Law 13617
Environmental Law Reporter 15050
Environmental News Digest 11086
Environmental Periodicals Bibliography 15362
Environmental Policy and Law 05998
The Environmental Professional 15856
Environmental Progress 13878
Environmental Protection Bulletin 12293
Environmental Report 14213
Environmental Resource Magazine 15392
Environmental Sanitation Abstracts 11082
Environmental Sanitation Reviews 11082
Environmental Science & Technology 13688
Environment Control in Biology 08256
Environment Features 04384

Environment Journal *14107*
EOQ Quality *10669*
EPA-Newsletter *08891*
epd Film *05772*
EPE Journal *01351*
Epeteris *03496*
Epetiris Etairias Byzantinin Spoudon: EEBS *06524*
The Ephemerist *11970*
EPI *01127*
Epidemiological Bulletin *16292, 16827*
Epilepsia *13801*
Epilepsie: Informationsblatt *10870*
Epilepsie-Selbsthilfe-Zeitung: Kontakte *10870*
Epilepsy News *06982*
Epilepsy Today *11558*
EPI Newsletter *16292*
Epistolodidaktika *08826*
Epitöanyog: Building Materials *06674*
EPPO-Bulletin *04479*
Equal Education Alert *16363*
Equity and Choice *15295*
ERA Technology News *11971*
Die Erde *05815*
Erdészeti Lapok: Forestry Bulletin *06669*
Erdöl Erdgas Kohle *00777*
Erdwissenschaftliche Forschung: Kommission für Erdwissenschaftliche Forschung *05050*
Eretz-Israel *07071*
Eretz Magazine *07111*
EREV-Schriftenreihe *05689*
Erfahrungsheilkunde *05789*
Erfinder und Neuheitendienst *05510*
Ergebnisse der Frobenius-Expeditionen *05765*
Ergon *06499*
Ergonomic Abstracts *11972*
Ergonomics *11972*
Ergonomics Society of Australia: Conference Proceedings *00393*
The Ergonomist *11972*
Ergotherapie *11006*
The Ericksonian Monographs *15741*
Erin *07028*
Erkenntnis *00679*
ernährung-nutrition *00778*
Ernährungs-Umschau: Organ der DGE *05346*
Ernest Bloch Society Bulletin *15054*
EROPA Bulletin *09518*
Erwachsenenbildung: Adult Education *06062*
Erwachsenenbildung: Berichte & Informationen der Erwachsenenbildung in Niedersachsen *06141*
Erziehungskunst *05225*
Erzmetall *05794*
Esalen Institute: Catalog *15055*
ESAMI Newsletter *11056*
ESAMRDC Newsletter *11057*
ESA Newsletter *15046*
ESARBICA Journal *08501*
ESARDA Bulletin *07681*
The Esatern Anthropologist *06758*
ESCAP/WMO Newsletter *09519*
Esercitazioni della Accademia Agraria di Pesaro *07125*
The ESOMAR Directory *08892*
ESOP Annual *15053*
Español Actual *10332*
Esperanto *09083*
Esperanto aktuell *05785*
Esperanto League for North America: Catalog *15057*
Espero Katolika *15428*
Essays in Ancient Greek Philosophy *16493*
Essays in Biochemistry *11440*
Essex Archaeological and Historical Congress: Newsletter *11976*
Essex Archaeological News *11977*
Essex Archaeology and History *11977*
Essex Journal *11976*
Essex Wildlife Magazine *11978*
The Establishment of an International Criminal Court *15463*
Estadística *09420*
Estadísticas Universitarias *03440*
Estadística Vitivinícola *00146*
Estatísticas Agrícolas *09673*
Estatísticas da Educação *09673*
Estatísticas Demográficas *09673*
Estatísticas Industriais *09673*
Estatísticas Monetárias e Financeiras *09673*
Estetika *03500*
Esthetics *13392*
Estuaries *15061*
Estuarine, Coastal and Shelf Science *11979*
Estuarine Research Federation: Newsletter *15061*
Estudio Económico de América Latina y el Caribe *03038*
Estudios CEDHAL *01623*
Estudios Clasicos *10386*
Estudios e Informes de la Cepal *03038*
Estudis escènics *10314*
E.T.A. Hoffmann-Gesellschaft e.V.: Mitteilungen *05644*
Etcetera *09269*
Ethica: Wissenschaft und Verantwortung *00656*
The Ethical Record *13082*
Ethics and Action *16805*
Ethiopian Library Association: Bulletin *03943*
Ethiopian Medical Journal *03944*
The Ethnic Reporter *15799*
L'Ethnie Française d'Europe *01123*
Ethnographia *06647*
L'Ethnographie *04722*
Ethnographie Albanaise *00003*
Ethnohistory *14060*
Ethnologia Scandinavica *10484*
Ethnologica Helvetica *10806*
Ethnologie Française *04723*
Ethnomedizin *05138*
Ethnomusicologie *04746*
Ethnomusicology *16520*
Ethnomusicology Newsletter *13883*
Ethnos *09661*
Ethnosciences *04746*
Ethos *16571*
Etnografia Shqiptare *00003*
Etruscan Foundation: Newsletter *15063*
Etruscans *15063*
ETT-European Transactions on Telecommunications and Related Technologies *07411*
Etude et Gestion des Sols *04247*
Etudes Celtiques *04382*
Etudes créoles *01895, 01948*
Etudes d'Education Comparée *04249*
Etudes ethnolinguistiques Maghreb-Sahara *04746*
Etudes et Sports Sous Marins *04526*
Etudes Hispano-Andalouses *11142*
Etudes Internationales *02177*
Etudes Juridiques *08560*
Etudes Mélanésiennes *09114*
Etudes Préhistoriques *04472*
Les Etudes Sociales *04651*
Etudes Tsiganes *04182*
ETUI Documentation Centre Bulletin *01353*
ETUI Newsletter *01353*
ETV-Mitteilungen *05639*
etz-Archiv *06351*
etz Elektrotechnische Zeitschrift *06351*
EUCARPIA Bulletin *08862*
Euclides *06510, 09013*
EUHOFA International Journal *10628*
EurACs *01289*
EURASIP Newsletter *10674*
EURO Bulletin *07253*
Euroclay *12223*
Euroinformazioni *07473*
Europa-Archiv: Zeitschrift für internationale Politik *05326*
Europäischer Verband für Produktivitätsförderung: News Release *05656*
Europäisches Zentrum für Wohlfahrtspolitik und Sozialforschung: Newsletter/Bulletin d'Information/Nachrichten *00585*
Europäische Vereinigung für Eigentumsbildung, Hermann-Lindrath-Gesellschaft e.V.: Pressedienst *05658*
Europa Medicophysica *01357*
Europa Orientalis: Studi e Ricerche sulle Culture dell'Est Europeo *07294*
Europarecht *06447*
European Anthropological Association: Newsletter *01270*
European Asphalt Magazine *08850*
European Association for Research on Plant Breeding: Proceedings of Section Meetings *08862*
European Association of Radiology: Newsletter *01296*
European Behavioural Therapists *11987*
European Bureau of Adult Education: Conference Reports *08873*
European Bureau of Adult Education: Newsletter *08873*
European Community Studies *03179*
European Contributions to American Studies *04480*
European Finance Association: Newsletter *01350*
European Grassland Federation: Proceedings *08889*
European Guide to Industrial Marketing Consultancy *12201*
European Healthcare Management Association: Newsletter *06947*
European Heart Journal *08893*
European Incoherent Scatter Scientific Association: Annual Report *10456*
European Journal for Gastroenterology and Hepatology *08853*
European Journal of Cardio-Thoracic Surgery *10672*
European Journal of Development Research *10677*
European Journal of Engineering Education *01460*
European Journal of Experimental Social Psychology *08866*
European Journal of Information Systems *12655*
European Journal of Internal Medicine *11997*
European Journal of Mineralogy *04819, 05471*
European Journal of Obstetrics, Gynecology and Reproductive Biology *08855*
European Journal of Operational Research *07253*
European Journal of Orthodontics *12036*
European Journal of Paediatric Surgery *12995*
European Journal of Phycology *11632*
European Journal of Physical Medicine and Rehabilitation *01357*
European Journal of Physics *12268*
European Journal of Plastic Surgery *04489*
European Journal of Political Research *12017*
European Journal of Population *08859*
European Journal of Social Psychology *01291*
European Journal of Soil Science *04247*
The European Journal of Surgery *09020*
European Journal of Surgical Oncology *11525*
European Journal of Teacher Education *01133*
European Journal of Toxicology *01126*
European Journal pf Entomology: Klapalekiana *03522*
European Marketing Research Board International: AB TGI *12032*
The European Movement: Facts *12034*
European Newsletter on Southeast Asian Studies *08964*
European Science Editing *12000*
European Society for Opinion and Marketing Research: Annual Congress Book of Papers *08892*
European Society for Opinion and Marketing Research: Newsbrief *08892*
European Society for Opinion and Marketing Research: The Monograph Series *08892*
European Studies Newsletter *14875*
European Taxation *08926*
European Transactions on Electrical Power Engineering *05640*
Europe de tradition orale *04746*
Europe en Péril? *04210*
Europhysics Letters *07874*
Europroductivity *01127*
Eurosocial Bulletin *00588*
Eurosocial Reports *00585*
Eurospace Bulletin *04556*
Euroturismo Progetto 2000 *07179*
EURYDICE Info *01260*
Euskera *10349*
EVAC Newsletter *02428*
Evangelische Akademie Bad Boll: Arbeitshilfen *05682*
Evangelische Akademie Bad Boll: Dokumente, Texte und Tendenzen *05682*
Evangelische Akademie Baden: Diskussionen: Zeitschrift für Akademiearbeit und Erwachsenenbildung *05683*
Evangelische Akademie Hofgeismar: Anstösse *05684*
Der Evangelische Buchberater *05544*
Evangelische Information *05772*
Evangelische Jugendhilfe *05689*
Evangelische Studiengemeinschaft e.V.: Texte und Materialien *05691*
Evelyn Waugh Newsletter *15067*
Everyman's Science *06813*
Evkönyv *06648*
Evolution *10744, 16616*
EWPOP-Bulletin *11022*
Excerpta Indonesica *08964*
Exchange Teacher: Annual Report *12443*
Executive Computing Newsletter *14421*
The Executive Counselor *13887*
The Executive Educator *16147*
Exeter Industrial Archaeology Group: Bulletin *12043*
Expanded Shale Clay and Slate Institute: Information Sheet *15069*
Expanded Shale Clay and Slate Institute: Membership Roster *15069*
Expanded Shale Clay and Slate Institute: Special Bulletin *15069*
Expansion *16690*
Expedition Report *11643*
Expedition Reports *13163*
Experiential Education *16155*
Experimental Aircraft Association: Chapter Bulletin *15070*
Experimental Mechanics *16523*
Experimental Medicine and Morphology *01777*
Experimental Physiology *12698*
Experimental Techniques *16523*
Expert Magazine: Intelligent Systems and their Applications *15317*
Explanatory Report *00505*
Explorations in Ethnic Studies *15799*
Explorations in Sights and Sounds *15799*
Explorer *13562*
Explorer News *15125*
The Explorers Club: Newsletter *15072*
Explorers Journal *15072*
Exportaciones Argentinas de Productos Vitivinícolas *00146*
Exposure *16570*
Extended Abstracts and Proceedings from Biennial Carbon Conferences *13677*
Extending Education *16275*
External Forestry *03201*
Extra Series *11888*
Eye *11796*

Faba Bean Abstracts *11720*
Fabian Discussion Papers *12045*
Fabian Pamphlet *12045*
Fabian Pamphlets *00290*
Fabian Research Series *12045*
Fabian Review *12045*
Face Facts *13393*
Facets *13932*
Facets of Fostoria *15118*
Fachzeitschrift für die Praxis in Wirtschaft und Verwaltung *05583*
Fachzeitschrift Heim *10725*
Facial Plastic Surgery Today *13393*
Facial Plastic Times *13393*
Facilities Management Magazine *14443*
Facsimile Series *03452*
Facts & Findings *15870*
Faculty Directory *14609*
Faculty Handbook *15309*
Faculty Newsletter *15174*
Faculty of Actuaries in Scotland: Transactions *12047*
Faculty of Actuaries in Scotland: Year Book *12047*
Faipar *06604*
Faith and Forum *15359*
Faith and Thought Bulletin *13182*
FAITS/FEITEN *01374*
Fakel: Magazine *01802*
Fakten, Daten, Zitate: Das Informationsangebot für Wissenschaft & Wirtschaft *00774*
Fall Harvest Catalog *16186*
Familia *10143*
Familia: The Ulster Genealogical Review *13153*
Familiengeschichte in Norddeutschland *05775*
Families *02778*
Famille Avertie *02593*
Famille et Développement *11103*
Family Advocate *13617*
Family History *12242*
Family History News & Digest *12065*
Family Law Quarterly *13617*
Family Medicine: Journal *16706*
Family Medicine Research Update *02354*
Family Planning Today *12056*
Family Relations *16020*
Family Safety and Health *16145*
Fanfare from the Royal Philharmonic Society *12839*
Fan va Turmush: Life and Science *16948*
Farbe und Lack *10936*
Farmaceutisk Tidende *03616*
Farmacia *09761, 09781*
Farmacija *01754*
Farmacja Polska *09565*
Il Farmaco *07774*
Farmakologia i Toksikologia: Pharmacology and Toxicology *09963*
Farm and Food News *12057*
Farmer and Rural World *11162*
Farming System Bulletin *08719*
Farmland Preservation Directory *16250*
Faro *07572*
Farumashia *08395*
The FASEB Journal *15081*
FAS Public Incerest Report *15080*
Fasti Archaeologici *07286*
Fast Job Listing Service *14414*
FBEDOC Informa *03062*
FCLA Teachers and their Areas of Expertise *15748*
FDI Dental World *12348*
Federal Bar Association: Membership Directory *15077*
Federal Bar News and Journal *15077*
Federal Funding Inquiry Reports *15901*
The Federal Veterinarian *15859*
Fédération Autonome de l'Education Nationale: Bulletin *04507*
Federation Bulletin *15090*
Federation for Unified Science Education: Proceedings of Annual Conference *15079*
Fédération Française de Sociétés des Sciences Naturelles: Bulletin *04521*
Fédération Française des Sociétés d'Amis de Musées: Bulletin *04523*
Fédération Française des Sociétés de Sciences Naturelles: Bulletin *04525*
Fédération Internationale des Universités Catholiques: News in Brief *04536*
Fédération Internationale pour l'Education des Parents: Lettre d'Information *04538*
Fédération Nationale des Syndicats Départementaux de Médecins Electro-Radiologistes Qualifiés: Bulletin *04542*
Federation News *15083*
Federation of Behavioral, Psychological and Cognitive Sciences: Annual Report *15083*
Federation of Government Information Processing Councils: Directory *15084*
Federation of Materials Societies: Conference Proceedings/Materials *15086*
Federation of Materials Societies: News *15086*
Federation of Prosthodontic Organizations: Directory *15088*
Federation of Prosthodontic Organizations: Newsletter *15088*
Federation of State Medical Boards of the United States: Handbook *15090*
Federation of Trainers and Training Programs in Psychodrama: Membership Directory *15091*
Federation of Trainers and Training Programs in Psychodrama: Newsletter *15091*
Fédération Québécoise des Directeurs et Directrices d'Ecole: Information *02461*
Federazione Medica *07695*
FedFacts *15084*
FEECA Bulletin *00586*
Feedback *00882, 14617, 15196*
Félagsblad K.I. *06693*
Feliciter *02189*
Fellow Directory *14776*
Fellowship of Australian Writers: Bulletin *00395*
Fellowship of Christian Writers: Newsheet *12072*
Felt and Damaging Earthquakes *12387*
F.E.N. HEBDO *04510*
Fennia *04047*
Fenxi Huaxue *03167*
FEPAFEM Informa: PAFAMS Newsletter *17006*
Fernseh-Dienst *06069*
Fernwärme International: District Heating/Chauffage Urbain *05098*
Ferrocement Abstracts *11082*
Fertility and Sterility *07834, 13811*
Fertility News *13811*
Fett-Wissenschaft, Technologie *05348*
Feuillets du Naturaliste *02332*
Feuillets Suisses de Pédagogie Musicale: Schweizer musikpädagogische Blätter *10900*
FGU-Dokumentationsbulletin *10696*
Fiber Society: Membership Directory *15092*
FICEMEA Bulletin *04532*
Fiches d'Identification du Plancton *03746*
Fiches Syndicales *04939*
Fichier Afrique *08065*
FID Directory *08895*
Fiddlehead Forum *13810*
Fidelity and Security News *13617*
FID News Bulletin *08895*
Field Crop Abstracts *11720*
Field Studies *12074*
Field Studies Council: Annual Report *12074*
Field Studies Council: Programmes of Courses *12074*
FIEP Bulletin *12062*
Fiji Society: Transactions *03947*
Fikrun wa fan (Geist und Leben) *06012*
Film Culture *14228*
film-dienst *06069*
Filmkritik *05701*
Filmkunst *00779*
Film Magazine *11560*
Filmovi Novini *01785*
Film/Video Manual *00779*
Filosofska i Sotsiologichna Dumka: Philosophy and Sociological Thought *11193*
Filozofický časopis *03500*
Filtration & Separation Magazine *12075*
Finanças e Desenvolvimento *01654*
Financial Journal *10009*
Financial Management *15093*
Financial Management Association: Membership/Professional Directory *15093*
Financial Management Collection *15093*
Financial News Analysis *10009*
Financial Plan *02640*
Financial Times Distributive Trades Survey *11837*
The Fine Line *16656*
The Finishing Line *14297*
The Finite String *14283*
FINNAM Newsletter *15095*
Finnish Economic Papers *04071*
The Finnish Journal of Education *04035*
Finnish Law I: Common Law *04045*
Finnish Law II: Public law *04045*
The Finnish Legal System: The English Language Publication *04045*
Finska Läkaresällskapets Handlingar *03961*
Finskt Museum *04051*
Fiorisce un Cenacolo *07170*
FIRA Bulletin *12096*
Fire Command *16051*
Fire Journal *16051*
Fire News *16051*
Fire Protection Reference Directory *16051*
Fire Surveyor *11319, 12192*
Fire Technology *16051*
Fire Technology – Calculations *12303*
Fire Technology – Chemistry & Combustion *12303*
FIR+IAW-Mitteilungen *05731*
First Break *08867*
First Destinations of Polytechnic Students *11804*
Fiscal Studies *12208*
Fish *12237*

Fisheries Society of Nigeria: Proceedings 09227
Fishery Bulletin 09227
Fitotecnia 08762
5 Minutes with ACHE 14286
FIW-Informationen 05732
Fizika Goreniya i Vzryva: Physics of Combustion and Explosion 09954
Fizika i Khimiya Obrabotki Materialov: Physics and Chemistry of Materials Processing 09954
Fizika i Khimiya Stekla: Physics and Chemistry of Glass 09954
Fizikai Szemle 06601
Fizika i Tekhnika Poluprovodnikov: Semiconductor Physics and Technology 09954
Fizika, Matematičko-Fizički List za Učenike Srednijih Škola 17079
Fizika Metallov i Metallovedenie: Physics of Metals and Metal Science 09954
Fizika Nizkikh Temperatur: Low-Temperature Physics 11193
Fizika Plazmy: Plasma Physics 09954
Fizika Tverdogo Tela: Solid State Physics 09954
Fiziko-khimicheskaya Mekhanika Materialov: Physical and Chemical Mechanics of Materials 11193
Fiziko-Tekhnicheskie Problemy Razrabotki Poleznykh Iskopaemykh: Physical and Technical Problems of Mineral Exploitation 09954
Fizik Tedavi Rehabilitasyon Dergisi 11145
Fiziologicheskii Zhurnal: Physiological Journal 09954, 11193
Fiziologiya Cheloveka: Human Physiology 09954
Fiziologiya i Biokhimiya Kulturnykh Rastenii: Physiology and Biochemistry of Cultivated Plants 11193
Fiziologiya Rastenii: Plant Physiology 09954
Fizko-himiceska Mehanika 01771
Fizyka Dielektryków i Radiospektroskopia 09612
The Flag Bulletin 15098
Flagmaster 12078
Flash: African Trade 10021
Flash Info 04518
Flash Market News 14257, 16183
Flat Earth News 15099
Flax, Hemp, Synthetic Fibre Industry Bulletin 06675
Fleet Safety Newsletter 16145
Flightline 15687
Flight Test News 16666
Flight Watch 15104
Flintshire Historical Society: Journal 12079
The Flock 14859
Flora de Colombia 03391
Flora Neotropica 16263
Flora og Fauna 03775
Floras 00107
Flower Essence Society: Directory 15101
The Flower Essence Society Newsletter 15101
Fluid Flow Measurement 11437
Fluid Flow Measurements Abstracts 11575
Fluid Power 11437
Fluid Power Abstracts 11575
Fluid Power Society: Newsletter 15102
Fluid Sealing 11437
Fluid Sealing Abstracts 11575
Fluids Engineering 14145
Flyer 14973
Flying Veterinarians Association: Directory 15106
Flying Veterinarians Association: Newsletter 15106
Flypast 12522
F & M: Feinwerktechnik & Meßtechnik 06331
Focus Newsletter 00251
Focus on Asian Studies 14253
Focus on Business Education 13065
Fodterapeuten 03763
Föld és Ég: Earth and Sky 06676
Földrajzi Közlemenyek: Geographical Review 06629
Földtani Közlöny 06635
Författaren 10589
Fogli di Informazione 07420
FOGRA Aktuell 05705
Fogra-Literaturdienst 05705
Fogra-Literatur-Profil 05705
Fogra-Mitteilungen 05705
Fogra-Patentschau 05705
Folha Mensal do Estado das Culturas e Previsão de Colheitas 09673
Folia Allergologica et Immunologica Clinica 07831
Folia Biologica 03500
Folia Entomologica Hungarica 06656
Folia Entomológica Mexicana 08759
Folia Geobotanica e Phytotaxonomica 03500
Folia Linguistica 00945
Folia Linguistica Historica 00945
Folia Microbiologica 03500, 03526
Folia Morphologica 03500
Folia Parasitologica 03500
Folia Pharmacologica Japonica 08273
Folia Zoologica 03500
Folk Directory 11961
Folkeskoten 03581
Folklore 06792, 12082
Folklore Americano 08740
Folklore Fellows' Communications 04000
Folklore Series 06717
Folklore suisse / Folclore svizzero 10856
Folk Music Journal 11961
Folletos 03009
Following Sea 16451
FOM-Instituut voor Atoom- en Molecuulfysica: Annual Report 08896
FOM-Instituut voor Plasmafysica Rijnhuizen: Annual Status Report 08897
Fondation Francqui: Rapport 01365
La Fonderia Italiana 07686
Fondo Nacional de las Artes: Informativo 00139
Fontes 09627
Fontes Artis Musicae 10472
Fonti Aragenesi 07228
Fonti e Testi 07178
Food and Chemical Toxicology 11439
Food and Nutrition Board: Activities Report 15108
Food and Nutrition Board: Directory 15108
Food Australia 00383
Food First News 15281
Food Research Quarterly 00377
Food Review 10179
Food Science and Technology Abstracts 12364
Food Science and Technology Today 12239
Food Science Profiles 12364
Food Technologist 09156
Food Technology 15321
Food Technology in Australia 00301
Footwear Business International 12886
For a Change 16565
Forecast 13785
Foreign Affairs Report 06789
Foreign Language Annals 13770
The Forensic Science Society: Journal 12085
Forest & Conservation History 15112
Forest History Cruiser 15112
Forest Industries Newsletter 16145
Forest Products Abstracts 11720
Forest Products Journal 15113
Forestry 12221
Forestry Abstracts 11720
Forestry Association of Nigeria: Proceedings of Annual Conferences 09228
Forestry Canada – Ontario Region: Program Review 02472
The Forestry Chronicle 02175
Forestry Newsletter 02472
Forestry Science and Technology 03201
Forest Science 16634
Forfatteren 03622
Formal Aspects of Computing 11547
Formation Sign 12526
Fornvännen 10493
Forrad Informazioni 08000
Forschungen und Beiträge zur Wiener Stadtgeschichte 00984
Forschungen und Berichte 05691
Forschungen zur Anthropologie und Religionsgeschichte 05726
Forschungen zur Antiken Sklaverei: Kommission für Geschichte des Altertums 05050
Forschungen zur deutschen Landeskunde 06463
Forschungen zur geschichtlichen Landeskunde der Steiermark 00650
Forschungen zur neueren Medizin- und Biologiegeschichte 05050
Forschung im Ingenieurwesen 06377
Forschung im Straßenwesen 05720
Forschungsarbeiten aus dem Strassenwesen 05720
Forschungsberichte 05364, 05575, 05750, 06303
Forschungsberichte: Reprint-Serie 01020
Forschungsberichte aus dem Gebiet der Luft- und Trocknungstechnik 05749
Forschungsberichte der Forschungsgemeinschaft Kalk und Mörtel e.V. 05713
Forschungsinstitut für Gesellschaftspolitik und beratende Sozialwissenschaft e.V.: Mitteilungsblatt 05729
Forskning Nu 10441
Forsttechnische Informationen: Mitteilungsblatt des KWF 06089
Fort 12087
Forthcoming International Scientific and Technical Conferences 11256, 11267
Forthcoming Scientific Congresses, Conferences and Symposia abroad and in Bulgaria, Series A: Natural and Mathematical Sciences 01775
Forthcoming Scientific Congresses, Conferences and Symposia abroad and in Bulgaria, Series B: Social Sciences 01775
Fortidsvern 09266
Fortschritte der Kieferorthopädie 05370
Fortschritte der Zoologie 05050
Fortschrittliche Betriebsführung und Industrial Engineering 06261
Fortschrittsberichte 05459
The Forum 15545
Forum der Schriftsteller: Forum des Ecrivains 10904
Forum E 06332
Forum för Ekonomi och Teknik 04075
The Forum for Health Care Planning: Membership Directory 15115
The Forum for Health Care Planning: Newsletter 15115
Forum Jugendhilfe 05104
Forum Loccum 05685
Forum Musikbibliothek 05570, 06003
Forum Newsletter 15554
Forum Pädagogik 06437
Forum Pädagogik: Zeitschrift für pädagogische Modelle und soziale Probleme 06010
Forum R 10898
Forum Ware: Die Ware und ihre Bedeutung für Mensch, Wirtschaft und Natur 00844
Fossils 08342
Fotointerpretacja w Geografii 09571
Foundation for Microbiology: Report 15129
Foundation for Research, Science and Technology: Annual Report 09125
Foundation for Science and the Handicapped: Newsletter 15132
Foundation for the Study of Infant Deaths: Newsletter 12089
Foundation for the Support of International Medical Training: Brochures 15134
Foundation for the Support of International Medical Training: Directory 15134
Foundation Forums 13337
Founders Letter 14278
FPRS Technical Newsletter 15113
Fra Fysikkens Verden 09314
Franca Esperantisto: Revue Française d'Espéranto 04980
Le Francais Aujourd'hui: Le Supplément au Français Aujourd'hui 04228
Le Français Moderne 04453
France Médecine 04514
Franchise Law Journal 13617
Franco-British Society: Newsletter 12091
Francophonie 11269
Frankfurter Geographische Hefte 05758
Fraser Forum 02475
Der Fraunhofer 05759
Fraunhofer-Gesellschaft zur Förderung der angewandten Forschung e.V.: Forschungsplan 05759
Fraunhofer-Gesellschaft zur Förderung der angewandten Forschung e.V.: Jahresbericht/Annual Report 05759
Fredeburger Hefte 05462
Der freie Arzt 06315
Freie Bildung und Erziehung 05251
Der Freie Zahnarzt 05761
Freight Cars Journal 16667
Freight Cars Journal Monograph 16667
Freiheit der Wissenschaft 05262
French Politics and Society 14825
French Review 13587
Fresh Morning of Life 13020
Freshwater Biological Association: Occasional Publications 12092
Freshwater Forum 12092
Fréttabréf 06706
Fréttabréf Laekna: Physicians News 06694
Freyr 06679
Friday Report 14672
Friends Historical Society: Journal 12093
Friends of Spannocchia 15063
Friends of Taliesin 15137
Friends of the National Libraries 12094
Fries Museum Bulletin 08898
Fritillary 11255
Frogpond 15184
Fronte Stomatologico 07357
Frontinus-Schriftenreihe 05766
Fruit Varieties Journal 14011
F.S.B.I. Newsletter 12077
FSMB Newsletter 15090
Fuel and Energy Abstracts 12235
Fugle 03660
Fuldaer Geschichtsblätter 05767
Functional and Developmental Morphology 03537
Functional Ecology 11554
Functional Orthodontist 13478
Fundación Juan March: Anales 10311
Fundamenta Informaticae 09585
Fundamental and Applied Toxicology 16708
Fundigao 09646
Funk-Korrespondenz 06069
Funktionsanalyse Biologischer Systeme 05050
Funktsionalnyi Analiz i Ego Prilozhenie: Functional Analysis and its Application 09954
Furniture History 12095
Fusión 17025
Fusion Technology 13960
Futura/Fer: Ergebnisse des Projektes Forschungspolitische Früherkennung des Schweizerischen Wissenschaftsrates 10924
Futures Canada 02032
Futures Research Quarterly 16927
Future Survey: A Monthly Abstract of Books, Articles and Reports concerning Forecasts, Trends and Ideas about the Future 16927
Futuribili 07727
The Futurist: A Journal of Forecasts, Trends and Ideas about the Future 16927
FWU-Magazin 05960
Fysiikka Tänään 04019
Fysik-Aktuellt 10550
Fysioterapeuten 09300
Fysioterapia 04018
Fysiovisie 09050

Gaceta Médica de Bilbao 10240
Gaceta Medica de Caracas 16962
Gaceta Médica de México 08690
Gaceta Numismatica 10278
Gaceta UDUAL 08777
Gaea 14365
Gallia 04382
Galton Institute: Symposium Proceedings 12099
Game Conservancy Annual Review 12100
Game Conservancy Newsletter 12100
Gamma 09010
GAMM-Mitteilungen 05798
Ganita 06741
Gann-Monograph on Cancer Research 08299
Ganzheitsmedizin: Zeitschrift für Regulationsmedizin 00868
Gaofenzi Tongbao 03167
Gaofenzi Xuebao 03167
The Garden 12810
Garden Today News 14992
Garten 00744
Garten und Landschaft 05353
Gas Abstracts 15322
Gas Engineering & Management 12304
Gas Research Institute Digest 15147
Gastroenterohepatološki Arhiv 17130
Gastroenterologia Japonica 08381
Gastroenterological Endoscopy 08263
Gastroenterology 13829
Gastrointestinal Endoscopy 14061
Gate 05430
The Gavel 13841
Gaz d'Aujourd'hui 04353
Gazeta Mathematica 09786
Gazette 00320, 02583, 04132, 06956, 12441, 13828
La Gazette des Archives 04166
Gazette des Mathématiciens 04890
Gazzetta Chimica Italiana 07774
GBO 01369
Gdańskie Studia Jezykoznawcze 09543
Gdańsk Wczesnośredniowieczny 09543
Der Geistig Schaffende 00958
Geldgeschichtliche Nachrichten 05833
Geliozekhnika: Helio Engineering 16948
Gelre: Bijdragen en Mededelingen 08900
Gemeinschaftsdiagnose: Die Lage der Weltwirtschaft und der westdeutschen Wirtschaft im Frühjahr und im Herbst eines jeden Jahres 05097
Gemmologian Työsaralta 04021
Gemologia 10261
Gendai no Toshokan 08271
Genealogical Institute of the Maritimes: Annual Report 02480
Genealogie 05293
The Genealogist 13674
Genealogists Magazine 13047
Genera et Species Animalium Argentinorum 00140
Genera et Species Plantarum Argentinorum 00140
General and Applied Entomology 00391
General Council of the Bar: Counsel 12103
General Engineering Transactions 00421
General Entomology 02436
General Insurance Seminar 00411
General Music Today 15763
General Register of Medical Practitioneers 07010
General Regulations & Examinations Timetable 13161
General Relativity and Gravitation 10756
General Semantics Bulletin 15323
General Systems Bulletin 15544
General Systems Yearbook 15544
General Test Practice Book 15170
Generations 02617
Genes and Development 12109
Genesis 14083
The Genetic Epistemologist 15596
Genetics 15151
Genetics Society of America: Membership Directory 15151
Genetics Society of Canada: Bulletin 02481
Genetics Society of Nigeria: Proceedings 09229
Genetika: Genetics 09954
Geneva Monitor-Disarmement 10640
The Geneva Papers on Risk and Insurance 10733
Gengo Kenkyu: Journal of the Linguistic Society of Japan 08202
The Genie 14242
Génie Rural: Column 04217
Genito-Urinary Medicine 12512
Genos: Suomen Sukututkimusseuran aikakauskirja 04060
Gens Nostra 08993
Genus: Rivista fondata da Corrado Gini edita sotto il Patrocinio del Consiglio Nazionale delle Ricerche semestrale 07616
Geoacta 00065
Geo-Archeologia 07279
Geobyte 13562
The Geochemical News 15152
Geochimica et Cosmochimica Acta 15152, 15723
Géochronique 04868
Geociencias 11088
Geodätische Arbeiten Österreichs für die Internationale Erdmessung 00862
Geodézia és Kartográfia 06605
Geofizicheskii Zhurnal: Geophysical Journal 11193
Geofysiikan päivät 03962
Geofysiske Publikasjoner 09250
Geografia 01612
Geografia Fisica e Dinamica Quaternaria: Bollettino del Comitato Claciologico Italiano 07614
La Geografia nelle Scuole 07293
Geografisch Tijdschrift 08965
Geografiska Annaler 10573
Geografiska Notiser 10464
Geografisk Tidsskrift 03759
Geografiya i Prirodnya Resursy: Geography and Natural Resources 09954
Geografski obzornik 10115
Geografski Razgledi: Geographical Surveys 08602
Geografski vestnik 10115
Geografski Vidik: Geographical Look 08602
Geografski zbornik: Acta Geographica 10105
Geographica Helvetica 10708, 10998
Geographical Association of Zimbabwe: Proceedings 17170
The Geographical Chronicles 03491
Geographical Education 00292
Geographical Education Magazine 17170
Geographical Journal 12806
Geographical Magazine 12806
Geographical Review 13831
Geographical Review of India 06760
Geographical Review of Japan: Chirigaku Hyoron 08289
The Geographical Society of China: Bulletin 03298
Géographie Physique et Quaternaire 02239
Geographische Gesellschaft Bern: Jahrbuch 10707
Geographische Gesellschaft: Mitteilungen 05776
Geographisches Jahrbuch Burgenland 00990
Geography 12110
Geokhimiya: Geochemistry 09954
Geolinguistics 14126
Geologian tutkimuskeskus, Opas 03963
Geologian tutkimuskeskus, Toimintakertomus 03963
Geologian tutkimuskeskus, Tutkimusraportti 03963
Geological Maps and Notes 00405
Geological Memoirs 00405
Geological, Mining and Metallurgical Society of India: Bulletin 06761
Geological, Mining and Metallurgical Society of India: Journal 06761
Geological, Mining and Metallurgical Society of Liberia: Bulletin 08563
Geological Papers 08643
Geological Review 03170
The Geological Society of America Bulletin 15153
Geological Society of Australia: Journal 00404
Geological Society of Australia: Meeting Abstracts 00404
Geological Society of Australia: Special Publications 00404
Geological Society of China: Memoir 03299
Geological Society of China: Proceedings 03299
Geological Society of New Zealand: Newsletter 09126
Geological Survey Bulletin 00505
Geological Survey of Finland, Bulletin 03963
Geological Survey of Finland, Special Paper 03963
Geological Survey of Ghana: Annual Report 06481
Geological Survey of India: Bulletin 06762
Geological Survey of India: Catalogue Series 06762
Geological Survey of India: Memoirs 06762
Geological Survey of India News 06762
Geological Survey of India: Records 06762
Geological Survey of India: Special Publications 06762
Geological Survey of Malawi: Bulletin 08622
Geological Survey of Malawi: Memoirs 08622

Geological Survey of Malawi: Records 08622
Geological Survey of Malaysia: Annual Report 08642, 08643
Geological Survey of Malaysia: Bulletin 08643
Geological Survey of Malaysia: Memoirs 08643
Geological Survey of Malaysia: Professional Papers (West) 08642
Geological Survey of Malaysia: Report 08643
Geological Survey of Malaysia: Reports and Bulletins (East) 08642
Geological Survey of New South Wales: Bulletin 00405
Geological Survey of New South Wales: Quarterly Notes 00405
Geological Survey of New South Wales: Records 00405
Geological Survey of Nigeria: Annual Report 09230
Geological Survey of Nigeria: Bulletin 09230
Geological Survey of Nigeria: Occasional Papers 09230
Geological Survey of Nigeria: Records 09230
Geological Survey of Western Australia: Bulletin 00406
Geological Survey of Western Australia: Information Pamphlets 00406
Geological Survey of Western Australia: Memoir 00406
Geological Survey of Western Australia: Record 00406
Geological Survey of Western Australia: Report 00406
Geological Survey Paper 00505
Geologicheskii Zhurnal: Geological Journal 11193
Géologie 01430
Geologie en Mijnbouw: International Journal of the Royal Geological and Mining Society of the Netherlands 08967
Geologische Rundschau 05781
Geologiska Föreningens i Stockholm Förhandlingar: GFF 10466
Geologists' Association Circular 12112
Geologists' Association: Proceedings 12112
Geologiya i Geofizika: Geology and Geophysics 09954
Geologiya Rudnykh Mestorozhdenii: Geology of Ore Deposits 09954
Geology 15153
Geomagnetizm i Aeronomiya: Geomagnetism and Aeronomy 09954
Géomètre 04609
Geomorfologiya: Geomorphology 09954
Geonews 06953
Geo-Nyt 03731
Geophysica 03962, 10491
Geophysical Journal 12772
Geophysical Journal International 05312
Geophysical Magazine 08147
Geophysical Prospecting 08867
Geophysical Research Letters 13833
Geophysics 16662
George Eliot Fellowship: Review 12113
Georgian Academy of Sciences: Bulletin 05011
Geoscience Canada 02482
Geosciences 01775
Geotechnical Abstracts 05345
Geotechnical Journal 14107
Geotechnical News 16831
Geotechnik 05345
Geotechnique 12294
Geotekhtonika: Geotectonics 09954
Geothermal Resources Council: Bulletin 15155
Geothermal Resources Council: Membership Roster 15155
Geotimes 13832
Gép: Machine 06606
Gépgyártástechnológia: Manufacturing Processes 06606
Gépipar: Machine Industry 06606
Geppo Haiku Journal 16943
Geriascope 02148
Germania 05568
The German Quarterly 13588
German Research 05305
German Teaching 11269
The Gerontologist 15158
Gerontology 07074
Gerontology: International Journal of Experimental and Clinical Gerontology 08741
Geschichte der Wasserversorgung 05766
Geschichte in Wissenschaft und Unterricht 06339
Geschichte, Politik und ihre Didaktik 06339
Geschichte und Politik in der Schule 06339
Geschiedkundige Vereniging Die Haghe: Yearbook 08906
Geschriften van de Vereniging voor Arbeidsrecht 09095
Die Gesellschaft 05791
Gesellschaft für die Geschichte und Bibliographie des Brauwesens e.V.: Jahrbuch 05811
Gesellschaft für Logotherapie und Existenzanalyse: Bulletin 00625
Gesellschaft für Naturkunde in Württemberg e.V.: Jahreshefte 05843
Gesellschaft für Rationelle Energieverwendung e.V.: Merkblätter 05856
Gesellschaft für Rechtsvergleichung e.V.: Mitteilungen 05857
Gesellschaft für Regionalforschung e.V.: Seminarberichte 05858
Gesellschaft für Salzburger Landeskunde: Mitteilungen 00632
Gesellschaft für Strategische Unternehmensführung: Informationsdienst 00634
Gesellschaft für Übernationale Zusammenarbeit e.V.: Dokumente: Zeitschrift für den deutsch-französischen Dialog 05865
Gesellschaft für Unternehmensgeschichte e.V.: Anno: Magazin für Unternehmensgeschichte 05866
Gesellschaft für vergleichende Felsbildforschung: Jahrbuch 00635
Gesellschaft für Versicherungswissenschaft und -gestaltung e.V.: Informationsdienst 05868
Gesellschaft für Versicherungswissenschaft und -gestaltung e.V.: Schriftenreihe 05868
Gesellschaft und Politik 00987
Gesellschaft zur Förderung der Lufthygiene und Silikoseforschung e.V.: Jahresberichte 05882
Gesnerus 10822
Gesta 15432
Gesundheits-Informations-Dienst 05257
Getah Asli 08651
Getreide, Mehl und Brot 05115
Get the Point! 13978
Gewerblicher Rechtsschutz und Urheberrecht (GRUR) 05614
GG-Nachrichten 05893
Ghalib 09365
Ghana Academy of Arts and Sciences: Proceedings 06482
Ghana Geographical Association: Bulletin 06485
Ghana Journal of Agricultural Science 06479
Ghana Journal of Science 06479
The Ghana Journal of Science 06489
Ghana Journal of Sociology 06490
Ghana Library Journal 06486
Ghana National Bibliography 06487
Ghaqda Bibljotekarji: Library Association Newsletter 08662
Ghaqda Bibljotekarji: Occasional Papers Series 08662
Il Giappone 07742
Il Giardino Fiorito 07821
Gibraltar Nature News 06495
Gidrobiologicheskii Zhurnal: Hydrobiological Journal 11193
Giessener Abhandlungen zur Agrar- und Wirtschaftsforschung des europäischen Ostens 06076
Giesserei 06376
Giessereiforschung 06376
Giesserei-Literaturschau 06376
Gießerei-Rundschau 00910
Gietwerk Perspektief 09006
Gift Catalog 15819
Ginecologia y Obstetricia de México 08702
La Ginnastica Medica 07884
Giornale Botanico Italiano 07773
Il Giornale degli Uccelli 07996
Il Giornale degli Urbanisti: Organo Ufficiale 07361
Il Giornale dei Poeti 07288
Giornale dell' A.A.B. 07261
Giornale della Arteriosclerosi 07882
Giornale dell'Accademia di Medicina di Torino 07169
Il Giornale della Filarmonica 07183
Il Giornale dell'Ingegnere 07374
Giornale dello Spettacolo: Notiziario FEDIC 07690
Giornale di Anestesia Stomatologica 07298
Giornale di Astronomica 07772
Giornale di Fisica 07874
Giornale di Gerontologia 07882
Giornale di Marketing 07333
Giornale Italiano di Chemioterapia 07849
Giornale Italiano di Senologia 07947
Giornale Storico della Lunigiana e del Territorio Lucense 07732
Giornalismo Emiliano-Romagnolo 07459
Gjuha jone 00003
GKSS-Information 05895
Glancoma 14114
Glareana 10710
GLARILEM Informations 00141
Glas: Review 17125
Glasgow Archaeological Journal 12118
Glasgow Archaeological Society: Bulletin 12118
Glasnik: Review 17080
Glass and Ceramics Newsletter 16145
Glass Art Society: Journal 15161
Glass Technology 13048
Glastechnische Berichte 05449
Glenbow 02486
Gli Uccelli d'Italia 07995
Global Biogeochemical Cycyles 13833
Global Electronics 16284
Global Gas Turbine News 14254
Global Learning Teacher Education Manual 15164
Global Missions 15437
Le Globe: Bulletin et Mémoires 10968
Globus 17127
Der Globusfreund 00668
Gloucestershire Record Series 11468
Gloucestershire Society for Industrial Archaeology: Journal 12122
Gloucestershire Society for Industrial Archaeology: Newsletter 12122
Goals & Strategies 16298
Godišnjak CANU 17080
Göttingische Gelehrte Anzeigen 05049
Going Right on 14771
Gold Papers 10755
Good Gardeners' Association: Newsletter 12123
Good Toy Guide 12704
Goodwin Series 10178
Gorjutsie Slantsö: Oil Shale 03933
Go Teach 12612
Governmental Research Association: Directory 15168
The Government Union Critique 16380
Government Union Review 16380
Governor's Journal 15491
Gower 12125
Graduate Education for Public Health 14498
Graduate Record Examinations Board: Board Newsletter 15170
Graduate Research Progress Reports 15040
Graduates of Italian Medical Schools: Membership Directory 15171
Graduates of Italian Medical Schools: Newsletter 15171
Graduate Woman 13595
Graeco-Roman Memoirs 11948
Grammy Pulse 15777
Grana 03573
Grants and Assistance News 16065
Grants and Awards Available to American Writers 16301
Graphic Communications Association: Newsletter 15172
Graphological Forum 15186
La Graphologie 04803
Grass Farmer 11570
Grass & Forage Science 11570
Grassland Society of Victoria: Conference Proceedings 00408
Grassland Society of Victoria: Newsletter 00408
Graue Literatur zur Orts-, Regional- und Landesplanung 05587
Gravestone Inscription Series 13154
Gravure Association of America: Membership Roster 15173
Gravure Association of America: Newsletter 15173
Gravure Environmental 15173
Great Lakes Commission: Membership List 15175
Great Lakes Commission: Minutes of Regular Meeting 15175
Great Lakes Research Checklist 15175
Great Plains Agricultural Council: Proceedings 15178
Great Plains Journal 15341
Great Western Echo 12127
Greece and Rome 11792
Greek Review 12128
Green Engineering 12307
The Green Sheet 15907
Gregoriusblad 09001
GRE Information Bulletin 15170
Grenzgebiete der Wissenschaft 00656
Grillparzer-Jahrbuch 00647
Groei & Bloei 08948
Groepsmedia 09076
Grønland 03709
Le Grotte d'Italia 08015
Ground Engineering 12294
Ground Engineering Yearbook 12294
Groundwater News 16298
Group Analysis 12129, 12240
Groupe Affairs 00382
Groupe Belge d'Etude de l'Arriération Mentale: Proceedings 01368
Groupe Phonétique de Paris: Report 04561
Group Health Association of America: Journal 15181
Group Health News 15181
Group Practice Journal 13835
Group Psychodrama 14128
Growing Edge 13318
GRS-Bericht 05799
GRS-Spektrum 05799
Grubensicherheitliche Kurzberichte 06431
Grüne Briefe 05945
Grupparbeten 10508
Gruppenpsychotherapie und Gruppendynamik 00882
Gruppe Olten, Schweizer Autorengruppe: Mitteilungsblatt 10724
Grus Americana 16892
GRV-Nachrichten 05855
GSA News & Information 15153
GSF-Bericht 05862
GST-Bulletin 10721
Guernsey Breeders' Journal 11963
Guidebook of Catholic Healthcare Facilities 14655
Guidelines for the Biological Evaluation of Pesticides 04479
Guidepost 13771
Guides to Forest and Conservation History of North America 15112
Guide to Children's Hospitals 15837
Guide to Christian Colleges 14744
Guide to College Film Courses 13813
Guide to County Museum Services in England and Wales 13041
The Guide to Film, Television and Communications Courses in Canada 02134
Guide to Graduate Departments of Sociology 14175
Guide to Independent Study through Correspondence Instruction 16175
Guide to Properties 12615
Guide to Research Libraries and Information Services in Finland 04064
Guide to Resources and Services 15583
Guide to the Graduate Record Examinations Program 15170
Guild of Church Musicians: Year Book 12133
Guild of Guide Lecturers: Member's Bulletin 12134
Guild of Guide Lecturers: Newsletter 12134
Guild of Pastoral Psychology: Bulletin 12136
Gujarat Research Society: Journal 06763
De Gulden Passer 01504
Gulf and Caribbean Fisheries Institute: Proceedings 15182
Gul Phul 09401
Gustav-Freytag-Blätter 05901
Guyana Library Association Bulletin 06577
gwa: Gas Wasser Abwasser 10919
gwf – Das Gas- und Wasserfach: Ausgaben Gas/Erdgas und Wasser/Abwasser 05556
gwf/Gas-Erdgas 06170
gwf/Wasser-Abwasser 06170
GwG-Zeitschrift 05872
Gymnasieskolen 03740
Gynecologic Investigation 16528
Gynécologie 04804
Gyógyszerészet 06633
Gypsy Lore Society: Membership Directory 15183
Gypsy Lore Society: Newsletter 15183

H20-Journal 09014
Haarlemse Voordrachten 08910
Haberdashery Technology 06675
Habitar 09464
Habitat 11851, 11969
Habitat Australia 00278
HAB-Journal 05912
Hadith Series 08502
Hadoar 15214
Haematologica 07867
Haemophilia Society: Bulletin 12140
Haemophilia Society: Update 12140
Hämophilie-Blätter 05453
Händel-Jahrbuch 05783
Haiku Journal 16943
Hakluyt Society – Extra Series 12141
Hakluyt Society – Second Series 12141
Haliotis 04806
Halle News 12143
Halle Year Book 12143
Haltung und Bewegung 05237
Hamatechet 07114
Hamburger Ärzteblatt 05035
Hamburger Zahnärzteblatt 06458
Hamdard-i-Sehat 09364
Hamdard Islamicus 09364
Hamdard Medicus 09364
Hamdard Naunehal 09364
Ha Mif'al: The Enterprise 07078
Hamizrah Hehadash: The New East 07084
Hampshire Field Club and Archaeological Society: Newsletter 12144
Hampshire Field Club and Archaeological Society: Proceedings 12144
Handassa 07046
Handbök Bånoa 06679
Handbook of Agricultural and Food Research 11213
Handbook of Austrian Development Aid 00743
Handbooks for the Identification of British Insects 12800
Handeln 06108
Handhaaf Newsletter 10141
hand in hand 06012
Handlingar och Tidskrift 10486
Hand Surgery Newsletter 13481
Hand to Hand 14535
Hankuk Munhunjungbohakhoeji: Journal of the Korean Library and Information Science Society 08535
Hannoversche Geographische Arbeiten 05779
Hansa 05904
Hardwood Research Council: Proceedings: Annual Hardwood Symposium 15187
Hardy Plant Society: Bulletin 12146
Hardy Plant Society: News Letter 12146
Harefuha 07082
Harita ve Kadastro Mühendisliği: Surveying Rngineering 11167
Hark 12155
Harry S. Truman Library Institute for National and International Affairs: Newsletter 15188
The Hartebeest 17174
Hartmannbund-Magazin 05909
Harvey Lectures 15190
Harz-Zeitschrift 05910
Hausarzt Bayern 06400
Der Hautarzt 06123
Havergal Brian Society: Newsletter 12150
Having Children 16568
Hawick Archaeological Society: Transactions 12151
Hawk and Owl Trust: Annual Report 12152
Hawk and Owl Trust: Newsletter 12152
HBNC 11436
Headache 13501
Headline Series 15110
Headmasters Association: Membership List 15192
Head Teachers Review 12567
Healing Hand 11936
Health Academy News 16378
Health Advocate 16071
Health and Beauty World 11305
Healthcare Education Dataline 14064
Healthcare Executive Magazine 13710
Healthcare Management: Gestion des Soins de Santé 02474
Health Care Newsletter 16145
Healthcare Protection Management 15383
Healthcare Risk Management Bulletin 12503
Health Device Alerts 15012
Health Devices 15012
Health Devices Sourcebook 15012
Health Directions Letter 15855
Health Education 13449, 14342
Health Education Quarterly 16573
Health Facts 14686
Health Freedom News 16070
Health & Hygiene 12819
Health Law Digest 16072
Health Law Vigil 13854
Health Lawyers News Report 16072
Health Letter 16374
Health Matters 11389
Health & Nutrition Update 02246
Health Physics Journal 15195
Health Physics Society: Membership Handbook 15195
Health Physics Society: Newsletter 15195
Health Progress: Official Journal 14655
Health Research Board: Annual Report 06954
Health Science Magazine 13955
Health Services Administration Education Directory 14529
Health Services Research 15231
Hearsay 02637
Heartbeat 02322, 10973, 15426, 15718
Heart Disease and Stroke 13844
Heather Society: Bulletin 12156
Heather Society: Yearbook 12156
Heat Transfer 14145
Hebrew Studies Journal 15882
Heemschut 08834
Hefte der Deutschen Gesellschaft für christliche Kunst 05336
Hegau 06386
Heidemijtijdschrift 08950
Heilbad und Kurort: Zeitschrift für das gesamte Bäderwesen 05502
Heilkunst: Forum anerkannter Naturheilverfahren 05249
Der Heilpraktiker 05694
Heimatkunde des Osnabrücker Landes in Einzelbeispielen 06387
Heimatleben 10935
Heimatschutz/Sauvegarde 10784
Heimen 09275
Hellenic Cardiological Review 06507
Helminthological Abstracts 11720
Helper 14042
Helvetica Chimica Acta 10774
Helvetica Chirurgica Acta 10983
Helvetica Physica Acta 10881
L'Hémophile 04230
Hemophilia World 16928
Henry Doubleday Research Association Newsletter 12158
Henry Williamson Society: Journal 12159
Henty Society: Bulletin 12160
Hepatology 13502
The Heraldic Register of America 13711
Héraldique et Généalogie 04515
Heraldisk Tidsskrift 03823
The Heraldry Gazette 12161
Heraldry in Canada 02499
Le Héraut 01416
Herbage Abstracts 11720
Herbert Hoover Presidential Library Association: Newsletter 15203
Herbertia 15425
Herb Gardens of the U.S. 13845
The Herd Book 11569
Heredity 12109
Heritage 13905
Heritage Australia 00281

Heritage Scotland 12615
Hermes 04382
Herpetologica 15205
Herpetological Circulars 16612
Herpetological Journal 11571
Herpetological Monographs 15205
Herpetological Review 16612
Hertfordshire Archaeology 11917, 12882
Hertfordshire Natural History Society and Field Club: Transactions 12165
Hertfordshire's Past 12164
Herzl Institute Bulletin 16777
Hesketh Hubbard Art Society: Annual Catalogue 12166
Hessische Beiträge zur Geschichte der Arbeiterbewegung 05923
Hessische Blätter für Volksbildung 05944
Hessisches Ärzteblatt 06092
Der Hessische Zahnarzt 06100
Der hessische Zahnarzt: DHZ 06047
HE-XTRA 14342
H.G. Wells Society: Newsletter 12167
Hid Islenska Fornleifafélag: Arbók 06688
Hidrológiai Közlöny 06634
Hidrológiai Tájékoztató 06634
High Energy Physics Index 05575
Higher Commission Newsletter 15736
Higher Education 12994
Higher Education in Europe 09750
Higher Education – National Affairs 13766
Higher Education Policy 04281
Higher Education Report 14450
Highlights in Space Technology 16812
High Performance Polymers 12268
Highway Digest 16436
Highways and Transportation 12305
Hikakuhô Konkyû: Comparative Law Journal 08131
Hilfsmittel für das Studium der finnisch-ugrischen Sprachen 04003
Himalayan Institute Quarterly 15211
Himalayan International Institute of Yoga Science and Philosophy of the U.S.A.: Research Bulletin 15211
Himalayan Research Bulletin 16191
Hindemith-Jahrbuch 10705
Híradástechnika: Telecommunication 06607
Hírlevél: Newsletter 06607
Hírlevél Magyar Könyvtárosok Egyesülete tagjaihoz 06644
Hispania 13590
Hispanic and Luso-Brazilian Council: Bulletin 12170
Hispanic Review 15213
Histochemical Society: Membership Roster 15216
Histoire des Sciences Médicales 04856
Historia Agriculturae 09075, 09075
Historia de Chile 03010
Historia Latinoamericana en Europa 11343
Historiales 10246
Historiallinen Aikakauskirja: Suuomen Historiallinen 03965
Historiallinen Arkisto 04027
Historiallinen Kirjasto 03965
Historiallisia Tutkimuksia 04027
The Historian 12172
Historian Aitta 03965
Historical Archaeology 16530
Historical Association of Oman: Bulletin 09355
Historical Booklets 02157
Historical Dance 11899
Historical Metallurgy 12256, 12262
Historical Papers 02157
Historical Records of Australian Science 00250
Historical Series 13154
Historical Society of Ghana: Transactions 06491
Historical Society of Nigeria: Journal 09231
Historical Society of Sierra Leone: Journal 10030
Historical Society of Washington, DC: Calendar of Events 15218
Historical Society of Washington, DC: Newsletter 15218
Historic Society of Lancashire and Cheshire: Transactions 12178
Historiens et Gèographes 04197
Historie & Samtid 03630
Historische Forschungen 05050
Historischer Verein der Pfalz e.V.: Mitteilungen 05920
Historischer Verein Dillingen an der Donau: Jahrbuch 05921
Historisches Jahrbuch 05896
Historisk-filosofiske Meddelelser 03710
Historisk-filosofiske Skrifter 03710
Historisk Tidskrift: Svensk Historisk Bibliografi 10554
Historisk Tidsskrift 03702
History, Archaeology and Ethnography 01775
History in Africa 13346
History News 00470, 13492
History News Dispatch 13492
History of Cajal Club 14631
History of Economics Society: Bulletin 15220
History of Education 15221
History of Education Society Bulletin 12179
History of Science Society: Newsletter 15222
The History Teacher 16531
Hitsaustekniikka-Svetsteknik 04028
HKLA Journal 06593
HLABC Forum 02497
HLH Heizung Lüftung/Klima Haustechnik 06327
HNO-Informationen 05359
HNO-Mitteilungen 05503
Hobart Papers 12233
Hochschule und Ausland 05496
Die Höhere Schule 05536
Die Höhle 00702
Die Höhle: Zeitschrift für Karst- und Höhlenkunde 00969, 06335
Hoehnea 01666
Hölderlin-Jahrbuch 05930
hörgeschädigte kinder 05448
Hörgeschädigtenpädagogik 05202
Hogaku Kyokai Zassi 08134
Hohenzollerische Heimat 05931
Hoja Informativa 10365
Hojas de Frijol 03360
Holistic Education Series 15688
Holistic Medicine 13849, 13850
Holistic Science 06865
Holzforschung und Holzverwertung: Mitteilungen 00791
Die Holzschwelle 06300
Home Economics Educator 15226
Home Economics Research Journal 13851
Home from Home 11749
The Homemaker Update 15800
Homeopathy Today 15944
Homeostasis Quarterly 13330
Hominid Remains: An Up-date 01469
Hommes et Destins 04105
Hommes et Fondérie 04350
Hommes et Terres du Nord 04657
The Homoeopath 13051
Homoeopathy 11572
Homøopatisk Tidsskrift 09320
Honeyguide 17176
Hong Kong Library Association: Newsletter 06593
Honors Catalog 16913
Honourable Society of Cymmrodorion: Transactions 12180
Hoppe-Seyler's Zeitschrift für Physiologische Chemie 05805
Horace Mann League of the U.S.A.: Newsletter 15228
Horisont 10569
Horizons 14779
Hornbill 06747
El Hornero 00092
Horological Journal 11573
Horticultural Abstracts 11720, 11723
Horticultural Bulletin 06710, 06711
Horticultural Research Institute: Research Letter 15230
HortScience 14066
HortTechnology 14066
Hoso Jiho 08137
Hospital 14065
Hospital Codes and Standards Update Letters 14067
Hospital & Community Psychiatry 14017
Hospital Engineering Bulletin 14067
Hospital & Health Services Administration 13710
Hospitality Educator 14920
Hospital Literature Index 13854
Hospital Podiatrist 13541
Hospital Technology Alerts 13854
Hospital Technology Series 13854
La Houille Blanche: Revue Internationale de l'Eau 04874
Hour Glass 16113
House Ear Institute: Research Bulletin 15232
Housing 12244
Housing and Planning Review 12593
Housing and Society 13542
Housman Society: Journal 12185
How Did It Start? 12303
HRD Documentation Bulletin 06724
HRD Newsletter 06724
HR Magazine 16535
HRMOB News 14456
Hrossaraektin 06679
HSRC Centre for Constitution Analysis 10147
HSRC/RGN In Focus 10147
Huaxue Jiaoyu 03167
Huaxue Tongbao 03167
Huaxue Tongxun 03167
Huaxue Xuebao 03167
Hudební věda 03500
Hugoku-Shakai to Bunka: China-Society and Culture 08433
Huguenot Society of Great Britain and Ireland: Proceedings 12186
Huisarts en Wetenschap 09052
Human Communication 02299
Human Communication Research 15444
Human Factors 15235
Human Factors Association of Canada: Annual Conference Proceedings 02508
Human Factors Association of Canada: Communiqué 02508
Human Factors Society: Bulletin 15235
Human Factors Society: Directory & Yearbook 15235
Humaniora 10491
Humanistic Judaism 16534
Humanistisk-Samhällsvetenskapliga Forskningsradet: Tvärsnitt 10468
Humanist News 11574
Humanities Education 15804
Humanities in the South 16726
Human Organization 16494
Human Power 15486
Human Resources 14452
Human Resources Administration 14065
Human Resources Research Organization: Bibliography 15240
Human Rights 13617
Human Rights Case Digest 11581
Human Science 06717
Humboldt 06012
Humboldt-Gesellschaft für Wissenschaft, Kunst und Bildung e.V.: Mitteilungen 05934
Humor: International Journal of Humor Research 15532
Hunan Agricultural Sciences 03204
The Hungarian PEN 06652
Hungarian Textile Engineering 06675
Hungarian Theatre Hungarian Drama 06658
Hunter Archaeological Society: Transactions 12187
Hunterian Society: Transactions 12188
Huntingdonshire Local History Society: Records 12189
Husflid 03632
Huxley Institute for Biosocial Research: Newsletter 15242
HVG-Mitteilungen 05932
Hychowanie Ojczyste 12708
Hydata News and Views 14209
Hydraulic Journal 14107
Hydrologic Bulletin 06571
Hydronymia Europaea 05050
Hydronymia Germaniae 05050
Hydroponic Society of America: Proceedings 15243
The Hydroponic/Soilless Grower 15243
Hydro Regulatory Report 16081
HYGIE: International Journal of Health Education 04588
The Hygienist 11616
Hyomen Gijutsu 08138
Hypatia 16629
Hyper Active Children's Support Group: Newsletter 12190
Hypertension 13844
Hypnotherapy in Review 13308
Hypnotherapy Today 13571
Hz: The Fylkingen Bulletin 10463

IAAAM Conference Abstract 15375
IABSE Congress Report 10742
IABSE Report 10742
IADC News 15404
IADR Newsletter 15379
IAEA Bulletin 00663
IAEA Newsbrief 00663
I.A.E.A. Newsletter 06768
IAEG Annual Review 06972
IAG Newsletter 08741
IAHR Bulletin 08918
ia Journal 12191
IAM-Journal 05994
The IAO Inspector 02534
IASA Journal 10473
IASC Bulletin: Bulletin de la SCAD 02515
IASCE Newsletter 15395
IASLIC Bulletin 06775
IASS Bulletin 10328
IASWR Conference Proceedings 15267
IATLIS Communication 06777
IATSS Research 08142
IATSS Review 08142
IATUL Newsletter 12334
IAU/UAI Information Bulletin 04971
IAWA Journal 08925
Iberian Studies 11754
IBI News 15347
Ibis 11626
ICAA News 10656
ICASE Newsletter 06594
ICA Today: Newsletter 15435
ICEA Bookmarks 15434
ICE-News Bulletin 12366
ICEP Coordinate Budget 12354
ICEP Flash 12354
ICES Oceanographic Data Lists and Inventories 03746
ICF Bugle 15462
Ichthyosis Focus 15097
ICID Bulletin 16819
ICLAS-News 03966
ICM Reference Book and List of Members 12226
ICOLD Technical Bulletin 04425
ICOM News 04458
Icomos News 04456
Icon: Cahier 01373
ICPE Newletter 15441
ICRU Report 15442
ICSSPE-Bulletin 03967
ICSSPE Sport Science Review 03967
ICSSR Journal of Abstracts and Reviews: Economics 06788
ICSSR Journal of Abstracts and Reviews: Geography 06788
ICSSR Journal of Abstracts and Reviews: Political Science 06788
ICSSR Journal of Abstracts and Reviews: Sociology and Social Anthropology 06788
ICSSR Newsletter 06788
ICSSR Research Abstracts 06788
ICSU Yearbook 04460
ICSW-Information 00664
Idea: Rivista di Cultura e di Vita Sociale 07602
Idea Exchange 15627
Ideas: Issues and Readings in Safety 13436
Ideas Newsletter 15253
Idées 01895
Identification Leaflets for Diseases and Parasites of Fish and Shellfish 03746
IDERA Calendor of Events 02551
IDERA Clipping Service 02551
IDF Bulletin 01395
IDF Newsletter 01395
El Idioma Quiché y su Grafía 06555
IDOC Internazionale 07467
Ido Dergi 11146
IDRC Communicator 15258
Idrotecnica 07284
IEA Newsletter 08922, 10470
IEA Readings 12233
IEC Bulletin 10649
IEC Yearbook 10649
IEE News 12296
IEE Proceedings 12296
IEE Review 12296
I.E.S. News Sheet 12302
I.E.S. Proceedings 12302
IFAP News 01557
IFAR Reports 15478
IFHP News Sheet 08932
IFIC Do it yourself Series 11082
IFIC Slide Presentation Series 11082
IFIC Specialized Bibliographies 11082
IFLA Annual 08934
IFLA Directory 08934
IFLA Journal 08934
IFLA Trends 08934
Ifo-Digest 05936
Ifo-Schnelldienst 05936
Ifo-Studien 05936
IfS-Information 00659, 05969
IFSW Newsletter 09270
IFUT News 06984
IFUW Newsletter 10702
Igaz Szó 09798
I Georgofili: Atti della Accademia Economico 07177
Iggeret 15882
Igiena 09767
IGIP-Report 00670
IHO Yearbook 08797
IIEP Newsletter 04573
IIP Monitor 00680
IIRA Bulletin 10750
IIRA Membership Directory 10750
IIRR Report 15490
IJO Newsletter 07714
Iker 10349
IKN Informationen Kerntechnische Normung 06183
ILAR News 15328
Il-Ballotra 08671
ILCO-Praxis: Organ der Deutschen Ileostomie-Colostomie-Urostomie-Vereinigung 05455
Iliria: Illyria 00003
Ilmailu 04030
Il-Merill 08669
ILSA Journal of International Law 16310
Image Communication 10674
Imagery Today: Newsletter 15488
Image Technology 11599
IMAG-publikaties 08912
IMAG-Research Reports 08912
IMA Journal of Applied Mathematics 12258
IMA Journal of Mathematical Control and Information 12258
IMA Journal of Mathematics Applied in Business and Industry 12258
IMA Journal of Mathematics Applied in Medicine and Biology 12258
IMA Journal of Numerical Analysis 12258
IMA News 06009
IMI Descriptions of Pathogenic Fungi and Bacteria 12378
IMI Distribution Maps of Plant Diseases 12378
IMLS Gazette 12261
IMM Abstracts 12309
Immigration History Newsletter 15249
IMM News 12257
IMMS Membership Directory 15700
Immunization Chart 15134
Immunologiya: Immunology 09952
Immunology 11657
Immunology and Allergy Practice 13697
I Monumenti d'Arte e di Storia del Canton Ticino 10775
Impact 10012
Implantologist: The International Journal of Oral Implantology 15447
Implant Update 13278
Imported Crude Oil and Petroleum Products 13995
Imprint 16165
The IMS Bulletin 15331
IMU Bulletin 01695
In ajutorul profesorului de geografie: A l'Aide du Professeur de Géographie 09784
Incite 00318
Income and Safety 15276
Incontri di Studio: Collana di Quaderni Meridionalistici 07583
Incorporated Society of Musicians: Yearbook 12195
Indépendance et Coopération 04572
Independência: Society's Review 09686
The Independent 15252, 16477
Independent Computer Consultants Association: Conference Proceedings 15252
Independent Energy: The Business Magazine of the Independent Energy Industry 13369
Independent Energy: The Business Magazine of the Independent Power Industry 13369
Independent Research Libraries Association: Directory 15254
Independent Scholars of Asia: Newsletter 15255
Independent Scholars of Asia: Proceedings 15255
Independent School 15868
The Independent Shavian 14578
Independent Study in the Humanities: Directory of Fellows 14870
Index AAI 01105
Index de la Documentation Économique, Scientifique et Technique 08805
The Indexer 14135
The Indexer: Micro Indexer 13052
Indexes to Testamentary Records 11639
Index Nominum 10885
Index of Current Research on Pigs 11720
Index of Fungi 11720, 12378
Index of JAMS 13297
Index of Periodicals 02581
Index Rétrospectifs Specialisés 08805
Index Seminum 12634
Index Series 15764
Index to Book Reviews in Religion: (IBRR) 14192
Index to Dental Literature 13779
Index to Foreign Legal Periodicals 13547
Index to Indian Legal Periodicals 06798
Index to NOLA Publications 15808
Index to Theses with Abstracts 11267
Index Veterinarius 11720
Indian Academy of Sciences: Proceedings 06767
The Indian Advocate 06739
Indian Anthropological Association: News Bulletin 06769
Indian Anthropologist 06769
Indian Association of Academic Librarians: Newsletter 06771
Indian Botanical Society: Journal 06779
Indian Ceramic Society: Transactions 06782
Indian Chemical Society: Journal 06783
Indian Council of Medical Research: Annual Report 06787
Indian Dairyman 06790
Indian Dissertation Abstracts 06788
Indian Economic Journal 06791
Indian Educational Review 06843
Indian Education Newsletter 16084
Indian Floras 06749
Indian Geohydrology 06773
The Indian Historical Review 06786
Indian Journal of Adult Education 06768
Indian Journal of Agricultural Economics 06815
Indian Journal of Biochemistry & Biophysics 06753
Indian Journal of Cancer 06781
Indian Journal of Chemistry, Section A 06753
Indian Journal of Chemistry, Section B 06753
Indian Journal of Criminology 06816
Indian Journal of Dairy Science 06790
Indian Journal of Experimental Biology 06753
Indian Journal of Fibre and Textile Research 06753
Indian Journal of Helminthology 06764
Indian Journal of History of Science 06804
Indian Journal of Marine Sciences 06753
Indian Journal of Mathematics 06713
Indian Journal of Pharmaceutical Sciences 06806
Indian Journal of Physics 06770
Indian Journal of Psychometry and Education 06810
Indian Journal of Public Health 06811
Indian Journal of Pure and Applied Mathematics 06804
Indian Journal of Pure and Applied Physics 06753
Indian Journal of Radio and Space Physics 06753
Indian Journal of Social Science 06788
Indian Journal of Surgery 06735
Indian Journal of Technology (Including Engineering) 06753
Indian Library Association: Bulletin 06799
Indian Library Science Abstracts 06775

Indian Linguistics *06833*
Indian Literature *06854*
Indian Medical Register *06835*
The Indian Mineralogist *06836*
Indian Minerals *06762*
Indian National Science Academy: Proceedings: Pt.A: Physical Sciences *06804*
Indian National Science Academy: Proceedings: Pt.B: Biological Sciences *06804*
Indian National Science Academy: Yearbook *06804*
The Indian PEN *06850*
Indian Phytopathology *06807*
Indian Psychological Abstracts *06788*
Indian Railway: Technical Bulletin *06853*
Indian Science Congress Association: Proceedings *06813*
Indian Society of Genetics and Plant Breeding: Journal *06818*
India Quarterly *06789*
Indice de Ciências Sociais *01714*
Indici e sussidi bibliografici della Biblioteca *07211*
Individual Psychology: The Journal of Adlerian Theory, Research & Practice *16224*
Indo-British Review *06823*
Indologica Taurinensia *04582*
Indo-Pacific Prehistory Association: Bulletin *00410*
Industria della Carta *07406*
L'Industria Italiana del Cemento *07352*
Industrial Accountant *09366*
Industrial Aerodynamics *11437*
Industrial Aerodynamics Abstracts *11575*
Industrial Archaeology Review *11266*
Industrial Corrosion *11437, 12229, 12295*
Industrial Development *15258*
Industrial Engineering *15325*
Industrial & Engineering Chemistry Research *13688*
Industrial Locomotive *12200*
Industrial Management *15325*
Industrial Marketing Research Association: Annual Conference Papers *12201*
Industrial Marketing Research Association: Symposium Papers *12201*
Industrial Mathematics *15259*
Industrial Property *11034*
Industrial Relations Research Association: Annual Research Volume *15260*
Industrial Relations Research Association: Newsletter *15260*
Industrial Research Institute: Annual Report *15261*
Industrial Setting Report *11437*
Industrial Society *12202*
Industrial Society: Directory of Sources *12202*
Industrial Therapy *11582*
Industrial Trends Survey *11837*
L'Industria Meccanica *07373*
Industria y Química *00096*
Industriebau *05119*
L'Industrie Nationale *04673*
Industry News *11388*
Industry Review *11467*
Infection Control Digest *13854*
Infectious Diseases Society of America: Membership Roster *15262*
Infeksion Dergisi *11166*
Infertility *07834*
Infoadmin *01388*
Infoagrar *05830*
Info ALS *01867*
Infoblatt *05080*
Infochange *14304*
In Focus *13823*
In Focus Newsletter *12542*
Info FIAV *15475*
Info Kinderphilosophie *00799*
InfoMAB *04621*
Info Pilote *04540*
Infor *02213*
Informaciones de Secretaría *03092*
Información Técnico-Económica *10248*
Informasi Pustaka Kebahasaan *06894*
Informatica *07343*
Informatika *17094*
Informatika dhe Matematika Llogaritese: Informatics and Computer Nathematics *00003*
Informatik-Spektrum *05829*
Die Information *05584*
Information Alert *16135*
Information and Control *03162*
Information Bulletin of the Coordinating Centre *09811*
L'Information du Technicien Biologiste *04279*
Informationen Bau-Rationalisierung ibp *06245*
Informationen der Gesellschaft für politische Aufklärung *00631*
Informationen des IfAPF *05948*
Informationen für den Biologie-Unterricht *06230*
Informationen für den Chemieunterricht *06230*
Informationen für den Geschichts- und Gemeinschaftskundelehrer *06339*
Informationen zur DDR-Pädagogik *06230*
Informationen zur Deutschdidaktik *00550*
Informationen zur modernen Stadtgeschichte *05587, 06389*
Information for Applicants to Schools and Colleges of Optometry *14495*
Information from the Secretariat *09985*
Information North *01871*
Information Pockets on Astronomy *14541*
Information Processing in Animals *05050*
Information Resources *01202*
Informationsblätter der GwG *05872*
Informationsblatt des ÖSLV *00902*
Informationsdienste Holz der Entwicklungsgemeinschaft Holzbau in der DGfH *05363*
Informationsdienst Textiltechnik Bekleidungstechnik Textilmaschinenbau *05692*
Informationsdienst Verpackung IV *06246*
Informationsdienst zur Ausländerarbeit *05970*
Informations Internationales *04597*
Informations Universitaires et Professionnelles Internationales *04283*
Information Systems Research *15330*
Information Technology and Libraries *15669*
Information Update *02301, 10147*
Information zur erziehungs- und bildungshistorischen Forschung (IZEBF) *05915*
Informativo INT *01690*
Informatore Botanico Italiano *07773*
Informazioni di Parapsicologia *07493*
Informe da Pesquisa *01738*
Informe Mensual Estadístico *00137*
Informes *00117, 00118*
Inforum *08717*
Infoscolaire *01907*
L'Ingegnere: Rivista Tecnica di Ingegneria e di Architettura *07374*
L'Ingegneria Ferroviaria *07607*
Ingegneria Nucleare *07374*
Ingegneria Sanitaria *07367, 07374*
Ingeniera Sanitaria *15348*
Ingeniería *13244*
Ingenieria Sanitaria *01582*
De Ingenieur *08963*
Ingénieur et Industrie *01150*
Ingénieurs et Architectes Suisses *10899*
De Ingenieurskrant *08963*
Ingenieursnieuws *08963*
Ingenieur-Werkstoffe *06328*
Ingeniøren *03635*
INHIGEO Newsletter *15443*
Iniziativa Giuridica *07482*
Inland Bird Banding Newsletter *15263*
Inland Seas *15176*
In-Natura *08671*
Inner Mongolian Social Science: Chinese Edition *03206*
Inner Mongolian Social Science: Mongolian Edition *03206*
Innovazione: Trimestrale di Informazione Tecnico-Scientifica *07725*
Innsbrucker Beiträge zur Kulturwissenschaft *00654*
Inorganic Chemistry *13688*
Inostrannaya Literatura *09989*
In Pharmation: B.C. Pharmacist *01986*
In Practice *11703*
Inquiry and Analysis *16147*
Inred *03023*
INRED: Indices y Resumenes en Educación *03064*
The IN-Report *16929*
Inroads *15798*
Inscape *11501*
Insecta Helvetica *10804*
Insect Molecular Biology *12800*
Insect Science and its Application *08463*
L'Insegnamento della Matematica e delle Scienze Integrate *07536*
Inside Housing *12244*
Inside ISHM *15533*
Inside MS *16111*
The Insiders *15276*
Insight *12655, 15250, 15505*
Insight Kommunikation *05523*
Insights *15612, 16146, 16289*
Institiuid Teangeolaiochta Eireann: Annual Report *06957*
Institut Africain pour le Développement Economique et Social: Bibliographies commented *08065*
Institut Africain pour le Développement Economique et Social: Listes d'acquisition *08065*
Institut Africain pour le Développement Economique et Social: Rapport d'activité *08065*
Institut Agrícola Català de Sant Isidre: Boletín *10313*
Institut Agrícola Català de Sant Isidre: Revista *10313*
Institut Archéologique du Luxembourg: Bulletin *01375*
Institut d'Egypte: Bulletin *03906*
The Institute *02525*
Institute for American Indian Studies: Research Report *15269*
Institute for Animal Health: Annual Report *12204*
Institute for Biblical Research: Newsletter *15270*
Institute for Certification of Computer Professionals: Newsletter *15271*
Institute for Educational Leadership: Newsletter *15278*
Institute for Expressive Analysis: Journal *15279*
Institute for Labor and Mental Health: Directory *15285*
Institute for Reality Therapy: Newsletter *15293*
Institute for Social Research: Newsletter *15296*
Institute for Social Research: The Research News *15296*
Institute for the Future: Perspectives *15303*
Institute for Theological Encounter with Science and Technology: Bulletin *15304*
Institute for Twenty-First Century Studies: Proceedings *15307*
Institute Insights *13865*
Institute of Actuaries: Journal *12211*
The Institute of Actuaries of Australia: Transactions *00411*
Institute of Architectural and Associated Technology: Newsletter *06959*
Institute of Australian Geographers: Newsletter *00412*
Institute of British Geographers: Transactions *12216*
Institute of Ceramics: Proceedings *12218*
Institute of Ceramics: Transactions and Journal *12218*
Institute of Chartered Accountants of Scotland: Annual Report *12220*
Institute of Concrete Technology: Convention Symposium Papers *12225*
Institute of Corrosion: Annual Report *12229*
Institute of Cost and Management Accountants of Pakistan: Professional Information Bulletin *09366*
Institute of Cultural Affairs: Highlights: Edges *15315*
Institute of Early American History and Culture: Newsletter *15316*
Institute of Economic Affairs: Occasional Papers *12233*
Institute of Engineers and Technicians: Journal *12236*
Institute of Environmental Sciences: Newsletter *15319*
Institute of Financial Education: Membership Newsletter *15320*
Institute of Food Research: Newsletter *12238*
Institute of Food Research: Report *12238*
Institute of Food Technologists: Membership Directory *15321*
Institute of General Semantics: Newsletter *15323*
Institute of Hospital Engineering: Institute Journal *12243*
Institute of Industrial Engineers: Transactions *15325*
Institute of Information Scientists: Inform *12246*
Institute of Jewish Affairs: Research Reports *12249*
Institute of Judicial Administration: Report *15327*
Institute of Marine Engineers: Bulletin *12255*
Institute of Materials Management: Members' Reference Book & Buyers' Guide *12257*
Institute of Mathematics and its Applications: Bulletin *12258*
Institute of Measurement and Control: Transactions *12259*
Institute of Nuclear Materials Management: Proceedings of Annual Meeting *15336*
Institute of Outdoor Drama: Newsletter *15337*
Institute of Packaging: Panorama *12265*
Institute of Physics, Singapore: Bulletin *10048*
Institute of Population Registration: Annual Report *12269*
Institute of Population Registration: Journal *12269*
Institute of Population Registration: Newsletter *12269*
Institute of Psychosexual Medicine Journal *12273*
Institute of Refrigeration: Proceedings *12277*
Institute of Transportation Engineers: Directory *15342*
Institute of Transportation Engineers: Journal *15342*
Institute of Wood Science: Journal *12291*
Institut für Angewandte Geodäsie: Mitteilungen *05947*
Institut für Wissenschaft und Kunst: Mitteilungen *00661*
Institut International de Recherches Betteravières: Proceedings *01387*
Institut International du Fer et de l'Acier: Bulletin *01389*
Institut International du Fer et de l'Acier: Conference Proceedings *01389*
Institut International du Froid: Bulletin *04575*
Institution of Electrical Engineers: Journal *09368*
Institution of Electrical Engineers: Newsletter *09368*
Institution of Electronics: Proceedings *12297*
Institution of Engineers amd Shipbuilders in Scotland: Transactions *12300*
Institution of Engineers – Australia: Conference Volumes *00421*
Institution of Engineers of Ireland: Monthly Journal *06968*
Institution of Engineers of Ireland: Transactions *06968*
Institution of Fire Engineers: Quarterly *12303*
Institution of Mechanical Engineers: Proceedings *12307*
Institution of Mining and Metallurgy: Transactions *12309*
Institution of Polish Engineers in Great Britain: Bulletin *12314*
Institution of Post Office Electrical Engineers: Journal *12315*
Institution of Railway Signal Engineers: Proceedings *12316*
Institution of Water and Environment Management: Newsletter *12318*
Institution of Water and Environment Management: Year Book *12318*
Institut National d'Archéologie et d'Art: Notes et Documents *11142*
Institut National Genevois: Bulletin *10730*
Institut National Genevois: Mémoires *10730*
Instituto Bonaerense de Numismática y Antigüedades: Boletín *00142*
Instituto de Botânica: Monographs *01666*
Instituto de Chile: Boletín *03053*
Instituto de España: Anuario *10318*
Instituto de Ingenieros de Chile: Anales *03055*
Instituto de Literatura: Boletín *00143*
Instituto de Literatura: Investigaciones *00143*
Instituto de Nutrición de Centro América y Panamá: Report *06570*
Instituto de Salud Publica de Chile: Boletín *03057*
Instituto Histórico e Geográfico Brasileiro: Revista *01675*
Instituto Histórico e Geográfico de São Paulo: Revista *01679*
Instituto Histórico e Geográfico do Maranhão: Revista *01681*
Instituto Histórico e Geográfico do Rio Grande do Sul: Revista *01684*
Institut Romand de Recherches et de Documentation Pédagogiques: Report *10731*
Institut Royal Belge du Pétrole: Annales/Annalen *01391*
Instituut voor Mechanisatie, Arbeid en Gebouwen: Annual Report *08912*
In-Stride *13972*
Instructional Innovator Magazine *14290*
Instrument Engineer's Yearbook *12259*
Insurance Counsel Journal *15404*
Insurance Facts *15346*
Insurance Review *15346*
INTECOL Newsletter *15380*
Integra *15582*
Integrated Manufacturing *16235*
Intelligence Report *13617*
Intelligent Buildings Institute: Directory *15347*
The Intelligent Enterprise *11267*
Intelligent Systems Engineering *12296*
Interact *10759*
Interacting with Computers *11547*
Interaction *14326*
Interactions *16638*
Inter-African Phytosanitary Bulletin *09244*
Inter-American Arbitration *15351*
Inter-American Association of Agricultural Librarians and Documentalists: Boletín Especial *03449*
Inter-American Association of Sanitary Engineering and Environmental Sciences: Newsletter *15348*
Inter-American Bar Association: Membership Directory *15349*
Inter-American Bar Association: Quarterly Newsletter *15349*
Inter-American College Association: Progress Bulletin *15350*
Inter-American Statistical Institute: Newsletter *09420*
Inter-American Tropical Tuna Comission: Annual Report *15355*
Inter-American Tropical Tuna Comission: Bulletin *15355*
Inter-American Tropical Tuna Comission: Special Report *15355*
Interauteurs *04438*
Interchange *16340*
Inter City Cost of Living Index *13687*
The Intercom *13294*
Intercom *14475, 14748, 16340*
Intercom Newsletter *16584*
Intercompany Announcements *16251*
Intercon *12319*
Interdisciplinaria *10491*
Intereconomics: Review of International Trade and Development *05935*
Interface *01875, 08713, 13288, 14505, 16454*
Interfaces *15330, 16253*
Interfaith Forum on Religion, Art and Architecture: Newsletter *15359*
Interior Design Educators Council: Record *15360*
Interlinguistika Informa Servo *05834*
Interlink *09490*
Interloc *13941*
Intermed *06650*
Internacia Jura Revuo *12350*
Internal Medicine *08243*
Internationaal Instituut voor Sociale Geschiedenis (Stichting): Annual Report *08916*
International Academy of Behavioral Medicine: Newsletter *15363*
International Academy of Chest Physicians and Surgeons: Journal *15364*
International Academy of Chest Physicians and Surgeons: Member Bulletin *15364*
International Academy of Chest Physicians and Surgeons: Membership Directory *15364*
International Academy of Nutrition and Preventive Medicine: Conference Proceedings *15367*
International Academy of Nutrition and Preventive Medicine: Directory *15367*
International Academy of Nutrition and Preventive Medicine: Journal *15367*
International Academy of Nutrition and Preventive Medicine: Professional Newsletter *15367*
International Aerospace Abstracts *13872*
International Affairs *12813*
International African Migratory Locust Organisation: Annual Report *08656*
International African Migratory Locust Organisation: Monthly Information Bulletin *08656*
International Airworthiness News *12353*
International and Comparative Law *11583*
International Association for Aquatic Animal Medicine: Newsletter *15375*
International Association for Comparative Research on Leukemia and Related Diseases: Symposium Proceedings *15376*
International Association for Computing in Education: Handbook and Directory *15378*
International Association for Computing in Education: Journal *15378*
International Association for Computing in Education: Newsletter *15378*
International Association for Cross-Cultural Psychology: Bulletin *02539*
International Association for Healthcare Security and Safety: Membership Directory *15383*
International Association for Healthcare Security and Safety: Newsletter *15383*
International Association for Hydraulic Research: Proceedings of Biennial Congresses *08918*
International Association for Media in Science: Newsletter *08919*
International Association for Personnel Women: Journal *15389*
International Association for Personnel Women: Membership Roster *15389*
International Association for Personnel Women: Newsletter *15389*
International Association for Research in Income and Wealth: Membership Directory *15391*
International Association for the Properties of Water and Steam: Releases and Guidelines *15394*
International Association Futuribles: Actualités prospectives *04579*
International Association of Agricultural Economists: Proceedings *15396*
International Association of Agricultural Information Specialists: Bulletin *04580*
International Association of Applied Psychology: Newsletter *08924*
International Association of Campus Law Enforcement Administrators: Conference Proceedings *15400*
International Association of Campus Law Enforcement Administrators: Membership Directory *15400*
International Association of Coroners and Medical Examiners: Newsletter *15402*
International Association of Counseling Services: Directory of Counseling Services *15403*
International Association of Defense Counsel: Committee Newsletter *15404*
International Association of Defense Counsel: Membership Directory *15404*
International Association of Educators for World Peace: Circulation Newsletter *15406*
International Association of Educators for World Peace: Directory *15406*
International Association of Forensic Toxicologists: Bulletin *15407*
International Association of Laryngectomees: Directory *15411*

International Association of Laryngectomees: News 15411
International Association of Sanskrit Studies: Newsletter 04582
International Association of School Librarianship: Annual Conference Proceedings 15417
International Association of School Librarianship: Newsletter 15417
International Association of Schools of Social Work: Newsletter 00662
International Association of Teachers of English as a Foreign Language: Newsletter 12333
International Association of University Professors of English: Bulletin 12335
International Bar News 12338
International Bibliography of the Social Sciences: Sociology, Political Science, Economics, Social and Cultural Anthropology 04409
International Biodeterioration 11441
International Bird Rescue Newsletter 15421
International Building Services Abstracts 11716
International Business Lawyer 12338
International Center for Research on Women: Series of Occasional Papers 15430
International Center of Medieval Art: Newsletter 15432
International Centre for Advanced Technical and Vocational Training: Bulletin 07709
International Centre for Theoretical Physics: Annual Report of Scientific Activities 07711
International Chemical Engineering 13878
International Childbirth Education Association: Membership Directory 15434
International Child Health 04261
International Chiropractors Association: Membership Directory 15435
International College of Dentists: Annual News Letter 15438
International College of Dentists: Roster 15438
International Commission for Plant-Bee Relationships: Reports of Meetings 04584
International Commission for the Prevention of Alcoholism and Drug Dependency: Quarterly Bulletin 15440
International Commission of Sugar Technology: Proceedings of General Assemblies 05986
International Commission on Occupational Health: Newsletter 10050
International Communications Association: Membership Directory 15444
International Communications Association: Newsletter 15444
International Communicator 13289
International Conference of Building Officials: Membership Roster 15445
International Conference of Building Officials: Newsletter 15445
International Conference on Mechanics in Medicine and Biology: Digest 15446
International Conference on Mechanics in Medicine and Biology: Directory of Conference Participants 15446
International Congress of Carboniferous Stratigraphy and Geology: Congress Proceedings 10329
International Congress of Oral Implantologists: Bulletin 15447
International Congress of Oral Implantologists: Membership Directory 15447
International Congress on Fracture: Proceedings 08143
International Congress on High-Speed Photography and Photonics: Proceedings 15448
International Copper Association: Report 15449
International Copyright Society: Yearbook 05987
International Council for Health, Physical Education and Recreation: Congress Proceedings 15452
International Council for the Exploration of the Sea: Bulletin Statistique 03746
International Council for the Exploration of the Sea: Rapports et Procès-Verbaux des Réunions 03746
International Council of Kinetography: Conference Proceedings 12347
International Council of Library Association Executives: Newsletter 15456
International Council of Museums: Notiziario 07713
International Council of Psychologists: Directory 15458
International Council on Archives: Archivum: International Review on Archives 04585
International Council on Education for Teaching: Newsletter 15460
International Council on Education for Teaching: Proceedings 15460
International Dental Journal 12348
International Design & Bulletin 11034
International Development Program of Australian Universities and Colleges: Newsletter 00425
International Development Program of Australian Universities and Colleges: Report 00425
International Development Research Centre: Annual Report 02552
International Digest of Health Legislation 11033
International Distribution & Handling Review 12601
International Dostoevsky Society: Newsletter 15467
International Dredging Abstracts 11575
Internationale Alpenschutzkommission: Info-Bulletin 08570
Internationale Briefe / International Journal / Revue Internationale 05995
International Economic Association: Newsletter 04586
International Education Magazine 02084
Internationale Gesellschaft für Getreidewissenschaft und -technologie: Newsletter 00669
Internationale Gesellschaft zur Erforschung und Förderung der Blasmusik: Mitteilungsblatt 00672
International Electronics Packaging Society: News 15469
International Electronics Packaging Society: Newsletter 15469
International Epidemiological Association: Membership Directory 15471
Die Internationale Politik 05326
Internationale Revue für soziale Sicherheit 10627
International Erosion Control Association: Membership Directory 15472
International Erosion Control Association: Newsletter 15472
Internationale Schönberg-Gesellschaft: Mitteilungen 00678
Internationales Dokumentations- und Studienzentrum für Jugendkonflikte: Cahier 05999
Internationales Institut für Öffentliche Finanzen: Proceedings 06000
Internationale Spectator 09053
Internationales Verkehrswesen 05623
Internationale Tagung der Historiker der Arbeiterbewegung: Auswahl der Konferenzbeiträge 00684
Internationale Tagung der Historiker der Arbeiterbewegung: Mitteilungsblatt 00684
Internationale Vereinigung gegen den Lärm: Informationsbulletin 10746
International Exchange of Information on Current Criminological Research Projects in Member States 04402
International Federation for Family Life Promotion: Message to Members 15473
International Federation for Family Life Promotion: Newsletter/Bulletin 15473
International Federation for Housing and Planning: Congress Papers Proceedings 08932
International Federation for Medical and Biological Engineering: Journal 02553
International Federation of Automatic Control: Newsletter 00686
International Federation of Free Teachers' Unions: Information Bulletin 08933
International Federation of Free Teachers' Unions: Newsletter 08933
International Federation of Multiple Sclerosis Societies: Annual Report 12357
International Federation of Multiple Sclerosis Societies: Update 12357
International Federation of Ophthalmological Societies: Proceedings 08935
International Fertilizer Correspondent/Corresponsal Internacional Agrícola 10739
International Forum of Information and Documentation 08895
International Gas Technology Highlights 15322
International Gas Turbine and Aeroengine Technology Report 14254
International Geographical Union: Bulletins 06007
International Graphic Arts Education Association: Membership Directory 15481
International Health Evaluation Association: Newsletter 15483
International Health Society: Bulletin 15484
International Health Society: Directory 15484
International Hospital Federation: Membership List 15485
International Humanist 08936
International Human Powered Vehicle Association: Newsletter 15486
International Huntington Association: Newsletter 08937
International Hydrographic Bulletin 08797
International Hydrographic Review 08797
International IAESTE Annual Report 14308
International Imagery Bulletin 15488
International Institute for Bioenergetic Analysis: Membership Directory 15489
International Institute of Public Finance: Papers and Proceedings 06008
International Insurance Society Yearbook 15491
International Joseph Diseases Foundation: Newsletter 15494
International Journal for Housing Science 15384
International Journal for Hybrid Microelectronics 15533
International Journal for Parasitology 00345, 00345
International Journal for Veterinary Homeopathy 07704
International Journal of Advertising 11205
International Journal of Africana Studies 13345
International Journal of Andrology 07834
International Journal of Anthropology 01270
International Journal of Biometeorology 02559
International Journal of Cancer 10732, 10993
International Journal of Childbirth Education 15434
International Journal of Chronobiology 15528
International Journal of Climatology 12832
International Journal of Clinical and Experimental Hypnosis 07684, 16508
International Journal of Cosmetic Science 13039
International Journal of Epidemiology 15471
International Journal of Experimental and Analytical Modal Analysis 16523
International Journal of Food Microbiology 04589
International Journal of Food Science and Technology 12239
International Journal of Group Psychotherapy 13836
International Journal of Group Tensions 15509
International Journal of Gynecology an Obstetrics 12356
International Journal of Hydrogen Energy 15385
International Journal of Japanese Sociology 08370
International Journal of Middle East Studies 15735
International Journal of Orthodontics 13415, 14092
International Journal of Pediatric Otorhinolaryngology 10159
International Journal of Physical Education 15452
International Journal of Project Management 10754
International Journal of Psycho-Analysis 11635, 12272
International Journal of Radiation Biology 01459
The International Journal of Radiation Oncology, Biology and Physics 14091
International Journal of Refrigeration 04575
International Journal of Rehabilitation Research 16406
International Journal of Remote Sensing 12749
International Journal of Science and Technology 14468
International Journal of Social Economics 12370
The International Journal of Social Psychiatry 11522
International Journal of Systematic Bacteriology 04589, 14073
International Journal of The Medical Defence Union 12503
International Journal of University Adult Education 02546
International Law Association: Conference Reports 12373
International Lawyer 13617
International Legal Practioner 12338
International Livestock Centre for Africa: Annual Report 03945
International Livestock Centre for Africa: Bulletin 03945
International Magnesium Association: Proceedings 15498
International Materials Review 12262
International Materials Reviews 14255
International Mathematical News 00867
International Metallographic Society: Directory 15501
The International Microform Journal of Legal Medicine and Forensic Sciences 15740
International Migration Review 14689
International Numismatic Commission: Compte rendu 12379
International Numismatic Newsletter 12379
International Numismatic Society Authentication Bureau: Newsletter 15505
International Omega Association: Proceedings 15506
International Organisation for the Study of the Endurance of Wire Ropes: Bulletin 08938
International Organization for Mycoplasmology: Congress Abstracts 15507
International Organization for Mycoplasmology: Congress Proceedings 15507
International Organization for Mycoplasmology: Newsletter 15507
International Organization for Septuagint and Cognate Studies: Bulletin 02555
International Orthopaedics 01462
International Packaging Abstracts 12701
International Pathology 15368
International Peat Journal 03969
International Personnel Management Association: Membership Directory 15511
International Pharmaceutical Abstracts 14132
International Political Science Abstracts 04409
International Poplar Commission: Session Report 07717
International Precious Metals Institute: Membership Directory 15515
International Professional Surrogates Association: Newsletter 15516
International Psychoanalytical Association: Bulletin 12385
International Psychologist 15458
International Public Relations Association: Members Manual 10755
International Red Locust Control Organisation for Central and Southern Africa: Annual Report 17153
International Red Locust Control Organisation for Central and Southern Africa: Quarterly Report 17153
International Reference Organization in Forensic Medicine and Sciences: Conference Proceeding 15519
International Reference Organization in Forensic Medicine and Sciences: Letter 15519
International Rehabilitation Review 16406
International Relations 08339
International Rescuer 15520
International Research Council on Biokinetics of Impacts: Proceedings 04587
International Research Group on Wood Preservation: Annual Report 10476
International Review of Administrative Sciences 01388
International Review of Applied Psychology 08924
International Review of Chiropractic 15435
International Review of Contemporary Law 01139
International Review of Psycho-Analysis 12272
International Review of Social History 08916
International Review of Trachoma 04614
International Rhinologic Society: Journal 01402
International Rubber Digest 12386
International Rubber Study Group: Proceedings of Annual Assembly 12386
International Schools Services: NewsLinks 15521
International Seaweed Association: Proceedings 01696
International Sharing 15490
International Social Security Review 10627
International Social Work 00662, 09270
International Society for Applied Ethology: Newsletter 12388
International Society for Astrological Research: Newsletter 15523
International Society for British Genealogy and Family History: Newsletter 15524
International Society for Labor Law and Social Security – U.S. National Branch: Proceedings of the Congress 15535
International Society for Philosophical Enquiry: Membership Roster 15538
International Society for Plant Molecular Biology: Directory of Members 15539
International Society for Research on Aggression: Bulletin 15540
International Society for Rock Mechanics: Journal 09662
International Society for Terrain-Vehicle Systems: Newsletter 15541
International Society for the Systems Sciences: Proceedings of Annual Meetings 15544
International Society of Air Safety Investigators: Membership Roster 15545
International Society of Arboriculture: Membership Roster 15546
International Society of Arboriculture: Yearbook 15546
International Society of Chemical Ecology: Proceedings of the Annual Meeting 15547
International Society of City and Regional Planners: The News Bulletin 08942
International Society of Differentiation: Conference Proceedings 15551
International Society of Differentiation: Newsletter 15551
International Society of Electrophysiological Kinesiology: Society Newsletter 02560
International Society of Hematology: Newsletter 15552
International Society of Soil Science: Bulletin 00689
International Sociology 10330
International Statistical Information 04278
International Statistical Review 04278, 08944
International Studies Newsletter 15558
International Studies Notes 15558
International Studies Quarterly 15558
International Surgery 15439
International Symposium Technical Proceedings 15533
International Test and Evaluation Association: Symposia Proceedings 15563
International Textile Calendar 13120
International Tsunami Information Center: Newsletter 15568
International Tsunami Information Center: Report 15568
International Turfgrass Society: Proceedings of Conference 15569
International Understanding at School 04143
International Union of Air Pollution Prevention and Environmental Protection Associations: Handbook 12400
International Union of Air Pollution Prevention and Environmental Protection Associations: Newsletter 12400
International Union of Pure and Applied Physics: News-Bulletin 02563
International Update 13524
International Update Newsletter 14818
International VAT Monitor 08926
International Veterinary Acupuncture Society: Newsletter 15570
International Wildlife 16180
International Wizard of Oz Club: Membership Directory 15574
International Women's Anthropology Conference: Bulletin 15575
International Work Group for Indigenous Affairs: Documents 03750
International Work Group for Indigenous Affairs: Newsletter 03750
International Work Group for Indigenous Affairs: Yearbook 03750
International Yearbook of Educational and Training Technology 11263
International Yearbook of Organization Studies 12030
International Yearbook on Teacher Education 15460
International Zoo Yearbook 13237
Internationella studier 10603
The Internist: Health Policy in Practice 14136
Internist's Intercom 14136
Internkontakt 09293
Inter NOS 15911
Interpretative Catalogue 14821
Interpro 13755
Inter-Society Color Council: Newsletter 15578
Interstate Conference on Water Policy: Annual Report 15580
Interstate Conference on Water Policy: Membership Directory 15580
Interstate Conference on Water Policy: News-In-Brief 15580
Interstate Conference on Water Policy: Proceedings 15580
Intertel: Membership List 15582
Inter-University Consortium for Political and Social Research: Annual Report 15583
Inter-University Consortium for Political and Social Research: Bulletin 15583
Inter-University Council for East Africa: Newsletter 11186
Inter-University Council for East Africa: Report 11186
The Interval 16645
In the Field of Building 07114
In Touch for Health 16786
Intractable Pain Society of Great Britain and Ireland: Forum 12409
Inventaire des Bibliothèques et Centres de Documentation Belges 01202
Inventories of Natural Gas Liquids and Liquefied Refinery Gases 13995
Inverse Problems 12268
Investigaciones y Ensayos 00050
Investigación y Progreso Agropecuario: Carillanca 03056
Investigación y Progreso Agropecuario: La Platina 03056
Investigación y Progreso Agropecuario: Quilamapu 03056
Investigación y Técnica del Papel 10248
Investigative Ophthalmology and Vision Science 14330

Investigative Radiology *14530*
In Your Hands *16836*
Inzhenerno-Fizicheskii Zhurnal: Engineerring Physics Journal *01088*
Inzhnernaya Geologiya: Engineering Geology *09954*
IPAC Bulletin *02531*
IPAC Quarterly *02513*
IPA Facts *00418*
IPA Newsletter *12385*
IPA Review *00418*
Ipar-Gazdaság: Industry-Economy *06673*
IPA Roster *12385*
IPB Geneva News *10640*
IPMA News *15511*
IPM Practitioner *14587*
IPO Annual Progress Report *08914*
IPORO Internacia Pedagogia Revuo *07716*
IPRA Newsletter *10755*
IPRA Review *10755*
IPS Bulletin *07086*
IP Statistics Service *12267*
Iqbaliat *09370*
Iqbaliat Farsi *09370*
Iqbal Review *09370*
Iran *06898, 11586*
Iranian Studies *16544*
Iraq Academy: Bulletin *06917*
IRED Forum *10662*
IRG Documents *10476*
IRI Research Institute: Bulletin *15584*
Irish Architect *07026*
Irish Association of Curriculum Development: Compass *06974*
Irish Birds *07006*
Irish Chemical News *06961*
Irish Chemical Newsletter *06961*
Irish Family Planning Association: Annual Report *06983*
Irish Family Planning Association: Newsletter *06983*
Irish Forestry *07033*
Irish Geography *06953*
Irish Georgian Society: Bulletin *06986*
Irish Journal of Earth Sciences *07028*
Irish Journal of Medical Science *07021*
Irish Medical Journal *06992*
Irish Motor Industry *07034*
The Irish Pharmacy Journal *07019*
The Irish Social Worker *06976*
Irish Studies in International Affairs *07028*
The Irish Sword *07014*
Irish Veterinary Journal *07004*
Irish Wildbird Conservancy News *07006*
The Iris Yearbook *11596*
Irodalomtörténet: Literary History *06638*
Iron and Steel Engineer *14449*
Ironic Blood: Newsletter *15586*
Ironmaking and Steelmaking *12256, 12262*
IRPA-Bulletin *08939*
Irrigation and Drainage Abstracts *11720, 11728*
Irrigation/Drainage Journal *14107*
L'Irrigazione *07478*
ISA Bulletin *10330, 10620*
ISAM Newsletter *15298*
I.S.A. Newsletter *07097*
ISCE Newsletter *15547*
ISCLT Newsletter *15529*
The ISC Newsletter *15548*
ISDS Newsletter *06011*
ISES News *07722*
ISG News *13096*
ISI Annual Report *06822*
ISI Bulletin *06822*
ISIG *07727*
ISI Handbook *06822*
ISIJ International *08392*
ISIM-Bulletin *10757*
ISIM Bulletin *10977*
Isis *15222*
Isis Critical Bibliography *15222*
Iskos *04051*
Island Naturalist *02674*
Islenzka bókmennatafélag: Annual Journal *06691*
Isle of Man Natural History and Antiquarian Society: Proceedings *12412*
Isle of Wight Birds *12413*
Isle of Wight Natural History and Archaeological Society: Proceedings *12413*
ISO 9000 News *10753*
ISO Bulletin *10753*
Isotope News *08186*
Israel Exploration Journal *07071*
Israel Geological Society: Abstracts of Annual Meeting *07073*
Israel Society for Microbiology: Newsletter *07094*
Israel Society of Special Libraries and Information Centres: Bulletin *07104*
ISSC Newsletter *04459*
ISS Directory of Overseas Schools *15521*
Issledovanie Zemli iz Kosmosa: Investigation of the Earth from Space *09954*
ISSMFE News *12390*
Issues in Health Care Technology *15012*
Istanbuler Mitteilungen *05568*
ISTA News Bulletin *10745*
Istituto Internazionale per l'Unificazione del Diritto Privato: NEWS Bulletin *07734*
Istituto Italiano degli Attuari: Giornale *07735*
Istituto Italiano di Paleontologia Umana: Memorie *07738*
Istituto per gli Studi di Politica Internazionale: Quaderni-Papers *07750*
Istituto per la Cooperazione Economica Internazionale e i Problemi di Sviluppo: Booklets and Documents *07751*
Istorija: History *08609*
Istoriki kai Ethnologiki Etaireia tis Ellados: Bulletin *06537*
Istoriko-Filologicheskii Zhurnal: Historical and Philological Journal *00203*
Istoriski glasnik *17084*
L'Italia e l'Europa: Rassegna di Diritto, Economia, Politica, Società *07500*
L'Italia Forestale e Montana *07194*
Italian American Librarians Caucus: Bulletin *15587*
Italian-American Newsletter *15588*
Italian-American Review *15588*
Italia Nostra *07757*
Italian Politics *14827*
Italia Ornitologica: Organo Ufficiale della Federazione Ornicoltori Italiani *07996*
Italica *13589*
It Beaken *08899*
Items *16485*
ITI Bulletin *12288*
IT in Nursing *11547*
IT Link *12246*
IT Link incorporating Automation Notes *11267*
It Novine *17134*
ITS-publikaties *08915*
ITS Today *15565*
IUCN Bulletin *10759*
IUFost Newsletter *06970*
IUFRO-NEWS *00677*
IUNS Directory *12403*
IUNS Newsletter *12403*
Iustitia *08046*
IVA-Newsletter *10469*
IVA-Nytt *10469*
IWAC Newsletter *15575*
iwd: Informationsdienst des Instituts der deutschen Wirtschaft *05943*
IWG-Berichte *05977*
IWG-Impulse *05977*
IWG-Mitteilungen *05977*
I.W.K.: Internationale wissenschaftliche Korrespondenz zur Geschichte der deutschen Arbeiterbewegung *05917*
IWL-Forum *05962*
IWL-Umweltbrief *05962*
IWRB News *12405*
IYYUN: The Jerusalem Philosophical Quarterly *07106*
Izvestiya AN Uzbekistana-Tekhnicheskie Nauki: Journal of the Uzbek Academy of Sciences: Engineering Sciences *16948*
Izvestiya Rossiiskoi Akademii Nauk: Bulletin of the Russian Academy of Sciences *09954*
Izvestiya Sibirskogo Otdeleniya Rossiiskoi Akademii Nauk: Bulletin of the Siberian Branch of the Russian Academy of Sciences *09954*

Jaarboek van de Maatschappij der Nederlandse Letterkunde te Leiden *08974*
JACT Bulletin *12422*
JACT Review *12422*
Jadeed Science *09399*
JAFTA Journal *10425*
Jagon *10349*
Jagoo aur Jagao *06768*
Jahrbuch Arbeit, Bildung, Kultur *05728*
Jahrbuch der Deutschen Akademie *05289*
Jahrbuch der Deutschen Bibliotheken *06373*
Jahrbuch der Deutschen Schiller-Gesellschaft *05571*
Jahrbuch der österreichischen Byzantinistik *00723*
Jahrbuch der Österreichischen Pädagogischen Gesellschaft *00876*
Jahrbuch der Schiffbautechnischen Gesellschaft *06275*
Jahrbuch der Schweizerischen Akademie der Naturwissenschaften *10788*
Jahrbuch der Schweizerischen Verkehrswirtschaft *10958*
Jahrbuch des Historischen Vereins für das Fürstentum Liechtenstein *08569*
Jahrbuch des Vereins für Geschichte der Stadt Wien *00984*
Jahrbuch des Vorarlberger Landesmuseumsvereins *01008*
Jahrbuch des Wiener Goethe-Vereins *01017*
Jahrbuch Dritte Welt: Daten, Übersichten, Analysen *05604*
Jahrbuch für Anthropologie und Religionsgeschichte *05726*
Jahrbuch für die Geschichte des Protestantismus Oesterreich *00617*
Jahrbuch für Exlibriskunst und Graphik *05302*
Jahrbuch für Landeskunde von Niederösterreich *00985*
Jahrbuch für Naturschutz und Landschaftspflege *05081*
Jahrbuch für österreichische Kulturgeschichte *00627*
Jahrbuch für Volkskunde *05896*
Jahrbuch für Westfälische Kirchengeschichte *06393*
Jahrbuch für Zeitgeschichte *00851*
Jahrbuch SGUF, Annuaire SSPA, Annuario SSPA *10853*
Jahresbericht der Gesellschaft Pro Vindonissa *10718*
Jahresberichte über die Tätigkeit des Stifterverbands *06295*
Jahresgabe der Johann-Joseph-Fux-Gesellschaft *00691*
JAMA: The Journal of the American Medical Association *13933*
Jamaica Library Association Annual Report *08083*
Jamaica Library Association Bulletin *08083*
Jamaican Geographical Society: Newsletter *08086*
James Joyce Journal *15592*
Jane Austen Society: Annual Report *12415*
Janus *04585*
Japanese Circulation Journal *08215*
Japanese Journalism Review *08377*
Japanese Journal of Allergology *08187*
Japanese Journal of Applied Entomology and Zoology *08355*
Japanese Journal of Applied Physics *08284*
Japanese Journal of Bacteriology, Microbiology and Immunology *08361*
Japanese Journal of Breeding *08210*
Japanese Journal of Cancer Research *08299*
Japanese Journal of Clinical Oncology *08184*
The Japanese Journal of Clinical Pathology *08250*
Japanese Journal of Crop Science *08252*
Japanese Journal of Ecology *08365*
The Japanese Journal of Educational Psychology *08239*
Japanese Journal of Educational Research *08237*
Japanese Journal of Genetics *08209*
Japanese Journal of Geriatrics *08251*
The Japanese Journal of Human Genetics *08316*
Japanese Journal of Hygiene *08197*
Japanese Journal of Industrial Health *08253*
Japanese Journal of Mathematics *08386*
Japanese Journal of Medical Mycology *08213*
Japanese Journal of Nuclear Medicine *08218*
Japanese Journal of Oral Surgery *08229*
Japanese Journal of Ornithology *08291*
Japanese Journal of Parasitology *08334*
Japanese Journal of Pharmacognosy *08389*
The Japanese Journal of Pharmacology *08273*
Japanese Journal of Psychology *08262*
Japanese Journal of Public Health *08343*
Japanese Journal of Tropical Medicine and Hygiene *08244*
The Japanese Journal of Urology *08305*
Japanese Journal of Zootechnical Science *08287*
Japanese Literature *08282*
Japanese Literature Today *08247*
Japanese Psychological Research *08262*
Japanese Scientific Monthly *08200*
Japan Oil Statistics Today *08412*
Japan Tappi *08153*
Jardins de France *04904*
Jarliboro *15428*
Járművek, Epitöipari és Mezógazdasági Gépek: Vehicles, Building and Agricultural Machines *06606*
Jazz Educators Journal *15409*
Jazzforschung: Jazz Research *00671*
J.B. Danquah Memorial Lectures *06482*
Jean Piaget Society: Symposium Proceedings *15596*
Jefferson Center for Character Education: Research Letter *15597*
Jef-Jel *10012*
Jefo Informa *04612*
Je Me Souviens *13826*
Jeofizik: Geophysics *11167*
Jeoloji Mühendisliği: Geological Engineering *11167*
Jersey at Home *12826*
Jessup Competition Compendium *16310*
Jesuit Secondary Education Association: News Bulletin *15601*
Jewish Academy of Arts and Sciences: Bulletin *15603*
Jewish Education *14878*
Jewish Education Directory *15605*
Jewish Educators Assembly: Yearbook *15606*
Jewish Historical Society of England: Transactions *12418*
Jewish Librarians Task Force: Newsletter *15608*
The Jewish Pharmacist *15609*
Jewish Science Interpreter *16675*
Jewish Studies Magazine *07119*
The Jews of Czechoslovakia *16596*
Jezik in Slovstvo *10104*
Jezyk Polski *09624*
JJAP *08402*
JLA News *08083*
JMC Administrator *14497*
Job Exchange *14291*
Jobline *14068*
Job Market Update *14206*
JobMart *14007*
Job Placement Bulletin *14775*
Jobs for Philosophers *13998*
Jogásnapló *06640*
Jogász Szövetségi Értekezések *06640*
Johann Strauß-Gesellschaft: Mitteilungen *00692*
John Burroughs Association: Bulletin *15611*
John Dewey Lectures *15612*
John Dewey Society: Current Issues *15612*
Johnson Society: Transactions *12420*
Joho Shori *08140*
Joho Shori Gakkai Ronbunshi: Transactions of the Information Processing Society of Japan *08140*
Joint Center for Political and Economic Studies: Focus *15615*
Joint Commission on Accreditation of Healthcare Organizations: Joint Commission Perspectives *15616*
Joint Liturgical Studies *11220*
Jojas de Arroz *03360*
The Jordanian National Bibliography *08452*
Jord og Viden *03605*
Jornal da Associação Paulista de Medicina *01606*
Jornal da Febab *01649*
Jornal de ABES *01588*
The Josephine Butler Society: News and Views *12423*
Jottings *12090*
Journal A.: Automatic Control *08963*
Journal A.: Benelux Quarterly Journal on Automatic Control *01380*
Journal and Transactions *13045*
Journal and Volume *11964*
Journal Arctic *01871*
Journal Arquitectos *09653*
Journal Asiatique *04631*
Journal Burns *15525*
Journal da Associação Médica Brasileira *01603*
Journal de Chimie Physique *04791*
Journal de Chimie physique et de physico-chimie biologique *04466*
Journal de Gynécologie, Obstétrique et Biologie de la Réproduction *04513*
Journal de Gynécologie Obstétrique et Biologie de la Reproduction *04770*
Journal de la Société de Statistique de Paris *04719*
Journal de la Société Finno-Ougrienne *04003*
Journal de Médecine Esthétique et de Chirurgie Dermatologique *04810*
Journal de Microscopie et de Spectroscopie Electronique *04818*
Journal de Mycologie Médicale *04821*
Journal de Physique *04836*
Journal de Recherche Océanographique *04974*
Journal des Africanistes *04691*
Journal des Diabétiques *10801*
Journal des Savants *04095*
Journal d'Ingénieur *01880*
Journal du Conseil *03746*
Journal d'Urologie *04865*
The Journal (E): English Edition *08353*
Journal Endocrinological Investigation *07868*
Journal ESA *04127*
Journal Européen de Radiothérapie *04844*
Journal for Continuing Medical Education *10151*
Journal for Inter-Regional Cooperation and Development *14344*
Journal for Legal Philosophy and Jurisprudence *09106*
Journal for Social Work Education in Africa *03941*
Journal for Specialists in Group Work *13771*
Journal for the Scientific Study of Religion *16610*
Journal for World Education *03561*
Journal für Ornithologie *05481*
Journal für Sozialforschung *00585, 00588*
Journal – Information Documentation *06909*
Journalism Association of Community Colleges: Directory *15627*
Journalism Association of Community Colleges: Newsletter *15627*
Journalism Educator *14292, 14497*
Journalism Monographs *14292*
Journal of Abnormal Psychology *14019*
Journal of Addictive Diseases *14096*
Journal of Adolescence *11276*
Journal of Adolescent Health Care *16490*
Journal of Advanced Transportation *13333*
Journal of Advertising *13380*
Journal of Advertising Research *13334*
Journal of Aesthetics and Art Criticism *01863*
Journal of AFES *06875*
Journal of African Meldicinal Plants *09244*
Journal of Agricultural and Food Chemistry *13688*
Journal of Agricultural Economics *11214*
Journal of Agronomic Education *14098, 14942*
Journal of Aircraft *13872*
Journal of Allied Health *14496*
Journal of Alloy Phase Diagrams *06796*
Journal of American College Health *13696*
Journal of American Ethnic History *15249*
Journal of American Folklore *13817*
Journal of American Studies *11485*
Journal of Anatomy *11227*
Journal of Andrology *07834*
Journal of Anesthesia *08241*
Journal of Animal Ecology *11554*
Journal of Animal Science *14100*
Journal of Antimicrobial Chemotherapy *11650*
Journal of AOA *13980*
Journal of Apicultural Research *12339*
Journal of Applied Arts, Science and Education *15792*
Journal of Applied Bacteriology *12964*
The Journal of Applied Behavioral Science *16234*
Journal of Applied Biomaterials *16502*
Journal of Applied Communications *13350*
Journal of Applied Cosmetology *07535*
Journal of Applied Crystallography *12401*
Journal of Applied Ecology *11554*
Journal of Applied Meteorology *13943*
Journal of Applied Physics: Microfiche Edition *13893*
Journal of Applied Physiology *14005*
Journal of Applied Psychology *14019*
Journal of Aquatic Animal Health *13816*
Journal of Aquatic Plant Management *14234*
Journal of Arboriculture *15546*
Journal of Architectural Education *14416*
Journal of Art & Design Education: NSEAD Newsletter *12608*
Journal of ASA *14182*
Journal of Asian Ecology *03238*
Journal of Asian Studies *14268*
Journal of Assam Research Society *06831*
Journal of Astrophysics & Astronomy *06767*
Journal of Atmospheric and Oceanic Technology *13943*
Journal of Bacteriology *14073*
Journal of Baltic Studies *14341*
Journal of Biblical Literature *16642*
Journal of Biocommunications *15196*
Journal of Biological Chemistry *14047*
Journal of Biological Education *12215*
Journal of Biological Standardization *12328*
Journal of Biomedical Engineering *11443*
Journal of Biomedical Materials Research *16502*
Journal of Biosciences *06767*
Journal of Black Psycholoy *14402*
Journal of BMHS *01074*
Journal of Bone and Joint Surgery *10213*
Journal of British Music Therapy *11659*
Journal of British Studies *16215*
Journal of Broadcasting & Electronic Media *14617*
Journal of Bryology *11535*
Journal of Building Structures *03115*
Journal of Burn Care and Rehabilitation *13673*
The Journal of Business Communication *14275*
Journal of Business & Economic Statistics *14182*
Journal of Canadian Petroleum Technology *02180*
Journal of Cancer Education *08851*
Journal of Cancer Research and Clinical Oncology *05461*
Journal of Cardiopulmonary Rehabilitation *13460*
Journal of Cataract and Refractive Surgery *14104*
Journal of Cell Biology *14049*
Journal of Chemical and Engineering Data *13688*
Journal of Chemical Ecology *15547*
Journal of Chemical Information and Computer Sciences *13688*
Journal of Chemical Physics: Microfiche Edition *13893*
Journal of Chemical Technology and Biotechnology *13035, 13035*
Journal of Child Psychiatry and Human Development *13575*
Journal of Child Psychology *11322*

Journal of Chinese Forestry *03268*
Journal of Chinese Studies *08529, 13473*
Journal of Chiropractic *13690*
Journal of Civil Procedure *08181*
Journal of Classical Studies *08366*
Journal of Climate *13943*
Journal of Clinical Endocrinology and Metabolism *15041*
Journal of Clinical Investigation *14050*
Journal of Clinical Microbiology *14073*
Journal of Clinical Neurophysiology *13793*
Journal of Clinical Oncology *14109*
Journal of Clinical Orthodontics *10159*
Journal of Clinical Pathology *11325*
Journal of Clinical Pharmacology *13702*
Journal of Coatings Technology *15089*
Journal of College and University Law *15839*
Journal of College Radio *15357*
Journal of College Reading and Learning *14778*
Journal of College Science Teaching *16151*
The Journal of College Student Personnel *13771*
Journal of Computer-Based Instructions *14350*
The Journal of Computer Information Systems *15377*
Journal of Computing *09140*
Journal of Conchology *11835*
Journal of Consulting and Clinical Psychology *14019*
Journal of Consumer Interest *13765*
Journal of Continuing Higher Education *14286*
Journal of Correctional Education *14865*
Journal of Cost Analysis *16652*
The Journal of Counseling & Development *13771*
Journal of Counseling Psychology *14019*
Journal of Craniomandibular Disorders Head and Neck Pain *13414*
Journal of Cranio-Maxillo-Facial Surgery *11989*
Journal of Criminal Justice Education *13286*
Journal of Cross-Cultural Psychology *02539*
Journal of Dairy Science *13775*
Journal of Dairy Technology *13042*
Journal of Dental Education *13528*
Journal of Dental Practice Administration *13389*
Journal of Dental Research *13477, 15379*
Journal of Dentistry for Children *14118*
Journal of Dermatologic Surgery and Oncology *14057*
The Journal of Dermatology *08303*
Journal of Developmental Medicine and Child Neurology *13376*
Journal of Development Communication *08634*
Journal of Diagnostic Medical Sonography *16654*
Journal of Documentation *11256, 11267*
Journal of Early Intervention *14979*
Journal of Eastern Religions *08297*
Journal of Ecology *11554*
Journal of Economic Entomology *15046*
Journal of Economic Issues *14295*
Journal of Economic Literature *13789*
Journal of Economics *08438*
Journal of Educational Equity and Leadership *16846*
Journal of Educational Measurement *16023*
Journal of Educational Psychology *14019*
Journal of Educational Sociology *08238*
Journal of Educational Statistics *13790, 14182*
The Journal of Educational Techniques and Technologies *15386*
Journal of Educational Technology Systems *16495*
Journal of Educational Television *11946*
Journal of Education for Library and Information Science *14313*
Journal of Education for Social Work *14932*
Journal of Egyptian Archaeology *11948*
Journal of Electrical and Electronics Engineering *00422*
Journal of Electrical & Electronics Engineering in Australia *00421*
Journal of Electronic Defense *14473*
Journal of Electronic Materials *15317, 15744*
The Journal of Electronic Packaging *14145*
Journal of Electrophysiological Technology *11954*
Journal of Employment Counseling *13771*
Journal of Endocrinological Investigation *07834*
Journal of Endocrinology *12971*
Journal of Engineering Thermophysics *03188*
Journal of Environmental Health *16043*
Journal of Environmental Horticulture *15230*
Journal of Environmental Quality *14098, 14942, 16717*
Journal of Environmental Sciences *15319*
Journal of Ethnobiology *16659*
Journal of European Integration *02105*
Journal of Experimental Botany *12973*
Journal of Experimental Psychology: Animal Behavior Processes *14019*
Journal of Experimental Psychology: Human Perception and Performance *14019*
Journal of Experimental Psychology: Learning, Memory and Cognition *14019*
Journal of Experiment Psychology: General *14019*
Journal of Extension *15858*
Journal of Extra-Corporal Technology *14122*
Journal of Fermentation and Bioengineering *08425*
Journal of Ferrocement *11082*
Journal of Field Ornithology *14437*
Journal of Film and Video *16847*
Journal of Finance *13814*
Journal of Fish Biology *12077*
Journal of Food Distribution Research *15109*
Journal of Food Science *15321*
Journal of Foot Surgery *13707*
Journal of Forensic Sciences *13396*
Journal of Forestry *16634*
Journal of Gemmology *12101*
Journal of General Physiology *16669*
Journal of General Virology *12975*
Journal of Genetics *06767, 06767*
Journal of Geography *08437, 15988*
Journal of Geological Education *15861*
Journal of Geomagnetism and Geo-Electricity *08288*
Journal of Geophysical Research *13833*
The Journal of Gerontology *15158*
Journal of Glaciology *12366*
The Journal of Graphoanalysis *15482*
Journal of Great Lakes Research *15382*
Journal of Guidance, Control and Dynamics *13872*
Journal of Hand Surgery *11669, 14087*
Journal of Hard Materials *12268*
Journal of Health Administration Education *14529*
Journal of Health and Physical Education *08390*
Journal of Health and Social Behavior *14175*
Journal of Healthcare Education and Training *14064*
Journal of Heat Treating *14255*
The Journal of Hellenic Studies and Supplement Archaeological Reports *13005*
Journal of Hepatology *08863*
Journal of Heredity *13830*
Journal of Herpetology *16612*
The Journal of Histochemistry and Cytochemistry *15216*
Journal of Home Economics *13851*
Journal of Hospitality Education *14920*
Journal of Humanistic Education *14303*
The Journal of Humanistic Education and Development *13771*
Journal of Humanistic Education and Development *14304*
Journal of Humanistic Psychology *14305*
Journal of Human Nutrition and Dietetics *11552*
Journal of Hydraulic Research *08918*
Journal of Hydrology (New Zealand) *09152*
Journal of Hyperbaric Medicine *16803*
Journal of Imaging Technology *16672*
Journal of Immunology *13543*
Journal of Indian and Buddhist Studies *08211*
Journal of Indian Education *06843*
Journal of Indian Museums *06837*
Journal of Individual Psychology *13362*
Journal of Indo-European Studies *15306*
Journal of Industrial Microbiology *16538*
Journal of Industrial Teacher Education *15869*
Journal of Infectious Diseases *15262*
Journal of Information Processing and Management: Joho Kanri *08216*
Journal of Information Resources Management Systems *14456*
Journal of Information Science *12246*
Journal of Inherited Metabolic Disease *13016*
Journal of Inner Asian Studies *14734*
Journal of Instructional Development *14290*
Journal of Interior Design Education and Research *15360*
Journal of International Business Studies *13293*
Journal of International Law and Diplomacy *08172*
Journal of International Marketing and Marketing Research *11803*
Journal of International Rhinology *14031*
Journal of Interpretation *15806*
The Journal of Investigative Dermatology *16543*
Journal of Islamic Studies *12667*
The Journal of Laboratory and Clinical Medicine *14716*
Journal of Laser Applications *15647*
Journal of Law and Human Behavior *14020*
Journal of Law and Technology *15496*
Journal of Legal Education *14382*
Journal of Legal Studies Education *13295*
Journal of Leisure Research *16173*
Journal of Library and Information Science *14736*
Journal of Life Insurance Medicine *13401*
Journal of Light and Visual Environment *08424*
Journal of Lightware Technology: Microfiche Edition *13893*
Journal of Lightwave Technology *15317, 16255*
Journal of Linguistics *12453*
Journal of Mammalogy *14142*
Journal of Management in Practice *14456*
Journal of Marine and Petroleum Geology *12111*
Journal of Marketing *13926*
Journal of Marketing Management *11763*
Journal of Marketing Research *13926*
Journal of Marriage and the Family *16020*
Journal of Materials Engineering *14255*
Journal of Materials Shaping Technology *14255*
Journal of Mechanical Engineering *03176*
Journal of Medical and Veterinary Mycology *04589*
Journal of Medical Education *14383*
Journal of Medical Entomology *15046*
Journal of Medical Microbiology *12682*
Journal of Medical Research *06787*
Journal of Medicinal Chemistry *13688*
Journal of Mental Imagery *15488*
Journal of Metals *15744*
Journal of Meteorological Research *08147*
Journal of Micromechanics and Microengineering *12268*
Journal of Microscopy *12833*
Journal of Microwave Power *15502*
Journal of Mineralogy, Petrology and Economic Geology *08300*
Journal of Molecular Endocrinology *12971*
Journal of Molecular Sciences *03167*
The Journal of Molluscan Studies *12478*
Journal of Mormon History *15755*
Journal of Multicultural Counseling & Development Guidance *13771*
Journal of Museum Education Anthology *15760*
Journal of Music Scores *16650*
Journal of Natural Products *14153*
Journal of Navigation *12814*
Journal of Near-Death Studies *15388*
Journal of Negro History *14356*
Journal of Nematology *16684*
Journal of Nephrology *07957*
Journal of Neurology, Neurosurgery and Psychiatry *13031*
Journal of Neurology, Psychiatry and Neurosurgery *01760*
Journal of Neuropathology and Experimental Neurology *13556*
Journal of Neurophysiology *14005*
The Journal of Neuroscience *16555*
Journal of Neurosurgery *13555*
Journal of Neurosurgical Sciences *07910*
Journal of Nuclear Materials Management *15336*
Journal of Nuclear Science and Technology *08203*
Journal of Nurse-Midwifery *13720*
Journal of Nutrition *13890*
Journal of Nutritional Science and Vitaminology *08281*
Journal of Nutrition Education *16557*
Journal of Occupational and Organizational Psychology *11636*
Journal of Oceanic Engineering *15317*
Journal of Ocular Therapy and Surgery *15413*
Journal of Offender Counseling *13771*
Journal of Operative Dentistry *13398*
Journal of Optics *06848*
Journal of Optometric Education *14495*
Journal of Optometric Vision Development *14776*
Journal of Oral and Maxillofacial Surgery *13557*
Journal of Oral Implantology *13399*
Journal of Oral Medicine *13413*
Journal of Oral Pathology and Medicine *10159*
The Journal of Organic Chemistry *13688*
Journal of Orgonomy *13724*
The Journal of Orthomolecular Medicine *02246*
Journal of Orthomolecular Psychiatry and Medicine *13559*
Journal of Orthopaedic Medicine *11585*
Journal of Orthopedic Research *16271*
Journal of Orthotics and Prosthetics *13976*
Journal of Osteopathic Sports Medicine *13979*
Journal of Otolaryngology *02289*
Journal of Otolaryngology of Japan *08313*
Journal of Pakistan Studies *09397*
Journal of Paleontology *16287*
Journal of Palestine Studies *15287*
Journal of Palliative Care *02327*
Journal of Palynology *06849*
Journal of Paralegal Education and Practice *13488*
Journal of Parametrics *15553*
The Journal of Parasitology *14152*
Journal of Parenteral and Enteral Nutrition *14076*
Journal of Pathology *12682*
Journal of Pediatric Psychology *16562*
Journal of Periodontology *13422*
Journal of Personality and Social Psychology *14019*
Journal of Personality Assessment *16566*
Journal of Petroleum Technology *16689*
Journal of Pharmaceutical Sciences *13996*
Journal of Pharmacology and Experimental Therapeutics *14077*
Journal of Pharmacy and Pharmacology *12838*
The Journal of Photographic Science *12841*
Journal of Physical and Chemical Reference Data *13688*
Journal of Physical and Reference Data *13893*
The Journal of Physical Chemistry *13688*
Journal of Physical Education, Recreation & Dance *13449*
Journal of Physical Oceanography *13943*
Journal of Physics A: Mathematical and General *12268*
Journal of Physics B: Atomic, Molecular and Optical Physics *12268*
Journal of Physics C: Condensed Matter *12268*
Journal of Physics D: Applied Physics *12268*
Journal of Physics G: Nuclear Physics *12268*
Journal of Physics of the Earth *08448*
The Journal of Physiology *12698*
Journal of Planning Education and Research *14417*
Journal of Plant Research *08383*
Journal of Plastic and Reconstructive Surgery *14157*
The Journal of Poetry Therapy *15810*
Journal of Polynesian Society *09189*
Journal of Popular Culture *16336*
Journal of Practical Nursing *15811*
Journal of Presbyterian History *16351*
Journal of Production Agriculture *14098, 14942, 16717*
Journal of Professional Nursing *13520*
Journal of Propulsion and Power *13872*
Journal of Prosthetic Dentistry *13290, 13404, 13435, 14016*
Journal of Prosthetic Dentisty *13395*
Journal of Prosthetics and Orthotics: C.P.O. *13417*
Journal of Protective Coatings and Linings *16746*
The Journal of Psychohistory *15290*
Journal of Psychology and Christianity *14743*
Journal of Public Budgeting and Finance *13469*
Journal of Public Health Dentistry *13528, 13576*
Journal of Quality Technology *14085*
Journal of Quantum Electronics *15317*
Journal of Radiation Research *08308*
Journal of Radiological Protection *12268, 12991*
Journal of Range Management *16574*
Journal of Rational-Emotive Therapy *15292*
Journal of Reading *15518*
Journal of Reading Behavior *16134*
Journal of Reality Therapy *15293*
Journal of Rehabilitation *16137*
Journal of Rehabilitation of the Deaf *13777*
The Journal of Religion and Psychical Research *13306*
Journal of Religion & the Applied Behavioral Sciences *14288*
Journal of Religious Studies *08266*
Journal of Research in Childhood Education *14279*
Journal of Research in Crime and Delinquency *16017*
Journal of Research in Music Education *15763*
Journal of Research in Reading *13167*
Journal of Research in Science Teaching *15812*
Journal of Rheology *16700*
Journal of Rhinology *01402*
The Journal of Risk and Insurance *14032*
Journal of Roman Studies *13008*
Journal of Roman Studies Monographs *13008*
Journal of Safety Research *16145*
Journal of School Health *14036*
Journal of Scientific and Industrial Research *06753*
Journal of Scientific and Statistical Computing *16536*
Journal of Sedimentary Paleontology *16578*
Journal of Seed Technology *14472*
The Journal of Sericultural Science of Japan *08254*
Journal of Sex Education & Therapy *13581*
The Journal of Sex Research *16611*
Journal of Shellfish Research *16153*
Journal of Ship Production *16683*
Journal of Ship Research *16683*
Journal of Slovene Studies *16579*
Journal of Small Animal Practice *11646*
Journal of Social and Behavioral Sciences *14502*
Journal of Social Issues *16608*
Journal of Socialist History *14701*
Journal of Soil and Water Conservation *16716*
Journal of Soil Science *11690*
Journal of Solid-State Circuits *15317*
The Journal of Southern History *16725*
Journal of Spacecraft and Rockets *13872*
Journal of Specialists in Group Work *14335*
Journal of Speech *16741*
Journal of Speech and Hearing Disorders *14178*
Journal of Speech and Hearing Research *14178*
Journal of Spelean History *14179*
Journal of Sport History *16223*
Journal of Sugar Beet Research *14168*
Journal of Supreme Court History *16755*
Journal of Surgical Research *14262*
Journal of Symbolic Logic *00679, 14338*
The Journal of Teacher Education *13519*
Journal of Technical and Vocational Education in South Africa *10180*
Journal of Technology Education *15560*
Journal of Technology Transfer *16769*
Journal of Terramechanics *15541*
Journal of Test and Evaluation *15563*
Journal of the AACAP *13383*
Journal of the AAHA *13451*
Journal of the AAVSO *13597*
Journal of the Academy of Ambulatory Foot Surgery *13280*
Journal of the Academy of Marketing Science *13297*
Journal of the Academy of Rehabilitative Audiology *13305*
Journal of the ACM *14285*
Journal of the Acoustical Society of America *13322*
Journal of the AES *14548*
Journal of the Air and Waste Management Association *13355*
Journal of the American Academy of Dermatology *13390*
Journal of the American Academy of Matrimonial Lawyers *13403*
Journal of the American Academy of Psychoanalysis *13432*
Journal of the American Bamboo Society *13615*
Journal of the American Ceramic Society *13685*
Journal of the American Ceramic Society incorporating Advanced Ceramic Materials *16093*
Journal of the American Chemical Society *13688*
Journal of the American Chestnut Foundation *13689*
Journal of the American College of Cardiology *13699*
Journal of the American Criminal Justice Association *13772*
Journal of the American Electroly Association *13794*
Journal of the American Institute of Homeopathy *13882*
Journal of the American Leather Chemists Association *13913*
Journal of the American Liszt Society *13918*
Journal of the American Mathematical Society *13929*
Journal of the American Society for Horticultural Science *14066*
Journal of the American Society for Psychical Research *14082*
Journal of the American Society of Echocardiography *14120*
Journal of the Anthropological Society of Nippon *08315*
Journal of the Anthropological Society of South Australia *00216*
Journal of the AOAC *14230*
Journal of the Arab Maritime Transport Academy *03874*
Journal of the ASBC *14103*
Journal of the Association of Food and Drug Officials *14439*
Journal of the Association of Teachers of Japanese *14514*
Journal of the Astronautical Sciences *13609*

Journal of the Atmospheric Sciences 13943
Journal of the Atomic Energy Society of Japan 08203
Journal of the Australian Entomological Society 00289
Journal of the Bahrain Medical Society 01058, 01058
Journal of the Balint Society 11420
Journal of the Bombay Natural History Society 06747
Journal of the British Astronomical Association 11530
Journal of the British Interplanetary Society 11595
The Journal of the Canadian Athletic 02078
Journal of the Catgut Acoustical Society 14650
Journal of the CCA 02095
Journal of the Ceramic Society of Japan 08369
Journal of the Chemical Society 12857
Journal of the China Coal Society 03131
Journal of the Chinese Foundrymen's Association 03269
Journal of the Chinese Institute of Engineers 03273
Journal of the Commonwealth Association for Development 11808
Journal of the Community Development Society 14810
Journal of the Confucius-Mencius Society 03292
Journal of the CTI 14858
Journal of the Czech Geological Society 03507
Journal of the Dental Association of South Africa 10134, 10159
The Journal of the East African Natural History Society 08497
The Journal of the Electrochemical Society of India 06757
Journal of the Foot and Ankle 13972
Journal of the Geodetic Society of Japan 08385
Journal of the Geological Society 12111
The Journal of the Geological Society of Jamaica 08079
Journal of the Herpetological Association of Africa 10146
The Journal of the Historical Metallurgy Society 12173
Journal of the History of Ideas 15542
Journal of the IAA 04583
Journal of the IAPD 12332
Journal of the I.E.S. 15248
Journal of the Indian Law Institute 06798
Journal of the Indian Mathematical Society: Mathematics Student 06800
Journal of the Indian Medical Association 06801
Journal of the Indian Musicological Society 06802
The Journal of the Indian Society of Oriental Art 06819
Journal of the Indian Society of Soil Science 06820
Journal of the Institute of Energy 12235
Journal of the Institute of Health Education 12241
Journal of the Institution of Water and Environment Management 12318
Journal of the International Association of Buddhist Studies 15399
Journal of the International Phonetic Association 12384
Journal of the Irish College of Physicians and Surgeons 07023
Journal of the Irish Dental Association 06981
The Journal of the Japan Diabetes Society 08393
Journal of the Japanese Stomatological Society 08231
Journal of The Japan Medical Association 08212
The Journal of the Japan Orthodontic Society 08240
Journal of the Japan Prosthodontic Society 08310
Journal of the Japan Society for Bioscience, Biotechnology and Agrochemistry 08348
Journal of the Kafka Society 15634
Journal of The Kansai Society of Naval Architects 08155
Journal of the Korean Chemical Society 08530
Journal of the Korean Forestry Society 08532
Journal of the Malaysian Branch of the Royal Asiatic Society 08646
Journal of the Marine Biological Association of the United Kingdom 12492
Journal of the Mathematical Society of Japan 08386
Journal of The Medical Defence Union 12503
Journal of the Meteorological Society of Japan 08335
Journal of the MPR Academy of Sciences 08798
Journal of the National Association of University Women 15913
Journal of the National Medical Association 16107
Journal of the Nepal Research Centre 05473
Journal of the Operational Research Society 12655
Journal of the Optical Society of America A 16255
Journal of the Optical Society of America B 16255
Journal of the Order of Indian Wars 16259
Journal of Theoretical Politics 09249
The Journal of the Otolaryngological Society of the Republic of China 03318
Journal of the Pembrokeshire Historical Society 12686
The Journal of the Perthshire Society of Natural Science 12689
Journal of the Philosophy of Sport 16313
Journal of the Physical Society of Japan 08284
Journal of the Printing Historical Society 12718
Journal of Therapeutic Horticulture 13853
Journal of Thermophysics and Heat Transfer 13872
Journal of the Royal Asiatic Society 12770
Journal of the Royal College of General Practitioners 12779
Journal of the Royal College of Physicians 12788
Journal of The Royal Historical Society of Queensland 00469
Journal of the Royal Institution of Cornwall 12821
Journal of The Royal Society of Antiquaries of Ireland 07030
Journal of the Royal Society of Western Australia 00481
Journal of the Royal Statistical Society, Series A: Statistics in Society 12866
Journal of the Royal Statistical Society, Series B: Methodological 12866
Journal of the Royal Statistical Society, Series C: Applied Statistics 12866
Journal of the Royal Statistical Society, Series D: The Statistician 12866
Journal of the SAH 16639
Journal of the Science of Food and Agriculture 13035, 13035
Journal of the Serbian Chemical Society 17129
Journal of the Singapore National Academy of Science 10056
Journal of the Society for Psychical Research 12989
Journal of the Society for Research in Asiatic Music 08441
Journal of the South African Institute of Mining and Metallurgy 10198
Journal of the South African Institution of Civil Engineers 10203
Journal of the South African Veterinary Association 10222
Journal of the Statistical and Social Inquiry Society of Ireland 07035
Journal of the Textile Institute 13120
Journal of the Tibet Society 16782
Journal of the Water Pollution Federation 16873
Journal of the World Aquaculture Society 16921
Journal of Thoracic and Cardiovascular Surgery 13503
Journal of Traditional Chinese Medicine 03127
Journal of Transportation Medicine 08233
The Journal of Turkish Phytopathology 11161
Journal of Ukrainian Studies 02186
Journal of Ultrasound in Medicine 13898
Journal of Urban Affairs 16850
Journal of Urology 14201
Journal of Vacuum Science and Technology A&B 13893
Journal of Vacuum Science and Technology (JVST) 14202
Journal of Value Inquiry 14094
Journal of Vascular Surgery 16626
Journal of Vegetation Science 06005
Journal of Vertebrate Paleontology 16712
Journal of Veterinary Diagnostic Investigation 13599
Journal of Veterinary Medical Education 14390
The Journal of Veterinary Medical Science 08214
Journal of Vinyl Technology 16692
Journal of Virology 14073
Journal of Vocational Education Research 14208
Journal of Weather Modification 16876
Journal of West African Education 11181
Journal of Youth Services in Libraries 14315
Journal of Zoo and Wildlife Medicine 13607
Journal of Zoology 13237
Journal on Applied Mathematics 16536
Journal on Computing 16536
Journal on Control 16536
Journal on Mathematical Analysis 16536
Journal on Numerical Analysis 16536
Journal on Selected Areas in Communications 15317
Journal on Transpersonal Psychology 14362
Journal SWA Scientific Society 08816
Journal, Transactions 08121
Journées de la Société de Legislation Comparée 04660
Judaica Librarianship 14451
The Judge's Journal 13617
Judicature 13908
Der Jugendpsychologe 11019
Jugoslovensko Voćarstvo: Journal of Yugoslav Pomology 17101
Junior Achievement Annual Report 15631
Junior Astronomical Society: Circular Newsletter 12424
Junta de Historia Eclesiástica Argentina: Boletín 00148
Jurimetrics: Journal of Law, Science and Technology 13617
Jurist 10109
Justice Directory of Services: Répertoire des services 02113
Justice for Whom? 12425
Justice Quarterly 13286
Justice Report: Actualités-Justice 02113
Justice System Journal 15947
Justiz und Recht 05518
Juvenile and Child Welfare Law Reporter 13617
Juvenile and Family Court Journal 16002
Juvenile and Family Court Newsletter 16002
Juvenile and Family Law Digest 16002

KAAD-Korrespondenz 06068
Kaafa 08565
Kabita 08817
Källa 10462
Käsikirjoja 04027
Kagaku Kisoron Kenkyu 08150
Kagaku to Kogyo 08318
Kagami: Japanischer Zeitschriftenspiegel 05844
Kairos 00679
Kakteen und andere Sukkulenten 05457
Kalevalaseuran vuosikirja 03972
Kali-Briefe/Potash Review/Revue de la Potassa/Revista de la Potasa 10739
Kalki: Studies in James Branch Cabell 15591
Kalori 00482
Kampsville Archaeological Center Research Series 14664
Kampsville Archaeological Center Technical Report 14664
Kanon: Jahrbuch der Gesellschaft für das Recht der Ostkirchen 00616
Kansantaloudellinen Aikakauskirja: The Finnish Economic Journal 03973
Kansatieteellinen Arkisto 04051
Kanser: The Turkish Journal of Cancer 11163
Karawan-e-Science 09384
Kardiologija 17130
Kardiologiya: Cardiology 09972
Karinthin 00712
Karl-August-Forster-Lectures: Mathematisch-Naturwissenschaftliche Klasse 05050
Karnatak Historical Review 06832
Karolinska Förbundets årsbok 10478
Karst und Höhle 06335
Karszt és Barlang 06642
Kartbladet 10479
Der Kartoffelbau: Fachzeitschrift über Züchtung, Vermehrung, Produktionstechnik, Verwertung und Ökonomik 05704
Kartographische Nachrichten 05369, 10826
Kartographische Schriftenreihe 10826
Karyogram 14424
Kasai 08220
Kassenärztliche Vereinigung Berlin: Mitteilungsblatt 06023
Kassenarzt-aktuell 06031
Katalog Fauny Pasozytniczej Polski 09598
Katholiek Documentatie Centrum: Jaarboeken 08946
Katholiek Documentatie Centrum: Scripta 08946
Katholische Bildung 06421
Kavnt: Quantum 09953
Kazakh Academy of Sciences: Izvestiya: Bulletin 08456
Kazakh Academy of Sciences: Vestnik: Herald 08456
Keats-Shelley Bulletin 07759
Keats-Shelley Journal 07759, 15636
The Keats-Shelley Review 12426
KEB-NordWest: Mitteilungsblatt für kath. Erwachsenenbildung 06063
Keel ja Kirjandus: Language and Literature 03933
Keeping Awake 02037
Keieigaku Ronshu 08326
Keikinzoku Gakkai: Journal 08156
Keizaigaku-shi Gakkai: Annual Bulletin 08159
Keizai-ho Gakkai: Journal 08160
Keizai Ronso: Economic Review 08178
Kekkaku 08329
Kelias News 08504
Kemisk Tidskrift 10555
Kent View 12429
The Kenya Farmer 08477
Kenya Historical Review 08502
The Kenya Journal of Sciences: Series A 08505
The Kenya Journal of Sciences: Series B 08505
The Kenya Journal of Sciences: Series C 08505
Keramicos 10185
Kerkime dhe Studime Hidraulike: Hydraulic Studies 00003
Kërkime/Istraživanja: Research 17076
Kernpunkte 10940
Keyboard Teacher 15637
Khimicheskaya Fizika: Chemical Physics 09954
Khimicheskaya Tekhnologiya: Chemical Technology 11193
Khimicheskii Zhurnal Armenii: Chemical Journal of Armenia 00203
Khimiya Geterotsiklicheskikh Soedinenii: Chemistry of Heterocyclic Compounds 08558
Khimiya i Tekhnologiya Topliv i Masel: Chemistry and Technology of Fuels and Oils 09954
Khimiya i Tekhnologiya Vody: Water Chemistry and Engineering 11193
Khimiya i Tekhnologiya Vody: Water Chemistry and Technology 09954
Khimiya i Zhizn: Chemistry and Life 09954
Khimiya Prirodnykh Soedinenii: Chemistry of Natural Compounds 09954, 16948
Khimiya Tverdogo Topliva: Solid Fuel Chemistry 09954
Khimiya Vysokikh Energii: High Energy Chemistry 09954
KHZ: Klassische hom. Zeitung 05566
Kibernetika, Kinematika i Fizika Nebesnykh Tel: Cybernetics, Kinematics and Physics of Heavenly Bodies 11193
Kidney International 14148
Kimya Mühendisliği: Chemical Engineering 11167
Das Kind: Halbjahresschrift für Montessori-Pädagogik 05472
Kind News 15803
Kiné Info 08579
Kinetika i Kataliz: Kinetics and Catalysis 09954
Kiniogenases 05050
Kinoizkustvo 01785
Kinoschriften 00951
The Kipling Journal 12431
Kipling Journal 15638
Kirchenmusikalisches Jahrbuch 05073, 05896
Kirjallisuudentutkijain Sauran Vuosikirja: The Yearbook of the Literary Research Society 03976
Kirjallisuudentutkijain Seuran vuosikirja: Yearbook of the Literary Research Society 04001
Kirjastolehti 04039
Kirkehistoriske Samlinger 03809
Kir ou Kirk 14247
Kisérletes Orvostudomány 06626
Kitab 09375
Kiyo 08298
KJEMI 09323
KLA Bulletin 08536
Klagenfurter Beiträge zur Sprachwissenschaft 00694
Das Kleine Ostpanorama 00628
Kleine Schriften der Cusanus-Gesellschaft 05280
Kleio 09089
Klinická biochemie a metabolismus: Clinical Biochemistry and Metabolism 03509
Klinická onkologie: Clinical Oncology 03509
Knight Letter 15665
Knijevni Jivot 09798
Knipselkrant 09063
Knitstats 12149
Knitwear Industry Review 06675
Knjižnica 10111
Knowledge is Power 03128
Kobunshi 08166
Kochel-Brief 05784
The Kodaly Envoy 16266
Köolaj és Földgáz: Oil and Gas 06670
Középiskolai Matematikai Lapok 06597
Kogyo Kayaku Kyokai: Journal 08168
Koho-Kenkyu 08227
Kokushigaku 08426
Koleopterologische Rundschau 01035
Kolloidny Zhurnal: Colloids Journal 09954
Kommunalpolitische Blätter 06080
Kommunalwirtschaft 06388
Kommunalwissenschaftliche Dissertationen 05587
Kompleksnoe Ispolzovanie Mineralnogo Syrya: Comprehensive Utilisation of Mineral Raw Materials 09954
Koncize 04612
Konferenz Schweizerischer Gymnasialrektoren: Colloquium 10761
Kongressberichte der Tagungen der NWDGIM 06145
Koninklijke Academie voor Geneeskunde van België: Verhandelingen 01407
Koninklijke Academie voor Nederlandse Taal- en Letterkunde: Verslagen en Mededelingen 01408
Koninklijke Academie voor Wetenschappen, Letteren en Schone Kunsten van België: Jaarboek 01409
Koninklijke Nederlandse Akademie van Wetenschappen: Akademie Nieuws 08952
Koninklijke Nederlandse Akademie van Wetenschappen: Medelingen: Nieuwe Reeks, Letterkunde 08952
Koninklijke Nederlandse Akademie van Wetenschappen: Proceedings: Series A: Mathematical Sciences; Series B: Palaeontology, Geology, Chemistry, Physics and Anthropology; Series C: Biological and Medical Sciences 08952
Koninklijke Nederlandse Akademie van Wetenschappen: Verhandelingen: Nieuwe Reeks, Letterkunde 08952
Koninklijke Nederlandse Akademie van Wetenschappen: Verhandelingen, Series I, II: Transactions 08952
Koninklijke Nederlandse Natuurhistorische Vereniging: Wetenschappelijke Mededelingen 08958
Koninklijk Genootschap voor Landbouwwetenschap: Adressenlijst: Address-List of all Dutch Agricultural Alumni 08962
Koninklijk Instituut van Ingenieurs: Yearbook 08963
Koninklijk Nederlandsch Genootschap voor Geslacht- en Wapenkunde: Series Publicaties 08966
Koninklijk Nederlandsch Genootschap voor Geslacht- en Wapenkunde: Series Werken 08966
Koninklijk Nederlands Geologisch Mijnbouwkundig Genootschap: Nieuwsbrief 08967
Koninklijk Oudheidkundig Genootschap Amsterdam: Annual Reports 08968
Konjunktur von Morgen: Brief fortnightly Survey of German and World Business Cycles and of the World's Commodity Marktets 05935
Konstruktion: Zeitschrift für Konstruktion und Entwicklung im Maschinen-, Apparate- und Gerätebau 06322
Konsulttidningen 10600
Kontakt 03772
Kontakten 09274
Kontakto 09083
Kontyu 08341
Konzerthaus-Nachrichten 01022
Koordinatsionnaya Khimiya: Coordination Chemistry 09954
Koreana: English Edition 08526
Koreana: Japanese Edition 08526
Korean Chemical Society: Bulletin 08530
Korean Economic Review 08531
Korea Newsreview 08526
Korean Journal of Clinical Psychology 08539
Korean Journal of Industrial Psychology 08539
The Korean Journal of Medicinal Chemistry 08530
The Korean Journal of Pharmacology 08540
Korean Journal of Psychology 08539
Korean Journal of Social Psychology 08539
Korean Medical Association: Journal 08537
Korean Medical Association: Newspaper 08537
Koroth 14003
Korrespondenz Abwasser 05029
Kortárs 06639
Kort Sagt!: Newsletter from D.L.A. 03576
Kosciuszko Foundation: Newsletter 15639
Kosmas 14947
Kosmicheskie Issledovaniya: Space Research 09954
Kosmos 09601, 10550, 15523
Krasoslovni zbornik: Acta Carsologica 10105
Kriminologisches Bulletin 10929
Kriobiologiya: Cryobiology 11193
Kristallografiya: Crystallography 09954
Kroeber Anthropological Society: Papers 15640
Kronika Stowarzyszenia Historyków Sztuki 09616
KSK Scanner 08165
Kujizuica Sipma 10098
Kuki-Chowa Eisei Kogakkai: Journal 08174
Kuki-Chowa Eisei Kogakkai: Transactions 08174
Kultura Popullore: Folk Culture 00003
kulturarbeit aktuell 05064
Kulturchronik 06012
Kulturminder 03810

Kulturno-Istorisko Nasledstvo na Makedonija: The Cultural and Historical Heritage of Macedonia 08606
Kulturpolitik 05241
Kulturrådet 10523
Kuml: Journal of the JAS 03754
Kunchong zhishi: Entomological Knowledge 03198
Kungliga Akademien för de fria Konsterna: Exhibition Catalogues 10482
Kungliga Fysiografiska Sällskapet i Lund: Arsbok 10483
Kungliga Humanistiska Vetenskaps-Samfundet i Uppsala: Arsbok 10485
Kungliga Humanistiska Vetenskaps-Samfundet i Uppsala: Skrifter: Acta 10485
Kungliga Humanistiska Vetenskaps-Samfundet i Uppsala: Yearbook 10485
Kungliga Musikaliska Akademien: Årsskrift 10487
Kungliga Vetenskaps- och Vitterhets-Samhället i Göteborg: Årsbok 10491
Kungliga Vetenskaps-Societeten i Uppsala: Arsbok 10492
Kunglig Samfundets Handlinger 10488
Kungl. Skogs- och Lantbruksakademiens Tidskrift 10489
Kunstchronik 06465
Kunstreisboek voor Nederland: Annual Report 08977
Kuratorium für Technik und Bauwesen in der Landwirtschaft e.V.: Schriften 06088
Kvant: Quantum 09954
Kvantovaya Elektronika: Quantum Electronics 09954
Kwant 03816
Kwantitatieve Methoden 09104
Kybernetika 03500, 03525
Kyrgyz Academy of Sciences: Izvestiya: Bulletin 08548

Laban/Bartenieff Institute of Movement Studies: Membership Directory 15642
Laban/Bartenieff Institute of Movement Studies: Newsletter 15642
Laboratory Animal Science 13484
Laboratory Hazards Bulletin 12857
Laboratory Medicine 14110
Labor Law Developments 16730
Labor Lawyer 13617
Labor und Medizin 10893
Lacio Drom: Bimestrale di Studi Zingari 07597
LACITO-Documents 04746
The Lactation Review 15236
Lääketieteellinen Aikakauskirja Duodecim 03997
The LAEE Membership Directory 15381
Laeknabladid: Icelandic Medical Journal 06694
Ländlicher Raum: Rundbrief der ASG 05044
Lakes Letter 15382
Lalit Kala 06840
Lamed Lĕšonēkha 07040
Lamishpacha 15214
Lam Lines 13897
Lancashire Dialect Society: Journal 12438
Land: Landscape Architectural News Digest 14139
Landmeter Expert Vastgoed 01493
Landscape Architectural Review 02285
Landscape Architecture 14139
Landscape Design 12440
Landtechnik: Fachzeitschrift für Agrartechnik und ländliches Bauen 06088
Land Use Digest 16801
Land Use Law and Zoning Digest 14007
Landwirtschaftliche Forschung 06354
Langenbecks Archiv für Chirurgie 05335
Langmuir 13688
Language 15674
Language Arts 16010
Language Learning Journal 11269
Language Sciences 08126
Language, Speech and Hearing Services in Schools 14178
Language World Newsheet 11269
Langues et cultures africaines 04746
Langues et cultures du Pacifique 04746
Langues et sociétés de l'Amérique traditionnelle 04746
Les Langues Modernes 04234
Larizza Informazioni 07575
Laryngologie, Rhinologie, Otologie 00789
Lasa Forum/Newsletter 15648
Lasers in Surgery and Medicine 14069
LAS Newsletter 10051
Lastentarha 03981
The Last Word 13755
Late Imperial China 16506
Latest Literature in Family Planning 12056
Lathaia 08144
Latin American Centre for Physics: Noticia 01698
Latin American Research Review 15648
Latin American Studies Association: Professional Journal 15648
Latinskaya Amerika: Latin America 09954
Latin Teaching 11290
LATIS: Landscape Architecture Technical Information Series 14139
Latomus: Revue d'Etudes Latines 01458
Lattice 15743
Latvijas Fizikas un Tehnisko Zinatnu Zurnals: Latvian Journal of Physics and Technical Sciences 08558
Latvijas Kimijas Zurnals: Latvian Chemical Journal 08558
Laubach Literacy Action: Directory 15649
Laudate 12133
Lauriston S. Taylor Lectures 16025
Lavori S.I.M. 07892
Lavra & Oficina 00034
Law and History Review 14070
Law Directory 06956
Law in Japan 15594
Law Librarian 11479
Law Library Journal 13547
Law, Medicine and Health Care 14140
Law of the Sea Institute: Occasional Papers 15654
Law of the Sea Institute: Proceedings 15654
Law Practice Management 13617
Law Reporter 14523, 14808
Laws 07037
Law School Admission Bulletin 15655
Law School Admission Council – Law School Admission Services: Annual Report 15655
Law Society Journal 00428
Law Society of Manitoba Communiqué 02579
Law Society of Saskatchewan: Journal 02581
Law Society of Scotland: Journal 12442
Law Society of Upper Canada: Communique 02583
Law & Society Review 15653
Law Talk 09162
Lawyers' Committee for Civil Rights under Law: Committee Report 15656
Lay Newsletter 15367
LBI News 15659
Lead and Zinc Statistics 12374
The Leader 02585
Leader in Action 13595
Leader Newsletter 16285
Leadership Reports 16147
Leadership Roster of Officers 14481
League of Canadian Poets: Newsletter 02584
Learn English in Britain 11383
Learning 02025
Die Lebenshilfe-Zeitung 05259
Lebensmittel-Technologie 10829
Lebensmittel-Wissenschaft und -Technologie: Food Science & Technology 10829
Lectura Petrarce 07655
Lecturas Matematicas 03409
Lectura Vida 15518
Lectures 12136
Lecture Series 09238, 12817
Lecture Series Notes 01512
Lecturi geografice: Des Lectures Géographiques 09784
Das Leder 06385
Leeds Philosophical and Literary Society: Proceedings 12444
Legal History Review 08136
Legalization Update 16083
Legal Memorandum 15892
Legal Plan Letter 16141
Legal Records 12939
Legal Studies 13062
Legal Studies Forum: A Journal of Interdisciplinary Legal Studies 13914
The Legend of Jennie Lee 14624
Legislative Report 15995, 16727
Legislative Watch 15895
Lehreragenda 11023
The Leicestershire Historian 12446
The Leisure Manager 12250
Leisure, Recreation and Tourism Abstracts 11720
Leisure Studies 12447
Leisure Studies Association: Conference Reports 12447
Lékař a technika: Biomedical Engineering 03509
Lembar Komunikasi 06894
Length of Stay, by Diagnosis, by Operation 14789
Leo Baeck Institute: Bulletin 15659
Leo Baeck Institute: Yearbook 15659
Leobener Grüne Hefte, Neue Folge 00708
Leopoldina: Mitteilungen der Deutschen Akademie der Naturforscher Leopoldina, Reihe 3 05285
Lepidoptera 03767
La Lepro 08356
Leprología 00175
Leprosy Review 11602
Lernen Fördern 06103
Leschetizky Association: News Bulletin 15662
Lesnictví 03501
Lĕšonénu: A Quarterly for the Study of the Hebrew Language and Cognate Subjects 07040
Lĕšonénu Laam 07040
Lesovedenie: Forestry Studies 09954
Lessing Yearbook 15663
Letopis Matice Srpske 17104
Letopis-Yearbook 10105
Lettera Circolare Qualità 07338
Lettere d'Affari 07525
Letter of the L.A.A. 02589
Lettre aux Générations 2000 04602
Lettre de l'ADBU 04130
Lettre de l'Ingénieur 04388
La Lettre de Psychiatrie Française 04223, 04927
Lettre de Syntec-Informatique 04388
Lettre d'Information des Historiens de l'Europe Contemporaine 01236
Lettre du Centre Français de Droit Comparé 04374
La Lettre du Président 04969
La Lettre Européenne du Progrés Technique 04305
Lettre Hebdomadaire des Médecins de Groupe 04950
Les Lettres Albanaises 00005
Lettres et Cultures de Langue Française 04180
Lettres et Journal de Physique 04836
Lettre Syndicale 04952
Leukemia Society of America: Society News 15664
Leusla 01682
Lexica Societatis Fenno-Ugricae 04003
Lexington Quarterly 15666
L-Ghasfur U L-Ambjent Naturali: The Bird and the Natural Habitat 08669
LGU-Mitteilungen 08571
Liaison: Saskatchewan's Heritage Review 02665
Liaison Bulletin 15896
Le Libéral de Chaine 04518
Le Libéral Dentaire 04518
Liberal Education 14378
Libërshënuesi/Spomenica: Diary 17076
Libra 08436
Librarium 10799
The Library 11438
Library Acquisitions Alert 16087
Library Administration and Management 15668
Library and Information Science Education Statistical Report 14313
Library and Information Technology Association: Newsletter 15669
Library & Archives News 15659
Library Association of Barbados: Bulletin 01087
Library Association of Barbados: Update: A Newsletter 01087
The Library Association of China: Bulletin 03305
Library Association of Trinidad and Tobago: Bulletin 11122
Library Association Record 12450
Library Association: Year Book 12450
Library Buildings Consultants List 15668
Library Bulletin 04384
Library Life 09164
Library Notes 12796
Library Resources and Technical Services 14314
Library Video Magazine 13916
Libro Almanaque Escuela para Todos 03446
Libro dei Programmi Annuali 07210
Libyan Studies 12978
Lichamelijke Opvoeding 08961
The Lichenologist 11604
Lidé a země 03500
Liebigs Annalen der Chemie 05793
Liechtensteiner Umweltbericht 08571
Life Insurance Marketing and Research Association: Proceedings 15671
Lifelines 14764
Lifelong Learning 13464
Lifespan 11664
Lighting Design + Application Magazine 15248
Lighting Journal 12306
Lighting Research and Technology 11767
Likovna Umetnost: Plastic Arts 08593
Likundoli 17146
Lilloa 00140
Limba si literatura romana 09782
Limbile moderne in scoala 09782
Limb. Tijdschrift voor Genealogie 08972
Limburgse Vogels 08979
L-Imnara 08663
Limnology and Oceanography 14141
Limosa 08999
The Linacre 16046
Lincoln Center Calendar of Events 15672
Lincoln Herald 15144
Lincoln Review 15673
Lincolnshire History and Archaeology 12979
Lincolnshire Past and Present 12979
Lines to Leaders 16793
Lingua e Vita 11592
Lingua Posnaniensis 09612
Lingüistica Española Actual 10332
The Linguist 12251
Linguistics and Literature 01775
Linguistic Series 06717
Linguistic Society of America: Annual Meeting Handbook 15674
Linguistic Society of America: Bulletin 15674
Linguistique Bibliographie 09060
Linguistique générale 04746
Link 11284, 12108
Linkletter 13354
The Linnean 12454
Le Linnéen 02933
Lipids 13964
Lippische Mitteilungen aus Geschichte und Landeskunde 06133
The Listener 15198
Listen to Norway 09331
List Forum für Wirtschafts- und Finanzpolitik 06104
Listing of Inter-American Graduate Programs 15350
List Lekar 17130
List of Accredited Programs in Architecture 15788
List of Building Courses 11761
List of Members and Officers 16554
List of Radix Teachers Worldwide 16390
Listy filologické 03500
The Liszt Society Journal 12455
The Liszt Society: Newsletter 12455
Literacy Advance 15649
Literarisches Österreich 00901
Das Literarische Wort 05982
Literary & Linguistic Computing 11271
Literary Review 03155
Literatura Ludowa: Folk Literature 09584
Literaturbericht 05708
Literaturberichte über Wasser, Abwasser und feste Abfallstoffe 06392
Literaturen Zbor: Literary Word 08612
Literaturkurzberichte Chemische Technik und Biotechnologie 05334
Literaturnaya Gazeta 09989
Literaturoznavstvo: Literary Studies 11193
Literaturschau Stahl und Eisen 06374
Literaturschnelldienst: Kunststoffe, Kautschuk, Fasern 05721
Literaturwissenschaftliches Jahrbuch 05896
Lithuanian Mathematical Journal 08573
Lithuanian Studies 15329
Lithuanistika: Lithuanian Studies 08573
Litigation 13617
Litir Nuachta 14217
Litologiya i Poleznye Iskopaemye: Lithology and Minerals 09954
Litteraria 09630
Livestock Production Science 07688
The Living Earth 13073
Living Historical Farms Bulletin 14316
Living Safety 02001
Living with Allergies 13448
Le Livre et l'Estampe 01486
Livsmedelsteknik 10563
Ljetopis 03459
LM & Rules 11503
LMS Bulletin 12462
LMS Journal 12462
LMS-Lingua 10515
LMS Proceedings 12462
Local Government Management 00416
Local Population Studies 12458
Loccumer Protokolle 05685
Locusta 08656
Łódzkie Studia Etnograficzne: Ethnographic Works and Materials 09584
The LOG 15767
The Log Analyst 16694
Logique et Analyse 01203
Logistics Focus 12252
Logistics Today 12257
Logistik Heute 05260
Logistik Spektrum: Logistik in Industrie, Handel und Dienstleistung 05376
Loisirs-Santé 04520
Loke 03819
Lok Niti 09500
London and Middlesex Archaeological Society: Newsletter 12460
London and Middlesex Archaeological Society: Transactions 12460
London Bird Report 12465
London Naturalist 12465
London's Infrastructure 12469
London Society: Journal 12468
London Topographical News 12470
Longevity Letter 13921
Longfellow Journal of Poetry 15677
Look-Listen Project Report 16171
The Loophole 02365
Lord Bossom Lectures 12050
Loreto 13 09642
Loss Prevention Bulletin 12293
Lotta contro la Tubercolosi e le Malattie Polmonari Sociali 07689
Lower Tideway Topics 12758
Lowveld Natural History Branch, Wildlife Society of Zimbabwe: Newsletter 17174
Lozanía: Acta Zoológica Colombiana 03391
LPS Supplements 12458
LRS 05256
LT-Journal 08962
Luceafărul 09798
Luce e Immagini 07385
Lucknow Librarian 06869
Lud: The People 09584
Luft- und Raumfahrt 05377
Lumen Vitae: International Review of Religious Education 01198
Lumen Vitae: Revue Internationale de la Formation Religieuse 01198
Lunar and Planetary Information Bulletin 16840
Lung Line 02785
Lung Line Letter 16098
Lungs Botaniska Förening Medlemsblad 10498
Luonnon Tutkija: The Naturalist 03993
Lutheran Educational Conference of North America: Papers and Proceedings 15681
Lutheran Higher Education Directory 15681
Lutheran Historical Conference: Minutes of Biennial Conference 15683
Lutheran Historical Conference: Newsletter 15683
Lutheran Libraries 15680
Lux: La Revue de l'Eclairage 04218
Lys 03769

Maal og Minne 09262
Maaperäkartta 03963
De Maasgouw 08972
Mabua 07047
Macedoniae Acta Archaeologica 08618
Machberes Hamenahel 15973
Macromolecules 13688
Madencilik: Mining 11167
Madrider Mitteilungen 05568
Maelkeritidende 03649
Il Maestro Sarto 07212
Magazine of Concrete Research 12294
Magazine Venture Inward 14328
The Maghreb Review 12476
Magnesium 15498
Magnetic Resonance Imaging 16550
Magnitnaya Gidrodinamika: Magnetic Hydrodynamics 08558
Magnitnyi Rezonans i ego Primenenie: Magnetic Resonance and its Application 09954
Magnolia 15686
The Magnolia Society: Membership Roster 15686
Magyar Biofizikai Társaság Ertesitője 06619
Magyar Biofizikai Társaság: Newsletter 06619, 06619
Magyar Epitöipar: Hungarian Building Industry 06602
Magyar Epületgépészet: Sanitary Engeneering 06602
Magyar Geofizika 06631
Magyar Grafika 06672
Magyar Jog 06640
Magyar Kémiai Folyóirat: Hungarian Journal of Chemistry 06643
Magyar Kémikusok Lapja: Hungarian Chemical Journal 06643
Magyar Napló 06639
Magyar Nyelv 06649
Magyarország Ujabbkori Forrásai 06659
Magyar Radiologia 06654
Magyar Sebészet 06657
Magyar Szinházi Hirek 06658
Magyar Traumatologia, Orthopaedia, Kézsebézet Plasztikai Sebészet 06660
Magyar Zene 06663
Mailings 13718
Mainstream 13159
Mainstream Magazine 14221
Maintenance 06846
Mainzer Naturwissenschaftliches Archiv 06263
Mainzer Reihe 05050
Mainzer Zeitschrift 06107
La Maison d'Hier et d'Aujourd'hui 01158
Maize Abstracts 11720
Majalah Ikatan Pustakawan Indonesia 06890
Majalah Insinyur Indonesia 06893
Majalah PPM 08649
Majallah Akitek 08648
Majallat al Majimma'al Ilmi al Iraqi 06917
Makednonski Veterinaren Pregled: Macedonian Veterinary Review 08611
Makedonika: History, Folklore, Archaeology of North Greece 06525
Makedonska Akademija na Naukite i Umetnostite: Letopis 08603
Makedonski Arhivist 08607
Makedonski Medicinski Pregled: Macedonian Medical Review 08605
Maktaba: Official Journal 08504
Malacological Review 16521
Malacological Society of Australia: Journal 00430
Malaria Risk Chart 15134
Malawi Library Association: Bulletin 08623
The Malayan Nature Journal 08644
Malaysia Dari Segi Sejarah 08647
Malaysia in History 08647
Malaysian Biochemical Society: Proceedings of Annual Conference 08645
Malaysian Rubber Research and Development Board: Annual Report 08651
Malaysian Rubber Review 08651
Il Malcontento 07130
Maledicta: The International Journal of Verbal Aggression 15499
Maleo 09685
Málfridur 06682
Mallorn 13136

Malone Society: Annual Report 12480
Malone Society: Bulletin 12480
Mammalian Species 14142
MAN 12768
Månadens Standard 10522
Managed Care-HMOs 15855
Management Accounting 11762
Management Accounting Research 11762
Management and Administration 14452
Management Decision 12375
Management France 04220
Management heute 05055
Management in Education 11555
Management in Engineering Journal 14107
Management in Nigeria 09239
Management Review 01100, 13372, 13925
Management Salaries Report 13329
Management Science 15330
Management Services 12254
Management Today 12253
Manakdoot 06822
Manav 06758
The Manchester Geographer 12483
Manchester Literary and Philosophical Society: Memoirs and Proceedings 12484
Manchester Medical Society: Annual Report 12485
Manchester Region Industrial Archaeology Society: Newsletter 12486
Manitoba Archaeological Quarterly 02599
Manitoba Association of School Trustees: Newsletter 02608
Manitoba Environmental Council: Report 02615
Manitoba Genealogical Society: Newsletter 02617
Manitoba Heritage Review 02603
Manitoba Naturalists Society: Bulletin 02629
The Manitoba Social Worker 02609
Manitoba Society of Occupational Therapists: Update 02636
Mankind 00214
Man-made Textiles India 06857
Manorial Society of Great Britain: Bulletin 12488
Manual de Signos Convencionales 01546
The Manual of Orthopaedic Surgery 13971
Manual of Textile Technology 13120
Manual Práctico Aduanero 13253
Manual Práctico del Contribuyente 13253
Manual Práctico del Exportador 13253
Manual Práctico del Importador 13253
Manual Práctico Sudamericano del Transporte Internacional por Carretera de Carga y Pasajeros 13253
Manual Series 06762
Manuel de Placement 04369
Manufacturing Engineer 12296
Manufacturing Review 14145
Manyo 08180
Map Bulletins 08642
MAPPS Capability Survey 15687
Marbacher Arbeitskreis für Geschichte der Germanistik, Mitteilungen 05571
Marine Conservation Society: Newsletter 12493
Marine Engineers Review 12255
Marinehistorisk Tidsskrift 03770
The Mariners' Mirror 12984
Marine Technology 16683
Marine Technology Society Journal 15691
Marine Technology Society: Membership Directory 15691
Marine Technology Society: Newsletter 15691
Marinews 11605
Maritime Provinces Higher Education Commission: Annual Report 02640
Maritime Research Bulletin 03874
Market Frontier News 14949
Marketing and Research Today 08892
Marketing Business 11763
Marketing News 13926
Marketing of Library & Information Services in India 06775
Marketing Science 15330
Marketing Science Institute: Newsletter 15693
Marketing Science Institute: Research Briefs 15693
Marketing Success 11763
Market Logic 15276
Market Profiles 16801
Market Update 16183
Das Markscheidewesen 05527
Marquandia Society: Newsletter 15697
Marquee 16773
Les Marques Internationales 11034
Marquetarian 12497
The Masonry Society: Journal 15698
MA Spezial 06108
Massage Journal 13927
Massenet Newsletter 15699
Mass Spectrometry Bulletin 12857
MAST 02608
Master Educator 13637
Masterstroke 12636
Matematica Balkanica 17079
Matematicheskii Sbornik: Mathematical Collection 09954
Matematicheskii Zametki: Mathematical Notes 09954
Matematički Bilten: Mathematical Bulletin 08610
Matematički List za Učenike Osnovnih Škola 17079
Matematički Vesnik 17087
Matematikai Lapok 06597
Matematisk-fysiske Meddelelser 03710
Matematyka Stosowana 09585
Material Culture 16321
Materialien aus dem Stiftungszentrum 06295
Materialien zum Internationalen Kulturaustausch 05952
Materialien zur Bildungspolitik 06295
Materialien zur Zeitgeschichte 00851
Materialoznanie i Technologija 01771
Materialprüfung 05446, 06328
Materialprüfung – Materials Testing – Matériaux Essais et Recherches 05549
Materials Australasia 00415
Materials Evaluation 14075
Materials Forum 00415
Materials Handling Outlook 15700
Materials in Civil Engineering Journal 14107
Materials Performance 15845
Materials Properties Council: Annual Report 15701
Materials Science and Technology 12256, 12262
Materialwissenschaft und Werkstofftechnik: Journal of Materials Technology and Testing 05334
Maternity Center Association: Special Delivery 15702
Mathematical and Physical Sciences 01775
The Mathematical Gazette 12498
Mathematical Geology 15387
Mathematical Physics: Microfiche Edition 13893
Mathematical Proceedings 11737
Mathematical Review 06510
Mathematical Reviews 13929
Mathematica-Revue d'Analyse Numérique et de Théorie del 'Approximation 09737
Mathematica Scandinavia 03645, 10564
Mathematics and Computer in Simulation: Applied Numerical Mathematics 01141
Mathematics and Mathematical Education 01786
Mathematics in School 12498
Mathematics Magazine 15703
Mathematics of Computation 13929
Mathematics of Operations Research 15330, 16253
Mathematics Teacher 16011
Mathematics Teaching 11397
Mathematikai Lapok 09786
Matica Srpska: Proceedings 17104
MAT-NYT 03645
Matrix 11644
The Matrix and Tensor 13119
Matsne: Herald 05011
The Matter 15241
Matthay News 13930
Matukio 11060
Der Maueranker 06423
MBA News 12492
McIlvainea 16219
MCMHA Newsletter 15725
MDU Nurse 12503
Measurement 06608
Measurement and Control 12259
Measurement and Evaluation in Counseling and Development 14269
Measurement and Evaluation in Counseling & Development 13771
Measurement Science and Technology 12268
La Meccanica Italiana 07368, 07686
Mechanical Design 14145
Mechanical Engineering 14145
Mechanical Engineering Transactions 00421
Mechanical Incorporated Engineer 12308
Mechanics 13405
Mechanics' Institute of Montréal: Annual Report 02642
Mechanics Research Communications 07470, 07710
Mechanika Teoretyczna i Stosowana: Journal of Theoretical and Applied Mechanics 09586
Meddelelser FRA Gymnasieskolernes Tysklaererforening 03741
Le Médecin de France 04435
Médecine Aéronautique et Spatiale 04807
Médecine Hospitaliére 04951
Médecine Légale 04670
Médecins de Groupe 04950
Le Médecin Spécialiste 01370
Le Médecin Vétérinaire du Québec 02396
Mededelingen van de VTB 09100
Mededelingen van het Wiskundig Genootschap 09110
Medelanden fra från Strindbergssällskapet 10527
Media-Analyse (MA) 05124
Mediacult Newsletter 00705
Mediaevalia 14687
Médiagaz 04353
Media Message 01901
Media Reporter 12501
Media Showcase 14601
Médiathèques Publiques 04326
Media-Trend-Journal 11007
Medica 06115
Medical and Dental Register 09232
Medical and Paediatric Oncology 08943
Medical Anthropology Quarterly 16551
Medical Archives 01799
Medical Association of Jamaica: Newsletter 08087
Medical Association of Thailand: Journal 11093
Medical & Biological Engineering & Computing 12296
Medical Directors Conference Proceedings 15181
Medical Education 11273, 11293
The Medical Executive 13550
Medical Journal 16799
Medical Journal of Australia 00322
Medical Journal of Malaysia 08650
Medical Knowledge Self-Assessment 13729
Medical Letter on Drugs and Therapeutics 15709
Medical Library Association: Bulletin 15710
Medical Mycological Society of the Americas: Bulletin 15711
Medical Mycological Society of the Americas: Directory 15711
Medical Officers of Schools Association: Proceedings and Report 12504
Medical Physics 13566, 13893
The Medical Register 12106
Medical Research Council: Annual Report 12508
Medical Research Council: Handbook 12508
Medical Research Council of Canada: Report of the President 02646
Medical Research Council of Ireland: Annual Report 07011
Medical Research Council of New Zealand: Research Review 09130
Medical Research Projects 00438
Medical Scientific Update 16098
Medical Society of the United States and Mexico: Directory 15713
Medical Staff Forum 13854
Medical & Veterinary Entomology 12800
Medical Woman 12514
Medical Women's International Association: Congress Report 06116
Medical World 11388, 12506
Medicina 00173, 03342
Medicina de España 10334
Medicina dello Sport 07693
Medicina dello Sport: Mensile di fisiopatologia dello sport 07686
La Medicina Estetica 07896
Medicina Fisica e de Reabilitação 09704
Medicina Geriatrica 07959
Medicina Interna 09770
Medicinal & Aromatic Plants Abstracts 06753
Medicina Psicosomatica 07901
Medicina Sociale: Organo della S.I.De.S. e della Società Italiana di Medicina Sociale 07975
Medicine and Science in Sports and Exercise 13737
Medicine-Biology-Environment 01383
Medicine, Science and the Law 11472
Medicinski pregled: Journal for General Medicine 17085
Medicinsk Revue 09278
Il Medico d'Italia 07695
Medico-Legal Journal 12515
Medico Legal Society of New South Wales: Proceedings 00433
Medicus Tropicus 09040
Medicus Universalis 06614
Mediendienst/Research News 05759
Medien-Journal 00803
medienpraktisch 05772
Medienspiegel 05759
Medieval Archaeology 12982
Medieval Settlement Research Group: Annual Report 12517
Mediterranean Survey 15286
Meditsinskaya Radiologiya: Medical Radiology 09952
medium 05772
Medium Aevum 13019
Medizin Heute: Zeitschrift für Patienten 05229
Medizinhistorisches Journal: Kommission für Geschichte der Medizin und der Naturwissenschaften 05050
Medizinische Forschung 05050
medizin populär: Patienteninformation der Österreichischen Ärztekammer 00721
Medizinsoziologie 05383
Medlemsnyt 03592
Med Tel International 10471
Medycyna Doswiadczalna i Mikrobiologia 09588
Medycyna Weterynaryjna 09591
Meetings on Atomic Energy 00663
Mehran 09401
Mehran Library Association: Newsletter 09374
Meida La Sefran 07080
Meieriposten 09303
Mein Leben 00737
Meinungen, Informationen, Nachrichten 06237
Mekhanika Kompozitnykh Materialov: Mechanics of Composite Materials 08558
Mekhanizatsiya i Elektrificatsiya Selskogo Khozyaistva: Mechanization and Electrification of Agriculture 09950
Mélanges Chinois et Bouddhiques 01381
Melita Historica 08666
Melville Society Extracts 15716
Membership Directory of ISBI 15525
me mikroelektronik 06351
Mémoires de l'Académie 04102
Mémoires de l'Académie Nationale de Metz 04119
Mémoires de la Classe des Beaux-Arts 01095
Mémoires de la Classe des Lettres 01095
Mémoires de la Classe des Sciences 01492
Mémoires de la Société Finno-Ougrienne 04003
Mémoires de la Société Géologique de France 04868
Mémoires de la Société Néophilologique: Monograph Series 04081
Mémoires de la Société Vaudoise des Sciences Naturelles 10990
Memoirs in Paleontology 16287
Memoranda Societatis pro Fauna et Flora Fennica 03995
Memoria Anual 00101, 00144, 03056
Memoria de El Colegio Nacional 08724
Memórias da Real Sociedade Arqueológica Lusitana 09668
Memorias de la Real Academia de Ciencias y Artes de Barcelona 10344
Memorie dell'Accademia delle Scienze di Torino 07150
Memorie dell'Accademia delle Scienze di Torino: Classe di Scienze morali 07150
Memorie della classe di scienze morali, storiche e filologiche 07211
Memorie della S.G.I.: Monografie scientifiche 07812
Memorie della Società Astronomica Italiana 07772
Memorie della Società Italiana di Scienze Naturali e del Museo Civico di Storia Naturale di Milano 07946
Memorie e Rendiconti 07175
Memorie lincee, matematica e applicazioni 07211
Memorie Storiche Forogiuliesi 07633
Memorie Valdarnesi 07246
Memory & Cognition 16373
Memo: To The President 13583
Mennonite Historical Bulletin 15217
Menntamálaráð: Almanak 06695
Men of the Stones: Yearbook & Directory 12518
Mensa Bulletin 13941
Mensa Research Journal 13941
Mensch und Arbeit 00556
Mensile di Informazione Sindacale sull'Istruzione Artistica 07767
Mental and Physical Disability Law Reporter 14787
Mental Handicap 11584
Mental Handicap Bulletin 11584
Mental Handicap Research 11584
Mental Health Matters 12635
Mental Research Institute: Newsletter 15720
Mercia Bioscope 12519
Mercury 14541
Mérés és Automatika: Measurement and Automation 06664
Meridian 08028
The Meridian 13325
Méridiens 04348
Der Merkurstab: Beiträge zu einer Entwicklung der Heilkunst 05788
Mesopotamia 07534
Le Message 05004
Message Line 13669
Messaggio Medico: Il Corriere di Roma 07233
Metabolic, Pediatric and Systemic Ophthalmology 15556
Metallofizika: Physics of Metals 11193
Metallogenic Maps and Notes 00405
Metallography 15501
La Metallurgia Italiana 07315, 07686
Metallurgical Transactions A 14255
Metallurgical Transactions B 14255
The Metallurgy of Iron and Steel 03274
Metal News 06796
Metals Abstracts 12524
Metals Abstracts Index 12524
Metals and Materials 12256, 12262
Metals & Materials 12524
Metals Newsletter 16145
Metals Science 12524
Metals Technology 12524
Metalurgija 01796
Metalurji: Metallurgy 11167
Metaphysical Society of America: Announcements 15722
Metaphysical Society of America: Membership List 15722
Metapsichica 07348
Meteoritics 15723
Meteor News 13944
Meteorological and Geoastrophysical Abstracts 13943
Meteorological Society of New Zealand: Newsletter 09131
Meteorologic Bulletin 06571
La Météorologie 04896
Meteorologische Zeitschrift 00809, 05469
Methods and Standards for the Production of Certified Milk 13549
Metodiky pro Zavádění Výzkumu do Praxe 03501
Metropolitan Association of Urban Designers and Environmental Planners: Newsletter 15724
Metsnierba da Technika 05011
Il Mezzo: Notiziario 07419
Mezzosecolo 07584
Miasto: The Town 09629
Micologia Italiana 08049
Microbiologia SEM 10391
Microbiological Review 14073
Microbiologie et Hygiène Alimentaire 11134
Microbiology 12975
Microbiology and Immunology 08272
Microchemical Journal 13945
Microfauna Marina: Mathematisch-Naturwissenschaftliche Klasse 05050
Microfiche Tax Data Base 16764
Microfilming Projects Newsletter 16461
Micro Magazine 15317
Micromath 11397
Microneurography Society: Directory 15728
Micronews 15727
Micronoticias 03030
Microscopy QMC Journal 12734
Microscopy Society of America: Bulletin 15729
Microscopy Society of America: Directory 15729
Microscopy Society of America: Proceedings of Annual Meeting 15729
Microwave World 15502
Mid-Continent Railway Historical Society: Membership Roster 15730
Middle Atlantic Planetarium Society: Constellation 15731
Middle East Geoscience News 11088
The Middle East Journal 15732
Middle East Librarians' Association: Notes 15733
Middle East Report 15734
Middle East Studies Association of North America: Abstracts of Papers Delivered at Annual Meeting 15735
Middle East Studies Association of North America: Bulletin 15735
Middle East Studies Association of North America: Newsletter 15735
Middle Ground 16110
Middle School Journal 16110
Middle States Association Newsletter 15736
Middle States Association of Colleges and Schools: Accredited Membership List 15736
The Middle Way 11715
Midemia Structures and Dimensions 12401
Midwives Chronicle 12780
MIE Közleményi 06637
Migraine Report 02654
Migration World 14689
Mijnbouwkundige Vereeniging: Jaarboek 08976
Mikhtav Lekhaver 07082
Mikrobiologiya: Microbiology 09954
mikroelektronik: Entwicklung und Produktion – Technik und Wirtschaft 06318
Mikroelektronika: Microelectronics 09954
Mikrographische Gesellschaft: Informationsblatt 00707
Militaert tidsskrift 03712
The Military Chest 11248
Military Collector and Historian 14815
Military Historical Society: Bulletin 12527
The Military Law and Law of War Review 01463
Military Medicine 14465
Military Uniforms in America 14815
Milton H. Erickson Foundation: Newsletter 15741
Milton Society of America: Bulletin 15742
Mimarlik: Architecture 11167
MIM Membership Directory 16709
MIM News Bulletin 16709
Mimos: Mitteilungen der SGIK 10850
Minas Tirith Evening Star 14197
Mind 12528
Mineral Industry 00405
Mineralogical Abstracts 12530, 15743
Mineralogical Association of Canada: Newsletter 02655
Mineralogical Magazine 12530
Mineralogical Society of Great Britain and Ireland: Monographs 12530
Mineralogicheskii Zhurnal: Mineralogical Journal 09954
Mineral Resources 00405

Minerals and Metallurgical Processing *16552*
Minerals, Metals and Materials Society: Membership Directory *15744*
Minerva *12824*
Minerva Anesthesiologica *07835*
Minerva Medico-Legale *07899*
Minerva Nipiologica *07916*
Minerva Pneumologica *07933*
Minerva Stomatologica *07919*
Mini-Beacon *14466*
Minicomputer Software Quarterly *14466*
Mining and Metallurgical Society of America: Bulletin *15745*
Mining and Metallurgical Society of America: News-Letter *15745*
The Mining Engineer *12310*
Mining Engineering *16552*
Mining Newsletter *16145*
Ministerial Formation *10665*
Minnestal: Obituaries *10491*
Mintek: Annual Report *10154*
Mintek Bulletin: Mintek Research Digest *10154*
Minutes of the Art Meetings *14492*
Mirovaya Ekonomika i Mezhdunarodnye Otnosheniya: World Economics and International Relations *09954*
Miscelanea Matematica: Boletín de la Sociedad Matematica Mexicana and Aportaciones Matematicas *08753*
Miscellaneous Publications *06762*
MIS Newsletter *12499*
MIS Quarterly *16540*
Mitekufat Haeven: Journal of the Israel Prehistoric Society *07088*
Mitglieder-Bulletin: à jour *10938*
Mitgliederzeitschrift der Österreichischen Hämophilie Gesellschaft *00858*
Mitteilungen aus Baltischem Leben *05164*
Mitteilungen aus dem Deutschen Kunststoff-Institut *05593*
Mitteilungen aus dem Forschungsinstitut für Wärmeschutz *05732*
Mitteilungen aus der Baunormung *06150*
mitteilungen aus der FfH *05738*
Mitteilungen aus der österreichischen Byzantinistik und Neogräzistik *00734*
Mitteilungen der Ärztekammer für Kärnten *00534*
Mitteilungen der Anthropologischen Gesellschaft in Wien *00544*
Mitteilungen der Antiquarischen Gesellschaft *10616*
Mitteilungen der Arbeitsgemeinschaft der Parlaments- und Behördenbibliotheken *05089*
Mitteilungen der Archäologischen Gesellschaft Steiermark *00568*
Mitteilungen der Deutschen Gesellschaft für allgemeine und angewandte Entomologie *05317*
Mitteilungen der Deutschen Malakozoologischen Gesellschaft *05465*
Mitteilungen der Deutschen Orient-Gesellschaft *05480*
Mitteilungen der Fränkischen Geographischen Gesellschaft *05757*
Mitteilungen der Geographischen Gesellschaft in Hamburg *05778*
Mitteilungen der Gesellschaft für Bibliothekswesen und Dokumentation des Landbaues *05804*
Mitteilungen der Gesellschaft für vergleichende Kunstforschung *00636*
Mitteilungen der Internationalen Stiftung Mozarteum *00683*
Mitteilungen der Mathematischen Gesellschaft in Hamburg *06111*
Mitteilungen der Münchner Entomologischen Gesellschaft *06124*
Mitteilungen der Mundartfreunde Österreichs *00978*
Mitteilungen der Naturforschenden Gesellschaft des Kantons Glarus *10766*
Mitteilungen der Naturforschenden Gesellschaft des Kantons Solothurn *10772*
Mitteilungen der Naturforschenden Gesellschaft Luzern *10770*
Mitteilungen der ÖGFKM *00779*
Mitteilungen der ÖGIST *00794*
Mitteilungen der ÖMG *00869*
Mitteilungen der Österreichischen Bodenkundlichen Gesellschaft *00733*
Mitteilungen der Österreichischen Exlibris-Gesellschaft *00741*
Mitteilungen der Österreichischen Geographischen Gesellschaft *00745*
Mitteilungen der Österreichischen Geologischen Gesellschaft *00747*
Mitteilungen der Österreichischen Gesellschaft für Chirurgie mit den assoziierten Fachgesellschaften *00769*
Mitteilungen der Österreichischen Numismatischen Gesellschaft *00872*
Mitteilungen der POLLICHIA *06238*
Mitteilungen der Rheinischen Naturforschenden Gesellschaft *06263*
Mitteilungen der Schweizerischen Entomologischen Gesellschaft/ Bulletin de la Société Entomologique Suisse *10804*
Mitteilungen der SGFF *10819*
Mitteilungen der Wilhelm-Busch-Gesellschaft *06442*
Mitteilungen des Badischen Landesvereins für Naturkunde und Naturschutz *05163*
Mitteilungen des Deutschen Archäologischen Instituts, Abteilung Kairo *05568*
Mitteilungen des Deutschen Germanistenverbandes *05513*
Mitteilungen des Dokumentationsarchives des österreichischen Widerstandes *00582*
Mitteilungen des Hochschulverbandes *05516*
Mitteilungen des Instituts für Angewandte Wirtschaftsforschung *05950*
Mitteilungen des Naturwissenschaftlichen Vereins für Steiermark *00713*
Mitteilungen des Verbandes der deutschen Höhlen- und Karstforscher e.V. *06335*
Mitteilungen des Vereins der Museumsfreunde in Wien *00979*
Mitteilungen des Vereins Freunde der Archäologie *00981*
Mitteilungen des Vereins für Forstliche Standortskunde und Forstpflanzenzüchtung *06384*
Mitteilungen für Lehrer slawischer Fremdsprachen *00962*
Mitteilungen und Forschungsbeiträge der Cusanus-Gesellschaft *05280*
Mitteilungsblätter der Deutschen MERU-Gesellschaft *05468*
Mitteilungsblatt der Berliner Zahnärzte *06041*
Mitteilungsblatt der Deutschen Gesellschaft für Orthopädie und Traumatologie *05392*
Mitteilungsblatt der GMDS *05382*
Mitteilungsblatt der IBG *00666*
Mitteilungsblatt der Wiener Beethoven-Gesellschaft *01012*
Mitteilungsblatt des Deutschen Akademikerinnenbundes e.V. *05495*
Mitteilungsblatt des Deutschen Altphilologenverbandes *05497*
Mitteilungsblatt des ÖVS *00905*
Mixing & Separation *11437*
MLA/ADE Job Information List *14427*
MLA/ADFL Job Information List *14428*
MLA Directory of Periodicals *15750*
MLA International Bibliography *15750*
MLA News *15710*
MMA Newsletter *08650*
Mnimeia Ellinikis Historias: Documents of Greek History *06498*
Mobil: Das Rheuma-Magazin *05515*
Modelling and Simulation in Materials Science and Engineering *12268*
Moderna Språk *10515*
Modern Austrian Literature *15374*
Modern Believing *12534*
Modern Casting *13824*
Moderne Sprachen *00961*
Modern Geriatric Topics *16064*
The Modern Language Journal *16047*
Modern Language Review *12535*
Modern Management *12285*
Modern Pain Control *14044*
Modern Steel Construction *13895*
Modus *12573*
Mokelas ir Tekhnika: Science and Technology *08573*
Mokuzai Gakkaishi: Journal of the Japan Wood Reseach Society *08344*
Moldovan Academy of Sciences: Izvestiya: Bulletin *08783*
Molecular Endocrinology *15041*
Molecular Pharmacology *14077*
Molecular Plant-Microbe Interactions *14006*
Molekulyarnaya Biologiya: Molecular Biology *09954*
Monatsberichte des Österreichischen Institutes für Wirtschaftsforschung *00912*
Monatshefte für Chemie *00637*
Monatsschrift für Kinderheilkunde *05371*
Le Monde Muselman Contemporain-Initiations *01205*
Mondes et Cultures *04105*
Mondo Aperto *07520*
Money Affairs *08712*
Mongolia Language and Literature *03206*
Mongolian Studies: Journal of the Mongolia Society *15752*
The Mongolia Society Bulletin *15752*
The Mongolia Society Newsletter *15752*
Monitor *00422, 14563, 14672*
Le Monitor des Architectes *04900*
Monografías de Economía del CIMMYT *08719*
Monografie Biochemiczne: Biochemical Monographs *09557*
Monografieën van de Nederlandse Entomologische Vereniging *08992*
Monografie parazytologiczne *09598*
Monografie Scientifiche del CNR *07628*
Monografies Mèdiques *10241*
Monographiae Botanicae *09558*
Monographs in Facial Plastic Surgery *11983*
Monographs of Allergy and Clinical Immunology *00153*
Monographs of Corporate and Institutional Histories *16194*
Monograph/Special Studies Series *15829*
Monograph Studies *13377*
Montana *16883*
Montaria *08712*
The Montessori Society A.M.I. (UK): Quarterly *12536*
Monthly Garden News Sheet *06710, 06711*
Monthly Oil Market Report *04126*
Monthly Weather Review *13943*
Monumental Brass Society: Bulletin *12537*
Monumental Brass Society: Transaction *12537*
Monumenta Musicae Svecicae *10575*
Monumenti Antichi *07211*
Monumenti Musicali Italiani *07909*
Morality Quarterly *03333*
Moreana *04132*
Moredun Research Institute: Scientific Report *11235*
Morgan Horse Magazine *11611*
Morgannwg *12116*
Morim Bulletin *15610*
Morphologia és Igazságügyi Orvosi Szemle *06651*
Morskoi Gidrofizicheskii Zhurnal: Marine Hydrophysical Journal *11193*
Morton Prince Digest of Hypnotherapy and Hypnoanalysis *15294*
Mosquito News Journal of the American Mosquito Control Association *13948*
Mosquito Systematics *13948*
MOTESZ Magazin *06650*
Motorcycle Safety Foundation: Annual Report *15756*
Mots *04543*
The Mountain World *10931*
Mouvement Français pour la Qualité: Newsletter *04601*
Mouvement Français pour la Qualité: Newsletter: Qualité en Mouvement *04601*
Movimento Operaio e Socialista *07514*
Movoznavstvo: Linguistics *11193*
Moznayim *07055*
MPG-Spiegel *06114*
MPT Metallurgical Plant and Technology International *06374*
MRC Annual Report *10206*
MRC News *10206, 12508*
MRC Newsletter *02646*
MRDA Membership Directory *15708*
MRDA News and Views *15708*
MRG Reports *12533*
MS Aktuell/SP Actuel/Attualitá SM *10873*
MS Biblioteksnyt *03772*
MSF: Bulletin *04600*
M.S. News *12541*
MS Research in Progress *12357*
MS Research Report *12357*
MS Revy *03772*
mta praxis *05554*
MTM Journal *15757*
Múanyag és Gumi: Plastics and Rubber *06606*
Müemlékvédelem: Care of Ancient Monuments *06676*
Mühendis ve Makina: Engineers and Machines *11167*
Münchener Medizinische Wochenschrift *00533*
Müszaki Magazin *06665*
Müvészettörténeti Értesítő *06655*
Multiphase Update *11437*
Multivariate Behavioral Research *16682*
Mundart: Forum des Vereins Schweizerdeutsch *11024*
Mundo Universitario *03348*
Munibe *10367*
Munibe (Antropologia-Arkeologia) *10371*
Munibe (Ciencias Naturales) *10371*
Municipal Engineer *06966, 11368*
Municipal Solid Waste News *16719*
Muse *02201*
Les Musées de Tunisie *11140*
Musées et Collections Publiques de France *04252*
Musei e Gallerie d'Italia *07363*
MUS'en *03651*
Museo *04052*
El Museo Canario *10372*
Museogramme *02201*
Museoviraston arkeologian julkaisuja *03983*
Museoviraston rakennushistorian osaston raportteja: Annual Report *03983*
Museoviraston työväenkulttuurijulkaisuja *03983*
Museum Anthropology *14880*
Museumbrief *01509*
Museum Helveticum: Schweizerische Zeitschrift für klassische Altertumswissenschaft *10939*
Museumleven *01509*
Museum News *13554*
Museum of the Fur Trade Quarterly *15758*
Museum of the Great Plains Newsletter *15341*
Museum Round-up *01983*
Museums and Galleries Commission: Annual Report *12544*
Museums Australia *00435*
Museumsführer *06263*
Museums Journal *12545*
Museums Journal of Pakistan *09390*
Museumskunde *05529*
Museums Newsletter *06837*
Museumsnytt *09302, 09305*
Museum Studies *08205, 09390*
Museums Yearbook *12545*
Museum Training Institute: News *12546*
Musica Domani: Trimestrale di Pedagogia Musicale *07967*
Musical Life *09966*
Musical Review *09966*
Musicanada *02203*
Musica Sacra *05073*
The Music Association of Ireland: Annual Report *07015*
Musica Sveciae *10487*
Music Cataloging Bulletin *15764*
Music Critics Association: Membership List *15762*
Music Critics Association: Newsletter *15762*
Music Educators Journal *15763*
Music Educators National Conference: Newsletter *15763*
Music in the Media: IMZ-Bulletin *00682*
Music Journal *12195*
Music Library Association: Newsletter *15764*
Music Library Association: Notes *15764*
Musicological Society of Australia: Newsletter *00436*
Musicology Australia *00436*
Music Therapy *13487*
Music USA-Industry Statistics *13950*
Musikalische Berufsstudien in der Schweiz: Etudes Musicales Professionnelles en Suisse *10962*
Musikalische Denkmäler *05050*
Musikerziehung: Zeitschrift der Musikerzieher Österreichs *00549*
Die Musikforschung *05842*
Musikforum: Referate und Informationen des Deutschen Musikrates *05530*
Musikfreunde *00612*
Musik i Sverige *10575*
Musik und Bildung *05543*
Musik und Bildung: Praxis Musikerziehung *06361*
Muskelreport *05339*
Mutisía: Acta Botánica Colombiana *03391*
Muttersprache: Zeitschrift zur Pflege und Erforschung der deutschen Sprache *05808*
Mutual Fund Forecaster *15276*
Muzeji *17118*
Muziek en Dans *08959*
Muzikalnaya Akademia *09966*
Muzikoznanie *01770*
Mycologia *15766*
Mycologia Memoirs *15766*
Mycological Research *11614*
Mycological Society of America: Newsletter *15766*
Mycologist *11614*
Mycopathologia *12378*
The Mycophile *16219*
Mykologické listy *03534*
Myrmecia *00289*
Mythic Circle *15768*
Mythic Society: Journal *06839*
Mythlore *15768*
Mythprint *15768*

NAACS News *14375, 16765*
NAAMM Membership Roster *16210*
NAAMM Newsletter *16210*
NAASR Newsletter *15793*
NABD-Mitteilungen *06152*
NABTE Review *15796*
NACADA Journal *15770*
NACADA Newsletter *15770*
NACCAS Review *15781*
NACDRAO Directory *15841*
NACF Magazine *12556*
Nachrichten aus Ärztlicher Mission *05581*
Nachrichten aus Chemie, Technik und Laboratorium *05793*
Nachrichten aus dem Karten- und Vermessungswesen *05947*
Nachrichtenblatt AGST *00568*
Nachrichtenblatt der Bayerischen Entomologen *06124*
Nachrichten-Bulletin SGO *10839*
Nachrichten der Gesellschaft für Natur- und Völkerkunde Ostasiens e.V.: Zeitschrift für Kultur und Geschichte Ost- und Südostasiens *05844*
Nachrichten Deutsche Geologische Gesellschaft *05311*
Nachrichten für Dokumentation *05341*
Nachrichtentechnische Zeitschrift: Communications Journal *05942*
Nachrichten zur Mahler-Forschung: News about Mahler Research *00673*
NACTA Journal *15842*
NADAP News/Report *15916*
NADSA Conference Directory *15851*
NADSA Update *15851*
NAE4-HA Membership Directory *15858*
NAEIR News *15819*
NAES Journal *15857*
NAFEO/AID Update *15798*
NaFo-nytt *09332*
NAH & WW Newsletter *15866*
Names *13952*
Namibiana *08816*
Namn och Bygd *10484*
NAMS Newsletter *15872*
NANASP News *15875*
Nano Technology *12268*
NAPT News Letter *15810*
NAREMCO Report *16135*
Narodna Tvorchist ta Etnografiya: Folk Art and Ethnography *11193*
NAS Avionics Newsletter *15918*
NASDA Letter *15895*
Naše jame *10101*
Naše řeč *03500*
NASPE News *16225*
NASPG Membership Directory *16222*
NA Sport Info *06206*
NAS-Proceedings *03780*
Nastava i va spitanje *17120*
NASW Newsletter *15891*
NATA News *00437*
NATE News *12564*
NATFHE Journal *12571*
Nathaniel Hawthorne Review *15769*
National *02081, 02586*
National Academic Advising Association: Proceedings of Annual Conference *15770*
National Academy of Arts: Bibliography *08542*
National Academy of Arts: Bulletin *08542*
National Academy of Arts: Journal *08542*
National Academy of Code Administration: Newsletter: Blueprint *15771*
National Academy of Design: Academy Bulletin *15772*
National Academy of Design: Academy Calendar *15772*
National Academy of Opticianry: Academy Newsletter *15775*
National Academy of Public Administration: Annual Report *15776*
National Academy of Recording Arts and Sciences: Program Book *15777*
National Academy of Sciences: Annual Number *06841*
National Academy of Sciences: Bibliography *08543*
National Academy of Sciences: Bulletin *08543*
National Academy of Sciences: Journal *08543*
National Academy of Sciences Letters *06841*
National Accounts *01664*
National Accrediting Agency for Clinical Laboratory Sciences: Newsletter *15780*
National Alliance of Black School Educators: Membership Roster *15784*
National Alliance of Black School Educators: News Briefs *15784*
National Alopecia Areata Foundation Newsletter *15786*
National Art Education Association: Newsletter *15790*
National Association for Bilingual Education: Journal *15794*
National Association for Bilingual Education: Newsletter *15794*
National Association for Biomedical Research: Report *15795*
National Association for Biomedical Research: Update *15795*
National Association for Family and Community Education: Handbook *15800*
National Association for Human Development: Digest *15802*
National Association for Industry-Education Cooperation: Journal *15805*
National Association for Industry-Education Cooperation: Newsletter *15805*
National Association for Outlaw and Lawman History: Newsletter *15808*
National Association for Outlaw and Lawman History: Quarterly *15808*
National Association for Physical Education in Higher Education: Newsletter *15809*
National Association for Research in Science Teaching: Abstracts of Papers Presented to Annual Meeting *15812*
National Association for Research in Science Teaching: Membership Directory *15812*
National Association for Research in Science Teaching: Newsletter *15812*
National Association for Search and Rescue: Update *15814*
National Association for the Exchange of Industrial Resources: Membership Directory *15819*
National Association for the Practice of Anthropology: Bulletin Series *15820*
National Association for Trade and Industrial Education: NewsNotes *15821*
National Association for Women in Education: Directory *15823*
National Association for Women in Education: Journal *15823*
National Association for Women in Education: Newsletter *15823*
National Association of Academies of Science: Directory and Proceedings *15826*
National Association of Academies of Science: Newsletter *15826*

National Association of Baptist Professors of Religion: Bibliographic Series *15829*
National Association of Biology Teachers: News + Views *15831*
National Association of Boards of Pharmacy: Newsletter *15834*
National Association of Bond Lawyers: Directory *15835*
National Association of Bond Lawyers: Newsletter *15835*
National Association of Children's Hospitals and Related Institutions: Newsletter *15837*
National Association of College and University Business Officers: Annual Report *15840*
National Association of College Deans, Registrars and Admissions Officers: Newsletter *15841*
National Association of College Deans, Registrars and Admissions Officers: Proceedings *15841*
National Association of College Wind and Percussion Instructors: Journal *15843*
National Association of County Engineers: Newsletter *15846*
National Association of County Planners: Roster *15847*
National Association of Dental Laboratories: Executive Information Series *15849*
National Association of Dramatic and Speech Arts: Newsletter *15851*
National Association of Emergency Medical Technicians: Newsletter *15854*
National Association of Emergency Medical Technicians: Perspective *15854*
National Association of Environmental Professionals: Newsletter *15856*
National Association of Episcopal Schools: Newsletter *15857*
National Association of Extension 4-H Agents: Membership Report *15858*
National Association of Extension 4-H Agents: News and Views *15858*
National Association of Flight Instructors: Newsletter *15860*
National Association of Geology Teachers: Membership Directory *15861*
National Association of Health Career Schools: Bulletin *15862*
National Association of Health Services Executives: Notes *15863*
National Association of Hebrew Day School Administrators: Directory *15864*
National Association of Industrial and Technical Teacher Educators: Directory *15869*
National Association of Industrial and Technical Teacher Educators: News and Views *15869*
National Association of Marine Surveyors: Annual Membership List *15872*
National Association of Nutrition and Aging Services Programs: Annual Report *15875*
National Association of Nutrition and Aging Services Programs: Monthly Membership Updates *15875*
National Association of Nutrition and Aging Services Programs: Special Bulletin *15875*
National Association of Principals of Schools for Girls: Proceedings *15879*
National Association of Private Schools for Exceptional Children: Directory *15880*
National Association of Private Schools for Exceptional Children: Newsbriefs *15880*
National Association of Private Schools for Exceptional Children: Newsletter *15880*
National Association of School Psychologists: Directory *15884*
National Association of School Psychologists: Journal Review *15884*
National Association of School Psychologists: Newsletter *15884*
National Association of School Safety and Law Enforcement Officers: Membership Directory *15885*
National Association of Schools of Art and Design: Directory *15887*
National Association of Schools of Art and Design: Handbook *15887*
National Association of Schools of Music: Directory *15888*
National Association of Schools of Music: Handbook *15888*
National Association of Schools of Music: Proceedings *15888*
National Association of Secondary School Principals: Bulletin *15892*
National Association of Secondary School Principals: Curriculum Report *15892*
National Association of State Archeologists: Directory *15893*
National Association of State Archeologists: Newsletter *15893*
National Association of State Directors of Teacher Education and Certification: Roster *15897*
National Association of State Supervisors of Vocational Home Economics: Directory *15905*
National Association of State Supervisors of Vocational Home Economics: Newsletter *15905*
National Association of State Universities and Land-Grant Colleges: Annual Report *15907*
National Association of Supervisors of Business Education: Newsletter *15909*
National Association of Teachers of Singing: Journal *15911*
National Association of Teachers of Singing: Membership Directory *15911*
National Association of Testing Authorities, Australia: Annual Report *00437*
National Association of University Women: Bulletin *15913*
National Association of University Women: Directory of Branch Presidents and Members *15913*
National Association of Women Lawyers: Membership Directory *15915*
National Biographies *13887*
National Black MBA Association Newsletter *15919*
The National Board Examiner *15926*
National Board for Respiratory Care: Annual Directory *15924*
National Board for Respiratory Care: Newsletter *15924*
National Board of Examiners in Optometry: Newsletter *15925*
National Board of Examiners in Optometry: Report & Examinations *15925*
National Board of Medical Examiners: Annual Report *15926*
National Bulletin *13587*
National Burn Information Exchange Newsletter *16088*
National Cancer Center: Annual Report *08184*
National Cancer Center: Collected Papers *08184*
National Catholic Business Education Association: Review *15933*
National Catholic Conference for Interracial Justice: Commitment *15934*
National Catholic Educational Association: Notes *15935*
National Catholic Educational Association: Update *15935*
National Center for Computer Crime Data: Annual Statistical Report *15941*
National Center for State Courts: Report and Master Calendar *15947*
National Character Laboratory: Newsletter *15952*
National Coalition News *15956*
National Coalition of Alternative Community Schools: Newsletter *15956*
National College of Foot Surgeons: Journal *15958*
National Committee on the Treatment of Intractable Pain: Newsletter *15965*
National Community Education Association: Membership Directory *15966*
National Conference of Commissioners on Uniform State Laws: Handbook and Proceedings *15969*
National Conference of Standards Laboratories: Conference Proceedings *15971*
National Conference of Standards Laboratories: Newsletter *15971*
National Conference of Yeshiva Principals: Newsletter *15973*
National Conference on Fluid Power *15974*
National Conference on Research in English: Directory *15975*
National Conference on Research in English: Newsletter *15975*
National Conference on the Advancement of Research: Proceedings *15976*
National Conference on Weights and Measures: Report *15977*
National Cooperative Highway Research Program Report *16790*
National Cooperative Transit Research and Development Program Reports *16790*
National Council for Accreditation of Teacher Education: Annual List of Accredited Institutions *15984*
National Council for Geographic Education: Perspective *15988*
National Council for Research and Planning: Newsletter *15990*
National Council for the Social Studies: Bulletin *15992*
National Council for Urban Economic Development: Commentary *15995*
National Council of Athletic Training: Directory *15996*
National Council of Athletic Training: Newsletter *15996*
National Council of Guilds for Infant Survival: Newsletter *16000*
National Council of Intellectual Property Law Associations: Chairman's Letter *16001*
National Council of Intellectual Property Law Associations: Newsletter *16001*
National Council of Teachers of Mathematics: News Bulletin *16011*
National Council of University Research Administrators: Newsletter *16013*
National Council on Alcoholism and Drug Dependence: Annual Report *16016*
National Council on Economic Education: Annual Report *16018*
National Council on Economic Education: Directory of Affiliated Councils and Centers *16018*
National Council on Family Relations: Newsletter *16020*
National Council on Health Laboratory Services: Membership Organization List *16022*
National Council on Radiation Protection and Measurements: Proceedings of the Annual Meeting *16025*
National Dental Association: Journal *16031*
National Dental Association: Newsletter *16031*
National Directory of Educational Programs in Gerontology *14299*
The National Directory of Internships *16155*
National Directory of Physician Assistant Programs *14482*
National Directory of Vocational Experts *13666*
National Eczema Society: Exchange *12586*
National Engineer *15878*
National Engineering Consortium: Conference Proceedings *16042*
National Federation of Abstracting and Information Services: Membership Directory *16045*
National Federation of Abstracting and Information Services: Report Series *16045*
National FFA Organization: Between Issues *16050*
National FFA Organization: Update *16050*
National Fire Codes *16051*
National Fire Protection Association: Yearbook *16051*
National Flags *15098*
National Foundation for Brain Research: Annual Report *16057*
National Foundation for Brain Research: Newsletter *16057*
National Foundation for Educational Research in England and Wales: Research Reports *12591*
National Future Farmer Magazine *16050*
National Genealogical Society: Newsletter *16062*
National Genealogical Society Quarterly *16062*
National Geographic *16063*
National Geographical Journal of India *06844*
National Geographic World *16063*
National Guild of Catholic Psychiatrists: Bulletin *16066*
National Guild of Community Schools of the Arts: Guildletter *16067*
The National Health *14022*
National Health and Medical Research Council: Council Session Reports *00438*
National Health Lawyers Association: Digest Index *16072*
National Health Lawyers Association: Register *16072*
National Hearing Aid Society: Confidential Report to Members *16073*
National Housing Conference: Newsletter *16078*
National Humanities Center: Newsletter *16080*
National Information Service for Earthquake Engineering: Current Abstract Update Service *16087*
National Institute Briefing Notes *12598*
National Institute Discussion Papers *12598*
National Institute Economic Review *12598*
National Institute for Certification in Engineering Technologies: Newsletter *16089*
National Institute for Compilation and Translation: News Bulletin *03312*
National Institute for Compilation and Translation: The Institute Periodical *03312*
National Institute of Science: Newsletter *16095*
National Institute of Science: Transactions *16095*
National Institute Report Series *12598*
Nationalities Papers *16470*
National Jewish Center for Immunology and Respiratory Medicine: Annual Report *16098*
National League of American PEN Women: Roster *16100*
The National Librarian *16102*
National Listing of Rolfers *16427*
National Marine Educators Association: Current: The Journal of Marine Education *16103*
National Mastitis Council: Annual Meeting Proceedings *16104*
National Materials Advisory Board: Newsletter *16105*
National Medical and Dental Association: Bulletin *16106*
National Medical Association Newsletter *16107*
National Medical Journal of China *03178*
National Mental Health Association: Focus *16109*
National Monuments Commission: Annual Report *17156*
National Monuments Commission: Research Publications *17156*
National Network Directory *16913*
National News *16048*
National News Report *16471*
Nationaløkonomisk Tidsskrift *03774*
National Optometric Association: Newsletter *16115*
National Orchestral Association: Fact Sheet *16116*
National Organization for Continuing Education of Roman Catholic Clergy: News Notes *16117*
National Organization of Minority Architects: Newsletter *16119*
National Organization on Legal Problems of Education: Notes *16120*
National Organization to Insure a Sound-Controlled Environment: Newsletter *16121*
National Osteopathic Guild Association: Newsletter *16123*
National Perinatal Association: Newsletter *16124*
National Perinatal Association: Proceedings *16124*
National Pest Control Association: Roster of Members *16125*
National Pharmaceutical Association: Journal *16126*
National Plant Board: Minutes *16129*
National Plant Board: Proceedings of the Annual Meeting *16129*
National Podriatic Medical Association: Annual Seminar Ad Book *16130*
National Podriatic Medical Association: Newsletter *16130*
National Psychological Association for Psychoanalysis: Bulletin *16132*
National Psychological Association for Psychoanalysis: News & Reviews *16132*
National PTA Bulletin *15865*
National Reading Conference: Yearbook *16134*
National Registry in Clinical Chemistry: Directory *16136*
National Rehabilitation Association: Newsletter *16137*
National Report on Human Resources *14093*
National Research Centre: Bulletin *03911*
National Research Council: News Report *16138*
National Roster of Black Elected Officials *15615*
National Safety Management Society: Focus *16146*
National Safety Management Society: Update Newsletter *16146*
National SAMPE Technical Conference Series *16589*
National Science *03314*
National Science Council: Abstracts of Research Papers *03314*
National Science Council: Proceedings *03314*
National Science Supervisors Association: Directory *16150*
National Science Supervisors Association: Newsletter *16150*
National Science Supervisors Association: Proceedings of Summer Conference *16150*
National Sculpture Review *16152*
National Sculpture Society: Newsletter *16152*
National Security Report *14038*
National Society for Clean Air: Proceedings of Annual Conferences and Seminars *12607*
National Society for the Preservation of Covered Bridges: Bulletin *16157*
National Society for the Preservation of Covered Bridges: Notices *16157*
National Society for the Study of Education: Yearbook *16158*
National Space Club: Newsletter *16161*
National Speleological Society: Membership List *16163*
National Speleological Society: Monthly News *16163*
National Student Speech Language Hearing Association: Journal *16166*
National Tax Journal *16168*
National Technical Association: Journal *16169*
National Technical Association: Newsletter *16169*
National Technical Services Association: Membership Roster *16170*
National Trust for Places of Historic Interest or Natural Beauty: Annual Report *12614*
National Trust for Places of Historic Interest or Natural Beauty: Members' and Visitors' Handbook *12614*
National Trust for Places of Historic Interest or Natural Beauty: Newsletter *12614*
National Tumor Registrars Association: Abstract *16174*
National Tumor Registrars Association: Membership Roster *16174*
National Tumor Registrars Association: Proceedings/Annual Report *16174*
National Union of Teachers: Annual Report *12616*
National University Continuing Education Association: Newsletter *16175*
National Update *16913*
National Waterline *16179*
National Water Resources Association Directory *16179*
National Waterways Transport Association: Newsletter *12617*
National Wetlands Newsletter *15050*
National Wildlife *16180*
National Workshop Proceedings *15806*
Natterjack *12879*
Náttúrufrådingurinn *06689*
Náttúrufraedingurinn *06692*
Natura *08958, 09695*
Natura: Rivista di Scienze Naturali *07946*
Natura Alpina *07789*
Natura e Montagna *08041*
Natural Environment Research Council: Annual Report *12619*
Natural History Bulletin *11099*
Natural History Society of Northumbria: Transactions *12620*
Naturalia *09695*
Les Naturalistes Belges: Bulletin *01414*
Natural Product Updates *12857*
Natural Rubber Technology *08651*
Natural Sciences and Engineering Research Council of Canada: Contact *02675*
Natural World: Annual Review *12850*
Natura & Montagna *07800*
Der Naturarzt *05531*
Natura Società *07698*
The Nature Conservancy Magazine *16188*
Nature Study *13956*
Naturforschende Gesellschaft in Bern: Mitteilungen *10768*
Naturforschende Gesellschaft in Zürich: Vierteljahresschrift *10769*
Naturforschende Gesellschaft Schaffhausen: Mitteilungen *10771*
Naturopa *04384*
Naturopa-Newsletter *04384*
Naturschutz heute *06132*
Naturschutz- und Naturparke *06422*
Die Naturstein-Industrie *05715*
Natur & Umwelt *05266, 05269*
Natur und Land *00898*
Natur und Mensch *06130, 10617*
Natur und Mensch: Schweizerische Blätter für Natur und Heimatschutz *10783*
Natur und Umwelt *05267*
Naturvetenskapliga Forskningsrådet: Annual Report *10504*
Naturvetenskapliga Forskningsrådet: Yearbook *10504*
Naturwissenschaftliche Gesellschaft Winterthur: Mitteilungen *10773*
Naturwissenschaftlicher Verein in Hamburg: Abhandlungen *06135*
Naturwissenschaftlicher Verein in Hamburg: Verhandlungen *06135*
Natuurbehoud *09086*
Natuur en Milieu *09073*
Natuur- en Stedeschoon *01411*
Natuurhistorisch Maandblad *08979*
Natuurkundige Voordrachten, Nieuwe Reeks *08949*
Nauchnoe Proborostroenie: Scientific Instrumentation *09954*
Nauka v Rossii: Science in Russia *09954*
Nauki v Uzbekistane: Social Sciences in Uzbekistan *16948*
Naunyn-Schmiedeberg's Archive of Pharmacology *05400*
Naunyn-Schmiedeberg's Archives of Pharmacology *05484*
Nautgriparaektin *06679*
Nautica Fennica *03983*
Nautologia *09592*
The Naval Architect incorporating Warship Technology *12823*
Naval Architecture and Ocean Engineering *08279*
Naval Engineers Journal *14147*
NAVA News *16227*
Navigation *08337, 15334*
Navigation News *12814*
Navy Records Society: Annual Report *12622*
Nawpa Pacha *15311*
NBS Handbook *15977*
NBTA News *02902*
NCALL News *16015*

NCA Quarterly 16228
NCBL Notes 15968
NCFL Newsletter 15936
NCHRP Synthesis of Highway Practice 16790
NCLC Energy Update 15983
NCLC Reports 15983
NCPI Hotline 16027
NCRP Commentaries 16025
NCRP Reports 16025
NCTRP Research Result Digest 16790
NCTRP Synthesis of Transit Practice 16790
NEAS Bulletin 16189
NEAS Newsletter 16189
NEA Today 16039
Nederland's Adelsboek 08838
Nederlands Archievenblad 09088
Nederlands Bosbouw Tijdschrift 08953
Nederlandsche Internisten Vereeniging: The Newsletter 08982
De Nederlandsche Leeuw 08966
Nederlandsche Vereniging voor Druk- en Boekkunst: Mededelingen 08984
Nederlandse Genealogieen 08966
Nederlandse Historische Bronnen 09051
Nederlandse Organisatie voor Wetenschappelijk Onderzoek: Jaarboek: Annual Report 08998
Nederlandse Vereniging voor Internationaal Recht: Mededelingen 09021
Nederlandse Vereniging voor Medisch Onderwijs: Bulletin: Officieel Blad 09026
Nederland's Patriciaat 08838
Nederlands Tandartsenblad: Dutch Dental Journal 08983
Nederlands Tijdschrift voor Anesthesiologie MTVA 09016
Nederlands Tijdschrift voor Fysiotherapie 09050
Nederlands Tijdschrift voor Manuele Therapie 09050
Nederlands Tijdschrift voor Obstetrie en Gynaecologie 09030
Nederlands Tijdschrift voor Opvoeding, Vorming en Onderwijs 09009
Nederlands-Zuidafrikaanse Vereniging: Annual Report 09057
Needlenews 13324
Neftekhimiya: Petrochemistry 09954
Negro History Bulletin 14356
NEHA-Bulletin 09084
Neige et Avalanches 04316
Neirofiziologiya: Neurophysiology 09954
Neirokhimiya: Neurochemistry 00203, 09954
Nematological Abstracts 11720
Nematology Newsletter 16684
Nenpo 08302
Nensho Kenkyu 08345
Nensho-Kenkyu 08346
Nentori: Review 00005
Neofilolog 09593
Neo-Hellenika 14690
Neo Metaphysical Digest 13056
Neometaphysical Series 13056
Neometaphysics & Current Affairs Series 13056
Nephrologische Nachrichten 00812
Néprajzi Hirek 06647
NERC News 12619
Nervenarzt 05410
Nestroyana: Blätter der Internationalen Nestroy-Gesellschaft 00675
Netherlands Journal of Agricultural Science 08962
Netherlands Journal of Cardiology 09017
The Netherlands Journal of Medicine 08982
Netherlands Journal of Zoology 08991
Netherlands Milk and Dairy Journal 08903
Netsu Shori: Journal of the Japan Society for Heat Treatment 08347
Network 12268
network 15075
The Network 16804
Network Circular Letter 03021
network en espanol 15075
network en français 15075
Network Exchange 13365
Network Magazine: The Magazine of Computer Communications 15317
Network News 00223, 16181
Network News Exchange 16531
Network Newsletter 02811
The Network: Newsletter 16192
Net Worth: Valeurs nettes 02197
Die Neue Hochschule 05928
Neue Horizonte 00870
Neue Musikzeitung 05143, 05543
Die Neue Ordnung 05961
Neue Studien zur Musikwissenschaft 05050
Neue Unterrichtspraxis 05852
Neujahrsblätter 10771
Neujahrsblatt 10769
Neuphilologische Mitteilungen 04081
Neurochirurgie 04674
Neurologia i Neurochirurgia Polska 09594
Neurologia Medico-Chirurgica I 08351
Neurologia Medico-Chirurgica II 08351
Neurology 13958
Neurophysiologie Clinique 04675
Neuropsychiatrie 00638
Neuroradiology 05680
Neuroscience News 08379
Neuroscience Newsletter 16555
Neuroscience Research 08379
Neuroscience Training Programs in North America 16555
Neurosurgery 14842
Neurosurgical Topics 13555
Neusprachliche Mitteilungen aus Wissenschaft und Praxis 05695
Newark and Nottinghamshire Agricultural Society: Catalogue 12624
New Brunswick Institute of Agrologists: Newsletter 02686
New Brunswick Naturalist: Le Naturaliste du New Brunswick 02680
New Brunswick Research and Productivity Council: Annual Report 02691
New Civil Engineer 06966, 12294
The Newcomen Society: Bulletin 12625
The Newcomen Society: Transactions 12625
New Dimensions 14978
New Dimensions in International Studies 15558
New Directions 15901, 16098
New Directors for Two-Year Colleges 15990
Newelectronics Journal 09145
New England 14782
New England Antiquities Research Association: Journal 16195
New England Association of Schools and Colleges: Membership Directory 16196
New England Association of Schools and Colleges: Newsletter 16196
New England Historical and Genealogical Register 16197
New English Art Club: Catalogue of Annual Exhibition 12626
The New Era 16925
New Era in Education 13218
New Europe Papers 12034
New Horizons 16807
New Invention Lists 12247
New Issues 15276
New Minerals Industry International 12309
New Moon 16595
New: Notes 13006
New Outlook 16342
New Paths 14221
New Political Science 14659
Newsalerts 16181
News and Views of NVATA 16177
The Newsboy 15229
News Brake 02001
The New Scholasticism 13681
New Series 11888
News for Teachers of Political Science 14010
News from CAST 14868
News from EMI 16040
News from ICFCYP 04380
News from ICTP 07711
News from Lightening Culture 09346
News from the Secretariat 02548
News in Physiological Sciences 14005
Newsleader 15892
Newsletter – APVA News 14353
Newsletter Energy & Environment Alert: Timely Environmental Pamphlets & Booklets 15987
Newsletter in Icelandic 06696
Newsletter of APLS 15425
Newsletter of the American Academy of History of Dentistry 13440
Newsletter of the American Committee on the History of the Second World War 16936
Newsletter of the American Liszt Society 13918
Newsletter of the Anthropological Society of South Australia 00216
Newsletter of the Asia Crime Prevention Foundation 08094
Newsletter of the Australian and New Zealand Society of Nuclear Medicine 00259
Newsletter of the Friends of the Maritime Trust 12495
Newsletter of the IAEG 04581
Newsletter of the Institute for URAM 02526
Newsletter of the NASPG 16222
Newsletter of the SCONUL Advisory Committee on Manuscripts 13096
Newsletter to Diplomates 13630, 13639
Newsletter to European Health Librarians 01276
Newsletter to Membership 16799
Newslook 14432
Newsmagazine Concern 16653
News of the International League 01399
News of the YIVO 16941
Newsounds 13361
Newspaper Techniques 05938
News Production and Inventary Management 14015
New Steel Construction 13100
New Swedish Technology 10529
New Testament Studies 12985
New Thought Quarterly 15504
New West Indian Guide 08964
New Writing Scotland 11283
New York Browning Society: Bulletin 16200
New York Browning Society: Directory 16200
New York Financial Writers' Association: Directory 16203
New York Genealogical and Biographical Record 16204
New York Pigment Club: Membership Roster 16205
The New Yugoslav Law 17135
New Zealand Agricultural Science 09153
New Zealand Association of Clinical Biochemists: Newsletter 09134
The New Zealand Author 09187
New Zealand Cartographic Journal 09139
New Zealand Council for Educational Research: Annual Report 09141
New Zealand Council for Educational Research: Newsletter 09141
New Zealand Dietetic Association: Journal 09143
New Zealand Engineering 09129
New Zealand Entomologist 09124
New Zealand Forestry 09157
New Zealand Geographer 09148
New Zealand Geographical Society: Proceedings 09148
New Zealand Geophysical Society: Newsletter 09149
New Zealand Historical Association: Newsletter 09150
New Zealand Institute of Surveyors: Journal 09161
New Zealand International Review 09158
New Zealand Journal of Archaeology 09133
New Zealand Journal of Geography 09148
New Zealand Leather and Shoe Research Association: Annual Report 09163
New Zealand Libraries 09164
New Zealand Limnological Society: Newsletter 09165
New Zealand Marine Sciences Society: Newsletter 09167
New Zealand Mathematical Society: Newsletter 09168
The New Zealand Medical Journal 09169
New Zealand Microbiological Society: Newsletter 09170
New Zealand Museum Journal 09116
The New Zealand National Society for Earthquake Engineering: Bulletin 09171
New Zealand Operational Research 09184
New Zealand Population Review 09190
New Zealand Psychological Society: Bulletin 09173
New Zealand Science Review 09135
New Zealand Society for Parasitology: Proceedings 09176
New Zealand Statistical Association: Newsletter 09181
New Zealand Veterinary Journal 09182
NFAIS Newsletter 16045
NGL blad 09048
NGT Geodesia 09019
NHA News from Washington 16081
NHC Policy and Resolutions 16078
NHES Quarterly Journal 16079
NHG-Standards for GP's 09052
NHM Info 06178
NIA 14402
NIAS-nytt: Nordic Newsletter of Asian Studies 03785
NIAS Report 03785
NIBR Notater 09321
NIBR Rapport 09321
NICOD: Annual Report 12628
Der Niedergelassene Arzt 06137
Niedersächsisches Ärzteblatt 05038
Niedersächsisches Zahnärzteblatt 06050, 06459
Nieuw Archief voor Wiskunde 09110
De Nieuwe Geografenkrant 08965, 08965
Nieuwe Verhandelingen 08831
Nieuws 01494
Nieuws van Archieven 09088
Nigerian Economic Society: Proceedings of Annual Conferences 09236
The Nigerian Entomologists' Magazine 09226
Nigerian Forum 09238
Nigerian Geographical Journal 09237
Nigerian Institute of International Affairs: Bulletin 09238
Nigerian Institute of International Affairs: Monograph Series 09238
Nigerian Journal for Microbiology 09241
Nigerian Journal of Economic and Social Studies 09236
Nigerian Journal of Entomology 09226
Nigerian Journal of Fisheries and Hydrobiology 09227
Nigerian Journal of Forestry 09228
Nigerian Journal of International Affairs 09238
Nigerian Libraries 09240
Nigerian Nutrition Newsletter 09243
Nigerian Veterinary Journal 09242
Nihon Furansu-go Furansu-bun-gaku-kai: Bulletin 08199
Nihon Gakujutsu Shinko-kai: Annual Report 08200
Nihon Gas Kyokai Chosabu Chosaka: Journal 08201
Nihon Hifuka Gakkai Zasshi 08303
Nihon Kasai Gakkai: Bulletin 08220
Nihon Kikai Gakkai: International Journal 08222
Nihon Kikai Gakkai: Journal 08222
Nihon Kikai Gakkai: Transactions 08222
Nihon Kikai Gakkai: Yearbook 08222
Nihon Kin Gakkai: News 08224
Nihon Kin Gakkai: Transactions 08224
Nihon Koku Eisei Gakkai: Journal 08230
Nihon Naika Gakkai: Journal 08243
Nihon no Sankotosho Shikiban 08271
Nihon no Shingaku: Theological Studies in Japan 08225
Nihon Reito Kyokai Ronbunshu: JAR Transactions 08248
Nihon Sen'i Kikai Gakkai: Journal: English Edition 08258
Nihon Sen'i Kikai Gakkai: Journal: Japanese Edition 08258
Nihon Shokaki-byo Gakkai Zasshi 08381
Nihon Shoni Geka Gakkai: Journal 08264
Nihon Tenmon Gakkai: Publications 08268
Nihon Tokei Gakkai: Journal 08270
Nihon Zosen Gakkai: Bulletin 08279
Nihon Zosen Gakkai: Journal 08279
Nikrobiologicheskii Zhurnal: Microbiological Journal 11193
El Niño en el Tropico 04376
Nippon Acta Radiologca 08311
Nippon Bunko Gakkai: Journal 08283
Nippon Byori Gakkai: Transactions 08286
Nippon Chishitsu Gakkai: Journal 08290
Nippon Denshi Kenbikyo Gakkai: Journal 08294
Nippon Dokumenesyon Kyokai: Informant: Microfiche-Editon 08296
Nippon Dokumenesyon Kyokai: Journal 08296
Nippon Geka Gakkai: Journal 08301
Nippon Hikaku Bungakukai: Bulletin 08304
Nippon Hikaku Bungakukai: Journal 08304
Nippon Hoi Gakkai: Journal 08306
Nippon Hotetsu Gakkai: Annual 08309
Nippon Junkatsu Gakkai: Journal 08317
Nippon Kagakukai: Bulletin 08318
Nippon Kagaku Kaishi 08318
Nippon Kagaku Kyoiku Gakukai: Journal 08319
Nippon Kagaku Kyoiku Gakukai: Letter 08319
Nippon Kaiho Gakkai: Report 08321
Nippon Kaiyo Gakkai: Journal 08323
Nippon Kazan Gakkai: Bulletin 08325
Nippon Keiho Gakkai: Journal 08327
Nippon Keizai Seisaku Gakkai: Annals 08328
Nippon Kessho Gakkai: Journal 08330
Nippon Kikan-Shokudo-ka Gakkai: Journal 08332
Nippon Kinzoku Gakkai: Bulletin 08333
Nippon Kinzoku Gakkai: Journal 08333
Nippon Kinzoku Gakkai: Transactions 08333
Nippon Kogakukai: Journal 08336
Nippon Kokai Gakkai: Journal 08337
Nippon Koku Gakkai: Journal 08338
Nippon Koku Gakkai: Transactions 08338
Nippon Koku Uchu Gakkai: Journal 08340
Nippon Koku Uchu Gakkai: Transactions 08340
Nippon Koseibutsu Gakkai: Special Papers 08342
Nippon Koseibutsu Gakkai: Transactions and Proceedings 08342
Nippon Onkyo Gakkai: Journal 08353
Nippon Ringakukai: Journal 08358
Nippon Ringakukai: Transactions of the Annual Meeting 08358
Nippon Rodo-ho Gakkai: Journal 08359
Nippon Rosiya Bungakkai: Bulletin 08360
Nippon Seikei Geka Gakkai: Journal 08363
Nippon Seiri Gakkai: Journal 08364
Nippon Shashin Gakkai: Journal 08371
Nippon Shashin Sokuryo Gakkai: Journal 08372
Nippon Shiho Gakkai: Journal 08373
Nippon Shika Hozon Gakkai: Journal 08375
Nippon Shika Igakkai: Journal 08376
Nippon Sugaku Kyoiku Gakkai: Journal 08387
Nippon Suisan Gakkai: Bulletin: Nippon Suisan Gakkaishi 08388
Nippon Teiinoshujutsu Kenkyukai: Summary of Annual Meeting 08391
Nippon Toshokan Gakkai: Annals 08394
Nippon Yukagaku Kyokai: Journal 08396
Nippon Yuketsu Gakkai: Journal 08397
NLA Newsletter 09240
NMC News 16112
NMEA News 16103
Noah Worcester Dermatological Society: Membership Roster 16207
NOA News 16116
NOA Newsletter 16114
NODA National News 12602
Nöroppsikiyatri Arsivi: Archives of Neuro-Psychiatry 11168
Nogyo-Doboku Gakkai: Journal 08398
Nogyo-Doboku Gakkai: Transactions 08398
Nogyo-ho Kenkyu 08399
Nogyokikai Gakkai: Journal 08400
Nogyo-Kisho: Journal of Agricultural Meteorology 08349
Noise and Vibration Control 03113
Noise Control Engineering Journal 15335
Noise/News 15335
Nomenclator Zoologicus 13237
Non-Destructive Testing 00294
Nonlinearity 12268, 12462
Noraforum 09335
Nord e Sud 07578
Das nordfriesische Jahrbuch 06423
Nordfriesland 06423
Nordiatrans Newspaper 09248
Nordic Concrete Research 04010
Nordic Journal of Latin American Studies 10510
Nordic Road Safety Council: Rapporte 09279
Nordinfo 03984
Nordinfo – Nytt 03984
Nordiska Afrikainstitutet: Annual Newsletter 10507
Nordiska Afrikainstitutet: Research Report 10507
Nordiska Afrikainstitutet: Seminar Proceedings 10507
Nordiska Institutet för Samhällsplanering: Arsrapport 10508
Nordiska Institutet för Samhällsplanering: Rapporter 10508
Nordiska Samarbetskommittén för Internationell Politik, inklusive Konflikt- och Fredsforskning: Newsletter: International Studies in the Nordic Countries 10509
Nordisk Byggedag: Congress Literature 03781
Nordiske Fortidsminder, Serie B – in quarto 03711
Nordisk Forening for Rettssociologi: Newsletter 03784
Nordisk Matematisk Tidskrift 09328, 10564
Nordisk Numismatisk Unions Medlemsblad 03790
Nordisk Statistik Arsbok: Yearbook of Nordic Statistics 03791
Nordisk Statistik Skriftserie 03791
Nordisk Tidskrift for Kriminalvidenskab 10557
NORDITA Preprint 03786
NORDITA Virksomhedsberetning: NORDITA Report 03786
Nord-Süd aktuell 05604
Norfolk Archaeology 12629
Norfolk Mardler 12836
Norges Geologiske Undersökelse: Bulletin 09283
Norges Geologiske Undersökelse: Skrifter 09283
Normas y Calidades 03386
normat 03645
Norsk arkivforum 09247
Norsk Forsikringsjuridisk Forening: Average 09312
Norsk Geografisk Tidsskrift 09259
Norsk Geologisk Tidsskrift 09317
Norsk Litterår Årbok 09262
Norsk Økonomisk Tidsskrift 09349
Norsk Skogbruk 09263
Norsk Skoleblad 09325
Norsk Tidsskrift for Logopedi 09326
Norsk Veterinaertidsskrift 09256
North American Association for Environmental Education: Monograph Series 16211
North American Association of Professors of Christian Education: Newsletter 16212
North American Bird Bander 15002, 15263, 16881
North American Clinical Dermatologic Society: Program 16214
North American Electric Reliability Council: Annual Report 16217
North American Journal of Fisheries Management 13816
North American Plant Propagator 15514
North American Simulation and Gaming Association: Conference Program 16221
North American Simulation and Gaming Association: Newsletter 16221
North American Society for Sport History: Newsletter 16223
North American Society for Sport History: Proceedings 16223
North American Society of Adlerian Psychology: Calendar-Newsletter 16224
North American Society of Adlerian Psychology: Membership List 16224
North American Thermal Analysis Society: Conference Proceedings 16226
North American Thermal Analysis Society: Membership Directory 16226

North American Thermal Analysis Society: Notes 16226
Northamptonshire Natural History Society and Field Club: Journal 12632
Northamptonshire Record Society: Journal 12633
Northeast Conference on the Teaching of Foreign Languages: Conference Program 16230
Northeast Conference on the Teaching of Foreign Languages: Newsletter 16230
Northeast Conference on the Teaching of Foreign Languages: Reports 16230
Northern Decisions 02023
The Northern Gardener 12634
Northern Journal of Applied Forestry 16634
Northern Libraries Bulletin 16231
Northern Perspectives 02023
Northern Studies 12929
Northline/Point Nord 01926
North Staffordshire Field Club: Transactions 12642
North Wales Arts Association: Report 12645
Northwest Association of Schools and Colleges: Directory of Accredited and Affiliated Institutions 16232
Northwest Association of Schools and Colleges: Newsletter 16232
North Western Naturalist 12646
Northwestern Naturalist 16715
North Western Naturalists' Union: Newsletter 12646
Norvegia Sacra 09273
Norway's Official Statistics (NOS) 09351
Norwegian-American Studies 16233
The Norwegian Cultural Fund 09341
Norwegian Tracks: Newsletter 16858
Notable Programs 16171
Notas de Población 03026
Notas Técnicas y Economicas 03452
Notatki Ornitologiczne 09608
Note di Informazioni Trimestrali e Note su Attività Tecnica 07650
Note di Tecnica Cinematografica 07407
Note Informative AISI 07346
Notenheft 11023
Notes from Unterground 14792
Notes Techniques Forestières 00018
Noticias COPANT: Catalogue of Standards 00122
Noticias Culturales 03377
Noticias de Galápagos 01212
Noticias del PEN 03399
Noticias de Seguridad 15354
Noticiero de la AMBAC 08699
Notiziario A.I.A.S. 07335
Notiziario AISM 07349
Notiziario Buiatrico 07846
Notiziario del Centro 07420
Notiziario del Circolo Speleologico Romano 07604
Notiziario del C.S.C. 07550
Notiziario dell'Accademia di Romania 07172
Il Notiziario dell'Accademia Internazionale d'Arte Moderna 07189
Notiziario della FAST 07686
Notiziario dell'AIRIEL L'Industria della Gomma/Elastica 07340
Notiziario del Museo Civico 07257
Notiziario d'Informazione del Gruppo Amici de La Nostra Famiglia 07354
Notiziario di Ortoflorofrutticoltura 07997
Notiziario e Bollettino 07922
Notiziario I.I.C.: Mensile d'Informazione 07730
Notiziario in La Salute Umana 07341
Notiziario S.I.E. 07869
Notiziario S.I.M. 07892
Notiziario SVP 08030
Notizie A.I.R.I. 07339
Notizie ARIPS 07274
Notizie C.S.E.I. 07556
Notizie degli scavi di antichità 07211
Nottinghamshire Historian 12650
Notulae Entomologicae 04029
Notulae Systematicae ad Floram Europaeam Spectantes 12080
Le Nouveau Biologiste 04973
Nouvelles CEQ 02324
Les nouvelles de l'Académie des Sciences 04564
Les Nouvelles de la F.I.P.E.S.O.: F.I.P.E.S.O. Newsletter 04535
Nouvelles de l'A.I.E.G.L. 04262
Nouvelles Etudes Pénales 04256
Nouvelles UMEC 08053
Nova 13984
Nova Acta 10492
Nova Acta Leopoldina: Abhandlungen der Deutschen Akademie der Naturforscher Leopoldina 05285
Nova et Vetera Iuris Gentium 09108
Nova Proizvodnja 10097
Nova Scotia Association of Optometrists: Newsletter 02723
Nova Scotia Dentist 02731
The Nova Scotia Genealogist 02479, 02870
Nova Scotia Law News 02725
Nova Scotia Library Association: Newsletter 02735
Nova Scotian Institute of Science: Proceedings 02740
Nova Scotia Research Foundation Corporation: Annual Report 02742
Nova Scotia School Boards Association: Newsletter 02744
Nova Scotia Society of Occupational Therapists: Info 02746
Novaya i Noveishaya Istoriya: Modern and Contemporary History 09954
Novedades CAFRA 11108
Novyi Mir 09989
Nový orient 03500
Nowiny Psychologiczne 09603
Now-World Publishing Monitor 12701
NPF Piano News 16128
NRCP Research Bulletin 09522
NRCP Technical Bulletin 09522
NRDC Newsline 16187
NSCA Members' Handbook 12607
The NSC Review 03314
NSC Special Publication 03314
NSC Symposium Series 03314
NSERC Awards Guide 02675
NSGF-Meldingsblad: Newsletter 09337
NSS Bulletin 16163
NSW Physiotherapy Bulletin 00329
NTRS Newsletter 16173
ntz-Archiv 06351
ntz Nachrichtentechnische Zeitschrift 06351
Nuclear Canada 02211
Nuclear Canada Yearbook 02211
Nuclear Energy 11620, 12294
The Nuclear Engineer 12311
Nuclear Europe 10685
Nuclear Industry 16821
Nuclear News 13960
Nuclear Science & Engineering 13960
Nuclear Science Journal 03250
Nuclear Technology 13960
Nucleotecnica 03037
The Nucleus 09380, 09535
Nuestras Aves 00092
Nuklearmedizin 05391
Numéros spéciaux 04746
Numismatic Chronicle 12837
Numismatic Literary Guild: Newsletter 16236
Numismatic Literature 13962
Numismatic Notes and Monographs 13962
Numismatic Report 03493
Numismatika Chronika 06512
Numismatisches Nachrichtenblatt 05477
Numismatische Zeitschrift 00872
numismatisk rapport 03656
The Numismatist 13961
Numizmatičke vijesti 03461
Numizmatika 03461
Numizmatikai Közlöny 06648
Il Nuovo Cimento A 07874
Il Nuovo Cimento B 07874
Il Nuovo Cimento C 07874
Il Nuovo Cimento D 07874
Nuovo Giappone 07512
Il Nuovo Saggiatore 07874
Nursing Bibliography 12782
Nursing Outlook: Journal of the American Academy of Nursing 13409
Nursing Review 07023
Nursing Standard 12782
Nursing the Elderly 12782
Nutrition Abstracts and Reviews: Human and Experimental 11720
Nutrition Abstracts and Reviews: Livestock Feeds and Feeding 11720, 11814
Nutrition Abstructs an Reviews: Human and Experimental 11814
Nutrition Action Healthletter 14699
Nutrition Education Materials Catalog 16029
Nutrition Forum 02264
Nutrition in Clinical Practice 14076
Nutrition News 16029, 17155
Nutrition Week 14814
Nuusbrief 10177
NVAG-Bulletin 09015
NVVK-Nieuws 09042
NVvL Jaarboek 09024
NWC Newsletter 16183
NWSA Action 16182
NWSA Journal 16182
Nyhetsbrev Europa Standardisering 10522
Nýmenntamál 06693
Ny Teknik 10588
Nyt fra Bibelselskabet 03704
Nyt fra Historien 03756
Nytt om Niotusen 10522
Nytt om Romfart 09292
NZEI Ralph Slade Memorial Lecture 09145
NZFMRA Annual Report 09146
NZ Journal of Psychology 09173

OAA News 16256
Oak Ridge Associated Universities: Annual Report 16239
Oberflächen-Werkstoffe 10910
Oberfläche/Surface 10836
Oberösterreichischer Musealverein: Jahrbuch: Teil I Abhandlungen, Teil II Berichte 00717
Oberösterreichischer Musealverein: Mitteilungen 00717
Oberösterreichischer Musealverein: Schriftenreihe 00717
Obol 03461
Obras de la Literatura Ecuatoriana 03839
Observations et Diagnostics Economiques 04543
Observations et Travaux 04633
Observer 13729, 15606
Observer's Handbook 02863
Obshchestvennye Nauki: Social Sciences 09954
Obstetrica si Ginecologia 09774
Obstetrics and Gynecology 13722
Obzornik za matematiko in fiziko 10098
OCA News 02765
Occasional Publications of the Entomological Society of Nigeria 09226
Occupancy Update 15945
Occupational Medicine 13058
Occupational Safety & Health 12852
Occupational Stress 15285
Ocean Challenge 11756
Oceanic Expeditions Brochure 16243
Oceanic Society: Newsletter 16242
Oceanographical Magazine 08147
Oceanologia et Limnologia 03181
Oceans Magazine 16243
ocg-Kommunikativ 00735
ocg-Schriftenreihe 00735
Odeo Olimpico 07225
Od-Maida 07068
Odontologia Uruguaya: Revista Científica Gremial e Informativa 13246
Odontostomatological Progress 06526
Odyssey Journal 16244
OEA Leader 16278
ÖAL-Richtlinien 00885
OEA News 16278
ÖAV-Informationen 00727
ÖGAI Journal 00762
ÖGOR-Nachrichten 00817
ÖGPW-Rundbrief 00824
Öko-Mitteilungen 06223
Ökosystemanalyse und Umweltforschung 05050
ÖNB-Kurier 00898
ÖNORM: Fachzeitschrift des ON Österreichisches Normungsinstitut 00914
Öntöde: Foundry 06670
ÖRG-Mitteilungen 00899
ÖROK-Schriftenreihe 00887
Österreich in Geschichte und Literatur (mit Geographie) 00658
Österreichische Ärztezeitung: Organ der Österreichischen Ärztekammer 00721
Österreichische Akademie der Wissenschaften: Sitzungsbericht I und II 00723
Österreichische Arbeitsgemeinschaft für Rehabilitation: Monatsbericht: Sozialpolitische Rundschau 00726
Der Österreichische Arzt 00996
Österreichische Dokumentation für Wohnen, Bauen und Planen (ODW) 00597
Der Österreichische Freiberufstierarzt 00574
Österreichische Geographische Gesellschaft, Zweigverein Innsbruck: Jahresbericht 00746
Österreichische Gesellschaft für Lungenerkrankungen und Tuberkulose: Jahresbericht der wissenschaftlichen Veranstaltungen 00807
Österreichische Gesellschaft für Musik: Beiträge 00811
Österreichische Gesellschaft für Raumforschung und Raumplanung: Schriftenreihe 00826
Österreichische Gesellschaft für Tropenmedizin und Parasitologie: Mitteilungen 00836
Österreichische Gesellschaft für Unfallchirurgie: Kongreßberichte: Hefte zur Unfallkunde 00837
Österreichische Gesellschaft für Wissenschaftsgeschichte: Mitteilungen 00849
Österreichische Ingenieur- und Architekten-Zeitschrift 00893
Die Österreichische Kunstforschung 00971
Österreichische Osthefte 00915
Österreichischer Himmelskalender 00886
Österreichischer Verband für Strahlenschutz: Tagungsberichte 00905
Österreichisches Archiv für Kirchenrecht 00801
Österreichisches Jahrbuch für Internationale Politik 00763
Österreichische Städtebibliographie 00884
Österreichisches Volksliedwerk: Jahrbuch 00922
Österreichische Tierärztezeitung 00576
Die Österreichische Volkshochschule 00974
Österreichische Zahnärzte-Zeitung 00736
Österreichische Zeitschrift für Physikalische Medizin 00823
Österreichische Zeitschrift für Pilzkunde 00871
Österreichische Zeitschrift für Politikwissenschaft; ÖZP 00824
Österreichische Zeitschrift für Soziologie 00832
Österreichische Zeitschrift für Statistika und Informatik 00919
Österreichische Zeitschrift für Statistik und Informatik 00834
Österreichische Zeitschrift für Vermessungswesen und Photogrammetrie 00907
Oesterreichische Zeitschrift für Volkskunde 00988
ÖVG-Spezial 00926
Offene Welt 06444
Office & Information Management International 12212
Office Job Evaluation Manual 12212
Office National d'Information sur les Enseignements et les Professions: Avenirs 04606
Office National d'Information sur les Enseignements et les Professions: Les Cahiers 04606
Office, Professional and Data Processing Report 13329
Officers and Directors 16121
Office Salaries Analysis 12212
Official ASLMS Newsletter 14069
Official Directory of Members 12220
Official International Membership Directory 16156
Official Journal 13709
Official List of Accredited Programs 13317
Official List of Shows and Sales 11390
Official Yearbook 15485
L'Officiel des Dermatologistes et Vénéréologistes 04969
Officiel des Mathématiques 04890
Offshore Engineer 12294
Offshore Engineering 11437
Offshore Engineer Yearbook 12294
Offshore Mechanics and Arctic Engineering 14145
Offshore Technology 12255
Oficina Regional de Educación de la UNESCO para América Latina y el Caribe: Contacto 03062
Oftalmologia 09775
O Glaso Romano 04410
Oikos 10513
Oil and Gas Reporter 16730
Oil Spill Intelligence Report 14700
OIML International Recommendations and International Documents 04615
L'Oiseau 04599
Okeanologiya: Oceanology 09954
Økonomi og Politik 03814
OLBG-INFO 05341
Old Cornwall 12067
Old English Newsletter 14687
Old Testament Abstracts 14652
Olea 10308
O Mentor 06499
Omnibus 12422
OmSLAget 03802
On Balance 02475
On Campus with Women 13595, 14378
Oncology Nursing Forum 16249
Oncology Nursing Society: Proceedings of Annual Congress 16249
One and All Magazine 12554
The O'Neill 13450
One-in-Ten 16406
Ongaku Gaku: Journal of the Musicological Society of Japan 08352
Onkologie 00863, 05358
Online 05077
Online Notes 11267
Onoma 01394
Onomasticon Vasconiae 10349
Ons Geestelijk Erf 01419
On Site 13864
ONS Newsletter 12662
ONS Occasional Papers 12662
Ontario Archaeology 02753
Ontario Association for Curriculum Development: Conference Report 02755
Ontario Conservation News 02389
Ontario Dentist 02773
The Ontario Genealogical Society: Newsleaf 02778
Ontario Gerontology Association: Newsletter 02779
Ontario Historical Society: Bulletin 02782
Ontario History 02782
Ontario Psychogeriatric Association: Newsletter 02789
Ontario Psychogeriatric Association: Proceeding 02789
Ontario Recycling Update 02855
Ontario Report 01819
Ontario Respiratory Care Society Update 02785
Ontario School Counsellor Association Reports 02794
Ontario Teachers' Superannuation Commission: Annual Report 02802
Ontario Teachers' Superannuation Commission: Exchange 02802
The Ontario Thoracic Review 02803
Ontario Thoracic Reviews 02785
Ontario Veterinary Association: Update 02804
On the Edge 16900
On the Line 13756
Ontogenez: Ontogenesis 09954
Openings, Job Opportunities for Scholars of Religion 13434
Open Mind 12529
Open Space 11806
Opera America: Bulletin 16251
Opera America: Membership Directory 16251
Opera Journal 16114
The Operational Geographer 02052
Operational Research Society: Newsletter 12655
Operation Enterprise Newsletter 16252
Operation Eyesight Universal: Newsletter 02806
Operations Forum 16873
Operations Research 16253
Operative Dentistry 13300
Ophthalmic and Physiological Optics 11544
Der Ophthalmologe 05478
Ophthalmological Society of Egypt: Annual Bulletin 03912
Ophthalmologie 04861
Ophthalmology 13410
Opinioni Libere 07570
Opticians Association of America: Guild Quarterly 16256
Optics and Spectroscpy 16255
Optics Letters 16255
Optics Letters: Microfiche Edition 13893
Optics News 13893, 16255
Optik 05321
Optika Atmosfery: Optics of Atmosphere 09954
Optika i Spektroskopiya: Optics and Spectroscopy 09954
Options 00687
Opto and Laser Europe 12268
Optometric Historical Society: Newsletter 16257
Optometrie 06450
Optometry 06805
Optometry and Vision Science 13411
Opus: Music Magazine 03652
Oracle 12281
Oral History 12656, 16381
Oral History Association: Annual Report and Membership Directory 16258
Oral History Association: Newsletter 16258
Oral History Review 16258
Oralité – Documents et Etudes 04746
Oral Surgery, Oral Medicine, Oral Pathology 13645
Orana 00318
Orbis Biblicus et Orientalis 10838
Orbit 06977
OR Briefing 12652
Die Orchidee 05479
Der Orchideenkurier 00874
The Orchid Review 12810
Ordem dos Médicos: Revista 09665
Order of the Indian Wars: Communique 16259
Orders & Medals 12657
Organ Handbook 16260
Organisation of Teachers of Transport Studies: Seminar & Annual Report 12660
The Organist 12784
The Organists' Review 12193
Organizational Dynamics 13372
Organizational Problems 09625
The Organization Development Journal 16262
Organization & Methods Newsletter 10199
Organizations and Change 16262
Organization Science 15330
Organometallics 13688
Orglet 03706
Oriens Antiquus 07754
Oriens Christianus 05896
Orient 05598
Oriente Moderno 07754
Original Contributions in English and Abstracts in English from the Transactions 08121
Oriolus 01412
Orion 10967
Orizont 09798
ORL-Verhandlungsberichte der wissenschaftlichen Frühjahrsversammlung 10840
Orman Mühendisliği: Forest Engineering 11167
Ornamental Horticulture 11720, 11723
Ornis Fennica 04046
Ornithological Monographs 13969
Ornithological Newsletter 13969
Ornithological Society of the Middle East: Bulletin 12663
Der Ornithologische Beobachter 10612
Orquidea 08703
Orria 10359
ORSA Journal on Computing 16253
Ortho-Briefs 13701
Orthodontic Bulletin 13558
Orthodox Theological Society in America: Bulletin 16269
Orthodox Theological Society in America: Membership Directory 16269
Orthomolecular Psychiatry 15242
Orthopädische Praxis 06412
Orthopedic Briefs 14911
Orthopedic Research Society: Proceedings of Annual Meeting 16271
Orthopedic Transactions 13825
Ortung und Navigation 05393
ORYX 12059
Osiris 15222
Osnabrücker Geschichtsquellen und Forschungen 06387

Osnabrücker Mitteilungen 06387
Osnabrücker Urkundenbuch 06387
OSNZ News 09185
Ost-Dokumentation Bildungswesen 00915
Ost-Dokumentation Wirtschaft 00915
Osteopathic College of Ophthalmology and Otorhinolaryngology: Newsletter 16273
Osteopathic Newsletter 12664
Osteoporosis: Bulletin for Physicians 02808
Osteuropa 05394
Osteuropa-Recht 05394
Osteuropastudien der Hochschulen des Landes Hessen: Reihe I 06076
Osteuropa-Wirtschaft 05394
Ostomy Quarterly 16814
Ostpanorama-Sonderausgabe 00628
Ostrich 10227
Otechestvennaya Istoriya: The Nation's History 09954
Otechestvennye Arkhivy: National Archives 09954
OT Line 01989
Otolaryngologia Polska: Polish Journal of Otolaryngology 09597
Oto Review 15232
L'Otorinolaringologia Pediatrica 07925
Oto-Rino-Laringologie 09776
Our Kids Magazine 13361
Outdoor News Bulletin 16899
Outlook 14869
Output 10941
Outreach 13854
Ouvertures 04303
Overseas Doctors Association in the UK: News Review 12666
Overseas Education Association: Journal 16278
Oversigt over Selskabets Virksomhed 03710
The Owl of Minerva 15201
Oxford Preservation Trust: Annual Report 12668
Oxoniensia 12669
Oyo-buturi 08402
Oz Trading Post 15574

PAA Affairs 16337
PABC Report 02818
PACE 16225
Pachyderm 08468
Pacific Dermatologic Association: Transactions 16280
Pacific Science Association: Information Bulletin 16282
Pacific Seabird Group: Bulletin 16283
Pacific Seabird Group: Membership Directory 16283
Padusa 07529
Padusa Notiziario 07529
paed 06064
Pädagogik und Information 05848
Paediatric Nursing 12782
Pädiatrie und Pädologie 00800
Page 11834
Pages Romandes 10860
Páginas de Contenido: Ciencias de la Información 03449
Pagine della Dante 07776
Paideuma 15074
Paideuma-Mitteilungen zur Kulturkunde 05765
Pain Control in Dentistry 14044
Paint Titles 12673
El Paisano 14966
PA Job Find 13425
Pakhto 09392
Pakistan Academy of Letters: Academy 09378
Pakistan Academy of Sciences: Monographs 09379
Pakistan Academy of Sciences: Proceedings 09379
Pakistan Academy of Sciences: Proceedings of Symposia 09379
The Pakistan Engineer 09369
Pakistan Historical Society: Journal 09385
Pakistan Horizon 09386
Pakistan Journal of Medical Research 09389
Pakistan Journal of Scientific and Industrial Research 09384
Pakistan Library Association: Conference Proceedings 09387
Pakistan Library Association: Journal 09387
Pakistan Library Association: Newsletter 09387
Pakistan Philosophical Journal 09391
PA-Kontakte 00681
The Palace Peeper 15160
Paläoklimaforschung 05050
Palaeontographical Society: Monographs 12674
Palaeontologica Indica 06762
Palaeontological Memoirs 00405
Paläontologische Zeitschrift 06231
Palaeontology 12675
Palaios 16578
Palavra Chave: Key Word 01605
Paleobiology 16287
Paleoceanography 13833
Paleontographica Americana 16286
Paleontologia Lombarda: Nuova Serie 07946
Paleontologicheskii Zhurnal: Palaeontological Journal 09954
Paleorient 04382
Palestine Exploration Quarterly 12676
Palmerston North Medical Research Foundation: Annual Report 09186
Palynott-nytt 03573
Památky archeologické 03500
Pan-American Biodeterioration Society: Conference Proceedings 16290
Pan-American Biodeterioration Society: Newsletter 16290
Panda 10280
Panduan Akitek 08648
Panel de Noticias 13243
Pan-Pacific Surgical Association: Newsletter 16294
Paper and Board Abstracts 12701
Papers and Proceedings of the Royal Society of Tasmania 00479
Papers in Meteorological Research 03307
Papers in Meteorology and Geophysics 08335
The Papers of George Catlett Marshall 15154
The Papers of Jefferson Davis 15598
Papers of Nathaniel Bacon 12630
Papers of the Seminar on the Acquisition of Latin American Library Materials 16461
Paper Technology 12701
Das Papier 06371
Papiripar 06672
El Papiro 03831
Papua New Guinea Scientific Society: Proceedings 09422
Paracelsusbrief: Mitteilungsblatt 00676
Paragraphs 16148
The Paralegal Educator 13488
Parametric World 15553
Parapsychological Association: Proceedings 16295
Parasitica 01151
Parasitología al Día 03079
Parassitologia 07926
Parazitologiya: Parasitology 09954
Le Parchemin 01416
Parents Guide-Special Needs-Special Provision 11914
Parergon 00257
Paris et Ile-de-France: Mémoires publiés par la Fédération des Sociétés Historiques et Archéologiques de Paris et de l'Ile de France 04516
The Parkinson Newsletter 12678
Parks and Recreation 16173
Parliamentary Affairs: A Journal of Comparative Politics 12145
Parliamentary Journal 13892
Parliaments, Estates and Representation 12342
Paroisses et Communes de France 04382
Parrot Society: Magazine 12680
Partners 15631
Partners in Hope 13367
Pashosh 07111
Pashtu Quarterly 00001
PAS Memo 14007
Passagen 10781
P.A.S.T.: Pioneer America Society Transactions 16321
Pastel Society: Annual Catalogue 12681
Pastoral Musician's Notebook 15877
Pastoral Music Magazine 15877
Patent Law Annual 16730
Pathologia 06651
Pathologica Japonica 08286
Pathology 00467
Pathways 16342
Patient Directory 14638
Patologia Polska 09599
Patologicheskiya Fiziologiya i Eksperimentalnaya Terapiya: Pathological Physiology and Experimental Therapy 09952
Patrimoine au Présent 04382
Patronato de Biología Animal: Revista 10335
Pattern Recognition 16299
Pattern Recognition and Artificial Intelligence 03162
Patterns of Prejudice 12249
Le Pays Bas-Normand: Revue 04619
PC Review 14421
PCT Gazette: Gazette of International Patent Applications 11034
Peace Chronicle 14849
Peace in the Sciences: Occasional Papers 00680
Peace Progress 15406
Peace Research 02221, 15658
Pe'amim: Studies in the Cultural Heritage of Oriental Jewry 07118
Peanut Research 13993
Peanut Science 13993
Pedagogical Bulletin 01170
Pedagogícja 17106
Pedagogic Reporter 15605
Pedagogika 03500
Pedagogika: Pedagogics 09953
Pédagogiques 01909
Pedagoška Biblioteka 17120
Pediatría 03412
Pediatria Polska 09600
Pediatric Cardiology 03951
Pediatric Dentistry 13420
Pediatric Research-Program Issue 13994, 16564
Pediatrics 13421
Pediatrie 09778
PEF Leader 16285
PEN: Public Economic Newsletter 00374
PEN American Center: Newsletter 16301
PEN-Brief 10661
PEN Club Argentino: Boletín 00151
PEN Club de Suisse Romande: Lettre de Information 10778
PEN Clube do Brasil-Associação Universal de Escritores: Boletim 01700
Le PEN Hongrois 06652
P.E.N. International 12382
PEN International (Sydney Centre): Newsletter 00445
PEN-Kwartaal 09059
P.E.N. Maison Internationale: Informations 04620
Penmens News Letter 02542
PEN-Nachrichten 00900
PEN-Tijdingen 01400
The PEN Woman 16100
People Plant Connection 13853
PEP Exchange 16297
Perception and Psychophysics 16373
Percussion News 16303
Percussive Notes 16303
Performance Instruction 16156
Performance of Constructed Facilities Journal 14107
Performing Arts Resources 16775
Perfusion Life 14122
Peribalticum 09543
Perinatal-Medizin 05398
Periodical Oud Holland 09066
Periodica Mathematica Hungarica 06597
Periodicum Biologorum 03462
Permanent International Association of Navigation Congresses – United States Section: Newsletter 16306
Personalhistorisk Tidsskrift 03797
Personeelbeleid 09034
Personnel Administrator 16535
Personnel Magazine 13372
Personnel Management 12266
Personnel Management Plus 12266
Personnel, Training and Education Journal 12451
Perspective in Philosophy 14692
Perspective on Consciousness and Psi Research 14328
Perspectives in Religious Studies 15829
Perspectives on Prevention 14516
Perspectives universitaires 01948
Persuasions 02568
Pesquisa em Andamento 01625, 01633, 01648
Pesticide Science 13035, 13035
Pest Management 16125
Peter Warlock Society: Newsletter 12690
Petroleum Newsletter 16145
Petroleum Review 12267
Pewter Review 11315
Pfälzer Heimat 06238
PGS Newsletter 16334
PHA Bulletin 09526
Phalanx 15739
Pharmaceutica Acta Helvetiae 10885
Pharmaceutical Historian 11672
The Pharmaceutical Journal 12838
Pharmaceutical Society of Ireland: Calendar 07019
Pharmaceutisch Weekblad 08956
Pharmacia Mediterranea 07804
Pharmacological Research 07871
Pharmacological Reviews 14077
The Pharmacologist 14077
Pharmacopeial Forum 16832
Pharmacy in History 13896
Pharmacy School Admission Requirements 13522
The Pharmacy Student 13996
Pharmacy Today 13996
Pharmakologiya i Toksikologiya: Pharmacology and Toxicology 09952
Pharmatimes 06806
Pharmazie in unserer Zeit 05485
Philippine Journal of Crop Science 09516
Philippine Journal of Nutrition 09523
Philippine Journal of Paediatrics 09530
Philippine Journal of Philosophy 09534
Philippine Library Association: Bulletin 09528
Philippine Library Association: Newsletter 09528
Philippine Pharmaceutical Association: Journal 09531
Philippine Surgeon 16691
Philological Monographs 13997
The Philological Society: Transactions 12693
Philologica Pragensia 03500
The Philosopher 12694
Philosophia Reformata 09097
Philosophical Research 03155
Philosophical Research Society: Journal 16312
Philosophical Society: Proceedings of Annual Conferences 10428
Philosophie-Österreich 00822
Philosophisches Jahrbuch 05896
Philosophy 12817
Philosophy and Phenomenological Research 15512
Philosophy and Political Action 16568
Philosophy, Morality and International Affairs 16568
Philosophy of Science 16314
Phlébologie 04831
Phlébologie 04984
Phlebologie und Proktologie 05401
Phoenix 11265
The Phoenix 16814
Photobiology Bulletin 11631
Photochemistry and Photobiology 14079
Photogrametric Engineering and Remote Sensing 14080
The Photogrammetric Record 12696
The Photographer 11588
Photographic Abstracts 12841
Photographic Canadiana 02816
Photographic Imaging Science 16672
The Photographic Journal 12841
Photographie und Gesellschaft 00630
Photonics Technology Letters 15317
Photo Roster 16273
Photosynthetica 03500
Physicalia 01442
Physicalia Magazine 01442
Physical Review 14000
Physical Review A, B, C, D: Microfiche Edition(s) 13893
Physical Review Abstracts 13893
Physical Review Letters 14000
Physical Review Letters: Microfiche Edition 13893
Physical Therapy 14001
Physical Therapy Resource and Buyer's Guide 14001
Physica Scripta 10490
Physicians Forum Bulletin 16317
Physicians Forum: Newsletter 16317
Physicians Referral List 16297
Physics Abstracts 12296
Physics and Chemistry of Glasses 13048
Physics Briefs 13893
Physics Briefs of Fluids: Microfiche Edition 13893
Physics Education 12268
Physics in Canada 02065
Physics in Medicine and Biology 12268
Physics in Medicine & Biology 13566
The Physics Teacher 13567
Physics Today: Microfiche Edition 13893
Physics Update 10048
Physics World 12268
Physikalische Berichte 05487
Physikalische Blätter 05487
Physikalische Medizin, Rehabilitationsmedizin, Kurortmedizin 06349
Physikalische Therapie in Theorie und Praxis 06367
Physiogeographica: Basler Beiträge zur Physiogeographie 10709
Physiologia Bohemoslovaca 03500
Physiological Entomology 12800
Physiological Measurement 12268
Physiological Reviews 14005
Physiologie Végétale 04835
The Physiologist 14005
Physiology Canada 02225
Der Physiotherapeut 10902
Physiotherapie 00957
Physiotherapy 11770
Phytomorphology 06827
Phytopathology 14006
Phyto-Protection 02922
Piceno 07455
Picture House 11786
Pierre Fauchard Academy: Membership Roster 16319
Pietismus und Neuzeit: Ein Jahrbuch zur Geschichte des neueren Protestantismus 05918
PI (European Poetry Quarterly) 01282
Pig News and Information 11720
Pilgrim Society Notes 16320
The Pioneer 14997
Pioneer America Society: Newsletter 16321
Pipeline 16325
Pipelines 11437
Pipe Roll Society: Annual Report 12700
PIPS: Public Information Paper 00374
Pira International: Printing Abstracts 12701
Placement Bulletin 14382
Plainsong and Mediaeval Music 12702
Plan Canada 02181
Planejamento 01651
Planetarian 15513
Planetary Report 16323
Planetology Papers 13833
Planned Parenthood Federation of America: Affiliates Directory 16324
Planned Parenthood Federation of America: Annual Report 16324
Planners Newsletter 02367
Planning Advisory Service Report 14007
Planning Bulletin 13137
Planning for Higher Education 16510
Planning Magazine 14007
Planta Medica 05802
Plant and Cell Physiology 08384
Plant Breeding Abstracts 11720
Plant Disease 14006
The Plant Engineer 12313
Plant Genetic Resources Newsletter 07708
Plant Growth Regulator Abstracts 11720
Plantinews 08631
The Plant Journal 12973
Plant Molecular Biology 15539
Plant Molecular Biology Reporter 15539
Plant/Operations Progress 13878
Plant Pathology 11663
Plant Physiology 14155
Plant Protection Society of Western Australia: Proceedings 00446
Plants 03121
Plant Science Bulletin 14610
The Plantsman 12810
Plant Variety Protection 10994
Plasma Physics and Controlled Fusion 12268
Plasma Sources Science and Technology 12268
Plastic and Reconstructive Surgeons combined Roster 14157
Plastics and Rubber International 12703
Plastics and Rubber Processing and Applications 12703
Plastics Engineering 16692
Plastics Institute of America: Institute Report 16325
Plastic Surgery News 14157
Plastic Surgery Research Council: Meeting Abstracts 16326
Platform 16875
Plating and Surface Finishing 13796
Platon: Deltion tes Hetaireias Hellenon Philologon 06534
Play and Parenting Connections 02070
Plowshare Press 14678
Plumbing Engineer 14158
Plus Lucis 01005
P.M.A. Newsletter 09529
PMLA 15750
Pneumologie 05407, 05608
PNGLA Librarian's Calendar 09421
PNGLA Nius 09421
Población y Desarrollo 03832
Pochvovedenie: Soil Science 09954
Pocket Charts 15277
The Poe Messenger 16329
Poetry Folio 12427
Poetry Magazine 15751
Poetry Review 12706
Poetry Society of America: Newsletter 16331
Pogrom: Zeitschrift für bedrohte Völker 05803
Point News: Historical Newssheets 12086
Point Papers Series 12086
Pokroky matematiky, fyziky a astronomie 03500, 03547
Polarforschung 05409
The Polar Times 14009
The Police Surgeon 11376
Policies and Procedures 13636
Policy Issues 00418
Policy Options 02525
Policy Statement 15580
Policy Studies Journal 16332
Policy Studies Review 16332
Polifónia 06663
Polish American Historical Association: Newsletter 16333
Polish American Studies 16333
Polish Review 16335
Polish Sociological Review 09605
Politica della Scuola 07380
Política e Estratégia: Revista de Política Internacional e Assuntos Militares 01707
Political Memo form COPE 14804
Political Science Quarterly 13303
Political Studies 12712
Political Studies Association of the United Kingdom: Newsletter 12712
Politica Meridionalista: Rivista Mensile di Cultura Economia Attualità 07583
Politická ekonomie 03500
Politics 00247, 12712
Politics Association Resources Bank 12713
Politiikka Géorges 04082
Politique de le Science 10861
Politische Vierteljahresschrift 05618
Politologen 10526
Polska Akademia Nauk: Bulletin 09549
Polski Przeglad Kartograficzny 09571
Polski Przeglag Radiologii 09550
Polygraphus 03795
Polymer Composites 16692
Polymer Engineering 16692
Polymer International 13035, 13035
Polymer Journal 08166
Polymer Preprints, Japan 08166
Polynesian Society: Memoirs 09189
Polytechnic Courses: A Guide to Full-Time and Sandwich Courses 11804
Polytechnic Courses Handbook 11804
Pomorskie Monografie Toponomastyczne 09543
Pomorski Zbornik: Maritime Annals 03457
Pomorze Gdańskie 09543
POMPI: Popular Music Periodicals Index 11603
Pontificia Accademia Romana di Archeologia: Memorie 16955
Pontificia Accademia Romana di Archeologia: Rendiconti 16955

Popular Astronomy 12424
Population and Development Review 16338
Population Bulletin 16340
Population Council: Annual Report 16338
Population Index 16337
Poradnik Bibliotekarza 09615
Poroshkovaya Metallurgiya: Powder Metallurgy 11193
Portfolio 12537
Il Porto di Genova agli Inizi del '600 07573
Portugiesische Forschungen 05896
Portuguese Studies 12535
Posebna izdanja: Monographs 17125
Positions Listing 14770
Positions Statement 14506
Positive Alternatives 14678
Postal History Journal 16341
Postcard 14318
Postcard Classics 14959
Post Convention Reports 13730
Postepy Astronautyki: Progress in Astronautics 09554
Postepy Fizyki: Advances in Physics 09569
Postępy Fizyki Medycznej: Progress in Medical Physics 09570
Postepy Mikrobiologii 09588
Postgraduate Medical Journal 12073
Post Harvest News and Information 11720
Postharvest News and Information 11723
Postopy Biochemii: Advances in Biochemistry 09557
Post-Secondary Programme Profile 02640
Potamon 08671
Potato Abstracts 11720
Poultry Abstracts 11720, 11730
Poultry Science 16345
Pour 04548
Poverkhnost: Fizika, Khimiya, Mekhanika: Surface: Physics, Chemistry, Mechanics 09954
Powder Metallurgy 12262, 12524
powder metallurgy international 05696
Power Engineering Journal 12296
Power Engineering Review 15317
Power Press and Forging Newsletter 16145
The Powys Journal 12714
Poznaj Świat 09571
Poznánskie Towarzystwo Przyjaciól Nauk: Bulletin: Série D: Sciences Biologiques 09612
Poznánskie Towarzystwo Przyjaciól Nauk: Sprawozdania 09612
P+P, Planung + Produktion: Fachzeitschrift für Organisation, Rationalisierung, Informationsverarbeitung, Produktionstechnik 10909
Prace Archeologiczne 09627
Prace Ethnologiczne: Ethnological Works 09584
Prace i Materiały Etnograficzne: Ethnographic Works and Materials 09584
Prace Komisji Archeologicznej 09612
Prace Komisji Automatyki 09612
Prace Komisji Biologicznej 09612
Prace Komisji Etnograficznej 09612
Prace Komisji Filologicznej 09612
Prace Komisji Filozoficnej 09612
Prace Komisji Geologiczno-Geograficznej 09612
Prace Komisji Historii Sztuki 09612
Prace Komisji Historycznej 09612
Prace Komisji Jezykoznawczej 09612
Prace Komisji Nauk Ekonomicznych 09612
Prace Komisji Nauk Rolniczych i Komisji Nauk Leśnych 09612
Prace Komisji Nauk Spolecznych 09612
Prace Monograficzne nad Przyroda Wielkopolskiego Parku Narodowego pod Poznaniem 09612
Prace Polonistyczne 09545
Prace Popularnonaukowe 09627
Prace Wydzialu Filologiczno-Filozoficznego 09627
Pracovní lékařství: Occupational Medicine 03509
Practice 11523
Practice Information 12811
Practicing Anthropology 16494
Practicing Architect 16637
Practicing Architect Magazine 16637
Practicus 03672
Practitioner 15892
Prähistorische Forschungen 00544
Präsentationshefter 03708
Pragmateiai: Papers 06498
Prajna 08817
Praktické zubní lékařství 03532
Praktické zubní lékařství: General Dentistry 03509
Praktický lékař: General Practitioner 03509
Praktika 06499
Praktika tes Akademias Athinon: Proceedings 06498
Der Praktiker 05551, 05753
Der Praktische Arzt 05201
Pramana: Journal of Physics 06767
PRA Newsletter 12673
Pratique 11895
Pravna misla 08616
Právník 03500
Pravni Život 17077
Praxis der Kinderpsychologie und Kinderpsychiatrie 05491
Pre-1855 Monumental Inscription Lists 12910
Precious Metals News amd Reviews 15515
Precision Measurements Association: Newsnotes 16350
Predškolsko Dete 17106
Prehistoric Society: Proceedings 12715
Přehled Lesnické a Myslivecké literatury 03501
Prelaw Handbook 15655
Premio Biella Poesia Junior: Biella Poetry Junior Price 07131
The Preparatory Schools Review 12194
Presek: list za mlade matematike, fizike in astronome 10098
Presentation 02871
Presenze Regionali 08040
Preservation Forum 16639
The President 13925
Presidential Address 15722
Presidential Studies Quarterly 14708
President's Bulletin 14930
The Presidents Journal 13887
President's Message 14105
President's Newsletter 14803, 15885
Presidents Newsletter 15915
President's Report 14435
Presidents Report 16269
Preslia 03500, 03504
Die Presserundschau 10814
Presseschau Ostwirtschaft 00915
Pressespiegel DESY 05575
La Presse Thermale et Climatique 04858
Pressure: Newsletter 16803
Pressure Vessel Technology 14145
Previsioni a Breve Termine 07525
Pribory i Tekhnika Eksperimenta: Instruments and Equipment for Experiments 09954
Price-Pottenger Nutrition Foundation: Membership Journal 16352
Prikladnaya Biokhimiya i Mikrobiologiya: Applied Biochemistry and Microbiology 09954
Prikladnaya Matematika i Mekhanika: Applied Mathematics and Mechanics 09954
Prikladnaya Mekhanika: Applied Mechanics 11193
Prilozi na Oddelenieto za Bioloski i Medicinski Nauki 08603
Prilozi na Oddelenieto za Matematicko-Tehnicki Nauki 08603
Prilozi na Oddelenieto za Opstestveni Nauki 08603
Primary Geographer 12110
Primary Health Care 12782
Primary Science Review 11282
Primate Eye 12717
Primers 14046
Prince Edward Island Association of Social Workers: Newsletter 02829
Principal 15853
Printing and Publishing Newsletter 16145
Printing History Journal 14014
Priroda 03462
Priroda: Nature 09954
The Prism 13516
Prism II 15079
Prison Information Bulletin 04402
Pritozi za Knjizevnost, Jezik, Istorija i Folklor 17090
Private Higher Education 13570
Private Investors Abroad: Problems and Solutions in International Business 16730
The Private Library 12719
Private Press Books: Annual Bibliography 12719
Private School Law Digest 15935
Prize Poems 16049
Probate and Property 13617
Probe 02944
The Probe 12105
Problemi attuali di scienza e di cultura 07211
Problemi dell'Informazione 07405
Problemi d'Oggi: Bimestrale di Scienze Umane e Formazione Sociale 07531
Problemi na balgarskija folklor 01770
Problemi na izkustvoto 01770
Problemi na Tehniceskata Kibernetika 01771
Problems of Ageing and Longevity 11189
Problems of Contemporary Theatre 09985
Problems of Obstetrics and Gynecology 01750
Problemy Dalnego Vostoka: Problems of the Far East 09954
Problemy Gastroenterologii: Problems of Gastroenterology 11045
Problemy Mashinostroeniya i Nadezhnosti Mashin: Problems of Engineering and Machine Reliability 09954
Problemy Medycyny i Farmacji 09631
Problemy Osvoyeniya Pustyn: Problems of Desert Development 11180
Problemy Peredachi Informatsii: Problems of Information Transmission 09954
Problemy Prochnosti: Problems of Strength 11193
Problemy Spetsialnoi Elektrometallurgii: Problems of Special Electrometallurgy 11193
Proceeding of IAMFE Conferences 10474
Proceedings A (Mathematics) 12858
Proceedings B (Biological Sciences) 12858
Proceedings (Chemical Sciences) 06767
Proceedings: civil engineering 12294
Proceedings (Earth and Planetary Sciences) 06767
Proceedings (Mathematical Sciences) 06767
Proceedings: Municipal Engineer 12294
Proceedings, NZFMRA Symposia 09146
Proceedings, NZFMRA Technical Conferences 09146
Proceedings of AIESEP 01137
Proceedings of Annual Riverside Conference on Stress, Emotions and Cancer 14639
Proceedings of Annual Science Conferences 09399
Proceedings of Annual Seminar-Workshop 09537
Proceedings of Annual Technical Meetings 12749
Proceedings of Canadian National Energy Forums 02205
Proceedings of CIESM Congress 08796
Proceedings of Conferences of the Institute of Lithuanian Studies: Lituanistikos Instituto Suvaziavimo Darbai 15329
Proceedings of Congress of Scientific Societies 07044
Proceedings of Engineering Thermophysics Conference 03188
Proceedings of European Conference 16939
Proceedings of History Week 08666
Proceedings of IIRA World Congresses 10750
Proceedings of International Congresses on Cybernetics 01135
Proceedings of International Discussion Forum 12386
Proceedings of International Peat Congresses 03969
Proceedings of IRRA Annual Meeting 15260
Proceedings of IRRA Spring Meeting 15260
Proceedings of Legal Conference 14689
Proceedings of Reciprocal Meat Conference 13931
The Proceedings of the Academy 09234
Proceedings of the Academy of Natural Sciences of Philadelphia 13299
Proceedings of the American Academy for Jewish Research 13377
Proceedings of the American Antiquarian Society 13454
Proceedings of the American Control Conference 13612
Proceedings of the American Mathematical Society 13929
Proceedings of the Annual Conference on Taxation 16168
Proceedings of the Annual Meeting of the Israel Mathematical Union 07081
Proceedings of the Annual Meeting on Information Science and Technology 08216
Proceedings of the Annual Veterinary Medical Forums 13739
Proceedings of the Association of Marine Laboratories of the Caribbean 09724
Proceedings of the Baltic Marine Biological Symposia 10438
Proceedings of the Chilean History of Medicine 03012
Proceedings of the Classical Association 11792
Proceedings of the Clinical Conferences 02764
Proceedings of the Convention of American Instructors of the Deaf 14857
Proceedings of the Desert Fishes Council 14965
Proceedings of the Dorset Natural History and Archaeological Society 11902
Proceedings of the Entomological Society of Ontario 02439
Proceedings of the Finnish Dental Society 04025
Proceedings of The Human Factors Society 15235
Proceedings of the International Commission on Trichinellosis 09544
Proceedings of the International Symposia 15467
Proceedings of the I.R.E.E. 00422
Proceedings of the Jesuit Philosophical Association 15600
Proceedings of the Meetings 07723
Proceedings of the New Zealand Society of Animal Production 09177
The Proceedings of The Nutrition Society 12651
Proceedings of the Physiological Society of New Zealand 09188
Proceedings of the Royal College of Physicians of Edinburgh 12787
Proceedings of the Royal Irish Academy 07028
Proceedings of the Saudi Biological Society 10005
Proceedings of the Società Italiana di Filosofia Giuridica e Politica 07873
Proceedings of the Society for Experimental Biology and Medicine 16522
Proceedings of the Society of Nutrition Physiology 05820
Proceedings of the Tanzania Veterinary Association Scientific Conferences 11063
Proceedings of the Virgil Society 13189
Proceedings of thr Nutrition Society of New Zealand 09183
Proceedings Sec. A: Physical Sciences 06841
Proceedings Sec. B: Biological Sciences 06841
Proceedings: structures and buildings 12294
Proceedings: Transport 12294
Proceedings U.B.S.S. 13175
Proceedings: Water, Maritime and Energy 12294
Process Studies 14693
Procès-Verbaux 15393
Procès-Verbaux des séances du CIPM 04363
Proche-Orient 08560
Pro Civitate Austriae 00884
Procurement Weekly 12274
Prodder Newsletter 10147
Prodeedings 06778
Productivity 06846
Productivity News 06846
Product Liability Law Reporter 14523
pro familia magazin: Sexualpädagogik und Familienplanung 06240
Professional Designers Newsletter 13877
Professional Documents on Social Development 03941
The Professional Energy Manager 14486
Professional Football Researches Association: Membership Directory 16356
Professional Freelance Directory 16183
The Professional Geographer 14379
The Professional Investigator 12271
The Professional Investor 12248
Professionalism 14562
Professional Manager 12253
The Professional Medical Assistant 13548
Professional Negligence Law Reporter 14523
Professional Psychology 14019
Professional Safety 14165
The Professional Sanitarian 16160
The Professional Statistician 12284
Professional Teacher 12724
Professional Women and Minorities: A Manpower Data Resource Service 14790
Professors of Curriculum: Directory 16359
Pro Forum 02846
Program-Automated Library and Information Systems 11267
Programmirovanie: Programming 09954
Progrès en Néonatologie 04823
Progress in Astronomy 03165
Progress in Rubber and Plastic Technology 12703
The Progressive Fish-Culturist 13816
Progress of Space Research 16812
Progress of Theoretical Physics 08284
Progress Report 13843, 13954, 14001, 16419
Progress Reports 16878
Project Management Journal 16362
Project Monograph: Urban Design Case Studies 15308
Project Progress Report 06550
Project Reference File 16801
Promeny 14947
Promoting Health 13854
Promyshlennaya Teplofizika: Industrial Thermal Physics 11193
Prontuario Economico del Turista: Spesa Giornaliera del Viaggiatore in 42 Paesi 07525
Properties of the National Trust 12614
La Propriété Industrielle 11034
The Prosecutor 16035
Prospettiva E.P. 07386
Prosthetics and Orthotics International 03748
Pro Technorama 10857
Protection 12312
Protection de la Nature 10890
Protestants-Christelijke Onderwijsvakorganisatie: Magazine 09062
Protetyka Stomatologiczna 09606
Proteus 10108
Protozoological Abstracts 11720
Proudh Shiksha 06768
Proust Research Association: Newsletter 16365
Provence Historique 04528
La Province du Maine 04871
PRP Activities Report 09666
Przeglad Biblioteczny 09615
Przeglad Dermatologiczny 09560
Przeglad Dermatologiczny: Polish Journal of Dermatology 09561
Przeglad Lubuski 09547
Przegląd Orientalistyczny: Oriental Review 09595
Przeglad Pediatryczny 09600
Przeglad Psychologiczny 09603
Przeglad Socjologiczny 09545
Przeglad Zoologiczny 09608
PS 14010
PSA 16322
PSE Symposia 12699
Psi Chi - The National Honor Society in Psychology: Handbook 16366
Psi Chi - The National Honor Society in Psychology: Newsletter 16366
Psicologia Italiana 07936
Psicosintesi Clinica 07939
Psicoterapie, Metodi e Tecniche 07504
Psihologija 01780
Psikhologicheskii Zhurnal: Psychological Journal 09954
The PSI Researcher 12989
PSNA Newsletter 16318
Psoriasis 12725
Psoriasis Research Association: Abstracts 16367
Psoriasis Research Association: Bulletin 16367
PS Quarterly 16371
Psych Discourse 14402
Psyche 07504
Psychiatria Polska 09602
Psychiatrie Française 04223
Psychiatrie Française 04927
Psychiatry and Art 14161
The Psychoanalytic Review 16132
Psychoanalytisch Forum 09056
Psychobiology 16373
Psychodrame 04726
Psychohistory News 15517
The Psychohistory Review 15180
Psychological Abstracts 14019
Psychological Bulletin 14019
Psychological Documents 14019
Psychological Review 14019
Psychologie 10844
Psychologie Française 04838
Psychologie für die Praxis 10932
Psychologie in Österreich 00575
Psychologie Médicale 04684
Psychologische Rundschau 05411
Psychologists Interested in Religious Issues: Newsletter 16370
Psychology and Pedagogics 01775
Psychology Society: Membership List 16371
Psychology Teaching 11286
Psychometrika 16372
Psychonomic Society: Bulletin 16373
Psychophysiology 16572
Psychosomatics 13304
Psychoterapia 09602
Psychotronics 00596
PsycSCAN: Applied Psychology 14019
PsycSCAN: Clinical Psychology 14019
PsycSCAN: Development Psychology 14019
PsycSCAN: LD/MR 14019
PsycSCAN Series from PsycINFO 14019
Psykologiumtiset 04056
Psykolog Nyt 03667
Pszichológiai Szemle: Psychological Review 06653
PTSM Series 15616
PTU-Nyt 03764
Pubblicazioni IEN 07728
Public Administration 12818
Public Administration and Development 12818
Public Administration Review 14084
Public Administration Times 14084
Publication of the ADS 13786
Publication Scientifique 15393
Publications in Librarianship 14414
Publications of the English Goethe Society 11962
Publications of the Mathematical Society of Japan 08386
Public Contract Law Journal 13617
Public Employee Newsletter 16145
Public Finance and Accountancy 11765
Public Focus 02001
Public Health 13061
Public Health Information 08343
Public Health News 11720
Public Justice Report 14694
Public Libraries 16377
Public Library Reporter 16377
Public Money and Management 11765
Public Personnel Management 15511
Public Responsibility in Medicine and Research: Conference Report 16379
Public Sphere 14702
Public Understanding of Science 12268
Public Utilities Newsletter 16145
Public Works Historical Society: Essay Series 16381

Public Works Historical Society: Newsletter 16381
Publikationen zu wissenschaftlichen Filmen 05956
Pulp and Paper Research Institute of Canada: Annual Report 02853
Pulse 13781
Pulvermetallurgie in Wissenschaft und Praxis 05696
Pump Abstracts 11575
Pumps 11437
Punime te Institutit te Fizikes Berthamore: Studies of the Institute of Nuclear Physics 00003
Purchasing & Supply Management 12274
Pure and Applied Chemistry 12404
Pure and Applied Optics 12268
PVS-Literatur 05618

Qadmoniot 07071
Quaderni AISI 07346
Quaderni ANIAI 07374
Quaderni CISCo 07508
Quaderni degli Studi Danteschi 07777
Quaderni del Centro 07432
Quaderni del Centro di Documentazione Dantesca e Medievale 07777
Quaderni dell'Accademia Arcadia 07199
Quaderni della Rivista Italiana di Musicologia 07909
Quaderni della Rubiconia Accademia dei Filopatridi 07144
Quaderni del Sapere Scientifico 07136
Quaderni di Filologia e Letteratura Siciliana 07792
Quaderni di Justitia 08046
Quaderni di Speleologia 07604
Quaderni d'italianistica 02261
Quaderni FAS 07372
Quaderni Scientifico-Tecnici 07233
Quadrant 14719
Quaestiones Mathematicae 10205
Quaker History 15140
Qualidade 09649
Qualità 07338
Qualität und Zuverlässigkeit: Zeitschrift für industrielle Qualitätssicherung 05414
Quality Circle 14324
Quality Circles Journal 14324
Quality Engineering 14085
Quality Outlook 15963
Quality Progress 14085
The Quality Review 14085
Quality Review Bulletin 15616
Quantum Optics 12268
Quarry Management 12275
The Quarterly 16184
Quarterly Journal of Engineering Geology 12111
Quarterly Journal of Experimental Psychology 12044
Quarterly of Humanistic Anthropology 16533
Quarterly Oil and Gas Statistics 04126
The Quarterly Report 02534
Quarto Series 12186
Quaternaria Nova: Atti – pubblicati nella rivista 07738
Quaternary International 12399
The Quaternary Research 08292
Québecensia 02931
Queensland Geographical Journal 00468
The Queensland Naturalist 00449
Quehacer cardiológico 08758
Quellen zur Theatergeschichte 01016
Quest 12733, 15809
La Questione Sarda-Autonomia 07271
Quête d'avenirs 02552
Quickening 13720
The Quill 14895
Química e Industria 03414, 10276
Quinzaine Universitaire 04946

Race and Class 12276
Race Relations News 10202
Race Relations Survey 10202
Rad 03459
The Radiance Technique Journal 16384
Radiation Research 16385
Radical Religion 14812
Radiobiologiya: Radiobiology 09954
Radio Communication 12737
The Radiographer 00313
Radio Graphics 16388
Radiography 11798
Radiography News 11798
Radioisotopes 08186
Radiokhimiya: Radiochemistry 09954
Radiología 10396
La Radiologia Medica 07321
Radiologie 09779
Il Radiologo 07321
Radiology 16388
Radionavigation Journal 16898
Radio Science 13833
Radius 05688
Radius-Magazine 12738
Radix Institute: Calendar of Events 16390
Radix Journal 16390
Radnorshire Society: Transactions 12739
Radyanske Pravo: Soviet Law 11193
Rädda Livet 10441
La Ragione 07365
RAIC Bulletin 02862
RAIC Practice Notes 02862
R A Illustrated 12760
Railroad Newsletter 16145
Railroad Station Historical Society: Monograph 16392
Railroad Station Historical Society: The Bulletin 16392
Railway Gazette 15730
Raising the Standard 11706
Rajasthan Academy of Science: Proceedings 06852
Rakennushistorian osaston julkaisuja 03983
R A Magazine 12760
RAM Magazine 12764
Rangelands 16574
Die Rapportryer 10142
Rapports Techniques CEBELCOR 01184
RAPRA News 12743
Rare Books and Manuscripts Librarianship 14414
Rare Earth Research Conference: Proceedings of Conference 16394
Rare Poultry Society: Newsletter 12744
RASD Update 16399
RASE Journal 12767
RASE News 12767
RAS Korea Branch Transactions 08527
Rassegna Amministrativa della Sanità 07581
Rassegna di Criminologia 07860
Rassegna di Medicina del Traffico 07894
Rassegna di Studi Dauni 07778
Rassegna di Studi Etiopici 07754
Rassegna di Studi Turistici 07377
Rassegna Frignanese: Rivista di Cultura e di Studi Regionali 07236
Rassegna Gallaratese di Storia e d'Arte 07811
Rassegna Italiana di Chirurgia Pediatrica 07856
Rassegna Italiana di Medicina Omeopatica 07193
Rassegna Medico-Chirurgica del Piceno 07207
Rassegna Storica del Risorgimento 07753
Rassegna Volterrana 07146
Rassid 11138
Rastitelnye Resursy: Plant Resources 09954
Rate Book 15015
RAT – Remscheider Arbeitshilfen und Texte 05064
Raumforschung und Raumordnung: Spatial Research and Spatial Management 05060
Raziskovalec: Researcher 10103
RCO Journal 12784
RCVS Newsletter 12793
R & D Focus 15497
R & D Preview 14873
Reaction 12610
Reading 13167
The Reading Informator 16397
Reading Matters 11460
Reading Research Quarterly 15518
Reading Teacher 15518
Real Academia de Bellas Artes de San Fernando: Boletín 10338
Real Academia de Ciencias Exactas, Físicas y Naturales: Anuario 10342
Real Academia de Ciencias Exactas, Físicas y Naturales: Memoria 10342
Real Academia de Ciencias Exactas, Físicas y Naturales: Revista 10342
Real Academia de Ciencias Morales y Políticas: Anales 10343
Real Academia de Córdoba de Ciencias, Bellas Letras y Nobles Artes: Boletín 10345
Real Academia de Farmacia: Anales 10346
Real Academia de la Historia: Boletín 10348
Real Academia de Nobles y Bellas Artes de San Luis: Boletín 10352
Real Academia Gallega: Boletín 10354
Real Academia Hispano-Americana: Boletín 10355
Real Academia Nacional de Medicina: Anales 10356
Real Academia Sevillana de Buenas Letras: Boletín 10357
Real Sociedad Española de Historia Natural: Actas 10362
Real Sociedad Fotográfica: Boletín 10364
Real Sociedad Geográfica: Boletín 10365
Real Sociedad Vascongada de los Amigos del País: Boletín 10367
Real World 00513
Recent Advances in Phytochemistry 16318
Recent Developments 14924
Recently Published Articles 13846
Recherche Sociale 04544
Recht, Bibliothek, Dokumentation 05105
Recht der Datenverarbeitung 05806
Rechtschreibung: Mitteilungen des Bundes für vereinfachte rechtschreibung 10637
Rechtsfilosofie en rechtstheorie 09106
Rechtsprechung in Strafsachen 10867
Recommandations Relatives à l' Eclairage 04218
Recommendations 09063
Recommended Practices 00493
Recorded Sound 14325
The Recorder 02320, 13902
Record Series 16650
Records Management Journal 11267
Records of Buckinghamshire 11243
Records of Columbia Historical Society 15218
Records of the American Society of Naturalists 14146
Recueil de l'Académie des Jeux Floraux 04096
Recueil des Notices et Mémoires 00023
Le Recueil Généalogique et Héraldique de Belgique 01416
Recusant History 11748
Red Double-Barred Cross 08163
Rééducation Orthophonique 04331
Reef Report 14662
REFA-Nachrichten 06261
Referátový výběr z anesteziologie a resuscitace: Reader's Digest – Anaesthesiology and Resuscitation 03517
The Referee 14230
The Reflector 14540
Reforma 16400
Reforma: National Association to Promote Library Services to the Spanish-Speaking: Membership Directory 16400
Reforma y Democracia: Revista del CLAD 16989
Regard sur la Biochimie 04645
Regard sur la Qualité 04601
Reggio Storia 08007
Regio Basiliensis: Basler Zeitschrift für Geographie/Revue de Géographie de Bâle 10709
Regional and Research Studies 00295
Regional Anesthesia 14164
Regional Arts and What's on Magazine 13231
Regional Catalogue of Earthquakes 12387
Regional Conference Proceedings 13606
Regional Guide to Further Education Courses in the North West 13161
Regional Institute of Social Welfare Research: Annual Report 16402
Regional List of Colleges in the North West 13161
Regional Memoirs 08642
Regional Natural Historics 00478
Regional Studies 12747
Regional Studies Association: Debates & Reviews 12747
Regional Studies Association: Newsletter 12747
Regional Studies Association: Special Issue 12747
Regionalwirtschaftliche Studien 05222
Register of Courses in European Studies in British Universities and Polytechnics 13174
Register of Current Research into European Integration 13174
Register of Dentists 06941
Register of Indexers 14135
The Register of Members and Guide for Buyers of Photography 11588
Register of Members' Interests 12910
Register of Patent Agents 11764
Register of Practices 12440
Register of Professional Sanitarians 13437
Register of Specialist Teachers 12195
Register of the Federation of Polish Medical Societies 09631
Registration and Clinical Statistics of Stomach Cancer in Japan 08184
Registri della Cancelleria Angioina 07228
Registry of Accredited Facilities and Certified Practitioners 13625
Regulatory Alert 15795
Regulus 08586
Rehabilitación 10397
Die Rehabilitation 05234
Rehabilitation 16406
Die Rehabilitation: Zeitschrift für alle Fragen der medizinischen, schulisch-beruflichen und sozialen Eingliederung 05612
Rehabilitation Bulletin 13490
Rehabilitation Counseling Bulletin 13771, 14028
Rehabilitation Information Round Table: Newsletter 16405
Rehabilitation Review 15883
Reito: Refrigeration 08248
Relaciones 00155
Relatively Speaking 01834
Relatório 01634
Relatório de Atividades 01738
Relatório Décimo Anual CPATU 01627
Relatório Técnico 01738
Relatório Técnico Anual 01633, 01648
Relazioni Internazionali 07750
Reliability Assessment 16217
Religion Index One: Periodicals 14192
Religion Index Two: Multi-Author Works: (RIT) 14192
Religious Studies 13434
Religious Studies Review 14903
Renaissance Monthly 03252
Renaissance Quarterly 16411
Rendiconti dell'Accademia Nazionale dei XL 07213
Rendiconti della classe di scienze morali, storiche e filologiche 07211
Rendiconti delle Adunanze Solenni 07211
Rendiconti lincei, scienze fisiche e naturali 07211
Renovatio: Zeitschrift für das interdisziplinäre Gespräch 06058, 06067
Répertoire Africain des Institutions de Recherche 17139
Répertoire des Peintures Flamandes du 15e Siècle 01196
Repertoire Survey 16251
Repertorio Boyacense 03335
Repertorio Histórico 03334
Repertorium Fontium Artis Historiae Portugaliae Instaurandum 09668
Report & Boara Activities 15925
Report, DK-edition 03557
Reporter 15168, 16170
Report Following Congresses 15416
Report from the Institute for Philosophy and Public Policy 15288
Reporting Service: Special Sheets for Quick Delivery of Urgent Information 04479
Report of Hematologic Neoplasmas Registration in Japan 08184
Report of the COPUOS 16812
Report of the International Congresses of Vexillology 15475
Report of the International Shakespeare Conference 12945
Report on Agricultural Research 11216
Report on International Symposium 08543
Report on Mathematical Education 08387
Report on Research in Iceland 06696
Report Psychologie 05204
Reports from International Scientific Congresses and Conferences, Series B: Social Sciences 01775
Reports from the Central Bureau of Statistics (REP) 09351
Reports of Spring and Autumn Meetings 08353
Reports on Mathematical Physics 09569
Reports on Progress in Physics 12268
Reports on Progress in Polymer Physics in Japan 08128
Reports on Rheumatic Diseases 11250
Reports to the Crown 12795
Report to Parliament 12804
Report & Transactions 11890
Report, UK-edition 03557
Representation 11950
RERIC Holdings List 11082
RERIC International Energy Journal 11082
RERIC Membership Directory 11082
RERIC News 11082
RES 16547
Research and Development in Federal Budget 13497
Research and Engineering Council of the Graphic Arts Industry: The Review: Recent Patents of Interest to the Graphic Arts Industry 16412
Research and Teaching Register 14825
Research Defence Society: Newsletter 12752
Research Edit 16303
Research & Industry 06753
Research in Ministry: (RIM) 14192
Research in Molecular Biology: Mathematisch-Naturwissenschaftliche Klasse 05050
Research in Nondestructive Evaluation 14075
Research in Social Psychology: Bulletin of the Japanese Society of Social Psychology 08260
Research in the Commonwealth 11808
Research into Lost Knowledge Organisation: Newsletter 12753
Research in Veterinary Science 11703
Research Journal 01065, 08390
Research Libraries Group Annual Report 16416
Research Libraries Group Directory 16416
Research Libraries Group News 16416
Research Management 15261
Research Management Review 16013
Research Methods Series 11392
Research Monograph 14306
Research Monographs 12233
Research Quarterly for Exercise and Sport 14327
Research Quarterly for Exercise & Sport 13449
Research Service Directory 15692
Research Society of Pakistan: Journal 09398
Research Topics 11457
Research to Prevent Blindness: Annual Report 16419
Research Work in Progress 11699
Resonance 04454, 15479
Resource 16535
Resource Directory 15161
Resource Hotline 13820
Resource Letters 15222
Resources 16117
Resources for Community-Based Economic Development 15978
Resources for Women in Science Series 14369
Resources Kit 02025
Resource Systems Institute Bulletin 15005
Respirer 04412
Response 15814
Ressources et Vous 02919
Restauro 05989
Resumenes de Cafe 03361
Resumenes Pastos Tropicales 03360
Resúmenes sobre Población Dominicana 03832
Resumos Informativos 01635
Retablo de Papel 08722
RE Today 11783
Reumatismo 07943
Review Ingenium 09664
Review of Agricultural Entomology 11720
Review of Applied Entomology, Series A (Agricultural), Series B (Medical and Veterinary) 11732
Review of Educational Research 13790, 13790
Review of Finnish Linguistics and Ethnology 04001
Review of Higher Education 14358
The Review of Income and Wealth 15391
Review of Medical and Veterinary Entomology 11720
Review of Medical and Veterinary Mycology 11720
Review of Plant Pathology 11720
Review of Progress in Coloration and Related Topics 13044
Review of Radio Science 01500
Review of Religious Research 16408
Review of Social Economy 14334
Review of the Bulgarian Geological Society 01749
Review of Tuberculosis for Public Health Nurses 08163
Reviews in Medical Microbiology 12682
Reviews of Geophysics and Space Physics 13833
Reviews of Infectious Diseases 15262
Reviews of Modern Physics 14000
Reviews of Modern Physics: Microfiche Edition 13893
Reviews of Scientific Instruments: Microfiche Edition 13893
Revista AIBDA 03449
Revista ALADEFE 03846
Revista Archivum 00148
Revista Argentina de Anestesiología 00134
Revista Argentina de Cirugía 00061
Revista Argentina de Endocrinologia y Metabolismo 00165
Revista Argentina del Toiax 00149
Revista Astronómica 00054
Revista Bio 01588
Revista Brasileira de Biblioteconomia e Documentação 01649
Revista Brasileira de Biologia 01563
Revista Brasileira de Economia 01654, 01664
Revista Brasileira de Entomologia 01709
Revista Brasileira de Estatística 01655
Revista Brasileira de Farmacia 01590
Revista Brasileira de Genética: Brazilian Journal of Genetics 01711
Revista Brasileira de Geociências 01713
Revista Brasileira de Geografia 01655
Revista Brasileira de Mandioca 01633
Revista Brasileira de Música 01629
Revista Brasileira de Tecnologia 01645
Revista CA 03029
Revista Cartográfica 08740
Revista Catalana de Psicoanàlisi 10395
Revista Centroamericana de Administración Pública 03437
Revista Chilena de Cirugía 03092
Revista Chilena de Educatión Quimica 03023
Revista Chilena de Entomología 03070
Revista Chilena de Historia Natural 03089, 03093
Revista Chilena de Historia y Geografía 03076
Revista Chilena de Ingeniería 03055
Revista Chilena de Obstetricia y Ginecología 03078
Revista Chilena de Pediatria 03081
Revista CIAF 03358
Revista Ciência e Trópico 01658
Revista Ciências da Terra 01713
Revista CODECI 03049
Revista Colombiana de Antropologia: Informes Antropologicos 03380
Revista Colombiana de Matemáticas 03409
Revista Colombiana de Psiquiatría 03413
Revista Colombina de Obstetricia y Ginecologia 03410
Revista Conjuntura Economia 01664
Revista Cubana de Ciencias Veterinarias 03474
Revista Cubana de Reproducciòn Animal 03474

Revista da Associação Médica Brasileira 01603
Revista da Sociedade Portuguesa de Oftalmologia 09708
Revista de Actualidad Odontoestomatológica Española 10299
Revista de Administraçao de Empresas 01654
Revista de Administração Pública 01654
Revista de Administración Pública 10287
Revista de Ciencias Agrarias 09679
Revista de Ciencias Biológicas 03478
Revista de Ciencias Económicas 00116
Revista de Ciencias Químicas 03478
Revista de Dialectologia y Tradiciones Populares 10322
Revista de Economía Política 10287
Revista de Educação AEC 01595
Revista de Educación 03023
Revista de Estudios Internacionales 10287
Revista de Estudios Políticos 10287
Revista de Filología Española 10322
Revista de filozofie: Journal of Philosophy 09736
Revista de Filozofie si Drept: Journal of Philosophy and Law 08783
Revista de Fizica si Chimie 09783
Revista de Guimarães 09687
Revista de Historia de America 08740
Revista de Instituciones Europeas 10287
Revista de Investigación y Desarrollo Pesquero 00144
Revista de istorie: Journal of History 09736
Revista de Istorie a Moldovei: Moldovan Historical Journal 08783
Revista de istorie şi teorie literară: Journal of Literary History and Theory 09736
Revista de la Asociación de Ciencias Naturales del Litoral 00078
Revista de la Asociación de Ortopedía y Traumatología 00070
Revista de la Asociación Geológica Argentina: RAGA 00082
Revista de la Cepal 03038
Revista de la Educación Superior 08705
Revista de la Sociedad Entomológica Argentina 00197
Revista de la Unión Matemática Argentina 00202
Revista del Centro de Estudios Constitucionales 10287
Revista de Lingvistica si Stiinta Literara: Journal of Linguistics and Study of Literature 08783
Revista del Instituto Histórico y Geográfico del Uruguay 13261
Revista de Literatura 10322
Revista de Logopedia 10263
Revista de Menorca 10285
Revista de Neurología, Neurocirugia y Psiquiatria 08768
Revista de Obstetricia y Ginecología de Venezuela 17011
Revista de Ortopedie si Traumatologie 09794
Revista de Planeación y Desarrollo 03371
Revista de Política Social 10287
Revista de Psihologie 09745
Revista de psihologie: Journal of Psychology 09736
Revista de Teatro 01704
Revista do I.G.H.B. 01672
Revista do I.H.G./R.N. 01683
Revista do IHGSC 01677
Revista do Instituto do Ceará 01669
Revista do Patrimônio Histórico e Artístico Nacional 01702
Revista El Campesino 03101
Revista Electrotécnica 00080, 00080
Revista El Niño Limitado 03023
Revista Española de Derecho Constitucional 10287
Revista Española de las Enfermedades del Aparato Digestivo y de la Nutrición 10394
Revista Estadística Española 10326
Revista Foresta Baracoa 03476
Revista Geofísica 08740
Revista Geográfica 08740
Revista Geográfica de Chile 03063
Revista Geológica de Chile 03063
Revista Hispánica Moderna 15212
Revista IADAP 03842
Revista Iberoamericana de Educación 10333
Revista Iberoamericana de Educación Superior a Distancia 10273
Revista ICAITI 06564
Revista INCAE 03436
Revista Interamericana de Educación de Adultos 08711
Revista Interamericana de Educación de Adultos 08722
Revista Internacional de Seguridad Social 10627
Revista Juridica de Cataluña 10288
Revista Matemática Iberoamericana 10366
Revista Médica de Chile 03099
Revista Medica de Rosario 00114
Revista Medicala Romana: Romanian Medical Journal 09789
Revista Medico-Chirurgicala: Medical Surgical Journal 09772
Revista Mexicana de Estudios Antropológicos 08756
Revista Mexicana de Fitopatología: El Vector 08763
Revista Mexicana de Pediatría 08770
Revista No. 107 01579
Revista Oftalmológica Venezolana 17027
Revista Paulista de Medicina 01606
Revista Pernambucana de Desenvolvimento 01668
Revista Peruana de Entomología 09476
Revista Physis 00060
Revista Plantas Medicinales 03474
Revista Portuguesa de Cardiologia: Portuguese Journal of Cardiology 09694
Revista Portuguesa de Ciências Veterinárias 09696
Revista Portuguesa de Farmcacia 09684
Revista Portuguesa de Química 09712
Revista Romana de Drept 09742
Revista Scientia 09414
Revista Venezolana de Urología 17034
Revitalized Signs Newsletter 15388
Revmatologhia 09931
Revue Africaine des Sciences de l'Education 17139
La Revue Agricole et Sucrière de l'Ile Maurice 08676
Revue Archéologie Médiévale 04382
Revue Archéologique de l'Est et du Centre-Est 04382
Revue Archéologique de Narbonnaise 04382
Revue ATIP 04354
Revue Belge d'Archéologie et d'Histoire de l'Art 01092
Revue Belge de Géographie 01473
Revue Belge de Musicology 01436
Revue Belge de Psychoanalyse 01444
Revue Belge du Feu 01149
Revue d'Archéométrie 04550
Revue de Bio-Mathématique 04876
Revue de Chirurgie Orthopédique 04792
La Revue de Droit International et de Droit Comparé 01377
Revue de Droit Rural: Tijdschrift voor Agrarisch Recht 01107
Revue d'Education Médicale 04441
Revue de Gemmologie 04216
Revue de Géographie de Lyon 04658, 04694
Revue de Géographie du Maroc 08802
Revue d'Egyptologie 04805
Revue de l'Académie Arabe de Damas 11036
Revue de l'ACELF 01873
La Revue de la Cinémathèque 02349
Revue de l'AICL 04267
Revue de la Saintonge et de l'Aunis 04517
Revue de la Soudure-Lastijdschrift 01378
La Revue de l'AUPELF 01895
Revue de Médecine Biotique 04437
Revue de Médecine Orthopédique 04812
Revue de Médecine Psychosomatique 04814
Revue de Metaphysique et de Morale 04830
Revue de Musicologie 04820
Revue de Neuropsychologie 04676
Revue de Physique Appliquée 04836
Revue des Archéologues et Historiens d'Art de Louvain 01116
La Revue des Echanges 01908
Revue des Etudes Juives 04744
Revue des Etudes Latines 04703
Revue des Etudes Sud-Est Européennes 09736
Revue des Questions Allemandes 04362
Revue des questions allemandes – Documents 05865
Revue des Questions Scientifiques 01490
Revue des Sciences Morales et Politiques: Nouvelle Série 04109
Revue des Sciences Sociales, Série de Psychologie 09745
Revue de Synthèse 04545
Revue d'Etudes Palestiniennes 15287
Revue d'Histoire de la Médecine Hébraique 04757
Revue d'Histoire de l'Amérique Française 02522
Revue d'Histoire de la Pharmacie 04758
Revue d'Histoire de l'Eglise de France 04767
Revue d'Histoire des Sciences 04545
Revue d'Histoire Diplomatique 04765
Revue d'Histoire du Théâtre 04761
Revue d'Imagerie Médicale 04839
Revue d'Odonto-Stomatologie 04907
Revue d'Odonto-Stomatologie Tropicale 04288
Revue Economique 04543
Revue Economique Française 04656
Revue Egyptienne de Droit International 03899
Revue E Tijdschrift 01478
Revue Française d'Administration Publique 04571
Revue Française d'Allergologie et d'Immunologie Clinique 04782
Revue Française de Droit Aérien et Spatial 04800
Revue Française de Science Politique 04227
Revue Française de Science Politique 04543
Revue Française de Service Social 04306
Revue Française de Transfusion et Hémobiologie 04903
Revue française d'héraldique et de sigillographie 04855
Revue Française d'Histoire d'Outre Mer 04857
Revue Française d'Histoire du Livre 04701
Revue Futuribles 04579
Revue Générale de Thermique 04846
Revue Générale du Froid 04239
Revue Générale Nucléaire 04824
Revue Géographique de l'Est 04184
Revue Hellénique de Droit International 06532
Revue Historique Ardennaise 04728
Revue historique vaudoise 10991
Revue Hospitalière de France 04529
Revue IAA 04173
Revue IBN 01379
Revue Internationale de Droit Comparé 04660
Revue Internationale de Droit Pénal 04256
Revue Internationale de Protection Civile 10776
Revue Internationale de Sécurité Sociale 10627
Revue Internationale des Services de Santé des Forces Armées: International Review of the Armed Forces Medical Services 01228
Revue Internationale pour l'Enseignment Commercial 03824
Revue Laitiers Sidérurgiques 04385
Revue Médecine et Travail 04970
Revue Médicale de l'Union des Médecins Arabes 11132
Revue Neurologique 04825
Revue Numismatique 04826
Revue Romaine de Physique 09737
Revue Roumaine de Biochimie 09737
Revue Roumaine de Biologie: Série de Biologie Animale 09737
Revue Roumaine de Biologie: Série de Biologie Végétale 09737
Revue Roumaine de Chimie 09737
Revue Roumaine de Géologie, Géophysique et Géographie 09737
Revue Roumaine de Linguistique 09737
Revue Roumaine de Mathématiques Pures et Appliquées 09737
Revue Roumaine des Science Sociales: Série des Sociologie/Série des Sciences Economiques 09736
Revue Roumaine des Sciences Sociales: Série de Psychologie 09736
Revue Roumaine des Sciences Techniques 09737
Revue Roumaine des Scienes Sociales: Série de Sciences Juridiques 09736
Revue Roumaine des Scienes Sociales: Série des Philosophie et Logique 09736
Revue Roumaine d'Etudes Internationales 09736
Revue Roumaine d'Études Internationales 09740
Revue Roumaine d'Histoire 09736
Revue Roumaine d'Histoire de l'Art: Série Théâtre, Musique, Cinéma 09736
Revue Roumaine d'Histoire de l'Art/Serie Beaux-Arts 09736
Revue Suisse de Viticulture, d'Arboriculture et d'Horticulture 10700
Revue Suisse de Zoologie: Annales d'Histoire Naturelle de Genève 10961
Revue Technique des Industries du Cuir 04232
Revue Trimestrielle 04757
Revue Valdôtaine d'Histoire Naturelle 08027
La Revuo Orienta: Esperanto 08198
Rheinisches Ärzteblatt 05039
Rheinisches Jahrbuch für Volkskunde 06264
Rheinisches Zahnärzteblatt 06051, 06460
Rheinisch-Westfälische Akademie der Wissenschaften: Abhandlungen 06265
Rheinisch-Westfälische Akademie der Wissenschaften: Jahrbuch 06265
Rheinisch-Westfälische Akademie der Wissenschaften: Jahresprogramm 06265
Rheinisch-Westfälische Akademie der Wissenschaften: Sitzungsberichte 06265
Rheinisch-Westfälische Zeitschrift für Volkskunde 06264, 06434
Rheologica Acta 05514
Rheology Bulletin 16700
Rhetoric Society Quarterly 16420
Rheumatism Review 13736
Rhododendrons 12810
RIA Digest 02945
RIAI Yearbook 07026
RIBA Journal 12812
The Ricardian 12755, 16421
The Ricardian Register 16421
Rice Abstracts 11720
Ricerca e Presenza: Mensile di Notizie ed Informazione Bibliografica 07427
Ricerca Operativa 07322
Ricerche Slavistiche 07294
Richard Jefferies Society: Annual Report and Bulletin 12756
RIIA Discussion Papers 12813
Riksbibliotektjenesten: Annual Report 09344
Riksbibliotektjenesten: Skrifter 09344
Rikusui-gaku Zasshi: Japanese Journal of Limnology 08357
Rimstock 12652
The Ring 09608
Ringing and Migration 11697
Rinnovare la Scuola 07381
Rinrigakukaironshu 08249
Rinsho Shinkei: Clinical Neurology 08378
RIPA Report 12818
Risalat Ul-Khalee Al-Arabi 09997
Risk Analysis Journal 16577
Rissalat al-Maktaba: Message of the Library 08452
RIT 06707
The Rivendell Review 13848
River Plate Shipping Guide 00110
Rivista Archeologica della Antica Provincia e Diocesi di Como 07770
Rivista della Ortoflorofrutticoltura Itáliana 08025
Rivista del Nuovo Cimento 07874
Rivista di Agricoltura Subtropicale e Tropicale 07724
Rivista di Agronomia 07830
Rivista di Diritto Penale e di Diritto della Guerra 07815
Rivista di Economia Agraria: Atti del Convegno di Studi 07864
Rivista di Emoterapia ed Immunoematologia 07827
Rivista di Ingegneria Agraria 07309
Rivista di Psicologia Analitica 07345
Rivista di Psicologia Individuale 07937
Rivista di Sessuologia 07495
Rivista di Storia dell'Agricoltura 07177
Rivista di Storia della Medicina 07952
Rivista di Studi Liguri 07732
Rivista di Teologia Morale 07408
Rivista di Tossicologia Sperimentale e Clinica 07955
Rivista Europea di Implantologia 07180
Rivista Geografica Italiana 07794
Rivista I.C.P. 07311
Rivista Ingauna e Intemelia 07732
Rivista Internazionale di Filosofia del Diritto 07873
Rivista Internazionale di Psicologia e Ipnosi 07816
Rivista Italiana della Saldatura 07736
Rivista Italiana di Agopuntura 07829
Rivista Italiana di Chirurgia della Mano 07853
Rivista Italiana di Economia, Demografia e Statistica 07865
Rivista Italiana di Geofisica e Scienze Affini 07881
Rivista Italiana di Geotecnica 07281
Rivista Italiana di Ipnosi Clinica e Sperimentale 07356
Rivista Italiana di Medicina e Igiene della Scuola 07895
Rivista Italiana di Medicina Sociale e Preventiva 07902
Rivista Italiana di Musicologia 07909
Rivista Italiana di Ornitologia 07946
Rivista Italiana di Pediatria 07932
Rivista Italiana di Stomatologia 07357
Rivista Italiana di Teosofia 08022
Rivista Rosminiana: Di Cultura e Filosofia 07474
Rivista Storica Calabrese (Nuova Serie) 07634
Rivista Storica del Mezzogiorno 08017
Rivista sull'Arte, la Storia e il Folklore di Somma Vesuviana 07548
Rivista Tecnica della Svizzera Italiana 10899
Rivista Ufficiale Annali Italiani di Medicina Interna 07898
RKW Kompass 06256
RKW-Mitteilungen 06248
RMF Newsletter 13320
Robert Burns Chronicle 14625
Robot and Automation 03162
Robotics Today Magazine 16423
Ročenka knihovnickej sekcie ZKI 10095
Rock Art 13746
Rocznik Chopinowski: Chopin Studies 09620
Rocznik Dendrologiczny 09558
Rocznik Gdański 09543
Roczniki Dziejów Spolecznych i Gospodarczych 09612
Roczniki Historyczne 09612
Roczniki TNT 09627
Rocznik Lubuski 09547
Rocznik Przemyski 09628
Rocznik Towarzystwa Literackiego im. Adama Mickiewicza 09622
Römische Historische Mitteilungen 00723
Römische Mitteilungen 05568
Römische Quartalschrift 05896
Römisches Österreich 00760
Roentgenological Briefs 14914
Röntgenpraxis 05198
Rogers Group: Newsletter 16426
Roll of Arms 14552
The Rolls of Dental Auxiliaries 12104
Romana Gens 07259
România Literaraí 09798
Romanian Alzheimer Newsletter 09795
Romanian Neurosurgery 09793
The Romanian Review of Psychiatry, Child Psychiatry and Clinical Psychiatry 09744
Romantique 04753
Romerike Berge 05184
Roopa Lekha 06714
Rosc 06937
Roshd: Educational Journal for Students and Teachers 06903
Rossiiskaya Arkheologiya: Russian Archaeology 09954
Rossiiskii Musei: Russian Museums 09954
Roster and Operations Manual 15830
Roster of ADI 13289
Roster of MESA Fellows 15735
Roster of Minority Firms 16119
Rostlinná Výroba 03501
Rostrum 11523, 16055
Royal Academy of Engineering: Annual Report 12763
Royal Academy of Engineering: Newsletter 12763
Royal Academy of Music: Prospectus 12764
The Royal African Society: Journal 12766
Royal Agricultural Society of New Zealand: Bulletin 09194
Royal Art Society of New South Wales: Newsletter 00454
Royal Asiatic Society, Hong Kong Branch: Journal 06595
Royal Asiatic Society of Sri Lanka: Journal 10418
Royal Association for Disability and Rehabilitation: Bulletin 12771
Royal Association for Disability and Rehabilitation: Contact 12771
Royal Astronomical Society: Monthly Notices 12772
Royal Astronomical Society of Canada: Journal 02863
Royal Astronomical Society of New Zealand: Newsletter 09195
Royal Astronomical Society: Quarterly Journal 12772
Royal Australasian College of Dental Surgeons: Annals 00455
Royal Australasian Ornithologists Union: Newsletter 00459
Royal Australian Historical Society: Journal 00464
Royal Choral Society: Prospectus 12777
The Royal College of Anaesthetists: Newsletter 12778
Royal College of Music Magazine 12781
Royal College of Obstetricians and Gynaecologists: Journal 12783
Royal College of Surgeons of Edinburgh: Journal 12791
The Royal College of Surgeons of England: Almanack 12792
The Royal College of Surgeons of England: Handbook 12792
Royal College of Veterinary Surgeons: Directory of Practices 12793
Royal College of Veterinary Surgeons: Register of Members 12793
Royal Commission on Historical Manuscripts: Annual Review 12795
Royal Commonwealth Society: Conference Reports 12796
Royal Commonwealth Society: Newsletter 12796
Royal Dublin Society Seminar Proceedings 07024
Royal Economic Society: Newsletter 12799
Royal Entomological Society of London: Symposium Volumes 12800
Royal Environmental Health Institute of Scotland: Annual Report 12801
Royal Forestry Society of England, Wales and Northern Ireland: Journal 12805
The Royal Highland News 12808
The Royal Historical Society of Queensland Bulletin 00469
Royal Historical Society of Victoria: Journal 00470
Royal Incorporation of Architects in Scotland: Prospect 12811
Royal Institute: Journal 11097
Royal Institute of Painters in Water Colours: Catalogue of Exhibitions 12816
Royal Institution of Great Britain: Proceedings 12822
Royal Institution of Great Britain: Record 12822
The Royal Institution of Naval Architects: Transactions 12823
Royalist Focus 12867
Royal Meteorological Society: Quarterly Journal 12832
Royal Microscopical Society: Proceedings 12833
The Royal Musical Association: Journal 12834

The Royal Musical Association: Research Chronicle *12834*
Royal Naval Bird Watching Society: News Bulletins *12835*
Royal Nova Scotia Historical Society: Collections *02870*
Royal Physical Society of Edinburgh: Proceedings *12842*
Royal Scientific Society: Current List of Periodical Holdings *08454*
Royal Scientific Society: Monthly Accession List *08454*
Royal Scottish Academy: Annual Report *12844*
The Royal Society: Annual Report *12848*
Royal Society for Asian Affairs: Journal *12849*
Royal Society for the Encouragement of Arts, Manufactures and Commerce: Journal *12851*
Royal Society Lectures *12050*
Royal Society of Arts and Sciences of Mauritius: Proceedings *08674*
Royal Society of British Artists: Annual Catalogue *12855*
Royal Society of Canada: Calendar: Annuaire *02871*
Royal Society of Canada: Proceedings and Transaction: Délibérations et Memoirs *02871*
Royal Society of Chemistry: Professional Bulletin *12857*
The Royal Society of Edinburgh: Annual Report *12858*
The Royal Society of Edinburgh: Year Book *12858*
The Royal Society of Health: Journal *12859*
Royal Society of Literature of the United Kingdom: Report *12860*
Royal Society of Literature of the United Kingdom: Transactions *12860*
The Royal Society of Medicine: Annual Report *12862*
The Royal Society of Medicine: Calendar *12862*
Royal Society of New South Wales: Journal *00476*
Royal Society of New South Wales: Newsletter *00476*
Royal Society of New South Wales: Proceedings *00476*
Royal Society of New Zealand: Bulletin *09196*
Royal Society of New Zealand: Journal *09196*
Royal Society of New Zealand: Proceedings *09196*
Royal Society of Portrait Painters: Catalogue of Annual Exhibition *12863*
Royal Society of Queensland: Newsletter *00477*
Royal Society of Queensland: Proceedings *00477*
Royal Society of Queensland: Symposia *00477*
Royal Society of South Africa: Transactions *10171*
Royal Society of South Australia: Transactions *00478*
Royal Society of Victoria: Proceedings *00480*
The Royal Society: Year Book *12848*
Royal Stuart Papers *12867*
Royal Stuart Review *12867*
Royal Welsh Agricultural Society: Annual Journal *12874*
Royal Western Australian Historical Society: Newsletter *00483*
Royal Zoological Society of Ireland: Annual Report *07031*
Royal Zoological Society of South Australia: Annual Report *00485*
Rozhledy v chirurgii: Surgical Review *03509*
Rozprawy Komisji Jeżykowej *09545*
Rozprawy Komisji Jezykowej *09630*
RPA News *12879*
RP Eye Research Foundation: Newsletter *02872*
RQ *16399*
RRS News *16385*
RSNA Today *16388*
RSS Newsletter *12749*
RTCA Digest *16389*
Rubber Developments *08651, 12479*
Rubber Statistical Bulletin *12386*
Ruch Filozoficzny: Philosophical Movement *09567*
Rudodobiv *01796*
Rundbrief des E.u.U.Z. *05641*
Rural Community Health Newsletter *15813*
Rural Development Abstracts *11720*
Rural Reconstruction Review *15490*
The Rural Sociologist *16430*
Rural Sociology *16430*
Rural Special Education Quarterly *13768*
Rural Wales Magazine *11741*
Rural Water News *16429*
Rusbogen *03795*
The Rushlight *16431*
Rusistika *11269*
Russell Society News *14579*
Russia Cristiana *07591*
Russian Academy of Agricultural Sciences: Doklady: Proceedings *09950*
Russkaya Literatura: Russian Literature *09954*
Russkaya Rech: Russian Speech *09954*
Russkii Yazyk i Natsionalnoi Shkole: Russian in the National School *09953*
RWI-Handwerksberichte *06267*
RWI-Konjunkturberichte *06267*
RWI-Konjunkturbriefe *06267*
RWI-Mitteilungen *06267*
RWI-Papiere *06267*
RX Ipsa Loquitur *14078*

SAAD Digest *12998*
SAAP Newsletter *16586*
S.A. Archaeological Bulletin *10178*
Saarländisches Ärzteblatt *05032*
Sabazia *08012*
Saber Leer *10311*
S.A.-Cancer Bulletin *10127*
SACO Tidningen Akademiker *10516*
Sacra Doctrina *08029*
Sadhana: Engineering Sciences *06767*
SAE-Australasia *00493*
Sächsische Akademie der Wissenschaften zu Leipzig: Abhandlungen *06273*
Sächsische Akademie der Wissenschaften zu Leipzig: Sitzungsberichte *06273*
SAE International: Handbook *16433*
SAE-News *00493*
Saerskrifter *03709*
SAE Update *16433*
Safe Association: Symposium Proceedings *16434*
Safe Cycling *15756*
Safe Driver *16145*
Safe Journal *16434*
Safe Multhull Cruising *12367*
Safety and Health *16145*
Safety Canada *02001*
Safety Education *12852*
Safetylines *02001*
Safety Management Monograph *16146*
Safe Worker *16145*
SAFR-Mitteilungen *10795*
Saga: Ný saga *06700*
Saga-Book *13185*
Sage Studies in International Sociology *10330*
Sagetrieb *15074*
Saggi: Rivista di Neuropsicologia Infantile, Psicopedagogia, Riabilitazione *07354*
Saggi e Studi di Pubblicistica *07739*
Saguaroland Bulletin *14964*
Saintpaulia and Houseplant Society: Bulletin *12884*
SAJM *10794*
SA Journal of Chemistry *10186*
SA Journal of Food Science and Nutrition *10158*
The SA Journal of Food Science & Nutrition *10179*
SALALM Bibliography and Reference Series *16461*
SALALM Newsletter *16461*
Salamandra: Zeitschrift für Herpetologie und Terrarienkunde *05361*
Salaries of Scientists, Engineers and Technicians *14790*
Sales & Marketing Management *12278*
The Sales Researcher *02876*
Salubritas *16926*
Salute *10206*
La Salute Umana *07615*
Salzburger Beiträge zum Sport unserer Zeit *01033*
Salzburger Beiträge zur Paracelsusforschung *00676*
Das Salzfaß *05927*
SAMAB: Southern African Museums Association Bulletin *10226*
Samadhi *12477*
SAM Advanced Management Journal *16491*
Samakaleena Bharateeya Sahitya *06854*
Samantix *10226*
Samferdsel: Communication *09338*
SAM Focus on Management *16491*
Samfundet til Udgivelse af Dansk Musik: Bulletin *03798*
Samlaren: Tidskrift för svensk litteraturvetenskaplig forskning *10562*
SAMPE Journal *16589*
SAMPE Quarterly *16589*
Samskrita Pratibha *06854*
Sananjalka *04037*
Sandgrouse *12663*
Sangeet Natak *06855*
San Martin News *16439*
SANTA Health Magazine *10210*
SANTA News *10210*
Santé et Société *04840*
Santé Publique *04840*
Sanvi Taleem *09393*
Sapporo Norin Gakkai: Journal *08407*
Saranukrom Thai *11097*
Sarascope Newsletter *16637*
S.A.S. Collections *13110*
Saskatchewan Bulletin *02905*
Saskatchewan Genealogical Society: Bulletin *02888*
Saskatchewan Institute of Agrologists: Newsletter *02892*
Saskatchewan Library Association: Forum *02893*
Sasmira Bulletin *06857*
Sasmira Technical Digest *06857*
Sasta Journal *00497*
Satapitaka Series *06826*
La Satellite *10020*
Saturn *13307*
SA Tydskrif vir Natuurwetenskappe en Tegnologie *10177*
Saudfjárraektin *06679*
Saudi Biological Society: Abstract and Programme of Annual Conference *10005*
SA Water Bulletin *10234*
S.A. Wild Flower Guide Series *10126*
SBAP-Bulletin *10888*
SCALA Conference Papers *13036*
SCALA Maintenance Expenature Revue *13036*
SCALA Newsletter *13036*
SCALA Study Day Papers *13036*
Scandinavian Actuarial Journal *10532*
Scandinavian Actuarial Journal: Published in cooperation with the societies of actuaries in Denmark, Finland and Sweden *09251*
The Scandinavian Economic History Review *03806*
Scandinavian Insurance Quarterly *10546*
Scandinavian Journal of Dental Research *03986*
Scandinavian Journal of Forest Research *10489*
Scandinavian Journal of Logopedics and Phoniatrics *03565*
Scandinavian Journal of Statistics *04065*
The Scandinavian Psychoanalytic Review *03666*
Scandinavian Refrigeration *03638*
Scandinavian Studies *16590*
SCA Newsletter *16648*
Scanner *13719, 14163*
Scarabée: Belgio *07463*
SCEH Newsletter *16508*
Schedule & Show Catalogue *12880*
Schizophrenia Association of Great Britain: Newsletter *12887*
Schleswig-Holsteinisches Ärzteblatt *05041, 06035*
Schmalenbachs Zeitschrift für betriebswirtschaftliche Forschung *06277*
School Board News *16147*
School Business Affairs *14494*
School Catalog *14734*
The School Counselor *13771*
School Directory *13947*
School Law Reporter *16120*
School Law Update *16120*
School Libraries in Australia *00336*
School Library Media *13579*
School Management Study Group: Newsletter *16442*
School Science *06843*
School Science and Mathematics *16443*
School Science and Mathematics Association: Convention Program *16443*
School Science and Mathematics Association: Newsletter *16443*
The School Science Review *11282*
School Security Journal *15885*
Schools in the Middle *15892*
School Technology News *15892*
Schopenhauer-Jahrbuch *06278*
Schriften der Deutschen Sektion *05574*
Schriften der Katholischen Akademie in Bayern *06059*
Schriften der Schweizerischen Gesellschaft für Theaterkultur *10850*
Schriftenreihe der FVS *00595*
Schriftenreihe der GDMB *05794*
Schriftenreihe der Katholischen Akademikerarbeit Deutschlands *06061*
Schriftenreihe des DRL *05538*
Schriftenreihe des Forschungsinstituts der Friedrich-Ebert-Stiftung *05764*
Schriftenreihe des RWI *06267*
Schriftenreihe des Vereins zur Förderung der Versicherungswissenschaft in München e.V. *06428*
Schriftenreihe Dokumentation *05952*
Schriftenreihe für Ländliche Sozialfragen *05044*
Schriftenreihe Wissenschaft *00999*
Schriftenreihe zum Konsumentenschutzrecht *11030*
Schrifttumsberichte zur deutschen Landeskunde *06463*
Schütz-Jahrbuch *05992*
Schulbibliothek Aktuell *05570*
Schule-Wirtschaft *05236*
Schutzgemeinschaft Alt Bamberg e.V.: Informationsheft *06279*
Schweissen und Schneiden *05551*
Schweißen und Schneiden: Welding and Cutting *05753*
Schweisstechnik *00829*
Schweißtechnik als Bestandteil der Technica *10920*
Der Schweizer Anwalt *10884*
Schweizer Archiv für Neurologie *10875*
Schweizer Archiv für Tierheilkunde *10721*
Schweizer Asiatische Studien *10798*
Schweizer Ingenieur und Architekt *10899*
Schweizerische Afrika-Gesellschaft: Newsletter *10786*
Schweizerische Akademie der medizinischen Wissenschaften: Bulletin *10787*
Schweizerische Akademie der Sozial- und Geisteswissenschaften: Bulletin *10789*
Schweizerische Apothekerzeitung *10885*
Schweizerische Blasmusikzeitung *10666*
Schweizerische Gesellschaft für Balneologie und Bioklimatologie: Congress report *10813*
Schweizerische Gesellschaft für Muskelkranke: Mitteilungsblatt *10835*
Schweizerische Gesellschaft für Soziologie: Bulletin *10848*
Schweizerische Hämophilie-Gesellschaft: Bulletin *10859*
Schweizerische Heilpädagogische Rundschau (SHR) *10860*
Schweizerische Hochschulkonferenz: Jahresbericht *10861*
Schweizerische Lehrerinnen- und Lehrer-Zeitung SLZ *10659*
Schweizerische Lichttechnische Gesellschaft: Mitteilung an die Mitglieder *10868*
Schweizerische Monatsschrift für Zahnmedizin *10959*
Schweizerische Musikforschende Gesellschaft: Jahrbuch *10874*
Schweizerische Numismatische Rundschau: Revue Suisse de Numismatique *10877*
Schweizerischer Altphilologen-Verband: Bulletin *10882*
Schweizerischer Münzkatalog *10877*
Schweizerischer Romanistenverband: Lettre Circulaire *10903*
Schweizerischer Verband der Ingenieur-Agronomen und der Lebensmittel-Ingenieure: Bulletin *10907*
Schweizerischer Verband für Frauenrechte: Contact *10913*
Schweizerischer Wissenschaftsrat: Jahresbericht *10924*
Schweizerisches Archiv für Volkskunde / Archives Suisses des Traditions Populaires *10856*
Schweizerische Schillerstiftung: Jahresbericht *10925*
Schweizerisches Institut für Kunstwissenschaft: Jahrbuch *10927*
Schweizerisches Institut für Kunstwissenschaft: Jahresbericht *10927*
Schweizerisches Jahrbuch für Internationales Recht *10947*
Schweizerische Technische Zeitschrift *10905*
Schweizerische Vereinigung der Versicherungsmathematiker: Mitteilungen *10937*
Schweizerische Vereinigung für Atomenergie: Bulletin *10940*
Schweizerische Vereinigung für Schiedsgerichtsbarkeit: Bulletin *10950*
Schweizerische Zeitschrift für Agrarwirtschaft und Agrarsoziologie *10809*
Schweizerische Zeitschrift für Forstwesen *10896*
Schweizerische Zeitschrift für Geschichte *10614*
Schweizerische Zeitschrift für kaufmännisches Bildungswesen *10827*
Schweizerische Zeitschrift für kaufmännisches Bildungswesen: Revue Suisse pour l'enseignement commercial *10897*
Schweizerische Zeitschrift für Psychologie *10844*
Schweizerische Zeitschrift für Soziologie: Revue Suisse de Sociologie *10848*
Schweizerische Zeitschrift für Volkswirtschaft und Statistik *10849*
Schweizer Kunst *10720*
Schweizer Landtechnik: Technique Agricole *10914*
Schweizer Münzblätter: Gazette Numismatique Suisse *10877*
Schweizer Musikdenkmäler *10874*
Schweizer Naturschutz *10890*
Schweizer Schule *10645*
Schweizer Theater-Jahrbuch *10850*
Schweizer Volkskunde: Korrespondenzblatt *10856*
Science *11098, 13497*
Science and Children *16151*
Science and Engineering Newsletter *15980*
Science and Engineering Research Council: Annual Report *12889*
Science and Life *09992*
Science and Technology Review *03128*
Science and Technology Studies *16580*
Science Bachchon Key Liye *09399*
Science Books and Films *13497*
Science Bulletin *03314*
Science Center Exhibitions Calendar *14500*
Science Chronicle *09384*
Science Education *03261*
Science Education News *13497*
Science Education Newsletter *13861*
Science for Peace Bulletin *02908*
Science for People *11667*
La Science Historique *04566*
Science in Parliament *12679*
Science Journal *01065*
Science-Ki-Duniya *06753*
Science Museum News *14499*
Science Name *09399*
The Science of Advaneed Materials + Process Engineering *16589*
Science of Food and Agriculture *14868*
Science Reporter *06753*
The Science Reporter *11316*
Science Scope *16151*
Sciences et Techniques de l'Eau *01955*
Sciences et Techniques Pharmaceutiques *04841*
Science Society of Thailand: Journal *11098*
The Science Teacher *16151*
Science Teaching *12888*
Science Technology *12279*
Science & the Public *11495*
Scientia Agriculturae Bohemoslovaca *03501*
Scientia Canadensis *02247*
Scientia Pharmaceutica *00878*
Scientific and Professional Meetings in Yugoslavia and Foreign Countries *17094*
Scientific Committee on Problems of the Environment: Newsletter *04623*
Scientific Communism, Philosophy, Sociology, Science of Science and Scientific Information *01775*
Scientific, Engineering, Technical Manpower Comments *14790*
Scientific Information Bulletin *08719*
Scientific Information serving in Natural, Mathematical and Social Sciences *01775*
Scientific Life *01799*
Scientific Papers *13339, 17153*
Scientific Proceedings *13898*
Scientific Publications *12092*
Scientific Report *01398*
Scientific Research in Israel *07108*
Scientific Strategy *12508*
Scientific World *05001, 16824*
Scientists Center for Aminal Welfare: Newsletter *16448*
SCNCER Newsletter *11071*
SCONULOG *13096*
Scope *11495*
SCOPE Series *04623*
Scottish Air News *11751*
The Scottish Association for Marine Sciemce: Annual Report *12895*
The Scottish Association for Marine Sciemce: Newsletter *12895*
The Scottish Banker *12214*
Scottish Bird News *12919*
Scottish Bird Report *12919*
Scottish Birds *12919*
Scottish Business Education Council: Annual Report *12901*
Scottish Business Education Council: Handbook *12901*
Scottish Church History Society: Records *12903*
Scottish Civil Law Reports *12442*
Scottish Criminal Case Reports *12442*
Scottish Educational Journal *11945*
Scottish Field Studies Association: Annual Report *12907*
Scottish Fly-Over Supplement *11751*
Scottish Forestry *12845*
The Scottish Genealogist *12910*
The Scottish Geographical Magazine *12846*
Scottish Industrial Heritage Society: Newsletter *12912*
Scottish Journal of Political Economy *12904*
Scottish Language *11283*
Scottish Libraries *12915*
Scottish Literary Journal *11283*
Scottish Literary Journal Supplements *11283*
Scottish Recreational Land Association: Newsletter *12923*
Scottish Rights of Way Society: Annual Report *12925*
Scottish Secondary Teachers' Association: Bulletin *12927*
Scottish Secondary Teachers' Association: Journal *12927*
Scottish Society for Crop Research: Bulletin *12928*
Scottish Society of the History of Medicine: Newsletter *12931*
Scottish Society of the History of Medicine: Report of Proceedings *12931*
Scottish Tartans Society: Proceedings *12932*
Scottish Text Society: Publications *12934*
Scottish Transport *12935*
Scottish Wildlife *12936*
Scrinium-Zeitschrift *00966*
Script Newsletter *15566*
Scriptum Geriatricum *00783*
Scritti in Favore dei Bambini Diabetici *07387*
The Scrivener *16450*
La Scuola l'Uomo *08042*
SCVIR Newsletter *16647*
SDA Newsletter *02887*
Seabird: Annual Journal *12937*
Seabird Group Newsletter *12937*
Sea Education Association: Annual Report *16451*

The Search 12542
Search Group: Report 16454
Searching 02552
Sea Shepherd Conservation Society: Newsletter 16455
Sea Swallow 12835
Seaways 12621
SEB Newsletter 16517
Secolul 20 09798
Secondary School Admission Test Board: Bulletin of Information 16456
Secondary School Admission Test Board: Newsletter 16456
The Secondary Teacher 06931
Section on Women in Legal Education of the AALS: Newsletter 16458
Sector 3 12908
SEDOS Bulletin 07647
Seed Abstracts 11720, 11723
SEEDBED 16592
Seedhead News 16186
Seedlisting 16186
Seed Science and Technology 10745
Seed Technologist News 02943
Sefarad 10322
SEFI-News 01460
SEG/SSE – Information 10806
S.E.H. 10388
Seibutsu Butsuri: Biophysics 08255
Seibutsu-kogaku Kaishi 08425
SEI Certified Product List 16435
Seisan Gijutsu Kenkyusho: Report 08408
Seishin Shinkeigaku Zasshi 08257
Seismological Society of America: Bulletin 16460
Seismologic Bulletin 06571
The Selborne Magazine 12938
Selected Essays 15045
Selection & Development Review 11636
Selektsiya i Semenovodstvo: Selection and Seed Science 09950
Selenology 13922
Selskab for Nordisk Filologi Arsberetning 03820
Selskokhozyaisstvennaya Biologiya: Agricultural Biology 09950
Semeia 16642
Semestral 10379
Semiannual Vehicle Forecasts 14558
Semiconductor Science and Technology 08129, 12268
Semina 04015
Seminar Books of Papers 08892
Seminario de Filología Vasca Julio de Urquijo: Anuario 10368
Semiotische Berichte: Zeitschrift für Semiotik 00830
The Sempervivum Society: Newsletter 12940
Semya i Shkola: Family and School 09953
Sen-i Gakkai: Journal 08413
Senior Nurse 12782
SEN-Science Education News 00488
SEPCR Newsletter 07682
The Sephardic Scholar 14167
SEPM Newsletter 16578
Septuagint and Cognate Studies (SCS) 02555
Sepu 03167
SERC Bulletin 12889
Serena Canada 02911
Serena West Newsletter 02910
Seria Bibliográfica 03027
Seriarte 07227
Seria Zródel 09543
Serie Bibliográfica 01536
Serie Carta de Suelos 00107
Série de Mécanique Appliquée 09737
Serie de Trabajos Varios: Monografias 10369
Serie Directorios 03027
Série Documento 01635
Série Documentos 01632
Série Encontros e Estudos 01647
Serie Estudios 03027
Série Estudos 01714
Serie Fitogeográfica 00107
Serie Información y Documentación 03027
Série Referência 01647
Série Técnica 01632
SER in Action Newsletter 16518
SERITEC 01809
Seritec 01810
Servicio de Información al Dia en Gestión Pública 16989
Servicio Nacional de Geología y Minería: Boletín 03063
SES-ESF tiedotuslehti 04011
Sessions des Comités Consultatifs (8): Electricité, Photométrie et Radiométrie, Thermométrie, Définition du Mètre, Définition de la Seconde, Etalons des Mesure des Reyonnement Ionisants, Unités, Masse et grandeurs apparentées 04363
Set: Research Information for Teachers 09141
La Settimana degli Ospedali 07957
Sexually Transmitted Diseases 14203
SF3 Locater List 16595
S.F.M. Bulletin 04815
SFPE Journal of Fire Protection Engineering 16664
SFS Catalogue 04059
SFS-tiedotus 04059
SGA-ASSPA-Bulletin 10812
SGFF-Jahrbuch 10819
SGF Rapport 10552
SGKM-Bulletin 10828
SGP Mitteilungen 10841
SGTS Publications 12909
SGU Bulletin 10852
S.H.A.A. Journal 13050
Shakaigaku Hyóron: Japanese Sociological Review 08370
Shakai-Keizai-Shigaku 08415
Shakespeare Association of America: Bulletin 16466
Shakespeare Association of America: Directory of Members 16466
Shakespeare Institue: Conference Report 12945
Shakespeare-Jahrbuch 05577
The Shavian 12947
Shavian Tracts 12947
Sheepherder 16594
Sheeraza 06830
Sheet Metal Industries: Trade Journal 12281
Sherlock Holmes Journal 12949
Sherwood Anderson Society: Bibliography 16469
Shevchenko Scientific Society: Memoirs 16470
Shevchenko Scientific Society: Newsletter 16470
Shevchenko Scientific Society: Proceedings 16470
Sheviley Hahinuch 14878
Shigaku-Kenkyu: Review of Historical Studies 08132
Shigaku-Zasshi 08417
Ship and Boat International 12823
Ship Repair and Conversion Technology 12823
The Shirin or the Journal of History 08418
Shokubai: Catalyst 08422
Short Book Reviews 08944
SHOT Newsletter 16598
Show Catalogue 11975, 12520
Show Times 13081
Shropshire Archaeological and Parish Register Society: Transactions 12951
SIA Journal 10053
SIAM Journal on Discrete Mathematics 16536
SIAM Journal on Matrix Analysis 16536
The Siam Society: Journal 11099
SIAS Monograph Series 03785
SIAS Occasional Papers 03785
Sibirskii Matematicheskii Zhurnal: Siberian Mathematical Journal 09954
Sibirskii Vestnik Selskokhozyaistvennoi Nauki: Siberian Agricultural Science Journal 01036
Sibrium: Collana di Studi e Documentazioni 07453
sicher aktuell 05867
Sicherheitswissenschaftliche Monographien 05859
Sicurezza Notizie 07439
SIECCAN Journal 02912
SIECCAN Newsletter 02912
SIECUS Newsletter 16464
SIECUS Report 16464
Sierra 16471
Sierra Leone Association of Archivists, Librarians and Information Scientists: Bulletin 10031
SIGART Newsletter 16737
Sight and Sound 11562
Sightlines 13812
Sigmund Freud House Bulletin 00944
Signes du Présent 08809
Signum 04064
The Sign Writer 14705
Silva Fennica 04050
Silver Lining Appeal Brochure 11598
The Simian 16475
Simian Society of America: Membership Roster 16475
Simulation 16512
Simulation: Games for Learning 12999
SINformation 16541
Singapore Institute of Architects: Yearbook 10053
Singapore Journal of Physics 10048
Singapore Libraries 10051
Singapore Medical Journal 10055
Sinhala Bauddhaya 10416
Sinh-Ly-Hoc 17061
Sino-American Relations 03252
Sinological Monthly 03252
Sinological Quarterly 03252
Sinopsis de Geriatría 00171
Sintesis Básica 00146
SIPE – Servizio Stampa Educazione e Sviluppo: Notiziario d'informazione 07752
Sir Frederick Hooper Essay Award 12088
Sir George Earle Memorial Lecture 12088
SIS-Bollettino 07951
SIS-Informazioni 07951
SIS Newsletter 16544
Sites et Monuments 04912
Sitzungsberichte der Physikalisch-Medizinischen Sozietät zu Erlangen 06236
Sitzungsberichte, Frankfurter Wissenschaftliche Beiträge 06446
Sivilingeniören 09306
SIWA News 12913
SJFR Eltgottråd 10520
Sjukgymnasten 10496
Skeptical Briefs 14798
The Skeptical Inquirer 14798
Skimir 06691
Skizze 05699
Skógraektarfélag Islands: Yearbook 06698
Skolebiblioteket 03586
Skolebiblioteksårbog 03587
Skoleforum 09354
Skolledaren 10521
Skolvärlden 10495
Skoop 09076
Skrifter fra NSGF: Conference Proceedings 09337
Skrifter utgivna av Svenska Litteratursällskapet i Finland 04067
Skriftserie 03716
Slavia Occidentalis 09612
Slavica Slovaca 10060
Slavic Review 13498
Slavistična Revija 10104
Slavonic and East European Review 12535
Slavyanovedenie: Slavonic Studies 09954
Sleep 16479
Sleep Research 16479
Sleutels 08946
Sliplines 15501
Slovenska Akademija Znanosti in Umetnosti: Razprave: Dissertations 10105
Slovenski biografski leksikon 10105
Small Town 16480
SMA Newsletter 10055
Smart Materials and Structures 12268
Smithsonian Contributions to Anthropology 16481
Smithsonian Contributions to Botany 16481
Smithsonian Contributions to Paleobiology 16481
Smithsonian Contributions to the Earth Sciences 16481
Smithsonian Contributions to the Marine Sciences 16481
Smithsonian Contributions to Zoology 16481
Smithsonian Folklife Studies 16481
Smithsonian Studies in History and Technology 16481
Smithsonian Studies to Air and Space 16481
SNALC-Info S. 1 04946
SNE Exchange 16557
SNV Bulletin 10876
Social Action and the Law 14697
Social and Economic Affairs 06480
Social Biology 16620
Social Biology and Human Affairs 11647
Social Care Update 12595
Social Compass: International Review of Sociology of Religion 01358
Social Economic Studies (SES) 09351
Social Education 15992
Social History of Medicine 13012
Social History Society of the United Kingdom: Bulletin 12955
Socialism & Health 12957
La Socialisto 00571
Socialist Review 14702
Socialpaedagogen 03822
Social Policy Review 12959
Social Problems 16621
Social Psychology 14175
Socialradgiveren 03691
Social Science Information 04459
Social Science Research Council: Annual Report 16485
Social Sciences and Humanities Research Council of Canada: Annual Report 02915
Social Sciences in China 03155
The Social Science Teacher 11295
Social Studies Professionals 15992
Social Welfare History Group: Newsletter 16487
Social Work Education Reporter 14932
Sociedad Argentina de Biología: Revista 00158
Sociedad Argentina de Botánica: Boletín 00159
Sociedad Argentina de Ciencias Neurológicas, Psiquiátricas y Neuroquirúrgicas: Revista 00162
Sociedad Argentina de Estudios Geográficos: Boletín 00167
Sociedad Argentina de Patología: Archivos 00180
Sociedad Científica del Paraguay: Revista 09434
Sociedad Colombiana de Cirugía: Revista 03406
Sociedad Colombiana de Economistas: Revista 03407
Sociedad de Amigos de Arqueología: Revista 13265
Sociedad de Pediatría de Madrid y Región Centro: Boletín 10375
Sociedade Anatómica Luso-Hispano-Americana: Actas dos Congressos 09673
Sociedade Anatómica Luso-Hispano-Americana: Anuário Estatístico: Continente e Ilhas Adjacentes 09673
Sociedade Broteriana: Anuário 09676
Sociedade Broteriana: Boletim 09676
Sociedade Broteriana: Memórias 09676
Sociedade de Geografia de Lisboa: Boletim 09682
Sociedade Geológica de Portugal: Boletim 09685
Sociedade Histórica da Independência de Portugal: Report 09686
Sociedade Portuguesa de Autores: Autores 09692
Sociedade Portuguesa de Ciências Naturais: Boletim 09695
Sociedade Portuguesa de Otorinolaringologia e Bronco-Esofagologia: Boletim 09710
Sociedad Española de Cerámica y Vidrio: Boletín 10380
Sociedad General de Autores de España: Boletín 10398
Sociedad General de Autores de la Argentina: Boletín 00198
Sociedad Mexicana de Geografía y Estadística: Boletín 08764
Sociedad Mexicana de Historia de la Ciencia y la Tecnología: Anales 08765
Sociedad Mexicana de Historia y Filosofia de la Medicina: Boletín 08767
Sociedad Peruana de Historia de la Medicina: Anales 09484
Sociedad Química del Perú: Boletín 09487
Sociedad Uruguaya de Patología Clínica: Revista 13270
Sociedad Venezolana de Ciencias Naturales: Boletín 17018
Sociedad Venezolana de Ciencias Naturales: Newsletter 17018
Sociedad Zoológica del Uruguay: Boletín 13272
Società Dalmata di Storia Patria: Atti e Memorie 07775
Società dei Naturalisti in Napoli: Memorie: Supplemento al Bollettino 07781
Società di Scienze Farmacologiche Applicate: Bolletino di Informazione 07788
Società di Studi Valdesi: Bollettino 07797
Società Domani, Dibattiti 07691
Società Entomologica Italiana: Bollettino 07801
Società Entomologica Italiana: Memorie: Supplemento al Bollettino 07801
Società Geologica Italiana: Bollettino 07813
Società Geologica Italiana: Memorie 07813
Società Istriana di Archeologia e Storia Patria: Atti e Memorie 07820
Società Italiana degli Economisti: Bollettino 07824
Società Italiana delle Scienze Veterinarie: Atti 07828
Società Italiana di Chirurgia: Archivio ed Atti 07850
Società Italiana di Chirurgia: Bollettino 07850
Società Italiana di Economia Demografia e Statistica: Collano di Studi e Monografie 07865
Società Italiana di Immunologia e di Immunopatologia: Proceedings 07887
Società Italiana di Mineralogia e Petrologia: Rendiconti 07908
Società Italiana di Patologia: Atti 07927
Società Italiana di Patologia: Bollettino 07927
Società Italiana di Patologia ed Allevamento dei Suini: Atti 07929
Società Italiana di Psichiatria: Bollettino 07934
Società Italiana di Studi sul Secolo XVIII: Bollettino 07953
Società Ligure di Storia Patria: Atti 07980
Società Medico-Chirurgica di Modena: Bollettino 07986
Società per le Belle Arti ed Esposizione Permanente 08002
Società Savonese di Storia Patria: Atti e Memorie 08012
Societatea Numismatica Romana: Buletinul 09791
Società Ticinese di Scienze Naturali: Bollettino 10965
Société Anatomique de Paris: Bulletins 04628
Société Archéologique d'Alexandrie: Bulletin 03913
Société Archéologique de Touraine: Bulletin 04629
Société Archéologique de Touraine: Mémoires 04629
Société Belge de Cardiologie: News Bulletin 01427
Société Belge de Photogrammétrie, de Télédétection et de Cartographie: Bulletin 01440
Société Belge des Auteurs, Compositeurs et Editeurs: Bulletin 01447
Société Belge d'Ophthalmologie: Bulletin 01451
Société Canadienne des Relations Publiques (Québec): Bulletin 02916
Société Centrale d'Architecture de Belgique: Bulletin Mensuel 01453
Société d'Anthropologie de Paris: Bulletins et Mémoires 04639
Société de Biogéographie: Compte Rendu des Séances 04642
Société de Biogéographie: Mémoires 04642
Société de la Flore Valdôtaine: Bulletin 08027
Société de Langue et de Littérature Wallonnes: Bulletin 01454
Société de l'Histoire de France: Annuaire-Bulletin 04661
Société de l'Histoire de France: Mémoires 04661
Société de l'Histoire de l'Ile Maurice: Bulletin 08675
Société de Médecine de Strasbourg: Bulletin 04667
Société d'Emulation Historique et Littéraire d'Abbéville: Bulletin 04671
Société de Mythologie Française: Bulletin 04672
Société de Pharmacie de Bordeaux: Bulletin 04680
Société des Américanistes: Journal 04693
Société des Amis du Louvre: Chronique 04697
Société des Auteurs et Compositeurs Dramatiques: Revue 04700
Société des Etudes Océaniennes: Bulletin 05003
Société des Océanistes: Journal 04711
Société des Océanistes: Publications 04711
Société des Poètes Français: Bulletin Trimestriel 04712
Société des Professeurs Français et Francophones en Amérique: Bulletin 16488
Société des Sciences Médicales du Grand-Duché de Luxembourg: Bulletin 08588
Société des Sciences Naturelles et Physiques du Maroc: Bulletin 08808
Société d'Etudes Dantesques: Bulletin 04729
Société d'Etudes et de Documentation Economiques, Industrielles et Sociales: Chroniques d'Actualité 04732
Société d'Etudes Jaurésiennes: Bulletin 04743
Société d'Histoire de la Suisse Romande: Mémoires et Documents 10971
Société d'Histoire et d'Archéologie: Bulletin 10972
Société d'Histoire et d'Archéologie: Mémoires et Documents 10972
Société d'Ophtalmologie de Paris: Bulletin 04774
Société Entomologique de France: Annales 04777
Société Entomologique de France: Bulletin 04777
Société Française d'Egyptologie: Bulletin 04805
Société Française de Numismatique: Bulletin 04826
Société Française de Pédagogie: Bulletin Trimestriel 04828
Société Française de Photogrammétrie et de Télédétection: Bulletin 04833
Société Française de Physique: Bulletin 04836
Société Française d'Ophthalmologie: Rapport 04861
Société Française d'Oto-Rhino-Laryngologie et de Pathologie Cervico-Faciale: Rapports Discutés au Congrès 04864
Société Généalogique Canadienne-Française: Mémoires 02928
Société Géologique de Belgique: Annales 01461
Société Géologique et Minéralogique de Bretagne: Bulletin 04869
Société Géologique et Minéralogique de Bretagne: Mémoires 04869
La Société Guernesiaise: Transactions 12961
Société Internationale de Criminologie: Newsletter 04877
Société Internationale de Psychopathologie de l'Expression: Newsletter 04879
Société Internationale de Psycho-Prophylaxie Obstétricale: Bulletin 04880
Société Internationale des Amis de Montaigne: Bulletin 04881
Société Internationale pour la Recherche sur les Maladies de Civilisation et sur l'Environnement: Congress Proceedings 01464
Société Jersiaise: Bulletin 12962
Société Mathématique de Belgique: Bulletin 01466
Société Mycologique de France: Bulletin Trimestriel 04897
Société Nationale Académique de Cherbourg: Mémoires 04898

Société Nationale des Sciences Naturelles et Mathématiques: Mémoires 04902
Société Neuchâteloise des Sciences Naturelles: Bulletin 10980
Société Royale Belge d'Entomologie: Bulletin et Annales 01476
Société Royale Belge d'Entomologie: Mémoires 01476
Société Royale d'Archéologie de Bruxelles: Annales 01480
Société Royale d'Archéologie de Bruxelles: Bulletin 01480
Société Royale de Chimie: Bulletin 01481
Société Royale des Sciences de Liège: Bulletin 01487
Société Suisse des Américanistes: Bulletin 10985
Société Zoologique de France: Bulletin 04923
Société Zoologique de France: Mémoires 04923
Society for Academic Emergency Medicine: Directory 16489
Society for American Archaeology: Bulletin 16492
Society for Applied Bacteriology: Symposium Series 12964
Society for Applied Learning Technology: Conference Proceedings 16495
Society for Applied Spectroscopy: Newsletter 16496
Society for Army Historical Research: Journal 12965
The Society for Asian Art Newsletter 16499
Society for Calligraphy: Newsletter 16504
Society for College and University Planning: Membership Roster 16510
Society for College and University Planning: News 16510
Society for Computer Simulation International: Transactions 16512
Society for Cooperation in Russian and Soviet Studies: Newsletter 12968
Society for Coptic Archaeology: Bulletin 03915
Society for Developmental Biology: Symposium Volume 16516
Society for Environmental Therapy: Newsletter 12972
Society for Epidemiologic Research: Abstracts 16519
Society for Epidemiologic Research: Newsletter 16519
Society for Ethnomusicology: Newsletter 16520
The Society for Experimental Biology: Seminar Series 12973
The Society for Experimental Biology: Symposia 12973
Society for French American Cultural Services and Educational Aid: Newsletter 16524
Society for General Microbiology Quarterly 12975
Society for German-American Studies: Newsletter 16527
Society for German-American Studies: Yearbook 16527
Society for Historians of American Foreign Relations: Newsletter 16529
Society for Historians of American Foreign Relations: Roster and Research List 16529
Society for Historical Archaeology: Newsletter 16530
Society for Humanistic Judaism: Newsletter 16534
Society for Industrial and Applied Mathematics: News 16536
Society for Industrial Archaeology: Bibliography 16537
Society for Industrial Archaeology: Journal 16537
Society for Industrial Archaeology: Newsletter 16537
Society for Information Display: Proceedings 16539
Society for Information Display: Quarterly Proceedings 16539
Society for Information Management: Member Forum 16540
Society for Information Management: Proceedings 16540
Society for International Development: Compass 08028
Society for Invertebrate Pathology: Abstracts of Symposia 16542
Society for Invertebrate Pathology: Newsletter 16542
Society for Italian Historical Studies: Membership List 16545
Society for Italian Historical Studies: Newsletter 16545
Society for Latin American Anthropology: Proceedings 16546
Society for Latin American Studies: Newsletter 12977
Society for Leukocyte Biology: Directory and Constitution 16547
Society for Neuroscience: Directory 16555
Society for New Language Study: Proceedings 16556
Society for Obstetric Anesthesia and Perinatology: Newsletter 16558
Society for Occlusal Studies: Newsletter 16559
Society for Occlusal Studies: Roster 16559
Society for Occupational and Environmental Health: Letter 16560
Society for Pediatric Dermatology: Newsletter 16561
Society for Pediatric Radiology: Membership Directory 16563
Society for Phenomenology and Existential Philosophy: Newsletter 16567
Society for Photographic Education: Membership Directory 16570
Society for Photographic Education: Newsletter 16570
Society for Post-Medieval Archaeology: Journal 12986
Society for Post-Medieval Archaeology: Newsheet 12986
Society for Psychical Research: Proceedings 12989
Society for Range Management: Mini-Directory 16574
Society for Research in Child Development: Monographs 16576
Society for Research into Higher Education: Abstracts 12994
Society for Research into Higher Education: International Newsletter 12994
Society for Research into Higher Education: Proceedings 12994
Society for Risk Analysis: Newsletter 16577
Society for Slovene Studies: Letter 16579
Society for Spanish and Portuguese Historical Studies Bulletin 16582
Society for Spanish and Portuguese Historical Studies: Membership Directory 16582
Society for Technical Communication: Proceedings of Annual Conference 16584
Society for the Advancement of American Philosophy: Membership Directory 16586
Society for the History of Czechoslovak Jews: Review I 16596
Society for the History of Czechoslovak Jews: Review II 16596
Society for the History of Natural History: Special Publications 13004
Society for the History of the Germans in Maryland: The Report: A Journal of German American History 16599
Society for the Investigation of Recurring Events: Abstract 16601
Society for the Investigation of Recurring Events: Bulletin 16601
Society for the Preservation of New England Antiquities: House Guide 16605
Society for the Preservation of New England Antiquities: Newsletter 16605
Society for the Protection of Old Fishes: Membership List 16607
Society for the Protection of Old Fishes: Newsletter 16607
Society for the Psychological Study of Social Issues: Newsletter 16608
Society for Theriogenology: Newsletter 16609
Society for Theriogenology: Proceedings of Annual Meeting 16609
Society for the Study of Architecture in Canada: Nouvelle: News 02940
Society for the Study of Evolution: Membership Directory 16616
Society for the Study of Human Biology: Symposia volumes 13015
Society for the Study of Social Problems: The Newsletter 16621
Society for the Study of Southern Literature: News-Letter 16622
Society for the Study of Symbolic Interaction: Journal 16623
Society for the Study of Symbolic Interaction: Newsletter 16623
Society for Visual Anthropology: Newsletter 16628
Society for Women in Philosophy – Eastern Division: Newsletter 16629
Society for Women in Philosophy – Pacific Division: Newsletter 16630
Society of Actuaries: Record 16631
Society of Actuaries: Transactions 16631
Society of Allied Weight Engineers: Newsletter 16632
Society of American Archivists: Membership Directory 16633
Society of American Archivists: Newsletter 16633
Society of Antiquaries of Scotland: Monographs 13024
Society of Antiquaries of Scotland: Proceedings 13024
Society of Architectural and Industrial Illustrators: Newsletter 13026
Society of Architectural and Industrial Illustrators: Yearbook 13026
Society of Architectural Historians: Newsletter 16639
Society of Architectural Historians of Great Britain: Newsletter 13027
Society of Archivists: Journal 13028
Society of Biological Chemists, India: Proceedings: Abstract of Papers Presented at the Annual Meeting 06858
Society of Carbide and Tool Engineers: Proceedings 16646
Society of Composers: Newsletter 16650
Society of Cosmetic Chemists: Journal 16651
Society of Cosmetic Chemists: Newsletter 16651
Society of Cost Estimating and Analysis: Newsletter 16652
Society of County Librarians: Information 13040
Society of Designer Craftsmen: News-Sheet 13043
Society of Diagnostics Medical Sonographers: Newsletter 16654
The Society of Dyers and Colourists: Journal 13044
Society of Economic Geologists: Membership List 16655
Society of Ethnic and Special Studies: Journal 16658
Society of Experimental Psychologists: Annual Report 16660
Society of Exploration Geophysicists: Roster 16662
Society of Federal Linguists: Membership Directory 16663
Society of Federal Linguists: Newsletter 16663
Society of Fellows Journal 11768
Society of Flavor Chemists: Newsletter 16665
Society of Flight Test Engineers: Membership Directory 16666
Society of Gastroenterology Nurses and Associates: Journal 16668
Society of Independent Professional Earth Scientists: Newsletter 16673
Society of Insurance Research: Research Review 16674
Society of Logistics Engineers: Annals 16676
Society of Logistics Engineers: Member's Handbook and Membership Directory 16676
Society of Logistics Engineers: Spectrum 16676
Society of Malawi: Journal 08624
Society of Medical Consultants to the Armed Forces: Newsletter 16678
Society of Medical Consultants to the Armed Forces: Roster 16678
Society of Medical Jurisprudence: Bulletin 16679
Society of Motion Picture and Television Engineers: Directory 16681
Society of Motion Picture and Television Engineers: Journal 16681
Society of Motion Picture and Television Engineers: News and Notes 16681
Society of Naval Architects and Marine Engineers: Transactions 16683
Society of Nematologists: Annual Meeting Presentations Abstracts 16684
Society of Neurosurgical Anesthesia and Critical Care: Newsletter 16686
Society of Obstetricians and Gynaecologists of Canada: Bulletin 02946
Society of Philippine Surgeons of America: Directory 16691
Society of Professional Archaeologists: Newsletter 16693
Society of Professional Well Log Analysts: Annual Transactions 16694
Society of Professors of Education: Membership Directory 16696
Society of Professors of Education: Monograph Series 16696
Society of Prospective Medicine: Membership Directory 16697
Society of Remedial Gymnasts and Recreational Therapies: Journal 13063
Society of Research Administrators: Journal 16699
Society of Research Administrators: Membership Directory 16699
Society of Research Administrators: Newsletter 16699
Society of Scribes and Illuminators: Journal 13064
Society of Soft Drink Technologists: Annual Meeting Proceedings 16701
Society of State Directors of Health, Physical Education and Recreation: Directory 16702
Society of State Directors of Health, Physical Education and Recreation: Newsletter 16702
Society of Teachers in Education of Professional Photography: Journal 16705
Society of Teachers in Education of Professional Photography: Membership Directory 16705
Society of Toxicology: Newsletter 16708
Society of USAF Flight Surgeons Newsletter 16710
Society of Vertebrate Paleontology: News Bulletin 16712
Society of Wildlife Artists: Annual Catalogue 13070
Society of Women Artists: Annual Catalogue 13071
Society of Wood Science and Technology: Newsletter 16714
Socijalističko Zemjodelstvo: Socialist Agriculture 08617
Sociolinguistique: systèmes de langues et interaction sociales et culturelles 04746
Sociológia 10060
Sociological Analysis 14355
Sociological Theory 14175
Sociology 17103
Sociology of Education 14175
Sociology of Law 08307
Sociology, Work, Employment and Society 11691
Socionomen 10595
Sodobna pedagogika 10116
Soemmering-Forschungen 05050
Software Engineering Journal 11547, 12296
Software Magazine 15317
Soil News 09180, 10160
Soils and Fertilizers 11720, 11728
Soils and Foundations 08123
Soil Science Society of America Journal 14942
Soil Use and Management 11690
SOJOURN: Social Issues in Southeast Asia 10049
SOK-medelingen 08979
Solanus: Bulletin of ACOSEEM 13096
Solar Bulletin 13597
Solar Energy 07722
Solar Energy Engineering 14145
Solar Progress incorporating South Wind 00260
Solar Today 14177
Soldiers of the Queen 13183
Soldiers Small Book 13183
Soletter 16676
Solidarnosc Bulletin 15657
Solid-Liquid Flow 11437
Solid Liquid Flow Abstracts 11575
Solos e Rochas 01592
Someni: Tanzania Library Association Journal 11060
Sondagem Conjuntural 01668
Sønderjyske Aarbøger 03742
Sønderjysk Maanedsskrift 03742
Sonderschriften des Frobenius-Instituts 05765
Sonnblick-Verein: Jahresbericht 00946
Sonneck Society: Membership Directory 16721
Sonneck Society: Newsletter 16721
Sonnenberg Internationale Briefe, International Journal, Revue Internationale: Rivista Trilingue (Tedesco, Inglese, Francese) 07485
Sonnenberg-News 05995
Sonnenenergie 05420
Soobshcheniya Byurakanskoi Observatorii: Report of the Byurakan Astrophysical Observatory 00203
SOPHE News and Views 16573
Sorghum and Millets Abstracts 11720
Sosialøkonomen 09349
Sot la Nape 07806
Sotsiologicheskie Issledovaniya: Sociological Research 09954
Soundings 16526
Soundings: An Interdisciplinary Journal 16625
Soundpost 07015
The Source 16432
South Africa Arts Calendar 10181
South African Academy of Science and Arts: Jaarverslag 10177
South African Archaeological Society: Monograph Series 10178
The South African Association of Physicists in Medicine and Biology: Congress Brochure 10183
The South African Banker 10148
South African Biological Society: Journal 10184
South African Ceramic Society: Conference Proceedings 10185
South African Forestry Journal 10225
South African Geographical Journal 10192
South African Institute for Librarianship and Information Science: Newsletter 10193
South African Institute of Assayers and Analysts: Bulletin 10196
The South African Institute of International Affairs: Bibliographical Series 10197
The South African Institute of International Affairs: Special Studies 10197
South African Journal for Librarianship and Information Science 10193
The South African Journal of Communication Disorders 10220
The South African Journal of Economics 10137
South African Journal of Geology 10144
South African Journal of International Affairs 10197
South African Journal of Photogrammetry, Remote Sensing and Cartography 10216
The South African Journal of Physiotherpy: Die Soid-Afrikaanse Tydskrif Fisioterapie 10219
South African Journal of Psychology 10167
South African Journal of Surgery 10124
South African Journal of Zoology 10237
South African Landscape Series 10192
South African Market Research Association: Report 10204
South African Medical Journal 10151, 10159
South African Medical News 10151
The South African Medical Research Council: Documentum 10206
The South African Numismatic Society: Bulletin 10211
South African Optometrist 10212
South African PEN Centre: Newsletter 10214
South African Statistical Journal 10221
South American Explorer 16722
South and West Asian Geoscience Newsletter 11088
South Asian Studies 12997
The South Australian Naturalist 00397
South Australian Ornithological Association: Newsletter 00496
The South Australian Ornithologist 00496
Southeast Asian Affairs 10049
Southerly 00390
Southern Africa Literature List 13341
Southern African Society of Aquatic Scientists: Journal 10228
Southern Africa Perspectives Series 13341
Southern Association of Colleges and Schools: Membership List 16723
Southern Association of Colleges and Schools: Newsletter 16723
Southern Association of Colleges and Schools: Proceedings 16723
The Southern Baptist Educator 14503
Southern Changes 16727
Southern Humanities Review 16726
Southern Journal of Applied Forestry 16634
Southern Society of Genealogists: Bulletin 16728
Southern Society of Genealogists: Directory 16728
Southern Speech Communication Journal 16729
Southern Stars 09195
South Wales Institute of Engineers: Proceedings 13084
Souvenir Catalogue 14598
Souvenir Festival Guide 11933
Sovetskaya Tyurkologiya: Soviet Turkology 00588
Soyabean Abstracts 11720
Soybean Rust Newsletter 03240
Sozialarbeit 10887
Der Sozialarbeiter 05504
Sozialarbeit & Suchtprobleme 10917
Soziale Arbeit: Deutsche Zeitschrift für soziale und sozialverwandte Gebiete 05607
Soziale Psychiatrie 05421
Soziale Welt: Zeitschrift für sozialwissenschaftliche Forschung und Praxis 05125
Sozialmedizinische Schriftenreihe 00929
Das sozialpolitische Forum 10951
Sozial- und Präventivmedizin 05422
Sozial- und Präventivmedizin: Médecine Sociale et Préventive 10847
Sozialwissenschaften und Berufspraxis 05205
Soziologie 05424
S.P.A.B. News 13010
Space Age Times: The Internntional Publication of Space News, Benefits and Education 16834
Space Explorer 11297
Spaceflight 11595
Spacereport 11297
Space Studies Institute: Update 16733
Space World 16162
Spanische Forschungen 05896
Speakers Forum 16297
Special Care in Dentistry 13288, 13540, 13779
Special Interest Group for Computer Personnel Research: Newsletter 16734
Special Interest Group for Symbolic and Algebraic Manipulation: Bulletin 16736
Special Interest Group for Symbolic and Algebraic Manipulation: Proceedings of Symposia 16736
Special Interest Group on Biomedical Computing: Newsletter 16738
Special Interest Group on Information Retrieval: Forum 16739
SpeciaList 16740
Specialist Periodical Reports 12857
Specializzazione 07195
Special Libraries 16740
Special Papers in Palaeontology 12675
SPEC Kit 16247
Spectra 15759, 16741
Spectroscopic Society of South Africa: Newsletter 10229
Spectrum der Augenheilkunde 00873
Spectrum Proceedings 15172
Speculum: A Journal of Medieval Studies 15714

SPE Drilling Engineering *16689*
Speech Communication *10674*
SPE Formation Evaluation *16689*
Speleologia *08015*
Spell/Binder *16604*
SPE Production Engineering *16689*
SPE Reservoir Engineering *16689*
Spina Bifida and Hydrocephalus Association of Ontario: Current: Pamphlet *02954*
Spinal Manipulation *15122*
Spiritual Fitness in Business *16355*
Spohr Journal *13087*
Spoken English *11966, 11966*
Spoletium *07240*
The Sporadical *16568*
Sport Aerobatics *15070*
Sportärzteverband Hessen - aktuell *06285*
Sportärztliche Mitteilungen *05180*
Sport Aviation *15070*
Sporterziehung in der Schule *10915*
Sports Medicine Bulletin *13737*
Sportsoziologie: Informationsschrift des Österreichischen Arbeitskreises für Soziologie des Sports und der Leibeserziehung *00883*
Sportwissenschaft *05240*
Spotlight *12953, 13525*
Spotlight on Children *11314*
Der Sprachdienst *05808*
Die Sprache: Zeitschrift für Sprachwissenschaft *01028*
Sprache im Technischen Zeitalter *06106*
Die Sprachheilarbeit *05425*
Der Sprachheilpädagoge *00833*
Sprachkunst: Beiträge zur Literaturwissenschaft *00723*
Sprachspiegel *10660*
Språklig Samling *09276*
Správy Slovenskej Parazitologickej Spoločnosti *10082*
Sprawozdania TNT *09627*
Sprawozdania z Czynności i Posiedzeń Naukowych /various/ *09545*
Sprawozdanie reczne, Notatki Płockie: Yearbook *09626*
Sprechstunde: Das Magazin der Ärzte für Ihre Gesundheit *11000*
S.P.R. Informations *04845*
Spring Floor *13079*
Spring Newsletter *12756*
SPSM Newsletter *16602*
SRC Home Record *16727*
SRF-Information *10594*
SRHE News *12994*
SRI Bulletin *08129*
Sri Lanka Association for the Advancement of Science: Proceedings *10419*
SRL-Information *06411*
SRM Notes *16574*
Srpska Akademija Nauka i Umetnosti: Godišnjak: Yearbook *17125*
Srpska Akademija Nauka i Umetnosti: Spomenik: Monument *17125*
Srpski arhiv za celokupno lekarstvo *17130*
Srpski etnografski zbornik: Serbian Ethnographic Collection *17125*
Srpsko Geografsko Društvo: Bulletin *17127*
SRRT Newsletter *16483*
SSATB Member Directory *16456*
SSC Newsletter *02960*
SSPC Members Directory *16746*
SSRC Annual Technical Session Proceedings *16748*
SSR-tidningen *10595*
Stadtbau-Informationen *05553*
Staffordshire Archaeological Studies *11790*
Staff Report *14529*
Stahlbau-Nachrichten *05541*
Stahlbau-Rundschau *00903*
Stahlbau-Rundschau-Mitteilungen *00903*
Stahl und Eisen *06374*
Stain Technology *14591*
Standard *11174*
Standardisering *09287*
Standards *03386*
Standards Action *13954*
Standards Association of Central Africa: Annual Report *17181*
Standards Canada *02301*
Standards Engineering *16744*
Standards Engineering Society: Membership Directory *16744*
Standards Engineering Society: Proceedings *16744*
Standards for the Certification of Teachers of the Hearing Impaired *14916*
Standards for the Evaluation of Programs for the Preparation of Teachers for the Hearing Impaired *14916*
Standards Report *14166*
Standards Worldover *06822*
Standing Committee on Commonwealth Forestry: Newsletter *13094*
The Standing Conference of National and University Libraries: Annual Review *13096*
Standort *05547*
Stannary Law Journal *11866*
Star and Furrow *11442*
Stars and Harp *13901*
Starts in the Classroom *16148*
State Board Forum *15398*
State Board Newsletter *15834*
State Court Journal *15947*
State Directors' Newsletter *15565*
State Education Indicators Report *14891*
State Education Leader *15027*
State Leadership Bulletin *16256*
State Librarian *11787*
Statens Råd för Byggnadsforskning: Documents: Research Reports in English *10524*
Statens Råd för Byggnadsforskning: Newsletter *10524*
Statens Råd för Byggnadsforskning: Rapporter: Research Reports in Swedish *10524*
State of Black America *16176*
State of Marketing *11763*
The State of the Child in the Arab World *03870*
State of the Nation *16871*
State Supervisors/Consultants of Trade and Industrial Education *15821*
Statistica *07267, 07425*
Statistical Analyses (SA) *09351*
Statistical and Economic Reports *01389*
Statistical Bulletin *00414*
Statistical Journal *10654*
Statistical Report *15015*
Statistical Science *15331*
Statistical Society of Australia: Newsletter *00499*
Statistical Summary *13990*
Statistical Theory and Method Abstracts *08944*
Statistical Yearbook *15015*
Statistica Neerlandica *09104*
Statistician *09181*
The Statistician *12284*
Statistic of Road Traffic Accidents in Japan *08142*
Statistics Bulletin *11716*
Statistics of Paper and Paperboard *13990*
Statistics on Social Work Education in the United States *14932*
Statsøkonomisk Tidsskrift *09352*
Status Report *16489*
Statute Law Review *13099*
Staub-Reinhaltung der Luft *06079*
Staying Well *15122*
Steaua *09798*
Steel Construction Yearbook *12294*
Steel in the USSR *12262*
Steel Research *06374*
Steel Structures Painting Bulletin *16746*
Steinbeck Monograph Series *15493*
Steinbeck Quarterly *15493*
Steine Sprechen *00772*
Steinschlag *00772*
Steirische Beiträge zur Hydrogeologie *00994*
Steirischer Naturschutzbrief *00898*
The Stewarts *13102*
ST-guides *03989*
Stichting voor de Technische Wetenschappen: Annual Report *09077*
Stichting voor Wetenschappelijk Onderzoek van de Tropen: Annual Report *09079*
Stiftelsen for Industriell og Teknisk Forskning ved Norges Tekniske Høgskole: Reports *09353*
Stiința Solului *09790*
S & T News *11059*
Stomatologie *09787*
Stomatološki glasnik Srbije *17130*
Stonehenge Viewpoint *16747*
Stopanski pregled: Economic Review *08613*
Storia della Storiografia *06598*
Strade Aperte *07576*
Strain *11668*
Strak of Gibraltar Bird Observation Report *06495*
Die Straße im Scheinwetter: Monatsberichte *00835*
Strassenverkehrstechnik *05720*
Strasse und Autobahn *05720*
Straße und Wirtschaft *05602*
Straße-Verkehr-Wirtschaft: (SVW)-Infodienst *05602*
Strategies *13449*
Stroke *13844*
Strolic *07806*
Strophes *16049*
The Structural Engineer *12317*
Structural Engineering Documents *10742*
Structural Engineering International *10742*
Structural Journal *14107*
Structure Reports *12401*
Die strukturpolitische Information *05878*
Strutture Ambientali *07484*
Student Acitivities Magazine *15892*
Student Lawyer *13617*
Students' Handbook *09366*
Students' Newsletter *11759*
Studia Albanica *00003*
Studia Archaeologica *08798*
Studia Diplomatica *01392*
Studia Ethnographica *08798*
Studia Fennica *04001*
Studia Folklorica *08798*
Studia Historica *04027, 08798*
Studia Islandica *06695*
Studia Iuridica *09627*
Studia Leibnitiana *05899*
Studia Leibnitiana Sonderhefte *05899*
Studia Leibnitiana Supplementa *05899*
Studia Mongolica *08798*
Studia Museologica *08798*
Studia Musicologica *06661*
Studia nad Historia Prawa Polskeigo *09612*
Studia Orientalia *04031*
Studia Philosophica *10880*
Studia Prawno-Ekonomiczne *09545*
Studia Scientiarium Mathematicarum *06661*
Studia Semiotyczne: Semiotic Studies *09604*
Studia Slavica *06661*
Studia Societatis Scientiarum Torunensis: Sectio C - geographia et geologia *09627*
Studia Societatis Scientiarum Torunensis: Sectio D - botanica *09627*
Studia Societatis Scientiarum Torunensis: Sectio E - zoologia *09627*
Studia Societatis Scientiarum Torunensis: Sectio F - astronomia *09627*
Studia Societatis Scientiarum Torunensis: Sectio G - physiologia *09627*
Studia Societatis Scientiarum Torunensis: Sectio H - medicina *09627*
Studi Bitontini *07537*
Studi Danteschi *07777*
Studi di Filologia Italiana: Bollettino dell'Accademia della Crusca *07147*
Studi di Grammatica Italiana *07147*
Studi di Lessicografia Italiana *07147*
Studi Economici e Sociali *07567*
Studi e Documenti *07213*
Studi e Materiali per la Storia della Cultura Popolare *07389*
Studien über Wirtschafts- und Systemvergleiche *01020*
Studien zu den Bogazköy-Texten *05050*
Studien zu den Fundmünzen der Antike *05050*
Studien zu Franz Schmidt *00598*
Studien zur Kinderpsychoanalyse *00920*
Studien zur Kulturkunde *05765*
Studien zur Musikwissenschaft *00645*
Studien zur Soziologie *00848*
Studier fra Sprog- og Oldtidsforskning *03725*
Studies and Reports in Hydrology *04577*
Studies for Trade Unionists *13215*
Studies in Art Education *15790*
Studies in Bibliography *14585*
Studies in Church History *11923*
Studies in Conservation *12369*
Studies in Educational Administration *00375*
Studies in Eighteenth-Century Culture *14058*
Studies in English Literature *08196*
Studies in Family Planning *16338*
Studies in French Language and Literature *08199*
Studies in Higher Education *12994*
Studies in Law *03155*
Studies in Moralogy *08183*
Studies in Stereoencephalotomy *16932*
Studies in the Education of Adults *12596*
Studies in the Japanese Language *08171*
Studies in the Philosophy of Education *08176*
Studies in Western History *08367*
Studies on Asian Topics *03785*
Studies on Chinese History *03155*
Studies on Economics *03155*
Studies on Latin America *10510*
Studi Genuensi *07732*
Studi Germanici *07741*
Studii clasice *09788*
Studii de literatura universala *09782*
Studii şi cercetări de istoria artei, seria Artă plastică: Studies and Researches in Art History/Fine Arts Series *09736*
Studii şi cercetări de istoria artei, seria Teatru, muzică, cinematografie: Studies and Researches in Art History/Theatre, Music, Cinematography Series *09736*
Studii şi cercetări de istorie veche şi arheologie: Studies and Researches in Ancient History and Archeology *09736*
Studii şi cercetări juridice: Juridical Studies and Researches *09736*
Studii si cercetari matematice: Studies and Research in Mathematics *09737*
Studi Maceratesi: Collana *07456*
Studime Filologjike: Philological Studies *00003*
Studime Gjeografike: Geographic Studies *00003*
Studime Historike: Historical Studies *00003*
Studime Meteorologjike e Hidrologjike: Meteorological and Hydrological Studies *00003*
Studime per Letersine Shqiptare: Studies on Albanian Literature *00003*
Studime/Studije: Studies *17076*
Studi Musicali *07221*
Studi Petrarcheschi *07124*
Studi Romagnoli *07795*
Studi Romanzi *07807*
Studi sul Petrarca *07655*
Study Handbook *12554*
Study Holidays *11749*
STUK-A-series *03989*
Stultifera Navis *10799*
Le Subiet *04739*
The Subterranean Sociology Newsletter *16751*
Sucht: Zeitschrift für Wissenschaft und Praxis *05429, 05454*
Sudan Notes and Records *10428*
Südosteuropa-Mitteilungen *06309*
Südwestdeutsche Blätter für Familien- und Wappenkunde *06383*
SUERF Papers on Monetary Policy and Financial Systems *09069*
SUERF Reprints *09069*
SUERF Series *09069*
SUERF Translations *09069*
Suffolk Archaeology & History Newsletter *13106*
Suffolk Institute of Archaeology and History: Proceedings *13106*
Sugaku *08386*
Sugar Industry Abstracts *11720*
Sugar Industry Technologists: Annual Technical Meeting Proceedings *02962*
Suggestion *14537*
Suicide and Life *13584*
Sulphur in Agriculture *16752*
Sumarski Pregled: Forester's Review *08615*
Sumi-E *16753*
Sumi-E Society of America: Catalogue *16753*
Sumi-E Society of America: Membership List *16753*
Summary Information on Master of Social Work Programs *14932*
Sun World *07722*
Suomalainen Teologisen Kirjallisuusseuran julkaisuja *03999*
Suomalainen Tiedeakatemia: Year Book *04000*
Suomalaiset kertojat *04038*
Suomen Autolehti *04008*
Suomen Eläinlääkärilehti: Finsk Veterinärtidskrift *04013*
Suomen geologinen kartta *03963*
Suomen Geologinen Seura: Bulletin *04022*
Suomen geologinen yleiskartta *03963*
Suomen Historian Lähteitä *04027*
Suomen Kirkkohistoriallisen Seuran Toimituksia *04040*
Suomen Kirkkohistoriallisen Seuran Vuosikirja: Yearbook *04040*
Suomen kirkot: Finland's Churches *03983*
Suomen Maataloustieteellinen Seura: Journal *04048*
Suomen Museo *04051*
Suomen museoliiton julkaisuja *04052*
Suomen Runotar *04038*
Suomen Sukututkimusseuran julkaisuja: Skrifter *04060*
Suomen Sukututkimusseuran vuosikirja: Årsskrift *04060*
Superconductor Science and Technology *12268*
Super Link *15561*
Super Trouper *15565*
Supervision Management *13372*
El Supervisor *15354*
Supplementa *10880*
Supplementary Issue *08387*
Supplement der Medizinischen Klinik *05367*
Supply and Demand for Scientists and Engineers *14790*
Supreme Court Historical Society: Annual Report *16755*
Supreme Court Historical Society: Newsletter *16755*
Surface Coating's International *12653*
Surface Engineering *12256, 12262*
The Surgeon *14095*
Surgery, Gynecology and Obstetrics *13738*
Surgical Association of the Republic of China: Journal *03331*
Surgical Forum *13738*
The Surgical Technologist *14511*
Surtsey Research Progress Reports *06701*
Survey Bulletin *02472*
Surveying and Mapping *13753*
Surveying Journal *14107*
Surveying World *12820*
Survey of Judicial Salaries *15947*
Survey of Korean Arts *08542*
Surveyors Historical Society: Membership Directory *16756*
Survey Research Center Monographs *15296*
Survey Review *11813*
Survey Statistician *04278*
Sussex Archaeological Collections *13113*
Sussex Archaeological Society: Newsletter *13113*
Sussex Industrial History *13114*
Sustainable Farming *13368*
SVAMPE *03737*
SVEB-Bulletin *10944*
Svejsning: Welding *03694*
Svenska Akademiens Handlingar *10530*
Svenska Läkaresällskapets Handlingar: Hygiea *10560*
Svenska Museer: Debate and Information on Museum Policies and Museology *10565*
Svenska Tekniska Vetenskapsakademien i Finland: Förhandlingar Proceedings *04068*
Svenska Tekniska Vetenskapsakademien i Finland: Meddelanden-Reports *04068*
Svensk Botanisk Tidskrift *10540*
Svensk Flyghistorisk Tidskrift *10578*
Svensk Kirurgi *10583*
Svensk Tidskrift för Musikforskning *10575*
Svensk Veterinärtidning *10598*
Sveriges Natur *10568*
Sverkhtverdye Materialy: Superhard Materials *11193*
Sverkhverdye Materialy: Superhard Materials *09954*
Svetsen *10599*
Sviluppo, Formazione, Economia, Cooperazione Internazionale *07395*
Svisa Esperanto-Societi Informas *10805*
Swara Magazine: Wildlife News *08498*
SWB-Information *10923*
SWEA Forum *16760*
The Swedenborg Society Magazine *13115*
The Swedish-American Historical Quarterly *16758*
Swedish Archaeology *10534*
Swedish Colonial Society: Directory *16759*
Swedish Dental Journal *10596*
Swedish Insurance Yearbook *10546*
Swiss-American Historical Society: Review *16761*
Swiss Papers in English Language and Literature *10791*
Swiss Studies in International Law *10947*
switec Information *10876*
SWS-Rundschau *00948*
Sylloge Nummorum Graecorum/The Collection of the American Numismatic Society *13962*
Sylwan *09583*
Sylwetki Łódzkich Uczonych *09545*
Symposium on Steel Production Technology *03274*
Synapse: A Canadian News Service for Biomedical Ethics *02327*
Synapse: Un service canadien d'information en éthique biomédicale *02327*
Syndicat National des Collèges et des Lycées: Bulletin *04943*
Synergy: Newsletter *02260*
Syn og Segn *09262*
Synopses *10524*
Synopsis *09344*
Synthese *00679*
Synthesis: Bulletin du Comité National de Littérature Comparée *09736*
Syöpä-Cancer *04061*
Systema Ascomycetum *12378*
Systematic Botany *14156*
Systematic Botany Monographs *14156*
Systematic Entomology *02436, 12800*
Systematic Zoology *16704*
Systems and Control *08314*
Systems Studies Monographs *03945*
Századok *06659*
Szene Schweiz/Scène suisse/Scena svizzera *10850*
Színházelméleti Füzetek *06658*
Színháztörténeti Könyvtár *06658*
Szlakami nauki *09545*

Tabelle Italiane Navali *07650*
Tähdet: Stars *04069*
Tähdet ja Avaruus: Stars and Space *04069*
Taiwan's Forestry Monthly *03268*
Tajik Academy of Sciences: Doklady: Report *11045*
Tajik Academy of Sciences: Izvestiya: Bulletin *11045*
Takra *06483*
Taleem-e-Tadrees *09393*
Taliesin *11198*
Talkback *12576*
Talking Point *11331*
Talking Politics incorporating Teaching Politics *12713*
Tally Sheet *11968*
The Tamarind Papers *16762*
Tamilchittu *06828*
Tamil Nilam *06828*
Tamil Olu *08655*
Tamil Pozhil *06864*
Tandlågebladet *03695*
Tandläkartidningen: The Journal of the S.D.A. *10596*
Tannlaeknabladid *06703*
Tanzania Commission for Science and Technology: Annual Report *11059*
Tanzania Notes and Records *11062*
Tanzania Veterinary Journal *11063*
Target *16110*
Tarikh *09231*
Tarim ve Mühendislik: Agriculture and Engineering *11167*
Tartans *12932*
Tasmanian Historical Research Association: Papers and Proceedings *00506*

1000 + 1 Buch 00681
La Tavola Rotonda 07190
Tax Analysts: Highlights and Documents 16764
The Tax Directory 16764
The Tax Lawyer 13617
Tax Notes 16764
TDES Newsletter 14990
Teach Abroad 11749
The Teacher 12616
Teacher Education Newsletter 15560
Teacher in Commerce: The National Journal of the Commercial Teaching Profession 12054
Teaching and Training 12572
Teaching Geography 12110
Teaching History 12172
Teaching Mathematics and its Applications 12258
Teaching Sociology 14175
Teaching Today 12570
Teangeolas 06957
Tea Research Foundation (Central Africa): Newsletter 08625
Tea Research Foundation (Central Africa): Report 08625
T.E.C. 04340
Tech Air 13055
Technical Association of the Graphic Arts: Newsletter 16767
Technical Association of the Graphic Arts: Proceedings 16767
Technical Bulletin 01626, 16012
Technical Committe Reports 16051
Technical Communication 16584
Technical Conference Reports 12255
Technical Documents 14067
Technical Documents in Hydrology 04577
Technical Guide Series 11623
Technical Handbook Series 11623
Technical Information 01626
Technical Marketing Society of America: Newsletter 16768
Technical Notes 01512
Technical Papers 13753, 15567, 16853
Technical Papers in Hydrology 04577
Technical Planner 13069
Technical Release 16125
Technical Report 00505, 17162
Technical Reports 07431, 13194, 15764
Technical Series 12964
Technical Surveys 01389
Technical Transactions 12255
Technica Manuel 16113
Technika i Nauka 12314
Technik i Tiden 10529
Technion Magazine 14088
Technion Research and Development Foundation Ltd: Research Report 07114
Technion USA Magazine 14088
technique de presse 05938
Technique Directory 13970
Techniques in Marine Environmental Sciences 03746
Techniques, Sciences, Methodes 04253
Technobrief 10133
Technology and Culture 16598
Technology and Economy 08146
Technology and Society Magazine 15317
Technology Digest 09384
Technology for Development in Southeast Asia and the Pacific 06887
Technology Teacher 15560
Technometrics 14085, 14182
Technonet Digest 10037
TECHNOS 13349
Tecnica Ospedaliera 07374
La Tecnica Professionale 07607
Tecniche Direzionali 07840
Tecnologie Alimentari: Sistemi per Produrre 07329
Tectonics 13833, 13833
TEE Newsletter 09209
Tehniceska Misal 01771
Tehnika 17133, 17134
TEHO 04080
Teilhard Perspective 14189
Teilhard Studies 14189
Tekhnicheskaya Diagnostika i Nerazrushayushchii Kontrol: Technical Diagnostics and Non-destructive Testing 11193
Tekhnicheskaya Elektrodinamika: Technical Electrodynamics 11193
Teknik & Standard 10522
Tekniske Rapporter 03791
Teknisk Ukeblad 09257, 09306
Tekstil ve Mühendis: Textile and Engineering 11167
Telecommunications 06125
Telemedium 16171
Telescope 15177
Teleuropa: Mensile di Cultura, Arte e Attualità 07245
Television 12868
Televizion 06002
Telicom 15538
Tellus 10551
TELMA 05385
Temars de Leprologia 00135
Temas Médicos 03342
Tenki 08335
Tenmon Kyositu: Astronomical Class 08139
The Tennyson Society: Annual Report 13118
The Tennyson Society: Bulletin 13118
Tensor 08428
Teoreticheskaya i Eksperimentalnaya Khimiya: Theoretical and Experimental Chemistry 11193
Teoreticheskaya i Matematicheskaya Fizika: Theoretical and Mathematical Physics 09954
Teoreticheskie Osnovy Khimicheskoi Tekhnologii: Theoretical Foundations of Chemical Technology 09954
Teoriya Veroyatnostei i ee Primenenie: Probability Theory and its Application 09954
Teploenergetika: Heat and Power Engineering 09954
Teplofizika Vysokikhk Temperatur: High Temperature Thermal Physics 09954
Terapia Familiare 07954
Teratology: The International Journal of Abnormal Development 16771
Természet Világa: World of Nature 06676
Termley 11397
La Termotecnica 07409, 07686
Te Roma Sequor 07272
Terra 04047, 09784
Terra America 07351
Terra Biodinamica 07390
Terrae Incognitae 16597
Terre et Hommes 17127
La Terre et la Via 04899
Tesisat Mühendisliği: Installation Engineering 11167
TESOL Newsletter 16766
TESOL Quarterly 16766
Tested Studies for Laboratory Teaching 14272
Test of the Month 13629
Tesugaku-Kenkyu: The Journal of Philosophical Studies 08179
Tetsugaku-zasshi 08429
Tetsu-to-Hagane 08392
Teva Ve'Aretz 07111
Texte zur Geschichte des Pietismus 05918
Textile Chemist and Colorist 13592
Textile Cleaning 06675
Textile Economy 06675
Textile Horizons 13120
Textile Progress 13120
Textiles 13120
Textile Technology Digest 15339
Textilforschung: Berichte des Forschungskuratoriums Gesamttextil 05733
Textil-Info für Unterricht & Bildung 05697
Textilnorm Mitteilungen 06209
Texts and Studies of the History of Cyprus 03496
Textus 07299
TGI 12032
Thanatology Abstracts 13886
Thambodala 10147
Theater in Österreich: Verzeichnis der Inszenierungen (theadok) 01016
Theaterjaarboek 09081
Theatre Crafts 13525
Theatre Design and Technology 16825
Theatre Historical Society: Directory 16773
Theatre Notebook 13000
Theatre Organ 14191
Theatre Survey 14090
Thematic Index of Music for Viols 13187
Thenmozhi 06828
Theodore Roosevelt Association: Journal 16776
Theodor Herzl Institute: Annual Season Preview 16777
Theological Education 14521
Theology 12987
Theoretical Computer Science 05665
The Theosophical Journal 13124
The Theosophical Movement: A Magazine devoted to the Living of the Higher Life 06867
Theosophy in Karachi 09372
Therapeutic Care and Education 11410
Therapeutic Claims in MS 12357
Therapeutic Recreation Journal 16173
Thérapie 04848
Therapists Association 02078
There Ought to be Free Choice 15956
Theriaca 03617
Thesaurus: Boletín Cuadrimestral 03377
Thinking Mission 13171
Third World Forum 14522
This Week's Law 02581
Thomas Paine Society: Bulletin 13127
Thomas Paine Society: Newsletter 13127
The Thomas Wolfe Review 16778
The Thoracic and Cardiovascular Surgeon 05431
Thoreau Society: Bulletin 16780
Thoresby Society: Publications 13130
Thorne Ecological Institute: Update 16781
Thraco-dacica 09736
Thrombosis and Hemostasis Journal 15557
Tibet Society Bulletin 16782
Tidal Gravity Corrections 08867
Tidbits 15807
Tidskrift 03954
Tidskrift för Documentation 10601
Tidskrift i Gymnastik & Idrott 10553
Tidskrift utgiven av Juridiska Föreningen i Finland 03971
Tidsskrift for landøkonomi 03760
Tidsskrift for Sukkersyge 03765
Tidsskrift Natur og Miljø 03583
Tietopalvelu 04077
Tijdschrift: Bulletin 01161
Tijdschrift van de Vereniging voor Nederlandse Muziekgeschiedenis 09102
Tijdschrift voor Diergeneeskunde 08957
Tijdschrift voor Economische en Sociale Geografie: Journal of Economic and Social Geography 08965
Tijdschrift voor Entomologie 08992
Tijdschrift voor Marketing 09054
Tikhookeanskaya Geologiya: Pacific Ocean Geology 09954
The Times 14933
Time to Learn 12596
Tin and its Use 16783
Tinnitus Today 14196
Tin Research Institute: Annual Report 16783
Tips 14992
Tips for Principals 15892
TLA Bulletin 11102
TMS Journal 13138
Toben Shinbun 08436
Tochi Seido Shigaku: The Journal of Agrarian History 08432
Today in Health Planning 13842
Today's Astrologer 13808
Today's Eduction 16039
Toegepaste Taalwetenschap: Publications in Applied Linguistics 09039
Tönnies-Forum 05699
Together 12606
Tohogaku 08434
Toimetised: Proceedings 03933
Toktok bilong haus buk 09421
Toktokki 10236
Toletum 10341
Tollways 15423
Toneel Theatraal 09081
Topical Problems of Science: A Survey with Subtitles corresponding to the Material included 01775
Topical Problems of Science and Science Policy 01775
Topologija 17079
The Torch 15114, 16502
Il Torchio Artistico e Letterario 07134
Torch: U.S. 15632
Torricelliana 08024
Tort and Insurance Law Journal 13617
Toshokan Zasshi 08271
Touch for Health Directory 16786
Touch for Health Journal 16786
Touch for Health Times 16786
Touchstone 15748
Tournaments Illuminated 16513
Towards a Better Life for the Arab Child 03870
Towarzystwo Naukowe Organizacji i Kierownictwa: Organization Review 09625
Town and Country Planning 13137
Toxicology and Applied Pharmacology 16708
Toxicology in Vitro 11439
The Toyoshi-Kenkyu: The Journal of Oriental History 08442
Trabajos de Estadística 10373
Trabajos de Investigación Operativa 10373
Trabalhos da Real Sociedade Arqueológica Lusitana 09668
Trabalhos de Antropologia e Etnologia 09691
The Tracker 16260
Tracts CJN 02332
Trade Monitor 15895
Trade Union Adviser 15645
Trade Union Studies Journal 13215
Traditiones 10105
Tradition Orale 04746
Traffic Safety 16145
Trainer's Workshop 13372
Training and Development 12287
Training Bulletin 15383
Training Information Directory 15971
Trance and Healing in Southeast Asia Today 15255
Transactional Analysis Journal 15566
Transaction on Biomedical Engineering 15246
Transactions A 15744
Transactions B 15744
Transactions: Earth Sciences 12858
Transactions of South Staffordshire Archaeological and Historical Society 13083
Transactions of the American Fisheries Society 13816
Transactions of the American Microscopical Society 13946
Transactions of the American Ophthalmological Society 13965
Transactions of the ASAE 14097
Transactions of the Buteshire Natural History Society 11718
Transactions of the Canadian Society for Mechanical Engineering 02435
Transactions of the College of Medicine of South Africa 10132
Transactions of the International Conference of Orientalists in Japan 08434
Transactions of the Manchester Statistical Society 12487
Transactions of the Ophthalmological Society of the Republic of China 03317
Transactions of the Oriental Ceramic Society 12661
Transactions of the Royal Historical Society 12809
Transactions of the Royal Society of Tropical Medicine and Hygiene 12864
Transactions of the Zimbabwe Sientific Association 17188
Transactions on Acoustics, Speech and Signal Processing 15317
Transactions on Aerospace and Electronic Systems 15317
Transactions on Antennas and Propagation 15317
Transactions on Automatic Control 15245, 15317
Transactions on Biomedical Engineering 15317
Transactions on Broadcasting 15317
Transactions on Circuits and Systems 15317
Transactions on Communications 15317
Transactions on Components, Hybrids and Manufacturing Technology 15317
Transactions on Computer-Aided Design of Integrated Circuits 15317
Transactions on Computers 15317
Transactions on Consumer Electronics 15317
Transactions on Education 15317
Transactions on Electrical Insulation 15317
Transactions on Electromagnetic Compatibility 15317
Transactions on Electron Devices 15317
Transactions on Energy Conversion 15317
Transactions on Engineering Management 15317
Transactions on Geoscience and Remote Sensing 15317
Transactions on Industrial Electronics 15317
Transactions on Industry Applications 15317
Transactions on Information Theory 15317
Transactions on Instrumentation and Measurement 15317
Transactions on Kowledge and Data Engineering 15317
Transactions on Magnetics 15317
Transactions on Medical Imaging 15317
Transactions on Microwave Theory and Techniques 15317
Transactions on Nuclear Science 15317
Transactions on Pattern Analysis and Machine Intelligence 15317
Transactions on Plasma Science 15317
Transactions on Power Delivery 15317
Transactions on Power Electronics 15317
Transactions on Power Systems 15317
Transactions on Professional Communication 15317
Transactions on Reliability 15317
Transactions on Robotics and Automation 15317
Transactions on Semiconductor Manufacturing 15317
Transactions on Software Engineering 15317
Transactions on Systems, Man and Cybernetics 15317
Transactions on the Association of American Physicians 14384
Transactions on Ultrasonics, Ferroelecrics and Frequency Control 15247
Transactions on Ultrasonics, Ferroelectrics and Frequency Control 15317
Transactions on Vehicular Technology 15317
Transcripts 14044
Transfusion 13511
Transfusion Today 04882
Transit Fact Book & Membership Directory 02313
Transit Topics 02313
Translation Journal on Magnetics in Japan 15317
Translation Review 13919
Transplantation 16788
Transplantation Proceedings 16788
Transport 11766
Transportation/Pipeline Journal 14107
Transportation Research Board: Special Report 16790
Transportation Research Circular 16790
Transportation Research Record 16790
La Trasfusione del Sangue 07886
Le Travail Humain 04690
Travail Social 10887
Traveaux 01208
Travel and Description Series 16233
Traveller Clinical Chart 15134
TREE 15253
Tree-Ring Bulletin 16791
Trees and Natural Resources 00440
Trees in South Africa 10232
The Tree Society of Southern Africa: Newsletter 10232
Treffpunkt Kindergarten/Forum Sozialpädagogik 06064
Trend 02853
Trends 15605
Trends and Techniques 15849
Trenie i Iznos: Friction and Wear 01088, 09954
TRI 08933
Trial 14523
Tribology 14145
Tribos 11437, 11575
La Tribune Psychique 04851
Trichologist 12289
Trinitarian Bible Society: Record 13143
Tripolitan 16792
Triss (Students) 10595
Tritsch-Tratsch 12419
Trolleybus 11696
Trolleybus Magazine 12613
Tropical Diseases Bulletin 11720
Tropical Grasslands 00512
Tropical Medicine and Hygiene News 14171
Tropical Science Center Occassional Papers 03452
Tropische und Subtropische Pflanzenwelt 05050
Trout Magazine 16793
Trout Unlimited: Chapter and Council Handbook 16793
Trudy Akademii Nauk Litvy: Bulletin of the Lithuanian Academy 08573
The Trumpeter 15630
The Trumpeter Swan Society: Conference Proceedings and Papers 16794
The Trumpeter Swan Society: Newsletter 16794
Trustee Digest 13917
Trustee Quarterly 14419
TSA Fact Sheets 13144
TSA&F Flyer 13134
TSA&F News 13134
TSA Newsletter 16787
TSA Scan 13144
Tsirkulyar Shemakhinskoi Astrofizicheskoi Observatorii: Newsletter of the Shemakha Astrophysics Observatory 01036
Tsitologiya: Cytology 09954
Tsitologiya i Genetika: Cytology and Genetics 11193
TSQUARED 16769
Tsuchi to Kiso 08123
La tua Città 07537
Tuarascail 06994
Tuberkulose und Lungenkrankheiten: Beilage zum Bulletin des Bundesamtes für Gesundheitswesen 10843, 10956
Türk Anesteziyoloji ve Reanimas yon Derneği Mecmuasi 11150
Türk Dili 11154
Türk Eczacilari Birligi Mecmuasi 11155
Türk Hukuk Lugati 11159
Türkiye Kurum Bülteni 11147
Türkiye Mühendislik Haberleri: News of Engineering in Turkey 11167
Türk Kütüphaneciliği 11165
Türk Mikrobiyoloji Cemiyeti Dergisi 11166
Türk Uroloji Dergisi: Turkish Journal of Urology 11177
TÜ-Technische Überwachung 06343
tum: technik und mensch 05255
Tuning In 13487
Tunnel 06302
Tunnelling and Underground Space Technology 04280
Turbomachinery 14145
Turkish American Physicians Association: Membership Roster 16795
Turkmen Academy of Sciences: Izvestiya: Bulletin 11180
Turner Society News 13145
Turrialba 03447
Turun Historiallinen Arkisto 04078
Tutorial Syllabus 13468
Tuttitalia 11269
Tutzinger Blätter 05687
Tutzinger Materialien 05687
The Twainian 15695
Twentieth Century Society: Journal 13146
Twins 07720, 07818
TWI: Research Bulletin 13147
Tydskrif vir Geesteswetenskappe 10177
Tyne & Tweed 11373

Uccelli 07996
Udenrigs 03715
Udenrigspolitiske Skrifter 03715
UEMS Bulletin 01496
L'Ufficio Moderno: Notiziario 07497
UFITA 05976
The Uganda Journal 11188
Ugandan Libraries 11187
Ugaritic Newsletter 02275
Ugeskrift for Jordbrug 03588, 03631
UICC Calendar of International Meetings on Cancer 10993
UICC News 10993
UIESP Newsletter 01404
UK Printing Industry Statistics 12701

Ukrainian Bibliographical Quarterly 16470
Ukrainian Engineers News 16797
Ukrainian Engineers' Society of America: Bulletin 16797
Ukrainian Institute of America: Newsletter 16798
Ukrainian Political Science Association in the United States: Newsletter 16800
Ukrainska Knyha: Ukrainian Book 16470
Ukrainskii Biokhimicheskii Zhurnal: Ukrainian Biochemical Journal 11193
Ukrainskii Botanichnyi Zhurnal: Ukrainian Botanical Journal 11193
Ukrainskii Fizicheskii Zhurnal: Ukrainiam Physics Journal 11193
Ukrainskii Istorichnyi Zhurnal: Ukrainian Historical Journal 11193
Ukrainskii Khimicheskii Zhurnal: Ukrainian Chemical Journal 11193
Ukrainskii Matematicheskii Zhurnal: Ukrainian Mathematical Journal 11193
UKSPA News 13168
UKSPA Science Park Directory 13168
Ulster Architectural Heritage Society: Newsletter 13150
Ulster Folklife 13152
Ulster Journal of Archaeology 13148
Ulster Local Studies 12061
Ulster Society in London: Newsletter 13157
Ultimas Adquisiciones 00101
Ultimate Reality and Meaning 01928
Ultimate Reality and Meaning: Interdisciplinary Studies in the Philosophy 02526
Umbau 00761
Umwelt 06377
Das Umweltgespräch: Schriftenreihe 05112
Umwelthygiene 05882
Umwelt-Report 00856
UNATEB Actualités 04993
UN-Common Group 16813
Underground 12294
Underground Services Directory 12294
Underground Water Resources Publication 06571
Undersea Biomedical Research: Journal 16803
Undersea & Hyperbaric Medicine: Abstracts from the Literature Abstract Journal 16803
Under the Window 15635
Underwater Naturalist 13920
Underwater Technology 13021
Une Lettre de la FIPF 04534
Unfallheilkunde: Traumatology 05432
Unicorn 00274
Unidade de Pesquisa de Ambito Estadual em Ponta Grossa: Circular 01738
L'Unificazione 07656
Unified List of United States Companies doing Business in South Africa 13341
Uniform Law Review 07734
Union Académique Internationale: Compte Rendu des Sessions 01492
Union Belge des Journalistes et Ecrivains du Tourisme: Nouvelles 01494
Union des Physiciens: Bulletin 04975
Unione Bolognese Naturalisti: Notiziario 08041
Unione Matematica Italiana: Bollettino 08048
Unione Matematica Italiana: Notiziario 08048
Unione Tecnica Italiana Farmacisti: Collegamento 08050
Union Internationale des Architectes: Bulletin d'Information 04985
Union Internationale pour la Liberté d'Enseignement: Congress Report 04990
Union Mondiale des Enseignants Catholiques: Proceedings of the World Congress 08053
The Union News 06910
Union of Rural Economics: Bjuletin-Vnedreni Novosti 01797
Union of Scientific Medical Societies in Bulgaria: Information Bulletin 01798
Union of Translators' in Bulgaria: News: Bulletin 01802
Union of Translators' in Bulgaria: Panorama: Magazine 01802
Union Radio-Scientifique Internationale: Proceedings of General Assemblies 01500
Union Technique de l'Automobile, du Motocycle et du Cycle: Bulletin Mensuel de Documentation 04997
Union Technique de l'Electricité: Bulletin 04998
Union Technique de l'Electricité: Catalogue 04998
Unitarian Historical Society: Transactions 13162
Unitarian Universalist Historical Society: Journal 16806
Unitarian Universalist Historical Society: Newsletter 16806
Unitarian Universalist Historical Society: Proceedings 16806
United Board for Christian Higher Education in Asia: Annual Report 16807
United Cancer Council: Coordinator 16808
United Kingdom Home Economics Federation: Bulletin: Conference Report 13165
United Kingdom Reading Association: Newsletter 13167
United New Conservationists: Membership Roster 16813
United Reformed Church History Society: Journal 13169
United Society for the Propagation of the Gospel: Quarterly Intercession Paper 13171
United Society of Artists: Annual Catalogue 13172
United States Committee of the International Association of Art: Information Bulletin 16818
United States Committee on Large Dams: Membership Directory 16820
United States Committee on Large Dams: Newsletter 16820
United States Council for Energy Awareness: INFO 16821
United States Federation for Culture Collections: Directory 16823
United States Federation for Culture Collections: Newsletter 16823
United States National Committee of the International Union of Radio Science: Program and Abstract of Meetings 16829
United States National Committee on Theoretical and Applied Mechanics: Proceedings 16830
United States Physical Therapy Association: Journal 16833
United States Space Education Association: Update 16834
Unit News in Anthropology Newsletter 13803
Univation 00362
Universal Serials and Book Exchange: Annual Report 16837
Universal Serials and Book Exchange: Newsletter 16837
Universe in the Classroom 14541
Universidades 08777
Universitas 01557, 16849
Universités 01895, 01948
L'Université Syndicaliste 04944
Universities Council on Water Resources: Newsletter 16838
Universities Council on Water Resources: Proceedings of Annual Meetings 16838
Universities Space Research Association: News and Notes 16840
University Affairs 01945
University Aviation Association: Newsletter 16843
University Aviation Association: Proceedings 16843
University Corporation for Atmospheric Research: Annual Report 16845
University Corporation for Atmospheric Research: Newsletter 16845
University Council for Educational Administration: Review 16846
University Film and Video Association: Digest 16847
University Film and Video Association: Membership Directory 16847
University Photographers Association of America: Newsletter 16848
University Today 17138
L'Universo 07729
Unsere Heimat 00985
Unsere Kunstdenkmäler/Nos Monuments d'Art et d'Histoire/I nostri Monumenti Storici 10716
Unterrichtsheft 11023
Unterrichtsjournal 11023
Die Unterrichtspraxis 13588
Untersuchungen zur Sprach- und Literaturgeschichte der Romanischen Völker 05050
Unyu-To-Keizai: Transportation & Economy 08443
Update in Medical Oncology 07317
Updating School Board Policies 16147
Upravlyayusshchie Sistemy i Mashiny: Control Systems and Computers 11193
Up the Gatineau! 02504
Uralica Linguistica 03933
Urban and Regional Information Systems Association: Annual Conference Proceedings 16851
Urban and Regional Information Systems Association: Directory 16851
Urban Design International 15308
Urban Design Update 15308
Urban Economic Developments: Newsletter 15995
Urban Focus 11791
Urbanisation and Health Newsletter 10206
Urbanistica 07749
Urbanistica Informazioni 07749
Urban Land Magazine 16801
The Urban Lawyer 13617
The Urban League News 16176
The Urban League Review 16176
Urban Libraries Exchange 16852
Urban Planning Journal 14107
URISA News 16851
Der Urologe 05199
Urology and Nephrology 09939
Ursa minor: Ursan jaostojentiedotuslehti 04069
URSI Information Bulletin 01500
URT 03595
USA Today 16588
La Usc di Ladins 08051, 08052
Use: Newsletter 16853
U.S. in World Affairs Journal 13494
USITT Newsletter 16825
USMA Newsletter 16826
USP DI Review 16832
U.S.P. Dispensing Information 16832
Uspekhi Fizicheskikh Nauk: Progress of Physics 09954
Uspekhi Fiziologicheskikh Nauk: Progress of Physiology 09954
Uspekhi Khimii: Progress in Chemistry 09954
Uspekhi Matematicheskikh Nauk: Progress in Mathematics 09954
Uspekhi Sovremennoi Biologii: Progress in Modern Biology 09954
USPG Yearbook 13171
US Population Data Sheet 16340
U.S. Section News 16925
Usted y su Familia 15354
U.S. Woman Engineer 16713
Ut de Smidte 08899
UTIAS Annual Report 02524
UTIAS Reports 02524
UTIAS Reviews 02524
UTIAS Technical Notes 02524
UTU News 13158
Utunk 09798
UWC Journal 13173
Uzbek Academy of Sciences: Doklady: Bulletin 16948
Uzbekskii Geologicheskii Zhurnal: Uzbek Geological Journal 16948
Uzbekskii Khimicheskii Zhurnal: Uzbek Chemical Journal 16948
Uzbek Tili va Adabieti: Uzbek Language and Literature 16948

Vaekst: Hedeselskabets Tidsskrift 03705
Världspolitikens dagsfrågor 10603
Växtekologiska Studier 10576
Vaglista 01796
Valóság: Reality 06676
Value 12290
The Valuer 00376
The Valuer and Land Economist 00316
Value World 16638
Vandteknik 03697
Vanguard 14810
Vantage Point 13760
VAÖ Mitteilungen 00956
Varilna Tehnika 10113
Varilna tehnika 17095
Városépites 06662
Vatra 09798
Vatten 10459
VBO 01369
vbo-informationen 06409
VBV/AFA Info-Bulletin 11013
VDD-Schriftenreihe 06316
VDI-Gesellschaft Agrartechnik: Mitglieder-Informationen 06319
VDI-Gesellschaft Fahrzeugtechnik: Mitglieder-Informationen 06323
VDI-Nachrichten 06377
VDI-Zeitschrift 06377
VDLUFA-Schriftenreihe 06354
Vector 11644
Veehouder en Dieren Arts 08957
Vegetable Grower 12720
The Vegon 13178
Vel: Collana Periodica di Psicanalisi 07399
Veld & Flora 10126
Venereology 00242
Venus 08217
Verband der Bibliotheken des Landes Nordrhein-Westfalen e.V.: Mitteilungsblatt 06334
Verband der Museen der Schweiz: Information 10997
Verband deutschsprachiger Schriftsteller in Israel: Rundschreiben 07116
Verband für medizinischen Strahlenschutz in Österreich: Informationsblatt 00964
Verband Jüdischer Lehrer und Kantoren der Schweiz: Bulletin 11002
Verband Schweizerischer Assistenz- und Oberärzte: Bulletin VSAO 11005
Verbindungstechnik in der Elektronik 05551, 05753
Verdi Newsletter 13871
Verein der Diplom-Bibliothekare an wissenschaftlichen Bibliotheken e.V.: Rundschreiben 06369
Verein Deutscher Emailfachleute e.V.: Mitteilungen 06375
Verein für Geschichte der Arbeiterbewegung: Archiv: Mitteilungsblatt 00983
Verein für Wasser-, Boden- und Lufthygiene e.V.: Schriftenreihe 06392
Vereinigung Schweizerischer Hochschuldozenten: Bulletin 11018
Verein Montandenkmal Altböckstein: Jahresberichte 00998
Verein Schweizerischer Geschichtslehrer: Bulletin 11025
Vereinte Nationen 05340
Verein zur Verbreitung naturwissenschaftlicher Kenntnisse: Schriften 01006
Vereniging Het Nederlands Kanker Instituut: Annual Report 09085
Verfassung und Recht in Übersee: Law and Politics in Africa, Asia and Latin America 05907
Vergilian Society: Newsletter 16856
Vergilius 16856
Verhaltenstherapie und Psychosoziale Praxis 05435
Verhandelingen van het KNGMG 08967
Verhandlungen der DPG 05487
Verhandlungen der Naturforschenden Gesellschaft in Basel 10767
Verhandlungen der Zoologisch-Botanischen Gesellschaft 01035
Verhandlungen des Österreichischen Juristentages 00894
Verkehrsannalen 00926
Verkfraedingafélag Islands: Arbók 06706
Vermessung, Photogrammetrie, Kulturtechnik 10921
Vermessung-Photogrammetrie-Kulturtechnik 11008
Vernacular Architecture 13179
Veröffentlichungen des INMM 05966
Veröffentlichungen des Tiroler Landesmuseums Ferdinandeum 01003
Veröffentlichungen zur Zeitgeschichte 00851
Versants 10648
Verschollene und Vergessene 05050
Versicherungs-Betriebswirt 06402
Die Versicherungsrundschau 00842
Verwaltungswissenschaftliche Informationen 05574
Vestis: Proceedings 08558
Věstník České Společnosti Zoologické 03531
Vestnik Dalnevostochnogo Otdeleniya Rossiiskoi Akademii Nauk: Journal of the Far Eastern Division of the Russian Academy of Sciences 09954
Vestnik Drevnei Istorii: Journal of Ancient History 09954
Vestnik Obshchestvennykh Nauk: Herald of Social Sciences 00203
Vestnik Rossiiskoi Akademii Meditsinskikh Nauk: Journal of the Russian Academy of Medical Sciences 09952
Vestnik Rossiiskoi Akademii Nauk: Journal of the Russian Academy of Sciences 09954
Vestnik Selskokhozyaistvennoi Nauki: Agricultural Science Journal 09950
Vestnik Zoologiya: Zoological Journal 11193
Vestsi: Bulletin 01088
Veterinaria Tropical 16991
Le Vétérinarius 02396
Veterinary and Human Toxicology 13638
Veterinary Bulletin 11720
Veterinary History 13181
Veterinary Pathology 13740
Veterinary Quarterly 08957
Veterinary Radiology 13741
Veterinary Record 11703
vfdb-Zeitschrift: Forschung und Technik im Brandschutz 06419, 06419
VGB Kraftwerkstechnik 06312
VGB-Kraftwerkstechnik 06432
VGS-Gesundheitsmagazin 11028
VHW-Mitteilungen 06365
Viaggiare in Liguria 07573
Viaggiatori Stranieri in Liguria 07573
Viaggiatori Tedeschi del '700 07573
Via Stellaris 16869
Viaţa Medicală 09754
Viata Româneascaí 09798
Vibes from the Libe 16298
Vibration and Acoustics 14145
The Victorian 16861
Victorian Artists' Society: Annual Report 00515
Victorian Artists' Society: News Letter 00515
The Victorian Naturalist 00396
Victorian Periodicals Review 16417
Victorian Review 02981
Victorian Society: Annual 13184
Victorian Society: Newsletter 13184
Vida Hispanica 11269
Le Vide, les Couches Minces 04866
Vidya Viyapathi 10419
La vie des sciences 04564
Vieilles Maisons Françaises 05000
Vienna Centre Newsletter 00591
Vierteljahresberichte des Forschungsinstituts der Friedrich-Ebert-Stiftung 05764
Vierteljahresschrift für wissenschaftliche Pädagogik 05896
Vierteljahrshefte zur Wirtschaftsforschung 05589
Vie Sociale 04369
Viewbox 13986
Viewfinder Magazine 11700
The View From Hyde Park 15136
Viewpoints on Digestive Diseases 13829
Views and Views 14435
Views and Vision 15066
Vigie info 04579
Vigyan Pragati 06753
Viitorul social 09746
Viiturul social: The Social Future, Journal of Sociology and Political Science 09736
Vijesti muzealaca i konzervatora Hrvatske 03466
VIM: Conference Proceedings 16862
Vingtième Siècle: Revue d'Histoire 04543
The Vintage Airplane 15070
Vintage Light Music 13186
Viola da Gamba Society: Newsletter 13187
VIP Gran Premio 07244
Virchow-Pirquet Medical Society: Proceedings 16863
Virginia Woolf Miscellany 16864
Virittäjä 03979
Virke 03948
Virus 08272
Visnyk Akademii Nauk Ukrainy: Journal of the Ukrainian Academy of Sciences 11193
Visual Arts Ontario: Agenda 02982
Visual Communications Journal 15481
Vita dell'Unione: Bollettino Mensile 08046
Vita Europea 07458
Vital Signs 16404
Vitamins 08281
Vitenskapelige Forhandlinger: Scientific Proceedings 09322
Vivre 04598
Vjesnik Bibliotekara Hrvatske 03460
Vjetari/Godisnjak: Annual Report 17076
Vjetar i Punimeve Shkencore te Qendres se Kerkimeve Biologjike: Yearbook of the Biological Research Centre 00003
Vlaamse Stam: Tijdschrift voor Familiegeschiedenis 01510
Vnitřní lékařství: Internal Medicine 03509
Vocational Education Journal 14206
Vocational Educator 14713
The Vocational Expert 13666
Vocational Industrial Clubs of America: Journal 16866
Vocational Industrial Clubs of America: Professional News 16866
La Voce dell'Unione: Atti Annuali 08044
Voce Francescana 07416
Vodni Problemi 01771
Vodnye Resursy: Water Resources 09954
Völkerkundliche Veröffentlichungen 00544
Voice 12777
Voice of Islam 09371
The Voice of Naprapathy 13953
The Voice of the Pharmacist 13698
Voices 13433
Voices from the Third World 09225
Les Voies de la Création Théatrale 04382
Volkacher Bote: Mitteilungsblatt 05286
Volkshochschule 05562
Volkshochschule: Zeitschrift des Deutschen Volkshochschul-Verbandes 06098
Volkskunde in Österreich: Nachrichtenblatt des Vereins für Volkskunde 00988
Volkskundliche Veröffentlichungen 00544
Volume of Essays 16567
Volumes 07539
Volunteer Committees of Art Museums of Canada and the United States: Conference Report 16868
Volunteer Committees of Art Museums of Canada and the United States: Directory 16868
Volunteer Committees of Art Museums of Canada and the United States: News 16868
Volunteer Work 11749
Voprosy Ekonomiki: Economic Questions 09954
Voprosy Filosofii: Questions of Philosophy 09954
Voprosy Ikhtiologii: Questions of Ichthyology 09954
Voprosy Literatury: Questions of Literature 09954
Voprosy Meditsinskoi Khimii: Problems of Medical Chemistry 09952
Voprosy Psikhologii: Problems of Psychology 09953
Voprosy Virusologii Khimii: Problems of Virology 09952
Voprosy Yazykoznaniya: Questions of Linguistics 09954
Vorgänge: Zeitschrift für Bürgerrechte und Gesellschaftspolitik 05933
Vorposy Istorii: Questions of History 09954
Vorposy Istorii Estestvoznaniya i Tekhniki: History of Natural Sciences and Technology 09954
Vorträge Betontag 05505
Vostok: East 09954
Vox Romanica 10648
V.R.B.-Informatie 01506
De Vrije Fries 08898
VS aktuell 05867
VSB Bulletin 10957
Vulkanologiya i Seismologiya: Vulcanology and Seismology 09954
Vuosikirja: Yearbook 04065
V.V.S.-Bulletin 09104
VVS danvak 03699

Vysokomolekulyarnye Soedineniya: High Molecular Compounds 09954

WA Bird Notes 00459
WAEC News 06493
Wakerobin 15611
Wald + Holz, La Foret: Organe des Waldwirtschaftsverbandes Schweiz 11029
Das Waldviertel: Zeitschrift für Heimat- und Regionalkunde des Waldviertels und der Wachau 01009
Walkerana-Proceedings 16521
Wallenberg Papers on International Finance 15496
Wall Paper 12590
Wandelhalle der Bücherfreunde 05790
Was: Zeitschrift für Kultur und Politik 01010
W.A. Shell Collector 00525
Washington Gleanings 15907
Washington Newsletter 13916, 14704
Washington Report 14375, 15580, 16875
Washington Update 14645, 15844, 16765
WAS Newsletter 16922
Wasser und Boden 05552
Wasserwirtschaft 05552
Waste, Recycling and Environmental Yearbook 12294
Watch It 16897
Water Environment Federation: Highlights 16873
Water International 15572
Waterlines 12320
Waterloo Historical Society 02984
Water Pollution Research Journal of Canada 02076
Water Quality International 12337
Water Research 12337
Water Research Foundation of Australia: Newsletter 00520
Water Resources Bulletin 14209
Water Resources Congress: Hotline 16875
Water Resources Journal 14107
Water Resources Research 13833
Water SA 10234
Water Science and Technology 12337
Watertek Newsletter 10133
Waterway/Port/Coastal Journal 14107
Watsonia 11464
WAVA-News 16923
Waves in Random Media 12268
Way of Life 12135
WCPS Quarterly 16914
WCRLA Newsletter 14778
Weather 12832
Weather and Forecasting 13943
Weather Modification Association: Newsletter 16876
Weed Abstracts 11720
Weed Research 05681
Weed Science 16877
Weed Science Society of America: Abstracts 16877
Weed Science Society of America: Newsletter 16877
Weed Science Society of South Australia: Newsletter 00522
Weeds Technology 16877
Weekly Epidemiological Record 11033
Weekly Report 14383
Weekly Wire 14544
Wehrtechnik 05440
Weight Engineering Journal 16632
Weizmann Now 13745
Welcome 12615
Weldasearch Industry News 13147
Welding Abstracts 13147
Welding and Metal Fabrication 11487
The Welding Institute News Video 13147
Welding Journal 14210
Welding Research Abroad 16878
Welding Research Council: Bulletin 16878
Welding Research Council: Yearbook 16878
Welding Research News 16878
Wellsian 12167
Weltbibliographie der sozialen Sicherheit 10627
The Wenner-Gren Center International Symposium Series 10605
Wenner-Gren Center Svenska Symposier 10605
Wenner-Gren Foundation for Anthropological Research: Annual Report 16879
Werbeforschung & Praxis 00928
Werbeforschung und Praxis 05625
Werk, Bauen und Wohnen 10638
Werken LGOG 08972
Werken NHG 09051
Werkstoffe und Korrosion 05123
Werkstoffe und Korrosion: Materials and Corrosion 05334
Werk und Zeit 05563
Wesley Historical Society: Proceedings 13205
West African Examinations Council: Annual Report 06493
West African Geoscience Newsletter 11088
West African Journal for History Teachers 06491
West African Journal of Agricultural Economics 09246
West African Science Association: Journal 06494
Westdeutsche Gesellschaft für Familienkunde e.V.: Mitteilungen 06438
Western American Literature 16884
Western Association of Map Libraries: Information Bulletin 16880
Western Canada Water and Wastewater Association: Bulletin 02990
Western College Association: Addresses and Proceedings 16882
Western Historical Quarterly 16883
Western Journal of Applied Forestry 16634
Western Society of Malacologists: Annual Report 16885
Western Society of Naturalists: Abstracts of Contributed Papers 16886
Western Society of Naturalists: Newsletter 16886
Western Surgical Association: Newsletter 16887
Western Surgical Association: Program 16887
Western Surgical Association: Transactions 16887
Western Veterinary Conference: Directory 16888
Westfälisches Ärzteblatt 05042
Weta 09124
Wetter und Leben: Zeitschrift für Angewandte Meteorologie 00809
WFSF Newsletter 08055
Whaka 09166
Whale Center Newsletter 16763
Whale News 13686
Whales Tales 16763
The Whalewatcher 13686
Whaling Museum Society: Annual Report 16890
Whaling Museum Society: Newsletter 16890
What's on in Computing 11547
What They Say About Forestry on the Hill 02139
Wheat, Barley and Triticale Abstracts 11720
Wheelchair Information Pack 11894
White Paper on Transportation Safety 08142
WHO Drug Information 11033
Whooping Crane Conservation Association: Membership Directory 16892
Who's Who in Special Libraries 16740
Who's Who in the Dental Laboratory Industry 15849
Whydah 08458
Wiadomości Botaniczne 09558
Wiadomości Chemiczne 09559
Wiadomosci Instytutu J. Pilsudskiego 15629
Wiadomości Matematyczne 09585
Wiadomości Numizmatyczne 09553
Wiadomości Parazytologiczne 09598
Wielewaal 01412
Wiener Arzt 00536
Wiener Beiträge zur Musikwissenschaft 00645
Wiener Bonbons 00692
Wiener Chopin-Blätter 00667
Wiener Geographische Schriften 00847
Wiener Geschichtsblätter 00984
Wiener Humanistische Blätter 01018
Wiener Institut für Entwicklungsfragen und Zusammenarbeit: Report Series 01019
Wiener Klinische Wochenschrift 00533, 00604
Wiener Naturschutznachrichten 00898
Wiener Rechtswissenschaftliche Studien 00827
Wiener Slavistisches Jahrbuch 00723
Wiener Slawistischer Almanach (WSA) 00643
Wiener Sprachblätter 00999
Wiener Studien: Zeitschrift für Klassische Philologie und Patristik 00723
Wiener Studien zur Tibetologie und Buddhismuskunde 00567
Wiener Tierärztliche Monatsschrift 00748
Wiener Völkerkundliche Mitteilungen (WVM) 00739
Wiener Zeitschrift für die Kunde Südasiens und Archiv für indische Philosophie 00723
Wiener Zeitschrift für Innere Medizin 00796
Wiener Zeitschrift für Nervenheilkunde 00825
Wild about Gwent 12138
Wilderness 16896
Wilderness Education Association: Newsletter 16894
Wilderness Medicine 16895
The Wilderness Society: Annual Report 16896
Wild Goose Association: Newsletter 16898
Wildlife Monographs 16901
Wildlifer 16901
Wildlife Society Bulletin 16901
Wild Rose Denturist 02415
Wilhelm Furtwängler Society: Newsletter 13207
Willem Mengelberg Society: Newsletter 16902
William and Mary Quarterly 15316
William Morris Society: Journal 13208
The Wilson Bulletin 16905
Wilson's Disease Association: Newsletter 16906
Wiltford Record Society: A Series of Historical Records of Local and National Interest 13210
Wiltshire Archaeological and Natural History Society: Magazine 13209
Windletter 14211
Winds of Change 13861
Winesburg Eagle 16469
Winrock International Institute for Agricultural Development: Annual Report 16907
wir produktiv 06253
Wirtschaft in der Praxis 01007
Wirtschaft & Produktivität 06247, 06254
Wirtschaftsdienst 05935
Wirtschaftsdienst: A Monthly Magazine on Economic Policies 05935
Wirtschaftskonjunktur 05936
Wirtschaftskurier 00975
Wirtschaft und Wissenschaft 06295
Wir Walser 11015
Wissenschaftliche Beihefte zur Zeitschrift Die Höhle 00969
Wissenschaftliche Berichte der Forschungsbereiche 05307
Wissenschaftliche Gesellschaft der Ärzte in der Steiermark: Abstracts über die wissenschaftlichen Sitzungen 01032
Wissenschaftlicher Bericht des jährlich abgehaltenen Seminars 00770
Wissenschaftliche Schriftenreihe 10943
Wissenschaftspolitik: Mitteilungsblatt der schweizerischen wissenschaftlichen Instanzen 10924
Wissenschaftsrat: Empfehlungen und Stellungnahmen 06451
WME Directory 16908
WMO Bulletin 10777
Wochenbericht des DIW 05589
Wohnbauforschung in Österreich (WBFÖ) 00597
Wohnmedizin 05442
Wolverton and District Archaeological Society: Newsletter 13212
Women and Mathematics Education: Newsletter 16908
Women and Mathematics Education of Girls and Women 16908
Women Educators: Annual Awards Report 16909
Women Educators: Newsletter 16909
Women Lawyers Journal 15915
Women's Caucus for Political Science: Membership Directory 16914
Women's Caucus for the Modern Languages: Concerns 16915
Women's Classical Caucus: Newsletter 16917
Women's Studies Newsletter 13215
Women's Studies Program Directory 16182
Wood and Fiber Science 16714
Wood Preserving Statistics 14213
Wool 12407
Wool and Wattles 13582
Woolhope Club Transactions 13213
Wool Industry Review 06675
Wool Questionnaire 12407
Word and Deed 13915
Word in Action 11478
The Wordsworth Circle 16920
Working Group Reports 04208
Working Holidays 11749
Workshop of Peace 01053
Workshop Proceedings 12607
World Affairs 13992
World Agric. Econ. & Rural Sociology Abstract 11720
World Aluminium Abstracts 05659
World Aquaculture 16921
World Aquaculture Society: Abstracts 16921
World Aquaculture Society: Directory 16921
World Archaeological Society: Special Publications 16922
World Association of Societies of Pathology (Anatomic and Clinical): News Bulletin 08445
World Birdwatch 11445
World Ceramics Abstracts 11755
World Congress Proceedings 12400
World Economy 03155
World Federation for Medical Education: Proceedings 13220
World Federation of Public Health Associations: Annual Report 16926
World Federation of Societies of Anaesthesiologists: Newsletter 13221
World Flow of Unwrought Aluminium 13217
World Flow of Unwrought Copper 13217
World Flow of Unwrought Lead 13217
World Flow of Unwrought Nickel 13217
World Flow of Unwrought Tin 13217
World Flow of Unwrought Zinc 13217
World Health 11033
World Health Forum 11033
World Health Magazine 13505
World Health Statistics Quarterly 11033
World Hospitals 15485
World Informo 06868
World Journal of Microbiology and Biotechnology 04589
The World Journal of Surgery 10976
World Journal of Surgery 14781
World Medical Journal 05002
World Metal Statistics 13217
World Metal Statistics Yearbook 13217
World of Aviation Maintenance 14562
The World of Books 11102
The World of Music: Intercultural Music Studies 06001
World Population Society: Proceedings of Annual Conferences 16930
World Ports & Harbours 11437
World Smoking and Health 13676
World's Poultry Science Association – U.S.A. Branch: Journal 16934
World's Poultry Science Journal 08887
World's Poultry Sciene Journal 13224
World Stainless Steel Statistics 13217
World Steel in Figures 01389
World Surface Coating Abstracts 12673
World Tin Statistics 13217
The World Today 12813
Worldwatch Papers 02389
World Water 12294
World Wrought Copper Statistics 13217
WRC Information 13194
Writings on Italian Americans 15587
Wroclawskie Towarzystwo Naukowe: Sprawozdania,: Reports A and B 09630
Wszechświat 09601
WT-Werkstattechnik 06326
WUCT Newsletter 08053
Württembergisch Franken 05926
Wuji Huaxue 03167
Wuli Huaxue 03167
Wydawnictwa Źródłowe Komisji Historycznej 09612
Wydział Języka i Literatury 09548
Wydział Nauk Historyczno-Społecznych 09548
Wydział Nauk Medycznych 09548
Wydział Nauk Przyrodniczych 09548

Xinjiang Agricultural Sciences 03224
XPLOR International: Proceedings of Worldwide Conference 16939
XVIIe Siècle 04727

Yadernaya Fizika: Nuclear Physics 09954
Yad-La-Kore 07080
Yakugaku Zasshi 08395
Y.A.S. Parish Register Series 13230
Y.A.S. Record Series 13230
Y.A.S. Wakefield Court Rolls Series 13230
Y Ddolen 13203
Year Book and List of Members 12300
Yearbook for Traditional Music 15454
Year Book of Adult Continuing Education 12596
Yearbook of Comparative and General Literature 13747
Yearbook of English Studies 12535
Yearbook of Iberian-Caucasian Linguistics 05011
Yearbook of Psychology 03888
Yearbook of the Swedish Linnaeus Society 10561
The Year-Rounder 15824
The Year's Work in English Studies 11960
The Year's Work in Modern Language Studies 12535
Yedyion Haaguda 07074
Yiddish 13572
Yiddish Studies Newsletter 13572
Yidishe Shprakh 16941
Yingyong Huaxue 03167
Ymer 10573
Yoga & Ayurveda 07479
Yoksa Hakbo 08534
Yorkshire Archaeological Journal 13230
Yorkshire Artscene 13231
Yorkshire Geological Society: Proceedings 13233
Yorkshire Philosophical Society: Annual Report 13234
Yorktech Directory 03000
Yorktech Newsletter 03000
Yosemite Association: Newsletter 16942
Yosetsu Gakkai: Journal 08446
Yosetsu Gakkai: Transactions 08446
Youji Huaxue 03167
Young Children 15818
Young Cinema International 04380
Young Citizen: Magazine 06964
Young Fabian Pamphlets 12045
Your Child 16836
Your Child and TV 02344
Your Consultant 14422
Youth Law News 15948
Youth TGI 12032
Yrke 09308
Yuki Teikei Haiku Society of the United States and Canada: Members Anthology 16943
YVL-guides 03989
Zacchia 07785
Zagadnienia Rodzajów Literackich 09545
Zagadnienia Wychowawcze a Zdrowie Psychiczne: Upbringing Problems and Mental 09576
Zahlentafeln der Physikalisch-Chemischen Untersuchungen des Rheinwassers 05993
Zahnärzteblatt Baden-Württemberg 06099
Zahnärzteblatt Westfalen-Lippe 06056, 06462
Zahnärztekammer Schleswig-Holstein: Mitteilungsblatt 06461
Zahnärztliche Informationen 06101
Zahnärztliche Mitteilungen 05261
Zahnärztliche Mitteilungen: ZM 06040
Zahnarzt aktuell: Informationsdienst für Mitglieder des Freien Verbandes Deutscher Zahnärzte e.V. 05761
Zambia Journal of Science and Technology 17154
Zambia Library Association: Journal 17158
Zambia Library Association: Newsletter 17158
Zambia Science Abstracts 17154
Zapiski Historyczne TNT 09627
Zapiski Mineralogo Obshchestva: Notes of the Mineralogical Society 09954
Zapiski Uzbekskogo Otdelenia Vsesoyuznogo Mineralohicheskogo Obshchestva: Notes of the Uzbek Division of the All-Union Mineralogy Society 16948
Zapisnici Srpskog Geološkog Društva 17128
ZAPP 03772
Zaschita Metallov: Protection of Metals 09954
Zavarivač 17095
Zavarivanje 17095
Zborniki geografskih zborovanj: Congress Proceedings 10115
Zborník INFOS 10095
ZDMG 05473
Zdrowie Psychiczne: Mental Health 09576
Zeitgeschichte 00679
Zeitschrift der Arbeitsgemeinschaft Österreichischer Entomologen 00560
Zeitschrift der Deutschen Gemmologischen Gesellschaft 05309
Zeitschrift der Deutschen Geologischen Gesellschaft 05311
Zeitschrift des Aachener Geschichtsvereins 05028
Zeitschrift des Bergischen Geschichtsvereins 05184
Zeitschrift des Deutschen Vereins für Kunstwissenschaft 05558
Zeitschrift des Historischen Vereins für Schwaben 05925
Zeitschrift des Historischen Vereins für Steiermark 00651
Zeitschrift des Instituts für Demographie: Demographische Informationen 00723
Zeitschrift Führung + Organisation 05847
Zeitschrift für Agrargeschichte und Agrarsoziologie 05796
Zeitschrift für Arbeitswissenschaft 05801
Zeitschrift für bayerische Kirchengeschichte 06379
Zeitschrift für Bibliothekswesen und Bibliographie: Offizielle Nachrichten 06373
Zeitschrift für die gesamte Versicherungswissenschaft 05560, 06006
Zeitschrift für die Geschichte der Saargegend 05922
Zeitschrift für die Geschichte des Oberrheins 06078
Zeitschrift für erziehungs- und sozialwissenschaftliche Forschung 05889
Zeitschrift für Ethnologie 05437
Zeitschrift für Flugwissenschaften und Weltraumforschung 05307
Zeitschrift für Fremdenverkehr 10747
Zeitschrift für Ganzheitsforschung 00619
Zeitschrift für Gastroenterologie 00781, 05434
Zeitschrift für Heereskunde: Wissenschaftliches Organ für die Kulturgeschichte der Streitkräfte, ihrer Bekleidung, Bewaffnung und Ausrüstung, für heeresmuseale Nachrichten und Sammler-Mitteilungen 05360
Zeitschrift für Historische Waffen- und Kostümkunde 05827
Zeitschrift für Hochschuldidaktik: Beiträge zu Studium, Wissenschaft und Beruf 00790
Zeitschrift für Hohenzollerische Geschichte 05931
Zeitschrift für Individualpsychologie 00906
Zeitschrift für Kardiologie: Verhandlungsbericht 05362
Zeitschrift für Kinder- und Jugendpsychiatrie 05372
Zeitschrift für Klinische Psychologie, Psychopathologie und Psychotherapie 05896
Zeitschrift für Kulturaustausch 05952
Zeitschrift für Lärmbekämpfung 05499

Zeitschrift für Lateinamerika 00913
Zeitschrift für Metallkunde 05381
Zeitschrift für Mund-, Kiefer- und Gesichtschirurgie 05386
Zeitschrift für öffentliche und gemeinwirtschaftliche Unternehmen 05846
Zeitschrift für Parlamentsfragen 05617
Zeitschrift für Physikalische Medizin, Balneologie und Medizinische Klimatologie 00765
Zeitschrift für Psychosomatische Medizin und Psychoanalyse 05491
Zeitschrift für Rechtsmedizin: Journal of Legal Medicine 05416
Zeitschrift für Rechtsvergleichung, Internationales Privatrecht und Europarecht 00827
Zeitschrift für Religions- und Geistesgeschichte 05823
Zeitschrift für Rheumatologie 05417, 10845
Zeitschrift für Säugetierkunde: International Journal of Mammalian Biology 05418
Zeitschrift für Sprachwissenschaft 05427
Zeitschrift für Stomatologie 00850
Zeitschrift für Unternehmensgeschichte 05866
Zeitschrift für Vermessungswesen 05559
Zeitschrift für Volkskunde 05439
Zeitschrift für Wirtschafts- und Sozialwissenschaften 05871
Zeitschrift für Württembergische Landesgeschichte 06078
Zeitschrift Physikalische Medizin, Rehabilitationsmedizin, Kurortmedizin 05404
Zeitungstechnik 05938
Zemědělská Ekonomika 03501
Zemědělská Informatika 03501
Zemědělská Technika 03501
Zemljišste i biljka 17100
Zemlya i Vselennaya: Earth and Universe 09954
Zentralblatt für Mathematik und Grenzgebiete 05911
Zhurnal Analiticheskoi Khimii: Journal of Analytical Chemistry 09954
Zhurnal Eksperimentalnoi i Teoreticheskoi Fiziki: Journal of Experimental and Theoretical Physics 09954
Zhurnal Evolyutsionnoi Biokhimii i Fiziologii: Journal of Evolutionary Biochemistry and Physiology 09954
Zhurnal Experimentalnoi i Klinicheskoi Medisiny: Journal of Experimental and Clinical Medicine 00203
Zhurnal Fizicheskoi Khimii: Journal of Physical Chemistry 09954
Zhurnal Nauchnoi i Prikladnoi Fotografii i Kinematografii: Journal of Scientific and Applied Photography and Cinematography 09954
Zhurnal Neorganicheskoi Khimii: Journal of Inorganic Chemistry 09954
Zhurnal Obshchei Biologii: Journal of General Biology 09954
Zhurnal Obshchei Khimii: Journal of General Chemistry 09954
Zhurnal Organicheskoi Khimii: Journal of Organic Chemistry 09954
Zhurnal Prikladnoi Khimii: Journal of Applied Chemistry 09954
Zhurnal Prikladnoi Mekhaniki i Tekhnicheskoi Fiziki: Journal of Applied Mechanics and Technical Physics 09954
Zhurnal Prikladnoi Spektroskopii: Journal of Applied Spectroscopy 01088
Zhurnal Strukturnoi Khimii: Journal of Structural Chemistry 09954
Zhurnal Tekhnicheskoi Fiziki: Journal of Technical Physics 09954
Zhurnal Vychislityelnoi Matematiki i Matematicheskoy Fiziki: Journal of Computational Mathematics and Mathematical Physics 09954
Zhurnal Vysshei Nervnoi Deyatelnosti: Journal of Higher Nervous Activity 09954
Zimbabwe Librarian 17186
Zimbabwe Prehistory 17179
Zimbabwe Science News 17188
Zimbabwe Veterinary Journal 17189
Zimbabwe Wildlife 17183
Zion 07056
Zirkular 00583
Zisin: Journal 08448
Zivilschutz Magazin: Fachzeitschrift für Zivilschutz, Katastrophenschutz und Selbstschutz 05252
Živočišná Výroba 03501
ZI – Ziegelindustrie International 05736
Zoning News 14007
Zoo 01410
Zoo Federation News 12069
Zoologica 10491
Zoological Journal 12454
Zoological Parks & Aquariums in the Americas 13606
Zoological Record 13237
Zoological Science 08295
Zoological Society of Glasgow and West of Scotland: Annual Report 13236
The Zoological Society of London: Symposia 13237
Zoologica Poloniae 09608
Zoologica Scripta 10490
Zoologicheskii Zhurnal: Zoological Journal 09954
Der Zoologische Garten 06364
Zootechnia 10274
Zootecnia 10402
Zootecnica e Nutrizione Animale 07401
Zootecnica International: Atti del Convegno Annuale 07928
Z Otchłani Wieków 09553
Zprávy ČAS: ČAS News 03502
Zprávy České Společnosti Parasitologické 03527
Zprávy ČSBS 03504
Zprávy SAD 03554
Züchtungskunde 05447
Zuid-Afrika 09057
Zukunftsforschung: Informationen über Zukunftsforschung, Planung und Zukunftsgestaltung 10955
ZUM: Zeitschrift für Urheber- und Medienrecht 05976
zur debatte: Themen der Katholischen Akademie in Bayern 06059
Zväz Slovenských Knihovníkov a Informatikov: Zväzový bulletin 10095

SOMEFI
SOCIEDAD MEXICANA DE FITOGENETICA, A.C.
MEXICAN SOCIETY OF PLANT GENETICS
Apartado Postal 21, Chapingo, México. 56230 MEXICO
Tel. (595) 4-22-00 ext. 5795 y 5362
Fax (595) 4-66-52

SOMEFI is in the agricultural field one of the earliest and more prestigious scientific society in Mexico. It was founded on February 1965, a century after Gregor Mendel published his article: "Experiments with plant hybrids", in which he developed the genetics mathematical basis.

Among several scientific activities, SOMEFI has organized Symposia on specific plant genetic issues, such as plant genetic resources, seed production, genotype environment interaction, and biotechnology of plants. The scientific papers presented in these meetings have been published in particular publications.

In addition, SOMEFI publishes the periodical scientific journal **Revista Fitotecnia Mexicana** and the scientific bulletin **Germen**, twice a year. "Revista Fitotecnia Mexicana" publishes scientific papers. "Germen" publishes literature reviews, papers describing the development of novelty research methods.

SOMEFI organizes a National Congress on Plant Breeding and Genetics every two years. The **XV Congress of Plant Genetics** will be hold at Monterrey City, Mexico, on September 25-30, 1994 and will host the plant genetics chapter of the **XI Latin American Congress of Genetics.**

A selection of technical dictionaries available from K.G. Saur Publishers

■ **Encyclopedic Dictionary of Electronics, Electrical Engineering and Information Processing**
Enzyklopädisches Wörterbuch der Elektronik, Elektrotechnik und Informationsverarbeitung
English-German/German-English
1990ff. 8 vols. c. 400pp per vol. HB.
Complete set: DM 1,920.00
ISBN 3-598-10680-7

■ **Dictionary of Electronics and Information Processing**
Wörterbuch der Elektronik und Informationsverarbeitung
English-German/German-English
Volume 1: A-K, Volume 2: L-Z
1990. 2 vols. Complete set: X,770pp. HB.
DM 296.00. ISBN 3-598-10885-0

■ **International Dictionary of Abbreviations and Acronyms of Electronics, Electrical Engineering, Computer Technology and Information Processing**
Internationales Verzeichnis der Abkürzungen und Akronyme der Elektronik, Elektrotechnik, Computertechnik und Informationsverarbeitung
English-German/German-English
1992. 2 vols. Complete set: 960pp. HB.
DM 336.00. ISBN 3-598-10977-6

If you would like further details on these titles by our author Peter Wennrich, please contact us at the number below.

K • G • Saur München • New Providence • London • Paris
A Reed Reference Publishing Company
K · G · Saur Verlag · Postfach 70 16 20 · D-81316 München · Tel. (089) 76902-0 · Fax (089) 76902-150

Yiddish Books from the Harvard College Library

Microfiche edition
1994. C. 5.500 fiches. 6 installments. Reader factor 24X
Diazo: DM 46,200.00*. ISBN 3-598-33176-2
Silver: DM 62,100.00*. ISBN 3-598-33184-3
(* prices subject to change without prior notice)

Like other European nations of the past millenium, the Jews of Central and Eastern Europe created an independent culture in their vernacular language. Because the development of Yiddish also coincided with the spread of printing, it generated a large popular literature. With the advent of the Enlightenment, Yiddish became simultaneously the vehicle of challenge to the traditional Jewish way of life and of its reinforcement. The high level of literacy among Jews made the printed page the main battleground and meeting point of their argument over modernity.
Modern Yiddish literature emerged as a dynamic expression of the transition from religious to secular culture.

The Yiddish Collection of the Harvard Microfiche Project has filmed non-copyright books and pamphlets of research value to scholars in every field of modern Jewish studies. These works were selected according to the criteria of scarcity, danger of disintegration, and primary research value. The collection includes materials in many areas of Jewish studies, e.g., Bible, Liturgy, Ethics, Jewish History, Philology. The largest section, Yiddish Literature, which includes poetry, drama, fiction, humour and satire, and miscellaneous prose, contains dozens of otherwise inaccessible 19th century popular novels, confiscated Soviet publications of the 1930's, and long since out-of-print literary anthologies and scholarly studies of the interwar period in Poland.

The collection reflects the complexity and diversity of Jewish culture throughout the period. In terms of date of publication, it extends from a Bible translation of Amsterdam 1749 to a posthumous collection of poems, Moscow 1970. In terms of imprint, it shows the reach of Yiddish publishing beyond the major centers of Vilna and Warsaw to Copenhagen, Czernowitz, Chicago. This microfiche collection greatly improves the possibilities of research and scholarship in a number of areas, making it a necessary complement to the collection of any library with a serious interest in modern Jewish studies.

Please inquire for more information

K·G·Saur Verlag
München · New Providence · London · Paris
Reed Reference Publishing
Postfach 70 16 20 · D-81316 München · Tel. (089) 7 69 02-0

Yiddish Books from the Harvard College Library